AMERICAN
MEN AND WOMEN
OF SCIENCE

AMERICAN MEN AND WOMEN OF SCIENCE

13TH EDITION

Edited by the Jaques Cattell Press

Volume 6 St–Z

R. R. BOWKER COMPANY
A Xerox Publishing Company
New York & London, 1976

Contents

Advisory Committee

Dr. Dael L. Wolfle, Chairman
Graduate School of Public Affairs
University of Washington

Dr. Randolph W. Bromery
Chancellor,
University of Massachusetts

Dr. Janet W. Brown
Program Head,
Office of Opportunities
in Science
American Association for the
Advancement of Science

Dr. Robert W. Cairns
Executive Director,
American Chemical Society

Dr. S. D. Cornell
Assistant to the President
National Academy of Sciences

Dr. Ruth M. Davis
Director,
Institute for Computer Science
and Technology
National Bureau of Standards

Dr. Carl D. Douglass
Deputy Director,
Division of Research Grants
National Institutes of Health

Dr. Richard G. Folsom
President Emeritus,
Rensselaer Polytechnic Institute

Dr. Robert E. Henze
Director,
Membership Division
American Chemical Society

Dr. Eugene L. Hess
Executive Director,
Federation of American Societies
for Experimental Biology

Dr. William C. Kelly
Executive Director,
Commission on Human Resources
National Research Council

Dr. Kenneth B. Raper
Department of Bacteriology
University of Wisconsin

Dr. A. L. Schawlow
Department of Physics
Stanford University

Dr. John F. Sherman
Vice President,
Association of American Medical
Colleges

Dr. Matthias Stelly
Executive Vice President,
American Society of Agronomy

Dr. John R. Whinnery
Department of Electrical Engineering
and Computer Sciences
University of California, Berkeley

Preface

The observance of an anniversary prompts reflection on the past, and it is natural to note here that the astounding growth in the dimension and stature of this 200 year old American nation has been influenced immeasurably by the achievements of her scientists. 1976 also marks the 70th anniversary of AMERICAN MEN & WOMEN OF SCIENCE as a chronicle of the lives and professional activities of those men and women most instrumental in affecting the shape and quality of science in America. The explosion in scientific activity over the past seven decades is clearly evident when one compares the 1906 edition, a single volume containing 4,000 entries, with the present edition, six volumes profiling nearly 110,000 men and women of importance in their fields.

This edition of AMERICAN MEN & WOMEN OF SCIENCE is a landmark in biographic achievement. The information contained in all seven volumes has been gathered, edited and compiled by the Jaques Cattell Press in the space of ten months. This is a radical, and beneficial, departure from the production of the 12th edition, which took three years to publish in its entirety. The acceleration was made possible by the use of a computerized printing method and more efficient production procedures.

The editors have not sacrificed quality in the interest of speed, however. The criteria were stringently applied in the selection of new entrants, all nominated by former biographees. The criteria follow.

Achievement, by reason of experience and training, of a stature in scientific work equivalent to that associated with the doctoral degree, coupled with presently continued activity in such work;

or

Research activity of high quality in science as evidenced by publication in reputable scientific journals; or, for those whose work cannot be published because of governmental or industrial security, research activity of high quality in science as evidenced by the judgment of the individual's peers;

or

Attainment of a position of substantial responsibility requiring scientific training and experience to the extent described for (1) and (2).

All data forms submitted by biographees were carefully edited and proofread. Those whose information was appearing for the first time received a proof before publication. Information on scientists who did not supply current material is included only if verification of present professional activity was established by our researchers. A former biographee whose current status could not be verified is given a reference to the 12th edition if the probability exists of a continued activity in science. References are also given to scientists who have died since publication of the last edition. Omitted are the names of those previously listed as retired and those who have entered fields or activities not covered by the scope of the directory. More than 20,000 nominations for inclusion were received and over 12,000 of those nominated returned information and were selected for listing. This is a significant increase in both nominations and selections over any previous edition and gives fuller representation to the developing scientist.

Geographic and discipline indexes make up the seventh volume of the set. The discipline index has been rearranged and is organized with major subject headings, providing easier access to this information.

Certain disciplines previously included in the directory are not represented in this edition. Engineering and economics will appear in separate directories with expanded criteria, to enable even broader coverage of these areas. Others, including sociology, psychology and political science were omitted because they are fully covered by membership directories in each field.

Great appreciation is expressed to the AMERICAN MEN & WOMEN OF SCIENCE Advisory Committee for their guidance in the planning of the 13th edition. Their efforts have contributed to an unusually good response to our requests for information and nominations, which has enhanced the value of the publication. Also to be thanked are the many scientific societies that provided membership lists for the use of our researchers or that published announcements in their bulletins and journals.

The staff of the Jaques Cattell Press deserves the highest accolade for their sustained interest, devotion and good will through the many hours of learning and implementing the new procedures necessary to the successful completion of this book. The overwhelming workload was shared by temporary employees who performed with the greatest diligence and responsibility. The job could not have been completed without their fine help. Everyone involved with this project gave outstanding service, but the contributions of Alice Smith, Pauline Stump, Joyce Howell and Ila Martin cannot go without mention. Special acknowledgement is given to Fred Scott, former general manager of the Jaques Cattell Press, who was responsible for initiating the formation of the advisory committee and for overseeing the planning and early stages of work on the 13th edition.

Comments and suggestions are invited and should be addressed to The Editors, American Men & Women of Science, Jaques Cattell Press, P.O. Box 25001, Tempe, Arizona, 85282.

Renee Lautenbach, Supervising Editor
Anne Rhodes, Administrative Managing Editor

JAQUES CATTELL PRESS

Desmond Reaney, Manager Book Editorial

R.R. BOWKER COMPANY

October, 1976

Abbreviations

AAAS—American Association for the Advancement of Science
abnorm—abnormal
abstr—abstract(s)
acad—academic, academy
acct—account, accountant, accounting
acoust—acoustic(s), acoustical
ACTH—adrenocorticotrophic hormone
actg—acting
activ—activities, activity
addn—addition(s), additional
Add—Address
adj—adjunct, adjutant
adjust—adjustment
Adm—Admiral
admin—administration, administrative
adminr—administrator(s)
admis—admission(s)
adv—adviser(s), advisory
advan—advance(d), advancement
advert—advertisement, advertising
AEC—Atomic Energy Commission
aerodyn—aerodynamic(s)
aeronaut—aeronautic(s), aeronautical
aerophys—aerophysical, aerophysics
aesthet—aesthetic(s)
AFB—Air Force Base
affil—affiliate(s), affiliation
agr—agricultural, agriculture
agron—agronomic, agronomical, agronomy
agrost—agrostologic, agrostological, agrostology
agt—agent
AID—Agency for International Development
Ala—Alabama
allergol—allergological, allergology
alt—alternate
Alta—Alberta
Am—America, American
AMA—American Medical Association
anal—analysis, analytic, analytical
analog—analogue
anat—anatomic, anatomical, anatomy
anesthesiol—anesthesiology
angiol—angiology
Ann—Annal(s)
ann—annual
anthrop—anthropological, anthropology
anthropom—anthropometric, anthropometrical, anthropometry
antiq—antiquary, antiquities, antiquity
antiqn—antiquarian

apicult—apicultural, apiculture
APO—Army Post Office
app—appoint, appointed
appl—applied
appln—application
approx—approximate(ly)
Apr—April
apt—apartment(s)
aquacult—aquaculture
arbit—arbitration
arch—archives
archaeol—archaeological, archaeology
archit—architectural, architecture
Arg—Argentina, Argentine
Ariz—Arizona
Ark—Arkansas
artil—artillery
asn—association
assoc(s)—associate(s), associated
asst(s)—assistant(s), assistantship(s)
Assyriol—Assyriology
astrodyn—astrodynamics
astron—astronomical, astronomy
astronaut—astronautical, astronautics
astronr—astronomer
astrophys—astrophysical, astrophysics
attend—attendant, attending
atty—attorney
audiol—audiology
Aug—August
auth—author
AV—audiovisual
Ave—Avenue
avicult—avicultural, aviculture

b—born
bact—bacterial, bacteriologic, bacteriological, bacteriology
BC—British Columbia
bd—board
behav—behavior(al)
Belg—Belgian, Belgium
bibl—biblical
bibliog—bibliographic, bibliographical, bibliography
bibliogr—bibliographer
biochem—biochemical, biochemistry
biog—biographical, biography
biol—biological, biology
biomed—biomedical, biomedicine
biomet—biometric(s), biometrical, biometry
biophys—biophysical, biophysics

bk(s)—book(s)
bldg—building
Blvd—Boulevard
Bor—Borough
bot—botanical, botany
br—branch(es)
Brig—Brigadier
Brit—Britain, British
Bro(s)—Brother(s)
bryol—bryology
Bull—Bulletin
bur—bureau
bus—business
BWI—British West Indies

c—children
Calif—California
Can—Canada, Canadian
cand—candidate
Capt—Captain
cardiol—cardiology
cardiovasc—cardiovascular
cartog—cartographic, cartographical, cartography
cartogr—cartographer
Cath—Catholic
CEngr—Corps of Engineers
cent—central
Cent Am—Central America
cert—certificate(s), certification, certified
chap—chapter
chem—chemical(s), chemistry
chemother—chemotherapy
chmn—chairman
citricult—citriculture
class—classical
climat—climatological, climatology
clin(s)—clinic(s), clinical
cmndg—commanding
Co—Companies, Company
coauth—coauthor
co-dir—co-director
co-ed—co-editor
coeduc—coeducation, coeducational
col(s)—college(s), collegiate, colonel
collab—collaboration, collaborative
collabr—collaborator
Colo—Colorado
com—commerce, commercial
Comdr—Commander
commun—communicable, communication(s)
comn(s)—commission(s), commissioned

ABBREVIATIONS

comnr—commissioner
comp—comparative
compos—composition
comput—computation, computer(s). computing
comt(s)—committee(s)
conchol—conchology
conf—conference
cong—congress, congressional
Conn—Connecticut
conserv—conservation, conservatory
consol—consolidated, consolidation
const—constitution, constitutional
construct—construction, constructive
consult(s)—consult, consultant(s), consultantship(s), consultation, consulting
contemp—contemporary
contrib—contribute, contributing, contribution(s)
contribr—contributor
conv—convention
coop—cooperating, cooperation, cooperative
coord—coordinate(d), coordinating, coordination
coordr—coordinator
corp—corporate, corporation(s)
corresp—correspondence, correspondent, corresponding
coun—council, counsel, counseling
counr—councilor, counselor
criminol—criminological, criminology
cryog—cryogenic(s)
crystallog—crystallographic, crystallographical, crystallography
crystallogr—crystallographer
Ct—Court
Ctr—Center
cult—cultural, culture
cur—curator
curric—curriculum
cybernet—cybernetic(s)
cytol—cytological, cytology
CZ—Canal Zone
Czech—Czechoslovakia

DC—District of Columbia
Dec—December
Del—Delaware
deleg—delegate, delegation
delinq—delinquency, delinquent
dem—democrat(s), democratic
demog—demographic, demography
demogr—demographer
demonstr—demonstrator
dendrol—dendrologic, dendrological, dendrology
dent—dental, dentistry
dep—deputy
dept—department(al)
dermat—dermatologic, dermatological, dermatology
develop—developed, developing, development, developmental
diag—diagnosis, diagnostic
dialectol—dialectological, dialectology
dict—dictionaries, dictionary
Dig—Digest
dipl—diploma, diplomate
dir(s)—director(s), directories, directory
dis—disease(s), disorders
Diss Abstr—Dissertation Abstracts
dist—district
distrib—distributed, distribution, distributive
distribr—distributor(s)
div—division, divisional, divorced

DNA—deoxyribonucleic acid
doc—document(s), documentary, documentation
Dom—Dominion
Dr—Drive

e—east
ecol—ecological, ecology
econ(s)—economic(s), economical, economy
economet—econometric(s)
ECT—electroconvulsive or electroshock therapy
ed—edition(s), editor(s), editorial
ed bd—editorial board
educ—education, educational
educr—educator(s)
EEG—electroencephalogram, electroencephalographic, electroencephalography
Egyptol—Egyptology
EKG—electrocardiogram
elec—electric, electrical, electricity
electrochem—electrochemical, electrochemistry
electrophys—electrophysical, electrophysics
elem—elementary
embryol—embryologic, embryological, embryology
emer—emeriti, emeritus
employ—employment
encour—encouragement
encycl—encyclopedia
endocrinol—endocrinologic, endocrinology
eng—engineering
Eng—England, English
engr(s)—engineer(s)
enol—enology
Ens—Ensign
entom—entomological, entomology
environ—environment(s), environmental
enzym—enzymology
epidemiol—epidemiologic, epidemiological, epidemiology
equip—equipment
ESEA—Elementary & Secondary Education Act
espec—especially
estab—established, establishment(s)
ethnog—ethnographic, ethnographical, ethnography
ethnogr—ethnographer
ethnol—ethnologic, ethnological, ethnology
Europ—European
eval—evaluation
evangel—evangelical
eve—evening
exam—examination(s), examining
examr—examiner
except—exceptional
exec(s)—executive(s)
exeg—exegeses, exegesis, exegetic, exegetical
exhib(s)—exhibition(s), exhibit(s)
exp—experiment, experimental
exped(s)—expedition(s)
explor—exploration(s), exploratory
expos—exposition
exten—extension

fac—faculty
facil—facilities, facility
Feb—February
fed—federal
fedn—federation
fel(s)—fellow(s), fellowship(s)
fermentol—fermentology

fertil—fertility, fertilization
Fla—Florida
floricult—floricultural, floriculture
found—foundation
FPO—Fleet Post Office
Fr—French
Ft—Fort

Ga—Georgia
gastroenterol—gastroenterological, gastroenterology
gen—general
geneal—genealogical, genealogy
geod—geodesy, geodetic
geog—geographic, geographical, geography
geogr—geographer
geol—geologic, geological, geology
geom—geometric, geometrical, geometry
geomorphol—geomorphologic, geomorphology
geophys—geophysical, geophysics
Ger—German, Germanic, Germany
geriat—geriatric(s)
geront—gerontological, gerontology
glaciol—glaciology
gov—governing, governor(s)
govt—government, governmental
grad—graduate(d)
Gt Brit—Great Britain
guid—guidance
gym—gymnasium
gynec—gynecologic, gynecological, gynecology

handbk(s)—handbook(s)
helminth—helminthology
hemat—hematologic, hematological, hematology
herpet—herpetologic, herpetological, herpetology
Hisp—Hispanic, Hispania
hist—historic, historical, history
histol—histological, histology
HM—Her Majesty
hochsch—hochschule
homeop—homeopathic, homeopathy
hon(s)—honor(s), honorable, honorary
hort—horticultural, horticulture
hosp(s)—hospital(s), hospitalization
hq—headquarters
HumRRO—Human Resources Research Office
husb—husbandry
Hwy—Highway
hydraul—hydraulic(s)
hydrodyn—hydrodynamic(s)
hydrol—hydrologic, hydrological, hydrology
hyg—hygiene, hygienic(s)
hypn—hypnosis

ichthyol—ichthyological, ichthyology
Ill—Illinois
illum—illuminating, illumination
illus—illustrate, illustrated, illustration
illusr—illustrator
immunol—immunologic, immunological, immunology
Imp—Imperial
improv—improvement
Inc—Incorporated
in-chg—in charge
incl—include(s), including
Ind—Indiana
indust(s)—industrial, industries, industry
inf—infantry
info—information
inorg—inorganic

ins—insurance
inst(s)—institute(s), institution(s)
instnl—institutional(ized)
instr(s)—instruct, instruction, instructor(s)
instrnl—instructional
int—international
intel—intelligence
introd—introduction
invert—invertebrate
invest(s)—investigation(s)
investr—investigator
irrig—irrigation
Ital—Italian

J—Journal
Jan—January
Jct—Junction
jour—journal, journalism
jr—junior
jurisp—jurisprudence
juv—juvenile

Kans—Kansas
Ky—Kentucky

La—Louisiana
lab(s)—laboratories, laboratory
lang—language(s)
laryngol—laryngological, laryngology
lect—lecture(s)
lectr—lecturer(s)
legis—legislation, legislative, legislature
lett—letter(s)
lib—liberal
libr—libraries, library
librn—librarian
lic—license(d)
limnol—limnological, limnology
ling—linguistic(s), linguistical
lit—literary, literature
lithol—lithologic, lithological, lithology
Lt—Lieutenant
Ltd—Limited

m—married
mach—machine(s), machinery
mag—magazine(s)
maj—major
malacol—malacology
mammal—mammalogy
Man—Manitoba
Mar—March
Mariol—Mariology
Mass—Massachusetts
mat—material(s)
mat med—materia medica
math—mathematic(s), mathematical
Md—Maryland
mech—mechanic(s), mechanical
med—medical, medicinal, medicine
Mediter—Mediterranean
Mem—Memorial
mem—member(s), membership(s)
ment—mental(ly)
metab—metabolic, metabolism
metall—metallurgic, metallurgical, metallurgy
metallog—metallographic, metallography
metallogr—metallographer
metaphys—metaphysical, metaphysics
meteorol—meteorological, meteorology
metrol—metrological, metrology
metrop—metropolitan
Mex—Mexican, Mexico
mfg—manufacturing

mfr(s)—manufacture(s), manufacturer(s)
mgr—manager
mgt—management
Mich—Michigan
microbiol—microbiological, microbiology
micros—microscopic, microscopical, microscopy
mid—middle
mil—military
mineral—mineralogical, mineralogy
Minn—Minnesota
Miss—Mississippi
mkt—market, marketing
Mo—Missouri
mod—modern
monogr—monograph
Mont—Montana
morphol—morphological, morphology
Mt—Mount
mult—multiple
munic—municipal, municipalities
mus—museum(s)
musicol—musicological, musicology
mycol—mycologic, mycology

n—north
NASA—National Aeronautics & Space Administration
nat—national, naturalized
NATO—North Atlantic Treaty Organization
navig—navigation(al)
NB—New Brunswick
NC—North Carolina
NDak—North Dakota
NDEA—National Defense Education Act
Nebr—Nebraska
nematol—nematological, nematology
nerv—nervous
Neth—Netherlands
neurol—neurological, neurology
neuropath—neuropathological, neuropathology
neuropsychiat—neuropsychiatric, neuropsychiatry
neurosurg—neurosurgical, neurosurgery
Nev—Nevada
New Eng—New England
New York—New York City
Nfld—Newfoundland
NH—New Hampshire
NIH—National Institutes of Health
NIMH—National Institute of Mental Health
NJ—New Jersey
NMex—New Mexico
nonres—nonresident
norm—normal
Norweg—Norwegian
Nov—November
NS—Nova Scotia
NSF—National Science Foundation
NSW—New South Wales
numis—numismatic(s)
nutrit—nutrition, nutritional
NY—New York State
NZ—New Zealand

observ—observatories, observatory
obstet—obstetric(s), obstetrical
occas—occasional(ly)
occup—occupation, occupational
oceanog—oceanographic, oceanographical, oceanography
oceanogr—oceanographer
Oct—October
odontol—odontology

OEEC—Organization for European Economic Cooperation
off—office, official
Okla—Oklahoma
olericult—olericulture
oncol—oncologic, oncology
Ont—Ontario
oper(s)—operation(s), operational, operative
ophthal—ophthalmologic, ophthalmological, ophthalmology
optom—optometric, optometrical, optometry
ord—ordnance
Ore—Oregon
org—organic
orgn—organization(s), organizational
orient—oriental
ornith—ornithological, ornithology
orthod—orthodontia, orthodontic(s)
orthop—orthopedic(s)
osteop—osteopathic, osteopathy
otol—otological, otology
otolaryngol—otolaryngological, otolaryngology
otorhinol—otorhinologic, otorhinology

Pa—Pennsylvania
Pac—Pacific
paleobot—paleobotanical, paleobotany
paleont—paleontological, paleontology
Pan-Am—Pan-American
parasitol—parasitology
partic—participant, participating
path—pathologic, pathological, pathology
pedag—pedagogic(s), pedagogical, pedagogy
pediat—pediatric(s)
PEI—Prince Edward Islands
penol—penological, penology
periodont—periodontal, periodontic(s), periodontology
petrog—petrographic, petrographical, petrography
petrogr—petrographer
petrol—petroleum, petrologic, petrological, petrology
pharm—pharmacy
pharmaceut—pharmaceutic(s), pharmaceutical(s)
pharmacog—pharmacognosy
pharmacol—pharmacologic, pharmacological, pharmacology
phenomenol—phenomenologic(al), phenomenology
philol—philological, philology
philos—philosophic, philosophical, philosophy
photog—photographic, photography
photogeog—photogeographic, photogeography
photogr—photographer(s)
photogram—photogrammetric, photogrammetry
photom—photometric, photometrical, photometry
phycol—phycology
phys—physical
physiog—physiographic, physiographical, physiography
physiol—physiological, physiology
Pkwy—Parkway
Pl—Place
polit—political, politics
polytech—polytechnic(al)
pomol—pomological, pomology
pontif—pontifical
pop—population
Port—Portugal, Portuguese
postgrad—postgraduate

ABBREVIATIONS

PQ—Province of Quebec
PR—Puerto Rico
pract—practice
practr—practitioner
prehist—prehistoric, prehistory
prep—preparation, preparative, preparatory
pres—president
Presby—Presbyterian
preserv—preservation
prev—prevention, preventive
prin—principal
prob(s)—problem(s)
proc—proceedings
proctol—proctologic, proctological, proctology
prod—product(s), production, productive
prof—professional, professor, professorial
Prof Exp—Professional Experience
prog(s)—program(s), programmed, programming
proj—project(s), projection(al), projective
prom—promotion
protozool—protozoology
prov—province, provincial
psychiat—psychiatric, psychiatry
psychoanal—psychoanalysis, psychoanalytic, psychoanalytical
psychol—psychological, psychology
psychomet—psychometric(s)
psychopath—psychopathologic, psychopathology
psychophys—psychophysical, psychophysics
psychophysiol—psychophysiological, psychophysiology
psychosom—psychosomatic(s)
psychother—psychotherapeutic(s), psychotherapy
Pt—Point
pub—public
publ—publication(s), publish(ed), publisher, publishing
pvt—private

Qm—Quartermaster
Qm Gen—Quartermaster General
qual—qualitative, quality
quant—quantitative
quart—quarterly

radiol—radiological, radiology
RAF—Royal Air Force
RAFVR—Royal Air Force Volunteer Reserve
RAMC—Royal Army Medical Corps
RAMCR—Royal Army Medical Corps Reserve
RAOC—Royal Army Ordnance Corps
RASC—Royal Army Service Corps
RASCR—Royal Army Service Corps Reserve
RCAF—Royal Canadian Air Force
RCAFR—Royal Canadian Air Force Reserve
RCAFVR—Royal Canadian Air Force Volunteer Reserve
RCAMC—Royal Canadian Army Medical Corps
RCAMCR—Royal Canadian Army Medical Corps Reserve
RCASC—Royal Canadian Army Service Corps
RCASCR—Royal Canadian Army Service Corps Reserve
RCEME—Royal Canadian Electrical & Mechanical Engineers
RCN—Royal Canadian Navy
RCNR—Royal Canadian Naval Reserve
RCNVR—Royal Canadian Naval Volunteer Reserve
Rd—Road

RD—Rural Delivery
rec—record(s), recording
redevelop—redevelopment
ref—reference(s)
refrig—refrigeration
regist—register(ed), registration
registr—registrar
regt—regiment(al)
rehab—rehabilitation
rel(s)—relation(s), relative
relig—religion, religious
REME—Royal Electrical & Mechanical Engineers
rep—represent, representative
repub—republic
req—requirements
res—research, reserve
rev—review, revised, revision
RFD—Rural Free Delivery
rhet—rhetoric, rhetorical
RI—Rhode Island
Rm—Room
RM—Royal Marines
RN—Royal Navy
RNA—ribonucleic acid
RNR—Royal Naval Reserve
RNVR—Royal Naval Volunteer Reserve
roentgenol—roentgenologic, roentgenological, roentgenology
RR—Railroad, Rural Route
rte—route
Russ—Russian
rwy—railway

s—south
SAfrica—South Africa
SAm—South America, South American
sanit—sanitary, sanitation
Sask—Saskatchewan
SC—South Carolina
Scand—Scandinavia(n)
sch(s)—school(s)
scholar—scholarship
sci—science(s), scientific
SDak—South Dakota
SEATO—Southeast Asia Treaty Organization
sec—secondary
sect—section
secy—secretary
seismog—seismograph, seismographic, seismography
seismogr—seismographer
seismol—seismological, seismology
sem—seminar, seminary
sen—senator, senatorial
Sept—September
ser—serial, series
serol—serologic, serological, serology
serv—service(s), serving
silvicult—silvicultural, silviculture
soc(s)—societies, society
soc sci—social science
sociol—sociologic, sociological, sociology
Span—Spanish
spec—special
specif—specification(s)
spectrog—spectrograph, spectrographic, spectrography
spectrogr—spectrographer
spectrophotom—spectrophotometer, spectrophotometric, spectrophotometry
spectros—spectroscopic, spectroscopy
speleol—speleological, speleology
Sq—Square

sr—senior
St—Saint, Street(s)
sta(s)—station(s)
stand—standard(s), standardization
statist—statistical, statistics
Ste—Sainte
steril—sterility
stomatol—stomatology
stratig—stratigraphic, stratigraphy
stratigr—stratigrapher
struct—structural, structure(s)
stud—student(ship)
subcomt—subcommittee
subj—subject
subsid—subsidiary
substa—substation
super—superior
suppl—supplement(s), supplemental, supplementary
supt—superintendent
supv—supervising, supervision
supvr—supervisor
supvry—supervisory
surg—surgery, surgical
surv—survey, surveying
survr—surveyor
Swed—Swedish
Switz—Switzerland
symp—symposia, symposium(s)
syphil—syphilology
syst(s)—system(s), systematic(s), systematical

taxon—taxonomic, taxonomy
tech—technical, technique(s)
technol—technologic(al), technology
tel—telegraph(y), telephone
temp—temporary
Tenn—Tennessee
Terr—Terrace
Tex—Texas
textbk(s)—textbook(s)
text ed—text edition
theol—theological, theology
theoret—theoretic(al)
ther—therapy
therapeut—therapeutic(s)
thermodyn—thermodynamic(s)
topog—topographic, topographical, topography
topogr—topographer
toxicol—toxicologic, toxicological, toxicology
trans—transactions
transl—translated, translation(s)
translr—translator(s)
transp—transport, transportation
treas—treasurer, treasury
treat—treatment
trop—tropical
tuberc—tuberculosis
TV—television
Twp—Township

UAR—United Arab Republic
UK—United Kingdom
UN—United Nations
undergrad—undergraduate
unemploy—unemployment
UNESCO—United Nations Educational Scientific & Cultural Organization
UNICEF—United Nations International Childrens Fund
univ(s)—universities, university
UNRRA—United Nations Relief & Rehabilitation Administration

UNRWA—United Nations Relief & Works Agency
urol—urologic, urological, urology
US—United States
USA—US Army
USAAF—US Army Air Force
USAAFR—US Army Air Force Reserve
USAF—US Air Force
USAFR—US Air Force Reserve
USAR—US Army Reserve
USCG—US Coast Guard
USCGR—US Coast Guard Reserve
USDA—US Department of Agriculture
USMC—US Marine Corps
USMCR—US Marine Corps Reserve
USN—US Navy
USNAF—US Naval Air Force
USNAFR—US Naval Air Force Reserve
USNR—US Naval Reserve
USPHS—US Public Health Service
USPHSR—US Public Health Service Reserve
USSR—Union of Soviet Socialist Republics
USWMC—US Women's Marine Corps

USWMCR—US Women's Marine Corps Reserve

Va—Virginia
var—various
veg—vegetable(s), vegetation
vent—ventilating, ventilation
vert—vertebrate
vet—veteran(s), veterinarian, veterinary
VI—Virgin Islands
vinicult—viniculture
virol—virological, virology
vis—visiting
voc—vocational
vocab—vocabulary
vol(s)—voluntary, volunteer(s), volume(s)
vpres—vice president
vs—versus
Vt—Vermont

w—west
WAC—Women's Army Corps

Wash—Washington
WAVES—Women Accepted for Voluntary Emergency Service
WHO—World Health Organization
WI—West Indies
wid—widow, widowed, widower
Wis—Wisconsin
WRCNS—Women's Royal Canadian Naval Service
WRNS—Women's Royal Naval Service
WVa—West Virginia
Wyo—Wyoming

yearbk(s)—yearbook(s)
YMCA—Young Men's Christian Association
YMHA—Young Men's Hebrew Association
yr(s)—year(s)
YWCA—Young Women's Christian Association
YWHA—Young Women's Hebrew Association

zool—zoological, zoology

AMERICAN
MEN AND WOMEN
OF SCIENCE

S

STAAB, FRANK WILLIAM, b Chicago, Ill, Nov 20, 18; m 41; c 4. ORGANIC CHEMISTRY. Educ: Northwestern Univ, BS, 40; Syracuse Univ, MS, 42. Prof Exp: Res chemist, E I du Pont de Nemours & Co, 42-43; instr, Syracuse Univ, 43-44; develop chemist, 44-47, head chem develop lab, 47-54, spec asst to pharmaceut prod mgr, 54, mgr parenteral drug prod, 54-66, mgr chem prod dept, 66-68, prod adminr, Chem Prod, Mat Control & Filling & Packaging Depts, 68-74, TECH SERV ENGR, BRISTOL LABS, 74- Mem: Am Chem Soc. Res: Design and testing of explosives in air and water; penicillin; streptomycin; tetracycline; novobiocin; kanamycin; semisynthetic penicillin production. Mailing Add: Bristol Labs PO Box 657 Syracuse NY 13201

STAAB, ROBERT ALLAN, analytical chemistry, see 12th edition

STAAL, GERARDUS BENARDUS, b Assen, Neth, Aug 8, 25; m 57; c 2. ENTOMOLOGY, BIOLOGY. Educ: State Agr Univ, Wageningen, Ing, 57, PhD(insect physiol), 61. Prof Exp: Sr res officer, Neth Orgn Appl Sci Res, 61-68; DIR BIOL RES, ZOECON CORP, 68- Concurrent Pos: Neth Orgn Pure Res fel biol, Harvard Univ, 62-63. Mem: AAAS; Entom Soc Am; Am Inst Biol Scientists; Am Soc Zoologists; Acridological Asn. Res: Application of insect hormones and analogs for control of insects; mode of action of insect hormones; comparative insect endocrinology. Mailing Add: Zoecon Corp 975 California Ave Palo Alto CA 94304

STAATS, PERCY ANDERSON, b Belleville, WVa, Feb 20, 21; m 44; c 4. PHYSICAL CHEMISTRY, PHYSICS. Educ: Marietta Col, AB, 43; Univ Minn, MS, 49. Hon Degrees: DS, Fisk Univ, 74. Prof Exp: Instr physics, Marietta Col, 43; tech supvr, Tenn Eastman Corp, 43-46; chemist, Rohm and Haas Co, Pa, 49-52; CHEMIST, OAK RIDGE NAT LAB, 52- Concurrent Pos: Guest lectr & lab dir, Infrared Spectros Inst, Fisk Univ, 57-; traveling lectr, Oak Ridge Inst Nuclear Studies, 59-60, 62-63 & 65-66. Mem: AAAS; Am Chem Soc; Sigma Xi. Res: Molecular structure by infrared spectroscopy; infrared spectra of gases as solids at low temperatures; inorganic ions in solid solution; isotopes, especially tritium; gas lasers. Mailing Add: 119 Manchester Rd Oak Ridge TN 37830

STAATZ, MORTIMER HAY, b Kalispell, Mont, Oct 20, 18; m 52; c 3. ECONOMIC GEOLOGY. Educ: Calif Inst Technol, BS, 40; Northwestern Univ, MS, 42; Columbia Univ, PhD(geol), 52. Prof Exp: Asst geol, Northwestern Univ, 41-42; GEOLOGIST, US GEOL SURV, 42-44 & 46- Mem: Geol Soc Am; Soc Econ Geologists; Mineal Soc Am. Res: Pegmatites of Colorado and South Dakota; geology of eastern Great Basin and Washington; beryllium, fluorspar and phosphate deposits; vein-type uranium and thorium deposits. Mailing Add: US Geol Surv Denver Fed Ctr Bldg 25 Denver CO 80225

STABA, EMIL JOHN, b New York, NY, May 16, 28; m 54; c 5. PHARMACOGNOSY. Educ: St John's Univ, NY, BS, 52; Duquesne Univ, MS, 54; Univ Conn, PhD(pharmacog), 57. Prof Exp: Prof pharmacog & chmn dept, Univ Nebr, Lincoln, 57-68; PROF PHARMACOG & CHMN DEPT, COL PHARM, UNIV MINN, MINNEAPOLIS, 68-, ASST DEAN, 74- Concurrent Pos: Consult var pharmaceut & govt agencies; NSF sr foreign scientist, Poland, Hungary & Czech, 69; Fulbright-Hays res fel, Ger, 70; Coun Sci & Indust Res vis scientist, India, 73; partic, US-Repub China Coop Sci Prog, Plant Cell & Tissue Culture, 74. Honors & Awards: Lunsford-Richardson Award, 58. Mem: AAAS; Am Soc Pharmacog (pres, 71-72); Am Chem Soc; Am Pharmaceut Asn; Soc Exp Biol & Med. Res: Cultivation, extraction and tissue culture of medicinal plants; literature retrieval. Mailing Add: Dept of Pharmacog Univ of Minn Col Pharm Minneapolis MN 55455

STABENFELDT, GEORGE H, b Shelton, Wash, June 26, 30; m 53; c 4. PHYSIOLOGY, ENDOCRINOLOGY. Educ: Wash State Univ, BA, 55, DVM, 56, MS, 62; Okla State Univ, PhD, 68. Prof Exp: Pvt pract, Ore, 56-57, Idaho, 57-58 & Wash, 58-60; instr vet path, Wash State Univ, 60-62; asst prof physiol, Okla State Univ, 62-68; assoc prof physiol & assoc res physiologist, Nat Ctr Primate Biol, 68-75, PROF PHYSIOL, DEPT REPROD, SCH VET MED, UNIV CALIF, DAVIS, 75- Mem: Am Vet Med Asn; Am Soc Vet Physiol & Pharmacol; Am Physiol Soc; Soc Study Fertility; Soc Study Reproduction. Res: Endocrinology of female reproductive cycle, including the estrous and menstrual cycles, pregnancy and parturition. Mailing Add: Dept of Clin Sci Sch Vet Med Univ of Calif Davis CA 95616

STABLEFORD, LOUIS TRANTER, b Meriden, Conn, July 30, 14; m 41; c 3. BIOLOGY. Educ: Univ Va, BS, 37; Yale Univ, PhD(zool), 41. Prof Exp: Lab asst, Univ Va, 34-37 & Yale Univ, 37-40; from instr to prof biol, 41-72, DANA PROF BIOL, LAFAYETTE COL, 72-, CHMN DEPT, 58- Honors & Awards: Christian &

Mary Lindbach Found Award, 65. Mem: Am Soc Zoologists; Soc Develop Biol. Res: Experimental embryology; early amphibian development; aging of connective tissue. Mailing Add: Dept of Biol Lafayette Col Easton PA 18042

STABLER, ROBERT MILLER, b Washington, DC, Oct 1, 04; m 29; c 1. ZOOLOGY. Educ: Swarthmore Col, AB, 27; Johns Hopkins Univ, ScD(parasitol), 31. Prof Exp: From asst instr to asst prof zool, Univ Pa, 29-47; asst protozool, Sch Hyg & Pub Health, Johns Hopkins Univ, 30-31; from assoc prof to prof, 47-74, head dept, 51-70, EMER PROF BIOL, COLO COL, 74- Concurrent Pos: Mem exped, Panama, 31. Mem: AAAS; Am Soc Parasitol (treas, 44-58, vpres, 59); Am Soc Zool; Am Micros Soc; Am Soc Trop Med & Hyg. Res: Cytology, life history and host-parasite relations of parasitic Protozoa; bird and reptile ecology; general parasitology. Mailing Add: Dept of Zool Colo Col Colorado Springs CO 80903

STABLER, TIMOTHY ALLEN, b Port Jervis, NY, Sept 27, 40. DEVELOPMENTAL BIOLOGY, ENDOCRINOLOGY. Educ: Drew Univ, BA, 62; DePauw Univ, MA, 64; Univ Vt, PhD(zool), 69. Prof Exp: Asst prof biol, Hope Col, 69-71; ASST PROF BIOL & HEALTH PROFESSIONS, IND UNIV NORTHWEST, 73- Concurrent Pos: NIH fel, Univ Minn, 68-69; Sigma Xi grant-in-aid, 71-72; mem, NSF workshop develop biol, Univ Calif, San Diego, 71; NIH trainee reproductive endocrinol, Sch Med, Boston Univ, 71-73; adj asst prof physiol, Northwest Ctr Med Educ, 74- Mem: AAAS; Am Soc Zool; Am Inst Biol Sci; NY Acad Sci. Res: Steroid biochemistry; competitive protein-binding of steroids as a method of examining the interaction between the protein ligand and steroid hormone. Mailing Add: Dept of Biol Ind Univ Northwest Gary IN 46408

STABY, GEORGE LESTER, b Greenwich, Conn, Apr 1, 44; m 66; c 2. HORTICULTURE, PLANT PHYSIOLOGY. Educ: Univ Conn, BS, 66; Mich State Univ, MS, 67, PhD(hort), 70. Prof Exp: ASST PROF HORT, OHIO STATE UNIV, 70- Concurrent Pos: Hatch-US Govt grant, Ohio State Univ, 71- Mem: Am Soc Hort Sci; Int Soc Hort Sci; Am Soc Plant Physiol. Res: Plant growth regulators, sterols and other terpenes. Mailing Add: Dept of Hort Ohio State Univ Columbus OH 43210

STACE-SMITH, RICHARD, b Creston, BC, May 2, 24; m 51. PLANT PATHOLOGY. Educ: Univ BC, BSA, 50; Ore State Col, PhD(plant path), 54. Prof Exp: Asst plant pathologist, 50-54, assoc plant pathologist, 54-58, PLANT PATHOLOGIST, CAN DEPT AGR, 58- Mem: Can Phytopath Soc; Agr Inst Can. Res: Rubus virus diseases; virus purification and properties. Mailing Add: Can Agr Res Sta 6660 NW Marine Dr Vancouver BC Can

STACEY, FRANCIS WILFRED, b Louisbourg, NS, Nov 19, 25; nat US; m 48; c 2. ORGANIC CHEMISTRY. Educ: Acadia Univ, BS, 46; McMaster Univ, MS, 50; Univ Chicago, PhD(chem), 53. Prof Exp: Instr chem, Acadia Univ, 46-48; res chemist, Cent Res Dept, 53-54 & 56-62, col rels rep, Employee Rels Dept, 62-70, COL RELS SUPVR, EMPLOYEE RELS DEPT, E I DU PONT DE NEMOURS & CO, 70- Mem: Am Chem Soc. Res: Free radical chemistry; radiation chemistry. Mailing Add: 1904 Longcom Dr Wilmington DE 19810

STACEY, JOHN SYDNEY, b June 15, 27; US citizen; m 54; c 3. PHYSICS. Educ: Univ Durham, BSc, 51; Univ BC, MASc, 58, PhD(physics), 62. Prof Exp: Engr, Marconi Wireless Tel Co, 51-55; lectr elec eng, Univ BC, 57-58; PHYSICIST, ISOTOPE GEOL BR, US GEOL SURV, 62- Mem: Am Geophys Union; Am Soc Mass Spectrometry. Res: Mass spectrometry in geologic studies and related data processing techniques; lead isotope studies in ore genesis. Mailing Add: Isotope Geol Br US Geol Surv Fed Ctr Denver CO 80225

STACEY, LARRY MILTON, b Greensboro, NC, July 30, 40. PHYSICS. Educ: Univ NC, Chapel Hill, BS, 62, PhD(physics), 68. Prof Exp: Fel physics, Rutgers Univ, New Brunswick, 67-70; chem engr, Calif Inst Technol, 70-71; ASST PROF PHYSICS, ST LOUIS UNIV, 71- Mem: Am Phys Soc; Am Asn Physics Teachers. Res: Applications of magnetic resonance to the study of solids and fluids; critical point phenomena. Mailing Add: Dept of Physics St Louis Univ St Louis MO 63103

STACEY, WESTON MONROE, JR, b US. REACTOR PHYSICS, APPLIED PHYSICS. Educ: Ga Inst Technol, BS, 59, MS, 63; Mass Inst Technol, PhD(nuclear eng), 66. Prof Exp: Nuclear engr, Knolls Atomic Power Lab, 62-64, mgr reactor kinetics, 66-69; sect head reactor theory, 69-72, assoc dir appl physics div, 72-73, DIR CTR PROG, ARGONNE NAT LAB, 73- Mem: Fel Am Nuclear Soc; Am Phys Soc. Res: Nuclear reactor theory; fusion reactor technology; plasma physics. Mailing Add: Appl Physics Div Argonne Nat Lab Argonne IL 60439

STACK, JOHN D, b Los Angeles, Calif, July 24, 38; m 63; c 1. THEORETICAL PHYSICS. Educ: Calif Inst Technol, BSc, 59; Univ Calif, Berkeley, PhD(physics), 65. Prof Exp: Actg asst prof physics, Univ Calif, Berkeley, 65-66; asst prof, 66-70, ASSOC PROF PHYSICS, UNIV ILL, URBANA, 70- Concurrent Pos: Vis assoc,

Calif Inst Technol, 69-70. Mem: Am Phys Soc. Mailing Add: Dept of Physics Univ of Ill Urbana IL 61801

STACK, STEPHEN M, b Monahans, Tex, Feb 12, 43; m 65; c 2. CYTOLOGY. Educ: Univ Tex, Austin, BAS, 65, PhD(cytol, bot), 69. Prof Exp: Asst prof, 69-74, ASSOC PROF BOT & PLANT PATH, COLO STATE UNIV, 74- Mem: AAAS; Bot Soc Am. Res: Somatic and premeiotic homologous chromosome pairing and development of the synaptinemal complex. Mailing Add: Dept of Bot & Plant Path Colo State Univ Ft Collins CO 80521

STACKELBERG, OLAF PATRICK, b Munich, Ger, Aug 2, 32; US citizen; m 54; c 3. MATHEMATICS. Educ: Mass Inst Technol, BS, 55; Univ Minn, MS, 60, PhD(math), 63. Prof Exp: Teaching asst math, Univ Minn, 58-63; asst prof, 63-68, ASSOC PROF MATH, DUKE UNIV, 68- Concurrent Pos: Alexander V Humboldt fel, Stuttgart Tech Univ, 65-66; vis assoc prof, Univ Ill, Urbana, 69-70; ed, Duke Math J, 71-74; vis assoc prof, Univ London, 74. Mem: Am Math Soc; Math Asn Am; Inst Math Statist. Res: Probability; metric number theory. Mailing Add: Dept of Math Duke Univ Durham NC 27706

STACKMAN, ROBERT W, b Dayton, Ohio, June 29, 35; m 61; c 3. ORGANIC CHEMISTRY, POLYMER CHEMISTRY. Educ: Univ Dayton, BS, 57; Univ Fla, PhD(org chem), 61. Prof Exp: Res asst cyclopolymers of silanes, Univ Fla, 58-61; res chemist, Summit Labs, 61-65, sr res chemist, Celanese Res Co, 65-72, res assoc, 72-74, RES SUPVR POLYMER FLAMMABILITY RES, CELANESE RES CO, CELANESE CORP, 74- Mem: AAAS; Am Chem Soc. Res: Condensation polymerization; emulsion polymerization; high temperature polymers; cyclopolymerization; organic synthesis; organosilicon compounds; addition polymerization; polymer modification and stabilization; flammability of polymers. Mailing Add: Celanese Res Co Box 1000 Summit NJ 07901

STACKPOLE, JOHN DUKE, b Boston, Mass, Dec 28, 35; m 60; c 3. METEOROLOGY. Educ: Amherst Col, BA, 57; Mass Inst Technol, MS, 59, PhD(meteorol), 64. Prof Exp: Res meteorologist, 64-73, SUPVRY RES METEOROLOGIST, NAT WEATHER SERV, NAT OCEANIC & ATMOSPHERIC ADMIN, 73- Res: Numerical weather prediction. Mailing Add: 7305 Kipling Pkwy District Heights MD 20028

STACY, CARL J, b Joplin, Mo, Jan 20, 29; m 51; c 3. POLYMER SCIENCE. Educ: Kans State Col Pittsburg, BA, 51; Purdue Univ, PhD(phys chem), 56. Prof Exp: Res fel starch, Purdue Univ, 55-56; res physicist, 56, RES PHYSICAL CHEMIST, PHILLIPS PETROL CO, 56- Mem: AAAS; Am Chem Soc; Sigma Xi. Res: Polymer and carbon black structure by light scattering, ultracentrifuge and other physical techniques. Mailing Add: 2929 Sheridan Rd Bartlesville OK 74003

STACY, EDNEY WEBB, mathematical statistics, see 12th edition

STACY, GARDNER WESLEY, b Rochester, NY, Oct 29, 21; m 67; c 5. ORGANIC CHEMISTRY. Educ: Univ Rochester, BS, 43; Univ Ill, PhD(org chem), 46. Prof Exp: Asst, Off Sci Res & Develop Proj, Ill, 43-46; fel biochem, Med Col, Cornell Univ, 46-48; from asst prof to assoc prof, 48-60, PROF CHEM, WASH STATE UNIV, 60- Concurrent Pos: Dir region VI, Am Chem Soc, 70-; chmn comt on educ & students, Am Chem Soc, 72- Honors & Awards: PRF Int Award, Australia & New Zealand, 63-64. Mem: AAAS; Am Chem Soc. Res: Sulfur-containing heterocyclic tautomeric systems and ring-chain tautomerism; prospective antimalarials. Mailing Add: Dept of Chem Wash State Univ Pullman WA 99163

STACY, PETER MICHAEL, physical chemistry, see 12th edition

STACY, RALPH WINSTON, b Middletown, Ohio, Feb 6, 20; m 43; c 2. PHYSIOLOGY, BIOENGINEERING. Educ: Miami Univ, BScE, 42; Ohio State Univ, MSc, 47, PhD(physiol), 48. Prof Exp: Instr physiol, Ohio State Univ, 47-48; asst prof, Univ Ky, 48-49; from asst prof to prof, Ohio State Univ, 49-62; prof biomath, NC State Univ, 62-65; prof bioeng & biomath, Univ NC, Chapel Hill, 65-69; scientist, Cox Heart Inst, 69-71; PROF PHYSIOL, SOUTHERN ILL UNIV, CARBONDALE, 71- Mem: Biophys Soc; Am Physiol Soc; Inst Elec & Electronics Engrs; Asn Comput Mach. Res: Physics of circulation and respiration; hemodynamics; biomedical uses of computers; medical electronics and instrumentation; physiological effects of environmental pollutants. Mailing Add: Dept of Physiol Southern Ill Univ Carbondale IL 62901

STACY, RICHARD D, organic chemistry, see 12th edition

STADEL, CHRISTOPH, b Donaueschingen, Ger, June 6, 38; m 67; c 3. URBAN GEOGRAPHY, POLITICAL GEOGRAPHY. Educ: Univ Freiburg, Politikum, 61; Univ Fribourg, DR(geog), 64. Prof Exp: Prof geog & Ger Inst Le Rosey, Switz, 64-67; teacher geog, Fr & Ger, Hillfield Col Ont, 67-68; asst prof, 68-74, ASSOC PROF GEOG, BRANDON UNIV, 74- Concurrent Pos: Guest prof, Salzburg Univ, Austria & res fel, Can Coun, 74-75. Mem: Am Asn Geog; Can Asn Latin Am Studies; Austrian Geog Soc; Swiss Geog Soc. Res: Human ecology of high mountain environments; urbanization in Columbia; settlement hierarchies in Manitoba; changing landscape of the Alps; squatter settlements. Mailing Add: Dept of Geog Brandon Univ Brandon MB Can

STADELBACHER, EARL A, entomology, see 12th edition

STADELBACHER, GLENN JOSEPH, horticulture, see 12th edition

STADELMAN, WILLIAM JACOB, b Vancouver, Wash, Aug 8, 17; m 42; c 2. FOOD SCIENCE. Educ: Wash State Univ, BS, 40; Pa State Univ, MS, 42, PhD(biochem), 48. Prof Exp: Asst poultry husb, Pa State Univ, 40-42; asst prof, Wash State Univ, 48-52, assoc prof, 52-55; from assoc prof to prof poultry sci, 55-62, PROF FOOD SCI, PURDUE UNIV, LAFAYETTE, 62- Concurrent Pos: Consult, Food Indust; mem, Tech Adv Comt, Poultry & Egg Inst Am, 57-; mem bd dirs, Res & Develop Assocs, Food & Container Inst, 66-69, 72-75; mem, Sci Adv Comt, Refrig Res Found, 67-; mem adv bd mil personnel supplies, Food Irradiation Comt, Nat Acad Sci, 67-69; mem, Tech Adv Comt, Nat Turkey Fedn, 71-, Tech Adv Comt, Am Egg Bd, 74- Honors & Awards: Christie Award, Poultry & Egg Nat Bd, 55; Res Award, Am Egg Bd, 75. Mem: Inst Food Technol (pres, 74-75); Am Poultry Sci Asn; Am Soc Heat, Refrig & Air-Conditioning Eng; World Poultry Sci Asn; Am Meat Sci Asn. Res: Effects of refrigeration and freezing on quality preservation of protein rich foods; poultry products quality evaluation and preservation; new product development. Mailing Add: Dept of Animal Sci Poultry Bldg Purdue Univ West Lafayette IN 47906

STADELMANN, EDUARD JOSEPH, b Graz, Austria, Sept 24, 20. PLANT PHYSIOLOGY, CELL PHYSIOLOGY. Educ: Innsbruck Univ, PhD(bot, philos), 53; Univ Freiburg, Venia Legendi, 57. Prof Exp: Asst bot, Freiburg Univ, 54-61, privat docent, 57, sr asst, 62-64; from asst prof to assoc prof, 64-72, PROF PLANT PHYSIOL, UNIV MINN, ST PAUL, 72- Concurrent Pos: Muellhaupt scholar biol, Ohio State Univ, 58-59; res assoc, Univ Minn, 63-64; US special prog, Humboldt Found, Bonn, Ger, 73-75. Mem: Ger Bot Soc; Swiss Bot Soc; Swiss Soc Natural Sci; Austrian Zool-Bot Soc; Am Inst Biol Sci. Res: Permeability; cytomorphology; salt resistance; protoplasmatology; radiation effects. Mailing Add: Dept of Hort Sci Univ of Minn St Paul MN 55108

STADLER, DAVID ROSS, b Columbia, Mo, May 24, 25; m 52; c 4. GENETICS. Educ: Univ Mo, AB, 48; Princeton Univ, MA, 50, PhD, 52. Prof Exp: Instr biol, Univ Rochester, 52; Gosney res fel genetics, Calif Inst Technol, 52-53, USPHS fel, 53-55; instr bot, 56-57, asst prof, 57-59, from asst prof to assoc prof genetics, 59-67, PROF GENETICS, UNIV WASH, 67- Concurrent Pos: Ed, Genetics, Genetics Soc Am, 73- Mem: Genetics Soc Am (treas, 69-71). Res: Genetics of microorganisms; mutation, recombination and development. Mailing Add: Dept of Genetics Univ of Wash Seattle WA 98195

STADLER, LOUIS BENJAMIN, b Monroe, Mich, Feb 26, 26; m 51; c 3. PHARMACEUTICAL CHEMISTRY, ANALYTICAL CHEMISTRY. Educ: Univ Mich, BS, 48, MS, 50, PhD(pharmaceut chem), 54. Prof Exp: Sr anal chemist, Parke, Davis & Co, 53-63, mgr anal stand, 63-64; asst head qual control, William S Merrell Co Div, Richardson-Merrell Inc, 64-65, head qual control, 66-71, Merrell Nat Labs, 71-73, Master Documents Admin, 73-75, MGR QO RECORDS, SYSTS & PLANNING, MERRELL-NAT LABS, 75- Concurrent Pos: Mem adv panel steroids, Nat Formulary, 60-65, comt specifications, 66-75, panel trypsin & chymotrypsin, 70-75; rev comt, US Pharmacopoeia, 70- Mem: Am Pharmaceut Asn; Am Chem Soc; Am Soc Qual Control. Res: Analytical methodology for testing drug substances, pharmaceutical dosage forms and associated standards; improved control techniques for pharmaceuticals; technical management. Mailing Add: Merrell-Nat Labs 110 E Amity Rd Cincinnati OH 45215

STADTHERR, LEON GREGORY, b New Ulm, Minn, Nov 27, 42. BIOINORGANIC CHEMISTRY. Educ: St John's Univ, Minn, BS, 65; Univ NDak, PhD(chem), 70. Prof Exp: Res assoc chem, Univ Va, 70-72; RES ASSOC CHEM, IOWA STATE UNIV, 73- Mem: Am Chem Soc; Sigma Xi. Res: Coordination compounds; stereochemistry of transition metal ion and lanthanide ion complexes; ion exchange. Mailing Add: Dept of Chem Iowa State Univ Ames IA 50011

STADTHERR, RICHARD JAMES, b Gibbon, Minn, Nov 24, 19. PLANT PHYSIOLOGY, PLANT BREEDING. Educ: Univ Minn, BS, 49, MS, 51, PhD(hort), 63. Prof Exp: Teaching asst, Univ Minn, 48-51, asst prof hort, Exten, 53-54, instr in charge of res turf & nursery crops, 54-61; res asst, Cornell Univ, 51-52; asst prof nursery crops res, Univ Mass, 52-53; assoc prof hort, Exten, NC State Univ, 63-67; assoc prof, 67-73, PROF HORT, LA STATE UNIV, BATON ROUGE, 73- Mem: Am Soc Hort Sci; Int Plant Propagators Soc; Am Forestry Asn; Am Hort Soc; Am Rose Soc. Res: Growth substances; plant propagation; testing-breeding of azaleas; groundcovers and woody ornamental trials. Mailing Add: Dept of Hort La State Univ Baton Rouge LA 70803

STADTMAN, EARL REECE, b Carrizozo, NMex, Nov 15, 19; m 43. BIOCHEMISTRY. Educ: Univ Calif, PhD(comp biochem), 49. Prof Exp: Asst food tech, Univ Calif, 43-46, sr lab technician, 48-49; AEC res fel, Mass Gen Hosp, 49-50; chemist, 50-57, chief sect enzymes, 57-62, CHIEF LAB BIOCHEM, NAT HEART & LUNG INST, 62- Concurrent Pos: Kresge-Hooker lectr, Wayne State Univ, 54; lectr, US USDA Grad Sch, 54-, Georgetown Univ, 56-58 & Univ Md, 59-; vis scientist, Max Planck Inst, Ger, 59-60 & Pasteur Inst, France, 60; ed, J Biol Chem, 60-; exec ed, Archives Biochem, Biophys, 60-; mem adv comt, Oak Ridge Nat Lab, 63-66; chmn biochem div, Found Advan Educ in Sci, 64-; NIH lectr, 66; lectr, Univ Ill, 66 & Univ Pisa, 66; mem coun, Am Soc Biol Chem, 74-77; del, US Nat Comt for Int Union of Biochem, 75-81. Honors & Awards: Lewis Award, Am Chem Soc, 53; Microbiol Award, Nat Acad Sci, 70. Mem: Am Chem Soc; Am Soc Biol Chem; Am Soc Microbiol; Am Acad Arts & Sci. Res: Microbial and intermediary metabolism; enzyme chemistry; biochemical function of vitamin B12 and ferredoxin; metabolic regulation of biosynthetic of biosynthetic pathways; membrane transport. Mailing Add: Lab of Biochem Nat Heart & Lung Inst NIH Bethesda MD 20014

STADTMAN, THRESSA CAMPBELL, b Sterling, NY, Feb 12, 20; m 43. BIOCHEMISTRY, ENZYMOLOGY. Educ: Cornell Univ, BS, 40, MS, 42; Univ Calif, PhD(microbiol), 49. Prof Exp: Asst nutrit, Agr Exp Sta, Cornell Univ, 42-43; res assoc food microbiol, Univ Calif, 43-46; asst, Harvard Med Sch, 49-50; BIOCHEMIST, NAT HEART INST, 50- Concurrent Pos: Whitney fel, Oxford Univ, 54-55; Rockefeller grant, Inst Cell Chem, Univ Munich, 59-60; French Govt fel, Inst Biol & Phys Chem, France, 60. Mem: Am Soc Microbiol; Am Soc Biol Chemists; Brit Biochem Soc. Res: Amino acid intermediary metabolism; one-carbon metabolism; methane formation; microbial biochemistry. Mailing Add: Nat Heart Inst Bethesda MD 20014

STAEBLER, ARTHUR E, b Detroit, Mich, May 3, 15; m 40; c 4. ORNITHOLOGY. Educ: Univ Mich, BS, 38, MS, 40, PhD(ornith), 48. Prof Exp: Asst prof conserv, Mich State Univ, 47-55; from asst prof to assoc prof, 55-66, PROF BIOL, CALIF STATE UNIV, FRESNO, 66- Mem: Am Ornith Union; Cooper Ornith Soc; Wilson Ornith Soc. Res: Bird migration; winter flocking behavior. Mailing Add: Dept of Biol Calif State Univ Fresno CA 93740

STAEBLER, DAVID LLOYD, b Ann Arbor, Mich, Apr 25, 40; m 61; c 2. OPTICS, SOLID STATE PHYSICS. Educ: Pa State Univ, BS, 62, MS, 63; Princeton Univ, MA, 67, PhD(elec eng), 70. Prof Exp: MEM TECH STAFF, RCA LABS, 63- Honors & Awards: Achievement Award, RCA Labs, 68, 72. Mem: Inst Elec & Electronics Eng; AAAS. Res: Photochromic and electrochromic phenomena; hologram storage in electro-optic materials; optical and electronic properties of dielectric materials. Mailing Add: RCA Labs Princeton NJ 08540

STAEBLER, GEORGE RUSSELL, forestry, silviculture, see 12th edition

STAEHELIN, LUCAS ANDREW, b Sydney, Australia, Feb 10, 39; m 65; c 3. CELL BIOLOGY, ELECTRON MICROSCOPY. Educ: Swiss Fed Inst Technol, DiplNatw, 63, PhD(biol), 66. Prof Exp: Res scientist, Dept Sci & Indust Res, NZ, 66-69; res fel cell biol, Harvard Univ, 69-70; asst prof, 70-73, assoc chmn dept molecular cell & develop biol, 72-73, ASSOC PROF CELL BIOL, UNIV COLO, BOULDER, 73- Concurrent Pos: Nat Inst Gen Med Sci grant, 71-76; consult, Nat Cell Biol Study Prog & mem, Ad Hoc Study Sect Pathol of Membranes, NIH, 74-75. Mem: AAAS; Am Soc Cell Biol; Am Soc Photobiology. Res: Structure and function of biological membranes; freeze-etch electron microscopy; photosynthesis. Mailing Add: Dept Molecular Cell Develop Biol Univ of Colo Boulder CO 80302

STAELIN, DAVID HUDSON, b Toledo, Ohio, May 25, 38; m 62; c 3. RADIO ASTRONOMY, METEOROLOGY. Educ: Mass Inst Technol, SB, 60, SM, 61, ScD(elec eng), 65. Prof Exp: From instr to asst prof, 65-69, ASSOC PROF ELEC ENG, MASS INST TECHNOL, 69- Concurrent Pos: Ford fel eng, 65-67; vis asst scientist, Nat Radio Astron Observ, 68-69; dir, Environ Res & Technol, Inc, 69-

Mem: AAAS; Am Astron Soc; Am Geophys Union; Inst Elec & Electronics Eng; Am Meteorol Soc. Res: Planetary atmospheres; pulsars; space-based and ground-based meteorological observations using passive microwave techniques; microwave and optical instrumentation; atmospheric sensing. Mailing Add: Dept of Elec Eng & Comput Sci Mass Inst Technol Cambridge MA 02139

STAETZ, CHARLES ALAN, b North Platte, Nebr, July 12, 45; m 68; c 2. ECONOMIC ENTOMOLOGY. Educ: Chadron State Col, BS, 67; Univ Nebr, MS, 72, PhD(entom), 75. Prof Exp: ENTOMOLOGIST, VELSICOL CHEM CORP, 75- Mem: Entom Soc Am. Res: Screening compounds for potential insecticidal activity; establishment and maintenance of several insect cultures; laboratory and field evaluations of candidate compounds. Mailing Add: Velsicol Agr Res Ctr PO Box 507 Woodstock IL 60098

STAFF, CHARLES HUBERT, b Barry, Ill, July 24, 32; m 59; c 2. FOOD SCIENCE, ORGANIC CHEMISTRY. Educ: Culver-Stockton Col, BA, 54; Southern Ill Univ, MS, 59. Prof Exp: Res chemist, Corn Prod Co, 59-63; sr res technologist, Pet Inc, 63-68; tech mgr, Pillsbury Co, 68-70; DIR RES, FAIRMONT FOODS CO, 70- Mem: Inst Food Technol; Am Asn Cereal Chem; Am Soc Quality Control. Res: Formulation, process and quality control of food products, particularly those foods produced from cereal, dairy and meat raw materials. Mailing Add: 10018 Rockbrook Rd Omaha NE 68124

STAFFELDT, EUGENE EDWARD, b Naperville, Ill, Feb 11, 26; m 44; c 2. PLANT PATHOLOGY. Educ: Northern Ill State Teachers Col, BS, 49; Iowa State Col, MS, 51, PhD(plant path), 53. Prof Exp: Asst, Iowa State Col, 49-53; plant pathologist, NMex State Univ, 53-55 & Int Coop Admin, Univ Sind, Pakistan, 55-57; from asst prof to assoc prof, 57-65, PROF BIOL, NMEX STATE UNIV, 65- Concurrent Pos: Consult, White Sands Missile Range & Jet Propulsion Lab, Calif Inst Technol. Mem: AAAS; Soc Indust Microbiol. Res: Soil microbiology, especially soil fungi; disease control; fungus metabolism. Mailing Add: Dept of Biol NMex State Univ Las Cruces NM 88003

STAFFIN, GERALD DAVID, organic chemistry, see 12th edition

STAFFORD, DARREL WAYNE, b Parsons, Kans, Mar 11, 35; m 57; c 3. ZOOLOGY, MOLECULAR BIOLOGY. Educ: Southwest Mo State Col, BA, 59; Univ Miami, Fla, PhD(cellular physiol), 64. Prof Exp: NIH fel, Albert Einstein Col Med, 64-65; asst prof, 65-70, ASSOC PROF ZOOL & BIOCHEM, UNIV N C, CHAPEL HILL, 70- Mem: AAAS. Res: Cell division and protein synthesis. Mailing Add: Dept of Zool 422 Wilson Hall Univ of NC Chapel Hill NC 27514

STAFFORD, EUGENE MARSHALL, entomology, see 12th edition

STAFFORD, FRED E, b New York, NY, Mar 31, 35; m 63. PHYSICAL INORGANIC CHEMISTRY. Educ: Cornell Univ, AB, 56; Univ Calif, Berkeley, PhD(chem), 59. Prof Exp: NSF fel, Free Univ Brussels, 59-61; from asst prof to assoc prof chem, Northwestern Univ, 61-74; prog officer sci develop progs, 74-75, PROG DIR SOLID STATE CHEMISTRY, NSF, 75- Concurrent Pos: Nat Res Coun Can-Nat Acad Sci Comt on High Temp Sci & Technol, 75- Mem: Am Chem Soc; Am Phys Soc; Am Soc Mass Spectrometry. Res: Mass spectrometry and spectroscopy of simple inorganic systems; high temperature chemistry; solid state chemistry. Mailing Add: Nat Sci Found Washington DC 20550

STAFFORD, GEORGE EWING, b Walnut, Kans, July 8, 06; m 34; c 2. PEDIATRICS, PUBLIC HEALTH. Educ: Univ Kans, AB, 28, BS, 30, MD, 32; Am Bd Pediat, dipl, 43. Prof Exp: Intern, Touro Infirmary, New Orleans, La, 32-33; resident pediat, Children's Mercy Hosp, Kansas City, Mo, 33-34; PROF CLIN PEDIAT, COL MED, UNIV NEBR, OMAHA, 54- Concurrent Pos: Attend physician, Lincoln Gen & Bryan Mem Hosps, 45-; consult allergy, Vet Admin Hosp, Lincoln. Mem: AMA; Am Acad Pediat; fel Am Col Allergists. Res: Allergy. Mailing Add: 1701 S 17th St Lincoln NE 68502

STAFFORD, HELEN ADELE, b Philadelphia, Pa, Oct 9, 22. PLANT PHYSIOLOGY. Educ: Wellesley Col, BA, 44; Conn Col, MA, 48; Univ Pa, PhD, 51. Prof Exp: Instr bot & res assoc biochem, Univ Chicago, 51-54; from asst prof to assoc prof, 54-65, PROF BOT, REED COL, 65- Concurrent Pos: Guggenheim fel, Harvard Univ, 58-59; NSF sr fel, Univ Calif, Los Angeles, 63-64. Mem: Bot Soc Am; Am Soc Plant Physiol; Am Soc Biol Chem; Brit Soc Exp Biol; Scandinavian Soc Plant Physiol. Res: Plant biochemistry; metabolism of phenolic compounds; regulation of and metabolism of phenolic compounds in higher plants. Mailing Add: Dept of Biol Reed Col Portland OR 97202

STAFFORD, HOWARD A, b West Chester, Pa, Sept 27, 33; c 2. ECONOMIC GEOGRAPHY. Educ: West Chester State Col, BS, 55; Univ Iowa, MA, 57, PhD(geog), 60. Prof Exp: Asst prof geog, Southern Ill Univ, 59-64; vis prof, Univ Wis, 64-65; assoc prof, 65-69, PROF GEOG, UNIV CINCINNATI, 69- Concurrent Pos: Grant, Southern Ill Univ, Carbondale, 61-62; Asn Am Geogr & NSF grant, Univ Cincinnati, 65-66; exchange lectr geog, Univ Wales, 66-67; Appalachian Regional Comn grant, Develop Dept, State of Ohio, 71-72. Res: Industrial location decision making; implications for regional economic development strategies. Mailing Add: Dept of Geog Univ of Cincinnati Cincinnati OH 45221

STAFFORD, OWEN LEROY, polymer chemistry, see 12th edition

STAFFORD, ROBERT OPPEN, b Milwaukee, Wis, Jan 28, 20; m 61; c 3. ENDOCRINOLOGY. Educ: Univ Wis, BA, 41, MA, 48, PhD(zool), 49. Prof Exp: Res scientist, 49-60, asst dir biol res, 60-62, biochem res, 62-68, asst to exec vpres pharmaceut div, 68-71, VPRES CORP PLANNING, UPJOHN CO, 71- Res: Pharmacology, virology, pathology; metabolic diseases; management of research. Mailing Add: Upjohn Co 7000 Portage Rd Kalamazoo MI 49001

STAFFORD, ROY ELMER, b Republic, Kans, Dec 28, 30; m 59; c 5. PLANT BREEDING. Educ: Kans State Univ, BS, 53; Univ Minn, MS, 62, PhD(genetics), 64. Prof Exp: Instr high sch, Nebr, 56-58; res asst, Univ Minn, 59-65; plant breeder, Am Crystal Sugar Co, Colo, 65-68; RES GENETICIST & PROJ LEADER, TEX A&M UNIV RES CTR, AGR RES SERV, USDA, 68- Concurrent Pos: Flax Develop Comt res fel; mem, Southern Regional Task Force New Crops & Minor Oilseeds, 74- Mem: Am Soc Agron; Crop Sci Soc Am; Am Soc Sugar Beet Technol. Res: Development of improved varieties of guar and sunflowers. Mailing Add: Tex A&M Univ Res Ctr USDA Box 1658 Vernon TX 76384

STAFFORD, STEWARD LAYNE, inorganic chemistry, see 12th edition

STAFIEJ, STANLEY FRANK, organic chemistry, see 12th edition

STAFSUDD, OSCAR M, JR, b Allison Park, Pa, Nov 10, 36; m 67; c 1. SOLID STATE SPECTROSCOPY. Educ: Univ Calif, Los Angeles, BA, 59, MS, 62, PhD(physics), 67. Prof Exp: Physicist, Atomics Int Div, NAm Aviation, Inc, 60-64 & Hughes Res Labs, 64-67; asst prof eng, 67-72, ASSOC PROF ENG, UNIV CALIF, LOS ANGELES, 72-, ASSOC PROF APPL SCI, 74- Concurrent Pos: Consult, Hughes Res Labs, 67- Mem: Am Phys Soc; Optical Soc Am; Am Soc Eng Educ. Res: Laser technology; crystal growth; solid state electronics. Mailing Add: Dept of Eng Univ of Calif Los Angeles CA 90024

STAGAT, ROBERT WALTER, theoretical physics, see 12th edition

STAGE, ALBERT RANDALL, forest management, statistics, see 12th edition

STAGE, GERALD IRVING, b Palo Alto, Calif, June 26, 35; m 58; c 2. SYSTEMATIC ENTOMOLOGY, ECOLOGY. Educ: Univ Calif, Berkeley, BS, 57, PhD(entom), 66. Prof Exp: Asst specialist entom, Exp Sta, Univ Calif, Berkeley, 64-66, asst res entomologist, 66; asst cur entom, Nat Mus Natural Hist, 66-70; ASST PROF ENTOM, UNIV CONN, 70- Concurrent Pos: Smithsonian Res Found grant, Smithsonian Inst, 67-70; Univ Conn Res Found grant, Univ Conn, 71-72. Mem: Soc Syst Zool (secy, 69-71); Entom Soc Am; Am Inst Biol Sci. Res: Biology and systematics of bees, especially melittids; pollination. Mailing Add: Biol Sci Group U-43 Univ of Conn Storrs CT 06268

STAGEMAN, PAUL JEROME, b Persia, Iowa, June 21, 16; m 37; c 1. BIOCHEMISTRY. Educ: Univ Omaha, AB, 39; Univ Iowa, MS, 50; Univ Nebr, PhD, 63. Prof Exp: Res chemist, Cudahy Packing Co, 39-41; from asst prof to assoc prof, 41-65, PROF CHEM, UNIV NEBR, OMAHA, 65- Mem: AAAS; Am Chem Soc; Am Inst Chem. Res: Ultracentrifugation; lipoproteins; atherosclerosis; plant pigments. Mailing Add: Dept of Chem Univ of Nebr 60th & Dodge Sts Omaha NE 68101

STAGER, CARL VINTON, b Kitchener, Ont, June 10, 35; m 62; c 4. SOLID STATE PHYSICS. Educ: McMaster Univ, BSc, 58; Mass Inst Technol, PhD(physics), 61. Prof Exp: Mem res staff, Francis Bitter Nat Magnet Lab, Mass Inst Technol, 60-63; from asst prof to assoc prof, 63-72, PROF PHYSICS, McMASTER UNIV, 72- Mem: Can Asn Physicists (treas, 64-68); Am Phys Soc. Res: Magnetism of insulating crystals; crystal fields spectra; electron paramagnetic resonance. Mailing Add: Dept of Physics McMaster Univ Hamilton ON Can

STAGER, HAROLD KEITH, b Gardena, Calif, Dec 5, 21; m 49; c 4. GEOLOGY. Educ: Univ Calif, Los Angeles, BA, 48. Prof Exp: Asst assayer, Golden Queen Mining Co, 41; geologist, Mineral Deposits Br, 48-63, for geol br, 63-64, base metals br, 64-65, FIELD OFFICER, OFF MINERALS EXPLOR, US GEOL SURV, 65- Mem: Geol Soc Am; Soc Econ Geol; Soc Mining Eng. Res: Mining geology; mineral deposits; strategic and rare metals. Mailing Add: US Geol Surv Menlo Park CA 94025

STAGER, KENNETH EARL, b Union City, Pa, Jan 28, 15; m 58; c 2. ZOOLOGY. Educ: Univ Calif, Los Angeles, AB, 40; Univ Southern Calif, MS, 53, PhD, 62. Prof Exp: Asst cur ornith & mammal, Los Angeles County Mus Natural Hist, 41-42, cur, 46-61, sr cur, 61-76; RETIRED. Concurrent Pos: Adj prof, Univ Southern Calif, 62-; consult, US Dept Interior, Geol Surv, 74- Honors & Awards: US Typhus Comn Medal; Pub Serv Award, US Dept Interior, 72; Commendation Award, Calif Dept Fish & Game, 73. Mem: Am Soc Mammal; Cooper Ornith Soc; Am Ornith Union. Res: Ornithology; behavior; ecology; mammalogy; distribution and taxonomical problems. Mailing Add: LA County Mus Natural Hist 900 Exposition Blvd Los Angeles CA 90007

STAGG, RONALD M, b Brooklyn, NY; m 52; c 5. PHYSIOLOGY, ENDOCRINOLOGY. Educ: Tusculum Col, BA, 50; Brooklyn Col, MA, 55; Rutgers Univ, PhD(endocrinol), 62. Prof Exp: Asst adminr, Willard F Greenwald, Med & Chem Consult, NY, 51-53; asst to med dir admin, Warner-Chilcott Lab Div, Warner-Lambert Pharmaceut Co, NJ, 53-56; instr zool, Drew Univ, 59-60; instr physiol, Med Col Va, 60-65; assoc prof, 65-74, PROF BIOL, HARTWICK COL, 74- Concurrent Pos: NIH grant, 63-65. Mem: Am Soc Zool. Res: Physiology, specifically mammalian; endocrinology, specifically the relationship between hormones and nutrition. Mailing Add: Dept of Biol Hartwick Col Oneonta NY 13820

STAGG, WILLIAM RAY, b Lexington, Ky, Sept 15, 37; m 62; c 2. PHYSICAL INORGANIC CHEMISTRY. Educ: Univ Ky, BS, 59; Iowa State Univ, PhD(chem), 63. Prof Exp: Res chemist, FMC Corp, NJ, 63-64; res assoc chem, Univ Ill, 66-67; asst prof, Colgate Univ, 67-71; ASSOC PROF CHEM, RANDOLPH-MACON WOMAN'S COL, 71- Mem: Am Chem Soc; The Chem Soc. Res: Complex equilibria of lanthanide elements; heteroatom ring systems of sulfur, nitrogen and phosphorous; environmental chemistry. Mailing Add: Dept of Chem Randolph-Macon Woman's Col Lynchburg VA 24504

STAHL, BARBARA JAFFE, b Brooklyn, NY, Apr 17, 30; m 51; c 4. COMPARATIVE ANATOMY, EVOLUTION. Educ: Wellesley Col, BA, 52; Radcliffe Col, AM, 53; Harvard Univ, PhD(biol), 65. Prof Exp: PROF BIOL, ST ANSELM'S COL, 54- Mem: AAAS; Soc Vert Paleont. Res: Evolution of holocephali and early vertebrates. Mailing Add: Dept of Biol St Anselm's Col Manchester NH 03102

STAHL, CHARLES JAY, III, b Philadelphia, Pa, Aug 5, 30; m 54; c 3. PATHOLOGY. Educ: Ursinus Col, BS, 52; Jefferson Med Col, MD, 56; Am Bd Path, dipl & cert anat & clin path, 61 & forensic path, 64. Prof Exp: From rotating gen intern to resident path, US Naval Hosp, Philadelphia, 56-61, asst pathologist, 61-62; resident forensic path, Armed Forces Inst Path, DC, 62-63; chief lab serv, US Naval Hosp, Guam, 63-65; chief forensic path br, Armed Forces Inst Path, 65-70, from asst chief to chief mil environ path div, 65-72, chief marine biopath br, 67-72, chief forensic sci div, 72-74, supvr residency training prog forensic path & registr forensic path, 65-75, chmn dept forensic sci, 74-75; CHMN, DEPT LAB MED, NAT NAVAL MED CTR, BETHESDA, MD, 75- & SUPVR RESIDENCY TRAINING PROG PATH, 75- Concurrent Pos: Dep med examr, Govt Guam, 63-65; consult, Prof Div, Bur Med & Surg, Dept Navy, DC, 70-, Guam Mem Hosp, Agana, 65 & Am Bd Path, 72-; prof lectr, George Washington Univ, 72-; assoc ed, J Forensic Sci, 71-72, ed, 72-74; clin course dir, Lab Schs, Naval Health Sci Educ & Training Command, Bethesda, 75- Mem: AMA; Col Am Path; Am Soc Clin Path; Am Acad Forensic Sci; Am Col Legal Med. Res: Drowning and scuba diving injuries; injuries by teargas pen guns; identification of unknown human remains; pathologic findings in poisoning; wound ballistics; toxic hazards in closed environments. Mailing Add: Nat Naval Med Ctr Bethesda MD 20014

STAHL, FRANKLIN WILLIAM, b Boston, Mass, Oct 8, 29; m 55; c 2. GENETICS. Educ: Harvard Univ, AB, 51; Univ Rochester, PhD(biol), 56. Hon Degrees: DSc, Oakland Univ. Prof Exp: Res fel biol, Calif Inst Technol, 55-58; assoc prof zool, Univ Mo, 58-59; assoc prof, 59-70, PROF BIOL, UNIV ORE, 70-, RES ASSOC, INST MOLECULAR BIOL, 59- Mem: Nat Acad Sci. Res: Genetics of bacteriophage. Mailing Add: Inst of Molecular Biol Univ of Ore Eugene OR 97403

STAHL, FRIEDA AXELROD, b Brooklyn, NY, May 27, 22; m 42; c 2. SOLID

STATE PHYSICS. Educ: Hunter Col, BA, 42; Hofstra Col, MA, 57; Claremont Grad Sch, PhD(educ), 69. Prof Exp: Jr physicist, US Army Signal Corps, NJ & Ala, 42-44 & Petty Labs, Petty Geophys Eng Co, Tex, 44-46; physicist, Hillyer Instrument Corp, NY, 46-48; sr physicist, Sylvania Res Labs, NY, 48-52; lectr physics, 58-59, from asst prof to assoc prof, 59-73, assoc dean acad planning, 70-75, PROF PHYSICS, CALIF STATE UNIV, LOS ANGELES, 73- Concurrent Pos: Res assoc physics, Harvey Mudd Col, 75-76. Mem: Am Phys Soc; Am Asn Physics Teachers; AAAS; Sigma Xi. Res: Ultrasound propagation in solid methane and deuteromethane as a function of temperature, with particular interest in the lambda-type phase transitions of these substances. Mailing Add: Dept of Physics Calif State Univ Los Angeles CA 90032

STAHL, JOHN BENTON, b Columbus, Ohio, Mar 28, 30. LIMNOLOGY. Educ: Iowa State Univ, BS, 51; Ind Univ, AM, 53, PhD(zool), 58. Prof Exp: Sessional lectr biol, Queen's Univ, Ont, 58-59; asst prof, Thiel Col, 59-63; asst prof, Wash State Univ, 63-66; asst prof, 66-72, ASSOC PROF BIOL, SOUTHERN ILL UNIV, 72- Mem: AAAS; Am Soc Limnol & Oceanog; Ecol Soc Am; Int Asn Theoret & Appl Limnol. Res: Chironomidae ans Chaoborus. Mailing Add: Dept of Zool Southern Ill Univ Carbondale IL 62901

STAHL, NEIL, b Sheridan, Ind, June 11, 42; m 67. MATHEMATICAL ANALYSIS. Educ: Ind Univ, Bloomington, AB, 64; Brown Univ, PhD(appl math), 70. Prof Exp: Asst prof math, 69-72, ASST PROF ECOSYSTS ANAL, UNIV WIS-GREENBAY, MARINETTE CAMPUS, 72- Mem: Soc Indust & Appl Math. Res: Differential equations and their applications; mathematical biology. Mailing Add: Marinette Campus Univ of Wis Marinette WI 54143

STAHL, PHILIP DAMIEN, b Wheeling, WVa, Oct 4, 41; m 68; c 1. PHYSIOLOGY, CELL BIOLOGY. Educ: WLiberty State Col, BS, 64; WVa Univ, PhD(pharmacol), 67. Prof Exp: ASST PROF PHYSIOL, MED SCH, WASH UNIV, 71- Concurrent Pos: Fel, Space Sci Res Ctr, Univ Mo, 67; Arthritis Found fel molecular biol, Vanderbilt Univ, 68-70. Mem: Brit Biochem Soc; Am Chem Soc. Res: Lysosomes and calcium homeostasis. Mailing Add: Dept of Physiol Wash Univ Med Sch St Louis MO 63110

STAHL, RALPH HENRY, b Berlin, Ger, Dec 29, 26; US citizen; m 55; c 3. EXPERIMENTAL PHYSICS. Educ: Harvard Univ, AB, 49, MA, 50, PhD(nuclear physics), 54. Prof Exp: Mem tech staff, Radiation Lab, Univ Calif, 54-56; mem tech staff, Gen Atomic Div, Gen Dynamics Corp, 56-68; secy-treas, Systs, Sci & Software, 68-72; MEM TECH STAFF, IRT CORP, 72- Concurrent Pos: Vis prof, Univ Ill, Urbana, 61. Mem: Am Phys Soc; Am Nuclear Soc. Res: Nuclear weapons effects on military systems; electronics and electronics components; nuclear reactor physics. Mailing Add: IRT Corp PO Box 80817 San Diego CA 92138

STAHL, ROLAND, EDGAR, b Northumberland, Pa, Sept 2, 25; m 55; c 2. ORGANIC CHEMISTRY. Educ: Bucknell Univ, BS, 50; Cornell Univ, PhD(chem), 54. Prof Exp: Asst org chem, Cornell Univ, 50-53; res chemist, Am Cyanamid Co, 54-56; res chemist, NY, 56-60, Tenn, 60-65, RES CHEMIST, E I DU PONT DE NEMOURS & CO, INC, DEL, 65- Mem: Am Asn Textile Chemists & Colorists. Res: Peroxide and radical chemistry; polymer applications; adhesives; textile finishing and applications. Mailing Add: Plastics Dept Chestnut Run Lab E I du Pont de Nemours & Co Wilmington DE 19898

STAHL, S SIGMUND, b Berlin, Ger, June 16, 25; nat; m 47; c 1. PERIODONTOLOGY, ORAL PATHOLOGY. Educ: Univ Minn, DDS, 47; Univ Ill, MS, 49; Am Bd Periodont, dipl. Prof Exp: Res assoc periodont, 49-50, from instr to prof, 50-71, PROF PERIODONT & ORAL MED & CHMN DEPT, COL DENT, NY UNIV, 71- Concurrent Pos: Attend, Vet Admin Hosp, Brooklyn, 50-53; consult, 58-; exec secy, Guggenheim Found Inst Dent Res, NY Univ, 64- Honors & Awards: Award, Int Asn Dent Res, 71. Mem: AAAS; Sigma Xi; Am Dent Asn; fel Am Med Writers Asn; fel Am Col Dent. Res: Effects of hormonal and dietary changes on oral structures; wound healing. Mailing Add: 67 Park Ave New York NY 10016

STAHL, SAUL, b Antwerp, Belg, Jan 23, 42; US citizen; m 72. MATHEMATICS. Educ: Brooklyn Col, BA, 63; Univ Calif, Berkeley, MA, 66; Western Mich Univ, PhD(math), 75. Prof Exp: Systs programmer, Int Bus Mach, 69-73; ASST PROF MATH, WRIGHT STATE UNIV, 75- Mem: Am Math Soc. Res: Graph theory. Mailing Add: Dept of Math Wright State Univ Dayton OH 45431

STAHL, WALTER BERNARD, b New York, NY, Aug 28, 28. PARASITOLOGY. Educ: Brooklyn Col, BA, 52; Columbia Univ, MS, 56, PhD(parasitol), 60. Prof Exp: Asst biol, Brooklyn Col, 52; asst entom, Univ Fla, 53-54; NIH asst, Biol Sta, Univ Mich, 56; lab instr parasitol, Columbia Univ, 56-59; instr med microbiol, Seton Hall Col Med & Dent, 59-61; res fel, Sch Trop Med, Univ PR, 61-62; NIH res fel, Sch Med, Keio Univ, Japan, 62-66; trainee trop dis, Sch Med, Nat Univ Mex, 66-67; res fel, Palo Alto Med Res Found, 67-68; res assoc trop med, Univ PR, San Juan, 68-69 & Trop Dis Clin, Dept Health, City of New York, 69-70; SR RES SCIENTIST, DEPT HEALTH, NY STATE, 70- Concurrent Pos: Fel trop med & parasitol, La State Univ, 60. Mem: AAAS; Am Soc Parasitologists; Am Soc Trop Med & Hyg; NY Acad Sci; Japanese Soc Parasitol. Res: Immunology of parasitic diseases; toxoplasmosis; schistosomiasis; oxyuriasis. Mailing Add: Div of Labs & Res NY State Dept of Health Albany NY 12201

STAHL, WILLIAM HERBERT, b Philadelphia, Pa, Mar 8, 13; m 39; c 2. CHEMISTRY. Educ: Pa State Univ, BS, 35; Mich State Univ, MS, 37, PhD(immunochem), 39. Prof Exp: Asst biochem, Mich State Univ, 35-37, asst chemist & USDA fel, Exp Sta, 37-41; chief chemist, Fairchild Bros & Foster, 41-45; group leader, Foster D Snell, Inc, NY, 45-47; sr biochemist, Pioneering Res Lab, US Army Qm Corps, 47-52, head anal sect, Qm Res & Develop Ctr, 52-57; RES MGR, McCORMICK & CO, INC, 57- Mem: AAAS; Am Chem Soc; Soc Appl Spectros; Inst Food Technologists; Am Soc Testing & Mat. Res: Microbiological degradation of keratins, cellulose and plasticizers; general instrument research in micro-analytical and macroanalytical methods; essential oil and oleoresin technology; correlation of objective and subjective methods of flavor analysis; physiology and psychology of flavor and odor perception and agronomy. Mailing Add: McCormich & Co Inc 204 Wright Ave Hunt Valley MD 21031

STAHL, WILLIAM J, b New York, NY, Jan 3, 39; m 64; c 1. BIOCHEMISTRY. Educ: Merrimack Col, AB, 60; Fordham Univ, MS, 61; St John's Univ, NY, PhD(biochem), 69. Prof Exp: Res asst biochem, Albert Einstein Med Ctr, 62-64; res chemist, Tenneco Chem Inc, 64-65; asst prof, 65-74, ASSOC PROF BIOCHEM, JOHN JAY COL CRIMINAL JUSTICE, 74- Mem: AAAS; Am Soc Microbiol; Am Chem Soc. Res: Enzymes associated with invasive microbes and biochemical intermediates; development of vaccine for heroin and related alkaloids. Mailing Add: John Jay Col Crim Justice 315 Park Ave S New York NY 10010

STAHL, WILLIAM LOUIS, b Glen Dale, WVa, Aug 2, 36; m 59; c 2. BIOCHEMISTRY, NEUROCHEMISTRY. Educ: Univ Notre Dame, BS, 58; Univ Pittsburgh, PhD(biochem), 63. Prof Exp: Res assoc biochem, NIH, 65-67; res asst prof neurochem, 67-71, ASSOC PROF NEUROCHEM, SCH MED, UNIV WASH, 71-; CHIEF NEUROCHEM, VET ADMIN HOSP, SEATTLE, 67- Concurrent Pos: United Cerebral Palsy Res & Educ Found fel, Biochem Dept, Inst Psychiat, Maudsley Hosp, Univ London, 63-65. Mem: Am Chem Soc; Brit Biochem Soc; Am Soc Neurochem; Int Soc Neurochem. Res: Chemistry and metabolism of the nervous system; energy linked transport systems; structure and function of biological membranes. Mailing Add: Div of Neurol Univ of Wash Sch of Med Seattle WA 98105

STAHLMAN, CLARENCE L, b Courtland, Kans, Apr 3, 21; m 43; c 1. DAIRY INDUSTRY. Educ: Kans State Univ, BS, 49; Hofstra Col, MS, 54. Prof Exp: Asst prof dairy indust, 50-61, assoc prof, 61-70, prof food processing technol, 70-72, PROF DAIRY & FOOD SCI & COORDR FOOD PROCESSING TECHNOL, STATE UNIV NY AGR & TECH COL FARMINGDALE, 72- Res: Dairy food products. Mailing Add: Dept of Food Processing Technol State Univ NY Agr & Tech Col Farmingdale NY 11735

STAHLMAN, MILDRED, b Nashville, Tenn, July 31, 22. PEDIATRICS, PHYSIOLOGY. Educ: Vanderbilt Univ, BA, 43, MD, 46; Am Bd Pediat, dipl, 54. Prof Exp: From instr to asst prof pediat, 51-64, from instr to asst prof physiol, 54-61, assoc prof pediat, 64-70, PROF PEDIAT, SCH MED, VANDERBILT UNIV, 70- Concurrent Pos: Lederle med fac award, 61-63; USPHS career develop award, 63-68; mem human embryol & develop study sect, USPHS, 64-68. Mem: Soc Pediat Res; Am Pediat Soc; Am Physiol Soc; Am Fedn Clin Res. Res: Newborn cardiorespiratory and fetal physiology; cardiology; rheumatic fever. Mailing Add: Dept of Pediat Vanderbilt Univ Sch of Med Nashville TN 37203

STAHLMAN, WILLIAM DUANE, b Clarion, Pa, May 27, 23; m 47; c 3. HISTORY OF SCIENCE. Educ: Mass Inst Technol, SB, 48; Amherst Col, MA, 50; Brown Univ, PhD(hist of math), 60. Prof Exp: Instr eng, Amherst Col, 48-50; asst prof hist sci, Mass Inst Technol, 56-60; vis lectr, Harvard Univ, 60-61; ASSOC PROF HIST OF SCI, UNIV WIS-MADISON, 61- Concurrent Pos: Vis asst prof, Brandeis Univ, 57. Mem: Hist Sci Soc (treas, 48, 56-). Res: History of the physical sciences, particularly astronomy, mathematics and physics. Mailing Add: Dept of Hist of Sci Univ of Wis Madison WI 53706

STAHLY, DONALD PAUL, b Columbus, Ohio, May 29, 37; m 59; c 2. MICROBIOLOGY. Educ: Ohio State Univ, BS, 59, MS, 61; Univ Ill, PhD(microbiol), 65. Prof Exp: NIH fel, Univ Minn, 65-66; asst prof, 66-72, ASSOC PROF MICROBIOL, UNIV IOWA, 72- Mem: AAAS; Am Soc Microbiol; Sigma Xi. Res: Bacterial sporulation and changes in amino acid metabolism during differentiation. Mailing Add: Dept of Microbiol Univ of Iowa Iowa City IA 52240

STAHLY, EDWARD ARTHUR, b Tiskilwa, Ill, Jan 31, 26; m 55; c 2. HORTICULTURE. Educ: Univ Ill, BS, 52; Univ Md, MS, 55, PhD, 59. Prof Exp: Horticulturist, Plant Indust Sta, 59-64, HORTICULTURIST, USDA, WASH, 64- Mem: Am Soc Hort Sci. Res: Fruit set and flower initiation of apple, pear and peach as influenced by synthetic and endogenous hormones; calcium nutrition of apple and pear. Mailing Add: Box 99 PO Annex 106 Wenatchee WA 98801

STAHLY, ELDON EVERETT, b Danvers, Ill, Feb 13, 08; m 46; c 2. ORGANIC CHEMISTRY. Educ: Blufton Col, BA, 29; Ohio State Univ, MA, 31; Pa State Col, PhD(org chem), 34. Prof Exp: Asst chem, Ohio State Univ, 29-31; asst chem, Pa State Col, 33-34; res chemist, Gen Elec Co, 34; res chemist, Standard Oil Co, La, 35-41; sr fel & admin fel, Mellon Inst, 41-48; dir chem res, Commercial Solvents Corp, 48-50; dir res, Burke Res Co, Mich & Fla, 50-63; proj mgr, Contract Res, W R Grace & Co, 63-72; ASSOC, TRIDENT ENG ASSOCS, 72-: CONSULT, POLYMER INST, UNIV DETROIT, 72- Concurrent Pos: Consult patents for chem processes, 72-; US ed, Flammability News Bulletin, 73-; registered prof eng & expert witness on fires & explosions, 73- Mem: AAAS; Am Chem Soc; Am Inst Chem; Am Inst Chem Eng; Brit Soc Chem Indust. Res: Resins; petroleum chemical processes; production of butadiene; manufacture of elastomers and their reinforcement; polymerizations; toxic chemical agents; flame-proofing combustibles; removal of carcinogens from tobacco smoke. Mailing Add: 2813 Deerfield Dr Ellicott City MD 21043

STAHMANN, MARK ARNOLD, b Spanish Fork, Utah, May 30, 14; m 41; c 2. BIOCHEMISTRY. Educ: Brigham Young Univ, BA, 36; Univ Wis, PhD(biochem), 41. Prof Exp: Asst chem, Rockefeller Inst, 42-44; res assoc org chem, Mass Inst Technol, 44-45; res assoc biochem, 46-47, from asst prof to assoc prof, 47-56, PROF BIOCHEM, UNIV WIS-MADISON, 56- Concurrent Pos: Guggenheim fel, Pasteur Inst, Paris, 55; Fulbright scholar, Nagoya, 67. Mem: AAAS; Am Chem Soc; Am Soc Biol Chem; Soc Exp Biol & Med; Am Phytopath Soc. Res: Anticoagulant 4-hydroxycoumarins: warfarin; biochemistry of plant diseases; synthetic polypeptides; polypeptidyl proteins; virus diseases; plant proteins; molecular pathology of atherosclerosis. Mailing Add: Dept of Biochem Univ of Wis Madison WI 53706

STAHNKE, HERBERT LUDWIG, b Chicago, Ill, June 10, 02; m 29; c 2. ARACHNOLOGY, SYSTEMATICS. Educ: Univ Chicago, SB, 28; Univ Ariz, MA, 34; Iowa State Univ, PhD(zool), 39. Prof Exp: Teacher pub schs, Ariz, 28-40; assoc prof sci, 41-47, prof zool & dir poisonous animals res lab, 47-72, head div life sci, 50-62, EMER PROF ZOOL & EMER DIR POISONOUS ANIMALS RES LAB, ARIZ STATE UNIV, 72- Mem: AAAS; Soc Syst Zool. Res: Scorpion taxonomy and biogeography. Mailing Add: 2016 E Laguna Dr Tempe AZ 85282

STAIB, JON ALBERT, b Toledo, Ohio, Mar 23, 40; m 67; c 1. COSMIC RAY PHYSICS. Educ: Univ Toledo, BS, 63; Case Western Reserve Univ, MS, 67, PhD(physics), 69. Prof Exp: Asst prof, 69-70, ASSOC PROF PHYSICS, MADISON COL, VA, 70- Mem: Am Phys Soc; AAAS; Int Soc Planetarium Educators. Res: Gamma ray astronomy; atmospheric gamma radiation. Mailing Add: Dept of physics Madison Col Harrisonburg VA 22801

STAIFF, DONALD C, b Everett, Wash, Feb 26, 36; m 59; c 2. PHARMACEUTICAL CHEMISTRY, BIONUCLEONICS. Educ: Univ Wash, BS, 59, PhD(pharmaceut chem), 63. Prof Exp: Asst prof pharmaceut chem, Ohio Northern Univ, 63-64; asst prof pharmaceut chem & bionucleonics, NDak State Univ, 64-67; anal res chemist, Western Pesticide Res Lab, Nat Commun Dis Ctr, USPHS, Wash, 67-72; CHIEF CHEMIST, WENATCHEE RES STA, ENVIRON PROTECTION AGENCY, 72- Concurrent Pos: Mead-Johnson Labs grant, 64-65; NSF inst grant, 65-66; Soc Sigma Xi grant-in-aid res, 65-66; guest lectr, Training Prog, Perrine Primate Res Lab, Environ Protection Agency, Fla. Mem: Am Pharmaceut Asn; Am Chem Soc; Health Physics Soc. Res: Conformational and configurational studies of some substituted phenyl-cyclohexane compounds by modern instrumental methods; metabolism studies including use of radiotracer techniques; effect of pesticides on health and persistence in the environment. Mailing Add: Wenatchee Res Sta Box 219 Environ Protection Agency Wenatchee WA 98801

STAIGER, JON CRAWFORD, b Milwaukee, Wis, May 13, 38; m 62; c 2. BIOLOGICAL OCEANOGRAPHY, ICHTHYOLOGY. Educ: Univ Mich, AB, 60;

Boston Univ, MA, 62; Univ Miami, PhD(marine sci), 70. Prof Exp: Res asst, 67-70, RES SCIENTIST BIOL OCEANOG, ROSENSTIEL SCH MARINE & ATMOSPHERIC SCI, UNIV MIAMI, 70- Mem: Am Soc Ichthyologists & Herpetologists; Am Soc Limnol & Oceanog; Am Fisheries Soc. Res: Ecology, distribution and systematics of deep-sea fishes. Mailing Add: Rosenstiel Sch of Marine & Atmospheric Sci Univ of Miami Miami FL 33149

STAIGER, ROGER POWELL, b Trenton, NJ, Nov 23, 21; m 44; c 1. ORGANIC CHEMISTRY. Educ: Ursinus Col, BS, 43; Univ Pa, MS, 48, PhD, 53. Prof Exp: From instr to assoc prof, 43-63, PROF & CHMN DEPT CHEM, URSINUS COL, 63- Concurrent Pos: Consult, Maumee Chem Co, 55-64; vis prof, Temple Univ, 63- & Alexandria Hosp, Nevis, 68- Mem: Am Chem Soc. Res: Synthesis of organic heterocyclic compounds. Mailing Add: 707 Chestnut St Collegeville PA 19426

STAIKOS, DIMITRI NICKOLAS, b Piraeus, Greece, Dec 18, 19; m 47; c 2. ELECTROCHEMISTRY. Educ: Nat Univ Athens, dipl, 42; Western Reserve Univ, MS, 50, PhD(chem), 51. Prof Exp: Off Naval Res asst, Western Reserve Univ, 49-50; res chemist, Pa, 50-56, res engr, Eng Res Lab, 56-60, sr res phys chemist, 60-64, SR RES PHYS CHEMIST, CENT RES DEPT, EXP STA, E I DU PONT DE NEMOURS & CO, INC, 64- Concurrent Pos: Instr, St Joseph's Col, Pa, 54-55. Mem: Am Chem Soc; Electrochem Soc. Res: Fundamentals of electrochemical processes; corrosion; electronic instrumentation; ultrasonics; surface tension; fused salts; electrochemistry in nonaqueous solutions; electroless deposition. Mailing Add: 1306 Quincy Dr Wilmington DE 19803

STAINER, DENNIS WILLIAM, b Liverpool, Eng, Aug 25, 32; m 57; c 2. BIOCHEMISTRY, MICROBIOLOGY. Educ: Univ Liverpool, BSc, 54, Hons, 55, PhD(biochem), 58. Prof Exp: RES MEM, CONNAUGHT LABS, LTD, 60- Concurrent Pos: Nat Res Coun Can fel biochem, Food & Drug Directorate, 57-59. Res: Production of diphtheria and tetanus toxoids and pertussis vaccine and studies on their immunogenicity; development of new bacteriological media and their application in bacterial fermentations; immunology. Mailing Add: 159 Driscoll Rd Richmond Hill ON Can

STAINER, HOWARD MARTIN, b London, Eng, Jan 19, 37; US citizen; m 61; c 4. COMPUTER SCIENCES. Educ: Queen's Col, NY, BS, 56; Calif Inst Technol, MS, 61; Univ Md, PhD(physics), 66. Prof Exp: PHYSICIST, APPL PHYSICS LAB, JOHNS HOPKINS UNIV, 66- Mem: Am Phys Soc; affil mem Inst Elec & Electronics Eng. Res: Advanced computer systems and techniques; software engineering and research; verification tools. Mailing Add: Appl Physics Lab Johns Hopkins Rd Johns Hopkins Univ Laurel MD 20810

STAINKEN, DENNIS M, b Staten Island, NY, Feb 9, 46. PHYSIOLOGICAL ECOLOGY, MARINE ZOOLOGY. Educ: Wagner Col, BS, 67; Rutgers Univ, MS, 70, PhD(zool), 75. Prof Exp: Sci asst biol, Richmond Col, 67-73; ASST PROF BIOL, COL MT ST VINCENT & MANHATTAN COL, 75- Concurrent Pos: Biologist, US Environ Protection Agency, 75- Mem: Wildlife Soc; Am Soc Limnol & Oceanog; Ecol Soc Am; Marine Biol Asn UK; AAAS. Res: Effects of pollutants on the physiology of marine animals and ecosystems; physiological ecology of marine animals. Mailing Add: 51 Coughlan Ave Staten Island NY 10310

STAINS, HOWARD JAMES, b Frenchtown, NJ, Apr 16, 24; m 48; c 3. ZOOLOGY. Educ: NC State Col, BS, 49, MS, 52; Univ Kans, PhD(zool), 55. Prof Exp: Lab instr econ zool, NC State Col, 48, res biologist, 49-51; res biologist, Univ Kans, 51-54, instr biol, 54-55; from asst prof to assoc prof, 55-71, PROF ZOOL, UNIV SOUTHERN ILL, 71- Mem: Wildlife Soc; Ecol Soc Am; Am Soc Mammal; Soc Study Evolution. Res: Furbearing mammals; osteology and ecology of mammals; wildlife techniques. Mailing Add: Dept of Zool Southern Ill Univ Carbondale IL 62901

STAINSBY, WENDELL NICHOLLS, b New York, NY, Nov 14, 28; m 52; c 4. PHYSIOLOGY. Educ: Bucknell Univ, AB, 51; Johns Hopkins Univ, ScD(physiol), 55. Prof Exp: From instr to assoc prof, 57-69, PROF PHYSIOL, COL MED, UNIV FLA, 69- Concurrent Pos: NIH res grant, 58- Mem: Am Physiol Soc; Am Col Sports Med. Res: Circulatory and muscle physiology; muscle metabolism and circulation; tissue gas transport. Mailing Add: Dept of Physiol Univ of Fla Col of Med Gainesville FL 32610

STAIR, ALVA TAYLOR, JR, molecular physics, see 12th edition

STAIRS, GERALD RAY, b Rupert, Idaho, Nov 18, 30; m 53; c 2. GENETICS, FORESTRY. Educ: Wash State Univ, BS, 58; Yale Univ, MF, 60, PhD(genetics), 63. Prof Exp: Forest mgt, Wash State Dept Natural Resources, 58-59; res collabr, Brookhaven Nat Lab, 61-63; asst prof forestry, State Univ NY Col Forestry, Syracuse, 63-65, assoc prof, 65-68; prof forestry & chmn dept, Univ Wis-Madison, 68-73; DEAN COL AGR, UNIV ARIZ, 73- Concurrent Pos: Res grants, USAEC & McIntire-Stennis Funds, 65-; res grant, NSF, 72-73. Mem: Soc Am Foresters. Res: Radiation, physiological and quantitative genetics; cytology; tree breeding. Mailing Add: Col of Agr Univ of Ariz Tucson AZ 85721

STAIRS, GORDON R, b Millville, NB, May 18, 32; m 54; c 4. INSECT PATHOLOGY. Educ: Univ NB, BSc, 54; McGill Univ, MSc, 58, PhD(entom), 63. Prof Exp: Res officer forest entom, Can Dept Agr, 54-58; res scientist, Insect Path Res Inst, 58-65; from asst prof to assoc prof, 65-73, PROF INSECT VIROL, OHIO STATE UNIV, 73- Res: Natural population biology and the effects of microorganisms on these populations; possible utilization of microorganisms in the control of pest populations. Mailing Add: Dept of Zool & Entom Ohio State Univ Columbus OH 43210

STAIRS, ROBERT ARDAGH, b Montreal, Que, June 10, 25; m 48; c 2. PHYSICAL CHEMISTRY. Educ: McGill Univ, BSc, 48; Univ Western Ont, MSc, 51; Cornell Univ, PhD(inorg chem), 55. Prof Exp: Instr chem, Cornell Univ, 53-55; lectr, Queen's Univ, Ont, 55-58; asst prof, 58-64, actg chmn dept, 66-67, assoc prof, 64-74, PROF CHEM, TRENT UNIV, 74- Mem: Chem Inst Can; AAAS; Am Chem Soc. Res: Physical properties of electrolyte solutions; surface chemistry; nonaqueous electrolyte solutions; metal-ammonia solutions; surface tension of solutions; viscosity of solutions. Mailing Add: Dept of Chem Trent Univ Peterborough ON Can

STAKE, PAUL ERIK, b Grandy, Minn, Jan 15, 44; m 66; c 3. ANIMAL NUTRITION. Educ: Univ Minn, BS, 68; SDak State Univ, MS, 71; Univ Ga, PhD(nutrit biochem), 74. Prof Exp: Lab supvr dairy sci, SDak State Univ, 68-71; res asst animal nutrit, Univ Ga, 71-74; ASST PROF NUTRIT SCI, UNIV CONN, 74- Mem: Am Dairy Sci Asn; Am Chem Soc; Am Soc Animal Sci; Nutrit Today Soc. Res: Comparative metabolism of dietary essential and non-essential trace elements in animals. Mailing Add: Dept of Nutrit Sci Univ of Conn Storrs CT 06268

STAKER, DONALD DAVID, b Wheelersburg, Ohio, Jan 16, 26; m 47; c 4. ORGANIC CHEMISTRY. Educ: Ohio Univ, BS, 47, MS, 48; Ohio State Univ, PhD(chem), 52. Prof Exp: Asst, Ohio Univ, 44-45, 47-48; asst, Ohio State Univ, 48-49; res chemist, Monsanto Chem Co, 52-56, res proj leader, 56-57, res group leader, 57-61; res sect leader, 61-67, RES SECT MGR, FATTY ACID DIV, EMERY INDUSTS, INC, 67- Mem: Am Chem Soc; Am Oil Chem Soc. Res: Fatty acid chemistry, production processes, utilization; reaction mechanisms; rubber chemicals; polymer properties; lubricant additives; process development; metalworking lubricants. Mailing Add: Res Lab 4900 Este Ave Emery Industs Inc Cincinnati OH 45232

STAKER, ROBERT DALE, b Newport, RI, July 3, 45. PHYCOLOGY. Educ: Univ Dayton, BS, 67; Univ Ariz, MS, 71, PhD(bot), 73. Prof Exp: Res asst phycol, Dept Biol, Univ PEI, 73-75; FEL PHYTOPLANKTOLOGY, NY OCEAN SCI LAB, 75- Mem: Int Soc Limnol; Sigma Xi. Res: Marine and freshwater algal taxonomy and ecology. Mailing Add: Dept of Microbiol NY Ocean Sci Lab Montauk NY 11954

STAKER, WILLIAM PAUL, b Aberdeen, SDak, Apr 9, 19; m 49; c 2. REACTOR PHYSICS. Educ: Ill State Univ, BS, 40; Univ Iowa, MS, 42; NY Univ, PhD(physics), 50. Prof Exp: Asst, Univ Iowa, 41-42; physicist & chief ballistic engr, Burnside Lab, E I du Pont de Nemours & Co, 42-46; res assoc, NY Univ, 46-50, proj dir, Res Div, Col Eng, 47-50; assoc physicist, Argonne Nat Lab, 50-52, group leader, 52; sr proj physicist, Eng Res Dept, Standard Oil Co, Ind, 52-56; sr physicist, Nuclear Div, 56-57, mgr exp physics, 57-59, mgr physics, Naval Reactors Div, 59-61, mgr eng & physics, 61-67, mgr fast breeder develop, Nuclear Power Dept, 67-73, mgr prod eng & develop, analysis, Nuclear Power Systs, 73-75, MGR, C-E/KWV COORD, NUCLEAR POWER SYSTS, COMBUSTION ENG, INC, 75- Mem: Am Phys Soc; Am Nuclear Soc. Res: Neutron physics; reactor engineering; tracer studies and radioisotope applications. Mailing Add: 64 Glenbrook Rd West Hartford CT 06107

STAKGOLD, IVAR, b Oslo, Norway, Dec 13, 25; nat US. APPLIED MATHEMATICS. Educ: Cornell Univ, BME, 45, MME, 46; Harvard Univ, MA, 48, PhD(appl math), 49. Prof Exp: From instr to asst prof appl math, Harvard Univ, 49-56; head math & logistics brs, US Off Naval Res, 56-59; assoc prof eng sci, Northwestern Univ, Evanston, 60-64, prof eng sci & math, 64-75, chmn dept eng sci, 69-75; PROF & CHMN DEPT MATH, UNIV DEL, 75- Concurrent Pos: Vis asst prof, Stanford Univ, 53-54; consult, Sylvania Elec Prod, Inc, 56-; lectr, Cath Univ Am, 57-58; liaison scientist, US Off Naval Res, London, 67-69; vis prof, Oxford Univ, 73-74; consult, Environ Protection Agency, 74- Mem: Am Math Soc Soc; Soc Indust & Appl Math. Res: Nonlinear boundary value problems. Mailing Add: Dept of Math Univ of Del Newark DE 19711

STAKLIS, ANDRIS A, b Valmiera, Latvia, Feb 4, 39; US citizen; m 61; c 1. ORGANIC CHEMISTRY. Educ: Univ Nebr, BS, 61, PhD(org chem), 65. Prof Exp: Asst head met develop sect, Manned Spacecraft Ctr, NASA, 65-67; sr scientist, 67-71, MGR MAT DEVELOP, NEW ORLEANS OPER, BELL AEROSPACE CO DIV, TEXTRON INC, 71- Mem: Am Chem Soc; Am Soc Test & Mat; Nat Asn Corrosion Eng; Am Soc Metals; Sigma Xi. Res: Direct materials development for air cushion ships; development and evaluation of nonmetallic materials; corrosion and stress corrosion protection of metallic materials; cavitation-erosion damage. Mailing Add: New Orleans Oper Box 29307 Bell Aerospace Co New Orleans LA 70189

STAKMAN, ELVIN CHARLES, b Algoma, Wis, May 17, 85; m 17. PLANT PATHOLOGY, MYCOLOGY. Educ: Univ Minn, AB, 06, AM, 10, PhD(plant path), 13. Hon Degrees: Dr Nat Sci, Univ Halle, 38; ScD, Yale Univ, 50, Univ RI, 53, Univ Minn, 54, Univ Wis, 54, Cambridge Univ, 64. Prof Exp: From instr to prof, 09-53, EMER PROF PLANT PATH, UNIV MINN, ST PAUL, 53- Concurrent Pos: Asst exp sta, Univ Minn, 09-10, asst plant pathologist, 10-13, head sect plant path, 13-40, chief, Div Plant Path & Bot, 40-53; pathologist & agent, USDA, 18-55, collab, 55-; guest prof, Univ Halle, 30-31 & Hitchcock prof, Univ Calif, 55; sci adv, Firestone Plantations Co, Liberia, 30; adv comt biol & med, AEC, 47-54, consult, 54-59; consult, Rockefeller Found, 53-72; ed-in-chief, Phytopath, 25-29 & Am ed, Phytopath Zeitschrift; mem div biol & agr, Nat Res Coun, 31-37, vchmn, 37, 47, mem exec comt, 32-34, 36-38, 48-50, war biol comt, 42-45, agr bd, 50-58, gov bd, 50-54; mem UNESCO Nat Comn, 50-56. Honors & Awards: Hansen Prize & Gold Medal, Copenhagen, 28; Appel Medal, 57; Cruz de Boyaca, 66. Mem: Nat Acad Sci; AAAS (pres, 49); fel Am Phytopath Soc (vpres, 20, pres, 22); Am Philos Soc; Bot Soc Am. Res: Epidemiology of cereal rusts; aerobiology; genetics of plant pathogens; plant disease resistance; agricultural improvement. Mailing Add: Dept of Plant Path Inst of Agr Univ of Minn St Paul MN 55101

STAKNIS, VICTOR RICHARD, b Bridgewater, Mass, June 14, 20; m 42. MATHEMATICS. Educ: Bridgewater Teachers Col, BS, 42; Mass Inst Technol, BS, 46; Boston Univ, MA, 50, PhD(math), 53. Prof Exp: Instr math, Ft Devens Br, Univ Mass, 46-49; lectr, Boston Univ, 51-52, instr, 52-53; asst prof, 53-59, ASSOC PROF MATH, NORTHEASTERN UNIV, 59- Mem: Am Math Soc. Res: Topology. Mailing Add: 90 Stoneleigh Dr Watertown MA 02172

STAKUTIS, VINCENT JOHN, b Boston, Mass, June 20, 20; m 47; c 4. ATMOSPHERIC PHYSICS, OPTICS. Educ: Boston Col, BS, 43; Brown Univ, MS, 50, PhD(physics), 54. Prof Exp: Lab instr optics, Boston Col, 42-43; instr & asst to headmaster, Marianapolis Prep Sch, Conn, 46-48; asst optics, photog, atomic physics & acoust, Brown Univ, 48-53; res physicist, Geophys Res Dir, Air Force Cambridge Res Labs, 53-57, supvry physicist, 57-58, br chief atmospheric optics, 58-60; mem tech staff, 60-63, MEM SR TECH STAFF, MITRE CORP, BEDFORD, 63- Mem: AAAS; Sigma Xi; Am Geophys Union; Optical Soc Am; Acoust Soc Am. Res: Ultrasonic attenuation in aqueous suspensions; electromagnetic propogation; scattering processes in the atmosphere; high altitude sky luminance and albedo measurements; vision in the atmosphere; satellite reconnaissance systems; nuclear weapon environmental effects. Mailing Add: 160 Grant St Lexington MA 02173

STALDER, ERNEST W, solid state physics, atmospheric physics, see 12th edition

STALEY, DAVID H, b Columbus, Ohio, Jan 30, 30; m 58; c 3. MATHEMATICS. Educ: Oberlin Col, AB, 52; Ohio Univ, MS, 54; Ohio State Univ, PhD(math), 63. Prof Exp: Instr math, Henry Ford Community Col, 56-57 & Oberlin Col, 60-61; from asst prof to assoc prof, 61-70, PROF MATH, OHIO WESLEYAN UNIV, 70- Mem: Math Asn Am; Nat Coun Teachers Math. Res: Commutability of operators in a topological space; mathematical models of insect populations. Mailing Add: Dept of Math Ohio Wesleyan Univ E Campus Delaware OH 43015

STALEY, DEAN ODEN, b Kennewick, Wash, Oct 18, 26; m 63; c 4. METEOROLOGY. Educ: Univ Wash, BS, 50, PhD(meteorol), 56; Univ Calif, Los Angeles, MA, 51. Prof Exp: From instr to asst prof, Univ Wis, 55-59; assoc prof & assoc meteorologist, 59-65, PROF & METEOROLOGIST, INST ATMOSPHERIC PHYSICS, UNIV ARIZ, 65- Mem: Am Meteorol Soc; Am Geophys Union. Res: Dynamic and synoptic meteorology; radiation and planetary atmospheres. Mailing Add: Inst of Atmospheric Physics Univ of Ariz Tucson AZ 85721

STALEY, JAMES TROTTER, b Brookings, SDak, Mar 14, 38; m 63; c 2. BACTERIOLOGY, MICROBIAL ECOLOGY. Educ: Univ Minn, Minneapolis, BA, 60; Ohio State Univ, MSc, 63; Univ Calif, Davis, PhD(bact), 67. Prof Exp: Instr microbiol, Mich State Univ, 67-69; asst prof environ sci & eng, Univ NC, Chapel Hill, 69-71; asst prof, 71-74, ASSOC PROF MICROBIOL, UNIV WASH, 74- Mem: Am Soc Microbiol; AAAS. Res: Biology of Ancalomicrobium, Prosthecomicrobium and other prosthecate bacteria; aquatic bacteriology; fresh water microbiology; microbial ecology. Mailing Add: Dept of Microbiol Univ of Wash Seattle WA 98195

STALEY, JOHN M, b Three Rivers, Mich, Sept 12, 29; m 53; c 3. PLANT PATHOLOGY, FORESTRY. Educ: Univ Mont, BS, 51; WVa Univ, MS, 53; Cornell Univ, PhD(plant path), 62. Prof Exp: Asst forest path, WVa Univ, 51-53; plant pathologist, Pac Northwest Forest & Range Exp Sta, 53, northeastern forest exp sta, 56-62, RES PLANT PATHOLOGIST, ROCKY MT FOREST & RANGE EXP STA, US FOREST SERV, 62- Concurrent Pos: Res asst, Cornell Univ, 56-62; affil prof, Grad Fac, Colo State Univ, 67- Mem: Am Phytopath Soc; Mycol Soc Am. Res: Complex diseases of forest trees; vascular wilt diseases of forest trees; foliage diseases of coniferous trees; rust diseases of cereals. Mailing Add: Rocky Mt Forest & Range Exp Sta 240 W Prospect Ft Collins CO 80521

STALEY, RALPH HORTON, b Boston, Mass, Mar 15, 45. PHYSICAL CHEMISTRY. Educ: Dartmouth Col, AB, 67; Calif Inst Technol, PhD(chem), 76. Prof Exp: Res physicist, Feldman Res Labs, Picatinny Arsenal, 68-71; ASST PROF PHYS CHEM, MASS INST TECHNOL, 75- Mem: Am Chem Soc; Am Phys Soc; Am Soc Mass Spectrometry; AAAS. Res: Gas-phase ion chemistry; determination of molecular structure and properties by spectroscopic techniques; reaction kinetics; chemical application of photoionization mass spectrometry, photoelectron spectroscopy and ion cyclotron resonance spectroscopy. Mailing Add: Dept of Chem Mass Inst Technol Cambridge MA 02139

STALEY, RAYMOND CLARENCE, physical oceanography, physical meteorology, see 12th edition

STALEY, ROBERT NEWTON, b Canova, SDak, Oct 15, 35; m 70. ORTHODONTICS, PHYSICAL ANTHROPOLOGY. Educ: Univ Minn, Minneapolis, BS, 57, DDS, 59; Univ Chicago, MA, 67; State Univ NY Buffalo, cert orthod, 69, MS, 70. Prof Exp: Intern dent, Zoller Mem Dent Clin, Univ Chicago Hosp & Clins, 59-60, mem staff, 62-65; asst prof, 70-73, ASSOC PROF ORTHOD, COL DENT, UNIV IOWA, 73- Concurrent Pos: Mem, Am Asn Dent Sch & Human Biol Coun. Mem: Int Asn Dent Res; Am Dent Asn; Am Asn Orthodont; Am Asn Univ Prof; Am Asn Phys Anthrop. Res: Dental morphology and genetics; craniofacial growth; biological evolution of human dentition. Mailing Add: Dept of Orthod Univ of Iowa Col of Dent Iowa City IA 52242

STALEY, STUART WARNER, b Pittsburgh, Pa, July 11, 38; m 63; c 2. ORGANIC CHEMISTRY. Educ: Williams Col, BA, 59; Yale Univ, PhD(phys org chem), 63. Prof Exp: Res assoc phys org chem, Univ Wis, 63-64; from asst prof to assoc prof, 64-74, PROF PHYS ORG CHEM, UNIV MD, COLLEGE PARK, 74- Concurrent Pos: Vis prof, Swiss Fed Inst Technol, 71-72. Mem: Am Chem Soc; The Chem Soc. Res: Thermodynamic studies of unsaturated hydrocarbons; organic reaction mechanisms; studies of hydrocarbon anions; thermal reorganization reactions. Mailing Add: Dept of Chem Univ of Md College Park MD 20742

STALEY, THEODORE EARNEST LEON, b Virginia City, Mont, Mar 10, 34; div; c 2. VETERINARY ANATOMY, VETERINARY PHYSIOLOGY. Educ: Carroll Col, Mont, BA, 58; Mich State Univ, DVM, 65, MS, 66. Prof Exp: Asst prof anat, Okla State Univ, 65-71; res pathologist, Vet Admin Hosp, Seattle, Wash, 71-72; ASSOC PROF ANAT-PHYSIOL, OKLA STATE UNIV, 72- Concurrent Pos: Fel, Mich State Univ, 65; consult, Okla State Univ & Univ San Carlos, Guatemala, 70; spec fel, Univ Wash, Vet Admin Hosp, Seattle, 71-72. Mem: Comp Gastroenterol Soc; Res Workers Animal Dis; Asn Gnotobiotics; Am Asn Vet Anatomists; World Asn Vet Anatomists. Res: Ultrastructural anatomy and pathophysiology of the neonatal intestine of domesticated animals as influenced by Escherichia coli and its enterotoxins. Mailing Add: Dept of Physiol Sci Col Vet Med Okla State Univ Stillwater OK 74074

STALFORD, HAROLD LENN, b Avery, Okla, July 22, 42; m 63; c 5. APPLIED MATHEMATICS. Educ: Okla State Univ, BS, 65; Univ Calif, Berkeley, MS, 66, PhD(appl mech), 70. Prof Exp: Asst res engr, Univ Calif, 70; opers res analyst, 70-71, MATH ANALYST, RADAR DIV, US NAVAL RES LAB, 71- Concurrent Pos: Sr analyst, Dynamics Res Corp, 76- Mem: Sigma Xi. Res: Control theory; game theory; computer science; statistics; system identification and classification. Mailing Add: Radar Div US Naval Res Lab Washington DC 20390

STALHEIM, OLE HENRY, b Garretson, SDak, Sept 23, 17; m 42; c 4. VETERINARY MICROBIOLOGY. Educ: Tex A&M Univ, DVM, 41; Univ SDak, MA, 61; Univ Wis, PhD(bact), 63. Prof Exp: Vet practitioner, Vermillion, SDak, 41-58; NIH fel, Univ Wis, 60-63; RES VET, NAT ANIMAL DIS CTR, AGR RES SERV, USDA, 63- Mem: Am Vet Med Asn; Am Soc Microbiol; Conf Res Workers Animal Dis; US Animal Health Asn. Res: Microbial nutrition, metabolism and virulence; immunity to bacterial infection; chemotherapy. Mailing Add: Nat Animal Dis Ctr PO Box 70 Ames IA 50010

STALICK, WAYNE MYRON, b Oregon City, Ore, Aug 24, 42; m 67. ORGANIC CHEMISTRY. Educ: Univ Ore, BA, 64; Northwestern Univ, PhD(org chem), 69. Prof Exp: Asst prof org chem, Calif State Univ, San Jose, 69-70; fel, Ohio State Univ, 70-72, lectr, 72; ASST PROF ORG CHEM, GEORGE MASON UNIV, 72- Mem: Am Chem Soc. Res: Base catalyzed reactions of hydrocarbons; heterocyclic, synthetic organic and organometallic chemistry. Mailing Add: Dept of Chem George Mason Univ Fairfax VA 22030

STALKER, ARCHIBALD MACSWEEN, b Montreal, Que, June 29, 24; m 51; c 4. GLACIAL GEOLOGY. Educ: McGill Univ, BA, 45, MSc, 48, PhD(geol), 50. Prof Exp: GEOLOGIST, GEOL SURV CAN, 50- Mem: Fel AAAS; fel Geol Soc Am; Geol Soc Can. Res: Glacial geology; geomorphology; relations of Cordilleran and Laurentide glaciations; preglacial drainage; early man in New World; climate, stratigraphy and mammals of Quaternary; Quaternary vertebrate paleontology. Mailing Add: Geol Surv of Can Ottawa ON Can

STALKER, HARRISON DAILEY, b Detroit, Mich, July 3, 15; m 41; c 1. GENETICS, EVOLUTIONARY BIOLOGY. Educ: Col Wooster, BA, 37; Univ Rochester, PhD(zool), 41. Prof Exp: Asst zool, Univ Rochester, 37-42; from instr to assoc porf, 42-56, PROF ZOOL, WASH UNIV, 56- Concurrent Pos: NSF sr fel, 61; mem genetics adv panel, NSF, 59-62. Mem: AAAS; Genetics Soc Am; Soc Study Evolution; Am Soc Nat; Am Soc Human Genetics. Res: Cytogenetics and isozymes of Drosphila species and populations; cytotaxonomy of Drosophila. Mailing Add: Dept of Biol Wash Univ St Louis MO 63130

STALL, ROBERT EUGENE, b Leipsic, Ohio, Dec 11, 31; m 52; c 2. PLANT PATHOLOGY. Educ: Ohio State Univ, BSc, 53, MSc, 54, PhD(bot, plant path), 57. Prof Exp: Res asst plant path, Ohio Agr Exp Sta, 57-63, assoc prof plant path & assoc plant pathologist, 63-69, PROF PLANT PATH, UNIV FLA, 69- Mem: AAAS; Am Phytopath Soc. Res: BActerial phytopathology. Mailing Add: Dept of Plant Path Univ of Fla Gainesville FL 32601

STALLARD, RICHARD E, b Eau Claire, Wis, May 30, 34. ANATOMY, PERIODONTOLOGY. Educ: Univ Minn, BS, 56, DDS, 58, MSD, 59, PhD(anat), 62. Prof Exp: Co-dir periodont res prog, EDent Dispensary, 62-65; from assoc prof to prof periodont & chmn dept, Univ Minn, Minneapolis, 65-68; asst dir, Eastman Dent Ctr, NY, 68-70; prof periodont & anat, Med Ctr & dir, Clin Res Ctr, Sch Grad Dent, Boston Univ, 70-74, asst dean, Sch Grad Dent, 72-74; DENT DIR & HEAD DEPT PERIODONT, GROUP HEALTH PLAN, INC, 75- Concurrent Pos: Consult, Cambridge State Sch & Hosp, 61-63; mem grad fac dent & grad fac anat, Univ Minn, Minneapolis, 65-68; consult, Vet Admin Hosp, 67-70, Univ Minn Hosp, 67-69 & consult, Wright-Patterson Air Force Base, 68-73; mem training grant comt, Nat Inst Dent Res, 69-73; consult, US Naval Hosp, Chelsea, Mass, 71-74, Boston Univ Hosp, 71-74 & Fairview Hosp, Minneapolis, 75- Mem: Fel AAAS; Am Acad Periodont (pres, 74); Am Dent Asn; Am Acad Oral Path; Int Col Oral Implantologists. Res: Microcirculation; occlusion; implantology and etiology of periodontal disease. Mailing Add: Group Health Med Ctr 8600 Nicollet Ave Bloomington MN 55420

STALLCUP, ODIE TALMADGE, b Paragould, Ark, Dec 2, 18; m 47; c 3. DAIRY HUSBANDRY. Educ: Univ Ark, BSA, 43; Univ Mo, AM, 47, PhD(dairy husb), 50. Prof Exp: Instr dairy husb, Univ Ark, 45-46; from asst instr to instr, Univ Mo, 46-50; assoc prof, 50-55, PROF DAIRY HUSB, UNIV ARK, FAYETTEVILLE, 55- Mem: Am Soc Animal Sci; Am Dairy Sci Asn; Soc Study Reproduction; Histochem Soc. Res: Ruminant nutrition; physiology of reproduction. Mailing Add: Dept of Animal Sci Univ of Ark Fayetteville AR 72701

STALLCUP, WILLIAM BLACKBURN, JR, b Dallas, Tex, Oct 18, 20; m 42; c 5. VERTEBRATE ZOOLOGY. Educ: Southern Methodist Univ, BS, 41; Univ Kans, PhD(zool), 54. Prof Exp: Instr biol, Southern Methodist Univ, 45-50; asst instr, Univ Kans, 50-53; from instr to assoc prof, 53-62, chmn dept, 63-67, assoc dean sch humanities & sci, 71-74, PROF BIOL, SOUTHERN METHODIST UNIV, 62-, ASSOC PROVOST, 74- Mem: AAAS; Am Soc Mammal; Am Ornith Union. Res: Comparative mycology and serology of birds; vertebrate natural history; general genetics. Mailing Add: Dept of Biol Southern Methodist Univ Dallas TX 75275

STALLEY, ROBERT DELMER, b Minneapolis, Minn, Oct 25, 24; m 50; c 4. MATHEMATICS. Educ: Ore State Col, BS, 46, MA, 48; Univ Ore, PhD(math), 53. Prof Exp: Instr math, Univ Ariz, 49-51; instr, Iowa State Col, 53-54, asst prof, 54-55; instr, Fresno State Col, 55-56; from asst prof to assoc prof, 56-66, PROF MATH, ORE STATE UNIV, 66- Mem: Am Math Soc; Math Asn Am. Res: Number density; recurring series; combinatorial analysis. Mailing Add: Dept of Math Ore State Univ Corvallis OR 97331

STALLING, DAVID LAURENCE, b Kansas City, Mo, Oct 24, 41; m 62; c 3. ANALYTICAL CHEMISTRY, ORGANIC CHEMISTRY. Educ: Mo Valley Col, BS, 62; Univ Mo-Columbia, MS, 64, PhD(biochem), 67. Prof Exp: Instr agr chem, Univ Mo-Columbia, 66-68; CHIEF CHEMIST, FISH-PESTICIDE RES LAB, 68- Concurrent Pos: Chem consult, Regis Chem Co, 66-69; cofounder, Anal Bicohem Labs, Inc, 67; assoc referee for contaminants in aquatic biota, Asn Official Anal Chemists, 75- Mem: AAAS; Am Chem Soc. Res: Development of rapid methods of analysis of biologically important compounds, especially amino acids, purine and pyrimidine bases, organic pollutants and pesticides by gas-liquid chromatography and combined gas chromatography-mass spectrometry computer techniques; biochemical effects of pesticides and organic contaminants on fish; gas-chromatography mass-spectrometry computer studies on environmental contaminants; organic residues in national pesticide monitoring programs. Mailing Add: Fish-Pesticide Res Lab Route 1 Columbia MO 65201

STALLINGS, CHARLES HENRY, b Durham, NC, Dec 28, 41; m 65; c 1. PLASMA PHYSICS. Educ: NC State Univ, BS, 63, MS, 64; Univ Wis-Madison, PhD(physics), 70. Prof Exp: SR PHYSICIST, PHYSICS INT CO, 70- Mem: Am Phys Soc; Inst Elec & Electronics Eng. Res: Generation and propagation of intense relativistic electron beams and their interaction with background on target plasma. Mailing Add: Physics Int Co 2700 Merced St San Leandro CA 94577

STALLINGS, EMMETT FRANCIS, b Carrabelle, Fla, Jan 10, 37; m 64; c 2. PHYSICAL GEOGRAPHY. Educ: Fla State Univ, BS, 59, MS, 61, PhD(geog), 70. Prof Exp: Instr geog, Utah State Univ, 64-66; asst prof, McNeese State Univ, 66-69; ASSOC PROF GEOG, UNIV SOUTHWESTERN LA, 69- Mem: Asn Am Geog; Am Acad Polit & Soc Sci; Nat Coun Geog Educ. Res: Contemporary geographic aphic thought; character of American geographical serials; water resources. Mailing Add: Box 3530 Univ of Southwestern La Lafayette LA 70501

STALLINGS, JAMES CAMERON, b Denton, Tex, Jan 16, 19; m 49; c 2. ORGANIC CHEMISTRY. Educ: Univ Tex, PhD(org chem), 49. Prof Exp: Prof chem, Sam Houston State Col, 49-57; sr res chemist, Celanese Corp Am, 57-59; PROF CHEM & DIR DEPT, SAM HOUSTON STATE UNIV, 59- Mem: Am Chem Soc. Res: Steric hindrance; synthetic hypnotics; free radicals in solution; organic synthesis; chemical education. Mailing Add: Dept of Chem Sam Houston State Univ Huntsville TX 77340

STALLINGS, JAMES HENRY, b Bryan, Tex, Sept 20, 92; m 23; c 2. SOIL CONSERVATION. Educ: Agr & Mech Col, Tex, BS, 14; Iowa State Col, MS, 17, PhD(soil fertility), 25. Prof Exp: Agronomist & farm dir high sch, Miss, 14-16; asst agronomist, Exp Sta, Ala, 16; asst prof soils, Iowa State Col, 19-20; assoc prof agron, Agr & Mech Col, Tex, 20-23, prof, 23-26; agronomist, J C Penny-Gwinn Corp Farms, Fla, 26-28; agronomist, Nat Fertilizer Asn, La, 29-34; regional dir, Soil Conserv Serv, NC, 34-37, sr soil conservationist, DC, 37-42, prin soil conservationist, War Food Admin, 42-44, soil conserv serv, 42-52, PRIN SOIL CONSERVATIONIST, AGR RES SERV, USDA, 53- Mem: AAAS; Am Soc Agron; Soil Sci Soc Am; NY Acad Sci. Res: Soil fertility; form of legume nitrogen when absorbed by non-legumes when the two are grown in association. Mailing Add: 5146 Nebraska Ave NW Washington DC 20008

STALLINGS, JOHN ROBERT, JR, b Morrilton, Ark, July 22, 35. TOPOLOGY. Educ: Univ Ark, BS, 56; Princeton Univ, PhD(math), 59. Prof Exp: NSF fel math, Oxford Univ, 59-60; from instr to assoc prof, Princeton Univ, 60-67; PROF MATH, UNIV CALIF, BERKELEY, 67- Concurrent Pos: Sloan Found fel, 62-65. Honors & Awards: Frank Nelson Cole Prize, Am Math Soc, 70. Mem: Am Math Soc. Res: Three-manifolds; geometric topology; group theory from topological and homological viewpoints. Mailing Add: Dept of Math Univ of Calif Berkeley CA 94720

STALLINGS, WILLIAM THOMAS, mathematics, see 12th edition

STALLKNECHT, GILBERT FRANKLIN, b Spooner, Minn, Sept 21, 35; m 58; c 4. PLANT PHYSIOLOGY, PLANT BIOCHEMISTRY. Educ: Univ Minn, BS, 62, MS, 66, PhD(plant physiol), 68. Prof Exp: Agr res technician, USDA Sugar Beet Invests, Univ Minn, St Paul, 63-67; ASST PROF PLANT PHYSIOL, UNIV IDAHO, 68-

Mem: Am Soc Plant Physiol; Japanese Soc Plant Physiol. Res: Fungus physiology; metabolic studies in host-parasite physiology; physiology of tuberization in potatoes; productivity and physiological age of potato tubers. Mailing Add: Res & Exten Ctr Univ of Idaho Aberdeen ID 83210

STALLMANN, FRIEDEMANN WILHELM, b Koenigsberg, Ger, July 29, 21; m 53; c 2. MATHEMATICS. Educ: Stuttgart Tech Univ, Dipl math, 49; Univ Giessen, Dr rer nat, 53. Prof Exp: Asst math, Univ Giessen, 53-55, lectr, 55-59; asst, Brunswick Tech Univ, 59-60; chief math sect, Spec Res Unit Med Electronic Data Processing, Vet Admin, DC, 60-64; assoc prof, 64-69, PROF MATH, UNIV TENN, KNOXVILLE, 69- Concurrent Pos: COnsult, Oak Ridge Nat Lab, 64- Mem: Am Math Soc. Res: Numerical analysis; conformal mapping and differential equations, lead field theory of electrocardiogram; computer analysis of electrocardiograms; complex variables. Mailing Add: Dept of Math Univ of Tenn Knoxville TN 37916

STALLONES, REUEL ARTHUR, b North Little Rock, Ark, Oct 10, 23; m 45; c 3. MEDICINE, EPIDEMIOLOGY. Educ: Western Reserve Univ, MD, 49; Univ Calif, MPH, 52; Am Bd Prev Med, dipl, 58. Prof Exp: Intern, Letterman Army Hosp, Med Corps, US Army, San Francisco, Calif, 49-50, resident pediat, 50, battalion surgeon, 50-51, prev med officer, Camp Pickett, Va, 52-54 & Ft Meade, Md, 54, asst chief dept epidemiol, Walter Reed Army Inst Res, DC, 54-56; lectr epidemiol, Univ Calif, Berkeley, 56-59, assoc prof pub health, 59-62, prof epidemiol, 62-68; DEAN UNIV TEX SCH PUB HEALTH HOUSTON, 68- Mem: Fel Am Pub Health Asn; fel Am Col Prev Med. Res: Epidemiology; cardiovascular disease. Mailing Add: Univ of Tex Sch of Pub Health PO Box 20186 Houston TX 77025

STALLWOOD, ROBERT ANTONY, b Oxbow, Sask, June 15, 25. NUCLEAR PHYSICS. Educ: Univ Toronto, BASc, 49, MA, 50; Carnegie Inst Technol, PhD(physics), 56. Prof Exp: Physicist, Nuclear Sci Sect, Gulf Res & Develop Co, 56-64; physicist, Gen Elec Space Sci Ctr, 64-67; ASSOC PROF PHYSICS, THIEL COL, 67- Mem: AAAS; Am Phys Soc. Res: High energy nuclear physics; gamma ray and x-ray spectroscopy; neutron physics. Mailing Add: Dept of Physics Thiel Col Greenville PA 16125

STALLWORTHY, WILSON BURNETT, b Westfield, NJ, Aug 28, 13; m 42; c 1. PHYSIOLOGY. Educ: Univ Toronto, BA, 35, PhD(physiol), 47. Prof Exp: Sci master, Upper Can Col, 40-42; teacher, Pickering Col, 45-46; asst prof physiol, Univ Maine, 46-53; from assoc prof to prof, 53-66, head dept, 56-69, RUGGLES GATES PROF BIOL, MT ALLISON UNIV, 66- Concurrent Pos: Davis exchange fel from Toronto to Ger; Wright res fel, Univ Toronto; bursary, Nat Res Coun Can. Res: Evaluation of environmental factors; stress in fish; oyster larvae. Mailing Add: Dept of Biol Mt Allison Univ Sackville NB Can

STALNAKER, CLAIR B, b Parkersburg, WVa, July 21, 38; m 63; c 2. FISH BIOLOGY, GENETICS. Educ: WVa Univ, BSF, 60; NC State Univ, PhD(zool), 66. Prof Exp: Res asst fisheries biol, NC State Univ, 60-66; asst prof, 66-72, ASSOC PROF FISHERIES BIOL, UTAH STATE UNIV, 72-, ASST UNIT LEADER, UTAH COOP FISHERY UNIT, 66- Mem: AAAS; Am Fisheries Soc; Wildlife Soc; Am Soc Ichthyol & Herpet; Soc Am Nat. Res: Physiological-genetic studies of fishes; aquatic environmental interactions. Mailing Add: Utah Coop Fishery Unit Utah State Univ Logan UT 84322

STALNAKER, NELSON DAVID, radiochemistry, physical chemistry, see 12th edition

STALTER, RICHARD, b Jan 16, 42; US citizen; m 68. BOTANY, PLANT ECOLOGY. Educ: Rutgers Univ, BS, 63; Univ RI, MS, 66; Univ SC, PhD(biol), 68. Prof Exp: Asst prof biol, Hihg Point Col, 68-69 & Pfeiffer Col, 69-70; asst prof, 71-74, ASSOC PROF BIOL, ST JOHN'S UNIV, NY, 75-, DIR ENVIRON STUDIES PROG, 75- Mem: Bot Soc Am; Ecol Soc Am; Sigma Xi. Res: Salt marsh ecology of the east coast of North America; flora of Long Island and Barrier Islands, South Carolina. Mailing Add: Dept of Biol St John's Univ Jamaica NY 11439

STAM, PAUL BOWMAN, physical chemistry, see 12th edition

STAMATELATOS, MICHAEL G, b Bucharest, Rumania, May 14, 41; m 70. NUCLEAR SCIENCE. Educ: Columbia Univ, BS, 64, MS, 65, DEngSc, 70. Prof Exp: ASST PROF PHYSICS, COOPER UNION, 69- Concurrent Pos: Consult, Consolidated Edison, NY, 71-72; Res Corp grant, Cooper Union, 72-73. Mem: Am Phys Soc; Am Nuclear Soc. Res: Nuclear physics, especially nuclear reaction theory and fast neutron physics; nuclear science and engineering, especially fast reactor physics and plama physics. Mailing Add: Dept of Physics Cooper Union Cooper Square New York NY 10003

STAMATIADOU, MARY NICKOLAS, biochemistry, biophysics, see 12th edition

STAMATOYANNOPOULOS, GEORGE, b Athens, Greece, Mar 11, 34; m 64; c 1. MEDICAL GENETICS, HEMATOLOGY. Educ: Nat Univ Athens, MD, 58, DSc, 60. Prof Exp: Asst med, Nat Univ Athens, 58-59, asst med & hemat, 61-64; res assoc med, 64-65, instr, 65-66, res asst prof, 65-69, assoc prof, 69-72, PROF MED, DIV MED GENETICS, UNIV WASH, 72- Concurrent Pos: Royal Hellenic Res Found fel, 61-64. Mem: Am Soc Human Genetics; Am Soc Clin Invest; Europ Soc Human Genetics; Genetics Soc Am. Res: Biochemical, developmental and human population genetics. Mailing Add: Div of Med Genet Dept of Med Univ of Wash SK-30 Seattle WA 98195

STAMBAUGH, JOHN EDGAR, JR, b Everrett, Pa, Apr 30, 40; m 61; c 4. ONCOLOGY, CLINICAL PHARMACOLOGY. Educ: Dickinson Col, BS, 62; Jefferson Med Col, MD, 66, Thomas Jefferson Univ, PhD(pharmacol), 68. Prof Exp: From instr to asst prof, 68-74, ASSOC PROF PHARMACOL, JEFFERSON MED COL, THOMAS JEFFERSON UNIV, 74- Concurrent Pos: AMA spec scholar, Thomas Jefferson Univ Hosp, 68-70, fel oncol, 70-72; resident med, Thomas Jefferson Univ Hosp, 68-70; staff physician, Cooper Hosp, Camden, NJ, 72- & Underwood Hosp, Woodbury, 73- Mem: Am Asn Cancer Res; Am Soc Clin Oncol; Am Soc Pharmacol & Exp Therapeut; Am Soc Clin Pharmacol. Res: Clinical 5drug metabolism and drug interactions; clinical oncology. Mailing Add: Dept of Pharmacol Jefferson Med Col Philadelphia PA 19107

STAMBAUGH, OSCAR FRANK, b Newport, Pa, Aug 4, 08; m 36; c 1. CHEMISTRY. Educ: Lebanon Valley Col, BS, 30; Pa State Univ, MS, 33, PhD(chem), 43. Prof Exp: Instr chem, Pa State Univ, 36-42 & Juniata Col, 42-43; chemist, Gulf Oil Corp, Pa, 43-46; prof chem & head dept, 46-73, actg dean, 66-67, EMER PROF CHEM, ELIZABETHTOWN COL, 73- Mem: Am Chem Soc. Res: Analytical chemistry. Mailing Add: Dept of Chem Elizabethtown Col Elizabethtown PA 17022

STAMBAUGH, RICHARD BULLA, b Boston, Mass, July 27, 13; m; c 2. PHYSICS. Educ: Kenyon Col, AB, 34; Mass Inst Technol, ScD(electronics), 41. Prof Exp: Res physicist, Goodyear Tire & Rubber Co, 39-52; supt, Develop Lab, Goodyear Atomic Corp, 52-63; MGR RES ANAL SERV, GOODYEAR TIRE & RUBBER CO, 63- Mem: AAAS; Am Phys Soc; Am Chem Soc. Res: Physical properties of rubber and plastics; analytical methods and instrumentation. Mailing Add: Res Dept Goodyear Tire & Rubber Co Akron OH 44316

STAMBAUGH, RICHARD L, b Mechanicsburg, Pa, Aug 16, 36; m 56; c 2. BIOCHEMISTRY. Educ: Albright Col, BS, 53; Univ Pa, PhD(biochem), 59. Prof Exp: Res assoc biochem, Philadelphia Gen Hosp, 58-59; dir biochem, Elwyn Res & Eval Ctr & instr biochem in pediat, Univ Pa, 59-63; sr res investr, Fels Res Inst & instr biochem, Sch Med, Temple Univ, 63-66; ASSOC PROF, DIV REPRODUCTIVE BIOL, SCH MED, UNIV PA, 66- Concurrent Pos: Consult, Penrose Res Lab, Zool Soc Philadelphia & Elwyn Res & Eval Ctr, 63-67. Mem: AAAS; Soc Study Reproduction; Fedn Am Socs Exp Biol. Res: Enzymology; biochemistry of reproduction. Mailing Add: Div of Reproductive Biol Univ of Pa Sch of Med Philadelphia PA 19104

STAMBAUGH, WILLIAM JAMES, b Allenwood, Pa, Dec 1, 27; m 52; c 3. FOREST PATHOLOGY. Educ: Pa State Univ, BS, 51, MS, 52; Yale Univ, PhD(forest path), 57. Prof Exp: Instr bot, Pa State Univ, 53-57, asst prof forest path, 57-61; from asst prof to assoc prof, 61-72, PROF FOREST PATH, DUKE UNIV, 72- Mem: Soc Am Foresters; Am Phytopath Soc. Res: Diseases of forest trees, with emphasis on epidemiology and control; microbiology of forest soils. Mailing Add: Sch of Forestry Duke Univ Durham NC 27706

STAMER, JOHN RICHARD, b Plankinton, SDak, May 19, 25; m 58; c 3. MICROBIOLOGY. Educ: Dakota Wesleyan Univ, BA, 50; SDak State Col, MS, 53; Cornell Univ, PhD(bact), 62. Prof Exp: Res assoc bact, Univ Ill, 53-56; jr scientist, Smith Kline & French Labs, 56-58; asst bact, Cornell Univ, 58-62, NIH fel, 62-64; asst prof, 64-70, ASSOC PROF BACT, NY STATE AGR EXP STA, CORNELL UNIV, 70- Mem: AAAS; Am Soc Microbiol. Res: Microbial physiology and nutrition. Mailing Add: Dept of Food Sci NY State Agr Exp Sta Geneva NY 14456

STAMER, PETER ERIC, b New York, NY, June 4, 39; m 68. PHYSICS. Educ: Stevens Inst Technol, BS, 61, MS, 63, PhD(physics), 66. Prof Exp: Instr physics, Upsala Col, 65-66; asst prof, 66-69, ASSOC PROF PHYSICS, SETON HALL UNIV, 69- Concurrent Pos: Jr res assoc physics, Stevens Inst Technol, 66-68, res assoc, 68- Mem: AAAS; Am Asn Physics Teachers; Am Phys Soc. Res: Experimental elementary particle physics; P-D interactions at 15.0 GeV/c and P-P interactions at 1.25-1.37 GeV/c. Mailing Add: Dept of Physics Seton Hall Univ South Orange NJ 07079

STAMEY, THOMAS ALEXANDER, b Rutherfordton, NC, Apr 26, 28; m 56; c 5. UROLOGY. Educ: Vanderbilt Univ, AB, 48; Johns Hopkins Univ, MD, 52; Am Bd Urol, dipl, 61. Prof Exp: Intern, Johns Hopkins Hosp, 52-53; mem, Brady Urol House Staff Residency Prog, Johns Hopkins Univ, 53-56; urol consult, US Armed Forces, UK, 56-58; from asst prof to assoc prof urol, Johns Hopkins Univ, 58-61; assoc prof surg, 61-64, PROF SURG, SCH MED, STANFORD UNIV, 64-, CHMN DIV UROL, 61- Concurrent Pos: Mem comt renal dis & urol training grants, NIH, 67-72, chmn, 71-72; mem sci adv bd, Nat Kidney Found, sci adv coun, Northern Calif Kidney Found & adv bd, Coop Study Pyelonephritis, USPHS; mem res comt, Northern Calif Kidney Found; mem sci adv comt, Hosp Sick Children, Toronto, Can; mem, Study of Res in Nephrology & Urol, NIH; consult, Santa Clara Valley Med Ctr, Palo Alto Vet Admin Hosp & Letterman Army Hosp, San Francisco; mem courtesy staff, El Camino Hosp; ed, Urol Dig; assoc ed, Campbell's Urol. Honors & Awards: Hugh Hampton Young Award, Am Urol Asn, 72. Mem: Am Urol Asn; AMA; Am Heart Asn; Am Surg Asn; Soc Univ Urol. Res: Renal physiology and disease and urinary tract infections; microbiology and hypertension. Mailing Add: Div of Urol Stanford Univ Med Ctr Stanford CA 94305

STAMEY, WILLIAM LEE, b Chicago, Ill, Oct 19, 22; m 45; c 3. MATHEMATICS. Educ: Univ Northern Colo, AB, 47; Univ Mo, MA, 49, PhD(math), 52. Prof Exp: From asst instr to instr math, Univ Mo, 47-52; asst prof, Ga State Col, 52-53; from asst prof to assoc prof, 53-62, assoc dean, 63-69, PROF MATH, KANS STATE UNIV, 62-, DEAN COL ARTS & SCI, 69- Mem: Am Math Soc; Math Asn Am. Mailing Add: Off of the Dean Col Arts & Sci Eisenhower Hall Kans State Univ Manhattan KS 66506

STAMIRES, DENNIS N, physical chemistry, see 12th edition

STAMLER, FREDERIC WILLIAM, b Muscatine, Iowa, Mar 28, 13; m 48; c 3. PATHOLOGY. Educ: Univ Iowa, BA, 39, MD, 43. Prof Exp: Asst, 46-47, from instr to assoc prof, 47-61, PROF PATH, COL MED, UNIV IOWA, 61- Mem: AAAS; Am Soc Clin Path; fel Am Col Path; Soc Exp Biol & Med. Res: Blood coagulation; vitamin K assay; endocrine factors involved in coagulation mechanism; extraction and assay of heparin; comparative pathology of pregnancy; nutritional factors in maternal toxemia and fetal and neonatal disease. Mailing Add: Dept of Path Univ of Iowa Col of Med Iowa City IA 52240

STAMLER, JEREMIAH, b New York, NY, Oct 27, 19; m 42; c 1. PREVENTIVE MEDICINE, PUBLIC HEALTH. Educ: Columbia Univ, AB, 40; State Univ NY, MD, 43. Prof Exp: Intern, Long Island Col Med Div, Kings County Hosp, 44; res assoc, Cardiovasc Dept, Med Res Inst, Michael Reese Hosp, Chicago, 49-55, asst dir dept, 55-58; dir heart dis control prog, Chicago Bd Health, 58-74, dir div adult health & aging, 63-74; from assoc to assoc prof, 59-71, PROF MED, SCH MED, NORTHWESTERN UNIV, CHICAGO, 71-, CHMN DEPT COMMUNITY HEALTH & PREV MED, 72-, DINGMAN PROF CARDIOL, MED SCH, 73-, CHMN DEPT COMMUNITY HEALTH & PREV MED, MEM HOSP, 73- Concurrent Pos: Fel path, Long Island Col Med, 47; res fel, Cardiovasc Dept, Med Res Inst, Michael Reese Hosp, Chicago, 48; Am Heart Asn estab investr, 52-58, fel coun arteriosclerosis, 63-64 & coun epidemiol, 64-66; dir chronic dis div, Chicago Bd Health, 61-63; chmn coun arteriosclerosis, Am Heart Asn, 63-64, mem exec comt, Coun Epidemiol, 64-66 & coun high blood pressure res; western hemisphere ed, Atherosclerosis, 63-; exec dir, Chicago Health Res Found, 63-72; consult, St Joseph Hosp, 64-, Presby-St Luke's Hosp, 64- & Atherosclerosis Cardiol Drug Lipid Coop Study & Cardiovasc Res Prog Eval Comt, Vet Admin, 65; prof lectr med, Div Biol Sci, Pritzker Sch Med, Univ Chicago, 70-; vis prof, Dept Internal Med, Rush Presby-St Luke's Med Ctr, 72-; attend physician, Northwest Mem Hosp, 72-; sponsor, Nat Health Educ Comt; mem, Worcester Found Exp Biol; specialist clin nutrit, Am Bd Nutrit; chmn coun epidemiol & prev, Int Soc Cardiol, 75- Honors & Awards: Med J Award, Lasker Found, 65; Blakeslee Awards, Am Heart Asn, 64, Award of Merit, 67; Citation, Inter-Soc Comn Heart Dis Resources, 68-71. Mem: Fel AAAS; fel Am Col Cardiol; Am Diabetes Asn; Am Fedn Clin Res; Asn Teachers Prev Med. Res: Cardiovascular physiology, medicine, epidemiology and preventive medicine, particularly atherosclerosis and hypertension; chronic disease, preventive medicine and public health. Mailing Add: Dept Community Health & Prev Med Northwestern Univ Med Sch Chicago IL 60611

STAMM, ALFRED JOAQUIN, b Los Angeles, Calif, Dec 29, 97; m. PHYSICAL CHEMISTRY, WOOD CHEMISTRY. Educ: Calif Inst Technol, BS, 21; Univ Wis,

MS, 23, PhD(colloid chem), 25. Prof Exp: Assoc chemist, Forest Prod Lab, US Forest Serv, 25-28, chemist, 28-36, sr chemist, 36-39, prin chemist, 39-45, chief div derived prod, 45-46, subj matter specialist, 46-59; Robertson distinguished prof wood & paper sci, Sch Forest Resources, 59-68, EMER ROBERTSON PROF WOOD & PAPER SCI, NC STATE UNIV, 68- Concurrent Pos: Rockefeller Found fel, Univ Uppsala, 28-29; chmn colloid div, Am Chem Soc, 34, Wis sect, 36, cellulose div, 53; Fulbright sr res fel, Div Forest Prod, Commonwealth Sci & Indust Res Orgn, Australia, 55-56; vis prof, Univ Calif, Berkeley, 69. Honors & Awards: Anselme Payen Award, Am Chem Soc, 68. Mem: Am Chem Soc; Forest Prod Res Soc; Tech Asn Pulp & Paper Indust; Soc Wood Sci & Technol; Int Acad Wood Sci. Res: Colloid chemistry; physics; derived products from wood; chemical processing of wood. Mailing Add: 3212 Rutherford Dr Raleigh NC 27609

STAMM, RALPH EUGENE, b Piedmont, SC, May 5, 32; m 61; c 2. ANALYTICAL CHEMISTRY, PETROLEUM CHEMISTRY. Educ: Wofford Col, BS, 54; Univ SC, PhD(phys chem), 62. Prof Exp: Sr chemist, Res & Tech Dept, 61-69, proj chemist, 69-73, SR PROJ CHEMIST, TEXACO INC, 73- Mem: Am Chem Soc. Res: Instrumental analysis; mass spectroscopy; computer sciences; spectroscopy. Mailing Add: 645 W Kitchen Port Neches TX 77651

STAMM, ROBERT FRANZ, b Mt Vernon, Ohio, Mar 28, 15; m 64. PHYSICAL CHEMISTRY, OPTICAL PHYSICS. Educ: Kenyon Col, AB, 37; Iowa State Univ, PhD(phys chem), 42. Prof Exp: Asst chem, Iowa State Univ, 37-42; res physicist, Am Cyanamid Co, 42-54, group leader, Basic Res Dept, 54-59 & Phys Res Dept, 59-61, res assoc, Chem Dept, 61-66, res fel, 66-72; SR SCI INVESTR, CLAIROL, INC, DIV BRISTOL MYERS CO, 73- Mem: Am Chem Soc; Am Phys Soc; Optical Soc Am. Res: Raman spectroscopy; light scattering; fluorescence; radiation chemistry and sterilization; neutron activation analysis; spectroscopy of triplet molecules and excited transients; photochromism; flash photolysis; kinetic spectroscopy; retroreflectors; optical properties of human hair fibers. Mailing Add: Clairol Res Labs 2 Blackley Rd Stamford CT 06902

STAMM, STANLEY JEROME, b Seattle, Wash, July 14, 24; m; c 3. MEDICINE. Educ: Seattle Univ, BS, 48; St Louis Univ, MD, 52; Am Bd Pediat, dipl, 58. Prof Exp: Intern, King County Hosp, Seattle, Wash, 52-53; resident pediat, Univ Wash, 53-55; instr, 57-58; dir cardiac diag lab, 58-59, co-dir dept cardiol & attend, 59-63, co-dir cystic fibrosis clin, 63, dir cardiopulmonary res lab, 62-67, dir cardiol dept, 67-70, DIR CARDIOPULMONARY DEPT, CHILDREN'S ORTHOP HOSP, 70- Concurrent Pos: NIH cardiac trainee, Children's Orthop Hosp, 55-56, hosp cardiac fel, 56-57; clin assoc prof, Univ Wash, 66-69. Mem: Fel Am Acad Pediat; Am Heart Asn; fel Am Col Chest Physicians; fel Am Col Angiol. Mailing Add: Children's Orthop Hosp & Med Ctr 4800 Sand Point Way NE Seattle WA 98105

STAMM, WALTER, b Meunster, Westphalia, Dec 14, 26; US citizen. CHEMISTRY. Educ: BSc, 51, MSc, 54, PhD, 57. Prof Exp: Res chemist, Adv Solvents Co, 57-59; sr res chemist, 59-61, group leader, 61-63, supvr, 63-68, spec projs mgr, 68-69, ASST TO VPRES RES, STAUFFER CHEM CO, 69- Mem: Am Chem Soc; NY Acad Sci. Res: Industrial chemistry and technology. Mailing Add: Stauffer Chem Co Westport CT 06880

STAMMELMAN, MORTIMER JACOB, b New York, NY, Dec 27, 97; m 35; c 1. PHYSICAL CHEMISTRY. Educ: Columbia Univ, AB, 20, AM, 21, PhD(phys chem), 23. Prof Exp: Chemist, Int Arms & Fuse Co, 17; chief chemist, Dusenberg Motors Corp, 18; asst chem, Columbia Univ, 18-22; pres, Atmos Prod Corp, 25-26, EMER PRES, ATMOS PROD CORP, 76- Concurrent Pos: Pres, Stammelman Realty Co, 28-32; secy, Hypodermic Medication, Pharmaceut & Biol Sect, C F Kirk Co, NJ, 32-34; humidification consult for tobacco indust. Honors & Awards: Tobacco Indust Distinguished Serv Award, 68. Mem: Fel AAAS; emer fel Am Inst Chem; Sigma Xi. Res: Hydrogen ions; humidification of tobacco and cigars; food preservation and activity coefficients of certain acids. Mailing Add: Atmos Prods Corp 151 W 28th St New York NY 10001

STAMMER, CHARLES HUGH, b Indianapolis, Ind, Apr 1, 25; m 47; c 2. ORGANIC CHEMISTRY. Educ: Univ Ind, BS, 48; Univ Wis, PhD(org chem), 52. Prof Exp: Res chemist, Merck & Co, Inc, NJ, 52-62; ASSOC PROF CHEM, UNIV GA, 62- Mem: Am Chem Soc. Res: Synthesis of peptides and potential antibiotics. Mailing Add: 718 Riverhill Dr Athens GA 30601

STAMMLER, MANFRED, inorganic chemistry, physical chemistry, see 12th edition

STAMPER, JAMES HARRIS, b Richmond, Ind, Sept 10, 38; m 59. THEORETICAL PHYSICS. Educ: Miami Univ, BA, 60; Yale Univ, MS, 62, PhD(physics), 65. Prof Exp: Asst prof physics, Elmira Col, 62-63, dir, 65-66; asst prof, Univ Fla, 67-70; PROF & CHMN DEPT PHYSICS, FLA SOUTHERN COL, 70- Concurrent Pos: Consult, Battelle Mem Inst, 69- Mem: Am Asn Physics Teachers. Res: Atomic and molecular physics. Mailing Add: Dept of Physics Fla Southern Col Lakeland FL 33802

STAMPER, JOHN ANDREW, b Middletown, Ohio, Mar 28, 30; m 59; c 1. PLASMA PHYSICS. Educ: Ohio State Univ, BS, 53; Univ Ky, MS, 58; Univ Md, PhD(physics), 68. Prof Exp: Mem tech staff semiconductor physics, Tex Instruments, Inc, 58-63; RES PHYSICIST, NAVAL RES LAB, DC, 68- Honors & Awards: E O Hulbert Award, US Naval Res Lab, 74. Mem: Am Phys Soc. Res: Experimental plasma physics, including physics of laser-matter interactions; semiconductor physics, including thermal, thermoelectric and thermomagnetic effects; laser-produced plasmas; interaction of laser radiation with plasmas. Mailing Add: Rt 2 Box 71-E Indian Head MD 20640

STAMPER, MARTHA C, b Dawson Springs, Ky, May 7, 25. ORGANIC CHEMISTRY. Educ: DePauw Univ, AB, 47; Univ Wis, PhD(org chem), 52. Prof Exp: ORG CHEMIST, PROCESS RES DIV, ELI LILLY & CO, 52- Mem: Am Chem Soc. Mailing Add: Process Res Div Eli Lilly & Co Indianapolis IN 46206

STAMPER, MAYNARD N, b Whitesburg, Ky, Sept 17, 09; m 43; c 5. INVERTEBRATE ZOOLOGY. Educ: Eastern Ky Univ, BS, 34; Univ Northern Colo, MA, 38; Ohio State Univ, PhD, 54. Prof Exp: Teacher high sch, Ky, 34-37; asst, Univ Northern Colo, 37-38; instr zool & chem, Scottsbluff Jr Col, Nebr, 38-42; asst prof zool & anat, Univ Denver, 46-50; instr gen zool, Ohio State Univ, 51-54; chmn dept biol, 54-68, prof, 54-74, EMER PROF BIOL SCI, UNIV NORTHERN COLO, 74- Concurrent Pos: Consult, Univ Dacca, 68-69. Mem: Nat Audubon Soc. Res: Fresh water Crustacea; nesting success of whiteface ibis in new nesting location. Mailing Add: Dept of Biol Univ of Northern Colo Greeley CO 80631

STAMPFER, JOSEPH FREDERICK, JR, b Dubuque, Iowa, Mar 15, 30; m 53; c 2. ATMOSPHERIC CHEMISTRY. Educ: Dartmouth Col, AB, 52; Univ NMex, PhD(chem), 58. Prof Exp: Staff mem, Los Alamos Sci Lab, 58-67; ASSOC PROF CHEM, UNIV MO-ROLLA, 67- Mem: AAAS; Am Chem Soc; Am Meteorol Soc; Air Pollution Control Asn. Res: Aerosols, generation and properties; distribution of atmospheric aerosols; surfactants; trace atmospheric constituents; surface to atmosphere exchange; aircraft sampling. Mailing Add: 106 Norwood Hall Cloud Physics Res Ctr Univ Mo Rolla MO 65401

STANABACK, ROBERT JOHN, b Weehawken, NJ, Dec 24, 30; m 57; c 1. ORGANIC CHEMISTRY. Educ: Rutgers Univ, BA, 53; Seton Hall Univ, MS, 64, PhD(chem), 66. Prof Exp: Assoc scientist, Warner-Lambert Res Inst, 56-67; SR CHEMIST, TENNECO CHEM, INC, PISCATAWAY, 67- Mem: Am Chem Soc. Res: Synthetic organic medicinals; thyroxine analogs; central nervous depressants; biocides; plasticizers; synthetic polymers; vinyl chloride technology and additives. Mailing Add: RD 1 Union Grove Rd Gladstone NJ 07934

STANACEV, NIKOLA ZIVA, b Milosevo, Yugoslavia, July 17, 28. BIOLOGICAL CHEMISTRY. Educ: Univ Zagreb, Chem E, 53, PhD(chem), 58. Prof Exp: Asst prof med, Univ Zagreb, 55-58; fel div biosci, Nat Res Coun Can, Ottawa, 58-59; res assoc chem & chem eng, Univ Ill, Urbana, 59-61; res assoc cell chem lab, Dept Biochem, Columbia Univ, 61-62; res assoc, Banting & Best Dept Med Res, Univ Toronto, 62-64, lectr, 64-65; res assoc biol chem, Harvard Med Sch, 65-67; assoc prof, 67-74, PROF PATH CHEM, UNIV TORONTO, 74- Res: Organic biochemistry; chemistry and biochemistry of lipids; isolation, determination of constitution and biosynthesis of complex lipids of animal and bacterial origin. Mailing Add: Banting Inst Univ of Toronto 100 College St Toronto ON Can

STANAITIS, OTONAS EDMUNDAS, b Gaisriai, May 28, 05; m 35; c 2. ANALYTICAL MATHEMATICS. Educ: Dipl, Kaunas State Univ, Lithuania, 30; Univ Würzburg, Ger, PhD(math), 32. Prof Exp: Sr asst math, Kaunas State Univ, Lithuania, 31-35, assoc prof, 35-40; assoc prof, Univ Vilnius, 40-44; from instr to prof, 49-74, EMER PROF MATH, ST OLAF COL, 74- Mem: Math Asn Am. Res: Special functions and equations of mathematical physics; infinite series. Mailing Add: Dept of Math St Olaf Col Northfield MN 55057

STANBROUGH, JESS HEDRICK, JR, b Ruston, La, May 1, 18; div; c 1. PHYSICS. Educ: Univ Tex, BS, 49, MA, 50. Prof Exp: Asst geophys, Univ Tex, 48-50, res physicist, Defense Res Lab, 50-60; exec secy, Undersea Warfare Res & Develop Planning Coun, 60-61, tech asst to dir, 61-67, exec secy joint oceanog insts deep earth sampling prog, 66-67, EXEC ASST & RES PHYSICIST, OCEAN ENG DEPT, WOODS HOLE OCEANOG INST, 67- Concurrent Pos: Tech asst, Comt Undersea Warfare, Nat Acad Sci, 56; consult, US Navy, 58-60. Mem: Acoust Soc Am; Marine Technol Soc. Res: Oceanography; underwater sound; navigation; instrumentation. Mailing Add: 36 Riddle Hill Rd Falmouth MA 02540

STANBURY, JOHN BRUTON, b Clinton, NC, May 15, 15; m 45; c 5. EXPERIMENTAL MEDICINE. Educ: Duke Univ, BA, 35; Harvard Med Sch, MD, 39; Am Bd Internal Med, dipl, 49. Prof Exp: House officer, Mass Gen Hosp, Boston, 40-41; asst in med, 46-49, asst clin prof, 56-60, assoc clin prof, 60-66, LECTR MED, HARVARD MED SCH, 66-; PROF EXP MED, MASS INST TECHNOL, 66- Concurrent Pos: Res fel pharmacol, Harvard Med Sch, 47-48; chief med resident, Mass Gen Hosp, 48-49, asst in med, 49-50, chief thyroid clin & lab, 49-66, from asst physician to physician, 50-66, consult physician, 66- Mem: Endocrine Soc; Am Soc Clin Invest; Asn Am Physicians; Am Soc Human Genetics; Am Thyroid Asn. Res: Endocrinology; metabolism; genetics; metabolic disease. Mailing Add: Mass Inst of Technol Cambridge MA 02139

STANCEL, GEORGE MICHAEL, b Chicago, Ill, Dec 29, 44; m 72. BIOCHEMISTRY, ENDOCRINOLOGY. Educ: St Thomas Col, BS, 66; Mich State Univ, PhD(biochem), 70. Prof Exp: ASST PROF PHARMACOL, UNIV TEX MED SCH HOUSTON, 72- Concurrent Pos: NIH fel endocrinol, Univ Ill, 71-72. Mem: Am Chem Soc; Tissue Cult Asn. Res: Biochemical endocrinology; steroid hormone action; hormone receptors; estrogen regulation of uterus and pituitary. Mailing Add: Dept of Pharmacol Univ of Tex Med Sch Houston TX 77025

STANCER, HARVEY C, b Toronto, Ont, Mar 6, 26; m 58; c 2. PSYCHIATRY, NEUROCHEMISTRY. Educ: Univ Toronto, BA, 50, PhD(path chem), 53, MD, 55; Royal Col Physicians Can, cert psychiat, 62, fel, 72. Prof Exp: Head neurochem, Toronto Psychiat Hosp, 62-66; assoc prof, 69-72, PROF PSYCHIAT, FAC MED, UNIV TORONTO, 72-, PROF PSYCHIAT RES, 74-, CHIEF CLIN INVEST UNIV & HEAD NEUROCHEM, CLARKE INST PSYCHIAT, 66- Concurrent Pos: McClean fel, Maudsley Inst, Univ London, 58-59; NY State Dept Hyg fel, Columbia Univ-Presby Med Ctr, 59-61; McLellan fel, Univ Toronto, 61-62; assoc, Med Res Coun Can, 62- Honors & Awards: Clarke Inst Prize, Univ Toronto, 70 & 74; McNeil Award, 72. Mem: Int Soc Neurochem; Neurochem Soc; Soc Biol Psychiat; Psychiat Res Soc; Can Endocrinol Soc & Metab. Res: Clinical psychiatric investigation and animal behavioral investigation of brain biogenic amines. Mailing Add: Clarke Inst of Psychiat Univ of Toronto Toronto ON Can

STANCL, MILDRED LUZADER, b Parkersburg, WVa,; m 66. TOPOLOGY. Educ: Marietta Col, AB, 49; Univ Ill, AM, 62, PhD(math), 69. Prof Exp: Systs analyst, Sperry-Rand Corp, 54-57 & Radio Corp Am, 57-60; teaching asst math, Univ Ill, 61-66; asst prof, Trenton State Col, 68-69 & Kans Univ, 69-72; assoc prof, 72-76, PROF MATH, NICHOLS COL, 76- Mem: Am Math Soc; Math Asn Am; Asn Teachers Quant Methods; Sigma Xi. Res: Isomorphisms of smooth manifolds surrounding polyhedra. Mailing Add: Dept of Math Nichols Col Dudley MA 01570

STANCLIFF, MERTON WESLEY, b Morristown, Vt, Apr 30, 30; m 56; c 2. ANTHROPOLOGY. Educ: Univ Vt, AB, 52; Columbia Univ, PhD(anthrop), 66. Prof Exp: Instr anthrop, Univ Man, 61-62, asst prof, 62-66; asst prof, Boston Univ, 66-71; ASSOC PROF ANTHROP, STATE UNIV NY COL PLATTSBURGH, 71- Mem: Fel Am Anthrop Asn; Am Ethnol Soc. Res: Cultural anthropology; peasant societies, European and general; basic values systems; culture change; acculturation; human cultural origins. Mailing Add: Dept of Anthrop State Univ NY Col Plattsburgh NY 12901

STANCYK, STEPHEN EDWARD, b Denver, Colo, Apr 8, 46. INVERTEBRATE ZOOLOGY, MARINE ECOLOGY. Educ: Univ Colo, BA, 68; Univ Fla, MS, 70, PhD(zool), 74. Prof Exp: Instr biol, Univ Fla, 74-75; ASST PROF BIOL & MARINE SCI, UNIV SC, 75- Mem: Am Soc Zoologists; AAAS; Ecol Soc Am; Sigma Xi. Res: Marine invertebrate ecology; life history patterns, general biology and reproductive and larval ecology of tropical and estuarine Atlantic benthic invertebrates, especially ophiuroids; marine community ecology. Mailing Add: Dept of Biol Univ of SC Columbia SC 29208

STANDAERT, FRANK GEORGE, b Paterson, NJ, Nov 12, 29; m 59; c 3. PHARMACOLOGY. Educ: Harvard Univ, AB, 51; Cornell Univ, MD, 55. Prof Exp: Intern med, Johns Hopkins Hosp, Baltimore, Md, 55-56; from instr to assoc prof pharmacol, Med Col, Cornell Univ, 59-67; PROF PHARMACOL & CHMN DEPT, SCH MED & DENT, GEORGETOWN UNIV, 67- Concurrent Pos: Res fel pharmacol, Med Col, Cornell Univ, 56-57; USPHS career develop award, 61-67; guest scientist, Naval Med Ctr, Bethesda, Ma; mem, Nat Res Coun & Am Asn Dent Schs;

chmn publ comt, Fedn Am Socs Exp Biol; mem comt toxicol, Nat Acad Sci-Nat Res Coun; neurol dis prog proj rev comt, NIH; ed, Neuropharmacol, J Pharmacol & Exp Therapeut; mem merit rev bd neurobiol, Vet Admin; mem basic pharmacol adv comt, Pharmaceut Mfrs Asn Found; mem bd dirs, Washington Heart Asn; secy, Asn Med Sch Pharmacol. Mem: AAAS; Am Soc Pharmacol & Exp Therapeut (secy-treas); Soc Exp Biol & Med; Sigma Xi; Am Soc Clin Pharmacol & Therapeut. Res: Pharmacology of neuromuscular transmission and cardiac rhythm; structure activity relationships of quaternary amines. Mailing Add: 8205 Stone Trail Dr Bethesda MD 20034

STANDAL, BLUEBELL R S'YIEM, b Shillong, India, Feb 8, 21; m 59; c 1. NUTRITION. Educ: Univ Calcutta, BSc, 42; Univ Calif, MS, 48, PhD, 52. Prof Exp: Res assoc path, Sch Med, Stanford Univ, 52-56; asst res officer, All-India Inst Med Sci, 57-58; asst prof & asst nutritionist, 58-68, assoc prof, 68-74, PROF FOODS & NUTRIT, UNIV HAWAII, 74-, ASSOC NUTRITIONIST, 68- Mem: AAAS. Res: Riboflavin deficiency and stress; metachromasy, polyploidy and basic proteins of tumor tissues; metabolites of estrone and progesterone; proteins and amino acids of foods of Hawaii. Mailing Add: Dept of Foods & Nutrit Univ of Hawaii Honolulu HI 96822

STANDEFER, JIMMY CLAYTON, b Stanton, Tex, Mar 2, 41; m 68; c 2. BIOCHEMISTRY, CLINICAL CHEMISTRY. Educ: Univ Kans, BA, 63, PhD(biochem), 67; Nat Registry Clin Chem, cert, 70. Prof Exp: Biochemist, Walter Reed Army Inst Res, 67-70; instr biochem, Sch Nursing, Univ Md, 69-70; ASST PROF CLIN CHEM, SCH MED, UNIV N MEX, 70- Mem: Am Asn Clin Chem; Am Acad Forensic Sci. Res: Membrane chemistry; isoenzymes; lipids. Mailing Add: Bernalillo Co Med Ctr Univ of NMex Sch of Med Albuquerque NM 87106

STANDEFORD, LEO VERN, b Terre Haute, Ind, June 12, 36. ASTRONOMY. Educ: Ind State Univ, Terre Haute, BS, 58, MS, 59; Univ Ill, Urbana, MS, 64, PhD(astron), 68. Prof Exp: ASSOC PROF ASTRON & MATH, MANKATO STATE COL, 68- Concurrent Pos: Consult astron, Aerial Phenomena Res Org, 69- Mem: AAAS; Am Astron Soc. Res: Solar system astronomy with attention to perturbing forces. Mailing Add: Dept of Math & Astron Mankato State Col Mankato MN 56001

STANDER, JOSEPH W, b Covington, Ky, Dec 2, 28. ALGEBRA. Educ: Univ Dayton, BS, 49; Cath Univ, MS, 57, PhD(math), 59. Prof Exp: Teacher, Hamilton Cath High Sch, 49-50; teacher, Colegio Ponceno, 50-55; asst prof, 60-65, ASSOC PROF MATH, UNIV DAYTON, 65-, DEAN GRAD STUDIES & RES, 69-, VPRES ACAD AFFAIRS & PROVOST, 75- Mem: Am Math Soc; Math Asn Am; Sigma Xi. Res: Matrix theory. Mailing Add: Univ of Dayton Dayton OH 45469

STANDER, RICHARD WRIGHT, b Grand Rapids, Mich, Oct 23, 22; m 43; c 4. OBSTETRICS & GYNECOLOGY. Educ: Univ Mich, MD, 51. Prof Exp: Instr obstet & gynec, Med Sch, Univ Mich, 56-57; instr, Col Med Evangelists, 57-58; from asst prof to assoc prof, Sch Med, Ind Univ, 58-67; PROF OBSTET & GYNEC & CHMN DEPT, COL MED, UNIV CINCINNATI, 67- Concurrent Pos: Vis prof, Kapiolani Hosp, 67; chmn test comt obstet & gynec, Nat Bd Med Exam; consult, Nat Inst Child Health & Human Develop. Mem: Am Col Obstet & Gynec; AMA; Am Gynec Soc; Am Asn Obstet & Gynec. Res: Obstetrics and gynecology; problems of human reproduction; uterine physiology. Mailing Add: Dept of Obstet & Gynec Univ of Cincinnati Col of Med Cincinnati OH 45267

STANDIFER, LEONIDES CALMET, JR, b Gulfport, Miss, Apr 24, 25; m 57; c 2. PLANT PHYSIOLOGY. Educ: Miss State Univ, BS, 50, MS, 54; Univ Wis, PhD(bot), 59. Prof Exp: Plant physiologist, Firestone Plantations Co, 54-61; from asst prof to assoc prof, 61-74, PROF BOT, LA STATE UNIV, BATON ROUGE, 74- Mem: Bot Soc Am; Am Soc Plant Physiol; Weed Sci Soc Am. Res: Patterns of plant recovery from flame injury; histological responses of certain plants to herbicides, and physiology of herbicidal action. Mailing Add: Dept of Hort La State Univ Baton Rouge LA 70803

STANDIFER, LONNIE NATHANIEL, b Itasca, Tex, Oct 28, 26; div. ENTOMOLOGY, PARASITOLOGY. Educ: Prairie View Agr & Mech Col, BS, 49; Kans State Col, MS, 51; Cornell Univ, PhD(med & vet entom & parasitol), 54. Prof Exp: Instr biol sci & supvr campus pest control, Tuskegee Inst, 51-52; asst livestock insect control, Cornell Univ, 53-54; asst prof biol sci, Southern Univ, 54-56; res scientist & apiculturist, 56-70, DIR BEE RES LAB, USDA, 70-, RES LEADER, HONEY BEE POLLINATION LAB, 72-, TECH ADV APICULTURE, WESTERN REGION, 73- Mem: AAAS; Entom Soc Am; Am Soc Parasitol; Am Beekeeping Fedn. Res: Medical and veterinary entomology and parasitology; control of insects of public health importance; insect physiology and nutrition; botany and plant pathology; honey bee physiology and nutrition, protein and lipids; pollen chemistry, fatty acids, sterols and hydrocarbons. Mailing Add: 8031 E 17th Place Tucson AZ 85710

STANDIL, SIDNEY, b Winnipeg, Man, Oct 19, 26; m 50; c 4. PHYSICS. Educ: Queen's Univ, Ont, BSc, 48, MSc, 49; Univ Man, PhD, 51. Prof Exp: From asst prof to assoc prof, 51-63, PROF PHYSICS, UNIV MAN, 63-, DEAN FAC GRAD STUDIES, 73- Mem: Am Phys Soc; Am Geophys Union; Can Asn Physicists. Res: Cosmic ray and space physics. Mailing Add: 772 Campbell St Winnipeg MB Can

STANDING, KEITH M, b Ogden, Utah, Aug 2, 28; m 56; c 7. VERTEBRATE ZOOLOGY. Educ: Brigham Young Univ, BS, 53, MS, 55; Wash State Univ, PhD(zool, bot), 60. Prof Exp: Assoc prof, 58-69, PROF & CHMN DEPT BIOL, CALIF STATE UNIV, FRESNO, 69- Concurrent Pos: NSF res grants, 64-67; mem, NSF Conf Histochem—Its Appl in Res & Teaching, Vanderbilt Univ, 65; mem gov bd, Moss Landing Marine Labs; bd dirs, Ctr Urban & Regional Studies; vis scholar, Univ Calif, Berkeley. Mem: AAAS; Am Soc Zool. Res: Histological analysis of reproductive organs of blue grouse; comparative histology of nephron units of kangaroo rats; isolation of nephron units of Dipodomys by various techniques; cytotaxonomy of Dipodomys; embryonic kidney development. Mailing Add: Dept of Biol Calif State Univ Fresno CA 93710

STANDING, KENNETH GRAHAM, b Winnipeg, Man, Apr 3, 25; m 61; c 4. NUCLEAR PHYSICS. Educ: Univ Man, BSc, 48; Princeton Univ, AM, 50, PhD(physics), 55. Prof Exp: From asst prof to assoc prof, 53-64, PROF PHYSICS, UNIV MAN, 64- Concurrent Pos: Nuffield Found Dom traveling fel, Wills Physics Lab, Bristol Univ, 58-59; Nat Res Coun Can sr res fel, Univ Grenoble, 67-68. Mem: Am Phys Soc; Can Asn Physicists. Res: Nuclear physics and applications; cosmic rays. Mailing Add: Dept of Physics Univ of Man Winnipeg MB Can

STANDISH, CHARLES JUNIOR, b Traingle, NY, Nov 10, 26. MATHEMATICS. Educ: Hamilton Col, NY, BA, 49; Johns Hopkins Univ, MA, 51; Cornell Univ, PhD(math), 54. Prof Exp: Instr math, Hamilton Col, NY, 51-52; asst, Cornell Univ, 52-54; asst prof, Union Univ, NY, 54-57; MATHEMATICIAN, IBM CORP, 57- Concurrent Pos: Vis assoc prof, NC State Univ, 60-61; vis lectr sch adv technol, State Univ NY Binghamton, 74- Mem: Am Math Soc; Soc Indust & Appl Math. Res: Measure theory; integral transforms; control theory. Mailing Add: RD 2 Greene NY 13778

STANDISH, E MYLES, JR, b Hartford, Conn, Mar 5, 39; m 68. ASTRONOMY. Educ: Wesleyan Univ, BA, 60, MA, 62; Yale Univ, PhD(astron), 68. Prof Exp: Asst prof astron, Yale Univ, 68-72; MEM TECH STAFF, JET PROPULSION LAB, 72- Mem: Am Astron Soc-Div Dynamic Astron. Res: Celestial mechanics; numerical analysis. Mailing Add: CPB-104 Jet Propulsion Lab Pasadena CA 91103

STANDISH, NORMAN WESTON, b Marion, Iowa, Apr 4, 30; m 56; c 3. ORGANIC CHEMISTRY. Educ: Beloit Col, BS, 52; Purdue Univ, MS, 57, PhD(org chem), 59. Prof Exp: Chemist, Selectron Div, Pittsburgh Plate Glass Co, 52-53; res assoc biochem, 60-66, tech dir plastics, Prophylactic Brush Div, 66-69, res supvr, Cleveland, 69-70, SUPVR DEVELOP, TECH SERV & POLYMERS, STANDARD OIL CO OHIO, 70- Mem: Am Chem Soc; Soc Plastics Engrs; Soc Nematol; fel Am Inst Chemists; Soc Plastics Indust. Res: Development of high nitrile barrier resins; design of plastic pressure vessels; high speed fabrication of plastic containers. Mailing Add: Standard Oil Co Ohio 4440 Warrensville Center Rd Cleveland OH 44128

STANDISH, SAMUEL MILES, b Campbellsburg, Ind, July 6, 23; m 49; c 2. ORAL PATHOLOGY. Educ: Ind Univ, DDS, 45, MS, 56; Am Bd Oral Path, dipl, 59. Prof Exp: Instr dent, 52-57, from asst prof to assoc prof oral path, 57-67, asst dean grad & postgrad educ, 69-74, PROF ORAL PATH, SCH DENT, IND UNIV, INDIANAPOLIS, 67-, ASSOC DEAN GRAD & POSTGRAD EDUC, 74- Mem: Am Dent Asn; fel Am Acad Oral Path; Int Asn Dent Res. Res: Salivary gland pathophysiology and experimental carcinogenesis; inflammatory mechanisms; striated muscle regeneration; muscle diseases; clinical oral pathology. Mailing Add: Dept of Oral Path Ind Univ Sch of Dent Indianapolis IN 46202

STANDLEE, WILLIAM JASPER, b Zybach, Tex, May 2, 29; m 58; c 3. POULTRY NUTRITION. Educ: Tex Tech Univ, BS, 54, MS, 55; Tex A&M Univ, PhD(poultry sci), 63. Prof Exp: Animal nutritionist, Standard Milling Co, Tex, 55-57; salesman, Van Waters & Rogers, 57-58; res asst, Tex Agr Exp Sta, 58-63; dir nutrit & res, Darragh Co, Ark, 63-65; dir nutrit, Burrus Feed Mills, Tex, 65-68; dir nutrit & res, Food Div, Valmac Industs, Inc, Ark, 68-71; NUTRITIONIST, B & D MILLS, 71- Mem: AAAS; Poultry Sci Asn. Res: Nutrition and feeding management of turkey breeders; broiler chicken breeders, market turkeys and broilers and egg production chickens. Mailing Add: 815 N Lucas Grapevine TX 76051

STANEK, JEAN CHAN, b Toyshan, China, Apr 24, 37; US citizen; m 60; c 2. MATHEMATICS. Educ: Univ Chicago, BS, 60, MS, 61; Univ Calif, Los Angeles, PhD(math), 71. Prof Exp: Asst prof math, Loyola Univ Los Angeles, 72-73; ASST PROF MATH, CALIF STATE COL, SONOMA, 73- Mem: Am Math Soc. Res: Convex sets. Mailing Add: Dept of Math Calif State Col Sonoma Rohnert Park CA 94928

STANEK, PETER, b Chicago, Ill, Dec 3, 37; m 60; c 2. MATHEMATICS. Educ: Univ Chicago, MS, 58, PhD(math), 61. Prof Exp: Mem, Inst Defense Anal, 61-62; analyst, Opers Eval Group, 62-63; asst prof math, Univ Southern Calif, 63-65; sr scientist, Jet Propulsion Lab, 65-68 & Lear Siegler Inc, 68-72; MEM STAFF, SYSTS APPLNS INC, 72- Mem: Am Math Soc; Soc Indust & Appl Math; Human Factors Soc; Am Inst Aeronaut & Astronaut. Res: Algebra; operations research; communications engineering; human factors engineering; computer applications. Mailing Add: 1281 Idylberry Rd San Rafael CA 94903

STANERSON, BRADFORD ROY, b Kanawha, Iowa, Jan 23, 07; m 32; c 3. CHEMISTRY. Educ: Iowa State Univ, BS, 30, MS, 31, PhD(inorg chem), 39. Prof Exp: Asst chem, Iowa State Univ, 30-31, instr, 36-40; instr, Dowling Col, 31-36; proj leader, Texaco, NY, 40-47; asst to exec secy, Am Chem Soc, DC, 47-54, asst secy, 54-57, dir membership affairs, 58-60, dep exec secy, 60-65, exec secy, 65-70; nat dir, Sigma Xi; 70-73; RETIRED. Concurrent Pos: Pres, Sci Manpower Comn, 59; secy-treas, InterSoc Comt Lab Serv Related to Health, 58-59; bd dirs, Am Chem Soc, 71- Mem: AAAS; Am Chem Soc; Am Inst Chem; Sigma Xi. Res: Adsorption of gases on catalysts; analytical methods in petroleum; analysis and production of butadiene, chemicals and high polymers from petroleum; manpower; administration. Mailing Add: 1216 Tanley Rd Silver Spring MD 20904

STANFIELD, JAMES AARMOND, b Covington, Ky, Aug 28, 17; m 42; c 3. ORGANIC CHEMISTRY. Educ: Eastern Ky State Col, BS, 40; Univ Tenn, MS, 42, PhD(phys org chem), 47. Prof Exp: Instr chem, Univ Ky, 41-42; instr, Univ Tenn, 42-46; from asst prof to assoc prof, 47-56, PROF CHEM, GA INST TECHNOL, 56-, ASST DIR SCH CHEM, 65-, RES ASSOC, RES INST, 51- Mem: Am Chem Soc. Res: Organic synthesis; catalytic hydrogenation kinetics; spirobarbituric acids; chemistry of uramil. Mailing Add: Sch of Chem Ga Inst Technol Atlanta GA 30332

STANFIELD, KENNETH CHARLES, b Los Angeles, Calif, Sept 21, 42; m 65; c 1. EXPERIMENTAL HIGH ENERGY PHYSICS. Educ: Univ Tex, BS, 64; Harvard Univ, AM, 67, PhD(physics), 69. Prof Exp: Res assoc physics, Univ Mich, 69-71; ASST PROF PHYSICS, PURDUE UNIV, 71- Concurrent Pos: Mem prog adv comt, Zero Gradient Synchrotron, Argonne Nat Lab, 75- Mem: Sigma Xi. Res: Experimental research, using electronic techniques, into the nature of elementary particle properties. Mailing Add: Dept of Physics Purdue Univ West Lafayette IN 47907

STANFIELD, MANIE K, b St Petersburg, Fla, Feb 15, 31. ORGANIC CHEMISTRY, BIOCHEMISTRY. Educ: Univ Chicago, BA, 54, MS, 57; Univ Calif, Los Angeles, PhD(org chem), 62. Prof Exp: Asst org chem, Mass Inst Technol, 62-63 & Rockefeller Univ, 63-65; ASST PROF BIOCHEM, SCH MED, TULANE UNIV, 65- Mem: AAAS; Am Chem Soc. Res: Organic syntheses of small biologically interesting molecules; inborn errors of amino acid metabolism; synthesis of central nervous system stimulating compounds, stability and polymerization of beta-lactam antibiotics. Mailing Add: Dept of Biochem Tulane Univ Med Sch New Orleans LA 70112

STANFORD, AUGUSTUS LAMAR, JR, b Macon, Ga, Jan 20, 31; m 52; c 2. SOLID STATE PHYSICS. Educ: Ga Inst Technol, BS, 52, MS, 57, PhD(physics), 58. Prof Exp: Sr staff consult, Sperry Rand Corp, 58-64; assoc prof, 64-74, PROF PHYSICS, GA INST TECHNOL, 74- Concurrent Pos: NASA res grant, 64- Mem: Am Phys Soc. Res: Nuclear spectroscopy; ferroelectrics; pyroelectrics; phonon in solids. Mailing Add: Sch of Physics Ga Inst of Technol Atlanta GA 30332

STANFORD, ERNEST HALL, b Bardolph, Ill, Dec 9, 11; m 42; c 2. PLANT BREEDING. Educ: Univ Minn, BS, 37; Univ Calif, PhD(genetics), 40. Prof Exp: Jr agronomist, 41-46, asst prof agron & asst agronomist, Exp Sta, 46-52, assoc prof & assoc agronomist, 52-59, PROF AGRON & AGRONOMIST, EXP STA, UNIV CALIF, DAVIS, 59- Mem: Am Soc Agron; Am Phytopath Soc; Genetics Soc Can. Res: Genetics and cytogenetics of alfalfa; breeding forage crops for disease and insect resistance. Mailing Add: Dept of Agron Univ of Calif Davis CA 95616

STANFORD, GEOFFREY BRIAN, b London, Eng, Mar 29, 16; m 59; c 3. ENVIRONMENTAL MANAGEMENT. Educ: Royal Col Physicians & Surgeons, dipl med radiol & LRCPS, 39. Prof Exp: Vis prof, Sch Archit & Environ Planning,

Calif State Polytech Col, San Luis Obispo, 70; adj prof, Environ Studies Ctr, Antioch Col, 71; vis prof & dir, Environics Ctr, St Edward's Univ, 71-72; biomed res scientist, Urban Health Module, Sch Pub Health, Univ Tex, Houston, 72-74; VIS PROF & RESOURCE RECOVERY PLANNING SPECIALIST, INST URBAN STUDIES, UNIV TEX, ARLINGTON, 74-; PRES, AGRO-CITY, INC, 74- Concurrent Pos: Adj prof, Sch Archit, Rice Univ, 72-74; mem environ & eco-systs planning comt, Prep Planning Group, UN Environ Prog for 1976 Habitat Conf, 74; UN Environ Prog deleg, Symp on Develop Patterns of Human Settlements in Developing Countries for Yr 2000, 75; proj dir res into effects of landmix, Environ Protection Agency; dir, Greenhills Environ Studies Ctr; trustee, Environic Found Int. Mem: Archit Asn Gt Brit; Am Soc Testing & Mat; fel Royal Photog Soc Gt Brit; Soil Asn; AAAS. Res: Use of municipal waste resources for restoring soil fertility and water characteristics; measurement of the effects on crop yield, water quality and climate; application to regional environmental management. Mailing Add: Agro City Inc Greenhills Rte 1 Box 861 Cedar Hill TX 75104

STANFORD, GEORGE, soil science, see 12th edition

STANFORD, GEORGE STAILING, b Halifax, NS, July 23, 28; m 56; c 3. REACTOR PHYSICS. Educ: Acadia Univ, BSc, 49; Wesleyan Univ, MA, 51; Yale Univ, PhD(nuclear energy levels), 56. Prof Exp: Proj engr infrared instrumentation, Perkin-Elmer Corp, 55-59; PHYSICIST, ARGONNE NAT LAB, 59- Mem: Am Nuclear Soc. Res: Experimental reactor physics. Mailing Add: Argonne Nat Lab D208 9700 S Cass Ave Argonne IL 60439

STANFORD, JACK ARTHUR, b Delta, Colo, Feb 18, 47; m 69; c 2. LIMNOLOGY. Educ: Colo State Univ, BS, 69, MS, 71; Univ Utah, PhD(limnol), 75. Prof Exp: Asst fish & wildlife, Colo State Univ, 65-69, asst zool, 69-72; res limnol, Univ Utah & Univ Mont, 72-74; ASST PROF LIMNOL & BIOL, N TEX STATE UNIV, 74- Concurrent Pos: Res biologist, Biol Sta, Univ Mont, 72-74; consult, 208 Planning Bd, Kalispell Mont, 75- Mem: Sigma Xi; Int Soc Theoret & Appl Limnol; Am Soc Limnol & Oceanog; Ecol Soc Am; NAm Benthological Soc. Res: All aspects of limnological study in lakes and streams with special interest in nutrient cycling by algae and heterotrophic bacteria; benthic ecology and life histories of the Plecoptera; applied biometrics. Mailing Add: Dept of Biol Sci NTex State Univ Denton TX 76203

STANFORD, JACK WAYNE, b Eldorado, Tex, Dec 21, 35; m 58; c 2. PLANT TAXONOMY. Educ: Baylor Univ, BA, 58; Tex Tech Univ, MS, 66; Okla State Univ, PhD(bot), 71. Prof Exp: Teacher jr high sch, Tex, 60-62, high sch, 62-66; asst prof biol, 66-68, assoc prof, 70-74, PROF BIOL, HOWARD PAYNE COL, 74- Res: Pollen morphology of the Mimosoideae; floristic studies of central Texas. Mailing Add: Dept of Biol Howard Payne Col Brownwood TX 76801

STANFORD, JOHN PERSHING, b Linden, Tex, Feb 24, 18; m 51. AGRONOMY. Educ: Agr & Mech Col, Tex, BS, 47; Pa State Col, PhD(agron, plant physiol), 51. Prof Exp: Agriculturist, Oldbury Electro-Chem Co, 51-56, agronomist, Hooker Chem Corp, 56-69; sr agr chem specialist, 69-75, MGR SODIUM CHLORATE PRODS, OCCIDENTAL CHEM CO, 75- Mem: Am Soc Agron; Am Soc Plant Physiol; Weed Sci Soc Am. Res: Turf maintenance and culture; weed eradication; adaptation of herbaceous materials for highway slope erosion control, pesticide sale and service. Mailing Add: 3038 Eagle Dr Memphis TN 38118

STANFORD, JOHN W, b Nashville, Tenn, Mar 22, 27; m 49; c 4. ORGANIC CHEMISTRY, BIOCHEMISTRY. Educ: Univ Md, BS, 52; Georgetown Univ, MS, 56, PhD(med & dent sci), 61. Prof Exp: Res assoc, Am Dent Asn, 52-66, dir div biophys, Res Inst, 70-73, SECY, COUN ON DENT MAT & DEVICES, AM DENT ASN, 66- Concurrent Pos: Guest scientist, Naval Med Res Inst, Nat Naval Med Ctr, Md, 59-61; mem dent study sect, Div Res Grants, NIH, 65-69; chmn rev & classification panel dent devices, Food & Drug Admin, 72-; chmn med devices, Stand Mgt Bd, Am Nat Stand Inst, 75- Mem: AAAS; Am Dent Asn; Am Col Dent; Int Dent Fedn; Am Inst Chem. Res: Basic properties of dental materials; development of test methods. Mailing Add: Am Dent Asn 211 E Chicago Ave Chicago IL 60611

STANFORD, LYLE MORRIS, b Livingston, Mont, Nov 26, 10; m 34; c 5. ZOOLOGY, BOTANY. Educ: Col Idaho, BA, 33; Univ Wash, MS, 37, PhD(zool), 42. Prof Exp: Teacher schs, Idaho, 32-40; mem staff chem & biol, 42-44, leader exped, 46-47, PROF BIOL & HEAD DEPT, COL IDAHO, 44- Concurrent Pos: Ford fel, Harvard Univ, 54-55; NSF sci fac fel biol conserv, US & Australia, 65-66; regional studies grant, 70-73; leader study exped Australia, 71; mem, Idaho State Bd Environ Protection & Health, 73. Res: Flora; petrified woods; history and biology of Snake River; fish of Snake River region; ecology. Mailing Add: Dept of Biol Col of Idaho Caldwell ID 83605

STANFORD, RICHARD HENRY, JR, crystallography, see 12th edition

STANG, LOUIS GEORGE, JR, b Portland, Ore, Oct 25, 19; m 43; c 3. RADIOCHEMISTRY. Educ: Reed Col, BA, 41. Frof Exp: Res chemist, Nat Defense Res Comt, Northwestern Univ, 42-43, Calif Inst Technol, 43, Clinton Labs, Tenn, 43-44, Metall Lab, Univ Chicago, 44-45, Monsanto Chem Co, Ohio, 45 & Universal Oil Prod Co, Ill, 45-47; CHEMIST & DIV HEAD, BROOKHAVEN NAT LAB, 47- Concurrent Pos: USAEC consult, Yugoslavia & Israel, 60 & rep, regional meetings utilization res reactors, Int Atomic Energy Agency, Manila, 63, Bombay, 64; Indian Atomic Energy Estab & Int Atomic Energy Agency lect prod radioisotopes, Bombay, 64. Honors & Awards: Distinguished Serv Award, Am Nuclear Soc, 69. Mem: AAAS; Am Chem Soc; Am Nuclear Soc. Res: Production of radioisotopes; spallation reactions; radionuclide generators; medical applications of radionuclides; design of radioactive laboratories and equipment; use of teeth as indicators of concentrations of trace elements in the human body. Mailing Add: Brookhaven Nat Lab Upton NY 11973

STANG, PETER JOHN, b Nürenberg, Ger, Nov 17, 41; US citizen; m 69. ORGANIC CHEMISTRY. Educ: DePaul Univ, BS, 63; Univ Calif, Berkeley, PhD(chem), 66. Prof Exp: NIH fel chem, Princeton Univ, 66-68, instr, 68-69; asst prof, 69-75, ASSOC PROF CHEM, UNIV UTAH, 75- Mem: Am Chem Soc; The Chem Soc; AAAS. Res: Kinetics and mechanisms of organic reactions; reactive intermediates; unsaturated cations and carbenes; molecules of medicinal and biological interest. Mailing Add: Dept of Chem Univ of Utah Salt Lake City UT 84112

STANGE, HUGO, b Elizabeth, NJ, June 24, 21; m 42; c 5. ORGANIC CHEMISTRY, INORGANIC CHEMISTRY. Educ: Northwestern Univ, BS, 42, PhD(chem), 50. Prof Exp: Chemist, Pa Ord Works, US Rubber Co, 42; chemist res dept, Olin Mathieson Chem Corp, 50-52, sect leader, 52-55; mgr org res, 55-60, mgr org & polymer res, 60-62, res mgr, 62-65, asst dir, Cent Res Dept, 65-72, dir cent tech dept, 72-74, DIR PRINCETON CTR TECH DEPT & MARCUS HOOK TECH SERVS, FMC CORP, 74- Mem: Am Chem Soc; The Chem Soc. Res: Thianaphthene and boron chemistry; industrial process and product development in organic and inorganic chemistry; agricultural pesticides; polymers; chemical research management. Mailing Add: 19 Hamilton Ave Princeton NJ 08540

STANGEL, HARVEY J, b Kewaunee, Wis, Oct 15, 23; m 51; c 12. AGRONOMY, PLANT PHYSIOLOGY. Educ: Univ Wis, BS, 49, MS, 50, PhD(agron, plant physiol), 53. Prof Exp: Asst agron, Univ Wis, 49-53; dist agronomist, Allied Chem Corp, 53-57, chief agronomist, 58-63, chief agriculturist, 64-66, mgr int mkt develop agr chem, 66-69, mgr mkt res, Planning Dept Agr Div, 69-70; CONSULT, HARVEY J STANGEL & ASSOCS, 70- Mem: Am Soc Agron; Soil Sci Soc Am; Am Soc Hort Sci; Am Mkt Asn; Fertilizer Soc. Res: Marketing management in chemicals specializing in agricultural chemicals; environmental assessment of projects. Mailing Add: 116 Lincoln Ave Highland Park NJ 08904

STANGER, PHILIP CHARLES, b Newark, NJ, Nov 11, 20; m 43; c 2. ASTRONOMY. Educ: Montclair State Teachers Col, AB, 42; Okla Agr & Mech Col, MS, 49; Ohio State Univ, MA, 54. Prof Exp: Instr math, Okla Agr & Mech Col, 46-48; instr, Ohio Univ, 48-50; from instr to asst prof, 52-59, from asst prof astron to assoc prof, 59-74, PROF ASTRON, OHIO WESLEYAN UNIV, 74- CHMN DEPT, 59- Mem: AAAS; Am Astron Soc. Res: Spectroscopic binaries; stellar atmospheres. Mailing Add: Dept of Astronomy Ohio Wesleyan Univ Delaware OH 43015

STANGHELLINI, MICHAEL EUGENE, b San Francisco, Calif, Mar 21, 40; m 66; c 2. PLANT PATHOLOGY. Educ: Univ Calif, Davis, BA, 63; Univ Hawaii, MS, 65; Univ Calif, Berkeley, PhD(plant path), 69. Prof Exp: Asst prof, 69-72, ASSOC PROF PLANT PATH, UNIV ARIZ, 72- Mem: Am Phytopath Soc. Res: Soil borne fungal pathogens. Mailing Add: Dept of Plant Path Univ of Ariz Tucson AZ 85721

STANIER, ROGER YATE, microbiology, see 12th edition

STANIFORTH, DAVID WILLIAM, b Esther, Alta, Sept 5, 19; m 45; c 3. AGRONOMY. Educ: Univ Sask, BSA, 44, MSc, 46; Iowa State Col, PhD(bot, plant physiol), 49. Prof Exp: Asst crop-breeding, Univ Sask, 41-44, asst crop ecol, 44-46; instr bot, 47-48, res assoc, 48-49, asst prof, 49-51, assoc prof bot & agron, 51-60, PROF BOT & AGRON, IOWA STATE UNIV, 60- Mem: Am Soc Plant Physiol; Weed Sci Soc Am; Am Soc Agron. Mem: Control of weeds in economic crops; physiology of herbicides; crop ecology; growth of crops and weeds. Mailing Add: Dept of Bot Iowa State Univ Ames IA 50010

STANIFORTH, RICHARD JOHN, b Sidmouth, Eng, Oct 2, 46; Can citizen; m 71. PLANT ECOLOGY. Educ: Univ Col NWales, BS, 68; Univ Col Western Ont, PhD(plant sci), 75. Prof Exp: Lectr biol, Univ Western Ont, 73-75; ASST PROF PLANT ECOL & TAXON, UNIV WINNIPEG, 75- Mem: Brit Ecol Soc; Can Bot Asn. Res: Comparative ecology of weed species of the genus Polygonum L; colonization and population dynamics of riverbank plant populations. Mailing Add: Dept of Biol Univ of Winnipeg Winnipeg MB Can

STANIFORTH, ROBERT ARTHUR, b Cleveland, Ohio, Oct 5, 17; m 44; c 3. INORGANIC CHEMISTRY. Educ: Case Western Reserve Univ, BA, 39; Ohio State Univ, MS, 42, PhD(inorg chem), 43. Prof Exp: Asst chem, Ohio State Univ, 39-43; res chemist, Monsanto Chem Co, Ohio, 44-46, group leader, 47, sect chief, AEC, Mound Lab, 48, res dir, 48-54, mgr chem develop, Inorg Chem Div, 54-59, asst dir develop, 59-62; mgr prod planning, 62-69, MGR COMMUN & INFO, MONSANTO INDUST CHEM CO, 69- Concurrent Pos: Res chemist, Gen Aniline & Film Corp, Pa, 43. Mem: Am Chem Soc; Electrochem Soc. Res: Ultramicrobalances; chelate compounds of the rare earth metals; radiochemistry; metals; semiconductors. Mailing Add: Monsanto Chem Co 800 N Lindbergh Blvd St Louis MO 63166

STANIN, THEODORE E, b Utica, NY, July 18, 11; m 36; c 3. ORGANIC CHEMISTRY. Educ: Univ Rochester, BA, 43. Prof Exp: Chemist, Eastman Kodak Co, 35-50; sr res chemist, 50-54, DEPT HEAD, TENN-EASTMAN CO, 54-, RES ASSOC, 67- Mem: Am Chem Soc. Res: Polymer chemistry; process and product development. Mailing Add: 1300 Greenfield Place Kingsport TN 37664

STANIONIS, VICTOR ADAM, b New York, NY, Dec 24, 38; m 60; c 2. APPLIED MATHEMATICS. Educ: Iona Col, BS, 60; NY Univ, MS, 64; Queen's Col, MA, 70; Columbia Univ, PhD(math), 75. Prof Exp: Instr math & physics, 61-66, ASST PROF PHYSICS, IONA COL, 66-, CHMN DEPT, 75- Concurrent Pos: Adj asst prof math, Queensborough Community Col, 69- Mem: Am Asn Physics Teachers; Math Asn Am; Nat Coun Teachers Math. Res: Effects of ultra-violet radiation; nature of problem solving in physics and mathematics. Mailing Add: Dept of Physics Iona Col New Rochelle NY 10804

STANISLAWSKI, DAN, b Bellingham, Wash, Apr 20, 03; m 34; c 2. HISTORICAL GEOGRAPHY, REGIONAL GEOGRAPHY. Educ: Univ Calif, Berkeley, AB, 37, PhD(geog), 44. Prof Exp: Lectr geog, Univ Calif, Berkeley, 40; instr, Syracuse Univ, 41-42; instr, Univ Calif, Berkeley, 42-45; asst prof, Univ Wash, 45-47; assoc prof, Univ Pa, 47-49; prof, Univ Tex, 49-61; head dept, 63-68, prof, 63-74, EMER PROF GEOG, UNIV ARIZ, 74- Res: Historical geography of the Mediterranean area and Latin America. Mailing Add: Dept of Geog & Area Develop Univ of Ariz Tucson AZ 85719

STANISLAWSKI, MICHAEL BARR, b Berkeley, Calif, June 17, 36; m 58. CULTURAL ANTHROPOLOGY, ARCHAEOLOGY. Educ: Stanford Univ, BA, 58; Univ Ariz, PhD(anthrop), 63. Prof Exp: Asst prof anthrop, Kans State Univ, 63-67; asst prof, Univ Ore, 67-70; assoc prof anthrop, Calif State Univ, Hayward, 70-75; WEATHERHEAD FEL, SCH AM RES, SANTA FE, 75- Mem: Am Anthrop Asn; Soc Am Archaeol; Am Ethnol Soc; AAAS; Am Asn Univ Profs. Res: Archaeological theory, and ethno-archaeology; material culture and craft arts; Southwestern, North American and Mesoamerican archaeology and ethnography; physical and cultural evolution. Mailing Add: Sch of Am Res PO Box 2188 Santa Fe NM 87501

STANITSKI, CONRAD LEON, b Shamokin, Pa, May 3, 39; m 63; c 2. INORGANIC CHEMISTRY. Educ: Bloomsburg State Col, BSEd, 60; State Col Iowa, MA, 64; Univ Conn, PhD(inorg chem), 71. Prof Exp: Teacher high sch, Pa, 60-63 & Goshen Cent Sch, 64-65; instr chem, Edinboro State Col, 65-67; teaching fel, Univ Conn, 70-71; asst prof, Ga State Univ, 71-75; ASSOC PROF CHEM, KENNESAW JR COL, 75- Mem: Am Chem Soc; Sigma Xi; AAAS. Res: Solid state hydride synthesis and reaction studies. Mailing Add: Natural Sci Div Kennesaw Jr Col Marietta GA 30062

STANKO, JOSEPH ANTHONY, b Wilkes-Barre, Pa, July 2, 41; m 62; c 1. INORGANIC CHEMISTRY. Educ: King's Col, BS, 62; Univ Ill, PhD(inorg chem), 66. Prof Exp: Asst prof chem, Pa State Univ, 66-73; ASSOC PROF CHEM, UNIV S FLA, TAMPA, 73- Mem: Am Chem Soc; The Chem Soc; Am Crystallog Asn. Res: Electronic and molecular structure of coordinations compounds of the transition metal elements; chemistry of platinum group metal complexes. Mailing Add: Dept of Chem Univ of SFla Tampa FL 33620

STANLEY, ALLAN JOHN, b Huntsville, Ala, Oct 25, 99; m 28; c 3. ENDOCRINOLOGY, EMBRYOLOGY. Educ: Simpson Col, AB, 22; Univ Iowa, MS, 35, PhD, 36. Hon Degrees: DSc, Simpson Col, 72. Prof Exp: Res assoc zool, Univ Iowa, 36-37; instr comp anat & physiol, La State Univ, 37-39, asst prof zool, 39-

43; from asst prof to prof physiol, 43-70, hon prof res anat, 69-70, EMER PROF PHYSIOL, BIOPHYS & RES ANAT, SCH MED, UNIV OKLA, 70- Concurrent Pos: Mem fac, St John's Col, Md, 42-43. Mem: Fel AAAS; Endocrine Soc; Nat Audubon Soc; Am Soc Zoologists; NY Acad Sci. Res: Embryology of Falconiformes, endocrines and allergy; genetics of susceptibility and resistance to cancer in rats; endocrinology, genetics and embryology of factors affecting fertility of male rats. Mailing Add: 612 NE 17th St Oklahoma City OK 73105

STANLEY, CHARLES WILLIAM, b Kansas City, Kans, Aug 11, 21; m 50; c 2. ANALYTICAL CHEMISTRY, PESTICIDE CHEMISTRY. Educ: Univ Chicago, BSc, 43. Prof Exp: Anal chemist, Carnegie-Ill Steel Corp, 42-43; res chemist, Metall Lab, Univ Chicago, 43-44, Clinton Labs, 44-46 & Los Alamos Sci Lab, 46-53; nuclear chemist, Walter Kidde Nuclear Labs, Inc, 53-55; sr nuclear chemist, Nuclear Sci & Eng Corp, 55-58; sr chemist, Midwest Res Inst, 58-69; SR RES CHEMIST, CHEMAGRO AGR DIV, MOBAY CHEM CORP, 69- Concurrent Pos: mem adv bd, J Agr & Food Chem, 73- Mem: AAAS; Am Chem Soc. Res: Metabolism of pesticides in crops and animals; development of analytical methods for pesticide residues in crops and animal tissues. Mailing Add: Chemagro Agr Div Mobay Chem Corp PO Box 4913 Kansas City MO 64120

STANLEY, DANIEL JEAN, b Metz, France, Apr 14, 34; US citizen; m 60; c 3. MARINE GEOLOGY. Educ: Cornell Univ, BSc, 56; Brown Univ, MSc, 58; Univ Grenoble, DSc, 61. Prof Exp: Res geologist, French Petrol Inst, 58-61; geologist, Pan-Am Petrol Corp, 61-62; asst to dir geol, US Army Engrs Waterways Exp Sta, 62-63; asst prof sedimentol, Univ Ottawa, 63-64; asst prof marine geol, Dalhousie Univ, 64-66; assoc cur sedimentol, 66-68, supvr div, 68-71, CUR SEDIMENTOL, SMITHSONIAN INST, 68-, GEOL OCEANOGR, 71- Concurrent Pos: Nat Res Coun Can travel award, USSR, 66; founder & ed, Maritime Sediments J, 64-66; Nat Acad Sci exchange award, Poland, 67; adj prof, Univ Maine, Orono, 74- Mem: AAAS; fel Geol Soc Am; Soc Econ Paleontologists & Mineralogists; Am Mineralogists; Am Asn Petrol Geologists; corresp mem Geol Soc Belg. Res: Sedimentology of flysch in Alps and Carpathians; modern slope, canyon and deep sea fan deposits; marine geology studies of Nova Scotian shelf, Bermuda, northwestern Atlantic and the Mediterranean. Mailing Add: Div of Sedimentol Smithsonian Inst Washington DC 20560

STANLEY, DAVID WARWICK, b Muncie, Ind, Oct 12, 39. FOOD SCIENCE. Educ: Univ Fla, BS, 62, MS, 63; Univ Mass, PhD(food sci), 67. Prof Exp: Res fel, Smith Col, 67-68; asst prof food sci, Univ Toronto, 68-70; asst prof, 70-74, ASSOC PROF FOOD SCI, UNIV GUELPH, 74- Mem: Inst Food Technol; Can Inst Food Sci & Technol. Res: Animal protein systems including muscle contraction and relaxation, meat tenderness and muscle protein biochemistry; cell membranes including isolation, composition, structure and function; plant protein systems including food uses of plant proteins; food analysis. Mailing Add: Dept of Food Sci Univ of Guelph Guelph ON Can

STANLEY, EDWARD ALEX, b New York, NY, Apr 7, 29; m 58; c 2. PALEOBIOLOGY. Educ: Rutgers Univ, BS, 54; Pa State Univ, MS, 56, PhD(geol), 60. Prof Exp: Res geologist palynologist, Pan-Am Petrol Corp, 60-62; res assoc, Univ Del, 62-64; asst prof paleont, 64-67, assoc prof paleont & polynology, 67-71, PROF PALEONT & PALYNOLOGY, UNIV GA, 71- Concurrent Pos: Nat Acad Sci vis scientist, Soviet Acad Sci, Moscow, 68-69 & 73. Mem: Fel AAAS; fel Geol Soc Am; Am Asn Petrol Geologists; Soc Econ Paleontologists & Mineralogists; Paleont Soc. Res: Upper Cretaceous-Tertiary plant micro-fossils, Dinoflagellates and Acritarchs; Holocene palynology, especially as applied to the study of ocean bottom sediments and cores; electron microscope study of fossil and extant pollen grains. Mailing Add: Dept of Geol Univ of Ga Athens GA 30602

STANLEY, EDWARD LIVINGSTON, b Orange, NJ, Sept 6, 19; m 43; c 2. CHEMISTRY. Educ: Princeton Univ, AB, 40, MA, 43, PhD(chem), 47. Prof Exp: Asst, Princeton Univ, 40-41, res assoc anal chem, Off Sci Res & Develop & Manhattan Dist Proj, 41-43; supvr, Anal Lab, 43-50, lab head, Res Div, 50-57, FOREIGN AREA SUPVR, ROHM AND HAAS CO, 57- Mem: AAAS; Am Chem Soc. Res: Analytical chemistry. Mailing Add: Rohm and Haas Co Independence Mall W Philadelphia PA 19105

STANLEY, EMILO JOSEPH, b Pozarevac, Yugoslavia, Jan 13, 24; US citizen; m 56; c 4. ENVIRONMENTAL STUDIES, GEOGRAPHY. Educ: Univ Mich, BA, 55, MA, 57, PhD(geog), 67. Prof Exp: Instr geog, San Diego State Col, 58-59; instr, Beloit Col, 59-60, asst prof, 60-64; asst prof polit geog, Pitzer Col & Claremont Grad Sch, 64-67; proj dir, Comput-Asst Educ Curric Prog, Univ Calif, Riverside & Irvine, 67-69; PROF GEOG & ENVIRON STUDIES, CALIF STATE POLYTECH UNIV, POMONA, 69- Concurrent Pos: US State Dept Bur Cult Affairs grant, Univs Belgrade, Zagreb, Skoplje, Sarajevo & Ljubljana, Yugoslavia, 71-72; fac fel int environ progs, Inst Multidisciplinary Progs, Calif State Polytech Univ, Pomona, 72-; vis prof, Rockford Col, 63-64; res consult, Foreign Demog Anal Div, US Bur Census, 65; lectr, Univ Calif, Riverside, 67-68. Mem: AAAS; Asn Comput Mach; Asn Am Geog; Am Geog Soc; World Future Soc. Res: Research, development and international applications of multidisciplinary environmental studies; political, regional and economic geography of Yugoslavia. Mailing Add: Int Environ Progs Inst for Multidisciplinary Progs Calif State Polytech Univ Pomona CA 91768

STANLEY, ERIC, b Liverpool, Eng, Feb 2, 24; m 59; c 2. CRYSTALLOGRAPHY. Educ: Univ Wales, BSc, 42 & 49, PhD(physics), 52; Univ Sask, DSc, 72. Prof Exp: Exp officer physics, Royal Naval Sci Serv, 42-47; res fel, Univ Southern Calif, 52-53; Nat Res Coun res fel, 53-55; lectr physics, Univ Manchester, 55-64; prof, Univ Sask, Regina, 65-72, dir acad serv, 69-72, dir acad serv & planning, 72; PROF PHYSICS & DEAN FAC, UNIV NB, ST JOHN, 73- Mem: Brit Inst Math & Appln; Am Crystallog Asn; Brit Inst Physics; Inst Pub Admin Can; fel Royal Soc Arts. Res: X-ray structural crystallography; theoretical and experimental studies in x-ray crystal structure determination; geometrical and physical optics. Mailing Add: Univ of NB in St John Tucker Park St John NB Can

STANLEY, EVAN RICHARD, b Sydney, Australia. CELL BIOLOGY, MEDICAL RESEARCH. Educ: Univ Western Australia, BSc, 67; Univ Melbourne, PhD(med biol), 70. Prof Exp: Fel med biol, Walter & Eliza Hall Inst Med Res, Melbourne, Australia, 70-72; lectr cell biol, Dept Med Biophys, Univ Toronto, 72-73, asst prof, 73-76; ASST PROF CELL BIOL, DEPTS MICROBIOL & IMMUNOL & CELL BIOL, ALBERT EINSTEIN COL MED, 76- Concurrent Pos: Mem sr sci staff, Ont Cancer Inst, 72-76. Res: Biochemical and genetic studies on humoral regulation of granulocyte and macrophage production and on the mechanism by which steroid hormone-dependent tumours become autonomous. Mailing Add: Dept of Microbiol & Immunol Albert Einstein Col of Med Bronx NY 10461

STANLEY, EVERETT MICHAEL, b Orangeburg, SC, Mar 7, 38; m 61; c 3. PHYSICAL OCEANOGRAPHY. Educ: The Citadel, BS, 60; Mass Inst Technol, MS, 64; Col William & Mary, PhD(phys oceanog), 76. Prof Exp: Physicist acoust, US Navy Engr Exp Sta, 60-62; phys oceanogr, US Coast & Geodetic Surv, 64-65; OCEANOGR, DAVID W TAYLOR NAVAL SHIP RES & DEVELOP CTR, US NAVY, 65- Res: Physical properties of seawater at pressure, numerical modeling of coastal and riverine dynamics, bubble formation and dynamics in open ocean waters, sulfide distribution and dynamics in natural waters. Mailing Add: 801 Chestnut Tree Dr Rt 9 Annapolis MD 21401

STANLEY, GLENN M, b Boise, Idaho, Feb 21, 22; m 42; c 3. GEOPHYSICS. Educ: Ore State Col, BS, 50, MS, 55. Prof Exp: Instr physics, Ore State Col, 49-51; physicist, Northwest Electrodevelop Lab, US Bur Mines, 50-52, US Navy Electronics Lab, 52-53, Geophys Inst, Univ Alaska, 53-58 & Arctic Aeromed Lab, Ft Wainwright, 58-62; prod mgr radio beaconry, Simmonds Precision Prod, Inc, NY, 62; pres, Dataco, Inc, Alaska, 62-63; assoc prof geophys, Geophys Inst, Univ Alaska, 63-66; sr physicist, Stanford Res Inst, Univ Calif, 66-69; assoc prof geophys, 69-74, PROF APPL SCI, GEOPHYS INST, UNIV ALASKA, 74- Concurrent Pos: NSF grant, 67-69; mem Int Union Radio Sci, 67. Res: Am Geophys Union; Inst Elec & Electronics Engrs. Res: Layered earth radio propagation at medium frequencies; application of search and rescue locator beaconry; ionospheric structures and morphology; biomedical techniques. Mailing Add: Geophys Inst Univ of Alaska Fairbanks AK 99701

STANLEY, HAROLD RUSSELL, b Salem, Mass, June 26, 23; m 46; c 3. ORAL PATHOLOGY. Educ: Univ Md, DDS, 48; Am Univ, BS, 52; Georgetown Univ, MS, 53; Am Bd Oral Path, dipl, 57. Prof Exp: Intern, Marine Hosp, USPHS, Baltimore, Md, 48-49; resident oral path, Armed Forces Inst Path, 51-53; mem staff, Nat Inst Dent Res, 53-66, clin dir, 66-68; CHMN DEPT ORAL MED, UNIV FLA, 70- Concurrent Pos: Hon prof, San Carlos, Univ Guatemala, 60- Mem: Hon fel Am Asn Endodont; Int Asn Dent Res; Am Dent Asn; Am Acad Oral Path (pres, 67). Res: Diseases of the human dental pulp; periodontium; oral mucous membranes. Mailing Add: Col of Dent Univ of Fla Gainesville FL 32601

STANLEY, HARRY EUGENE, b Norman, Okla, Mar 28, 41. STATISTICAL MECHANICS, SOLID STATE PHYSICS. Educ: Wesleyan Univ, BA, 62; Harvard Univ, PhD(physics), 67. Prof Exp: Staff mem solid state theory group, Lincoln Lab, Mass Inst Technol, 67-68; fel physics, Miller Inst Basic Res Sci, Univ Calif, Berkeley, 68-69; asst prof, 69-71, ASSOC PROF PHYSICS, MASS INST TECHNOL, 71-, HERMANN VON HELMHOLTZ ASSOC PROF HEALTH SCI & TECHNOL, 73- NSF fel theoret physics, 67-68; consult, Lincoln Lab, Mass Inst Technol, 69-71; vis prof physics, Osaka Univ, 75. Honors & Awards: Choice Award for Outstanding Acad Bk of 1971, Am Asn Acad Bk Publ, 72. Mem: AAAS; fel Am Phys Soc; Biophys Soc; NY Acad Sci. Res: Phase transitions and critical phenomena; magnetism; biomedical physics, especially cooperative phenomena in proteins, membrane transport, vision, and biological structure as deduced by laser Raman spectroscopy. Mailing Add: Dept of Physics Rm 13-2114 Mass Inst of Technol Cambridge MA 02139

STANLEY, HARRY EUGENE, b Duncan, Okla, Aug 28, 21; m 40; c 2. TEXTILE PHYSICS. Educ: Univ Okla, BS, 42, MS, 43; Mass Inst Technol, PhD(phys chem), 49. Prof Exp: Res chemist, Textile Fibers Dept, E I du Pont de Nemours & Co, 43-46; asst, Mass Inst Technol, 46-49; res chemist, 49-55, sr res chemist, 55-57, res supvr, 57-61, RES ASSOC, TEXTILE FIBERS DEPT, E I DU PONT DE NEMOURS & CO, INC, 61- Mem: Am Chem Soc; Fiber Soc. Res: Fundamental science of textile fibers and fabrics. Mailing Add: Chestnut Run Textile Res Lab E I du Pont de Nemours & Co Inc Wilmington DE 19898

STANLEY, HUGH P, b Modesto, Calif, July 14, 26; m 59; c 3. ELECTRON MICROSCOPY, CELL BIOLOGY. Educ: Univ Calif, Berkeley, BA, 51; Ore State Univ, MA, 58, PhD(zool), 61. Prof Exp: NIH fel zool, Zool Sta, Naples, Italy, 61-63 & Cornell Univ, 63 & sr fel biol struct, Univ Wash, 63-65; asst prof anat, Univ Minn, 65-66; asst prof zool, 66-68, ASSOC PROF ZOOL, UTAH STATE UNIV, 68- Mem: AAAS; Electron Micros Soc Am; Am Soc Cell Biol; Am Soc Zool. Res: Ultrastructure of developing cell systems, especially vertebrate spermatid differentiation. Mailing Add: Dept of Zool Utah State Univ Logan UT 84322

STANLEY, JOHN, b London, Eng, June 25, 05; m 30; c 1. ZOOLOGY. Educ: Univ BC, BA, 27; Univ Minn, AM, 29, PhD, 31. Prof Exp: Asst zool, Univ Minn, 27-28, asst entom, 28-30; asst, Exp Sta, Asn Hawaiian Pineapple Canners, 30-32; from asst prof to assoc prof biol, Queen's Univ, Ont, 32-46; prof zool, McGill Univ, 46-65, chmn dept, 47-63, chmn biol sci group, 48-63; prof pop dynamics, 65-70; CONSULT MED STATIST & PROG, 70- Concurrent Pos: Jr res investr, Nat Res Coun Can, 32. Mem: Zool Soc London. Res: Biomathematics; mathematical theory of population growth; computer programming. Mailing Add: 2575 W King Edward Ave Vancouver BC Can

STANLEY, JOHN PEARSON, b Washington, DC, Dec 17, 15; m 41; c 7. BIOCHEMISTRY. Educ: Cath Univ Am, BS, 37, MS, 39, PhD(biochem), 42. Prof Exp: Chemist & spec agt, Fed Bur Invest, Washington, DC, 41-44; res chemist, Gelatin Prod Co, Mich, 44, head bact labs, 44-46, asst chief control dept & res & develop labs, 46; dir res & develop, 46-68, TECH DIR, R P SCHERER CORP, 68- Mem: Fel AAAS; Am Chem Soc; Am Pharmaceut Asn; fel Am Inst Chemists; NY Acad Sci. Res: Soft gelatin capsules; gelatin; vitamins; nutrition; research and development administration; product development; technical service. Mailing Add: R P Scherer Corp 9425 Grinnell Ave Detroit MI 48213

STANLEY, JON G, b Edinburg, Tex, Oct 28, 37; m 65; c 3. COMPARATIVE PHYSIOLOGY. Educ: Univ Mo, AB, 60, BS, 63, PhD(zool), 66. Prof Exp: Asst prof biol, DePaul Univ, 66-69; from asst prof to assoc prof, Univ Wis-Milwaukee, 69-72; FISHERIES BIOLOGIST, MAINE COOP FISHERY RES UNIT, UNIV MAINE, ORONO, 72- Mem: AAAS; Am Soc Zoologists; Am Fisheries Soc. Res: Osmoregulation in lower vertebrates; metabolic studies of fish; gynogenesis and breeding of freshwater fish; environmental effects and physiology of grass carp. Mailing Add: Maine Coop Fishery Res Unit Univ of Maine Orono ME 04473

STANLEY, KENNETH EARL, b Auburn, NY, Nov 7, 47; m 71. STATISTICS. Educ: Alfred Univ, BA, 69; Bucknell Univ, MA, 70; Univ Fla, PhD(statist), 74. Prof Exp: Teaching asst statist, Univ Fla, 70-74; RES ASST PROF STATIST, STATE UNIV NY BUFFALO, 75- Concurrent Pos: Coord statistician, Working Party for Ther of Lung Cancer, 75- Mem: Am Statist Asn. Res: Nonparametric statistical techniques applicable for clinical trials in cancer research. Mailing Add: Statist Lab 4230 Ridge Lea Rd Amherst NY 14226

STANLEY, KENNETH OLIVER, b Santa Maria, Calif, Sept 23, 41; m 67. GEOLOGY. Educ: Univ Calif, Los Angeles, AB, 64, MA, 66; Univ Wis-Madison, PhD(geol), 69. Prof Exp: Asst prof, 69-72, ASSOC PROF GEOL, UNIV NEBR, LINCOLN, 72- Mem: AAAS; Geol Soc Am; Soc Econ Paleont & Mineral. Res: Sedimentary petrology and petrography; sedimentology. Mailing Add: Dept Geol 433 Morrill Hall Univ of Nebr Lincoln NE 68508

STANLEY, LESTER NELSON, b Worcester, Mass, Mar 15, 10; m 35; c 2. ORGANIC CHEMISTRY. Educ: Tufts Col, BS, 31; Mass Inst Technol, MS, 32, PhD(org chem),

36. Prof Exp: Res chemist, Monsanto Chem Co, 35-41 & Esso Labs, Standard Oil Develop Co, 41-43; sr chemist, Gen Aniline & Film Corp, 43-49, leader napthols sect, 49-67, res specialist, Process Res & Develop Dept, GAF Corp, 67-70, tech assoc, 70-73; RETIRED. Honors & Awards: Awarded 67 US Patents. Mem: Am Chem Soc; Am Asn Textile Chemists & Colorists. Res: Oxidation; dehydrogenation; polymerization; organic synthesis and halogenation; dye intermediates; azoic dyes; ultraviolet light absorbers; synthetic fiber dyes; diazotypy. Mailing Add: 40 Roweland Ave Delmar NY 12054

STANLEY, MALCOLM MCCLAIN, b Henderson, Ky, Mar 2, 16; m 43; c 2. MEDICINE. Educ: Centre Col, AB, 37; Univ Louisville, MD, 41; Am Bd Internal Med & Am Bd Gastroenterol, dipl. Prof Exp: From intern to asst resident med, Gallinger Munic Hosp, DC, 41-43; from asst resident to chief resident, Evans Mem Hosp, Boston, 43-46; from instr to assoc prof, Med Sch, Tufts Univ, 46-57; prof exp med, Sch Med, Univ Louisville, 57-59, prof med, 59-62; prog dir, Clin Res Ctr, 62-70, SECT CHIEF GASTROENTEROL, HINES VET ADMIN HOSP, 70-; PROF MED, UNIV ILL COL MED, 62- Concurrent Pos: Am Cancer Soc fel, Joseph H Pratt Diag Hosp, Boston, 46-48; res assoc, Pratt Diag Hosp, Boston, 48, mem staff, 49-57. Mem: Am Soc Clin Invest; Am Physiol Soc; Am Gastroenterol Asn; Am Fedn Clin Res. Res: Gastroenterology, especially intestinal absorption and secretion; bile salt metabolism. Mailing Add: Vet Admin Hosp PO Box 23 Hines IL 60141

STANLEY, MELISSA SUE MILLAM, b South Bend, Wash, June 23, 31; m 58. EXPERIMENTAL ZOOLOGY. Educ: Univ Ore, BS, 53, MA, 59; Univ Utah, PhD(zool, entom), 65. Prof Exp: Med technologist, Hosps & Labs, 53-57; teaching asst biol, Univ Ore, 57-58; med technologist, Hosps & Labs, 58-59; instr, Westminster Col, Utah, 59-61, asst prof, 61-63, actg chmn dept, 59-60, chmn dept, 60-63; res assoc, Pioneering Lab Insect Path, USDA, 65-67; from asst prof to assoc prof biol, 67-74, prog coordr, 68-69, PROF BIOL, GEORGE MASON UNIV, 74- Mem: Entom Soc Am; Tissue Cult Asn; Am Inst Biol Sci. Res: Arthropod tissue culture. Mailing Add: Dept of Biol George Mason Univ Fairfax VA 22030

STANLEY, NATHANIEL RICHARD, b New York, NY, Aug 17, 16; m 51; c 2. MATHEMATICS. Educ: City Col New York, BS, 36; NY Univ, MS, 50, PhD(math), 59. Prof Exp: Mem staff, Gen Precision Labs, 50-51; mathematician, Reeves Instrument Corp, 51-52; asst to dir behav models proj, Columbia Univ, 53; instr math, Rutgers Univ, 53-56; instr, NY Univ, 56-59, asst prof, 59-60; sr mathematician, Univac, 60-62 & Perkin-Elmer Corp, 62-63; econ engr, Am Can Co, 63-64; staff mathematician, Vitro Labs, 64-65 & Univac, 65-66; assoc prof, 66-72, PROF MATH, UNIV LI, 72- Mem: Am Math Soc. Res: Exponential sums; Green's functions; ordinary linear boundary-value problems; applied probability; stochastic processes. Mailing Add: Dept of Math Long Island Univ Brooklyn NY 11201

STANLEY, NORMAN FRANCIS, b Rockland, Maine, May 6, 16; m 63; c 2. CHEMISTRY. Educ: Res chemist, Algin Corp Am, 40-53, res dir, 53-59; asst tech dir, 59-64, RES CHEMIST, MARINE COLLOIDS, INC, 64- Mem: AAAS; Am Chem Soc; Soc Rheol; Am Inst Aeronaut & Astronaut. Res: Polysaccharide chemistry; chemistry and technology of marine algae, algal products and watersoluble gums; design and analysis of experiments. Mailing Add: Marine Colloids Inc PO Box 308 Rockland ME 04841

STANLEY, RAYMOND WALLACE, b Allahabad, United Provinces, India, Oct 9, 16; US citizen; m 49; c 1. GEOGRAPHY OF LATIN AMERICA, POLITICAL GEOGRAPHY. Educ: Univ Chicago, AB, 41, MA, 47; Univ Calif, Los Angeles, PhD(geog), 54. Prof Exp: Asst prof geog, George Peabody Col, 47-50; asst, Univ Calif, Los Angeles, 50-52; asst prof, Fresno State Col, 53-54; asst prof, Bradley Univ, 54-56; from asst prof to assoc prof, 56-70, PROF GEOG, SAN JOSE STATE UNIV, 70- Mem: Asn Am Geog; Nat Coun Geog Educ; Am Geog Soc. Mailing Add: Dept of Geog San Jose State Univ San Jose CA 95192

STANLEY, RICHARD PETER, b New York, NY, June 23, 44; m 71; c 1. ALGEBRA. Educ: Calif Inst Technol, BS, 66; Harvard Univ, PhD(math), 71. Prof Exp: Miller fel math, Miller Inst Basic Res Sci, 71-73; Moore instr, 70-71, asst prof, 73-75, ASSOC PROF MATH, MASS INST TECHNOL, 75- Concurrent Pos: Res scientist & consult, Jet Propulsion Lab, 65-; consult, Bell Tel Labs, 73- Honors & Awards: Polya Prize, Soc Indust & Appl Math, 75. Mem: Am Math Soc; Math Asn Am; Soc Indust & Appl Math. Res: Development of a unified foundation to combinatorial theory; applications of commutative algebra to combinatorics. Mailing Add: Dept of Math Mass Inst of Technol Cambridge MA 02139

STANLEY, RICHARD W, b Milesburg, Pa, Dec 16, 28; m 52; c 3. NUTRITION, BIOCHEMISTRY. Educ: Pa State Univ, BS, 56, MS, 58, PhD(dairy sci), 61. Prof Exp: Instr nutrit, Pa State Univ, 57-61; from asst prof to assoc prof, 61-70, PROF NUTRIT, COL TROP AGR, UNIV HAWAII, 70-, CHMN DEPT ANIMAL SCI, 68- Concurrent Pos: Fel, Univ Mo, 67-68. Mem: Am Dairy Sci Asn. Res: Ruminant nutrition, including utilization of metabolites formed in the rumen as they influence the productive performance of domestic cattle. Mailing Add: Dept of Animal Sci Univ of Hawaii Col Trop Agr Honolulu HI 96822

STANLEY, ROBERT GRUNBLATT, forestry, physiology, deceased

STANLEY, ROBERT LAUREN, b Seattle, Wash, Dec 30, 21; m 47; c 2. MATHEMATICS. Educ: Univ Wash, BS, 43, MA, 47; Harvard Univ, PhD, 51. Prof Exp: Guest lectr philos, Univ BC, 51-52, guest lectr math, 52-54; asst prof, Univ SDak, 54-57; asst prof, Wash State Univ, 57-61; assoc prof, 61-66, PROF MATH, PORTLAND STATE UNIV, 66- Mem: Am Math Soc; Asn Symbolic Logic; Math Asn Am. Res: Mathematical logic and foundations; logical analysis in philosophy of science. Mailing Add: Dept of Math Portland State Univ Portland OR 97207

STANLEY, ROBERT LEE, JR, b Dodge Co, Ga, Mar 7, 40; m 68; c 2. AGRONOMY. Educ: Univ Ga, BSA, 63, PhD(agron), 69; Clemson Univ MSA, 64. Prof Exp: Asst prof, 68-74, ASSOC PROF AGRON, UNIV FLA, 74- Mem: Am Soc Agron; Crop Sci Soc Am; Soc Range Mgt. Res: Forage crops management and utilization. Mailing Add: PO Box 470 Quincy FL 32351

STANLEY, ROBERT WEIR, b Ann Arbor, Mich, July 18, 25; m 48; c 5. PHYSICS. Educ: Auburn Univ, BS, 48; Johns Hopkins Univ, PhD(physics), 53. Prof Exp: From asst prof to assoc prof physics, Clarkson Tech Univ, 52-56; asst prof, 56-68, ASSOC PROF PHYSICS, PURDUE UNIV, WEST LAFAYETTE, 68- Concurrent Pos: Exchange prof, Univ Paris, 66; dean sch sci, Free Univ of the Congo, 69-71. Mem: Fel Optical Soc Am; Am Phys Soc; Am Asn Physics Teachers. Res: Optics; spectroscopy; interferometry. Mailing Add: Dept of Physics Purdue Univ West Lafayette IN 47907

STANLEY, ROLFE S, b Brooklyn, NY, Nov 4, 31; m 52; c 4. GEOLOGY. Educ: Williams Col, BA, 54; Yale Univ, MS, 55, PhD(geol), 62. Prof Exp: Geologist, Shell Oil Co, 57-59; NSF fel & lectr geol, Yale Univ, 62-64; from asst prof to assoc prof, 64-72, instnl res grant, 64-66, PROF GEOL, UNIV VT, 72-, CHMN DEPT, 64- Concurrent Pos: Res assoc, Ctr Technophysics, Tex A&M Univ, 71 & 72. Mem: Fel Geol Soc Am. Res: Structural analysis and structural petrology. Mailing Add: Dept of Geol Univ of Vt Burlington VT 05401

STANLEY, RONALD ALWIN, b Edinburg, Tex, June 18, 39; m 63; c 3. PLANT PHYSIOLOGY. Educ: Univ Ark, BS, 61, MS, 63; Duke Univ, PhD(plant physiol), 70. Prof Exp: Botanist, Tenn Valley Authority, 64-75; ASST PROF BIOL, UNIV S DAK, SPRINGFIELD, 75-; PLANT PHYSIOLOGIST, ENVIRON PROTECTION AGENCY, 76- Mem: AAAS; Am Inst Biol Sci; Am Soc Plant Physiologists; Asn Aquatic Vascular Plant Biologists; Sigma Xi. Res: Synergistic interactions of chemical, physical and biological factors in the environment with aquatic macrophytes; adaptation to the aquatic environment. Mailing Add: Off of Pesticide Progs Environ Protection Agency WH 568 Crystal Mall Bldg 2 Arlington VA 22202

STANLEY, STEVEN MITCHELL, b Detroit, Mich, Nov 2, 41; m 69. PALEONTOLOGY, EVOLUTION. Educ: Princeton Univ, AB, 63; Yale Univ, PhD(geol), 68. Prof Exp: Asst prof paleont, Univ Rochester, 67-69; from asst prof to assoc prof, 69-74, PROF PALEOBIOL, JOHNS HOPKINS UNIV, 74- Mem: AAAS; Paleont Soc; Geol Soc Am; Soc Study Evolution. Res: Functional morphology and evolution of bivalve mollusks and other taxa with fossil records; rates, trends and patterns of evolution; benthic marine ecology and paleoecology; biomechanics. Mailing Add: Dept of Earth & Planetary Sci Johns Hopkins Univ Baltimore MD 21218

STANLEY, THOMAS WILFRED, chemistry, mathematics, see 12th edition

STANLEY, WENDELL MEREDITH, JR, b New York, NY, Nov 9, 32; m 58; c 3. MOLECULAR BIOLOGY, BIOCHEMISTRY. Educ: Univ Calif, Berkeley, AB, 57; Univ Wis-Madison, MS, 59, PhD(biochem), 63. Prof Exp: From instr to asst prof biochem, Sch Med, NY Univ, 65-67; asst prof, 67-70, ASSOC PROF BIOCHEM, UNIV CALIF, IRVINE-CALIF COL MED, 70- Concurrent Pos: USPHS grant, Sch Med, NY Univ, 63-65. Mem: AAAS; Am Soc Biol Chem. Res: Control of protein biosynthesis in eukaryotes; RNA-containing tumor viruses. Mailing Add: Dept of Molecular Biol & Biochem Univ of Calif Irvine CA 92717

STANLEY, WILLARD FRANCIS, b Ames, Iowa, May 11, 01; m 27, 37; c 3. ZOOLOGY. Educ: NCent Col, AB, 27; Univ Ill, AM, 28, PhD(zool), 32. Prof Exp: Asst zool, Univ Ill, 28-31; prof biol, NDak State Teachers Col, Minot, 31-35; head dept sci, 35-62, prof biol, 35-71, chmn dept, 62-66, EMER PROF BIOL & CUR, WILLARD F STANLEY MUS, STATE UNIV NY COL FREDONIA, 71- Concurrent Pos: Ed-publ, Bio-Philately. Mem: AAAS; Am Soc Zoologists; Am Soc Ichthyologists & Herpetologists; Cooper Ornith Soc; assoc Am Ornithologists Union. Res: Genetics of Drosophila; herpetology; ornithology; field biology; effect of temperature upon wing size in Drosophila melanogaster with temperature effective periods. Mailing Add: Dept of Biol State Univ of NY Fredonia NY 14063

STANLEY, WILLIAM GORDON, b Wichita, Kans, Aug 4, 23; m 46, c 5. PHYSICAL CHEMISTRY. Educ: Southwestern Col, Kans, BA, 48; Kans State Univ, MS, 50, PhD(phys chem), 52. Prof Exp: Chemist, 51-56, group leader, Amoco Chem Corp, 56-63, dir propellants develop, 63-65, mgr info & commun, Res Dept, Am Oil Co, 66-70, DIR INFO SERV, RES DEPT, AMOCO RES CTR, STANDARD OIL CO (IND), 70- Mem: Am Chem Soc; Am Soc Info Sci. Res: Mechanism of combustion and catalysis of solid propellants; system management of rocket systems; information retrieval systems. Mailing Add: Amoco Res Ctr Standard Oil Co Ind Box 400 Naperville IL 60540

STANLEY, WILLIAM LYONS, b Teh Chou, China, May 30, 16; US citizen; m 41; c 3. ORGANIC CHEMISTRY. Educ: Marietta Col, AB, 39; Univ Calif, PhD(chem), 48. Prof Exp: Chemist & asst group leader, Carbide & Carbon Chem Corp, WVa, 39-45; res chemist, Western Regional Res Lab, USDA, 48-51 & Res & Develop Ctr, Union Oil Co, 51-54; prin chemist, Fruit & Veg Chem Lab, USDA, 54-60, chief fruit lab, Western Utilization Res & Develop Div, Agr Res Serv, Calif, 60-68; UN develop prog citrus res technologist, Food Inst, Centre Indust Res, Haifa, Israel, 68-69; CHIEF RES CHEMIST, FRUIT & VEG PROCESSING LAB, WESTERN REGIONAL RES LAB, AGR RES SERV, USDA, 70- Concurrent Pos: Mem comt fruit & veg prod, Adv Bd Mil Personnel Supplies, Nat Res Coun-Nat Acad Sci. Honors & Awards: Superior Serv Award, USDA, 62. Mem: Fel Am Chem Soc; Inst Food Technologists; Phytochem Soc NAm. Res: Synthetic organic chemistry; petrochemicals; chemistry of natural products; flavor components of citrus fruits; immobilized enzymes in food processing. Mailing Add: Fruit & Veg Processing Lab 800 Buchanan St Albany CA 94710

STANNARD, JAMES NEWELL, b Owego, NY, Jan 2, 10; m 35; c 1. RADIOBIOLOGY, TOXICOLOGY. Educ: Oberlin Col, AB, 31; Harvard Univ, AM, 34, PhD(biol), 35. Prof Exp: Asst physiol, Harvard Univ, 32-34 & biol, 34-35; instr physiol, Sch Med & Dent, Univ Rochester, 35-39; from instr to asst prof pharmacol, Emory Univ, 39-41; from pharmacologist to sr pharmacologist, NIH, 41-47, prin physiologist, 47; from asst prof to assoc prof radiation biol, 48-59, sect head, Atomic Energy Proj, 48-59, assoc dir educ, 59-69, assoc dean grad studies, 59-75, prof, 59-75, EMER PROF RADIATION BIOL, BIOPHYS & PHARMACOL, SCH MED & DENT, UNIV ROCHESTER, 75- Concurrent Pos: Chmn, Am Inst Biol Sci adv comt educ & training to US AEC, 57-61; mem subcomt relative biol effects, Nat Comt Radiation Protection; mem subcomt inhalation hazards & subcomt internal emitters, Biol Effects Atomic Radiation Comt, Nat Acad Sci-Nat Res Coun; consult, USPHS, mem nat adv comt radiation; consult, AEC, Energy Res & Develop Agency, Electric Power Res Inst, Bettis Atomic Power Lab & Environ Protection Agency; chmn sci comt, 34, Nat Coun Radiation Protection & Measurements, 69- Mem: AAAS; Am Physiol Soc; Am Soc Pharmacol & Exp Therapeut; Radiation Res Soc; Health Physics Soc. Res: Cellular respiration; metabolism of microorganisms, including parasites; radiation biology; muscle and aviation physiology; toxicology of radioactive materials. Mailing Add: Dept of Radiation Biol & Biophys Univ Rochester Sch of Med & Dent Rochester NY 14642

STANNARD, LEWIS JUDSON, JR, b Philadelphia, Pa, Apr 15, 18; m 54; c 1. ENTOMOLOGY. Educ: Pa State Col, BS, 46; Univ Ill, MS, 47, PhD, 52. Prof Exp: From asst taxonomist to assoc taxonomist, 47-61, TAXONOMIST, ILL STATE NATURAL HIST SURV, 61-; PROF AGR ENTOM, UNIV ILL, URBANA, 69- Concurrent Pos: Guggenheim fel, 54-55. Mem: Fel Entom Soc Am; fel Royal Entom Soc London. Res: Taxonomy and evolution of Thysanoptera. Mailing Add: Ill Natural Hist Surv Urbana IL 61801

STANNARD, WILLIAM A, b Whitefish, Mont, Sept 5, 31; m 51; c 4. MATHEMATICS. Educ: Univ Mont, BA, 53; Stanford Univ, MA, 58; Mont State Univ, EdD, 66. Prof Exp: Instr math, Northern Mont Col, 60-62; instr, Mont State Univ, 62-66; assoc prof, 66-74, PROF MATH, EASTERN MONT COL, 74-, CHMN DEPT, 68- Concurrent Pos: Partic NSF Math Inst, Rutgers Univ, 67. Res: Teaching undergraduate mathematics. Mailing Add: Dept of Math Eastern Mont Col Billings MT 59101

STANNETT, VIVIAN THOMAS, b Langley, Eng, Sept 1, 17; nat US; m 46; c 1. POLYMER CHEMISTRY. Educ: London Polytech Inst, BSc, 39; Polytech Inst Brooklyn, PhD(chem), 50. Prof Exp: Plant chemist, Brit Celanese Co, 39-41; chief chemist, Utilex, Ltd, 44-47, dir, 50-51; asst group leader polymers, Koppers Co, 51-52; asst prof polymer chem, State Univ NY Col Forestry, Syracuse Univ, 52-56, prof, 56-61; assoc dir, Camille Dreyfus Lab, Res Triangle Inst, 61-67; prof, 67-69, CAMILLE DREYFUS PROF CHEM ENG, NC STATE UNIV, 69-, VPROVOST & DEAN GRAD SCH, 75- Concurrent Pos: Res assoc, Mellon Inst, 51. Honors & Awards: Borden Award & Payen Award, Am Chem Soc, 74. Mem: Am Chem Soc; Tech Asn Pulp & Paper Indust; fel NY Acad Sci; Soc Chem Indust. Res: Physical chemistry and engineering properties of plastics; cellulosic plastics; plastics-paper combinations; radiation chemistry of polymers. Mailing Add: Dept of Chem Eng NC State Univ Raleigh NC 27607

STANOJEVIC, CASLAV V, b Belgrade, Yugoslavia, June 23, 28; US citizen; m 70; c 1. MATHEMATICS. Educ: Univ Belgrade, BS, 52, MS, 54, PhD(math), 55. Prof Exp: From asst prof to assoc prof math, Univ Belgrade, 58-61; from assoc prof to prof math, Univ Detroit, 62-68; PROF MATH, UNIV MO-ROLLA, 68- Concurrent Pos: Vis prof, Ohio State Univ, 67-68 & La State Univ, New Orleans, 71-52. Mem: Am Math Soc; Math Asn Am; Inst Math Statist. Res: Integrability of some cosine series; geometry of quantum states and normed linear spaces; applied probability; analysis; theory of probability. Mailing Add: Dept of Math Univ of Mo Rolla MO 65401

STANONIS, DAVID JOSEPH, b Louisville, Ky, Mar 19, 26. ORGANIC CHEMISTRY. Educ: Univ Ky, BS, 45; Northwestern Univ, PhD(chem), 50. Prof Exp: Instr chem, Northwestern Univ, 48-49; asst prof, Clark Univ, 49-50 & Loyola Univ, Ill, 50-54; RES CHEMIST, SOUTHERN REGIONAL RES LAB, USDA, 56- Mem: Am Chem Soc; Sigma Xi; Am Inst Chem; Fiber Soc; Am Asn Textile Chemists & Colorists. Res: Stereochemistry; mechanisms of organic reactions; cellulose chemistry. Mailing Add: 3406 Canal Apt E New Orleans LA 70119

STANONIS, FRANCIS LEO, b Louisville, Ky, July 9, 31; m 52; c 2. MINERALOGY, PETROLOGY. Educ: Univ Ky, BS, 51, MS, 56; Pa State Univ, PhD(mineral, petrol), 58. Prof Exp: Geologist, Carter Oil Co, 58-60 & George A Hoffman Co, 60-62; pres, Mitchell & Stanonis Inc, 62-67 & Int Pollution Control, Inc, 67-69; ASSOC PROF GEOL & GEOG, IND STATE UNIV, EVANSVILLE, 69-; PRES ENVIRO-SCI CORP, 70- Concurrent Pos: Owner, Red Banks Oil & Gas Co, 60- Mem: Geol Soc Am; Am Inst Prof Geol; Am Inst Mining, Metall & Petrol Eng. Res: Industrial waste disposal; petroleum and natural gas exploration; industrial water systems. Mailing Add: 142 N Arlington PO Box 150 Henderson KY 42420

STANOVICK, RICHARD PAUL, b Buffalo, NY, Apr 6, 33; m 55; c 4. SOIL MICROBIOLOGY, BIOCHEMISTRY. Educ: Univ Ga, BS, 58, MS, 60. Prof Exp: Chemist, Niagara Chem, FMC Corp, 60-63, group leader chem, 63-68; mgr, 68-69, mgr prog, 69-72, DIR, HAZELTON LABS, 72- Mem: Am Chem Soc. Res: Agricultural pesticides; chemistry of metabolism; development of analytical procedures for agricultural pesticides. Mailing Add: Hazleton Labs 9200 Leesburg Pike Vienna VA 22180

STANSBERRY, KENT GARDNER, b Joplin, Mo, June 11, 43; m 67; c 2. SPACE PHYSICS. Educ: Iowa State Univ, BS, 65, MS, 67; Univ Calif, Riverside, PhD(physics), 72. Prof Exp: Mem tech staff missile flight test anal, 67-70, MEM TECH STAFF COMMAND CONTROL COMMUN ANAL, AEROSPACE CORP, 72- Mem: Am Inst Physics. Res: Theoretical study of the space environment near Jupiter with particular focus on the energy and particle sources and losses in the Jovian Magnetosphere. Mailing Add: Aerospace Corp PO Box 92957 Los Angeles CA 90009

STANSBERY, DAVID HONOR, b Upper Sandusky, Ohio, May 5, 26; m 48; c 4. ZOOLOGY. Educ: Ohio State Univ, BS, 50, MS, 53, PhD, 60. Prof Exp: Fel, Stone Lab Hydrobiol, Ohio State Univ, 53-55, asst instr gen zool, 56-57, instr animal ecol, 58-60, from asst prof to assoc prof, 61-71, PROF ZOOL & DIR MUS ZOOL, OHIO STATE UNIV, 71- Concurrent Pos: Cur natural hist, Ohio State Mus, 61-72. Mem: AAAS; Ecol Soc Am; Soc Syst Zool; Soc Study Evolution; Am Malacol Union. Res: Zoogeography, ecology, evolution and taxonomy of freshwater forms, especially bivalve molluscs and decapod crustaceans. Mailing Add: Mus of Zool Ohio State Univ 1813 N High St Columbus OH 43210

STANSBREY, JOHN JOSEPH, b St Louis, Mo, Dec 30, 18; m 63; c 2. PHYSICAL CHEMISTRY. Educ: Washington Univ, AB, 41, MS, 43, PhD(chem), 47. Prof Exp: Asst chem, Washington Univ, 41-45, lectr, US Army Training Prog, 43-44; res chemist, Am Can Co, Ill, 45-47; res physicist, Anheuser-Busch, Inc, 47-53; mem tech staff, Bell Tel Labs, Inc, 53-56; MEM GROUP STAFF, RES DEVELOP & MFG DIV, NCR CORP, 56- Concurrent Pos: Instr, Webster Col, 44-45; instr exten div, Univ Cincinnati, 56-57. Mem: Am Chem Soc; Am Statist Asn; Opers Res Soc Am; Inst Mgt Sci. Res: Colloid chemistry; thixotropy; electrochemistry; conductance; electrophoresis; ultracentrifuge; light scattering and viscosity of protein solutions; emission and absorption spectra of biological materials; solid state physics and transistors; photochemistry; systems and operations research; stochastic processes; queuing theory. Mailing Add: Res Develop & Mfg Div NCR Corp Main & K Sts Dayton OH 45479

STANSBURY, EDWARD JAMES, b Oakville, Ont, Aug 1, 27; m 52; c 2. PHYSICS. Educ: Univ Toronto, BA, 49, MA, 50, PhD(physics), 52. Prof Exp: Res assoc physics, Univ Toronto, 52-53; mem tech staff, Bell Tel Labs, 53-56; lectr physics, 56-57, from asst prof to assoc prof, 57-69, vdean fac arts & sci, 68-69, dean, 69-71, PROF PHYSICS, McGILL UNIV, 69-, DEAN FAC SCI, 71- Mem: Can Asn Physicists. Res: Meteorological physics; nucleation of ice; thunderstorm electricity. Mailing Add: Fac of Sci Box 6070 McGill Univ Montreal PQ Can

STANSBURY, HARRY ADAMS, JR, b Morgantown, WVa, Sept 14, 17; m 44; c 4. CHEMISTRY. Educ: WVa Univ, AB, 40, MS, 41; Yale Univ, PhD(org chem), 44. Prof Exp: Group leader, Union Carbide Corp, WVa, 44-71; STATE DIR COMPREHENSIVE HEALTH PLANNING, GOV OFF, 71- Mem: Am Chem Soc. Res: Development of new pesticides; registration of pesticides; synthetic organic chemistry; residue analyses. Mailing Add: Gov Off Charleston WV 25305

STANSBURY, MACK FULTON, b Perry, La, Nov 5, 18; m 44; c 4. AGRICULTURAL CHEMISTRY. Educ: Southwestern La Univ, BS, 38; La State Univ, MS, 40. Prof Exp: Instr chem, Southwestern La Univ, 41; chemist, Southern Regional Res Lab, Bur Agr & Indust Chem, 41-53, chemist, Southern Utilization Res Br, Agr Res Serv, 53-59, prin analyst, 59-61, asst to dir, 61-70, asst to dir, Southern Mkt & Nutrit Res Div, 70-72, ASST TO DIR, SOUTHERN REGIONAL RES CTR, AGR RES SERV, USDA, 73- Mem: Am Chem Soc; Am Oil Chemists Soc; Sigma Xi; fel Am Inst Chemists. Res: Plant pigments; analytical methods; composition and utilization of agricultural commodities; chemistry of vegetable oils; cotton textile chemistry. Mailing Add: Southern Regional Res Ctr USDA 1100 Robert E Lee Blvd New Orleans LA 70179

STANSBY, MAURICE EARL, b Cedar Rapids, Iowa, Apr 25, 08; m 38; c 1. MARINE CHEMISTRY. Educ: Univ Minn, BChem, 30, MSc, 33. Prof Exp: Jr chemist, US Bur Commercial Fisheries, Mass, 31-35, Md, 35-37 & Wash, 38-40, technologist chg fishery prod lab, Alaska, 40-42, dir tech lab, 42-66, dir food sci, Pioneer Res Lab, Wash, 66-71, dir environ conserv div, 72-75, SCI CONSULT CONTAMINANTS RES, NORTHWEST FISHERIES CTR, NAT MARINE FISHERIES SERV, 75- Concurrent Pos: Lectr, Col Fisheries, Univ Wash, 38- Honors & Awards: Distinguished Serv Award, US Dept Interior, 66. Mem: Am Chem Soc; Inst Food Technologists; Am Oil Chemists Soc. Res: Analysis, preservation and processing of fish; chemistry of fish oils; effects of contaminants in the environment upon fish. Mailing Add: Northwest Fisheries Ctr Nat Marine Fisheries Serv Seattle WA 98112

STANSEL, JAMES WILBERT, b Angleton, Tex, Apr 8, 34; m 54; c 2. PLANT BREEDING, PLANT GENETICS. Educ: Tex A&M Univ, BS, 56, MS, 59; Purdue Univ, PhD(plant breeding & genetics), 65. Prof Exp: Asst geneticist, 60-66, asst prof genetics, 66-70, ASST PROF AGRON, AGR RES & EXTEN CTR BEAUMONT, TEX AGR EXP STA, TEX A&M UNIV, 70-, ASSOC PROF GENETICS & ENVIRON & SCIENTIST IN CHG, WESTERN DIV, 72- Mem: Am Soc Agron; Soil Sci Soc Am; Am Asn Cereal Chemists; Am Genetics Asn. Res: Increasing rice yields by developing through genetic and cultural manipulation morphologically and physiologically superior rice plants which more effectively utilize available sunlight in intensified cultural practice systems. Mailing Add: Western Div Beaumont Res & Exten Ctr Tex A&M Univ Eagle Lake TX 77434

STANSFIELD, BARRY LIONEL, b Toronto, Ont, June 10, 42; m 67; c 3. PLASMA PHYSICS. Educ: Univ Toronto, BASc, 65; Univ BC, MASc, 67, PhD(plasma physics), 71. Prof Exp: Fel, 71-72, PROF PLASMA PHYSICS, INST NAT SCI RES ENERGY.xENERGY, 72- Mem: Can Asn Physicists; Am Phys Soc. Res: Confinement of plasmas using electric as well as magnetic fields; development of plasma diagnostics; laser-produced plasmas and laser diagnostics. Mailing Add: Inst Nat Sci Res Energy CP 1020 Varennes PQ Can

STANSFIELD, CHARLES ARTHUR, b Philadelphia, Pa, Apr 16, 37; m 61; c 3. RESOURCE GEOGRAPHY. Educ: West Chester State Col, BS, 59; Pa State Univ, MS, 61; Univ Pittsburgh, PhD(geog), 65. Prof Exp: Instr geog, Frostburg State Col, 61-62; asst prof, Univ Del, 65-66; PROF GEOG & CHMN DEPT, GLASSBORO STATE COL, 66- Mem: Asn Am Geog; Nat Coun Geog Educ; Nat Recreation & Park Asn. Res: Geography of resorts; geography of recreation resource development; geography of population; urban geography. Mailing Add: Dept of Geog Glassboro State Col Glassboro NJ 08028

STANSFIELD, ROGER ELLIS, b Sanford, Maine, July 16, 26; m 51; c 2. ORGANIC CHEMISTRY. Educ: Northwestern Univ, BS, 50; Carnegie Inst Technol, PhD(chem), 55. Prof Exp: Fel, Duke Univ, 54-56; from asst prof to assoc prof, 56-65, chmn dept, 71-74, PROF CHEM, BALDWIN-WALLACE COL, 65- Concurrent Pos: Vis prof, Forman Christian Col, W Pakistan, 64-66. Mem: AAAS; Am Chem Soc. Res: Synthesis of peptides; alkaloids of nicotiana; esterification; chemistry of pyroles. Mailing Add: Dept of Chem Baldwin-Wallace Col Berea OH 44017

STANSFIELD, WILLIAM D, b Los Angeles, Calif, Feb 7, 30; m 53; c 3. GENETICS. Educ: Calif State Polytech Col, BS, 52, MA, 58; Univ Calif, Davis, MS, 61, PhD(animal breeding), 63. Prof Exp: INSTR BIOL, CALIF POLYTECH STATE UNIV, SAN LUIS OBISPO, 63- Mem: Genetics Soc Am; Am Genetic Asn; Sigma Xi. Res: Immunogenetics. Mailing Add: Dept of Biol Sci Calif Polytech State Univ San Luis Obispo CA 93407

STANSLOSKI, DONALD WAYNE, b Big Rapids, Mich, June 22, 39; m 59; c 4. MEDICINAL CHEMISTRY, CLINICAL PHARMACY. Educ: Ferris State Col, BS, 61; Univ Nebr-Lincoln, MS, 70, PhD(pharmaceut sci), 71. Prof Exp: Asst prof pharm, Col Pharm, Univ Nebr-Lincoln, 70-72; CLIN COORDR, RAABE COL PHARM, OHIO NORTHERN UNIV, 72- Concurrent Pos: Consult, Vet Admin Hosp, Lincoln, Nebr, 71 & Ohio Dept Pub Welfare, 74- Mem: Am Pharmaceut Asn. Res: Chemistry of mesoionic compounds and role of pharmacist in the provision of drug therapy. Mailing Add: Rte 2 Box 329 Ada OH 45810

STANSLY, PHILIP GERALD, b New York, NY, July 23, 12; m 37; c 2. BIOCHEMISTRY, MICROBIOLOGY. Educ: Cornell Univ, BS, 33; NY Univ, MSc, 36; Univ Minn, PhD(biochem), 44. Prof Exp: Asst, Virus Lab, NY Univ, 33-37; asst bacteriologist, Biol Labs, E R Squibb Co, 37-39; instr chem, Northwestern Inst Med Technol, 40-41; asst, Vet Div, Univ Minn, 41-43, instr, 43-44; sr biologist, Am Cyanamid Co, 44-52; res assoc, Detroit Inst Cancer Res, 54-66; sr res scientist, Mich Cancer Found, 66-67; dir microbiol, Mason Res Inst, 67-68; PROG DIR VIRAL ONCOL, DIV CANCER RES RESOURCES & CTRS, NAT CANCER INST, 68- Concurrent Pos: Fel, Inst Enzyme Res, Univ Wis, 52-54; assoc prof, Sch Med, Wayne State Univ, 54-67. Mem: AAAS; Am Soc Biol Chem; Am Asn Cancer Res; Soc Exp Biol & Med. Res: Antibiotics; enzymology; experimental oncology; eperythrozoonosis; cell biology; tumor virology. Mailing Add: Div of Cancer Res Resour & Ctrs Nat Cancer Inst Bethesda MD 20014

STANTON, ALFRED HODGIN, psychiatry, see 12th edition

STANTON, CHARLES MADISON, b San Diego, Calif, July 2, 42; m 71. ANALYTICAL MATHEMATICS. Educ: Wesleyan Univ, BS, 64; Stanford Univ, PhD(math), 69. Prof Exp: Lectr, 68-69, ASST PROF MATH, WESLEYAN UNIV, 69- Mem: Am Math Soc; Math Asn Am. Res: Complex analysis; algebras of analytic funtions; Riemann surfaces. Mailing Add: Dept of Math Wesleyan Univ Middletown CT 06457

STANTON, GARTH MICHAEL, b Cleveland, Ohio, June 15, 33; m 57; c 5. ORGANIC CHEMISTRY. Educ: Univ Detroit, BS, 56, MS, 58; Purdue Univ, PhD(chem), 62. Prof Exp: Res chemist, 62-69, SR RES CHEMIST, CHEVRON RES CO, STANDARD OIL CO CALIF, 69- Mem: AAAS; Am Chem Soc. Mailing Add: Chevron Res Co 576 Standard Ave Richmond CA 94804

STANTON, GEORGE EDWIN, b Danville, Pa, Mar 28, 44; m 65; c 2. AQUATIC ECOLOGY, INVERTEBRATE ECOLOGY. Educ: Bucknell Univ, BS, 66; Univ Maine, Orono, PhD(zool, entom), 69. Prof Exp: USDA res asst, Univ Maine, Orono, 66-69; asst prof, 69-72, ASSOC PROF BIOL, COLUMBUS COL, 72- Concurrent Pos: Adj assoc prof, Ga State Univ, 72-73. Mem: Am Inst Biol Sci; Ecol Soc Am; Entom Soc Am; Am Soc Oceanog & Limnol. Res: Ecology of small watershed systems and their impoundments; dynamics and succession of macroinvertebrate communities associated with carrion decomposition. Mailing Add: Dept of Biol Columbus Col Columbus GA 31907

STANTON, HENRY EDMUND, b Washington, DC, July 22, 10; m 40; c 2. PHYSICS. Educ: George Washington Univ, BS, 32; Univ Chicago, PhD(physics), 44. Prof Exp: Engr, Potomac Elec Power Co, DC, 33-35; res engr, Dept Terrestrial Magnetism, Carnegie Inst, Peru, 35-39; tech asst math biophys, Univ Chicago, 40-44; patent counr, Argonne Nat Lab, 44-52, engr res & develop, 46-51, assoc physicist, 52-75; RETIRED. Concurrent Pos: Teacher, Univ Chicago, 42-44. Mem: AAAS; Am Phys Soc; Sigma Xi. Res: Cosmic rays; mass spectroscopy. Mailing Add: 9402 S Sproat Ave Oak Lawn IL 60453

STANTON, HUBERT COLEMAN, b Orofino, Idaho, May 3, 30; m 50; c 2. PHARMACOLOGY. Educ: Idaho State Col, BS, 51; Ore State Col, MS, 53; Univ Iowa, PhD(pharmacol), 58. Prof Exp: Asst pharmacol, Univ Iowa, 55-58; instr, Sch Med, Univ Colo, 58-60; sr pharmacologist & group leader, Mead Johnson & Co, 60-65; asst prof pharmacol, Col Med, Baylor Univ, 65-68; HEAD DEPT ANIMAL PHYSIOL, BIOL SCI RES CTR, SHELL DEVELOP CO, 68- Mem: AAAS; Soc Exp Biol & Med; Am Soc Pharmacol & Exp Therapeut; Am Soc Animal Sci. Res: Autonomic pharmacology; neonatal physiology. Mailing Add: Biol Scis Res Ctr Shell Develop Co Modesto CA 95352

STANTON, MEARL FREDRICK, b Staunton, Ill, Aug 14, 22; m 51; c 4. MEDICINE. Educ: St Louis Univ, MD, 48. Prof Exp: Sr instr path, Sch Med, St Louis Univ, 50-54; PATHOLOGIST, NAT CANCER INST, 56-, ED-IN-CHIEF JOUR, 68- Mem: AMA; Am Soc Exp Path Res: Host-parasite relationships of obligate intracellular parasites; experimental a[illegible] clinical cancer research; pulmonary diseases. Mailing Add: Lab of Path Nat Cance[illegible] Bet[illegible]sda MD 20014

STANTON, NANCY [illegible]N, b San Francisco, Calif, Mar 23, 48; m 71. MATHEMATICS. Edu[illegible] Univ, BS, 69; Mass Inst Technol, PhD(math), 73. Prof Exp: Instr math, Mas[illegible] Technol, 73-74; lectr, Univ Calif, Berkeley, 74-76; RITT ASST PROF MATH[illegible]UMBIA UNIV, 76- Mem: Am Math Soc; Math Asn Am. Res: Spectrum of the Laplacian on complex manifolds, geometry of complex manifolds with boundary. Mailing Add: Dept of Math Columbia Univ New York NY 10027

STANTON, NOEL RUSSELL, b Dover, NJ, Dec 29, 37; m 62; c 1. HIGH ENERGY PHYSICS. Educ: Rutgers Univ, BA, 60; Cornell Univ, PhD(exp physics), 65. Prof Exp: Res assoc exp high energy physics, Univ Mich, 65-68; asst prof, 68-73, ASSOC PROF PHYSICS, OHIO STATE UNIV, 73- Mem: Am Phys Soc. Res: Strong interactions of elementary particles at high energy. Mailing Add: Dept of Physics Ohio State Univ Columbus OH 43210

STANTON, RICHARD EDMUND, b Brooklyn, NY, Aug 31, 31; m 57; c 5. THEORETICAL CHEMISTRY. Educ: Niagara Univ, BS, 52; Univ Notre Dame, PhD(phys chem), 57. Prof Exp: Fel, Cath Univ Am, 56-57; from asst prof to assoc prof, 57-69, PROF PHYS CHEM, CANISIUS COL, 69- Concurrent Pos: Consult, Union Carbide Res Inst, 61-63; Sloan fel, 69-71; vis prof, Univ Manchester, 70-71. Mem: Am Chem Soc; Am Phys Soc. Res: Quantum chemistry, especially electron correlation, isoelectronic sequence theory and self-consistency problems; new applications of group theory to chemical kinetics. Mailing Add: Dept of Chem Canisius Col Buffalo NY 14208

STANTON, ROBERT JAMES, JR, b Los Angeles, Calif, June 17, 31; m 53; c 2. GEOLOGY, PALEONTOLOGY. Educ: Calif Inst Technol, BS, 53, PhD(geol), 60; Harvard Univ, MA, 56. Prof Exp: Res geologist, Shell Develop Co, 59-67; assoc prof, 67-72, PROF GEOL, TEX A&M UNIV, 72- Mem: AAAS; Geol Soc Am; Soc Econ Paleontologists & Mineralogists; Paleont Soc. Res: Paleoecology; Cenozoic paleontology of Pacific Coast; ichnology. Mailing Add: Dept of Geol Tex A&M Univ College Station TX 77843

STANTON, WILLIAM ALEXANDER, b Washington, DC, Sept 9, 15; m 42; c 3. ORGANIC CHEMISTRY. Educ: Univ Md, BS, 36, PhD(org chem), 41. Prof Exp: Res chemist, Tech Div, Photo Prod Dept, NJ, 41-45, group leader, 46-49, chief supvr, Plant Process Dept, Prod Div, 49-50, plant process supt, 50-52, prod supt, 53-56, asst plant mgr, NY, 57-58, dir, Parlin Res Lab, 58-63, mgr prod mkt, 64-65, dir printing & indust sales, 66, DIR FOREIGN OPERS, E I DU PONT DE NEMOURS & CO, INC, 67- Honors & Awards: J Award, Soc Motion Picture & TV Engrs. Mem: AAAS; Am Chem Soc; Soc Motion Picture & TV Engrs; Soc Photog Scientists & Engrs; Am Inst Chemists. Res: Natural products; photographic emulsions and processing solutions; synthetic color-forming polymers for photographic emulsions. Mailing Add: Photo Prod Dept E I du Pont de Nemours & Co Inc Wilmington DE 19898

STAPELFELDT, HEINZ EWALD, analytical chemistry, photographic science, see 12th edition

STAPF, ROBERT JOSEPH, b Pittsburgh, Pa, Nov 21, 23; m 49; c 3. FOODS. Educ: Univ Pittsburgh, BSc, 44, PhD(chem), 50. Prof Exp: Lab instr chem, Univ Pittsburgh, 46-48, asst, 48-50; asst, Kraft Foods Co, 50-53, group leader, 53-56, tech serv mgr, 56-58; tech serv mgr, Res & Develop Div, Nat Dairy Corp, 58-63; vpres, Sanna Div, 70-74, TECH DIR, SANNA DAIRIES DIV, BEATRICE FOODS CO, 63-, PRES, 74- Mem: Inst Food Technologists. Res: Soybean oil; margarine; shortening; salad dressing; fats and oil; toppings; dairy; drying; analytical chemistry; administration; specification; technical sales; aseptic processing. Mailing Add: Sanna Dairies Div Beatrice Foods Co Box 1587 Madison WI 53701

STAPFER, CHRISTIAN HENRI, organometallic chemistry, polymer chemistry, see 12th edition

STAPLE, PETER HUGH, b Tonbridge, Eng, Oct 15, 17; m 52; c 2. PHYSIOLOGY, DENTISTRY. Educ: Univ London, BDS, 40, BSc, 49, PhD(sci, histochem), 52. Prof Exp: Mem sci staff, Med Res Coun, Eng, 51-57; lectr physiol, Univ Birmingham, 57; res assoc, Med Res Labs, Charing Cross Hosp, Univ London, 57-59; instr pharmacol, Univ Ala, 59-60, assoc prof, 60-63, assoc prof dent, Med Col & Sch Dent, 59-63; assoc prof oral biol, 63-72, PROF ORAL BIOL, SCH DENT, STATE UNIV NY BUFFALO, 72- Concurrent Pos: Hon vis assoc prof, Univ BC, 71. Mem: Microcirc Soc; Histochem Soc; Int Asn Dent Res. Res: Histochemistry in relation to dental disease; dilantin sodium and gingival hyperplasia; microcirculation in the gingiva. Mailing Add: Dept Oral Biol State Univ of NY at Buffalo Sch of Dent Buffalo NY 14226

STAPLE, TOM WEINBERG, b Hamburg, Ger, May 6, 31; US citizen; m 64; c 1. MEDICINE, RADIOLOGY. Educ: Univ Ill, Chicago, BS, 53, MD, 55. Prof Exp: From instr to assoc prof, 62-73, PROF RADIOL, MALLINCKRODT INST RADIOL, SCH MED, WASH UNIV, 73- Concurrent Pos: Consult radiol, H G Phillips Hosp, St Louis, Mo. Mem: Asn Univ Radiol. Res: Bone growth and arthrography. Mailing Add: Mallinckrodt Inst of Radiol Barnes Hosp St Louis MO 63110

STAPLES, ALBERT FRANKLIN, b Sanford, Maine, Sept 3, 22; m 53; c 4. ORAL SURGERY, PHYSIOLOGY. Educ: Tufts Univ, DMD, 51; Baylor Univ, BS in Dent, 54, PhD(physiol), 70. Prof Exp: From asst prof to assoc prof oral surg, Col Dent, Baylor Univ, 61-67; PROF ORAL SURG & CHMN DEPT, COL DENT, UNIV OKLA, 70- Concurrent Pos: Consult oral surg, US Naval Hosp, Corpus Christi, Tex, 59-61 & Vet Admin Hosp, Oklahoma City, Okla, 70- Mem: Int Asn Dent Res. Res: Estrogenic hormones in gingiva; relationship of cyclic adenosine monophosphate to levels of estrogenic hormones. Mailing Add: Dept of Oral Surg Univ of Okla Health Sci Ctr Oklahoma City OK 73190

STAPLES, BERT ROLAND, b Hazleton, Pa, June 17, 35; m 57. PHYSICAL INORGANIC CHEMISTRY. Educ: Univ Buffalo, BA, 57; Univ Md, MS, 65, PhD(phys chem), 67. Prof Exp: Anal chemist, Carborundum Co, 57-58; chemist, Nat Bur Standards, 61; NIH asst phys chem, Univ Md, 61-65; RES CHEMIST, NAT BUR STANDARDS, 67- Mem: Am Chem Soc; The Chem Soc. Res: Experimental and theoretical studies concerning properties and behavior of electrolytes in aqueous, nonaqueous and mixed solvents to understand fundamental processes involving ions; establish scales and reference materials for precise measurement of pH and acidity; critical evaluation of electrolyte solution data; microcalorimetric studies of biologically important reactions. Mailing Add: Nat Bur of Standards Washington DC 20234

STAPLES, GEORGE EMMETT, b Kanosh, Utah, Nov 2, 18; m 48; c 6. VETERINARY MEDICINE, ANIMAL NUTRITION. Educ: Utah State Univ, BS, 47; SDak State Univ, MS, 49; Colo State Univ, DVM, 54. Prof Exp: Instr animal husb, SDak State Univ, 47-48, asst animal husbandman, 48-50; pvt pract, 54-56; vet livestock inspector & area vet, Animal Dis Eradication Div, Agr Res Serv, USDA, 56-60, res vet, Animal Dis & Parasite Div, Colo, 60; res vet, Theracon, Inc, Kans, 60-64; ASSOC PROF VET MED & ANIMAL DIS RES, NDAK STATE UNIV, 65- Mem: Am Vet Med Asn; Am Soc Animal Sci. Res: Digestibility studies with prairie hay at different maturity stages; designing and testing therapeutic diets; neonatal diseases of farm animals therapy and prophylaxis; serological investigations; bovine chromosome investigations. Mailing Add: Dept of Vet Sci NDak State Univ Fargo ND 58102

STAPLES, JON T, b Waterville, Maine, Sept 14, 38; m 62; c 3. ORGANIC CHEMISTRY, POLYMER CHEMISTRY. Educ: Bowdoin Col, AB, 61; Univ NC, Chapel Hill, PhD(org chem), 66. Prof Exp: Res assoc org chem, Mass Inst Technol, 66-67; SR RES CHEMIST, EASTMAN KODAK LABS, ROCHESTER, 67- Mem: Am Chem Soc. Res: Protecting group chemistry; peptide synthesis; monomer and polymer synthesis; photographic science. Mailing Add: Eastman Kodak Co Res Lab 1669 Lake Ave Rochester NY 14650

STAPLES, LLOYD WILLIAM, b Jersey City, NJ, July 9, 08; m 41; c 3. GEOLOGY. Educ: Columbia Univ, AB, 29; Univ Mich, MS, 30; Stanford Univ, PhD(mineral), 35. Prof Exp: With Mich State Geol Surv, 30-31; instr geol, Mich Col Mining & Technol, 31-33; res assoc mineral, Stanford Univ, 35-36; instr geol, Ore State Col, 36-37; from instr to prof, 39-74, head dept, 58-68, EMER PROF GEOL, UNIV ORE, 74- Concurrent Pos: Chief geologist, Horse Heaven Mines, Sun Oil Co, 37-41 & Cordero Mining Co, Nev, 41-45; Guggenheim fel, Mex, 60-61; consult, UNESCO, Paris, 68-71. Mem: Fel Geol Soc Am; fel Mineral Soc Am; Am Inst Mining, Metall & Petrol Engrs. Res: Mineralogy and crystallography; economic geology of quicksilver; microchemistry of minerals; mineral determination by microchemical methods; field and x-ray study of zeolites. Mailing Add: Dept of Geol Univ of Ore Eugene OR 97403

STAPLES, MILFRED LAWSON, b Victoria Road, Ont, Sept 14, 12; m 38; c 2. TEXTILE CHEMISTRY, TEXTILE TECHNOLOGY. Educ: Univ Toronto, BA, 34, MA, 35. Prof Exp: Res asst chem, 34-36, res fel, Textile Dept, 37-46, sr res fel, 47-49, asst dir textile dept, 50-67, dir, 67-73, DIR DEPT TEXTILES, CLOTHING & FOOTWEAR, ONT RES FOUND, 73- Concurrent Pos: Mem comt textile test methods, Can Govt Specifications Bd, 52-, chmn, 69, chmn comt standardization of garment sizes, 53-, mem comt care labeling of textiles, 63-, chmn, 69; mem Can adv comt, Tech Comt Textiles, Inst Standardization Org, 54-, leader Can deleg, Eng, 54, 61, 65 & 70, NY, 56, France, 67 & 71, Mass, 71 & Ger, 71; vpres, Inst Textile Sci Can, 63, pres, 64. Honors & Awards: Award of Merit, Can Govt Specif Bd, 75. Mem: AAAS; Fiber Soc; Am Soc Testing & Mat; Am Asn Textile Chemists & Colorists; fel Brit Soc Dyers & Colourists. Res: Functional finishes by chemical modification of textile fibers; consumer serviceability standards; systems for standardizing sizes of clothing; moisture relations of textiles; laboratory test methods; fiber analysis; flammability of textiles. Mailing Add: Textiles Clothing & Footwear Ont Res Found Sheridan Park Mississauga ON Can

STAPLES, RICHARD CROMWELL, b Hinsdale, Ill, Jan 29, 26; m 54; c 3. PHYTOPATHOLOGY. Educ: Colo State Univ, BS, 50; Columbia Univ, AM, 53, PhD(plant physiol), 57. Prof Exp: Fel biochem, 52-57, from asst biochemist to assoc biochemist, 57-64, PLANT BIOCHEMIST, BOYCE THOMPSON INST PLANT RES, INC, 64-, PROG DIR PHYSIOL OF PARASITISM, 66- Concurrent Pos: Adj prof biol, Manhattanville Col, 74- Mem: Am Soc Plant Physiologists; Am Phytopath Soc; Am Chem Soc; Am Soc Microbiol; Am Inst Biol Sci. Res: Developmental physiology of fungi and fungal plant parasites. Mailing Add: Boyce Thompson Inst 1086 N Broadway Yonkers NY 10701

STAPLES, ROBERT, b Philadelphia, Pa, Dec 9, 16; m 43; c 3. ENTOMOLOGY. Educ: Univ Mass, BS, 40; Cornell Univ, PhD, 48. Prof Exp: Asst entomologist, Conn Agr Exp Sta, 49-50; assoc prof entom, 50-74, PROF ENTOM, UNIV NEBR, LINCOLN, 74-, ASSOC ENTOMOLOGIST, 50- Mem: Entom Soc Am. Res: Arthropod transmission of plant viruses; economic entomology. Mailing Add: Dept of Entom Univ of Nebr Lincoln NE 68503

STAPLES, ROBERT EDWARD, b Cobourg, Ont, Dec 5, 31; m 57; c 2. TERATOLOGY. Educ: Univ Sask, BSA, 54, MSc, 56; Cornell Univ, PhD(reproductive physiol), 61. Prof Exp: Asst animal husb, Univ Sask, 54-56 & Cornell Univ, 56-61; sect head, Endocrinol Dept, William S Merrell Co, 61-63; staff scientist, Worcester Found Exp Biol, 63-67; head, Unit Teratology & Reproduction, Merck Inst Therapeut Res, 67-71; head sect exp teratology, 71-75, ASST BR CHIEF TERATOLOGY, ENVIRON TOXICOL BR, NAT INST ENVIRON HEALTH SCI, 75- Concurrent Pos: Prin investr, USPHS grant, 64-67; consult, William S Merrell Co, 63-67. Mem: AAAS; Soc Study Reproduction; Soc Toxicol; Teratology; Europ Teratology Soc. Res: Early embryo development in mammals, including the influence of environmental factors; animal reproductive physiology; identification and delineation of environmental factors most likely to constitute and hazard for embryo development in the human. Mailing Add: Nat Inst Environ Health Sci PO Box 12233 Research Triangle Park NC 27709

STAPLETON, GEORGE EDWARD, biology, see 12th edition

STAPLETON, HARVEY JAMES, b Kalamazoo, Mich, Dec 22, 34; m 57; c 3. PHYSICS. Educ: Univ Mich, BS, 57; Univ Calif, Berkeley, PhD(physics), 61. Prof Exp: From asst prof to assoc prof, 61-69, PROF PHYSICS, UNIV ILL, URBANA, 69- Mem: AAAS; Am Phys Soc. Res: Paramagnetic resonance; electron spin-lattice relaxation; nuclear orientation; low temperature solid state physics. Mailing Add: Dept of Physics Univ of Ill Urbana IL 61801

STAPLETON, JAMES H, b Royal Oak, Mich, Feb 8, 31; m 63; c 1.

MATHEMATICAL STATISTICS. Educ: Eastern Mich Univ, AB, 52; Purdue Univ, MS, 54, PhD(math statist), 57. Prof Exp: Statistician, Gen Elec Co, 57-58; from asst prof to assoc prof statist, 58-73, chmn dept statist & probability, 69-75, PROF STATIST, MICH STATE UNIV, 73- Concurrent Pos: NSF fac sci fel, Univ Calif, Berkeley, 66-67. Mem: Am Math Asn; Inst Math Statist; Am Statist Asn. Res: Nonparametric statistics. Mailing Add: Dept of Statist & Probability Mich State Univ East Lansing MI 48823

STAPLETON, JOHN F, b Brooklyn, NY, Jan 25, 21; m 50; c 5. MEDICINE. Educ: Fordham Univ, AB, 42; Georgetown Univ, MD, 45; Am Bd Internal Med, dipl, 53; Am Bd Cardiovasc Dis, dipl, 57. Prof Exp: Intern, Providence Hosp, Washington, DC, 45-46; resident med, Georgetown Univ, 49-51, clin instr med, Hosp, 52-54, from instr to asst prof, 54-65; assoc prof med & chief cardiol, Woman's Med Col Pa, 65-67; PROF MED & ASSOC DEAN, SCH MED, GEORGETOWN UNIV, 67-, MED DIR, UNIV HOSP, 67- Concurrent Pos: Nat Heart Inst res fel, Georgetown Univ, 51-52; dir med educ, St Vincent Hosp, Worcester, Mass, 54-65; attend physician, Philadelphia Vet Admin Hosp, 65-67; consult, Vet Admin Hosp, Wilmington, Del, 65-67. Mem: Am Heart Asn; Am Col Physicians; AMA. Res: Clinical cardiology; medical education; hospitals. Mailing Add: Georgetown Univ Med Ctr 3800 Reservoir Rd NW Washington DC 20007

STAPLEY, EDWARD OLLEY, b Brooklyn, NY, Sept 25, 27; m 49; c 3. MICROBIOLOGY. Educ: Rutgers Univ, BS, 50, MS, 54, PhD(microbiol), 59. Prof Exp: From jr microbiologist to sr microbiologist, Merck Sharp & Dohme Res Labs, 50-66, res fel microbiol, 66-69; asst dir basic microbiol res, 69-74, DIR MICROBIAL CHEMOTHERAPEUT, MERCK INST THERAPEUT RES, 74- Concurrent Pos: Am Inst Biol Sci vis biologist, 69-72. Mem: AAAS; Am Acad Microbiol; Am Soc Microbiol; Soc Indust Microbiol (vpres, 74-75, pres, 76-77); NY Acad Sci. Res: Isolation of microorganism; mutation; fermentation; ergosterol production by yeasts; microbial transformations of steroids; isolation and utility of antibiotic-resistant microorganisms; detection, characterization and evaluation of new antibiotics; microbial transformations of sulfur. Mailing Add: Merck Sharp & Dohme Res Labs Rahway NJ 07065

STAPLIN, FRANK LYONS, b Santa Fe, NMex, Oct 5, 23; m 46; c 4. GEOLOGY, PALEONTOLOGY. Educ: Univ Tex, BS, 49, MA, 50; Univ Ill, PhD(geol), 53. Prof Exp: RES GEOLOGIST & HEAD PALEONT GROUP, IMP OIL, LTD, 56- Mem: Am Asn Petrol Geologists. Res: Palynology; organic debris in sediments; geothermal gradients; freshwater ostracods. Mailing Add: 339-50 Ave SE Calgary AB Can

STAPP, HENRY P, b Cleveland, Ohio, Mar 23, 28. PHYSICS. Educ: Univ Mich, BS, 50; Univ Calif, MA, 52, PhD, 55. Prof Exp: Theoret physicist, Lawrence Berkeley Lab, Univ Calif, 55-58 & Inst Theoret Physics, Swiss Fed Inst Technol, 58; THEORET PHYSICIST, LAWRENCE BERKELEY LAB, UNIV CALIF, 59- Res: Elementary particle physics. Mailing Add: Lawrence Berkeley Lab Univ of Calif Berkeley CA 94720

STAPP, JOHN PAUL, b Bahia, Brazil, July 11, 10; US citizen; m 57. BIOPHYSICS. Educ: Baylor Univ, BA, 31, MA, 32; Univ Tex, PhD(biophys), 40; Univ Minn, BM & MD, 44; Am Bd Prev Med, dipl, 56. Hon Degrees: DSc, Baylor Univ, 56. Prof Exp: Instr zool, Decatur Col, 32-34; proj officer, Aero Med Lab, Wright Field, US Air Force, 46-53, chief, Aero Med Field Lab, Holloman Air Force Base, 53-58, chief, Aero Med Lab, Wright Air Force Base, 58-60, asst to comdr aerospace med, Aerospace Med Ctr, Brooks Air Force Base, 60-65, resident in biophys, Armed Forces Inst Path, 65-67; prin med scientist, Nat Hwy Safety Bur, 67-72; ADJ PROF & CONSULT, SAFETY & SYSTS MGT CTR, UNIV SOUTHERN CALIF, 72- Concurrent Pos: Vpres, Int Astron Fedn, 60; consult, Nat Acad Sci, Nat Traffic Safety Agency, Gen Serv Admin & Nat Bur Stand; permanent chmn, Annual Stapp Car Crash Conf, Soc Automotive Engrs, 55- Honors & Awards: Cheney Award Valor, 54; Gorgas Medal, Mil Surg Asn, 57; Cresson Medal, Franklin Inst, 73; Excalibur Award, Safety Adv Coun, US Dept Transp, 75. Mem: Fel Am Inst Aeronaut & Astronaut (pres, 59); Soc Automotive Eng; Aerospace Med Asn (vpres, 57); AMA; Civil Aviation Med Asn (pres, 68). Res: Aerospace and industrial medicine; biodynamics of crashing and ditching; impact injury; medical biophysics. Mailing Add: Safety Systs Mgt Ctr Univ of Southern Calif Los Angeles CA 90007

STAPP, PAUL R, organic chemistry, see 12th edition

STAPP, WILLIAM B, b Cleveland, Ohio, June 17, 29; m 55; c 3. CONSERVATION. Educ: Univ Mich, Ann Arbor, BA, 51, MA, 58, PhD(conserv), 63. Prof Exp: Instr sci, Cranbrook Sch Boys, Mich, 51-52, instr biol, 54-58; conservationist, Aullwood Audubon Ctr, Ohio, 58-59; instr conserv, 59-61, lectr, 63-64, from asst prof to assoc prof, 64-72, PROF NATURAL RESOURCES & CHMN ENVIRON EDUC & OUTDOOR RECREATION PROG, 72- Concurrent Pos: Res assoc, Cranbrook Inst Sci, Mich, 55-57; consult conserv, Ann Arbor Pub Schs, 61-68, youth progs, Nat Audubon Soc, New York, 65-66; consult, Int Film Bur, Chicago, Kalamazoo Nature Ctr, Mich & Creative Visuals, Tex, 66-68; consult, environ educ prog, University City Pub Schs, Mo, 66-67, DeKalb Pub Schs, Ill, 67-68, Grand Haven Pub Schs, Mich, 67-, Raleigh County Sch Syst, WVa, 68-69, Toledo Bd Educ, Ohio, 70-, State of Alaska, 72-; consult, Nat Youth Movement Natural Beauty & Conserv, Washington, DC, 67-69, NJ Environ Educ Prog, 67-69, High Rock Interpretive Ctr, NY, 68-70, Seven Ponds Nature Ctr, 69-, Tapes Unlimited, Div Educ Unlimited Corp, 72-; consult environ interpretive ctr, Dept of Interior, 67-69 & environ educ, Dept Health, Educ & Welfare, 68, div col support, Off Educ, 70-; consult audio-cassette series ecol, Am Soc Ecol Educ, 72- & proj man & environ, Nat TV Learning Systs, Miami, Fla, 72-; mem conserv comt, Mich Dept Pub Instr, 65-68; mem bd dirs, Drayton Plains Interpretive Ctr, Pontiac, Mich, 67-; prog dir, Ford Found grant, 68-70; mem working comt, Ann Arbor Environ Interpretive Ctr, 69-; mem bd dirs, Mich Pesticide Coun, 69-71; mem comn educ, Int Union Conserv Nature & Natural Resources, 69-; mem bd adv, Gill Inst Environ Studies, NJ, 70-; mem, Ecol Ctr Commun Coun, Washington, DC, 71- & Educ Resources Info Ctr Sci, Math & Environ Educ, Columbus, Ohio, 71-; fac adv, Ecol Ctr Ann Arbor, 71-, vpres & mem bd dirs, 72-; mem, Pub Sanit Systs, Los Angeles, Calif, 72- & Mich Pop Coun, 72-; chmn Gov Task Force Develop State Environ Educ Plan, 72-; mem task force estab guidelines environ educ elem & sec schs, Mich Dept Educ, 72- Honors & Awards: Samuel Trask Dana Award Conserv, 62; Key Man Award, Conserv Educ Asn, 71; Conserv Educator of Year Award, State of Mich, 71. Mem: Am Nature Study Soc (vpres, 66-67, pres, 69-70); Conserv Educ Asn; Nat Audubon Soc; Asn Interpretive Naturalists (vpres, 69-71). Res: Environmental education and ecology; programs directed at helping urbanized man to develop a fuller understanding of environmental resource problems and his role in helping to resolve them. Mailing Add: Samuel T Dana Bldg Sch of Natural Resources Univ of Mich Ann Arbor MI 48104

STAPPER, CHARLES HENRI, b Amsterdam, Neth, Mar 27, 34; US citizen; m 58. ELECTRICAL ENGINEERING, SOLID STATE PHYSICS. Educ: Mass Inst Technol, BS, 59, MS, 60; Univ Minn, Minneapolis, PhD(elec eng, physics), 67. Prof Exp: Coop student elec eng, Gen Radio Co, 57-59; elec engr, IBM Corp, 60-65; teaching assoc elec eng, Univ Minn, Minneapolis, 66-67; engr-mgr elec eng & physics, 67-69, DEVELOP ENGR, IBM CORP, 70- Mem: Inst Elec & Electronics Engrs; Sigma Xi. Res: Application of mathematical theory to practical engineering and physics problems; statistical models for semiconductor devices and semiconductor manufacturing processes. Mailing Add: Dept N83 Bldg 967 IBM Corp PO Box A Essex Junction VT 05452

STAR, AURA E, b New York, NY, Mar 15, 30; m 50; c 2. BOTANY. Educ: Hunter Col, BA, 49; Mt Holyoke Col, MA, 51; Rutgers Univ, PhD(cytogenetics), 67. Prof Exp: Chemist, Baltimore Light & Power Co, Md, 51-52; instr biol, Morgan State Col, 52-53; from asst prof to assoc prof, 67-75, PROF BIOL, TRENTON STATE COL, 75- Concurrent Pos: Sigma Xi res grant, 63. Mem: AAAS; Bot Soc Am; Am Inst Biol Sci; Phytochem Soc; Torrey Bot Club. Res: Biochemical systematics of ferns and grasses; flavonoid chemistry; chemical biogeography of Pityrogramma; phytochemistry of marijuana. Mailing Add: Dept of Biol Trenton State Col Trenton NJ 08625

STARACE, ANTHONY FRANCIS, b New York, NY, July 24, 45; m 68. ATOMIC PHYSICS. Educ: Columbia Univ, AB, 66; Univ Chicago, MS, 67, PhD(physics), 71. Prof Exp: Res assoc physics, Imp Col, Univ London, 71-72; asst prof, 73-75, ASSOC PROF PHYSICS, UNIV NEBR-LINCOLN, 75- Concurrent Pos: Alfred P Sloan Found fel, 75-77. Mem: Am Phys Soc; Brit Inst Physics. Res: Theory of atomic photoabsorption and photoionization processes. Mailing Add: Behlen Lab of Physics Univ of Nebr Lincoln NE 68588

STARBIRD, MICHAEL PETER, b Los Angeles, Calif, July 10, 48. TOPOLOGY. Educ: Pomona Col, BA, 70; Univ Wis-Madison, MA, 73, PhD(math), 74. Prof Exp: ASST PROF MATH, UNIV TEX, AUSTIN, 74- Concurrent Pos: NSF res grant, 75- Mem: Am Math Asn; Math Asn Am. Res: Geometric and general topology. Mailing Add: Dept of Math Univ of Tex Austin TX 78712

STARBUCK, WESLEY CURTIS, b Denver, Colo, June 22, 34; div; c 4. PHARMACOLOGY, CHEMISTRY. Educ: Univ Colo, BS, 56, MS, 58; Baylor Univ, PhD(pharmacol), 62. Prof Exp: From instr to asst prof, 61-69, ASSOC PROF PHARMACOL, BAYLOR COL MED, 69- Mem: AAAS; Am Pharmaceut Asn; Am Chem Soc; Am Asn Cancer Res; Am Asn Clin Chemists. Res: Protein chemistry; isolation, purification, structure and function of basic nuclear proteins; histones and their role in cancer; dietary modification of foods for special diets. Mailing Add: Dept of Pharmacol Baylor Col of Med Houston TX 77025

STARCHER, BARRY CHAPIN, b Los Angeles, Calif, Dec 1, 38; m 60; c 3. NUTRITION, BIOCHEMISTRY. Educ: Univ Calif, Davis, BS & MS, 62; NC State Univ, PhD(biochem), 65. Prof Exp: Asst mem biochem, Inst Biomed Res, 66-70; asst prof path, Univ Colo Med Ctr, 70-72; asst prof biochem, Med Ctr, Univ Ala, Birmingham, 72-74; RES ASST PROF, PULMONARY DIV, SCH MED, WASHINGTON UNIV, 74- Mem: Am Nutrit Soc. Res: Studies on the biochemistry of copper and zinc metabolism; enzyme induction in relation to stress, and the chemistry of the crosslinking amino acids in elastin; connective tissue components of lung; the role of elastin in calcification and arteriosclerosis. Mailing Add: Pulmonary Div Washington Univ Sch of Med St Louis MO 63110

STARCHER, GEORGE WILLIAM, b Ripley, WVa, Jan 14, 06; m 28; c 1. MATHEMATICS. Educ: Ohio Univ, AB, 26; Univ Ill, MA, 27, PhD(math), 30. Hon Degrees: DHum, Ohio Univ, 71; LLD, Univ NDak, 75. Prof Exp: From instr to prof math, Ohio Univ, 30-43, actg dean col arts & sci & grad col, 43-45, prof math, 45-46, actg dean univ col, 46-47, dean 47-51, dean col arts & sci, 51-54; pres, 54-71, EMER PRES, UNIV NDAK, 71- Concurrent Pos: Fel, Harvard Univ, 45-46. Mem: Am Math Soc; Math Asn Am. Res: Q-difference equations, especially solution; functional equations; interpretation in number theory. Mailing Add: 3605 Jaffa Dr Sarasota FL 33579

STARCHER, PAUL SPENCER, organic chemistry, see 12th edition

STARCHMAN, DALE EDWARD, b Wallace, Idaho, Apr 16, 41; m 69; c 2. MEDICAL PHYSICS. Educ: Kans State Col, BS, 63; Univ Kans, MS, 65, PhD(radiation biophys), 68. Prof Exp: Chief health physicist, Ill Inst Technol Res Inst & radiol phyicist, Inst Radiation Therapy Mercy Hosp & Med Ctr, Chicago, 68-71; HEAD, MED PHYSICS SERV, 71- Concurrent Pos: Consult, Aultman Hosp & Timken Mercy Hosp, Canton, Ohio & Northeast Ohio Conjoint Radiation Oncol Ctr, 71-; mem bd, Mideast Region Radiol Physics Ctr Bd Adv & instr, Univ Akron, 73-; chmn, Physics Curric Develop Comt, Northeastern Ohio Univ Col Med, 74-; mem, Adv Staff, Akron Gen Med Ctr, 75- Mem: Am Asn Physicists Med; Soc Nuclear Med; Health Physics Soc; Am Asn Therapeut Radiologists. Res: Electron beam perturbation by cavities; radiation dosimetry; post irradiation atrophic changes of bone; information optimization with dose minimization in diagnostic radiology; radiation oncology treatment development. Mailing Add: Med Physics Serv 5942 Easy Pace Circle NW Canton OH 44718

STARE, FREDRICK JOHN, b Columbus, Wis, Apr 11, 10; m 59; c 3. NUTRITION. Educ: Univ Wis, BS, 31, MS, 32, PhD(biochem), 34; Univ Chicago, MD, 41. Hon Degrees: DSc, Suffolk Univ, 64 & Trinity Col, Dublin, 65. Prof Exp: Asst biochem, Univ Wis, 31-34; Nat Res Coun fel, Sch Med, Washington Univ, 34-35; Gen Educ Bd fel, Cambridge Univ, 35-36, Univ Szeged, 36 & Univ Zurich, 37; asst biochem, Univ Wis, 37-39; intern med, Barnes Hosp, St Louis, 41-42; from asst prof to assoc prof, 42-47, PROF NUTRIT, HARVARD UNIV, 47-, HEAD DEPT, 42- Concurrent Pos: Jr med assoc, Peter Bent Brigham Hosp, 42-44, assoc, 44-; ed, Nutrit Rev, 42-69; dir, Continental Can Co, 65- Honors & Awards: Goldberger Award, 64; Shattuck Award Pub Health, 69; Elvehjem Award Nutrit, 69. Mem: Am Chem Soc; Am Soc Biol Chemists; Am Soc Clin Invest; fel Am Pub Health Asn; fel AMA. Res: Application of nutrition to problems of medicine and public health, particularly cardio-vascular disease, obesity, osteoporosis and tooth decay; professional and lay education in nutrition. Mailing Add: Dept of Nutrit Harvard Univ Boston MA 02115

STARFIELD, BARBARA HOLTZMAN, b Brooklyn, NY, Dec 18, 32; m 55; c 4. PEDIATRICS. Educ: Swarthmore Col, BA, 54; State Univ NY Downstate Med Ctr, MD, 59; Johns Hopkins Univ, MPH, 63. Prof Exp: Teaching asst anat, State Univ NY Downstate Med Ctr, 54-57; from intern to asst resident pediat, Harriet Lane Home, Johns Hopkins Hosp, 59-62; from instr to asst prof pediat, Sch Med, 63-73, instr pub health admin, Sch Hyg & Pub Health, 65-66, from asst prof to assoc prof med care & hosps, 66-75, PROF HEALTH CARE ORGN, SCH HYG & PUB HEALTH, JOHNS HOPKINS UNIV, 75-, ASSOC PROF PEDIAT, 73- Concurrent Pos: Nat Ctr Health Serv Res & Develop res scientist develop award; med dir, Community Nursing Proj, Dept Pediat & dir, Pediat Med Care Clin, Johns Hopkins Hosp, 63-66, asst dir community health, Comprehensive Child Care Proj & mem, Comt Planning & Develop, 65-67, pediatrician, dir, Pediat Clin Scholars Prog, 71-; mem spec rev comt, Exp Med Care Rev Orgns, Dept Health, Educ & Welfare, mem, Health Serv Res Study Sect, consult, Spec Res & Develop Projs Div, Nat Ctr Health Serv Res & Develop; mem proj adv comt, Chart Audit Study, Nat Bd Med Exam; consult, Proj Expanded Health Serv Children, Am Acad Pediat; consult, Proj Regionalized Perinatal Care. Honors & Awards: Award, Enuresis Found, 67. Mem:

Sigma Xi; Am Pub Health Asn; Am Med Women's Asn; Soc Pediat Res; Int Epidemiol Asn. Mailing Add: Sch of Hyg & Pub Health Johns Hopkins Univ Baltimore MD 21205

STARK, BARBARA LOUISE, b Washington, DC, July 9, 44. ANTHROPOLOGY. Educ: Rice Univ, BA, 66; Yale Univ, MPh, 70, PhD(anthrop), 74. Prof Exp: Instr anthrop, Upsala Col, 71-72; ASST PROF ANTHROP, ARIZ STATE UNIV, 72- Mem: AAAS; Soc Am Archaeol; Am Anthrop Asn. Res: The origin and development of civilization in mesoamerica, especially in the Gulf Coast area; settlement pattern and cultural ecology in Mesoamerica. Mailing Add: Dept of Anthrop Ariz State Univ Tempe AZ 85281

STARK, BETTY SALZBERG, b Denver, Colo, Jan 19, 44; m 64; c 1. ALGEBRA. Educ: Univ Calif, Los Angeles, BA, 64; Univ Mich, MA, 66, PhD(math), 71. Prof Exp: ASST PROF MATH, NORTHEASTERN UNIV, 71- Mem: Am Math Soc; Math Asn Am. Res: Geometry of root-type groups in finite groups of Lie type. Mailing Add: Dept of Math Northeastern Univ Boston MA 02115

STARK, CHARLES R, b Detroit, Mich, Apr 2, 34; m 59; c 3. PEDIATRICS, EPIDEMIOLOGY. Educ: Univ Mich, MD, 60, DrPH, 63. Prof Exp: Med officer, Nat Cancer Inst, 63-68; epidemiologist, Nat Inst Child Health & Human Develop, 68-71, chief, Epidemiol Br, 71-75; GEM MED OFFICER, US COAST GUARD ACAD, NEW LONDON, CT, 75- Concurrent Pos: Instr, Sch Med, Georgetown Univ, 68- Res: Epidemiology of childhood neoplasms and congenital anomalies; use of electronic computers and mathematics in epidemiological research; cause and prevention of prematurity and infant mortality. Mailing Add: US Coast Guard Acad New London CT 06335

STARK, DENNIS MICHAEL, b Baltimore, Md, May 16, 42. IMMUNOLOGY. Educ: Univ Ga, DVM, 66; Cornell Univ, PhD(immunol), 69. Prof Exp: Res asst immunol, Cornell Univ, 66-69; asst prof, C W Post Col, Long Island Univ, 69-73; ASSOC PROF PATH & DIR ANIMAL FACIL, MED CTR, NY UNIV, 73- Mem: AAAS; Am Soc Microbiol; Am Asn Lab Animal Sci; Am Vet Med Asn; NY Acad Sci. Res: Immunochemistry of streptococcus species; autoimmunity of systemic lupus. Mailing Add: Berg Inst NY Univ Med Ctr 550 First Ave New York NY 10016

STARK, EGON, b Vienna, Austria, Sept 28, 20; nat US; m 48; c 3. MICROBIOLOGY. Educ: Univ Man, BS, 47, MS, 48; Purdue Univ, PhD(microbiol), 51. Prof Exp: Asst org chem, Univ Man, 44-47, asst microbiol, 45-48; asst bact, Purdue Univ, 48-51, Purdue Res Found Indust fel & res assoc microbiol, 51-53; consult microbiol, Univ Chicago, 53-54; sr res scientist, Joseph E Seagram & Sons, Ky, 54-66; PROF BIOL, ROCHESTER INST TECHNOL, 66- Mem: AAAS; Am Soc Microbiol; Am Chem Soc. Res: Microbiology of bacteria, yeasts, fungi; taxonomy, physiology, enzymology, ecology, fermentations, water pollution, waste disposal; process of producing a heat-stable bacterial amylase. Mailing Add: Dept of Biol Rochester Inst of Technol Rochester NY 14623

STARK, ERIC WALTER, b Ger, Mar 31, 10; nat US; m 36; c 3. WOOD TECHNOLOGY. Educ: Purdue Univ, BS, 32; State Univ NY, MS, 34, PhD, 52. Prof Exp: Asst wood tech, State Univ NY Col Forestry, Syracuse Univ, 34-38; asst prof forestry, Univ Idaho, 38-40; dir forest prod res, State Forest Serv, Tex, 40-43; assoc prof, 43-59, PROF FORESTRY, PURDUE UNIV, LAFAYETTE, 59- Mem: Soc Am Foresters. Res: Minute anatomy of woods native to the United States. Mailing Add: Dept of Forestry Purdue Univ West Lafayette IN 47906

STARK, FORREST OTTO, b Bay City, Mich, Mar 31, 30; m 57; c 4. CHEMISTRY. Educ: Univ Pittsburgh, BS, 55; Pa State Univ, University Park, PhD(chem), 62. Prof Exp: Res chemist, 55-57, 62-70, mgr tech serv & develop, Med Prod, 70-75, MGR RESINS & CHEM RES, DOW CORNING CORP, 75- Mem: Am Chem Soc; Sigma Xi. Mailing Add: 5311 Sunset Dr Midland MI 48640

STARK, FRANCIS C, JR, b Drumright, Okla, Mar 19, 19; m 41; c 2. HORTICULTURE. Educ: Okla State Univ, BS, 40; Univ Md, MS, 41, PhD(hort), 48. Prof Exp: From asst prof to prof veg crops, 45-64, prof hort & head dept, 64-74, chmn food sci fac, 66-73, PROVOST, DIV AGR & LIFE SCI, UNIV MD, COLLEGE PARK, 74- Concurrent Pos: Chmn, Gov Comn on Migratory Labor, Md, 63-; trustee, Lynchburg Col, 70-; dir, Coun Agr Sci & Technol, 74- Mem: Fel AAAS; fel Am Soc Hort Sci; Am Inst Biol Sci. Res: Nutrition, physiology, breeding and culture of vegetable crops. Mailing Add: Div of Agr & Life Sci Univ of Md College Park MD 20742

STARK, FRANK LOUIS, JR, plant pathology, see 12th edition

STARK, GEORGE ROBERT, b New York, NY, July 4, 33; m 56; c 2. BIOCHEMISTRY. Educ: Columbia Univ, BA, 55, MA, 56, PhD(chem), 59. Prof Exp: Res assoc biochem, Rockefeller Inst, 59-61, asst prof, 61-63; from asst prof to assoc prof, 63-71, PROF BIOCHEM, SCH MED, STANFORD UNIV, 71- Concurrent Pos: Guggenheim fel, 70-71. Mem: Am Soc Biol Chem; Am Chem Soc. Res: Chemistry and reactions of proteins; control of mammalian gene expression; structure-function relationships of enzymes; proteins of DNA tumor viruses. Mailing Add: Dept of Biochem Stanford Univ Sch of Med Stanford CA 94305

STARK, HAROLD EMIL, b San Diego, Calif, July 26, 20; m 44, 71; c 4. ENTOMOLOGY. Educ: San Diego State Col, BA, 43; Univ Utah, MS, 48; Univ Calif, PhD, 65. Prof Exp: Asst, Univ Utah, 46-48; jr entomologist, USPHS, 48-49, med entomologist, 50-63, trainin officer health mobilization, 63-64, ecol & chief vert-vector unit, Commun Dis Ctr, Ga, 64-68, entomologist, Walter Reed Army Inst Res, US Army Med Component/SEATO, Bangkok, Thailand, 68-70, res entomologist, Ecol Invest, Ctr Dis Control, USPHS, 70-73; RES ZOOLOGIST, ENVIRON & ECOL BR, DUGWAY PROVING GROUND, UTAH, 73- Mem: Entom Soc Am; Wildlife Dis Asn; Ecol Soc Am. Res: Systematics of Siphonaptera; ecology of small wild rodents and fleas in relation to natural occurrence of plague; preparation of training literature and audiovisuals for vector-borne diseases; preparation of environmental impact assessments and statements. Mailing Add: 520 Bonafin Dugway UT 84022

STARK, HAROLD MEAD, b Los Angeles, Calif, Aug 6, 39; m 64. NUMBER THEORY. Educ: Calif Inst Technol, BS, 61; Univ Calif, Berkeley, MA, 63, PhD(math), 64. Prof Exp: From instr to asst prof math, Univ Mich, Ann Arbor, 64-66; asst prof, Univ Mich, Dearborn Ctr, 66-67; from asst prof to assoc prof, Univ Mich, Ann Arbor, 67-68; assoc prof, 69-72, PROF MATH, MASS INST TECHNOL, 72- Concurrent Pos: Off Naval Res fel, 67-68; Sloan fel, 68-70. Mem: Am Math Soc; Math Asn Am. Res: Analytic and elementary number theory with emphasis on zeta functions and applications to quadratic fields. Mailing Add: Dept of Math Mass Inst of Technol Cambridge MA 02139

STARK, JAMES CORNELIUS, b Port Jefferson, NY, Sept 1, 41; m 63; c 3. ORGANIC CHEMISTRY, BIOCHEMISTRY. Educ: Eastern Nazarene Col, BS, 63; Purdue Univ, Lafayette, PhD(org chem), 69. Prof Exp: ASSOC PROF CHEM, EASTERN NAZARENE COL, 68- Mem: Am Chem Soc. Res: Preparation and reactions of polyhalo-organic compounds. Mailing Add: Dept of Chem Eastern Nazarene Col Quincy MA 02170

STARK, JEREMIAH MILTON, b Norfolk, Va, Apr 1, 22; m 49, 62; c 3. MATHEMATICS. Educ: US Coast Guard Acad, BS, 44; NTex State Col, BS, 46; Mass Inst Technol, SM, 49, PhD(math), 54. Prof Exp: Instr math, Mass Inst Technol, 49-52, mathematician instrumentation lab, 54-56; PROF MATH & HEAD DEPT, LAMAR UNIV, 56- Concurrent Pos: NSF sci fac fel, Stanford Univ, 63-64. Mem: AAAS; Am Math Soc; Math Asn Am; Soc Indust & Appl Math. Res: Analysis; complex variables. Mailing Add: Dept of Math Lamar Univ Beaumont TX 77705

STARK, JOEL, b New York, NY, Nov 18, 30; m 50; c 2. SPEECH PATHOLOGY. Educ: Long Island Univ, BA, 50; Columbia Univ, MA, 51; NY Univ, PhD(speech), 56. Prof Exp: Instr speech, Long Island Univ, 51-54; asst prof, City Col New York, 54-65; assoc prof speech path, Sch Med, Stanford Univ, 65-68; assoc prof commun arts & sci, 68-72, PROF COMMUN ARTS & SCI & DIR SPEECH & HEARING CTR, QUEENS COL, NY, 72- Concurrent Pos: Nat Inst Neurol Dis & Blindness fel, 62-64; mem coun except children. Mem: Am Speech & Hearing Asn; Am Asn Ment Deficiency. Res: Communications disorders; language development and disorders in children. Mailing Add: Speech & Hearing Ctr Queens Col Flushing NY 11367

STARK, JOHN BENJAMIN, biochemistry, see 12th edition

STARK, JOHN FREDERICK, b Olean, NY, May 11, 27; m 53; c 3. PHARMACEUTICAL CHEMISTRY. Educ: Grove City Col, BS, 50; Univ Buffalo, BS, 53; Purdue Univ, MS, 55, PhD(pharmaceut chem), 57. Prof Exp: Sr res scientist, Eaton Labs Div, 57-61, asst dir, Pharmaceut Res Div, 61-65, dir, 65-75, DIR RES ACQUISITIONS & LICENSES, NORWICH PHARMACAL CO, 75- Mem: AAAS; Am Chem Soc; Am Pharmaceut Asn. Res: Pharmaceutical product development; medicinal chemistry; application of radioisotopes to invitro drug behavioral studies; manufacturing pharmacy; research administration. Mailing Add: RD 2 E River Rd Norwich NY 13815

STARK, JOHN HOWARD, b Port Jefferson, NY, Sept 1, 41; m 63; c 3. CARBOHYDRATE CHEMISTRY. Educ: Eastern Nazarene Col, BS, 63; Purdue Univ, MS, 65, PhD(biochem), 69. Prof Exp: SR RES ASSOC CARBOHYDRATE CHEM, INT PAPER CO, 69- Mem: Am Chem Soc; Am Heart Asn. Res: Medical applications of polysaccharide derivatives and other polymers; cellulose derivatives and chemicals. Mailing Add: Int Paper Co Tuxedo NY 10987

STARK, LARRY GENE, b Abilene, Kans, Dec 31, 38; m 60; c 2. PHARMACOLOGY. Educ: Univ Kans, BS, 61, MS, 63; Stanford Univ, PhD(pharmacol), 68. Prof Exp: ASST PROF PHARMACOL, SCH MED, UNIV CALIF, DAVIS, 69- Concurrent Pos: NIH fel, Univ Chicago, 68-69. Mem: Soc Neurosci; Am Soc Pharmacol & Exp Therapeut. Res: Anticonvulsant drugs and animal models of epilepsy; correlations between the distribution of drugs in the brain and their neuropharmacological properties. Mailing Add: Dept of Pharmacol Univ of Calif Sch of Med Davis CA 95616

STARK, LAWRENCE, b New York, NY, Feb 21, 26; m 49; c 3. BIOENGINEERING. Educ: Columbia Univ, AB, 45; Albany Med Col, MD, 48; Am Bd Psychiat & Neurol, dipl, 57. Prof Exp: Intern, US Naval Hosp, St Albans, 48-49; res asst biochem, Oxford Univ & mem, Trinity Col, 49-50; res asst physiol, Univ Col, London, 50-51; asst prof physiol & pharmacol & res assoc neurophysiol & neuromuscular physiol, NY Med Col, 51; from instr to asst prof neurol & assoc physician, Yale Univ, 55-60; head neurol sect, Ctr Commun Sci, Res Lab Electronics & Electronic Systs Lab, Mass Inst Technol, 60-65; prof bioeng, neurol & physiol & chmn biomed eng dept, Univ Ill, Chicago Circle, 65-68; PROF PHYSIOL OPTICS & ENG SCI, UNIV CALIF, BERKELEY, 68- Concurrent Pos: Fel neurol & EEG, Neurol Inst, Columbia Univ & Presby Hosp, 51-52; fel neurol, Sch Med, Yale Univ, 54-55; fel, Mass Inst Technol, 60-65; fel neurol, Mass Gen Hosp, 60-65; Guggenheim fel, 68-70; vis app, Univ Col, London, 50-51, Nobel Inst Neurophysiol, Stockholm, 57, Harvard Univ, 63-65, Univ Calif, Los Angeles, 65 & Stanford Univ, 68 & 75; dir, Biosysts, Inc, Cambridge, 62-67 & Biocontacts, Inc, Berkeley, 69-73; chmn neurosci work session, Math Concepts of Cent Nerv Syst, 64, Gordon Conf Biomath, 65 & bioeng training comt, Nat Inst Gen Med Sci, 67-68; consult, NIH, NSF & var indust companies; assoc ed or ed bd mem, Math Biosci, Inst Elec & Electronics Eng-SMC, Brain Res, J Appl Physiol, J Neurosci & Comput in Biol & Med. Mem: Am Physiol Soc; Biophys Soc; Am Acad Neurol; fel Inst Elec & Electronics Eng; Asn Comput Mach. Res: Application of communication and information theory to neurophysiology; normal and abnormal neurological control systems; cybernetics; pattern recognition and artificial intelligence; information flow in biological evolution and economic theory. Mailing Add: 226 Minor Hall Univ of Calif Berkeley CA 94720

STARK, MARVIN MICHAEL, b Mich, Mar 14, 21; m; c 3. DENTISTRY. Educ: Univ Calif, Los Angeles, AB, 48, DDS, 52. Prof Exp: PROF OPER DENT & ORAL BIOL, SCH DENT, UNIV CALIF, SAN FRANCISCO, 53-; CHIEF DENT OFFICER, STATE OF CALIF DEPT HEALTH, 75- Concurrent Pos: Res fel dent med, Sch Dent Med, Harvard Univ, 52-53. Mem: Am Asn Endodont; fel Int Col Dent; fel Am Col Dent; Am Dent Asn; Int Asn Dent Res. Mailing Add: Dept Oper Dent & Oral Biol Univ of Calif Sch of Dent San Francisco CA 94122

STARK, NELLIE MAY, b Norwich, Conn, Nov 20, 33; m 62. PLANT ECOLOGY. Educ: Conn Col, BA, 56; Duke Univ, MA, 58, PhD(plant ecol, bot), 62. Prof Exp: Botanist, Pac Southwest Forest & Range Exp Sta, US Forest Serv, 58-64; res assoc, Lab Atmospheric Physics, Desert Res Inst, Univ Nev, Reno, 64-72; ASSOC PROF FORESTRY, UNIV MONT, 72- Concurrent Pos: Mem, Alph Helix Res Exped for Desert Res Inst, Brazil & Peru, 67; Int Biol Prog grants, 72-73; co-chmn comt hemispheric coop, Inst Ecol, 73-74. Mem: Ecol Soc Am; Int Soc Trop Ecol; Bot Soc Am; Am Inst Biol Sci; Asn Trop Biol. Res: Nutrient cycling and soil ecology in tropical and temperate forests; fire and logging ecology; applied concept of the biological life of a soil to land use management. Mailing Add: Sch of Forestry Univ of Mont Missoula MT 59801

STARK, PAUL, b Philadelphia, Pa, Feb 1, 29; m 52; c 3. PHARMACOLOGY, PHYSIOLOGY. Educ: McGill Univ, BSc, 49; Univ Rochester, PhD(pharmacol), 63. Prof Exp: Chief prod biochem, Gerber Prod Co, 52-55; res technician, Stromberg-Carlson Co, 55-57; res chemist, Allerton Chem, 57-60; sr pharmacologist, 63-66, res scientist, 67-72, RES ASSOC, ELI LILLY & CO, 72- Concurrent Pos: Assoc prof, Sch Med, Ind Univ, Indianapolis. Mem: Int Col Neuropsychopharmacol; Am Soc Pharmacol & Exp Therapeut; Am Physiol Soc; Soc Neurosci. Res: Neuropharmacological and psychopharmacological techniques in the study of neuro-transmitters within the central nervous system. Mailing Add: MC304 Lilly Res Labs 307 E McCarty St Indianapolis IN 46206

STARK, PHILIP HERALD, b Iowa City, Iowa, Mar 2, 36. GEOLOGY. Educ: Univ

Okla, BS, 58; Univ Wis, MS, 61, PhD(geol), 63. Prof Exp: Explor geologist, Mobil Oil Corp, 63-65, sr explor geologist, 65-66, regional comput coordr, 66-69; mgr geol applns, 69-74, DIR TECH APPLNS, PETROL INFO CORP, 74- Mem: Geol Soc Am; Soc Econ Paleontologists & Mineralogists; Am Asn Petrol Geologists. Res: Stratigraphy and micropaleontology of Paleozoic flysch facies; computer applications in geology for petroleum exploration. Mailing Add: Petrol Info Corp PO Box 2612 Denver CO 80202

STARK, RICHARD B, b Conrad, Iowa, Mar 31, 15; m 67. PLASTIC SURGERY. Educ: Stanford Univ, AB, 36; Cornell Univ, MD, 41; Am Bd Plastic Surg, dipl, 52. Prof Exp: Intern, Peter Bent Brigham & Children's Hosps, Boston, Mass, 41-42; resident, Children's Hosp, 42; plastic surgeon, Northington Gen Hosp, Tuscaloosa, Ala, 45-46 & Percy Jones Gen Hosp, Battle Creek, Mich, 46; surgeon, Kingsbridge Vet Hosp & NY Hosp, 47-50; from asst prof clin surg to assoc prof, 55-73, PROF CLIN SURG, COL PHYSICIANS & SURGEONS, COLUMBIA UNIV, 73- Concurrent Pos: Fel, Med Sch, Stanford Univ, 46-47; plastic surgeon, NY Hosp, 48; attend surgeon chg plastic surg, St Luke's Hosp, 58-; vis prof, Univ Tex, 65, Univ Mich, 66, Walter Reed Med Ctr, 70 & Univ Man, 71; from vpres to pres, Am Bd Plastic Surg, 66-68. Honors & Awards: Res Prize, Am Soc Plastic & Reconstruct Surg Found, 51; Medal of Honor, Vietnam, 67 & 69; Order of San Carlos, Colombia, 69. Mem: AAAS; Soc Univ Surg; AMA; Am Asn Plastic Surg; Am Col Surgeons. Res: Circulation in skin grafts; homologous transplants of skin; pathogenesis of harelip and cleft palate. Mailing Add: 115 E 67th St New York NY 10021

STARK, RICHARD HARLAN, b Ozawkie, Kans, Dec 5, 16; m 42; c 4. COMPUTER SCIENCE. Educ: Univ Kans, AB, 38; Northwestern Univ, MS, 42, PhD(math), 46. Prof Exp: Jr physicist, US Naval Ord Lab, 42-43 & Los Alamos Sci Lab, 44-45; instr math, Northwestern Univ, 46-48; mathematician, Los Alamos Sci Lab, 48-51; mgr math anal, Knolls Atomic Power Lab, Gen Elec Co, 52-56, mgr math & comput oper, Atomic Power Equip Dept, 56-61, consult analyst, Comput Dept, 62-64; from assoc prof to prof math & info sci, Wash State Univ, 64-69; PROF COMPUT SCI, N MEX STATE UNIV, 69- Mem: Soc Indust & Appl Math; Asn Comput Mach. Res: Proofs of program validity. Mailing Add: Dept of Comput Sci NMex State Univ Las Cruces NM 88001

STARK, RONALD WILLIAM, b Can, Dec 4, 22; nat US; m 44; c 2. FOREST ENTOMOLOGY. Educ: Univ Toronto, BScF, 48, MA, 51; Univ BC, PhD(forest entom), 58. Prof Exp: Agr res officer, Div Forest Biol, Sci Serv, Can Dept Agr, 48-59; asst prof entom & asst entomologist, Agr Exp Sta, Univ Calif, Berkeley, 59-61, from assoc prof to prof, 61-70, vchmn dept entom & parasitol, 68-70, entomologist, 61-70; PROF FORESTRY & ENTOM, COORDR RES & DEAN GRAD SCH, UNIV IDAHO, 70- Concurrent Pos: NSF sr fel, 67-68; collabr, Pac Southwest Forest & Range Exp Sta, US Forest Serv; Am rep & chmn working group forest entom, Int Union Forest Res Orgn; proj leader, Pest Mgt Prog Pine Back Beetle Ecosyst, Int Biol Prog. Mem: AAAS; Soc Am Foresters; Entom Soc Am; Ecol Soc Am; Entom Soc Can. Res: Population dynamics; pest management. Mailing Add: Grad Sch Morrill Hall Univ of Idaho Moscow ID 83843

STARK, ROYAL WILLIAM, b Wellington, Ohio, Apr 30, 37; m 62; c 2. SOLID STATE PHYSICS, LOW TEMPERATURE PHYSICS. Educ: Case Inst Technol, BS, 59, MS, 61, PhD(physics), 62. Prof Exp: Res assoc solid state physics, Case Inst Technol, 62; from instr to prof, Univ Chicago & Inst Study Metals, 63-72; PROF PHYSICS, UNIV ARIZ, 72- Concurrent Pos: Alfred P Sloan res fel, 64-70. Mem: Am Phys Soc. Res: Electronic properties of metals; magnetic breakdown; Fermi surface and band structure; ferromagnetism; plasma effects. Mailing Add: Dept of Physics Univ of Ariz Tucson AZ 85721

STARK, WILLIAM HUBBARD, bacteriology, biochemistry, see 12th edition

STARK, WILLIAM MAX, b Ind, June 14, 14; m 46; c 3. INDUSTRIAL MICROBIOLOGY. Educ: Ind State Univ, BS, 39; Purdue Univ, MS, 49, PhD(bact), 52. Prof Exp: Instr bact, Purdue Univ, 50-51; sr microbiologist, 51-60, sr scientist, 60-61, res assoc, 61-72, RES ADV, ELI LILLY & CO, 72- Mem: AAAS; Am Soc Microbiol; Am Chem Soc; Soc Indust Microbiol (pres, 75-76); Am Inst Biol Sci. Res: Biosynthesis of fermentation products. Mailing Add: Lilly Res Labs Eli Lilly & Co Indianapolis IN 46206

STARK, WILLIAM RICHARD, b Lexington, Ky, Apr 28, 45; m 68; c 1. MATHEMATICAL LOGIC. Educ: Univ Ky, BS, 68; Univ Wis-Madison, PhD(math), 75. Prof Exp: Asst math, Univ Wis-Madison, 69-74; INSTR MATH, UNIV TEX, AUSTIN, 74- Mem: Am Math Soc; Asn Symbolic Logic. Res: Forcing methods in recursive set theory and their application to completeness and compactness for large languages; independence proof of a Moore space problem; building models by Henkin's method in computers. Mailing Add: Dept of Math Univ of Tex Austin TX 78712

STARKE, ALBERT CARL, JR, b Cleveland, Ohio, Jan 14, 16; m 41; c 3. ORGANIC CHEMISTRY. Educ: Fla Southern Col, BS, 36; Northwestern Univ, PhD(chem), 40. Prof Exp: Lab instr, Dent Sch, Northwestern Univ, 37-40, Nat Defense Res Comt fel, 40-42; res chemist, GAF Corp, 42-46, patent searcher, 46-47, patent liaison, 48-50, patent agt, 50-55, supvr tech info serv, 55-66, mgr tech info, 66-72; RES SPECIALIST, UNIV CONN, 72-; INFO MGR, NEW ENG RES APPLN CTR, 72- Concurrent Pos: Mem, Franklin Inst, Chem Abstr communicator, 66- Mem: Fel AAAS; Am Chem Soc; Am Soc Info Sci; fel Am Inst Chemists. Res: Physiological and synthetic organic chemistry; color photography; patent soliciting and prosecution; warfare agents; polymers; storage and retrieval of technical information; computer systems design; documentation; computer searching. Mailing Add: PO Box 172 Mansfield Center CT 06250

STARKE, EMORY POTTER, b New York, NY, Jan 31, 96; m 23; c 1. MATHEMATICS. Educ: Columbia Univ, AB, 16, MA, 17, PhD(math), 27. Prof Exp: Instr math, City Col New York, 17-18; from asst prof to prof & chmn dept, 19-61, EMER PROF MATH, RUTGERS UNIV, 61- Concurrent Pos: Assoc ed, Am Math Monthly, 46-; prof, Bloomfield Col, 61-66; consult, NSF, 62-67. Mem: Am Math Soc; Math Asn Am. Res: Functions of complex variable; theory of numbers; uniform functions of rational functions. Mailing Add: 1000 Kensington Ave Plainfield NJ 07060

STARKEY, EUGENE EDWARD, b Yakima, Wash, July 14, 26; m 54; c 3. DAIRY SCIENCE. Educ: Calif State Polytech Col, BS, 52; Univ Wis, MS, 54, PhD(dairy husb, genetics), 58. Prof Exp: Asst dairying, Univ Wis, 52-55; dairy husbandman, Dairy Husb Res Br, Agr Res Serv, USDA, 55-57; asst prof dairy prod, Utah State Univ, 57-60; PROF DAIRY PROD, UNIV WIS-MADISON, 60- Mem: Am Dairy Sci Asn. Res: Dairy cattle breeding sire selection and evaluation of environmental influences on production. Mailing Add: 5126 Loruth Terr Madison WI 53711

STARKEY, JOHN, b Manchester, Eng, Aug 11, 36; m 58; c 2. STRUCTURALGEOLOGY. Educ: Univ Liverpool, BSc, 57, PhD(geol), 60. Prof Exp: NATO & Dept Sci & Indust Res Gt Brit res fels geol, Inst Crystallog & Petrol, Swiss Fed Inst Technol, 60-62; Miller res fel, Univ Calif, Berkeley, 62-65; asst prof, 65-69, ASSOC PROF GEOL, UNIV WESTERN ONT, 69- Concurrent Pos: Leverhulme Europ fel, 60-61; Royal Soc bursary, Imp Col, Univ London, 71-72. Mem: Brit Mineral Soc; Am Mineral Soc. Res: Petrofabric analysis of rocks, primarily by x-ray techniques; crystallography of plagioclase feldspars and their twinning; crystal chemistry of rock forming minerals. Mailing Add: Dept of Geol Univ of Western Ont London ON Can

STARKEY, OTIS PAUL, b Buffalo, NY, Apr 14, 06; m 60. ECONOMIC GEOGRAPHY, GEOGRAPHY OF ANGLO-AMERICA. Educ: Columbia Univ, BS, 27, MA, 30, PhD(geog), 39. Prof Exp: Instr econ, Washington & Jefferson Col, 27-28; ed asst & personal asst to Prof J Russell Smith, Columbia Univ, 28-31; from instr to asst prof geog, Univ Pa, 31-42; geogr, Mil Intel, War Dept, 42-45; dep dir planning div, Off Foreign Liquidation Comn, US Dept State, 45-46; prof geog, 46-70, EMER PROF GEOG, IND UNIV, 70-; LECTR GEOG, CALIF STATE COL, SAN BERNARDINO, 74- Concurrent Pos: Prin investr, Off Naval Res Contract, 48-61. Honors & Awards: Meritorious Civilian Serv Award, Off Chief of Staff, War Dept, 46. Mem: Asn Am Geog; Am Geog Soc; Nat Coun Geog Educ. Res: Historical, economic and cultural factors in the settlement and development of North American regions. Mailing Add: 18801 Via San Marco Irvine CA 92664

STARKEY, PAUL EDWARD, b Fultonham, Ohio, Dec 9, 20; m 42; c 4. PEDODONTICS, DENTISTRY. Educ: Ind Univ, DDS, 43; Am Bd Pedodont, dipl, 58. Prof Exp: Instr pedodontics, Ohio State Univ, 55-56; assoc prof, 59-63, chmn clin div, 61 & dept pedodontics, 68, PROF PEDODONTICS, SCH DENT, IND UNIV, INDIANAPOLIS, 63- Concurrent Pos: Pvt pract, 46-59; exam mem, Am Bd Pedodont, 69. Honors & Awards: Frederick Bachman Lieber Distinguished Teaching Award, Ind Univ, 68. Mem: Am Soc Dent for Children (pres, 67); Am Acad Pedodont. Res: Clinical children's dentistry; educational research. Mailing Add: Dept of Pedodont Ind Univ Sch of Dent Indianapolis IN 46202

STARKEY, ROLAND JOSEPH, JR, microbiology, organic chemistry, see 12th edition

STARKLOFF, GENE B, b St Louis, Mo, July 21, 14; m 38; c 2. SURGERY. Educ: Wash Univ, AB, 35, MD, 39; Am Bd Surg, dipl, 48. Prof Exp: From instr surg to assoc prof clin surg, 59-67, PROF CLIN SURG, SCH MED, ST LOUIS UNIV, 67- Concurrent Pos: Mem surg staff, St Mary's Hosp, 49-; chief staff, St Louis State Hosp, 53-; consult, St Louis City Hosp & Vet Admin Hosp, 53- Mem: Fel Am Col Surg. Res: Cancer surgery. Mailing Add: Dept of Surg St Louis Univ Sch of Med St Louis MO 63104

STARKOVSKY, NICOLAS ALEXIS, b Alexandria, Egypt, Jan 15, 22; US citizen; m 59; c 2. ORGANIC CHEMISTRY. Educ: Univ Cairo, BS, 46, MS, 54, PhD(chem), 56. Prof Exp: Res chemist, Memphis Chem Co, Egypt, 51-60; fel, Columbia Univ, 60-61; res chemist, Dow Chem Co, 61-64; dir res & develop, Collab Res, Inc, Mass, 64-70; dir proj develop, Ortho Res Found, NJ, 71-72; RES ASST, PRINCETON PROD CO DIV, CARTER WALLACE, INC, PRINCETON, 73- Mem: AAAS; Am Chem Soc; NY Acad Sci. Res: Immunology; clinical medicine. Mailing Add: 788 W Foothill Rd Bridgewater NJ 08807

STARKS, AUBRIE NEAL, JR, b Dermott, Ark, Aug 20, 46. ANALYTICAL CHEMISTRY. Educ: Southern Ill Univ, Carbondale, BA, 67; Univ Ark, Fayetteville, PhD(chem), 75. Prof Exp: Instr chem, Univ Ark, Fayetteville, 73-74; intern, Hendrix Col, 74-75; ASST PROF CHEM, THIEL COL, 75- Mem: Am Chem Soc. Res: Photovoltammetric investigation of transition metal complexes for photocurrents generated in optically-shielded electrode systems. Mailing Add: Dept of Chem Thiel Col Greenville PA 16125

STARKS, CHARLES MASTERSON, chemistry, see 12th edition

STARKS, FRED W, b Milford, Ill, Aug 16, 21; m 46; c 3. ORGANIC CHEMISTRY. Educ: Univ Ill, BS, 43, MS, 47; Univ Nebr, PhD, 50. Prof Exp: Lab supvr, US Rubber Co, 43-44; res chemist, E I du Pont de Nemours & Co, 50-57, group leader, 57; PRES & DIR RES, STARKS ASSOCS, INC, 57- Mem: Am Chem Soc; fel Am Inst Chemists; NY Acad Sci. Res: Chemistry of hydrogen cyanide, chlorinated hydrocarbons, activated carbon and surfaces; synthesis of metabolite analogs; patent development. Mailing Add: 742 Highland Ave Kenmore NY 14223

STARKS, KENNETH JAMES, b Ft Worth, Tex, July 27, 24; m 51; c 2. ENTOMOLOGY. Educ: Univ Okla, BS, 50, MS, 51; Iowa State Col, PhD(entom), 54. Prof Exp: Asst prof, Univ Ky, 53-61; mem staff, USDA, Uganda, 61-69; prof entom, Okla State Univ, 69; RES & LOCATION LEADER, AGR RES SERV, USDA, 69- Mem: Entom Soc Am. Res: Grain insects investigations. Mailing Add: USDA Agr Res Serv Dept of Entom Okla State Univ Stillwater OK 74074

STARKS, PAUL (BRECKENRIDGE), b Hardin, Ky, May 11, 17; m 42; c 1. ANIMAL SCIENCE. Educ: Western Ky Univ, BS, 41; Univ Ky, MS, 47; Univ Ill, PhD(animal sci), 53. Prof Exp: Teacher high sch, NC, 41-42 & Ky, 47-50; assoc prof agr, Univ Tenn, 53-66; assoc prof, 66-72, PROF ANIMAL SCI, SOUTHWEST MO STATE UNIV, 72- Mem: Am Soc Animal Sci. Res: Value of sodium sulfate; elemental sulfur; methionine and cystine in the nutrition of growing lambs. Mailing Add: 646 Katella Lane Springfield MO 65804

STARKS, THOMAS HAROLD, b Owatonna, Minn, Aug 19, 30; m 59; c 2. STATISTICS. Educ: Mankato State Col, BA, 52; Purdue Univ, MS, 54; Va Polytech Inst, PhD(statist), 59. Prof Exp: Spec serv engr, E I du Pont de Nemours & Co, 59-61; ASSOC PROF MATH, SOUTHERN ILL UNIV, CARBONDALE, 61- Mem: AAAS; Am Statist Asn; Math Asn Am; Inst Math Statist. Res: Design of experiments; statistical inference. Mailing Add: Dept of Math Southern Ill Univ Carbondale IL 62901

STARKWEATHER, GARY KEITH, b Lansing, Mich, Jan 9, 38; m 61; c 2. PHYSICAL OPTICS, ELECTROOPTICS. Educ: Mich State Univ, BS, 60; Univ Rochester, MS, 66. Prof Exp: Engr, Bausch & Lomb, Inc, 62-64; AREA MGR OPTICAL SYSTS, XEROX PALO ALTO RES CTR, 64- Concurrent Pos: Instr optics, Monroe Community Col, 68-69. Mem: Optical Soc Am. Res: Optics and electronics and their specific system interaction, involving display and hard copy image systems. Mailing Add: Xerox Palo Alto Res Ctr 3333 Coyote Hill Rd Palo Alto CA 94304

STARKWEATHER, HOWARD WARNER, JR, b Cambridge, Mass, July 20, 26; m 48; c 3. PHYSICAL CHEMISTRY. Educ: Haverford Col, AB, 48; Harvard Univ, AM, 50; Polytech Inst Brooklyn, 50-52, PhD(chem), 53. Prof Exp: Chemist, Rayon Dept, 47, res chemist, Ammonia Dept, 48-49 & Plastics Dept, 52-57, sr res chemist, 57-66, RES ASSOC, PLASTICS DEPT, E I DU PONT DE NEMOURS & CO, INC, 66- Mem: Am Chem Soc. Res: Polymer chemistry; polymerization kinetics; polymer properties and molecular structure; polymer crystallography. Mailing Add: Plastics Dept E I du Pont de Nemours & Co Inc Wilmington DE 19898

STARLING, JAMES HOLT, b Troy, Ala, June 28, 12; m; c 1. BIOLOGY, ZOOLOGY. Educ: Univ Ala, AB, 33, MA, 37; Duke Univ, PhD(zool), 42. Prof Exp: High sch instr, 33-39; asst gen zool, Duke Univ, 39-42; instr, 42-43, from asst prof to assoc prof gen biol & comp anat, 46-51, PROF GEN BIOL & COMP ANAT, WASHINGTON & LEE UNIV, 51-, COORDR PRE-MED STUDIES, 64- Concurrent Pos: Glenn grant, Natural Hist Mus, London, Eng, 53; partic, NSF Insts, Univ NC, 59; consult, NSF Insts, Appalachian State Col, 60 & 63; partic, Oak Ridge Inst Nuclear Studies, 61. Mem: AAAS; Sigma Xi. Res: Human intestinal parasites such as malaria, micro-filaria and schistosomes; ecological and taxonomical studies of microscopic animals of the soil. Mailing Add: 207 Paxton St Lexington VA 24450

STARLING, JAMES LYNE, b Henry Co, Va, Aug 16, 30. PLANT BREEDING, STATISTICS. Educ: Va Polytech Inst, BS, 51; Pa State Univ, MS, 55, PhD(agron), 58. Prof Exp: Asst forage crop breeding, 54-57, from instr to assoc prof agron, 57-69, PROF AGRON & HEAD DEPT, PA STATE UNIV, UNIVERSITY PARK, 69- Mem: AAAS; Am Soc Agron. Res: Forage crop breeding; genetics and cytogenetics of forage crop species; experimental design. Mailing Add: Pa State Univ 119 Tyson Bldg University Park PA 16802

STARLING, THOMAS MADISON, b Loneoak, Va, Aug 12, 23; m 61; c 2. PLANT BREEDING. Educ: Va Polytech Inst, BS, 44; Iowa State Univ, MS, 47, PhD, 55. Prof Exp: From asst agronomist to assoc agronomist, 44-60, prof agron, 60-70, assoc dean grad sch, 70-71, PROF AGRON, VA POLYTECH INST & STATE UNIV, 71- Mem: Fel Am Soc Agron; Crop Sci Soc Am. Res: Plant breeding and genetics of winter barley and wheat. Mailing Add: 618 Woodland Dr Blacksburg VA 24060

STARMER, C FRANK, b Greensboro, NC, Sept 4, 41; m 63; c 3. COMPUTER SCIENCE, MEDICINE. Educ: Duke Univ, BSEE, 63, MSEE, 65; Rice Univ, 65-66; Univ NC, PhD(biomath, bioeng), 68. Prof Exp: Res assoc med, 63-65, assoc biomath, 66-68, asst prof community health sci, 68-71, ASST PROF MED, MED CTR, DUKE UNIV, 71- Concurrent Pos: NIH career develop award, Duke Univ, 72-77, NIH res grant, 72-75. Mem: AAAS; Biomet Soc; Asn Comput Mach; Am Heart Asn; Inst Elec & Electronics Eng. Res: Computer science and pattern recognition in heart disease; applied statistics. Mailing Add: Dept of Med Duke Univ Med Ctr Durham NC 27710

STARNES, ORDWAY, b Lancaster, SC, Oct 22, 19; m 48; c 2. ENTOMOLOGY. Educ: Clemson Univ, BS, 42; Rutgers Univ, PhD(entom), 49. Prof Exp: Assoc specialist entom, Col Agr, Rutgers Univ, 49-53, prof agr res, 56-65, asst dir, NJ Agr Exp Sta, 53-56, assoc dir, 56-61, dir, 61-65; dir agr res, E African Agr & Forestry Res Orgn, 65-71, dir Indian agr prog, 71-74, AGR SCI REP IN AFRICA, ROCKEFELLER FOUND, 74-; ASSOC DIR, INT LAB RES ANIMAL DIS, 74- Mem: Entom Soc Am; Soc Nematol. Res: Economic entomology and pesticides. Mailing Add: Rockefeller Found 1133 Ave of the Americas New York NY 10036

STARNES, PAUL KISER, b Cherokee Falls, SC, June 7, 20; m 43; c 2. ORGANIC CHEMISTRY, ANALYTICAL CHEMISTRY. Educ: Wofford Col, BS, 43; Univ NC, MA, 48. Prof Exp: Chemist, 48-55, proj leader anal res, 55-62, sr chemist, 62-64, sect head anal res, 64-69, SR SECT HEAD, ANAL & INFO DIV, ESSO RES & ENG CO, 69- Concurrent Pos: Mem proj nitrogen constituents petrol, Am Petrol Inst, 64-65, chmn, 65-66, mem proj composition heavy ends petrol, 66-; adv bd chem technol, Union County Tech Inst, 65- Mem: Am Chem Soc. Res: Analytical research as related to pilot plant, engineering and products development research. Mailing Add: Anal & Info Div Esso Res & Eng Co PO Box 121 Linden NJ 07036

STARNES, WILLIAM HERBERT, JR, organic chemistry, see 12th edition

STAROSTKA, RAYMOND WALTER, b Silver Creek, Nebr, Nov 27, 23; m 51; c 4. AGRONOMY. Educ: Univ Nebr, BS, 47; Univ Wis, PhD(soils), 50. Prof Exp: Asst soil chem, Purdue Univ, 47-48; assoc soil scientist, Soil & Water Conserv Res Br, USDA, 50-54, soil scientist, 54-56; supvr agr chem res, W R Grace & Co, 56-62, mgr agr res, 62-65; assoc dir agr res, Int Minerals & Chem Corp, Ill, 65-67; PRES, AGROSERV, INC, 67- Mem: AAAS; Am Chem Soc; Am Soc Agron; Soil Sci Soc Am. Res: Soil fertility; plant physiology; agronomic evaluation of fertilizers; inorganic chemistry, especially of phosphate compounds; agricultural chemicals; management of agricultural systems research. Mailing Add: Agroserv Inc Box 36 Silver Creek NE 68663

STARR, ALBERT, b New York, NY, June 1, 26; m 55; c 2. THORACIC SURGERY. Educ: Columbia Col, BA, 46; Columbia Univ, MD, 49. Prof Exp: Asst surg, Columbia Univ, 56-57; from instr to assoc prof surg, 57-64, PROF CARDIOPULMONARY SURG, MED SCH, UNIV ORE, 64-, HEAD DIV, 63- Honors & Awards: Sci Achievement Award, Am Heart Asn, 63; Rene Le Riche Award Cardiovasc Surg, 65. Mem: Am Surg Asn; Am Asn Thoracic Surg; Am Col Cardiol; Int Cardiovasc Soc. Res: Prosthetic values for cardiac surgery. Mailing Add: Div of Cardiopulmonary Surg Univ of Ore Med Sch Portland OR 97201

STARR, DAVID WRIGHT, b Anna, Tex, Dec 8, 12; m 37, 61; c 2. MATHEMATICS. Educ: Southern Methodist Univ, AB, 33; Univ Ill, AM, 37, PhD(math), 40. Prof Exp: High sch instr, Tex, 33-37, prin, 34; asst math, Univ Ill, 38-40; instr ground sch aviation, 40-43, from instr to assoc prof math, 40-48, coordr war training serv, Civil Aeronaut Admin, 41-44, PROF MATH, SOUTHERN METHODIST UNIV, 48-, CHMN DEPT, 63- Mem: Am Math Soc; Math Asn Am. Res: Analysis; Schrodinger wave equation from the point of view of singular integral equations. Mailing Add: Dept of Math Southern Methodist Univ Dallas TX 75222

STARR, DONALD DEE, chemistry, see 12th edition

STARR, DUANE FRANK, b Pasadena, Calif, Oct 20, 42; m 65; c 1. PHYSICAL CHEMISTRY. Educ: Wesleyan Univ, BA, 64; Ore State Univ, PhD(phys chem), 73. Prof Exp: Resident res assoc lasers, Naval Res Lab, Nat Res Coun, 73-75; STAFF CHEMIST PROPELLANT CHEM, ALLEGANY BALLISTICS LAB, HERCULES INC, 75- Mem: Am Chem Soc; Am Phys Soc. Res: Cure chemistry in crosslinked double base propellants; effects of solids distribution on propellant rheology; fogging properties of solid propellants. Mailing Add: Allegany Ballistics Lab Hercules Inc PO Box 210 Cumberland MD 21502

STARR, JAMES LEROY, b Almont, Mich, Aug 14, 39; m 60; c 2. SOIL PHYSICS, MICROBIOLOGY. Educ: Mich State Univ, BS, 61, MS, 70; Eastern Baptist Theol Sem, MA, 66; Univ Calif, Davis, PhD(soil sci), 73. Prof Exp: Voc Agr teacher, Carson City Community Schs, Mich, 61-62 & Mayville Community Schs, 62-64; teaching asst, Mich State Univ, 67-68, instr soils & dir audiotutorial lab, 68-70; res asst, Univ Calif, 70-72, staff res assoc, 72-74; ASST SCIENTIST SOIL PHYSICS, CONN AGR EXP STA, 74- Mem: Soil Sci Soc Am; Int Soc Soil Sci. Res: Reaction and transport of nitrogenous materials through soil profiles to underground water supplies. Mailing Add: Conn Agr Exp Sta 123 Huntington St New Haven CT 06504

STARR, JASON LEONARD, b Chelsea, Mass, Aug 13, 28; m 51; c 3. ONCOLOGY, BIOCHEMISTRY. Educ: Harvard Univ, AB & AM, 49, MD, 53. Prof Exp: Intern med, Beth Israel Hosp, Boston, Mass, 53-54; asst prof, Med Sch, Northwestern Univ, 61-65; from assoc prof to prof, Col Med, Univ Tenn, Memphis, 65-72; prof med, Sch Med, Univ Calif, Los Angeles, 73-74; CLIN PROF MED, UNIV TENN, MEMPHIS, 74-; DIR ONCOL, BAPTIST MEM HOSP, MEMPHIS, 74- Concurrent Pos: Fel, Mayo Found, Univ Minn, 56-58; fel biochem, Sch Med, Western Reserve Univ, 58-61; Am Cancer Soc fel, 61-63 & scholar, 71-72; sr investr, Arthritis Found, 63-68. Mem: Am Soc Clin Invest; Am Asn Immunol; Am Chem Soc. Res: Genetic control of antibody synthesis; biochemistry of neoplastic cells; cancer chemotherapy; immuno-oncology. Mailing Add: Baptist Mem Hosp 920 Madison Ave Memphis TN 38103

STARR, JOHN EDWARD, b St Louis, Mo, July 12, 39; m 61; c 2. PHOTOGRAPHY. Educ: Colo Col, BA, 61; Stanford Univ, PhD(org chem), 65. Prof Exp: Sr chemist, 65-69, LAB HEAD, RES LABS, EASTMAN KODAK CO, 69- Mem: Am Chem Soc. Res: Spectral sensitization of photographic emulsions by sensitizing dyes. Mailing Add: Res Labs Eastman Kodak Co Kodak Park Rochester NY 14650

STARR, JOHN THORNTON, JR, b Baltimore, Md, Aug 4, 39; m 63; c 2. TRANSPORTATION GEOGRAPHY, URBAN GEOGRAPHY. Educ: Johns Hopkins Univ, BA, 61, MAT, 62; Univ Chicago, PhD(geog), 72. Prof Exp: Geogr, Md State Planning Dept, 64-66; ASST PROF GEOG, UNIV MD, BALTIMORE COUNTY, 67- Concurrent Pos: Mem, Transp Res Forum, 67-; co-dir, NSF In-Serv Teachers Training Prog, Morgan State Col, 70-71; consult, Dept Planning, City Baltimore, 72-; mem panel future port requirements of US Maritime Transp Res Bd, Nat Acad Sci, 73- Mem: Asn Am Geog; Am Geog Soc. Res: Intermodal competition in the transport industries; coordination of land-water transport systems; bulk freight transport; impact of changing transport patterns on urban centers; urban mass transit. Mailing Add: Dept of Geog Univ Md Baltimore Co Baltimore MD 21228

STARR, LEON, b Bronx, NY, May 2, 37; m 60. ORGANIC CHEMISTRY. Educ: Polytech Inst Brooklyn, BS, 58; Univ Mo, PhD(org chem), 62. Prof Exp: Sr res chemist, Mobil Chem Co, 62-66, group leader process res, 66-67; res assoc, Celanese Plastics Co, Clark, 67-69, proj mgr, Newark, 69-71, lab dir, Summit, 71-72, tech mgr, 73-74; DIR RES, CELANESE CHEM CO, 74- Mem: AAAS; Am Chem Soc; The Chem Soc. Res: Exploratory polymerization using interfacial polycondensation; photochemistry research; liquid phase oxidation and transition metal catalysis; polymer synthesis and development; polyacetals. Mailing Add: Celanese Chem Co Tech Ctr PO Box 9077 Corpus Christi TX 78408

STARR, MERLE ARTHUR, b Portland, Ore, Apr 8, 11. ELECTRONICS. Educ: Reed Col, BA, 33; Univ Calif, MA & PhD(physics), 37. Prof Exp: Asst physics, Washington Univ, 37-39; instr, Univ Ore, 39-41; mem staff, Radiation Lab, Mass Inst Technol, 41-46; asst prof physics, Univ Ore, 46-47; asst prof, 47-50, ASSOC PROF PHYSICS, UNIV PORTLAND, 50- Mem: Am Phys Soc; Am Asn Physics Teachers. Res: Cosmic radiations by means of cloud chamber and Geiger counters; measurement of atomic electron velocity; secondary emission of electrons; circuits and screens for cathode ray tubes; production of cosmic-ray showers in lead. Mailing Add: Dept of Physics Univ of Portland Portland OR 97203

STARR, MORTIMER PAUL, b New York, NY, Apr 13, 17; m 44; c 3. MICROBIOLOGY, PHILOSOPHY OF BIOLOGY. Educ: Brooklyn Col, BA, 38; Cornell Univ, MS, 39, PhD(bact, biochem, plant path), 43. Prof Exp: Tutor biol, Brooklyn Col, 39-44, from instr to asst prof, 44-47; from asst prof bact & asst bacteriologist to assoc prof bact & assoc bacteriologist, 47-58, spec asst to chancellor res grants & contracts, 63-67, PROF & BACTERIOLOGIST, EXP STA, UNIV CALIF, DAVIS, 58- Concurrent Pos: Nat Res Coun fel, Hopkins Marine Sta, Stanford Univ, 44-46; cur, Int Collection of Phytopathogenic Bacteria, 47-; vis specialist fac agron, Nat Univ Colombia, 49; NIH spec fels, Cambridge & Ghent Univs, 53-54, Plant Dis Div, Univ Auckland, NZ, 62; Guggenheim Mem Found fel, Max Planck Inst Med Res, 58, guest, 59; vis prof, Chile, 66; Guggenheim Mem Found fel, Swiss Fed Inst Technol, Zurich, 68-69; vis prof, Univ Hamburg, 72-73; partic, Int Cong Microbiol, Rome, 53, Stockholm, 58, Montreal, 62, Moscow, 66, Mexico, 70; partic, Int Bot Cong, Paris, 54, Montreal, 59 & Edinburgh, 64; mem, Int Conf Sci Probs Plant Protection, Budapest, 60, Int Conf Cult Collections, Toronto, 62 & Tokyo, 68; mem, Conf Global Impacts Appl Microbiol, Stockholm, 63, Addis Ababa, 67, Bombay, 69; mem, Int Conf Phytopathogenic Bacteria, Harpenden, 64, Lisbon, 67; assoc ed, Ann Rev Microbiol, 58-72, ed, 72- Honors & Awards: Bernardo O'Higgins Medal, First Class, Repub of Chile, 67; Silver Medal, Purkyne Univ, Brno, 68. Mem: AAAS; Electron Micros Soc Am; Am Phytopath Soc; Am Soc Microbiol; fel Am Acad Microbiol. Res: Biochemistry, metabolism, genetics, ecology and taxonomy of phytopathogenic bacteria; microbial pigments; pectin metabolism; philosophical grounds of taxonomy and ecology; bacterial morphogenesis; bacterial diversity and ecology; international aspects of microbiology; university administration. Mailing Add: Dept of Bact Univ of Calif Davis CA 95616

STARR, NORMAN, b Scranton, Pa, Apr 14, 33; m 65. MATHEMATICAL STATISTICS. Educ: Univ Mich, BA, 55, MA, 60; Columbia Univ, PhD(math statist), 65. Prof Exp: Assoc math, Evans Res & Develop Corp, 61-65; asst prof statist, Univ Minn, 65-66; from asst prof to assoc prof, Carnegie-Mellon Univ, 66-68; assoc prof, 68-73, PROF MATH, DEPT STATIST, UNIV MICH, ANN ARBOR, 68- Mem: Inst Math Statist; Am Math Soc; Am Statist Asn. Res: Sequential analysis; optimal stopping; statistical allocation and theory; applied probability. Mailing Add: Dept of Statist Univ of Mich Ann Arbor MI 48104

STARR, NORTON, b Kansas City, Mo, June 18, 36; m 59; c 2. MATHEMATICS. Educ: Harvard Univ, AB, 58; Mass Inst Technol, PhD(math), 64. Prof Exp: Instr math, Mass Inst Technol, 64-66; asst prof, 66-71, ASSOC PROF MATH, AMHERST COL, 71- Concurrent Pos: Vis asst prof, Univ Waterloo, 72-73. Mem: Am Math Soc; Math Asn Am. Res: Operator limit theory. Mailing Add: Dept of Math Amherst Col Amherst MA 01002

STARR, PATRICIA RAE, b Hood River, Ore, Feb 28, 35. MICROBIOLOGY. Educ: Ore State Univ, BS, 57, MS, 62; Univ Ore, PhD(microbiol), 68. Prof Exp: Instr microbiol, Dent Sch, Univ Ore, 68-69; assoc, Ore State Univ, 69-71; asst prof, Univ Ill, Urbana, 71-75; RES ASSOC, PROVIDENCE HOSP, 75- Mem: AAAS; Am Soc Microbiol. Res: Relationship of sterol synthesis to respiratory adaptation in yeast; amino acid uptake systems in yeast; white blood cell function. Mailing Add: Dept of Med Educ Providence Hosp Portland OR 97213

STARR, PHILLIP HENRY, b Poland, Nov 16, 20; nat US; m 49; c 3. PSYCHIATRY. Educ: Univ Toronto, MD, 44. Prof Exp: Asst prof neuropsychiat & pediat, Sch Med, Wash Univ, 52-55, dir community child guid clin, 52-56; ASSOC PROF NEUROL & PSYCHIAT, SCH MED, UNIV NEBR, OMAHA, 57- Concurrent Pos: Chief psychiat consult, St Louis Children's Hosp, 51-56; chief children's outpatient serv, Nebr Psychiat Inst, 56-64, consult, 64- Mem: Fel Am Psychiat Asn; fel AMA; Int Asn Child Psychiat; fel Am Acad Child Psychiat. Res: Child and adult psychiatry. Mailing Add: Dept of Psychiat Univ of Nebr Sch of Med Omaha NE 68105

STARR, RICHARD CAWTHON, b Greensboro, Ga, Aug 24, 24. PHYCOLOGY.

Educ: Ga Southern Col, BS, 44; George Peabody Col, MA, 47; Vanderbilt Univ, PhD(biol), 52. Prof Exp: From instr to assoc prof, 52-60, PROF BOT, IND UNIV, BLOOMINGTON, 60- Concurrent Pos: Guggenheim fel, 59. Mem: AAAS; Bot Soc Am; Phycol Soc Am; Brit Phycol Soc. Res: Morphology and cultivation of green algae; genetics and development of Volvox. Mailing Add: Dept of Plant Sci Ind Univ Bloomington IN 47401

STARR, ROBERT BREWSTER, b Buffalo, NY, Mar 18, 25; m 59; c 1. OCEANOGRAPHY. Educ: Cornell Univ, BCE, 50, MS, 55. Prof Exp: Phys oceanogr, US Naval Oceanog Off, 55-61; phys oceanogr, US Coast & Geod Surv, 61-62, chief oceanog sect, 62-63, chief oceanog anal br, 63-65; phys oceanogr, Inst Oceanog, Environ Sci Serv Admin, 65-67; res oceanogr, 67-73, CHIEF DATA SERV & QUAL CONTROL GROUP, ATLANTIC OCEANOG & METEOROL LABS, NAT OCEANIC & ATMOSPHERIC ADMIN, 73- Mem: Geol Soc Am; Am Soc Limnol & Oceanog; Am Geophys Union. Res: Physical oceanography and marine geology of the Antarctic; physical oceanography, marine geology and geophysics of the Caribbean Sea and adjacent Atlantic Ocean; marine ecosystems analysis; New York bight. Mailing Add: Atlantic Oceanog & Meteorol Labs 15 Rickenbacker Causeway Miami FL 33149

STARR, ROBERT I, b Laramie, Wyo, Dec 11, 32; m 56; c 2. CHEMISTRY, ENVIRONMENTAL SCIENCES. Educ: Univ Wyo, BS, 56, MS, 59, PhD(plant physiol), 72. Prof Exp: Res biochemist, US Fish & Wildlife Serv, Colo, 60-63; plant physiologist, Colo State Univ, 63-64; anal chemist, US Food & Drug Admin, 64-65; chemist, Colo State Univ, 65-69; res chemist, Bur Sport Fisheries & Wildlife, US Dept Interior, 69-74; ENVIRON SCIENTIST, US GEOL SURV, 74- Concurrent Pos: Res biochemist, Wildlife Res Ctr, US Fish & Wildlife Serv, Colo, 68-69. Mem: AAAS; Am Chem Soc; fel Am Inst Chemists; Am Soc Plant Physiologists. Res: Pesticide chemistry as related to plants and soils, including method development studies; plant, soil, and chemistry matters relating to mining operations. Mailing Add: US Geol Surv Bldg 25 Fed Ctr Denver CO 80225

STARR, SELIG, mathematics, see 12th edition

STARR, THEODORE JACK, b Plainfield, NJ, Aug 22, 24; m 54; c 3. MICROBIOLOGY. Educ: City Col New York, BS, 49; Univ Mass, MS, 51; Univ Wash, PhD(microbiol), 53. Prof Exp: Res assoc microbiol, Haskins Labs, 48-49; instr biol, Univ Ga, 54-55; fishery res biologist, US Fish & Wildlife Serv, 55-57; McLaughlin fel virol, Med Br, Univ Tex, 57-60; assoc res scientist, Lab Comp Biol, Kaiser Found Res Inst, 60-62; assoc prof microbiol, Univ Notre Dame, 62-68; prof biol sci & assoc head dept, Col Arts & Sci, Univ Ill, Chicago Circle, 68-70; PROF BIOL SCI & HEAD DEPT, STATE UNIV NY COL BROCKPORT, 70-, ASST VPRES ACAD AFFAIRS, GRAD STUDIES & RES, 75- Concurrent Pos: Fac exchange scholar, State Univ NY, 75. Mem: Am Soc Microbiol; Soc Exp Biol & Med. Res: Cytochemistry; virology; marine and space biology; gnotobiology. Mailing Add: Dept of Biol Sci State Univ of NY Brockport NY 14420

STARR, WALTER LEROY, b Portland, Ore, Feb 9, 24; m 44; c 2. PHYSICS. Educ: Univ Southern Calif, BS, 50; Calif Inst Technol, MS, 51. Prof Exp: Physicist, US Naval Missile Test Ctr, 50; res physicist, US Naval Civil Eng Lab, 51-55; res scientist, Phys Sci Lab, Lockheed Missiles & Space Co, 55-67; PHYSICIST, AMES RES CTR, NASA, 67- Mem: AAAS; Am Phys Soc; Am Geophys Union. Res: Atomic and molecular physics; atmospheric processes; ionization and excitation; absorption cross sections. Mailing Add: Planetary Environ Br Ames Res Ctr NASA Moffett Field CA 94035

STARRATT, ALVIN NEIL, b Paradise, NS, Sept 18, 36; m 67; c 2. NATURAL PRODUCTS CHEMISTRY. Educ: Acadia Univ, BSc, 59; Univ Western Ont, PhD(org chem), 63. Prof Exp: Brit Petrol Co res fel, Imp Col, Univ London, 63-64; res fel, Res Inst Med & Chem, Mass, 64-65; RES SCIENTIST, RES INST, CAN AGR, 65- Concurrent Pos: Hon lectr, Univ Western Ont, 69- Mem: Am Chem Soc; The Chem Soc; fel Chem Inst Can. Res: Natural products influencing the behavior of insects. Mailing Add: Res Inst Can Agr University Sub PO London ON Can

STARRETT, ANDREW, b Greenwich, Conn, Mar 18, 30; m 51; c 3. MAMMALOGY. Educ: Univ Conn, BS, 51; Univ Mich, MS, 55, PhD(zool), 58. Prof Exp: Instr zool, Univ Mich, 56-57; from instr to asst prof biol, Univ Southern Calif, 57-64; asst prof, Northeastern Univ, 64-65; assoc prof, 65-69, PROF BIOL, CALIF STATE UNIV, NORTHRIDGE, 69- Concurrent Pos: Res assoc, Los Angeles County Mus Natural Hist. Mem: AAAS; Soc Syst Zool; Soc Study Evolution; Am Soc Mammalogists. Res: Vertebrate and mammalian evolution and distribution; mammalian morphology and systematics, particularly Chiroptera. Mailing Add: Dept of Biol Calif State Univ Northridge CA 91324

STARRETT, AUSTIN LEROY, b China, Maine, Sept 18, 07; m 36; c 3. MATHEMATICS. Educ: Dartmouth Col, AB, 29; Harvard Univ, AM, 30. Prof Exp: Instr math, Hobart Col, 31-33; assoc prof, 35-56, prof, 56-74, EMER PROF MATH, GA INST TECHNOL, 74- Mem: Math Asn Am. Res: Third order differential operators; algebraic numbers; Laplace and Fourier transforms. Mailing Add: Dept of Math Ga Inst of Technol Atlanta GA 30332

STARRFIELD, SUMNER GROSBY, b Los Angeles, Calif, Dec 29, 40; m; c 2. THEORETICAL ASTROPHYSICS. Educ: Univ Calif, Berkeley, BA, 62; Univ Calif, Los Angeles, PhD(astron), 69. Prof Exp: Lectr astron, Yale Univ, 67-69, asst prof, 69-71; scientist, Thomas J Watson Res Ctr, IBM Corp, 71-72; asst prof, 72-75, ASSOC PROF ASTROPHYS, ARIZ STATE UNIV, 75- Concurrent Pos: NSF res grant, 74; consult, Los Alamos Sci Lab, 74- Mem: Royal Astron Soc; Int Astron Union; Am Astron Soc. Res: Stellar structure and evolution; hydrodynamical studies of novae; star formation. Mailing Add: Dept of Physics Ariz State Univ Tempe AZ 85281

STARSHAK, ALBERT JOSEPH, b Chicago, Ill, June 20, 33; m 67; c 3. STATISTICS, DATA PROCESSING. Educ: Xavier Univ, Ohio, LitB, 57; Case Inst Technol, MS, 65; Ill Inst Technol, PhD(chem), 70. Prof Exp: Res asst chem, Case Inst Technol, 60-64; sr instr math, St Ignatius Prep Sch, 64-65; asst chemist, IIT Res Inst, 65-68; res assoc chem, Ill Inst Technol, 68-70; sr scientist data anal, 70-72, SR SCIENTIST PROD DEVELOP, TISSUE/TOWEL DEVELOP, AM CAN CO, 72- Mem: Am Chem Soc; Am Soc Qual Control. Res: Statistical design of experiments; cost analysis; market research; new product development. Mailing Add: Am Can Co 1915 Marathon Ave Neenah WI 54956

STARZAK, MICHAEL EDWARD, b Woonsocket, RI, Apr 21, 42; m 67; c 1. BIOPHYSICAL CHEMISTRY. Educ: Brown Univ, BS, 63; Northwestern Univ, PhD(chem), 68. Prof Exp: Actg instr chem & grant, Univ Calif, Santa Cruz, 68-69, actg asst prof, 69-70; ASST PROF CHEM, STATE UNIV NY, BINGHAMTON, 70- Concurrent Pos: Corp mem, Marine Biol Lab, 74- Mem: AAAS; Am Chem Soc; Am Phys Soc; Biophys Soc. Res: Excitable membrane phenomena; photochemistry; stochastic processes; energy transfer. Mailing Add: Dept of Chem State Univ of NY Binghamton NY 13901

STARZL, THOMAS E, b Le Mars, Iowa, Mar 11, 26; m 54; c 3. SURGERY. Educ: Westminister Col, Mo, BA, 47; Northwestern Univ, MA, 50, PhD(anat) & MD, 52. Hon Degrees: Dr, Westminister Col, Mo, New York Med Col, Univ Wyo & Westmar Col. Prof Exp: Intern srug, Johns Hopkins Hosp, 52-53, asst resident, 55-56; resident, Sch Med, Univ Miami, 56-58; resident & instr, Northwestern Univ, 58-59, assoc, 59-61, asst prof, 61; assoc prof, 62-64, PROF SURG, SCH MED, UNIV COLO, DENVER, 64-, CHMN DEPT, 72- Honors & Awards: Int Soc Surg Medal, 65; Eppinger Prize (Freiburg), 70; Brookdale Award in Med, 74; Middleton Award, 68; Mod Med Distinguished Achievement Award, 69. Mem: Am Col Surg; fel Am Acad Arts & Sci; Soc Univ Surg; Am Surg Asn; Soc Vascular Surg. Res: General and thoracic surgery; neurophysiology, cardiac physiology; transplantation of tissues and organs. Mailing Add: Dept of Surg Univ of Colo Sch of Med Denver CO 80220

STARZYK, MARVIN JOHN, b Chicago, Ill, Feb 3, 35; m 58; c 4. MICROBIOLOGY. Educ: Loyola Univ, Chicago, BS, 57; Univ Wis-Madison, PhD(microbiol), 62. Prof Exp: Asst prof natural sci, Northern Ill Univ, 61-64; group leader microbiol, Res Dept, Brown & Williamson Tobacco Corp Ky, 64-65, sect leader biol sci, 65-66; ASSOC PROF MICROBIOL, NORTHERN ILL UNIV, 66- Concurrent Pos: Consult, Brown & Williamson Tobacco Corp, 66-67. Mem: Am Soc Microbiol; Can Soc Microbiol; Int Asn Water Pollution Res. Res: Water microbiology, the ecology of microorganisms associated with pure and polluted waters. Mailing Add: Dept of Biol Sci Northern Ill Univ De Kalb IL 60115

STASHEFF, JAMES DILLON, b New York, NY, Jan 15, 36; m 59; c 2. MATHEMATICS. Educ: Univ Mich, BA, 56; Princeton Univ, MA, 58, PhD(math), 61; Oxford Univ, DPhil(math), 61. Prof Exp: Moore instr math, Mass Inst Technol, 60-62; from asst prof to prof, Univ Notre Dame, 62-70; PROF MATH, TEMPLE UNIV, 70- Concurrent Pos: NSF grants, 64-; mem Inst Advan Study, 64-65, Sloan fel, 69-70; vis prof, Princeton Univ, 68-69. Mem: AAAS; Am Math Soc. Res: Algebraic topology, especially homotopy theory; fibre space, H-spaces and characteristic classes. Mailing Add: Dept of Math Temple Univ Philadelphia PA 19122

STASIW, ROMAN OREST, b Ukraine, Russia, May 3, 41; US citizen; m 68; c 1. CLINICAL CHEMISTRY, BIOCHEMISTRY. Educ: Univ Rochester, BS, 63; State Univ NY Buffalo, PhD(inorg chem), 68. Prof Exp: Analyst, E I du Pont de Nemours & Co, summers 62 & 63; asst scientist, Cancer Res Ctr, Columbia, Mo, 68-73; SR BIOCHEMIST, TECHNICON, 73- Mem: Am Chem Soc; Am Asn Clin Chem. Res: Inorganic and synthetic organic chemistry; enzymology; clinical automation. Mailing Add: 98 N Grant Ave Congers NY 10920

STASKIEWICZ, BERNARD ALEXANDER, b Monessen, Pa, Aug 20, 24; m 49; c 5. PHYSICAL CHEMISTRY. Educ: Washington & Jefferson Col, AB, 46; Carnegie Inst Technol, MS, 50, PhD(chem), 53. Prof Exp: Instr chem, Washington & Jefferson Col, 46-51; res chemist, Esso Res & Eng Co, Standard Oil Co, NJ, 53-56 & Rayonier, Inc, 56-58; assoc prof, 58-62, PROF CHEM, WASHINGTON & JEFFERSON COL, 62-, CHMN DEPT, 67- Mem: Am Chem Soc. Res: Thermodynamics; cellulose chemistry; automotive lubricants. Mailing Add: Dept of Chem Washington & Jefferson Col Washington PA 15301

STASKO, AIVARS B, b Riga, Latvia, May 22, 37; Can citizen; m 63; c 3. AQUATIC ECOLOGY, FISHERIES. Educ: Univ Toronto, BASc, 60, PhD(zool), 69. Prof Exp: Res assoc limnol, Univ Wis-Madison, 67-70; RES SCIENTIST, BIOL STA, FISHERIES & MARINE SERV, 70- Concurrent Pos: Ed, Underwater Telemetry Newslett, 71- Mem: AAAS; Animal Behav Soc; Am Fisheries Soc; Can Soc Zoologists. Res: Crab and lobster biology and fisheries; underwater biotelemetry; responses of fish to environmental factors. Mailing Add: Biol Sta Fisheries & Marine Serv St Andrews NB Can

STASZAK, DAVID JOHN, b Milwaukee, Wis, Mar 29, 44; m 65; c 2. ANIMAL PHYSIOLOGY, BIOCHEMISTRY. Educ: Iowa State Univ, BS, 66, MS, 68, PhD(physiol), 71. Prof Exp: Res asst insect physiol, Iowa State Univ, 66-68, teaching asst human physiol, 68-69, res asst insect physiol, 69-71; asst prof biochem, Ill Col, 71-72; ASST PROF PHYSIOL, DEPT BIOL, COL MILLEDGEVILLE, 72- Concurrent Pos: USDA grant, Iowa State Univ, 66-71; asst prof, Dept Biol, MacMurray Col, 71-72; consult, Biochem Sect, Res Dept, Regional Ment Health Ctr, Cent State Hosp, 75- Mem: Am Inst Biol Sci; Sigma Xi; AAAS; Entom Soc Am. Res: Influence of low temperature on animals; chill-coma; thermal acclimation; hypertension. Mailing Add: Dept of Biol Ga Col Milledgeville GA 31061

STATE, DAVID, b London, Ont, Nov 13, 14; nat US; m 45; c 5. SURGERY. Educ: Univ Western Ont, BA, 36, MD, 39; Univ Minn, MS, 45, PhD(surg), 47; Am Bd Surg, dipl, 46; Bd Thoracic Surg, dipl, 52. Prof Exp: Intern, Victoria Hosp, Ont, 39-40; from intern to sr resident surg, Univ Minn Hosp, 41-45, res asst, Univ, 45-46, from instr to assoc prof, 46-52, dir cancer detection ctr, Univ Hosp, 48-52; clin assoc prof surg, Sch Med, Univ Southern Calif, 52-58; prof, Albert Einstein Col Med, 58-71, chmn dept, 59-71; PROF SURG & VCHMN DEPT, UNIV CALIF, LOS ANGELES & CHMN DEPT SURG, HARBOR GEN HOSP, 71- Concurrent Pos: Fel path, St Luke's Hosp, Chicago, 40-41; dir surg, Cedars of Lebanon Hosp, Los Angeles, Calif, 53-58, Bronx Munic Hosp Ctr, New York, 59-71 & Hosp of Albert Einstein Col Med, 66-71. Mem: AAAS; Soc Exp Biol & Med; Am Thoracic Soc; Soc Univ Surg; AMA. Res: General, thoracic and open heart surgery; gastrointestinal physiology. Mailing Add: Harbor Gen Hosp Dept of Surg 1000 W Carson St Torrance CA 90509

STATE, HAROLD M, b Washington, Mo, Apr 15, 10; m 39; c 1. ANALYTICAL CHEMISTRY, INORGANIC CHEMISTRY. Educ: Cent Col, Mo, AB, 32; Princeton Univ, AM, 35, PhD(chem), 36. Prof Exp: Asst prof chem, Culver-Stockton Col, 36-37; from instr to prof, 37-75, EMER PROF CHEM, ALLEGHENY COL, 75- Concurrent Pos: Vis asst prof, Univ Ill, 41-42. Mem: Am Chem Soc. Res: Application of Werner complexes to analysis; higher valent complexes of nickel; chemistry of coordination compounds. Mailing Add: RD 1 Saegertown PA 16433

STATEN, RAYMOND DALE, b Stillwater, Okla, May 17, 22; m 46; c 4. AGRONOMY, BOTANY. Educ: Okla State Univ, BS, 47; Univ Nebr, MS, 49, PhD(agron), 51. Prof Exp: Asst prof agron, Univ Ark, 51-56; asst prof, 56-60, ASSOC PROF AGRON, TEX A&M UNIV, 60- Mem: Am Soc Agron. Res: Forage crop breeding and improvement; pasture management; grain and fiber crops production; morphology. Mailing Add: Dept of Soil & Crop Sci Tex A&M Univ College Station TX 77843

STATES, JACK STERLING, b Laramie, Wyo, Nov 6, 41; m 65; c 2. MICROBIAL ECOLOGY, MYCOLOGY. Educ: Univ Wyo, BAEd, 64, MSc, 66; Univ Alta, PhD(bot), 69. Prof Exp: Res assoc bot, Univ Wyo, 69-70; ASST PROF BIOL, NORTHERN ARIZ UNIV, 70- Concurrent Pos: High sch instr, Wyo, 69-70. Mem: Mycol Soc Am; Am Inst Biol Sci; Bot Soc Am; Torrey Bot Club. Res: Soil microfungi; ecological studies and effects of industrial pollutants; developmental studies of high Basidiomycetes. Mailing Add: Dept of Biol Northern Ariz Univ Flagstaff AZ 86001

STATES, JAMES BRUCE, b Laramie, Wyo, Nov 6, 41; m 65; c 3. ECOLOGY. Educ: Univ Wyo, BA, 64, MS, 68; Ore State Univ, PhD(behav ecol), 73. Prof Exp: High sch teacher, Sheridan Pub Schs, Wyo, 65-66; SECT MGR TERRESTRIAL ECOL, ECOL CONSULTS INC, 73- Mem: Sigma Xi; Ecol Soc Am. Res: Species adaptation in ecotones, particularly ponderosa pine transitions; ongoing morphological and behavioral evolution in yellow pine chipmunks; improving impact assessments, mitigation plans and monitoring designs for proposed industrial activities. Mailing Add: Ecol Consults Inc Box 1057 Ft Collins CO 80521

STATTON, GARY LEWIS, b New Brighton, Pa, Nov 4, 37; m 58; c 3. POLYMER CHEMISTRY. Educ: Geneva Col, BS, 59; Univ Fla, PhD(org chem), 64. Prof Exp: Res chemist, 64-66, SR RES CHEMIST, ATLANTIC RICHFIELD CO, 66- Mem: Am Chem Soc; The Chem Soc. Res: Organometallics; polymers. Mailing Add: 214 W Rose Valley Rd Wallingford PA 19086

STATTON, WILLIAM OSBORNE, b Monte Vista, Colo, Feb 3, 21; m 45; c 2. POLYMER PHYSICS. Educ: Univ Calif, AB, 43; Mass Inst Technol, PhD(inorg chem), 50. Prof Exp: Anal supvr, Tenn Eastman Co, 45-46; res assoc, Mass Inst Technol, 48-50; res chemist, RCA Labs, 50-51; res chemist, E I du Pont de Nemours & Co, 51-58, sr res chemist, 58-63, res fel, 63-69; prof mat sci & eng, Univ Utah, 69-76; CONSULT, 76- Concurrent Pos: Vis res fel, Univ Manchester Inst Sci & Technol, 75-76. Mem: Am Chem Soc; Am Crystallog Asn; Am Phys Soc; Fiber Soc. Res: X-ray studies of fibers; low angle scattering of x-rays; nuclear magnetic and electron paramagnetic resonance of fibers; polymer deformation and fracture. Mailing Add: 157 Mano Dr Kula HI 96790

STATZ, HERMANN, b Herrenberg, Ger, Jan 9, 28; nat US; m 53; c 2. PHYSICS. Educ: Stuttgart Tech Univ, MS, 49, Dr rer nat(physics), 51. Prof Exp: Res assoc, Max Planck Inst Metal Res, Ger, 49-50; Ger Res Asn fel physics, Stuttgart Tech Univ, 51-52; mem solid state & molecular theory group, Mass Inst Technol, 52-53; group leader, 53-58, asst gen mgr, 58-69, ASST GEN MGR & TECH DIR, RAYTHEON RES DIV, RAYTHEON CO, 69- Mem: Fel Am Phys Soc. Res: Semiconductor physics, surfaces and devices; ferromagnetism; paramagnetic resonance; exchange interactions in solids; masers and lasers. Mailing Add: Raytheon Res Div 28 Seyon St Waltham MA 02154

STATZ, JOYCE ANN, b Minn, July 21, 47. COMPUTER SCIENCES. Educ: Col St Benedict, Minn, BA, 69; Syracuse Univ, MA, 71, PhD(comput sci), 73. Prof Exp: Res assoc comput sci, Syracuse Univ, 72-73; ASST PROF COMPUT SCI, BOWLING GREEN STATE UNIV, 73- Concurrent Pos: Ed, Sgcue Bull, Asn Comput BOWLING GREEN STATE UNIV, 73- Concurrent Pos: Ed, SIGCUE Bull, Asn Comput Mach, 74- Mem: Asn Comput Mach; Nat Coun Teachers Math. Res: Computer education; logo. Mailing Add: Dept of Comput Sci Bowling Green State Univ Bowling Green OH 43403

STAUB, HERBERT WARREN, b Brooklyn, NY, Aug 31, 27; m 55; c 1. BIOCHEMISTRY. Educ: Syracuse Univ, AB, 49; Rutgers Univ, MS, 57, PhD(biochem, physiol), 60. Prof Exp: Asst, Rutgers Univ, 57-60; SR RES SPECIALIST NUTRIT, TECH CTR, GEN FOODS CORP, 60- Concurrent Pos: Mem coun arteriosclerosis, Am Heart Asn; adj prof nutrit, Pac Univ Westchester, 75- Mem: AAAS; Am Chem Soc; Soc Nutrit Educ; Inst Food Technologists; NY Acad Sci. Res: Nutritional biochemistry; atherosclerosis; proteins; carbohydrates; enzymology; relationship of dietary carbohydrates to metabolic activity; protein quality evaluation and protein nutrition. Mailing Add: Gen Foods Tech Ctr 250 North St White Plains NY 10625

STAUB, LAURENCE CHARLES, chemistry, see 12th edition

STAUB, NORMAN CROFT, b Syracuse, NY, June 21, 29; m 53; c 5. PHYSIOLOGY. Educ: Syracuse Univ, AB, 50; State Univ NY, MD, 53. Prof Exp: Intern, Walter Reed Army Med Ctr, Washington, DC, 54; instr physiol, Grad Sch Med, Univ Pa, 57-58; vis asst prof, 58-59, asst res physiologist, 59-60, from asst prof to assoc prof physiol, 60-70, PROF PHYSIOL, CARDIOVASC RES INST, MED CTR, UNIV CALIF, SAN FRANCISCO, 70-, MEM SR STAFF, 58- Concurrent Pos: Res fel physiol, Grad Sch Med, Univ Pa, 56-58. Mem: AAAS; Am Physiol Soc; Microcirc Soc; Int Soc Lymphology. Res: Pulmonary physiology; pulmonary structure-function relations; kinetics of reaction of oxygen and hemoglobin; diffusion of oxygen and carbon monoxide; pulmonary capillary bed; pulmonary edema and blood flow; pulmonary lymph and lymphatics. Mailing Add: Cardiovasc Res Inst Univ of Calif San Francisco CA 94143

STAUB, ROBERT J, b Chicago, Ill, Jan 29, 22. ECOLOGY, BOTANY. Educ: St Mary's Col, Minn, BS, 43; Univ Minn, Minneapolis, MS, 49, PhD(ecol), 66. Prof Exp: Teacher high schs, Mo, Minn, Tenn & Ill, 43-50; instr biol, 50-53, assoc prof, 59-61, PROF BIOL, CHRISTIAN BROS COL, 70-, CHMN DEPT, 61- Concurrent Pos: Res grants, Dept of Interior, 67-69 & Environ Protection Agency, 70-72. Mem: Ecol Soc Am; Bot Soc Am; Am Water Resources Asn; Am Inst Biol Sci; Am Bryol & Lichenological Soc. Res: Plant variation; water pollution and its effects on phytoplankton; aquatic ecology. Mailing Add: Dept of Biol Christian Brothers Col 650 E Parkway S Memphis TN 38104

STAUBER, LESLIE ALFRED, parasitology, deceased

STAUBER, WILLIAM TALIAFERRO, b East Orange, NJ, June 15, 43; m 70; c 1. PHYSIOLOGY. Educ: Ithaca Col, BS, 67; Rutgers Univ, MS, 69, PhD(physiol), 72. Prof Exp: Tech lab asst physiol chem, Rutgers Univ & Univ Louvain, 69-70; NSF fel, 72-73, MUSCULAR DYSTROPHY ASN FEL PHYSIOL, UNIV IOWA, 74- Concurrent Pos: John Polachek Found Med Res grant, 75. Mem: Sigma Xi; AAAS. Res: Physiology-pathology of skeletal muscle protein breakdown as related to organelle function and the involvement of lysosomes, peroxisomes and sarcoplasmic reticulum in autophagy. Mailing Add: Dept of Physiol & Biophys Basic Sci Bldg Univ Iowa Iowa City IA 52242

STAUBITZ, WILLIAM JOSEPH, b Buffalo, NY, Mar 19, 15; m 44; c 4. MEDICINE, UROLOGY. Educ: Gettysburg Col, AB, 38; Univ Buffalo, MD, 42. Prof Exp: Chmn urol, Roswell Park Mem Inst, 49-60; PROF UROL & CHMN DEPT, SCH MED, STATE UNIV NY COL BUFFALO, 60- Concurrent Pos: Chmn dept urol, Buffalo Gen Hosp, Buffalo Children's Hosp & Edward J Meyer Mem Hosp, 60-; consult, Roswell Park Mem Inst, 60-; consult & mem dean's comt, Vet Admin Hosp, 65-; mem, Residency Rev Comt Urol, 68- Mem: Can Urol Asn; Am Urol Asn; Am Col Surg; Am Acad Pediat; Am Asn Genito-Urinary Surg. Res: Carcinoma of the prostate; carcinoma of the testes; urinary tract infections. Mailing Add: Dept of Urol State Univ of NY Sch of Med Buffalo NY 14215

STAUBUS, JOHN REGINALD, b Cissna Park, Ill, Mar 21, 26; m 51; c 1. DAIRY SCIENCE. Educ: Univ Ill, BS, 50, MS, 56, PhD(dairy sci), 59. Prof Exp: Asst dairy sci, Univ Ill, 54-59, res assoc, 59-60; from asst prof to assoc prof, 60-69, PROF DAIRY SCI, OHIO STATE UNIV, 69-, EXTEN SPECIALIST, 60- Mem: AAAS; Am Dairy Sci Asn; Sigma Xi. Res: Nutrition in dairy science; ruminant nutrition and physiology; bacteriology of silage; forage plant physiology and composition. Mailing Add: Dept of Dairy Sci Ohio State Univ Columbus OH 43210

STAUCH, JOHN EDWARD, b Ann Arbor, Mich, Nov 4, 20; m 46; c 3. BACTERIOLOGY. Educ: Univ Mich, BS, 43, MS, 50, PhD(bact), 58. Prof Exp: Clin bacteriologist, Hosp, Lackland AFB, Tex, 50-52, dep comdr, 494th Clin Lab, London, Eng, 52-55, actg chief, Epidemiol Lab, Lackland AFB, 57-58, dep chief, 58-59, clin microbiologist, Hosp, Andrews AFB, Md, 60-65; asst prof med, Sch Med, George Washington Univ, 65-72; dir lab serv, Clinton Community Hosp, Md, 67-74; DIR, OXON HILL DIAG CTR, 72- Concurrent Pos: Rep, Bact & Mycol Study Sect, Grants Div, NIH. Mem: Am Soc Microbiol; Am Soc Clin Path; Am Asn Bioanalysts; fel Asn Clin Sci; Royal Soc Med. Res: Immunochemistry; host response to endotoxic materials; host-parasite relationships. Mailing Add: Oxon Hill Diag Ctr 52 Albert Dr Oxon Hill MD 20022

STAUD, MARGARET C, physiology, zoology, see 12th edition

STAUDENMAYER, RALPH, b July 28, 42; US citizen. METALLURGICAL CHEMISTRY. Educ: Univ Calif, Los Angeles, BS, 66; Univ Ariz, MS, 68; Univ Ark, PhD(chem), 73. Prof Exp: CHIEF CHEMIST METALL, TRW INC, WENDT SONIS, 73- Res: Surface chemistry and gas deposition on cemented carbides. Mailing Add: 1616 Hammond St Fayetteville AR 72701

STAUDER, JACK, b Pueblo, Colo, Mar 2, 39. CULTURAL ANTHROPOLOGY. Educ: Harvard Univ, BA, 62; Cambridge Univ, MA, 64, PhD(anthrop), 68. Prof Exp: Lectr social anthrop, Harvard Univ, 68-71; asst prof anthrop, Northeastern Univ, 71-73; ASSOC PROF ANTHROP, SOUTHEASTERN MASS UNIV, 73- Mem: Fel Am Anthrop Asn. Res: Cultural ecology; tribal societies; peasant societies; capitalist and socialist societies; imperialism and underdevelopment; Marxism and revolutionary change. Mailing Add: Dept of Sociol Southeastern Mass Univ North Dartmouth MA 02747

STAUDER, M FRANCIS BORGIA, b Witt, Ill, Apr 7, 11. MATHEMATICS. Educ: St Louis Univ, AB, 37; Univ Notre Dame, MS, 43, PhD(math), 47. Prof Exp: High sch teacher, Mo, 37-45; prof math, Le Clerc Col, 47-49; PROF MATH, NOTRE DAME COL, MO, 54- Mem: Math Asn Am. Res: Projective geometry; projective generalizations of metric geometry. Mailing Add: Dept of Math Notre Dame Col St Louis MO 63125

STAUDER, WILLIAM, b New Rochelle, NY, Apr 23, 22. GEOPHYSICS, SEISMOLOGY. Educ: St Louis Univ, AB, 43, MS, 48; Univ Calif, PhD(geophys), 59. Prof Exp: Instr, Marquette Univ High Sch, 48-49; res asst geophys, Univ Calif, 57-59; from instr to assoc prof geophys, 60-66, chmn dept earth & atmospheric sci, 72-75, PROF GEOPHYS, ST LOUIS UNIV, 66-, DEAN GRAD SCH/UNIV RES ADMINR, 75- Concurrent Pos: Mem geophys adv panel, Air Force Off Sci Res, 61-71; mem panel seismol, Comt Alaska Earthquake, Nat Acad Sci-Nat Res Coun, 64-72; mem adv panel, Nat Ctr Earthquake Res, 66-; mem ad hoc comt triggering of earthquakes, AEC, 69-72. Mem: Fel Am Geophys Union; Seismol Soc Am (vpres, 64, pres, 65). Res: Focal mechanism of earthquakes; crustal structure in central United States; earth tides and long period seismic waves; seismicity of southeastern Missouri. Mailing Add: Grad Sch St Louis Univ St Louis MO 63103

STAUDINGER, WILBUR LEONARD, b Molalla, Ore, May 16, 31; m 53; c 3. PLANT PATHOLOGY, MICROBIOLOGY. Educ: Ore State Univ, BS, 57; Iowa State Univ, MS, 59, PhD(plant path), 61. Prof Exp: Instr bot, Iowa State Univ, 61-62; from asst prof to assoc prof biol, Cornell Col, 62-68; PROF BOT & CHMN DEPT BIOL, NEBR WESLEYAN UNIV, 68- Concurrent Pos: Consult, Natur's Way, 74-75. Mem: Am Phytopath Soc. Res: Production of bovine and human antisecretory immunoglobulin A. Mailing Add: Dept of Biol Nebr Wesleyan Univ Lincoln NE 68504

STAUFFER, ALLAN DANIEL, b Kitchener, Ont, Mar 11, 39; m 62; c 2. ATOMIC PHYSICS. Educ: Univ Toronto, BSc, 62; Univ London, PhD(appl math), 66. Prof Exp: Asst lectr math, Royal Holloway Col, 64-66; fel, 66-67, asst prof, 67-71, ASSOC PROF PHYSICS, YORK UNIV, 71- Concurrent Pos: Vis prof, Royal Holloway Col, London, 74-75. Mem: Am Phys Soc; Royal Astron Soc; Brit Inst Physics; Can Asn Physicists. Res: Theoretical atomic collisions; atomic structure problems. Mailing Add: Dept of Physics York Univ Downsview ON Can

STAUFFER, CHARLES HENRY, b Harrisburg, Pa, Apr 17, 13; m 39; c 3. PHYSICAL CHEMISTRY. Educ: Swarthmore Col, AB, 34; Harvard Univ, AM, 36, PhD(chem), 37. Prof Exp: Lab asst org chem, Harvard Univ, 34-36, from instr to assoc prof chem, Worcester Polytech Inst, 37-58; prof & head dept, St Lawrence Univ, 58-65; PROF CHEM & CHMN DIV NATURAL SCI, BATES COL, 65- Mem: Am Chem Soc. Res: Enolization of unsymmetrical ketones; gaseous formation and decomposition of tertiary alkyl halides; reaction kinetics; experimental and theoretical calculations of rates of reaction in gas and liquid phases. Mailing Add: 10 Champlain Ave Lewiston ME 04240

STAUFFER, CLYDE E, b Duluth, Minn, Nov 8, 35; m 58; c 2. BIOCHEMISTRY. Educ: NDak State Univ, BS, 56, MS, 58; Univ Minn, PhD(biochem), 63. Prof Exp: RES CHEMIST, PROCTER & GAMBLE CO, 63- Mem: AAAS; Am Chem Soc; Am Soc Biol Chemists; Sigma Xi. Res: Protein biophysical chemistry; enzymology; surface and interfacial adsorption from solution; immunochemistry. Mailing Add: Miami Valley Lab Procter & Gamble Co Cincinnati OH 45239

STAUFFER, DALE ADRIAN, organic chemistry, see 12th edition

STAUFFER, EDWARD KEITH, b Logan, Utah, July 6, 41; m 65; c 2. MEDICAL PHYSIOLOGY. Educ: Utah State Univ, BS, 64, MS, 69; Univ Ariz, PhD(physiol), 74. Prof Exp: Assoc, Col Med, Univ Ariz, 74-75; ASST PROF PHYSIOL, SCH MED, UNIV MINN, DULUTH, 75- Mem: Am Physiol Soc; Soc Neurosci; AAAS. Res: Neurophysiological studies of motor control with emphasis afferent, central and efferent mechanisms found in the spinal cord. Mailing Add: Sch of Med Univ of Minn 2205 E Fifth St Duluth MN 55812

STAUFFER, GARY DEAN, b Wenatchee, Wash, Feb 26, 44; m 68; c 1. FISHERIES. Educ: Univ Wash, BS, 66, MS, 69, PhD(fisheries & statist), 73. Prof Exp: Fishery biologist salmon res, Quinault Resource Develop Proj, Quinault Tribal Coun, 71-72; FISHERY BIOLOGIST RECREATIONAL FISHERIES, NAT MARINE FISHERY SERV, 73- Mem: Am Fishery Soc. Res: Stock assessment and fishery evaluation of southern California recreational-commercial fisheries for developing management information. Mailing Add: Southwest Fishery Ctr PO Box 271 La Jolla CA 92038

STAUFFER, GEORGE FRANKLIN, b Hanover, Pa, Oct 23, 07; m 31. ASTRONOMY. Educ: Millersville State Col, BS, 32; Univ Pa, MS, 38, EducD, 63. Prof Exp: Teacher pub sch, Pa, 26-27 & high schs, 29-57; PROF ASTRON,

MILLERSVILLE STATE COL, 57- Mem: AAAS; Am Astron Soc; Nat Sci Teachers Asn. Mailing Add: Millersville State Col Millersville PA 17551

STAUFFER, HOWARD BOYER, b Philadelphia, Pa, Aug 10, 41. APPLIED MATHEMATICS. Educ: Williams Col, BA, 64; Univ Calif, Berkeley, PhD(math), 69. Prof Exp: Fel, Univ BC, 69-70; ASST PROF MATH, CALIF STATE UNIV, HAYWARD, 70- Concurrent Pos: Fulbright prof, Nat Univ Malaysia, 74-75; res fel, Pac Forest Res Ctr, Victoria, BC, 75-76. Res: Mathematical models in population dynamics and population genetics. Mailing Add: Dept of Math Calif State Univ Hayward CA 94542

STAUFFER, JAMES, b Butler, Pa, Apr 27, 02; m; c 2. GENETICS. Educ: Univ Mich, AB, 22, AM, 28; Cornell Univ, PhD(plant morphol), 34. Prof Exp: Instr biol, Berea Col, 27-29; asst prof, Colgate Univ, 30-48; assoc prof, 48-52, prof, 52-74, EMER PROF BIOL, LEWIS & CLARK COL, 74- Mem: AAAS; Bot Soc Am. Res: Human genetics; morphology of Labiatae. Mailing Add: Dept of Biol Lewis & Clark Col Palatine Hill Rd Portland OR 97219

STAUFFER, JOHN FREDERICK, b Vance, Miss, July 24, 07; m 32; c 2. PLANT PHYSIOLOGY. Educ: Miss State Col, BS, 28, MS, 30; Univ Wis, PhD(bot), 33. Prof Exp: From instr to asst prof bot, Miss State Col, 28-30; asst, 30-34, from instr to prof, 34- 73, chmn dept, 47-65, EMER PROF BOT, UNIV WIS-MADISON, 73- Mem: AAAS; Am Soc Plant Physiol; Am Chem Soc; Soc Gen Physiol; Bot Soc Am. Res: Photosynthesis and respiration; quantum efficiency of photosynthesis; biological effects of radiation; natural and induced variability in antibiotic-producing fungi. Mailing Add: Dept of Bot Univ of Wis Madison WI 53706

STAUFFER, JOSEPH, biochemistry, see 12th edition

STAUFFER, MEL R, b Edmonton, Alta, July 16, 37; m 58; c 3. STRUCTURAL GEOLOGY. Educ: Univ Alta, BSc, 60, MSc, 61; Australian Nat Univ, PhD(geol), 64. Prof Exp: From asst prof to assoc prof, 65-75, PROF STRUCT GEOL, UNIV SASK, 75- Concurrent Pos: Vis lectr, Univ Alta, 64-65; Nat Res Coun fel, Univ BC, 65-66. Mem: Geol Asn Can. Mailing Add: Dept of Geol Sci Univ of Sask Saskatoon SK Can

STAUFFER, ROBERT CLINTON, b Cleveland, Ohio, May 26, 13. BIOLOGY, HISTORY OF SCIENCE. Educ: Dartmouth Col, BA, 34; Harvard Univ, MA, 39, PhD(hist), 48. Prof Exp: Field asst, Minn State Geol Surv, 31-32; instr biol, Dartmouth Col, 35-36; asst, Harvard Univ, 37-39, fel hist of sci & hist, 39-41; teaching fel, Radcliffe Col, 41-42; asst prof, 47-65, ASSOC PROF HIST SCI, UNIV WIS-MADISON, 65- Concurrent Pos: Fel, Woods Hole Oceanog Inst, 36-38; prin investr, NSF proj, Cambridge Univ, 60-61. Mem: AAAS; Hist Sci Soc; Brit Soc Hist Sci. Res: History of modern natural science; history of ecology; Darwin. Mailing Add: Dept of Hist of Sci Univ of Wis Madison WI 53706

STAUFFER, ROBERT ELIOT, b Chicago, Ill, June 9, 13; m 34; c 4. PHYSICAL CHEMISTRY. Educ: Mt Union Col, BA, 32; Harvard Univ, MA, 34, PhD(phys chem), 36. Hon Degrees: DSc, Mt Union Col, 58. Prof Exp: Asst electrochem, Harvard Univ, 34-36; res assoc, 36-54, from assoc head to head Emulsion Res Div, 54-71, ASST DIR RES, EMULSION RES DIV, EASTMAN KODAK CO, 71- Concurrent Pos: Instr, Rochester Inst Technol, 47-49. Mem: AAAS; Am Chem Soc; NY Acad Sci. Res: Theory of electrolytes; protein electrochemistry and composition; photographic emulsions; viscosities of strong electrolytes; photographic chemistry. Mailing Add: Eastman Kodak Co Rochester NY 14650

STAUFFER, THOMAS MIEL, b Edmore, Mich, June 24, 26; m 54; c 2. FISH BIOLOGY. Educ: Mich State Univ, Lansing, BS, 49, MS, 66. Prof Exp: From fisheries technol fish res to supvr sea lamprey res, Mich Dept Conserv, 50-64; BIOLOGIST IN CHARGE FISH RES, MARQUETTE FISHERIES RES STA, 64- Concurrent Pos: Head, Great Lakes Res, Mich Dept Natural Resources, 64-72, anadromous fisheries res, 72-75. Mem: Am Fisheries Soc; Am Inst Fisheries Res Biologists. Res: Determination of the cause of reproductive failure of planted lake trout and assessment of reproduction by coho and chinook salmon in the Great Lakes. Mailing Add: Marquette Fisheries Res Sta 484 Cherry Creek Rd Marquette MI 49855

STAUFFER, TRUMAN PARKER, SR, b Illmo, Mo, May 29, 19; m 45; c 1. PHYSICAL GEOGRAPHY. Educ: Univ Kansas City, BA, 61; Univ Mo, Kansas City, MA, 64; Univ Nebr, PhD(geog), 72. Prof Exp: From teacher geog to admin supt aide, Ft Osage Sch Dist, 61-68; asst prof, 70-75, ASSOC PROF GEOG, UNIV MO, KANSAS CITY, 75- Concurrent Pos: Coun mem, Underground Construct Res Coun, Am Soc Civil Engrs, 74-; consult, Union Carbide of AEC, 75. Mem: Asn Am Geogr; fel Geog Soc Am; Sigma Xi; Nat Coun Geog Educ; Int Conf Bldg Off. Res: Utilization and economic development of underground space for the conservation of space and energy by planned excavation and conversion of mined areas preserving the qualities of the surface. Mailing Add: Dept of Geosci Univ of Mo 5100 Rockhill Rd Kansas City MO 64110

STAUGAARD, BURTON CHRISTIAN, b Paterson, NJ, Aug 6, 29; m 53; c 4. EMBRYOLOGY. Educ: Brown Univ, AB, 50; Univ RI, MS, 54; Univ Conn, PhD(embryol), 64. Prof Exp: Dir med photog dept, RI Hosp, 50-52; sales rep, x-ray dept, Gen Elec Co, 54-58; instr zool, Univ NH, 61-64, asst prof, 64-67; res fel anat, Sch Med, Vanderbilt Univ, 67-68, asst prof, 68-70; ASSOC PROF BIOL, UNIV NEW HAVEN, 70- Mem: AAAS; Am Soc Zoologists; Zool; Biol Photog Asn. Res: Morphological and functional changes of the developing mammalian embryo in preparation for independent existence; developmental biochemistry and anatomy of kidney, liver and placenta. Mailing Add: Dept of Biol Univ of New Haven PO Box 1306 New Haven CT 06505

STAUM, MUNI M, b New York, NY, Oct 30, 21; m 46; c 2. RADIOCHEMISTRY, PHARMACEUTICAL CHEMISTRY. Educ: City Col New York, BS, 42; Columbia Univ, BS, 51; Univ Fla, PhD(pharmaceut chem), 61. Prof Exp: Develop chemist, Am Cyanamid Co, 53-57; sr res scientist, Olin Mathieson Chem Corp, 61-67; ASST PROF RADIOL, SCH MED, UNIV PA, 67- Concurrent Pos: Am Found Pharmaceut Educ fel. Mem: Am Chem Soc; Am Pharmaceut Asn; Soc Nuclear Med. Res: Organic reaction mechanisms; pharmaceutical drug development; development of radioactive pharmaceuticals for diagnostic nuclear medicine. Mailing Add: Dept of Radiology Hosp-Univ of Pa Philadelphia PA 19174

STAUSS, GEORGE HENRY, b East Orange, NJ, Mar 25, 32; m 59; c 2. PHYSICS. Educ: Princeton Univ, AB, 53; Stanford Univ, MS, 58, PhD(physics), 61. Prof Exp: PHYSICIST, US NAVAL RES LAB, 61- Mem: Am Phys Soc. Res: Nuclear magnetic resonance, principally in magnetically ordered compounds and alloys. Mailing Add: Code 6451 US Naval Res Lab Washington DC 20390

STAVCHANSKY, SALOMON AYZENMAN, b Mexico City, Mex, May 7, 47; m 70; c 2. PHARMACY, PHARMACEUTICS. Educ: Nat Univ Mex, BS, 69; Univ Ky, PhD(pharmaceut sci), 74. Prof Exp: Anal chemist, Nat Med Ctr, Mex, 68-69; develop pharmacist, Syntex Labs, Mex, 69-70; vis scientist, Sloan Kettering Inst Cancer Res, 74; ASST PROF PHARM, UNIV TEX, AUSTIN, 74-, BIOPHARMACEUT COORDR, DRUG DYNAMICS INST, 75- Concurrent Pos: Consult, Alcon Labs, 75- & Dept Health, Educ & Welfare, 76- Mem: Am Pharmaceut Asn; Am Chem Soc; Mex Pharmaceut Asn. Res: Analytical chemistry of pharmaceutical systems; protein binding; application of short lived isotopes for the identification of neoplastic tumors. Mailing Add: Col of Pharm Univ of Tex Austin TX 78712

STAVE, UWE, b Hamburg, Ger, Feb 4, 23; m 46; c 4. PEDIATRICS, PREVENTIVE MEDICINE. Educ: Univ Hamburg, 50. Prof Exp: Res asst pharmacol, Med Sch, Univ Hamburg, 50-53; resident pediat, Med Sch, Univ Marburg, 53-59, asst prof, 59-61; res assoc pediat, Fels Res Inst, 61-75; ASSOC PROF & PROG DIR MAILMAN CTR, SCH MED, UNIV MIAMI, 75- Mem: Soc Pediat Res; Ger Soc Pediat; Int Orgn Study Human Develop; Int Acad Prev Med. Res: Physiological and functional development of renal and liver function; amino acid metabolism; child development; metabolism of the new born; hypoxia; perinatal physiology; bone metabolism. Mailing Add: Mailman Ctr Univ of Miami PO Box 520006 Miami FL 33152

STAVELY, HOMER EATON, biochemistry, deceased

STAVELY, JOSEPH RENNIE, b Wilmington, Del, May 28, 39; m 65; c 1. PLANT PATHOLOGY. Educ: Univ Del, BS, 61; Univ Wis-Madison, MS, 63, PhD(plant path, bot), 65. Prof Exp: Fel plant path, Univ Wis-Madison, 65-66; RES PLANT PATHOLOGIST, TOBACCO LAB, PLANT GENETICS & GERMPLASM INST, AGR RES SERV, USDA, 66- Mem: AAAS; Am Phytopath Soc. Res: Disease resistance in Nicotiana; tobacco leaf diseases, their epiphytology, histology, effect on tobacco physiology and quality, and relationship to leaf age to their development; diseases of cigar tobaccos. Mailing Add: Beltsville Agr Res Ctr West Agr Res Serv USDA Beltsville MD 20705

STAVER, ALLEN ERNEST, b Scribner, Nebr, Dec 5, 23; m 65; c 4. DYNAMIC METEOROLOGY. Educ: Univ Nebr, Omaha, BGen Ed, 56; NY Univ, MS, 59; Univ Wis-Madison, PhD(meteorol), 69. Prof Exp: Weather officer, Air Weather Serv, US Air Force, 43-67; asst prof, 69-72, ASSOC PROF METEOROL, NORTHERN ILL UNIV, 72- Mem: Am Meteorol Soc; Sigma Xi. Res: Dynamics and synoptic meteorology utilizing satellite data; modeling and computer programming of air pollution transport and diffusion. Mailing Add: Dept of Geog Northern Ill Univ De Kalb IL 60115

STAVINOHA, WILLIAM BERNARD, b Temple, Tex, June 11, 28; m 56, 67; c 3. PHARMACOLOGY, TOXICOLOGY. Educ: Univ Tex, BS, 51, MS, 54, PhD(pharmacol), 59. Prof Exp: From instr to asst prof pharmacol & toxicol, Med Br, Univ Tex, 58-60; chief toxicol res, Civil Aeromed Inst, Fed Aviation Agency, Okla, 60-68; assoc prof pharmacol, 68-72, PROF PHARMACOL, UNIV TEX MED SCH SAN ANTONIO, 72- Concurrent Pos: Asst res prof, Med Ctr, Univ Okla, 60, adj prof, 62. Mem: Soc Toxicol; Soc Neurochem. Res: Neurochemistry; insecticides; adaptive mechanisms. Mailing Add: 3910 Tupelo San Antonio TX 78229

STAVITSKY, ABRAM BENJAMIN, b Newark, NJ, May 14, 19; m 42; c 2. IMMUNOLOGY. Educ: Univ Mich, AB, 39, MS, 40; Univ Minn, PhD(bact, immunol), 43; Univ Pa, VMD, 46. Prof Exp: Asst bact, Med Sch, Univ Minn, 42; bacteriologist, Dept Pediat, Univ Pa, 44-46; asst prof immunol, 47-49, from asst prof to assoc prof microbiol, 49-63, PROF MICROBIOL, SCH MED, CASE WESTERN RESERVE UNIV, 63- Concurrent Pos: Res fel immunochem, Calif Inst Technol, 46-47; NSF fel, Nat Inst Med Res, Eng, 58-59; bacteriologist, State Dept Health, Minn, 42 & Children's Hosp, Philadelphia, Pa, 44-46; estab investr, Am Heart Asn, 54-59; mem microbiol fel panel, USPHS, 60-63; expert comts immunochem & teaching immunol, WHO, 63-; ed, J Cellular Physiol, Wistar Inst, 66-; mem microbiol test comt, Nat Bd Med Examr, 70-; ed, Immunochem, J Immunol Methods & Lab Animal Sci. Mem: AAAS; Am Soc Microbiol; Am Asn Immunol; Brit Soc Immunol. Res: Induction and regulation of cellular and humoral immunity. Mailing Add: Dept of Microbiol Case Western Reserve Univ Cleveland OH 44106

STAVN, ROBERT HANS, b Palo Alto, Calif, July 30, 40. ECOLOGY. Educ: San Jose State Col, BA, 63; Yale Univ, MS, 65, PhD(ecol), 69. Prof Exp: Lectr biol, City Univ New York, 67-70, instr, 70-71; ASST PROF BIOL, UNIV NC, GREENSBORO, 71- Concurrent Pos: Grant-in-aid, Univ NC, Greensboro, Res Coun, 71-76; res grant, NC Bd Sci & Technol, 74-75. Mem: AAAS; Am Inst Biol Sci; Ecol Soc Am; Biomet Soc; Am Soc Zoologists. Res: Aquatic ecology; physiological ecology; optical properties of lakes; theory of the ecological niche. Mailing Add: Dept of Biol Univ of NC Greensboro NC 27412

STAVRIC, BOZIDAR, b Skopje, Yugoslavia, Oct 31, 26; Can citizen; m 58; c 2. TOXICOLOGY, PHARMACOLOGY. Educ: Univ Zagreb, BSc, 50, PhD(org chem), 58. Prof Exp: Lectr org chem, Univ Zagreb, 50-63, asst prof, 63; res scientist biochem, Health Protection Br, 65-72, RES SCIENTIST TOXICOL, FOOD DIRECTORATE, HEALTH PROTECTION BR, HEALTH & WELFARE CAN, 72- Concurrent Pos: Nat Res Coun Can fel biosci, 63-65. Mem: AAAS; Am Chem Soc; Soc Exp Biol & Med; Can Biochem Soc. Res: Drug interactions; isolation and identification of drug metabolites and/or impurities in drugs or food additives; experimentally induced hyperuricemia in animals for studies in the fields of hyperuricemia and hyperuricosuria. Mailing Add: Toxicol Res Div Hlth Protect Br Health & Welfare Can Ottawa ON Can

STAVRIC, STANISLAVA, biochemistry, microbiology, see 12th edition

STAVROPOULOS, WILLIAM SPYROS, clinical chemistry, medicinal chemistry, see 12th edition

STAVROUDIS, ORESTES NICHOLAS, b New York, NY, Feb 22, 23; m 49; c 2. MATHEMATICS, OPTICS. Educ: Columbia Univ, AB, 48, MA, 49; Imp Col, dipl & Univ London, PhD, 59. Prof Exp: Asst math, Rutgers Univ, 50-51; mathematician, US Dept Navy, 51; mathematician, Nat Bur Stand, 51-54, in chg lens anal & design, 57-67; PROF, OPTICAL SCI CTR, UNIV ARIZ, 67- Mem: Fel AAAS; Optical Soc Am; Am Math Soc; Soc Indust & Appl Math; Math Asn Am. Res: Geometric and physical optics; differential equations; differential geometry; variational calculus; time-sharing computers. Mailing Add: Optical Sci Ctr Univ of Ariz Tucson AZ 85721

STAY, BARBARA, b Cleveland, Ohio, Aug 31, 26. INSECT MORPHOLOGY. Educ: Vassar Col, AB, 47; Radcliffe Col, MA, 49, PhD(biol), 53. Prof Exp: Asst biol, Harvard Univ, 52; Fulbright Scholar, Commonwealth Sci & Indust Res Orgn, Australia, 53-54; entomologist, Qm Res & Eng Ctr, US Dept Army, 54-59; Lalor fel, Harvard Univ, 59; vis asst prof zool, Pomona Col, 60; asst prof biol, Univ Pa, 61-67; ASSOC PROF ZOOL, UNIV IOWA, 67- Mem: Entom Soc Am; Am Soc Zoologists; Am Soc Cell Biol. Res: Histochemistry of blowfly during metamorphosis and larval blowfly midgut; histology of scent glands, physiology and fine structure of accessory reproductive glands in cockroaches; reproductive behavior of cockroaches. Mailing Add: Dept of Zool Univ of Iowa Iowa City IA 52242

STEAD, EUGENE ANSON, JR, b Atlanta, Ga, Oct 6, 08; m 40; c 3. MEDICINE. Educ: Emory Univ, BS, 28, MD, 32. Prof Exp: Intern med, Peter Bent Brigham Hosp, Boston, 32-33, intern surg, 34-35; instr, Univ Cincinnati, 35-37; asst, Harvard Med Sch, 37-39, instr, 39-41, assoc, 41-42; prof, Sch Med, Emory Univ, 42-46, dean, 45-46; FLORENCE McALISTER PROF MED, SCH MED, DUKE UNIV, 47- Concurrent Pos: Fel, Harvard Univ, 33-34; from asst resident to resident, Cincinnati Gen Hosp, 35-37; resident physician, Thorndike Mem Lab & asst, Boston City Hosp, 37-39; assoc med, Peter Bent Brigham Hosp, 39-42, actg physician-in-chief, 42; physician-in-chief, Univ Div, Grady Hosp, 42-46 & Duke Hosp, 47-67. Mem: Am Soc Clin Invest (secy, 46-48); Am Fedn Clin Res; Asn Am Physicians (secy, 62-67, pres, 71-72). Res: Cardiovascular studies. Mailing Add: Box 3910 Duke Hosp Durham NC 27710

STEAD, WILLIAM WHITE, b Decatur, Ga, Jan 4, 19; m 75; c 1. INTERNAL MEDICINE, PULMONARY DISEASES. Educ: Emory Univ, AB, 40, MD, 43. Prof Exp: Resident med, Emory Univ, 44-45, Univ Cincinnati, 46-47 & Univ Minn, 48-49; chief of serv pulmonary dis, Vet Admin Hosp, Minneapolis, Minn, 54-57; assoc prof, Col Med, Univ Fla, 57-60; prof, Med Col Wis, 60-72; chief pulmonary dis, Vet Admin Hosp, Little Rock, Ark, 72-73; DIR TUBERC PROG, ARK DEPT HEALTH, 73-; PROF PULMONARY DIS, UNIV ARK, LITTLE ROCK, 72- Concurrent Pos: Fel cardiol, Univ Cincinnati, 47-48; med dir, Muirdale Sanatorium, Milwaukee, 60-72; consult, Dept Med, Vet Admin Hosp, Little Rock, 73- & Arthur D Little Co, Mass, 74. Mem: AAAS; Am Soc Clin Invest; Am Fedn Clin Res (secy, 55-58, vpres, 58-59, pres, 59-60); Am Thoracic Soc; Am Col Chest Physicians. Res: Pulmonary physiology, development of spirometers; clinical and public health aspects of tuberculosis. Mailing Add: Ark Dept of Health 4815 W Markham St Little Rock AR 72201

STEADMAN, JAMES ROBERT, b Cleveland, Ohio, Feb 7, 42; m 64; c 3. PLANT PATHOLOGY. Educ: Hiram Col, BA, 64; Univ Wis-Madison, MS, 68, PhD(plant path), 69. Prof Exp: Asst prof bot & plant path, 69-74, ASSOC PROF PLANT PATH, UNIV NEBR-LINCOLN, 75- Concurrent Pos: US Dept Interior water resources res grant, 72-78. Mem: Am Phytopath Soc. Res: Epidemiology; vegetable diseases; pathogen dissemination in water; white mold disease; soil-borne pathogens; plant disease and microclimate interaction; disease resistance; fungal sclerotia dormancy and germination. Mailing Add: Dept of Plant Path Univ of Nebr Lincoln NE 68583

STEADMAN, JOHN WILLIAM, b Cody, Wyo, Oct 13, 43; m 64; c 2. ELECTRICAL ENGINEERING, BIOENGINEERING. Educ: Univ Wyo, BS, 64, MS, 66; Colo State Univ, PhD(elec eng), 71. Prof Exp: Res engr life sci res, Convair Div, Gen Dynamics Corp, Calif, 66-68; ASST PROF BIOENG & ELEC ENG, UNIV WYO, 71- Mem: Inst Elec & Electronics Engrs; Aerospace Med Asn. Res: Machine analysis of electroencephalograms; information processing in the nervous system. Mailing Add: Dept of Elec Eng Univ of Wyo Laramie WY 82070

STEADMAN, ROBERT GEORGE, b Sydney, NSW, Nov 8, 39; m 70; c 1. TEXTILES. Educ: Univ New South Wales, BSc, 61, PhD(textile physics), 65. Prof Exp: Officer in charge, Cotton Fiber Lab, NSW Dept Agr, 64-66; textile mgr, Australian Wool Testing Authority, 66-68; assoc prof clothing & textiles, Univ Man, 68-71; exec vol, Can Exec Serv Overseas, Nigeria, 71-72; ASST PROF CLOTHING & TEXTILES, TEX TECH UNIV, 72- Concurrent Pos: Nat Res Coun Can fel, Univ Man, 69-71. Mem: Am Asn Textile Chemists & Colorists; Brit Textile Inst; Metric Asn. Res: Textiles as thermal insulators and application to human biometeorology; economics of clothing and textiles; consumer problems in textiles. Mailing Add: Dept Clothing & Textiles Tex Tech Univ PO Box 4170 Lubbock TX 79409

STEADMAN, THOMAS REE, organic chemistry, see 12th edition

STEAR, ADRIAN N, b Bristol, Eng, Aug 6, 38; m 61; c 3. INORGANIC CHEMISTRY, MANAGEMENT SCIENCE. Educ: Univ Birmingham, BSc, 60; Cambridge Univ, PhD(chem), 63; NY Univ, MBA, 71. Prof Exp: Res fel chem, Univ Tex, 63; chemist, Plastics Tech Ctr, Shell Chem Co, NJ, 64-67, technologist, Plastics & Resins Div, NY, 67-71, mgr scientist, Shell Int Chem Co, Eng, 71-72; PLANNING DIR, ATLANTIC RICHFIELD CO, 72- Mem: The Chem Soc. Res: Financial information systems. Mailing Add: 314 S Highland Ave Los Angeles CA 90036

STEARN, COLIN WILLIAM, b Bishops Stortford, Eng, July 16, 28; m 53; c 3. PALEONTOLOGY, STRATIGRAPHY. Educ: McMaster Univ, BSc, 49; Yale Univ, PhD(geol), 52. Prof Exp: From asst prof to prof geol, 52-68, asst dean fac grad studies & res, 60-63, chmn dept geol sci, 69-74, LOGAN PROF GEOL, McGILL UNIV, 68- Mem: Geol Soc Am; Paleont Soc; Am Asn Petrol Geol; Geol Asn Can; Royal Soc Can. Res: Lower Paleozoic stratigraphy and paleontology; historical geology; fossil stromatoporoids; growth of West Indian reefs. Mailing Add: Dept of Geol Sci McGill Univ Box 6070 Montreal PQ Can

STEARN, JOSEPH LEONARD, b NY, Feb 1, 08; m 33; c 2. GEODESY. Educ: City Col New York, BS, 30, MS, 34. Prof Exp: Mathematician, US Coast & Geod Surv, 31-65; asst prof math & head dept, Northern Va Community Col, 66-73; RETIRED. Honors & Awards: Medal, Dept Com, 59. Mem: Am Geophys Union; Int Asn Geod. Res: Matrix theory and applications; numerical analysis; theory of errors; mathematical geodesy; matrix algebra. Mailing Add: 3511 Inverrary Dr Apt 108 Lauderhill FL 33319

STEARNER, SIGRID PHYLLIS, b Chicago, Ill, Jan 10, 19. CARDIOVASCULAR PHYSIOLOGY, RADIOBIOLOGY. Educ: Univ Chicago, BS, 41, MS, 42, PhD(zool), 46. Prof Exp: BIOLOGIST, DIV BIOL & MED RES, ARGONNE NAT LAB, 46- Mem: AAAS; Radiation Res Soc; NY Acad Sci; Am Soc Cell Biol; Electron Micros Soc Am. Res: Late effects of ionizing radiations on the microcirculation, physiological and ultrastructural studies; other physiological effects of radiations on biological systems; pigmentation changes. Mailing Add: Div of Biol & Med Res Argonne Nat Lab Argonne IL 60439

STEARNS, BRENTON FISK, b Chicago, Ill, July 28, 28; m; c 2. ENERGY CONVERSION, SURFACE PHYSICS. Educ: Pomona Col, BA, 49; Wash Univ, PhD(physics), 56. Prof Exp: Asst prof physics, Univ Ark, 54-57; from asst prof to assoc prof, Tufts Univ, 57-68; chmn dept physics, 68-74, assoc provost, 74-75, PROF PHYSICS, HOBART & WILLIAM SMITH COLS, 68- Mem: AAAS; Am Phys Soc; Am Asn Physics Teachers. Res: Applications of energy storage; interactions at surfaces. Mailing Add: Dept of Physics Hobart & William Smith Cols Geneva NY 14456

STEARNS, CHARLES EDWARD, b Billerica, Mass, Jan 20, 20; m 42; c 6. GEOLOGY. Educ: Tufts Univ, AB, 39; Harvard Univ, MA, 42, PhD(geol), 50. Hon Degrees: LLD, Southeastern Mass Technol Inst, 62. Prof Exp: Asst geol, Tufts Univ, 41, instr, 41-42, 45, 46-48, asst prof, 48-51; asst prof, Harvard Univ, 51-54; assoc prof, 54-57, dean col lib arts, 54-67 & 68-69, actg provost, 66-67, PROF GEOL, TUFTS UNIV, 57- Mem: AAAS; Geol Soc Am. Res: Pleistocene stratigraphy; shoreline geomorphology. Mailing Add: 381 Boston Rd Billerica MA 01821

STEARNS, CHARLES R, b McKeesport, Pa, May 2, 25. METEOROLOGY. Educ: Univ Wis, BS, 50, MS, 52, PhD(meteorol), 67. Prof Exp: Asst meteorol, Univ Wis, 55-56; chief physicist, Winzen Res, Inc, 56-57; res assoc meteorol, 57-65, asst prof, 65-69, chmn, Inst Environ Studies, 72-74, ASSOC PROF METEOROL, UNIV WIS-MADISON, 69- Concurrent Pos: Consult, Aberdeen Proving Ground, Md, 69- Mem: Am Meteorol Soc. Res: Micrometeorology, particularly boundary layer problems; evaporation from lakes; diffusion from power plants. Mailing Add: Dept of Meteorol Univ of Wis Madison WI 53706

STEARNS, DAVID WINROD, b Muskegon, Mich, Mar 23, 29; m 48; c 5. STRUCTURAL GEOLOGY. Educ: Univ Notre Dame, BS, 53; SDak Sch Mines & Technol, MS, 55; Tex A&M Univ, PhD, 69. Prof Exp: Geologist, Shell Oil Co, 55-56 & Shell Develop Co, 56-66; assoc prof, 67-71, PROF GEOL & HEAD DEPT, TEX A&M UNIV, 71- Concurrent Pos: Consult, Amoco Prod Co, 70- Mem: Am Geophys Union; Geol Soc Am. Res: Structural and mechanical relationships of laramide deformation in the Rocky Mountain forelands. Mailing Add: Dept of Geol Tex A&M Univ College Station TX 77843

STEARNS, EDWIN IRA, b Matawan, NJ, Sept 3, 11; m 34; c 3. PHYSICAL CHEMISTRY. Educ: Lafayette Col, BS, 32; Rensselaer Polytech Inst, MS, 33; Rutgers Univ, PhD(phys chem), 45. Prof Exp: Physicist, Am Cyanamid Co, NJ, 33-43, chief physicist, 44-45, asst dir physics res, 45-51, mgr prod improv, Dyestuff Dept, 52-54, asst mgr, Midwest Territory, 54-59, tech mgr, Dyes Dept, 59-63, mgr sales develop, Dyes & Textile Chem Dept, 64-69, res assoc, 69-72; HEAD DEPT TEXTILE SCI, CLEMSON UNIV, 72- Concurrent Pos: Instr, Cooper Union, 38-39 & Adult Sch, Bound Brook, 40. Honors & Awards: Olney Medal, Am Asn Textile Chemists & Colorists, 67; Godlove Medal, Inter-Soc Color Coun, 67. Mem: Am Chem Soc; Am Asn Textile Chemists & Colorists (pres, 71-72); Tech Asn Pulp & Paper Indust; Inter-Soc Color Coun; Am Asn Textile Technol. Res: Photochemistry; phase rule; visual and infrared spectrophotometry; instrumentation; optical properties of pigments; spectrophotometer improvements; instrumentation in chemical processes. Mailing Add: 321 Woodland Way Clemson SC 29631

STEARNS, EUGENE MARION, JR, b Evanston, Ill, May 3, 32; m 55; c 3. BIOCHEMISTRY. Educ: Denison Univ, BA, 54; Purdue Univ, West Lafayette, MS, 61, PhD(biochem), 65. Prof Exp: Res fel, 65-67, res assoc lipid biochem, 67-70, ASST PROF LIPID BIOCHEM, HORMEL INST, UNIV MINN, 70- Mem: AAAS; Am Chem Soc; Am Oil Chem Soc; Am Inst Biol Sci; Tissue Cult Asn. Res: Plant growth regulators; lipid synthesis and metabolism; plant biochemistry, plant tissue culture. Mailing Add: Hormel Inst Univ of Minn 801 16th Ave NE Austin MN 55912

STEARNS, FOREST, b Milwaukee, Wis, Sept 10, 18; m 43, 56; c 4. ECOLOGY, BOTANY. Educ: Harvard Univ, AB, 39; Univ Wis, PhM, 40, PhD(bot), 47. Prof Exp: Asst bot, Univ Wis, 40-42, 46-47; instr bot exp sta, Purdue Univ, 47-49, asst prof, 49-57; botanist, Vicksburg Res Ctr, US Forest Serv, 57-60, proj leader forest wildlife habitat res, NCent Forest Exp Sta, 61-68; PROF BOT, UNIV WIS-MILWAUKEE, 68- Mem: AAAS; Ecol Soc Am; Bot Soc Am; Wildlife Soc. Res: Autecology of trees and shrubs; seed germination; early succession and productivity; wetland and urban ecology and phenology. Mailing Add: Dept of Bot Univ of Wis Milwaukee WI 53201

STEARNS, HAROLD THORNTON, b Wallingford, Conn, Aug 25, 00. GEOLOGY. Educ: Wesleyan Univ, BS, 21; George Washington Univ, PhD(geol), 26. Prof Exp: Asst geol, Wesleyan Univ, 19-21; mineral examiner, Gen Land Off, 21-23; asst geologist, US Geol Surv, 23-26, assoc geologist, 26-29, geologist, 29-36, sr geologist, 36-46, in charge Hawaiian groundwater invests & Pac invests, 43-46; CONSULT ENG GEOLOGIST, 46-; NSF RES ASSOC, HAWAIIAN INST GEOPHYS, 64- Concurrent Pos: Pres & dir, Waipio Land Co; consult, Armed Forces at Pac Bases, 41-45. Honors & Awards: Medal for Merit. Mem: Geol Soc Am. Res: Volcanology; groundwater; coral reefs; geology of dam and reservoir sites. Mailing Add: Box 158 Hope ID 83836

STEARNS, MARTIN, b Philadelphia, Pa, Aug 16, 16; m 48; c 2. PHYSICS. Educ: Univ Calif, Los Angeles, BA, 43; Cornell Univ, PhD(physics), 52. Prof Exp: Res assoc, Carnegie Inst Technol, 52-57; staff scientist, Gen Atomic Div, Gen Dynamics Corp, 57-60; prof physics & chmn dept, 60-62, DEAN COL LIBERAL ARTS, WAYNE STATE UNIV, 62- Concurrent Pos: Consult, Ramo-Wooldridge Corp, 55. Mem: Fel Am Phys Soc. Res: High energy nuclear physics; plasma physics; bremsstrahlung and pair production; photoproduction of mesons; mesonic x-rays; nuclear reactors. Mailing Add: Col of Liberal Arts Wayne State Univ Detroit MI 48202

STEARNS, MARY BETH GORMAN, b Minneapolis, Minn, Feb 5, 25; m 48; c 2. PHYSICS. Educ: Univ Minn, BS, 46; Cornell Univ, PhD(physics), 52. Prof Exp: Asst physics, Cornell Univ, 47-51; res physicist, Carnegie Inst Technol, 52-56, Univ Pittsburgh, 57 & Gen Atomic Div, Gen Dynamics Corp, 58-60; SR SCIENTIST, SCI LAB, FORD MOTOR CO, 60- Mem: Fel Am Phys Soc. Res: Photonuclear reactions; meson spectroscopy; thermoelectricity; solids; low energy nuclear physics; magnetism; Mössbauer effect and pulsed nuclear magnetic resonance studies; electron scattering. Mailing Add: Ford Res Lab PO Box 2053 Dearborn MI 48121

STEARNS, RICHARD EDWIN, b Caldwell, NJ, July 5, 36; m 63; c 2. MATHEMATICS, COMPUTER SCIENCE. Educ: Carleton Col, BA, 58; Princeton Univ, PhD(math), 61. Concurrent Pos: Mathematician, Res Lab, 61-65, mathematician, Res & Develop Ctr, 65-71, MATHEMATICIAN, CORP RES & DEVELOP, GEN ELEC CORP, 71- Mem: Am Math Soc; Math Asn Am; Asn Comput Mach. Res: Game theory; computer science; automata theory. Mailing Add: Corp Res & Develop Gen Elec Corp Schnectady NY 12345

STEARNS, RICHARD GORDON, b Buffalo, NY, Apr 28, 27; m 50; c 2. GEOLOGY. Educ: Vanderbilt Univ, AB, 48, MS, 49; Northwestern Univ, PhD(geol), 53. Prof Exp: Asst state geologist, State Div Geol, Tenn, 53-61; from asst prof to assoc prof geol, 61-68, PROF GEOL, VANDERBILT UNIV, 68-, CHMN DEPT, 67- Mem: AAAS; Geol Soc Am; Am Asn Petrol Geol; Am Geophys Union. Res: Stratigraphy; structure; paleogeography; hydrogeology. Mailing Add: Box 1615 Sta B Vanderbilt Univ Nashville TN 37235

STEARNS, RICHARD S, b St Paul, Minn, May 10, 19; m 53; c 1. PHYSICAL CHEMISTRY. Educ: Univ Wis, BS, 41; Univ Chicago, PhD(phys chem), 47. Prof Exp: Res assoc, Univ Chicago, 42-47; chemist, Firestone Tire & Rubber Co, 47-59; chemist, 59-62, MGR NEW PROD & PROCESS DEVELOP, CORP RES DEPT, SUN OIL CO, 62- Mem: AAAS; Am Chem Soc. Res: Development of new products, processes and applications for materials capable of being synthesized or obtained from crude petroleum sources; new refining processes; new petrochemical processes; new materials development and application. Mailing Add: 27 Harvey Ln Malvern PA 19355

STEARNS, ROBERT INMAN, b Atlanta, Ga, Feb 26, 32; m 66; c 2. INORGANIC CHEMISTRY. Educ: Loyola Univ, La, BS, 53; Tulane Univ, MS, 55, PhD(inorg chem), 58. Prof Exp: Res specialist, Cent Res Dept, Monsanto Co, Mo, 59-68; DIR RES, LORVIC CORP, 68- Concurrent Pos: Asst prof chem, Eve Div, Univ Mo-St Louis, 66-72, assoc prof, 72- Res: Physical chemistry of fluorides in preventative dentistry; dental materials, cements and polymers; semiconductor materials research, particularly vapor phase depositon of single crystal thin films. Mailing Add: Lorvic Corp 8810 Frost Ave St Louis MO 63134

STEARNS, ROBERT L, b New Haven, Conn, July 28, 26; m 58; c 2. PHYSICS. Educ: Wesleyan Univ, BA, 50; Case Inst Technol, MS, 52, PhD(physics), 55. Prof Exp: Instr physics, Case Inst Technol, 52-55 & Queens Col, 55-58; from asst prof to assoc prof, 58-68, chmn dept, 62-64 & 66-69, PROF PHYSICS, VASSAR COL, 68-, DEAN FRESHMEN, 74- Concurrent Pos: Vis assoc physicist, Brookhaven Nat Lab, 64-65; vis scientist, Europ Orgn Nuclear Res, Geneva, 70-71. Mem: Am Phys Soc; Am Asn Physics Teachers. Res: Neutron physics; scattering of cold neutrons from cyrstals and liquids; nuclear structure physics using high energy proton scattering; nuclear structure-mesic atoms. Mailing Add: Dept of Physics Vassar Col Poughkeepsie NY 12601

STEARNS, THOMAS W, b New York, NY, June 17, 09; m 37; c 2. BIOCHEMISTRY. Educ: Univ Fla, BS, 34, MS, 37; Univ Minn, PhD(biochem), 40. Prof Exp: Asst, Univ Minn, 38-40; asst prof vet res, Iowa State Col, 40-46; asst prof chem, 46-49, assoc prof agr chem, 49-55, PROF CHEM & ASST CHMN DEPT, UNIV FLA, 55- Mem: Am Chem Soc. Res: Physical chemistry bacteria; biochemistry foods; biosynthesis riboflavin. Mailing Add: 1731 NW 12th Rd Gainesville FL 32605

STEBBINGS, RONALD FREDERICK, b London, Eng, Mar 20, 29; m 52; c 3. PHYSICS. Educ: Univ Col, Univ London, BSc, 52, PhD(atomic physics), 56. Prof Exp: Scientist, Atomic Physics Lab, San Diego, Calif, 58-65; reader physics, Univ Col, Univ London, 65-68; chmn dept space sci, 69-74, PROF PHYSICS & SPACE SCI, RICE UNIV, 68- Mem: Am Phys Soc; Am Geophys Union. Res: Experimental atomic physics, particularly as it relates to problems of astrophysical or aeronomic interest. Mailing Add: Dept Space Physics & Astron Rice Univ Houston TX 77001

STEBBINGS, WILLIAM LEE, b Orange Co, Calif, Mar 1, 45; m 68; c 1. STRUCTURAL CHEMISTRY. Educ: Iowa State Univ, BS, 66; Univ Wis, PhD(org chem), 72. Prof Exp: SR CHEMIST, 3 M CO, 72- Mem: Am Chem Soc; Am Soc Mass Spectrometry. Res: Applications of mass spectrometry in analytical chemistry. Mailing Add: 3M Co 3M Ctr Bldg 201-BW St Paul MN 55133

STEBBINS, DEAN WALDO, b Billings, Mont, Jan 14, 13; m 37; c 1. PHYSICS, ACADEMIC ADMINISTRATION. Educ: Mont State Col, BS, 35; Iowa State Univ, PhD(appl physics), 38. Prof Exp: Instr physics, State Col Wash, 38-39 & Agr & Mech Col Tex, 39-41; asst prof, Lehigh Univ, 46-47; from assoc prof to prof, Iowa State Univ, 47-60; physicist, Rand Corp, Calif, 60-63; prof physics & head dept, 63-65, dean fac, 65-66, VPRES ACAD AFFAIRS, MICH TECHNOL UNIV, 66- Concurrent Pos: Consult, Opers Anal Off, Hq, US Dept Air Force, 50-60, Radiation Lab, Univ Calif, 56-58, Westinghouse Elec Corp, Pa, 57 & Ramo-Wooldridge Corp, 59. Mem: AAAS; Am Phys Soc; Am Asn Physics Teachers; Am Soc Eng Educ; Am Geophys Union. Res: Classical physics; geophysics; presence and distribution of matter; interplanetary and interstellar space. Mailing Add: Mich Technol Univ Houghton MI 49931

STEBBINS, GEORGE LEDYARD, b Lawrence, NY, Jan 6, 06; m 31, 58; c 3. BOTANY. Educ: Harvard Univ, AB & AM, 28, PhD(biol), 31. Hon Degrees: Dr, Univ Paris, 62. Prof Exp: Asst bot, Harvard Univ, 29-31; instr biol, Colgate Univ, 31-35; jr geneticist, 35-39, from asst prof to prof, 39-73, EMER PROF GENETICS, UNIV CALIF, DAVIS, 73- Concurrent Pos: Jesup lectr, Columbia Univ, 46; Guggenheim fels, 54 & 60; secy gen, Int Union Biol Sci, 59-64; fac res lectr, Univ Calif, Davis, 62. Honors & Awards: Lewis Prize, Am Philos Soc, 60. Mem: Nat Acad Sci; Am Soc Naturalists (pres, 69); Bot Soc Am (pres, 62); Soc Study Evolution (vpres, 47, pres, 48); Am Philos Soc. Res: Cytogenetics of parthenogenesis in the higher plants; production of hybrid and polyploid types of forage grasses; natural selection, developmental genetics and morphogenesis of higher plants; mechanisms of evolution. Mailing Add: Dept of Genetics Univ of Calif Davis CA 95616

STEBBINS, RICHARD GILBERT, chemistry, see 12th edition

STEBBINS, ROBERT CYRIL, b Chico, Calif, Mar 31, 15; m 41; c 3. ZOOLOGY. Educ: Univ Calif, MA, 41, PhD(zool), 43. Prof Exp: From instr to assoc prof, 44-58, PROF ZOOL, UNIV CALIF, BERKELEY, 58-, CURATOR HERPERT, MUS VET ZOOL, 48- Concurrent Pos: Guggenheim fel, 49; ed, Am Soc Ichthyol & Herpet Jour, 55; NSF sr fel, 58-59. Mem: Soc Syst Zool; Am Soc Ichthyol & Herpet; fel Am Acad Zool. Res: Natural history and factors in the evolution of amphibians and reptiles; population studies of amphibians and reptiles; function of pineal apparatus; research development of biological science topics for schools; scientific illustrations. Mailing Add: Mus of Vertebrate Zool Univ of Calif Berkeley CA 94720

STEBBINS, WILLIAM COOPER, b Watertown, NY, June 6, 29; m 53; c 3. BIOACOUSTICS. Educ: Yale Univ, BA, 51; Columbia Univ, MA, 54, PhD(psychol), 57. Prof Exp: Res assoc otol, NY Univ Med Ctr, 57; asst prof psychol, Hamilton Col, 57-61; fel neurophsyiol, Med Sch, Univ Wash, 61-63; from asst prof to assoc prof psychol & otorhinolaryngol, 63-70, PROF PSYCHOL & OTORHINOLARYNGOL, MED SCH, UNIV MICH, ANN ARBOR, 70- Concurrent Pos: Sigma Xi res grants, 60-61; NIH res grants, 60-61 & 64-; fel, Univ Wash, 61-63; NSF res grants, 74- Mem: AAAS; Acoust Soc Am; Psychonomic Soc; Int Primatol Soc; Asn Res Otolaryngol. Res: Comparative bioacoustics and the evolution of hearing, methodology in animal psychophysics, hearing and auditory perception in nonhuman primates. Mailing Add: Kresge Hearing Res Inst Univ of Mich Med Sch Ann Arbor MI 48104

STEBELSKY, IHOR, b Krakow, Poland, Sept 6, 39; Can citizen; m 63; c 3. ENVIRONMENTAL SCIENCES, EARTH SCIENCES. Educ: Univ Toronto, BA, 62, MA, 64; Univ Wash, PhD(geog), 67. Prof Exp: Res asst geog, Univ Wash, 65-67, res assoc, 68; asst prof, 68-72, ASSOC PROF GEOG, UNIV WINDSOR, 72- Concurrent Pos: Russian & Far Eastern Inst res assoc, Univ Wash & Moscow & Lenningrad, USSR, 68; Ont Dept Univ Affairs grant, Univ Windsor, 70-71; Can Coun res grant, 74. Mem: Am Asn Advan Slavic Studies; Am Geog Soc; Asn Am Geog; Can Asn Geog; Can Asn Slavists. Res: Geography of agricultural resources. Mailing Add: Dept of Geog Univ of Windsor Windsor ON Can

STEBEN, JOHN D, b Hinsdale, Ill, Feb 27, 36; m 59; c 2. PHYSICS. Educ: Univ Ill, BS, 58, MS, 59, PhD(physics), 65. Prof Exp: Physicist, Midwest Univs Res Asn, 65-67; physicist phys sci lab, Univ Wis-Madison, 67-74, lectr nuclear eng, 70-74; MEM STAFF, STEIN RES CTR, THOMAS JEFFERSON UNIV, 74- Mem: Am Phys Soc. Res: Nuclear physics, particle accelerator physics; plasma and medical physics, computer methods in these areas. Mailing Add: Stein Res Ctr Thomas Jefferson Univ 920 Chancellor St Philadelphia PA 19107

STECHER, EMMA DIETZ, b Brooklyn, NY, 1905; m 44. ORGANIC CHEMISTRY. Educ: Columbia Univ, BA, 25, MA, 26; Bryn Mawr Col, PhD(chem), 29. Prof Exp: Res chemist, Harvard Univ, 29-34; Am Asn Univ Women Berliner fel, Univ Munich, 34-35; res chemist, Exp Sta, Hercules Powder Co, Del, 35-37; lectr, Moravian Col Women, 38-41; res chemist, Merck & Co, Inc, NJ, 41; asst prof, Conn Col, 41-43; res chemist, Gen Aniline & Film Corp, Pa, 43-45; from instr to prof org chem, 45-71, EMER PROF ORG CHEM, BARNARD COL, COLUMBIA UNIV, 71- Concurrent Pos: Adj prof, Pace Univ, 71- Mem: AAAS; Sigma Xi; Am Chem Soc; NY Acad Sci. Res: Microanalysis; diazotype paper; chlorophyll; unsaturated ketoacids and lactones; synthesis and oxidation potentials of benzanthraquinones. Mailing Add: 423 W 120th St New York NY 10027

STECHER, MILTON, b US, June 7, 13; m 43; c 2. PHYSICS. Educ: City Col New York, BS, 35, MS, 36; NY Univ, MS, 50. Prof Exp: Physicist degaussing sect, US Dept Navy, 41-43; instr math & physics, Drew Univ, 43-45; tutor physics, City Col New York, 45-46; instr, 48-51, from asst prof to assoc prof, 51-70, PROF PHYSICS, COOPER UNION, 70- Mem: Am Asn Physics Teachers. Res: Science education; diffraction of light due to ultrasonic waves; physics of musical instruments; geometrical and physical optics; moire phenomena. Mailing Add: Dept of Physics Cooper Union Cooper Square New York NY 10003

STECHER, THEODORE P, b Kansas City, Mo, Dec 15, 30; m 56; c 4. ASTRONOMY. Educ: Univ Iowa, BA, 53, MS, 56. ASTRONOMER, NASA GODDARD SPACE FLIGHT CTR, 59-, HEAD, OBSERV ASTRON BR, 72- Concurrent Pos: Mem space sci sub-comt astron, NASA, 68-70; independent res fel, Goddard Space Flight Ctr, 71-72; vis fel, Joint Inst Lab Astrophys, Univ Colo/Nat Bur Standards, 71-72; US proj scientist, Astron Netherlands Satellite. Honors & Awards: John C Lindsay Mem Award, NASA, 66. Mem: Am Astron Soc; Int Astron Union; fel Royal Astron Soc. Res: Ultraviolet stellar spectrophotometry from rockets; stellar physics; interstellar grains and molecules; space instrumentation; gum nebula; gaseous nebulae. Mailing Add: Code 672 Goddard Space Flight Ctr Greenbelt MD 20771

STECHSCHULTE, AGNES LOUISE, b Owosso, Mich, Jan 9, 24. BIOLOGY, MICROBIOLOGY. Educ: Siena Heights Col, BS, 47; Detroit Univ, MS, 53; Cath Univ, PhD(biol), 61. Prof Exp: Teacher, St Dominic Elem Sch, 43-46 & St Agatha Elem Sch, 46-48, St Ambrose High Sch, 48-53 & Aquinas High Sch, 53-57; from instr to asst prof, 60-70, PROF BIOL, BARRY COL, 70-, CHMN DEPT, 61- Concurrent Pos: NIH res grant, 62-65. Mem: AAAS; Am Soc Microbiol; Nat Asn Biol Teachers; NY Acad Sci. Res: Lysozyme resistant mutants. Mailing Add: Dept of Biol Barry Col 11300 NE 2nd Ave Miami FL 33161

STECK, EDGAR ALFRED, b Philadelphia, Pa, Dec 24, 18; m 47; c 2. ORGANIC CHEMISTRY. Educ: Temple Univ, AB, 39; Univ Pa, MS, 41, PhD(org chem), 42. Prof Exp: Asst bact, Temple Univ, 36-39; sr res org chemist, Winthrop Chem Co, 42-46; assoc mem, Sterling-Winthrop Res Inst, 46-56, mem, 56-58; med res group leader, Res Ctr, Johnson & Johnson, 58-60; dir res, Wilson Labs, 60-61; sr scientist, Nalco Chem Co, 61-65; dir res, McKesson Labs, 65-67; PROJ DIR, WALTER REED ARMY INST RES, 67- Mem: Am Soc Trop Med & Hyg; Am Chem Soc; The Chem Soc; Royal Soc Trop Med & Hyg; Swiss Chem Soc. Res: Antiparasitic agents; nitrogen heterocyclic compounds; effects of drugs on central nervous system; chemotheraphy of parasitic diseases. Mailing Add: Walter Reed Army Inst Res Walter Reed Army Med Ctr Washington DC 20012

STECK, THEODORE LYLE, b Chicago, Ill, May 3, 39; m 61; c 2. BIOCHEMISTRY. Educ: Lawrence Col, BS, 60; Harvard Univ, MD, 64. Prof Exp: Intern med, Beth Israel Hosp, Boston, 64-65; res fel, Sch Med, Harvard Univ, 65-66, 68-70 & Mass Gen Hosp, 68-70; res assoc, Nat Cancer Inst, 66-68; asst prof med, 70-73, asst prof biochem, 73-74, ASSOC PROF MED, UNIV CHICAGO & PRITZKER SCH MED, 74- Concurrent Pos: Schweppe Found fel, 71; fac res award, Am Cancer Soc, 75, mem adv comt biochem & chem carcinogenesis, 75-78. Mem: AAAS; Am Soc Biol Chemists. Res: Membrane biochemistry; molecular basis of membrane structure and function, especially in the erythrocyte and cellular slime mold. Mailing Add: 920 E 58th St Chicago IL 60637

STECK, WARREN FRANKLIN, organic chemistry, plant biochemistry, see 12th edition

STECKEL, JOSEPH ERIS, agronomy, see 12th edition

STECKEL, THOMAS FRIER, organic chemistry, see 12th edition

STECKER, FLOYD WILLIAM, b New York, NY, Aug 12, 42; m 65; c 1. PHYSICS, ASTRONOMY. Educ: Mass Inst Technol, SB, 63; Harvard Univ, AM, 65, PhD(astron), 68. Prof Exp: Res assoc astrophys, NASA-Nat Res Coun, 67-68; astrophysicist, Lab Theoret Studies, 68-71, ASTROPHYSICIST, LAB SPACE PHYSICS, GODDARD SPACE FLIGHT CTR, NASA, 71- Honors & Awards: NASA Med Exceptional Sci Achievement, 73. Mem: Am Astron Soc; fel Am Phys Soc. Res: High-energy astrophysics; cosmic-ray physics; gamma-ray astronomy and cosmology. Mailing Add: Theoret Studies Group Goddard Space Flight Ctr NASA Greenbelt MD 20771

STECKER, HERBERT CHRISTIAN, organic chemistry, physiological chemistry, see 12th edition

STECKLER, BERNARD MICHAEL, b Hebron, NDak, Jan 23, 32; m 54; c 4. ORGANIC CHEMISTRY, HISTORY OF SCIENCE. Educ: St Martins Col, BS, 53; Univ Wash, PhD(org chem), 57. Prof Exp: Chemist, Nat Bur Standards, Washington, DC, 53, Northwest Labs, 54 & Shell Develop Co, 57-61; assoc prof, 61-75, PROF CHEM, SEATTLE UNIV, 75- Mem: AAAS; Am Chem Soc. Res: History and philosophy of science; interdisciplinary approaches to teaching physical science; integration of humanities and science disciplines; non-traditional studies curriculum development; phosphorus in delocalized pi-electron systems. Mailing Add: Dept of Chem Seattle Univ Seattle WA 98122

STECKLER, ROBERT, b Vienna, Austria, Nov 27, 14; nat US; m 49; c 1. CHEMISTRY. Educ: Univ Vienna, PhD, 38. Prof Exp: Asst, Graz Univ, 36; res mgr resins & plastics div, Arco Co, 40-45; mgr, R Steckler Labs, Ohio, 45-68; pres, Permacryl, Inc, Ohio, 67-68; VPRES RES, ALCOLAC INC, 68- Mem: Fel AAAS; fel Am Inst Chem; Asn Consult Chemists & Chem Eng (pres, 61-62). Res: Acrylics, epoxies, polyesters; polymers; plastics; protective coatings; functional monomers; surfactants. Mailing Add: Alcolac Inc 3440 Fairfield Rd Baltimore MD 21226

STEDMAN, DONALD HUGH, b Dundee, Scotland, Feb 8, 43; m 64; c 3. ATMOSPHERIC CHEMISTRY. Educ: Cambridge Univ, BA, 64; Univ EAnglia, MSc, 65, PhD(chem), 67. Prof Exp: US Dept Health Educ & Welfare grant, Kans State Univ, 67-69; sr res scientist air pollution chem, Sci Res Labs, Ford Motor Co, 69-72; vis lectr atmospheric chem, Inst Environ Qual, 72-73, ASST PROF CHEM & ATMOSPHERIC & OCEANIC SCI, UNIV MICH, ANN ARBOR, 73- Mem: AAAS; The Chem Soc; Am Chem Soc; Am Phys Soc. Res: Gas phase chemical kinetics and spectroscopy of small molecules, particularly as related to aeronomy, atmospheric chemistry and air pollution; trace analysis of atmospheric pollutants. Mailing Add: Dept of Chem Univ of Mich Ann Arbor MI 48104

STEDMAN, EARL DAVID, biochemistry, molecular biology, see 12th edition

STEDMAN, ROBERT JOHN, b Marlow, Eng, Jan 28, 29; m 60; 2. ORGANIC CHEMISTRY. Educ: Cambridge Univ, BA, 49, MA & PhD(chem), 52. Prof Exp: Fel chem, Nat Res Coun Can, 52-54; res assoc, Med Col, Cornell Univ, 54-56; res assoc, Banting Inst, Univ Toronto, 57-58; res chemist, Chas Pfizer & Co, Conn, 58-60 & Smith Kline & French Labs, Pa, 60-69; ASSOC PROF PHYS ORG CHEM, SCH PHARM, TEMPLE UNIV, 69- Mem: AAAS; Am Chem Soc; The Chem Soc. Res: Natural products and medicinals; nuclear magnetic resonance spectroscopy. Mailing Add: Health Sci Ctr Temple Univ Sch Pharm Philadelphia PA 19140

STEED, GUY PERCY F, b Singapore, Malaysia, Oct 20, 38; m 67; c 1. ECONOMIC GEOGRAPHY. Educ: McGill Univ, BA, 61; Univ Wash, PhD(geog), 66. Prof Exp: Lectr econ geog, Queen's Univ Belfast, 64-65; asst prof geog, 66-71, ASSOC PROF GEOG, SIMON FRASER UNIV, 71- Concurrent Pos: Brit Coun grant, 67; Can Coun grant, 67-68, 69-71 & 73-74. Mem: Asn Am Geog; Am Econ Asn; Regional Sci Asn; Can Asn Geog. Res: Geography of business enterprise; manufacturing; regional planning. Mailing Add: Dept of Geog Simon Fraser Univ Burnaby BC Can

STEEG, CARL W, JR, b Indianapolis, Ind, Aug 17, 22; m 46; c 2. APPLIED MATHEMATICS. Educ: DePauw Univ, AB, 43; Mass Inst Technol, PhD(math), 52. Prof Exp: Instr math, Mass Inst Technol, 47-52, res mathematician, 52-56; mgr anal & simulation, Airborne Systs Lab, Radio Corp Am, 56-59, systs engr, Missile Electronics & Controls Div, 59-61, systs engr Saturn projs, Aerospace Communications & Controls, 61-63; dir tech planning, Indust Labs Div, Int Tel & Tel Corp, 63, dir prod develop, 63-69; PROF ELEC ENG, PURDUE UNIV, FT WAYNE, 69- Concurrent Pos: Lectr, St Francis Col, 69- Mem: Fel AAAS; Am Math Soc; Am Chem Soc. Res: Applications of statistical theory to the optimization of electronic and optical filters and predictors and in non-linear, adaptive and sampled-data control systems for video and other electronic signal enhancement. Mailing Add: Dept of Elec Eng Purdue Univ Ft Wayne IN 46805

STEEGMANN, ALBERT THEODORE, JR, b Cleveland, Ohio, Aug 15, 36; m 63; c 2. BIOLOGICAL ANTHROPOLOGY, PHYSICAL ANTHROPOLOGY. Educ: Univ Kans, BA, 58; Univ Mich, MA, 61, PhD(anthrop), 65. Prof Exp: From instr to asst prof anthrop, Univ Mo-Columbia, 64-66; from asst prof to assoc prof, 68-74, PROF ANTHROP, STATE UNIV NY BUFFALO, 74- Concurrent Pos: Vis colleague, Univ Hawaii, 67-68; NSF res grants, 67-70 & 73-75; State Univ NY Buffalo fac res grants, 69-72; res assoc, Royal Ont Museum, 70- Mem: AAAS; Am Anthrop Asn; Am Asn Phys Anthrop; Soc Study Human Biol; Soc Study Evolution. Res: Human cold response, physiological and physical; cranio-facial evolution; American sub-arctic. Mailing Add: Dept of Anthrop State Univ NY Buffalo Amherst NY 14226

STEEL, COLIN, b Aberdeen, Scotland, Feb 7, 33; m 58; c 3. PHYSICAL CHEMISTRY. Educ: Univ Edinburgh, BSc, 55, PhD(chem), 58. Prof Exp: Res assoc chem, State Univ NY Col Forestry, Syracuse Univ, 58-59; res assoc, Brandeis Univ, 59-60; asst prof, Univ Toronto, 60-61; res scientist, Itek Corp, 61-63; asst prof chem, 63-66, ASSOC PROF CHEM, BRANDEIS UNIV, 66- Mem: Am Chem Soc; The Chem Soc. Res: Reaction kinetics and photochemistry. Mailing Add: Dept of Chem Brandeis Univ Waltham MA 02154

STEEL, COLIN GEOFFREY HENDRY, b London, Eng, June 15, 46; m 70. INVERTEBRATE PHYSIOLOGY, COMPARATIVE ENDOCRINOLOGY. Educ: Univ Cambridge, BA, 67, MA, 71; Queen's Univ, PhD(zool), 71; Univ London, DIC, 75. Prof Exp: Fel insect physiol, Imp Col, Univ London, 71-72, res fel, 72-75; RES ASSOC, DEPT BIOL, YORK UNIV, 75- Mem: Fel Royal Entom Soc London; Soc Exp Biol; Europ Soc Comp Endocrinol; Soc Endocrinol. Res: Neurosecretion in invertebrates, especially insects and crustacea; nervous and hormonal mechanisms controlling development; photoperiodic regulation of the endocrine system; control of polymorphism. Mailing Add: Dept of Biol York Univ Downsview ON Can

STEEL, HOWARD HALDEMAN, b Philadelphia, Pa, Apr 17, 21; m 64; c 4. ORTHOPEDIC SURGERY. Educ: Colgate Univ, BA, 42; Temple Univ, MD, 45, MS, 51; Am Bd Orthop Surg, 52, dipl, 52. Prof Exp: Resident, Temple Univ Hosp & Shriners Hosp Crippled Children, 48-52; PROF ORTHOP SURG, MED CTR, TEMPLE UNIV, 65-; CHIEF ORTHOP SURGEON, SHRINERS HOSP CRIPPLED CHILDREN, 65- Concurrent Pos: Staff surgeon, Med Ctr, Temple Univ, 51-; assoc prof, Div Grad Med, Univ Pa, 55-; clin prof, Med Sch, Univ Wash, 64-65; consult, Vet Admin Hosp, Philadelphia, 67- & Walson Army Hosp, Ft Dix, NJ, 67-; attend surgeon, St Christopher's Hosp for Children. Mem: Orthop Res Soc; AMA; Am Acad Orthop Surg; Am Orthop Asn; Am Fedn Clin Res. Res: Clinical investigation of the developmental defects in the child; clinical and bacteriological investigations of nosocomial infections. Mailing Add: Dept of Orthop Surg Temple Univ Philadelphia PA 19140

STEEL, R KNIGHT, b New York, NY, Dec 1, 39; m 65; c 1. INTERNAL MEDICINE. Educ: Yale Univ, BA, 61; Columbia Univ, MD, 65. Prof Exp: From intern to chief resident med, Univ NC, Chapel Hill, 65-71, asst prof, 71-72; ASST PROF MED, UNIV ROCHESTER, 72-; ASSOC DIR, MONROE COMMUNITY HOSP, 72- Res: Medical education; geriatrics; health care delivery. Mailing Add: Monroe Community Hosp Rochester NY 14603

STEEL, ROBERT, b Winnipeg, Man, Mar 17, 23; m 52; c 2. MICROBIOLOGY. Educ: Univ Man, BS, 49, MS, 51; Univ Manchester, PhD(microbiol, biochem), 56. Prof Exp: Jr res officer, Div Appl Biol, Nat Res Coun Can, 51-54; Imp Chem Industs res fel, Univ Manchester, 55-58; res assoc, 58-71, HEAD MICROBIOL & CHEM SERV, UPJOHN CO, 71- Mem: Am Soc Microbiol; Can Soc Microbiol; Am Chem Soc. Res: Steroid bioconversions; utilization of agricultural wastes by fermentation; production of 2, 3-butanediol, citric acid; biochemical engineering; agitation aeration studies in fermentation; mixing and scale-up of antibiotic fermentations. Mailing Add: Upjohn Co 301 Henrietta Kalamazoo MI 49001

STEEL, ROBERT GEORGE DOUGLAS, b St John, NB, Sept 2, 17; m 41; c 1. STATISTICAL ANALYSIS. Educ: Mt Allison Univ, BA, 39, BSc, 40; Acadia Univ, MA, 41; Iowa State Univ, PhD(statist), 49. Prof Exp: Asst prof math, Univ Wis & statistician, Agr Exp Sta, 49-52; assoc prof biol statist, Cornell Univ, 52-60; PROF STATIST & GRAD ADMINR, NC STATE UNIV, 60- Concurrent Pos: Mem math res ctr, US Dept Army, Univ Wis, 58-59. Mem: Fel Am Statist Asn; Biomet Soc. Res: Nonparametric statistics; experimental design; data analysis. Mailing Add: Dept of Statist NC State Univ PO Box 5457 Raleigh NC 27607

STEEL, THOMAS B, JR, b Berkeley, Calif, Mar 2, 29; m 52; c 2. COMPUTER SCIENCE. Educ: Univ Calif, Berkeley, AB, 49, MA, 51. Prof Exp: Assoc mathematician, Rand Corp, 55-57; comput syst specialist, Syst Develop Corp, 57-61, sr res leader comput sci, 61-65, prin scientist, 65-70; independent consult, 70-71; SR TECH CONSULT, EQUITABLE LIFE ASSURANCE SOC OF US, 71- Concurrent Pos: Mem data processing panel, Winter Study Group, US Air Force, 60 & mem data processing & display panel, Proj Forecast, 63; US rep tech comt prog lang, Int Fedn Info Processing, 63-, chmn, 69-; mem exec bd, Share, 65-66; chmn, Am Nat Standards Inst Info Processing Standards Planning & Requirements Comt, 69- Honors & Awards: Silver Core, Int Fedn Info Processing, 74. Mem: AAAS; Asn Comput Mach; Asn Symbolic Logic; Sigma Xi. Res: Formal tools for analysis and description of both syntactic and semantic elements of formal languages, especially computer languages. Mailing Add: Equitable Life Tech Serv 1285 Ave of Americas New York NY 10019

STEEL, WARREN G, b New York, NY, Feb 16, 20; m 43; c 2. GEOLOGY. Educ: Univ NC, BS, 46, MS, 49. Prof Exp: Asst prof geol, NC State Col, 48-55; prof geol, 55-72, E B ANDREWS PROF NATURAL SCI, MARIETTA COL, 72-, CHMN DEPT GEOL, 55- Concurrent Pos: Geologist, US Geol Surv, 47-48 & 51 & NC Dept Conserv & Develop, 49-50 & 52; geologist-petrogr, Rare Minerals Br, US Bur Mines, 53-55; consult geologist, 55- Mem: Geol Soc Am; Nat Asn Geol Teachers. Res: Petrography; structural geology; geomorphology. Mailing Add: Dept of Geol Marietta Col Marietta OH 45750

STEELE, ARTHUR BURNS, b Racine, Wis, Mar 6, 20; m 43; c 4. CHEMISTRY. Educ: Univ Wis, BS, 41; Univ Pittsburgh, PhD(org chem), 49. Prof Exp: Indust fel, Mellon Inst Sci, 41-49, sr fel, 49-55; mgr tech serv dept & dir tech serv lab, Union Carbide Chem Co, 55-62, dir res & develop, Chem Div, Union Carbide Corp, 62-64, sales mgr, 64-66, OPERS MGR, COATING INTERMEDIATES DIV, UNION CARBIDE CORP, 67- Mem: Am Chem Soc; Am Soc Testing & Mat; NY Acad Sci; Soc Chem Indust. Res: Surface active agents; formulation of agricultural chemicals; extractive solvent technology; chemistry of polyketones; technology of gas purification. Mailing Add: Union Carbide Corp 270 Park Ave New York NY 10017

STEELE, CLELLIE TRUMAN, b Hurley, Mo, July 10, 13; m 38; c 1. CHEMISTRY. Educ: Drury Col, BS, 34; Univ Ark, MS, 35; Univ Mo, PhD(chem), 38. Prof Exp: Instr chem, Univ Ark, 36-37; res chemist, Esso Standard Oil Co, La, 38-42, sect head, Chem Prod Lab, 42-46, asst head petrol tech serv, 46-50, asst chief chemist, 50-56; coordr petrochem res, Res Dept, Imp Oil, Ltd, 56-67, mgr chem div, 67-75; ASSOC DIR, HYDROCARBON RES CTR, UNIV ALTA, 75- Mem: Am Chem Soc; Am Inst Chem Eng; Chem Inst Can; Sigma Xi. Res: Kinetics of benzoylation of aromatic amines; conversion of petroleum to chemical raw materials and olefins and diolefins to polymers and oxygenated compounds. Mailing Add: Apt 806 12141 Jasper Ave Edmonton AB Can

STEELE, DONALD HAROLD, b London, Ont, Nov 5, 32; m 59; c 1. ZOOLOGY. Educ: Univ Western Ont, BSc, 54; McGill Univ, MSc, 56, PhD(zool), 61. Prof Exp: Technician, Biol Sta, St Andrews, NB, 55-56; lectr biol, Sir George Williams Univ, 60-62, asst prof, 62; from asst prof to assoc prof, 62-75, PROF BIOL, MEM UNIV NFLD, 75- Mem: AAAS; Ecol Soc Am; Brit Ecol Soc; Int Asn Ecol; Can Soc Zool; Arctic Inst NAm. Res: Marine ecology; zoogeography; systematics of marine amphipoda. Mailing Add: Dept of Biol Mem Univ of Nfld St John's NF Can

STEELE, FRANCES M, b Hartford, Conn, Feb 11, 36. VIROLOGY, MICROBIOLOGY. Educ: Boston Univ, BA, 59; Univ Conn, MS, 62; Yale Univ, PhD(epidemiol), 66. Prof Exp: USPHS trainee, Yale Univ, 66-67 & Albert Einstein Col Med, 67-68; MICROBIOLOGIST, LAB DIV, CONN STATE DEPT HEALTH, 68- Mem: Am Soc Microbiol. Res: Effects of 2-thiouracil on poliovirus; protein composition of the poliovirus capsid; development of a pediatric virology service fo Connecticut hospitals. Mailing Add: Conn State Dept Health PO Box 1689 Hartford CT 06101

STEELE, GRANT, b Salt Lake City, Utah, June 8, 24; m 45; c 3. STRATIGRAPHY, PALEONTOLOGY. Educ: Univ Utah, BS, 49; Univ Wash, PhD(geol), 59. Prof Exp: Instr geol, Syracuse Univ, 52-53; div paleontologist, Gulf Oil Corp, Tulsa, 53-56, div stratigr, Denver, 56-63, sr geologist, Gulf Res & Develop Co, Pittsburgh, 63-66, GEOL SUPVR, HOUSTON TECH SERV CTR, GULF RES & DEVELOP CO, 66- Mem: Am Asn Petrol Geologists; Paleont Soc; Soc Econ Paleontologists & Mineralogists. Res: New applications of biostratigraphy and lithostratigraphy. Mailing Add: Houston Tech Serv Box 36506 Gulf Res & Develop Co Houston TX 77036

STEELE, IAN MCKAY, geology, crystallography, see 12th edition

STEELE, JACK, b Indianapolis, Ind, Jan 22, 42; m 68; c 3. INORGANIC CHEMISTRY, PHYSICAL CHEMISTRY. Educ: DePauw Univ, BA, 64; Univ Ky, PhD(inorg chem), 68. Prof Exp: Am Chem Soc Petrol Res Fund grant & teaching intern, Wash State Univ, 68-70; asst prof, 70-75, ASSOC PROF CHEM, ALBANY STATE COL, 75- Concurrent Pos: NSF col sci improv prog mem, Albany State Col, 72-73, minority sch biomed support prog mem, 72-76. Mem: Am Chem Soc; Sigma Xi. Res: Etiologic factors of disease; clinical chemistry; stereochemistry of metal chelates of biologically important compounds; optical activity and absolute configuration of metal chelates. Mailing Add: Dept of Chem Albany State Col Albany GA 31705

STEELE, JAMES HARLAN, b Chicago, Ill, Apr 3, 13; m 41; c 2. VETERINARY MEDICINE. Educ: Mich State Univ, DVM, 41; Harvard Univ, MPH, 42; Am Bd Vet Pub Health, dipl. Prof Exp: With, State Health Dept, Ohio, 42-43; vet, USPHS, 43-45, chief vet pub health, Nat Commun Dis Ctr, 45-68, asst surgeon gen vet affairs, 68-71; PROF ENVIRON HEALTH, INST ENVIRON HEALTH, UNIV TEX SCH PUB HEALTH HOUSTON, 71- Concurrent Pos: Consult, Pan-Am Sanit Bur, 44-, US Dept State, 48-, USDA, 50- & White House Comt Consumer Protection; secy, Am Bd Vet Pub Health, 50-52; consult, WHO, 50-, pub health consult, Philippines, 72 & Somoa & Fiji, 73; consult, Food & Agr Orgn, UN, 60, pub health consult, Cyprus, 75-76; chmn, WHO-Food & Agr Orgn Expert Comt Zoonoses, 3rd Report, 67. Honors & Awards: Meritorious Medal, USPHS, 63, Bronfman Award, 71; Centennial Award, Am Pub Health Asn, 72; Honorary Dipl, XX World Vet Cong, 75. Mem: Am Soc Trop Med & Hyg; Asn Mil Surg US; Am Vet Med Asn; Am Vet Epidemiol Soc (pres, 75-77); Animal Health Asn. Res: Veterinary public health; epidemiology of zoonoses and chronic diseases common to animals and man; cost benefits of international veterinary public health programs. Mailing Add: Inst Environ Health Univ Tex Sch Pub Health Houston TX 77025

STEELE, JAMES PATRICK, b Louisville, Ky, Dec 6, 19; m 44; c 2. RADIOLOGY. Educ: Univ Louisville, MD, 43; Am Bd Radiol, dipl, 50. Prof Exp: PROF RADIOL, SCH MED, UNIV S DAK, VERMILLION, 49- Concurrent Pos: Radiologist, Sacred Heart Hosp, Yankton, 49-; prof, Sch Med, Univ Nebr; pres, SDak Health Res Inst. Mem: Am Roentgen Ray Soc; Radiol Soc NAm; AMA; Am Col Radiol. Res: Night vision as related to fluoroscopy; psychological effects of total body radiation. Mailing Add: Sacred Heart Hosp PO Box 650 Yankton SD 57078

STEELE, JOHN A, b Raeford, NC, Mar 26, 28; m 61; c 1. ORGANIC CHEMISTRY,

ANALYTICAL CHEMISTRY. Educ: JC Smith Univ, BS, 51; Howard Univ, MS, 63. Prof Exp: Res chemist, Tenn Valley Authority, 53 & NIH, 56-63; forensic anal chemist, US Internal Revenue Serv, 63-69, RES ANAL CHEMIST, BUR ALCOHOL, TOBACCO & FIREARMS, DEPT TREAS, 69- Mem: AAAS; fel Am Inst Chemists. Res: Partial synthesis and degradation of steroidal compounds; preparation and purification of specific working groups for nucleotide synthesis; tobacco and tobacco products; anthrasteroid rearrangement mechanisms. Mailing Add: Bur of Alcohol Tobacco & Firearms Rm 7575 IRS Bldg Washington DC 20226

STEELE, JOHN EARLE, b St John's, Nfld, Jan 29, 32; m 57; c 3. ZOOLOGY, ENDOCRINOLOGY. Educ: Dalhousie Univ, BSc, 54; Univ Western Ont, MSc, 56; Univ Sask, PhD(biol), 59. Prof Exp: Res officer, Can Dept Agr, 59-64; from asst prof to assoc prof, 64-75, PROF ZOOL, UNIV WESTERN ONT, 75- Mem: Can Soc Zool; Can Physiol Soc. Res: Hormonal control of metabolism, particularly in the insects; hormonal control of growth and development. Mailing Add: Dept of Zool Univ of Western Ont London ON Can

STEELE, JOHN WISEMAN, b Motherwell, Scotland, May 27, 34; m 58; c 4. PHARMACEUTICAL CHEMISTRY. Educ: Glasgow Univ, BSc, 55, PhD(pharmaceut chem), 59. Prof Exp: Lectr, 58-59, from asst prof to assoc prof, 59-68, PROF PHARMACEUT CHEM, FAC PHARM, UNIV MAN, 68- Concurrent Pos: Fel, Chelsea Col Sci & Technol, 65-66; vis scientist, Med-Chem Inst, Univ Bern, 72-73. Mem: Can Pharmaceut Asn; assoc Royal Inst Chem; The Chem Soc; Asn Faculties Pharm Can (pres, 75-76). Res: Drug metabolism, especially of anabolic steroids and other drugs likely to be abused by athletes; methods of drug analysis, including gas-liquid chromatography and gas-liquid chromatography/Masr spectral analysis. Mailing Add: Fac of Pharm Univ of Man Winnipeg MB Can

STEELE, KENNETH F, b Statesville, NC, Jan 16, 44; m 62; c 2. GEOLOGY, GEOCHEMISTRY. Educ: Univ NC, Chapel Hill, BS, 62, PhD(geol), 71. Prof Exp: Instr, 70-72, ASST PROF GEOL, UNIV ARK, FAYETTEVILLE, 72- Mem: Geochem Soc; Geol Soc Am; Soc Environ Geochem & Health; Nat Asn Geol Teachers; Int Asn Earth & Environ Sci. Res: Major and trace element geochemical investigations concerned with the origin of rocks and the distribution of elements; geochemical studies of water related to rock. Mailing Add: Dept of Geol Univ of Ark Fayetteville AR 72701

STEELE, LEON, b Ill, Apr 8, 15; m 41; c 3. PLANT BREEDING. Educ: Ill Wesleyan Univ, BS, 40, DSc, 67. Prof Exp: Res assoc, Michael-Leonard Seed Co, 36-40; mgr dept res, 40-52, assoc res dir, 52-57, RES DIR, FUNK SEEDS INT, 57-, VPRES, 63-, PRES, FUNK SEEDS INT CAN, LTD, 63- Mem: AAAS; Am Soc Agron; Bot Soc Am; Genetics Soc Am; Am Genetic Asn. Res: Commercial and hybrid corn breeding; physiology of corn plant; corn diseases and their control through breeding for resistance. Mailing Add: Funk Seeds Int Inc Bloomington IL 61702

STEELE, MARTIN CARL, b New York, NY, Dec 25, 19; m 41; c 4. SOLID STATE ELECTRONICS. Educ: Cooper Union, BChE, 40; Univ Md, MS, 49, PhD, 52. Prof Exp: Physicist & chief cryomagnetics sect, US Naval Res Lab, 47-55; res physicist, Res Lab, Radio Corp Am, 55-58, head semiconductor res group, 58-60, dir res labs, Japan, 60-63, head solid state electron physics group, 63-72; HEAD SEMICONDUCTOR MAT & DEVICE RES, RES LABS, GEN MOTORS CORP, 72- Concurrent Pos: Vis lectr, Princeton Univ, 65-66. Mem: Fel Am Phys Soc; Electrochem Soc; Inst Elec & Electronic Engrs. Res: Solid state physics; superconductivity; galvanomagnetic effects in metals and semiconductors; high electric field effects in semiconductors; solid state plasma effects; microwave devices; infrared detection; integrated circuits; MOS devices; semiconductor surfaces. Mailing Add: Res Labs Gen Motors Corp Warren MI 48090

STEELE, MARY PHILIP, b Sioux City, Iowa, Nov 6, 05. MATHEMATICS. Educ: Rosary Col, BA, 27; Univ Wis, MA, 40; Cath Univ Am, PhD(math), 43. Prof Exp: Teacher, Bethlehem Acad, Minn, 30-37; teacher, High Sch, Wis, 37-40; instr math, Cath Univ Am, 42-43; from asst prof to prof, 43-72, chmn dept, 56-67, EMER PROF MATH, ROSARY COL, 72- Mem: AAAS; Am Math Soc; Soc Indust & Appl Math. Res: Abstract algebra; group theory; geometic interpretation and some applications of dihedral group G-16; analysis; probability and statistics. Mailing Add: Dept of Math Rosary Col River Forest IL 60305

STEELE, RICHARD, b Charlotte, NC, Sept 6, 21; m 49; c 2. CHEMISTRY. Educ: Univ NC, SB, 42; Princeton Univ, MA, 48, PhD(chem), 49. Prof Exp: Res chemist, Rohm and Haas Co, 42-46; res chemist & head phys org chem sect, Textile Res Inst, 50-53; lab head, Rohm and Haas Co, 53-65; dir appln & prod develop, NY, 65-66, vpres & tech dir, 67-71, SR VPRES TECHNOL & ADMIN, CELANESE FIBERS MKT CO, 71- Honors & Awards: Olney Medal, Am Asn Textile Chemists & Colorists, 63. Mem: AAAS; Am Chem Soc; Am Asn Textile Chemists & Colorists; Brit Textile Inst. Res: Structure of natural and synthetic fibers; chemistry of textile wet-finishing processes; cellulose chemistry. Mailing Add: Celanese Fibers Co PO Box 1414 Charlotte NC 28232

STEELE, RICHARD HAROLD, b Buffalo, NY, Aug 1, 19; m 52; c 3. BIOCHEMISTRY. Educ: Univ Ala, BS, 48; Tulane Univ, PhD(biochem), 53. Prof Exp: Vis investr, Inst Muscle Res, Marine Biol Lab, 54-57; PROF BIOCHEM, TULANE UNIV, 57- Concurrent Pos: Lederle Med Fac Award, 57-60; NIH sr res fels, 60 & 65. Mem: Am Chem Soc; Am Soc Biol Chem. Res: Energy generation and transfer; spectroscopy; chemiluminescence and bioluminescence. Mailing Add: Dept of Biochem Tulane Univ Sch Med New Orleans LA 70112

STEELE, ROBERT, biochemistry, deceased

STEELE, ROBERT, b Scotland, Jan 16, 29; m 55; c 2. PREVENTIVE MEDICINE, EPIDEMIOLOGY. Educ: Univ Edinburgh, DPH, 56; Univ Sask, MD, 60. Prof Exp: Asst prof prev med, Col Med, Univ Sask, 58-62; med officer, Scottish Health Dept, 62-64; assoc prof prev med, Fac Med & dir res, 64-68, PROF COMMUNITY HEALTH & EPIDEMIOL & HEAD DEPT, FAC MED, QUEEN'S UNIV, ONT, 68- Concurrent Pos: Consult, Kingston Gen Hosp, Ont. Mem: Asn Teachers Prev Med; fel Am Pub Health Asn; Royal Med Soc; Can Asn Teachers Social & Prev Med. Res: Epidemiology of sudden death in infants; cancer; growth and development in children; medical care; community health. Mailing Add: Dept of Community Health Queen's Univ Fac of Med Kingston ON Can

STEELE, ROBERT L, b Jane Lew, WVa, Nov 30, 30; m 53; c 2. DAIRY SCIENCE. Educ: WVa Univ, BS, 53, MS, 57. Prof Exp: Instr dairy sci, WVa Univ, 55-56, asst exten dairyman, 56, assoc exten dairyman, 57-60; mgr dairy processing, Producers Coop Dairy, WVa, 60; dir commodities & mkt, Pa Farmers Asn, 60-67; dairy specialist, 67-70, dir appl res, 70-72, mgr dairy-livestock develop, 72-73, DIR FIELD RES & DEVELOP, AGWAY, INC, 73- Mem: Am Dairy Sci Asn; Animal Nutrit Res Coun; Am Soc Animal Sci. Res: Dairy, swine, beef feeding and management; crops varieties, fertility and cultural practices and manure management. Mailing Add: Agway Inc PO Box 1333 Syracuse NY 13201

STEELE, ROBERT WILBUR, b Denver, Colo, Aug 13, 20; m 42, 61; c 6. FOREST MANAGEMENT. Educ: Colo State Univ, BSF, 42; Univ Mich, MSF, 49; Univ Mont, PhD(forest fire sci), 75. Prof Exp: Forest guard, US Forest Serv, Ore, 42-43, forester, Pac Northwestern Exp Sta, 46-55; forest mgr, SDS Lumber Co, 55-56; asst prof, 56-67, ASSOC PROF FORESTRY, UNIV MONT, 67- Concurrent Pos: Consult, 56- Mem: Soc Am Foresters; Am Meteorol Soc. Res: Forest fire control; development of techniques and machinery for fire detection and control; use and effects of prescribed fire in the forest; forest fire science; land surveying. Mailing Add: Sch of Forestry Univ of Mont Missoula MT 59801

STEELE, ROGER L, b Fostoria, Ohio, Feb 12, 24; m 42; c 4. AIR POLLUTION. Educ: Univ Iowa, BS, 48; Pa State Univ, MS, 51. Prof Exp: Instr mech eng, Pa State Univ, 48-51; engr, E I du Pont de Nemours & Co, 51-54; sr engr, Consol Papers, Inc, 54-58, res supvr, 58-61; instr mech eng, Exten Div, Univ Wis, 58; assoc prof thermodyn, Colo State Univ, 61-67, prof & dir atmospheric simulation lab, 66-71; PROF ATMOSPHERIC PHYSICS & PROG MGR AIR RESOURCES, ENERGY & ATMOSPHERIC ENVIRON CTR, DESERT RES INST, UNIV NEV SYST, 71- Concurrent Pos: NSF res grants, 63- Mem: Am Meteorol Soc; Air Pollution Control Asn. Res: Weather modification; heterogeneous nucleation and cloud precipitation processes; environmental assessment, fossil fueled generating stations; waste heat utilization; solar/wind energy development. Mailing Add: Desert Res Inst Sage Bldg Univ of Nev Reno NV 89507

STEELE, SIDNEY RUSSELL, b Toledo, Ohio, June 30, 17; m 44; c 2. CHEMISTRY. Educ: Univ Toledo, BS, 39; Ohio State Univ, PhD(chem), 43. Prof Exp: Res chemist, Girdler Corp, Ky, 43-47; assoc prof, 47-60, PROF CHEM, EASTERN ILL UNIV, 60-, HEAD DEPT, 67- Mem: Am Chem Soc. Res: Polarography; abnormal diffusion currents; water gas-shift catalysis; methanation of carbon monoxide. Mailing Add: Dept of Chem Eastern Ill Univ Charleston IL 61920

STEELE, TIMOTHY DOAK, b Muncie, Ind, Apr 12, 41; m 65; c 2. HYDROLOGY, RESOURCE MANAGEMENT. Educ: Wabash Col, AB, 63; Stanford Univ, MS, 65, PhD(hydrol), 68; USDA Grad Sch, advan cert acct, 73. Prof Exp: Res hydrologist, Water Resources Div, Washington, DC, 66-68, res hydrologist, Systs Lab Group, 68-72, hydrologist, Qual Water Br, 72-74, PROJ CHIEF HYDROLOGIST, YAMPA RIVER BASIN ASSESSMENT STUDY, US GEOL SURV, COLO, 75- Concurrent Pos: Water qual specialist, US AID, Pakistan, 72. Mem: Am Geophys Union; Int Asn Sci Hydrol. Res: Design of hydrologic data-collection networks; statistical analysis of data; hydrologic simulation and modeling; water resources planning and systems analysis; hydrogeochemistry; water quality; assessments of environmental impacts of energy-resource development. Mailing Add: US Geol Surv MS 415 Colo Dist Denver Fed Ctr Box 25046 Lakewood CO 80225

STEELE, VERNON EUGENE, b Blairsville, Pa, July 23, 46; m 68; c 1. RADIOBIOLOGY. Educ: Bucknell Univ, BS, 68; Univ Rochester, MS, 74, PhD(radiation biol), 75. Prof Exp: INVESTR CARCINOGENESIS, BIOL DIV, OAK RIDGE NAT LAB, 75- Res: Effects of radiation, chemical carcinogens and promoters on cell and tissue kinetics, morphology and physiology. Mailing Add: Bldg 9211 Biol Div Oak Ridge Nat Lab Box Y Oak Ridge TN 37830

STEELE, VLADISLAVA JULIE, b Prague, Czech, July 8, 34; m 59; c 1. INVERTEBRATE PHYSIOLOGY, HISTOLOGY. Educ: McGill Univ, BSc, 57, MSc, 59, PhD(zool), 65. Prof Exp: Lectr histol, McGill Univ, 60-61; vis lectr, 62-63, lectr histol & embryol, 63-65, asst prof, 65-72, ASSOC PROF HISTOL & EMBRYOL, MEM UNIV NFLD, 72- Mem: Am Inst Biol Sci; Can Soc Zool; Nutrit Today Soc. Res: Photoperiod and neurosecretion in marine amphipods; influence of environmental factors on the reproduction of boreo-arctic intertidal amphipods. Mailing Add: Dept of Biol Mem Univ Nfld St John's NF Can

STEELE, WARREN CAVANAUGH, b Pocatello, Idaho, Oct 25, 29; m 55. PHYSICAL CHEMISTRY. Educ: Ore State Col, BA, 51, PhD(phys chem), 56. Prof Exp: Res chemist, Dow Chem Co, 56-58; res assoc chem, Tufts Univ, 58-60 & 62-64; res fel, Harvard Univ, 60-62; sr staff scientist, Space Systs Div, Avco Corp, Wilmington, 64-75; SR SCIENTIST, ENERGY RESOURCES CO, INC, CAMBRIDGE, 75- Mem: AAAS; Am Chem Soc. Res: Mass spectrometry; gas-surface reaction kinetics; high temperature thermochemistry; environmental chemistry. Mailing Add: 18 Old Dee Rd Cambridge MA 02138

STEELE, WILLIAM A, b St Louis, Mo, June 4, 30; m 55; c 2. PHYSICAL CHEMISTRY. Educ: Wesleyan Univ, BA, 51; Univ Wash, PhD(phys chem), 54. Prof Exp: Fel, Cryogenic Lab, 54-55, from asst prof to assoc prof, 55-66, PROF PHYS CHEM, PA STATE UNIV, UNIVERSITY PARK, 66- Concurrent Pos: NSF fel, 57-58, sr fel, 63-64; mem comt colloid & surface chem, Nat Acad Sci-Nat Res Coun, 66-72. Mem: AAAS; Am Chem Soc; Am Phys Soc. Res: Thermodynamics and statistical mechanics of liquids and physical adsorption of gases on solids. Mailing Add: Dept of Chem Pa State Univ University Park PA 16802

STEELE, WILLIAM F, b Quincy, Mass, Mar 14, 20; m 54; c 2. MATHEMATICS. Educ: Boston Univ, AB, 51, MA, 52; Univ Pittsburgh, PhD(math), 61. Prof Exp: From instr to assoc prof, 52-63, PROF MATH, HEIDELBERG COL, 63- Concurrent Pos: Lectr, Univ Pittsburgh, 58-; NSF vis scholar, Mass Inst Technol, 71-72. Mem: Am Math Soc; Math Asn Am; Soc Indust & Appl Math. Res: Summability of sequences by matrix methods. Mailing Add: Dept of Math Heidelberg Col Tiffin OH 44883

STEELE, WILLIAM JOHN, b Philadelphia, Pa, Mar 31, 29; m 54; c 3. BIOCHEMICAL PHARMACOLOGY. Educ: Univ Pa, AB, 51, PhD(biochem), 58. Prof Exp: Res assoc cancer, Univ Pa, 57-60, Chester Beatty Res Inst, Royal Cancer Hosp, 60-62; asst prof, Col Med, Baylor Univ, 62-67; from asst prof to assoc prof pharmacol, 67-74, PROF PHARMACOL, COL MED, UNIV IOWA, 74- Mem: Am Asn Cancer Res; Am Chem Soc; Am Soc Pharmacol & Exp Therapeut; Am Soc Cell Biol; Brit Biochem Soc. Res: Biochemical basis of drug and carcinogen induced alterations in gene expression; chromosome structure and function; separation of subcellular components. Mailing Add: Dept of Pharmacol Univ of Iowa Col of Med Iowa City IA 52242

STEELE, WILLIAM KENNETH, b Ft Wayne, Ind, Nov 2, 42; m; c 1. GEOLOGY, GEOPHYSICS. Educ: Case Western Reserve Univ, BS, 65, PhD(geol), 70. Prof Exp: ASST PROF GEOL, EASTERN WASH STATE COL, 70- Mem: AAAS. Res: Paleomagnetism, x-ray emission spectroscophy, general geophysics, especially gravity and magnetic surveys, local environmental problems. Mailing Add: Dept of Geol Eastern Wash State Col Cheney WA 99004

STEELHAMMER, JOE CHARLES, physical chemistry, radiation physics, see 12th edition

STEELINK, CORNELIUS, b Los Angeles, Calif, Oct 1, 22; m 49; c 2. ORGANIC CHEMISTRY. Educ: Calif Inst Technol, BS, 44; Univ Southern Calif, MS, 50; Univ

Calif, Los Angeles, PhD(chem), 56. Prof Exp: Lectr chem, Univ Southern Calif, 49-50 & Orange Coast Col, 50-53; asst, Univ Calif, Los Angeles, 53-56; res fel, Univ Liverpool, 56-57; from asst prof to assoc prof chem, 57-70, PROF CHEM, UNIV ARIZ, 70- Mem: Am Chem Soc. Res: Structure of lignin; electron spin resonance studies on naturally-occurring compounds; isolation and structural elucidation of plant terpenoids; enzymatic degradation of lignin; forensic chemistry. Mailing Add: Dept of Chem Univ of Ariz Tucson AZ 85721

STEELMAN, CARROL DAYTON, b Vernon, Tex, Dec 9, 38; m 60; c 1. MEDICAL ENTOMOLOGY, VETERINARY ENTOMOLOGY. Educ: Okla State Univ, BS, 61, MS, 63, PhD(entom), 65. Prof Exp: From asst prof to assoc prof, 65-75, PROF MED & VET ENTOM, LA STATE UNIV, BATON ROUGE, 75- Concurrent Pos: USDA res grant, 67-69. Mem: Entom Soc Am; Am Mosquito Control Asn. Res: External parasites of domestic animals and man; disease-vector-host biological, ecological and control relationships; effects of insect parasites on animal hosts. Mailing Add: Dept of Entom La State Univ Life Sci Bldg Baton Rouge LA 70803

STEELMAN, SANFORD LEWIS, b Hickory, NC, Oct 11, 22; m 45; c 2. CLINICAL PHARMACOLOGY. Educ: Lenoir-Rhyne Col, BS, 43; Univ NC, PhD(biol chem), 49. Prof Exp: Biochemist, Armour & Co Labs, 49-50, head endocrinol sect, 51-53, head dept biochem res, 53-56; assoc prof biochem, Baylor Col Med, 56-58; dir endocrinol, Merck Inst Therapeut Res, 68-70, SR CLIN ASSOC, MERCK SHARP & DOHME RES LABS, RAHWAY, 70- Concurrent Pos: Assoc prof, Postgrad Sch Med, Univ Tex, 56-58. Mem: AAAS; Am Chem Soc; Soc Exp Biol & Med; Endocrine Soc; Am Soc Biol Chem. Res: Isolation and biological and physicochemical properties of protein and peptide hormones; physiology, pharmacology and bioassay of steroidal hormones; clinical pharmacology. Mailing Add: Dept of Clin Pharmacol Merck Sharp & Dohme Res Lab Rahway NJ 07065

STEEN, EDWIN BENZEL, b Wheeling, Ind, July 23, 01; m 27; c 2. PARASITOLOGY. Educ: Wabash Col, AB, 23; Columbia Univ, AM, 26; Purdue Univ, PhD(zool), 38. Prof Exp: Instr zool, Wabash Col, 23-25, actg head dept, 26-27; asst biol, NY Univ, 25-26; instr zool, Univ Cincinnati, 27-31; instr zool, Purdue Univ, 31-34, instr fish & game, Univ & Exp Sta, 34-36; tutor biol, City Col New York, 38-40; from asst prof to prof, 41-72, head dept, 64-65, EMER PROF BIOL, WESTERN MICH UNIV, 72- Mem: AAAS; Sigma Xi; Am Inst Biol Sci. Res: Mammalian anatomy and physiology; medical and biological lexicography; zoology. Mailing Add: 2011 Greenlawn Ave Kalamazoo MI 49007

STEEN, FREDERICK HENRY, b Brooklyn, NY, Nov 26, 07; m 37; c 5. MATHEMATICS. Educ: Colgate Univ, AB, 29; Harvard Univ, AM, 31, PhD(math), 34. Prof Exp: Asst instr physics, Colgate Univ, 28-29; instr math, Harvard Univ, 31-34; from instr to asst prof, Ga Inst Technol, 34-42; from asst prof to assoc prof math, 42-44, secy to faculty, 49-69, PROF MATH, ALLEGHENY COL, 44- Mem: Am Math Soc; Math Asn Am. Res: Analysis; analytic geometry; differential equations; heuristics; statistics. Mailing Add: Dept of Math Allegheny Col Meadville PA 16335

STEEN, JAMES SOUTHWORTH, b Vicksburg, Miss, Oct 26, 40; m 60; c 2. BIOLOGY, IMMUNOLOGY. Educ: Delta State Col, BS, 62; Univ Miss, MS, 64, PhD(biol), 68. Prof Exp: Asst prof, 68-74, ASSOC PROF BIOL, DELTA STATE COL, 74- Mem: Am Soc Microbiol. Res: Carbohydrate metabolism in bacteria; tissue transplantation and immunosuppression as related to the enhancement phenomenon in inbred strains of mice. Mailing Add: Dept of Sci Delta State Col Cleveland MS 38732

STEEN, JOHN CARL, b Detroit Lakes, Minn, May 20, 41; m 64; c 3. PHYSICAL ANTHROPOLOGY. Educ: Univ Ore, BS, 65, MA, 69, PhD(anthrop), 74. Prof Exp: Sr instr, Dept Anthrop, Ore State Univ, 71-73; teaching asst anthrop, 65-67, res asst, 67-69, instr, 69-71, RES ASSOC, CHILD STUDY CLIN, HEALTH SCI CTR, UNIV ORE, 75- Concurrent Pos: Consult anthrop, Soc Sci Curric, WAlbany High Sch, Ore, 72-73; consult primate behav, Child Study Biophys Lab, Sch Dent, Univ Ore, 70-74; consult forensic anthrop, Dist Attys Off, Linn County, Ore, 74, Multnomah County Med Examr & Ore State Med Examrs Off, 75- Mem: Am Asn Phys Anthropologists. Res: Child growth; human morphology; dental development; dental histology; quantitative change in human morphology during growth and dental histology techniques for use in paleopathology. Mailing Add: Child Study Clin Univ of Ore Health Sci Ctr Portland OR 97201

STEEN, LYNN ARTHUR, b Chicago, Ill, Jan 1, 41; m 63. MATHEMATICS. Educ: Luther Col, BA, 61; Mass Inst Technol, PhD(math), 65. Prof Exp: From asst prof to assoc prof, 65-75, PROF MATH, ST OLAF COL, 75- Concurrent Pos: NSF sci faculty fel, Mittag-Leffler Inst, Sweden, 71-72; assoc ed, Am Math Monthly, 70-; ed, Math Mag, 76- Honors & Awards: Lester R Ford Award, Math Asn Am, 73 & 75. Mem: Am Math Soc; Math Asn Am. Res: Analysis; function algebras; mathematical logic; general topology. Mailing Add: Dept of Math St Olaf Coll Northfield MN 55057

STEEN, STEPHEN N, b London, Eng, Sept 6, 23; US citizen; m. MEDICINE. Educ: Mass Inst Technol, SB, 43; Univ Geneva, ScD(med biochem), 51, MD, 52; Am Bd Anesthesiol, dipl, 60. Prof Exp: Intern Abbot Hosp, Minneapolis, Minn, 53-54; res anesthesiol, Columbia-Presby Ctr, 54-56; instr, Albert Einstein Med Sch, 56-60; from asst prof to assoc prof, State Univ NY Downstate Med Ctr, 61-69; physician-in-chief anesthesia res, Cath Med Ctr Brooklyn & Queens, Inc, NY, 69-71; PROF ANESTHESIOL, SCH MED, UNIV CALIF, LOS ANGELES, 71-, DIR TRAINING & RES, HARBOR GEN HOSP, 71- Concurrent Pos: Actg dir, Delafield Hosp, NY, 56, attend anesthesiologist, 56-61; instr, Bronx Munic Hosp Ctr, 56-60 & Columbia Univ, 57-61; physician, Beth Israel Hosp, 57-61; attend anesthesiologist, Cent Islip State Hosp & St Francis Hosp, 58-, Misericordia Hosp, 58-62, Vet Admin 62- & St Barnabas Hosp, Bronx, NY, 56-61, dir anesthesiol, 60-61; vis attend anesthesiologist, Brooklyn Vet Admin Hosp, 61-66; from assoc vis anesthesiologist to vis anesthesiologist, Kings County Hosp Ctr, 61- Mem: Fel Am Col Anesthesiol; sr mem Am Chem Soc; AMA; Am Soc Anesthesiol; Asn Am Med Cols. Res: Anesthesiology. Mailing Add: Dept of Anesthesiol Univ of Calif Sch of Med Los Angeles CA 90024

STEEN, WILLIAM CHARLES, soil microbiology, soil biochemistry, see 12th edition

STEEN, WILSON D, b Hugo, Okla, July 21, 23; m; c 2. PREVENTIVE MEDICINE, PUBLIC HEALTH. Educ: Univ Okla, BS, 49, PhD(prev med, pub health), 63; Columbia Univ, MSPH, 52. Prof Exp: Sanitarian & actg dir, Kiowa County Health Dept, Okla, 49-50; health educ consult, Okla State Dept Health, 52-53; exec secy, Okla Health Coun, 53-54; field rep, Okla Div, Am Cancer Soc, 54-56; exten specialist, Exten Div, 56-61, dir health studies, Col Continuing Educ & res assoc, Dept Prev Med & Pub Health, 61-63, from asst prof to assoc prof prev med & pub health, Col Med, 63-69, assoc prof health admin, Col Health, 67-69, assoc dean, Col Health, 68-73, PROF FAMILY PRACT, COMMUNITY MED & DENT, COLS MED & DENT, VCHMN COMMUNITY MED & DENT & PROF HEALTH ADMIN, COL HEALTH, UNIV OKLA, 69- Concurrent Pos: Mem rev panel, Nat Heart & Lung Inst. Mem: Fel Am Pub Health Asn. Res: Measuring and evaluating community resources for the control of heart disease; continuing education for the health professions. Mailing Add: Health Sci Ctr Univ of Okla Oklahoma City OK 73190

STEENBERG, NEIL RICHARD, nuclear physics, high energy physics, see 12th edition

STEENBERGEN, JAMES FRANKLIN, b Glasgow, Ky, May 11, 39; m 67. MICROBIOLOGY. Educ: Western Ky State Col, BS, 62; Ind Univ, MA, 65, PhD(microbiol), 68. Prof Exp: Res assoc marine microbiol, Ore State Univ, 68-69, vis asst prof microbiol, 69-70; asst prof, 70-75, ASSOC PROF MICROBIOL, SAN DIEGO STATE UNIV, 75- Mem: Am Soc Microbiol. Res: Microbial ecology; diseases of wildlife; diseases of crustaceans and fish; invertebrate immunology. Mailing Add: Dept of Microbiol San Diego State Univ San Diego CA 92115

STEENBURG, RICHARD WESLEY, b Aurora, Nebr, Feb 3, 25; m 50; c 2. SURGERY. Educ: Harvard Univ, MD, 48; Am Bd Surg, dipl. Prof Exp: Assoc prof surg, Johns Hopkins Univ, 65-69; PROF SURG, COL MED, UNIV NEBR, OMAHA, 69- Concurrent Pos: Surgeon in chief, Baltimore City Hosps, 67-69. Mem: Soc Univ Surg. Res: General surgery; surgical endocrinology; vascular disease; renal physiology. Mailing Add: Dept of Surg Univ of Nebr Col of Med Omaha NE 68105

STEEN-MCINTYRE, VIRGINIA CAROL, b Chicago, Ill, Dec 3, 36; m 67. QUATERNARY GEOLOGY, TEPHROCHRONOLOGY. Educ: Augustana Col, Ill, 59; Wash State Univ, MS, 65; Univ Idaho, PhD(geol), 76. Prof Exp: Asst geologist, George H Otto, Consult Geologist, Chicago, 59-61; jr geologist, Lab Anthrop, Wash State Univ, 64-66; PHYS SCI TECHNICIAN, US GEOL SURV, DENVER, 70- Concurrent Pos: Corresp mem, Int Asn Quaternary Res Comn Tephrochronology, 73-77; tephrochronologist, Valsequillo Early Man Proj, Mex, NSF, 66-, El Salvador Protoclassic Proj, 75- Mem: Geol Soc Am; Int Asn Quaternary Res; Am Asn Quaternary Res; Asn Field Archaeology. Res: Volcanic ash chronology; archaeologic site stratigraphy; petrography of friable Pleistocene deposits; tephra hydration dating; weathering of volcanic ejecta. Mailing Add: Box 1167 Idaho Springs CO 80452

STEENSEN, DONALD H J, b Clinton, Iowa, Apr 26, 29; m 54. FOREST ECONOMICS. Educ: Iowa State Univ, BS, 58; Duke Univ, MF, 60, PhD(forest econ), 65. Prof Exp: Asst prof forest econ & sampling, Auburn Univ, 60-65; asst prof forest econ & mensuration, 65-73, ASSOC PROF FOREST ECON & MENSURATION, SCH FOREST RESOURCES, NC STATE UNIV, 73- Mem: Soc Am Foresters. Res: Forest mensuration. Mailing Add: Dept Forest Mgt-Wood Sci & Tech NC State Univ Sch Forest Resour Raleigh NC 27607

STEEPLES, DONALD WALLACE, b Hays, Kans, May 15, 45; m 67; c 2. GEOPHYSICS. Educ: Kans State Univ, BS, 69, MS, 70; Stanford Univ, MS, 74, PhD(geophys), 75. Prof Exp: Geophysicist seismol, US Geol Surv, 72-75; RES ASSOC GEOPHYS, STATE GEOL SURV KANS, 75- Mem: Am Geophys Union; Soc Explor Geophysicists; Seismol Soc Am. Res: Crust and upper mantle structure of central North America; Kimberlites of Kansas; Pleistocene drainage of Kansas; passive seismic methods in exploration for geothermal energy. Mailing Add: Kans Geol Surv Univ of Kans Lawrence KS 66044

STEER, ARTHUR, pathology, see 12th edition

STEER, CHARLES MELVIN, b Midvale, NJ, Feb 13, 13; m 45; c 2. OBSTETRICS & GYNECOLOGY. Educ: Columbia Univ, AB, 33, MD, 37, MedScD, 42. Prof Exp: Instr obstet & gynec, 46-50, from asst prof to assoc prof clin obstet & gynec, 50-69, actg chmn dept, 68-70, PROF CLIN OBSTET & GYNEC, COL PHYSICIANS & SURGEONS, COLUMBIA UNIV, 69- Concurrent Pos: From asst attend obstetrician & gynecologist to attend, Presby Hosp, 46- Mem: Am Col Surgeons; fel Am Col Obstet & Gynec. Res: Mechanical and physiological problems of labor; x-ray pelvimetry; electrohysterography; fetal mortality and wastage. Mailing Add: Dept of Obstet Columbia Univ New York NY 10032

STEER, MAX DAVID, b New York, NY, June 14, 10; m 42. SPEECH & HEARING SCIENCES. Educ: Long Island Univ, 32, LLD, 57; Univ Iowa, MA, 33, PhD(psychol), 38. Prof Exp: Asst speech path, Univ Iowa, 33-35; from instr to prof speech sci, 35-70, dir speech & hearing clin, 46-70, head dept audiol & speech sci, 63-70, HANLEY DISTINGUISHED PROF AUDIOL & SPEECH SCI, PURDUE UNIV, WEST LAFAYETTE, 70- Concurrent Pos: Consult, State of Ind Hearing Comn, 66-, US Off Educ, NIH, NSF & Ind State Training Sch Ment Retarded; consult, Neurol & Sensory Dis Control Prof, USPHS, Pan-Am Health Orgn, 72-, Latin Am Fedn Logopedics, Phoniatrics, Audiol, 72- & Univ Bogota, 74; mem nat res adv comt, Bur Educ Handicapped, US Off Educ, 72-; consult & vis lectr, Nat Rehab Inst, Panama, 74. Mem: AAAS; Acoust Soc Am; Am Psychol Asn; fel Am Speech & Hearing Asn (vpres, 49, pres, 51); Int Asn Logopedics & Phoniatrics (vpres, 63-65, 71-77). Res: Speech disorders and acoustics; audiology; clinical psychology; neurology; physiology and psychology of communication. Mailing Add: Dept of Audiol & Speech Sci Purdue Univ West Lafayette IN 47907

STEER, RONALD PAUL, b Regina, Sask, Mar 7, 43; m 64; c 2. PHYSICAL CHEMISTRY. Educ: Univ Sask, BA, 64, PhD(chem), 68. Prof Exp: USPHS fel, Univ Calif, Riverside, 68-69; asst prof, 69-73, ASSOC PROF CHEM, UNIV SASK, 73- Concurrent Pos: Vis fel, Univ Southampton, 75-76. Mem: Chem Inst Can. Res: Photochemistry and excited states of small ring organic molecules; importance of excited states of small molecules in formation of photochemical smog. Mailing Add: Dept of Chem Univ of Sask Saskatoon SK Can

STEERE, RICHARD C, b Kansas City, Mo, Mar 5, 09; m 36; c 3. METEOROLOGY, PHYSICAL OCEANOGRAPHY. Educ: US Naval Acad, BS, 31; Mass Inst Technol, SM, 41. Prof Exp: Prog mgr, John I Thompson & Co, 61-75; WITH TRACO-SITKO INC, 75- Concurrent Pos: Chmn & mem joint oceanog subcomt, Joint Chiefs of Staff, 44-45; sr asst to dir meteorol, Naval Weather Serv, 44-46, fac mem meteorol & oceanog, Armed Forces Staff Col, 46-48. Mem: Am Meteorol Soc; Marine Technol Soc. Res: Marine meteorology; development and application of methods for wave and surf forecasting.

STEERE, RUSSELL LADD, b Ann Arbor, Mich, Aug 23, 17; m 43; c 4. VIROLOGY, BIOPHYSICS. Educ: Univ Mich, BA, 41, MA, 47, PhD(bot), 50. Prof Exp: Asst physics, Univ Mich, 46-47, res assoc, 47-50; asst, Rockefeller Inst Med Res, 50-51; res asst, Virol Lab, Univ Calif, 51-59; BOTANIST & LEADER, PLANT VIROL LAB, PLANT PROTECTION INST, AGR RES SERV, USDA, 59- Honors & Awards: Cert of Achievement, Walter Reed Inst Res & Superior Serv Award, USDA, 69; Ruth Allen Award, Am Phytopath Soc, 72. Mem: Am Phytopath Soc; Bot Soc Am; Electron Micros Soc Am (pres, 70). Res: Purification of viruses, subcellular particles and macromolecules; structure and multiplication of viruses; development of new techniques for preparation of biological materials for electron microscopy. Mailing Add: Plant Protection Inst USDA Plant Indust Sta Beltsville MD 20705

STEERE, WILLIAM CAMPBELL, b Muskegon, Mich, Nov 4, 07; m 27; c 3.

BOTANY. Educ: Univ Mich, BS, 29, MA, 31, PhD(bot), 32. Hon Degrees: DSc, Univ Montreal, 59, Univ Mich, 62. Prof Exp: Instr bot, Temple Univ, 29-31; from instr to prof, Univ Mich, 31-50, chmn dept, 47-50; prof, Stanford Univ, 50-58, dean grad div, 55-58; dir, 58-68, exec dir, 68-70, pres, 70-72, EMER PRES & SR SCIENTIST, NY BOT GARDEN, 72-; PROF, COLUMBIA UNIV, 58- Concurrent Pos: Botanist, Univ Mich-Carnegie Inst Exped, Yucatan, 32; ed in chief, The Bryologist, 38-54 & Am J Bot, 53-57; epxhange prof, Univ PR, 39-40; sr botanist, Foreign Econ Admin, Colombia, 42-43 & Ecuador, 43-44; mem exped, Great Bear Lake, Can, 48; botanist, US Geol Surv, Alaska, 49; botanist, Arctic Res Lab, Pt Barrow, 51-53, investr, 60, 61, 63 & 65; pres, Bryol Sect, Int Bot Cong, Paris, 54; prog dir, NSF, DC, 54-55; trustee, Biol Abstr, 61-66 & 67-73, pres, 64-65; mem exped, McMurdo Sound, Antarctica, 65; Phi Beta Kappa vis scholar, 66-67. Honors & Awards: Liberty Hyde Medal, Am Hort Soc; Medal, Int Bot Cong, Montreal, 59; Mt Steere named in his honor, Antarctica, 68; Mary Soper Pope Medal, Cranbrook Inst Sci, 70; Imperial Order of the Sacred Treasure, Japan, 72. Mem: AAAS (vpres, 48, 66 & 67); Bot Soc Am (pres, 59); fel Am Geog Soc; Am Soc Naturalists (pres, 57); Am Asn Mus (vpres, 67, pres, 68-69). Res: Cytology, ecology, geography and taxonomy of bryophytes; communication of scientific information. Mailing Add: NY Bot Garden Bronx NY 10458

STEERS, ARTHUR WALTER, b Ketchikan, Alaska, May 27, 13; m 40; c 3. PHARMACEUTICAL CHEMISTRY. Educ: Univ Wash, BS, 38, MS, 39, PhD(pharmaceut chem), 46. Prof Exp: Chemist, San Francisco Procurement Dist, Chem Warfare Serv, US Dept Army, 42-44; chemist, US Food & Drug Admin, 44-57, chief chemist, Seattle Dist, 57-67, dir, Nat Ctr Drug Anal, Mo, 67-74; RETIRED. Honors & Awards: Award of Merit, US Food & Drug Admin. Mem: Acad Pharmaceut Sci; Am Pharmaceut Asn; Am Chem Soc; Asn Off Anal Chem. Res: Drug standardization and analysis; automated methods of drug analysis. Mailing Add: 4970 Brunston Dr St Louis MO 63128

STEERS, EDWARD, b Bethlehem, Pa, July 15, 10; m 30; c 3. MICROBIOLOGY, CHEMISTRY. Educ: Moravian Col, BS, 32; Lehigh Univ, MS, 37; Univ Pa, PhD(microbiol), 49. Prof Exp: Prof biol, Moravian Col, 39-45; assoc prof bact, Sch Med, Univ Md, 49-56; assoc prof, Sch Med, Univ Pa, 56-63; prof microbiol, New York Med Col, 64-66; PROF PATH, STATE UNIV NY, 66- Concurrent Pos: Consult clin microbiol & infectious dis, Community Med Ctr, Moses-Taylor Hosp & Mercy Hosp, Scranton, Pa & Brooklyn-Cumberland Med Ctr, 70- Mem: Soc Am Microbiol; Am Acad Microbiol. Res: Bacterial metabolism; clinical microbiology. Mailing Add: Box 199C Paupack PA 18451

STEERS, EDWARD, JR, b Bethlehem, Pa, May 21, 37; m 57; c 3. BIOLOGICAL CHEMISTRY. Educ: Univ Pa, AB, 59, PhD(biol), 63. Prof Exp: Staff fel biochem, 63-65, RES BIOLOGIST, LAB CHEM BIOL, NAT INST ARTHRITIS, METABOLISM & DIGESTIVE DIS, 65- Concurrent Pos: Lectr, George Washington Univ, 66- Mem: Am Soc Biol Chem; Am Soc Cell Biol. Res: Protein chemistry; relationship of structure to function in proteins; cell organelles. Mailing Add: Bldg 10 Rm 9N307 NIH Bethesda MD 20014

STEEVES, HARRISON ROSS, III, b Birmingham, Ala, July 2, 37; m 57; c 4. HISTOCHEMISTRY, TAXONOMY. Educ: Univ of the South, BS, 58; Univ Va, MS, 60, PhD(biol), 62. Prof Exp: Instr zool, Univ Va, 62; fel histol, Med Ctr, Univ Ala, 62-65, instr, 65-66; asst prof histol & histochem, 66-68, ASSOC PROF ZOOL, VA POLYTECH INST & STATE UNIV, 68- Honors & Awards: Andrew Fleming award biol res, 61. Mem: AAAS. Res: Invertebrate histochemistry; taxonomy, ecology and physiology of cave crustaceans and histochemistry of digestion in these forms. Mailing Add: Dept of Biol Va Polytech Inst & State Univ Blacksburg VA 24061

STEEVES, LEA CHAPMAN, b New Westminster, BC, Nov 4, 15; m 42; c 5. MEDICINE. Educ: Mt Allison Univ, BA, 36; McGill Univ, MD, CM, 40; FRCP(C), 47. Hon Degrees: LLD, Mt Allison Univ, 69; DSc, Mem Univ Nfld, 74. Prof Exp: From intern to resident, Royal Victoria Hosp, 40-43; from asst prof to assoc prof med, 48-57, dir postgrad div, 57-69, asst dean, 64-69, ASSOC DEAN FAC MED, DALHOUSIE UNIV, 69-, PROF MED, 57- Concurrent Pos: Fel cardiol, Royal Victoria Hosp, Montreal, 46-47; specialist, Camp Hill Vet Hosp, Halifax, NS, 48-; from asst attend physician to attend, Victoria Gen Hosp, 48-; consult, Can Serv & Halifax Children's Hosps, 48-70. Mem: Fel Am Col Physicians; Can Med Asn; Can Cardiovasc Soc. Res: Cardiovascular diseases; internal medicine; continuing medical and intern-resident education. Mailing Add: Fac of Med Dalhousie Univ Halifax NS Can

STEEVES, MARGARET WOLFE, paleobotany, see 12th edition

STEEVES, RICHARD ALLISON, b Fredericksburg, Va, Feb 2, 38; m 65; c 3. VIROLOGY, ONCOLOGY. Educ: Univ Western Ont, MD, 61; Univ Toronto, PhD(med biophys), 66. Prof Exp: From sr cancer res scientist to assoc cancer res scientist, Roswell Park Mem Inst, 67-72; ASSOC PROF DEVELOP BIOL & CANCER, ALBERT EINSTEIN COL MED, 72- Concurrent Pos: Nat Cancer Inst Can fel, Dept Biol, McMaster Univ, 66-67. Mem: Am Asn Cancer Res; Am Soc Microbiol. Res: Defectiveness of spleen focus-forming virus in the Friend leukemia virus complex; envelope antigens of murine leukemia viruses. Mailing Add: Dept of Develop Biol & Cancer Albert Einstein Col of Med Bronx NY 10461

STEEVES, TAYLOR ARMSTRONG, b Quincy, Mass, Nov 29, 26; m 56; c 3. BOTANY. Educ: Univ Mass, BS, 47; Harvard Univ, AM, 49, PhD(biol), 51. Prof Exp: Jr fel, Soc Fels, Harvard Univ, 51-54, asst prof bot, Biol Labs, 54-59; assoc prof biol, 59-64, PROF BIOL, UNIV SASK, 64- Concurrent Pos: Lalor Found Award, 58. Mem: Bot Soc Am; Am Soc Plant Physiol; Soc Exp Biol & Med; Soc Develop Biol; Can Bot Asn (pres, 72-73). Res: Morphogenesis of vascular plants; plant tissue culture and growth hormones. Mailing Add: Dept of Biol Univ of Sask Saskatoon SK Can

STEFANCSIK, ERNEST ANTON, b Brooklyn, NY, Sept 10, 23. ORGANIC CHEMISTRY. Educ: St John's Univ, NY, BS, 43; NY Univ, MS, 47, PhD(chem), 53. Prof Exp: Chemist, Am Cyanamid Co, 43-46; teaching fel, NY Univ, 47-49, asst, 50-52; res chemist, 52-70, SR RES CHEMIST, PIGMENTS DEPT, E I DU PONT DE NEMOURS & CO, INC, NEWARK, 70- Mem: AAAS; Am Chem Soc. Res: Physics and chemistry of organic pigments. Mailing Add: RD 6 Box 595 Flemington NJ 08822

STEFANI, ANDREW PETER, b Island of Cyprus, July 10, 26; US citizen; m 55; c 1. ORGANIC CHEMISTRY, PHYSICAL CHEMISTRY. Educ: Mich State Univ, BA, 56; Univ Colo, PhD(chem), 60. Prof Exp: Res assoc chem, State Univ NY Col Forestry, Syracuse Univ, 60-62; asst prof, Purdue Univ, 62-63; asst prof, 63-64, ASSOC PROF CHEM, UNIV MISS, 64- Concurrent Pos: Res grants, Res Corp, 64, Petrol Res Fund & NSF, 64-68. Mem: AAAS; Am Chem Soc. Res: Chemical kinetics; free radical reactivity; solvent effects in electroneutral reactions; high pressure chemistry. Mailing Add: Dept of Chem Univ of Miss University MS 38677

STEFANI, STEFANO, b Trieste, Italy, Apr 19, 29; US citizen; m 58; c 2. RADIOTHERAPY, RADIOBIOLOGY. Educ: Univ Trieste, BS, 48; Univ Perugia, MD, 54. Prof Exp: Intern, Univ Trieste, 54-55; resident radiol, Med Sch, Univ Padua, 55-56, asst prof, 56-57; resident radiother, G Roussy Inst, Paris, 57-59; asst prof, Med Sch, Univ Paris, 59-60; resident, Univ Md Hosp, 61-62; res radiobiologist, Med Sch, Northwestern Univ, 62-63; from instr to asst prof radiol, Stritch Sch Med, Loyola Univ Chicago, 63-69; CHIEF THERAPEUT RADIOL SERV, VET ADMIN HOSP, HINES, 69- Concurrent Pos: Fels radiother, Curie Found, Paris, 57 & radiobiol, Nat Inst Nuclear Sci & Technol, Saclay, France, 58-59; consult, WHO, Geneva, 61, Argonne Nat Lab, AEC, 64-65; clin investr, Vet Admin Hosp, Hines, 64-69, staff physician, 67-68; abstractor, J Surg, Gynec & Obstet, 66-; consult radiother, Weiss Mem Hosp, 72. Mem: AAAS; AMA; Radiation Res Soc; Transplantation Soc; Radiol Soc NAm. Res: Radiation potentiators; total body irradiation and radioprotection; physiology and pathology of lymphocytes; bone marrow radiosensitivity and transfusion. Mailing Add: Therapeut Radiol Serv Vet Admin Hosp Hines IL 60141

STEFANIAK, JEROME JOSEPH, microbiology, see 12th edition

STEFANINI, MARIO, b Chieri, Italy, June 11, 18; nat US; m 49; c 2. PATHOLOGY. Educ: Univ Rome, MD, 39; Univ Freiburg, PhD, 40; Univ Calcutta, dipl, 43; Marquette Univ, MS, 47; Am Bd Path, dipl. Prof Exp: Asst prof med, Univ Rome, 45-46; instr biochem, Sch Med, Marquette Univ, 46-49; instr med, Med Sch, Tufts Univ, 49-51, assoc prof, 51-52, prof lab med, 52-61; DIR LABS & PATH, ST ELIZABETH'S HOSP, 61- Concurrent Pos: USPHS sr res fel biochem, Sch Med, Marquette Univ, 46-49; Damon Runyon Fund sr res fel med, Tufts Univ, 49-51; physician, New Eng Ctr Hosp, Boston, 50-54; physician-in-chief, Clin Lab, Boston Dispensary, 51-54; lectr, Univ PR, 52; estab investr, Am Heart Asn, 52-58; dir res labs & hematologist, St Elizabeth's Hosp, 54-61. Honors & Awards: Piccinini Prize, Italy, 46; Award, Am Asn Blood Banks, 54 & 56. Mem: AAAS; Soc Exp Biol & Med; Am Physiol Soc; Am Soc Clin Invest; Am Immunol Asn. Res: Tropical diseases; liver function; hemorrhagic diseases and physiology of blood coagulation. Mailing Add: St Elizabeth's Hosp 600 Sager Ave Danville IL 61832

STEFANO, GEORGE BOGDON, b New York, NY, Sept 11, 45; m 68. NEUROBIOLOGY. Educ: Wagner Col, BS, 67; Fordham Univ, MS, 69, PhD(physiol), 73. Prof Exp: ASST PROF PHYSIOL, CITY UNIV NEW YORK, 72-; RES ASSOC NEUROPHYSIOL, FORDHAM UNIV, 73- Concurrent Pos: Consult neurophysiol, Albert Einstein Col Med, 74- Mem: Am Soc Zoologists; NY Acad Sci; AAAS. Res: Histofluorescent localization of monamines in invertebrate nervous systems; microspectrofluorometric analysis of monoamines in such nervous systems; the effect of mono and divalent ions on monoamine activity. Mailing Add: Dept of Biol Sci Fordham Univ Bronx NY 10458

STEFANOU, HARRY, b New York, NY, June 16, 47; m 69; c 1. POLYMER PHYSICS. Educ: City Col New York, BS, 69; City Univ New York, PhD(phys chem), 73. Prof Exp: Res assoc polymer physics, Princeton Univ, 72-73; SR RES CHEMIST POLYMER PHYSICS, PENNWALT CORP, 73- Mem: AAAS; Am Chem Soc; Am Phys Soc. Res: The areas of polymer crystallization kinetics; polymer viscoelasticity and solution thermodynamics. Mailing Add: Pennwalt Corp 900 First Ave King of Prussia PA 19406

STEFANOVIC, MOMCILO, biochemistry, see 12th edition

STEFANSKI, RAYMOND JOSEPH, b Buffalo, NY, July 22, 41; m 69; c 1. PHYSICS. Educ: State Univ NY Buffalo, BA, 63; Yale Univ, PhD(physics), 69. Prof Exp: Res assoc physics, Yale Univ, 68-69; PHYSICIST I, FERMI NAT ACCELERATOR LAB, 69- Mem: Am Phys Soc. Res: High energy physics and cosmic rays; hyperon beams; hyperon decays and interactions; K-meson decays; neutrino beams; cosmic ray muon distributions; heavy stable particles in cosmic rays. Mailing Add: Neutrino Area Fermi Nat Accelerator Lab PO Box 500 Batavia IL 60510

STEFANSSON, BALDUR ROSMUND, b Vestfold, Man. PLANT BREEDING. Educ: Univ Man, BSA, 50, MS, 52, PhD(plant sci), 66. Prof Exp: Res assoc, 52-66, assoc prof, 66-74, PROF PLANT SCI, UNIV MAN, 74- Honors & Awards: Royal Bank Award, Royal Bank of Can, 75. Mem: Fel Agr Inst Can. Res: Plant breeding and related research with oilseed crops, formerly soybeans, currently rapeseed; compositional changes in rapeseed, including low erucic acid, low glucosinolate and low fibre content, induced by breeding. Mailing Add: Dept of Plant Sci Univ of Man Winnipeg MB Can

STEFFAN, WALLACE ALLAN, b St Paul, Minn, Aug 10, 34; m 66; c 1. ENTOMOLOGY. Educ: Univ Calif, Berkeley, BS, 61, PhD(entom), 65. Prof Exp: ENTOMOLOGIST, BERNICE P BISHOP MUS, 64-, HEAD DIPTERA SECT, 68- Concurrent Pos: NIH grant, 67-72; partic island ecosyst int res prog, US Int Biol Prog, 70; Grace H Griswold lectr, Cornell Univ, 70; mem affiliated grad faculty, Univ Hawaii. Mem: AAAS; Am Entom Soc; Am Mosquito Control Asn. Res: Systematics of Culicidae and Sciaridae; ecology of Culicidae; cytogenetics of Sciaridae. Mailing Add: Entom Dept Bishop Mus PO Box 6037 Honolulu HI 96818

STEFFEE, CLYDE HAROLD, b South Bend, Ind, Aug 23, 20; m 42; c 3. PATHOLOGY. Educ: Univ Chicago, SB, 43, SM, 46, PhD(path), & MD, 49; Am Bd Path, dipl, 51. Prof Exp: Asst, Univ Chicago, 43-46, instr, 46-49; from intern to resident path, USPHS Hosp, New Orleans, La, 49-51, pathologist, Lab Path, Nat Cancer Inst, 51-55; pathologist, Oak Ridge Hosp, Tenn, 56-60; dir labs, Methodist Hosp, Memphis, 60-65; dir path, Cape Fear Valley Hosp & Highsmith-Rainey Mem Hosp, Fayetteville, 65-70 & NC Sanitarium, McCain, 70- Mem: Am Soc Exp Path; Soc Nuclear Med; fel Am Soc Clin Path; AMA; Am Asn Cancer Res. Res: Protein nutrition and metabolism; pulmonary carcinogenesis; biological effects of chromates; rare earths; radioisotopes; epidemiology of staphylococcal infections; clinical pathology. Mailing Add: Pinehurst Path Ctr PO Box 3000 Pinehurst NC 28374

STEFFEK, ANTHONY J, b Milwaukee, Wis, Aug 6, 35; m 60; c 3. DENTISTRY, PHARMACOLOGY. Educ: Marquette Univ, DDS, 62, MS, 63, PhD(physiol), 65. Prof Exp: Res assoc pharmacol, Nat Inst Dent Res, 65-70; ASST PROF & DIR DIV DEVELOP BIOL, WALTER G ZOLLER MEM DENT CLIN, UNIV CHICAGO, 75-, RES ASSOC, DEPT ANAT, 75- Concurrent Pos: Am Dent Asn grant pharmacol, Nat Inst Dent Res, 65-; vis prof pharmacol, Charles Univ, Prague, Czechoslovakia. Mem: Am Dent Asn; Am Physiol Soc. Res: Metabolism and pharmacological disposition of environmental agents producing experimentally-induced oral-facial malformations, specifically cleft lip and palate; teratology. Mailing Add: Am Dent Asn Health Found 211 E Chicago Ave Chicago IL 60611

STEFFEN, ALBERT HARRY, b Menomonee Falls, Wis, May 24, 14; m 39; c 2. FOOD CHEMISTRY. Educ: Univ Wis, BS, 40. Prof Exp: Anal food chemist fats & oils, 40-46, res food chemist, 46-52, fat & oil researcher, 52-56, asst prod mgr, Refining Div, 56-57, tech dir & qual control mgr, Lookout Oil & Refining Co Div, 57-67, prod mgr by-prod & mem staff, Cent Qual Control, Nebr, 67-68, mem cent qual assurance staff, Ill, 68-71, TECH SPECIALIST, QUAL ASSURANCE STAFF, ARMOUR & CO, ARIZ, 71- Mem: Am Oil Chem Soc; Am Soc Qual Control; Inst

Food Technol. Res: Meat and meat by-products; fats and oils. Mailing Add: 4105 W Mission Ln Phoenix AZ 85021

STEFFEN, ROLF MARCEL, b Basel, Switz, June 17, 22; m 49; c 2. NUCLEAR PHYSICS. Educ: Cantonal Col, Zurich, BS, 41; Swiss Fed Inst Technol, MS, 46, PhD, 48. Prof Exp: From asst prof to assoc prof, 49-56, PROF PHYSICS, PURDUE UNIV, WEST LAFAYETTE, 56- Concurrent Pos: Asst, Swiss Fed Inst Technol, 49-54; Sigma Xi Res Award, 64. Mem: Fel Am Phys Soc; Swiss Phys Soc. Res: Nuclear spectroscopy; angular correlations of nuclear radiation; influence of extranuclear fields on angular correlation; beta decay; hyperfine fields; heavy ion nuclear reactions. Mailing Add: Dept of Physics Purdue Univ West Lafayette IN 47907

STEFFENS, GEORGE LOUIS, b Bryantown, Md, June 13, 30; m 59; c 2. PLANT PHYSIOLOGY. Educ: Univ Md, BS, 51, MS, 53, PhD(agron), 56. Prof Exp: Agron biochemist, Gen Cigar Co, Inc, Pa, 58-61; plant physiologist, Coastal Plain Exp Sta, 61-63, PLANT PHYSIOLOGIST, PLANT HORMONE & GROWTH REGULATORS LAB, PLANT PHYSIOL INST, AGR RES SERV, USDA, 63- Honors & Awards: Philip Morris Award, 72. Mem: AAAS; Am Soc Plant Physiologists; Am Chem Soc; Scand Soc Plant Physiologists. Res: Growth regulating chemicals and their development and effect on plants, especially tobacco. Mailing Add: Plant Physiol Inst Agr Res Serv USDA Beltsville MD 20705

STEFFENS, JAMES JEFFREY, b Lakewood, Ohio, Nov 1, 42; m 66; c 2. BIO-ORGANIC CHEMISTRY, DRUG METABOLISM. Educ: Amherst Col, BA, 64; Mass Inst Technol, PhD(chem), 71. Prof Exp: Fel, Pa State Univ, 71-74; SR RES CHEMIST, DRUG METAB, MERCK & CO, INC, 74- Res: Reaction mechanisms of enzymes involved in the metabolism of xenobiotics. Mailing Add: Dept of Drug Metab & Radiochem Merck & Co Inc Rahway NJ 07065

STEFFENSEN, DALE MARRIOTT, b Salt Lake City, Utah, Apr 17, 22; m 50; c 3. GENETICS. Educ: Univ Calif, Los Angeles, AB, 48; Univ Calif, Berkeley, PhD(genetics), 52. Prof Exp: Res geneticist asst, Univ Calif, 52; from assoc geneticist to geneticist, Brookhaven Nat Lab, 52-61; PROF BOT, UNIV ILL, URBANA, 61- Concurrent Pos: USPHS spec fel, Naples, Italy, 67-68. Mem: Genetics Soc Am; Am Soc Cell Biol; Soc Develop Biol. Res: Chromosome structure, genetics and cell biology; isotopic chemistry and biochemistry of nucleus and related structures. Mailing Add: Dept of Bot Univ of Ill Urbana IL 61801

STEFFEY, ORAN DEAN, b Billings, Okla, May 19, 21; m 43; c 2. INDUSTRIAL HYGIENE. Educ: Phillips Univ, AB, 48; Okla State Univ, MS, 50, PhD(bot), 53. Prof Exp: Asst zool, Phillips Univ, 42, asst bot & zool, 47-48; res assoc, Res Found, 50-53, from instr to asst prof bot & plant path, 53-56; radiol health physics officer, Radiation Lab, 56-72, INDUSTRIAL HYGIENIST, MED DEPT, CONTINENTAL OIL CO, 72- Concurrent Pos: Mem, Okla State Radiation Adv Comt, 72- Mem: Health Physics Soc; Sigma Xi; Am Indust Hygiene Asn. Res: Variations of the fruits of Quercus Macrocarpa; development of an autoradiographic technique for use in botanical investigations; cyto-morphogenetic studies in sorghum. Mailing Add: 2017 N Sixth Ponca City OK 74601

STEFFGEN, FREDERICK WILLIAMS, b San Diego, Calif, Nov 21, 26; m 49; c 2. FUEL SCIENCE, SURFACE CHEMISTRY. Educ: Stanford Univ, BS, 47; Northwestern Univ, Evanston, MS, 49; Univ Del, PhD(org chem), 53. Prof Exp: Res chemist, Cutter Labs, Calif, 49-50; res chemist, Esso Res & Eng Co, Standard Oil Co (NJ), La, 52-56; sr res chemist, Union Oil Res, Union Oil Co Calif, 56-60; res assoc, Richfield Res Ctr, Atlantic Richfield Co, 60-69; mgr process develop, Antox, Inc, WVa, 70-71; supvr res chemist, Pittsburgh Energy Res Ctr, US Bur Mines, 71, RES SUPVR CHEM, PITTSBURGH ENERGY RES CTR, US ENERGY RES & DEVELOP ADMIN, 75- Mem: Am Chem Soc; Am Inst Chem Engr; Catalysis Soc. Res: Heterogeneous catalysis, processes and catalyst preparation; conversions to produce low sulfur, clean fuels from coal, petroleum and organic wastes. Mailing Add: Pittsburgh Energy Res Ctr USERDA 4800 Forbes Ave Pittsburgh PA 15213

STEFKO, PAUL LOWELL, b Middletown, Conn, June 29, 15; m 40; c 1. EXPERIMENTAL SURGERY, PHARMACOLOGY. Educ: Univ Va, BS, 44. Prof Exp: Tech surg asst, NY Hosp, New York, 38-43; exp surgeon & pharmacologist, 47-61, SR PHARMACOLOGIST & EXP SURGEON, HOFFMANN-LA ROCHE, INC, 61- Concurrent Pos: Exp surg consult, USAEC & Off Naval Res & Develop, 40-43, cert, 45. Mem: AAAS; Sigma Xi; Am Indust Hyg Asn; Am Asn Lab Animal Sci; NY Acad Sci. Res: Vascular surgery; blood vessel transposition; gastroenterology; spasmolytics; antitussives. Mailing Add: Hoffmann-La Roche Inc Nutley NJ 07110

STEFL, EUGENE PAUL, organic chemistry, see 12th edition

STEGELMANN, ERICH J, b Lütjenburg, Ger, Mar 28, 14; m 51; c 3. ELECTROOPTICS. Educ: Tech Univ, Berlin, Dipl Ing, 38, Dr Ing, 39. Prof Exp: Develop engr, Rheinmetall-Borsig, Ger, 44-45; scientist, Technisches Büro 11 der USSR, Berlin, 46-48; sr res engr, Can Aviation Electronics, Montreal, 52-54; design & res specialist, Lockheed Calif Corp, 54-60, sr res scientist, 60-63, mem tech staff, Northrop Space Lab, 63-64; sr tech specialist, Space Div, NAm Aviation, Inc, Calif, 64-68, mem tech staff, Autonetics Div, NAm Rockwell Corp, 68-70; MEM TECH STAFF, ELECTRO-OPTICAL LABS, HUGHES AIRCRAFT CO, CULVER CITY, 72- Mem: Optical Soc Am. Res: Space physics; propagation of light in atmosphere. Mailing Add: 18559 Chatsworth St Northridge CA 91324

STEGEN, GILBERT ROLLAND, b Long Beach, Calif, Aug 19, 39; m 66. FLUID MECHANICS. Educ: Mass Inst Technol, BS, 61; Stanford Univ, MS, 62, PhD(aeronaut, astronaut), 67. Prof Exp: Develop engr, United Tech Ctr, United Aircraft Corp, 62-63; vis sr res fel, Col Aeronaut Eng, 67; asst res engr, Univ Calif, San Diego, 67-69; asst prof civil eng, Colo State Univ, 69-70; asst prof geol & geophys sci, Princeton Univ, 70-74; MEM STAFF, FLOW RES INC, 74- Mem: Am Phys Soc; Am Geophys Union. Res: Experimental fluid mechanics; experimental studies in atmospheric and oceanic turbulence. Mailing Add: Flow Res Inc Suite 72 1819 S Central Ave Kent WA 98031

STEGENGA, DAVID ALLAN, b Chicago, Ill, Aug 20, 46; m 71. MATHEMATICAL ANALYSIS. Educ: Purdue Univ, BS, 68; Univ Wis, MA, 71, PhD(math), 73. Prof Exp: Teaching asst math, Univ Wis, 68-73; ASST PROF MATH, IND UNIV, 73- Concurrent Pos: Vis mem, Inst Advan Study, 73-74. Mem: Am Math Soc. Res: The field of Fourier analysis, particularly the interplay between function theoretic and functional analytic methods of mathematical analysis. Mailing Add: Dept of Math Ind Univ Bloomington IN 47401

STEGER, THEODORE ROOSEVELT, JR, b Jacksonville, Fla, May 17, 46; m 70. POLYMER PHYSICS. Educ: Univ Fla, BS, 67; Mass Inst Technol, PhD(physics), 74. Prof Exp: SR RES PHYSICIST POLYMER PHYSICS, MONSANTO CO, 74- Mem: Am Phys Soc; Am Crystallog Asn. Res: Physical properties of solids and fluids. Mailing Add: Monsanto Co 800 N Lindergh Blvd St Louis MO 63166

STEGINK, LEWIS D, b Holland, Mich, Feb 8, 37; m 62; c 2. BIOLOGICAL CHEMISTRY. Educ: Hope Col, BA, 58; Univ Mich, MS & PhD(biol chem), 63. Prof Exp: Fel biochem, 63-65, asst prof pediat, 65-71, asst prof pediat & biochem, 68-71, ASSOC PROF PEDIAT & BIOCHEM, UNIV IOWA, 71- Mem: AAAS; Am Chem Soc; Am Inst Nutrit; Am Soc Biol Chemists; NY Acad Sci. Res: Biochemistry of normal and abnormal growth and development; amino acids; parental nutrition; acetylated proteins. Mailing Add: Dept of Pediat Univ of Iowa Iowa City IA 52240

STEGNER, ROBERT W, b Fairmount, NDak, Aug 21, 14; m 41; c 4. BIOLOGY. Educ: NDak State Univ, BS, 36; Univ Colo, MEd, 47, MS, 60, PhD(biol), 62. Prof Exp: Teacher, Pub Schs, Minn, 36-42 & Colo, 47-61; from instr to asst prof biol, Univ Colo, 61-63; from asst prof to assoc prof biol & educ, 63-71, PROF BIOL & EDUC & CO-DIR POP CURRICULUM STUDY, UNIV DEL, 71-, DIR, MARINE ENVIRON CURRICULUM STUDY, 75- Concurrent Pos: Univ Res Found grant, Univ Del, 65-66; partic, NSF Inst Biol for Teachers, 66-67; consult on course of study in biol of human pop, UNESCO; grants, du Pont, Christiana Found, US Off Educ, C S May, Pop Coun & Nat Oceanic & Atmospheric Admin. Honors & Awards: Lindback Award, 67. Mem: AAAS; Am Inst Biol Sci; Nat Sci Teachers Asn; Nat Asn Biol Teachers; Sigma Xi. Res: Physiological characteristics of albino maize; preparation of science teachers; development of learning experiences in population-environment studies. Mailing Add: Hall Bldg Univ of Del Newark DE 19711

STEGUN, IRENE ANNE, b Yonkers, NY, Feb 9, 19. NUMERICAL ANALYSIS. Educ: Col Mt St Vincent, BA, 40; Columbia Univ, MA, 41. Prof Exp: MATHEMATICIAN, NAT BUR STAND, 43 Honors & Awards: Gold Medal, Nat Bur Stand, 65. Mem: Am Math Soc; Soc Indust & Appl Math; Asn Comput Mach. Res: Automatic computation. Mailing Add: Nat Bur of Stand Washington DC 20234

STEHBENS, WILLIAM ELLIS, b Australia. PATHOLOGY. Educ: Univ Sydney, MB & BS, 50, MD, 62; Oxford Univ, PhD(path), 60; FRCPath(Australia), 61; FRCPath(E), 63; FRACP. Prof Exp: From lectr to sr lectr path, Univ Sydney, 53-62; from assoc prof to prof, Wash Univ, 66-68; prof path, Albany Med Col, 68-74; PROF PATH, WELLINGTON CLIN SCH, UNIV OTAGO, NZ, 74- Concurrent Pos: Teaching fel path, Univ Sydney, 52; sr res fel, Australian Nat Univ, 62-66, sr fel, 66; pathologist-in-chief & dir dept path & lab med, Jewish Hosp St Louis, 66-68; dir electron micros unit, Vet Admin Hosp, Albany, 68-74. Honors & Awards: R T Hall Res Prize, Cardiac Soc Australia & NZ, 66. Res: Relationship of intimal thickening, thrombosis and hemodynamics to the pathogenesis of atherosclerosis and cerebral aneurysms. Mailing Add: Dept of Path Wellington Clin Sch Wellington New Zealand

STEHL, RUDOLPH HERMAN, analytical chemistry, see 12th edition

STEHLE, PHILIP MCLELLAN, b Philadelphia, Pa, Mar 3, 19; m 42; c 3. QUANTUM ELECTRONICS. Educ: Univ Mich, AB, 40, AM, 41; Princeton Univ, PhD(physics), 44. Prof Exp: Asst, Univ Mich, 40-41; instr physics, Princeton Univ, 41-44; instr, Harvard Univ, 46-47; asst prof, 47, PROF PHYSICS, UNIV PITTSBURGH, 47- Mem: Am Phys Soc. Res: Quantum theory; quantum optics. Mailing Add: Dept of Physics Univ of Pittsburgh Pittsburgh PA 15260

STEHLE, RANDALL GEMMER, physical pharmacy, see 12th edition

STEHLI, FRANCIS GREENOUGH, b Montclair, NJ, Oct 16, 24; m 48; c 4. GEOLOGY. Educ: St Lawrence Univ, BS, 49, MS, 50; Columbia Univ, PhD(geol), 53. Prof Exp: Asst geol, St Lawrence Univ, 49-50; asst, Columbia Univ, 50-52; asst prof invert paleont, Calif Inst Technol, 53-56; res engr res ctr, Pan Am Petrol Corp, 57, tech group supvr, 58-60; chmn dept, 61-73, prof geol, 60-74, SAMUEL ST JOHN PROF EARTH SCI, CASE WESTERN RESERVE UNIV, 74-, ACTG DEAN SCI, 75- Concurrent Pos: Mem, NSF Earth Sci Adv Panel, 67-69, chmn, 69. Mem: Paleont Soc; Geol Soc Am; Geochem Soc. Res: Paleozoic brachiopods; Mesozoic stratigraphy of Gulf Coast and western Mexico; paleocology; carbonate rock formation and diagenesis; continental drift, polar wondering and paleoclimatology. Mailing Add: Dept of Earth Sci Case Western Reserve Univ Cleveland OH 44106

STEHLY, DAVID NORVIN, b Bethlehem, Pa, Oct 3, 33; m 61; c 2. INORGANIC CHEMISTRY. Educ: Moravian Col, BS, 59; Lehigh Univ, MS, 62, PhD(inorg chem), 67. Prof Exp: Instr chem, 60-67, from asst prof to assoc prof, 67-75, PROF CHEM, MUHLENBERG COL, 75- Mem: AAAS; Am Chem Soc. Res: Metal chelates of substitued amides; trace metal ions in surface waters. Mailing Add: Dept of Chem Muhlenberg Col Allentown PA 18104

STEHNEY, ANDREW FRANK, b Chicago, Ill, May 4, 20; m 43; c 3. RADIOCHEMISTRY, RADIOBIOLOGY. Educ: Univ Chicago, BS, 42, PhD(chem), 50. Prof Exp: Instr chem, Univ Chicago, 49-50; assoc chemist, 50-71, SR CHEMIST, ARGONNE NAT LAB, 71- Concurrent Pos: Lectr, Univ Chicago, 56-61; vis scientist, Europ Orgn Nuclear Res, 61-63; mem subcomt radiochem, Nat Res Coun, 69- Mem: AAAS; Am Chem Soc; Am Phys Soc. Res: Toxicity of radium and other internal emitters; applications of radiochemistry to environmental studies; nuclear reactions. Mailing Add: Radiol & Environ Res Div Argonne Nat Lab Argonne IL 60439

STEHNEY, ANN KATHRYN, b Oak Ridge, Tenn, June 20, 46. PURE MATHEMATICS. Educ: Bryn Mawr Col, AB, 67; State Univ NY, Stony Brook, MA, 69, PhD(math), 71. Prof Exp: ASST PROF MATH, WELLESLEY COL, 71- Concurrent Pos: Vis scholar, Enrico Fermi Inst, Univ Chicago, 74-75. Mem: Am Math Soc; Math Asn Am; Asn Women Math. Res: Differential geometry. Mailing Add: Dept of Math Wellesley Col Wellesley MA 02181

STEHOUWER, DAVID MARK, b Grand Rapids, Mich, July 14, 43; m 68; c 2. ORGANIC CHEMISTRY. Educ: Hope Col, BA, 65; Univ Mich, Ann Arbor, MS, 67, PhD(org chem), 70. Prof Exp: SR CHEMIST RES LABS, TEXACO INC, 70- Mem: Am Chem Soc. Res: Lubricant additive chemistry; mechanisms of lubricant performance; new lubricant technology; corrosion mechanisms and inhibition. Mailing Add: Texaco Res Labs PO Box 509 Beacon NY 12508

STEHR, FREDERICK WILLIAM, b Athens, Ohio, Dec 23, 32; m 59; c 2. ENTOMOLOGY. Educ: Univ Ohio, BS, 54; Univ Minn, MS, 58, PhD(entom), 64. Prof Exp: Res fel entom, Univ Minn, 62-65; asst prof, 65-71, ASSOC PROF ENTOM, MICH STATE UNIV, 71- Honors & Awards: Karl Jordan Medal, Lepidop Soc, 74. Mem: Entom Soc Am; Soc Syst Zool; Lepidop Soc; Ecol Soc Am; Entom Soc Can. Res: Systematics of Lepidoptera and immature insects; biological control; ecology. Mailing Add: Dept of Entom Mich State Univ East Lansing MI 48823

STEHSEL, MELVIN LOUIS, b Long Beach, Calif, Oct 3, 24; m 57; c 2. ORGANIC CHEMISTRY, PLANT PHYSIOLOGY. Educ: Univ Calif, Berkeley, BS, 45, MS, 47, PhD(plant physiol), 50. Prof Exp: Off Naval Res fel biochem genetics, 50-51; French Govt fel, 51; res chemist, Socony Mobil Oil Co, 51-56; chief phys lab, Aerojet-Gen Corp, 56-65; assoc prof biol, 65-72, PROF BIOL, PASADENA CITY COL, 72- Mem: AAAS; Bot Soc Am; Am Chem Soc. Res: High temperature mechanical properties of graphite for nuclear rockets; action of natural and synthetic plant growth hormones; stimulation and cause of cell differentiation in plant tissue culture; plant biochemistry. Mailing Add: 1570 E Colorado Blvd Pasadena City Col Pasadena CA 91106

STEIB, MICHAEL LEE, b May 2, 41; US citizen; m 66. MATHEMATICAL ANALYSIS. Educ: Univ Tex, BA, 63, MA, 65, PhD(math), 66. Prof Exp: ASST PROF MATH, UNIV HOUSTON, 66- Mem: Am Math Soc. Res: Integration theory; questions concerning conditions which guarantee integrability. Mailing Add: 3717 Nottingham Houston TX 77005

STEIB, RENE J, b Vacherie, La, July 30, 18; m 44; c 2. PLANT PATHOLOGY. Educ: La State Univ, PhD(sugarcane dis), 49. Prof Exp: Asst gen mgr, Am Sugarcane League, La, 49-51; from asst prof to assoc prof, 51-64, PROF PLANT PATH, AGR EXP STA, LA STATE UNIV, BATON ROUGE, 64- Concurrent Pos: Consult to govt, indust & orgn, 64- Mem: Am Phytopath Soc; Am Soc Sugarcane Technologists; Inst Soc Sugarcane Technologists. Res: Fungal and virus diseases of sugarcane, especially in Louisiana; testing of new varieties for disease resistance. Mailing Add: Dept of Plant Path La State Univ Baton Rouge LA 70803

STEICHEN, RICHARD JOHN, b Wichita, Kans, July 1, 44; m 68; c 2. ANALYTICAL CHEMISTRY. Educ: Rockhurst Col, BA, 66; Univ Kans, PhD(anal chem), 71. Prof Exp: Asst prof chem, Rockhurst Col, 71-73; SR RES CHEMIST, GOODYEAR TIRE & RUBBER CO, 73- Mem: Am Chem Soc. Res: Analytical methods development for industrial applications including process control, Occupational Safety and Health Administration and Food and Drug Administration compliance. Mailing Add: 142 Goodyear Blvd Akron OH 44316

STEIGER, FRED HAROLD, b Cleveland, Ohio, May 11, 29; m 52; c 2. APPLIED CHEMISTRY, COSMETIC CHEMISTRY. Educ: Univ Pa, BA, 51; Temple Univ, MA, 56. Prof Exp: Res chemist, Rohm and Haas Co, Pa, 51-60; group leader sanit protection, 60-62, sr res chemist, 62-68, SR RES SCIENTIST, PERSONAL PROD CO DIV, JOHNSON & JOHNSON, 68- Concurrent Pos: Abstractor, Chem Abstr, 53-68, ed textile sect, 68-; vpres prof affairs, NJ Inst Chemists, 75- Mem: Am Chem Soc; Am Asn Textile Chem & Colorists; fel Am Inst Chem; Am Soc Test & Mat. Res: Textile chemistry; fibers and polymers; sanitary protection; absorption of liquids. Mailing Add: Res & Develop Div Personal Prod Co Milltown NJ 08850

STEIGER, LEONARD WILLIAM, b Jersey City, NJ, Jan 16, 98; m 21; c 1. PHARMACEUTICAL CHEMISTRY. Educ: Columbia Univ, PhG, 17. Prof Exp: Chemist, Lederle Antitoxin Labs, Am Cyanamid Co, 17-18; instr chem col pharm, Columbia Univ, 19-20; chemist, Maywood Chem Works Div, 20-48, dir res, 48-59, tech coordr, 59-63, CONSULT, STEPAN CHEM CO, 63- Mem: Am Chem Soc; Am Pharmaceut Asn; NY Acad Sci. Res: Electrochemistry; lithium products; syndets. Mailing Add: RR 2 Green Pond Newfoundland NJ 07435

STEIGER, ROGER ARTHUR, b Potosi, Wis, Dec 29, 39; m 67; c 3. HIGH TEMPERATURE CHEMISTRY. Educ: Wis State Univ-Platteville, BS, 61; Univ Iowa, PhD(phys chem), 67. Prof Exp: SR RES CHEMIST INDUST CHEM DIV, PPG INDUSTS INC, 67- Mem: AAAS; Am Chem Soc; Am Ceramic Soc. Res: Thermodynamics of refractory compounds; plasma chemical reactions; powder metallurgy of fine-grained cemented carbides; preparation and fabrication of sinterable submicron ceramic powders. Mailing Add: Indust Chem Div PPG Indust Inc PO Box 31 Barberton OH 44203

STEIGER, RONALD PAUL, chemistry, see 12th edition

STEIGER, WALTER RICHARD, b Colo, Sept 4, 23; m 46; c 2. PHYSICS. Educ: Mass Inst Technol, BS, 48; Univ Hawaii, MS, 50; Univ Cincinnati, PhD(physics), 53. Prof Exp: Asst physics, Univ Hawaii, 48-50; instr, Univ Cincinnati, 51-52; from asst prof to assoc prof, 53-65, PROF PHYSICS, UNIV HAWAII, 65-, CHMN DEPT PHYSICS & ASTRON, 72- Concurrent Pos: Vis researcher high altitude observ, Univ Colo, 59-60; Fulbright res scholar, Tokyo Astron Observ, 66-67. Mem: AAAS; Am Phys Soc; Am Asn Physics Teachers; Am Geophys Union; Int Astron Union. Res: Upper atmosphere physics, ionosphere; airglow. Mailing Add: Dept of Physics & Astron Univ of Hawaii Honolulu HI 96822

STEIGERT, FREDERICK EDWARD, b New York, NY, Sept 11, 28; m 50; c 4. NUCLEAR PHYSICS. Educ: Union Col, BS, 49; Univ Ind, MA, 50, PhD(physics), 53. Prof Exp: Asst metal res lab, Gen Elec Co, 49; asst physics, Univ Ind, 49-53; instr, Yale Univ, 53-56, asst prof, 56-62; ASSOC PROF, UNIV CONN, 62- Mem: Am Phys Soc. Res: Energy levels of light nuclei; elastic and ineleastic scattering from light nuclei; heavy ion reactions. Mailing Add: Dept of Physics Univ of Conn Storrs CT 06268

STEIGERWALD, BERNARD J, environmental health, see 12th edition

STEIGMAN, ALEX J, b Philadelphia, Pa, July 23, 16; m 43. PEDIATRICS. Educ: Temple Univ, BS, 34, MD, 38, MSc, 47. Hon Degrees: ScD, Temple Univ, 56. Prof Exp: Intern, Temple Univ, 38-40; asst instr pediat, Univ Pa, 40-41; instr pediat & prev med, Harvard Univ, 42-43; asst prof pediat, Temple Univ, 46-47; consult clin epidemiol, Nat Found Infantile Paralysis, Inc, 49-50; prof child health, Sch Med, Univ Louisville, 50-62, chmn dept pediat, 54-62; planning dir, Joint Pediat Res Educ Pract Comt, Evanston, Ill, 62-63; vis prof pediat, Univ Chicago, 63-64 & Rockefeller Found Spec Staff for India, 64-66; PROF PEDIAT, MT SINAI SCH MED, 66- Concurrent Pos: Fel, Harvard Univ, 41-42; Nat Res Coun sr fel, Children's Hosp Res Found, Ohio, 47-49. Mem: Infectious Dis Soc Am; Soc Pediat Res; Am Pediat Soc; Soc Exp Biol & Med; Am Acad Pediat. Res: Infectious disease; medical education. Mailing Add: Mt Sinai Sch of Med 11 E 100th St New York NY 10029

STEIGMAN, GARY, b New York, NY, Feb 23, 41; m 71. ASTROPHYSICS. Educ: City Col New York, BS, 61; NY Univ, MS, 63, PhD(physics), 68. Prof Exp: Vis fel, Inst Theoret Astron, Cambridge, Eng, 68-70; res fel, Calif Inst Technol, 70-72; ASST PROF ASTRON, YALE UNIV, 72- Concurrent Pos: Vis scientist, Nat Radio Astron Observ, 75. Res: Theoretical and atomic astrophysics; cosmology. Mailing Add: Dept of Astron Yale Univ New Haven CT 06520

STEIGMAN, JOSEPH, chemistry, see 12th edition

STEIGMANN, FREDERICK, b Austria, Apr 25, 05; nat US; m 37; c 3. MEDICINE. Educ: Univ Ill, BS, 28, MD, 30, MS, 38. Prof Exp: Asst, 33-34, from instr to assoc prof, 34-72, CLIN PROF MED, UNIV ILL COL MED, 72- Concurrent Pos: Pvt pract, 33-; dir depts therapeut & gastroenterol & attend physician, Cook County Hosp, 40- Mem: Fel Soc Exp Biol & Med; fel Am Soc Pharmacol & Exp Therapeut; fel Am Fedn Clin Res; fel Am Col Physicians; Am Gastroenterol Asn. Res: Gastroenterology; liver; vitamin A; protein metabolism. Mailing Add: 30 N Michigan Ave Chicago IL 60602

STEILA, DONALD, b Cleveland, Ohio, Sept 26, 39; m 68; c 1. CLIMATOLOGY. Educ: Kent State Univ, BS, 65, MA, 66; Univ Ga, PhD(geog), 71. Prof Exp: Instr geog, Univ Ga, 70-71; asst prof, Univ Ariz, 71-72; asst prof, 72-75, ASSOC PROF GEOG, E CAROLINA UNIV, 75- Mem: Asn Am Geogr; Am Meteorol Soc; Soil Sci Soc Am; Am Soc Agron. Res: Quantitative identification of drought intensity; the impact of drought upon vegetation and the temporal and spatial patterns of drought occurrence. Mailing Add: Dept of Geog ECarolina Univ Greenville NC 27834

STEIMAN, HENRY ROBERT, b Winnipeg, Man, Aug 2, 38; m 73; c 1. PHYSIOLOGY, DENTISTRY. Educ: NDak State Univ, BS, 64; Wayne State Univ, MS, 67, PhD(physiol), 69; Univ Detroit, DDS, 73. Prof Exp: Asst prof, 69-73, ASSOC PROF PHYSIOL & DIR DIV BIOL SCI, DENT SCH, UNIV DETROIT, 73-, CHMN DEPT PHYSIOL, 70- Concurrent Pos: Mich Asn Regional Med Progs grant, Dent Sch, Univ Detroit, 74-; consult, Hypertension Coordinating & Planning Comt, Southeastern Mich, 74- Mem: Am Dent Asn; Am Physiol Soc. Res: Hypertension screening and detection; role of dental personnel. Mailing Add: Dept of Physiol Univ of Detroit Sch of Dent Detroit MI 48207

STEIN, ABRAHAM MORTON, b Chicago, Ill, Aug 9, 23; m 43; c 1. BIOCHEMISTRY. Educ: Univ Calif, Los Angeles, AB, 49, MA, 51; Univ Southern Calif, PhD(biochem), 57. Prof Exp: NIH res fel biochem, Brandeis Univ, 57-59; res assoc, Univ Pa, 59-65, sr res investr, 65-67; assoc prof col med, Univ Fla, 67-71; chmn dept biol sci, 71-75, PROF BIOL SCI, FLA INT UNIV, 71- Concurrent Pos: Adj prof col med, Univ Miami, 71- Mem: AAAS; Am Soc Biol Chem; Am Chem Soc; NY Acad Sci. Res: Enzymology; flavoproteins; metabolic control; pyridine nucleotides; porphyrin synthesis; developmental biochemistry. Mailing Add: Dept of Biol Sci Fla Int Univ Miami FL 33199

STEIN, ALLAN RUDOLPH, b Edmonton, Alta, Nov 14, 38. PHYSICAL ORGANIC CHEMISTRY. Educ: Univ Alta, BSc, 60; Univ Ill, Urbana, PhD(org chem), 64. Prof Exp: Asst org chem, Univ Ill, Urbana, 60-61; res scientist, Domtar Cent Res Labs, Senneville, Que, 64-65; from asst prof to assoc prof chem, 65-75, PROF CHEM, MEM UNIV, NFLD, 75- Concurrent Pos: Nat Res Coun Can res grants, 65-69 & 75-78; vis prof, King's Col, Univ London & Univ Umea, Sweden, 72-73; res grant, Swedish Res Coun, 73. Mem: Am Chem Soc; Brit Chem Soc; Chem Inst Can. Res: Organic reaction mechanism studies, especially reactions of ambident ions, the isonitriles and of phenol alkylation; organic polarographic and carbene chemistry studies; ion-pair mechanism of nucleophilic displacement. Mailing Add: Dept of Chem Mem Univ St John's NF Can

STEIN, ALVIN, b Brooklyn, NY, Nov 24, 25; m 75; c 2. POLYMER CHEMISTRY. Educ: City Col New York, BS, 50; Brooklyn Polytech Inst, PhD(org chem), 55. Prof Exp: Chemist adhesives, Manhattan Paste & Glue Co, 49-52; group leader polymers, Monsanto Co, 55-65; res dir chem, Springdale Labs, Time Inc, 65-70; ASST VPRES CHEM, PHILIP A HUNT CHEM CO, 70 Mem: Am Chem Soc. Res: Photopolymers; dispersion polymerization; polymer characterization. Mailing Add: 10 Ridgewood Rd Barrington RI 02806

STEIN, ARTHUR, b New York, NY, Aug 5, 18; m 41; c 2. MATHEMATICAL STATISTICS. Educ: City Col New York, BS, 38; Columbia Univ, MA, 44. Prof Exp: Statistician, Ballistic Res Labs, Aberdeen Proving Ground, US Dept Army, 41-44, mathematician, Ord Ballistic Team, 44-46, supvry ballistician, 46-51, dir qual control eng sect, Ord Ammunition Command, 52-55; prin engr & asst head, Opers Res Dept, 55-65, head, Systs Res Dept, 65-70, ASSOC DIR, APPL TECHNOL GROUP, CORNELL AERONAUT LAB, INC, 70- Concurrent Pos: Lectr, Univ Chicago, 53-55; lectr, Univ Buffalo, 56. Mem: Opers Res Soc; Inst Math Statist; Am Statist Asn; Am Soc Qual Control; Am Ord Asn. Res: Operations research; weapons effectiveness; aircraft vulnerability; terminal ballistics; sample inspection and quality control; environmental systems analysis; reliability; probability; arms control; applied mathematics; mathematical modeling. Mailing Add: 30 Chapel Woods Williamsville NY 14221

STEIN, ARTHUR A, b Toronto, Ont, Mar 12, 22; nat US; m 48; c 3. MEDICINE, PATHOLOGY. Educ: Univ Toronto, MD, 45; Am Bd Path, dipl, 52. Prof Exp: Intern, Victoria Hosp, London, Ont, 45-46; asst resident med, Jewish Hosp, St Louis, Mo, 47; asst resident path, City Hosp, 47-48; resident, Sch Trop Med, Univ PR, 48-49; instr path & bact, 49-52, from asst prof to assoc prof path, 52-59, dir res inst exp path & toxicol, 65-69, PROF PATH, ALBANY MED COL, 59- Concurrent Pos: Surg pathologist, Albany Hosp, 55-; sci adv, Ky Tobacco & Health Bd, Commonwealth of Ky, 71-; dir Micros Biol Res, Inc, 67- Mem: Am Soc Clin Path; AMA; Col Am Path; Am Acad Clin Toxicol; Soc Toxicol. Res: Toxicology. Mailing Add: Dept of Path Albany Med Col Albany NY 12201

STEIN, ARTHUR HENRY, JR, b St Louis, Mo, June 11, 24; m 46; c 1. ORTHOPEDIC SURGERY. Educ: Amherst Col, AB, 46; Wash Univ, MD, 48; Am Bd Orthop Surg, dipl. Prof Exp: Resident orthop surg, Barnes Hosp, 53-55, asst, 55-56; from instr to assoc prof, 56-72, PROF ORTHOP SURG, SCH MED, WASH UNIV, 72- Concurrent Pos: Fel orthop surg, Barnes Hosp, 50-51. Mem: AMA; Clin Orthop Soc; Am Orthop Asn; Asn Orthop Chmn; Am Col Surg. Res: Blood supply to the bones; clinical orthopedic problems. Mailing Add: 4989 Barnes Hosp Plaza Queeny Towers St Louis MO 63110

STEIN, BARRY EDWARD, b New York, NY, Dec 3, 44; m 68. NEUROPHYSIOLOGY, DEVELOPMENTAL PHYSIOLOGY. Educ: Queens Col, BA, 66, MA, 69; City Univ New York, PhD(neuropsychol), 70. Prof Exp: Fel neurophysiol & neuroanat, Univ Calif, Los Angeles, 70-72, asst res anatomist, 72-75; ASST PROF PHYSIOL, MED COL VA, VA COMMONWEALTH UNIV, 75- Mem: Sigma Xi; Int Brain Res Orgn; AAAS; Am Psychol Asn; Soc Neurosci. Res: The ontogenesis of sensory systems; neurophysiological, neuroanatomical and behavioral changes during early life. Mailing Add: Dept of Physiol Med Col Va Commonwealth Univ Richmond VA 23298

STEIN, BARRY FRED, b Philadelphia, Pa, Nov 2, 37; m 62; c 3. SOLID STATE PHYSICS. Educ: Univ Pa, BA, 59, MS, 61, PhD(physics), 65. Prof Exp: PRIN PHYSICIST, UNIVAC DIV, SPERRY RAND CORP, 65- Mem: Am Phys Soc; Am Vacuum Soc. Res: Magnetic susceptibility and nuclear magnetic resonance in transition metal monophosphides; electrical properties of gallium arsenide; chemical vapor deposition and magnetic and optical properties of gadolinium iron garnet; liquid phase epitaxial growth of garnets used as bubble memory elements. Mailing Add: Univac Div Sperry Rand Corp Box 500 Blue Bell PA 19422

STEIN, BENNETT M, b New York, NY, Feb 2, 31; m 55; c 2. NEUROSURGERY. Educ: Dartmouth Col, BA, 52; McGill Univ, MD, 55; Am Bd Neurol Surg, cert, 66. Prof Exp: Intern, US Naval Hosp, St Albans, NY, 58-59; surg resident, Columbia Presby Hosp, New York, 59-60, from asst resident to chief resident neurosurg, 60-64; asst prof, Neurol Inst, Columbia Univ, 68-71; PROF NEUROSURG & CHMN DEPT, NEW ENG MED CTR, TUFTS UNIV, 71- Concurrent Pos: Fulbright scholar neurol, Nat Inst, Queens Sq, London, Eng, 58-59; NIH spec fel neuroanat, Columbia

Univ, 64-66; consult, US Naval Hosp, Chelsea, Lemuel Shattuck Hosp, Boston & Vet Admin Hosp, Boston, 71- Mem: Fel Am Col Surg; Am Asn Anat; Soc Neurol Surg; Cong Neurol Surg; Am Acad Neurol Surg. Res: Cerebrovascular reactions, specifically cerebrovasospasm in response to subarachnoid hemorrhage; neuroanatomical problems. Mailing Add: 171 Harrison Ave Boston MA 02111

STEIN, BERNARD REUBIN, science policy, physical chemistry, see 12th edition

STEIN, CHARLES W C, b Philadelphia, Pa, Apr 28, 14; m 38. ORGANIC CHEMISTRY. Educ: Univ Pa, BS, 36, MS, 39, PhD(org chem), 42. Prof Exp: Asst instr chem, Lehigh Univ, 37; asst instr, Drexel Inst, 37-42; res chemist, Gen Aniline & Film Corp, 42-46; process develop chemist, Calco Div, Am Cyanamid Co, 46-51; TECH ASSOC, GAF CORP, 51- Mem: AAAS; Am Chem Soc. Res: Dyestuff chemistry; pigments; organic synthesis; plastics. Mailing Add: 910 Summit Ave Westfield NJ 07090

STEIN, ELIAS M, b Antwerp, Belg, Jan 13, 31; nat US; m 59; c 2. MATHEMATICS. Educ: Univ Chicago, AB, 51, MS, 53, PhD, 55. Prof Exp: Instr, Mass Inst Technol, 56-58; mem fac, Univ Chicago, 58-62, assoc prof math, 61-62; mem, Inst Advan Study, 62-63; chmn dept, 68-71, PROF MATH, PRINCETON UNIV, 63- Concurrent Pos: Sloan Found res fel, 61-63; NSF sr fel, 62-63 & 71-72; sr vis fel, Sci Res Coun Gt Brit, 68. Mem: Nat Acad Sci; Am Math Soc. Res: Topics in harmonic analysis related to the Littlewood-Paley theory; singular integrals and differentiality properties of functions. Mailing Add: 132 Dodds Lane Princeton NJ 08540

STEIN, FRANK S, b Lancaster, Pa, Jan 11, 21; m 47; c 3. PHYSICS. Educ: Franklin & Marshall Col, BS, 42; Columbia Univ, MA, 47; Univ Buffalo, PhD(physics), 51. Prof Exp: Res physicist, Manhattan Proj, Columbia Univ, 44-46; instr physics, Univ Buffalo, 48-51; res physicist res labs, Westinghouse Elec Corp, 51-55, mgr semiconductor dept electronic tube div, 55, mgr power devices develop sect semiconductor dept, 55-60; mgr eng semiconductor div, Gen Instrument Corp, NJ, 60-61, dir res appl res lab, 61-63; sr scientist semiconductor dept, Delco Radio Div, 63-66, mgr semiconductor res & eng dept, 66-70, chief engr solid state prod, Delco Electronics Div, 70-75, CHIEF ENGR ADVAN ENG, DELCO ELECTRONICS DIV, GEN MOTORS CORP, 75- Concurrent Pos: Chmn joint electron device eng coun, Electronic Industs Asn & Nat Electronics Mfrs Asn Semiconductor Device Coun, 69-70 & 73-75; mem, Nat Acad Sci/Nat Acad Eng/Nat Res Coun Eval Panel Electronic Technol Div, Nat Bur Standards, 75- Mem: Inst Elec & Electronics Eng. Res: Mass spectrometry; properties of semiconductors and semiconductor devices; solid state electronics; microelectronics. Mailing Add: Delco Electronics Div Gen Motors Corp PO Box 1104 Kokomo IN 46901

STEIN, FREDERICK MAX, b Wyaconda, Mo, Feb 17, 19; m 43; c 2. MATHEMATICS. Educ: Iowa Wesleyan Univ, AB, 40; Univ Iowa, MS, 47, PhD(math), 55. Prof Exp: Instr high schs, Iowa, 40-43; instr math, Univ Iowa, 43-44, asst, 46-47; assoc prof, Iowa Wesleyan Univ, 47-53; instr math, Univ Iowa, 53-54; assoc prof, 55-63, PROF MATH, COLO STATE UNIV, 63- Mem: Am Math Soc; Soc Indust & Appl Math; Math Asn Am. Res: Approximation; orthogonal functions; differential and integro-differential equations; Sturm-Liouville systems. Mailing Add: Dept of Math Colo State Univ Ft Collins CO 80523

STEIN, GARY S, b Brooklyn, NY, July 30, 43; m 66. BIOCHEMISTRY, CELL BIOLOGY. Educ: Hofstra Univ, BA, 65, MA, 66; Univ Vt, PhD(cell biol), 69. Prof Exp: Res assoc biochem, Temple Univ, 71-72; asst prof, 72-75, ASSOC PROF BIOCHEM, SCH MED, UNIV FLA, 75- Concurrent Pos: NIH fel, Sch Med, Temple Univ, 69-71; Damon Runyon Mem Fund cancer res grant, Sch Med, Univ Fla, 71-; Am Cancer Soc, NSF & NIH grants, 74- Mem: AAAS; Am Soc Biol Chem; Am Asn Cancer Res; Am Soc Cell Biol; Am Soc Zool. Res: Role of chromosomal proteins in the control of cell division and gene expression. Mailing Add: Dept of Biochem Univ of Fla Sch of Med Gainesville FL 32601

STEIN, GEORGE J, medical microbiology, malariology, see 12th edition

STEIN, GEORGE NATHAN, b Philadelphia, Pa, Aug 11, 17; m 48; c 3. RADIOLOGY. Educ: Univ Pa, BA, 38; Jefferson Med Col, MD, 42; Am Bd Radiol, dipl, 49. Prof Exp: Intern, Jewish Hosp, Philadelphia, 43; resident radiol, Grad Hosp, 47-49, assoc radiologist, 49-51, from instr to assoc prof, Div Grad Med, 51-61, assoc dir dept, Ctr, 67-71, dir dept radiol, Presby-Univ Pa Med Ctr, 71, PROF RADIOL, SCH MED, UNIV OF PA, 61- Concurrent Pos: Hon prof, Pontif Univ Javeriana, Colombia, 60. Mem: Radiol Soc NAm; Am Roentgen Ray Soc; fel Am Col Radiol. Res: Gastrointestinal radiology. Mailing Add: Dept of Radiol Presby-Univ of Pa Med Ctr Philadelphia PA 19104

STEIN, GRETCHEN HERPEL, b Asbury Park, NJ, Mar 27, 45; m 66. CELL BIOLOGY. Educ: Brown Univ, AB, 65; Stanford Univ, PhD(molecular biol), 71. Prof Exp: NIH fel cell biol, Sch Med, Stanford Univ, 71-73, res fel cell biol & molecular biol, 73-74; RES ASSOC CELL BIOL & MOLECULAR BIOL, UNIV COLO, 74- Mem: Biophys Soc. Res: Study of regulation of cellular reproduction in normal and transformed human cells; role of DNA-binding proteins and nature of mechanisms controlling the initiation of DNA replication. Mailing Add: Molecular Cellular & Develop Biol Univ of Colo Boulder CO 80302

STEIN, GUSTAV ALBERT, b Mannheim, Ger, Sept 13, 99; nat US; m 29. PHARMACEUTICAL CHEMISTRY. Educ: Univ Heidelberg, PhD(chem), 24. Prof Exp: Factory chemist, Chem Factory Loerrach, Ger, 24-25; res chemist, Nat Bur Standards, 25-27; res chemist, Thiophene Co, Ill, 28-29; sr res chemist, Merck & Co, Inc, 29-64; RETIRED. Honors & Awards: Co-recipient Patent Award, NJ Coun Res & Develop, 71. Mem: Am Chem Soc. Res: Tetrazine dicarbonic acid; medicinal agents. Mailing Add: 29 D Cardinal Cedar Glen W Lakehurst NJ 08733

STEIN, HARVEY PHILIP, b Brooklyn, NY, May 4, 40; m 65; c 3. ORGANIC CHEMISTRY. Educ: Queens Col, NY, BS, 61; Mass Inst Technol, PhD(org chem), 67. Prof Exp: Chemist, Stamford Res Labs, Am Cyanamid Co, summer 61; asst, Mass Inst Technol, 61-65; instr chem, Pa State Univ, 65-67, asst prof, 67-68; asst prof, Trenton State Col, 68-71; prof, Western Col, 71-75; SR SCIENTIST, USPHS, 75- Mem: Am Chem Soc. Res: Reaction mechanisms; use of isotopes; chemical education; public health especially occupational cancer. Mailing Add: 419 College Pkway Pkwy Rockville MD 20850

STEIN, HERMAN H, b Chicago, Ill, May 27, 30; m 51; c 2. BIOCHEMISTRY, MEDICINAL CHEMISTRY. Educ: Univ Ill, BS, 51; Univ Minn, MS, 53; Northwestern Univ, PhD(chem), 56. Prof Exp: Lab asst, Northwestern Univ, 53-54; res chemist, Toni Co Div, Gillette Co, 56-61; sr res chemist, Abbott Labs, 61-66, group leader, 66, sect head, 67-72; ASSOC RES FEL, PHARMACOL DEPT, ABBOTT LABS, 72- Mem: Am Chem Soc; Am Soc Pharmacol & Exp Therapeut. Res: Enzymology; automated metabolic and enzymic analyses; neurochemistry; electroanalytical chemistry; antianginal agents; cyclic adenosine monophosphate metabolism; pharmacology of nucleosides; beta-adrenergic blocking agents. Mailing Add: Pharmacol Dept Abbott Labs 1400 Sheridan Rd North Chicago IL 60064

STEIN, HOWARD JAY, b Baltimore, Md, May 28, 33; m 57; c 5. PLANT PHYSIOLOGY. Educ: Temple Univ, BA, 54; Univ Mich, MA, 58, PhD(bot), 61. Prof Exp: From asst prof to assoc prof biol, Kans State Col, Pittsburgh, 60-65; assoc prof, 65-71, chmn dept, 66-68 & 71-75, PROF BIOL, GRAND VALLEY STATE COL, 71- Concurrent Pos: NSF grant, 62-64; staff biologist, Off Biol Educ, Am Inst Biol Sci, 69-71, vis biologist, 71-72, curric consult bur, 71-74. Mem: AAAS; Am Soc Plant Physiol; Bot Soc Am; Nat Asn Biol Teachers; Am Inst Biol Sci. Res: Amino acid metabolism in plant roots; metabolism in plant mitochondria. Mailing Add: Dept of Biol Grand Valley State Col Allendale MI 49401

STEIN, IRVIN, b Fayetteville, NC, Oct 17, 06; m; c 3. ORTHOPEDIC SURGERY. Educ: Univ NC, AB, 26; Jefferson Med Col, MD, 30. Prof Exp: Intern gen surg, Sinai Hosp, 30-31; resident orthop surg, Philadelphia Orthop Hosp, Johns Hopkins Hosp & Childrens Hosp Sch, Baltimore, Md, 31-34; asst chief, Orthop Hosp, 34-41; asst prof orthop surg, 41-67, ASSOC PROF ORTHOP SURG, SCH MED, UNIV PA, 67- Concurrent Pos: Asst chief orthop surg, Methodist Hosp, 38-41; vis orthop surgeon, Children's Seashore Home, Atlantic City, NJ, 39-57; chief orthop surg, Doctors Hosp, 41- & Philadelphia Gen Hosp, 41-; chmn dept orthop surg, Albert Einstein Med Ctr; mem attend staff, Univ Hosp, 41-; consult, Eagleville Sanatorium. Mem: Fel Geront Soc; Orthop Res Soc; fel Am Rheumatism Asn; fel Am Col Surgeons; fel Int Col Surg. Res: Physiology and biochemistry of bone; circulation of bone using isotopes and histologic techniques. Mailing Add: 220 W Rittenhouse Sq Philadelphia PA 19103

STEIN, IRVING F, JR, b Chicago, Ill, July 6, 18; m 50; c 2. SURGERY. Educ: Dartmouth Col, AB, 39; Northwestern Univ, MS, 41, MD, 43, PhD, 51; Am Bd Surg, dipl, 50. Prof Exp: ASST PROF SURG, MED SCH, NORTHWESTERN UNIV, 54- Concurrent Pos: Attend surgeon, Cook County Hosp; chief surg, Highland Park Hosp. Mem: AAAS; fel AMA; fel Am Col Surg; Am Fedn Clin Res. Res: Gastrointestinal and surgical research. Mailing Add: 625 Roger Williams Ave Highland Park IL 60035

STEIN, JAMES DEWITT, JR, b New York, NY, Aug 29, 41. MATHEMATICS. Educ: Yale Univ, BA, 62; Univ Calif, Berkeley, MA & PhD(math), 67. Prof Exp: Asst prof math, Univ Calif, Los Angeles, 67-74, NSF grant, 70-74; MEM FAC, CALIF STATE UNIV, LONG BEACH, 74- Mem: Am Math Soc. Res: Banach algebras; continuity and boundedness problems in Banach spaces; measure theory. Mailing Add: Calif State Univ Long Beach Long Beach CA 90840

STEIN, JANET RUTH, b Denver, Colo. BOTANY. Educ: Univ Colo, BA, 51; Wellesley Col, MA, 53; Univ Calif, PhD(bot), 57. Prof Exp: Lab technician & cur, Univ Calif, 57-59; from instr to assoc prof bot, biol & phycol, 59-71, PROF BOT, UNIV BC, 71- Concurrent Pos: Vis assoc prof, Univ Calif, 65; dir, Western Bot Serv, Ltd, 72-75; ed, J Phycol, Phycological Soc Am, 75-; Kilham sr fel, Univ BC, 75. Honors & Awards: Darbaker Award, Bot Soc Am, 60. Mem: Bot Soc Am; Phycol Soc Am (ed, News Bull, 60-64, Newsletter, 65-66, pres, 65); Can Bot Asn (vpres, 69-70, pres, 70-71); Brit Phycol Soc; Int Phycol Soc. Res: Morphology, physiology, ecology and distribution of algae. Mailing Add: Dept of Bot Univ of BC Vancouver BC Can

STEIN, JEROME D, JR, b Brooklyn, NY, Aug 23, 19; m 47; c 2. PHYSIOLOGY. Educ: Univ Calif, AB, 51, MA, 52, PhD(physiol), 55. Prof Exp: Res assoc physiol, Fairmont Alameda County Hosp, San Leandro, Calif, 55-57; res assoc, Ortho Res Found, Johnson & Johnson, 57-60, res info coordr, Ortho Pharmaceut Co, NJ, 60-68; MGR TECH INFO DEPT, WARNER-LAMBERT CO, 68- Mem: AAAS; Am Soc Info Sci; Am Chem Soc; NY Acad Sci. Res: Physiology of the reproductive system; endocrinology of reproduction; information science, especially systems analysis with respect to storage and retrieval of scientific information storage and retrieval. Mailing Add: Warner-Lambert Co 170 Tabor Rd Morris Plains NJ 07950

STEIN, JEROME EVERETT, biology, chemistry, see 12th edition

STEIN, JOHN MICHAEL, b Vienna, Austria, May 29, 35; US citizen; m 69; c 4. SURGERY. Educ: Harvard Univ, AB, 57, MD, 61; Am Bd Surg, dipl, 68. Prof Exp: Assoc surg, New York Hosp-Cornell Med Ctr, 62-63; asst instr, Albert Einstein Col Med, 66-67; chief burn study br, US Army Inst Surg Res, Tex, 67-69; ASSOC PROF SURG & BURNS, ALBERT EINSTEIN COL MED, 69- Concurrent Pos: NIH fel surg, Albert Einstein Col Med, 64-65. Mem: Am Burn Asn; NY Surg Soc; fel Am Col Surg; Am Trauma Soc; Asn Acad Surg. Res: Surgical training and research, expecially metabolic care of surgical and burned patients. Mailing Add: Dept of Surg Albert Einstein Col of Med Bronx NY 10461

STEIN, JOLYON ADAM, food technology, see 12th edition

STEIN, JUNIOR, b Orange, Calif, Dec 31, 40. MATHEMATICS. Educ: Calif State Col, Long Beach, BS, 62, MA, 63; Univ Calif, Los Angeles, CPhil, 70, PhD(math), 71. Prof Exp: ASST PROF MATH, UNIV TOLEDO, 71- Mem: Am Math Soc; Math Asn Am. Res: Calculus of variations; optimal control theory; numerical analysis; differential equations; index theory of quadratic forms. Mailing Add: Dept of Math Univ of Toledo Toledo OH 43606

STEIN, JUSTIN JOHN, b Haskell, Tex, Oct 18, 07; m 36; c 1. SURGERY, CANCER. Educ: Baylor Univ, MD, 33; Am Bd Radiol, dipl. Prof Exp: Intern, Cincinnati Gen Hosp, Ohio, 34-35; resident gen & tumor surg & radiol, Edward Hines, Jr Hosp, Ill, 35-41; asst clin prof surg, Col Med Evangelists, 48-52; prof radiol, 52-75, EMER PROF RADIOL, SCH MED, UNIV CALIF, LOS ANGELES, 75-; CHIEF RADIATION THERAPY, VET ADMIN HOSP, LONG BEACH, 75- Concurrent Pos: Del Amo fel, Univ Madrid, 58 & 60; consult radiol, path & oncol, var hosps & med progs, 38-; attend surg staff, Calif Hosp, Los Angeles, 45-; tumor surgeon & radiotherapist, Los Angeles Tumor Inst, 46-52; attend surgeon & mem tumor bd, Malignancy Serv, Los Angeles County Hosp, 48-49; attend staff, Desert Hosp, Palm Springs, 50-60, Santa Monica Hosp & Univ Calif Hosp, Los Angeles, 52 & Crenshaw Hosp, 53-60, lectr radiol, US Naval Hosp, San Diego, 50-; chmn, Gov Emergency Med Adv Comt, Calif, 50-67; mem tumor bd & attend radiologist, Los Angeles County Harbor Gen Hosp, 52-60; mem, Calif Bd Med Examr, 52-60, pres, 56, vpres, 66; exchange prof, Med Sch, Nat Univ Mex, 53 & Ministry Higher & Specialized Educ, Moscow & Leningrad, USSR, 63; dir, Calif Inst Cancer Res, 55-70; mem radiother subcomt bladder cancer, adjuvant study grant chemother-radiother, NIH, 64-; mem, Adv Comt Radiol, Vet Admin, DC, 65-, Sci Adv Bd, Inst Laryngol & Voice Disorders, Inc, Los Angeles, 66- & US Civil Defense Coun; dir-at-large, Am Cancer Soc, chmn med & sci comt, 71- Mem: AAAS; fel Am Col Surg; fel Am Col Radiol; AMA; Am Radium Soc (secy, 62-64, pres, 65-66). Res: Tumor surgery and therapy. Mailing Add: Long Beach Vet Hosp Radiation Ther 114T Long Beach CA 90801

STEIN, KURT, chemical engineering, organic chemistry, see 12th edition

STEIN, LARRY, b New York, NY, Nov 10, 31; m 60. PHYSIOLOGICAL PSYCHOLOGY. Educ: NY Univ, BA, 52; Univ Iowa, MA, 53, PhD, 55. Prof Exp: Res psychologist, Walter Reed Army Inst Res, 55-57; res psychologist, Vet Admin Res Labs Neuropsychiat, 57-59; sr res scientist, 59-64, MGR DEPT PSYCHOPHARMACOL, WYETH LABS, 64- Concurrent Pos: Res assoc, Bryn Mawr Col, 61-72, adj prof, 72-; adj prof med sch, Univ Pa. Honors & Awards: Bennett Award, Soc Biol Psychiat, 61. Mem: Am Psychol Asn; Am Physiol Soc; Am Soc Pharmacol & Exp Therapeut; Soc Neurosci. Res: Behavior; the brain and drugs. Mailing Add: Wyeth Labs PO Box 8299 Philadelphia PA 19101

STEIN, LAWRENCE, b Hampton, Va, June 21, 22; m 52; c 2. INORGANIC CHEMISTRY. Educ: George Washington Univ, BS, 48; Univ Wis, PhD(chem), 52. Prof Exp: Anal chemist, Nat Bur Standards, 48; asst, Univ Wis, 50-51; CHEMIST, ARGONNE NAT LAB, 51- Concurrent Pos: Adv panelist comt biol effects atmospheric pollutants, Nat Res Coun, 71-72; consult, Nat Inst Occupational Safety & Health, 73. Mem: AAAS; Am Chem Soc. Res: Fluorine chemistry; interhalogen compounds; chemistry of noble gases, particularly radon; environmental radiation; biologic effects of atmospheric pollutants; infrared spectroscopy; chemistry of actinide elements. Mailing Add: Argonne Nat Lab 9700 S Cass Ave Argonne IL 60439

STEIN, MARJORIE LEITER, b New York, NY, Dec 8, 46. MATHEMATICS. Educ: Barnard Col, AB, 68; Princeton Univ, MA, 71, PhD(math), 72. Prof Exp: Res assoc math, Univ Wis-Madison, 72-73, lectr comput sci, 73; res assoc, Nat Bur Standards, 73-75; SR MATH STATISTICIAN, US POSTAL SERV, 75- Concurrent Pos: Res assoc, Nat Res Coun-Nat Acad Sci, 73- Mem: Am Math Soc; Math Asn Am; Math Prog Soc; Asn Women Math. Res: Combinatorial theory; networks; linear programming. Mailing Add: US Postal Serv 475 L'Enfant Plaza West SW Washington DC 20260

STEIN, MARTIN EDWARD, pharmaceutical chemistry, see 12th edition

STEIN, MARVIN, b St Louis, Mo, Dec 8, 23; m 50; c 3. PSYCHIATRY. Educ: Wash Univ, BS, 45, MD, 49. Prof Exp: Intern, St Louis City Hosp, 49-50; asst resident psychiat, Sch Med, Wash Univ, 50-51; asst instr, Sch Med, Univ Pa, 53-54, res assoc, 54-56, from asst prof to assoc prof, 56-63; prof, Med Col, Cornell Univ, 63-66; prof & chmn dept, State Univ NY Downstate Med Ctr, 66-71; PROF PSYCHIAT & CHMN DEPT, MT SINAI SCH MED & PSYCHIATRIST-IN-CHIEF, 71- Concurrent Pos: USPHS fel clin sch, Sch Med, Univ Pittsburgh, 51-53; fel psychiat, Sch Med, Univ Pa, 53-54; ment health career investr, NIMH, 56-61, mem ment health fels rev panel, 61-64, ment health res career award comt, 63-65, chmn, 65-67. Mem: Am Psychiat Asn. Res: Investigation of respiratory and psychological variables; bronchial asthma; central nervous system mechanisms and immune processes. Mailing Add: Dept of Psychiat Mt Sinai Sch of Med New York NY 10029

STEIN, MARVIN L, b Cleveland, Ohio, July 15, 24; m 44; c 3. MATHEMATICS, COMPUTER SCIENCE. Educ: Univ Calif, Los Angeles, BA, 47, MA, 49, PhD, 51. Prof Exp: Asst math, Univ Calif, Los Angeles, 47-48, mathematician inst numerical anal, 48-52; sr res engr, Consol-Vultee Corp, 52-55; from asst prof to assoc prof math, 55-61, dir univ comput ctr, 58-70, actg head comput info & control sci dept, 70-71, PROF MATH, UNIV MINN, MINNEAPOLIS, 61- Concurrent Pos: Lectr, Univ Calif, Los Angeles, 54-55; Guggenheim fel, 63-64; vis prof, Tel Aviv Univ & Hebrew Univ, Jerusalem, 71-72. Mem: Am Math Soc; Soc Indust & Appl Math; Asn Comput Mach. Res: Numerical analysis; applications of digital computers; computer systems, machine arithmetic. Mailing Add: Comput Info & Control Sci Dept Univ of Minn Minneapolis MN 55455

STEIN, MICHAEL ROGER, b Milwaukee, Wis, Mar 21, 43; m 67; c 1. MATHEMATICS. Educ: Harvard Univ, BA, 64; Columbia Univ, PhD(math), 70. Prof Exp: Asst prof math, 70-74, ASSOC PROF MATH, NORTHWESTERN UNIV, EVANSTON, 74- Concurrent Pos: Fel, Hebrew Univ, Jerusalem, 72-73. Mem: Am Math Soc. Res: Algebraic K-theory; algebra. Mailing Add: Dept of Math Northwestern Univ Evanston IL 60201

STEIN, MYRON, b East Boston, Mass, May 27, 25; m 53; c 4. PHYSIOLOGY, MEDICINE. Educ: Dartmouth Col, BA, 48; Tufts Univ, MD, 52. Prof Exp: Instr med, Harvard Med Sch, 57-64, assoc, 64-65; from assoc prof to prof med sci, Brown Univ, 69-73; PROF MED, UNIV CALIF, LOS ANGELES, 73-; DIR PULMONARY DIV, BROTMAN MEM HOSP, 73- Concurrent Pos: Consult, Mass Rehab Comt, 60-65, Vet Admin Hosps, West Roxbury, 63- & Davis Park, RI, 65- Mem: Am Fedn Clin Res; Am Physiol Soc. Res: Pulmonary physiologic effects of pulmonary embolism; relationship of acid-base states and thyroid hormone transport; physiologic studies in clinical lung diseases. Mailing Add: Brotman Mem Hosp 3828 Delmas Terr Culver City CA 90230

STEIN, NELSON, b New York, NY, July 23, 26; m 58; c 3. NUCLEAR PHYSICS. Educ: City Col New York, BS, 57; Univ Ill, MS, 58, PhD(physics), 63. Prof Exp: Res assoc nuclear physics, Univ Wash, 63-66; res assoc, Yale Univ, 66-68, asst prof physics, 68-72; PHYSICIST, LOS ALAMOS SCI LAB, UNIV CALIF, 72- Concurrent Pos: Vis staff mem, Los Alamos Sci Lab, 69-72; NATO sr fel sci, Ctr Nuclear Res, Strasbourg, France, 72. Mem: Am Phys Soc. Res: Structure of medium and heavy nuclei; nuclear reaction mechanisms; nuclear models. Mailing Add: Physics Div MS 458 PO Box 1663 Los Alamos Sci Lab Los Alamos NM 87544

STEIN, NORMAN, b New York, NY, Dec 28, 33; m 63; c 2. MATHEMATICS. Educ: Cornell Univ, AB, 54, PhD(math), 57. Prof Exp: Instr math, Yale Univ, 57-59; J F Ritt instr, Columbia Univ, 59-62; assoc prof, State Univ NY, Stony Brook, 62-64; assoc prof, Haverford Col, 64-66; assoc prof, 66-75, PROF MATH, UNIV ROCHESTER, 75- Mem: Am Math Soc. Res: Topology of manifolds; homotopy theory; differential geometry. Mailing Add: Dept of Math Univ of Rochester Rochester NY 14627

STEIN, OTTO LUDWIG, b Augsburg, Ger, Jan 14, 25; nat US; m 58; c 4. PLANT MORPHOGENESIS. Educ: Univ Minn, BS, 49, MS, 52, PhD(bot), 54. Prof Exp: Asst bot, Univ Minn, 48-53; instr, Univ Mo, 55; USPHS res fel, Brookhaven Nat Lab, 55-58; from asst prof to assoc prof bot, Mont State Univ, 58-64; assoc prof, 64-70, head dept, 70-74, PROF BOT, UNIV MASS, AMHERST, 70- Concurrent Pos: Res collab, Brookhaven Nat Lab, 58-70; vis asst prof, Univ Calif, 61-62; sr NATO res fel, Imp Col, Univ London, 71-72. Mem: Bøt Soc Am; Soc Study Develop Biol; Am Soc Cell Biol; Soc Exp Biol & Med; Am Genetic Asn. Res: Genetics; cytology; developmental anatomy of apical meristems and their derivatives. Mailing Add: Dept of Bot Univ of Mass Amherst MA 01002

STEIN, PAUL CARY, biology, parasitology, see 12th edition

STEIN, PHILIP, b New York, NY, Apr 28, 32. PHYSIOLOGY. Educ: Brooklyn Col, BA, 53; George Washington Univ, MS, 54; Columbia Univ, MA, 59; Univ Geneva, PhD(biochem), 61. Prof Exp: Instr chem, Brooklyn Col, 60-62; instr, New York Community Col, 62-64; asst prof biol, Fairleigh Dickinson Univ, 64-68; ASSOC PROF BIOL, STATE UNIV NY COL, NEW PALTZ, 68- Concurrent Pos: USPHS grant, 62-64; NSF grant, 64-; consult biochemist, Nat Sugar Industs; consult, Sugar Refinery, Pepsi Cola, 72- Mem: Fel Am Inst Chem; Am Chem Soc; Am Soc Biol Chem. Res: Chemical composition of the thyrotropic hormone secreted by the anterior pituitary gland; role of hypothalamus in regulation of thyroid function. Mailing Add: Dept of Biol State Univ of NY Col New Paltz NY 12561

STEIN, PHILIP, b New York, NY, June 16, 24; m 46; c 4. MEDICAL PHYSICS, PHYSICS. Educ: Adelphi Univ, BS, 61, MA, 70. Prof Exp: Instr physics & math, RCA Inst, 45-61, head dept physics, 61-74; VPRES, INST AUDIO RES, INC, 74-; HEAD MED-PHYSICS DIV, MAIMONIDES MED CTR, 67- Concurrent Pos: Proj engr, Lewyt Corp, 44-46 & 49-50; pres, Writers & Designers Inc, 51-53; consult & tech writer, Jos Racker Co, 53-56 & Caldwell-Clements Co, 56-57; dir, RCA Inst-IBM Coop Training Prog, 57-59; chmn curric rev comt, RCA Inst, 61-62; instr, Adelphi Univ, 61-71; dir audiometry clin, Brooklyn Jewish Geriat Ctr, 73- Mem: AAAS; Am Inst Physics; Inst Elec & Electronics Eng. Res: Application of magnetic fields to generate movement in malfunctioning smooth, skeletal and cardiac muscle and to control compression of hollow viscera such as urethra and colostomy and ileostomy stoma. Mailing Add: Maimonides Med Ctr 4802 Tenth Ave Brooklyn NY 11219

STEIN, REINHARDT P, b New York, NY, Dec 19, 35; m 62; c 2. ORGANIC CHEMISTRY. Educ: Rensselaer Polytech Inst, BS, 58; Ohio State Univ, PhD(org chem), 63. Prof Exp: Res chemist, Dow Chem Co, 63-64; RES CHEMIST, WYETH LABS, INC, 64- Mem: Am Chem Soc. Res: Total synthesis of natural products; new totally synthetic steroids; structural elucidation of natural products; synthesis of new drugs. Mailing Add: Wyeth Labs Inc Box 8299 Philadelphia PA 19101

STEIN, RICHARD BERNARD, b New Rochelle, NY, June 14, 40; m 62; c 2. NEUROPHYSIOLOGY, BIOPHYSICS. Educ: Mass Inst Technol, BS, 62; Oxford Univ, MA & DPhil(physiol), 66. Prof Exp: Res fel med res, Exeter Col, Oxford Univ, 65-68; assoc prof physiol, 68-72, PROF PHYSIOL, UNIV ALTA, 72- Concurrent Pos: USPHS fel, 66-68. Mem: Brit Physiol Soc; Can Physiol Soc; Neurosci Soc. Res: Motor control; information processing by nerve cells; sensory feedback; neural models. Mailing Add: Dept of Physiol Univ of Alta Edmonton AB Can

STEIN, RICHARD JAMES, b Palmerton, Pa, Aug 10, 30; m 53. POLYMER CHEMISTRY. Educ: Pa State Univ, BS, 58; Univ Akron, MS, 60, PhD(polymer chem), 67. Prof Exp: Res chemist, Goodyear Tire & Rubber Co, 60-63, sr res chemist, 66-71; POLYMERS SPECIALIST, INSULATING MAT DEPT, GEN ELEC CO, 71- Mem: Am Chem Soc. Res: Polymer synthesis and properties. Mailing Add: 239 Pinewood Dr Schenectady NY 12303

STEIN, RICHARD STEPHEN, b Far Rockaway, NY, Aug 21, 25; m 51; c 4. POLYMER CHEMISTRY. Educ: Polytech Inst Brooklyn, BS, 45; Princeton Univ, MA, 48, PhD(phys chem), 49. Prof Exp: Asst, Polytech Inst Brooklyn, 45; asst, Princeton Univ, 49-50; asst prof chem, 50-57, from assoc prof to prof, 57-61, COMMONWEALTH PROF CHEM & DIR POLYMER RES INST, UNIV MASS, AMHERST, 61- Concurrent Pos: Fulbright vis prof, Kyoto Univ, 68. Honors & Awards: Int Award, Soc Plastics Eng, 69; Borden Award, Am Chem Soc, 72; Bingham Medal, Soc Rheol, 72. Mem: Am Chem Soc; Am Phys Soc; Soc Rheol. Res: Molecular structure; light scattering; mechanical and optical properties of high polymers. Mailing Add: Dept of Chem Univ of Mass Amherst MA 01002

STEIN, ROBERT A, organic chemistry, see 12th edition

STEIN, ROBERT CARRINGTON, b Brooklyn, NY, Nov 16, 31; m 54; c 1. ORNITHOLOGY, BIOACOUSTICS. Educ: St Olaf Col, BA, 52; Cornell Univ, MS, 54, PhD(ornith), 57. Prof Exp: Asst prof biol, Ursinus Col, 56-60; res supvr lab ornith, Cornell Univ, 60-65; assoc prof biol, 65-66, PROF BIOL, STATE UNIV NY COL, BUFFALO, 66- Concurrent Pos: Res assoc, Cornell Univ, 58-60; Am Philos Soc res grant, 58 & 67-68; NSF res grant, 59 & 66-68; sci fac fel, Australia, 71-72. Mem: Am Soc Zool; Am Ornith Union. Res: Study of animal sounds, their role in behavior and evolution; syringeal anatomy and function; passerine birds, especially Empidonax and Parulidae. Mailing Add: Dept of Biol State Univ of NY Col Buffalo NY 14222

STEIN, ROBERT FOSTER, b New York, NY, Mar 4, 35; m 58; c 2. ASTROPHYSICS. Educ: Univ Chicago, BS, 57; Columbia Univ, PhD(physics), 66. Prof Exp: Res fel astrophys, Carnegie Inst Wash Mt Wilson & Palomar Observs, 66-67; res fel, Harvard Observ, 67-69; ASST PROF ASTROPHYS, Brandeis UNIV, 69-; CONSULT, SMITHSONIAN ASTROPHYS OBSERV, 69- Concurrent Pos: Vis fel joint inst lab astrophys, Nat Bur Standards & Univ Colo, 73-74. Mem: Am Astron Soc; Int Astron Union. Res: Astrophysical fluid dynamics; cloud formation in interstellar medium and heating of solar chromosphere and corona; galaxy formation. Mailing Add: Dept of Physics Brandeis Univ Waltham MA 02154

STEIN, ROBERT GEORGE, b Milwaukee, Wis, May 21, 38; m 62; c 2. AGRICULTURAL CHEMISTRY, MEDICINAL CHEMISTRY. Educ: Univ Wis-Milwaukee, BS, 61; Loyola Univ, MS, 65. Prof Exp: Sr chemist, Aldrich Chem Co, 66-69; chemist, 61-66, GROUP LEADER AGR CHEM, ABBOTT LABS, 69- Mem: Am Chem Soc; Int Soc Heterocyclic Chem. Res: Design and synthesis organic compounds to be utilized as ecologically safe biodegradable pesticides; synthesis plant growth regulators, such as abscission agents and ethylene producing compounds. Mailing Add: Abbott Labs North Chicago IL 60064

STEIN, ROBERT PRESTON, physics, see 12th edition

STEIN, SAMUEL C, b Philadelphia, Pa, Apr 11, 12; m 45; c 1. MEDICINE. Educ: Villanova Univ, BS, 33; Jefferson Med Col, MD, 37; Univ Pa, MMedSc, 50; Am Bd Internal Med, dipl, 49, cert pulmonary dis, 50. Prof Exp: Chief clins, Henry Phipps Inst, 48-60, ASST PROF CLIN & COMMUNITY MED, SCH MED & DIV GRAD MED, UNIV PA, 53- Concurrent Pos: Chief serv, Philadelphia Gen Hosp, 46-60, consult, 60-; med dir, Union Health Ctr, Int Ladies Garment Workers Union, 54-; chief tuberc control sect, City Dept Health, Philadelphia, 58-70, chief chronic respiratory dis, 70- Mem: Am Thoracic Soc; fel Am Pub Health Asn; fel Am Col Physicians; NY Acad Sci; fel Royal Soc Health. Res: Tuberculosis; chronic diseases. Mailing Add: William Penn House 1919 Chestnut St Philadelphia PA 19103

STEIN, SAMUEL H, b New York, NY, Jan 6, 37; m 57; c 2. PHOTOGRAPHIC CHEMISTRY, PHYSICAL ORGANIC CHEMISTRY. Educ: City Col New York, BS, 57; Boston Univ, PhD(org chem), 67. Prof Exp: Res chemist, Nat Cash Register Co, Ohio, 57-59; res chemist, Itek Corp, summer 60, sr res chemist & proj leader org chem, Lexington Res Labs, 65-68, mgr paper develop group, Lexington Develop Labs, 68-69, mgr sci staff negative lithographic plate group, 69-71, mgr positive lithographic plate group, 71-72, mgr lithographic systs res dept, Lithographic Technol Lab, 72-74; MGR EMULSION RES, CHEMCO PHOTOPROD, 74- Mem: AAAS; Am Chem Soc; Soc Photog Sci & Eng. Res: Photochemistry, including unconventional photographic process and silver halide photo processes; silver halide emulsions for positive and negative systems. Mailing Add: 36 Preston St Huntington NY 11743

STEIN, SEYMOUR, b Brooklyn, NY, Apr 4, 28; m 54; c 2. PHYSICS. Educ: Harvard Univ, SM, 50, PhD(appl physics), 55. Prof Exp: Eng specialist, Sylvania Elec Prods, Inc, 53-56; mem tech staff, Hermes Electronic Co, 56-59; sr scientist & head commun res dept appl res lab, Sylvania Electronic Systs, Gen Tel & Electronics Corp, Mass, 59-64, assoc dir, 64-66, dir commun systs labs, 66-68; PRES, STEIN ASSOCS, INC, 69- Concurrent Pos: Vis prof, Polytech Inst Brooklyn, 62; vis lectr, Harvard Univ, 66; vpres & dir, Adams-Russell Co, 74- Mem: Fel Inst Elec & Electronics Eng. Res: Communication theory; ditigal signal processing. Mailing Add: Stein Assocs Inc 280 Bear Hill Rd Waltham MA 02154

STEIN, SEYMOUR NORMAN, b Chicago, Ill, Nov 23, 13; m 36; c 1. PHYSIOLOGY. Educ: Univ Ill, BS, 41, MD, 43. Prof Exp: From res asst to res assoc neurophysiol, Univ Ill, 46-49, asst prof & res physiologist, 49-51; head neurophysiol br & submarine & diving med br, Naval Med Res Inst, 52-60, head physiol div, 53-57; dep bio-sci officer, Pac Missile Range, US Dept Navy, 61-63; chief life sci officer, 64-65, CHIEF MED OFFICER, NASA-AMES RES CTR, 66- Concurrent Pos: Consult, Surgeon Gen, US Dept Army, 47-, NIMH, 53-; mem panel underwater swimmers, Nat Res Coun, 54-55; ed, J Biol Photog Asn, 56. Mem: Fel AAAS; Am Physiol Soc; Soc Exp Biol & Med; Biol Photog Asn; assoc fel Am Inst Aeronaut & Astronaut. Res: Basic and clinical studies on convulsions; effects of acute and chronic exposure to hypernormal amounts of carbon dioxide and oxygen; space medicine; bioinstrumentation. Mailing Add: NASA-Ames Res Ctr Moffett Field CA 94035

STEIN, SHERMAN KOPALD, b Minneapolis, Minn, Aug 11, 26; m 50; c 3. MATHEMATICS. Educ: Calif Inst Technol, BSc, 46; Columbia Univ, MA, 47, PhD(math), 52. Hon Degrees: DH, Marietta Col, 75. Prof Exp: PROF MATH, UNIV CALIF, DAVIS, 53- Concurrent Pos: Mem, Math Panel Calif Teacher Prep & Cert, 73- Honors & Awards: L R Ford Award, Math Asn Am, 75. Mem: Am Math Soc; Math Asn Am. Res: Algebraic applications to geometry; convex bodies. Mailing Add: Dept of Math Univ of Calif Davis CA 95616

STEIN, SIDNEY J, b New York, NY, Oct 9, 21; m 45; c 4. PHYSICAL CHEMISTRY, MICROELECTRONICS. Educ: Brooklyn Col, AB, 42; Polytech Inst Brooklyn, MS, 46, PhD(phys chem), 51. Prof Exp: Chief chem operator, Amecco Chem Co, 42-43; from res chemist to group leader, Manhattan Proj, Kellex Corp & Carbide & Carbon Chem Corp, 43-46; res assoc, Polytech Inst Brooklyn, 47-48; tech dir & partner, Maybunn Chem Co, 48-49; sr res chemist, Int Resistance Co, 49-50, asst res dir, 50-52, dir res, 52-58, res & eng, 58-60, vpres, 60-62; PRES & FOUNDER, ELECTRO-SCI LABS, INC, 62- Concurrent Pos: Dir, Process & Instruments Inc, NY, 61-, Sci Capital Corp, Pa, 61-65; dir & chmn bd, Microstate Electronics Corp, NJ, 62-64; dir, Ala Binder & Chem Co, 63-66; exec vpres, Apollo Industs Inc, Pa, 63-67, Universal Petrochem Inc, NJ, 65-; mem, Conf Elec Insulation, Nat Res Coun; bus mgr, Electro-Sci Labs, GmbH, West Ger, 72-; managing dir, ESL et Compagnie, France, 75- Mem: AAAS; Am Chem Soc; fel Am Inst Chem; Inst Elec & Electronics Eng; Int Soc Hybrid Microelectronics. Res: Polymers and resins; rheological measurements; evaporation of metals; high vacuum and hermetic sealing; glass and plastics to metal adhesion; conductive, resistive, dielectric and insulating glazes; ceramics; new glasses; precious metals technology; materials for hybrid microelectronic circuits. Mailing Add: Electro-Sci Labs Inc 1601 Sherman Ave Pennsauken NJ 08110

STEIN, STEPHEN ELLERY, b Manhattan, NY, Dec 13, 48; m 74. PHYSICAL CHEMISTRY, CHEMICAL KINETICS. Educ: Univ Rochester, BS, 69; Univ Wash, PhD(phys chem), 74. Prof Exp: Res assoc, 74-75, PHYS CHEMIST, STANFORD RES INST, 75- Mem: Am Chem Soc. Res: Elementary chemical processes; unimolecular reactions; reaction rate theory; gas-surface reactions and catalysis; computer modeling of complex reacting systems; coal gasification; combustion; ignition processes. Mailing Add: Stanford Res Inst Menlo Park CA 94025

STEIN, T PETER, b London, Eng, Apr 27, 41; m 67. BIOCHEMISTRY. Educ: Univ London, BSc, 62, MSc, 63; Cornell Univ, PhD(chem), 67. Prof Exp: Asst chem, Cornell Univ, 63-67; instr res surg & biochem, 69-72, ASSOC SURG & PHYSIOL, SCH MED, UNIV PA, 72- Concurrent Pos: NIH fel biochem, Univ Calif, Los Angeles, 67-69. Mem: AAAS; Am Chem Soc; The Chem Soc. Res: Oxygen-18 techniques; analytical biochemistry; nitrogen-15 metabolism and rates of protein synthesis in man; lung biochemistry; oxygen-18 exchange reactions. Mailing Add: Dept of Physiol Univ of Pa Sch of Med Philadelphia PA 19104

STEIN, TALBERT SHELDON, b Detroit, Mich, Jan 6, 41; m 63; c 1. PHYSICS. Educ: Wayne State Univ, BS, 62; Brandeis Univ, MA, 64, PhD(physics), 68. Prof Exp: Res assoc physics, Univ Wash, 67-70; asst prof, 70-75, ASSOC PROF PHYSICS, WAYNE STATE UNIV, 75- Mem: Am Phys Soc; Am Asn Physics Teachers. Res: Experimental atomic physics including low energy positron-atom interactions; precision measurement of the g factor of the free electron; studies of effects of electric fields on neutral atoms. Mailing Add: Dept of Physics Wayne State Univ Detroit MI 48202

STEIN, WAYNE ALFRED, b Minneapolis, Minn, Dec 6, 37. ASTROPHYSICS. Educ: Univ Minn, BPhys, 59, PhD(physics), 64. Prof Exp: Res assoc astrophys, Princeton Univ, 64-66; asst res physicist, 66-69, asst prof, 69-73, ASSOC PROF ASTROPHYS, UNIV CALIF, SAN DIEGO, 73- Concurrent Pos: Alfred P Sloan Found fel, Univ Minn, Minneapolis & Univ Calif, San Diego, 69-; from asst prof to assoc prof, Univ Minn, Minneapolis, 69- Mem: Am Astron Soc. Res: Infrared astronomy. Mailing Add: Dept of Physics Univ of Calif at San Diego La Jolla CA 92037

STEIN, WILLIAM EARL, b Rochester, NY, May 30, 24; m 47; c 4. ELECTRON PHYSICS, NUCLEAR PHYSICS. Educ: Univ Va, BEE, 46; Stanford Univ, MS, 50; Univ NMex, PhD(physics), 62. Prof Exp: PHYSICIST, LOS ALAMOS SCI LAB, 49- Mem: Am Phys Soc. Res: Electron-photon interactions and production of nearly monochromatic soft x-rays by the inverse Compton effect. Mailing Add: P-DOR MS 458 Los Alamos Sci Lab Los Alamos NM 87545

STEIN, WILLIAM HOWARD, b New York, NY, June 25, 11; m 36; c 3. BIOCHEMISTRY. Educ: Harvard Univ, BS, 33; Columbia Univ, PhD(biochem), 38. Hon Degrees: DSc, Columbia Univ & Albert Einstein Col Med, Yeshiva Univ, 73. Prof Exp: Asst chem, 39-43, assoc, 43-49, assoc mem, 49-52, PROF CHEM & MEM, ROCKEFELLER UNIV, 52- Concurrent Pos: Am Swiss Found fel, 56; vis prof, Univ Chicago, 60; vis prof, Harvard Univ, 64; consult, Qm Corps, US Army; trustee, Montefiore Hosp, New York; sci counselor, Nat Inst Neurol Dis & Blindness, 61-66; chmn US nat comt, Int Union Biochem, 65-68. Honors & Awards: Nobel Prize in Chem, 72; Kaj-Linderstrom-Lang Gold Medal & Prize, 72; Chromatography & Electrophoresis Award, Am Chem Soc, 64, Theodore William Richards Medal, Northeast Sect, 72. Mem: Nat Acad Sci; AAAS; Am Soc Biol Chem (ed-in-chief, 68-71); Am Chem Soc; Harvey Soc; fel Am Acad Arts & Sci. Res: Chemistry and biochemistry of amino acids, peptides and proteins; mechanism of enzyme action; chromatographic separation processes. Mailing Add: Rockefeller Univ 66th St & York Ave New York NY 10021

STEIN, WILLIAM IVO, b Wurzburg, Ger, July 22, 22; nat US; m 48; c 12. FORESTRY. Educ: Pac Col, BS, 43; Ore State Univ, BS, 48; Yale Univ, MF, 52, PhD, 63. Prof Exp: Forester timber mgt res, 48-52, timber mgt asst, summer 47, forester forest admin, summer 48, res forester, 52-63, PRIN PLANT ECOLOGIST, PAC NORTHWEST FOREST & RANGE EXP STA, US FOREST SERV, 63- Mem: Soc Am Foresters. Res: Forest ecology; regeneration research. Mailing Add: Forestry Sci Lab 3200 Jefferson Way Corvallis OR 97331

STEIN, WILLIAM WARNER, anthropology, see 12th edition

STEIN, ZENA, b Durban, SAfrica, July 7, 22; m 49; c 3. EPIDEMIOLOGY. Educ: Univ Cape Town, MA, 42; Univ Witwatersrand, MB, BCh, 50. Prof Exp: Med officer social med, Alexandria Univ Health Ctr, Univ Witwatersrand, 51-55; assoc prof, 66-72, PROF EPIDEMIOL, COLUMBIA UNIV, 73-, DIR RES UNIT EPIDEMIOL OF MENT RETARDATION, NY STATE DEPT MENT HYG, 68- Concurrent Pos: Med Res Coun res fel, Univ Manchester, 56-65; Ment Health Res Fund Gt Brit sr fel, 59-62; Med Res Coun Gt Brit grant, 63-66. Mem: Am Pub Health Asn; Int Epidemiol Asn; Am Asn Ment Deficiency. Res: Socio-medical research; epidemiology of mental illness; mental retardation, malnutrition and psychological disturbances. Mailing Add: 600 W 168th St New York NY 10032

STEINBACH, ALAN B, physiology, biophysics, see 12th edition

STEINBACH, HENRY BURR, b Dexter, Mich, Oct 7, 05; m 34; c 4. ZOOLOGY, PHYSIOLOGY. Educ: Univ Mich, BA, 28; Brown Univ, MA, 30; Univ Pa, PhD(zool), 33. Prof Exp: Asst physiol & biochem, Brown Univ, 28-30; instr zool, Univ Pa, 30-33; Nat Res Coun fel, Univ Chicago, 33-34; fel, Univ Rochester, 34-35; instr zool, Univ Minn, 35-36, asst prof, 36-38; asst prof, Columbia Univ, 38-42; assoc prof, Wash Univ, St Louis, 42-46, prof, 46-47; prof, Univ Minn, 47-57; prof, Univ Chicago, 57-73, chmn dept, 57-68, dir marine biol lab, 66-68; PRES, OCEANIC INST, 73- Concurrent Pos: Dean grad sch, Woods Hole Oceanog Inst, Mass, 68-73, mem corp, trustee & mem coun, Marine Biol Lab; asst dir, NSF, 53-54; chmn div biol & agr, Nat Res Coun, 58-62; managing ed, Biol Bull, 42-50. Mem: Am Soc Zool (pres, 57-58); Am Physiol Soc; Soc Gen Physiol (pres, 56-57). Res: Bioelectric phenomena; injury potentials; sodium potassium equilibrium of cells; ontogenesis of enzyme systems in chicks; enzyme systems of granular componenets. Mailing Add: Oceanic Inst Makapuu Point Waimanalo HI 96795

STEINBACH, HOWARD LYNNE, b Pittsburgh, Pa, Sept 17, 18; m 51; c 3. RADIOLOGY. Educ: Univ Calif, Los Angeles, AB, 40; St Louis Univ, MS, 43. Prof Exp: Intern, Los Angeles County Gen Hosp, 44; from asst resident to resident, Univ Calif Hosp, 45-48; pvt pract, 58-59; from instr to prof radiol, 49-72, CLIN PROF RADIOL, SCH MED, UNIV CALIF, SAN FRANCISCO, 72-; CLIN PROF RADIOL, STANFORD UNIV, 72- Concurrent Pos: Consult, Gen Hosp, 61- & Letterman Gen Hosp, 65- Mem: Radiol Soc NAm; AMA; Am Col Radiol; Asn Univ Radiol; Am Roentgen Ray Soc. Res: Roentgen manifestations of metabolic and endocrine abnormalities of the skeletal system; arthritis and rheumatism; pediatric radiology; endocrine and metabolic diseases; gastroenterology. Mailing Add: 4141 Geary Blvd San Francisco CA 94118

STEINBACH, LEONARD, b New York, NY, Feb 26, 27; m 51; c 3. ORGANIC CHEMISTRY. Educ: City Col New York, BS, 47; Polytech Inst Brooklyn, MS, 54. Prof Exp: Group leader org res, 56-62, prod mgr, 63-64, dir res & develop, 65, dir corp develop, 66-67, gen mgr, 67-69, VPRES, INT FLAVORS & FRAGRANCES, INC, UNION BEACH, 69- Concurrent Pos: Mem chm fac, Monmouth Col, 56-61. Mem: AAAS; Am Chem Soc; NY Acad Sci. Res: Organic synthesis and development in aromatic chemicals, perfumes and flavor materials. Mailing Add: 10 Ramsgate Rd Cranford NJ 07016

STEINBACH, WAYNE ROBERT, b Green Bay, Wis, Mar 22, 47; m 69; c 1. MOLECULAR SPECTROSCOPY. Educ: Lawrence Univ, BA, 69; Duke Univ, PhD(physics), 74. Prof Exp: Res asst molecular spectros, Dept Physics, Grad Sch Arts & Sci, Duke Univ, 71-74; PROF MGR SOLID STATE SCI, AIR FORCE OFF SCI RES, AIR FORCE SYSTS COMMAND, 74- Mem: Am Phys Soc; Am Asn Physics Teachers. Res: High frequency, high resolution microwave submillimeter rotational molecular spectroscopy. Mailing Add: Air Force Off of Sci Res Bolling AFB Washington DC 20332

STEINBECK, HERBERT D, b Cape Girardeau, Mo, Jan 19, 25; m 46; c 4. MEDICINE. Educ: St Louis Univ, MD, 47. Prof Exp: Pvt pract, 48-50 & 52-64; SR STAFF ADV BIOMED DATA PROCESSING SYSTS DEVELOP, ADVAN SYSTS DEVELOP DIV, IBM CORP, 64- Res: Biomedical data processing systems development. Mailing Add: IBM Corp 2651 Strang Blvd Yorktown Heights NY 10598

STEINBECK, KLAUS, b Munich, Ger, Dec 11, 37; US citizen; m 60; c 3. FORESTRY, PHYSIOLOGY. Educ: Univ Ga, BSF, 61, MS, 63; Mich State Univ, PhD(forestry), 65. Prof Exp: Res plant physiologist, US Forest Serv, 65-68; asst prof, 69-72, ASSOC PROF FOREST RESOURCES, UNIV GA, 72- Mem: Soc Am Foresters. Res: Mineral nutrition and growth of forest trees. Mailing Add: Sch of Forest Resources Univ of Ga Athens GA 30601

STEINBERG, ARTHUR GERALD, b Port Chester, NY, Feb 27, 12; m 39; c 2. MEDICAL GENETICS, HUMAN GENETICS. Educ: City Col New York, BSc, 33; Columbia Univ, MA, 34, PhD(zool), 41. Prof Exp: Lectr genetics, McGill Univ, 40-44; mem opers res group, Off Sci Res & Develop, US Dept Navy, 44-46; assoc prof genetics, Antioch Col & chmn dept genetics, Fels Res Inst, 46-48; consult div biomet & med statist, Mayo Clinic, 48-52; geneticist, Children's Cancer Res Found & res assoc, Children's Hosp, Boston, Mass, 52-56; prof biol, 56-72, from asst prof to assoc prof dept prev med, 56-70, prof human genetics, Dept Reproductive Biol, 70-75, FRANCIS HOBART HERRICK PROF BIOL, CASE WESTERN RESERVE UNIV, 72-, PROF HUMAN GENETICS DEPT MED, 75- Concurrent Pos: Consult permanent comt int human genetics cong, NIH, 66-71; chmn med adv bd, Nat Genetics Found, 68-, sr ed, Progress in Med Genetics; consult ed, Transfusion; contrib ed, Vox Sanguinis; consult, WHO. Mem: AAAS; Am Soc Human Genetics (ed, J Human Genetics, pres, 64); Genetics Soc Am; Am Asn Immunol; hon mem Japanese Soc Human Genetics. Res: Immunogenetics; study of genetic control of human immunoglobulins; population genetics; genetics of diabetes. Mailing Add: Dept of Biol 403N Millis Sci Ctr Case Western Reserve Univ Cleveland OH 44106

STEINBERG, BERNARD ALBERT, b New York, NY, Oct 2, 24; m 46; c 3. MICROBIOLOGY. Educ: NY Univ, AB, 47; Univ Ill, MS, 48, PhD(bact), 50. Prof Exp: Head sect bact & mycol, Squibb Inst Med Res, NJ, 50-56; head virus & cancer res, Wm S Merrell Co, 56-63; GROUP LEADER CHEMOTHER SECT, STERLING-WINTHROP RES INST, 63- Mem: AAAS; Am Soc Microbiol; Sigma Xi. Res: Antiviral chemotherapy; upper respiratory viruses; veterinary viruses; virus vaccines and immunology of viruses; viral etiology of cancer. Mailing Add: Chemother Dept Sterling-Winthrop Res Inst Rensselaer NY 12144

STEINBERG, BERNHARD, b New York, NY, June 18, 97; m 31; c 2. PATHOLOGY. Educ: Boston Univ, MD, 22. Prof Exp: House officer, Hosp Div, Med Col Va, 22-23; asst physician, Boston Psychopath Hosp, 23-24; dir labs & res, Toledo Hosp, Ohio, 27-64, dir inst med res, 43-64; ASSOC PROF RES IN PATH, SCH MED, LOMA LINDA UNIV, 64- Concurrent Pos: Nat Res Coun fel, Sch Med, Western Reserve Univ, 24-26, Crile res fel, 26-27; pvt pract oncol & hemat, Pomona, 65- Honors & Awards: Am Soc Clin Path Silver Medal, 37; Cincinnati Proctol Soc Cancer Res Award, 50. Mem: Am Soc Clin Path; AMA; Am Asn Path & Bact; Am Soc Exp Path; Am Soc Hemat. Res: Peritoneal infections; lung diseases; leukemias; hematopoiesis; cancer. Mailing Add: PO Box 1016 Pacific Palisades CA 90272

STEINBERG, DANIEL, b Windsor, Ont, July 21, 22; US citizen; m 46; c 3. BIOCHEMISTRY. Educ: Wayne State Univ, BS, 42, MD, 44; Harvard Univ, PhD(biochem), 51. Prof Exp: Intern internal med, Boston City Hosp, 44-45; resident, Detroit Receiving Hosp, 45-46; instr physiol, Med Sch, Boston Univ, 47-48; res scientist, Sect Cellular Physiol, Nat Heart Inst, 51-54, from actg chief to chief sect metab, 54-68; PROF MED, HEAD DIV METAB DIS & PROG DIR BASIC SCI IN MED, SCH MED, UNIV CALIF, SAN DIEGO, 68- Concurrent Pos: Vis scientist, Carlsberg Labs, Copenhagen, 52-53; pres, Found Advan Educ in Sci, 59; ed-in-chief, J Lipid Res, 61-64; chmn coun arteriosclerosis, Am Heart Asn, 68-69. Mem: AAAS; Am Soc Biol Chem; Am Chem Soc; Am Oil Chem Soc; Soc Exp Biol & Med. Res: Mechanisms of hormone action; biochemistry of lipid and lipoprotein metabolism and its relation to atherosclerosis. Mailing Add: Div of Metab Dis Univ of Calif Sch of Med La Jolla CA 92037

STEINBERG, DANIEL J, b Washington, DC, Feb 13, 35; m 59; c 2. THERMODYNAMICS, HYDRODYNAMICS. Educ: Johns Hopkins Univ, AB, 55; Harvard Univ, MA, 57, PhD(physics), 61. Prof Exp: SR PHYSICIST, LAWRENCE LIVERMORE LAB, UNIV CALIF, 61- Mem: Am Phys Soc. Res: Thermodynamics and equations of state; production of multi-megagauss magnetic fields utilizing high explosives. Mailing Add: Lawrence Livermore Lab L-24 Lawrence CA 94550

STEINBERG, DAVID H, b Bronx, NY, Nov 24, 29; m 52; c 4. ORGANIC CHEMISTRY. Educ: Yeshiva Univ, BA, 51; NY Univ, MS, 56, PhD(org chem), 60. Prof Exp: Res asst biochem, Montefiore Hosp, NY, 52-53; res assoc org chem res div, NY Univ, 53-59; res chemist res div, Geigy Chem Corp, 59-72, RES ASSOC PLASTICS & ADDITIVES RES, CIBA-GEIGY CHEM CORP, 72- Mem: Am Chem Soc; The Chem Soc. Res: Organic chemistry encompassing synthesis, reaction mechanism, stereochemistry and structure-activity relationships. Mailing Add: Ciba-Geigy Chem Corp 444 Saw Mill River Rd Ardsley NY 10502

STEINBERG, ELIOT, b New York, NY, June 5, 23; m 47; c 3. ORGANIC CHEMISTRY, RESEARCH ADMINISTRATION. Educ: Polytech Inst Brooklyn, BS, 43, MS, 47. Prof Exp: Res chemist, Johnson & Johnson, NJ, 43-44; res chemist, Chilcott Labs Div, Maltine Co, 47-52; res adminr, Warner-Chilcott Labs, 52-58, DIR RES ADMIN, WARNER-LAMBERT RES INST, 58- Mem: AAAS; Am Chem Soc; Am Inst Chem; Drug Info Asn; NY Acad Sci. Res: Information retrieval. Mailing Add: Warner-Lambert Res Inst 170 Tabor Rd Morris Plains NJ 07950

STEINBERG, ELLIS PHILIP, b Chicago, Ill, May 26, 20; m 44; c 3. CHEMISTRY. Educ: Univ Chicago, SB, 41, PhD(chem), 47. Prof Exp: Jr chemist, Elwood Ord Plant, US War Dept, Ill, 41-43; jr chemist Manhattan dist metall lab, Univ Chicago, 43-46; consult, AEC, 46-47, assoc chemist & asst group leader, 47-58, sr chemist & group leader, 58-74, SECT HEAD NUCLEAR & INORG CHEM, ARGONNE NAT LAB, 74- Concurrent Pos: Guggenheim fel, 57-58; mem sci adv comt space radiation effects lab accelerator, Col William & Mary; mem subcomt nuclear instruments & tech, Nat Acad Sci-Nat Res Coun, 48-59 & subcomt radiochem, 59-70; Am Chem Soc rep, Am Nat Standards Inst Subcomt Nuclear Med, 70-74. Mem: AAAS; Am Chem Soc; Am Phys Soc; Sigma Xi. Res: Nuclear and radiochemistry; nuclear fission; high energy nuclear reactions. Mailing Add: Chem Div Argonne Nat Lab 9700 S Cass Ave Argonne IL 60439

STEINBERG, GEORGE MILTON, organic chemistry, see 12th edition

STEINBERG, GUNTHER, b Cologne, Ger, Apr 14, 24; nat US; m 49; c 2. SURFACE CHEMISTRY, PHYSICAL CHEMISTRY. Educ: Univ Calif, Los Angeles, BS, 48, MS, 50, PhD(physiol chem), 56. Prof Exp: Res assoc, Scripps Metab Clin, Calif, 50; biochemist atomic energy proj, Univ Calif, Los Angeles, 50-56; res chemist, Martinez Res Lab, Shell Oil Co, 56-61, res chemist, Shell Develop Co, 61-64; res chemist, Stanford Res Inst, 64-67; SR STAFF CHEMIST, MEMOREX CORP, 67- Mem: Am Chem Soc. Res: Surface and physical chemistry of polymer composites; physical and chemical measurements; magnetic tape development; electrical contact phenomena; heterogeneous catalysis; radiotracer applications. Mailing Add: Memorex Corp 1200 Memorex Dr Santa Clara CA 95052

STEINBERG, HERBERT AARON, b Bronx, NY, Sept 19, 29; m 55; c 2. MATHEMATICS. Educ: Cornell Univ, BA, 50; Yale Univ, MA, 51, PhD(math), 55. Prof Exp: Aerodynamicist, Repub Aviation Corp, 55-56; digital systs engr, Sperry Gyroscope Co, 56; sr mathematician, TRG Div, Control Data Corp, 56-68; DIR SCI SERV, MATH APPLNS GROUP, INC, 68- Mem: Soc Indust & Appl Math; Asn Comput Mach; Inst Elec & Electronics Eng; Math Asn Am; Am Math Soc. Res: Systems engineering; signal processing; Monte Carlo methods; computer simulation; radiation transport; random noise; stochastic processes; numerical analysis. Mailing Add: Math Applns Group Inc 3 Westchester Plaza Elmsford NY 10523

STEINBERG, HOWARD, b Chicago, Ill, Aug 23, 26; m 46; c 3. ORGANIC CHEMISTRY. Educ: Univ Ill, BS, 48; Univ Calif, Los Angeles, PhD(chem), 51. Prof Exp: AEC fel reaction mechanisms, Mass Inst Technol, 51-52; res chemist, Aerojet-Gen Corp, 52; res assoc org synthesis, Univ Calif, Los Angeles, 52-53; collabr natural prod, USDA, 53-54; res chemist, Pac Coast Borax Co, 54, mgr org res, US Borax Res Corp, 55-58, asst dir chem res, 58, from assoc dir to dir, 59-63, vpres, 63-69, PRES, US BORAX RES CORP, 69-, VPRES, US BORAX & CHEM CORP, 69- Mem: Am Chem Soc; The Chem Soc. Res: Boron and synthetic organic chemistry; reaction mechanisms; kinetics. Mailing Add: US Borax & Chem Corp 3075 Wilshire Blvd Los Angeles CA 90010

STEINBERG, ISRAEL VICTOR, organic chemistry, polymer chemistry, see 12th edition

STEINBERG, JAMES, b Winnipeg, Man, Aug 4, 35; m 58; c 2. MEDICINE, BIOCHEMISTRY. Educ: Univ Man, MD & BSc, 59, MSc, 62. Prof Exp: From intern to asst resident med, Winnipeg Gen Hosp, 59-61; sr resident, Boston City Hosp, 62-63; resident renal metab, Boston Vet Admin Hosp, 63-64; asst med, Peter Bent Brigham Hosp, 64-67; vis scientist, Strangeways Res Lab, Cambridge, Eng, 67-69; res assoc, Cancer Res Inst, New Eng Deaconess Hosp, 69-74; ASST PROF MED, HARVARD MED SCH, 74- Concurrent Pos: Clin fel endocrinol, Winnipeg Gen Hosp, 61-62; res fel, Harvard Med Sch, 64-69, Med Found Inc fel, 67-70; instr, Dept Med, Harvard Med Sch, 69-73; asst, Boston City Hosp, 65-67; assoc med, Peter Bent Brigham Hosp, 74-; res assoc, Robert B Brigham Hosp, 74-; endocrinologist, West Roxbury Vet Admin Hosp, 74- Mem: AAAS; Am Fedn Clin Res; Endocrine Soc; Brit Biochem Soc; Brit Bone & Tooth Soc. Res: Metabolic bone disease; disorders of connective tissue; cellular control mechanisms; tissue culture. Mailing Add: Robert B Brigham Hosp 125 Parker Hill Ave Boston MA 02120

STEINBERG, JOHN CHRISTIAN, b Lakota, Iowa, June 21, 95; m 27; c 2. UNDERWATER ACOUSTICS. Educ: Coe Col, BS, 16, MS, 17; Univ Iowa, PhD(physics), 22. Prof Exp: Mem tech staff, Western Elec Co, 22-25; mem tech staff, Bell Tel Labs, Inc, 25-57; prof marine sci, Univ Miami, 60-72; SR RES SCIENTIST, INST ACOUST RES, MIAMI DIV, PALISADES GEOPHYS INST, 72- Mem: AAAS; Am Phys Soc; Acoust Soc Am (vpres, 44-46, pres, 47-49). Res: Physical characteristics of speech, music and noise; characteristics of hearing; speech transmission with reduced frequency range; underwater sound propagation, devices and systems; relationships between variations of underwater sound waves in the ocean and variations in the medium. Mailing Add: 8730 SW 48th St Miami FL 33165

STEINBERG, JOSEPH, b New York, NY, Mar 22, 20; m 49; c 2. MATHEMATICAL STATISTICS. Educ: City Col New York, BS, 39. Prof Exp: Statistician pop div, US Bur Census, 40-42; math statistician, Social Security Bd, 42-44; statistician statist res div, US Bur Census, 44-46, chief statist methods br pop & housing div, 46-59, statist methods officer, 59-60, officer statist methods div, 60-63; chief math statistician, Social Security Admin, 63-72; dir off surv methods res & asst comnr surv design, Bur Labor Statist, 72-75; PRES, SURV DESIGN, INC, 75- Concurrent Pos: Lectr, USDA Grad Sch, 42-; consult, Orgn Am States, Chile, 65 & 67; vis prof surv res ctr, Inst Social Res, Univ Mich, 68, 69, 70 & 72; mem assembly behav & soc sci, Nat Acad Sci-Nat Res Coun, 72-, mem comt energy consumption measurement, 75-; mem, Inter-Am Statist Inst. Honors & Awards: Meritorious Serv Award, Dept Com, 55; Comnr Citation, Soc Security Admin, 65; Distinguished Serv Award, Dept Health, Educ & Welfare, 68. Mem: Hon fel AAAS; hon fel Am Statist Asn; Am Soc Qual Control; Inst Math Statist; Int Asn Surv Statisticians. Res: Sample survey design and statistical analysis; evaluation of non-sampling errors; response variance and bias; data linkage; computer analysis; quality control; operations research; cost functions and optimization. Mailing Add: 962 Wayne Ave Suite 422 Silver Spring MD 20910

STEINBERG, MALCOLM SAUL, b New Brunswick, NJ, June 1, 30; m 56; c 4. DEVELOPMENTAL BIOLOGY, EMBRYOLOGY. Educ: Amherst Col, BA, 52; Univ Minn, MA, 54, PhD(zool), 56. Prof Exp: Instr zool, Univ Minn, 55; fel embryol, Carnegie Inst, 56-58; from asst prof to assoc prof biol, Johns Hopkins Univ, 58-66; prof, 66-75, HENRY FAIRFIELD OSBORN PROF BIOL, PRINCETON UNIV, 75- Concurrent Pos: Instr-in-charge embryol course, Woods Hole Marine Biol Lab, 67-72, trustee, 69- Mem: Fel AAAS; Am Soc Cell Biol; Am Soc Zool; Soc Develop Biol (Secy, 70-73); Int Soc Develop Biol. Res: Regeneration in coelenterates; mechanisms of morphogenesis; biological self-organization; tissue reconstruction by dissociated cells; physics, chemistry and physiology of cell adhesion. Mailing Add: Dept of Biol Princeton Univ Princeton NJ 08540

STEINBERG, MARIA WEBER, mathematics, see 12th edition

STEINBERG, MARSHALL, b Pittsburgh, Pa, Sept 18, 32; m 62; c 3. PHARMACOLOGY, TOXICOLOGY. Educ: Georgetown Univ, BS, 54; Univ Pittsburgh, MS, 56; Univ Tex Med Br Galveston, PhD(pharmacol, toxicol), 66; Nat Registry Clin Chem, cert, 70. Prof Exp: Asst dir trop testing, Univ Pittsburgh, 55-56; Med Serv Corps, US Army, 57-, chief clin path lab, 97th Gen Hosp, Frankfurt, Ger, 56-60, chief biochem & toxicol div, 4th Army Med Lab, San Antonio, Tex, 61-63, chief toxicol div, Environ Hyg Agency, 66-71, dir lab serv US Army environ Hyg Agency, 72-75; CONSULT TO SURGEON GEN, LAB SCI, 75- Concurrent Pos: Liaison mem, Armed Forces Pest Control Bd, 67-75; mem pesticide monitoring panel, Fed Working Group Pesticide Mgt, President's Cabinet Comt on Environ, 71-72 & safety panel, 72- Mem: Soc Toxicol; Am Soc Pharmacol & Exp Therapeut; Am Conf Govt Indust Hygienists; Am Indust Hyg Asn. Res: Applied research in industrial and environmental toxicology, insect repellants, pesticides and fire extinguishants, particularly in regard to hazards owing to skin penetration or irritation as well as toxic effects due to inhalation. Mailing Add: HQ Dept of the Army Office of the Surgeon Gen Washington DC 20310

STEINBERG, MARTIN, b Chicago, Ill, Apr 18, 20; m 42; c 3. PHYSICAL CHEMISTRY. Educ: Univ Ill, BS, 41; Univ Chicago, PhD(chem), 49. Prof Exp: Res chemist, Continental Carbon Co, 41-45; instr chem, Wilson Jr Col, 48; res assoc stable isotopes inst nuclear studies, Univ Chicago, 49-51; res chemist, Gen Elec Co, 51-56; sr scientist, Armour Res Found, 56-61; HEAD CHEM PHYSICS, DELCO ELECTRONICS DIV, GEN MOTORS CORP, GOLETA, 61- Mem: AAAS; Am Chem Soc; Am Phys Soc; Combustion Inst. Res: Carbon black formation and properties in rubber; electrochemistry and polarography in fused salts; stable isotope equilibria; mass spectrometry; combustion; reentry physics; high temperature kinetics; atom recombination on surfaces; gaseous radiation and spectroscopy; air pollution chemical lasers. Mailing Add: 345 N Ontare Rd Santa Barbara CA 93105

STEINBERG, MARVIN PHILLIP, b Philadelphia, Pa, Oct 4, 22; m 46; c 3. FOOD TECHNOLOGY. Educ: Univ Minn, BS, 43, MS, 49; Univ Ill, PhD(food technol), 53. Prof Exp: Asst chem eng, Univ Minn, 45-49; asst food tech, 49-52, from instr to assoc prof, 52-63, PROF FOOD TECH, UNIV ILL, URBANA, 63- Mem: Am Chem Soc; Inst Food Technol. Res: Application of chemical engineering to problems in food processing; heat transfer and fluid flow to dehydration, freezing, canning and fermentation. Mailing Add: Dept of Food Sci Univ of Ill Urbana IL 61801

STEINBERG, MAYNARD ALBERT, b Winthrop, Mass, Aug 8, 19; m 52; c 2. FOOD SCIENCE, FISHERIES. Educ: Univ Mass, BS, 47, PhD(food technol), 55; Ore State Col, MS, 48. Prof Exp: Biochemist, Polytech Inst Brooklyn, 48-52; biochemist, DCA Food Industs, 55-57; biochemist, 57-67, DIR, PAC UTILIZATION RES CTR, US FISH & WILDLIFE SERV, 66- Mem: Am Chem Soc; Inst Food Technol. Res: Animal proteins; enzymes; lipids. Mailing Add: Pac Utilization Res Ctr Nat Marine Fish Serv NOAA 2725 Montlake Blvd E Seattle WA 98112

STEINBERG, MELVIN SANFORD, b Canton, Ohio, Mar 28, 28; m 54; c 2. THEORETICAL PHYSICS. Educ: Univ NC, BS, 49, MS, 51; Yale Univ, PhD(physics), 55. Prof Exp: Asst physics, Yale Univ, 54-55; asst prof, Stevens Inst Technol, 55-59; assoc prof, Univ Mass, 59-62; ASSOC PROF PHYSICS, SMITH COL, 62- Concurrent Pos: Res assoc, Woods Hole Oceanog Inst, 56-58 & 62; res assoc, Air Force Cambridge Res Lab, 60-61; NSF sci fac fel, 66-67. Mem: Am Phys Soc. Res: Theory of solids; acoustics; electrodynamics. Mailing Add: Dept of Physics Smith Col Northampton MA 01060

STEINBERG, MERL, physiology, biochemistry, see 12th edition

STEINBERG, MITCHELL IRWIN, b Philadelphia, Pa, Jan 22, 44; m 66; c 2. PHARMACOLOGY. Educ: Philadelphia Col Pharm & Sci, BSc, 66; Univ Mich, PhD(pharmacol), 70. Prof Exp: Res fel, Univ Conn Health Ctr, 70-71, instr pharmacol, 71-72; SR PHARMACOLOGIST, ELI LILLY RES LABS, 72- Mem: AAAS. Res: Interactions of pharmacological agents with excitable membranes of mammalian cardiac & neuronal tissues. Mailing Add: Dept of Pharmacol MC 304 Eli Lilly Res Labs Indianapolis IN 46206

STEINBERG, PHILLIP HENRY, b Cincinnati, Ohio, Dec 11, 31; m 61; c 3. EXPERIMENTAL HIGH ENERGY PHYSICS. Educ: Univ Cincinnati, BS, 54; Northwestern Univ, PhD(physics), 60. Prof Exp: From asst prof to assoc prof physics, 59-75, PROF PHYSICS, UNIV MD, COLLEGE PARK, 75- Concurrent Pos: NSF fel, 65-66. Mem: Am Phys Soc. Res: Experimental elementary particle physics; hypernuclei; leptonic K-meson decay; meson-nucleon scattering. Mailing Add: Dept of Physics & Astron Univ of Md College Park MD 20740

STEINBERG, ROBERT, b Stykon, Rumania, May 25, 22; m 52. MATHEMATICS. Educ: Univ Toronto, PhD(math), 48. Prof Exp: Lectr math, Univ Toronto, 47-48; from instr to assoc prof, 48-62, PROF MATH, UNIV CALIF, LOS ANGELES, 62- Mem: Am Math Soc; Math Asn Am. Res: Group representations; algebraic groups. Mailing Add: Dept of Math 6364 Math Sci Bldg Univ of Calif Los Angeles CA 90024

STEINBERG, ROY HERBERT, b New York, NY, Dec 9, 35; m 59; c 1. NEUROPHYSIOLOGY. Educ: Univ Mich, BA, 56, MA, 57; NY Med Col, MD, 61; McGill Univ, PhD(neurophysiol), 65. Prof Exp: Intern med, Mass Mem Hosp, 61-62; USPHS fel, NIMH, 62-65; head Neurophysiol Br, Neurol Sci Div, Naval Aerospace Med Inst, Fla, 68-69; asst res physiologist, 69-72, ASSOC PROF PHYSIOL, SCH MED, UNIV CALIF, SAN FRANCISCO, 72- Concurrent Pos: USPHS career develop award, 71-; William C Bryant, Bernard C Spiegel & Harpuder Awards, NY Med Col. Mem: AAAS; Asn Res Vision & Ophthal; Am Phys Soc. Res: Physiology of the nervous system; vision, especially physiology of the retina. Mailing Add: Dept of Physiol Univ of Calif Med Ctr San Francisco CA 94122

STEINBERG, SHELDON A, veterinary medicine, neurology, see 12th edition

STEINBERG, STANLY, b Traverse City, Mich, Mar 10, 40; m 60. APPLIED MATHEMATICS. Educ: Mich State Univ, BS, 62; Stanford Univ, PhD(math), 68. Prof Exp: Asst prof math, Purdue Univ, Lafayette, 67-74; ASST PROF MATH, UNIV NMEX, 74- Res: Partial differential equations. Mailing Add: Dept of Math Univ of NMex Albuquerque NM 87131

STEINBERG, STUART ALVIN, b Chicago, Ill, Feb 3, 41; m 66; c 1. MATHEMATICS. Educ: Univ Ill, Urbana, BS, 63, PhD(math), 70; Univ Chicago, MS, 65. Prof Exp: Asst prof math, Univ Mo-St Louis, 70-71; asst prof, 71-75, ASSOC PROF MATH, UNIV TOLEDO, 75- Mem: Am Math Soc. Res: Algebra, especially ring theory and ordered algebraic structures. Mailing Add: Dept of Math Univ of Toledo Toledo OH 43615

STEINBERG, WILLIAM, b Brooklyn, NY, Apr 25, 41; m 69; c 3. MICROBIOLOGY. Educ: Brooklyn Col, BS, 62; Univ Wis, MS, 65, PhD(microbiol), 67. Prof Exp: NSF fel, Ctr Molecular Genetics, Gif-sur-Yvette, France, 67-68, Europ Molecular Biol Orgn fel, 68-69; ASST PROF MICROBIOL, MED SCH, UNIV VA, 69- Mem: Am Soc Microbiol. Res: Bacterial spore formation and germination; enzyme structure and function; amino-acyl transfer DNA synthetases; genetics of Bacillus. Mailing Add: Dept of Microbiol Univ of Va Sch Med Charlottesville VA 22903

STEINBERGER, ANNA S, b Radom, Poland, Jan 1, 28; US citizen; m 50; c 2. CELL BIOLOGY, IMMUNOLOGY. Educ: Univ Iowa, MS, 52; Wayne State Univ, PhD(microbiol), 62. Prof Exp: Bacteriologist, Univ Iowa, 53-55; res virologist, Parke, Davis & Co, Mich, 55-56 & 58-59; asst mem, Albert Einstein Med Ctr, 61-71; PROF REPRODUCTIVE BIOL & ENDOCRINOL, UNIV TEX MED SCH HOUSTON, 71- Concurrent Pos: USPHS res grant, Albert Einstein Med Ctr, 61-71. Mem: Endocrine Soc; Soc Study Reproduction; Tissue Cult Asn; NY Acad Sci; Am Soc Andrology. Res: Spermatogenesis and steroidogenesis in male mammalian gonads; hormonal control of spermatogenesis; secretion of gonadotropins; cytogenetics; tissue culture. Mailing Add: 6400 W Cullen St Univ of Tex Med Sch Houston TX 77025

STEINBERGER, JACK, b Bad Kissingen, Ger, May 25, 21; nat US; m 61; c 2. PHYSICS. Educ: Univ Chicago, BS, 42, PhD, 48. Prof Exp: Mem, Inst Advan Study, Princeton, 48-49; res asst, Univ Calif, Berkeley, 49-50; HIGGINS PROF PHYSICS, COLUMBIA UNIV, 50-; PHYSICIST, EUROP CTR NUCLEAR RES, 68- Concurrent Pos: Mem, Inst Advan Study, Princeton, 59-60. Mem: Nat Acad Sci. Res: Mesons, spin, parity and other properties of pions; particle, spins, other properties of strange particles; two neutrinos; CP violating properties of kaons. Mailing Add: Europ Ctr for Nuclear Res Geneva 23 Switzerland

STEINBRECHER, LESTER, b Philadelphia, Pa, Sept 17, 27; m 51; c 3. INORGANIC CHEMISTRY. Educ: Temple Univ, BA, 50; Drexel Inst, MS, 57. Prof Exp: Chemist, Socony Mobil Oil Corp, 52-58; res chemist, 58-70, DIR RES, AMCHEM PROD INC, AMBLER, 70- Mem: Nat Asn Corrosion Eng; Am Soc Electroplaters. Res: Solid state reactions; inorganic metallic coatings; analytical chemistry; coatings for metals. Mailing Add: Amchem Prod Inc Ambler PA 19002

STEINBRENNER, ARTHUR H, b New York, NY, July 23, 17; m 48; c 2. MATHEMATICS EDUCATION. Educ: Columbia Univ, AB, 40, AM, 41, PhD, 55. Prof Exp: Instr math & physics, Graham-Eckes Sch, Fla, 41-46; asst math, Teachers Col, Columbia Univ, 46-48; asst prof, US Naval Acad, 48-53; from instr to asst prof, 53-58, from asst prof math & educ to assoc prof math & educ, 58-70, coordr, Sch Math Study Group, 58-61, PROF MATH & EDUC, UNIV ARIZ, 70- Concurrent Pos: Fulbright lectr, Australia, 63; coordr Ariz Ctr Minn Math & Sci Teaching Proj. Mem: Math Asn Am. Res: Mathematics curricula experiments. Mailing Add: Dept of Math Univ of Ariz Tucson AZ 85721

STEINBRENNER, EUGENE CLARENCE, b St Paul, Minn, Sept 3, 21; m 44; c 4. FORESTRY, SOILS. Educ: Univ Minn, BS, 49; Univ Wis, MS, 51; Univ Wash, Seattle, PhD(forestry), 54. Prof Exp: Asst, Univ Wis, 49-51; asst, State Conserv Dept, Wis, 49-50; FOREST SOILS SPECIALIST, FORESTRY RES CTR, WEYERHAEUSER CO, 52- Concurrent Pos: Bullard fel, Harvard Univ, 68. Mem: Soil Sci Soc Am; Soc Am Foresters. Res: Soil classification and nutrition; productivity; soil management for site protection, rehabilitation and improvement; nursery soils. Mailing Add: Forestry Res Ctr Weyerhaeuser Co Box 420 Centralia WA 98531

STEINDLER, MARTIN JOSEPH, b Vienna, Austria, Jan 3, 28; nat US; m 52; c 2. INORGANIC CHEMISTRY, RESEARCH ADMINISTRATION. Educ: Univ Chicago, PhB, 47, BS, 48, MS, 49, PhD(chem), 52. Prof Exp: Res asst, US Navy Inorg Proj, Univ Chicago, 48-52, consult, 53; assoc chemist, 53-74, SR CHEMIST, ARGONNE NAT LAB, 74- Concurrent Pos: Mem, Atomic Safety & Licensing Bd Panel, 72-; consult, Adv Comt Reactor Safeguards, 67- Mem: Am Chem Soc; Am Nuclear Soc; The Chem Soc. Res: Nuclear fuel cycle; radiological safety; fission product disposal; fluorine chemistry of the actinide elements and fission product elements; reactor fuel reprocessing; non-aqueous inorganic kinetics. Mailing Add: Argonne Nat Lab Bldg 205 RmC227 9700 S Cass Ave Argonne IL 60439

STEINER, ANDRE LOUIS, b Haguenau, France, June 21, 28. ANIMAL BEHAVIOR. Educ: Univ Strasbourg, BSc, 52; Univ Paris, DSc(animal behav), 60. Prof Exp: Res assoc animal biol, Nat Res Coun, Paris, France, 52-56; teaching asst zool, Univ Montpellier, 56-59, teaching asst psychophysiol, 59-61, asst prof psychophysiol & animal behav, 61-65; vis prof zool & animal behav, Univ Montreal, 65-66; assoc prof animal behav, 66-72, PROF ANIMAL BEHAV, UNIV ALTA, 72- Concurrent Pos: Mem, Int Union Conserv Nature & Natural Resources, Switz, 67. Honors & Awards: Medal, Nat Res Coun, France, 63; Cert Merit, Dict Int Biog, 70. Mem: AAAS; Am Soc Mamma; Animal Behav Soc; Can Soc Zool; Ecol Soc Am. Res: Behavior, ecology and distribution of solitary wasps and of some vertebrates; ecology, especially behavioral aspects and field studies; animal distribution and dispersion; cine-photo analysis of behavior. Mailing Add: Dept of Zool Biosci Bldg Univ of Alta Edmonton AB Can

STEINER, ANNE KERCHEVAL, b Warrensburg, Mo, Aug 5, 36. MATHEMATICS. Educ: Univ Mo, AB, 58, MA, 63; Univ NMex, PhD(math), 65. Prof Exp: Eng asst, Am Tel & Tel Corp, 58-59; asst prof math, Tex Tech Col, 65-6;; asst prof, Univ NMex, 66-69; assoc prof, 69-72, PROF MATH, IOWA STATE UNIV, 72- Concurrent Pos: Vis assoc prof, Univ Alta, 70-71. Mem: Am Math Soc; Math Asn Am; Sigma Xi. Res: Point set topology. Mailing Add: Dept of Math Iowa State Univ Ames IA 50011

STEINER, ARNOLD BYRON, b Duluth, Minn, Nov 29, 10; m 39; c 3. ORGANIC POLYMER CHEMISTRY. Educ: Stanford Univ, AB, 33, ChE, 34, PhD(phys chem), 37. Prof Exp: Chemist, Shell Oil Co, 37-38; res chemist, Kelco Co, San Diego, 39-44, tech dir, 44-54, vpres, 54-70, sr vpres, 71-74; RETIRED. Mem: Am Chem Soc; Soc Rheol; Inst Food Technol; Am Inst Chem; Soc Photog Sci & Eng. Res: Photochemistry; colloidal chemistry; petroleum, algin and kelp products; biofermentation gums; vegetable gums and related hydrophilic colloids. Mailing Add: PO Box 166 Rancho Santa Fe CA 92067

STEINER, BRUCE, b Oberlin, Ohio, May 14, 31; m 60; c 1. PHYSICAL CHEMISTRY. Educ: Oberlin Col, AB, 53; Princeton Univ, PhD, 56. Prof Exp: Res assoc physics, Univ Chicago, 58-61; PHYSICIST, NAT BUR STANDARDS, 61- Mem: AAAS; Am Phys Soc; Am Chem Soc; Optical Soc Am; Am Inst Chem. Res: Optical radiation measurement; breakdown of molecules under impact; electron detachment phenomena in ions and molecules. Mailing Add: Nat Bur of Standards Washington DC 20234

STEINER, DIETER, b Menziken, Switz, Sept 21, 32; m 58; c 3. GEOGRAPHY. Educ: Univ Zurich, MSc, 57, PhD(geog), 60. Prof Exp: Res asst geog, Univ Zurich, 57-62; instr, Univ Chicago, 63-64; asst prof, Univ Zurich, 64-68, assoc prof, 68; assoc prof, 68-70, PROF GEOG, UNIV WATERLOO, 70- Concurrent Pos: Chmn comn interpretation of aerial photographs, 64-68 & secy comn geog data sensing & processing, 68, Int Geog Union. Mem: Can Asn Geog; Regional Sci Asn; Am soc Photogram. Res: Photo interpretation; remote sensing; application of quantitative methods to geography; computer mapping. Mailing Add: Dept of Geog Univ of Waterloo Waterloo ON Can

STEINER, DONALD FREDERICK, b Lima, Ohio, July 15, 30. BIOCHEMISTRY, ENDOCRINOLOGY. Educ: Univ Cincinnati, BS, 52; Univ Chicago, MS & MD, 56. Hon Degrees: Dr, Royal Univ Umea, Sweden, 73. Prof Exp: Intern, King County Hosp, Seattle, Wash, 56-57; asst med, Univ Wash, 57-60, med resident, 59-60; from asst prof to prof biochem, 60-70, actg chmn dept biochem, 72-73, A N PRITZKER PROF BIOCHEM & MED, UNIV CHICAGO, 70-, CHMN DEPT BIOCHEM, 73-, DIR DIABETES-ENDOCRINOL CTR, 74- Concurrent Pos: Res fel med, Univ Wash, 57-59; mem metab study sect, USPHS, 65-70; Jacobaeus lectr, Nordisk Insulin Fund, Copenhagen, 70; E F F Copp mem lect, La Jolla, Calif, 71. Honors & Awards: Ernst Oppenheimer Award, Endocrine Soc, 70; Hans Christian Hagedorn Medal, Steensen Mem Hosp, Copenhagen, 70; Lilly Award, Am Diabetes Asn, 69; Gairdner Award, Gairdner Found, Can, 71; 50th Anniversary Medallion, Am Diabetes Asn, 72; Diaz-Cristobal Award, Span Soc Study Diabetes, 73. Mem: Nat Acad Sci; AAAS; Am Soc Biol Chem; Biochem Soc; Am Diabetes Asn. Res: Discovery, isolation, structural analysis and biosynthesis of proinsulin; mechanism of conversion of proinsulin to insulin; insulin binding to tissues and mechanism of action; evolutionary development of insulin and related hormones. Mailing Add: Dept of Biochem Univ of Chicago Chicago IL 60637

STEINER, EDWIN CHARLES, organic chemistry, physical chemistry, see 12th edition

STEINER, ERICH E, b Thun, Switz, Apr 9, 19; nat US; m 44; c 3. GENETICS. Educ: Univ Mich, BS, 40; Ind Univ, PhD(bot), 50. Prof Exp: From instr to asst prof bot, 50-58, assoc prof, 58-61, chmn dept bot, 68-71, PROF BOT, UNIV MICH ANN ARBOR, 61-, DIR MATTHAEI BOT GARDENS, 71- Concurrent Pos: NSF sr fel, 60-61. Mem: Bot Soc Am; Genetics Soc Am; Soc Study Evolution; Am Soc Naturalists; Soc Econ Bot. Res: Genetics and evolutionary biology of Oenothera; genetics of incompatability. Mailing Add: Dept of Bot Univ of Mich Ann Arbor MI 48104

STEINER, EUGENE FRANCIS, b St Louis, Mo, July 15, 34; m 63; c 1. MATHEMATICS. Educ: Univ Mo, BS, 56, MA, 60, PhD(math), 63. Prof Exp: Asst prof physics, Southwestern La Inst, 56-57; eng physicist, McDonnell Aircraft Corp, 58-59; asst prof math, Univ NMex, 63-65; assoc prof, Tex Tech Col, 65-66; assoc prof, Univ NMex, 66-68; assoc prof, 68-72, PROF MATH, IOWA STATE UNIV, 72- Concurrent Pos: Vis prof, Univ Alta, 70-71. Mem: Am Math Soc; Math Asn Am. Res: General topology. Mailing Add: Dept of Math Iowa State Univ Ames IA 50010

STEINER, GARY W, b Rexburg, Idaho, Sept 21, 35; m 63; c 3. PLANT PATHOLOGY. Educ: Univ Idaho, BS, 61, MS, 64; Mich State Univ, PhD(plant path), 68. Prof Exp: Assoc pathologist, 68-75, PATHOLOGIST EXP STA, HAWAIIAN SUGAR PLANTERS' ASN, 75- Concurrent Pos: Fel, Mont State Univ, 70. Mem: Am Phytopath Soc. Res: Reaction of fungal spores in soil; root diseases of sugar cane; ecology of cane soils; quarantines; sugar cane diseases. Mailing Add: Hawaiian Sugar Planters Asn PO Box 2450 Honolulu HI 96804

STEINER, GEORGE, b Czech, Mar 11, 36; Can citizen; m 66; c 2. MEDICAL RESEARCH. Educ: Univ BC, BA, 56, MD, 60; FRCP(C), 65. Prof Exp: Resident med, Royal Victoria Hosp, McGill Univ, 60-62; fel, Harvard Med Sch, Peter Bent Brigham Hosp & Joslin Clin Res Lab, 62-64; resident med & endocrinol, Royal Victoria Hosp, McGill Univ, 64-66; from lectr to asst prof, 66-73, ASSOC PROF MED & PHYSIOL, UNIV TORONTO, 73-; DIR LIPID RES CLIN & DIABETES CLIN, TORONTO GEN HOSP, 66- Concurrent Pos: Med Res Coun Can scholar, 67-72; sci panel mem, Can Heart Found, 74-; mem coun atherosclerosis, Am Heart Asn. Mem: Am Diabetes Asn; Am Fedn Clin Res; Am Physiol Soc; Endocrine Soc; Can Soc Clin Invest. Res: Interaction of carbohydrate and lipid metabolism;

hyperlipemia and atherosclerosis. Mailing Add: Dept of Med Rm 7302 Med Sci Bldg Univ of Toronto Toronto ON Can

STEINER, GILBERT, b Moscow, USSR, Jan 19, 37; US citizen; m 71. MATHEMATICS. Educ: Univ Mich, BS, 58, MS, 59; Univ Calif, Berkeley, PhD(math), 62. Prof Exp: Instr math, Reed Col, 62-64; asst prof, Dalhousie Univ, 64-68; asst prof, 68-72, ASSOC PROF MATH, FAIRLEIGH DICKINSON UNIV, TEANECK CAMPUS, 72- Mem: Am Math Soc; Math Asn Am. Res: Functional analysis. Mailing Add: Dept of Math Fairleigh Dickinson Univ Teaneck NJ 07666

STEINER, HARVEY, biochemistry, toxicology, see 12th edition

STEINER, HERBERT M, b Goppingen, Ger, Dec 8, 27; nat US. PARTICLE PHYSICS. Educ: Univ Calif, Berkeley, BS, 51, PhD(physics), 56. Prof Exp: Lectr, 57-60, from asst prof to assoc prof, 61-67, PROF PHYSICS, UNIV CALIF, BERKELEY, 67-, PHYSICIST, LAWRENCE BERKELEY LAB, 53- Concurrent Pos: Guggenheim fel, 60-61. Mem: Am Phys Soc. Res: High energy physics; elementary particle interactions. Mailing Add: Dept of Physics Univ of Calif Berkeley CA 94720

STEINER, JOHANN, b Vienna, Austria, May 1, 26; Can citizen; m 57; c 1. GEOLOGY. Educ: Univ Alta, BSc, 61, MSc, 62; Australian Nat Univ, PhD(geol), 67. Prof Exp: Geologist, Shell Can Ltd, 67-69; ASST PROF GEOL, UNIV ALTA, 69- Mem: AAAS; Geol Soc Am; Geol Asn Can; Am Asn Petrol Geol; Int Asn Sedimentol. Res: Litho-stratigraphy and sedimentation; coal geology; geological history and its relationship to the dynamics of the Milky Way Galaxy. Mailing Add: Dept of Geol Univ of Alta Edmonton AB Can

STEINER, JOHN F, b Milwaukee, Wis, July 21, 08; m 49. ELECTROCHEMISTRY. Educ: Univ Wis, BS, 29, MS, 32, PhD(chem), 33. Prof Exp: Instr chem, Univ Wis, 29-33, Alumni Asn fel, 33-34; develop chemist, Milwaukee Gas Specialty Co, 35-36; res chemist, Globe-Union, Inc, Wis, 36-37; res chemist, Miner Labs, Ill, 38-53; head food lab, Guardite Corp, 53-56; DIR, CHEM RES LABS, 56- Concurrent Pos: Dir develop, Sound Recording Serv, 47-; mem staff, Univ Ill & Wilson Col, 59-; consult, Ill Dept Revenue, 64- & US Dept Internal Revenue, 65- Mem: AAAS; Am Chem Soc; Electrochem Soc. Res: Limnology; ceramics; dentifrices; cereals; food technology; sound recording; information storage-retrieval. Mailing Add: Chem Res Labs 2748 S Superior St Milwaukee WI 53207

STEINER, KIM CARLYLE, b Alton, Ill, Nov 21, 48; m 70; c 2. FOREST GENETICS. Educ: Colo State Univ, BS, 70; Mich State Univ, MS, 71, PhD(forest genetics), 75. Prof Exp: ASST PROF FOREST GENETICS, PA STATE UNIV, 74- Mem: Sigma Xi. Res: Genetic improvement of amenity trees for densely populated areas; genetic adaptation of trees to the environment including genecology; taxonomy, distribution and geographic variation of forest trees. Mailing Add: Sch of Forest Resources Pa State Univ University Park PA 16802

STEINER, LISA AMELIA, b Vienna, Austria, May 12, 33; US citizen. IMMUNOLOGY. Educ: Swarthmore Col, BA, 54; Radcliffe Col, MA, 56; Yale Univ, MD, 59. Prof Exp: Asst prof biol, 57-70, ASSOC PROF BIOL, MASS INST TECHNOL, 70- Concurrent Pos: Helen Hay Whitney res fel microbiol med sch, Wash Univ, St Louis, 62-65; Am Heart Asn res fel immunol, Wright-Fleming Inst, London, 65-67; mem biochem & molecular biol fels rev comt, NIH, 69-70, mem biochem training comt, 71-73, mem allergy & immunol study sect, 74- Mem: Am Asn Immunol; Am Soc Biol Chem; Biophys Soc. Res: Structure and function of immunoglobulins; antigen-antibody reactions; protein chemistry. Mailing Add: Mass Inst of Technol 77 Massachusetts Ave Cambridge MA 02139

STEINER, LOREN FRANKLIN, b Bluffton, Ohio, July 26, 04; m 27; c 1. ENTOMOLOGY. Educ: Ohio State Univ, BS, 26, MS, 27. Hon Degrees: DSc, Ohio State Univ, 67. Prof Exp: Asst entomologist exp sta, Purdue Univ, 27-30; asst entomologist bur entom & plant quarantine, Ga, 30-34, sr entomologist in charge fruit insect lab, Ind, 34-49, res entomologist, Hawaii Fruit Fly Invest, Entom Res Div, Agr Res Serv, 49-68, in charge chem control res, 49-57, tech adv, Fla Mediter Fruit Fly Eradication Prog, 56-57, invest leader & sta head, Hawaii, 57-68, invests leader, Carribean Fruit Fly Invests, US Plant Introduction Sta, Fla, 68-72, CONSULT & COLLABR, USDA, 72- Concurrent Pos: Consult fruit fly eradication methods, USDA, AID, Int Atomic Energy Agency, UN Food & Agr Orgn & several for govts, including Cent Am, Europe, NAfrica, Port Azores & Australia. Honors & Awards: Super Serv Award, USDA, 57, Unit Awards, 53, 59 & 64; J Everett Bussart Honor Award, Entom Soc Am, 66. Mem: AAAS; Entom Soc Am. Res: Deciduous and sub-tropical fruit insect research; biology, behavior, control, new tropical fruit fly detection and eradication methods with lures, insecticides, and sterile insect releases. Mailing Add: 2650 Pearce Dr Apt 301 Clearwater FL 33520

STEINER, MARION ROTHBERG, b New York, NY, May 23, 41; m 63; c 2. BIOCHEMISTRY, ANIMAL VIROLOGY. Educ: Smith Col, BA, 62; Univ Ky, PhD(biochem), 68. Prof Exp: Instr biochem, Univ Ky, 68-70; fel, 71-73, ASST PROF VIROL, BAYLOR COL MED, 73- Concurrent Pos: Nat Cancer Inst spec fel, 74-76. Mem: Am Chem Soc; Am Soc Microbiologists; Sigma Xi. Res: Examination of the structure and function of the surface membrane of oncogenic cells, including studies of glycolipid and glycoprotein composition and metabolism. Mailing Add: Dept of Virol & Epidemiol Baylor Col of Med Tex Med Ctr Houston TX 77030

STEINER, MORRIS, b Shenandoah, Pa, Mar 1, 04; m 29; c 2. PEDIATRICS. Educ: NY Univ, MD, 28. Prof Exp: From asst clin prof to assoc clin prof, 53-66, from assoc prof to prof, 66-74, EMER PROF PEDIAT, STATE UNIV NY DOWNSTATE MED CTR, 74- Concurrent Pos: NY Tuberc & Health Asn grant, State Univ NY Downstate Med Ctr, 66-70; consult, Jewish Hosp, Brooklyn, 69- & King's County Hosp Med Ctr, 69- Mem: Am Acad Pediat; Am Thoracic Soc. Res: Tuberculosis; asthma, chronic broncho-pulmonary disease in children. Mailing Add: Dept of Pediat State Univ NY Downstate Med Ctr Brooklyn NY 11203

STEINER, PINCKNEY ALSTON, b Athens, Ga, Apr 5, 38; m 60; c 2. SOLID STATE PHYSICS, MAGNETIC RESONANCE. Educ: Univ Ga, BS, 59; Duke Univ, PhD(physics), 65. Prof Exp: Res fel chem physics, H C Orsted Inst, Copenhagen Univ, 64-66; ASST PROF PHYSICS, CLEMSON UNIV, 66- Mem: Am Phys Soc; Sigma Xi. Res: Electron paramagnetic resonance applied to various problems in solid state physics and in biophysics. Mailing Add: Dept of Physics & Astron Clemson Univ Clemson SC 29631

STEINER, RAY PHILLIP, b Bronx, NY, Apr 28, 41; m 72. NUMBER THEORY. Educ: Univ Ariz, BSEE, 63, MS, 65; Ariz State Univ, PhD(math), 68. Prof Exp: Asst prof, 68-72, ASSOC PROF MATH, BOWLING GREEN STATE UNIV, 72- Mem: Fibonacci Asn; Am Math Soc; Math Asn Am; Nat Coun Teachers Math. Res: Finding the units and class numbers of algebraic number fields by linear programming techniques, Diophantine equations and Fibonacci numbers. Mailing Add: Dept of Math Bowling Green State Univ Bowling Green OH 43403

STEINER, ROBERT FRANK, b Manila, Philippines, Sept 29, 26; US citizen; m 56; c 2. PHYSICAL BIOCHEMISTRY. Educ: Princeton Univ, AB, 47; Harvard Univ, PhD(phys chem), 50. Prof Exp: Fel, US Naval Med Res Inst, 50-51, phys chemist, 51-70; PROF CHEM, UNIV MD, BALTIMORE COUNTY, 70-, CHMN DEPT, 74- Concurrent Pos: Jewett fel, 50-51; lectr, Georgetown Univ, 57-58 & Howard Univ, 58-59 & 60-61; mem, US Civil Serv Bd Exam, 58-; mem molecular biol panel, NSF, 67-70; ed, Res Commun Chem Path & Pharmacol, 70- & Biophys Chem, 73- Mem: Am Chem Soc; Biophys Soc; Am Soc Biol Chem; NY Acad Sci. Res: Light scattering; fluorescence; protein interactions; nucleic acids; synthetic polynucleotides; statistical thermodynamics. Mailing Add: 2609 Turf Valley Rd Ellicott City MD 21043

STEINER, ROBERT HENRY, plastics chemistry, see 12th edition

STEINER, RODNEY, b Los Angeles, Calif, June 11, 28; m 53; c 3. PHYSICAL GEOGRAPHY. Educ: Univ Calif, Los Angeles, BA, 50, MA, 51; Univ Wash, PhD(geog), 54. Prof Exp: PROF GEOG, CALIF STATE UNIV, LONG BEACH, 56- Mem: Asn Am Geog; Am Geog Soc. Res: California water supply; metropolitan peripheral land use; rural land ownership, regional geography of California. Mailing Add: Dept of Geog Calif State Univ Long Beach CA 90840

STEINER, RUSSELL IRWIN, b Lebanon, Pa, July 21, 27; m 56; c 4. INDUSTRIAL ORGANIC CHEMISTRY. Educ: Lebanon Valley Col, BS, 49; Univ Conn, MS, 52, PhD(chem), 55. Prof Exp: Res chemist, Nat Aniline Div, Allied Chem Corp, 55-63, group leader, 63-65, res supvr indust chem div, 65-67; res assoc, 67-74, GROUP LEADER DISPERSE DYES, DYES & CHEM DIV, CROMPTON & KNOWLES CORP, 74- Mem: Am Chem Soc; Am Asn Textile Chem & Colorists. Res: Textile dyes; food colors. Mailing Add: Res & Develop Dept Dyes Chem Div Crompton & Knowles Corp Reading PA 19603

STEINER, SHELDON, b Bronx, NY, Apr 23, 40; m 63; c 2. VIROLOGY. Educ: Drew Univ, BA, 61; Univ Ky, MS, 64, PhD(microbiol), 67. Prof Exp: Instr, 71-73, ASST PROF VIROL, BAYLOR COL MED, 73- Concurrent Pos: NIH grant, Univ Ky, 69-71; NIH spec fel, Baylor Col Med, 72-75. Mem: AAAS; Am Soc Microbiol. Res: Structure and function of procaryotic and eucaryotic membranes; biochemical characterization of membranes of malignant cells. Mailing Add: Dept of Virol Baylor Col of Med Houston TX 77025

STEINER, WERNER DOUGLAS, b Milwaukee, Wis, Oct 8, 32; m 61; c 3. ORGANIC CHEMISTRY. Educ: Univ Karlsruhe, BA, 56, MA, 58; Univ Pa, PhD(org chem), 64. Prof Exp: Res & develop chemist, Org Chem Dept, Jackson Lab, 63-65, process develop chemist, 67-68 & dyes & intermediates, 68-75, SR CHEMIST, E I DU PONT DE NEMOURS & CO, INC, DEEPWATER, NJ, 62- Mem: AAAS; Am Chem Soc. Res: Research and development of dyes for natural and man-made fibers; research and development of fluorine compounds for use as aerosol propellants, hydraulic fluids, instrument fluids, convective coolants and working fluids; process development of dyes and fluorine compounds; dye process development, manufacture, intermediates Mailing Add: 7 Crenshaw Dr Wilmington DE 19810

STEINERT, LEON ALBERT, theoretical physics, see 12th edition

STEINETZ, BERNARD GEORGE, JR, b Germantown, Pa, May 30, 27; m 49; c 3. ENDOCRINOLOGY. Educ: Princeton Univ, AB, 49; Rutgers Univ, PhD(zool), 54. Prof Exp: Asst org res chemist, Irving Varnish & Insulator Co, 47; asst, Rutgers Univ, 52-54; sr scientist physiol res, Warner-Lambert Res Inst, 54-62, sr res assoc, 62-67; head reproductive physiol & fel, Ciba Res, Ciba Pharmaceut Co, NJ, 67-71, MGR ENDOCRINOL & METAB, PHARMACEUT DIV, CIBA-GEIGY CORP, ARDSLEY, NY, 71- Concurrent Pos: Res assoc prof physiol, NY Univ Sch Med, 74- Mem: Endocrine Soc; Am Physiol Soc; Am Soc Zool; NY Acad Sci; Soc Exp Biol & Med. Res: Reproductive physiology; hormone metabolism; hormones and connective tissue; hormones and aging; hormone interactions. Mailing Add: 336 Longbow Dr Franklin Lakes NJ 07417

STEINFELD, JEFFREY IRWIN, b Brooklyn, NY, July 2, 40. PHYSICAL CHEMISTRY. Educ: Mass Inst Technol, BS, 62; Harvard Univ, PhD(chem), 65. Prof Exp: NSF fel, 65-66; asst prof chem, 66-70, ASSOC PROF CHEM, MASS INST TECHNOL, 70- Concurrent Pos: Alfred P Sloan res fel, 69-71; John Simon Guggenheim fel, Univ Calif, Berkeley & Kammerlingh-Onnes Lab, Leiden, 72-73; mem, Nat Acad Sci-Nat Res Coun Comt Basic Res Adv to US Army Res Off, Durham, 72-75; consult, Aerospace Corp, 74- & A D Little Inc, 73- Mem: AAAS; Am Phys Soc; Fedn Am Sci. Res: Molecular spectroscopy; energy transfer in molecular collisions; applications of lasers to chemical kinetics, isotope separation and atmospheric monitoring. Mailing Add: Dept of Chem Room 2-221 Mass Inst of Technol Cambridge MA 02139

STEINFELD, JESSE LEONARD, b West Aliquippa, Pa, Jan 6, 27; m 53; c 3. CANCER, MEDICINE. Educ: Univ Pittsburgh, BS, 45; Western Reserve Univ, MD, 49; Am Bd Internal Med, dipl, 58. Hon Degrees: LLD, Gannon Col, 72. Prof Exp: Instr med, Univ Calif, 52-54 & George Washington Univ, 54-58; asst dir, Blood Hosp, City of Hope, Duarte, Calif, 58-69; from asst prof to prof med, Sch Med, Univ Southern Calif, 59-68, cancer coordr, 66-68; dep dir, Nat Cancer Inst, 68-69; dep asst secy health & sci affairs, Dept Health, Educ & Welfare, 69-72; surgeon gen, USPHS, 69-73; prof med, Univ Calif, Irvine, 74-76; chief med serv, Long Beach Vet Admin Hosp, 74-76; DEAN, SCH MED, MED COL VA, 76- Concurrent Pos: AEC fel med, Univ Calif, 52-53; clin investr, Nat Cancer Inst, 54-58, mem, Krebiozen Rev Comt, 63- & Clin Studies Panel, 64-66, consult, 66-, mem chemother adv comt, 67 & cancer spec prog adv comt, 67; consult, Vet Admin Hosp, Long Beach, Calif, 59-, City of Hope Med Ctr, 60- & Kern County Gen Hosp, Bakersfield, 64-; mem, Calif State Cancer Adv Coun, 61-68; prof med & dir dept oncol, Mayo Clin & Mayo Med Sch, Rochester, Minn, 73-74. Mem: Soc Nuclear Med; Am Asn Cancer Res; AMA; fel Am Col Physicians; Am Fedn Clin Res. Res: Cancer chemotherapy; hematology; health administration. Mailing Add: Med Col of Va Richmond VA 23298

STEINFELD, LEONARD, b New York, NY, Nov 16, 25; m 65; c 4. MEDICINE. Educ: Hofstra Col, BA, 49; State Univ NY Downstate Med Ctr, MD, 53. Prof Exp: Intern, Los Angeles County Gen Hosp, 53-54; resident pediat, Mt Sinai Hosp, New York, NY, 54-56, instr pediat, Col Physicians & Surgeons, Columbia Univ, 58-68; assoc prof, 66-69, PROF PEDIAT, MT SINAI SCH MED, 69- Concurrent Pos: NIH fel pediat cardiol, Mt Sinai Hosp, New York, 56-57, NY Heart Asn fel, 57-58; from asst attend pediatrician to assoc attend pediatrician, Mt Sinai Hosp, 59-69, attend pediatrician, 69-; consult, USPHS Hosp, Staten Island, 60- & Perth Amboy Gen Hosp, NY, 70- Mem: Am Acad Pediat; Am Col Cardiol; Am Heart Asn; NY Acad Sci. Res: Heart disease in infants and children. Mailing Add: Dept of Pediat Mt Sinai Sch of Med New York NY 10029

STEINFINK, HUGO, b Vienna, Austria, May 22, 24; nat US; m 48; c 2. PHYSICAL CHEMISTRY. Educ: City Col New York, BS, 47; Columbia Univ, MA, 48; Polytech Inst Brooklyn, PhD(chem), 54. Prof Exp: Res chemist, Shell Develop Co, 48-51 & 54-60; assoc prof chem eng, 60-63, PROF CHEM ENG, UNIV TEX, AUSTIN, 63-

Mem: Am Chem Soc; Mineral Soc Am; Am Crystallog Asn. Res: Crystal structures of silicate minerals and silicate-organic complexes; crystal chemistry and physical properties of semiconductor materials; materials science research. Mailing Add: Dept of Chem Eng Univ of Tex Austin TX 78712

STEINGISER, SAMUEL, b Springfield, Mass, June 6, 18; m 46; c 3. CHEMISTRY. Educ: City Col New York, BS, 38; Polytech Inst Brooklyn, MS, 41; Univ Conn, PhD, 49. Prof Exp: Chemist, Rockefeller Inst, 41-42; asst atom bomb proj s a m labs, Columbia Univ, 42-43; asst, Carbide & Carbon Chems Corp, NY, 44-46; res assoc metall lab, Univ Chicago, 43-44; group leader phys chem, Publicker Industs, Pa, 46; res assoc, US Off Naval Res, Conn, 46-50; group leader & res scientist cent res dept, Monsanto Chem Co, 50-54; group leader, Mobay Chem Co, 54-59, asst res dir, 59-65; scientist, Monsanto Res Corp, Ohio, 66-70, RES MGR & SCI FEL, LOPAC PROJ, MONSANTO CO, 70- Concurrent Pos: Mem mat adv bd, Nat Res Coun, 61-63; mem, Int Standardization Orgn. Mem: Fel AAAS; Am Chem Soc; Soc Plastics Indust; Am Soc Test & Mat; Soc Rheol. Res: Mass spectroscopy; corrosion; electrolysis; nuclear chemistry; high-vacuum phenomena; magneto-optics; magnetic susceptibility; high polymer physics; mechanical properties; advanced composites; instrumentation design; plastics processing and development. Mailing Add: 10 Fox Chase Rd Bloomfield CT 06002

STEINHARDT, CHARLES KENDALL, b Milwaukee, Wis, Mar 1, 35. ORGANIC CHEMISTRY. Educ: Univ Wis, BS, 57; Univ Ill, PhD(org chem), 63. Prof Exp: Instr chem, Univ Wash, Seattle, 63-65; mem staff, Lubrizol Corp, Ohio, 65-71; CHEMIST, NAPKO CORP, 73- Mem: Am Chem Soc. Mailing Add: 11427 Dunlap Dr Houston TX 77035

STEINHARDT, JACINTO, b New York, NY, May 20, 06; m 24; c 2. PHYSICAL CHEMISTRY, BIOCHEMISTRY. Educ: Columbia Univ, AB, 27, AM, 28, PhD(biophys), 34. Prof Exp: Asst instr biol, City Col New York, 27-28; asst instr zool, Columbia Univ, 28-29, instr biophys, 29-34; Nat Res Coun fel, Univ Copenhagen, Univ Upsala, Cambridge Univ & Harvard Univ, 34-36; Rockefeller Found fel, Harvard Univ, 37-38; phys chemist res lab textile found, Nat Bur Standards, 38-42; res assoc div war res, Columbia Univ, 42-43; mem opers eval group & naval warfare anal group, Mass Inst Technol, US Dept Navy, DC, 45-62; PROF CHEM, GEORGETOWN UNIV, 62- Honors & Awards: Medal of Freedom, 45; Presidential Cert Merit, 46. Mem: Am Soc Biol Chem; Am Chem Soc. Res: Physical chemistry of proteins and other macromolecules; operations research. Mailing Add: Dept of Chem Rm 306 Sci Ctr Georgetown Univ Washington DC 20015

STEINHARDT, RALPH GUSTAV, JR, b Newark, NJ, Sept 15, 18; m 46; c 3. CHEMISTRY, CHEMICAL PHYSICS. Educ: Lehigh Univ, BS, 40, MS, 41, PhD(chem), 50; Va Polytech Inst, BS, 44. Prof Exp: Chemist, Los Alamos Sci Lab, 44-46; res assoc, Lehigh Univ, 50-53, asst prof chem, 53-54; assoc prof, Va Polytech Inst, 54-56; chmn dept, 56-71, PROF CHEM, HOLLINS COL, 56- Concurrent Pos: Res partic, Oak Ridge Nat Lab, 56-; vis lectr, Stanford Univ, 63-64; res collabr, Brookhaven Nat Lab, 70- Mem: Fel AAAS; Am Chem Soc; fel Am Inst Chemists. Res: Molecular spectroscopy of condensed phases; stereospecific biochemical reactions of psychological interest; x-ray photoelectron spectroscopy of solids. Mailing Add: Dept of Chem Hollins College VA 24020

STEINHARDT, RICHARD ANTONY, b Washington, DC, Sept 23, 39. NEUROBIOLOGY, CELL BIOLOGY. Educ: Columbia Univ, AB, 61, PhD(biol sci), 66. Prof Exp: NSF fel, Agr Res Coun Inst Animal Physiol, Babraham & Plymouth Marine Sta, Eng, 66-67; asst prof zool, 67-73, ASSOC PROF ZOOL, UNIV CALIF, BERKELEY, 73- Mem: AAAS; Am Soc Zool; Soc Develop Biol. Res: Cellular physiology related to the development and function of the nervous system; biophysics of fertilization; ion transport; membrane permeability. Mailing Add: Dept of Zool Univ of Calif Berkeley CA 94720

STEINHART, CAROL ELDER, b Cleveland, Ohio, May 27, 35; m 58; c 3. PLANT PHYSIOLOGY. Educ: Albion Col, AB, 56; Univ Wis, PhD(bot), 60. Prof Exp: Biologist lab gen & comp biochem, NIMH, 61-66, sci analyst div res grants, NIH, Md, 66-68, biologist, 68-70; SCI WRITER & ED, 70- Mem: Am Inst Biol Sci. Res: Growth, differentiation and nutrition of plant tissue cultures; hormonal control of enzyme synthesis in plants; ecology and environmental problems. Mailing Add: 104 Lathrop St Madison WI 53705

STEINHART, JOHN SHANNON, b Chicago, Ill, June 3, 29; m 58; c 2. GEOPHYSICS, SCIENCE POLICY. Educ: Harvard Univ, AB, 51; Univ Wis, PhD, 60. Prof Exp: Proj leader, Woods Hole Oceanog Inst, 56; NSF fel, Carnegie Inst Wash Dept Terrestrial Magnetism, 60-61, mem staff, 61-68; tech asst, Resources & Environ, Off Sci & Technol, Exec Off of the Pres, 68-70; PROF GEOPHYS & ENVIRON STUDIES, UNIV WIS-MADISON, 70-, ASSOC DIR, MARINE STUDIES CTR, 73- Mem: Soc Explor Geophys; Seismol Soc Am; Am Geophys Union. Res: Physics of solid earth; terrestrial heat flow; environmental policy research; science and public policy research and teaching. Mailing Add: Marine Studies Ctr Univ of Wis 1225 W Dayton St Madison WI 53706

STEINHART, WILLIAM LEE, biochemistry, molecular genetics, see 12th edition

STEINHAUER, ALLEN LAURENCE, b Winnipeg, Man, Oct 17, 31; US citizen; m 58; c 2. ENTOMOLOGY, ECOLOGY. Educ: Univ Man, BSA, 53; Ore State Univ, MS, 55, PhD(entom), 58. Prof Exp: From asst prof entom to assoc prof entom, Univ Md, 58-66; assoc prof, Ohio State Univ, 66-69; assoc prof, 69-71, PROF ENTOM, UNIV MD, COLLEGE PARK, 71- Concurrent Pos: Entom specialist, Ohio State Univ-US Agency Int Develop, Brazil, 66-69; ed, Environ Entom. Mem: AAAS; Entom Soc Am; Int Orgn Biol Control; Brazilian Entom Soc. Res: Applied ecology; forage crop insects; biological control; pest management; insect behavior; graduate training. Mailing Add: Dept of Entom Univ of Md College Park MD 20742

STEINHAUS, DAVID WALTER, b Neillsville, Wis, July 29, 19; m 49; c 4. ATOMIC PHYSICS, SPECTROCHEMISTRY. Educ: Lake Forest Col, AB, 41; Johns Hopkins Univ, PhD(physics), 52. Prof Exp: Physicist, Cent Sci Co, 41-45, Cenco Indust fel, 41-42; res asst inst coop res, Johns Hopkins Univ, 46-52; mem staff & physicist, 52-73, SECT LEADER, LOS ALAMOS SCI LAB, 73- Concurrent Pos: Mem comt line spectra of the elements, Nat Acad Sci-Nat Res Coun, 66-70. Mem: AAAS; Optical Soc Am; Am Asn Physics Teachers; Soc Appl Spectros. Res: Instrument development; visible and ultraviolet spectroscopy; time resolution of spectra from spark discharges; high resolution spectroscopy; optical spectra of the heavy elements; spectrochemical analysis. Mailing Add: 1342 Sage Loop Los Alamos NM 87544

STEINHAUS, JOHN EDWARD, b Omaha, Nebr, Feb 23, 17; m 43; c 5. ANESTHESIOLOGY. Educ: Univ Nebr, BA, 40, MA, 41; Univ Wis, MD, 45, PhD(pharmacol), 50; Am Bd Anesthesiol, dipl, 59. Prof Exp: Asst prof pharmacol, Marquette Univ, 50-51; assoc prof, Univ Wis, 51-54, asst prof anesthesiol, 54-58; assoc prof, 58-59, PROF ANESTHESIOL & CHMN DEPT, SCH MED, EMORY UNIV, 59- Mem: Am Soc Anesthesiol (pres, 70); Asn Univ Anesthetists (pres, 71); Am Soc Pharmacol & Exp Therapeut; AMA. Res: Drug reactions and intoxications; antiarrythmic agents; cough mechanism and suppression; depression of respiratory reflexes. Mailing Add: Dept of Anesthesiol Emory Univ Sch of Med Atlanta GA 30322

STEINHAUS, RALPH K, b Sheboygan, Wis, June 21, 39; m 65; c 2. ANALYTICAL CHEMISTRY. Educ: Wheaton Col, BS, 61; Purdue Univ, PhD(anal chem), 66. Prof Exp: Asst prof chem, Wis State Univ-Oshkosh, 65-68; asst prof, 68-73, ASSOC PROF CHEM, WESTERN MICHIGAN UNIV, 73- Concurrent Pos: Res assoc, Ohio State Univ, 75-76. Mem: Am Chem Soc; Sigma Xi. Res: Kinetics and mechanisms of transition metal chelates; factors affecting stability of metal chelates. Mailing Add: Dept of Chem Western Mich Univ Kalamazoo MI 49001

STEINHAUSER, FREDRIC R, b Lamberton, Minn, July 16, 18; m 49; c 4. GEOGRAPHY. Educ: Mankato State Col, BS, 49; Univ Chicago, MA, 51; Univ Minn, PhD(geog), 60. Prof Exp: Teaching asst geog, Mankato State Col, 47-49; teaching asst, 54-57, from instr to assoc prof, 57-70, PROF GEOG, UNIV MINN, MINNEAPOLIS, 70- Concurrent Pos: State coordr, Nat Coun Geog Educ, 60-; geog consult, Univ Minn Proj Social Studies, 65-67; Hill Found grant community study, 65; univ study grants community study, Univ Minn, Minneapolis, 65-66; consult & study dir, St Anthony Park Community, 65-; pres & dir, St Paul Asn Communities, 70; chmn, Mayor's Finance Comt, 70; chmn, St Paul Sch Bond Comt, 71. Mem: Nat Coun Geog Educ. Res: Urban geography, especially city and community planning; geography education, especially junior college teaching. Mailing Add: Dept of Geog Gen Col Univ of Minn Minneapolis MN 55455

STEINHOFF, HAROLD WILLIAM, b Ft Morgan, Colo, Mar 9, 19; m 44; c 2. WILDLIFE BIOLOGY. Educ: Colo Agr & Mech Col, BS, 41; Syracuse Univ, MS, 47, PhD(wildlife biol), 57. Prof Exp: Timber sale asst, US Forest Serv, 40, timber estimator, 41; asst, State Univ NY Col Forestry, Syracuse, 41-42; asst prof wildlife mgt, Colo State Univ, Ft Collins, 47-54, from assoc prof to prof, 54-70, actg head dept, 64-66, actg asst dean grad sch, 66-67, centennial prof, 70-74, SOUTHWEST REGIONAL ADMINR, COLO STATE UNIV, 74- Mem: Wildlife Soc. Res: Wildlife population dynamics; zoogeography; wildlife, bitterbrush and snow ecology; bio-economic and multiple resource use models; small mammal populations; elk. Mailing Add: Southwest Regional Off Rm 102A Ft Lewis Col Durango CO 81301

STEINHOFF, RAPHAEL JOHN, forestry, genetics, see 12th edition

STEINHOFF, RAYMOND OAKLEY, b Hart, Mich, Apr 22, 25; m 52; c 1. GEOLOGY. Educ: Southern Methodist Univ, BS & MS, 48; Tex A&M Univ, PhD(geol), 65. Prof Exp: Instr geol, Tex A&M Univ, 48-51; geologist, Humble Oil & Refining Co, 52-57; from asst prof to assoc prof geol, Tulane Univ, 57-70, chmn dept, 69-70; PROF GEOL & HEAD DEPT, STEPHEN F AUSTIN STATE UNIV, 70- Mem: Am Asn Petrol Geol. Res: Structural and petroleum geology. Mailing Add: Dept of Geol Stephen F Austin State Univ Nacogdoches TX 75961

STEININGER, DONALD HARRY, physics, see 12th edition

STEINITZ, MICHAEL OTTO, b New York, NY, June 12, 44; m 65; c 2. SOLID STATE PHYSICS, MATERIALS SCIENCE. Educ: Cornell Univ, BE, 65; Northwestern Univ, Evanston, PhD(mat sci), 70. Prof Exp: Coop student comput logic, Philco Corp, 62-63, student radio propagation, 63; Nat Res Coun Can fel dept physics, Univ Toronto, 70-72, proj scientist solid state physics dept metall, 72-73, instr, Scarborough Col, 70-72; ASST PROF PHYSICS, ST FRANCIS XAVIER UNIV, 73- Mem: Am Phys Soc; Can Asn Physicists. Res: Magnetic properties of non-ferromagnetic transition metals; thermal expansion; magnetostriction; neutron diffraction; ultrasonic attenuation; phase transitions; chromium; defect structures in oxides. Mailing Add: PO Box 154 Dept of Physics St Francis Xavier Univ Antigonish NS Can

STEINKE, FREDERICH H, b Wilmington, Del, Nov 26, 35; m 60; c 2. POULTRY NUTRITION. Educ: Univ Del, BS, 57, MS, 59; Univ Wis, PhD(biochem), 62. Prof Exp: Asst mgr poultry nutrit, 62-63, mgr turkey res, 63-72, MGR NUTRIT RES, CENT RES LAB, RALSTON-PURINA CO, 72- Mem: Poultry Sci Asn. Res: Poultry nutrition including broilers, laying hens and turkeys; nutrition evaluation; human nutrition. Mailing Add: 8818 Brookview Dr Crestwood MO 63126

STEINKE, PAUL KARL WILLI, b Friedeberg, Ger, July 13, 21; nat US; m 44; c 3. AGRICULTURAL MICROBIOLOGY. Educ: Univ Wis, BS, 47, MS, 48, PhD(agr bact), 51. Prof Exp: Asst agr bact, Univ Wis, 47-51; food bacteriologist, Grain Belt Co, 51-56; bacteriologist, Paul-Lewis Labs, Inc, 56-61; mgr tech serv, Chas Pfizer & Co, Inc, NY, 61-63, mgr com develop, 64-69, prod mkt mgr, 70-71, TECH DIR, PFIZER, INC, MILWAUKEE, 72- Mem: Am Soc Microbiol; Am Soc Brewing Chem; Master Brewers Asn Am; Am Dairy Sci Asn. Res: Microbiology of meat products; food poisoning bacteria; thermal death time of food spoilage organisms; dehydration of foods; brewing bacteriology and technology; chemistry of hops; cheese cultures and media. Mailing Add: 6060 N Kent Ave Whitefish Bay WI 53217

STEINKER, DON COOPER, b Seymour, Ind, Oct 6, 36; m 59. PALEONTOLOGY. Educ: Ind Univ, BS, 59; Univ Kans, MS, 61; Univ Calif, Berkeley, PhD(paleont), 69. Prof Exp: Res Paleontologist, Univ Calif, Berkeley, 62-63, teaching asst paleont, 62-65, instr, 66-67; asst prof geol, San Jose State Col, 65-66; lectr, Univ Calif, Davis, summer 66; asst prof, 67-75, ASSOC PROF GEOL, BOWLING GREEN STATE UNIV, 75- Mem: Paleont Soc; Nat Asn Geol Teachers. Res: Foraminiferal biology and ecology; paleobiology; primate evolution. Mailing Add: Dept of Geol Bowling Green State Univ Bowling Green OH 43403

STEINKRAUS, KEITH HARTLEY, b Bertha, Minn, Mar 15, 18; m 41; c 5. MICROBIOLOGY. Educ: Univ Minn, BA, 39; Iowa State Col, PhD(bact), 51. Prof Exp: Chemist, Am Crystal Sugar Co, Minn, 39; microbiologist, Jos Seagram & Sons, Inc, 42; instr electronics, US Army Air Force Tech Training Sch, SDak, 42-43; res microbiologist, Gen Mills, Inc, Minn, 43-47; res microbiologist, Pillsbury Mills, Inc, 51; from asst prof to assoc prof bact, 52-62, PROF BACT, NY STATE COL AGR, CORNELL UNIV, 62- Concurrent Pos: Food & agr specialist, US Nutrit Surv, Ecuador & Vietnam, 59, Burma, 61 & 64, Malaya, Thailand & Korea, 64; spec consult interdept comt nutrit for nat defense, NIH, 59-60, spec consult off int res, 67-68; vis prof microbiol col agr, Univ Philippines, 67-69 & Polytech South Bank, London, 72-73; vis prof, UNESCO/Int Cell Res Orgn/UN Environ Prog Training Course Appl Microbiol, Inst Technol, Bandung, Indonesia, 74. Mem: Fel AAAS; Fel Am Acad Microbiol; Am Soc Microbiol; Inst Food Technol. Res: Biochemical and microbial changes in fermented protein-rich foods; biological and chemical transformations by yeasts, molds and bacteria; biological control of insects with parasitic spore-forming bacteria; extraction of plant proteins. Mailing Add: NY State Agr Exp Sta Cornell Univ Geneva NY 14456

STEINLAGE, RALPH CLETUS, b St Henry, Ohio, July 2, 40; m 62; c 3. MATHEMATICS. Educ: Univ Dayton, BS, 62; Ohio State Univ, MS, 63, PhD(math),

66. Prof Exp: Asst prof math, 66-71, ASSOC PROF MATH, UNIV DAYTON, 71- Mem: Math Asn Am; Am Math Soc. Res: Measure theory and topology; Haar measure on locally compact Hausdorff spaces; function spaces; conditions related to equicontinuity; non-standard analysis; fuzzy topological spaces. Mailing Add: Dept of Math Univ of Dayton Dayton OH 45469

STEINLE, EDMUND CHARLES, JR, b Scranton, Pa, Feb 7, 24; m 47; c 3. ORGANIC CHEMISTRY. Educ: ePauw Univ, AB, 47; Univ Iowa, MS, 49, PhD(org chem), 52. Prof Exp: Chemist, Ethyl Corp Res Lab, 52-55; CHEMIST, PLASTICS & CHEM RES & DEVELOP LAB, UNION CARBIDE CORP, 55- Mem: AAAS; Am Chem Soc; Am Oil Chemists Soc; fel The Chem Soc. Res: Fatty alchohols; biodegradable surfactants; surfactant intermediates. Mailing Add: Chems & Plastics Res & Dev Lab Union Carbide Corp South Charleston WV 25303

STEINMAN, CHARLES ROBERT, b New York, NY, Aug 3, 38. RHEUMATOLOGY, BIOCHEMISTRY. Educ: Princeton Univ, AB, 59; Columbia Univ, MD, 63. Prof Exp: From intern to resident med, Presby Hosp, Columbia Univ, 63-65 & 68-69; assoc biochem, Nat Inst Arthritis, Metab & Digestive Dis, NIH, 65-67; fel rheumatology, Presby Hosp, Columbia Univ, 67-68 & 69-70; vis fel rheumatology, The London Hosp, 70; ASST PROF RHEUMATOLOGY, MT SINAI SCH MED, 70- Concurrent Pos: Asst attend physician, Mt Sinai Hosp, 70-, Beth Israel Hosp, 71- & Bronx Vet Admin Hosp, 71- Mem: Am Rheumatism Asn; AAAS. Res: Study of rheumatoid arthritis; systemic lupus erythematosus and related disorders to determine their pathogenesis by biochemical, immunological and microbiological approaches. Mailing Add: One E 100 St New York NY 10029

STEINMAN, HARRY GORDON, b Trenton, NJ, Jan 5, 13; m 36; c 1. ORGANIC CHEMISTRY. Educ: Mass Inst Technol, BS, 33; Rutgers Univ, MS, 36; Columbia Univ, PhD(chem), 42. Prof Exp: From biochemist to head sect biochem, Lab Clin Invest, Nat Inst Allergy & Infectious Dis, 38-66, chief viral reagents, Nat Cancer Inst, 66-69, MEM STAFF, OFF RES SAFETY, NAT CANCER INST, 69- Mem: Am Chem Soc; Soc Exp Biol & Med; Am Soc Biol Chem; Am Soc Microbiol; Tissue Cult Asn. Res: Cell-mediated immunity; chemotherapy of infectious diseases; penicillins and penicillinases; drug-protein interactions; viral oncology. Mailing Add: Nat Cancer Inst Off Res Safety 9000 Rockville Pike Bethesda MD 20014

STEINMAN, IRVIN DAVID, b New York, NY, Nov 7, 24; m 54; c 2. MICROBIOLOGY. Educ: Brooklyn Col, AB, 48; Univ Chicago, MS, 49; Rutgers Univ, PhD(microbiol), 58. Prof Exp: Instr microbiol col dent, NY Univ, 51-54; asst instr biol, Rutgers Univ, 54-58; mem fac, Monmouth Jr Col, NJ, 58-59; dir biol serv, US Testing Co, 59-61; dir prof serv, White Labs, 61-67; head med commun, Bristol Labs Int Corp, 67-68; clin res assoc, Ciba Pharm Co, 68-69; mem staff, 69-74, ACTG ASST DIR DIV SCI OPINION, BUR CONSUMER PROTECTION, FED TRADE COMN, 74- Mem: AAAS; Am Soc Microbiol; Am Med Asn; Asn Mil Surg US; fel Am Acad Microbiol. Res: Microbial genetics; metabolism; pigmentation and production of antibiotics; scientific evaluation of false and misleading advertising claims; health sciences administration. Mailing Add: 14601 Notley Rd Silver Spring MD 20904

STEINMAN, MARTIN, b Passaic, NJ, Feb 16, 37; m 61; c 2. MEDICINAL CHEMISTRY, ORGANIC CHEMISTRY. Educ: Rutgers Univ, BS, 58, MS, 62; Univ Kans, PhD(med chem), 65. Prof Exp: From chemist to sr chemist, 65-70, prin scientist, 70-73, SECT LEADER, SCHERING CORP, 73- Mem: Am Chem Soc; NY Acad Sci. Res: Stereochemistry; new heterocyclic ring systems; synthesis of potentially useful medicinal agents; structure-activity relationships. Mailing Add: Schering Corp 60 Orange St Bloomfield NJ 07003

STEINMAN, RALPH R, b Asheville, NC, Nov 23, 10; m 38; c 3. DENTISTRY. Educ: Emory Univ, DDS, 38; Univ Mich, MS, 53. Prof Exp: From instr to assoc prof caries control, 53-65, PROF ORAL MED, SCH DENT, LOMA LINDA UNIV, 65- Mem: AAAS; Int Asn Dent Res. Res: Pathology of dental caries, especially as related to nutrition; hypothalamic-parotid endocrine axis and its relation to dental caries. Mailing Add: Loma Linda Univ Sch of Dent Loma Linda CA 92354

STEINMAN, ROBERT, b New York, NY, Mar 30, 18; m 39; c 4. INORGANIC CHEMISTRY, ORGANIC CHEMISTRY. Educ: Carnegie Inst Technol, BS, 39; Univ Ill, MS, 40, PhD(chem), 42. Prof Exp: Res chemist, Owens-Corning Corp, 42-47; dir res, Waterway Projs, 47-48; pres, Garan Chem Corp, Calif, 48-67; VPRES RES & DEVELOP, WHITTAKER CORP, 67- Honors & Awards: Ord Award, US Dept Navy, 44. Mem: AAAS; Am Chem Soc; Soc Plastics Eng; Soc Plastics Indust. Res: Phosphorus nitrogen chemistry; surface chemistry of glass; reinforced plastics. Mailing Add: Whittaker Corp 210 E Alondro Blvd PO Box 192 Gardena CA 90248

STEINMEIER, ROBERT C, b Glendale, Calif, Apr 16, 43; m 68; c 2. BIOCHEMISTRY Educ: Univ Nebr, Lincoln, BS, 65, PhD(biochem), 75. Prof Exp: Immunochemist, Hyland Div, Travenol Labs, 67-69; RES ASSOC BIOENERGETICS, STATE UNIV NY BUFFALO, 74-, NAT CANCER INST FEL, 75- Res: Studies on proteins of the mitochondrial inner membrane, including illucidation of structure-function relationships utilizing various physical and chemical approaches; rapid reaction kinetic studies on hemoproteins and enzymes. Mailing Add: Bioenergetics Lab 168 Acheson Hall State Univ NY Buffalo NY 14214

STEINMETZ, CHARLES, JR, b Brooklyn, NY, Feb 15, 31; m 53; c 1. ZOOLOGY. Educ: Univ Conn, BA, 54; Univ Mich, MS, 55, PhD(fisheries), 61. Prof Exp: Asst, State Bd Fish & Game, Conn, 53; lab asst, US Fish & Wildlife Serv, 55; instr zool, Mich State Univ, 55-56; aide conserv inst fisheries res, Univ Mich, 56-58; asst prof zool, 59-70, PROF ZOOL, SOUTHERN CONN STATE COL, 70- Mem: Am Fisheries Soc; Am Soc Ichthyol & Herpet. Res: Fish hybridization, predation, behavior and management; aquatic biology; larval marine fishes. Mailing Add: Dept of Biol 501 Crescent St Southern Conn State Col New Haven CT 06515

STEINMETZ, CHARLES HENRY, b Logansport, Ind, Oct 5, 29; m 71; c 4. OCCUPATIONAL HEALTH, MEDICAL ADMINISTRATION. Educ: Ind Univ, AB, 50, PhD(comp physiol), 53; Univ Cincinnati, MD, 60; Johns Hopkins Univ, MPH, 72; Am Bd Prev Med, cert gen prev med, 73. Prof Exp: Asst physiol, Ind Univ, 49-50, anat, 50-51 & zool, 51-53; asst chief space biol, Aero Med Field Lab, Holloman AFB, NMex, 53-56; epidemiologist, Off of Dir, Robert A Taft Sanit Eng Ctr, USPHS, Ohio, 58-60; intern, Staten Island Marine Hosp, NY, 60-61; asst gen mgr, Life Sci Opers, NAm Aviation, Inc, 62-68; dir, Systemed Corp, Md, 68-72; vpres, Nat Health Serv, 72-74; ASST CORP MED DIR, MOBIL OIL CORP, 74-; CLIN INSTR PUB HEALTH, MED COL, CORNELL UNIV, 75- Mem: Fel Aerospace Med Asn; fel Am Acad Occup Med; fel Am Occup Med Asn; fel Am Col Prev Med; fel NY Acad Med. Res: General preventive medicine; occupational health; systems analysis. Mailing Add: Mobil Oil Corp Rm 1425 150 E 42nd St New York NY 10017

STEINMETZ, RICHARD, b Cuxhaven, Ger, Aug 3, 22; US citizen; m 54; c 3. GEOLOGY. Educ: Princeton Univ, AB, 54; Pa State Univ, MS, 57; Northwestern Univ, PhD(geol), 62. Prof Exp: Geologist, Plymouth Oil Co, 56-58; geologist, Pan Am Petrol Corp, 61-64, sr scientist, 64-67; asst prof geol, Tex Christian Univ, 67-71; STAFF RES SCIENTIST, AMOCO PROD CO, 71- Mem: Am Asn Petrol Geol; Soc Econ Paleont & Mineral; Int Asn Sedimentol. Res: Sedimentary petrology; sedimentology; petroleum geology. Mailing Add: Amoco Prod Co Res Ctr PO Box 591 Tulsa OK 74102

STEINMETZ, WALTER EDMUND, b Washington, DC, Jan 12, 21; m 47; c 2. ORGANIC CHEMISTRY. Educ: St Ambrose Col, BS, 43; Univ Iowa, MS, 47, PhD(org chem), 49. Prof Exp: Res chemist, Nalco Chem Co, 48-49, group leader org chem, 49-53, dir, 53-57, sr tech adv, 57-60; sect leader, El Paso Natural Gas Prod Co, 60-65; sect leader, 65-69, mgr chem res div, 69-75, SR TECH ADV, PENNZOIL CO, 75- Mem: Am Chem Soc. Res: Herbicides; oil treatment chemicals; petrochemicals. Mailing Add: Chem Res Div Pennzoil Co PO Box 6199 Shreveport LA 71106

STEINMETZ, WAYNE EDWARD, b Huron, Ohio, Feb 16, 45. PHYSICAL CHEMISTRY, MOLECULAR SPECTROSCOPY. Educ: Oberlin Col, AB, 67; Harvard Univ, AM, 68, PhD(chem), 73. Prof Exp: Instr phys sci, St Peter's Boys' High Sch, 69-70; lab instr, Oberlin Col, 70-71; ASST PROF CHEM, POMONA COL, 73- Mem: Am Chem Soc; AAAS. Res: Molecular structure and spectroscopy; the application of low resolution microwave spectroscopy to conformational analysis. Mailing Add: Seaver Chem Lab Pomona Col Claremont CA 91711

STEINMETZ, WILLIAM JOHN, b Wheeling WVa, Nov 14, 39; m 71. APPLIED MATHEMATICS. Educ: St Louis Univ, BS, 60; Ga Inst Technol, MS, 62; Rensselaer Polytech Inst, PhD(math), 70. Prof Exp: Res scientist aeronaut eng, NASA Ames Res Ctr, 61-66; asst prof, 70-75, ASSOC PROF MATH, ADELPHI UNIV, 75- Mem: AAAS; Soc Indust & Appl Math. Res: Applied mathematics, in particular singular perturbations of ordinary and partial differential equations; stochastic differential equations. Mailing Add: Dept of Math Adelphi Univ Garden City NY 11530

STEINMULLER, DAVID, b New York, NY, June 17, 34. TRANSPLANTATION BIOLOGY. Educ: Swarthmore Col, BA, 56; Univ Pa, PhD(zool), 61. Prof Exp: Mus cur, Wistar Inst Anat & Biol, 61-62; lectr embryol, Fac Med, Univ Valencia, 62-63; res instr med genetics, Sch Med, Univ Wash, 63-65, res instr path, 65-66; from instr to asst prof, Sch Med, Univ Pa, 66-68; asst prof, 68-71, ASSOC PROF PATH & SURG, COL MED, UNIV UTAH, 71- Concurrent Pos: Res fel, Wistar Inst Anat & Biol, 61-62; Fulbright fel, Univ Valencia, 62-63; res assoc, Div Path, Inst Cancer Res, 66-68; mem dent res insts & prog adv comt, Nat Inst Dent Res, 71- Mem: Transplantation Soc; Am Asn Cancer Res; Am Asn Immunol. Res: Immunology and biology of tissue and organ transplants; carcinogenesis and tumor immunity; cellular immunology. Mailing Add: Dept of Path Univ of Utah Med Ctr Salt Lake City UT 84132

STEINRAUF, LARRY KING, b St Louis, Mo, June 8, 31; m 68; c 1. BIOCHEMISTRY, PHYSICAL CHEMISTRY. Educ: Univ Mo, BS & MA, 54; Univ Wash, PhD(biochem), 57. Prof Exp: Asst prof phys chem, Univ Ill, Urbana, 59-64; assoc prof biochem, 64-68, PROF BIOCHEM & BIOPHYS, SCH MED, IND UNIV, INDIANAPOLIS, 68- Concurrent Pos: Res fel chem, Calif Inst Technol, 57-58; USPHS fel crystallog, Cavendish Lab, Cambridge, 58-59. Mem: Am Crystallog Asn. Res: Relation of molecular structure to biological activity. Mailing Add: 825 E 80th St Indianapolis IN 46240

STEINSCHNEIDER, ALFRED, b Brooklyn, NY, June 11, 29; m 50; c 2. PEDIATRICS. Educ: NY Univ, BA, 50; Univ Mo, MA, 52; Cornell Univ, PhD(psychol), 55; State Univ NY, MD, 61. Prof Exp: Asst psychol, Univ Mo, 50-52; asst, Cornell Univ, 52-54; engr, Advan Electronics Ctr, Gen Elec Co, NY, 54-57; res assoc, 58-64, asst prof, 64-68, ASSOC PROF PEDIAT, STATE UNIV NY UPSTATE MED CTR, 68- Mem: Am Psychosom Soc; Soc Res Child Develop; Soc Psychophysiol Res; Am Acad Pediat. Res: Sudden infant death syndrome; child development; psychophysiology. Mailing Add: Dept of Pediat State Univ of NY Upstate Med Ctr Syracuse NY 13210

STEINWACHS, DONALD MICHAEL, b Boise, Idaho, Sept 9, 46; m 72. OPERATIONS RESEARCH, PUBLIC HEALTH ADMINISTRATION. Educ: Univ Ariz, BS, 68, MS, 70; Johns Hopkins Univ, PhD(opers res), 73. Prof Exp: RES MGR, HEALTH SERV RES & DEVELOP CTR, JOHNS HOPKINS UNIV, 72-, ASST PROF, DEPT PUB HEALTH ADMIN, 73- Mem: AAAS; Opers Res Soc Am; Inst Mgt Sci; Am Pub Health Asn. Mailing Add: Health Serv Res & Develop Ctr Johns Hopkins Univ Baltimore MD 21205

STEINWAND, PAUL JOSEPH, organic chemistry, see 12th edition

STEITZ, JOAN ARGETSINGER, b Minneapolis, Minn, Jan 26, 41; m 66. BIOCHEMISTRY, MOLECULAR BIOLOGY. Educ: Antioch Col, BS, 63; Harvard Univ, MA, 67, PhD(biochem, molecular biol), 68. Prof Exp: Asst prof, 70-74, ASSOC PROF MOLECULAR BIOPHYS & BIOCHEM, YALE UNIV, 74- Concurrent Pos: NSF fel, Med Res Coun Lab Molecular Biol, Cambridge, Eng, 68-69, Jane Coffin Childs Fund med res fel, 69-70. Mem: Am Soc Biol Chem. Res: Control of transcription and translation; RNA sequence analysis; bacteriophage. Mailing Add: Dept Molec Biophys & Biochem Yale Univ New Haven CT 06510

STEITZ, THOMAS ARTHUR, b Milwaukee, Wis, Aug 23, 40; m 66. MOLECULAR BIOLOGY. Educ: Lawrence Univ, BA, 62; Harvard Univ, PhD(molecular biol, biochem), 66. Prof Exp: NIH grant, Harvard Univ, 66-67; Jane Coffin Childs Mem Fund Med Res fel, Med Res Coun Lab Molecular Biol, Cambridge Univ, 67-70; asst prof biochem, Univ Calif, Berkeley, 70; asst prof, 70-74, ASSOC PROF MOLECULAR BIOPHYS & BIOCHEM, YALE UNIV, 74- Mem: Am Crystallog Asn; Am Soc Biol Chemists. Res: X-ray crystallographic structure determination of biological macromolecules; relation of enzyme structure and mechanism; structure studies of protein-nucleic acid interaction. Mailing Add: Dept of Molecular Biophys & Biochem Yale Univ New Haven CT 06520

STEJSKAL, EDWARD OTTO, b Chicago, Ill, Jan 19, 32; m 57. PHYSICAL CHEMISTRY. Educ: Univ Ill, BS, 53, PhD(chem), 57. Prof Exp: Asst phys chem, Univ Ill, 53-56; NSF fel, Harvard Univ, 57-58; from instr to asst prof phys chem Univ Wis, 58-64, Wis Alumni Res Found fel, 58-59; res specialist, 64-67, SR RES SPECIALIST, MONSANTO CO, 67- Mem: AAAS; Am Phys Soc; Am Chem Soc. Res: Molecular structure; spectroscopy; nuclear magnetic resonance; molecular motion in solids, liquids and gases; nuclear spin relaxation and diffusion phenomena; fluid rheology; lubrication. Mailing Add: Corp Res Dept Monsanto Co 800 N Lindbergh Blvd St Louis MO 63166

STEJSKAL, RUDOLF, b Budejovice, Czech, Apr 16, 31; m 68; c 2. PATHOLOGY. Educ: Charles Univ, Czech, DDS, 56; Univ Chicago, MS, 69; Chicago Med Sch-Univ Health Sci, PhD(path), 72. Prof Exp: Trainee path, Univ Chicago, 67-69; res assoc exp path, Mt Sinai Hosp Med Ctr, Chicago, 70-72; res assoc & asst prof path, Univ

Chicago, 72-73; sr res investr, 73-75, SR RES SCIENTIST PATH, SEARLE LABS, 75- Concurrent Pos: Clin assoc path, Chicago Med Sch, 70-72. Mem: Reticuloendothelial Soc; Soc Pharmacol & Environ Pathologists; Int Acad Path; Am Soc Exp Path; NY Acad Sci. Res: Fate of blood group substances in human cancer; drug-induced carcinogenesis. Mailing Add: Searle Labs PO Box 5110 Chicago IL 60680

STEKEL, FRANK D, b Hillsboro, Wis, Aug 26, 41; m 67. SCIENCE EDUCATION, PHYSICS. Educ: Univ Wis-La Crosse, BS, 63, Univ Wis-Madison, MS, 65; Ind Univ, Bloomington, EdD(sci educ), 70. Prof Exp: From instr to asst prof 65-70, ASSOC PROF PHYSICS, UNIV WIS-WHITEWATER, 70- Mem: AAAS; Am Asn Physics Teachers; Nat Asn Res Sci Teaching; Nat Sci Teachers Asn. Res: Development, implementation and evaluation of physical science instruction at all levels, from the elementary school up to the college and university level. Mailing Add: Dept of Physics Univ of Wis Whitewater WI 53190

STEKIEL, WILLIAM JOHN, b Milwaukee, Wis, Jan 1, 28; m 55; c 2. BIOPHYSICS. Educ: Marquette Univ, BS, 51; Johns Hopkins Univ, PhD(biophys), 57. Prof Exp: USPHS fel & instr physiol, 57-60, asst prof biophys, 60-66, ASSOC PROF BIOPHYS, MED COL WIS, 66- Concurrent Pos: Res exchange prof, Med Univ, Budapest, 63. Mem: AAAS; Biophys Soc; Am Physiol Soc; Inst Elec & Electronics Engrs; Microcirc Soc. Res: Electrophysiology; circulatory physiology; neurophysiology and neurochemistry. Mailing Add: Dept of Physiol Med Col of Wis 561 N 15th St Milwaukee WI 53233

STELCK, CHARLES RICHARD, b Edmonton, Alta, May 20, 17; m 45; c 4. GEOLOGY. Educ: Univ Alta, BSc, 37, MSc, 41; Stanford Univ, PhD(geol), 50. Prof Exp: Lab asst, Univ Alta, 37-41; well site geologist, BC Dept Mines, 40-42; field geologist, Canol Proj, 42-44 & Imp Oil Co, 44-48; lectr, 48-53, assoc prof, 54-57, PROF GEOL, UNIV ALTA, 58- Mem: Geol Soc Am; Paleont Soc; fel Royal Soc Can; Geol Asn Can. Res: Stratigraphic paleontology of western Canada. Mailing Add: Dept of Geol Univ of Alta Edmonton AB Can

STELL, GEORGE ROGER, b Glen Cove, NY, Jan 2, 33. PHYSICS, MATHEMATICS. Educ: Antioch Col, BS, 55; NY Univ, PhD(math), 61. Prof Exp: Instr physics, Univ Ill, Chicago, 55-56; assoc res scientist, Inst Math Sci, NY Univ, 61-64; assoc res scientist, Belfer Grad Sch Sci, Yeshiva Univ, 64-65; from asst prof physics to assoc prof physics, Polytech Inst Brooklyn, 65-68; assoc prof mech, 68-70, PROF MECH, STATE UNIV NY STONY BROOK, 70- Concurrent Pos: Consult, Lawrence Radiation Lab, Univ Calif, 63-66; visitor, Lab Theoret & High Energy Physics, Nat Ctr Sci Res, France, 67-68; prin investr, NSF grant, 71- Mem: Am Math Soc; Am Phys Soc. Res: Statistical mechanics; especially molecular theory of fluids and lattice systems; mathematics associated with statistical mechanics, especially graph theory; generating functionals and non-linear integral equations; theory of critical phenomena and thermodynamic perturbation theory; theory of critical properties of fluids and magnetic materials. Mailing Add: Dept of Mech State Univ of NY Stony Brook NY 11790

STELL, WILLIAM KENYON, b Syracuse, NY, Apr 21, 39. VISUAL PHYSIOLOGY, NEUROBIOLOGY. Educ: Swarthmore Col, BA, 61; Univ Chicago, PhD(anat), 66, MD, 67. Prof Exp: Staff assoc, Lab Neurophysiol, Nat Inst Neurol Dis & Stroke, 67-68, staff assoc neurocytol, Lab Neuropath & Neuroanat Sci, 68-69 & Lab of the Dir, 69, sr staff fel, Off Dir Intramural Res, 69-71 & Lab Neurophysiol, Sect Cell Biol, 71-72; ASSOC PROF OPHTHAL, JULES STEIN EYE INST, UNIV CALIF, LOS ANGELES, 72-, ASSOC PROF ANAT, 75- Honors & Awards: E Gellhorn Award, Univ Chicago, 66. Mem: AAAS; Am Soc Cell Biol; Am Asn Anat; Assoc Am Physiol Soc; Asn Res Vision & Ophthal. Res: Neurocytology, ultrastructure and functional interconnections in vertebrate retina. Mailing Add: Jules Stein Eye Inst Med Ctr Univ of Calif Los Angeles CA 90024

STELLA, VALENTINO JOHN, b Melbourne, Australia, Oct 27, 46; m 69. PHARMACY. Educ: Victorian Col Pharm, Melbourne, BPharm, 68; Univ Kans, PhD(anal pharmaceut chem, pharmaceut), 71. Prof Exp: Pharmacist, Bendigo Base Hosp, Australia, 67-68; asst prof pharm, Univ Ill, Med Ctr, 71-73; ASST PROF PHARMACEUT CHEM, SCH PHARM, UNIV KANS, 73- Concurrent Pos: Consult, Inter Res Corp, 73- Honors & Awards: Lederle Award, Lederle Labs, 72 & 75. Mem: Am Chem Soc; Am Pharmaceut Asn; Victorian Pharmaceut Soc; Acad Pharmaceut Sci. Res: Physical pharmacy; pro-drugs and drug latentiation; drug stability; ionization kinetics; biopharmaceutics and pharmacokinetics. Mailing Add: Dept of Pharmaceut Chem Univ Kans Sch Pharm Lawrence KS 66045

STELLAR, ELIOT, b Boston, Mass, Nov 1, 19; m 45; c 2. PHYSIOLOGICAL PSYCHOLOGY. Educ: Harvard Univ, AB, 41; Brown Univ, MSc, 42, PhD(psychol), 47. Prof Exp: From instr to asst prof psychol, Johns Hopkins Univ, 47-54; from assoc prof to prof physiol psychol, 54-65, dir, Inst Neurol Sci, 65-73, PROVOST, SCH MED, UNIV PA, 73- Concurrent Pos: Mem, Nat Comn Protection Human Subj Biomed & Behav Res, 74-76. Mem: Nat Acad Sci; AAAS; Am Psychol Asn; Am Acad Arts & Sci. Res: Motivation; learning; physiology of motivation. Mailing Add: Univ of Pa Sch of Med Philadelphia PA 19104

STELLER, KENNETH EUGENE, b Lancaster, Pa, Mar 14, 41; m 64; c 2. ORGANIC CHEMISTRY. Educ: Franklin & Marshall Col, BS, 63; Northwestern Univ, PhD(org chem), 67. Prof Exp: Res chemist, 67-74, SR RES CHEMIST, HERCULES INC, 74- Mem: Am Chem Soc. Res: Polyether polymerization; elastomers; polymer modification; thermoplastic elastomers; peroxide research and development. Mailing Add: Hercules Res Ctr Hercules Inc Wilmington DE 19899

STELLMACHER, KARL L, b Brandenburg, Ger, Mar 24, 09. ANALYTICAL MATHEMATICS. Educ: Univ Gottingen, PhD, 35, Habil, 48. Prof Exp: Privat-docent math, Univ Gottingen, 48-54; mem staff, Inst Fluid Dynamics & Appl Math, 55-56, PROF MATH, UNIV MD, 56- Res: Mathematical physics; hyperbolic differential equations; differential geometry; general relativity. Mailing Add: Dept of Math Univ of Md College Park MD 20742

STELLO, PHYLLIS GREENE, b Salt Lake City, Utah, June 15, 20; m 46; c 1. SOLID STATE PHYSICS. Educ: Univ Calif AB, 43. Prof Exp: Sci aide, Western Regional Lab, USDA, 42-43, asst physicist, 43-44; Nat Bur Stand, 44-46; res engr, NAm Aviation, Inc, 49-52; mem tech staff, Hughes Aircraft Co, 52-54, dept head single crystals optical mat, 54-60, mem tech staff & group head, 60-75; RETIRED. Res: Thin film processing techniques to improve stability and achieve new material characteristics; multilayer processing of large scale I integrated subsystems; failure analysis. Mailing Add: 8443 Vicksburg Ave Los Angeles CA 90045

STELLWAGEN, EARLE C, b Joliet, Ill, June 14, 33; m 58; c 4. BIOCHEMISTRY. Educ: Elmhurst Col, BS, 55; Northwestern Univ, MS, 58; Univ Calif, Berkeley, PhD(biochem), 63. Prof Exp: NIH res fel biochem, Univ Vienna, 63-64; from asst prof to assoc prof, 64-72, PROF BIOCHEM, UNIV IOWA, 72- Concurrent Pos: NIH career develop award, 67-72; mem biophysics & biophysical study sect B, NIH, 70-, chmn, 72-; vis scientist, Bell Labs, 71-72. Mem: Am Soc Biol Chem. Res: Relationship of structure of proteins to their biological function. Mailing Add: Dept of Biochem Univ of Iowa Iowa City IA 52242

STELLWAGEN, ROBERT HARWOOD, b Joliet, Ill, Jan 6, 41; m 63; c 2. BIOCHEMISTRY. Educ: Harvard Univ, AB, 63; Univ Calif, Berkeley, PhD(biochem), 68. Prof Exp: Res biochemist, Univ Calif, Berkeley, 68; staff fel molecular biol, Nat Inst Arthritis & Metab Dis, 68-69; asst prof, 70-74, ASSOC PROF BIOCHEM, SCH MED, UNIV SOUTHERN CALIF, 74- Concurrent Pos: USPHS fel biochem, Univ Calif, San Francisco, 69-70. Mem: AAAS; Am Soc Biol Chem. Res: Biochemical control mechanisms in animal cells; enzyme induction by hormones; cellular differentiation. Mailing Add: Dept of Biochem Univ of Southern Calif Sch Med Los Angeles CA 90033

STELLY, MATTHIAS, b Arnaudville, La, Aug 7, 16; c 4. SOIL FERTILITY. Educ: Univ Southwestern La, BS, 37; La State Univ, MS, 39; Iowa State Univ, PhD(soil fertil), 42. Prof Exp: Asst agronomist, Exp Sta, La State Univ, 42-43; from assoc prof to prof soils, Univ Ga, 46-57; prof in chg soil testing serv, La State Univ, 57-59, prof soil chem, 59-61; exec secy & treas, 61-70, EXEC VPRES & ED IN CHIEF, ASA PUBLS, AM SOC AGRON, 70- Concurrent Pos: Soil specialist, USDA, E Africa, 52-53; Am secy, Comn Soil Fertil & Plant Nutrit, 7th Cong, Int Soc Soil Sci, 58-60; prog dir transl & printing, Soviet Soil Sci; ed, Agron J, 61- Mem: Fel AAAS; fel Am Soc Agron; Crop Sci Soc Am; Soil Sci Soc Am; Int Soc Soil Sci. Res: Soil chemistry; identification of inorganic soil phosphorus compounds; radioactive materials as plant stimulants; crop rotation; radioactive phosphorus; fertilizer requirements and chemical composition of Bermuda grasses; methodology of soil testing; chemical and mineralogical investigations of Louisiana soils. Mailing Add: Am Soc of Agron 677 S Segoe Rd Madison WI 53711

STELMACH, HONORATE (ANN), pharmaceutical chemistry, see 12th edition

STELOS, PETER, b Lowell, Mass, May 17, 23. IMMUNOLOGY, IMMUNOCHEMISTRY. Educ: Berea Col, AB, 48; Univ Chicago, PhD(microbiol), 56. Prof Exp: Res assoc microbiol, Univ Chicago, 56-57; cancer res scientist, Roswell Park Mem Inst, 59-60, sr cancer res scientist, 60-63, assoc cancer res scientist, 63-65; ASSOC PROF MICROBIOL, HAHNEMANN MED COL, 65- Concurrent Pos: Fel chem, Yale Univ, 57-59. Mem: AAAS; Am Asn Immunol; Am Chem Soc. Res: Separation and properties of immuno-globulins. Mailing Add: Dept of Microbiol Hahnemann Med Col Philadelphia PA 19102

STELSON, HUGH EUGENE, b Elmdale, Kans, Jan 16, 03; m 25; c 6. MATHEMATICS. Educ: Kans Wesleyan, AB, 23; Northwestern Univ, AM, 24; Univ Iowa, PhD(math), 30. Prof Exp: Head dept math, Hedding Col, 24-26; asst & instr, Iowa State Col, 26-28; assoc prof, Marshall Col, 29-30; from asst prof to prof, Kent State Univ, 30-47; from assoc prof to prof, 47-71, EMER PROF MATH, MICH STATE UNIV, 71- Concurrent Pos: Vis prof, Univ Hawaii, 49-50. Mem: Am Math Soc; Math Asn Am. Res: Mathematical analysis. Mailing Add: Dept of Math Mich State Univ East Lansing MI 48823

STELSON, PAUL HUGH, b Ames, Iowa, Apr 9, 27; m 50; c 4. PHYSICS. Educ: Purdue Univ, BS, 47, MA, 48; Mass Inst Technol, PhD(physics), 50. Prof Exp: Asst physics, Purdue Univ, 47; mem staff div indust coop, Nuclear Sci & Eng Lab, Mass Inst Technol, 50-52; assoc dir, 52-73, DIR PHYS DIV, OAK RIDGE NAT LAB, 73- Concurrent Pos: Ford Found prof physics, Univ Tenn. Mem: Fel Am Phys Soc. Res: Nuclear physics; accelerators. Mailing Add: Oak Ridge Nat Lab Oak Ridge TN 37830

STELTENKAMP, ROBERT JOHN, organic chemistry, see 12th edition

STELTS, MARION LEE, b Oregon City, Ore, Mar 28, 40; m 61; c 2. NUCLEAR PHYSICS. Educ: Univ Ore, BA, 61; Univ Calif, Davis, MS, 70, PhD(appl sci), 75. Prof Exp: Physicist, Lawrence Livermore Lab, 61-75; PHYSICIST NUCLEAR PHYSICS, BROOKHAVEN NAT LAB, 75- Mem: Am Phys Soc. Mailing Add: Brookhaven Nat Lab Physics Dept Upton NY 11973

STELZER, LORIN ROY, b Bloomer, Wis, Mar 6, 31; m 58; c 2. RESEARCH ADMINISTRATION, ENTOMOLOGY. Educ: Wis State Col, Whitewater, BEd, 53; Univ Wis, MS, 55, PhD(entom), 57. Prof Exp: Field res specialist, 57-62, supvr, Southern Field Res Sta, Fla, 62-69, East East/South Field Res Sta, NJ, 69-71, MGR TECH COORD RES & DEVELOP, ORTHO DIV, CHEVRON CHEM CO, CALIF, 71- Mem: Entom Soc Am. Res: Field research, evaluation and development of chemicals for agricultural and home use. Mailing Add: Ortho Div Chevron Chem Co 940 Hensley Richmond CA 94804

STELZNER, HARRY DONALD, biochemistry, nutrition, see 12th edition

STEMBRIDGE, GEORGE EUGENE, b Canton, Ga, Sept 27, 36; m 61; c 2. POMOLOGY. Educ: Clemson Univ, BS, 58; Univ Md, MS, 59, PhD(hort), 62. Prof Exp: Asst prof, 61-62 & 64-67, assoc prof, 67-73, PROF HORT, CLEMSON UNIV, 73- Mem: Am Soc Hort Sci; Int Soc Hort Sci. Res: Use of growth regulators in fruit production; culture of fruit trees. Mailing Add: Dept of Hort Clemson Univ Clemson SC 29631

STEMBRIDGE, VERNIE ALBERT, b El Paso, Tex, June 7, 24; m 44; c 3. PATHOLOGY. Educ: Tex Col Mines, BA, 43; Univ Tex, MD, 48; Am Bd Path, cert anat & clin path, 53. Prof Exp: Intern, Marine Hosp, Norfolk, Va, 48-49; resident path, Med Br, Univ Tex, 49-52, assoc dir clin labs & from asst prof to assoc prof, 52-56; chief aviation path sect, Armed Forces Inst Path, 56-59; assoc prof, 59-61, PROF PATH, UNIV TEX HEALTH SCI CTR DALLAS, 61-, CHMN DEPT, 66- Concurrent Pos: Consult, Vet Admin Hosp, Dallas, 59-, Civil Air Surgeon, Fed Aviation Agency, DC, 59-70 & Surgeon Gen, US Air Force, 62-70; trustee, Am Bd Path, 69-; mem sci adv bd, Armed Forces Inst Path, DC, 71-75. Mem: AMA; Am Asn Path & Bact; Am Soc Exp Path; Col Am Path; Am Soc Clin Path. Res: Neoplasms. Mailing Add: Dept of Path Univ of Tex Health Sci Ctr Dallas TX 75235

STEMKE, GERALD W, b Watseka, Ill, Oct 7, 35; m 61; c 2. IMMUNOCHEMISTRY. Educ: Ill State Univ, BS, 57; Univ Ill, PhD(chem), 63. Prof Exp: Teaching asst, Univ Ill, 59-60; trainee microbiol, NY Univ Med Ctr, 60-62, res assoc, 62-63; NIH fel, Pasteur Inst, Paris, 63-64, asst prof biol, Univ Pittsburgh, 65-66, asst prof microbial & molecular biol, 66-67, assoc prof biophys & microbiol, 67-70; PROF MICROBIOL, UNIV ALTA, 70- Concurrent Pos: Nat Res Coun Can res grant, 70-74. Mem: AAAS; Am Chem Soc; Am Soc Microbiol; Am Asn Immunol; Brit Biochem Soc. Res: Structure and function of rabbit immunoglobulins; immunogenetics. Mailing Add: Dept of Microbiol Univ of Alta Edmonton AB Can

STEMLER, ALAN JAMES, b Chicago, Ill, July 29, 43; m 70. PLANT PHYSIOLOGY. Educ: Mich State Univ, BS, 65; Univ Ill, Urbana, PhD(plant physiol), 74. Prof Exp: Instr plant physiol, Univ Ill, Urbana, 74-75; FEL, DEPT PLANT BIOL, CARNEGIE INST WASH, 75- Mem: Am Soc Plant Physiologists. Res: Primary photochemical events of photosynthesis. Mailing Add: 2061 Liberty Park Ave Menlo Park CA 94025

STEMM, ROBERT MARVIN, b Oklahoma City, Okla, Sept 3, 23; m 46; c 3. ORTHODONTICS. Educ: Univ Nebr, BS, 52, DDS, 54, MS, 61. Prof Exp: Instr crown & bridge, 55-61, asst prof, 61-64, assoc prof orthod, 64-70, PROF ORTHOD, COL DENT, UNIV NEBR-LINCOLN, 70-, CHMN DEPT, 71- Concurrent Pos: Fels, Col Dent, Univ Nebr-Lincoln, 67-; state coordr oral-plastic prog, Nebr Serv for Crippled Children, 65- Mem: Int Asn Dent Res; Am Dent Asn; Am Asn Orthod; Am Cleft Palate Asn. Res: Oral-facial displasias. Mailing Add: Dept of Orthod Univ Nebr Col of Dent Lincoln NE 68503

STEMMER, EDWARD ALAN, b Cincinnati, Ohio, Jan 20, 30; m 54; c 5. THORACIC SURGERY. Educ: Univ Chicago, BA, 49, MD, 53; Am Bd Surg, dipl, 62; Am Bd Thoracic Surg, dipl, 63. Prof Exp: Intern med, Univ Chicago Clins, 53-54, res asst surg, Sch Med, 54-55, from asst resident to sr resident, 54-60, instr, 59-60; chief resident, Stanford Univ, 60-61, clin teaching asst, 61; chief resident, Palo Alto Vet Admin Hosp, 61-62, asst chief surg serv, 62-64; asst prof surg, Univ Utah, 64-65; asst prof in residence, 66-70, ASSOC PROF SURG, UNIV CALIF, IRVINE, 70-; CHIEF SURG SERV, LONG BEACH VET ADMIN HOSP, 65- Concurrent Pos: Responsible investr, Vet Admin Hosp, 62-; attend surgeon, Salt Lake County Hosp, Utah, 64-65; prin investr, NIH, 64-70. Mem: Fel Am Col Surg; Am Asn Thoracic Surg; Am Col Chest Physicians; Soc Thoracic Surg; Am Surg Asn. Res: Vascular surgery; metabolism of plasma proteins; myocardial functions; control of regional blood flow. Mailing Add: Long Beach Vet Admin Hosp 5901 E Seventh St Long Beach CA 90801

STEMMERMANN, GRANT N, b Bronx, NY, Oct 28, 18; m 44; c 2. PATHOLOGY. Educ: Trinity Col, Conn, 35-37; McGill Univ, MD, 43. Prof Exp: Lab dir path, Hilo Hosp, Hawaii, 51-58; LAB DIR PATH, KUAKINI HOSP, 58-; CLIN PROF PATH, SCH MED, UNIV HAWAII, MANOA, 66- Concurrent Pos: Prin investr, Japan-Hawaii Cancer Study. Mem: AAAS; Col Am Path; Am Asn Path & Bact; Am Soc Clin Path; Int Acad Path. Res: Geographic pathology, particularly neoplastic and cardiovascular diseases in migrants; biology of elastic tissue and disease patterns in the feral mongoose. Mailing Add: Kuakini Hosp 347 N Kuakini St Honolulu HI 96817

STEMNISKI, JOHN ROMAN, b Nanticoke, Pa, Apr 29, 33; m 59; c 3. ORGANIC CHEMISTRY. Educ: Fordham Univ, BS, 55; Carnegie Inst Technol, MS, 59, PhD(org chem), 60. Prof Exp: Res chemist, Monsanto Res Corp, 59-65; staff chemist, 65-69, prin chemist polymer prod, 69-73, CHIEF MAT & PROCESS CONTROL LAB, C S DRAPER LAB, MASS INST TECHNOL, 73- Mem: Am Chem Soc; The Chem Soc; Am Soc Testing and Mat. Res: Synthesis of medicinal compounds; synthesis of oil additives; lubricant systems; oxidation mechanism; high density fluids; gel permeation chromatography; fluorine containing fluids, materials science studies. Mailing Add: C S Draper Lab, 68 Albany St Cambridge MA 02139

STEMNISKI, MICHAEL ANDREW, b Wilkes-Barre, Pa, Oct 4, 40; m 67. ORGANIC CHEMISTRY. Educ: Villanova Univ, BS, 62; PhD(org chem), Fordham Univ, 67. Prof Exp: Res chemist exp sta, E I du Pont de Nemours & Co, Inc, Del, 67-70; SCI TEACHER, BRANDYWINE SPRINGS JR HIGH SCH, 70-; CHEM TEACHER, THOMAS McKEAN HIGH SCH, 72- Concurrent Pos: Part time instr org chem, Univ Del, 73- Mem: AAAS; Am Chem Soc; The Chem Soc. Res: Beckmann rearrangement and Schmidt reaction on methyl substituted 1-hydrindanones. Mailing Add: RD 1 Box 105-B Rolling Ridge Hockessin DE 19707

STEMPAK, JEROME G, b Chicago, Ill, Dec 24, 31; m 61; c 1. ANATOMY. Educ: Roosevelt Univ, BS, 58; Univ Ill, MS, 60, PhD(teratology), 62. Prof Exp: Instr, 63-66, ASST PROF, ANAT, STATE UNIV NY DOWNSTATE MED CTR, 66- Concurrent Pos: USPHS fel, 62-63. Mem: AAAS; Am Asn Anat; Am Soc Cell Biol. Res: Electron microscopy of differentiating cells and tissues; electron microscopic and biochemical investigations on generation of cell organelles. Mailing Add: Dept of Anat State Univ NY Downstate Med Ctr Brooklyn NY 11203

STEMPEL, ARTHUR, b Brooklyn, NY, June 8, 17; m 50; c 2. CHEMISTRY. Educ: City Col New York, BS, 37; Columbia Univ, MA, 39. PhD(chem), 42. Prof Exp: Asst, Col Physicians & Surgeons, Columbia Univ, 38-39, org chem, Col Pharm, 40-41, res assoc, Chem Labs, 42-43; tutor & fel chem, Queens Col, 41-42; sr chemist, 43-68, RES GROUP CHIEF, CHEM RES DEPT, HOFFMANN-LA ROCHE, INC, 68- Mem: Am Chem Soc; NY Acad Sci. Res: Isolation of natural products of animal and plant origin; antibiotics; organic synthesis of pharmaceuticals; investigations on loco weeds; synthesis of benzodiazepines; cholinesterase inhibitors; anticurare compounds. Mailing Add: Chem Res Dept Hoffman-La Roche Inc 340 Kingsland St Nutley NJ 07110

STEMPEL, EDWARD, b Brooklyn, NY, Mar 7, 26; m 59; c 1. PHARMACY. Educ: L I Univ, BS, 49; Columbia Univ, MS, 52, MA, 55, EdD, 56. Prof Exp: From instr to assoc prof, 49-64, PROF PHARM & CHMN DEPT, BROOKLYN COL PHARM, 64- Mem: Am Pharmaceut Asn; Am Asn Col Pharm. Res: Dispensing pharmacy; long-active dosage forms. Mailing Add: Brooklyn Col of Pharm 600 Lafayette Ave Brooklyn NY 11216

STEMPEN, HENRY, b Phila, Pa, May 10, 24; m 53; c 4. MICROBIOLOGY. Educ: Phila Col Pharm, BS, 45; Univ Pa, PhD(microbiol), 51. Prof Exp: Res asst cytol, Lankenau Hosp Res Inst, Phila, Pa, 45-46; from instr to asst prof, bact, Jefferson Med Col, 50-57, assoc prof microbiol, 57-62; asst prof biol, 62-63, ASSOC PROF MICROBIOL, RUTGERS UNIV, CAMDEN, 63- Concurrent Pos: NIH grants, 59-60 & 64-66. Mem: AAAS; Am Soc Microbiol; Mycol Soc Am. Res: Bacterial cytology and genetics; morphogenesis of myxomycetes. Mailing Add: Dept of Biol Rutgers Univ 311 N Fifth St Camden NJ 08102

STEMPIEN, MARTIN F, JR, b New Britain, Conn, Sept 2, 30. BIOCHEMISTRY. Educ: Yale Univ, BS, 52, MS, 53, PhD(org chem), 57; Cambridge Univ, PhD(org chem), 60. Prof Exp: Res assoc & sr investr bio-org chem, 60-66, ASST TO DIR & BIO-ORG CHEMIST, OSBORN LABS, MARINE SCI, NY AQUARIUM, NY ZOOL SOC, 66- Concurrent Pos: Consult, Off Naval Res, 68; fel, NY Zool Soc, 68- Mem: AAAS; Am Chem Soc; NY Acad Sci; Am Soc Zool; The Chem Soc. Res: Physiologically active materials from extracts of marine invertebrates, particularly Porifera and Echinodermata; bio-chemical taxonomy of Phylum Porifera. Mailing Add: Osborn Labs of Marine Sci NY Aquarium Coney Island Brooklyn NY 11224

STEMPLE, JOEL G, b Brooklyn, NY, Feb 3, 42; m 68. MATHEMATICS. Educ: Brooklyn Col, BS, 62; Yale Univ, MA, 64, PhD, 66. Prof Exp: Asst prof math, 66-70, ASSOC PROF MATH, QUEEN'S COL (NY), 70- Res: Theory of graphs. Mailing Add: Dept of Math Queen's Col Flushing NY 11367

STENBACK, WAYNE ALBERT, b Brush, Colo, June 12, 29; m 54; c 1. MICROBIOLOGY, ELECTRON MICROSCOPY. Educ: Univ Colo, BS, 55; Univ Denver, MS, 57; Univ Mo, PhD(microbiol), 62. Prof Exp: Instr exp biol, 62-66, asst prof exp biol, Dept Surg, Baylor Col Med, 66-75; ELECTRON MICROSCOPIST, DEPT PATH, TEX CHILDREN'S HOSP, 75- Concurrent Pos: Mem, Int Asn Comp Res on Leukemia & Related Dis. Mem: AAAS; Am Soc Microbiol; Am Asn Cancer Res; Electron Micros Soc Am. Res: Density gradient studies of Newcastle disease virus; control of endemic microorganisms in mouse colonies; electron microscopy, biological and biophysical studies of viruses associated with neoplasms. Mailing Add: Dept of Path Tex Children's Hosp Houston TX 77025

STENBERG, VIRGIL IRVIN, b Grygla, Minn, May 18, 35; m 56; c 4. ORGANIC CHEMISTRY. Educ: Concordia Col, Moorhead, Minn, BA, 56; Iowa State Univ, PhD(org chem), 60. Prof Exp: From asst prof to assoc prof, 60-67, PROF CHEM, UNIV NDAK, 67- Concurrent Pos: Res grants, Res Corp, 60-62, Petrol Res Fund, 61-63, NIH, 61-64, 68-71 & 74-76, career develop award, 70-75; NSF res grant, 66-68; vis prof, Imp Col, Univ London, 71-72; Energy Res Develop Admin contract, 75-78. Mem: Am Chem Soc; Sigma Xi. Res: Photochemistry, particularly nitrogen compounds and charge transfer complexes; photochemistry of natural products; catalyst development. Mailing Add: Dept of Chem Univ of NDak Grand Forks ND 58201

STENCHEVER, MORTON ALBERT, b Paterson, NJ, Jan 25, 31; m 55; c 3. REPRODUCTIVE BIOLOGY, HUMAN GENETICS. Educ: NY Univ, AB, 51; Univ Buffalo, MD, 56; Am Bd Obstet & Gynec, 65. Prof Exp: Intern med, Mt Sinai Hosp, NY, 56-57; resident, Columbia-Presby Med Ctr, 57-60; from instr to assoc prof obstet & gynec, Case Western Reserve Univ, 64-70, dir tissue cult lab, 64-70, assoc, Dept Med Educ, 68-70; PROF OBSTET & GYNEC & CHMN DEPT, UNIV UTAH, 70- Concurrent Pos: NIH res training fel genetics, 62-64; Oglebey res fel, Case Western Reserve Univ, 64; chief obstet & gynec serv, Malmstrom AFB Hosp, Mont. Mem: AAAS; AMA; fel Am Col Obstet & Gynec; fel Am Gynec Soc; Soc Cryobiol. Res: Human and cell genetics; human reproduction. Mailing Add: Dept of Obstet & Gynec Univ of Utah Salt Lake City UT 84112

STENDELL, REY CARL, b San Francisco, Calif, Aug 12, 41; m 66; c 2. POLLUTION BIOLOGY. Educ: Univ Calif, Santa Barbara, BA, 63, MA, 67; Univ Calif, Berkeley, PhD(zool), 72. Prof Exp: Res biologist, 72-73, RES COORDR POLLUTION BIOL, US FISH & WILDLIFE SERV, PATUXENT WILDLIFE RES CTR, 73- Mem: Am Ornithologists Union; Cooper Ornithol Soc; Am Soc Mammalogists; Ecol Soc Am; Raptor Res Found. Res: Evaluation of the effects of environmental pollutants on wildlife, particularly birds. Mailing Add: Patuxent Wildlife Res Ctr Laurel MD 20811

STENERSON, RICHARD OWEN, physics, see 12th edition

STENESH, JOCHANAN, b Magdeburg, Ger, Dec 19, 27; US citizen; m 57. BIOCHEMISTRY. Educ: Univ Ore, BS, 53; Univ Calif, Berkeley, PhD(biochem), 58. Prof Exp: Res assoc biochem, Weizmann Inst, 58-60 & Purdue Univ, 60-63; from asst prof to assoc prof, 63-71, PROF CHEM, WESTERN MICH UNIV, 71- Concurrent Pos: Res grants, Nat Inst Allergy & Infectious Dis, 64-70, Am Cancer Soc, 65-67. Mem: AAAS; Am Chem Soc; Am Asn Biol Chemists; Am Soc Microbiol. Res: Physical biochemistry; enzymology; ribosomes and protein synthesis; molecular biology of proteins and nucleic acids. Mailing Add: Dept of Chem Western Mich Univ Kalamazoo MI 49001

STENGER, FRANK, b Veszprem, Hungary, July 6, 38; m 61; c 2. APPLIED MATHEMATICS. Educ: Univ Alta, BSc, 61, MSc, 63, PhD(math), 65. Prof Exp: Guest worker, Nat Bur Stand, Washington, DC, 63-64; asst prof comp sci, Univ Alta, 65-66; asst prof math, Univ Mich, 66-69; assoc prof, 69-73, PROF MATH, UNIV UTAH, 74- Concurrent Pos: NSF grant, Univ Utah, 70-71; vis math res ctr, Univ Montreal, 71-72; US Army res grant, Univ Utah, 74-76. Mem: Am Math Soc; Can Math Cong; Soc Indust & Appl Math; Asn Comput Mach. Res: Quadrature; numerical solution of differential and integral equations; asymptotic approximation of integrals; asymptotic solution of differential equations. Mailing Add: Dept of Math Univ of Utah Salt Lake City UT 84112

STENGER, RICHARD J, b Cincinnati, Ohio, Dec 13, 27; m 51; c 2. PATHOLOGY, ELECTRON MICROSCOPY. Educ: Col of the Holy Cross, AB, 49; Univ Cincinnati, MD, 53. Prof Exp: Intern, Cincinnati Gen Hosp, 53-54; resident path, Mass Gen Hosp, 54-55, 57-58, fel 58-59, asst, 59-60; asst prof, Col Med, Univ Cincinnati, 60-64; assoc prof path, Case Western Reserve Univ, 65-68; prof, New York Med Col, 68-72; PROF PATH, MT SINAI SCH MED, 72-; DIR PATH & LABS, BETH ISRAEL MED CTR, 72- Concurrent Pos: Nat Inst Arthritis & Metab Dis res grant & Nat Inst Gen Med Sci career develop award, 64-68; instr, Harvard Med Sch, 59-60; attend pathologist, Cincinnati Gen Hosp, 60-64; assoc pathologist, Cleveland Metrop Gen Hosp, 64-68. Mem: AAAS; Am Soc Exp Path; Am Asn Path & Bact; Int Acad Path. Res: Light and electron microscopic studies of the liver, especially with regards to toxic effects. Mailing Add: Beth Israel Med Ctr New York NY 10003

STENGER, ROBERT ARTHUR, organic chemistry, polymer chemistry, see 12th edition

STENGER, VERNON ARTHUR, b Minneapolis, Minn, June 11, 08; m 33; c 5. ANALYTICAL CHEMISTRY. Educ: Univ Denver, BS, 29, MS, 30; Univ Minn, PhD(anal chem), 33. Hon Degrees: DSc, Univ Denver, 71. Prof Exp: Asst, Eastman Kodak Co, 29-30; anal chemist, Univ Minn, 33-35; anal res chemist, 35-53, dir spec serv lab, 53-61, anal scientist, 61-73, CONSULT, DOW CHEM CO, 73- Honors & Awards: Anachem Award, 70. Mem: Am Chem Soc; Sigma Xi; Geochem Soc; NY Acad Sci; fel Am Inst Chem. Res: Purity of reagents; technology of bromine and its compounds; instrumentation for water analysis; analytical methods for environmental analysis. Mailing Add: 1108 E Park Dr Midland MI 48640

STENGER, VICTOR JOHN, b Bayonne, NJ, Jan 29, 35; m 62; c 2. PHYSICS. Educ: Newark Col Eng, BS, 56; Univ Calif, Los Angeles, MS, 58, PhD(physics), 63. Prof Exp: Mem tech staff, Hughes Aircraft Co, 56-59; asst physics, Univ Calif, Los Angeles, 59-63; assoc prof, 63-74, PROF PHYSICS & ASTRON, UNIV HAWAII, 74- Mem: Am Phys Soc. Res: Elementary physics. Mailing Add: Dept of Physics Univ of Hawaii Honolulu HI 96822

STENGER, WILLIAM, b Bayonne, NJ, Jan 25, 42; m 68; c 2. MATHEMATICS. Educ: Stevens Inst Technol, BS, 63; Univ Md, PhD(math), 67. Prof Exp: Asst comput, Davidson Lab, Stevens Inst Technol, 62-63; res asst, Univ Md, 66-67; asst prof math, American Univ, 67-68 & Georgetown Univ, 68-69; asst prof, 69-72, PROF MATH, AMBASSADOR COL, 72- Concurrent Pos: Res grants, Air Force Off Sci Res, 67-68 & NSF, 68-69. Mem: Math Asn Am; Am Math Soc; Soc Indust Appl Math. Res: Variational theory of eigenvalues and related inequalities. Mailing Add: Dept of Math Ambassador Col 300 W Green St Pasadena CA 91123

STENGLE, JAMES MARSHALL, b Wilkinsburg, Pa, Aug 7, 17. INTERNAL MEDICINE, HEMATOLOGY. Educ: Oberlin Col, AB, 42; Northwestern Univ, MD, 46. Prof Exp: From intern to resident, Evanston Hosp, Ill, 45-49; clin investr, Nat Microbiol Inst, 53-55; sr investr, Nat Cancer Inst, 55-59, exec secy hemat study sect,

Div Res Grants, NIH, 59-61, chief, Training Grants & Fels Br, Nat Heart & Lung Inst, 61-63, chief, Off Prog Planning & Eval, Extramural progs, 63-67, chief nat blood resource prog, 67-74, DEP DIR MED AFFAIRS, LISTER HILL NAT CTR BIOMED COMMUN, NAT LIBR MED, 74- Concurrent Pos: Fel hemat, Cook County Hosp, Chicago, 49-50; Nuffield fel, Oxford Univ, 51-52; chmn blood prog med adv comt, Am Red Cross, 65-66; mem task force intravascular thrombosis, Nat Res Coun, 65-66; mem exec comt, Coun Thrombosis, Am Heart Asn, 72-; chmn med & sci adv comt, Nat Hemophilia Found, 73- Mem: Am Soc Hemat; Int Soc Thrombosis & Haemostasis (secy & treas, 70-76); Int Soc Hemat. Res: Clinical and laboratory hematology. Mailing Add: Nat Libr of Med 9000 Rockville Pike Bethesda MD 20014

STENGLE, THOMAS RICHARD, b Lancaster, Pa, Nov 25, 29. PHYSICAL CHEMISTRY. Educ: Franklin & Marshall Col, BS, 51; Univ Mich, MS, 53, PhD(chem), 61. Prof Exp: From instr to assoc prof, 59-73, PROF CHEM, UNIV MASS, AMHERST, 73- Mem: Am Chem Soc; Am Phys Soc. Res: Application of nuclear magnetic resonance techniques to fast reactions, inorganic reaction mechanisms, molecular biology and solute-solvent interactions. Mailing Add: Dept of Chem Univ of Mass Amherst MA 01002

STENGLE, WILLIAM BERNARD, b Lancaster, Pa, Feb 21, 23; m 48; c 3. WOOD CHEMISTRY. Educ: Franklin & Marshall Col, BS, 43; State Univ NY Col Forestry, Syracuse, MS, 49. Prof Exp: Chemist, Animal Trap Co Am, 47; asst, State Univ NY Col Forestry, Syracuse, 49; res chemist, Crossett Co, Ark, 49-58, asst tech serv dir, Paper Mill Div, 58-62; TECH DIR, TENN RIVER PULP & PAPER CO, COUNCE, 62- Mem: Am Chem Soc; Tech Asn Pulp & Paper Indust; Air Pollution Control Asn. Res: Hydroxy acids; alkaline pulping; bleaching; paper manufacture; tall oil; corrugated paperboard; water and air pollution abatement. Mailing Add: 1109 Cypress St Savannah TN 38372

STENLUND, MILTON HARVEY, biology, see 12th edition

STENN, FREDERICK, b Chicago, Ill, Dec 25, 08; m 36; c 3. MEDICINE. Educ: Univ Chicago, BS, 27, MS, 36; Rush Med Col, MD, 32; Am Bd Internal Med, dipl, 50. Prof Exp: Clin instr med, Col Med, Univ Ill, 50-55; asst prof, 55-63, ASSOC PROF MED, MED SCH, NORTHWESTERN UNIV, CHICAGO, 63- Concurrent Pos: Mem staff, Chicago Wesley Mem Hosp. Mem: Fel Am Col Physicians. Res: Medical ethics and history; social medicine. Mailing Add: Northwestern Univ Med Sch 301 E Chicago Ave Chicago IL 60611

STENSAAS, LARRY J, b Nov 13, 32; US citizen; m 62; c 1. NEUROANATOMY. Educ: Univ Calif, Berkeley, BA, 55, MA, 57; Univ Calif, Los Angeles, PhD(neuroanat), 65. Prof Exp: Stratigrapher, Richmond of Columbia, 57-58; head paleont sect, Cuba Calif Oil Co, 58-59; micropaleontologist, Standard of Calif, 59-60; asst prof, 68-72, ASSOC PROF PHYSIOL, UNIV UTAH, 72- Concurrent Pos: Cerebral Palsy Educ & Res Found fel, 65-67; NIH fel, 67-68. Mem: AAAS; Am Asn Anat; Soc Neurosci. Res: Light and electron microscopy of normal and regenerating central nervous tissue. Mailing Add: 4C202 Univ Med Ctr Univ of Utah Salt Lake City UT 84112

STENSAAS, SUZANNE SPERLING, b Oakland, Calif, Mar 15, 39; m 62; c 1. NEUROANATOMY. Educ: Pomona Col, BA, 59; Univ Calif, Los Angeles, MA, 62; Univ Utah, PhD(anat), 75. Prof Exp: Instr, 71-75, ASST PROF ANAT, COL MED, UNIV UTAH, 76- Mem: Soc Neurosci; AAAS. Res: Biological compatability of materials with the brain; development of neuroprostheses; effects of electrical stimulation on the brain; problems of development and regeneration of the central nervous system. Mailing Add: Dept of Anat Col of Med Univ of Utah Salt Lake City UT 84132

STENSETH, RAYMOND EUGENE, b Ludlow, SDak, Aug 5, 31; m 65; c 1. ORGANIC CHEMISTRY, PHARMACY. Educ: Univ Mich, BS, 53, MS, 57, PhD(pharmaceut chem), 61. Prof Exp: Sr res chemist, 60-68, RES SPECIALIST, MONSANTO CO, 68- Mem: Am Chem Soc. Res: Organic syntheses; nitrogen heterocycles; organophosphorus chemistry; food chemicals; pharmaceuticals; bacteriostats; fungistats; chemical processes. Mailing Add: Monsanto Co 800 N Lindbergh Blvd St Louis MO 63166

STENSRUD, HOWARD LEWIS, b Minneapolis, Minn, Nov 16, 36; m 60; c 2. GEOLOGY, GEOCHEMISTRY. Educ: Univ Minn, Minneapolis, BA, 58; Univ Wyo, MA, 63; Univ Wash, PhD(geol), 70. Prof Exp: Explor geologist, Humble oil & Ref Co, 60; asst prof geol, Univ Minn, Morris, 61-66; ASST PROF GEOL, CALIF STATE UNIV, CHICO, 70- Mem: Geol Soc Am; Mineral Soc Am. Res: Metamorphic petrology. Mailing Add: Dept of Geol & Phys Sci Calif State Univ First & Normal St Chico CA 95926

STENSTROM, RICHARD CHARLES, b Elkhorn, Wis, June 19, 36; m 60; c 1. GEOLOGY. Educ: Beloit Col, BS, 58; Univ Chicago, MS, 62, PhD(geophys sci), 64. Prof Exp: Am Chem Soc res fel geophys sci, Univ Chicago, 64-65; from instr to asst prof, 65-71, ASSOC PROF GEOL, BELOIT COL, 71- Mem: AAAS; fel Geol Soc Am; Mineral Soc Am; Geochem Soc; Am Geophys Union. Res: Diffusion rates through sediments; hydrologic effects of urbanization; environmental geologic hazards. Mailing Add: 2531 E Ridge Rd Beloit WI 53511

STENT, GUNTHER SIEGMUND, b Berlin, Ger, Mar 28, 24; nat US; m 51; c 1. MOLECULAR BIOLOGY, NEUROBIOLOGY. Educ: Univ Ill, PhD(phys chem), 48. Prof Exp: Asst chem, Univ Ill, 44-48; Nat Res Coun Merck fel biol, Calif Inst Technol, 48-50; Am Cancer Soc fel, Copenhagen & Pasteur Inst, Paris, 50-52; asst res biochemist & lectr bact, 52-56, assoc prof bact, 57-58, prof bact & virol, 59-63, PROF MOLECULAR BIOL, UNIV CALIF, BERKELEY, 63- Concurrent Pos: Consult, Field Info Agencies Technol, 46-47; mem genetics study sect, NIH, 59-64; mem genetic biol panel, NSF, 65-69; external mem, Max Planck Inst Molecular Genetics, 67-; Guggenheim fel, Med Sch, Harvard Univ, 69-70. Mem: Am Acad Arts & Sci. Res: Nervous control of behavior; cellular regulation. Mailing Add: Dept of Molecular Biol Univ of Calif Berkeley CA 94720

STENZEL, GEORGE, b Richmond, NY, Nov 26, 10; m 46. FOREST MANAGEMENT, FOREST ENGINEERING. Educ: Univ NH, BS, 38; Yale Univ, MF, 39. Prof Exp: Mapper & crew chief, US Forest Serv, 39-40, proj forester, 40-42; asst chief control, Woods Dept, Brown Co, 46-49; from instr to assoc prof forestry, 49-62, PROF LOGGING ENG, UNIV WASH, 62- Mem: Soc Am Foresters. Res: Timber harvesting and control; forest road development and construction; logging engineering; cost and production control. Mailing Add: Col of Forest Resources AR-10 Univ of Wash Seattle WA 98195

STENZEL, KURT HODGSON, b Stamford, Conn, Nov 3, 32; m 57; c 3. NEPHROLOGY, MEDICINE. Educ: NY Univ, AB, 54; Cornell Univ, MD, 58. Prof Exp: Intern med & nephrology, Second Cornell Med Div, Bellevue Hosp, NY, 58-59, asst resident, 59-60, cardio-renal resident, 62-63; res fel, 63-64, from instr to assoc prof, 64-75, PROF SURG & BIOCHEM, MED COL, CORNELL UNIV, 75- Concurrent Pos: NY Heart Asn res fel, 63-66 & sr res fel, 66- Mem: Biophys Soc; fel Am Col Physicians; Am Fedn Clin Res; Am Soc Artificial Internal Organs. Res: Dialysis; biomaterials; transplantation immunology; collagen immunochemistry, amino acid metabolism. Mailing Add: Cornell Univ Med Col 525 E 68th St New York NY 10021

STENZEL, WOLFRAM G, b Berlin, Ger, May 24, 19; nat US; m 43; c 1. PHYSICS. Educ: City Col New York, BS, 39, MS, 41. Prof Exp: Asst biochem, Warner Inst Therapeut Res, 42; res scientist phys chem, Nuodex Prod Co, Inc, 46-48; instr physics & math, Bloomfield Col, 48-51; physicist, B G Corp, 51-55; prin engr physics, Ford Instrument Co Div, 55-69, SR ENGR, SYSTS MGT DIV, SPERRY RAND CORP, 69- Concurrent Pos: Lectr, Adelphi Col, 57-60. Mem: Am Phys Soc; Am Nuclear Soc. Res: Nuclear reactors and effects; mathematical physics; thermionics; traffic control; rocket trajectories. Mailing Add: 77-21 250th St Bellerose NY 11426

STEPAKOFF, GERALD LIONEL, physical chemistry, see 12th edition

STEPAN, ALFRED HENRY, b St Paul, Minn, Jan 2, 20; m 44; c 5. ORGANIC CHEMISTRY. Educ: Col St Thomas, BS, 42; Univ Nebr, MS, 45, PhD(org chem), 51. Prof Exp: Chemist, Tenn Eastman Corp, 44-46; sr chemist, Continental Oil Co, 48-56; sr res chemist, 56-60, supvr, 60-63, MGR, MINN MINING & MFG CO, 63- Mem: Am Chem Soc. Res: Development of synthetic leather. Mailing Add: 3M Co 3M Ctr Bldg 235 St Paul MN 55101

STEPENUCK, STEPHEN JOSEPH, JR, b Salem, Mass, Oct 12, 37; m 68; c 3. ENVIRONMENTAL CHEMISTRY. Educ: Merrimack Col, BS, 59; Col of the Holy Cross, MS, 61; Univ NH, PhD(phys chem), 71. Prof Exp: Instr chem, Merrimack Col, 61-65; teaching fel, Univ NH, 66-68, instr, 69-70; asst prof, 70-73, ASSOC PROF CHEM, KEENE STATE COL, 73- Mem: Am Chem Soc. Res: Radiation chemistry; clinical and diagnostic analyses; environmental analyses; marine chemistry, occupational health. Mailing Add: Dept of Chem Keene State Col Keene NH 03431

STEPHANY, EDWARD O, b Rochester, NY, July 6, 16; m 41; c 1. MATHEMATICS, STATISTICS. Educ: Univ Rochester, AB, 37, AM, 38; Syracuse Univ, PhD(statist), 56. Prof Exp: PROF MATH & CHMN DEPT, STATE UNIV NY COL, BROCKPORT, 47- Mem: Math Asn Am; Am Math Soc. Res: Technique for comparing factors obtained through factor analysis; mathematics education. Mailing Add: Dept of Math State Univ of NY Col Brockport NY 14420

STEPHAS, PAUL, b New York, NY, Aug 31, 29; m 59; c 1. PHYSICS. Educ: Univ Wash, Seattle, BS, 56; Rensselaer Polytech Inst, MS, 59; Univ Ore, PhD(physics), 66. Prof Exp: Asst res lab, Gen Elec Co, 56-58; metallurgist & ceramist, Vallecitos Atomic Lab, 58-62; asst prof physics, Univ BC, 66-69; ASSOC PROF PHYSICS, EASTERN ORE COL, 69- Mem: Am Phys Soc; Am Asn Physics Teachers; AAAS. Res: Atomic effects associated with beta decay; low energy nuclear physics; special relativistic mechanics. Mailing Add: Dept of Physics Eastern Ore State Col La Grande OR 97850

STEPHEN, CHARLES RONALD, b Montreal, Que, Mar 16, 16; nat US; m 41; c 3. ANESTHESIOLOGY. Educ: McGill Univ, BSc, 38, MD & CM, 40; Royal Col Surgeons, fel fac anesthetists, 64. Prof Exp: Chief dept anesthesia, Montreal Neurol Inst, 46-47; asst prof anesthesia, McGill Univ, 48-50; prof, Sch Med & chief div, Univ Hosp, Duke Univ, 50-66; prof, Univ Tex Southwest Med Sch Dallas, 66-71; MALLINCKRODT PROF ANESTHESIOL & CHMN DEPT, SCH MED, WASH UNIV, 71-; HEAD DEPT ANESTHESIA, BARNES HOSP, ST LOUIS, 71- Concurrent Pos: Chief dept anesthesia, Children's Mem Hosp, Montreal, 47-50; ed, Surv Anesthesiol, 56-; anesthesiologist, Parkland Mem Hosp, Dallas, 66-71; chief anesthesia, Children's Med Ctr, Dallas, 66-71. Mem: Am Soc Anesthesiol; Acad Anesthesiol; Inst Anesthesia Res Soc. Res: Pharmacology and physiology of clinical anesthesiology. Mailing Add: Dept of Anesthesiol Wash Univ Sch of Med St Louis MO 63110

STEPHEN, KEITH H, b Ft Wayne, Ind, Jan 3, 34; m 59; c 2. INORGANIC CHEMISTRY. Educ: Wabash Col, BA, 59; Northwestern Univ, PhD(inorg chem), 65. Prof Exp: SR RES CHEMIST, EASTMAN KODAK CO, 64- Mem: Am Chem Soc; Soc Photog Scientists & Engrs. Res: Coordination compounds; electron transfer reactions; photographic chemistry. Mailing Add: Eastman Kodak Co Res Labs 1669 Lake St Rochester NY 14650

STEPHEN, MICHAEL JOHN, b Johannesburg, SAfrica, Apr 7, 33; m 66. PHYSICS. Educ: Univ Witwatersrand, BS, 52, MS, 54; Oxford Univ, PhD(phys chem), 56. Prof Exp: Ramsey fel, 56-58; res assoc chem, Columbia Univ, 58-60; Imp Chem Industs res fel math, Oxford Univ, 60-62; from asst prof to assoc prof, Yale Univ, 62-68; PROF PHYSICS, RUTGERS UNIV, 68- Concurrent Pos: Consult, Bell Tel Labs, 64-66; vis prof, Mass Inst Technol, 67-68. Mem: Am Phys Soc. Res: Low temperature, solid state and molecular physics. Mailing Add: Dept of Physics Busch Campus Rutgers Univ New Brunswick NJ 08903

STEPHEN, WILLIAM ARCHIBALD, b Cedarville, Ont, July 19, 06; m 39; c 1. APICULTURE. Educ: Ont Agr Col, BSA, 35; Univ Toronto, MSA, 41. Prof Exp: Teacher elem sch, Ont, 25-27, prin, 29-31; asst, Can Dept Agr, 35-47; exten beekeeper, NC State Col, 47-63; prof apicult & exten specialist, 63-72, EMER PROF APICULT, OHIO STATE UNIV, 73- Concurrent Pos: Apiculturist, Bur Entom & Plant Quarantine, Div Bee Cult, USDA, Wis, 44-46; apicult specialist, supvr honey proj, Church World Serv, Sparta & Thessaloniki, Greece, 72- Mem: Fel AAAS; Entom Soc Am; Bee Res Asn; Int Fedn Beekeepers Asns. Res: Physical, chemical and biological properties pertaining to honey storage, removal of moisture from honey; Nosema apis infection of honey bees; population densities of honey bee colonies as determined by management practices; educational program to increase income from beekeeping in Greece. Mailing Add: 34 Orchard Dr Worthington OH 43085

STEPHEN, WILLIAM FRANCIS, JR, biochemistry, see 12th edition

STEPHEN, WILLIAM PROCURONOFF, b St Boniface, Man, June 6, 27; nat US; m 52; c 4. ENTOMOLOGY. Educ: Univ Man, BSA, 48; Univ Kans, PhD, 52. Prof Exp: From asst entomologist to assoc entomologist, Sci Serv, Can Dept Agr, 48-53; from asst prof entom & asst entomologist to assoc prof entom & assoc entomologist, 53-63, PROF ENTOM, ORE STATE UNIV, 63- Concurrent Pos: Consult, Orgn Am States, Chile, 70-71 & Food & Agr Orgn, UN, Arg, 72; mem, Bee Res Inst. Mem: AAAS; Entom Soc Am; Animal Behav Soc; Soc Study Evolution; Soc Syst Zool. Res: Insect behavior; systematic zoology; development physiology; population genetics; pollination. Mailing Add: Dept of Entom Ore State Univ Corvallis OR 97331

STEPHENS, ARTHUR BROOKE, b Dinuba, Calif, July 6, 42; m 68. NUMERICAL ANALYSIS. Educ: Univ Colo, BA, 64; Univ Md, PhD(math), 69. Prof Exp: Asst prof math, Univ Hawaii, 69-71; ASST PROF MATH, MT ST MARY'S COL, 73- Mem: Am Math Soc; Math Asn Am. Res: Optimization problems in functional analysis;

estimates for complex eigenvalues of positive matrices. Mailing Add: Dept of Sci & Math Mt St Mary's Col Emmitsburg MD 21727

STEPHENS, BOBBY GENE, b Glendale, SC, Mar 8, 35; m 57; c 4. ANALYTICAL CHEMISTRY. Educ: Wofford Col, BS, 57; Clemson Univ, MS, 61, Nat Defense Educ Act fel & PhD(anal chem), 64. Prof Exp: From asst prof to assoc prof, 63-74, DEAN COL CHEM, WOFFORD COL, 72-, PROF, 74- Concurrent Pos: USPHS res grants, 65-68 & 68-73; co-dir study mission, Czech, 69; grant proj dir, NSF Col Sci Improv Prog, 70-73. Honors & Awards: Jefferson Award, SC Acad Sci, 69. Mem: Am Chem Soc. Res: Light absorption spectrophotometry, solvent extraction; complexometric titrations; classical and modern methods of chemical analysis and primary standards; trace methods for determining atmospheric contaminants. Mailing Add: Wofford Col Spartanburg SC 29301

STEPHENS, CHARLENE BARR, b Jackson, Miss, Sept 2, 36. OTOLARYNGOLOGY. Educ: Univ Miss, BA, 58, MS, 66, PhD(physiol), 70. Prof Exp: Therapist, Pub Schs, Miss, 58-59; teacher, Pub Schs, Ala, 59-60; therapist, Pub Schs, Tenn, 60-62 & Miss, 62-64; speech pathologist & audiologist, 64-67, instr, 67-70, ASST PROF OTOLARYNGOL, MED CTR, UNIV MISS, 70- Concurrent Pos: Mem grad fac, Univ Miss, 72. Mem: AAAS; Asn Res Otolaryngol; Am Speech & Hearing Asn. Res: Embryology, physiology, anatomy and pathology of the ear and larynx; genetics of head and neck lesions; physiology and pathology of voice, speech and language. Mailing Add: Div of Otolaryngol Univ of Miss Med Ctr Jackson MS 39216

STEPHENS, CHARLES ARTHUR LLOYD, JR, b Brooklyn, NY, Apr 4, 17; m 39, 64; c 2. INTERNAL MEDICINE. Educ: Cornell Univ, AB, 38, MD, 42; Am Bd Internal Med, dipl, 52. Prof Exp: DIR RES, SOUTHWESTERN CLIN & RES INST, INC, UNIV ARIZ, 46-, PRES, 71-, ASSOC INTERNAL MED, COL MED, 70- Concurrent Pos: Sr consult, Tucson Med Ctr, St Joseph's & Palo Verde Hosps, Ariz; pvt pract; chmn med adv comt, Southern Chap, Am Red Cross; mem bd dirs, Southwest Chap, Arthritis Found; vis prof agr biochem, Univ Ariz, adj prof microbiol, 76- Mem: AMA; fel Am Col Physicians; Am Rheumatism Asn; NY Acad Sci; Am Heart Asn. Res: Tissue culture research in rheumatic diseases. Mailing Add: Holbrook-Hill Med Group 5100 E Grant Rd Tucson AZ 85712

STEPHENS, CHARLES ROBERT, JR, organic chemistry, see 12th edition

STEPHENS, CLARENCE FRANCIS, b Gaffney, SC, July 24, 17; m 42; c 2. MATHEMATICS. Educ: J C Smith Univ, BS, 38; Univ Mich, MS, 39, PhD(math), 43. Hon Degrees: DSc, J C Smith Univ, 54. Prof Exp: Instr math, Prairie View Col, 40-42, prof, 46-47; prof & head dept, Morgan State Col, 47-62; prof, State Univ NY Col Geneseo, 62-69; PROF MATH & CHMN DEPT, STATE UNIV NY COL POTSDAM, 69- Concurrent Pos: Ford fel & mem, Inst Advan Study, 53-54. Mem: Assoc Am Math Soc; assoc Math Asn Am. Res: Non-linear difference equations analytic in a parameter. Mailing Add: 81 Pierrepont Ave Potsdam NY 13676

STEPHENS, DALE NELSON, b Los Angeles, Calif, Dec 20, 41; m 64; c 2. ORGANIC CHEMISTRY. Educ: Westmont Col, BA, 63; Univ Ariz, PhD(org chem), 67. Prof Exp: Res chemist, Corn Prod Co, 67-68; asst prof, 68-71, ASSOC PROF CHEM, BETHEL COL, MINN, 71- Mem: AAAS; Am Chem Soc. Res: Urethanes; epoxies; natural product synthesis; x-ray crystallography; organic synthesis. Mailing Add: Dept of Phys Sci Bethel Col Arden Hills MN 55113

STEPHENS, EDGAR RAY, b Detroit, Mich, Aug 26, 24; m 66. PHYSICAL CHEMISTRY. Educ: Carnegie Inst Technol, BS, 45; Princeton Univ, MS, 49, PhD(chem), 51. Prof Exp: Res chemist, Shawinigan Resins Corp, 45-47; asst phys chem, Princeton Univ, 47-50; res chemist, Franklin Inst, 50-59 & Scott Res Labs, Inc, 59-63; lectr plant path, 63-71, RES CHEMIST, UNIV CALIF, RIVERSIDE, 63-, PROF ENVIRON SCI, 71- Mem: AAAS; Am Chem Soc; Air Pollution Control Asn. Res: Chemical kinetics; air pollution; photochemistry. Mailing Add: Air Pollution Res Ctr Univ of Calif Riverside CA 92502

STEPHENS, FRANK SAMUEL, b Ind, June 30, 31; m 59. NUCLEAR CHEMISTRY. Educ: Oberlin Col, AB, 52; Univ Calif, PhD(chem), 55. Prof Exp: RES CHEMIST, LAWRENCE RADIATION LAB, UNIV CALIF, BERKELEY, 55- Concurrent Pos: Ford Found grant, Inst Theoret Physics, Copenhagen, Denmark, 59-60; guest prof, Physics Sect, Univ Munich, 70-71. Mem: Am Phys Soc. Res: Coulomb excitation; nuclear structure; heavy ion physics. Mailing Add: Lawrence Radiation Lab Univ of Calif Berkeley CA 94704

STEPHENS, GEORGE ROBERT, JR, forestry, see 12th edition

STEPHENS, GROVER CLEVELAND, b Oak Park, Ill, Jan 12, 25; m 49; c 3. ZOOLOGY. Educ: Northwestern Univ, BS, 48, MA, 49, PHD(biol), 52. Prof Exp: Asst philos, Northwestern Univ, 48-49, asst zool, 49-51; instr biol, Brooklyn Col, 52-53; asst zool, Univ Minn, Minneapolis, 53-55, from asst prof to prof, 55-64; prof organismic biol, 64-69, chmn dept organismic biol, 64-69, PROF DEVELOP & CELL BIOL, UNIV CALIF, IRVINE, 69- Concurrent Pos: Mem corp, Marine Biol Lab, Woods Hole, 53-, instr, 53-60, in-charge invert zool, 58-60; NSF sr fel, 59-60. Mem: AAAS; Am Soc Zool; Am Soc Limnol & Oceanog. Res: Invertebrate physiology; biological rhythms; feeding mechanisms in invertebrates; algal physiology; amino acid transport. Mailing Add: Dept of Develop & Cell Biol Univ of Calif Irvine CA 92664

STEPHENS, HAROLD W, b Trenton, NJ, Mar 16, 19; m 46; c 1. MATHEMATICS. Educ: Trenton State Col, BS, 41; Columbia Univ, MA, 44, EdD, 64. Prof Exp: Head dept math, Farragut Naval Acad, 44-46; instr, Univ Fla, 46-48, Univ Md, 48-49 & McCoy Col, Johns Hopkins Univ, 49-50; asst prof, Ball State Univ, 52-55 & Univ Tenn, 55-60; asst prof, 60-64, PROF MATH, MEMPHIS STATE UNIV, 64- Concurrent Pos: Lectr, NSF insts & workshops, Univ Tenn, Memphis State, Murray State Univ & Southwestern Univ, Memphis, 57- Mem: Math Asn Am; Am Math Soc. Res: Mathematics courses for teacher training; algebra. Mailing Add: Dept of Math Memphis State Univ Memphis TN 38152

STEPHENS, HENRY LEROY, b Dodgeville, Wis, Sept 13, 98; m 34; c 2. PLANT PHYSIOLOGY. Educ: Univ Wis, BS, 24, MS, 26, PhD(plant chem), 32. Prof Exp: From instr to prof, 27-73, EMER PROF BIOL, WESTERN KY UNIV, 73- Mem: Bot Soc Am. Res: Light effects and nitrate assimilation in plants; nitrogen relation of fungi. Mailing Add: Dept of Biol Western Ky Univ Bowling Green KY 42101

STEPHENS, HOWARD L, b Akron, Ohio, Oct 9, 19; m 45; c 1. POLYMER CHEMISTRY. Educ: Univ Akron, BS, 49, MS, 50, PhD(polymer chem), 60. Prof Exp: Res chemist rubber res labs, 49-52, inst rubber chem, 53-56 & inst rubber res, 56-57, admin asst, 57-65, from instr to assoc prof chem, 57-73, PROF CHEM & POLYMER SCI, UNIV AKRON, 73-, MGR APPL RES, INST POLYMER SCI, 66- Mem: Am Chem Soc. Res: Polymer oxidation; preparation and structure of graph polymers; emulsion polymerization; vulcanization. Mailing Add: Dept of Polymer Sci Univ of Akron Akron OH 44325

STEPHENS, HOWARD PAGE, physical chemistry, see 12th edition

STEPHENS, JAMES BRISCOE, b San Francisco, Calif, Mar 5, 36; m 67; c 2. ATMOSPHERIC PHYSICS, SPACE PHYSICS. Educ: Univ Okla, BS, 64, MS, 66, PhD(eng physics), 71. Prof Exp: Scientist remote sensing, 66-69, scientist statist physics, 70-72, TASK TEAM LEADER TERRESTRIAL DIFFUSION, MARSHALL SPACE FLIGHT CTR, NASA, 73- Concurrent Pos: Mem, Atmos Effects Panel, NASA Hq, 74- Mem: Am Phys Soc; Sigma Xi. Res: The environmental effects from aerospace effluents. Mailing Add: Aerospace Environ Div NASA Marshall Space Flight Ctr Huntsville AL 35812

STEPHENS, JAMES FRED, b Lexington, Tenn, Sept 29, 32; m 72; c 2. POULTRY SCIENCE, MICROBIOLOGY. Educ: Univ Tenn, BS, 54, MS, 59, PhD(bact), 64. Prof Exp: Asst poultry, Univ Tenn, 56-62; from asst prof to assoc prof poultry sci, Clemson Univ, 62-68; ASSOC PROF POULTRY SCI, OHIO STATE UNIV, 68- Mem: Poultry Sci Asn; Am Soc Microbiol; World Poultry Sci Asn. Res: Poultry disease; nutrition relationships; pathogenecity of Salmonellae; physiological effects of coccidiosis in chickens; drug resistance in enteric bacteria. Mailing Add: Dept of Poultry Sci Ohio State Univ 674 W Lane Ave Columbus OH 43210

STEPHENS, JAMES REGIS, b Pittsburgh, Pa, Mar 16, 25; m 55; c 4. ORGANIC POLYMER CHEMISTRY. Educ: St Vincent Col, BS, 47; Univ Pittsburgh, MS, 49; Northwestern Univ, PhD(chem), 53. Prof Exp: Res chemist, Sinclair Ref Co, 50 & Am Cyanamid Co, 53-57; sr res scientist, 57-67, group leader res dept, 67-70, SECT LEADER, RES & DEVELOP, AMOCO CHEM CORP, 70- Mem: Am Chem Soc. Res: Organic nitrogen compounds; stereochemistry of dioxanes; applications of aromatic acids; alkyd baking enamels; heterocyclic and aromatic polymers for high temperature service; magnet wire enamels. Mailing Add: Res & Devel Amoco Chem Corp PO Box 400 Naperville IL 60540

STEPHENS, JESSE JERALD, b Oklahoma City, Okla, June 3, 33; m 55; c 3. PHYSICAL METEOROLOGY. Educ: Univ Tex, BS, 58, MA, 61; Tex A&M Univ, Nat Defense Educ Act fel & PhD(meteorol), 66. Prof Exp: Meteorologist, US Weather Bur, 57-59; lectr meteorol, Univ Tex, 59-61, from instr to asst prof, 64-66; assoc prof, Univ Okla, 66-67; assoc prof, 67-71, PROF METEOROL, FLA STATE UNIV, 71-, CHMN DEPT, 75- Mem: Am Meteorol Soc; Optical Soc Am; Res: Scattering processes in the atmosphere; geophysical data processing. Mailing Add: Dept of Meteorol Fla State Univ Tallahassee FL 32306

STEPHENS, JOHN ARNOLD, b Council Bluffs, Iowa, Aug 11, 21; m 44; c 1. ORGANIC CHEMISTRY. Educ: Univ Omaha, Munic, BA, 43; Univ Nebr, MA, 49, PhD(chem), 51. Prof Exp: Res chemist, 50-57, res group leader, 57-62, mgr res & develop, Agr Res Dept, 62-75, DIR BIOL EVAL, MONSANTO AGR PROD CO, 75- Mem: Am Chem Soc. Res: Nitrogen heterocycles; industrial synthesis; agricultural chemicals. Mailing Add: Agr Res Dept Monsanto Co 400 N Lindbergh St Louis MO 63166

STEPHENS, JOHN FIRTH, b Covington, Ky, Dec 22, 10; m 39. INDUSTRIAL HYGIENE. Educ: Univ Tenn, BS, 35; dipl, Am Acad Indust Hyg. Prof Exp: Coop eng student, Tenn Eastman Corp, 30-34; lab asst, Tenn Valley Authority, 34; sales engr, Warren Candies, Inc, 35-36; chemist, Am Bemberg Corp, 37-42; line foreman, Hercules Powder Co, 42-44; sr chemist, Clinton Eng Works-Tenn Eastman Corp, 44-46; indust hygienist, Carbide & Carbon Chem Corp, 46-49; asst indust hygienist, City Bur Indust Hyg, Detroit, Mich, 49-51; sr indust hygienist, Gen Motors Corp, 51-67, res engr, Indust Hyg Dept, 67-75; CONSULT, 76- Mem: Am Chem Soc; Am Indust Hyg Asn; Air Pollution Control Asn; fel Am Inst Chem. Res: Ammonium sulphate leaching of low-grade manganese ores of east Tennessee; chemical analysis and production supervision; technical supervision in an atomic energy installation; industrial hygiene and air pollution control engineering. Mailing Add: 14970 Lindsay Detroit MI 48227

STEPHENS, JOHN JAMES, III, b Brooklyn, NY, May 16, 32; m 57; c 2. VERTEBRATE PALEONTOLOGY. Educ: Lafayette Col, AB, 54; Univ Mich, MA, 56, PhD(geol), 59. Prof Exp: Res asst geol, Univ Mich, 58; instr geol & cur, Ohio State Univ, 59-61, asst prof, 61-70, asst dean col arts & sci, 62-65, off acad affairs, 65-66, col biol sci, 66-70, assoc for col develop, 69-70; acad officer, Bd Regents, State of Kans, 70-71; consult environ educ, Univ Kans, 71; asst to chancellor, Univ Nebr, Lincoln, 72, exec asst to chancellor, 72-74; EXEC ASST TO ASST SECY EDUC, DEPT HEALTH, EDUC & WELFARE, 74- Concurrent Pos: NSF res grant paleont, Mex, 62; Am Coun Educ fel, Acad-Admin Internship Prog, 68-69. Mem: Fel Geol Soc Am; Paleont Soc; Soc Vert Paleont. Res: Fossil mammals of Pliocene and Pleistocene age; rodents. Mailing Add: Off of Asst Secy for Educ 400 Maryland Ave SW Washington DC 20202

STEPHENS, JOHN STEWART, JR, b Los Angeles, Calif, May 12, 32; m 53; c 1. MARINE BIOLOGY, FISH BIOLOGY. Educ: Stanford Univ, BS, 54; Univ Calif, Los Angeles, MA, 57, PhD, 60. Prof Exp: Asst zool, Univ Calif, Los Angeles, 54-58, assoc biol, Santa Barbara, 58-69; from instr to prof biol, 59-74, JAMES IRVINE PROF ENVIRON BIOL, OCCIDENTAL COL, 74- Concurrent Pos: Dir, Vantuna Oceanog Prog, Occidental Col, 69- Mem: Am Soc Ichthyol & Herpet; Soc Syst Zool; Am Fisheries Soc; Am Inst Fishery Res Biologists. Res: Systematics and distribution of blenniod fishes; especially Chaenopsidae; osteology of tropical blennies; ecology of nearctic fishes of California, including effects of pollution and habitat destruction. Mailing Add: Dept of Biol Occidental Col Los Angeles CA 90041

STEPHENS, KENNETH S, b Kutztown, Pa, Dec 11, 32; m 53; c 3. APPLIED STATISTICS, MATHEMATICAL STATISTICS. Educ: LeTourneau Inst, BS, 55; Rutgers Univ, MS, 60, PhD(appl & math statist), 66. Prof Exp: Qual control engr, Western Elec Co, Inc, Pa, 55-58 & 59-60; asst & instr math, Rutgers Univ, 58-59; chief qual control eng dept, 60-62, chief eng personnel & staff & Kans City coord dept, 62-63, res leader appl math & statist group systs res & develop, NJ, 63-67; prof & coordr math & indust eng, Letourneau Col, 67-72; LECTR, SCH INDUST & SYSTS ENG, GA INST TECHNOL & RES CONSULT, ECON DEVELOP LAB, ENG EXP STA, 74- Concurrent Pos: Mem, UN team statist qual control, India, 62-63 & UN Indust Develop Orgn adv standardization & qual control, Thai Indust Stand Inst, Bangkok, Thailand, 72-74. Mem: Fel Am Soc Qual Control; Am Statist Asn; Europ Orgn Qual Control; Indian Asn Qual & Reliability; Sigma Xi. Res: Industrial statistics; quality control systems; standardization and certification; industrial engineering. Mailing Add: Sch Indust & Systs Eng Ga Inst of Technol Atlanta GA 30332

STEPHENS, LAWRENCE JAMES, b Chicago, Ill, Aug 11, 40; m 64; c 3. ORGANIC CHEMISTRY, SCIENCE EDUCATION. Educ: Loyola Univ Chicago, BS, 63; Univ Nebr, Lincoln, PhD(org chem), 69. Prof Exp: Res assoc, Stanford Univ, 68-69; asst prof chem, Findlay Col, 69-73; ASSOC PROF CHEM, ELMIRA COL, 73- Mem: Am Chem Soc; Nat Sci Teachers Asn. Res: Curriculum development in the natural sciences; synthesis of terpenoids. Mailing Add: Dept of Chem Elmira Col Elmira NY 14901

STEPHENS, LEE BISHOP, JR, b Atlanta, Ga, Oct 22, 25; m 58; c 3. EMBRYOLOGY. Educ: Morehouse Col, BS, 47; Atlanta Univ, MS, 50; Univ Iowa, PhD, 57. Prof Exp: Instr biol, Dillard Univ, 50-53; instr, NC Col Durham, 53-54; assoc prof, Southern Univ, 57-62; from asst prof to assoc prof, 62-70, PROF BIOL, CALIF STATE UNIV, LONG BEACH, 70- Mem: Am Soc Zool; Am Micros Soc. Res: Neuroembryology; regeneration; endocrinology and development of the nervous system. Mailing Add: Dept of Biol Calif State Univ Long Beach CA 90801

STEPHENS, MARVIN WAYNE, b Grand Rapids, Mich, Mar 24, 43; m 66; c 1. BIOCHEMISTRY. Educ: Cedarville Col, BS, 65; Univ Nebr, Lincoln, PhD(chem), 72. Prof Exp: Asst prof, 69-75, ASSOC PROF CHEM, MALONE COL, 75- Mem: AAAS. Res: Genetic and biochemical aspects of pigmentation in the fowl; pitcher plant biochemistry. Mailing Add: Dept of Chem Malone Col Canton OH 44709

STEPHENS, MAYNARD MOODY, b Connersville, Ind, Apr 25, 08; m 31; c 6. PETROLEUM ENGINEERING, ECONOMIC GEOLOGY. Educ: Univ Minn, BA, 30, MA, 31, PhD(geol), 34; Pa State Univ, PE, 43. Prof Exp: Res geologist, Minn Geol Surv & Univ Minn, 31-36; supvr petrol & natural gas exten, Pa State Univ, 36-41; partner, Ryder Scott Co, 41-52; prof geol, Midwestern Univ, 52-60, dean sch petrol & phys sci, 57-60; consult, Maynard M Stephens Co, 57-68; indust specialist, 68-72, MGR VULNERABILITY STUDIES, OFF OIL & GAS, US DEPT INTERIOR, 72- Concurrent Pos: Geologist & engr, Drill Well Oil Co, 53-55 & Assoc Petrol Engrs, 55-57. Mem: Am Asn Petrol Geologists; Geol Soc Am; Am Inst Mining, Metall & Petrol Eng; Nat Soc Prof Engrs; Soc Petrol Eng; Am Soc Indust Security. Res: Mineralography; secondary recovery of oil by water flooding; polished surface of ores; photomicrography; petroleum and natural gas engineering; petroleum production; security and national vulnerability as related to oil and gas. Mailing Add: R652 South Fed Bldg New Orleans LA 70130

STEPHENS, MICHAEL A, b Bristol, Eng, Apr 26, 27; m 62. MATHEMATICAL STATISTICS, APPLIED STATISTICS. Educ: Bristol Univ, BSc, 48; Harvard Univ, AM, 49; Univ Toronto, PhD(math), 62. Prof Exp: Instr math, Tufts Col, 49-50; lectr, Woolwich Polytech, Eng, 52-53 & Battersea Col Technol, 53-56; instr, Case Western Reserve Univ, 56-59; lectr, Univ Toronto, 59-62, asst prof, 62-63; from asst prof to prof, McGill Univ, 63-70; prof, Univ Nottingham & Univ Grenoble, 70-72; PROF MATH, McMASTER UNIV, 72- Concurrent Pos: Consult, Can Packers Ltd, 62-63 & various Montreal res Drs, 63-67. Mem: Fel Am Statist Asn; Can Statist Soc; Can Statist Sci Asn; Inst Math Statist; Can Math Cong. Res: Mathematical statistics, distributions on a circle or a sphere; goodness of fit statistics. Mailing Add: Dept of Appl Math McMaster Univ Hamilton ON Can

STEPHENS, NEWMAN LLOYD, b Kanth, India, Feb 28, 26; Can citizen; m 67; c 2. PHYSIOLOGY, BIOSTATISTICS. Educ: Univ Lucknow, MB & BS, 50, DM, 53. Prof Exp: Resident med officer, King George's Med Col, Univ Lucknow, 50-53; head sect med, Clara Swain Hosp, Bareilly, India, 55-58; med registr, Univ Col Hosp, London, Eng, 59-61; from asst prof to assoc prof, 67-73, PROF PHYSIOL, FAC MED, UNIV MAN, 73- Concurrent Pos: Res fel cardiol, Res & Educ Hosp, Univ Ill, 62-64; res fel physiol, Sch Hyg, Johns Hopkins Univ, 64-65; res fel med, Winnipeg Gen Hosp, Man, 65-66; Can Heart Found scholar & Med Res Coun Can grant, Fac Med, Univ Man, 67- Mem: Fel Royal Soc Med; Am Physiol Soc; Biophys Soc; Can Physiol Soc; Can Soc Clin Invest. Res: Smooth muscle, biophysics, biochemistry and ultrastructure of normal muscle, effects of acidosis and hypoxia on these parameters. Mailing Add: Dept of Physiol Univ of Man Fac of Med Winnipeg MB Can

STEPHENS, NOEL, JR, b Richmond, Ky, Dec 27, 28; m 58; c 4. ANIMAL SCIENCE, ANIMAL NUTRITION. Educ: Univ Ky, BS, 55, MS, 56, PhD(animal sci), 64. Prof Exp: From instr to assoc prof, 56-69, PROF ANIMAL SCI, BEREA COL, 69- Mem: Am Soc Animal Sci. Res: Amino acids and trace mineral research in swine nutrition. Mailing Add: Dept of Animal Sci Berea Col Berea KY 40403

STEPHENS, PHILIP J, b West Bromwich, Eng, Oct 9, 40; m 62; c 1. THEORETICAL CHEMISTRY. Educ: Oxford Univ, BA, 62, DPhil(chem), 64. Prof Exp: Res fel chem, Univ Copenhagen, 64-65; res fel, Univ Chicago, 65-67; asst prof, 67-70, ASSOC PROF CHEM, UNIV SOUTHERN CALIF, 70- Concurrent Pos: Sci Res Coun fel, 64-66; Alfred P Sloan res fel, 68-70. Mem: Am Chem Soc; The Chem Soc; Am Phys Soc. Res: Magneto-optical and spectroscopic properties of matter. Mailing Add: Dept of Chem Univ of Southern Calif Los Angeles CA 90007

STEPHENS, RAYMOND EDWARD, b Pittsburgh, Pa, Mar 5, 40. CELL BIOLOGY, PROTEIN CHEMISTRY. Educ: Geneva Col, BS, 62; Univ Pittsburgh, MS, 63; Dartmouth Med Sch, PhD(molecular biol), 65. Prof Exp: Fel, Univ Hawaii, 66; NIH fel, Harvard Univ, 66-67; asst prof, 67-70, ASSOC PROF BIOL, BRANDEIS UNIV, 70-; INVESTR, MARINE BIOL LAB, 70- Concurrent Pos: Mem cell biol study sect, NIH, 71-75. Mem: AAAS; Am Chem Soc; Soc Gen Physiol; Am Soc Cell Biol. Res: Protein subunit association; chemistry of cell division and cell movement. Mailing Add: Marine Biol Lab Woods Hole MA 02543

STEPHENS, RAYMOND WEATHERS, JR, b Marietta, Ga, Apr 20, 28; m 51; c 2. PETROLEUM GEOLOGY. Educ: Univ Ga, BS, 51; La State Univ, MS, 56, PhD(geol), 60. Prof Exp: Geologist, Shell Oil Co, 59-66; dist geologist, Pubco Petrol Co, 66-72; asst prof, 72-74, ASSOC PROF EARTH SCI, UNIV NEW ORLEANS, 74- Mem: Am Asn Petrol Geologists. Res: Stratigraphic and paleontologic geology. Mailing Add: Dept of Earth Sci Univ of New Orleans New Orleans LA 70122

STEPHENS, ROGER, b Manchester, Eng, Apr 6, 45; m 70. CHEMISTRY. Educ: Cambridge Univ, BA, 67; Bristol Univ, MSc, 69; Univ London, PhD(chem), 71. Prof Exp: ASST PROF CHEM, DALHOUSIE UNIV, 71- Concurrent Pos: Nat Res Coun Can res grant, 71- Mem: The Chem Soc; Royal Inst Chemists; Spectros Soc Can. Res: Analytical spectroscopy; flame systems; spectroscopic sources. Mailing Add: Dept of Chem Dalhousie Univ Halifax NS Can

STEPHENS, STANLEY GEORGE, b Dudley, Eng, Sept 2, 11; nat US; m 38; c 2. EVOLUTIONARY BIOLOGY. Educ: Cambridge Univ, BA, 33, MA, 36; Univ Edinburgh, PhD(bot), 41. Prof Exp: Asst plant breeder, Scottish Plant Breeding Sta, Univ Edinburgh, 36-38; asst geneticist, Empire Cotton Corp, BWI, 38-44, asst prof genetics, McGill Univ, 44-45; res assoc, Cold Spring Harbor, Carnegie Inst Technol, 45-47; prof genetics & cotton cytogeneticist, Tex A&M Univ, 47-49; res prof agron, 49-51, William Neal Reynolds prof genetics, 51-74, EMER PROF GENETICS, NC STATE UNIV, 75- Concurrent Pos: Guggenheim fel, NC State Univ, 58. Honors & Awards: Award, Nat Cotton Coun, 62. Mem: Nat Acad Sci; Soc Study Evolution; Soc Am Archaeol. Res: Cytogenetics of cotton; mechanisms of species differentiation; evolution and dispersal of cultivated plants. Mailing Add: 3219 Darien Dr Raleigh NC 27607

STEPHENS, STANLEY LAVERNE, b Niagara Falls, NY, Apr 23, 43; m 64; c 1. MATHEMATICS. Educ: Anderson Col, BA, 65; Lehigh Univ, MS, 67, PhD(math), 72. Prof Exp: Instr, Moravian Col, 68-71; ASST PROF MATH, ANDERSON COL, 71- Mem: Am Math Soc. Res: Prime power groups, particularly the automorphism group of p-groups. Mailing Add: Dept of Math Anderson Col Anderson IN 46011

STEPHENS, TIMOTHY LEE, b Bellingham, Wash, June 27, 44; m 69; c 1. MOLECULAR PHYSICS, ATMOSPHERIC PHYSICS. Educ: Calif Inst Technol, BS, 66; Harvard Univ, AM, 67, PhD(physics), 71. Prof Exp: Res fel physics, Smithsonian Astrophys Observ, 69-70; PHYSICIST, GEN ELEC CO-TEMPO, 70- Res: Optical emission processes of atoms and molecules; energy partitioning during high energy electron deposition in air; chemical and hydrodynamic properties of the atmosphere; environmental effects of nuclear weapons. Mailing Add: Gen Elec-TEMPO PO Drawer QQ Santa Barbara CA 93102

STEPHENS, VERLIN CLARK, pharmaceutical chemistry, see 12th edition

STEPHENS, WILLIAM D, b Paris, Tenn, Nov 17, 32; div; c 3. ORGANIC CHEMISTRY. Educ: Western Ky State Col, BS, 54; Vanderbilt Univ, PhD(chem), 60. Prof Exp: Group leader high energy oxidizers, Thiokol Chem Corp, 59-61, sect chief org chem, 61-63; group leader basic mat, Goodyear Tire & Rubber Co, 63-66; PRIN CHEMIST, THIOKOL CHEM CORP, 66- Concurrent Pos: Adj assoc prof, Univ Ala, Huntsville, 70- Mem: Am Chem Soc. Res: Solid propellant research; explosives, burning-rate catalysts; organometallic, organic nitrogen, sulfur and cyclic compounds; polymer chemistry; adhesives; bonding agents; urethane catalysts; antioxidants. Mailing Add: 12002 Branscomb Rd Huntsville AL 35803

STEPHENS, WILLIAM EDWARDS, b St Louis, Mo, May 29, 12; m 42; c 2. PHYSICS. Educ: Wash Univ, AB, 32, MS, 34; Calif Inst Technol, PhD(physics), 38. Prof Exp: Lectr, Univ Pittsburgh, 39; fel, Res Labs, Westinghouse Elec Corp, 39-40; instr physics, Stanford Univ, 40-41; from instr to assoc prof, 41-48, chmn dept physics, 63-68, vprovost & actg dean col, 68-69, dean & vprovost, Col Arts & Sci, 69-74, PROF PHYSICS, UNIV PA, 48-, DIR TANDEM ACCELERATOR, 62- Concurrent Pos: Nat Defense Res Comt tech rep, 44-45; vis prof, Univ Zurich, 57 & 69. Mem: Fel Am Phys Soc; Am Astron Soc; fel AAAS. Res: Nuclear physics; photonuclear reactions; mass spectroscopy; nucleosynthesis. Mailing Add: Rittenhouse Lab Univ of Pa Philadelphia PA 19174

STEPHENS, WILLIAM LEONARD, b Covington, Ky, Apr 19, 29; m 57. MICROBIOLOGY. Educ: Sacramento State Col, BS, 57; Univ Calif, Davis, PhD(microbiol), 63. Prof Exp: Res asst bact, Univ Calif, Davis, 57-63; from asst prof to assoc prof, 63-70, chmn dept, 68-74, PROF BACT, CHICO STATE COL, 70- Mem: Am Soc Microbiol. Res: Carotenoid pigments of bacteria. Mailing Add: Dept of Biol Sci Chico State Col Chico CA 95926

STEPHENS-NEWSHAM, LLOYD G, b Saskatoon, Sask, Apr 30, 21; m 50; c 2. BIOPHYSICS. Educ: Univ Sask, BA, 43; McGill Univ, PhD(nuclear physics), 48. Prof Exp: Asst prof physics, Dalhousie Univ, 48-51; from asst prof to assoc prof, Fac Med, McGill Univ, 52-66; from assoc prof to prof physiol, 66-74, PROF, FAC PHARM, UNIV ALTA, 74- Concurrent Pos: Consult, Victoria Gen Hosp, 48-51; radiation physicist, Royal Victoria Hosp, Montreal, 52-66. Mem: Can Physiol Soc; Can Asn Physicists; Biophys Soc; Sigma Xi. Res: Effects of ionizing radiation on active substances in tissues. Mailing Add: Fac Pharm & Pharmaceut Sci Univ of Alta Edmonton AB Can

STEPHENSON, ALFRED BENJAMIN, b Unity, Va, May 24, 12; m 41; c 3. POULTRY HUSBANDRY. Educ: Va Polytech Inst, BS, 33; Rutgers Univ, MS, 34; Iowa State Col, PhD, 49. Prof Exp: Asst poultry breeding, Iowa State Col, 46-49; from asst prof to assoc prof, Utah State Agr Col, 49-53; assoc prof, 53-58, PROF POULTRY BREEDING, UNIV MO-COLUMBIA, 58- Mem: Poultry Sci Asn. Res: Quantitative inheritance in poultry breeding. Mailing Add: Dept Poultry Bldg T-14 Univ of Mo Columbia MO 65201

STEPHENSON, BETTY ANN, b Dallas, Tex, Nov 11, 43; m 71. BIO-ORGANIC CHEMISTRY. Educ: Univ Calif, Berkeley, BS, 66; Stanford Univ, PhD(org chem), 71. Prof Exp: Smith, Kline & French fel pharmacol, Stanford Univ, 70-72; fel org chem, Syva Co, 72- & Stanford Res Inst, 72-73; asst prof chem, Univ Santa Clara, 73-75; LECTR, DEPT CHEM, CASE WESTERN RESERVE UNIV, 75- Mem: Am Chem Soc. Res: Organic stereochemistry; microbial oxidations. Mailing Add: Dept of Chem Case Western Reserve Univ Cleveland OH 44106

STEPHENSON, CHARLES BRUCE, b Little Rock, Ark, Feb 9, 29; m 52. ASTRONOMY. Educ: Univ Chicago, BS, 49, MS, 51; Univ Calif, PhD(astron), 58. Prof Exp: Asst astron, Dearborn Observ, Northwestern Univ, 51-53 & Univ Calif, 56-57; from instr to assoc prof, 58-68, PROF ASTRON, CASE WESTERN RESERVE UNIV, 68- Mem: AAAS; Am Astron Soc; Int Astron Union. Res: Stellar spectra; galactic structure; observational astrophysics. Mailing Add: Dept of Astron Case Western Reserve Univ Cleveland OH 44106

STEPHENSON, CHARLES V, b Centerville, Tenn, Oct 1, 24; m 48; c 3. SOLID STATE PHYSICS. Educ: Vanderbilt Univ, BA, 48, MA, 49, PhD(physics), 52. Prof Exp: Res physicist, Sandia Corp, 52-56; asst prof physics, Ala Polytech Inst, 56-58; head physics sect, Southern Res Inst, 58-62; chmn dept, 67-74, PROF ELEC ENG, VANDERBILT UNIV, 62- Concurrent Pos: Consult, Sandia Corp, 56-58. Mem: AAAS; fel Am Phys Soc; Acoust Soc Am; Am Asn Physics Teachers; sr mem Inst Elec & Electronics Eng. Res: Molecular spectroscopy; solid state physics. Mailing Add: Dept of Elec Eng Vanderbilt Univ Nashville TN 37235

STEPHENSON, CLARK CONKLING, b Augusta, Kans, Dec 28, 11; m 42; c 1. PHYSICAL CHEMISTRY. Educ: Univ Kans, AB, 32; Univ Calif, PhD(chem), 36. Prof Exp: Fel, Mellon Inst, 36-37; from instr to assoc prof, 37-60, PROF CHEM MASS INST TECHNOL, 60- Mem: Am Acad Arts & Sci. Res: Thermodynamics; low temperature physics. Mailing Add: Dept of Chem Mass Inst Technol Cambridge MA 02139

STEPHENSON, DANNY LON, b Ft Worth, Tex, Nov 7, 37; m 63; c 1. ORGANIC CHEMISTRY, SPECTROSCOPY. Educ: Tex Christian Univ, BA, 59, MA, 60; Rice Univ, PhD(org chem), 64. Prof Exp: Res chemist, Phillips Petrol Co, 64-65; PROF CHEM, HOWARD PAYNE COL, 65-, HEAD DEPT & DEAN, SCH SCI & MATH, 74- Mem: Am Chem Soc. Res: Mechanistic study of various condensation reactions with zinc chloride as the catalyst; natural products. Mailing Add: 3700 Second St Brownwood TX 76801

STEPHENSON, DAVID A, b Moline, Ill, July 1, 36; m 58; c 2. HYDROLOGY, GEOLOGY. Educ: Augustana Col, Ill, BA, 58; Wash State Univ, MS, 61; Univ Ill, PhD(ground water geol), 65. Prof Exp: Res assoc hydrogeol, Desert Res Inst, Reno, Nev, 62-64; res asst, Ill State Geol Surv, 64-65; asst prof, 65-70, ASSOC PROF HYDROGEOL & DIR WATER RESOURCES, UNIV WIS-MADISON, 70- Concurrent Pos: Chmn water resources mgt prog, Univ Wis-Madison, 68-, mem exec comt, Inst Environ Studies, 72- Mem: AAAS; Geol Soc Am; Am Geophys Union. Res: Hydrogeology; investigation of geologic controls to quality and quantity of water resource; problems of pollution control in lakeshore environments; importance of water as geologic factor in engineering studies. Mailing Add: Sci Hall Univ of Wis Madison WI 58706

STEPHENSON, DAVID ALLEN, b Denver, Colo, Nov 23, 42; m 63; c 2. MOLECULAR PHYSICS. Educ: NMex State Univ, BS, 64; Univ Mich, Ann Arbor, MS, 65, PhD(physics), 68. Prof Exp: Nat Res Coun-Environ Sci Serv Admin res fel, Environ Sci Serv Admin Res Labs, Colo, 68-70; ASSOC SR RES PHYSICIST, GEN MOTORS RES LABS, 70- Mem: Optical Soc Am. Res: Raman spectroscopy of gases; gas phase reactions. Mailing Add: Physics Dept Gen Motors Res Labs Warren MI 48090

STEPHENSON, EDWARD LUTHER, b Calhoun, Tenn, May 5, 23; m 47; c 2. ANIMAL NUTRITION. Educ: Univ Tenn, BS, 46, MS, 47; State Col Wash, PhD(poultry nutrit), 52. Prof Exp: From asst prof to assoc prof, 49-57, PROF ANIMAL NUTRIT, COL AGR, UNIV ARK, FAYETTEVILLE, HEAD DEPT, 64- Mem: Am Soc Animal Sci; Soc Exp Biol & Med; fel Poultry Sci Asn; Am Inst Nutrit. Res: Poultry nutrition. Mailing Add: Dept of Animal Sci Univ Ark Col Agr Fayetteville AR 72701

STEPHENSON, ELIZABETH WEISS, b Newark, NJ, Apr 1, 27; m 46; c 3. PHYSIOLOGY. Educ: Univ Chicago, BS, 47; George Washington Univ, PhD(physiol), 64. Prof Exp: Res asst, Ill Neuropsychiat Inst, 47-49; from instr to asst prof physiol, Sch Med, George Washington Univ, 64-71; SR STAFF FEL, LAB PHYS BIOL, NAT INST ARTHRITIS, METAB & DIGESTIVE DIS, 71- Mem: AAAS; Am Physiol Soc; Biophys Soc. Res: Ion transport across cellular and intracellular membranes. Mailing Add: Lab of Phys Biol Nat Inst of Arthritis Metab & Digestive Dis Bethesda MD 20014

STEPHENSON, FRANCIS CREIGHTON, b Brantford, Ont, Mar 24, 24; m 48; c 4. PHYSICS. Educ: Univ Toronto, BASc, 49, MA, 51, PhD(physics), 54. Prof Exp: Res physicist, Lamp Develop Dept, Gen Elec Co, 53-65; asst prof, 65-67, ASSOC PROF PHYSICS, CLEVELAND STATE UNIV, 67- Mem: Am Asn Physics Teachers. Res: Molecular spectroscopy; incandescent radiation; vibration; gas discharge. Mailing Add: Dept of Physics 1983 E 24th St Cleveland State Univ Cleveland OH 44115

STEPHENSON, GERARD J, JR, b Yonkers, NY, Mar 4, 37; m 60; c 1. THEORETICAL NUCLEAR PHYSICS. Educ: Mass Inst Technol, BS, 59, PhD(physics), 64. Prof Exp: Res fel physics, Calif Inst Technol, 64-66; from asst prof to assoc prof physics, Univ Md, College Park, 69-74; STAFF MEM, LOS ALAMOS SCI LAB, 74- Concurrent Pos: Guggenheim Mem fel, Los Alamos Sci Lab, 72-73. Mem: AAAS; Am Phys Soc. Res: Theoretical studies of nuclear structure and of low and intermediate energy nuclear reactions, particularly the use of the few body reactions to investigate the nucleon-nucleon interaction. Mailing Add: Medium Energy Theory Group T-5 Los Alamos Sci Lab Los Alamos NM 87545

STEPHENSON, HAROLD PATTY, b Angier, NC, Dec 22, 25; m 56; c 2. MOLECULAR SPECTROSCOPY. Educ: Duke Univ, BSME, 47, MA, 49, PhD(physics), 52. Prof Exp: Instr physics, Duke Univ, 48-49 & 51-52; asst, Appl Physics Lab, Johns Hopkins Univ, 51; assoc prof physics, Ill Wesleyan Univ, 52-53, prof & chmn dept, 53-57; assoc prof mech eng, Duke Univ, 57-60; assoc prof, 60-63, PROF PHYSICS, PFEIFFER COL, 63-, HEAD DEPT, 60- Mem: Am Asn Physics Teachers. Res: Near ultraviolet absorption spectra of poly-atomic molecules; thermodynamics; mechanics. Mailing Add: Dept of Physics Pfeiffer Col Misenheimer NC 28109

STEPHENSON, HUGH EDWARD, JR, b Columbia, Mo, June 1, 22; m 64; c 2. THORACIC SURGERY, CARDIOVASCULAR SURGERY. Educ: Univ Mo, AB & BS, 43; Wash Univ, MD, 45; Am Bd Surg, dipl, 53; Bd Thoracic Surg, dipl, 63. Prof Exp: Instr surg, Sch Med, NY Univ, 51-53; from asst prof to assoc prof, 53-55, chmn dept, 56-60, PROF SURG, SCH MED, UNIV MO-COLUMBIA, 56-, CHIEF, GEN SURG SECT, 76- Concurrent Pos: Chief of surg, Univ Hosps, Univ Mo-Columbia, 56-60. Mem: Soc Vascular Surg; AMA; Am Asn Surg of Trauma; Am Col Surg; Am Col Chest Physicians. Res: Cardiovascular research. Mailing Add: 807 Stadium Blvd Columbia MO 65201

STEPHENSON, JOHN, b Chichester, Eng, 1939; m 65; c 3. THEORETICAL PHYSICS. Educ: Univ London, BSc, 61, PhD(theoret physics), 64. Prof Exp: Lectr math, Univ Adelaide, 65-68; res assoc physics, 68-70, vis asst prof, 70-71, asst prof, 71-74, ASSOC PROF PHYSICS, UNIV ALTA, 74- Mem: Can Asn Physicists; Am Phys Soc. Res: Statistical mechanics and critical phenomena in fluids and magnetic systems. Mailing Add: Dept of Physics Univ of Alta Edmonton AB Can

STEPHENSON, JOHN LESLIE, b Farmington, Maine, Dec 4, 21; m 46; c 3. BIOPHYSICS. Educ: Harvard Univ, BS, 43; Univ Ill, MD, 49. Prof Exp: Asst theoret physics, Metall Lab, Univ Chicago, 43-45; physicist, US Naval Ord Lab, 45; intern, Staten Island Marine Hosp, NY, 49-50; from res assoc to asst prof, Univ Chicago, 52-54; scientist, 54-73, HEAD SECT THEORET BIOPHYS, NAT HEART & LUNG INST, 73- Concurrent Pos: USPHS fel anat, Univ Chicago, 50-52; vis prof, Inst Fluid Dynamics & Appl Math, Univ Md, College Park, 73-74. Mem: Am Phys Soc; Am Physiol Soc; Int Soc Nephrology; Soc Math Biol; Biophys Soc. Res: Mathematical theory of transport in biological systems; theory of renal function. Mailing Add: Bldg 31 Rm 9A Nat Heart & Lung Inst Bethesda MD 20014

STEPHENSON, JOHN LOUIS, organic chemistry, inorganic chemistry, see 12th edition

STEPHENSON, LEE PALMER, b Fresno, Calif, Oct 21, 23; m 48; c 3. GEOPHYSICS. Educ: Fresno State Col, AB, 47; Univ Ill, MS, 49, PhD(physics), 53. Prof Exp: Asst physics, Univ Ill, 47-53; res physicist, Calif Res Corp, Stand Oil Co, Calif, 53-57, group supvr, 57-59, res assoc geophys, 59-63, SR RES ASSOC, CHEVRON OIL FIELD RES CO, 63- Mem: AAAS; Am Phys Soc; Am Chem Soc; Soc Explor Geophys. Res: Exploration seismology; seismic signal detection, data processing and interpretation; physical properties of earth materials; compaction and cementation of clastic sediments; optics; astronomy; astronomical instrumentation. Mailing Add: Chevron Oil Field Res Co PO Box 446 La Habra CA 90631

STEPHENSON, LEONARD MERRIMAN, organic chemistry, see 12th edition

STEPHENSON, MARY LOUISE, b Brookline, Mass, Feb 23, 21. MOLECULAR BIOLOGY. Educ: Conn Col, AB, 43; Radcliffe Col, PhD(biochem), 56. Prof Exp: Res fel, 56-59, asst biochemist, 59-66, assoc biol chem, 66-74, ASSOC BIOCHEMIST, MASS GEN HOSP, HARVARD MED SCH, 74-, PRIN RES ASSOC, HARVARD MED SCH, 69- Mem: Am Asn Biol Chemists; Am Asn Cancer Res; Am Soc Cell Biol. Res: Biosynthesis of proteins. Mailing Add: Huntington Lab Mass Gen Hosp Boston MA 02114

STEPHENSON, NORMAN ROBERT, b Toronto, Ont, Mar 15, 17; m 43; c 3. BIOCHEMISTRY, SCIENCE ADMINISTRATION. Educ: Univ Toronto, BA, 38, MA, 40, PhD, 42; Carleton Univ, DPA, 71. Prof Exp: Asst chemist, Insulin Comt Lab, Univ Toronto, 38-39, asst med res, Banting Inst, 39-42; res chemist, Stand Brands, Ltd, 45-48 & Consumers Res Labs, 48-50; res chemist, Physiol & Hormones Sect, Food & Drug Labs, 50-66, head anti-cancer sect & sci adv, Div Med & Pharmacol, Drug Adv Bur, 66-73, CHIEF, HEALTH CARE PROD DIV, PLANNING & EVAL DIRECTORATE, HEALTH PROTECTION BR, DEPT NAT HEALTH & WELFARE, 73- Mem: Can Physiol Soc; Can Pharmacol Soc; fel Chem Inst Can; Can Asn Res Toxicol. Res: Endocrinology; biological and chemical assays of hormones and adrenal corticosteroids; estrogens, androgens; thyroid hormone; anterior pituitary hormones; insulin. Mailing Add: 44 Kilbarry Crescent Ottawa ON Can

STEPHENSON, PAUL BERNARD, b Jena, La, Dec 16, 37; m 59; c 3. PHYSICS. Educ: La Polytech Inst, BS, 60, MS, 61; Duke Univ, PhD(physics), 66. Prof Exp: Engr, Tex Instruments Inc, 61-62; res asst physics, Duke Univ, 62-66; from asst prof to assoc prof, 66-75, PROF PHYSICS, LA TECH UNIV, 75- Mem: Am Asn Physics Teachers; Am Phys Soc. Res: Solid state physics, particularly luminescence of organic crystals. Mailing Add: Dept of Physics La Tech Univ Ruston LA 71270

STEPHENSON, RICHARD ALLEN, b Cleveland, Ohio, June 8, 31; m 52; c 3. GEOMORPHOLOGY, MARINE SCIENCES. Educ: Kent State Univ, BA, 59; Univ Tenn, MS, 61; Univ Iowa, PhD(geog, geol), 67. Prof Exp: Assoc prof phys geog, ECarolina Univ, 62-67; asst prof phys geog & earth sci, Univ Ga, 67-71; assoc prof phys geog, 71-74, PROF & DIR, INST COASTAL & MARINE SOURCES, ECAROLINA UNIV, 74- Mem: Asn Am Geog; Geol Soc Am; Sigma Xi. Res: Geomorphology; water resources; environmental resources; hydrology. Mailing Add: Inst Coastal & Marine Sources ECarolina Univ Sta Greenville NC 27834

STEPHENSON, ROBERT CHARLES, b Oxford, Ohio, Dec 27, 16; m 42; c 3. GEOLOGY. Educ: Miami Univ, AB, 38; Johns Hopkins Univ, PhD(geol), 43. Prof Exp: Geologist & mining engr, Titanium Div, MacIntyre Develop, Nat Lead Co, 42-43; field geologist, Union Mines Develop Corp, Colo, 43-44 & Rocky Mt Div, Union Oil Co, Calif, 44-45; sr geologist, State Topog & Geol Surv, Pa, 45-46, asst state geologist, 46-52; geologist, Woodward & Dickerson, Inc, 52-55; exec dir, Am Geol Inst, DC, 55-63; prof geol & exec dir res found, Ohio State Univ, 63-72; dir, Ctr Marine Resources, 72-74, PROF MGT & SPEC PROGS ADMINR, OFF UNIV RES, TEX A&M UNIV, 74- Concurrent Pos: Assoc Univ Progs, Div Univ Progs, Energy Res & Develop Admin, Washington, DC, 75-77. Mem: AAAS; fel Geol Soc Am; Soc Econ Geologists; Am Asn Petrol Geol; Am Inst Mining, Metall & Petrol Eng; Nat Asn Geol Teachers. Res: Economic geology of metalliferous deposits; petroleum geology; geology of nonmetallic mineral resources; management of marine and coastal resources. Mailing Add: Off Univ Res Tex A&M Univ College Station TX 77843

STEPHENSON, ROBERT LLOYD, b Portland, Ore, Feb 18, 19; m 46. ANTHROPOLOGY, ARCHAEOLOGY. Educ: Univ Ore, BA, 40, MA, 42; Univ Mich, PhD(anthrop), 56. Prof Exp: Supvr archaeol lab, Univ Tex, Works Progress Admin, 40-41; asst prof archaeol excavation, Washington & Jefferson Col, 41; field dir river basin surv, Smithsonian Inst, Tex, 46-51, dir, Missouri Basin, Lincoln, Nebr, 52-63, dir river basin surv, Washington, DC, 63-66; coordr, Nev Archaeol Surv, Univ Nev, 66-68; DIR INST ARCHAEOL & ANTHROP, UNIV SC, 68- Concurrent Pos: Assoc prof, Univ Nebr, 61-63; asst ed, Soc Am Archaeol, 61-63; fel archaeol study settlement patterns in Missouri Valley, Smithsonian Inst & Missouri River Basin, 62-64; res prof, Univ Nev, 66-68; adj prof, Univ SC, 68-, dir univ mus, 71-74; state archaeologist, SC, 68-; consult, Tenn Valley Authority, 72- NSF fel, Univ SC, 72-73. Mem: Fel AAAS; fel Am Anthrop Asn; Soc Am Archaeol; Soc Vert Paleont; Soc Hist Archaeol. Res: Prehistory of North America, particularly the Archaic and early Ceramic periods; historical American archaeology; administration of archaeological research programs. Mailing Add: Inst of Archaeol & Anthrop Univ of SC Columbia SC 29208

STEPHENSON, ROBERT MOFFATT, JR, b Atlanta, Ga, Dec 25, 40; m 65; c 1. MATHEMATICS. Educ: Vanderbilt Univ, BA, 62; Tulane Univ, MS, 65, PhD(math), 67. Prof Exp: Asst prof math, Univ NC, Chapel Hill, 67-73; ASSOC PROF MATH & COMPUT SCI, UNIV SC, 73- Mem: Am Math Soc. Res: General topology. Mailing Add: Dept of Math & Comput Sci Univ of SC Columbia SC 29208

STEPHENSON, SAMUEL EDWARD, JR, b Bristol, Tenn, May 16, 26; m 50; c 3. MEDICINE, ORGANIC CHEMISTRY. Educ: Univ SC, BS, 46; Vanderbilt Univ, MD, 50; Am Bd Surg & Bd Thoracic Surg, dipl, 57. Prof Exp: Asst surg, Sch Med, Vanderbilt Univ, 53-55; from instr to asst prof, 55-61, assoc prof, Sch Med & Assoc dir, Clin Res Ctr, 61-67, dir, S R Light Lab Surg Res, 59-62; PROF SURG, UNIV FLA, 67-; CHMN DEPT SURG, UNIV HOSP JACKSONVILLE, 67-; CHMN DEPT SURG, UNIV HOSP JACKSONVILLE, 67- Concurrent Pos: Consult, Regional Respiratory & Rehab Ctr, 58 & Thayer Vet Admin Hosp, 59. Res: Medical electronics, especially physiological control of respiration and cardiac rate; experimental atherogenesis; malignant disease; cardiovascular surgery and neoplasms. Mailing Add: Dept of Surg Univ Hosp of Jacksonville Jacksonville FL 33209

STEPHENSON, STEPHEN NEIL, b Hayden Lake, Idaho, Feb 3, 33; m 53; c 4. BOTANY, ECOLOGY. Educ: Idaho State Univ, BS, 55; Rutgers Univ, MS, 63, PhD(bot), 65. Prof Exp: Park ranger, Nat Park Serv, 57-61; instr bot, Douglass Col, Rutgers Univ, 62-63; asst prof, 65-72, ASSOC PROF BOT, MICH STATE UNIV, 72- Concurrent Pos: Mem eastern deciduous forest biome coord comt, Int Biol Prog, 68- Mem: AAAS; Am Soc Mammal; Ecol Soc Am; Am Inst Biol Sci. Res: Community structure and organization; biosystematics of Gramineae; biogeography of North America, especially arid and semiarid regions. Mailing Add: Dept of Bot & Plant Path Mich State Univ East Lansing MI 48823

STEPHENSON, THOMAS EDWIN, economic geology, see 12th edition

STEPHENSON, WILLIAM KAY, b Chicago, Ill, Apr 6, 27; m 51; c 3. PHYSIOLOGY. Educ: Knox Col, AB, 50; Univ Minn, PhD, 55. Prof Exp: Phys chemist, Nat Bur Stand, 49-50; asst zool, Univ Minn, 50-53, instr, 54; from asst prof to assoc prof biol, 54-64, PROF BIOL & CHMN DEPT, EARLHAM COL, 64- Mem: AAAS; Am Physiol Soc; Am Soc Zool; Am Soc Cell Biol. Res: Ion distribution; active transport; bioelectric phenomena; cnidarian behavior. Mailing Add: Dept of Biol Earlham Col Richmond IN 47374

STEPKA, WILLIAM, b Veseli, Minn, Apr 13, 17; m 48; c 1. PLANT PHYSIOLOGY, PLANT BIOCHEMISTRY. Educ: Univ Rochester, AB, 46; Univ Calif, PhD, 51. Prof Exp: Asst, Univ Rochester, 46-47 & Univ Calif, 48-49; res assoc bot, Univ Pa, 51-54, asst prof, 54-55; PLANT PHYSIOLOGIST IN CHG, RADIOL NUTRICULTURE LAB, MED COL VA, VA COMMONWEALTH UNIV, 55-, PROF PHARMACOG, HEALTH SCI DIV, UNIV, 68- Concurrent Pos: Plant physiologist & biochemist, Am Tobacco Co, 55-68. Mem: Am Soc Plant Physiol; Am Inst Biol Sci; Am Soc Pharmacognosy; Am Asn Cols Pharm. Res: Biosynthesis of radioactively labeled compounds; discovery and isolation of cardioactive, hypotensive and contraceptive compounds from natural sources. Mailing Add: Dept Pharm Hlth Sci Div Va Cmmnwlth Univ Box 666 MCV Sta Richmond VA 23298

STEPLEMAN, ROBERT SAUL, b New York, NY, Nov 2, 42; m 67; c 1. MATHEMATICS. Educ: State Univ NY Stony Brook, BS, 64; Univ Md, College Park, PhD(math), 69. Prof Exp: Res assoc numerical anal, Inst Fluid Dynamics & Appl Math, Univ Md, College Park, 69; asst prof appl math & comput sci, Sch Eng & Appl Sci, Univ Va, 69-73; MEM TECH STAFF, DAVID SARNOFF RES CTR, RCA, 73- Mem: Math Asn Am; Asn Comput Mach; Soc Indust & Appl Math. Res: Numerical analysis; convergence of numerical methods; matrix theory and its application to boundary value problems for ordinary and partial differential equations. Mailing Add: David Sarnoff Res Ctr RCA Princeton NJ 08540

STEPNICZKA, HEINRICH EDUARD, chemistry, see 12th edition

STEPONKUS, PETER LEO, b Chicago, Ill, Sept 18, 41; m 62; c 4. PLANT PHYSIOLOGY, HORTICULTURE. Educ: Colo State Univ, BSc, 63; Univ Ariz, MSc, 64; Purdue Univ, PhD(plant physiol), 66. Prof Exp: Asst prof hort & asst horticulturist, Univ Ariz, 66-68; asst prof, 68-72, ASSOC PROF HORT, CORNELL UNIV, 72- Honors & Awards: Kenneth Post Award, Soc Cryobiol, 71. Mem: AAAS; Am Soc Hort Sci; Am Soc Plant Physiol; Scand Soc Plant Physiol; Soc Cryobiol. Res: Stress physiology; biochemical mechanisms of cold acclimation; freezing injury; hormonal controls in high temperature injury and senescence post-harvest physiology of cut flowers. Mailing Add: Dept Floricult & Ornmntl Hort Cornell Univ Ithaca NY 14850

STEPP, JAMES WILSON, mathematics, see 12th edition

STEPPLER, HOWARD ALVEY, b Morden, Man, Nov 8, 18; m 45; c 1. AGRONOMY. Educ: Univ Man, BSA, 41; McGill Univ, MSc, 48, PhD, 55. Prof Exp: Lectr plant sci, Univ Man, 41-42; plant breeder, Dom Exp Farm, 48-49; from asst prof to assoc prof, 49-57, chmn dept, 55-73, PROF AGRON, MACDONALD COL, MCGILL UNIV, 57- Concurrent Pos: Nuffield traveling fel, 55; vis prof, Univ West Indies, 64; exec mem, Nat Soil Fertil Comt, 58; mem coord comt, Can Forage Seeds Proj, 58; mem, Nat Comt Agrometeorol, 59; chmn adv comt agr, World's Fair, Montreal, 67; agr adv, Can Int Develop Agency, 70-71; mem bd trustees, Int Ctr Trop Agr, Colombia, 72- Honors & Awards: Merit cert, Am Grassland Coun, 64. Mem: AAAS; Am Soc Agron; Can Soc Agron (pres, 58); Biomet Soc; fel Agr Inst Can (vpres, 55-57, pres, 64-65). Res: Plant breeding; forage crops; grassland management, especially ecology, physiology, climatology and morphology in relation to growth and response under animal grazing. Mailing Add: Dept of Agron MacDonald Col St Anne de Bellevue PQ Can

STEPTO, ROBERT CHARLES, b Chicago, Ill, Oct 6, 20; m 42; c 2. OBSTETRICS & GYNECOLOGY, PATHOLOGY. Educ: Northwestern Univ, BS, 41; Howard Univ, MD, 44; Univ Chicago, PhD(path), 48. Prof Exp: Asst, Col Med, Univ Ill, 42; clin instr obstet & gynec, Stritch Sch Med, Loyola Univ Chicago, 50-60; from clin asst prof to clin assoc prof, Chicago Med Sch, 60-70, chmn dept, 70-74, prof obstet & gynec, 70-75; PROF OBSTET & GYNEC, RUSH MED COL, 75- Concurrent Pos: USPHS fel, Inst Res, Michael Reese Hosp, 48-50; chmn, Dept Obstet & Gynec, Provident Hosp, 53-63 & Mt Sinai Hosp & Med Ctr, 70-; dir obstet & gynec, Cook County Hosp, 72-75; mem, Food & Drug Adv Comt Obstet & Gynec, 72-; mem, Family Planning Coord Coun; mem maternal & preschool nutrition comt, Nat Acad Sci. Mem: Fel AMA; Am Col Obstet & Gynec; Am Col Surg; Int Col Surg Am; Am Fertil Soc. Res: Endocrine pathology; sex hormones influence on tissue synthesis. Mailing Add: 5201 S Cornell Ave Chicago IL 60615

STERBENZ, FRANCIS JOSEPH, b Queens, NY, May 11, 24; m 56; c 3. BIOCHEMICAL PHARMACOLOGY. Educ: St John's Univ, NY, BS, 50, MS, 52; NY Univ, PhD, 57. Prof Exp: Asst bacteriologist, New York City Dept Hosps, 51-52; res assoc protozool, St John's Univ, NY, 52-56; instr physiol, NJ Col Med & Dent, 56-58, instr microbiol, 58-59; sr res microbiologist, Squibb Inst Med Res, 59-65; asst dept head biochem, Bristol-Myers Co, 65-70, DEPT HEAD, RES & DEVELOP LAB, BRISTOL-MYERS PROD, HILLSIDE, 70- Mem: AAAS; Soc Protozool; Am Soc Microbiol; NY Acad Sci. Res: Bio-availability and biochemistry of analgesic, sedative and related drugs; immunochemistry; mechanisms involved in microbial pathogenicity; chemotherapy; nutritional physiology of microorganisms. Mailing Add: 60 Drake Rd Somerset NJ 08873

STERE, ATHLEEN JACOBS, b Boston, Mass, Feb 1, 21; m 43; c 3. BIOLOGY. Educ: Bryn Mawr Col, AB, 41; Radcliffe Col, MA, 42; Pa State Univ, University Park, PhD(biol), 71. Prof Exp: Res asst immunol, Sch Med, Boston Univ, 44-46; res asst microbiol, Res Div, Albert Einstein Med Ctr, Philadelphia, 59-63; res asst, 63-71, ASST PROF BIOL, PA STATE UNIV, UNIVERSITY PARK, 71- Mem: AAAS. Res: Histochemistry; effect of oxygen deprivation on cellular metabolism. Mailing Add: Dept of Biol Pa State Univ University Park PA 16802

STERGIS, CHRISTOS GEORGE, b Greece, Dec 22, 19; US citizen; m 48; c 2. PHYSICS. Educ: Temple Univ, AB, 42, AM, 43; Mass Ins Technol, PhD(physics), 48. Prof Exp: Physicist, Radiation Lab, Mass Inst Technol, 44-45, res asst physics, 45-48; asst prof, Temple Univ, 48-51; physicist, 51-58, chief, Space Physics Lab, 59-63, CHIEF UPPER ATMOSPHERE PHYSICS, AIR FORCE CAMBRIDGE RES LABS, 63- Concurrent Pos: Hon res asst, Univ Col, Univ London, 64-65; mem, Comt High Altitude Rocket & Balloon Res, Nat Acad Sci, 63-66, Comt Int Quiet Sun Yr, 63-67, Comt Solar-Terrestrial Res, 71-, Aeronomy Panel, Interdept Comt Atmospheric Sci, 71- Mem: Am Phys Soc; Am Geophys Union; Sigma Xi. Res: Structure of the earth's upper atmosphere by means of rockets and satellites; scattering of solar radiations by the atoms and molecules of the upper atmosphere. Mailing Add: Aeronomy Lab AF Cambridge Res Labs LG Hanscom Field Bedford MA 01730

STERIADE, MIRCEA, b Bucharest, Romania, Aug 20, 24; c 1. NEUROPHYSIOLOGY. Educ: Col Culture, Bucharest, BA, 44; Fac Med, Bucharest, MD, 52; Inst Neurol, Acad Sci, Bucharest, DSc(neurophysiol), 55. Prof Exp: Sr scientist, Inst Neurol, Acad Sci, Bucharest, 55-62, head lab, 62-68; assoc prof, 68-69, PROF PHYSIOL, FAC MED, UNIV LAVAL, 69- Honors & Awards: Claude Bernard Medal, Univ Paris, 65. Mem: AAAS; Int Brain Res Orgn; Can Physiol Soc; Fr Neurol Soc; Fr Asn Physiol. Res: Neuronal circuitry of thalamic nuclei and cortical areas; responsiveness of thalamic and cortical relay cells; thalamic and cortical inhibitory mechanisms during sleep and waking; inhibitory interneurons. Mailing Add: Dept of Physiol Univ Laval Fac of Med Quebec PQ Can

STERKEN, GORDON JAY, b Ft Atkinson, Wis, Mar 17, 30. ORGANIC CHEMISTRY. Educ: Hope Col, AB, 51; Mich State Univ, PhD(org chem), 60. Prof Exp: Chemist, Stauffer Chem Co, 59-66 & Sinclair Oil Co, 66-69; CHEMIST, HILTON-DAVIS CHEM CO DIV, STERLING DRUG CO, 69- Mem: Am Chem Soc. Res: Synthesis of fluoran dyes, pigments and pharmaceutical intermediates; product and process development in these areas. Mailing Add: Hilton-Davis Chem Co Langdon Farm Rd Cincinnati OH 45237

STERLING, ANNE, b New York, July 30, 44; m. DEVELOPMENTAL GENETICS. Educ: Univ Wis, BA, 65; Brown Univ, PhD(genetics), 70. Prof Exp: Instr med sci, 71-72, ASST PROF BIOMED SCI, BROWN UNIV, 72- Mem: AAAS; Soc Develop Biol. Res: Oogenesis and early development in Drosophila melanogaster. Mailing Add: Dept of Med Sci Brown Univ Providence RI 02912

STERLING, CLARENCE, b Millville, NJ, Mar 25, 19; m 42; c 5. BOTANY. Educ: Univ Calif, AB, 40, PhD(bot), 44. Prof Exp: Asst forestry, Univ Calif, 40-41 & bot, 44; res asst, Univ Ill, 46-47; instr, Univ Wis, 47-50; from asst prof to assoc prof food sci & technol, 50-60, PROF FOOD SCI & TECHNOL, UNIV CALIF, DAVIS, 60- Concurrent Pos: Fulbright grant, 56-57; Guggenheim fels, 56-57 & 63-64. Mem: AAAS; Bot Soc Am; Linnean Soc London. Res: Plant anatomy and morphology; cytology; submicroscopic structure of gels; crystal structure. Mailing Add: Dept of Food Sci & Technol Univ of Calif Davis CA 95616

STERLING, DANIEL J, b New York, NY, July 18, 32; m 56; c 3. MATHEMATICS. Educ: St Lawrence Univ, BS, 53; Columbia Univ, MA, 55; Univ Wis, PhD(math), 62. Prof Exp: Teacher, Barnard Sch Boys, 56-57; instr math & physics, C W Post Col, Long Island, 57-59; from instr math to asst prof math, Bowdoin Col, 62-68; ASSOC PROF MATH, COLORADO COL, 68- Concurrent Pos: NSF grant, 66-68. Mem: AAAS; Am Math Soc; Math Asn Am. Res: Lie algebras; algebraic and Lie groups; cohomology of groups. Mailing Add: Dept of Math Colorado Col Colorado Springs CO 80903

STERLING, HAROLD MELVIN, b Sioux City, Iowa, Dec 3, 21; m 56; c 4. PHYSICAL MEDICINE, PEDIATRICS. Educ: Univ Iowa, BA, 47; Yale Univ, MD, 51. Prof Exp: Med dir, Ill Children's Hosp Sch, 54-55; instr phys med & rehab, Med Sch, Univ Minn, 56-59; med dir, Joseph P Kennedy Jr Mem Hosp, 59-62; from asst prof to prof phys med & rehab & head div, Sch Med, Tufts Univ, 62-68, prof pediat, 66-68; prof maternal & child health, Sch Pub Health, Univ Calif, Berkeley, 68-70; PROF PEDIAT & PHYS MED & REHAB, SCH MED, UNIV CALIF, DAVIS, 70- Res: Rehabilitation. Mailing Add: Univ of Calif Sch of Med Davis CA 95616

STERLING, HENRY SOMERS, b New York, NY, Mar 4, 05; m 30; c 2. GEOGRAPHY. Educ: Columbia Univ, BA, 27, AM, 34; Univ Wis, PhD(geog), 39. Prof Exp: From instr to assoc prof, 37-57, chmn dept, 62-63 & 65-66, prof, 57-73, EMER PROF GEOG, UNIV WIS-MADISON, 73- Concurrent Pos: Mem staff res & anal br, Off Strategic Serv, DC & Europe, 41-45; Soc Sci Res Coun fel, 46-47; dir interdisciplinary res team, Venezuelan Andes, 52-53; consult, Ministry of Agr, Venezuela, 57; NDEA fel, Brazil, 61. Honors & Awards: Medal of Freedom, 46; Atwood Medal, Pan-Am Inst Geog & Hist, 59. Mem: Asn Am Geog; Am Geog Soc; Conf Latin Am Geog. Res: Trends in agricultural settlement and land use, colonization and agrarian reform in Latin American tropics. Mailing Add: 427 Virginia Terr Madison WI 53705

STERLING, JOHN DEO, b Latham, Mo, Aug 3, 21; m 45; c 2. CHEMISTRY. Educ: Cent Col, Mo, BA, 42; Univ Md, PhD(org chem), 49. Prof Exp: Res chemist, Univ Md, 43-45; res chemist, 48-53, res supvr, 53-62, DIV HEAD, JACKSON LAB, E I DU PONT DE NEMOURS & CO, INC, 62- Mem: Am Chem Soc. Res: Antimalarial drugs; methoxylated organic compounds; organic fluorine compounds; dyes; petroleum chemicals; lithium batteries. Mailing Add: Jackson Lab E I du Pont de Nemours & Co Inc Wilmington DE 19898

STERLING, NICHOLAS J, b Cooperstown, NY, Nov 7, 34; m 62; c 2. MATHEMATICS. Educ: Williams Col, BA, 56; Syracuse Univ, MS, 61, PhD(math), 66. Prof Exp: Asst prof, 66-70, ASSOC PROF MATH, STATE UNIV NY BINGHAMTON, 70- Mem: Am Math Soc. Res: Non-associative ring theory. Mailing Add: Dept of Math State Univ NY Binghamton NY 13901

STERLING, PETER, b New York, NY, June 28, 40; m 61; c 2. NEUROANATOMY, NEUROPHYSIOLOGY. Educ: Western Reserve Univ, PhD(bbiol), 66. Prof Exp: Asst prof, 69-74, ASSOC PROF NEUROANAT, SCH MED, UNIV PA, 74- Concurrent Pos: NSF fel, Med Sch, Harvard Univ, 66-68, NIH fel, 68-69; NSF grant, Sch Med, Univ Pa, 69-71, NIH grant, 71- Honors & Awards: C Judson Herrick Award, Am Asn Anat, 71. Mem: Am Asn Anat; Soc Neurosci. Res: Relation between form of nerve cells and their physiological functioning, particularly in the visuo-motor system. Mailing Add: Dept of Anat Univ of Pa Sch of Med Philadelphia PA 19104

STERLING, REX ELLIOTT, b Eldorado, Kans, Sept 5, 24; m 48; c 2. BIOCHEMISTRY. Educ: Cent Mo State Col, BS, 48; Univ Ark, MS, 49; Univ Colo, PhD(biochem), 53. Prof Exp: Clin biochemist, Los Angeles County Gen Hosp, 53-71; from instr to assoc prof, 53-72, PROF BIOCHEM, UNIV SOUTHERN CALIF, 72-; HEAD CLIN BIOCHEMIST, LOS ANGELES COUNTY GEN HOSP, 71- Mem: Am Asn Clin Chem. Res: Diabetes and carbohydrate in cataract formation; carbohydrate metabolism and adrenal cortical function; prophyrins and porphyria; clinical biochemical methodology and automation. Mailing Add: Labs & Path LAC-USC Med Ctr Los Angeles CA 90033

STERLING, ROBERT FILLMORE, b Toledo, Ohio, Aug 19, 19; m 47; c 2. ORGANIC POLYMER CHEMISTRY. Educ: Carnegie Inst Technol, BS, 42; Univ Pittsburgh, cert bus mgt, 62. Prof Exp: Res metallurgist, Magnesium Div, Dow Chem Co, 42-44; res engr, Res Labs, 46-49, mat & process engr, 50-52, res engr, Res Labs, 52-55, adv scientist, Atomic Power Dept, 55-57, mgr chem sect, 57-62, FEL ENGR, MISSILE LAUNCHING & HANDLING DEPT, MARINE DIV, WESTINGHOUSE ELEC CORP, 62- Mem: Am Chem Soc. Res: Thermal insulation materials; synthetic resins of improved electrical properties; chemistry and materials of nuclear reactors; reliability, research and development of missile launching and handling equipment; biological oceanography and agricultural biochemistry. Mailing Add: 1457 Hollenbeck Ave Sunnyvale CA 94087

STERLING, STEWART ALLEN, b Rochester, NY, Mar 1, 41; m 75; c 1. SOLID STATE PHYSICS. Educ: Mass Inst Technol, BS, 64; Univ Calif, Berkeley, PhD(physics), 70. Prof Exp: STAFF MEM PHYS CHEM, GEN ATOMIC CO, 69- Mem: Am Phys Soc. Res: Experimental characterization of diffusional kinetics fission products at high temperatures; development of models describing thermally activated creep in modern super alloys. Mailing Add: Gen Atomic Co Box 81608 San Diego CA 92138

STERLING, THEODOR DAVID, b Vienna, Austria, July 3, 23; nat US; m 48; c 2. COMPUTER SCIENCE, BIOMETRY. Educ: Univ Chicago, AB, 49, MA, 53; Tulane Univ, PhD, 55. Prof Exp: Instr math, Univ Ala, 54-55, asst prof statist, 55-57; asst prof statist, Mich State Univ, 57-58; asst prof prev med, Col Med, Univ Cincinnati, 58-66, assoc prof biostatist, 61-63, prof & dir med ctr, 63-66; prof comput sci, Wash Univ, 66-72; PROF COMPUT SCI & DIR DEPT, SIMON FRASER UNIV, 72- Concurrent Pos: Consult, NSF, 67, Environ Protection Agency, 71 & Fed Trade Comn, 72. Mem: Asn Comput Mach; Am Statist Asn. Res: Humanizing effects of automation; errors and foibles in investigations, especially medical. Mailing Add: Dept of Comput Sci Simon Fraser Univ Burnaby BC Can

STERLING, WARREN MARTIN, b Chicago, Ill, Jan 4, 47. INFORMATION SCIENCE, ELECTRONIC ENGINEERING. Educ: Univ Ill, BS, 68; Carnegie-Mellon Univ, MS, 70, PhD(elec eng), 74. Prof Exp: Engr numerical control, Westinghouse Elec Corp, 68-74; ENGR ELEC ENG, XEROX CORP, EL SEGUNDO, CALIF, 74- Concurrent Pos: Res instr elec eng, Carnegie-Mellon Univ, 72. Mem: Inst Elec & Electronics Engrs. Res: Office information systems, including advanced display processors and high speed local communications; digital image processing; optical computing. Mailing Add: PARC/ADL Xerox Corp MS C3-50 701 Aviation Blvd El Segundo CA 90245

STERLING, WINFIELD LINCOLN, b Edinburg, Tex, Sept 18, 36; m 61; c 3. ENTOMOLOGY. Educ: Pan Am Col, BA, 62; Tex A&M Un- iv, MS, 66, PhD(entom), 69. Prof Exp: Res assoc entom, 64-66, asst prof, 69-74, ASSOC PROF ENTOM, TEX A&M UNIV, 74- Concurrent Pos: AID consult, Univ Calif, Berkeley, 74-75; post-doctoral fel, Univ Queensland, 75-76. Mem: Entom Soc Am; Entom Soc Can; Am Inst Biol Sci. Res: Insect ecology, pest management and population dynamics. Mailing Add: Dept of Entom Tex A&M Univ College Station TX 77843

STERMAN, MAURICE B, b St Paul, Minn, Dec 31, 35; m 64; c 2. NEUROPSYCHOLOGY. Educ: Univ Calif, Los Angeles, AB, 58, PhD(neuropsychol), 63. Prof Exp: Fel neurophysiol, Brain Res Inst, Univ Calif, Los Angeles, 60-62; res staff psychologist, Vet Admin Hosp, 62-68, CHIEF NEUROPSYCHOL RES, VET ADMIN HOSP, 68-; PROF ANAT & PSYCHIAT, UNIV CALIF, LOS ANGELES, 75- Concurrent Pos: Lectr, Yale Univ, 64-65; asst prof anat & physiol, Univ Calif, Los Angeles, 65-70, assoc prof anat, 70-75; consult-res assoc, Los Angeles Co Newborn Serv, Univ Southern Calif Med Ctr, 74-; assoc ed, J Biofeedback & Self Regulation. Mem: Am Psychol Asn; Am Physiol Soc; Am Asn Anat; Asn Psychophysiol Study Sleep. Res: Neurophysiology of sleep; integrated inhibition of behavior; biofeedback and epilepsy. Mailing Add: Neuropsychol Res Vet Admin Hosp Sepulveda CA 91343

STERMAN, MELVIN DAVID, b Brooklyn, NY, Sept 19, 30; m 56; c 4. COLLOID CHEMISTRY, POLYMER CHEMISTRY. Educ: City Col New York, BS, 51; Purdue Univ, PhD(phys chem), 55. Prof Exp: Teaching asst, Iowa State Univ, 51-52; from res chemist to sr res chemist, 55-62, RES ASSOC, RES LABS, EASTMAN KODAK CO, 63- Mem: Am Chem Soc. Res: Characterization of polymers by physical chemical techniques; electrical properties of polymers; chemistry of cross-linking of polymers; properties of cross-linked polymer works; polymer adsorption on surfaces; preparation and stability of lyophobic colloids in non-aqueous solvents; electrophoretic mobility of colloidal particles. Mailing Add: Res Labs Eastman Kodak Co 1999 Lake Ave Rochester NY 14650

STERMAN, SAMUEL b Buffalo, NY, June 6, 18; m 53; c 4. PHYSICAL CHEMISTRY. Educ: Univ Buffalo, BS, 39. Prof Exp: Develop chemist, Nat Carbon Co Div, Union Carbide Corp, 40-45, res chemist, Linde Co Div, 45-50, supvr spec prod develop, Silicone Div, 50-66, asst dir res & develop, Chem & Plastics, 66-73, ASSOC DIR RES & DEVELOP, UNION CARBIDE TECH CTR, UNION CARBIDE CORP, 73- Honors & Awards: Award, Soc Plastics Indust, 61. Mem: Am Chem Soc; Soc Plastics Indust; AAAS. Res: Textile chemicals; protective coatings; water repellants; surface active agents; composites; interface bonding; organofunctional silanes; urethane foam; high temperature polymers; surface chemistry; elastomers; patent management. Mailing Add: Res & Develop Union Carbide Tech Ctr Tarrytown NY 10591

STERMITZ, FRANK, b Thermopolis, Wyo, Dec 3, 28; m 54; c 5. ORGANIC CHEMISTRY. Educ: Univ Notre Dame, BS, 50; Univ Colo, MS, 51, PhD(chem), 58. Prof Exp: Res chemist, Merck & Co, Inc, NJ, 51-53 & Lawrence Radiation Lab, Univ Calif, Berkeley, 58-61; from asst prof to assoc prof chem, Utah State Univ, 61-67; assoc prof, 67-69, CENTENNIAL PROF CHEM, COLO STATE UNIV, 69- Concurrent Pos: USPHS res career develop award, 63-67; vpres, Elars Biores Labs, 74- Mem: Am Chem Soc; Am Soc Pharmacog. Res: Alkaloid and other natural product isolation, structure proof and biosynthesis; medicinal chemistry; organic photochemistry; chemotaxonomy. Mailing Add: Dept of Chem Colo State Univ Ft Collins CO 80521

STERN, AARON MILTON, b Detroit, Mich, Aug 12, 20; m 55; c 2. PEDIATRICS, CARDIOLOGY. Educ: Univ Mich, AB, 42, MD, 45; Am Bd Pediat, dipl, 52. Prof Exp: Intern, Saginaw Gen Hosp, Mich, 45-46; from asst resident to resident, Univ Hosp, Univ Mich, Ann Arbor, 46-48; resident, Saginaw Gen Hosp, Mich, 48-49; from instr to assoc prof, 49-70, PROF PEDIAT & COMMUN DIS, MED SCH, UNIV MICH, ANN ARBOR, 70- Concurrent Pos: Consult, Alpena County Rheumatic Fever Clin, Mich, 51-56. Mem: Am Heart Asn; Am Acad Pediat. Res: Contrast studies of cardiovascular lesions and effects of cardiac surgery on physiologic and mental status of patients. Mailing Add: Dept of Pediat Univ of Mich Med Ctr Ann Arbor MI 48104

STERN, ADOLPH JOHN, b Nuermberg, Ger, Feb 12, 00; nat US; m 38; c 1. CHEMISTRY. Educ: Munich Tech Univ, dipl, 23, Dr Ing, 25, Dr habil, 33. Prof Exp: Asst org chem, Munich Tech Univ, 25-28, asst org phys chem, 31-33, privatdocent, 33-35, docent, 35-37; res assoc, Children's Fund, Detroit, Mich, 38-42; from assoc prof to prof chem, Wagner Col, 42-69, distinguished prof, 69-70, chmn dept, 50-58, dean sch, 51-69, spec asst to the pres, 66-69; ADJ PROF CHEM, STATEN ISLAND COMMUNITY COL, 70- Honors & Awards: Grand Silver Cross, Austrian Govt, 69. Mem: AAAS; Am Chem Soc; hon fel Am Inst Chemists. Res: Organic synthetic and organic physical chemistry; absorption and Raman spectra; polarography; calorimetry; fluorescence; optical activity. Mailing Add: 16 Lloyd Ct Staten Island NY 10310

STERN, ALBERT VICTOR, b New York, NY, Apr 26, 23; div; c 2. ASTRONOMY, SYSTEMS ENGINEERING. Educ: Univ Calif, Berkeley, AB, 47, PhD(astron), 50. Prof Exp: Sect head digital subsysts, Hughes Aircraft Co, Fullerton, 54-57, sr scientist, 57-59, lab mgr adv systs, 59-66, asst div mgr, 66-69, chief scientist, Syst Div, 69-73, PROG MGR, MISSILE SYSTS GROUP, HUGHES AIRCRAFT CO, CANOGA PARK, 73- Res: Celestial mechanics; weapons systems analysis; systems engineering. Mailing Add: 279 Ravenna Dr Long Beach CA 90803

STERN, ALFRED, b Würsburg, Ger, Dec 31, 35; US citizen; m 60; c 3. INDUSTRIAL ORGANIC CHEMISTRY. Educ: City Col New York, BS, 57; Brandeis Univ, MA, 59; Polytech Inst Brooklyn, PhD(chem) 64. Prof Exp: Teaching asst, Brandeis Univ, 59; res chemist, Nat Starch & Chem Corp, 63-65; res chemist chem div, Universal Oil Prod Co, NJ, 65-68, group leader org res, UOP Chem Co Div, 68-69; GROUP LEADER, MAKHTESHIM CHEM WORKS, LTD, ISRAEL, 69- Concurrent Pos: Lectr, Univ of the Negev, Israel. Mem: Am Chem Soc. Res: Organometallic chemistry; aroma products; specialty chemicals; monomer-polymer synthesis; organic reactions on polymers; high pressure-temperature reactions; pesticides; intermediates. Mailing Add: Makhteshim Chem Works Ltd PO Box 60 Beersheva Israel

STERN, ARTHUR IRVING, b New York, NY, Dec 8, 30; m 62; c 3. PLANT PHYSIOLOGY, PHOTOBIOLOGY. Educ: City Col New York, BS, 53; Brandeis Univ, PhD(biol), 62. Prof Exp: NIH fel develop biol, Brandeis Univ, 62-63; Kettering fel photosynthesis, Weizmann Inst, 63-64, USPHS fel, 64-65; asst prof, 65-70, ASSOC PROF BOT, UNIV MASS, AMHERST, 70- Concurrent Pos: NSF grant, 66-67. Mem: AAAS; Am Soc Plant Physiol; Am Soc Photobiol. Res: Correlation of chloroplast structure and function; chloroplast development; photophosphorylation; pellicle structure of algae. Mailing Add: Dept of Bot Univ of Mass Amherst MA 01002

STERN, ARTHUR MARVIN, bacteriology, see 12th edition

STERN, CURT, b Hamburg, Ger, Aug 30, 02; nat US; m 31; c 3. ZOOLOGY. Educ: Univ Berlin, PhD(zool), 23. Hon Degrees: DSc, McGill Univ, 58; Dr rer nat, Univ Munich, 72. Prof Exp: Investr, Kaiser Wilhelm Inst, 23-33; res assoc zool, Univ Rochester, 33-35, from asst prof to assoc prof, 35-41, prof exp zool & chmn div biol sci, 41-47; prof zool, 47-70, prof genetics, 58-70, EMER PROF ZOOL & GENETICS, UNIV CALIF, BERKELEY, 70- Concurrent Pos: Int Ed Bd fel, 24-26; privatdocent, Univ Berlin, 28-33; vis prof, Case Western Reserve Univ, 32; Rockefeller Found fels, Univ Ill, Calif Inst Technol, 32-33 & Univ Paris, 51; mem adv comt biol & med, AEC, 50-55; Guggenheim Found fel, 55 & 63. Honors & Awards: Kimber Genetics Award, Nat Acad Sci, 63. Mem: Nat Acad Sci; Am Acad Arts & Sci; Am Soc Zoologists (pres, 62); Genetics Soc Am (pres, 50); Soc Human Genetics (vpres, 50, pres, 57). Res: Cyto and developmental genetics of Drosophila and man. Mailing Add: Dept of Zool Univ of Calif Berkeley CA 94720

STERN, DANIEL HENRY, b Richmond, Va, June 18, 34; m 63; c 1. LIMNOLOGY, ECOLOGY. Educ: Univ Richmond, BS, 55, MS, 59; Univ Ill, PhD(zool), 64. Prof Exp: Asst prof biol, Tenn Technol Univ, 64-66 & La State Univ, New Orleans, 66-69; from asst prof to assoc prof, 69-75, PROF BIOL, UNIV MO-KANSAS CITY, 75- Mem: Ecol Soc Am; Am Soc Limnol & Oceanog; Soc Protozool; Micros Soc Am; NAm Benthological Soc. Res: Ecology of Protozoa, Algae and Micrometazoa; invertebrate ecology; applied ecology and environmental impacts. Mailing Add: Dept of Biol Univ of Mo Kansas City MO 64110

STERN, DAVID P, b Decin, Czech, Dec 17, 31; US citizen; m 61; c 3. SPACE PHYSICS. Educ: Hebrew Univ, Israel, MSc, 55; Israel Inst Technol, DSc(physics), 59. Prof Exp: Res assoc physics, Univ Md, 59-61; Nat Acad Sci-Nat Res Coun resident res assoc, 61-63, PHYSICIST, GODDARD SPACE FLIGHT CTR, NASA, 63- Mem: Am Phys Soc; Am Geophys Union; Am Asn Physics Teachers. Res: Terrestrial and interplanetary magnetic fields, their structure; description by Euler potentials and motion of charged particles in them; theory of magnetospheric tail, boundary and electric field. Mailing Add: Code 602 NASA Goddard Space Flight Ctr Greenbelt MD 20771

STERN, DONALD, physics, see 12th edition

STERN, EDWARD ABRAHAM, b Detroit, Mich, Sept 19, 30; m 55; c 3. SOLID STATE SCIENCE. Educ: Calif Inst Technol, BS, 51, PhD(physics), 55. Prof Exp: Res fel solid state physics, Calif Inst Technol, 55-57; from asst prof to prof, Univ Md, 57-66; PROF PHYSICS, UNIV WASH, 66- Concurrent Pos: Guggenheim fel, 63-64; NSF sr res fel, 70-71. Mem: Am Phys Soc. Res: Electronic structure of metals and alloys; collective effects; magnetism; atomic structure of amorphous and biological matter. Mailing Add: Dept of Physics Univ of Wash Seattle WA 98195

STERN, ELIZABETH, b Cobalt, Ont, Sept 19, 15; US citizen; m 40; c 3. EXPERIMENTAL PATHOLOGY, EPIDEMIOLOGY. Educ: Univ Toronto, MD, 39; Am Bd Path, dipl, 58. Prof Exp: Resident path, Cedars of Lebanon Hosp-Good Samaritan Hosp, Los Angeles, 42-46; assoc pathologist, Cedars of Lebanon Hosp, 46-49; dir lab & res, Cancer Detection Ctr, Los Angeles, 50-60; res coordr cancer proj, Med Sch, Univ Southern Calif, 61; lectr path & chief cytol lab, Sch Med, 61-63, assoc res pathologist, Sch Pub Health, 63-65, PROF PUB HEALTH, SCH PUB HEALTH, UNIV CALIF, LOS ANGELES, 65- Mem: AAAS; fel Col Am Path; Am Soc Cytol; Am Pub Health Asn; AMA. Res: Epidemiology of cancer. Mailing Add: Sch of Pub Health Univ of Calif Ctr for Health Sci Los Angeles CA 90024

STERN, ERIC WOLFGANG, b Vienna, Austria, Nov 4, 30; nat US; m 60; c 1. CHEMISTRY. Educ: Syracuse Univ, BS, 51; Northwestern Univ, PhD(chem), 54. Prof Exp: Res chemist, Texaco, Inc, 54-57; res chemist, M W Kellogg Co Div, Pullman, Inc, NJ, 58-63, res assoc, 63-70; SECT HEAD RES & DEVELOP, ENGELHARD INDUSTS DIV, ENGELHARD MINERALS & CHEM CORP, EDISON, 70- Concurrent Pos: Co-adj prof, Rutgers Univ, 65- Mem: AAAS; Am Chem Soc; Catalysis Soc; NY Acad Sci; The Chem Soc. Res: Homogeneous catalysis; coordination chemistry; reaction mechanisms; molecular structure; heterogeneous catalysis; catalyst characterization. Mailing Add: 234 Oak Tree Rd Mountainside NJ 07092

STERN, ERNEST, b Wetter, Ger, June 5, 28; US citizen; m 53; c 4. SOLID STATE PHYSICS, ACOUSTICS. Educ: Columbia Univ, BS, 53. Prof Exp: Sr engr, Sperry Gyroscope Co, 55-57; mem staff, Electronics Lab, Gen Elec Co, 58-62; vpres, Microwave Chem Lab, 62-64; staff mem, 64-68, GROUP LEADER ACOUSTICS, LINCOLN LAB, MASS INST TECHNOL, 68- Concurrent Pos: Co-chmn, Gorden Conf Ultrasonics, 70. Mem: Sigma Xi; Am Phys Soc; Inst Elec & Electronics Engrs. Res: Gyromagnetic phenomena; microwave frequencies; nonlinear magnetic phenomena; surface acoustics and acousto-electric phenomena; components and devices; x-ray iithography. Mailing Add: 31 Oxbow Rd Concord MA 01742

STERN, FRANK, b Koblenz, Ger, Sept 15, 28; nat US; m 55; c 2. THEORETICAL SOLID STATE PHYSICS. Educ: Union Col, NY, BS, 49; Princeton Univ, PhD(physics), 55. Prof Exp: Physicist, US Naval Ord Lab, Md, 53-62; res staff mem, Thomas J Watson Res Ctr, 62-65, Zurich Res Lab, 65-66, RES STAFF MEM, THOMAS J WATSON RES CTR, IBM CORP, 66- Concurrent Pos: Lectr, Univ Md, 55-58, part-time prof, 59-62. Mem: AAAS; fel Am Phys Soc; Sigma Xi. Res: Cohesive energy of iron; semiconductors; injection lasers; optical properties; quantum effects in inversion layers; solid state theory. Mailing Add: Thomas J Watson Res Ctr IBM Corp Yorktown Heights NY 10598

STERN, HERBERT, b Can, Dec 22, 18; m 53; c 3. CELL BIOLOGY. Educ: McGill Univ, BSc, 40, MSc, 42, PhD, 45. Prof Exp: Royal Soc Can fel, Univ Calif, 46-48; lectr cell physiol, Med Sch, Univ Witwatersrand, 48-49; assoc, Rockefeller Inst Med Res, 49-55; head biochem cytol, Plant Res Inst, Can Dept Agr, 55-60; prof bot, Univ Ill, Urbana, 60-65; PROF BIOL, UNIV CALIF, SAN DIEGO, 65-, CHMN DEPT, 74- Concurrent Pos: Mem develop biol panel, NSF; mem cell biol panel, NIH; mem spec subcomt cellular & subcellular struct & function, Nat Acad Sci. Mem: Am Soc Plant Physiol; Soc Develop Biol (pres, 64-65); Am Soc Cell Biol; fel Am Soc Biol Chem; Genetics Soc Am. Res: Cell biology and biochemistry. Mailing Add: Dept of Biol PO Box 109 Univ of Calif at San Diego La Jolla CA 92037

STERN, IRVING B, b New York, NY, Sept 12, 20; m; c 3. DENTISTRY, CELL BIOLOGY. Educ: City Col New York, BS, 41; NY Univ, DDS, 46; Columbia Univ, cert, 56. Prof Exp: Lectr periodont, Sch Dent, Univ Wash, 59-60, from asst prof to prof, 60-75; PROF PERIODONT & CHMN DEPT, SCH DENT MED, TUFTS UNIV, 75- Concurrent Pos: Spec res fel anatomy, Sch Dent, Univ Wash, 61-62; USPHS grant. Mem: AAAS; Am Dent Asn; Am Asn Periodont; Am Soc Cell Biol; Int Asn Dent Res. Res: Ultrastructure and biology of oral epithelium and epithelial derivatives; ultrastructure of dento-gingival junction and cementum. Mailing Add: Dept of Periodont Tufts Univ Sch Dent Med Boston MA 02111

STERN, IVAN J, b Chicago, Ill, Jan 5, 30; m 55; c 2. BIOCHEMISTRY, BACTERIOLOGY. Educ: Univ Ill, BSc, 51, MSc, 53; Ore State Univ, PhD(bact), 58. Prof Exp: Sr res biochemist, Norwich Pharmacal Co, 58-62; sr res biochemist, 62-72, head drug metab, 72-75, MGR BIOCHEM SECT, BAXTER LABS, 75- Concurrent Pos: Lectr med, Univ Ill, 66-; adj asst prof, Stritch Sch Med, Loyola Univ Chicago. Mem: AAAS; Am Soc Pharmacol & Exp Therapeut. Res: Biochemical mechanisms of drug toxicity; enzymic degradation of mucopolysaccharides and glycoproteins; immunoassay; drug metabolism; pharmacokinetics. Mailing Add: Baxter Labs 6301 N Lincoln Morton Grove IL 60053

STERN, JACK TUTEUR, JR, b Chicago, Ill, Jan 18, 42; m 67; c 1. BIOMECHANICS. Educ: Univ Chicago, PhD(anat), 69. Prof Exp: Instr anat, Univ Chicago, 69-70, asst prof, 70-74; ASSOC PROF ANAT, STATE UNIV NY STONY BROOK, 74- Concurrent Pos: USPHS res career develop award, 73. Mem: Am Asn Phys Anthrop; Am Asn Anat; Int Primatological Soc. Res: Functional anatomy of primates; evolution of erect posture; biomechanics and evolution of muscles. Mailing Add: Dept of Anat Sci State Univ NY Stony Brook NY 11794

STERN, JOEL R, b Newark, NJ, Jan 31, 23; m 54; c 2. BIOCHEMISTRY. Educ: Rutgers Univ, PhD(biochem), 49. Prof Exp: Res assoc nutrit, Wash State Univ, 49-52; res assoc, Michael Reese Hosp, 52-60; dir biochem lab, Edgewater Hosp, Chicago, 60-66; DIR BIOCHEM LAB, LUTHERAN GEN HOSP, 66- Mem: Fel Am Inst Chem; Am Chem Soc; Am Asn Clin Chem; Soc Nuclear Med. Res: Chemical changes in disease; diagnostic use of radioisotopes. Mailing Add: Biochem Lab Lutheran Gen Hosp 1775 Dempster St Park Ridge IL 60068

STERN, JOHN HANUS, b Brno, Czech, May 21, 28; nat US; m 49. PHYSICAL CHEMISTRY. Educ: Univ Calif, BS, 53; Univ Wash, MS, 54, PhD(chem), 58. Prof Exp: From asst prof to assoc prof, 58-67, PROF CHEM, CALIF STATE UNIV, LONG BEACH, 67- Concurrent Pos: Am Chem Soc-Petrol Res Found int fac fel, Univ Florence, 64-65; vis prof, Hebrew Univ, Jerusalem, 71-72. Mem: Am Chem Soc. Res: Thermodynamics of electrolytes and non-electrolytes in aqueous solutions. Mailing Add: Dept of Chem Calif State Univ Long Beach CA 90840

STERN, JOSEPH AARON, b New York, NY, Apr 24, 27; m 50; c 3. MICROBIOLOGY. Educ: Mass Inst Technol, SB, 49, SM, 50, PhD(food technol), 53. Prof Exp: Food technologist, Davis Bros Fisheries, Mass, 48-49; asst food technol, Mass Inst Technol, 50-53; from asst prof to assoc prof fisheries technol, Univ Wash, 53-58; chief biochem unit, Space Med Sect, Boeing Co, 58-59, res prog dir, Bioastronaut Sect, 59-61, mgr adv space prog, 61-65 & Voyager prog planetary quarantine, 65-66, adv interplanetary explor prog, 66; sterilization group supvr, Environ Requirements Sect, Jet Propulsion Lab, Calif, 66-67, asst sect mgr sterilization, 67-69; PRES, BIONETICS CORP, 69- Concurrent Pos: Mem comt animal food prod, NSF-Nat Res Coun, 58-60. Mem: AAAS; Am Inst Aeronaut & Astronaut; NY Acad Sci. Res: Spoilage and preservation of food products; biochemical systems in space flight and extraterrestrial missions; planetary quarantine; spacecraft sterilization. Mailing Add: Bionetics Corp 18 Research Dr Hampton VA 23366

STERN, JOSEPH RICHARD, biochemistry, deceased

STERN, JUDITH S, b Brooklyn, NY, Apr 25, 43; m 64. NUTRITION. Educ: Cornell Univ, BS, 64; Harvard Univ, MS, 66, ScD, 70. Prof Exp: From res assoc to asst prof, Rockefeller Univ, 69-74; ASST PROF NUTRIT, UNIV CALIF, DAVIS, 75- Mem: Am Inst Nutrit; Am Dietetic Asn; AAAS; Sigma Xi. Res: Studies of some critical factors involved in the development of obesity which include adipose cellularity, food intake, diet composition, exercise, hyperinsulinemia and tissue resistance in muscle and adipose. Mailing Add: Dept of Nutrit Univ of Calif Davis CA 95616

STERN, KARL, psychiatry, deceased

STERN, KINGSLEY ROWLAND, b Port Elizabeth, SAfrica, Oct 30, 27; nat US; m 56; c 2. TAXONOMIC BOTANY. Educ: Wheaton Col, Ill, BS, 49; Univ Mich, MA, 50; Univ Minn, PhD(bot), 59. Prof Exp: Asst, Univ Mich, 49-51 & Univ Ill, 54-55; asst, Univ Minn, 55-56 & 57-58, instr bot, 58-59; instr biol, Hamline Univ, 57-58; from asst prof to assoc prof, 59-68, PROF BOT, CALIF STATE UNIV, CHICO, 68- Concurrent Pos: NSF res grants, 59, 60 & 63-71. Mem: Am Soc Plant Taxon; Bot Soc Am; Int Soc Plant Taxon. Res: Taxonomy of vascular plants, especially pollen grains, anatomy, cytology and morphogenesis. Mailing Add: Dept of Biol Sci Calif State Univ Chico CA 95929

STERN, KURT, b Vienna, Austria, Apr 3, 09; nat US; m 39; c 3. PATHOLOGY, CANCER. Educ: Univ Vienna, MD, 33. Prof Exp: Instr biochem, Inst Med Chem, Univ Vienna, 30-33, res assoc, 33-38; jr physician, State Inst Study & Treatment Malignant Dis, Buffalo, 43-45; asst pathologist, Mt Sinai Hosp, 45-48, from asst to assoc dir, Mt Sinai Med Res Found, 48-60, dir blood ctr, 50-60; prof path & pathologist, Res & Educ Hosp, 60-70, EMER PROF PATH, UNIV ILL COL MED, 70-; PROF LIFE SCI, BAR-ILAN UNIV, ISRAEL, 69- Concurrent Pos: Res fel, New York Cancer Hosp & Div Cancer, Bellevue Hosp, 39-40; from assoc to assoc prof, Chicago Med Sch, 49-60; sci ed, Bull Am Asn Blood Banks, 60. Mem: Fel Am Soc Clin Path; Soc Exp Biol & Med; Am Soc Exp Path; Am Asn Immunol; Am Asn Cancer Res. Res: Experimental cancer research; immunology; experimental pathology; blood groups and immunohematology; physiopathology of reticulo-endothelial system. Mailing Add: Dept of Life Sci Bar-Ilan Univ Ramat-Gan Israel

STERN, KURT HEINZ, b Vienna, Austria, Dec 26, 26; nat US; m 60; c 2. PHYSICAL INORGANIC CHEMISTRY. Educ: Drew Univ, AB, 48; Univ Mich, MS, 50; Clark Univ, PhD(chem), 53. Prof Exp: Asst, Univ Mich, 50; instr chem, Clark Univ, 50-52; from instr to assoc prof, Univ Ark, 52-60; res chemist, Electrochem Sect, Nat Bur Stand, 60-68; sect head high temperature electrochem, 68-74, CONSULT, NAVAL RES LAB, 74- Concurrent Pos: Res assoc, Nat Acad Sci-Nat Res Coun, 59-60; mem fac, Grad Sch, NIH, 63- Honors & Awards: Turner Prize, Electrochem Soc, 51, Blum Award, 71. Mem: Am Chem Soc; Electrochem Soc; The Chem Soc; AAAS. Res: High temperature electrochemistry; molten salts; vaporization and thermal decomposition of inorganic salts. Mailing Add: Inorg Chem Br Naval Res Lab Washington DC 20375

STERN, LEO, b Montreal, Que, Jan 20, 31; m 55; c 4. PERINATAL BIOLOGY, CLINICAL PHARMACOLOGY. Educ: McGill Univ, BSc, 51; Univ Man, MD, 56; FRCPS(C), 64. Prof Exp: Demonstr, McGill Univ, 62-66, lectr, 66-67, from asst prof to assoc prof pediat, 67-73; PROF PEDIAT & CHMN SECT REPRODUCTIVE & DEVELOP MED, BROWN UNIV, 73- Concurrent Pos: Mead Johnson res fel, Karolinska Inst, Sweden, 58-59, Nat Res Coun Can med res fel, 59-60; Queen Elizabeth II scientist for res in dis of children, McGill Univ, 66-72; mem, Comn Study Perinatal Mortality, Prov of Que, 67-73; dir dept newborn med, Montreal Childrens Hosp, 69-73; pediatrician-in-chief, RI Hosp, 73- Honors & Awards: Queen Elizabeth II Res Scientist Award, 66. Mem: Perinatal Res Soc; Soc Pediat Res; Am Pediat Soc; Am Soc Clin Nutrit; Am Soc Clin Pharmacol & Therapeut. Res: Development pharmacology; perinatal biology, adaptation to extrauterine life, thermoregulation and bilirubin metabolism in the new born, respiratory adaptation in the normal and abnormal newborn infant. Mailing Add: Dept of Pediat RI Hosp Providence RI 02902

STERN, MARTIN, b New York, NY, Jan 9, 33; m 69; c 1. ORAL SURGERY. Educ: Harvard Univ, DMD, 56; Am Bd Oral Surg, dipl, 63. Prof Exp: ATTEND SURGEON IN CHG ORAL SURG, LONG ISLAND JEWISH MED CTR/QUEENS HOSP CTR AFFILIATION, 67-, ASSOC DIR DENT, 72-; PROF ORAL SURG, SCH DENT MED, STATE UNIV NY STONY BROOK, 71- Concurrent Pos: Asst clin prof, Sch Dent, Columbia Univ, 68-70. Mem: Am Dent Asn; Am Soc Oral Surg; Int Asn Dent Res. Mailing Add: Dept Oral Surg Queens Hosp Ctr 82-68 164th St Jamaica NY 11432

STERN, MARTIN OSCAR, physics, see 12th edition

STERN, MARVIN, b New York, NY, Jan 6, 16; m 42; c 3. PSYCHIATRY. Educ: City Col New York, BS, 35; NY Univ, MD, 39. Prof Exp: From fel to assoc prof, 40-62, PROF PSYCHIAT, SCH MED, NY UNIV, 62-, ATTEND PSYCHIATRIST, UNIV HOSP, 52- Concurrent Pos: Consult, US Vet Admin Regional Off, Brooklyn, 51-66 & Manhattan Vet Admin Hosp, 66-; assoc vis neuropsychiatrist, Bellevue Hosp, 52-62, vis neuropsychiatrist, 62- Mem: Psychosom Soc; Asn Res Nerv & Mental Dis; Am Psychopath Asn; Am Psychiat Asn. Res: Psychosomatic medicine; altered brain function in organic disease. Mailing Add: 184 Rugby Rd Brooklyn NY 11226

STERN, MARVIN JOSEPH, physical chemistry, deceased

STERN, MAX HERMAN, b Sioux City, Iowa, Mar 23, 20; m 46; c 2. ORGANIC CHEMISTRY. Educ: Morningside Col, BA, 41; Univ Wis, MS, 43, PhD(org chem), 45. Prof Exp: Asst, Univ Wis, 42-45; res chemist, Distillation Prod Industs, 45-65, RES ASSOC, RES LABS, EASTMAN KODAK CO, 65- Mem: Fel Am Inst Chemists; Am Chem Soc. Res: Vitamins A and E; carotenoids; soysterols; terpenes; photochemistry; photographic addenda. Mailing Add: Res Labs Eastman Kodak Co Rochester NY 14650

STERN, MELVIN ERNEST, b New York, NY, Jan 22, 29; m 56; c 2. HYDRODYNAMICS. Educ: Cooper Union, BEE, 50; Ill Inst Technol, MS, 51; Mass Inst Technol, PhD(meteorol), 56. Prof Exp: From res assoc meteorol to physicist, Woods Hole Oceanog Inst, 51-64; PROF OCEANOG, GRAD SCH, UNIV RI, 64- Concurrent Pos: Guggenheim fel, 70-71. Mem: Fel Am Acad Arts & Sci, 75. Res: Oceanic circulation and turbulence; non-linear stability theory. Mailing Add: Grad Sch of Oceanog Univ of RI Kingston RI 02881

STERN, MICHELE SUCHARD, b Chicago, Ill, Mar 17, 43; m 63. PLANT PHYSIOLOGY, BIOCHEMISTRY. Educ: Univ Ill, Urbana, BS, 64; Tenn Technol Univ, MS, 66; Tulane Univ, PhD(biol), 69. Prof Exp: Asst prof, 69-75, ASSOC PROF BIOL, UNIV MO-KANSAS CITY, 75- Mem: Am Soc Plant Physiol; Am Chem Soc; Ecol Soc Am; Am Soc Limnol & Oceanog; Am Inst Biol Sci. Res: Isolation and characterization of plant proteases and their relationship to senescence; aquatic entomology and limnology; water pollution. Mailing Add: Dept of Biol Univ of Mo Kansas City MO 64110

STERN, PAULA HELENE, b New Brunswick, NJ, Jan 20, 38; m 59. PHARMACOLOGY. Educ: Univ Rochester, BA, 59; Univ Cincinnati, MS, 61; Univ Mich, PhD(pharmacol), 63. Prof Exp: Instr pharmacol, Univ Mich, 65-66; asst prof, 66-72, ASSOC PROF PHARMACOL, MED SCH, NORTHWESTERN UNIV, 72- Concurrent Pos: Fel pharmacol, Rochester Univ, 63-64 & Marine Biol Lab, Woods Hole, Mass, 64; res career develop award, NIH; consult, Food & Drug Admin; mem, Gen Med B Study Sect, NIH. Mem: AAAS; Am Soc Pharmacol & Exp Therapeut; Endocrine Soc; Soc Exp Biol & Med. Res: Calcium metabolism; mechanisms of action of drugs and hormones on bone. Mailing Add: Dept of Pharmacol Northwestern Univ Med Sch Chicago IL 60611

STERN, RICHARD, b Paterson, NJ, Nov 27, 29; m 58. ACOUSTICS. Educ: Univ Calif, Los Angeles, BA, 52, MS, 56, PhD(physics), 64. Prof Exp: Asst res physicist, 64-65, asst prof eng, 66-71, ASSOC PROF ENG, UNIV CALIF, LOS ANGELES, 71- Concurrent Pos: Exchange fel, Imp Col, Univ London, 64-65. Mem: Am Phys Soc; fel Acoust Soc Am; Inst Elec & Electronics Engrs. Res: Experimentation in physical, engineering and medical acoustics. Mailing Add: Sch of Eng & Appl Sci Univ of Calif Los Angeles CA 90024

STERN, RICHARD CECIL, b New York, NY, Jan 4, 42; m 64. CHEMICAL PHYSICS. Educ: Cornell Univ, AB, 63; Harvard Univ, AM, 65, PhD(chem), 68. Prof Exp: From asst prof to assoc prof chem, Columbia Univ, 68-74; CHEMIST, LAWRENCE LIVERMORE LAB, UNIV CALIF, 74- Mem: Am Chem Soc; Am Phys Soc. Res: Laser isotope separation; photochemical kinetics; scattering and chemical reactions of low energy electrons; molecular beam and time-of-flight technology. Mailing Add: Lawrence Livermore Lab Univ of Calif Livermore CA 94550

STERN, RICHARD M, b New York, NY, Aug 20, 33; m 62; c 1. SOLID STATE PHYSICS. Educ: Cornell Univ, BA, 54; Brown Univ, MSc, 57; Aachen Technol Univ, Dr rer nat(physics), 61. Prof Exp: Scientist, Aachen Technol Univ, 59-61; sr researcher electron micros, Cambridge Univ, 61-62; res assoc electron diffraction, Cornell Univ, 62-63; from asst prof to assoc prof, 63-70, PROF PHYSICS, POLYTECH INST NY, 70- Mem: Am Phys Soc; Acoust Soc Am. Res: Ultrasonic attenuation in solids; radiation damage in metals; electron microscopy; low energy electron diffraction; surface physics; optical properties of solids; photoelectric and electron spectroscopy. Mailing Add: Dept of Physics Polytech Inst Brooklyn Brooklyn NY 11201

STERN, ROBERT, b Bad Kreuznach, Ger, Feb 11, 36; US citizen; m 63; c 3. MEDICINE, BIOCHEMISTRY. Educ: Harvard Univ, BA, 57; Univ Wash, MD, 62. Prof Exp: USPHS officer, Nat Inst Dent Res, 63-65; SR SCIENTIST, NAT INST DENT RES, 67-; RESIDENT ANAT PATH, NAT CANCER INST, 74- Concurrent Pos: Nat Cancer Inst spec fel, Weizmann Inst Sci, 65-67. Mem: AAAS; Am Soc Biol Chem; Am Soc Microbiol. Res: Transcriptional, translational controls in animal cells; anatomic pathology; translation of collagen messenger RNA; animal viruses and their effect on animal cells. Mailing Add: 8102 Maple Ridge Rd Bethesda MD 20014

STERN, ROBERT LOUIS, b Newark, NJ, Apr 10, 35; m 58; c 3. ORGANIC CHEMISTRY. Educ: Oberlin Col, AB, 57; Johns Hopkins Univ, MA, 59, PhD(org chem), 64. Prof Exp: Asst prof chem, Northeastern Univ, 62-65, assoc prof, 65-68; ASSOC PROF CHEM, OAKLAND UNIV, 68-, CO-CHMN DEPT, 74- Mem: AAAS; Am Chem Soc. Res: Organoanalytical chemistry; organic reaction mechanisms; biosynthesis; separation mechanisms of structurally related organic molecules; organic photochemistry. Mailing Add: Dept of Chem Oakland Univ Rochester MI 48063

STERN, ROBERT MALCOLM, b Milwaukee, Wis, June 7, 15; m 55; c 3. BACTERIOLOGY. Educ: Univ Wis, BS, 37, MS, 38, PhD(bact), 40. Prof Exp: Asst agr bact, Univ Wis, 37-40, consult chemist, 40; res biochemist, Nat Canners Asn, DC, 40-46, Nat Res Coun, 46 & Pabst Brewing Co, 46-59; PRES & DIR RES, GREAT LAKES BIOCHEM CO, INC, 59- Concurrent Pos: Mem Nutrit Found, 43. Mem: AAAS; Am Chem Soc; Animal Nutrit Res Coun. Res: Bacteriology; vitamins; nutrition; fermentation; water chemistry. Mailing Add: Great Lakes Biochem Co Inc 6120 W Douglas Ave Milwaukee WI 53218

STERN, RONALD JOHN, b Chicago, Ill, Jan 20, 47; div. TOPOLOGY. Educ: Knox Col, BA, 68; Univ Calif, Los Angeles, MA, 70, PhD(math), 73. Prof Exp: Mem, Inst Advan Study, 73-74; INSTR MATH, UNIV UTAH, 74- Mem: Am Math Soc. Res: Geometrical topology emphasizing the structure of topological manifolds. Mailing Add: Dept of Math Univ of Utah Salt Lake City UT 84112

STERN, SAMUEL T, b Buffalo, NY, May 27, 28; m 57; c 3. MATHEMATICS. Educ: Univ Buffalo, BA, 57, MA, 60; State Univ NY Buffalo, PhD(math), 62. Prof Exp: Instr math, Univ Buffalo, 58-62; asst prof, 62-65, PROF MATH, STATE UNIV NY COL BUFFALO, 65- Concurrent Pos: State Univ NY fac res fel, 67 & 70. Mem: Math Asn Am. Res: Noncommutative number theory; modern algebra; theory of skew groups and skew rings. Mailing Add: Dept of Math State Univ NY Col Buffalo NY 14222

STERN, SIDNEY CHARLES, b Brooklyn, NY, Oct 30, 14; m 42; c 2. ATMOSPHERIC PHYSICS, ENVIRONMENTAL SCIENCES. Educ: City Col New York, BS, 36. Prof Exp: Chemist, NJ Stern, NY, 37-41; micrometeorologist, Biol Labs, US Dept Army, Ft Detrick, 49-56; sr tech specialist, Appl Sci Div, Litton Industs, Inc, 55-66; vpres, Cambridge Technol, Mass, 66-70; iro/tech pres, Enviro/Tech Sci, Inc, Newton Upper Falls, 70-74; CONSULT, 74- Concurrent Pos: Consult, McGraw Hill Info Systs, 70-, Arthur D Little Inc, 74- & FRL Corp, Albany, 74- Mem: Sci Res Soc Am. Res: Micrometeorology; aerosol physics; atmospheric electricity; electrostatics; sampling and detection of atmospheric pollutants; filtration and impaction theory; atmospheric diffusion; air pollution; control. Mailing Add: 10 Debra Lane Framingham MA 01701

STERN, SILVIU ALEXANDER, b Bucharest, Romania, June 18, 21; US citizen; m 73; c 2. PHYSICAL CHEMISTRY, CHEMICAL ENGINEERING. Educ: Israel Inst Technol, BS, 45; Ohio State Univ, MS, 48, PhD(phys chem), 52. Prof Exp: Res assoc chem eng, Ohio State Univ, 52-55; res engr chem eng, Linde Div, Union Carbide Corp, NY, 55-58, group leader phys chem, 58-60, res supvr, 60-67; PROF CHEM ENG, SYRACUSE UNIV, 67- Mem: Am Chem Soc; Am Inst Chem Engrs; AAAS. Res: Transport phenomena in polymers; separation processes, particularly membrane separation processes; surface phenomena; biomedical engineering. Mailing Add: Dept of Chem Eng & Mat Sci 312 Hinds Hall Syracuse Univ Syracuse NY 13210

STERN, THEODORE, b Briarcliffe, NY, July 27, 17; m 42; c 2. ANTHROPOLOGY. Educ: Bowdoin Col, AB, 39; Univ Pa, AM, 41, PhD(anthrop), 48. Prof Exp: Asst prof anthrop, Univ Ore, 48-56, assoc prof, 56-64, chmn dept, 69-72, PROF ANTHROP, UNIV ORE, 64- Concurrent Pos: Soc Sci Res Coun fel, Klamath Reservation, Ore, 50, Am Coun Learned Soc fel, 51; Am Philos Soc fel, Umatilla Reservation, Ore, 53 & 55; Fulbright res fel, Univ Rangoon, 54-55; NSF fel, Thailand, 64-65. Mem: Fel Am Anthrop Asn; Am Ethnol Soc; Am Folklore Soc; Burma Res Soc. Res: Cultural change; ethnolinguistics; folklore; ethnohistory; American Indian; Southeast Asia. Mailing Add: Dept of Anthrop Univ of Ore Eugene OR 97403

STERN, THOMAS WHITAL, b Chicago, Ill, Dec 12, 22; m 55; c 1. GEOCHRONOLOGY. Educ: Univ Chicago, SB, 47; Univ Tex, MA, 48. Prof Exp: Geologist, 48-68, chief isotope geol br, 68-71, GEOLOGIST, US GEOL SURV, 71- Mem: AAAS; Geol Soc Am; Mineral Soc Am; Am Geophys Union; Geochem Soc. Res: Geochemistry; mineralogy; lead-uranium age determinations; isotope geology; autoradiography. Mailing Add: US Geol Surv 929 Nat Ctr Reston VA 22092

STERN, VERNON MARK, b Sykeston, NDak, Mar 28, 23; m 47; c 2. ENTOMOLOGY. Educ: Univ Calif, Berkeley, BS, 49, PhD, 52. Prof Exp: Res asst entom, Univ Calif, Berkeley, 49-52; entomologist, Producers Cotton Oil Co, Ariz, 52-56; asst entomologist, 56-62, assoc prof, 62-68, PROF ENTOM, UNIV CALIF, RIVERSIDE, 68-, ASSOC RES ENTOMOLOGIST, LAB NUCLEAR MED & RADIATION BIOL, LOS ANGELES, 66- Concurrent Pos: Collabr, USDA, 53-56; vpres, Ariz State Bd Pest Control, 53-56; coordr, Producers Agr Found, 54-56; NSF res grants, 61-69; Cotton Producers Inst res grant, 63-69; consult, US AEC, 65-66 & UN Food & Agr Orgn, 66-; Cotton Inst res grant, 69-; USDA res grant, 71-; int biol prog, NSF res grant, 72- Mem: Entom Soc Am; Ecol Soc Am. Res: Insect ecology; integrated control of arthropod pests; environmental radiation; radioecology; arthropods; population dynamics; insect migration and biology of insects. Mailing Add: Dept of Entom Univ of Calif Riverside CA 92502

STERN, W EUGENE, b Portland, Ore, Jan 1, 20; m 46; c 4. SURGERY. Educ: Univ Calif, AB, 41, MD, 43. Prof Exp: Clin instr neurol surg, 51-52, from asst prof to assoc prof surg, 52-59, PROF SURG, SCH MED, UNIV CALIF, LOS ANGELES, 59-, CHIEF NEUROSURG DIV, 67- Concurrent Pos: Consult, Los Angeles Vet Admin Hosp, 52- Mem: AMA; Am Surg Asn; Am Asn Neurol Surg; Soc Neurol Surg; Neurosurg Soc Am. Res: Cerebral swelling; intracranial circulatory dynamics and intracranial mass dynamics. Mailing Add: Neurosurg Div Univ of Calif Sch of Med Los Angeles CA 90024

STERN, WILLIAM LOUIS, b Paterson, NJ, Sept 10, 26; m 49; c 2. PLANT ANATOMY. Educ: Rutgers Univ, BS, 50; Univ Ill, MS, 51, PhD(bot), 54. Prof Exp: From instr to asst prof wood anat, Sch Forestry, Yale Univ, 53-60; cur, Samuel James Record Mem Collection, 53-60; cur, Div Plant Anat, Smithsonian Inst, 60-64, chmn dept bot, 64-67; PROF BOT, UNIV MD, COLLEGE PARK, 67-, CUR HERBARIUM, 73- Concurrent Pos: Ed, Trop Woods, 53-60, Plant Sci Bull, 61-64 & Biotropica, 68-73; expert, UN Food & Agr Orgn, Philippines, 63-64; mem sci adv comt, Pac Trop Bot Garden, 69 & H P du Pont Winterthur Mus, 73; ed, Memoirs, Torrey Bot Club, 72-75; mem comt, Visit Arnold Arboretum, Harvard Univ, 72, vchmn comt, 73. Mem: Bot Soc Am; Am Soc Plant Taxon; Torrey Bot Club; Int Asn Wood Anat; Int Asn Plant Taxon. Res: Plant anatomy and its relationship to taxonomic botany; plant morphology and phylogeny; tropical dendrology. Mailing Add: Dept of Bot Univ of Md College Park MD 20742

STERNBACH, DANIEL DAVID, b Montclair, NJ, May 28, 49. ORGANIC CHEMISTRY. Educ: Univ Rochester, BS, 71; Brandeis Univ, PhD(org chem), 76. Prof Exp: SWISS NAT SCI FOUND RES ASST, SWISS FED INST TECHNOL, 76- Mem: Am Chem Soc. Res: Synthesis of interesting and biologically significant organic compounds and investigation of new synthetic methods. Mailing Add: Org Chem Lab Swiss Fed Inst Technol 8006 Zurich Switzerland

STERNBACH, LEO H, b Abbazia, Austria, May 7, 08; nat US; m 41; c 2. MEDICINAL CHEMISTRY. Educ: Krakow, Poland, MPharm, 29; Jagiellonian Univ, PhD(org chem), 31. Hon Degrees: DrTechSc, Vienna Inst Technol, 71. Prof Exp: Res assoc, Krakow, 31-37; Wislicki Found fel, Inst Med Colloid Chem, Vienna, 37; fel, Swiss Fed Inst Technol, 38, res assoc, 38-40; res chemist, Hoffmann-La Roche, Inc, Switz, 40-41, NJ, 41-59, group chief, 59-65, sect chief, 65-67, dir med chem, Nutley, 67-73; CHEM CONSULT, 73- Mem: Am Chem Soc; NY Acad Sci; Swiss Chem Soc; fel The Chem Soc. Res: Structure determination of natural products; degradation of antibiotics; synthesis of vitamins and pharmaceuticals; biotin and homologs; heterocyclic compounds; quinazoline N-oxides; benzodiazepines; anticholinergics; psychotherapeutic agents; chlordiazepoxide, Librium; diazepam, Valium; flurazepam, Dalmane; nitrazepam, Mogadon; clonazepam; Clonepin. Mailing Add: 10 Woodmont Rd Upper Montclair NJ 07043

STERNBACH, RICHARD ALAN, physiological psychology, see 12th edition

STERNBERG, CHARLES MORTRAM, b Lawrence, Kans, Sept 18, 85; Can citizen; m 11; c 3. VERTEBRATE PALEONTOLOGY. Hon Degrees: LLD, Univ Alta, 60; DSc, Carleton Univ, 74. Prof Exp: Collector & preparator vert fossils, Geol Surv-Nat Mus Can, 12-51, cur, 29-51, EMER CUR, VERT PALEONT, NAT MUS CAN, 51- Mem: Soc Vert Paleont; fel Royal Soc Can. Res: Vertebrate paleontology, especially upper Cretaceous dinosaurs. Mailing Add: 169 Holmwood Ave Ottawa ON Can

STERNBERG, DAVID, b Brooklyn, NY, May 18, 22; m 61. PHYSICS, OCEANOGRAPHY. Educ: City Col New York, BS, 42; Columbia Univ, MA, 49, PhD(physics), 57. Prof Exp: Res physicist, Hudson Labs, Columbia Univ, 53-63, assoc dir, 63-67, sr scientist, 67-69; analyst, Ctr Naval Anal, Va, 69-70; PROF PHYSICS, SETON HALL UNIV, 70-, CHMN DEPT, 75- Concurrent Pos: Exec secy, Undersea Warfare Res & Develop Planning Coun, 62-63; consult, Ctr Naval Anal, Va, 70-; consult, Ocean & Atmospheric Systs, NY, 71- Mem: AAAS; Am Phys Soc; Acoust Soc Am; Am Math Soc; Am Asn Physics Teachers. Res: Classical and atomical scattering; wave propagation; acoustical vibrations. Mailing Add: Dept of Physics Seton Hall Univ South Orange NJ 07079

STERNBERG, ELI, b Vienna, Austria, Nov 13, 17; nat US; m 56; c 2. MECHANICS. Educ: NC State Univ, BCE, 41; Ill Inst Technol, MSc, 43, PhD(mech), 45. Hon Degrees: DSc, NC State Univ, 63. Prof Exp: Instr mech, 43-45, from asst prof to prof, Ill Inst Technol, 45-56; vis prof, Delft Univ Technol, 56-57; prof appl math, Brown Univ, 57-63; vis prof, Keio Univ Japan, 63-64; prof appl mech, 64-70, PROF MECH, CALIF INST TECHNOL, 70- Concurrent Pos: Fulbright award, 56 & 70; Guggenheim award, 63. Mem: Am Acad Arts & Sci; Am Soc Mech Eng; Am Math Soc; Soc Natural Philos; Nat Acad Eng. Res: Applied mathematics; continuum mechanics; elasticity and viscoelasticity theories. Mailing Add: Div of Eng & Appl Sci Calif Inst of Technol Pasadena CA 91109

STERNBERG, HEINZ WALTER, b Vienna, Austria, Apr 25, 11; nat US; m 45. CHEMISTRY. Educ: Univ Vienna, PhD(chem), 35. Prof Exp: Asst, Univ Vienna, 35-38; res chemist, Wilmington Chem Corp, 38-46; group leader, Paraffine Cos, Inc, 46-49; assoc chemist, Atlantic Refinery Co, 49-50; supvry res chemist, US Bur Mines, 50-75; CONSULT, 75- Mem: AAAS; Philos Sci Asn; Brit Soc Philos Sci; The Chem Soc; Am Chem Soc. Res: Quantitative organic elementary analysis; organic and coal chemistry; chemistry of metal carbonyls and metalo-organic complexes; electro-organic chemistry. Mailing Add: 1100 Ptarmigan Dr No 6 Walnut Creek CA 94595

STERNBERG, HILGARD O'REILLY, b Rio de Janeiro, Brazil, July 5, 17; m 42; c 5. GEOGRAPHY OF BRAZIL & THE HUMID TROPICS. Educ: Fed Univ Rio de Janeiro, Bacharel, 40, Lic, 41; La State Univ, PhD(geog), 56; Fed Univ Rio de Janeiro, Dr(geog), 58. Hon Degrees: Dr, Univ Toulouse, 62. Prof Exp: Prof geog, Cath Univ Rio de Janeiro, 41-44; prof, Fed Univ Rio de Janeiro, 44-64; PROF GEOG, UNIV CALIF, BERKELEY, 64- Concurrent Pos: Mem nat textbk comn, Nat Educ Coun Brazil, 46-47; prof, Foreign Serv Inst, Ministry Foreign Affairs, 46-55; mem, Brazilian Govt Cult Mission to Uruguay, 49; consult, Nat Geog Coun Brazil, 49, Ministry Educ rep, Exec Bd, 53-57; dir, Ctr Res Brazilian Geog, Fed Univ Rio de Janeiro, 51-64; vpres, Int Geog Union, 52-56, 1st vpres, 56-60; mem adv comt arid zone res, UNESCO, 55-56, chmn, 56, rep, Mission to Egypt, 57; vis prof, Ind Univ, 59, Univ Heidelberg, 61, Stockholm Sch Econ, 61, Univ Fla, 63, Univ Calif, Los Angeles, 63, McGill Univ, 63 & Columbia Univ, 63-64; Brazil rep prep comt, UN Conf Human Environ, 70. Honors & Awards: Comendador, Nat Order Merit Brazil, 56 & Order Rio Branco, 67. Mem: Leopoldina Ger Acad Res Natural Sci; Brazilian Acad Sci (gen secy, 57-58); hon mem Serbian Geog Soc; hon mem Royal Geog Soc; corresp mem Mex Soc Geog & Statist. Res: Pioneer settlements. Mailing Add: Dept of Geog 466 Michigan Ave Univ of Calif Berkeley CA 94707

STERNBERG, HYMAN M, applied mathematics, physics, see 12th edition

STERNBERG, JAMES CONSTANTINE, physical chemistry, see 12th edition

STERNBERG, MOSHE, b Marculesti, Rumania, Sept 3, 29; m 55; c 2. FOOD SCIENCE. Educ: Parhon Univ, Bucharest, Rumania, 52, PhD(org chem), 61. Prof Exp: Lab chief, Chem Pharmaceut Res Inst, Bucharest, Rumania, 55-57; res fel, Israel Inst Technol, 61-62; DIR, FUNDAMENTAL PROTEIN RES, MILES LABS, INC, 62- Mem: Am Chem Soc; Am Asn Cereal Chemists; Inst Food Technologists; AAAS. Res: Separation of industrial enzymes; proteins separation and characterization. Mailing Add: Miles Labs Inc 1127 Myrtle Elkhart IN 46514

STERNBERG, RICHARD WALTER, b Mt Pleasant, Iowa, Nov 21, 34; m 57; c 3. GEOLOGICAL OCEANOGRAPHY, MARINE SEDIMENTATION. Educ: Univ Calif, Los Angeles, BA, 58; Univ Wash, MSc, 61, PhD(oceanog), 65. Prof Exp: Assoc oceanog, Univ Wash, 63-65, res asst prof, 65-66; fel, Geomorphol Lab, Uppsala Univ, 66; res geophysicist, Univ Calif, San Diego, 67-68; asst prof, 68-73, ASSOC PROF OCEANOG, UNIV WASH, 73-, ADJ ASSOC PROF ENVIRON STUDIES, 70- Mem: Soc Econ Paleontologists & Mineralogists. Res: Geological oceanography, especially processes of sediment transport and boundary-layer flow near the seafloor. Mailing Add: Dept of Oceanog Univ of Wash Seattle WA 98105

STERNBERG, ROBERT LANGLEY, b Newark, NJ, Apr 9, 22; m 50; c 3. APPLIED MATHEMATICS, OCEAN ENGINEERING. Educ: Northwestern Univ, BS, 46, MA, 48, PhD(math), 51. Prof Exp: Asst instrument engr, Clinton Labs, Manhattan Proj, 44-45, jr physicist, 45-46; asst math, Northwestern Univ, 44-50, lectr, 50-51; mathematician, Lab for Electronics, Inc, 51-63; prof staff mem, Inst Naval Studies, 63-66; chief adv study group & staff scientist, Res Dept, Elec Boat Div, Gen

Dynamics Corp, 66-71; PROF MATH, UNIV RI, 71-, LECTR ENG, 73- Concurrent Pos: Consult, Army Res Off, 63, McGill Univ, 65 & Sanders Assocs, 66; mathematician, Naval Underwater Systs Ctr, 72-73 & 74-; lectr math, Univ New Haven, 74-76; lectr statist, Univ Conn, 75-76. Mem: Am Math Soc; Brit Interplanetary Soc. Res: Systems of differential equations; applied mathematics; Bennett functions; microwave lens antennas; ocean resources; operations research; astronautics; thermo-nuclear deterrence; applied mathematics and ocean engineering. Mailing Add: 113 Seneca Dr Noank CT 06340

STERNBERG, ROLF, b Jever, Ger, Aug 28, 26; US citizen; m 53. URBAN GEOGRAPHY, GEOGRAPHY OF LATIN AMERICA. Educ: Ursinus Col, BA, 53; Clark Univ, MA, 56; Syracuse Univ, PhD(geog), 71. Prof Exp: Jr planner, Worcester, Mass, 55-56 & Syracuse, NY, 56-57; lectr geog, City Col New York, 61-70; ASST PROF GEOG, MONTCLAIR STATE COL, 70- Mem: Asn Am Geogrs; Am Geog Soc; Latin Am Studies Asn; Conf Latin Am Geogrs. Res: Central places and their growth processes in historical perspectives; Santa Fe Province, Argentina, 1887-1914; Latin America, land redistribution and the future. Mailing Add: Dept of Geog & Urban Studies Montclair State Col Upper Montclair NJ 07043

STERNBERG, WILLIAM HOWARD, b Brooklyn, NY, Aug 16, 13; m 42; c 2. PATHOLOGY. Educ: Cornell Univ, AB, 33, MD, 37. Prof Exp: From instr to assoc prof, 46-54, PROF PATH, SCH MED, TULANE UNIV, 54- Mem: Am Asn Path & Bact; Col Am Path; Int Acad Path; Am Soc Cytol. Res: Endocrine and gynecologic pathology; cytogenetics. Mailing Add: Dept of Path Tulane Univ Sch of Med New Orleans LA 70112

STERNBERGER, LUDWIG AMADEUS, b Munich, Ger, May 26, 21; nat US; m 62. MEDICINE. Educ: Am Univ, Beirut, MD, 45. Prof Exp: Sr med bacteriologist, Div Labs & Res, State Dept Health, NY, 50-52, sr med biochemist, 52-53; asst prof med, Med Sch & assoc dir, Allergy Res Lab, Northwestern Univ, 53-55; chief path br, US Army Chem Res & Develop Labs, 55-67, CHIEF BASIC SCI DEPT, MED RES LABS, ARMY CHEM CTR, 67-; ASST PROF MICROBIOL, SCH MED, JOHNS HOPKINS UNIV, 66- Concurrent Pos: Fel exp path, Mem Cancer Ctr, New York, 48-50; assoc surg res, Sinai Hosp Baltimore, 66-; consult, Univ Iowa. Honors & Awards: Paul A Siple Award, 72. Mem: Soc Exp Biol & Med; Am Asn Immunol; Am Acad Allergy; Histochem Soc; Venezuela Soc Electron Micros. Res: Immunocytochemistry. Mailing Add: Basic Sci Dept Med Res Labs Edgewood Arsenal MD 21010

STERNBURG, JAMES GORDON, b Chicago, Ill, Feb 22, 19; m 54; c 3. ENTOMOLOGY. Educ: Univ Ill, AB, 49, MS, 50, PhD(entom), 52. Prof Exp: Res assoc entom, 52-54, from asst prof to assoc prof, 54-63, PROF ENTOM, UNIV ILL, URBANA, 63- Mem: Entom Soc Am; Lepidop Soc. Res: Insect physiology and toxicology of insecticides; enzymatic detoxication of dichloro-diphenyl-trichloroethane by resistance house flies; effects of insecticides on neuroactivity in insects; behavior of nearctic and neotropical Lepidoptera. Mailing Add: Dept of Entom Univ of Ill Urbana IL 61801

STERNER, JAMES HERVI, b Bloomsburg, Pa, Nov 14, 04; m 32, 71; c 3. MEDICINE. Educ: Pa State Univ, BS, 28, Harvard Univ, MD, 32; Am Bd Prev Med, dipl, 55; Am Bd Indust Hyg, dipl. Prof Exp: Intern, Lankenau Hosp, Philadelphia, 32-34, chief resident physician, 34-35; dir lab indust med, Eastman Kodak Co, NY, 36-49, from assoc med dir to med dir, 49-68; assoc dean, 68-71, PROF ENVIRON/OCCUP HEALTH, UNIV TEX SCH PUB HEALTH, HOUSTON, 68- Concurrent Pos: Instr indust med, Univ Rochester, 40-50, assoc prof med, Sch Med & Dent, 51-58, clin assoc prof, 58-68, clin assoc prof prev med, 59-61, clin assoc prof prev med & community health, 61-68; med consult, Holston Ord Works, Tenn, 41-45; med dir, Clinton Eng Works, Tenn Eastman Co, 43-45; mem interim med adv bd, Manhattan Proj, AEC, 45-47, mem radiol safety sect & medico-legal bd, Oper Crossroads, 46, consult, Off Oper Safety, AEC, 48-, mem adv comt biol & med, 60-66 & gen adv comt, 71-74; mem comt toxicol, Nat Acad Sci-Nat Res Coun, 47-55 & 71-74; mem comt environ physiol, 65-68; mem expert adv comt social & occup health, WHO, 51-, chmn expert comt med supvn in radiation work, 59; vis lectr, Harvard Med Sch, 52-56; mem comt occup health & safety, Int Labor Off, 52-; chief indust med staff, Rochester Gen Hosp, NY, 55-62, consult, 63-68; trustee, Am Bd Prev Med, 55-67, vchmn occup med, 59-60, chmn 62-67; mem main comt, Nat Coun Radiation Protection & Measurements, 55-68; spec consult, Cancer Control Comt, Nat Cancer Inst, 57-61; mem adv comt, Nat Health Surv, 57-61; spec consult & chmn comt radiation studies, USPHS, 57-61, consult, Nat Ctr Health Statist, 66-; mem, Gen Adv Comt Atomic Energy, NY, 58-65; sr assoc physician, Strong Mem Hosp, NY, 58-68; mem, Am Adv Bd, Am Hosp, Paris, 58-; chmn forum occup health, Nat Health Coun, 59, pres, 61; mem, Permanent Comn & Int Asn Occup Health, 60-; mem environ health panel, Exec Off Sci & Technol, 61-65; mem, Nat Environ Health Comt, 64-67; chmn, Nat Air Conserv Comn, 67; mem, Nat Adv Dis Prev & Environ Control Coun, 67-68; actg city health dir, Houston, Tex, 70; mem comn toxicol & indust hyg, Int Union Chem. Honors & Awards: Cummings Award, Am Indust Hyg Asn, 55; Knudsen Award, Indust Med Asn, 57; Award, Am Acad Occup Med, 59. Mem: fel Am Pub Health Asn; Am Indust Hyg Asn (pres, 48); Am Col Prev Med (pres, 59-60); Am Acad Occup Med (pres, 52-53); fel Royal Soc Health. Res: Clinical and experimental toxicology in industrial hygiene and environmental health. Mailing Add: Univ of Tex Sch of Pub Health PO Box 20186 Houston TX 77025

STERNER, JOHN, b London, Eng, Oct 26, 12; nat US; m 55; c 2. PHYSICS. Educ: Mass Inst Technol, BS, 33, DSc(physics), 50. Prof Exp: Jr physicist, Watertown Arsenal, US Dept Army, 35-40; vpres, Baird-Atomic, Inc, 40-55; dir flight test opers, Space Tech Labs, Inc, 55-59; EXEC VPRES, CORDIS CORP, 59- Mem: Am Phys Soc; Optical Soc Am; Am Inst Aeronaut & Astronaut; Am Ord Asn. Res: Medical instrumentation; rocket flight testing; spectrochemistry; interferometry. Mailing Add: 3915 Biscayne Blvd Miami FL 33137

STERNFELD, LEON, b Brooklyn, NY, June 15, 13; m 34; c 2. MEDICAL ADMINISTRATION, RESEARCH ADMINISTRATION. Educ: Univ Chicago, SB, 32, MD, 36, PhD(biochem), 37; Columbia Univ, MPH, 43. Prof Exp: Intern pediat, Johns Hopkins Univ, 38-39 & Sydenham Hosp, 39-40; asst res, Jewish Hosp, Brooklyn, 40-41; epidemiologist-in-training, State Dept Health, NY, 41-42, jr epidemiologist, 42, asst dist state health officer, 43-44, dir med rehab, 44-50, dist health officer, 50-51; asst dir, Tuberc Div, State Dept Pub Health, Mass, 51-52; assoc dir, Field Training Unit, Harvard Univ, 52-53, lectr, Sch Pub Health, 53-57, from asst clin prof to assoc clin prof maternal & child health, 58-70, vis lectr maternal & child health, 70-74; MED DIR, UNITED CEREBRAL PALSY ASNS, INC, 71- Concurrent Pos: Chief Pub Health Admin, Korean Civil Asst Command, 53-55; chief prev med, Ft Devons, US Army, 55; City health comnr, Cambridge, Mass, 55-61; lectr, Simmons Col, 56-69; assoc physician, Children's Med Ctr, Boston, 57-69; dep health comnr, Mass Dept Pub Health, 61-69. Mem: Am Pub Health Asn; AMA; Am Acad Cerebral Palsy. Res: Chemical properties of essential bacterial growth factor; essential fructosuria; pathophysiology; medical and public health aspects of cerebral palsy and mental retardation; public health methodology and community health. Mailing Add: 1385 York Ave New York NY 10021

STERNFELD, MARVIN, organic chemistry, see 12th edition

STERNGLANZ, ROLF, b Sewell, Chile, May 18, 39; US citizen; m 64. MOLECULAR BIOLOGY, BIOCHEMISTRY. Educ: Oberlin Col, AB, 60; Harvard Univ, PhD(phys chem), 67. Prof Exp: NIH res fel biochem, Sch Med, Stanford Univ, 66-68; ASST PROF BIOCHEM, STATE UNIV NY STONY BROOK, 69- Concurrent Pos: Am Cancer Soc res grants, State Univ NY Stony Brook, 69- Mem: AAAS. Res: Mechanism of DNA replication; physical chemistry of DNA. Mailing Add: Dept of Biochem State Univ of NY Stony Brook NY 11790

STERNGLASS, ERNEST JOACHIM, b Berlin, Ger, Sept 24, 23; nat US; m 57; c 2. PHYSICS. Educ: Cornell Univ, BEE, 44, MS, 51, PhD(eng physics), 53. Prof Exp: Asst physics, Cornell Univ, 44, res assoc, 51-52; physicist, US Naval Ord Lab, 46-52; res physicist, Res Labs, Westinghouse Elec Co, Pa, 52-60, adv physicist, 60-67; PROF RADIATION PHYSICS, UNIV PITTSBURGH, 67- Concurrent Pos: Assoc, George Washington Univ, 46-47; Westinghouse Res Lab fel, Inst Henri Poincare, Paris, 57-58; vis prof, Inst Theoret Physics, Stanford Univ, 66-67. Mem: AAAS; fel Am Phys Soc; Am Astron Soc; Fedn Am Sci; Am Asn Physicists in Med. Res: Secondary electron emission; physics of electron tubes; electron and elementary particle physics; electronic imaging devices for astronomy and medicine; radiation physics; biological effects of radiation. Mailing Add: 1417 Shady Ave Pittsburgh PA 15217

STERNHEIMER, RUDOLPH MAX, b Saarbruecken, Ger, Apr 26, 26; nat US; m 52. ATOMIC PHYSICS. Educ: Univ Chicago, BS, 43, MS, 46, PhD, 49. Prof Exp: Jr scientist, Div War Res, Metall Lab, Columbia Univ, 45-46; instr physics, Univ Chicago, 46-48; asst & instr, Yale Univ, 48-49; mem staff, Los Alamos Sci Lab, 49-51; from assoc physicist to physicist, 52-65, SR PHYSICIST, BROOKHAVEN NAT LAB, 65- Mem: Fel Am Phys Soc. Res: Atomic and nuclear physics; theory of solids; theory of nuclear quadrupole coupling; theory of ionization loss and Cerenkov radiation; focusing magnets; polarization of nucleons; theory of meson production; electronic polarizabilities of ions. Mailing Add: Dept of Physics Brookhaven Nat Lab Upton NY 11973

STERNICK, EDWARD SELBY, b Cambridge, Mass, Feb 10, 39; m 60; c 3. MEDICAL PHYSICS. Educ: Tufts Univ, BS, 60; Boston Univ, MA, 63; Univ Calif, Los Angeles, PhD(med physics), 68. Prof Exp: Res scientist biophys, Nat Aeronaut & Space Admin, 63-64; instr radiol, 68-72, ASST PROF CLIN MED, DARTMOUTH-HITCHCOCK MED CTR, 72- Concurrent Pos: Prof, NH Voc Tech Inst, 72-; adj asst prof bioeng, Thayer Sch Eng, Dartmouth Univ, 73-; consult, Vet Admin Hosp, 74- Honors & Awards: First Place Sci Exhibit Award, Am Asn Physicists in Med, 74. Mem: Am Asn Physicists in Med; Soc Nuclear Med; Health Physics Soc. Res: Application of computer technology to radiation medicine. Mailing Add: Dept of Therapeut Radiol Dartmouth-Hitchcock Med Ctr Hanover NH 03755

STERNLIEB, IRMIN, b Czernowitz, Rumania, Jan 11, 23; US citizen; m 53. GASTROENTEROLOGY, ELECTRON MICROSCOPY. Educ: Univ Geneva, MSc, 49, MD, 52. Prof Exp: Intern, Morrisania City Hosp, Bronx, NY, 52-53; resident internal med, Bronx Munic Hosp Ctr, NY, 55-57; asst instr, 56-57, instr & assoc, 57-61, from asst prof to assoc prof, 61-72, PROF MED, COL MED, ALBERT EINSTEIN COL MED, 72- Concurrent Pos: Fel internal med & gastroenterol, Mt Sinai Hosp, New York, 53-55; USPHS fel, 57-60 & spec fel, Lab Atomic Synthesis & Proton Optics, Ivry, France, 64-65. Mem: Am Soc Clin Invest; Int Asn Study Liver; Am Gastroenterol Asn; Am Asn Study Liver Dis; Am Soc Cell Biol. Res: Clinical, genetic, biochemical, diagnostic and morphologic aspects of human inherited copper toxicity; electron microscopy of human liver. Mailing Add: Albert Einstein Col of Med 1300 Morris Park Ave Bronx NY 10461

STERNLIGHT, HIMAN, b New York, NY, May 31, 36; m 58; c 3. PHYSICAL CHEMISTRY. Educ: Columbia Univ, BA, 57, BS, 58; Calif Inst Technol, PhD(chem), 63. Prof Exp: Mem tech staff, Bell Tel Labs, NJ, 63-65; asst prof chem, Univ Calif, Berkeley, 65-70; MEM TECH STAFF, BELL LABS, INC, 70- Concurrent Pos: NIH res grant, 66-69. Mem: Am Phys Soc; Am Chem Soc. Res: Magnetic resonance studies, including small and macromolecular systems. Mailing Add: Bell Labs Inc Mountain Ave Murray Hill NJ 07974

STERNSTEIN, MARTIN, b Chicago, Ill, Apr 25, 45; m 70. MATHEMATICS. Educ: Univ Chicago, BS, 66; Cornell Univ, PhD(math), 71. Prof Exp: Asst prof, 70-75, ASSOC PROF MATH, ITHACA COL, 75-, CHMN DEPT, 72- Mem: Am Math Soc; Math Asn Am. Res: Algebraic topology. Mailing Add: Dept of Math Ithaca Col Ithaca NY 14850

STERNSTEIN, SANFORD SAMUEL, b New York, NY, June 19, 36; m 58; c 2. POLYMER PHYSICS, POLYMER ENGINEERING. Educ: Univ Md, BS, 58; Rensselaer Polytech Inst, PhD(chem eng), 61. Prof Exp: From asst prof to prof polymers, 61-73, WILLIAM WEIGHTMAN WALKER PROF POLYMER ENG, RENSSELAER POLYTECH INST, 73- Concurrent Pos: NSF & Inst Paper Chem Pioneering Res grants, 63-65; Nat Inst Dent Res grant, 65-70; NSF res grants, 73-78. Mem: Am Inst Chem Eng; Soc Rheol; Am Phys Soc; Am Chem Soc. Res: Rheology; fracture; dynamic mechanical properties of polymers; polymer-solvent interactions and crazing; polymer network mechanics and rubber elasticity. Mailing Add: Mat Div Rensselaer Polytech Inst Troy NY 12181

STERRETT, ANDREW, b Pittsburgh, Pa, Apr 3, 24; m 48; c 2. MATHEMATICS. Educ: Carnegie Inst Technol, 48; Univ Pittsburgh, MS, 50, PhD(math), 56. Prof Exp: Lectr math, Univ Pittsburgh, 48-50; instr, Univ Ohio, 50-53; from asst prof to assoc prof, 53-64, chmn dept math, 60-63 & 65-68, dir comt undergrad prog in math, 70-72, PROF MATH, DENISON UNIV, 64-, DEAN COL, 73- Concurrent Pos: NSF fac fel statist, Stanford Univ, 59-60. Mem: Am Math Soc; Math Asn Am; Am Statist Asn. Mailing Add: Dept of Math Denison Univ Granville OH 43023

STERRETT, FRANCES SUSAN, b Vienna, Austria, Sept 25, 13; nat US; m 39; c 2. ENVIRONMENTAL CHEMISTRY. Educ: Univ Vienna, PhD(chem), 38. Prof Exp: Res chemist, Lab, France, 38-39; asst biochem, Med Ctr, Columbia Univ, 39-40; res chemist, van Ameringen & Haebler, Inc, NJ, 40-41, Woburn Degreasing Co, 43 & Fritzsche Bros, Inc, NY, 43-49; lectr chem, 53-57, from instr to assoc prof, 57-73, PROF CHEM, HOFSTRA UNIV, 73- Concurrent Pos: Lectr biochem sec high sch teachers, NSF, 65-67. Mem: AAAS; Am Chem Soc; fel Am Inst Chemists; NY Acad Sci. Res: Chemistry and chemical reactions in the environment in reference to the atmosphere, hydrosphere, lithosphere and biosphere; microanalysis; aromatic chemicals; essential oils; inorganic and qualitative chemistry; organic chemistry; quantitative analysis; environmental problems in reference to water pollution on Long Island, New York. Mailing Add: Dept of Chem Hofstra Univ Hempstead NY 11550

STERRETT, JOHN KENNETH, b Manhattan, Kans, July 29, 11; m 37; c 3. MATHEMATICS. Educ: Washburn Col, BS, 35; Univ Kans, MA, 38; Univ Pittsburgh, PhD(math), 53. Prof Exp: Teacher high sch, Kans, 37-38; instr math, Dodge City Jr Col, 38-40; instr, Hutchinson Jr Col, 40-42; head measurements anal sect, Comput Lab, Ballistics Res Lab, US Dept Army, 50-52; head opers res sect,

Anal & Method Br, Systs Div, US Naval Res Lab, 52-54; chief aeroballistics br, Air Res & Develop Command, US Dept Air Force, 54-55; opers control systs group, Off Naval Res, 55-57; head opers math br, Off Qm Gen, US Dept Army, 57-59; staff scientist, Off Space Flight Opers, NASA, 59-61; SCI ADV, DEP CHIEF STAFF, N AM AIR DEFENSE COMMAND, ENT AFB, 61- Mem: Am Math Soc; Soc Indust & Appl Math; Math Asn Am; Sigma Xi; Opers Res Soc Am. Res: Projective geometry; numerical analysis and data adjustment; ballistics; high speed computers. Mailing Add: 1311 N Iowa Ave Colorado Springs CO 80909

STERRETT, JOHN PAUL, b Springfield, Ohio, Dec 14, 24; m 49; c 2. PLANT PHYSIOLOGY. Educ: Univ WVa, BS, 50; Va Polytech Inst, MS, 61, PhD(plant physiol), 66. Prof Exp: County forester, WVa Conserv Comn, 50-53; forester, Bartlett Tree Expert Co, 53-59; res asst plant physiol, Va Polytech Inst, 59-61, asst prof, 61-69; plant physiologist, Veg Control Div, Ft Detrick, US Army, 69-74; PLANT PHYSIOLOGIST, WEED PHYSIOL & GROWTH REGULATOR RES LAB, AGR RES SERV, USDA, 74- Mem: Soc Am Foresters; Weed Sci Soc Am; Am Soc Plant Physiol. Res: Plant growth regulators for abscission, bud break, height and yield. Mailing Add: PO Box 1209 Frederick MD 21701

STERRETT, KAY FIFE, b McKeesport, Pa, May 20, 31; m 60; c 3. GEOPHYSICS, PHYSICAL CHEMISTRY. Educ: Univ Pittsburgh, BS, 53, PhD(phys chem), 57. Prof Exp: Asst, Univ Pittsburgh, 53-57; phys chemist, Nat Bur Stand, 57-61; mem res staff, Northrop Space Labs, 62-64, head, Space Physics & Chem Lab, 64-66; head phys chem lab, Northrop Space Labs, 66-67; CHIEF, RES DIV, US ARMY COLD REGIONS RES & ENG LAB, 67- Concurrent Pos: Neth Govt fel, Kamerlingh Onnes Lab, Univ Leiden, 57-58; mem exten teaching staff, Univ Calif, Los Angeles, 65-66; mem Army res coun, Dept Army, 67-68. Mem: Fel AAAS; Am Chem Soc; Am Phys Soc; The Chem Soc; Sigma Xi. Res: Thermodynamics of solids; high pressure physics; planetary interiors; low temperature physics; cold regions environment; physics of snow, ice and frozen soil; research management. Mailing Add: US Army Col Reg Res & Eng Lab PO Box 282 Hanover NH 03755

STERZER, FRED, b Vienna, Austria, Nov 18, 29; nat US; m 64. PHYSICS. Educ: City Col New York, BS, 51; NY Univ, MS, 52, PhD(physics), 55. Prof Exp: Dir microwave appl res lab, Radio Corp of Am, 54-69, mgr advan technol lab, 69-72, DIR MICROWAVE TECHNOL CTR, RCA LABS, 72- Mem: Fel Inst Elec & Electronics Engrs. Res: Microwave spectroscopy, tubes and solid state devices; photovoltaic devices. Mailing Add: Microwave Technol Ctr RCA Labs Princeton NJ 08540

STESSEL, GEORGE JOHN, plant pathology, see 12th edition

STETLER, DAVID ALBERT, b Pasadena, Calif, June 17, 35; m 64; c 2. PLANT CYTOLOGY. Educ: Univ Southern Calif, BSc, 59; Univ Calif, Berkeley, PhD(bot), 67. Prof Exp: Asst prof bot, Univ Minn, 67-69; asst prof biol, Dartmouth Col, 69-73; ASST PROF BOT, VA POLYTECH INST & STATE UNIV, 73- Mem: Am Soc Plant Physiologists; Bot Soc Am; Tissue Cult Asn; Int Asn Plant Tissue Cult. Res: Organelle development in plant cells; ultrastructure of plant tissues. Mailing Add: Dept of Biol Va Polytech Inst & State Univ Blacksburg VA 24061

STETSON, ALVIN RAE, b San Diego, Calif, July 23, 26; m 47; c 2. PHYSICAL CHEMISTRY. Educ: San Diego State Col, AB, 48. Prof Exp: Anal chemist, 48-50, phys chemist, 50-53, from staff engr to sr res staff engr, 53-66, CHIEF PROCESS RES, SOLAR DIV, INT HARVESTER CO, 66- Mem: Am Chem Soc; Nat Asn Corrosion Eng; Am Soc Metals. Res: Fused salt plating; high temperature metallic and ceramic protective coatings; reaction of materials at high temperatures; reentry and gas turbine environment simulation; braze joining of dissimilar metals; plasma arc testing and spraying; materials research supervision. Mailing Add: Solar Div Int Harvester Co 2200 Pac Hwy San Diego CA 92112

STETSON, CHANDLER ALTON, b Boston, Mass, Nov 3, 21; m 48; c 4. MEDICINE. Educ: Bowdoin Col, BS, 41; Harvard Med Sch, MD, 44. Prof Exp: House officer, Children's Hosp, Boston, Mass, 44-45 & Maine Gen Hosp, Portland, 45-47; vis investr, Rockefeller Inst, 48-51; from assoc prof to prof path, Sch Med, NY Univ, 55-72, chmn dept, 58-72; DEAN COL MED, UNIV FLA, 72-, VPRES HEALTH AFFAIRS, 74- Concurrent Pos: Fel pediat, Johns Hopkins Hosp, 47-48; res fel, Univ Minn, 53-55. Mem: Asn Am Physicians; Am Soc Clin Invest; Harvey Soc; Am Asn Immunol; Am Soc Exp Path. Res: Experimental pathology; bacteriology; immunology. Mailing Add: Off of the Dean Univ of Fla Col of Med Gainesville FL 32601

STETSON, MILTON H, b Springfield, Mass, Nov 25, 43; m 66; c 2. REPRODUCTIVE ENDOCRINOLOGY, BIOLOGICAL RHYTHMS. Educ: Cent Conn State Col, BA, 65; Univ Wash, MS, 68, PhD(zool), 70. Prof Exp: NIH fel, 66-70; res fel reproduction, Univ Tex, Austin, 71-73; ASST PROF BIOL SCI, UNIV DEL, 73- Mem: Am Soc Zool; AAAS; Soc Study Reproduction; Am Asn Univ Profs. Res: Role of the circadian system in the timing of reproductive events; neural and neuroendocrine generation of female reproductive cyclicity. Mailing Add: Dept of Biol Sci Univ of Del Newark DE 19711

STETSON, RICHARD PRATT, b Dorchester, Mass, Apr 24, 98; m 28; c 2. MEDICINE. Educ: Dartmouth Col, AB, 22; Harvard Med Sch, MD, 26; Am Bd Internal Med, dipl, 47. Prof Exp: Intern med, Mass Gen Hosp, Boston, 27-28; asst instr, Sch Med, Yale Univ, 28-30; from asst to assoc, Harvard Med Sch, 30-55, asst clin prof, 55-63, clin prof, 63-64; chief of staff, Vet Admin Hosp, West Roxbury, 63-69, LIAISON FOR EDUC, VET ADMIN REGION I, 69- Concurrent Pos: Asst resident & assoc physician, New Haven Hosp, Conn, 28-30; mem staff, Harvard Med Serv & Thorndike Mem Lab, Boston City Hosp, 30-64, physician-in-chief, 2nd Med Serv, Hosp, 48-59, vis physician, Hosp, 48-64, consult, 64-; physician, New Eng Deaconess Hosp, Boston, 36-57, consult, 57-; assoc vis physician, Mass Mem Hosps, Boston, 51-55, physician, 55-61, consult, 62-; lectr, Sch Med, Boston Univ, 56-64; mem spec med adv group, Vet Admin, 57-59, area dir prof serv, 59-63; mem, Arthritis Found. Consult, Vet Admin Hosp, West Roxbury, 46-49, USPHS, 57-69, Morton Hosp, Taunton & Addison Gilbert Hosp, Gloucester. Mem: AAAS; AMA; Am Clin & Climat Asn; Am Diabetes Asn; Am Col Physicians. Res: Hematology, especially pernicious anemia; diabetes. Mailing Add: 235 Woodland Rd Chestnut Hill MA 02167

STETSON, ROBERT FRANKLIN, b Lewiston, Maine, Apr 17, 32; m 67. PLASMA PHYSICS. Educ: Bates Col, BS, 54; Wesleyan Univ, MA, 56; Univ Va, PhD(physics), 59. Prof Exp: Asst physics, Wesleyan Univ, 54-56; instr, Univ Va, 56-58; asst prof, Univ Fla, 59-64; assoc prof, 64-69, PROF PHYSICS, FLA ATLANTIC UNIV, 69-, DIR FAC SCHOLARS PROG, 71- Concurrent Pos: Proj scientist, Air Force Off Sci Res, 62-64; consult, Col Entrance Exam Bd. Mem: Am Phys Soc; Am Asn Physics Teachers. Res: Angular correlation of gamma rays; neutron scattering; non-traditional higher education; computer simulation of plasma and thermodynamic problems. Mailing Add: Dept of Physics Fla Atlantic Univ Boca Raton FL 33432

STETTEN, DEWITT, JR, b New York, NY, May 31, 09; m 41; c 4. MEDICINE. Educ: Harvard Univ, AB, 30; Columbia Univ, MD, 34, PhD(biochem), 40. Hon Degrees: DSc, Wash Univ, 74. Prof Exp: Intern & resident, 3rd Med Serv, Bellevue Hosp, New York, 34-37; asst biochem, Columbia Univ, 38-39, from instr to asst prof, 41-47; asst prof, Harvard Med Sch, 47-48; chief div nutrit & physiol, Pub Health Res Inst, New York, NY, 48-54; dir intramural res, Nat Inst Arthritis & Metab Dis, 54-62; dean, Rutgers Med Sch, 62-70; dir, Nat Inst Gen Med Sci, 70-74, DEP DIR SCI, NIH, 74- Concurrent Pos: Assoc, Peter Bent Brigham Hosp, Boston, Mass, 47-48; mem sci adv comt, Okla Med Res Found, 63-67, chmn, 66; exec comt mem, Div Med Sci, Nat Res Coun, 66; chmn nat adv comt, Roche Inst Molecular Biol, 67-70. Honors & Awards: Smith Prize, 43; Alvarenga Prize, 54; Banting Medal, Am Diabetes Asn, 57; Superior Serv Honor Award, Dept Health, Educ & Welfare, 73; Gold Medal, Distinguished Achievement in Med, Asn Alumni, Columbia Univ, Col Physicians & Surgeons, 74. Mem: Nat Acad Sci; AAAS; Am Soc Biol Chem; Soc Exp Biol & Med; Harvey Soc. Res: Intermediary metabolism of fats and carbohydrates; experimental diabetes; evaluation of rates of chemical processes in the intact animal; metabolic defects in gout and muscular dystrophy; insulin metabolism. Mailing Add: NIH 9000 Rockville Pike Bethesda MD 20014

STETTEN, MARJORIE ROLOFF, b New York, NY, July 13, 15; m 41; c 4. BIOCHEMISTRY. Educ: Rutgers Univ, BS, 37; Columbia Univ, PhD(biochem), 44. Prof Exp: Asst biochem, Col Physicians & Surgeons, Columbia Univ, 40-47; fel biol chem, Harvard Med Sch, 47-48; assoc div nutrit & physiol, Pub Health Res Inst, New York, 48-54; biochemist, Nat Inst Arthritis & Metab Dis, 54-63; res prof exp med, Sch Med, Rutgers Univ, New Brunswick, 63-71; BIOCHEMIST, NAT INST ARTHRITIS, METAB & DIGESTIVE DIS, NIH, 71- Concurrent Pos: Mem, Marine Biol Lab, Woods Hole. Mem: Am Soc Biol Chem. Res: Intermediary metabolism of amino acids and carbohydrates; particulate enzymes of liver and kidney. Mailing Add: Bldg 10-9B-02 Nat Inst of Health Bethesda MD 20014

STETTENHEIM, PETER, b New York, NY, Dec 27, 28; m 65; c 2. ORNITHOLOGY. Educ: Haverford Col, BS, 50; Univ Mich, MA, 51, PhD(zool), 59. Prof Exp: Res zoologist, USDA, Avian Anat Proj, 58-69; book rev ed, Wilson Bull, 70-74; ED, CONDOR, 74- Concurrent Pos: Panelist NSF-Am Ornith Union workshop on status of ornith, 75- Honors & Awards: Co-recipient, Tom Newman Mem Int Award, Brit Poultry Breeders & Hatcheries Asn, 73. Mem: Wilson Ornith Soc; Cooper Ornith Soc; Am Ornith Union; Brit Ornith Union; Ger Ornith Soc. Res: Descriptive and functional anatomy of avian integument, particularly feather structure. Mailing Add: Meriden Rd Lebanon NH 03766

STETTER, JOSEPH ROBERT, b Buffalo, NY, Dec 15, 46; m 72. PHYSICAL CHEMISTRY, SURFACE CHEMISTRY. Educ: State Univ NY Buffalo, BA, 69, PhD(phys chem), 75. Prof Exp: Res asst, Linde Div, Union Carbide Corp, 66-68, chemist, 69; teaching asst chem, State Univ NY Buffalo, 70-71, res asst, Dept Chem, 71-74; SR RES CHEMIST, ENER GETICS SCI INC, 74- Mem: Am Chem Soc; Am Inst Physics; Am Vacuum Soc. Res: Physical adsorption, chemisorption, heterogeneous catalytic systems, and the physical, electronic and chemical characterization of solid surfaces and the solid-gas interface. Mailing Add: RD 2 Aqueduct Rd Continental Village Peekskill NY 10566

STETTLER, JOHN DIETRICH, b Cleveland, Ohio, Mar 15, 34; m 57; c 5. THEORETICAL PHYSICS. Educ: Univ Notre Dame, BS, 56; Mass Inst Technol, PhD(physics), 62. Prof Exp: Res asst physics, Mass Inst Technol, 57-60; asst prof, Univ Mo-Rolla, 60-64; RES PHYSICIST, US ARMY MISSILE COMMAND, 64- Concurrent Pos: Adj prof physics, Univ Ala, Huntsville, 70- Mem: AAAS; Am Phys Soc; Am Asn Physics Teachers. Res: Theory of electron-phonon interactions at defects in ionic crystals; theory of intermolecular energy exchange in gases. Mailing Add: 410 Cumberland Dr Huntsville AL 35803

STETTLER, REINHARD FRIEDERICH, b Steckborn, Switz, Dec 27, 29; m 55; c 1. FOREST GENETICS. Educ: Swiss Fed Inst Technol, dipl, 55; Univ Calif, Berkeley, PhD(genetics), 63. Prof Exp: Res officer silvicult, Res Div, BC Forest Serv, Can, 56-58; res assoc forest mgt, Fed Inst Forest Res, Switz, 58-59; from asst prof to assoc prof, 63-74, PROF FOREST GENETICS, UNIV WASH, 74- Concurrent Pos: Alexander von Humboldt fel, Inst Forest Genetics, Schmalenbeck, Ger, 69-70. Mem: AAAS; Bot Soc Am; Genetics Soc Am. Res: Genetic control of morphogenesis in higher plants; reproductive physiology of forest trees; factors affecting crossability in poplars; induction of haploid parthenogenesis. Mailing Add: Col of Forest Resources Univ of Wash Seattle WA 98195

STEUBER, WALTER, physics, see 12th edition

STEUER, MALCOLM F, b Marion, SC, Dec 16, 28; m 58; c 3. NUCLEAR PHYSICS. Educ: US Merchant Marine Acad, BS, 50; Clemson Col, MS, 54; Univ Va, PhD(physics), 57. Prof Exp: Res fel, Univ Va, 57-58; asst prof, 58-63, ASSOC PROF PHYSICS, UNIV GA, 63- Concurrent Pos: NSF sci fac fel, Univ Wis, 64-65. Mem: Am Phys Soc. Res: Interactions of neutrons with nuclei. Mailing Add: Dept of Physics Univ of Ga Athens GA 30601

STEUNENBERG, ROBERT KEPPEL, b Caldwell, Idaho, Sept 18, 24; m 47. INORGANIC CHEMISTRY. Educ: Col Idaho, BA, 47; Univ Wash, PhD(chem), 51. Prof Exp: Mem staff, 51-67, SR CHEMIST, ARGONNE NAT LAB, 67- Mem: Am Chem Soc; Sigma Xi; Am Nuclear Soc; Electrochem Soc. Res: Fluorocarbons; interhalogen compounds; pyrometallurgical methods for processing nuclear reactor fuels; nuclear technology; high-temperature batteries; energy conversion; molten salt chemistry. Mailing Add: Argonne Nat Lab Bldg D-205 Rm W-109 9700 S Cass Ave Argonne IL 60439

STEVEN, THOMAS AUGUST, b Dryden, Ore, Oct 14, 17; m 45; c 2. GEOLOGY. Educ: San Jose State Col, AB, 39; Univ Calif, Los Angeles, PhD(geol), 50. Prof Exp: GEOLOGIST, US GEOL SURV, 42- Mem: AAAS; Soc Econ Geologists; Geol Soc Am. Res: Economic and general geology. Mailing Add: US Geol Surv Bldg 25 Denver Fed Ctr Denver CO 80225

STEVENS, ALAN DOUGLAS, b Nashua, NH, Aug 17, 26; m 49; c 5. OCCUPATIONAL HEALTH, VETERINARY MEDICINE. Educ: Cornell Univ, DVM, 47. Prof Exp: Private practice, Ga, 47-50; prog officer res grants, Div Environ Eng & Food Protection, US Pub Health Serv, 63-66, chief res grants, 66-67, chief res & training grants rev, Nat Ctr Urban & Indust Health, 67-68, environ control admin, 68-69, chief res grants, Bur Safety & Occup Health, Environ Control Admin, 69-70, ASST DIR & DIR EXTRAMURAL PROGS, NAT INST OCCUP SAFETY & HEALTH, CTR DIS CONTROL, DEPT HEALTH, EDUC & WELFARE, 70- Mem: AAAS; Am Inst Chemists; Am Conf Govt Indust Hygienists. Res: Food chemistry and microbiology; irradiation of foods; virology; laboratory animal medicine; research administration; information retrieval. Mailing Add: 100 Silver Ave Ft Mitchell KY 41017

STEVENS, ALFRED LYMAN, b Brown City, Mich, Apr 28, 35; m 56; c 2. ENERGY CONVERSION, SOLID MECHANICS. Educ: Mich State Univ, BS, 60, MS, 61, PhD(appl mech), 68. Prof Exp: Tech staff mem, Sandia Labs, 61-64; instr solid mech,

Mich State Univ, 66-67; tech staff mem, 68-76, SUPVR, OIL SHALE PROGS DIV, SANDIA LABS, 76- Mem: Soc Exp Stress Anal; Am Phys Soc. Res: Shock wave physics; continuum mechanics; stress analysis. Mailing Add: 7105 Lantern Rd NE Albuquerque NM 87109

STEVENS, ANN REBECCA (LARKIN), b Huntington, WVa, July 22, 39; m 65; c 2. BIOCHEMISTRY, CELL BIOLOGY. Educ: Univ Ala, Tuscaloosa, BS, 61; Univ Colo, Denver, PhD(biochem), 66. Prof Exp: Res assoc cell biol, Inst Cellular, Molecular & Develop Biol, Univ Colo, Boulder, 66-68; asst prof, 68-74, ASSOC PROF BIOCHEM, COL MED, UNIV FLA, 74-; RES INVESTR & DIR ELECTRON MICROS LABS, VET ADMIN HOSP, 68- Concurrent Pos: Nat Inst Allergy & Infectious Dis res grants, 70-76; Fulbright awardee, Pasteur Inst, Lille, France. Mem: AAAS; Am Soc Cell Biol; Soc Exp Biol & Med; NY Acad Sci. Res: Aspects of nucleic acid metabolism during growth and differentiation in pathogenic and nonpathogenic strains of Acanthamoeba and Naegleria; chemotherapy of pathogenic free-living amoebae. Mailing Add: Cell Biol Labs Vet Admin Hosp Gainesville FL 32601

STEVENS, AUDREY L, b Leigh, Nebr, July 21, 32; m 64; c 2. BIOCHEMISTRY, MICROBIOLOGY. Educ: Iowa State Univ, BS, 53; Western Reserve Univ, PhD(biochem), 58. Prof Exp: NSF fel, 58-60; instr pharmacol, Sch Med, Univ St Louis, 60-62, asst prof, 62-63; from asst prof to assoc prof biochem, Sch Med, Univ Md, Balitmore City, 63-66; MEM RES STAFF, BIOL DIV, OAK RIDGE NAT LAB, 66- Concurrent Pos: NSF res grants, 60-66. Mem: Am Soc Biol Chem. Res: Nuclear acid biosynthesis. Mailing Add: Biol Div Oak Ridge Nat Lab Oak Ridge TN 37831

STEVENS, BRIAN, b South Elmsall, Eng, July 22, 24; m 53; c 2. PHOTOCHEMISTRY. Educ: Oxford Univ, BA & MA, 50, DPhil(phys chem), 53. Prof Exp: Fel, Nat Res Coun Can, 53-55; res asst chem, Princeton Univ, 55-56, res assoc, 56-57; res assoc tech off chem eng, Esso Res & Eng Co, NJ, 57-58; lectr chem, Univ Sheffield, 58-65, reader photochem, 65-67; PROF CHEM, UNIV S FLA, 67- Mem: Am Chem Soc; The Chem Soc. Res: Molecular luminescence and electronic energy transfer in complex molecules; photosensitized peroxidation of unsaturated molecules and reduction of dyes. Mailing Add: Dept of Chem Univ of South Fla Tampa FL 33620

STEVENS, CALVIN H, b Sheridan, Wyo, Apr 3, 34; m 61; c 2. PALEOECOLOGY, STRATIGRAPHY. Educ: Univ Colo, AB, 56, MA, 58; Univ Southern Calif, PhD(geol), 63. Prof Exp: Geologist, Res Lab, Humble Oil Co, 58-60; asst prof geol, San Jose State Col, 63-65 & Univ Colo, 65-66; from asst prof to assoc prof, 66-72, PROF GEOL, SAN JOSE STATE UNIV, 72- Mem: Am Asn Petrol Geol; fel Geol Soc Am; Soc Econ Paleont & Mineral. Res: Late Paleozoic paleoecology and paleontology; Great Basin geology and stratigraphy. Mailing Add: Dept of Geol San Jose State Univ San Jose CA 95114

STEVENS, CALVIN LEE, b Edwardsville, Ill, Nov 3, 23; m 47; c 1. CHEMISTRY. Educ: Univ Ill, BS, 44; Univ Wis, PhD(org chem), 47. Prof Exp: Asst org chem, Univ Wis, 44-47; Du Pont fel, Mass Inst Technol, 47-48; from asst prof to assoc prof, 48-54, PROF ORG CHEM, WAYNE STATE UNIV, 54- Concurrent Pos: Guggenheim fel, Univ Paris, 55-56; sci liaison officer, Off Naval Res, London, 59-60; Fulbright fels, Sorbonne, 64-65, Univ Paris, 71-72. Mem: Am Chem Soc; The Chem Soc; Swiss Chem Soc; Chem Soc France. Res: Organic chemistry; epoxyethers and nitrogen analogs of ketenes; natural products; amino-sugars; amino-ketone rearrangements. Mailing Add: Dept of Chem 221 Chem Bldg Wayne State Univ Detroit MI 48202

STEVENS, CARL MANTLE, II, b Washington, DC, Oct 31, 15; m 49; c 4. BIOCHEMISTRY. Educ: Am Univ, BA, 37; Univ Ill, PhD(biochem), 41. Prof Exp: Lab asst, Int Paper Co, Maine, 37; asst biochem, Univ Ill, 38 & Med Col, Cornell Univ, 41-45; from asst prof to assoc prof, 46-51, actg chmn dept chem, 60-61, chmn, 61-71, PROF BIOCHEM, WASH STATE UNIV, 51- Concurrent Pos: Merck sr fel, Calif Inst Technol, 54-55; vis lectr, Univ Ill, 57; vis sr res, Sir Wm Dunn Sch Path, Univ Oxford, 73-74. Mem: AAAS; Am Chem Soc; Harvey Soc; Am Soc Biol Chemists. Res: Biochemical genetics; biosynthetic pathways in microorganisms; amino acid metabolism. Mailing Add: Dept of Chem Wash State Univ Pullman WA 99163

STEVENS, CHARLES DAVID, b Pittsburgh, Pa, Feb 1, 12; m 37; c 5. BIOCHEMISTRY. Educ: Univ Cincinnati, AB, 33, MSc, 34, PhD(biochem), 37. Prof Exp: Res assoc biochem, Cardiac Lab, Sch Med, Univ Cincinnati, 38-42 & Lab Aviation Med, 42-45; biochemist, Dow Chem Co, 45-46; res assoc biochem, Gastric Lab, Univ Cincinnati, 46-50, asst prof, Dept Prev Med & Indust Health, 50-65; assoc prof, 65-68, PROF BIOMET, EMORY UNIV, 68- Concurrent Pos: Fel, Med Col Va, 61-62. Mem: Fel AAAS; Soc Exp Biol & Med; Am Asn Cancer Res. Res: Selective localization of chemicals in acidic cancer tissue; respiration; biomathematics; synthesis and metabolism of organolead compounds. Mailing Add: Dept of Statist & Biomet Emory Univ Atlanta GA 30322

STEVENS, CHARLES EDWARD, b Minneapolis, Minn, June 5, 27; m 52; c 7. VETERINARY PHYSIOLOGY. Educ: Univ Minn, BS, 51, DVM & MS, 55, PhD(vet physiol & pharmacol), 58. Hon Degrees: Hon Prof, San Marcos Univ, Peru, 72. Prof Exp: Instr vet anat, Univ Minn, 51-52, asst vet physiol, 52-55, res assoc, 58-60; vet physiologist, Agr Res Serv, US Dept Agr, 60-61; assoc prof, 61-66, PROF VET PHYSIOL, NY STATE VET COL, CORNELL UNIV, 66-, CHMN DEPT PHYSIOL, BIOCHEM & PHARMACOL, 73- Concurrent Pos: NIH spec res fel, 62-63 & res grant, 65-, dir training prog comp gastroenterol, 71-; field rep grad physiol, Cornell Univ, 68-70; mem gen med study sect, NIH, 69-73; Fulbright lectr, 72. Mem: Am Soc Vet Physiol & Pharmacol (pres, 67-68); Am Physiol Soc; Am Vet Med Asn; Conf Res Workers Animal Dis; Comp Gastroenterol Soc. Res: Comparative physiology of the stomach and large intestine; microbial digestion and mechanisms of secretion, absorption and digesta passage. Mailing Add: Dept of Physiol NY State Vet Col Cornell Univ Ithaca NY 14850

STEVENS, CHARLES F, b Chicago, Ill, Sept 1, 34; m 56; c 3. NEUROPHYSIOLOGY. Educ: Harvard Univ, BA, 56; Yale Univ, MD, 60; Rockefeller Univ, PhD, 64. Prof Exp: From asst prof to prof physiol & biophys, Sch Med, Univ Wash, 63-75; PROF PHYSIOL, SCH MED, YALE UNIV, 75- Res: Neural integration; synaptic physiology; properties of excitable membranes. Mailing Add: Dept of Physiol Yale Univ Sch Med 383 Cedar St New Haven CT 06510

STEVENS, CHARLES LE ROY, b Chicago, Ill, Aug 8, 31; m 52; c 4. BIOPHYSICAL CHEMISTRY. Educ: Valparaiso Univ, BA, 53; Univ Pittsburgh, MS, 60, PhD(biophys), 62. Prof Exp: Physicist, US Army Biol Labs, 55-56; NIH res fel, 62-64; asst prof, 64-67, ASSOC PROF BIOPHYS, UNIV PITTSBURGH, 67- Concurrent Pos: NATO sr fel, Univ Uppsala, 69. Mem: AAAS; NY Acad Sci; Biophys Soc. Res: Physical chemistry of proteins and nucleic acids; the role of water in the structure of biological macromolecules; self-association of proteins. Mailing Add: Dept of Biophys & Microbiol Univ of Pittsburgh Pittsburgh PA 15260

STEVENS, DALE JOHN, b Ogden, Utah, June 27, 36; m 62; c 4. PHYSICAL GEOGRAPHY. Educ: Brigham Young Univ, BA, 61; Ind Univ, MA, 63; Univ Calif, Los Angeles, PhD(geog), 69. Prof Exp: Instr geog, Univ Wyo, 63-64; ASSOC PROF GEOG, BRIGHAM YOUNG UNIV, 66- Concurrent Pos: Univ develop grant, Brigham Young Univ, 71-72. Mem: Asn Am Geog; Asn Pac Coast Geog. Res: Morphometric analysis of land forms, especially Karst and natural arches; noncommercial agricultural geography; geography of micro-climatology. Mailing Add: Dept of Geog Brigham Young Univ 167 HGB Provo UT 84602

STEVENS, DAVID ARTHUR, b Salem, NJ, June 22, 16; m 39; c 1. BACTERIOLOGY, VIROLOGY. Educ: Univ Pa, BA, 43. Prof Exp: Assoc res virol & bact, Sch Med, Western Reserve Univ, 48-61; res assoc, Pub Health Res Inst New York, 61-67; asst cur of viruses, 67-71, actg cur of viruses, 71-74, HEAD DEPT VIROL, AM TYPE CULT COLLECTION, 74- Mem: Am Soc Microbiol. Res: Epidemiology of infectious diseases; characterization and classification of viruses. Mailing Add: Am Type Cult Collection 12301 Parklawn Dr Rockville MD 20852

STEVENS, DEAN FINLEY, b Derby, Conn, Oct 19, 23; m 51; c 3. ZOOLOGY, CELL BIOLOGY. Educ: Boston Univ, AB, 49, AM, 50; Clark Univ, PhD(cell biol), 64. Prof Exp: Res asst biol, Boston Univ, 50-51; teaching master, Mt Hermon Sch Boys, 51-54; staff scientist, Worcester Found Exp Biol, 54-67; ASSOC PROF ZOOL, UNIV VT, 67- Concurrent Pos: Fels, USPHS, Am Cancer Soc & NIH; Ortho Res Found spec grant. Mem: Am Soc Cell Biol; Am Asn Cancer Res. Res: Vascular physiology; cancer; mechanisms of cell division. Mailing Add: Dept of Zool Univ of Vt Burlington VT 05401

STEVENS, DONALD KEITH, b Troy, NY, July 30, 22; m 45, 65, 74; c 2. SOLID STATE PHYSICS. Educ: Union Col, BS, 43; Univ NC, PhD(chem), 53. Prof Exp: Physicist, US Naval Res Lab, 43-49; consult radiation effects in solids, Oak Ridge Nat Lab, 49-51, physicist, 53-57; chief metall & mat br, Div Res, US AEC, 57-60, asst dir res metall & mat progs, 60-74, ASST DIR MAT SCI, DIV PHYS RES, US ENERGY RES & DEVELOP ADMIN, 74- Concurrent Pos: Mem mat adv bd, Nat Acad Sci-Nat Res Coun, 59-62. Mem: AAAS; Sci Res Soc Am; Am Phys Soc. Res: Radiation effects in solids. Mailing Add: Div of Phys Res US Energy Res & Develop Admin Washington DC 20545

STEVENS, E PHILIP, pharmaceutical chemistry, see 12th edition

STEVENS, ERNEST DONALD, b Calgary, Alta, July 5, 41; m 64; c 3. PHYSIOLOGY, ZOOLOGY. Educ: Victoria Univ, BSc, 63; Univ BC, MSc, 65, PhD(zool), 68. Prof Exp: Assoc prof zool, Univ Hawaii, 68-75; MEM FAC ZOOL, UNIV GUELPH, 75- Mem: AAAS; Can Soc Zool; Soc Exp Biol & Med. Res: Physiology, primarily of fish; mechanisms of respiration and circulation, especially as they are affected by muscular exercise. Mailing Add: Dept of Zool Univ of Guelph Guelph ON Can

STEVENS, EVELYN VICTORIA, b London, Eng, May 9, 40. ENZYMOLOGY. Educ: Univ Buenos Aires, MS, 65, PhD(biochem), 71. Prof Exp: Teaching & res asst, Dept Plant Biochem, Sch Pharm & Biochem, Univ Buenos Aires, 66-71; res assoc biochem, Rockefeller Univ, 71-73; RES ASSOC ENZYMOL, DEPT CHEM, UNIV DEL, 73- Mem: AAAS; Am Chem Soc. Res: Mechanism of enzyme function; mechanisms of catalysis and control; structure structure-function relationships; structure of multi-subunit enzymes; chemical modification of essential amino acid residues. Mailing Add: Dept of Chem Univ of Del Newark DE 19711

STEVENS, FRANK JOSEPH, b Peru, Ill, Mar 9, 19; m; c 2. CHEMISTRY. Educ: Univ Ill, BS, 41; Iowa State Col, PhD(bio-org chem), 47. Prof Exp: Instr chem, Iowa State Col, 42-47; from asst prof to assoc prof, 47-59, PROF CHEM, AUBURN UNIV, 59- Concurrent Pos: Chmn, Premed-Predent Adv Comt, Auburn Univ, 69- Mem: AAAS; Am Chem Soc; NY Acad Sci. Res: Organic and pharmaceutical chemistry; plant growth regulators; indole and pyridazine derivatives. Mailing Add: Dept of Chem Saunders Lab Auburn Univ Auburn AL 36830

STEVENS, FRITS CHRISTIAAN, b Ghent, Belg, Sept 18, 38; m 65; c 2. BIOCHEMISTRY, PROTEIN CHEMISTRY. Educ: Univ Ghent, Lic chem, 59; Univ Calif, Davis, PhD(biochem), 63. Prof Exp: Asst biochem, Pharmaceut Inst, Univ Ghent, 59-60 & Univ Calif, Davis, 60-63; sr researcher, Univ Brussels, 63-64; fel, Univ Calif, Los Angeles, 65-67; asst prof, 67-71, ASSOC PROF BIOCHEM, UNIV MAN, 71- Mem: Can Biochem Soc; Belg Biochem Soc; Am Chem Soc; Am Soc Biol Chem. Res: Structure-function relationships in proteins. Mailing Add: Dept of Biochem Univ of Man Fac of Med Winnipeg MB Can

STEVENS, GEORGE PUTNAM, b Hammond, Ind, Nov 1, 18; m 44; c 1. ECONOMIC GEOGRAPHY. Educ: Ind Univ, BS, 41; Univ Wis, PhM, 45, PhD(geog), 62. Prof Exp: Instr geog, Okla State Univ, 46-49; PROF GEOG, NORTHERN ILL UNIV, 50- Mem: Asn Am Geogr; Am Geog Soc; Am Cong Surv & Mapping. Res: Middle East petroleum and agriculture; urban land use in northern Illinois. Mailing Add: Dept of Geog Northern Ill Univ De Kalb IL 60115

STEVENS, GEORGE RICHARD, b Norfolk, Va, May 28, 31; m; c 4. STRUCTURAL GEOLOGY, TECTONICS. Educ: Johns Hopkins Univ, AB, 54, MA, 55, PhD(geol, tectonics), 59. Prof Exp: Asst prof geol, Lafayette Col, 57-66, asst acad dean, 64-65; PROF GEOL & HEAD DEPT, ACADIA UNIV, 66- Concurrent Pos: Consult, 55-67. Mem: AAAS; fel Geol Soc Am; Am Geophys Union. Res: Petrology; metamorphic recrystallization; fabric of deformed rock and of igneous rock. Mailing Add: Dept of Geol Acadia Univ Wolfville NS Can

STEVENS, HAROLD, b Salem, NJ, Oct 18, 11; m 38; c 2. NEUROLOGY. Educ: Pa State Univ, BS, 33; Univ Pa, AM, 34, PhD, 37, MD, 41. Prof Exp: PROF NEUROL, SCH MED, GEORGE WASHINGTON UNIV, 54- Concurrent Pos: Consult pediat neurol, DC Health Dept, 46-; sr attend neurologist, Children's Hosp, 51-; consult, Vet Hosp, 54- & Walter Reed Hosp & NIH, 57-; nat consult to Surg Gen, US Air Force; consult, FDA. Mem: Am Asn Neurol Surg; Am Neurol Asn; Am Electroencephalog Soc; fel Am Col Physicians; fel Am Acad Neurol. Res: Clinical and pediatric neurology. Mailing Add: 3301 New Mexico Ave NW Washington DC 20016

STEVENS, HARRY NELSON, b Port Huron, Mich, May 6, 14; m 37; c 3. RUBBER CHEMISTRY. Educ: Yale Univ, BS, 35. Prof Exp: Chemist, B F Goodrich Co, 35-40, patent attorney, 40-42, res coordr, 42-52, dir res, 52-58, secy res & develop policy coun, 58-64, tech dir, 65-73; RETIRED. Mem: Am Chem Soc. Res: Plastics; synthetic rubber; fiber forming polymers. Mailing Add: 9135 Highland Dr Brecksville OH 44141

STEVENS, HENRY CONRAD, b Vienna, Austria, Apr 17, 18; nat US; m 41; c 2. ORGANIC CHEMISTRY. Educ: Columbia Univ, BS, 41; Western Reserve Univ, MS, 49, PhD(chem), 51. Prof Exp: Res chemist, H Kohnstamm & Co, 41-42; res supvr, Chem Div, Pittsburgh Plate Glass Co, 42-72, RES SUPVR, CHEM DIV, PPG INDUSTS, INC, BARBERTON, 72- Mem: Am Chem Soc. Res: Chemistry of phosgene derivatives; free radical polymerization; polycarbonate resins; cycloadditions; tropolone syntheses; peroxides; epoxides; phase transfer catalysis. Mailing Add: 149 Birdwood Rd Akron OH 44313

STEVENS, JACK G, physics, see 12th edition

STEVENS, JACK GERALD, b Port Angeles, Wash, Nov 3, 33; m 57; c 1. VIROLOGY, EXPERIMENTAL PATHOLOGY. Educ: Wash State Univ, DVM, 57; Colo State Univ, MS, 59; Univ Wash, PhD(virol), 62. Prof Exp: Asst prof microbiol, Wash State Univ, 62-63; from asst prof to assoc prof med microbiol & immunol, 63-73, PROF MED MICROBIOL & IMMUNOL, SCH MED, UNIV CALIF, LOS ANGELES, 73- Mem: AAAS; Am Soc Microbiol; Infectious Dis Soc Am; Am Asn Immunologists; Am Soc Exp Pathologists. Res: Viral pathogenesis, particularly latent infections; diseases of the nervous system; neoplasms. Mailing Add: Dept of Med Microbiol & Immunol Univ of Calif Sch of Med Los Angeles CA 90024

STEVENS, JANICE R, b Portland, Ore; m 45; c 2. NEUROLOGY, PSYCHIATRY. Educ: Reed Col, BA, 44; Boston Univ, MD, 49. Prof Exp: Intern med, Mass Mem Hosp, 49-50; resident neurol, Boston City Hosp, 50-51; fel neurol & assoc physician, Sch Med, Yale Univ, 51-54; resident, 54-55, from instr to assoc prof, 55-71, PROF NEUROL, MED SCH, UNIV ORE, 71-, ASSOC PROF PSYCHIAT, 74- Concurrent Pos: Vis prof psychiat, Harvard Med Sch & Mass Gen Hosp, 71-73; NIMH guest worker, St Elizabeth's Hosp, Washington, DC, 75-76. Mem: AAAS; Am Acad Neurol; Am Electroencephalog Soc (pres, 73-74); Am Epilepsy Soc. Res: Neurology and electroencephalography of behavior and epilepsy. Mailing Add: Depts of Neurol & Psychiat Univ of Ore Med Sch Portland OR 97201

STEVENS, JERRY BRUCE, b Browerville, Minn, Apr 1, 32; m 51; c 6. VETERINARY PATHOLOGY, BIOCHEMISTRY. Educ: Univ Minn, BS, 59, DVM, 61, PhD(vet path), 67. Prof Exp: USPHS fel, Univ Minn, 61-67, asst prof vet path, 67-69; med scientist, Brookhaven Nat Lab, 69; PROF VET PATH, COL VET MED, UNIV MINN, ST PAUL, 69- Mem: Am Soc Vet Clin Path. Res: Blood cell kinetics; clinical biochemistry; metabolic and cellular profile testing of dairy cattle. Mailing Add: Dept of Path Univ of Minn Col of Vet Med St Paul MN 55108

STEVENS, JOHN CHARLES, b Ft Collins, Colo, July 19, 46; m 67; c 3. MAGNETOHYDRODYNAMICS, X-RAY ASTRONOMY. Educ: Calif Inst Technol, BS, 68, PhD(physics), 72. Prof Exp: DESIGN PHYSICIST, LAWRENCE LIVERMORE LAB, 72- Res: Magnetohydrodynamics of low temperature, high density, high beta plasmas. Mailing Add: Lawrence Livermore Lab Univ of Calif PO Box 808 Livermore CA 94550

STEVENS, JOHN GEHRET, b Mount Holly, NJ, Dec 16, 41; m 63; c 3. PHYSICAL CHEMISTRY. Educ: NC State Univ, BS, 64, PhD(chem), 69. Prof Exp: Asst prof, 63-73, ASSOC PROF CHEM, UNIV NC, ASHEVILLE, 73- Concurrent Pos: Mem ad hoc comt Mössbauer spectros data & conv, Nat Acad Sci, 70-73; ed Mössbauer Effect Data Index, Univ NC & Nat Bur Standards, 70-; res assoc, Max Planck Inst Solid State Physics, 73; mem exec comt, Int Comn Applns of Mössbauer Spectros, 75- Mem: AAAS; Am Chem Soc; Am Phys Soc; Fedn Am Sci. Res: Mössbauer spectroscopy; antimony chemistry; information sciences; evaluation of data. Mailing Add: Dept of Chem Univ of NC Asheville NC 28804

STEVENS, JOHN I, organic chemistry, see 12th edition

STEVENS, JOHN JOSEPH, b London, Eng, July 16, 41. CANCER. Educ: Univ Buenos Aires, MD, 64. Prof Exp: Res physician, Inst Biol & Exp Med, Buenos Aires, 64-67, Nat Acad Med, Arg, 65-67; res fel, 67-70, RES ASSOC, RES INST, HOSP JOINT DIS, NEW YORK, 70- Concurrent Pos: Instr, Dept Biochem, Mt Sinai Sch Med, City Univ New York, 70-73, res asst prof, 73-; spec fel, Leukemia Soc Am, 74-76, scholarship, 76-81. Mem: Endocrine Soc; Am Asn Cancer Res. Res: Mechanism of steroid hormone action; studies on glucocorticoid-induced lymphocytolysis of malignant lymphocytes; chemotherapy of cancer. Mailing Add: Res Inst of Hosp for Joint Dis 1919 Madison Ave New York NY 10035

STEVENS, JOSEPH ALFRED, b Cleveland, Ohio, Jan 3, 27. MEDICAL MYCOLOGY, MEDICAL MICROBIOLOGY. Educ: Univ Dayton, BS, 49; Mich State Univ, MS, 53, PhD(microbiol), 57. Prof Exp: Res instr & fel, Mich State Univ, 57-59, res assoc, 59-61; from instr to assoc prof, 61-71, actg chmn dept, 70-74, PROF MICROBIOL, CHICAGO COL OSTEOP MED, 71- Mem: Am Soc Microbiol; Mycol Soc Am; NY Acad Sci; Med Mycol Soc Am; Am Inst Biol Sci. Res: Immune-deficiency diseases; physiology of fungi. Mailing Add: Dept of Microbiol Chicago Col of Osteop Med Chicago IL 60615

STEVENS, JOSEPH CHARLES, b Grand Rapids, Mich, Feb 28, 29. EXPERIMENTAL PSYCHOLOGY. Educ: Calvin Col, AB, 50; Mich State Univ, 53; Harvard Univ, PhD(psychol), 57. Prof Exp: Res fel psychol, Harvard Univ, 57-59, from instr to asst prof, 59-66; LECTR & RES ASSOC, YALE UNIV, 66-, FEL, JOHN B PIERCE FOUND LAB, SCH MED, 66-, JONATHAN EDWARDS COL, 71- Mem: AAAS; Acoust Soc Am; Optical Soc Am; Psychonomic Soc; Soc Neurosci. Res: Psychophysics; psychology of sensory and perceptual processes. Mailing Add: John B Pierce Found Lab 290 Congress Ave New Haven CT 06519

STEVENS, KENNETH LLOYD, b Fresno, Calif, Mar 12, 37; m 59; c 4. NATURAL PRODUCTS CHEMISTRY. Educ: Fresno State Col, BS, 59, MA, 60; Univ Wash, Seattle, PhD(org chem), 63. Prof Exp: CHEMIST, US DEPT AGR, 63- Mem: Am Chem Soc; Phytochem Soc NAm. Res: Isolation, identification and syntheses of natural occurring aquatic herbicides. Mailing Add: 800 Buchanan St Albany CA 94710

STEVENS, LAURENCE GUY, b Detroit, Mich, Apr 22, 32; m 54; c 3. INORGANIC CHEMISTRY, EXTRACTIVE METALLURGY. Educ: Alma Col, BS, 54; Mich Technol Univ, MS, 58; Wayne State Univ, PhD(inorg chem), 62. Prof Exp: From asst prof to assoc prof chem, Mich Technol Univ, 61-65; sr proj scientist, Calumet Div, Calumet & Hecla, Inc, 65-67, dir res & develop, 67-69; res coordr, 69-74, DIR TECH DEVELOP MINERAL SCI DIV, UNIVERSAL OIL PROD CO, 74- Concurrent Pos: Consult, Calumet & Hecla, Inc, 64-65; Petrol Res Fund grant, 64-66. Mem: Am Inst Mining, Metall & Petrol Eng; Am Chem Soc. Res: Extractive metallurgy of copper, nickel and cobalt; hydrometallurgy; industrial copper chemistry; organogallium compounds; high vacuum techniques. Mailing Add: Minerals Sci Div UOP Inc 999 E Touhy Ave Des Plaines IL 60018

STEVENS, LEROY CARLTON, JR, b Kenmore, NY, June 5, 20; m 42; c 3. DEVELOPMENTAL BIOLOGY. Educ: Cornell Univ, BS, 42; Univ Rochester, PhD, 52. Prof Exp: Asst, Univ Rochester, 48-52, instr, Univ Sch, 51-52; res fel, 52-55, res assoc, 55-57, staff scientist, 57-67, SR STAFF SCIENTIST, JACKSON LAB, 67- Concurrent Pos: Guggenheim fel, Exp Embryol Lab, Col of France, 61-62. Mem: Am Soc Zool; Am Asn Anat; Am Asn Cancer Res: Soc Develop Biol. Res: Experimental embryology; cancer; mammalian embryology and teratocarcinogenesis. Mailing Add: Jackson Lab Bar Harbor ME 04609

STEVENS, LEWIS AXTELL, b Butte, Mont, Nov 17, 13; m 35; c 3. BIOPHYSICS. Educ: San Jose State Col, AB, 50. Prof Exp: Meteorol aid, Sci Serv Div, US Weather Bur, 51-52; physicist, Aviation Ord Dept, US Naval Ord Test Sta, 52-56, electronic scientist, 56-57, electronic scientist, Fuze Eval Div, Test Dept, 57-60; gen engr & head measurements br, 60-61, gen proj engr, 62, res physicist, Explosives & Pyro-Tech Div, Propulsion Develop Dept, 62-70; CONSULT PHYSICIST, 70- Concurrent Pos: Mem fuze field tests subcomt, Joint Army-Navy-Air Force, 61-64. Mem: AAAS; Am Phys Soc; Am Inst Aeronaut & Astronaut. Res: Development of new medical tools and techniques; nuclear physics; meteorological aspects of health physics; technology of high speed aerial tow targets; technology of soft lunar landings. Mailing Add: LASTEV Lab 725 Randall St Ridgecrest CA 93555

STEVENS, LLOYD WEAKLEY, b Philadelphia, Pa, Jan 14, 14; m 41; c 3. SURGERY. Educ: Univ Pa, AB, 33, MD, 37; Am Bd Surg, dipl, 44. Prof Exp: Assoc surg, Grad Sch Med, 46-49, assoc prof clin surg, Sch Med, 53-60, PROF CLIN SURG, SCH MED, UNIV PA, 60- Concurrent Pos: Assoc prof, Women's Med Col Pa, 46-49; dir surg, Presby-Univ Pa Med Ctr & assoc surgeon, Univ Hosp; chief surg, Philadelphia Gen Hosp. Mem: Am Col Surgeons; Soc Surg Alimentary Tract. Res: Acute cholecystitis; peptic ulcer; ulcerative colitis. Mailing Add: 316 Mill Creek Rd Haverford PA 19041

STEVENS, MALCOLM PETER, b Birmingham, Eng, Apr 3, 34; US citizen; m 60; c 2. ORGANIC POLYMER CHEMISTRY. Educ: San Jose State Col, BS, 57; Cornell Univ, PhD(org chem), 61. Prof Exp: Res chemist, Chevron Res Co, Standard Oil Co Calif, 61-64; asst prof chem, Robert Col, Istanbul, 64-67; asst prof, Univ Hartford, 67-68; from asst prof to assoc prof, Am Univ Beirut, 68-71; ASSOC PROF CHEM, UNIV HARTFORD, 71- Mem: Am Chem Soc. Res: Photopolymerization; condensation polymerization; polymer degradation; reaction mechanisms. Mailing Add: Dept of Chem Univ of Hartford 200 Bloomfield Ave West Hartford CT 06117

STEVENS, MERWIN ALLEN, b Mt Carmel, Utah, Aug 12, 35; m 60; c 2. PLANT GENETICS. Educ: Utah State Univ, BS, 57, MS, 61; Ore State Univ, PhD(hort, genetics), 67. Prof Exp: Soil scientist, Soil Conserv Serv, US Dept Agr, 60-61; county exten agent, Exten Serv, Ore State Univ, 61-64, asst hort, Univ, 64-67; res assoc, Campbell Inst Agr Res, NJ, 67-70; asst geneticist, 70-74, ASSOC GENETICIST, DEPT VEG CROPS, UNIV CALIF, DAVIS, 74- Honors & Awards: Nat Canners Asn Award, 68; Asgrow Award, 71; Campbell Award, 73. Mem: AAAS; Am Soc Hort Sci; Inst Food Technol. Res: Genetics and chemistry of tomato quality; genetics and physiology of processes limiting yield and quality in tomatoes; breeding tomatoes for processing. Mailing Add: Dept of Veg Crops Univ of Calif Davis CA 95616

STEVENS, MICHAEL FRED, b Urbana, Ill, May 17, 41; m 65; c 2. INDUSTRIAL CHEMISTRY. Educ: Eastern Ill Univ, BS, 64; Univ Ill, Urbana, MS, 66; Univ Nebr, Lincoln, PhD(org chem), 70. Prof Exp: Res assoc, 70-73, SR RES ASSOC ORG CHEM, APPLETON PAPERS INC, NAT CASH REGISTER CO, 73- Mem: Am Chem Soc. Res: Applied research in the fields of microencapsulation and coacervation; coating technology. Mailing Add: Lawe St Lab Res Dept Appleton Papers Inc Appleton WI 54911

STEVENS, MICHAEL THOMAS, b London, Eng, Aug 28, 35; US citizen; m 65; c 2. PHYSICS. Educ: Queens Col, NY, BS, 58; Polytech Inst Brooklyn, MS, 61, PhD(physics), 69. Prof Exp: PHYSICIST, ZENITH RADIO CORP, 69- Mem: Am Phys Soc. Res: Solid state physics, specifically opto-electronic properties of materials. Mailing Add: Zenith Radio Corp 6001 W Dickens Ave Chicago IL 60639

STEVENS, NELSON PIERCE, b Haverhill, Mass, May 23, 12; m 42; c 2. GEOCHEMISTRY. Educ: Univ Mass, BS, 35, MS, 37. Prof Exp: Instr chem, Univ Mass, 37-38; res assoc, Magnolia Petrol Co, 40-56; mgr explor res div, Field Res Lab, Mobil Oil Corp, Tex, 56-66, asst to west coast region explor mgr, 66-68, EXPLOR & PROD RES ADV, MOBIL RES & DEVELOP CORP, 68- Mem: Am Chem Soc; Geochem Soc; Am Asn Petrol Geol. Res: Petroleum geochemistry and exploration. Mailing Add: Mobil Res & Develop Corp 150 E 42nd St New York NY 10017

STEVENS, RAYMOND SAWTELL, b Nashua, NH, Apr 15, 94; m 20; c 2. RESEARCH ADMINISTRATION. Educ: Mass Inst Technol, BS, 17. Prof Exp: Asst supt, Isko Co, Ill, 19-20; from serv mgr to vpres, Arthur D Little, Inc, 20-56, pres, 56-60, 61-62, chmn exec comt, 60-62, chmn bd, 62-64, consult, 64-74; RETIRED. Concurrent Pos: Dir surv res in indust, Nat Res Coun-Nat Resources Planning Bd, 40-41; consult, War Prod Bd, 42; chmn comt qm probs, Nat Res Coun-Off Qm Gen, 43-58, chmn, 55-58; mem corp, Mass Inst Technol, 50-55; trustee, Gordon Res Conf, 55-58, chmn, 57-58; pres corp, Woods Hole Oceanog Inst, 55-61; pres, Mass Small Bus Investment Co, 63,64, chmn, 65-; Honors & Awards: Award, Am Inst Chem, 56; Outstanding Civilian Serv Medal, US Dept Army, 60. Mem: AAAS; Am Chem Soc; Am Inst Chem; Inst Food Technol; Soc Chem Indust. Res: Research administration and policy. Mailing Add: 100 Memorial Dr Apt 5-3B Cambridge MA 02142

STEVENS, RICHARD EDWARD, b Minneapolis, Minn, Mar 5, 31; m 53; c 2. GEOGRAPHY. Educ: Concordia Teachers Col, Nebr, BS, 52; Univ Colo, MA, 58; Univ Kans, PhD(geog), 62. Prof Exp: Teacher parochial sch, Tex, 52-54; asst prof geog, Nebr State Teachers Col, 61-63; from asst prof to assoc prof, 63-75, PROF GEOG, UNIV COLO, DENVER, 75- Concurrent Pos: Fulbright lectr, Univ Botswania, Lesotho & Swaziland, 67-68. Mem: AAAS; Asn Am Geog. Res: Agricultural geography, especially land tenure and agricultural practices among primitive peoples; microclimatology and production of crops. Mailing Add: Dept of Geog Univ of Colo Denver CO 80202

STEVENS, RICHARD EDWARD, b Washington, DC, Oct 30, 32; m 57; c 3. WATER CHEMISTRY. Educ: Washington Col, BS, 54; Pa State Univ, MS, 56; Rensselaer Polytech Inst, PhD(phys chem), 75. Prof Exp: Asst chemist, Pa State Univ, 54-56 & Univ Colo, 56-59; scientist-chemist, Rocky Flats Div, Dow Chem Co, Colo, 59-62, develop chemist, 62-64; res chemist, Am Cyanamid Co, 64-67; microscopist, Ernest F Fullam, Inc, NY, 67-74; sr res microscopist, Walter C McCrone Assocs, 74-76; SR CHEMIST, NALCO CHEM CO, 76- Mem: AAAS; Am Chem Soc; Electron Micros Soc Am. Res: Chemical and electron microscopy; crystallography; ultramicroanalysis; correlation of physical properties of solids with performance; the microscope in archaeology and art conservation; nondestructive analytical methods; water quality. Mailing Add: 747 S Lombard Ave Oak Park IL 60304

STEVENS, RICHARD JOSEPH, b Rochester, NY, Oct 31, 41; m 65; c 2. NEUROSCIENCES. Educ: Univ Rochester, BS, 63; Univ Ill, Urbana, MS, 65, PhD(biophysics), 69. Prof Exp: Aerospace technologist, NASA-Lewis Res Ctr, 63; res asst, Dept Physics, Univ Ill, 64-65, teaching asst human & cellular physiol, 66-67; fel neuroanat, Dept Anat, Brain Res Inst, Univ Calif, Los Angeles, 69-70; asst prof, 70-75, ASSOC PROF HUMAN ADAPTABILITY, COL HUMAN BIOL, UNIV WIS-GREEN BAY, 75- Concurrent Pos: President's teaching improvement grant, Univ

Wis, 72; consult, Green Bay Childbirth Educ Asn, 75- Mem: AAAS; Sigma Xi. Res: Neurophysiology of vision, neuro-behavioral aspects of environmental contaminants; neuro-behavioral aspects of pain perception; inovative teaching of biology. Mailing Add: Col of Human Biol Univ of Wis Green Bay WI 54302

STEVENS, RICHARD S, b Cranston, RI, Mar 22, 25; m 52; c 3. MARINE GEOLOGY, PHYSICAL OCEANOGRAPHY. Educ: Brown Univ, AB, 50. Prof Exp: Oceanogr, US Naval Oceanog Off, 52-62, OCEANOGR, OFF NAVAL RES, 62- Mem: Am Geophys Union; Am Soc Limnol & Oceanog; Sigma Xi. Res: Geological oceanography; ocean circulation, ocean bottom processes and their effects on sediment distribution and structure. Mailing Add: Off of Naval Res 715 Broadway New York NY 10003

STEVENS, ROBERT E, b Medford, Ore, Jan 19, 28; m 51; c 2. FOREST ENTOMOLOGY. Educ: Ore State Univ, BS, 51; Univ Calif, MS, 58, PhD(entom), 65. Prof Exp: Entomologist, Ore State Bd Forestry, 51-52 & Bur Entom & Plant Quarantine, Calif, 52-54; entomologist, Pac Southwest Forest & Range Exp Sta, 54-65, asst dir forest insect res, Washington, DC, 65-68, ENTOMOLOGIST & PROJ LEADER, ROCKY MT FOREST & RANGE EXP STA, US FOREST SERV, 68- Mem: Entom Soc Am. Res: Ecology and control of forest insects. Mailing Add: Rocky Mt Forest & Range Exp Sta US Forest Serv 240 W Prospect St Ft Collins CO 80521

STEVENS, ROBERT R, b Orleans, Ind, Aug 31, 35; m 64. MATHEMATICS. Educ: Purdue Univ, BSEE, 58, MS, 60; Univ Ariz, PhD(math), 65. Prof Exp: Asst prof math, Calif State Col Hayward, 65-66 & Colo State Univ, 66-67; asst prof, 67-73, ASSOC PROF MATH, UNIV MONT, 73- Concurrent Pos: Mem staff, NASA, 69. Mem: Am Math Soc; Math Asn Am. Res: Qualitative theory of differential equations; partial differential equations; functional analysis. Mailing Add: Dept of Math Univ of Mont Missoula MT 59801

STEVENS, ROBERT VELMAN, b Mason City, Iowa, Mar 24, 41; m 63; c 1. ORGANIC CHEMISTRY. Educ: Iowa State Univ, BS, 63; Ind Univ, PhD(org chem), 66. Prof Exp: From asst prof to assoc prof, 66-72, PROF ORG CHEM, RICE UNIV, 72- Concurrent Pos: A P Sloan fel; consult, Merck Sharpe & Dohme Res Labs, 74-; consult, Nat Inst Health, 75-79. Mem: Am Chem Soc; The Chem Soc. Res: Synthesis of complex natural products and development of new synthetic methods. Mailing Add: Dept of Chem Rice Univ Houston TX 77001

STEVENS, ROBERT WILLIAM CLARK, genetics, see 12th edition

STEVENS, ROSEMARY ANNE, b Bourne, Eng, Mar 18, 35; US citizen. PUBLIC HEALTH. Educ: Oxford Univ, BA, 57, MA, 61; Univ Manchester, dipl social admin, 59; Yale Univ, MPH, 63, PhD(epidemiol), 68. Prof Exp: Res asst, 62-65, res assoc, 66-68, from asst prof to assoc prof, 68-74, PROF PUB HEALTH, SCH MED, YALE UNIV, 74- Concurrent Pos: Hon res officer, London Sch Econ & Polit Sci, 62-63, vis lectr, 63-64; lectr, Sch Pub Health, Johns Hopkins Univ, 67-68; guest scholar, Brookings Inst, 67-68; consult, Dept Health, Educ & Welfare, 68- Mem: Am Pub Health Asn; Soc Social Hist Med. Res: History of medicine; comparative studies in health care policy; physician migration to the United States. Mailing Add: Dept of Epidemiol & Pub Health Yale Univ Sch of Med New Haven CT 06510

STEVENS, ROY WHITE, b Troy, NY, Sept 4, 34; m 56; c 2. MEDICAL MICROBIOLOGY, IMMUNOLOGY. Educ: State Univ NY Albany, BS, 56, MS, 58; Albany Med Col, PhD(microbiol), 65. Prof Exp: Bacteriologist, 58-61, sr bacteriologist, 62-65, assoc bacteriologist, 65-67, sr assoc prin res scientist immunol, 67-73, PRIN RES SCIENTIST IMMUNOL, NY STATE DEPT HEALTH, 73- Mem: AAAS; Am Soc Microbiol. Res: Diagnostic immunology, serology; medical microbiology. Mailing Add: Div of Labs & Res NY State Dept of Health Albany NY 12201

STEVENS, RUSSELL BRADFORD, b Washington, DC, Oct 31, 15; m 49; c 3. PLANT PATHOLOGY. Educ: Univ Va, BS, 37; Univ Wis, PhD(bot), 40. Prof Exp: Asst prof biol, Birmingham-Southern Col, 40-42 & Univ Louisville, 46; assoc prof bot, Auburn Univ, 46-47, Univ Tenn, 47-51 & US Govt, 51-54; exec secy biol coun, Nat Res Coun, 54-57; prof bot, George Washington Univ, 57-66; EXEC SECY DIV BIOL SCI, NAT RES COUN, 64- Mem: AAAS; Mycol Soc Am; Bot Soc Am; Am Phytopath Soc; Soc Econ Bot. Res: Phytopathology; general mycology and botany; epidemiology. Mailing Add: Div of Biol Sci Nat Res Coun 2101 Constitution Washington DC 20418

STEVENS, SANDRA, b Philadelphia, Pa, Sept 6, 47. SYNTHETIC ORGANIC CHEMISTRY. Educ: Dickinson Col, BS, 69; Princeton Univ, MA, 71, PhD(chem), 73. Prof Exp: RES CHEMIST, JACKSON LAB, E I DU PONT DE NEMOURS & CO, INC, 73- Mem: Am Chem Soc. Res: Synthesis and development of organic dyes. Mailing Add: Jackson Lab Org Chem E I du Pont de Nemours & Co Wilmington DE 19898

STEVENS, STANLEY SMITH, psychophysics, deceased

STEVENS, SUE CASSELL, b Roanoke, Va. BIOCHEMISTRY. Educ: Goucher Col, BA, 30; Columbia Univ, MA, 31, PhD(chem), 40. Prof Exp: Res biochemist, NY Skin & Cancer Hosp, New York, 32-35; biochemist, Fifth Ave Hosp, 35; res chemist, Col Physicians & Surgeons, Columbia Univ, 35-39, NY Orthop Hosp, 40-41 & Calif Milk Prod Co, 41-43; res dairy chemist, Golden State Co, Ltd, 43-46 & Swift & Co, 46-47; dir res & qual control, Steven Candy Kitchens, 47-48; assoc prof chem & biol, MacMurray Col, 48-49; chief biochemist, US Vet Admin Ctr, Dayton, Ohio, 49-52, res biochemist, 52-56, supvr res lab Hosp, Lincoln, Nebr, 56-65; DIR DIV ENDOCRINE CHEM, JEWISH HOSP ST LOUIS, 65- Concurrent Pos: Asst prof path, Sch Med, Wash Univ, 67- Mem: Fel AAAS; fel Am Inst Chem; Am Soc Qual Control; NY Acad Sci; Am Chem Soc. Res: Clinical chemistry methods; electrolytes in biological fluids; steroids; hormones; automation. Mailing Add: PO Box 4854 Field Station St Louis MO 63108

STEVENS, THOMAS MCCONNELL, b Plainfield, NJ, May 25, 27; m 54; c 4. VIROLOGY, ENTOMOLOGY. Educ: Haverford Col, BA, 50; Rutgers Univ, MS, 55, PhD(entom), 57. Prof Exp: Fel microbiol, St Louis Univ, 57-58, from instr to asst prof, Sch Med, 58-63; asst prof, Med Sch, Rutgers Univ, New Brunswick, 63-66; assoc prof exp med & from assoc dir to dir teaching labs, 66-72, PROF MICROBIOL, RUTGERS MED SCH, COL MED & DENT NJ, 72-, ASST DEAN, 72- Mem: Am Soc Microbiol. Res: Physical and chemical nature of the togaviruses using dengue virus as a model. Mailing Add: Rutgers Med Sch Col of Med & Dent of NJ Piscataway NJ 08854

STEVENS, TRAVIS EDWARD, b Leigh, Nebr, Dec 22, 27; m 61; c 4. ORGANIC CHEMISTRY. Educ: Wayne State Col, AB, 51; Iowa State Col, PhD(chem), 55. Prof Exp: Asst, Iowa State Col, 51-53; SR RES CHEMIST, ROHM AND HAAS CO, 55- Concurrent Pos: Vis prof, Ind Univ, 65. Mem: Am Chem Soc; The Chem Soc. Res: Synthesis and properties of high-energy compounds; molecular rearrangements; polymer synthesis; paper chemicals; coatings and textile chemistry. Mailing Add: Rohm and Haas Co Spring House PA 19477

STEVENS, VERNON CECIL, b Greenup, Ky, Aug 11, 35; m 65; c 2. PHYSIOLOGY. Educ: Colo State Univ, BS, 57; Ohio State Univ, MSc, 58, PhD, 62. Prof Exp: Dir bioassay labs, 62-64, DIR DIV REPROD BIOL, DEPT OBSTET & GYNEC, OHIO STATE UNIV, 64- Concurrent Pos: Coordr task force pop & fertil control, WHO, 72-; consult contraception, Battelle Mem Inst, 74- Mem: Endocrine Soc; Soc Study Reprod; Brit Soc Endocrinol; Am Fertility Soc; World Pop Soc. Res: Reproduction; endocrinology; immunology; development of methods for human fertility control. Mailing Add: Dept of Obstet & Gynec Ohio State Univ Hosps Columbus OH 43210

STEVENS, VERNON LEWIS, b Tacoma, Wash, Oct 10, 30; m 54; c 4. BIOCHEMISTRY, ANALYTICAL CHEMISTRY. Educ: Cent Wash State Col, BS, 57; Ore State Univ, MS, 60. Prof Exp: Res asst & biochemist, William S Merrell Co Div, Richardson-Merrell, Inc, 59-67; head anal chem, Enzomedic Lab, Inc, Wash, 67-69; CHEMIST, PUGET SOUND PLANT, TEXACO INC, 69- Mem: Am Chem Soc. Res: Development of analytical procedures for gas-liquid and thin layer chromatography, autoanalyzer, radioisotopes and spectronic equipment; lipid synthesis in animals; nucleotides; clinical, environmental and petroleum chemistry. Mailing Add: Puget Sound Plant Texaco Inc PO Box 622 Anacortes WA 98221

STEVENS, VINCENT LEROY, b Boston, Mass, July 14, 30; m 58. BIOCHEMISTRY. Educ: Univ Calif, Berkeley, AB, 53, PhD(biochem), 57. Prof Exp: Jr res biochemist, Med Ctr, Univ Calif, San Francisco, 57-59; asst prof chem & biochem, 59-62, assoc prof chem, 62-67, PROF CHEM, EASTERN WASH STATE COL, 67-, CHMN DEPT, 70-, DEAN, DIV HEALTH SCI, 74- Concurrent Pos: Consult, Deaconess Hosp, Spokane, 62- Mem: Am Chem Soc. Res: Organic and physical chemistry of nucleic acids and their derivatives. Mailing Add: Dept of Chem Eastern Wash State Col Cheney WA 99004

STEVENS, WALTER, b Salt Lake City, Utah, Dec 6, 33; m 55; c 4. ANATOMY, RADIOBIOLOGY. Educ: Univ Utah, BS, 56, PhD(anat, radiobiol), 62. Prof Exp: From instr to assoc prof, 62-74, PROF ANAT, UNIV UTAH, 74-, HEAD CHEM GROUP, ANAT & RADIOBIOL DIV, 70-, DIR, NAT INST GEN MED SCI TRAINING GRANT, 74- Mem: Endocrine Soc; Am Asn Anatomists; Soc Neurosci; Am Physiol Soc; Radiation Res Soc. Res: Mechanism of action of glucocorticoids in lymphoid tissues, central nervous system and lung; interaction of transuranic elements with biological systems. Mailing Add: Dept of Anat 2C110 Med Ctr Univ of Utah Salt Lake City UT 84132

STEVENS, WALTER JOSEPH, b Atlantic City, NJ, Apr 29, 44; m 66; c 2. THEORETICAL CHEMISTRY, CHEMICAL PHYSICS. Educ: Drexel Univ, BS, 67; Ind Univ, Bloomington, PhD(chem physics), 71. Prof Exp: NSF fel, Argonne Nat Lab, 71-72, lab fel, 72-74; MEM STAFF, TIME & ENERGY DIV, NAT BUR STANDARDS, 74- Mem: AAAS; Am Chem Soc; Am Phys Soc. Res: Quantum chemistry; ab initio calculation of molecular wavefunctions and properties; computer applications to chemical education. Mailing Add: Time/Energy Div Radio Bldg 3540 Nat Bur of Standards Boulder CO 80302

STEVENS, WILLIAM CLARK, b Richland, Tex, Mar 24, 21; m 44; c 2. MICROBIOLOGY, CELL PHYSIOLOGY. Educ: Harding Col, BS, 49; Univ Ark, MA, 51; Vanderbilt Univ, PhD(biol), 56. Prof Exp: Instr sci, Beebe Jr Col, 47-49; asst prof biol, Harding Col, 50-52, prof, 55-66; instr, Vanderbilt Univ, 54-55; PROF & HEAD DEPT BIOL, ABILENE CHRISTIAN COL, 66- Concurrent Pos: NIH res fel, 62-63. Res: Bacterial physiology and biochemistry; animal virology; microbiology of water. Mailing Add: 902 Scott Place Abilene TX 79601

STEVENS, WILLIAM GEORGE, b Champaign, Ill, Sept 20, 38; m 61; c 3. ELECTROCHEMISTRY, ANALYTICAL CHEMISTRY. Educ: Mass Inst Technol, BS, 61; Univ Wis-Madison, PhD(chem), 66. Prof Exp: Sr chemist, Corning Glass Works, 66-69; res specialist nonaqueous batteries, 69-72, RES SPECIALIST ANAL CHEM, RES & DEVELOP DIV, WHITTAKER CORP, 72- Mem: Am Chem Soc; Electrochem Soc; Inst Elec & Electronics Engrs. Res: Polymer characterization and physical properties of materials. Mailing Add: Res & Develop Div Whittaker Corp 3540 Aero Ct San Diego CA 92123

STEVENS, WILLIAM HARMER, b London, Ont, Apr 12, 18; m 42; c 5. ANALYTICAL CHEMISTRY. Educ: Queen's Univ, Ont, BSc, 40, MSc, 41; McMaster Univ, PhD(chem), 52. Prof Exp: Res officer, Dept Nat Defense, Royal Mil Col, 42-45; instr chem, Queen's Univ, 45-47; from asst res officer to assoc res officer, 47-55, SR RES OFFICER & HEAD GEN CHEM BR, ATOMIC ENERGY CAN, LTD, 55- Mem: Fel Chem Inst Can; Can Soc Chem Eng; The Chem Soc. Res: Physical organic chemistry; analytical chemistry of nuclear power development; infrared spectroscopy; radiation and isotope chemistry; electrochemistry; catalysis. Mailing Add: Gen Chem Br Nuclear Lab Atomic Energy of Can Ltd Chalk River ON Can

STEVENSON, ALDEN, b Santiago, Chile, Mar 4, 28; US citizen; m 53; c 4. SOLID STATE PHYSICS. Educ: Swarthmore Col, BA, 50; Univ Va, PhD(physics), 53. Prof Exp: Mem tech staff semiconductors, Honeywell Res Ctr, Minn, 53-56; Pac Semiconductors, Inc, Calif, 56-57, dir mat res, 57-59, res, 59-61; dir appl res solid state physics, Litton Systs, Inc, Calif, 61-72; dir eng, Semiconductor Prod Div, Motorola, Inc, 72-75; MGR SEMICONDUCTOR ENG, SIEMENS COMPONENTS GROUP, 75- Mem: Electrochem Soc. Res: Microelectronic research and development in semiconductors; thin-films; advanced computer design and large-scale integrated circuit design; lasers; batch-process memories; materials and process engineering; semiconductor research and development; memory technology. Mailing Add: Siemens Components Group 8700 E Thomas Rd Scottsdale AZ 85252

STEVENSON, ARTHUR CHARLES, b Coldwater, Mich, Aug 26, 11; m 34; c 2. RUBBER CHEMISTRY. Educ: Eastern Mich Univ, AB, 34; Univ Mich, MS, 40, PhD(chem), 42. Prof Exp: Teacher chem, Eastern Mich Univ, 33-34 & high sch, Mich, 34-39; res chemist, 42-48, res supvr, 48-52, div head neoprene res, 52-54, asst dir, ELASTOMERS LAB, 54-59, dir, 59-68, mkt mgr Adiprene, Viton & chem, 68-71, ASST TO DIR SALES, E I DU PONT DE NEMOURS & CO, INC, 71- Mem: AAAS; Am Chem Soc; Am Inst Chem; Soc Chem Indust. Res: Hormones; dyes; intermediates; synthetic rubber. Mailing Add: Elastomers Chem Dept E I du Pont de Nemours & Co Inc Wilmington DE 19898

STEVENSON, CHARLES EDWARD, b Mt Vernon, NY, Feb 26, 13; m 43; c 3. ORGANIC CHEMISTRY. Educ: Pa State Univ, BS, 34, MS, 37, PhD(org chem), 41. Prof Exp: Asst, Pa State Univ, 34-42; res chemist, Standard Oil Develop Corp, NJ, 42-45; chemist, Diamond Glass Co, Pa, 45-47; assoc dir, Chem Eng Div, Argonne Nat Lab, 47-53; tech dir, Idaho Chem Process Plant, Phillips Petrol Co, 54-60; supvr exp breeder reactor II, Fuel Cycle Facil, 60-65, proj mgr, 65-69, SR SCIENTIST, ARGONNE NAT LAB, 69- Concurrent Pos: Chmn comt nuclear fuel cycle, Am Nat

Standards Inst; vis scientist, Regulatory Stand Directorate, US Nuclear Regulatory Comn, 73-75. Mem: Am Chem Soc; Am Nuclear Soc; Am Inst Chem Eng. Res: Methylamines; lubrication oils; chemical engineering with radioactive substances; nuclear fuel processing and waste disposal; nuclear standards. Mailing Add: Chem Eng Div Argonne Nat Lab 9700 S Cass Ave Argonne IL 60439

STEVENSON, CHRIS G, b Edmonton, Alta, Mar 15, 07; US citizen; m 47; c 1. CHEMISTRY. Educ: Univ Wash, Seattle, BS(chem) & BS(librarianship), 40. Prof Exp: Asst dir, Statewide Library Proj, 40-42; county librn, Clark County Library, 43-47; mgr tech info, Hanford Atomic Prod Oper, Gen Elec Co, 47-64 & Pac Northwest Labs, Battelle Mem Inst, 64-70; ASSOC DIR PARTNERS OF THE AMERICAS PROG, CENT WASH STATE COL, 70- Concurrent Pos: Mem tech info panel, US Atomic Energy Comn, 48-70. Res: Utilization of data processing equipment in the management of technical information. Mailing Add: 2210 Humphrey St Richland WA 99352

STEVENSON, D RICHARD, b Windsor, Ont, Dec 11, 31; m 63; c 3. PHYSICS. Educ: Univ Toronto, BASc, 53; Univ Mich, MSE, 54; Mass Inst Technol, MechE, 56, ScD(mech eng), 57. Prof Exp: Lectr physics, Univ Western Ont, 57-59; from asst prof to assoc prof, 59-68, PROF PHYSICS & DIR MAGNET & CRYOGENICS LABS, McGILL UNIV, 68- Concurrent Pos: Can del, comt experts on res high magnetic fields, Orgn Econ Coop & Develop, 65-67; consult, Nat Magnet Lab, Mass Inst Technol, 67; pres, Can Superconductor & Cryogenics Co Ltd, 71- Mem: Am Phys Soc; Can Asn Physicists; hon foreign fel, Indian Cryogenics Coun. Res: Magnetic properties of solids; high pressure physics. Mailing Add: Magnet Lab McGill Univ 151 Boulevard Industriel Longueuil PQ Can

STEVENSON, DAVID, b Farnworth, Eng, Apr 27, 44; m 70. ORGANIC CHEMISTRY, BIOCHEMISTRY. Educ: Oxford Univ, BA, 66, MA & DPhil(org chem), 68. Prof Exp: Jr demonstr org chem, Dyson Perrins Lab, Oxford Univ, 66-68; res assoc, Univ Wash, 68-70; Welch res fel, Tex Tech Univ, 70-72; vis asst prof biochem & exp endocrinol, Univ Ala, Birmingham, 72-73; proj engr, Schwarz-Mann Div, Orangeburg, NY, 73-75, SR PROJ CHEMIST, CLAY-ADAMS DIV, BECTON, DICKINSON & CO, PARSIPPANY, NJ, 75- Mem: Am Chem Soc; The Chem Soc. Res: Improved methods for synthesis and purification of biologically active polypeptides; enzyme isolation and purification; development of improved reagents for clinical analyses; aliphatic, aromatic and heterocyclic syntheses especially related to dyes. Mailing Add: 5 Kristoffersen Ct Suffern NY 10901

STEVENSON, DAVID AUSTIN, b Albany, NY, Sept 6, 28; m 58; c 3. MATERIALS SCIENCE. Educ: Amherst Col, BA, 50; Mass Inst Technol, PhD(phys chem), 54. Prof Exp: Res assoc metall, Mass Inst Technol, 53-54, asst prof, 55-58; Fulbright scholar, Univ Munich, 54-55; PROF MAT SCI, STANFORD UNIV, 58- Concurrent Pos: Fulbright sr res fel, Max Planck Inst Phys Chem, 68, 69. Mem: Am Soc Metals; Am Inst Mining, Metall & Petrol Eng; Electrochem Soc. Res: Synthesis and properties of semiconducting materials and device applications; solid state electrochemistry; diffusion in compound semiconductors. Mailing Add: Dept of Mat Sci Stanford Univ Stanford CA 94305

STEVENSON, DAVID STUART, b Virden, Man, Jan 23, 24; m 46; c 2. SOIL PHYSICS. Educ: Univ BC, BSA, 51; Ore State Univ, MSc, 56, PhD(soils), 63. Prof Exp: Res officer, Dom Exp Farm, Can Dept Agr, Sask, 56-57, Agr Res Sta, Alta, 62-66, RES SCIENTIST SUMMERLAND RES STA, CAN DEPT AGR, 66- Mem: AAAS; Am Soc Agron; Soil Sci Soc Cam. Res: Irrigation; soil-water-plant growth relationship. Mailing Add: Can Agr Res Sta Summerland BC Can

STEVENSON, DENNIS A, b Mt Holly, NJ, Jan 25, 44; m 66; c 3. BIOPHYSICS. Educ: Gettysburg Col, BA, 66; Univ Del, MS, 68, PhD(physics), 72. Prof Exp: Teaching res asst physics, Univ Del, 66-72; res assoc biophys, Univ Pittsburgh, 72-73; ASST PROF PHYSICS, NORTHEAST LA UNIV, 73- Mem: Biophys Soc; Am Phys Soc. Res: Physical studies of biologically important macromolecules, particularly protein-nucleic acid interactions, virology, enzymes, and hemoglobin including effects of various ionizing radiations on the macromolecules as well as other physicochemical studies. Mailing Add: Dept of Physics Northeast La Univ Monroe LA 71201

STEVENSON, DONALD THOMAS, b Washington, DC, Sept 8, 23; m 46; c 3. SOLID STATE PHYSICS. Educ: Cornell Univ, AB, 44; Mass Inst Technol, PhD(physics), 50. Prof Exp: Asst physics, 49-50, res assoc, 50-51, mem staff, Lincoln Lab, 51-53, asst group leader solid state physics, 53-57, group leader, 57-61, ASST DIR, FRANCIS BITTER NAT MAGNET LAB, MASS INST TECHNOL, 60- Mem: AAAS; Am Phys Soc. Res: Semiconductors; high magnetic fields. Mailing Add: Francis Bitter Nat Magnet Lab NW 14-3218 Mass Inst of Technol Cambridge MA 02139

STEVENSON, ELMER CLARK, b Pine City, Wash, Aug 20, 15; m 39; c 6. HORTICULTURE. Educ: Univ Md, BS, 37; Univ Wis, PhD(agron, plant path), 42. Prof Exp: Asst plant path, Univ Wis, 38-42; from asst plant pathologist to plant pathologist, Drug Plant Invests, US Dept Agr, 42-48; from assoc prof to prof hort, Purdue Univ, 48-67, head dept, 58-67; PROF HORT, ASSOC DEAN AGR & DIR RESIDENT INSTRUCT, ORE STATE UNIV, 67- Concurrent Pos: Consult, US Dept Agr & Univ Ky, 58, US Agency Int Develop, Brazil, 62 & US Dept Agr & Miss State Univ, 64. Mem: Fel Am Soc Hort Sci. Res: Corn diseases and breeding; diseases of medicinal and special crops; mint breeding and production; vegetable breeding and genetics. Mailing Add: Sch of Agr Ore State Univ Corvallis OR 97331

STEVENSON, ENOLA L, b Feb 20, 39; US citizen. PLANT PHYSIOLOGY. Educ: Southern Univ, BS, 60; Univ NH, MS, 62, PhD(plant physiol), 68. Prof Exp: Res asst plant physiol, Univ NH, 60-62, 67-68; instr bot, Southern Univ, 62-64; asst prof, 68-72, ASSOC PROF BIOL, ATLANTA UNIV, 72- Res: Effects of light quality and intensity on plant growth and metabolism. Mailing Add: Dept of Biol Atlanta Univ Atlanta GA 30314

STEVENSON, EUGENE HAMILTON, b Chicago Heights, Ill, June 29, 19; m 43; c 3. ORGANIC CHEMISTRY, NUTRITION. Educ: Cornell Col, BS, 42; Ill Inst Technol, MS, 44. Prof Exp: Chemist, Swift & Co, Ill, 46-48; asst secy, Coun Foods & Nutrit, Am Med Asn, 48-55, actg secy, 55-57, assoc secy, 57-60; asst to dir, Div Nutrit, US Food & Drug Admin, 60-66, asst dir, 66-68; DIR NUTRIT PROD INFO, RES CTR, MEAD JOHNSON & CO, 68- Mem: AAAS; Am Chem Soc; Am Pub Health Asn; Inst Food Technol; Am Inst Chem. Res: Human nutrition; nutrient requirements; composition of foods; diet and nutritional status surveys; federal food regulations. Mailing Add: Res Ctr Mead Johnson & Co Evansville IN 47721

STEVENSON, EVERETT E, b Buffalo, NY, Jan 14, 23; m 45; c 3. MATHEMATICS. Educ: State Univ NY Col Buffalo, BS, 44; Univ Houston, MEd, 52; Ohio State Univ, PhD(math, math ed), 61. Prof Exp: From instr to assoc prof math, US Air Force Acad, 56-66, chief enrichment br, 66-67; fac mem, Indust Col Armed Forces, 67-68; PROF & ASSOC CHMN DEPT MATH SCI, MEMPHIS STATE UNIV, 69- Mem: Am Math Soc; Math Asn Am. Res: Mathematics education; complex variables; differential equations. Mailing Add: Dept of Math Sci Memphis State Univ Memphis TN 38152

STEVENSON, FORREST FREDERICK, b Kismet, Kans, Nov 12, 16; m 47; c 1. PLANT MORPHOLOGY. Educ: Cent Mo State Col, BS, 46; Univ Mo, MA, 48; Univ Mich, PhD(bot), 56. Prof Exp: Instr biol, Univ Kans City, 48-50 & McCook Jr Col, 50-51; from asst prof to assoc prof, 55-65, PROF BIOL, BALL STATE UNIV, 65- Mem: Bot Soc Am; Am Bryol & Lichenological Soc. Res: Experimental plant morphology. Mailing Add: Dept of Biol Ball State Univ Muncie IN 47306

STEVENSON, FRANK JAY, b Logan, Utah, Aug 2, 22; m 52; c 3. SOILS. Educ: Brigham Young Univ, BS, 49; Ohio State Univ, PhD(agron), 52. Prof Exp: From asst prof to assoc prof, 53-62, PROF SOIL BIOL, UNIV ILL, 62- Mem: Soil Sci Soc Am; Am Soc Agron; Geochem Soc. Res: Biochemical properties of soils; chemistry of soil organic matter. Mailing Add: Dept of Agronomy Univ of Ill Urbana IL 61801

STEVENSON, FRANK ROBERT, b Brooklyn, NY, Aug 29, 31; m 56; c 6. SOLID STATE PHYSICS. Educ: Polytech Inst Brooklyn, BS, 53, MS, 59. Prof Exp: Physicist, Sperry Gyroscope Co, NY, 53-56 & Curtis Wright Corp, Pa, 56-58; res assoc, RIAS Div, Martin Co, Md, 58-63; physicist, Lewis Res Ctr, NASA, Ohio, 63-71; ENVIRON SCIENTIST, FORD MOTOR CO, 71- Concurrent Pos: Lectr, Goucher Col, 61-62 & John Carroll Univ, 64-65. Mem: Am Phys Soc. Res: Radiation damage; microwave electronics; low temperature physics; accelerators; air pollution; industrial noise control; energy conservation; water pollution management. Mailing Add: Ford Motor Co PO Box 9898 Cleveland OH 44142

STEVENSON, GEORGE FRANKLIN, b St Thomas, Ont, Sept 13, 22; US citizen; m 45; c 2. PATHOLOGY. Educ: Univ Western Ont, BA, 44, MD, 45. Prof Exp: Prof clin path, dean sch allied health sci & dir med technol prog, Med Col SC, 66-71; dep comnr med technol, 65-67, comnr continuing educ, 67-72, exec vpres, 71-74, SCI DIR, AM SOC CLIN PATHOLOGISTS, 74-; PROF PATH, MED SCH, NORTHWESTERN UNIV, CHICAGO, 72- Mem: Asn Am Med Cols; Am Asn Pathologists & Bacteriologists; Col Am Pathologists; Am Soc Clin Pathologists; Asn Clin Sci (pres, 59-60). Res: Administrative medicine. Mailing Add: Am Soc of Clin Pathologists 2100 W Harrison St Chicago IL 60612

STEVENSON, GEORGE WILLIAM, b Salt Lake City, Utah, Dec 13, 24; m 47; c 4. TOXICOLOGY. Educ: Occidental Col, AB, 45; Univ Calif, MD, 48, MS, 53. Prof Exp: Intern, USPHS Hosp, San Francisco, Calif, 48-49, asst surgeon, Outpatient Dept, San Diego, 49-50; USPHS fel, Morphine Metab, Univ Calif, Berkeley, 53-55, asst prof pharmacol & toxicol, Sch Med, Univ Calif, Los Angeles, 55-61, lectr, Sch Pub Health, 58-61; chief toxicol div, Bio-Sci Labs, 61-67; DIR, BIO-ANAL INC, 67- Concurrent Pos: Consult, Beckman Instruments, Inc, Calif, 59-61. Mem: Am Chem Soc: Am Acad Forensic Sci; Am Asn Clin Chemists. Res: Chemical analysis of biological and other samples for drugs, poisons, elements and biochemicals; clinical and forensic toxicology; drug metabolism; relation of chemical, physical and biological properties of substances. Mailing Add: Bio-Anal Inc 1701 Berkeley St Santa Monica CA 90404

STEVENSON, HARLAN QUINN, b Waynesboro, Pa, Apr 1, 27; m 60; c 2. CYTOGENETICS, RADIOBIOLOGY. Educ: Pa State Univ, BS, 50; Univ Fla, PhD(radiation biol), 63. Prof Exp: Asst bot, Pa State Univ, 50-51 & Cornell Univ, 51-56; res assoc biol, Brookhaven Nat Lab, 56-60; asst prof, Univ Fla, 63-64; from asst prof to assoc prof, 64-72, PROF BIOL, SOUTHERN CONN STATE COL, 72-, CHMN DEPT, 75- Mem: AAAS; Genetics Soc Am; Soc Study Evolution; Am Inst Biol Sci; Am Soc Human Genetics. Res: Chemical and radiation induced chromosomal aberrations; genetic and radiation effects in plant tumors; evolutionary and practical significance of multiple allopolyploidy; cytotaxonomy and sex determination; genetic counseling; bioethics. Mailing Add: Dept of Biol Southern Conn State Col New Haven CT 06515

STEVENSON, HEBER JOHN RICHARDS, b Salt Lake City, Utah, May 6, 16; m 49; c 4. BIOPHYSICS, SCIENCE ADMINISTRATION. Educ: Univ Utah, AB, 41; Calif Inst Technol, MS, 42; Stanford Univ, MA, 48; Univ Cincinnati, ScD(indust health), 65. Prof Exp: Biophysicist, Chem Corps, US Army, 49-51 & US Pub Health Serv, 51-74; RETIRED. Res: The understanding and control through the use of physical sciences and engineering of conditions affecting living systems in community and clinical environments Mailing Add: 217 Cecilia Ct St Augustine Shores FL 32084

STEVENSON, HENRY MILLER, b Birmingham, Ala, Feb 25, 14; m 39; c 4. ORNITHOLOGY. Educ: Birmingham-Southern Col, AB, 35; Univ Ala, MS, 39; Cornell Univ, PhD(ornith), 43. Prof Exp: Lab asst geol, Birmingham-Southern Col, 35-36; lab asst biol, Univ Ala, 38-39; lab asst bot, Vanderbilt Univ, 40-41; lab asst ornith, Cornell Univ, 42-43; actg assoc prof biol, Univ Miss, 43-44; assoc prof, Emory & Henry Col, 44-46; from asst prof to prof, 46-75, EMER PROF ZOOL, FLA STATE UNIV, 75-; RES FEL, TALL TIMBERS RES STA, 75- Concurrent Pos: Consult, Conserv Consults, Inc, 73-; ed, Fla Field Naturalist. Mem: Wilson Ornith Soc; Nat Audubon Soc; Am Ornith Union. Res: Avian taxonomy; quantitative field studies of birds; geographical distribution and migration of birds. Mailing Add: Tall Timbers Res Sta Tallahassee FL 32303

STEVENSON, IAN, b Montreal, Que, Oct 31, 18; nat US; m 47. MEDICINE, PSYCHIATRY. Educ: McGill Univ, BSc, 40, MD, CM, 43. Prof Exp: Intern & asst resident med, Royal Victoria Hosp, Montreal, 44-45; from intern to resident, St Joseph's Hosp, Phoenix, Ariz, 45-46; fel internal med, Ochsner Med Found, New Orleans, La, 46-47; Commonwealth fel med, Med Col, Cornell Univ, 47-49; asst prof med & psychiat, Sch Med, La State Univ, 49-52, assoc prof psychiat, 52-57; prof psychiat & chmn dept neurol & psychiat, 57-67, CARLSON PROF PSYCHIAT, SCH MED, UNIV VA, 67- Concurrent Pos: Consult, New Orleans Parish Sch Bd, 49-52, State Dept Pub Welfare, 50-52 & Southeast La State Hosp, Mandeville, 52-57; vis physician, Charity Hosp, New Orleans, 52-57; hon mem staff, DePaul Hosp, 52-57; psychiatrist-in-chief, Univ Va Hosp, 57-67. Mem: AAAS; Am Psychosom Soc; Am Soc Psychical Res; Am Psychiat Asn; AMA. Res: Experimental psychoses; psychotherapy; parapsychology. Mailing Add: Div of Parapsychol Univ of Va Med Ctr Box 152 Charlottesville VA 22901

STEVENSON, IAN LAWRIE, b Hamilton, Ont, Dec 28, 26; m 53; c 2. AGRICULTURAL MICROBIOLOGY, CYTOLOGY. Educ: Ont Agr Col, BSA, 49; Univ Toronto, MSA, 51; Univ London, PhD(microbiol), 55. Prof Exp: From bacteriologist to sr bacteriologist, Microbiol Res Inst, 51-67, head physiol & nutrit unit, 59-67, head cytol & physiol unit, Chem & Biol Res Inst, Ont, 67-72, assoc dir res sta, Lethbridge, Alta, 72-74, PRIN RES SCIENTIST, CHEM & BIOL RES INST, CAN DEPT AGR, 74- Concurrent Pos: Lectr, Univ Ottawa, 55-59; vis scientist, Nat Inst Med Res, Mill Hill, 63-64. Mem: Am Soc Microbiol; Can Soc Microbiol; Brit Soc Gen Microbiol. Res: Physiology and growth of micro-organisms; microbiology of the soil; electron microscopy; cytology. Mailing Add: Chem & Biol Res Inst Can Dept of Agr Ottawa ON Can

STEVENSON, IRA MORLEY, b Ont, Can, Mar 27, 20; m 42; c 5. GEOLOGY. Educ:

McGill Univ, BSc, 49, MSc, 51, PhD(geol), 54. Prof Exp: GEOLOGIST, DEPT ENERGY, MINES & RESOURCES CAN, 54- Mem: Geol Asn Can. Mailing Add: 27 Madawaska Dr Ottawa ON Can

STEVENSON, IRONE EDMUND, JR, b Linthicum, Md, Apr 21, 30; m 60; c 2. BIOCHEMISTRY. Educ: Univ Md, BS, 53; Univ Pa, PhD(biochem), 61. Prof Exp: Asst instr biochem, Univ Pa, 54-58; asst zool, Yale Univ, 60-63; RES CHEMIST, E I DU PONT DE NEMOURS & CO, INC, 63- Mem: Am Soc Biol Chem; Entom Soc Am. Res: Degradation of cholesterol by mammalian enzymes; intermediary metabolism of insects. Mailing Add: Cent Res & Develop Dept E I du Pont de Nemours & Co Inc Wilmington DE 19898

STEVENSON, ISAAC GLENN, b Labette, Kans, July 18, 15; m 41; c 4. CHEMISTRY, BACTERIOLOGY. Educ: Bethel Col, Kans, AB, 37; Univ Kans, MA, 40, PhD(chem), 42. Prof Exp: Lab asst, Bethel Col, 36-37; lab asst, Univ Kans, 37-41, purchasing agent, Dept Chem, 41-42; res chemist & tech mgr, GAF Corp, 42-68; mgr advan technol, Gen Elec Co, 68-72; PRIN SCIENTIST, NY STATE ASSEMBLY SCI STAFF, 72- Mem: Am Chem Soc; Am Soc Qual Control. Res: Organic synthesis; photographic chemistry; process development and control; application of science and technology to public affairs and problems of society. Mailing Add: 226 Executive Dr Guilderland NY 12084

STEVENSON, JAMES CAMERON, b Kerrobert, Sask, Can, Jan 16, 18; m 44; c 2. MARINE BIOLOGY. Educ: Univ Sask, BA, 39, MA, 42; Univ Toronto, PhD, 55. Prof Exp: Limnologist, Univ Sask, 38-42; fisheries biologist, Fisheries Res Bd Can, 43-48, head herring invests, 48-53, asst dir, 54-59; asst area dir Pac, Can Dept Fisheries, 59-62; DIR SCI & TECH PUBL, FISHERIES & MARINE SERV, CAN DEPT OF ENVIRON, 62- Mem: Coun Biol Ed; Am Fisheries Soc (pres, 75-76); Can Conf Fisheries Res (secy-treas, 71-); Can Soc Zool; Can Soc Environ Biol. Res: Aquatic science, primarily from viewpoint of scientific editing and publishing. Mailing Add: Dept of the Environ 116 Lisgar St Fisheries & Marine Serv Ottawa ON Can

STEVENSON, JAMES HAROLD, b Volant, Pa, Mar 22, 14; m 40; c 2. FISH BIOLOGY. Educ: Westminster Col, Pa, BS, 35; Oberlin Col, MA, 37; Okla State Univ, PhD, 50. Prof Exp: Metallurgist, Carnegie-Ill Steel Corp, 39-42; instr chem, Westminster Col, 42-46; instr zool, Little Rock Univ, 46-48; res biologist, State Game & Fish Comn, Ark, 50-56; chief fish farming exp sta, Bur Sport Fisheries & Wildlife, US Fish & Wildlife Serv, 60-65; DEAN, SCH SCI, ARK STATE UNIV, 65- Concurrent Pos: Chmn sci div, Little Rock Univ, 50-60; consult, Pakistan, 59-60. Mem: Am Fisheries Soc. Res: Ecology; productivity of large impoundments; fish culture. Mailing Add: Ark State Univ State University AR 72467

STEVENSON, JAMES RUFUS, b Trenton, NJ, May 19, 25; m 55; c 3. SURFACE PHYSICS. Educ: Mass Inst Technol, SB, 50; Univ Mo, PhD(physics), 58. Prof Exp: Res participant, Oak Ridge Nat Lab, 55; asst prof, 55-62, actg dir, 68-69, ASSOC PROF PHYSICS, GA INST TECHNOL, 62-, DIR SCH PHYSICS, 69- Concurrent Pos: Physicist, US Naval Res Lab, DC, 58, consult, 60-; Fulbright-Hays vis prof, Univ Sci & Technol, Ghana, 65-66; mem comt applns physics, Am Phys Soc, 75-78. Mem: Am Phys Soc; Am Asn Physics Teachers; Optical Soc Am; Am Soc Eng Educ. Res: Synchrotron radiation, Auger spectroscopy and optical surface studies of metals, metal oxides, and semiconductors with applications to corrosion. Mailing Add: Sch of Physics Ga Inst of Technol Atlanta GA 30332

STEVENSON, JEAN MOORHEAD, b Circleville, Ohio, Oct 2, 04; m 40; c 3. SURGERY. Educ: Miami Univ, AB, 26; Univ Cincinnati, MB, 30, MD, 31. Prof Exp: Resident surg, Cincinnati Gen Hosp, 31-33 & 34-37 & Univ Calif, 33-34; from instr to assoc prof, 37-61, PROF SURG, COL MED, UNIV CINCINNATI, 61- Mem: Soc Univ Surgeons; Soc Clin Surgeons; AMA; Am Col Surgeons; Int Soc Surg. Res: Wound healing; development of technics for the management of wounds of violence; care of tissues in all surgical wounds. Mailing Add: Dept of Surg Univ of Cincinnati Med Ctr Cincinnati OH 45267

STEVENSON, JOHN CRABTREE, b Everett, Wash, Feb 24, 37; m 60; c 3. MATHEMATICS. Educ: NY Univ, BA, 63, MS, 63; Adelphi Univ, PhD(math), 70. Prof Exp: From instr to assoc prof, 68-74, PROF MATH, C W POST COL, LONG ISLAND UNIV, 74-, CHMN DEPT, 72- Concurrent Pos: C W Post Col grant, dept physics, Imp Col, Univ London, 70-71. Mem: AAAS; Math Asn Am; Am Math Soc; Soc Indust & Appl Math. Res: Numerical solution of hyperbolic partial differential equations; plasma physics in the solar atmosphere and magnetosphere; multiple pool analysis of metabolic pathways. Mailing Add: Dept of Math C W Post Col Long Island Univ Greenvale NY 11548

STEVENSON, JOHN SINCLAIR, b New Westminster, BC, Sept 21, 08; m 35; c 2. MINERALOGY, GEOLOGY. Educ: Univ BC, BA, 29, BASc, 30; Mass Inst Technol, PhD(econ geol), 34. Prof Exp: Instr & asst geol, Mass Inst Technol, 31-34; engr chg, Longacre Long Lac Gold Mines, Ont, 34-35; from asst res mining engr to mining engr, Dept Mines, Victoria, 35-50; from assoc prof to prof mineral, 50-72, chmn dept geol sci, 66-68, DAWSON PROF GEOL, McGILL UNIV, 72- Concurrent Pos: Can travel fel, Guggenheim Found, 47-48; consult, 50- & Int Nickel Co Can, Ltd, 65- Mem: Fel Geol Soc Am; Soc Econ Geologists; fel Mineral Soc Am; Geochem Soc; Sigma Xi. Res: Mining geology; investigation of gold, mercury, tungsten and molybdenum deposits of British Columbia; varieties of Coast Range intrusives of British Columbia and types of related metallization; mineralogy of urinary calculi; uranium and columbium mineralization; medical mineralogy; origin of Sudbury, Ontario, irruptive; geology of sulfide nickel. Mailing Add: Dept of Geol Sci McGill Univ Montreal PQ Can

STEVENSON, JOSEPH ROSS, b Canton, China, Sept 4, 31; US citizen; m 54; c 3. DEVELOPMENTAL BIOLOGY, ENDOCRINOLOGY. Educ: Oberlin Col, BA, 53; Northwestern Univ, MS, 55, PhD, 60. Prof Exp: Asst zool & chem, Oberlin Col, 52-53; asst biol, Northwestern Univ, 53-55; instr, Chatham Col, 56-59; res assoc zool, Univ Wash, 59-60; from instr to assoc prof, 60-71, assoc dean grad col, 73-74, PROF ZOOL, KENT STATE UNIV, 71- Concurrent Pos: Jacques Loeb assoc, Rockefeller Univ, 63-64. Mem: AAAS; Am Soc Zool; Soc Develop Biol; Am Inst Biol Sci; Am Soc Cell Biol. Res: Developmental physiology; control of growth processes; endocrinology of arthropods; physiology and biochemistry of crustacean growth and molting. Mailing Add: Dept of Biol Sci Kent State Univ Kent OH 44242

STEVENSON, KENNETH EUGENE, b Modesto, Calif, June 3, 42; m 69; c 1. FOOD MICROBIOLOGY. Educ: Univ Calif, Davis, BS, 64, PhD(microbiol), 70. Prof Exp: Instr biol, Napa Col, Calif, 71; res microbiologist, Univ Calif, Davis, 71; ASST PROF FOOD MICROBIOL, MICH STATE UNIV, 71- Concurrent Pos: Sci adv, US Food & Drug Admin, Detroit, 72- Mem: Soc Indust Microbiol; Am Soc Microbiol; Inst Food Technol. Res: Microbiological analyses of foods; food poisoning microorganisms; microbiological aspects of plant sanitation and waste disposal; use of microorganisms in the production of food. Mailing Add: Dept of Food Sci & Human Nutrit Mich State Univ East Lansing MI 48824

STEVENSON, KENNETH JAMES, b Calgary, Alta, Apr 16, 41; m 64; c 2. PROTEIN CHEMISTRY. Educ: Univ Alta, BSc, 62, PhD(biochem), 66. Prof Exp: Med Res Coun fel, Lab Molecular Biol, Cambridge Univ, 66-67; Killam fel, Univ BC, 67-69; asst prof, 69-75, ASSOC PROF BIOCHEM, UNIV CALGARY, 75- Mem: Can Biochem Soc; Brit Biochem Soc. Res: Structure and function of proteolytic enzymes; techniques in protein chemistry. Mailing Add: Dept of Chem Univ of Calgary Calgary AB Can

STEVENSON, KENNETH KNOX, organic chemistry, see 12th edition

STEVENSON, KENNETH LEE, b Fort Wayne, Ind, Aug 1, 39; m 59; c 2. PHOTOCHEMISTRY. Educ: Purdue Univ, BS, 61, MS, 65; Univ Mich, PhD(phys chem), 68. Prof Exp: Teacher high schs, Ind & Mich, 61-65; asst prof, 68-73, ASSOC PROF CHEM, PURDUE UNIV, FT WAYNE, 73- Concurrent Pos: Fel, Chem Dept, NMex State Univ, 75-76. Mem: AAAS; Am Chem Soc; Am Asn Univ Profs. Res: Photochemistry of coordination compounds; kinetics and thermodynamics of ligand exchange reactions; induction of optical activity using light; photochemical conversion of solar energy. Mailing Add: Dept of Chem Purdue Univ Ft Wayne IN 46805

STEVENSON, L HAROLD, b Bogalusa, La, Mar 18, 40; m 61; c 1. MICROBIOLOGY. Educ: Southeastern La Col, BS, 62; La State Univ, MS, 64, PhD(microbiol), 67. Prof Exp: Asst prof, 67-71, ASSOC PROF BIOL, UNIV SC, 71-, MARINE SCI, 73- Mem: AAAS; Am Soc Microbiol. Res: Bacterial ecology; distribution, activity and taxonomy of estuarine bacteria. Mailing Add: Dept of Biol Univ of SC Columbia SC 29208

STEVENSON, LOUISE STEVENS, b Seattle, Wash, July 28, 12; Can citizen; m 35; c 2. MINERALOGY. Educ: Univ Wash, BS, 32; Radcliffe Col, AM, 33. Prof Exp: Res asst climat, US Weather Bur, Seattle, 34; lectr geol & geog, Victoria Col, BC, 48-49; mus assoc geol, 51-57, CUR GEOL, REDPATH MUS, McGILL UNIV, 57- Concurrent Pos: Convener sect 17, Int Geol Cong, 70-72. Mem: Fel Geol Asn Can; Mineral Soc Am; Mineral Asn Can; Nat Asn Geol Teachers; Sigma Xi. Res: Petrogenesis of rare minerals; mineralogy applied to medicine and dentistry; petrology of siliceous lavas; adult education in mineralogy; geological education through university museums. Mailing Add: Redpath Mus McGill Univ Montreal PQ Can

STEVENSON, MERLON LYNN, b Salt Lake City, Utah, Oct 31, 23; m 48; c 5. PARTICLE PHYSICS. Educ: Univ Calif, AB, 48, PhD(physics), 53. Prof Exp: From asst to lectr, 48-58, from asst prof to assoc prof, 58-64, PROF PHYSICS, UNIV CALIF, BERKELEY, 64-, PHYSICIST, LAWRENCE BERKELEY LAB, 51- Concurrent Pos: NSF sr fel & vis prof physics, Inst High Energy Physics, Univ Heidelberg, 66-67. Res: Neutrino physics and new particle search at Fermi Lab. Mailing Add: Lawrence Berkeley Lab Univ of Calif Berkeley CA 94720

STEVENSON, MERRITT RAYMOND, b Chicago, Ill, Feb 5, 36; m 58; c 3. PHYSICAL OCEANOGRAPHY. Educ: Ohio State Univ, BSc, 58, MA, 61; Ore State Univ, PhD(oceanog), 66. Prof Exp: SR SCIENTIST, INTER-AM TROP TUNA COMN, 66- Mem: AAAS; Am Geophys Union. Res: Circulation in the eastern tropical Pacific; direct current measurements; upwelling; remote sensing; science education. Mailing Add: Inter-Am Trop Tuna Comn Scripps Inst of Oceanog La Jolla CA 92037

STEVENSON, NANCY ROBERTA, b Vinton, Iowa, Feb 14, 38; m 73. PHYSIOLOGY, NUTRITION. Educ: Univ Northern Iowa, BS, 60; Rutgers Univ, MS, 63, PhD(nutrit), 69. Prof Exp: Nat Inst Arthritis, Metab & Digestive Dis fel, 69-71, instr physiol, 71-72, ASST PROF PHYSIOL, RUTGERS MED SCH, COL MED & DENT NJ, 72- Mem: AAAS; Am Gastroenterol Asn; Am Dietetic Asn; Am Physiol Soc. Res: Gastrointestinal digestion and absorption; diurnal rhythms. Mailing Add: Rutgers Med Sch Dept Physiol Col of Med & Dent of NJ Piscataway NJ 08854

STEVENSON, NELL ELIZABETH, b San Antonio, Tex, Jan 24, 44. TOPOLOGY. Educ: Univ Tex, Austin, BA, 65, MA, 68, PhD(math), 69. Prof Exp: Res assoc math, State Univ NY Binghamton, 69-70, asst prof, 70-71; tech rep, Gen Elec Info Serv Bus Div, 73-74; CONSULT & SYSTS ANALYST, ADVAN COMPUT TECHNIQUES CORP, 75- Mem: Math Asn Am. Res: Point set topology and fundamental questions concerning the proper approach to the foundations of mathematics and mathematical logic. Mailing Add: Apt B-2 960 Bloomfield Ave Glen Ridge NJ 07028

STEVENSON, PETER COOPER, b New York, NY, Apr 8, 24; m 46; c 2. RADIOCHEMISTRY. Educ: Princeton Univ, AB, 44, PhD(chem), 50. Prof Exp: Res chemist, 50-63, radiochem div leader, 63-69, asst dept head chem, 69-73, NUCLEAR CHEMIST, LAWRENCE LIVERMORE LAB, UNIV CALIF, 73- Mem: AAAS; NY Acad Sci. Res: Formation and growth of colloidal particles; separation techniques useful in radiochemistry; fission phenomenon; on-line isotope separation. Mailing Add: 4352 Emory Way Livermore CA 94550

STEVENSON, PHILIP ERIK, theoretical chemistry, see 12th edition

STEVENSON, RALPH GIRARD, JR, b Jersey Shore, Pa, Feb 14, 25; m 52; c 4. MINERALOGY, PETROLOGY. Educ: Univ NMex, BS, 49, MS, 50; Ind Univ, PhD(geol), 65. Prof Exp: Geol engr, Water Resources Div, US Geol Surv, 50-51; geologist, Skelly Oil Co, 51-55 & Shell Develop Co, 55-61; res geologist, Gulf Res & Develop Co, 64-66, staff geologist, Tech Serv Ctr, Gulf Oil Corp, 66-68; asst prof, 68-75, ASSOC PROF GEOL, UNIV S FLA, 75- Mem: AAAS; Mineral Soc Am; Clay Minerals Soc; Am Inst Mining, Metall & Petrol Eng; Mineral Asn Can. Res: Mineralogy and petrology of polymetamorphic and sedimentary rocks; crystal chemistry and geochemistry of phosphate minerals and clay minerals at atmospheric conditions. Mailing Add: Dept of Geol Univ of South Fla Tampa FL 33620

STEVENSON, RICHARD, biology, see 12th edition

STEVENSON, ROBERT EDWIN, b Columbus, Ohio, Dec 2, 26. MICROBIOLOGY, SCIENCE ADMINISTRATION. Educ: Ohio State Univ, BSc, 47, MSc, 50, PhD(bact), 54; Am Bd Microbiol, dipl. Prof Exp: Res assoc, US Pub Health Serv, 52-54, virologist, 54-58; head tissue cult div, Tissue Bank Dept, US Naval Med Sch, 58-60; head cell cult & tissue mat sect, Virol Res Resources Br, Nat Cancer Inst, 60-62, actg chief, 62-63, chief, 63-66, chief etiology, Viral Carcinogenesis Br, 66-67; mgr biol sci, Develop Dept, Union Carbide Res Inst, NY, 67-71; VPRES, LITTON BIONETICS & GEN MGR, NAT CANCER INST-FREDERICK CANCER RES CTR, 72- Concurrent Pos: Mem, Nat Inst Allergy & Infectious Dis, bd virus reference reagent, 63-65; cell cult comt, Int Asn Microbiol Soc, 63-67; comt transplantation, Nat Acad Sci-Nat Res Coun, 66-70; chmn cell cult comt, Am Type Cult Collection, 71-75, mem bd trustees, 72- Mem: AAAS; Am Soc Microbiol; Soc Cryobiol; Tissue Cult Asn. Res: Viral oncology; biomedical instrumentation; biological standardization. Mailing Add: Frederick Cancer Res Ctr PO 21 Frederick MD 21701

STEVENSON, ROBERT EVANS, b Des Moines, Iowa, May 5, 16; m 48; c 2. GEOLOGY. Educ: Univ Hawaii, BS, 39; State Col Wash, MS, 42; Lehigh Univ,

PhD(geol), 50. Prof Exp: Asst, Univ Hawaii, 39; lab asst, State Col Wash, 39-42; geologist, Wash State Div Geol, 42-44; field geologist, Venezuelan Atlantic Ref Co, 44-46; instr geol, Lehigh Univ, 46-50; geologist, State Geol Surv, SDak, 50-51; from asst prof to prof geol, 51-71, chmn dept geol, 57-67, PROF EARTH SCI, UNIV S DAK, 71-, CUR GEOL, MUS, 73- Concurrent Pos: Geologist, NY State Sci Serv, 47-48. Mem: AAAS; Geol Soc Am; Am Asn Petrol Geol; Paleont Soc. Res: Stratigraphy, sedimentation and paleontology of South Dakota; paleoecology. Mailing Add: Dept of Earth Sci-Physics Univ of SDak Vermillion SD 57069

STEVENSON, ROBERT EVERETT, b Fullerton, Calif, Jan 15, 21; m 63; c 2. OCEANOGRAPHY. Educ: Univ Calif, AB, 46, AM, 48; Univ Southern Calif, PhD(marine geol), 54. Prof Exp: Instr geol, Compton Col, 47-49; lectr, Univ Southern Calif, 49-51, dir inshore res oceanog, Hancock Found, 53-59, 60-61; res scientist, Off Naval Res, London, 59; dir marine lab, Tex A&M Univ, 61-63; assoc prof meteorol & geol, Fla State Univ, 63-65; res oceanogr & asst dir biol lab, US Bur Commercial Fisheries, 65-70; SCI LIAISON OFFICER, OFF NAVAL RES, SCRIPPS INST OCEANOG, 70- Mem: Fel Geol Soc Am; Am Meteorol Soc; Am Geophys Union. Res: Coastal oceanography, meteorology and climatology; application of spacecraft sensors to analyses of surface-layer oceanography. Mailing Add: Sci Liaison Officer Off Naval Res Scripps Inst of Oceanog La Jolla CA 92093

STEVENSON, ROBERT FINDLAY, b Lowell, Mass, Mar 11, 23; m 60. ANTHROPOLOGY. Educ: Columbia Col, AB, 48; Columbia Univ, PhD(anthrop), 65. Prof Exp: Assoc anthrop, Columbia Univ, 63; lectr, Hunter Col, 64-65; asst prof, 66-68, ASSOC PROF ANTHROP, STATE UNIV NY STONY BROOK, 68- Concurrent Pos: NIMH res fel & grant, 65-66; chmn sem ecol systs & cult evolution, Columbia Univ, 69- Mem: AAAS; fel Am Anthrop Asn; fel African Studies Asn; fel Am Geog Soc; Int African Inst. Res: Human ecology; comparative political systems and evolution of the state; development of anthropological theory; social organization. Mailing Add: Dept of Anthrop State Univ NY Stony Brook NY 11790

STEVENSON, ROBERT LOVELL, b Long Beach, Calif. ANALYTICAL CHEMISTRY. Educ: Reed Col, BA, 63; Univ Ariz, PhD(chem), 66. Prof Exp: Sr chemist, Shell Develop Co, 66-69; sr chemist, Varian Aerograph, 69-75, MGR LIQUID CHROMATOGRAPHY-RES & DEVELOP, VARIAN ASSOCS, 75- Mem: Am Chem Soc. Res: Managing a research and development group developing high speed liquid chromatographs and accessories. Mailing Add: 2700 Mitchell Dr Walnut Creek CA 94598

STEVENSON, ROBERT THOMAS, b Washington, DC, July 23, 16; m 43; c 4. BIOLOGY. Educ: Am Univ, BA, 38; Univ Wis, MPh, 40, PhD(zool), 43. Prof Exp: Asst, Univ Wis, 38-43; asst prof biol & physiol, Univ Utah, 46-48; assoc prof, 48-68, head dept sci, 62-68, PROF BIOL, SOUTHWEST MO STATE UNIV, 68-, HEAD DEPT LIFE SCI, 68- Concurrent Pos: Parasitologist, Neiman Stephenson Co, 41-42. Res: Parasitology; comparative histology and ultrastructure. Mailing Add: Dept of Life Sci Southwest Mo State Univ Springfield MO 65802

STEVENSON, ROBERT WILLIAM, b Philadelphia, Pa, Oct 22, 30; m 55; c 3. ORGANIC CHEMISTRY. Educ: Univ Pa, BS, 54; Ga Inst Technol, PhD(org chem), 58. Prof Exp: Res chemist, Celanese Corp Am, 58-61; sr res chemist, Mobil Chem Co, 62-69; chmn dept phys sci, 74-75, PROF SCI, CHEYNEY STATE COL, 69- Mem: Am Chem Soc; Sigma Xi. Res: Synthesis of linear polyamides and polyesters, polyacetals and polyolefins; modification of fats. Mailing Add: Dept of Phys Sci Cheyney State Col Cheyney PA 19319

STEVENSON, STUART SHELTON, b Bridgeport, Conn, Nov 11, 14. PEDIATRICS. Educ: Yale Univ, BA, 35, MD, 39; Harvard Univ, MPH, 44. Prof Exp: Instr pediat, Sch Med, Yale Univ, 41-43; Rockefeller fel, Harvard Univ, 43, asst maternal & child health, Sch Pub Health, 44, assoc child health, 46-47, asst prof, 47-49; res prof pediat, Sch Med, Univ Pittsburgh, 49-59; prof & chmn dept, Seton Hall Col Med, 59-64; clin prof, Col Physicians & Surgeons, Columbia Univ, 64-72, prof pediat, 72-74; dir pediat & attend pediatrician, St Luke's Hosp Ctr, 64-74. Concurrent Pos: Staff health comnr, Rockefeller Found, 44-46. Honors & Awards: Order of the Nile, Egypt, 47. Mem: Soc Pediat Res; Am Pediat Soc; fel Am Acad Pediat; NY Acad Med. Res: The newborn, especially carbonic anhydrase, hyaline membrane, thyroid, congenital malformations, nutrition, growth failure and prematurity. Mailing Add: 2 Fifth Ave New York NY 10011

STEVENSON, THOMAS DICKSON, b Columbus, Ohio, Sept 23, 24; m 52; c 3. MEDICINE. Educ: Ohio State Univ, BA, 45, MD, 48; Am Bd Internal Med, dipl, 56; Am Bd Path, dipl, 64. Prof Exp: Intern med, Johns Hopkins Hosp, 48-49; asst resident, Univ Minn Hosps, 50-51; from asst resident to resident, Ohio State Univ Hosps, 51-53; investr clin gen med & exp therapeut, Nat Heart Inst, 53-55; asst prof med & assoc dir div hemat, Sch Med, Univ Louisville, 55-61; assoc prof, 61-71, PROF PATH, COL MED, OHIO STATE UNIV, 71- Mem: Am Fedn Clin Res; Am Soc Clin Path; Am Soc Cytol. Res: Biochemical aspects of erythropoiesis; vitamin B-12 metabolism. Mailing Add: Dept of Path Col of Med Ohio State Univ Hosp Columbus OH 43210

STEVENSON, WILLIAM CAMPBELL, b Brooklyn, NY, Jan 22, 31; m 55; c 3. BIOCHEMISTRY. Educ: St John's Col, BS, 52. Prof Exp: Chemist, Quaker Maid Co, 54-57 & Nat Biscuit Co, 57-58; chemist, 58-67, SR SCIENTIST, MEAD JOHNSON CO, 67- Mem: Am Chem Soc; NY Acad Sci. Res: Isolation of natural products; enzymes; proteins. Mailing Add: 3520 Laurell Lane Evansville IN 47712

STEVENSON, WILLIAM HENRY, b Philadelphia, Pa, May 11, 27; m 51; c 4. FISHERIES MANAGEMENT. Educ: Univ Del, BA, 51. Prof Exp: Resident mgr, Marine Biol Lab, Del, 51-54; explor fishing & gear res, Fish Prod Co, Del, 54-58; fishery consult, 58-60; fishing gear res, Smith Res & Develop Co, 60-61; fishery methods & equip specialist, Bur Com Fisheries, 61-66, foreign fisheries adv, Agency Int Develop, 66-69, chief gear experimentation sect, 69-70, chief div explor fishing & gear res, 70-74, mgr fisheries eng lab, 74, REGIONAL DIR, SOUTHEAST REGION, NAT MARINE FISHERIES SERV, NOAA, DEPT COM, 74- Res: Commercial fisheries development; biological resource survey and analysis; resource harvesting methods improvement; marine biological and fisheries engineering; fisheries remote sensing; fisheries administration and management. Mailing Add: Nat Marine Fisheries Serv NOAA Duval Bldg 9450 Gandy Blvd St Petersburg FL 33702

STEVER, H GUYFORD, b Corning, NY, Oct 24, 16; m 46; c 4. AERONAUTICS, ASTRONAUTICS. Educ: Colgate Univ, AB, 38; Calif Inst Technol, PhD(physics), 41. Hon degrees: Numerous degrees from Am univs. Prof Exp: Mem staff radiation lab & instr, Army-Navy Officers Radar Sch, Mass Inst Technol, 41-42;sci liaison officer, London Mission, Off Sci Res & Develop, 42-45; exec officer, Guided Missiles Prog, Mass Inst Technol, 46-48, from asst prof to assoc prof aeronaut eng, 46-55, prof aeronaut & astronaut, 56-65, assoc dean eng, 56-59, head depts aeronaut & naval archit &marine eng, 61-65; pres, Carnegie-Mellon Univ, 65-72; dir, NSF, 72-76; HEAD, WHITE HOUSE OFF SCI & TECHNOL, 76- Concurrent Pos: Mem guided missiles tech eval group, Res & Develop Bd, 46-48; mem sci adv bd to chief staff, US Dept Air Force, 47-69, from vchmn to chmn, 56-69, chief scientist, 55-56, mem air defense syst eng comt, 49-51; mem tech adv panel ord, Asst Secy Defense, 54-56, mem steering comt tech adv panel aeronaut, 56-62; chmn spec comt space technol, Nat Adv Comt Aeronaut, 58, res adv comt missile & spacecraft aerodyn, NASA, 59-65, mem, Defense Sci Bd; mem adv panel, Comt Sci & Astronaut, US House Rep, 60-72; mem systs command bd of visitors, US Air Force, 64-72; mem, President's Comn Patent Syst, 65-67; mem exec comt, Div Eng & Indust Res, Nat Acad Sci-Nat Res Coun; mem, Nat Sci Bd, 70-72. Honors & Awards: Presidential Cert Merit, 48; Civilian Serv Award, US Dept Air Force, 56; Scott Gold Medal, Am Ord Asn, 60; Distinguished Pub Serv Medal, Dept Defense, 68 Mem: Nat Acad Eng; AAAS; fel Am Phys Soc; Am Inst Aeronaut & Astronaut (vpres, Inst Aerospace Sci, 58, pres, 60); fel Am Acad Arts & Sci. Res: Gas discharge; Geiger counters; cosmic rays; radar guided and ballistic missiles; hypersonic aerodynamics; shock tubes; transonic aircraft; nuclear propulsion of aircraft; condensation in high speed flow; space flight. Mailing Add: The White House Washington DC 20500

STEVERDING, BERNARD, b Stadtlohn, Ger, Aug 3, 26; US citizen; m 59. THEORETICAL PHYSICS. Educ: Univ Münster, BS, 48, PhD(physics), 51; Aachen Tech Univ, DrEng, 56. Prof Exp: Asst prof phys chem, Aachen Tech Univ, 54-56; res assoc metall, Deutsche Edelstahl Werke, 56-57; res assoc electronics, Signal Corps, US Army, 57-58; asst dir mat sci, 58-67, TECH DIR NUCLEAR EFFECTS, US ARMY MISSILE COMMAND, REDSTONE ARSENAL, 67- Concurrent Pos: Consult, Nat Acad Sci, 67-68; Dept Defense fel & vis scholar, Univ Calif, Berkeley, 71-72. Mem: Ger Soc Scientists & Physicians. Res: Fracture mechanics; shock waves; material sciences, radiation physics; transport phenomena; mass and heat transfer; ablation; applied mathematics; non-linear optics; lasers; high temperature phenomena plasma physics of dense matter; equation of state. Mailing Add: Phys Sci Div Army Missile Command Redstone Arsenal AL 35809

STEWARD, FREDERICK CAMPION, b London, Eng, June 6, 04; m 29; c 1. BOTANY, CELL BIOLOGY. Educ: Univ Leeds, BSc, 24, PhD(bot), 26. Hon Degrees: DSc, Univ London, 36. Prof Exp: Demonstr bot, Univ Leeds, 26-27, asst lectr, 29-33; Rockefeller fel, Cornell Univ & Univ Calif, 27-29; Rockefeller fel, Univ Calif & Carnegie Inst Technol, 33-34; reader, Univ London, 34-47; dir aircraft equip, Ministry of Aircraft Prod, 40-45; res assoc, Univ Chicago, 45-46; vis prof bot & chmn dept, Univ Rochester, 46-50; prof, 50-65, Alexander prof biol, 65-72, CHARLES A ALEXANDER EMER PROF BIOL SCI, CORNELL UNIV, 72- Concurrent Pos: Guggenheim fel, 64. Honors & Awards: Merit Award, Bot Soc Am, 61; Stephen Hales Award, Am Soc Plant Physiol, 64. Mem: Bot Soc Am; Am Soc Plant Physiol; fel Royal Soc; fel Am Acad Arts & Sci. Res: Plant physiology and biochemistry; respiration; salt intake; metabolism; protein synthesis; chromatography of amino acids; morphogenesis. Mailing Add: 1612 Inglewood Dr Charlottesville VA 22901

STEWARD, JAMES GORDON, b North Fond du Lac, Wis, Aug 13, 23; m 43; c 3. MATHEMATICS, COMPUTER SCIENCE. Educ: Lawrence Col, BS, 47; Kans State Col, MS, 49. Prof Exp: Mathematician, Ballistics Res Lab, Aberdeen Proving Ground, Md, 49-51; mathematician, Dept Appl Mech, Southwest Res Inst, Tex, 51-52; mathematician, Dept Biomet, US Air Force Sch Aviation Med, 52-55; sr analyst, Standard Oil Co, Tex, 55-59; comput supvr, 59-62, res assoc, 62-66, COMPUT RES SUPVR, RES CTR, AMOCO PROD CO, OKLA, 66- Concurrent Pos: Consult, Seeligson Eng, Tex, 52-54; lectr, Trinity Univ, 52-55. Mem: Asn Comput Mach; Soc Indust & Appl Math. Res: Computers in data display and man-machine interaction; supervision of computing research dealing primarily with new applications and equipment for petroleum problems. Mailing Add: Res Ctr Amoco Prod Co PO Box 591 Tulsa OK 74102

STEWARD, JOHN P, b Huntington Park, Calif, Oct 9, 27. MEDICAL MICROBIOLOGY, IMMUNOLOGY. Educ: Stanford Univ, AB, 48, MD, 55. Prof Exp: Nat Inst Allergy & Infectious Dis fel med microbiol, Sch Med, 58-60, from instr basic med sci to asst prof exp med, 60-70, asst dean, Sch Med, 64-65, asst dir, Fleischmann Labs Med Sci, 64-70, sr lectr med microbiol, Sch Med & actg dir, Fleischmann Labs Med Sci, 70-74, ADJ PROF MED MICROBIOL, SCH MED, STANFORD UNIV, 74-, ASSOC DEAN SCH MED, 71- Concurrent Pos: Lectr, Sch Pub Health, Univ Calif, Berkeley, 63. Mem: Am Asn Immunologists; Am Soc Microbiol. Res: Host-parasite relationship between enterobacteriaceae and experimental animals. Mailing Add: Stanford Univ Sch of Med Stanford CA 94305

STEWARD, KERRY KALAN, b Skowhegan, Maine, June 2, 30; m 56; c 1. PLANT PHYSIOLOGY. Educ: Univ Conn, BS, 58, MS, 62, PhD(bot), 66. Prof Exp: RES PLANT PHYSIOLOGIST, AGR RES SERV, USDA, 66- Concurrent Pos: Nat Park Serv grant, Everglades Nat Park, US Forest Serv, 72-73. Mem: Am Soc Plant Physiol; Am Inst Biol Sci; Bot Soc Am; Weed Sci Soc Am. Res: Physiology of aquatic plants; mineral nutrition of aquatic plants; effects of eutrophication on growth of aquatic plants. Mailing Add: Agr Res Serv USDA 3205 SW 70th Ave Ft Lauderdale FL 33314

STEWARD, OMAR WADDINGTON, b Woodbury, NJ, May 28, 32; m 58; c 3. ORGANOMETALLIC CHEMISTRY, INORGANIC CHEMISTRY. Educ: Univ Del, BS, 53; Pa State Univ, PhD(chem), 57. Prof Exp: Proj leader fluorine & organosilicon chem, Dow Corning Corp, 57-62; NSF fel, Univ Leicester, 62-63; instr inorg chem, Univ Ill, 63; asst prof, Southern Ill Univ, 63-64; from asst prof to assoc prof, 64-72, PROF INORG CHEM, DUQUESNE UNIV, 72- Mem: Am Chem Soc; Am Inst Chem; The Chem Soc; Sigma Xi. Res: Structure, bonding and reaction mechanisms of group IVb organometallic compounds; bonding and reaction mechanisms of coordination compounds. Mailing Add: Dept of Chem Duquesne Univ Pittsburgh PA 15219

STEWARD, ROBERT F, b Springboro, Pa, June 2, 23; m 46; c 2. MATHEMATICS. Educ: Wheaton Col, Ill, BS, 47; Rutgers Univ, MS, 49; Auburn Univ, PhD(math), 61. Prof Exp: Instr math, Va Mil Inst, 49-53; asst prof, Drexel Inst Technol, 53-57; instr, Auburn Univ, 57-58, 59-60; assoc prof, Western Carolina Col, 60-61, prof & chmn dept, 61-63; chmn dept, 67-73, PROF MATH, R I COL, 63- Mem: Math Asn Am. Res: Numerical analysis. Mailing Add: Dept of Math Rhode Island Col Providence RI 02908

STEWARD, VINCENT WILLIAM, b Bristol, Eng, Nov 6, 27. NEUROPATHOLOGY. Educ: Univ London, MB, BS, 53. Prof Exp: Demonstr anat, London Hosp Med Col, 55-56, lectr histol, 56-57; instr anat, Emory Univ, 58-59; demonstr-lectr, Oxford Univ, 63-64; from registr to sr registr path, Nat Hosps for Nerv Dis, London, 64-67; asst prof path & dir neuropath lab, 67-74, RES ASSOC PROF PATH, DEPTS RADIOL & SURG, UNIV CHICAGO, 74- Concurrent Pos: Med Res Coun Eng scholar cytol & histochem, London Hosp Med Col, Eng, 54-55; Lederle med fac fel, Emory Univ, 59-60; clin & teaching fel neurol, Harvard Med Sch, Mass Gen Hosp, Boston, 60-61 & neuropath, 61-62, res & teaching fel neurol, 62; Wellcome Trust spec traveling fel, Inst Neurobiol, Gothenburg, Sweden, 66-67. Mem: Am Asn Path & Bact; Am Heart Asn; fel Royal Soc Med; fel Royal Micros Soc; Inst Elec & Electronics Eng. Res: Development and application of automatic methods for the quantitative morphologic analysis of neural tissue; development and application of proton beam and heavy ion radiography and densitometry for diagnosis. Mailing Add: Dept of Path Univ of Chicago Chicago IL 60637

STEWART, ALBERT BURNS, b Greensburg, Pa, Mar 23, 19; m 41; c 4. EXPERIMENTAL ATOMIC PHYSICS. Educ: Antioch Col, BS, 42; Johns Hopkins Univ, PhD(physics), 48. Prof Exp: Asst, C F Kettering Found, 39-42; jr instr physics, Johns Hopkins Univ, 42-44, 46-47, instr, 47-48; from asst prof to assoc prof, 48-56, dean of fac, 66-67, Harvard Proj Physics, 67-68, PROF PHYSICS, ANTIOCH COL, 56- Concurrent Pos: Vis fel, Princeton Univ, 51-52; fel, Univ Va, 56-57; Am consult, Physics Inst, Univ Mysore, 65, 69; staff mem, Carnegie Corp Study Future of Lib Arts Col, 65-67. Mem: AAAS; Am Phys Soc; Am Asn Physics Teachers; Hist Sci Soc. Res: Spectroscopy; gaseous electronics; history and philosophy of science. Mailing Add: Dept of Physics Antioch Col Yellow Springs OH 45387

STEWART, ALBERT CLIFTON, b Detroit, Mich, Nov 25, 19; m 49. RADIATION CHEMISTRY. Educ: Univ Chicago, SB, 42, SM, 48; St Louis Univ, PhD(chem), 51. Prof Exp: Chemist, Sherwin-Williams Paint Co, Ill, 43-44; asst inorg chem, Univ Chicago, 47-49; instr & res assoc, St Louis Univ, 49-51; sr chemist, Oak Ridge Nat Lab, 51-56; group leader, Res Lab, Nat Carbon Co Div, Union Carbide Corp, 56-59, asst dir res, Consumer Prod Div, 60-63, asst develop dir, 63-65, planning mgr new mkt develop, 65-66, mkt develop mgr, Chem & Plastics Develop Div, 66-69, mkt mgr rubber chem, Mkt Area, 69-71, mkt mgr chem coatings solvents, 71-73, INT BUS MGR, CHEM & PLASTICS DIV, UNION CARBIDE CORP, 73- Concurrent Pos: Prof, Knoxville Col, 53-56; lectr, John Carroll Univ, 56-63; consult, Pub Affairs Div, Ford Found, 63; adminstr officer, NASA, 63 & Agency Int Develop, 64-69; treas, NY State Dormitory Authority, 71- Honors & Awards: Cert Merit, Soc Chem Professions Cleveland, 62. Mem: AAAS; Am Chem Soc; Am Nuclear Soc; Radiation Res Soc. Res: Physical inorganic and radiation chemistry; research, development and general administration; marketing management. Mailing Add: 50 E 89th St Apt 5A New York NY 10028

STEWART, ALEC THOMPSON, b Can, 25; m 60; c 3. PHYSICS. Educ: Dalhousie Univ, BSc, 46, MSc, 49; Cambridge Univ, PhD(physics), 52. Prof Exp: From asst res officer to assoc res officer, 52-57; from asst prof to assoc prof physics, Dalhousie Univ, 53-60; from assoc prof to prof, Univ NC, Chapel Hill, 60-68; PROF PHYSICS & HEAD DEPT, QUEEN'S UNIV, ONT, 68- Concurrent Pos: J S Guggenheim fel & Kenan travelling prof, 65-66. Mem: Fel Am Phys Soc; Can Asn Physicists (pres, 72-73); fel Royal Soc Can. Res: Solid state by positron annihilation in matter and neutron inelastic scattering. Mailing Add: Dept of Physics Queen's Univ Kingston ON Can

STEWART, ALLAN GREENWOOD, b Kingston, Ont, Aug 11, 20; m 43; c 2. BIOCHEMISTRY. Educ: Univ Western Ont, PhD(med res), 52. Prof Exp: Asst biochem, Ont Vet Col, 45-53; from asst prof biochem to assoc prof pediat, Univ Alta, 53-69; ASST PROF PATH, DALHOUSIE UNIV, 69-; BIOCHEMIST, IZAAK WALTON KILLAM HOSP CHILDREN, 69- Mem: NY Acad Sci; Can Biochem Soc; Can Physiol Soc; Can Soc Clin Chemists; Can Inst Chemists. Res: Lipids; tumor host relationships; bilirubin metabolism; red cell enzymes. Mailing Add: Div of Clin Chem Izaak W Killam Hosp for Children Halifax NS Can

STEWART, ALVA THEODORE, JR, b Beckley, WVa, July 4, 29; m 53; c 2. ORGANIC CHEMISTRY, SCIENCE ADMINISTRATION. Educ: Duke Univ, BS, 50, MA, 52, PhD(chem), 54. Prof Exp: Fel, Mellon Inst Sci, 54-55; chemist, Shell Develop Co, 59-62; chemist, Avco Corp, 63-65; mem tech staff, Northrop Corp, 66-69; head chem & plastics res & sr scientist, 69-71, CHIEF SCIENTIST, SCI DEVELOP, BECHTEL CORP, 71- Mem: AAAS; Am Chem Soc; Am Ord Asn; Am Inst Chem. Res: New technology assessment; bioprotein; nutrition; food technology. Mailing Add: Bechtel Corp 50 Beale St San Francisco CA 94119

STEWART, ANNE MARIE, b Stoneham, Mass, June 12, 36; m 58; c 4. ANIMAL BEHAVIOR, ANIMAL PHYSIOLOGY. Educ: Simmons Col, BS, 58; Univ Mass, MA, 61; Tufts Univ, PhD(biol), 67. Prof Exp: Instr zool, Univ Mass, 61-62; res asst histol, McLain Hosp, 62-63; asst prof biol, Cape Cod Community Col, 65-66; asst prof, 65-66, ASSOC PROF BIOL, WINDHAM COL, 66- Mem: AAAS; Animal Behav Soc; Am Soc Zoologists. Res: Photic behavior of animals; field biology. Mailing Add: Dept of Biol Windham Col Putney VT 05346

STEWART, ARTHUR VAN, b Buffalo, NY, July 25, 38; m 65; c 3. DENTISTRY. Educ: Univ Pittsburgh, BS, 60, MEd, 64, DMD, 68, PhD(educ admin), 73. Prof Exp: Instr pedodont, dent auxiliary utilization & restorative dent, Univ Pittsburgh, 68-70; prof community & prev dent & chmn dept, Sch Dent, Fairleigh Dickinson Univ, 70-75, asst dean, dir learning resources & chmn dept continuing educ, 73-75; PROF, DEPT COMMUNITY DENT, SCH DENT, UNIV LOUISVILLE HEALTH SCI CTR, 75-, ASST DEAN ACAD AFFAIRS, 75- Concurrent Pos: Consult, NJ Dent Asn, 71-75, Northside Family Health Ctr, 71-75, Patrick House Family Health Ctr, 71-75, North Bergen Dent Health Ctr, 72-75, Headstart Progs, 72-75 & Paterson Bd Educ Dent Health Ctr, 73-75. Mem: Fel Am Col Dent; Am Dent Asn; Am Asn Dent Schs; Am Sch Health Asn; World Future Soc. Res: Dental education; health manpower; preventive dentistry; community health; manpower development in dental education. Mailing Add: Sch of Dent Univ of Louisville Health Sci Ctr Louisville KY 40201

STEWART, BABETTE TAYLOR, cell biology, see 12th edition

STEWART, BOBBY ALTON, b Erick, Okla, Sept 26, 32; m 56; c 3. SOIL FERTILITY, SOIL CHEMISTRY. Educ: Okla State Univ, BS, 53, MS, 57; Colo State Univ, PhD(soil sci), 61. Prof Exp: Res assoc soils, Soil & Water Res Div, 53-57, res soil scientist, 57-68, DIR & RES SOIL SCIENTIST, AGR RES SERV, US DEPT AGR, 68- Concurrent Pos: Instr agron, Okla State Univ, 53-57; fac affil, Colo State Univ, 57-68, spec lectr, 62. Mem: Soil Conserv Soc Am; fel Am Soc Agron; Soil Sci Soc Am; Int Soil Sci Soc. Res: Nitrogen-sulfur relationships in plant tissue, plant residues and soil organic matter; animal waste management; movement of fertilizer nitrogen through soil profiles. Mailing Add: Southwest Gr Plains Res Ctr US Dept of Agr Bushland TX 79012

STEWART, BONNIE MADISON, b Loveland, Colo, July 10, 14; m 40; c 2. MATHEMATICS. Educ: Univ Colo, BA, 36; Univ Wis, PhM, 37, PhD, 40. Prof Exp: Asst math, Univ Wis, 38-40; instr, Mich State Univ, 40-42; asst prof, Denison Univ, 42-43; from asst prof to assoc prof, 43-53, PROF MATH, MICH STATE UNIV, 53- Mem: Am Math Soc; Math Asn Am. Res: Matrix theory; number theory; graph theory; Euclidian geometry. Mailing Add: 4494 Wausau Rd Okemos MI 48864

STEWART, C GORDON, b Chatham, Ont, June 22, 17; m 48; c 2. BIOCHEMISTRY, MEDICINE. Educ: Univ Toronto, BA, 39, MD, 43, MA, 48; McGill Univ, dipl trop med, 46. Prof Exp: Nat Cancer Inst fel biochem, Univ Toronto, 49-54; head med res br, 55, DIR MED DIV, CHALK RIVER NUCLEAR LABS, ATOMIC ENERGY CAN LTD, 55-, CHIEF MED OFFICER, 66- Concurrent Pos: Mem, Int Comn Radiol Protection, 62-, vchmn, 65-69, chmn, 69-, mem comt II, 56- Mem: Health Physics Soc. Res: Toxicology; radiation dosimetry; radiochemistry; radiation protection. Mailing Add: Chalk River Nuclear Labs Atomic Energy of Canada Ltd Chalk River ON Can

STEWART, CARLETON C, b Schenectady, NY, July 13, 40; m 63; c 2. IMMUNOLOGY, BIOPHYSICS. Educ: Hartwick Col, BA, 62; Univ Rochester, MS, 64, PhD(radiation), 67. Prof Exp: Atomic Energy Proj res asst & instr radiation physics, Univ Rochester, 62-67; instr immunol, Univ Pa, 67-69; sr scientist, Smith Kline & French Labs, 69-70; ASST PROF CANCER BIOL IN RADIOL, SCH MED, WASHINGTON UNIV, 70- Concurrent Pos: USPHS fel & grant, 67-69; Am Cancer Soc grant, 68-69; Cancer Ctr grant, Nat Cancer Inst. Mem: AAAS; Am Asn Exp Pathologists; NY Acad Sci; Reticuloendothelial Soc. Res: Cellular immunology; tumor immunology. Mailing Add: Sect of Cancer Biol Washington Univ Sch of Med St Louis MO 63110

STEWART, CECIL R, b Monmouth, Ill, Mar 11, 37; m 58; c 2. PLANT PHYSIOLOGY. Educ: Univ Ill, BS, 58; Cornell Univ, MS, 63, PhD(plant physiol), 67. Prof Exp: NIH fel plant physiol, Purdue Univ, 66-68; asst prof 68-71, ASSOC PROF BOT, IOWA STATE UNIV, 71- Mem: AAAS; Am Soc Plant Physiol; Am Inst Biol Sci. Res: Plant metabolism. Mailing Add: Dept of Bot & Plant Path Iowa State Univ Ames IA 50011

STEWART, CHARLES EDWARD, mathematics, see 12th edition

STEWART, CHARLES JACK, b Rawlins, Wyo, June 17, 29; m 56; c 3. BIOCHEMISTRY. Educ: San Diego State Col, BA, 50; Ore State Univ, MS, 52, PhD(biochem), 55. Prof Exp: Fulbright grant biochem, Inst Org Chem, Univ Frankfurt, 54-55; from instr to assoc prof, 55-65, PROF CHEM, SAN DIEGO STATE UNIV, 65- Concurrent Pos: NIH res grant, 62-, spec fel, 63-64; guest prof chem, Max Planck Inst Med Res, Heidelberg, Ger, 75-76. Mem: AAAS; Am Soc Biol Chemists; Am Chem Soc. Res: Mechanism of enzymes and antimetabolites; synthesis and enzymatic properties of coenzyme A analogs. Mailing Add: Dept of Chem San Diego State Univ San Diego CA 92182

STEWART, CHARLES NEIL, b Albany, NY, May 6, 45; m 68; c 1. SURFACE PHYSICS. Educ: Union Col, NY, BS, 67; Univ Ill, Urbana-Champaign, MS, 69, PhD(physics), 74. Prof Exp: Asst physics, Dudley Observ, NY, 65-67; lab instr, Union Col, NY, 67; asst, Univ Ill, Urbana-Champaign, 67-73; RES PHYSICIST, LAMP PHENOMENA RES LAB, GEN ELEC CO, 73- Mem: Am Phys Soc; Sigma Xi. Res: Gas-solid interactions, particularly chemisorption, physisorption, and reactive processes; field emission. Mailing Add: Gen Elec Lamp Phenomena Res Lab Nela Park Cleveland OH 44112

STEWART, CHARLES RANOUS, b La Crosse, Wis, Aug 6, 40. MICROBIAL GENETICS. Educ: Univ Wis, BS, 62; Stanford Univ, PhD(genetics), 67. Prof Exp: Am Cancer Soc fel biochem, Albert Einstein Col Med, 67-69; asst prof, 69-74, ASSOC PROF BIOL, RICE UNIV, 74- Mem: Am Soc Microbiol. Res: Genetics and biochemistry of Bacillus subtilis and its virulent bacteriophages. Mailing Add: Dept of Biol Rice Univ Houston TX 77001

STEWART, CHARLES WINFIELD, b Wilmington, Del, Jan 27, 40; m 62; c 3. THEORETICAL CHEMISTRY. Educ: Univ Del, BS, 62, PhD(chem), 66. Prof Exp: RES CHEMIST, ELASTOMER CHEM DEPT, E I DU PONT DE NEMOURS & CO, INC, WILMINGTON, 66- Mem: Am Chem Soc. Res: Polymer physics. Mailing Add: 42 Millbrook Rd Fireside Park Newark DE 19711

STEWART, CHESTER BRYANT, b Norboro, PEI, Dec 17, 10; m 42; c 2. MEDICINE, EPIDEMIOLOGY. Educ: Dalhousie Univ, BSc, 36, MD, CM, 38; Johns Hopkins Univ, MPH, 46, DrPH, 53; FRCP(C), 62. Hon Degrees: LLD, Univ Prince Edward Island, 73. Prof Exp: Asst secy assoc comt med res, Nat Res Coun Can, 38-40; dean med, 54-71, PROF EPIDEMIOL, DALHOUSIE UNIV, 46-, VPRES HEALTH SCI, 71- Concurrent Pos: Res assoc, Johns Hopkins Univ, 51-52. Honors & Awards: Centennial Medal, 67; Officer, Order of Can, 72. Mem: Fel Am Pub Health Asn; Asn Can Med Cols (pres, 61-63); Can Physiol Soc; Can Pub Health Asn (pres, 68); Can Med Asn. Res: Aviation medicine; decompression sickness and anoxia; Bacillus Calmette-Guerin vaccination; tuberculosis; immunity; delivery of health care; hospital and medical insurance. Mailing Add: Sir Charles Tupper Med Bldg Dalhousie Univ Halifax NS Can

STEWART, CLARE A, JR, organic chemistry, polymer chemistry, see 12th edition

STEWART, CLYDE EVERETT, agricultural economics, see 12th edition

STEWART, D K R, b St John, NB, Sept 22, 27; m 66. ORGANIC CHEMISTRY. Educ: Dalhousie Univ, BSc & Ba, 48, MSc, 50; Univ NB, PhD(org chem), 55. Prof Exp: Nat Res Coun Can res fel, 55-57; RES SCIENTIST, KENTVILLE RES STA, CAN DEPT AGR, 57- Mem: Chem Inst Can. Res: Structure of peptides and alkaloids; pesticide residues in soils and crops. Mailing Add: Res Sta Can Dept of Agr Kentville NS Can

STEWART, DAVID BENJAMIN, b Springfield, Vt, July 18, 28; m 52; c 2. GEOLOGY, MINERALOGY. Educ: Harvard Univ, AB, 51; AM, 52, PhD(petrol), 56. Prof Exp: GEOLOGIST, US GEOL SURV, 51- Concurrent Pos: Guest prof, Univ Toronto, 68 & Swiss Fed Inst Technol, 71; mem, Lunar Sample Rev Bd, 70-72; prin investr, lunar feldspar Apollo 11-15, 69-72, lunar metamorphism, 73-76. Honors & Awards: Award, Mineral Soc Am, 66. Mem: Fel Geol Soc Am; fel Mineral Soc Am; Am Geophys Union. Res: Crystal chemistry and phase relations of feldspar, silica and rock-forming silicates; thermal expansion measurements by x-ray diffractometry; metamorphic recrystallization of lunar minerals; continental edge tectonics Maine Devonian coastal volcanic belt. Mailing Add: Nat Ctr 959 US Geol Surv Reston VA 22092

STEWART, DAVID BRADSHAW, b Winnipeg, Man, June 18, 16; m 41; c 4. OBSTETRICS & GYNECOLOGY, ZOOLOGY. Educ: Univ Man, BSc, 36, MD, 41. Prof Exp: Consult obstet & gynec, Regional Hosp Bd, Aberdeen, Scotland, 51-52; prof, Univ West Indies, 53-70, vdean fac med, 58-62, dean, 63-64; PROF ZOOL, BRANDON UNIV, 70- Concurrent Pos: Mem, Order of Brit Empire, 46; fel, Royal Col Physicians & Surgeons, Can, 55; fel, Royal Col Obstet & Gynec, 58; Carnegie Found traveling fel, African Med Schs, 60; vis prof, Inst Obstet & Gynec, Univ London, 61 & Queen's Univ, Belfast, 67; consult, WHO, India, 69. Mem: Fel Royal Soc Med; Soc Obstet & Gynec Can; Can Soc Zool. Res: Comparative physiology of reproduction; incidence and effects of chlamydial infections in migrant birds. Mailing Add: Dept of Zool Brandon Univ Brandon MB Can

STEWART, DAVID PERRY, b Summersville, WVa, Mar 14, 16; m 43; c 2. GEOMORPHOLOGY. Educ: WVa Univ, AB, 38; Mich State Univ, MS, 48; Syracuse Univ, PhD(geol), 54. Prof Exp: Instr phys sci, Mich State Univ, 46-49; assoc prof geol, Marshall Col, 49-56; from asst prof to assoc prof, 56-70, PROF GEOL, MIAMI

UNIV, 70- Mem: Geol Soc Am; Nat Asn Geol Teachers. Res: Glacial geology and geology of the United States; glacial geology of New York and Vermont; glacial history of New England and eastern United States. Mailing Add: Dept of Geol Miami Univ Oxford OH 45056

STEWART, DAVID WYLIE, chemistry, see 12th edition

STEWART, DONALD BORDEN, b Sask, Can, Mar 15, 17; nat US; m 52; c 2. ENVIRONMENTAL MANAGEMENT. Educ: Univ Wash, BS, 39. Prof Exp: Chemist, B F Goodrich Co, 39-41, mgr, Gen Chem Lab, 41-42, oper mgr, Res Div, 42-48, opers mgr, Res Ctr, 48-56; bus mgr, Cent Labs, Gen Foods Corp, 56-57, dir admin serv, Res Ctr, 57-61; vpres, Sterling Forest Corp, 61-66; admin officer, 66-67, supt, 67-74, ASST GEN MGR, PALISADES INTERSTATE PARK COMN, 74- Mem: AAAS; Am Chem Soc; fel Am Inst Chem. Res: Analytical methods; industrial safety and hygiene. Mailing Add: Palisades Interstate Park Comn Bear Mountain NY 10911

STEWART, DONALD CHARLES, b Salt Lake City, Utah, Dec 15, 12; m 48; c 2. RADIOCHEMISTRY. Educ: Univ Calif, Los Angeles, AB, 35; Univ Southern Calif, MS, 40; Va Polytech Inst & State Univ, BS, 44; Univ Calif, Berkeley, PhD(biochem), 50. Prof Exp: Chemist, Knudsen Creamery, 35-42; group leader, Metall Lab, Univ Chicago, 44-45, asst sec chief, 45-46; chemist, Radiation Lab, Univ Calif, 46-52; asst dir, 52-54, assoc chemist, 54-59, ASSOC DIR CHEM DIV, ARGONNE NAT LAB, 59- Concurrent Pos: Co-ed, Prog in Nuclear Energy, Series IX, Pergamon Press, 63-74. Mem: Fel AAAS; Am Chem Soc; Sigma Xi. Res: Analytical and inorganic chemistry of rare earth and actinide elements. Mailing Add: Chem Div Argonne Nat Lab 9700 S Cass Ave Argonne IL 60439

STEWART, DONALD GEORGE, b Pocatello, Idaho, Jan 9, 33. MATHEMATICS. Educ: Univ Utah, BA, 59, MS, 61; Univ Tenn, PhD, 63. Prof Exp: Asst prof math, Univ Tenn, 63-64; asst prof, 64-72, ASSOC PROF MATH, ARIZ STATE UNIV, 72- Mem: Am Math Soc. Res: Point-set topology. Mailing Add: Dept of Math Ariz State Univ Tempe AZ 85281

STEWART, DONALD IRVING, operations research, see 12th edition

STEWART, DONALD MARTIN, b Rembrandt, Iowa, Jan 20, 08; m 38; c 2. PLANT PATHOLOGY. Educ: Univ Minn, BS, 31, PhD, 53. Prof Exp: Asst instr forest path, Univ Minn, 32-33; techician white pine blister rust control, USDA, 33-35, dist leader northern Minn, 36-50, res plant pathologist, Coop Rust Lab, Univ Minn & assoc prof plant path, 51-70; LIAISON OFFICER & ACTG PROJ MGR, EGYPT, FOOD & AGR ORGN, UN, 70- Concurrent Pos: Fulbright res fel & lectr, Romania, 65. Mem: Am Phytopath Soc; Mycol Soc Am; Brit Mycol Soc. Res: Physiologic specialization and epidemiology of cereal rusts in wheat and oats. Mailing Add: c/o UN Develop Progs PO Box 982 Cairo Egypt

STEWART, DORIS MAE (MRS FELIX POWELL), b Sandsprings, Mont, Dec 12, 27; m 56; c 2. ZOOLOGY, PHYSIOLOGY. Educ: Univ Puget Sound, BS, 48, MS, 49; Univ Wash, PhD(zool), 53. Prof Exp: NIH fel, Univ Wash, 54; from instr to asst prof zool, Univ Mont, 54-57; asst prof biol, Univ Puget Sound, 57-58; head dept sci, Am Col Girls, Istanbul, 58-62; res asst prof zool, Univ Wash, 63-67, res assoc prof, 67-69; assoc prof biol, Cent Mich Univ, 70-72; res assoc prof zool, Univ Wash, 72-73; ASSOC PROF, SCI DEPT, UNIV BALTIMORE, 73- Mem: AAAS; Am Physiol Soc. Res: Muscle atrophy and hypertrophy and circulation and molting physiology in the spider. Mailing Add: Dept of Sci Univ of Baltimore Baltimore MD 21201

STEWART, ELMO JOSEPH, b Salt Lake City, Utah, Nov 2, 13; m 42; c 1. MATHEMATICS. Educ: Univ Utah, BS, 37, MS, 39; Rice Inst, PhD(math), 53. Prof Exp: Instr math, Univ Utah, 46-53; mathematician, Bendix Corp, 53-54; asst prof math, Calif State Polytech Col, 54-55; assoc prof, 55-61, PROF MATH, NAVAL POSTGRAD SCH, 61- Concurrent Pos: Consult, Bendix Corp, 54-55, Firestone Tire & Rubber Co, 57-58 & Tech Opers, Inc, 58-59. Mem: Math Asn Am. Mailing Add: Dept of Math Naval Postgrad Sch Monterey CA 93940

STEWART, ELWIN LYNN, b Ellensburg, Wash, July 22, 40; m 64; c 1. MYCOLOGY. Educ: Eastern Wash State Col, BA, 69; Ore State Univ, PhD(mycol), 74. Prof Exp: Fel mycol, Dept Bot & Plant Path, Ore State Univ, 74-75; ASST PROF MYCOL, DEPT PLANT PATH, UNIV MINN, ST PAUL, 75- Mem: Mycol Soc Am; Am Phytopath Soc; Brit Mycol Soc; Int Res Group Wood Prescrv. Res: Utilization of mycorrhizal fungi for increasing crop production and the systematics of mycorrhizal fungi. Mailing Add: Dept of Plant Path 304 Stakman Hall of Plant Path St Paul MN 55108

STEWART, FRANK EDWIN, b Dallas, Tex, July 9, 41; m 72. PHYSICS, CHEMISTRY. Educ: Univ Tex, Arlington, BS, 61; Tex A&M Univ, MS, 64, PhD(physics), 66. Prof Exp: Instr physics, Tex A&M Univ, 64-66; Nat Acad Sci resident res assoc chem physics, Jet Propulsion Lab, Univ Calif, 66-67; asst prof physics, Northeast La Univ, 67-71; PROF MATH, PHYSICS & ASTRON & DIR PLANETARIUM, COOKE COUNTY JR COL, 71- Mem: Am Asn Physics Teachers; Am Phys Soc; Sigma Xi. Res: Electron paramagnetic resonance; charge-transfer complexes. Mailing Add: Dept of Math-Physics Box 815 Cooke County Jr Col Gainesville TX 76240

STEWART, FRANK MOORE, b Beirut, Lebanon, Dec 27, 17; US citizen; m 46; c 1. MATHEMATICS. Educ: Princeton Univ, AB, 39; Harvard Univ, MA, 41, PhD(math), 47. Prof Exp: From instr to assoc prof, 47-61, PROF MATH, BROWN UNIV, 61- Mem: AAAS; Am Math Soc; Math Asn Am. Res: Differential equations; population genetics. Mailing Add: Dept of Math 79 Waterman St Brown Univ Providence RI 02912

STEWART, FRANKLIN BURTON, b Sparta, Tenn, Aug 17, 22; m 44. SOIL FERTILITY. Educ: Tenn Polytech Inst, BS, 46; Univ Tenn, MS, 47; Univ Md, PhD(soil chem), 55. Prof Exp: SOIL SCIENTIST, VA TRUCK & ORNAMENTALS RES STA, 55- Mem: Soil Sci Soc Am; Am Soc Agron; Am Hort Soc. Res: Vegetable crop fertilization; soil testing; minor elements; plant nutrition; turf management. Mailing Add: Va Truck & Ornamentals Res Sta PO Box 2160 Norfolk VA 23501

STEWART, GARY FRANKLIN, b Okmulgee, Okla, Apr 3, 35; m 56; c 3. GEOLOGY. Educ: Okla State Univ, BS, 57; Univ Okla, MS, 63; Univ Kans, PhD(geol), 73. Prof Exp: Geologist, Humble Oil & Refining Co, 58-60; asst prof, Kans State Geol Surv, Univ Kans, 62-71; asst prof, 71-73, ASSOC PROF GEOL, OKLA STATE UNIV, 73- Concurrent Pos: Consult, Oak Ridge Nat Lab, 70-72. Mem: Geol Soc Am. Res: Environmental geology; geomorphology; stratigraphy; geologic mapping for environmental purposes; depositional environments of sedimentary rocks. Mailing Add: 1102 N Payne St Stillwater OK 74074

STEWART, GEORGE FRANKLIN, b Mesa, Ariz, Feb 22, 08; m 33; c 3. FOOD TECHNOLOGY. Educ: Univ Chicago, BS, 30; Cornell Univ, PhD(biochem), 33. Prof Exp: Res chemist, Omaha Cold Storage Co, 33-38; from assoc prof to prof, Iowa State Univ, 38-48, assoc dir, Exp Sta, 48-51; prof poultry husb, chmn dept & biochemist, Exp Sta, 51-59, prof food sci & technol, 59-74, dir food protection & toxicol ctr, 64-70, chmn agr toxicol & residue res lab, 66-70, EMER PROF FOOD SCI & TECHNOL, UNIV CALIF, DAVIS, 74- Concurrent Pos: Fulbright fel, Australia, 57-58. Honors & Awards: Christie Award, 50; Int Award, Inst Food Technol, 66. Mem: AAAS; Am Dairy Sci Asn; Poultry Sci Asn; Am Pub Health Asn; Inst Food Technol (pres, 66-67). Res: Poultry, meat and agricultural products; poultry plant sanitation; public health problems associated with poultry products. Mailing Add: Dept of Food Sci & Technol Univ of Calif Davis CA 95616

STEWART, GEORGE HAMILL, b Chambersburg, Pa, July 6, 25; m 50; c 5. BIOMEDICAL ENGINEERING. Educ: Union Col, NY, BSEE, 50; Drexel Univ, MS, 62; Temple Univ, PhD(med physics), 68. Prof Exp: ASSOC PROF MED PHYSICS, SCH MED, TEMPLE UNIV, 50- Concurrent Pos: Mem coun basic sci, Am Heart Asn. Mem: Inst Elec & Electronics Engrs; Asn Advan Med Instrumentation. Res: Biomedical instrumentation; applications of computers to biomedical research; cardiovascular dynamics. Mailing Add: Dept of Med Physics Temple Univ Sch of Med Philadelphia PA 19140

STEWART, GEORGE HUDSON, b Brooklyn, NY, May 13, 25; m 58; c 5. PHYSICAL CHEMISTRY. Educ: Univ Calif, Berkeley, BS, 49; Univ Utah, PhD(phys chem), 58. Prof Exp: Res asst chem, Univ Utah, 58-59; from instr to asst prof, Gonzaga Univ, 59-64, assoc prof chem & chmn dept chem & chem eng, 64-70, dean grad sch, 67-70; PROF CHEM & CHMN DEPT, TEX WOMAN'S UNIV, 70- Mem: AAAS; Am Chem Soc. Res: Physical chemistry of chromatography; dynamics of gas-liquid interface; flow in porous media. Mailing Add: 2003 W Oak St Denton TX 76201

STEWART, GERALD WALTER, b Hamilton, Ohio, Oct 8, 44; m 65; c 2. PHYSICAL CHEMISTRY. Educ: Wilmington Col, BS, 65; SDak Sch Mines & Technol, MS, 67; Univ Idaho, PhD(phys chem), 71. Prof Exp: Res assoc phys chem, Washington Univ, 71-73 & Mass Inst Technol, 73-74; ASST PROF PHYS CHEM, WVA UNIV, 74- Mem: Am Chem Soc. Res: Chemical kinetics; ion molecule reactions by ion-cyclotron resonance spectrometry, hot atom chemistry, molecular beams, chemiluminescence, reactions of excited intermediates. Mailing Add: Dept of Chem WVa Univ Morgantown WV 26506

STEWART, GLENN ALEXANDER, b Ellensburg, Wash, Jan 14, 41; m 62; c 2. SOLID STATE PHYSICS, SURFACE PHSYICS. Educ: Amherst Col, BA, 62; Univ Wash, MSE, 65, PhD(physics), 69. Prof Exp: Fel physics, Univ Wash, 69-70; res fel physics, Calif Inst Technol, 70-72; ASST PROF PHYSICS, UNIV PITTSBURGH, 72- Mem: Am Phys Soc. Res: Phase transitions in surface films, particularly in physically absorbed noble gas monolayers. Mailing Add: Dept of Physics Univ of Pittsburgh Pittsburgh PA 15260

STEWART, GLENN RAYMOND, b Riverside, Calif, Feb 7, 36; m 63; c 2. VERTEBRATE ZOOLOGY, NATURAL HISTORY. Educ: Calif State Polytech Col, BS, 58; Ore State Univ, MA, 60, PhD(zool), 64. Prof Exp: From asst prof to assoc prof, 63-73, PROF ZOOL, CALIF STATE POLYTECH UNIV, 73- Mem: AAAS; Am Soc Ichthyol & Herpet; Am Soc Mammal; Am Inst Biol Sci; Soc Study Amphibians & Reptiles. Res: Ecology, taxonomy and behavior of reptiles, amphibians and mammals; status of endangered and rare species. Mailing Add: Dept of Biol Sci Calif State Polytech Univ Pomona CA 91768

STEWART, GLENN WILLIAM, b East Rochester, NH, Jan 8, 14; m 39; c 2. GEOLOGY. Educ: Univ NH, BS, 35; Syracuse Univ, MS, 37; Harvard Univ, MA, 50. Prof Exp: Instr geol, Purdue Univ, 39-41; asst prof, 41-58, ASSOC PROF GEOL, UNIV NH, 58-; STATE GEOLOGIST, NH, 63- Concurrent Pos: Mem staff, US Geol Surv, 44-45; Fed Off Water Resources res grant, 73. Mem: Fel AAAS; fel As Am Geogrs; Mineral Soc Am; Geochem Soc; Nat Asn Geol Teachers. Res: Nonmetallic minerals; structural geology; water in crystalline rocks. Mailing Add: Dept of Geol James Hall Univ of NH Durham NH 03824

STEWART, GORDON ARNOLD, b Denver, Colo, July 2, 34; m 62; c 1. DAIRY SCIENCE. Educ: Univ Mo, BS, 56, MS, 58, PhD(dairy), 60. Prof Exp: Assoc prof agr, Southeast Mo State Col, 60-65; asst prof, 65-68, ASSOC PROF DAIRY SCI, LA TECH UNIV, 68- Concurrent Pos: Assoc prof animal sci, Col of Agr, Haile Sellassie Univ, Alemaya, Ethiopia, 73-75. Mem: Nat Asn Cols & Teachers Agr (treas, 65-73); Am Dairy Sci Asn; Am Soc Animal Sci. Res: Dairy cattle nutrition, breeding and management. Mailing Add: Dept of Animal Indust La Tech Univ PO Box 5618 Ruston LA 71270

STEWART, GORDON ERVIN, b San Bernardino, Calif, June 25, 34; m 59; c 3. MICROWAVE PHYSICS, PLASMA PHYSICS. Educ: Univ Calif, Los Angeles, BS, 55, MS, 57; Univ Southern Calif, PhD(elec eng), 63. Prof Exp: Mem tech staff, Hughes Aircraft Co, 55-62; mem tech staff, Plasma Res Lab, 62-70, SECT HEAD, AEROSPACE CORP, 70- Concurrent Pos: Asst prof, Univ Southern Calif, 62-66; mem comn 6, Int Union Radio Sci. Mem: Inst Elec & Electronic Engrs; Am Phys Soc. Res: Radar scattering; antenna theory; wave propagation in plasmas; acoustic holography. Mailing Add: Aerospace Corp 2350 E El Segundo Blvd El Segundo CA 90061

STEWART, GORDON LEROY, b Aurora, Utah, Oct 7, 33; m 57; c 4. SOIL PHYSICS, HYDROLOGY. Educ: Utah State Univ, BS, 55, MS, 57; Wash State Univ, PhD(soil physics), 62. Prof Exp: Res asst soil physics, Utah State Univ, 55-56 & Wash State Univ, 58-62; proj chief isotopic hydrol, Water Resources Div, US Geol Surv, 62-68; ASSOC PROF SOIL PHYSICS, UNIV MASS, AMHERST, 68- Concurrent Pos: Comt mem, US-Int Hydrol Decade Work Group Nuclear Tech in Hydrol, 66-69; consult. Honors & Awards: Blue Ribbon Award, Soil Sci Soc Am, 62. Mem: Am Geophys Union; Soil Sci Soc Am; Am Soc Agron. Res: Saturated and unsaturated water flow in porous media; isotopic techniques to trace water and pollutant movement and to study clay-water interaction, movement and behavior of potential pollutants in soils. Mailing Add: Dept of Plant & Soil Sci Univ of Mass Amherst MA 01002

STEWART, HAROLD BROWN, b Chatham, Ont, Can, Mar 9, 21; m 50. BIOCHEMISTRY. Educ: Univ Toronto, MD, 44, PhD, 50; Cambridge Univ, PhD, 55. Prof Exp: Assoc prof, 55-60, chmn dept, 65-72, PROF BIOCHEM, UNIV WESTERN ONT, 60-, DEAN FAC GRAD STUDIES, 72- Concurrent Pos: Med Res Coun vis scientist, Cambridge Univ, 71-72. Mem: Am Soc Biol Chem; Can Physiol Soc; Can Biochem Soc; Brit Biochem Soc. Res: Intermediary metabolism in animals and microorganisms. Mailing Add: 118 Base Line Rd E London ON Can

STEWART, HAROLD JULIAN, medicine, cardiology, deceased

STEWART, HARRIS BATES, JR, b Auburn, NY, Sept 19, 22; m 59; c 2. OCEANOGRAPHY. Educ: Princeton Univ, AB, 48; Univ Calif, MS, 52, PhD(oceanog), 56. Prof Exp: Hydrographer, US Naval Hydrographic Off, 48-50; instr,

Hotchkiss Sch, Conn, 50-51; res asst oceanog, Scripps Inst Oceanog, Univ Calif, 51-56; chief oceanogr, US Coast & Geod Surv, 57-65; dir inst oceanog, Environ Sci Serv Admin, 65-70, DIR ATLANTIC OCEANOG & METEOROL LABS, NAT OCEANIC & ATMOSPHERIC ADMIN, 70- Concurrent Pos: Chmn ocean surv panel, Interagency Comt Oceanog, 62-66 & int progs panel, 65-67; US nat coordr, Coop Invest of Caribbean & Adjacent Regions, 68-; chmn adv coun, Dept Geol & Geophys Sci, Princeton Univ, 73- Honors & Awards: Silver Medal, US Dept Com, 60, Gold Medal, 65; Almirante Padilla Decoration, Repub Colombia, 72. Mem: Fel AAAS; fel Geol Soc Am; Am Geophys Union; Marine Technol Soc (vpres, 74-); Int Oceanog Found (vpres). Res: Coastal lagoons; marine geology; physical oceanography. Mailing Add: Atlantic Oceanog & Meteorol Labs Nat Ocean & Atmospher Adm Va Key Miami FL 33149

STEWART, HERBERT, b Stanton, Ky, July 18, 28; m 53; c 3. SCIENCE EDUCATION, BIOLOGY. Educ: Univ Conn, BA, 54, MS, 56; Columbia Univ, EdD, 58. Prof Exp: Teacher pub sch, Ky, 51-53; instr bot, Univ Conn, 54-56; instr biol, Teachers Col, Columbia Univ, 57-58; prof sci educ & biol, Md State Teachers Col, Towson, 58-59; asst prof biol, Sch Com, NY Univ, 59-60; asst prof sci educ, Rutgers Univ, 60-61; assoc prof, Univ SFla, 61-67; PROF SCI EDUC, FLA ATLANTIC UNIV, 67- Concurrent Pos: Sci Manpower fel, Columbia Univ, 59-60; dir, NSF In-Serv Inst, 66-67. Mem: AAAS; Nat Asn Res Sci Teaching. Res: Curriculum research and development in science education at elementary and secondary school levels; application of biological bases of learning to education. Mailing Add: Dept of Sci Educ Fla Atlantic Univ Boca Raton FL 33432

STEWART, HUGH BARNES, b New Holland, Ohio, Dec 17, 16; m 41; c 2. PHYSICS. Educ: Kent State Univ, BS, 38; Ohio State Univ, MS, 39, PhD(physics), 47. Prof Exp: Asst, Ohio State Univ, 39-41; res engr, Allison Div, Gen Motors Corp, Ind, 41-45; res assoc, Knolls Atomic Power Lab, Gen Elec Co, 47-59, mgr reactor physics, SIR Proj, 55-57, DIG Proj, 57-59; leader reactor physics, HTGR Proj, Gen Atomic Div, Gen Dynamics Corp, 59-61, chmn, Nuclear Anal & Reactor Physics Dept, 61-63, asst lab dir, 63-69; VPRES HIGH TEMPERATURE GAS-COOLED REACTOR FUELS, GEN ATOMIC CO, 69- Mem: Am Phys Soc; Am Nuclear Soc. Res: Mechanical vibrations; infrared spectroscopy; critical experiments; reactor physics; nuclear fuel engineering. Mailing Add: Gen Atomic Co PO Box 81608 San Diego CA 92138

STEWART, IVAN, b Stanton, Ky, July 24, 22; m 47; c 3. PLANT CHEMISTRY. Educ: Univ Ky, BS, 48, MS, 49; Rutgers Univ, PhD(soils), 51. Prof Exp: From asst biochemist to assoc biochemist, 51-61, BIOCHEMIST, CITRUS EXP STA, UNIV FLA, 61- Res: Mineral nutrition of plants. Mailing Add: Inst of Food & Agr Sci Agr Res & Educ Ctr Univ of Fla PO Box 1088 Lake Alfred FL 33850

STEWART, JACK LAUREN, b Covington, Okla, Apr 3, 24; m 48; c 4. DENTISTRY. Educ: Univ Kansas City, DDS, 52. Prof Exp: Pvt pract, 52-62; asst prof, 63-67, assoc prof & coordr res, 67-70, PROF DENT & ASST DEAN RES & CONTINUING EDUC, SCH DENT, UNIV MO-KANSAS CITY, 70- Concurrent Pos: Investr, US Army res contract, 63-67, co-responsible investr, 67-70; consult, Wadsworth Vet Admin Ctr, 64- & Kansas City Vet Admin Hosp, 67-; abstractor, Oral Res Abstr, Am Dent Asn; chmn sect comput appln, Am Asn Dent Schs, 68-69, from vchmn to chmn sect learning resources, 69-72. Mem: AAAS; Am Dent Asn; Int Asn Dent Res; Am Asn Lab Animal Sci. Res: Research administration; maxillofacial injuries; oral lesions. Mailing Add: 650 E 25th St Kansas City MO 64108

STEWART, JAMES ALLEN, b Pembroke, Ont, Can, Jan 7, 27; m 54; c 3. PHYSICAL CHEMISTRY. Educ: Queen's Univ, Ont, BA, 51, MA, 53; Univ Ottawa, PhD(phys chem), 59. Prof Exp: Anal res chemist, Dept Nat Health & Welfare, Can, 54-59; from asst prof to assoc prof, 59-69, PROF PHYS CHEM, UNIV NDAK, 69- Mem: Am Chem Soc. Res: Chemical kinetics of hydrolytic enzyme systems, ester hydrolyses and excited alkali metal reactions; solvent isotope effects on reaction rates. Mailing Add: Dept of Chem Univ of NDak Grand Forks ND 58201

STEWART, JAMES ANTHONY, b Manchester, NH, Aug 2, 38; m 63; c 3. BIOCHEMISTRY. Educ: St Anselm's Col, BA, 63; Univ Conn, PhD(biochem), 67. Prof Exp: Investr, Biol Div, Oak Ridge Nat Lab, 67-68; asst prof, 68-73, ASSOC PROF BIOCHEM, UNIV NH, 73- Mem: AAAS; Soc Develop Biol; Am Inst Biol Sci; Am Soc Biol Chemists. Res: Regulation of protein and nucleic acid synthesis during development and differentiation of the mouse central nervous system. Mailing Add: Dept of Biochem Univ of NH Durham NH 03824

STEWART, JAMES COLLIER, b Memphis, Tenn, Mar 30, 13. MATHEMATICS. Educ: Univ Miss, BA, 34, MA, 36; La State Univ, MS, 40; Univ Ill, PhD(math), 46. Prof Exp: Teacher high sch, Miss, 36-38; instr math, La State Univ, 42-46; from asst prof to assoc prof, 46-64, PROF MATH, LAWRENCE UNIV, 64- Concurrent Pos: NSF fac fel, Harvard Univ, 60-61. Mem: Math Asn Am. Res: Differential geometry; geodesic correspondences between surfaces of revolution. Mailing Add: Dept of Math Lawrence Univ Appleton WI 54911

STEWART, JAMES DREWRY, b Toronto, Ont, Mar 29, 41. MATHEMATICAL ANALYSIS. Educ: Univ Toronto, BSc, 63, PhD(math), 67; Stanford Univ, MS, 64. Prof Exp: Nat Res Coun Can fel, Univ London, 67-69; asst prof, 69-74, ASSOC PROF MATH, MCMASTER UNIV, 74- Mem: Am Math Soc; Math Asn Am; Can Math Cong. Res: Abstract harmonic analysis, functional analysis, history of mathematics. Mailing Add: Dept of Math McMaster Univ Hamilton ON Can

STEWART, JAMES EDWARD, b Anyox, BC, Aug 3, 28; m 67; c 2. BACTERIAL PHYSIOLOGY. Educ: Univ BC, BSA, 52, MSA, 54; Univ Iowa, PhD, 58. Prof Exp: Scientist, Fisheries Res Bd Can, 58-74, SCIENTIST, RES & DEVELOP DIRECTORATE, FISHERIES SERV CAN, 74- Mem: AAAS; Soc Invert Path; Am Soc Microbiol; Can Soc Microbiol. Res: Microbial oxidation of hydrocarbons; enzymes; bacterial metabolism; defense mechanisms and diseases of marine animals. Mailing Add: Halifax Lab R&D Directorate Fisheries Serv Can PO Box 429 Halifax NS Can

STEWART, JAMES EUGENE, chemical physics, see 12th edition

STEWART, JAMES LLOYD, b Chengtu, China, Jan 5, 18; nat US; m 45; c 1. PHYSICS. Educ: Univ Sask, BA, 38, MA, 40; Johns Hopkins Univ, PhD(physics), 43. Prof Exp: Lab asst physics, Univ Sask, 35-38, technician, Radon Plant, Sask Cancer Comn, 38-40; jr instr physics, Johns Hopkins Univ, 40-43; sci officer, Ballistics Lab, Can Armament Res & Develop Estab, 43-45; lectr physics, Queen's Univ, Ont, 45-46; asst prof, Rutgers Univ, 46-51; physicist, US Navy Electronics Lab, 51-67; PHYSICIST, NAVAL UNDERSEA CTR, 67- Mem: Sr mem Inst Elec & Electronics Eng; fel Acoust Soc Am. Res: Slow neutrons; ballistics; ultrasonics; underwater acoustics; signal processing theory. Mailing Add: 2941 Aber St San Diego CA 92117

STEWART, JAMES MCDONALD, b Taft, Tenn, Sept 22, 41; m 64; c 2. PLANT PHYSIOLOGY. Educ: Okla State Univ, BS, 63, PhD(plant physiol), 68. Prof Exp: PLANT PHYSIOLOGIST COTTON PHYSIOL, AGR RES SERV, USDA, 68-; ASST PROF PLANT & SOIL SCI, UNIV TENN, KNOXVILLE, 68- Mem: Sigma Xi; Am Soc Plant Physiologists; Bot Soc Am; Agron Soc Am; AAAS. Res: Basic and applied research on cotton fiber and seed development and maturation with correlated studies concerning the effects of environment and heritable traits theron. Mailing Add: Dept of Plant & Soil Sci Univ of Tenn Knoxville TN 37916

STEWART, JAMES RAY, b Beeville, Tex, Aug 5, 37; m 68. CELL BIOLOGY, MICROBIOLOGY. Educ: NTex State Univ, BS, 59; Univ Ala, Tuscaloosa, MS, 65; Univ Tex, Austin, PhD(biol sci), 70. Prof Exp: NIH fel, Univ Tex Med Sch, San Antonio, 70-74; MEM FAC, DEPT BIOL, TYLER STATE COL, 74- Mem: Am Soc Microbiol; Phycol Soc Am. Res: Taxonomic studies of myxobacteria; studies of enzymes involved in morphogenesis of myxobacteria and Acanthamoeba. Mailing Add: Dept of Biol Tyler State Col Tyler TX 75701

STEWART, JAMES RUSSELL, physics, see 12th edition

STEWART, JAMES T, b Birmingham, Ala, Dec 1, 38; m 63; c 3. PHARMACEUTICAL CHEMISTRY. Educ: Auburn Univ, BS, 60, MS, 63; Univ Mich, PhD(pharmaceut chem), 67. Prof Exp: Asst prof, 67-71, ASSOC PROF PHARMACEUT CHEM, UNIV GA, 71- Concurrent Pos: Mead-Johnson res grant, 67-68; NIH biomed sci grant, 68-69. Mem: Am Chem Soc; Am Pharmaceut Asn. Res: Fluorometric analysis of pharmaceuticals; gas chromatography; synthesis of fluorescent tagging agents. Mailing Add: Sch of Pharm Univ of Ga Athens GA 30601

STEWART, JAY JUNIOR, organic chemistry, see 12th edition

STEWART, JENNIFER KEYS, b Rome, Ga, May 15, 47. ENDOCRINOLOGY. Educ: Emory Univ, BS, 68, MS, 69, PhD(physiol), 75. Prof Exp: Instr biol, Mercer Univ, 69-71; RES FEL ENDOCRINOL, HARBORVIEW MED CTR, 75- Res: Endocrine regulation of protein metabolism; mechanism and control of hormone secretion; mechanism of action of hormones. Mailing Add: Dept of Endocrinol Harborview Med Ctr 325 Ninth Ave Seattle WA 98104

STEWART, JOHN ALDEN, biology, see 12th edition

STEWART, JOHN ALLAN, b Saskatoon, Sask, Feb 18, 24; m 48; c 5. SOIL SCIENCE. Educ: Univ BC, BSA, 50, MSA, 53; Univ Wis, PhD, 64. Prof Exp: Lectr soils, Univ BC, 50-51 & 52-54; plant physiologist, Can Dept Agr, 54-65; agronomist, Can, 65-67, res agronomist, Ill, 67-68, mgr fertilizers & cropping systs res, Res & Develop Div, 68-70, mgr agr res, 70-72, DIR RES & DEVELOP, INT MINERALS & CHEM CORP, 72- Concurrent Pos: Mem, Coun Agr Sci & Technol. Mem: Am Soc Agron; Can Soc Soil Sci; Agr Inst Can; AAAS. Res: Mineral nutrition of agricultural crops; environmental aspects of agricultural technology; industrial minerals applications. Mailing Add: 1328 Redwood Libertyville IL 60048

STEWART, JOHN CONYNGHAM, b New York, NY, Feb 10, 30; m 56; c 2. GEOLOGY. Educ: Trinity Col, Conn, BA, 52; Princeton Univ, MA, 56, PhD(geol), 57. Prof Exp: Geologist, Harvard NSF res proj, Dordogne dist, France, 57; geophys interpreter, Mobil Oil Co, Venezuela, 58-61; from instr to assoc prof, 61-74, PROF GEOL, BROOKLYN COL, 74-, CHMN DEPT, 68- Concurrent Pos: Danforth assoc. Mem: AAAS; Geol Soc Am; Soc Econ Paleont & Mineral; Am Asn Petrol Geol; Am Geophys Union. Res: Sedimentology; stratigraphy. Mailing Add: 34 Westcott Rd Princeton NJ 08540

STEWART, JOHN DOUGLAS, b Hayward, Calif, May 14, 27; m 63; c 3. PSYCHOPHYSICS. Educ: Univ Calif, Berkeley, BS, 50, MS, 53. Prof Exp: Eng intern, Ames Res Lab, Nat Adv Comt Aeronaut, 51-52; engr, Marquardt Aircraft Co, 53; RES SCIENTIST, AMES RES CTR, NASA, 54- Mem: Am Inst Aeronaut & Astronaut. Res: Pilots perception and use of angular acceleration motion cues, vestibular-visual interactions, and the effects of vibration on the pilot's opinion and performance in closed loop control situations. Mailing Add: Ames Res Ctr NASA Mail Stop 239-3 Moffett Field CA 94035

STEWART, JOHN HARRIS, b Berkeley, Calif, Aug 7, 28; m 62; c 2. GEOLOGY. Educ: Univ NMex, BS, 50; Stanford Univ, PhD, 61. Prof Exp: GEOLOGIST, US GEOL SURV, 51- Mem: Geol Soc Am. Res: Stratigraphy; sedimentology; regional stratigraphy of Triassic rocks in Utah, Colorado, Nevada, Arizona and New Mexico and of late Precambrian and Cambrian in Nevada and California; regional and local mapping in Nevada; compilation of geologic map of Nevada. Mailing Add: US Geol Surv 345 Middlefield Rd Menlo Park CA 94025

STEWART, JOHN MATHEWS, b Vermillion, SDak, Apr 5, 20; m 43; c 1. ORGANIC CHEMISTRY. Educ: Univ Mont, BA, 41; Univ Ill, PhD(org chem), 44. Prof Exp: Res chemist, War Prod Bd, Univ Ill, 43-45 & Calif Res Corp, 45-46; from asst prof to assoc prof, 46-56, chmn dept, 59-67, PROF CHEM, UNIV MONT, 56-, DEAN GRAD SCH, 68-, ACTG ACAD VPRES, 75- Mem: Am Chem Soc. Res: Reactions of olefin sulfides; additions to unsaturated nitriles; participation of cyclopropane rings in conjugation, ring-opening reactions of cyclopropanes; use of diazomethane in synthesis of heterocyclic compounds. Mailing Add: Grad Sch Univ of Mont Missoula MT 59801

STEWART, JOHN MORROW, b Guilford Co, NC, Oct 31, 24; m 49; c 3. BIOCHEMISTRY, PHARMACOLOGY. Educ: Davidson Col, BS, 48; Univ Ill, MS, 50, PhD(org chem), 52. Prof Exp: Instr chem, Davidson Col, 48-49; asst, Rockefeller Univ, 52-57, from asst prof to assoc prof biochem, 57-68; PROF BIOCHEM, MED SCH, UNIV COLO, DENVER, 68- Mem: Am Chem Soc; Am Soc Pharmacol & Exp Therapeut; Harvey Soc; NY Acad Sci; Am Soc Biol Chemists. Res: Chemistry and pharmacology of peptide hormones, methods of peptide synthesis, antimetabolites, lipid biosynthesis, amino acids; synthetic organic chemistry. Mailing Add: Dept of Biochem Univ of Colo Med Sch Denver CO 80220

STEWART, JOHN WESTCOTT, b New York, NY, Nov 15, 26; m 54; c 1. PHYSICS. Educ: Princeton Univ, AB, 49; Harvard Univ, MA, 50, PhD(physics), 54. Prof Exp: Res fel, 54-56, asst prof, 56-60, ASSOC PROF PHYSICS, UNIV VA, 60-, ASST DEAN COL, 70- Mem: Fel Am Phys Soc; Am Asn Physics Teachers. Res: Properties of matter under combined field of high pressure and low temperature; meteorology. Mailing Add: Dept of Physics Univ of Va Charlottesville VA 22903

STEWART, JOHN WILEY, biochemistry, see 12th edition

STEWART, JOHN WRAY BLACK, b Coleraine, NIreland, Jan 16, 36; Can citizen; m 65; c 2. SOIL SCIENCE, CHEMISTRY. Educ: Queen's Univ, Belfast, BSc, 58, BAgr, 59, PhD(soil sci), 63. Prof Exp: From sci officer to sr sci officer soil sci, Chem Res Div, Ministry Agr, NIreland, 59-64; fel, 64-65, asst prof, 65-70, ASSOC PROF SOIL SCI, UNIV SASK, 70- Concurrent Pos: Tech expert, Int Atomic Energy Agency, Vienna, 71-72. Mem: Agr Inst Can; Brit Soc Soil Sci; Am Soc Agron. Res: Soil chemistry and fertility; cycling of macro and micro nutrients and heavy metals in the soil plant system. Mailing Add: Dept of Soil Sci Univ of Sask Saskatoon SK Can

STEWART, JOSEPH KYLE, b Oceana, WVa, Mar 24, 06; m 33; c 1. MATHEMATICS. Educ: Marshall Col, AB, 27; WVa Univ, MS, 31, PhD(math), 34. Prof Exp: Teacher high schs, WVa, 27-30; instr, 32, 34-37, from asst prof to prof, 37-74, dir, Kanawha Valley Grad Ctr Sci & Eng, 58-60, chmn dept math, 60-65, asst grad dean, 61-65, EMER PROF MATH, W VA UNIV, 74- Mem: Am Math Soc; Math Asn Am. Res: Geometry of hyperspace; electronics; radar. Mailing Add: Dept of Math WVa Univ Morgantown WV 26506

STEWART, JOSEPH LETIE, b Salida, Colo, Aug 2, 27; m 50; c 3. AUDIOLOGY, SPEECH PATHOLOGY. Educ: Univ Denver, BA, 49, MA, 50; Univ Iowa, PhD(speech path, audiol, anthrop), 59. Prof Exp: Res assoc speech path, Univ Iowa, 58-59; asst prof audiol & dir hearing ctr, Univ Denver, 59-65; consult audiol & speech path, Neurol & Sensory Dis Control Prog, Nat Ctr Chronic Dis Control, USPHS, 65-70, CHIEF SENSORY DISABILITIES PROG, INDIAN HEALTH SERV, USPHS, 70- Res: Cultural variation of incidence of stuttering; auditory deprivation in children; inhibition and facilitation of competing sensory stimuli; otitis media prevalence. Mailing Add: Indian Health Serv USPHS 2701 Frontier Pl NE Albuquerque NM 87131

STEWART, KENNETH C, b Warren, Vt, Oct 19, 17; m 41; c 2. ACOUSTICS, ENVIRONMENTAL SCIENCE. Educ: Univ Vt, BSEE, 50; Univ Pittsburgh, MS, 52. Prof Exp: Asst prof, 57-63, ASSOC PROF INDUST HYG, GRAD SCH PUB HEALTH, UNIV PITTSBURGH, 63-, ADJ ASSOC PROF AUDIOL, 66- Concurrent Pos: Career develop award math, Univ Vt, 57-59; govt & indust consult, 62-; prog dir acoust environ control, NIH Training grant, 70- Mem: Acoust Soc Am. Res: Physical acoustics and noise control. Mailing Add: Grad Sch Pub Health Univ of Pittsburgh Pittsburgh PA 15261

STEWART, KENNETH MALCOLM, b Tecumseh, Nebr, June 16, 16; m 60; c 2. ANTHROPOLOGY, ETHNOLOGY. Educ: Univ Calif, Berkeley, AB, 38, MA, 40, PhD(anthrop), 46. Prof Exp: Asst prof anthrop, Fresno State Col, 47; assoc prof, 47-58, PROF ANTHROP, ARIZ STATE UNIV, 58- Concurrent Pos: Fac res grant, Ariz State Univ & Colo River Reservation, 70-71. Mem: Am Anthrop Asn; Am Ethnol Soc; Am Soc Ethnohist; Sigma Xi. Res: Ethnography and ethnohistory of Southwestern Indian tribes, with emphasis on the Mohave. Mailing Add: Dept of Anthrop Ariz State Univ Tempe AZ 85281

STEWART, KENNETH WILSON, b Walters, Okla, Mar 5, 35; m 56; c 3. ENTOMOLOGY, AQUATIC ECOLOGY. Educ: Okla State Univ, BS, 58, MS, 59, PhD(entom, zool), 63. Prof Exp: Entomologist, Rocky Mt Forest & Range Exp Sta, US Forest Serv, 58-59; head dept biol, Coffeyville Col, 60-61; from instr to assoc prof, 61-74, PROF BIOL, N TEX STATE UNIV, 74-, FAC RES GRANTS, 63- Concurrent Pos: NIH res grant, 66-68; consult investr, US Corps Engrs. Mem: Entom Soc Am; Am Entom Soc. Res: Southwest United States stream benthos community structure and dynamics; Southwest Plecoptera; passive dispersal of Algae and Protozoa by aquatic insects; food habits and life histories of aquatic insects and spiders. Mailing Add: Dept of Biol Sci NTex State Univ Denton TX 76203

STEWART, KENT KALLAM, b Omaha, Nebr, Sept 5, 34; m 56; c 4. BIOCHEMISTRY. Educ: Univ Calif, Berkeley, AB, 56; Fla State Univ, PhD(chem), 65. Prof Exp: USPHS guest investr biochem, Rockefeller Univ, 65-67, res assoc, 67-68, asst prof, 68-69; res chemist, 70-75, LAB CHIEF, NUTRIENT COMPOS LAB, NUTRIT INST, AGR RES SERV, USDA, 75- Mem: AAAS; Am Chem Soc; NY Acad Sci; Inst Food Technologists. Res: Protein chemistry as related to structure and function; separation chemistry; naturally occuring protease inhibitors; automated analyses; nutrient composition of foods. Mailing Add: 220 Hillsboro Dr Silver Spring MD 20902

STEWART, KENTON M, b Withee, Wis, Aug 28, 31; m 54; c 3. ZOOLOGY, LIMNOLOGY. Educ: Wis State Univ, Stevens Point, BS, 55; Univ Wis-Madison, MS, 59, PhD(zool), 65. Prof Exp: Asst limnol, ecol & invert zool, Univ Wis-Madison, 58-61, asst limnol, 61-65, fel, 65-66; asst prof, 66-71, ASSOC PROF BIOL, STATE UNIV NY BUFFALO, 71- Mem: Am Soc Limnol & Oceanog; Ecol Soc Am; Int Asn Theoret & Appl Limnol; Int Asn Gt Lakes Res. Res: Physical limnology and eutrophication; comparative limnology of Finger Lakes of New York. Mailing Add: Dept of Biol State Univ of NY Buffalo NY 14214

STEWART, KIRKLAND BRUCE, b Tacoma, Wash, July 24, 22. APPLIED STATISTICS. Educ: Col Puget Sound, BA, 48, MS, 50. Prof Exp: Engr, Boeing Co, 52-54; sr statistician, Gen Elec Co, 55-65; res engr, Boeing Co, 65-67; SR RES SCIENTIST APPL STATIST, PAC NORTHWEST LAB, BATTELLE MEM INST, 67- Concurrent Pos: Mem subcomt statist, Am Nat Stand Inst, 75- Mem: Inst Nuclear Mat Mgt; AAAS. Res: Application of statistical, probabalistic and game theory techniques to the mass balance accounting of nuclear materials. Mailing Add: Pac Northwest Lab Battelle Mem Inst Box 999 Richland WA 99352

STEWART, LAURA CHRISTINE, biochemistry, see 12th edition

STEWART, LELAND TAYLOR, b San Francisco, Calif, Nov 24, 28; m 56; c 2. STATISTICS. Educ: Stanford Univ, BS, 51, MS, 57, PhD(statist), 65. Prof Exp: Res engr, Autonetics Div, NAm Aviation, Inc, 51-56 & Electronic Defense Labs, Sylvania Elec Prod, Inc, 58-61; statistician, C-E-I-R, Inc, 61-65; STAFF SCIENTIST, LOCKHEED MISSILES & SPACE CO, 65- Mem: Am Statist Asn. Res: Bayesian statistics and decision theory. Mailing Add: Res Lab Lockheed Missiles & Space Co 3251 Hanover Palo Alto CA 94300

STEWART, LEVER F, b Philadelphia, Pa, May 10, 24; m 52; c 2. NEUROLOGY. Educ: Princeton Univ, AB, 49; Univ Pa, MD, 49; McGill Univ, MSc, 56. Prof Exp: Asst prof, 58-63, ASSOC PROF NEUROL, SCH MED, UNIV VA, 63-, DIR EEG LAB, UNIV HOSP, 58- Mem: Am Electroencephalog Soc; AMA; Am Acad Neurol. Res: Electroencephalography; epilepsy. Mailing Add: 1935 Blue Ridge Rd Charlottesville VA 22903

STEWART, LYNN MARTIN, b Kansas City, Kans, Nov 13, 33; m 62; c 1. BIOCHEMISTRY, ENZYMOLOGY. Educ: Univ Kans, BS, 55; Southwestern Baptist Theol Sem, MRE, 58; Kans State Univ, PhD(biochem), 62. Prof Exp: Asst nuclear engr, Gen Dynamics/Convair, Tex, 57-58; asst chem, Kans State Univ, 58-60 & biochem, 60-62; NIH fel chem, Ind Univ, 62-64; asst prof biochem, Meharry Med Col, 64-66; Fulbright lectr, Univ Nangrahar, Afghanistan, 66-67; asst prof, 67-72, ASSOC PROF BIOCHEM, MEHARRY MED COL, 72- Mem: AAAS; fel Am Inst Chemists; Am Chem Soc; Am Sci Affiliation; NY Acad Sci. Res: Catalytic function of specific amino acyl residues of proteins in function and action patterns of carbohydrate enzymes; coupled enzyme assays. Mailing Add: Dept of Biochem Meharry Med Col Nashville TN 37208

STEWART, MARGARET MCBRIDE, b Greensboro, NC, Feb 6, 27; m 69. VERTEBRATE ECOLOGY, HERPETOLOGY. Educ: Univ NC, AB, 48, MA, 51; Cornell Univ, PhD(vert zool), 56. Prof Exp: Lab instr anat & physiol, Woman's Col, Univ NC, 50-51; instr biol, Catawba Col, 51-53; asst bot & taxon, Cornell Univ, 53-56; from asst prof to assoc prof, 56-65, PROF VERT BIOL, STATE UNIV NY ALBANY, 65- Concurrent Pos: Res Found grant-in-aid, 58-61, 65-71 & 73-74. Mem: Ecol Soc Am; Asn Trop Biol; Am Soc Ichthyol & Herpet; Soc Study Amphibians & Reptiles. Res: Competition in tropical and temperate frogs; pattern polymorphism; population dynamics of Adirondack frogs; amphibians of Jamaica; small mammal ecology; amphibians of Malawi. Mailing Add: Dept of Biol Sci State Univ of NY Albany NY 12222

STEWART, MARK ARMSTRONG, b Yeovil, Eng, July 23, 29; US citizen; m 55; c 3. BIOCHEMISTRY, PSYCHIATRY. Educ: Cambridge Univ, BA, 52; Univ London, LRCP & MRCS, 56. Prof Exp: Asst psychiat, Sch Med, Wash Univ, 57-61, instr, 61-63, asst prof psychiat & pediat, 63-67, from assoc prof to prof psychiat, 67-72, assoc prof pediat, 68-72; IDA P HALLER PROF CHILD PSYCHIATRY, COL MED, UNIV IOWA, 72- Concurrent Pos: NIMH res career develop award, 61-71; dir psychiat, St Louis Children's Hosp. Mem: Asn Res Nerv & Ment Dis; Am Soc Biol Chem; Am Soc Neurochem; Psychiat Res Soc; Soc Res Child Develop. Res: Relationship of temperament and its biochemical correlates to children's behavior problems. Mailing Add: 500 Newton Rd Iowa City IA 52240

STEWART, MELBOURNE GEORGE, b Detroit, Mich, Sept 30, 27; m 54; c 3. PHYSICS. Educ: Univ Mich, AB, 49, MS, 50, PhD(physics), 55. Prof Exp: Instr physics, Univ Mich, 54-55; res assoc, Iowa State Univ, 55-56, from asst prof to assoc prof physics, 56-63; chmn dept, 63-73, PROF PHYSICS, WAYNE STATE UNIV, 63-, ASSOC PROVOST, 73- Mem: Am Phys Soc. Res: Nuclear structure. Mailing Add: Mackenzie Hall Wayne State Univ Detroit MI 48202

STEWART, NORMAN REGINALD, b Los Angeles, Calif, Oct 1, 28; m 55. CULTURAL GEOGRAPHY. Educ: Univ Calif, Los Angeles, BA, 50, MA, 55, PhD(geog), 61. Prof Exp: Teaching asst geog, Univ Calif, Los Angeles, 54-55; teaching asst, La State Univ, Baton Rouge, 55-56; teaching asst, Univ Calif, Los Angeles, 56-58 & 59-60; instr, Los Angeles State Col, 60-61; prof, Univ Huamanga, Peru, 61-62; from asst prof to assoc prof, Univ Nebr, Lincoln, 62-68; ASSOC PROF GEOG, UNIV WIS-MILWAUKEE, 68- Concurrent Pos: Fulbright-Hays grant, Peru, 61-62; Soc Sci Res Coun-Am Coun Learned Socs foreign area fel prog & Woods fel, Ecuador, 66-67. Mem: AAAS; Asn Am Geog; Am Geog Soc; Latin Am Studies Asn. Res: Latin America; pioneer settlement in the lowland tropics. Mailing Add: Dept of Geog Univ of Wis Milwaukee WI 53201

STEWART, OMER CALL, b Provo, Utah, Aug 17, 08; m 36; c 4. APPLIED ANTHROPOLOGY, ETHNOHISTORY. Educ: Univ Utah, AB, 33; Univ Calif, Berkeley, PhD(anthrop), 39. Prof Exp: Instr anthrop, Univ Tex, Austin, 39; instr, Univ Minn, Minneapolis, 41-42; from asst prof to prof, 45-74, chmn dept soc sci, 52-54, chmn dept anthrop, 54-59, EMER PROF ANTHROP, UNIV COLO, BOULDER, 74- Concurrent Pos: Soc Sci Res Coun fel, Univ Minn & Zuni Indian Pueblo, 40-41; expert witness for Chippewa, Shoshone, Paiute & Calif Indians, US Indians Claims Comn, DC, 51-59; Nat Inst Ment Health grant, Reservation Town, Colo, 59-64; dir, Tri-Ethnic Res Proj, Univ Colo-Nat Inst Ment Health, 59-64 & Univ Colo Study Abroad Prog, Bordeaux, France, 67-69; vis prof anthrop, Univ Bordeaux, 68-69; Am Coun Learned Socs grant in aid hist Peyote relig var US reservations & arch, 72. Mem: AAAS; Am Anthrop Asn; Soc Am Archaeol; Soc Appl Anthrop (pres, 65-66). Res: Peyote religion; values and behavior of a tri-ethnic community; aboriginal fires and their influence on vegetation; American Indian land use; territory of tribes of the Great Basin Indians. Mailing Add: Dept of Anthrop Univ of Colo Boulder CO 80302

STEWART, P BRIAN, b Aug 7, 22; Can citizen; m 49; c 3. IMMUNOLOGY, PHYSIOLOGY. Educ: Univ London, MB, BS, 50. Prof Exp: House physician, Middlesex & Brompton Hosps, Eng, 50-52; gen pract, Barnstaple, Eng, 52-54; med res assoc, 56-57, ASSOC PROF MED, McGILL UNIV, 57-; RES DIR, PHARMA-RES CAN LTD, 62- Concurrent Pos: Lilly res fel physiol, McGill Univ, 55-56, Nat Res Coun Can res assoc med, 56-57; med dir, Geigy Can Ltd, 57-62; assoc physician, Royal Victoria Hosp Can, 57- Mem: Am Physiol Soc; Am Asn Immunologists; Can Soc Immunologists; Pharmacol Soc Can; Can Med Asn. Res: Allergy. Mailing Add: Pharma-Res Can Ltd 250 Hymus Blvd Pointe Claire PQ Can

STEWART, PAUL ALVA, b Leetonia, Ohio, June 24, 09; m 47; c 2. ECOLOGY, ORNITHOLOGY. Educ: Ohio State Univ, BS, 52, MS, 53, PhD(zool), 57. Prof Exp: Admin asst, State Dept Conserv, Ind, 58-59; wildlife res biologist, US Bur Sport Fisheries & Wildlife, 59-65; res entomologist, Agr Res Serv, USDA, 65-73; WILDLIFE CONSULT, 73- Concurrent Pos: Consult blackbird nuisance prob, Nat Audubon Soc, 75- Mem: Am Soc Mammal; Wildlife Soc; Cooper Ornith Soc; Wilson Ornith Soc; Am Ornith Union. Res: Ecology and management of the wood duck; ecology of blackbird congregations; biological control of insect pests. Mailing Add: 203 Mooreland Dr Oxford NC 27565

STEWART, PAUL RICH, biology, deceased

STEWART, PETER ARTHUR, b Sask, May 12, 21; m 52; c 2. BIOPHYSICS, PHYSIOLOGY. Educ: Univ Man, BSc, 43; Univ Minn, MS, 49, PhD(biophys), 51. Prof Exp: Instr physiol, Univ Ill, 51-53, asst prof neurophysiol, Neuropsychiat Inst, Col Med, 53-54; from asst prof to assoc prof physiol, Emory Univ, 54-65, assoc prof physics, 61-65; PROF MED SCI, BROWN UNIV, 65- Concurrent Pos: Markle scholar, 56-61. Mem: AAAS; Am Physiol Soc; Biophys Soc; Biomed Eng Soc; Am Asn Physics Teachers. Res: Electrical parameters nerve membrane; electroencephalographic analysis; protoplasmic movement and structure in slime molds; biological control systems analysis; computers in biomedicine; theoretical biology. Mailing Add: Div of Biol & Med Sci Brown Univ Providence RI 02912

STEWART, RALPH RANDLES, b West Hebron, NY, Apr 15, 90; m 16; c 2. SYSTEMATIC BOTANY. Educ: Columbia Univ, AB, 11, AM, 15, PhD(bot), 16. Hon Degrees: DSc, Punjab Univ, India, 52; LLD, Alma Col, 63. Prof Exp: Prof biol, 11-60, prin, 34-55, EMER PRIN, GORDON COL, PAKISTAN, 55-; RES ASSOC BOT, UNIV MICH, ANN ARBOR, 60- Concurrent Pos: Asst, Columbia Univ, 14-16, Cutting fel, 16-17; actg cur, NY Bot Garden, 42-44; NSF grants, 62-69. Honors & Awards: Kaiser i Hind Gold Medal, 38; Sitara i Imtiaz, Govt Pakistan, 61. Res: Botany of the western Himalaya, especially northern Pakistan and Kashmir; ferns of Nepal; meaning of all fern generic names; dictionary of generic and specific names of flora Indica and flora Iranica areas. Mailing Add: Herbarium North Univ Bldg Univ of Mich Ann Arbor MI 48104

STEWART, RAY EDWARD, b Eugene, Ore, Nov 29, 42; m 64; c 2. PEDODONTICS, MEDICAL GENETICS. Educ: Univ Ore, DMD, 68, MS, 71. Prof Exp: Fel med genetics, Med Sch, Univ Ore, 68-71; asst prof pediat dent, Univ Calif, Los Angeles, 71-74, ASSOC PROF, DIV MED GENETICS, DEPT PEDIAT & ASSOC PROF

PEDIAT DENT, HARBOR GEN HOSP-UNIV CALIF, LOS ANGELES, 74-, DENT COORDR, CRANIOFACIAL ANOMALIES CTR, 72- Concurrent Pos: Consult pediat dent, Wadsworth Vet Admin Hosp, 73- Mem: Sigma Xi; Am Cleft Palate Asn; Soc Craniofacial Geneticists; Am Acad Dent for Handicapped. Res: Clinical and laboratory research on pathogenesis and etiology of craniofacial malformations; development of animal models for human chondro-dystrophies. Mailing Add: Harbor Gen Hosp 1000 W Carson St Torrance CA 90509

STEWART, REGINALD BRUCE, b Moose Jaw, Sask, May 30, 28; m 50; c 1. ANALYTICAL CHEMISTRY. Educ: Univ Man, BSc, 50. Prof Exp: Res chemist, Hudson Bay Mining & Smelting Co, 50-55; asst chief chemist, Noranda Mines Ltd, 55-62; res off anal chem, 62-69, HEAD ANAL SCI BR, ATOMIC ENERGY CAN, 69- Mem: Chem Inst Can. Res: Analytical sciences, particularly nuclear power research and development. Mailing Add: Anal Sci Br Atomic Energy of Can Pinawa MB Can

STEWART, RICHARD CUMMINS, microbiology, see 12th edition

STEWART, RICHARD DONALD, b Lakeland, Fla, Dec 26, 26; m 52; c 3. INTERNAL MEDICINE, TOXICOLOGY. Educ: Univ Mich, AB, 51, MD, 55, MPH, 62; Am Bd Internal Med, dipl, 74. Prof Exp: Resident internal med, Med Ctr, Univ Mich, 59-62; staff physician, Med Dept, Dow Chem Co, 56-59, dir med res sect, Biochem Res Lab, 62-66; asst prof internal med & assoc prof prev med & toxicol, 66-69, PROF ENVIRON MED IN INTERNAL MED & TOXICOL & ENVIRON MED, SCH MED, MED COL WIS, 69-, CHMN DEPT ENVIRON MED, 66- Concurrent Pos: Corp med adv, S C Johnson & Son, 71- Mem: Fel Am Col Physicians; Am Soc Artificial Internal Organs; Soc Toxicol; Am Acad Clin Toxicol; fel Am Occup Med Asn. Res: Human toxicology; development of the hollow fiber artificial kidney; air pollution epidemiological studies; experimental human exposures to artificial environments. Mailing Add: Dept of Environ Med Med Col of Wis Milwaukee WI 53226

STEWART, RICHARD JOHN, b Duluth, Minn, May 30, 42; m 67. GEOLOGY. Educ: Univ Minn, BA, 65; Stanford Univ, PhD(geol), 70. Prof Exp: ASST PROF GEOL, UNIV WASH, 69- Concurrent Pos: Geologist, Olympia Mts, US Geol Surv, 70; sedimentologist deep sea drilling proj, NSF, 71, res grant, Univ Wash, 72-73. Mem: Mineral Soc Am; Geol Soc Am; Am Asn Petrol Geologists; Soc Econ Paleontologists & Mineralogists; Am Geophys Union. Res: Sedimentary petrology; structural geology; geological and tectonic history of the northeast Pacific Ocean and its continental margin. Mailing Add: Dept of Geol Sci Univ of Wash Seattle WA 98195

STEWART, RICHARD WILLIS, b Atlanta, Ga, Dec 27, 36; m 64; c 2. ATMOSPHERIC PHYSICS. Educ: Univ Fla, BS, 60; Columbia Univ, MA, 63, PhD(physics), 67. Prof Exp: Res assoc atmospheric physics, 67-69, STAFF SCIENTIST, GODDARD INST SPACE STUDIES, 69- Concurrent Pos: Asst prof, Rutgers Univ, 68 & City Col New York, 69- Mem: Am Meteorol Soc; Am Geophys Union. Res: Aeronomy; planetary ionospheres. Mailing Add: Goddard Inst for Space Studies 2880 Broadway New York NY 10025

STEWART, ROBERT A, b St Louis, Mo, May 8, 13; m 40; c 2. NUTRITION. Educ: Butler Univ, AB, 35; Pa State Univ, MS, 38, PhD(nutrit), 40. Prof Exp: Biochemist, Quaker Oats Co, 40-46; res chemist, 46-47, nutrit res mgr, 47-63, RES DIR, GERBER PROD CO, 63- Concurrent Pos: Mem adv comt, Nutrit Found. Mem: Am Inst Nutrit. Res: Food technology; biochemistry; infant nutrition; products to meet the needs of mass feeding. Mailing Add: Res Ctr Gerber Prods Co Fremont MI 49412

STEWART, ROBERT ARCHIE, II, b Houston, Miss, Jan 23, 42. PALYNOLOGY, PLANT ECOLOGY. Educ: Miss State Univ, BS, 65, MS, 67; Ariz State Univ, PhD(bot), 71. Prof Exp: Partic, NSF advan seminar trop bot, Univ Miami, 68; ASST PROF BIOL, DELTA STATE UNIV, 70- Mem: AAAS; Bot Soc Am; Am Asn Stratig Palynologists; Ecol Soc Am. Res: Cenozoic palynology; ecology. Mailing Add: Dept of Biol Sci Delta State Univ Cleveland MS 38732

STEWART, ROBERT BLAYLOCK, b Stilwell, Okla, Feb 10, 26; m 48; c 2. PLANT PATHOLOGY. Educ: Okla State Univ, BS, 50; Tex A&M Univ, MS, 53, PhD(plant path), 57. Prof Exp: Instr agron, Okla State Univ, 54-56; asst prof plant path, Tex A&M Univ, 56-58; from assoc prof to prof bot, 58-74, PROF BIOL, SAM HOUSTON STATE UNIV, 74- Concurrent Pos: Assoc prof from Okla State Univ, Imp Ethiopian Col Agr & Mech Arts, 59. Mem: AAAS; Mycol Soc Am; Soc Econ Bot. Res: Paleoethnobotany. Mailing Add: Dept of Biol Sam Houston State Univ Huntsville TX 77340

STEWART, ROBERT BRUCE, b Toronto, Ont, May 2, 26; m 45; c 2. ANIMAL VIROLOGY. Educ: Mt Allison Univ, BSc, 49; Queen's Univ, Ont, MA, 51, PhD(bact), 55. Prof Exp: Res officer, Defence Res Bd, Can, 51-55; from instr to asst prof bact, Sch Med & Dent, Univ Rochester, 55-63; assoc prof, 63-66, PROF BACT, QUEEN'S UNIV, ONT, 66-, HEAD DEPT, 71- Honors & Awards: Browncroft Pediat Res Found Award, 58. Mem: Am Soc Microbiol; Can Soc Microbiol. Res: Virology; regulation of virus growth; infectious disease. Mailing Add: Dept of Microbiol & Immunol New Med Bldg Queen's Univ Kingston ON Can

STEWART, ROBERT CLARENCE, b Sharon, Pa, Sept 23, 21; m 59. MATHEMATICS. Educ: Washington & Jefferson Col, BA, 42, MA, 44; Yale Univ, MA, 48. Prof Exp: Instr math, Washington & Jefferson Col, 42-44, 45-46; asst, Yale Univ, 46-50; from instr to assoc prof, 50-67, PROF MATH, TRINITY COL, CONN, 67- Mem: Am Math Soc; Math Asn Am. Res: Modern algebra; matrix theory; differential equations. Mailing Add: Dept of Math Trinity Col Hartford CT 06106

STEWART, ROBERT DANIEL, b Salt Lake City, Utah, June 15, 23; m 47; c 4. PHYSICAL CHEMISTRY, METALLURGICAL CHEMISTRY. Educ: Univ Utah, BS, 50; Univ Wash, PhD(phys chem), 54. Prof Exp: Sr res chemist, Am Potash & Chem Corp, 54-56, group leader, 56-58, sect head, 58-68; group leader, Garrett Res & Develop Co, 68-75; GROUP LEADER, OCCIDENTAL RES CORP, 75- Mem: Am Chem Soc; Electrochem Soc; Am Inst Mining, Metall & Petrol Engrs. Res: Gas phase kinetics; thermodynamics; inorganic polymers; extractive hydrometallurgy. Mailing Add: 17052 El Cajon Ave Yorba Linda CA 92686

STEWART, ROBERT EARL, b Kansas City, Mo, Apr 16, 13; m 36; c 4. WILDLIFE ECOLOGY, POPULATION ECOLOGY. Educ: Univ Iowa, BS, 36; Univ Mich, MS, 37. Prof Exp: Biologist, US Civil Serv Comn, 38-40; wildlife res biologist, Patuxent Wildlife Res Ctr, 40-60, WILDLIFE RES BIOLOGIST, NORTHERN PRAIRIE WILDLIFE RES CTR, US BUR SPORT FISHERIES & WILDLIFE, 61- Mem: Fel Am Ornith Union; Ecol Soc Am; Wildlife Soc; Wilson Ornith Soc; Cooper Ornith Soc. Res: Habitat and population investigations of birds, including waterfowl; biotic communities; biogeographical distribution of birds; plant ecology. Mailing Add: Northern Prairie Wildlife Rs Ctr Jamestown ND 58401

STEWART, ROBERT F, b Seattle, Wash, Dec 31, 36; m 59; c 2. PHYSICAL CHEMISTRY. Educ: Carleton Col, AB, 58; Calif Inst Technol, PhD(chem), 63. Prof Exp: NIH fel, Univ Wash, 62-64; fel, 64-69, ASSOC PROF CHEM, CARNEGIE-MELLON UNIV, 69- Concurrent Pos: Alfred P Sloan fel, 70-72. Res: Ultraviolet absorption of single crystals; valence structure from x-ray scattering; x-ray diffraction. Mailing Add: Carnegie-Mellon Univ 5000 Forbes Ave Pittsburgh PA 15213

STEWART, ROBERT FRANCIS, b Birmingham, Ala, Oct 31, 26; m 58; c 4. NUCLEAR CHEMISTRY, FUEL TECHNOLOGY. Educ: Univ Ala, BS, 49, MS, 50. Prof Exp: Jr chemist, Nat Southern Prod Corp, 50-52; head radioisotope lab, 63-69, RES CHEMIST, US BUR MINES, 54-, RES SUPVR, MORGANTOWN ENERGY RES CTR, 69- Mem: Am Chem Soc; Am Soc Testing & Mat; Instrument Soc Am. Res: Nuclear methods of continuous analysis of bulk materials for process control based on neutron interactions in matter. Mailing Add: Rte 8 Box 228E Morgantown WV 26505

STEWART, ROBERT LEWIS, b Sullivan, Ind, July 11, 19; m 63; c 1. PSYCHIATRY. Educ: Ind Univ, BS, 41, MD, 44; Chicago Inst Psychoanal, cert, 60. Prof Exp: From assoc prof to prof psychiat & exec dir cent psychiat clin, Col Med, Univ Cincinnati, 61-72; TRAINING & SUPV ANALYST, CINCINNATI PSYCHOANAL INST, 73- Concurrent Pos: Mem fac, Chicago Inst Psychoanal, 60-73, training analyst, 68-73; clin prof psychiat, Col Med, Univ Cincinnati, 72- Mem: Int Psychoanal Asn; Am Psychiat Asn; Am Psychoanal Asn; AMA. Res: Psychoanalytic education; supervision; technique; application to community problems. Mailing Add: 264 William Howard Taft Rd Cincinnati OH 45219

STEWART, ROBERT MALCOLM, b Washington, DC, Oct 17, 25; wid; c 2. APPLIED PHYSICS, NEURO-TECHNOLOGY. Educ: Calif Inst Technol, BS, 47; George Washington Univ, MS, 50. Prof Exp: Instr physics, George Washington Univ, 48-49; aerodynamicist, Ryan Aeronaut Co, 49-50; sr res engr in charge syst anal group, Jet Propulsion Lab, Calif Inst Technol, 50-58; mgr res & adv tech lab, Space Electronics Corp, 58-60, lab dir, Ctr Res & Educ, Space Gen Corp, 60-68; PRES & CHMN BD DIRS, LEVEL-SEVEN INC, 68- Concurrent Pos: Lectr commun theory, Calif Inst Technol, 52-53; adj prof neurol & eng, Univ Southern Calif, 66- Res: Electrochemical computer research; electrodynamics and functional organization of brain tissue; neurodynamic models; research and development in the relationship between technology and normal development and functioning of the human nervous system, especially as directed toward enhanced nonverbal learning and perception. Mailing Add: 15720 Morrison St Encino CA 91316

STEWART, ROBERT MURRAY, JR, b Washington, DC, May 6, 24; m 45; c 2. COMPUTER SCIENCE. Educ: Iowa State Col, BS, 45, PhD(physics), 54. Prof Exp: Instr elec eng, 46-48, asst physics, 48-50, res assoc, 50-54, from asst prof to assoc prof, 54-58, engr in charge cyclone comput lab, 56-67, assoc prof physics & elec eng, 58-60, PROF PHYSICS & ELEC ENG, IOWA STATE UNIV, 60-, ASSOC DIR COMPUT CTR, 67-, CHMN DEPT COMPUT SCI, 69- Concurrent Pos: Sr physicist, Ames Lab, AEC; mem bd ed consults, Electronic Assocs, Inc; consult, Midwest Res Inst; mem educ comt, Am Fedn Info Processing Socs, 68-; vchmn, Comput Sci Conf Bd, 75- Mem: Am Phys Soc; Asn Comput Mach; Inst Elec & Electronics Eng; AAAS. Res: Design of logical control systems and digital computer systems; pattern recognition and adaptive logic. Mailing Add: 3416 Oakland St Ames IA 50010

STEWART, ROBERT N, b Sioux Falls, SDak, Sept 18, 16; m 41; c 3. PHYTOCHEMISTRY. Educ: Univ SDak, AB, 39; Univ Md, College Park, MS, 42, PhD(bot), 46. Prof Exp: Asst prof bot, Univ Md, College Park, 46 & Barnard Col, Columbia Univ, 46-48; res geneticist, 48-55, RES HORTICULTURIST, AGR RES SERV, USDA, 55- Concurrent Pos: Assoc prof lectr, George Washington Univ, 61-67. Mem: AAAS; Am Genetic Asn; Bot Soc Am; Genetics Soc Am; Phytochem Soc. Res: Genetic basis of somatic variability; use of chimeras in studying ontogeny; poinsettia breeding; investigation at the cellular level of factors influencing flower color. Mailing Add: Rm 108 Bldg 004 Beltsville Agr Res Ctr USDA Beltsville MD 20705

STEWART, ROBERT WILLIAM, b Boston, Mass, Dec 24, 18. GEOLOGY. Educ: Mass Inst Technol, BS, 40, MS, 46, PhD, 51. Prof Exp: Instr geol, Hofstra Col, 47-48; geologist, Pure Oil Co, Mont, 48-49 & Texaco, Inc, 51-56; consult geologist, 56-66; consult geologist, Phelps Dodge Corp, Ariz, 66-68; dist explor geologist, Cominco Am Inc, Mo, 68-70; sr staff geologist, 70-74, TECH & ADMIN COORDR, DRESSER MINERALS DIV, DRESSER INDUSTS, INC, 74- Mem: Am Asn Petrol Geol; Geol Soc Am; Am Inst Mining, Metall & Petrol Eng; Soc Econ Paleont & Mineral; fel Inst Mining & Metall. Res: Mining and mining exploration. Mailing Add: Explor Dept Dresser Minerals Div Dresser Industs Inc PO Box 6504 Houston TX 77005

STEWART, ROBERT WILLIAM, b Smoky Lake, Alta, Aug 21, 23; div; c 3. PHYSICS. Educ: Queen's Univ, Ont, BSc, 45, MSc, 47; Cambridge Univ, PhD(physics), 52. Hon Degrees: DSc, McGill Univ, 72. Prof Exp: Lectr physics, Queen's Univ, Ont, 46; defence sci serv officer, Pac Naval Lab Can, 50-55; from assoc prof to prof, 55-70, HON PROF PHYSICS, UNIV BC, 70-; DIR-GEN, PAC REGION, OCEAN & AQUATIC SCI, CAN DEPT ENVIRON, 70- Concurrent Pos: Vis prof, Dalhousie Univ, 60-61 & Harvard Univ, 64; distinguished vis prof, Pa State Univ, 64; Commonwealth vis prof, Cambridge Univ, 67-68; chmn joint organizing comt, Global Atmospheric Res Prog, 72- Mem: Fel Royal Soc Can; fel Royal Soc; Can Asn Physicists; Can Meteorol Soc. Res: Turbulence; physical oceanography; air-sea interaction. Mailing Add: Dept of the Environ 5th Floor 1230 Government St Victoria BC Can

STEWART, ROBERTA A, b Rochester, NH, Aug 24, 23. ORGANIC CHEMISTRY. Educ: Univ NH, BS, 44; Smith Col, MA, 46, PhD, 49. Prof Exp: Res assoc chem, Smith Col, 48-49; instr, Wellesley Col, 49-53; from asst prof to assoc prof, 53-70, chmn dept chem, 63-66 & 69-73, chmn div natural sci & math, 67-73, asst to pres, 69-75, PROF CHEM, HOLLINS COL, 70-, DEAN COL, 75- Concurrent Pos: NSF fel, Radcliffe Col, 60-61; Am Coun on Educ fel acad admin, Univ Del, 66-67. Mem: AAAS; fel Am Inst Chem; Am Chem Soc. Res: Synthetic experiments in direction of morphine; some reactions of beta tetralone; derivatives of cyclohexanone. Mailing Add: Box 9685 Hollins College VA 24020

STEWART, ROBIN KENNY, b Ayr, Scotland, May 22, 37; m 64; c 3. ANIMAL ECOLOGY. Educ: Glasgow Univ, BS, 61, PhD(entom), 66. Prof Exp: Asst lectr agr zool, Glasgow Univ, 63-66; asst prof, 66-70, coordr biol sci div, 72-75, ASSOC PROF ENTOM & ZOOL, MACDONALD COL, MCGILL UNIV, 70-, CHMN DEPT ENTOM, 75- Concurrent Pos: Consult, Can Pac Investment, 74-75 & UN Develop Proj, 75- Mem: Can Entom Soc; British Ecol Soc. Res: Agricultural zoology; ecology; entomology; photoperiodism; integrated control. Mailing Add: Dept of Entom Macdonald Campus of McGill Univ Ste Anne de Bellevue PQ Can

STEWART, ROGER MALCOLM, geophysics, see 12th edition

STEWART, ROLLAND KEITH, b Cheyenne, Wyo, Sept 1, 32; m 62; c 3. AQUATIC BIOLOGY. Educ: Univ Wyo, BS, 58, MS, 59. Prof Exp: Fisheries res biologist, SDak

Dept Game Fish & Parks, 59-64; aquatic biologist, USPHS, 64-66; aquatic biologist, Fed Water Pollution Control Admin, 66-71, AQUATIC BIOLOGIST, US ENVIRON PROTECTION AGENCY, 71- Mem: Am Soc Limnol & Oceanog. Res: Water pollution ecology. Mailing Add: 6706 Foothill Circle Anchorage AK 99504

STEWART, ROSS, b Vancouver, BC, Mar 16, 24; m 46; c 2. PHYSICAL ORGANIC CHEMISTRY. Educ: Univ BC, BA, 46, MA, 48; Univ Wash, PhD, 54. Prof Exp: Lectr chem, Univ BC, 47-49; lectr, Can Serv Col, Royal Roads, 49-51, from asst prof to assoc prof, 51-55; from asst prof to assoc prof, 55-62, PROF CHEM, UNIV BC, 62- Mem: Royal Soc Can; Am Chem Soc; fel Chem Inst Can. Res: Organic oxidation mechanisms; protonation of weak organic bases; general acid catalysis; ionization of weak acids in strongly basic solution. Mailing Add: Dept of Chem Univ of BC Vancouver BC Can

STEWART, RUTH CAROL, b Englewood, NJ, Dec 18, 28. MATHEMATICS. Educ: Rutgers Univ, AB, 50, MA, 63, EdD(math educ), 69. Prof Exp: Dir music, high sch, NJ, 50-53; instr, Wiesbaden Am High Sch, Ger, 54-57; chmn dept math, Frankfurt Am High Sch, 58-62 & 63-64; PROF MATH, MONTCLAIR STATE COL, 64- Mem: Math Asn Am. Res: Mathematics in areas of algebra and analysis; mathematics education in areas of curriculum and instruction. Mailing Add: Dept of Math Montclair State Col Upper Montclair NJ 07043

STEWART, SARAH ELIZABETH, b Tecalitian, Mex, Aug 16, 06; US citizen. BACTERIOLOGY. Educ: NMex Agr Col, BS, 27; Mass State Col, MS, 30; Univ Chicago, PhD(bact), 39; Georgetown Univ, MD, 49. Prof Exp: Asst bacteriologist, Exp Sta, Univ Colo, 30-33; from asst bacteriologist to bacteriologist, NIH, 36-44; instr bact, 44-47, asst prof, 48-55, PROF PATH, SCH MED, GEORGETOWN UNIV, 55- ; SECT CHIEF, VIRAL BIOL BR, NAT CANCER INST, 60- Concurrent Pos: From sr asst surgeon to sr surgeon, Nat Cancer Inst, 50-60. Mem: AAAS; Am Asn Cancer Res. Res: Oncogenic viruses; polyoma viruses which cause multiple tumors in rodents. Mailing Add: Dept of Path Georgetown Univ Sch of Med Washington DC 20007

STEWART, SHEILA FRANCES, b Halifax, NS, Mar 16, 27. REPRODUCTIVE ENDOCRINOLOGY. Educ: Univ BC, BA, 48, MS, 63; Rutgers Univ, PhD(endocrinol), 69. Prof Exp: Scientist, 68-75, SR SCIENTIST ENDOCRINOL, WARNER LAMBERT RES INST, 75- Res: Effects of underfeeding, sex, hemigonadectomy, drug and hormone treatment on pituitary secretions of gonadatropins; prostaglandin effects on the pseudopregnant ovary and pituitary secretions. Mailing Add: Warner Lambert Res Inst 170 Tabor Rd Morris Plains NJ 07920

STEWART, SHELTON E, b Sanford, NC, Oct 1, 34; m 65; c 1. BOTANY, ZOOLOGY. Educ: ECarolina Col, BS, 56; Univ NC, MA, 59; Univ Ga, PhD(bot), 66. Prof Exp: Teacher, Pine Forest High Sch, 56-57; prof sci, Ferrum Jr Col, 59; prof biol, Lander Col, 59-63; instr bot, Univ Ga, 64-65; PROF BIOL, LANDER COL, 66- Mem: AAAS. Res: Plant taxonomy; plant biosystematics. Mailing Add: Dept of Biol Lander Col Greenwood SC 29646

STEWART, THOMAS BONNER, b Sao Paulo, Brazil, Nov 24, 24; US citizen; m 56; c 4. ZOOLOGY, PARASITOLOGY. Educ: Univ Md, BS, 49; Auburn Univ, MS, 53; Univ Ill, Urbana, PhD(vet med sci), 63. Prof Exp: Parasitologist, USDA, Ala, 50-53 & Ga Coastal Plain Exp Sta, 53-60; fel, Univ Ill, Urbana, 60-62; res parasitologist, 63-64, DIR, ANIMAL PARASITE RES LAB, GA COASTAL PLAIN EXP STA, USDA, 64- Concurrent Pos: Mem grad fac & coun mem, Fac Parasitol, Univ Ga. Mem: AAAS; Am Soc Parasitol; Soc Protozool; Am Micros Soc. Res: Life history of Cooperia punctata; gastrointestinal parasites of cattle; eradication of the kidneyworm of swine; beetles as intermediate hosts of nematodes; Strongyloides ransomi of swine; ecology of swine parasites; trans-uterine and trans-milk infection of host by nematode parasites; anthelmintics for swine and cattle parasites. Mailing Add: Animal Parasite Res Lab Ga Coastal Plain Exp Sta Tifton GA 31794

STEWART, THOMAS DALE, b Delta, Pa, June 10, 01; m 32, 52; c 1. ANTHROPOLOGY. Educ: George Washington Univ, AB, 27; Johns Hopkins Univ, MD, 31. Hon Degrees: DSc, Nat Univ Cuzco, 49. Prof Exp: Aide phys anthrop, Nat Mus Natural Hist, DC, 27-31; from asst cur to cur, 31-61, head cur anthrop, 61-62, dir, 62-65, sr phys anthropologist, 65-71, EMER PHYS ANTHROPOLOGITST, NAT MUS NATURAL HIST, WASHINGTON, DC, 71- Concurrent Pos: Mem, Smithsonian Inst field trips, Alaska, 27, Mex, 39, Peru, 41, Guatemala, 47 & 49, Iraq, 57, 60 & 62; vis prof, Sch Med, Washington Univ, 43 & Nat Sch Anthrop, Mex, 45; ed jour, Am Asn Phys Anthrop, 43-48; res consult, Am Graves Regist Serv, Japan, 54; mem lang develop adv comt, Off Educ, 63-65; bd sci adv, Delta Primate Res Ctr, Tulane Univ, 64-73. Mem: Nat Acad Sci; AAAS; Am Asn Phys Anthrop (pres, 50-51, secy-treas, 60-64); Inst Human Paleont (pres, 55-61); Am Acad Forensic Sci. Res: Comparative human osteology; paleopathology; human identification. Mailing Add: 1191 Crest Lane McLean VA 22101

STEWART, THOMAS HENRY MCKENZIE, b Hertfordshire, Eng, Aug 17, 30; Can citizen; m 60; c 4. INTERNAL MEDICINE, IMMUNOLOGY. Educ: Univ Edinburgh, MB, ChB, 55; FRCP(C), 62. Prof Exp: House officer surg, All Saints Hosp, Chatham, Eng, 55-56; house officer med, Eastern Gen Hosp, Edinburgh, Scotland, 56 & Westminster Hosp, London, Ont, 58-60; resident, Ottawa Gen Hosp, 61-62; lectr nuclear med, Univ Mich, Ann Arbor, 63-64; lectr, 64-66, asst prof, 66-72, ASSOC PROF MED, UNIV OTTAWA, 72- Concurrent Pos: Teaching fel, Path Inst, McGill Univ, 60-61; res fel hemat, Univ Ottawa, 62-63. Mem: Can Soc Immunol; Can Soc Clin Invest; Soc Nuclear Med; Am Asn Cancer Res. Res: Immunology of cancer; host-tumor relationships; immunochemotherapy of human cancer; immunology of inflammatory bowel disease. Mailing Add: 1 Mt Pleasant Ave Ottawa ON Can

STEWART, THOMAS WILLIAM WALLACE, b London, Ont, June 22, 24; m 55; c 4. ACOUSTICS, ELECTRONICS. Educ: Univ Western Ont, BSc, 53, MSc, 55; Univ London, PhD(physics), 62. Prof Exp: Res assoc physics, Univ Col, Univ London, 55-57; res asst, 58-59, instr, 59-60, lectr, 60-62, from asst prof to assoc prof, 62-74, PROF PHYSICS, UNIV WESTERN ONT, 74- Mem: AAAS; Acoust Soc Am; Can Asn Physicists. Res: Hearing; musical and architectural acoustics. Mailing Add: Dept of Physics Univ of Western Ont London ON Can

STEWART, VINCENT EVANS, analytical chemistry, see 12th edition

STEWART, WELLINGTON BUEL, b Chicago, Ill, June 18, 20; m 45; c 3. PATHOLOGY. Educ: Univ Notre Dame, BS, 42; Univ Rochester, MD, 45. Prof Exp: Intern path, Strong Mem Hosp, Rochester, NY, 45-46; Rockefeller Found fel, Univ Rochester, 48-49, Veteran fel, 48-50; assoc, Columbia Univ, 50-51, asst prof, 51-54; assoc prof, Col Physicians & Surgeons, 54-60; prof & chmn dept, Col Med, Univ Ky, 60-70; dir med comput ctr, 70-75, PROF PATH, UNIV MO-COLUMBIA, 70-, DIR LABS, 75- Concurrent Pos: Asst pathologist, Presby Hosp, New York, 50-54, assoc attend pathologist, 54-60; chmn, Bd Registry Med Technol, 62-67. Mem: Am Soc Exp Pathologists; Soc Exp Biol & Med; Harvey Soc; Am Asn Pathologists & Bacteriologists; Col Am Pathologists. Res: Iron metabolism; fatty liver; red cell physiology; computers in medicine. Mailing Add: Dept of Path Univ of Mo Sch of Med Columbia MO 65201

STEWART, WILLIAM ALLAN, mechanical engineering, see 12th edition

STEWART, WILLIAM ANDREW, b Liberty Center, Ohio, Apr 6, 33; m 60; c 5. MEDICINE. Educ: Miami Univ, AB, 54; Ohio State Univ, MD, 58. Prof Exp: Intern, 58-65, ASST PROF NEUROSURG, STATE UNIV NY UPSTATE MED CTR, 67- Concurrent Pos: Reader neurosurg, Fac Health Sci, Univ Ife, Nigeria, 74-75. Mem: Cong Neurol Surgeons; fel Am Col Surgeons; Am Asn Neurol Surgeons. Res: Trigeminal nerve; head injury. Mailing Add: 725 Irving Ave Syracuse NY 13210

STEWART, WILLIAM CHARLES, b Guam, Jan 10, 26; m 54; c 3. STATISTICAL ANALYSIS, OPERATIONS RESEARCH. Educ: Cornell Univ, BSME, 46; Drexel Inst Technol, MBA, 50, MME, 53; Univ Pa, PhD(statist), 65. Prof Exp: Engr, Philco Corp, 46-48; engr, Schutte & Koerting Co, 48-51; prod engr, Esterbrook Pen Co, 51-57; prod supt, SKF Industs, 57-60; prod mgr, J Bishop Co, 60-61; assoc prof statist, Drexel Inst Technol, 61-67; PROF STATIST & CHMN DEPT, TEMPLE UNIV, 67- Concurrent Pos: Consult, Drexel Inst Technol, 64-67, Naval Air Develop Ctr, Warminster, Pa, 65-, USDA, Philadelphia, 66-70, Dept Educ Admin, Sch Educ, Temple Univ, 66- & Einstein Med Ctr, 67. Mem: Am Statist Asn. Res: Application of statistics to engineering problems; operations research methods particularly for scheduling problems. Mailing Add: Dept of Statist Temple Univ Philadelphia PA 19122

STEWART, WILLIAM CHRISTOPHER, b Dunnville, Ont, Apr 22, 19; m 45; c 1. PHYSIOLOGY, PHARMACOLOGY. Educ: Univ Toronto, BA, 39; Queen's Univ, Ont, MA, 42, MD, CM, 46. Prof Exp: Assoc prof physiol & pharmacol, Univ Alta, 46-52; defence scientist, 52-71, DIR RES, DEFENCE RES ESTAB SUFFIELD, 71- Mem: Can Physiol Soc; Pharmacol Soc Can; Can Med Asn. Res: Toxicology of war gases; therapy of anticholinesterase poisoning; clinical pharmacology and toxicology. Mailing Add: Defence Res Estab Suffield Ralston AB Can

STEWART, WILLIAM HOGUE, JR, b Mullins, SC, Dec 25, 36; m 59; c 3. SOLID STATE PHYSICS. Educ: The Citadel, BSEE, 59; Univ Cincinnati, MS, 61; Clemson Univ, PhD(physics), 64. Prof Exp: Res physicist, 66-71, sr res physicist, 71-74, RES ASSOC, DEERING MILLIKEN RES CORP, 74- Mem: Am Phys Soc. Res: Fiber physics; static electricity and ion physics; paramagnetic resonance spectroscopy; solid state diffusion; x-ray diffraction and spectroscopy; nonlinear servomechanisms. Mailing Add: PO Box 1927 Spartanburg SC 29301

STEWART, WILLIAM HUFFMAN, b Minneapolis, Minn, May 19, 21; m 46; c 2. MEDICINE. Educ: Univ Minn, 39-41; La State Univ, MD, 45; Am Bd Pediat, dipl, 52. Prof Exp: Resident pediatrician, Charity Hosp, New Orleans, 48-50; pvt pract, 50-51; epidemiologist, Commun Dis Ctr, USPHS, 51-53, actg chief heart dis control prog, 54-55, chief, 55-56, asst dir, Nat Heart Inst, 56-57, asst to surgeon gen, 57-58, chief div pub health methods, Off Surgeon Gen, 57-61 & div community health serv, 61-63, asst to spec asst to secy health & med affairs, 63-65, dir, Nat Heart Inst, 65, surgeon gen, 66-69; chancellor, 69-73, PROF PEDIAT, PREV MED & PUB HEALTH, MED CTR, LA STATE UNIV, NE W ORLEANS, 73- Concurrent Pos: Mem tech adv bd, Milbank Fund & adv med bd, Leonard Wood Mem. Mem: AAAS; Am Pub Health Asn; Am Acad Pediat; Am Heart Asn; Am Social Health Asn. Res: Epidemiology; medical administration. Mailing Add: La State Univ Med Ctr 1542 Tulane Ave New Orleans LA 70112

STEWART, WILLIAM SHELDON, b San Diego, Calif, Nov 14, 14; m 40; c 3. PLANT PHYSIOLOGY. Educ: Univ Calif, Los Angeles, AB, 36, AM, 37; Calif Inst Technol, PhD(plant physiol), 39. Prof Exp: Asst, Univ Calif, Los Angeles, 36-37; from asst plant physiologist to plant physiologist & mem emergency rubber proj, Bur Plant Indust, USDA, Mex, 39-45; from asst plant physiologist to assoc plant physiologist, Citrus Exp Sta, Univ Calif, 45-50; head dept plant physiol, Pineapple Res Inst, Hawaii, 50-53; horticulturist & head dept hort, Citrus Exp Sta, Univ Calif, 53-55; dir, Los Angeles State & County Arboretum, 55-70; dir, Pac Trop Bot Garden, 70-75; RETIRED. Concurrent Pos: Res assoc, Calif Inst Technol, 55-; Fulbright lectr, Univs Western Australia & Tasmania, 59 & B-R Col, Agra, 64; US-SAfrica leader exchangee, 62. Mem: AAAS; Am Soc Plant Physiol; Bot Soc Am; Am Soc Hort Sci. Res: Application of plant growth regulators to agriculture. Mailing Add: PO Box 6301 Santa Barbara CA 93111

STEWART, WILLIAM THOMAS, b Globe, Ariz, Sept 16, 15; m 41; c 3. ORGANIC CHEMISTRY. Educ: Univ Ariz, BS, 37, MS, 38; Calif Inst Technol, PhD(org chem), 41. Prof Exp: Res chemist, Chevron Res Co Div, Standard Oil Co Calif, 41-50, sr res chemist, 50-56, supvr res chemist, 56, supvr, 56-66, sr res assoc, 66-70; RES SCIENTIST, DIV ENVIRON RES & ENG, PHELPS DODGE CORP, 70- Mem: Am Chem Soc; Am Inst Mining, Metall & Petrol Engrs. Res: Exploratory and development research; lubricating oil additives; synthetic oils; engine fuel research and development; radiation effects on organic compounds; structure and properties of heavy petroleum fractions; environmental engineering and research of urban, rural and smelter areas. Mailing Add: Phelps Dodge Corp Suite 607 Tucson Fed Tower 32 NStone Ave Tucson AZ 85701

STEWART, WILSON NICHOLS, b Madison, Wis, Dec 7, 17; m 41; c 2. BOTANY. Educ: Univ Wis, BA, 39, PhD(bot), 47; Univ Ill, MA, 46. Prof Exp: From instr to prof bot, Univ Ill, Urbana, 47-65; PROF BOT, UNIV ALTA, 66- Mem: Bot Soc Am; Am Inst Biol Sci; Can Bot Asn. Res: Plant morphology; paleobotany; study of the pteridosperms; morphology of the Isoetales and their fossil ancestors. Mailing Add: Dept of Bot Univ of Alta Edmonton AB Can

STEWART, WOLCOTT ELMER, animal physiology, see 12th edition

STEYERMARK, AL, b St Louis, Mo, Jan 29, 04; m 41. MICROCHEMISTRY. Educ: Washington Univ, BS & MS, 28, PhD, 30. Prof Exp: Org res chemist, Thomas & Hochwalt Labs, Inc, Ohio, 31-33 & Res Lab, SC Hooker, NY, 33-36; org res chemist, Sci Dept, Hoffmann-La Roche, Inc, 36-38, head microchem dept, 38-63, chem res group chief, Res Div, 63-67, asst to vpres chem res, 67-69; VIS PROF CHEM, RUTGERS UNIV, 69- Concurrent Pos: Mem comt anal chem, Nat Acad Sci-Nat Res Coun, 57-61; chmn comn microchem technol, Int Union Pure & Appl Chem, 58-65; ed-in-chief, Microchem J, 62- Honors & Awards: Fritz Pregl Plaquette, Austrian Asn Microchem & Anal Chem, 61. Mem: Fel AAAS; NY Acad Sci; fel Asn Off Anal Chemists; hon mem Am Microchem Soc (pres, 43-44); Sigma Xi. Res: Organic analysis; quantitative organic microanalysis. Mailing Add: 115 Beech St Nutley NJ 07110

STEYERMARK, JULIAN ALFRED, b St Louis, Mo, Jan 27, 09; m 37. TAXONOMIC BOTANY. Educ: Wash Univ, AB, 29, MS, 30, PhD(bot), 33; Harvard Univ, MA, 32. Prof Exp: Asst to Dr R E Woodson, Jr, Mo Bot Garden, 33-34; instr high sch, Mo, 35-37; from asst cur to cur herbarium, Chicago Natural Hist Mus, 37-58; botanist, 59-74, taxonomist, 69-74, CUR HERBARIUM, INST BOT, MINISTRY AGR &

ANIMAL BREEDING, 74- Concurrent Pos: Mem bot exped, Western Tex, 31; collector, US Forest Serv Exped, 32-35 & 36-38 & Washington Univ-Mo Bot Garden Exped, Panama, 34-35; taxonomist & ecologist, Plant Surv, US Forest Serv, Mo, 36; collector, Field Mus Natural Hist Exped, Guatemala, 39-40 & 41-42; botanist, Cinchona Mission, Ecuador, 43 & Venezuela, 43-44; mem exped, Venezuela, 43-45 & 53-55; hon res assoc, Mo Bot Garden, 48-; bot consult, Eli Lilly & Co, 56-; vis prof, Southern Ill Univ, 59; vis cur, NY Bot Gardens, 61-65, 68-69 & 71; mem comn preserv natural areas Venezuela, Acad Phys Sci, Math & Natural Sci, 75. Honors & Awards: Order Quetzal, Guatemala; distinguished Serv Award, Washington Univ, 55; Distinguished Serv Award, NY Bot Gardens, 65; Amigos de Venezuela, Venezuelan Orgn, 73; Order of Andres Bello & Order of Merito de Trabajo, Venezuelan Govt, 74. Mem: Am Soc Plant Taxon; Asn Trop Biol; Int Asn Plant Taxon; hon mem Venezuelan Soc Natural Sci; hon mem Ecuadorian Inst Natural Sci. Res: Rubiaceae of northern South America; flora of Missouri, Guatemala, Ecuador and Venezuela; general flora of Venezuela; Piperaceae of Venezuela. Mailing Add: Ministry of Agr & Animal Breeding Bot Inst Apartado 2156 Caracas Venezuela

STEYERT, WILLIAM ALBERT, b Allentown, Pa, Sept 20, 32; m 61; c 2. PHYSICS. Educ: Mass Inst Technol, BS, 54; Calif Inst Technol, MS, 56, PhD(physics), 60. Prof Exp: Res asst physics, Calif Inst Technol, 54-60; res assoc, Univ Ill, 60-61; STAFF MEM, LOS ALAMOS SCI LAB, UNIV CALIF, 61- Concurrent Pos: NSF fel, Univ Tokyo, 69; adj prof, Utah State Univ. Mem: Fel Am Phys Soc. Res: Nuclear physics at very low temperatures; very low temperature solid state research; low temperature refrigeration. Mailing Add: Los Alamos Sci Lab Box 1663 Los Alamos NM 87544

STIBBS, GERALD DENIKE, b Schreiber, Ont, Apr 25, 10; m 55; c 3. DENTISTRY. Educ: Univ Ore, BS & DMD, 31. Prof Exp: Chmn dept oper dent, dir dent operatory & clin coordr, Sch Dent, 48-70, exec officer, Dept Fixed Partial Dentures, 50-57, chmn dept oper dent grad prog, 50-70, prof oper dent, Sch Dent, 48-70, prof fixed partial dentures, 54-70, spec asst to dean, 70-73, PROF RESTORATIVE DENT, SCH DENT, UNIV WASH, 73-, MEM STAFF, GRAD SCH, 50- Concurrent Pos: Mem assoc comt dent res, Nat Res Coun Can, 45-48; consult, Madigan Army Hosp, 48-52, coun dent health, Am Dent Asn, 54-55 & Pac Northwest Labs, Battelle Mem Inst, 70-74. Mem: Am Dent Asn; Am Acad Restorative Dent; Am Acad Gold Foil Opers; Int Asn Dent Res; Acad Operative Dent. Res: Restorative dentistry; dental materials. Mailing Add: 6227 51st Ave NE Seattle WA 98115

STIBITZ, GEORGE ROBERT, b York, Pa, Apr 30, 04; m 30; c 2. MEDICAL RESEARCH. Educ: Denison Univ, PhB, 26; Union Col, NY, MS, 27; Cornell Univ, PhD(physics), 30. Hon Degrees: DSc, Denison Univ, 66. Prof Exp: Res mathematician, Bell Tel Labs, 30-41; tech aide, Nat Defense Res Comt, 41-45; math consult, 45-66; prof, 66-73, EMER PROF PHYSIOL, DARTMOUTH MED SCH, 73- Honors & Awards: Harry Goode Award, Am Fedn Info Processing Socs. Res: Computing devices; automatic control and stability; dynamic testing; logical design of computers; electronic music; mathematical and computer models of biomedical systems; computer programs for radiation therapy dosage; mathematical and computer models of physiological systems; passive electrical properties of cardiac cells. Mailing Add: Dept of Physiol Dartmouth Med Sch Hanover NH 03755

STICH, HANS F, b Prague, Czech, Dec 24, 27; Can citizen; m 55; c 1. CELL BIOLOGY, GENETICS. Educ: Univ Würzburg, PhD(zool), 49. Prof Exp: Res assoc cell biol, Max Planck Inst Marine Biol, 50-57; assoc prof cancer res unit, Nat Cancer Inst Can, 57-60; from assoc prof to prof genetics, Queen's Univ, Ont, 60-66; prof biol & chmn dept, McMaster Univ, 66-70; PROF ZOOL, UNIV BC, 70- Concurrent Pos: Fulbright grants, Univ Wis & Western Reserve Univ, 55-57; vis prof, Med Ctr, Stanford Univ, 58, Univ Tex MD Anderson Hosp & Tumor Inst, 61 & Roswell Park Mem Inst, 64. Mem: Am Soc Cell Biol; Am Soc Human Genetics; Am Asn Cancer Res; NY Acad Sci; Can Soc Cell Biol (pres, 66). Res: Cytogenetic studies of virus, x-ray and chemically induced neoplastic cells; immunological studies on virus induced neoplastic cells using fluorescein and ferritin conjugated antibodies; regulation of gene activity; DNA repair; in vitro bioassays for carcinogens. Mailing Add: Cancer Res Ctr Univ of BC Vancouver BC Can

STICHT, FRANK DAVIS, b Plattsburg, Miss, June 14, 19; m 41; c 2. PHARMACOLOGY. Educ: Univ Miss, BS Pharm, 48; Baylor Univ, DDS, 56; Univ Tenn, Memphis, MS, 65. Prof Exp: From instr to asst prof, 61-73, ASSOC PROF PHARMACOL, UNIV TENN CTR HEALTH SCI, MEMPHIS, 73- Res: Autonomic and cardiovascular pharmacology; influence of drugs on the blood pressure within the tooth pulp. Mailing Add: Dept of Pharmacol Univ of Tenn Ctr for Health Sci Memphis TN 38163

STICKEL, DELFORD LEFEW, b Falling Waters, WVa, Dec 12, 27; m 52; c 1. SURGERY. Educ: Duke Univ, AB, 49, MD, 53; Am Bd Surg, dipl, 63; Bd Thoracic Surg, dipl, 63. Prof Exp: Asst surg, 57-59, from instr to assoc prof, 59-72, PROF SURG, MED CTR, DUKE UNIV, 72-, ASSOC DIR HOSP, 72- Concurrent Pos: Markle scholar & NIH career develop award, 62; attend physician, Durham Vet Admin Hosp, 65-66, chief surg serv, 66-68, chief of staff, 70-72; consult, Watts Hosp, Durham, 66- & NC Eye & Human Tissue Bank, 69. Mem: AAAS; AMA; Am Fedn Clin Res; Soc Cryobiol; Am Col Surgeons. Res: Clinical renal transplantation. Mailing Add: Dept of Surg Duke Univ Med Ctr Durham NC 27710

STICKEL, LUCILLE FARRIER, b Mich, Jan 11, 15; m 41. ZOOLOGY. Educ: Eastern Mich Univ, BA, 36; Univ Mich, MS, 38, PhD, 49. Hon Degrees: DSc, Eastern Mich Univ, 74. Prof Exp: Biologist, 43-47 & 61-72, DIR PATUXENT WILDLIFE RES CTR, US FISH & WILDLIFE SERV, 72- Honors & Awards: Aldo Leopold Award, Wildlife Soc, 74. Res: Vertebrate population ecology; ecology and pharmacotoxicology of environmental pollution. Mailing Add: Patuxent Wildlife Res Ctr US Fish & Wildlife Serv Laurel MD 20811

STICKEL, WILLIAM HENSON, b Terre Haute, Ind, Nov 8, 12; m 41. WILDLIFE RESEARCH, POLLUTION BIOLOGY. Educ: Univ Mich, BS, 34, MA, 35. Prof Exp: Asst zool, Univ Mich, 35-40; jr biologist, US Civil Serv Comn, DC, 40-41; RES BIOLOGIST, PATUXENT RES CTR, US FISH & WILDLIFE SERV, 41- Concurrent Pos: Ed, Wildlife Rev, 52-59. Mem: Wildlife Soc; Am Soc Mammal. Res: Effects of pesticides and pollutants on wildlife and habitat. Mailing Add: Patuxent Res Ctr US Fish & Wildlife Serv Laurel MD 20811

STICKER, ROBERT EARL, b New York, NY, Mar 4, 30; m 57; c 3. ORGANIC CHEMISTRY. Educ: Cornell Univ, AB, 53; Columbia Univ, AM, 57; Univ Kans, PhD(pinane chem), 65. Prof Exp: Res chemist, Eastman Kodak Co, 57-60; RES CHEMIST, NIAGARA CHEM DIV, FMC CORP, 65- Mem: Am Chem Soc. Res: Terpenes; heterocycles; surfactants; pesticides; herbicides. Mailing Add: Res & Develop Niagara Chem Div FMC Corp 100 Niagara St Middleport NY 14105

STICKLE, GENE P, b New Castle, Pa, Apr 11, 29; m 54; c 2. CHEMICAL ENGINEERING, BIOCHEMISTRY. Educ: Univ Tenn, BS, 53, MS, 54. Prof Exp: Res engr, Squibb Inst Med Res, Olin Mathieson Chem Corp, 54-61, sect head, Fermentation Mfg Dept, E R Squibb & Sons, 61-62, dept head microbiol develop pilot plant, Squibb Inst Med Res, 62-66, dept head chem develop pilot plant, 67-69, asst dept dir chem develop, 69-74, ANTIBIOTICS MFG DEVELOP MGR, E R SQUIBB & SONS, INC, 74- Mem: Am Chem Soc; Am Inst Chem Eng; Sigma Xi. Res: Fermentation technology, scaleup, process design and development; plant start-up and manufacture of antibiotics, steroids, vitamins and enzymes; technical liaison; organic synthetics manufacture. Mailing Add: PO Box 4000 Princeton NJ 08540

STICKLE, RALPH, JR, organic chemistry, see 12th edition

STICKLER, FRED CHARLES, b Villisca, Iowa, Dec 11, 31; m 55; c 3. AGRONOMY, CROP ECOLOGY. Educ: Iowa State Univ, BS, 53, PhD, 58; Kans State Univ, MS, 55. Prof Exp: From asst prof to assoc prof agron, Kans State Univ, 58-64; res agronomist, 64-71, prod planner, 71-73, MGR AGR EQUIP PLANNING, DEERE & CO, 73- Res: Ecological aspects of crop production and management; crop management for improved mechanization. Mailing Add: Prod Planning Dept Deere & Co Moline IL 61265

STICKLER, WILLIAM CARL, b Stuttgart, Ger, Jan 25, 18; nat US; m 42, 58, 68; c 2. ORGANIC CHEMISTRY. Educ: Columbia Univ, AB, 41, AM, 44, PhD(chem), 47. Prof Exp: Asst chem, Columbia Univ, 41-44, 46-47; from asst prof to assoc prof, 47-63, actg chmn dept, 71-72, PROF CHEM, UNIV DENVER, 63- Concurrent Pos: Instr, Sarah Lawrence Col, 43-44 & Hofstra Col, 46-47; vis prof & lectr, Univ Munich & Munich Tech Univ, 56-58; NSF fac fel, 57-58. Mem: Am Chem Soc; The Chem Soc; Soc German Chem. Res: Stereochemistry; reaction mechanisms; natural products and physiologically important compounds. Mailing Add: Dept of Chem Univ of Denver Denver CO 80210

STICKLEY, ELMER EUGENE, b Brackenridge, Pa, Jan 23, 15; m 39, 73; c 1. RADIOLOGY, PHYSICS. Educ: Carnegie Inst Technol, BS, 37; Univ Pittsburgh, MS, 40, PhD(physics), 42. Prof Exp: Asst physics, Univ Pittsburgh, 37-42; indust fel, Mellon Inst, 42-43; physicist, Glass Res Div, Pittsburgh Plate Glass Co, 43-51; med physicist, Med Dept, Brookhaven Nat Lab, 51-61; assoc prof radiol, Col Physicians & Surgeons, Columbia Univ, 61-66; PROF RADIOL SCI, MED COL VA, VA COMMONWEALTH UNIV, 66- Concurrent Pos: Instr, Pa Col Women, 38-42 & Defense Training Exten, Pa State Col, 40-42. Mem: AAAS; fel Am Col Radiol; Am Asn Physicists in Med; Health Physics Soc; Am Phys Soc. Res: Acoustics; x-ray diffraction; properties and ultimate structure of glass; neutron dosimetry; nuclear reactor and accelerator utilization in biological and medical research; application of radionuclides to therapeutic, diagnostic and tracer experiments; diagnostic x-ray physics. Mailing Add: Med Col of Va Richmond VA 23298

STICKNEY, ALDEN PARKHURST, b Providence, RI, Sept 7, 22; m 51; c 2. MARINE ECOLOGY. Educ: Univ RI, BSc, 48; Harvard Univ, MA, 51. Prof Exp: Fishery aide, US Fish & Wildlife Serv, 51-53; res asst, Stirling Sch Med, Yale Univ, 53-54; chief Atlantic salmon invest, US Fish & Wildlife Serv, 54-60, fishery res biologist, Fishery Biol Lab, Bur Com Fisheries, Maine, 60-72 & Biol Lab, Nat Marine Fisheries Serv, 72-73; MARINE RESOURCES SCIENTIST, MAINE DEPT MARINE RESOURCES, 74- Mem: Ecol Soc Am; Am Fisheries Soc. Res: Estuarine ecology; shellfish biology; physiology of larval shellfish; behavior and ecology of sea herring and pandalid shrimp. Mailing Add: Biol Lab Maine Dept of Marine Resources Boothbay Harbor ME 04575

STICKNEY, JAMES MINOTT, b Nashville, Tenn, Nov 2, 07; m 37; c 4. MEDICINE. Educ: Univ Chicago, PhB, 29; Rush Med Col, MD, 33; Univ Minn, MS, 39. Prof Exp: From instr to assoc prof med, 39-65, prof clin med, 65-76, EMER PROF CLIN MED, MAYO GRAD SCH MED, UNIV MINN, 76-, EMER PROF MED, MAYO MED SCH, 76- Concurrent Pos: Consult med, Mayo Clin, 39-75. Mem: AMA; fel Am Col Physicians. Res: Hematology; clinical medicine; leukemia. Mailing Add: Dept of Med Mayo Clin Rochester MN 55901

STICKNEY, JANICE LEE, b Tallahassee, Fla, July 21, 41. PHARMACOLOGY, CARDIOVASCULAR PHYSIOLOGY. Educ: Oberlin Col, AB, 62; Univ Mich, PhD(pharmacol), 67. Prof Exp: Acad Senate grants, Univ Calif, San Francisco, 68 & 69, from instr to asst prof pharmacol, Sch Med, 69-72; asst prof, 72-75, ASSOC PROF PHARMACOL, MICH STATE UNIV, 75- Concurrent Pos: Training fel, 67-68; Bay Area Heart Asn grant, Univ Calif, San Francisco, 69-70; Nat Heart & Lung Inst, Nat Inst Drug Abuse & Mich Heart Asn grants. Mem: AAAS; Pharmacol Soc Can; NY Acad Sci; Am Soc Pharmacol & Exp Therapeut. Res: Cardiovascular pharmacology, especially role of the sympathetic nervous system in the cardiac arrhythmias produced by large doses of digitalis, with emphasis on the mechanisms by which cardiac glycosides produce sympathetic effects; general cardiovascular effects of narcotic analgesics; antiarrhythmic drugs. Mailing Add: Dept of Pharmacol Mich State Univ East Lansing MI 48824

STICKNEY, JOHN CLIFFORD, b Vancouver, Wash, July 15, 09. PHYSIOLOGY, BIOPHYSICS. Educ: Wheaton Col, BS, 33; Univ Wash, MS, 36; Univ Minn, PhD(physiol), 40. Prof Exp: Lab asst gen zool, Wheaton Col, 31-33, instr zool, 35-36; asst physiol, Univ Minn, 37-40; from instr to prof physiol, Sch Med, WVa Univ, 40-74, prof biophys, 67-74; RETIRED. Mem: AAAS; Soc Exp Biol & Med; Am Physiol Soc. Res: Potassium balance; hypoxia; gastrointestinal tract. Mailing Add: Dept of Physiol & Biophys WVa Univ Med Ctr Morgantown WV 26506

STICKNEY, PALMER BLAINE, b Columbus, Ohio, Nov 1, 15; m 37; c 4. POLYMER CHEMISTRY. Educ: Ohio State Univ, AB, 38, PhD(phys chem), 49. Prof Exp: Res engr, Battelle Mem Inst, 40-42, 46, 49-52, asst chief, Rubber & Plastics Div, 52-60, chief polymer res, 60-68, tech adv, 68-73; PROF, WILBERFORCE UNIV, 73- Mem: Fel Am Inst Chemists; Soc Plastics Engrs; Am Chem Soc; Am Asn Univ Profs. Res: Polymerization and processing of polymers. Mailing Add: 2870 Halstead Rd Columbus OH 43221

STICKSEL, PHILIP RICE, b Cincinnati, Ohio, Feb 15, 30; m 53; c 2. METEOROLOGY. Educ: Univ Cincinnati, BS, 52; Fla State Univ, MS, 59, PhD(meteorol), 66. Prof Exp: Jr develop engr, Goodyear Aircraft Corp, 53-54; res assoc meteorol, Fla State Univ, 64-65; res meteorologist, Environ Sci Serv Admin, 65-69 & Nat Air Pollution Control Admin, 67-69; SR METEOROLOGIST, BATTELLE MEM INST, 69- Mem: Am Meteorol Soc; Air Pollution Control Asn. Res: Air pollution meteorology and education; upper atmosphere ozone; air quality management; ozone transport; visible emissions. Mailing Add: 563 Mohican Way Westerville OH 43081

STIDD, BENTON MAURICE, b Bloomington, Ind, June 30, 36; m 58; c 4. PALEOBOTANY. Educ: Purdue Univ, BS, 58; Emporia Kans State Col, MS, 63; Univ Ill, PhD(bot), 68. Prof Exp: Teacher, Wheatland High Sch, 58-62; partic, NSF Acad Year Inst & Res Participation Prog Teachers, Kans State Teachers Col, 62-63; teacher, NKnox High Sch, 63-64; asst prof anat, morphol & paleobot, Univ Minn, Minneapolis, 68-70; ASST PROF BIOL SCI, WESTERN ILL UNIV, 70- Concurrent Pos: Univ Minnesota Grad Sch res grant, 68-69; Sigma Xi res grants-in-aid, 69-70; res coun grant, Western Ill Univ, 71-72; NSF res grant, 74. Mem: Bot Soc Am; Int Orgn

Paleobot; Int Asn Plant Taxon. Res: Paleozoic paleobotany, especially Carboniferous coal ball plants. Mailing Add: Dept of Biol Sci Western Ill Univ Macomb IL 61455

STIDD, CHARLES KETCHUM, b Independence, Ore, Aug 12, 18; m 42; c 2. METEOROLOGY. Educ: Ore State Univ, BS, 41. Prof Exp: Res forecaster, US Weather Bur, 47-55; self employed, 55-62; res assoc meteorol & hydrol, Univ Nev, Reno, 62-71; SPECIALIST METEOROL, SCRIPPS INST OCEANOG, UNIV CALIF, SAN DIEGO, 71- Concurrent Pos: Long-range forecaster, Calif Dept Water Resources, 75- Mem: Am Meteorol Soc; Am Geophys Union. Res: Rainfall and climatic probabilities; general circulation of the atmosphere; moisture, energy and momentum balances; mechanics of rainfall; orthogonal methods in air-sea interaction and long-range forecasting. Mailing Add: Scripps Inst of Oceanog La Jolla CA 92093

STIDHAM, HOWARD DONATHAN, b Memphis, Tenn, Sept 14, 25; div. PHYSICAL CHEMISTRY. Educ: Trinity Col, BS, 50; Mass Inst Technol, PhD, 55. Prof Exp: Spectroscopist, Dewey & Almy Chem Co, 55-56; asst prof, 56-71, ASSOC PROF CHEM, UNIV MASS, 71- Mem: AAAS; Optical Soc Am; Am Chem Soc; Am Phys Soc. Res: Molecular spectroscopy; statistical mechanics. Mailing Add: Dept of Chem Univ of Mass Amherst MA 01002

STIDHAM, SUE NICKERSON, computer science, see 12th edition

STIDWORTHY, GEORGE H, b Viborg, SDak, May 28, 24; m 48; c 4. BIOCHEMISTRY. Educ: Univ SDak, BA, 49, MA, 51; Univ Okla, PhD(biochem), 61. Prof Exp: Chemist, Rayonier Corp, 46-47; res asst biochem, Okla Med Res Found, 51-53; asst chief, Gen Med Res Lab, Vet Admin Hosp, Oklahoma City, 53-57, chief biochemist, 57-59; chief biochemist, Cancer Res Lab, Vet Admin Hosp, Martinsburg, WVa, 59-64; supvr, Med Res Lab, 64-72, CHIEF GEN MED RES LAB, VET ADMIN HOSP, 72-; ASST PROF BIOCHEM, SCH MED, BOSTON UNIV, 64- Concurrent Pos: Nat Cancer Inst grant, Vet Admin Hosp, Martinsburg, WVa, 61-65. Mem: AAAS; Geront Soc; Tissue Cult Asn; Am Aging Asn. Res: Aging effects upon connective tissues; chemistry and biology of the intracellular matrix; in vitro aging of cells in culture; environmental effects in cell metabolism. Mailing Add: Res Lab Vet Admin Hosp 200 Springs Rd Bedford MA 01730

STIEF, LOUIS J, b Pottsville, Pa, July 26, 33; m 63; c 2. PHOTOCHEMISTRY, ASTROCHEMISTRY. Educ: La Salle Col, BA, 55; Cath Univ, PhD(chem), 60. Prof Exp: Asst chem, Cath Univ, 55-59; Nat Acad Sci-Nat Res Coun res fel, Nat Bur Standards, 60-61; NATO fel, Univ Sheffield, 61-62, Dept Sci & Indust Res fel, 62-63; sr chemist, Res Div, Melpar, Inc, 63-65, sr scientist, 65-68; Nat Acad Sci-Nat Res Coun sr res fel, 68-69, aerospace technol chemist, 69-74, HEAD, SPACE CHEM SECT, GODDARD SPACE FLIGHT CTR, NASA, 74- Concurrent Pos: Adj prof, Dept Chem, Cath Univ Am, DC, 75- Mem: Am Phys Soc; Am Chem Soc; The Chem Soc; fel Am Inst Chem; Am Astron Soc. Res: Vacuum-ultraviolet photochemistry; flash photolysis; interstellar molecules; planetary atmospheres; chemical kinetics. Mailing Add: Code 691 NASA/Goddard Space Flight Ctr Greenbelt MD 20771

STIEFEL, EDWARD I, inorganic chemistry, see 12th edition

STIEGLER, JAMES HAROLD, soil science, see 12th edition

STIEGLITZ, RONALD DENNIS, b Milwaukee, Wis, Aug 25, 41; m 65; c 3. GEOLOGY. Educ: Univ Wis-Milwaukee, BS, 63; Univ Ill, Urbana, MS, 67, PhD(geol), 70. Prof Exp: Teaching asst geol, Univ Wis-Milwaukee, 63-64 & Univ Ill, 64-69; lectr, Univ Wis-Milwaukee, 71-72; geologist, Ohio State Geol Surv, 72-74, HEAD, REGIONAL GEOL SECT, OHIO DNR DIV, GEOL SURV, 74- Mem: AAAS; Soc Econ Paleont & Mineral; Sigma Xi. Res: Economic mineral and land capability investigations; subsurface stratigraphy; sedimentology of fine-grained sedimentary deposits; application of scanning electron microscopy to sedimentologic problems. Mailing Add: Ohio DNR Div Geol Surv Fountain Square Columbus OH 43224

STIEHL, ROY THOMAS, JR, b Hay Springs, Nebr, Jan 27, 28; m 54; c 3. POLYMER CHEMISTRY. Educ: Univ Nebr, BS, 50, MS, 51; Univ Ill, PhD(org chem), 53. Prof Exp: Asst, Univ Nebr, 49-51 & Off Rubber Reserv, Univ Ill, 51-53; res chemist, 53-68, SR RES CHEMIST, E I DU PONT DE NEMOURS & CO, INC, 68- Mem: Am Chem Soc. Res: Spandex chemistry; butadiene copolymerization; vinyl monomer synthesis; organophosphorous compounds; textile compounds; textile chemistry. Mailing Add: 400 Ridge Circle Waynesboro VA 22980

STIEHLER, ROBERT DANIEL, b Springfield, NY, July 16, 10; m 49; c 2. CHEMISTRY. Educ: Johns Hopkins Univ, PhD(chem), 33. Prof Exp: Nat Res Coun fel, Calif Inst Technol, 33-34; Lewisohn fel, Wilmer Inst, Sch Med, Johns Hopkins Univ, 34-36, asst ophthal, 36-38; mem staff rubber res, B F Goodrich Co, 39-42; sr chemist, Qm Corps, US Dept Army, Mass, 42-43; tech asst, Off Rubber Reserve, 43-46; mem staff, 46-48, chief, Testing & Specifications Sect, Org & Fibrous Mat Div, 48-64 & Eval Criteria Sect, 64-69, mgr eng standards, US Metric Study, 69-71, CONSULT, NAT BUR STANDARDS, 71- Honors & Awards: Silver Medal, Dept Com, 63; Award of Merit, Am Soc Testing & Mat, 68. Mem: Fel AAAS; fel Am Soc Testing & Mat; Am Soc Eng Educ; Am Chem Soc; fel Brit Plastics & Rubber Inst. Res: Oxidation reduction; thermochemistry of purines; biochemistry of secretion of intraocular and spinal fluids; rubber chemistry; vulcanization kinetics; test methods for rubber and tires; standardization of testing procedures; engineering standards. Mailing Add: 3234 Quesada St NW Washington DC 20015

STIEHM, E RICHARD, b Milwaukee, Wis, Jan 22, 33; m 58; c 3. PEDIATRICS, IMMUNOLOGY. Educ: Univ Wis-Madison, BS, 54, MD, 57. Prof Exp: USPHS fels, physiol chem, Univ Wis, 58-59 & pediat immunol, Univ Calif, San Francisco, 63-65; from asst prof to assoc prof pediat, Med Sch, Univ Wis, 65-69; assoc prof, 69-72, PROF PEDIAT, SCH MED, UNIV CALIF, LOS ANGELES, 72- Concurrent Pos: Markle scholar acad med, 67. Honors & Awards: Ross res award pediat res, 71; E Mead Johnson Award pediat res, 74. Mem: Am Acad Pediat; Soc Pediat Res; Am Asn Immunologists; Am Pediat Soc. Res: Pediatric immunology, immunodeficiency disease; newborn defense mechanisms; human gamma globulin; immunology of malnutrition. Mailing Add: Dept of Pediat Univ of Calif Ctr Health Sci Los Angeles CA 90024

STIEL, EDSEL FORD, b Los Angeles, Calif, Dec 19, 33. MATHEMATICS. Educ: Univ Calif, Los Angeles, AB, 55, MA, 59, PhD(math), 63. Prof Exp: Math analyst, Douglas Aircraft Co, Calif, 55-56, comput analyst, 57-59; sr math analyst, Lockheed Missiles & Space Co, 60; from asst prof to assoc prof, 62-72, chmn dept, 62-74, PROF MATH, CALIF STATE UNIV, FULLERTON, 72- Mem: Am Math Soc; Math Asn Am. Res: Isometric immersions of Riemannian manifolds; differential geometry. Mailing Add: Dept of Math Calif State Univ 800 N State Col Fullerton CA 92634

STIEN, HOWARD M, b Montevideo, Minn, Apr 11, 26; m 47; c 2. ZOOLOGY, PHYSIOLOGY. Educ: Northwestern Col, Minn, BA, 56; Macalester Col, MA, 58; Univ Wyo, PhD(physiol), 63. Prof Exp: Instr biol, Pepperdine Col, 58-60; asst prof zool, Univ Wyo, 61-64; assoc prof biol, Northwestern Col, Minn, 64-65; assoc prof, 65-72, PROF BIOL, WHITWORTH COL, WASH, 72-, CHMN DEPT, 65- Mem: AAAS. Res: Immunogenetics, especially the ontogeny of molecular individuality. Mailing Add: Dept of Biol Whitworth Col Spokane WA 99251

STIENING, RAE FRANK, b Pittsburgh, Pa, May 26, 37. HIGH ENERGY PHYSICS. Educ: Mass Inst Technol, SB, 58, PhD(physics), 62. Prof Exp: Physicist, Lawrence Berkeley Lab, Univ Calif, 63-71; PHYSICIST, FERMI NAT ACCELERATOR LAB, 71- Res: Weak interactions of K-mesons; particle accelerators. Mailing Add: Fermi Nat Accelerator Lab PO Box 500 Batavia IL 60510

STIENSTRA, WARD CURTIS, b Holland, Mich, June 19, 41; m 63; c 2. PLANT PATHOLOGY. Educ: Calvin Col, ABGen, 63; Mich State Univ, MS, 66, PhD(plant path), 70. Prof Exp: Asst prof, 70-75, ASSOC PROF PLANT PATH, UNIV MINN, ST PAUL, 75- Mem: Am Phytopath Soc; Am Inst Biol Sci. Res: Diseases of turf and ornamentals; soil borne diseases. Mailing Add: Dept of Plant Path Univ of Minn St Paul MN 55101

STIER, ELIZABETH FLEMING, b Riverside, NJ, Nov 24, 25; m 47; c 3. BIOCHEMISTRY. Educ: Rutgers Univ, BS, 47, MS, 49, PhD(biochem), 51. Prof Exp: Asst, 47-51, res assoc, 51-59, from asst prof to assoc prof, 59-72, PROF FOOD SCI, RUTGERS UNIV, NEW BRUNSWICK, 72- Mem: AAAS; Inst Food Technol; NY Acad Sci. Res: Methodology of flavor evaluation; flavor evaluation as a tool on pesticide treated fruits and vegetables; objective flavor techniques. Mailing Add: Dept of Food Sci Rutgers Univ New Brunswick NJ 08901

STIER, HOWARD LIVINGSTON, b Delmar, Del, Nov 28, 10; m 40; c 5. RESEARCH ADMINISTRATION. Educ: Univ Md, BS, 32, MS, 37, PhD(plant physiol), 39. Prof Exp: Agent potato breeding, USDA, 33-35; res asst hort, Univ Md, 35-39, asst prof, 39-41, prof mkt & head dept, 46-51; dir div statist & mkt res, Nat Canners Asn, 51-61; dir qual control, United Fruit Co, 61-71, dir develop & prod supporting serv, 72-73, vpres qual control, 73-74, VPRES RES DEVELOP & QUAL CONTROL, UNITED BRANDS CO, 74- Concurrent Pos: Dir prog anal, War Assets Admin, 46, consult, 46-47; prof lectr, George Washington Univ, 57-61; mem res adv comt, USDA, 62-64; mem adv comt, Dept Defense, 63; mem task group statist qual control of foods, Nat Acad Sci, 64. Mem: Fel AAAS; fel Am Soc Qual Control (vpres, 66-68, pres elec, 73-74, pres, 74-75); Am Statist Asn; Biomet Soc; Inst Food Tech. Res: Plant breeding; factors affecting quality and growth and development of horticultural crops, food processing, statistical control of quality. Mailing Add: United Brands Co Prudential Ctr Boston MA 02199

STIER, PAUL MAX, b Eden, NY, Aug 18, 24; m 47; c 4. PHYSICS. Educ: Univ Buffalo, BS, 44; Cornell Univ, PhD(physics), 52. Prof Exp: Sr physicist, Oak Ridge Nat Lab, 50-55; group leader chem physics, Union Carbide Corp, 55-60, asst dir, Res Lab, Union Carbide Nuclear Co, 60-65, mgr, Nucleonics Res Lab, 65-66, prog mgr phys sci, Tarrytown Tech Ctr, 66-72, OPERS MGR, CORP RES DEPT, STERLING FOREST LAB, UNION CARBIDE CORP, 72- Mem: Am Phys Soc. Res: Stopping of heavy ions; energy range 10-250 kilo-electron-volts; energy loss; ionization; charge exchange; field emission; field ionization microscopy; chemisorption; surface diffusion; radiation damage in solids. Mailing Add: Corp Res Dept Union Carbide Corp PO Box 324 Tuxedo NY 10987

STIERWALT, DONALD L, b Fremont, Ohio, Sept 20, 26; m 53; c 2. OPTICS. Educ: Univ Toledo, BS, 50; Syracuse Univ, MS, 53, PhD(physics), 61. Prof Exp: Res physicist, Naval Ord Lab, Corona, 58-70, RES PHYSICIST, ELECTRONIC MAT SCI DIV, NAVAL ELECTRONICS LAB CTR, 70- Mem: Optical Soc Am. Res: Low temperature infrared spectral emittance and transmittance of optical materials and components. Mailing Add: Code 4600 Naval Electronics Lab Ctr San Diego CA 92157

STIFEL, FREDERICK BENTON, b St Louis, Mo, Jan 30, 40; m 63; c 2. BIOCHEMISTRY, NUTRITION. Educ: Iowa State Univ, BS, 62, PhD(biochem, nutrit), 67. Prof Exp: Lab supvr & res biochemist, US Army Med Res & Nutrit Lab, Fitzsimons Gen Hosp, Denver, 67-74; LAB SUPVR & RES CHEMIST, LETTERMAN ARMY INST RES, SAN FRANCISCO, 74- Honors & Awards: Meritorious Sci Achievement Awards, US Army-Sci Conf, 70 & 72. Mem: AAAS; Am Inst Nutrit; Am Fedn Clin Res; Am Soc Clin Nutrit. Res: Effects of diet, vitamins, drugs and hormones upon jejunal and hepatic enzyme regulation; relationship of enzyme regulation to human disease; nature of metabolic regulation in mammals. Mailing Add: 8 Rubicon Court San Rafael CA 94903

STIFEL, PETER BEEKMAN, b Wheeling, WVa, Feb 9, 36; c 2. GEOLOGY. Educ: Cornell Univ, BA, 58; Univ Utah, PhD(geol), 64. Prof Exp: Res asst, Univ Utah, 60-63; ASSOC PROF GEOL, UNIV MD, COLLEGE PARK, 66- Mem: AAAS; Geol Soc Am; Soc Econ Paleont & Mineral; Am Malacol Union. Res: Paleontology, stratigraphy and sedimentation. Mailing Add: Dept of Geol Univ of Md College Park MD 20742

STIFF, ROBERT H, b Pittsburgh, Pa, Apr 23, 23; m 45; c 3. DENTISTRY. Educ: Univ Pittsburgh, BS, 43, DDS, 45, Med, 53. Prof Exp: Instr oper dent, 45-58, asst prof oral med, 56-57, assoc prof oral genetic path & microbiol, 58-59, assoc prof oral med, 60-65, PROF ORAL MED & CHMN DEPT, SCH DENT, UNIV PITTSBURGH, 65- Concurrent Pos: Consult, USPHS, 62-66 & Dent Div, Pa Dept Health, 64-66. Mem: Am Acad Oral Path; fel Am Col Dent. Res: Task analysis of dental practice; dental education; dental treatment for the handicapped; attitudes of dental students towards treatment of chronically ill and aged; caries inhibiting effectiveness of a stannous fluoride-insoluble sodium metaphosphate dentifrice in children. Mailing Add: Dept of Oral Med Univ of Pittsburgh Sch of Dent Pittsburgh PA 15213

STIFFLER, DANIEL F, b Los Angeles, Calif, Nov 27, 42; m 67; c 2. PHYSIOLOGY, ZOOLOGY. Educ: Univ Calif, Santa Barbara, BA, 68; Ore State Univ, MS, 70, PhD(physiol), 72. Prof Exp: NIH-USPHS trainee physiol, Health Sci Ctr, Univ Ore, 72-74; lectr animal physiol, Univ Calif, Davis, 74-75; lectr, 75-76, ASST PROF PHYSIOL & ZOOL, CALIF STATE POLYTECH UNIV, POMONA, 76- Mem: AAAS; Am Physiol Soc; Am Soc Zoologists. Res: Renal physiology; epithelial transport physiology; comparative physiology of osmotic and ionic regulation. Mailing Add: Dept of Biol Sci Calif State Polytech Univ Pomona CA 91768

STIFFLER, PAUL W, b Buffalo, NY, June 24, 43; m 67; c 1. MEDICAL MICROBIOLOGY. Educ: Bowling Green State Univ, BA, 65, MA, 67; Mich State Univ, PhD(microbiol), 72. Prof Exp: Res assoc microbiol, Michael Reese Hosp & Med Ctr, Chicago, 72-73; USPHS fel clin microbiol, Dept Med, Sect Infectious Dis, Pritzker Sch Med, Univ Chicago, 73-75; MICROBIOLOGIST, MASON-BARRON LABS, INC, 75- Mem: Am Soc Microbiol; Asn Practitioners Infection Control. Res: Antibiotic resistance determinants and their mode of transfer inter- and intra-generically. Mailing Add: Mason-Barron Labs Inc 4720 W Montrose Ave Chicago IL 60641

STIGERS, CHARLES ARTHUR, physics, see 12th edition

STIGGALL, DIANA LEE, physical organic chemistry, see 12th edition

STIGLER, ROBERT LEATH, JR, anthropology, see 12th edition

STIGLER, STEPHEN MACK, b Minneapolis, Minn, Aug 10, 41; c 2. STATISTICS. Educ: Carleton Col, BA, 63; Univ Calif, Berkeley, PhD(statist), 67. Prof Exp: Asst prof, 67-71, ASSOC PROF STATIST, UNIV WIS-MADISON, 71- Mem: AAAS; Am Statist Asn; Inst Math Statist. Res: Order statistics; experimental design; history of statistics. Mailing Add: Dept of Statist Univ of Wis 1210 W Dayton St Madison WI 53706

STIGLIAMI, WILLIAM MICHAEL, b Stamford, Conn, Jan 9, 45. PHYSICAL CHEMISTRY. Educ: Univ Conn, BA, 66; Princeton Univ, MA, 69, PhD(chem), 71. Prof Exp: Fel chem, Bristol Univ, Eng, 71-73; fel, 73-74, INSTR CHEM, PRINCETON UNIV, 74- Mem: Am Chem Soc; AAAS. Res: Study of intramolecular rearrangement mechanisms of molecules containing large organic polycyclic systems; vibrational motions of carbon skeleton are studied utilizing vibrational and pure rotational spectroscopy. Mailing Add: Frick Chem Lab Princeton Univ Princeton NJ 08540

STILES, DAVID A, b Harrow, Eng, Apr 28, 38; m 66; c 3. ANALYTICAL CHEMISTRY, ENVIRONMENTAL CHEMISTRY. Educ: Univ Birmingham, BSc, 60, PhD(electron spin resonance spectros), 63. Prof Exp: Asst prof chem, Univ Calgary, 63-64; univ fel, Univ Alta, 64-66; asst prof, 66-72, ASSOC PROF CHEM, ACADIA UNIV, 72- Mem: Chem Inst Can; The Chem Soc. Res: Agricultural pollution; fate of pesticides and heavy metals in sandy soils; applications of molecular emission cavity analysis. Mailing Add: Dept of Chem Acadia Univ Wolfville NS Can

STILES, JOHN I, (JR), b Ft Wayne, Ind, July 6, 48; m 70. MOLECULAR GENETICS. Educ: Ind Univ, Bloomington, BA, 70; Cornell Univ, PhD(plant physiol), 76. Prof Exp: NIH FEL MOLECULAR GENETICS, SCH MED, UNIV ROCHESTER, 75- Mem: AAAS; Am Soc Plant Physiologists. Res: Isolation of messenger RNA for yeast cytochrome c and investigation of its regulation. Mailing Add: Dept of Radiation Biol & Biophys Univ of Rochester Sch of Med Rochester NY 14627

STILES, MARTIN, b Huntington, WVa, Aug 25, 27; m 54. ORGANIC CHEMISTRY. Educ: Ohio State Univ, BSc, 50; Harvard Univ, MA, 51, PhD(chem), 54. Prof Exp: Res assoc, 53-55, from instr to assoc prof, 55-64, PROF CHEM, UNIV MICH, ANN ARBOR, 64- Concurrent Pos: Alfred P Sloan fel, 61-65; John Simon Guggenheim Mem fel, 62-63; ed, J Am Chem Soc, 69-75. Mem: Am Chem Soc. Res: Organic reaction mechanisms. Mailing Add: Dept of Chem Univ of Mich Ann Arbor MI 48104

STILES, PHILIP GLENN, b Terre Haute, Ind, Nov 24, 31; m 56; c 1. FOOD TECHNOLOGY, POULTRY SCIENCE. Educ: Univ Ark, BS, 53; Univ Ky, MS, 56; Mich State Univ, PhD(food tech), 58. Prof Exp: Assoc prof food tech, Univ Conn, 59-69; PROF POULTRY SCI & FOOD TECHNOL, ARIZ STATE UNIV, 69- Concurrent Pos: Consult, Nixon Baldwin Div, Tenneco Co, 67- Mem: Inst Food Technol; Poultry Sci Asn; World Poultry Asn. Res: Food technology as applied to poultry products and food packaging. Mailing Add: Div of Agr Ariz State Univ Tempe AZ 85281

STILES, PHILLIP JOHN, b Manchester, Conn, Oct 31, 34; m 56; c 5. SOLID STATE PHYSICS. Educ: Trinity Col, Conn, BS, 56; Univ Pa, PhD(physics), 61. Prof Exp: Fel & res assoc physics, Univ Pa, 61-62; NSF fel, Cambridge Univ, 62-63; mem res staff, Thomas J Watson Res Ctr, Int Bus Mach Corp, NY, 63-70; PROF PHYSICS, BROWN UNIV, 70-, CHMN DEPT, 74- Honors & Awards: Outstanding Contrib Invention Achievement, IBM Corp, 66 & 70. Mem: AAAS; Am Phys Soc; Am Astron Soc; Acoust Soc Am. Res: Solid state and low temperature physics; electronic properties of metals and semiconductors. Mailing Add: Dept of Physics Brown Univ Providence RI 02912

STILES, RAEBURN BRACKETT, b Middlebury, Vt, Mar 14, 15; m 41; c 1. APPLIED MATHEMATICS. Educ: Middlebury Col, AB, 38; Peabody Col, MA, 49; Vanderbilt Univ, BE, 58. Prof Exp: Instr high sch, NY, 38-41 & La, 41-42; from instr to assoc prof appl math, Vanderbilt Univ, 46-64; engr dir comput serv div, Tenn Dept Hwy, 64-72; regional dir comput serv, Cols & Univs, State of Tenn, 72-74; DIR MGT SYSTS, DEPT OF TREAS, STATE OF TENN, 74- Mem: Am Soc Eng Educ. Res: Statistics; applications of computer to engineering education; numerical analysis. Mailing Add: 3911 Trimble Rd Nashville TN 37215

STILES, ROBERT NEAL, b Mar 15, 33; m 59; c 3. PHYSIOLOGY. Educ: Univ Mo, BS, 59, MA, 63; Northwestern Univ, PhD, 66. Prof Exp: Asst prof zool & physiol, Butler Univ, 66-68; asst prof physiol & biophys, 68-75, ASSOC PROF PHYSIOL & BIOPHYS, UNIV TENN CTR HEALTH SCI, MEMPHIS, 75- Concurrent Pos: USPHS grant, Univ Tenn Ctr Health Sci, Memphis, 69- Mem: AAAS; Sigma Xi; Am Physiol Soc; Soc Neurosci. Res: Human limb tremor; muscle mechanics; motor control system. Mailing Add: Univ of Tenn Ctr for Health Sci Memphis TN 38163

STILES, WARREN CRYDER, b Dias Creek, NJ, June 16, 33; m 55; c 4. HORTICULTURE. Educ: Rutgers Univ, BS, 54, MS, 55; Pa State Univ, PhD(hort), 58. Prof Exp: Asst prof pomol, Rutgers Univ, 58-63; assoc prof, 63-69, PROF POMOL, UNIV MAINE, 69-, EXTEN FRUIT SPECIALIST, 63-, SUPT, HIGHMOOR FARM, 66- Mem: AAAS; Am Soc Hort Sci. Res: Nutrition and post-harvest physiology of fruit. Mailing Add: Highmoor Farm Maine Agr Exp Sta Monmouth ME 04259

STILES, WILBUR J, b Suffern, NY, Jan 12, 32; m 56; c 2. MATHEMATICS. Educ: Lehigh Univ, BS, 54; Ga Inst Technol, BS, 60, MS, 62, PhD(math), 65. Prof Exp: ASSOC PROF MATH, FLA STATE UNIV, 65- Mem: Am Math Soc; Math Asn Am; Can Math Cong. Res: Functional analysis, geometry of Banach spaces. Mailing Add: Dept of Math Fla State Univ Tallahassee FL 32306

STILES, WILLIAM WHITFIELD, b Washington, DC, May 26, 08; m 40; c 3. PREVENTIVE MEDICINE, PUBLIC HEALTH. Educ: Univ Colo, BS, 30; Univ Rochester, MD, 39; Univ Calif, MPH, 47. Prof Exp: PROF PUB HEALTH, SCH PUB HEALTH, UNIV CALIF, BERKELEY, 47- Concurrent Pos: Consult, Environ Sanit Technician Sch, US Navy, 48- Honors & Awards: Taylor Instrument Award, 40; Bosch & Lomb Award, 41. Mem: AMA; fel Am Pub Health Asn; Asn Teachers Prev Med; Am Col Prev Med. Res: Clinical and epidemiological investigations of leptospirosis. Mailing Add: Sch of Pub Health Univ of Calif Berkeley CA 94720

STILL, CECIL CALVERT, plant biochemistry, physiology, see 12th edition

STILL, CHARLES NEAL, b Richmond, Va, Apr 15, 29; m 58; c 3. NEUROLOGY. Educ: Clemson Univ, BS, 49; Purdue Univ, MS, 51; Med Col SC, MD, 59. Prof Exp: Instr chem, Clemson Univ, 51-52 & US Mil Acad, 53-55; intern, Univ Chicago Clins, 59-60; resident neurol, Baltimore City Hosps & Johns Hopkins Hosp, 60-63; CHIEF NEUROL SERV, WILLIAM S HALL PSYCHIAT INST, 65-; ASSOC PROF NEUROL, SCH MED, UNIV SC, 76- Concurrent Pos: Fel neurol med, Sch Med, Johns Hopkins Univ, 60-63; Nat Inst Neurol Dis & Blindness spec res fel neuropath, Res Lab, McLean Hosp & Harvard Med Sch, 63-65; fel neurol, Seizure Unit, Children's Hosp Med Ctr, Boston, 66; assoc clin prof neurol, Med Univ SC, 73-; chmn grants rev bd, SC Dept Ment Health, 73-; mem, Huntington's Chorea Res Group, World Fedn Neurol. Mem: Fel Am Acad Neurol; Am Chem Soc; fel Am Geriat Soc; fel Am Inst Chemists; Geront Soc. Res: Clinical neurology; behavioral neurology; aging; bioethics. Mailing Add: Neurol Serv William S Hall Psychiat Inst Columbia SC 29202

STILL, EDWIN TANNER, b Monroe, Ga, Nov 2, 35; m 59; c 2. RADIOBIOLOGY. Educ: Univ Rochester, MS, 64; Univ Ga, DVM, 59. Prof Exp: Res scientist, Sch Aerospace Med, US Air Force, Brooks AFB, Tex, 64-67 & Naval Radiol Defense Lab, Calif, 67-69; res contracts adminr, Div Biol & Med, US AEC, 69-75; CHMN, RADIATION BIOL DEPT, ARMED FORCES RADIOBIOL RES INST, 75- Mem: Sigma Xi. Res: Low-level radiation effects; beneficial applications of radiation. Mailing Add: Armed Forces Radiobiol Res Inst Bethesda MD 20014

STILL, EUGENE UPDYKE, physiology, see 12th edition

STILL, GERALD G, b Seattle, Wash, Aug 13, 33; m 54; c 3. BIOCHEMISTRY, ORGANIC CHEMISTRY. Educ: Wash State Univ, BS, 59; Ore State Univ, MS, 63, PhD(biochem), 65. Prof Exp: RES BIOCHEMIST, RADIATION & METAB RES LAB, AGR RES SERV, USDA, 65- Mem: Am Soc Plant Physiol; Am Chem Soc. Res: Metabolism of pesticides; isolation and characterization of pesticide metabolites from the agricultural environment; development of chromatographic methodology for purifications of polar metabolic products. Mailing Add: Radiation & Metab Res Lab USDA State Univ Sta NDak State Univ Fargo ND 58102

STILL, IAN WILLIAM JAMES, b Rutherglen, Scotland, July 5, 37; m 64; c 2. ORGANIC CHEMISTRY. Educ: Glasgow Univ, BSc, 58, PhD(chem), 62. Prof Exp: Res assoc, Univ Toronto, 62-63; sci officer, Allen & Hanburys Ltd, Eng, 63-64; from asst lectr to lectr chem, Huddersfield Col Tech Eng, 64-65; asst prof, 65-70, ASSOC PROF CHEM, UNIV TORONTO, 70- Mem: Am Chem Soc; Chem Inst Can; fel The Chem Soc. Res: Chemical and photochemical reactions, especially of organic sulfur compounds; C-13 nuclear magnetic resonance. Mailing Add: Erindale Col Univ of Toronto Mississauga ON Can

STILL, W CLARK, JR, b Augusta, Ga, Aug 31, 46; m 67. SYNTHETIC ORGANIC CHEMISTRY. Educ: Emory Univ, BS, 69, PhD(org chem), 72. Prof Exp: IBM fel theoret org chem, Princeton Univ, 72-73; fel synthetic org chem, Columbia Univ, 73-75; ASST PROF ORG CHEM, VANDERBILT UNIV, 75- Mem: Am Chem Soc. Res: Organic synthesis; new synthetic methods. Mailing Add: Dept of Chem Vanderbilt Univ Nashville TN 37235

STILL, WILLIAM JAMES SANGSTER, b Aberdeen, Scotland, Sept 16, 23; m 51; c 2. PATHOLOGY. Educ: Univ Aberdeen, MB, ChB, 51, MD, 60. Prof Exp: Lectr path, Univ London, 57-60; asst prof, Sch Med, Washington Univ, 60-62; sr lectr, Univ London, 62-65; assoc prof, 65-70, PROF PATH, MED COL VA, 70- Concurrent Pos: Fel coun arteriosclerosis, Am Heart Asn, 65. Mem: Col Am Path; Path Soc Gt Brit & Ireland. Res: Cardiovascular disease, particularly arterial disease. Mailing Add: Dept of Path Med Col of Va Richmond VA 23219

STILLE, JOHN KENNETH, b Tucson, Ariz, May 8, 30; m 58; c 2. ORGANIC CHEMISTRY, POLYMER CHEMISTRY. Educ: Univ Ariz, BS, 52, MS, 53; Univ Ill, PhD(org chem), 57. Prof Exp: Asst, Univ Ariz, 51-53; from instr to assoc prof, 57-65, PROF ORG CHEM, UNIV IOWA, 65- Concurrent Pos: Consult, Dunlop Res, 61- & E I du Pont de Nemours & Co, 64-; vis prof, Royal Inst Technol, Sweden, 68; mem comt org polymer characterization of mat adv bd, Nat Acad Sci; mem eval panel for polymers div, Inst Mat Res, Nat Bur Standards. Mem: Am Chem Soc; The Chem Soc. Res: Physical organic chemistry; organometallic reactions and mechanisms; asymmetric synthesis catalyzed by transition metals polymer synthesis and reaction mechanisms. Mailing Add: Dept of Chem Univ of Iowa Iowa City IA 52242

STILLER, MARY LOUISE, b Salem, Ohio, Nov 29, 31. PLANT PHYSIOLOGY, BIOCHEMISTRY. Educ: Purdue Univ, BS, 54, MS, 56, PhD(plant physiol), 59. Prof Exp: NSF fel biochem, Univ Chicago, 58-60, USPHS trainee, 60-61; fel, Univ Pa, 61-62; asst prof, 62-66, ASSOC PROF BIOL SCI, PURDUE UNIV, LAFAYETTE, 66- Concurrent Pos: NIH career develop award, 65- Mem: AAAS; Am Soc Plant Physiol. Res: Biochemistry of photosynthesis, photoreduction and respiration. Mailing Add: Dept of Biol Sci Purdue Univ West Lafayette IN 47906

STILLER, RICHARD L, b New York, NY, Feb 15, 33; m 70. BIOCHEMISTRY. Educ: Hunter Col, AB, 59; St John's Univ, NY, MS, 70, PhD(biol chem), 72. Prof Exp: RES SCIENTIST BIOCHEM, NY PSYCHIAT INST, 61-, HEAD ANAL SERV SECT, NEUROTOXICOL RES UNIT, 75- Concurrent Pos: Adj asst prof, Queensborough Community Col, 72- Mem: AAAS; Am Chem Soc. Res: Synthesis and biosynthesis of sphingolipids; neurochemistry of brain and nerve tissue; lipid chemistry; drug effects on central nervous system; methods for neuropsychotropic agent detection in biological medium. Mailing Add: NY Psychiat Inst Neurotoxicol Res Unit 722 W 168th St New York NY 10032

STILLERMAN, MAXWELL, b Brooklyn, NY, Jan 25, 09; m 45; c 3. PEDIATRICS. Educ: State Univ NY Downstate Med Ctr, MD, 32; Am Bd Pediat, dipl, 42. Prof Exp: Fel, Manhattan Convalescent Serum Lab, 38-41; assoc, Meningitis Div, City Health Dept, New York, 41-42; from asst prof to assoc prof clin pediat, Med Col, Cornell Univ, 57-72; PROF CLIN PEDIAT, STATE UNIV NY STONY BROOK, 72- Concurrent Pos: Nat Found Infantile Paralysis investr, 39-40; attend pediatrician, North Shore Hosp, 54- & Jewish Hosp, 55-; NIH & NY Heart Asn grants, 56-59. Mem: AMA; Am Acad Pediat; Infectious Dis Soc Am; NY Acad Sci. Res: Beta-hemolytic streptococcal infections. Mailing Add: 20 Polo Rd Great Neck NY 11023

STILLEY, JERRY LEE, b Albany, La, Oct 9, 41; m 71. PHYSICS. Educ: La State Univ, Baton Rouge, BS, 63, PhD(physics), 70. Prof Exp: Systs eng asst comput, IBM Corp, 69-70; asst prcf, 70-74, ASSOC PROF PHYSICS, MCNEESE STATE UNIV, 74- Mem: Am Phys Soc. Res: Atomic collision theory. Mailing Add: Dept of Physics McNeese State Univ Lake Charles LA 70601

STILLINGER, FRANK HENRY, theoretical chemistry, see 12th edition

STILLINGS, BRUCE ROBERT, b Portland, Maine, May 18, 37; m 59; c 4.

NUTRITION. Educ: Univ Maine, BS, 58; Pa State Univ, MS, 60, PhD(animal nutrit), 63. Prof Exp: NIH fel, Cornell Univ, 63-66; supvry res chemist & dep lab dir, US Nat Marine Fisheries Serv, 66-74; DIR RES ACTIVITIES, RES CTR, NABISCO INC, 74- Mem: AAAS; Am Inst Nutrit; Inst Food Technologists. Res: Nutritional studies on metabolism and utilization of minerals and amino acids; nutritive value of food-proteins and protein concentrates; factors affecting protein quality of foods. Mailing Add: Nabisco Inc Res Ctr 2111 Rt 208 Fair Lawn NJ 07410

STILLINGS, ROBERT ALMON, b Frazee, Minn, Sept 12, 16; m 39; c 2. CHEMISTRY. Educ: Univ Mont, BS, 37; Lawrence Univ, MS, 39, PhD(chem), 41. Prof Exp: Res chemist, 41-45, tech supt, Lakeview Mill, 45-48, staff control supt wadding, 49-52, asst to mgr mills, 52-55, prod supt, Neenah Mill, 55-57, mgr tissue sect, Res & Develop, 57-65, mgr consumer prod, Res & Eng, Paper Prod & Process Develop, 65-68, dir res & eng, Consumer Prod Div, 68-75; RETIRED. Mem: Tech Asn Pulp & Paper Indust. Res: Degradation of cellulose; new pulping processes; bleaching and testing of wood pulp; action of ultraviolet light upon cellulose; statistical quality control. Mailing Add: 1323 Oakcrest Court Appleton WI 54911

STILLIONS, MERLE C, b Bedford, Ind, Feb 15, 29; m 53; c 5. LABORATORY ANIMAL SCIENCE, NUTRITION. Educ: Purdue Univ, BS, 57, MS, 58; Rutgers Univ, PhD(nutrit), 62. Prof Exp: Instr nutrit, Rutgers Univ, 58-62, chmn, Dairy Dept, Chico State Univ, 62-63; dir nutrit, Morris Res Lab, 63-72; RES DIR, AGWAY INC, 72- Concurrent Pos: Mem, Equine Comt, Nat Res Coun, 69-73. Mem: Am Soc Animal Sci; Am Asn Lab Animal Soc. Res: Laboratory animal and fish nutrition and feed control programs. Mailing Add: Agway Tech Ctr 777 Warren Rd Ithaca NY 14850

STILLMAN, IRVING MAYER, b Queens, NY, Oct 15, 33; m 58. BIOPHYSICS, MEDICINE. Educ: Queens Col, NY, BS, 55; Washington Univ, MD, 59; Polytech Inst Brooklyn, PhD(phys chem), 68. Prof Exp: Med intern, Jersey City Med Ctr, NJ, 59-60; NIH fel, 62-65; RES ASSOC BIOPHYS, NAT INST NEUROL DIS & STROKE, 65- Mem: AAAS; fel Am Inst Chemists; Am Chem Soc; Am Phys Soc; NY Acad Sci. Res: Organic semiconductors; electrochemistry; neurophysiology; membrane biophysics; quantum chemistry and biology. Mailing Add: 5374 Fallriver Row Ct Columbia MD 21044

STILLMAN, RICHARD ERNEST, b Grand Island, Nebr, Dec 6, 29; m 56; c 2. MATHEMATICS, CHEMICAL ENGINEERING. Educ: Univ Kans, BS, 51, MS, 56; Pa State Univ, University Park, PhD(chem eng), 61. Prof Exp: Staff engr process control, Res Div, 58-63, adv engr, Systs Develop Div, 64-65, SR ENGR, DATA PROCESSING DIV, IBM CORP, PALO ALTO, 66- Mem: Am Inst Chem Engrs. Res: Formulation of mathematical models of chemical processes; numerical methods for solving partial and ordinary differential equations; gradient optimization procedures and multicomponent distillation calculations. Mailing Add: 1659 Fairorchard Ave San Jose CA 95125

STILLO, HORATIO SERAFINO, physical chemistry, see 12th edition

STILLWAY, LEWIS WILLIAM, b Casper, Wyo, Feb 27, 39; m 59; c 2. BIOCHEMISTRY. Educ: Col Idaho, BS, 62; Univ Idaho, MS, 65, PhD(biochem), 68. Prof Exp: Fel, Inst Marine Sci, Univ Miami, 68-69; assoc chem, 69-71, ASST PROF BIOCHEM, MED UNIV SC, 71- Mem: AAAS. Res: Natural products; growth factors; lipid chemistry and metabolism; marine biology; enzymes; marine lipids and nutrition. Mailing Add: Dept of Biochem Med Univ of SC Charleston SC 29401

STILLWELL, EDGAR FELDMAN, b Staten Island, NY, Nov 2, 29. PHYSIOLOGY. Educ: Wagner Mem Lutheran Col, BS, 51; Duke Univ, MA, 53, PhD(zool), 57. Prof Exp: Asst zool, Duke Univ, 52-56, res assoc, 56-57; asst prof biol, Longwood Col, 57-60 & Univ SC, 60-61; assoc prof zool, ECarolina Univ, 61-68; ASSOC PROF BIOL, OLD DOM UNIV, 68- Concurrent Pos: NASA-Am Soc Eng Educ fac res fel, Langley Res Ctr, 69-70; NASA res grant, 71-72. Mem: AAAS; Am Soc Cell Biol. Res: Mitogenetic control mechanisms in central nervous system neurons in tissue culture. Mailing Add: Dept of Biol Old Dom Univ Norfolk VA 23508

STILLWELL, EPHRAIM POSEY, JR, b Sylva, NC, Aug 29, 34; m 60; c 2. SOLID STATE PHYSICS. Educ: Wake Forest Col, BS, 56; Univ Va, MS, 58, PhD(physics), 60. Prof Exp: From asst prof to assoc prof, 60-69, head dept, 71-74, PROF PHYSICS, CLEMSON UNIV, 69- Concurrent Pos: US Air Force Off Sci Res grant, 63-69. Mem: Am Asn Physics Teachers; Am Phys Soc; AAAS. Res: Magnetoresistance in metals; superconductivity. Mailing Add: Dept of Physics Clemson Univ Clemson SC 29631

STILLWELL, GEORGE KEITH, b Moose Jaw, Sask, July 11, 18; m 43; c 2. PHYSICAL MEDICINE. Educ: Univ Sask, BA, 39; Queen's Univ, Ont, MD, CM, 42; Univ Minn, PhD(phys med & rehab), 54; Am Bd Phys Med & Rehab, dipl, 52. Prof Exp: Instr, Univ Minn, 50-54, from instr to assoc prof, Mayo Grad Sch Med, Univ Minn, 55-73, PROF PHYS MED & REHAB, MAYO MED SCH, UNIV MINN, 73-, CHMN DEPT, 73- Concurrent Pos: Consult, Mayo Clin, 54-73. Mem: AAAS; Cong Rehab Med; Am Acad Phys Med & Rehab. Res: Rehabilitation; physiologic effects of therapeutic procedures; edema of peripheral origin. Mailing Add: Dept of Phys Med & Rehab Mayo Clin Rochester MN 55901

STILLWELL, HAROLD DANIEL, b Staten Island, NY, Mar 21, 31. PHYSICAL GEOGRAPHY, BIOGEOGRAPHY. Educ: Duke Univ, BS, 52, MF, 54; Mich State Univ, PhD, 61. Prof Exp: Forestry aid, US Forest Serv, NC, 52; asst, Ore Forest Res Ctr, 54-57; asst geog, Mich State Univ, 57-59; asst prof, Eastern Mich Univ, 60-61 & Univ Tex, 61-62; assoc prof, ECarolina Univ, 62-71; PROF GEOG, APPALACHIAN STATE UNIV, 71- Mem: Asn Am Geographers; Sigma Xi. Res: Forestry; wood technology; landforms; geography of national parks and mountains. Mailing Add: Dept of Geog Appalachian State Univ Boone NC 28607

STILLWELL, RICHARD NEWHALL, b Princeton, NJ, Nov 22, 35. ORGANIC CHEMISTRY, COMPUTER SCIENCE. Educ: Princeton Univ, BA, 57; Harvard Univ, MA, 59, PhD(chem), 64. Prof Exp: From instr to asst prof, 63-74, ASSOC PROF CHEM, BAYLOR COL MED, 74- Mem: Am Chem Soc; Am Soc Mass Spectrometry; Asn Comput Mach. Res: Chemistry of natural products; chemical modelling; analytical systems. Mailing Add: Inst Lipid Res Baylor Col of Med 12 Moursund Ave Houston TX 77025

STILLWELL, WILLIAM DUNCAN, b Tillamook, Ore, Dec 3, 07; m 33; c 2. APPLIED CHEMISTRY. Educ: San Diego State Teachers Col, AB, 29; Tulane Univ, MS, 31; Western Reserve Univ, PhD(inorg chem), 33. Prof Exp: Res chemist, Harshaw Chem Co, 33-38 & tech serv, 38-40, supvr prod records, War Priorities & Control Measures, Control Off, 40-43, chief, Manhattan Dist Contracts, 43-47, supvr costs, 47-49, asst to exec vpres, 50-51, mgr oper statist, 51-52, dir field res, 52-55, prod mgr catalysts, 55-57, dir com develop, 57-61; adminstr dept chem, Univ Calif, San Diego, 61-73, exec officer, 73-75; RETIRED. Mem: Am Chem Soc. Res: Fluorochemistry; general inorganic chemistry; ceramic products; catalysts. Mailing Add: 205 Ocean View Del Mar CA 92014

STILSON, WALTER LESLIE, b Sioux Falls, SDak, Dec 13, 08; m 33; c 3. RADIOLOGY. Educ: Columbia Union Col, BA, 29; Loma Linda Univ, MD, 33; Am Bd Radiol, dipl, 38. Prof Exp: Intern, White Mem Hosp, Los Angeles, 34; instr roentgenol, 35-41, instr radiol, 40-41, asst clin prof, 41-45, assoc prof, 45-53, dir, Sch X-ray Technol, 41-66, exec secy dept, 45-50, head dept, 50-62, chmn dept, 62-69, CHIEF DIAG RADIOL, MED CTR & PROF RADIOL, SCH MED, LOMA LINDA UNIV, 53-, CHMN DEPT RADIOL TECHNOL, SCH ALLIED HEALTH PROF, 66- Concurrent Pos: Resident, Los Angeles County Gen Hosp, 34-36, sr attend phys radiologist, 56-67; mem, Int Cong Radiol, 37 & 53; radiologist, White Mem Hosp, 37-39, dir dept radiol, 39-67; consult, Loma Linda Sanitarium, 37-42 & Los Angeles County Health Dept, 43-67; assoc radiologist, Glendale Sanitarium, 46-50, consult, 50-; assoc radiologist, Behren's Mem Hosp, 46-56. Mem: AAAS; Radiol Soc NAm; Am Cancer Soc; fel AMA; fel Am Col Radiol. Res: Medicine. Mailing Add: Dept of Radiol Loma Linda Univ Med Ctr Loma Linda CA 92354

STILWELL, DONALD LONSON, b Detroit, Mich, Dec 29, 18. ANATOMY. Educ: Wayne State Univ, AB, 41, MD, 44. Prof Exp: Intern, Harper Hosp, Detroit, 44-45, resident surg, 45-56; from instr to asst prof, 49-59, asst dean, 64-65, ASSOC PROF ANAT, SCH MED, STANFORD UNIV, 60- Concurrent Pos: Fel anat, Wayne State Univ, 58-59. Mem: AAAS; Am Asn Anat. Res: Anatomy; experimental pathology; vascularization of vertebral column; innervation of hand, foot, joints, spine and eye; blood supply of brain. Mailing Add: Dept of Anat Stanford Univ Stanford CA 94305

STILWELL, KENNETH JAMES, b Poughkeepsie, NY, Apr 4, 34; m 56; c 3. MATHEMATICS. Educ: Bob Jones Univ, BS, 56; Ariz State Univ, MA, 59; Univ Ariz, MS, 64; Hunter Col, MA, 65; Univ Northern Colo, EdD(math educ), 71. Prof Exp: Instr high schs, Ariz, 57-64; asst prof math, King's Col, NY, 65-66; assoc prof, 66-74, PROF MATH, NORTHEAST MO STATE UNIV, 74- Mem: Math Asn Am. Res: Mathematics education; effect of video-tape and critique on attitude of pre-service mathematics teachers. Mailing Add: Dept of Math Northeast Mo State Univ Kirksville MO 63501

STIMLER, MORTON, b New York, NY, Sept 14, 24; m 64. PHYSICS, ENGINEERING. Educ: City Col New York, BS, 48; Univ Md, MS, 63. Prof Exp: Engr, US Naval Ord Lab, 50-56, scientist, 56-61, physicist, 61-66, RES PHYSICIST, US NAVAL SURFACE WEAPONS CTR, 66- Concurrent Pos: Mem mine develop comt, US Naval Ord Lab, 56-57, head exp laser br, 61-63; US Naval Surface Weapons Ctr invited speaker lasers, Various Univs, 62-65. Mem: Am Phys Soc; Optical Soc Am. Res: Acoustic and magnetic mines; magnetic amplifiers; radar; missiles; communication systems; photon momentum; electron injection; optics; lasers; imaging; electro optics; opto-acoustics; acoustic lenses. Mailing Add: 19 Watchwater Way Rockville MD 20850

STIMLER, SUZANNE STOKES, b Aberdeen, SDak, Sept 25, 28; m 64. PHYSICAL CHEMISTRY. Educ: Univ Colo, BA, 50; Mt Holyoke Col, MA, 54; Univ Rochester, PhD(phys chem), 58. Prof Exp: Chemist, Shell Oil Co, 51-52; instr chem, Wellesley Col, 57-58; res chemist, US Navl Res Lab, 58-68; health scientist adminr, Nat Inst Child Health & Human Develop, 68-71, HEALTH SCIENTIST ADMINR, DIV RES RESOURCES, NIH, 71- Mem: AAAS; Am Chem Soc. Res: Molecular electronic spectroscopy, particularly absorption and emission; infrared absorption spectroscopy; photodegradation of polymers. Mailing Add: Div of Res Resources Nat Inst of Health Bethesda MD 20014

STIMPERT, FRED DEWEY, bacteriology, see 12th edition

STIMPFLING, JACK HERMAN, b Denver, Colo, June 11, 24; m 50; c 4. GENETICS. Educ: Univ Denver, BS, 49, MS, 50; Univ Wis, PhD(genetics), 57. Prof Exp: Asst yeast genetics, Southern Ill Univ, 51-52; immunogenetics, Univ Wis, 52-57; fel, Jackson Mem Lab, 57-59, assoc staff scientist, 59-61, staff scientist, 61-64; res assoc, 65-68, DIR, McLAUGHLIN RES INST, 68- Concurrent Pos: Mem study sect B Allergy & infectious dis, NIH. Mem: Genetics Soc Am; Am Asn Immunologists. Res: Immunogenetics; inheritance of cellular antigens; cellular antigens in tissue transplantation. Mailing Add: McLaughlin Res Inst Columbus Hosp Great Falls MT 59401

STIMPSON, EDWIN GREENWOOD, biochemistry, see 12th edition

STIMSON, MIRIAM MICHAEL, b Chicago, Ill, Dec 24, 13. ORGANIC CHEMISTRY. Educ: Siena Heights Col, BS, 36; Inst Divi Thomae, MS, 39, PhD(chem), 48. Prof Exp: Head res lab, Siena Heights Col, 36-39, instr chem, 39-46, asst prof, 46-50, prof natural sci & head div, 50-69; chmn dept, 69-74, PROF CHEM, KEUKA COL, 69- Honors & Awards: Charles Williams Award, 42. Mem: Am Chem Soc; Am Phys Soc; Coblentz Soc; NY Acad Sci. Res: Infrared and ultraviolet absorption in the solid state by potassium bromide disks; effect of irradiation on pyrimidines in the solid state. Mailing Add: Dept of Chem Keuka Col Keuka Park NY 14478

STINCHCOMB, THOMAS GLENN, b Tiffin, Ohio, Sept 12, 22; m 45; c 4. RADIATION PHYSICS, NUCLEAR PHYSICS. Educ: Heidelberg Col, BS, 44; Univ Chicago, MS, 48, PhD(physics), 51. Prof Exp: From instr to asst prof physics, State Col Wash, 51-54; from assoc prof to prof & head dept, Heidelberg Col, 54-61; res physicist, Nuclear & Radiation Physics Sect, IIT Res Inst, 61-65, sr physicist & group leader, 65-68; PROF PHYSICS & CHMN DEPT, DEPAUL UNIV, 68- Concurrent Pos: Actg mgr, Nuclear & Radiation Physics Sect, IIT Res Inst, 66-67. Mem: AAAS; Am Asn Physics Teachers; Am Nuclear Soc; Am Phys Soc; Am Geophys Union. Res: Nuclear radiation physics. Mailing Add: Dept of Physics DePaul Univ 1215 WFullerton Ave Chicago IL 60614

STINCHFIELD, FRANK E, b Warren, Minn, Aug 12, 10; m 30; c 2. ORTHOPEDIC SURGERY. Educ: Northwestern Univ, MD, 34; Am Bd Orthop Surg, dipl, 46. Hon Degrees: DSc, Carleton Col, 60. Prof Exp: PROF ORTHOP SURG & CHMN DEPT, COL PHYSICIANS & SURGEONS, COLUMBIA UNIV, 56- Concurrent Pos: Attend orthop surgeon, Columbia-Presby Med Ctr, 51-, consult, Neurol Inst, 47-; Dept Defense & Dept Air Force orthop surg consult, Asst Secy Defense, 65-, tour Vietnam & Far East installations, 66; pres, Am Bd Orthop Surg. Honors & Awards: Centennial Award, Northwestern Univ, 59. Mem: Am Surg Asn; Am Asn Surg of Trauma; Am Acad Orthop Surg; Am Orthop Asn (treas); NY Acad Med. Res: Effect of anticoagulant therapy on bone repair; osteogenesis of bone isolated from soft tissue blood supply. Mailing Add: 161 Ft Washington Ave New York NY 10032

STINE, CAWLEY RICHARD, analytical chemistry, see 12th edition

STINE, CHARLES MAXWELL, b Osceola Mills, Pa, Mar 4, 25; m 51; c 3. FOOD SCIENCE. Educ: Pa State Univ, BS, 51, MS, 52; Univ Minn, PhD(dairy tech), 57. Prof Exp: Assoc prof, 57-68, PROF FOOD SCI, MICH STATE UNIV, 68- Mem:

Am Oil Chem Soc; Am Dairy Sci Asn. Res: Lipid oxidation in food products; spray dried foods. Mailing Add: Dept of Food Sci Mich State Univ East Lansing MI 48823

STINE, GERALD JAMES, b Johnstown, Pa, May 29, 35; m 62; c 2. MICROBIAL GENETICS. Educ: Southern Conn State Col, BS, 61; Dartmouth Col, MA, 63; Univ Del, PhD(genetics), 66. Prof Exp: Biologist, Oak Ridge Nat Lab, 66-68; asst prof microbial genetics, Univ Tenn, Knoxville, 68-72; ASSOC PROF NATURAL SCI, UNIV N FLA, 72- Concurrent Pos: Union Carbide fel, 66-68; consult, Oak Ridge Nat Lab, 68- Mem: Genetics Soc Am; NY Acad Sci; Am Microbial Soc; assoc Inst Soc Ethics & Life Sci. Res: Isolation of specific gene fragments of Escherichia coli after transfer of the Escherichia coli chromosome into Proteus mirabilis; studies on development in Neurospora crassa. Mailing Add: Dept of Natural Sci Univ of NFla Jacksonville FL 32216

STINE, HARRISON M, physical chemistry, see 12th edition

STINE, JAMES BRYAN, b Pecos, Tex, June 27, 11; m 34; c 2. FOOD MICROBIOLOGY. Educ: Tex Tech Col, BS, 33; Iowa State Col, MS, 34, PhD(dairy bact), 36. Prof Exp: Res, Kraft Foods Co Div, Nat Dairy Prod Corp, 34-39, prod mgr, 39-47, dir cheese res, 47-53, prod mgr, 53-62, nat prod mgr, 62-69, VPRES QUAL STAND & REGULATORY COMPLIANCE, KRAFT FOODS DIV, KRAFTCO CORP, 69- Concurrent Pos: Mem US del, Codex Alimentarius. Mem: Fel Am Dairy Sci Asn. Res: Chemistry and bacteriology of cheese; milk products; flavoring products of butter; defects of cheese spreads; Roquefort type cheese; Swiss cheese; manufacture of malted milk; food standards and technology. Mailing Add: Qual Stand & Reg Compliance Dept Kraftco Corp 500 Peshtigo Ct Chicago IL 60690

STINE, OSCAR CEBREN, b Washington, DC, June 11, 27; m 51; c 5. PEDIATRICS, PUBLIC HEALTH. Educ: Oberlin Col, BA, 50; George Washington Univ, MD, 54; Johns Hopkins Univ, DrPH, 60; Am Bd Pediat, dipl, 59. Prof Exp: Intern pediat, Rochester Gen Hosp, NY, 54-55, resident, 55-57; instr pub health admin, Sch Hyg & Pub Health, Johns Hopkins Univ, 57-62, asst prof maternal & child health, 62-66, instr pediat, Sch Med, 57-66, dir maternal & child health clin, 62-66; ASSOC PROF PEDIAT, SCH MED, UNIV MD, BALTIMORE CITY, 67-; CHIEF AMBULATORY CARE, GTR BALTIMORE MED CTR, 75- Concurrent Pos: Sch physician, Baltimore City Health Dept, 59-62; mem, Baltimore County Bd Health, 63-66; consult, Proj Head Start, Am Acad Pediat, 66-; consult, Nat Found Birth Defects Ctrs, 66-72. Mem: Fel Am Acad Pediat; fel Am Pub Health Asn; Asn Teachers Prev Med; fel Am Sch Health Asn. Res: Children, growth and development; utilization of health services. Mailing Add: 6701 N Charles St Baltimore MD 21204

STINE, WILLIAM R, b Schenectady, NY, Dec 14, 38. ORGANIC CHEMISTRY. Educ: Union Col, BS, 60; Syracuse Univ, PhD(chem), 66. Prof Exp: Asst prof, 65-69, ASSOC PROF CHEM, WILKES COL, 69- Mem: Am Chem Soc. Res: Structure of pentavalent phosphorus compounds; reactions of tertiary phosphines with positive halogen compounds. Mailing Add: Dept of Chem Wilkes Col Wilkes-Barre PA 18703

STINEBRING, WARREN RICHARD, b Niagara Falls, NY, July 31, 24; m 48; c 3. MEDICAL MICROBIOLOGY. Educ: Univ Buffalo, BA, 48; Univ Pa, MS, 50, PhD(bact), 51. Prof Exp: From asst instr bact to instr microbiol, Univ Pa, 48-53, assoc microbiol, 53-55; McLaughlin fel, Univ Tex Med Br, 55-57; assoc res prof, Inst Microbiol, Rutgers Univ, 57-60; asst res prof, Sch Med, Univ Pittsburgh, 60-61, from asst prof to assoc prof, 61-65; prof & chmn dept, Univ Calif-Calif Col Med, Irvine, 65-67; PROF MED MICROBIOL & CHMN DEPT, COL MED, UNIV VT, 67- Mem: AAAS; Am Soc Cell Biol; Am Soc Microbiol; Reticuloendothelial Soc; Tissue Cult Asn. Res: Interferon production by bacteria and endotoxin; delayed hypersensitivity; intracellular growth of brucellae and pleuropneumonia-like organisms; nonantibody resistance mechanisms. Mailing Add: Dept of Med Microbiol Univ of Vt Col of Med Burlington VT 05401

STINGELIN, RONALD WERNER, b New York, NY, May 29, 35. GEOLOGY, REMOTE SENSING. Educ: City Col New York, BS, 57; Lehigh Univ, MS, 59; Pa State Univ, PhD(geol), 65. Prof Exp: Res geologist, 65-67, sr res geologist, 67-68, mgr, Environ Sci Br, 68-72, PRIN GEOLOGIST, ENVIRON ANAL DEPT, HRB SINGER, INC, STATE COLLEGE, PA, 72- Concurrent Pos: NSF-Am Soc Photogram vis scientist, 68-71. Mem: Fel Geol Soc Am. Res: Application of remote sensing to environmental problems; energy and resources studies with emphasis on fossil fuels; subsidence and seam interaction in coal mining; surface mine reclamation and acid mine drainage abatement. Mailing Add: 120 Ronan Dr State College PA 16801

STINGL, HANS ALFRED, b Eger, Czech, Oct 13, 27; US citizen; m 54; c 2. INDUSTRIAL ORGANIC CHEMISTRY. Educ: Univ Erlangen, dipl, 54, PhD(org chem), 56. Prof Exp: Res assoc org chem, Univ Ill, Urbana, 56-58; res & develop chemist, 58-75, RES ASSOC, TOMS RIVER CHEM CORP, 75- Mem: Fel Am Inst Chem; Am Chem Soc. Res: Organometallics; senecio alkaloids; organic dyestuffs and intermediates. Mailing Add: 852 Ocean View Dr Toms River NJ 08753

STINI, WILLIAM ARTHUR, b Oshkosh, Wis, Oct 9, 30; m 50; c 3. HUMAN BIOLOGY, PHYSICAL ANTHROPOLOGY. Educ: Univ Wis, BBA, 60, MS, 67, PhD(human biol), 69. Prof Exp: From asst prof to assoc prof anthrop, Cornell Univ, 68-73; ASSOC PROF ANTHROP, UNIV KANS, 73- Mem: AAAS; Am Asn Phys Anthrop; NY Acad Sci; Am Anthrop Asn; Brit Soc Study Human Biol. Res: Effects of stress on human development including growth and maturation as measured by gross morphological and serological parameters; evaluation of stress as evolutionary force. Mailing Add: Dept of Anthrop Univ of Kans Lawrence KS 66044

STINNER, RONALD EDWIN, b New York, NY, July 27, 43; m 63; c 2. POPULATION ECOLOGY. Educ: NC State Univ, BS, 65; Univ Calif, Berkeley, PhD(entom), 70. Prof Exp: Res assoc entom, Tex A&M Univ, 70; res assoc, 70-73, ASST PROF ENTOM, NC STATE UNIV, 73- Mem: Entom Soc Am; Entom Soc Can; Japanese Soc Pop Ecologists; Int Orgn Biol Control. Res: Modeling of population dynamics of agricultural pest insects and pathogens; studies on effects of behavior and host interactions on system dynamics. Mailing Add: Dept of Entom NC State Univ Raleigh NC 27607

STINNETT, JAMES LEBARON, b Washington, DC, Aug 1, 38; m 63; c 3. PSYCHIATRY. Educ: Princeton Univ, AB, 60; Univ Pa, MD, 65. Prof Exp: Intern med, Hosp Univ Pa, 65-66, resident psychiat, 66-69, chief resident, 69-70; chief psychiat, Munson Army Hosp, Ft Leavenworth, Kans, 70-72; asst chmn psychiat, Univ Pa, 72-73; CHIEF ALCOHOL TREATMENT UNIT, PHILADELPHIA VET ADMIN HOSP, 73- Mem: Am Psychiat Asn; Am Psychosomatic Soc. Res: Biochemical determinants to alcoholism; behavior therapy in alcoholism. Mailing Add: Dept of Psychiat Hosp of Univ of Pa Philadelphia PA 19174

STINSON, EDGAR ERWIN, organic chemistry, see 12th edition

STINSON, GLEN MONETTE, b Sarnia, Ont, Dec 27, 39; m 62; c 3. EXPERIMENTAL NUCLEAR PHYSICS. Educ: Univ Toronto, BASc, 61; Univ Waterloo, MSc, 62; McMaster Univ, PhD(nuclear physics), 66. Prof Exp: Fel physics, 66-68, res assoc, Tri-Univ Meson Facility, 68-69, asst res physicist, 69-71, ASST PROF PHYSICS, TRI-UNIV MESON FACILITY, UNIV ALTA, 71- Concurrent Pos: Lectr, Univ Alta, 66-68. Mem: Am Phys Soc; Can Asn Physicists. Res: Proton induced reactions; design and use of high precision magnetic spectrometers; design of charged particle beam transport systems. Mailing Add: Tri-Univ Meson Facility Univ of Alta Edmonton AB Can

STINSON, HARRY THEODORE, JR, b Newport News, Va, Oct 26, 26; m 49; c 3. GENETICS. Educ: Col William & Mary, BS, 47; Ind Univ, PhD(cytogenetics), 51. Prof Exp: Asst prof biol, Col William & Mary, 51-52; res asst genetics, Conn Agr Exp Sta, 52-53, res assoc, 53-60, chief geneticist, 60-62; chmn dept bot, 64-65, PROF GENETICS, CORNELL UNIV, 62-, CHMN SECT GENETICS, DEVELOP & PHYSIOL, 65- Mem: AAAS; Soc Study Evolution; Bot Soc Am; Genetics Soc Am; Am Soc Nat (treas, 63-66). Res: Cytology. Mailing Add: Sect of Genet Develop & Physiol Cornell Univ Ithaca NY 14850

STINSON, JAMES ROBERT, b Bakersfield, Calif, Mar 24, 21; m 51. METEOROLOGY. Educ: Univ Calif, Santa Barbara, BA, 48; St Louis Univ, MS, 55, PhD(geophys, meteorol), 58. Prof Exp: Jr res meteorologist, Univ Calif, Los Angeles, 49-51; asst prof geophys & geophys eng, St Louis Univ, 51-60; asst prof earth sci, Northern Ill Univ, 60-62; sr scientist, Meteorol Res Inc, chief res div, Navy Weather Res Facil, 64-66; assoc chief off atmospheric water resources, Bur Reclamation, 66-69; v pres develop & sr scientist, Meteorol Res Inc, 69-74; PROF GEOG, CALIF STATE COL, DOMINGUEZ HILLS, 74- Concurrent Pos: Instr, Okla State Univ, 51-52. Mem: AAAS; Am Meteorol Soc; Am Geophys Union. Res: General meteorology; weather modification; cloud physics; environmental pollution; severe local storms. Mailing Add: Calif State Col Geog Dept 1000 Victori Dominguez Hills CA 90247

STINSON, PERRI JUNE, US citizen. OPERATIONS RESEARCH, STATISTICS. Educ: Univ Calif, Santa Barbara, AB, 48; Okla State Univ, MS, 52, PhD(statist), 55. Prof Exp: Asst prof health orgn res, St Louis Univ, 58-60; asst prof math, Northern Ill Univ, 60-62; biostatistician, Vet Admin Res Support Ctr, 62-64; mathematician & statistician, US Naval Aviation Safety Ctr, 64-65; head statist & math systs res, Douglas Aircraft Co, 65-67; prof environ eng, Univ Denver, 67-69; PROF OPERS RES & STATIST, CALIF STATE UNIV, LONG BEACH, 69- Concurrent Pos: Fac res grant, Univ Denver, 67-69; fac res grant, Calif State Univ, Long Beach, 69-72; consult, US Off Educ, 70-, Tex Water Develop Bd, 71-72 & Meteorol Res, Inc, 71- Mem: Opers Res Soc Am; Inst Mgt Sci; Soc Gen Systs Res; Am Statist Asn; Inst Math Statist. Res: Statistical and operations research applications to problems in the medical sciences, atmospheric pollution, public administration, and human resources. Mailing Add: Dept of Quant Systs Calif State Univ Long Beach CA 90840

STINSON, RICHARD FLOYD, b Cleveland, Ohio, Feb 4, 21; m 54. FLORICULTURE. Educ: Ohio State Univ, BS, 43, MS, 47, PhD, 52. Prof Exp: Instr floricult, State Univ NY Sch Agr Alfred, 47-48; asst prof, Univ Conn, 48-55; from asst prof to assoc prof hort, Mich State Univ, 55-67; assoc prof, 67-73, PROF AGR EDUC & HORT, PA STATE UNIV, UNIVERSITY PARK, 73- Mem: Am Soc Hort Sci; Am Voc Asn; Sigma Xi. Res: Algae control on clay flower pots; application of infrared heating to greenhouse crops; horticultural and natural resources instruction material in agricultural education. Mailing Add: Dept of Agr Educ Pa State Univ University Park PA 16802

STINSON, ROBERT ANTHONY, b Hamilton, Ont, Sept 30, 41; m 64. BIOCHEMISTRY. Educ: Univ Toronto, BScA, 64; Univ Alta, PhD(plant biochem), 68. Prof Exp: Med Res Coun Can fel molecular enzym, Bristol Univ, 68-71; asst prof path, 71-74, ASSOC PROF PATH, UNIV ALTA, 74- Res: Enzyme mechanisms and kinetics; isoenzymes and clinical applications. Mailing Add: Med Lab Sci Clin Sci Bldg Univ of Alta Edmonton AB Can

STINSON, ROBERT HENRY, b Toronto, Ont, Sept 17, 31; m 54; c 3. BIOPHYSICS, PHYSICS. Educ: Univ Toronto, BSA, 53, MSA, 57; Univ Western Ont, PhD(biophys), 60. Prof Exp: Mem faculty physics dept, Ont Agr Col, 53-63; prof physics, State Univ NY Col Potsdam, 63-67; ASSOC PROF PHYSICS, UNIV GUELPH, 67- Mem: AAAS; Biophys Soc; Sigma Xi. Res: X-ray and neutron diffraction of connective tissue. Mailing Add: Dept of Physics Univ of Guelph Guelph ON Can

STINSON, STEPHEN CHARLES, b Elmhurst, Ill, Nov 25, 36; m 65. ORGANIC CHEMISTRY. Educ: Rutgers Univ, BS, 59; Univ Iowa, MS, 62, PhD(org chem), 64. Prof Exp: Asst prof chem, Purdue Univ, 64-65 & Univ Toledo, 65-69; ed asst, Chem & Eng News, DC, 69-70, asst ed, 70-71; ASSOC ED, PLASTICS TECHNOL, BILL COMMUN, INC, 71- Mem: AAAS; Am Chem Soc; Soc Plastics Eng. Mailing Add: Bill Commun Inc 633 Third Ave New York NY 10017

STINSON, WILLIAM SICKMAN, JR, food technology, horticulture, see 12th edition

STIPANOVIC, BOZIDAR J, b Zagreb, Yugoslavia, Jan 9, 33; m 59. CHEMISTRY. Educ: Univ Belgrade, BS, 60, PhD(org chem), 65. Prof Exp: Teaching asst org chem, Univ Belgrade, 61-65; fel, Ipatieff High Pressure & Catalytic Lab, Northwestern Univ, 66-69; vis assoc prof org chem, Cent Univ Venezuela, 69-70; DIR RES & DEVELOP, CORAL CHEM CO, WAUKEGAN, 70- Mem: Am Chem Soc. Res: Organic catalytic reactions; base catalyzed alkylations; conversion and chemical coatings on metals; corrosion inhibitors. Mailing Add: 1458 Crowe Ave Deerfield IL 60051

STIPANOVIC, ROBERT DOUGLAS, b Houston, Tex, Oct 28, 39; m 63; c 3. NATURAL PRODUCT CHEMISTRY. Educ: Loyola Univ, La, BS, 61; Rice Univ, PhD(chem), 66. Prof Exp: Res assoc chem, Stanford Univ, 66-67; asst prof, Tex A&M Univ, 67-71; RES CHEMIST, USDA, 71- Mem: Am Chem Soc; The Chem Soc. Res: Natural product synthesis and structure determination; mass spectroscopy structure determination and reaction mechanisms; nuclear magnetic resonance studies. Mailing Add: Nat Cotton Path Res Lab PO Drawer JF College Station TX 77840

STIPANOWICH, JOSEPH J, b Canton, Ill, Apr 14, 21; m 47; c 2. MATHEMATICS. Educ: Western Ill Univ, BS, 46; Univ Ill, MS, 47; Northwestern Univ, EdD(math), 56. Prof Exp: Head dept, 58-68, PROF MATH, WESTERN ILL UNIV, 47- Mem: Math Asn Am. Res: Basic mathematics. Mailing Add: Dept of Math Western Ill Univ Macomb IL 61455

STIPE, CLAUDE EDWIN, b Calexico, Calif, May 13, 26; m 52; c 2. ANTHROPOLOGY. Educ: Wheaton Col, BA, 52; Univ Calif, Los Angeles, MA, 55; Univ Minn, PhD(anthrop), 68. Prof Exp: From instr to asst prof anthrop, Ft Wayne Bible Col, 55-59; from instr to assoc prof, Bethel Col, 59-68; asst prof, 68-71, ASSOC PROF ANTHROP, MARQUETTE UNIV, 71- Concurrent Pos: Lectr exten div, Univ Minn, 63-68. Mem: Am Anthrop Asn; Am Soc Ethnohist; Soc Appl Anthrop; Am Sci

Affil. Res: Comparative religious systems; religion and culture change. Mailing Add: Dept of Sociol & Anthrop Marquette Univ Milwaukee WI 53233

STIPE, JOHN GORDON, JR, b Oxford, Ga, Jan 1, 14; m 43; c 2. PLANETARY SCIENCES. Educ: Emory Univ, AB, 33, MS, 38; Princeton Univ, PhD(physics), 45; US Dept Army-US Dept Navy, Cert, 47. Prof Exp: Engr, Claude Neon Southern Corp, Ga, 35-38; instr physics, Emory Univ, 38-40; asst exp physicist, Comt Fortification Design, Nat Res Coun, NJ, 40-45; assoc physicist, Nat Defense Res Comt, 45-56; assoc prof physics & head dept, Allegheny Col, 46-47; prof & head dept, Randolph-Macon Woman's Col, 47-58; assoc prof, 58-63, PROF PHYSICS, BOSTON UNIV, 63- Concurrent Pos: Carnegie intern, Harvard Univ, 56-57. Mem: AAAS; Am Geophys Union; Am Asn Physics Teachers. Res: Terminal ballistics; cratering; administration; solid earth geophysics and planetary physics. Mailing Add: 12 Partridge Hill Rd Southborough MA 01772

STIPPES, MARVIN C, b Chicago, Ill, Aug 8, 22; m 46; c 1. APPLIED MECHANICS. Educ: Univ Ill, Urbana, BS, 43; Univ Wash, MS, 46; Va Polytech Inst & State Univ, PhD(appl mech), 57. Prof Exp: Asst prof math, Mont State Col, 48-49; asst prof mech, Univ Ill, Urbana, 49-51; assoc prof mech, Wash Univ, 51-53; from asst prof to assoc prof, 53-59, PROF MECH, UNIV ILL, URBANA, 59- Res: Static and dynamic classical elasticity. Mailing Add: Dept of Theoret & Appl Mech Univ of Ill Urbana IL 61803

STIREWALT, EDWARD NEALE, b Hartsville, SC, Nov 29, 18; m 47; c 3. ORGANIC CHEMISTRY. Educ: High Point Col, AB, 38; Univ NC, MA, 42. Prof Exp: Chemist, Tenn Eastman Co, 42-44; US Naval Bur Aeronaut fel, Univ Md, 47-48; sci asst, US AEC, 48-53; sci asst, US Naval Res Lab, 53-57; chief tactical br, Anal Serv Inc, 58-62; prin scientist, Booz-Allen Appl Res, Inc, 64-68; consult, US Marine Corps Develop Ctr, 69-74; PRES, STIREWALT & ASSOCS, INC, 71- Mem: AAAS; Am Chem Soc. Res: Organic synthesis; systems analysis. Mailing Add: PO Box 584 Herndon VA 22070

STIREWALT, MARGARET AMELIA, b Hickory, NC, Jan 18, 11; m 53. MEDICAL PARASITOLOGY. Educ: Randolph-Macon Woman's Col, BA, 31; Columbia Univ, MA, 35; Univ Va, PhD(zool), 38; Am Bd Microbiol, Cert pub health & med lab parasitol, 65. Prof Exp: Teacher high sch, Va, 31-33; asst zool, Univ Va, 38-40; asst prof biol, Flora Macdonald Col, 40-42; head helminth, Naval Med Res Inst, 45-58, head parasitol, Naval Med Sch, 58-70, dir div parasitol, Naval Med Res Inst, 70-71; PARASITOLOGIST, AM FOUND BIOL RES, 72- Concurrent Pos: Va Acad Sci grant, 38-40; instr, Woman's Col, Univ NC, 39. Mem: AAAS; Soc Syst Zool; Am Soc Parasitol (vpres, 72); Am Soc Trop Med & Hyg; fel Royal Soc Trop Med & Hyg. Res: Taxonomy, morphology and physiology of Turbellaria; bionomics and control of human schistosomes. Mailing Add: Am Found for Biol Res 12111 Parklawn Dr Rockville MD 20852

STIRLING, CHARLES E, b Havelock, NC, Nov 30, 33; m 62; c 3. PHYSIOLOGY, BIOPHYSICS. Educ: George Washington Univ, BA, 61; State Univ NY, PhD(physiol), 67. Prof Exp: Instr physiol, State Univ NY Upstate Med 66-67; asst prof, 68-74, ASSOC PROF PHYSIOL, UNIV WASH, 74- Mem: Am Physiol Soc; Physcolog Soc Am; Biophys Soc; Am Soc Cell Biol. Res: Active transport. Mailing Add: Dept of Physiol & Biophys Univ of Wash Seattle WA 98105

STIRLING, JAMES HEBER, b Los Angeles, Calif, Aug 14, 21; c 5. ANTHROPOLOGY. Educ: Walla Walla Col, BTh, 44; Andrews Univ, MA, 55; Univ Calif, Berkeley, MA, 64; Univ Calif, Los Angeles, PhD(anthrop), 68. Prof Exp: From asst prof to assoc prof anthrop, 64-73, assoc head dept sociol & anthrop, 72-74, PROF ANTHROP, LOMA LINDA UNIV, 73- Concurrent Pos: Asst prof, Redlands Univ, 67-69; vis prof, Towson State Col, 75-76; lectr, Johns Hopkins Univ Eve Col, 75-76. Mem: AAAS; fel Am Anthrop Asn; Soc Appl Anthrop; Soc Am Archaeol; Soc Med Anthrop. Res: Culture change; immigration and cultural adjustments; archaeology and prehistory of Central America and the Near East. Mailing Add: Dept of Sociol & Anthrop Loma Linda Univ Loma Linda CA 92354

STIRLING, WILLIAM LEAKE, physics, see 12th edition

STIRN, RICHARD J, b Milwaukee, Wis, Dec 5, 33; m 67; c 3. ENERGY CONVERSION. Educ: Univ Wis, BS, 61; Purdue Univ, MS, 63, PhD(physics), 66. Prof Exp: MEM TECH STAFF, JET PROPULSION LAB, CALIF INST TECHNOL, 66- Mem: Am Phys Soc. Res: Properties of energy barriers in semiconductors, including p-n junctions and Schottky barriers, photovoltaics, laser energy conversion, transport properties and band structure in III-V semiconductors. Mailing Add: Jet Propulsion Lab Calif Inst of Technol Pasadena CA 91103

STIRRAT, JAMES HILL, b Johnstone, Scotland, Sept 17, 13; Can citizen; m 39; c 1. PATHOLOGY, BACTERIOLOGY. Educ: Glasgow Univ, BSc, 35, MB, ChB, 38, MD, 45; Univ Alta, cert path, 64; FRCP, 63. Prof Exp: Intern & resident med, surg & obstet, Glasgow Munic Hosps, 38-40, asst pathologist & bacteriologist, 40-45; assoc prof bact, Univ Alta, 48-52, assoc prof path & asst prof pathologist, 52-56, prof path, 57-60; dir dept virol & pathologist, State Pub Health Lab Serv, Western Australia, 60; assoc pathologist, Gen Hosp, Calgary, Alta, 61-63; dir dept lab med, 63-73, BACTERIOLOGIST, PATHOLOGIST & INFECTION CONTROL OFFICER, MISERICORDIA HOSP, 73- Concurrent Pos: Adv, Sch Med Technol, Northern Alta Inst Technol, 63-73; consult pathologist, Grand Prairie Munic Hosp, High Prairie & Sisters of Providence Hosp, Alta, 63-68, Edmonton Gen Hosp, 63-75, McLennan Munic Hosp, Alta & Devon Munic Hosp, Alta, 63-; asst prof path, Univ Alta, 64- Mem: Fel Col Am Pathologists; Can Med Asn; Can Asn Pathologists; Can Soc Forensic Sci; Brit Med Asn. Res: Forensic medicine; spread of carcinoma in the human body; typhus group of fevers in India; bacteria associated with insects parasitic on livestock. Mailing Add: 601 Valleyview Manor 12207 Jasper Ave Edmonton AB Can

STITCH, MALCOLM LANE, b Apr 23, 23; m 65, 75; c 2. PHYSICS, ENGINEERING. Educ: Southern Methodist Univ, BA & BS, 47; Columbia Univ, PhD(physics), 53. Prof Exp: Res asst radiation lab, Columbia Univ, 48-51; instr, Cooper Union Eng Col, 51-52; res asst radiation lab, Columbia Univ, 52-53; res physicist, Varian Assocs, 53-56; res physicist & head molecular beams group, Res Labs, Hughes Aircraft Co, 56-60, sr staff physicist, 59-60, head laser develop sect, Res & Develop Div, 61-62, mgr laser develop dept, 62-65, asst mgr high frequency lab, Res & Develop Div, 65-67, chief scientist, 67-68; asst gen mgr, Korad Dept, Union Carbide Corp, Calif, 68-73; sr scientist & consult, Ctr Laser Studies, Naval Res Lab, Univ Southern Calif, 73-74; MGR ELECTRO-OPTICAL OPERS, EXXON NUCLEAR RES & TECHNOL CTR, 74- Concurrent Pos: Instr, Sarah Lawrence Col, 49-51; mem comt safe use lasers, Am Nat Standards Inst, 72-; adj prof elec eng, Univ Southern Calif, 74- Mem: AAAS; Am Phys Soc; fel Inst Elec & Electronics Engrs; NY Acad Sci; Soc Photo-Optical Instrumentation Engrs. Res: Quantum electronics; laser technology and applications; laser isotope separation; coherent optics; accusto and electro-optics; laser dye chemistry. Mailing Add: Exxon Nuclear Res & Technol Ctr 2955 George Washington Way Richland WA 99352

STITELER, WILLIAM MERLE III, b Kane, Pa, July 30, 42; m 64; c 1. STATISTICS, FORESTRY Educ: Pa State Univ, BS, 64, MS, 66, PhD(statist), 70. Prof Exp: Asst prof statist, Pa State Univ, University Park, 70-75; MEM FAC, DEPT FORESTRY, STATE UNIV NY COL FORESTRY, 75- Mem: Inst Math Statist; Am Statist Asn; Int Asn Ecol; Biomet Soc. Res: Spatial patterns in ecological populations; modeling and simulation of biological populations. Mailing Add: Dept of Forestry State Univ NY Col of Forestry Syracuse NY 13210

STITES, JOSEPH GANT, JR, inorganic chemistry, physical chemistry, see 12th edition

STITH, LEE S, b Tulia, Tex, Aug 30, 18; m 47; c 1. PLANT BREEDING. Educ: NMex State Univ, BS, 40; Univ Tenn, MS, 42; Iowa State Univ, PhD(crop breeding), 55. Prof Exp: Asst county supvr, Farm Security Admin, DeBaca County, NMex, 40-41; res asst agr econ, Univ Tenn, 41-42; agronomist, El Paso Valley Substa 17, Tex A&M Univ, 46-47, cotton breeder, 47-55; PLANT BREEDER & PROF PLANT BREEDING, UNIV ARIZ, 55-, DIR HYBRID COTTON RES PROJ, 72- Mem: Am Soc Agron. Res: Crop breeding; plant pathology and physiology; statistics; cotton and grain sorghum. Mailing Add: Dept of Plant Sci Col of Agr Univ of Ariz Tucson AZ 85721

STITH, REX DAVID b Hominy, Okla, Dec 11, 42; m 64; c 2. ENDOCRINOLOGY. Educ: Okla State Univ, BS, 64, MS, 66; Purdue Univ, PhD(physiol), 71. Prof Exp: Instr biol, Southeast Mo State Col, 66-68; vet physiol, Purdue Univ, 68-71; res assoc pharmacol, Univ Mo-Columbia, 71-72; asst prof, 72-75, ASSOC PROF PHYSIOL, HEALTH SCI CTR, UNIV OKLA, 75- Mem: AAAS; Sigma Xi; Soc Exp Biol & Med; Am Physiol Soc; Endocrine Soc. Res: Mechanisms of action of glucocorticoids; interactions of steroids in target cells; effects of glucocorticoids on intracellular functions; mechanisms of action of steroid hormones on target tissues, especially brain tissues and biochemical interactions of steroids in these tissues. Mailing Add: Dept Phys & Biophys Box 26901 Univ of Okla Health Sci Ctr Oklahoma City OK 73190

STITH, WILLIAM JOSEPH, b Oklahoma City, Okla, Feb 7, 42; m 66; c 3. BIOCHEMISTRY. Educ: Phillips Univ, BA, 64; Univ Okla, PhD(biochem), 72. Prof Exp: Chief microbiol, US Naval Hosp, Philadelphia, 66-67, chief blood bank & serol, 67-68; asst officer in-chg, Armed Serv Whole Blood Processing Lab, McGuire AFB, NJ, 68-69; fel human biol chem & genetics, Univ Tex Med Br, Galveston, 72-73; SCIENTIST PROD EXPLOR, FENWAL DIV, BAXTER LABS, INC, 73- Mem: Am Chem Soc; Sigma Xi. Res: Development and evaluation of column procedures for removal and collection of leukocytes from whole blood; development and evaluation of solutions and materials for storage of blood in liquid or frozen state. Mailing Add: Fenwal Labs Div Baxter Labs Rte 120 & Wilson Rd Round Lake IL 60073

STITT, JAMES HARRY, b Sellersville, Pa, Dec 13, 39; m 64; c 2. GEOLOGY, PALEONTOLOGY. Educ: Rice Univ, BA, 61; Univ Tex, Austin, MA, 64, PhD(geol), 68. Prof Exp: ASSOC PROF GEOL, UNIV MO-COLUMBIA, 68- Mem: Paleont Soc; Geol Soc Am. Res: Late Cambrian and early Ordovician trilobites; invertebrate paleontology and biostratigraphy; carbonate petrology. Mailing Add: Dept of Geol Univ of Mo Columbia MO 65201

STITT, JOHN THOMAS, b Belfast, Northern Ireland, Nov 7, 42; m 66; c 1. PHYSIOLOGY, NEUROPHYSIOLOGY. Educ: Queens Univ Belfast, BSc, 65; Queens Univ, Ont, MSc, 67, PhD(physiol), 69. Prof Exp: Can Med Res Coun fel, Med Sch, 69-72, asst fel, 69-73, ASSOC FEL PHYSIOL, JOHN B PIERCE FOUND LAB, YALE UNIV, 73-, ASST PROF ENVIRON PHYSIOL, MED SCH, 72- Mem: Am Physiol Soc; Can Physiol Soc; Soc Neurosci. Res: Physiology of thermoregulation in mammals; role of the hypothalamus in the homeostasis of body temperature; neurophysiological mechanisms of fever. Mailing Add: John B Pierce Found Lab Yale Univ Sch of Med New Haven CT 06519

STITZEL, ROBERT ELI, b New York, NY, Feb 22, 37; m 61; c 1. PHARMACOLOGY. Educ: Columbia Univ, BS, 59, MS, 61; Univ Minn, PhD(pharmacol), 64. Prof Exp: Res asst pharmacol, Univ Minn, 61-64; asst prof, WVa Univ, 65-66; Swed Med Res Coun fel, 66-67; from asst prof to assoc prof, 67-73, PROF PHARMACOL & DIR GRAD STUDIES, DEPT PHARMACOL, W VA UNIV, 73- Concurrent Pos: USPHS fel, 64-65; USPHS res career develop award; vis prof, Univ Adelaide, Australia, 73. Mem: AAAS; Am Soc Pharmacol & Exp Therapeut; Int Soc Biochem Pharmacol; Am Soc Neurochem. Res: Physiological and pharmacological factors affecting catecholamine release; relationship between stress and drug metabolism. Mailing Add: Dept of Pharmacol WVa Univ Morgantown WV 26506

STIVALA, SALVATORE SILVIO, b New York, NY, June 23, 23; m 50; c 2. PHYSICAL CHEMISTRY. Educ: Columbia Univ, AB, 48; Stevens Inst Technol, MSChE, 52, MS, 58; Univ Pa, PhD(chem), 60. Hon Degrees: MEng, Stevens Inst Technol, 64. Prof Exp: Res engr, US Testing Co, 49-50; mat engr, Picatinny Arsenal, 50-51, 54-57; NSF sci fac fel, Univ Pa, 57-59; from asst prof to assoc prof phys & polymer chem, 59-64, instr chem eng, 52-57, PROF PHYS & POLYMER CHEM, STEVENS INST TECHNOL, 64- Concurrent Pos: Consult, 52-57, 59- Honors & Awards: Ottens Res Award, 68. Mem: Am Chem Soc; Soc Plastics Eng; Am Inst Chem Eng; NY Acad Sci. Res: Physical chemistry of high polymers; solution properties; kinetics of polymer degradation; physico-chemical aspects of biopolymers. Mailing Add: Dept of Chem & Chem Eng Stevens Inst of Technol Castle Point Hoboken NJ 07030

STIVEN, ALAN ERNEST, b St Stephen, NB, Nov 12, 35; m 72; c 3. ECOLOGY, POPULATION BIOLOGY. Educ: Univ NB, BSc, 57; Univ BC, MA, 59; Cornell Univ, PhD(ecol), 62. Prof Exp: From asst prof to assoc prof zool, 62-71, chmn dept, 67-72, PROF ZOOL, UNIV NC, CHAPEL HILL & CHMN ECOL CURRICULUM, 71- Concurrent Pos: NSF res grants, 63-; USPHS ecol training grant, 66-69; res fel, Univ BC, 70; res grant, Sea Grant, Nat Oceanog & Atmos Asn, 74-76. Mem: Ecol Soc Am; Am Soc Naturalists; Japanese Soc Pop Ecol. Res: Population ecology; secondary productivity in aquatic systems; population energetics; stream and salt marsh ecology. Mailing Add: Dept of Zool Univ of NC Chapel Hill NC 27514

STIVER, JAMES FREDERICK, b Elkhart, Ind, Jan 27, 43; m 65; c 4. MEDICINAL CHEMISTRY, BIONUCLEONICS. Educ: Purdue Univ, BS, 66, MS, 68, PhD(med chem, bionucleonics), 70. Prof Exp: Asst prof, 69-73, ASSOC PROF PHARMACEUT CHEM & BIONUCLEONICS, COL PHARM, N DAK STATE UNIV, 73-, RADIOL SAFETY OFFICER, 70- Mem: AAAS; Am Chem Soc; Am Pharmaceut Asn; Am Pub Health Asn; Health Physics Soc. Res: Radioisotope labeling synthesis of organic compounds and drugs; radioisotope tracer techniques and tracer methodology development; metabolism of drug and toxic chemicals; radioactive nuclide levels in soil due to fallout. Mailing Add: Col of Pharm NDak State Univ Fargo ND 58102

STIVERS, RUSSELL KENNEDY, b Marshall Co, Ill, May 9, 17; m 47; c 3. SOIL FERTILITY. Educ: Univ Ill, BS, 39; Purdue Univ, MS, 48, PhD, 50. Prof Exp: Assoc agronomist, Va Polytech Inst, 50-55; ASSOC PROF AGRON, PURDUE UNIV,

LAFAYETTE, 55- Mem: Am Soc Agron; Soil Sci Soc Am. Res: Soil testing. Mailing Add: Dept of Agron Purdue Univ West Lafayette IN 47906

STIX, THOMAS HOWARD, b St Louis, Mo, July 12, 24; m 50; c 2. PLASMA PHYSICS. Educ: Calif Inst Technol, BS, 48; Princeton Univ, PhD(physics), 53. Prof Exp: Res asst, 53-54, res assoc, 54-56, assoc head exp div, 56-61, CO-HEAD EXP DIV, PLASMA PHYSICS LAB, PRINCETON UNIV, 61-, PROF ASTROPHYS SCI, 62- Concurrent Pos: NSF sr fel, 60-61; mem adv comt, thermonuclear div, Oak Ridge Nat Lab, 66-68; John Simon Guggenheim Mem Found fel, 69-70; assoc ed, Int J Eng & Sci, 69- Mem: Fel Am Phys Soc. Res: Controlled fusion; waves and instabilities; plasma heating and confinement. Mailing Add: Plasma Physics Lab Princeton Univ Princeton NJ 08540

STJERNHOLM, RUNE LEONARD, b Stockholm, Sweden, Apr 25, 24; nat US; m 53; c 2. BIOCHEMISTRY. Educ: Stockholm Tech Inst, BS, 44; Western Reserve Univ, PhD(biochem), 58. Prof Exp: From asst prof to prof biochem, Case Western Reserve Univ, 58-71; PROF BIOCHEM, MED SCH, TULANE UNIV, 71- Mem: Am Chem Soc; Am Soc Microbiol; The Chem Soc; Swed Chem Soc. Res: Intermediary metabolism of microorganisms; carbohydrate metabolism in leukocytes. Mailing Add: Dept of Biochem Tulane Univ Med Sch New Orleans LA 70112

STOB, MARTIN, b Chicago, Ill, Feb 20, 26. ANIMAL SCIENCE. Educ: Purdue Univ, PhD(physiol), 53. Prof Exp: Asst, 49-53, asst prof animal husb, 53-58, assoc prof animal sci, 58-63, PROF ANIMAL SCI, PURDUE UNIV, WEST LAFAYETTE, 63- Mem: AAAS; Am Soc Animal Sci; Endocrine Soc; Soc Study Fertil; Soc Study Reproduction. Res: Hormonal regulation of growth; occurrence of compounds with estrogenic activity in plant material; microbiological synthesis and metabolism of estrogens; reproductive physiology. Mailing Add: Dept of Animal Sci Purdue Univ West Lafayette IN 47906

STOBAUGH, ROBERT EARL, b Humboldt, Tenn, June 24, 27; m 56. ORGANIC CHEMISTRY. Educ: Southwestern at Memphis, BS, 47; Univ Tenn, MS, 49, PhD(chem), 52. Prof Exp: Res assoc, Ohio State Univ, 52-54; from asst ed to sr assoc ed, 54-61, from asst dept head to dept head, 61-65, tech adv registry div, 65-67, MGR CHEM INFO SCI, CHEM ABSTR, 67- Mem: Am Soc Info Sci; Am Chem Soc. Res: Steroids; chemical literature; chemical information storage and retrieval; chemical structural data; chemical information science. Mailing Add: Chem Abstr Serv Ohio State Univ Columbus OH 43210

STOBBE, ELMER HENRY, b Matsqui, BC, Jan 26, 36; m 62; c 3. AGRONOMY, WEED SCIENCE. Educ: Univ BC, BSA, 61, MSA, 65; Ore State Univ, PhD(crop sci), 69. Prof Exp: Asst prof, 68-72, ASSOC PROF WEED SCI, UNIV MAN, 72- Mem: Weed Sci Soc Am; Agr Inst Can. Res: Weed science, especially chemical and cultural weed control, physiology of herbicides, weed-crop ecology, biology of weeds and mechanics of herbicide application; agronomy, especially zero tillage research and effect of cultivation on crop yield. Mailing Add: Dept of Plant Sci Univ of Man Winnipeg MB Can

STOBER, HENRY CARL, b Brooklyn, NY, June 20, 35; m 61; c 2. ANALYTICAL CHEMISTRY, PHARMACEUTICAL CHEMISTRY. Educ: City Col New York, BS, 58; Seton Hall Univ, MS, 69, PhD(chem), 71. Prof Exp: Res asst biol chem, Letterman Army Hosp, US Army, 58-60; chemist, Ciba Pharmaceut Co, 60-66, supvr anal chem, 66-70; teaching asst chem, Seton Hall Univ, 70-71; sr scientist, 71-74, SR STAFF SCIENTIST, CIBA-GEIGY CORP, SUFFERN, NY, 74- Mem: Am Chem Soc. Res: Analysis and solid state characterization of pharmaceuticals and related chemicals. Mailing Add: 84 Hamilton St Madison NJ 07940

STOBER, QUENTIN JEROME, b Billings, Mont, Mar 25, 38; m 65; c 2. AQUATIC ECOLOGY, WATER POLLUTION. Educ: Mont State Univ, BS, 60, MS, 62, PhD(zool), 68. Prof Exp: Aquatic biologist, Southeast Water Lab, Div Water Supply & Pollution Control, USPHS, Ga, 62-65; res asst prof estuarine ecol, 69-72, RES ASSOC PROF ESTUARINE & STREAM ECOL, FISHERIES RES INST, UNIV WASH, 72- Concurrent Pos: Consult, Atomic Safety & Licensing Bd Panel, US Nuclear Regulatory Comn, 74- Honors & Awards: W F Thompson Award, Am Inst Fisheries Res Biol, 71. Mem: AAAS; Am Fisheries Soc; Am Soc Limnol & Oceanog; Am Inst Fisheries Res Biol; Sigma Xi. Res: Fisheries problems related to hydro and thermal nuclear energy production; estuarine ecology of effects of municipal and industrial wastes; stream ecology and water shed management. Mailing Add: Fisheries Res Inst Univ Wash Col of Fisheries Seattle WA 98195

STOBO, JOHN DAVID, b Somerville, Mass, Sept 1, 41; m 64; c 3. IMMUNOLOGY. Educ: Dartmouth Col, AB, 63; State Univ NY Buffalo, MD, 68. Prof Exp: Res assoc immunol, NIH, 70-72; chief resident med, Johns Hopkins Hosp, 72-73; ASSOC PROF IMMUNOL, MAYO MED SCH & FOUND, 73- Concurrent Pos: Sr investr, Am Arthritis Asn, 73; rep, Nat Heart Asn, 75-77. Mem: Am Asn Immunologists; Am Arthritis Asn. Res: Cellular immunology; forces involved in the regulation of cell mediated and humoral immune responses. Mailing Add: Dept of Immunol Mayo Clin & Found Rochester MN 55901

STOCK, CHARLES CHESTER, b Terre Haute, Ind, May 19, 10; m 36. CHEMOTHERAPY. Educ: Rose-Hulman Inst Technol, BS, 32; Johns Hopkins Univ, PhD(physiol chem), 37; NY Univ, MS, 41. Hon Degrees: ScD, Rose-Hulman Inst, 54. Prof Exp: Instr bact, Col Med, NY Univ, 37-41; vol worker, Rockefeller Hosp Med Res, 41-42; tech aide, Comt Treatment Gas Casualties, Div Med Sci, Nat Res Coun, 42-45, exec secy, Insect Control Comt, 45-46, chmn chem coding panel chem-biol, Coord Ctr, 46-52; assoc, 46-50, chief div exp chemother, 47-72, assoc dir inst, 57-60, sci dir, 60-61, vpres, 61-72, MEM, SLOAN-KETTERING INST CANCER RES, 50-, DIR WALKER LAB, RYE, 60-, VPRES INST AFFAIRS, CANCER CTR, 74-; EMER PROF BIOCHEM, SLOAN-KETTERING DIV, MED COL, CORNELL UNIV, 76- Concurrent Pos: Prof biochem, Sloan-Kettering Div, Med Col, Cornell Univ, 51-75; mem comt tumor nomenclature & statist, Int Cancer Res Comn, 52-54; chmn screening panel, Cancer Chemother Nat Serv Ctr, NIH, 55-58 & drug eval panel, 58-; mem sci adv bd, Roswell Park Mem Inst, 57-66; mem chemother rev bd, Nat Adv Cancer Coun, 58-59; mem US nat comt, Int Union Against Cancer, 67-, chmn, 75-; mem bd dirs, Am Cancer Soc, 72- Honors & Awards: Alfred P Sloan Award, 65. Mem: Am Chem Soc; Am Soc Biol Chemists; Soc Exp Biol & Med; Am Asn Cancer Res. Res: Enzymes; hypertension; experimental chemotherapy of cancer. Mailing Add: Sloan-Kettering Inst Cancer Res 410 E 68th St New York NY 10021

STOCK, DAVID ALLEN, b Elyria, Ohio, Feb 8, 41; m 64; c 1. MICROBIOLOGY, GENETICS. Educ: Mich State Univ, BS, 63; NC State Univ, MS, 66, PhD(genetics), 68. Prof Exp: Instr microbiol, Sch Med, Univ Miss, 67-68; USDA fel, Baylor Col Med, 68-69, NIH fel, 69-70; ASST PROF BIOL, STETSON UNIV, 70- Mem: Am Soc Microbiol; Am Phytopath Soc. Res: Cytogenetics, physiology and pathogenesis of Candida albicans; metabolism of heavy metals in aquatic ecosystems, nucleic acid metabolism of mycoplasma, genetics of fungi. Mailing Add: Dept of Biol Stetson Univ De Land FL 32720

STOCK, JOHN JOSEPH, b Oakville, Ont, June 9, 20; m 45; c 2. MICROBIOLOGY, MEDICAL MYCOLOGY. Educ: Univ Toronto, BSA, 44; McGill Univ, MSc, 49, PhD(bact, immunol), 51. Prof Exp: Tech supvr antibiotic fermentations, Merck & Co, Que, 44-45; prod, control & res bacteriologist, F W Horner Co, 45-47; asst prof bact & immunol, 51-61, assoc prof microbiol, 61-72, PROF MICROBIOL, UNIV BC, 73- Mem: Am Soc Microbiol; Med Mycol Soc of the Americas; Can Soc Microbiol. Res: Factors influencing the virulence of dermatophytes, as well as studies on their composition, metabolism and immunology; investigations of microbial enzymes and their immunospecificities. Mailing Add: Dept of Microbiol Univ BC Vancouver BC Can

STOCK, JOHN THOMAS, b Margate, Eng, Jan 26, 11; nat US; m 45; c 1. ANALYTICAL CHEMISTRY. Educ: Univ London, BSc, 39 & 41, MSc, 45, PhD(chem), 49, DSc, 65. Prof Exp: Sci off chem, Ministry Supply, Gt Brit, 40-44; actg chief chemist, Fuller's Ltd, 44-46; lectr chem, Norwood Tech Col, 46-51, head dept, 51-56; assoc prof, 56-59, PROF CHEM, UNIV CONN, 59- Concurrent Pos: London County Coun Blair fel, Univ Minn, 53-54; consult, 50- Mem: Fel Am Chem Soc; fel The Chem Soc; Soc Chem Indust; Royal Inst Chem; Electrochem Soc. Res: Design of microchemical and general scientific apparatus; history of chemistry. Mailing Add: Dept of Chem Univ of Conn Storrs CT 06268

STOCK, JURGEN, astronomy, see 12th edition

STOCK, LEON M, b Detroit, Mich, Oct 15, 30; m 61; c 2. ORGANIC CHEMISTRY. Educ: Univ Mich, BS, 52; Purdue Univ, PhD, 59. Prof Exp: From instr to assoc prof, 58-70, PROF CHEM, UNIV CHICAGO, 71- Concurrent Pos: Consult, Phillips Petrol Co, 64- Mem: Am Chem Soc; The Chem Soc. Res: Electrophilic aromatic substitution reactions; influences of structure and solvents on reactivity; models for evaluation of inductive influences of substituents; electron paramagnetic resonance spectra of organic radicals; the structure of coal. Mailing Add: Dept of Chem Univ of Chicago Chicago IL 60637

STOCK, WERNER, b Berlin, Ger, May 16, 16; US citizen; m 51; c 2. BIOCHEMISTRY. Prof Exp: Tech, Visking Co, Ill, 47-50, chemist, 51-62; mgr new appln food prod div, 62-64, mgr prod develop, 65-67, VPRES INT DEPT, FILMS-PACKAGING DIV, UNION CARBIDE CORP, CHICAGO, 67- Mem: Inst Food Technol. Res: Function of semipermeable membranes in food processing; protective packaging of perishable foods. Mailing Add: 232 Thomas St Park Forest IL 60466

STOCKBRIDGE, ROBERT R, b Worcester, Mass, Aug 21, 10; m 37; c 1. ANIMAL HUSBANDRY, POULTRY HUSBANDRY. Educ: Univ Mass, BVA, 34; Hofstra Col, MS, 46. Prof Exp: From instr to assoc prof, 38-60, PROF POULTRY SCI, STATE UNIV NY AGR & TECH COL FARMINGDALE, 60-, CHMN AGR DEPT, 60- Mem: Poultry Sci Asn; World Poultry Sci Asn. Mailing Add: Dept of Agr State Univ of NY Agr & Tech Col Farmingdale NY 11735

STOCKBURGER, GEORGE JOSEPH, b Philadelphia, Pa, May 23, 27; m 61; c 2. INDUSTRIAL ORGANIC CHEMISTRY. Educ: St Joseph's Col, Pa, BS, 50; Univ Pa, MS, 52, PhD, 55. Prof Exp: SR RES CHEMIST, ICI US INC, 55- Mem: Am Chem Soc. Res: Reaction kinetics and mechanism; catalysis. Mailing Add: ICI US Inc Wilmington DE 19897

STOCKDALE, FRANK EDWARD, b Long Beach, Calif, Mar 15, 36. DEVELOPMENTAL BIOLOGY, ONCOLOGY. Educ: Yale Univ, AB, 58; Univ Pa, MD & PhD(develop biol), 63. Prof Exp: Intern internal med, Univ Hosps, Western Reserve Univ, 63-64; staff assoc, Nat Inst Arthritis & Metab Dis, 64-66; sr resident, Univ Hosp, 66-67, from instr to asst prof med, 68-74, asst prof biol, 70-74, ASSOC PROF MED & BIOL SCI, SCH MED, STANFORD UNIV, 74- Mem: Am Soc Clin Invest; AAAS; Soc Develop Biol; Int Soc Develop Biol; Am Soc Cell Biol. Res: Medical oncology; mechanisms for control of cell differentiation and growth during embryogenesis; hormonal control of cell function and growth. Mailing Add: Dept of Med Stanford Univ Sch of Med Stanford CA 94305

STOCKDALE, HAROLD JAMES, b Aplington, Iowa, Dec 3, 31; m 51; c 2. ECONOMIC ENTOMOLOGY. Educ: Iowa State Univ, BS, 58, MS, 59, PhD(entom), 64. Prof Exp: EXTEN ENTOMOLOGIST, IOWA STATE UNIV, 61- Mem: Entom Soc Am; Am Mosquito Control Asn. Res: Field crop insect management; household and structural insect control. Mailing Add: Insectary Bldg Iowa State Univ Ames IA 50011

STOCKDALE, JOHN ALEXANDER DOUGLAS, b Ipswich, Australia, Mar 15, 36; m 57; c 3. PHYSICS. Educ: Univ Sydney, BSc, 57, MSc, 60; PhD(physics), Univ Tenn, 69. Prof Exp: Res scientist, Australian AEC, 58-66; PHYSICIST, HEALTH PHYSICS DIV, OAK RIDGE NAT LAB, 66- Concurrent Pos: Vis prof, NY Univ, 75-76. Mem: Am Phys Soc. Res: Atomic and molecular physics. Mailing Add: 907 W Outer Dr Oak Ridge TN 37830

STOCKEL, RICHARD F, b Jersey City, NJ, July 22, 35; m 59; c 3. ORGANIC CHEMISTRY. Educ: Univ Iowa, BS, 57; Univ Vt, MS, 59; Clemson Univ, PhD(org chem), 62; Fairleigh Dickinson Univ, MBA, 68. Prof Exp: Res chemist, Am Cyanamid Co, Bound Brook, 62-71; mgr indust applns, 71, VPRES & DIR RES & DEVELOP, HYDRON LABS, INC DIV, NAT PATENT DEVELOP CORP, 71- Mem: Am Chem Soc; Am Mgt Asn. Res: Organic fluorine, organic phosphorous, cellulosic and polymer chemistry; hydrophilic polymers for medical, dental, coatings and consumer products; polymer research in the areas of enzyme immobilization and gel permeation chromatography. Mailing Add: 783 Jersey Ave New Brunswick NJ 08902

STOCKELL-HARTREE, ANNE, b Nashville, Tenn, Jan 11, 26; m 59; c 2. BIOCHEMISTRY. Educ: Vanderbilt Univ, BA, 46, MS, 49; Univ Utah, PhD(biochem), 56. Hon Degrees: MA, Cambridge Univ, 62. Prof Exp: Asst biochem, Vanderbilt Univ, 46-51; asst, Univ Utah, 51-54; fel USPHS Johnson Res Found, Univ Pa, 56-58; Med Res Coun Unit Molecular Biol, Cavendish Lab, 58-59, Jane Coffin Childs Fund, 59-60, RES WORKER DEPT BIOCHEM, CAMBRIDGE UNIV, 60- Concurrent Pos: Res fel, Girton Col, 62-65; external mem sci staff, Med Res Coun, 70- Mem: AAAS; Brit Biochem Soc; Brit Soc Endocrinol; Am Soc Biol Chemists. Res: Amino acid analysis; enzyme kinetics; protein structure and function; pituitary proteins. Mailing Add: Dept of Biochem Cambridge Univ Cambridge England

STOCKER, BRUCE ARNOLD DUNBAR, b Haslemere, Eng, May 26, 17; m 56; c 2. MICROBIAL GENETICS, MEDICAL MICROBIOLOGY. Educ: Univ London, MB & BS, 40, MD, 47. Prof Exp: House physician, Royal United Hosp, Bath, Eng, 40-41; supernumerary pathologist, Westminster Hosp Med Sch, 46-47; lectr bact, London Sch Hyg & Trop Med, Univ London, 48-53, head Guinness-Lister Res Unit, Lister Inst Prev Med, 53-65, Guinness prof microbiol, 63-65; PROF MED MICROBIOL, SCH MED, STANFORD UNIV, 66- Concurrent Pos: Commonwealth Fund fel microbiol, Col Med, NY Univ, 51-52; lectr, Am Soc Microbiol Found for Microbiol, 69-70; Guggenheim fel microbiol, Australian Nat Univ, 71. Honors & Awards: Paul

Ehrlich Prize, Munich, Ger, 65. Mem: Fel Royal Soc; Am Soc Microbiol; Brit Soc Gen Microbiol; Brit Genetical Soc; Path Soc Gt Brit & Ireland. Res: Bacterial genetics, especially Salmonella typhimurium; genetics of lipopolysaccharide structure, motility and flagellar characters, virulence, plasmids, transduction. Mailing Add: Dept of Med Microbiol Stanford Univ Sch of Med Stanford CA 94305

STOCKER, FRED BUTLER, b Kenyon, Minn, Jan 31, 31; m 53; c 2. ORGANIC CHEMISTRY. Educ: Hamline Univ, BS, 53; Univ Minn, MS, 55; Univ Colo, PhD(org chem), 58. Prof Exp: Assoc prof, 58-69, PROF CHEM, MACALESTER COL, 69-, CHMN DEPT, 70- Concurrent Pos: Consult, 59- Mem: Am Chem Soc. Res: Imidazole derivatives. Mailing Add: Dept of Chem Macalester Col St Paul MN 55101

STOCKER, HANS J, solid state physics, see 12th edition

STOCKER, JACK HUBERT, b Detroit, Mich, May 3, 24; m 64; c 2. ORGANIC CHEMISTRY. Educ: Olivet Col, BS, 44; Ind Univ, MA, 47; Tulane Univ, PhD(org chem), 55. Prof Exp: Control chemist, R P Scherer Corp, Mich, 48-50; control chemist, Atlas Pharmaceut Co, 50; Fullbright traveling fel, Heidelburg Univ, 55-56; assoc prof chem, Univ Southern Miss, 56-58; assoc prof, 58-71, admin asst to dean col sci, 65-67, PROF CHEM, UNIV NEW ORLEANS, 71- Concurrent Pos: Res partic, Oak Ridge Inst Nuclear Studies, 59; consult, Food & Drug Admin; res assoc, Gulf South Res Inst; vis prof chem, Univ Lund, Sweden, 74-75. Mem: Am Chem Soc; The Chem Soc; Electrochem Soc. Res: Acetals and ketals; organometallics; stereoselective reactions; c14 techniques; organic photochemistry and electrochemistry. Mailing Add: Dept of Chem Univ of New Orleans Lakefront LA 70122

STOCKER, RICHARD LOUIS, b Honolulu, Hawaii, Apr 22, 41. GEOPHYSICS. Educ: Lehigh Univ, BA, 64; Yale Univ, MS, 66, PhD(geophys), 73. Prof Exp: Asst prof geol, Lehigh Univ, 73-75; ASST PROF GEOL, ARIZ STATE UNIV, 75- Mem: Sigma Xi; Am Geophys Union. Res: Transport properties of minerals and rocks, particularly rheology and atomistic diffusion; point defect chemistry of minerals. Mailing Add: Dept of Geol Ariz State Univ Tempe AZ 85281

STOCKHAMMER, KARL ADOLF, b Ried, Austria, July 19, 26; m 56; c 3. ZOOLOGY, ENTOMOLOGY. Educ: Graz Univ, PhD(zool), 51. Prof Exp: Res assoc zool, Univ Munich, 51-58; instr, Univ Göttingen, 58-59; asst prof, 59-67, ASSOC PROF ENTOM & PHYSIOL, UNIV KANS, 67- Mem: Entom Soc Am. Res: Detection of e-vector of polarized light in insects; behavioral and physiological aspects of nesting in native bees. Mailing Add: Dept of Entom Univ of Kans Lawrence KS 66045

STOCKHAUSEN, JOHN HARVEY, physics, see 12th edition

STOCKING, CLIFFORD RALPH, b Riverside, Calif, June 22, 13; m 37; c 2. PLANT PHYSIOLOGY. Educ: Univ Calif, BS, 37, MS, 39, PhD(plant physiol), 42. Prof Exp: Asst plant physiol, Univ Calif, 38-39, assoc bot, 39-42; food chemist, Puccinelli Packing Co, 42-45; assoc, Exp Sta, Univ Calif, Davis, 45-46, asst prof bot, Univ & asst botanist, Exp Sta, 46-52, assoc prof & assoc botanist, 52-58, actg chmn dept bot, 66-67, chmn dept, 68-74, PROF & BOTANIST, EXP STA, UNIV CALIF, DAVIS, 58- Concurrent Pos: Merck sr fel biochem, Univ Wis, 55-56; NSF fel, Imp Col, Univ London, 63-64, sr fel King's Col, 70-71. Mem: Fel AAAS; Bot Soc Am; Am Soc Plant Physiol; Scandinavian Soc Plant Physiol; Japanese Soc Plant Physiol. Res: Biochemistry of chloroplasts; plant water relations; intracellular distribution of enzymes and phosynthetic products. Mailing Add: Dept of Bot Univ of Calif Davis CA 95616

STOCKING, GORDON GARY, b Axin, Mich, Jan 12, 24; m 47; c 2. VETERINARY MEDICINE. Educ: Mich State Univ, DVM, 46. Prof Exp: Res vet, Upjohn Farms, 46-49, from asst vet to assoc vet, 49-57, dir vet div, 57-64, asst dir agr div, 65-69, prod mgr, 69-73, SR STAFF VET AGR DIV, UPJOHN CO, 73- Concurrent Pos: Ranch mgr & vet, Kellogg Ranch, Calif State Polytech Col, 51-52. Mem: Am Vet Med Asn; Indust Vet Asn; US Animal Health Asn; Am Asn Lab Animal Sci; Am Asn Equine Practr. Res: Equine reproduction and disease. Mailing Add: Agr Div Upjohn Co Kalamazoo MI 49001

STOCKING, HOBART EBEY, b Clarendon, Tex, Nov 16, 06; m 34; c 2. GEOLOGY. Educ: John Hopkins Univ, MS, 38; Univ Chicago, PhD(geol), 49. Prof Exp: Asst dist engr, Tex Pipeline Co, 28-29; topog engr, Angola Petrol Co, 29-31; field asst, US Geol Surv, 33-34; geologist, Shell Petrol Corp, 36-38; instr geol, Univ WVa, 40-41, asst prof, 41-43; dist geologist, Petrol Admin War, 43-45; US State Dept vis prof, Costa Rica, 45-46; geologist, NMex Bur Mines, 46; prof geol, Okla State Univ, 46-51; from asst chief geologist to chief geologist, USAEC, 52-58; prof geol, 59-75, EMER PROF GEOL, OKLA STATE UNIV, 75- Concurrent Pos: Hon prof, Costa Rica, 52; consult AEC, Argentina, 58. Mem: AAAS; Geol Soc Am. Res: Economic geology. Mailing Add: Dept of Geol Okla State Univ Stillwater OK 74074

STOCKING, KENNETH MORGAN, b Riverside, Calif, Jan 4, 11; m 35; c 2. PLANT TAXONOMY, ECOLOGY. Educ: Univ Southern Calif, PhD(bot), 50. Prof Exp: Teacher gen biol & bot, San Joaquin Delta Col, 39-58; from assoc prof to prof bot & sci educ, Univ of the Pac, 49-63; chmn dept biol, 64-68, PROF BIOL, CALIF STATE COL, SONOMA, 63-, PROVOST, SCH ENVIRON STUDIES, 72- Mem: AAAS; Am Soc Plant Taxon. Res: Taxonomy and ecology of echinocystis, marah and echinopepon. Mailing Add: 3567 Green Hill Dr Santa Rosa CA 95404

STOCKLAND, ALAN EUGENE, b Huron, SDak, July 18, 38; m 68; c 2. MICROBIOLOGY. Educ: Univ Nebr, BS & BA, 61; Mich State Univ, MS, 67, PhD(microbiol), 70. Prof Exp: Teacher secondary sch, Malaysia, 62-64; ASSOC PROF MICROBIOL, WEBER STATE COL, 70- Res: Microbiological control of insect pests. Mailing Add: Dept Microbiol Weber State Col 3750 Harrison Blvd Ogden UT 84403

STOCKLAND, WAYNE LUVERN, b Lake Lillian, Minn, May 4, 42; m 71; c 2. NUTRITION. Educ: Univ Minn, BS, 64, PhD(nutrit), 69. Prof Exp: Res asst nutrit, Univ Minn, 69, res fel, 69-70; RES NUTRITIONIST & STATIST MGR, INT MULTIFOODS CORP, 70- Mem: Am Soc Animal Sci; Am Dairy Sci Asn; Poultry Sci Asn; Am Inst Nutrit. Res: Swine, poultry and ruminant nutrition and management, especially the protein and amino acid requirements and the effect of energy level, temperature and other nutrients on these requirements. Mailing Add: Supersweet Res Farm PO Box 117 Courtland MN 56021

STOCKMAN, DAVID LYLE, b Lansing, Mich, Aug 12, 36; m 58; c 4. PHYSICAL CHEMISTRY. Educ: Eastern Ill Univ, BS, 57; Univ Minn, PhD(phys chem), 61. Prof Exp: Scientist, electronics lab, Gen Elec Co, NY, 61-65; scientist res labs, 65-69, mgr explor photoconductor systs br, 69-71, mgr photoreceptor technol, 71-73, MGR ALLOY PHOTORECEPTOR TECHNOL CTR, XEROX CORP, 73- Mem: AAAS; fel Am Inst Chem; Am Phys Soc; Am Chem Soc. Res: Photophysics and photochemistry of polyatomic molecules, especially with respect to excited state phenomena; dye laser development; photoconductivity. Mailing Add: Xerox Corp J C Wilson Ctr for Technol Rochester NY 14644

STOCKMAN, GAIL DIANE, b Richmond, Calif, July 24, 44. IMMUNOLOGY. Educ: Univ Tex, Austin, BA, 66; Univ Calif, Berkeley, MA, 67; Baylor Col Med, PhD(immunol), 71. Prof Exp: ASST PROF IMMUNOL, BAYLOR COL MED, 71- & UNIV TEX M D ANDERSON HOSP & TUMOR INST, HOUSTON, 71- Mem: Soc Exp Hemat. Res: Effects of cyclophosphamide on immune responsiveness; immunosuppression in cardiac transplantation; role of lymphocyte subpopulations in in vitro and in vivo immune responses; immunological monitoring of cancer patients. Mailing Add: Dept of Obstet & Gynec Baylor Col of Med Houston TX 77025

STOCKMANN, VOLKER ERWIN, b Stuttgart, Ger, Feb 10, 40; US citizen; m 64; c 2. SOLID MECHANICS, POLYMER PHYSICS. Educ: Univ Hamburg, MS, 64, PhD(physics), 68. Prof Exp: Scientist wood physics, Fed Orgn Forestry, Reinbek, 65-68; physicist paper physics, Forest Prod Lab, Madison, Wis, 68-74; SECT HEAD FIBER PHYSICS, WEYERHAEUSER CO, TACOMA, 74- Mem: Tech Asn Pulp & Paper Indust; Soc Wood Sci & Technol. Res: Paper and fiber physics; cell wall structure; technology of papermaking. Mailing Add: Weyerhaeuser Co 3400 13th Ave SW Seattle WA 98166

STOCKMAYER, WALTER HUGO, b Rutherford, NJ, Apr 7, 14; m 38; c 2. PHYSICAL CHEMISTRY. Educ: Mass Inst Technol, SB, 35, PhD(chem), 40; Oxford Univ, BSc, 37. Hon Degrees: Dr, Univ Louis Pasteur, 72. Prof Exp: Instr chem, Mass Inst Technol, 39-41 & Columbia Univ, 41-43; asst prof, Mass Inst Technol, 43-46, from assoc prof to prof phys chem, 46-61; chmn dept chem, 63-67 & 73-76, PROF CHEM, DARTMOUTH COL, 61- Concurrent Pos: Consult, E I du Pont de Nemours & Co, Inc, 45-; Guggenheim fel, 54-55; trustee, Gordon Res Conf, 63-66. Honors & Awards: Award, Mfg Chem Asn, 60; Award, Am Chem Soc, 66, Peter Debye Award phyPhys Chem, 74, High Polymer Physics Prize, 75. Mem: Nat Acad Sci; Am Chem Soc; fel Am Phys Soc; fel Am Acad Arts & Sci; The Chem Soc. Res: High polymers; applied statistical mechanics; dynamics and statistical mechanics of macromolecules. Mailing Add: Dept of Chem Dartmouth Col Hanover NH 03755

STOCKMEYER, LARRY JOSEPH, b Evansville, Ind, Nov 13, 48. COMPUTER SCIENCE. Educ: Mass Inst Technol, SB & SM, 72, PhD(comput sci), 73. Prof Exp: MEM RES STAFF, THOMAS J WATSON RES CTR, IBM CORP, 74- Mem: Asn Comput Mach. Res: Computational complexity; analysis of algorithms; information storage and retrieval. Mailing Add: IBM Thomas J Watson Res Ctr PO Box 218 Yorktown Heights NY 10598

STOCKMEYER, PAUL KELLY, b Detroit, Mich, May 1, 43; m 66; c 2. MATHEMATICS. Educ: Earlham Col, AB, 65; Univ Mich, Ann Arbor, MA, 66, PhD(math), 71. Prof Exp: ASST PROF MATH, COL WILLIAM & MARY, 71- Mem: Math Asn Am; Am Math Soc. Res: Combinatorial analysis; graph theory. Mailing Add: Dept of Math Col of William & Mary Williamsburg VA 23185

STOCKNER, JOHN G, b Kewanee, Ill, Sept 17, 40; m 62; c 2. LIMNOLOGY, ECOLOGY. Educ: Augustana Col, Ill, 62; Univ Wash, PhD(zool), 67. Prof Exp: Fel phytoplankton ecol, Windermere Lab, Freshwater Biol Asn, Eng, 67-68; limnologist, Freshwater Inst, 68-71, LIMONOLOGIST, PAC ENVIRON INST, FISHERIES RES BD CAN, 71- Mem: Am Soc Limnol & Oceanog; Int Asn Theoret & Appl Limnol. Res: Phytoplankton ecology and paleolimnology; marine plankton ecology and benthic algal and phytoplankton production. Mailing Add: 2614 Mathers Ave West Vancouver BC Can

STOCKS, DOUGLAS ROSCOE, JR, b Dallas, Tex, Sept 4, 32; m 51; c 3. MATHEMATICS. Educ: Univ Tex, BA, 58, MA, 60, PhD(math), 64. Prof Exp: Spec instr math, Univ Tex, 60-64; from asst prof to assoc prof, Univ Tex, Arlington, 64-69; ASSOC PROF MATH, UNIV ALA, BIRMINGHAM, 69- Concurrent Pos: Mathematician, US Navy Electronics Lab, 63; Tex Col & Univ Syst res grant, 66-67. Mem: Am Math Soc; Math Asn Am. Res: Lattice paths and graph theory; foundations of mathematics; geometry; topology; point set theory. Mailing Add: Dept of Math Univ of Ala Birmingham AL 35294

STOCKTON, DORIS S, b New Brunswick, NJ, Feb 9, 24; m 48; c 2. MATHEMATICS. Educ: Rutgers Univ, BSc, 45; Brown Univ, MSc, 47, PhD(math), 58. Prof Exp: Instr, Brown Univ, 52-54; asst prof, 58-73, ASSOC PROF MATH, UNIV MASS, AMHERST, 73- Mem: Am Math Soc; Math Asn Am. Res: Functional analysis. Mailing Add: Dept of Math Univ of Mass Amherst MA 01002

STOCKTON, GERALD WILLIAM, b Stockport, Eng, Dec 16, 44; Can citizen; m 69; c 1. MOLECULAR BIOPHYSICS, NUCLEAR MAGNETIC RESONANCE. Educ: NStaffordshire Polytech, RIC, 68; Univ Alta, PhD(chem), 73. Prof Exp: Exp officer, Pharmaceut Div, Imp Chem Indust, 63-68; fel, 73-75, RES ASSOC MOLECULAR BIOPHYS, DIV BIOL SCI, NAT RES COUN CAN, 75- Res: Development and application of nuclear magnetic resonance methods for the study of the molecular dynamics of biological membranes; deuterium NMR of specifically labelled lipids. Mailing Add: Div of Biol Sci Nat Res Coun of Can Ottawa ON Can

STOCKTON, JACK JENKS, b St Paris, Ohio, July 9, 20; m 42; c 4. MICROBIOLOGY, PARASITOLOGY. Educ: Ohio State Univ, DVM, 43; Mich State Univ, MS, 50; Univ Mich, PhD(epidemiol sci), 62. Prof Exp: From instr to prof microbiol, Mich State Univ, 47-68, chmn dept, 60-65; assoc dean, 68-70, actg dean, 70-71, DEAN SCH VET SCI & MED, PURDUE UNIV, 71- Concurrent Pos: Chief party, Mich State Univ group & vis prof, Univ Ryukyus, 65-67; mem vet med rev comt, Dept Health, Educ & Welfare, 71-74. Mem: Am Vet Med Asn; Asn Am Vet Parasitol; Am Soc Parasitol; Am Soc Microbiol. Res: Nutrition of parasitic Protozoa, particularly the venereal trichomonads; host-parasite interactions emphasizing the effect of the host immune response on the parasite. Mailing Add: Sch of Vet Sci & Med Purdue Univ West Lafayette IN 47907

STOCKTON, JOHN RICHARD, b Jarrell, Tex, Feb 19, 17; m 53; c 6. RESEARCH MANAGEMENT. Educ: Univ Tex, BSc, 38, MA, 41, PhD(microbiol), 51. Prof Exp: Pharmacist, Baylor Hosp, 38-39; tutor pharm, Univ Tex, 39-41, asst prof, 41-46; dir res, Hyland Labs, 46-50; tech asst to dir biol prod, Merck Sharp & Dohme, 50-54; mgr res & qual control, Pillsbury Co, 54-62, dir res & develop, 62-66, dir sci activities, 66-68; mgr res & develop Corn Prod Food Technol Inst, 68-70, assoc dir res & qual control, Best Foods, CPC Int, Inc, 70-73; vpres res & develop, Nutri Co, 73-74; PRES, MGT CATALYSTS, 74- Mem: AAAS; Inst Food Technol; Am Chem Soc; Am Soc Microbiol; NY Acad Sci. Res: Antimicrobial agents; physicochemical characteristics of drugs; biological products; food science; product development; nutrition; research management. Mailing Add: Mgt Catalysts PO Box E Ship Bottom NJ 08008

STOCKTON, MARY ROSE, b Cincinnati, Ohio, Dec 8, 08. ORGANIC CHEMISTRY. Educ: Athenaeum of Ohio, BS, 37; Univ Cincinnati, MA, 42, PhD(org chem), 43. Prof Exp: Teacher high sch, Ind, 31-32, teacher high sch, Ohio, 32-41; from asst prof to assoc prof, 43-57, PROF CHEM, MARIAN COL, IND, 57-, HEAD DEPT, 43- Concurrent Pos: Abstractor, Chem Abstr, 56-; res corp grant, 58-59; NSF lectr grant, 59-61, res grants, 60 & 64. Mem: Fel Am Inst Chem; Am Chem Soc. Res: Pyridine and boron chemistry. Mailing Add: Dept of Chem Marian Col 3200 Cold Springs Rd Indianapolis IN 46222

STOCKTON, WILLIAM DENIS, b Oakley, Tenn, May 21, 19; m 51; c 4. ENTOMOLOGY. Educ: Univ Akron, BS, 50; Cornell Univ, MS, 54, PhD(entom), 56. Prof Exp: Aquarium attend, US Fish & Wildlife Serv, Mass, 50; res assoc oceanog, US Off Naval Res, 51-53; lab asst gen biol, Cornell Univ, 54-55, mus asst, 56-57; from asst prof to assoc prof biol, 57-65, PROF ENTOM, CALIF STATE UNIV, LONG BEACH, 65- Concurrent Pos: Mem, Int Document Ctr Arachnology. Res: Taxonomy, ecology and biology of Arachnida; theoretical astronomy; education of children in science. Mailing Add: Dept of Biol Calif State Univ 6101 E Seventh St Long Beach CA 90840

STOCKWELL, CHARLES WARREN, b Port Angeles, Wash, Dec 31, 40; m 66; c 1. NEUROSCIENCES. Educ: Western Wash State Col, BA, 64; Univ Ill, MA, 66, PhD(psychol), 68. Prof Exp: Res psychologist, Naval Aerospace Med Ctr, 69-71; ASST PROF OTOLARYNGOL, COL MED, OHIO STATE UNIV, 72- Mem: Psychonomic Soc; Asn Res Otolaryngol; Soc Neurosci. Res: Vestibular function. Mailing Add: Dept of Otolaryngol Ohio State Univ Col of Med Columbus OH 43210

STOCKWELL, CLIFFORD HOWARD, b Estevan, Sask, Sept 26, 97; m 35; c 3. GEOLOGY. Educ: Univ BC, BASc, 24; Univ Wis, PhD(geol), 30. Prof Exp: Asst geologist, 27-29, assoc geologist, 29-39, geologist, 39-53, sr geologist, 53-57, RES SCIENTIST, GEOL SURV CAN, 57- Concurrent Pos: Vis prof, McGill Univ, 47-48; consult, NJ Zinc Co, 51-52. Mem: Fel Geol Soc Am; fel Royal Soc Can; Can Inst Mining & Metall. Res: Precambrian and structural geology. Mailing Add: Geol Surv of Can Ottawa ON Can

STODDARD, ALONZO EDWIN, JR, b Huntington, WVa, Feb 6, 26; m 49; c 2. PHYSICS. Educ: Univ Mich, BS, 48, MS, 49, PhD(physics), 53. Prof Exp: Physicist, Calif Res & Develop Co, 53-55, physicist, Chevron Res Corp, 55-60; PROF PHYSICS, HARVEY MUDD COL, 60- Concurrent Pos: Physicist, US Geol Surv, 66-67. Mem: Am Phys Soc. Res: Color centers and luminescence in solids; nuclear spectroscopy; geophysics. Mailing Add: Dept of Physics Harvey Mudd Col Claremont CA 91711

STODDARD, CARL C, b Clan Harbor, NS, July 9, 12; m 43; c 5. ANESTHESIOLOGY. Educ: Dalhousie Univ, MD, CM, 38; Royal Col Physicians & Surgeons Can, cert specialist, 43. Prof Exp: Head dept anesthesia, Royal Can Naval Hosp, Halifax, NS, 43-46; lectr, 46-49, PROF ANESTHESIA, DALHOUSIE UNIV, 49-; HEAD DEPT ANESTHESIA, VICTORIA GEN HOSP, 47- Concurrent Pos: Mem, Adv Comt Anesthesia, Royal Col Physicians & Surgeons Can, 44-, regional mem, 67; consult, Halifax Infirmary, Halifax children's Hosp, Vet Hosp Halifax & Armed Forces Hosp, 49. Mem: Am Soc Anesthesiol; fel Am Col Anesthesiol; Can Anaesthetists Soc (vpres, 52); Int Anesthesia Res Soc; fel Int Col Anesthetists. Res: Anesthesia teaching and administration. Mailing Add: Dept of Anesthesia 10-032 Victoria Gen Hosp Halifax NS Can

STODDARD, DAN WARREN, b Richmond, Utah, Nov 20, 25; m 47; c 3. MATHEMATICS. Educ: Utah State Univ, MS, 52. Prof Exp: Instr math, Utah State Univ, 49-52; staff mem, Sandia Corp, 52-56; asst prof math, Brigham Young Univ, 56-57; RES SCIENTIST, KAMAN SCI CORP, 57- Res: Systems, operations and statistical analysis; weapon systems; applied mathematics. Mailing Add: Kaman Sci Corp Garden of the Gods Rd Colorado Springs CO 80907

STODDARD, DAVID LEE, plant pathology, see 12th edition

STODDARD, GEORGE EDWARD, b Boise, Idaho, July 15, 21; m 46; c 5. ANIMAL NUTRITION. Educ: Univ Idaho, BS, 43; Univ Wis, MS, 48, PhD(dairy husb), 50. Prof Exp: Asst, Univ Wis, 46-49; asst prof dairy husb, Iowa State Col, 49-52; assoc prof, 52-55, PROF DAIRY HUSB, UTAH STATE UNIV, 55-, HEAD DEPT DAIRY SCI, 60- Concurrent Pos: Pres, Agri-Mgt Consults, Inc, 72-; staff collabr, Agriserv Found, 75- Mem: AAAS; Am Soc Animal Sci; Am Dairy Sci Asn. Res: Rumen physiology; effect of feeds on milk composition; nutrient absorption; feed value of silage; hay and grain substitutions; grain feeding on pasture; hay quality and harvest methods; fluorosis in dairy animals; insecticides on hay fed to dairy animals; dietary fats and urinary metabolites of rats; costs of milk production. Mailing Add: Dept of Dairy Sci Utah State Univ Logan UT 84322

STODDARD, JAMES H, b Saginaw, Mich, June 17, 30; m 57; c 2. MATHEMATICS. Educ: Univ Mich, BS, 52, PhD(math), 61. Prof Exp: Instr math, Univ Mich, 60-61; asst prof, Oakland Univ, 61-62; asst prof, Syracuse Univ, 62-66; asst prof, Univ of the South, 66-67; assoc prof, Kenyon Col, 67-70; prof, Upsala Col, 70-72; PROF MATH, MONTCLAIR STATE COL, 72- Concurrent Pos: Mem col level exam comt, Educ Testing Serv, NJ, 65-69. Mem: Am Math Soc; Math Asn Am; Asn Comput Mach. Res: Topology; application to social sciences. Mailing Add: Dept of Math Montclair State Col Upper Montclair NJ 07043

STODDARD, LELAND DOUGLAS, b Hillsboro, Ill, Mar 15, 19; m 46. PATHOLOGY. Educ: DePauw Univ, AB, 40; Johns Hopkins Univ, MD, 43. Prof Exp: Asst & asst resident path, Sch Med, Duke Univ, 47-48, instr & resident, 49-50, assoc, 50-51; from asst prof to assoc prof, Med Sch, Univ Kans, 51-54; chmn dept, 54-73, PROF PATH, EUGENE TALMADGE MEM HOSP, MED COL GA, 54- Concurrent Pos: Chief staff, Eugene Talmadge Mem Hosp, 64-65; chief res path, Atomic Bomb Casualty Comn Japan, 61-62; vis prof, Med Sch, Osaka Univ, 66; mem path training comt, Vet Admin, 69-72; mem, Intersoc Path Coun, Int Coun Socs Path, US Nat Comt & Sci Adv Bd of Consult, Armed Forces Inst Path, 70-75. Mem: Am Asn Cancer Res; Asn Hist Med; Am Asn Pathologists & Bacteriologists; Am Soc Exp Pathologists; Int Acad Path (secy-treas, US-Can div coun, 70-, vpres, Acad, 74-). Res: Cervical carcinoma; experimental and human renal diseases; medical education. Mailing Add: Dept of Path Med Col of Ga Augusta GA 30902

STODDARD, ROBERT HUGH, b Auburn, Nebr, Aug 29, 28; m 55; c 3. GEOGRAPHY. Educ: Nebr Wesleyan Univ, BA, 50; Univ Nebr, MA, 60; Univ Iowa, PhD(geog), 66. Prof Exp: From instr to asst prof geog, Nebr Wesleyan Univ, 61-67; asst prof, 67-70, ASSOC PROF GEOG, UNIV NEBR, LINCOLN, 70- Concurrent Pos: Fulbright-Hays lectureship grant to Nepal, 75-76. Mem: Asn Am Geog; Asn Asian Studies; Nat Coun Geog Educ. Res: Geography of social problems, methods of spatial analysis and geography of South Asia. Mailing Add: Dept of Geog Univ of Nebr Lincoln NE 68508

STODDARD, THEODORE LOTHROP, anthropology, see 12th edition

STODOLSKY, MARVIN, microbiology, see 12th edition

STOEBER, WERNER, b Göttingen, Ger, May 8, 25; m 55; c 1. PHYSICAL CHEMISTRY, BIOPHYSICS. Educ: Univ Göttingen, Dipl, 53, Dr rer nat, 55. Prof Exp: Sci asst, Med Res Inst, Max Planck Soc, 55-61; res fel aerosol sci, Calif Inst Technol, 61-63; sci asst, Inst Med Physics, Univ Münster, 64-65, docent, 65-66; assoc prof, 66-70, PROF RADIATION BIOL & BIOPHYS, MED SCH, UNIV ROCHESTER, 70- Concurrent Pos: USPHS int res fel, 61-62; prof, Univ Münster, 68. Res: Surface chemistry; silicosis; aerosol. Mailing Add: Sch of Med & Dent Med Ctr Univ of Rochester 601 Elmwood Ave Rochester NY 14642

STOECKELER, ERNEST GEORGE, photo interpretation, deceased

STOECKENIUS, WALTHER, b Giessen, Ger, July 3, 21; m 52; c 3. CYTOLOGY. Educ: Univ Hamburg, MD, 51. Prof Exp: Intern, Pharmacol Inst, Univ Hamburg, 51, intern internal med, 51 & obstet & gynec, 52, researcher virol, Inst Trop Med, 52-54, res asst path, 54-58, privat docent, 58; guest investr, Rockefeller Inst, 59, from asst prof to assoc prof cytol, 59-67; PROF CELL BIOL, SCH MED, UNIV CALIF, SAN FRANCISCO, 67- Res: Fine structure of cells at the molecular level; energy transducing membranes; photobiology. Mailing Add: Dept of Biochem & Biophys Univ of Calif San Francisco CA 94122

STOECKER, ROBERT EUGENE, biology, ecology, see 12th edition

STOECKLE, JOHN DUANE, b Highland Park, Mich, Aug 17, 22; m 47; c 4. MEDICINE. Educ: Antioch Col, BS, 48; Harvard Med Sch, MD, 48; Am Bd Internal Med, dipl, 58. Prof Exp: Intern med, Mass Gen Hosp, Boston, 48-49, asst resident, 49-50, resident, 51-52; panel dir med aspects of atomic energy, Comt Med Sci, Res & Develop Bd, Dept Defense, 52-54; instr, 54-58, assoc, 58-67, asst prof, 67-69, ASSOC PROF MED, HARVARD MED SCH, 69-; CHIEF MED CLIN, MASS GEN HOSP, 54- Concurrent Pos: Physician, Mass Gen Hosp, 69-; mem comt on coal miner's safety & health, Dept Health, Educ & Welfare, 71-73. Mem: Fel Am Pub Health Asn; assoc mem Am Sociol Asn; fel Am Anthrop Asn; fel Am Psychosom Soc; Soc Appl Anthrop. Res: Medical care administration and health and illness behavior; longitudinal study of occupational lung diseases. Mailing Add: Mass Gen Hosp 32 Fruit St Boston MA 02114

STOECKLEY, THOMAS ROBERT, b Ft Wayne, Ind, Dec 6, 42. ASTRONOMY. Educ: Mich State Univ, BS, 64; Cambridge Univ, PhD(astron), 67. Prof Exp: Asst prof, 67-74, ASSOC PROF ASTRON, MICH STATE UNIV, 67- Mem: Am Astron Soc; Am Inst Physics; Royal Astron Soc. Res: Stellar rotation. Mailing Add: Dept of Astron & Astrophys Mich State Univ East Lansing MI 48823

STOECKLY, ROBERT E, b Schenectady, NY, June 9, 38; m 69. PHYSICS, ASTRONOMY. Educ: Princeton Univ, PhD(astrophys sci), 64. Prof Exp: Res fel astron, Mt Wilson & Palomar Observs, 64-65; asst prof physics & astron, Rensselaer Polytech Inst, 65-72; PHYSICIST, MISSION RES CORP, 72- Mem: Am Astron Soc. Res: Fluid dynamics; plasma physics; astrophysics. Mailing Add: Mission Res Corp Drawer 719 Santa Barbara CA 93102

STOEHR, HENRY ARTHUR, b Chicago, Ill, May 21, 10; m 40. FORESTRY. Educ: Univ Minn, BS, 33; Duke Univ, MF, 46; Univ Mich, PhD, 54. Prof Exp: Tech foreman, US Forest Serv, 33-36, farm planner, soil conserv serv, 37-43; asst prof forest mensuration & mgt, Mich State Univ, 46-55; assoc prof forest mensuration, Univ Ga, 55-65; chief forester, Pulp & Paper Div, Brown & Root, Inc, 65-69; resource analyst, Champion Int Corp, 69-75; RETIRED. Concurrent Pos: Mem contract team, US Foreign Aid Prog, Cambodia, 60-62. Mem: Soc Am Foresters. Res: Growth rate of trees; tree stem form; tropical forestry; feasibility studies for domestic and foreign proposed pulp and paper mills. Mailing Add: 107 Crossbow Circle Winterville GA 30683

STOELTING, VERGIL KENNETH, b Freelandville, Ind, Feb 10, 12; m 36; c 2. MEDICINE. Educ: Ind Univ, BS & MD, 36. PROF ANESTHESIOL, SCH MED & DIR DEPT, UNIV HOSPS, IND UNIV, INDIANAPOLIS, 47- Concurrent Pos: Consult, Vet Admin, Methodist & Community Hosps, Indianapolis. Mem: Am Soc Anesthesiol. Res: Anesthesiology as related to physiology and pharmacology. Mailing Add: Dept of Anesthesiol Ind Univ Sch of Med Indianapolis IN 46207

STOENNER, HERBERT GEORGE, b Levasy, Mo, June 17, 19; m 46; c 4. BACTERIOLOGY, VIROLOGY. Educ: Iowa State Col, DVM, 43. Prof Exp: Asst scientist, Commun Dis Ctr, USPHS, Ga, 47-50, sr asst vet, 50-52, vet, 52-56, sr vet, 56-61, asst dir lab, 62-64, DIR ROCKY MOUNTAIN LAB, NAT INST ALLERGY & INFECTIOUS DIS, 64-, VET OFFICER DIR, USPHS, 61- Concurrent Pos: Fac affil, Univ Mont. Honors & Awards: Distinguished Serv Medal, Dept Health, Educ & Welfare, 71; K F Meyer Gold Headed Canc Award, 74. Mem: Am Vet Med Asn; Am Pub Health Asn; US Animal Health Asn; Conf Res Workers Animal Dis. Res: Zoonoses; leptospirosis; rickettsioses; brucellosis; psittacosis. Mailing Add: 1102 S Second St Hamilton MT 59840

STOENNER, RAYMOND WILLIAM, analytical chemistry, see 12th edition

STOERK, HERBERT CARL, b Vienna, Austria, Mar 2, 08; nat US; m 55. PATHOLOGY. Educ: Univ Vienna, MD, 38. Prof Exp: Asst path, Univ Paris, 38-40; instr, Columbia Univ, 40-46; instr bact, Harvard Med Sch, 46-47; dir exp path, Merck Inst Therpaeut Res, 48-63; from asst prof to assoc prof, 60-68, PROF PATH, COL PHYSICIANS & SURGEONS, COLUMBIA UNIV, 68- Concurrent Pos: Claude Bernard prof, Univ Montreal; consult, Vet Admin Hosp, 63-; ed, Ergebnisse der allgem Pathologie, 65. Mem: AAAS; Soc Exp Biol & Med; Am Soc Exp Pathologists. Res: Experimental pathology; regression of lympho-sarcoma implants in pyridoxine deficient mice; morphologic and functional changes in acute pyridoxine deficiency; influence of endocrines upon lymphoid tissue and immunity; immunity to homografts; physiology of adrenals and parathyroids. Mailing Add: Col of Physicians & Surgeons Columbia Univ 630 W 168th St New York NY 10032

STOERMER, EUGENE F, b Webb, Iowa, Mar 7, 34; m 60; c 3. ALGOLOGY, LIMNOLOGY. Educ: Iowa State Univ, BS, 58, PhD(bot), 63. Prof Exp: NIH fel phycol, Iowa State Univ, 63-65; assoc res algologist, 66-71, RES ALGOLOGIST, GREAT LAKES RES DIV, UNIV MICH, ANN ARBOR, 71-, LECTR BOT, BIOL STA, 69- Concurrent Pos: McHenry fel, Acad Natural Sci, Philadelphia, Pa, 59. Mem: AAAS; Am Soc Limnol & Oceanog; Bot Soc Am; Phycol Soc Am. Res: Taxonomy and ecology of Bacillariophyta and Laurentian Great Lakes algal flora. Mailing Add: Great Lakes Res Div Univ of Mich IST Bldg Ann Arbor MI 48104

STOESSER, ALBERT V, b St Paul, Minn, Nov 28, 01. PEDIATRICS. Educ: Univ Minn, BS, 22, MD, 25, PhD, 29. Prof Exp: From instr to assoc prof pediat, Med Sch, Univ Minn, Minneapolis, 29-46, clin prof, 46-70; CHIEF ALLERGY CLIN, US ARMY HOSP, FT ORD, 70- Concurrent Pos: Consult dir, Pediat Allergy Clin, Univ Hosp, Univ Minn; pediatrician & allergist, St Barnabas Hosp, Minn. Mem: AAAS; Soc Pediat Res; Soc Exp Biol & Med; Am Asn Immunol; Am Col Allergists. Res: Value of vaccine therapy in the control of infections in allergic children; relationship between chronic headaches in children and allergy; role of adrenocorticotrophic hormone and cortisone-derivatives in the treatment of severe allergic manifestations in children; origin of so-called emphysema in children. Mailing Add: Allergy Clin Med Arts Ctr Silas B Hays Army Hosp Ft Ord CA 93941

STOESSL, ALBERT, b Linz, Austria, Feb 24, 24; Can citizen; m 54; c 1. NATURAL PRODUCTS CHEMISTRY. Educ: Univ London, BSc, 55, PhD(chem), 60. Prof Exp: Res fel org chem, Univ Western Ont, 60-61; res scientist, 61-73, SR RES SCIENTIST, RES INST, CAN DEPT AGR, 73- Mem: The Chem Soc; Can Inst Chem; Phytochem Soc NAm. Res: Chemistry of natural products; chemical basis of disease resistance in plants. Mailing Add: Res Inst Can Dept of Agr Univ Sub PO London ON Can

STOETZEL, MANYA BROOKE, b Houston, Tex, Apr 11, 40; m 62; c 2. ENTOMOLOGY. Educ: Univ Md, College Park, BS, 66, MS, 70, PhD(entom), 72. Prof Exp: Entomologist, First US Army Med Lab, Ft Meade, Md, 66-68; Presidential intern, 73-74, RES ENTOMOLOGIST, SYST ENTOM LAB, AGR RES SERV, USDA, 74- Mem: Entom Soc Am; Am Mosquito Control Asn. Res: Morphology and taxonomy of aphids, aleyrodid, psyllids and armored scale insects. Mailing Add: Syst Entom Lab Agr Res Serv USDA Bldg 004 Rm 6 Beltsville MD 20705

STOEVER, EDWARD CARL, JR, b Milwaukee, Wis, Mar 13, 26; m 54; c 2. Educ: Purdue Univ, BS, 48; Univ Mich, MS, 50, PhD(geol), 59. Prof Exp: Res geologist, Int Minerals & Chem Corp, 52-54; from asst prof to assoc prof geol, 56-69, assoc dir undergrad studies, Sch Geol & Geophys, 70-72, PROF GEOL & GEOPHYS, UNIV OKLA, 69- Concurrent Pos: Dir Okla Geol Camp, 64-69; dir inst earth sci, NSF, 65-69 & 70-72; assoc dir, Earth Sci Curriculum Proj, 69; assoc prog dir teacher educ sect, NSF, 69-70; sr staff consult, Earth Sci Teacher Prep Proj, 70-72; dir, Okla Earth Sci Educ Proj, 72-74. Mem: Nat Asn Geol Teachers (vpres, 74-75, pres, 75-76); AAAS; fel Geol Soc Am; Am Asn Petrol Geologists; Nat Sci Teachers Asn. Mailing Add: Sch of Geol & Geophys Rm 107 830 Van Vleet Oval Norman OK 73069

STOEWSAND, GILBERT SAARI, b Chicago, Ill, Oct 20, 32; m 57; c 2. ANIMAL NUTRITION, TOXICOLOGY. Educ: Univ Calif, Davis, BS, 54, MS, 58; Cornell Univ, PhD, 64. Prof Exp: Res assoc poultry sci, Cornell Univ, 58-61; res nutritionist, US Army Natick Lab, Mass, 63-66; res assoc, Inst Exp Path & Toxicol, Albany Med Col, 66-67; asst prof, 67-73, ASSOC PROF TOXICOL, EXP STA, NY STATE COL AGR & LIFE SCI, CORNELL UNIV, 73- Mem: AAAS; Inst Food Technol; Am Inst Nutrit; Am Chem Soc. Res: Heavy metal toxicology; effect of nutrition on toxicity; food additives; natural food toxicants. Mailing Add: Dept of Food Sci & Technol NY State Agr Exp Sta Cornell Univ Geneva NY 14456

STOFFER, JAMES OSBER, b Homeworth, Ohio, Oct 16, 35; m 57; c 2. ORGANIC CHEMISTRY. Educ: Mt Union Col, BS, 57; Purdue Univ, PhD(org chem), 61. Prof Exp: Res asst, Purdue Univ, 57-59; res assoc, Cornell Univ, 61-63; asst prof, 63-66, ASSOC PROF CHEM, UNIV MO-ROLLA, 66- Mem: Am Chem Soc. Res: Beta deuterium isotope effects; separation of optical isomers by gas chromatography; small ring compounds. Mailing Add: Dept of Chem Univ of Mo Rolla MO 65401

STOFFER, ROBERT LLEWELLYN, b North Georgetown, Ohio, Sept 16, 27; m 51; c 3. INDUSTRIAL HYGIENE, ANALYTICAL CHEMISTRY. Educ: Ashland Col, AB, 50; Ohio State Univ, PhD(anal chem), 54. Prof Exp: Anal chemist, 54-56, from asst proj chemist to sr proj chemist, 56-72, INDUST HYG CHEMIST, RES DEPT, STANDARD OIL CO (IND), 72- Mem: Am Chem Soc. Res: Development of new analytical methods in the industrial hygiene field. Mailing Add: Amoco Res Ctr PO Box 400 Naperville IL 60540

STOFFEY, DONALD G, organic chemistry, see 12th edition

STOFFOLANO, JOHN GEORGE, JR, b Gloversville, NY, Dec 31, 39; m 65; c 2. ENTOMOLOGY, NEUROBIOLOGY. Educ: State Univ NY Col Oneonta, BS, 62; Cornell Univ, MS, 67; Univ Conn, PhD(entom), 70. Prof Exp: ASST PROF ENTOM, UNIV MASS, AMHERST, 69- Concurrent Pos: NSF fel neurobiol, Princeton Univ, 70-71; NIH fel, Univ Mass, Amherst, 72-75. Mem: AAAS; Entom Soc Am; Am Inst Biol Sci; Soc Nematol. Res: Integrative studies on the ecology, neurobiology and physiology of diapausing, nondiapausing and aging flies of the genus Musca and Phormia. Mailing Add: Dept of Entom Univ of Mass Amherst MA 01002

STOFFYN, PIERRE JULES, b Ixelles, Belgium, May 10, 19; US citizen; m 50; c 2. CHEMISTRY, BIOCHEMISTRY. Educ: Univ Brussels, MS, 41, PhD(chem), 49. Prof Exp: Res fel med, Mass Gen Hosp, Harvard Med Sch, 52-54, res assoc med, 56-61, res assoc biol chem, 61-63, LECTR, MASS GEN HOSP, HARVARD MED SCH, 63-, SR ASSOC BIOL CHEM, 73- Concurrent Pos: Res fel, Am Heart Asn, 57-59, adv res fel, 59-61, estab investr, 61-66; from asst biochemist to biochemist, McLean Hosp, Belmont, 61-74; biochemist, Eunice Kennedy Shriver Ctr & Mass Gen Hosp, 74- Mem: Am Chem Soc; Am Soc Biol Chem; Belgian Soc Biochem; Int Soc Neurochem; Am Soc Neurochem. Res: Microchemistry; organic chemistry; carbohydrates; lipids; neurochemistry. Mailing Add: E K Shriver Ctr Waltham MA 02154

STOHLER, RUDOLF, b Basel, Switz, Dec 5, 01; nat US; m 29; c 5. ZOOLOGY. Educ: Univ Basel, MA & PhD(zool), 28. Prof Exp: Inst Student Exchange fel, 28-30, res assoc zool, 32-34, instr, 34-35, res assoc, 35-41, prin lab tech, 41-69, instr zool & biol, Exten, 35-69, assoc res zoologist, 55-66, res zoologist, 66-69, EMER RES ZOOLOGIST, UNIV CALIF, BERKELEY, 69- Mem: Am Malacol Union; Swiss Soc Natural Sci; hon mem Swiss Zool; Sigma Xi. Res: Cytology of toads; sex reversal in fish; genetics of human twinning; Gastropoda of California coast; laboratory techniques. Mailing Add: 1584 Milvia St Berkeley CA 94709

STOHLMAN, FREDERICK, JR, hematology, deceased

STÖHR, WALTER B, b Vienna, Austria, Aug 20, 28. GEOGRAPHY. Educ: Univ Commerce, Vienna, Dilpomkaufmann, 51, Doktor(econ geol), 63. Prof Exp: Chief economist, Austrian Inst Urban & Regional Planning, Vienna, 55-64; sr regional planning adv, Ford Found, 64-69; PROF GEOG, McMASTER UNIV, 69- Concurrent Pos: Regional planning adv, UN, 69-; corresp mem comn regional aspects econ develop, Int Geog Union, 71- Mem: Regional Sci Asn; Inter-Am Planning Soc. Res: Interurban and regional development. Mailing Add: Dept of Geog McMaster Univ Hamilton ON Can

STÖHRER, GERHARD, b Heidelberg, Ger, May 28, 39. ORGANIC CHEMISTRY, BIOCHEMISTRY. Educ: Univ Heidelberg, dipl chem, 62, PhD(chem), 65. Prof Exp: ASSOC BIOCHEM, SLOAN-KETTERING INST CANCER RES, 66- Mem: Am Chem Soc; Ger Chem Soc. Res: Molecular biology of oncogenesis. Mailing Add: Walker Lab 145 Boston Post Rd Sloan-Kettering Inst Cancer Res Rye NY 10580

STOHS, SIDNEY JOHN, b Ludell, Kans, May 24, 39; m 60; c 1. PHARMACOGNOSY, BIOCHEMISTRY. Educ: Univ Nebr, BS, 62, MS, 64; Univ Wis, PhD(biochem), 67. Prof Exp: From asst prof to assoc prof pharmacog, 67-74, chmn dept, 68-71, chmn dept med chem & pharmacog, 72, PROF MED CHEM & PHARMACOG & CHMN DEPT, UNIV NEBR, LINCOLN, 74- Concurrent Pos: Nebr Heart Asn grants, 68-69, 71-72 & 72-73; equip grants, Smith Kline & French Labs & Nebr Res Coun, 68-; NSF grants, 69-73; NIH grants, 69-72. Mem: AAAS; Am Soc Pharmacog; Am Pharmaceut Asn; Int Plant Tissue Cult Asn; Am Chem Soc. Res: Phytosterol biosynthesis; alkaloid biosynthesis; cardiac glycoside biosynthesis and metabolism; microsomal oxidation. Mailing Add: Col of Pharm Univ of Nebr Lincoln NE 68508

STOIBER, RICHARD EDWIN, b Cleveland, Ohio, Jan 28, 11; m 41; c 2. ECONOMIC GEOLOGY, VOLCANOLOGY. Educ: Dartmouth Col, AB, 32; Mass Inst Technol, PhD(econ geol), 37. Prof Exp: Instr, 35-36, 37-50, asst prof, 40-48, prof, 48-71, FREDERICK HALL PROF GEOL, DARTMOUTH COL, 71- Concurrent Pos: Part-time with Nfld Geol Surv & US Geol Surv; govt sponsored res, Cent Am; consult, UN; pvt consult. Mem: Mineral Soc Am; Soc Econ Geol; Geol Soc Am; Am Inst Mining, Metall & Petrol Eng. Res: Ore deposits; volcanoes; optical crystallography. Mailing Add: Dept of Earth Sci Dartmouth Col Hanover NH 03755

STOICHEFF, BORIS PETER, b Bitol, Yugoslavia, June 1, 24; nat Can; m 54; c 1. LASERS, MOLECULAR SPECTROSCOPY. Educ: Univ Toronto, BASc, 47, MA, 48, PhD(physics), 50. Prof Exp: McKee-Gilchrist fel, Univ Toronto, 50-51; fel, Nat Res Coun Can, 52-53, res officer, Div Pure Physics, 53-64; PROF PHYSICS, UNIV TORONTO, 64-, CHMN ENG SCI, 72- Concurrent Pos: Vis scientist, Mass Inst Technol, 63-64. Honors & Awards: Centennial Medal of Can, 67; Gold Medal, Can Asn Physicists, 74. Mem: Fel Am Phys Soc; hon fel Indian Acad Sci; fel Royal Soc, London; fel Royal Soc Can; fel Optical Soc Am (pres-elect, 75, pres, 76). Res: Molecular spectroscopy and structure; Rayleigh, Brillouin and Raman scattering; lasers and their applications in spectroscopy; stimulated scattering processes and two photon absorption; elastic constants of rare gas single crystals. Mailing Add: Dept of Physics Univ of Toronto Toronto ON Can

STOJANOVIC, BORISLAV JOVAN, b Zajecar, Yugoslavia, Nov 29, 19; US citizen; m 52. MICROBIOLOGY, BIOCHEMISTRY. Educ: Univ Bonn, BS, 48, Dr Agr, 50; Cornell Univ, MS, 55, PhD(soil microbiol), 56. Prof Exp: From asst prof to assoc prof, 56-67, PROF SOIL MICROBIOL, MISS STATE UNIV, 67- Mem: Am Soc Agron; Am Soc Microbiol. Res: Microbial transformations of soil proteinaceous materials and biocides. Mailing Add: Dept of Agron Miss State Univ Mississippi State MS 39762

STOKELY, ERNEST MITCHELL, b Greenwood, Miss, Mar 26, 37; m 64; c 2. BIOMEDICAL ENGINEERING. Educ: Miss State Univ, BSEE, 59; Southern Methodist Univ, MSEE & EE, 68, EE, 71, PhD(biomed eng), 73. Prof Exp: Sr elec engr, Tex Instruments, Inc, 59-69; ASST PROF RADIOL, UNIV TEX HEALTH SCI CTR DALLAS, 73- Concurrent Pos: Adj prof, Southern Methodist Univ, 73- & Univ Tex, Arlington, 73- Mem: Inst Elec & Electronic Engrs; Nat Asn Biomed Engrs. Res: Medical image and signal processing; biological system modeling. Mailing Add: Dept of Radiol Univ of Tex Health Sci Ctr Dallas TX 75235

STOKER, HOWARD STEPHEN, b Salt Lake City, Utah, Apr 16, 39; m 64; c 5. INORGANIC CHEMISTRY, ENVIRONMENTAL CHEMISTRY. Educ: Univ Utah, BA, 63; Univ Wis, PhD(chem), 68. Prof Exp: ASSOC PROF INORG CHEM, WEBER STATE COL, 68- Mem: Am Chem Soc; Sigma Xi. Res: Air and water pollution; interhalogen compounds. Mailing Add: Dept of Chem Weber State Col Ogden UT 84408

STOKER, JAMES JOHNSTON, b Dunbar, Pa, Mar 2, 05; m 28; c 4. MATHEMATICS, MECHANICS. Educ: Carnegie Inst Technol, BS, 27, MS, 31; Tech Hochsch Zurich, DrMath, 36. Prof Exp: Instr mech, Carnegie Inst Technol, 28-31, asst prof, 31-37; from asst prof to assoc prof, 37-45, PROF MATH, NY UNIV, 45-, DIR COURANT INST MATH SCI & HEAD ALL-UNIV DEPT MATH, 58- Concurrent Pos: Res mathematician, Appl Math Panel, Nat Defense Res Comt, 43-45. Honors & Awards: Heineman Prize, Am Phys Soc. Mem: AAAS; Am Math Soc. Res: Differential geometry; elasticity; vibration theory; hydrodynamics. Mailing Add: Courant Inst of Math Sci NY Univ 251 Mercer St New York NY 10012

STOKES, ALLEN WOODRUFF, b Philadelphia, Pa, Sept 16, 14; m 44; c 2. ANIMAL BEHAVIOR. Educ: Haverford Col, BS, 36; Harvard Univ, MA, 40; Univ Wis, PhD(zool, wildlife mgt), 52. Prof Exp: From asst prof to assoc prof, 52-61, PROF WILDLIFE MGT, UTAH STATE UNIV, 61- Concurrent Pos: NSF sr fel, 58. Honors & Awards: Award, Wildlife Soc, 54. Mem: Animal Behav Soc (pres, 71); Wildlife Soc; Am Ornith Union; Cooper Ornith Soc; Am Soc Mammal. Res: Behavioral ecology. Mailing Add: Dept of Wildlife Sci Utah State Univ Logan UT 84322

STOKES, ARNOLD PAUL, b Bismarck, NDak, Jan 24, 32; m 57; c 6. PURE MATHEMATICS. Educ: Univ Notre Dame, BS, 55, PhD(math), 59. Prof Exp: Staff mathematician, Res Inst Adv Study, 58-60; NSF fel math, Johns Hopkins Univ, 60-61; from asst prof to assoc prof, Catholic Univ, 61-65; chmn dept, 67-70, PROF MATH, GEORGETOWN UNIV, 65- Concurrent Pos: Sr res assoc, Nat Res Coun-Nat Acad Sci, Goddard Space Flight Ctr, NASA, 74-75. Mem: Am Math Soc. Res: Nonlinear differential equations. Mailing Add: Dept of Math Georgetown Univ Washington DC 20057

STOKES, CHARLES SOMMERS, b Philadelphia, Pa, Apr 24, 29; m 54; c 2. PHYSICAL CHEMISTRY. Educ: Ursinus Col, BS, 51; MA, 53. Prof Exp: Res chemist, 53-56, res assoc, 56-61, mgr test site, 61-72, VPRES, GERMANTOWN LABS, INC, 72- Mem: Am Chem Soc; Am Inst Aeronaut & Astronaut; Am Inst Chem; Combustion Inst. Res: Fluorine chemistry; propellents; high temperatures; plasma jet chemistry. Mailing Add: 127 Madison Rd Willow Grove PA 19090

STOKES, DAVID KERSHAW, JR, b Camden, SC, Feb 3, 27; m 50; c 3. FAMILY MEDICINE. Educ: Clemson Univ, BS, 48; Univ Ga, MS, 52; Tex A&M Univ, PhD, 56; Med Univ SC, MD, 57. Prof Exp: Intern, 58, DIR FAMILY PRACT RESIDENCY PROG, SPARTANBURG GEN HOSP, 72-; ASSOC PROF FAMILY PRACT, MED UNIV SC, 72- Mem: AMA; Am Heart Asn; Am Acad Family Physicians; Am Rheumatism Asn. Mailing Add: Family Pract Residency Prog Spartanburg Gen Hosp Spartanburg SC 29301

STOKES, DONALD EUGENE, b Andalusia, Ala, Aug 25, 31; m; c 5. NEMATOLOGY. Educ: Univ Fla, BS, 55, MA, 63, PhD(nematol), 72. Prof Exp: Nematologist III, 56-75, CHIEF NEMATOL, DIV PLANT INDUST, FLA DEPT AGR & CONSUMER SERV, 75- Concurrent Pos: Courtesy Fac Appl & Grad Fac Status, Univ Fla, 75. Mem: Soc Nematologists; Orgn Trop Am Nematologists; Europ Soc Nematologists. Res: Regulatory aspects of nematology, which include pathogenicity, taxonomy and response to chemicals of the various plant parasitic nematodes. Mailing Add: PO Box 1269 Gainesville FL 32602

STOKES, GEORGE ALWIN, b Winnfield, La, Dec 29, 20; m 45; c 2. GEOGRAPHY. Educ: La State Norm Col, AB, 42; La State Univ, MS, 48, PhD(geog), 54. Prof Exp: From asst prof to assoc prof, 49-61, PROF GEOG, NORTHWESTERN STATE UNIV, 61-, DEAN, COL LIB ARTS, 63- Mailing Add: Col of Lib Arts Northwestern State Univ Natchitoches LA 71457

STOKES, GRANVILLE WOOLMAN, b Anderson, Ind, Aug 1, 20; m 41; c 1. CROP BREEDING, PLANT PATHOLOGY. Educ: Purdue Univ, BS, 49; Univ Wis, MS, 51, PhD(genetics & plant path), 53. Prof Exp: Asst plant pathologist, 53-57, asst prof plant path & assoc pathologist, 57-60, assoc prof, 60-61, PROF PLANT PATH, UNIV KY, 61-, ASSOC DEAN, COL AGR, 69- Concurrent Pos: Lectr, Univ Calif, Berkeley, 60-61; assoc dir, Agr Exp Sta, Univ Ky, 66-68, dir tobacco & health res prog, 67-68. Mem: AAAS. Res: Cytogenetics and genetics of Nicotiana tabacum; breeding for disease resistance and quality of tobacco. Mailing Add: Agr Sci Ctr Rm S129 Col of Agr Univ of Ky Lexington KY 40506

STOKES, ILEY EDGAR, agronomy, see 12th edition.

STOKES, JACOB LEO, b Warsaw, Poland, Sept 27, 12; US citizen; m 42; c 2. MICROBIOLOGY. Educ: Rutgers Univ, BS, 34, PhD(microbiol), 39; Univ Ky, MS, 36. Prof Exp: Asst bact, Univ Ky, 34-36; asst marine bact, Rutgers Univ, 36-37, soil microbiol, 37-39; from microbiologist to head sect microbiol metab, Res Labs, Merck & Co, Inc, 39-47; res assoc, Hopkins Marine Sta, Stanford Univ, 48-50; assoc prof bact, Ind Univ, 50-53; bacteriologist, Western Utilization Res Br, USDA, 53-59; chmn dept bact & pub health, 59-68, PROF BACT & PUB HEALTH, WASH STATE UNIV, 59- Mem: AAAS; Am Soc Microbiol. Res: Relation of algae to other microorganisms in nature; antibiotics; iron bacteria; nutrition, physiology and biochemistry of microorganisms; psychrophilic microorganisms. Mailing Add: Dept of Bact & Pub Health Wash State Univ Pullman WA 99163

STOKES, JIMMY CLEVELAND, b Cochran, Ga, Nov 29, 44; m 72; c 1. SCIENCE EDUCATION. Educ: Univ Ga, BS, 66, MEd, 67, EdD(chem educ), 69. Prof Exp: Instr chem, Univ Ga, 69-70; asst prof, Clayton Jr Col, 70-74; ASST PROF CHEM, W GA COL, 74- Concurrent Pos: Consult, Wadsworth Publ Co, Calif, 73- Mem: Am Chem Soc; Nat Sci Teachers Asn; Asn Educ Commun & Technol. Res: Construction and evaluation of self paced and individualized programs of instruction in chemistry; development of drug education materials and programs. Mailing Add: Dept of Chem W Ga Col Carrollton GA 30117

STOKES, JOSEPH, III, b Philadelphia, Pa, Dec 2, 24; m 65; c 4. INTERNAL MEDICINE. Educ: Haverford Col, AB, 46; Harvard Med Sch, MD, 49; Am Bd Internal Med, dipl. Prof Exp: Intern med, Osler Serv, Hopkins Hosp, Johns Hopkins Univ, 49-50, asst resident, Serv & asst, Sch Med, Univ, 50-51; res fel prev med, Harvard Med Sch, 51-53, teaching fel med, 53-54, instr prev med, 54-56, assoc, 56-61, chief family health prog, 54-61; dir, Hawaii Cardiovasc Study, 61-64; dean sch med, 64-67, chmn dept community med, 67-75, PROF COMMUNITY MED, SCH MED, UNIV CALIF, SAN DIEGO, 64- Concurrent Pos: Sr asst surgeon, Framingham, Mass Heart Dis Epidemiol Study, Nat Heart Inst, 51-53, med officer, 55-60; asst, Peter Bent Brigham Hosp, Boston, 51-53; resident, Mass Gen Hosp, 53-54, clin asst, 54-55, asst, 55-61, actg chief infectious dis unit, 60-61; heart dis control officer, Hawaii State Health Dept, 61-62; asst chief med, Queen's Hosp & consult, Tripler Army Hosp, Honolulu, 63-64; ed, New Eng J Med Progress Series, Little, Brown & Co, 64-75; spec consult, Manpower & Training Comt, Nat Libr Med, 66-68; mem, Harvard Med Alumni Coun, 66-69; mem res comt, Asian-Pac Cong Cardiol; mem adv comt cardiovasc dis, SPac Comn; mem couns clin cardiol & epidemiol, Am Heart Asn; mem, Epidemiol & Dis Control Study Sect, 74- & adv comt, Pac Southwest Regional Med Libr Serv, 74- Mem: Am Fedn Clin Res; Am Soc Human Genetics; Am Cancer Soc; Am Col Physicians; Am Soc Internal Med. Res: Preventive medicine; epidemiology of cardiovascular disease. Mailing Add: Dept of Community Med M-007 Univ of Calif at San Diego La Jolla CA 92093

STOKES, JOSEPH FRANKLIN, b Havana, Ark, Feb 27, 34; m 59; c 2. MATHEMATICS. Educ: Univ Ark, Fayetteville, BS, 56, MA, 57; George Peabody Col, PhD(math), 72. Prof Exp: Instr math, Kans State Univ, 58-61; instr, Auburn Univ, 61-62; ASSOC PROF MATH, WESTERN KY UNIV, 62- Mem: Math Asn Am. Res: Importance sampling. Mailing Add: Dept of Math Western Ky Univ Bowling Green KY 42101

STOKES, LEE W, b Boise, Idaho, May 19, 39; m 60; c 2. BIOLOGY, WATER POLLUTION. Educ: Univ Idaho, BS, 61; Univ Minn, MPH, 66, PhD(environ biol), 69. Prof Exp: Microbiologist, Mont Bd Health, 61-62; microbiologist, Idaho Dept Health, 63-65, consult limnologist, 69-70, chief water pollution control sect, 71-74, ADMINR DIV ENVIRON, IDAHO DEPT HEALTH & WELFARE, 74- Concurrent Pos: Lectr, Boise State Col, 70- Mem: Am Soc Limnol & Oceanog; Water Pollution Control Fedn. Res: Primary productivity of periphytic algae. Mailing Add: Idaho Dept of Health & Welfare Div of Environ Statehouse Boise ID 83720

STOKES, PETER E, b Haddonfield, NJ, Aug 27, 26; m 56; c 3. MEDICINE, PSYCHIATRY. Educ: Trinity Col, BS, 48; Cornell Univ, MD, 52; Am Bd Internal Med, dipl, 59; Am Bd Psychiat & Neurol, dipl, 71; Am Bd Radiol, dipl & cert nuclear med, 72. Prof Exp: Intern med, New York Hosp, 52-53, asst resident, 53-54, asst resident med & endocrinol 54-55; NIH trainee fel, 55-57; instr med in endocrinol, 57-59, asst prof med in psychiat, 59-63, ASSOC PROF PSYCHIAT & MED, MED COL, CORNELL UNIV, 63- Concurrent Pos: Physician, Outpatient Clin, New York Hosp, 55-56, from asst attend to assoc attend physician, 61-; dir clin res labs, Payne Whitney Psychiat Clin, New York Hosp-Cornell Med Ctr, 57-, dir psychobiol study unit, 67, assoc attend psychiatrist, 69; assoc vis physician, Cornell Div, Bellevue Hosp, 58-; pvt pract consult, 59- Mem: Endocrine Soc; Am Soc Nuclear Med; Am Fedn Clin Res; NY Acad Sci; fel Am Col Physicians. Res: Neuroendocrine function in emotional disorders; hypothalamic pituitary adrenocortical function control systems in animals; alcoholism and the effects of alcohol on neuroendocrine function; calcium, strontium and creatinine metabolism; problems in growth thyroid; adrenal and ovarian function; clinical psychiatric problems, especially affective disorders; lithium metabolism. Mailing Add: Payne Whitney Clin New York Hosp-Cornell Med Ctr New York NY 10021

STOKES, RICHARD HIVLING, b Troy, Ohio, Apr 30, 21; m 56. EXPERIMENTAL PHYSICS. Educ: Case Univ, BS, 42; Iowa State Univ, PhD(physics), 51. Prof Exp: Staff mem, Underwater Sound Ref Lab, Nat Defense Res Comt, 42-44; res assoc, Inst Atomic Res, Iowa State Col, 46-51; staff mem, 51-67, GROUP LEADER, LOS ALAMOS SCI LAB, 67- Concurrent Pos: US del, Conf Peaceful Uses Atomic Energy, Geneva, 58; lectr, Univ Minn, 60-61. Mem: Fel Am Phys Soc. Res: Nuclear physics; accelerator research; fission; spectroscopy of light nuclei; nuclear detectors; heavy ion accelerators; heavy ion reactions. Mailing Add: 2450 Club Rd Los Alamos NM 87544

STOKES, ROBERT ALLAN, b Richmond, Ky, June 25, 42; m 63. ASTROPHYSICS. Educ: Univ Ky, BS, 64; Princeton Univ, MA, 66, PhD(physics), 68. Prof Exp: Asst

prof physics & astron, Univ Ky, 68-72; sr scientist, 72-74, MGR SPACE SCI, PAC NORTHWEST LABS, BATTELLE MEM INST, 74-; ASSOC PROF PHYSICS & ASTRON, UNIV KY, 73- Concurrent Pos: Adv comt, Geophys Inst, Univ Alaska, 75- Mem: AAAS; Am Phys Soc; Am Astron Soc. Res: Experimental cosmology and relativity; planetary astrophysics; radiative transfer theory. Mailing Add: Battelle Observ Battelle-Northwest Richland WA 99352

STOKES, ROBERT MITCHELL, b Vandalia, Ill, May 21, 36; m 59; c 3. COMPARATIVE PHYSIOLOGY. Educ: Mich State Univ, BS, 58, MS, 59, PhD(physiol), 63. Prof Exp: Asst prof, 63-70, ASSOC PROF BIOL SCI, KENT STATE UNIV, 70- Concurrent Pos: Consult, Great Lakes Basin Comn Water Qual Task Force, 68-70. Mem: Am Soc Zool. Res: Biochemical and biophysical aspects of membrane transport phenomena in fish, especially glucose transport by intestine; fish physiology, metabolism and toxicology. Mailing Add: Dept of Biol Sci Kent State Univ Kent OH 44242

STOKES, RUSSELL AUBREY, b Preston, Miss, May 1, 22; m 59. MATHEMATICS. Educ: Miss State Univ, BS, 48; Univ Miss, MA, 51; Univ Tex, PhD(math), 63. Prof Exp: From asst prof to assoc prof, 56-66, PROF MATH, UNIV MISS, 66- Mem: Am Math Soc; Math Asn Am. Res: Measure and integration. Mailing Add: Dept of Math Univ of Miss University MS 38677

STOKES, WILLIAM GLENN, b Corsicana, Tex, Dec 26, 21; m; c 2. MATHEMATICS. Educ: Sam Houston State Col, BS, 46, MA, 47; Peabody Col, PhD(math), 57. Prof Exp: Head dept math, Navarro Jr Col, 47-53; instr appl math, Vanderbilt Univ, 54-55; head dept math, Austin Peay State Col, 55-57; assoc prof, Northwestern State Col, 57-59 & East Tex State Univ, 59-60; assoc prof, 60-74, PROF MATH, AUSTIN PEAY STATE UNIV, 74- Mem: Math Asn Am; Am Math Soc. Mailing Add: Dept of Math Austin Peay State Univ Clarksville TN 37040

STOKES, WILLIAM LEE, b Hiawatha, Utah, Mar 27, 15; m 39; c 4. STRATIGRAPHY. Educ: Brigham Young Univ, BS, 37, MS, 38; Princeton Univ, PhD(geol), 41. Prof Exp: Asst, Princeton Univ, 41-42; from jr geologist to asst geologist, US Geol Surv, 42-46; from asst prof to assoc prof geol, 47-54, chmn dept, 54-68, PROF GEOL, UNIV UTAH, 54-; ASSOC GEOLOGIST, US GEOL SURV, 46- Concurrent Pos: Consult, USAEC, 52-54. Mem: AAAS; fel Geol Soc Am; Soc Vert Paleont; Am Asn Petrol Geol; Am Geophys Union. Res: Mesozoic stratigraphy; dinosaurs; sedimentary structures; uranium deposits; textbook writing and popularization of earth science. Mailing Add: 1354 Second Ave Salt Lake City UT 84103

STOKES, WILLIAM MOORE, b Cleveland, Ohio, Sept 18, 21. ORGANIC CHEMISTRY. Educ: Franklin & Marshall Col, BS, 44; Yale Univ, PhD(chem), 52. Hon Degrees: MA, Providence Col, 61. Prof Exp: Chemist, Hamilton Watch Co, 44-46; lab asst, Yale Univ, 46-48, 49-51; from asst prof to assoc prof med res, 51-59, PROF CHEM & DIR MED RES LAB, PROVIDENCE COL, 59- Mem: Fel AAAS; Am Chem Soc; Am Oil Chem Soc; NY Acad Sci. Res: Neurochemistry; isolation of natural products; steroid metabolism; correlation of optical activity with molecular structure. Mailing Add: Dept of Chem Providence Col Providence RI 02918

STOKINGER, HERBERT ELLSWORTH, b Boston, Mass, June 19, 09. TOXICOLOGY. Educ: Harvard Univ, AB, 30; Columbia Univ, PhD(biochem), 37. Prof Exp: Instr chem, City Col New York, 32-39; res assoc bact, Sch Med & Dent, Univ Rochester, 39-43, chief indust hyg sect, AEC, 43-51, from asst prof to assoc prof pharm & toxicol, 45-51; CHIEF TOXICOLOGIST, NAT INST OCCUP SAFETY & HEALTH, USPHS, 51- Concurrent Pos: Res assoc, Col Physicians & Surgeons, Columbia Univ, 37-39; res assoc, Atomic Bomb Test, Bikini, 46; mem subcomt toxicol, Nat Res Coun, 46, chmn comt, 70-73; chmm subcomt toxicol, USPHS Drinking Water Stas, 58-70; chmn threshold limits comt, Am Conf Govt Indust Hygenists, 62- Honors & Awards: Award of Merit, Am Conf Govt Indust Hygienists, 58, Meritorious Achievement Award, 65; Donald E Cummings Mem Award for outstanding contrib to knowledge & pract of indust hygiene, Am Indust Hygiene Asn, 69; Eminent Chemist Award, Cincinnati Am Chem Soc, 70; Super Serv Award, Dept Health, Educ & Welfare, 72; Award of Merit, Soc of Toxicol, 73; S C Weisfeld Mem Lect Award, 75. Mem: AAAS; Soc Exp Biol & Med; Am Chem Soc; Am Indust Hyg Asn; Am Asn Immunol. Res: Pharmacology and toxicology of atomic energy materials; bacteriological chemistry of gonococcus; toxins; chemotherapy of sulfonamides and arsenicals; prophylaxis of industrial poisons; industrial, water and air pollution toxicology. Mailing Add: Nat Inst Occup Safety & Health 1014 Broadway Cincinnati OH 45202

STOKOWSKI, STANLEY E, b Lewiston, Maine, Dec 28, 41; m 64; c 3. SOLID STATE PHYSICS. Educ: Mass Inst Technol, SB, 63; Stanford Univ, PhD(physics), 68. Prof Exp: Physicist, Nat Bur of Stand, 68-70; mem tech staff, Bell Tel Labs, NJ, 70-72; PRIN INVESTR, RES INST ADVAN STUDIES DIV, MARTIN MARIETTA CORP, 72- Mem: Am Phys Soc. Res: Crystal field theory; phase transitions; ferroelectricity; color centers; optical properties of crystals; infrared detectors. Mailing Add: Res Inst for Advan Studies Martin Marietta Corp Baltimore MD 21227

STOKSTAD, EVAN LUDVIG ROBERT, b China, Mar 6, 13; US citizen; m 34; c 2. BIOCHEMISTRY. Educ: Univ Calif, BS, 34, PhD(animal nutrit), 37. Prof Exp: Biochemist, Western Condensing Co, 37-39; biochemist, Golden State Co, Ltd, 39-40; Lalor fel, Calif Inst Technol, 40-41; chemist, Lederle Labs Div, Am Cyanamid Co, 41-63; actg chmn dept nutrit sci, 68-69, PROF NUTRIT, UNIV CALIF, BERKELEY & BIOCHEMIST, AGR EXP STA, 63- Concurrent Pos: Mem food & nutrit bd, Div Biol & Agr, Nat Acad Sci-Nat Res Coun, 68-72, comt food standards & fortification policy, 72. Honors & Awards: Borden Award, Poultry Sci Asn; Mead-Johnson Award, Am Inst Nutrit. Mem: Soc Exp Biol & Med; Am Soc Biol Chemists; Am Chem Soc; Poultry Sci Asn; Am Inst Nutrit (treas, 70-73, pres-elect, 75). Res: Water soluble vitamin requirements for chicks; antibiotics in animal nutrition; bacterial nutrition; chemistry and biochemistry of thioctic acid; dental nutrition and mineral metabolism; chemistry and metabolism of folic acid and vitamin B12. Mailing Add: Dept of Nutrit Sci Univ of Calif Berkeley CA 94720

STOLAR, MORRIS EMMANUEL, b Alexandria, Egypt, Mar 28, 29; US citizen; m 61; c 2. PHARMACY, PHARMACOLOGY. Educ: Columbia Univ, BS, 53, MS, 55; Philadelphia Col Pharm & Sci, PhD(pharm), 58. Prof Exp: Sr res assoc pharm & pharmacol, Warner-Lambert Res Inst, 58-64; dir pharm develop, Dome Labs Div, Miles Labs, Inc, 64-72; DIR PHARMACEUT DEVELOP, ABIC LTD, ISRAEL, 71- Mem: Am Chem Soc; Am Pharmaceut Asn; Acad Pharmaceut Sci; Soc Cosmetic Chem; NY Acad Sci. Res: Effect of various ointment bases on the percutaneous absorption of salicylates through intact rabbit skin; evaluation of certain factors influencing oil deposition on skin after immersion in an oil bath; bath oils. Mailing Add: ABIC Ltd 3 Hayozma St Ramat Gan Israel

STOLARSKY, KENNETH B, b Chicago, Ill, May 9, 42; m 69. MATHEMATICS. Educ: Calif Inst Technol, BS, 63; Univ Wis-Madison, MS, 65, PhD(math), 68. Prof

Exp: Fel math, Inst Advan Study, 68-69; asst prof, 69-73, ASSOC PROF MATH, UNIV ILL, URBANA, 73- Mem: Am Math Soc; Math Asn Am. Res: Number theory; combinatorics; geometric inequalities. Mailing Add: Dept of Math Univ of Ill at Urbana-Champaign Urbana IL 61801

STOLBACH, LEO LUCIEN, b Geneva, Switz, Feb 25, 33; US citizen; m 61; c 2. ONCOLOGY, INTERNAL MEDICINE. Educ: Harvard Univ, BA, 54; Univ Rochester, MD, 58; Am Bd Internal Med, cert, 67. Prof Exp: Intern med, Univ Hosp, Cleveland, 58-59, resident, 59-60; clin assoc, Endocrinol Br, NIH, 60-62, sr investr, 62-63; resident, Univ Hosp, Cleveland, 63-64; res physician oncol, Lemuel Shattuck Hosp, Boston, 64-70; CHIEF MED, PONDVILLE HOSP, 70- Concurrent Pos: From instr to asst prof, Sch Med, Tufts Univ, 64-73, assoc prof, 73- Mem: Am Asn Cancer Res; Am Soc Clin Oncol; fel Am Col Physicians; Am Asn Cancer Educ. Res: Clinical chemotherapy; tumor immunology; tumor markers. Mailing Add: Pondville Hosp Box 111 Walpole MA 02081

STOLBERG, HAROLD JOSEF, b San Juan, PR, Aug 4, 40; m 71. MATHEMATICS. Educ: Univ PR, Rio Piedras, BS, 62; Cornell Univ, PhD(math), 69. Prof Exp: Asst prof math, Ithaca Col, 67-68; fel, Carnegie-Mellon Univ, 68-69, asst prof, 69-71; ASSOC PROF MATH & CHMN DEPT, UNIV PR, RIO PIEDRAS, 71- Concurrent Pos: Consult, Col Entrance Exam Bd, PR, 71-74. Mem: Am Math Soc; Math Asn Am; AAAS. Res: Commutative algebra; flat modules; mathematics education. Mailing Add: Dept of Math Univ of PR Rio Piedras PR 00931

STOLBERG, MARVIN ARNOLD, b New York, NY, Oct 29, 25; m 49; c 2. ORGANIC CHEMISTRY. Educ: Columbia Univ, BS, 50; Univ Del, MS, 54, PhD(chem), 56. Prof Exp: Org chemist, Chemother Br, US Army Chem Ctr, Md, 50-53, asst br chief, 54-56; head chem dept, Tracerlab, Inc, Mass, 56-60; tech dir-vpres, 60-72, PRES & CHIEF EXEC OFF, NEW ENGLAND NUCLEAR CORP, 72- Mem: Am Chem Soc; Soc Nuclear Med. Res: Organic and inorganic synthesis with radioactive isotopes; radioactive pharmaceuticals and assay of labeled compounds; tracer techniques for solving problems concerning food and drug acceptability criteria; product evaluation; organic reaction mechanisms. Mailing Add: New England Nuclear Corp 575 Albany St Boston MA 02118

STOLC, VIKTOR, b Bratislava, Czech, Oct 5, 32; m 73; c 2. ENDOCRINOLOGY, HEMATOLOGY. Educ: Univ Commenius Bratislava, RNDr, 56; Slovak Acad Sci, Czech, PhD(biochem), 63. Prof Exp: Biochemist, Endocrine Sta, Inst Health, Czech, 56-57; independent scientist, Inst Endocrinol, Slovak Acad Sci, 57-68; res assoc, 65-66 & 68-70, asst res prof, 70-71, RES ASSOC PROF PATH, SCH MED, UNIV PITTSBURGH, 71- Mem: AAAS; Endocrine Soc; Am Soc Biol Chemists; Reticuloendothelial Soc. Res: Development of pituitary-thyroid axis in postnatal rats; regulation of thyroid hormone biosynthesis in isolated thyroid cells, endocrine factors in normal and leukemic leukocytes. Mailing Add: Dept of Path Univ of Pittsburgh Sch of Med Pittsburgh PA 15261

STOLDT, STEPHEN HOWARD, organic chemistry, see 12th edition

STOLEN, JOANNE SIU, b Chicago, Ill, June 22, 43; m 72. IMMUNOLOGY. Educ: Univ Mich, BS, 65; Seton Hall Univ, MS, 68; Rutgers Univ, PhD(biochem), 72. Prof Exp: Res intern immunol, Inst Microbiol, Rutgers Univ, 69-72; fel, Dept Serol & Bact, Univ Helsinki; RES IMMUNOL, SANDY HOOK LAB & UNIV HELSINKI, 74- Mem: Sigma Xi. Res: Cellular immunology; thymus derived and bone marrow derived cell function in the mouse and presently in lower animals such as the fish. Mailing Add: Mid Atlantic Fisheries Ctr Sandy Hook Lab Highlands NJ 07732

STOLEN, ROGERS HALL, b Madison, Wis, Sept 18, 37. SOLID STATE PHYSICS. Educ: St Olaf Col, BA, 59; Univ Calif, Berkeley, PhD(physics), 65. Prof Exp: Fel, Univ Toronto, 64-66; MEM TECH STAFF SOLID STATE OPTICS, BELL LABS, 66- Mem: Am Phys Soc. Res: Optical wave guides; light scattering in glass. Mailing Add: Bell Labs 4B-421 Holmdel NJ 07733

STOLER, DAVID, b Brooklyn, NY, Aug 21, 36; m 58; c 1. ELEMENTARY PARTICLE PHYSICS, QUANTUM OPTICS. Educ: City Col New York, BS, 58; Yeshiva Univ, PhD(physics), 66. Prof Exp: Instr physics, Staten Island Community Col, 59-60; sr res asst, Microwave Res Inst, 60-66; asst prof, 66-73, ASSOC PROF PHYSICS, POLYTECH INST NEW YORK, 73- Concurrent Pos: Instr, Brooklyn Col, 60-61; consult liquid crystal displays, Riher-Maxson Corp, 72. Mem: Am Phys Soc. Res: Electromagnetic properties of hadrons; quantum field theory; quantum theory of coherence; nonlinear optics. Mailing Add: Dept of Physics Polytech Inst of NY 333 Jay St Brooklyn NY 11201

STOLER, PAUL, b Brooklyn, NY, June 8, 38; m 66. EXPERIMENTAL NUCLEAR PHYSICS. Educ: Brooklyn Col, BS, 60; Rutgers Univ, MS, 62, PhD(physics), 66. Prof Exp: ASSOC PROF PHYSICS, RENSSELAER POLYTECH INST, 66- Mem: Am Phys Soc. Res: Nuclear structure physics using techniques of proton scattering, photonuclear reactions and neutron cross section measurements; photopion production from complex nuclei. Mailing Add: Dept of Physics Rensselaer Polytech Inst Troy NY 12181

STOLFI, JULIUS E, b New York, NY, Feb 3, 16; m 43; c 3. INTERNAL MEDICINE. Educ: City Col New York, BS, 36; NY Univ, MS, 38; St Louis Univ, MD, 42; Am Bd Internal Med, dipl, 50. Prof Exp: Clin instr med, 47-55, asst clin prof, 55-61, clin assoc prof, 61-67, clin prof, 67-71, PROF ADMIN MED, COL MED, STATE UNIV NY DOWNSTATE MED CTR, 71-, ASSOC DEAN CLIN AFFAIRS, 68-, VPRES, HOSP AFFAIRS, 72-, DIR, STATE UNIV HOSP, 71- Concurrent Pos: Consult med, Long Island Col Hosp, Brooklyn Vet Admin Hosp, Brooklyn Cumberland Med Ctr, Caledonian & Lutheran Med Ctr. Mem: AAAS; fel Am Col Physicians (vpres). Res: Endocrinology; arthritis; thyroid. Mailing Add: 7902 Colonial Rd Brooklyn NY 11209

STOLFI, ROBERT LOUIS, b Brooklyn, NY, Sept 16, 38; m 68. IMMUNOCHEMISTRY, IMMUNOLOGY. Educ: Brooklyn Col, BS, 60; Univ Miami, PhD(microbiol), 67. Prof Exp: Bact technician, Jewish Hosp Brooklyn, 60-61; asst bacteriologist, Bellevue Hosp, 61-62; res assoc immunochem, Howard Hughes Med Inst, 63-67; from instr to asst prof microbiol, Sch Med, Univ Miami, 67-71; DIR TRANSPLANTATION IMMUNOL, DEPT SURG, CATH MED CTR BROOKLYN & QUEENS INC, 69- Concurrent Pos: Res assoc immunochem, Variety Children's Res Found, Fla, 67-71. Mem: AAAS; Am Asn Immunologists; Am Soc Microbiol. Res: Therapeutic methods for the alteration of immunological reactivity in the tumor-bearing host, or in the recipient of a histoincompatible normal tissue transplant. Mailing Add: St Anthony's Hosp 89-15 Woodhaven Blvd Woodhaven NY 11421

STOLINE, MICHAEL ROSS, bJefferson, Iowa, Sept 17, 40; m 60; c 4. STATISTICAL ANALYSIS. Educ: Univ Iowa, BA, 62, MA, 64, PhD(statist), 67. Prof Exp: ASSOC PROF MATH, WESTERN MICH UNIV & STATIST CONSULT, COMPUT CTR, 67- Concurrent Pos: On leave, Upjohn Co & Univ Calif, Berkeley, 75-76. Mem: Am Statist Asn; Inst Math Statist. Res: Problems in the analysis of variance and regression; multiple comparisons. Mailing Add: 3306 Everett Tower Dept of Math Western Mich Univ Kalamazoo MI 49001

STOLK, JON MARTIN, b Englewood, NJ, Oct 15, 42; m 73. PSYCHIATRY, PHARMACOLOGY. Educ: Middlebury Col, AB, 64; Dartmouth Col, PhD(pharmacol), 69; Stanford Univ, MD, 72. Prof Exp: Fel psychiat, Sch Med, Stanford Univ, 69-71; asst prof pharmacol, 72-73, asst prof psychiat, 74-76, ASSOC PROF PSYCHIAT, DARTMOUTH MED COL, 76- Mem: Am Soc Pharmacol & Exp Therapeut; Int Soc Psychoneuroendocrinol. Res: Biogenic amines and behavior; psychopharmacology; central nervous system polypeptides; neurochemistry. Mailing Add: Dept of Psychiat Dartmouth Med Sch Hanover NH 03755

STOLL, ALICE MARY, b New York, NY, Aug 25, 17. MEDICAL BIOPHYSICS. Educ: Hunter Col, BA, 38; Cornell Univ, MS, 48. Prof Exp: Asst allergy, metab & infrared spectrophotog, New York Hosp & Med Col, Cornell Univ, 38-43, temperature regulation, 46-48, asst physiol, res assoc environ thermal radiation, Med Col & instr sch nursing, 48-53; physiologist med res dept, 53-56, spec tech asst, 56-60, head thermal lab, 60-64, head biophys & bioastronaut div, 64-70, HEAD BIOPHYS LAB, CREW SYSTS DEPT, US NAVAL AIR DEVELOP CTR, 70- Concurrent Pos: Consult, Arctic Aero-Med Lab, Ladd AFB, Alaska, 52-53. Honors & Awards: Fed Civil Serv Award, 65; Achievement Award, Soc Women Engrs, 69; Paul Bert Award, Aerospace Med Asn, 72. Mem: AAAS; Biophys Soc; Am Physiol Soc; fel Aerospace Med Asn; Am Soc Mech Eng. Res: Environmental thermal radiation; temperature regulation and special instrumentation; heat transfer and thermal protection principles. Mailing Add: Crew Systems Dept US Naval Air Develop Ctr Warminster PA 18974

STOLL, MANFRED, b Calw, Ger, Aug 24, 44; US citizen; m 66; c 1. MATHEMATICAL ANALYSIS. Educ: State Univ NY Albany, BS, 67; Pa State Univ, MA, 69, PhD(math), 71. Prof Exp: ASST PROF MATH, UNIV SC, 71- Mem: Am Math Soc. Res: Harmonic, holomorphic and plurisubharmonic function theory on bounded symmetric domains and generalized half planes; spaces and algebras of holomorphic functions of one and several complex variables. Mailing Add: Dept of Math & Comput Sci Univ of SC Columbia SC 29208

STOLL, ROBERT ROTH, b Pittsburgh, Pa, May 19, 15; m 37; c 3. MATHEMATICS. Educ: Univ Pittsburgh, BS, 36, MS, 37; Yale Univ, PhD(math), 43. Prof Exp: Instr math, Rensselaer Polytech Inst, 37-39; instr, Yale Univ, 39-42; asst prof, Williams Col, 42-46; from asst prof to assoc prof, Lehigh Univ, 46-52; prof, Oberlin Col, 52-71, actg chmn dept, 53-54, 60-61 & 66-67; PROF MATH & CHMN DEPT, CLEVELAND STATE UNIV, 71- Concurrent Pos: Nat Res Coun fel, Ind Univ, 45-46; consult, NSF, 52 & 70; NSF fac fel, Calif Inst Technol, 58-59; fel Mass Inst Technol, 67-68; vis Fulbright lectr, Am Univ Beirut, 64-65. Mem: Am Math Soc; Math Asn Am. Res: Theory of semi-groups; foundation of mathematics. Mailing Add: Dept of Math Cleveland State Univ Cleveland OH 44115

STOLL, WILHELM, b Freiburg, Ger, Dec 22, 23; m 55; c 4. MATHEMATICS. Educ: Univ Tübingen, Dr rer nat, 53, Dr habil, 54. Prof Exp: Asst math, Univ Tübingen, 53-59, docent, 54-60, appl prof, 60; head dept math, 66-68, PROF MATH, UNIV NOTRE DAME, 60- Concurrent Pos: Vis lectr, Univ Pa, 54-55; mem, Inst Adv Study, 57-59; vis prof, Stanford Univ, 68-69 & Tulane Univ, 73. Mem: Am Math Soc; Math Asn Am; Ger Math Asn. Res: Complex analysis; value distribution in several variables, modifications mero- morphic maps; families of divisors; continuation of analytic sets and maps; algebraic dependence of meromorphic functions. Mailing Add: Dept of Math Univ of Notre Dame Notre Dame IN 46556

STOLL, WILLIAM FRANCIS, b Lamoni, Iowa, July 21, 32; m 54; c 2. FOOD SCIENCE. Educ: Iowa State Univ, BS, 55, MS, 57; Univ Minn, St Paul, PhD(dairy sci), 66. Prof Exp: Asst prof dairy sci, SDak State Univ, 57-67; SR FOOD SCIENTIST, PROD DEVELOP DEPT, GREEN GIANT CO, 67- Mem: Am Asn Cereal Chemists; Inst Food Technologists; Am Dairy Sci Asn. Res: Use of physical, chemical, microbiological principles for design and fabrication of new food products. Mailing Add: Prod Develop Dept Green Giant Co Le Sueur MN 56058

STOLL, WILLIAM RUSSELL, b Los Angeles, Calif, July 8, 31; m 55; c 2. PHARMACOLOGY, CHEMISTRY. Educ: Union Univ, NY, BS, 52; Univ Rochester, PhD(pharmacol), 56. Prof Exp: From instr to assoc prof, 56-70, RES ASSOC PHARMACOL, ALBANY MED COL, 64- Mem: Am Chem Soc. Res: Chemical nature of sodium and carbonate in bone mineral; autonomic pharmacology; nature of 2-halo-2-phenethylamines; applications of nucleonics in biological research. Mailing Add: Dept of Pharmacol Albany Med Col Albany NY 12208

STOLLAR, BERNARD DAVID, b Saskatoon, Sask, Aug 11, 36; m 56; c 3. IMMUNOLOGY. Educ: Univ Sask, BA, 58, MD, 59. Prof Exp: Res fel biochem, Brandeis Univ, 60-62; dep chief biol sci div, Air Force Off Sci Res, 62-64; asst prof pharmacol, 64-67, from asst prof to assoc prof biochem, 67-74, PROF BIOCHEM, SCH MED, TUFTS UNIV, 74- Concurrent Pos: NSF grant, 64-; consult, Biol Sci Div, Air Force Off Sci Res, 66-69; sr fel, Weizmann Inst Sci, 71-72. Mem: AAAS; Am Asn Immunologists. Res: Immunochemistry of nucleic acids and nucleoprotein, especially in relation to auto-immune disease; complement fixation properties of immunoglobulin classes. Mailing Add: Dept of Biochem & Pharmacol Tufts Univ Sch of Med Boston MA 02111

STOLLAR, VICTOR, b Saskatoon, Sask, Dec 6, 33; m 67; c 1. MICROBIOLOGY, VIROLOGY. Educ: Queen's Univ, Ont, MDCM, 56. Prof Exp: Fel, Brandeis Univ, 58-62; fel Weizman Inst Sci, 62-65; from asst prof to assoc prof, 65-75, PROF MICROBIOL, RUTGERS MED SCH, COL MED & DENT NJ, 75- Mem: AAAS; Am Soc Microbiol. Res: Replication of toga viruses in vertebrate and in insect cells. Mailing Add: Dept Microbiol Rutgers Med Sch Col of Med & Dent of NJ Piscataway NJ 08854

STOLLBERG, ROBERT, b Toledo, Ohio, May 27, 15; m 43; c 4. SCIENCE EDUCATION. Educ: Univ Toledo, BS, 35, BEd, 36; Columbia Univ, MA, 40, EdD(sci ed, electronics), 47. Prof Exp: Instr, Rossford High Sch, Ohio, 36-39; asst prof physics, Wabash Col, 46-47; ed, Purdue Univ, 47-49; assoc prof physics, 49-59, chmn dept interdisciplinary phy sci, 67-69, PROF PHYSICS, SAN FRANCISCO STATE UNIV, 59-, ASSOC DEAN, SCH NATURAL SCI, 69, 71- Concurrent Pos: Mem, Harvard Univ Conf Prob Sci Ed, 53; chmn, Nat Conf Prob High Sch Sci, 59; mem, President's Comt Develop Scientists & Engrs; sci adv, Columbia Univ team, India, 65-66; Columbia Univ-USAID contract lectr, Inst Educ, Makerere Univ, Uganda, 69-71; vis prof, Columbia Univ. Mem: Fel AAAS; Am Asn Physics Teachers; Nat Sci Teachers Asn (pres, 55-56); Am Inst Physics. Mailing Add: Off Assoc Dean Sch Natural Sci San Francisco State Univ San Francisco CA 94132

STOLLER, BENJAMIN BORIS, b Apr 10, 07; m 40. AGRICULTURAL MICROBIOLOGY, BIOCHEMISTRY. Educ: Iowa State Col, MS, 36; Univ Wis, PhD(biochem), 45. Prof Exp: Dir res, L F Lambert, Inc, Pa, 36-43; dir res, Lenny's Food, Inc, Minn, 45-47; head malting res, Pabst Brewing Co, Wis, 49-50; dir res,

West Foods, Calif, 55-56; OWNER, STOLLER RES CO, 57- Mem: AAAS; Am Chem Soc; Am Soc Microbiol. Res: Growth of plants under industrial conditions; mushroom and bean sprout culture; malting; insecticides; fungicides. Mailing Add: PO Box 1071 Santa Cruz CA 95060

STOLLER, EDWARD W, b McCook, Nebr, Jan 9, 37; m 60; c 2. PLANT PHYSIOLOGY, WEED SCIENCE. Educ: Univ Nebr, BS, 58; Purdue Univ, MS, 62; NC State Univ, PhD(soil fertil), 66. Prof Exp: PLANT PHYSIOLOGIST, N CENT REGION, AGR RES SERV, USDA, 65- Mem: AAAS; Am Soc Plant Physiol; Am Soc Agron; Weed Sci Soc Am. Res: Weed physiology and control. Mailing Add: Dept of Agron 215 Davenport Hall Univ Ill Urbana IL 61801

STOLLERMAN, GENE HOWARD, b New York, NY, Dec 6, 20; m 45; c 3. MEDICINE. Educ: Dartmouth Col, AB, 41; Columbia Univ, MD, 44; Am Bd Internal Med, dipl, 52. Prof Exp: From intern to chief med resident, Mt Sinai Hosp, New York, 44-49; res fel microbiol, Col Med, NY Univ, 49-50, instr med, 51-55; instr, Col Med, State Univ NY Downstate Med Ctr, 50-51; from asst prof to prof, Med Sch, Northwestern Univ, 55-64; PROF MED & CHMN DEPT, COL MED, UNIV TENN, MEMPHIS, 65- Concurrent Pos: Med dir, Irvington Hosp, New York, 51-55; prin investr, Sackett Found Res Rheumatic Fever & Allied Dis, 55-64; physician-in-chief, City Memphis Hosps, 65-; consult, Memphis Vet Admin Hosp, 65-; chmn coun rheumatic fever & congenital heart dis, Am Heart Asn, 65-; mem, Am Bd Internal Med, 67-73, chmn written exam comt, 70- & exec comt, 72-73; ed, Advan Internal Med, 68-; attend physician, Vet Admin Res Hosp & Passavant Mem Hosp, New York; mem, Expert Comt Cardiovasc Dis, WHO; pres, Cent Soc Clin Res, 74-75. Mem: Asn Profs Med (pres, 75-76); Asn Am Physicians; fel Am Col Physicians; Am Soc Clin Invest; Am Asn Immunologists. Res: Infectious and rheumatic diseases; biology of streptococcus; etiology of rheumatic fever. Mailing Add: 951 Court Ave Rm 339M Memphis TN 38163

STOLMAN, ABRAHAM, b New York, NY, Oct 6, 08; m 44. TOXICOLOGY. Educ: Rutgers Univ, BSc, 32; Univ Edinburgh, PhD(toxicol), 48. Prof Exp: Chemist, Rockefeller Inst, 30-31; asst, Med Sch, Columbia Univ, 32-41; biochemist, Med Sch NY Univ, 41-43; toxicologist, City of Kingston Lab, NY, 50; STATE LEGAL TOXICOLOGIST, STATE DEPT HEALTH, CONN, 50- Concurrent Pos: Asst prof, Med Sch, Univ Conn. Mem: AAAS; Am Chem Soc; Am Acad Forensic Sci; NY Acad Sci. Res: Forensic toxicology. Mailing Add: State Dept of Health Box 1689 Hartford CT 06101

STOLOFF, IRWIN LESTER, b Philadelphia, Pa, May 9, 27; m 52; c 3. MEDICINE. Educ: Jefferson Med Col, MD, 51; Am Bd Internal Med, dipl, 58. Prof Exp: From intern to resident med, Jefferson Med Col Hosp, 51-53; resident, Baltimore City Hosps, Md, 53-54 & Mt Sinai Hosp, New York, 56-57; fel med, 57-58, from instr to asst prof, 59-69, ASSOC PROF MED & PREV MED, JEFFERSON MED COL, 69- Mem: Fel Am Col Physicians; fel Am Col Chest Physicians. Res: Immunology; autoimmune diseases; cancer immunology; epidemiology of chronic lung disease. Mailing Add: Dept of Med Jefferson Med Col 1025 Walnut St Philadelphia PA 19107

STOLOFF, LEONARD, b Boston, Mass, Mar 24, 15; m 40. BIOCHEMISTRY. Educ: Mass Inst Technol, BS, 36. Prof Exp: Chemist, Granada Wines, Inc, 36-37; self employed, 38-41; chemist, US Dept Navy, 41; chemist, Consumer's Union, 42; res chemist, Agar Substitute Prog, US Fish & Wildlife Serv, 42-44; res chemist, Krim-Ko Corp, 44-51; res dir, Seaplant Chem Corp, 51-59; asst tech dir, Marine Colloids, Inc, 59-63; chief mycotoxins & enzymes sect, Div Food Chem, 63-71, NATURAL TOXICANTS SPECIALIST, BUR FOODS, FOOD & DRUG ADMIN, 71- Mem: AAAS; Am Chem Soc; Inst Food Technol. Res: Chemistry, toxicology and occurrence of natural poisins in foods. Mailing Add: 13208 Bellevue St Silver Spring MD 20904

STOLOV, HAROLD L, b New York, NY, May 27, 21. PHYSICS, METEOROLOGY. Educ: City Col New York, BS, 42; Mass Inst Technol, MS, 47; NY Univ, PhD(physics, meteorol), 53. Prof Exp: Asst radio & electronics, Signal Corps, US Dept Army, 42; instr physics, City Col New York, 47-50; lectr, Hunter Col, 50-51; res assoc, NY Univ, 50-53; instr, Douglass Col, Rutgers Univ, 51-53, asst prof, 53-59; from asst prof to assoc prof, 59-70, PROF PHYSICS, CITY COL NEW YORK, 70- Concurrent Pos: Consult, Martin Co, 56-59; consult, Res & Adv Develop Div, Avco Corp, 60-61; Nat Acad Sci-Nat Res Coun sr res associateship, Inst Space Studies, New York, 65-67. Mem: Am Phys Soc; Am Meteorol Soc; Am Asn Physics Teachers; Am Geophys Union. Res: Physics of the upper atmosphere; tidal oscillations; physics education; magnetosphere; solar-terrestrial physics. Mailing Add: 2575 Palisade Ave New York NY 10463

STOLOVY, ALEXANDER, b Brooklyn, NY, Nov 21, 26; m 55; c 3. NUCLEAR PHYSICS. Educ: Brooklyn Col, BS, 48; Calif Inst Technol, MS, 50; NY Univ, PhD(physics), 55. Prof Exp: Nuclear physicist, Brookhaven Nat Lab, 53-54; NUCLEAR PHYSICIST, RADIATION TECHNOL DIV, NAVAL RES LAB, 55- Concurrent Pos: Partic guest scientist, Lawrence Livermore Lab, 74-75. Mem: Am Phys Soc; Sigma Xi. Res: Slow neutron spectroscopy; neutron capture gamma ray studies using time-of-flight techniques. Mailing Add: Radiation Technol Div Naval Research Lab Washington DC 20375

STOLOW, NATHAN, b Montreal, Que, May 4, 28; m 50; c 2. CHEMISTRY. Educ: McGill Univ, BSc, 49; Univ Toronto, MA, 52; Univ London, PhD(conserv), 56. Prof Exp: Res chemist, Nat Res Coun Can, 49-50; vis lectr chem & physics, Sir John Cass Col, Univ London, 52-55, res assoc conserv, Courtauld Inst Art, 55-56; dir conserv, Nat Gallery Can, 56-72, dir Can Conserv Inst, 72-75, SPEC ADV CONSERV, NAT MUS CAN, 76- Concurrent Pos: Carnegie travel grant, Nat Gallery Can, 56-57; rapporteur, Comt Conserv, Int Coun Mus, 64-, chmn Can nat comt, 70-; mem coun, Int Inst Conserv Hist & Artistic Works, 72- Honors & Awards: Can Medal, Govt Can, 67. Mem: Fel Chem Inst Can. Res: Museum conservation; problems of deterioration in works of art related to conservation; solution of museological problems by chemical and physical approaches; interaction of art history and scientific research; exhibition conservation research. Mailing Add: Spec Adv Conserv Nat Mus of Can Ottawa ON Can

STOLOW, ROBERT DAVID, b Boston, Mass, Mar 9, 32; m 53; c 3. ORGANIC CHEMISTRY. Educ: Mass Inst Technol, SB, 53; Univ Ill, PhD(chem), 56. Prof Exp: From instr to asst prof, 58-64, ASSOC PROF ORG CHEM, TUFTS UNIV, 64- Concurrent Pos: Fel, Calif Inst Technol, 67-68. Mem: Am Chem Soc; The Chem Soc. Res: Physical organic chemistry; stereochemistry; conformational analysis; the hydrogen bond. Mailing Add: Dept of Chem Tufts Univ Medford MA 02155

STOLPE, STANLEY GEORGE, b DeKalb, Ill, May 7, 12; m 53; c 3. ENDOCRINOLOGY. Educ: Northern Ill State Teachers Col, BE, 34; Univ Iowa, MS, 38, PhD(zool), 48. Prof Exp: Instr mammalian physiol, 48-53, asst prof physiol & anat, 53-60, ASSOC PROF PHYSIOL, UNIV ILL, URBANA, 60- Mem: AAAS. Res: Ovary and adrenal transplantation studies; 17-keto-steroids in nutritional and climatic stress; mammalian sex determination; ergonomics. Mailing Add: Dept of Physiol & Biophys Univ of Ill Urbana IL 61801

STOLTEN, HANS JOSEPH, b New York, NY, June 25, 22; m 58; c 2. PHYSICAL CHEMISTRY. Educ: NY Univ, PhD(chem), 51. Prof Exp: Asst chem, NY Univ, 47-51; res analyst, 51-58, MGR ANAL DEPT, CENT RES LAB, GAF CORP, 58- Mem: Am Chem Soc. Res: Instrumental methods of analysis; analytical chemistry. Mailing Add: RD 1 Bangor PA 18013

STOLTENBERG, CARL H, b Monterey, Calif, May 17, 24; m 49; c 5. FOREST ECONOMICS. Educ: Univ Calif, BS, 48, MF, 49; Univ Minn, PhD(agr econ), 52. Prof Exp: Instr forestry, Univ Minn, 49-51; asst prof forest econ, Duke Univ, 51-56; head resource econ res, Northwest Forest Exp Sta, US Forest Serv, Pa, 56-58, chief div forest econ res, 58-60; prof forestry & head dept, Iowa State Univ, 60-67; PROF FORESTRY, DEAN SCH FORESTRY & DIR FOREST RES LAB, ORE STATE UNIV, 67- Concurrent Pos: Mem forestry res comn, Nat Acad Sci, 63-65; mem, Nat Adv Bd Coop Forestry Res, 62-66, Ore Bd Forestry, 67-, chmn, 74-; mem Secy of Agr State & Pvt Forestry Adv Comn, 70-74; chmn, Ore & Calif Adv Bd, Bur Land Mgt, 72- Mem: Fel Soc Am Foresters; Am Econ Asn; Forest Prod Res Soc; Sigma Xi; AAAS. Res: Economic analysis of forest management alternatives; forest policy; resource allocation in forestry; natural resource policy. Mailing Add: Sch of Forestry Ore State Univ Corvallis OR 97331

STOLTMAN, JAMES BERNARD, b Minneapolis, Minn, Feb 6, 35; m 60; c 3. ANTHROPOLOGY, ARCHAEOLOGY. Educ: Univ Minn, BA, 57, MA, 62; Harvard Univ, PhD(anthrop), 67. Prof Exp: Instr anthrop, Tulane Univ La, 65-66; from asst prof to assoc prof, 66-75, PROF ANTHROP, UNIV WIS-MADISON, 75- Concurrent Pos: Pres, Wis Archaeol Surv, 71-73; gov appointee, Hist Preserv Rev Bd, Wis, 72- Mem: AAAS; Soc Am Archaeol; fel Am Anthrop Asn; fel Current Anthrop. Res: North American and Mesoamerican prehistory; environmental archaeology. Mailing Add: Dept of Anthrop Univ of Wis Madison WI 53706

STOLTZ, LEONARD PAUL, b Kankakee, Ill, Dec 5, 27; m 53; c 5. HORTICULTURE. Educ: Agr & Mech Col, Tex, BS, 55; Ohio State Univ, MS, 56; Purdue Univ, PhD(hort), 65. Prof Exp: Res assoc floricult, Rutgers Univ, 57-60; asst prof hort, Univ RI, 60-62; asst prof, 65-70, ASSOC PROF HORT, UNIV KY, 70- Concurrent Pos: USDA res grant, 66-70. Honors & Awards: Kenneth Post Award, Cornell Univ, 67; L M Ware Award, Am Soc Hort Sci, Southern Region, 68, 72. Mem: Am Soc Hort Sci; Int Plant Propagators Soc. Res: Plant propagation; naturally occurring chemical components of plants; tissue and embryo culture; marketing of ornamental plants; turf management. Mailing Add: Dept of Hort Univ of Ky Lexington KY 40506

STOLTZ, ROBERT LEWIS, b Bakersfield, Calif, May 15, 45; m 71. ENTOMOLOGY. Educ: Univ Calif, Davis, BS, 67; Univ Calif, Riverside, PhD(entom), 73. Prof Exp: Fel entom, Univ Calif, Riverside, 73-74 & Univ Mo-Columbia, 74-75; EXTEN SPECIALIST ENTOM, COOP EXTEN SERV, UNIV IDAHO, 75- Mem: Entom Soc Am; Biol Res Inst Am; Sigma Xi. Res: Insect control, particularly in potatoes, sugar beets, beans, peas, and alfalfa hay; black fly control; black plant bug resistance in grasses. Mailing Add: Univ of Idaho Coop Exten Serv 634 Addison Ave W Twin Falls ID 83301

STOLTZFUS, CONRAD MARTIN, biochemistry, virology, see 12th edition

STOLTZFUS, JOSEPH CHRISTIAN, b Ft Dodge, Iowa, Aug 9, 33; m 56; c 3. NUCLEAR PHYSICS. Educ: Goshen Col, BA, 53; Univ Iowa, MS, 57, PhD(physics), 61. Prof Exp: Asst prof physics, Va Polytech Inst, 60-63; asst prof, Mich State Univ, 63-65; asst prof, 65-71, ASSOC PROF PHYSICS, BELOIT COL, 71- Concurrent Pos: Resident assoc, Argonne Nat Lab, 68-69, consult, 70-72; sci guest, Max Planck Inst Nuclear Physics, 72-73. Mem: Am Phys Soc. Res: Instrumentation for research in nuclear physics. Mailing Add: Dept of Physics Beloit Col Beloit WI 53511

STOLTZFUS, WILLIAM BRYAN, b Martinsburg, Pa, Apr 25, 32; m 57; c 5. ENTOMOLOGY. Educ: Goshen Col, BS, 57; Kent State Univ, MS, 66; Iowa State Univ, PhD(entom), 74. Prof Exp: Instr biol, Eastern Mennonite Col, 66-70 & entom, Iowa State Univ, 73-74; ASSOC PROF BIOL & CHMN DEPT, WILLIAM PENN COL, 74- Mem: Entom Soc Am. Res: Life history of fruitflies as it relates to their taxonomy. Mailing Add: Dept of Biol William Penn Col Oskaloosa IA 52577

STOLWIJK, JAN ADRIANUS JOZEF, b Amsterdam, Netherlands, Sept 29, 27; nat US; m 57. BIOPHYSICS. Educ: State Agr Univ, Wageningen, MS, 51, PhD, 55. Prof Exp: Cabot res fel biol, Harvard Univ, 55-57; from assoc fel to fel biol, 57-74, ASSOC DIR, JOHN B PIERCE FOUND, 74-; PROF EPIDEMIOL, SCH MED, YALE UNIV, 75- Concurrent Pos: From asst prof to assoc prof, Sch Med, Yale Univ, 63-75. Mem: Aerospace Med Asn; Int Soc Biometeorol; Am Physiol Soc; Biophys Soc. Res: Body temperature regulation; regulatory systems in physiology; thermal receptor structures; radiant heat exchange with environment construction and application of mathematical models for study of complex physiological systems; environmental physiology. Mailing Add: John B Pierce Found 290 Congress Ave New Haven CT 06519

STOLZ, HAL FISHER, b Columbia, SC, Nov 21, 34; m 58; c 2. MEDICAL RESEARCH. Educ: Univ Ga, DVM, 58; Univ Rochester, MS, 68; Am Bd Vet Pub Health, dipl, 75. Prof Exp: Asst post vet, US Army, Ft Campbell, Ky, 58-60, area vet, Area Serv Command, SKorea, 61-62, post vet, US Army, Ft Gordon, Ga, 62-64, instr, US Army Qm Sch, Ft Lee, Va, 64-67, chief med effects div, Defense Atomic Support Agency, Washington, DC, 68-71, action officer med res mgt, Off Chief Res & Develop, Dept Army, 72-74, res mgt action officer, US Army Med Res & Develop Command Res Planning Off, 74-76, CHMN BEHAV SCI DEPT, ARMED FORCES RADIOBIOL RES INST, 76- Mem: Am Vet Med Asn. Res: Management of a broad range of military related medical research and planning a coordinated program. Mailing Add: 8115 Langbrook Rd Springfield VA 22152

STOLZBERG, RICHARD JAY, b Winthrop, Mass, Feb 5, 48. ANALYTICAL CHEMISTRY. Educ: Tufts Univ, BS, 69; Mass Inst Technol, PhD(anal chem), 73. Prof Exp: RES ASSOC & PRIN INVESTR, HAROLD EDGERTON RES LAB, NEW ENG AQUARIUM, 73- Mem: Am Chem Soc; Sigma Xi. Res: Characterization of trace metal-organic interactions in natural waters; effect of metal speciation on bioavailability; organism response to trace metal stress; chromatography; electrochemistry; spectroscopy. Mailing Add: Harold Edgerton Res Lab New Eng Aquarium Cent Wharf Boston MA 02110

STOLZE, ROBERT GARDNER, b St Louis, Mo, Oct 31, 27; m 52; c 1. SYSTEMATIC BOTANY. Educ: Univ Notre Dame, BS, 49. Prof Exp: Herbarium asst, 64-66, CUSTODIAN PTERIDOPHYTE HERBARIUM, FIELD MUS NATURAL HIST, 67- Mem: Am Fern Soc (treas, 72-73); Bot Soc Am. Res: Neotropical systematic botany, specializing in Pteridophyta, especially Central American ferns and neo-tropical tree ferns. Mailing Add: Dept of Bot Field Mus of Natural Hist Chicago IL 60605

STOLZENBERG, GARY ERIC, b Southampton, NY, Dec 1, 39; m 69. PESTICIDE CHEMISTRY. Educ: Rensselaer Polytech Inst, BS, 62; Kans State Univ, PhD(biochem), 68. Prof Exp: Asst biochem, Kans State Univ, 62-68; RES CHEMIST, METAB & RADIATION RES LAB, USDA, 68- Mem: Am Chem Soc. Res: Pesticide metabolism in plants; herbicide formulation agents; surfactant analysis. Mailing Add: Metab & Radiation Res Lab US Dept of Agr Fargo ND 58102

STOLZENBERG, SIDNEY JOSEPH, b New York, NY, Nov 30, 27; m 58; c 2. REPRODUCTIVE PHYSIOLOGY, ENDOCRINOLOGY. Educ: NY Univ, BA, 50; Univ Mo, MS, 54; Cornell Univ, PhD(reproductive physiol), 66. Prof Exp: Biochemist, Lederle Labs Div, Am Cyanamid Co, 54-59, agr div, 59-72; ENDOCRINOLOGIST, LIFE SCI DIV, STANFORD RES INST, 72- Mem: AAAS; Am Soc Animal Sci; Am Physiol Soc; Soc Study Reproduction; Am Inst Biol Sci; NY Acad Sci. Res: Reproductive physiology in female and male; includes work on contraceptive activity of new synthetic compounds, intrauterine devices, studies on blastocyst implantation and corpus luteum function. Mailing Add: Life Sci Div Stanford Res Inst Menlo Park CA 94025

STOLZY, LEWIS HAL, b Mich, Dec 11, 20; m 47; c 3. SOIL PHYSICS. Educ: Mich State Col, BS, 48, MS, 50, PhD, 54. Prof Exp: Actg proj supvr, Soil Conserv Serv, USDA, 50-52; asst, Mich State Univ, 52-54; asst irrig engr, 54-61, assoc soil physicist, 61-66, ASSOC PROF SOIL PHYSICS, UNIV CALIF, RIVERSIDE, 66- Concurrent Pos: Fulbright sr res scholar, Univ Adelaide, 64-65. Mem: Soil Sci Soc Am; Am Geophys Union; Am Soc Agron; Am Phytopath Soc; Am Soc Hort Sci. Res: Soil moisture and aeration. Mailing Add: Dept of Soils & Plant Nutrit Univ of Calif Riverside CA 92502

STOMBAUGH, TOM ATKINS, b Vancouver, Wash, Aug 22, 21; m 44; c 4. BIOLOGY. Educ: Ill State Norm Univ, BEd, 41; Univ Ill, MS, 46; Ind Univ, PhD, 53. Prof Exp: Asst prof zool, Eastern Ill State Col, 48-50; PROF BIOL, SOUTHWEST MO STATE UNIV, 53- Mem: Am Soc Mammal. Res: Taxonomy of the voles of sub-genus Pedomys; mammalian taxonomy and ecology. Mailing Add: Dept of Sci Southwest Mo State Univ Springfield MO 65802

STOMBLER, MILTON PHILIP, b New York, Ny, Dec 19, 39; m 67; c 3. EXPERIMENTAL SOLID STATE PHYSICS. Educ: Univ Md, College Park, BS, 62; Univ SC, MS, 66, PhD(physics), 69. Prof Exp: Asst engr, Aerospace Div, Westinghouse Elec Corp, 64-65; fel, Univ Del, 69-71; asst prof physics, 71-73, DIR SPONSORED RES, STATE UNIV NY COL POTSDAM, 73- Mem: Am Phys Soc. Res: Electron paramagnetic resonance. Mailing Add: State Univ of NY Col at Potsdam Potsdam NY 13676

STOMMEL, HENRY MELSON, b Wilmington, Del, Sept 27, 20. OCEANOGRAPHY. Educ: Yale Univ, BS, 42. Hon Degrees: DSc, Gothenburg Univ, 64, Yale Univ, 70, Univ Chicago, 70. Prof Exp: Instr math & astron, Yale Univ, 42-44; res assoc phys oceanog, Oceanog Inst, Woods Hole, 44-59; prof oceanog, Mass Inst Technol, 59-60; prof, Harvard Univ, 60-63; PROF OCEANOG, MASS INST TECHNOL, 63- Mem: Nat Acad Sci; Am Astron Soc; Am Soc Limnol & Oceanog; Am Acad Arts & Sci; Am Geophys Union. Res: Dynamics of ocean currents. Mailing Add: Rm 54-1416 Mass Inst of Technol Cambridge MA 02139

STONE, ADOLF, geography, geography of Europe, see 12th edition

STONE, ALBERT MORDECAI, b Boston, Mass, Dec 24, 13; m 41, 68; c 3. PLASMA PHYSICS, MICROWAVE PHYSICS. Educ: Harvard Univ, AB, 34; Mass Inst Technol, PhD(physics), 38. Prof Exp: Res assoc, Mass Inst Technol, 38-39, staff mem, Radiation Lab, 42-46; instr, Middlesex Col, 36-38; physicist, US Naval Torpedo Sta, 40-41; from asst prof to assoc prof physics, Mont State Col, 41-46; sci liaison officer, US Embassy, London, Eng, 46-48; assoc mem comt electronics & comt basic phys sci, Res & Develop Bd, US Dept Defense, 48-49; tech asst to dir appl physics lab, 49-72, DIR ADVAN RES PROJS, APPL PHYSICS LAB, JOHNS HOPKINS UNIV, 72-, HEAD TECH INFO DIV, 62- Honors & Awards: Cert of Appreciation, Off of Naval Res, 48. Mem: Fel AAAS; fel Am Phys Soc. Res: Electronics; gaseous discharges; radar signal thresholds; guided missiles; countermeasures; controlled thermonuclear plasmas; nuclear effects. Mailing Add: Appl Physics Lab Johns Hopkins Rd Laurel MD 20810

STONE, ALEXANDER GLATTSTEIN, b Hungary, Jan 30, 16; nat US; m 58; c 2. MATHEMATICS. Educ: Debrecen Univ Med, Dr Laws, 40; George Washington Univ, MS, 61. Prof Exp: Mathematician, Repub Aviation Corp, 55-56; mathematician appl physics lab, Johns Hopkins Univ, 56-69, supvr programmers digital comput, 66-69, lectr, univ, 59-60, instr, eve col, 66-67; MEM TECH STAFF, JET PROPULSION LAB, CALIF INST TECHNOL, 69- Mem: Math Asn Am; Asn Comput Mach. Res: Programming for automatic digital computers; Boolean algebra. Mailing Add: Jet Propulsion Lab Calif Inst of Technol Pasadena CA 91103

STONE, ALEXANDER PAUL, b West New York, NJ, June 28, 28; m 60; c 1. MATHEMATICS. Educ: Columbia Univ, BS, 52; Newark Col Eng, MS, 56; Univ Ill, Urbana, PhD(math), 65. Prof Exp: Engr, Western Elec Co, 52-56; instr elec eng, Manhattan Col, 56-58; asst prof physics, Dickinson Col, 58-60; asst prof math, Univ Ill, Chicago Circle, 65-69, assoc prof, 69-70; ASSOC PROF MATH, UNIV NMEX, 70- Mem: Am Math Soc. Res: Differential geometry; applied mathematics. Mailing Add: Dept of Math & Statist Univ of NMex Albuquerque NM 87131

STONE, ARTHUR HAROLD, b London, Eng, Sept 30, 16; m 42; c 2. PURE MATHEMATICS. Educ: Cambridge Univ, BA, 38; Princeton Univ, PhD(math), 41. Prof Exp: Mem, Inst Advan Study, 41-42; instr math, Purdue Univ, 42-44; math physicist, Geophys Lab, Carnegie Inst, 44-46; fel, Trinity Col, Cambridge Univ, 46-47; lectr math, Univ Manchester, 47-56, sr lectr, 56-61; PROF MATH, UNIV ROCHESTER, 61- Mem: Am Math Soc; Math Asn Am. Res: Point-set topology; aerodynamics; graph theory; general topology; descriptive set theory. Mailing Add: Dept of Math Univ of Rochester Rochester NY 14627

STONE, AUDREY LARACK, biochemistry, see 12th edition

STONE, BENJAMIN CLEMENS, III, b Shanghai, China, July 26, 33; nat US; m. SYSTEMATIC BOTANY. Educ: Pomona Col, BA, 54; Univ Hawaii, PhD, 60. Prof Exp: Res asst syst bot, Smithsonian Inst, 60-61; from assoc prof to head dept, Col Guam, 62-65; READER BOT & CUR HERBARIUM, UNIV MALAYSIA, 65- Concurrent Pos: Ed, Micronesica, J Col Guam, 64- Mem: Fel AAAS; Bot Soc Am; Am Soc Plant Taxon; fel Linnaean Soc London. Res: Systematic, morphologic amd geographic studies of vascular plants, especially of Pacific Islands and Southeast Asia; family Pandanaceae, Rutaceae and Araliaceae. Mailing Add: Div of Bot Sch Biol Sci Univ of Malaya Kuala Lumpur Malaysia

STONE, BENJAMIN P, b Dover, Tenn, Aug 28, 35; m 56; c 1. PLANT PHYSIOLOGY. Educ: Austin Peay State Univ, BS, 59; Univ Tenn, Knoxville, MS, 61, PhD(bot), 68. Prof Exp: Asst prof biol, Austin Peay State Univ, 61-65; res partic radiation biol, Cornell Univ, 65-66; asst prof plant physiol, Purdue Univ, West Lafayette, 69; assoc prof biol, 69-72, PROF BIOL, AUSTIN PEAY STATE UNIV, 72- Concurrent Pos: Fel hort, Purdue Univ, West Lafayette, 69. Mem: Bot Soc Am; Am Soc Plant Physiol. Res: Nucleic acid; protein synthesis. Mailing Add: Dept of Biol Austin Peay State Univ Clarksville TN 37040

STONE, BOBBIE DEAN, b Paulton, Ill, June 11, 27; m 50; c 2. SOLID STATE CHEMISTRY, INORGANIC CHEMISTRY. Educ: Univ Southern Ill, BS, 49; Northwestern Univ, PhD(chem), 52. Prof Exp: Res chemist, Mound Lab, 52-53, cent res dept & res & eng div, 53-62 & inorg chem div, 62-65, res group leader, Semiconductor Mat Dept, 65-69, silicon res mgr, 69-72, sr res specialist, Electronics Prod Div, 72-74, ENG FEL, ELECTRONICS PROD DIV, MONSANTO CO, 74- Mem: Am Chem Soc; Am Soc Crystal Growth; AAAS. Res: Semiconductor grade silicon; neutron transmutation duping; III-V compounds. Mailing Add: Electronics Prod Div Monsanto Co PO Box 8 St Peters MO 63376

STONE, CHARLES DEAN, b Athens, Ga, Sept 6, 26; m 50; c 1. FOOD SCIENCE, BIOCHEMISTRY. Educ: Univ Ga, BS, 49, PhD(food sci), 64; Fla State Univ, MS, 59. Prof Exp: Partner, Stone's Ideal Bakery, 50-53; instr baking sci & mgt food serv bakery, Fla State Univ, 53-59, asst prof baking sci & mgt, 59-61; sect mgr food res, Quaker Oats Co, 64-69, mgr cereal res, 69-72, mgr cereals, mixes & corn goods res, 72-73; SR RES SCIENTIST, M&M/MARS, 73- Mem: Am Asn Cereal Chemists; Am Soc Bakery Eng; Inst Food Technologists. Res: Cereal and confectionery; flavor, nutrition, rheology, structural, crystallization, stability, cariogenicity, new products; fermentation; ion-protein interactions as affected by fermentation. Mailing Add: M&M/Mars High St Hackettstown NJ 07840

STONE, CHARLES JOEL, b Los Angeles, Calif, July 13, 36; m 66; c 2. MATHEMATICS, STATISTICS. Educ: Calif Inst Technol, BS, 58; Stanford Univ, PhD(math statist), 61. Prof Exp: Instr math, Princeton Univ, 61-62; asst prof, Cornell Univ, 62-64; from asst prof to assoc prof, 64-69, PROF MATH, UNIV CALIF, LOS ANGELES, 69-, PROF BIOMATH, 75- Concurrent Pos: NSF grant, Univ Calif, Los Angeles, 64-; consult, Rand Corp, 66-67, Planning Res Corp, 66-68, Gen Elec Tech Mil Planning Oper, 68-70, Fed Aviation Admin, 70-74, Consol Anal Ctr Inc, 71-74 & Urban Inst, 75- Mem: Fel Inst Math Statist; Am Math Soc; Am Statist Asn. Res: Probability and statistics, including random walks, birth and death, diffusion and infinitely divisible processes; renewal theory; infinite particle systems; nonparametric estimation and regression. Mailing Add: Dept of Math Univ of Calif Los Angeles CA 90024

STONE, CHARLES PORTER, b Owatonna, Minn, Sept 16, 37; m 59; c 4. WILDLIFE RESEARCH, WILDLIFE ECOLOGY. Educ: Univ Minn, BA, 60; Colo State Univ, MS, 63; Ohio State Univ, PhD(zool), 73. Prof Exp: Res biologist, Patuxent Wildlife Res Ctr, 63-66; asst leader, Ohio Coop Wildlife Res Unit, 66-70; lectr, Ohio State Univ, 70; res biologist, Denver Wildlife Res Ctr, 71-73, asst dir, 73-75, SUPVRY WILDLIFE BIOLOGIST, DENVER WILDLIFE RES CTR, US FISH & WILDLIFE SERV, 73-, CHIEF WILDLIFE ECOL PUB LANDS, 75- Concurrent Pos: Instr wildlife biol, Ohio State Univ, 66-70; mem adj fac, Colo State Univ, 73- Mem: AAAS; Wildlife Soc; Am Ornithologists Union; Am Soc Mammalogists; Wilson Ornith Soc. Res: Effects of energy development, forest and range management practices, and other land disturbances upon wildlife abundance, distribution and behavior; ecology of animal damage to crops. Mailing Add: US Fish & Wildlife Ctr Bldg 16 Denver Wildlife Res Ctr Denver CO 80225

STONE, CLEMENT A, b Hastings, Nebr, May 23, 23; m 52; c 3. PHARMACOLOGY. Educ: Univ Nebr, BSc, 46; Univ Ill, MS, 48; Boston Univ, PhD(physiol), 52. Prof Exp: From instr to asst prof physiol, Sch Med, Boston Univ, 51-54; res assoc, Res Div, Sharp & Dohme, Inc, 54-57, dir pharmacodynamics, Merck Inst Therapeut Res, 56-57, from assoc dir to dir, 58-66, exec dir, 66-71, VPRES, MERCK SHARP & DOHME RES LABS, 71- Concurrent Pos: Lectr, St Andrews, 53. Mem: AAAS; Am Soc Pharmacol & Exp Therapeut; Am Chem Soc; Soc Exp Biol & Med; NY Acad Sci. Res: Pharmacology of adrenergic, ganglionic blocking drugs and antihypertensive agents. Mailing Add: Merck Sharp & Dohme Res Labs West Point PA 19486

STONE, DANIEL BOXALL, b Gravesend, Eng, May 15, 25; US citizen; m 49; c 2. INTERNAL MEDICINE, ENDOCRINOLOGY. Educ: Univ London, BS & MD, 48, dipl psychiat, 50. Prof Exp: Intern & resident internal med, Univ London, 48-56; from asst prof to prof, Col Med, Univ Iowa, 59-71, exec assoc dean, 67-71; MILLARD PROF MED, UNIV NEBR MED CTR, OMAHA, 71- Concurrent Pos: Fel internal med, Univ London, 48-56; fel internal med & endocrinol, Col Med, Univ Iowa, 57-59; Markle scholar acad med, 60-; consult, Vet Admin Hosp, Iowa City, Iowa, 63-71 & Coun Drugs, AMA, 66- Mem: Fel Am Col Physicians; Am Diabetes Asn; Am Heart Asn; Endocrine Soc; Royal Soc Med. Res: Influence of diet on serum lipids; geographic pathology of diabetes; metabolism of adipose tissue; influence of hypoglycemic drugs on lipolysis in adipose tissue. Mailing Add: 530 Doctors Bldg 4239 Farnam Omaha NE 68131

STONE, DANIEL JOSEPH, b Passaic, NY, Dec 19, 18; m 50; c 3. MEDICINE. Educ: Johns Hopkins Univ, BA, 39; George Washington Univ, MD, 43; Am Bd Internal Med, dipl, 51. Prof Exp: Fel internal med, New York Med Col, 46-47; asst chief, Pulmonary Dis Serv, 49-54, assoc, Cardiopulmonary Lab, 50-54, CHIEF PULMONARY DIS SERV & DIR RESPIRATION LAB, BRONX VET ADMIN HOSP, 54-, CHMN INHALATION THER COMT, 60-; PROF MED, NY MED COL, 75-, DIR PULMONARY SECT, 75- Concurrent Pos: Adv ed of res, Handbk of Biol Sci, Nat Acad Sci, 59; assoc prof, Mt Sinai Sch Med, 68. Mem: Am Physiol Soc; fel Am Col Physicians; fel AMA; Am Fedn Clin Res; fel Am Thoracic Soc. Res: Pulmonary diseases; lung mechanics and the mechanisms of pulmonary failure. Mailing Add: Pulmonary Dis Sect Vet Admin Hosp Bronx NY 10468

STONE, DAVID, b Eng, Mar 28, 19; nat US; m 43; c 2. BIOCHEMISTRY, GENETICS. Educ: Univ Manchester, BSc, 40; Yale Univ, PhD, 51. Prof Exp: Nat Nutrit res fel biochem, Yale Univ, 50-52; SR SCIENTIST, WORCESTER FOUND EXP BIOL, 52-; ASSOC RES DIR, COUN TOBACCO RES, USA, INC, 74- Concurrent Pos: NSF sr fel, Cambridge Univ, 57-58; dir biol res, Worcester State Hosp, Mass. Mem: Am Soc Biol Chemists; Am Physiol Soc; Soc Develop Biol; Tissue Cult Asn; NY Acad Sci. Res: Carcinogenesis; influences of nucleic acids on broken cell systems and cells in culture; behavioral genetics. Mailing Add: Coun Tobacco Res USA Inc 110 E 59th St New York NY 10022

STONE, DAVID B, b Guernsey, UK, Sept 14, 33; m 60; c 3. GEOPHYSICS. Educ: Univ Keele, BA, 56; Univ Newcastle, PhD(geophys), 63. Prof Exp: Sr demonstrator geophys, Univ Newcastle, 63-66; ASSOC PROF GEOPHYS, GEOPHYS INST, UNIV ALASKA, 66- Mem: Fel Royal Astron Soc; Am Geophys Union. Res: Geomagnetism; paleomagnetism; continental drift; geotectonics. Mailing Add: Geophys Inst Univ of Alaska College AK 99701

STONE, DAVID ROSS, b Little Rock, Ark, Aug 30, 42; m 65; c 1. ALGEBRA. Educ: Ga Inst Technol, BS, 64; Univ SC, PhD(math), 68. Prof Exp: ASSOC PROF MATH, GA SOUTHERN COL, 68- Mem: Am Math Soc; Math Asn Am. Res: Rings and modules; torsion theory. Mailing Add: Dept of Math Ga Southern Col Statesboro GA 30458

STONE, DEBORAH BENNETT, b Portchester, NY, Oct 26, 38; m 65; c 2. PHYSICAL BIOCHEMISTRY, BIOCHEMICAL PHARMACOLOGY. Educ: Smith Col, BA, 60; Yale Univ, PhD(pharmacol), 65. Prof Exp: Res assoc pharmacol, Sch Med, Stanford Univ, 64-66; USPHS trainee phys biochem, Cardiovasc Res Inst, 66-68, ASST RES BIOCHEMIST, CARDIOVASC RES INST, UNIV CALIF, SAN FRANCISCO, 68-, LECTR PHYSIOL, 68- Concurrent Pos: USPHS res career develop award, 68-73. Mem: Biophys Soc. Res: Serotonin metabolism in the developing rat brain; regulation of phosphofructokinase activity; molecular mechanisms in muscle contraction. Mailing Add: Cardiovasc Res Inst Univ of Calif San Francisco CA 94143

STONE, DONALD EUGENE, b Eureka, Calif, Dec 10, 30; m 52; c 3. BOTANY, GENETICS. Educ: Univ Calif, AB, 52, PhD(bot), 57. Prof Exp: Asst cytol, biosysts & gen bot, Univ Calif, 54-57; from instr to asst prof bot, Tulane Univ, 57-63; from asst prof to assoc prof, 63-70, PROF BOT, DUKE UNIV, 70- Concurrent Pos: Assoc prog dir syst biol, NSF, 68-69. Mem: Bot Soc Am; Soc Study Evolution; Am Soc Naturalists; Am Soc Plant Taxonomists (secy, 73-75); Orgn Trop Studies. Res: Biosystematics of temperate and tropical families. Mailing Add: Dept of Bot Duke Univ Durham NC 27706

STONE, DORIS ZEMURRAY, b New Orleans, La, Nov 19, 09; m 30; c 2. ANTHROPOLOGY, ARCHAEOLOGY. Educ: Radcliffe Col, BA, 30. Hon Degrees: LLD, Tulane Univ La, 57; LittD, Union Col, NY, 73. Prof Exp: ASSOC ARCHAEOL, MID AM RES INST, TULANE UNIV LA, 30-; RES ASSOC CENT AM ARCHAEOL & ETHNOL, PEABODY MUS, HARVARD UNIV, 42- Concurrent Pos: Mem, Andean Inst; mem permanent coun, Int Cong Americanists, 58- Honors & Awards: Comendador, Order of Ruben Dario, Nicaragua, 55; Hon Citizen, Honduras, 56; Comendador, Order of Francisco Morazan, Honduras, 57; Caballero, Order of Vasco Nunez de Balboa, Panama, 57; Chevalier, Legion de Honor, France, 58. Mem: Am Anthrop Soc; fel Soc Am Archaeol; Am Ethnol Soc; Royal Anthrop Inst Gt Brit & Ireland; Mex Anthrop Soc. Res: Central American archaeology and ethnology; Costa Rican archaeology.

STONE, DOROTHY MAHARAM, b Parkersberg, WVa, July 1, 17; m 42; c 2. MATHEMATICS. Educ: Carnegie Inst Technol, BSc, 37; Bryn Mawr Col, PhD(math), 40. Prof Exp: Asst lectr math, Univ Manchester, 52-61; PROF MATH, UNIV ROCHESTER, 61- Concurrent Pos: NSF fel math, 65-66. Mem: Am Math Soc. Res: Measure theory; ergodic theory; probability; linear operators. Mailing Add: Dept of Math Univ of Rochester Rochester NY 14627

STONE, EARL LEWIS, JR, b Phoenix, NY, July 12, 15; m 41; c 3. FOREST SOILS. Educ: State Univ NY, BS, 38; Univ Wis, MS, 40; Cornell Univ, PhD(soils), 48. Prof Exp: Field asst & jr forester, Southern Forest Exp Sta, US Forest Serv, 40-41; from asst prof to assoc prof forest soils, 48-62, CHARLES LATHROP PACK PROF FOREST SOILS, CORNELL UNIV, 62-, DIV BIOL SCI, 66- Concurrent Pos: Collabr & consult, Southern Forest Exp Sta, US Forest Serv, 47-48 & 52; soil scientist, Pac Sci Bd, Nat Acad Sci, Marshall Islands, 50; Am-Swiss Found fel, 54-55; vis prof, Philippines, 58-60; Fulbright res fel, Forest Res Inst, New Zealand, 62; ed, Forest Sci, Soc Am Foresters, 65-71; Bullard fel, Harvard Univ, 69-70; consult, Biotrop, Indonesia, 70; mem adv panel ecol, NSF, 70-; mem adv comn, Ecol Sci Div, Oak Ridge Nat Lab, 71-; mem, Int Union Forest Res Orgns. Honors & Awards: Barrington Moore Award, Soc Am Foresters, 73. Mem: Fel AAAS; fel Soc Am Foresters; Soil Sci Soc Am; Ecol Soc Am; fel Am Soc Agron. Res: Forest nutrition; ecology; Pacific tropics. Mailing Add: Dept of Agron Cornell Univ Ithaca NY 14850

STONE, EDWARD, b Fall River, Mass, Dec 7, 32; m 56; c 4. ORGANIC CHEMISTRY, POLYMER CHEMISTRY. Educ: Southeastern Mass Tech Inst, BS, 55; Univ Md, PhD(org chem), 62. Prof Exp: Teaching asst, Univ Md, 55-57; chemist, Metals & Controls Corp, 56; anal res chemist, Nat Inst Drycleaning, 57-61, consult, 61; sr res chemist, Tex-US Chem Co, NJ, 61-65; TECH MGR NEW PROD RES & DEVELOP, INMONT CORP, 65- Mem: Am Chem Soc; The Chem Soc. Res: Radiation curing of inks and coatings; polymer research and development; block and graft copolymers; polyurethanes, polyesters, polyolefins; structure-property correlations; organic synthesis; instrumental analysis. Mailing Add: 4 Inwood Rd Morris Plains NJ 07950

STONE, EDWARD CARROLL, JR, b Knoxville, Iowa, Jan 23, 36; m 62; c 2. PHYSICS. Educ: Univ Chicago, SM, 59, PhD(physics), 64. Prof Exp: Res fel, 64-67, sr res fel, 67, asst prof, 67-71, ASSOC PROF PHYSICS, CALIF INST TECHNOL, 71- Concurrent Pos: Mem particles & fields adv comt, NASA, 69-71, consult, 71-; Alfred P Sloan res fel, Calif Inst Technol, 71-73. Mem: AAAS; Am Phys Soc; Am Geophys Union. Res: Solar and galactic cosmic rays; magnetosphere; interplanetary medium; modulation of cosmic rays; geomagnetically trapped particles; satellite and balloon instrumentation. Mailing Add: Dept Physics Calif Inst Technol 1201 E California Blvd Pasadena CA 91109

STONE, EDWARD CURRY, b Ill, Nov 28, 17; m 41; c 2. PLANT PHYSIOLOGY. Educ: Univ Calif, BS, 40, PhD(plant physiol), 48. Prof Exp: Plant physiologist, Calif Forest & Range Exp Sta, US Forest Serv, 48-49; PROF FOREST ECOL & SILVICULTURIST, AGR EXP STA, UNIV CALIF, BERKELEY, 49- Concurrent Pos: Fulbright res scholar, Univ NZ, 59-60; Guggenheim fel, 60. Mem: Ecol Soc Am; Soc Am Foresters; Am Soc Plant Physiol; Bot Soc Am; Scand Soc Plant Physiol. Res: Forest physiology; dormancy; root growth; drought resistance; fire response; cone production; nutritional requirements of forest vegetation. Mailing Add: 145 Mulford Hall Univ of Calif Berkeley CA 94720

STONE, EDWARD JOHN, b Minersville, Pa, July 27, 30; m 56; c 2. ORGANIC CHEMISTRY, BIOCHEMISTRY. Educ: Pa State Univ, BS, 53, MS, 56, PhD(biochem), 62. Prof Exp: Res chemist, Campbell Soup Co, 59-62, SR RES CHEMIST, CAMPBELL INST FOOD RES, 62- Mem: NY Acad Sci; Am Chem Soc. Res: Flavor chemistry; radiochemistry. Mailing Add: Campbell Inst for Food Res Campbell Pl Camden NJ 08102

STONE, ELLEN ROSE, b Cambridge, Eng, June 12, 47. TOPOLOGY. Educ: Univ Rochester, BA, 65; Cornell Univ, MA, 68, PhD(math), 72. Prof Exp: ASST PROF MATH, MICH STATE UNIV, 75- Mem: Am Math Soc. Res: Smooth symmetric structures on manifolds. Mailing Add: Dept of Math Mich State Univ East Lansing MI 48823

STONE, ERIKA MARES, b Prague, Czech, Jan 26, 38; US citizen; m; c 1. MATHEMATICS. Educ: Pa State Univ, BA, 60, MA, 62, PhD(math), 64. Prof Exp: Instr math, Swarthmore Col, 64-65; sr res mathematician, HRB-Singer, Inc, Pa, 65-68, lectr dept comput sci, Pa State Univ, 68; vis asst prof, Dept Math & Comput Sci, Univ SC, 73-75. Mem: Am Math Soc. Res: Structure theory of semiperfect rings and the generalization of theory for modules. Mailing Add: 2106 Strebor Rd Durham NC 27705

STONE, ERNEST COBB, veterinary physiology, see 12th edition

STONE, FRED WILBUR, organic chemistry, see 12th edition

STONE, FREDERICK LOGAN, b Biloxi, Miss, Mar 31, 15; m 38; c 2. BIOLOGY, PUBLIC HEALTH. Educ: Middlebury Col, BS, 37; Univ Rochester, MS, 42, PhD(biol), 48. Prof Exp: Instr biol, Univ Rochester, 47-48; chief res fel br, Div Res Grants, NIH, 48-51, chief extramural progs, Nat Inst Neurol Dis & Blindness, 51-54; asst to vchancellor, Schs Health Professions, Univ Pittsburgh, 54-55; dir med & sci dept, Nat Multiple Sclerosis Soc, NY, 55-56; asst to assoc dir, NIH, 56-57, asst chief div res grants & chief res training br, 57-58, asst chief div gen med sci & chief res training br, 58-62, actg chief div, 62, chief div res facil & resources, 62-64, dir, Nat Inst Gen Med Sci, 64-70; prof commun & prev med & pres, NY Med Col, Flower & Fifth Ave Hosps, 70-72; from interim dep adminr to dep adminr, Health Serv & Ment Health Admin, Dept Health, Educ & Welfare, 72-73, dep adminr, Health Serv Admin, 73-74; DEP DIR, BOSTON UNIV MED CTR, 74- Concurrent Pos: Mem med adv bd, Nat Multiple Sclerosis Soc, 54-; consult, Excerpta Medica Found, New York & Amsterdam, 58-71; mem gov bd sci info exchange, Smithsonian Inst, 61-65; mem panel on facil, President's Comn on Heart Dis, Cancer & Stroke, 64-65; mem jury, Albert Lasker Med Res Awards, 61-67; mem curriculum comt, Med Sch, Univ Pa, 67-70; consult, Arctic Biol Res Inst, Fairbanks, Alaska, 69-70, Nat Inst Dent Res, 70-71 & Health Serv & Ment Health Admin, Dept Health, Educ & Welfare, 70-71; pres, Westchester Med Ctr Found, NY, 70-71; consult neurosci res prog, Mass Inst Technol, 70-72; mem-at-lg pub affairs comt, Fedn of Am Socs Exp Biol, 70-72; mem adv coun dept biol, Princeton Univ, 70-73. Honors & Awards: Superior Serv Award, Dept Health, Educ & Welfare, 64, Secy Spec Citation, 66. Mem: Asn Am Med Cols; Am Acad Neurol; Am Soc Ichthyol & Herpet; Asn Res Nerv & Ment Dis; Asn Res Vision & Ophthal. Res: Research administration. Mailing Add: Boston Univ Med Ctr 720 Harrison Ave Boston MA 02118

STONE, GILBERT C H, b New York, NY, Jan 16, 04; m 34; c 2. BIOCHEMISTRY. Educ: City Col New York, BS, 25; Columbia Univ, AM, 27, PhD(biochem), 33. Prof Exp: Mem staff, Rockefeller Inst, 25-27; res chemist, Nestle's Food Co, 27-29; tutor biochem, City Col New York, 29-34, from instr to assoc prof, 34-66; DIR LABS, E LEITZ, INC, 46- Mem: AAAS; Am Chem Soc; Am Soc Biol Chemists; Am Asn Clin Chemists. Res: Physical chemistry of proteins; effect of proteins on activity of salts; dialysis of protein solutions; spatial relation of polar groups in protein molecule; aliphatic disulfonic acids; chemistry of allergy; photoelectric colorimeters and methods. Mailing Add: 511 W 232nd St New York NY 10463

STONE, GORDON EMORY, b Sioux City, Iowa, July 12, 33; m 55; c 2. CELL BIOLOGY. Educ: Univ Iowa, BA, 56, MSc, 58, PhD(zool), 61. Prof Exp: Res fel, NIH, 61-63 & AEC, 63-64; from asst prof to assoc prof anat, Sch Med, Univ Colo, 64-72; PROF BIOL SCI & CHMN DEPT, UNIV DENVER, 72- Concurrent Pos: NIH career develop award, 65-70. Mem: AAAS; Am Soc Cell Biol; Soc Protozool; Am Soc Zool. Res: Cytochemical studies on cell growth and division, especially the sequential macromolecular events during the interdivision interval leading to division with emphasis on microtubule protein synthesis. Mailing Add: Dept of Biol Sci Univ of Denver Denver CO 80210

STONE, HENRY OTTO, JR, b Spartanburg, SC, Apr 10, 36; m 60. VIROLOGY, BIOCHEMISTRY. Educ: Wofford Col, BS, 59; Duke Univ, PhD(zool), 64. Prof Exp: Am Cancer Soc fel biochem, Duke Univ, 64-66; res chemist, E I du Pont de Nemours & Co, Inc, Del, 66-70; NIH spec res fel animal virol, St Jude Children's Res Hosp, 70-72; ASST PROF MICROBIOL, UNIV KANS, 72- Mem: Am Soc Microbiol; Soc Gen Microbiol. Res: Paramyxoviruses; viral RNA synthesis; genome transcription; viral proteins. Mailing Add: Dept of Microbiol Univ of Kans Lawrence KS 66044

STONE, HERBERT, b Washington, DC, Sept 14, 34; m 64; c 2. NUTRITION, ORGANIC CHEMISTRY. Educ: Univ Mass, BSc, 55, MSc, 58; Univ Calif, Davis, PhD(nutrit), 62. Prof Exp: Specialist, Exp Sta, Univ Calif, Davis; food scientist, Stanford Res Inst, 62-64, dir dept food & plant sci, 67-74; PRES, TRAGON CORP, 74- Mem: AAAS; Sigma Xi; Am Soc Enol; Inst Food Technol; Am Soc Testing & Mat. Res: Basic mechanisms of the olfactory response of man; techniques for odor measurement; taste perception; psychophysics of taste and odor measurement; technoeconomic and management services in consumer products industries, sensory measurement of product acceptance; food and cosmetic product acceptance; nutrition and food acceptance. Mailing Add: Tragon Corp PO Box 783 Palo Alto CA 94302

STONE, HERMAN, b Munich, Ger, Nov 3, 24; nat US; m 49; c 6. ORGANIC CHEMISTRY. Educ: Bethany Col, WVa, BSc, 44; Ohio State Univ, PhD(chem), 50. Prof Exp: Anal chemist, Nat Aniline Div, Allied Chem Corp, 44-45, anal res chemist, 51-53, res chemist, 53-61, group leader appln res chem, 61-63, mgr chem res, Indust Chem Div, 63-68, dir res, Specialty Chem Div, 68-69, res assoc, Corp Chem Res Lab, 69-72; DIR CHEM RES, MALDEN MILLS INC, 72- Mem: AAAS; Am Chem Soc; Soc Plastics Engrs. Res: Analytical and exploratory research on polymer intermediates; urethane polymer technology. Mailing Add: Malden Mills Inc 48 Safford St Lawrence MA 01841

STONE, HOMER EDWARD, mathematics, see 12th edition

STONE, HOWARD ANDERSON, b Claremont, NH, Nov 21, 40; m 62; c 2. GENETICS, VIROLOGY. Educ: Univ NH, BS, 62, MS, 65; Mich State Univ, PhD(poultry), 72. Prof Exp: Asst poultry genetics, Univ NH, 62-65; RES GENETICIST, AGR RES SERV, USDA, 65- Mem: Poultry Sci Asn. Res: Investigations of the genetic control of Marek's disease and lymphoid leukosis in chickens; maintenance and development of highly inbred lines of chickens. Mailing Add: USDA Agr Res Serv 3606 E Mount Hope Rd East Lansing MI 48823

STONE, HRANT H, b Philadelphia, Pa, Feb 17, 17; m 42; c 4. MEDICINE. Educ: Univ Pa, BS, 39, MD, 43; Am Bd Anesthesiol, dipl, 48. Prof Exp: From instr to assoc prof, 46-58, PROF ANESTHESIA, SCH MED, UNIV PA, 58-, DIR ANESTHESIA DEPT, GRAD HOSP, 46- Concurrent Pos: Examr, Am Bd Anesthesiol, 48; consult, US Army, 49-51; from assoc prof to prof, Med Col Pa, 51-; consult, Vet Serv, 51- Mem: AAAS; Soc Anesthesiol; AMA. Res: Cardiovascular changes during anesthesia; drug effects in anesthesia cerebral blood flow, shock hypothermia and cardiopulmonary by-pass. Mailing Add: Dept of Anesthesia Grad Hosp 19th & Lombard St Philadelphia PA 19146

STONE, HUBERT LOWELL, b Baton Rouge, La, July 22, 36; m 57; c 3. PHYSIOLOGY. Educ: Rice Univ, BS, 58; Univ Ill, Urbana, MS, 59, PhD(physiol), 61. Prof Exp: Asst, Univ Ill, 58-61; from instr to asst prof physiol & biophys, Med Sch, Univ Miss, 61-64; res physiologist, US Air Force Sch Aerospace Med, 64-71;

PROF PHYSIOL, UNIV TEX MED BR GALVESTON & CHIEF CARDIOVASC CONTROL SECT, MARINE BIOMED INST, 71- Concurrent Pos: USPHS fel, Med Sch, Univ Miss, 61-63. Honors & Awards: Outstanding Certs, Dept Defense & US Air Force, 68 & 70. Mem: NY Acad Sci; Am Heart Asn; Soc Exp Biol & Med; Soc Neurosci; Am Physiol Soc. Res: Cardiovascular physiology. Mailing Add: Marine Biomed Inst Univ of Tex Med Br Galveston TX 77550

STONE, IRVING CHARLES, JR, b Chicago, Ill, Dec 18, 30; m 55; c 3. FORENSIC SCIENCE. Educ: Iowa State Univ, BS, 52; George Washington Univ, MS, 61, PhD(geochem), 67. Prof Exp: Spec agt-microscopist, Fed Bur Invest, 55-61; res chemist, Res Div, W R Grace & Co, 61-63, proj leader, 63-64, res supvr, 64-68; dir, Geochem Surv, Tex, 68-72; criminalist, 72-74, CHIEF PHYS EVIDENCE SECT, INST FORENSIC SCI, 74- Concurrent Pos: Lectr police sci, Montgomery Jr Col, 67-70; instr forensic sci, Univ Tex Health Sci Ctr Dallas, 72-; adj asst prof, Univ Tex, Arlington, 73- Mem: Sigma Xi; Am Soc Firearms & Toolmark Examrs; Am Soc Crime Lab Dir. Res: Analytical chemistry, especially x-ray diffraction, spectrometry, light microscopy; applied research in forensic sciences, specifically glass, firearm residues and instrumental analytical applications. Mailing Add: Inst of Forensic Sci Box 35728 Dallas TX 75235

STONE, IRWIN, b New York, NY, Nov 18, 07; m 31; c 1. BIOCHEMISTRY, MEDICAL RESEARCH. Prof Exp: Chemist, Pease Labs, 24-34; from res chemist to head brewing res sect, Wallerstein Co, Travenol Labs, Inc, NY, 34-71; dir res, Megascorbic Res, Inc, 72-75; INDEPENDENT RES, 75- Mem: Am Soc Brewing Chem (pres, 62-63); fel Am Inst Chem; fel Int Acad Preventive Med; fel Australasian Col Bio-Med Scientists. Res: Enzyme, ascorbate chemistry; genetics of scurvy; description of human genetic liver-enzyme disease, hypoascorbemia; elimination of chronic subclinical scurvy as a widespread mammalian disease; megascorbic therapy. Mailing Add: 1331 Charmwood Sq San Jose CA 95117

STONE, JAY D, b Littlefield, Tex, Oct 14, 44; m 65; c 2. ENTOMOLOGY. Educ: West Tex State Univ, BS, 68; Iowa State Univ, MS, 70, PhD(entom), 73. Prof Exp: Res assoc entom, Iowa State Univ, 72-73; ASST PROF ENTOM, KANS STATE UNIV, 73- Mem: Entom Soc Am. Res: Biology, economic thresholds, and control of arthropod pests of field corn in Kansas. Mailing Add: Kans Agr Exp Sta Garden City KS 67846

STONE, JOE THOMAS, b Miami, Okla, June 25, 41; m 63; c 2. PHYSICAL ORGANIC CHEMISTRY, PHOTOGRAPHIC CHEMISTRY. Educ: Harvey Mudd Col, BS, 63; Univ Wash, PhD(org chem), 67. Prof Exp: NIH res fel, Univ Wash, 67-68; SR RES CHEMIST, EASTMAN KODAK CO, 68- Mem: The Chem Soc; Am Chem Soc. Res: Organic reaction mechanisms; application of physical-organic techniques to biological processes, enzyme kinetics and mechanism; homogeneous and heterogeneous catalysis and reaction kinetics. Mailing Add: Eastman Kodak Co Res Labs Kodak Park Rochester NY 14650

STONE, JOHN ARTHUR, textile chemistry, organic chemistry, see 12th edition

STONE, JOHN AUSTIN, b Paintsville, Ky, Nov 30, 35; m 61; c 3. NUCLEAR CHEMISTRY. Educ: Univ Louisville, BS, 55; Univ Calif, Berkeley, PhD(nuclear chem), 63. Prof Exp: Chemist, 63-73, staff chemist, 73-74, RES STAFF CHEMIST, SAVANNAH RIVER LAB, E I DU PONT DE NEMOURS & CO, INC, 74- Mem: Am Phys Soc; Am Chem Soc. Res: Mössbauer spectroscopy; solid state and chemical properties of the actinides; radioactive waste management. Mailing Add: Savannah River Lab E I du Pont de Nemours & Co Inc Aiken SC 29801

STONE, JOHN BRUCE, b Forfar, Ont, Sept 23, 30; m 54; c 4. ANIMAL SCIENCE. Educ: Ont Agr Col, BSA, 53, MSA, 54; Cornell Univ, PhD, 59. Prof Exp: Asst prof animal husb, Ont Agr Col, 54-62; asst prof animal sci, Cornell Univ, 62-66; PROF ANIMAL SCI, UNIV GUELPH, 66- Mem: Am Dairy Sci Asn; Agr Inst Can; Am Soc Animal Sci. Res: Dairy cattle nutrition; forages for dairy cattle rations; calf-raising programs; systems analyses for dairy production. Mailing Add: Dept of Animal & Poultry Sci Univ of Guelph Guelph ON Can

STONE, JOHN ELMER, b Montgomery, Ala, Aug 12, 31; m 59; c 2. GEOLOGY. Educ: Ohio Wesleyan Univ, BA, 53; Univ Ill, MS, 58, PhD(geol), 60. Prof Exp: Asst prof geol, Univ Tex, 60-62; geologist, Minn Geol Surv, 62-67; PROF GEOL & HEAD DEPT, OKLA STATE UNIV, 67- Concurrent Pos: Res grants, Univ Tex Excellence Fund, 61 & grad sch, Univ Minn, 62, 67; NSF summer grants, 68-72, sci equip grant, 69; Okla State Univ Res Found res grant, 70. Mem: AAAS; fel Geol Soc Am; Nat Asn Geol Teachers; Am Quaternary Asn; Soc Econ Paleont & Mineral. Res: Glacial and engineering geology. Mailing Add: Dept of Geol Okla State Univ Stillwater OK 74074

STONE, JOHN ERNEST, b London, Eng, May 29, 21; m 44; c 2. PHYSICAL CHEMISTRY. Educ: Univ London, BSc, 42, PhD(phys chem), 48. Prof Exp: Res officer, Nat Res Coun Can, 48-52; res assoc, Inst Paper Chem, 53-57; sr chemist, Pulp & Paper Res Inst Can, 57-69, head fiber chem, 69-70; prog coordr, Forest Prod, 70-75, DIR FORESTRY RELS & TECHNOL TRANSFER, CAN FORESTRY SERV, DEPT ENVIRON CAN, 75- Concurrent Pos: Chmn, Can/USSR Working Group Forest Based Industs, 75- Mem: Tech Asn Pulp & Paper Indust; Can Pulp & Paper Asn. Res: Structure of the plant cell wall. Mailing Add: Can Forestry Serv Int Forestry Dept Environ Ottawa ON Can

STONE, JOHN FLOYD, b York, Nebr, Oct 13, 28; m 53; c 4. SOIL PHYSICS. Educ: Univ Nebr, BSc, 52; Iowa State Univ, MS, 55, PhD, 57. Prof Exp: Res assoc, Dept Agron & Inst Atomic Res, Iowa State Univ, 56-57; from asst prof to assoc prof agron, 57-69, PROF AGRON, OKLA STATE UNIV, 69- Concurrent Pos: Assoc ed, Soil Sci Soc Am, 68-75. Mem: Am Soc Agron; Soil Sci Soc Am; Am Geophys Union; Int Soc Soil Sci. Res: Water conservation, evapotranspiration; water flow in plants; electronic instrumentation. Mailing Add: Dept of Agron Okla State Univ Stillwater OK 74074

STONE, JOHN GROVER, II, b Pueblo, Colo, Aug 6, 33; m 64; c 5. GEOLOGY. Educ: Yale Univ, BS, 55; Stanford Univ, PhD(geol), 58. Prof Exp: Staff geologist, 58-69, ASST CHIEF GEOLOGIST, HANNA MINING CO, 69- Mem: Geol Soc Am; Soc Econ Geologists. Res: Genesis of ore deposits. Mailing Add: Hanna Mining Co 100 Erieview Plaza Cleveland OH 44114

STONE, JOHN PATRICK, b Algood, Tenn, Sept 5, 39; m 64; c 3. ENDOCRINOLOGY, RADIOBIOLOGY. Educ: Wayne State Univ, BS, 61, PhD(biol), 72; Purdue Univ, Lafayette, MS, 64. Prof Exp: Teaching asst bionucleonics, Purdue Univ, Lafayette, 62-64; teaching asst endocrinol & radiobiol, Wayne State Univ, 65-68; fel radiobiol, Div Biol & Med Res, Argonne Nat Lab, 72-74; asst scientist, 74-75, ASSOC SCIENTIST, RADIOBIOL DIV, MED DEPT, BROOKHAVEN NAT LAB, 76- Mem: Radiation Res Soc; Am Soc Zoologists. Res: Hormonal control of neutron induced mammary tumorigenesis; effects of chronic gamma irradiation upon endocrine and hematopoietic systems; pigment cell biochemistry and physiology. Mailing Add: Med Dept Brookhaven Nat Lab Upton NY 11973

STONE, JOSEPH, b Holyoke, Mass, June 3, 20. BIOCHEMISTRY, PHARMACOLOGY. Educ: Mass Col Pharm, BS, 47; Univ Colo, PhD(pharmacol), 54. Prof Exp: Proj assoc, McArdle Mem Labs, Univ Wis, 54-56; staff pharmacologist, Vet Admin Hosp, Chicago, 56; asst prof, 57-63, ASSOC PROF PHARMACOL, MED CTR, UNIV ARK, LITTLE ROCK, 63- Mem: AAAS; NY Acad Sci. Res: Diffusion respiration; testing systems for DNA antimetabolites; central nervous system biochemistry and pharmacology. Mailing Add: Dept of Pharmacol Univ of Ark Med Ctr Little Rock AR 72201

STONE, JOSEPH LOUIS, b Claremont, NH, Jan 25, 18; m 43; c 2. MEDICAL BACTERIOLOGY. Educ: Univ NH, BS, 40; MS, 41; Boston Univ, PhD, 48. Prof Exp: Sr bacteriologist, Biol Lab, State Dept Pub Health, Mass, 48-56; sr res investr, Wyeth Labs, Inc, Marietta, 56-72, SR RES INVESTR, WYETH LABS, INC, RADNOR, 72- Mem: AAAS; Am Geog Soc. Res: Antibiotics; tetanus toxin and toxoid; virus vaccines. Mailing Add: 717 Croton Rd Wayne PA 19087

STONE, JULIAN, b New York, NY, Apr 12, 29; m 51; c 3. PHYSICS. Educ: City Col New York, BS, 50; NY Univ, MS, 51, PhD(physics), 58. Prof Exp: Electronic scientist, Naval Mat Lab, 52; tutor physics, City Col New York, 52-53; res scientist, Hudson Lab, Columbia Univ, 53-69, assoc dir physics, 66-69; MEM TECH STAFF, BELL LABS, 69- Concurrent Pos: Tutor physics, City Col New York, 56-57. Mem: Am Phys Soc. Res: Lasers; spectroscopy; underwater sound propagation. Mailing Add: Bell Labs Room R113 Box 400 Holmdel NJ 07733

STONE, KIRK HASKIN, b Bay Village, Ohio, Apr 27, 14; m 36; c 1. GEOGRAPHY. Educ: Univ Mich, PhD(geog), 49. Prof Exp: Asst prof geog, Univ Toledo, 38-42; geogr, Off Strategic Serv, DC & Ceylon, 42-46; prof geog, Univ Wis-Madison, 47-65; RES PROF GEOG, UNIV GA, 65- Concurrent Pos: Geogr, US Bur Land Mgt, 41, 45-46 & 48; Univ Wis Grad Res Comt grants, 47-65; consult, Bur Census, 49 & 59, Dept Defense, 51-53 & Bur Land Mgt, 54; Arctic Inst NAm res grant, 51; geogr, Lincoln Lab, Mass Inst Technol, 52-53; Fulbright sr res scholar, Univ Oslo, 56-57; Off Naval Res grant, Univ Wis, Scandinavia & Finland, 59-63; guest lectr, Advan Inst Remote Sensing, Univ Mich, 68, Univ Tenn, 69 & Univ Denver, 71; Am Coun Learned Soc-Soc Sci Res Coun res grant, 70; Am Philos Soc res grant, 72. Honors & Awards: Citation for Meritorious Serv, Asn Am Geogr, 65; Michael Award, Univ Ga, 68. Mem: AAAS; Asn Am Geogr; Int Geog Union; Pop Asn Am; Int Union Sci Study Pop. Res: Planning for new rural settling throughout the world; world population geography. Mailing Add: Dept of Geog Univ of Ga Athens GA 30602

STONE, LAWRENCE DAVID, b St Louis, Mo, Sept 2, 42; m 67. MATHEMATICS, OPERATIONS RESEARCH. Educ: Antioch Col, BS, 64; Purdue Univ, West Lafayette, MS, 66, PhD(math), 67. Prof Exp: From assoc to sr assoc, 67-74, VPRES, DANIEL H WAGNER ASSOCS, 74- Concurrent Pos: Off Naval Res grant, 69-76. Mem: Am Math Soc; Inst Math Statist; Oper Res Soc; Math Prog Soc. Res: Theory of search for stationary objects; constrained extremal problems; threshold crossing problems for markov and semi-markov processes; optimal stochastic control of semi-markov processes. Mailing Add: Daniel H Wagner Assocs Paoli PA 19301

STONE, LOYD RAYMOND, b Prague, Okla, Jan 6, 45; m 65; c 2. SOIL PHYSICS. Educ: Okla State Univ, BS, 67, MS, 69; SDak State Univ, PhD(agron), 73. Prof Exp: ASST PROF AGRON, EVAPOTRANSPIRATION LAB, KANS STATE UNIV, 73- Mem: Am Soc Agron; Soil Sci Soc Am; Sigma Xi; Soil Conserv Soc Am. Res: Management of the soil-water-crop environment for efficient use of water; irrigation water use efficiency and plant root systems analysis. Mailing Add: Evapo-transpiration Lab Kans State Univ Manhattan KS 66506

STONE, MARGARET HODGMAN, b Cleveland, Ohio, May 20, 14; m 41; c 3. PLANT TAXONOMY. Educ: Western Reserve Univ, BA, 36, MA, 37, PhD(bot), 40. Prof Exp: Dir hort, Garden Ctr Gtr Cleveland, 40-41; agr res dir bot, State Univ NY Col Agr, Cornell Univ, 41-42; instr, Western Reserve Univ, 43-45; agr res dir State Univ NY Col Agr, Cornell Univ, 45-46; prof lectr, Col Agr, Philippines, 58-60; SR CUR TAXON BOT, L H BAILEY HORTORIUM, CORNELL UNIV, 65- Concurrent Pos: Mercer res fel, Arnold Arboretum, Harvard Univ, 69-70. Mem: AAAS; Int Asn Plant Taxon; Am Hort Soc. Res: Hydrogen-ion concentration of soil in relation to distribution of flora; dormant buds in Pinus; taxonomy of cultivated plants. Mailing Add: L H Bailey Hortorium Cornell Univ Ithaca NY 14850

STONE, MARSHALL HARVEY, b New York, NY, Apr 8, 03; m 27, 62; c 4. MATHEMATICS. Educ: Harvard Univ, BA, 22, MA, 24, PhD(math), 26. Hon Degrees: ScD, Kenyon Col, 39, Amherst Col, 54 & Colby Col, 59; Dr, Univ San Marcos, Peru, 43, Univ Buenos Aires, 47 & Univ Athens, 54; ScD, Univ Mass, 66. Prof Exp: Instr math, Harvard Univ, 22-23 & 27-28, asst prof, 28-31; assoc prof, Yale Univ, 31-33; from assoc prof to prof, 33-46, Harvard Univ, chmn dept, 42; instr, Columbia Univ, 25-27; Andrew Macleish Distinguished Serv prof, 46-68, chmn dept, 46-52, EMER PROF MATH, UNIV CHICAGO, 68-; PROF MATH, UNIV MASS, AMHERST, 73- Concurrent Pos: Guggenheim fel, Inst Advan Study, 36-37; Ames lectr, Univ Wash, 42; vis prof, Univ Buenos Aires, 43, Univ Brazil, 47, Tata Inst Fundamental Res, India, 49-50, 55-56 & Col France, 53; vis lectr, Univs, Japan, 49, 56, Australia, 59 & Univ Islamabad, WPakistan, 69-70; external examr, Univ Malaya, 58-60, Middle East Tech Univ, Turkey, 63, Res Inst Math Sci, Madras, India, 63-64, 65, Kyoto, Japan, 65, Univ Geneva, 64, Int Math Union & Acad Sci, Pakistan, 64, Univ Hong Kong, 65, Thai Math Soc, 65, Univs Taiwan, 65 & Cern, Geneva, 66; George David Birkhoff prof, Univ Mass, Amherst, 68-73. Ed, Math Rev, 45-50; vchmn div math & phys sci, Nat Res Coun, 46-52, div math, 51-52; pres, Int Math Union, 52-54; pres, Int Comn Math Instruct, 59-62; pres, Royaumont Cong, Orgn Europ Econ Coop, 59, mem expert group, Yugoslavia, 60; mem panel elem sch math, Sch Math Study Group, 60; mem comt social thought, Univ Chicago, 62-68; mem inter-union comn sci teaching, Int Coun Sci Unions, 62-65; mem, Inter-Am Comt Math Educ, 61- Mem: Nat Acad Sci; AAAS; Am Math Soc (pres, 43-44); Am Geog Soc; Am Philos Soc. Res: Analysis; general topology; algebra of logic; Hilbert space theory; foundations of mathematics and physics. Mailing Add: 260 Lincoln Ave Amherst MA 01002

STONE, MARTIN L, b New York, NY, June 11, 20; m 43; c 1. OBSTETRICS & GYNECOLOGY. Educ: Columbia Univ, BS, 41; New York Med Col, MD, 44, MMSc, 49; Am Bd Obstet & Gynec, dipl, 52. Prof Exp: PROF OBSTET & GYNEC & CHMN DEPT, NEW YORK MED COL, 56-, VPRES HOSP AFFAIRS, 72- Concurrent Pos: Attend, Flower & Fifth Ave Hosps, Metrop & Bird S Coler Hosps, 56-; consult, Southampton Hosp, 60- & Deepdale Gen Hosp, 66- Mem: Fel Am Gynec Soc; fel Am Pub Health Asn; fel Am Asn Obstetricians & Gynecologists; assoc fel Royal Soc Med; fel Am Col Obstetricians & Gynecologists (secy, 71-). Mailing Add: Dept of Obstet & Gynec New York Med Col New York NY 10029

STONE, MAX WENDELL, b Petersburg, Tenn, Mar 6, 29; m 50. COMPUTER SCIENCES. Educ: Union Univ, Tenn, BS, 49; Peabody Col, MA, 50. Prof Exp: High sch teacher, Ark, 50-51; scientist & supvr comput & data reduction, Rohm and Haas Co, Redstone Res Labs, 53-64; mgr corp data processing, Sci Systs, Inc, 64-73; HEAD COMPUT OPERS & ANAL SUPPORT, SYST DEVELOP CORP, 73- Concurrent Pos: Teacher eve div, Univ Ala, Huntsville Ctr, 54-65. Mem: Am Defense Preparedness Asn. Res: Digital computer applications to problems in engineering, science and business, including management information systems, accounting functions, inventory control and data reduction; solid rocket propellant grain design; operation of large computers. Mailing Add: 826 Jacqueline Dr SE Huntsville AL 35802

STONE, MICHAEL GATES, b Midland, Tex, Oct 9, 38; m 63; c 1. MATHEMATICS. Educ: Wesleyan Univ, BA, 60; La State Univ, Baton Rouge, MS, 62; Univ Colo, Boulder, PhD(math), 69. Prof Exp: Asst prof, 69-74, ASSOC PROF MATH, UNIV CALGARY, 74- Mem: Am Math Soc; Can Math Cong. Res: Universal algebra; lattice theory; model theory. Mailing Add: Dept Math Statist & Comput Sci Univ of Calgary Calgary AB Can

STONE, NEWTON C, b Burnt Prairie, Ill, Jan 21, 11; m 35; c 3. METEOROLOGY, ATMOSPHERIC SCIENCES. Educ: Southern Ill Univ, BEd, 34; Univ Ill, MS, 35; Calif Inst Technol, BS, 40, MS, 41. Prof Exp: Teacher high sch, Ill, 35-39; instr meteorol, Calif Inst Technol, 40-43, asst prof, 43-48; weather engr, Krick Assocs, Inc, 48-52, dir meteorol, Krick Assocs, Inc, 52-54, V PRES, IRVING P KRICK ASSOCS, 54-; WATER RESOURCES DEVELOP CORP, 54- Concurrent Pos: Dir meteorol, Water Resources Develop Corp, 52-54; vpres, Am Inst Aerological Res, 54-; dir, Krick Assocs, Can, Ltd & Krick, Inc, Tex, 59-; consult, AROWA, 54-55. Mem: Am Meteorol Soc; Am Geophys Union; Nat Soc Prof Engrs. Res: Short and long range weather forecasting; northern hemisphere research; weather modification. Mailing Add: Water Resources Develop Corp 68392 Kings Rd Palm Springs CA 92262

STONE, ORVILLE JOSEPH, b Ottawa, Ill, Mar 3, 31; m 52; c 4. DERMATOLOGY. Educ: Univ Ill, Urbana-Champaign, BA, 53, BS, 54; Univ Ill Col Med, MD, 56. Prof Exp: Internship, St Francis Hosp, Peoria, Ill, 56-57; resident, 59-62, from instr to prof dermat, Univ Tex Med Br Galveston, 62-73; PROF MED, UNIV CALIF, IRVINE-CALIF COL MED, 73- Concurrent Pos: Consult, John Sealy Hosp, Galveston, Tex, 61-73, Brooke Gen Hosp, Ft Sam Houston, 67-73, USPHS Hosp, Galveston, 68-73, Shrine Burn Inst, 69-73, Long Beach Vet Admin Hosp, 73-, Orange County Med Ctr, 73- & US Naval Hosp, San Diego, Calif, 73- Mem: Fel Royal Soc Trop Med & Hyg; Am Acad Dermat; Soc Invest Dermat. Res: Inflammation; leukocyte abscesses; hookworm; nails; parasitology; iodides; psoriasis. Mailing Add: Dept of Med Div of Dermat Univ of Calif Irvine CA 92664

STONE, PETER HUNTER, dynamic meteorology, applied mathematics, see 12th edition

STONE, PHILIP M, b Wilkinsburg, Pa, Nov 23, 33; m 55; c 3. ATOMIC PHYSICS, PLASMA PHYSICS. Educ: Univ Mich, BSE, 55, MSE, 56, PhD(nuclear eng), 62. Prof Exp: Staff mem, Los Alamos Sci Lab, 56-63, mem advan study prog, 59-60, grad thesis prog, 60-62; staff mem, Sperry Rand Res Ctr, 63-68, head radiation sci dept, 67-68; assoc prof, Div Interdisciplinary Studies, State Univ NY Buffalo, 68-69; liaison scientist, Sperry Rand Res Ctr, 69-71, mgr systs studies dept, 71-75; PHYSICIST, ENERGY RES & DEVELOP ADMIN, 75- Concurrent Pos: Fel, Univ Col, Univ London, 65-66; vis assoc prof, Univ Pittsburgh, 67-68; vis scientist, Ctr d'Etudes Nucleaires de Saclay, 74. Mem: AAAS; Am Phys Soc; Sigma Xi; NY Acad Sci; Inst Elec & Electronic Engrs. Res: Theoretical atomic and plasma physics; electron-atom scattering, photoabsorption, recombination, line shapes and intensities, nonequilibrium populations, microwave radiation from plasmas; signal processing and system analysis. Mailing Add: Div of CTR Energy Res & Develop Admin Washington DC 20545

STONE, RICHARD L, b Blackfoot, Idaho, Aug 9, 33. ANTHROPOLOGY. Educ: Idaho State Col, BA, 60; Univ Hawaii, MA, 65, PhD(anthrop), 70. Prof Exp: Instr hist & anthrop, Ateneo de Manila Univ, 61-62, res assoc anthrop, 64-65; instr, Univ Hawaii, 65-66; proj dir, Ateneo de Manila Univ-Pa State Univ Basic Res Proj, 66-68; asst prof anthrop, Univ Hawaii, 68; asst prof, Calif State Univ, Long Beach, 68-70; asst prof behav sci, John Jay Col Criminal Justice, 70-71; asst prof, 71-74, ASSOC PROF ANTHROP, CALIF STATE UNIV, LOS ANGELES, 74-, DIR CTR ASIAN STUDIES, 72- Concurrent Pos: Consult, Peace Corps Training Prog, Hawaii, 66 & 68, Cent Inst Training Resettlement Urban Squatters, Repub of Philippines Social Welfare Agency, 67 & Asia Training Ctr, Honolulu, 68. Mem: Philippine Sociol Soc; Soc Appl Anthrop. Res: Legal systems; culture change; urbanization; southeast Asia political anthropology. Mailing Add: Dept of Anthrop Calif State Univ 5151 State University Dr Los Angeles CA 90032

STONE, RICHARD O'NEILL, b Los Angeles, Calif, May 4, 20. GEOMORPHOLOGY. Educ: Colo Sch Mines, GeolE, 41; Univ Southern Calif, MS, 53, PhD, 56. Prof Exp: From asst prof to assoc prof, 56-67, PROF GEOL, UNIV SOUTHERN CALIF, 67-, CHMN DEPT, 72- Mem: Geol Soc Am; Am Geol Inst; Sigma Xi; AAAS. Res: Desert geology, morphology and terrain; eolian studies; sedimentation; quantitative geomorphology. Mailing Add: Dept of Geol Sci Univ of Southern Calif Los Angeles CA 90007

STONE, RICHARD SPILLANE, b Huntington, NY, Sept 14, 25; m 48; c 3. PHYSICS. Educ: Rensselaer Polytech Inst, BS, 49, MS, 50, PhD(physics), 52. Prof Exp: Asst physics, Rensselaer Polytech Inst, 49-52; res assoc nuclear physics instrumentation, Knolls Atomic Power Lab, Gen Elec Co, 52-57; physicist in chg, TRIGA Proj & sect mgr, HTGR Proj, Gen Atomic Div, Gen Dynamics Corp, 57-63; vpres eng & sales, Tech Measurement Corp, 63-64; HEAD PHYSICS SECT, ARTHUR D LITTLE, INC, 64- Mem: Am Phys Soc; Am Nuclear Soc. Res: Underwater acoustics; computer applications; system analysis instrumentation design and development; reactor physics; design, construction and operation of experimental reactors; critical assemblies and in-pile experiments; nuclear power plant analysis and test. Mailing Add: Arthur D Little Inc 15 Acorn Park Cambridge MA 02140

STONE, ROBERT EDWARD, JR, b Spokane, Wash, Feb 20, 37; m 62; c 3. SPEECH & HEARING SCIENCES, SPEECH PATHOLOGY. Educ: Whitworth Col, BS, 60; Univ Ore, MEd, 64; Univ Mich, PhD(speech path, speech sci), 71. Prof Exp: Teacher math & sci, Oswego Pub Schs, Ore, 60-63; speech therapist, Portland Pub Schs, 63-64; instr speech, Ore State Syst Higher Educ, 64-66; speech pathologist & res asst otorhinolaryngol, 70-71, INSTR OTORHINOLARYNGOL, UNIV MICH, ANN ARBOR, 71- Mem: Am Speech & Hearing Asn; Asn Res in Otorhinolaryngol; assoc Acoust Soc Am. Res: Laryngeal physiology and effects of aberrant production of voice. Mailing Add: Kresge Hearing Res Inst 1301 E Ann St Ann Arbor MI 48104

STONE, ROBERT GILBERT, physics, radio astronomy, see 12th edition

STONE, ROBERT LOUIS, b Frankfort, Ky, Dec 30, 21. MICROBIOLOGY. Educ: Univ Ky, BS, 47, MS, 52; Ind Univ, PhD, 59. Prof Exp: Sr microbiologist, 49-55 & 58-72, RES MICROBIOLOGIST, LILLY RES LABS, ELI LILLY & CO, 72- Concurrent Pos: Lectr, Butler Univ, 68- Mem: AAAS; Am Soc Microbiol; NY Acad Sci. Res: Bacteriophage; immunology; chemotherapy; medical virology. Mailing Add: 5611 Brendon Way Ct Indianapolis IN 46226

STONE, ROBERT MARION, b Cleveland, Ohio, June 18, 30; m 72; c 4. PHYSICAL CHEMISTRY. Educ: Vanderbilt Univ, BA, 53, PhD(chem), 59. Prof Exp: Res chemist, 59-63, res supvr, 63-69, sr supvr, 69-71, RES & DEVELOP MGR, EXP STA LAB, E I DU PONT DE NEMOURS & CO, SEAFORD, DEL, 71- Concurrent Pos: Prof, Chattanooga Col, 60. Mem: Am Chem Soc; Am Asn Textile Chemists & Colorists. Res: Reaction kinetics and mechanisms; rocket fuels and propellants; textile chemistry; thermodynamics of high energy fuels; polymer chemistry; kinetics; polymer spinning technology; carpet fiber technology. Mailing Add: 602 Sussex Ave Seaford DE 19973

STONE, ROBERT SAMUEL, pathology, see 12th edition

STONE, ROBERT WILLIAM, b Redland, Ore, Apr 17, 10; m 35; c 3. MICROBIOLOGY. Educ: Ore State Col, BS, 32; Iowa State Univ, PhD(bact), 36. to prof, 37-74, head dept bact, 48-70, chmn div biol sci, 60-64, EMER PROF BACT, PA STATE UNIV, 74- Concurrent Pos: Nat Heart Found fel, Inst Enzyme Res, Univ Wis, 50-51; mem adv comt, Microbiol Sect, Off Naval Res, 55-60 & Div Biol & Agr, Nat Res Coun, 60-70; trustee, Am Type Cult Collection, 66-72. Mem: Acad Microbiol; Soc Indust Microbiol (vpres, 65-66); fel NY Acad Sci. Res: Oxidative metabolism of microorganisms; iron-oxidizing bacteria; mine drainage. Mailing Add: Dept of Microbiol Pa State Univ University Park PA 16802

STONE, SAMUEL ARTHUR, b Keene, NH, Sept 10, 14; m 40; c 2. MATHEMATICS. Educ: Univ NH, BS, 36, MS, 37; Boston Univ, PhD(physics), 53. Prof Exp: Instr math, Univ NH, 37-40, Mass Nautical Acad, 40-41 & Northeastern Univ, 41-45; asst lens designer, Polaroid Corp, Mass, 45-48; instr math, 48-53, asst prof math & physics & in charge depts, 53-57, prof & head depts, 57-63, dean col arts & sci, 63-69, COMMONWEALTH PROF MATH, SOUTHEASTERN MASS UNIV, 69- Mem: Am Math Soc; Math Asn Am; Am Asn Physics Teachers. Res: Mathematical analysis of non-linear stretch; vibration spectra of ethylene oxide. Mailing Add: Dept of Math Southeastern Mass Univ North Dartmouth MA 02747

STONE, SANFORD HERBERT, b New York, NY, Sept 9, 21; m 53; c 3. IMMUNOLOGY. Educ: City Col New York, BS, 47; Univ Paris, DSc, 51. Prof Exp: NIH res fel, Sch Med, Johns Hopkins Univ, 51-52; res asst immunol, New York Med Col, 53; asst, Appl Immunol Div, Pub Health Res Inst, New York, 54-57; head sect natural & acquired resistance, Lab Immunol, 57-62, head sect allergy & hypersensitivity, 63-69, head immunol sect, Lab Microbiol, 70-74, HEAD IMMUNOL SECT, OSD, NAT INST ALLERGY & INFECTIOUS DIS, 74- Concurrent Pos: Prof lectr, Howard Univ, 61-75. Mem: AAAS; Am Asn Immunologists; Am Soc Exp Pathologists; Reticuloendothelial Soc; Fr Soc Microbiol. Res: Tissue antigens and antibodies; mechanism of hypersensitivity; autoimmunity; transplantation immunity. Mailing Add: Bldg 5 Rm 210 Nat Inst of Allergy & Infectious Dis Bethesda MD 20014

STONE, SIDNEY NORMAN, b Rochester, NY, May 11, 22; m 51; c 2. OPTICAL PHYSICS, ASTROPHYSICS. Educ: Univ Calif, BA, 51, MA, 52, PhD(astron), 57. Prof Exp: Physicist, Ballistic Res Lab, Aberdeen Proving Ground, Md, 44-49; asst astron, Univ Calif, 53-54, STAFF MEM, LOS ALAMOS SCI LAB, UNIV CALIF, 57- Mem: Am Astron Soc. Res: Spectroscopy; optical instrumentation; spectroscopic binary stars; physics of the upper atmosphere; photographic sensitometry; high speed photography; radiation dosimetry. Mailing Add: Los Alamos Sci Lab Univ Calif MS 664 Box 1663 Los Alamos NM 87545

STONE, SOLON WALLINGFORD, b Meadville, Pa, Aug 19, 16; m 41; c 3. GEOLOGY. Educ: Allegheny Col, 38; Syracuse Univ, MS, 40; Harvard Univ, PhD(geol), 51. Prof Exp: Geologist, Magnolia Petrol Co, 41-42; instr geol, Trinity Col, Conn, 47-51; geologist, Carter Oil Co, 51-61; STAFF GEOLOGIST, EXXON CORP, 61- Concurrent Pos: Instr, Centenary Col, 52-55. Mem: Geol Soc Am; Soc Econ Paleontologists & Mineralogists; Am Asn Petrol Geologists. Res: Structure and stratigraphy of Northwest Vermont; petroleum geology of northern Rocky Mountains and Gulf Coast. Mailing Add: 9214 Riddlewood Houston TX 77025

STONE, STANLEY S, b Old Forge, Pa, Apr 4, 21; m 50; c 4. BIOCHEMISTRY, IMMUNOCHEMISTRY. Educ: Loyola Col, Md, BS, 50; Georgetown Univ, MS, 53, PhD(biochem), 57. Prof Exp: Biochemist, USDA, 49-52; biochemist, NIH, 52-57; biochemist, Plum Island Animal Dis Lab, USDA, 57-62 & 64-66, EAfrican Vet Res Lab, 62-64, EAfrican Vet Res Orgn, 66-70 & Plum Island Animal Dis Lab, 70-71, HEAD BIOCHEM BIOPHYS, NAT ANIMAL DIS CTR, USDA, 72- Mem: Am Chem Soc; Am Asn Immunologists; Am Soc Microbiol; NY Acad Sci; Reticuloendothelial Soc. Res: Isolation and characterization of immunoglobulins from farm domestic animals, particularly immunoglobulins of the exocrine secretions; reactions of immunoglobulins with viral antigens. Mailing Add: USDA Nat Animal Dis Ctr Ames IA 50010

STONE, WARREN KENNETH, b Ogden, Utah, Mar 5, 13; m 40; c 3. FOOD SCIENCE. Educ: Utah State Univ, BS, 38, MS, 41; Univ Wis, PhD(dairy & food indust), 51. Prof Exp: Asst dairy tech, Utah State Univ, 39-40; asst chief sanitarian, City Dept Health, Salt Lake City, Utah, 41-43; operating engr, US Dept Army, 43-44; foods inspector, State Dept Agr, Utah, 44-47; asst dairy mfg res, Univ Ill, 47-48 & Univ Wis, 48-51; food technologist, Qm Food & Container Inst, Ill, 51-55; prof dairy sci, 55-68, PROF FOOD SCI & TECHNOL, VA POLYTECH INST & STATE UNIV, 68- Mem: Fel AAAS; fel Am Inst Chem; Am Chem Soc; Am Dairy Sci Asn; Inst Food Technol. Res: Chemistry and microbiology of dairy and food products; food plant sanitation; new processes and product development. Mailing Add: Dept of Food Sci & Technol Va Polytech Inst & State Univ Blacksburg VA 24061

STONE, WILLIAM ELLIS, b Colton, Calif, Jan 22, 11; m 41; c 3. PHYSIOLOGY. Educ: Calif Inst Technol, BS, 33; Univ Minn, PhD(physiol chem), 39. Prof Exp: Coxe Mem fel, Lab Neurophysiol, Yale Univ, 39-40; res assoc surg, Col Med, Wayne Univ, 40-47, from instr to asst prof physiol chem, 41-47; from asst prof to assoc prof, 47-64, PROF PHYSIOL, UNIV WIS-MADISON, 64- Mem: Am Physiol Soc; Am Epilepsy Soc. Res: Chemical physiology fo the brain. Mailing Add: Dept of Physiol Univ of Wis Madison WI 53706

STONE, WILLIAM HAROLD, b Boston, Mass, Dec 15, 24; m 48; c 2. GENETICS, IMMUNOLOGY. Educ: Brown Univ, AB, 48; Univ Maine, MS, 49; Univ Wis, PhD(genetics, biochem), 53. Prof Exp: Lab asst biol, Brown Univ, 46-47; asst bact & genetics, Jackson Mem Lab, 47-48; asst, 48-49, from instr to assoc prof, 49-62, PROF GENETICS & MED GENETICS, UNIV WIS-MADISON, 62- Concurrent Pos: Consult, Wis Alumni Res Found, 54; Bell Tel Co & Am Inst Biol Sci Films; NIH res fel, 60; mem expert panel blood group scientists, Food & Agr Orgn, 63-65; ed, Immunogenetics Letter, 66 & J Transfusion, 69-; NIH res fel, Univ Barcelona, 70; mem comt vet med sci, Nat Res Coun, 75- Honors & Awards: Ivanov Medal, USSR Acad Sci, 74. Mem: AAAS; Genetics Soc Am; Soc Human Genetics; Am Asn

Immunologists; Transplantation Soc. Res: Immunogenetics and immunochemical studies of blood groups; immunology of fertility and sterility; transplantation and tolerance. Mailing Add: Lab of Genetics Univ of Wis Madison WI 53706

STONE, WILLIAM JACK HANSON, b Pearland, Tex, Dec 28, 32; m 53; c 2. PLANT PATHOLOGY, WEED SCIENCE. Educ: Tex A&M Univ, BS, 55, MS, 57; Purdue Univ, PhD(plant path), 63. Prof Exp: Res plant pathologist, USDA, 57-59; asst plant pathologist, Univ Ariz, 63-66; res & develop pathologist, 66, HEAD FLA SUBSTA, UPJOHN CO, 66- Mem: Am Soc Plant Physiol; Am Phytopath Soc. Res: Pesticides for agricultural uses. Mailing Add: Agr Prod Div Unjohn Co PO Box P Delray Beach FL 33444

STONE, WILLIAM MATTHEWSON, b Oregon City, Ore, Feb 4, 15; m 61; c 2. MATHEMATICS. Educ: Willamette Univ, BA, 38; Ore State Col, MA, 40, PhD(math), 47. Prof Exp: Asst prof math, Ore State Col, 47-51, assoc prof, 53-59; res engr, Boeing Co, 51-53, 59-61; PROF MATH, ORE STATE UNIV, 61- Concurrent Pos: Fac assoc, Boeing Co, 54-63; univ assoc, Marathon Oil Co, 67. Mem: AAAS; Am Math Soc; Math Asn Am; Soc Indust & Appl Math; Inst Math Statist. Res: Applied mathematics; information theory; stochastic processes. Mailing Add: Dept of Math Ore State Univ Corvallis OR 97331

STONE, WILLIAM MORGAN, b Bloomfield, Ky, Feb 16, 21; m 47. PARASITOLOGY. Educ: Transylvania Col, AB, 47; Univ Ky, MS, 50. Prof Exp: Asst, Narcotics Res Lab, USPHS, 47-48; chemist, Trop Dis Div, NIH, 48-49; asst parasitologist, Dept Vet Sci, Univ Ky, 50-51; asst parasitol, Univ Fla, 51-62; parasitologist, Vet Diag & Invests Lab, Col Vet Med, Univ Ga & Ga Dept Agr, 62-75; RETIRED. Mem: Am Soc Parasitol. Res: Host-parasite relationships and helminth control; investigation of the role of milk-borne nematode larvae in infection of mammalian hosts; study of milk-borne infection and somatic migratory larvae. Mailing Add: 237 Carolina Dr Tifton GA 31794

STONE, WILLIAM SPENCER, b Ogden, Utah, Mar 20, 02; m 28; c 2. TROPICAL MEDICINE. Educ: Univ Idaho, BS, 24, MS, 25; Univ Louisville, MD, 29. Hon Degrees: PhD, Univ Louisville, 48. Prof Exp: Chief protozool & virus dis sect, Army Med Sch, US Army, 34-38, mem, Army Med Res Bd, CZ, 38-39, bacteriologist, Bd Health Lab, Panama, 39-40, chief sanit & labs div, Off Surgeon Gen, 41-43, consult & chief prev med, 43-45, chief prev med div, 45-46, chmn med res & develop bd, 46-50, commandant, Army Med Sci Grad Sch, 50-54; dean sch med, 54-70, DIR MED EDUC & RES, UNIV MD, BALTIMORE CITY, 54-, PROF PATH & MED, 70- Mem: Am Soc Trop Med & Hyg; fel AMA. Res: Bacteriology; immunity; preventive medicine. Mailing Add: Dept of Med Univ of Md Sch of Med Baltimore MD 21201

STONE, WINFIELD S, b Binghamton, NY, Aug 19, 12; m 41; c 4. VETERINARY MEDICINE. Educ: Cornell Univ, DVM, 35, MS, 36, PhD(path), 41. Prof Exp: Instr path, Cornell Univ, 36-41; asst dir livestock dis control, Dept Agr, Mass, 41-44; asst animal indust, Dept Agr, State NY, 44-59, assoc vet, Dept Health, 59-64; prof agr, 64-70, ASST TO PRES, STATE UNIV NY AGR & TECH COL, DELHI, 70- Concurrent Pos: Lectr, Animal Med Ctr, 65-; dir pilot prog, Fed Off Educ, 65-69; mem, Comt Tech Educ, Nat Acad Sci, 66-68. Mem: Am Vet Med Asn; Am Asn Lab Animal Sci. Res: Infectious diseases of animals; laboratory animal medicine; zoonoses. Mailing Add: Off of the Pres State Univ NY Agr & Tech Col Delhi NY 13753

STONEBRAKER, PETER MICHAEL, b Glendale, Calif, Apr 18, 45; m 74. CHEMISTRY. Educ: Whitworth Col, Spokane, Wash, BS, 67; Univ Wash, PhD(org chem), 73. Prof Exp: RES CHEMIST, CHEVRON RES CO, STAND OIL CALIF, 73- Mem: Am Chem Soc. Res: Lubricating oil additives; reaction mechanisms of heterocycles. Mailing Add: Chevron Res Co 576 Standard Ave Richmond CA 94804

STONECYPHER, ROY W, b Atlanta, Ga, Mar 20, 33; m 54; c 3. FORESTRY, GENETICS. Educ: NC State Univ, BS, 59, PhD(forestry), 66. Prof Exp: Proj leader, Int Paper Co, 63-67, res forester, 67-70; assoc prof forest genetics, Okla State Univ, 70-72; QUANT GENETICIST, FORESTRY RES CTR, WEYERHAUSER CO, 72- Concurrent Pos: Adj asst prof, NC State Univ, 67-70; affil assoc prof, Univ Wash, 73- Mem: Soc Am Foresters. Res: Quantitative genetics work in pine populations; applied forest tree breeding; statistical analyses of forestry related research using electronic computers. Mailing Add: Weyerhauser Co Forestry Res Ctr 505 N Pearl St Centralia WA 98531

STONEHAM, RICHARD GEORGE, b Chicago, Ill, Feb 22, 20. MATHEMATICS. Educ: Ill Inst Technol, BSc, 42; Brown Univ, ScM, 44; Univ Calif, PhD(math), 52. Prof Exp: Instr math, Univ Ill, 46-47; asst, Univ Calif, 47-49, Off Naval Res Proj, 51, jr res mathematician, 52, lectr, univ, 51-52, instr math, 52, asst prof, 55-58, mathematician, radiation lab, 54; asst prof math, San Diego State Col, 53-54; lectr math, 59-61, from asst prof to assoc prof, 61-73, PROF MATH, CITY COL, CITY UNIV NEW YORK, 74- Concurrent Pos: Mathematician, Ramo-Wooldridge Corp, 54-55. Mem: Am Math Soc; Math Asn Am; Math Soc France. Res: Applied mathematics; mathematical theory of elasticity; partial differential equations; number theory; normal numbers, especially uniform distributions, exponential sums. Mailing Add: Dept of Math City Col of City Univ New York New York NY 10031

STONEHILL, DAVID L, computer science, see 12th edition

STONEHILL, ROBERT BERRELL, b Philadelphia, Pa, Feb 14, 21; m 44; c 3. INTERNAL MEDICINE. Educ: Temple Univ, BA, 42, MD, 45; Am Bd Internal Med, dipl, 56, re-cert, 74, cert pulmonary dis, 65; Am Bd Prev Med, dipl, 57. Prof Exp: Chief pulmonary physiol lab, Samson AFB Hosp, NY, 55-56, chief pulmonary dis serv, Wilford Hall, US Air Force Hosp, Lackland AFB, Tex, 56-61, chmn dept med, 61-67; PROF MED, SCH MED, IND UNIV, INDIANAPOLIS, 67- Concurrent Pos: Rep to Surgeon Gen, Combined Vet Admin Armed Forces Comt Pulmonary Physiol, 56-57; rep to Surg Gen & mem exec comt, Vet Admin Armed Forces Coccidioidomycosis Coop Study Group, 58-65. Mem: Fel Am Col Physicians; fel Am Col Chest Physicians; fel Royal Soc Med; fel Am Col Prev Med. Res: Pulmonary physiology; aerospace medicine. Mailing Add: Ind Univ Sch of Med 1100 W Michigan St Indianapolis IN 46202

STONEHOUSE, ALBERT JAMES, b Burlington, Iowa, Jan 2, 29; m 50; c 2. INORGANIC CHEMISTRY, PHYSICAL CHEMISTRY. Educ: Iowa State Univ, BS, 49. Prof Exp: Res engr, Brush Beryllium Co, Cleveland, 49-55, mgr chem dept, 55-69, mgr mat sci, 69-71, mgr mat sci, Brush Wellman, Inc, 71-72, DIR BERYLLIUM RES, BRUSH WELLMAN, INC, CLEVELAND, 72- Mem: Electrochem Soc; Am Inst Mining, Metall & Petrol Engrs; Metall Soc. Res: Transition element beryllides; physical metallurgy of beryllium and beryllium corrosion. Mailing Add: 4985 N Sedgewick Lyndhurst OH 44124

STONEHOUSE, HAROLD BERTRAM, b Eng, Apr 13, 22; nat US; m 50; c 4. GEOCHEMISTRY. Educ: Univ London, BSc, 43; Univ Toronto, PhD, 52. Prof Exp: Geologist, Ex-Lands, Nigeria, WAfrica, 43-45; Geologist, Brit Guiana Consol Goldfields, 45-46; geologist & mgr, Can-Guiana Mines, 47-48; assoc geologist, State Geol Surv, Univ Ill, 54-55; from asst prof to assoc prof, 55-74, PROF GEOL, MICH STATE UNIV, 74- Mem: Geol Soc Am. Res: Mineralogy; economic geology; hydrogeology. Mailing Add: Dept of Geol Mich State Univ East Lansing MI 48823

STONEMAN, DAVID MCNEEL, b Madison, Wis, Oct 11, 39; m 64; c 2. MATHEMATICAL STATISTICS. Educ: Univ Wis, BS, 61, MS, 63, PhD(statist), 66. Prof Exp: Math statistician, Forest Prod Lab, Forest Serv, USDA, 64-66; asst prof math & statist, 66-69, assoc prof math, 69-75, PROF MATH, UNIV WIS-WHITEWATER, 75- Mem: Math Asn Am; Am Statist Asn. Res: Experimental design. Mailing Add: Dept of Math Univ of Wis Whitewater WI 53190

STONEMAN, ELVYN ARTHUR, geography, see 12th edition

STONER, ADAIR, b Oklahoma City, Okla, Oct 15, 28; m 53; c 2. ENTOMOLOGY. Educ: Okla State Univ, BS, 56, MS, 60. Prof Exp: Entomologist, Pest Control Div, 58-60, entomologist, Western Cotton Insect Invests, Cotton Insect Br, Entom Res Div, 60-67, res entomologist, 67-72, res entomologist, Western Cotton Res Lab, 72-75, RES ENTOMOLOGIST, HONEY BEE DIS UNIT, AGR RES SERV, USDA, 75- Mem: Entom Soc Am. Res: Effects of pesticides on honey bees. Mailing Add: USDA Honey Bee Lab Univ Sta Box 3168 Laramie WY 82071

STONER, ALLAN K, b Muncie, Ind, July 6, 39; m 62; c 2. HORTICULTURE. Educ: Purdue Univ, West Lafayette, BS, 61, MS, 63; Univ Ill, PhD(hort), 65. Prof Exp: Horticulturist, Crops Res Div, Plant Indust Sta, USDA, Md, 65-71, HORTICULTURIST, VEG LAB, PLANT GENETICS & GERMPLASM INST, AGR RES CTR-W, USDA, 71- Mem: Am Soc Hort Sci. Res: Vegetable breeding and production; breeding of tomatoes. Mailing Add: Plant Genetics & Germplasm Inst USDA Agr Res Ctr-W Beltsville MD 20705

STONER, ALLAN WILBUR, b Tipton, Ind, Sept 15, 31; m 58; c 3. PHYSICAL CHEMISTRY. Educ: Ind Univ, BS, 53; Univ Calif, PhD(chem), 56. Prof Exp: Res scientist, Res Ctr, 56-58, res scientist, Indust Reactor Labs, 58-60, fiber develop mgr, Fiber & Textile Div, 60-66, plant mgr, NC, 66-69, mgr res & develop, Plastic Prod, Ind, 69-71, dir res & develop, Plastic & Indust Prod Div, Oxford Mgt & Res Ctr, 71-73, DIR MKT, INDUST PROD DIV, OXFORD MGT & RES CTR, UNIROYAL, INC, 73- Mem: Am Chem Soc. Res: Nuclear physics and spectroscopy; radiochemistry; application of tracers; radiation and polymer chemistry; physical properties of fibers and elastomeric materials. Mailing Add: Oxford Mgt & Res Ctr Uniroyal Inc Middlebury CT 06749

STONER, CLINTON D, biochemistry, see 12th edition

STONER, ELAINE CAROL BLATT, b New York, NY, Dec 31, 39; m 65; c 2. PHYSICAL CHEMISTRY. Educ: Brooklyn Col, BS, 61; Univ Calif, Berkeley, PhD(chem), 64. Prof Exp: NIH fel, Univ Wis, 64-65; asst ed electrochem & anal chem elec phenomena, 65-68, assoc ed elec phenomena, 68-69, SR ASSOC INDEXER ELEC PHENOMENA, CHEM ABSTR SERV, AM CHEM SOC, 69- Mem: Am Chem Soc. Res: Nuclear magnetic resonance of exchange rates of ligands in coordination complexes; solid state physics; semiconductors; superconductors; electric phenomena. Mailing Add: Chem Abstr Serv Dept 57 Columbus OH 43210

STONER, GARY DAVID, b Bozeman, Mont, Oct 25, 42; m 69; c 2. MICROBIOLOGY. Educ: Mont State Univ, BS, 64; Univ Mich, MS, 68, PhD(microbiol), 70. Prof Exp: Asst res scientist, 70-75, ASSOC RES SCIENTIST, DEPT COMMUNITY MED, SCH MED, UNIV CALIF, SAN DIEGO, 75- Concurrent Pos: Nat Heart & Lung Inst young investr grant pulmonary res, 74. Mem: Am Asn Cancer Res; Am Tissue Cult Asn; Am Soc Microbiol. Res: Chemical and viral carcinogenesis; differentiation of cultured mammalian lung cells. Mailing Add: Univ of Calif Sch of Med PO Box 109 La Jolla CA 92037

STONER, GEORGE GREEN, b Wilkinsburg, Pa, Jan 29, 12; m 40; c 3. ORGANIC CHEMISTRY. Educ: Col Wooster, AB, 34; Ohio State Univ, AM, 36; Princeton Univ, PhD(org chem), 39. Prof Exp: Res chemist, Wallace Labs, 39 & Columbia Chem Div, Pittsburgh Plate Glass Co, 39-42; res engr, Battelle Mem Inst, 42-48; res chemist, Gen Aniline & Film Corp, 48-50, group leader, 50-51, res fel, 52-56; mgr prod develop, Avon Prod, Inc, 56-65; supvr patent liaison, 65-70, MGR INFO SERV, J P STEVENS & CO, INC, GARFIELD, NJ, 70- Concurrent Pos: Instr, Ohio State Univ, 47-48. Mem: Fel AAAS; Am Chem Soc; fel Am Inst Chemists. Res: Fat acids; tall oil; plasticizers; organic sulfur chemistry; alkyl diselenides; Reppe chemistry; photosensitizing dyes; allyl resins; polyethylene; hydrocarbon oxidation; peroxides; technical writing; information processing; management; leather preservation; cosmetics; fiber modification. Mailing Add: 2 Parkside Dr Suffern NY 10901

STONER, GLENN EARL, b Springfield, Mo, Oct 26, 40; m 62; c 3. ELECTROCHEMISTRY. Educ: Univ Mo-Rolla, BS, 62, MS, 63; Univ Pa, PhD(chem), 68. Prof Exp: Sr scientist mat sci, Sch Eng, Univ Va, 68-71; vis assoc prof chem, Univ Mo-Rolla, 71; lectr chem eng, Univ Va, 71-72, sr scientist mat sci, 72-73; vis assoc prof electrochem, Fac Sci, Univ Rouen, France, 73-74; RES ASSOC PROF MAT SCI, UNIV VA, 74- Concurrent Pos: Consult, Owens-Ill, Inc, 75- Honors & Awards: Cert Recognition, NASA, 75. Mem: Electrochem Soc. Res: Applied research in bioelectrochemistry and biomaterials research; interaction with industry towards development of innovative concepts. Mailing Add: Dept of Mat Sci Univ of Va Charlottesville VA 22901

STONER, GRAHAM ALEXANDER, b Saginaw, Mich, June 13, 29; m 55; c 4. ANALYTICAL CHEMISTRY, AGRICULTURAL CHEMISTRY. Educ: Univ Mich, BS, 51, MS, 52; Tulane Univ La, PhD(chem), 55. Prof Exp: Chemist, Dow Chem Co, 55-58, prof leader, 58-60; chemist, Ethyl Corp, 60-62; mgr anal chem, Bioferm Div, Int Minerals & Chem Corp, Calif, 62-64, assoc dir anal labs, Chem Div, 64-67; dir anal & tech serv, IMC Growth Sci Ctr, Ill, 67-69; plant mgr, Infotronics Corp, Tex, 69-70; dir res & develop spec prod div, Kennecott Copper Corp, 70-74; vpres technol, 74-75, VPRES MKT & DEVELOP, KOCIDE CHEM CORP, 75- Concurrent Pos: Guest scientist, Brookhaven Nat Lab, 56-57. Mem: AAAS; Am Chem Soc; Sigma Xi; Hyacinth Control Soc; NY Acad Sci. Res: Pesticide research, testing, registration, formulation; governmental regulations; enzymatic methods of analysis; automated analysis; radiochemistry. Mailing Add: 12701 Almeda Rd Houston TX 77045

STONER, JOHN CLARK, b Toledo, Ohio, Feb 26, 33; m 54; c 2. VETERINARY MEDICINE. Educ: Ohio State Univ, DVM, 60. Prof Exp: Clin res vet, Agr Div, Am Cyanamid Co, NJ, 60-63; sr res vet, Ciba Res Farm, 63-69; asst dir animal health documentation, Squibb Agr Res Ctr, 69-71, DIR VET PROFESSIONAL & REGULATORY AFFAIRS, E R SQUIBB & SONS, INC, 71- Mem: Am Vet Med Asn; Indust Vet Asn. Res: Clinical research for animal health pharmaceuticals and drug regulatory affairs. Mailing Add: E R Squibb & Sons Inc PO Box 4000 Princeton NJ 08540

STONER, JOHN OLIVER, JR, b Milton, Mass, Oct 4, 36; m 60; c 4. ATOMIC SPECTROSCOPY. Educ: Pa State Univ, BS, 58; Princeton Univ, MA, 59,

PhD(physics), 64. Prof Exp: Res assoc physics, Univ Wis, 63-66, asst prof, 66-67; ASST PROF PHYSICS, UNIV ARIZ, 67- Concurrent Pos: Mem comt line spectra, Nat Res Coun, 75- Mem: Am Phys Soc; AAAS; Optical Soc Am. Res: Atomic and beam-foil spectroscopy; atomic beam and other techniques for producing narrow spectral lines. Mailing Add: Dept of Physics Univ of Ariz Tucson AZ 85721

STONER, MARSHALL ROBERT, b Kenesaw, Nebr, Sept 24, 38. ORGANIC CHEMISTRY. Educ: Hastings Col, BA, 60; Iowa State Univ, PhD(chem), 64. Prof Exp: Asst prof, 64-70, ASSOC PROF CHEM, UNIV SDAK, 70- Mem: AAAS; Am Chem Soc; The Chem Soc. Res: Synthesis and rearrangements of bicyclic compounds; intramolecular photochemical cycloaddition reactions in bicyclic compounds; photochemical reactions of alcohols with unsaturated acids. Mailing Add: Dept of Chem Univ of SDak Vermillion SD 57069

STONER, MARTIN FRANKLIN, b Pasadena, Calif, Jan 19, 42; m 63. PLANT PATHOLOGY, MYCOLOGY. Educ: Calif State Polytech Col, BS, 63; Wash State Univ, PhD(plant path), 67. Prof Exp: From asst prof to assoc prof, 67-75, PROF BOT, CALIF STATE POLYTECH UNIV, POMONA, 75- Mem: AAAS; Am Phytopath Soc; Am Soc Plant Physiol; Bot Soc Am; Mycol Soc Am. Res: Host-parasite interactions; soil-borne fungi; microbial ecology; general plant pathology and mycology. Mailing Add: Dept of Biol Sci Calif State Polytech Univ Pomona CA 91768

STONER, RICHARD DEAN, b Newhall, Iowa, Mar 29, 19; m 45; c 2. IMMUNOLOGY. Educ: Univ Iowa, BA, 40, PhD(zool), 50. Prof Exp: Jr scientist, 50-52, assoc med bacteriologist, 52-54, from asst scientist to scientist, 52-62, SR SCIENTIST, MED DEPT, BROOKHAVEN NAT LAB, 62- Concurrent Pos: Consult, Off Surgeon Gen & Dep Dir Comn on Radiation & Infection, Armed Forces Epidemiol Bd, 63- Mem: Am Inst Biol Sci; Am Soc Microbiol; Radiation Res Soc; Am Soc Parasitol; Am Soc Exp Pathologists. Res: Radiation effect upon immune mechanisms; antibody formation; cellular defense mechanism; anaphylaxis; immunity to parasitic infections. Mailing Add: Med Dept Brookhaven Nat Lab Upton NY 11973

STONER, RICHARD GRIFFITH, b Buffalo, NY, June 13, 19; m 46; c 3. PHYSICS. Educ: Princeton Univ, AB, 41, MA, 46, PhD(physics), 47. Prof Exp: Asst physics, Columbia Univ, 41-42; asst physicist, Nat Defense Res Comt, Princeton Univ, 42-46, instr physics, Univ, 47-48; from asst prof to assoc prof, Pa State Univ, 48-60, actg head dept, 60-61, prof, 60-63; chmn dept, 63-74, PROF PHYSICS, ARIZ STATE UNIV, 63- Concurrent Pos: Proj scientist, Air Force Off Sci Res, Wash, DC, 62-63. Mem: AAAS; Am Phys Soc; Am Asn Physics Teachers. Res: Spectroscopy; fluid dynamics; attenuation of spherical shock waves in air; refraction of shock waves. Mailing Add: Dept of Physics Ariz State Univ Tempe AZ 85281

STONER, RONALD EDWARD, b Indianapolis, Ind, Nov 25, 37; m 60; c 2. SOLID STATE PHYSICS. Educ: Wabash Col, BA, 59; Purdue Univ, MS, 61, PhD(physics), 66. Prof Exp: From asst prof to assoc prof, 66-74, PROF PHYSICS, BOWLING GREEN STATE UNIV, 74- Mem: Am Phys Soc; Sigma Xi. Res: Astrophysics; computational physics; theoretical physics. Mailing Add: Dept of Physics Bowling Green State Univ Bowling Green OH 43403

STONER, WARREN NORTON, b Iowa City, Iowa, Apr 12, 22; m 43; c 4. VIROLOGY, ENTOMOLOGY. Educ: Univ Calif, BS, 43, PhD(entom), 49. Prof Exp: Asst entom, Univ Calif, 46-49; asst plant pathologist, Exp Sta, Univ Fla, 49-53; entomologist & plant pathologist, Ministry Agr, Venezuela, 54; entomologist & plant pathologist, Exp Sta, Univ RI, 57-58; virologist, Minister Agr, Nigeria, 59-60; asst vis prof zool, Univ RI, 60; RES ENTOMOLOGIST, NORTHERN GRAIN INSECT RES LAB, AGR RES SERV, USDA, 61- Concurrent Pos: Prof, Grad Div, SDak State Univ. Mem: Entom Soc Am; Am Phytopath Soc. Res: Insect transmission of plant pathogens; insect ecology; economic entomology; zoogeography. Mailing Add: Northern Grain Insect Res Lab Agr Res Serv USDA Brookings SD 57006

STONER, WILLIAM WEBER, b Columbus, Ohio, June 4, 44. OPTICAL PHYSICS, RADIOLOGICAL PHYSICS. Educ: Union Col, NY, BS, 66; Princeton Univ, PhD(physics), 75. Prof Exp: Scientist radiol, Machlett Labs, Raytheon, Inc, Stamford, Conn, 73-75, SCIENTIST NUCLEAR MED, RAYTHEON RES DIV, 75- Mem: Optical Soc Am. Res: Development of imaging devices for medical diagnosis in the fields of ultrasound, radiology, and nuclear medicine. Mailing Add: Raytheon Res Div 28 Seyon St Waltham MA 02154

STONES, ROBERT C, b Portland, Ore, May 19, 37; m 57; c 8. ENVIRONMENTAL PHYSIOLOGY. Educ: Brigham Young Univ, BS, 59, MS, 60; Purdue Univ, West Lafayette, PhD(environ physiol), 64. Prof Exp: From asst prof to assoc prof, 64-70, PROF PHYSIOL, MICH TECHNOL UNIV, 70-, HEAD DEPT BIOL SCI, 70- Concurrent Pos: Mem, Hibernation Info Exchange, 64-; NSF res grants, 67-71. Mem: Am Asn Higher Educ; Nat Asn Biol Teachers; Am Forestry Asn; Am Soc Mammal; Australian Soc Mammal. Res: Thermal regulation of hibernating species of bats; comparative and animal physiology; comparative anatomy. Mailing Add: Dept of Biol Sci Mich Technol Univ Houghton MI 49931

STONEY, SAMUEL DAVID, JR, b Charleston, SC, Dec 20, 39; m 59; c 2. NEUROSCIENCE. Educ: Univ SC, BS, 62; Tulane Univ, PhD(physiol), 66. Prof Exp: From instr to asst prof physiol, New York Med Col, 66-70; NIH res grant, 70, asst prof, 70-74, ASSOC PROF PHYSIOL, MED COL GA, 74- Mem: Am Physiol Soc. Res: Electrophysiological studies of the organization of motor sensory cortex and pyramidal motor systems. Mailing Add: Dept of Physiol Med Col of Ga Augusta GA 30902

ST-ONGE, DENIS ALDERIC, b Ste-Agathe, Man, May 11, 29; m 55; c 2. GEOMORPHOLOGY. Educ: St-Boniface Col, Man, BA, 51; Cath Univ, Louvain, LicSc, 57, DocSc(geog), 62. Prof Exp: Teacher elem sch, Sask, 51-52; teacher high sch, Ethiopia, 53-55; teacher, Col Jean de Brebeuf, Montreal, 57-58; geographer, Geog Br, Dept Mines & Technol Surv, 58-65; res scientist, Geol Surv Can, 65-68; prof geomorphol, Univ Ottawa, 68-70; res scientist, Geol Surv Can, 70-74; CHMN DEPT GEOG, UNIV OTTAWA, 74- Concurrent Pos: Nat Res Coun Can-NATO fel, 61-62; part-time prof, Dept Geog, Univ Ottawa, 70-74. Mem: Geol Asn Can; Can Asn Geog; corresp mem Int Union Geod & Geophys. Res: Urban geology; geology and planning; erosion control in oil development regions. Mailing Add: Dept of Geog Univ of Ottawa 78 Laurier Est Ottawa ON Can

STONIER, TOM TED, b Hamburg, Ger, Apr 29, 27; nat US; m 53; c 5. SCIENCE POLICY, CELL PHYSIOLOGY. Educ: Drew Univ, AB, 50; Yale Univ, MS, 51, PhD, 55. Prof Exp: Asst, Yale Univ, 51-52; jr res assoc biol, Brookhaven Nat Lab, 52-54; vis investr, Rockefeller Inst, 54-57, res assoc, 57-62; assoc prof biol, Manhattan Col, 62-71, prof biol & dir peace studies prog, 71-75; PROF SCI & SOCIETY, UNIV BRADFORD, 75- Concurrent Pos: USPHS fel, 54-56; Damon Runyon Mem fel, 56-57; consult, Living Sci Labs, 61-62, Hudson Inst, 65-69, MacMillan Co, 68, Environ Defense Fund, 68-70 & Drew Univ, 69-71; instr, New Sch Social Res, 68-70 & State Univ NY Col Purchase, 72. Mem: AAAS; Am Soc Plant Physiol; Fedn Am Sci (secy, 66-67); NY Acad Sci; Scand Soc Plant Physiol. Res: Impact of science and technology on society; cell physiology of plant growth, cancer and ageing. Mailing Add: Sch of Sci & Society Univ of Bradford Bradford England

STOODLEY, GERALD RALPH, mathematics, computer science, see 12th edition

STOOKEY, GEORGE K, b Waterloo, Ind, Nov 6, 35; m 54; c 4. DENTISTRY. Educ: Ind Univ, AB, 57, MS, 62, PhD, 71. Prof Exp: Dir lab res, 63-64, asst dir, Prev Dent Res Inst, 69-72, exec secy, Oral Health Res Inst, 72-74, ASST PROF PREV DENT, SCH DENT, IND UNIV-PURDUE UNIV, INDIANAPOLIS, 64-, ASSOC DIR, ORAL HEALTH RES INST, 74- Mem: AAAS; Int Asn Dent Res; Am Dent Asn; Am Soc Prev Dent. Res: Metabolism of fluoride and other trace elements in experimental animals and humans; various types of dental caries preventive measures, including fluorides and various aspects of nutrition. Mailing Add: Oral Health Res Inst Ind Univ-Purdue Univ Indianapolis IN 46202

STOOKEY, STANLEY DONALD, b Hay Spring, Nebr, May 23, 15; m; c 3. PHYSICAL CHEMISTRY. Educ: Coe Col, AB, 36; Mass Inst Technol, PhD(phys chem), 40. Prof Exp: Res chemist, 40-58, mgr fundamental chem res, 58-62, DIR FUNDAMENTAL CHEM RES, CORNING GLASS WORKS, 62- Mem: Am Ceramics Soc; Am Chem Soc. Res: Glass composition; photosensitive, photochronic and opal glasses; glass ceramics. Mailing Add: Res & Develop Labs Corning Glass Works Corning NY 14830

STOOL, JOSEPH A, medicine, deceased

STOOLMILLER, ALLEN CHARLES, b Battle Creek, Mich, Nov 3, 40; m 61; c 2. BIOLOGICAL CHEMISTRY. Educ: Western Reserve Univ, AB, 61; Univ Mich, MA, 63, PhD(biochem), 66. Prof Exp: Fel, Chicago & Ill Heart Asns, 66-68; instr pediat, Univ, 68-69, ASST PROF PEDIAT & RES ASSOC BIOCHEM, DEPT PEDIAT & LA RABIDA INST, UNIV CHICAGO, 69- Mem: Am Chem Soc. Res: Mucopolysaccharide biosynthesis, control and regulation; tissue culture; subcellular fractionation and membrane biochemistry. Mailing Add: Dept of Pediat Univ of Chicago Chicago IL 60637

STOOPS, JAMES KING, b Charleston, WVa, Sept 15, 37; m 62; c 2. BIOCHEMISTRY. Educ: Duke Univ, BS, 60; Northwestern Univ, Evanston, PhD(chem), 66. Prof Exp: Sr demonstr biochem, Univ Queensland, 66-67, Australian Res Comt grants fel biochem, 67-70; NIMH fel, Duke Univ, 70-71; ASST PROF BIOCHEM, BAYLOR COL MED, 71- Mem: Am Chem Soc. Res: Enzymology and protein chemistry. Mailing Add: Dept of Biochem Baylor Col of Med Houston TX 77025

STOPFORD, WOODHALL, b Jersey City, NJ, Feb 25, 43; m 66. INTERNAL MEDICINE, CLINICAL TOXICOLOGY. Educ: Dartmouth Col, BA, 65; Dartmouth Med Sch, BMS, 67; Harvard Univ, MD, 69. Prof Exp: ASST PROF COMMUNITY HEALTH SCI, DUKE MED CTR, DUKE UNIV, 73- Mem: Am Acad Occup Med; Am Occup Med Asn; Am Indust Hyg Asn. Res: Clinical toxicologic studies of heavy metal and chlorinated hydrocarbon exposures; pharmacokinetics of heavy metals in man. Mailing Add: Duke Med Ctr PO Box 2914 Durham NC 27710

STOPHER, EMMET CARSON, b Noblesville, Ind, June 5, 10; m 36; c 4. MATHEMATICS. Educ: Miami Univ, AB, 32; Kent State Univ, BS, 32; Univ Iowa, MS, 33, PhD(math), 37. Prof Exp: From asst prof to assoc prof math, Ashland Col, 37-41; asst prof math & sci, State Univ NY Teachers Col Brockport, 41-47; asst prof math, Miami Univ, 47-49; prof & head dept, Ft Hays Kans State Col, 49-57; PROF MATH, STATE UNIV NY COL OSWEGO, 57- Mem: Am Math Soc; Math Asn Am. Res: Point set theory; objective tests; topology. Mailing Add: Dept of Math State Univ of NY Col Oswego NY 13126

STOPKIE, ROGER JOHN, b Perth Amboy, NJ, July 17, 39; m 62; c 1. MICROBIOLOGY, BIOCHEMISTRY. Educ: St Lawrence Univ, BS, 61; St Louis Univ, PhD(microbial physiol), 68. Prof Exp: Res asst biochem & microbiol, Merck & Co, 62-64; res biochemist, 69-75, SR RES & INFO SCIENTIST, ICI US INC, 75- Mem: AAAS; Am Soc Microbiologists; Am Chem Soc; Sigma Xi; Leukemia Soc Am. Res: Biology of mycoplasma; information systems; enzyme regulation; inflammation biochemistry; neurochemistry. Mailing Add: ICI US Inc Wilmington DE 19897

STOPPER, WILLIAM W, b Toronto, Can, Apr 21, 18; US citizen; m 40; c 2. POULTRY HUSBANDRY, AGRICULTURAL ECONOMICS. Educ: Pa State Univ, BS, 39; Cornell Univ, MS, 54. Prof Exp: Teacher poultry, State Univ NY Agr & Tech Col Alfred, 46-54; consult, Agency Int Develop, Israel, 54-56; head agr dept, State Univ NY Agr & Tech Co Alfred, 56-60; consult poultry, Agency Int Develop, India, 60-62; CHMN AGR DIV, STATE UNIV NY AGR & TECH COL ALFRED, 62- Res: Extension education. Mailing Add: Div of Agr State Univ of NY Agr & Tech Col Alfred NY 14802

STOPPS, GORDON JAMES, b London, Eng, Feb 12, 26; m 54; c 3. INTERNAL MEDICINE, PHYSIOLOGY. Educ: Univ London, MB, BS, 50. Prof Exp: Asst path, New York Hosp-Cornell Med Ctr, 52-53; asst resident internal med, Shaughnessy Hosp, Vancouver, BC, 53-54; sr intern pediat, Hosp Sick Children, Toronto, Ont, 55-56; med officer, Int Nickel Co, 56-58; asst resident internal med, Royal Victoria Hosp, Montreal, Que, 58-59; physiologist, 59-64, chief physiol sect, 64-69, ASST DIR, HASKELL LAB, E I DU PONT DE NEMOURS & CO, INC, 69- Mem: AAAS; Am Pub Health Asn; Am Indust Hyg Asn; AMA. Res: Application of physiology to ergonomics and toxicology. Mailing Add: Haskell Lab E I du Pont de Nemours & Co Inc Newark DE 19711

STORAASLI, JOHN PHILLIP, b St Paul, Minn, Jan 28, 21; m 50; c 2. MEDICINE. Educ: Univ Minn, BS, 44, MB, 45, MD, 46. Prof Exp: Res asst, AEC Proj, Sch Med, 47-48, resident, Univ Hosp, 48-50, from instr to assoc prof, 50-61, PROF RADIOL, SCH MED, CASE WESTERN RESERVE UNIV, 61-, RES ASSOC, 56-, ASSOC RADIOLOGIST, HOSP, 50- Mem: Radiation Res Soc; Radiol Soc NAm; Am Roentgen Ray Soc; Am Soc Therapeut Radiol; fel Am Col Radiol. Res: Clinical therapeutic radiology; biological effects of ionizing radiation; diagnostic and therapeutic uses of radioactive isotopes. Mailing Add: 2065 Adelbert Rd Cleveland OH 44106

STORB, URSULA, b Stuttgart, Ger. IMMUNOBIOLOGY. Educ: Univ Tübingen, MD, 60. Prof Exp: Asst prof, 72-75, ASSOC PROF MICROBIOL, UNIV WASH, 75- Concurrent Pos: NIH res grants, 72- Mem: Am Asn Immunol; Am Soc Cell Biol; Reticuloendothelial Soc; Asn Women in Sci. Res: Organization of immunoglobulin genes; control of antibody gene expression. Mailing Add: Microbiol & Immunol SC-42 Univ of Wash Seattle WA 98195

STORCH, RICHARD HARRY, b Evanston, Ill, Mar 16, 37; m 63; c 2. ENTOMOLOGY. Educ: Carleton Col, BA, 59; Univ Ill, MS, 61, PhD(entom), 66.

Prof Exp: Temporary asst prof entom, USDA, 65-66, asst prof, 66-69, ASSOC PROF ENTOM, UNIV MAINE, ORONO, 69- Mem: Entom Soc Am; Entom Soc Can. Res: Embryonic and postembryonic development of cervicothoracic structure and musculature; behavior and ecology of Coccinellidae. Mailing Add: Dept of Entom 209 Deering Hall Univ of Maine Orono ME 04473

STORCK, ROGER LOUIS, b Brussels, Belgium, Feb 22, 23; US citizen; m 49; c 1. MICROBIOLOGY, BIOCHEMISTRY. Educ: Indust Fermentation Inst, Brussels, MS, 46; Univ Ill, Urbana, PhD(microbiol), 60. Prof Exp: Asst microbiol, Indust Fermentation Inst, Brussels, 46-48, instr, 48-54 & 55-57; res assoc bact, Univ Ill, Urbana, 57-60, fel microbiol, 60-61; from asst prof to assoc prof, Univ Tex, 61-66; PROF BIOL, RICE UNIV, 66- Mem: AAAS; Am Soc Microbiol; Biophys Soc; Mycol Soc Am; Am Soc Biol Chemists. Res: Bacterial metabolism; enzymes localization; ribosomes and nucleic acids in morphogenesis; systematics and phylogeny of fungi; biochemistry and genetics of fungal morphogenesis. Mailing Add: Dept of Biol Rice Univ Houston TX 77001

STORER, EDWARD HAMMOND, b Rockland, Maine, Mar 1, 21; m 44; c 3. SURGERY. Educ: Univ Chicago, SB, 43, MD, 45; Am Bd Surg, dipl, 56. Hon Degrees: MA, Yale Univ, 70. Prof Exp: Asst physiol, Univ Chicago, 43-44, from res asst to res assoc surg, 48-50; res assoc, Univ Wash, 51-55; from asst prof to prof, Col Med, Univ Tenn, Memphis & dir res, 55-69, assoc prof physiol, 63-69; PROF SURG, SCH MED, YALE UNIV, 69-, ASSOC DEAN, SCH MED, 74-; CHIEF SURG SERV, VET ADMIN HOSP, WEST HAVEN, 69-, CHIEF STAFF, 74- Concurrent Pos: Former consult, Vet Admin Hosp & US Naval Hosps, Memphis, Tenn & Blytheville AFB Hosp. Honors & Awards: Gold Medal, AMA, 50. Mem: Am Physiol Soc; Soc Univ Surgeons; Am Col Surgeons; Am Gastroenterol Asn; Am Surg Asn. Res: Surgical physiology of the gastrointestinal tract. Mailing Add: Dept of Surg Vet Admin Hosp W Spring St West Haven CT 06516

STORER, JOHN B, b Rockland, Maine, Oct 16, 23; m 45; c 4. RADIOBIOLOGY. Educ: Univ Chicago, MD, 47. Prof Exp: Intern, Mary Imogene Bassett Hosp, Cooperstown, NY, 47-48; USPHS res fel path, Univ Chicago, 48-49, res assoc, Toxicity Lab, 49-50; staff mem, Biomed Res Group, Los Alamos Sci Lab, 50-58; staff scientist, Jacksom Mem Lab, 58-67; dep dir div biol & med, AEC, Md, 67-69; sci dir path & immunol, 69-75, DIR BIOL DIV, OAK RIDGE NAT LAB, 75- Concurrent Pos: Alt leader, Biomed Res Group & Leader Radiobiol Sect, Los Alamos Sci Lab, 52-58; mem subcomt relative biol effectiveness, Nat Coun Radiation Protection, 57-62; consult, Argonne Nat Lab, 59-67; mem, Radiation Study Section, NIH, 62-66 & 71-75, chmn, 72-75; mem adv comt, Atomic Bomb Casualty Comn, 69-74; mem subcomt radiobiol, Nat Coun Radiation Protection & Measurements, 69-, mem bd dirs, 75-; mem adv comt biol & med to AEC Sci Secy, 69-73; mem sci adv bd, Nat Ctr Toxicol Res, 72-75. Honors & Awards: E O Lawrence Award, 68. Mem: Radiation Res Soc; Am Soc Exp Path; Am Asn Cancer Res; Geront Soc; Soc Exp Biol & Med. Res: Late effects of ionizing radiation; aging. Mailing Add: Biol Div Oak Ridge Nat Lab Oak Ridge TN 37830

STORER, KEN RODGER, agronomy, see 12th edition

STORER, ROBERT WINTHROP, b Pittsburgh, Pa, Sept 20, 14; m 55; c 2. ZOOLOGY. Educ: Princeton Univ, AB, 36; Univ Calif, MA, 42, PhD(zool), 49. Prof Exp: Tech asst, Mus Vert Zool, Univ Calif, 41-42, mus technician, 48-49, assoc, Div Entom & Parasitol, Exp Sta, 45, asst zool, Univ, 46-48; from instr to assoc prof zool, Univ, 49-63, asst cur birds, Mus Zool, 49-56, PROF ZOOL, UNIV MICH, ANN ARBOR, 63-, CUR BIRDS, MUS ZOOL, 56- Concurrent Pos: Ed, The Auk, Am Ornith Union, 53-57, ed, Ornith Monogr, 63-70; mem comt, Int Ornith Cong, 58- Mem: Soc Study Evolution; Wilson Ornith Soc; Cooper Ornith Soc (vpres, 70); fel Am Ornith Union (pres, 70-72); Brit Ornith Union. Res: Avian morphology; systematics; distribution; paleontology; avian behavior. Mailing Add: Mus of Zool Univ of Mich Ann Arbor MI 48104

STORER, THOMAS, US citizen. MATHEMATICS. Educ: Univ Calif, Los Angeles, BA, 59; Univ Southern Calif, PhD, 64. Prof Exp: Mem, Inst Advan Study, 64-65; ASSOC PROF MATH, UNIV MICH, ANN ARBOR, 65- Res: Easy mathematics. Mailing Add: Dept of Math Univ of Mich Ann Arbor MI 48104

STOREY, ARTHUR THOMAS, b Sarnia, Ont, July 8, 29; m 64; c 3. PHYSIOLOGY, ORTHODONTICS. Educ: Univ Toronto, DDS, 53; Univ Mich, MS, 60, PhD(physiol), 64. Prof Exp: From instr to asst prof orthod, Sch Dent & Physiol & Sch Med, Univ Mich, 62-66; assoc prof, Fac Dent & asst prof physiol, Fac Med, 66-70, PROF DENT, FAC DENT & ASSOC PROF PHYSIOL, FAC MED, UNIV TORONTO, 70- Mem: Am Physiol Soc; Am Asn Orthod; Can Dent Asn; Int Asn Dent Res. Res: Oral, pharyngeal and laryngeal receptors and reflexes. Mailing Add: Fac of Dent Univ of Toronto Toronto ON Can

STOREY, BAYARD THAYER, b Boston, Mass, July 13, 32; m 58; c 4. CELL PHYSIOLOGY, PHYSICAL BIOCHEMISTRY. Educ: Harvard Univ, AB, 52, PhD(phys org chem), 58; Mass Inst Technol, MS, 55. Prof Exp: Res chemist, Ion Exchange Lab, Rohm and Haas Co, 58-60, head ion exchange synthesis lab, 60-65; Nat Inst Gen Med Sci spec fel, Johnson Res Found, 65-67, asst prof phys biochem, Univ, 67-73, ASSOC PROF OBSTET & GYNEC, PHYSIOL & PHYS BIOCHEM, UNIV PA, 73- Mem: Am Chem Soc; Am Soc Plant Physiol; Soc Study Reproduction; Am Fertil Soc; Am Soc Biol Chemists. Res: Energy metabolism in spermatozoa; energy conservation mechanism in mitochondrial membranes. Mailing Add: Dept of Obstet & Gynec Univ of Pa Philadelphia PA 19174

STOREY, JAMES BENTON, b Avery, Tex, Oct 25, 28; m 48; c 2. POMOLOGY, PLANT PHYSIOLOGY. Educ: Tex A&M Univ, BS, 49, MS, 53; Univ Calif, Los Angeles, PhD(bot sci), 57. Prof Exp: Asst county agr agt, Tex Agr Exten Serv, 49-52, asst hort, 52-53; asst plant physiol, Univ Calif, Los Angeles, 53-57; from asst prof to assoc prof pomol, 57-74, PROF POMOL, TEX A&M UNIV, 74- Concurrent Pos: Ed, Pecan Quart & Tex Horticulturist; exec dir, Tex Pecan Producer's Bd. Mem: Am Soc Hort Sci; Am Soc Plant Physiol; Int Soc Hort Sci; Am Pomol Soc. Res: Control of vegetative and fruiting responses in pecans; nutrition, salinity and post-harvest studies in pecans; coordinator pecan research program in Texas. Mailing Add: Hort Sect Tex A&M Univ College Station TX 77843

STOREY, PATRICK BRENDAN, b Co Wicklow, Ireland, Apr 9, 24; nat US; m 47; c 4. MEDICINE. Educ: Georgetown Univ, MD, 47; Am Bd Internal Med, dipl, 55. Prof Exp: Intern, DC Gen Hosp, 47-48, resident pulmonary dis, 48-49, asst resident internal med, 49-50, chief resident, 50-51, assoc med officer, Div Pulmonary Dis, 51-53; staff physician, Baltimore Vet Admin Hosp, 55-56, chief med serv, 56-58, dir prof serv, 58-60; from asst prof to assoc prof med, Univ Md, 56-64; dir dept postgrad progs, AMA, 64-66; prof commun med & head dept, Hahnemann Med Col, 66-72; PROF MED & COMMUN MED, SCH MED, UNIV PA, 72-, ASSOC DEAN SCH MED, 72-; MED DIR, PHILADELPHIA GEN HOSP, 75- Concurrent Pos: Asst prof, Johns Hopkins Univ, 59-60; dir, Pa Urban Health Serv Ctr & med dir, Grad Hosp, Univ Pa, 72-75. Mem: Am Thoracic Soc; Am Heart Asn; fel Am Col Physicians; fel Am Col Chest Physicians; fel Am Col Prev Med. Res: Internal medicine; pulmonary diseases. Mailing Add: Philadelphia Gen Hosp 700 Civic Center Blvd Philadelphia PA 19104

STOREY, THEODORE GEORGE, b Fresno, Calif, Sept 6, 23; m 46; c 5. FORESTRY. Educ: Univ Calif, BS, 48, MS, 68. Prof Exp: Forester, Hammond Lumber Co, 48-49; forester forest influences res, 49-52, forester fire res, 52-59, forester, Southern Forest Fire Lab, 59-62, FORESTER, RIVERSIDE FOREST FIRE LAB, US FOREST SERV, 62- Mem: Soc Am Foresters; Am Geophys Union. Res: Fire management and control systems; fire behavior; forest and urban fire and blast damage from nuclear weapons; watershed management. Mailing Add: 1520 Ransom Pl Riverside CA 92506

STOREY, WILLIAM BICKNELL, genetics, plant breeding, see 12th edition

STORFER, STANLEY J, b Brooklyn, NY, July 31, 30; m 56; c 2. ORGANIC CHEMISTRY. Educ: Polytech Inst Brooklyn, BS, 54, PhD(org chem), 60. Prof Exp: Jr chemist, Am Cyanamid Co, 54-56; chemist, Esso Res & Eng Co, 60-63, from sr chemist to sr res chemist, 63-73, RES ASSOC, EXXON CHEM-TECHNOL DEPT, 73- Mem: Am Chem Soc. Res: Rheology of water-soluble polymer solutions; new product applications; statistical design of experiments; solvents technical service and market development. Mailing Add: 24 Ten Eyck Pl Edison NJ 08817

STORHOFF, BRUCE NORMAN, b Lanesboro, Minn, Jan 2, 42. INORGANIC CHEMISTRY. Educ: Luther Col, Iowa, BA, 64; Univ Iowa, PhD, 69. Prof Exp: Asst prof, 68-69, 70-75, ASSOC PROF CHEM, BALL STATE UNIV, 75- Concurrent Pos: Fel, Ind Univ, 69-70. Mem: Am Chem Soc. Res: Organic derivatives of transition metals; chemistry of carboranes. Mailing Add: Dept of Chem Ball State Univ Muncie IN 47306

STORK, DANIEL FRANKLIN, mathematics, see 12th edition

STORK, DONALD HARVEY, b Minn, Mar 22, 26; m 48; c 6. NUCLEAR PHYSICS. Educ: Carleton Col, BA, 48; Univ Calif, PhD(phyiscs), 53. Prof Exp: Asst physics, Univ Calif, 48-51, Lawrence Radiation Lab, 51-53, res assoc, 53-56, asst prof, Univ Calif, Los Angeles, 56-59, assoc prof, 59-64, PROF PHYSICS, UNIV CALIF, LOS ANGELES, 64- Mem: Fel Am Phys Soc. Res: Pions; K mesons; hyperons and antiprotons; production; beams; interactions; decay. Mailing Add: Dept of Physics Univ of Calif Los Angeles CA 90024

STORK, GILBERT JESSE, b Brussels, Belgium, Dec 31, 21; nat US; m 44; c 4. SYNTHETIC ORGANIC CHEMISTRY. Educ: Univ Fla, BS, 42; Univ Wis, PhD(chem), 45. Hon Degrees: DSc, Lawrence Col, 61. Prof Exp: Sr res chemist, Lakeside Labs, Inc, 45-46; instr chem, Harvard Univ, 46-48, asst prof, 48-53; from assoc prof to prof, 53-67, chmn dept, 73-76, EUGENE HIGGINS PROF CHEM, COLUMBIA UNIV, 67- Concurrent Pos: Consult, NSF, 58-61, US Army Res Off, 66-69, NIH, 67-71 & Sloane Found, 74-; various lectureships & professorships, US & abroad, 58-; Guggenheim fel, 59; mem adv bd, Petrol Res Fund, 63-66; mem comt org chem & comt postdoctoral fels, Nat Res Coun, 59-62. Honors & Awards: Am Chem Soc Awards, 57 & 67; Baekeland Medal, 61; Harrison Howe Award, 62; Franklin Mem Award, 66; Synthetic Org Chem Mfg Asn Gold Medal, 71. Mem: Nat Acad Sci; Am Acad Arts & Sci; Am Chem Soc (chmn org div, 66-67); Brit Soc Chem Indus(; The Chem Soc. Res: Total synthesis of complex structure; design of new synthetic reactions and reaction mechanisms. Mailing Add: Columbia Univ Dept of Chem Box 666 Havemeyer Hall New York NY 10027

STORLAZZI, JOSEPH JORDAN, b New Haven, Conn, Apr 19, 08; m 31; c 4. BIOLOGY, BACTERIOLOGY. Educ: Dickinson Col, BS, 31; Univ Pisa, ScD, 35. Prof Exp: Lab asst biol, Dickinson Col, 30-31; lab instr, Univ Pisa, 33-35; asst prof, 35-40, PROF BIOL & CHMN SCI DIV, WIDENER COL, 40- Concurrent Pos: Bacteriologist, City of Morton, 44- & City of Chester, 61-; dir, Chester Clin Labs, 45- Mem: AAAS; Am Soc Med Technol; Am Nuclear Soc; Am Genetic Asn; Nat Asn Biol Teachers. Res: Comparative studies on Baloenoptera and on fossil Cervus. Mailing Add: Sci Div Widener Col Chester PA 19013

STORM, CARLYLE BELL, b Baltimore, Md, Mar 2, 35; m 57; c 3. BIOINORGANIC CHEMISTRY. Educ: Johns Hopkins Univ, BA, 61, MA, 63, PhD(chem), 65. Prof Exp: NIH res fel chem, Stanford Univ, 65-66; staff fel biochem, NIMH, 66-68; from asst prof to assoc prof chem, 68-73, PROF CHEM, HOWARD UNIV, 73- Concurrent Pos: NIH res career develop award, 73-78; sr visitor, Inorg Chem Lab, Oxford Univ, 74-75. Mem: Am Chem Soc; The Chem Soc. Res: Mechanism of action and structure of metal containing enzymes; inorganic reagents in biochemistry. Mailing Add: Dept of Chem Howard Univ Washington DC 20059

STORM, DANIEL RALPH, b Hawarden, Iowa, June 21, 44; m 66; c 3. BIOCHEMISTRY. Educ: Univ Wash, BS, 66, MS, 67; Univ Calif, Berkeley, PhD(biochem), 71. Prof Exp: Res asst biochem, Univ Calif, Berkeley, 67-71, NIH res fel, Harvard Univ, 71-72, NSF fel, 72-73; ASST PROF BIOCHEM, UNIV ILL, URBANA, 73- Concurrent Pos: Indust consult, Pharmaco Inc, 75- Mem: Am Chem Soc; Am Soc Biol Chemists; Am Soc Microbiol. Res: Structure and function of biological membranes; mechanism of enzymatic catalysis; membrane active antibiotics and molecular pharmacology at the membrane level. Mailing Add: Dept of Biochem Univ of Ill Urbana IL 61801

STORM, DAVID LYNN, organic chemistry, biochemistry, see 12th edition

STORM, LEO EUGENE, b Valeda, Kans, Aug 29, 28. COMPUTER SCIENCE, STATISTICS. Educ: Okla Agr & Mech Col, BA, 53. Prof Exp: Seismic engr, Seismic Eng Co, 53-54; meteorologist, US Weather Bur, 54-55; mathematician, Northwestern Univ, 55; qual control engr, Metro Bottle Glass Co, 55-56; jr engr, US Testing Co, 56; assoc staff mem, Gen Precision Lab, Inc, 56-57; sr statistician, Nuclear Fuel Oper, Olin Mathieson Chem Corp, 57-61; statist qual control supvr, United Nuclear Corp, 61-62; opers analyst, United Aircraft Corp Systs Ctr, 62-67; sr sci programmer, NY Med Col, 67-70; syst analyst, Texaco Inc, 70-71; programmer analyst, Data Develop, Inc, 71-73; SR SYST ANALYST, NABISCO, 73- Mem: AAAS; Am Meteorol Soc; Am Soc Qual Control; Am Statist Asn. Res: Simulation of communication systems and airport configurations; digital computer programming for management systems; statistical sample surveys; war gaming; programming analysis for statistical accounting and biomedical applications. Mailing Add: 37-14 89th St Jackson Heights NY 11372

STORM, MARTIN LEE, physics, see 12th edition

STORM, PAUL CARLYLE, organic chemistry, deceased

STORM, ROBERT MACLEOD, b Calgary, Alta, July 9, 18; US citizen; m 43, 59; c 6. ZOOLOGY. Educ: Northern Ill State Teachers Col, BE, 39; Ore State Col, MS, 41, PhD(zool), 48. Prof Exp: From instr to assoc prof, 48-62, PROF ZOOL, ORE STATE UNIV, 62- Mem: Assoc Am Soc Ichthyologists & Herpetologists. Res:

Natural history of cold-blooded land vertebrates. Mailing Add: Dept of Zool Ore State Univ Corvallis OR 97331

STORMER, JOHN CHARLES, JR, b Englewood, NJ, Oct 28, 41; m 63; c 2. PETROLOGY, GEOCHEMISTRY. Educ: Dartmouth Col, BA, 63; Univ Calif, Berkeley, PhD(geol), 71. Prof Exp: Asst geologist, Climax Molybdenum Co, Colo, 67; ASST PROF GEOL, UNIV GA, 71- Concurrent Pos: Vis prof, Inst Geosci, Univ Sao Paulo, 73. Mem: Mineral Soc Am; Geochem Soc; Am Geophys Union; Brazilian Geol Soc. Res: Mineralogy and geochemistry of igneous rocks as applied to petrology; thermochemical data and methods of investigating the origin of igneous rocks, and applications to various rock suites and petrographic provinces. Mailing Add: Dept of Geol Univ of Ga Athens GA 30602

STORMONT, CLYDE J, b Viola, Wis, June 25, 16; m 40; c 5. GENETICS, IMMUNOLOGY. Educ: Univ Wis, BA, 38, PhD(genetics), 47. Prof Exp: Instr genetics, Univ Wis, 46-47, lectr, 47, asst prof, 48; Fulbright scholar, Univ NZ, 49-50; asst prof, 50-54, assoc prof vet med & assoc serologist, Exp Sta, 54-59, PROF IMMUNOGENETICS, UNIV CALIF, DAVIS, 59- Concurrent Pos: E B Scripps fel, San Diego Zool Soc, 56-57 & 66-67. Mem: Genetics Soc Am; Am Asn Immunol; Soc Exp Biol & Med; Int Soc Animal Blood Group Res; Am Soc Nat. Res: Blood groups; animal blood groups and biochemical polymorphisms; genetic markers in animal blood. Mailing Add: Dept of VM Reprod Sch of Vet Med Univ of Calif Davis CA 95616

STORMONT, ROBERT TULLOCH, b LaCrosse, Wis, Apr 28, 12; m 38; c 2. PHARMACOLOGY. Educ: Univ Wis, BS, 34, PhD(pharmacol), 39; Univ Chicago, MD, 42. Prof Exp: Asst pharmacol, Univ Wis, 34-39, instr, 38-39; instr, Univ Chicago, 39-42; intern med, US Marine Hosp, Chicago, 42-43; chief pharmacol, Toxicol Sect, US Naval Med Res Inst, 45-46; med officer, Food & Drug Admin, 46-47, med dir, 47-50; dir div ther & res & secy coun pharm & chem, AMA, 50-55, dir dept ther & res, 55-59; vpres, 59-69, DIR RES LIAISON, RICHARDSON-MERRELL, INC, 69- Concurrent Pos: Prof lectr, Med Sch, Georgetown Univ, 47-50. Mem: AAAS; Am Soc Pharmacol & Exp Therapeut; Am Fedn Clin Res. Res: Respiratory alkalosis and anesthesia; motion sickness; antimalarial drugs. Mailing Add: Richardson-Merrell Inc 110 E Amity Rd Cincinnati OH 45215

STORMS, EDMUND KUGLER, radiochemistry, see 12th edition

STORMS, LOWELL H, b Schenectady, NY, Feb 14, 28; m 55; c 3. NEUROPSYCHOLOGY. Educ: Univ Minn, BA, 50, MS, 51, PhD(clin psychol), 56. Prof Exp: Psychologist, Hastings State Hosp, Minn, 54-56; Fulbright grant, Inst Psychiat, Univ London, 56-57; from instr to prof clin psychol, Neuropsychiat Inst, Univ Calif, Los Angeles, 57-71; PROF CLIN PSYCHOL, SCH MED, UNIV CALIF, SAN DIEGO, 71- Concurrent Pos: Consult, Vet Admin, 64-71 & Encounters Unlimited, 68-71; psychologist, Vet Admin Hosp, San Diego. Mem: AAAS; Am Psychol Asn. Res: Behavior of schizophrenics; behavior therapy; clinical psychology. Mailing Add: Sch of Med Dept of Psychiat Univ of Calif at San Diego La Jolla CA 92037

STORMSHAK, FREDRICK, b Enumclaw, Wash, July 4, 36; m 63; c 2. REPRODUCTIVE ENDOCRINOLOGY. Educ: Wash State Univ, BSc, 59, MSc, 60; Univ Wis, PhD(endocrinol), 65. Prof Exp: Res physiologist, USDA, 65-68; asst prof physiol, 68-72, actg head dept animal sci, 74, ASSOC PROF PHYSIOL, ORE STATE UNIV, 72- Concurrent Pos: Postdoctoral trainee endocrinol, Univ Wis, 74-75; sect ed, J Animal Sci, 75-77. Mem: Am Soc Animal Sci; Soc Study Fertility; Endocrine Soc; Soc Study Reproduction. Res: Quantitative measurement of steroid hormones of ovarian and adrenal origin; factors affecting the regression and maintenance of the corpus luteum; pituitary, ovarian and uterine interrelationships in reproduction. Mailing Add: Dept of Animal Sci Ore State Univ Corvallis OR 97331

STORR, JOHN FREDERICK, b Ottawa, Ont, Aug 17, 15; m 42; c 1. AQUATIC ECOLOGY. Educ: Queen's Univ, Ont, BA, 42; Columbia Univ, MA, 48; Cornell Univ, PhD(marine ecol), 55. Prof Exp: Instr biol, Queen's Col, Bahamas, 42-45; asst prof physiol, Adelphi Col, 47-52; res asst prof marine ecol, Univ Miami, 55-58; ASSOC PROF ECOL & INVERT ZOOL, STATE UNIV NY BUFFALO, 58- Concurrent Pos: US Fish & Wildlife Serv grant, 55-57; limnol consult, Niagara Mohawk Power Corp, 63-; aquatic ecol consult, Rochester Gas & Elec Corp, 68- Mem: Ecol Soc Am; Int Asn Theoret & Appl Limnol; Am Fisheries Soc. Res: Coral reef zonation; ecology of sponges of Gulf of Mexico and of benthic organisms in Lake Erie and Lake Ontario of New York. Mailing Add: Biol Div State Univ of NY Buffalo NY 14214

STORROW, HUGH ALAN, b Long Beach, Calif, Jan 13, 26; m 53; c 3. PSYCHIATRY. Educ: Univ Southern Calif, AB, 46, MD, 50; Am Bd Psychiat & Neurol, dipl, 55. Prof Exp: Intern, USPHS Hosp, Baltimore, 49-50; resident, Sheppard & Enoch Pratt Hosp, Towson, Md, 50-51; staff psychiatrist, US Penitentiary Hosp, Leavenworth, Kans, 51-52; resident, USPHS Hosp, Lexington, Ky, 52-53, staff psychiatrist, 53-54; resident, Brentwood Vet Admin Hosp, Los Angeles, 54-55; instr psychiat, Sch Med, Yale Univ, 55-56; asst prof, Sch Med, Univ Calif, Los Angeles, 56-60, attend psychiatrist, Med Ctr, 57-60; assoc prof, Col Med, Univ Ky, 60-65; prof psychiat, Univ Minn, 65-66; PROF PSYCHIAT, COL MED, UNIV KY, 66- Concurrent Pos: Attend psychiatrist, Brentwood Vet Admin Hosp, Calif, 57-60; consult, United Cerebral Palsy Asn, Los Angeles County, Calif, 57-60, USPHS & Vet Admin Hosps, Ky, 60- Mem: Am Psychiat Asn; AMA; Asn Am Med Cols. Res: Behavior modification; teaching methods for psychiatry. Mailing Add: Dept of Psychiat Univ of Ky Col of Med Lexington KY 40506

STORRS, CHARLES LYSANDER, b Shaowu, Fukien, China, Oct 25, 25; US citizen; m 57; c 3. NUCLEAR PHYSICS. Educ: Mass Inst Technol, BS, 49, PhD, 52. Prof Exp: Mem staff, Aircraft Nuclear Propulsion Dept, Gen Elec Co, 52-56, supvr initial engine test opers, 56-59, flight engine test opers, 59-61, mgr reactor test opers, Nuclear Propulsion Dept, 61, SL-1 Proj, Nuclear Mat & Propulsion Opers, 61-62, 710 Proj, 62-65; dir heavy water organic cooled reactor, Atomics Int-Combustion Eng Joint Venture, NAm Aviation, Inc, Calif, 65-67; asst dir advan reactor eng, 67-69, dir advan reactor develop, 69-71, DIR PROJS, NUCLEAR POWER DEPT, COMBUSTION ENG, INC, 71- Mem: AAAS; Am Nuclear Soc; Am Phys Soc. Res: Engineering, design and development of technology leading to the application of nuclear energy to power generation and desalination; reactor test operations; management of technical enterprises. Mailing Add: Nuclear Power Dept Combustion Eng Inc Windsor CT 06095

STORRS, ELEANOR EMERETT, b Cheshire, Conn, May 3, 26; m 63; c 2. BIOCHEMISTRY, MEDICAL RESEARCH. Educ: Univ Conn, BS, 48; NY Univ, MS, 58; Univ Tex, PhD(biochem), 67. Prof Exp: Asst, Boyce Thompson Inst Plant Res, 48-59, asst biochemist, 59-62; res scientist, Clayton Found Biochem Inst, Univ Tex, 62-65; res chemist, Pesticides Res Lab, USPHS, Fla, 65-67; res chemist, 67-71, DIR DEPT COMP BIOCHEM, GULF SOUTH RES INST, 71- Mem: Am Chem Soc; Bot Soc Am; fel Am Inst Chemists; Reticuloendothelial Soc; Int Soc Trop Dermat. Res: Armadillo in biomedical research; leprosy; biochemical individuality; analytical methods for biochemical, environmental and residue analyses; drug metabolism; mode of fungicidal, insecticidal action. Mailing Add: Dept of Biochem Gulf South Res Inst New Iberia LA 70560

STORTS, RALPH WOODROW, b Zanesville, Ohio, Feb 5, 33; m 60; c 3. VETERINARY PATHOLOGY. Educ: Ohio State Univ, DVM, 57, PhD(vet path), 66; Purdue Univ, West Lafayette, MSc, 62. Prof Exp: Instr vet microbiol, Purdue Univ, West Lafayette, 57-60; instr vet path, Ohio State Univ, 61-66; from asst prof to assoc prof, 66-73, PROF VET PATH, TEX A&M UNIV, 73- Mem: Am Vet Med Asn; Am Col Vet Path; Int Acad Path; Conf Res Workers Animal Dis. Res: Veterinary neuropathology including cytology of normal and infected cultures of nervous tissue. Mailing Add: Dept of Vet Path Tex A&M Univ College Station TX 77843

STORTZ, CLARENCE B, b Marlette, Mich, July 23, 33; m 56; c 5. MATHEMATICS. Educ: Wayne State Univ, BS, 55; Univ Miami, MS, 58; Univ Mich, Ann Arbor, DEd(math), 68. Prof Exp: Asst prof math, Northern Mich Univ, 63-66 & Cent Mich Univ, 66-68; assoc prof, 68-72, PROF & HEAD DEPT MATH, NORTHERN MICH UNIV, 72- Res: General topology; history of mathematics. Mailing Add: Dept of Math Northern Mich Univ Marquette MI 49855

STORVICK, CLARA A, b Emmons, Minn, Oct 31, 06. NUTRITION. Educ: St Olaf Col, AB, 29; Iowa State Univ, MS, 33; Cornell Univ, PhD(nutrit, biochem), 41. Prof Exp: Instr chem, Augustana Acad, 30-32; asst, Iowa State Univ, 32-34; nutritionist, Fed Emergency Relief Admin, Minn, 34-36; asst prof nutrit, Okla State Univ, 36-38; asst, Cornell Univ, 38-41; asst prof nutrit, Univ Wash, 41-45; from assoc prof to prof, 45-72, head home econ res, 55-72, dir nutrit res inst, 65-72, EMER PROF NUTRIT, ORE STATE UNIV, 72- Concurrent Pos: Sabbatical leaves, Chem Dept, Columbia Univ & Inst Cytophysiol, Denmark, 52, Lab Nutrit & Endocrinol, NIH, 59 & Div Clin Oncol, Med Sch, Univ Wis, 66. Honors & Awards: Borden Award, Am Home Econ Asn, 52. Mem: Am Home Econ Asn; Am Dietetic Asn; fel Am Pub Health Asn; Am Inst Nutrit; fel AAAS. Res: Calcium, phosphorus, nitrogen, ascorbic acid, thiamine and riboflavin metabolism; nutrition and dental caries; vitamin B-6. Mailing Add: 124 NW 29th St Corvallis OR 97330

STORVICK, DAVID A, b Ames, Iowa, Oct 24, 29; m 52; c 3. MATHEMATICS. Educ: Luther Col, Iowa, AB, 51; Univ Mich, MA, 52, PhD(math), 56. Prof Exp: From instr to asst prof math, Iowa State Univ, 55-57; from asst prof to assoc prof, 57-66, PROF MATH, UNIV MINN, MINNEAPOLIS, 66-, ASSOC HEAD SCH MATH, 64- Concurrent Pos: Res assoc, US Army Math Res Ctr, Wis, 62-63. Mem: Am Math Soc; Math Asn Am. Res: Complex function theory. Mailing Add: Sch of Math Univ of Minn Minneapolis MN 55455

STORVICK, WALDEMAR O, b Albert Lea, Minn, Oct 8, 31; m 56; c 3. BIOCHEMISTRY. Educ: Luther Col, BA, 57; Va Polytech Inst, MS, 58, PhD(biochem, nutrit), 61. Prof Exp: Sr biochemist, 61-66, dept head, 66-70, mgr insulin mfg opers, 70-73, MGR BIOCHEM MFG OPERS, ELI LILLY & CO, 73- Mem: Am Chem Soc. Res: Production and product development management; radioimmunoassay procedures development and automation. Mailing Add: Dept M410 Eli Lilly & Co Indianapolis IN 46206

STORY, DEE ANN, b Houston, Tex, Dec 12, 31; m 61. ANTHROPOLOGY, ARCHAEOLOGY. Educ: Univ Tex, Austin, BA, 53, MA, 56; Univ Calif, Los Angeles, PhD(anthrop), 63. Prof Exp: Mus technician, Univ Calif, Los Angeles, 56-57; archaeologist, Univ Utah, 58-60; cur anthrop, Tex Mem Mus, 60-62, archaeol asst dir, Tex Archaeol Salvage Proj, 62-64, ASST PROF ANTHROP & EXEC DIR TEX ARCHAEOL RES LAB, UNIV TEX, AUSTIN, 65- Concurrent Pos: NSF res grant, 65-66. Mem: Fel Am Anthrop Asn; Soc Am Archaeol. Res: North American archaeology, especially prehistory of Texas and adjacent region, also paleoecology of this area. Mailing Add: Dept of Anthrop Univ of Tex Austin TX 78712

STORY, HAROLD S, b Catskill, NY, Oct 5, 27; m 51; c 2. SOLID STATE PHYSICS, NUCLEAR MAGNETIC RESONANCE. Educ: NY State Col Teachers, BA, 49, MA, 50; Univ Maine, Orono, MS, 52; Case Inst Technol, PhD(physics), 57. Prof Exp: Mem tech staff, Bell Tel Labs, NJ, 56-59; assoc prof, 59-63, PROF PHYSICS, STATE UNIV NY ALBANY, 63- Mem: Am Phys Soc; Sigma Xi; Am Asn Physics Teachers. Res: Structure, defects and conduction processes in superionic conductors, utilizing nuclear magnetic resonance. Mailing Add: Dept of Physics State Univ of NY Albany NY 12222

STORY, JIM LEWIS, b Alice, Tex, July 30, 31; m 58; c 4. NEUROSURGERY. Educ: Tex Christian Univ, BS, 52; Vanderbilt Univ, MD, 55. Prof Exp: From instr to asst prof neurosurg, Med Sch, Univ Minn, 61-67; PROF NEUROSURG, UNIV TEX HEALTH SCI CTR SAN ANTONIO, 67- Concurrent Pos: Univ fels neurol surg, Univ Minn, 56-59 & 60-61; USPHS fels anat, Univ Calif, Los Angeles, 59-60 & Univ Minn, 60-62. Mem: Am Asn Neurol Surgeons; Soc Neurol Surgeons; Am Col Surgeons; Neurosurg Soc Am; Am Acad Neurol Surgeons. Res: Intracranial pressure monitoring; etiology of brain tumors. Mailing Add: Div of Neurol Surg Univ of Tex Health Sci Ctr San Antonio TX 78229

STORY, JON ALAN, b Odebolt, Iowa, Apr 7, 46; m 69; c 2. BIOCHEMISTRY, NUTRITION. Educ: Iowa State Univ, BS, 68, MS, 70, PhD(zool), 72. Prof Exp: Instr zool, Iowa State Univ, 71-72; trainee lipid metab, 72-74, RES ASSOC LIPID METAB, WISTAR INST ANAT & BIOL, 74- Mem: Sigma Xi; Am Chem Soc. Res: Investigation into the effects of several dietary components and age on cholesterol and bile acid metabolism as involved in development of experimental atherosclerosis. Mailing Add: Wistar Inst of Anat & Biol 36th & Spruce Sts Philadelphia PA 19104

STORY, PAUL RICHARD, organic chemistry, see 12th edition

STORY, TROY LEE, b Montgomery, Ala, Nov 11, 40. CHEMICAL PHYSICS. Educ: Morehouse Col, BS, 62; Univ Calif, Berkeley, PhD(chem), 68. Prof Exp: Mem staff & fel chem, Univ Calif, Berkeley, 69-70; fel physics, Chalmers Univ Technol, Sweden, 70-71; ASST PROF CHEM, HOWARD UNIV, 71- Mem: Am Chem Soc; Am Phys Soc. Res: Experimental determination of dipole moments using molecular beam resonance and deflection techniques; theoretical quantum mechanical model for the analysis of rotational and vibrational distributions for reactive scattering experiments; topological analysis of nonrigid bodies. Mailing Add: Dept of Chem Howard Univ Washington DC 20059

STORZ, JOHANNES, b Hardt/Schramberg, Ger, Apr 29, 31; US citizen; m 59; c 3. VIROLOGY, MICROBIOLOGY. Educ: Vet Col, Hannover, dipl, 57; Univ Munich, DrMedVet, 58; Univ Calif, Davis, PhD(comp path), 61; Am Col Vet Microbiol, dipl, 69. Prof Exp: Res assoc, Fed Res Inst Viral Dis Animals, Tübingen, Ger, 57-58; lectr vet microbiol, Univ Calif, Davis, 58-61; from asst prof to assoc prof vet virol, Utah State Univ, 61-65; assoc prof, 65-68, PROF VET VIROL, COLO STATE UNIV, 68- Concurrent Pos: USPHS res grant, Utah State Univ, 62-65; USPHS res grant, Colo

State Univ, 66-72. WHO grant, 72-77; vis scientist, Univ Giessen, 71-72; consult, WHO, Geneva, 71. Honors & Awards: Andrew G Clark Award, Colo State Univ, 75. Mem: AAAS; Am Soc Microbiol; Am Vet Med Asn; Conf Res Workers Animal Dis; World Asn Buiatrics. Res: Chlamydiology; pathogenic mechanisms in intrauterine viral and chlamydial infections; chlamydial polyarthritis; intestinal viral and chlamydial infections; multiple viral infections; parvoviruses; cell biology of chlamydial infections. Mailing Add: Dept of Microbiol Colo State Univ Ft Collins CO 80521

STOSICK, ARTHUR JAMES, b Milwaukee, Wis, Dec 1, 14; m 37; c 3. CHEMISTRY. Educ: Univ Wis, BS, 36; Calif Inst Technol, PhD(struct chem), 39. Prof Exp: Fel, Calif Inst Technol, 39-40, instr gen chem, 40-41, Nat Defense Res Comt res assoc, 41-43, res chemist, Jet Propulsion Lab, 44-46, chief rockets & mat div, 50-56; res chemist, Aerojet Eng Corp, 43-44; assoc prof phys chem, Iowa State Col, 46-47; prof, Univ Southern Calif, 47-50; asst dir, Union Carbide Res Inst, 56-59; staff scientist, Gen Atomic Div, Gen Dynamics Corp, 59-60; sr scientist, Aerojet-Gen Corp, 60-71; SR SCIENTIST, UNITED TECHNOL CORP, 72- Concurrent Pos: Mem rocket eng subcomt, Nat Adv Comt Aeronaut, 52-56; asst to vpres, Adv Res Proj Agency, Off Secy Defense, 58-59. Mem: Am Chem Soc; Am Phys Soc; Am Crystallog Asn. Res: Molecular structures by diffraction; physical chemistry as related to molecular structures; propellants; high temperature chemistry; metallurgy. Mailing Add: 1153 Lime Dr Sunnyvale CA 94087

STOSKOPF, N C, b Mitchell, Ont, June 11, 34; m 60; c 2. CROP BREEDING. Educ: Univ Toronto, BSA, 57, MSA, 58; McGill Univ, PhD(agron), 62. Prof Exp: Lectr agron, Ont Agr Col, 58-59; instr & exten specialist, Kemptville Agr Sch, 59-60; asst prof crop sci, Ont Agr Col, 62-66, assoc prof, 66-69, PROF CROP SCI, UNIV GUELPH, 69-, DIR DIPL PROG AGR, 74- Mem: Agr Inst Can. Res: Winter wheat breeding given a physiological basis with yield as main objective; plants selected for upright leaves to achieve a high optimum leaf area, a high net assimilation rate and a long period of grain filling; cereal physiology. Mailing Add: Dept of Crop Sci Univ of Guelph Guelph ON Can

STOTHERS, JOHN BAILIE, b London, Ont, Apr 16, 31; m 53; c 2. ORGANIC CHEMISTRY. Educ: Univ Western Ont, BSc, 53, MSc, 54; McMaster Univ, PhD(phys org chem), 57. Prof Exp: Res chemist, Res Dept, Imp Oil, Ltd, 57-59; lectr chem, 59-61, from asst prof to assoc prof, 61-67, PROF CHEM, UNIV WESTERN ONT, 67- Concurrent Pos: Merck, Sharp & Dohme lect award, 71. Mem: Am Chem Soc; fel Chem Inst Can; The Chem Soc. Res: Nuclear magnetic resonance spectroscopy; applications of deuterium and carbon-13 nuclear magnetic resonance to organic structural, stereochemical and mechanistic problems; deuterium exchange processes and molecular rearrangements. Mailing Add: Dept of Chem Univ of Western Ont London ON Can

STOTSKY, BERNARD A, b New York, NY, Apr 8, 26; m 52; c 5. PSYCHOLOGY, PSYCHIATRY. Educ: City Col New York, BS, 48; Univ Mich, MA, 49, PhD(psychol), 51; Western Reserve Univ, MD, 62. Prof Exp: Staff psychologist, Ment Hyg Clin, Vet Admin, Detroit, 51-53; instr & assoc, Boston Univ, 54-56, asst prof psychol, 56-57; asst prof, Duke Univ, 57-58; staff psychologist, Vet Admin Hosp, Brockton, Mass, 58-61; intern, George Washington Univ Hosp, 62-63; fel psychiat & resident psychiat, Mass Ment Health Ctr, 63-65 & Boston State Hosp, 65-66; lectr, 64-67, assoc prof, 67-68, head dept psychol, 72-73, PROF PSYCHIAT, BOSTON STATE COL, 68- Concurrent Pos: Chief counseling psychologist, Vet Admin Hosp, Brockton, Mass, 53-56, consult, 56-57 & 70- & chief psychologist, Durham, NC, 57-58; consult, Brockton Family Serv, 54, Hayden Goodwill Inn, 54-57, Mass Dept Pub Health, 67 & Boston State Hosp, 63-65, 68-75; prin investr & lectr, Northeastern Univ, 64-; assoc psychiat, Tufts Univ, 66-67, asst prof, 67-; lectr, Clark Univ, 69-70 & Mt Sinai Sch Med, 69-72; consult, Food & Drug Admin, 72-75 & Nat Inst Child Health & Human Develop, 73-; prof psychiat & behav sci, Univ Wash, 73-; dir outpatient psychiat clin, St Elizabeth's Hosp Boston, 73, assoc dir psychiat educ, 74-; consult campus sch, Boston Col, 74- Mem: Fel Am Psychol Asn; Am Psychiat Asn. Res: Psychopharmacology; pathology of mental disease; personality and organic factors in rehabilitation of chronically ill patients; geriatrics. Mailing Add: 246 Clark Rd Brookline MA 02146

STOTT, DONALD FRANKLIN, b Reston, Man, Apr 30, 28; m 60; c 3. GEOLOGY. Educ: Univ Manitoba, BSc, 53, MSc, 54; Princeton Univ, AM, 56, PhD, 58. Prof Exp: Asst geol, Princeton Univ, 54-55; head regional geol subdiv, Inst Sedimentary & Petrol Geol, 72-73, GEOLOGIST, GEOL SURV CAN, 57-, DIR, INST SEDIMENTARY & PETROL GEOL, 73- Mem: Fel Geol Soc Am; Can Soc Petrol Geologists; Geol Asn Can; Soc Econ Paleontologists & Mineralogists. Res: Physical stratigraphy and sedimentation, particularly of Cretaceous system of Rocky Mountain foothills, Canada. Mailing Add: Inst of Sedimentary & Petrol Geol 3303 33rd St NW Calgary AB Can

STOTT, GERALD H, b Kanosh, Utah, Mar 7, 24; m 47; c 6. DAIRY SCIENCE. Educ: Utah State Univ, BS, 51, MS, 52; Univ Wis, PhD, 56. Prof Exp: Asst prof dairy sci, Univ Ga, 56-57; assoc prof dairy sci & dairy physiologist, 57-63, PROF & HEAD DEPT DAIRY SCI, UNIV ARIZ, 63-, DAIRY SCIENTIST, 74- Mem: AAAS; Am Dairy Sci Asn. Res: Genetics; animal physiology; nutrition; parathyroid activity; calcium and phosphorous metabolism; reproduction and nutrition under high climatic temperatures. Mailing Add: Dept of Dairy Sci Univ of Ariz Tucson AZ 85721

STOTTER, PHILIP LEWIS, organic chemistry, see 12th edition

STOTTLEMEYER, QUAYTON RAY, physical chemistry, see 12th edition

STOTZ, ELMER HENRY, b Boston, Mass, July 29, 11; m 36; c 5. BIOCHEMISTRY. Educ: Mass Inst Technol, BS, 32; Harvard Univ, PhD(biochem), 36. Prof Exp: Instr biochem, Univ Pittsburgh, 36-37, Univ Chicago, 37-38 & Harvard Med Sch, 38-42; prof agr & biochem & head dept, Cornell Univ, 43-47; PROF BIOCHEM & CHMN DEPT, SCH MED & DENT, UNIV ROCHESTER, 47- Concurrent Pos: Dir labs, McLean Hosp, 38-43; mem nat comt biochem, Nat Acad Sci-Nat Res Coun, 54-; treas, Int Union Biochem, 55-; trustee, Assoc Univs, Inc, 57-; mem div chem & chem technol, Nat Res Coun, 57- Mem: Am Chem Soc; Am Soc Biol Chemists (secy, 50-53); Biol Stain Comn (treas, 45-). Res: Cytochromes; analytical methods; biological and fatty acid oxidation; stain chemistry. Mailing Add: Dept of Biochem Univ of Rochester Med Ctr Rochester NY 14642

STOTZ, ROBERT WILLIAM, b Monroe, Mich, July 18, 42; m 71; c 1. RESEARCH ADMINISTRATION. Educ: Univ Toledo, BS, 64, MS, 66; Univ Fla, PhD(inorg chem), 70. Prof Exp: Teaching asst inorg chem, Univ Toledo, 64-66; res asst, Univ Fla, 66-70; res assoc, Mich State Univ, 70-71; instr chem, Eastern Mich Univ, 71-72; asst prof, Mercer Univ, 72-73; asst prof, Tri-State Col, 73-74; SUPVR INORG ANAL RES, INST GAS TECHNOL, 74- Mem: Am Chem Soc. Res: Development and modification of various wet chemical and instrumental methods for determination of inorganic constituents in complex inorganic systems. Mailing Add: Inst of Gas Technol 3424 S State St Chicago IL 60616

STOTZKY, GUENTHER, b Leipzig, Ger, May 24, 31; nat US; m 58; c 3. MICROBIAL ECOLOGY. Educ: Calif State Polytech Col, BS, 52; Ohio State Univ, MS, 54, PhD(agron & microbiol), 56. Prof Exp: Res asst soil biochem & microbiol, Ohio State Univ, 53-56; res assoc bot & plant nutrit, Univ Mich, 56-58; head soil microbiol, Cent Res Labs, United Fruit Co, 58-63; microbiologist & chmn, Kitchawan Res Lab, Brooklyn Bot Garden, 63-68; assoc prof biol, 68-70, adj assoc prof, 67-68, PROF BIOL & CHMN DEPT, NY UNIV, 70- Concurrent Pos: Spec scientist, Argonne Nat Lab, 55; mem, Am Inst Biol Sci-NASA Regional Coun, 65-68; regional ed, J Soil Biol & Biochem, 69-; assoc ed, Appl Microbiol, 71- & Can J Microbiol, 71-75; mem ad hoc comt rev biomed & ecol effects of extremely low frequency radiation, Bur Med & Surg, Dept Navy, 72-; vis prof, Inst Advan Studies, Polytech Inst, Mexico, 73; mem bd trustees, NY Ocean Sci Lab, Affil Cols & Univs, Inc, 73-; mem comn human resources, Nat Res Coun, 75- Mem: AAAS; Am Soc Microbiologists; Bot Soc Am; Am Inst Biol Sci; Am Soc Agron. Res: Soil microbiology and biochemistry; plant-microbe relations; soil-borne plant and animal diseases; clay mineralogy; soil, air and water pollution; seed germination; cell surfaces. Mailing Add: Dept of Biol NY Univ New York NY 10003

STOUB, KENNETH PAUL, b Chicago, Ill, Jan 18, 43; m 64; c 2. ANALYTICAL CHEMISTRY. Educ: Calvin Col, BS, 64. Prof Exp: Res chemist anal chem, Swift & Co, 64-66; res chemist, 66-71, group leader, 71-74, SECT HEAD ANAL CHEM, HUNT-WESSON FOODS, INC, 74- Mem: Am Chem Soc; Am Oil Chemists Soc. Res: Chemistry of food products, mainly fats and tomatoes, particularly reaction mechanisms of processes and development of improved process techniques. Mailing Add: 1645 W Valencia Dr Fullerton CA 92634

STOUDT, HARRY NATHANIEL, b Bernville, Pa, Dec 27, 08; m 33; c 1. BOTANY. Educ: Temple Univ, BS, 31, MEd, 33; Johns Hopkins Univ, PhD(biol), 39. Prof Exp: Asst biol, Temple Univ, 31-33, instr, 33-36, 37-38; asst bot, Johns Hopkins Univ, 36-37, 38-39; teacher high sch, NJ, 41-43, head sci dept, 43-46; instr biol, Temple Univ, 46-48, asst prof, 48-61; assoc prof, 61-64, PROF BIOL, GLASSBORO STATE COL, 64- Concurrent Pos: Mem bot exped, Jamaica, 36. Mem: AAAS; Bot Soc Am; Int Soc Plant Morphol. Res: Floral morphology; phylogeny; gametogenesis in angiosperms. Mailing Add: Dept of Sci Glassboro State Col Glassboro NJ 08028

STOUDT, HOWARD WEBSTER, b Pittsburgh, Pa, May 13, 25; m 53; c 2. MEDICAL EDUCATION, PUBLIC HEALTH. Educ: Harvard Univ, AB, 49, SM, 62; Univ Pa, AM, 53, PhD(anthrop), 59. Prof Exp: Res asst, Sch Pub Health, Harvard Univ, 52-55; res & educ specialist, Air Univ, 55-57; res assoc phys anthrop, Sch Pub Health, Harvard Univ, 57-66, asst prof, 66-73; PROF & CHMN DEPT COMMUNITY MED, COL OSTEOP MED, MICH STATE UNIV, 73- Concurrent Pos: Consult, Nat Health Exam Surv, USPHS Ctr Dis Control; res investr, Normative Aging Study, Vet Admin, Boston. Mem: AAAS; Am Asn Phys Anthrop; Am Pub Health Asn; Am Anthrop Asn; Asn Behav Sci in Med Educ. Res: Behavioral sciences in medical education; health care delivery; epidemiology of noninfectious disease; constitutional and environmental correlates of health and performance; applied physical anthropology; human factors engineering. Mailing Add: Dept of Community Med Mich State Univ Col Osteop Med East Lansing MI 48824

STOUDT, THOMAS HENRY, b Temple, Pa, Apr 6, 22; m 43; c 3. MICROBIOLOGY. Educ: Albright Col, BS, 43; Rutgers Univ, MS, 44; Purdue Univ, West Lafayette, PhD(org chem), 49. Prof Exp: Asst chem, Rutgers Univ, 43-44 & Purdue Univ, West Lafayette, 46-47; sr microbiologist, 49-58, from sect head to sr sect head, 58-69, dir appl microbiol, 69-75, SR DIR APPL MICROBIOL & NAT PROD ISOLATION, MERCK & CO, INC, 75- Mem: Am Chem Soc; Soc Indust Microbiol; Int Asn Dent Res; Am Soc Microbiologists; NY Acad Sci. Res: Microbial transformations; microbial biosyntheses; antibiotics; microbial physiology and genetics, rumen microbiology; oral microbiology; microbial enzymology; vaccines and immunology; scale-up of industrial fermentation processes. Mailing Add: Merck & Co Inc Rahway NJ 07065

STOUFER, ROBERT CARL, b Ashland, Ohio, Nov 3, 30; m 54; c 2. INORGANIC CHEMISTRY. Educ: Otterbein Col, BA & BS, 52; Ohio State Univ, PhD, 59. Prof Exp: ASSOC PROF INORG CHEM, UNIV FLA, 58- Mem: Am Chem Soc. Res: Preparation and characterization of inorganic complexes. Mailing Add: Dept of Chem Univ of Fla Gainesville FL 32601

STOUFFER, JAMES L, b Harrisburg, Pa, Sept 25, 35; m 68. AUDIOLOGY, PSYCHOACOUSTICS. Educ: State Univ NY Col Buffalo, BSc, 64; Pa State Univ, MSc, 66, PhD(audiol, statist), 69. Prof Exp: Clin audiologist, Pa State Univ, 64-69; asst prof audiol, Univ Western Ont, 69-70, DIR COMMUN DIS & CHIEF SPEECH & HEARING SERV, UNIV WESTERN ONT & UNIV HOSP, LONDON, ONT, 72- Concurrent Pos: NIH fel psychoacoust, Commun Sci Lab, Univ Fla, 69-70; consult, Oxford County Ment Health Centre, 70-; ed asst, Ont Speech & Hearing Asn, 71- Mem: Am Speech & Hearing Asn; Acoust Soc Am. Res: Physiological responses to auditory signals of normals versus abnormals; underwater acoustics. Mailing Add: 909 Maitland London ON Can

STOUFFER, JAMES RAY, b Glen Elder, Kans, Jan 12, 29; m 55; c 2. ANIMAL SCIENCE. Educ: Univ Ill, BS, 51, MS, 53, PhD(meats), 56. Prof Exp: Asst prof animal husb, Univ Conn, 55-56; ASST PROF ANIMAL HUSB, CORNELL UNIV, 56- Mem: Am Soc Animal Sci; Inst Food Technol; Am Meat Sci Asn. Res: Carcass evaluation of meat animals, particularly the relationship of live animal and carcass characteristics. Mailing Add: Dept of Animal Sci Cornell Univ Ithaca NY 14853

STOUFFER, JOHN EMERSON, b Sioux City, Iowa, Dec 4, 25; m 55; c 2. BIOCHEMISTRY. Educ: Northwestern Univ, BS, 49; Boston Univ, PhD(org chem), 57. Prof Exp: Res assoc biochem, Med Col, Cornell Univ, 57-59, instr, 59-61; asst prof, 61-66, ASSOC PROF BIOCHEM, BAYLOR COL MED, 67- Mem: Am Chem Soc; Am Oil Chemists Soc; Am Soc Biol Chemists; Endocrine Soc; Am Inst Chemists. Res: Thyroid hormones; structure function relationships of hormones and mechanism of action; membrane receptor sites. Mailing Add: dept of Biochem Baylor Col of Med Houston TX 77025

STOUFFER, RICHARD FRANKLIN, b Welch, WVa, July 3, 32; m 57; c 2. PLANT PATHOLOGY, VIROLOGY. Educ: Vanderbilt Univ, BA, 54; Cornell Univ, PhD(plant path), 59. Prof Exp: Asst, Cornell Univ, 54-59; asst res prof, Univ RI, 59-60; asst virologist, Univ Fla, 61-65; asst prof plant path, 65-71, ASSOC PROF PLANT PATH, FRUIT RES LAB, PA STATE UNIV, 71- Mem: Am Phytopath Soc; Brit Asn Appl Biol. Res: Plant virology. Mailing Add: Pa State Univ Fruit Res Lab Box 309 Biglerville PA 17307

STOUGHTON, RAYMOND WOODFORD, b Tehachapi, Calif, Aug 6, 16; m 41; c 4. PHYSICAL CHEMISTRY, NUCLEAR CHEMISTRY. Educ: Univ Calif, BS, 37, PhD(chem), 40. Prof Exp: Asst chem, Univ Calif, 37-40, res chemist, Radiation Lab, 41-43; instr, Agr & Mech Col, Tex, 40-41; RES CHEMIST, UNION CARBIDE NUCLEAR CO, OAK RIDGE NAT LAB, 43- Concurrent Pos: Res chemist, Metall Lab, Univ Chicago, 43. Mem: AAAS; Am Chem Soc; Am Phys Soc; Am Nuclear

Soc; NY Acad Sci. Res: Reaction kinetics; application of computers to chemical and physical problems; solution thermodynamics; solution chemistry of heavy elements; radiochemistry; process development; neutron cross sections. Mailing Add: Oak Ridge Nat Lab Box X Oak Ridge TN 37830

STOUGHTON, RICHARD BAKER, b Duluth, Minn, July 4, 23; m 46; c 1. DERMATOLOGY. Educ: Univ Chicago, SB, 45, MD, 47; Am Bd Dermat, dipl, 52. Prof Exp: From instr to asst prof dermat, Univ Chicago, 50-56; assoc prof med dir dermat, Sch Med, Case Western Reserve Univ, 57-67; HEAD DEPT DERMAT, SCRIPPS CLIN & RES FOUND, 67-; PROF DERMAT & CHIEF DIV, UNIV CALIF, SAN DIEGO, 75- Concurrent Pos: Consult, US Army Chem Ctr, Md, 54-58; mem subcomt dermat, Nat Res Coun, 59-; ed-in-chief, Soc Invest Dermat, 67; dir, Am Bd Dermat. Mem: Soc Invest Dermat; Am Acad Dermat. Res: Histochemistry, percutaneous absorption and pathologic anatomy of dermatology. Mailing Add: Dermat Div Scripps Clin & Res Found La Jolla CA 92037

STOUT, BENJAMIN BOREMAN, b Parkersburg, WVa, Mar 2, 24; m 45; c 3. FOREST ECOLOGY. Educ: WVa Univ, BSF, 47; Harvard Univ, MF, 50; Rutgers Univ, PhD, 67. Prof Exp: Forester, Pond & Moyer Co, 47-49; silviculturist, Harvard Black Rock Forest, 50-55, supvr, 55-59; from asst prof to assoc prof forestry, 59-68, PROF FORESTRY, RUTGERS UNIV, NEW BRUNSWICK, 68-, CHMN DEPT BIOL SCI, 74- Mem: Ecol Soc Am; Soc Am Foresters; Sigma Xi. Res: Ways and means of quantifying vegetations response to environment. Mailing Add: 27 Fairview Ave East Brunswick NJ 08816

STOUT, CHARLES ALLISON, b Beaumont, Tex, Sept 20, 30. ORGANIC CHEMISTRY. Educ: Rice Inst, BA, 52, MA, 53; Ohio State Univ, PhD(chem), 62. Prof Exp: Chemist, Goodyear Tire & Rubber Co, Ohio, 53-54; NIH res fel photochem, Calif Inst Technol, 62-63; res chemist, Chevron Oil Field Res Co, 64-72; RES CHEMIST, DIVERSIFIED CHEM CORP, 72- Mem: Am Chem Soc; Soc Petrol Engrs. Res: Physical organic chemistry, reaction mechanisms; molecular orbital treatments of condensed aromatic systems; photochemical reactions in solutions and solids, photosensitization; structure of interfacial films. Mailing Add: Diversified Chem Corp 8100 Electric Ave Stanton CA 90680

STOUT, EDGAR LEE, b Grants Pass, Ore, Mar 13, 38; m 58; c 1. MATHEMATICS. Educ: Ore State Col, BA, 60; Univ Wis, MA, 61, PhD(math), 64. Prof Exp: Instr math, Yale Univ, 64-65, asst prof, 65-69, Off Naval Res res assoc, 67-68; ASSOC PROF MATH, UNIV WASH, 69- Mem: Math Asn Am; Am Math Soc. Res: Functions of one or several complex variables; function algebras. Mailing Add: Dept of Math Univ of Wash Seattle WA 98105

STOUT, EDWARD IRVIN, b Washington Co, Iowa, Mar 2, 39; m; c 3. ORGANIC CHEMISTRY. Educ: Iowa Wesleyan Col, BS, 60; Bradley Univ, MS, 68; Univ Ariz, PhD(org chem), 74. Prof Exp: Chemist, Lever Bros Co, 61-62; RES CHEMIST, NORTHERN REGIONAL RES CTR, USDA, 62- Concurrent Pos: Instr org chem, Bradley Univ, 70-75. Mem: Am Chem Soc (secy, 75). Res: Preparation and characterization of starch derivatives including starch graft copolymers. Mailing Add: Northern Regional Res Lab USDA 1815 N University Peoria IL 61604

STOUT, ERNEST RAY, b Boone, NC, Oct 31, 38; m 61; c 3. MOLECULAR BIOLOGY, BIOCHEMISTRY. Educ: Appalachian State Univ, BS, 61; Univ Fla, PhD(bot, biochem), 65. Prof Exp: Nat Cancer Inst fel biochem genetics, Univ Md, 65-67; asst prof molecular biol, 67-72, ASSOC PROF MOLECULAR BIOL, VA POLYTECH INST & STATE UNIV, 72- Mem: AAAS; Am Soc Plant Physiol. Res: Mechanism of nucleic acid synthesis in higher plants; control of nucleic acid synthesis; plant growth and differentiation. Mailing Add: Dept of Biol Va Polytech Inst & State Univ Blacksburg VA 24061

STOUT, FLOYD M, animal nutrition, genetics, see 12th edition

STOUT, GEORGE HUBERT, b St Louis, Mo, Sept 30, 32; m 55; c 1. NATURAL PRODUCTS CHEMISTRY, X-RAY CRYSTALLOGRAPHY. Educ: Harvard Univ, AB, 53, MS, 54, PhD(chem), 56. Prof Exp: NSF fel, Swiss Fed Inst Technol, 56-57; from asst prof to assoc prof chem, 57-69, PROF CHEM, UNIV WASH, 69- Mem: Am Chem Soc; Am Crystallog Asn; The Chem Soc. Res: Structure determination and synthesis of natural products, especially oxygen heterocycles; x-ray crystallography as applied to the determination of organic structures; nonheavy atom-x-ray phasing methods. Mailing Add: Dept of Chem Univ of Wash Seattle WA 98105

STOUT, GLENN EMANUEL, b Fostoria, Ohio, Mar 23, 20; m 42; c 2. METEOROLOGY. Educ: Findlay Col, BS, 42; Univ Chicago, cert, 43. Hon Degrees: DSc, Findlay Col, 73. Prof Exp: Asst math, Findlay Col, 39-42; instr meteorol, Univ Chicago, 42-43 & US War Dept, Chanute AFB, Ill, 46-47; asst engr, Ill State Water Surv, 47-52, head atmospheric sci sect, 52-71, asst to chief, 71-74, DIR WATER RESOURCES CENTER, ILL STATE WATER SURV, UNIV ILL, URBANA, 73-, PROF METEOROL, INST ENVIRON STUDIES, UNIV, 75- Concurrent Pos: Consult, Crop-Hail Ins Actuarial Assoc, Ill, 61-69; prog coordr, Nat Ctr Atmospheric Res, NSF, 69-71. Honors & Awards: Einstein Award, Findlay Col, 42. Mem: Am Meteorol Soc; Am Geophys Union; Am Water Resources Asn; AAAS; Int Asn Water Resources. Res: Hail climatology; weather modification; water resources; environmental science; environmental management. Mailing Add: Univ of Ill Water Resources Ctr 2535 Hydrosysts Lab Urbana IL 61801

STOUT, ISAAC JACK, b Clarksburg, WVa, July 20, 39; m 64; c 2. ECOLOGY. Educ: Ore State Univ, BS, 61, Va Polytech Inst & State Univ, MS, 67; Wash State Univ, PhD(zool), 72. Prof Exp: Wildlife Mgt Inst fel waterfowl ecol, Va Coop Wildlife Res Univ, 64-65; field ecologist, Old Dominion Univ, 65-67; USPHS fel appl ecol, Wash State Univ, 67-69; ASST PROF BIOL SCI, FLA TECHNOL UNIV, 72- Concurrent Pos: Mem, Environ Effect & Fate Solid Rocket Emission Prod, NASA, Kennedy Space Ctr, 75. Mem: Ecol Soc Am; Brit Ecol Soc; Wildlife Soc; Am Soc Mammalogists; Sigma Xi. Res: Population and community ecology; tick-host relations; applied ecology. Mailing Add: Dept of Biol Sci Fla Technol Univ Box 25000 Orlando FL 32816

STOUT, JOHN FREDERICK, b Takoma Park, Md, Jan 20, 36; m 56; c 2. ETHOLOGY, NEUROBIOLOGY. Educ: Columbia Union Col, BA, 57; Univ Md, PhD(zool), 63. Prof Exp: Instr biol, Walla Walla Col, 62, asst prof, 63-65, dir marine sta, 64-69, assoc prof biol, 66-69; assoc prof, 69-70, PROF BIOL, ANDREWS UNIV, 70- Concurrent Pos: USPHS spec fel & vis researcher, Univ Cologne, 69-70; guest res prof, Max Planck Inst Behav Physiol, 75-76. Honors & Awards: Alexander von Humboldt Sr US Scientist Award, 75. Mem: AAAS; Am Soc Zoologists; Am Physiol Soc. Res: Communication during social behavior; neurobiology of acoustic communication; behavioral physiology. Mailing Add: Dept of Biol Andrews Univ Berrien Springs MI 49104

STOUT, JOHN WILLARD, b Seattle, Wash, Mar 13, 12; m 48; c 1. PHYSICAL CHEMISTRY, CHEMICAL PHYSICS. Educ: Univ Calif, BS, 33, PhD(phys chem), 37. Prof Exp: Instr chem, Univ Calif, 37-38, Lalor fel, 38-39; instr chem, Mass Inst Technol, 39-41; investr, Nat Defense Res Comt, Univ Calif, 41-44, group leader, Manhattan Dist, Los Alamos Sci Lab, 44-46; assoc prof chem, 46-54, PROF CHEM, UNIV CHICAGO, 54- Concurrent Pos: Ed, J Chem Physics, 59- Mem: AAAS; Am Chem Soc; fel Am Phys Soc. Res: Thermodynamics; calorimetry; crystal spectra; cryogenics; paramagnetism and antiferromagnetism. Mailing Add: Dept of Chem Univ of Chicago Chicago IL 60637

STOUT, LANDON CLARKE, JR, b Kansas City, Mo, Feb 20, 33; m 54; c 5. PATHOLOGY, INTERNAL MEDICINE. Educ: Univ Md, MD, 57. Prof Exp: Resident internal med, Med Ctr, Univ Okla, 58-61, asst prof, 63-72, dir inst comp path, 65-69, resident path, 66-67, from asst prof to assoc prof, 68-72, interim chmn dept, 70-72; assoc prof, 72-74, PROF PATH, UNIV TEX MED BR GALVESTON, 74- Concurrent Pos: Nat Heart Inst spec fel, Univ Okla, 67-68; consult, Okla Med Res Found, 71-72; mem coun epidemiol, Am Heart Asn. Mem: Am Asn Path & Bact; Am Col Physicians; Am Gastroenterol Asn; Am Soc Exp Path. Res: Comparative pathology; atherosclerosis; gastroenterology; diabetes. Mailing Add: Dept of Path Univ of Tex Med Br Galveston TX 77550

STOUT, MARTIN LINDY, b N Hollywood, Calif, Feb 11, 34; m 56; c 2. GEOLOGY. Educ: Occidental Col, BA, 55; Univ Wash, MS, 57, PhD(geol), 59. Prof Exp: Asst geol, Occidental Col, 53-55 & Univ Wash, 55-59; from asst prof to assoc prof, 60-71, PROF GEOL, CALIF STATE UNIV, LOS ANGELES, 71-; SR ENG GEOLOGIST, MOORE & TABER ENGRS, GEOLOGISTS, 63- Concurrent Pos: Econ geologist, Aerogeophysics, Inc, 55; Geol Soc Am Penrose res grant, 58; partic, Int Field Inst, Scandinavia, Am Geol Inst, 63; NSF grant, 65, fel, Iceland & Norway, 66-67. Mem: AAAS; Asn Eng Geol; Geol Soc Am; Am Asn Geol Teachers. Res: Geochemical studies of basalts in Washington, Iceland, Norway; gravity movements and rates of tectonism in southern California; distribution and mechanism of failure of landslides in the capistrano formation of southern California; engineering properties of volcanic rocks. Mailing Add: Dept of Geol Calif State Univ 5151 State College Dr Los Angeles CA 90032

STOUT, MASON GARDNER, b Salt Lake City, Utah, Apr 30, 35; m 60; c 4. ORGANIC CHEMISTRY. Educ: Univ Utah, BS, 58, PhD(org chem), 60. Prof Exp: Res chemist, Lasdon Found Res Inst Chemother, 60-64; investr nucleotides & coenzymes, Pabst Labs, Pabst Brewing Co, 64-65, dir res & develop, P-L Biochem, Inc, 65-66; fel synthesis nucleosides, Univ Utah, 66-69; res chemist, 69-71, head res admin, 71-73, HEAD CHEM DIV, ICN PHARMACEUT INC, 73- Mem: Am Chem Soc. Res: Synthesis of potential mgdicinals, both flavanoids and nucleosides. Mailing Add: 27 Bethany Dr Irvine CA 92715

STOUT, MINARD WILLIAM, research administration, see 12th edition

STOUT, PAUL RICHARD, b Hamilton, Ohio, Nov 21, 18; m 41; c 3. APPLIED STATISTICS, RESEARCH MANAGEMENT. Educ: Miami Univ, AB, 39. Prof Exp: Org chemist, John Stuart Res Labs, Quaker Oats Co, 40-45, group leader chem, 46-50, statistician, 51-53, sect leader, Statist Serv, 54-65, MGR MATH APPLN, JOHN STUART RES LABS, QUAKER OATS CO, BARRINGTON, 65- Honors & Awards: Peters Medal, Quaker Oats Co. Mem: Am Statist Asn; Am Soc Qual Control. Res: Philosophy of science. Mailing Add: 150 N Main St Lombard IL 60148

STOUT, PERRY ROBERT, agricultural chemistry, deceased

STOUT, THOMPSON MYLAN, b Big Springs, Nebr, Aug 16, 14; m 40. GEOLOGY, VERTEBRATE PALEONTOLOGY. Educ: Univ Nebr, Lincoln, BSc, 36, MSc, 37. Prof Exp: Res asst vert paleont, State Mus, 33-38, from instr to assoc prof, 38-52, PROF GEOL UNIV NEBR, LINCOLN, 52-, ASSOC CUR, STATE MUS, 57- Concurrent Pos: Res assoc, Frick Lab, Am Mus Natural Hist, New York, NY, 38-; studies of fossil rodents & geol in Europ museums, 48-69; corresp, Nat Mus Natural Hist, Paris, 66. Mem: Geol Soc Am; Soc Vet Paleont; Paleont Soc; Am Soc Mammal; NY Acad Sci. Res: Stratigraphy and vertebrate paleontology, with special reference to the Tertiary and Quaternary and to intercontinental correlations in connection with revisionary studies of fossil rodents; cyclic sedimentation and geomorphology. Mailing Add: Dept of Geol 433 Morrill Hall Univ of Nebr Lincoln NE 68503

STOUT, VIRGIL L, b Emporia, Kans, Mar 14, 21; m 46; c 2. PHYSICS. Educ: Univ Mo, PhD(physics), 51. Prof Exp: Res assoc, Stanford Res Inst, 51-52; physicist, Gen Elec Co Res Labs, 52-57, mgr phys electronics br, Develop Ctr, 57-58, mgr solid state physics lab, 68-72, MGR SOLID STATE & ELECTRONCIS LAB, GEN ELEC CO, 72- Mem: Am Phys Soc. Res: Experimental investigations of electronic properties of surfaces. Mailing Add: Solid State & Elec Lab Gen Elec Res & Develop Ctr PO Box 8 Schenectady NY 12301

STOUT, VIRGINIA FALK, b Buffalo, NY, Jan 5, 32; m 55; c 1. ENVIRONMENTAL CHEMISTRY. Educ: Cornell Univ, AB, 53; Harvard Univ, AM, 55; Univ Wash, PhD(org chem), 61. Prof Exp: RES CHEMIST, PAC UTILIZATION RES CTR, US DEPT COM, 61- Concurrent Pos: Affil assoc prof, Col Fisheries, Univ Wash, 72- Mem: Am Chem Soc. Res: Pesticide residues in fishery products; organic contaminants and trace metals in fishery products. Mailing Add: Pac Utilization Res Ctr 2725 Montlake Blvd E Seattle WA 98112

STOUT, WALTER CLAY, b Ennis, Tex, Sept 2, 08; m 41; c 2. DENTISTRY. Educ: Emory Univ, DDS, 33. Prof Exp: From assoc prof to prof periodont, 52-73, chmn dept pedodontics, 60-73, asst dean, 69-73, EMER PROF PERIODONT, BAYLOR COL DENT, 73- Concurrent Pos: Hon staff mem, Children's Med Ctr, 60- Mem: Am Col Dentists; Am Acad Periodont; Am Acad Hist Dent; Int Asn Dent Res. Res: Treatment of infectious diseases of the oral cavity and cervial sensitivity. Mailing Add: Box 219 Ennis TX 75119

STOUT, WILLIAM F, b Wilkensburg, Pa, July 3, 40; m 65. MATHEMATICS. Educ: Pa State Univ, BS, 62; Purdue Univ, MS, 64, PhD(probability), 67. Prof Exp: Asst prof, 67-73, ASSOC PROF MATH, UNIV ILL, URBANA-CHAMPAIGN, 73- Mem: Am Statist Asn; Am Math Soc; Inst Math Statist. Res: Probability limit theorems. Mailing Add: 310 Altgeld Hall Univ of Ill Dept of Math Champaign IL 61820

STOUTAMIRE, DONALD WESLEY, b Roanoke, Va, Mar 10, 31; m 56; c 3. ORGANIC CHEMISTRY. Educ: Roanoke Col, BS, 52; Univ Wis, PhD(org chem), 57. Prof Exp: CHEMIST, SHELL DEVELOP CO, 57- Mem: Am Chem Soc; AAAS. Res: Agricultural chemicals; animal health products. Mailing Add: Shell Develop Co Box 4248 Modesto CA 95352

STOUTAMIRE, WARREN PETRIE, b Salem, Va, July 5, 28; m 63; c 2. PLANT TAXONOMY, EVOLUTION. Educ: Roanoke Col, BS, 49; Univ Ore, MS, 50; Ind Univ, PhD(taxon), 54. Prof Exp: Botanist, Cranbrook Inst Sci, 56-66; ASSOC PROF BIOL, UNIV AKRON, 66- Concurrent Pos: Collabr, Bot Garden, Univ Mich, 57-67. Mem: AAAS; Am Hort Soc; Am Soc Plant Taxon; Bot Soc Am; Asn Trop Biol. Res: Evolution of the genus Gaillardia; physiology of orchid seed germination; pollination of terrestrial orchid species. Mailing Add: Dept of Biol Univ of Akron Akron OH 44304

STOUTER, VINCENT PAUL, b Jersey City, NJ, Apr 28, 24; m 53; c 5. ZOOLOGY, NEUROENDOCRINOLOGY. Educ: Spring Hill Col, BS, 49; Fordham Univ, MS, 51; Univ Buffalo, PhD(biol), 59. Prof Exp: Instr biol, physiol & genetics, Canisius Col, 51-52; instr biol, anat & genetics, Gannon Col, 52-53; instr gen chem, D'Youville prof to assoc prof, 59-69, chmn dept biol, 59-71, PROF BIOL, PHYSIOL & ANAT, CANISIUS COL, 69-, CHMN COMT PRE-MED & PRE-DENT QUALIFICATION, 62- Mem: NY Acad Sci; Asn Am Med Cols. Res: Hypothalamic neurosecretion; electrolyte and salt balance in mammals. Mailing Add: Dept of Biol Canisius Col Buffalo NY 14208

STOVER, BETSY JONES, b Salt Lake City, Utah, May 13, 26; m 50; c 2. RADIOBIOLOGY, PHARMACOLOGY. Educ: Univ Utah, AB, 47; Univ Calif, PhD(chem), 50. Prof Exp: Asst, Univ Calif, 47-49, asst, Radiation Lab, 48-50; asst res prof chem, Univ Utah, 50-58, assoc res prof, 58-70; ASSOC PROF PHARMACOL, SCH MED, UNIV NC, CHAPEL HILL, 70-, DIR GRAD TRAINING PROG, 74- Concurrent Pos: Chemist, Radiobiol Lab, Univ Utah, 50-70, adj assoc res prof anat, 70-75, adj res prof, 75-; mem panel eval NSF Grad Fel Applns, Nat Res Coun, 73-; mem adv comt, Health Physics Div, Oak Ridge Nat Lab, 75-77. Mem: Am Phys Soc; Am Chem Soc; Radiation Res Soc; Am Soc Pharmacol & Exp Therapeut; AAAS. Res: Toxicology of radionuclides; rate processes in biology. Mailing Add: Dept of Pharmacol Univ of NC Chapel Hill NC 27514

STOVER, JAMES ANDERSON, JR, b Hayesville, NC, June 9, 37; m 63. OPERATIONS RESEARCH, SYSTEMS SCIENCE. Educ: Univ Ga, BS, 59; Univ Ala, MA, 66, PhD(math), 69. Prof Exp: Physicist, US Army Missile Command, Redstone Arsenal, Ala, 60-62; control systs engr, Marshall Space Flight Ctr, NASA, 62-65; consult systs anal, Anal Serv, Inc, Va, 69; asst prof math, Memphis State Univ, 69-74; PRIN STAFF, OPERS RES INC, APPL RES LAB, PA STATE UNIV, 74- Res: Automatic control systems; stability theory; underwater sound propagation. Mailing Add: Pa State Univ Appl Res Lab Box 30 State College PA 16801

STOVER, LEON EUGENE, b Lewistown, Pa, Apr 9, 29; m 56. CULTURAL ANTHROPOLOGY. Educ: Western Md Col, BA, 50; Columbia Univ, MA, 52, PhD(anthrop), 62. Prof Exp: Instr anthrop, Am Mus Natural Hist, 55-57; from instr to asst prof, Hobart & William Smith Cols, 57-65; assoc prof, 65-74, PROF ANTHROP, ILL INST TECHNOL, 74- Concurrent Pos: Vis asst prof cult anthrop, Univ Tokyo, 63-65, Grad Sch, 64-65; Human Ecol Fund fel, Japan, 63-64, NIH fel, 64-65. Mem: Asn Asian Studies. Res: Cultural ecology of China; Chinese behavior; science fiction of H G Wells. Mailing Add: Dept of Social Sci Ill Inst of Technol Chicago IL 60616

STOVER, LEWIS EUGENE, b Philadelphia, Pa, Apr 12, 25; m 51; c 3. PALEONTOLOGY, PALYNOLOGY. Educ: Dickinson Col, BSc, 51; Univ Rochester, PhD, 56. Prof Exp: Res geologist, Esso Prod Res Co, Tex, 56-65, res assoc, 66-71, mem staff, Esso Standard Oil Ltd, 71-74, MEM STAFF, EXXON PROD RES CO, HOUSTON, 74- Mem: Geol Soc Am; Paleont Soc; Int Asn Plant Taxon. Res: Geology; fossil spore; pollen; microplankton; small calcareous fossils; acid-in-soluble microfossils. Mailing Add: Exxon Prod Res Co Box 2159 Houston TX 77001

STOVER, RAYMOND WEBSTER, b Pittsburgh, Pa, Mar 20, 38; m 60; c 2. PHYSICS. Educ: Lehigh Univ, BS, 60; Syracuse Univ, MS, 62, PhD(physics), 67. Prof Exp: SR SCEINTIST, XEROX CORP, 66- Concurrent Pos: Adj fac mem, Rochester Inst Technol, 74- Mem: Am Asn Physics Teachers; Soc Photog Scientists & Engrs; Asn Am Scientists. Res: Search for an electron-proton charge difference; xerographic development process; electrostatics; triboelectricity; small particle physics. Mailing Add: Xerographic Technol Dept Xerox Corp Webster NY 14580

STOVER, ROBERT HARRY, b Chatham, Ont, Dec 2, 26; m 49. PLANT PATHOLOGY. Educ: Ont Agr Col, BSA, 47; Univ Toronto, PhD(plant path), 50. Prof Exp: Plant pathologist, Dominion Lab Plant Path, Ont, 47-51; plant pathologist, Dept Trop Res, United Brands Co, 51-57, head dept plant path, 57-61, asst sci dir, Dept Trop Res, 61-75, DIR, VINING C DUNLAP LABS, UNITED BRANDS CO, 75- Mem: Am Phytopath Soc; Can Phytopath Soc. Res: Diseases of tropical crops; soil microbiology. Mailing Add: Vining C Dunlap Labs United Brands Co La Lima Honduras

STOVER, SAMUEL LANDIS, b Bucks Co, Pa, Nov 19, 30; m; c 3. PEDIATRICS. Educ: Goshen Col, BA, 52; Jefferson Med Col, MD, 59; Am Bd Pediat, dipl, 69; Am Bd Phys Med & Rehab, dipl, 71. Prof Exp: Intern, St Luke's Hosp, Bethlehem, Pa, 59-60; gen pract in Ark, 60-61 & Indonesia, 61-64; resident pediat, Children's Hosp, Philadelphia, 64-66; asst med dir, Children's Seashore House, Atlantic City, NJ, 66-67; resident phys med & rehab, Univ Pa, 67-69; ASSOC PROF PEDIAT & PROF PHYS MED & REHAB, UNIV ALA, BIRMINGHAM, 69-, CHIEF PEDIAT UNIT, SPAIN REHAB CTR, UNIV HOSP & CLINS, 69- Concurrent Pos: Chief pediat unit, Children's Hosp, Birmingham, 69- Mem: Am Acad Pediat; Am Acad Phys Med & Rehab; Am Cong Rehab Med; AMA. Mailing Add: Spain Rehab Ctr 1717 Sixth Ave S Birmingham AL 35233

STOVER, STEPHEN LEECH, b McPherson, Kans, Aug 10, 19; m 44; c 5. GEOGRAPHY. Educ: McPherson Col, AB, 40; Univ Kans, MA, 41; Univ Wis, MS, 55, PhD, 60. Prof Exp: Instr high sch, Kans, 41-42, asst prin, 46-54; instr soc sci & asst dean, Garden City Jr Col, 46-54; asst geog, Univ Wis, 55-56, instr, Univ Wis-Milwaukee, 57-60, asst prof, 60-64; asst prof, 64-69, ASSOC PROF GEOG, KANS STATE UNIV, 69- Concurrent Pos: Fulbright vis lectr, Univ Auckland, 66-67; coordr, Nat Coun Geog Educ, Kans, 69-; ed, Kans Geogr, 69-; NSF travel grant, NZ, 74; assoc ed, Transactions, 75- Mem: Asn Am Geogrs; Nat Coun Geog Educ; Agr Hist Soc. Res: Historical geography of the United States; agricultural geography; Australia, New Zealand. Mailing Add: Dept of Geog Kans State Univ Manhattan KS 66506

STOW, GEORGE CLIFFORD, organic chemistry, see 12th edition

STOW, RICHARD W, b Medina, Ohio, June 13, 16; m 45; c 4. BIOPHYSICS. Educ: Mich State Univ, BS, 37; Pa State Univ, MS, 40; Univ Minn, PhD(biophys), 53. Prof Exp: Instr physics, William Penn Col, 46-47; asst biophys, Mayo Grad Sch Med, Univ Minn, 47-53; from asst prof med to assoc prof phys med & physiol, 53-72, PROF PHYS MED & PHYSIOL, COL MED, OHIO STATE UNIV, 72- Concurrent Pos: Fulbright lectr physiol, Iran, 60-61. Mem: Biophys Soc. Res: Ciculatory physiology; instrumentation in biology. Mailing Add: Dept of Phys Med Ohio State Univ Col of Med Columbus OH 43210

STOW, STEPHEN HARRINGTON, b Oklahoma City, Okla, Sept 18, 40; m 65. GEOCHEMISTRY, ENVIRONMENTAL GEOLOGY. Educ: Vanderbilt Univ, BA, 62; Rice Univ, MA, 65, PhD(geochem), 66. Prof Exp: Res scientist, Plant Foods Res Div, Continental Oil Co, 66-69; asst prof, 69-74, ASSOC PROF GEOL, UNIV ALA, TUSCALOOSA, 74- Concurrent Pos: Consult, Ala Geol Surv, 69- Mem: Mineral Soc Am; Geochem Soc; Geol Soc Am; Am Asn Petrol Geologists; Nat Asn Geol Teachers. Res: Element distribution in igneous and metamorphic rocks, hydrothermal alteration of igneous rocks, geology and geochemistry of phosphates; thorium and uranium geochemistry; geochronology; environmental geology; land-use planning. Mailing Add: Dept of Geol & Geog Univ of Ala University AL 35486

STOWE, BRUCE BERNOT, b Neuilly-sur-Seine, France, Dec 9, 27; US citizen; m 51; c 2. PLANT PHYSIOLOGY, BIOCHEMISTRY. Educ: Calif Inst Technol, BS, 50; Harvard Univ, MA, 51, PhD(biol), 54. Hon Degrees: MA, Yale Univ, 71. Prof Exp: NSF fel, Univ Col NWales, 54-55; instr biol, Harvard Univ, 55-58, lectr bot, 58-59, tutor biochem sci, 56-58; asst prof bot, 59-63, assoc prof biol, 63-71, PROF BIOL, YALE UNIV, 71-, DIR, MARSH BOT GARDENS, 75- Concurrent Pos: Mem, Metab Biol Panel, NSF, 60-61, Subcomt Plant Sci Planning & Comt Sci & Pub Policy, Nat Acad Sci, 64-66; Guggenheim fel, Nat Ctr Sci Res, France, 65-66; vis prof, Univ Osaka Prefecture, Japan, 72 & Waite Agr Res Inst, Univ Adelaide, 72-73. Mem: Am Soc Biol Chemists; Am Soc Plant Physiologists (secy, 63-65); Bot Soc Am; Soc Develop Biol; Phytochem Soc NAm. Res: Biochemistry and physiology of plant hormones, especially auxins, gibberellins and their relations to lipids and membrane structure. Mailing Add: Kline Biol Tower Yale Univ New Haven CT 06520

STOWE, CLARENCE M, b Brooklyn, NY, Mar 19, 22; m 46; c 3. PHARMACOLOGY, VETERINARY MEDICINE. Educ: NY Univ, BS, 44; Queens Col, NY, BS, 46; Univ Pa, VMD, 50; Univ Minn, PhD, 55. Prof Exp: Instr, 50-55, assoc prof, 55-57, prof & asst dean, 57-60, head dept, 60-71, PROF PHARMACOL, COL VET MED, UNIV MINN, ST PAUL, 60- Concurrent Pos: Spec appointee, Rockefeller Found, Columbia Univ, 64-65; prof, Nat Univ Columbia, 64-65; chmn comt vet drug efficacy, Nat Acad Sci-Nat Res Coun; mem tox study sect, NIH, 65-69; vis scholar, Univ Cambridge, 71-72; mem coun biol & therapeut agents, Am Vet Med Asn. Mem: Soc Exp Biol & Med; Am Soc Pharmacol & Exp Therapeut; Am Soc Vet Physiol & Pharmacol; Am Vet Med Asn; Am Dairy Sci Asn. Res: Drug distribution and excretion, muscle relaxants, chemotherapy of large domestic and wild animals. Mailing Add: Dept of Clin Sci Col of Vet Med Univ of Minn St Paul MN 55101

STOWE, HOWARD DENISON, b Greenfield, Mass, Mar 31, 27. PATHOLOGY. Educ: Univ Mass, BS, 48; Mich State Univ, MS, 56, DVM, 60, PhD(vet path), 62. Prof Exp: Instr animal husb & dairy prod, Bristol County Agr Sch, Mass, 49-53; asst animal husb, anat & vet path, Mich State Univ, 55-60; assoc prof vet sci & chief nutrit sect, Univ Ky, 63-68; from asst prof to assoc prof path, Sch Med, Univ NC, Chapel Hill, 68-74; ASSOC PROF PATH, SCH VET MED, AUBURN UNIV, 74- Concurrent Pos: Vis researcher, Dunn Nutrit Lab, Cambridge Univ, 62 & Dept Nutrit & Biochem, Denmark Polytech Inst, Copenhagen, 62; pathologist, Div Lab Animal Med, Sch Med, Univ NC, Chapel Hill. Mem: Am Vet Med Asn; Conf Res Workers Animal Dis; Am Asn Lab Animal Sci; Am Inst Nutrit. Res: Effects of lead upon reproduction in rats; cadmium toxicity in rabbits; canine and avian lead toxicity. Mailing Add: Dept of Path & Parasitol Auburn Univ Sch of Vet Med Auburn AL 36830

STOWE, KEITH S, b Midland, Mich, Feb 16, 43; m 67; c 2. ELEMENTARY PARTICLE PHYSICS. Educ: Ill Inst Technol, BS, 65; Univ Calif, San Diego, MS, 67, PhD(physics), 71. Prof Exp: Lectr physics, 71-74, ASST PROF PHYSICS, CALIF POLYTECH STATE UNIV, SAN LUIS OBISPO, 74- Mem: Am Phys Soc. Res: Elementary particle theory; ether theory. Mailing Add: Dept of Physics Calif Polytech State Univ San Luis Obispo CA 93407

STOWE, MARY EVELYN, chemistry, see 12th edition

STOWE, ROBERT ALLEN, b Kalamazoo, Mich, July 26, 24; m 47; c 4. SURFACE CHEMISTRY. Educ: Kalamazoo Col, BA, 48; Brown Univ, PhD(chem), 53. Prof Exp: Res chemist, 52-58, res & develop lab, Ludington Div, 58-64, sr res chemist, 64-69, hydrocarbons & monomers res lab, 69-71, assoc scientist, 71-74, ASSOC SCIENTIST, HYDROCARBONS & ENERGY RES LAB, DOW CHEM USA, 74- Honors & Awards: Victor J Azbe Lime Award, Nat Lime Asn, 64. Mem: Am Chem Soc; Am Inst Chemists; Sigma Xi; Catalysis Soc. Res: Heterogeneous catalysis; hydrocarbon processes; inorganic chemistry; organic fluorine chemistry; statistics; carbon monoxide methanation; Fisher-Tropsch synthesis. Mailing Add: Hydrocarbons & Energy Res Lab 677 Bldg Dow Chem USA Midland MI 48640

STOWELL, ELLERY CORY, (JR), b Washington, DC, Feb 24, 19; m 41. BIOCHEMISTRY. Educ: Calif Inst Technol, BS, 40; Univ Calif, MA, 43, PhD(biochem), 48. Prof Exp: Res assoc, Univ Wash, 47-50; from instr to asst prof biochem, Univ Southern Calif, 50-57; RES CHEMIST, MED RES PROGS, VET ADMIN HOSP, LONG BEACH, 57- Concurrent Pos: Consult, Los Angeles County Gen Hosp, 55-; adj asst prof, Univ Southern Calif, 58-; trainee, Inst Advan Educ Dent Res, 64. Mem: Fel AAAS; Soc Nuclear Med; assoc mem Am Dent Asn; NY Acad Sci; Int Asn Dent Res. Res: Permeability of dental enamel and fillings; fluoride biochemistry; iontophoresis in teeth; hydroxyapatite adsorption of salivary constituents; chemistry and metabolism of sulfur amino acids; thyroxine analogs; respiratory function; tubercle bacilli metabolism. Mailing Add: Dent Res Lab Vet Admin Hosp Long Beach CA 90801

STOWELL, EWELL ADDISON, b Ashland, Ill, Sept 2, 22; m 53. PLANT PATHOLOGY. Educ: Ill State Norm Univ, BEd, 43; Univ Wis, MS, 47, PhD(bot), 55. Prof Exp: Asst bot, Univ Wis, 46-47; instr, Univ Wis, Milwaukee, 47-49, asst, 49-53; from instr to assoc prof, 53-66, PROF BOT, ALBION COL, 66-, CHMN DEPT BIOL, 72- Concurrent Pos: Vis lectr, Univ Wis, 63; assoc prof, Univ Mich, 64. Mem: Bot Soc Am; Mycol Soc Am, Am Inst Biol Sci. Res: Taxonomy and morphology of Ascomycetes. Mailing Add: Dept of Biol Albion Col Albion MI 49224

STOWELL, JOHN CHARLES, b Passaic, NJ, Sept 10, 38; m 64; c 2. ORGANIC CHEMISTRY. Educ: Rutgers Univ, New Brunswick, BS, 60; Mass Inst Technol, PhD(org chem), 64. Prof Exp: Res specialist, Cent Res Lab, 3M Co, 64-69; NIH fel org chem, Ohio State Univ, 69-70; asst prof, 70-73, ASSOC PROF ORG CHEM, UNIV NEW ORLEANS, 73- Concurrent Pos: Res Corp & Petrol Res Fund grants, Univ New Orleans, 71-73. Mem: Am Chem Soc. Res: Organic synthesis; small ring heterocyclic compounds; sterically hindered compounds. Mailing Add: Dept of Chem Univ of New Orleans New Orleans LA 70122

STOWELL, ROBERT EUGENE, b Cashmere, Wash, Dec 25, 14; m 45; c 2. PATHOLOGY. Educ: Stanford Univ, AB, 36, MD, 41; Wash Univ, PhD(path), 44. Prof Exp: From asst to assoc prof path, Sch Med, Wash Univ, 42-48; prof path & oncol, Sch Med & dir cancer res, Med Ctr, Univ Kans, 48-59, chmn dept oncol, 48-51, path & oncol, 51-59, pathologist-in-chief, 51-59; sci dir, Armed Forces Inst Path, Washington, DC, 59-67; mem nat adv comt, Nat Ctr Primate Biol, 67-68, dir, 69-71, chmn dept path, 67-69, asst dean, Sch Med, 67-71, PROF PATH, SCH MED, UNIV CALIF, DAVIS, 67- Concurrent Pos: Commonwealth Fund advan med study & res fel, Inst Cell Res, Stockholm, Sweden, 46-47; mem morphol & genetics study sect,

NIH, 49-53 & path study sect, 54-55, chmn, 55-57, mem path training comt, div gen med sci, 58-61, chmn animal resources adv comt, Div Res Resources, 70-74; mem, Intersoc Comt Res Potential in Path, 56-, pres, 57-60; vis prof, Sch Med, Univ Md, 60-67; mem fedn bd, Fedn Am Soc Exp Biol, 63-66; mem subcomt comp path, Comt Path, Nat Acad Sci-Nat Res Coun, 63-69, subcomt manpower needs in path, 64-69 & US Nat comt, Int Coun Socs Path, 66-, chmn, 72-; mem, Intersoc Comt Path Info, 65-69, chmn, 66-67; mem adv med bd, Leonard Wood Mem, 65-69; mem div biol & agr, Nat Res Coun, 65-68 & comt doc data anat & clin path, 66-68; mem bd dirs, Coun Biol Sci Info, Nat Acad Sci, 67-; ed, Lab Invest, Int Acad Path, 67-72; mem med adv comt & consult, Vet Admin Hosp, Martinez, Calif, 69-; mem, Int Coun Socs Path, 70-; mem bd dirs, Univ Asn Res & Educ Path, 74- Honors & Awards: Except Civilian Serv Award, US Dept Army & Meritorious Civilian Serv Award, 63. Mem: AAAS; Am Soc Clin Path; Am Asn Path & Bact (vpres, 69-70, pres, 70-71); Am Soc Exp Path (vpres, 63-64, pres, 64-65); Int Acad Path (vpres, 57-58, pres elect, 58-59, pres, 59-60). Res: Cancer; experimental pathology; comparative pathology. Mailing Add: Dept of Path Univ of Calif Sch of Med Davis CA 95616

STOWENS, DANIEL, b New York, NY, Oct 27, 19; m 44; c 2. PATHOLOGY. Educ: Columbia Univ, AB, 41, MD, 43; Am Bd Pediat, dipl, 51; Am Bd Path, dipl, 54. Prof Exp: Chief, Sect Pediat Path, Armed Forces Inst Path, US Army, Washington, DC, 54-58; assoc prof path, Univ Southern Calif, 58-61 & Univ Louisville, 61-65; PATHOLOGIST, ST LUKE'S MEM HOSP CTR, 66- Concurrent Pos: Registr, Am Registry Pediat Path, 54-58; consult, Walter Reed Army Hosp, 56-58; pathologist, Children's Hosp, Los Angeles, 58-61; dir labs & chief prof servs, Children's Hosp, Louisville, 61-65. Mem: Fel Am Soc Clin Path; Soc Pediat Res; Am Asn Path & Bact; Int Acad Path. Res: Pediatric pathology, especially pathophysiology of fetus and mechanisms of development. Mailing Add: St Luke's Mem Hosp Ctr Utica NY 13503

STOY, WILLIAM S, b New York, NY, Sept 23, 25; m 49; c 2. CHEMISTRY. Educ: Queens Col, NY, BS, 45. Prof Exp: Chemist paint ptod res & develop, Socony Mobil Oil Co, Ind, 45-50, supvr, 50-58; sr chemist, Res Labs, Columbian Carbon Co, 58-64, mgr plastics applns, 64-71, MGR COATINGS, PLASTICS & INKS, PETROCHEM RES, CITIES SERV CO, CRANBURY, NJ, 71- Mem: Am Chem Soc; Soc Plastics Engrs; Am Soc Testing & Mat. Res: Plastics development; coatings and inks; pigment syntheses and applications; flame retardants; polymer chemistry. Mailing Add: 221 Herrontown Rd Princeton NJ 08540

STOYLE, JUDITH, b Quincy, Mass, Aug 26, 28. APPLIED STATISTICS. Educ: Univ Mass, BS, 50; Pa State Univ, MS, 57, PhD(agr econ, statist), 62. Prof Exp: Secy, Fuerst Stock Farm, Pine Plains, NY, 50-52; off mgr, Md Aberdeen Angus Asn, Towson, 52-54; partner, Nat Aberdeen-Angus Sales Serv, 54-55; res asst agr econ, Pa State Univ, 55-57, asst prof statist, 57-66; assoc prof, 66-74, PROF STATIST, TEMPLE UNIV, 74- Concurrent Pos: Consult statist, Head Start Eval Ctr, 67-70; trustee, Morris Animal Found, Denver, 71- Mem: Am Statist Asn; Am Inst Decision Sci. Res: Mobility of Pennsylvania rural youth; characteristics of inner city children; influence of Head Start on inner city children; incomes of Actors' Equity members; factors related to cat mortality. Mailing Add: Dept of Statist Temple Univ Philadelphia PA 19122

ST-PIERRE, CLAUDE, b Montreal, Que, Jan 7, 32; m 54; c 2. NUCLEAR PHYSICS. Educ: Univ Montreal, BSc, 54, MSc, 56, DSc(physics), 59. Prof Exp: Sci officer, Defence Res Bd Can, 58-61; Nat Res Coun Can fel, Ctr Nuclear Res, Strasbourg, France, 61-62; from asst prof to assoc prof, 62-70, PROF PHYSICS, LAVAL UNIV, 70-, CHMN DEPT, 73- Mem: Can Asn Physicists; Am Phys Soc. Res: Nuclear spectroscopy. Mailing Add: Dept of Physics Laval Univ Quebec PQ Can

ST-PIERRE, JACQUES, b Trois-Rivieres, PQ, Aug 30, 20; c 6. APPLIED STATISTICS. Educ: Univ Montreal, LSc, 45 & 48, MSc, 51; Univ NC, PhD(math statist), 54. Prof Exp: From asst prof to assoc prof math, 47-60, vdean, Fac Sci, 61-64, head dept comput sci, 66-69, dir, Comput Ctr, 64-72, PROF MATH STATIST, UNIV MONTREAL, 60-, VPRES PLANNING, 72- Concurrent Pos: Consult, Inst Microbiol & Hyg, 54- Mem: Can Asn Univ Teachers (pres, 65-66); Inst Math Statist; Am Statist Asn; Biometrics Soc; Asn Inst Res. Res: Statistical methods relative to public health and epidemiology. Mailing Add: Off of VPres Planning Univ of Montreal CP 6128 Montreal PQ Can

STRAAT, HAROLD WILLIAM, physics, optics, see 12th edition

STRAAT, PATRICIA ANN, b Rochester, NY, Mar 28, 36. BIOCHEMISTRY, ENZYMOLOGY. Educ: Oberlin Col, BA, 58; Johns Hopkins Univ, PhD(biochem), 64. Prof Exp: Lab instr biol, Johns Hopkins Univ, 58-59, USPHS res fel radiol sci, 64-67, res assoc, 67-68, asst prof, 68-70; sr res biochemist, 70-75, RES COORDR, BIOSPHERICS, INC, 75- Concurrent Pos: Lectr, Dept Radiol Sci, Sch Hyg & Pub Health, Johns Hopkins Univ, 70-72. Mem: AAAS; Am Inst Biol Sci; Am Chem Soc; NY Acad Sci. Res: Electron transport and inorganic nitrogen metabolism; mechanisms of nucleic acid replication; extraterrestial life detection; biological and chemical aspects of water pollution. Mailing Add: Biospherics Inc 4928 Wyaconda Rd Rockville MD 20852

STRAATSMA, BRADLEY RALPH, b Grand Rapids, Mich, Oct 29, 27; c 3. MEDICINE. Educ: Yale Univ, MD, 51. Prof Exp: Intern, New Haven Hosp, Yale Univ, 51-52; vis scholar, Col Physicians & Surgeons, Columbia Univ, 52, asst resident, 55-58; spec clin trainee, Nat Inst Neurol Dis & Blindness, 58-59; assoc prof surg & ophthal, 59-63, chief div ophthal, Sch Med, 59-68, PROF SURG & OPHTHAL, SCH MED, UNIV CALIF, LOS ANGELES, 63-, DIR, JULES STEIN EYE INST, 64-, CHMN DEPT OPHTHAL, UNIV, OPHTHALMOLOGIST IN CHIEF, HOSP & MEM BD, UNIV HOSP CHAPLAINCY SERV, INC, 68- Concurrent Pos: Resident, Inst Ophthal, Presby Hosp, New York, 55-58; fel ophthalmic path, Armed Forces Inst Path, Walter Reed Army Med Ctr, DC, 58-59; fel ophthal, Wilmer Inst, Johns Hopkins Univ, 58-59; mem vision res training comt, Nat Inst Neurol Dis & Blindness, 59-63 & neurol & sensory dis prog proj comt, 64-68; consult to Surgeon Gen, USPHS, 59-68; ophthal examr, aid to blind progs, Calif Dept Social Welfare, 59-; mem med adv comt, Nat Coun Combat Blindness, 60-; consult, Vet Admin Hosp, Long Beach, Calif, 60-; attend physician, Vet Admin Ctr, Wadsworth Gen Hosp, Los Angeles, 60-; attend physician & consult, Los Angeles County Harbor Gen Hosp, Torrance, 60-; vis consult, St John's Hosp, Santa Monica, 60-; mem courtesy staff, Santa Monica Hosp, 60- & St Vincent's Hosp, Los Angeles, 60-; mem sensory dis serv panel, Bur States Serv, USPHS, 63-65; trustee, John Thomas Dye Sch, Los Angeles, 67-72; prof, New Orleans Acad Ophthal, 68-; mem med adv bd, Int Eye Found, 70-; mem nat adv comt, Pan-Am Cong Ophthal, 71-72 & bd dirs, Conrad Berens Int Eye Film Libr, 71-; mem, Am Bd Ophthal, 72- & Pan-Am Ophthal Found. Honors & Awards: William Warren Hoppin Award, NY Acad Med, 56; co-recipient, Silver Award, Cert of Merit, AMA, 57, Knapp Award, 61 & Cert of Appreciation, Bd Trustees, 68; co-recipient, Silver Award, Am Soc Clin Path & Col Am Pathologists, 62; co-recipient, Conrad Berens Award, Int Eye Film Festival, 65; Award of Merit, Am Acad Ophthal & Otolaryngol, 67. Mem: Am Acad Ophthal & Otoleryngol; AMA; Am Asn Ophthal; Am Ophthal Soc; Asn Res Vision & Ophthal. Mailing Add: Jules Stein Eye Inst Univ of Calif Sch of Med Los Angeles CA 90024

STRACHAN, ALEC RONALD, b Colorado Springs, Colo, May 23, 33; m 60; c 2. ENVIRONMENTAL BIOLOGY. Educ: Calif State Col, Long Beach, BS, 62. Prof Exp: Seasonal aide, Calif Dept Fish & Game, 62-64, aquatic biologist, 64-65, asst marine biologist, 65-71; marine biologist, 71-74, SR MARINE BIOLOGIST, SOUTHERN CALIF EDISON CO, 74- Mem: Am Fisheries Soc; Am Nuclear Soc. Res: Development, conducting and analyzing of studies of the effect of proposed and existing generating stations upon the marine environments, including studies of water quality, cooling water discharges and the biological community. Mailing Add: So Calif Edison Co 2244 Walnut Grove Ave Rosemead CA 91770

STRACHAN, DONALD STEWART, b Highland Park, Mich, 32; div; c 4. HISTOLOGY, ORAL BIOLOGY. Educ: Wayne State Univ, BA, 54; Univ Mich, DDS, 60, MS, 62, PhD(anat), 64. Prof Exp: From instr to assoc prof oral biol, Sch Dent & Anat, Sch Med, 63-73, PROF DENT & ORAL BIOL, SCH DENT, UNIV MICH, ANN ARBOR, 73-, ASST DEAN, 69- Concurrent Pos: USPHS res career award, 63-68; consult Vet Admin Hosp, DC, 66-68 & coun dent educ, Am Dent Asn. Mem: Am Dent Asn; Int Asn Dent Res. Res: Histochemistry of esterase isoenzymes; lactic dehydrogenase in developing teeth and healing bone; data analysis and programming in the analysis of gel electrophoretic patterns; educational research; computer assisted instruction; self instructional media development. Mailing Add: Sch of Dent Univ of Mich Ann Arbor MI 48104

STRACHAN, WILLIAM MICHAEL JOHN, organic chemistry, see 12th edition

STRACHER, ALFRED, b Albany, NY, Nov 16, 30; m 54. BIOCHEMISTRY. Educ: Rensselaer Polytech Inst, BS, 52; Columbia Univ, MA, 54, PhD, 56. Prof Exp: From asst prof to assoc prof, 59-68, PROF BIOCHEM, COL MED, STATE UNIV NY DOWNSTATE MED CTR, 68-, CHMN DEPT, 72- Concurrent Pos: Nat Found Infantile Paralysis fel biochem, Rockefeller Inst, 56-58 & Carlsberg Lab, Copenhagen, 58-59; Guggenheim fel, 73-74. Mem: Am Soc Biol Chemists. Res: Relationship of structure of muscle proteins to mechanism of muscular contraction; relationship of protein structure to biological activity; contractility in non-muscle systems. Mailing Add: Dept of Biochem State Univ NY Downstate Med Ctr Brooklyn NY 11203

STRACK, CHARLES MILLER, b Grundy Center, Iowa, Aug 18, 15; m 40; c 4. GEOGRAPHY. Educ: Univ Iowa, PhD(geog), 50. Prof Exp: Instr soc sci, Univ Iowa, 46-50; dean, Gen Col, 54-71, PROF GEOG, HENDERSON STATE UNIV, 50-, DEAN ACAD SERV, 71- Concurrent Pos: Inst dir, Title IV Civil Rights Act, 66-68. Mem: Asn Am Geogrs; Nat Coun Geog Educ. Res: General education programs and curricula; economic geography; spatial and environmental perception. Mailing Add: Henderson State Univ Off of Acad Dean Arkadelphia AR 71923

STRADA, SAMUEL JOSEPH, b Kansas City, Mo, Oct 6, 42; m 72. PHARMACOLOGY, NEUROBIOLOGY. Educ: Univ Mo-Kansas City, BSPharm, 64, MS, 66; Vanderbilt Univ, PhD(pharmacol), 70. Prof Exp: Asst pharmacol, Univ Mo-Kansas City, 64-66; NIMH staff fel pharmacol, St Elizabeth's Hosp, Washington, DC, 70-72; ASST PROF PHARMACOL, UNIV TEX MED SCH, HOUSTON, 72- Concurrent Pos: Assoc fac, Univ Tex Grad Sch Biomed Sci, Houston, 72- Mem: AAAS; Soc Neurosci; Tissue Cult Asn. Res: Role of cyclic nucleotides in the nervous system; release of neurotransmitters and synaptic transmission; relation of the nervous system to hormone release mechanisms. Mailing Add: Pharmacol Prog Univ of Tex Med Sch Houston TX 77025

STRADFORD, H TODD, b Birmingham, Ala, Jan 8, 15; m 42; c 3. ORTHOPEDIC SURGERY, PATHOLOGY. Educ: Univ Chicago, BS, 36, MD, 38; Am Bd Orthop Surg, dipl, 54. Prof Exp: Fel orthop, Hosp for Ruptured & Crippled, 42 & Fracture Serv, Presby Hosp, New York, 43; resident children's orthop, NC Orthop Hosp Crippled Children, 48-49; resident adult orthop, Philadelphia Naval Hosp, 51-52; sr instr orthop path, Armed Forces Inst Path, 58-61; asst prof, 61-67, ASSOC PROF ORTHOP SURG, SCH MED & DENT, UNIV ROCHESTER, 67- Concurrent Pos: Consult, Highland Hosp, Rochester, NY, Genesee Hosp & Rochester Gen Hosp, 67-; NIH fel, Univ Rochester, 69- Mem: Fel Am Acad Orthop Surgeons; NY Acad Sci. Res: Bone pathology; clinical and laboratory investigation of metabolic diseases of bone and of bone tumors. Mailing Add: Div of Orthop Sug Univ Rochester Sch of Med & Dent Rochester NY 14642

STRADLING, SAMUEL STUART, b Hamilton, NY, Dec 11, 37; m 63; c 3. ORGANIC CHEMISTRY. Educ: Hamilton Col, AB, 59; Univ Rochester, PhD(org chem), 64. Prof Exp: From asst prof to assoc prof, 63-74, PROF CHEM, ST LAWRENCE UNIV, 74- Concurrent Pos: NSF acad year exten grant, 65-67. Mem: AAAS; Am Chem Soc; Sigma Xi. Res: Reaction mechanisms; natural product chemistry. Mailing Add: Dept of Chem St Lawrence Univ Canton NY 13617

STRAETER, TERRY ANTHONY, b St Louis, Mo, June 12, 42; m 64; c 2. APPLIED MATHEMATICS. Educ: William Jewell Col, AB, 64; Col William & Mary, MA, 66; NC State Univ, PhD(appl math), 71. Prof Exp: Instr math, William Jewell Col, 66-67; MATHEMATICIAN, LANGLEY RES CTR, NASA, 67- Concurrent Pos: Lectr, Christopher Newport Col, 71- & George Washington Univ, 71- Mem: Soc Appl & Indust Math; Am Math Soc; Am Inst Aeronaut & Astronaut; Math Asn Am. Res: Optimal control theory; optimization techniques theory and application; computational mathematics. Mailing Add: NASA Langley Res Ctr Ms 125 Hampton VA 23365

STRAF, MIRON L, b New York, NY, Apr 13, 43. STATISTICS. Educ: Carnegie-Mellon Univ, BS, 64, MS, 65; Univ Chicago, PhD(statist), 69. Prof Exp: Asst prof statist, Univ Calif, Berkeley, 69-74; RES ASSOC, COMT NAT STATIST & STAFF OFFICER, PANEL TO REV STATIST ON SKIN CANCER, NAT ACAD SCI-NAT RES COUN, 74-, STAFF DIR, STUDY GROUP ON ENVIRON MONITORING, 75- Mem: Inst Math Statist; Am Statist Asn; AAAS. Res: Applied and mathematical statistics; analysis and evaluation of environmental data on pollution and its effects; analysis of effects of ultraviolet radiation on skin cancer; applications of statistics to public policy. Mailing Add: Comt on Nat Statist Nat Acad Sci 2101 Constitution Ave Washington DC 20418

STRAFFON, RALPH ATWOOD, b Croswell, Mich, Jan 4, 28; m 54; c 5. MEDICINE, UROLOGY. Educ: Univ Mich, MD, 53; Am Bd Urol, dipl, 62. Prof Exp: Intern, Univ Hosp, Ann Arbor, Mich, 53-54, from asst resident to resident gen surg, 54-55, resident surg, 56-57, from jr clin instr to sr clin instr, 57-59; staff mem, 59-63, HEAD, CLEVELAND CLIN FOUND, 63- Concurrent Pos: Res fel med, Renal Lab, Peter Bent Brigham Hosp, Boston, 56. Mem: AMA; fel Am Col Surg; Am Urol Asn; Am Asn Genito-Urinary Surg; fel Am Acad Pediat. Mailing Add: 9500 Euclid Ave Cleveland OH 44106

STRAFUSS, ALBERT CHARLES, b Princeton, Kans, Jan 24, 28; m 54; c 5. COMPARATIVE PATHOLOGY, ONCOLOGY. Educ: Kans State Univ, BS & DVM, 54; Iowa State Univ, MS, 58; Univ Minn, PhD(comp path), 63. Prof Exp: Practitioner, Hastings, Nebr, 54-56; instr, Iowa State Univ & pathologist, Iowa Vet Med Diag Lab, 56-59; instr path, Col Vet Med, Univ Minn, 59-63; assoc prof path, Col Vet Med, Iowa State Univ & pathologist, Vet Med Res Inst, 63-64; assoc prof path, Sch Vet Med, Univ Mo-Columbia, 64-68; ASSOC PROF PATH, COL VET MED, KANS STATE UNIV, 68- Mem: Am Vet Med Asn; Electron Micros Soc Am; Conf Res Workers Animal Diseases. Res: Pathologic and epidemiologic studies of animal neoplasms; ultrastructure studies on the pathogenesis of morphological tissue alterations. Mailing Add: Dept of Vet Path Col of Vet Med Kans State Univ Manhattan KS 66506

STRAHILEVITZ, MEIR, b July 13, 35; Israeli citizen. PSYCHIATRY, MEDICAL RESEARCH. Educ: Hadassah Hebrew Univ, MD, 63. Prof Exp: Intern med, Hadassah Univ Hosp, 62-63; res asst exp biol & immunol, Weitzman Inst Sci, Israel, 64; resident psychiat, Ohio State Univ, 65-66 & Univ Wash, 66-69; resident grad physician, Galesburg State Res Hosp, 69-70; fel immunol, Univ Man, 70-71; staff psychiatrist, Rochester State Hosp, Minn, 71-72; clin dir training & res, Malcolm Bliss Ment Health Ctr, 72-74; ASSOC PROF PSYCHIAT, SCH MED, SOUTHERN ILL UNIV, 74- Concurrent Pos: Res fel, Tudichum Brain Res Lab, 69-70; asst prof, Wash Univ, 72-74. Mem: Am Psychiat Asn; AMA; Royal Col Physicians & Surgeons Can; Soc Biol Psychiat. Res: Biological research in psychiatric illness; the application of immunological methods to psychiatric diagnosis and treatment. Mailing Add: Sch of Med Southern Ill Univ PO Box 3926 Springfield IL 62708

STRAHL, ERWIN OTTO, b New York, NY, July 2, 30; m 52; c 4. MINERALOGY, PETROLOGY. Educ: City Col, New York, BS, 52; Pa State Univ, PhD(mineral), 58. Prof Exp: Res asst mineral, Pa State Univ, 52-58; mineralogist, Mineral Resources Dept, Kaiser Aluminum & Chem Corp, 58-59, Metals Div, Res Lab, 59-69, HEAD X-RAY & ELECTRON OPTICS LAB, ANAL RES DEPT, KAISER CTR TECHNOL, 69- Mem: Mineral Soc Am. Res: Mineralogy and petrology of soils and sedimentary rocks; x-ray diffraction analysis of inorganic oxides and hydroxides; quantitative x-ray diffraction and spectographic analysis; phase equilibria studies of aluminum oxide systems. Mailing Add: Anal Res Dept Kaiser Ctr Tech Box 870 Pleasanton CA 94566

STRAHM, NORMAN DALE, b Toronto, Kans, Feb 22, 40. PHYSICS, ELECTRICAL ENGINEERING. Educ: Mass Inst Technol, SB, 62, SM & EE, 64, PhD(elec sci & eng), 69. Prof Exp: Mem tech staff, Lincoln Lab, Mass Inst Technol, 69-70; VIS ASST PROF PHYSICS, UNIV ILL, CHICAGO CIRCLE, 70- Mem: AAAS; Am Phys Soc; Inst Elec & Electronics Engrs. Res: Light scattering; crystal lattice dynamics; quantum optics and quantum electronics. Mailing Add: Dept of Physics Univ of Ill at Chicago Circle Chicago IL 60680

STRAHS, GERALD, b New York, NY, May 26, 38; m 60; c 3. INORGANIC CHEMISTRY. Educ: Cooper Union Univ, BChE, 60; Univ Ill, MS, 62, PhD(phys chem), 65. Prof Exp: Assoc chem, Univ Calif, San Diego, 65-68; asst prof biochem, NY Med Col, 68-71; chemist, Crime Lab Sect, New York Police Dept, 72; chemist, US Assay Off, New York, 72-73; CHIEF CHEMIST, CONSOLIDATED REFINING CO, 73- Concurrent Pos: Am Cancer Soc fel, 65-68; chemist, US Assay Off, 73-75. Mem: AAAS; Am Chem Soc; Am Crystallog Asn. Res: Precious metals, refining, assaying, recovery. Mailing Add: 130-09 230 St Laurelton NY 11413

STRAIGHT, RICHARD COLEMAN, b Rivesville, WVa, Sept 8, 37; m 63; c 3. PHOTOBIOLOGY. Educ: Univ Utah, BA, 61, PhD(molecular biol), 67. Prof Exp: Asst dir radiation biol summer inst, Univ Utah, 61-63; SUPVRY CHEMIST, MED SERV, VET ADMIN HOSP, 65- Mem: AAAS; Am Chem Soc; Biophys Soc; Am Soc Photobiol; Int Solar Energy Soc. Res: Photodynamic action of biomonomers and biopolymers; tumor immunology; effect of antigens on mammary adenocarcinoma of C3H mice; ageing; biochemical changes in ageing; venom toxicology; mechanism of action of psychoactive drugs. Mailing Add: Res Dept VA Hospital 500 Foothill Dr Salt Lake City UT 84113

STRAILE, WILLIAM EDWIN, b Beaver, Pa, Mar 22, 31; m 53; c 4. BIOLOGICAL SCIENCE, NEUROSCIENCES. Educ: Westminster Col, AB, 53; Brown Univ, ScM, 55, PhD(biol), 57. Prof Exp: Sr cancer res scientist, Springville Labs, Roswell Park Mem Inst, 61-65; assoc prof anat & head cell res sect, Med Sch, Temple Univ, 66-75; GRANTS ASSOC, DIV RES GRANTS, NIH, 75- Concurrent Pos: Nat Cancer res fel, Univ London, 57-58; res fel, Brown Univ, 58-61; asst res prof, Grad Sch, State Univ NY Buffalo, 62-65. Res: Electron microscopy and electrophysiology of nerve endings; neurotransmitter chemicals in the control of neuronal functions, blood flow and cell division; neural elements in melanotic and epidermal neoplasia. Mailing Add: Dept of Dermat Temple Univ Sch of Med Philadelphia PA 19140

STRAIN, BOYD RAY, b Laramie, Wyo, July 19, 35; m 58; c 2. PHYSIOLOGICAL ECOLOGY. Educ: Black Hills State Col, BS, 60; Univ Wyo, MS, 61; Univ Calif, Los Angeles, PhD(plant sci), 64. Prof Exp: Asst prof bot & plant ecol, Univ Calif, Riverside, 64-69; ASSOC PROF BOT, DUKE UNIV, 69- Concurrent Pos: Mem, Panel Ecol Sect, NSF, 72-75 & Comt Mineral Resources & Environ, Nat Res Coun, 73-74. Mem: Ecol Soc Am; Bot Soc Am; Am Inst Biol Sci. Res: Physiological adaptations of plants to extreme environments; ecosystems analysis. Mailing Add: Dept of Bot Duke Univ Durham NC 27706

STRAIN, FRANKLIN, chemistry, see 12th edition

STRAIN, JEROME CHAMBERLAIN, b Marysville, Calif, Oct 24, 03; m 29; c 2. DENTISTRY. Educ: Univ Calif, DDS, 27. Prof Exp: Pvt pract, Calif, 27-41; from assoc clin prof to prof dent prosthetics, 46-69, EMER PROF DENTURE PROSTHESIS, SCH DENT, UNIV CALIF, SAN FRANCISCO, 69- Mem: Am Dent Asn; Asn Mil Surgeons US; fel Am Col Dentists. Res: Denture prosthetics; carcinogenic and toxic factors of color matter used to color denture base materials; met- alizing of denture bases. Mailing Add: Dept of Dent Univ of Calif Sch of Dent San Francisco CA 94122

STRAIN, JOHN HENRY, b Worcester, Eng, Oct 28, 22; Can citizen; m 49; c 3. POULTRY SCIENCE. Educ: Univ Sask, BSAgr, 49; Iowa State Univ, MS, 60, PhD(poultry breeding), 61. Prof Exp: Hatcheryman, Swift Can Co, 49-50; res off, 50-60, scientist poultry genetics, 60-70, HEAD ANIMAL SCI SECT, RES BR, CAN DEPT AGR, 70- Mem: Genetics Soc Can; Poultry Sci Asn; Can Soc Animal Sci. Res: Poultry genetics, mainly selection and genotype-environment interaction studies; dwarf broiler breeding management systems. Mailing Add: Res Sta Can Dept of Agr Brandon MB Can

STRAIN, WILLIAM HENRY, b Darien, Ga, June 27, 03; m 37; c 3. CHEMISTRY. Educ: State Univ NY, BS, 21; Northwestern Univ, MS, 23; Mass Inst Technol, PhD(chem), 31. Prof Exp: Chemist, Western Elec Eng Dept, Bell Tel Labs, Inc, 23-24; instr, Williams Col, 27-30; Mass Inst Technol travel fel, Munich Tech Univ, 31-32; fel biochem & exp path, Sch Med & Dent, Univ Rochester, 32-36, res assoc radiol, 36-43, assoc, 43-70; consult surg, 70-74, ASST, SCH MED & DENT, CASE WESTERN RESERVE UNIV, 70-; DIR TRACE ELEMENT LAB, CLEVELAND METROP GEN HOSP, 70- Concurrent Pos: Chemist, Eastman Kodak Co, 37-43; asst & Hoover fel, Northwestern Univ; asst, Mass Inst Technol. Mem: AAAS; Am Chem Soc; Am Asn Neurol Surg; Newcomen Soc NAm; Am Heart Asn. Res: Animal pigments; steroids; azo dyes; radiologic diagnostic agents; wound healing; zinc chelates; zinc therapy for atherosclerosis; trace elements. Mailing Add: Cleveland Metrop Gen Hosp 3395 Scranton Rd Cleveland OH 44109

STRAIN, WILLIAM SAMUEL, b Alpine, Tex, July 2, 09; m 36; c 1. VERTEBRATE PALEONTOLOGY. Educ: West Tex State Univ, BS, 32; Univ Okla, MS, 37; Univ Tex, PhD(geol), 64. Prof Exp: Instr geol & dir mus, 37-46, from asst prof to prof geol, 46-74, chmn dept geol sci, 71-74, EMER PROF GEOL SCI, UNIV TEX, EL PASO, 74- Mem: Geol Soc Am; Am Asn Petrol Geol; Am Inst Mining, Metall & Petrol Eng; Soc Vert Paleont; Nat Asn Geol Teachers. Res: Stratigraphy. Mailing Add: Dept of Geol Sci Univ of Tex El Paso TX 79968

STRAIT, PEGGY, b Canton, China, Apr 20, 33; US citizen; m 55; c 2. MATHEMATICS. Educ: Univ Calif, Berkeley, BA, 53; Mass Inst Technol, MS, 57; NY Univ, PhD(math), 65. Prof Exp: Programmer math, Liverpool Radiation Lab, Univ Calif, 54-55 & Lincoln Lab, Mass Inst Technol, 55-57; res assoc, G C Dewey Corp, NY, 57-62; lectr, 64-65, from asst prof to assoc prof, 65-72, PROF MATH, QUEENS COL, NEW YORK, 76- Concurrent Pos: Lincoln lab assoc staff fel, Mass Inst Technol, 56-57; res assoc fel, NY Univ, 62-64; NSF sci fac fel, 71-72. Mem: Am Math Soc. Res: Stochastic processes; probability theory and applications. Mailing Add: Dept of Math Queens Col Flushing NY 11367

STRAKA, LEORA E, organic chemistry, see 12th edition

STRAKA, WILLIAM CHARLES, b Phoenix, Ariz, Oct 21, 40; m 66. ASTROPHYSICS. Educ: Calif Inst Technol, BS, 62; Univ Calif, Los Angeles, MA, 65, PhD(astron), 69. Prof Exp: Teacher astron & phys sci, Long Beach City Col, 66-70; asst prof astron, Boston Univ, 70-74; ASST PROF ASTRON, JACKSON STATE UNIV, 74- Mem: Am Astron Soc; AAAS. Res: Structure and evolution of small mass stars; galactic nebulae; dynamics of supernova shells. Mailing Add: Dept of Physics Jackson State Univ Jackson MS 39217

STRALEY, H W, III, b Mercer Co, WVa, 06; m 28; c 2. GEOLOGY, GEOPHYSICS. Educ: Univ Chicago, PhD(geol), 37; Univ NC, Chapel Hill, PhD(geophysics), 38. Prof Exp: Geol engr, Seneca Oil Co, WVa, 26-28; geol engr, Pocatello Coal & Coke Co, 30-31; instr geol eng, Univ NC, 34-39; assoc prof geol & head dept, Baylor Univ, 39-42; geol engr, Foreign Econ Admin, 42-46; lectr, Am Univ, 46; geol engr, Int Bank Reconstruct & Develop, 47-48; spec lectr, Catholic Univ, 48-49; prof geol, Ga Inst Technol & res assoc, Eng Exp Sta, 50-66; sr geologist, Dames & Moore, 68-72; CONSULT GEOLOGIST, 57- Concurrent Pos: Indust consult, 33-; res grants, Smith Fund, 35-38, AAAS, 36-37, Sigma Xi, 55-57 & NSF, 65-67; partner, Johnson & Straley, 36-42; mem tech mission, Cuba, 42-43; vpres, H W Straley, Ltd, 50-58, pres, 58; deleg, Int Geol Cong, Copenhagen, 60 & Montreal, 72; prof structural & econ geol, Morehead State Univ, 69-71. Mem: Am Asn Petrol Geol; Am Geophys Union; Am Inst Mining, Metall & Petrol Eng; Geol Soc Am; Soc Explor Geophys. Res: Economic and structural geology; world mineral resources and structure; seismicity and structure of Appalachians and Atlantic Coastal Plain; cause and occurrence of damaging earthquakes. Mailing Add: 5910 Riverwood Dr NW Atlanta GA 30328

STRALEY, JAMES MADISON, b Madison, Wis, Nov 12, 10; m 36; c 3. ORGANIC CHEMISTRY. Educ: Univ Ill, BS, 32; Kans State Teachers Col, MS, 33; Iowa State Col, PhD(org chem), 36. Prof Exp: Res chemist, Eastman Kodak Co, NY, 36-42; res chemist, Gen Aniline & Film Corp, 43-45; res chemist, 46-50, res assoc, 50-62, sr res assoc, 62-68, RES FEL, RES DEPT, TENN EASTMAN CO, 68- Mem: Am Inst Chemists; Am Chem Soc; Am Asn Textile Chemists & Colorists. Res: Organometallics; physiological action of organic compounds; diazo-types; synthetic fibers; leuco derivatives of vat dyes; dye application; synthesis of dyes. Mailing Add: Res Dept Tenn Eastman Co Kingsport TN 37662

STRALEY, JOSEPH WARD, b Paulding, Ohio, Oct 6, 14; m 39; c 3. SPECTROSCOPY. Educ: Bowling Green State Univ, BSEd, 36; Ohio State Univ, MSc, 37, PhD(physics), 41. Prof Exp: Asst, Ohio State Univ, 37-38, 40-41; actg instr physics, Heidelberg Col, 38-39; instr, Univ Toledo, 41-42, asst prof, 42-44, actg head dept, 43-44; from asst prof to assoc prof, 44-58, PROF PHYSICS, UNIV NC, CHAPEL HILL, 58- Concurrent Pos: Guggenheim fel, 56-57. Mem: Am Phys Soc; Am Asn Physics Teachers. Res: Spectroscopy; research in learning and teaching physics. Mailing Add: Dept of Physics 232 Phillips Hall Univ of NC Chapel Hill NC 27514

STRALEY, TINA, b New York, NY, Sept 4, 43; c 1. MATHEMATICS. Educ: Ga State Univ, BA, 65, MS, 66; Auburn Univ, PhD(math), 71. Prof Exp: Teacher math, Miami Beach Sr High Sch, 66-67; instr, Spelman Col, 67-68 & Auburn Univ, 71-73; ASST PROF MATH, KENNESAW JR COL, 73- Mem: Am Math Soc. Res: Embeddings, extensions and automorphisms of Steiner systems. Mailing Add: Dept of Natural Sci & Math Kennesaw Jr Col Marietta GA 30061

STRALKA, ALBERT R, b Wilkes-Barre, Pa, Jan 18, 40; m 65; c 2. MATHEMATICS. Educ: Wilkes Col, AB, 61; Pa State Univ, MA, 64, PhD(math), 67. Prof Exp: Instr math, Wilkes Col, 61-62 & Pa State Univ, 66-67; asst prof, 67-72, ASSOC PROF MATH, UNIV CALIF, RIVERSIDE, 72- Mem: Am Math Soc. Res: Topological semigroups and lattices. Mailing Add: Dept of Math Univ of Calif Riverside CA 92502

STRAMPP, ALICE, immunochemistry, microbiology, see 12th edition

STRANAHAN, GEORGE S, b Toledo, Ohio, Nov 5, 31; m 54; c 5. PHYSICS. Educ: Calif Inst Technol, BS, 53; Carnegie Inst Technol, MS, 57, PhD(physics), 62. Prof Exp: Res assoc physics, Purdue Univ, 62-65; from asst prof to assoc prof, Mich State Univ, 65-72, chmn bd, Aspen Ctr Physics, 68-72. Res: Theoretical physics; field theory; many-body theory, particularly first order phase transitions. Mailing Add: Box 125 Woody Creek CO 81656

STRAND, FLEUR LILLIAN, b Bloemfontein, SAfrica, Feb 24, 28; m 46; c 1. BIOLOGY. Educ: NY Univ, AB, 48, MS, 50, PhD(biol), 52. Prof Exp: Instr biol, Brooklyn Col, 51-57; NIH fel, Physiol Inst, Free Univ Berlin, 57-59; from asst prof to assoc prof, 61-73; PROF BIOL, NY UNIV, 73- Mem: AAAS; Am Physiol Soc; Soc Neurosci; Soc Exp Biol & Med; NY Acad Sci. Res: Neurohormonal integration; effect of hormones on nerve and muscle. Mailing Add: Dept of Biol New York Univ New York NY 10003

STRAND, JOHN A, III, b Red Bank, NJ, July 22, 38; m 63; c 3. POLLUTION BIOLOGY. Educ: Lafayette Col, AB, 60; Lehigh Univ, MS, 62; Univ Wash, PhD(fisheries biol), 75. Prof Exp: Fisheries biologist, NJ Bur Fisheries Lab, 62-63; res scientist, US Naval Radiol Defense Lab, 64-69; SR RES SCIENTIST AQUATIC ECOL, PAC NORTHWEST LABS, BATTELLE MEM INST, 69- Concurrent Pos: Tech merit reviewer, Environ Protection Agency, 72. Mem: Am Fisheries Soc. Res: Aquatic radioecology; biological accumulation of radioisotopes in biological systems and their effects; effects and fate of petroleum residues in biological systems;

mariculture. Mailing Add: Ecosysts Dept Pac NW Labs PO Box 999 Richland WA 99352

STRAND, KAJ AAGE, b Hellerup, Denmark, Feb 27, 07; nat US; m 43, 49; c 2. ASTRONOMY. Educ: Univ Copenhagen, BA & MSc, 31, PhD(astron), 38. Prof Exp: Geodesist, Geod Inst, Univ Copenhagen, 31-33; asst to dir observ, Univ Leiden, 33-38; res assoc astron, Swarthmore Col, 38-42, res astronr, 46, Am-Scand Found fel, 38-39, Danish Rask-Orsted Found fel, 39-40; assoc prof astron, Univ Chicago, 46-47, res assoc, 47-67; prof astron, Northwestern Univ & dir, Dearborn Observ, 47-58; dir astrometry & astrophys, 58-63, SCI DIR, US NAVAL OBSERV, 63- Concurrent Pos: Guggenheim fel, 46; consult, NSF, 53-56. Honors & Awards: Distinguised Serv Award, US Dept Navy, 73. Mem: Int Astron Union; Am Astron Soc; Netherlands Astron Soc; Royal Danish Acad. Res: Photographic observations of double stars; stellar parallaxes; orbial motion in double and multiple systems; instrumentation. Mailing Add: 3203 Rowland Pl NW Washington DC 20008

STRAND, OLIVER ERIC, b Boyceville, Wis, Oct 9, 22; c 3. AGRONOMY, PLANT PHYSIOLOGY. Educ: Univ Mich, BS, 54; Univ Minn, MS, 66, PhD(agron & plant physiol), 69. Prof Exp: Soil conserv agent, Exten Serv, 56-59, agr agent, 59-66, EXTEN AGRONOMIST, UNIV MINN, ST PAUL, 66- Mem: Am Soc Agron; Weed Sci Soc Am. Res: Weed control in field crops. Mailing Add: Dept of Agron Univ of Minn St Paul MN 55101

STRAND, ROBERT CHARLES, b Newark, NJ, Sept 22, 25; m 51; c 3. ORGANIC POLYMER CHEMISTRY. Educ: Union Col, NY, BS, 51; Stevens Inst Technol, MS, 55; State Univ NY Buffalo, PhD(chem), 63. Prof Exp: Assoc chemist, Allied Chem Co, 51-59; res chemist, Sinclair Res, Inc, Ill, 61-62, group leader polymers, 62-68; chief chemist, NY Labs, Stein Hall Co, Inc, Long Island City, NY, 68-69, dir polymers res, 69-73; ASSOC DIR, BARRINGTON RES LAB, AM CAN CO, ILL, 74- Mem: Am Chem Soc. Res: Polymer applications, synthesis and characterization; adhesives; hot melts; specialty coatings; coatings for rigid and flexible packaging; polyacrylamides. Mailing Add: 4709 Valerie Dr Crystal Lake IL 60014

STRAND, ROBERT FENTON, b Ft Lewis, Wash, May 9, 28; m 53; c 4. FOREST ECOLOGY, FOREST SOILS. Educ: Univ Wash, BS, 51, MF, 57; Ore State univ, PhD(forest ecol), 64. Prof Exp: Res forester, Cent Res, 56-64, RES SUPVR, CENT RES, CROWN ZELLERBACH CORP, 64- Mem: Soc Am Foresters. Res: Forest fertilization, especially nutritional status correlation to responses; nitrogen loss; growth response; forest biometrics; application quality; stand simulation. Mailing Add: Cent Res Crown Zellerbach Corp Camas WA 98607

STRANDBERG, GERALD WILLIAM, b Oak Park, Ill, Oct 13, 39; m 60; c 3. BACTERIOLOGY, BIOCHEMISTRY. Educ: Loyola Univ, Ill, BS, 61; Univ Wis-Madison, MS, 63, PhD(bact), 66. Prof Exp: RES MICROBIOLOGIST, NORTHERN UTILIZATION RES & DEVELOP DIV, AGR RES SERV, USDA, 66- Mem: Am Soc Microbiol. Res: Carbohydrate and polyol metabolism; microbial physiology. Mailing Add: Northern Regional Res Lab 1815 University St Peoria IL 61604

STRANDBERG, MALCOM WOODROW PERSHING, b Box Elder, Mont, Mar 9, 19; m 47; c 4. SOLID STATE PHYSICS. Educ: Harvard Univ, SB, 41; Mass Inst Technol, PhD(physics), 48. Prof Exp: Res assoc, Mass Inst Technol, 41-42, mem staff, Off Sci Res & Develop, 42-43, microwave develop, 43-45, res assoc, 45-48, from asst prof to assoc prof, 48-60, PROF PHYSICS, MASS INST TECHNOL, 60- Concurrent Pos: Fulbright lectr, Univ Grenoble, 61-62. Mem: Fel Am Phys Soc; fel Inst Elec & Electronics Engrs; fel Am Acad Arts & Sci; NY Acad Sci. Res: Design of microwave components, radio transmitters and receivers; microwave physics. Mailing Add: Mass Inst of Technol 26-353 Cambridge MA 02139

STRANDJORD, PAUL EDPHIL, b Minneapolis, Minn, Apr 5, 31; m 53; c 2. CLINICAL CHEMISTRY, LABORATORY MEDICINE. Educ: Univ Minn, BA, 51, MA, 52; Stanford Univ, MD, 59. Prof Exp: Intern med, Sch Med, Univ Minn, 59-60, from instr to assoc prof lab med, 63-69; PROF LAB MED & CHMN DEPT, SCH MED, UNIV WASH, 69- Concurrent Pos: USPHS med fel, Univ Minn, 61-63. Honors & Awards: Borden Res Award, Stanford Univ, 59; C J Watson Res Award, Univ Minn, 63. Mem: AAAS; Acad Clin Lab Physicians & Sci; Am Chem Soc; Am Fedn Clin Res; Am Asn Clin Chem. Res: Diagnostic enzymology; diagnosis of liver disease; pattern recognition; computer assisted diagnosis. Mailing Add: Dept of Lab Med Univ of Wash Sch of Med Seattle WA 98105

STRANDNESS, DONALD EUGENE, JR, b Bowman, NDak, Sept 22, 28; m 57; c 3. MEDICINE, SURGERY. Educ: Pac Lutheran Univ, BA, 50; Univ Wash, MD, 54. Prof Exp: From instr to assoc prof, 62-70, PROF SURG, SCH MED, UNIV WASH, 70- Concurrent Pos: Res fel, Nat Heart Inst, 59-60; NIH career develop award, 65-; clin investr, Vet Admin, 62-65. Mem: Soc Vascular Surg; Am Inst Ultrasonics in Med; Am Col Surg; Am Surg Asn; Int Cardiovasc Soc. Res: Peripheral vascular disease and physiology. Mailing Add: Dept of Surg Univ of Wash Sch of Med Seattle WA 98195

STRANDTMANN, RUSSELL WILLIAM, b Maxwell, Tex, Apr 9, 10; m 36; c 2. TAXONOMY, ZOOLOGY. Educ: Southwestern Tex State Col, BS, 35; Tex A&M Univ, MS, 37; Ohio State Univ, PhD(entom), 44. Prof Exp: Instr sci & math, ETex State Univ, 37-42; field entomologist, Bur Entom & Plant Quarantine, USDA, 43; asst prof entom, Med Br, Univ Tex, 43-48; prof, 48-75, EMER PROF INVERT ZOOL, TEX TECH UNIV, 75- Concurrent Pos: Acarologist, Bernice P Bishop Mus, Honolulu, Hawaii, 67-68. Res: Biology and systematics of free living prostigmatic acarines of the polar regions. Mailing Add: Dept of Biol Tex Tech Univ Lubbock TX 79409

STRANG, ROBERT M, b Gt Brit, 26; m 53; c 5. FOREST ECOLOGY. Educ: Univ Edinburgh, BSc, 50; Univ London, PhD(ecol), 65. Prof Exp: Res off forestry, Colonial Develop Corp, Swaziland, Nyasaland & Tanganyika, 50-57 & Rhodesian Wattle Co, Ltd, 57-62; forest ecologist, Northern Forest Res Ctr, Forestry Serv, Can Dept Environ, 65-73, biologist, Northern Natural Resources & Environ Br, Arctic Land Use Res, Can Dept Indian & Northern Affairs, 73-74, HEAD ENVIRON STUDIES SECT, NORTHERN NATURAL RESOURCES & ENVIRON BR, DEPT INDIAN & NORTHERN AFFAIRS, CAN, 74- Concurrent Pos: Hon res assoc, Univ NB, 67-71; secy, Conserv Coun NB, 69-71. Mem: Commonwealth Forestry Asn; Can Bot Asn; Can Inst Forestry. Res: Resource and land management. Mailing Add: Northern Natural Res & Environ Br Dept Indian & Northern Affairs Ottawa ON Can

STRANG, RUTH HANCOCK, b Bridgeport, Conn, Mar 11, 23. PEDIATRICS, CARDIOLOGY. Educ: Wellesley Col, BA, 44; New York Med Col, MD, 49. Prof Exp: Intern, Flower & Fifth Ave Hosps, New York, 49-50, resident pediat, 50-52; from instr to asst prof bact, New York Med Col, 52-57, instr pediat, 52-56, asst clin prof, 56-57; from asst prof to assoc prof, 62-70, PROF PEDIAT, UNIV MICH, ANN ARBOR, 70- Concurrent Pos: Fel cardiol, Babies Hosp, New York, 56-57 & Hopkins Hosp, Baltimore, 57-59; res fel, Children's Hosp, Boston, 59-62. Mem: Fel Am Acad Pediat. Res: Congenital heart disease; effect on growth; ventricular performance; echocardiography. Mailing Add: Dept of Pediat Univ of Mich Hosp Ann Arbor MI 48104

STRANG, WILLIAM GILBERT, b Chicago, Ill, Nov 27, 34; m 58; c 3. MATHEMATICS. Educ: Mass Inst Technol, SB, 55; Oxford Univ, BA, 57; Univ Calif, Los Angeles, PhD(math), 59. Prof Exp: Moore instr, 59-61, from asst prof to assoc prof, 62-69, PROF MATH, MASS INST TECHNOL, 69- Concurrent Pos: Fels, NATO, 61-62 & Sloan Found, 65-67. Mem: Am Math Soc; Math Asn Am. Res: Partial difference and differential equations; matrix analysis. Mailing Add: Dept of Math Mass Inst of Technol Cambridge MA 02139

STRANGE, HAROLD OTTO, organic chemistry, see 12th edition

STRANGE, JOHN PHILLIP, b Canonsburg, Pa, Dec 26, 15; m 39; c 2. PHYSICS. Educ: Waynesburg Col, BS, 37. Prof Exp: Res physicist, Mine Safety Appliances Co, 37-48; sr res engr, Stanolind Oil & Gas Co, Standard Oil Co, Ind, 48-50; chief physicist, 50-57, mgr appl res & eng, 57-61, ASSOC DIR RES & ENG, MINE SAFETY APPLIANCES CO, 61- Mem: Am Phys Soc; fel Instrument Soc Am; Am Soc Testing & Mat; Am Ord Asn; Air Pollution Control Asn. Res: Instrumentation for process stream analysis and monitoring atmospheres for combustible or toxic contaminants; breathing apparatus for high altitude flight and oxygen deficient atmospheres; personal protective equipment. Mailing Add: Mine Safety Appliances Co 100 N Braddock Ave Pittsburgh PA 15208

STRANGE, JOHN RUBLE, b Knoxville, Tenn, Feb 25, 43; m 65; c 2. DEVELOPMENTAL PHYSIOLOGY, ENVIRONMENTAL PHYSIOLOGY. Educ: ETenn State Univ, BS, 65, MA, 67; Univ Tenn, PhD(zool), 70. Prof Exp: ASST PROF BIOL, GAS INST TECHNOL, 70- Mem: AAAS; Am Soc Zoologists; Am Inst Biol Sci; Soc Develop Biol. Res: Teratogenic agents; effects of radiation on the prenatal organism; effects of phosphate on aquatic environment. Mailing Add: Sch of Biol Ga Inst of Technol Atlanta GA 30332

STRANGE, RONALD STEPHEN, b Covington, Ky, Nov 18, 43; m 70; c 2. INORGANIC CHEMISTRY. Educ: Univ Ky, BS, 65; Univ Ill, Urbana, MS, 67, PhD(inorg chem), 71. Prof Exp: Instr chem, Ill Inst Technol, 70-71; ASST PROF CHEM, FARLEIGH DICKINSON UNIV, FLORHAM-MADISON CAMPUS, 71-, CHMN DEPT, 75- Concurrent Pos: Fac res grant-in-aid, Farleigh Dickinson Univ, 72-73. Mem: Am Chem Soc; Sigma Xi. Res: Transition metal base chemistry; semi empirical self consistent field molecular orbital calculations. Mailing Add: Dept of Chem Farleigh Dickinson Univ Florham-Mad Campus Madison NJ 07940

STRANGE, WILLIAM ERNEST, b Meridian, Miss, Aug 28, 07; m 36; c 3. MATHEMATICS. Educ: Univ Miss, BA, 30; Duke Univ, MEd, 36. Prof Exp: Coach, Newton High Sch, Miss, 30-34 & Louisville High Sch, 34-38; registr, Pearl River Jr Col, 38-41; accountant, Flintkote Co, 41-43; from assoc prof to prof, 43-70, T H STANLEY PROF MATH, MISS COL, 70-, HEAD DEPT, 45- Res: Analysis; geometry. Mailing Add: Dept of Math Miss Col Clinton MS 39056

STRANGWAY, DAVID W, b Simcoe, Ont, June 7, 34; m 57; c 2. GEOPHYSICS. Educ: Univ Toronto, BA, 56, MA, 58, PhD(physics), 60. Prof Exp: Sr geophysicist, Dominion Gulf Co, 56; chief geophysicist, Ventures Ltd, Ont, 56-57; res geophysicist, Kennecott Copper Corp, Colo, 60-61; asst prof geol, Univ Colo, 61-64; asst prof geophys, Mass Inst Technol, 65-68; assoc prof physics, 68-71, PROF GEOL, UNIV TORONTO, 71- Concurrent Pos: Consult, Kennecott Copper Corp, Anaconda Co, UN, Alyeska Pipelines & NASA; chief geophys br, NASA-Manned Spacecraft Ctr, 70-71. Mem: Soc Explor Geophys; Am Geophys Union; Europ Asn Explor Geophys; Soc Terrestrial Magnetism & Elec Japan; fel Royal Astron Soc. Res: History of the earth's magnetic field; studies of ancient reversals of the field; changes in direction and intensity and secular variation; exploration using electromagnetic techniques. Mailing Add: Dept of Geol Univ of Toronto Toronto ON Can

STRANO, ALFONSO J, b Ambridge, Pa, Apr 7, 27; m 57; c 1. VIROLOGY, PATHOLOGY. Educ: Hiram Col, BA, 50; Duquesne Univ, MS, 53; Univ Okla, PhD(path), 57; Univ Tex, MD, 60. Prof Exp: From instr to asst prof path, Univ Tex Med Br Galveston, 62-67; pathologist, Armed Forces Inst Path & chief, Viro-Path Br, 67-73; CLIN PROF PATH, SCH MED, SOUTHERN ILL UNIV, 73- Concurrent Pos: Am Cancer Soc res fel, 60-62. Mem: AMA; Col Am Path; Reticuloendothelial Soc; Int Acad Path. Res: Immunologic aspects of infectious disease, cellular immunity; histologic reaction to viral infections. Mailing Add: St Johns Hosp Labs Springfield IL 62702

STRANSKY, JOHN JANOS, b Budapest, Hungary, Sept 2, 23; nat US; m 47; c 2. SILVICULTURE. Educ: Univ Munich, BF, 47; Harvard Univ, MS, 54. Prof Exp: Plant propagator, Bussey Inst, Harvard Univ, 54-57; RES FORESTER, SOUTHERN FOREST EXP STA, US FOREST SERV, 57- Concurrent Pos: Lectr, Sch Forestry, Stephen F Austin State Univ. Mem: Soc Am Foresters; Wildlife Soc. Res: Silvicultural aspects of combining timber production with wildlife habitat practices in southern forests. Mailing Add: Wildlife Habitat & Silvicult Lab US Forest Serv Box 7600 Nacogdoches TX 75961

STRASBERG, MURRAY, b New York, NY, Aug 11, 17; m 45. ACOUSTICS. Educ: City Col New York, BS, 38; Cath Univ, MS, 48, PhD, 56. Prof Exp: Patent examr, US Patent Off, 38-42; physicist, David Taylor Model Basin, 42-49 & 52-58; noise consult, US Bur Ships, 49-52; sci liaison officer, Off Naval Res, London, 58-60; PROJ COORDR, NAVAL SHIP RES & DEVELOP CTR, MD, 60- Concurrent Pos: Fulbright lectr, Tech Univ Denmark, 63; adj prof, Am Univ, 64-; vis prof, Cath Univ, 74- Mem: Fel Acoust Soc Am (pres, 74-75); Am Phys Soc. Res: Underwater acoustics; hydrodynamics; cavitation; hydrodynamic noise; electroacoustic instrumentation; mechanical vibrations. Mailing Add: Naval Ship Res & Dev Ctr 1901 Bethesda MD 20034

STRASBURG, DONALD WISHART, b Benton Harbor, Mich, Sept 13, 24; m 52. FISH BIOLOGY. Educ: US Naval Acad, BS, 45; Univ Hawaii, PhD(marine zool), 53. Prof Exp: Asst zool, Univ Hawaii, 50-53; instr zool, Duke Univ & asst to dir, Marine Lab, 53-55; fishery res biologist, Bur Com Fisheries, US Fish & Wildlife Serv, 55-67; spec tech asst to mgr underwater technol, Elec Boat Div, Gen Dynamics Corp, Conn, 67-70; asst mgr, Fisheries Eng Lab, Nat Marine Fisheries Serv, 70-72; HEAD BR MARINE BIOL, NAVAL RES LAB, 72- Concurrent Pos: Asst marine zoologist, Pac Sci Bd Exped, Arno Atoll, Marshall Islands, 50, asst geologist, Onotoa Atoll, Gilbert Islands, 51; ichthyologist, Eniwetok Marine Biol Lab, Marshall Islands, 55; mem grad fac, Univ Hawaii, 57-67. Mem: Am Soc Ichthyol & Herpet; Lepidop Soc; Am Inst Fishery Res Biologists; Marine Technol Soc. Res: Biology and systematics of blennies, sharks, remoras and marlins; use of submarines for research; remote sensing of marine resources; biological problems of naval interest. Mailing Add: Code 8350 Naval Res Lab Washington DC 20375

STRASDINE, GEORGE ALFRED, b Edmonton, Alta, Sept 7, 33; m 69; c 7. MICROBIOLOGY, BIOCHEMISTRY. Educ: Univ BC, BSc, 56, MSc, 58,

PhD(microbiol), 61. Prof Exp: Fel, Nat Res Coun Can, Ottawa, 61-63; prof bact, Univ Sask, 63-65; RES SCIENTIST MICROBIOL, VANCOUVER TECHNOL LAB, FISHERIES RES BD CAN, 65- Mem: Can Soc Microbiol. Res: Microbial food poisoning; industrial waste treatment and utilization; research management. Mailing Add: Vancouver Technol Lab Fisheries Res Bd of Can Vancouver BC Can

STRASSBURG, ROGER WILLIAM, organic chemistry, see 12th edition

STRASSENBURG, ARNOLD ADOLPH, b Victoria, Minn, June 8, 27; m 49; c 3. PHYSICS. Educ: Ill Inst Technol, BS, 51; Calif Inst Technol, MS, 53, PhD(physics), 55. Prof Exp: From asst prof to assoc prof physics, Univ Kans, 55-66; prof, State Univ NY Stony Brook, 66-75; HEAD, MAT & INSTR DEVELOP SECT, NSF, 75- Concurrent Pos: Staff physicist, Comn Col Physics, 63-65; dir, Div Educ & Manpower, Am Inst Physics, 66-72; exec officer, Am Asn Physics Teachers, 72- Honors & Awards: Millikan Lectr Award, Am Asn Physics Teachers, 72. Mem: AAAS; Am Asn Physics Teachers; Nat Sci Teachers Asn. Res: High energy physics; fundamental particles; measurement of educational outcomes resulting from the application of alternative instructional materials and modes. Mailing Add: Div of Higher Educ in Sci Nat Sci Found Washington DC 20550

STRASSER, ELVIRA RAPAPORT, b Hungary; US citizen; wid; c 2. MATHEMATICS. Educ: Washburn Univ, BS, 43; Smith Col, MS, 51; NY Univ, PhD(math), 56. Prof Exp: Off Naval Res fel, 59-60; lectr math, Hunter Col, 61; from asst prof to assoc prof, Polytech Inst Brooklyn, 61-67; PROF MATH, STATE UNIV NY STONY BROOK, 67- Mem: Am Math Soc. Res: Group theory; graph theory; combinatorial problems. Mailing Add: Dept of Math State Univ of NY Stony Brook NY 11790

STRATBUCKER, ROBERT A, b Omaha, Nebr, Jan 29, 31; m 54; c 1. PHYSIOLOGY, BIOENGINEERING. Educ: Univ Omaha, BA, 55; Univ Nebr, MD, 60. Prof Exp: Instr physiol, 61-64, asst prof elec eng, 64-69, ASSOC PROF ELEC ENG, UNIV NEBR-LINCOLN, 69-; ASSOC PROF PHYSIOL & BIOPHYS & ASST PROF INTERNAL MED, UNIV NEBR MED CTR, OMAHA, 67- Concurrent Pos: Nat Heart Inst spec fel, Univ Wash, 66-67. Mem: Inst Elec & Electronics Engrs. Res: Cardiovascular physiology and instrumentation; electrocardiography; cellular electrophysiology. Mailing Add: Dept of Elec Eng Univ of Nebr Lincoln NE 68508

STRATFORD, EUGENE SCOTT, b Waterloo, Iowa, June 18, 42; m 66; c 1. MEDICINAL CHEMISTRY. Educ: Idaho State Univ, BSPharm, 66; Ohio State Univ, PhD(med chem), 70. Prof Exp: ASST PROF MED CHEM, SCH PHARM, UNIV CONN, 70- Concurrent Pos: Nat Heart & Lung Inst res grant, Univ Conn, 72- Mem: Am Chem Soc; Am Asn Cols Pharm; The Chem Soc. Res: Structure-activity relationships of biologically active organic compounds, particularly those with hypolipemic properties as potentially valuable for the treatment of atherosclerosis. Mailing Add: Dept of Med Chem Univ of Conn Sch of Pharm Storrs CT 06268

STRATFORD, JOSEPH, b Brantford, Ont, Sept 5, 23; m 52; c 2. NEUROSURGERY. Educ: McGill Univ, BSc, 45, MD, CM, 47, MSc, 51, dipl neurosurg, 54; FRCS(C), 56. Prof Exp: Lectr neurosurg, McGill Univ, 55-56; from asst prof to prof surg, Univ Sask, 56-62; assoc prof, 62-72, PROF NEUROSURG, McGILL UNIV, 72- Concurrent Pos: Dir div neurosurg, Montreal Gen Hosp. Mem: Am Asn Neurol Surg; fel Am Col Surgeons; Am Acad Neurol; Cong Neurol Surg; fel Royal Soc Med. Mailing Add: Montreal Gen Hosp Montreal PQ Can

STRATHDEE, GRAEME GILROY, b Edinburgh, Scotland, June 29, 42; Can citizen; m 67; c 1. SURFACE CHEMISTRY, INORGANIC CHEMISTRY. Educ: McGill Univ, BSc, 63, PhD(chem), 67. Prof Exp: ASSOC RES OFFICER CHEM, WHITESHELL NUCLEAR RES ESTAB, ATOMIC ENERGY CAN, LTD, 67- Mem: Chem Inst Can; Am Chem Soc. Res: Homogeneous catalysis; catalytic activation of small molecules; hydrogen isotope exchange reactions; enrichment of deuterium; adsorption phenomena; foaming and antifoaming. Mailing Add: Atomic Energy of Can Ltd Whiteshell Nuclear Res Estab Pinawa MB Can

STRATHMANN, RICHARD RAY, b Pomona, Calif, Nov 25, 41; m 64; c 2. MARINE BIOLOGY, ZOOLOGY. Educ: Pomona Col, BS, 63; Univ Wash, MS, 66, PhD(zool), 70. Prof Exp: NIH training grant, Univ Calif, Los Angeles, 70; NSF fel, Univ Hawaii, 70-71; asst prof zool, Univ Md, College Park, 71-73; ASST PROF ZOOL, UNIV WASH & RESIDENT ASSOC DIR, FRIDAY HARBOR LABS, 73- Mem: Am Soc Naturalists; Am Soc Limnol & Oceanog; Am Soc Zoologists; Marine Biol Asn UK. Res: Population biology, form and function of marine invertebrates; biology of invertebrate larvae; biology of suspension feeding. Mailing Add: Friday Harbor Labs Univ of Wash Friday Harbor WA 98250

STRATMAN, FREDERICK WILLIAM, b Dodgeville, Wis, Nov 26, 27; m 51; c 1. CHEMISTRY. Educ: Univ Wis, BS, 50, MS, 57, PhD(animal-dairy husb, biochem), 61. Prof Exp: Res asst animal husb, Univ Wis, 57-61; researcher, Wis Alumni Res Found, 62; res assoc, Univ Wis, 62-63, NIH fel reprod physiol, 63-65; asst prof animal sci & biochem, Univ Ife, Nigeria, 65-67; asst prof animal sci, 67-68, proj assoc, Inst Enzyme Res, 68-70, proj assoc & Babcock fel, 70-71, NIH spec fel, 71-73, ASST RES PROF, UNIV WIS-MADISON, 71- Mem: Am Soc Biol Chemists. Res: Hormonal regulation of protein synthesis, particularly sulfhydryls, polyamines, methylation, phosphorylation, muscle, liver, tumors, perfusions, testosterone, somatomedin; hormonal regulation of gluconeogenesis; hepatocytes. Mailing Add: Inst for Enzyme Res Univ of Wis Madison WI 53706

STRATMEYER, MELVIN EDWARD, b Peoria, Ill, Aug 30, 42; m 66. RADIOBIOLOGY, BIOCHEMISTRY. Educ: Purdue Univ, Lafayette, BS, 65, MS, 66, PhD(bionucleonics), 69. Prof Exp: Lab technician, Miles Labs, 62-64; lab technician, Hort Dept, Purdue Univ, 64-65, teaching asst bionucleonics & health physics, 65-66; RES CHEMIST, BUR RADIOL HEALTH, US FOOD & DRUG ADMIN, 69- Mem: AAAS. Res: Ionizing radiation effects on nucleic acid and protein metabolism; ionizing radiation effects on mitochrondrial systems; microwave and ultrasound effects on nucleic acid and protein metabolism. Mailing Add: Div of Biol Effects Bur Radiol Health US Food & Drug Admin Rockville MD 20852

STRATOPOULOS, GEORGE, b Methoni-Messinia, Greece, May 25, 31; US citizen; m 60. MATHEMATICS. Educ: Univ Utah, BA, 60, MS, 63, PhD(math), 66. Prof Exp: Chmn dept, 72-74, ASSOC PROF MATH, US INT UNIV, WEST CAMPUS, 72- Mailing Add: Dept of Math US Int Univ 10455 Pomerado Rd San Diego CA 92131

STRATTA, JULIUS JOHN, physical chemistry, polymer chemistry, see 12th edition

STRATTON, CEDRIC, b Langley, Eng, Apr 26, 31; US citizen; m 61; c 1. INORGANIC CHEMISTRY, ANALYTICAL CHEMISTRY. Educ: Univ Nottingham, BSc, 53; Univ London, PhD(inorg chem), 63. Prof Exp: Qual control chemist, Richard Klinger, Ltd, Eng, 53-55; develop chemist, Small & Parkes, Ltd, 55-56; sci officer anal res, Brit Insulated Callender's Cables, 57-61; NSF res fel, Univ Fla, 63-65; assoc prof, 65-72, PROF INORG & ANAL CHEM, ARMSTRONG STATE COL, 72- Mem: Am Chem Soc; The Chem Soc. Res: Chemistry of group V elements, their heterocyclic derivatives; concentration of minerals in local well-water; legal consultancy. Mailing Add: Dept of Chem Armstrong State Col 11935 Albercorn St Savannah GA 31406

STRATTON, CHARLES ABNER, b Canyon, Tex, Mar 28, 16; m 51; c 3. COLLOID CHEMISTRY, SCIENCE EDUCATION. Educ: WTex State Col, BS, 36; Univ Southern Calif, MS, 50, PhD(chem), 53. Prof Exp: Chemist, Borger Refinery, Phillips Petrol Co, 39-47; asst chem, Univ Southern Calif, 47-51; CHEMIST, RES DIV, PHILLIPS PETROL CO, 52- Concurrent Pos: Part-time instr chem, Bartlesville Wesleyan Col, 74- Mem: Am Chem Soc. Res: Chemical treatment of kerosene, gasoline and liquified petroleum gases; compounding of greases with inorganic thickeners; drilling mud chemicals; water-soluble polymers; water-flood chemicals. Mailing Add: 1233 N Wyandotte Dewey OK 74029

STRATTON, CHARLOTTE DIANNE, b Brooklyn, NY, Mar 7, 29. ORGANIC CHEMISTRY. Educ: Bucknell Univ, BS, 51; Pa State Univ, MS, 52. Prof Exp: From asst res chemist to assoc res chemist, 52-70, RES CHEMIST, PARKE, DAVIS & CO, WARNER-LAMBERT CO, INC, 70- Mem: Am Chem Soc. Res: Medicinal chemistry, especially natural products isolation and organic synthesis of cardiovascular drugs. Mailing Add: Parke Davis & Co Res Lab 2800 Plymouth Rd Ann Arbor MI 48106

STRATTON, DONALD BRENDAN, b Escanaba, Mich, Jan 6, 41; m 67. PHYSIOLOGY. Educ: Northern Mich Univ, BS, 63, MA, 64; Southern Ill Univ, PhD(physiol), 71. Prof Exp: ASST PROF BIOL, DRAKE UNIV, 71- Mem: Neurosci Soc. Res: Electroanesthesia; electrosleep; mechanisms underlying fading during electrically induced sleep and anesthesia. Mailing Add: Dept of Biol Drake Univ Des Moines IA 50311

STRATTON, EVERETT FRANKLIN, b Cambridge City, Ind, July 1, 07; m 28; c 1. GEOLOGY, GEOPHYSICS. Educ: DePauw Univ, AB, 28; Harvard Univ, AM, 31, ScD(geol), 32. Prof Exp: Instr geol, Harvard Univ, 28-32; res engr, Delco-Remy Corp, Gen Motors Corp, 33-36; res engr, Schlumberger Well Surv Corp, 36-58, exec vpres, 58-63, vpres, Schlumberger Ltd, 63-72, pres, Schlumberger Tech Corp, 66-72; RETIRED. Concurrent Pos: Mem, Bd Dirs, Rocky Mountain Natural Gas Co. Mem: AAAS; Soc Explor Geophys; Am Asn Petrol Geologists; Am Inst Mining, Metall & Petrol Engrs; Am Geophys Union. Res: Mineralogy; x-ray and chemical analysis; economic and structural geology; electrical well logging; geophysical methods for geological study and exploration. Mailing Add: 3435 Westheimer Houston TX 77027

STRATTON, JAMES FORREST, b Chicago Heights, Ill, Nov 29, 43. PALEONTOLOGY. Educ: Ind State Univ, Terre Haute, BS, 65; Ind Univ, Bloomington, MAT, 67, AM, 72, PhD(paleont), 75. Prof Exp: Instr geol, Shippensburg State Col, 67-70; ASST PROF GEOL, EASTERN ILL UNIV, 75- Mem: Soc Econ Paleontologists & Mineralogists; Int Bryozool Asn; Am Asn Petrol Geologists; Brit Palaeont Asn. Res: Quantitative analysis of morphological and structural characters of Fenestellidae for the study of taxonomy and functional morphology. Mailing Add: Dept of Geol Eastern Ill Univ Charleston IL 61920

STRATTON, JULIUS ADAMS, b Seattle, Wash, May 18, 01; m 35; c 3. PHYSICS. Educ: Mass Inst Technol, SB, 23, SM, 26; Swiss Fed Inst Technol, ScD(math, physics), 27. Hon Degrees: DEng, NY Univ, 55, LHD, Hebrew Union Col, 62, Oklahoma City Univ, 63, Jewish Theol Sem Am, 65; LLD, Northeastern Univ, 57, Union Col, NY, 58, Harvard Univ, 59, Brandeis Univ, 59, Carleton Col, 60, Univ Notre Dame, 61, Johns Hopkins Univ, 62; ScD, St Francis Xavier Univ, 57, Col William & Mary, 64, Carnegie Inst Technol, 65, Univ Leeds, 67. Prof Exp: Res assoc commun, 24-26, asst prof elec eng, 28-31, from asst prof to prof physics, 31-51, mem staff, Radiation Lab, 40-45, dir, Res Lab Electronics, 45-49, provost, 49-56, vpres, 51-56, chancellor, 56-59, actg pres, 57-59, pres, 59-66, EMER PRES, MASS INST TECHNOL, 66- Concurrent Pos: Expert consult, Secy War, 42-46; chmn, Comt Electronics, Res & Develop Bd, 46-49; chmn bd, Ford Found, 66-71; chmn, Comn Marine Sci, Eng & Resources, 67-69; mem, Nat Adv Comt Oceans & Atmosphere, 71-73; mem corp, Mass Inst Technol; trustee, Boston Mus Sci; mem, Coun Foreign Rels. Honors & Awards: Medal Merit, 46; Cert Award, US Dept Navy, 57; Medal Hon, Inst Radio Eng, 57; Faraday Medal, Brit Inst Elec Engrs, 61; Officer, French Legion Hon, 61; Orden de Boyaca, Govt Colombia, 64; Boston Medal Distinguished Achievement, 66; Knight Commander, Order Merit, Fed Repub Ger, 66. Mem: Nat Acad Sci; Nat Acad Eng; fel Am Phys Soc; fel Inst Elec & Electronics Engrs; fel Am Acad Arts & Sci. Res: Electromagnetic theory. Mailing Add: Mass Inst of Technol Cambridge MA 02139

STRATTON, LEWIS PALMER, b West Chester, Pa, Aug 22, 37; m 60; c 2. BIOCHEMISTRY. Educ: Juniata Col, BS, 59; Univ Maine, MS, 61; Fla State Univ, PhD(chem), 67. Prof Exp: Asst prof, 67-74, ASSOC PROF BIOL, FURMAN UNIV, 74- Mem: AAAS; Am Soc Microbiol; Asn Southeastern Biologists; Sigma Xi. Res: Comparative protein biochemistry. Mailing Add: Dept of Biol Furman Univ Greenville SC 29613

STRATTON, PAUL OSWALD, b Rawlins, Wyo, Aug 18, 23; m 48; c 2. ANIMAL SCIENCE. Educ: Univ Wyo, BS, 47, MS, 50; Univ Minn, PhD, 52. Prof Exp: Supply instr animal prod, Univ Wyo, 49-50; asst animal breeding, Univ Minn, 50-52; asst prof, 52-58, PROF ANIMAL SCI & HEAD DEPT, COL AGR, UNIV WYO, 58- Mem: Am Soc Animal Sci. Res: Improvement of beef cattle and sheep through breeding methods. Mailing Add: Animal Sci Div Univ of Wyo Laramie WY 82070

STRATTON, ROBERT, b Vienna, Austria, Aug 14, 28; US citizen; m 53; c 2. THEORETICAL PHYSICS. Educ: Univ Manchester, BSc, 49, PhD(theoret physics), 52. Prof Exp: Res physicist, Metrop Vickers Elec Co, Ltd, Eng, 52-59; mem tech staff, 59-63, dir, Physics Res Lab, 63-71, assoc dir, Cent Res Lab, 71-72, dir semiconductor res & develop labs, 72-75, DIR CENT RES LABS, TEX INSTRUMENTS, INC, 75- Mem: Fel Am Phys Soc; sr mem Inst Elec & Electronics Engrs; fel Brit Inst Physics & Phys Soc. Res: Solid state theory, including field emission, space charge barriers, thermoelectricity, high electric fields, thermal conductivity, dielectric breakdown and surface energies of solids. Mailing Add: Cent Res Labs Tex Instruments PO Box 5936 MS 136 Dallas TX 75222

STRATTON, ROBERT ALAN, b Selma, Ala, Feb 4, 36; m 61; c 4. POLYMER CHEMISTRY. Educ: Univ Nev, BS, 58; Univ Wis, PhD(chem), 62. Prof Exp: Sr res chemist, Mobil Chem Co, 62-69; ASSOC PROF CHEM, INST PAPER CHEM, 69- Mem: Soc Rheol; Am Chem Soc; Tech Asn Pulp & Paper Indust. Res: Rheology of polymer melts and solutions; dilute solution properties of polymers; flocculation of colloids; use of polymers in papermaking and waste water treatment. Mailing Add: Inst of Paper Chem Appleton WI 54911

STRATTON, THOMAS FAIRLAMB, physics, see 12th edition

STRATTON, WILLIAM R, b River Falls, Wis, May 15, 22; m 52; c 3. PHYSICS.

Educ: Univ Minn, PhD(physics), 52. Prof Exp: Res assoc, Univ Minn, 52; MEM STAFF, LOS ALAMOS SCI LAB, UNIV CALIF, 52- Concurrent Pos: US del, Int Conf Peaceful Uses Atomic Energy, 58 & Fast Reactor Prog, Cadarache, France, 65-66; mem, Adv Comt Reactor Safeguards, AEC, 66- Mem: Am Phys Soc; Am Nuclear Soc. Res: Scattering and reaction in nuclear physics; nuclear forces; reactor physics. Mailing Add: Los Alamos Sci Lab Univ of Calif Los Alamos NM 87545

STRATTON, WILMER JOSEPH, b Newark, NJ, June 4, 32; m 55; c 3. CHEMISTRY. Educ: Earlham Col, AB, 54; Ohio State Univ, PhD(chem), 58. Prof Exp: Asst prof chem, Ohio Wesleyan Univ, 58-59 & Earlham Col, 59-64; vis lectr, Univ Ill, 64-65; assoc prof, 65-70, chmn dept, 65-68, PROF CHEM, EARLHAM COL, 70- Mem: Am Chem Soc. Res: Metal coordination compounds, including synthesis of new polydentate chelates and bonding in chelate systems. Mailing Add: Dept of Chem Earlham Col Richmond IN 47374

STRAUB, DAREL K, b Titusville, Pa, May 17, 35. INORGANIC CHEMISTRY. Educ: Allegheny Col, BS, 57; Univ Ill, PhD(inorg chem), 61. Prof Exp: Instr, 61-62, asst prof, 62-68, ASSOC PROF CHEM, UNIV PITTSBURGH, 68- Mem: Am Chem Soc; AAAS. Res: Iron porphyrins; complexes of sulfur-containing ligands; Mössbauer spectroscopy. Mailing Add: Dept of Chem Univ of Pittsburgh Pittsburgh PA 15260

STRAUB, HARALD WALTER, physics, optics, see 12th edition

STRAUB, LESLIE ELLEN, anthropology, see 12th edition

STRAUB, WILLIAM ALBERT, b Philadelphia, Pa, June 21, 31; m 58; c 2. ANALYTICAL CHEMISTRY. Educ: Univ Pa, BA, 53; Cornell Univ, PhD, 58. Prof Exp: Technologist, 57-67, sr res chemist, 67-75, ASSOC RES CONSULT, US STEEL CORP, 75- Mem: Am Chem Soc. Res: Effluent gas, process solution analysis. Mailing Add: US Steel Corp Research Lab Monroeville PA 15146

STRAUB, WOLF DETER, b Boston, Mass, Apr 27, 27; m 61; c 2. SOLID STATE PHYSICS. Educ: Yale Univ, BS, 50; Univ Mich, MS, 52. Prof Exp: Staff mem solid state physics, Res Div, Raytheon Co, 52-65; physicist, Electronics Res Ctr, NASA, 65-70 & M/K Systs, Inc, Mass, 70-72; MGR ANAL LAB, COULTER INFO SYSTS, INC, BEDFORD, 72- Mem: Am Phys Soc. Res: Galvanometric properties of semiconductors and semimetals; radiation damage and studies of microwave generation in semiconductors; electrical and mechanical properties of dielectric thin films; problems in electrophotography; surface physics. Mailing Add: 158 Barton Dr Sudbury MA 01776

STRAUBE, ROBERT LEONARD, b Chicago, Ill, Sept 16, 17; m 44; c 2. RADIOBIOLOGY. Educ: Univ Chicago, BS, 39, PhD(physiol), 55. Prof Exp: Asst path, Univ Chicago, 43-46; prof radiobiol, Assoc Cols Midwest, 63-64; assoc scientist, Argonne Nat Lab, 47-65; EXEC SECY RADIATION STUDY SECT, DIV RES GRANTS, NIH, 65- Mem: AAAS; Radiation Res Soc; Am Physiol Soc; Soc Exp Biol & Med; Am Asn Cancer Res. Res: Nature of radiation effects and their modification by chemical agents; growth processes in neoplastic cells. Mailing Add: Div of Res Grants Nat Inst of Health Bethesda MD 20014

STRAUCH, FRED, b Windsor, Colo, Apr 20, 24; m 48; c 2. SOIL SCIENCE, AGRONOMY. Educ: Colo State Univ, BS, 48. Prof Exp: Soil scientist, Soil Conserv Serv, USDA, 48-58, Agr Res Serv, 58-72, ASST TO DEP ADMINR, WESTERN REGION, AGR RES SERV, USDA, 72- Concurrent Pos: Agr Res Serv rep, Colo Conserv Needs Comt, 63-, Land Use & Mgt Work Group, Upper Colo River Basin Interagency Surv Team, 67- Honors & Awards: Merit Awards, Soil Conserv Serv, 56 & Agr Res Serv, 68; President's Citation, Soil Conserv Soc Am, 69, Commendation Award, 71. Mem: Am Soc Agron; Soil Sci Soc Am; fel Soil Conserv Soc Am. Res: Soil survey; land classification; soil morphology and genesis; soil and water conservation; agricultural research administration. Mailing Add: Agr Res Serv USDA 2850 Telegraph Ave Berkeley CA 94705

STRAUCH, KARL, b Giessen, Ger, Oct 4, 22; nat US; m 51. PARTICLE PHYSICS. Educ: Univ Calif, AB, 43, PhD(physics), 50. Prof Exp: Soc Fels jr fel, 50-53, from asst prof to assoc prof, 53-62, PROF PHYSICS, HARVARD UNIV, 62- Concurrent Pos: Dir, Cambridge Electron Accelerator, Harvard Univ, 67-74. Mem: Am Phys Soc; Am Acad Arts & Sci. Res: High energy reactions; elementary particles. Mailing Add: Dept of Physics Harvard Univ Cambridge MA 02138

STRAUCH, RALPH EUGENE, b Springfield, Mass, May 14, 37; m 58; c 2. MATHEMATICS, STATISTICS. Educ: Univ Calif, Los Angeles, AB, 59, Univ Calif, Berkeley, MA, 64, PhD(statist), 65. Prof Exp: Consult, 64-65, MATHEMATICIAN, RAND CORP, 65- Mem: AAAS; Inst Math Statist. Res: Dynamic programming; statistical decision theory; national security policy; policy analysis methodology. Mailing Add: Rand Corp 1700 Main St Santa Monica CA 90406

STRAUGHAN, ISDALE (DALE) MARGARET, b Pittsworth, Australia, Nov 4, 39; m 62. BIOLOGY, ECOLOGY. Educ: Queensland Univ, BSc, 60, Hons, 62, PhD(zool), 66. Prof Exp: Demonstr zool, Queensland Univ, 66; sr demonstr, Univ Col, Townsville, 66-67; asst prof & res assoc, Allan Hancock Found, 69-74, RES SCIENTIST, ALLAN HANCOCK FOUND, UNIV SOUTHERN CALIF, 74- Concurrent Pos: Consult biologist, Northern Elec Authority, Queensland, 66-68; Am Asn Univ Women fel, 68-69. Mem: AAAS. Res: Ecology and taxonomy of serpulids and other marine and estuarine invertebrates dealing mainly with problems of fouling and pollution; comparison of man-induced change to natural biological fluctuations. Mailing Add: Allan Hancock Found Univ of Southern Calif Los Angeles CA 90007

STRAUGHN, WILLIAM RINGGOLD, JR, b Dubois, Pa, May 21, 13; m 41; c 4. BACTERIOLOGY. Educ: Mansfield State Col, BS, 35; Cornell Univ, MS, 40; Univ Pa, PdD(bact), 58. Prof Exp: Teacher high sch, Pa, 35-36 & NY, 36-38; asst bact, Univ NC, 40-42; instr math & chem, Md State Teachers Col, Salisbury, 42-44; from instr to assoc prof, 44-69, PROF BACT, SCH MED, UNIV NC, CHAPEL HILL, 69- Mem: Am Soc Microbiol. Res: Bacterial physiology and metabolism; antibacterial agents; enzyme synthesis; amino acid decarboxylases-mechanisms of formation and action; bacterial membranes and transport mechanisms. Mailing Add: Dept of Bact Univ of NC Sch of Med Chapel Hill NC 27514

STRAUMANIS, JOHN JANIS, JR, b Riga, Latvia, Apr 22, 35; US citizen; m 59; c 2. PSYCHIATRY. Educ: Univ Iowa, BA, 57, MD, 60, MS, 64. Prof Exp: Intern med, Georgetown Univ Hosp, 60-61; resident psychiat, Univ Iowa, 61-64; asst prof psychiat & Nat Inst Ment Health res career develop grant, 66-71, ASSOC PROF PSYCHIAT, TEMPLE UNIV, 71- Mem: Am Psychiat Asn; Soc Biol Psychiat; Am Psychopath Asn; Am Electroencephalographic Soc; Eastern Asn Electroencephalographers. Res: Electrophysiology pf psychiatric disorders. Mailing Add: Eastern Pa Psych Inst 3300 Henry Ave Philadelphia PA 19129

STRAUMFJORD, JON VIDALIN, JR, b Portland, Ore, Feb 23, 25; m 47; c 2. MEDICINE, CLINICAL PATHOLOGY. Educ: Willamette Univ, BA, 48; Univ Ore, MS & MD, 53; Univ Iowa, PhD(biochem), 58. Prof Exp: Res fel biochem, Univ Iowa, 54-58; resident path & consult, Providence Hosp, Portland, Ore, 58-60; asst prof path, Univ Miami, 60-62; assoc prof, Med Col Ala, 62-65, prof clin path & chmn dept, 65-70, dir clin labs, 62-65, clin pathologist in chief, Univ Hosp, 65-70; PROF PATH & CHMN DEPT, MED COL WIS, 70- Concurrent Pos: Asst pathologist, Div Clin Path, Jackson Mem Hosp, Miami, Fla, 60-62; dir labs, Milwaukee City Gen Hosp, Wis, 70-; mem surg adv bd, Shrine Burn Units. Mem: AAAS; Am Soc Clin Path; Am Asn Clin Chem; Col Am Pathologists; NY Acad Sci. Res: Surface characteristics of cells; clinical chemical screening procedures. Mailing Add: Dept of Path Med Col of Wis 8700 W Wisconsin Ave Milwaukee WI 53226

STRAUS, ALAN EDWARD, b Berkeley, Calif, May 14, 24; m 53; c 2. ORGANIC CHEMISTRY. Educ: Univ Calif, Berkeley, BS, 49. Prof Exp: From asst res chemist to res chemist, 49-67, SR RES CHEMIST, CHEVRON RES CO, 67- Mem: Am Chem Soc. Res: Petrochemicals; hydrocarbon oxidation; surface active agents; hydrocarbon pyrolysis; organic synthesis; polymers; heterogeneous catalysis. Mailing Add: Chevron Res Co 576 Standard Ave Richmond CA 94802

STRAUS, BERNARD, b New York, NY, July 30, 11; m 35; c 2. INTERNAL MEDICINE. Educ: NY Univ, BS, 31; Long Island Col Med, MD, 35; Am Bd Internal Med, dipl, 47. Prof Exp: Chief med serv, Vet Admin Hosp, Bronx, NY, 46-54; assoc prof clin med, Albert Einstein Col Med, 54-64; prof internal med, NY Med Col, 64-71; PROF MED, MT SINAI SCH MED, 71-; DIR MED, BETH ISRAEL MED CTR, 71- Concurrent Pos: Former dir med, Ctr Chronic Dis, Bird S Coler Hosp, New York. Mem: Fel Am Col Physicians; Asn Am Med Cols. Res: Diagnosis; liver disease; malariology; lymphomas and sleep; medical education. Mailing Add: Dept of Med Beth Israel Med Ctr New York NY 10003

STRAUS, DAVID BRADLEY, b Chicago, Ill, July 26, 30; m 55; c 3. BIOCHEMISTRY. Educ: Reed Col, BA, 53; Univ Chicago, PhD(biochem), 60. Prof Exp: Asst biochem, Med Sch, Univ Ore, 53-54; asst, Univ Chicago & Argonne Cancer Res Hosp, 55-60; res assoc chem, Princeton Univ, 60-64, res staff mem, 64-65; asst prof biochem, State Univ NY Buffalo, 65-72; ASSOC PROF CHEM, STATE UNIV NY COL NEW PALTZ, 73- Mem: AAAS; Am Chem Soc; NY Acad Sci. Res: Chemical synthesis of polynculeotides; nucleic acid enzymology and chemistry; protein-nucleic acid interactions. Mailing Add: Dept of Chem State Univ of NY Col New Paltz NY 12561

STRAUS, DAVID CONRAD, b Evansville, Ind, Apr 27, 47; m 75. MEDICAL MICROBIOLOGY. Educ: Wright State Univ, BS, 70; Loyola Univ Chicago, PhD(microbiol), 74. Prof Exp: Teaching asst microbiol, Sch Med, Loyola Univ Chicago, 70-74; fel, Med Ctr, Univ Cincinnati, 74-75; INSTR MICROBIOL, UNIV TEX HEALTH SCI CTR, SAN ANTONIO, 75- Concurrent Pos: Instr microbiol, Ill Col Podiatric Med, 72-73. Mem: Am Soc Microbiol; Sigma Xi. Res: Study of mechanisms of bacterial pathogenicity and host response; study of bacterial exotoxins; study of nutrilite membrane transport in phagocytic cells. Mailing Add: Dept of Microbiol Univ of Tex Health Sci Ctr 7703 Floyd Curl Dr San Antonio TX 78284

STRAUS, ERNST GABOR, b Munich, Ger, Feb 25, 22; m 44; c 2. MATHEMATICS. Educ: Columbia Univ, MA, 42, PhD(math), 48. Prof Exp: Lectr math, Columbia Univ, 42-44; asst to Prof Albert Einstein, Inst Advan Study, 44-48; from instr to assoc prof, 48-60, PROF MATH, UNIV CALIF, LOS ANGELES, 60- Mem: Am Math Soc. Res: Number theory; geometry; analysis; algebra; relativity theory. Mailing Add: Dept of Math Univ of Calif Los Angeles CA 90024

STRAUS, FRANCIS HOWE, II, b Chicago, Ill, Mar 16, 32; m 55; c 4. PATHOLOGY. Educ: Harvard Univ, AB, 53; Univ Chicago, MD, 57, MS, 64. Prof Exp: Intern, Clins, 57-58, resident path, 58-62, chief resident, 62-63, from instr to asst prof, 62-71, ASSOC PROF PATH, SCH MED, UNIV CHICAGO, 71- Concurrent Pos: Am Cancer Soc advan clin fel, 65-68. Mem: AAAS; Am Soc Clin Path; Am Asn Pathologists & Bacteriologists; Int Acad Path; NY Acad Sci. Res: Morphology in surgical pathology as it relates to diagnosis and prognosis of clinical disease; cellular aspects of host-tumor interaction; endocrine pathology. Mailing Add: Dept of Path Univ of Chicago Sch of Med Chicago IL 60637

STRAUS, HELEN LORNA PUTTKAMMER, b Chicago, Ill, Feb 15, 33; m 55; c 4. ANATOMY, BIOLOGY. Educ: Radcliffe Col, AB, 55; Univ Chicago, MS, 60, PhD(anat), 62. Prof Exp: Fel anat, 62-63, res assoc, 63-64, instr anat & biol, 64-67, asst prof biol & asst dean undergrad students, 67-71, ASSOC PROF BIOL, UNIV CHICAGO, 73-, DEAN UNDERGRAD STUDENTS, 71-, DEAN ADMISSIONS, 75- Mem: Am Asn Anatomists; AAAS. Res: Histochemistry, histology and cytology of secretory process. Mailing Add: Dept of Anat Univ of Chicago Chicago IL 60637

STRAUS, JOE MELVIN, b Dallas, Tex, May 27, 46; m 71. ATMOSPHERIC PHYSICS. Educ: Rice Univ, BA, 68; Univ Calif, Los Angeles, MS, 69, PhD(planetary, space physics), 72. Prof Exp: MEM TECH STAFF, SPACE SCI LAB, THE AEROSPACE CORP, 73- Mem: Sigma Xi; Am Geophys Union. Res: Theoretical studies of atmospheric physics; aeronomy; convection in atmospheres, oceans, stars, planetary interiors; geophysical fluid dynamics. Mailing Add: Space Sci Lab The Aerospace Corp PO Box 92957 Los Angeles CA 90009

STRAUS, JOZEF, b Velke Kapusany, Czech, July 18, 46; Can citizen. EXPERIMENTAL SOLID STATE PHYSICS. Educ: Univ Alta, BS, 69, PhD(physics), 74. Prof Exp: Fel, Univ Alta, 69-74; MEM SCI STAFF & NAT RES COUN CAN FEL, BELL NORTHERN RES LTD, 74- Res: Fabrication and study of physical properties of light emitting diodes and of solid state lasers; fiber optics, fiber optics communication; electron tunneling in normal and superconducting metals, Josephson tunneling. Mailing Add: Bell Northern Res Ltd Dept 5C23 PO Box 3511 Sta C Ottawa ON Can

STRAUS, MARC J, b New York, NY, June 2, 43; m 64; c 2. ONCOLOGY, CHEMOTHERAPY. Educ: Franklin & Marshall Col, AB, 64; State Univ NY Downstate Med Ctr, MD, 68; Am Bd Internal Med, dipl & cert med oncol, 75. Prof Exp: CHIEF MED ONCOL, MED CTR, BOSTON UNIV, 74-, ASSOC PROF MED, SCH MED, 75- Concurrent Pos: Prin investr, Eastern Coop Oncol Group, Med Ctr, Boston Univ, 75-; consult oncol, Framingham Union Hosp, Newton-Wellesley Hosp & Boston Vet Admin Hosp, 75- Mem: Am Soc Clin Oncol; Am Fedn Clin Res; Working Party Ther Lung Cancer; Am Asn Cancer Res. Res: Application of cellular kinetics in animal and human tumors to the design of clinical cancer treatment programs; clinical cancer chemotherapy. Mailing Add: 75 Newton St Boston MA 02118

STRAUS, NEIL ALEXANDER, b Kitchener, Ont, Apr 29, 43; m 66; c 2. MOLECULAR BIOLOGY. Educ: Univ Toronto, BSc Hons, 66, MSc, 67, PhD(molecular biol), 70. Prof Exp: Fel biophys, Carnegie Inst, Washington, DC, 70-72; ASST PROF MOLECULAR BIOL, UNIV TORONTO, 72- Concurrent Pos: Nat Res Coun Can, Med Res Coun Can & Nat Cancer Inst Can res grants, 72- Mem: Am Soc Cell Biol; Can Soc Cell Biol. Res: Chromosomes structure; gene regulation; DNA sequence relationships in eukaryotes. Mailing Add: Dept of Bot Univ of Toronto Toronto ON Can

STRAUS, THOMAS MICHAEL, b Berlin, Ger, Oct 25, 31; US citizen; m 57; c 3. APPLIED PHYSICS. Educ: Univ Mich, BS, 52; Harvard Univ, MA, 56, PhD(appl physics), 59. Prof Exp: Mem tech staff, microwaves & lasers, Hughes Aircraft Co, 59-64, sr staff mem, 64-69, sr staff engr, Laser Dept, 69-74; SR SCIENTIST, THETA-COM, 74- Concurrent Pos: Lectr, Eng Exten, Univ Calif, Los Angeles, 62-67. Mem: Inst Elec & Electronics Engrs. Res: Development of microwave and laser components and systems. Mailing Add: Theta-Com 2216 W Peoria Phoenix AZ 85029

STRAUS, WERNER, b Offenbach, Ger, June 5, 11; nat US. BIOCHEMISTRY. Educ: Univ Zurich, PhD(chem), 38. Prof Exp: Res assoc path, Long Island Col Med, 47-50; asst prof, State Univ NY Downstate Med Ctr, 50-58; vis scientist, Cath Univ Louvain & Free Univ Brussels, 59-61 & Univ NC, 62-63; ASSOC PROF BIOCHEM, CHICAGO MED SCH, 64- Mem: AAAS; Am Soc Cell Biol; Histochem Soc. Res: Intracellular localization of enzymes; lysosomes and phagosomes; cell biology; immuno-cytochemistry. Mailing Add: Dept of Biochem Chicago Med Sch 2020 W Ogden Ave Chicago IL 60612

STRAUSE, STERLING FRANKLIN, b Summit Station, Pa, Jan 4, 31; m 56; c 4. ORGANIC CHEMISTRY. Educ: Lebanon Valley Col, BS, 52; Univ Del, MS, 53, PhD(chem), 55. Prof Exp: Develop chemist, Chem Develop Dept, Gen Elec Co, 55-57, spec process develop, 57-58, qual control engr, 58-60, mgr, 60-65, qual control, 65-68, mgr polycarbonate res & develop, 68-71; dir, 71-74, VPRES RES & DEVELOP, W H BRADY CO, MILWAUKEE, 74- Mem: Am Chem Soc; Am Soc Qual Control. Res: Polymeric peroxide and free radical chemistry; polymer processes. Mailing Add: 7716 W Bonniwell Rd Mequon WI 53092

STRAUSER, WILBUR ALEXANDER, b Charleroi, Pa, June 15, 24; m 45; c 2. PHYSICS, MATHEMATICS. Educ: Washington & Jefferson Col, AB, 45. Prof Exp: Physicist, Manhattan Eng Dist, Tenn, 46-47; assoc physicist, Oak Ridge Nat Lab, 47-50; sci analyst & chief declassification br, AEC, 50-55, asst to mgr, San Francisco Opers Off, 55-56, dep dir, Div Classification, 56-63, asst dir safeguards, Div Int Affairs, 63-70, CHIEF WEAPONS BR, DIV CLASSIFICATION, AEC, 70- Mem: AAAS; Am Phys Soc; Am Asn Physics Teachers. Res: Neutron diffraction; security classification; safeguards. Mailing Add: 11816 Charles Rd Silver Spring MD 20906

STRAUSS, AARON SOLOMON, b Pittsburgh, Pa, July 6, 39; m 63. MATHEMATICS. Educ: Case Western Reserve Univ, BS, 61; Univ Wis, MS, 62, PhD(math), 64. Prof Exp: From asst prof to assoc prof, 64-74, PROF MATH, UNIV MD, COLLEGE PARK, 74- Concurrent Pos: Consult, NASA, 65-66; NSF fel, Univ Florence, 66-67. Mem: Am Math Soc. Res: Stability theory of ordinary differential equations; optimal control theory. Mailing Add: Dept of Math Univ of Md College Park MD 20742

STRAUSS, BELLA S, b Camden, NJ, May 28, 20. MEDICINE. Educ: Columbia Univ, BA, 42; Western Reserve Univ, MD, 53; Am Bd Internal Med, dipl, 61. Prof Exp: Intern med, First Div, Bellevue Hosp, New York, 53-54; asst resident path, Univ Hosps, Med Ctr, Univ Mich, Ann Arbor, 54-55; asst resident med, Manhattan Vet Admin Hosp, New York, 55-56; asst resident First Div, Bellevue Hosp, 56-57, chief resident, Chest Serv Div, 57-58; career scientist, Health Res Coun New York, 62-66; vis specialist, Care/Medico, Avicenna Hosp, Kabul, Afghanistan, 66-67; staff physician, Maine Coast Mem Hosp, Ellsworth, 67-68; ASSOC PROF MED, DARTMOUTH MED SCH, 68- Concurrent Pos: NY Tuberc & Health Asn Miller fel, Col Physicians & Surgeons, Columbia Univ, 58-60; guest investr, Rockefeller Inst, 62-64; asst prof, Col Physicians & Surgeons, Columbia Univ, 64-66. Honors & Awards: Career Scientist Award, Health Res Coun New York, 62. Mem: Fel Am Col Physicians. Res: Internal and chest medicine; pathophysiology; training of paramedical personnel. Mailing Add: Dept of Med Dartmouth Med Sch Hanover NH 03755

STRAUSS, BERNARD, b Odessa, Russia, Apr 10, 04; nat US; m 64. MEDICINE. Educ: State Univ NY, MD, 27; Am Bd Urol, dipl, 43. Prof Exp: Instr urol, Sch Med, Stanford Univ, 39-42; asst prof, Sch Med, Loma Linda Univ, 56-65; assoc prof, 65-72, EMER ASSOC CLIN PROF UROL, DEPT SURG, SCH MED, UNIV SOUTHERN CALIF, 72- Mem: Am Urol Asn; fel Am Col Surgeons; corresp mem Belg Soc Urol. Res: Urology. Mailing Add: 2080 Century Park East Los Angeles CA 90067

STRAUSS, BERNARD S, b New York, NY, Apr 18, 27; m 49; c 3. MOLECULAR BIOLOGY, CELL BIOLOGY. Educ: City Col New York, BS, 47; Calif Inst Technol, PhD(biochem), 50. Prof Exp: Hite fel cancer res & biochem genetics, Univ Tex, 50-52; from asst prof to assoc prof, Syracuse Univ, 52-60; assoc prof, 60-64, PROF MICROBIOL, UNIV CHICAGO, 64-, CHMN DEPT, 69- Concurrent Pos: Fulbright & Guggenheim fels, Osaka Univ, 58; mem genetics training comt, NIH, 62-68 & 70-74, chmn, Genetics Comt, 62-76; vis prof, Univ Sydney, 67. Mem: Am Soc Biol Chemists; Genetics Soc Am; Am Soc Microbiol; Mutation Res Soc. Res: Chemical mutagenesis; DNA repair and replication in mammalian cells; lymphocyte transformations. Mailing Add: Dept of Microbiol Univ of Chicago Chicago IL 60637

STRAUSS, BRUCE PAUL, b Elizabeth, NJ, Aug 19, 42; m 64; c 2. CRYOGENICS, LOW TEMPERATURE PHYSICS. Educ: Mass Inst Technol, SB, 64, ScD(solid state physics), 67; Univ Chicago, MBA, 72. Prof Exp: Prin res engr, Avco-Everett Res Lab, Avco Corp, 67-68; physicist, Argonne Nat Lab, 68-69; ENGR, NAT ACCELERATOR LAB, BATAVIA, IL, 69- Concurrent Pos: Vis scientist, Univ Wis-Madison, 71- Mem: Am Phys Soc; Am Soc Metals; Am Inst Mining, Metall & Petrol Engrs; Cryogenic Soc Am. Res: Cryogenic magnet systems; optimization of materials and performance. Mailing Add: 624 62nd St Downers Grove IL 60515

STRAUSS, CARL RICHARD, b Chicago, Ill, May 18, 36; m 59; c 3. POLYMER CHEMISTRY. Educ: Univ Ill, BS, 58; Univ Akron, MS, 65, PhD(polymer chem), 70. Prof Exp: Plant engr chlorinated organics, Pittsburgh Plate Glass Chem Div, 58-63, res chemist reinforcement elastomers, 63-69; advan scientist polyesters, 69-72, sr scientist phenolic binder, 72-74, SR SCIENTIST RESINS & BINDERS, OWENS-CORNING FIBERGLAS CORP, 74- Concurrent Pos: Instr, Cent Ohio Tech Col, 73-75. Mem: Am Chem Soc. Res: Cure and mechanical properties of organic binders; glass-binder interaction and mechanical performance of fiberglass composites; binder development. Mailing Add: Owens-Corning Fiberglas Corp Tech Ctr Granville OH 43023

STRAUSS, CHARLES MICHAEL, b Providence, RI, Oct 18, 38; m 61; c 2. COMPUTER SCIENCE, APPLIED MATHEMATICS. Educ: Harvard Col, AB, 60; Brown Univ, ScM, 66, PhD(appl math), 69. Prof Exp: ASST PROF APPL MATH, BROWN UNIV, 68- Mem: AAAS; Asn Comput Mach; Soc Indust & Appl Math; Math Asn Am. Res: Computer graphics; numerical analysis. Mailing Add: Div of Appl Math Brown Univ Providence RI 02912

STRAUSS, ELLEN GLOWACKI, b New Haven, Conn, Sept 25, 38; m 69. MOLECULAR GENETICS, VIROLOGY. Educ: Swarthmore Col, BA, 60; Calif Inst Technol, PhD(biochem), 66. Prof Exp: NIH fel biochem, Univ Wis, 66-68, fel, 68-69; res fel biol, 69-73, SR RES FEL BIOL, CALIF INST TECHNOL, 73- Mem: Sigma Xi; Am Soc Microbiologists. Res: Molecular biology of the replication of togaviruses, particularly alphavirus Sindbis, primarily through isolation and characterization of conditional lethal mutants. Mailing Add: Div of Biol Calif Inst of Technol Pasadena CA 91125

STRAUSS, ELLIOTT WILLIAM, b Brooklyn, NY, Jan 25, 23; m 51; c 3. PATHOLOGY, ANATOMY. Educ: Columbia Univ, AB, 44; NY Univ, MD, 49. Hon Degrees: MSc, Brown Univ, 72. Prof Exp: Asst med, Peter Bent Brigham Hosp, 57-59; res fel med, Harvard Med Sch, 57-59, res fel anat, 59-61, res assoc path, 61-65; asst prof, Univ Colo Med Ctr, Denver, 65-70; ASSOC PROF MED SCI, BROWN UNIV, 70- Concurrent Pos: USPHS career develop award, 61-65; NIH grants, 61-66 & 67-72. Mem: AAAS; Am Gastroenterol Asn; Am Soc Cell Biol; Am Asn Pathologists & Bacteriologists; Am Soc Exp Path; Am Soc Zoologists. Res: Electron microscopy; lipid chemistry; vitamin B-12; normal and abnormal mechanisms for adsorption and transport by intestine and vessels. Mailing Add: Div of Biol & Med Sci Brown Univ Providence RI 02912

STRAUSS, FREDERICK BODO, b Bad Wildungen, Ger, Feb 24, 31; US citizen; m 54; c 3. MATHEMATICS. Educ: Univ Calif, Los Angeles, BA, 59, MA, 62, PhD(math), 64. Prof Exp: Asst prof math, Univ Hawaii, 64-68; ASSOC PROF MATH, UNIV TEX, EL PASO, 68- Mem: AAAS; Am Math Soc; Math Asn Am. Res: Linear algebra and functional analysis; matrix Lie algebras; theory of rings. Mailing Add: Dept of Math Univ of Tex El Paso TX 79902

STRAUSS, GEORGE, b Vienna, Austria, Nov 27, 21; US citizen; m 54; c 2. BIOPHYSICAL CHEMISTRY. Educ: Univ London, BSc, 50; Lehigh Univ, PhD(chem), 55. Prof Exp: Chemist, A S Harrison & Co, 45-52; Colgate fel, 55-57, from asst prof to assoc prof, 57-66, res assoc, Inst Microbiol, 57-64, PROF CHEM, RUTGERS UNIV, NEW BRUNSWICK, 66- Concurrent Pos: Rutgers Univ fac fel & USPHS fel, Univ Sheffield, 64-65. Mem: AAAS; Am Chem Soc; Biophys Soc. Res: Photochemical kinetics; electronic absorption and emission spectroscopy; excited states and energy transfer in biological systems; photosynthesis; complexes of nucleic acids with organic molecules; interactions in lipid membranes. Mailing Add: Dept of Chem Douglass Col Rutgers Univ New Brunswick NJ 08903

STRAUSS, HERBERT L, b Aachen, Ger, Mar 26, 36; US citizen; m 60; c 3. PHYSICAL CHEMISTRY. Educ: Columbia Univ, AB, 57, MA, 58, PhD(chem), 60. Prof Exp: Ramsey fel from Univ Col, London Univ & NSF fel, Oxford Univ, 60-61; from asst prof to assoc prof, 61-73, PROF CHEM, UNIV CALIF, BERKELEY, 73- Concurrent Pos: Sloan res fel, 66-68; vis prof, Indian Inst Technol, Kampur, 68; NSF res grants. Mem: Am Chem Soc; Am Phys Soc. Res: Experimental and theoretical spectroscopy; far infrared; light scattering; configuration of ring compounds; coupling of various types of molecular motion. Mailing Add: Dept of Chem Univ of Calif Berkeley CA 94720

STRAUSS, JAMES HENRY, b Galveston, Tex, Sept 16, 38; m 69. BIOCHEMISTRY, VIROLOGY. Educ: St Mary's Univ, Tex, BS, 60; Calif Inst Technol, PhD, 67. Prof Exp: NSF fel, Albert Einstein Col Med, 66-67, res fel, 66-69; asst prof, 69-75, ASSOC PROF BIOL, CALIF INST TECHNOL, 75- Mem: AAAS; Am Soc Microbiol; Sigma Xi. Res: Structure and replication of animal viruses; cell surface modification and RNA replication during arbovirus infection; biogenesis of cell plasma membranes. Mailing Add: Div of Biol Calif Inst of Technol Pasadena CA 91125

STRAUSS, JOHN STEAVEN, b Cleveland, Ohio, Aug 18, 32; m 61; c 2. PSYCHIATRY. Educ: Swarthmore Col, BA, 54; Yale Univ, MD, 59. Prof Exp: Res psychiatrist, NIMH, 64-68, chief psychiat assessment sect, 68-72; DIR CLIN PSYCHIAT RES PROG, MED SCH, UNIV ROCHESTER, 72- & PROF PSYCHIAT, 76- Concurrent Pos: Collabr investr, Int Pilot Study Schizophrenia, WHO, 66- & consult psychiat res, 67-; consult, Comt Nomenclature & Statistics, Am Psychiat Asn, 74- Mem: Soc Life Hist Res Psychopath; fel Am Psychiat Asn; Psychiat Res Soc; Am Psychopath Soc; Am Psychosomatic Soc; AAAS; Res: Diagnostic and prognostic issues in psychopathology; family functioning human development; cross cultural aspects of psychiatric disorder. Mailing Add: Dept of Psychiat Sch of Med Univ of Rochester Rochester NY 14642

STRAUSS, JOHN STEINERT, b New Haven, Conn, July 15, 26; m 50; c 2. MEDICINE. Educ: Yale Univ, BS, 46, MD, 50. Prof Exp: Instr dermat, Univ Pa, 57; from asst prof to assoc prof, 57-66, PROF DERMAT, SCH MED, BOSTON UNIV, 66- Concurrent Pos: Fel dermat, Univ Pa, 51-52 & 54-55, USPHS fel, 55-57; asst mem, Evans Mem Dept Clin Res, Univ Hosp, 58-63, assoc mem, 63-, assoc chief dermat, 59-; consult, Lemuel Shattuck & Boston Vet Admin Hosps; assoc dir, Dept Dermat, Boston City Hosp, 74- Mem: Soc Invest Dermat; AMA; Am Acad Dermat; Am Dermat Asn; Am Fedn Clin Res. Res: Dermatology; pilosebaceous physiology and pathology. Mailing Add: Dept of Dermat Boston Univ Sch of Med Boston MA 02118

STRAUSS, LOTTE, b Nuremberg, Ger, Apr 15, 13; nat US. PATHOLOGY. Educ: Univ Siena, MD, 37. Prof Exp: Res asst bact, Beth Israel Hosp, New York, 38-41 & Mt Sinai Hosp, 47-49; asst pathologist, Lebanon Hosp, 50-52; assoc pathologist, Mt Sinai Hosp, 53-66, PROF PATH, MT SINAI SCH MED, 66- Concurrent Pos: Fel path, Mt Sinai Hosp, 44-47; asst prof, Col Physicians & Surgeons, Columbia Univ, 58-66; consult, Dept Path, Elmhurst City Hosp. Mem: Int Acad Path; Am Asn Path & Bact; Col Am Path; NY Acad Sci. Res: Pediatric pathology. Mailing Add: Dept of Path Mt Sinai Hosp 11 E 100th St New York NY 10029

STRAUSS, MARK A, physical chemistry, operations research, see 12th edition

STRAUSS, MARY JO, b Columbus, Ohio, June 10, 27; m 57; c 2. PHYSICAL CHEMISTRY. Educ: Bowling Green State Univ, BS, 49; Mich State Univ, PhD(phys chem), 55. Prof Exp: Phys chemist, US Naval Res Lab, 54-59, pvt consult, 60-71; RES ASSOC, COL GEN STUDIES, GEORGE WASHINGTON UNIV, 72- Mem: Fel Am Inst Chemists; Am Coun Consumer Interests; Sigma Xi; AAAS; Am Asn Women Sci. Res: Physical and chemical properties of ammonium amalgam; physical chemistry of the iron-oxygen-water system, particularly corrosion mechanisms; environmental studies. Mailing Add: 4506 Cedell Pl Camp Springs MD 20031

STRAUSS, MAURICE BENJAMIN, medicine, deceased

STRAUSS, MONTY JOSEPH, b Tyler, Tex, Aug 26, 45. MATHEMATICS. Educ: Rice Univ, BA, 67; NY Univ, PhD(math), 71. Prof Exp: Asst prof, 71-75, ASSOC PROF, TEX TECH UNIV, 75- Concurrent Pos: NSF res grant, 75. Mem: Am Math Soc; Math Asn Am. Res: Partial differential equations, particularly the theoretical aspects of existence and uniqueness of solutions and several complex variables. Mailing Add: Dept of Math Tex Tech Univ Lubbock TX 79409

STRAUSS, NORMAN, microbiology, see 12th edition

STRAUSS, PHYLLIS R, b Worcester, Mass, Mar 19, 43. CELL PHYSIOLOGY. Educ: Brown Univ, BA, 64; Rockefeller Univ, PhD(life sci), 71. Prof Exp: Res fell cell physiol, Harvard Med Sch, 71-73; ASST PROF CELL PHYSIOL, NORTHEASTERN UNIV, 73- Mem: AAAS; Am Soc Protozoologists; NY Acad Sci; Soc Am Cell Biologists. Res: Regulation of plasma membrane transport by macrophages and normal and leukemic lymphocytes; substrates are amino acids, sugars, purines and pyrimidines. Mailing Add: Dept of Biol Northeastern Univ 360 Huntington Ave Boston MA 02115

STRAUSS, ROBERT R, b Chelsea, Mass, Nov 4, 29; m 51; c 3. BIOCHEMISTRY, MICROBIOLOGY. Educ: Univ Pa, BA, 54; Hehnemann Med Col, MS, 56, PhD(microbiol), 58. Prof Exp: Sr res scientist, Nat Drug Co, Div Richardson-Merrell, Inc, 58-61, dir biochem res, 61, dir biochem & bact res labs, 61-67; res microbiologist, St Margaret's Hosp, Boston, 67-73; ASSOC DIR MICROBIOL, ALBERT EINSTEIN MED CTR, 73-; RES ASSOC PROF, SCH MED, TEMPLE UNIV, 73- Concurrent Pos: Asst prof, Sch Med, Tufts Univ. Mem: AAAS; Am Soc Exp Path; Reticuloendothelial Soc; Am Soc Microbiol. Res: Biochemistry of inflammation; virus purification; biochemistry of phagocytosis. Mailing Add: Dept of Microbiol Albert Einstein Med Ctr Philadelphia PA 19141

STRAUSS, ROGER WILLIAM, b Buffalo, NY, Sept 23, 27; m 50; c 4. PAPER TECHNOLOGY. Educ: State Univ NY Col Forestry, Syracuse, BS, 49, MS, 50, PhD(chem), 61. Prof Exp: Develop engr, Bauer Bros, Ohio, 50-52; paper sales develop engr, Hammermill Paper Co, Pa, 52-55; instr paper sci, State Univ NY Col Forestry, Syracuse, 55-60; mgr res, Nekoosa-Edwards Paper Co, Wis, 60-66; prof paper sci & eng, State Univ NY Col Environ Sci & Forestry, 66-75; DIR RES & SCI SERV, BOWATER INC, CONN, 75- Mem: Tech Asn Pulp & Paper Indust. Res: Pulping and bleaching of wood pulp; paper production and coating. Mailing Add: Bowater Inc 1500 E Putnam Ave Old Greenwich CT 06807

STRAUSS, RONALD GEORGE, b Mansfield, Ohio, Nov 29, 39; m 62; c 3. PEDIATRICS, HEMATOLOGY. Educ: Capital Univ, BS, 61; Univ Cincinnati, MD, 65; Am Bd Pediat, dipl, 70, cert pediat hemat-oncol, 74. Prof Exp: Intern pediat, Boston City Hosp, 65-66; from jr resident to chief resident, Children's Hosp, Cincinnati, 66-69; pediatrician, David Grant US Air Force Med Ctr, 69-71; fel pediat hemat, Children's Hosp Res Found, Cincinnati, 71-73, asst prof pediat, Col Med, Univ Cincinnati, 73-74; ASST PROF PEDIAT, COL MED, UNIV TENN, MEMPHIS, 74-; ASST MEM HEMAT-ONCOL, ST JUDE CHILDREN'S RES HOSP, 74- Mem: Reticuloendothelial Soc; Am Soc Hemat. Res: Leukocyte physiology and function. Mailing Add: St Jude Children's Res Hosp Memphis TN 38101

STRAUSS, SIMON WOLF, b Poland, Apr 15, 20; nat US; m 57; c 2. CHEMISTRY. Educ: Polytech Inst Brooklyn, BS, 44, MS, 47, PhD(chem), 50. Prof Exp: Inorg chemist, Nat Bur Standards, 51-55; phys chemist, US Naval Res Lab, 55-57, head chem metall sect, 57-63; STAFF CHEMIST, HQ, AIR FORCE SYSTS COMMAND, 63- Mem: Fel AAAS; Am Chem Soc; fel Am Inst Chemists. Res: Solid state reactions; structure and electrical properties of glass; nature and structure of liquid metals; technical management. Mailing Add: 4506 Cedell Pl Camp Springs MD 20031

STRAUSS, STEVEN, b Czech, Dec 4, 30; US citizen; m 59; c 3. PHARMACY. Educ: Brooklyn Col Pharm, BS, 55, MS, 65; Univ Pittsburgh, PhD(pharm), 70. Prof Exp: Asst prof, 65-71, alumni dir, 65-70, ASSOC PROF PHARM ADMIN, BROOKLYN COL PHARM, 71-; FIELD DIR, MKT MEASURES, 72-, IMS AM, LTD, 73- Concurrent Pos: Dir continuing educ, Brooklyn Col Pharm, 72- Mem: Am Pharmaceut Asn; assoc AMA; fel Am Col Apothecaries. Res: Pharmacy administration; marketing; market research. Mailing Add: Brooklyn Col of Pharm 600 Lafayette Ave Brooklyn NY 11216

STRAUSS, ULRICH PAUL, b Frankfort, Ger, Jan 10, 20; nat US; m 50; c 4. PHYSICAL CHEMISTRY. Educ: Columbia Univ, AB, 41; Cornell Univ, PhD(chem), 44. Prof Exp: Sterling fel, Yale Univ, 46-48; from asst prof to assoc prof chem, 48-60, dir sch chem, 65-71, PROF PHYS CHEM, RUTGERS UNIV, NEW BRUNSWICK, 60-, CHMN DEPT & DIR GRAD PROG CHEM, 74- Concurrent Pos: NSF sr fel, 61-62; Guggenheim fel, 71-72. Mem: AAAS; Am Chem Soc; Am Inst Chemists; NY Acad Sci. Res: Experimental and theoretical investigations of high polymers and colloidal electrolytes. Mailing Add: Sch of Chem Rutgers Univ New Brunswick NJ 08903

STRAUSS, WALTER, b Nürnberg, Ger, Nov 6, 23; US citizen m 59; c 2. PHYSICS, ELECTRICAL ENGINEERING. Educ: City Col New York, BEE, 48; Columbia Univ, PhD(physics), 61. Prof Exp: Tutor elec eng, City Col New York, 48-53; asst physics, Columbia Radiation Lab, 53-59; lectr elec eng, City Col New York, 59-60; MEM TECH STAFF, BELL TEL LABS, 60- Mem: Am Phys Soc. Res: Magnetic domain devices; magnetoelastic properties of yttrium iron garnet; magnetic materials; piezoelectricity; magnetron oscillators; microwave delay lines. Mailing Add: Bell Tel Labs Rm 2D-264 Murray Hill NJ 07974

STRAUSS, WALTER A, b Aachen, Ger, Oct 28, 37; US citizen. MATHEMATICS. Educ: Columbia Univ, AB, 58; Univ Chicago, MS, 59; Mass Inst Technol, PhD(math), 62. Prof Exp: NSF fel, Mass Inst Technol & Univ Paris, 62-63; vis asst prof math, Stanford Univ, 63-66; assoc prof, 66-71, PROF MATH, BROWN UNIV, 71- Concurrent Pos: Guggenheim fel, 71; vis scientist, Univ Tokyo, 72. Mem: Am Math Soc. Res: Nonlinear partial differential equations; scattering theory; abstract analysis. Mailing Add: Dept of Math Brown Univ Providence RI 02912

STRAUSSER, HELEN R, b New York, NY, Oct 31, 22; m 43; c 2. PHYSIOLOGY, ZOOLOGY. Educ: Hunter Col, AB, 46; Univ Pa, MS, 49; Rutgers Univ, PhD(zool), 58. Prof Exp: Instr physiol, Hunter Col, 49-50; from instr to assoc prof, 59-70, PROF PHYSIOL, RUTGERS UNIV, 70- Mem: AAAS; Am Physiol Soc; Am Soc Zoologists. Res: Immunology, particularly humoral and cell-mediated immune defense in ageing; autoimmunity and tumor immunity; endotoxins; endocrinology, particularly endocrine effects on cells and receptor sites for hormones. Mailing Add: Dept of Zool & Physiol Rutgers Univ Newark NJ 07102

STRAUSZ, OTTO PETER, b Miskolc, Hungary, 24; Can citizen; c 1. CHEMISTRY. Educ: Eötvös Lorand Univ, Hungary, MSc, 52; Univ Alta, PhD(chem), 62. Prof Exp: Res asst, 62-63, from asst prof to assoc prof, 63-71, PROF CHEM, UNIV ALTA, 71-, DIR HYDROCARBON RES CTR, 74- Mem: Fel Chem Inst Can; Am Chem Soc; AAAS; NY Acad Sci. Res: Mechanism and kinetics of chemical reactions induced photochemically or thermally and the chemistry of atoms, free radicals and reactive intermediates; chemical composition, analytical chemistry and the origin of petroleum. Mailing Add: Dept of Chem Univ of Alta Edmonton ON Can

STRAUTZ, ROBERT LEE, b Savanna, Ill, Jan 25, 35. HISTOLOGY, PHYSIOLOGY. Educ: Am Univ, BS, 63; Univ Md, PhD(anat), 66; Howard Univ, MD, 73. Prof Exp: Dir lab drug res, Hazelton Labs, Va, 61-62; instr biol, Am Univ, 65-66, asst prof biol & physiol, 66-70, assoc prof biol, 70-75; resident pathologist, George Washington Univ Hosp, 73-75; RESIDENT PATHOLOGIST, HARBOR GEN HOSP, TORRANCE, CALIF, 75- Concurrent Pos: USPHS grant. Mem: AAAS; Endocrine Soc; Am Physiol Soc; Am Asn Anatomists; Am Soc Pathologists. Res: Transplantation of pancreatic islets in diabetes. Mailing Add: Dept of Path Harbor Gen Hosp Torrance CA 90509

STRAW, HARRY ARTHUR, b US, Apr 22, 24; m 45; c 2. ORGANIC CHEMISTRY. Educ: Yale Univ, BS, 45, PhD(org chem), 50. Prof Exp: Instr chem, Yale Univ, 46-50; res chemist, Polychem Dept, 50-69, MGR INDUST DIAMONDS, E I DU PONT DE NEMOURS & CO, 69- Mem: Am Chem Soc; Sigma Xi. Res: Thiophane derivatives; Willgerodt reaction; polymer characterization; market research; plastics product and market development. Mailing Add: 8 Pinecrest Dr Wilmington DE 19810

STRAW, JAMES ASHLEY, b Farmville, Va, Apr 12, 32; m 54; c 2. PHARMACOLOGY. Educ: Univ Fla, BS, 58, PhD(physiol), 63. Prof Exp: From asst prof to assoc prof, 65-75, PROF PHARMACOL, SCH MED, GEORGE WASHINGTON UNIV, 75- Concurrent Pos: NIH fel physiol, Univ Fla, 63-64 & res grant, 64-65. Mem: Am Soc Pharmacol & Exp Therapeut. Res: Drug metabolism; action of drugs on endocrine glands; steroid biosynthesis; physiological disposition of anticancer drugs. Mailing Add: Dept of Pharmacol George Washington Univ Sch of Med Washington DC 20037

STRAW, RICHARD MYRON, b St Paul, Minn, July 25, 26; m 49; c 5. POPULATION BIOLOGY, SYSTEMATIC BOTANY. Educ: Univ Minn, BA, 49; Claremont Col, PhD(bot), 55. Prof Exp: Asst prof biol, Deep Springs Col, 55-56; from asst prof to assoc prof, 56-68, assoc dean acad affairs, Sch Letters & Sci, 70-75, PROF BIOL, CALIF STATE UNIV, LOS ANGELES, 68-, ASSOC DEAN ACAD PLANNING, 75- Concurrent Pos: NSF res grants, 57-59 & 60-62; consult, Children's Hosp, Los Angeles, 57-66; Fulbright lectr biol, Peru, 63-64; dir curric planning, Calif State Univ, 65-66, coordr biol interdept prog, 69-70; consult sci & math educ, US Peace Corps, Kuala Lumpur, Malaysia, 66-68. Mem: AAAS; Soc Study Evolution; Am Soc Naturalists; Am Soc Plant Taxonomists. Res: Evolution; genetics; biosystematics; population ecology and genetics; pollination mechanisms. Mailing Add: Dept of Biol Calif State Univ Los Angeles CA 90032

STRAW, ROBERT NICCOLLS, b Burlington, Iowa, Aug 24, 38; m 61; c 2. PHARMACOLOGY. Educ: Univ Iowa, BS, 60, MS, 65, PhD(pharmacol), 67. Prof Exp: RES ASSOC PHARMACOL, UPJOHN CO, 67- Mem: AAAS; Am Soc Pharmacol & Exp Therapeut. Res: Neuropharmacology; neurophysiology; pathophysiology of convulsions; mechanism of action of anticonvulsant drugs; effect of drugs on electroencephalogram. Mailing Add: Cent Nerv Syst Dis Res Upjohn Co 301 Henrietta St Kalamazoo MI 49001

STRAW, THOMAS EUGENE, b St Paul, Minn, Nov 20, 36; m 57; c 3. AQUATIC BIOLOGY. Educ: Univ Minn, St Paul, BS, 65, PhD(biochem), 69. Prof Exp: Asst prof, 68-73, ASSOC PROF BIOL, UNIV MINN, MORRIS, 73- Mem: AAAS; Am Soc Limnol & Oceanog; Sigma Xi. Res: Biochemical limnology; phosphorus metabolism in aquatic environments. Mailing Add: Dept of Sci & Math Univ of Minn Morris MN 56267

STRAW, WILLIAM THOMAS, b Griffin, Ind, Sept 29, 31; m 56; c 3. GEOLOGY. Educ: Ind Univ, BS, 58, MA, 60, PhD(geol), 68. Prof Exp: Geologist, Humble Oil & Refining Co, 60-65; lectr geol, Ind Univ, 67-68; asst prof, 68-70, actg chmn dept, 71, chmn, 71-74, ASSOC PROF GEOL, WESTERN MICH UNIV, 70- Concurrent Pos: Actg assoc dir, Geol Field Sta, Ind Univ, 70-75; geologist, Ind Geol Surv, 70. Mem: Am Asn Petrol Geologists; Geol Soc Am. Res: Glacial geology, hydrogeology and geomorphology; fluvial sedimentation; geology of valley trains. Mailing Add: Dept of Geol Western Mich Univ Kalamazoo MI 49001

STRAWBRIDGE, DENNIS WINSLOW, b Reading, Pa, Oct 10, 20; m 46. ECOLOGY. Educ: Univ Chicago, PhD(zool), 53. Prof Exp: Instr zool, Univ Pa, 53-55; from asst prof to assoc prof, 56-63, PROF NATURAL SCI, MICH STATE UNIV, 64- Res: Population ecology; biometrics; history and philosophy of science. Mailing Add: Dept of Natural Sci Mich State Univ East Lansing MI 48823

STRAWCUTTER, RICHARD, b Kane, Pa, Oct 8, 27; m 48; c 4. VERTEBRATE ZOOLOGY. Educ: Pa State Teachers Col, Indiana, BS, 50; Columbia Univ, AM, 56. Prof Exp: Teacher jr high sch, Conn, 50-53 & high sch, 54-56; from asst prof to assoc prof, 57-68, PROF BIOSCI, INDIANA UNIV, PA, 68-, CHMN DEPT ALLIED HEALTH PROFESSIONS. Res: Vertebrate zoology and allied health professions education. Mailing Add: Dept of Allied Health Prof Indiana Univ of Pa Indiana PA 15701

STRAWDERMAN, WAYNE ALAN, b Wakefield, RI, Oct 11, 36; m 58; c 2. APPLIED MECHANICS. Educ: Univ RI, BS, 58, MS, 61; Univ Conn, PhD(appl mech), 67. Prof Exp: Res engr, E I du Pont de Nemours & Co, 58-59; teaching asst mech eng, Univ RI, 59-61; mech engr, Elec Boat Div, Gen Dynamics Corp, 61-63; RES MECH ENGR, NEW LONDON LAB, NAVAL UNDERWATER SYSTS CTR, 63- Concurrent Pos: Lectr, Univ Conn, 67-69. Mem: Acoust Soc Am; Am Acad Mech. Res: Response of coupled mechanical-acoustical systems to random excitation, turbulence induced noise, random vibrations, acoustics. Mailing Add: Homestead Rd Ledyard CT 06339

STRAWN, ROBERT KIRK, b De Land, Fla, May 26, 22; m; c 3. ICHTHYOLOGY. Educ: Univ Fla, BS, 47, MS, 53; Univ Tex, PhD(zool), 57. Prof Exp: Asst malaria control, USPHS, 42-43; lab asst biol, Univ Fla, 46-48; instr, Southwestern Univ, 55-56; asst prof, Lamar State Col, 56-59; asst prof wildlife mgt, Agr & Mech Col, Tex, 59-60; from asst prof to assoc prof zool, Univ Ark, 60-66; assoc prof, 66-69, PROF WILDLIFE SCI, TEX A&M UNIV, 69- Mem: Am Fisheries Soc; Am Soc Limnol & Oceanog; Am Soc Ichthyol & Herpet; Ecol Soc Am; Soc Study Evolution. Res: Ecology and speciation of fishes. Mailing Add: Dept of Wildlife Sci Tex A&M Univ College Station TX 77843

STRAZDINS, EDWARD, b More, Latvia, Sept 19, 18; US citizen; m 43; c 2. PHYSICAL CHEMISTRY, POLYMER CHEMISTRY. Educ: Darmstadt Tech Univ, MS, 49. Prof Exp: Mill chemist, Baltic Wood Pulp & Paper Mills, 41-42, supvr, 43-44; res chemist, Am Cyanamid Co, 49-62, sr res chemist, 63-66, proj leader chem res, Cent Res Div, 67-68, res assoc, 68-74, PRIN RES SCIENTIST, CHEM RES DIV, AM CYANAMID CO, 74- Honors & Awards: Sci Achievement Award, Am Cyanamid Co, 74. Mem: Am Chem Soc; Tech Asn Pulp & Paper Indust; fel Am Inst Chemists. Res: Paper chemistry; polyelectrolytes; sizing; retention; flocculation aids; theoretical aspects of paper making process; ecology; electrokinetic phenomena; polymer research and surface chemistry. Mailing Add: Cent Res Div Am Cyanamid Co 1937 W Main St Stamford CT 06902

STREAMS, FREDERICK ARTHUR, b Mercer, Pa, Sept 8, 33; m 56; c 3. INSECT ECOLOGY. Educ: Indiana State Col, Pa, BS, 55; Cornell Univ, MS, 60, PhD(entom), 62. Prof Exp: Teacher pub schs, NY, 57-58; entomologist, Entom Res Div, USDA, 62-64; asst prof entom, 64-69, assoc prof biol, 69-74, PROF BIOL, UNIV CONN, 74-

Mem: Fel AAAS; Entom Soc Am; Ecol Soc Am; Entom Soc Can. Res: Ecology and evolution of populations; biological control of insects; preditor-prey interactions in insects. Mailing Add: Ecol Sect Biol Sci Group Univ of Conn Storrs CT 06268

STREBE, DAVID DIEDRICH, b Tonawanda, NY, Oct 6, 18; m 42; c 2. MATHEMATICS. Educ: State Univ NY Teachers Col, Buffalo, BS, 40; Univ Buffalo, MA, 49, PhD(math), 52. Prof Exp: Instr math, LeTourneau Tech Inst, 46-47 & Univ Buffalo, 47-54; assoc prof, Univ SC, 54-57; prof, State Univ NY Col, Oswego, 57-58, Univ SC, 58-70 & Westmont Col, 70-71; PROF MATH, COLUMBIA COL, S C, 71- Mem: Am Math Soc; Math Asn Am. Res: Set theoretic topology. Mailing Add: Dept of Math Columbia Col Columbia SC 29203

STRECKER, GEORGE EDISON, b Ft Collins, Colo, Feb 25, 38; m 60; c 2. MATHEMATICS. Educ: Univ Colo, Boulder, BS(elec eng) & BS(bus), 61; Tulane Univ, La, PhD(math), 66. Prof Exp: Instr math, Univ Colo, 58-60; teaching asst, Tulane Univ, La, 64-65; res assoc, Univ Amsterdam, 65-66, Fulbright fel, 66; fel, Univ Fla, 66-67, asst prof, 67-71; assoc prof, Univ Pittsburgh, 71-72; ASSOC PROF MATH, KANS STATE UNIV, 72- Mem: Am Math Soc; Math Asn Am. Res: General topology; compactifications and extensions; relationship to category theory. Mailing Add: Dept of Math Kans State Univ Manhattan KS 66506

STRECKER, HAROLD ARTHUR, b Marietta, Ohio, June 11, 18; m 42; c 5. INORGANIC CHEMISTRY. Educ: Cornell Univ, AB, 40, PhD(chem), 48. Prof Exp: Chemist, Marietta Dyestuff Co, 40-41 & Nat Defense Comn, 42-45; res assoc, 47-58, supvr process res, 58-60, sr res assoc, 60-68, SUPVR SPECTROS & MICROS, STANDARD OIL CO, OHIO, 69- Mem: Am Chem Soc; Soc Appl Spectros. Res: Catalysis; reaction kinetics; atomic spectroscopy; x-ray fluorescence and diffraction; electron microscopy. Mailing Add: 7131 Rotary Dr Walton Hills OH 44146

STRECKER, HAROLD JOSEPH, biochemistry, deceased

STRECKER, JOSEPH LAWRENCE, b Kansas City, Mo, Mar 30, 32; m 60; c 2. THEORETICAL PHYSICS. Educ: Rockhurst Col, BS, 55; Johns Hopkins Univ, PhD(physics), 61. Prof Exp: Jr instr physics, Johns Hopkins Univ, 55-58, res asst, 58-61, sr res scientist, Gen Dynamics, Ft Worth, 61-66; assoc prof physics, Univ Dallas, 66-68; ASSOC PROF PHYSICS, WICHITA STATE UNIV, 68- Concurrent Pos: Adj prof, Tex Christian Univ, 62-67. Mem: Am Phys Soc. Res: Superconductivity; quantum field theory and application to solid state phenomena; statistical mechanics, especially phase transitions. Mailing Add: Dept of Physics Wichita State Univ Wichita KS 67208

STRECKER, ROBERT LOUIS, b Marietta, Ohio, Feb 10, 25; m; c 3. ECOLOGY, MAMMALOGY. Educ: Marietta Col, BA, 46; Univ Wis, MA, 47, PhD(zool), 51. Prof Exp: From asst prof to assoc prof zool, Miami Univ, Ohio, 51-65; mem fac, San Diego City Col, 66-75, chmn dept biol, 70-75; MEM FAC BIOL, SAN DIEGO EVE COL, 75- Concurrent Pos: Partic, Pac Islands Rat Ecol Proj, Nat Res Coun, 56-58. Mem: Am Soc Mammalogists; Ecol Soc Am; Am Soc Zoologists. Res: Vertebrate ecology; rodent ecology and behavior; limnology. Mailing Add: Dept of Biol San Diego Eve Col 3375 Camino del rio S San Diego CA 92108

STRECKFUSS, JOSEPH LARRY, b Shirley, Mo, Feb 23, 31; m 52; c 3. MICROBIOLOGY, IMMUNOLOGY. Educ: Southern Ill Univ, Carbondale, BA, 58, MA, 61, PhD(virol, immunol), 68. Prof Exp: Dir diag microbiol, Holden Hosp, Carbondale, Ill, 57-66; asst res prof oral microbiol, 68-74, ASSOC PROF MICROBIOL, DEPT PATH & ASSOC PROF IN RESIDENCE, DENT SCI INST, UNIV TEX DENT BR, HOUSTON, 74- Concurrent Pos: Comt mem curric, Univ Tex Grad Sch Biomed Sci, Houston, 71- Res: Oral microbiology; study of morphologic variation of oral filamentous organisms with emphasis on the mechanism of variation, calcification of these organisms with interests on the ecologic contribution to periodontal diseases. Mailing Add: Dept of Path Univ of Tex Dent Sci Inst Houston TX 77025

STRECOK, ANTHONY J, b Chicago, Ill, June 5, 31. MATHEMATICS. Educ: Northwestern Univ, BS, 53, MS, 54. Prof Exp: Asst mathematician, 55-68, ASSOC COMPUT SCIENTIST, ARGONNE NAT LAB, 68- Mem: Am Math Soc; Soc Indust & Appl Math. Res: Differential equations; integration techniques; difference equations; computer programming. Mailing Add: Argonne Nat Lab 9700 S Cass Ave Argonne IL 60439

STREET, DANA MORRIS, b New York, NY, May 7, 10; m 40; c 4. ORTHOPEDIC SURGERY. Educ: Haverford Col, BS, 32; Cornell Univ, MD, 36. Prof Exp: Chief orthop sect, Kennedy Vet Admin Hosp, Memphis, Tenn, 46-59; prof orthop surg, Sch Med, Univ Ark, 59-62; PROF SURG IN RESIDENCE, SCH MED, UNIV CALIF, LOS ANGELES, 62-; HEAD ORTHOP DIV, HARBOR GEN HOSP, 62- Concurrent Pos: Mem staff, Orthop Hosp, Los Angeles, St Mary's Hosp & Mem Hosp, Long Beach, Calif. Mem: AMA; Am Acad Orthop Surgeons. Res: Fracture treatment by use of medullary nail, particularly in femur and forearm; treatment and rehabilitation of the paraplegic. Mailing Add: Orthop Div Harbor Gen Hosp Torrance CA 90509

STREET, EVAN HOSKINS, JR, physical inorganic chemistry, see 12th edition

STREET, JABEZ CURRY, b Opelika, Ala, May 5, 06; m 39; c 2. HIGH ENERGY PHYSICS. Educ: Ala Polytech Inst, BS, 27; Univ Va, MS, 30, PhD(physics), 31. Hon Degrees: AM, Harvard Univ, 42. Prof Exp: Fel, Bartol Res Found, 31-32; from instr to prof, 32-70, chmn dept, 56-60, MALLINCKRODT PROF PHYSICS, HARVARD UNIV, 70-, ASST TO DEAN SCI, 66- Concurrent Pos: Mem, Carnegie Cosmic Ray Exped, Peru, 33; res assoc, Radiation Lab, Mass Inst Technol, 40-45; trustee, Assoc Univs, Inc, 70- Mem: Nat Acad Sci; Am Acad Arts & Sci; fel Am Phys Soc. Res: High energy particle physics; cosmic rays; electronic circuits; electrical discharges in gases; experiments using Geiger counters; ionization and cloud chambers; circuit development; radar; bubble chambers. Mailing Add: 56 Fletcher Rd Belmont MA 02178

STREET, JAMES STEWART, b Chicago, Ill, July 26, 34; m 57; c 3. GEOLOGY. Educ: Univ Ill, Urbana, BS, 58; Syracuse Univ, MS, 63, PhD(geol), 66. Prof Exp: Geologist, Texaco Inc, La, 65-66; asst prof, 66-72, ASSOC PROF GEOL, ST LAWRENCE UNIV, 72- Concurrent Pos: Dir, NY State Tech Serv Prog, St Lawrence Univ, 66- Mem: Geol Soc Am; Am Asn Geol Teachers; Int Asn Quaternary Res. Res: Geomorphology and glacial geology. Mailing Add: Dept of Geol St Lawrence Univ Canton NY 13617

STREET, JOHN MALCOLM, b McIntosh, SDak, May 28, 24; m 60; c 2. BIOGEOGRAPHY. Educ: Univ Calif, Berkeley, BA, 48, PhD(geog), 60. Prof Exp: Lectr geog, Far East Prog, Univ Calif, 55-56; lectr, Far East Prog, Univ Md, 56-57; asst prof, Univ Calif, Davis, 60; PROF GEOG, UNIV HAWAII, 60- Concurrent Pos: NSF res grants, Univ Hawaii & Govt New Guinea, 64 & 67 & Peru, 69-70; fel, East-West Ctr, Univ Hawaii, 68-69; Univ Hawaii Found grant, Southeast Asia, 71; NSF grant, Nat Ctr Sci Res, France, 73. Res: Impact of man on the tropical biosphere, especially processes and consequences of the creation and maintenance of grasslands derived from forest. Mailing Add: Dept of Geog Univ of Hawaii Honolulu HI 96822

STREET, JOSEPH CURTIS, b Bozeman, Mont, Aug 30, 28; m 49; c 4. TOXICOLOGY, PESTICIDE CHEMISTRY. Educ: Mont State Col, BS, 50, MS, 52; Okla State Univ, PhD(chem), 54. Prof Exp: Instr agr chem, Okla State Univ, 53; from instr to assoc prof, 53-67, PROF ANIMAL SCI, CHEM & BIOCHEM, UTAH STATE UNIV, 67- Mem: AAAS; Am Chem Soc; fel Am Col Vet Toxicol; Am Dairy Sci Asn; Soc Toxicol. Res: Mammalian metabolism of pesticides; food chemicals and natural toxicants; biochemistry of lipids and animal nutrition. Mailing Add: Dept of Animal Sci Utah State Univ Logan UT 84321

STREET, KENNETH, JR, nuclear chemistry, see 12th edition

STREET, PAUL FRANKLIN, plant pathology, see 12th edition

STREET, ROBERT ELLIOTT, b Belmont, NY, Dec 11, 12; m 41, 69; c 3. AERODYNAMICS, NUMERICAL ANALYSIS. Educ: Rensselaer Polytech Inst, BS, 33; Harvard Univ, AM, 34, PhD(math, physics), 39. Prof Exp: Instr math, Rensselaer Polytech Inst, 37-41; asst physicist, Nat Adv Comt Aeronaut, Langley Field, Va, 41-43; asst prof physics, Dartmouth Col, 43-44; engr, Gen Elec Co, NY, 44-47; assoc prof physics, Univ NMex, 47-58; assoc prof, 48-55, PROF AERONAUT & ASTRONAUT, UNIV WASH, 55- Mem: Am Inst Aeronaut & Astronaut; Math Asn Am; Soc Indust & Appl Math. Res: Numerical fluid mechanics. Mailing Add: Dept of Aeronaut & Astronaut Univ of Wash Seattle WA 98195

STREETEN, DAVID HENRY PALMER, b Bloemfontein, SAfrica, Oct 3, 21; nat US; m 52; c 3. INTERNAL MEDICINE. Educ: Univ Witwatersrand, MB, BCh, 46; Oxford Univ, DPhil(pharmacol), 51. Prof Exp: Intern med & surg, Gen Hosp, Johannesburg, SAfrica, 47; jr lectr med, Univ Witwatersrand, 48; Nuffield demonstr pharmacol, Oxford Univ, 48-51; asst med, Peter Bent Brigham Hosp, Boston, 51-53, jr assoc, 53; from instr to asst prof internal med, Univ Mich Hosp, 53-60; assoc prof, 60-64, PROF MED, STATE UNIV NY UPSTATE MED CTR, 64- Concurrent Pos: Rockefeller traveling fel & res fel med, Harvard Univ, 51-52; investr, Howard Hughes Found, 55-61; consult, Vet Admin Hosp, Syracuse, 61-, Crouse Irving Mem Hosp, 61-, St Joseph's Hosp, 64- & Utica State Hosp, 65- Mem: Endocrine Soc; Am Fedn Clin Res. Res: Physiology and pathology of adrenal cortex, especially effects of its secretions on water and electrolyte metabolism and their role in causation of disease. Mailing Add: Dept of Med State Univ of NY Hosp Syracuse NY 13210

STREETER, JOHN GEMMIL, b Ellwood City, Pa, Feb 25, 36; m 60; c 1. PLANT PHYSIOLOGY, AGRONOMY. Educ: Pa State Univ, BS, 58, MS, 64; Cornell Univ, PhD(bot), 69. Prof Exp: Asst prof, 69-74, ASSOC PROF AGRON, OHIO AGR RES & DEVELOP CTR, 74- Mem: AAAS; Am Soc Agron; Am Soc Plant Physiol. Res: Nitrogen metabolism in plants; amino acid biosynthesis; mechanisms of nitrogen transport in plant tissues and cells; symbiotic nitrogen fixation. Mailing Add: Dept of Agron Ohio Agr Res & Develop Ctr Wooster OH 55691

STREETER, ROBERT GLEN, b Madison, SDak, Feb 1, 41; m 64; c 2. WILDLIFE CONSERVATION, WILDLIFE RESEARCH. Educ: SDak State Univ, BS, 63; Va Polytech Inst & State Univ, MS, 65; Colo State Univ, PhD(wildlife biol, physiol), 69. Prof Exp: Asst biologist avian depredation res, SDak State Dept Game, Fish & Parks, 63; res asst elk range ecol, Va Coop Wildlife Res Unit, US Fish & Wildlife Serv, 63-65; res asst bighorn sheep ecol & mgt, Colo Coop Wildlife Res Unit, 65-69; res physiologist, US Air Force Sch Aerospace Med, Brooks AFB, Tex, 69-72; wilflife biologist & res asst leader, Colo Coop Wildlife Res Unit, Colo State Univ & US Fish & Wildlife Serv, 72-73; head coop wildlife units, Div Res, Washington, DC, 73-75, COAL PROJ RES MGR, WESTERN ENERGY & LAND USE TEAM, US FISH & WILDLIFE SERV, 75- Concurrent Pos: Vis mem grad fac, Tex A&M Univ, 71-72; asst prof, Colo State Univ, 72-73. Honors & Awards: Hibbs Award, Colo State Univ & Colo State Div Wildlife, 67. Mem: Sigma Xi; Wildlife Soc; Am Soc Mammalogists. Res: Effects of coal extraction, conversion, transportation and related social developments on fish and wildlife populations, development of mitigation options and management decision alternatives. Mailing Add: 208 Federal Bldg Ft Collins CO 80521

STREETMAN, JOHN ROBERT, b Ft Worth, Tex, Apr 12, 30; m 51; c 2. PHYSICAL CHEMISTRY. Educ: Baylor Univ, BS, 51; Univ Tex, MA, 53, PhD(phys chem), 55. Prof Exp: Aeronaut res scientist, Nat Adv Comt Aeronaut, 55-56; sr nuclear engr, Gen Dynamics/Convair, 56-59; MEM STAFF, LOS ALAMOS SCI LAB, UNIV CALIF, 59- Mem: Am Phys Soc. Res: Monte Carlo neutron and gamma transport; quantum mechanics; solid state physics. Mailing Add: Los Alamos Sci Lab Univ of Calif Los Alamos NM 87545

STREETMAN, WILLIAM EDWARD, organic chemistry, see 12th edition

STREETS, RUBERT BURLEY, b Helena, Mont, May 22, 95; m 22; c 3. PLANT PATHOLOGY. Educ: Mont State Col, BS, 18; Univ Wis, MS, 22, PhD, 24. Prof Exp: Asst plant pathologist, Off Cereal Invests, USDA, 18-20; asst plant path, Univ Wis, 20-22, instr, 22-24; from asst prof to prof, 24-70, head dept, 52-60, EMER PROF PLANT PATH, UNIV ARIZ, 70- Mem: AAAS; Am Phytopath Soc. Res: Flax wilt; brown rot of stone fruits; Texas root rot; diseases of date palm, citrus and subtropical plants; guar and roses; antibiotics for plant disease control; serology of citrus viruses. Mailing Add: Dept of Plant Path Univ of Ariz Tucson AZ 85721

STREETT, JAMES CLARK, JR, b St Louis, Mo, Aug 13, 13; m 44; c 2. BIOLOGY. Educ: Princeton Univ, AB, 36, AM, 38, PhD, 39. Prof Exp: From instr to asst prof biol, Tex Christian Univ, 39-42, asst prof, 45-46; assoc prof, Univ Miss, 46-47; bacteriologist, Swift & Co, 47-50; assoc prof, 50-56, PROF BIOL, TEX WESLEYAN COL, 56-, CHMN DIV SCI, 74- Mem: AAAS. Res: Experimental embryology; regeneration of viscera in amphibia. Mailing Add: Dept of Biol Sci Tex Wesleyan Col Ft Worth TX 76105

STREETT, WILLIAM BERNARD, b Lake Village, Ark, Jan 27, 32; m 55; c 4. PHYSICAL CHEMISTRY, HIGH PRESSURE PHYSICS. Educ: US Mil Acad, BS, 55; Univ Mich, MS, 61, PhD(mech eng), 63. Prof Exp: Instr astron & astronaut, US Mil Acad, 61-62 & 63-64, asst prof, 64-65, NATO res fel low temperature chem, Oxford Univ, 66-67, ASST DEAN ACAD RES & DRI SCI RES LAB, US MIL ACAD, WEST POINT, 67- Concurrent Pos: Guggenheim fel, Oxford Univ, 74-75; Colonel, US Army. Mem: The Chem Soc. Res: Experimental measurements of physical and thermodynamic properties of fluids and fluid mixtures at low temperatures and high pressures; computer simulations of liquids. Mailing Add: Sci Res Lab US Mil Acad West Point NY 10996

STREEVER, RALPH L, b Schenectady, NY, June 7, 34; m 64; c 2. SOLID STATE PHYSICS. Educ: Union Col, NY, BS, 55; Rutgers Univ, PhD(physics), 60. Prof Exp: Physicist, Nat Bur Standards, 60-66; PHYSICIST, US ARMY ELECTRONICS

COMMAND, 66- Mem: Am Phys Soc. Res: Nuclear magnetic resonance studies of magnetic properties, hyperfine interactions and relaxation mechanisms in ferromagnetic metals, alloys and compounds. Mailing Add: US Army Electronics Command Ft Monmouth NJ 07703

STREHLER, ALLEN FREDERICK, b Detroit, Mich, Apr 9, 21. MATHEMATICS. Educ: Oberlin Col, BA, 42; Ohio State Univ, MA, 44; Univ Wis, PhD(math), 49. Prof Exp: Instr math, Univ Chicago, 49-51; asst prof, Haverford Col, 51-53; res mathematician, Lovelace Found, 53-55; from asst prof to assoc prof math, 55-70, assoc dean, 65-70, DEAN GRAD STUDIES, CARNEGIE-MELLON UNIV, 70-, SR LECTR MATH, 74- Concurrent Pos: Consult, Sandia Corp, 50-53 & Baldwin-Whitehall Pub Sch, 58-59; vis assoc prof, Western Reserve Univ, 60. Mem: Am Math Soc; Math Asn Am. Res: Abstract algebra; classical number theory; graduate education. Mailing Add: Off of Grad Studies Carnegie-Mellon Univ 500 Forbes Ave Pittsburgh PA 15213

STREHLER, BERNARD LOUIS, b Johnstown, Pa, Feb 21, 25; m 48; c 3. BIOLOGY, BIOCHEMISTRY. Educ: Johns Hopkins Univ, BS, 47, PhD, 50. Prof Exp: Biochemist, Oak Ridge Nat Lab, 50-53; asst prof biochem, Univ Chicago, 53-56; chief cellular & comp physiol sect, Geront Res Ctr, Nat Inst Child Health & Human Develop, 56-67; PROF BIOL SCI, UNIV SOUTHERN CALIF, 67- Concurrent Pos: Dir aging res satellite lab, Vet Admin Hosp, Baltimore, Md, 64-67. Honors & Awards: Karl August Forster Prize, Ger Acad Sci & Lett, 75. Mem: Am Soc Biol Chemists; Soc Develop Biol; Geront Soc; Am Soc Naturalists. Res: Bioluminescence; photosynthesis; aging; bioenergetics. Mailing Add: Dept of Biol Univ of Southern Calif University Park Los Angeles CA 90007

STREHLOW, CLIFFORD DAVID, b Mineola, NY, July 10, 40; m 63; c 2. ENVIRONMENTAL HEALTH. Educ: Muhlenberg Col, BS, 62; Mass Inst Technol, MS, 64; NY Univ, PhD(environ health), 72. Prof Exp: Res scientist, Inst Environ Med, Med Ctr, NY Univ, 64-72; CONSULT, INT LEAD ZINC RES ORGN & RES FEL, ST MARY'S HOSP MED SCH, ENG, 72- Concurrent Pos: Lectr environ health, St Mary's Hosp Med Sch, 75- Mem: Am Chem Soc; Brit Occup Hyg Soc. Res: Lead metabolism ; epidemiology; trace element analysis; air and water pollution; radiochemistry. Mailing Add: Pediatric Unit St Mary's Hosp Med Sch London England

STREHLOW, RICHARD ALAN, b Chicago, Ill, Sept 20, 27; m 50; c 1. CHEMISTRY. Educ: Univ Chicago, SB, 48; Univ Ill, PhD(chem), 57. Prof Exp: RES STAFF MEM, UNION CARBIDE NUCLEAR CO, OAK RIDGE NAT LAB, 56- Mem: Am Chem Soc; Am Vacuum Soc; Am Soc Testing & Mat; Am Nuclear Soc; Sigma Xi. Res: Catalysis and surface chemistry, coal conversion; electro-organic and fused salt chemistry; high vacuum research; mass spectrometry; fusion reactor design; graphite fabrication research. Mailing Add: Oak Ridge Nat Lab 4500-S PO Box X Oak Ridge TN 37830

STREHLOW, ROGER ALBERT, b Milwaukee, Wis, Nov 25, 25; m 48; c 2. PHYSICAL CHEMISTRY, FLUID DYNAMICS. Educ: Univ Wis, BS, 47, PhD(chem), 50. Prof Exp: Phys chemist, Ballistic Res Lab, Aberdeen Proving Ground, 50-58, chief physics br, Interior Ballistics Lab, 59-61; PROF AERONAUT & ASTRONAUT ENG, UNIV ILL, URBANA, 61- Concurrent Pos: Ford Found vis prof, Univ Ill, Urbana, 60-61; consult, Los Alamos Sci Lab, 63-72, Aro Inc, Arnold Air Force Sta, 65-70, Weapon Command, Rock Island Arsenal, Ill, 69-71, Environ Protection Agency, 71 & Brookhaven Nat Lab, 75-; dep ed, Combustion & Flame, 75- Mem: AAAS; Am Chem Soc; fel Am Phys Soc; assoc fel Am Inst Aeronaut & & Astronaut; Combustion Inst. Res: Reactive gas dynamics; combustion. Mailing Add: 101 Transportation Bldg Univ of Ill Urbana IL 61801

STREHLOW, WOLFGANG HANS, b Rathenow, Ger, July 26, 37; m 66; c 2. PHYSICAL CHEMISTRY, RESEARCH ADMINISTRATION. Educ: Free Univ Berlin, Vordiplom, 61; Univ Frankfurt, Diplom, 63, Dr phil nat (physics), 66. Prof Exp: Ger Res Asn fel, Inst Phys Chem, Frankfurt, 65-66; sr physicist & res specialist solid state physics, 66-70, supvr quantum electronics, 70-73, mgr electronics group, 73-74, LAB MGR, ELECTRONICS & OPTICS DEPT, CENT RES LAB, 3M CO, 74- Res: Nuclear quadrupole and nuclear magnetic resonance; thin film and crystal growth; optical properties of semiconductors; surface properties of solids; quantum electronics; optical memories; holography. Mailing Add: 3M Co Cent Res Lab PO Box 33221 St Paul MN 55133

STREIB, JOHN FREDRICK, b Avalon, Pa, Mar 21, 15; m 46, 54; c 2. PHYSICS. Educ: Calif Inst Technol, BS, 36, PhD(physics), 41. Prof Exp: Asst physicist, Carnegie Inst Technol, 41; asst physicist, Nat Bur Standards, 41-42, assoc physicist, 42-43, physicist, 43; scientist, Los Alamos Sci Lab, 43-46; mem tech staff, Bell Tel Labs, Inc, NY, 46; asst prof physics, Univ Colo, 46-47; asst prof, 47-60, ASSOC PROF PHYSICS, UNIV WASH, 60- Mem: Am Phys Soc; Am Asn Physics Teachers. Res: Nuclear physics; fluorine plus proton reactions; positron absorption. Mailing Add: Dept of Physics FM 17 Univ of Wash Seattle WA 98105

STREIB, WILLIAM E, b New Salem, NDak, Mar 16, 31; m 61; c 2. PHYSICAL CHEMISTRY. Educ: Jamestown Col, BS, 53; Univ NDak, MS, 55; Univ Minn, PhD(phys chem), 62. Prof Exp: Fel, Harvard Univ, 62-63; instr phys chem, 63-64, asst prof chem, 64-68, assoc dir labs, 68, DIR LABS, DEPT CHEM, IND UNIV, BLOOMINGTON, 68- Mem: Am Chem Soc; AAAS; Am Cyrstallog Asn. Res: X-ray crystallography; crystal and molecular structure; low temperature x-ray diffraction techniques. Mailing Add: Dept of Chem Ind Univ Bloomington IN 47401

STREICHER, EUGENE, b New York, NY, Oct 25, 26; m 51. NEUROPHYSIOLOGY. Educ: Cornell Univ, BA, 47, MA, 48; Univ Chicago, PhD(physiol), 53. Prof Exp: Physiologist, US Army Chem Ctr, Md, 48-50; neurophysiologist, Sect Aging, NIMH, 54-62, physiologist, Nat Inst Neurol Dis & Stroke, SCIENTIST ADMINR, NAT INST NEUROL DIS & STROKE, 64- Mem: AAAS; Soc Exp Biol & Med; Am Asn Neuropath. Res: Physiological chemistry of central nervous system. Mailing Add: Nat Inst of Neurol Dis & Stroke Bethesda MD 20901

STREIFER, WILLIAM, b Poland, Sept 13, 36; US citizen; m 58; c 3. APPLIED MATHEMATICS. Educ: City Col New York, BEE, 57; Columbia Univ, MS, 59; Brown Univ, PhD(elec eng), 62. Prof Exp: Res engr, Heat & Mass Flow Analyzer Lab, Columbia Univ, 58-59; from asst prof to prof elec eng, Univ Rochester, 62-72; PRIN SCIENTIST, XEROX CORP, 72- Concurrent Pos: Lectr, City Col New York, 57-59; consult lectr, Eastman Kodak Co, 65-68; consult, Xerox Corp, 68-72; vis assoc prof, Stanford Univ, 69-70. Mem: AAAS; Inst Elec & Electronics Eng; Optical Soc Am. Res: Electromagnetic theory; optics; mathematical ecology. Mailing Add: Xerox Palo Alto Res Ctr 3333 Coyote Hill Rd Palo Alto CA 94304

STREIFF, ANTON JOSEPH, b Jackson, Mich, Apr 1, 15; m 41; c 4. PETROLEUM CHEMISTRY. Educ: Univ Mich, BS, 36, MS, 37. Prof Exp: Res assoc, Nat Bur Stand, 37-50; sr res chemist, 50-71, asst chmn dept chem, 66-74, DIR AM PETROL INST RES PROJ, 58, 60-, PRIN RES CHEMIST, CARNEGIE-MELLON UNIV, 72-, ADMIN OFFICER, DEPT CHEM, 74- Mem: AAAS; fel Am Inst Chem; Am Chem Soc. Res: Fractionation, purification, purity and analysis of hydrocarbons; American Petroleum Institute standard reference materials. Mailing Add: Dept of Chem Carnegie-Mellon Univ Pittsburgh PA 15213

STREIFF, RICHARD REINHART, b Highland, Ill, June 1, 29; m 59; c 3. MEDICINE, HEMATOLOGY. Educ: Wash Univ, AB, 51; Univ Basel, MD, 59. Prof Exp: Intern med, Harvard Med Serv, Boston City Hosp, 59-60; intern, Mt Auburn Hosp, Cambridge, Mass, 60-61, resident, 61-62; resident, Harvard Med Serv, Boston City Hosp, 62-63; instr med, Harvard Med Sch, 66-68; from asst prof to assoc prof, 68-74, PROF MED, COL MED, UNIV FLA, 74- CHIEF HEMAT UNIT, VET ADMIN HOSP, GAINESVILLE, 72-, CHIEF MED SERV, 73- Concurrent Pos: Res fel hemat, Thorndike Med Lab, Harvard Med Sch, 63-68; clin investr, Vet Admin, 69-71. Mem: Am Soc Hemat; Am Fedn Clin Res; Am Soc Clin Nutrit; Am Inst Nutrit; Fedn Am Socs Exp Biol. Res: Vitamin B-12 and folic acid deficiency anemias; metabolism and biological function of vitamin B-12 and folic acid. Mailing Add: Dept of Med Univ of Fla Col of Med Gainesville FL 32610

STREILEIN, JACOB WAYNE, b Johnstown, Pa, June 19, 35; m 57; c 3. IMMUNOLOGY, GENETICS. Educ: Gettysburg Col, AB, 56; Univ Pa, MD, 60. Prof Exp: Intern, Univ Hosp, Univ Pa, 60-61, resident internal med, 61-63, from asst prof to assoc prof med genetics, Sch Med, 66-71; PROF CELL BIOL & ASSOC PROF MED, UNIV TEX SOUTHWESTERN MED SCH, DALLAS, 72- Concurrent Pos: Fel allergy & immunol, Univ Pa, 63-64; fel transplantation immunity, Wistar Inst Anat & Biol, 64-65; Markle scholar acad med, 68-74. Mem: Am Asn Immunol; Transplantation Soc; Soc Exp Hemat. Res: Transplantation immunobiology with special reference to cellular immunity, immunoregulation, graft-versus-host disease, immunogenetic disparity. Mailing Add: Dept of Cell Biol Univ of Tex Southwestern Med Sch Dallas TX 75235

STREIPS, LAIMONS VALDEMARS, analytical chemistry, see 12th edition

STREIPS, ULDIS NORMUNDS, b Riga, Latvia, Feb 1, 42; US citizen. MICROBIOLOGY. Educ: Valparaiso Univ, BA, 64; Northwestern Univ, PhD(microbiol), 69. Prof Exp: ASST PROF MICROBIAL GENETICS, SCH MED, UNIV LOUISVILLE, 72- Concurrent Pos: Damon Runyon Mem Fund Cancer res grant microbial genetics, Scripps Clin & Res Found, 69-70 & Sch Med & Dent, Univ Rochester, 70-72. Mem: AAAS; Am Soc Microbiol. Res: Genetic transformation in Bacillus subtilis; DNA excretion; restriction endonucleases; mutagenesis assays; cell fusion in eukaryotic systems. Mailing Add: Sch Med Dept Microbiol & Immunol Univ of Louisville Med Ctr Louisville KY 40201

STREISINGER, GEORGE, b Budapest, Hungary, Dec 27, 27; nat US; m 49; c 2. GENETICS. Educ: Cornell Univ, BS, 50; Univ Ill, PhD(bact), 54. Prof Exp: Res fel biophys, Calif Inst Technol, 53-56; mem staff, Carnegie Inst Washington Genetics Res Unit, 56-60; assoc prof biol, 60-63, co-chmn dept, 68-71, PROF BIOL, UNIV ORE, 63- Concurrent Pos: Nat Found Infantile Paralysis sr fel, Med Res Coun Unit Molecular Biol, Eng, 57-58; instr, Winter Sch Molecular Biol, Tata Inst Fundamental Res, India, 67; Guggenheim fel, 71-72. Mem: Nat Acad Sci. Res: Behavioral genetics of lower vertebrates; genetic and developmental studies of zebrafish; behavioral genetics of zebrafish; molecular mechanism of mutation in bacteriophage T4. Mailing Add: Inst Molecular Biol Univ of Ore Eugene OR 97403

STREITFELD, MURRAY MARK, b New York, NY, Sept 16, 22; c 1. MICORBIOLOGY, CHEMOTHERAPY. Educ: City Col New York, BS, 43; McGill Univ, MS, 48; Univ Calif, Los Angeles, PhD(microbiol), 52. Prof Exp: Asst chemist toxicol, Off Chief Med Examr, New York, 43; instr, Med Lab, Beaumont Gen Hosp, Tex, 45-46; teaching asst bact, McGill Univ, 47-48 & Univ Calif, Los Angeles, 51-52; from instr to asst prof, 53-66, ASSOC PROF MICROBIOL, SCH MED, UNIV MIAMI, 66- Concurrent Pos: Instr med lab, Brookes Med Ctr, Tex, 45-46; res bacteriologist, Nat Children's Cardiac Hosp, Miami, Fla, 52-57; asst dir res bact lab, Variety Children's Hosp, 57-59, res assoc, Variety Res Found, 60-; resident attend, Vet Admin Hosp, Coral Gables, 60- Mem: Am Soc Microbiol; fel Am Acad Microbiol; Sigma Xi. Res: Bacteriology; rheumatic fever; antibiotics; prophylaxis of dental infection; streptococcal and staphylococcal epidemiology; pseudomonas and gonorrhea immunity; gamma globulin; staphyloccocal toxins; gonococcal cellular immunity; streptococcal virulence; antibiotics. Mailing Add: Dept Microbiol Univ Miami Sch Med Box 875 Biscayne Annex Miami FL 33152

STREITWIESER, ANDREW, JR, b Buffalo, NY, June 23, 27; m 67; c 2. PHYSICAL ORGANIC CHEMISTRY. Educ: Columbia Univ, AB, 48, MA, 50, PhD(chem), 52. Prof Exp: From instr to assoc prof, 52-63, PROF CHEM, UNIV CALIF, BERKELEY, 63- Concurrent Pos: Sloan Found fel, Univ Calif, Berkeley, 58-62; NSF faculty fel, 59-60; Guggenheim fel, 69- Honors & Awards: Award, Am Chem Soc, 67. Mem: Nat Acad Sci; AAAS; Am Chem Soc; The Chem Soc. Res: Theoretical organic chemistry; molecular orbital theory; reaction mechanisms; isotope effects; acidity and basicity; rare earth organometallic chemistry. Mailing Add: Dept of Chem Univ of Calif Berkeley CA 94720

STREJAN, GILL HENRIC, b Galati, Romania, Sept 24, 30; m 63. IMMUNOLOGY. Educ: Univ Bucharest, MS, 53; Hebrew Univ Jerusalem, PhD(immunol), 65. Prof Exp: Instr bact & immunol, Hebrew Univ Jerusalem, 63-65; asst prof, 68-73, ASSOC PROF IMMUNOCHEM, UNIV WESTERN ONT, 73- Concurrent Pos: Res fel, NIH training grant & Fulbright travel grant immunochem, Calif Inst Technol, 65-68. Mem: Am Asn Immunol; Can Soc Immunol. Res: Immunochemistry of reagin-mediated hypersensitivity. Mailing Add: Dept of Bact & Immunol Univ of Western Ont London ON Can

STREM, JOSEPH GEORGE, organic chemistry, physical chemistry, see 12th edition

STREM, MICHAEL EDWARD, b Pittsburgh, Pa, Apr 1, 36; m 67. ORGANIC CHEMISTRY. Educ: Brown Univ, AB, 58; Univ Pittsburgh, MS, 61, PhD(chem), 64. Prof Exp: PRES, STREM CHEM INC, 64- Mem: Am Chem Soc. Res: Organometallic chemistry, including its use in organic synthesis. Mailing Add: 150 Andover St Danvers MA 01923

STRENG, ALEX G, b Bachmut, Russia, Feb 25, 07; nat US; m 36; c 1. PHYSICAL CHEMISTRY. Educ: Chem Inst, Leningrad, ChE, 32; Chem Inst, Moscow, PhD(chem), 36. Prof Exp: Chief chemist & consult, Olschieferverwertung, Ger, 47-49; chief chemist & lab supvr, Vereinigte Farben & Lack-Fabriken, 48-50; res assoc & proj dir, Res Inst, Temple Univ, 50-69; RES ASSOC & PROJ DIR, GERMANTOWN LABS, INC, 69- Concurrent Pos: Abstractor, Chem Abstr, 57- Mem: AAAS; Am Ord Asn; Am Chem Soc; NY Acad Sci; Combustion Inst. Res: Analytical, cryogenic and high temperature chemistry; paints, fuels, industrial explosives, fire damp, high energy propellants, combustion and detonation studies; ozone, fluorine, fluorine compounds, noble gas compounds, synthesis and characterization of inorganic compounds. Mailing Add: 5229 N 16th St Philadelphia PA 19141

STRENG, WILLIAM HAROLD, b Milwaukee, Wis, Mar 6, 44; m 67; c 2. PHYSICAL CHEMISTRY. Educ: Carroll Col, Wis, BS, 66; Mich Technol Univ, MS, 68, PhD(phys chem), 71. Prof Exp: Res assoc theoret chem, Clark Univ, 72-73; SR CHEMIST, MERRELL-NAT LABS, RICHARDSON-MERRELL, INC, 73- Mem: Am Chem Soc; Am Pharmaceut Asn. Res: Elucidation of interactions in electrolyte solution from both theoretical and experimental considerations. Mailing Add: Merrell-Nat Labs 110 Amity Rd Cincinnati OH 45215

STRENGTH, DELPHIN RALPH, b Brewton, Ala, May 24, 25; m 46; c 4. BIOCHEMISTRY. Educ: Auburn Univ, BS, 48, MS, 50; Cornell Univ, PhD(biochem), 52. Prof Exp: Res assoc biochem, Cornell Univ, 52-53; instr, St Louis Univ, 53-54, sr instr, 54-56, from asst prof to assoc prof, 56-61; assoc prof animal sci, 61-65, PROF BIOCHEM & NUTRIT, AUBURN UNIV, 65- Mem: AAAS; Am Chem Soc; Am Soc Biol Chem; Soc Exp Biol & Med; Am Inst Nutrit. Res: Enzyme chemistry; nutrition; phospholipids; proteins. Mailing Add: Dept of Animal & Dairy Sci Auburn Univ Auburn AL 36830

STRENKOSKI, LEON FRANCIS, b Shamokin, Pa, Mar 26, 41; m 64; c 2. CLINICAL MICROBIOLOGY. Educ: King's Col, Pa, BS, 62; Cath Univ Am, PhD(microbiol), 68. Prof Exp: Asst biol, Cath Univ Am, 62-67; asst prof microbiol, Rensselaer Polytech Inst, 67-70; mem staff, 70-73, SUPVR AMES MICROBIOL PROD RES & DEVELOP, MILES LABS, INC, 73- Honors & Awards: Specialist Microbiologist, Am Acad Microbiol, 74. Mem: Am Soc Microbiol. Res: Bacterial metabolism and genetics; inorganic nitrogen assimilation in Hydrogenomonas eutropha, a facultative-autotrophic bacterium. Mailing Add: Miles Labs Inc 1127 Myrtle St Elkhart IN 46514

STRENZWILK, DENIS FRANK, b Rochester, NY, Oct 27, 40. SOLID STATE PHYSICS. Educ: Le Moyne Col, NY, BS, 62; Clarkson Col Technol, MS, 65, PhD(physics), 68. Prof Exp: RES PHYSICIST, US ARMY BALLISTIC RES LABS, 68- Concurrent Pos: Teacher, Exten Sch, Univ Del, Aberdeen Proving Ground, Md, 70. Mem: Am Asn Physics Teachers; Am Phys Soc. Res: Magnetism, effective field theory for yttrium iron garnet, lattice dynamics. Mailing Add: US Army Ballistics Res Labs Aberdeen Proving Ground MD 21005

STRETCH, ALLAN WESLEY, plant pathology, see 12th edition

STREU, HERBERT THOMAS, b Elizabeth, NJ, May 16, 27; m; c 1. ENTOMOLOGY, ZOOLOGY. Educ: Rutgers Univ, BS, 51, MS, 59, PhD(entom), 60. Prof Exp: Nematologist, Agr Res Serv, USDA, 60-61; assoc res prof, 61-70, RES PROF ENTOM, RUTGERS UNIV, 70- Mem: Fel AAAS; Entom Soc Am; Soc Nematol; Acarological Soc Am; Ecol Soc Am. Res: Ecology and control of arthropods in turfgrass; biology and control of insects, nematodes and other economic arthropod pests attacking ornamental crops. Mailing Add: Dept of Entom & Econ Zool Rutgers Univ New Brunswick NJ 08903

STREULI, CARL ARTHUR, b Bronxville, NY, May 7, 22; m 50; c 3. ANALYTICAL CHEMISTRY. Educ: Lehigh Univ, BS, 43; Cornell Univ, AM, 50, PhD(anal chem), 52. Prof Exp: Chemist, Foster D Snell, Inc, 43-44 & 46-47; res assoc, Cornell Univ, 52-53; res chemist, Co, 53-57, group leader, 57-63, res assoc, 63-69, GROUP LEADER, LEDERLE LABS, AM CYANAMID CO, 69- Concurrent Pos: Lectr, Univ Conn, Stamford, 63-66; fel, Purdue Univ, 66-67. Mem: Am Chem Soc. Res: Analytical chemistry in nonaqueous solvents; acid base theory; electroanalytical chemistry; gas-liquid and liquid-liquid chromatography. Mailing Add: 102 Ken Ct Stamford CT 06907

STRIANSE, SABBAT JOHN, b US, Nov 16, 13; m 42; c 3. PHARMACY. Educ: Long Island Univ, BS, 41. Prof Exp: Pharmacist, Walgreen Drug Co, NY, 41-42; res chemist, George W Luft Co, Inc, 45-48; sr res chemist, Warner-Hudnut, Inc, 48-50; tech dir, Sofskin Div, Vick Chem Co, NJ, 50-56; dir res, Shulton, Inc, 56-59; DIR RES, YRADLEY & CO LTD, 59-, EXEC DIR, CO, 65-, VPRES IN CHARGE RES, YARDLEY OF LONDON, INC, TOTOWA, 67-, CORP DIR, 69- Concurrent Pos: Mem, Textile Res Inst; pres, Int Fedn Socs Cosmetic Chem, 63-64; sci controller, Brit-Am Cosmetics Ltd, 71- Honors & Awards: Medal Award, Soc Cosmetic Chemists, 73. Mem: AAAS; Soc Cosmetic Chemists (pres, 57); Am Chem Soc; Am Pharmaceut Asn; NY Acad Sci. Res: Cosmetics and pharmaceuticals; emulsion and solubilization technology; surfactants general; polymer chemistry as applied to cosmetics; cosmetic manufacture and analysis. Mailing Add: 40 Westville Ave Caldwell NJ 07006

STRIBLEY, REXFORD CARL, b Kent, Ohio, Mar 12, 18; m 45; c 2. ORGANIC CHEMISTRY. Educ: Kent State Univ, BS, 39. Prof Exp: Chemist, 40-41, res chemist, Mason Lab, 45-50 & 52-55, chief res & develop, 55-70, TECH DIR NUTRIT DIV, WYETH LABS, INC DIV, AM HOME PROD CORP, 70- Concurrent Pos: US indust adv, Comt Food for Special Dietary Uses, UN Codex Alimentorius Comn, 72-74. Mem: Am Chem Soc; Am Oil Chem Soc; Am Dairy Sci Asn; Inst Food Technol. Res: Chemistry and development of infant formulas; infant nutrition; milk chemistry; dairy manufacturing technology and engineering; special dietary food products. Mailing Add: Wyeth Labs Inc Mason MI 48854

STRICHARTZ, ROBERT STEPHEN, b New York, NY, Oct 14, 43; m 68; c 1. MATHEMATICS. Educ: Dartmouth Col, BA, 63; Princeton Univ, MA, 65, PhD(math), 66. Prof Exp: NATO fel, Fac Sci, Orsay, France, 66-67; C L E Moore instr math, Mass Inst Technol, 67-69; asst prof, 69-71, ASSOC PROF MATH, CORNELL UNIV, 71- Mem: Am Math Soc. Res: Harmonic analysis and partial differential equations. Mailing Add: Dept of Math Cornell Univ Ithaca NY 14850

STRICK, ELLIS, b Pikeville, Ky, Mar 19, 21. GEOPHYSICS, OCEANOGRAPHY. Educ: Va Polytech Inst & State Univ, BS, 42; Purdue Univ, West Lafayette, PhD(theoret physics), 50. Prof Exp: Physicist radio eng, US Naval Res Lab, DC, 42-46; asst prof physics, Univ Wyo, 50-51; res assoc theoret seismol, Shell Explor & Prod Res Lab, Tex, 51-68; ASSOC PROF GEOPHYS, UNIV PITTSBURGH, 68- Concurrent Pos: Lectr physics, Univ Houston, 51-67; NSF grant, Univ Pittsburgh, 70-71. Mem: Soc Explor Geophysicists; Seismol Soc Am. Res: Anelastic wave propagation in solids at low frequencies. Mailing Add: Dept of Earth & Planetary Sci 506 Langley Hall Univ Pittsburgh 4200 Fifth Ave Pittsburgh PA 15213

STRICKBERGER, MONROE WOLF, b Brooklyn, NY, July 3, 25; m 57; c 2. EVOLUTION, GENETICS. Educ: NY Univ, BA, 49; Columbia Univ, MA, 59, PhD(genetics), 62. Prof Exp: Res fel genetics, Univ Calif, Berkeley, 62-63; from asst prof to assoc prof biol, St Louis Univ, 63-68; assoc prof, 68-71, PROF BIOL, UNIV MO-ST LOUIS, 71- Concurrent Pos: NIH res grant, 63-69. Mem: AAAS; Genetics Soc Am; Am Eugenics Soc; Am Genetic Asn; Am Soc Naturalists. Res: Evolution of fitness in Drosophila populations; induction of sexual isolation. Mailing Add: Dept of Biol Univ Mo 8001 Natural Bridge Rd St Louis MO 63121

STRICKER, EDWARD MICHAEL, b New York, NY, May 23, 41; m 64; c 2. NEUROPSYCHOLOGY. Educ: Univ Chicago, BS, 60, MS, 61; Yale Univ, PhD(psychol), 65. Prof Exp: Fel, Med Ctr, Univ Colo, 65-66 & Inst Neurol Sci, Med Sch, Univ Pa, 66-67; from asst prof to assoc prof psychol, McMaster Univ, 67-71; ASSOC PROF PSYCHOL & BIOL, UNIV PITTSBURGH, 71- Concurrent Pos: Consult ed, J Comp & Physiol Psychol, 72- Res: Physiological and behavioral mechanisms that maintain water and electrolyte balance, body temperature and energy metabolism; the neurochemical basis for recovery of function following brain damage; central controls of motivated behavior. Mailing Add: Dept of Psychol Univ of Pittsburgh Pittsburgh PA 15260

STRICKHOLM, ALFRED, b New York, NY, July 3, 28; m 52; c 3. PHYSIOLOGY, BIOPHYSICS. Educ: Univ Mich, BS, 51; Univ Minn, MS, 56; Univ Chicago, PhD(physiol), 60. Prof Exp: Fel biophys, Physiol Inst, Univ Uppsala, 60-61; asst prof physiol, Sch Med, Univ Calif, San Francisco, 61-66; assoc prof, 66-72, PROF ANAT & PHYSIOL, CTR NEURAL SCI, IND UNIV, BLOOMINGTON, 72- Concurrent Pos: USPHS grant, 62- Mem: AAAS; Am Physiol Soc; Soc Neurosci; Soc Gen Physiol; Biophys Soc. Res: Biophysics of the cell; contraction coupling in muscle; permeability, active transport, and excitation; structure and function of cell membranes. Mailing Add: Dept of Anat-Physiol Ind Univ Bloomington IN 47401

STRICKLAND, BARNEY RALPH, organic chemistry, see 12th edition

STRICKLAND, ERASMUS HARDIN, b Spartanburg, SC, May 18, 36; m 66; c 1. BIOPHYSICS. Educ: Pa State Univ, BS, 58, MS, 59, PhD(biophys), 61. Prof Exp: Chief phys chem sect, US Army Med Res Lab, Ft Knox, 61-63; from asst prof to assoc prof biophys, 63-70, ASSOC RES BIOPHYSICIST, RADIATION BIOL LAB, UNIV CALIF, LOS ANGELES, 69- Mem: AAAS; Biophys Soc. Res: Circular dichroism and absorption spectroscopy of biological molecules. Mailing Add: Lab Nuclear Med & Radiation Biol Univ of Calif Los Angeles CA 90024

STRICKLAND, JAMES SHIVE, b Harrisburg, Pa, Nov 18, 29; m 55; c 3. EXPERIMENTAL PHYSICS. Educ: Franklin & Marshall Col, BS, 51; Mass Inst Technol, PhD(physics), 57. Prof Exp: staff physicist, Phys Sci Study Comt, Mass Inst Technol, 57-58; staff physicist, Educ Develop Ctr, Inc, Mass, 58-72; vis scientist, Mass Inst Technol, 72-73; PROF PHYSICS & CHMN DEPT, GRAND VALLEY STATE COL, 73- Mem: AAAS; Am Asn Physics Teachers; Am Phys Soc. Res: Development of new materials for science education. Mailing Add: Dept of Physics Grand Valley State Col Allendale MI 49401

STRICKLAND, JOHN CLAIBORNE, b Petersburg, Va, Feb 3, 15. BIOLOGY. Educ: Univ Richmond, BA, 37; Univ Va, MA, 39, PhD(biol), 43. Prof Exp: Actg asst prof biol, Col William & Mary, 43-46; from asst prof to assoc prof, 46-58, chmn dept, 57-64, PROF BIOL, UNIV RICHMOND, 58- Concurrent Pos: Res assoc, Va Inst Sci Res, Va Acad Sci, 47-48. Mem: AAAS; Bot Soc Am; Am Soc Plant Taxonomists; Phycol Soc Am. Res: Taxonomy, culture and physiology of Myxophyceae; plant anatomy; botany. Mailing Add: Dept of Biol Univ of Richmond Richmond VA 23173

STRICKLAND, JOHN WILLIS, b Wichita, Kans, Mar 23, 25; m 47; c 4. PETROLEUM GEOLOGY. Educ: Univ Okla, BS, 46. Prof Exp: Geologist, Skelly Oil Co, 47-50; res geologist, 51-55, div geologist, 55-61, explor mgr, Ireland, 61-64, coordr explor res, 64-66, dir adv geol, 66-67, chief geologist, 67-76, MGR, AFRICA & LATIN AM, CONTINENTAL OIL CO, 76- Mem: Am Asn Petrol Geol. Res: Petroleum geology of world; factors controlling generation and distribution of hydrocarbons. Mailing Add: Continental Oil Co PO Box 2197 Houston TX 77001

STRICKLAND, KENNETH PERCY, b Loverna, Sask, Aug 19, 27; m 48; c 3. BIOCHEMISTRY. Educ: Univ Western Ont, BSc, 49, MSc, 50, PhD(biochem), 53. Prof Exp: Nat Res Coun Can fel chem path, Guy's Hosp Med Sch, Univ London, 53-55; from asst prof to assoc prof, 55-66, PROF BIOCHEM, UNIV WESTERN ONT, 66- Concurrent Pos: Lederle med fac award, 55-58; res assoc, Med Res Coun Can, 58- Mem: AAAS; Am Soc Biol Chemists; Can Biochem Soc; Can Physiol Soc; Chem Inst Can. Res: Biochemistry of central nervous system, especially biosynthesis of lipid components; biochemistry of muscle, especially relating to degenerating diseases. Mailing Add: Dept of Biochem Univ of Western Ont London ON Can

STRICKLAND, WALTER NICHOLAS, b Shamva, Rhodesia, Feb 15, 30; US citizen. GENETICS. Educ: Univ Natal, BSc, 51; Glasgow Univ, PhD(genetics), 57. Prof Exp: Asst lectr genetics, Glasgow Univ, 56-57; res assoc, Stanford Univ, 57-61; res assoc, Dartmouth Col, 61-64; from asst prof to assoc prof, Univ Utah, 64-69; SR LECTR BIOCHEM, UNIV CAPE TOWN, 69- Res: Genetic control of protein structure and function; structure and function of histones and basic proteins. Mailing Add: Univ of Cape Town Dept Biochem Cape Province P/BAG Rondebosch Republic of South Africa

STRICKLAND, WILLIAM ALEXANDER, JR, b Doyer, Ark, July 25, 23; m 46; c 3. PHARMACY. Educ: Univ Tenn, BS, 44; Univ Wis, MS, 52, PhD, 55. Prof Exp: From assoc prof to prof pharm, Sch Pharm, Univ Ark, 55-67; dean, Sch Pharm, 67-73, PROF PHARM & MED, UNIV MO-KANSAS CITY, 74- Concurrent Pos: Consult, Parke, Davis & Co, 58-; dir, Outreach, Western Mo Area Health Educ Ctr, 74- Mem: AAAS; Am Pharmaceut Asn; Am Asn Cols Pharm. Res: Physical pharmacy; tableting and aerosol technology; physical reactions of drug molecules as in complex formation and hydration. Mailing Add: 9959 Goddard Overland Park KS 66214

STRICKLER, DWIGHT JOHNSTON, b Newell, Pa, Jan 5, 06; m 32; c 2. GENETICS. Educ: Olivet Nazarene Col, AB, 29; Mich State Univ, MS, 40. Hon Degrees: DSc, Olivet Nazarene Col, 71. Prof Exp: Chmn dept biol sci, 44-70, dir visual aids, 40-51, PROF BIOL SCI, OLIVET NAZARENE COL, 30-, HON CHMN DEPT, 70- Concurrent Pos: Prin, High Sch, Ill, 33-47. Res: Postnatal development of the mouse. Mailing Add: Dept of Biol Olivet Nazarene Col Kankakee IL 60901

STRICKLER, HERBERT SHARPLESS, b Columbia, Pa, Aug 18, 08; m 42; c 2. CHEMISTRY. Educ: Carnegie Inst Technol, BS, 30, MS, 32, ScD(phys chem), 36. Prof Exp: Instr, Carnegie Inst Technol, 35-36; from instr to asst prof phys chem, Sch Med, Univ Pittsburgh, 36-46; asst prof chem, Albright Col, 46-47; res chemist, Res Labs, Westinghouse Elec Corp, 47-50; res chemist, Singer Inst, Allegheny Gen Hosp, 50-75; RETIRED. Concurrent Pos: Res chemist, Elizabeth Steel Magee Hosp, 36-46; adj assoc prof, Pa State Univ, 64- Mem: AAAS; Am Chem Soc; Am Asn Clin Chem. Res: Peat bacteriology; coal ash fusibility; thermodynamics of metallic systems; determination of blood iodine; hormone derivations in urine; technology of powders; phosphorous metabolism in cancer; automation of clinical chemistry analyses. Mailing Add: 180 W Hutchinson Ave Pittsburgh PA 15218

STRICKLER, PAUL DONOVAN, b Knox City, Mo, July 5, 28; m 48; c 2. ORGANIC CHEMISTRY. Educ: Culver-Stockton Col, BA, 49; Univ Mo, MA, 52. Prof Exp: Res chemist, Spencer Chem Co, 51-59, staff specialist, 59-62, group leader, 62-65; sect supvr, 65-67, dir agr chem div, 67-69, div dir res, Petrol Prod Dept, New Prod Div, 69-71, dir, Prod Technol Div, 71-75, DIR, INDUST PROD DIV, GULF RES & DEVELOP CO, 75- Res: Am Chem Soc. Res: Organic synthesis in agricultural pesticides, especially carbamates and quaternary ammonium compounds; synthesis of hydroxylamine; utilization of organic derivatives. Mailing Add: Gulf Res & Dev Co Indust Prod Div PO Drawer 2038 Pittsburgh PA 15230

STRICKLER, PAUL MEREDITH, physics, mathematics, see 12th edition

STRICKLER, STEWART JEFFERY, b Mussoorie, India, July 12, 34; US citizen; m 59; c 2. PHYSICAL CHEMISTRY. Educ: Col Wooster, BA, 56; Fla State Univ, PhD(phys chem), 61. Prof Exp: Chemist, Radiation Lab, Univ Calif, 61; res assoc, Rice Univ, 61-62, lectr chem, 62-63; from asst prof to assoc prof, 63-73, PROF CHEM, UNIV COLO, BOULDER, 73-, CHMN DEPT, 74- Mem: Am Chem Soc; Am Phys Soc. Res: Molecular spectroscopy; photochemistry; quantum chemistry. Mailing Add: Dept of Chem Univ of Colo Boulder CO 80302

STRICKLER, THOMAS DAVID, b Ferozepur, India, Nov 11, 22; US citizen; m 48; c 4. ATOMIC PHYSICS. Educ: Col Wooster, BA, 47; Yale Univ, MS, 48, PhD(physics), 53. Prof Exp: Instr physics, Yale Univ, 52-53; from asst prof to assoc prof, 53-61, CHARLES F KETTERING PROF PHYSICS & CHMN DEPT, BEREA COL, 61- Concurrent Pos: NSF faculty fel, 60-61; consult, NSF Physics Inst, Chandigarh, India, 66, Gauhati Univ, India, 67 & Calcutta, India, 68; Fulbright lectr, Comt Int Exchange Persons, 73-74. Mem: AAAS; Am Asn Physics Teachers. Res: Neutron and gamma ray scattering; health physics; gaseous electronics. Mailing Add: 114 Van Winkle Grove Berea KY 40403

STRICKLIN, BUCK, b Clovis, NMex, Dec 30, 22; m 42; c 1. ORGANIC CHEMISTRY. Educ: Tex Tech Col, BS, 48; Univ Colo, PhD(org chem), 52. Prof Exp: Asst, Univ Colo, 48-52; res mgr, 52-69, TECH DIR, PAPER PROD DIV, MINN MINING & MFG CO, 69- Mem: Am Chem Soc; Tech Asn Pulp & Paper Indust. Res: Fluorocarbons; fluoroethers; chlorination; photochemistry; photoconductivity; polymers. Mailing Add: 454 Hilltop Ave St Paul MN 55113

STRICKLIN, FRED LEE, JR, geology, see 12th edition

STRICKLING, EDWARD, b Woodsfield, Ohio, Oct 20, 16; m 41; c 3. SOILS. Educ: Ohio State Univ, BS, 37, PhD, 49. Prof Exp: Instr, High Sch, 37-42; PROF SOILS, UNIV MD, COLLEGE PARK, 50- Mem: Soil Sci Soc Am; Am Soc Agron. Res: Soil physics, especially soil structure and evapotranspiration. Mailing Add: 6904 Calverton Dr Hyattsville MD 20782

STRICKMEIER, HENRY BERNARD, JR, b Galveston, Tex, Sept 28, 40. MATHEMATICS EDUCATION. Educ: Tex Lutheran Col, BS, 62; Univ Tex, Austin, MA, 67, PhD(math educ), 70. Prof Exp: Teacher high sch, Tex, 62-65; ASSOC PROF MATH, CALIF POLYTECH STATE UNIV, SAN LUIS OBISPO, 70- Mem: Math Asn Am; Am Educ Res Asn. Res: Evaluation of mathematics curricula; analysis of mathematics teaching. Mailing Add: Dept of Math Calif Polytech State Univ San Luis Obispo CA 93401

STRICKON, ARNOLD, b New York, NY, July 19, 30; m 58. CULTURAL ANTHROPOLOGY. Educ: City Col New York, BA, 52; Columbia Univ, MA, S4, PhD(anthrop), 60. Prof Exp: Asst prof social anthrop, Univ Nev, 60-61; asst prof anthrop, Brandeis Univ, 61-65; assoc prof anthrop, 65-72, PROF ANTHROP & INTEGRATED LIB STUDIES, UNIV WIS-MADISON, 72- Mem: AAAS; Am Anthrop Asn; Am Ethnol Soc; Soc Appl Anthrop; Latin-Am Studies Asn. Res: Cultural anthropology and application of its methods and theories to Latin America, especially modern communities in South America. Mailing Add: Dept of Anthrop Univ of Wis-Madison Madison WI 53706

STRICOS, DAVID PETER, b Pittsfield, Mass, Jan 12, 32; m 55; c 4. ANALYTICAL CHEMISTRY. Educ: Union Col, NY, BS, 53, MS, 60; Rensselaer Polytech Inst, PhD(chem), 63. Prof Exp: Anal chemist, State Div Standards & Purchase, NY, 55-56; anal chemist, 55-60, RADIOCHEMIST, KNOLLS ATOMIC POWER LAB, 63- Concurrent Pos: Prin nuclear power analyst, NY State Dept Pub Serv, 73- Mem: Am Chem Soc; fel Am Inst Chem; Am Nuclear Soc. Res: Analysis of energy research and development; polarography; rheology of polymers; radiochemistry; nuclear measurements. Mailing Add: 14 Oak Tree Ln Schenectady NY 12309

STRIDDE, GEORGE ERICH, organic chemistry, see 12th edition

STRIDER, DAVID LEWIS, b Salisbury, NC, Feb 12, 29; m 54; c 4. PLANT PATHOLOGY. Educ: NC State Col, MS, 57, PhD(plant path), 59. Prof Exp: Res asst prof, 59-64, assoc prof, 64-70, PROF PLANT PATH, NC STATE UNIV, 70- Concurrent Pos: Mem, NC State Univ-US AID Mission, Peru, 70-71. Mem: Am Phytopath Soc. Res: Control of horticultural crops diseases; disease control of greenhouse floral crops. Mailing Add: Dept of Plant Path NC State Univ Raleigh NC 27607

STRIEDER, WILLIAM, b Erie, Pa, Jan 19, 38; m 67; c 2. PHYSICAL CHEMISTRY, CHEMICAL ENGINEERING. Educ: Pa State Univ, BS, 59; Case Inst Technol, PhD(phys chem), 63. Prof Exp: Res fel irreversible thermodyn, Free Univ Brussels, 63-65; res fel statist mech, Univ Minn, 65-66; asst prof eng sci, 66-70, ASSOC PROF CHEM ENG, UNIV NOTRE DAME, 70- Mem: Am Chem Soc; Am Phys Soc. Res: Molecular theory of transport processes; flow through random porous media; transport phenomena; thermodynamics. Mailing Add: Dept of Chem Eng Univ of Notre Dame Notre Dame IN 46556

STRIER, MURRAY PAUL, b New York, NY, Oct 19, 23l m 55; c 3. PHYSICAL CHEMISTRY, ORGANIC CHEMISTRY. Educ: City Col New York, BChE, 44; Emory Univ, MS, 47; Univ Ky, PhD(chem), 52; Am Inst Chemists, cert. Prof Exp: Asst & instr, Univ Ky, 48-50; res chemist & proj leader, Reaction Motors, Inc, 52-56; sr chemist & head polymers sect, Air Reduction Co, Inc, 56-58; chief chemist, Fulton-Irgon Corp, 58-59; group leader fiber res, Rayonier, Inc, 59-60, suprvr develop res, 60-61; res chemist, T A Edison Res Lab, 61-64; sr res scientist, Douglas Aircraft Co, Inc, 64-67, chief fuel cell & battery res sect, 67-69; res assoc, Hooker Res Ctr, 69-71; prin chem engr, Cornell Aeronaut Lab, 71-72; CHEMIST, ENVIRON PROTECTION AGENCY, 72- Concurrent Pos: Consult, NSF, 73-75. Mem: AAAS; Am Chem Soc; Am Inst Chemists; Am Soc Testing & Mat; Electrochem Soc. Res: Organic polarography; physical chemistry of rocket propellants; physical properties of organic coatings, plastics and fibers; viscose chemistry; fuel cells and batteries; environmental science; industrial water pollution. Mailing Add: 10 Surry Ct Rockville MD 20850

STRIETELMEIER, JOHN HENRY, b Columbus, Ind, Feb 9, 20; m 43; c 3. CULTURAL GEOGRAPHY, GEOGRAPHY OF GREAT BRITAIN. Educ: Valparaiso Univ, AB, 42; Northwestern Univ, MA, 47. Hon Degrees: LittD, Concordia Sem, Mo, 63. Prof Exp: From instr to assoc prof, 43-67, PROF GEOG, VALPARAISO UNIV, 67- Concurrent Pos: Managing ed, Cresset, Valparaiso Univ, 49-69, univ ed, 58-67; chmn comt publ, Nat Coun Geog Educ, 50-53; Davidsmeyer lectr, Concordia Theol Sem, Ill, 61; consult, Southeastern Dist, Lutheran Church-Mo Synod, 65. Mem: Asn Am Geogrs. Res: The geography of religions. Mailing Add: Dept of Geog Valparaiso Univ Valparaiso IN 46383

STRIETER, GREDERICK JOHN, b Davenport, Iowa, Sept 14, 34; m 57; c 2. PHYSICAL CHEMISTRY. Educ: Augustana Col, Ill, AB, 56; Univ Calif, Berkeley, PhD, 60. Prof Exp: Asst chem, Univ Calif, 56-57; asst crystallog, Lawrence Radiation Lab, Univ Calif, 57-59; mem tech staff, 59-75, CIRCUITS DEVELOP PILOT LINE MGR, TEX INSTRUMENTS, INC, 75- Mem: Am Crystallog Asn; Electrochem Soc (treas, 73-); Inst Elec & Electronic Engr. Res: Semiconductor device process technology; impurity diffusion in semiconductors; ion implantation of impurities in semiconductors; electron beam pattern definition. Mailing Add: 7814 Fallmeadow Ln Dallas TX 75240

STRIFFLER, DAVID FRANK, b Pontiac, Mich, Oct 24, 22; m 49; c 2. PUBLIC HEALTH, DENTISTRY. Educ: Univ Mich, DDS, 47, MPH, 51; Am Bd Dent Pub Health, dipl, 55. Prof Exp: Consult dent, Dearborn Pub Schs, Mich, 50-51, dir sch health, 51-53; dir div dent health, State Dept Pub Health NMex, 53-61; assoc prof pub health, Sch Pub Health & assoc prof dent, Sch Dent, 61-65, chmn dept community dent, Sch Dent, 63-67, PROF PUB HEALTH DENT, SCH PUB HEALTH & PROF DENT, SCH DENT, UNIV MICH, ANN ARBOR, 65-, DIR PROG DENT PUB HEALTH, SCH PUB HEALTH, 62- Concurrent Pos: Mem, Pub Health Res Study Sect, NIH, 59-60, Health Serv Res Study Sect, 60-62, Dis Control Study Sect, 64-65 & Prev Med & Dent Rev Comt, 70-74; mem, Nat Adv Comt Pub Health Training, USPHS, 60-61 & Health Serv Res Training Comt, 66-68; ed, J Pub Health Dent, 75- Mem: Sigma Xi; Am Dent Asn; fel Am Pub Health Asn; fel Am Col Dent; Fel Int Col Dent. Res: Fluoridation; epidemiology of periodontal diseases; delivery of dental health services. Mailing Add: 2217 Vinewood Blvd Ann Arbor MI 48104

STRIFFLER, WILLIAM D, b Oberlin, Ohio, July 10, 29; m 56; c 4. FOREST HYDROLOGY. Educ: Mich State Univ, BS & BSF, 52; Univ Mich, MF, 57, PhD(forest hyrdol), 63. Prof Exp: Res forester & proj leader groundwater hydrol & steambank erosion, Lake States Forest Exp Sta, Mich, 57-63; hydrologist, Stripmined Areas Restoration Res Proj, Northeastern Forest Exp Sta, Ky, 64-66; from asst prof to assoc prof watershed mgt, 66-75, PROF EARTH RESOURCES, COLO STATE UNIV, 75- Mem: Am Geophys Union; Am Water Resource Asn. Res: Wildland hydrology; land use hydrology; erosion and sedimentation processes; water quality; grassland hydrology; instrumentation. Mailing Add: Dept of Earth Resources Colo State Univ Ft Collins CO 80521

STRIGHT, I LEONARD, b Mercer Co, Pa, May 7, 16; m 40; c 3. MATHEMATICS. Educ: Allegheny Col, AB, 35, AM, 38; Western Reserve Univ, PhD(math), 46. Prof Exp: Instr, High Schs, Pa, 35-42; assoc prof math, Baldwin-Wallace Col, 42-46; prof & head dept, Northern Mich Univ, 46-47; prof, Indiana Univ Pa, 47-57, dean grad sch, 57-71; ACAD VPRES, OHIO NORTHERN UNIV, 71- Concurrent Pos: Fel, Univ Chicago, 63. Mem: Math Asn Am. Res: Mathematics education; applied statistics; tests and measurements. Mailing Add: Off Acad VPres Ohio Northern Univ Ada OH 45810

STRIGHT, PAUL LEONARD, b St Paul, Minn, May 12, 30; m 60; c 2. ORGANIC CHEMISTRY. Educ: Grinnell Col, BA, 51; Univ Minn, Minneapolis, PhD(org chem), 56. Prof Exp: Res chemist, Esso Res & Eng Co, 56-59; res chemist, Allied Chem Corp, 59-65, res supvr, 65-68, res assoc, 68-70; asst prof org chem, Univ Minn, Morris, 70-71; assoc scientist neurochem, Univ Minn, Minneapolis, 71-73; INSTR CHEM, LAKE MICH COL, 73- Mem: Am Chem Soc. Mailing Add: Lake Mich Col 2755 E Napier Ave Benton Harbor MI 49022

STRIKE, DONALD PETER, b Mt Carmel, Pa, Oct 24, 36; m 72. PHARMACEUTICAL CHEMISTRY. Educ: Philadelphia Col Pharm & Sci, BS, 58; Iowa State Univ, MS, 61, PhD(org chem), 63. Prof Exp: NIH fel, Univ Southampton, 63-64; res chemist, 65-69, RES GROUP LEADER, WYETH LABS, AM HOME PROD CORP, 69- Mem: Am Chem Soc. Res: Natural products; steroids; prostaglandins. Mailing Add: Wyeth Labs Res & Develop Lancaster Pike & Morehall Rd Radnor PA 19088

STRILKO, PETER SHEPARD, physical organic chemistry, see 12th edition

STRIMLING, WALTER EUGENE, b Minneapolis, Minn, Jan 6, 26; m 57; c 3. MATHEMATICS, ELECTRICAL ENGINEERING. Educ: Univ Minn, BPhys & MA, 45, PhD(math), 53. Prof Exp: Instr math & educ, Col St Catherine, 45-46; asst math, Univ Minn, 49-53; engr, Raytheon Co, 53-55; PRES, US DYNAMICS, 55- Mem: Am Math Soc; Am Phys Soc; Math Asn Am; Inst Elec & Electronics Engrs. Res: Theoretical physics; chemistry. Mailing Add: 63 Westcliff Rd Weston MA 02193

STRIMPLE, HARRELL LEROY, b Yates Center, Kans, Jan 7, 12; m 34; c 1. INVERTEBRATE PALEONTOLOGY, STRATIGRAPHY. Prof Exp: Mem staff, Phillips Petrol Co, 32-59; curator, Geol Enterprises, 60-61; consult paleontologist, 61-62; RES ASSOC GEOL & CURATOR, UNIV IOWA, 62- Mem: Paleont Res Inst; fel Geol Soc Am; Paleont Soc. Res: Paleozoic Echinodermata; stratigraphy. Mailing Add: Dept of Geol Univ of Iowa Iowa City IA 52242

STRINGALL, ROBERT WILLIAM, b San Francisco, Calif, Dec 12, 33; c 2. MATHEMATICS EDUCATION. Educ: San Jose State Col, BA, 59; Univ Wash, MS, 63, PhD(math), 65. Prof Exp: ASSOC PROF MATH, UNIV CALIF, DAVIS, 65- Concurrent Pos: Consult, Elem & Sec Educ Act Title III, 67-68, Proj Sem, 70- Res: Algebra. Mailing Add: Dept of Math Univ of Calif Davis CA 95616

STRINGAM, ELWOOD WILLIAMS, b Alberta, Can, Dec 10, 17; m 44; c 6. ANIMAL SCIENCE, AGRICULTURE. Educ: Univ Alta, BSc, 40, MSc, 42; Univ Minn, PhD(agr), 48. Prof Exp: Asst, Dom Range Exp Sta, Alta, 40; fieldman, Livestock Prod Serv, 41-42; instr animal husb, Univ Minn, 46-48; assoc prof animal sci, Univ Man, 48-51; prof animal husb, Ont Agr Col, 51-54; head dept animal sci, 54-73, PROF ANIMAL SCI, UNIV MAN, 54- Concurrent Pos: Mem, Nat Animal Breeding Comt, Can, 58-63; adv comt, Western Vet Col, 66-; dir nat adv comt for agr, World's Fair, 67; mem, Nat Genetic Adv Comt Cattle Importations, 69- Mem: AAAS; Am Soc Animal Sci; Am Genetic Asn; Can Soc Animal Prod; fel Agr Inst Can (vpres, 62-63, pres, 66-67). Res: Agricultural education; animal genetics and physiology; farm animal production and management; beef cattle production. Mailing Add: Dept of Animal Sci Univ of Man Winnipeg MB Can

STRINGAM, GARY RICE, b Cardston, Alta, May 24, 37; m 61; c 6. CYTOGENETICS. Educ: Brigham Young Univ, BS, 61; Univ Minn, MS, 65, PhD(genetics), 66. Prof Exp: Res asst genetics, Univ Minn, 61-63; fel radiation physiol, Univ Hawaii, 66-67; RES SCIENTIST CYTOGENETICS, CAN DEPT AGR, 67- Mem: Genetics Soc Can; Am Genetic Asn; Agr Inst Can. Res: Tissue culture; plant breeding. Mailing Add: Agr Can Res Sta 107 Science Crescent Saskatoon SK Can

STRINGFELLOW, DALE ALAN, b Ogden, Utah, Sept 13, 44; m 66; c 2. VIROLOGY, IMMUNOBIOLOGY. Educ: Univ Utah, BS, 67, MS, 70, PhD(microbiol), 72. Prof Exp: NIH fel & instr microbiol, Univ Utah, 72-73; RES SCIENTIST VIROL, UPJOHN CO, 73- Mem: Am Soc Microbiol. Res: Antiviral agents; interrelationship between host defense systems and virus infection; pathogenesis of virus infection; mechanisms modulating nonspecific immunity; viral ecology. Mailing Add: Exp Biol Dept Upjohn Co Kalamazoo MI 49001

STRINGFELLOW, FRANK, b Cheriton, Va, Oct 27, 40; m 68. ZOOLOGY. Educ: St Louis Univ, BS, 62; Drake Univ, MA, 64; Univ SC, PhD(biol), 67. Prof Exp: Asst gen biol, Drake Univ, 63-64; instr anat & physiol, Univ SC, 64-65; ZOOLOGIST, ANIMAL PARASITOL INST, AGR RES CTR, US DEPT AGR, 67- Mem: Am Micros Soc; Am Soc Parasitol; Sigma Xi. Res: Pathobiology of gastrointestinal parasites of cattle; mechanisms of tissue invasion of Entamoeba histolytica; analysis of taxonomic characters; histochemistry of nematodes; special histological staining techniques; physiology and biochemistry of parasites. Mailing Add: Animal Parasitol Inst Agr Res Ctr USDA Beltsville MD 20705

STRINGFELLOW, GERALD B, b Salt Lake City, Utah, Apr 26, 42; m 62; c 3. MATERIALS SCIENCE, SEMICONDUCTORS. Educ: Univ Utah, BS, 64; Stanford Univ, MS, 66, PhD(mat sci), 67. Prof Exp: Mem tech staff, Solid State Physics Lab, 67-71, PROJ MGR, HEWLETT PACKARD LABS, 71- Mem: Am Phys Soc; Electrochem Soc; Am Inst Mining, Metallurgical & Petroleum Engrs. Res: Electrical and optical properties of alloys between III-V compound semiconductors; luminescence and photoconductivity of II-VI and III-V compounds; crystal growth and thermodynamics in ternary III-V systems. Mailing Add: Solid State Physics Lab Hewlett Packard Labs 1501 Page Mill Rd Palo Alto CA 94304

STRINGFIELD, VICTOR TIMOTHY, b Franklinton, La, Sept 10, 02; m 29; c 2. HYDROGEOLOGY. Educ: La State Univ, BS, 25; Wash Univ, MS, 27. Prof Exp: Asst geol & geog, Wash Univ, 25-27; instr geol, Okla Agr & Mech Col, 27-28; asst prof, NMex Sch Mines, 28-30; asst geologist, US Geol Surv, 30-36, assoc geologist, 36-39, geologist in charge ground water invests, Southeastern States, 39-42, sr geologist in charge ground water invests, Eastern States, 42-47, prin geologist & chief sect ground water geol, 47-57, chief sect radiohydrol, 57-60, staff geol specialist, 60-75; RETIRED. Concurrent Pos: Geologist, NMex Bur Mines, 28-30; consult hydrol, Food & Agr Orgn, UN, Jamaica, WI, 65- Mem: Fel Geol Soc Am; Soc Econ Geol; Am Asn Petrol Geol; Am Geophys Union. Res: Ground water geology; radiohydrology; ground water hydrology; radioisotopes in soils and water; hydrogeology of carbonate terranes. Mailing Add: 4208 50th St NW Washington DC 20016

STRINGHAM, REED MILLINGTON, JR, b Salt Lake City, Utah. PHYSIOLOGY, ORAL BIOLOGY. Educ: Northwestern Univ, Evanston, DDS, 58; Univ Utah, BS, 64, PhD(molecular & genetic biol), 68. Prof Exp: Nat Inst Dent Res fel, 65-68, RES ASSOC PLASTIC SURG, MED SCH, UNIV UTAH, 68-; PROF OCCUP HEALTH & ZOOL & DEAN SCH ALLIED HEALTH SCI, WEBER STATE COL, 69- Concurrent Pos: Resource person, Intermountain Regional Med Prog, 69, consult, Oral Cancer Screening Proj, 71- Mem: Am Dent Asn. Res: Salivary gland physiology; health manpower. Mailing Add: Sch of Allied Health Sci Weber State Col Ogden UT 84403

STRITAR, JEFFREY ALLAN, inorganic chemistry, see 12th edition

STRITTMATTER, CORNELIUS FREDERICK, b Philadelphia, Pa, Nov 16, 26; m 55; c 1. BIOCHEMISTRY. Educ: Juniata Col, BS, 47; Harvard Univ, PhD(biol chem), 52. Prof Exp: Instr biol chem, Harvard Med Sch, 52-54, assoc, 55-58, asst prof, 58-61; ODUS M MULL PROF BIOCHEM & CHMN DEPT, BOWMAN GRAY SCH MED, 61- Concurrent Pos: USPHS res fel, Oxford Univ, 54-55; USPHS sr res fel, Harvard Med Sch, 61; consult, New Eng Deaconess Hosp, 58-61; mem fel comt, NIH, 67-70 & 74; consult, NC Alcoholism Res Auth, 74-76. Mem: Am Chem Soc; Am Soc Biol Chemists; Soc Develop Biol; Soc Exp Biol & Med; Am Soc Zool. Res: Enzymic differentiation and control mechanisms during embryonic development and aging; characterization of electron transport systems; cellular control mechanisms in metabolism; comparative biochemistry; mechanisms in enzyme systems. Mailing Add: Dept of Biochem Bowman Gray Sch of Med Winston-Salem NC 27103

STRITTMATTER, PETER ALBERT, b Bexleyheath, Eng, Sept 12, 39; m 67; c 2. ASTRONOMY. Educ: St John's Col, Cambridge Univ, BA, 64, PhD(math), 66. Prof Exp: Mem staff astron, Inst Theoret Astron, Cambridge, 67-68; res assoc, Mt Stromlo & Siding Spring Observ, 69; mem staff, Inst Theoret Astron, Cambridge, 70; res physicist, Univ Calif, San Diego, 71; assoc prof, 71-73, PROF ASTRON, STEWARD OBSERV, UNIV ARIZ, 73- Concurrent Pos: Consult astron adv panel, NSF, 75. Mem: Am Astron Soc; Royal Astron Soc; Int Astron Union; Ger Astron Soc. Res: Quasistellar objects; Seyfert galaxies; radio sources; white dwarfs; novae. Mailing Add: Steward Observ Univ of Ariz Tucson AZ 85721

STRITTMATTER, PHILIPP, b Philadelphia, Pa, July 13, 28; m 56; c 2. BIOCHEMISTRY, ENZYMOLOGY. Educ: Harvard Univ, PhD, 54. Prof Exp: From instr to prof biochem, Wash Univ, 54-68; chmn dept, 68-74, PROF BIOCHEM, UNIV CONN, STORRS, 68- Res: Oxidative enzyme mechanisms. Mailing Add: Dept of Biochem Univ of Conn Storrs CT 06268

STRITZEL, JOSEPH ANDREW, b Cleveland, Ohio, June 11, 22; m 50; c 9. AGRONOMY. Educ: Iowa State Univ, BS, 49, MS, 53, PhD(soil fertil, prod econ), 58. Prof Exp: Exten soil fertil specialist, 50-63, PROF SOILS, IOWA STATE UNIV, 63- Mem: Am Soc Agron. Res: Soil fertility; production economics. Mailing Add: Dept of Soils Iowa State Univ Ames IA 50010

STRITZKE, JIMMY FRANKLIN, b South Coffeyville, Okla, Sept 9, 37; m 59; c 2. AGRONOMY. Educ: Okla State Univ, BS, 59, MS, 61; Univ Mo, PhD(field crops), 67. Prof Exp: Res scientist weed control, Agr Res Serv, USDA, 61-66; asst prof agron, SDak State Univ, 66-70; ASST PROF AGRON, OKLA STATE UNIV, 70- Mem: Am Soc Agron; Weed Sci Soc Am; Soc Range Mgt. Res: Weed and brush control in pasture and rangelands, herbicide residue and translocation. Mailing Add: Dept of Agron Okla State Univ Stillwater OK 74074

STRNISTE, GARY F, b Springfield, Mass, May 31, 44; m 75. MOLECULAR BIOLOGY, BIOLOGICAL CHEMISTRY. Educ: Univ Mass, BS, 66; Pa State Univ, MS, 69, PhD(biophys), 71. Prof Exp: Fel molecular radiobiol, Los Alamos Sci Lab, 71-73; fel molecular biol, City Univ New York, 73-74; MEM STAFF, LOS ALAMOS SCI LAB, 75- Mem: AAAS; Biophys Soc; Am Soc Microbiol. Res: Nucleic acid and protein interactions; structure and function of chromatin, mutagenic and carcinogenic induced modifications in chromatin; in vivo and in vitro repair of radiation induced changes in DNA. Mailing Add: Cellular & Molecular Biol Los Alamos Sci Lab Los Alamos NM 87545

STROBACH, DONALD ROY, b St Louis, Mo, Jan 10, 33; m 60; c 2. ORGANIC CHEMISTRY, BIOCHEMISTRY. Educ: Wash Univ, AB, 54, PhD(chem), 59. Prof Exp: NIH fels, 60-63; res chemist, Cent Res Dept, 63-69, RES CHEMIST, FREON PROD LAB, E I DU PONT DE NEMOURS & CO, INC, 70- Res: Carbohydrates; synthesis and structure determination; synthesis of oligonucleotides; aerosol technology and product development. Mailing Add: Freon Prod Lab E I du Pont de Nemours & Co Inc Wilmington DE 19898

STROBEL, CHARLES WILLIAM, b Bertrand, Mo, Oct 31, 29; m 57; c 4. POLYMER CHEMISTRY. Educ: Southeast Mo State Col, BS, 51; Univ Mo, MA, 54, PhD, 56. Prof Exp: Sr res chemist, Phillips Petrol Co, Tex, 56-69; SR RES CHEMIST, DeSOTO INC, DES PLAINES, 69- Res: Polymer synthesis; synthesis of elastomer, impact resistant plastics, sealants, electrocoat vehicles, vehicles for water base coatings Mailing Add: Des Plaines IL

STROBEL, DARRELL FRED, b Fargo, NDak, May 13, 42; m 68. PLANETARY SCIENCES. Educ: NDak State Univ, BS, 64; Harvard Univ, AM, 65, PhD(appl physics), 69. Prof Exp: Res assoc planetary astron, Kitt Peak Nat Observ, 68-70, asst physicist, 70-72, assoc physicist, 72-73; RES PHYSICIST, NAVAL RES LAB, 73- Mem: AAAS; Am Astron Soc; Am Geophys Union; Am Meteorol Soc. Res: Physics of planetary atmospheres; planetary aeronomy; ionospheric physics. Mailing Add: Code 7750 Naval Res Lab Washington DC 20375

STROBEL, GARY A, b Massillon, Ohio, Sept 23, 38; m 63; c 2. PLANT PATHOLOGY. Educ: Colo State Univ, BS, 60; Univ Calif, Davis, PhD(plant path), 63. Prof Exp: From asst prof to assoc prof, 63-72, PROF BOT, MONT STATE UNIV, 72- Concurrent Pos: Investr, coop state res serv, USDA, 64-68, 72-77; NSF res grants, fac assoc, 64-67, prin investr, 65-67 & 65-68; NIH career develop award, 69-74. Mem: AAAS; Am Phytopath Soc; Am Soc Plant Physiologists; Am Soc Microbiol; Am Soc Biol Chemists. Res: Plant disease physiology; biochemistry of fungi and bacteria that cause plant diseases; phytotoxic glycopeptides; metabolic regulation in diseased plants; nature and mechanism of action of host specific toxins; cyanide metabolism by microbes. Mailing Add: Dept of Bot & Microbiol Mont State Univ Bozeman MT 59715

STROBEL, GEORGE L, b Pratt, Kans, May 26, 37; m 57; c 2. THEORETICAL NUCLEAR PHYSICS, OPTICS. Educ: Kans State Univ, BS, 58; Univ Pittsburgh, MS, 61; Univ Southern Calif, PhD(physics), 65. Prof Exp: Scientist physics, Westinghouse Bettis Atomic Power Lab, 58-61 & Douglas Aircraft Co, 61-64; res assoc, Univ Southern Calif, 65 & Univ Calif, Davis, 65-67; ASSOC PROF PHYSICS, UNIV GA, 67- Concurrent Pos: Vis prof, Nuclear Res Ctr, Jülich, WGer, 71-72. Mem: Am Phys Soc. Mailing Add: Dept of Physics & Astron Univ of Ga Athens GA 30601

STROBEL, HOWARD AUSTIN, b Bremerton, Wash, Sept 5, 20; m 53; c 3. ANALYTICAL CHEMISTRY, PHYSICAL CHEMISTRY. Educ: State Col Wash, BS, 42; Brown Univ, PhD(phys chem), 47. Prof Exp: Jr res chemist, Manhattan Dist, Brown Univ, 43-45, res assoc, Univ, 47-48; from instr to assoc prof chem, 48-64, asst dean, Trinity Col, 56-64, assoc dean, 64-66, PROF CHEM, DUKE UNIV, 64-, DEAN BALDWIN RESIDENTIAL FEDN, 72- Mem: Am Chem Soc; The Chem Soc. Res: Solute-solvent interactions in mixed media; ion exchange phenomena; chemical instrumentation. Mailing Add: Dept of Chem Duke Univ Durham NC 27706

STROBEL, JAMES WALTER, b Steubenville, Ohio, Oct 31, 33; m 55; c 2. PLANT PATHOLOGY. Educ: Ohio Univ, AB, 55; Wash State Univ, PhD, 59. Prof Exp: Asst plant path, Wash State Univ, 55-59; from asst plant pathologist to assoc plant pathologist, Univ Fla, 59-68, prof plant path & plant pathologist, 68-74, chmn ornamental hort-agr exp stas, 70-74, dir agr & res ctr, Brandenton, 68-70; CHMN DEPT HORT SCI, NC STATE UNIV, 74- Mem: Am Phytopath Soc. Res: Etiology, epidemiology, and control of vegetable diseases, particularly control of verticillium wilt of tomato and strawberry by breeding for resistance. Mailing Add: Dept of Hort Sci NC State Univ Raleigh NC 27607

STROBEL, RUDOLF G K, b Kiessling, Ger, Feb 7, 27; m 58; c 4. BIOCHEMISTRY. Educ: Univ Munich, Dipl, 53, Dr rer nat, 58. Prof Exp: Asst, Max Planck Inst Protein & Leather Res, Ger, 56-58; RES BIOCHEMIST, RES DIV, PROCTER & GAMBLE CO, 58- Mem: AAAS; Am Chem Soc. Res: Histochemistry; histology; protein composition and structure; enzymology; natural products; microbiology; flavor research. Mailing Add: 7305 Thompson Rd Cincinnati OH 45247

STROBELL, JOHN DIXON, JR, b Newark, NJ, Dec 28, 17; m 49; c 2. GEOLOGY. Educ: Yale Univ, AB, 39, MS, 42, PhD(geol), 56. Prof Exp: GEOLOGIST, US GEOL SURV, 42- Mem: Geol Soc Am; AAAS. Res: Geology of deposits of copper, uranium and vanadium; geology of Colorado Plateau province. Mailing Add: US Geol Surv 601 E Cedar Ave Flagstaff AZ 86001

STROBOS, ROBERT JULIUS, b The Hague, Netherlands, July 2, 21; nat US; m 47, 67; c 5. NEUROLOGY. Educ: Univ Amsterdam, BS, 41, MD, 45; Am Bd Psychiat & Neurol, dipl, 55. Prof Exp: Asst resident neurol, Montefiore Hosp, New York, 50, resident, 52-53; asst neurol, Columbia Univ, 52-54; asst prof, Bowman Gray Sch Med, Wake Forest Univ, 54-60, assoc physiol & pharmacol, 56-60; assoc prof, 60-64, PROF NEUROL & CHMN DEPT, NEW YORK MED COL, 64- Concurrent Pos: Res fel neurosurg, Neurol Inst, New York, 51-52; asst neurol, Nat Hosp, London, Eng, 50; asst resident, NY Hosp, White Plains, 53-54. Mem: Am Epilepsy Soc; Asn Res Nerv & Ment Dis; AMA; Am Acad Neurol; Am Electroencephalog Soc. Res: Convulsive disorders; brain physiology and behavior; electroencephalography. Mailing Add: Dept of Neurol New York Med Col Flower & Fifth Ave Hosps New York NY 10029

STROCK, HERMAN, b New York, NY, Mar 14, 09; m 46; c 2. FOOD SCIENCE, FOOD TECHNOLOGY. Educ: NY Univ, BS, 31; Columbia Univ, MA, 33; Rutgers Univ, PhD(food sci), 63. Prof Exp: Res chemist, Sardic Food Prod Corp, NY, 33-40; commissary officer, US Navy, 40-46, officer chg res & develop foods & logistics, Navy Supply Corps Sch NJ, 46-47, liaison officer, Qm Food & Container Inst, Ill, 48-50, ship supply officer, Korean Campaign, 50-52, dir res food, Food Serv & Logistics, Bur Supplies & Acct, 52-55, dir supply admin, Naval Shipyard, Pa, 55-57, asst to chief inventory control, Defense Clothing & Textile Supply Agency, Pa, 57-58, comptroller, Navy Regional Accts Off, Va, 59-60, officer chg food sci & eng & logistics eng, Naval Supply Res & Develop Facil, NJ, 60-65, officer in chg, Cheatham Annex, 65-68; PROF HOTEL, RESTAURANT & INSTNL MGT & CHMN DEPT, NORTHERN VA COMMUNITY COL, 68- Mem: Inst Food Technologists. Res: Food science and food service systems, including administration, preparation, equipment and food controls. Mailing Add: Hotel Restaurant & Instnl Mgt Northern Va Community Col Annandale VA 22003

STRODT, WALTER CHARLES, b Brooklyn, NY, Mar 11, 15; m 41; c 2. MATHEMATICS. Educ: Columbia Univ, AB, 36, MA, 37, PhD(math), 39. Prof Exp: Asst math, Columbia Univ, 36-37; Fine instr, Princeton Univ, 38-40; from instr to prof math, Columbia Univ, 40-69; PROF MATH, ST LAWRENCE UNIV, 69-

Concurrent Pos: Asst prof, State Col Wash, 44; Nat Res Coun fel, Harvard Univ, 46-47; vis prof, Math Res Ctr, US Army, Univ Wis, 63-64. Mem: Am Math Soc; Soc Indust & Appl Math. Res: Differential equations. Mailing Add: Dept of Math St Lawrence Univ Canton NY 13617

STROEBEL, CHARLES FREDERICK, III, b Chicago, Ill, May 25, 36; m 59; c 2. PSYCHOPHYSIOLOGY, NEUROPHYSIOLOGY. Educ: Univ Minn, BA, 58, PhD, 61; Yale Univ, MD, 73. Prof Exp: Res asst biophys, Mayo Clin, 55-58; res asst, Psychiat Animal Res Labs, Univ Minn, 58-61, actg dir labs & lectr, Univ, 62; DIR LABS PSYCHOPHYSIOL, INST LIVING HOSP, 62-, DIR CLINS, 74- Concurrent Pos: Adj prof, Univ Hartford, 64-72, res prof, 72-; adj prof, Trinity Col, Conn, 72-; lectr psychiat, Sch Med, Yale Univ, 73- Mem: AAAS; Biophys Soc; Soc Psychophysiol Res; Asn Res Nerv & Ment Dis; Am Psychol Asn. Res: Physiologic and behavioral mechanisms of stress and drugs; biologic rhythms; biofeedback; biostatistics; neurophysiology of learning and emotion. Mailing Add: Inst of Living Hosp 400 Washington St Hartford CT 06106

STROEHLEIN, JACK LEE, b Cobden, Ill, Dec 22, 32; m 65. SOIL SCIENCE. Educ: Southern Ill Univ, BS, 54; Univ Wis, MS, 58, PhD(soils), 62. Prof Exp: Asst prof, 62-67, ASSOC PROF AGR CHEM & SOILS, UNIV ARIZ, 67-, ASSOC AGR CHEMIST, AGR EXP STA, 74- Concurrent Pos: Adv soils & soil fertil, Brazil Prog, AID, 70-71. Mem: Am Soc Agron. Res: Soil-plant-water relationships; soil testing; fertilization and fertilizer use. Mailing Add: Dept of Soils Water & Eng Univ of Ariz Tucson AZ 85721

STROH, WILLIAM RICHARD, b Sunbury, Pa, May 5, 23. PHYSICS. Educ: Harvard Univ, SB, 46, AM, 50, PhD(appl physics), 57. Prof Exp: Instr physics, Bucknell Univ, 46-49; res fel acoustics, Harvard Univ, 57-58; from asst prof to assoc prof elec eng, Univ Rochester, 58-62; assoc prof, 62-68, PROF PHYSICS, GOUCHER COL, 68-, CHMN DEPT PHYSICS & ASTRON, 64- Mem: Inst Elec & Electronics Engr; Am Phys Soc; Acoust Soc Am; Am Asn Physics Teachers; Sigma Xi. Res: Acoustics; instrumentation. Mailing Add: Dept of Physics Goucher Col Towson MD 21204

STROHBEHN, JOHN WALTER, b San Diego, Calif, Nov 21, 36; m 58; c 3. BIOMEDICAL ENGINEERING, RADIOPHYSICS. Educ: Stanford Univ, BS, 58, MS, 59, PhD(elec eng), 64. Prof Exp: From asst prof to assoc prof, 63-74, PROF ENG, DARTMOUTH COL, 74- Concurrent Pos: Partic, Nat Acad Sci-Acad Sci USSR Exchange Prog, 67; mem comn II, Int Sci Radio Union; Inter-Union Comt Radio Meteorol; consult, McGraw-Hill, Inc & Avco Corp. Honors & Awards: Distinguished Authorship Award, Nat Oceanic & Atmospheric Admin, 74. Mem: AAAS; Am Geophys Union; Inst Elec & Electronics Engrs; Am Soc Eng Educ; fel Optical Soc Am. Res: Radio propagation in the troposphere; optical propagation through a turbulent medium; image processing and tomography. Mailing Add: Radiophysics Lab Dartmouth Col Hanover NH 03755

STROHECKER, HENRY FREDERICK, b Macon, Ga, Oct 15, 05; m 34; c 3. BIOLOGY. Educ: Mercer Univ, AB, 26; Univ Chicago, PhD(zool), 36. Prof Exp: Teacher, High Sch, Ga, 26-33; asst, Univ Chicago, 33-36; asst prof zool, Univ Miami, 36-37; asst prof biol, Kenyon Col, 37-44; vis asst prof, Wayne State Univ, 44-45; prof zool, NMex Highlands Univ, 45-46; from assoc prof to prof, 46-72, EMER PROF ZOOL, UNIV MIAMI, 72- Mem: Am Soc Zool; Soc Syst Zool. Res: Ecology; insect physiology and taxonomy. Mailing Add: Dept of Zool Univ of Miami Coral Gables FL 33124

STROHL, EVERETT LEE, surgery, deceased

STROHL, GEORGE RALPH, JR, b Ardmore, Pa, Oct 19, 19; m 46; c 2. MATHEMATICS. Educ: Haverford Col, BA, 41; Univ Pa, MA, 47; Univ Md, PhD(math), 56. Prof Exp: From instr to assoc prof, 47-63, chmn dept, 70-76, PROF MATH, US NAVAL ACAD, 63- Mem: Am Math Soc; Am Soc Eng Educ. Res: Topology and analysis. Mailing Add: Dept of Math US Naval Acad Annapolis MD 21402

STROHL, JOHN HENRY, b Forest City, Ill, Oct 2, 38; m 60; c 2. ANALYTICAL CHEMISTRY. Educ: Univ Ill, BS, 59; Univ Wis, PhD(chem), 64. Prof Exp: Asst chem, Univ Wis, 59-64; asst prof, 64-70, ASSOC PROF CHEM, WVA UNIV, 70- Mem: Am Chem Soc. Res: Preparative electrochemistry and continuous electrolysis. Mailing Add: Dept of Chem WVa Univ Morgantown WV 26506

STROHL, WILLIAM ALLEN, b Bethlehem, Pa, Nov 1, 33; m 57; c 2. VIROLOGY. Educ: Lehigh Univ, AB, 55; Calif Inst Technol, PhD(biol), 60. Prof Exp: Instr microbiol, Sch Med, St Louis Univ, 59-63; res assoc, 64-66, asst prof, 66-70, ASSOC PROF MICROBIOL, RUTGERS MED SCH, COL MED & DENT NJ, 70- Concurrent Pos: Nat Found fel, 59-61. Mem: AAAS; Am Soc Microbiol; NY Acad Sci. Res: Animal viruses; viral oncogenesis. Mailing Add: Rugers Med Sch Col of Med & Dent of NJ Piscataway NJ 08854

STROHM, JERRY LEE, b West Union, Ill, Jan 9, 37; m 57; c 4. GENETICS, PLANT BREEDING. Educ: Univ Ill, BS, 59; Univ Minn, PhD(genetics), 66. Prof Exp: PROF BIOL, UNIV WIS-PLATTEVILLE, 64-, HEAD DEPT, 68- Res: Soybean genetics. Mailing Add: Dept of Biol Univ of Wis-Platteville Platteville WI 53818

STROHM, PAUL F, b Pennsauken, NJ, Jan 13, 35; m 57; c 5. AGRICULTURAL CHEMISTRY, ORGANIC CHEMISTRY. Educ: La Salle Col, BA, 56; Temple Univ, PhD(org chem), 61. Prof Exp: Res chemist, Atlantic Refining Co, 60-62 & Houdry Labs, Air Prod & Chem Inc, 62-68; res chemist, 68-72, GROUP LEADER, AGR CHEM DIV LAB, AMCHEM PROD, INC, 72- Mem: Am Chem Soc. Res: Organic synthesis: agricultural chemicals, especially herbicides and plant growth regulators; kinetics of urethane reactions; mechanism of epoxy curing reactions; bicyclic amine chemistry. Mailing Add: Agr Chem Div Lab Amchem Prod Inc Ambler PA 19002

STROHM, WALTER WILLIAM, nuclear physics, see 12th edition

STROHMAN, RICHARD CAMPBELL, b New York, NY, May 5, 27; m; c 2. ZOOLOGY. Educ: Columbia Univ, PhD, 58. Prof Exp: Instr zool & cell physiol, Columbia Univ, 55-56; from asst prof to assoc prof, 58-70, PROF ZOOL, UNIV CALIF, BERKELEY, 70-, CHMN DEPT, 73- Mem: Soc Gen Physiol. Res: Physiology and biochemistry of muscle growth and development. Mailing Add: Dept of Zool Univ of Calif Berkeley CA 94720

STROHMAYER, HERBERT FRANZ, organic chemistry, see 12th edition

STROHMEIER, GUSTAV H, b Nuremberg, Ger, July 26, 09; US citizen; m 42. PHYSICS. Educ: Munich Technol Univ, BS, 31; Univ Jena, Dr phil nat, 35. Prof Exp: Asst physics, Univ Jena, 33-36; res physicist, Res Lab, Siemens-Halske, 36-37; actg sec gen, Lilienthal Soc Aviation Res, 37-45; mem staff, Ger Acad Aerospace Sci, 41-45, actg secy gen, 44-45; scientist, Wright Field, US Air Force, 46-48; res consult to chief res & develop procurement, Air Materiel Command, 48-54; dir defense res, Armour Res Found, 55-64, V PRES, IIT RES INST, 64- Concurrent Pos: Mem, Bd Dirs, Growth Indust Shares, Inc. Mem: Newcomen Soc; Inst Aeronaut & Astronaut. Mailing Add: IIT Res Inst 10 W 35th St Chicago IL 60616

STROIKE, JAMES EDWARD, b Enid, Okla, May 1, 42; m 61; c 3. PLANT BREEDING, PLANT GENETICS. Educ: Okla State Univ, BS, 64, MS, 67; Univ Nebr, Lincoln, PhD(plant breeding & genetics), 72. Prof Exp: Assoc secy-treas seed cert, Okla Crop Improv Asn, 66-67; from instr to asst prof agron, Univ Nebr-Lincoln, 71-75; PLANT GENETICIST, ROHM AND HAAS CHEM CO, 75- Mem: Am Soc Agron; Crop Sci Soc Am; Sigma Xi. Res: Chemical male gametocides for hybrid seed production; wheat. Mailing Add: Res Lab Rohm and Haas Chem Co Spring House PA 19477

STROJNY, EDWIN JOSEPH, b Chicago, Ill, Jan 1, 26; m 55; c 2. INDUSTRIAL CHEMISTRY. Educ: Ill Inst Technol, BS, 51; Univ Ill, PhD(chem), 55. Prof Exp: Asst instr gen chem, Univ Ill, 51-52; org chemist, G D Searle & Co, 54-57; ORG CHEMIST, DOW CHEM CO, 57- Mem: AAAS; Am Chem Soc. Res: Phenolic compounds and derivatives; aromatic chemistry; heterogeneous and homogeneous catalysis; oxidation of organic compounds by oxygen; reaction mechanisms. Mailing Add: 3713 Orchard Dr Midland MI 48640

STROKE, GEORGE W, b Zagreb, Yugoslavia, July 29, 24; US citizen. PHYSICAL OPTICS. Educ: Univ Montpellier, BSc, 42; Univ Paris, Ing Dipl, 49, Dr es sc(physics), 60. Prof Exp: Res assoc, Spectros Lab, Mass Inst Technol, 52-56, mem staff, Res Lab Electronics, 55-58, mem defense res staff, Instrumentation Lab, 57-58, res assoc, Spectros Lab, 58-59, lectr elec eng, 60-63; prof elec eng & head electro-optical sci lab, Univ Mich, 63-67; PROF ELEC SCI & MED BIOPHYS, STATE UNIV NY STONY BROOK, 67- Concurrent Pos: Consult, Jarrell-Ash Co, 54-59 & 60-66, Jobin et Yvon, Paris, 59-60, Lincoln Lab, Mass Inst Technol, 61, Perkin-Elmer Corp, 61-63, NASA Electronics Res Ctr, 66-67, Am Cancer Soc, 72- & NSF, 72-; asst res prof, Boston Univ, 56-57; adv, US Air Force Systs Command Laser Task Force, 64; Nat Acad Sci rep, Comn I, Int Sci Radio Union, 65-; vis prof, Harvard Med Sch, 70-73; mem spec task force med ultrasonic imaging, NSF, 73- Mem: Fel Optical Soc Am; fel Inst Elec & Electronics Engr; Am Astron Soc; fel Am Phys Soc; Fr Phys Soc. Res: Holography and its application to biophysics; electromagnetic diffraction theory and ruling of optical diffraction gratings; coherent optics and laser applications; digital and optical image storage and retrieval; optical information processing, computing and communications. Mailing Add: Dept of Elec Sci State Univ NY Stony Brook NY 11790

STROKE, HINKO HENRY, b Zagreb, Yugoslavia, June 16, 27; US citizen; m 56; c 2. ATOMIC SPECTROSCOPY, NUCLEAR PHYSICS. Educ: Newark Col Eng, BS, 49; Mass Inst Technol, MS, 52, PhD, 54. Prof Exp: Consult, Atomic Instrument Co, 50; consult, Sci Translation Serv, Mass Inst Technol, 51-52; Nat Res Coun Can res fel, 54; res assoc physics, Princeton Univ, 54-57; lectr & res staff mem, Mass Inst Technol, 57-63; assoc prof, 63-68, PROF PHYSICS, NY UNIV, 68- Concurrent Pos: Consult, TRG, Inc, 59, Air Force, Cambridge Res Ctr, 63, Laser, Inc, Am Optical Co, 63-69 & Int Tel & Tel Fed Labs, 66; vis prof, Univ Paris, 69-70; ed, Comments Atomic & Molecular Physics, 73-; consult, Nat Aeronaut & Space Admin, 75; sr fel sci, NATO, 75. Mem: AAAS; fel Am Phys Soc; Optical Soc Am; Fr Phys Soc; Europ Phys Soc. Res: Hyperfine structure and isotope shifts of stable and radioactive atoms by magnetic resonance and optical spectroscopy; nuclear moments; charge and magnetization distribution; coherence in atomic radiation; solar spectra; spectroscopic instrumentation; laser systems. Mailing Add: Dept of Physics NY Univ New York NY 10003

STROM, EDWIN THOMAS, b Des Moines, Iowa, June 11, 36; m 58; c 2. PHYSICAL ORGANIC CHEMISTRY, GEOCHEMISTRY. Educ: Univ Iowa, BS, 58; Univ Calif, Berkeley, MS, 61; Iowa State Univ, PhD(phys org chem), 64. Prof Exp: Res technologist, 64-67, SR RES CHEMIST, FIELD RES LAB, MOBIL RES & DEVELOP CORP, 67- Concurrent Pos: Vis lectr, Dallas Baptist Col, 69-70, El Centro Community Col, 70-72 & Univ Tex, Dallas, 74. Mem: Am Chem Soc. Res: Organic geochemistry free radicals, magnetic resonance, mass spectrometry, aromatic substituent effects, organic molecular orbital theory. Mailing Add: Mobil Res & Develop Corp Field Res Lab PO Box 900 Dallas TX 75221

STROM, ROBERT GREGSON, b Long Beach, Calif, Oct 1, 33; m 55; c 1. ASTROGEOLOGY. Educ: Univ Redlands, BS, 55; Stanford Univ, MS, 57. Prof Exp: Geologist, Stand Vacuum Oil Co, 57-60; asst res geologist, Univ Calif, Berkeley, 61-63; asst prof, Lunar & Planetary Lab, 63-72, ASSOC PROF, DEPT PLANETARY SCI, UNIV ARIZ, 72- Concurrent Pos: Mem, Apollo Lunar Oper Working Group, 68-69, Imaging Sci Team, Mariner Venus/Mercury Mission, 69-75, Lunar Sci Inst, Lunar Sci & Cartog Comn, 74-, NASA Comet Working Group, Jet Propulsion Lab Jupiter Orbiter Sci Working Group & NASA Mercury Geol Mapping Prog, 75-; rep, Planetary Prog, Jet Propulsion Lab Imaging Syst Instrument Develop Prog, 75- Honors & Awards: Pub Serv Group Achievement Award Mariner 10 TV Exp, NASA, 74. Mem: Am Geophys Union; Int Astron Union. Res: Lunar and planetary geology; origin and evolution of lunar and planetary surfaces; space craft imaging of planetary surfaces. Mailing Add: Dept of Planetary Sci Univ of Ariz Tucson AZ 85721

STROM, STEPHEN, b Bronx, NY, Aug 12, 42; m 60; c 4. ASTRONOMY, ASTROPHYSICS. Educ: Harvard Univ, AB, 62, AM & PhD(astron), 64. Prof Exp: Astrophysicist, Smithsonian Astrophys Observ, 62-69; lectr astron, Harvard Univ, 64-69; assoc prof physics & earth & space sci, State Univ NY Stony Brook, 69-71, prof astron, 71-72, coordr astron & astrophys, 69-72; ASTRONR, KITT PEAK NAT OBSERV, 72-, CHMN GALACTIC & EXTRAGALACTIC PROG, 75- Concurrent Pos: Alfred P Sloan Found res fel, 70-72; mem at large, Assoc Univs Res Astron, 71-; res assoc, Smithsonian Astrophys Observ. Honors & Awards: Bart J Bok Prize Astron, Harvard Univ, 71. Mem: Am Astron Soc; Int Astron Union. Res: Evolution of stars; structure and evolution of galaxies. Mailing Add: Kitt Peak Nat Observ Tucson AZ 85717

STROMAN, DAVID WOMACK, b Corpus Christi, Tex, June 1, 44; m 65; c 1. BIOCHEMISTRY, MICROBIOLOGY. Educ: Bethany Nazarene Col, BS, 66; Univ Okla, PhD(biochem), 70. Prof Exp: NIH fel microbiol, Sch Med, Washington Univ, 70-72; RES SCIENTIST ANTIBIOTIC DEVELOP, UPJOHN CO, 72- Mem: Am Chem Soc; Am Soc Microbiol. Res: Molecular biology and biochemical genetics with special emphasis on the ribosome and cellular control mechanisms. Mailing Add: Infectious Dis Res Upjohn Co 7000 Portage Rd Kalamazoo MI 49001

STROMATT, ROBERT WELDON, b Muskogee, Okla, Mar 27, 29; m 56; c 3. ANALYTICAL CHEMISTRY. Educ: Emporia State Teachers Col, BS, 54; Kans State Univ, PhD(chem), 58. Prof Exp: Res chemist, Hanford Labs, Gen Elec Co, 57-66; sr res scientist, Pac Northwest Lab, Battelle Mem Inst, 66-70; FEL SCIENTIST, WESTINGHOUSE-HANFORD, 70- Mem: Am Chem Soc; Sigma Xi. Res: Electroanalytical chemistry and general methods development. Mailing Add: 411 Franklin Richland WA 99352

STROMBERG, KARL ROBERT, b Modoc, Ind, Dec 1, 31; m 68; c 3. MATHEMATICS. Educ: Univ Ore, BA, 53, MA, 54; Univ Wash, PhD(math), 58. Prof Exp: Res assoc & Off Naval Res fel math, Yale Univ, 58-59; res lectr, Univ Chicago, 59-60; from asst prof to assoc prof, Univ Ore, 60-68; PROF MATH, KANS STATE UNIV, 68- Concurrent Pos: Vis res prof, Univ York, Eng, 74-75. Mem: Am Math Soc; Math Asn Am. Res: Measure and integration theory; real variable theory; harmonic analysis; topological groups; functional analysis. Mailing Add: Dept of Math Kans State Univ Manhattan KS 66506

STROMBERG, KURT, b Albuquerque, NMex, Mar 3, 39. PATHOLOGY. Educ: Amherst Col, BA, 61; Univ Colo, MD, 66; Am Bd Path, dipl, 74. Prof Exp: Intern path, Yale-New Haven Hosp, 66-67; res assoc, Nat Cancer Inst, 68-72; resident path, Columbia Univ, 72-74; mem res staff, Inst Cancer Res, Delafield Hosp, New York, 72-74; STAFF INVESTR, NAT CANCER INST, 72- Res: Viral oncology. Mailing Add: Nat Cancer Inst Bldg 37 Rm 2E-08 Bethesda MD 20014

STROMBERG, LAWAYNE ROLAND, b Minneapolis, Minn, Nov 18, 29; m 54; c 3. SURGERY, NUCLEAR MEDICINE. Educ: Univ Calif, Berkeley, BA, 51; Univ Calif, Los Angeles, MD, 55; Univ Rochester, MS, 63. Prof Exp: Med Corps, US Army, 56-, intern & resident gen surg, Sch Med, Univ Calif, Los Angeles, 55-58, resident, Vet Admin Hosp, Los Angeles, 58-60, cmndg officer & surgeon, 11th Evacuation Hosp, Korea, 61-62, nuclear med res officer, Walter Reed Army Inst Res, 65-68, DIR ARMED FORCES RADIOBIOL RES INST, NAT NAVAL MED CTR, 68- Concurrent Pos: Res fel radiation biol, Walter Reed Army Inst Res, 63-65. Res: Effect of radiation on response to trauma. Mailing Add: Armed Forces Radiobiol Res Inst Nat Naval Med Ctr Bethesda MD 20014

STROMBERG, MELVIN WILLARD, b Quamba, Minn, Nov 2, 25; m 48; c 6. GROSS ANATOMY, NEUROANATOMY. Educ: Univ Minn, BS, 53, DVM, 54, PhD(vet anat), 57. Prof Exp: Asst vet anat, Univ Minn, 53-54, from instr to assoc prof, 54-60; assoc prof, 60-62, PROF VET ANAT, PURDUE UNIV, WEST LAFAYETTE, 62-, HEAD DEPT, 63- Concurrent Pos: NIH spec fel anat, Karolinska Inst, Sweden, 70-71. Mem: Am Asn Vet Anat (pres, 66-67); Am Asn Anat; World Asn Vet Anat. Res: Veterinary nuerology; neuroanatomy of domestic animals; acupuncture. Mailing Add: Dept Anat Sch Vet Med Purdue Univ West Lafayette IN 47906

STROMBERG, ROBERT REMSON, b Buffalo, NY, Feb 2, 25; m 47; c 4. POLYMER SCIENCE. Educ: Univ Buffalo, BA, 48, PhD(phys chem), 51. Prof Exp: Asst phys chem, Univ Buffalo, 48-50; phys chemist, 51-62, chief phys chem br, Off Saline Water, 62, phys chemist, 62-67, chief polymer interface sect, 67-75, DEP CHIEF POLYMERS DIV, NAT BUR STANDARDS, 69- Concurrent Pos: Chmn, Gordon Res Conf Sci Adhesion, 66; US Dept Com fel sci & technol, 75-76. Honors & Awards: US Dept Com Silver Medal, 73. Mem: Am Chem Soc. Res: Polymer adsorption; interfaces; ellipsometry; surface properties; biomaterials; medical devices. Mailing Add: Nat Bur Standards Washington DC 20234

STROMBERG, THORSTEN FREDERICK, b Aberdeen, Wash, Aug 13, 36. LOW TEMPERATURE PHYSICS. Educ: Reed Col, BA, 48; Iowa State Univ, PhD(physics), 65. Prof Exp: Res fel, Los Alamos Sci Lab, 65-67; asst prof, 67-74, ASSOC PROF PHYSICS, N MEX STATE UNIV, 74- Mem: Am Phys Soc. Res: Thermal and magnetic properties of superconductors, particularly type-II superconducting materials. Mailing Add: Dept of Physics NMex State Univ Las Cruces NM 88001

STROMBERG, VERNER L, JR, b Salt Lake City, Utah, May 6, 20; m 43; c 2. ORGANIC CHEMISTRY. Educ: Univ Utah, BS, 41, MS, 43; Univ Wis, PhD(org chem), 49. Prof Exp: Chemist, WVa Ord, 43 & Gen Chem Co, 43; res chemist, Hercules Powder Co, 44, lab supvr, Badger Ord Works, 45; head dept phys sci, Weber Col, 49-51; res chemist, Nat Heart Inst, 51-54; mgr corrosion & additive res, Petrolite Corp, 54-65; mgr food sci, Pet, Inc, 65-69, DIR CONTECH LAB, PET MILK INC, 69- Concurrent Pos: Mem, Nutrit Found. Mem: Am Chem Soc. Res: Natural products; foods; sterilization and preservation techniques; instrumental analysis; synthetic organic chemistry. Mailing Add: Pet Milk Inc Louis Latzer Dr Greenville IL 62246

STROME, FORREST C, JR, b Kalamazoo, Mich, May 19, 24; m 45; c 2. SOLID STATE PHYSICS, LASERS. Educ: Univ Ill, BS, 45; Univ Mich, MS, 48, PhD(physics), 54. Prof Exp: Test engr, Gen Elec Co, 46-47; proj physicist, Apparatus & Optical Div, 53-60, sr res physicist, Res Labs, 60-68, RES ASSOC, EASTMAN KODAK CO, 68- Mem: Am Phys Soc; Optical Soc Am. Res: Dye lasers and molecular photophysics. Mailing Add: Eastman Kodak Res Labs 1669 Lake Ave Rochester NY 14650

STROMER, MARVIN HENRY, b Readlyn, Iowa, Sept 1, 36; m 60; c 1. CELL BIOLOGY, BIOCHEMISTRY. Educ: Iowa State Univ, BS(animal sci) & BS(agr educ), 59, PhD(cell biol), 66. Prof Exp: Foreman prod develop, George A Hormel & Co, Minn, 59-62; res asst biochem, Iowa State Univ, 62-66; fel, Mellon Inst, 66-68; ASSOC PROF ANIMAL SCI, IOWA STATE UNIV, 68- Concurrent Pos: Humboldt fel, 74; vis scientist, Max Planck Inst Med Res, Heidelberg, WGer, 74-75. Mem: Electron Micros Soc Am; Am Heart Asn; Am Soc Cell Biol; Biophys Soc. Res: Ultrastructure and biochemistry of muscle and connective tissues. Mailing Add: Dept of Animal Sci Iowa State Univ Ames IA 50011

STROMINGER, NORMAN LEWIS, b New York, NY, June 1, 34; m 57; c 3. NEUROANATOMY. Educ: Univ Chicago, AB, 55, BS, 56, PhD(biopsychol), 61. Prof Exp: Trainee neuroanat, Columbia Univ, 62-65; from asst prof to assoc prof, 65-74, PROF ANAT, ALBANY MED COL, 74- Mem: AAAS; Am Asn Anat; Soc Neurosci. Res: Psychophysiological and neuroanatomical studies of auditory and motor systems. Mailing Add: Dept of Anat Albany Med Col Albany NY 12208

STROMMEN, DENNIS PATRICK, b Milwaukee, Wis, Sept 2, 38; m 68. INORGANIC CHEMISTRY, SPECTROSCOPY. Educ: Wis State Univ-Whitewater, BA, 66; Cornell Univ, PhD(chem), 71. Prof Exp: Res assoc spectros, Ctr Mat Res, Univ Md, 70-71; ASST PROF CHEM, CARTHAGE COL, 71- Mem: Soc Appl Spectros; Am Chem Soc. Res: Characterization of compounds through vibrational analysis, especially with regard to their Raman spectra; isolation of reactive molecular species in frozen gas matrices. Mailing Add: Dept of Chem Carthage Col Kenosha WI 53140

STROMMEN, NORTON D, meteorology, see 12th edition

STROMSTA, COURTNEY PAUL, b Muskegon, Mich, Apr 25, 22; m 50; c 2. SPEECH PATHOLOGY, AUDIOLOGY. Educ: Western Mich Univ, BS, 48; Ohio State Univ, MA, 51, PhD(speech & hearing sci), 56. Prof Exp: Audiol trainee, Vet Admin-Walter Reed Hosp & New York City Regional Off, 51; dir speech & hearing clin, ECarolina Univ, 54-56; prof speech & hearing sci, Ohio State Univ, 56-68; PROF SPEECH & HEARING SCI, WESTERN MICH UNIV, 68- Concurrent Pos: Nat Inst Neurol Dis & Blindness res grant, Ohio State Univ, 57-67 & US Off Educ res grant, 65-67; NIH spec res fel, Karolinska Inst, Sweden, 71-72; consult, Electronic Teaching Labs, Washington, DC, 61-65; guest prof, Univ Zagreb, 65-66. Mem: AAAS; Am Speech & Hearing Asn; Acoust Soc Am. Res: Cybernetic relationship of speech and hearing with emphasis on stuttering and acoustically-impaired children; effects of shaping acoustical signals on perception of speech by hearing impaired children. Mailing Add: Dept of Speech Path & Audiol Western Mich Univ Kalamazoo MI 49008

STRONCK, DAVID RICHARD, b Chicago, Ill, Feb 3, 31, m 69; c 3. SCIENCE EDUCATION. Educ: St Patrick's Col, Calif, BA, 53; Ore State Univ, MS, 66, PhD(sci educ), 68. Prof Exp: Teacher chem, Serra High Sch, San Mateo, Calif, 58-65 & 67-68; instr biol, Sacramento State Col, 68-69; asst prof sci educ, Univ Tex, Austin, 69-71; asst prof biol, 71-73, ASSOC PROF BIOL, WASH STATE UNIV, 73- Concurrent Pos: Consult, 70-; dir, NSF grants, 71 & 73-75, co-dir, NSF grants, 72-75, judge proposals, NSF, 73 & 74; dir, Nat Dairy Coun grant, 73. Mem: Nat Sci Teachers Asn; Nat Asn Res Sci Teaching; Nat Asn Biol Teachers; Am Inst Biol Sci. Res: Behavioral modification of students in dietary preferences and in environmental attitudes, especially through use of modern media; concept formation through laboratory experiments and modularized instruction, including diagnostic pretesting and advising. Mailing Add: Prog in Gen Biol Wash State Univ Pullman WA 99163

STRONG, ALAN EARL, b Boston, Mass, May 30, 41; m 66; c 2. OCEANOGRAPHY. Educ: Kalamazoo Col, BA, 63; Univ Mich, MS, 65, PhD(oceanog), 68. Prof Exp: RES METEOROLOGIST-OCEANOGR, NAT ENVIRON SATELLITE SERV, NAT OCEANIC & ATMOSPHERIC ADMIN, 68- Mem: AAAS; Am Meteorol Soc; Am Geophys Union; Int Asn Gt Lakes Res; Sigma Xi. Res: Lake and sea breezes, air-sea interface; marine meteorology; spacecraft oceanography; remote sensensing—infrared, microwave, visible. Mailing Add: Nat Environ Satellite Serv NOAA-FOB4 Stop D Suitland MD 20233

STRONG, CAMERON GORDON, b Vegreville, Alta, Sept 18, 34; m 59; c 2. INTERNAL MEDICINE, NEPHROLOGY. Educ: Univ Alta, MD, 58; McGill Univ, MS, 66. Prof Exp: Resident internal med & path, Queens Hosp, Honolulu, 59-61; resident internal med, Mayo Grad Sch Med, 61-64; from instr to asst prof, 67-74, ASSOC PROF MED, MAYO MED SCH, 74-; CONSULT, DIV NEPHROLOGY & INTERNAL MED, MAYO CLIN & FOUND, 66-, CHMN DIV, 73- Concurrent Pos: Fel nephrology, Hotel Dieu Montreal, 64-66; res assoc hypertension, Dept Physiol, Univ Mich, Ann Arbor, 66-67; mem med adv bd, Coun High Blood Pressure Res, 71- Mem: Fel Am Col Physicians; fel Am Col Cardiol. Res: Hypertension; renal disease; vascular smooth muscle physiology; prostaglandins; renin-angiotensin system. Mailing Add: Div Nephrology & Internal Med Mayo Clin Rochester MN 55901

STRONG, DOROTHY HUSSEMANN, b Peoria, Ill, Feb 15, 08; m 57. FOOD SCIENCE. Educ: Univ Ill, BS, 28, MS, 29, PhD(food bact), 46. Prof Exp: Asst, Univ Ill, 29-30; instr, 30-40, from asst prof to prof foods, 40-70, chmn dept foods & nutrit, 51-68, prof food sci, 68-70, EMER PROF FOOD SCI, UNIV WIS-MADISON, 70- Honors & Awards: Borden Res Award, 70. Mem: Am Home Econ Asn; Inst Food Technol. Res: Cookery temperature as related to food poisoning microorganisms; microbiological flora of frozen foods; staphylococcus food poisoning; thermal relationships of Salmonella; Clostridium perfringens as a food poisoning agent. Mailing Add: 625 Anthony Ln Madison WI 53711

STRONG, DOUGLAS C, agricultural economics, resource economics, see 12th edition

STRONG, FRANK EDWARD, entomology, see 12th edition

STRONG, FRANK MORGAN, b Brewerton, NY, Sept 7, 08; m 28; c 5. BIOCHEMISTRY. Educ: Syracuse Univ, AB, 28, MA, 29; Univ Wis, PhD(org chem), 32. Hon Degrees: ScD, Syracuse Univ, 59. Prof Exp: Asst, 29-32, asst biochem, 32-33, instr & res assoc, 33-38, from asst prof to prof, 38-72, EMER PROF BIOCHEM, UNIV WIS-MADISON, 72- Concurrent Pos: Rockefeller fel, Univ Zurich & State Univ Utrecht, 35-36; consult, Northern & Western Utilization Res & Develop Divs, Agr Res Serv, USDA; chmn study group smoking & health, NIH, 56-57, mem, Biochem Study Sect, 56-59, mem, Med Chem Study Sect, 59-64; mem food protection comt, Nat Acad Sci-Nat Res Coun, 67-72. Honors & Awards: Borden Award, Am Inst Nutrit, 56. Mem: AAAS; Am Chem Soc; Am Soc Biol Chem; Am Inst Nutrit; Venezuelan Asn Advan Sci. Res: Microbiological determination and chemistry of B-vitamins; distribution of B-vitamins and amino acids in foods; synthesis of unsaturated fatty acids; isolation and chemistry of physiologically active natural products, natural pigments, coenzyme A, rat lathyrism factor, antimycin A, cytokinins, trichothecenes and mycotoxins. Mailing Add: 625 Anthony Ln Madison WI 53711

STRONG, FREDERICK CARL, III, b Denver, Colo, Nov 17, 17; m 41; c 2. ANALYTICAL CHEMISTRY. Educ: Swarthmore Col, BA, 39; Lehigh Univ, MS, 41; Bryn Mawr Col, PhD, 54. Prof Exp: Chief chemist, Superior Metal Co, 40-42; res chemist, Lea Mfg Co, 42-43; asst, Wesleyan Univ, 43-45; res chemist, Enthone Co, 45; instr chem, Cedar Crest Col, 45-47; asst prof, Villanova Col, 47-51; from asst prof to assoc prof chem & chem eng, Stevens Inst Technol, 51-60; prof chem & chmn dept, Inter-Am Univ PR, 60-63 & Univ Bridgeport, 63-68; prof chem, Nat Tsing Hua Univ, Taiwan, 69 & Univ El Salvador, 70-72; tech expert, UN Indust Develop Orgn, Asuncion, Paraguay, 72-73; TITULAR PROF, UNIV ESTADUAL DE CAMPINAS, BRAZIL, 73- Concurrent Pos: Ed-in-chief, Appl Spectros, 55-60; Leverhulme fel, Aberdeen Univ, 64-65; Fulbright-Hays lectr, Tribhuvan Univ, Nepal. Mem: Fel Am Inst Chemists; Soc Appl Spectros; Am Chem Soc; Coblentz Soc; Sigma Xi. Res: Spectrochemical analysis; qualitative analysis; food analysis, copper complexes of carbohydrates. Mailing Add: Fac Eng Alimentos Univ Estadual de Campinas Campinas SP Brazil

STRONG, HERBERT MAXWELL, b Wooster, Ohio, Sept 30, 08; m 35; c 2. HIGH PRESSURE PHYSICS, PHYSICAL OPTICS. Educ: Univ Toledo, BS, 30; Ohio State Univ, MS, 32, PhD(physics), 36. Prof Exp: Asst, Ohio State Univ, 31-35; res physicist, Bauer & Black Div, Kendall Co, Ill, 35-45 & Kendall Mills Div, Mass, 45-46; res assoc, Res Lab, Gen Elec Co, 46, physicist, Res & Develop Ctr, 46-73; RES ASSOC PHYSICS, UNION COL, NY, 73- Concurrent Pos: Consult, Gen Elec Res & Develop Ctr, 73-74; consult technol use of diamond, Lazar Kaplan & Sons, 73- Honors & Awards: Award, Soc Mfg Eng, 62; Modern Pioneers Award, Nat Asn Mfrs, 65. Mem: AAAS; fel Am Phys Soc; Sigma Xi. Res: Technological and industrial applications of diamonds; physical optical studies of rocket motor flames extreme high pressure techniques; measurements and phase equilibria; synthesis of gem diamond. Mailing Add: 1165 Phoenix Ave Schenectady NY 12308

STRONG, IAN B, b Cohoes, NY, July 11, 30; m 60; c 1. PHYSICS, ASTRONOMY. Educ: Glasgow Univ, BSc, 53; Pa State Univ, PhD(physics), 63. Prof Exp: Mem tech staff, Bell Tel Labs, Inc, 53-55; staff mem, Ord Res Lab, 55-57; MEM STAFF, LOS ALAMOS SCI LAB, 61- Mem: AAAS; Am Phys Soc; Am Geophys Union; Am Astron Soc. Res: Acoustics, transmission through solids and liquids, ultrasonics; nuclear physics, particle detection, passage of radiation through matter, multiple scattering; astrophysics, interplanetary medium, high energy astronomy; history;

philosophy; sociology of science. Mailing Add: 229 Rio Bravo Dr Los Alamos NM 87544

STRONG, JACK PERRY, b Birmingham, Ala, Apr 27, 28; m 51; c 4. PATHOLOGY. Educ: Univ Ala, BS, 48; La State Univ, MD, 51; Am Bd Path, dipl, 57 & 58. Prof Exp: Intern, Jefferson Hillman Hosp, Birmingham, Ala, 51-52; asst, 52-53, from instr to assoc prof, 55-64, PROF PATH, SCH MED, UNIV NEW ORLEANS, 64-, HEAD DEPT, 66- Concurrent Pos: USPHS fel, 57; consult, Southwest Found Res & Educ, 54-55; sabbatical leave, Social Med Res Unit, Med Res Coun, London, Eng, 62-63; mem path A study sect, USPHS, 65-69, chmn, 67-69; mem sci adv bd consult, Armed Forces Inst Path, 71-; mem coun arteriosclerosis, Am Heart Asn; mem epidemiol & biomet adv comt, Nat Heart & Lung Inst, NIH. Mem: Am Asn Path & Bact (asst secy, 59-62); Am Soc Exp Path; Am Soc Clin Path; Col Am Path; Int Acad Path. Res: Pathology of cardiovascular diseases; atherosclerosis in the human and the experimental animal and in primates; epidemiology; geographic pathology and pathogenesis of atherosclerosis. Mailing Add: Dept of Path La State Univ Med Ctr New Orleans LA 70112

STRONG, JERRY GLENN, b Dawson, NMex, Nov 12, 41; m 71. PESTICIDE CHEMISTRY. Educ: Austin Col, BA, 63; Northwestern Univ, PhD(org chem), 68. Prof Exp: SR RES CHEMIST, RES & DEVELOP LABS, MOBIL CHEM CO, 67- Mem: Am Chem Soc. Res: Synthesis and development of new crop chemicals. Mailing Add: Mobil Chem Co R&D Labs Box 240 Edison NJ 08817

STRONG, JOHN (DONOVAN), b Riverdale, Kans, Jan 15, 05; m 28; c 2. PHYSICS. Educ: Univ Kans, AB, 26; Univ Mich, MS, 28, PhD(physics), 30. Hon Degrees: DSc, Southwestern at Memphis, 62. Prof Exp: Instr chem, Univ Kans, 25-27; instr physics, Univ Mich, 27-29, asst investr eng res, 29-30; Nat Res Coun fel physics, Calif Inst Technol, 30-32, fel, Astrophys Observ, 32-37, asst prof physics, 37-42; spec fel, Harvard Univ, 42-45; prof exp physics & dir, Lab Astrophys & Phys Meteorol, Johns Hopkins Univ, 45-67; PROF PHYSICS & ASTRON, UNIV MASS, AMHERST, 67- Concurrent Pos: Consult, Libbey-Owens-Ford Glass Co, Ohio. Honors & Awards: Longstreth & Levy Medals, Franklin Inst; Ives Medal, Optical Soc Am, 59. Mem: Fel Am Phys Soc; fel Optical Soc Am (pres, 59); fel Am Acad Arts & Sci; corresp mem Royal Belg Soc Sci; Int Acad Astronaut. Res: Experimental physics; evaporation in vacuum; infrared spectroscopy; meteorology; optics; astrophysical observations from high altitudes. Mailing Add: Dept of Phys & Astron Univ Mass Astron Res Facility Amherst MA 01002

STRONG, JUDITH ANN, b Cooperstown, NY, June 19, 41. PHYSICAL CHEMISTRY. Educ: State Univ NY Albany, BS, 63; Brandeis Univ, MA, 66, PhD(phys chem), 70. Prof Exp: Asst prof, 69-73, ASSOC PROF CHEM, MOORHEAD STATE UNIV, 73- Mem: Am Chem Soc. Res: Nonaqueous solutions; sulfur in ammonia and amines. Mailing Add: Dept of Chem Moorhead State Univ Moorhead MN 56560

STRONG, LAURENCE EDWARD, b Kalamazoo, Mich, Sept 3, 14; m 38; c 4. PHYSICAL CHEMISTRY. Educ: Kalamazoo Col, AB, 36; Brown Univ, PhD(chem), 40. Prof Exp: Asst phys chem, Harvard Med Sch, 40-41, res assoc, 41-43, assoc dir pilot plant, 43-46; from assoc prof to prof chem, Kalamazoo Col, 46-52; head dept, 52-65, PROF CHEM, EARLHAM COL, 52- Concurrent Pos: Dir, UNESCO Pilot Proj, Asia, 65-66; vis prof chem, Macquarie Univ, Australia, 71-72. Honors & Awards: SAMA Award for Chem Educ, Am Chem Soc, 71. Mem: Fel AAAS; Am Chem Soc; Sigma Xi. Res: Electrical properties of solutions; fractionation of proteins; thermodynamics of acid ionization. Mailing Add: Dept of Chem Earlham Col Richmond IN 47374

STRONG, LOUISE CONNALLY, b San Antonio, Tex, Apr 23, 44; m 70; c 2. HUMAN GENETICS, CANCER. Educ: Univ Tex, Austin, BA, 66; Univ Tex Med Br Galveston, MD, 70. Prof Exp: Fel med genetics, Tex Res Inst Ment Sci & Univ Tex Grad Sch Biomed Sci Houston, 70-72, res assoc cancer genetics, 72-73; ASST PROF BIOL, UNIV TEX HEALTH SCI CTR HOUSTON, 73-; DIR MED GENETICS CLIN & CONSULT MED GENETICS, UNIV TEX SYST CANCER CTR, M D ANDERSON HOSP & TUMOR INST, HOUSTON, 73- Concurrent Pos: Consult med genetics, Nat Large Bowel Cancer Proj, 76. Mem: Am Soc Human Genetics; AAAS. Res: Clinical cancer genetics; etiology and epidemiology of cancer. Mailing Add: Sect of Med Genetics Univ of Tex Syst Cancer Ctr M D Anderson Hosp & Tumor Inst Houston TX 77030

STRONG, MERVYN STUART, b Kells, Ireland, Jan 28, 24; nat US; m 50; c 2. OTOLARYNGOLOGY. Educ: Trinity Col, Dublin, BA, 45; Univ Dublin, MD, 47; FRCS(I), 49; FRCS (Eng), 50. Prof Exp: Registr otolaryngol, Royal Infirmary, Edinburgh, 49-50; instr, 52-56, PROF OTOLARYNGOL, SCH MED, BOSTON UNIV, 56- Concurrent Pos: Fel otolaryngol, Lahey Clin, Boston, 50-52; asst, Boston Univ Hosp, 52-56, chief serv, 56-; chief otolaryngol, Boston Vet Admin Hosp, 65- Mem: Fel Am Soc Head & Neck Surg; AMA; fel Am Col Surgeons; Soc Univ Otolaryngol (pres, 73-74); fel Am Acad Opthal & Otolaryngol. Res: Multicentric origins of carcinoma of oral cavity and pharynx. Mailing Add: 75 E Newton St Boston MA 02118

STRONG, PHILIP LEON, organic chemistry, see 12th edition

STRONG, ROBERT LYMAN, b Hemet, Calif, May 30, 28; m 51; c 4. PHYSICAL CHEMISTRY. Educ: Univ Calif, BS, 50; Univ Wis, PhD(chem), 54. Prof Exp: Res fel chem, Nat Res Coun Can, 54-55; from asst prof to assoc prof phys chem, 55-62, PROF PHYS CHEM, RENSSELAER POLYTECH INST, 62- Concurrent Pos: NSF sci faculty fel, 62-63. Mem: AAAS; Am Chem Soc. Res: Photochemistry and flash photolysis; atom recombination in gas and solution systems; halogen atom charge-transfer complexes; optical rotary dispersion of excited states and intermediate species in photochemical processes. Mailing Add: Dept of Chem Rensselaer Polytech Inst Troy NY 12181

STRONG, ROBERT STANLEY, b Sargent, Nebr, May 4, 24; m 50; c 5. ANALYTICAL CHEMISTRY. Educ: Cent Wash State Col, BA, 51; Ore State Univ, MS, 57; Univ of the Pac, PhD(org chem), 69. Prof Exp: Teacher high schs, Wash, 51-57; instr chem, Columbia Basin Col, 57-64, chmn div sci, 60-64; prof chem, Univ SDak, Springfield, 65-73; ASST PROF CHEM, FITCHBURG STATE COL, 73- Concurrent Pos: Dean col, Univ SDak, 67-72, dir instnl res, 72-73. Mem: Am Chem Soc. Res: D-galactosamine and its derivatives; thin layer chromatography and its applications; chemical instrumentation. Mailing Add: Dept of Chem Fitchburg State Col Fitchburg MA 01420

STRONG, RUDOLPH GREER, b Utica, Miss, Nov 5, 24; m 58; c 4. ENTOMOLOGY. Educ: Miss State Univ, BS, 46, MS, 48; Cornell Univ, PhD(econ entom), 56. Prof Exp: Asst & instr zool & entom, Miss State Univ, 46-49; from asst entomologist to assoc entomologist, Agr Exten Serv, La State Univ, 49-51; tech & sales rep, Stauffer Chem Co, 51-52; asst entom, Cornell Univ, 53-56; from asst entomologist to assoc entomolgist, 56-75, ENTOMOLOGIST, UNIV CALIF, RIVERSIDE, 75- Mem: Entom Soc Am. Res: Biology, ecology and prevention of stored-product insects and other urban-industrial pests; insecticide development. Mailing Add: Dept of Entom Univ of Calif Riverside CA 92502

STRONG, STANLEY STERLING, b Sopris, Colo, Feb 23, 23; m 47; c 2. NUCLEAR PHYSICS, SPACE PHYSICS. Educ: Northwestern State Col, BS, 48; Univ Okla, MS, 51. Prof Exp: Engr, Chance Vought Aircraft, Inc, 51-52; nuclear engr, Convair, Ft Worth, 52-55, sr nuclear engr, 55-59; mem tech staff, Space Tech Labs, Inc, Thomson-Ramo-Wooldridge, Inc, 59-60; mgr booster off, 60-62, SYSTS ENG DIR, AEROSPACE CORP, 62- Mem: Am Phys Soc; Am Inst Aeronaut & Astronaut. Res: Research, development, design and operation of neutron and gamma dosimeters and gamma spectrometers; application of computer techniques to large scale data systems; planning, development and operation of earth orbiting vehicles for various uses. Mailing Add: Aerospace Corp Rm 1425 Bldg 115 PO Box 92957 Los Angeles CA 90009

STRONG, WALKER ALBERT, b Baltimore, Md, Apr 3, 18; m 61; c 3. CHEMISTRY. Educ: Dickinson Col, BS, 40; Pa State Univ, MS, 42, PhD(org chem), 44. Prof Exp: Res asst, Univ Ill, 44-46; res chemist, Wm S Merrell Co, Ohio, 46-47; sr res chemist, 47-64, admin asst to dir res, 64-71, SR RES CHEMIST, CHEM DIV, PPG INDUSTS, INC, 71- Mem: Am Chem Soc. Res: Organo-silicon compounds; streptomycin and other antibiotics; organic sulfur and chlorine compounds; phosgene derivatives; peroxycarbonates; polyurethanes; peroxide stabilization; stone-scripts resembling ancient Phoenician found in Pennsylvania. Mailing Add: Tech Ctr Chem Div PPG Indust Inc Barberton OH 44203

STRONG, WILLIAM J, b Idaho Falls, Idaho, Jan 1, 34; m 59; c 6. ACOUSTICS. Educ: Brigham Young Univ, BS, 58, MS, 59; Mass Inst Technol, PhD(physics), 64. Prof Exp: Asst prof, 67-71, ASSOC PROF PHYSICS, BRIGHAM YOUNG UNIV, 71- Mem: Acoust Soc Am; Inst Elec & Electronic Engr. Res: Physics of musical instruments; analysis and synthesis of instrumental tones and of speech; machine recognition of speech; speech aids for the deaf. Mailing Add: Dept of Physics Brigham Young Univ Provo UT 84602

STRONGIN, MYRON, b New York, NY, July 27, 36; m 57; c 2. LOW TEMPERATURE PHYSICS. Educ: Rensselaer Polytech Inst, BS, 56; Yale Univ, MS, 57, PhD(physics), 62. Prof Exp: Mem staff, Lincoln Labs, Mass Inst Technol, 61-63; from asst physicist to assoc physicist, 63-67, physicist, 67-74, SR PHYSICIST, BROOKHAVEN NAT LAB, 74- Concurrent Pos: Adj prof, City Univ New York. Mem: Fel Am Phys Soc. Res: Properties of superconducting materials; analysis of surfaces and influence of surfaces on superconducting properties; epitaxy of films and superconductivity of films. Mailing Add: Physics Dept Brookhaven Nat Lat Upton NY 11973

STROSBERG, ARTHUR MARTIN, b Albany, NY, Sept 16, 40. PHARMACOLOGY. Educ: Siena Col, NY, BS, 62; Univ Calif, San Francisco, PhD(pharmacol), 70. Prof Exp: Pharmacologist, 70-72, sect head cardiovasc pharmacol, 72-75, PRIN SCIENTIST, SYNTEX RES, 75- Mem: AAAS; Sigma Xi. Res: Cardiovascular pharmacology, cardiotonic agents, antianginal agents, antiarrhythmic agents; contractile properties of cardiac muscle; cardiac muscle contraction mechanisms, cardiac muscle ultrastructure and oscillations. Mailing Add: Dept of Exp Pharmacol Syntex Res Stanford Indust Park Palo Alto CA 94304

STROSS, FRED HELMUT, b Alexandria, Egypt, Aug 22, 10; nat US; m 36; c 2. PHYSICAL CHEMISTRY, ARCHAEOLOGY. Educ: Case Inst Technol, BS, 34; Univ Calif, PhD(chem), 38. Prof Exp: Chemist, Shell Develop Co, 38-52, supvr res, 52-70; res asst anthrop, Univ Calif, Berkeley, 70-75; GUEST SCIENTIST, LAWRENCE BERKELEY LAB, UNIV CALIF, 75- Concurrent Pos: Consult, Lowie Mus & Univ Art Mus, Berkeley, 70-; chmn Nat Res Coun subcomt gas chromatography group, Int Union Pure & Appl Chem. Mem: AAAS; Am Chem Soc. Res: Photochemistry; asphalt technology; catalytic industrial processes; physical chemistry of solids; gas chromatography; analytical physical chemistry, including applications to characterization of polymers; archaeometry, the application of physical sciences to archaeology. Mailing Add: 44 Oak Dr Orinda CA 94563

STROSS, RAYMOND GEORGE, b St Charles, Mo, July 2, 30; m 64; c 3. ECOLOGY. Educ: Univ Mo, BS, 52; Univ Idaho, MS, 55; Univ Wis, PhD(zool), 58. Prof Exp: Res asst, Univ Wis, 54-58; NIH fel, Oceanog Inst, Woods Hole, 58-59; from asst prof to assoc prof zool, Univ Md, 59-67; ASSOC PROF ZOOL, STATE UNIV NY ALBANY, 67- Mem: AAAS; Am Soc Zool; Ecol Soc Am; Am Soc Limnol & Oceanog. Res: Arthropod diapause; biological rhythms of plankton populations; biological limnology; experimental ecology; arctic ecology. Mailing Add: Dept of Biol Sci State Univ NY Albany NY 12222

STROTHER, ALLEN, b Sweetwater, Tex, Feb 20, 28; m 57; c 2. PHARMACOLOGY, BIOCHEMISTRY. Educ: Tex Tech Col, BS, 55; Univ Calif, Davis, MS, 57; Tex A&M Univ, PhD(biochem, nutrit), 63. Prof Exp: Assoc animal sci, Univ Calif, Davis, 56-57; asst to trustee, Burnett Estate, Ft Worth, Tex, 58; dir nutrit res, Uncle Johnny Feed Mills, Houston, 59; res biochemist, Food & Drug Admin, DC, 63-65; from asst prof to assoc prof, 65-75, PROF PHARMACOL, SCH MED, LOMA LINDA UNIV, 75- Mem: Am Chem Soc; Am Soc Pharmacol & Exp Therapeut; Am Soc Animal Sci; Poultry Sci Asn. Res: Large and small animal nutrition; dietary energy levels; mineral requirements; drug and pesticide metabolism. Mailing Add: Dept of Pharmacol Loma Linda Univ Sch of Med Loma Linda CA 92354

STROTHER, CORNEILLE OSBURN, b Columbus, Ga, Apr 7, 08; m 41; c 3. INDUSTRIAL CHEMISTRY. Educ: Williams Col, AB, 30; Oberlin Col, MA, 31; Princeton Univ, MA, 32, PhD(chem), 33. Prof Exp: Int exchange fel, Univ Munich, 33-34; Loomis fel, Princeton Univ, 34-35; res chemist, Carbide & Carbon Chems Corp, WVa, 35-38 & Mellon Inst, Pa, 38-39; res chemist, Res Lab, Linde Co, 39-43, head div phys chem, 43-47, asst to supt, 47-51, asst supt, 51-53, asst dir res, 53-56, asst mgr res admin, Union Carbide Corp, 56-58, vpres res, Union Carbide Nuclear Co, 58-64, vpres nuclear div, Union Carbide Corp, 64-65, dir univ rels, 66-73; RETIRED. Concurrent Pos: Mem sci adv panel, US Dept Army, 58-65; mem, NY State Gen Adv Comt Atomic Energy, 61- Honors & Awards: Schoellkopf Medal, Am Chem Soc, 51. Mem: Am Chem Soc; AAAS. Res: Heterogeneous catalysis; ultraviolet absorption spectroscopy; kinetics of vinyl polymerizations; chemical reactions at very high pressures; reactions of metallo-organic compounds. Mailing Add: 197 Clinton Ave Dobbs Ferry NY 10522

STROTHER, GREENVILLE KASH, b Huntington, WVa, July 27, 20; m 50; c 3. BIOPHYSICS. Educ: Va Polytech Inst, BS, 43; George Washington Univ, MS, 54; Pa State Univ, PhD(physics), 57. Prof Exp: Asst prof physics, 57-61, assoc prof biophys, 61-72, PROF BIOPHYS, PA STATE UNIV, UNIVERSITY PARK, 72- Mem: Biophys Soc. Res: Microspectrophotometry of cellular systems; biophysical instrumentation. Mailing Add: Dept Physics 104 Davey Lab Pa State Univ University Park PA 16802

STROTHER, WAYMAN L, b US, Apr 19, 23; div; c 2. MATHEMATICS. Educ: Ala State Teachers Col, BS, 43; Univ Chicago, MS, 49; Tulane Univ, PhD(math), 51. Prof Exp: Asst prof math, Univ Miami, 48; instr, Ill Inst Technol, 49; asst prof, Univ Ala, 51; from asst prof to assoc prof, Univ Miami, 52-59; Buckingham prof & chmn dept, Miami Univ, 59-64; head dept, 64-72, PROF MATH, UNIV MASS, AMHERST, 64- Concurrent Pos: Mathematician, US Naval Ord Testing Sta, Calif, 56; sr res scientist, Missile & Space Div, Lockheed Aircraft Corp, 56-59, consult, Lockheed Missile & Space Corp, 59-61. Mem: Am Math Soc; Math Asn Am. Res: Topology; applied mathematics. Mailing Add: Dept of Math Univ of Mass Amherst MA 01002

STROTHMANN, RUDOLPH OTTO, b Milwaukee, Wis, Sept 13, 22; m 60. FORESTRY. Educ: Univ Mich, BSF, 50, MF, 51, PhD(forestry), 64. Prof Exp: RES FORESTER, US FOREST SERV, 53- Mem: Soc Am Foresters. Res: Silviculture. Mailing Add: US Forest Serv 1550 B St Arcata CA 95521

STROTTMAN, DANIEL, b Sumner, Iowa, Apr 15, 43; m 66; c 1. NUCLEAR PHYSICS. Educ: Univ Iowa, BA, 64; State Univ NY Stony Brook, MA, 66, PhD(physics), 69. Prof Exp: Niels Bohr fel nuclear physics, Niels Bohr Inst, Copenhagen, Denmark, 69-70; res officer, Oxford Univ, 70-74; ASST PROF PHYSICS, STATE UNIV NY STONY BROOK, 74- Concurrent Pos: Vis Nordita prof, Physics Inst, Univ Oslo, 73-74. Res: Group theory applications. Mailing Add: Dept Physics State Univ NY Stony Brook NY 11790

STROUBE, EDWARD W, b Hopkinsville, Ky, Apr 2, 27; m 54; c 3. AGRONOMY. Educ: Univ Ky, BS, 51, MS, 59; Ohio State Univ, PhD(agron), 61. Prof Exp: Agr exten agent, Univ Ky, 54-57, res asst agron, 57-58; res asst, 58-60, from instr to assoc prof, 60-70, PROF AGRON, OHIO STATE UNIV & OHIO AGR RES & DEVELOP CTR, 70- Mem: Am Soc Agron; Weed Sci Soc Am. Res: Weed control of field crops involving the evaluations of herbicides, tillage practices, flaming and crop rotations; soil and crop residue studies involving herbicides. Mailing Add: Dept of Agron Ohio State Univ Columbus OH 43210

STROUBE, WILLIAM HUGH, b Sturgis, Ky, June 24, 24; m 50; c 3. PLANT SCIENCE. Educ: Murray State Col, BSA, 49; Univ Ky, MSA, 51; La State Univ, PhD(plant path), 53. Prof Exp: Asst plant path, La State Univ, 51-53, asst plant pathologist, 53-54; agronomist, Agr Exp Sta, Univ Ky, 55-56, assoc agronomist, 56-66, actg chmn dept agron, 65-66; asst dean col sci & technol, 69-70, PROF AGR, WESTERN KY UNIV, 66- ASSOC DEAN COL SCI & TECHNOL, 71- Concurrent Pos: Collabr, Field Crops Sect, Agr Res Serv, USDA, 52-54. Mem: Am Phytopath Soc; Am Soc Agron. Res: Crop production and management. Mailing Add: Col of Sci & Technol Western Ky Univ Bowling Green KY 42101

STROUD, CARLOS RAY, b Owensboro, Ky, July 9, 42; m 62; c 3. QUANTUM OPTICS. Educ: Centre Col Ky, AB, 63; Wash Univ, PhD(physics), 69. Prof Exp: Asst prof, 70-75, ASSOC PROF OPTICS, UNIV ROCHESTER, 75- Concurrent Pos: NSF grant, Univ Rochester, 70- Mem: Am Phys Soc. Res: Foundations of quantum theory; quantum and semiclassical radiation theory; interactions of electromagnetic fields with matter; high resolution dye laser spectroscopy. Mailing Add: Inst of Optics Univ of Rochester Rochester NY 14627

STROUD, F AGNES NARANJO SCHMINK, b Albuquerque, NMex, July 23, 22; m 50, 66; c 1. RADIOBIOLOGY. Educ: Univ NMex, BS, 45; Univ Chicago, PhD, 66. Prof Exp: Res technician hemat, Los Alamos Sci Lab, NMex, 45-46; assoc cytologist, Argonne Nat Lab, 46-69; dir, Dept Tissue Cult, Pasadena Found Med Res, 69-70; sr res cytogeneticist, Image Processing Group, Sci Data Anal Sect, Jet Propulsion Lab, 70-75; STAFF CYTOGENETICIST, HEALTH RES DIV, LOS ALAMOS SCI LAB, 75- Honors & Awards: C Morrison Prize, NY Acad Sci, 55; Dipl Hon, Pan-Am Cancer Cytol Cong, 57. Mem: Radiation Res Soc; Am Soc Cell Biol; Biophys Soc; Tissue Cult Asn (corresp secy, 58); NY Acad Sci. Res: Tissue culture; automation of chromosome analysis by computers; effects of radiation on animal tumors; effects of ionizing radiation in vitro and in vivo; cell kinetics; mammalian radiation biology; chromosome analysis. Mailing Add: Health Res Div MS-880 Los Alamos Sci Lab Los Alamos NM 87545

STROUD, JACKSON SWAVERLY, b Cabarrus Co, NC, June 1, 31; m 61; c 2. EXPERIMENTAL SOLID STATE PHYSICS, ENGINEERING. Educ: Union Col, BS, 53; Ohio State Univ, MS, 57. Prof Exp: Physicist, Corning Glass Works, 57-67; PHYSICIST, BAUSCH & LOMB, INC, 67- Mem: Am Phys Soc; Am Ceramic Soc; Optical Soc Am; Inst Elec & Electronics Eng. Res: Solid state physics with specialized knowledge of glass; radiation chemistry; glass tank design optical properties of solids. Mailing Add: Bausch & Lomb Inc 1400 N Goodman St Rochester NY 14602

STROUD, JUNIUS BRUTUS, b Greensboro, NC, June 9, 29; m 55; c 3. ALGEBRA. Educ: Davidson Col, BS, 51; Univ Va, MA, 62, PhD(math), 65. Prof Exp: Instr math & sci, Fishburne Mil Sch, 53-57; teacher, High Sch, 57-58; from instr to asst prof math, 60-67, ASSOC PROF MATH, DAVIDSON COL, 67- Concurrent Pos: Vis lectr, Sec Schs, 62-63 & 65-66. Mem: Am Math Asn. Res: Simple Jordan algebras of characteristic two; finitely generated modules over a Dedekind ring. Mailing Add: Dept of Math Davidson Col Davidson NC 28036

STROUD, MALCOLM HERBERT, b Birmingham, Eng, May 17, 20; m 49; c 3. MEDICINE. Educ: Univ Birmingham, MB, ChB, 45; FRCS, 52; Am Bd Otolaryngol, dipl, 60. Prof Exp: From asst prof to assoc prof, 57-72, PROF MED, SCH MED, WASH UNIV, 72- Mem: Am Acad Ophthal & Otolaryngol. Res: Otology. Mailing Add: Dept of Otolaryngol Wash Univ Sch of Med St Louis MO 63110

STROUD, RICHARD HAMILTON, b Dedham, Mass, Apr 24, 18; m 43; c 2. ZOOLOGY. Educ: Bowdoin Col, BS, 39; Univ NH, MS, 42. Prof Exp: Asst bot, Bowdoin Col, 39; asst zool, Univ NH, 40-42; jr aquatic biologist, Tenn Valley Authority, 42, aquatic biologist, 46-48; chief aquatic biologist, Mass Dept Conserv, 49-53; asst exec vpres, 53-55, EXEC V PRES, SPORT FISHING INST, 55- Concurrent Pos: Consult, Calif Fish & Game Dept, 65-66, Ark Game & Fish Comn, 69, Iowa Conserv Comn, 70-71 & Tenn Valley Authority, 71-72; vpres, Sport Fishery Res Found, 62-; mem, World Panel Fishery Experts, Food & Agr Orgn, UN, Ocean Fisheries & Law of Sea Adv Comts, Dept State, Exec Comt, Natural Resources Coun Am, NAm Atlantic Salmon Coun & Marine Fisheries Adv Comt, Dept Com; fishery expert adv to Sen Select Comt Govt Opers. Mem: Am Fisheries Soc; Am Soc Ichthyologists & Herpetologists; Am Soc Limnol & Oceanog; Am Inst Fishery Res Biologists. Res: Fish population dynamics, behavior, ecology and life history. Mailing Add: Sport Fishing Inst 608 13th St NW Suite 801 Washington DC 20005

STROUD, ROBERT CHURCH, b Oakland, Calif, Jan 5, 18; m 47. PHYSIOLOGY. Educ: Princeton Univ, AB, 40; Univ Rochester, MS, 50, PhD(physiol), 52. Prof Exp: Chemist, Calco Chem Div, Am Cyanamid Co, 40-44 & Lederle Labs Div, 44-48; instr physiol, Grad Sch Med, Univ Pa, 52-53; assoc med physiol, Brookhaven Nat Lab, 53-54; asst prof pharmacol & res assoc aviation physiol, Ohio State Univ, 54-55; asst prof physiol, Med Col SC, 55-56; supvr physiologist, US Naval Med Res Lab, 56-61; pulmonary physiologist, Occup Health Res & Training Facil, USPHS, 61-62; chief res prog mgr life sci, Ames Res Ctr, NASA, 62-64; chief sci rev sect, Health Res Facil Br, NIH, 64-69, chief health res facil br, Div Educ & Res Facil, 69-70, chief, Training Grants & Awards Br, Nat Heart & Lung Inst, 70-73, EXEC SECY, REV BR, NAT HEART & LUNG INST, 73- Concurrent Pos: Lectr, Stanford Univ, 63-64. Honors & Awards: Lederle Med Fac Award, 55. Mem: Am Physiol Soc. Res: Cardiopulmonary and respiratory physiology; physiology of adaptation to high altitudes and submarine environments; physiology of diving; aerospace physiology. Mailing Add: Nat Heart & Lung Inst Bethesda MD 20014

STROUD, ROBERT MALONE, b St Louis, Mo, Mar 12, 31; m 55; c 2. IMMUNOLOGY. Educ: Harvard Univ, BS, 52, MD, 56. Prof Exp: Intern med, Cook County Hosp, Chicago, Ill, 56-57; resident, Barnes Hosp, St Louis, Mo, 59-61; dir rheumatology, Ga Warm Springs Found, 65-66; from asst prof to assoc prof med, 66-71, ASSOC PROF MICROBIOL & PROF MED, MED SCH, UNIV ALA, BIRMINGHAM, 71- Concurrent Pos: USPHS fel, Med Sch, Johns Hopkins Univ, 61-63, Helen Hay Whitney fel, 63-65. Mem: Am Asn Immunol; Am Rheumatism Asn; Am Soc Clin Invest; Am Acad Allergy. Res: Complement components and their relationship to inflammation. Mailing Add: Div of Clin Immunol & Rheumatol Univ of Ala Med Sch Birmingham AL 35294

STROUD, ROBERT WAYNE, b Jonesboro, Ark, May 24, 29; m 57; c 2. TEXTILE CHEMISTRY. Educ: Ark Col, BS, 50; Ga Inst Technol, MSCh, 54; Univ Tex, Austin, PhD(org chem), 63. Prof Exp: Teacher, Pub Schs, Ark, 49-50; chemist, Carbide & Carbon Chem Co, Tex, 53-54 & 56-58; res chemist, 62-75, SR RES CHEMIST, E I DU PONT DE NEMOURS & CO, INC, 75- Mem: Am Chem Soc. Res: Textile fibers chemistry and engineering. Mailing Add: 188 S Crest Rd Chattanooga TN 37404

STROUD, THOMAS WILLIAM FELIX, b Toronto, Ont, Apr 7, 36; m 62; c 4. STATISTICS. Educ: Univ Toronto, BA, 56, MA, 60; Stanford Univ, PhD(statist), 68. Prof Exp: Asst prof math, Acadia Univ, 60-64; asst prof, 68-75, ASSOC PROF MATH, QUEEN'S UNIV, ONT, 75- Concurrent Pos: Res asst, Sch Educ, Stanford Univ, 66-68; Nat Res Coun Can res grant, Queen's Univ, Ont, 69-; Educ Testing Serv vis res fel, Educ Testing Serv, 72-73. Mem: Am Statist Asn; Inst Math Statist; Psychomet Soc. Res: Multivariate analysis; large sample theory; data analysis; statistical applications in education and psychology; linear models; forecasting theory and methodology. Mailing Add: Dept of Math Queen's Univ Kingston ON Can

STROUP, RICHARD EUGENE, physics, see 12th edition

STROUSE, CHARLES EARL, b Ann Arbor, Mich, Jan 29, 44; m 72. CHEMISTRY. Educ: Pa State Univ, University Park, BS, 65; Univ Wis-Madison, PhD(phys chem), 69. Prof Exp: AEC fel, Los Alamos Sci Lab, 69-71; ASST PROF CHEM, UNIV CALIF, LOS ANGELES, 71- Mem: AAAS; Am Chem Soc; Am Crystallog Asn. Res: Structural chemistry. Mailing Add: Dept of Chem Univ of Calif Los Angeles CA 90024

STROUT, GEORGE M, b Mexico, Maine, Mar 1, 26; m 49; c 3. APPLIED PHYSICS. Educ: Harvard Univ, BS, 48; Univ NH, MEd, 55; Mich State Univ, PhD, 70. Prof Exp: Instr math & physics, Monson Acad, Mass, 47-50; instr physics, NH Tech Inst, 50-55; assoc prof, Hudson Valley Community Col, 55-63; dir, NH Tech Inst, 64-72; DIR PROG & FACILITIES DEVELOP, POST-SEC DIV, NH STATE DEPT EDUC, 72- Mem: Am Soc Eng Educ; Am Asn Physics Teachers; Am Asn Higher Educ. Res: Various aspects of vocational technical education at the two year college level. Mailing Add: Gilmanton Iron Works NH 03837

STROUT, RICHARD GOOLD, b Auburn, Maine, Nov 11, 27; m 50; c 2. ZOOLOGY, PARASITOLOGY. Educ: Univ Maine, BS, 50; Univ NH, MS, 54, PhD(parasitol), 61. Prof Exp: Instr poultry sci, 54-60, from asst prof to assoc prof, 60-68, PROF PARASITOL, UNIV NH, 68-, PARASITOLOGIST, 63- Concurrent Pos: Fel, Sch Med, La State Univ, 67. Mem: AAAS; Am Soc Parasitol. Res: In vitro culture and pathogenicity of blood parasites of fish and birds and avian coccidiosis. Mailing Add: Dept of Animal Sci Rm 404 Kendall Hall Univ of NH Durham NH 03824

STROUTH, MARY CLARUS, inorganic chemistry, organic chemistry, see 12th edition

STROVINK, MARK WILLIAM, b Santa Monica, Calif, July 22, 44; m 65; c 2. EXPERIMENTAL HIGH ENERGY PHYSICS. Educ: Mass Inst Technol, BS, 65; Princeton Univ, PhD(physics), 70. Prof Exp: From instr to asst prof physics, Princeton Univ, 70-73; ASST PROF PHYSICS, UNIV CALIF, BERKELEY, 73- Concurrent Pos: Vis asst prof physics, Cornell Univ, 71-72; mem subpanel future facilities, High Energy Physics Adv Panel, US Energy Res & Develop Agency, 74- Res: Muon interactions and production at high energy; principles of invariance to changes of energy scale and charge-parity inversion. Mailing Add: Lawrence Berkeley Lab 50-139 Univ of Calif Berkeley CA 94720

STROZIER, JOHN ALLEN, JR, b Miami, Fla, June 3, 34; m 62; c 3. SURFACE PHYSICS. Educ: Cornell Univ, BEP, 58; Univ Utah, PhD(physics), 66. Prof Exp: Instr physics, Univ Utah, 66-67; res assoc mat sci, Cornell Univ, 67-69; sr res assoc, State Univ NY Stony Brook, 69-71, asst prof, 71-74; PHYSICIST, BROOKHAVEN NAT LAB, 74- Mem: Am Phys Soc. Res: Surface physics; low energy electron diffraction, catalysis. Mailing Add: Dept of Physics Brookhaven Nat Lab Upton NY 11973

STRUBE, RICHARD ERNST, organic chemistry, see 12th edition

STRUBLE, DEAN L, b Wawota, Sask, Aug 29, 36; m 57; c 2. SYNTHETIC ORGANIC CHEMISTRY. Educ: Univ Sask, BA & MA, 61, PhD(org chem), 65. Prof Exp: Develop chemist, Du Pont of Can, Ont, 62-63; Nat Res Coun Can overseas fels, Imp Col, Univ London, 65-66 & Univ Adelaide, 66-67; RES SCIENTIST, CAN DEPT AGR, 68- Mem: Chem Inst Can. Res: Sterochemistry of free radical and nucleophilic addition reactions to activated olefinic compounds; sex pheromones of insects, particularly Lepidoptera and Coleoptera; chemical behavior of organophosphorus pesticides. Mailing Add: Res Sta Can Dept Agr Lethbridge AB Can

STRUBLE, GEORGE W, b Philadelphia, Pa, July 6, 32; m 55; c 3. COMPUTER SCIENCE. Educ: Swarthmore Col, AB, 54; Univ Wis, MS, 57, PhD(math), 61. Prof Exp: Proj supvr, Numerical Anal Lab, Univ Wis, 60-61; from asst prof to assoc prof math, 61-69, res assoc, 61-65, assoc dir statist lab & comput ctr, 65-69, dir comput ctr, 69-74, ASSOC PROF COMPUT SCI, UNIV ORE, 69- Concurrent Pos: Consult, Computer Mgt Serv, Inc, Portland, Ore, 74- Mem: Data Processing Mgt Asn; Asn Comput Mach. Res: Business data processing. Mailing Add: Dept of Computer Sci Univ of Ore Eugene OR 97403

STRUBLE, GORDON LEE, b Cleveland, Ohio, Mar 7, 37; m 61; c 4. NUCLEAR CHEMISTRY. Educ: Rollins Col, BS, 60; Fla State Univ, PhD(chem), 64. Prof Exp: Fel, Lawrence Radiation Lab, Univ Calif, 64-66; asst prof chem, Univ Calif, Berkeley,

66-71; STAFF CHEMIST, LAWRENCE LIVERMORE LAB, 71- Concurrent Pos: Prof physics, Univ Munich, 75. Mem: AAAS; Am Chem Soc; Am Phys Soc. Res: Experimental and theoretical low energy nuclear structure and reaction physics, determination of characteristics of low energy excitations in nuclei by nuclear reactions and decay processes and their description by theoretical many body techniques. Mailing Add: Lawrence Livermore Lab Dept Chem PO Box 808 Livermore CA 94550

STRUBLE, RAIMOND ALDRICH, b Forest Lake, Minn, Dec 10, 24; m 46; c 5. MATHEMATICS. Educ: Univ Notre Dame, PhD(math), 51. Prof Exp: Aerodynamicist, Douglas Aircraft Co, 51-53; asst prof math, Ill Inst Technol, 53-58; asst prof, 58-60, PROF MATH, NC STATE UNIV, 60- Concurrent Pos: Consult, Armour Res Found, 54-58. Res: Fourier analysis and almost periodic functions; nonlinear differential equations; applied mathematics. Mailing Add: Dept of Math NC State Univ Raleigh NC 27607

STRUCHTEMEYER, ROLAND AUGUST, b Wright City, Mo, Jan 4, 18; m 40; c 2. SOILS. Educ: Univ Mo, BS, 39, MA, 41; Ohio State Univ, PhD(agron), 52. Prof Exp: Asst soils, Univ Mo, 39-40 & Ohio State Univ, 40-42; explosive chemist, Certainteed Corp, Tex, 42-43; head dept soils, 46-71, PROF SOILS, UNIV MAINE, ORONO, 46- Concurrent Pos: Agron fel, 65; soil fertility specialist, IRI Res Inst, Brazil, 66. Mem: Am Soc Agron; Soil Sci Soc Am; Int Soc Soil Sci. Res: Plant nutrition. Mailing Add: Dept Plant & Soil Sci Univ of Maine Orono ME 04473

STRUCK, CHARLES WILLIAM, physical chemistry, see 12th edition

STRUCK, JACOB, JR, b Paterson, NJ, Mar 13, 27; m 51; c 2. BIOCHEMISTRY. Educ: Rutgers Univ, BS, 51; Mass Inst Technol, PhD(biochem), 58. Prof Exp: Jr biochemist, Takamine Lab, Inc, 51-53; asst, Mass Inst Technol, 53-58; res biochemist, Miles Chem Co, 58-61; sr scientist, 61-69, dir div biochem, 69-70, ASST DIR RES DIAG, ORTHO RES FOUND, 70- Mem: AAAS; Am Chem Soc; Am Asn Clin Chem; NY Acad Sci. Res: Enzyme chemistry; metabolism of amino and hydroxy acids; clincial chemistry. Mailing Add: Ortho Res Found Raritan NJ 08869

STRUCK, ROBERT FREDERICK, b Pensacola, Fla, Jan 9, 32; m 63. ORGANIC CHEMISTRY. Educ: Auburn Univ, BS, 53, MS, 57, PhD(org chem), 61. Prof Exp: Assoc scientist, Southern Res Inst, 57-58; org chemist, Fruit & Veg Prod Lab, Agr Res Serv, USDA, 61; res scientist, 61-64, SR SCIENTIST, SOUTHERN RES INST, 64- Mem: Am Chem Soc; Am Asn Cancer Res. Res: Metabolism of anticancer drugs; organophosphorous and natural products chemistry; synthesis in organic heterocyclic chemistry. Mailing Add: Southern Res Inst 2000 Ninth Ave S Birmingham AL 35205

STRUCK, WILLIAM ANTHONY, b Paterson, NJ, Mar 17, 20; m 43; c 3. ANALYTICAL CHEMISTRY. Educ: Calvin Col, AB, 40; Univ Mich, MS, 62, PhD, 63. Prof Exp: Microanalyst, 41-48, head chem res anal, 48-62, mgr phys & anal chem res, 62-68, from asst dir to dir supportive res, 68-74, DIR, INT PHARMACEUT RES & DEVELOP, UPJOHN CO, 74- Mem: AAAS; Am Chem Soc. Res: Organic electrochemistry; organic analysis; optical rotatory dispersion. Mailing Add: 2102 Waite Ave Kalamazoo MI 49008

STRUCKMEYER, BURDEAN ESTHER, b Cottage Grove, Wis, May 25, 12. HORTICULTURE. Educ: Univ Wis, BA, 35, MA, 36, PhD, 39. Prof Exp: From instr to assoc prof, 39-65, PROF HORT, UNIV WIS-MADISON, 65- Mem: Fel Am Soc Hort Sci; Bot Soc Am; Am Soc Plant Physiologists; Int Soc Hort Sci; Int Soc Plant Morphol. Re Horticultural plants, especially their anatomical structure as influenced by mineral nutrition, physiological and pathological diseases and growth substances; investigations on flowering of plants. Mailing Add: Dept of Hort Univ of Wis-Madison Madison WI 53706

STRUEMPLER, ARTHUR W, b Lexington, Nebr, Dec 12, 20; m 50; c 2. ANALYTICAL CHEMISTRY. Educ: Univ Nebr, BS, 50, MS, 55; Iowa State Univ, PhD, 57. Prof Exp: Asst prof, Chico State Col, 57-60; fel biochem, Univ Calif, Davis, 60-62; opers analyst, Strategic Air Command Hq, Nebr, 62-65; HEAD DIV SCI & MATH, CHADRON STATE COL, 65- Mem: Am Chem Soc; Sigma Xi; AAAS. Res: Weather modification studies relating to element concentrations in precipitation; geochemical studies. Mailing Add: Div of Sci & Math Chadron State Col Chadron NE 69337

STRUEVER, STUART, b Peru, Ill, Aug 4, 31; m 56; c 2. ANTHROPOLOGY, ARCHAEOLOGY. Educ: Dartmouth Col, BA, 53; Northwestern Univ, MA, 60; Univ Chicago, PhD(anthrop), 68. Prof Exp: Instr anthrop, Univ Chicago, 64-65; from instr to assoc prof, 65-71, PROF ANTHROP, NORTHWESTERN UNIV, EVANSTON, 71- Concurrent Pos: Chmn bd, Found Ill Archaeol, 65-; lectr, Univ Chicago, 68-; Wenner-Gren Found Anthrop Res res grant, 68-69; NSF res grants, 68-70; mem, Hist Sites Adv Coun, Ill, 69-; ed, Memoirs, Soc Am Archaeol, 69- Mem: Soc Am Archaeol; Am Anthrop Asn. Res: Intensive reconstruction of prehistoric lifeways in the central Mississippi Valley from 8000 BC to European arrival; development of archaeological field strategy and methods using the long-term program in the central Mississippi Valley as a field laboratory. Mailing Add: Dept of Anthrop Northwestern Univ Evanston IL 60201

STRUHSAKER, JEANNETTE ADAIR WHIPPLE, b Mt Vernon, Wash, Jan 5, 36; div. FISH BIOLOGY, ENVIRONMENTAL PHYSIOLOGY. Educ: Western Wash Col, BA, 58; Univ Hawaii, PhD(zool), 66. Prof Exp: Instr zool, Duke Univ, 63-65, res assoc marine zool, Marine Lab, 65-67; asst marine biologist, Hawaii Inst Marine Biol, Univ Hawaii, 67-70, affil staff, Dept Oceanog, 71-72; FISHERIES BIOLOGIST, NAT MARINE FISHERIES SERV, TIBURON, SOUTHWEST FISHERIES CTR, 72- Concurrent Pos: Res coinvestr, Marine Lab, Duke Univ, NSF grant, 65-67; proj leader, Fisheries Prog-baitfish, Hawaii Inst Marine Zool, 67-70, primary investr, NSF grant, 67-69, coordr larval rearing, Sea Grant prog, 70-72 & coinvestr, NSF, grant, 71-72, chief physiol invest, Nat Marine Fisheries Serv, Tiburon, Southwest Fisheries Ctr, 73- Honors & Awards: Merit Award, Nat Marine Fisheries Serv, Tiburon Lab, 74. Res: Marine ecology; population ecology mollusks; fisheries biology, physiology; biology and rearing of marine larvae; aquaculture; effects of oil pollutants on fishes and marine food chains. Mailing Add: Nat Marine Fisheries Serv Southwest Fisheries Ctr Tiburon CA 94920

STRUHSAKER, PAUL JAMES, b Lansing, Mich, July 16, 35. MARINE ZOOLOGY, FISHERIES. Educ: Mich State Univ, BS, 58; Univ Hawaii, MS, 66, PhD(zool), 73. Prof Exp: FISHERIES BIOLOGIST, NAT MARINE FISHERIES SERV, 59-65 & 69- Mem: Am Soc Ichthyologists & Herpetologists. Res: Systematics and ecology of marine fishes; fisheries development. Mailing Add: Nat Marine Fisheries Serv PO Box 3830 Honolulu HI 96812

STRUIK, RUTH REBEKKA, b Mass, Dec 15, 28; div; c 3. MATHEMATICS. Educ: Swarthmore Col, BA, 49; Univ Ill, MA, 51; NY Univ, PhD(math), 55. Prof Exp: Digital comput programmer, Univ Ill, 50-51; asst, Univ Chicago, 52; lectr math, Sch Gen Studies, Columbia Univ, 55; asst prof, Drexel Inst Technol, 56-57; lectr, Univ BC, 57-61; actg asst prof, 61-62, asst prof, 62-63, 64-65, ASSOC PROF MATH, UNIV COLO, BOULDER, 65- Mem: Math Asn Am; Am Math Soc. Res: Groups; modern algebra. Mailing Add: Dept of Math ESCR 2-38 Univ of Colo Boulder CO 80302

STRUMEYER, DAVID HYMAN, b Brooklyn, NY, Oct 11, 34; m 57. BIOCHEMISTRY. Educ: Brooklyn Col, BS, 55; Harvard Univ, MA, 56, PhD(biochem), 59. Prof Exp: Am Cancer Soc res fel, Univ Calif, Berkeley, 59-61; res assoc biochem, Brookhaven Nat Lab, 61-63; sr res scientist, Bristol-Meyers Co, 63-64; asst prof biochem & microbiol, 64-68, ASSOC PROF BIOCHEM & MICROBIOL, RUTGERS UNIV, 68- Mem: Am Chem Soc; Am Soc Biol Chemists. Res: Protein chemistry; mechanism of enzyme action especially of proteolytic and amylolytic enzymes; plant phenolics, interaction with proteins; automated analyses; immobilized enzyme technology; amylese inhibitors; celiac disease; amino acid analysis. Mailing Add: Dept of Biochem & Microbiol Lipman Hall Cook Col Rutgers Univ New Brunswick NJ 08903

STRUMPF, ALBERT, b New York, NY, June 17, 23; m 53; c 2. APPLIED MATHEMATICS. Educ: Polytech Inst Brooklyn, BCE, 44; Stevens Inst Technol, MS, 53, PhD(appl math), 64. Prof Exp: Res engr, exp towing tank, 44-56, staff scientist, 56-60, DIV CHIEF STABILITY & CONTROL, DAVIDSON LAB, STEVENS INST TECHNOL, 60- Concurrent Pos: Mem hydroballistics adv comt, Bur Ord, 60-63; mem monitor activities stability & control & mem hydroballistics adv comt, Bur Naval Weapons, 63-66 & Naval Ord Systs Command, 66-; assoc prof ocean eng, Stevens Inst Technol, 70-73. Res: Dynamics and control of vehicles that move through fluids; submersibles, towed bodies, rockets, surface ships, missiles, automobiles and airships. Mailing Add: Davidson Lab Stevens Inst of Technol 711 Hudson St Hoboken NJ 07030

STRUMWASSER, FELIX, b Trinidad, BWI, Apr 16, 34; nat US. PHYSIOLOGY, NEUROBIOLOGY. Educ: Univ Calif, Los Angeles, BA, 53, PhD(zool), 57. Prof Exp: Asst, Univ Calif, Los Angeles, 56-57; from asst scientist to sr asst scientist, Lab Neurophysiol, NIMH, 57-60; res assoc neurophysiol, Walter Reed Inst Res, 60-64; assoc prof, 64-69, PROF BIOL, CALIF INST TECHNOL, 69- Concurrent Pos: Res assoc, Washington Sch Psychiat, 60-64; Penn lectr, Univ Pa, 67; Carter-Wallace lectr, Princeton Univ, 68; mem fel comt, NIH, 68-70; mem biochronometry comt, NSF, 69. Mem: AAAS; Am Soc Zoologists; Soc Gen Physiol; Am Physiol Soc; NY Acad Sci. Res: Neurophysiology; neurocellular basis of behavior, sleep-waking, reproductive; mechanisms of circadian rhythms in nervous systems; long-term studies on single identifiable neurons in organ culture; integrative mechanisms of the neuron; comparative neurophysiology; pacemaker mechanisms in neurons. Mailing Add: Div of Biol Calif Inst of Technology Pasadena CA 91109

STRUNK, DUANE H, b Irene, SDak, Mar 14, 20; m 45; c 2. ANALYTICAL CHEMISTRY. Educ: Univ SDak, BA, 42; Univ Louisville, MS, 51. Prof Exp: Res supvr, 42-43, prod supvr, 43-44, maintenance supvr, 43-45, res chemist, 46-65, CONTROL LABS ADMINSTR, JOSEPH E SEAGRAM & SONS, INC, 66- Mem: Am Chem Soc; Am Water Works Asn. Res: Microanalytical methods, especially colorimetric, flame spectrophotometry and atomic absorption methods for copper and magnesium; water, food and wood chemistry; high accuracy particle counter. Mailing Add: 3104 Gambriel Ct Louisville KY 40205

STRUNK, RICHARD JOHN, b Jamaica, NY, July 6, 41; m 68; c 2. ORGANIC CHEMISTRY. Educ: Gettysburg Col, AB, 63; State Univ NY Albany, PhD(org chem), 67. Prof Exp: RES SCIENTIST, RES CTR, UNIROYAL INC, 67- Mem: Am Chem Soc; Sigma Xi. Res: Organometallic and free radical chemistry; organometallic and pesticide chemistry. Mailing Add: Uniroyal Inc Oxford Mgt & Res Ctr Middlebury CT 06749

STRUNK, WILLIAM G, chemistry, see 12th edition

STRUTHERS, PAUL HERBERT, soil chemistry, see 12th edition

STRUTHERS, ROBERT CLAFLIN, b Syracuse, NY, June 2, 28; m 52; c 6. COMPARATIVE ANATOMY, DEVELOPMENTAL ANATOMY. Educ: Syracuse Univ, BA, 50, MA, 52; Univ Rochester, PhD(biol), 56. Prof Exp: Asst comp & develop anat, Syracuse Univ, 50-56; from instr to asst prof anat, Ohio State Univ, 56-61; assoc prof, 61-62, PROF ANAT, WHEELOCK COL, 62- Res: Morphology of early embryonic stages of vertebrate animals, particularly on the pharynx and its derivatives. Mailing Add: Dept of Sci & Math Wheelock Col 200 Riverway Boston MA 02215

STRUTT, JOSEPH RAYMOND ARNOLD, mathematics, see 12th edition

STRUVE, WILLIAM GEORGE, b Milwaukee, Wis, Mar 19, 38; c 2. BIOCHEMISTRY, CHEMISTRY. Educ: Lake Forest Col, BA, 62; Northwestern Univ, PhD, 66. Prof Exp: Res chemist, Am Cyanamid Co, 66-68; ASSOC PROF BIOCHEM, UNIV TENN, MEMPHIS, 68- Concurrent Pos: Vis scholar, Stanford Univ, 69-70. Res: Biological membranes; molecular biology of the central nervous system; acetylocholinesterase. Mailing Add: Rm 218 Nash Bldg Dept of Biochem Univ of Tenn Memphis TN 38163

STRUVE, WILLIAM SCOTT, b Utica, NY, May 1, 15; m 39; c 3. ORGANIC CHEMISTRY. Educ: Univ Mich, BS, 37, MS, 38, PhD(org chem), 40. Prof Exp: Du Pont fel, Univ Mich, 40-41; chemist, Jackson Lab, 41-49, res supvr, 49-58, lab dir, Color Res Lab, 58-73, DIR NEWARK LAB & MGR COLORS RES & DEVELOP, E I DU PONT DE NEMOURS & CO, 73- Mem: AAAS; Am Chem Soc. Res: Carcinogenic hydrocarbons; organic fluorine compounds; dyestuffs; pigments. Mailing Add: 29 Dellwood Ave Chatham NJ 07928

STRUZYNSKI, RAYMOND EDWARD, b Jersey City, NJ, Dec 10, 37; m 65; c 2. LOW TEMPERATURE PHYSICS. Educ: Stevens Inst Technol, BEng, 59, MS, 61, PhD(physics), 65. Prof Exp: From instr to asst prof, 64-70, ASSOC PROF PHYSICS, BROOKLYN COL, 70- Concurrent Pos: Res scientist, Hudson Labs, Columbia Univ, 65-67. Mem: Am Phys Soc. Res: Radiative beta decay; liquid helium; quantum mechanics of many boson systems. Mailing Add: Dept of Physics Brooklyn Col Bedford Ave & Ave H Brooklyn NY 11210

STRYCKER, STANLEY JULIAN, b Goshen, Ind, Aug 30, 31; m 52; c 4. ORGANIC CHEMISTRY. Educ: Goshen Col, AB, 53; Univ Ill, PhD(org chem), 56. Prof Exp: Res chemist, 56-63, sr res chemist, 63-68, group leader, 68-69, res mgr pharmaceut sci, 69-74, DIR CHEM & PROD DEVELOP, DOW CHEM CO, 74- Concurrent Pos: Sabbatical, Col Med, Univ Iowa, 67-68. Mem: AAAS; Am Chem Soc; Sigma Xi. Res: Organic synthesis; heterocyclics; medicinal chemistry. Mailing Add: R2 Box 100-53 Zionsville IN 46077

STRYER, LUBERT, b Tientsin, China, Mar 2, 38; US citizen; m 58; c 2. BIOCHEMISTRY, BIOPHYSICS. Educ: Univ Chicago, BS, 57; Harvard Univ, MD,

61. Prof Exp: From asst prof to assoc prof biochem, Stanford Univ, 63-69; PROF MOLECULAR BIOPHYS & BIOCHEM, YALE UNIV, 69- Concurrent Pos: Helen Hay Whitney fel, Harvard Univ & Med Res Coun Lab Molecular Biol, Cambridge, Eng, 61-63; consult, NIH, 67-71; assoc ed, Ann Rev Biophys & Bioeng, 70; mem biol vis comt, Brookhaven Nat Lab, 76- Honors & Awards: Eli Lilly Award, Am Chem Soc, 70. Mem: Am Soc Biol Chemists; Biophys Soc; fel Am Acad Arts & Sci. Res: Protein structure and function; visual excitation; excitable membranes; spectroscopy; neutron diffraction. Mailing Add: Yale Univ Box 1937 Yale Sta New Haven CT 06520

STRYK, ROBERT A, physics, see 12th edition

STRYKER, HARRY KANE, b Arkansas City, Kans, Feb 5, 21; m 42; c 2. POLYMER CHEMISTRY, APPLIED STATISTICS. Educ: Kans State Col, Pittsburg, BS, 42, MS, 50. Prof Exp: From analyst to sr analyst chem, Spencer Chem Co, 47-49, staff asst chem res, 49-52, staff assoc, 52-53, from staff mem to sr staff mem, 53-58, staff specialist, 58-62, sr res chemist, Spencer Chem Co & Gulf Oil Co, 62-66, sr res chemist, Gulf Res & Develop Co, 66-74, RES ASSOC, GULF OIL CHEM CO, 74- Mem: Am Chem Soc; fel Am Inst Chemists. Res: Oxidation, reduction and polymerization by flow reactions; emulsion and high pressure polymerization of ethylene; stability and applications of polyethylene latexes; polymer characterization by techniques of gel permeation chromatography; column elution; differential scanning calorimetry. Mailing Add: Gulf Oil Chem Co Houston Res Lab PO Box 79070 Houston TX 77079

STRYKER, LYNDEN JOEL, b Stamford, NY, Feb 19, 43; m 71. COLLOID CHEMISTRY, SURFACE CHEMISTRY. Educ: Clarkson Col Technol, BS, 64, PhD(chem), 69. Prof Exp: Asst, Clarkson Col Technol, 67-68; lectureship, Brunel Univ, 69-70; RES CHEMIST, WESTVACO CORP, 70- Mem: Am Chem Soc. Res: Emulsion technology; coagulation of colloidal dispersions; surface characterization of solids; air and water pollution control; cement chemistry. Mailing Add: 1257 Wappetaw Pl Mt Pleasant SC 29464

STRYKER, WALTER ALBERT, b Grand Rapids, Mich, Dec 4, 10; m 40; c 3. PATHOLOGY. Educ: Univ Mich, AB, 33; Univ Chicago, MD & PhD(path), 40. Prof Exp: Asst path, Med Sch, Univ Chicago, 36-40; from instr to asst prof, Univ Mich, 42-46; PATHOLOGIST & DIR LAB, WYANDOTTE GEN HOSP, 46- Concurrent Pos: Asst prof path, Wayne State Univ, 47-53, assoc prof, 53- Mem: Am Soc Clin Path; Am Asn Pathologists & Bacteriologists; Col Am Pathologists. Res: Experimental general pathology; histopathology. Mailing Add: Dept of Path Wyandotte Gen Hosp Wyandotte MI 48192

STRYLAND, JAN CORNELIS, b Doorn, Neth, Oct 11, 14; m 43; c 3. MOLECULAR PHYSICS. Educ: Univ Amsterdam, PhD(physics), 53. Prof Exp: Conservator, Van der Waals Lab, Amsterdam, 43-54; PROF PHYSICS, UNIV TORONTO, 54- Mem: Am Phys Soc; Can Asn Physicists; Neth Phys Soc. Res: Intermolecular forces; investigation of physical phenomena at high pressures. Mailing Add: Dept of Physics Univ of Toronto Toronto ON Can

STUART, ALFRED HERBERT, b Farmville, Va, 13; m 44; c 2. PHOTOGRAPHIC CHEMISTRY. Educ: Hampden-Sydney Col, BS, 33; Univ Va, PhD(org chem), 37. Prof Exp: Res fel, Univ Va, 37-39; res chemist, Schieffelin & Co, NY, 39-43, dir chem res, 43-47; develop mgr, Charles Bruning Co, Ill, 47-65, tech dir, 65-71, CHIEF CHEMIST, BRUNING DIV, ADDRESSOGRAPH-MULTIGRAPH CORP, GUILFORD, 71- Res: Diazotype and electrostatic copying processes. Mailing Add: 11-D Harbour Village Branford CT 06405

STUART, ALFRED WRIGHT, b Pulaski, Va, Nov 16, 32; m 60; c 4. URBAN GEOGRAPHY. Educ: Univ SC, BS, 55; Emory Univ, MS, 56; Ohio State Univ, PhD(geog), 66. Prof Exp: Glaciologist, US Antarctic Res Prog, 58-60; res assoc glaciol, Inst Polar Studies, Ohio State Univ, 60-61, asst instr geog, 61-62, community planner, City of Roanoke, Va, 63-64; asst prof geog, Univ Tenn, 64-69; assoc prof, 69-73, PROF GEOG, UNIV NC, CHARLOTTE, 73-, CHMN DEPT, 69- Concurrent Pos: Urban planner, US Bur Mines, 68-69; mem ed bd, Southeastern Geogr, 68-; consult, Urban Decentralization Study, Oak Ridge Nat Lab, 69; co-ed, NC Atlas Proj, 72- Mem: Asn Am Geog. Res: Suburbanization of manufacturing within urban areas; rural industrialization in relation to population change; urban and industrial structure of the Piedmont South. Mailing Add: Dept of Geog & Earth Sci Univ of NC Charlotte NC 28223

STUART, ANN ELIZABETH, b Harrisburg, Pa, Oct 5, 43. NEUROBIOLOGY. Educ: Swarthmore Col, BA, 65; Yale Univ, MS, 67, PhD(physiol), 69. Prof Exp: Res fel neurophysiol, Dept Physiol, Univ Calif, Los Angeles Med Ctr, 71-73; res fel neurochem, 69-71, ASST PROF, DEPT NEUROBIOL, HARVARD MED SCH, 73- Mem: Soc Neurosci; Soc Gen Physiologists; Int Brain Res Orgn. Res: Synaptic mechanisms; integration and synaptic transmission between single cells of small populations of neurons; integration in invertebrate central nervous systems. Mailing Add: Dept of Neurobiol Harvard Med Sch 25 Shattuck St Boston MA 02115

STUART, ARCHIBALD PAXTON, organic chemistry, see 12th edition

STUART, DAVID EDWARD, b Anniston, Ala, Jan 9, 45; m 71. ANTHROPOLOGY, RESEARCH ADMINISTRATION. Educ: WVa Wesleyan Col, BA, 67; Univ NMex, MA, 70, PhD(anthrop), 72. Prof Exp: Asst prof anthrop, Eckerd Col, 72-74; RES & PROJ COORDR ANTHROP, OFF CONTRACT ARCHAEOL, UNIV N MEX, 74- Concurrent Pos: Instr, Div Continuing Educ, Univ NMex, 74-, asst prof anthrop, 75- Mem: Am Anthrop Asn; Royal Anthrop Inst Gt Brit; Am Soc Ethnohist. Res: Cultural ecology of nonagricultural populations with specific emphasis on the adaptive nature of subsistence behavior, population distribution and mobility, and social organization among hunter-gatherers. Mailing Add: Dept of Anthrop Univ of NMex Albuquerque NM 87131

STUART, DAVID GORDON, b Hollis, Maine, Apr 7, 36; m 56; c 1. MICROBIOLOGY. Educ: Gordon Col, BA, 61, BS, 63; Univ NH, PhD(microbiol), 68. Prof Exp: Asst prof, 67-72, ASSOC PROF MICROBIOL, MONT STATE UNIV, 72- Mem: AAAS; Soil Conserv Soc Am; Am Soc Microbiol. Res: Bacterial physiology; water quality. Mailing Add: Dept of Bot & Microbiol Mont State Univ Bozeman MT 59715

STUART, DAVID MARSHALL, b Ogden, Utah, May 20, 28; m 51; c 5. PHARMACY, CHEMISTRY. Educ: Univ Utah, BS, 51; Univ Wis, PhD(pharmaceut chem), 55. Prof Exp: Asst prof pharmaceut chem, Univ Tex, 55-S7 & Ore State Col, 57-60; coordr sci info, Neisler Labs, 60-64; PROF PHARMACEUT CHEM, COL PHARM, OHIO NORTHERN UNIV, 64- Concurrent Pos: Pharm consult, Ohio Dept Pub Welfare, 71-; pres, Pharm Health & Related Mgt, Inc, 74- Mem: Am Pharmaceut Asn; NY Acad Sci. Res: Scientific literature research and writing; biochemistry. Mailing Add: Dept of Pharmaceut Chem Col of Pharm Ohio Northern Univ Ada OH 45810

STUART, DAVID W, b Lafayette, Ind, June 15, 32; m 64; c 2. METEOROLOGY. Educ: Univ Calif, Los Angeles, BA, 55, MA, 57, PhD(meteorol), 62. Prof Exp: Res meteorologist, Univ Calif, Los Angeles, 55-61, teaching asst meteorol, 57-61; asst prof, 62-66, ASSOC PROF METEOROL, FLA STATE UNIV, 66-, ASSOC CHMN DEPT, 67- Concurrent Pos: Assoc prof, Naval Postgrad Sch, 66-67. Mem: Am Meteorol Soc; Am Geophys Union. Res: Synoptic meteorology and numerical weather prediction, especially diagnostic studies. Mailing Add: Dept of Meteorol Fla State Univ Tallahassee FL 32306

STUART, DERALD ARCHIE, b Bingham Canyon, Utah, Nov 9, 25; m 48; c 2. SOLID STATE PHYSICS. Educ: Univ Utah, BS, 47, MS, 48, PhD(physics), 50. Prof Exp: Asst physics, Univ Utah, 47-50; asst prof eng mat, Cornell Univ, 50-52, assoc prof eng mech & mat, 52-58; propulsion staff mgr & resident rep to Aerojet-Gen Corp, 58-59, asst to Polaris Missile syst mgr & Polaris resident rep to Aerojet-Gen Corp, 59-61, mgr propulsion staff, Missile Systs Div, 61-62, dir propulsion systs, 62-64, asst chief engr, 64-66, asst gen mgr eng & develop, 66-67, vpres & asst gen mgr, 67-70, VPRES CORP & VPRES & GEN MGR MISSILE SYSTS DIV, LOCKHEED MISSILES & SPACE CO, LOCKHEED AIRCRAFT CORP, 70- Concurrent Pos: Consult, Cornell Aeronaut Lab, 52-54, Allegany Ballistics Lab, 52-58, Lincoln Lab, Mass Inst Technol & Ramo-Wooldridge Corp, 54-56. Honors & Awards: Meyer Award, Am Ceramic Soc, 54; Meritorious Pub Serv Award & Cert Commendation, US Navy, 61; Montgomery Award, Nat Soc Aerospace Prof, 64. Mem: Assoc fel Am Inst Aeronaut & Astronaut. Res: Glassy state; plastic behavior of materials; solid fuel rockets. Mailing Add: Dept 80-01 Bldg 153 PO Box 504 Lockheed Missiles & Space Co Inc Sunnyvale CA 94088

STUART, DOUGLAS GORDON, b Casino, NSW, Australia, Oct 5, 31; US citizen; m 57; c 4. PHYSIOLOGY. Educ: Mich State Univ, BS, 55, MA, 56; Univ Calif, Los Angeles, PhD(physiol), 61. Prof Exp: Res fel anat, Univ Calif, Los Angeles, 61-63, asst prof physiol in residence, 63-65, assoc prof, Univ Calif, Davis, 65-67; PROF PHYSIOL, COL MED, UNIV ARIZ, 70- Mem: Am Physiol Soc. Res: Neural control of posture and locomotion. Mailing Add: Dept of Physiol Univ of Ariz Col of Med Tucson AZ 85724

STUART, EDWARD E, veterinary medicine, see 12th edition

STUART, GEORGE WALLACE, b New York, NY, Apr 5, 24; m 48; c 3. THEORETICAL PHYSICS. Educ: Rensselaer Polytech Inst, BEE, 49, MS, 50; Mass Inst Technol, PhD(physics), 53. Prof Exp: Head theoret physics unit, Hanford Atomic Prod Oper, Gen Electric Co, 52-57; res adv spec nuclear effects lab, Gen Atomic Div, Gen Dynamics Corp, Calif, 58-67; sr res scientist, Systs, Sci & Software, 67-72; STAFF SCIENTIST, SCI APPLNS INC, 72- Mem: Am Phys Soc. Res: Nuclear reactors; plasma theory; atomic physics. Mailing Add: Sci Applns Inc 1250 Prospect La Jolla CA 92037

STUART, JEANNE JONES, b Atlanta, Ga, Dec 22, 42; m 61; c 3. ZOOLOGY. Educ: Jacksonville State Univ, BS, 65, MS, 66; Auburn Univ, PhD(zool), 72. Prof Exp: Teacher french, Calhoun County Bd Educ, Ala, 65-67; instr biol, Jacksonville State Univ, 67-68; teaching asst zool, Auburn Univ, 68-72, res assoc biochem, 72-73; ASSOC PROF BIOL & HEAD DEPT, BELMONT ABBEY COL, 73- Mem: AAAS; Sigma Xi; Am Soc Parasitologists; Am Inst Biol Sci. Res: Immunological phenomena associated with trichostrongylid parasitism. Mailing Add: Dept of Biol Belmont Abbey Col Belmont NC 28012

STUART, JOE DON, b Brownsboro, Tex, Feb 28, 32; m 63; c 2. ENVIRONMENTAL SCIENCES, COMPUTER SCIENCE. Educ: Univ Tex, BA, 57, PhD(physics), 63. Prof Exp: Sr physicist appl physics lab, Johns Hopkins Univ, 63-65; sr scientist, Tracor, Inc, 65-72; SR SCIENTIST, RADIAN CORP, 72- Mem: Opers Res Soc Am. Res: Air, surface and ground water pollution control operations research. Mailing Add: 4009 Knollwood Dr Austin TX 78731

STUART, LAURENCE COOPER, b Dubois, Pa, July 9, 07; m 31; c 1. ZOOLOGY. Educ: Univ Mich, BS, 30, MS, 31, PhD(zool), 33. Prof Exp: From instr to prof zool, 33-69, EMER PROF ZOOL, UNIV MICH, ANN ARBOR, 70- Concurrent Pos: Res assoc, Mus Zool, Univ Mich, 33-47 & Lab Vert Zool, 38-47, from asst biologist to assoc biologist, 47-56; res assoc, Cranbrook Inst Sci, Mich, 39-48; exchange prof, Nat Univ Mex, 45; consult, Off Strategic Serv, 43-45; mem sci adv panel, US Dept Army, 60-65; mem expeds, Cent Am & US. Mem: Distinguished fel Am Soc Ichthyol & Herpet; fel Royal Geog Soc; corresp mem Soc Geog & Hist Guatemala. Res: Herpetology, especially systematics; ecology and zoogeography of Neotropica. Mailing Add: Chalet 2945 Panajachel Solola Guatemala

STUART, MERRILL M, b Wabash, Ind, Aug 14, 39; m 61; c 2. GEOGRAPHY. Educ: Carroll Col, Wis, BA, 61; Univ Hawaii, MA, 63; Columbia Univ, EdD(geog), 68. Prof Exp: Instr geog, Bowling Green State Col, 64-65; asst prof, 67-71, ASSOC PROF GEOG, CALIF STATE UNIV, FRESNO, 71- Mem: Asn Am Geogrs; Am Geog Soc; Nat Coun Geog Educ. Res: Economic transportation and Anglo-American geography. Mailing Add: Dept of Geog Calif State Univ Shaw & Cedar Ave Fresno CA 93710

STUART, RICHARD NORWOOD, b Medford, Ore, Apr 23, 26; m 58; c 4. PHYSICS. Educ: Univ Calif, AB, 43, PhD, 52. Prof Exp: Lectr nuclear eng, 58-62, THEORET PHYSICIST, LAWRENCE LIVERMORE LAB, UNIV CALIF, 52- Res: Reactor and solid state physics. Mailing Add: PO Box 293 Diablo CA 94528

STUART, RICHARD ROY, zoology, see 12th edition

STUART, ROBERT LEE, b Cincinnati, Ohio, June 16, 18; m 48; c 2. RADIOCHEMISTRY. Educ: Univ Cincinnati, BS, 49. Prof Exp: Chemist, Berghausen Chem Co, Cincinnati, 49-52; radiochemist, Aircraft Nuclear Propulsion Dept, Gen Elec Co, 52-69, RADIOCHEMIST, KNOLLS ATOMIC POWER LAB, GEN ELEC CO, 69- Mem: Am Soc Testing & Mat. Res: Development of a fuel scanner to eliminate the need for x-ray equipment, utilizing isotopes with gamma energies favorably located on the fuel absorption edge. Mailing Add: Gen Elec Co PO Box 1072 Knolls Atomic Power Lab Schenectady NY 12301

STUART, RONALD S, b Tingley, NB, Mar 26, 19; m 46; c 4. ORGANIC CHEMISTRY. Educ: Univ NB, BA, 40; Univ Toronto, MA, 41, PhD(org chem), 44. Prof Exp: Demonstr chem, Univ Toronto, 40-42; res assoc, Nat Res Coun Can, 43-45; asst dir res, Dom Tar & Chem, 45-48; mgr chem & biol control, Merck & Co, Ltd, 48-53, mgr sci develop, 53-60, mgr tech & prod opers, 60-63, dir res, Merck Sharp & Dohme Can, 63-65, Charles E Frosst & Co, 65-68, DIR RES, MERCK FROSST LABS, 68- Mem: AAAS; Am Chem Soc; NY Acad Sci; Chem Inst Can; Can Res Mgt Asn. Res: Medicinal chemistry. Mailing Add: Merck Frosst Labs Box 1005 Pointe Claire Dorval PQ Can

STUART, SARAH ELIZABETH, b Arlington, Va, Jan 23, 48; m 75. MOLECULAR BIOLOGY. Educ: Westhampton Col, BS, 69; Med Col Va, Va Commonwealth Univ, PhD(microbiol), 72. Prof Exp: RES ASSOC BIOCHEM, MICH STATE UNIV, 73-

Concurrent Pos: NIH fel biochem, 74- Mem: AAAS; Sigma Xi; Am Soc Microbiol. Res: Nucleo-cytoplasmic communication; post transcriptional modification and processing of RNA. Mailing Add: Dept of Biochem Mich State Univ East Lansing MI 48824

STUART-ALEXANDER, DESIREE ELIZABETH, b London, Eng, Apr 6, 30; US citizen. GEOLOGY. Educ: Westhampton Col, BA, 52; Stanford Univ, MSc, 59, PhD(geol), 67. Prof Exp: Geologist explor dept, Utah Construct & Mining Co, 58-60; asst prof geol, Haile Selassie Univ, 63-65; GEOLOGIST, US GEOL SURV, 66- Concurrent Pos: Prog scientist, NASA Hq, 74-75. Mem: AAAS; Geol Soc Am. Res: Lunar and Martian geology including studies based on remote sensing data and petrology of lunar rocks; terrestrial studies of metamorphic problems and lunar analogs. Mailing Add: Dept of Astrogeol US Geol Surv 345 Middlefield Rd Menlo Park CA 94025

STUBBE, JOHN SUNAPEE, b New York, NY, Feb 21, 19; m 43; c 4. MATHEMATICS. Educ: Univ NH, BS, 41; Brown Univ, MS, 42; Univ Cincinnati, PhD(math), 45. Prof Exp: Instr math, army specialized training prog, Univ Cincinnati, 43-44; instr, Univ Ill, 45-47; asst prof, Univ NH, 47-49; asst prof, 49-53, dir comput ctr, 64-70, ASSOC PROF MATH, CLARK UNIV, 53- Concurrent Pos: Prof, Worcester Polytech Inst, 69-72. Mem: Am Math Soc; Math Asn Am. Res: Summability; Fourier series. Mailing Add: Dept of Math Clark Univ Worcester MA 01610

STUBBEMAN, ROBERT FRANK, b Midland, Tex, May 9, 35; m 56; c 3. PHYSICAL CHEMISTRY. Educ: Austin Col, BA, 57; Univ Tex, MA, 61, PhD(chem), 64. Prof Exp: Res chemist, Esso Res & Eng Co, 63-66; sr res chemist, 66-71, GROUP LEADER, SPECTROS LAB, CELANESE CHEM CO, 71- Mem: Am Chem Soc; fel Am Inst Chemists. Res: Shock tube kinetics; mass spectrometry, including qualitative and quantitative low and high resolution; process research. Mailing Add: Res Dept Celanese Chem Co Box 9077 Corpus Christi TX 78408

STUBBINGS, ROBERT LAMB, b Chicago, Ill, Apr 3, 21; m 42; c 1. PHYSICAL CHEMISTRY, BIOCHEMISTRY. Educ: Lehigh Univ, BS, 41, MS, 46, PhD(chem), 49. Prof Exp: Res assoc chem, Lehigh Univ, 49-53; dir, Inst Leather Technol, Milwaukee Sch Eng, 53-71; V PRES RES, DEVELOP & PROCESS, FRED RUEPING LEATHER CO, 71- Honors & Awards: Alsop Award, Am Leather Chem Asn, 58. Mem: Instrument Soc Am; Am Leather Chem Asn (pres, 61); fel Am Inst Chemists. Res: Leather chemistry and technology; inorganic complex ions; protein chemistry and structure; instrumentation. Mailing Add: 2109 Rienzi Rd Fond du Lac WI 54935

STUBBINS, JAMES FISKE, b Honolulu, Hawaii, Feb 19, 31; m 59; c 3. MEDICINAL CHEMISTRY. Educ: Univ Nev, BS, 53; Purdue Univ, MS, 58; Univ Minn, PhD(pharmaceut chem), 65. Prof Exp: Asst prof pharmaceut chem, Univ Fla, 62-63; asst prof, 63-68, ASSOC PROF PHARMACEUT CHEM, MED COL VA, 68- Mem: Am Chem Soc; Am Pharmaceut Asn; Acad Pharmaceut Sci; Am Asn Univ Prof; Am Asn Col Pharm Coun Fac. Res: Synthesis of medicinal agents; pharmacology of drugs in the autonomic and central nervous systems; antimetabolite theory and chemotherapy; drugs acting on blood cells; contraceptive agents. Mailing Add: Sch of Pharm Med Col of Va Va Commonwealth Univ Richmond VA 23298

STUBBINS, WARREN FENTON, physics, see 12th edition

STUBBLEFIELD, BEAUREGARD, b Navasota, Tex, July 31, 23; m 50; c 5. TOPOLOGY. Educ: Prairie View State Col, BS, 43, MA, 45; Univ Mich, MS, 51, PhD(math), 59. Prof Exp: Asst math, Prairie View State Col, 43-44; prof & head dept, Univ Liberia, 52-56; lectr & NSF fel, Univ Mich, 59-60; supvr anal sect, Int Elec Corp, 60-61; assoc prof math, Oakland Univ, 61-67; vis prof & vis scholar, Tex Southern Univ, 68-69; sr prog assoc, Inst Serv Educ, 69-71; PROF MATH, APPALACHIAN STATE UNIV, 71- Concurrent Pos: Consult, Detroit Arsenal, 57-59; asst prof, Stevens Inst Technol, 60-61; vis prof, Prairie View A&M Col, 70. Mem: AAAS; Am Math Soc; Math Asn Am; Soc Indust & Appl Math. Mailing Add: Dept of Math Appalachian State Univ Boone NC 28607

STUBBLEFIELD, BONNIE MCGREGOR, b Fitchburg, Mass, June 13, 42; m 75. OCEANOGRAPHY. Educ: Tufts Univ, BS, 64; Univ RI, MS, 67; Univ Miami, PhD(marine sci), 75. Prof Exp: Res asst oceanog, Univ RI, 65-70, res assoc, 70-72; OCEANOGR, ATLANTIC OCEANOG & METEOROL LABS, NAT OCEANOG & ATMOSPHERIC ADMIN, 72- Mem: Am Geophys Union; Am Geol Inst. Res: Processes responsible for the evolution of the sea floor; sedimentary framework and the processes which determine the sediment stability of a passive continental margin. Mailing Add: Atlantic Oceanog & Meteorol Lab NOAA 15 Rickenbacker Cswy Miami FL 33149

STUBBLEFIELD, CEDRIC TAYLOR, b Navasota, Tex, Sept 25, 21; m 52; c 2. PHYSICAL CHEMISTRY. Educ: Tex Southern Univ, BS, 42; Prairie View Agr & Mech Col, MS, 47; Univ Iowa, PhD(chem), 54. Prof Exp: Instr chem, Prairie View Agr & Mech Col, 47-48; asst prof, Tex Southern Univ, 48-51; ASSOC PROF CHEM, PRAIRIE VIEW A&M UNIV, 54- Concurrent Pos: Res Corp grant, 54-55; Robert A Welch Found grants, 56-69; Minnie Stevens Piper Prof, 68; NASA grants, 69-73; mem rev panel proposals, col scisci improv prog D, NSF, 72. Mem: AAAS; Am Chem Soc. Mailing Add: Dept of Chem Prairie View A&M Univ Prairie View TX 77445

STUBBLEFIELD, CHARLES BRYAN, b Viola, Tenn, Sept 14, 31; m 60; c 2. ANALYTICAL CHEMISTRY. Educ: Mid Tenn State Univ, BS, 53; Univ Tenn, MS, 60. Prof Exp: Res chemist, PPG Corp, 60-65, group leader anal chem, 65-67; res chemist, 67-72, DIR QUAL CONTROL, LITHIUM CORP AM, GULF RESOURCES & CHEM CORP, 72- Mem: Am Chem Soc; Sigma Xi; Am Inst Chemists; Am Soc Qual Control; Am Soc Testing & Mat. Res: Chemical and instrumental analytical methods development for improving analytical procedures; implementation of better analytical methods to improve process and product reliability. Mailing Add: Lithium Corp of Am PO Box 795 Bessemer City NC 28016

STUBBLEFIELD, FRANK MILTON, b Hillsboro, Ill, June 25, 11; m 32; c 2. ORGANIC CHEMISTRY. Educ: Univ Ill, AB, 32, MS, 36, PhD(chem), 42. Prof Exp: Chemist, Univ Ill, 35-37 & 39-42; chemist, Swift & Co, 37-39; res chemist, Weldon Spring Ord Works, Atlas Powder Co, Mo, 42-43; prof chem & head dept, Davis & Elkins Col, 43-47; assoc prof, Univ Ill, 47-56; CHEMIST, US GOVT, 56- Concurrent Pos: Consult, 43-46; chief, Chem Div, Chem & Radiol Labs, Chem Corps, 51-53. Mem: Fel Am Inst Chemists; Am Chem Soc. Res: Organophosphorous chemistry-phosphonates mechanism of action and antidotes; toxicology of compounds of high physiological activity, with special emphasis on heterocyclic nitrogen compounds. Mailing Add: 1030 Dead Run Dr McLean VA 22101

STUBBLEFIELD, ROBERT DOUGLAS, b Decatur, Ill, Mar 4, 36; m 58; c 3. ANALYTICAL CHEMISTRY. Educ: Eureka Col, BS, 59. Prof Exp: Anal chemist, 59-62, chemist, 62-64, RES CHEMIST, NORTHERN REGIONAL RES LAB, AGR RES SERV, USDA, 64- Concurrent Pos: Co-dir int collab study aflatoxin M methods milk, Int Union Pure & Appl Chem-Asn Off Anal Chemists, 72-73. Mem: Am Chem Soc. Res: Identification, preparation, and determination of known and unknown toxic compounds produced by the action of molds on agricultural commodities and products. Mailing Add: Northern Regional Res Lab USDA 1815 N University Peoria IL 61604

STUBBLEFIELD, ROBERT LEE, b Carbon, Tex, Jan 17, 20; m 42; c 2. PSYCHIATRY. Educ: Univ Tex, BA, 40, MD, 42; Am Bd Psychiat & Neurol, dipl, 49. Prof Exp: Intern, St Louis City Hosp, 43, resident psychiat, 45-46; resident psychiat, Med Ctr, Univ Colo, 46-47, resident child psychiat, 47-49, staff psychiatrist, Sch Med, 47-51, assoc prof psychiat & dir psychiat clin, 54-57; clin assoc prof, La State Univ, 52; clin asst prof, Sch Med, Tulane Univ, 52; staff psychiatrist, NIMH, 53-54; prof psychiat & chmn dept, Univ Tex Southwestern Med Sch Dallas, 57-71; assoc dir psychiat, Western Interstate Comn Higher Educ, 71-74; MED DIR & ADMINR, SILVER HILL FOUND, NEW CANAAN, CONN, 74- Concurrent Pos: Consult, US Dept Air Force, 57-, USPHS, 57- & Vet Admin Hosp, 57-; mem Tex del, White House Conf Children & Youth, 60; clin prof, Sch Med, Univ Colo, Denver, 71-74; clin prof, Dept Psychiat, Med Sch, Yale Univ & Yale Child Study Ctr, New Haven, 75- Mem: Fel Am Psychiat Asn; fel Am Orthopsychiat Asn; AMA; Am Acad Child Psychiat; fel Am Col Psychoanalysts. Res: Language development in children; chronic asthma in children; effects of steroids on behavior. Mailing Add: Silver Hill Found PO Box 1177 New Canaan CT 06840

STUBBLEFIELD, TRAVIS ELTON, b Austin, Tex, May 27, 35; m 57; c 2. CELL BIOLOGY. Educ: NTex State Univ, BS, 57; Univ Wis, MS, 59, PhD(exp oncol), 61. Prof Exp: NSF fel animal virol, Max Planck Inst Virus Res, Tübingen, WGer, 62-63; asst prof, 65-68, ASSOC PROF BIOL, UNIV TEX GRAD SCH BIOMED SCI HOUSTON, 68- Concurrent Pos: Asst biologist, Univ Tex M D Anderson Hosp & Tumor Inst Houston, 63-68, assoc biologist, 68-73. Mem: Am Soc Cell Biol. Res: Structure and physiology of mammalian chromosomes; synchronized cell culture; cell differentiation; structure and function of centrioles. Mailing Add: Dept of Biol Health Sci Ctr Univ Tex Grad Sch Biomed Sci 6414 Fannin Houston TX 77025

STUBBLEFIELD, VERNON SHAW, organic chemistry, forensic science, see 12th edition

STUBBS, DONALD WILLIAM, b Seguin, Tex, Sept 26, 32; m 53; c 4. PHYSIOLOGY. Educ: Tex Lutheran Col, BA, 54; Univ Tex, MA, 56, PhD(physiol), 64. Prof Exp: Instr zool, Auburn Univ, 56-60; from instr to assoc prof, 63-75, PROF PHYSIOL, UNIV TEX MED BR GALVESTON, 75- Concurrent Pos: NIH res grant, 65-71; multidisciplinary res grant ment health, 74- Mem: AAAS; Am Physiol Soc. Res: Biosynthesis of ascorbic acid; hormonal induction of enzyme activities. Mailing Add: Dept of Physiol Univ of Tex Med Br Galveston TX 77551

STUBBS, JOHN DORTON, b Cape Girardeau, Mo, Oct 9, 38; m 62; c 2. MOLECULAR BIOLOGY, BIOCHEMISTRY. Educ: Wash Univ, BA, 60; Univ Wis-Madison, MA, 62, PhD(biochem), 65. Prof Exp: USPHS fel genetics, Univ Wash, 65-67; from asst prof to assoc prof molecular biol, Calif State Univ, San Francisco, 68-75, PROF MOLECULAR BIOL, SAN FRANCISCO STATE UNIV, 75-, CHMN DEPT CELL & MOLECULAR BIOL, 72- Concurrent Pos: Brown-Hazen grant, Calif State Univ, San Francisco, 69-70; USPHS res grant, 70-72. Mem: AAAS. Res: Regulation of gene expression; transcriptional and translational control of the tryptophan operon in E coli; molecular mechanisms of membrane assembly; developmental biochemistry. Mailing Add: Dept of Cell & Molecular Biol San Francisco State Univ San Francisco CA 94132

STUBBS, MORRIS FRANK, b Sterling, Kans, May 25, 98; m 23; c 1. CHEMISTRY. Educ: Sterling Col, AB, 21; Univ Chicago, MS, 25, PhD(chem), 31. Hon Degrees: DSc, Sterling Col, 60. Prof Exp: Teacher chem & physics, Elgin Jr Col, 21-23; prof chem & physics & head dept phys sci, Tenn Wesleyan Col, 23-42, dean, 31-42; prof chem & head dept, Carthage Col, 42-44, Tenn Polytech Inst, 44-46 & NMex Inst Mining & Technol, 46-63, dir col div, 62-63; prof chem, Tex Tech Univ, 63-68; PROF CHEM, CHMN DEPT CHEM & DIR DIV NATURAL SCI & MATH, UNIV ALBUQUERQUE, 68- Concurrent Pos: Off Naval Res grant, 50-54. Honors & Awards: Clark Medal, Am Chem Soc, 65. Mem: Fel AAAS; Am Chem Soc. Res: Chemistry of indium; geochemical tests; general chemistry and qualitative analysis. Mailing Add: Dept of Chem Univ of Albuquerque Albuquerque NM 87140

STUBBS, ULYSSES SIMPSON, JR, b Kingstree, SC, Dec 11, 11; m 35. CHEMISTRY. Educ: Claflin Col, BS, 35; Columbia Univ, MA, 39; NY Univ, PhD(chem), 58. Prof Exp: Instr chem & physics, Hampton Inst, 39-42, 43-47; from asst prof to assoc prof, 47-68, PROF CHEM, MORGAN STATE COL, 68- Mem: AAAS; Am Chem Soc. Res: Reduction of organic compounds by inorganic methods. Mailing Add: Dept of Chem Morgan State Col Baltimore MD 21212

STUBER, CHARLES WILLIAM, b St Michael, Nebr, Sept 19, 31; m 53; c 1. GENETICS. Educ: Univ Nebr, BS, 52, MS, 61; NC State Univ, PhD(genetics, exp statist), 65. Prof Exp: Instr high sch, Nebr, 56-59; from asst prof to assoc prof genetics, 65-75, PROF GENETICS, NC STATE UNIV, 75-; RES GENETICIST, AGR RES SERV, USDA, 62- Mem: AAAS; Genetics Soc Am; Am Soc Agron; Crop Sci Soc Am. Res: Quantitative genetics; inheritance of quantitative traits and correlated biochemical traits. Mailing Add: Dept of Genetics NC State Univ Raleigh NC 27607

STUBER, FRED A, b Paris, France, Dec 7, 33; m 64; c 1. PHYSICAL ORGANIC CHEMISTRY. Educ: Univ Zurich, ChemEng, 57, DSc(nuclear chem), 61. Prof Exp: Res assoc mass spectros, Inst Reactor Res, Switz, 61-62; fel, Univ Notre Dame, 62-63 & Mellon Inst, 63-64; res chemist, Ciba, Switz, 64-67; res chemist, 67-75, MGR, D S GILMORE RES LAB, UPJOHN CO, NORTH HAVEN, 75- Res: Ionization and appearance potentials; analysis of nuclear magnetic resonance spectra; photopolymers. Mailing Add: 65 Chapel Hill Rd North Haven CT 06473

STUCK, BARTON W, b Detroit, Mich, Oct 25, 46; m 75. APPLIED MATHEMATICS. Educ: Mass Inst Technol, BS & MS, 69, ScD(elec eng), 72. Prof Exp: MEM TECH STAFF, MATH & STATIST RES CTR, BELL LABS, AM TEL & TEL CO, 72- Mem: Soc Indust & Appl Math; Math Asn Am; Inst Elec & Electronics Engrs. Res: Mathematical physics; applied probability theory. Mailing Add: Bell Labs 600 Mountain Ave Murray Hill NJ 07974

STUCKER, JOSEPH BERNARD, b Chicago, Ill, Feb 28, 14; m 41; c 3. CHEMISTRY. Educ: Univ Chicago, BS, 35. Prof Exp: Chemist, Pure Oil Co, 35-41, asst supt grease plant, 41-43, group leader prod develop, 43-50, sect supvr, 50-52, div dir, 52-65, sr res assoc, Res Dept, Union Oil Co, Calif, 65-68, mgr prod develop, Pure Oil Div, 68-69 & Union 76 Div, 69-71, MGR PROD QUAL, REFINING DIV, UNION OIL CO CALIF, 71- Mem: Soc Automotive Eng; Am Soc Testing & Mat; Am Petrol Inst. Res: Product development of lubricants, particularly greases, gear and crankcase oils and industrial lubricants. Mailing Add: Refining Div Union Oil Co of Calif 200 E Golf Rd Palatine IL 60067

STUCKER, ROBERT EVAN, b Burlington, Iowa, Jan 28, 36; m 56; c 3. PLANT BREEDING, STATISTICS. Educ: Iowa State Univ, BS, 59; Purdue Univ, MS, 61; NC State Univ, PhD(genetics), 66. Prof Exp: Res geneticist, Forage & Range Br, Crops Res Div, Agr Res Serv, USDA, 65-68; asst prof agron & plant genetics, 68-72, ASSOC PROF AGRON & PLANT GENETICS, UNIV MINN, ST PAUL, 72- Mem: Am Soc Agron; Crop Sci Soc Am. Res: Corn quantitative genetics; statistical design of experiments in agronomy; application of quantitative genetics in plant breeding. Mailing Add: Dept of Agron & Plant Genetics Univ of Minn St Paul MN 55101

STUCKEY, BEN NELSON, organic chemistry, food technology, see 12th edition

STUCKEY, IRENE HAWKINS, b Griffin, Ga, Apr 6, 11. PLANT PHYSIOLOGY. Educ: Vanderbilt Univ, AB, 32; Cornell Univ, PhD(cytol, plant physiol), 36. Prof Exp: From asst to assoc prof, 37-65, PROF PLANT PHYSIOL, UNIV RI, 65- Honors & Awards: Graphic Arts Award, Printing Industs Am, 68; Outstanding Educr Am Award, 75. Mem: Am Soc Plant Physiol; Bot Soc Am; Am Soc Agron; Soil Sci Soc Am. Res: Conservation of native plants; salt marsh ecology. Mailing Add: Dept of Plant & Soil Sci Univ of RI Kingston RI 02881

STUCKEY, JACKSON H, b China, Tex, Jan 12, 16. SURGERY. Educ: Univ Tex, AB, 39; Yale Univ, MD, 42; Am Bd Surg, dipl, 57. Prof Exp: Assoc prof, 58-64, PROF SURG, STATE UNIV NY DOWNSTATE MED CTR, 64- Concurrent Pos: Attend surgeon, Univ Hosp, 66-; chief thoracic surg, State Univ-Kings County Health Ctr, 71-; consult thoracic-cardiovasc surg, Lutheran Med Ctr, 72-; attend, Kings County Hosp; consult, Brooklyn Vet Hosp. Mem: Fel Am Col Surgeons. Res: Cardiovascular research. Mailing Add: Dept of Surg State Univ NY Downstate Med Ctr Brooklyn NY 11203

STUCKEY, JAMES MORLAN, physical chemistry, see 12th edition

STUCKEY, JOHN EDMUND, b Stuttgart, Ark, Dec 6, 29; m 55; c 3. PHYSICAL CHEMISTRY, INORGANIC CHEMISTRY. Educ: Hendrix Col, BA, 51; Univ Okla, MS, 53, PhD(chem), 57. Prof Exp: Res chemist, Oak Ridge Nat Lab, Union Carbide Corp, 57; asst prof chem, La Polytech Inst, 57-58; PROF CHEM, HENDRIX COL, 58- Mem: Am Chem Soc. Res: Preparations and properties of monofluorophosphate compounds; solution chemistry in the critical temperature region; x-ray crystallography. Mailing Add: Dept of Chem Hendrix Col Conway AR 72032

STUCKEY, RONALD LEWIS, b Bucyrus, Ohio, Jan 9, 38. BOTANY. Educ: Heidelberg Col, BS, 60; Univ Mich, MA, 62, PhD(bot), 65. Prof Exp: Instr bot, Univ Mich, 65; asst prof, 65-70, cur herbarium, 67-75, ASSOC PROF BOT, OHIO STATE UNIV, 70- Mem: AAAS; Am Soc Plant Taxon; Int Asn Plant Taxon; Bot Soc Am. Res: Taxonomy and distribution of angiosperms; history of American botany; monographic studies in the Cruciferae, particularly Rorippa; Ohio vascular plant flora and phytogeography; history of plant taxonomy in North America; taxonomy and distribution of angiosperms, particularly aquatic and marsh flora. Mailing Add: Dept of Bot Ohio State Univ 1735 Neil Ave Columbus OH 43210

STUCKEY, WALTER JACKSON, JR, b Fairfield, Ala, Mar 6, 27; m 52; c 3. INTERNAL MEDICINE, HEMATOLOGY. Educ: Univ Ala, BS, 47; Tulane Univ, MD, 51. Prof Exp: Intern, Charity Hosp La, New Orleans, 51-52, resident internal med, 55, Vet Admin Hosp, New Orleans, 55-56 & Charity Hosp La, 56-57; from instr to assoc prof, 58-68, PROF MED, SCH MED, TULANE UNIV, 68-, CHIEF HEMAT SECT, 63- Concurrent Pos: Sr vis physician, New Orleans Charity Hosp; consult, Baptist, East Jefferson, Hotel Dieu, Mercy, Methodist, Sara Mayo, Touro & USPHS Hosps, New Orleans, La, Vet Admin Hosps, New Orleans & Alexandria, Huey P Long Hosp Pineville & Lallie Kemp Hosp, Independence. Mem: Am Asn Cancer Educ; Am Soc Clin Oncol; fel Am Col Physicians; Am Soc Internal Med; Am Soc Hemat. Res: Bone marrow function in health and disease. Mailing Add: Sect of Hematol Tulane Univ Sch of Med New Orleans LA 70112

STUCKEY, WAYNE K physical chemistry, see 12th edition

STUCKI, JACOB CALVIN, b Neillsville, Wis, Nov 30, 26; m 48; c 3. ENDOCRINOLOGY, RESEARCH ADMINSTRATION. Educ: Univ Wis, BS, 48, MS, 51, PhD(zool, physiol), 54. Prof Exp: Res asst, Univ Wis, 50-54; endocrinologist, Wm S Merrell Co, 54-57; res assoc, 57-60, dept head endocrinol, 60-61, mgr pharmacol res, 61-68, DIR RES PLANNING & ADMIN, PHARMACEUT RES & DEVELOP, UPJOHN CO, 68- Mem: AAAS; Soc Exp Biol & Med; Endocrine Soc. Res: Reproduction; inflammation; pharmacology; research management. Mailing Add: Upjohn Co Kalamazoo MI 49001

STUCKI, WILLIAM PAUL, b Neillsville, Wis, Sept 28, 31; m 55; c 5. BIOCHEMISTRY. Educ: Univ Wis, BS, 57, MS, 59, PhD(biochem), 62. Prof Exp: Asst prof Univ Puerto Rico & scientist, PR Nuclear Ctr, Mayaguez, 62-63; asst prof, Antioch Col & assoc biochem, Fels Res Inst, 63-67; res biochemist, 67-75, SR RES BIOCHEMIST, PARKE DAVIS & CO, 75- Mem: Am Chem Soc; Am Soc Parasitol; Am Soc Trop Med & Hyg; NY Acad Sci. Res: Protein and amino acid nutrition and metabolism; plant biochemistry; biochemistry of natural products; chemotherapy, immunology and immunochemistry; atherothrombotic disease. Mailing Add: Pharmacol Dept Parke Davis & Co 2800 Plymouth Rd Ann Arbor MI 48106

STUCKWISCH, CLARENCE GEORGE, b Seymour, Ind, Oct 13, 16; m 42; c 5. CHEMISTRY. Educ: Ind Univ, AB, 39; Iowa State Col, PhD(org chem), 43. Prof Exp: Res assoc, Iowa State Col, 43; asst prof chem, Wichita State Univ, 43-44; res chemist, Eastman Kodak Co, 44-4S; from asst prof to assoc prof chem, Wichita State Univ, 45-60; assoc prof, NMex Highlands Univ, 60-62, prof & head dept, 62-64, dir inst sci res, 63-64; prof chem & exec officer dept, State Univ NY Buffalo, 64-68; chmn dept, 68-74, PROF CHEM, UNIV MIAMI, 68-, DEAN GRAD SCH, 72- Mem: Am Chem Soc. Res: Organometallic, psychopharmacological and organosulfur compounds; anthraquinone dyes; synthesis of amino acids. Mailing Add: 210 Ferre Bldg Univ of Miami Coral Gables FL 33124

STUCKY, DONALD J, agronomy, see 12th edition

STUCKY, GALEN DEAN, b McPherson, Kans, Dec 17, 36; m 61; c 2. INORGANIC CHEMISTRY. Educ: McPherson Col, BS, 57; Iowa State Univ, PhD(chem), 62. Prof Exp: Fel physics, Mass Inst Technol, 62-63; NSF fel, Quantum Chem Inst, Fla, 63-64; from asst prof to assoc prof chem, 64-72, PROF CHEM, UNIV ILL, URBANA, 72- Concurrent Pos: Consult div univ & col, Argonne Nat Labs, 66; vis prof physics, Univ Uppsula, Sweden, 71. Mem: Am Chem Soc; Am Crystallog Asn. Res: Crystal structural analysis; organometallic chemistry of electron deficient sites; solid state chemistry. Mailing Add: 263 Noyes Lab Dept of Chem Univ of Ill Urbana IL 61801

STUCKY, GARY LEE, b Murdock, Kans, May 18, 41; m 72. BIOINORGANIC CHEMISTRY. Educ: Bethel Col, Kans, AB, 63; Kans State Univ, PhD(inorg chem), 67. Prof Exp: Instr chem, Halstead Sch Nursing, Kans, 62-63 & Kans State Univ, 63-65; res scientist bioinorg chem, Miles Labs, Ind, 67-70 & Kivuvu Inst Med Evangel, Kimpese, Zaire, 71; asst prof, 72, ASSOC PROF BIOINORG CHEM, EASTERN MENNONITE COL, 72- Concurrent Pos: Am Leprosy Mission res grant, Eastern Mennonite Col, 72. Mem: AAAS; Am Chem Soc; The Chem Soc. Res: Electrochemistry of leprosy; inorganic synthesis; ion-selective electrodes; hot-atom chemistry. Mailing Add: Dept of Chem Eastern Minnonite Col Harrisonburg VA 22801

STUDDEN, WILLIAM JOHN, b Timmins, Ont, Sept 30, 35; m. MATHEMATICAL STATISTICS. Educ: McMaster Univ, BSc, 58; Stanford Univ, PhD(statist), 62. Prof Exp: Res assoc math, Stanford Univ, 62-64; from asst prof to assoc prof statist, 64-71, PROF STATIST, PURDUE UNIV, LAFAYETTE, 71- Concurrent Pos: NSF fels, Purdue Univ, Lafayette, 68-71, 72-75. Mem: Fel Inst Math Statist. Res: Optimal designs; Tchebycheff systems. Mailing Add: Dept of Statist Purdue Univ Lafayette IN 47906

STUDEBAKER, GERALD A, b Freeport, Ill, July 22, 32; m 55; c 4. AUDIOLOGY. Educ: Ill State Univ, BS, 55; Syracuse Univ, MS, 56, PhD(audiol), 60. Prof Exp: Supvr clin audiol, Vet Admin Hosp, DC, 59-61, chief audiol & speech path serv, Syracuse, NY, 61-62; supvr clin audiol, Med Ctr, Univ Okla, 62-66, from asst prof to assoc prof audiol & consult, Dept Otorhinolaryngol, 62-72, res audiologist, 66-72; RES PROF AUDIOL, MEMPHIS STATE UNIV, 72- Concurrent Pos: Mem commun sci study sect, NIH, 70-74. Mem: AAAS; fel Am Speech & Hearing Asn; Acoust Soc Am. Res: Bone-conduction hearing thresholds, auditory masking; loudness estimation procedures and adaptation; speech discrimination; ear mold acoustics. Mailing Add: Memphis Speech & Hearing Ctr Memphis State Univ Memphis TN 38105

STUDEBAKER, JOEL FRANKLIN, biophysics, biochemistry, see 12th edition

STUDEBAKER, MERTON LELAND, b McPherson, Kans, Nov 8, 13; m 43; c 4. RUBBER CHEMISTRY, COLLOID CHEMISTRY. Prof Exp: Res chemist, Gen Atlas Carbon Co, Tex, 37-42 & War Prod Bd, Washington, DC, 42-43; supvry engr, 43-73, CARBON BLACK CONSULT, PHILLIPS CHEM CO, 73- Mem: AAAS; Am Chem Soc; Plastics & Rubber Inst. Res: Carbon black; reinforcement of rubber by carbon black; vulcanization and oxidation of rubber. Mailing Add: Phillips Chem Co 1501 Commerce Dr Stow OH 44224

STUDIER, EUGENE H, b Dubuque, Iowa, Mar 16, 40; div; c 2. PHYSIOLOGICAL ECOLOGY. Educ: Univ Dubuque, BS, 62; Univ Ariz, PhD(zool), 6S. Prof Exp: From asst prof to assoc prof biol, NMex Highlands Univ, 65-72; assoc prof, 72-74, PROF BIOL, UNIV MICH-FLINT, 74- Concurrent Pos: USPHS grant, 66-67; Sigma Xi res grant-in-aid, 69; Am Philos Soc grant, 69. Mem: Fel AAAS; Am Soc Mammalogists; Am Soc Zoologists. Res: Mammalian physiology; physiological adaptation in bats and rodents. Mailing Add: Dept of Biol Univ of Mich-Flint Flint MI 48503

STUDIER, FREDERICK WILLIAM, b Waverly, Iowa, May 26, 36; m 62; c 2. BIOPHYSICS. Educ: Yale Univ, BS, 58; Calif Inst Technol, PhD(biophys), 63. Prof Exp: NSF fel biochem, Med Ctr, Stanford Univ, 62-64; from asst biophysicist to biophysicist, 64-74, SR BIOPHYSICIST, BIOL DEPT, BROOKHAVEN NAT LAB, 74- Mem: AAAS; Biophys Soc; Am Soc Biol Chemists. Res: Physical and chemical properties of nucleic acids; genetics and physiology of bacteriophage T7. Mailing Add: Biol Dept Brookhaven Nat Lab Upton NY 11973

STUDIER, MARTIN HERMAN, b Leola, SDak, Nov 10, 17; m 44; c 4. CHEMISTRY. Educ: Luther Col, BA, 39; Univ Chicago, PhD(chem), 47. Prof Exp: Asst, Iowa State Col, 39-41, instr, 41-42; res chemist, 43-36, SR CHEMIST, ARGONNE NAT LAB, 46- Mem: AAAS; Am Chem Soc; Am Phys Soc; Sigma Xi. Res: Nuclear chemistry of the heavy elements; mass spectrometry; organic matter in meteorites. Mailing Add: Argonne Nat Lab 9700 S Cass Ave Argonne IL 60439

STUDNICKA, BENEDICT JOSEPH, organic chemistry, inorganic chemistry, see 12th edition

STUDT, WILLIAM LYON, b Ypsilanti, Mich, Mar 12, 47; m 66; c 3. MEDICINAL CHEMISTRY. Educ: Eastern Mich Univ, BA, 69; Univ Mich, PhD(org chem), 73. Prof Exp: Res fel org chem, Yale Univ, 73-74; SECT HEAD RES MED CHEM, RORER-AMCHEM INC, 74- Res: The organic synthesis of natural products and biologically active compounds. Mailing Add: William H Rorer Inc 500 Virginia Dr Ft Washington PA 19034

STUDZINSKI, GEORGE P, b Poznan, Poland, Oct 30, 32; m 59; c 4. EXPERIMENTAL PATHOLOGY, CELL BIOLOGY. Educ: Glasgow Univ, BS, 55, MB, 58, PhD(exp path), 62. Prof Exp: Brit Empire Cancer Campaign res fel path, Glasgow Royal Infirmary, 59-60, resident, 60-62; from instr to assoc prof, 62-73, PROF PATH, JEFFERSON MED COL, 73- Mem: Tissue Cult Asn; Am Soc Exp Path; Histochem Soc; Am Asn Cancer Res; Am Soc Cell Biol. Res: Study of effect of cancer chemotherapeutic agents on cultured diploid and aneuploid mammalian cells by a combinatiôn of cytochemical and biochemical methods. Mailing Add: Dept of Path Jefferson Med Col Philadelphia PA 19107

STUEBE, CARL, b Cleveland, Ohio, June 20, 15; m 44; c 1. ORGANIC CHEMISTRY. Educ: Fenn Col, BSc, 41; Case Sch Appl Sci, MSc, 48; Western Reserve Univ, PhD(chem), 54. Prof Exp: Chemist, 41-56, group leader lubricant detergents, 56-65, RES CHEMIST, LUBRIZOL CORP, 65- Mem: AAAS; Am Chem Soc; Am Soc Lubrication Eng. Res: Organic chemistry applied to lubricant additives, especially detergents, dispersants and extreme pressure agents; chemistry of organic sulfonates, carboxylates, amides, phosphorous nitrogen and sulfur compounds; thermal analysis of polymers and monomers. Mailing Add: Lubrizol Corp Euclid Sta Cleveland OH 44117

STUEBEN, EDMUND BRUNO, b Cuxhaven, Ger, Apr 22, 20; nat US; m 4S; c 4. PARASITOLOGY. Educ: NY Univ, BS, 41; Baylor Univ, MA, 49; Univ Fla, PhD(zool), 53. Prof Exp: Instr, biol lab & parasitol lab, Univ Fla, 50-53; asst prof biol, Arlington State Col, 53; assoc prof, 54-63, PROF ZOOL & PHYSIOL, UNIV SOUTHWESTERN LA, 63- Res: Larval development of filariae in arthropods; physiology of filaria larva; transmission of infective state filaria larva; antihistamine effect on coronary circulation. Mailing Add: Dept of Biol Univ of Southwestern La Lafayette LA 70501

STUEBEN, EDWIN FRANK, b Chicago, Ill, Sept 22, 36; m 72. MATHEMATICS. Educ: Ill Inst Technol, BS, 58, MS, 60, PhD(math), 63. Prof Exp: From instr to asst prof, 62-70, ASSOC PROF MATH, ILL INST TECHNOL, 70- Mem: Am Math Soc; Math Asn Am. Res: Number theory; point set topology. Mailing Add: Dept of Math Ill Inst of Technol Chicago IL 60616

STUEBEN, KENNETH CHARLES, b New York, NY, Aug 24, 31; m 53; c 2. ORGANIC CHEMISTRY, POLYMER CHEMISTRY. Educ: Brooklyn Col, BS, 56; Polytech Inst Brooklyn, PhD(org chem), 60. Prof Exp: Res technician, Colgate-Palmolive Co, NJ, 49-55; res chemist, Union Carbide Plastics Co, 59-65, proj scientist, 65-70, RES SCIENTIST, PLASTICS DIV, UNION CARBIDE CORP, 70- Mem: Am Chem Soc. Res: High temperature polymers; oxidation kinetics; smoke generation in polymer systems; adhesives; photocure. Mailing Add: Plastics Div Union Carbide Corp River Rd Bound Brook NJ 08805

STUEBER, ALAN MICHAEL, b St Louis, Mo, Apr 18, 37. GEOCHEMISTRY. Educ: Wash Univ, BS, 58, MA, 61; Univ Calif, San Diego, PhD(earth sci), 65. Prof Exp: Res assoc earth sci, Wash Univ, 65-66; fel geochem, Carnegie Inst Washington, 66-67; asst prof, 67-72, ASSOC PROF GEOL, MIAMI UNIV, 72- Mem: Geochem Soc. Res: Earth sciences; strontium isotope studies. Mailing Add: Dept of Geol Miami Univ Oxford OH 45056

STUEBING, EDWARD WILLIS, b Cincinnati, Ohio, Sept 9, 42; m 61; c 2. CHEMICAL PHYSICS. Educ: Univ Cincinnati, BS, 65; Johns Hopkins Univ, MA, 69, PhD(chem physics), 70. Prof Exp: Res physicist, 70-74, RES CHEMIST, PITMAN-DUNN LABS, US ARMY FRANKFORD ARSENAL, 74- Concurrent Pos: Adj asst prof chem, Drexel Univ, 74- Honors & Awards: Army Res & Develop Achievement Award, US Army, 74. Mem: Am Phys Soc; Asn Comput Mach; Int Soc Quantum Biol; Am Chem Soc. Res: Aerosol light scattering; theoretical study of molecular structure, excited states and energy transfer; interaction of matter with laser light at high power density; mathematical modeling and operations research analyses. Mailing Add: Pitman-Dunn Labs US Army Frankford Arsenal Philadelphia PA 19137

STUECEK, GUY LINSLEY, b New Haven, Conn, Jan 22, 42; m 65; c 2. PLANT PHYSIOLOGY. Educ: Univ Conn, BS, 63, PhD(plant physiol), 68; Yale Univ, MF, 65. Prof Exp: Nat Res Coun Can fel, Forest Prod Lab, BC, 68-69; asst prof biol, 69-72, ASSOC PROF BIOL, MILLERSVILLE STATE COL, 72- Mem: AAAS; Am Soc Plant Physiol; Scan Soc Plant Physiol; Ecol Soc Am. Res: Influence of mechanical stress on plant growth and development; phloem transport and mineral nutrition. Mailing Add: Dept of Biol Millersville State Col Millersville PA 17551

STUEDEMANN, JOHN ALFRED, b Clinton, Iowa, Oct 3, 42; m 67. ANIMAL NUTRITION. Educ: Iowa State Univ, BS, 64; Okla State Univ, MS, 67, PhD(ruminant nutrit), 70. Prof Exp: RES PHYSIOLOGIST, SOUTHERN PIEDMONT CONSERV RES CTR, AGR RES SERV, USDA, 70- Mem: Am Soc Animal Sci. Res: Ruminant nutrition; forage production and utilization; waste disposal and land fertilization; health problems of beef cattle; forage finishing of cattle. Mailing Add: Southern Piedmont Conserv Res Ctr USDA PO Box 555 Watkinsville GA 30677

STUEHR, JOHN EDWARD, b Aug 30, 35; US citizen; m 62; c 4. BIOPHYSICAL CHEMISTRY. Educ: Western Reserve Univ, BA, 57, MS, 59, PhD(chem), 61. Prof Exp: NIH res fel chem, Max Planck Inst, Göttingen, WGer, 62-63; from asst prof to assoc prof, 64-74, PROF CHEM, CASE WESTERN RESERVE UNIV, 74- Mem: Am Chem Soc; Sigma Xi; Fedn Biol Chemists. Res: Reaction kinetics of fast processes in solution; relaxation spectroscopy; metal complexing; biochemical kinetics; elementary steps in enzyme kinetics. Mailing Add: Dept of Chem Case Western Reserve Univ Cleveland OH 44106

STUELPNAGEL, JOHN CLAY, b Houston, Tex, Nov 12, 36; m 59; c 3. MATHEMATICS. Educ: Yankton Col, BA, 55; Johns Hopkins Univ, PhD(math), 62. Prof Exp: Fel math, Res Inst Advan Studies, Martin-Marietta Corp, 61-64; sr engr, 64-66, FEL ENGR, AEROSPACE DIV, WESTINGHOUSE ELEC CORP, 66- Mem: Soc Indust & Appl Math; Math Asn Am. Res: Lie groups; linear algebra; differential equations; computation and computer design. Mailing Add: MS 450 Westinghouse Elec Corp Box 746 Baltimore MD 21203

STUESSY, TOD FALOR, b Pittsburgh, Pa, Nov 18, 43; m 68. SYSTEMATIC BOTANY. Educ: DePauw Univ, BA, 65; Univ Tex, Austin, PhD(bot), 68. Prof Exp: Asst prof, 68-74, ASSOC PROF BOT, OHIO STATE UNIV, 74- Concurrent Pos: Maria Moors Cabot res fel, Gray Herbarium, Harvard Univ, 71-72. Mem: AAAS; Am Inst Biol Sci; Am Soc Plant Taxon; Bot Soc Am; Soc Study Evolution. Res: Systematics and evolution of Compositae. Mailing Add: Dept of Bot Ohio State Univ 1735 Neil Ave Columbus OH 43210

STUETZER, OTMAR MICHAEL, b Nuernberg, Ger, Apr 30, 12; nat US; m 38; c 3. PHYSICS. Educ: Univ Munich, MA, 36; Munich Inst Technol, Dr rer tech, 38, Dr Habil, 43. Prof Exp: Dir gen sci & microwave depts, Flight Res Inst, Ger, 40-45; res physicist electronic components lab, Wright Air Develop Ctr, 46-49, chief tube tech sect, 49-53, adv develop br, 53-55; mgr physics & electronics res, Mech Div, Gen Mills, Inc, 55-60, from dir develop to tech dir, 61-63; MGR EXPLOR INSTRUMENTATION DEPT, SANDIA CORP, 63- Concurrent Pos: Chief radar sect, German Res Coun, 43-44 & German Air Force, 41-45. Mem: Fel Inst Elec & Electronics Eng; fel Am Phys Soc; Am Inst Aeronaut & Astronaut; Am Ord Asn; Metric Asn. Res: Electronic physics; electrohydrodynamics. Mailing Add: 708 Lamp Post Circle SE Albuquerque NM 87123

STUEWER, FREDERICK WILLIAM, mammalogy, wildlife ecology, see 12th edition

STUEWER, REINHOLD F, chemistry, see 12th edition

STUEWER, ROGER HARRY, b Sept 12, 34; US citizen; m 60; c 2. HISTORY OF PHYSICS. Educ: Univ Wis, BS, 58, MS, 64, PhD(hist sci & physics), 68. Prof Exp: Instr physics, Heidelberg Col, 60-62; from asst prof to assoc prof hist physics, Univ Minn, Minneapolis, 67-71; assoc prof hist sci, Boston Univ, 71-72; assoc prof hist physics, 72-74, PROF HIST PHYSICS, UNIV MINN, MINNEAPOLIS, 74- Concurrent Pos: Mem adv panel hist & philos sci, NSF, 70-72, res support, 70-; Am Coun Learned Soc fel, 74-75,consult, Franklin Mint, 74- Mem: Hist Sci Soc (secy, 72-); Philos Sci Asn; Sigma Xi; Brit Soc Hist Sci; Brit Soc Philos Sci. Res: History of nineteenth and twentieth century physics, especially optics, quantum theory, and nuclear physics; Compton effect as a turning point in physics. Mailing Add: Sch of Physics & Astron Univ of Minn Minneapolis MN 55455

STUFFLEBEAM, CHARLES EDWARD, b St Louis, Mo, Feb 22, 33; m 52; c 3. ANIMAL PHYSIOLOGY, BIOCHEMISTRY. Educ: Univ Mo, BS, 58, MS, 61, PhD, 64. Prof Exp: Asst county agent, Exten Div, Univ Mo, 59-61, instr animal husb & agr biochem, 62-64; assoc prof range animal sci, Sul Ross State Col, 64-65; assoc prof animal sci, Northwestern State Univ, 65-69; ASSOC PROF ANIMAL SCI, SOUTHWEST MO STATE UNIV, 69- Mem: Am Soc Animal Sci. Res: Genetics; biochemistry, physiology and nutrition of domestic animals and their application to agriculture. Mailing Add: Dept of Agr Southwest Mo State Univ Springfield MO 65802

STUGARD, FREDERICK, b Kosmosdale, Ky, Feb 15, 19; m 44; c 3. GEOLOGY. Educ: Yale Univ, PhD(geol), 50. Prof Exp: Geologist, US Geol Surv, Colo, 48-54; sr geologist, Phillips Petrol Co, Alexandria, Egypt, 54-69; exec geologist, Int Resources Ltd, London, Eng, 69-70; PETROL & MINERAL CONSULT, 70- Mem: Am Asn Petrol Geol; Soc Econ Geol; Geol Soc London. Res: Geology of oil and gas; research and exploration; well log interpretation; petroleum geology of North Sea region, North Africa and Middle East; geology of uranium deposits; finding water. Mailing Add: Durris Cottage Durris Banchory Kincardineshire Scotland

STUHLINGER, ERNST, b Niederrimbach, Ger, Dec 19, 13; nat US; m 50; c 3. PHYSICS. Educ: Univ Tübingen, PhD(physics), 36. Prof Exp: Asst prof, Berlin Inst Technol, 36-41; res asst guid & control, Rocket Develop Ctr, Peenemuende, 43-45; asst res & develop, Ord Corps, US Army, Ft Bliss, Tex & White Sands Proving Ground, NMex, 46-60, aeronaut res adminr & supvry phys scientist, Ballistic Missile Agency, Redstone Arsenal, Ala, 50-60; dir, Space Sci Lab, 60-68, ASSOC DIR SCI, MARSHALL SPACE FLIGHT CTR, NASA, 68- Honors & Awards: Distinguished Civilian Serv Award, US Army, 58; Galabert Prize, Paris, 62; Herman Oberth Award, 62 & Medal, 64; Propulsion Award, Am Inst Aeronaut & Astronaut, 60. Mem: Fel Am Inst Aeronaut & Astronaut; fel Am Astronaut Soc; Ger Phys Soc; Ger Soc Rockets & Space Flight; Brit Interplanetary Soc. Res: Feasibility and design studies of electrical propulsion systems for space ships; scientific satellites and space probes. Mailing Add: Code AD-S NASA Marshall Space Flight Ctr Huntsville AL 35812

STUHLMAN, ROBERT AUGUST, b Cincinnati, Ohio, Apr 9, 39; m 61; c 3. LABORATORY ANIMAL MEDICINE, MEDICAL RESEARCH. Educ: Ohio State Univ, DVM, 68; Univ Mo, Columbia, MS, 71; Am Col Lab Animal Med, dipl, 74. Prof Exp: Res asst vet clin, Ohio State Univ, 66-68; res assoc vet med & surg, Univ Mo, Columbia, 68-71 & instr path & asst dir lab animal med, Med Ctr, 71-75; DIR LAB ANIMAL RESOURCES & COORDR INTERDISCIPLINARY TEACHING LABS, WRIGHT STATE UNIV, 75- Concurrent Pos: Vet med officer, Vet Admin Hosp, Columbia, Mo, 72-75; consult gen med res, Vet Admin Ctr, Dayton, Ohio, 75- Mem: Am Vet Med Asn; Am Asn Lab Animal Sci; Am Col Lab Animal Med; Am Soc Lab Animal Practitioners. Res: Diabetes mellitus, especially development of the animal model; diagnosis; pathogenesis; establishment of secondary complications; therapy; inheritance patterns. Mailing Add: Lab Animal Resources Wright State Univ Sch of Med Dayton OH 45431

STUHT, JOHN NEAL, b Munising, Mich, Dec 23, 39; m 69; c 2. WILDLIFE DISEASES. Educ: Northern Mich Univ, BS, 62; Univ Mich, MPH, 65, PhD(wildlife mgt), 72. Prof Exp: Lab asst epidemiol, Sch Pub Health, Univ Mich, 64-65; biol technician, Fish & Wildlife Serv, US Dept Interior, 68; WILDLIFE RES BIOLOGIST, WILDLIFE DIV, MICH DEPT NATURAL RESOURCES, 70- Mem: Wildlife Soc; Wildlife Dis Asn. Res: Specific diseases and parasites of wild animals and birds to determine their effects on wildlife populations and their relationships to man, domestic animals and the total environment. Mailing Add: Rose Lake Wildlife Res Ctr Wildlife Path Lab Rte 1 East Lansing MI 48823

STUIVER, MINZE, b Vlagtwedde, Neth, Oct 25, 29; m 56; c 2. EARTH SCIENCE. Educ: State Univ Groningen, MSc, 53, PhD(biophys), 58. Prof Exp: Res assoc & fel geol, Yale Univ, 59-62, sr res assoc geol & biol & dir radiocarbon lab, 62-69; PROF GEOL & ZOOL, UNIV WASH, 69- Res: Biophysics of sense organs; low level counting techniques; geophysical implications of variations in atmospheric radiocarbon content; Pleistocene geology, oceanography and limnology. Mailing Add: Quaternary Res Ctr Univ of Wash Seattle WA 98195

STUKUS, PHILIP EUGENE, b Braddock, Penn, Oct 22, 42; m 66; c 3. MICROBIOLOGY, MOLECULAR BIOLOGY. Educ: St Vincent Col, BA, 64; Cath Univ Am, MS, 66, PhD(microbiol), 68. Prof Exp: ASST PROF BIOL, DENISON UNIV, 68- Mem: AAAS; Am Soc Microbiol. Res: Autotrophic and heterotrophic metabolism of hydrogen bacteria; degradation of detergent additives and pesticides by soil microorganisms. Mailing Add: Dept of Biol Denison Univ Granville OH 43023

STULA, EDWIN FRANCIS, b Colchester, Conn, Jan 3, 24; m 55; c 1. VETERINARY PATHOLOGY. Educ: Univ Conn, BS, 50, PhD(animal path), 63; Univ Toronto, DVM, 55. Prof Exp: Instr vet med & exten vet, Univ Conn, 55-62; CHIEF RES PATHOLOGIST, HASKELL LAB TOXICOL & INDUST MED, E I DU PONT DE NEMOURS & CO, INC, 63- Mem: Am Vet Med Asn; Soc Pharmacol & Environ Pathologists; NY Acad Sci; Int Acad Path; Am Asn Lab Animal Sci. Res: Bovine vibriosis, mastitis and infertility; leptospirosis in chincillas and guinea pigs; spontaneous diseases of laboratory animals; industrial medicine; pathologic effects in animals exposed to various chemicals by various routes; morphologic effects using both light and electron microscopes; carcinogenicity and embryotoxicity. Mailing Add: 235 Mercury Rd Newark DE 19711

STULBERG, CYRIL SIDNEY, b Chicago, Ill, Apr 11, 19. TISSUE CULTURE, VIROLOGY. Educ: Univ Minn, BA, 43, MS, 45, PhD, 47. Prof Exp: Virologist, Children's Fund Mich, 49-52; SR RES ASSOC, CHILD RES CTR MICH, 54-; PROF IMMUNOL & MICROBIOL, SCH MED, WAYNE STATE UNIV, 61- Concurrent Pos: Affil, Lab Med, Children's Hosp Mich, 53-; consult, Beaumont Hosp, Royal Oak, Mich & Sinai Hosp, Detroit, Mich. Mem: Am Acad Microbiol; Am Asn Cancer Res; Am Asn Immunol; Am Soc Microbiol; Tissue Cult Asn. Res: Inter and intraspecies of cultured animal cells; viruses and cancer. Mailing Add: Child Res Ctr of Mich 3901 Beaubien Blvd Detroit MI 48201

STULBERG, MELVIN PHILIP, b Duluth, Minn, May 17, 25; m 55; c 3. BIOCHEMISTRY. Educ: Univ Minn, BS, 49, MS, 55, PhD(biochem), 58. Prof Exp: Res assoc, Biol Div, Oak Ridge Nat Lab, 58-59, biochemist, 59-61; biochemist, AEC, 61-63; BIOCHEMIST, BIOL DIV, OAK RIDGE NAT LAB, 63- Mem: Am Chem Soc; Am Soc Biol Chemists. Res: Protein biosynthesis; isolation and function of transfer RNA; enzyme mechanisms. Mailing Add: Biol Div Oak Ridge Nat Lab PO Box Y Oak Ridge TN 37830

STULL, DANIEL RICHARD, b Columbus, Ohio, May 28, 11; m 36; c 2. THERMOCHEMISTRY. Educ: Baldwin-Wallace Col, BS, 33; Johns Hopkins Univ, PhD(chem), 37. Prof Exp: Instr chem, Johns Hopkins Univ, 35-37; res chemist, US Indust Chem Co, 37; prof chem, E Carolina Teachers Col, 37-40; phys res chemist, Res Lab, 40-51, dir thermal lab, 51-69, RES SCIENTIST, DOW CHEM CO, 69- Concurrent Pos: Lectr, Mich State Univ, 52. Mem: AAAS; Am Chem Soc; Am Inst Chemists. Res: Evaluation and tabulation of vapor pressure and thermodynamic properties for chemical species; evaluation of reactivity and hazard associated with chemical species (or their mixtures) which are capable of self decomposition. Mailing Add: 1113 W Park Dr Midland MI 48640

STULL, ELISABETH ANN, b Fayette, Mo, Jan 7, 43. LIMNOLOGY, PHYCOLOGY. Educ: Lawrence Univ, BA, 65; Univ Ga, MS, 69; Univ Calif, Davis, PhD(ol PhD(zool), 72. Prof Exp: Asst prof biol sci, 71-75, ASST PROF ECOL & EVOLUTIONARY BIOL, UNIV ARIZ, 75- Mem: Am Soc Limnol & Oceanog; Int Soc Limnol; Phycol Soc Am; AAAS. Res: Energetics and trophic ecology of unicellular taxa; primary productivity and plankton dynamics; algal floristics; regional and geographical patterns in limnology and water quality. Mailing Add: Dept of Ecol & Evolutionary Biol Univ of Ariz Tucson AZ 85721

STULL, JAMES TRAVIS, b Ashland, Ky, Feb 7, 44; m 66; c 2. PHARMACOLOGY, BIOCHEMISTRY. Educ: Southwestern at Memphis, BS, 66; Emory Univ, PhD(pharmacol), 71. Prof Exp: Adj asst prof biol chem, Sch Med, Univ Calif, Davis, 73-74; ASST PROF PHARMACOL, SCH MED, UNIV CALIF, SAN DIEGO, 74- Concurrent Pos: Damon Runyon Mem Fund Cancer Res fel, Univ Calif, Davis, 71-73; estab investr, Am Heart Asn, 73. Mem: AAAS. Res: Protein phosphorylation reactions in regulation of muscle metabolism, contraction and responses to hormones and adrenergic drugs. Mailing Add: Dept of Med Univ of Calif at San Diego La Jolla CA 92037

STULL, JOHN LEETE, b Dansville, NY, June 2, 30; m 52; c 2. PHYSICS, CERAMICS. Educ: Alfred Univ, BS, 52, MS, 54, PhD(ceramics), 58. Prof Exp: Res assoc ceramics, 52-58, from asst prof to assoc prof physics, 58-68, chmn dept, 72-75, PROF PHYSICS, ALFRED UNIV, 68-, DIR OBSERV, 68- Mem: AAAS; Am Asn Physics Teachers. Res: Astronomy; development of physics teaching apparatus; teaching films. Mailing Add: Dept of Physics Alfred Univ Alfred NY 14802

STULL, JOHN WARREN, b Benton, Ill, Nov 23, 21; m 45; c 5. DAIRY SCIENCE. Educ: Univ Ill, BS, 42, MS, 47, PhD(food technol), 50. Prof Exp: Asst dairy technol, Univ Ill, 46-49; from asst prof to assoc prof dairy sci, 49-58, actg head dept, 56-57, PROF DAIRY & FOOD SCI, COL AGR, UNIV ARIZ, 58-, DAIRY & FOOD SCIENTIST, AGR EXP STA, 74- Mem: Am Chem Soc; Am Dairy Sci Asn; Inst Food Technologists. Res: Factors related to the oxidative deterioration of the constituents of milk; food value of milk; chemical residues in milk; biochemistry of milk and food lipids. Mailing Add: Dept of Dairy & Food Sci Col of Agr Univ of Ariz Tucson AZ 85721

STULL, VINCENT ROBERT, b Pottsville, Pa, July 25, 31; m 58; c 3. APPLIED PHYSICS, SYSTEMS ANALYSIS. Educ: Univ Scranton, BS, 53; Univ Pittsburgh, PhD(physics), 58. Prof Exp: Physicist, US Naval Air Sta, Pa, 53; res asst & lectr physics, Univ Pittsburgh, 53-58; res scientist, Aeronutronic Div, Ford Motor Co, 58-62; res scientist, Gen Res Corp, 62-71; RES SCIENTIST, SCI SPECTRUM, INC, 71- Mem: AAAS; Am Phys Soc; Am Geophys Union; Optical Soc Am. Res: Atomic and molecular physics; radiative transfer; atmospherics; electromagnetic scattering; space physics; radiative phenomena in flames and hypersonic aerodynamics; microwave interaction with plasmas and solids; instrumentation development; surface radiative emission. Mailing Add: 1526 Monte Vista Rd Santa Barbara CA 93108

STULL, WILLIAM DEMOTT, b Madison, NJ, July 3, 12; m 41; c 2. ZOOLOGY. Educ: Middlebury Col, BS, 34, MS, 36; Univ Md, PhD(zool), 40. Prof Exp: From instr to assoc prof biol & geol, Cent Col, Mo, 40-45; assoc prof, 45-51, PROF ZOOL, OHIO WESLEYAN UNIV, 51- Concurrent Pos: Res assoc, Stanford Univ, 65. Mem: Wilson Ornith Soc; Am Ornithologists Union; Am Soc Zoologists. Res: Physiology of the response of frogs to low temperature; ecological relationships of two competing species of salamanders; taxonomy and distribution of salamanders; behavior of frogs; life history of birds. Mailing Add: Dept of Zool Ohio Wesleyan Univ Delaware OH 43015

STULLKEN, DONALD EDWARD, b Sullivan, Ill, Apr 11, 20; m 42; c 3. PHYSIOLOGY. Educ: DePauw Univ, BA, 41; Purdue Univ, MS, 42, PhD, 50. Hon Degrees: DSc, DePauw Univ, 67. Prof Exp: Instr physiol, Purdue Univ, 46-50, asst prof, 53-54; aviation physiologist, US Navy Sch Aviation & mem staff, Chief of Naval Air Training, 54-62; chief recovery opers br, Manned Spacecraft Ctr, 62-71, chief flight opers & recovery, 72-75, CHIEF TRAINING DEVELOP & INTEGRATION, JOHNSON SPACE CTR, NASA, 75- Honors & Awards: Chief Naval Air Training Commendation, 59; NASA Superior Achievement Award, 69, Except Serv Medal, 73. Mem: AAAS; US Naval Inst. Res: Space flight operations; environmental physiology; aerospace management. Mailing Add: NASA Johnson Space Ctr Code CG-3 Houston TX 77058

STULTZ, ROBERT LEE, JR, b Washington, DC, July 24, 26; m 48; c 5. PHYSICAL CHEMISTRY. Educ: Univ WVa, BA, 49. Prof Exp: Instr chem, Wyo Sem, 49-53; res chemist, Union Carbide Chem Co, 53-58, group leader, 58-63; mgr dyeing & finishing labs, 63-64, textile process develop, 64-65, prod dir fortrel & nylon, 65-67, dir retail & consumer mkt, 67-68, vpres, 68-73, SR VPRES, CELANESE FIBERS MKT CO, 73- Res: Non-cellulosic synthetic fibers. Mailing Add: PO Box 1414 Charlotte NC 28232

STULTZ, WALTER ALVA, b St John, NB, Mar 14, 04; nat US; m 31, 50; c 5. ANATOMY. Educ: Acadia Univ, BA, 27; Yale Univ, PhD(zool, anat), 32. Prof Exp: Prin sch, NB, Can, 22-24; asst biol, Yale Univ, 27-30; instr, Spring Hill Sch, Conn, 30-31; instr & actg head dept, Trinity Col, Conn, 31-32; prof, Mt Union Col, 32-33; asst, Yale Univ, 33-34; fel anat, Sch Med, Univ Ga, 34-35, fel histol & embryol, 35-36; instr anat, Med Col SC, 36-37; from asst prof to assoc prof, 37-69, EMER PROF ANAT, COL MED, UNIV VT, 69- Concurrent Pos: Sr lectr gross anat & histol, Sch Med, Univ Calif, San Diego, 69- Mem: Am Asn Anat. Res: Experimental embryology of Amblystoma; relations of symmetry in fore and hind limbs; interrelationships between limbs and nervous system. Mailing Add: RR 2 Williston VT 05495

STUMBO, CHARLES RAYMOND, b Monmouth, Kans, July 25, 14; m 36; c 3. MICROBIOLOGY, FOOD SCIENCE. Educ: Kans State Univ, BS, 36, MS, 37, PhD(bact), 41. Prof Exp: Instr microbiol, SDak State Col, 37-39; agent, USDA, 39-41; asst dir res, John Morell & Co, 41-44; biochemist, Mich Dept Health & res prof biochem & microbiol, Mich State Col, 44-45; res supvr microbiol, Owens Ill Glass Co, 45-47; chief bacteriologist, Food Mach & Chem Corp, 47-50; assoc prof food microbiol, Wash State Col, 50-53; head bact res, H J Heinz Co, 53-59; consult & dir prod develop, Producers Creamery Co, 59-63; assoc prof, 63-65, PROF FOOD SCI & NUTRIT, UNIV MASS, AMHERST, 65- Mem: Am Soc Microbiol; Inst Food Technologists. Res: Microbiology, particularly as it relates to food preservation; thermobacteriology and other phases of sterilization. Mailing Add: Dept of Food Sci & Nutrit Univ of Mass Amherst MA 01002

STUMP, ALEXANDER BELL, b Emmorton, Md, Jan 13, 05; m 42; c 4. PROTOZOOLOGY. Educ: Univ Va, BS, 30, MS, 31, PhD(protozool), 34. Prof Exp: Res assoc, Univ Va, 34-35; prof biol, Flora Macdonald Col, 36-47; prof, 47-72, EMER PROF, PRESBY COL, SC, 72- Mem: Soc Protozoologists. Res: Mitosis in the testacea; cytology of protozoa. Mailing Add: Rt 1 Box 57 Cross Hill SC 29332

STUMP, BILLY LEE, b Morristown, Tenn, Jan 11, 30; m 58; c 3. PHYSICAL CHEMISTRY, ORGANIC CHEMISTRY. Educ: Carson-Newman Col, BS, 52; Univ Tenn, PhD(chem), 59. Prof Exp: Res chemist, Carson-Newman Col, 52-53; res technician, Oak Ridge Inst Nuclear Studies, 53-54; chemist, Redstone Div, Thiokol Chem Corp, 59-60; res chemist, Film Res Lab, E I du Pont de Nemours & Co, 60-62 & Spruance Film Res & Develop Lab, 62-63; assoc prof chem, Carson-Newman Col, 63-66; ASSOC PROF CHEM, VA COMMONWEALTH UNIV, 66- Mem: Am Chem Soc. Res: Kinetics and reaction mechanisms; catalysis and kinetics of catalytic hydrogenation; polymer chemistry. Mailing Add: Dept of Chem 901 W Franklin St Va Commonwealth Univ Richmond VA 23220

STUMP, EUGENE CURTIS, JR, b Charleston, WVa, May 19, 30; m 58; c 3. ORGANIC CHEMISTRY, FLUORINE CHEMISTRY. Educ: WVa Univ, BS, 52; Columbia Univ, MA, 53; Univ Fla, PhD(org chem), 60. Prof Exp: Dir contract res, 60-70, VPRES CONTRACT RES DIV, PCR, INC, 70- Mem: Am Chem Soc; The Chem Soc. Res: Synthesis of containing compounds, particularly ethers, olefins, nitroso and difluoramine compounds; synthesis of fluorine-containing polymers as low temperature elastomers; synthesis of thermally and oxidatively stable fluids. Mailing Add: Contract Res Div PCR Inc PO Box 1466 Gainesville FL 32601

STUMP, JOHN EDWARD, b Galion, Ohio, June 3, 34; m 55; c 2. VETERINARY ANATOMY. Educ: Ohio State Univ, DVM, 58; Purdue Univ, PhD, 66. Prof Exp: Private vet pract, Ohio, 58-61; from instr to assoc prof vet anat, 61-76, PROF VET ANAT, PURDUE UNIV, WEST LAFAYETTE, 76- Mem: AAAS; Am Vet Med Asn; World Asn Vet Anat; Am Asn Vet Anat; Asn Am Vet Med Cols. Res: Electron microscopy of epidermal tissue, especially the equine hoof under normal and disease conditions; gross anatomy of domestic animals. Mailing Add: Dept of Vet Anat Purdue Univ West Lafayette IN 47907

STUMP, JOHN M, b Charleston, WVa, June 26, 38; m 64. PHARMACOLOGY. Educ: WVa Univ, BS, 61, MS, 62, PhD(pharmacol), 64. Prof Exp: Res pharmacologist, 64-72, SR RES PHARMACOLOGIST, STINE LAB, PHARMACEUT RES DIV, E I DU PONT DE NEMOURS & CO, INC, 72- Res: Autonomic and cardiovascular pharmacology. Mailing Add: Stine Lab E I du Pont de Nemours & Co Inc Newark DE 19711

STUMP, JOSEPH HARRY, JR, organic chemistry, see 12th edition

STUMP, ROBERT, b Indianapolis, Ind, Oct 16, 21; m 43; c 4. PHYSICS. Educ: Butler Univ, BA, 42; Univ Ill, MS, 48, PhD(physics), 50. Prof Exp: From asst prof to assoc prof, 50-60, PROF PHYSICS, UNIV KANS, 60- Concurrent Pos: Consult, Aeronaut Radio, Inc, DC, 52-53; vis scientist, Midwestern Univs Res Asn, Wis, 59, Europ Orgn Nuclear Res, Geneva, 63-64 & Polytech Sch, Paris, 70-71; vis physicist, Brookhaven Nat Lab, 62-63. Mem: AAAS; fel Am Phys Soc. Res: Experimental nuclear and elementary particle physics; angular correlations; low temperature effects; hydrogen and heavy liquid bubble chamber experiments. Mailing Add: 2418 Orchard Lane Lawrence KS 66044

STUMP, WILLIAM LESTER, physical chemistry, see 12th edition

STUMPF, FOLDEN BURT, b Lansing, Mich, Aug 18, 28; m 54; c 2. PHYSICS. Educ: Kent State Univ, BS, 50; Univ Mich, MS, 51; Ill Inst Technol, PhD(physics), 56. Prof Exp: From asst prof to assoc prof, 56-66, PROF PHYSICS, OHIO UNIV, 66- Mem: Acoust Soc Am; Am Asn Physics Teachers. Res: Ultrasonic transducers; application of ultrasonics to liquids. Mailing Add: Dept of Physics Ohio Univ Athens OH 45701

STUMPF, HENRY JOHN, applied mechanics, physics, see 12th edition

STUMPF, PAUL KARL, b New York, NY, Feb 23, 19; m 47; c 5. LIPID BIOCHEMISTRY. Educ: Harvard Univ, AB, 41; Columbia Univ, PhD(biochem), 45. Prof Exp: Chemist, Div War Res, Columbia Univ, 44-46; instr epidemiol, Sch Pub Health, Univ Mich, 46-48; asst prof plant nutrit, Univ Calif, Berkeley, 48-52, from assoc prof to prof plant biochem, 52-58, chmn dept, 53-57, PROF PLANT BIOCHEM & CHMN DEPT BIOCHEM, UNIV CALIF, DAVIS, 58- Concurrent Pos: NIH sr fel, 54-55; NSF sr fel, 61-62, 68; Guggenheim fel, 62-69; ed, J Phytochem, 60- & Archives Biochem & Biophys, 60-65, exec ed, 65-; ed, J Lipid Res, 63-66, Anal Biochem, 69- & Biochem Prep; mem physiol chem study sect, NIH, 60-64; chmn subcomt biochem nomenclature comn, Nat Acad Sci-Nat Res Coun, 61-62; metab biol panel, NSF, 65-68; mem, City of Davis Planning Comn, 67-69; mem div biol & agr, Nat Res Coun, 67-; vis scientist, Commonwealth Sci & Indust Res Orgn, Canberra ACT, Australia, 75-76; sr US Sci fel, Von Humboldt Fedn, Ger, 76. Honors & Awards: Stephen Hales Award, Am Soc Plant Physiologist, 74; Lipid Chem Prize, Am Oil Chemists Soc, 74. Mem: Am Soc Plant Physiol; Am Oil Chem Soc; Am Soc Biol Chem; Brit Biochem Soc; foreign mem Royal Danish Acad Arts & Sci. Res: Lipid biochemistry of higher plants; photobiosynthesis; developmental biochemistry. Mailing Add: Dept of Biochem & Biophys Univ of Calif Davis CA 95616

STUMPF, WALTER ERICH, b Oelsnitz, Ger, Jan 10, 27; m 61; c 4. NEUROENDOCRINOLOGY, PHARMACOLOGY. Educ: Univ Berlin, MD, 52, cert neurol & psychiat, 57; Univ Chicago, PhD(pharmacol), 67. Prof Exp: Intern, Charite Hosp, Univ Berlin, 52-53, resident neurol & psychiat, 53-57; sci asst, Univ Marburg, 58-60 & Lab Radiobiol & Isotope Res, 61-62; res assoc pharmacol, Univ Chicago, 63-67, asst prof, 67-70; assoc prof anat & pharmacol & mem labs for reproductive biol, 70-73, PROF ANAT & PHARMACOL, UNIV NC, CHAPEL HILL, 73- Concurrent Pos: Trainee, psychother & psychoanal, Inst Psychother, WBerlin, 54-56; lectr clin neurol, Charite Hosp, Univ Berlin, 56-57; vis psychiatrist, Maudsley Hosp, London, 59; consult, Microtome-Cyrostats, 69-; mem, Neurolbiol Prog, 70-; assoc, Carolina Pop Ctr, 72-; res scientist, Biol Sci Res Ctr, 72-; consult, Life Sci Inst, Research Triangle Park, NC, 73- Mem: AAAS; Am Soc Zoologists; Histochem Soc; Int Brain Res Orgn; Am Asn Anatomists. Res: Development of histochemical techniques; low temperature sectioning and freeze-drying; dry-mount autoradiography for the localization of hormones and drugs in the brain and other tissues. Mailing Add: Labs for Reproductive Biol Univ of NC at Chapel Hill Chapel Hill NC 27514

STUMPFF, HOWARD KEITH, b Holden, Mo, May 26, 30; m 57; c 3. MATHEMATICS. Educ: Cent Mo State Col, BSEd, 51; Univ Mo, AM, 53; Univ Kans, PhD(math ed), 68. Prof Exp: Asst instr math, Univ Mo, 51-53; instr, Univ NMex, 57-63; from asst prof to assoc prof, 63-72, PROF MATH, CENT MO STATE COL, 72-, HEAD DEPT, 69- Concurrent Pos: Asst instr, Univ Kans, 61-62. Mem: Am Math Soc; Math Asn Am. Res: Mathematics education. Mailing Add: Dept of Math WCM 222 Cent Mo State Col Warrensburg MO 64093

STUNKARD, ALBERT J, b New York, NY, Feb 7, 22. PSYCHIATRY. Educ: Yale Univ, BS, 43; Columbia Univ, MD, 45. Prof Exp: Resident physician, Johns Hopkins Hosp, 48-51; fel psychiat, 51-52; res fel med, Col Physicians & Surgeons, Columbia Univ, 52-53; Commonwealth fel med, Med Col, Cornell Univ, 53-56, asst prof, 56-57; from assoc prof to prof psychiat & chmn dept, Sch Med, Univ Pa, 62-73; PROF PSYCHIAT, STANFORD UNIV, 73- Concurrent Pos: Fel, Ctr Advan Study Behav Sci, Calif, 71-72; vis prof, Inst Neurol Sci, Univ Pa, 75-76. Mem: Am Psychosom Soc; Am Psychiat Asn; Am Fedn Clin Res; fel NY Acad Sci; Psychiat Res Soc. Res: Obesity and regulation of energy balance. Mailing Add: Dept of Psychiat Stanford Univ Stanford CA 94305

STUNKARD, JIM A, b Sterling, Colo, Jan 25, 35; m 67; c 1. LABORATORY

ANIMAL MEDICINE, VETERINARY MICROBIOLOGY. Educ: Colo State Univ, BS, 57, DVM, 59; Tex A&M Univ, MS, 66; Am Col Lab Animal Med, dipl. Prof Exp: Vet, Glasgow Animal Hosp, Ky, 59-61; vet in charge, Sentry Dog Procurement, Training & Med Referral Ctr, US Air Force Europe, 61-64 & Lab Animal Colonies & Zoonoses Control Ctr, 61-64, resident lab animal med, sch aerospace med, Brooks AFB, Tex, 64-66, dir vet med sci dept, Naval Med Res Inst, Nat Naval Med Ctr, Md, 66-71. Concurrent Pos: Vet consult, Turkish Sentry Dog Prog, US Air Force Europe, 61-63, Can Air Force Europe, 62-64, Bur Med & Surg, US Navy, Washington, DC, Navy Toxicol Unit, Md, 66-71 & AEC, 68-71; consult to dean vet med, Colo State Univ, 69-71; US Navy rep ad hoc comt, Dept Defense, 66-67 & Inst Lab Animal Resouces, Nat Acad Sci-Nat Res Coun, Washington, DC, 66-71. Mem: Am Vet Med Asn. Res: Veterinary medicine, dentistry and surgery, especially all phases of laboratory animal medicine. Mailing Add: Bowie Animal Hosp 3428 Crain Hwy Bowie MD 20715

STUNTZ, CALVIN FREDERICK, b Buffalo, NY, Aug 6, 18; m 51; c 3. CHEMISTRY. Educ: Univ Buffalo, BA, 39, PhD(chem), 47. Prof Exp: Teacher high sch, NY, 39-40; anal chemist, Linde Air Prods Co, Union Carbide & Carbon Corp, 40-41; asst, Univ Buffalo, 41-43, 45-46; from asst prof to assoc prof chem, 46-69, PROF CHEM, UNIV MD, COLLEGE PARK, 69- Mem: Am Chem Soc. Res: Quantitative analysis; chemical microscopy. Mailing Add: Dept of Chem Univ Of Md College Park MD 20742

STUNTZ, DANIEL ELLIOT, b Milford, Ohio, Mar 15, 09. MYCOLOGY. Educ: Univ Wash, BS, 35; Yale Univ, PhD, 40. Prof Exp: From instr to assoc prof, 40-59, PROF BOT, UNIV WASH, 59- Mem: Bot Soc Am; Mycol Soc Am. Res: Morphology; taxonomy of Basidiomycetes, Ascomycetes. Mailing Add: Dept of Bot Univ of Wash Seattle WA 98195

STUPEGIA, DONALD CHARLES, physical chemistry, see 12th edition

STUPIAN, GARY WENDELL, b Alhambra, Calif, Oct 17, 39. SOLID STATE PHYSICS, SURFACE PHYSICS. Educ: Calif Inst Technol, BS, 61; Univ Ill, Urbana, MS, 63, PhD(physics), 67. Prof Exp: Res asst physics, Univ Ill, Urbana, 61-67; res assoc mat sci, Cornell Univ, 67-69; MEM TECH STAFF, AEROSPACE CORP, 69- Mem: AAAS; Am Phys Soc. Res: Nuclear magnetic resonance; Auger spectroscopy; heterogeneous catalysis; analytical instrumentation. Mailing Add: Bldg 120/12431 Chem Physics Dept Aerospace Corp PO Box 92957 Los Angeles CA 90009

STUPP, EDWARD HENRY, b Brooklyn, NY, Dec 10, 32; m 54; c 2. SOLID STATE PHYSICS. Educ: City Col New York, BS, 54; Syracuse Univ, MS, 58, PhD(physics), 60. Prof Exp: Asst physics, Columbia Univ, 54-55, Watson Lab, 55-56 & Syracuse Univ, 56-69; staff physicist, Thomas J Watson Res Ctr, 59-62; sr physicist, 62-66, STAFF PHYSICIST & SUPVR INFRARED IMAGING & IMAGE TUBES, PHILIPS LABS DIV, N AM PHILIPS CO, 66-, SUPVR PHOTOEMIS- SION, 69- Concurrent Pos: Consult infrared imaging, Philips Broadcast Equip Cor, 72-73. Mem: Am Phys Soc; Inst Elec & Electronics Eng. Res: Experimental solid state physics, especiallly the interaction of radiation with solids; photoemission; visible and infrared camera tubes, photodetectors, electron multiplication, image tubes; cold cathodes. Mailing Add: Philips Labs 345 Scarborough Rd Briarcliff NY 10510

STURBAUM, BARBARA ANN, b Cleveland, Ohio, June 10, 36. PHYSIOLOGY. Educ: Marquette Univ, BS, 59, MS, 61; Univ NMex, PhD(zool), 72. Prof Exp: From asst prof to assoc prof biol & earth sci, St John Col, Ohio, 61-75; ASST PROF ANAT BIOL & PHYSIOL, ORAL ROBERTS UNIV, 75- Concurrent Pos: Consult radionuclide metab & toxicity, Lovelace Found Med Educ & Res, 74-75. Mem: Radiation Res Soc; Am Soc Zoologists; Am Inst Biol Sci; AAAS. Res: Environmental physiology and radiobiology, particularly behavioral a nd physiological responses of animals to environmental factors and the effects of radiation on animals. Mailing Add: Dept of Natural Sci Oral Roberts Univ Tulsa OK 74102

STURCH, CONRAD RAY, b Cincinnati, Ohio, Nov 5, 37; m 62; c 1. ASTRONOMY. Educ: Miami Univ, BA, 58, MS, 60; Univ Calif, Berkeley, PhD(astron), 65. Prof Exp: Res asst astron, Lick Observ, Univ Calif, 65; from instr to asst prof, Univ Rochester, 65-73; vis asst prof, Univ Western Ont, 73-74; VIS ASST PROF ASTRON, CLEMSON UNIV, 74- Mem: Am Astron Soc; Int Astron Union. Res: Variable stars; stellar populations; interstellar reddening; galactic structure. Mailing Add: Dept of Physics & Astron Clemson Univ Clemson SC 29631

STURDIVANT, HARWELL PRESLEY, b West Point, Ga, Dec 15, 02; m 28; c 1. ZOOLOGY. Educ: Emory Univ, BS, 25, MA, 26; Columbia Univ, PhD(zool), 34. Prof Exp: Asst, Emory Univ, 25-26, instr, 26-27; instr, Wash Sq Col, NY Univ, 27-29, Columbia Univ, 28-31 & Emory Univ, 31-32; prof biol, Union Col, Ky, 32-44; prof biol & head dept sci, Middle Tenn State Col, 44-46; prof biol, Millsaps Col, 46-48; PROF BIOL & HEAD DEPT, WESTERN MD COL, 48- Concurrent Pos: Prof, Simpson Col, 41-42; US AID consult, India, 65; sci consult, India, 68. Mem: Fel AAAS. Res: Cytology; cytogenetics; the centriole; nonflagellated. Mailing Add: 712 Washington Rd Westminster MD 21157

STURDY, ROBERT ALLAN, b Woodson, Ill, Nov 14, 14; m 38; c 2. VETERINARY MEDICINE. Educ: Univ Ill, AB, 36; Ohio State Univ, DVM, 45. Prof Exp: Res chemist, Moorman Mfg Co, 37-42; control chemist feed lab, State Dept Agr, Ohio, 43-45; res vet, 45-68, DIR FED & STATE RELS, MOORMAN MFG CO, 68- Mem: Am Vet Med Asn; Entom Soc Am. Res: Animal nutrition and nutritional diseases; animal anthelmintics and insecticides; systematic animal insecticides. Mailing Add: Moorman Mfg Co 1000 N 30th St Quincy IL 62301

STURGE, MICHAEL DUDLEY, b Bristol, Eng, May 25, 31; m 56; c 4. EXPERIMENTAL SOLID STATE PHYSICS. Educ: Cambridge Univ, BA, 52, PhD(physics), 57. Prof Exp: Mem staff, Mullard Res Lab, 56-58; sr res fel, Royal Radar Estab, 58-61; MEM TECH STAFF, BELL LABS, 61- Concurrent Pos: Res assoc, Stanford Univ, 65; vis scientist, Univ BC, 69; exchange visitor, Philips Res Labs, Eindhoven, Neth, 73-74; vis lectr physics, Drew Univ, 75. Mem: Fel Am Phys Soc. Res: Magnetic insulators; optical properties of solids; semiconductor luminescence. Mailing Add: Bells Labs Murray Hill NJ 07974

STURGEON, EDWARD EARL, b Irving, Ill, Apr 28, 16; m 43; c 2. FORESTRY. Educ: Univ Mich, PhD(forestry), 54. Prof Exp: Instr forest policy, Univ Mich, 50-51; assoc prof forestry & head dept forestry & biol, Mich Technol Univ, 51-59; assoc prof forestry, Humboldt State Col, 59-66, coordr dept, 60-66; head dept, 66-73, PROF FORESTRY, OKLA STATE UNIV, 66- Mem: AAAS; Soc Am Foresters; Am Forestry Asn. Res: Public-private balance in forest land ownership; forest environment and administration. Mailing Add: Dept of Forestry Okla State Univ Stillwater OK 74074

STURGEON, GEORGE DENNIS, b Sioux Falls, SDak, Sept 21, 37. SOLID STATE CHEMISTRY, HIGH TEMPERATURE CHEMISTRY. Educ: Univ NDak, BS, 59; Mich State Univ, PhD(chem), 64. Prof Exp: Instr chem, Mich State Univ, 64; asst prof, 64-73, ASSOC PROF CHEM, UNIV NEBR, LINCOLN, 73- Mem: AAAS; Am Chem Soc; Sigma Xi. Res: Chemistry of refractory materials; high-temperature thermodynamics; chemistry of complex fluorides. Mailing Add: Dept of Chem Univ of Nebr Lincoln NE 68588

STURGEON, MYRON THOMAS, b Salem, Ohio, Apr 27, 08; m 46; c 2. PALEONTOLOGY, STRATIGRAPHY. Educ: Mt Union Col, AB, 31; Ohio State Univ, AM, 33, PhD(paleont), 36. Prof Exp: Found inspector, US Corps Engrs, Ohio, 34; from asst to assoc prof geol, Mich State Norm Col, 37-46; from asst prof to assoc prof, 46-54, PROF GEOL, OHIO UNIV, 54- Mem: AAAS; Paleont Soc; assoc Soc Econ Paleont & Mineral; fel Geol Soc Am; Am Ornith Union. Res: Stratigraphy and invertebrate paleontology of the Pennsylvanian system of eastern Ohio. Mailing Add: Dept of Geol Porter Hall Ohio Univ Athens OH 45701

STURGEON, PHILLIP, b Los Angeles, Calif, Feb 17, 18; m 41; c 3. HEMATOLOGY. Educ: Univ Calif, Los Angeles, BS, 39; Univ Southern Calif, MD, 43; Am Bd Pediat, dipl, 52. Prof Exp: Resident path, pediat & hemat, Los Angeles Children's Hosp, 46-50; from asst prof to prof pediat, Sch Med, Univ Southern Calif, 51-60; assoc dir res, Am Red Cross, 60-66; PROF PEDIAT, SCH MED, UNIV CALIF, LOS ANGELES, 66- Concurrent Pos: Res fel hemat, Los Angeles Children's Hosp, 50-52; Markle scholar, 52-57. Mem: Int Soc Hemat; Int Soc Blood Transfusion. Res: Pediatrics; immunology; genetics. Mailing Add: Apt 406P 201 Ocean Ave Santa Monica CA 90402

STURGEON, ROY V, JR, b Wichita, Kans, July 1, 24; m 50; c 2. PLANT PATHOLOGY. Educ: Okla State Univ, BS, 61, MS, 64; Univ Minn, Minneapolis, PhD(plant path), 67. Prof Exp: Instr bot & plant path, 63-65, from instr to assoc prof, Col Arts & Sci & Agr Exten, 67-74, PROF BOT & PLANT PATH, COL ARTS & SCI & AGR EXTEN, OKLA STATE UNIV, 74-, EXTEN PLANT PATHOLOGIST, FED EXTEN SERV, 67- Concurrent Pos: Private Plant Health Consult Serv, Okla, 67- Mem: Am Phytopath Soc; Soc Nematol; Am Soc Agron. Res: Evaluation of chemicals for disease control, especially fungicides and nematicides. Mailing Add: Dept of Plant Path Okla State Univ 115 Life Sci E Stillwater OK 74074

STURGES, WILTON, III, b Dothan, Ala, July 21, 35; m 57; c 3. PHYSICAL OCEANOGRAPHY. Educ: Auburn Univ, BS, 57; Johns Hopkins Univ, MA, 63, PhD(oceanog), 66. Prof Exp: Res asst phys oceanog, Johns Hopkins Univ, 63-66; from asst prof to assoc prof, Univ RI, 66-72; ASSOC PROF PHYS OCEANOG, FLA STATE UNIV, 72- Concurrent Pos: Instr, US Naval Res Off Sch, 63-66; mem ocean-wide surv panel, Comt Oceanog, Nat Acad Sci, 68-71, Buoy Technol Assessment Panel Marine Bd, Nat Acad Eng, 72-, Ocean Sci Comt, Nat Res Coun, 75- & Comn Marine Geodesy, Am Geophys Union, 74-; assoc chmn, Joint FGGE Adv Panel, Ocean Sci Comn, US Global Atmospheric Res Prog, 74. Mem: Am Geophys Union. Res: Analysis of serial oceanographic data; slope of sea level; ocean circulation; mixing in bottom currents. Mailing Add: Dept of Oceanog Fla State Univ Tallahassee FL 32306

STURGESS, JENNIFER MARY, b Nottingham, Gt Brit, Sept 26, 44; m 66; c 2. MICROBIOLOGY, PATHOLOGY. Educ: Bristol Univ, BSc, 65; Univ London, PhD(path), 70. Prof Exp: Res asst microbiol, Agr Col Norway, 64 & Clin Res Unit, Med Res Coun Eng, 65-66; lectr exp path, Inst Dis Chest, Brompton Hosp, Univ London, 66-70; INVESTR & ASST PROF PATH, UNIV TORONTO, 71- Concurrent Pos: Res fel path, Hosp for Sick Children, Toronto, 70-71, Med Res Coun Can term grant & scholar, 71- Mem: Am Soc Cell Biol; Electron Micros Soc Am; Micros Soc Can. Res: Glycoprotein biosynthesis and secretion; role of golgi complex as membrane site of control of glycoprotein in normal cell; abnormal glycoprotein in disease. Mailing Add: Hosp for Sick Children 555 University Ave Toronto ON Can

STURGILL, BENJAMIN CALEB, b Wise Co, Va, Apr 27, 34; m 55; c 2. MEDICINE, PATHOLOGY. Educ: Berea Col, BA, 56; Univ Va, MD, 60. Prof Exp: Intern med, New York, Hosp-Cornell Med Ctr, 60-61; resident path, Univ Va, 61-62; clin assoc, NIH, 62-64; from instr to asst prof, 64-71, ASSOC PROF PATH, SCH MED, UNIV VA, 71-, ACTG CHMN DEPT, 74- Mem: Int Acad Path; Am Asn Path & Bact; Am Soc Exp Path. Res: Immunopathology and renal diseases. Mailing Add: Sch of Med Univ of Va Charlottesville VA 22903

STURGIS, BERNARD MILLER, b Butler, Ind, Nov 27, 11; m 36; c 2. PETROLEUM CHEMISTRY. Educ: DePauw Univ, AB, 33; Mass Inst Technol, PhD(org chem), 36. Prof Exp: Res chemist, Jackson Lab, 36-42, group leader auxiliary chem sect, Elastomer Div, 42-46, head petrol chem div, 46-51 & combustion & scavenging div, Petrol Lab, 51-53, from asst dir to dir, 53-62, mgr mid-continent region, Petrol Chem Div, 62-64, MGR PATENTS & CONTRACTS DIV, ORG CHEM DEPT, E I DU PONT DE NEMOURS & CO, INC, 64- Honors & Awards: Horning Mem Award, Soc Automotive Eng, 56; Rector Award, 58. Mem: Am Chem Soc; Soc Automotive Eng; Combustion Inst. Res: Synthetic organic and rubber chemicals; accelerators; antioxidants; sponge blowing agents; peptizing agents; nonsulfur vulcanization of rubber; petroleum additives; tetraethyl lead; combustion; lead scavenging from engines. Mailing Add: E I du Pont de Nemours & Co Inc Patents & Contracts Div Org Chem Dept Wilmington DE 19898

STURGIS, HOWARD EWING, b Pasadena, Calif, June 1, 36; m 57; c 2. COMPUTER SCIENCE. Educ: Calif Inst Technol, BS, 58; Univ Calif, Berkeley, PhD(comput sci), 73. Prof Exp: RES SCIENTIST COMPUT SCI, XEROX PALO ALTO RES CTR, 72- Mem: Asn Comput Mach. Res: Architecture of computer operating systems; semantics of programming languages. Mailing Add: Xerox Palo Alto Res Ctr 3333 Coyote Hill Rd Palo Alto CA 94304

STURGUL, JOHN ROMAN, b Hurley, Wis, Jan 3, 40; m 65; c 2. GEOPHYSICS, ENGINEERING SCIENCE. Educ: Mich Technol Univ, BS, 61; Univ Ariz, MS, 63; Univ Ill, PhD(eng), 66. Prof Exp: Asst prof eng & dir seismol observ, Univ Miss, 66-68; assoc prof geophys, Univ Ariz, 68-76; ASSOC PROF MINING ENG & HEAD DEPT MINING & PETROL ENG, NMEX INST MINING & TECHNOL, 76- Concurrent Pos: Fulbright-Hays fel, Univ Queensland, 72; vis lectr, Univ Melbourne, 75. Mem: AAAS; Am Geophys Union; Soc Explor Geophysicists; Seismol Soc Am; Am Inst Mining, Metall & Petrol Engrs. Res: Mining engineering; seismology; geodynamics; geomorphology; applied mathematics; computer applications in mining and geophysics. Mailing Add: Dept of Mining & Petrol Eng NMex Inst Mining & Technol Socorro NM 87801

STURKIE, PAUL DAVID, b Proctor, Tex, Sept 18, 09; m 40, 64; c 2. PHYSIOLOGY. Educ: Tex A&M Univ, BS, 33, MS, 36; Cornell Univ, PhD(genetics, physiol), 39. Prof Exp: Res asst, Tex A&M Univ, 34-36 & Cornell Univ, 36-39; assoc prof, Auburn Univ, 39-44; assoc prof, 44-50, PROF PHYSIOL, RUTGERS UNIV, NEW BRUNSWICK, 71- Concurrent Pos: Guest reseacher, Agr Res Coun, 60. Honors & Awards: Poultry Sci Res Award, 47; Borden Award, 56; Linback Res Award, Rutgers Univ, 74. Mem: Fel AAAS; Am Physiol Soc; Poultry Sci Asn; Am Heart Asn; Microcirc Soc. Res: Physiology of reproduction, heart and circulation of birds. Mailing Add: Dept of Environ Physiol Rutgers Univ New Brunswick NJ 08903

STURLEY, ERIC AVERN, b Dibden Hants, Eng, June 9, 15; nat US; m 47; c 3. MATHEMATICS. Educ: Yale Univ, BA, 37, MA, 39; Univ Grenoble, cert, 45; Columbia Univ, EdD, 56. Prof Exp: Instr, Berkshire Sch, 41-42 & Lawrenceville Sch, 46-47; from instr to assoc prof math, Allegheny Col, 47-57; instr, 57-61, actg head div sci & math, 58-60, asst dean grad sch, 62-64, PROF MATH, SOUTHERN ILL UNIV, EDWARDSVILLE, 61-, CHIEF ACAD ADV, 59-, COORDR DEANS COL, 67- Concurrent Pos: Consult, Talon, Inc, Pa, 55-57; chief party, Southern Ill Univ Contract Team, Mali, WAfrica, 64-67, Nepal, 70-71. Mem: Am Math Soc; Math Asn Am. Res: Statistics; history of mathematics. Mailing Add: Dept of Math Southern Ill Univ Edwardsville IL 62025

STURM, EDWARD, US citizen; m 50; c 3. GEOLOGY, MINERALOGY. Educ: NY Univ, BA, 48; Univ Minn, MSc, 50; Rutgers Univ, PhD(geol), 57. Prof Exp: Res geologist, Hebrew Univ Jerusalem, 51-52; asst res specialist crystallog, Bur Eng Res, Rutgers Univ, 56-58; asst prof geol, Tex Technol Col, 58-63; from asst prof to assoc prof, 63-74, PROF GEOL, BROOKLYN COL, 74- Mem: Geol Soc Am; Mineral Soc Am; Am Crystallog Asn. Res: Clay mineralogy; crystallography of silicates; preferred orientation studies; geochemistry of solids. Mailing Add: Dept of Geol Brooklyn Col Brooklyn NY 11210

STURM, WILLIAM JAMES, b Marshfield, Wis, Sept 10, 17; m 51; c 2. NUCLEAR PHYSICS, APPLIED PHYSICS. Educ: Marquette Univ, BS, 40; Univ Chicago, MS, 42; Univ Wis, PhD(physics), 49. Prof Exp: Asst nuclear physics, Manhattan Proj, Metall Lab, Univ Chicago, 42-43, jr physicist, 43-46; assoc physicist & group leader, Argonne Nat Lab, 46-47; consult physicist, 49-51; from physicist to sr physicist, Oak Ridge Nat Lab, 51-56; assoc physicist, Int Inst Nuclear Sci & Eng, Argonne Nat Lab, 56-59; assoc physicist, Int Sch Nuclear Sci & Eng, 59-60, Int Inst Nuclear Sci & Eng, 60-65 & Off Col & Univ Coop, 65-67, ASST DIR APPL PHYSICS DIV, ARGONNE NAT LAB, 67- Mem: AAAS; Am Phys Soc; Am Nuclear Soc. Res: Neutron cross sections and diffraction; nuclear reactions, reactor physics and absolute nuclear particle energies; irradiation effects in solids; subcritical and critical reactor studies; reactor safety. Mailing Add: Appl Physics Div Argonne Nat Lab 9700 S Cass Ave Argonne IL 60439

STURMAN, JOHN ANDREW, b Hove, Eng, Aug 10, 41. BIOCHEMISTRY, NUTRITION. Educ: Univ London, BSc, 62, MSc, 63, PhD(biochem), 66. Prof Exp: ASSOC RES SCIENTIST, DEPT PEDIAT RES, INST BASIC RES MENT RETARDATION, 67- Concurrent Pos: Res study grant red cell metab, King's Col Hosp, Med Sch, Univ London, 63-67. Mem: AAAS; Am Inst Nutrit; Brit Biochem Soc; Am Soc Neurochem; Int Soc Neurochem. Res: Sulfur amino acid metabolism in normal and vitamin B-6 deficiency, in fetal, neonatal and adult tissue and in inborn errors of metabolism. Mailing Add: Dept of Pediat Res Inst Basic Res Ment Retardation Staten Island NY 10314

STURMAN, LAWRENCE STUART, b Detroit, Mich, Mar 13, 38; m 59; c 3. VIROLOGY. Educ: Northwestern Univ, BS, 57, MS & MD, 60; Rockefeller Univ, PhD(virol), 68. Prof Exp: Intern, Hosp Univ Pa, 60-61; staff assoc virol, Nat Inst Allergy & Infectious Dis, 68-70; RES PHYSICIAN VIROL, DIV LABS & RES, NY STATE DEPT HEALTH, 70- Concurrent Pos: Adj asst prof microbiol, Dept of Microbiol & Immunol, Albany Med Col, 73- Mem: Am Soc Microbiol, Soc Gen Microbiol. Res: Coronaviruses; viral biosynthesis; host dependent differences in virus replication; pathogenesis of viral disease. Mailing Add: Div of Labs & Res NY State Dept of Health New Scotland Ave Albany NY 12201

STURMER, DAVID MICHAEL, b Norfolk, Va, July 27, 40; m 64; c 2. PHYSICAL ORGANIC CHEMISTRY, PHOTOGRAPHIC CHEMISTRY. Educ: Stanford Univ, BS, 62; Ore State Univ, PhD(org chem), 66. Prof Exp: NSF fel chem, Yale Univ, 66-67; sr chemist, 67-72, RES ASSOC CHEM, EASTMAN KODAK CO RES LABS, 72- Mem: Am Chem Soc; Sigma Xi; Soc Photog Scientists & Engrs. Res: Molecular orbital calculations; heterocyclic dye synthesis; spectral sensitization of silver halides; solid state photochemistry; radiotracer methods. Mailing Add: Eastman Kodak Co Res Labs 1669 Lake Ave Rochester NY 14650

STURR, JOSEPH FRANCIS, b Syracuse, NY, Apr 29, 33; m 60; c 4. PSYCHOPHYSIOLOGY. Educ: Wesleyan Univ, BA, 55; Fordham Univ, MA, 57; Univ Rochester, PhD(psychol), 62. Prof Exp: Asst exp psychol, Fordham Univ, S6-57 & Univ Rochester, 57-58, asst vision res lab, 58-61; USPHS res fel psychophysiol lab, Ill State Psychiat Inst, 61-64; from asst prof to assoc prof, 64-72, PROF PHYSIOL PSYCHOL, SYRACUSE UNIV, 72- Concurrent Pos: Consult, Vet Admin Hosp, Syracuse, 6S- Mem: AAAS; Am Psychol Asn; Optical Soc Am; Asn Res Vision & Ophthal. Res: Vision; psychophysics and electrophysiology; binocular vision, spatio-temporal factors; flicker, increment thresholds; target detection; visual masking and excitability; sensitivity; cerebral factors in vision. Mailing Add: Collendale Psychol Labs Syracuse Univ Syracuse NY 13210

STURROCK, PETER ANDREW, b Grays, Eng, Mar 20, 24; US citizen; m 63; c 3. ASTROPHYSICS, PLASMA PHYSICS. Educ: Cambridge Univ, BA, 45, MA, 48, PhD(math), 51. Prof Exp: Harwell sr fel, Atomic Energy Res Estab, Eng, 51-53; fel, St John's Col, Cambridge Univ, S3-55; res assoc microwaves, Stanford Univ, 55-58; Ford fel plasma physics, Europ Orgn Nuclear Res, Switz, 58-59; res assoc, 59-60, prof eng sci & appl physics, 61-66, chmn inst plasma res, 64-74, PROF SPACE SCI & ASTROPHYS, STANFORD UNIV, 66- Concurrent Pos: Consult, Varian Assocs, Calif, 57-64 & NASA Ames Res Ctr, 62-64; dir, Enrico Fermi Summer Sch Plasma-Astrophys, Varenna, Italy, 66. Honors & Awards: Gravity Found Prize, 67, 71. Mem: AAAS; Inst Astron Union; fel Am Phys Soc; Am Geophys Union; Am Astron Soc. Res: Plasma processes of solar physics; solar activity; pulsars; radio galaxies; quasars. Mailing Add: Inst for Plasma Res Stanford Univ Via Crespi Stanford CA 94305

STURROCK, PETER EARLE, b Miami, Fla, Dec 6, 29; m 58. ANALYTICAL CHEMISTRY, ELECTROCHEMISTRY. Educ: Univ Fla, BS, 51, BA, 51; Stanford Univ, MS, 54; Ohio State Univ, PhD(chem), 60. Prof Exp: Asst prof, 60-65, ASSOC PROF CHEM, GA INST TECHNOL, 65- Concurrent Pos: Res grant, Petrol Res Fund, 64-66. Mem: AAAS; Am Chem Soc. Res: Instrumental chemical analysis; equilibria of complex ions; kinetics of electrode reactions. Mailing Add: Dept of Chem Ga Inst of Technol Atlanta GA 30332

STURROCK, THOMAS TRACY, b Havana, Cuba, Dec 9, 21; US citizen; m 48; c 5. HORTICULTURE, PLANT MORPHOLOGY. Educ: Univ Fla, BSA & MSA, 43, PhD(fruit crops), 61. Prof Exp: Partner, Sturrock Trop Fruit Nursery, 46-56; inspector, State Plant Bd Fla, 56-57; teacher high sch, 57-58; res asst fruit crops, Univ Fla, 58-60; instr biol, Palm Beach Jr Col, 60-64; from asst prof to assoc prof bot, 64-74, PROF BOT, FLA ATLANTIC UNIV, 74-, ASST DEAN COL SCI, 71- Mem: AAAS; Am Soc Hort Sci. Res: Tropical horticulture; factors influencing fertilization, embryological development and fruit-set of the mango. Mailing Add: Dept of Biol Sci Fla Atlantic Univ Boca Raton FL 33432

STURTEVANT, FRANK MILTON, b Evanston, Ill, Mar 8, 27; m 50; c 2. PHARMACOLOGY. Educ: Lake Forest Col, BA, 48; Northwestern Univ, MS, 50, PhD(biol), 51. Prof Exp: Asst, Northwestern Univ, 50-51; sr investr, G D Searle & Co, 51-58; sr pharmacologist, Smith Kline & French Labs, 58-60; dir sci & regulatory affairs, Mead Johnson & Co, 60-72; lectr genetics, Univ Evansville, 72; ASSOC DIR RES & DEVELOP, G D SEARLE & CO, 72- Mem: Drug Info Asn; Soc Exp Biol & Med; Am Soc Pharmacol & Exp Therapeut; Am Fertil Soc; Int Soc Chronobiol. Res: Hypertension; pharmacokinetics; biochemorphology; glucoregulation; genetics; reproduction; central nervous system; chronobiology. Mailing Add: G D Searle & Co PO Box 5110 Chicago IL 60680

STURTEVANT, JULIAN MUNSON, b Edgewater, NJ, Aug 9, 08; m 29; c 2. BIOPHYSICAL CHEMISTRY. Educ: Columbia Univ, AB, 27; Yale Univ, PhD(chem), 31. Hon Degrees: ScD, Ill Col, 62. Prof Exp: From instr to asst prof chem, Yale Univ, 31-43; staff mem, Radiation Lab, Mass Inst Technol, 43-46; from assoc prof to prof chem, 46-63, chmn dept, 59-62, assoc dir Sterling Chem Lab, 50-59, PROF CHEM, MOLECULAR BIOPHYS & BIOCHEM, YALE UNIV, 62- Concurrent Pos: Consult, Mobil Oil Co, 46-66; Guggenheim fel & Fulbright scholar, Cambridge Univ, 55-56; Fulbright scholar, Univ Adelaide, 62-63; vis prof, Univ Calif, San Diego, 66-67 & 69-70; vis fel, Seattle Res Ctr, Battelle Mem Inst, 72-73; mem, US Nat Comt Data Sci & Technol, 74-; vis prof, Stanford Univ, 75-76. Honors & Awards: Huffman Award, Calorimetry Conf US, 67. Mem: AAAS; Am Chem Soc; Nat Acad Sci; fel Am Acad Arts & Sci. Res: Thermochemistry; thermodynamics; physical chemistry of natural and synthetic macromolecules; macromolecular kinetics; mechanism of enzyme action. Mailing Add: Kline Chem Lab Yale Univ New Haven CT 06520

STURTEVANT, RUTHANN PATTERSON, b Rockford, Ill, Feb 7, 28; m 50; c 2. GROSS ANATOMY, BIOLOGICAL RHYTHMS. Educ: Northwestern Univ, Evanston, BS, 49, MS, 50; Univ Ark, Little Rock, PhD(anat), 72. Prof Exp: From instr to asst prof life sci, Ind State Univ, Evansville, 65-74; adj asst prof, Sch Med, Ind Univ, 72-74; lectr, Sch Med, Northwestern Univ, 74-75; ASST PROF ANAT & SURG, STRITCH SCH MED, LOYOLA UNIV, CHICAGO, 75- Concurrent Pos: Grad Women Sci fel, 73. Mem: Am Asn Anatomists; Int Soc Chronobiol; Sigma Xi; AAAS; Am Inst Biol Sci. Res: Chronobiology; chronopharmacokinetics. Mailing Add: Loyola Univ Stritch Sch of Med Dept of Anat 2160 S First Ave Maywood IL 60153

STURTEVANT, WILLIAM CURTIS, b Morristown, NJ, July 26, 26; m 52; c 3. ANTHROPOLOGY. Educ: Univ Calif, Berkeley, BA, 49; Yale Univ, PhD(anthrop), 55. Prof Exp: Res assoc tri-instnl Pac prog ling, Yale Univ, 53-54, instr anthrop & asst cur, 54-56; from ethnologist to gen anthropologist, Bur Am Ethnol, 56-65, GEN ANTHROPOLOGIST & CUR DEPT ANTHROP, NAT MUS NATURAL HIST, SMITHSONIAN INST, 65- Concurrent Pos: NSF res grants, Fla, 59 & Burma, 63-64; Am Philos Soc res grant, Europe, 60 & Okla, 62; Fulbright lectr, Inst Social Anthrop, Oxford Univ, 67-68; Wenner-Gren Found grant, Kashmir, 68; mem, Ctr Study Man, Smithsonian Inst, 68-; adj prof anthrop, Johns Hopkins Univ, 74-78; coun del, AAAS, 74-76, mem comt coun affairs, 74-75. Mem: Am Anthrop Asn; AAAS; Royal Anthrop Inst Gt Brit & Ireland; Societe des Americanistes de Paris; Am Soc Ethnohist (pres, 65-66). Res: Ethnography, ethnology and culture history of North American Indians, especially Eastern; historical ethnography 16th century Antilles; Burmese ethnography; art, technology and museum studies; ethnoscience; ethnohistory; anthropological linguistics. Mailing Add: Dept Anthrop Nat Mus Natural Hist Smithsonian Inst Washington DC 20560

STURZENEGGER, AUGUST, b Switz, May 3, 21; nat US; m 55; c 3. ORGANIC CHEMISTRY, CHEMICAL ENGINEERING. Educ: Swiss Fed Inst Technol, MS, 45, PhD, 48. Prof Exp: Chemist, Royal Dutch Shell Co, Holland, 48; chemist, Steinfels, Inc Switz, 49; chemist, 49-59, DIR ADVAN TECHNOL, HOFFMANN-LA ROCHE INC, 59- Mem: Am Chem Soc; Am Astronaut Soc; Am Inst Chem Eng; Swiss Chem Soc; Am Phys Soc. Res: Process development; detergents; petroleum chemistry; pharmaceuticals; systems analysis and automation; multidisciplinary interactions. Mailing Add: 25 Rensselaer Rd Essex Fells NJ 07021

STUSHNOFF, CECIL, b Saskatoon, Sask, Aug 12, 40; m 63; c 2. HORTICULTURE. Educ: Univ Sask, BSA, 63, MSc, 64; Rutgers Univ, PhD(hort, embryol), 67. Prof Exp: Res asst interspecific hybridization in vaccinium, Rutgers Univ, 64-67; from asst prof to assoc prof fruit breeding, 67-75, PROF HORT SCI & LANDSCAPE ARCHIT, UNIV MINN, ST PAUL, 75- Concurrent Pos: Consult, North-Gro, Inc. Honors & Awards: Paul Howe Shepard Award; Joseph Harvey Gourley Award. Mem: Am Soc Hort Sci; Int Soc Hort Sci; Am Pomol Soc. Res: Fruit and ornamentals breeding and genetics; cold tolerance breeding; mutation breeding; fruit culture. Mailing Add: Dept of Hort Sci Univ of Minn St Paul MN 55101

STUSNICK, ERIC, b Edwardsville, Pa, Aug 18, 39; m 67; c 1. ACOUSTICS. Educ: Carnegie-Mellon Univ, BS, 60; NY Univ, MS, 62; State Univ NY Buffalo, PhD(physics), 71. Prof Exp: Asst prof physics, Niagara Univ, 69-72; assoc physicist, Cornell Aeronaut Lab, Inc, 72-73; RES PHYSICIST, CALSPAN CORP, 73- Concurrent Pos: Lectr, Niagara Univ, 72- Mem: AAAS; Am Phys Soc; Am Asn Physics Teachers; Acoust Soc Am; Sigma Xi. Res: Environmental acoustics; acoustic simulation and modeling; digital signal processing and analysis; noise source identification techniques; diesel engine noise. Mailing Add: Calspan Corp PO Box 235 Buffalo NY 14221

STUTE, FRANCIS BERNARD, analytical chemistry, spectroscopy, see 12th edition

STUTEVILLE, DONALD LEE, b Okeene, Okla, Sept 7, 30; m 52; c 3. PLANT PATHOLOGY. Educ: Kans State Univ, BS, 59, MS, 61; Univ Wis, PhD(plant path), 64. Prof Exp: Res asst plant path, Univ Wis, 61-64; asst prof, 64-69, ASSOC PROF PLANT PATH, KANS STATE UNIV, 69-, RES FORAGE PATHOLOGIST, AGR EXP STA, 74- Mem: Am Phytopath Soc. Res: Diseases of forage crops; improving resistance in forage crops, particularly alfalfa to diseases caused by bacteria, fungi and viruses. Mailing Add: Dept of Plant Path Kans State Univ Manhattan KS 66506

STUTH, CHARLES JAMES, b Greenville, Tex, Jan 9, 32; m 53; c 3. ALGEBRA. Educ: East Tex State Univ, 51, MEd, 53; Univ Kans, PhD(math), 63. Prof Exp: Instr math, East Tex State Univ, 56-58; asst instr, Univ Kans, 58-62; from asst prof to prof, East Tex State Univ, 62-66; asst prof, Univ Mo-Columbia, 66-70; CHMN DEPT, STEPHENS COL, 70- Mem: Am Math Soc; Math Asn Am. Res: Group theory; theory of semigroups. Mailing Add: Dept of Math Stephens Col Columbia MO 65201

STUTHMAN, DEON DEAN, b Pilger, Nebr, May 7, 40; m 62; c 2. PLANT GENETICS, PLANT BREEDING. Educ: Univ Nebr, BSc, 62; Purdue Univ, MSc, 64, PhD(genetics of alfalfa), 67. Prof Exp: Asst prof, 66-71, ASSOC PROF OAT GENETICS & BREEDING, UNIV MINN, ST PAUL, 71- Mem: Am Soc Agron; Am Crop Sci Soc. Res: Breeding and genetics of oats. Mailing Add: Inst of Agr Agron & Plant Genetics Univ of Minn St Paul MN 55101

STUTMAN, LEONARD JAY, b Boston, Mass, Apr 8, 28; m 51; c 4. HEMATOLOGY, CARDIOLOGY. Educ: Mass Inst Technol, BS, 48; Boston Univ, MA, 49; Univ Rochester, MD, 53. Prof Exp: Intern & resident internal med, 4th Med Div, Bellevue

Hosp, New York, 53-56; instr clin med, Post-Grad Med Sch, NY Univ, 56-61, asst prof path, Sch Med, 61-65; HEAD COAGULATION RES LAB, DEPT MED, ST VINCENT'S HOSP & MED CTR, 65- Concurrent Pos: Lillia-Babbit-Hyde res fel metab dis, Sch Med, NY Univ, 56-57, Nat Heart Inst spec advan res fel, 59-61; Ripple Found coagulation res grant, 66; John A Polacheck Found fel, 66-67; fel coun arteriosclerosis, Am Heart Asn, 57-; attend physician, Nyack Hosp, 59-; co-investr, Nat Heart Inst grants, 60-65; assoc attend physician, St Vincent's Hosp, 65-; med dir, Presidential Life Ins Co, Nyack, NY, 65-; dir, Ford Found-Vera Inst Cardiovasc Epidemiol Proj, 71- Mem: AAAS; Am Col Physicians; assoc fel Am Col Cardiol; Am Fedn Clin Res. Res: Blood coagulation proteins in normal and pathologic states, including biochemistry and biophysics of cellular lipoproteins; epidemiology of cardiovascular disease; high altitude physiology, including effects on erythrocytes; biochemical genetics in clotting disorders. Mailing Add: Coagulation Res Lab St Vincent's Hosp & Med Ctr New York NY 10011

STUTMAN, OSIAS, b Buenos Aires, Arg, June 4, 33. IMMUNOLOGY, PATHOLOGY. Educ: Univ Buenos Aires, MD, 57. Prof Exp: Lectr, Inst Med Res, Univ Buenos Aires, 57-63; mem res staff physiol, Inst Biol & Exp Med, Buenos Aires, 63-66; from instr to assoc prof path, Med Sch, Univ Minn, Minneapolis, 66-72; MEM & SECT HEAD, SLOAN-KETTERING INST CANCER RES, 73-; PROF BIOL, CORNELL UNIV GRAD SCH MED SCI, 75- Concurrent Pos: USPHS res fel, Med Sch, Univ Minn, Minneapolis, 66-69; Am Cancer Soc res assoc, 69-74. Mem: Am Asn Immunol; Am Soc Exp Path; Am Asn Cancer Res; Transplantation Soc. Res: Development of immune functions in mammals, especially role of thymus and mechanisms of cell-mediated immunity in relation to normal functions and as defense against tumor development. Mailing Add: Sloan-Kettering Inst Cancer Res 410 E 68th St New York NY 10021

STUTSMAN, PAUL SNELL, b Chicago, Ill, Sept 15, 10; m 35; c 1. ORGANIC POLYMER CHEMISTRY, PETROLEUM CHEMISTRY. Educ: Univ Ill, BS, 34; Univ Wis, PhD(org chem), 38. Prof Exp: Res chemist, Texaco, Inc, 38-42; asst to dir res, 43-52, supvr chem res dept, 53-54; from asst mgr res & develop to pres, Tex-US Chem Co, 55-67; assoc prof chem & dir gen chem labs, 67-73, EMER PROF CHEM, ARIZ STATE UNIV, 73- Mem: AAAS; Am Chem Soc; Am Inst Chem Eng. Res: Petroleum refining processes and products; industrial chemistry; petrochemicals; elastomers and elastomer intermediates; general chemistry curricula and teaching methods. Mailing Add: 10350 Salem Dr Sun City AZ 85351

STUTTE, CHARLES AUTEN, b Wapanucka, Okla, July 19, 33; m 55; c 3. PLANT PHYSIOLOGY, AGRONOMY. Educ: Southeastern State Col, Durant, BS, 55; Okla State Univ, MS, 61, PhD(bot, plant physiol), 67. Prof Exp: Teacher, Prof Exp: Teacher high schs, Okla, 55-64; instr biol & ecol, ECent State Col, 64-65; adv plant physiol, forest physiol & gen plant physiol, Okla State Univ, 65-67; asst prof, 67-72, ASSOC PROF PLANT NUTRIT & BEN J ALTHEIMER CHAIR SOYBEAN RES, UNIV ARK, FAYETTEVILLE, 72- Concurrent Pos: Partic, NSF Acad Year Inst, 60-61 & res participation prog for col teachers, 65. Mem: AAAS; Am Soc Plant Physiologists; Crop Sci Soc Am; Am Soc Agron. Res: Physiological stress and growth regulator responses observed in wheat and soybean plants in relation to changes in isozyme systems and their intercellular molecular variations resulting from different environmental growing conditions. Mailing Add: Dept of Agron Univ of Ark Fayetteville AR 72701

STUTTS, ENSEL C, nutritional biochemistry, see 12th edition

STUTZ, CONLEY I, b Currie, Minn, Aug 18, 32; m 55; c 1. PHYSICS. Educ: Wayne State Col, BSE, 57; Univ NMex, MSE, 60; Univ Nebr, PhD(physics), 68. Prof Exp: High sch teacher, Iowa, 59; assoc prof physics, Pac Univ, 60-64; asst prof, 68-74, PROF PHYSICS, BRADLEY UNIV, 74- Concurrent Pos: Partic, NSF Acad Year Inst, Univ NMex, 60-61. Mem: Am Inst Physics; Am Asn Physics Teachers. Res: Study of the apprach to equilibrium of quantum mechanical systems; statistical mechanics; optics. Mailing Add: Dept of Physics Bradley Univ Peoria IL 61606

STUTZ, FREDERICK PAUL, b Chattanooga, Tenn, Mar 14, 44; m 66; c 1. URBAN GEOGRAPHY, ECONOMIC GEOGRAPHY. Educ: Valparaiso Univ, BA, 66; Northwestern Univ, MS, 68; Mich State Univ, PhD(geog), 70. Prof Exp: Teaching asst geog, Northwestern Univ, 66-68; res assoc geog & comput consult, Mich State Univ, 68-70; ASST PROF GEOG, CALIF STATE UNIV, SAN DIEGO, 70- Concurrent Pos: Consult, Jim Daughtry Assoc, Transp Planners; Ford Found res grant, Comprehensive Planning Orgn Sand Diego, Calif, 72; consult, Western Behav Sci Inst, 74- & Calif Dept Transp, 74-75; prin investr, Environ Planning Contract, US Air Force, 74-76; chmn, Asn Am Geographers Transp Comn, 75- Mem: Asn Am Geog; Int Geog Union; Regional Sci Asn. Res: Neighborhood identification and environmental management; inter-residential social travel patterns and urban transportation planning; spatial decision making and trip making. Mailing Add: Dept of Geog 5204 College Ave Calif State Univ La Jolla CA 92115

STUTZ, HOWARD COOMBS, b Cardston, Alta, Aug 24, 18; nat US; m 40; c 7. GENETICS. Educ: Brigham Young Univ, BS, 40, MS, 51; Univ Calif, PhD, 56. Prof Exp: Prin, high sch, Utah, 42-44; chmn dept biol, Snow Col, 46-51; asst prof, 56-67, PROF BOT, BRIGHAM YOUNG UNIV, 67- Concurrent Pos: Guggenheim fel, 60; vis prof, Am Univ Beirut, 67. Mem: Genetics Soc Am; Soc Study Evolution. Res: Cytogenetic studies of Secale L and related grasses; phyllogenetic studies of western browse plants; origin of cultivated rye; dominance-penetrance relationships; phyllogenetic studies within the grass tribe Hordeae. Mailing Add: Dept of Bot Brigham Young Univ Provo UT 84601

STUTZ, ROBERT EUGENE, b Corvallis, Ore, Jan 5, 21; m 45; c 5. BIOCHEMISTRY. Educ: Ore State Col, BS, 43; Univ Wis, MS, 46, PhD(biochem), 50. Prof Exp: Assoc biochemist, Biomed Div, Argonne Nat Lab, 49-56; res chemist, Western Pine Asn, Ore, 56-60; mgr res & develop, Chapman Chem Co, 60-66; self-employed, 66-68; CONSULT, KOPPERS CO, 68- Mem: AAAS; Am Inst Chemists; Am Chem Soc; Am Soc Plant Physiologists; Forest Prod Res Soc. Res: Plant biochemisty; indole-3-acetic acid 2-carbon-14; plant peroxidases; enzymatic and fungal stains in lumber; new fungicides and mechanism of fungicidal action. Mailing Add: 25310 Elena Rd Los Altos Hills CA 94022

STUTZ, ROBERT L, b Kansas City, Kans, Aug 1, 31; m 60; c 2. ORGANIC CHEMISTRY, PHARMACEUTICAL CHEMISTRY. Educ: Univ Kans, BA, 53, MS, 57, PhD(org chem), 61. Prof Exp: Asst chemist, Stand Oil Co, Ind, 56-57; sr chemist, Minn Mining & Mfg Co, 61-64; res chemist, C J Patterson Co, Kansas City, Mo, 64-65, head chem sect, 65-73; PRES, VANGUARD SYSTS, INC, 63- Concurrent Pos: Frederick Gardner Cottrell grant, 58-59. Mem: Am Chem Soc; Am Oil Chem Soc; fel Am Inst Chem; The Chem Soc; NY Acad Sci. Res: Surfactants; food emulsifiers; microbial inhibitors; specialty chemicals. Mailing Add: 5630 Belinder Rd Shawnee Mission KS 66205

STUTZMAN, JACOB WILLIAM, b Berlin, Pa, Jan 17, 17; m 46; c 1. PHARMACOLOGY. Educ: Franklin & Marshall Col, BS, 37; Univ Wis, PhD(med physiol), 41, MD, 43. Prof Exp: Asst med physiol, Univ Wis, 37-40, from instr to asst prof, 40-47; assoc prof pharmacol, Sch Med, Boston Univ, 47-50; head pharmacol sect, Smith Kline & French Labs, 50-52; vpres res & develop, Riker Labs, 52-61, vpres & gen mgr, 61-62, pres, 62-66, chmn bd, 66-70, PRES, RIKER LABS, INC, 3M CO, 70-; DIR, DART INDUSTS INC, 72- Concurrent Pos: Vpres ethical drug group, Rexall Drug & Chem Co, 66-70. Mem: AAAS; Am Physiol Soc; Am Soc Clin Pharmacol & Exp Therapeut; Am Soc Pharmacol & Exp Therapeut; Soc Exp Biol & Med. Res: Effects of anesthetics on the cardiovascular system; autonomic drugs. Mailing Add: Riker Labs Inc 3M Ctr St Paul MN 55101

STUTZMAN, LEON, b Wheeling, WVa, June 8, 24; m 48; c 2. INTERNAL MEDICINE, ONCOLOGY. Educ: WVa Univ, AB, 44, BS, 45; Wash Univ, MD, 47; Am Bd Internal Med, dipl, 58, cert med oncol, 73. Prof Exp: ASSOC CHIEF DEPT MED, ROSWELL PARK MEM INST, 57-; RES PROF MED, SCH MED, STATE UNIV NY BUFFALO, 63- Mem: Fel Am Col Physicians; Am Soc Clin Oncol; Am Asn Cancer Res. Res: Cancer chemotherapy; pathophysiology of lymphomas and chronic leukemia. Mailing Add: Dept of Med B Roswell Park Mem Inst Buffalo NY 14263

STUX, PAUL, b Vienna, Austria, Jan 17, 21; nat US; m 55; c 3. ORGANIC CHEMISTRY. Educ: Ill Inst Technol, BS, 44; Univ Chicago, MS, 48, PhD, 51. Prof Exp: Res chemist, Edwal Labs, Inc, Ill, 41-44 & Emulsol Chem Corp, 45; res group leader, Carlisle Chem Works, Inc, Ohio, 51-58; supvr & info & lit scientist, Chem Div, Pittsburgh Plate Glass Co, 58-71; PATENT CONSULT, 71- Mem: Am Chem Soc. Res: Free radical reactions; light-sensitive diazo compounds; chemistry of lubrication and chlorine compounds; chemical production and pilot plants; information retrieval; digital computers. Mailing Add: 584 Moreley Ave Akron OH 44320

STUY, JOHAN HARRIE, b Bogor, Indonesia, Jan 17, 25; m 52; c 2. BACTERIOLOGY. Educ: State Univ Utrecht, Bachelor, 48, Drs, 52, PhD(microbiol), 61. Prof Exp: Mem res staff radiobiol, N V Philips Labs, Netherlands, 52-65; assoc prof biol, 65-74, PROF BIOL SCI, FLA STATE UNIV, 74- Concurrent Pos: Fel, biol dept, Brandeis Univ, S7-58, biol div, Oak Ridge Nat Lab, 58-59 & biophys dept, Yale Univ, 59-60; vis prof, Fla State Univ, 62-63; US Atomic Energy Comn grant, 68- Mem: Brit Soc Gen Microbiol. Res: Recombination in bacteria and bacteriophages at the DNA level. Mailing Add: Dept of Biol Sci Fla State Univ Tallahassee FL 32306

STWALLEY, WILLIAM CALVIN, b Glendale, Calif, Oct 7, 42; m 63; c 2. PHYSICAL CHEMISTRY, ATOMIC PHYSICS. Educ: Calif Inst Technol, BS, 64; Harvard Univ, PhD(phys chem), 68. Prof Exp: From asst prof to assoc prof, 68-75, PROF CHEM, UNIV IOWA, 75- Concurrent Pos: A P Sloan fel, 72-75; assoc prog dir quantum chem, NSF, 75-76. Mem: Am Chem Soc; Am Phys Soc. Res: Intermolecular forces; gas phase chemical reaction kinetics; molecular beams; atomic and molecular scattering and spectroscopy. Mailing Add: Dept of Chem Univ of Iowa Iowa City IA 52242

STYLES, ERNEST DEREK, b Canterbury, Eng, Oct 19, 26; m 65; c 2. GENETICS. Educ: Univ BC, BSA, 60; Univ Wis, PhD(genetics), 65. Prof Exp: Res asst genetics, Univ Wis, 60-64, from proj asst to proj assoc, 64-66; asst prof, 66-71, ASSOC PROF GENETICS, UNIV VICTORIA, 71- Mem: AAAS; Genetics Soc Am; Genetics Soc Can; Am Genetics Asn. Res: Maize genetics; genetic control of flavonoid biosynthesis; paramutation. Mailing Add: Dept of Biol Univ of Victoria Victoria BC Can

STYLES, SALMA MAHMOUD, b Antioch, Syria, Sept 13, 36; m 65; c 2. GENETICS. Educ: Damascus Univ, BSc, 59; Univ Wis-Madison, MSc, 62, PhD(genetics), 65. Prof Exp: Res assoc, dept genetics, Univ Wis-Madison, 65-66; GENETICIST, SCIENTIST & HEAD DEPT MED GENETICS, LABS, ROYAL JUBILEE HOSP, VICTORIA, BC, 68- Mem: Genetics Soc Am; Genetics Soc Can. Res: Human cytogenetics; chromosomal abnormalities as contributors to etiology of genetic disease; tissue culturing techniques as applied to prenatal diagnosis of genetic disease and to diagnosis of disease in general; genetic counseling. Mailing Add: Dept of Med Genetics Royal Jubilee Hosp Labs Victoria BC Can

STYLES, TWITTY JUNIUS, b Prince Edward Co, Va, May 18, 27; m 62; c 2. PARASITOLOGY, BIOLOGY. Educ: Va Union Univ, BS, 48; NY Univ, MS, 57, PhD(biol), 63. Prof Exp: Jr bacteriologist, New York City Health Dept, 53-54, jr scientist, State Univ NY Downstate Med Ctr, 55-64; fel parasitol, Nat Univ Mex, 64-65; asst prof, 65-69, ASSOC PROF BIOL, UNION COL, NY, 69- Concurrent Pos: Lectr, City Col New York, 64; consult off higher educ planning, NY State Educ Dept, 70-71; lectr, Narcotics Addiction Control Comn, NY State, 71-72; NSF course histochem, Vanderbilt Univ, 72; sabbatical, Dept Vet Microbiol & Immunol, Univ Guelph, 72. Mem: Am Soc Parasitologists; Soc Protozool; Am Soc Microbiol; Nat Asn Biol Teachers; NY Acad Sci. Res: Effect of marine biotoxins on parasitic infections; effect of endotoxin of Trypanosoma lewisi infections in rats and Plasmodium berghei infections in mice. Mailing Add: Dept of Biol Sci Union Col Schenectady NY 12308

STYRIS, DAVID LEE, b Pomona, Calif, Apr 21, 32; m 58; c 1. EXPERIMENTAL PHYSICS. Educ: Pomona Col, BA, 57; Univ Ariz, MS, 62, PhD(physics), 67. Prof Exp: Dynamics engr, Airframe Design, Convair-Pomona, 56-58; res physicist, Weapons Testing, Edgerton, Germeshausen & Grier, 58-60; res assoc field ion micros, Cornell Univ, 67-69; asst prof physics, Wash State Univ, 69-74; SR RES SCIENTIST, BATTLELLE NORTHWEST LAB, 74- Mem: AAAS; Int Solar Energy Soc. Res: Radiation damage of materials related to controlled thermonuclear reactor systems, solar energy. Mailing Add: Battelle Northwest Lab PO Box 999 Richland WA 99352

STYRON, CHARLES WOODROW, b New Bern, NC, Nov 6, 13; m 39; c 2. MEDICINE. Educ: NC State Univ, BS, 34; Duke Univ, MD, 38; Am Bd Internal Med, dipl. Prof Exp: Intern pediat, Duke Univ, 38; intern & resident med, Boston City Hosp, 38-40; assoc, 50-65, ASST PROF MED, MED CTR, DUKE UNIV, 65- Concurrent Pos: Fel, Joslin Clin, New Eng Deaconess Hosp, Boston, 40-42; pvt pract, 46-; mem coun foods & nutrit, AMA, 69-76; chmn, NC Gov Comt Health Care Delivery, 71-73. Mem: Fel Am Col Physicians; AMA; Am Diabetes Asn; fel Am Heart Asn; Am Soc Internal Med. Res: Internal medicine; diabetes mellitus and endocrinology. Mailing Add: 615 St Mary's St Raleigh NC 27605

STYRON, CLARENCE EDWARD, JR, b Washington, NC, Sept 14, 41; m 69; c 1. ECOLOGY. Educ: Davidson Col, BS, 63; Emory Univ, MS, 65, PhD(biol), 67. Prof Exp: ASST PROF BIOL, ST ANDREWS PRESBY COL, 69- Concurrent Pos: Consult, Oak Ridge Nat Lab, 69-; res radiobiologist, Inst Marine Biomed Res, Univ NC, 73- Honors & Awards: Outstanding Educr Am, 72, 74, 75. Mem: Ecol Soc Am; Am Inst Biol Sci; Am Soc Limnol & Oceanog; Marine Biol Asn UK. Res: Ecology of invertebrate communities; effects of radioactive fallout on terrestrial systems; systems modeling; transport of heavy metals in marine benthic systems. Mailing Add: Sci Div St Andrews Presby Col Laurinburg NC 28352

SU, AARON CHUNG LIONG, organic chemistry, organometallic chemistry, see 12th edition

SU, CHAU-HSING, b Fukien, China, Nov 23, 35; m 60; c 4. FLUID MECHANICS, PLASMA PHYSICS. Educ: Nat Taiwan Univ, BS, 56; Univ Minn, MS, 59; Princeton Univ, PhD(eng), 64. Prof Exp: Asst prof eng, Mass Inst Technol, 63-66; res assoc plasma physics, Princeton Univ, 66-67; ASSOC PROF APPL MATH, BROWN UNIV, 67- Concurrent Pos: Ford fels, Mass Inst Technol, 64-66. Mem: AAAS; Am Phys Soc. Res: Dynamics of a system of vortices. Mailing Add: Dept of Appl Math Brown Univ Providence RI 02912

SU, CHE, b Taipei, Taiwan, June 12, 32; m 56; c 3. PHARMACOLOGY. Educ: Nat Taiwan Univ, BS, 55, MS, 60; Univ Calif, Los Angeles, PhD(pharmacol), 65. Prof Exp: Lectr pharmacol, Nat Taiwan Univ, 60-63; pharmacologist, Riker Labs, Inc, 64-67; asst prof, 67-72, ASSOC PROF PHARMACOL, UNIV CALIF, LOS ANGELES, 72- Mem: Int Soc Toxicol; Am Soc Pharmacol & Exp Therapeut; Microciro Soc; Japanese Pharmacol Soc. Res: Pharmacology of snake venoms; neuromuscular blocking agents; vascular smooth muscle electrophysiology and pharmacology; sympathetic transmission mechanisms in blood vessels; pharmacology of blood vessels. Mailing Add: Dept of Pharmacol Univ of Calif Sch of Med Los Angeles CA 90024

SU, CHEH-JEN, b Taipei, Taiwan, June 11, 34; US citizen; m 67; c 3. POLYMER CHEMISTRY, PAPER CHEMISTRY. Educ: Taipei Inst Technol, Taiwan, BS, 55; NC State Univ, BS, 60; State Univ NY Col Forestry, Syracuse Univ, MS, 63. Prof Exp: Chem engr, Taiwan Pulp & Paper Co, 55-59; res chemist, Owens-Ill, Inc, Ohio, 65-67; sr res scientist II paper & polymers, 67-75, SR RES SCIENTIST I POLYMER & FOREST PROD, CONTINENTAL CAN CO, INC, 75- Mem: Am Chem Soc. Res: Characterization of polymers and plastic molded articles; chemicals and materials from renewable sources. Mailing Add: Continental Can Co Inc 7622 S Racine Ave Chicago IL 60620

SU, GEORGE CHUNG-CHI, b Amoy, China, Aug 8, 39; m 71. ORGANIC CHEMISTRY. Educ: Hope Col, AB, 62; Univ Ill, MS, 64, PhD(org chem), 66. Prof Exp: Res chemist plastics dept, E I du Pont de Nemours & Co, 66-69; NIH fel, Dept Biochem, Mich State Univ, 69-70, res assoc, Pesticide Res Ctr, 70-72; biochemist, Pesticide Sect, Bur Labs, Mich Dept Pub Health, 72-74; CHIEF TECH SERV, AIR POLLUTION CONTROL, DEPT NATURAL RESOURCES, STATE MICH, 74- Mem: Am Chem Soc; The Chem Soc; NY Acad Sci. Res: Organic reaction mechanisms; air monitoring techniques; pesticide photochemistry; analytical techniques for isolation, detection, identification and quantitation of submicrogram quantities of environmental pollutants; toxicology and enzymology. Mailing Add: 4391 Cherrywood Dr Okemos MI 48864

SU, HELEN CHIEN-FAN, b Nanping, China, Dec 26, 22; nat US. ORGANIC CHEMISTRY. Educ: Hwa Nan Col, China, BA, 44; Univ Nebr, MS, 51, PhD(chem), 53. Prof Exp: Asst chem, Hwa Nan Col, 44-47, instr, 47-49; prof, Lambuth Col, 53-55; res asst, Res Found, Auburn Univ, 55-57; res chemist, Borden Chem Co, 57-63; assoc scientist, Lockheed-Ga Co, 63-65, res scientist, 65-68; RES CHEMIST, STORED PROD INSECTS RES & DEVELOP LAB, AGR RES SERV, USDA, 68- Mem: AAAS; fel Am Inst Chem; Am Chem Soc; NY Acad Sci. Res: Heterocyclic nitrogen and sulfur compounds; unsaturated aliphatic compounds; natural products; naturally occurring pesticides; insect pheromones; insect repellents and attractants. Mailing Add: 12209 Bedford Dr Savannah GA 31406

SU, JIN-CHEN, b Anhwei, China, Dec 30, 32; US citizen; m 60; c 3. TOPOLOGY. Educ: Nat Taiwan Univ, BS, 55; Univ Pa, PhD(math), 61. Prof Exp: Asst prof math, Univ Va, 61-64; math mem, Inst Advan Study, 64-66; assoc prof, 66-72, PROF MATH, UNIV MASS, AMHERST, 72- Mem: Am Math Soc. Res: Transformation groups. Mailing Add: Dept of Math Univ of Mass Amherst MA 01002

SU, JUDY YA-HWA LIN, pharmacology, see 12th edition

SU, KENNETH SHYAN-ELL, b Taipei, Taiwan, Nov 26, 41; US citizen; m 70; c 1. PHARMACEUTICS. Educ: Taipei Med Col, BS, 65; Univ Wis, MS, 69, PhD(pharmaceut), 71. Prof Exp: Res fel biochem, US Naval Med Res Unit 2, 64-65; pharmaceut chemist, William S Merrell Co, 71; SR PHARMACEUT CHEMIST, ELI LILLY & CO, 71- Mem: Am Pharmaceut Asn. Res: Studies of disperse systems with particular interest in particle interactions and surface phenomena, adsorption at liquid and solid interfaces, and chemical kinetics at interfaces. Mailing Add: Lilly Res Labs Eli Lilly & Co Indianapolis IN 46206

SU, KWEI LEE, b Ping Tong, Taiwan, Mar 18, 42; m 68; c 1. LIPID BIOCHEMISTRY. Educ: Nat Taiwan Univ, BS, 64; Univ Minn, PhD(biochem), 71. Prof Exp: Instr pharmacog, Col Pharm, Nat Taiwan Univ, 64-66; Hormel fel, Hormel Inst, Univ Minn, 70-72, from res fel to res assoc lipid chem, 72-74, asst prof, 74-75; RES ASST PROF NEUROCHEM, SINCLAIR COMP MED RES FARM, UNIV MO, COLUMBIA, 75- Mem: Am Oil Chemists Soc. Res: Isolation, structural determination, biosynthesis and function of ether lipids in mammals; effects of neurotransmitters on lipid metabolism in brain subcellular membranes. Mailing Add: Sinclair Comp Med Res Farm Univ of Mo Rte 3 Columbia MO 65201

SU, LAO-SOU, b Kaohsung, Taiwan, Dec 13, 32; m 66; c 2. PHYSICAL CHEMISTRY. Educ: Taiwan Norm Univ, BS, 57; Ind Univ, MS, 63, PhD(phys chem), 67. Prof Exp: Fel, Univ Mich 67-69 & Ind Univ, 69; SR RES CHEMIST, S C JOHNSON & SON, INC, 69- Mem: Am Chem Soc. Res: Corrosion study of aerosol products; elemental analysis by means of x-ray fluorescence spectrometry; electron diffraction study of molecular structure. Mailing Add: S C Johnson & Soc Inc 1525 Howe St Racine WI 53403

SU, RUTH WOLF, b Tulsa, Okla, June 29, 43; m 67; c 3. ANALYTICAL MATHEMATICS. Educ: Coe Col, BA, 65; Mich State Univ, MS, 67, PhD(math), 74. Prof Exp: PROGRAMMER & ANALYST ENG APPLNS, MICH DEPT STATE HWYS & TRANSP, 72- Mailing Add: 4849 Kenmore Dr Okemos MI 48864

SU, SHIN-YI, b Taipei, Taiwan, China, July 18, 40; m 70; c 1. SPACE PHYSICS. Educ: Nat Taiwan Univ, BS, 63; Dartmouth Col, PhD(eng sci), 70. Prof Exp: Res asst space physics, Dartmouth Col, 65-69; postdoctoral fel, Univ Calgary, 70-72; resident res assoc at Johnson Space Ctr, Nat Acad Sci-Nat Res Coun, 72-74; PRIN SCIENTIST SPACE PHYSICS, LOCKHEED ELECTRONICS CO INC, 74- Mem: Am Geophys Union. Res: Study of the ultra low frequency wave phenomenon and the substorm injection process from the low energy proton data taken from ATS-6 satellite. Mailing Add: C-23 Lockheed Electronics Co Inc 16811 El Camino Real Houston TX 77058

SU, STANLEY Y W, b Fukien, China, Feb 18, 40; US citizen; m 65; c 2. COMPUTER SCIENCE. Educ: Tamkang Col Arts & Sci, BA, 61; Univ Wis, MS,65, PhD(comput sci), 68. Prof Exp: Proj asst syst prog, Comput Ctr, Univ Wis, 64-67, res asst regional Am English proj, 67, res asst natural lang processing, 67-68; mathematician comput ling, Rand Corp, 68-70; asst prof, Dept Elec Eng & Commun Sci Lab, 70-74, ASSOC PROF COMPUT & INFO ENG, DEPT ELEC ENG & INST ADVAN STUDY COMMUN PROCESSES, UNIV FLA, 74- Concurrent Pos: Mem, Spec Interest Group Operating Syst & Spec Interest Group Mgt Data, Asn Comput Mach, 73-; consult, Creativity Ctr Consortium, 73-74, Fla Keys Community Col, 74-75 & Cent Fla Community Col, 74-; staff consult, Queueing Systs, Inc, 74-75; lectr continuing educ, George Washington Univ, 75- Mem: Asn Comput Mach; Asn Comput Ling. Res: Associative processing systems; computer architecture for data base management; data base translation and program conversion; data base semantics; application of microprocessor network to non-numeric processing; man-machine communications. Mailing Add: 227 Larsen Hall Univ of Fla Gainesville FL 32601

SU, TAH-MUN, b Taiwan, July 22, 39; m; c 2. ORGANIC CHEMISTRY, BIOENGINEERING. Educ: Chen Kung Univ, Taiwan, BSc, 62; Univ Nev, MS, 65; Princeton Univ, PhD(chem), 70. Prof Exp: Res fel geochem, Biodyn Lab, Univ Calif, Berkeley, 69-70; res fel chem, Union Carbide Res Inst, 70-71; STAFF SCIENTIST CHEM & BIOENG, CORP RES & DEVELOP CTR, GEN ELEC CO, 72- Mem: Am Chem Soc. Res: Single cell protein from cellulosic fiber; enzymatic, saccharification of cellulose; mechanism of organic chemical reactions. Mailing Add: Corp Res & Develop Gen Elec Co River Rd Schenectady NY 12301

SU, YAO SIN, b Ping-tung, Taiwan, Oct 17, 29; US citizen; m 54; c 2. ANALYTICAL CHEMISTRY. Educ: Taiwan Univ, BS, 52; Univ Pittsburgh, PhD(chem), 62. Prof Exp: Chemist, Union Res Inst, Taiwan, 53-58; sr res chemist, 63-73, RES SUPVR, CORNING GLASS WORKS, 73- Mem: Am Chem Soc. Res: Inorganic chemical analysis; electroanalysis; classical wet methods. Mailing Add: 197 Cutler Ave Corning NY 14830

SUAREZ, KENNETH ALFRED, b Queens, NY, June 27, 44; m 68. PHARMACOLOGY, TOXICOLOGY. Educ: Univ RI, BS, 67, MS, 70, PhD(pharmacol), 72. Prof Exp: Instr, 72-74, ASST PROF PHARMACOL, CHICAGO COL OSTEOP MED, 74- Mem: AAAS. Res: Halogenated hydrocarbon induced hepatic injury. Mailing Add: Dept of Pharmacol Chicago Col of Osteop Med Chicago IL 60615

SUAREZ, THOMAS H, b Temperley, BsAs, Arg, Dec 7, 36; m 61; c 3. RESOURCE MANAGEMENT. Educ: Univ Buenos Aires, MS, 59, PhD(phys org chem), 61. Prof Exp: Teaching asst org chem, Univ Buenos Aires, 59-61, head lab course, 61; res chemist, Textile Fibers Dept, Dacron Mfg Div, 61-66, anal res supvr, 66-69, supvr process develop, 69-71, tech supt, Polymer Intermediates Dept, 71-74, PLANNING MGR, POLYMER INTERMEDIATES DEPT, E I DU PONT DE NEMOURS & CO, 74- Mem: AAAS; Am Chem Soc; Arg Chem Asn. Res: Nucleophylic aromatic substitution; reaction kinetics and mechanisms; polymer chemistry; melt spining synthetic fibers; physical and chemical characterization of polymers. Mailing Add: E I du Pont de Nemours & Co Polymer Intermediates Dept Wilmington DE 19898

SUBBARAO, MATHUKUMALLI VENKATA, b Yazali, India, May 4, 21; m 45; c 2. MATHEMATICS. Educ: Univ Madras, MA, 42, MSc, 45, PhD(math), 53. Prof Exp: Lectr math, Univ Madras & Andhra States Educ Serv, 45-55; reader, Sri Venkateswara Univ, India, 56-60; assoc prof, Univ Mo, 60-63 & Univ Alta, 63-66; prof math, Univ Kerala, 66-67 & Univ Mo, 67-68; PROF MATH, UNIV ALTA, 68- Concurrent Pos: NSF fel, 62-63; Nat Res Coun Can fels, 63-67 & 68- Mem: Am Math Soc; Math Asn Am; Can Math Cong; Indian Math Soc. Res: Number theory; functional analysis in number theory; arithmetic functions. Mailing Add: Dept of Math Univ of Alta Edmonton AB Can

SUBCASKY, WAYNE JOSEPH, chemistry, see 12th edition

SUBDEN, RONALD ERNEST, biochemical genetics, see 12th edition

SUBERKROPP, KELLER FRANCIS, b Wamego, Kans, Apr 12, 43; m 71; c 4. MYCOLOGY, PHYSIOLOGY. Educ: Kans State Univ, BS, 65, MS, 67; Mich State Univ, PhD(bot), 71. Prof Exp: Res assoc microbial ecol, Kellogg Biol Sta, Mich State Univ, 71-75; ASST PROF BIOL SCI, IND UNIV-PURDUE UNIV, FT WAYNE, 75- Mem: Mycol Soc Am; Brit Mycol Soc; Am Soc Microbiol. Res: Role of fungi in decomposition of leaf litter in aquatic habitats; effects of environmental factors on growth and sporulation of these fungi. Mailing Add: Dept of Biol Sci Ind Univ-Purdue Univ Ft Wayne IN 46805

SUBLETT, AUDREY J, b San Antonio, Tex, Apr 27, 37. PHYSICAL ANTHROPOLOGY. Educ: Univ Ariz, BA, 61; Univ Toronto, MA, 63; State Univ NY Buffalo, PhD(anthrop), 66. Prof Exp: NSF res assoc anthrop, State Univ NY Buffalo, 66-67; asst prof, 67-70, ASSOC PROF ANTHROP, FLA ATLANTIC UNIV, 70- Mem: Fel AAAS; Am Asn Phys Anthrop; fel Am Anthrop Asn; Soc Am Archaeol. Res: Human osteology; microevolutionary changes; skeletal anomalies. Mailing Add: Dept of Anthrop Fla Atlantic Univ Boca Raton FL 33432

SUBLETT, BOBBY JONES, b Paintsville, Ky, Aug 27, 31; m 56; c 4. ORGANIC CHEMISTRY, POLYMER CHEMISTRY. Educ: Eastern Ky State Col, BS, 58; Univ Tenn, MS, 60. Prof Exp: From res chemist to sr res chemist, 60-75, RES ASSOC, TENN EASTMAN CO, 75- Mem: Am Chem Soc. Res: Reaction mechanisms; tobacco smoke analysis; condensation polymers; textile chemicals; adhesives. Mailing Add: 1205 Jerry Lane Kingsport TN 37664

SUBLETT, MICHAEL DEAN, b Kansas City, Mo, Jan 5, 43. GEOGRAPHY. Educ: Univ Mo-Columbia, BA, 66, MA, 67; Univ Chicago, PhD(geog), 74. Prof Exp: ASST PROF GEOG, ILL STATE UNIV, 70- Mem: Am Geog Soc; Asn Am Geogr; Agr Hist Soc. Res: Extent of fan territories for professional sports teams; local self-guided instructional field trips; evolution of Illinois counties; transport of human remains prior to burial . Mailing Add: Dept of Geog & Geol Ill State Univ Normal IL 61761

SUBLETT, ROBERT L, b Columbia, Mo, Apr 10, 21; m 46; c 3. CHEMISTRY. Educ: Univ Mo, AB, 43, PhD, 50; Ga Inst Technol, MS, 48. Prof Exp: Instr chem, Ga Inst Technol, 47; res chemist, Chemstrand Corp, 52-55; assoc prof chem, Ark State Teachers Col, 55-56; assoc prof, 56-70, PROF CHEM, TENN TECHNOL UNIV, 70-, CHMN DEPT, 72- Mem: Am Chem Soc. Res: High polymers; Friedels-crafts; organic and high polymer analytical chemistry; instrumental analysis. Mailing Add: Dept of Chem Tenn Technol Univ Cookeville TN 38501

SUBLETTE, JAMES EDWARD, b Healdton, Okla, Jan 19, 28; m 50; c 4. ZOOLOGY. Educ: Univ Ark, BS, 48, MS, 50; Univ Okla, PhD(zool), 53. Prof Exp: Biologist, Corps Engrs, US Dept Army, 49-51; asst prof zool, Southwestern La Inst, 51-53 & Henderson State Teachers Col, 53; from asst prof to assoc prof, Northwestern State Col, 53-60; assoc prof, Tex Western Col, 60-61; assoc prof, 61-66, PROF BIOL & DEAN SCH GRAD STUDIES, EASTERN N MEX UNIV, 66- Mem: Soc Syst Zool; Entom Soc Am; Int Asn Theoret & Appl Limnol. Res: Taxonomy and ecology of aquatic insects, particularly Chironomidae; limnology. Mailing Add: Sch of Grad Studies Eastern NMex Univ Portales NM 88130

SUBLUSKEY, LEE ANTHONY, organic chemistry, see 12th edition

SUBRAHMANIAM, KATHLEEN, b Pittsburgh, Pa, Mar 31, 38; m 62; c 2. STATISTICS. Educ: Muskingum Col, BS, 60; Johns Hopkins Univ, MS, 63, DSc(statist), 69. Prof Exp: Lectr math, Muskingum Col, 62-63; consult statist, Dept Community Med, Univ Western Ont, 65-70; sessional lectr statist, 67-69, asst prof, 69-73, ASSOC PROF, UNIV MAN, 73- Mem: Biomet Soc; Am Statist Asn. Res: Applied multivariate analysis and analysis of discrete data. Mailing Add: Dept of Statist Univ of Man Winnipeg MB Can

SUBRAHMANIAM, KOCHERLAKOTA, b Bangalore, India, Feb 3, 35; m 62; c 2. MATHEMATICAL STATISTICS. Educ: Univ Col Sci, Benares, India, BSc, 54, MSc, 57; Inst Agr Res Statist, dipl, 63; Johns Hopkins Univ, DSc(biostatist), 64. Prof Exp: Jr res fel statist, Inst Agr Res Statist, 57-58, sr res fel, 58-59; investr, Rockefeller Found, India, 59-60; asst prof math, Univ Western Ont, 64-66; assoc prof, 66-70, PROF STATIST, UNIV MAN, 70- Concurrent Pos: Nat Res Coun operating grants pure & appl math, 66- Mem: Inst Math Statist; Am Statist Asn; fel Royal Statist Soc. Res: Multivariate analysis; distribution theory; statistical tests of significance; applied probability theory; non-normality. Mailing Add: Dept of Statist Univ of Man Winnipeg MB Can

SUBRAMANIAN, GOPAL, b Madras, India, Apr 4, 37; m 66; c 2. NUCLEAR MEDICINE, CHEMICAL ENGINEERING. Educ: Univ Madras, BSc, 58 & 60; Johns Hopkins Univ, MSE, 64; Syracuse Univ, PhD(chem eng), 70. Prof Exp: Chem engr, Prod Dept, E Asiatic Co (India) Pvt, Ltd, 60-62; res assoc radiochem, Med Insts, Johns Hopkins Univ, 64-65; res assoc radiopharmaceut, 65-68, from instr to asst prof, 68-72, ASSOC PROF RADIOL, STATE UNIV NY UPSTATE MED CTR, 72- Concurrent Pos: NIH grant, State Univ NY Upstate Med Ctr, 69-75; consult, New Eng Nuclear Corp, 68- & Am Nat Stand Inst, 71-; mem, adv panel radiopharmaceut, US Pharmacoepia, 71-; asst prof, Syracuse Univ, 71- Honors & Awards: Gold Medal, Soc Nuclear Med, 72. Mem: AAAS; Soc Nuclear Med; fel Am Inst Chem; Am Inst Chem Eng. Res: Radiochemistry; radiopharmaceuticals; fluid dynamics as applied to chemical engineering. Mailing Add: Dept of Nuclear Med State Univ of NY Upstate Med Ctr Syracuse NY 13210

SUBRAMANIAN, PALLATHERI MANACKAL, b Ottapalam, Kerala, India, Jan 10, 31; m 66. ORGANIC CHEMISTRY. Educ: Univ Madras, BSc, 50; Univ Bombay, MSc, 58; Wayne State Univ, PhD(org chem), 64. Prof Exp: Chemist, Godrej Soaps, India, 50-58; res chemist, Electrochem Dept, 64-70, SR RES CHEMIST, PLASTICS DEPT, CHESTNUT RUN LABS, E I DU PONT DE NEMOURS & CO, INC, 70- Mem: Am Chem Soc; The Chem Soc; Sigma Xi. Res: Physical organic chemistry; kinetics of elimination reactions in organic bicyclic systems; nuclear magnetic resonance spectroscopy of organic compounds; synthetic organic high polymers; adhesives and coatings; synthesis and process of plastics. Mailing Add: 2710 Tanager Dr Wilmington DE 19808

SUBRAMANIAN, RAVANASAMUDRAM VENKATACHALAM, b Kalakad, India, Jan 16, 33; m 53; c 2. POLYMER CHEMISTRY, POLYMER SCIENCE. Educ: Presidency Col, Madras, India, BSc, 53; Loyola Col, Madras, India, MSc, 54; Univ Madras, PhD(polymer chem), 57. Prof Exp: Jr res fel polymer chem, Nat Chem Lab, Poona, India, 57, Coun Sci & Indust Res India sr res fel, 57-59, jr sci officer, 59-63; res assoc chem, Case Inst Technol, 63-66 & Inst Molecular Biophys, Fla State Univ, 66; pool officer, Dept Phys Chem, Madras Univ, 66-67; asst prof, Harcourt Butler Tech Inst, Kanpur, India, 67-69; res chemist, Mat Chem Sect, Col Eng Res Div, 69-73, ASSOC PROF MAT SCI, WASH STATE UNIV, 73-, HEAD POLYMER MAT SECT, 74- Concurrent Pos: NSF fel, Case Inst Technol, 63-66; AEC fel, Inst Molecular Biophys, Fla State Univ, 66; Coun Sci & Indust Res India grant, 68-69. Mem: Fel Am Inst Chemists; NY Acad Sci; Am Chem Soc; Soc Plastics Engrs. Res: Kinetics and mechanisms of polymerization; polymer structure and proper properties; electropolymerization; interface modification in carbon fiber reinforced composites; basalt fibers; organotin monomers and polymers. Mailing Add: Dept of Mat Sci Wash State Univ Pullman WA 99163

SUCHANNEK, RUDOLF GERHARD, b Hindenburg, Ger, Oct 17, 21. EXPERIMENTAL ATOMIC PHYSICS. Educ: Univ Hamburg, dipl(physics), 58; Univ Alaska, PhD(physics), 74. Prof Exp: Engr, Westinghouse Elec Corp, 58-62; eng specialist, Microwave Comp Lab, Sylvania Co, 62-64; physicist, Unified Sci Asn Inc, 64-66; sr res asst atomic collision, Geophys Inst, Univ Alaska, 66-73, fel, 74; RES ASSOC ATOMIC PHYSICS, RES LAB ELECTRONICS, MASS INST TECHNOL, 75- Mem: Am Phys Soc; Inst Elec & Electronic Engrs. Res: Excitation transfer collisions of laser excited atoms and molecules; charge exchange collisions of protons with atoms and molecules. Mailing Add: Rm 26-244 Res Lab of Electronics Mass Inst of Technol Cambridge MA 02139

SUCHARD, STEVEN NORMAN, b Chicago, Ill, Feb 8, 44; m 64; c 4. LASERS. Educ: Univ Calif, Berkeley, BS, 65; Mass Inst Technol, PhD(chem physics), 69. Prof Exp: Res asst, Lawrence Berkeley Lab, 64-65; teaching asst chem, Mass Inst Technol, 65-69; lectr, Univ Calif, Berkeley, 69-70; ASSOC DEPT HEAD CHEM PHYSICS, AEROSPACE CORP, 70- Concurrent Pos: Fel chem physics, Univ Calif, Berkeley, 69-70. Mem: Am Phys Soc. Res: Effect of system variables on the output of pulsed and continuous wave chemical lasers; flash photolysis; energy transfer in molecular systems; laser optics in high gain media; determination of the feasibility of producing new chemically and electrically pumped electronic transition lasers; measurement of molecular and kinetic parameters effecting optical gain of potential laser systems. Mailing Add: Aerospace Corp 2350 E El Segundo Blvd El Segundo CA 90245

SUCHESTON, MARTHA ELAINE, b Bowling Green, Ky, June 17, 39; m 68; c 2. DEVELOPMENTAL ANATOMY. Educ: Western Ky Univ, BSc, 60; Ohio State Univ, MSc, 61, PhD(anat), 65. Prof Exp: Asst prof gross anat & embryol, Ohio State Univ, 67-68; asst prof gross anat, Stanford Univ, 68-69; asst prof, 70-74, ASSOC GROSS ANAT & EMBRYOL, OHIO STATE UNIV, 75- Concurrent Pos: Bremer Found Fund fel, Ohio State Univ, 71-73; vis assoc prof gross anat, Univ BC, 74-75. Mem: AAAS; Am Asn Anat; Teratology Soc; Pan-Am Asn Anat. Res: Pre and postnatal development; histochemical demonstration of certain enzymes; electron microscopy of the adrenal cortex of the Mongolian gerbil. Mailing Add: Dept of Anat Ohio State Univ Col of Med Columbus OH 43210

SUCHMAN, DAVID, b New York, NY, June 23, 47; m 70; c 1. METEOROLOGY. Educ: Rensselaer Polytech Inst, BS, 68; Univ Wis-Madison, MS, 70, PhD(meteorol), 74. Prof Exp: Proj assoc meteorol, 74-76, ASST SCIENTIST METEOROL, SPACE SCI & ENG CTR, UNIV WIS-MADISON, 76- Mem: Am Meteorol Soc; Sigma Xi. Res: Application of geostationary satellite data to the study of the dynamics of tropical and mid-latitude mesoscale systems. Mailing Add: Space Sci & Eng Ctr Univ of Wis Madison WI 53706

SUCHOW, LAWRENCE, b New York, NY, June 24, 23; m 68. SOLID STATE CHEMISTRY, INORGANIC CHEMISTRY. Educ: City Col New York, BS, 43; Polytech Inst Brooklyn, PhD(chem), 51. Prof Exp: Anal chemist, Aluminum Co Am, 43-44; res chemist, Baker & Co, Inc, 44, Manhattan Proj, Oak Ridge, Tenn, 45-46, Baker & Co, Inc, 46-47, Signal Corps Eng Labs, US Dept Army, 50-54 & Francis Earle Labs, Inc, 54-58; sr res chemist, Westinghouse Elec Corp, 58-60; mem res staff, Watson Res Ctr, Int Bus Mach Corp, 60-64; from asst prof to assoc prof, 64-70, PROF CHEM, NJ INST TECHNOL, 70- Concurrent Pos: NSF grants, 67-74; sabbatical leave, Imp Col, Univ London, 74. Mem: Am Chem Soc; Sigma Xi; fel NY Acad Sci. Res: High temperature inorganic reactions; physical properties of solids; x-ray crystallography; crystal growth; phosphors; rare earths. Mailing Add: Dept of Chem Eng & Chem NJ Inst of Technol Newark NJ 07102

SUCHSLAND, OTTO, b Jena, Ger, June 18, 28; US citizen; m 56; c 2. WOOD TECHNOLOGY. Educ: Univ Hamburg, BS, 52, Dr nat sci(wood technol), 56. Prof Exp: Res engr, Swed Forest Prod Lab, Stockholm, 52-55; tech dir, Elmendorf Res, Inc, Calif, 55-57; from asst prof to assoc prof forest prod, 57-71, PROF FORESTRY, MICH STATE UNIV, 71- Mem: Forest Prod Res Soc. Res: Adhesives; gluing of wood; technology of composite wood products. Mailing Add: 138 Natural Resources Bldg Mich State Univ East Lansing MI 48823

SUCIU-FOCA, NICOLE M, b Romania, Dec 6, 38; stateless; m 63. IMMUNOLOGY, HISTOLOGY. Educ: Univ Bucharest, BS, 59, MS, 60, PhD(transplantation immunol), 65. Prof Exp: Res assoc transplantation immunol, Inst Oncol, Bucharest, Romania, 61-64; asst prof immunol & dir immunol lab, Fac Med, Univ Bucharest, 65-70; vis res fel transplantation immunol, Sloan-Kettering Inst Cancer Res, 69-70; res assoc, 71-73, ASST PROF PATH, COL PHYSICIANS & SURGEONS, COLUMBIA UNIV, 73-, DIR IMMUNOL LAB, FRANCIS DELAFIELD HOSP, 71- Concurrent Pos: Int Union Against Cancer Eleanor Roosevelt fel, 69. Mem: Transplantation Soc; Am Fedn Clin Res; Am Asn Clin Histocompatibility Testing. Res: Cellular immune responsiveness in organ allograft recipients; patients with cancer and autoimmune disease. Mailing Add: Francis Delafield Hosp Columbia Univ Dept of Surg New York NY 10032

SUCOFF, EDWARD IRA, b NJ, Nov 17, 31; PLANT PHYSIOLOGY. Educ: Univ Mich, BS, 55, MS, 56; Univ Md, PhD(bot), 60. Prof Exp: Res forester, US Forest Serv, 56-60; assoc prof, 60-71, PROF FORESTRY, UNIV MINN, ST PAUL, 71- Mem: AAAS; Am Soc Plant Physiologists; Soc Am Foresters. Res: Tree growth. Mailing Add: Col of Forestry Univ of Minn St Paul MN 55101

SUD, GIAN CHAND, b Noormahal, Punjab, India, Nov 6, 35; m 66. ZOOLOGY, BIOCHEMISTRY. Educ: Panjab Univ, India, BSc, 53, MSc, 54; Univ Wis, PhD(zool, biochem), 64. Prof Exp: Lectr zool & bot, GMM Col, India, 54-58; instr biol & chem, Stout State Col, 58-59; asst prof biol, Wis State Univ, Oshkosh, 64-66; ASSOC PROF BIOL, WESTERN MICH UNIV, 66- Concurrent Pos: Consult, Wis State Univ, Oshkosh & Northern Mich Univ, 64-66; admin assoc, Teachers Col, Columbia Univ, 70-71. Res: Histochemistry and cytochemistry with the aid of an electron microscope of a muscle tissue from vitamin E deficient animals; biology as a conceptual tool in nursing, health sciences, health education and allied health sciences. Mailing Add: Dept of Biol Western Mich Univ Kalamazoo MI 49001

SUDAN, RAVINDRA NATH, b Kashmir, India, June 8, 31; m 59; c 2. PLASMA PHYSICS. Educ: Panjab Univ, India, BA, 48; Indian Inst Sci, dipl, 52; Univ London, DIC & PhD(elec eng), 55. Prof Exp: Elec engr, Brit Thomson Houston Co, Eng, 55-57; instruments engr, Imp Chem Industs, Ltd, India, 57-58; res assoc, 58-59, from asst prof to assoc prof elec eng, 59-68, mem ctr radiophysics & space res, 63-68, chmn exec comt lab plasma studies, 68-74, PROF ELEC ENG & APPL PHYSICS, CORNELL UNIV, 68-, DIR, LAB PLASMA PHYSICS, 74- Concurrent Pos: Vis scientist, Int Ctr Theoret Physics, Trieste, 65-66, 70; vis res physicist, plasma physics lab, Princeton Univ, 66-67; head theoret plasma physics, Naval Res Lab, DC, 70-71; consult, Lawrence Radiation Lab, Univ Calif & Naval Res Lab, DC. Mem: AAAS; fel Am Phys Soc; sr mem Inst Elec & Electronics Engrs. Res: Plasma, electro-discharge and space physics; magnetohydrodynamics; electrodynamics; energy conversion. Mailing Add: Sch of Eng & Appl Physics Clarke Hall Cornell Univ Ithaca NY 14850

SUDARSANAN, KESAVAN, b Trivandrum, Kerala, India, July 5, 28; m 57; c 2. X-RAY CRYSTALLOGRAPHY, COMPUTER SCIENCE. Educ: Univ Kerala, BSc, 48, MSc, 50; Univ Paris, DSc, 65. Prof Exp: Lectr physics, Univ Col, Trivandrum, India, 51-60; jr researcher, Mineral Lab, Sorbonne, 60-61, researcher physics, 62-65; fel, 66-68, RES SCIENTIST PHYSICS, PHYS SCI DIV, ENG EXP STA, GA INST TECHNOL, 68- Mem: Am Crystallog Asn. Res: Inorganic structures by x-ray and neutron diffraction; computer programming. Mailing Add: Phys Sci Div Eng Exp Sta Ga Inst of Technol Atlanta GA 30332

SUDARSHAN, ENNACKEL CHANDY GEORGE, b Kottayam, India, Sept 16, 31; m 54; c 3. THEORETICAL PHYSICS. Educ: Univ Madras, BSc, 51, MA, 52; Univ Rochester, PhD(physics), 58. Hon Degrees: DSc, Univ Wis-Milwaukee, 66 & Univ Delhi, 73. Prof Exp: Demonstr physics, Christian Col, Madras, 51-52; res asst, Tata Inst Fundamental Res, 52-55 & Univ Rochester, 55-57; res fel, Harvard Univ, 57-59; from asst prof to assoc prof, Univ Rochester, 59-64; prof, Syracuse Univ, 64-69; PROF PHYSICS, CTR PARTICLE THEORY, UNIV TEX, AUSTIN, 69-, DIR CTR, 70- Concurrent Pos: Guest prof, Univ Bern, 63-64; vis prof, Brandeis Univ, 64; Sir C V Raman distinguished vis prof, Univ Madras, 70-71; prof & dir, Ctr Theoret Studies, Indian Inst Sci, Bangalore, 72- Honors & Awards: Honor Award for Outstanding Achievement, Nat Asn Indians Am, 75. Mem: Fel Indian Nat Acad Sci; fel Am Phys Soc; fel Indian Acad Sci. Res: Quantum field theory; elementary particles; high energy physics; classical mechanics; quantum optics; Lie algebras and their application to particle physics; foundations of physics; philosophy and history of contemporary physics. Mailing Add: PMA Bldg 9.328 Ctr for Particle Theory Univ of Tex Austin TX 78712

SUDBOROUGH, IVAN HAL, b Royal Oak, Mich, Dec 19, 43; m 69; c 2. INFORMATION SCIENCE. Educ: Calif State Polytech Col, BS, 66, MS, 67; Pa State Univ, PhD(comput sci), 71. Prof Exp: ASST PROF COMPUT SCI, NORTHWESTERN UNIV, 71- Concurrent Pos: NSF res grant, 74. Mem: Asn Comput Mach; Soc Indust & Appl Math. Res: Computational complexity; formal languages; automata theory; theory of computation. Mailing Add: Dept of Comput Sci Technol Inst Northwestern Univ Evanston IL 60201

SUDBURY, JOHN DEAN, b Natchitoches, La, July 29, 25; m 47; c 3. PHYSICAL CHEMISTRY. Educ: Univ Tex, BS, 44, MS, 47, PhD(phys chem), 49. Prof Exp: Sr res chemist, Develop & Res Dept, Continental Oil Co, 49-56, supv res chemist, 56-66, dir petrochem res div, Okla, 66-69, gen mgr, C/A Nuclear Fuels Div, Calif, 69-70, asst to vpres res, NY, 70-72, ASST DIR RES, CONOCO COAL DEVELOP CO, 72- Honors & Awards: Speller Award, Nat Asn Corrosion Engrs, 66. Mem: Am Chem Soc; Nat Asn Corrosion Engrs. Res: Advanced systems for liquified natural gas, arctic transport; development of conversion processes to get coal into more desirable energy sources; conversion of coal to liquids and gases; removal of sulfur from combustion products of coal. Mailing Add: Res Div Conoco Coal Develop Co Library PA 15129

SUDDARTH, STANLEY KENDRICK, b Westerly, RI, Oct 22, 21; m 51. FORESTRY. Educ: Purdue Univ, BSF, 43, MS, 49, PhD(agr econ forestry), 52. Prof Exp: Assoc dir bomb effectiveness res, US Dept Air Force Proj, Res Found, 51-54, from asst prof to assoc prof forestry, 54-60, PROF WOOD ENG, AGR EXP STA, PURDUE UNIV, WEST LAFAYETTE, 60- Concurrent Pos: Consult home mfg indust, 55- & US Forest Prod Lab, Madison, Wis, 70-72; tech adv, Am Inst Timber Construct & Truss Plate Inst. Honors & Awards: Res Award, Truss Plate Inst, 70; Markwardt Eng Res Award, Forest Prod Res Soc, 71; Markwardt Award, Am Soc Testing & Mat, 72. Mem: Forest Prod Res Soc; Int Acad Wood Sci; Am Soc Agr Engrs; Am Soc Civil Engrs. Res: Applied mathematics in engineering and economic problems; engineering properties and uses of wood. Mailing Add: Dept Forestry & Natural Resources Purdue Univ Agr Exp Sta West Lafayette IN 47907

SUDDATH, FRED LEROY, (JR), b Macon, Ga, May 6, 42; m 65; c 2. BIOLOGICAL STRUCTURE, X-RAY CRYSTALLOGRAPHY. Educ: Ga Inst Technol, BS, 65, PhD(chem), 70. Prof Exp: NIH fel, Mass Inst Technol, 70-72, res assoc biol, Lab Molecular Struct, 72-75; ASST PROF BIOCHEM, INVESTR INST DENT RES & ASSOC SCIENTIST COMPREHENSIVE CANCER CTR, MED CTR, UNIV ALA, BIRMINGHAM, 75- Concurrent Pos: Am Cancer Soc fel, Mass Inst Technol, 72-73. Mem: AAAS; Am Chem Soc; The Chem Soc; Am Crystallog Asn. Res: Structure and function of transfer RNA; structural studies of nucleic acids and proteins; correlation of molecular structure and biological function; experimental methods development. Mailing Add: Inst of Dent Res Univ of Ala Med Ctr University Station Birmingham AL 35294

SUDDERTH, WILLIAM DAVID, b Dallas, Tex, Apr 29, 40; m 62; c 2. MATHEMATICS, MATHEMATICAL STATISTICS. Educ: Yale Univ, BS, 63; Univ Calif, Berkeley, MS, 65, PhD(math), 67. Prof Exp: Asst prof statist, Univ Calif, Berkeley, 67-68; asst prof math, Morehouse Col, 68-69; asst prof statist, 69-71, ASSOC PROF STATIST, UNIV MINN, MINNEAPOLIS, 71- Mem: Am Math Soc; Inst Math Statist. Res: Probability, espeically the study of finitely additive probability measures and abstract gambling theory, which is also known as dynamic programming and stochastic control. Mailing Add: Sch of Statist Univ of Minn Minneapolis MN 55455

SUDDICK, RICHARD PHILLIPS, b Omaha, Nebr, Feb 3, 34; m 55; c 4. PHYSIOLOGY. Educ: Creighton Univ, BS, 58, MS, 59, DDS, 61; Univ Iowa, PhD(physiol), 67. Prof Exp: Instr physiol, Univ Iowa, 63-65; asst prof biol sci, Creighton Univ, 65-68, from assoc prof to prof oral biol & head dept, 68-74, assoc prof physiol, Sch Med, 70-74; ASSOC PROF PHYSIOL & ASST DEAN RES, COL DENT MED, MED UNIV SC, 74- Mem: AAAS; Int Asn Dent Res; Am Physiol Soc. Res: Physiology of exocrine secretion, primarily secretion of saliva; function of the saliva in the oral cavity and the alimentary tract and its relationship to normal and diseased states. Mailing Add: Col of Dent Med Med Univ of SC 80 Barre St Charleston SC 29401

SUDDS, RICHARD HUYETTE, JR, b State College, Pa, Feb 13, 27; m 52; c 2. PARASITOLOGY. Educ: Univ Conn, BA, 50, MA, 51; Univ NC, MSPH, 54, PhD(parasitol), 59. Prof Exp: From asst prof to assoc prof, 58-72, PROF MICROBIOL, STATE UNIV NY COL PLATTSBURGH, 72- Mem: Am Soc Trop Med & Hyg; Am Pub Health Asn; Am Soc Parasitologists. Res: Host-parasite relationships. Mailing Add: Dept of Biol Sci State Univ of NY Col Plattsburgh NY 12901

SUDERMAN, HAROLD JULIUS, b Myrtle, Man, July 24, 21; m 47; c 3. BIOCHEMISTRY. Educ: Univ Man, BSc, 49, MSc, 52, PhD, 62. Prof Exp: Demonstr biochem, Univ Man, 51-52, lectr, 52-56, asst prof, 56-63; asst prof, Ont Agr Col, 63-65, asst prof, 65-74, ASSOC PROF BIOCHEM, UNIV GUELPH, 74- Mem: AAAS; Can Biochem Soc; NY Acad Sci. Res: Molecular properties, structure and function of proteins; comparative biochemistry of hemoglobins. Mailing Add: Dept of Chem Univ of Guelph Guelph ON Can

SUDIA, THEODORE WILLIAM, b Ambridge, Pa, Oct 10, 25; m 49; c 3. ENVIRONMENTAL PHYSIOLOGY. Educ: Kent State Univ, BS, 50; Ohio State Univ, MS, 51, PhD(bot), 54. Prof Exp: Asst prof biol sci, Winona State Col, 55-58; res fel plant physiol, Univ Minn, St Paul, 58-59, res assoc physiol ecol, 59-61, asst prof, 61-63, assoc prof plant path & bot, 63-67; assoc dir, Am Inst Biol Sci, Washington, DC, 67-69; chief ecol serv, Off Natural Sci, 69-73, CHIEF SCIENTIST, US NAT PARK SERV, 73- Mem: AAAS; Ecol Soc Am; Am Soc Plant Physiologists; Bot Soc Am; NY Acad Sci. Res: Research administration. Mailing Add: Off of Chief Scientist Nat Park Serv US Dept Interior 18th & D Sts NW Washington DC 20240

SUDIA, WILLIAM DANIEL, b Ambridge, Pa, Aug 19, 22; m 49; c 2. ENTOMOLOGY, VIROLOGY. Educ: Univ Fla, BS, 49; Ohio State Univ, MS, 50, PhD(entom), 58. Prof Exp: Entomologist, Med Entom Unit, Ctr Dis Control, USPHS, 51-53, asst chief virus vector unit, 53-60, asst chief arbovirus vector lab, 60-65, lab consult & develop sect, 66, CHIEF ARBOVIRUS ECOL LAB, CTR DIS CONTROL, USPHS, 67- Honors & Awards: Meritorious Serv Medal, USPHS. Mem: Sigma Xi; Am Soc Trop Med & Hyg; Am Mosquito Control Asn. Res: Ecology of arthropod-borne encephalitis viruses; mosquito vectors and vertebrate hosts. Mailing Add: Ctr Dis Control LD4 Bldg 6 Rm 291 Lab Training Br C Atlanta GA 30333

SUDMEIER, JAMES LEE, b Minneapolis, Minn, Feb 14, 38; m 62; c 2. ANALYTICAL CHEMISTRY. Educ: Carleton Col, BA, 59; Princeton Univ, MA, 61, PhD(chem), 66. Prof Exp: Actg asst prof chem, Univ Calif, Los Angeles, 65-66, asst prof, 66-70; asst prof, 70-71, ASSOC PROF CHEM, UNIV CALIF, RIVERSIDE, 71- Mem: Am Chem Soc. Res: Nuclear magnetic resonance studies of coordination compounds and metal binding to biopolymers. Mailing Add: Dept of Chem Univ of Calif Riverside CA 92502

SUDO, SARA ZEECE, b New Orleans, La, Jan 15, 44; m 66; c 1. MICROBIOLOGY, BIOCHEMISTRY. Educ: Univ Minn, BA, 65, MS, 68, PhD, 70. Prof Exp: NIH fel, Univ Calif, San Diego, 70-72; asst prof microbiol, 72-73, ASST PROF DENT, SCH DENT, UNIV MINN, MINNEAPOLIS, 73- Mem: AAAS; Am Soc Microbiol. Res: Microbial attachment to solid supports; microbial interactions. Mailing Add: Dept of Periodont Univ of Minn Sch of Dent Minneapolis MN 55455

SUDWEEKS, EARL MAX, b Richfield, Utah, Dec 27, 33; m 60; c 8. ANIMAL NUTRITION, DAIRY NUTRITION. Educ: Utah State Univ, BS, 60, MS, 62; NC State Univ, PhD(nutrit biochem), 72. Prof Exp: Res assoc animal nutrit, Utah State Univ, 62-65, from asst prof to assoc prof exten, 65-68; res asst animal nutrit, NC State Univ, 68-72; ASST PROF ANIMAL NUTRIT, UNIV GA, 72- Mem: Am Dairy Sci Asn; Am Soc Animal Sci; Sigma Xi. Res: The role of roughages in rumen physiology, energy utilization, feed conversion and animal longevity of beef and dairy cattle. Mailing Add: Univ of Ga Ga Sta Experiment GA 30212

SUDWEEKS, WALTER BENTLEY, b Buhl, Idaho, May 22, 40; m 65; c 2. INDUSTRIAL CHEMISTRY, EXPLOSIVES. Educ: Brigham Young Univ, BS, 65, PhD(org chem), 70. Prof Exp: RES CHEMIST, POLYMER INTERMEDIATES DEPT, E I DU PONT DE NEMOURS & CO, INC, 69- Mem: Am Chem Soc. Res: Organic synthesis; hydrometallurgical processes; explosives research. Mailing Add: Potomac River Develop Labs Polymer Intermediates Dept E I du Pont de Nemours & Co Inc Martinsburg WV 25401

SUELTENFUSS, ELIZABETH ANNE, biology, see 12th edition

SUELTER, CLARENCE HENRY, b Lincoln, Kans, Dec 15, 28; m 55; c 3. ENZYMOLOGY. Educ: Kans State Univ, BS, 51, MS, 53; Iowa State Univ, PhD(biochem), 59. Prof Exp: From asst prof to assoc prof, 61-69, PROF BIOCHEM, MICH STATE UNIV, 69- Concurrent Pos: USPHS fel, Univ Minn, 59-61, res career develop award, Mich State Univ, 65-75; consult, NIH Title IV Grad Fels, 63-67; mem enzyme nomenclature comt, Nat Acad Sci, 70-74. Mem: AAAS; Am Chem Soc; Soc Exp Biol & Med. Res: Monovalent cation activation of enzymes; structure and function of enzymes. Mailing Add: Dept of Biochem Mich State Univ East Lansing MI 48824

SUEOKA, NOBORU, b Kyoto, Japan, Apr 12, 29; m 57; c 1. GENETICS. Educ: Kyoto Univ, BS, 53, MS, 55; Calif Inst Technol, PhD(biochem genetics), 59. Prof Exp: Res fel biochem genetics, Harvard Univ, 58-60; asst prof microbiol, Univ Ill, 60-62; from assoc prof to prof biol, Princeton Univ, 62-72; PROF BIOL, UNIV COLO, BOULDER, 72- Concurrent Pos: Fulbright grant, 55-56. Mem: Am Soc Biol Chemists; Am Soc Microbiol; Genetics Soc Am. Res: Biochemical genetics; molecular biology, particularly genetic aspects of biological macromolecules, nucleic acids and protein. Mailing Add: Dept Molecular Cell Develop Biol Univ of Colo Boulder CO 80302

SUESS, GENE GUY, b Beaver, Okla, Apr 16, 41; m 68; c 1. MEAT SCIENCE. Educ: Tex Tech Univ, BS, 63; Univ Wis-Madison, MS 66, PhD(meat sci, animal sci), 68. Prof Exp: Res technologist, 68-73, NEW PROD DEVELOP SUPVR, OSCAR MAYER & CO, 73- Mem: Inst Food Technologists. Res: Meats processing. Mailing Add: Res Dept Oscar Mayer & Co PO Box 1409 Madison WI 53701

SUESS, HANS EDUARD, b Vienna, Austria, Dec 16, 09; nat US; m 42; c 2. CHEMISTRY. Educ: Univ Vienna, PhD(chem), 36; Univ Hamburg, Dr habil, 39. Prof Exp: Demonstr, Univ Vienna, 34-36; res assoc, Univ Hamburg, 37-48, assoc prof, 49-50; res fel, Univ Chicago, 50-51; chemist, US Geol Surv, 51-55; PROF CHEM, UNIV CALIF, SAN DIEGO, 55- Honors & Awards: V M Goldschmidt Medal, Geochem Soc, 74. Mem: Nat Acad Sci; Austrian Acad Sci; fel Am Acad Arts & Sci. Res: Chemical kinetics; nuclear hot atom chemistry; cosmic abundances of nuclear species; nuclear shell structure; geologic age determinations; carbon-14 dating. Mailing Add: Dept of Chem Univ of Calif at San Diego La Jolla CA 92093

SUESS, JAMES FRANCIS, b Rock Island, Ill, Nov 27, 19; m 46; c 3. PSYCHIATRY. Educ: Northwestern Univ, BS, 50, MD, 52. Prof Exp: Resident psychiat, Warren State Hosp, Warren, Pa, 53-56, clin dir, 56-62; PROF PSYCHIAT, SCH MED, UNIV MISS, 62- Concurrent Pos: Fel psychiat, Med Sch, Univ Pa, 53; exchange teaching fel, Med Sch, Univ Pittsburgh, 55; fel, Col Physicians & Surgeons, Columbia Univ, 58; consult, Vet Admin, 62- & Gov Drug Coun, Miss, 72- Mem: Am Psychiat Asn; Asn Am Med Cols; AMA. Res: Medical education in psychiatry; use of television and videotape in medical education; programmed teaching with television. Mailing Add: Dept of Psychiat Univ of Miss Med Sch Jackson MS 39216

SUESS, STEVEN TYLER, b Los Angeles, Calif, Aug 4, 42; m 64; c 2. FLUID DYNAMICS. Educ: Univ Calif, Berkeley, AB, 64, PhD(planetary, space sci), 69. Prof Exp: Nat Acad Sci-Nat Res Coun res assoc, Environ Sci Serv Admin Res Labs, Boulder, Colo, 69-71, PHYSICIST, SPACE ENVIRON LAB, NAT OCEANIC & ATMOSPHERIC ADMIN, 71- Concurrent Pos: Guest worker, Max Planck Inst Aeronomy, 75. Mem: Am Geophys Union; Am Astronom Soc; Sigma Xi; Int Astron Union. Res: Dynamics of the sun and stars; oscillations of stars; stellar winds; magnetohydrodynamics of rotating fluids. Mailing Add: Space Environ Lab Nat Oceanic & Atmospheric Admin Boulder CO 80302

SUFFET, IRWIN HENRY, b Brooklyn, NY, May 11, 39; m 62; c 2. ANALYTICAL CHEMISTRY, ENVIRONMENTAL CHEMISTRY. Educ: Brooklyn Col, BS, 61; Univ Md, College Park, MS, 64; Rutgers Univ, New Brunswick, PhD(environ sci), 69. Prof Exp: Asst prof, 69-73, ASSOC PROF CHEM & ENVIRON SCI, DREXEL UNIV, 74- Concurrent Pos: Consult, Western Elec Co, 70-72; grants, Western Elec Co, 70-72, Environ Protection Agency, 71-74, NSF, 72-73 & City Philadelphia Water Dept, 72-75. Mem: Am Chem Soc; Am Water Works Asn; Am Pub Health Asn. Res: Fate of pollutants and pesticides in the environment; analytical environmental analysis of trace organics; chemical nature of water and wastes. Mailing Add: Environ Studies Inst Drexel Univ Philadelphia PA 19104

SUGA, NOBUO, b Japan, Dec 17, 33; m 63; c 2. PHYSIOLOGY. Educ: Tokyo Metrop Univ, PhD(physiol), 63. Prof Exp: NSF fel hearing physiol, Harvard Univ, 63-64; res zoologist, Brain Res Inst, Univ Calif, Los Angeles, 65; res neuroscientist, Sch Med, Univ Calif, San Diego, 66-68; ASSOC PROF BIOL, WASH UNIV, 69- Mem: AAAS; Am Soc Zoologists; Am Physiol Soc; Acoust Soc Am. Res: Auditory physiology. Mailing Add: Dept of Biol Wash Univ St Louis MO 63130

SUGAR, JACK, b Baltimore, Md, Dec 22, 29; m 56; c 3. ATOMIC SPECTROSCOPY. Educ: Johns Hopkins Univ, BA, 56, PhD(physics), 60. Prof Exp: PHYSICIST, NAT BUR STAND, 60- Concurrent Pos: Fulbright res traveling grant, 66-67. Honors & Awards: Silver Medal, US Dept Com, 71. Mem: Fel Optical Soc Am. Res: Spectra of solids; atomic spectra and energy levels; nuclear moments; ionization energies. Mailing Add: Nat Bur of Stand A167 Physics Bldg Washington DC 20236

SUGAR, OSCAR, b Washington, DC, July 9, 14; m 44; c 3. PHYSIOLOGY, NEUROSURGERY. Educ: Johns Hopkins Univ, AB, 34; George Washington Univ, MA, 37, MD, 42; Univ Chicago, PhD(physiol), 40. Prof Exp: Asst physiol, Univ Chicago, 36-38; clin asst, 46-48, from instr to assoc prof, 48-58, PROF NEUROL SURG, UNIV ILL COL MED, 58-, HEAD DEPT, 71- Concurrent Pos: Sci consult, Nat Inst Neurol Dis & Stroke, 72-75. Mem: Am Asn Neurol Surg; Am Neurol Soc; Soc Neurol Surg. Res: Degeneration and regeneration of the peripheral and central nervous system; effects of oxygen lack on cells and electrical activity of the nervous system; visualization of the blood supply and vascular anomalies of the brain. Mailing Add: Dept of Neurol Surg Univ of Ill Col of Med Chicago IL 60612

SUGAR, ROBERT LOUIS, b Chicago, Ill, Aug 20, 38; m 66. THEORETICAL PHYSICS. Educ: Harvard Univ, AB, 60; Princeton Univ, PhD(physics), 64. Prof Exp: Res assoc physics, Columbia Univ, 64-66; from asst prof to assoc prof, 66-73, PROF PHYSICS, UNIV CALIF, SANTA BARBARA, 73- Mem: Am Phys Soc. Res: High energy physics. Mailing Add: Dept of Physics Univ of Calif Santa Barbara CA 93106

SUGARMAN, MEYER LOUIS, JR, b Atlanta, Ga, Aug 4, 17; m 66; c 4. RESEARCH ADMINISTRATION. Educ: Univ Fla, BS, 39; Ohio State Univ, MS, 40. Prof Exp: Res chemist, Kryptar Corp, 45-49; sr chemist, Eastman Kodak Co, 49-52; res engr, RCA Labs, RCA Corp, 52-56; dir res, Apeco Corp, 56-64; chmn & tech dir photog, Opto/Graphics Inc, 64-71; MGR RES LAB ELECTRONICS, ZENITH RADIO

CORP, 71- Mem: Am Chem Soc; Soc Photog Scientists & Engrs. Res: Color television picture tubes and display devices. Mailing Add: 4000 Dundee Rd Northbrook IL 60062

SUGARMAN, NATHAN, b Chicago, Ill, Mar 3, 17; m 40; c 2. NUCLEAR CHEMISTRY. Educ: Univ Chicago, BS, 37, PhD(phys chem), 41. Prof Exp: Am Philos Soc fel, Univ Chicago, 41-42, res assoc, Metall Lab, 42-43, sect chief, 43-45; group leader, Los Alamos Sci Lab, Univ Calif, 45-46; from asst prof to assoc prof, 46-52, PROF CHEM, ENRICO FERMI INST, UNIV CHICAGO, 52- Mem: AAAS; Am Chem Soc; fel Am Phys Soc; Fedn Am Scientists. Res: Nuclear reactions; fission studies; recoil experiments. Mailing Add: Enrico Fermi Inst Univ of Chicago Chicago IL 60637

SUGATHAN, KANNETH KOCHAPPAN, b Palliport, India, Mar 23, 26; m 56; c 4. ORGANIC CHEMISTRY. Educ: Univ Kerala, BSc, 51, PhD(terpene chem), 67; Univ Saugar, MSc, 53. Prof Exp: Demonstr chem, SKV Col, Trichur, India, 53-54, lectr, 54-64; lectr, Univ Kerala, 64-68; Nat Res Coun-Agr Res Serv fel, USDA Naval Stores Lab, Olustee, Fla, 68-70; Am Cancer Soc res fel, Univ Miss, 71-72; RES CHEMIST, CROSBY CHEM, INC, 72- Concurrent Pos: Sr demonstr, Christian Med Col, Vellore, India, 61-64. Mem: AAAS; Am Chem Soc; Am Oil Chemists' Soc. Res: Terpene chemistry and resin acid chemistry; development of novel industrial products from terpenes and rosin, particularly polyterpenes, terpene phenolics, terpene phenol formaldehydes, polyamides, polyalcohols, polyesters, and terpene-petroleum hydrocarbon copolymers. Mailing Add: 313 S Steele Ave Picayune MS 39466

SUGERMAN, ABRAHAM ARTHUR, b Dublin, Ireland, Jan 20, 29; nat US; m 60; c 4. PSYCHIATRY. Educ: Univ Dublin, BA, 50, MB, BCh & BAO, 52; Royal Col Physicians & Surgeons, dipl psychol med, 58; State Univ NY, MedDSc(psychiat), 62; Univ Newcastle, dipl, 66; Am Bd Psychiat & Neurol, dipl, 69. Prof Exp: House officer, Meath Hosp, Dublin, Ireland, 52-53 & St Nicholas Hosp, London, 53; sr house physician, Brook Gen Hosp, 54; registr psychiat, Kingsway Hosp, Derby & Med Sch, King's Col, Newcastle, 55-58; clin psychiatrist, Trenton State Hosp, NJ, 58-59; chief sect invest psychiat & dir clin invest unit, NJ Neuropsychiat Inst, 61-73; dir outpatient serv, 72-74, MED DIR, CARRIER CLIN, 74- Concurrent Pos: Res fel psychiat, State Univ NY Downstate Med Ctr, 59-61; res consult, Trenton Psychiat Hosp, 64-; res consult & assoc psychiatrist, Carrier Clin, NJ, 68-72, res dir, Carrier Clin Found, 72-; clin assoc prof, Rutgers Med Sch, 72-; consult, Med Ctr, Princeton, 72-; prof, Grad Sch Appl & Prof Psychol, Rutgers Univ, 74- Mem: fel AAAS; fel Am Psychiat Asn; fel Am Col Clin Pharmacol; AMA; fel Am Col Neuropsychopharmacol. Res: Evaluation of new psychiatric drugs; nosology; psychology and prognosis in schizophrenia and alcoholism; quantitative analysis of the electroencephalogram. Mailing Add: 125 Roxboro Rd Lawrenceville NJ 08648

SUGG, WINFRED LINDLEY, b Snow Hill, NC, May 8, 32; m 63; c 3. THORACIC SURGERY, CARDIOVASCULAR SURGERY. Educ: Univ NC, Chapel Hill, BS, 54, MD, 57. Prof Exp: Intern, Barnes Hosp, St Louis, Mo, 57-58, resident, 58-62; asst prof thoracic & cardiovasc surg, Univ Tex Southwestern Med Sch Dallas, 66-69; assoc prof, Sch Med, Univ NC, Chapel Hill, 69-70; assoc prof thoracic & cardiovasc surg & chmn div, Southwestern Med Sch, Univ Tex Health Sci Ctr Dallas, 70-74, CLIN ASSOC PROF THORACIC & CARDIOVASC SURG, UNIV TEX SOUTHWESTERN MED SCH, 74- Concurrent Pos: Res grants & award, Dallas Heart Asn, 67-69, Tex Heart Asn, 67-69, 71-72, NIH, 67-69, 70-71 & Southwestern Med Found, 71-73; co-investr, NIH res grants, 68-75. Mem: Am Asn Thoracic Surg; Am Col Chest Physicians; Am Col Surg; Am Soc Artificial Internal Organs; fel Am Col Cardiol. Res: Cardiac research; artificial heart and lung assist devices; pediatric cardiac surgery. Mailing Add: 6161 Harry Hines Blvd Dallas TX 75235

SUGGITT, ROBERT MURRAY, b Toronto, Ont, June 24, 25; nat US; m 59; c 3. PHYSICAL CHEMISTRY. Educ: Univ Toronto, BA, 47; Univ Mich, MS, 48, PhD(chem), 52. Prof Exp: Chemist, Texaco, Inc, 52-57, group leader, 57-68, res assoc, 68-75, TECH ASSOC, TEXACO DEVELOP CORP, 75- Res: Catalysis, petroleum and petrochemical processing, lubrication, petroleum production, patents. Mailing Add: Texaco Develop Corp 135 E 42nd St New York NY 10017

SUGGS, JOSEPH ELLIS, biochemistry, see 12th edition

SUGGS, MORRIS TALMAGE, JR, b Ft Myers, Fla, June 17, 27; m 52; c 3. MICROBIOLOGY. Educ: Wake Forest Col, BS, 50; Fla State Univ, MS, 57; Univ NC, MPH, 65, DrPH, 67. Prof Exp: Teacher, Fla Pub Sch, 52-54; microbiologist, Ala, 57, asst chief tissue cult unit, 58-59, res asst, 60-62, res asst virol training unit, 62-68, spec asst biol reagents sect, Lab Prog, 68, DIR BIOL PROD DIV, LAB BUR, CTR DIS CONTROL, USPHS, 68- Mem: Am Soc Microbiologists, Sigma Xi; Am Pub Health Asn; Conf of State & Prov Pub Health Lab Dirs. Res: Standardization and quality assurance of in vitro diagnostic products. Mailing Add: Biol Prod Div Lab Bur Ctr for Dis Control 1600 Clifton Rd Atlanta GA 30333

SUGIHARA, JAMES MASANOBU, b Las Animas, Colo, Aug 6, 18; m 44; c 2. ORGANIC CHEMISTRY, ACADEMIC ADMINISTRATION. Educ: Univ Calif, BS, 39; Univ Utah, PhD(chem), 47. Prof Exp: From instr to prof chem, Univ Utah, 43-64; dean col chem & physics, 64-73, dean col sci & math, 73, PROF CHEM, N DAK STATE UNIV, 64-, DEAN GRAD SCH & DIR RES ADMIN, 74- Concurrent Pos: Fel, Ohio State Univ, 48. Mem: Am Chem Soc; Am Inst Chemists; Geochem Soc Am. Res: Reaction mechanisms; porphyrin chemistry; origin of petroleum. Mailing Add: Off of Dean NDak State Univ Grad Sch Fargo ND 58102

SUGIHARA, THOMAS TAMOTSU, b Las Animas, Colo, June 14, 24; m 52; c 2. NUCLEAR CHEMISTRY. Educ: Kalamazoo Col, AB, 45; Univ Chicago, SM, 51, PhD(phys chem), 52. Prof Exp: Res assoc, Mass Inst Technol, 52-53; from asst prof to prof chem, Clark Univ, 53-67, chmn dept, 63-66; PROF CHEM, TEX A&M UNIV, 67-, DIR CYCLOTRON INST, 71- Concurrent Pos: Assoc scientist, Woods Hole Oceanog Inst, 54-67; mem subcomt nuclear ship waste disposal, Nat Acad Sci-Nat Res Coun, 58, mem surv panel nuclear chem, 64, mem subcomt low level contamination mat, 65-66; Guggenheim fel, Univ Oslo, 61-62. Mem: AAAS; Am Chem Soc; Am Phys Soc. Res: Structure of transition nuclei; conversion-electron spectroscopy; high-span states; heavy-ion reactions. Mailing Add: Cyclotron Inst Tex A&M Univ College Station TX 77843

SUGIMOTO, ROY, b Los Angeles, Calif, Jan 21, 17; m 55; c 3. ORGANIC CHEMISTRY. Educ: Univ Calif, Los Angeles, AB, 39, MA, 41; Purdue Univ, PhD, 49. Prof Exp: Proj leader, Ethyl Corp, 48-52, asst res supvr, 52-54, supvr process develop, 54-60, proj mgr antioxidants, 60-64, sr prod mgr com develop, Mich, 64-68; TECH DIR, HOUSTON CHEM CO, PPG INDUSTS, INC, 68- Mem: Am Chem Soc; Air Pollution Control Asn; Com Develop Asn; Soc Automotive Engrs. Res: Gasoline and fuel additives; auto exhaust emission control systems; auto exhaust particulate traps; glycol antifreeze. Mailing Add: Houston Chem Co Box 4026 Corpus Christi TX 78408

SUGIOKA, KENNETH, b Hollister, Calif, Apr 19, 20; m 47, 66; c 5. ANESTHESIOLOGY. Educ: Univ Denver, BS, 45, Wash Univ, MD, 49; Am Bd Anesthesiol, dipl, 55. Prof Exp: Intern, Univ Iowa Hosp, 49-50, resident anesthesiol, 50-52, instr, 52; actg chief anesthesiol, Vet Admin Hosp, Des Moines, 52; resident & instr, Vet Admin Hosp, Iowa City, 52; from asst prof to assoc prof, 54-63, PROF ANESTHESIOL, SCH MED, UNIV NC, CHAPEL HILL, 63-, CHMN DEPT, 69- Concurrent Pos: NIH spec res fel, 62; consult, Vet Admin Hosp, Fayetteville, NC; vis prof, Inst Physiol, Univ Göttingen, 62 & Med Sch, King's Col, Univ London, 63. Mem: AAAS; Am Soc Anesthesiol; Soc Exp Biol & Med; Am Physiol Soc; Asn Univ Anesthetists. Res: Application of electronic instrumentation to physiological measurements; electrochemical methods of biological analysis; cation sensitive glass electrodes; respiratory physiology. Mailing Add: Dept of Anesthesiol Univ of NC Sch of Med Chapel Hill NC 27514

SUGITA, EDWIN T, b Honolulu, Hawaii, Feb 1, 37; m 59; c 2. PHARMACEUTICS. Educ: Purdue Univ, BS, 59, MS, 62, PhD(pharmaceut), 63. Prof Exp: From asst prof to assoc prof, 64-72, PROF PHARM, PHILADELPHIA COL PHARM & SCI, 72- Concurrent Pos: Res grant, Smith, Kline & French Labs, 65-67; NIH grant, 65-71 & 74-77. Mem: Acad Pharmaceut Sci; Sigma Xi. Res: Mild acid catalysis of penicillins; immunogenic properties of degradation products of penicillins; gastrointestinal absorption of charged drugs; drug dosing regimens in humans. Mailing Add: Philadelphia Col of Pharm & Sci 43rd St & Kingsessing Ave Philadelphia PA 19104

SUGIURA, MASAHISA, b Tokyo, Japan, Dec 8, 25; m 62; c 1. SPACE PHYSICS. Educ: Univ Tokyo, MS, 49; Univ Alaska, PhD, 55. Prof Exp: Asst prof geophys res, Geophys Inst, Univ Alaska, 55-57, assoc prof geophys, 57-62, prof, 62; Nat Acad Sci sr assoc, NASA, 62-64, MEM STAFF, GODDARD SPACE FLIGHT CTR, 64- Concurrent Pos: Gugge nheim fel, 59; prof, Univ Wash, 66-67. Mem: Am Geophys Union; Am Phys Soc. Res: Magnetospheric physics; geophysics. Mailing Add: Lab for Planetary Atmospheres NASA Goddard Space Flight Ctr Greenbelt MD 20771

SUGIYAMA, HIROSHI, b Alameda, Calif, Nov 10, 16; m 51; c 2. BACTERIOLOGY, PHARMACOLOGY. Educ: Univ Calif, AB, 39; Univ Chicago, PhD(bact), 50. Prof Exp: Res assoc, Food Res Inst, Univ Chicago, 51-54, from asst prof to assoc prof, Univ, 55-66; PROF BACT, FOOD RES INST, UNIV WIS-MADISON, 66- Mem: Am Soc Microbiol; Am Acad Microbiol; Soc Exp Biol & Med. Res: Bacterial food poisoning; clostridium botulinum toxin. Mailing Add: Food Res Inst Univ of Wis Madison WI 53706

SUH, JOHN TAIYOUNG, b Seoul, Korea, Aug 9, 27; US citizen; m 58; c 4. MEDICINAL CHEMISTRY, ORGANIC CHEMISTRY. Educ: Butler Univ, BS, 53; Univ Wis, MS, 56, PhD(med chem, org chem), 58. Prof Exp: Teaching asst, Univ Wis, 55; sr res chemist, Res Ctr, Johnson & Johnson, 55-58 & McNeil Labs, Inc, 58-63; group leader res, 63-65, SECT HEAD RES, DEPT MED CHEM, LAKESIDE LABS DIV, COLGATE-PALMOLIVE CO, 65- Concurrent Pos: Vis lectr chem, Marquette Univ, 72- Mem: Am Chem Soc. Res: Medicinal chemistry in areas of cardiovascular, psychopharmacological and hematinic agents; organic chemistry in areas of stereochemistry, natural products, heterocyclic and organometallic chemistry; pulmonary and allergy research. Mailing Add: 3709 W Scenic Ave Mequon WI 53092

SUH, TAE-IL, b Chungdo, Korea, June 1, 28; m 55; c 3. ALGEBRA. Educ: Kyung-Pook Nat Univ, Korea, BS, 52; Yale Univ, PhD(math), 61. Prof Exp: Asst prof math, Kyung-Pook Nat Univ, Korea, 61-63; assoc prof, Sogang Univ, Korea, 63-65; assoc prof, 65-68, PROF MATH, E TENN STATE UNIV, 68- Mem: Am Math Soc; Math Asn Am; Math Soc Japan. Res: Non-associative algebras. Mailing Add: Dept of Math ETenn State Univ 2648 University Sta Johnson City TN 37601

SUHADOLNIK, ROBERT J, b Forest City, Pa, Aug 15, 25; m 49; c 5. BIOCHEMISTRY. Educ: Pa State Univ, BS, 49, PhD, 56; Iowa State Univ, MS, 53. Prof Exp: Res assoc biochem, Univ Ill, 56-57; asst prof, Okla State Univ, 57-61; res mem, Albert Einstein Med Ctr, 61-70, head dept bio-org chem, 68-70; PROF BIOCHEM & RES PROF MICROBIOL, SCH MED, TEMPLE UNIV, 70- Mem: Am Chem Soc; Am Soc Biol Chemists. Res: Alkaloid biogenesis; metabolism of allose and allulose; biosynthesis and biochemical properties of nucleoside antibiotics. Mailing Add: Dept of Biochem Temple Univ Sch of Med Philadelphia PA 19140

SUHAYDA, JOSEPH NICHOLAS, b Flint, Mich, Feb 23, 44; m 66; c 2. OCEANOGRAPHY. Educ: Calif State Univ, Northridge, BS, 66; Univ Calif, San Diego, PhD(phys oceanog), 72. Prof Exp: Asst prof, 72-75, ASSOC PROF MARINE SCI, COASTAL STUDIES INST, LA STATE UNIV, BATON ROUGE, 75- Mem: Am Geophys Union; Am Shore & Beach Preserv Asn. Res: Coastal oceanography, primarily nearshore processes on beaches and reefs, and the influence of storm waves on sediment on the continental shelf. Mailing Add: Coastal Studies Inst La State Univ Baton Rouge LA 70803

SUHL, HARRY, b Leipzig, Ger, Oct 18, 22; nat US; wid. PHYSICS. Educ: Univ Wales, BSc, 43; Oxford Univ, PhD(theoret physics), 48. Prof Exp: Exp officer, Admiralty Signal Estab, Eng, 43-46; mem tech staff, Bell Tel Labs, Inc, 48-60; vis lectr, 60, PROF PHYSICS, UNIV CALIF, SAN DIEGO, 61- Mem: Nat Acad Sci; Fel Am Phys Soc. Res: Theoretical solid state physics. Mailing Add: Dept of Physics Univ of Calif at San Diego La Jolla CA 92037

SUHM, RAYMOND WALTER, b Springfield, Mass, June 9, 41; m 64; c 1. STRATIGRAPHY. Educ: Southeast Mo State Col, BS, 63; Southern Ill Univ, Carbondale, MS, 65; Univ Nebr-Lincoln, PhD(geol), 70. Prof Exp: Instr geol, Southern Ill Univ, Carbondale, 65; geophysicist, Humble Oil Co, Calif, 65-67; teaching asst geol, Univ Nebr-Lincoln, 67-69; asst prof, 70-75, ASSOC PROF GEOL, TEX A&I UNIV, 75- Concurrent Pos: Consult, Cockrell Corp, 72. Mem: Am Asn Petrol Geologists; Soc Econ Paleontologists & Mineralogists; assoc Sigma Xi. Res: Ordovician stratigraphy and paleontology; Ozark geology; coastal sedimentation and geomorphology; historical geology. Mailing Add: Dept of Geog & Geol Tex A&I Univ Kingsville TX 78363

SUHOVECKY, ALBERT J, b Youngstown, Ohio, May 3, 26; m 51; c 3. PLANT PATHOLOGY. Educ: Western Reserve Univ, BS, 50; Kent State Univ, MA, 51; Ohio State Univ, PhD(agron), 55. Prof Exp: Res asst field pesticides, Res Found, Ohio State Univ, 52-53; res asst exten field crops, Ohio Agr Exp Sta, 53-55; plant pathologist, Monsanto Chem Co, 55-60; sr cereal chemist, 60-70, ASST QUAL CONTROL DIR, FALSTAFF BREWING CORP, 70- Concurrent Pos: Mem comt hops res, US Brewers Asn & mem tech comt, Malting Barley Improv Asn, 61- Mem: Am Asn Cereal Chemists; Inst Sanit Mgt; Entom Soc Am; Am Soc Brewing Chemists; Am Soc Agron. Res: Development and evaluation of organic chemicals as agricultural fungicides; quality control of barley, malt and adjuncts; sanitation control. Mailing Add: 50 St Eugene Lane Florissant MO 63033

SUHR, NORMAN HENRY, b Chicago, Ill, June 13, 30; m 53; c 4. SPECTROSCOPY, GEOCHEMISTRY. Educ: Univ Chicago, AB, 50, MS, 54. Prof Exp: Spectroscopist & mineralogist, Heavy Minerals Co, Vitro Corp Am, 56-58; spectroscopist, Labs, 58-65, asst dir, 65-70, res assoc, Univ, 63-67, asst prof, 67-69, ASSOC PROF GEOCHEM, PA STATE UNIV, 69-, DIR MINERAL CONST LABS, 70- Mem: AAAS; Soc Appl Spectros; Geochem Soc. Res: X-ray and emission spectroscopy and atomic absorption, primarily in the fields of earth sciences. Mailing Add: Mineral Const Labs 311 Mineral Sci Bldg Pa State Univ University Park PA 16802

SUHRLAND, LEIF GEORGE, b Schroon Lake, NY, Apr 9, 19; m 50; c 3. HEMATOLOGY, ONCOLOGY. Educ: Cornell Univ, BS, 42; Univ Rochester, MD, 50. Prof Exp: Bacteriologist, USPHS, 42-43; intern & jr asst med, Univ Hosps, Cleveland, 50-54; instr, Western Reserve Univ, 57-59, asst prof med & asst clin pathologist, 59-67; PROF MED & VCHMN DEPT, COL HUMN MED, MICH STATE UNIV, 67- Concurrent Pos: Am Cancer Soc fel, Univ Hosps, Cleveland, 54-56; Howard M Hanna & Anna Bishop fels, Sch Med, Western Reserve Univ, 56-57. Mem: Am Fedn Clin Res; Am Soc Hemat. Res: Host tumor relationships. Mailing Add: B-220 Life Sci Bldg Mich State Univ Col of Human Med East Lansing MI 48823

SUICH, JOHN EDWARD, b Bridgeport, Conn, Sept 28, 36; m 57; c 2. INFORMATION SCIENCE. Educ: Harvard Univ, BA, 58; Mass Inst Technol, PhD(nuclear eng), 63. Prof Exp: Sr physicist, Savannah River Lab, 63-65, res supvr, 65-66, res mgr appl math, 66-68, dir comput sci sect, 68-71, mgr telecommun planning, Gen Servs Dept, 71-72, asst mgr comput sci div, Cent Systs & Servs Dept, Del, 72, mgr com systs div mgt sci, 72-75, RES ASSOC, SAVANNAH RIVER LAB, E I DU PONT DE NEMOURS & CO, INC, 75- Res: Geophysical data bases and numerical transport models. Mailing Add: Savannah River Lab E I du Pont de Nemours & Co Inc Aiken SC 29801

SUICH, RONALD CHARLES, b Cleveland, Ohio, Nov 16, 40; m 62; c 3. STATISTICS. Educ: John Carroll Univ, BSBA, 62; Case Western Reserve Univ, MS, 64, PhD(statist), 68. Prof Exp: Mkt researcher, Cleveland Elec Illum Co, 62-64; from instr to asst prof statist, Case Western Reserve Univ, 64-70; ASST PROF MATH, UNIV AKRON, 70- Mem: Am Statist Asn. Res: Sequential tests. Mailing Add: Dept of Math Univ of Akron Akron OH 44325

SUIE, TED, b Akron, Ohio, June 20, 23. MICROBIOLOGY, IMMUNOLOGY. Educ: Univ Akron, BSc, 48; Ohio State Univ, MSc, 49, PhD(bacteriol), 53. Prof Exp: From instr to assoc prof, 53-72, PROF OPHTHAL, OHIO STATE UNIV, 72- Mem: Am Acad Ophthal & Otolaryngol; hon mem Am Soc Mil Ophthal. Res: Infections of the eye and immunology of the eye. Mailing Add: Dept of Ophthal Ohio State Univ Columbus OH 43210

SUIT, HERMAN DAY, b Houston, Tex, Feb 8, 29. RADIOTHERAPY. Educ: Univ Houston, AB, 48; Baylor Univ, SM & MD, 52; Oxford Univ, DrPhil(radiobiol), 56. Prof Exp: Intern, Jefferson Davis Hosp, Houston, Tex, 52-53, resident radiol, 53-54; house surgeon radiother, Churchill Hosp, Oxford, Eng, 54, res asst radiobiol lab, 54-56, registr radiother, 56-57; sr asst surgeon, Radiation Br, Nat Cancer Inst, 57-59; asst radiotherapist, Univ Tex M D Anderson Hosp & Tumor Inst Houston, 59-63, assoc radiotherapist, 63-68, radiotherapist, 68-71, chief sect exp radiother, 62-70; PROF RADIATION THER, HARVARD MED SCH, 70-; HEAD DEPT RADIATION MED, MASS GEN HOSP, 71- Concurrent Pos: Nat Cancer Inst res career develop award, Univ Tex M D Anderson Hosp & Tumor Inst Houston, 64-68; gen fac assoc, Univ Tex Grad Sch Biomed Sci, 65-70, prof radiation ther, 68-71; staff mem, NASA Manned Spacecraft Ctr, 69-71; subcomt radiation biol, Nat Acad Sci. Mem: AAAS; Am Col Radiol; Am Soc Therapeut Radiol (secy, 70-72); AMA; Am Asn Cancer Res. Mailing Add: Dept of Radiation Med Mass Gen Hosp Boston MA 02114

SUIT, JOAN C, b Ontario, Ore, Apr 14, 31; m 60. MICROBIOLOGY. Educ: Ore State Col, BS, 53; Stanford Univ, MA, 55, PhD(med microbiol), 57. Prof Exp: Res assoc biochem, Biol Div, Oak Ridge Nat Lab, 57-59; res assoc sect molecular biol, Univ Tex M D Anderson Hosp & Tumor Inst, Houston, 59-66, assoc biologist & assoc prof biol, Univ Tex Grad Sch Biomed Sci, Houston, 66-73; RES ASSOC BIOL, MASS INST TECHNOL, 73- Mem: Am Soc Microbiol; Am Soc Cell Biol. Res: Microbial genetics; DNA replication; microbial growth. Mailing Add: Dept of Biol Rm 56-417A Mass Inst of Technol Cambridge MA 02139

SUITS, CHAUNCEY GUY, b Oshkosh, Wis, Mar 12, 05; m 31; c 2. PHYSICS. Educ: Univ Wis, AB, 27; Swiss Fed Inst Technol, DSc, 29. Hon Degrees: DSc, Union Col, NY, 44, Hamilton Col, 46, Drexel Inst Technol, 55 & Marquette Univ, 59; DEng, Rensselaer Polytech Inst, 50. Prof Exp: Consult physics, Forest Prod Lab, US Forest Serv, Wis, 29-30; res physicist, Gen Elec Co, 30-40, asst to dir res, 40-45, vpres & dir res, 45-65; CONSULT INDUST RES MGT, 65- Concurrent Pos: Fel, Inst Int Educ Mem Div 14 & chief div 15, Nat Defense Res Comt, Off Sci Res & Develop, 42-46; mem sci adv bd, Nat Security Agency; mem ord comt, Res & Develop Bd, US Dept Defense, 49-50, mem comt electronics, 52-54, mem tech adv panel electronics, 54-; mem spec tech adv group, Joint Chiefs of Staff & Res & Develop Bd, 50-53; chmn, Naval Res Adv Comt, 58-61; chmn adv comt corp assocs, Am Inst Physics, 58-62, dir-at-large gov bd, 61-64; mem bd dirs, NY State Sci & Technol Found, 65-67, vchmn, 68-; mem res mgt adv panel, US House of Rep Comt Sci & Astronaut. Honors & Awards: Presidential Medal for Merit & King's Medal Serv in Cause of Freedom, Gt Brit, 48; Proctor Prize, Sigma Xi, 58; Distinguished Serv Award, Am Mgt Asn, 59; Medal, Indust Res Inst, 62; Charles M Schwab Mem Lectr, Am Iron & steel Inst, 63; Medal for Advan Res, Am Soc Metals, 66; Frederik Philips Award, Inst Elec & Electronics Engrs, 74. Mem: Nat Acad Sci; Nat Acad Eng; Sigma Xi; fel Am Phys Soc; Am Mgt Asn (vpres, 56). Res: Nonlinear electronic circuits; high pressure arcs. Mailing Add: Crosswinds Pilot Knob NY 12844

SUITS, GWYNN HALYBURTON, physics, see 12th edition

SUITS, JAMES CARR, b Schenectady, NY, May 29, 32; m 54; c 3. PHYSICS. Educ: Yale Univ, BS, 54; Harvard Univ, PhD(appl physics), 60. Prof Exp: STAFF PHYSICIST, RES LAB, IBM CORP, SAN JOSE, 60- Mem: Fel Am Phys Soc; Inst Elec & Electronics Engrs. Res: Magnetism, ultra-high vacuum evaporated thin films, magneto-optics, discovery and development of novel magnetic materials, garnet film growth. Mailing Add: IBM Res Lab Monterey & Cottle Rd San Jose CA 95193

SUJISHI, SEI, b San Pedro, Calif, Nov 9, 21; m 55; c 1. INORGANIC CHEMISTRY. Educ: Wayne State Univ, BS, 46, MS, 48; Purdue Univ, PhD(chem), 49. Prof Exp: From instr to assoc prof chem, Ill Inst Technol, 49-59; assoc prof, 59-65, PROF CHEM, STATE UNIV NY STONY BROOK, 65-, CHMN DEPT, 74- Mem: Am Chem Soc. Res: Chemistry of silicon and germanium hydrides. Mailing Add: Dept of Chem State Univ of NY Stony Brook NY 11790

SUK, WADI NAGIB, b Khartoum, Sudan, Oct 26, 34; US citizen; m 67; c 4. INTERNAL MEDICINE, NEPHROLOGY. Educ: Am Univ Beirut, BS, 55, MD, 59. Prof Exp: Resident internal med, Parkland Mem Hosp, Dallas, 61-63; from instr to assoc prof, 65-71, PROF INTERNAL MED, BAYLOR COL MED, 71-, CHIEF RENAL SECT, 68- Concurrent Pos: Res fel exp med, Univ Tex Southwestern Med Sch Dallas, 59-61, USPHS res fel nephrology, 63-65, Dallas Heart Asn res grant, 67-68; Nat Heart & Lung Inst res grant, Baylor Col Med, 68-72, training grant, 71-76, Nat Inst Arthritis, Metab & Digestive Dis res grant, 74-77, Nat Inst Allergy & Infectious Dis contract, 74-78, Nat Aeronaut & Space Admin contract, 75; attend physician, Ben Taub Gen Hosp, 68-; consult, Vet Admin Hosp, 68- & Wilford Hall, USAF Med Ctr, 72-; chief renal sect, Methodist Hosp, 69-; pres med adv bd, Kidney Found Houston & Greater Gulf Coast, 69-71; chmn nat med adv coun, Nat Kidney Found, 71-73, trustee-at-large, 71-; mem exec comt, Coun Kidney in Cardiovasc Dis, Am Heart Asn, 71-74; mem gen med B study sect, NIH, 75- Mem: Am Fedn Clin Res; Am Soc Clin Invest; Am Physiol Soc; fel Am Col Physicians; Am Soc Nephrology. Res: Renal, fluid and electrolyte physiology and pathophysiology; renal disease, dialysis and transplantation. Mailing Add: Renal Sect Dept of Med Baylor Col of Med Houston TX 77025

SUKAVA, ARMAS JOHN, b Elma, Man, Mar 1, 17; m 50; c 2. PHYSICAL CHEMISTRY. Educ: Univ Man, BSc, 46, MSc, 49; McGill Univ, PhD, 55. Prof Exp: Lectr chem, Univ Man, 47-49; res & develop chemist, Consol Mining & Smelting Co, 49-50; lectr chem, Univ BC, 50-51 & Univ Alta, 51-52; instr, 54-55, lectr, 55-56, asst prof, 56-59, ASSOC PROF CHEM, UNIV WESTERN ONT, 59- Mem: Electrochem Soc; Chem Inst Can. Res: Cathode overpotential and surface-active additives; physicochemical properties of electrolyte systems. Mailing Add: Dept of Chem Univ of Western Ont London ON Can

SUKER, JACOB ROBERT, b Chicago, Ill, Oct 17, 26; m 56; c 3. INTERNAL MEDICINE. Educ: Northwestern Univ, Evanston, BS, 47; Northwestern Univ, Chicago, MS, 54, MD, 56. Prof Exp: Dir med educ, Chicago Wesley Mem Hosp, 64-68; asst dean grad educ, 68-70, ASSOC PROF MED, MED SCH, NORTHWESTERN UNIV, CHICAGO, 68-, ASSOC DEAN GRAD EDUC, 70- Concurrent Pos: Assoc attend physician, Chicago Wesley Mem Hosp, 64- Res: Parathyroid disease. Mailing Add: Dept of Med Northwestern Univ Med Sch Chicago IL 60611

SUKHATME, BALKRISHNA VASUDEO, b Poona, India, Nov 3, 24; m 56; c 1. STATISTICS. Educ: Univ Delhi, BA, 45, MA, 47; Inst Agr Res Statist, New Delhi, dipl, 49; Univ Calif, Berkeley, PhD(statist), 55. Prof Exp: Sr res statistician, Indian Coun Agr Res, 55-58, prof statist, 58-65, dep statist adv, 62-63, sr prof statist, 65-67; assoc prof, 67-68, PROF STATIST, IOWA STATE UNIV, 68- Concurrent Pos: Vis assoc prof statist, Mich State Univ, 59-60; ed jour, Indian Soc Agr Statist, 59-67; consult, FAO, Rome, 65; mem, Int Statist Inst, The Hague, 72- Mem: Inst Math Statist; fel Am Statist Asn; Int Asn Surv Statisticians; Indian Soc Agr Statist (joint secy, 56-58). Res: Sampling theory and its applications; nonparametric tests for scale and randomness and asymptotic theory of order statistics and generalized U-statistics; planning, organization and conduct of large-scale sample surveys. Mailing Add: Dept of Statist Iowa State Univ Ames IA 50011

SUKHATME, SHASHIKALA BALKRISHNA, b Karad, Maharashtra, India; c 1. MATHEMATICAL STATISTICS. Educ: Univ Poona, BSc, 53, Hons, 54, MSc, 55; Mich State Univ, PhD(statist), 60. Prof Exp: Lectr statist, Univ Delhi, 63-67; ASST PROF STATIST, IOWA STATE UNIV, 67- Concurrent Pos: Univ Grants Comn, India Fel, Univ Delhi, 61-63. Mem: Inst Math Statist. Res: Nonparametric statistical theory. Mailing Add: Statist Lab Iowa State Univ Ames IA 50010

SUKMAN, EDWIN LOUIS, organic chemistry, see 12th edition

SUKOW, WAYNE WILLIAM, b Merrill, Wis, Dec 9, 36; m 59; c 2. MOLECULAR BIOPHYSICS, BIOPHYSICAL CHEMISTRY. Educ: Univ Wis-River Falls, BA, 59; Case Inst Technol, MS, 63; Wash State Univ, PhD(chem physics), 74. Prof Exp: ASSOC PROF PHYSICS, UNIV WIS- RIVER FALLS, 61- Concurrent Pos: Vis prof physics, Macalester Col, 67; physicist, 3M Co, 67; NSF sci fac fel, Wash State Univ, 70-72. Mem: Biophys Soc; Am Asn Physics Teachers. Res: Protein-ligand binding, particularly the mechanism of detergent binding to membrane proteins; conformational changes of proteins monitored by electron paramagnetic resonance using spin probe molecules; photoelectron microscopy of biological materials. Mailing Add: Dept of Physics Univ of Wis River Falls WI 54022

SUKOWSKI, ERNEST JOHN, b Chicago, Ill, Nov 17, 32; div; c 3. PHYSIOLOGY, PHARMACOLOGY. Educ: Loyola Univ Chicago, BS, 54; Univ Ill, MS, 58, PhD(physiol, pharmacol), 62. Prof Exp: ASSOC PROF PHYSIOL, CHICAGO MED SCH, 63- Mem: NY Acad Sci; Am Physiol Soc; Am Heart Asn. Res: Cardiac and liver metabolism; cardiovascular physiology; hypertension; sub-cellular physiology. Mailing Add: Dept of Physiol & Biophys Chicago Med Sch 2020 W Ogden Ave Chicago IL 60612

SULAK, LAWRENCE RICHARD, b Columbus, Ohio, Aug 29, 44; m 70. ELEMENTARY PARTICLE PHYSICS, EXPERIMENTAL HIGH ENERGY PHYSICS. Educ: Carnegie-Mellon Univ, BS, 66; Princeton Univ, AM, 68, PhD(physics), 70. Prof Exp: Res physicist, Univ Geneva, 70-71; asst prof, 72-75, ASSOC PROF PHYSICS, HARVARD UNIV, 75- Concurrent Pos: Vis scientist, Europ Orgn Nuclear Res, 70-71; vis physicist, Fermi Nat Acceleratory Lab, 71-; guest assoc physicist, Brookhaven Nat Lab, 74- Mem: Am Phys Soc. Res: Experimental K-meson physics; neutrino physics, particularly deep inelastic scattering, tests of scaling and studies of neutral currents and dimuons. Mailing Add: 252 Jefferson Lab Harvard Univ Cambridge MA 02138

SULAKHE, PRAKASH VINAYAK, b Nov 18, 41; Indian citizen; m 73. PHYSIOLOGY. Educ: Bombay Univ, BS, 62, MS, 65; Univ Man, PhD(physiol), 71. Prof Exp: Lectr physiol, Topiwala Nat Med Col, India, 66; sci officer med div, Bhaha Atomic Res Ctr, India, 66-67; demonstr & teaching fel physiol, Univ Man, 68-71; Med Res Coun Can fel pharmacol, Univ BC, 71-73; ASST PROF PHYSIOL, UNIV SASK, 73- Mem: Can Physiol Soc; Int Study Group Res Cardiac Metab & Struct; NY Acad Sci; AAAS. Res: Regulation and metabolism and function of contractile tissues and brain; cyclic nucleotides, calcium ions and membranes. Mailing Add: Dept of Physiol Col of Med Univ of Sask Saskatoon SK Can

SULAVIK, STEPHEN B, b New Britain, Conn, Aug 11, 30; m 55; c 8. MEDICINE. Educ: Providence Col, BS, 52; Georgetown Univ, MD, 56. Prof Exp: Asst chief chest dis, Vet Admin Hosp, Bronx, NY, 61-62; clin instr, 62-63, from instr to asst prof, 63-69, ASSOC CLIN PROF MED, SCH MED, YALE UNIV & ASSOC PROF MED & ACTG HEAD PULMONARY DIV, UNIV CONN, 69-; CHMN DEPT MED, ST FRANCIS HOSP, HARTFORD, 69- Concurrent Pos: Mem med adv comt, Dept Health, Educ & Welfare. Mem: Am Thoracic Soc; AMA. Res: Anatomy and physiology of intrathoracic lymphatic system. Mailing Add: St Francis Hosp 114 Woodland St Hartford CT 06105

SULENTIC, JACK WILLIAM, b Waterloo, Iowa, Apr 10, 47; m 75. ASTRONOMY. Educ: Univ Ariz, BS, 69; State Univ NY Albany, PhD(astron), 75. Prof Exp: HALE OBSERV FEL ASTRON, HALE OBSERV, 75- Mem: Am Astron Soc; Royal Astron Soc. Res: Application of optical and radio observations to understanding the origin and evolution of galaxies. Mailing Add: Hale Observ 813 Santa Barbara St Pasadena CA 91101

SULERUD, RALPH L, b Fargo, NDak, June 6, 32. GENETICS, ZOOLOGY. Educ: Concordia Col, Moorhead, Minn, BA, 54; Univ Nebr, MS, 58, PhD(zool), 68. Prof Exp: Instr biol, St Olaf Col, 58-59; from instr to asst prof, 64-68, ASSOC PROF BIOL, AUGSBURG COL, 68-, CHMN DEPT, 74- Mem: AAAS; Genetics Soc Am; Am Genetic Asn; Am Inst Biol Sci; Soc Study Evolution. Res: Taxonomy; genetics of Drosophila. Mailing Add: Dept of Biol Augsburg Col 8th St at 21st Ave S Minneapolis MN 55404

SULING, WILLIAM JOHN, b New York, NY, June 12, 40; m 65; c 3. MEDICAL MICROBIOLOGY, CHEMOTHERAPY. Educ: Manhattan Col, BS, 62; Duquesne Univ, MS, 65; Cornell Univ, PhD(microbiol), 75. Prof Exp: Res asst cancer chemother, Sloan-Kettering Inst Cancer Res, 64-70; sr bacteriologist, Biol Lab, Mass Dept Pub Health, 74-75; RES MICROBIOLOGIST, SOUTHERN RES INST, 75- Mem: NY Acad Sci; Am Soc Microbiol. Res: Folate metabolism and its inhibition by folate analogues; the biochemistry of antimicrobial drug resistance and the use of microorganisms for studies involving cancer chemotherapy. Mailing Add: Southern Res Inst 2000 Ninth Ave S Birmingham AL 35205

SULKIN, NORMAN M, anatomy, deceased

SULKIN, SIMON EDWARD, microbiology, see 12th edition

SULKOWSKI, EUGENE, b Plonsk, Poland, May 22, 34; m 64; c 2. BIOCHEMISTRY. Educ: Univ Warsaw, MS, 56, PhD(biochem), 60. Prof Exp: Res asst biochem, Inst Biochem & Biophys, Polish Acad Sci, 56-60; exchange scientist, Univ Sorbonne, 62-63; res asst, Polish Acad Sci, 63-65; ASSOC CANCER RES SCIENTIST, ROSWELL PARK MEM INST, 65- Concurrent Pos: Res fel, Marquette Univ, 60-62. Mem: AAAS; Am Soc Biol Chemists; Polish Biochem Soc. Res: Enzymology; human interferon. Mailing Add: Med Viral Oncol Roswell Park Mem Inst 666 Elm St Buffalo NY 14263

SULKOWSKI, THEODORE SYLVESTER, organic chemistry, see 12th edition

SULLENGER, DON BRUCE, b Richmond, Mo, Feb 8, 29; m 64; c 3. SOLID STATE CHEMISTRY. Educ: Univ Colo, AB, 50; Cornell Univ, PhD, 69. Prof Exp: Trainee Chemet Prog, Chem Div, Gen Elec Co, 50-51 & 53, res chemist, Res Lab, 53-54; SR RES CHEMIST, MOUND LAB, MONSANTO RES CORP, 62- Mem: AAAS; Am Inst Chemists; Am Chem Soc; Am Crystallog Asn. Res: X-ray crystallographic structure determination of inorganic substances, principally theory of structure determinations via the direct methods; solid state chemistry of inorganic materials. Mailing Add: Monsanto Res Corp Mound Lab Miamisburg OH 45342

SULLIVAN, ALFRED DEWITT, b New Orleans, La, Feb 2, 42; m 62; c 2. FORESTRY, BIOMETRICS. Educ: La State Univ, BS, 64, MS, 66; Univ Ga, PhD(forest biomet), 69. Prof Exp: Asst prof statist, Va Polytech Inst & State Univ, 69-73; ASSOC PROF FORESTRY, MISS STATE UNIV, 73- Mem: Soc Am Foresters; Biomet Soc. Res: Prediction of forest growth and yield; application of statistical methodology to natural resource problems. Mailing Add: Dept of Forestry PO Drawer FD Miss State Univ Mississippi State MS 39762

SULLIVAN, ANDREW JACKSON, b Birmingham, Ala, Mar 3, 26; m 53. BIOCHEMISTRY, FOOD CHEMISTRY. Educ: Univ Richmond, BS, 47; Univ Mo, PhD(bot), 52. Prof Exp: USPHS fel, Univ Pa, 52-53; res chemist, Campbell Soup Co, 53-55 & 57-59, head div flavor biochem res, 59-71, div head environ sci & chem technol, Campbell Inst for Food Res, 71-76, DIR SCI RESOURCES, CAMPBELL INST FOR FOOD RES, 76- Mem: Inst Food Technologists; NY Acad Sci; fel Am Inst Chemists. Res: Chemistry of microorganisms; food and flavor chemistry. Mailing Add: Campbell Inst for Food Res Campbell Place Camden NJ 08101

SULLIVAN, ANN CLARE, b Tillamook, Ore, June 3, 43; m 68. BIOCHEMISTRY. Educ: Col Notre Dame Md, BA, 65; Northwestern Univ, MS, 67; NY Univ, PhD(biochem), 73. Prof Exp: Res assoc, Sci & Eng Inc, 66-68; RES GROUP CHIEF, HOFFMANN-LA ROCHE INC, 69- Concurrent Pos: Lectr nutrit, Sch Med, Columbia Univ, 76. Mem: NY Acad Sci; AAAS; Am Chem Soc; Am Oil Chemists Soc; Asn Study Obesity. Res: Regulation of lipid and carbohydrate metabolism, control of appetite and energy balance, and metabolic aspects of obesity and hyperlipidemia. Mailing Add: Hoffmann-La Roche Inc Dept of Biochem Nutrit Nutley NJ 07110

SULLIVAN, BETTY, b Minneapolis, Minn, May 31, 02. BIOCHEMISTRY. Educ: Univ Minn, BS, 22, PhD(biochem), 35. Prof Exp: Asst chemist, Russell Miller Milling Co, 22-24, chief chemist, 27-47, vpres & dir res, 47-55, vpres & dir res, Peavey Co Flour Mills, 55-67; from vpres to pres, 67-73, V CHMN BD, EXPERIENCE, INC, 73- Honors & Awards: Thomas Burr Osborne Award, Am Asn Cereal Chemists, 48; Garvan Medal, Am Chem Soc, 54. Mem: AAAS; Am Chem Soc; Am Asn Cereal Chemists (vpres, 43, pres, 44). Res: Chemistry of wheat and flour; fermentation. Mailing Add: 4825 Queen Ave S Minneapolis MN 55410

SULLIVAN, CHARLES IRVING, b Milwaukee, Wis, Nov 18, 18; m 48; c 2. ORGANIC POLYMER CHEMISTRY. Educ: Boston Univ, AB, 43. Prof Exp: Chemist, UBS Chem Co Div, A E Staley Mfg Co, 43-46, supvr indust chem & develop sect, 46-58, mgr res & develop, 58-67, res assoc, 67-69; sr scientist, 69-70, RES ASSOC, POLAROID CORP, 70- Mem: AAAS; Am Chem Soc; fel Am Inst Chemists. Res: Emulsion polymerization; paints; floor finishes; paper coatings and binders; wood coatings; adhesives; textile backings; aqueous polymer research and development related to membrane-like structures, functional coatings, binders and colloids. Mailing Add: 148 Bellevue Ave Melrose MA 02176

SULLIVAN, CHARLES RAYMOND, b Los Angeles, Calif, Sept 27, 34. PHYSICS. Educ: Univ Tex, El Paso, BS, 60; Vanderbilt Univ, PhD(physics), 66. Prof Exp: Res assoc physics, Vanderbilt Univ, 66-67; res assoc, 67-68, ASST PROF PHYSICS, CASE WESTERN RESERVE UNIV, 68- Mem: Am Phys Soc. Res: High energy experimental physics. Mailing Add: Dept of Physics Case Western Reserve Univ Cleveland OH 44106

SULLIVAN, CHARLOTTE MURDOCH, b St Stephen, NB, Dec 18, 19. ANIMAL PHYSIOLOGY. Educ: Dalhousie Univ, BSc, 41, MSc, 43; Univ Toronto, PhD(zool), 49. Prof Exp: Nat Res Coun Can overseas fel, Cambridge Univ, 49-50; from lectr to asst prof, 50-61, ASSOC PROF ZOOL, UNIV TORONTO, 61- Res: Physiology of animal behavior. Mailing Add: Dept of Zool Univ of Toronto Toronto ON Can

SULLIVAN, DAN ALLEN, b Cairo, Ill, Sept 26, 33; m 63; c 2. GEOLOGY. Educ: Northwestern Univ, BA, 55; Wash Univ, MA, 60, PhD(geol), 66. Prof Exp: Asst prof eng geol, Univ Valle, Colombia, 66-67; assoc prof geol, Murray State Univ, 67-68; ASSOC PROF GEOL, DePAUW UNIV, 68- Concurrent Pos: Rockefeller Found grant univ develop in Latin Am & consult, City of Cali, Colombia & Corp Autonoma del Valle del Cauca, 66-67; consult opers geologist, Bolt Beranek & Newman, Inc, 73-75. Mem: Am Inst Mining, Metall & Petrol Engrs; NY Acad Sci; Am Asn Petrol Geologists; Am Inst Prof Geologists. Res: Environmental geology; stratigraphy; sedimentation; Pacific-margin tertiary basin, Gulf of Alaska. Mailing Add: Dept of Earth Sci DePauw Univ Sci Ctr Greencastle IN 46135

SULLIVAN, DANIEL JOSEPH, b New York, NY, Apr 22, 28. ENTOMOLOGY, ANIMAL BEHAVIOR. Educ: Fordham Univ, BS, 50, MS, 58; Univ Vienna, cert Ger, 58; Univ Strasbourg, cert French, 62; Univ Innsbruck, cert theol, 62; Univ Calif, Berkeley, PhD(entom), 69. Prof Exp: Teacher, NY High Sch, 55-57; asst prof, 69-74, ASSOC PROF ZOOL, FORDHAM UNIV, 74- Mem: AAAS; Entom Soc Am; Ecol Soc Am; Animal Behav Soc; Am Inst Biol Sci. Res: Biological control of insect pests, with special reference to the primary parasites and hyperparasites of aphids; ecology and behavior of aphids and parasites. Mailing Add: Dept of Biol Sci Fordham Univ Bronx NY 10458

SULLIVAN, DANIEL RICHARD, b Chicago, Ill, Nov 27, 41; m 68; c 2. ORGANIC CHEMISTRY, LIPID CHEMISTRY. Educ: St Mary's Col, Minn, AB, 63; Xavier Univ, Ohio, MS, 70; Loyola Univ, Chicago, PhD(chem), 73. Prof Exp: SCIENTIST LIPID CHEM, CENT SOYA CO INC, 73- Mem: Am Chem Soc; Sigma Xi; Am Oil Chemists Soc. Res: Research and development of edible and industrial vegetable oils and emulsifiers. Mailing Add: Cent Soya Co Inc 1825 N Laramie Ave Chicago IL 60639

SULLIVAN, DARRELL THORNTON, b Steele, Mo, Jan 4, 17; m 42; c 3. HORTICULTURE. Educ: Miss State Col, BS, 39; Univ Ga, MS, 40; Ohio State Univ, PhD(hort), 54. Prof Exp: Asst prof hort, Univ Ga, 40-42, 46-51; PROF HORT, N MEX STATE UNIV, 54- Mem: Am Soc Hort Sci; Am Pomol Soc. Res: Pomological research in post harvest physiology and growth regulators. Mailing Add: Dept of Hort NMex State Univ Las Cruces NM 88003

SULLIVAN, DAVID JOHN, veterinary pathology, see 12th edition

SULLIVAN, DAVID L, organic chemistry, see 12th edition

SULLIVAN, DAVID THOMAS, b Salem, Mass, Mar 20, 40; m 66; c 1. DEVELOPMENTAL GENETICS. Educ: Boston Col, BS, 61, MS, 63; Johns Hopkins Univ, PhD(biol), 67. Prof Exp: USPHS res fel biochem, Calif Inst Technol, 67-69; asst prof, 70-74, ASSOC PROF GENETICS, SYRACUSE UNIV, 74- Mem: Genetics Soc Am; Soc Develop Biol. Res: Genetic and biochemical control of animal development. Mailing Add: Dept of Biol Syracuse Univ Syracuse NY 13210

SULLIVAN, DONALD, b Merthyr Tydfil, Wales, Mar 23, 36; m 61. MATHEMATICS. Educ: Univ Wales, BSc, 57, PhD(appl math), 60. Prof Exp: Asst lectr math, Univ Col, Univ Wales, 60-61; asst prof, 61-66, ASSOC PROF MATH, UNIV NB, 66- Concurrent Pos: Mem, Inst Math & Its Appln, 67. Mem: Can Math Cong; London Math Soc. Res: Fluid mechanics; phase plane analysis of differential equations; functional equations. Mailing Add: Dept of Math Univ of NB Fredericton NB Can

SULLIVAN, DONALD BARRETT, b Phoenix, Ariz, June 13, 39; m 59; c 3. LOW TEMPERATURE PHYSICS. Educ: Tex Western Col, BS, 61; Vanderbilt Univ, MA, 63, PhD(physics), 65. Prof Exp: Res assoc physics, Vanderbilt Univ, 65; physicist & br chief, Radiation Physics Br, US Army Nuclear Defense Lab, 65-67; Nat Res Coun assoc, 67-69, PHYSICIST, CRYOGENICS DIV, NAT BUR STAND, 69- Mem: Am Phys Soc. Res: Josephson effect and quantum interference in superconductors; development of measurement instruments using these and other low temperature phenomena. Mailing Add: Cryogenics Div Nat Bur of Stand Boulder CO 80302

SULLIVAN, DONITA B, b Marlette, Mich, Feb 11, 31. PEDIATRICS. Educ: Siena Heights Col, BS, 52; St Louis Univ, MD, 56; Am Bd Pediat, dipl, 61. Prof Exp: Intern, Henry Ford Hosp, Detroit, Mich, 56-57; resident pediat, Children's Hosp of Mich, 57-59, sr resident, 59; res assoc, Sch Med, Wayne State Univ, 59; clin instr, 59-62, asst prof pediat & dir birth defects treatment ctr, 62-69, ASSOC PROF PEDIAT & DIR PEDIAT REHAB & RHEUMATOLOGY SECT, MED SCH, UNIV MICH, ANN ARBOR, 69- Concurrent Pos: Pediat consult, Wayne County Gen Hosp, 62-, Field Clins, Mich Crippled Children's Comn, 64- & Cath Social Servs, 66-; mem med adv comt, Washtenaw County Chapters, Nat Found & Nat Cystic Fibrosis Res Found, 60-74; prog consult, Nat Found, 66-69; bd trustees, Siena Heights Col, 70-75. Res: Handicapped children; children with birth defects; clinical and immunologic aspects of connective tissue disease in children. Mailing Add: Dept of Pediat Univ of Mich Med Ctr Ann Arbor MI 48104

SULLIVAN, EDWARD AUGUSTINE, b Salem, Mass, July 5, 29; m 59; c 6. INORGANIC CHEMISTRY. Educ: Col Holy Cross, BS, 50; Mass Inst Technol, MS, 52. Prof Exp: Asst, Sugar Res Found, 50-52; res chemist, Metal Hydrides, Inc, 52-63, sr res chemist, Metal Hydrides Div, Ventron Corp, 63-71, TECH MGR, RES CHEM, VENTRON CORP, 71- Mem: Am Chem Soc; Tech Asn Pulp & Paper Indust. Res: Chemistry of hydrides; inorganic synthesis; industrial applications of hydrides. Mailing Add: Chem Div Ventron Corp 12-24 Congress St Beverly MA 01915

SULLIVAN, EDWARD FRANCIS, b Portland, Maine, Sept 16, 20; m 48; c 4. AGRONOMY. Educ: Univ Maine, BS, 49; Cornell Univ, MSA, 51, PhD(agron), 53. Prof Exp: Asst agron, Cornell Univ, 50-53; asst prof, Southern Ill Univ, 53-56 & Pa State Univ, 56-61; from agronomist to sr agronomist, 61-75, MGR CROP ESTAB & PROTECTION, AGR RES CTR, GREAT WESTERN SUGAR CO, 75- Concurrent Pos: Asst prof, Univ Ill, 54-56. Mem: Am Soc Agron; Weed Sci Soc Am; Am Soc Sugar Beet Technologists. Res: Weed control; plant growth regulators; crop production. Mailing Add: Great Western Agr Res Ctr Sugarmill Rd Longmont CO 80501

SULLIVAN, EDWARD T, b Flushing, NY, June 28, 20; m 54; c 1. FOREST ECONOMICS. Educ: NC State Col, BSF, 46; Duke Univ, MS, 47, DF, 53. Prof Exp: Acct, southern woodlands dept, WVa Pulp & Paper Co, 47-50; vis instr forest econ, sch forestry, Duke Univ, 50-51; asst prof forestry, Univ Minn, 54-59; ASSOC PROF FOREST ECON, UNIV FLA, 59-, ASSOC FORESTER, 71- Mem: Am Econ Asn; Soc Am Foresters. Res: Marketing of forest products; demand for pulpwood. Mailing Add: Sch of Forestry Univ of Fla Gainesville FL 32601

SULLIVAN, FRANCIS JOSEPH, physiology, see 12th edition

SULLIVAN, GEORGE ALLEN, b Bronxville, NY, Dec 1, 35; m 60; c 2. INFORMATION SCIENCE, STATISTICS. Educ: Grinnell Col, AB, 57; Univ Rochester, AM, 59; Univ Nebr, PhD(solid state physics), 64. Prof Exp: Res asst solid state physics, Rensselaer Polytech Inst, 64-66; sr physicist, Electronic Res Div, Clevite Corp, Cleveland, 66-69; asst prof elec eng, Air Force Inst Technol, Wright Patterson AFB, 69-70; vis scientist, Physics Inst, Chalmers Univ Technol, Sweden, 70-71; criminal justice systs planner IV, 71-74, actg dir, 74-75, EVAL COORDR, RES &

STATIST DIV, PA BD PROBATION & PAROLE, 75- Mem: Am Phys Soc; Nat Speleol Soc; Nat Coun Crime & Delinquency. Res: Point defects and diffusion in metals; thermal mass transport and electromigration in solids; semiconductors; properties of materials; physics of solar cells; criminal justice information statistics systems; social rehabilitation programs; social research; operations and program planning. Mailing Add: Res & Statist Div Pa Bd of Probation & Parole Box 1661 3101 N Front St Harrisburg PA 17120

SULLIVAN, GERALD, b Magazine, Ark, Aug 11, 34; m 58; c 2. PHARMACY, PHARMACOGNOSY. Educ: Wash State Univ, BS & BPharm, 57, MS, 63; Univ Wash, PhD(pharm), 66. Prof Exp: Asst prof, 66-71, ASSOC PROF PHARMACOG, COL PHARM, UNIV TEX, AUSTIN, 71- Mem: Am Soc Pharmacog; Am Pharmaceut Asn; Acad Pharmaceut Sci. Res: Biosynthesis of secondary metabolites; fermentative processes; isolation and identification of plant toxins and secondary constituents; chemotaxonomy. Mailing Add: Col of Pharm Univ of Tex Austin TX 78712

SULLIVAN, HARRIS MARTIN, b Graves Co, Ky, Mar 12, 09; m 35; c 2. INSTRUMENTATION, MATERIALS SCIENCE. Educ: Univ Ky, BS, 31, MS, 33; Pa State Univ, PhD(physics, phys chem), 38. Prof Exp: Asst physics, Univ Ky, 31-33, instr, 33-35; asst, Pa State Univ, 35-37, instr, 37-38; asst prof phys metall & metall eng, Rensselaer Polytech Inst, 38-42; chief metallurgist, Adirondack Foundries & Steel, 42-45; dir res & develop, Cent Sci Co, 45-50, vpres & dir res, 50-55; mgr electronics lab, Gen Elec Co, 55-57, mgr adv semiconductor lab, 57-62; dir res, Johnson Controls, Inc, 62-74; CONSULT, 74- Concurrent Pos: Mem war metall comt, Watertown Arsenal, 42-45. Mem: AAAS; Am Phys Soc; Am Chem Soc; Am Soc Metals; Inst Elec & Electronic Engrs. Res: Chemical spectroscopy; x-ray crystal analysis; x-ray metallography; vacuum systems and equipment; ferrous metallurgy; physical metallurgy; electrochemistry; electronic instruments; analytical and process instruments; solid state physics; semiconductor physics; research administration. Mailing Add: Stone Haven Rd RFD 1 Belgium WI 53004

SULLIVAN, HARRY MORTON, b Winnipeg, Man, Apr 14, 21; m 49; c 3. PHYSICS. Educ: Queen's Univ, Ont, BSc, 45; Carleton Univ, BSc, 50; McGill Univ, MSc, 54; Univ Sask, PhD(upper atmosphere physics), 62. Prof Exp: Chemist, Can Civil Serv, 45-47; engr, Canadair Ltd, 54-56; physicist, Can Civil Serv, 56-59; physicist, Nat Ctr Sci Res, France, 62-64; asst prof, 64-69, ASSOC PROF PHYSICS, UNIV VICTORIA, 69- Mem: Can Asn Physicists. Res: Upper atmosphere; airglow and related phenomena, particularly twilight glow and day glow sodium; rarer constituents of upper atmosphere; photometer calibration techniques; standard radiation sources. Mailing Add: Dept of Physics Univ of Victoria Victoria BC Can

SULLIVAN, HERBERT J, b Ebbw Vale, Gt Brit, May 20, 33; m 58; c 3. GEOLOGY, PALYNOLOGY. Educ: Univ Sheffield, BSc, 54, PhD(geol), 59. Prof Exp: Dept Sci & Indust Res fel, 59-61; from asst lectr to lectr geol, Univ Sheffield, 61-64; sr res scientist, Res Ctr, Pan Am Petrol Corp, Okla, 64-71; SR STAFF GEOLOGIST, AMOCO CAN PETROL CO LTD, 71- Mem: Brit Geol Soc; Brit Paleont Asn; Am Asn Stratig Palynologists; Paleont Soc. Res: Paleozoic palynology, especially its stratigraphical applications. Mailing Add: Amoco Can Petrol Co Ltd 444 Seventh Ave SW Calgary AB Can

SULLIVAN, HUGH D, b Butte, Mont, June 16, 39; m 61; c 4. MATHEMATICS. Educ: Univ Mont, BA, 62, MA, 64; Wash State Univ, PhD(math), 68. Prof Exp: Teaching asst math, Univ Mont, 62-64 & Wash State Univ, 64-67; asst prof, 67-70, ASSOC PROF MATH & CHMN DEPT, EASTERN WASH STATE COL, 70- Mem: Am Math Soc; Math Asn Am. Res: Abstract systems theory; topology; probability and statistics. Mailing Add: Dept of Math Eastern Wash State Col Cheney WA 99004

SULLIVAN, HUGH R, b Indianapolis, Ind, Apr 8, 26; m 48; c 5. DRUG METABOLISM. Educ: Univ Notre Dame, BS, 48; Temple Univ, MA, 54. Prof Exp: Assoc res chemist, Socony-Vacuum Oil Co, 48-51; from assoc res chemist to sr res chemist, 51-69, res scientist, 69-72, RES ASSOC, RES LABS, ELI LILLY & CO, INC, 72- Mem: Am Chem Soc; Am Soc Pharmacol & Exp Therapeut; Am Soc Mass Spectrometry. Res: Mechanism of drug action and detoxication; analgesics; antibiotics; pharmocokinetics; quantitative mass fragmentography. Mailing Add: Eli Lilly & Co Inc Lilly Res Labs Indianapolis IN 46206

SULLIVAN, JAMES BOLLING, b Rome, Ga, Mar 19, 40; m 63; c 3. BIOCHEMISTRY, ZOOLOGY. Educ: Cornell Univ, AB, 62; Univ Tex, Austin, PhD(zool), 66. Prof Exp: ASST PROF BIOCHEM, DUKE UNIV, 70- Concurrent Pos: USPHS fel biochem, Duke Univ, 67-70. Mem: AAAS; Soc Study Evolution; Am Soc Biol Chemists; Lepidop Soc. Res: Comparative protein chemistry. Mailing Add: Marine Lab Duke Univ Beaufort NC 28516

SULLIVAN, JAMES DOUGLAS, b Chicago, Ill, Oct 27, 40; m 67; c 2. PHYSICS, SPACE SCIENCE. Educ: Univ Chicago, SB, 62, SM, 64, PhD(physics), 70. Prof Exp: Physicist, Enrico Fermi Inst, Univ Chicago, 70-71; asst res physicist, Univ Calif, Berkeley, 71-74; MEM STAFF, CTR SPACE RES, MASS INST TECHNOL, 74- Mem: Am Phys Soc; Am Geophys Union. Res: Space physics, solar particles and magnetospheric physics; nuclear physics, high energy heavy ion reactions; astrophysics, cosmic rays and origin of gamma rays. Mailing Add: Ctr for Space Res Mass Inst of Technol Cambridge MA 02139

SULLIVAN, JAMES F, b Peoria, Ill, Feb 17, 24; m; c 7. MEDICINE. Educ: Eureka Col, BS, 49; St Louis Univ, MD, 51. Prof Exp: Instr med, Sch Med, St Louis Univ, 55-59, asst prof clin med, 59-61; assoc prof, 61-64, PROF MED, SCH MED, CREIGHTON UNIV, 64-; CHIEF MED, OMAHA VET ADMIN HOSP, 72- Mem: AMA; fel Am Col Physicians; Am Fedn Clin Res; Am Soc Clin Nutrit. Res: Lipid, alcohol and trace metal metabolism. Mailing Add: Sch of Med Creighton Univ Omaha NE 68131

SULLIVAN, JAMES MICHAEL, b Butte, Mont, July 1, 34. NEUROANATOMY. Educ: Carroll Col, BA, 56; Univ Ore, MS, 64; St Louis Univ, PhD(anat), 73. Prof Exp: From instr to asst prof biol, Carroll Col, 61-69, assoc prof, 73-74; fel & res assoc, 74-75, instr, 75-76, ASST PROF ANAT, SCH MED, ST LOUIS UNIV, 76- Mem: Sigma Xi. Res: Anatomy, physiology and pharmacology of the autonomic nervous system with special emphasis on the cardiovascular system of man. Mailing Add: Dept of Anat Sch of Med St Louis Univ 1402 SGrand Blvd St Louis MO 63108

SULLIVAN, JAMES THOMAS, JR, b Seekonk, Mass, May 30, 28; m 55; c 4. PHYSICAL CHEMISTRY. Educ: Providence Col, BS, 50; Cath Univ Am, PhD, 55. Prof Exp: Asst, Cath Univ Am, 53-54, res assoc, 54-55; from asst prof to assoc prof, 55-67, PROF CHEM, UNIV ST THOMAS, TEX, 67-, EXEC ASST ACAD AFFAIRS, 72- Mem: AAAS; Am Chem Soc. Res: Chemical kinetics; teaching. Mailing Add: Dept of Phys Sci Univ of St Thomas Houston TX 77006

SULLIVAN, JAY MICHAEL, b Brockton, Mass, Aug 3, 36; m 64; c 3. CARDIOVASCULAR DISEASES. Educ: Georgetown Univ, BS, 58, MD, 62. Prof Exp: House officer med, Peter Bent Brigham Hosp, 62-63, resident, 63-67; res assoc biochem, Harvard Med Sch, 67-69, from instr to asst prof med, 69-74; PROF MED & CHMN DIV CIRCULATORY DIS, COL MED, UNIV TENN, MEMPHIS, 74- Concurrent Pos: Nat Heart Inst res fel med, Harvard Med Sch, 64-66; res fel, Med Found, 67-69; mem, Coun High Blood Pressure Res, 70; dir hypertension unit, Peter Bent Brigham Hosp, 70-74; physician-in-chief, Boston Hosp Women, 73-74; consult, Nat Heart & Lung Inst, 74 & Vet Admin Hosp, Memphis, 74; fel, Coun Circulation, Am Heart Asn, 75. Mem: AAAS; fel Am Col Cardiol; Am Fedn Clin Res; fel Am Col Physicians. Res: Hypertension, regulation of the circulation, pharmacology of antihypertensive and anti-platelet drugs. Mailing Add: 951 Court Ave Memphis TN 38163

SULLIVAN, JEREMIAH B, b Fitchburg, Mass, Aug 18, 31; m 55; c 3. PHARMACEUTICAL CHEMISTRY, ORGANIC CHEMISTRY. Educ: Mass Col Pharm, BS, 53, MS, 55; Univ Wash, PhD, 63. Prof Exp: Asst prof, Fordham Univ, 60-61; res instr & USPHS grant, Univ Wash, 63; supvr pharmaceut chem, 63-64, res mgr, Chem-Bioeng Dept, 64-65, from assoc dir to dir life systs div, 65-69, dir sci develop, 69-72, sci adv to pres, 72-74, V PRES, HAZLETON LABS AM, INC, 74-, DIR PROD SAFETY DIV, 75- Concurrent Pos: Mem, Therapeut Res Found. Mem: Am Chem Soc; Am Pharmaceut Asn; Europ Soc Toxicol. Res: Synthetic organic chemistry and instrumental analysis; analytical method development; biochemical studies; in-vitro and in-vivo metabolism; distribution and elimination using radioactive materials; evaluation of pharmaceutical formulations. Mailing Add: 9301 Schubert Ct Vienna VA 22180

SULLIVAN, JERRY STEPHEN, b Havre, Mont, July 17, 45; m 67; c 2. SOLID STATE ELECTRONICS. Educ: Univ Colo, Boulder, BSc, 67, MSc, 69, PhD(physics), 70. Prof Exp: Res scientist solid state devices, N V Philips Gloeilampenfabrieken, Eindhoven, 71-75, RES SCIENTIST SOLID STATE ELECTRONICS, PHILIPS LABS, 75- Mem: Inst Elec & Electronics Engrs; AAAS; Am Phys Soc; Europ Phys Soc; Am Asn Physics Teachers. Res: Numerical analysis; network theory; application of computers to semiconductor device modeling and integrated circuit analysis; electron paramagnetic and nuclear magnetic resonance and exchange interactions of ion pairs; microcomputer architecture and microprocessor design. Mailing Add: Philips Labs 345 Scarborough Rd Briarcliff Manor NY 10510

SULLIVAN, JOHN DENNIS, b Lake Forest, Ill, June 17, 28; m 55; c 3. FORESTRY. Educ: Univ Idaho, BS, 52, MS, 54; Mich State Univ, PhD(wood technol), 58. Prof Exp: Asst wood technol, Mich State Univ, 54-56, instr, 56-58; from asst prof to assoc prof wood sci, Sch Forestry, Duke Univ, 58-68; prin wood scientist, 68-69, agr adminr, 69-71, DEP ADMINR, COOP STATE RES SERV, USDA, 71- Mem: Am Soc Testing & Mat; Soc Wood Sci & Technol; Forest Prod Res Soc; Soc Am Foresters; fel Brit Inst Wood Sci. Res: Cellulose morphology at a molecular level, especially high resolution electron microscopy; adhesion and wood; liquid interactions. Mailing Add: Coop State Res Serv US Dept of Agr Washington DC 20250

SULLIVAN, JOHN FRANCIS, physical chemistry, biochemistry, see 12th edition

SULLIVAN, JOHN FRANCIS, b Dorchester, Mass, May 17, 11; m 39; c 2. NEUROLOGY. Educ: Boston Univ, AB, 33, MD, 37. Prof Exp: Asst resident neurol, Pratt Diagnostic Hosp, Boston, Mass, 47; resident, Boston City Hosp, 48; INSTR NEUROL, HARVARD MED SCH, 49-; PROF, SCH MED, TUFTS UNIV, 55- Concurrent Pos: Fel, Mass Gen Hosp, Boston, 49; vis neurologist, St Elizabeth's Hosp, 53-57, consult neurologist, 57-; physician, Boston Floating Hosp, 54-; assoc prof, Tufts Univ, 54-55; neurologist-in-chief & chmn dept neurol, New Eng Med Ctr Hosp, Boston, 54-, chmn hosp coun, 58-61; neurologist-in-chief, Boston Dispensary, 55; asst vis neurologist, Boston City Hosp, 55, consult in neurol, 68-; mem med adv bd, Nat Multiple Sclerosis Soc, 56-, chmn, 69-71, mem fel comt, 60-; mem training grant comt, Nat Inst Neurol Dis & Blindness, consult, 56-62; adv comt psychiat, neurol & psychol serv, Vet Admin, DC, 63-65. Mem: Fel AMA; Am Neurol Asn; Asn Res Nerv & Ment Dis; fel Am Acad Neurol. Res: Epilepsy; clinical neurology. Mailing Add: Neurology 171 Harrison Ave Boston MA 02111

SULLIVAN, JOHN HENRY, b New Haven, Conn, May 18, 19; m 47; c 2. PHYSICAL CHEMISTRY. Educ: Calif Inst Technol, PhD(chem), 50. Prof Exp: CHEMIST, LOS ALAMOS SCI LAB, UNIV CALIF, 50- Mem: Am Chem Soc. Res: Kinetics of gaseous reactions. Mailing Add: 3536-A Arizona St Los Alamos NM 87544

SULLIVAN, JOHN JOSEPH, b New York, NY, Mar 28, 35; m 63; c 2. REPRODUCTIVE PHYSIOLOGY. Educ: Rutgers Univ, BS, 57, PhD(dairy sci, physiol), 63; Univ Tenn, MS, 59. Prof Exp: Res assoc, 63-65, ASSOC DIR LABS, AM BREEDERS SERV, INC, W R GRACE & CO, 65- Mem: Am Dairy Sci Asn; Am Soc Animal Sci; Soc Cryobiol; Soc Study Reproduction. Res: Physiology of reproduction and related fields; artificial insemination of domestic animals; low temperature biology. Mailing Add: Rte 2 Stevenson Dr Poynette WI 53955

SULLIVAN, JOHN M, b Philadelphia, Pa, June 21, 32; m 56; c 6. ORGANIC CHEMISTRY. Educ: Dartmouth Col, AB, 54; Univ Mich, MS, 56, PhD(org chem), 60. Prof Exp: PROF CHEM, EASTERN MICH UNIV, 58- Mem: Am Chem Soc. Res: Heterocyclics; conformational analysis. Mailing Add: Dept of Chem Eastern Mich Univ Ypsilanti MI 48197

SULLIVAN, JOHN W, b Fargo, NDak, Nov 1, 32; m 64; c 3. CEREAL CHEMISTRY. Educ: NDak State Univ, BS, 54, MS, 58; Kans State Univ, PhD(cereal chem), 66. Prof Exp: Proj leader food res, John Stuart Res Labs, Quaker Oats Co, 61-68, sect mgr food res, 68-74; DIR RES, ROMAN MEAL CO, 74- Mem: Am Chem Soc; Am Asn Cereal Chemists; Inst Food Technologists. Res: Physical and chemical changes in starch and associated carbohydrates; enzyme changes in physical structure of starches and proteins. Mailing Add: 2101 S Tacoma Way Tacoma WA 98409

SULLIVAN, JOSEPH ARTHUR, b Boston, Mass, June 5, 23; m 46; c 4. MATHEMATICS. Educ: Boston Col, AB, 44; Mass Inst Technol, SM, 47; Ind Univ, PhD(math), 50. Prof Exp: From instr to assoc prof math, Univ Notre Dame, 50-60; PROF MATH, BOSTON COL, 60- Mem: Am Math Soc; Math Asn Am. Res: Mathematical analysis. Mailing Add: Dept of Math Boston Col Chestnut Hill MA 02167

SULLIVAN, JULIA CHRISTINE, b Haverhill, Mass, Dec 21, 06. MICROBIOLOGY. Educ: Emmanuel Col, AB, 27; Mass Inst Technol, MPH, 42. Prof Exp: Technician, City Health Dept, Haverhill, Mass, 27-40; res asst prev med, Harvard Med Sch, 41; jr bacteriologist, State Dept Pub Health, Mass, 42; assoc bacteriologist, Md, 43; res asst immunol, Harvard Med Sch, 43-46; asst immunologist, State Dept Pub Health, Mass, 46-53; immunologist, Protein Found, Inc, 53-55; independent researcher, Mass Gen Hosp, Boston, 55-56; asst path, Harvard Med Sch, 56-58, tech asst prev med, 58-63; res assoc pediat, Sch Med, Stanford Univ, 63-67; RES ASSOC PEDIAT, CASE

WESTERN RESERVE UNIV, 67- Mem: AAAS; Am Soc Microbiol; Am Soc Cell Biol; Tissue Cult Asn; NY Acad Sci. Res: Bacterial and viral diseases; immunology of human gamma globulin; growth requirements and metabolic control mechanisms of human cell cultures. Mailing Add: Dept of Pediat Res Bldg 346 Cleveland Metrop Gen Hosp Cleveland OH 44109

SULLIVAN, LAWRENCE PAUL, b Hot Springs, SDak, June 16, 31; m 55; c 3. PHYSIOLOGY. Educ: Univ Notre Dame, BS, 53; Univ Mich, MS, 56, PhD(physiol), 59. Prof Exp: Asst physiol, Univ Mich, 55-59, instr, 59-60; asst prof, George Washington Univ, 60-61; from asst prof to assoc prof, 61-69, PROF PHYSIOL, MED CTR, UNIV KANS, 69- Concurrent Pos: USPHS career develop award, 65-70; vis prof, Sch Med, Yale Univ, 69-70. Mem: Am Soc Nephrol; Am Physiol Soc. Res: Renal physiology. Mailing Add: Dept of Physiol Univ of Kans Med Ctr Kansas City KS 66103

SULLIVAN, LLOYD JOHN, b Lowell, Ariz, Sept 6, 23; m 48; c 4. BIOCHEMISTRY, PHYSICAL ORGANIC CHEMISTRY. Educ: Univ Ariz, BS, 50; Univ Pittsburgh, MS, 54. Prof Exp: Res asst phys chem, Mellon Inst Sci, 50-52, res assoc, 52-54, jr fel, 54-56, fel, 56-58; mem tech staff, Cent Res Lab, Tex Instruments, 58-60, sect head & dir energy conversion, Apparatus Div, 60-61; fel phys & anal chem, 61-67, SR FEL PHYS & ANAL CHEM & HEAD CHEM & BIOCHEM SECT, CHEM HYG FEL, MELLON INST SCI, 67- Mem: Fel AAAS; fel Am Inst Chemists; NY Acad Sci; Am Chem Soc. Res: Physical properties of organic materials; separation and purification of organic compounds, particularly natural products; gas, liquid and thin layer chromatography; analytical biochemistry; metabolism of organic compounds in vivo and by tissue culture techniques. Mailing Add: Chem Hyg Fel Mellon Inst Sci Pittsburgh PA 15213

SULLIVAN, LOUIS WADE, b Atlanta, Ga, Nov 3, 33; m 55; c 2. INTERNAL MEDICINE. Educ: Morehouse Col, BS, 54; Boston Univ, MD, 58; Am Bd Internal Med, dipl, 66. Prof Exp: Intern, NY Hosp-Cornell Med Ctr, 58-59, resident med, 59-60; resident gen path, Mass Gen Hosp, 60-61; instr, Harvard Med Sch, 63-64; asst prof, NJ Col Med, 64-66, asst attend physician, 64-65, assoc attend physician, 65-66; asst prof med, 66-70, ASSOC PROF MED & PHYSIOL, SCH MED, BOSTON UNIV, 70- Concurrent Pos: Res fel med, Thorndike Mem Lab, Boston City Hosp & Harvard Med Sch, 61-63; res assoc, Thorndike Mem Lab, Boston City Hosp, 63-64; USPHS res career develop award, 65-66, 67-71; asst vis physician, Boston Univ Hosp, 66- Mem: AAAS; Soc Exp Biol & Med; Am Fedn Clin Res; Am Soc Hemat; Am Soc Clin Nutrit. Res: Metabolism of vitamin B-12 and folic acid in man. Mailing Add: Dept of Med Boston Univ Sch of med Boston MA 02118

SULLIVAN, M HELEN, mathematics, see 12th edition

SULLIVAN, MARGARET P, b Lewistown, Mont, Feb 7, 22. PEDIATRICS, MEDICINE. Educ: Rice Inst, BA, 44; Duke Univ, MD, 50; Am Bd Pediat, dipl, 56. Prof Exp: Pediatrician, Atomic Bomb Casualty Comn, Japan, 53-55; from asst pediatrician to assoc pediatrician, 56-73, PROF PEDIAT & PEDIATRICIAN, UNIV TEX M D ANDERSON HOSP & TUMOR INST HOUSTON, 73- Mem: AAAS; Am Asn Cancer Res; Am Acad Pediat; AMA; Am Med Women's Asn (vpres, 72, pres, 74). Res: Pediatric oncology. Mailing Add: Dept of Pediat Univ of Tex M D Anderson Hosp & Tumor Inst Houston TX 77025

SULLIVAN, MARY LOUISE, b Butte, Mont, Oct 3, 06. INORGANIC CHEMISTRY. Educ: St Mary Col, Kans, BS, 35; St Louis Univ, MS, 39, PhD(chem), 47. Prof Exp: Instr chem, 46-53, dean, 53-74, DIR INST RES, ST MARY COL, KANS, 74- Mem: Am Asn Physics Teachers; Nat Sci Teachers Asn. Res: Analytic chemistry, especially trace analysis for metals using dithizone. Mailing Add: St Mary Col Leavenworth KS 66048

SULLIVAN, MAURICE FRANCIS, b Butte, Mont, Feb 15, 22; m 51; c 6. PHARMACOLOGY. Educ: Mont State Col, BS, 50; Univ Chicago, PhD(pharmacol), 55. Prof Exp: Scientist, Hanford Labs, Gen Elec Co, 55-56, sr scientist, 56-65; mgr physiol sect, 65-71, STAFF SCIENTIST, PAC NORTHWEST LABS, BATTELLE MEM INST, 71- Concurrent Pos: NIH fel, Med Res Coun, Harwell, Eng, 61- Mem: Fel AAAS; Am Physiol Soc; Radiation Res Soc. Res: Biological effects of radiation, especially on the gastrointestinal tract; biochemistry; physiology; pathology; pharmacological effects of slow-release administration of narcotic antagonists and the long term smoking of marihuana. Mailing Add: Biol Dept Pac Northwest Labs Battelle Mem Inst Box 999 Richland WA 99352

SULLIVAN, MICHAEL FRANCIS, b New Paltz, NY, Aug 20, 42; m 67; c 3. PHOTOGRAPHIC CHEMISTRY. Educ: St Lawrence Univ, BS, 63; Univ NC, Chapel Hill, PhD(chem), 68. Prof Exp: Sr chemist, 67-73, RES LAB HEAD, EASTMAN KODAK CO, 73- Mem: Nat Micrographics Asn; Nat Microfilm Asn. Res: Organometallic chemistry as applied to photographic science; applied and exploratory research directed toward design of new photographic materials. Mailing Add: 341 Brooksboro Dr Webster NY 14580

SULLIVAN, MICHAEL JOSEPH, JR, b Chicago, Ill, Jan 8, 42; m 64; c 4. MATHEMATICS. Educ: DePaul Univ, BS, 63; Ill Inst Technol, MS, 64, PhD(math), 67. Prof Exp: Teaching asst math, Ill Inst Technol, 62-65; asst prof, 65-70, ASSOC PROF MATH, CHICAGO STATE UNIV, 70- Concurrent Pos: Am Coun Educ acad admin internship, 70-71. Mem: Am Math Soc; Math Asn Am. Res: Differential geometric aspects of dynamics; polygenic functions; applications of mathematics in the management and behavioral sciences; mathematics education. Mailing Add: Dept of Math Chicago State Univ 95th at King Dr Chicago IL 60628

SULLIVAN, MILES VINCENT, b Fargo, NDak, Aug 19, 17; m 48; c 3. PHYSICAL CHEMISTRY. Educ: Wabash Col, AB, 41; Purdue Univ, MS, 42, PhD(phys chem), 48. Prof Exp: Res chemist, US Naval Res Lab, 42-46; ENGR, BELL LABS, 48- Mem: Am Chem Soc. Res: Flammability; lubricating oils; wire insulations; heats of combustions; solid state materials and process development; photolithography. Mailing Add: Bell Labs Murray Hill NJ 07974

SULLIVAN, NICHOLAS, b Philadelphia, Pa, Dec 20, 27. SPELEOLOGY. Educ: Cath Univ Am, BS, 50; Univ Pittsburgh, MS, 54; Univ Notre Dame, PhD(ecol), 60. Hon Degrees: DSc, Univs Kyoto & San Carlos & Notre Dame Col, 63. Prof Exp: Asst prof biol, 62-66, ASSOC PROF EARTH SCI, LA SALLE COL, 67-, ASST TO PRES, 69- Concurrent Pos: Ed, Int J Speleol, 65- Mem: AAAS; Nat Speleol Soc (pres, 57-62); Nat Asn Biol Teachers (vpres, 61); Am Soc Mammalogists; Am Soc Zoologists. Res: Biospeleology; terrestrial ecology. Mailing Add: Dept of Earth Sci La Salle Col Philadelphia PA 19141

SULLIVAN, PATRICIA ANN NAGENGAST, b New York, NY, Nov 22, 39; m 66. BIOLOGY. Educ: Notre Dame Col Staten Island, AB, 61; NY Univ, MS, 64, PhD(biol), 67. Prof Exp: Part-time instr, Notre Dame Col Staten Island, 64-67; asst prof biol, Wagner Col, 67-68; NIH trainee anat & cell biol, State Univ NY Upstate Med Ctr, 68-69 & fel cell biol, 69-70; asst prof, 70-74, ASSOC PROF BIOL, WELLS COL, 74-, CHAIRWOMAN DIV LIFE SCI, 75- Mem: AAAS; NY Acad Sci; Am Soc Hemat. Res: Bioethics and implications of biological research; hematopoiesis and its regulation; chromatin structure and function. Mailing Add: Dept of Biol Wells Col Aurora NY 13026

SULLIVAN, PAUL JOSEPH, b Morrickville, Ont, Mar 2, 39; m 62; c 2. APPLIED MATHEMATICS, ENGINEERING. Educ: Univ Waterloo, BSc, 64, MSc, 65; Cambridge Univ, PhD(appl math), 68. Prof Exp: ASST PROF APPL MATH & ENG, UNIV WESTERN ONT, 68- Concurrent Pos: Res fel, Calif Inst Technol, 69-70. Mem: Am Acad Mech; Can Soc Mech Eng. Res: Dispersion within turbulent fluid flow; convection phenomenon in fluids; turbulent fluid flow generally. Mailing Add: Dept of Appl Math Univ of Western Ont London ON Can

SULLIVAN, PETER KEVIN, b San Francisco, Calif, June 14, 38. POLYMER PHYSICS. Educ: Univ San Francisco, BS, 60; Cornell Univ, MS, 63; Rensselaer Polytech Inst, PhD(phys chem), 65. Prof Exp: Res fel, Nat Bur Stand, 65-67, chemist, 67-73; PHYS CHEMIST, CELANESE RES CO, 74- Mem: Am Phys Soc; Am Chem Soc. Res: Physics, chemistry and mechanical properties of polymers. Mailing Add: Celanese Res Co Box 1000 Summit NJ 07901

SULLIVAN, RAYMOND, b Ebbw Vales, Wales, Oct 27, 34; m 62; c 1. GEOLOGY. Educ: Univ Sheffield, BSc, 57; Glasgow Univ, PhD(geol), 60. Prof Exp: Demonstr geol, Glasgow Univ, 57-60; paleontologist, Shell Oil Co Can, 60-62; from asst prof to assoc prof, 62-74, assoc dean natural sci, 69-72, PROF GEOL, CALIF STATE UNIV, SAN FRANCISCO, 74- Concurrent Pos: NSF res grant, 66-67. Mem: Paleont Soc; Soc Econ Paleontologists & Mineralogists; Nat Asn Geol Teachers; Geol Soc Am. Res: Upper Paleozoic biostratigraphy; sedimentary petrology of carbonate and clastic rocks. Mailing Add: Dept of Geol Calif State Univ 1600 Holloway Ave San Francisco CA 94132

SULLIVAN, RICHARD FREDERICK, b Olathe, Colo, Dec 26, 29; m 68; c 2. PHYSICAL CHEMISTRY. Educ: Univ Colo, BA, 51, PhD(chem), 56. Prof Exp: Res chemist, Calif Res Corp, 55-68, SR RES CHEMIST, CHEVRON RES CO, 68- Mem: Am Chem Soc. Res: Photochemistry; mechanisms of hydrocarbon reactions; catalysis; petroleum process research and development. Mailing Add: Chevron Res Co 576 Standard Ave Richmond CA 94802

SULLIVAN, RICHARD WARD, mathematics, see 12th edition

SULLIVAN, ROBERT EMMETT, b Sioux City, Iowa, May 28, 32; m 61. DENTISTRY, PEDODONTICS. Educ: Morningside Col, BA, 54; Univ Nebr, DDS, 61, MSD, 63; Am Bd Pedodont, dipl, 67. Prof Exp: From instr to assoc prof pedodont, 63-72, ASSOC PROF PEDIAT, UNIV NEBR-LINCOLN, 69-, PROF PEDODONT, 72- Concurrent Pos: Consult, Omaha-Douglas County Dent Pub Health, 62- & Omaha-Douglas County Children & Youth Proj, 68- Mem: Am Dent Asn; Am Soc Dent for Children. Res: Vital staining of teeth; mechanism of action of dental preventative materials. Mailing Add: 1201 Piedmont Rd Lincoln NE 68510

SULLIVAN, ROBERT LITTLE, b Chicago, Ill, Oct 27, 28. GENETICS. Educ: Univ Del, AB, 50; NC State Univ, MS, 53, PhD, 56. Prof Exp: Res assoc entom, Univ Kans, 56-61; asst prof biol, Washburn Univ, 61-62; ASST PROF BIOL, WAKE FOREST UNIV, 62- Mem: Genetics Soc Am; Soc Study Evolution. Res: Insect genetics, including radiation studies on the genetics of the house fly. Mailing Add: Dept of Biol Wake Forest Univ Winston-Salem NC 27109

SULLIVAN, SERAPHIN A, b Lawrence, Mass, Sept 2, 23. PHYSICS. Educ: St Bonaventure Col, BA, 46; St Bernardine of Siena Col, BS, 54, MS, 55; Cath Univ Am, PhD(physics), 62. Prof Exp: Instr, NY Parochial High Sch, 50-52; instr phys sci, St Bonaventure Univ, 52-53; instr, NY Parochial High Sch, 55-56; asst prof physics, St Bonaventure Univ, 62-68; ASSOC PROF PHYSICS, SIENA COL, NY, 68- Mem: Optical Soc Am. Res: Emission and infrared absorption spectroscopy; interferometry. Mailing Add: Dept of Physics Siena Col Loudonville NY 12211

SULLIVAN, THOMAS DONALD, b Fair Haven, Vt, Feb 16, 12. CYTOLOGY. Educ: St Michael's Col, Vt, AB, 34; Cath Univ Am, MA, 39; Fordham Univ, PhD(cytol), 47. Prof Exp: Instr Latin & Greek, 39-43, acad dean, 42-44, head dept biol, 47-67, dir res unit, 52-58, vpres, Col, 58-64, trustee, 67-70, PROF BIOL, ST MICHAEL'S COL, VT, 47- Mem: AAAS; Torrey Bot Club. Res: Cytogenetics of petunia; plant growth substances; mouse ascites tumors. Mailing Add: St Michael's Col St Michaels VT 05404

SULLIVAN, THOMAS FREDERICK, b Covington, Ky, Nov 13, 30; m 61. ORGANIC CHEMISTRY, NUCLEAR CHEMISTRY. Educ: Univ Notre Dame, BS, 52; Northwestern Univ, Evanston, PhD(chem), 58. Prof Exp: Res chemist, Glidden Co, 56-58; head radioisotope lab, Armour & Co, 58-60; head radioisotope lab, Abbott Labs, 60-61; mgr chem prod & serv, Tracer Lab, 61-63; vpres res prod, 63-72, CLERK OF CORP, NEW ENG NUCLEAR CORP, 72- Concurrent Pos: Armour & Co fel, Radiocarbon Lab, Univ Ill, 58-59; mem comt specif & criteria of nucleotides & related compounds, Nat Res Coun, 71- Mem: Am Chem Soc. Res: Biochemistry; chemical management. Mailing Add: New Eng Nuclear Corp 549 Albany St Boston MA 02118

SULLIVAN, THOMAS WESLEY, b Rover, Ark, Sept 30, 30; m 55. POULTRY NUTRITION, BIOCHEMISTRY. Educ: Okla State Univ, BS, 51; Univ Ark, MS, 56; Univ Wis, PhD, 58. Prof Exp: Instr, Ark Pub Schs, 51-52; res asst poultry nutrit, Univ Ark, 54-55; res asst, Univ Wis, 55-58, from asst prof to assoc prof, 58-65; PROF POULTRY SCI, UNIV NEBR-LINCOLN, 65- Concurrent Pos: Consult feed mfg & ingredients. Honors & Awards: Res Award, Nat Turkey Fedn, 68. Mem: Poultry Sci Asn; Am Inst Nutrit; Soc Exp Biol & Med. Res: Nutrient interrelationships; evaluation of poultry feed additives and mineral sources. Mailing Add: Dept of Poultry & Wildlife Sci Univ of Nebr Lincoln NE 68583

SULLIVAN, VICTORIA I, b Avon Park, Fla, Nov 14, 41. BIOSYSTEMATICS. Educ: Univ Miami, BA, 63; Fla State Univ, PhD(biol), 72. Prof Exp: Naturalist, Everglades Nat Park, Nat Park Serv, 63-64; tech asst, Fairchild Trop Garden, Miami, 67-68; instr, Iowa State Univ, 71-72; botanist, Trustees Internal Improv Trust Fund, State of Fla, 72-73; ASST PROF BIOL, UNIV SOUTHWESTERN LA, 73- Mem: AAAS; Am Soc Plant Taxonomists; Soc Study Evolution; Am Soc Naturalists; Sigma Xi. Res: Biosystematics of Eupatorium compositae species and hybrids, including pollination, karyotypes, phytochemistry and breeding systems; niche partitioning of polyploid and diploid races; primarily productivity of Louisiana marshes, burned versus unburned. Mailing Add: Dept of Biol Univ of Southwestern La Lafayette LA 70501

SULLIVAN, WALTER JAMES, b New York, NY, Apr 27, 25; m 51; c 3. PHYSIOLOGY, BIOPHYSICS. Educ: Manhattan Col, BS, 46; Cornell Univ, MD, 51. Prof Exp: Instr physics, Manhattan Col, 46-57; intern med, Univ Va, 51-52; instr, Cornell Univ, 54-55; vis investr, Rockefeller Inst, 55-56; group leader, Lederle Labs, 56-63; from asst prof to assoc prof, 63-73, PROF PHYSIOL, SCH MED, NY UNIV,

73-, ACTG CHMN DEPT, 71- Concurrent Pos: Fel physiol, Cornell Univ, 52-54. Mem: Am Physiol Soc; Biophys Soc; Soc Gen Physiol; Am Soc Nephrology. Res: Renal physiology; micropuncture study of renal ion transport. Mailing Add: Dept of Physiol & Biophys NY Univ Sch of Med New York NY 10016

SULLIVAN, WALTER SEAGER, b New York, NY, Jan 12, 18; m 50; c 3. SCIENCE WRITING, JOURNALISM. Educ: Yale Univ, BA, 40. Hon Degrees: LHD, Yale Univ, 69, Newark Col Eng, 74; DS, Hofstra Univ, 75. Prof Exp: Mem staff, 40-48, foreign correspondent, 48-56, sci news ed, 60-63, SCI ED, NEW YORK TIMES, 64- Concurrent Pos: Mem bd gov, Arctic Inst NAm, 59-65; Bromley lectr, Yale Univ, 65, mem univ coun, 70-75; mem adv comt pub rels, Am Inst Physics, 65-; partic, Sem Technol & Social Change, Columbia Univ & mem adv coun, Dept Geol & Geophys Sci, Princeton Univ, 71- Mem exped, Arctic, 35 & 46 & Antarctic, 46, 54, 56 & 57. Honors & Awards: George Polk Mem Award in Jour, 59; Nat Book Awards Jurist in Sci, 60 & 69; Westinghouse-AAAS Writing Awards, 63, 68 & 72; Int Nonfiction Book Prize, Frankfurt Fair, Ger, 65; Grady Award, Am Chem Soc, 69; Am Inst Physics-US Steel Found Award in Physics & Astron, 69; Washburn Award, Boston Mus Sci, 72; Daly Medal, Am Geog Soc, 73; Ralph Coats Roe Medal, Am Soc Mech Engrs, 75. Mem: Fel AAAS; fel Arctic Inst NAm; Am Geog Soc; Am Geophys Union. Mailing Add: The New York Times 229 W 43 St New York NY 10036

SULLIVAN, WILLIAM ALBERT, JR, b Nashville, Tenn, Apr 6, 24; m 49; c 2. SURGERY. Educ: Tulane Univ, MD, 47; Univ Minn, MS, 56; Am Bd Surg, dipl, 58. Prof Exp: Asst prof surg, 56-61, asst dean, Med Sch, 68-73, assoc prof surg, Med Sch, 61-68, DIR DEPT CONTINUATION MED EDUC, UNIV MINN, MINNEAPOLIS, 58-, ASSOC DEAN ADMIS & STUDENT AFFAIRS, MED SCH, 73- Mem: AMA; Am Col Surgeons. Res: Asymptomatic detection of cancer; medical education. Mailing Add: Dept of Surg Univ of Minn Med Sch Minneapolis MN 55455

SULLIVAN, WILLIAM DANIEL, b Boston, Mass, Nov 18, 18. BIOCHEMISTRY. Educ: Boston Col, AB, 44, MA, 45; Fordham Univ, MS, 48; Cath Univ Am, PhD(biol), 57. Prof Exp: Teacher, Cranwell Prep Sch, 45-46, Fairfield Prep Sch, 46-47 & Cheverus High Sch, 52-53; asst prof bact, Fairfield Univ, 57-58; from asst prof to assoc prof, 58-65, chmn dept, 58-69, PROF BIOL, BOSTON COL, 65- Concurrent Pos: Dir cancer res inst, Boston Col; consult, Sta WGBH TV. Mem: AAAS; Am Soc Microbiol; Am Soc Parasitol; Soc Protozool; Electron Micros Soc Am. Res: Biochemistry of protozoa and cancer cells, especially effect of radiation on enzymatic activities; protein synthesis; electron microscopy with autoradiography of macromolecules in protozoan and cancer cells. Mailing Add: Cancer Res Inst Boston Col Chestnut Hill MA 02167

SULLIVAN, WILLIAM FRANCIS, b Newark, NJ, Dec 20, 14; m 44; c 2. PHYSICAL INORGANIC CHEMISTRY. Educ: Univ Mich, BS, 37, MS, 39; Oak Ridge Inst Reactor Technol, cert, 52; Rutgers Univ, PhD, 53. Prof Exp: Res chemist, 39-47, asst supvr, 47-62, sect head, 62-69, asst mgr, New Prod Dept, 69-72, RES ASSOC, TITANIUM PIGMENT DIV, N L INDUSTS, 73- Mem: AAAS; Am Chem Soc; Am Inst Chemists. Res: Preparation and properties of titanium dioxide pigments; application of radioisotopes to research and industrial problems; multiple scattering of light by transport theory. Mailing Add: 25 Ridge Rd Roseland NJ 07068

SULLIVAN, WILLIAM RICHARD, b Niles, Ohio, Oct 22, 16; m 41. CHEMISTRY. Educ: Western Reserve Univ, AB, 37, MA, 39; Univ Wis, PhD(biochem), 42. Prof Exp: Asst biol, Western Reserve Univ, 38-39; asst biochem, Univ Wis, 39-42; chemist, 42-45, asst to dir chem res, 45-56, gen secy res, 56-59, res coordr, 59-63, DIR RES SERV, HOFFMANN-LA ROCHE INC, 63- Mem: AAAS; Am Chem Soc; Am Inst Chemists; NY Acad Sci. Res: Medicinal chemistry; research administration; information science. Mailing Add: Res Div Hoffmann-La Roche Inc Nutley NJ 07110

SULLWOLD, HAROLD H, JR, b St Paul, Minn, Dec 22, 16; m 40; c 2. GEOLOGY. Educ: Univ Calif, Los Angeles, BA, 39, MA, 40, PhD(geol), 59. Prof Exp: Geologist, Wilshire Oil Co, 41, US Geol Surv, 42-44 & W R Cabeen & Assocs, 44-52; instr, Univ Calif, Los Angeles, 52-58; consult geologist, 58-60; CONSULT, GEORGE H ROTH & ASSOCS, 60- Concurrent Pos: Adj prof, Univ SC, 70-71. Mem: Geol Soc Am; Am Asn Petrol Geologists. Res: Petroleum geology. Mailing Add: 560 Concha Loma Dr Carpinteria CA 93013

SULSER, FRIDOLIN, b Grabs, Switz, Dec 2, 26; m 55; c 4. PHARMACOLOGY. Educ: Univ Basel, MD, 55. Prof Exp: Asst prof pharmacol, Univ Berne, 56-58; head dept pharmacol, Wellcome Res Labs, NY, 63-65; prof pharmacol, Sch Med, Vanderbilt Univ, 65-74; DIR, TENN NEUROPSYCHIAT INST, 74- Concurrent Pos: Int Pub Health Serv fel, NIH, 59-62. Mem: AAAS; Am Soc Pharmacol; Am Col Neuropsychopharmacol; Am Fedn Clin Res; NY Acad Sci. Res: Pharmacology of psychotropic drugs; neurochemistry; biochemical mechanisms of drug action. Mailing Add: 4621 Tara Dr Nashville TN 37215

SULSKI, LEONARD C, b Buffalo, NY, Mar 3, 36; m 66; c 1. MATHEMATICS. Educ: Canisius Col, BS, 58; Univ Notre Dame, PhD(math), 63. Prof Exp: Instr math, Univ Notre Dame, 63-64; lectr, Univ Sussex, 64-65; asst prof, Col of the Holy Cross, 66-68; lectr, Univ Sussex, 68-70; ASSOC PROF MATH & CHMN DEPT, COL OF THE HOLY CROSS, 70- Mem: Am Math Soc; Math Asn Am; London Math Soc; Soc Indust & Appl Math. Res: Analysis; functional analysis; theory of distributions and differential equations. Mailing Add: Dept of Math Col of the Holy Cross Worcester MA 01610

SULTZER, BARNET MARTIN, b Union City, NJ, Mar 24, 29; m 56; c 1. MICROBIOLOGY. Educ: Rutgers Univ, BS, 50; Mich State Univ, MS, 51, PhD(bact), 58. Prof Exp: Asst bact, Mich State Univ, 56-58; res assoc microbiol, Princeton Labs, Inc, 58-64; asst prof, 64-66, ASSOC PROF MICROBIOL & IMMUNOL, STATE UNIV NY DOWNSTATE MED CTR, 66- Concurrent Pos: Vis scientist, Karolinska Inst, Sweden, 71-72. Mem: AAAS; Am Asn Immunologists; Am Soc Microbiol; NY Acad Sci; Harvey Soc. Res: Lymphocyte activation, mitogenic properties of tuberculin, host responses to microbial endotoxins. Mailing Add: Dept of Microbiol State Univ NY Downstate Med Ctr Brooklyn NY 11203

SULYA, LOUIS LEON, b North Monmouth, Maine, Aug 17, 11; m 37; c 2. BIOCHEMISTRY. Educ: Col Holy Cross, BS, 32, MS, 33; St Louis Univ, PhD(org chem), 39. Prof Exp: Instr chem, Spring Hill Col, 34-36; res chemist, Reardon Co, Mo, 40-42; assoc prof, 45-50, PROF BIOCHEM & CHMN DEPT, SCH MED, UNIV MISS, 50- Mem: AAAS; Endocrine Soc; Am Chem Soc; Am Soc Biol Chemists; Int Soc Nephrol. Res: Endocrinology and comparative biochemistry. Mailing Add: 1076 Parkwood Pl Jackson MS 39206

SULZBACHER, WILLIAM LOUIS, b Braddock, Pa, Sept 11, 13; m 41; c 1. FOOD SCIENCE. Educ: Univ Pittsburgh, BS, 36, MS, 38. Prof Exp: Instr bact, Univ Pittsburgh, 36-38; bacteriologist, Clyde H Campbell Consult Lab, Pa, 37-41; ord engr, Philadelphia Naval Shipyard, US Dept Navy, 41-45; bacteriologist, Animal Husb Div, Bur Animal Indust, USDA, 45-54, bacteriologist, Washington Utilization Res Br, 54-55, chief meat lab, Eastern Utilization Res & Develop Div, Agr Res Serv, 55-71, mem bd civil serv exam, 48-60; CONSULT, 71- Concurrent Pos: Visxlectr,xUniv Concurrent Pos: Chmn meat processing comt, Reciprocal Meat Conf, 57-60; co-chmn, Gordon Res Conf Chem & Psychophysiol Odor & Flavor, 66; vis lectr, Univ Md, 67-; chmn meat microbiol comt, Am Soc Testing & Mat, 73- Honors & Awards: Signal Serv Award, Am Meat Sci Asn, 68. Mem: Fel AAAS; Am Soc Microbiol; Inst Food Technologists; Brit Soc Appl Bact; fel Am Inst Chemists. Res: Microbiology of meats; physiology and taxonomy of psychrophilic microorganisms; chemistry of muscle proteins and meat flavors; rancidity and fat oxidation; meat technology; mechanical engineering. Mailing Add: 8527 Clarkson Dr Fulton MD 20759

SULZBERG, THEODORE, b New York, NY, May 28, 36; m 57; c 3. ORGANIC POLYMER CHEMISTRY. Educ: City Col New York, BS, 57; Brooklyn Col, MA, 59; Mich State Univ, PhD(org chem), 62. Prof Exp: Res fel org chem, Ohio State Univ, 62-63; res chemist, Chem & Plastics Div, Union Carbide Corp, 63-70; RES GROUP LEADER, CORP LAB, SUN CHEM CORP, 70- Mem: Am Chem Soc; The Chem Soc; Am Asn Textile Chemists & Colorists. Res: Low temperature condensation polymerization; charge-transfer complexes; impact modification of addition polymers; carbonium ions; chemical finishing of textiles; flame retardants; rosin derivatives; hydrocarbon polymers. Mailing Add: Corp Lab Sun Chem Corp Carlstadt NJ 07072

SULZER, ALEXANDER JACKSON, b Emmett, Ark, Feb 13, 22; m 42; c 2. MEDICAL PARASITOLOGY. Educ: Hardin-Simmons Univ, BA, 49; Emory Univ, MSc, 60, PhD(parasitol), 62. Prof Exp: Med parasitologist, Commun Dis Ctr, 52-62, res parasitologist, 62-74, RES MICROBIOLOGIST, CTR DIS CONTROL, 74- Mem: Sigma Xi; Am Soc Trop Med & Hyg; Am Soc Parasitol; fel Royal Soc Trop Med & Hyg; fel Am Acad Microbiol. Res: Immunoparasitology; diagnostic serology of parasitic diseases of man and animals; tagged systems in serology especially fluorescent tagged materials; fine structure of protozoan parasites. Mailing Add: Ctr for Dis Control 1600 Clifton Rd Atlanta GA 30333

SUMARTOJO, JOJOK, b Surabaya, Indonesia, July 5, 37; Australian citizen; m 66; c 2. SEDIMENTARY PETROLOGY, ECONOMIC GEOLOGY. Educ: Bandung Inst Technol, BS, 61; Univ Ky, MS, 66; Univ Cincinnati, PhD(geol), 74. Prof Exp: From instr to lectr geol, Bandung Inst Technol, 59-62; lectr geol, Univ Pajajaran, 68-69; instr geol, Univ Adelaide, 69-75; ASST PROF GEOL, VANDERBILT UNIV, 75- Mem: Sigma Xi; Geol Soc Australia; assoc Geoscientists Int Develop; Int Asn Math Geol. Res: Trace and major-element geochemistry and mineralogy of fine-grained detrital sedimentary rock, especially black shales and red-beds. Mailing Add: Dept of Geol Vanderbilt Univ Nashville TN 37235

SUMBERG, DAVID ALLAN, b Utica, NY, June 28, 42; m 64; c 1. EXPERIMENTAL PHYSICS. Educ: Utica Col, BA, 64; Mich State Univ, MS, 66, PhD(physics), 72. Prof Exp: Physicist, Eastman Kodak Co, 72-74; ASST PROF PHYSICS, ST JOHN FISHER COL, 74- Mem: Am Phys Soc. Res: Development of techniques useful in Mössbauer spectroscopy. Mailing Add: 105 Seymour Rd Rochester NY 14609

SUMERFORD, WOOTEN TAYLOR, b Valdosta, Ga, Mar 4, 09; m 36; c 2. PHARMACEUTICAL CHEMISTRY, ORGANIC CHEMISTRY. Educ: Univ Ga, BS, 30, MS, 33; Univ Md, PhD(chem), 39. Prof Exp: From instr to prof pharm, Univ Ga, 31-45, actg dean, 45-46; assoc prof org chem, La State Univ, 46-48; chemist, USPHS, 48-50, in charge toxicol lab, 50-52; dir pharmaceut org chem div, Mead Johnson & Co, 52-66; PROF CHEM, VALDOSTA STATE COL, 66- Mem: AAAS; Am Chem Soc; Am Pharmaceut Asn. Res: Chemistry of medicinal agents and economic poisons. Mailing Add: Dept of Chem Valdosta State Col Valdosta GA 31601

SUMMA, A FRANCIS, b Waterbury, Conn, Jan 28, 33; m 59; c 4. PHARMACEUTICAL CHEMISTRY. Educ: St John's Univ, NY, BS, 54; Univ Conn, MS, 57, PhD(pharmaceut chem), 60. Prof Exp: Anal chemist, Bur Sci Res, Div Pharmaceut Chem, US Food & Drug Admin, 59-65; ASST SECT HEAD ANAL CHEM, VICK CHEM CO, RICHARDSON-MERRELL INC, 65- Mem: Am Pharmaceut Asn; Am Chem Soc. Res: Analytical method development; gas chromatography; polarography. Mailing Add: Vick Chem Co 1 Bradford Rd Mt Vernon NY 10557

SUMMER, GEORGE KENDRICK, b Cherryville, NC, May 8, 23; m 52; c 2. BIOCHEMISTRY, NUTRITION. Educ: Univ NC, BS, 44; Harvard Med Sch, MD, 51. Prof Exp: From instr to asst prof pediat, 57-65, from asst prof to assoc prof biochem & nutrit, 65-72, PROF BIOCHEM & NUTRIT, UNIV NC, CHAPEL HILL, 72-, ASSOC CLIN PROF PEDIAT, 66-, RES SCIENTIST, CHILD DEVELOP INST, 71- Concurrent Pos: Fel pediat metab, Univ NC, Chapel Hill, 54-57; NIH res career develop award, 65-70; vis scientist, Galton Lab, Univ Col, London & Med Res Coun human biochem genetics res unit & dept biochem, King's Col, London, 62-63. Mem: AAAS; Soc Exp Biol & Med; Am Inst Nutrit; Tissue Cult Asn; Am Chem Soc; Am Acad Pediat. Res: Biochemistry of cells in tissue culture; study of inborn errors of metabolism; pathophysiology of disease; pediatrics. Mailing Add: Dept Biochem & Nutrit Div Health Affairs Univ of NC Chapel Hill NC 27514

SUMMERFELT, ROBERT C, b Chicago, Ill, Aug 2, 35; m 60; c 3. FISH BIOLOGY. Educ: Univ Wis-Stevens Point, BS, 57; Southern Ill Univ, MS, 59, PhD(zool), 64. Prof Exp: Lectr zool, Southern Ill Univ, 62-64; asst prof, Kans State Univ, 64-66; assoc prof, 66-71, PROF ZOOL & LEADER OKLA COOP FISHERY RES UNIT, OKLA STATE UNIV, 71- Concurrent Pos: Grants, Bur Com Fisheries, 66-69, Nat Marine Fisheries Serv, Off Water Resources Res, 70-71 & 74-75, Environ Protection Agency, 71-72 & Bur Reclamation, 72 & 74-75. Honors & Awards: Spec Achievement Award, US Fish & Wildlife Serv, 71; Commendation, Sport Fishing Inst, 75. Mem: Soc Invert Path; Ecol Soc Am; Am Fisheries Soc; Wildlife Dis Asn; Am Inst Fishery Res Biologists. Res: Biology of fishes, especially trophic relationships and reproductive biology; microsporidan parasites of fishes; fish hematology. Mailing Add: Okla Coop Fishery Res Unit Okla State Univ Stillwater OK 74074

SUMMERFIELD, MARTIN, b New York, NY, Oct 20, 16; m 45; c 1. PHYSICS. Educ: Brooklyn Col, BS, 36; Calif Inst Technol, MS, 37, PhD(physics), 41. Prof Exp: Res engr, jet propulsion proj, Calif Inst Technol, 40-42, asst chief engr, 42-43, chief rockets & mat div, Jet Propulsion Lab, 45-49; chief liquid rocket develop dept, Aerojet Eng Corp, Calif, 43-45; ed, aeronaut publ prog, 49-52, PROF AERONAUT ENG, PRINCETON UNIV, 51- Concurrent Pos: Mem subcomt combustion, NASA, 49- Mem: AAAS; Am Phys Soc; Am Soc Mech Engrs; Am Inst Aeronaut & Astronaut; Int Acad Astronaut. Res: Infrared spectroscopy; soil erosion; rocket propellants; combustion; jet engines; heat transfer. Mailing Add: Dept of Aeronaut Eng Princeton Univ Princeton NJ 08540

SUMMERLIN, LEE R, b Sumiton, Ala, Apr 15, 34; m 58; c 4. CHEMISTRY, SCIENCE EDUCATION. Educ: Samford Univ, AB, 55; Birmingham Southern Col, MS, 60; Univ Md, College Park, PhD(sci educ), 71. Prof Exp: Chemist, Southern Res Inst, 56-59; teacher, Fla High Sch, 59-61; asst prof chem, Fla State Univ, 62-71; asst prof sci educ, Univ Ga, 71-72; ASSOC PROF CHEM & SCI EDUC, UNIV ALA,

BIRMINGHAM, 72- Concurrent Pos: Chemist, US Pipe & Foundry Co, 53-55; consult, Chem Educ Mat Study, 63-70, Cent Treaty Orgn, 64-, US Agency Int Develop, 66-68 & India Proj, NSF, 68-; teaching assoc, Univ Md, 70-71. Honors & Awards: James Conant Award, Am Chem Soc, 69. Mem: AAAS; Am Chem Soc; Am Inst Chemists; Nat Asn Res Sci Teaching; Nat Sci Teachers Asn. Res: Computer assisted instruction; developing chemistry material for nonscience majors; autotutorial and individualized instruction in chemistry. Mailing Add: 1786 Cornwall Rd Birmingham AL 35226

SUMMERS, CHARLES EUGENE, b Murray, Ky, Apr 30, 33; m 51; c 2. ANIMAL NUTRITION. Educ: Univ Ky, BS, 55, MS, 56; Iowa State Univ, PhD(ruminant nutrit), 59. Prof Exp: Asst, Univ Ky, 55-56; asst, Iowa State Univ, 56-59, asst prof animal husb & animal husbandman, Exten, 59-63; field res rep, Eli Lilly & Co, Ind, 63-72; dir tech sales, Quali-Tech Prod Co, Minn, 72-73; DIR MKT, CENT SOYA CO, INC, 73- Res: Basic requirements of ruminant nutrition. Mailing Add: 10102 Coverdale Rd Ft Wayne IN 46809

SUMMERS, CHARLES GEDDES, b Ogden, Utah, Dec 24, 41. ECONOMIC ENTOMOLOGY. Educ: Utah State Univ, BS, 64, MS, 66; Cornell Univ, PhD(entom), 70. Prof Exp: Res fel entom, Utah State Univ, 64-66; res asst, Cornell Univ, 66-70; asst entomologist, 70-75, ASSOC ENTOMOLOGIST, UNIV CALIF, BERKELEY, 75- Mem: Entom Soc Am; Entom Soc Can; Crop Sci Soc Am; AAAS. Res: Biology and population dynamics of arthropods associated with field crops; host-plant resistance to arthropods attacking alfalfa and economic threshold levels of arthropods attacking alfalfa and cereal crops. Mailing Add: Dept of Entom Sci Univ of Calif Berkeley CA 94720

SUMMERS, CHARLES GENE, organic chemistry, see 12th edition

SUMMERS, DENNIS BRIAN, b Natrona Heights, Pa, Aug 4, 43; m 63; c 2. PLANT BREEDING. Educ: Ind Univ, Pa, BS, 66, MEd, 68; Pa State Univ, PhD(bot), 73. Prof Exp: Teacher pub sch, Pa, 66-67; mem staff & fac chem & biol warfare, US Army Chem Ctr & Sch, 67-69; PLANT BREEDER, ASGROW SEED CO, SUBSID UPJOHN CO, 74- Mem: Nat Sweet Corn Breeders Asn; Am Seed Trade Asn; Am Soc Hort Sci. Res: Breeding for disease resistance at the cell culture level. Mailing Add: 125 E Ave Nevada IA 50201

SUMMERS, DONALD BALCH, b Maplewood, NJ, Oct 18, 02; m 43; c 1. CHEMISTRY. Educ: Wesleyan Univ, BS, 24, MA, 26; Princeton Univ, AM, 29; Columbia Univ, PhD(org chem), 32. Prof Exp: Asst, Wesleyan Univ, 24-26; instr chem, Amherst Col, 26-27 & Princeton Univ, 27-29; res & develop chemist, Thomas A Edison Co, 29-30 & Chas Pfizer & Co, 30-33; instr chem, High Sch, Columbia Univ, 33-42 & 46-57; proj asst, Fund Adv Educ Chem Film Proj, Univ Fla, 57-58; prof chem, Glassboro State Col, 58-60; Olin chem teacher, Alton Sr High Sch, Ill, 60-67; prof, 67-73, EMER PROF CHEM, N MEX STATE UNIV, 73- Concurrent Pos: Lectr, South Orange-Maplewood Adult Sch, West Orange Adult Sch & Pelham Adult Sch, NJ, 37-42; consult, 34- Mem: AAAS; Am Chem Soc; fel Am Inst Chemists. Res: Catalysis; electro-organic chemistry; corrosion; photography; chemical education. Mailing Add: 1710 Altura Ave Las Cruces NM 88001

SUMMERS, FRANCIS MARION, b Fredericktown, Mo, June 8, 06; m 40; c 1. ENTOMOLOGY. Educ: Univ Calif, AB, 30; Columbia Univ, MA, 31, PhD(protozool), 35. Prof Exp: Tutor, City Col New York, 31-34; tutor & instr biol, Bard Col, 34-37; tutor & instr anat, Med Col, Cornell Univ, 37-38; instr biol, City Col New York, 38-43; asst entomologist, 44-52, from assoc prof to prof, 52-73, EMER PROF ENTOM, UNIV CALIF, DAVIS, 73- Res: Economic entomology. Mailing Add: Dept of Entom Univ of Calif Davis CA 95616

SUMMERS, GEOFFREY P, b London, Eng. SOLID STATE PHYSICS. Educ: Oxford Univ, BA, 65, PhD(physics), 69. Prof Exp: Res asst physics, Univ NC, 70-72; ASST PROF PHYSICS, OKLA STATE UNIV, 73- Mem: Am Phys Soc. Res: Experimental investigation of the electronic structure of defects in solids by means of photoconductivity, luminescence, optical absorption and lifetime studies. Mailing Add: Dept of Physics Okla State Univ Stillwater OK 74074

SUMMERS, JAMES THOMAS, b Nashville, Tenn, Nov 4, 38; m 58; c 3. INORGANIC CHEMISTRY. Educ: Vanderbilt Univ, AB, 60; Fla State Univ, PhD(inorg chem), 63. Prof Exp: Fel, Fla State Univ, 64 & Univ Tex, 64-65; res chemist, 65-71, SR RES CHEMIST, TEXTILE FIBERS DEPT, E I DU PONT DE NEMOURS & CO, INC, 71- Mem: Am Chem Soc; The Chem Soc. Res: Coordination chemistry; polymer chemistry. Mailing Add: 823 Kay Circle Chattanooga TN 37421

SUMMERS, JOHN CLIFFORD, b Chicago, Ill, Dec 4, 36. PESTICIDE CHEMISTRY. Educ: Augustana Col, Ill, BA, 58; Univ Ill, PhD(org chem), 63. Prof Exp: RES CHEMIST, EXP STA, E I DU PONT DE NEMOURS & CO, INC, 63- Mem: Am Chem Soc. Mailing Add: Exp Sta Bldg 324 E I du Pont de Nemours & Co Inc Wilmington DE 19898

SUMMERS, JOHN D, agriculture, see 12th edition

SUMMERS, LAWRENCE, b Bevier, Mo, June 21, 14; div; c 2. ORGANIC CHEMISTRY. Educ: Iowa State Col, BS, 39, PhD(chem), 50; Utah State Agr Col, MS, 41. Prof Exp: Jr chemist, US Bur Mines, 41-42; res chemist, Remington Arms Co, E I du Pont de Nemours & Co, Inc, 42-45; res chemist, Columbia Chem Div, Pittsburgh Plate Glass Co, 46-47; from asst prof to assoc prof, 50-56, honors prog coordr, 57-71, PROF CHEM, UNIV N DAK, 56- Concurrent Pos: Res assoc, Georgetown Univ, 60-61. Mem: Am Chem Soc. Res: Organometallic chemistry; organic chlorine compounds; linguistics; machine translation. Mailing Add: Dept of Chem Univ of NDak Grand Forks ND 58201

SUMMERS, MURRAY RUDULPH, biochemistry, see 12th edition

SUMMERS, PETER WILLIAM, b Birmingham, Eng, Sept 18, 29; m 59; c 2. METEOROLOGY. Educ: Univ Nottingham, BSc, 51; Imp Col, Univ London, dipl, 52; McGill Univ, PhD(meteorol), 64. Prof Exp: Meteorologist, NZ Meteorol Serv & Can Meteorol Serv, 52-58; consult meteorologist, Weather Eng Corp Can, Ltd, 58-62; res asst meteorol, McGill Univ, 62-63; res meteorologist, Res Coun Alta, 64-71, sr res officer & head hail & cloud physics sect, 71-73; CHIEF ATMOSPHERIC DISPERSION DIV, ATMOSPHERIC ENVIRON SERV, CAN DEPT ENVIRON, 73- Honors & Awards: Prize in Appl Meteorol, Can Meteorol Soc, 72. Mem: Am Meteorol Soc; Air Pollution Control Asn; Can Meteorol Soc; fel Royal Meteorol Soc. Res: Weather modification and hailstorm research; hail suppression experiments; development of cloud seeding techniques; chemistry of precipitation; air pollution and the urban heat island effect; environmental impact research. Mailing Add: Atmospheric Environ Serv 4905 Dufferin St Toronto ON Can

SUMMERS, ROBERT GENTRY, JR, b Sonora, Calif, Jan 10, 43; m 70. ZOOLOGY. Educ: Univ Notre Dame, BS, 65, MS, 67; Tulane Univ, PhD(anat), 71. Prof Exp: Fel, 70-71, asst prof, 71-75, ASSOC PROF ZOOL, UNIV MAINE, ORONO, 75- Mem: Am Soc Zoologists; Am Asn Anatomists; Am Soc Cell Biol; Soc Develop Biol. Res: Developmental biology, particularly invertebrate embryology and morphogenesis; electron microscopy of developing invertebrates and their gametes; control systems in growth and regeneration. Mailing Add: Dept of Zool Univ of Maine Orono ME 04473

SUMMERS, ROBERT MILTON, organic chemistry, biochemistry, see 12th edition

SUMMERS, ROBERT WENDELL, b Lansing, Mich, July 28, 38; m 61; c 3. GASTROENTEROLOGY. Educ: Mich State Univ, BS, 61; Univ Iowa, MD, 65. Prof Exp: NIH gastroenterol res fel, 68-70, asst prof, 71-74, ASSOC PROF GASTROENTEROL, DEPT MED, DIV GASTROENTEROL, UNIV IOWA HOSPS, 74- Concurrent Pos: Assoc dir, Gastroenterol Prog, Gastroenterol Div, Dept Med, Vet Admin Hosp, Iowa City, 70-71, consult, 71- Mem: Am Gastroenterol Asn; Am Col Physicians; Am Soc Gastrointestinal Endoscopy; Am Fedn Clin Res. Res: Interrelationships of electrical and motor activity of the small and large intestine, and intestinal flow as modulated by physiologic, pharmacologic, and pathologic influence; the therapy of Crohns Disease. Mailing Add: Dept of Internal Med Univ Hosp Iowa City IA 52242

SUMMERS, SELBY EDWARD, b Toronto, Ont, Aug 26, 24; US citizen; m 50; c 4. CHEMISTRY. Educ: State Univ NY Albany, BA, 50, MA, 51. Prof Exp: Engr, Gen Eng Lab, Gen Elec Co, 51-55, appln engr, X-ray Dept, 55-58; x-ray microscopist, Ernest F Fullam Inc, 58-59, lab mgr, 59-62, vpres, 63-72; mat analyst, RRC Int, Inc, 72-74; SR SCIENTIST, SPRAGUE ELEC CO, 74- Mem: Am Chem Soc; Electron Micros Soc Am; Am Soc Metals; Electron Probe Anal Soc Am. Res: Electron optics; x-ray microscopy; x-ray emission and electron microprobe analysis. Mailing Add: Res & Develop Ctr Sprague Elec Co North Adams MA 01247

SUMMERS, THOMAS EUGENE, b Senoia, Ga, June 7, 19; m 43, 61; c 4. ENTOMOLOGY, PLANT PATHOLOGY. Educ: Univ Ga, BSA, 41; Iowa State Univ, MS, 47, PhD, 50. Prof Exp: Asst plant path, Agr Exp Sta, Iowa State Univ, 46-50; plant pathologist, 50-65, ENTOMOLOGIST, ENTOM RES DIV, AGR RES SERV, USDA, 65- Concurrent Pos: Consult, Agency Int Develop, Cuba, Guatemala & Colombia. Mem: AAAS. Res: Economic entomology; biological control of sugarcane borer; wireworms; diseases of stem and leaf fiber plants, particularly jute, kenaf, sisal, Sansevieria and ramie; parasitic nematodes fiber plants and sugarcane; white grub control. Mailing Add: PO Box 584 Canal Point FL 33438

SUMMERS, WILLIAM ALLEN, SR, b Gary, Ind, Apr 22, 14; m 40; c 3. PARASITOLOGY, MEDICAL MICROBIOLOGY. Educ: Univ Ill, AB, 35, MS, 36; La State Univ, Tulane Univ, PhD(trop dis), 40. Prof Exp: Sr parasitologist pub health lab, Fla State Bd Health, 40-42; parasitologist malaria control, US Army, 42-45; asst prof bact & parasitol, Med Col Va, 45-47; from asst prof to assoc prof, 47-64, PROF MICROBIOL, SCH MED, IND UNIV, INDIANAPOLIS, 64- Concurrent Pos: USPHS res grants, Sch Med, Ind Univ, 48-; fel trop med, La State Univ, 58; vis scientist, Gorgas Mem Lab, Panama, 66-67. Mem: Am Soc Trop Med & Hyg; Am Soc Parasitol; Am Soc Microbiol; Tissue Cult Asn; Sigma Xi. Res: Amebiasis, toxoplasmosis, anaplasmosis, malaria; helminth parasites, especially diagnosis, research into growth and chemotherapy; electron and light microscopy; fluorescent microscopy. Mailing Add: Dept of Microbiol Ind Univ Sch of Med Indianapolis IN 46202

SUMMERS, WILLIAM ALLEN, JR, b Atlanta, Ga, Dec 4, 44; m 67; c 2. INDUSTRIAL ORGANIC CHEMISTRY, PHYSICAL ORGANIC CHEMISTRY. Educ: Wabash Col, AB, 66; Northwestern Univ, PhD(chem), 71. Prof Exp: Res assoc photochem, Okla Univ, 70-72, vis asst prof, 70-71; prog dir & group leader nucleotide synthesis, Ash Steven Inc, 72-73; sr res assoc photochem, Okla Univ, 73-74; DEVELOP CHEMIST HETEROGENEOUS CATALYSIS, IMC CHEM GROUP, 74- Mem: Am Chem Soc; Am Soc Photobiol; Sigma Xi. Res: Research and development involving heterogeneous catalysis particularly catalytic reduction and photochemistry of nitroparaffins and their derivatives; applications of and design of new reduction reactions. Mailing Add: IMC Chem Group 1331 S First St Terre Haute IN 47808

SUMMERS, WILLIAM CLARKE, b Corvallis, Ore, Sept 13, 36; m 64; c 2. MARINE ECOLOGY. Educ: Univ Minn, BME, 59, PhD(zool), 66. Prof Exp: Res assoc & investr squid biol, Marine Biol Lab, 66-72, NIH contract squid ecol, 69-72; ASSOC PROF ECOL, HUXLEY COL ENVIRON STUDIES, WESTERN WASH STATE COL & DIR, SHANNON POINT MARINE CTR, 72- Mem: AAAS; Am Soc Mech Engrs; Am Soc Zoologists. Res: Physiological ecology of aquatic mollusks; life history, autecology and population biology of the squid, Loligo pealei; similar studies on Puget Sound octopus species; marineecology; aquatic biology. Mailing Add: Huxley Col of Environ Studies Western Wash State Col Bellingham WA 98225

SUMMERS, WILLIAM COFIELD, b Janesville, Wis, Apr 17, 39. MOLECULAR BIOLOGY, BIOCHEMISTRY. Educ: Univ Wis-Madison, BS, 61, MS, 63, PhD(molecular biol)& MD, 67. Prof Exp: Asst prof radiobiol, 68-70, ASSOC PROF RADIOBIOL, MOLECULAR BIOPHYS & BIOCHEM, YALE UNIV, 70-, ASSOC PROF HUMAN GENETICS, 72- Concurrent Pos: NSF fel, Mass Inst Technol, 67-68. Mem: AAAS; Am Soc Microbiol; Am Soc Biol Chemists. Res: Regulation of gene expression in normal and virus infected cells and in the course of development and differentiation. Mailing Add: 333 Cedar St New Haven CT 06510

SUMMERS, WILLIAM FRANCIS, b St John's, Nfld, Sept 13, 19; m 50; c 7. GEOGRAPHY. Educ: Dalhousie Univ, BSc, 48; McGill Univ, MSc, 49, PhD(geog), 57. Prof Exp: From lectr to asst prof geog, McGill Univ, 50-59; assoc prof, 60-65, PROF GEOG, MEM UNIV NFLD, 65- Concurrent Pos: Mem, Can Coun Urban & Regional Res, 64-67 & Nat Adv Coun Geog Res, 67- Res: Atlas of Newfoundland; natural and recreational resources; resource use; population analysis of Newfoundland. Mailing Add: Dept of Geog Mem Univ of Nfld St John's NF Can

SUMMERS, WILLIAM HUNLEY, b Dallas, Tex, Feb 5, 36. MATHEMATICS. Educ: Univ Tex, Arlington, BS, 61; Purdue Univ, West Lafayette, MS, 63; La State Univ, Baton Rouge, PhD(math), 68. Prof Exp: Asst prof, 68-73, ASSOC PROF MATH, UNIV ARK, FAYETTEVILLE, 73- Concurrent Pos: NSF res grants, 69-74; Univ Ark travel grant, Brazil, 72; Fulbright-Hayes travel grant, Brazil, 73; vis prof, Inst Math, Fed Univ Rio de Janeiro, 73-75. Mem: Am Math Soc; Math Soc France. Res: Weighted approximation theory. Mailing Add: Dept of Math Univ of Ark Fayetteville AR 72701

SUMMERS, WILMA POOS, b Richmond, Ind, Dec 8, 37; m 65. BIOCHEMICAL GENETICS. Educ: Ohio Univ, BS, 59; Univ Wis-Madison, PhD(oncol), 66. Prof Exp: Fel oncol, McArdle Lab Cancer Res, Univ Wis-Madison, 66-67; fel biochem, Harvard Med Sch, 67-68; fel pharmacol, 68-69, res assoc, 69-72, RES ASSOC VIROL, DEPT THERAPEUT RADIOL, SCH MED, YALE UNIV, 72- Concurrent Pos: Consult, Nat Cancer Inst, NIH, 73-74. Mem: Am Soc Biol Chemists. Res: Biochemical genetics of herpes simplex virus; the isolation and characterization of herpes virus mutants to be used in the development of suppressor genetics in mammalian cell systems. Mailing Add: Radiobiol Lab Sch of Med Yale Univ 333 Cedar St New Haven CT 06510

SUMMERS-GILL, ROBERT GEORGE, b Sask, Can, Dec 22, 29. EXPERIMENTAL NUCLEAR PHYSICS. Educ: Univ Sask, BA, 50, MA, 52; Univ Calif, PhD(physics), 56. Prof Exp: From asst prof to assoc prof, 56-66, PROF PHYSICS, McMASTER UNIV, 66- Mem: Am Phys Soc; Optical Soc Am; Can Asn Physicists; Brit Inst Physics. Res: Atomic beam resonances; nuclear spectroscopy; direct reactions; nuclear shell model calculations; heavy ion reactions. Mailing Add: Dept of Physics McMaster Univ Hamilton ON Can

SUMMERSKILL, WILLIAM HEDLEY JOHN, b London, Eng, Jan 8, 26; nat US; m 51; c 1. GASTROENTEROLOGY. Educ: Oxford Univ, BA, 47, BM, BCh, 49, MA, 51, DM, 55, DSc, 75; FRCP. Prof Exp: Intern & resident, Affil Hosps, Med Sch, Univ London, 50-51; registr & tutor, Post-Grad Med Sch, Univ London & Hammersmith Hosp, 53-55; Rockefeller traveling fel, Med Res Coun, Thorndike Mem Lab, Harvard Med Sch, 55-56; sr registr, Dept Gastoenterol, Cent Middlesex Hosp, 57-58; from asst prof to assoc prof, 59-67, PROF MED, MAYO GRAD SCH & MED SCH, UNIV MINN, 67-, DIR GASTROENTEROL UNIT, MAYO CLIN, 63- Concurrent Pos: Consult, NIH, mem gastroenterol & nutrit training grants comt, 71-74; consult, Vet Admin Res Eval, 66-68; mem gastroenterol drug comt, Food & Drug Admin, 70-, chmn, 74- Mem: Fel Am Col Physicians; Asn Am Physicians; Am Gastroenterol Asn; Soc Exp Biol & Med; Am Asn Study Liver Dis (pres, 74). Res: Clinical and metabolic aspects of liver disease and liver failure; pathophysiology of digestion. Mailing Add: Gastroenterol Unit Mayo Clin Rochester MN 55901

SUMMERSON, CHARLES HENRY, b Catlettsburg, Ky, Nov 15, 14; m 44; c 3. GEOLOGY. Educ: Univ Ill, BS, 38, MS, 40, PhD(geol), 42. Prof Exp: Asst geol, Univ Ill, 38-42; asst geologist, US Geol Surv, 43-45; asst prof geol, Mo Sch Mines, 46-47; from asst prof to assoc prof, 47-72, asst to dir res found, 58-65, PROF GEOL, OHIO STATE UNIV, 72- Concurrent Pos: Staff assoc, NSF, 65-66. Mem: AAAS; Soc Econ Paleont & Mineral; Am Asn Petrol Geol; Geol Soc Am; Int Asn Sedimentol. Res: Stratigraphy; sedimentary petrography; paleontology; micropaleontology; photogeology. Mailing Add: Dept of Geol Ohio State Univ Columbus OH 43210

SUMMERVILLE, RICHARD MARION, b Shippenville, Pa, May 20, 38; m 62; c 2. MATHEMATICS. Educ: Clarion State Col, BS, 59; Wash Univ, AM, 65; Syracuse Univ, PhD(math), 69. Prof Exp: Instr math, Clarion State Col, 60-64; asst prof, State Univ NY Col Oswego, 68-69; res assoc, Syracuse Univ, 69-70; assoc prof math & chmn dept, 70-74, PROF MATH & HEAD DEPT MATH & COMPUT SCI, ARMSTRONG STATE COL, 74- Mem: AAAS; Math Asn Am; Am Math Soc. Res: Complex analysis; conformal and quasiconformal mapping; Schlicht functions. Mailing Add: Dept of Math & Comput Sci Armstrong State Col Savannah GA 31406

SUMMITT, ROBERT L, b Knoxville, Tenn, Dec 23, 32; m 55; c 3. PEDIATRICS, MEDICAL GENETICS. Educ: Univ Tenn, MD, 55, MS, 62; Am Bd Pediat, dipl, 62. Prof Exp: Intern, Mem Res Ctr & Hosp, Univ Tenn, Knoxville, 56; asst resident pediat, Col Med, Univ Tenn, Memphis, 59-60, chief resident, 60-61, USPHS trainee pediat endocrine & metab dis, Univ, 61-62; fel med genetics, Univ Wis, 63; from instr to assoc prof, 64-71, PROF PEDIAT & ANAT, COL MED, UNIV TENN, MEMPHIS, 71- Concurrent Pos: NIH, Children's Bur, State Ment Health Dept & Nat Found res grants; mem pediat staff, City of Memphis, Le Bonheur Children's Hosp; pediat consult, Baptist Mem, St Josph's, Methodist, US Naval, St Jude Children's Res & Arlington Hosps. Mem: AAAS; Am Acad Pediat; Am Soc Human Genetics; NY Acad Sci; Soc Pediat Res. Res: Clinical genetics and cytogenetics. Mailing Add: Dept of Peidat Univ of Tenn Col of Med Memphis TN 38163

SUMNER, DARRELL DEAN, b Kansas City, Mo, Jan 1, 41; m 63; c 3. MEDICINAL CHEMISTRY, METABOLISM. Educ: Univ Kans, AB, 63, PhD(med chem), 68. Prof Exp: Sr scientist drug metab, McNeil Labs, Inc, 68-71; sr anal chemist, 71-75, PROJ SCIENTIST, AGR DIV, CIBA-GEIGY CORP, 75- Concurrent Pos: Vis asst prof, Univ of NC Greensboro, 75-76. Mem: AAAS; Am Chem Soc. Res: Fate of pesticides in plants, animals and the environment. Mailing Add: Ciba-Geigy Corp Agr Div 410 Swing Rd Greensboro NC 27409

SUMNER, DONALD RAY, b Studley, Kans, Sept 20, 37; m 68. PLANT PATHOLOGY. Educ: Kans State Univ, BS, 59; Univ Nebr, MS, 64, PhD(plant path), 67. Prof Exp: Plant pathologist, Green Giant Co, Minn, 67-69; ASST PROF PLANT PATH, COASTAL PLAIN EXP STA, UNIV GA, 69- Mem: Am Phytopath Soc. Res: Root and foliage diseases of cucurbits, curcifers and other vegetables caused by fungi; ecology of soil-borne pathogens. Mailing Add: Dept of Plant Path Univ of Ga Coastal Plain Exp Sta Tifton GA 31794

SUMNER, DONOVAN BRADSHAW, b Cathcart, SAfrica, Aug 11, 10; nat Can; m 37; c 3. MATHEMATICS. Educ: Univ Witwatersrand, BA, 31 & 33, MA, 34, PhD, 47; Cambridge Univ, MSc, 40. Prof Exp: From lectr to sr lectr math, Univ Witwatersrand, 37-45; vis asst prof, La State Univ, 47-48; vis lectr, Univ Toronto, 48-49; from asst prof to prof, McMaster Univ, 50-70; PROF PURE MATH, UNIV WATERLOO, 70- Mem: Am Math Soc; Math Asn Am; Can Math Cong. Res: Integral transforms; operators. Mailing Add: Dept of Math Univ of Waterloo Waterloo ON Can

SUMNER, EDWARD D, b Spartanburg, SC, Mar 21, 25; m 47; c 2. PHARMACY. Educ: Wofford Col, BS, 48; Med Col SC, BS, 50; Univ NC, MS, 64, PhD(pharm), 66. Prof Exp: Instr pharm, Univ NC, 61-65; from asst prof to assoc prof, Univ Ga, 66-75; PROF PHARM COL PHARM, MED UNIV SC, 75- Concurrent Pos: Consult pharm internship, Vet Admin Hosp, Augusta, Ga, 68- Mem: Am Soc Hosp Pharmacists; Am Asn Cols Pharm; Sigma Xi; Am Pharmaceut Asn. Res: Physics of tablet compression and drug release from tablets; drug utilization in nursing homes and mental retardation centers; medication studies on the senior citizen such as compliance to physician's orders and factors determing when, what, and where drugs are purchased. Mailing Add: Col of Pharm 80 Barre St Med Univ of SC Charleston SC 29401

SUMNER, GEORGE GARDNER, b Canton, Ohio, June 13, 29; m 57; c 3. PHYSICAL CHEMISTRY, CRYSTALLOGRAPHY. Educ: Ohio State Univ, BSc, 51, MSc, 54; Univ Pittsburgh, PhD(phys chem), 61. Prof Exp: Res assoc crystallog, Mellon Inst, 54-61; MEM TECH STAFF, TEX INSTRUMENTS INC, 61- Mem: Am Crystallog Asn. Res: Epitaxial growth of thin metal and semiconductor films; structure and physical properties of crystal surfaces; preparation and properties of epitaxial magnetic garnet films. Mailing Add: Tex Instruments Inc MS-134 PO Box 5936 Dallas TX 75222

SUMNER, JOHN RANDOLPH, b Corpus Christi, Tex, Aug 28, 44; m 69; c 2. GEOPHYSICS. Educ: Univ Ariz, BS, 66; Stanford Univ, MS, 68, PhD(geophys), 71. Prof Exp: Asst prof earth sci, Univ Calif, Santa Cruz & fel, Crown Col, 71-72; ASST PROF GEOPHYS, LEHIGH UNIV, 72- Mem: AAAS; Geol Soc Am; Soc Explor Geophys; Am Geophys Union. Res: Solid earth geophysics; gravity and magnetic fields of geologic structures. Mailing Add: Dept of Geol Sci Lehigh Univ Bethlehem PA 18015

SUMNER, JOHN STEWART, b Bozeman, Mont, June 24, 21; m 43; c 3. EXPLORATION GEOPHYSICS, GEOLOGY. Educ: Univ Minn, BS, 47 & 48; Univ Wis, PhD, 55. Prof Exp: Staff geophysicist, Jones & Laughlin Steel Corp, 54-55; asst prof geophys, Western State Col, Colo, 55-56; mgr, McPharm Geophys, Inc, 56-57; chief geophysicist, Phelps Dodge Corp, 57-63; prof geophys, Col Mines, 63-72, PROF GEOPHYS, COL EARTH SCI, UNIV ARIZ, 72- Concurrent Pos: Consult, Mining Co. Mem: Soc Explor Geophys; Am Inst Mining, Metall & Petrol Eng; Am Geophys Union. Res: Mining geophysics including electrical resistivity and induced polarization methods and gravity, magnetic, and seismic exploration techniques. Mailing Add: 728 N Sawtelle Tucson AZ 85716

SUMNER, KENNETH, b New York, NY, Sept 29, 42; m 69; c 1. BIOINORGANIC CHEMISTRY. Educ: Union Col, NY, BS, 64; State Univ NY Upstate Med Ctr, PhD(biochem), 69. Prof Exp: USPHS fel, Univ Chicago, 69-71; asst prof chem, 71-74, ASSOC PROF CHEM, BRIDGEWATER STATE COL, 74- Concurrent Pos: Environ chem consult, Environ Protection Agency, 75-76. Mem: Am Chem Soc. Res: Trace metals; stability constants of lead-amino acid complexes. Mailing Add: Dept of Chem Bridgewater State Col Bridgewater MA 02324

SUMNER, LOWELL, b New York, NY, Dec 7, 07; m 30; c 1. BIOLOGY. Educ: Pomona Col, BA, 28; Univ Calif, MA, 33. Prof Exp: Econ biologist, State Div Fish & Game, Calif, 30-34; biologist, Nat Park Serv, 35-42, park planner, NMex, 45, biologist, 46-54, biologist, Sequoia & Kings Canyon Nat Parks, 54-58, regional res biologist, Calif, 58-60, prin biologist, 60-63, res biologist, 63-67; ENVIRON CONSULT, 68- Concurrent Pos: Sr ed tech staff, Desert Bighorn Coun. Mem: Wildlife Soc. Res: Ecology of game animals with relation to effects of predators. Mailing Add: Box 278 Glenwood NM 88039

SUMNER, RICHARD LAWRENCE, b Albany, NY, May 28, 38; m 64; c 2. HIGH ENERGY PHYSICS. Educ: State Univ NY Albany, BS, 59; Univ Chicago, SM, 65, PhD(physics), 75. Prof Exp: Res assoc physics, 72-73, MEM RES STAFF PHYSICS, PRINCETON UNIV, 73- Res: High energy elementary particle studies; direct muon production and particle production at large transverse momentum. Mailing Add: Dept of Physics Princeton Univ Princeton NJ 08540

SUMNER, ROBERT JOCELYN, b Springfield, Mass, Oct 26, 15; m 40; c 4. NATURAL PRODUCTS CHEMISTRY. Educ: Hobart Col, BA, 37; Cornell Univ, PhD, 41. Prof Exp: Asst, NY Exp Sta, 38-41; indust fel, Mellon Inst, 41-45; dir res, C J Patterson Co, 45-47; head res dept, Am Inst Baking, 47-48; asst res dir, 48-52, dir cent res dept, 52-67, MGR BAKERY PRODUCTS PROD, ANHEUSER-BUSCH, INC, 60- Mem: Indust Res Inst; Am Soc Bakery Eng; Am Asn Cereal Chem. Res: Bread staling, processing and microbiology; enzyme chemistry; cereal carbohydrates and amylases; yeast and starch technology; corn wet milling. Mailing Add: Anheuser-Busch Inc 9 Pestalozzi St St Louis MO 63118

SUMNER, ROGER D, b Hammond, Ind, Apr 18, 34; m 60; c 2. GEOPHYSICS. Educ: Rice Inst, BA, 56; Calif Inst Technol, MS, 58; Univ Wis, PhD(geophys), 65. Prof Exp: RES GEOPHYSICIST, EXPLOR DIV, GULF RES & DEVELOP CO, 65- Mem: Seismol Soc Am; Am Geophys Union. Res: Seismic wave propagation. Mailing Add: 216 Falconhurst Dr Pittsburgh PA 15238

SUMNER, THOMAS, b Akron, Ohio, Apr 16, 26. ORGANIC CHEMISTRY. Educ: Yale Univ, BS, 46, PhD(chem), 51. Prof Exp: Instr chem, Yale Univ, 46-50; from instr to prof chem, 50-57, Columbia-Southern prof, 57-61, from actg head to head dept chem, 52-61, from actg dean to dean col lib arts, 60-62, DEAN GEN COL, UNIV AKRON, 62-, PROF CHEM, 74- Mem: Am Chem Soc. Res: Organic synthesis. Mailing Add: 484 Hampshire Rd Akron OH 44313

SUMNICHT, RUSSELL WILLIAM, b Bingham Canyon, Utah, May 26, 16; m 38; c 3. DENTISTRY. Educ: Univ Southern Calif, DDS, 40; Johns Hopkins Univ, MPH, 56; Am Bd Dent Pub Health, dipl. Prof Exp: Dent officer, Dent Corps, US Army, 41-51, dept dir dent div, Army Med Serv Grad Sch, 51-55, commanding officer, 86th Med Detachment, Europe, 57-60, dent adv, Med Field Serv Sch, 60-61, chief prev dent br, Off Surgeon Gen, Dept Army, 61-66; asst dean, 66-67, actg dean, 67-68, ASSOC DEAN, SCH DENT, UNIV MO-KANSAS CITY, 68- Concurrent Pos: Consult, Surgeon Gen, US Army, 60- & Vet Admin Hosps, Kansas City & Wadsworth, Kans, 66-; mem deans comt, Vet Admin Hosp, Kansas City; mem, Am Asn Dent Schs; mem, Mo Health Manpower Planning Task Force, 74- Mem: Am Dent Asn; fel Am Pub Health Asn; Int Asn Dent Res. Res: Military dental service; role ot the dentist in mass disaster; preventive dentistry; dental public health. Mailing Add: Sch of Dent Univ of Mo Kansas City MO 64110

SUMPTION, LAVON JOHN, animal breeding, see 12th edition

SUMRALL, H GLENN, b Macon, Miss, Nov 8, 42; m 63; c 1. PLANT PHYSIOLOGY, MICROBIOLOGY. Educ: Southeastern La Col, BS, 64; La State Univ, MS, 66, PhD(plant path), 69. Prof Exp: Asst prof biol, Cornell Col, 69-73; ASSOC PROF BIOL, LIBERTY BAPTIST COL, 73- Mem: Am Phytopath Soc. Res: Fecal coliform pollution in streams and lakes; physiology of plant disease. Mailing Add: PO Box 1111 Liberty Baptist Col Lynchburg VA 24501

SUMRELL, GENE, b Apache, Ariz, Oct 7, 19. ORGANIC CHEMISTRY. Educ: Eastern NMex Col, AB, 42; Univ NMex, BS, 47, MS, 48; Univ Calif, PhD(chem), 51. Prof Exp: Asst, Univ NMex, 47-48; asst, Univ Calif, 48-51; asst prof, Eastern NMex Univ, 51-53; res chemist, J T Baker Chem Co, 53-58; sr org chemist, Southwest Res Inst, 58-59; proj leader chem & plastics div, Food Mach & Chem Corp, 59-61; res sect leader, El Paso Natural Gas Prod Co, Tex, 61-64; sr chemist, Southern Utilization Res & Develop Div, 64-67, head invests, 67-73, RES LEADER OILSEED & FOOD LAB, SOUTHERN REGIONAL RES CTR, AGR RES SERV, USDA, 73- Mem: AAAS; Am Chem Soc; fel Am Inst Chem; Am Oil Chem Soc; Am Asn Textile Chem & Colorists. Res: Fatty acids in tubercle bacilli; organic synthesis; pharmaceuticals; branched-chain compounds; triazoles; petrochemicals; fats and oils; monomers and polymers; cellulose chemistry; cereal chemistry. Mailing Add: Southern Regional Res Ctr PO Box 19687 New Orleans LA 70179

SUN, ALBERT YUNG-KWANG, b Amoy, Fukien, China, Oct 13, 32; m 64; c 1. BIOCHEMISTRY, NEUROCHEMISTRY. Educ: Nat Taiwan Univ, BS, 57; Ore State Univ, PhD(biochem), 67. Prof Exp: AEC fel & res assoc biochem, Case Western Reserve Univ, 67-68; res assoc lab neurochem, Cleveland Psychiat Inst, 68-72, proj dir, 72-74; PROF, SINCLAIR COMP MED RES FARM, UNIV MO, 74- Concurrent Pos: NIH gen res support grant, Cleveland Psychiat Inst, 68-73; Cleveland Diabetic Fund res grant, 72; alcohol grant, Nat Inst Alcohol Abuse & Alcoholism, 74; aging grant, Nat Inst Neurol Commun Dis & Strokes, 75. Mem: AAAS; Am Chem Soc; Biochem Soc; Am Soc Neurochem; Soc Neurosci. Res: Functional and structural relationship of the central nervous system membranes; active transport mechanism; drug effects on synaptic transmission of the central nervous system. Mailing Add: Sinclair Res Farm Univ of Mo Columbia MO 65201

SUN, ALEXANDER SHIHKAUNG, b Feb 21, 39; US citizen; m 68; c 1. BIOCHEMISTRY, CELL BIOLOGY. Educ: Taiwan Normal Univ, BS, 63; Univ Calif, Berkeley, PhD(biochem), 71. Prof Exp: Guest investr, Dept Biochem Cytol, Rockefeller Univ, 71-72; asst res physiologist, Dept Physiol & Anat, Univ Calif, Berkeley, 72-75; ASST PROF PATH, MT SINAI SCH MED, 75- Mem: Am Soc Cell Biol; Am Soc Photobiol. Res: Biochemistry of human aging, especially the effects of oxidative damage generated by subcellular organelles on cell aging, the mechanism controlling the different lifespans of normal and tumor cells in vitro. Mailing Add: Dept of Path Mt Sinai Sch of Med 5th Ave at 100th St New York NY 10029

SUN, BERNARD CHING-HUEY, b Nanking, China, Aug 23, 37; m 64; c 1. FOREST PRODUCTS, PULP CHEMISTRY. Educ: Nat Taiwan Univ, BSA, 60; Univ BC, MS, 67, PhD(wood-pulp sci), 70. Prof Exp: Jr specialist, Nat Taiwan Univ Res Forest, 61-63; demonstr wood pulp sci, Univ BC, 68-70; asst prof wood sci, 70-75, ASSOC PROF WOOD SCI DEPT FORESTRY, MICH TECHNOL UNIV, 75- Mem: Forest Prod Res Soc; Tech Asn Pulp & Paper Indust; Soc Wood Sci & Technol. Res: Wood products; wood-pulp relationships; wood pulping: developing a new pulping process. Mailing Add: Dept of Forestry Mich Technol Univ Houghton MI 49931

SUN, CHAO NIEN, b Hopeh, China, Dec 4, 14; m 46; c 2. EXPERIMENTAL PATHOLOGY. Educ: Nat Peking Univ, BSc, 40; Univ Okla, MSc, 50; Ohio State Univ, PhD, 53. Prof Exp: Asst bot, Nat Peking Univ, 40-45, lectr, 45-48; asst, Univ Okla, 48-50; asst, Ohio State Univ, 50-51; asst biophys, St Louis Univ, 52-53, res assoc biol & biophys, 54-57; fel anat, Wash Univ, 57-62; from asst prof to assoc prof path, St Louis Univ, 62-67; res assoc prof, Baylor Univ, 67-69; assoc prof path, Sch Med & electron microscopist, Hosp, 69-73, PROF PATH, SCH MED, UNIV ARK, LITTLE ROCK & CHIEF ELECTRON MICROSCOPE LAB, VET ADMIN HOSP, 73- Concurrent Pos: Fel, Inst Divi Thomae Found, 53-54. Mem: Am Soc Cell Biol; Electron Micros Soc Am. Res: Experimental virology; histochemistry; differentiation; tissue culture; biological and pathological ultrastructure; electron microscopy. Mailing Add: 2104 Gunpowder Rd Little Rock AR 72207

SUN, CHIH-REE, b Hsu-Chen, China, May 6, 23; m 56; c 3. HIGH ENERGY PHYSICS, NUCLEAR PHYSICS. Educ: Univ Calcutta, BSc, 47; Univ Calif, Los Angeles, MS, 51, PhD(physics), 56. Prof Exp: Teacher, Overseas Chinese High Sch, India, 47-49; mem res staff physics, Princeton Univ, 56-62; asst prof, Northwestern Univ, 62-65; assoc prof, Queens Col, NY, 65-68; ASSOC PROF PHYSICS, STATE UNIV NY ALBANY, 68- Concurrent Pos: Consult, Princeton Univ, 66- Mem: Am Phys Soc. Res: High energy experiments with bubble chambers study of elementary particles; accelerator; nuclear instrumentation. Mailing Add: Dept of Physics State Univ of NY Albany NY 12203

SUN, DAVID CHEN HWA, b Shanghai, China, Feb 18, 22; nat US; m 67. MEDICINE. Educ: St John's Univ, BS, 44, MD, 46; Univ Pa, MSc, 51, DSc(med), 55; Temple Univ, MD, 53. Prof Exp: Intern, St John's Univ Hosp, 45-46; fel, Hosp Univ Pa, 48-49 & Grad Hosp, 50-51; res assoc, Sch Med, Temple Univ, 51-54, instr med, 54-57, assoc, 57-59, asst prof, 59-60; dir gastroenterol & res lab, Vet Admin Hosp, Washington, DC, 60-65; DIR, INST GASTROENTEROL, GOOD SAMARITAN HOSP, 67- Concurrent Pos: Assoc clin prof med, Sch Med, George Washington Univ, 60-65; consult gastroenterol, Vet Admin Hosp, Phoenix, 65-73, assoc chief of staff, 67-73; consult gastroenterol, Good Samaritan Hosp, 67- Mem: Am Physiol Soc; Am Gastroenterol Asn; AMA; Am Fedn Clin Res; Am Col Physicians. Res: Gastrointestinal physiology; nutrition; pathogenesis of peptic ulcer disease and pancreatic function test; intestinal absorption studies. Mailing Add: 926 E McDowell Rd Phoenix AZ 85006

SUN, FRANK F, b Kiangshi, China, July 26, 38; m; c 2. BIOCHEMISTRY. Educ: Tunghai Univ, Taiwan, BS, 59; Tex Tech Univ, MS, 63; Univ Tex, Austin, PhD(biochem), 66. Prof Exp: NIH fel biochem, Purdue Univ, Lafayette, 66-68; RES BIOCHEMIST DEPT EXP BIOL, UPJOHN CO, 68- Mem: Am Chem Soc. Res: Lipid biochemistry; prostaglandin analysis and metabolism; drug metabolism; bioenergetics. Mailing Add: Dept of Exp Biol Res Lab Upjohn Co Kalamazoo MI 49001

SUN, HUGO SUI-HWAN, b Hong Kong, Oct 19, 40; m 67; c 2. ALGEBRA, NUMBER THEORY. Educ: Univ Calif, Berkeley, BA, 63; Univ Md, College Park, MA, 66; Univ NB, Fredericton, PhD(math), 69. Prof Exp: Asst prof math, Univ NB, Fredericton, 69-70; asst prof, 70-74, ASSOC PROF MATH, CALIF STATE UNIV, FRESNO, 74- Mem: Am Math Soc; Math Asn Am; NY Acad Sci. Res: Group theory, number theory, combinatorial analysis, and finite geometry. Mailing Add: Dept of Math Calif State Univ Fresno CA 93740

SUN, JAMES MING-SHAN, b China, May 10, 18; nat US; m 53; c 2. MINERALOGY, GEOPHYSICS. Educ: Nat Cent Univ, China, BS, 40; Univ Chicago, MS, 47; La State Univ, PhD(geol), 50. Prof Exp: Res fel, Columbia Univ, 50-51; mineralogist, State Bur Mines & Mineral Resources, NMex Inst Mining & Technol, 51-62; resident res fel, Jet Propulsion Lab, Calif Inst Technol, 62-64; res physicist, Air Force Weapons Lab, 64-69; res & writing, 69-74; PROF GEOPHYS SCI, NAT CENT UNIV, TAIWAN, REPUB CHINA, 74- Mem: AAAS; fel Mineral Soc Am; Am Geophys Union; Geochem Soc. Res: X-ray crystallography and fluorescent spectroscopy; volcanic rocks; authigenic minerals; minerals of New Mexico; physics of high pressure; hypervelocity impact and cratering; digital simulation techniques. Mailing Add: 7704 Sierra Azul NE Albuquerque NM 87110

SUN, KUAN-HAN, b Shaohsing, Chekiang, June 7, 14; US citizen; m 46; c 3. NUCLEAR SCIENCE, GLASS TECHNOLOGY. Educ: Chekiang Univ, BS, 36; Univ Pittsburgh, MS, 38, PhD(chem), 40. Prof Exp: Fel chem, Univ Pittsburgh, 40-42; res chemist, Eastman Kodak Co, 42-46; res physicist, 46-55, MGR RADIATION & NUCLEONICS, RES LABS, WESTINGHOUSE ELEC CORP, 55- Concurrent Pos: Lectr, Univ Pittsburgh 46-50; consult, Bausch & Lomb Co, 52-56; Fulbright prof, Nat Tsing Hua Univ, 59; dir, Inst Nuclear Sci Taiwan, 59-60, Int Atomic Energy Agency vis prof, 62. Honors & Awards: Meyer Award, Am Ceramic Soc, 48; Award, Chinese Inst Eng, 60; achievement cert, US Educ Found China, 60. Mem: Fel Am Phys Soc; fel Am Inst Chem; fel Am Ceramic Soc; Am Chem Soc. Res: Nuclear reaction; nuclear detection; radiation effects; nuclear application and technology; glass composition and properties; space radiation. Mailing Add: Res Labs Westinghouse Elec Corp Pittsburgh PA 15235

SUN, NAI CHAU, b Shanghai, China, June 15, 36; m 64. GENETICS, CELL BIOLOGY. Educ: Nat Taiwan Univ, BSc, 60; Univ Man, MSc, 65; Iowa State Univ, PhD(genetics, cell biol), 70. Prof Exp: NIH fel aging study, Biol Div, Oak Ridge Nat Lab, 70-72; RES ASSOC SOMATIC CELL GENETICS, MED SCH, UNIV MICH, ANN ARBOR, 72- Mem: Am Soc Microbiol. Res: Auxotrophic mutants induction and selection in cultured Chinese hamster cells in the study of pyrimidine pathway and glycolysis. Mailing Add: Dept of Human Genetics Univ of Mich Med Sch Ann Arbor MI 48104

SUN, PAUL LUN-FANG, plant breeding, plant genetics, see 12th edition

SUN, SIAO FANG, b Shaoshing, China, Feb 19, 22; m 51; c 3. PHYSICAL CHEMISTRY. Educ: Nat Chengchi Univ, Taiwan, LLB, 45; Univ Utah, MA, 50; Loyola Univ Chicago, MS, 56; Univ Chicago, PhD, 58; Univ Ill, PhD, 62. Prof Exp: Prof math, Northland Col, 60-64; asst prof chem, 64-70, ASSOC PROF CHEM, ST JOHN'S UNIV, NY, 70- Mem: Am Math Soc; Am Chem Soc; Am Phys Soc. Res: Theoretical molecular kinetics; physical chemistry of macromolecules. Mailing Add: Dept of Chem St John's Univ Jamaica NY 11432

SUN, SUNG HUANG, b Hongchow, China, June 17, 23; nat US; m 47; c 1. MEDICAL MYCOLOGY. Educ: St John's Univ, BS, 44; Mich State Col, BS, 50, PhD, 53. Prof Exp: Res asst, Yale Univ, 53; res asst, Univ Ga, 53-54; res assoc dermat dept med, Univ Chicago, 54-63; sr engr, NAm Aviation, 63-65; microbiologist, Vet Admin Hosp, San Fernando, 65-72, MICROBIOLOGIST, VET ADMIN HOSP, LONG BEACH, 72- Mem: AAAS; Mycol Soc Am; Bot Soc Am. Res: Morphological and physiological studies of actinomycetes; mechanism of airborne infections; various aspects of biological studies of Coccidioides immitis. Mailing Add: 151 Mycol Res Vet Admin Hosp 5901 E Seventh St Long Beach CA 90801

SUN, YUN PEI, b China, June 20, 10; nat US; m 38; c 2. INSECT TOXICOLOGY. Educ: Nat Cheking Univ, China, BS, 32; Univ Minn, MS, 41, PhD(entom), 43. Prof Exp: Jr chemist in charge insecticides lab, Nat Agr Res Bur, China, 35-39; res fel, Cornell Univ, 44-48; res entomologist & dir insecticides text lab, Julius Hyman & Co, 48-52; asst mgr in charge entom & residue anal labs, Shell Develop Co, 52-54, mgr entom dept, 54-57, chief entomologist biol sci res ctr, 57-70; CONSULT, 70- Concurrent Pos: Consult, Sino-US Joint Comn Rural Reconstruct, 71 & 72-74; prof, Nat Taiwan Univ, 72; dir, Plant Protection Ctr, Taiwan, 72-74, hon adv, 74- Mem: AAAS; Entom Soc Am; Sigma Xi. Res: Bioassay of insecticides and their residues; correlation between chemical structure and insect toxicity; joint action of insecticides; dynamics of pesticide toxicology; pesticide regulations. Mailing Add: 1918 La Villa Rose Court Modesto CA 95350

SUNAHARA, FRED AKIRA, b Vancouver, BC, Jan 22, 24; m 52; c 5. PHARMACOLOGY. Educ: Univ Western Ont, BSc, 48, PhD(physiol), 52. Prof Exp: Fel physiol, Univ Western Ont, 52-53; sci officer aviation med, Defence Res Med Lab, Toronto, 53-61; sr pharmacologist, Ayerst Res Lab, Montreal, 61-64; PROF PHARMACOL, UNIV TORONTO, 64- Concurrent Pos: Assoc physiol, Univ Toronto, 59-61; Ont Heart Found res grants, 65- Mem: Am Physiol Soc; Can Physiol Soc; Pharmacol Soc Can. Res: Cardiovascular and respiratory physiology with reference to experimental anemia and to hypobaric environment; autonomic and cardiovascular pharmacology; interrelationship of autonomic drugs and prostaglandin group of substances. Mailing Add: Dept of Pharmacol Univ of Toronto Fac of Med Toronto ON Can

SUND, ELDON H, b Plentywood, Mont, June 6, 30; m 57; c 4. ORGANIC CHEMISTRY. Educ: Univ Ill, Urbana, BS, 52; Univ Tex, PhD(org chem), 60. Prof Exp: Res chemist, E I du Pont de Nemours & Co, Del, 59-66; asst prof chem, Ohio Northern Univ, 66-67; from asst prof to assoc prof, 67-73, PROF CHEM, MIDWESTERN STATE UNIV, 73- Concurrent Pos: Hardin prof, Hardin Found, 75-76. Mem: AAAS; Am Chem Soc. Res: Synthesis of heterocyclic compounds. Mailing Add: Dept of Chem Midwestern State Univ Wichita Falls TX 76308

SUND, KENNETH A, plant physiology, agronomy, see 12th edition

SUND, PAUL N, b Thief River Falls, Minn, Nov 13, 32; m 56; c 2. OCEANOGRAPHY. Educ: Univ Calif, Santa Barbara, BA, 54; Univ Wash, Seattle, MA, 56. Prof Exp: Biol oceanogr, Inter-Am Trop Tuna Comn, 56-63; oceanogr, 63-67, asst chief br marine fisheries, 67-71, NAT COORDR, PLATFORMS OF OPPORTUNITY PROGS, NAT MARINE FISHERIES SERV, 71- Concurrent Pos: Mem plankton adv comt, Smithsonian Inst, 64-; adj prof, Fla Atlantic Univ, 66-68; fel, Nat Inst Pub Affairs, 69. Mem: AAAS; fel Am Inst Fishery Res Biol; Marine Technol Soc; Am Soc Limnol & Oceanog; Marine Biol Asn UK. Res: Fishery and zooplankton ecology; chaetognath taxonomy and ecology; biological oceanography relative to biological indicators of water mass; ecology and life history of tropical tunas; aerial remote sensing of marine mammals. Mailing Add: Nat Marine Fisheries Serv Tiburon Lab 3150 Paradise Dr Tiburon CA 94920

SUND, RAYMOND EARL, b Capac, Mich, Dec 14, 32; m 61; c 4. NUCLEAR PHYSICS. Educ: Univ Mich, BSE(physics) & BSE(math), 55, MS, 56, PhD(physics), 60. Prof Exp: Assoc res physicist, Univ Mich Res Inst, 60-61; staff assoc, Gulf Radiation Technol, 61-63, STAFF MEM, GEN ATOMIC CO, SAN DIEGO, 63- Mem: Am Phys Soc; Am Nuclear Soc. Res: Investigation of decay schemes of radioactive isotopes; studies of photonuclear reactions and of prompt and delayed gamma rays from fission; afterheat and shielding studies for reactors. Mailing Add: 2635 Shalimar Cove Del Mar CA 92014

SUND, ROBERT B, b Los Angeles, Calif, Jan 24, 25. SCIENCE EDUCATION. Educ: Reed Col, BA, 50; Stanford Univ, MA, 52, EdD, 59; Ore State Univ, MS, 60. Prof Exp: Teacher pub high schs, Calif, 52-59; assoc prof sci educ, 60-70, PROF SCI EDUC, UNIV NORTHERN COLO, 70- Concurrent Pos: Lectr, Col Guam, 66 & 68 & Univ Hawaii, 68-69; mem bd dir, Sci Math Coun, Univ Northern Colo, 74-75; consult, Metro Teacher Ctr, Nashville, Tenn, 74-75. Mem: Fel AAAS; fel Asn Res Sci Teaching; Nat Asn Biol Teachers; Nat Sci Teachers Asn; Jean Piaget Soc. Res: Inquiry science teaching; science behavioral objectives; questioning techniques and applications of Piaget's theory to education; Piagetian applications to education and humanistic education. Mailing Add: Dept of Sci Educ Ross Hall Univ of Northern Colo Greeley CO 80639

SUNDARALINGAM, MUTTAIYA, b Taiping, Malaysia, Sept 21, 31; nat US; m 66; c 1. CRYSTALLOGRAPHY. Educ: Ceylon Univ, BSc, 56; Univ Pittsburgh, PhD(chem), 61. Prof Exp: Res assoc crystallog sch med, Univ Wash, Seattle, 62-65; res assoc lab molecular biol, Children's Cancer Res Found, Boston, Mass & Harvard Med Sch, 65-66; assoc prof chem, Case Western Reserve Univ, 66-69; PROF BIOCHEM, UNIV WIS-MADISON, 69- Concurrent Pos: John Simon Guggenheim fel lab molecular biophys, Dept Zool, Oxford Univ, 75-76. Mem: AAAS; Am Crystallog Asn; Am Chem Soc; Biophys Soc; The Chem Soc. Res: X-ray diffraction investigation of biological structures; conformational analysis; nucleic acids transfer ribonucleic acids; proteins; nucleic acid-protein complexes; membrane phospholipids; carbohydrates. Mailing Add: Dept of Biochem Univ of Wis Madison WI 53706

SUNDARAM, SWAMINATHA, b Arnipatty, India, Oct 22, 24; m 46; c 2. PHYSICS. Educ: Annamalai Univ, Madras, BSc, 45, MA, 47, PhD, 57, DSc(physics), 60; Ill Inst Technol, MS, 60. Prof Exp: Lectr physics, Annamalai Univ, Madras, 45-57; instr, Ill Inst Technol, Technol, 57-59; res assoc, Univ Chicago, 59-60; mem staff, BC Res

Coun, 60-61; assoc prof mech eng, Univ Sask, 61-62; assoc prof physics, Ill Inst Technol, 62-65; actg head dept, 67-68, PROF PHYSICS, UNIV ILL, CHICAGO, 65-, HEAD DEPT, 68- Mem: Fel Am Phys Soc. Res: Molecular and solid state spectroscopy; optics; thermodynamics; transport properties; atomic physics; radiative transfer; astrophysics. Mailing Add: Dept of Physics Univ of Ill Chicago IL 60680

SUNDARESAN, PERUVEMBA RAMNATHAN, b Madras, India, Aug 11, 30; m 70; c 1. NUTRITIONAL BIOCHEMISTRY. Educ: Univ Banaras, BSc, 50, MSc, 53; Indian Inst Sci, Bangalore, PhD(biochem), 58. Prof Exp: Res asst biochem, Coun Sci & Indust Res, New Delhi, 56-58; res asst, Indian Inst Sci, Bangalore, 58-59; sr res fel, Coun Sci & Indust Res, 59-61; res assoc nutrit biochem radio carbon lab, Univ Ill, Urbana, 61-62; res assoc, Mass Inst Technol, 62-64; Nat Acad Sci-Nat Res Coun res assoc environ biochem, US Army Res Inst Environ Med, Mass, 64-66, res biochemist, 66-68; CHIEF LIPIDS LAB RES INST, ST JOSEPH HOSP, 68- Concurrent Pos: NIH res grants dept animal sci, Univ Ill, 60-61 & dept nutrit & food sci, Mass Inst Technol, 61-64; res consult, Millersville State Col, 72-; consult biochem, Vet Admin Hosp, 73- Mem: AAAS; Am Inst Nutrit; Brit Biochem Soc; Am Oil Chem Soc. Res: Biochemical function and metabolism of vitamin A. Mailing Add: Lipids Lab Res Inst St Joseph Hosp Lancaster PA 17604

SUNDBERG, MICHAEL WILLIAM, b Battle Creek, Mich. PHYSICAL CHEMISTRY. Educ: Albion Col, BA, 69; Stanford Univ, PhD(phys chem), 73. Prof Exp: RES CHEMIST, EASTMAN KODAK CO RES LABS, 73- Honors & Awards: Von Hevesy Prize Nuclear Med, Soc Nuclear Med, 74. Mem: Am Chem Soc; AAAS. Mailing Add: Eastman Kodak Co Res Labs 1669 Lake Ave Kodak Park Rochester NY 14650

SUNDBERG, RICHARD J, b Sioux Rapids, Iowa, Jan 6, 38; m 63; c 2. ORGANIC CHEMISTRY. Educ: Univ Iowa, BS, 59; Univ Minn, Minneapolis, PhD(org chem), 62. Prof Exp: From asst prof to assoc prof, 64-74, PROF CHEM, UNIV VA, 74- Concurrent Pos: NIH res fel, Stanford Univ, 71-72. Mem: Am Chem Soc. Mailing Add: Dept of Chem Univ of Va Charlottesville VA 22903

SUNDBERG, ROBERT LEE, b Sterling, Ill, Feb 23, 18; m 42; c 3. ORGANIC CHEMISTRY, RESEARCH ADMINISTRATION. Educ: Knox Col, AB, 40; Univ Iowa, MS, 42, PhD(chem), 44. Prof Exp: Asst, Univ Iowa, 40-44, asst, Nat Defense Res Comt, 42-44; res fel, Gen Aniline & Film Corp, 44-55; DIR PHARMACEUT RES, JOHNSON & JOHNSON, 55- Mem: AAAS; Am Chem Soc. Res: Chemistry of surface actives; carbohydrate chemistry; ethylene oxide reactions; pharmaceutical product and process development. Mailing Add: Johnson & Johnson New Brunswick NJ 08903

SUNDBERG, RUTH DOROTHY, b Chicago, Ill, July 29, 15; div. ANATOMY. Educ: Univ Minn, BS, 37, MA, 39, PhD(anat), 43, MD, 53; Am Bd Path, dipl, 60. Prof Exp: Technician anat, Univ Minn, 37-39; instr path, Wayne Univ, 39-41; asst, Univ, 41-43, from instr to assoc prof anat, 43-60, dir hemat labs, Hosps, 45-74, prof lab med, Univ, 63-73, PROF ANAT, UNIV MINN, MINNEAPOLIS, 60-, PROF LAB MED & PATH, 73-, HEMATOLOGIST, UNIV HOSPS, 42 & 45-, CO-DIR HEMAT LABS, 74- Mem: Soc Exp Biol & Med; Am Soc Cell Biol; Am Soc Path & Bact; Am Soc Hemat; Am Asn Anatomists. Res: Morphologic hematology; diagnosis by aspiration or trephine biopsy of marrow; lymphocytogenesis in human lymph nodes; histopathology of lesions in the bone marrow; agnogenic myeloid metaplasia; sideroblastic anemia and hemochromatosis; fatty acid deficiency; laboratory medicine. Mailing Add: Dept Lab Med & Path Univ Minn Hosp Box 198 Mayo Bldg Minneapolis MN 55455

SUNDBERG, WALTER JAMES, b San Francisco, Calif, Sept 16, 39; m 64; c 2. MYCOLOGY. Educ: San Francisco State Col, BA, 62, MA, 67; Univ Calif, Davis, PhD(bot), 71. Prof Exp: Lectr bot, Univ Calif, Davis, 71-72; ASST PROF BOT, SOUTHERN ILL UNIV, CARBONDALE, 72- Mem: Mycol Soc Am; Brit Mycol Soc; NAm Mycol Asn. Res: Ultrastructure, taxonomy and ecology of fungi, emphasis Basidiomycetes. Mailing Add: Dept of Bot Southern Ill Univ Carbondale IL 62901

SUNDE, MILTON LESTER, b Volga, SDak, Jan 7, 21; m 46; c 3. POULTRY NUTRITION. Educ: SDak State Col, BS, 47; Univ Wis, MS, 49, PhD, 50. Prof Exp: From asst prof to assoc prof poultry sci, 51-57, PROF POULTRY SCI, UNIV WIS-MADISON, 57- Concurrent Pos: Res scientist, Rockefeller Found, Colombia, SAm, 60; mem animal nutrit comt, Nat Res Coun, 70; Int Feed Ingredient Asn travel grant, 71. Honors & Awards: Res Award, Am Feed Mfrs Asn, 61; teaching award, Poultry Sci Asn, 62. Mem: Am Chem Soc; Soc Exp Biol & Med; Am Inst Nutrit; fel Poultry Sci Asn (2nd vpres, 65; 1st vpres, 66, pres, 67-68); NY Acad Sci. Res: Unidentified factors; vitamins; amino acids; energy for chickens, turkey and pheasants. Mailing Add: Animal Sci Bldg Univ of Wis Madison WI 53706

SUNDEEN, JOSEPH EDWARD, b Manchester, NH, Nov 5, 43; m 64; c 2. ORGANIC CHEMISTRY. Educ: Rensselaer Polytech Inst, BS, 64; Purdue Univ, PhD(chem), 68. Prof Exp: Fel org chem, Syntex Res Div, Syntex Corp, 68-69; INVESTR CARDIOVASC RES, SQUIBB INST MED RES, 69- Mem: Am Chem Soc; NY Acad Sci. Res: Cardioactive medicinals. Mailing Add: Squibb Inst for Med Res Princeton NJ 08540

SUNDELIN, KURT GUSTAV RAGNAR, b Pitea, Sweden, Dec 21, 37; US citizen; m 63; c 4. ORGANIC CHEMISTRY, MEDICINAL CHEMISTRY. Educ: Idaho State Univ, BS, 62 & 65, MS, 65; Univ Kans, PhD(med chem), 69. Prof Exp: CHEMIST BIOL SCI RES CTR, SHELL DEVELOP CO, 69- Mem: Am Chem Soc. Res: Organic chemical synthesis of biologically active agents in area of animal health, nutrition and pesticides. Mailing Add: Biol Sci Res Ctr Shell Develop Co PO Box 4248 Modesto CA 95352

SUNDELIN, RONALD M, b New York, NY, Oct 20, 39; m 67; c 1. ELEMENTARY PARTICLE PHYSICS. Educ: Mass Inst Technol, BS, 61; Carnegie Inst Technol, MS, 63, PhD(physics), 67. Prof Exp: Res physicist, Carnegie-Mellon Univ, 67-69; res assoc elem particle physics, Wilson Lab, 69-75, SR RES ASSOC ELEM PARTICLE PHYSICS, NEWMAN LAB, CORNELL UNIV, 75- Mem: Am Phys Soc. Res: Medium energy experimental physics, especially muon physics; high energy experimental physics; accelerator physics. Mailing Add: Newman Lab Cornell Univ Ithaca NY 14850

SUNDELIUS, HAROLD WESLEY, b Escanaba, Mich, July 6, 30; m 55; c 2. GEOLOGY. Educ: Augustana Col, AB, 52; Univ Wis, MS, 57, PhD(geol), 59. Prof Exp: Geologist mil geol br, US Geol Surv, 59-61, regional geologist eastern br, 61-65; from asst to assoc prof geol, Wittenberg Univ, 65-74, assoc dean col, 71-75, prof, 74-75; VPRES ACAD AFFAIRS & DEAN COL, AUGUSTANA COL, ILL, 75- Mem: Geol Soc Am; Soc Econ Geol; Nat Asn Geol Teachers. Res: Appalachian geology, especially Piedmont; economic geology and mineral economics; military geology; geology of the Carolina slate belt; Precambrian geology of the Lake Superior region; massive sulfide deposits in greenstone belts. Mailing Add: Augustana Col Rock Island IL 61201

SUNDERLIN, CHARLES EUGENE, b Reliance, SDak, Sept 28, 11; m 36; c 4. ORGANIC CHEMISTRY. Educ: Univ Mont, AB, 33; Oxford Univ, BA, 35; Univ Rochester, PhD(chem), 39. Prof Exp: Instr chem, Union Col, NY, 38-41; instr, US Naval Acad, 41-43, from asst prof to assoc prof, 45-46; sci liaison officer, US Off Naval Res, London, 46-47, from dept sci dir to sci dir, 48-51; dep dir, NSF, 51-57; dep dir, Union Carbide Europ Res Assocs, SA, Belg, 57-62, res mgr defense & space systs dept, Union Carbide Corp, 62-65; spec asst to pres, Nat Acad Sci, 65-69; VPRES & SECY, ROCKEFELLER UNIV, 69- Concurrent Pos: US del gen assembly, Int Coun Sci Unions, Amsterdam, 52 & Oslo, 55; US del, Dirs Nat Res Ctrs, Milan, 55; mem, Comt Experts Scientists' Rights, Paris, 53; treas, Engrs & Scientists Comt, Inc, People-to-People Prog, 66- Mem: AAAS; Am Chem Soc; The Chem Soc; Royal Inst Gt Brit; Brit Soc Chem Indust. Res: Research administration and management; international cooperation in science and technology. Mailing Add: Rockefeller Univ 1230 York Ave New York NY 10021

SUNDERMAN, DONALD W, plant genetics, see 12th edition

SUNDERMAN, DUANE NEUMAN, b Wadsworth, Ohio, July 14, 28; m 53; c 3. RESEARCH ADMINISTRATION. Educ: Univ Mich, AB, 49, MS, 54, PhD(chem), 56. Prof Exp: Res chemist, Argonne Nat Lab & E I du Pont de Nemours & Co, 51-52; res chemist, Savannah River Proj, 52-54; res asst, Univ Mich, 54-55; prin chemist, 56, proj leader, 56-58, asst div chief, 58-59, chief chem physics div, 59-65, assoc mgr physics dept, 65-69, coordr basic res, 67-69, asst dir, 69-70, mgr soc & mgt systs dept, 70-73, assoc dir, 73-74, DIR TECH DEVELOP, BATTELLE MEM INST, 74- Concurrent Pos: Partic prog mgt develop, Harvard Bus Sch, 69; trustee, Columbus Area Leadership Prog, 75- Mem: Am Chem Soc; Am Nuclear Soc; hon mem Am Soc Testing & Mat. Res: Nuclear fuel development; nuclear reactor chemistry; industrial applications of radioisotopes; environmental radioactivity; research management; coal conversion. Mailing Add: 2011 Pevensey Ct Columbus OH 43220

SUNDERMAN, FREDERICK WILLIAM, b Altoona, Pa, Oct 23, 98; m 25; c 2. INTERNAL MEDICINE, CLINICAL PATHOLOGY. Educ: Gettysburg Col, BS, 19; Univ Pa, MD, 23, MS, 27, PhD(res med), 29; Am Bd Internal Med, dipl, 37; Am Bd Path, dipl, 44; Am Bd Clin Chem, dipl, 53. Hon Degrees: ScD, Gettysburg Col, 52. Prof Exp: Instr, Gettysburg Acad, 19; asst dermat, Univ Pa, 23, from instr to asst prof res med, 25-47, lectr, 34-47, ward physician, Univ Hosp, 34-40, from assoc to chief chem div, Wm Pepper Lab, 34-47; prof clin path, Sch Med, Temple Univ & dir lab clin med, Univ Hosp, 47-48; dir clin res, Univ Tex M D Anderson Hosp & Tumor Inst, 48-50; prof clin med, Emory Univ, 50-51; prof clin med & dir div metab res, Jefferson Med Col, 51-65; DIR, INST CLIN SCI, 65-; PROF PATH, HAHNEMANN MED COL, 70-, DIR, INST CLIN SCI, 74- Concurrent Pos: Resident physician, Pa Hosp, 23-25, chief chem lab, 29-33, chief metab & diabetic clins, 29-46, physician, 39-47; med dir explosives res lab, US Bur Mines, Carnegie Inst Technol, 43-45; actg med dir & med consult, Brookhaven Nat Lab, 47-48; consult, Los Alamos Sci Lab, 47-48, US Army Ord, Redstone Arsenal, Ala, Abington Mem & Vet Admin Hosps, 53-66; mem staff, Cleveland Clin Found, 48-49; trustee & vpres, Am Bd Path, 48-51, life trustee, 61-; prof, Post-Grad Sch Med, Univ Tex, 49; mem & chmn bd trustees, Gettysburg Col, 67-; med adv, Rohm and Haas Co; mem corp, Bermuda Biol Sta; mem, Pa Governor's Task Force Environ Health, 68-; ed in chief, Annals Clin Lab Sci, 71- Mem: Fel Am Soc Clin Path (pres, 50); fel Am Soc Clin Invest; fel Am Chem Soc; Asn Clin Scientists (pres, 56-58); fel Col Am Path. Res: Serum electrolytes; hazards of nickel exposure; metabolism; clinical chemistry; research medicine. Mailing Add: 1833 Delancey Pl Philadelphia PA 19103

SUNDERMAN, FREDERICK WILLIAM, JR, b Philadelphia, Pa, June 23, 31; m 63; c 3. CLINICAL PATHOLOGY, EXPERIMENTAL PATHOLOGY. Educ: Emory Univ, BS, 52; Jefferson Med Col, MD, 55. Prof Exp: Intern, Jefferson Med Col Hosp, 55-56, instr med, 60-63, assoc, 63-64; from assoc prof to prof path & dir clin lab, Col Med, Univ Fla, 64-68; PROF LAB MED & HEAD DEPT, SCH MED, UNIV CONN, 68- Mem: Am Asn Cancer Res; Am Asn Clin Chem; Am Asn Path & Bact; Am Col Physicians; Asn Clin Scientists (pres, 64-65). Res: Experimental carcinogenesis and trace metal metabolism; clinical biochemistry. Mailing Add: Dept of Lab Med Univ of Conn Health Ctr Farmington CT 06032

SUNDERMAN, HARVEY C, geology, see 12th edition

SUNDERWIRTH, STANLEY GEORGE, b El Dorado Springs, Mo, Aug 12, 30; m 52; c 4. ORGANIC CHEMISTRY. Educ: Tarkio Col, AB, 51; Ohio State Univ, PhD(org chem), 55. Prof Exp: From instr to assoc prof chem, Colo State Univ, 55-64; prof & chmn dept, Kans State Col, Pittsburg, 64-72; DEAN SCI & MATH, METROP STATE COL, 72- Concurrent Pos: Fulbright fel, Uruguay, 65, 68 & 70; NSF-AID consult, India, 67 & 69; Fulbright-Hays lectr, India, 75. Mem: AAAS; Am Chem Soc; Sigma Xi. Res: Organic mechanisms. Mailing Add: Metrop State Col 250 W 14th Ave Denver CO 80204

SUNDET, SHERMAN ARCHIE, b Litchville, NDak, Sept 25, 18; m 44; c 5. POLYMER CHEMISTRY. Educ: Concordia Col, BS, 39; Univ Idaho, MS, 41; Univ Minn, PhD(org chem), 48. Prof Exp: Chemist, B F Goodrich Co, Ohio, 42-45; instr org chem, Univ Calif, Los Angeles, 48-50; res chemist textile fibers dept pioneering res div, 50-54, res supvr, 54-70, RES ASSOC PLASTICS DEPT, E I DU PONT DE NEMOURS & CO, INC, 70- Mem: AAAS; Am Chem Soc; Sigma Xi. Res: Structure, properties and applications of polymers. Mailing Add: Plastics Dept Exp Sta E I du Pont de Nemours & Co Inc Wilmington DE 19898

SUNDFORS, RONALD KENT, b Santa Monica, Calif, June 3, 32; m 63; c 3. SOLID STATE PHYSICS. Educ: Stanford Univ, BS, 54, MS, 55; Cornell Univ, PhD(exp physics), 63. Prof Exp: Res assoc, 63-65, asst prof physics, 65-69, ASSOC PROF PHYSICS, WASH UNIV, ST LOUIS, 69- Mem: AAAS; Am Phys Soc; Am Asn Physics Teachers. Res: Nuclear magnetic resonance; low temperature physics; semiconductor research; ultrasonics; acoustic coupling to nuclear spins. Mailing Add: Dept of Physics Wash Univ St Louis MO 63130

SUNDHARADAS, GNANASIGAMONI, b Palliyadi, India, Mar 18, 36; m 69; c 2. BIOCHEMISTRY, IMMUNOLOGY. Educ: Univ Kerala, India, BS, 57; Princeton Univ, PhD(biochem), 66. Prof Exp: Res assoc biochem, Princeton Univ, 66-67 & 70-71; res assoc, Yale Univ, 67-68; asst prof, All India Inst Med Sci, 68-70; res assoc immunol, 71-73, ASST PROF MED MICROBIOL, UNIV WIS-MADISON, 73- Res: Biochemistry of membranes; anti-inflammatory effects of tumor cells. Mailing Add: Dept of Med Microbiol Univ of Wis-Madison Madison WI 53706

SUNDHEIM, BENSON ROSS, physical chemistry, see 12th edition

SUNDHOLM, NORMAN KARL, b Ely, Minn, Feb 3, 14; m 44; c 2. ORGANIC CHEMISTRY. Educ: Univ Minn, BChem, 41; Univ Ill, PhD(org chem), 44. Prof Exp: Sr res chemist, 44-56, group leader rubber chem res, 56-65, RES SCIENTIST, UNIROYAL CHEM DIV, 65- Mem: Am Chem Soc. Res: Agricultural fungicides; restricted rotation in aryl amines; rubber vulcanizing agents; accelerators; antioxidants and antiozonants; plastics curing agents; tire cord adhesives. Mailing Add: Bayberry Terr Middlebury CT 06762

SUNDICK, ROY, b Brooklyn, NY, May 8, 44; m 69. IMMUNOLOGY. Educ: Harpur Col, BA, 65; State Univ NY Buffalo, MA, 69, PhD(microbiol), 72. Prof Exp: Austrian Res Coun fel, Inst Gen & Exp Path, Univ Vienna, 71-73; ASST PROF IMMUNOL & MICROBIOL, SCH MED, WAYNE STATE UNIV, 74- Mem: Ger Soc Immunol; Sigma Xi. Res: Pathogenesis of autoimmune disease and mechanism of self-recognition. Mailing Add: Dept Immunol & Microbiol Wayne State Univ Sch Med Detroit MI 48201

SUNDSTEN, JOHN WALLIN, b Seattle, Wash, Jan 16, 33; m 63; c 5. ANATOMY. Educ: Univ Calif, Los Angeles, AB, 56, PhD(anat), 61. Prof Exp: Asst anat, Sch Med, Univ Calif, Los Angeles, 57-59; NSF fel, 61-62; from instr to asst prof, 62-70, ASSOC PROF ANAT, SCH MED, UNIV WASH, 70- Concurrent Pos: Vis scientist, USPHS, 64-66; USPHS res grant, 64-; NIH spec fel, Bristol, Eng, 68-69; vis prof, Univ Malaya, 73-74. Mem: AAAS; Am Asn Anatomists. Res: Neuroendocrinology; hypothalamic regulatory mechanisms; psychophysiology; neurophysiology. Mailing Add: Dept of Biol Structure Univ of Wash Sch of Med Seattle WA 98105

SUNG, CHENG-PO, b Hsinchu, Taiwan, Oct 21, 35; m 65; c 2. BIOCHEMISTRY. Educ: Chung Hsing Univ, Taiwan, BSc, 59; McGill Univ, PhD(biochem), 67. Prof Exp: Fel biochem, McGill Univ, 66-67; res scientist, Food & Drug Directorate, Dept Nat Health & Welfare, Can, 67-68; res assoc dept pharmacol, Univ Wis, 68-69; assoc sr investr, Smith Kline & French Labs, 69-75, ASSOC SR INVESTR BIOCHEM, SMITH-KLINE LABS, 75- Mem: AAAS; NY Acad Sci; Biochem Soc; Am Chem Soc. Res: Enzyme related to membrane transport; cyclic nucleotide research; biological aspect of immediate hypersensitivity. Mailing Add: Smith-Kline Labs F10 1500 Spring Garden St Philadelphia PA 19101

SUNG, CHI CHING, b Nanking, China, Mar 5, 36; m 68. THEORETICAL PHYSICS. Educ: Taiwan Nat Univ, BS, 57; Univ Calif, Berkeley, PhD(physics), 65. Prof Exp: Res assoc physics, Ohio State Univ, 65-67, lectr, 67-68, asst prof, 68-72; ASSOC PROF PHYSICS, UNIV ALA, HUNTSVILLE, 72- Res: Superconductivity; transport theory; magnetism. Mailing Add: Dept of Physics Univ of Ala Huntsville AL 35807

SUNG, CHIEN-BOR, b Shanghai, China, Feb 1, 25; nat US; m 53; c 2. RESEARCH ADMINISTRATION, INTERDISCIPLINARY SCIENCES. Educ: Chiao-Tung Univ, China, BS, 45; Mass Inst Technol, SM, 48; Harvard Univ, MBA, 50. Prof Exp: From engr to dept chief, Nanking-Shanghai Rwy Systs Admin, China, 45-47; develop engr instrumentation, Ruge-De Forest, Inc, 50-52; engr, Res Labs, Bendix Corp, 52-62, asst gen mgr, 62-64, gen mgr & dir, 64-67, vpres eng & res, 67-69, vpres & group exec advan technol, 69-74; PRES & CHIEF EXEC, CMA, INC, 74- Concurrent Pos: Mem adv panel nat metric study, Secy Com, 69-71; mem vis bd, Col Eng, Oakland Univ, 74-; mem vis comt, Carnegie-Mellon Univ, 74-; mem bd dirs, Codata Corp, ETEC Corp & Galileo Electro-Optics Corp, 75-; chmn bd, Airborne Mfg Co, 75-; mem indust comt, Col Eng, Univ Mich, 75- Mem: Sigma Xi. Res: Provide consultation to companies engaged in fluid control technology, communication systems utilizing carrier-current and digital technologies, electro-optics utilizing channel-electron technology, fiber-optic data transmission, scanning electron microscopes and electron-beam lithography. Mailing Add: CMA Inc PO Box 6347 Cleveland OH 44101

SUNG, JOO HO, b Korea, Feb 18, 27; US citizen; m 59; c 3. PATHOLOGY, NEUROPATHOLOGY. Educ: Yonsei Univ, Korea, MD, 52. Prof Exp: Resident path, Newark Beth Israel Hosp, 54-57; fel neuropath, Col Physicians & Surgeons, Columbia Univ, 57-61, asst prof, 61-62; from asst prof to assoc prof, 62-69, PROF NEUROPATH, MED SCH, UNIV MINN, MINNEAPOLIS, 69- Concurrent Pos: Nat Inst Neurol Dis & Stroke fel, Columbia Univ, 59-61; consult, Minneapolis Vet Admin Hosp, 62-68; mem, NIH Neurol Sci Res Training Comt, 70-73; vis prof, Med Col, Yonsei Univ, Korea, 72. Mem: Am Asn Neuropath; Am Acad Neurol; Asn Res Nerv & Ment Dis. Res: Aging changes in the nervous system; x-radiation effects on the nervous system. Mailing Add: Neuropath Lab Univ of Minn Med Sch Minneapolis MN 55455

SUNG, SHAN-CHING, biochemistry, neurochemistry, see 12th edition

SUNIER, JULES WILLY, b Saint-Imier, Switz, Nov 10, 34; m 58; c 2. EXPERIMENTAL NUCLEAR PHYSICS. Educ: Swiss Fed Inst Technol, dipl, 57, Dr sc nat(nuclear physics), 62. Prof Exp: Res physicist, Swiss Fed Inst Technol, 57-64; from asst prof to assoc prof physics, Univ Calif, Los Angeles, 64-72; MEM STAFF PHYSICS DIV, LOS ALAMOS SCI LAB, UNIV CALIF, 72- Mem: Am Phys Soc; Swiss Phys Soc. Res: Nuclear structure; few nucleon transfer and charge exchange reactions; on line computers and data acquisition systems. Mailing Add: Physics Div Los Alamos Sci Lab Univ of Calif Los Alamos NM 87545

SUNLEY, JUDITH S, b Detroit, Mich, July 26, 46. NUMBER THEORY. Educ: Univ Mich, BS, 67, MS, 68; Univ Md, PhD(math), 71. Prof Exp: Asst prof math, 71-75, ASSOC PROF MATH, DEPT MATH, STATIST & COMPUT SCI, AM UNIV, 75- Mem: Am Math Soc. Res: Eisenstein series of Siegel Modular Group; generalized prime discriminants in totally real fields; class numbers of totally imaginary quadratic extensions of totally real fields. Mailing Add: Dept of Math Statist & Comput Sci Am Univ Washington DC 20016

SUNSHINE, IRVING, b New York, NY, May 17, 16; m 39; c 2. CLINICAL CHEMISTRY. Educ: NY Univ, BS, 37, MA, 41, PhD, 50. Prof Exp: Instr chem, Newark Col Eng, 41-47; asst prof, NJ State Teachers Col, 47-50; toxicologist & clin chemist, City of Kingston Lab, NY, 50-51; sr instr path & pharmacol, 51-54, from asst prof to assoc prof, 54-73, PROF PATH & PHARMACOL, CASE WESTERN RESERVE UNIV, 73-; TOXICOLOGIST, CUYAHOGA COUNTY CORONERS LAB, 51- Concurrent Pos: Asst biochemist, Univ Hosp, Univ Ohio. Honors & Awards: Ames Award, Am Asn Clin Chemists, 73. Mem: Am Chem Soc; Am Asn Clin Chemists; Am Acad Forensic Sci; Am Asn Poison Control Ctrs. Res: Alcohol; barbiturates; toxicology methodology; poison prevention programming; drugs of abuse. Mailing Add: 2121 Adelbert Rd Cleveland OH 44106

SUNSHINE, MELVIN GILBERT, b Chicago, Ill, Oct 14, 36; m 70; c 1. BACTERIAL GENETICS. Educ: Univ Ill, BS, 58; Univ Southern Calif, PhD(bact), 68. Prof Exp: Res microbiologist, San Diego State Col, 67-68; Jane Coffin Childs Mem Fund Med Res fel, Karolinska Inst, Sweden, 68-70; USPHS trainee, Dept Molecular Biol & Virus Lab, Univ Calif, Berkeley, 70-72; vis asst prof microbiol, Sch Med, Univ Southern Calif, 72-73; res assoc, 73-75, RES SCIENTIST MICROBIOL, UNIV IOWA, 76- Mem: Am Soc Microbiol. Res: Bacterial genetics; genetics of the temperate bacterial viruses P2 and P4; host factors associated with phages P2 and P4. Mailing Add: Dept of Microbiol Univ of Iowa Sch of Med Iowa City IA 52242

SUNTHARALINGAM, NAGLINGAM, b Jaffna, Ceylon, June 18, 33; m 61; c 3. RADIOLOGICAL PHYSICS. Educ: Univ Ceylon, BSc, 55; Univ Wis, MS, 66, PhD(radiol sci), 67. Prof Exp: Asst lectr physics, Univ Ceylon, 55-58; from instr radiol physics to assoc prof radiol, 62-72, PROF RADIOL & RADIATION THERAPY, JEFFERSON MED COL, 72- Concurrent Pos: Vis lectr, Grad Sch Med, Univ Pa, 67-, consult dept physics, 68- Mem: Am Asn Physicists in Med; Health Physics Soc; Soc Nuclear Med; Am Col Radiol. Res: Radiation dosimetry; thermoluminescence dosimetry; clinical dosimetry. Mailing Add: Dept of Radiol Jefferson Med Col Philadelphia PA 19107

SUNYAR, ANDREW WILLIAM, b Henderson, Mich, Sept 9, 20; m 43; c 3. NUCLEAR PHYSICS. Educ: Albion Col, AB, 42; Univ Ill, MS, 44, PhD(physics), 49. Prof Exp: Asst, Univ Ill, 42-43, 46-49, res assoc, 49; assoc physicist, 49-53, physicist, 53-60, SR PHYSICIST, BROOKHAVEN NAT LAB, 60- Concurrent Pos: NSF sr fel, Inst Theoret Physics, Denmark, 60-61. Mem: Fel Am Phys Soc. Res: Nuclear isomerism; nuclear disintegration schemes; heavy ion-induced reactions; nuclear magnetic moments; Coulomb excitation. Mailing Add: Brookhaven Nat Lab Upton NY 11973

SUOMI, VERNER EDWARD, b Eveleth, Minn, Dec 6, 15; m 41; c 3. METEOROLOGY. Educ: Winona State Col, BE, 38; Univ Chicago, PhD(meteorol), 53. Prof Exp: Teacher, pub schs, Minn, 38-42; res assoc meteorol, Univ Chicago, 44-48; from asst prof to assoc prof, 48-53, PROF METEOROL & SOILS, UNIV WIS-MADISON, 53-, DIR SPACE SCI & ENG CTR, 66- Concurrent Pos: Assoc prog dir atmospheric sci, NSF, DC, 62; chief scientist, US Weather Bur, 64-65; chmn comt adv to Nat Oceanic & Atmospheric Admin, Nat Acad Sci, 66-69, mem comt atmospheric sci, 66-, chmn US comt, Global Atmospheric Prog, 71-74; mem Nat Adv Comt Oceans & Atmosphere, 71-72. Honors & Awards: Meisinger Award, Am Meteorol Soc, 61, Rossby Res Medal, 68; Presidential Citation, 70; Robert M Losey Award, Am Inst Aeronaut & Astronaut, 71. Mem: Nat Acad Eng; AAAS; Am Meteorol Soc (pres, 68-69); Am Geophys Union; foreign mem Finnish Acad Sci & Lett. Res: Atmospheric radiation; meteorological satellites; environmental observation systems. Mailing Add: Space Sci & Eng Ctr Univ of Wis-Madison Madison WI 53706

SUPER, ARLIN B, b Flensburg, Minn, Nov 3, 38; m 60; c 1. METEOROLOGY. Educ: St John's Univ, Minn, BS, 60; Univ Wis, MS, 62, PhD(meteorol), 65. Prof Exp: Proj asst meteorol, Univ Wis, 60-65; assoc prof, 65-71, PROF METEOROL, MONT STATE UNIV, 71- Mem: AAAS; Am Geophys Union; Am Meteorol Soc. Res: Air-water interface phenomena; weather modification; hydrometeorology; mountain meteorology; limnology. Mailing Add: Dept of Meteorol Mont State Univ Bozeman MT 59715

SUPLINSKAS, RAYMOND JOSEPH, b Hartford, Conn, Aug 29, 39; m S9; c 3. PHYSICAL CHEMISTRY. Educ: Yale Univ, BS, 61; Brown Univ, PhD(chem), 65. Prof Exp: Mem tech staff, Bell Tel Labs, 64-65; from asst prof to assoc prof chem, Yale Univ, 65-72; ASSOC PROF CHEM & CHMN DEPT, SWARTHMORE COL, 72- Res: Theory of molecular collision processes; statistical mechanics of liquids. Mailing Add: Dept of Chem Swarthmore Col Swarthmore PA 19081

SUPPE, FREDERICK (ROY), b Los Angeles, Calif, Feb 22, 40. PHILOSOPHY OF SCIENCE, HISTORY OF SCIENCE. Educ: Univ Calif, Riverside, AB, 62; Univ Mich, AM, 64, PhD(philos), 67. Prof Exp: Instr philos, Univ Mich, 64-67; asst prof, Univ Ill, Urbana, 67-73; ASSOC PROF PHILOS, UNIV MD, COLLEGE PARK, 73-, CHAIRPERSON COMT HIST & PHILOS SCI, 75- Concurrent Pos: Educ adv, Indo-Am Prog, USAID, Kanpur, India, 65-67; NSF res grant, 73; Am Coun Learned Soc int travel award, 74; mem adv bd, Nat Workshop Teaching Philos, 74- Honors & Awards: Amicus Poloniae Award, Poland, 75. Mem: AAAS; Philos Sci Asn; Hist Sci Soc; Asn Symbolic Logic; Am Philos Asn. Res: Nature of scientific knowledge, including structure of theories and models, explanation, facts and scientific observation; growth of scientific knowledge; history of the philosophy of science; automata theory; sexual morality. Mailing Add: Comt Hist & Philos Sci 1131 Skinner Hall Univ of Md College Park MD 20742

SUPPE, JOHN EDWARD, b Los Angeles, Calif, Nov 30, 42; m 65; c 2. GEOLOGY, GEOPHYSICS. Educ: Univ Calif, Riverside, BA, 65; Yale Univ, PhD(geol), 69. Prof Exp: Assoc res geologist, Yale Univ, 69; NSF fel geol, Univ Calif, Los Angeles, 69-71; ASST PROF GEOL, PRINCETON UNIV, 71- Concurrent Pos: Assoc ed, Am J Sci, 75-81. Mem: Geol Soc Am; Am Geophys Union. Res: Tectonics; regional structural geology. Mailing Add: Dept of Geol & Geophys Sci Princeton Univ Princeton NJ 08540

SUPPLE, JEROME HENRY, b Boston, Mass, Apr 27, 36; m 64. ORGANIC CHEMISTRY. Educ: Boston Col, BS, 57, MS, 59; Univ NH, PhD(org chem), 63. Prof Exp: Res chemist, Univ Calif, Berkeley, 63-64; asst prof org & gen chem, 64-69, assoc dean arts & sci, 72-73, actg assoc provost, State Univ NY Cent Admin, 74-75, ASSOC PROF ORG & GEN CHEM, STATE UNIV NY COL, FREDONIA, 69-, ASSOC VPRES ACAD AFFAIRS, 73-, CHMN DEPT CHEM, 75- Concurrent Pos: NSF sci fac fel, Univ E Anglia, 70-71. Mem: AAAS; Am Chem Soc. Res: Heterocyclic chemistry; natural products; stereochemistry; organic spectroscopy; conformational studies in the heterocyclic systems; narcotic antagonists; homogeneous catalysis. Mailing Add: 4804 Berry Rd Fredonia NY 14063

SUPRAN, MICHAEL KENNETH, b New York, NY, Feb 13, 39; m 67; c 2. FOOD SCIENCE. Educ: Univ Ga, BS, 61, MS, 63, PhD(food sci), 68. Prof Exp: Scientist, Nutrit Prod Div, Mead Johnson & Co, 63-65; fel food sci, USPHS, 65-68; Fulbright fel food sci, Danish Meat Res Inst, 68-69; ASSOC DIR FOOD RES, THOMAS J LIPTON, INC, 69- Mem: Inst Food Technologists; Am Chem Soc; Am Oil Chemists Soc; Dairy Sci Asn; Sigma Xi. Res: Research and development of new food concepts, systems and products; administration of sensory testing of food products. Mailing Add: Thomas J Lipton Inc 800 Sylvan Ave Englewood Cliffs NJ 07632

SUPRUNOWICZ, KONRAD b Pulkovnikov, Siberia, Mar 3, 19; US citizen; m 52; c 1. MATHEMATICS. Educ: Univ Nebr, BSc, 52, MA, 53, PhD(math), 60. Prof Exp: Instr physics, Minot State Col, 54-55; instr math, Univ Nebr, 57-60; asst prof, Univ Idaho, 60-61; assoc prof, 61-69, PROF MATH, UTAH STATE UNIV, 69- Concurrent Pos: Vis assoc prof, Univ Nebr, 62-63. Mem: Asn Symbolic Logic; Math Asn Am; Am Math Soc. Res: Application of symbolic logic to the study of relational systems; methodology of science. Mailing Add: Dept of Math Utah State Univ Logan UT 84321

SURAK, JOHN GODFREY, b Milwaukee, Wis, July 13, 48; m 71. TOXICOLOGY. Educ: Univ Wis-Madison, BS, 71, MS, 72, PhD(toxicol), 74. Prof Exp: Res asst food sci, Univ Wis-Madison, 70-74; ASST PROF TOXICOL, UNIV FLA, 74- Mem: AAAS; Am Chem Soc; Inst Food Technologists; Sigma Xi. Res: Food toxicology as related to the analysis, metabolism, mode of action and excretion of food additive, natural products and pesticides. Mailing Add: Pesticide Res Lab Dept of Food Sci Univ of Fla Gainesville FL 32611

SURANYI, PETER, b Budapest, Hungary, Jan 31, 35; m 60; c 2. HIGH ENERGY PHYSICS. Educ: Eötvös Lorand Univ, Budapest, BS, 58; Acad Sci, USSR, PhD(physics), 64. Prof Exp: Jr res fel cosmic ray physics, Cent Res Inst Physics,

Budapest, Hungary, 58-61; res fel theoret physics, Joint Inst Nuclear Studies, Moscow, 61-65; sr res fel theoret high energy physics, Cent Res Inst Physics, Budapest, Hungary, 65-69; vis lectr physics, Johns Hopkins Univ, 69-70, res assoc, 70-71; assoc prof, 71-74, PROF PHYSICS, UNIV CINCINNATI, 74- Honors & Awards: Schmidt Award, Hungarian Phys Soc, 68. Mem: Am Phys Soc. Res: High energy behavior of strong interactions; group theoretic methods in elementary particle physics. Mailing Add: Dept of Physics Univ of Cincinnati Cincinnati OH 45221

SURAPANENI, CHALAPATHI RAO, b Unguturu, India, Jan 1, 41; m 70; c 1. SYNTHETIC ORGANIC CHEMISTRY. Educ: Andhra Univ, India, BS, 60, MS, 64; Rensselaer Polytech Inst, PhD(org chem), 69. Prof Exp: Demonstr chem, Hindu Col, Andhra Univ, Masulipatam, India, 61-62, demonstr, W Godavar Bhimavaram Col, 62-63; jr res fel org chem, Cent Leather Res Inst Adyar, 64-65; res asst, Rensselaer Polytech Inst, 65-69; asst prof chem, 69-75, ASSOC PROF CHEM, HUDSON VALLEY COMMUNITY COL, 75- Mem: Am Chem Soc. Res: Heterocyclic chemistry; synthesis and structural determination of heterocyclic compounds; cycloaddition reactions of heterocyclic compounds. Mailing Add: Dept of Chem Hudson Valley Community Col Troy NY 12180

SURATT, EDGAR CECIL, organic chemistry, see 12th edition

SURAWICZ, BORYS, b Moscow, Russia, Feb 11, 17; nat US; m 46; c 3. INTERNAL MEDICINE, CARDIOLOGY. Educ: Stefan Batory Univ, Poland, MD, 39; Am Bd Internal Med, dipl; Am Bd Cardiovasc Dis, dipl. Prof Exp: Instr cardiol, Sch Med, Univ Pa, 54-55; instr med, Col Med, Univ Vt, 55-57, asst prof exp & clin med, 56-62; assoc prof, 62-66, PROF MED, COL MED, UNIV KY, 66-, DIR CARDIOVASC DIV, 62- Concurrent Pos: Fel coun clin cardiol, Am Heart Asn. Mem: Fel Am Col Physicians; fel Am Col Cardiol; AMA; Am Physiol Soc; Asn Univ Cardiologists. Res: Electrocardiology; role of electrolytes in cardiac arrhythmias. Mailing Add: 806 Overbrook Ct Lexington KY 40502

SURBEY, DONALD LEE, b North Canton, Ohio, July 19, 40; m 61; c 2. ORGANIC CHEMISTRY. Educ: Manchester Col, BS, 61; Univ Notre Dame, PhD(org chem), 68. Prof Exp: Control chemist, Miles Labs, Inc, 61-63; RES CHEMIST, LUBRIZOL CORP, 67- Mem: Am Chem Soc. Res: Organic chemistry as related to process and product development in field of polymer chemistry and lubricant additives. Mailing Add: Lubrizol Corp PO Box 17100 Cleveland OH 44117

SURDY, TED E, b Wheeling, WVa, Jan 25, 25; m 50; c 4. BACTERIOLOGY, BIOCHEMISTRY. Educ: Purdue Univ, BS, 58, MS, 59, PhD(bact), 62. Prof Exp: Instr bact, Purdue Univ, 59-61; assoc prof bact & cell physiol, Kans State Teachers Col, 62-67; res assoc biol, Educ Res Coun Greater Cleveland, 67-68; chmn dept biol, 68-74, PROF BIOL, SOUTHWEST MINN STATE COL, 74- Concurrent Pos: NSF grant, 65-68. Mem: AAAS; Am Soc Microbiol; Soc Indust Microbiol; Nat Asn Biol Teachers. Res: Lytic reactions of gram-negative bacterial cell walls; membrane permeability of gram-negative bacteria; effect of lipids on lysis of gram-negative bacteria; audio-tutorial bacteriology. Mailing Add: Dept of Biol Southwest Minn State Col Marshall MN 56258

SURGALLA, MICHAEL JOSEPH, b Nicholson, Pa, May 12, 20; m 48; c 4. MEDICAL MICROBIOLOGY. Educ: Univ Scranton, BS, 42; Univ Chicago, PhD(bact), 46. Prof Exp: Bacteriologist, E R Squibb & Sons, NJ, 46-48; res assoc, Univ Chicago, 48-54; bacteriologist, Biol Sci Lab, Dept of Army, Ft Detrick, Md, 54-71; DIR CLIN MICROBIOL, ROSWELL PARK MEM INST, 71- Mem: AAAS; Am Soc Microbiol; Am Acad Microbiol; NY Acad Sci; Am Soc Clin Pathologists. Res: Medical bacteriology; staphylococcus food poisoning; influenza virus; experimental plague; bacterial virulence; pathogenic mechanisms; host resistance; endotoxins; fibrinolysis; opportunist pathogens. Mailing Add: Roswell Park Mem Inst 666 Elm St Buffalo NY 14263

SURGENOR, DOUGLAS MACNEVIN, b Hartford, Conn, Apr 7, 18; m 46; c 5. BIOCHEMISTRY. Educ: Williams Col, AB, 39; Mass State Col, MS, 41; Mass Inst Technol, PhD(org chem), 46. Prof Exp: Mem staff, Div Indust Coop, Mass Inst Technol, 42-45; res assoc phys chem, Harvard Med Sch, 45-50, asst prof, 50-55, asst prof biol chem, 55-60; sr investr, Protein Found, 56-60; head dept biochem, 60-64, dean, 62-68, provost, Fac Health Sci, 67-70, PROF BIOCHEM, SCH MED, STATE UNIV NY BUFFALO, 60-, RES PROF, SCH MGT, 71- Concurrent Pos: Assoc mem lab phys chem related to med & pub health, Harvard Univ, 50-54; consult, Vet Admin Hosp, Buffalo, 60-, chmn dean's comt, 62-68; mem med bd, Buffalo Gen & Buffalo Children's Hosps, 62-68; mem med coun, NY State Educ Dept, 62-68; bd sci counr, Div Biol Stand, NIH, 63-68; mem, Nat Heart Inst Prog Proj Comt B, 65-68; consult med bd, Millard Fillmore Hosp, 66-70; mem, Int Comt Thrombosis & Haemostasis, 63-, chmn, 70-72; mem nat blood resource prog adv comt, Nat Heart & Lung Inst, 69-73, chmn, 70-73; mem med adv comt, Am Nat Red Cross, 70-; pres, Ctr Blood Res, Boston, 72-; mem bd trustees, Children's Hosp Med Ctr, Boston, 75- Mem: AAAS; Am Soc Biol Chemists; Am Heart Asn; Am Soc Hemat; Int Soc Thrombosis & Haemostasis. Res: Blood biochemistry; blood and public policy; blood coagulation. Mailing Add: Sch of Med State Univ of NY Buffalo NY 14214

SURI, BALWANT RAI, food technology, see 12th edition

SURIA, AMIN, b Dhoraji, India, Aug 24, 42; m 74; c 1. NEUROPHARMACOLOGY. Educ: Univ Karachi, Pakistan, BS, 63, MS, 64; Vanderbilt Univ, Nashville, PhD(pharmacol), 71. Prof Exp: Chemist, United Paints Ltd, Karachi, Pakistan, 64; chemist & in-chg lab, Textile Dyes & Auxiliary Dept, Hoechst Pharmaceut Co, Ltd, Karachi, Pakistan, 64-66; vis fel, Lab Clin Pharmacol, Nat Heart & Lung Inst, NIH, Bethesda, Md, 71-72; fel, Lab Preclin Pharmacol, NIMH, St Elizabeth's Hosp, Washington, DC, 72-74, staff fel, 74-75; ASST PROF PHARMACOL, GEORGE WASHINGTON UNIV, WASHINGTON, DC, 75- Concurrent Pos: Guest worker, Lab Preclin Pharmacol, St Elizabeth's Hosp, NIMH. Mem: Am Soc Pharmacol & Exp Therapeut; Soc Neurosci; NY Acad Sci; AAAS. Res: Molecular mechanisms by which anti-anxiety, anticonvulsant, and antidepressant drugs exert their actions on complex neuronal pathways, research entails using electrophysiological and biochemical techniques. Mailing Add: Dept Pharmacol Med Sch George Washington Univ 2300 I St NW Washington DC 20037

SURIANI, ERNESTO, organic chemistry, physical chemistry, see 12th edition

SURKAN, ALVIN JOHN, b Drumheller, Alta, June 5, 34; m 67; c 2. PHYSICS, APPLIED MATHEMATICS. Educ: Univ Alta, BSc, 54; Univ Toronto, MA, 56; Univ Western Ont, PhD(physics), 59. Prof Exp: Sr demonstr geophys, Univ Western Ont, 58-59; Nat Res Coun Can fel, Univ Alta, 59-61; sci officer marine physics, Can Defence Res Bd, 61-62; fac mem physics, Univ BC, 62-63; staff consult geophys comput, res & develop ctr, IBM Corp, 64-65, mem res staff environ sci group, phys sci dept, Watson Res Ctr, 65-69; mem inst water resources res, 69-73, PROF COMPUT SCI, UNIV NEBR, LINCOLN, 69- Concurrent Pos: Consult geophys comput & resident visitor physics dept, Bell Tel Labs, 72. Mem: Am Geophys Union; Soc Explor Geophysicists; Am Inst Physics; Inst Elec & Electronics Engrs. Res: Magnetic and seismic methods in exploration geophysics; mathematical modeling in hydrology, geomorphology and educational psychology; algorithms for data interpretation; nonlinear optimization and symbolic computation in geophysics and chemical engineering. Mailing Add: Dept of Comput Sci Univ of Nebr Lincoln NE 68508

SURKO, CLIFFORD MICHAEL, b Sacramento, Calif, Oct 11, 41; m 65; c 2. PHYSICS. Educ: Univ Calif, Berkeley, AB, 64, PhD(physics), 68. Prof Exp: Res assoc physics, Univ Calif, Berkeley, 68-69; MEM TECH STAFF PHYSICS, BELL LABS, 69- Mem: AAAS; Am Inst Physics. Res: Experimental research in low temperature physics and in gas plasma physics; study of atoms, condensed matter and plasmas with light scattering. Mailing Add: Room 1C352 Bell Labs Murray Hill NJ 07974

SURKO, PAMELA TONI, b Britton, SDak, June 15, 42; m 65; c 2. ELEMENTARY PARTICLE PHYSICS. Educ: Univ Calif, Berkeley, AB, 63, PhD(physics), 70. Prof Exp: Res assoc, 70-71, instr, 71-72, ASST PROF PHYSICS, PRINCETON UNIV, 72- Mem: Am Phys Soc. Res: Strangeness-changing neutral currents; muon-induced events at high energy. Mailing Add: Dept of Physics Princeton Univ Princeton NJ 08540

SURLS, JOSEPH PLEAS, JR, chemistry, see 12th edition

SURMATIS, JOSEPH D, b Dickson City, Pa, Mar 22, 13; m 45; c 1. ORGANIC CHEMISTRY. Educ: Pa State Univ, BS, 36, MS, 37, PhD(org chem), 42. Prof Exp: Instr org chem, Pa State Univ, 40-42; consult, G J Esselen, Inc, Mass, 42-45; fel, 45-74, tech fel, 64, SR FEL, HOFFMANN-LA ROCHE, INC, 74- Concurrent Pos: Chmn, Int Carotenoid Symp, Cluj, Romania, 72. Mem: Fel AAAS; fel Am Inst Chem; Am Chem Soc; Sigma Xi. Res: Vitamin A chemistry; carotenoid chemistry; antibiotics; vitamin E. Mailing Add: 4 Sunset Rd West Caldwell NJ 07006

SURPURIYA, VIJAY B, b Poona, India, Oct 26, 41; US citizen; m 70; c 1. PHARMACEUTICAL CHEMISTRY, PHYSICAL PHARMACY. Educ: Ferguson Col, Poona, BS, 63; Philadelphia Col Pharm & Sci, BS, 69; Univ Mich, Ann Arbor, PhD(pharmaceut chem), 73. Prof Exp: SR PHARMACIST, AYERST LABS, 73- Res: Development of oral solid and liquid dosage forms and basic research in this area. Mailing Add: Ayerst Labs 64 Maple Ave Rouses Point NY 12979

SURREY, ALEXANDER ROBERT, b New York, NY, Mar 13, 14; m 39; c 1. ORGANIC CHEMISTRY. Educ: City Col New York, BS, 34; NY Univ, PhD(chem), 40. Prof Exp: Nat Defense Res Comt fel, Cornell Univ, 40-41; res chemist, 41-57, sect head, 57-60, asst dir chem res, 60-64, sr res fel & dir new prod, 64-67, dir develop res, 67-72, VPRES RES & DEVELOP, STERLING-WINTHROP RES INST, 72- Concurrent Pos: Lectr & adj prof, Rensselaer Polytech Inst, 58-64. Mem: Am Chem Soc; fel NY Acad Sci; fel The Chem Soc. Res: Medicinals. Mailing Add: 15 Harvard Ave Albany NY 12208

SURREY, KENNETH, b India, Dec 6, 22; nat US; m 52; c 4. PHYTOCHEMISTRY. Educ: Univ Punjab, India, BSc, 46, MA, 52; Univ Mo, MA, 55, PhD(phytochem), 57. Prof Exp: Lab instr chem, Forman Christian Col, Pakistan, 49-53; asst bot, Univ Mo, 54-57; asst plant physiologist, Argonne Nat Lab, 57-66; HEALTH SCIENTIST ADMINR, NIH, 66- Res: Histochemistry of protein constituents by azo-coupling reactions in plants; metabolic responses of regenerating meristems and germinating seeds as influenced by visible and ionizing radiation; action and interaction of red and far-red radiation on metabolic processes of developing seedlings; physiological bases for morphological development. Mailing Add: Nat Inst of Neurol & Commun Disorders & Stroke Westbard Ave Bethesda MD 20014

SURVANT, WILLIAM G, b Owensboro, Ky, Aug 26, 07; m 36. SOIL SCIENCE. Educ: Univ Ky, BS, 31, MS, 45; Ohio State Univ, PhD(soil sci), 51. Prof Exp: Teacher high sch, Ky, 31-38; soil conservationist soil conserv serv, USDA, 38-42; exten soil conservationist, 42-47, asst prof soils, 47-50, from assoc prof to prof agron, 51-74, EMER PROF AGRON, UNIV KY, 74- Mem: Am Soc Agron; Soil Sci Soc Am. Res: Soil conservation. Mailing Add: 120 Tahoma Rd Lexington KY 40503

SURVER, WILLIAM MERLE, JR, b Altoona, Pa, June 26, 43. DEVELOPMENTAL GENETICS. Educ: St Francis Col, BA, 66; Univ Notre Dame, PhD(genetics), 72. Prof Exp: Instr biol, Univ Notre Dame, 71-72; ASST PROF ZOOL, UNIV RI, 72- Mem: AAAS; Am Soc Zool; Entom Soc Am; Genetics Soc Am. Res: Genetic effects on developing systems; genetics of kelp fly, Coelopa frigida; genetics of complex loci in Drosophila. Mailing Add: Dept of Zool Biol Sci Ctr Univ of RI Kingston RI 02881

SURWILLO, WALTER WALLACE, b Rochester, NY, Nov 25, 26; m 55. PSYCHOPHYSIOLOGY. Educ: Wash Univ, St Louis, BA, 51, MA, 53; McGill Univ, PhD(psychol), 55. Prof Exp: Asst psychol, Wash Univ, St Louis, 50-53; asst, McGill Univ, 53-55; res assoc psychophysiol, Allan Mem Inst Psychiat, 55-57; res psychophysiologist gerontol br, NIH, 57-65; assoc prof psychiat, 65-70, PROF PSYCHIAT SCH MED, UNIV LOUISVILLE, 70-, ASSOC PROF PSYCHOL, GRAD SCH, 71- Concurrent Pos: Mem, NIH Exp Psychol Study Sect, 70-74; consult, NSF, 73- Mem: Am Psychol Asn; Soc Psychophysiol Res; NY Acad Sci; Soc Neurosci. Res: Nervous system function and its relation to behavior; psychophysiological and electrophysiological methods of investigation; instrumentation; central nervous system and behavioral changes with development and senescence. Mailing Add: Dept of Psychiat & Behav Sci Univ of Louisville Sch of Med Louisville KY 40201

SURYARAMAN, MARUTHUVAKUDI GOPALASASTRI, b Madras, India, Mar 2, 25; m 52; c 3. ANALYTICAL CHEMISTRY, PHYSICAL CHEMISTRY. Educ: Univ Madras, BSc, 46, MS, 52; Univ Colo, PhD(chem), 61. Prof Exp: Demonstr chem, Madras Christian Col, 46-49 & Vivekananda Col, Madras, 52; lectr, Sri Venkateswara Univ Cols, Andhra, 52-57; asst, Univ Colo, 57-59, 60-61; sr res chemist, Monsanto Co, Mo, 61-66; asst prof, 66-70, ASSOC PROF CHEM, HUMBOLDT STATE UNIV, 70- Mem: Am Chem Soc; fel Royal Inst Chem; fel Indian Chem Soc. Res: Analytical chemistry, electrochemistry and ion exchange; general inorganic chemistry. Mailing Add: Dept of Chem Humboldt State Univ Arcata CA 95521

SUSALLA, ANNE A, b Parisville, Mich. PLANT ANATOMY. Educ: Madonna Col, BA, 62; Univ Detroit, MS, 67; Ind Univ, Bloomington, PhD(bot), 72. Prof Exp: Asst prof, 72-75, ASSOC PROF BIOL, ST MARY'S COL, 75- Mem: Bot Soc Am; Am Inst Biol Sci; AAAS. Res: Ultrastructure of plastids in phenotypically green leaf tissue of a genetic albino strain of Nicotiana; tissue culture work is being employed to study the developmental stages of these plastids. Mailing Add: Dept of Biol Sci Hall St Mary's Col Notre Dame IN 46556

SUSDORF, DIETER HANS, b Neustadt, Ger, Aug 16, 30; nat US; m 54; c 3. IMMUNOLOGY. Educ: Univ Mo, BA, 52; Univ Chicago, PhD(microbiol), 56. Prof Exp: Logan fel, Univ Chicago, 57-58; resident res assoc biol, Argonne Nat Lab, 58-59; res fel immunochem, Calif Inst Technol, 59-61; res immunochemist, NIH, 61-63; asst prof, 64-72, ASSOC PROF MICROBIOL, MED COL & GRAD SCH MED SCI,

CORNELL UNIV, 72- Concurrent Pos: Author & consult, Scott, Foresman & Co, 59- Honors & Awards: David Anderson-Berry Prize, 61. Res: Cellular mechanism of immunocompetence: function of thymus and non-thymic tissues in humoral and cellular immunity. Mailing Add: Dept of Microbiol Cornell Univ Med Col New York NY 10021

SUSI, FRANK ROBERT, b Boston, Mass, Dec 10, 36. ANATOMY, ORAL PATHOLOGY. Educ: Boston Col, BS, 58; Harvard Univ, DMD, 62, cert, 65; Tufts Univ, PhD(anat), 67. Prof Exp: Instr anat, Sch Med, 67-68, from asst prof to assoc prof oral path, 67-74, PROF ORAL PATH, SCH DENT MED, TUFTS UNIV, 74-, DIR DIV, 73-, DIR BASIC HEALTH SCI, 70-, ASST PROF ANAT, SCH MED, 68- Concurrent Pos: Fel anat, McGill Univ, 68-69. Mem: Am Dent Asn; Am Asn Anat; Histochem Soc; Am Acad Oral Path; Int Asn Dent Res. Res: Histochemistry; autoradiography; electron microscopy; keratinization, carcinogenesis; spermiogenesis. Mailing Add: Dept Oral Path Sch Dent Med Tufts Univ One Kneeland St Boston MA 02111

SUSI, HEINO, physical chemistry, see 12th edition

SUSI, PETER VINCENT, b Philadelphia, Pa, Apr 26, 28; m 54; c 2. ORGANIC CHEMISTRY. Educ: Univ Pa, BA, 50; Univ Del, MS, 51, PhD(chem), 57. Prof Exp: From res chemist to sr res chemist, 56-63, GROUP LEADER, AM CYANAMID CO, 63- Mem: AAAS; Am Chem Soc. Res: Synthetic organic chemistry in fields of plastics additives, light stabilizers, antioxidants, antistatics; ultraviolet and infrared absorbers; flame retardants. Mailing Add: Chem Res Div Am Cyanamid Co Bound Brook NJ 08805

SUSINA, STANLEY V, b Berwyn, Ill, Apr 14, 23; m 48; c 3. PHARMACY, PHARMACOLOGY. Educ: Univ Ill, BS, 48, MS, 51, PhD(pharmacol), 55. Prof Exp: Asst pharm, Univ Ill, 48-50, from instr to asst prof, 50-61, assoc prof & actg head dept, 61-62; PROF PHARM & CHMN DEPT, SAMFORD UNIV, 62- Mem: Acad Pharmaceut Sci; Am Asn Cols Pharm; Am Pharmaceut Asn. Res: Antihistamines; neuromuscular blocking agents; local anesthetics; radioactive isotopes. Mailing Add: Dept of Pharm Samford Univ 800 Lakeshore Dr Birmingham AL 35209

SUSKIND, RAYMOND ROBERT, b New York, NY, Nov 29, 13; m 44; c 2. MEDICAL SCIENCE, HEALTH SCIENCES. Educ: Columbia Univ, AB, 34; State Univ NY, MD, 43; Am Bd Dermat & Syphil, dipl, 49. Prof Exp: Resident dermat & syphil, Cincinnati Gen Hosp, 44-46; res fel dermat & syphil, Col Med, Univ Cincinnati, 48-49, from asst prof to assoc prof prev med & indust health, 49-62, asst prof dermat & syphil, 50-52, dir dermat res, Kettering Lab, 48-62; prof dermat & head div environ med, Med Sch, Univ Ore, 62-69; PROF ENVIRON HEALTH & MED, CHMN DEPT ENVIRON HEALTH & DIR, KETTERING LAB, COL MED, UNIV CINCINNATI, 69- Concurrent Pos: Attend dermatologist, Hosps, Cincinnati; mem comt cutaneous syst, Nat Res Coun, 58-65; attend physician, Univ Ore Hosp, 62-69; consult, US Army Med Res & Develop Command, 65-71 & Pan Am Health Orgn, 66-; mem comt biol effects air pollutants, Panel Polycyclic Org Mat, Nat Acad Sci, 70-73 & ad hoc comt enzyme detergents, 71; mem nat air qual adv comt, Environ Protection Agency, 70-73, consult & mem, Mercury Adv Comt, Pesticide Regulation Div, 70-72, consult, Toxicol Res Ctr, 73-75; fellow, Grad Sch, Univ Cincinnati. Mem: Fel Am Col Physicians; AAAS; fel Am Acad Dermat; Soc Invest Dermat; Am Indust Hyg Asn. Res: Environmental medicine and dermatology; percutaneous absorption; cutaneous hypersensitivity; effects of physical environment on skin reactions to irritants and allergens; environmental cancer. Mailing Add: Dept of Environ Health Univ Cincinnati Col Med Cincinnati OH 45267

SUSKIND, SIGMUND RICHARD, b New York, NY, June 19, 26; m 51; c 3. MICROBIOLOGY. Educ: NY Univ, AB, 48; Yale Univ, PhD(microbiol), 54. Prof Exp: Asst microbiol, Yale Univ, 50-54; USPHS fel, NY Univ, 54-56; from asst prof to assoc prof biol, 56-65, PROF BIOL, McCOLLUM-PRATT INST, JOHNS HOPKINS UNIV, 65-, DEAN ACAD PROGS, 71- Concurrent Pos: Consult, Am Inst Biol Sci, 58-59; spec consult, USPHS, 66-70; head molecular biol sect, NSF, 70-71; consult, Coun Grad Schs & Mid States Asn Cols & Sec Schs, 73- & NIH, 66-70. Mem: Am Soc Microbiol; Genetics Soc Am; Am Asn Immunol; Am Soc Biol Chem. Res: Immunochemistry and biochemistry of mechanism of gene action in microorganisms. Mailing Add: McCollum-Pratt Inst Johns Hopkins Univ Baltimore MD 21218

SUSKIND, STUART PAUL, organic chemistry, polymer chemistry, see 12th edition

SUSMAN, MILLARD, b St Louis, Mo, Sept 1, 34; m 57; c 2. GENETICS. Educ: Wash Univ, AB, 56; Calif Inst Technol, PhD(genetics), 62. Prof Exp: NIH fel, Med Res Coun Microbial Genetics Res Unit, Hammersmith Hosp, London, Eng, 61-62; from asst prof to assoc prof, 62-73, chmn dept, 71-75, PROF GENETICS, LAB GENETICS, UNIV WIS-MADISON, 72- Mem: AAAS; Genetics Soc Am. Res: Bacteriophage genetics and developmental genetics; effects of acridines on bacteriophage growth, recombination and mutation; role of the host cell in phage growth. Mailing Add: Lab of Genetics Univ of Wis Madison WI 53706

SUSSER, MERVYN W, b Johannesburg, SAfrica, Sept 26, 21; m 49; c 3. EPIDEMIOLOGY, SOCIAL MEDICINE. Educ: Univ Witwatersrand, MB, BCh, 50; FRCP(E), 70. Prof Exp: Med officer, Alexandria Health Ctr & Univ Clin, Johannesburg, 51-55; from lectr to reader social med, Univ Manchester, 57-65; Asn Aid Crippled Children Belding scholar, 65-66; PROF EPIDEMIOL, COLUMBIA UNIV, 66- Concurrent Pos: Clin tutor med, Univ Witwatersrand, 51-55; John Simon Guggenheim fel, 72-73; mem comt, Sect Epidemiol & Community Psychiat, World Psychiat Asn. Mem: Am Pub Health Asn; Am Sociol Asn; Soc Epidemiol Res. Res: Social and cultural factors in human development and disease. Mailing Add: Div of Epidemiol Columbia Univ 600 W 168th St New York NY 10032

SUSSEX, IAN MITCHELL, b Auckland, NZ, May 4, 27. BOTANY. Educ: Univ NZ, BS, 48, MSc, 50; Manchester Univ, PhD, 52. Prof Exp: Asst lectr bot, Victoria Univ Col, 54-55; asst prof, Univ Pittsburgh, 55-60; assoc prof, 60-73, PROF BOT, YALE UNIV, 73- Concurrent Pos: Fel, Ezra Stiles Col. Mem: AAAS; Soc Develop Biol; Bot Soc Am; Int Soc Plant Morphol; Am Soc Cell Biol. Res: Plant morphogenesis; tissue culture. Mailing Add: Dept of Biol Yale Univ New Haven CT 06520

SUSSEX, JAMES NEIL, b Northcote, Minn, Oct 2, 17; m 43; c 4. PSYCHIATRY. Educ: Univ Kans, AB, 39, MD, 42. Prof Exp: Resident psychiat, US Naval Hosp, Mare Island, Calif, 46-49; asst clin prof psychiat, Sch Med, Georgetown Univ, 53-55; assoc prof, Med Col Ala, 55-59, prof & chmn dept, 59-68; prof, 68-70, CHMN DEPT PSYCHIAT, SCH MED, UNIV MIAMI, 70- Concurrent Pos: Fel child psychiat, Philadelphia Child Guid Clin, Univ Pa, 49-51; dir, Am Bd Psychiat & Neurol, 75-79; consult, NIMH; consult, Vet Admin, mem, Ment Adv Coun; pres, Serv Children, 72-74; dir ment health serv div, Jackson Mem Hosp, Miami, 70- Mem: AMA; Am Psychiat Asn; Am Col Psychiat; Am Acad Child Psychiat. Res: Child psychiatry; child development in cross-cultural perspective; atypical culture-bound syndromes; dissociative states. Mailing Add: Jackson Mem Hosp Miami FL 33136

SÜSSKIND, CHARLES, b Prague, Czech, Aug 19, 21; nat US; m 45; c 3. BIOENGINEERING, HISTORY OF TECHNOLOGY. Educ: Calif Inst Technol, BS, 48; Yale Univ, MEng, 49, PhD(elec eng), 51. Prof Exp: Res assoc, Stanford Univ, 51-55, lectr elec eng, univ & asst dir, Microwave Lab, 53-55; from asst prof to assoc prof elec eng, 55-64, asst dean eng, 64-68, PROF ENG SCI, UNIV CALIF, BERKELEY, 64- Concurrent Pos: Coordr acad affairs, Statewide Univ, 69-74. Honors & Awards: Clerk Maxwell Premium, Brit Inst Electronic & Radio Eng, 52. Mem: Biomed Eng Soc; Hist Sci Soc; Int Soc Technol Assessment; fel Inst Elec & Electronics Eng; Brit Inst Electronic & Radio Eng. Res: Bioelectronics; electron optics; history of technology; higher education. Mailing Add: Col of Eng Univ of Calif Berkeley CA 94720

SUSSMAN, ALFRED SHEPPARD, b Portsmouth, Va, July 4, 19; m 48; c 3. MYCOLOGY. Educ: Univ Conn, BS, 41; Harvard Univ, AM, 48, PhD(biol), 49. Prof Exp: Instr microbiol, Mass Gen Hosp, 48-49; instr bot, 50-52, from asst prof to assoc prof, 53-61, chmn dept bot, 63-68, assoc dean col lit, sci & arts, 68-70, actg dean, 70-71, assoc dean, H H Rackham Sch Grad Studies, 72-74, PROF BOT, UNIV MICH, ANN ARBOR, 61-, DEAN, H H RACKHAM SCH GRAD STUDIES, 74- Concurrent Pos: Nat Res Coun fel, Univ Pa; Lalor Found fel, 56; NSF sr fel, Calif Inst Technol, 59-60; consult panel develop biol, NSF, 63-65, mem steering comt & comt innovation in lab instr, Biol Sci Curric Study, comnr comn undergrad educ biol sci, 66-69; chmn comt educ, Am Inst Biol Sci; mem biol comt, Argonne Univ Asn, 69-71, chmn, 70-71, mem comt bio & med, 72- & trustee, 74- Mem: Bot Soc Am; Am Soc Microbiol; Soc Develop Biol; Am Acad Microbiol; Am Soc Biol Chem. Res: Physiological mycology; microbial physiology and development; dormancy in microorganisms. Mailing Add: 1615 Harbal Dr Ann Arbor MI 48105

SUSSMAN, HOWARD H, b Portland Ore, Oct 21, 34. BIOCHEMISTRY, MEDICINE. Educ: Univ Ore, BS, 57, MS & MD, 60. Prof Exp: NIH fel chem pharmacol, 61-63, fel biochem, NIH, 64-68, staff scientist, 65-68; ASST PROF PATH, SCH MED, STANFORD UNIV, 68- Mem: AAAS. Res: Enzymology; developmental biochemistry. Mailing Add: Dept of Path Stanford Univ Sch of Med Stanford CA 94305

SUSSMAN, IRVING, b New York, NY, Feb 12, 08; m 40; c 2. MATHEMATICS. Educ: Columbia Univ, BS, 43; Johns Hopkins Univ, MA, 47; Univ Calif, PhD(math), 53. Prof Exp: Instr math & physics, Ricker Jr Col, 43-44; instr math, Johns Hopkins Univ, 44-46; asst prof, Univ Dayton, 47-48; asst prof math & physics, St Mary's Col, Calif, 48-51; asst prof math, Univ San Francisco, 51-54; prof, Calif State Polytech Col, 54-56; prof, 56-73, chmn dept, 56-58, actg chmn, 72, EMER PROF MATH, UNIV SANTA CLARA, 73- Concurrent Pos: Dir insts math teachers, NSF, 59-; Ford Found & US AID prof, Latin Am, 64; spec lectr, Calif State Univ, San Jose, 73-75. Mem: Am Math Soc; Math Asn Am. Res: Abstract algebra; generalized Boolean rings; analysis. Mailing Add: 19174 Montara Lane Los Gatos CA 95030

SUSSMAN, KARL EDGAR, b Baltimore, Md, May 29, 29; m 55; c 2. MEDICINE, ENDOCRINOLOGY. Educ: Johns Hopkins Univ, BA, 51; Univ Md, MD, 55. Prof Exp: From instr to assoc prof med, 62-72, head div endocrinol, 69-72, PROF MED, UNIV COLO MED CTR, DENVER, 73- Concurrent Pos: Chief med serv, Denver Vet Admin Hosp, 72- Mem: Am Col Physicians; Am Diabetes Asn; Am Fedn Clin Res; Am Physiol Soc; Endocrine Soc. Res: Factors controlling insulin secretion in isolated perfused rat pancreases; hormonal control of carbohydrate-lipid metabolism; relationship of intermediary metabolism to insulin secretion. Mailing Add: Univ of Colo Med Ctr 4200 E Ninth Ave Denver CO 80220

SUSSMAN, MAURICE, b New York, NY, Mar 2, 22; m 48; c 3. DEVELOPMENTAL BIOLOGY, MOLECULAR BIOLOGY. Educ: City Col New York, BS, 42; Univ Minn, PhD(bact), 49. Prof Exp: USPHS fel & instr bact, Univ Ill, 49-50; instr biol sci, Northwestern Univ, 50-53, from asst prof to assoc prof, 53-58; assoc prof, Brandeis Univ, 58-60, prof, 60-73; prof inst life sci, Hebrew Univ Jerusalem, 73-76; PROF & CHMN DEPT LIFE SCI, UNIV PITTSBURGH, 76- Concurrent Pos: Instr, Marine Biol Lab, Woods Hole, 56-60 & 67-70. Honors & Awards: NIH career develop award, 66. Mem: Am Soc Microbiol; Soc Gen Physiol; Soc Develop Biol; Am Soc Biol Chem; Brit Soc Gen Microbiol. Res: Cellular differentiation and morphogenesis; molecular genetics. Mailing Add: Dept of Life Sci Univ of Pittsburgh Pittsburgh PA 15213

SUSSMAN, MYRON MAURICE, b Trenton, NJ, Oct 7, 45; m 70. NUMERICAL ANALYSIS. Educ: Mass Inst Technol, SB, 67; Carnegie-Mellon Univ, MS, 68, PhD(math), 75. Prof Exp: Instr math, Carnegie-Mellon Univ, 68-69 & Robert Morris Col, 69-71; MATHEMATICIAN, BETTIS ATOMIC POWER LAB, WESTINGHOUSE ELEC CO, 75- Mem: Soc Indust & Appl Math; Sigma Xi; Am Math Soc. Res: Numerical analysis of partial differential equations and iterative solution of large linear systems of algebraic equations. Mailing Add: 5026 Belmont Ave Bethel Park PA 15102

SUSSMAN, RAQUEL ROTMAN, b Arg, Oct 22, 21; nat US; m 48; c 3. MICROBIOLOGY. Educ: Univ Chile, BS, 44; Univ Ill, PhD(bact), 52. Prof Exp: Asst viruses, Inst Bact Chile, 44-48; asst microbiol, Univ Minn, 48-49 & Univ Ill, 49-50; res assoc, Northwestern Univ, 50-58 & Brandeis Univ, 58-74; mem staff, Dept Molecular Biol, Hadassah Med Sch, Hebrew Univ, Israel, 74-76; MEM STAFF, DEPT LIFE SCI, UNIV PITTSBURGH, 76- Res: Cell biology, chiefly genetics and differentiation. Mailing Add: Dept of Life Sci Univ of Pittsburgh Pittsburgh PA 15213

SUSSMAN, SIDNEY, b Brooklyn, NY, Aug 29, 14; m 39; c 3. WATER CHEMISTRY, CORROSION. Educ: Polytech Inst Brooklyn, BS, 34; Mass Inst Technol, PhD(org chem), 37. Prof Exp: Res chemist, E I du Pont de Nemours & Co, 37-40; res chemist, Permutit Co, 40-44, chief res chemist, 44-46; chief chemist, Liquid Conditioning Corp, NJ, 46-49; chief chemist, Water Serv Labs, Inc, 49-61, tech dir, 61-69, vpres, 66-69; tech dir, Olin Water Serv Labs, 69-71, tech dir water treatment dept, 71-75, TECH DIR, OLIN WATER SERV, OLIN CORP, 75- Concurrent Pos: Mem nat tech adv comt water qual for indust uses, Fed Water Pollution Control Admin, 67-68; mem panel water qual criteria for indust uses, Nat Acad Sci, 71-72. Mem: Am Soc Testing & Mat; Am Chem Soc; Am Water Works Asn; fel Am Inst Chem; Soc Chem Indust. Res: Ion exchange resin synthesis and applications; water conditioning; desalting seawater; industrial water treatment including cooling water, boilerwater and process water. Mailing Add: Olin Water Serv Olin Corp 120 Long Ridge Rd Stamford CT 06904

SUTCLIFFE, CHARLES HERBERT, b Salamanca, NY, Nov 2, 25; m 50; c 5. PHYSICS. Educ: Grove City Col, BS, 49; Univ Pa, PhD(physics), 53. Prof Exp: From asst to instr physics, Univ Pa, 49-53; physicist, Philco Corp, 53-55, sect mgr res, 55-58, mgr, Spec Components Dept, 58-60, gen mgr, Spec Prod Oper, 60-62, dir eng, 62-63, gen mgr, Semiconductor Oper, 63-64; vpres & gen mgr, Rectifier Div, Gen Instruments Corp, 64-66; mgr complex arrays, Fairchild Semiconductor Div, 66-68, instrument group mgr, Fairchild Camera & Instrument Corp, 68-69; pres, Four-Phase Systs, Inc, 69-74; OWNER, SUTCLIFFE ASSOCS, 74- Mem: Sr mem Inst Elec & Electronics Engrs. Res: Solid state physics; semiconductor devices and materials;

infrared and microwave fields. Mailing Add: Sutcliffe Assocs 15474 Via Vanquero Monte Sereno CA 95030

SUTCLIFFE, WILLIAM GEORGE, b Detroit, Mich, Nov 25, 37; m 60; c 4. PHYSICS. Educ: Univ Mich, BS, 60; Univ Del, PhD(physics), 69. Prof Exp: Instr physics, US Naval Nuclear Power Sch, 62-64; PHYSICIST, LAWRENCE LIVERMORE LAB, 68- Mem: Am Asn Physics Teachers. Res: Design and development of large computer codes, including hydrodynamics, radiation transport and neutroincs. Mailing Add: Lawrence Livermore Lab L-71 PO Box 808 Livermore CA 94550

SUTCLIFFE, WILLIAM HUMPHREY, JR, b Miami, Fla, Nov 8, 23; m 45, 49, 64; c 5. ZOOLOGY. Educ: Emory Univ, BA, 45; Duke Univ, MA, 47, PhD(zool), 50. Prof Exp: Instr zool, Duke Univ, 49-50; investr marine biol, NC Inst Fisheries Res, 50-51, 50-51; staff biologist lobster invests, Bermuda Biol Sta, 51-53, dir, 53-69; SR RES SCIENTIST, BEDFORD INST OCEANOG, 67-, HEAD FISHERIES OCEANOG SECT, 75- Concurrent Pos: Assoc, Woods Hole Oceanog Inst, 57; dir marine sci ctr, Lehigh Univ, 64-67. Mem: AAAS; Am Soc Zool; Am Soc Limnol & Oceanog; Marine Biol Asn UK. Res: Dynamics of plankton populations; air-sea interaction; marine food chains. Mailing Add: Bedford Inst of Oceanog Dartmouth NS Can

SUTER, DANIEL B, b Hinton, Va, Apr 25, 20; m 41; c 4. HUMAN ANATOMY, PHYSIOLOGY. Educ: Bridgewater Col, BA, 47; Vanderbilt Univ, MA, 48; Med Col Va, PhD(anat), 63. Prof Exp: Asst prof biol, 48-60, assoc prof, 62-63, PROF BIOL, EASTERN MENNONITE COL, 63-, CHMN DIV NATURAL SCI & MATH, 64-, CHMN DEPT LIFE SCI, 72- Concurrent Pos: NIH fel, Univ Calif, Davis, 70-71. Mem: Asn Am Med Cols; Am Asn Anat; Am Sci Affiliation. Res: Effects of radiation and pesticides, especially organophosphates, on the central nervous system. Mailing Add: Dept of Biol Eastern Mennonite Col Harrisonburg VA 22801

SUTER, EMANUEL, b Basel, Switz, Feb 7, 18; m 54. MEDICAL EDUCATION, IMMUNOLOGY. Educ: Univ Basel, MD, 44. Prof Exp: Res fel bact, Univ Basel, 45-48; asst, Rockefeller Inst, 49-52; assoc bact & immunol, Harvard Med Sch, 52-53, asst prof, 53-56; prof microbiol & head dept, Col Med, Univ Fla, 56-65, dean, 65-72; DIR DIV INT MED EDUC & DIV EDUC RESOURCES, ASN AM MED COLS, 72- Res: Experimental tuberculosis; pathogenesis of infectious diseases and immunology; health profession education. Mailing Add: Div of Int Med Educ Asn of Am Med Cols Washington DC 20036

SUTER, M ST AGATHA, b Philadelphia, Pa, Feb 19, 97. BIOLOGY. Educ: Inst Divi Thomae, PhD(biochem), 50. Prof Exp: Teacher, parochial schs, 20-41; instr biol, 41-50, prof 50-74, EMER PROF BIOL, IMMACULATA COL, PA, 74- Mem: AAAS; Nat Asn Biol Teachers; Am Physiol Soc; Nat Sci Teachers Asn; Am Chem Soc. Res: Cancer research. Mailing Add: Dept of Biol Immaculata Col Immaculata PA 19345

SUTER, ROBERT WINFORD, b Warren, Ohio, Aug 3, 41; m 63; c 2. CHEMISTRY. Educ: Bluffton Col, BA, 63; Ohio State Univ, MS, 66, PhD(chem), 69. Prof Exp: ASSOC PROF CHEM, BLUFFTON COL, 69- Mem: Am Chem Soc. Res: Inorganic chemistry, particularly nonmetals; solution phenomena. Mailing Add: Dept of Chem Box 907 Marbeck Ctr Bluffton Col Bluffton OH 45817

SUTER, STUART ROSS, b Harrisonburg, Va, Apr 1, 41; m 63; c 2. ORGANIC CHEMISTRY. Educ: Bridgewater Col, BA, 63; Univ Mich, Ann Arbor, MS, 65; Univ Va, PhD(org chem), 71. Prof Exp: Assoc chemist, Smith Kline & French Labs, 65-67; app, Univ Vt, 70-71; patent chemist, Smith Kline & French Labs, 71-75, PATENT CHEMIST SMITH KLINE CORP, 75- Mem: Am Chem Soc. Res: Aryl nitrenes; synthetic organic chemistry; medicinal chemistry. Mailing Add: Patent Dept Smith Kline Corp 1500 Spring Garden St Philadelphia PA 19101

SUTFIN, DUANE, b Chicago, Ill, Aug 27, 29; m 51; c 4. PHYSIOLOGY. Educ: Hope Col, AB, 50; Univ Iowa, MS, 52, PhD(physiol), 56. Prof Exp: Instr physiol, Univ Iowa, 54-55; instr, Univ Ill, Urbana, 55-57, asst prof, 57-62; prin res scientist, Honeywell, Inc, 62-64, prod mgr med, 64-72, MGR HEALTH CARE INDUST MKT, HONEYWELL INFO SYST, INC, 72- Mem: Soc Advan Med Systs; Asn Advan Med Instrumentation. Res: Physiology and pharmacology of the cardiovascular system. Mailing Add: 2605 E Dartmouth Denver CO 80210

SUTHERLAND, ANGUS JOHNSTON, b NS, Mar 16, 15; nat US; m 49. CHEMISTRY. Educ: McGill Univ, BSc, 37, MSc, 38; Dalhousie Univ, dipl, 46. Prof Exp: Anal chemist, NS Agr Col, 38-41 & 47-51; spectrographer, Limestone Prod Corp Am, 51-57; SPECTROGRAPHER ANAL CHEM SECT, RES CTR, JOHNSMANVILLE CORP, 57- Mem: Soc Appl Spectros. Res: Spectrography; determination of vitamin A in fish oils, carotene in milk and cobalt in herbage; tolerance of growing chickens for dietary copper. Mailing Add: 13061 W Ohio Ave Lakewood CO 80228

SUTHERLAND, BILL, b Sedalia, Mo, Mar 31, 42. PHYSICS. Educ: Wash Univ, AB, 63; State Univ NY Stony Brook, MA, 65, PhD, 67. Prof Exp: Res assoc, State Univ NY Stony Brook, 67-69; asst res physicist, Univ Calif, Berkeley, 69-70; ASST PROF PHYSICS, UNIV UTAH, 70- Res: Statistical mechanics. Mailing Add: Dept of Physics Univ of Utah Salt Lake City UT 84112

SUTHERLAND, CHARLES F, b Camp Grant, Ill, Oct 1, 21; m 44; c 4. FORESTRY ECONOMICS. Educ: Univ Idaho, BS, 48, MFor, 54; Univ Mich, PhD(forestry econ), 61. Prof Exp: Res forester, Potlatch Forests Inc, 48-53; forest economist, Lake States Forest Exp Sta, US Forest Serv, St Paul, Minn, 56-58; asst prof forestry econ, 59-68, ASSOC PROF FORESTRY ECON, SCH FORESTRY, ORE STATE UNIV, 68- Res: Biology and management in private industry; marketing and production problems in forestry; taxation and forest protection problems. Mailing Add: Sch of Forestry Ore State Univ Corvallis OR 97331

SUTHERLAND, CLAUDIA SEBESTE, b Johnstown, NY, Nov 29, 24; m. PHARMACOLOGY. Educ: Univ Md, BS, 44, MS, 49; Georgetown Univ, PhD, 52. Prof Exp: Pharmacologist, Div Pharmacol, US Food & Drug Admin, 48-54 & New Drug Off, Bur Med, 54-59; mem staff, Psychopharmacol Serv Ctr, NIMH, 59-60, exec secy, Cancer Chemother Study Sect, Div Res Grants, 60-63; asst to dir med affairs, 63-67, DIR OFF SPONSORED RES, SCH MED, VANDERBILT UNIV, 67- Concurrent Pos: Asst res prof, Sch Med, George Washington Univ, 57-59. Mem: Am Chem Soc; Am Soc Pharmacol & Exp Therapeut. Res: Medicinal chemistry; metabolism of sulfur compounds. Mailing Add: Off of Sponsored Res Vanderbilt Univ Sch of Med Nashville TN 37232

SUTHERLAND, DAVID M, b Bellingham, Wash, Oct 5, 40. PLANT TAXONOMY. Educ: Western Wash State Col, BA, 63; Univ Wash, PhD(bot), 67. Prof Exp: Asst prof, 67-74, ASSOC PROF BIOL, UNIV NEBR, OMAHA, 74- Mem: AAAS; Int Asn Plant Taxonomists; Bot Soc Am; Am Soc Plant Taxon. Res: Biosystematics of larkspurs; floristics of Great Plains. Mailing Add: Dept of Biol Univ of Nebr Omaha NE 68101

SUTHERLAND, DONALD JAMES, b Chelsea, Mass, Oct 5, 29. PHYSIOLOGY, BIOCHEMISTRY. Educ: Tufts Univ, BS, 51; Univ Mass, MS, 57; Rutgers Univ, PhD(entom), 60. Prof Exp: From asst prof to assoc prof, 60-67, PROF ENTOM, RUTGERS UNIV, NEW BRUNSWICK, 67- Mem: Entom Soc Am; Am Mosquito Control Asn. Mailing Add: Dept of Entom Rutgers Univ New Brunswick NJ 08903

SUTHERLAND, DONALD RALPH, b Boston, Mass, Nov 24, 41; m 63; c 1. ANTHROPOLOGY. Educ: Dartmouth Col, AB, 63; Tulane Univ, PhD(anthrop), 71. Prof Exp: Instr, 67-70, ASST PROF ANTHROP, UNIV SC, 70- Concurrent Pos: Univ SC Comt Res & Prod Scholar res grant, Colombia, 72. Mem: Soc Am Archaeol; Inst Andean Studies. Res: Archaeology, especially of northwestern South America and southeastern United States. Mailing Add: Dept of Anthrop & Sociol Univ of SC Columbia SC 29208

SUTHERLAND, EARL WILBUR, JR, pharmacology, deceased

SUTHERLAND, GEORGE LESLIE, b Dallas, Tex, Aug 13, 22; m 47; c 3. ORGANIC CHEMISTRY. Educ: Univ Tex, BS, 43, MA, 47, PhD(org chem), 50. Prof Exp: Lilly fel, Univ Tex, 49-51; chemist, 51-57, group leader, 58-62, mgr metab & anal res, 62-66, dir prod develop & govt registr, 66-69, asst dir res & develop, 69-70, dir res & develop, 70-73, VPRES RES & DEVELOP, LEDERLE LABS DIV, AM CYANAMID CO, 73- Mem: AAAS; Am Chem Soc; NY Acad Sci; The Chem Soc. Res: Organometallic compounds; anticonvulsants; microbiological growth factors; hypotensive agents; anticoccidials; chemical process development; formulation, metabolism and analysis of agricultural chemicals; discovery and development of pharmaceuticals. Mailing Add: Lederle Labs Div Am Cyanamid Co Pearl River NY 10965

SUTHERLAND, GERALD BONAR, b Winnipeg, Man, July 31, 19; m 48; c 2. PHYSIOLOGY. Educ: Univ BC, BA, 48; Stanford Univ, PhD(physiol), 54. Prof Exp: Asst, Stanford Univ, 52-53; fel, Calif Inst Technol, 53-56; asst physiol, Univ Southern Calif, 55-56; instr, Univ Kans, 56-58, asst prof, 58-60; assoc prof, 60-68, PROF PHYSIOL, UNIV SASK, 68- Mem: NY Acad Sci; Can Physiol Soc; Can Soc Immunol. Res: Immunology; body fluids; effect of antiserum on mosquitoes. Mailing Add: Dept of Physiol Univ of Sask Saskatoon SK Can

SUTHERLAND, JAMES HENRY RICHARDSON, b Can, July 6, 23; nat US; m 43; c 3. PHARMACOLOGY. Educ: Univ Calif, AB, 48, PhD(physiol), 56. Prof Exp: Trainee cardiovasc res prog, 55-56, from asst prof to assoc prof pharmacol, 56-64, chmn dept, 64-68, dir div health commun, 68-74, PROF PHARMACOL, MED COL GA, 64- Mem: AAAS; Health Sci Commun Asn (pres, 71-72). Res: Cardiovascular pressure and flow; influence of the adrenergic agents on the cardiovascular system; pharmacology of the adrenergic receptor; educational communications. Mailing Add: Dept of Pharmacol Med Col of Ga Augusta GA 30902

SUTHERLAND, JAMES MCKENZIE, b Chicago, Ill, Aug 8, 23; m 53; c 2. PEDIATRICS. Educ: Univ Chicago, SB, 47, MD, 50; Am Bd Pediat, dipl, 55. Prof Exp: Intern, Cincinnati Gen Hosp, 50-51; resident pediat, Children's Hosp, 51-54; instr, Col Med, Univ Cincinnati, 53-54; fel, Harvard Med Sch, 54-56; from asst prof to assoc prof, 56-67, PROF PEDIAT, COL MED, UNIV CINCINNATI, 67- Concurrent Pos: NIH fel, Children's Med Ctr, Boston, 54-56; attend pediatrician, Children's Hosp, Cincinnati, 56-; dir newborn servs & clinician, Out-Patient Dept, Cincinnati Gen Hosp, 57-; res assoc & head div newborn physiol, Children's Hosp Res Found. Mem: AAAS; Soc Pediat Res; Am Pediat Soc; AMA. Res: Physiology of normal and abnormal respiration in infants. Mailing Add: Dept of Pediat Univ of Cincinnati Cincinnati OH 45229

SUTHERLAND, JEFFREY C, b Rochester, NY, June 18, 39; m 62; c 1. GEOCHEMISTRY, HYDROLOGY. Educ: Cornell Univ, AB, 62; Syracuse Univ, PhD(geol), 68. Prof Exp: Asst prof geol, Slippery Rock State Col, 67-68, assoc prof, 68-70; GEOCHEMIST & HYDROLOGIST, 70-; DEPT HEAD & PROJ ADV, WILLIAMS & WORKS, INC, 72- Concurrent Pos: Consult, Cent NY Onondaga Lake Study, 68-70. Mem: AAAS; Geol Soc Am; Geochem Soc; Soc Econ Paleont & Mineral; Am Inst Prof Geologists; Am Geophys Union. Res: Applied and basic research in low temperature mineral-water reactions and their bearing on quality of inland and estuarine waters. Mailing Add: Williams & Works Inc 611 Cascade W Parkway SE Grand Rapids MI 49503

SUTHERLAND, JOHN CLARK, b New York, NY, Sept 2, 40; m 65. BIOPHYSICS. Educ: Ga Inst Technol, BS, 62, MS, 64, PhD(physics), 67. Prof Exp: Biophysicist, Walter Reed Res Inst, 67-69; res fel, Lab Chem Biodyn, Univ Calif, Berkeley, 69-72, USPHS fel, 69-71; res fel chem, Univ Southern Calif, 72-73; ASST PROF PHYSIOL, CALIF COL MED-UNIV CALIF, IRVINE, 73- Mem: Am Chem Soc; Am Phys Soc; Biophys Soc. Res: Optical spectroscopy and photochemistry of biological materials; vacuum ultraviolet spectroscopy of metals. Mailing Add: Dept of Physiol Calif Col of Med Univ of Calif Irvine CA 92664

SUTHERLAND, JOHN PATRICK, b Salem, Ore, Oct 1, 42; m 64; c 1. MARINE ECOLOGY. Educ: Univ Wash, Seattle, BS, 64; Univ Calif, Berkely, PhD(zool), 69. Prof Exp: Asst prof marine ecol, 69-75, ASSOC PROF MARINE ECOL, MARINE LAB, DUKE UNIV, 75- Concurrent Pos: Off Naval Res & NSF grants, 72- Mem: Ecol Soc Am. Res: Comparative studies on the dynamics and bioenergetics of marine inverte- brates; structure and function of subtidal, epibenthic fouling communities. Mailing Add: Duke Univ Marine Lab Beaufort NC 28516

SUTHERLAND, JUDITH ELLIOTT, b Clovis, NMex, June 6, 24; m 47; c 2. POLYMER CHEMISTRY. Educ: Univ Tex, BSChem, 45; Univ Conn, MA, 68; Univ Mass, PhD(polymer chem), 72. Prof Exp: Res biochemist, Tex Agr Exp Sta, College Station, 45-46; res chemist biochem, Clayton Found, Biochem Inst, Austin, Tex, 46-48; chemist, Stamford Chem Co, Conn, 60-61 & Am Cyanamid Co, Conn, 61-69; RES CHEMIST POLYMERS, EASTMAN KODAK CO RES LABS, 72- Mem: Am Chem Soc; Am Phys Soc. Res: Polymer synthesis and characterization. Mailing Add: Eastman Kodak Co Res Labs 1669 Lake Ave Rochester NY 14650

SUTHERLAND, LESLIE HUNT, chemistry, see 12th edition

SUTHERLAND, PATRICK KENNEDY, b Dallas, Tex, Feb 17, 25. PALEOBIOLOGY, STRATIGRAPHY. Educ: Univ Okla, BSc, 46; Cambridge Univ, PhD(geol), 52. Prof Exp: Geologist, Phillips Petrol Co, 46-49, 52-53; asst prof geol, Univ Houston, 53-57; assoc prof, 57-64, PROF GEOL, SCH GEOL & GEOPHYS, UNIV OKLA, 64- Concurrent Pos: Nat Acad Sci vis exchange fel, USSR, 71. Mem: Fel Geol Soc Am; Paleont Soc; Soc Econ Paleont & Mineral; Am Asn Petrol Geol; Soc Syst Zool. Res: Paleobiology; biostratigraphy; paleoecology; carboniferous Rugose corals and brachiopods. Mailing Add: Sch of Geol & Geophysics Univ of Okla Norman OK 73069

SUTHERLAND, ROBERT CARVER, b Whitefish, Mont, Apr 27, 30; m 50; c 5. PLANT PHYSIOLOGY, PLANT MORPHOLOGY. Educ: Univ SDak, BS, 53,

MNS, 59; Tex A&M Univ, PhD(plant physiol), 66. Prof Exp: Mem fac bot, 59-63, chmn dept biol, 66-74, PROF BIOL, WAYNE STATE COL, 66- Res: Development of foliar embryos in the genus Kalanchoe. Mailing Add: Dept of Biol Wayne State Col Wayne NE 68787

SUTHERLAND, ROBERT MELVIN, b Moncton, NB, Oct 21, 40; m 62; c 2. CELL BIOLOGY, RADIATION BIOLOGY. Educ: Acadia Univ, BSc, 61; Univ Rochester, PhD(radiation biol), 66. Prof Exp: Fel radiation biol, Norsk Hydro's Inst Cancer Res, Oslo, Norway, 66-67; fel radiation biol, 67-68, RADIOBIOLOGIST, ONT CANCER FOUND, LONDON CLIN, VICTORIA HOSP, LONDON, ONT, 68-; ASSOC PROF THERAPEUT RADIOL & BIOPHYS, UNIV WESTERN ONT, 72- Concurrent Pos: Radiation res fel, James Picker Found, 66-68; hon lectr biophys, Univ Western Ont, 67-68, lectr therapeut radiol & asst prof biophys, 68-72. Mem: Radiation Res Soc; Can Soc Cell Biol; Am Asn Cancer Res. Res: Radio-sensitivity of cultured mammalian cells; in vitro tumor models; lymphocyte transformation; control of cell proliferation. Mailing Add: Ont Cancer Res Found London ON Can

SUTHERLAND, RONALD GEORGE, b Belfast, Northern Ireland, May 4, 35; m 60; c 2. CHEMISTRY. Educ: Univ Strathclyde, BSc, 59; Univ St Andrews, PhD(org chem), 62. Prof Exp: Res assoc, Columbia Univ, 62-63; fel, Calif Inst Technol, 63-64; Imp Chem Industs res fel chem, Queen's Col, Dundee, 64; from asst prof to assoc prof, 64-73, PROF CHEM, UNIV SASK, 73- Concurrent Pos: Sci Res Coun UK sr vis fel, Edinburgh Univ, 75-76. Mem: The Chem Soc; Royal Inst Chem; Am Chem Soc. Res: Organometallic chemistry; organic photochemistry; chemical decomposition of pesticides. Mailing Add: Dept of Chem Univ of Sask Saskatoon SK Can

SUTHERLAND, STEPHEN MERTON, geography, see 12th edition

SUTHERLAND, VIOLETTE CUTTER, b Chicago, Ill, Jan 23, 15; wid; c 1. PHARMACOLOGY. Educ: Univ Kans, AB, 40; Stanford Univ, MA, 47; Univ Calif, San Francisco, PhD(pharmacol), 55. Prof Exp: Assoc pharmacol, 53-54, jr res pharmacologist, 53-56, lectr dent pharmacol, 54-56, asst res pharmacologist, 56-59, from asst prof to assoc prof, 56-68, PROF PHARMACOL, UNIV CALIF, SAN FRANCISCO, 68- Concurrent Pos: Vis scientist, Nat Inst Alcohol Abuse & Alcoholism, NIMH, 70-71. Mem: Fel AAAS; Am Soc Pharmacol & Exp Therapeut; Am Heart Asn; NY Acad Sci; Brit Biochem Soc. Res: Effect of alcohol and other neurotropic agents on neurometabolism in vivo and in vitro. Mailing Add: Dept of Pharmacol Univ of Calif Med Ctr San Francisco CA 94143

SUTHERLAND, WILLIAM HARRISON, b South Norwalk, Conn, Aug 11, 15; m 46; c 3. OPERATIONS RESEARCH. Educ: Yale Univ, BE, 36; Stevens Inst Technol, MS, 44. Prof Exp: Proj engr exp towing tank, Stevens Inst Technol, 41-51; opers res analyst, Opers Res Off, Johns Hopkins Univ, 51-61; opers res analyst, Res Anal Corp, McLean, Va, 61-72; OPERS RES ANALYST, OPERS RES INC, MD, 74- Concurrent Pos: Consult comt undersea warfare, Nat Res Coun, 53-54. Honors & Awards: UN Serv Medal, 52. Mem: AAAS; Opers Res Soc Am; Brit Oper Res Soc. Res: Program evaluation; program assessment in research and development; cost uncertainty analysis; cost benefit analysis; operations research in transportation and naval architecture. Mailing Add: 2901 McComas Ave Kensington MD 20795

SUTHERLAND, WILLIAM NEIL, b Linden, Iowa, Aug 10, 27; m 49; c 4. SOIL FERTILITY, AGRICULTURAL ECONOMICS. Educ: Iowa State Univ, BS, 50, MS, 53, PhD(soil fertil), 60. Prof Exp: Soil conserv agent exten serv, Univ Minn, 54-56; res assoc, Iowa State Univ, 56-60; AGRICULTURIST, TENN VALLEY AUTH, UNIV NEBR, LINCOLN, 60- Mem: Am Soc Agron; Soil Conserv Soc Am. Res: Evaluation of Tennessee Valley Authority's experimental fertilizers agronomically and in fertilizer use systems in cooperation with land grant universities and fertilizer industry firms; educational programs to encourage efficient fertilizer use. Mailing Add: Tenn Valley Auth 308 Filley Hall Univ of Nebr Lincoln NE 68503

SUTHERLAND, WILLIAM ROBERT, b Hastings, Nebr, May 10, 36; m 57; c 3. COMPUTER SCIENCE, ELECTRICAL ENGINEERING. Educ: Rensselaer Polytech Inst, BEE, 57; Mass Inst Technol, SM, 63, PhD(elec eng), 66. Prof Exp: Assoc group leader comput graphics & comput-aided design & mem tech staff, Lincoln Lab, Mass Inst Technol, 64-69; mgr interactive systs dept, Bolt Beranek & Newman, Inc, 69-72, div vpres & dir comput sci div, 72-75; MGR SYSTS SCI LAB, XEROX PALO ALTO RES CTR, 75- Mem: Asn Comput Mach. Res: Distributed computing; graphics. Mailing Add: Xerox Palo Alto Res Ctr 3333 Coyote Hill Rd Palo Alto CA 94304

SUTHERLAND-BROWN, ATHOLL, b Ottawa, Ont, June 20, 23; m 48; c 1. ECONOMIC GEOLOGY. Educ: Univ BC, BASc, 50; Princeton Univ, PhD(geol), 54. Prof Exp: Geologist econ geol, 51-71, dep chief geol, 71-75, CHIEF GEOLOGIST ECON GEOL, MINERAL RESOURCES BR, BC DEPT MINES & PETROLEUM RESOURCES, 75- Concurrent Pos: Ed, Can Inst Mining & Metall, 74-76; mem, Geol Soc Am Del, Am Comn Stratig Nomenclature, 69-72; mem, Can Nat Comn, Int Geol Correlation Prog, 74-; del, Int Union Geol Sci-UNESCO Meeting Govt Experts, Int Geol Cong, Paris, 71; mem, Nat Orgn Comt, 24th Int Geol Cong, 69-72; exec mem, Nat Adv Comn, Res Geol Sci, 69-73. Mem: Geol Soc Am; Geol Asn Can; Can Inst Mining & Metall; Soc Econ Geologists. Res: Morphology, classification, distribution and tectonic setting of porpityry deposits; metallogeny and distribution of metals in deposits and background , particularly in Canadian Cordillera; geology and tectonics of Queen Charlotte Islands and insular tectonic belt. Mailing Add: Dept of Mines & Petrol Resources Victoria BC Can

SUTHERS, RODERICK ATKINS, b Columbus, Ohio, Feb 2, 37. PHYSIOLOGY. Educ: Ohio Wesleyan Univ, BA, 60; Harvard Univ, AM, 61, PhD(biol), 64. Prof Exp: Res fel biol, Harvard Univ, 64-65; from asst prof to assoc prof anat & physiol, 65-74, PROF ANAT & PHYSIOL, UNIV IND, BLOOMINGTON, 74-, PROF, CTR NEUROL SCI, 75- Mem: AAAS; Am Soc Zool; Am Ornith Union; Cooper Ornith Soc; Acoust Soc Am. Res: Sensory physiology and behavior; comparative sensory physiology of animal sonar systems; neural processing of visual and auditory information in echolating animals; cross modal interaction; laryngeal mechanisms of vocalization by echolocating bats. Mailing Add: Dept of Anat & Physiol Ind Univ Bloomington IN 47401

SUTIN, JEROME, b Albany, NY, Mar 12, 30; m 56; c 2. NEUROANATOMY, NEUROPHYSIOLOGY. Educ: Siena Col, BS, 51; Univ Minn, MS, 53, PhD(anat), 54. Prof Exp: Asst anat, Univ Minn, 52-53; hon res asst, Univ Col, Univ London, 53-54; asst, Univ Minn, 54; jr res anatomist, Univ Calif, Los Angeles, 55-56; from instr to assoc prof anat, Sch Med, Yale Univ, 56-66; PROF ANAT & CHMN DEPT, EMORY UNIV, 66- Concurrent Pos: Vis investr, Autonomics Div, Nat Phys Lab, Middlesex, Eng; vis prof, Inst Psychiat, Maudsley Hosp, London; Nat Found Infantile Paralysis fel anat, Univ Calif, Los Angeles, 55-56. Mem: Am Asn Anatomists; Am Physiol Soc; Am Soc Zool; Soc Neurosci; Am Acad Neurol. Res: Hypothalamic organization; basal ganglia and motor function. Mailing Add: Dept of Anat Emory Univ Atlanta GA 30322

SUTIN, NORMAN, b SAfrica, Sept 16, 28; nat US; m 58; c 2. PHYSICAL INORGANIC CHEMISTRY. Educ: Univ Cape Town, BSc, 48, MSc, 50; Cambridge Univ, PhD(chem), 53. Prof Exp: Imp Chem Industs fel, Durham Univ, 54-55; res assoc, 56-57, from assoc chemist to chemist, 58-66, SR CHEMIST, BROOKHAVEN NAT LAB, 66- Concurrent Pos: Affil, Rockefeller Univ, 58-62; vis fel, Weizmann Inst, 64; vis prof, State Univ NY Stony Brook, 68, Columbia Univ, 68 & Tel Aviv Univ, Israel, 73. Mem: Am Chem Soc; The Chem Soc. Res: Kinetics and mechanisms of inorganic reactions; bioinorganic chemistry; photochemistry of transition metal complexes. Mailing Add: Dept of Chem Brookhaven Nat Lab Upton NY 11973

SUTLIFF, THOMAS MARK, organic chemistry, see 12th edition

SUTLIFF, WHEELAN DWIGHT, b Rhinelander, Wis, June 20, 00; m 38; c 2. INTERNAL MEDICINE. Educ: Univ Wis, AB, 21, MS, 22; Cornell Univ, MD, 24. Prof Exp: Intern, Bellevue Hosp, NY, 24-26; asst med, Cornell Univ, 26-28, traveling fel, 28; asst med, Harvard Med Sch, 29-30, instr, 31-34; res instr, Univ Chicago, 34-37; asst dir bur labs, City Dept Health, New York, 37-46; from asst prof to assoc prof med, 46-67, prof, 68-70, EMER PROF MED COL MED, UNIV TENN, MEMPHIS, 71- Concurrent Pos: Staff physician, WTenn Chest Dis Hosp; consult, Vet Admin & Baptist Mem Hosps, Memphis. Mem: Am Thoracic Soc; AMA; Am Col Physicians; Infectious Dis Soc Am; Med Mycol Soc Am. Res: Infectious disease; systemic mycoses: etiologic diagnosis; theory. Mailing Add: 286 Windover Rd Memphis TN 38111

SUTMAN, FRANK X, b Newark, NJ, Dec 20, 27; m 56; c 3. CHEMISTRY. Educ: Montclair State Col, AB, 49, AM, 52; Columbia Univ, EdD, 56. Prof Exp: Instr, Pub Schs, NJ, 49-51; instr, High Schs, 51-55; asst prof sci, Paterson State Col, 55-57; assoc prof natural sci & chmn div, Inter-Am Univ, PR, 57-58; PROF SCI, SCH EDUC & DIR, NSF INSTS, TEMPLE UNIV, 62- Concurrent Pos: NSF-AID lectr, India, 67; observer, Orgn Am States Coun Sci Educ & Cult, 71; vis prof, Hebrew Univ, Jerusalem, Israel, 73; consult, Israel Environ Protection Serv, 75. Honors & Awards: Recognition Award, Internal US Dept State AID, 72. Mem: Fel AAAS; Am Chem Soc; Nat Asn Res Sci Teaching (pres, 72); Nat Sci Teachers Asn. Res: Chemical education research. Mailing Add: Dept of Educ 341 Ritter Hall Temple Univ Philadelphia PA 19122

SUTNICK, ALTON IVAN, b Trenton, NJ, July 6, 28; m 58; c 2. INTERNAL MEDICINE. Educ: Univ Pa, AB, 50, MD, 54. Prof Exp: From intern to resident anesthesiol & med, Hosp Univ Pa, 54-57, USPHS fel, Univ, 56-57; from resident to chief resident med, Marion County Gen Hosp, Indianapolis, Ind, 57-61; USPHS fel, Temple Univ, 61-63, instr med, 63-64, assoc, 64-65; res physician, 65-72, ASSOC DIR, INST CANCER RES, FOX CHASE CANCER CTR, 72-; ASSOC PROF MED, SCH MED, UNIV PA, 71- Concurrent Pos: Consult, Coun Drugs, AMA, 62-64; mem US nat comt, Int Union Against Cancer, 69-72; mem ed bd, Res Commun Chem Path Pharmacol, 69-; vis prof med, Med Col Pa, 71-; asst ed, Ann Internal Med, 72-; mem, Nat Cancer Control Planning Conf, Nat Cancer Inst, 73, consult, Diag Res Adv Group, 74-; sect ed for med, Int J Dermat, 74- Mem: Am Asn Cancer Res; Am Fedn Clin Res; Am Col Physicians; Am Cancer Soc; AMA. Res: Cancer epidemiology; susceptibility to cancer; Australia antigen; hepatitis; pulmonary surfactant. Mailing Add: Inst for Cancer Res 7701 Burholme Ave Fox Chase Philadelphia PA 19111

SUTTER, DAVID FRANKLIN, b Ft Wayne, Ind, Nov 21, 35; m 59; c 3. APPLIED PHYSICS, INSTRUMENTATION. Educ: Purdue Univ, BS, 58; Cornell Univ, MS, 67, PhD(physics), 69. Prof Exp: Asst, Cornell Univ, 62-69; physicist, Fermi Nat Accelerator Lab, 69-75; PHYSICIST, DIV PHYS RES, ENERGY RES & DEVELOP ADMIN, 75- Mem: Am Phys Soc; AAAS. Res: Computer monitoring and control of accelerators; digital and analog instrumentation; theory of operation of accelerators; electron beam optics; development of superconducting magnet systems. Mailing Add: Div Phys Res MS J-309 ERDA Washington DC 20545

SUTTER, GERALD RODNEY, b Fountain City, Wis, Sept 20, 37; m 58; c 3. ENTOMOLOGY. Educ: Winona State Col, BA, 60; Iowa State Univ, MS, 63, PhD(entom), 65. Prof Exp: Res entomologist, European Corn Borer Lab, Arkeny, Iowa, 65, Northern Grain Insect Res Lab, 65-73, RES LEADER ENTOM, ENTOM RES DIV, NORTHERN GRAIN INSECT RES LAB, AGR RES SERV, USDA, 73- Mem: Entom Soc Am; Soc Invert Path. Res: Utilization of microorganisms in the biological control of insects. Mailing Add: Northern Grain Insect Res Lab RR 3 Brookings SD 57006

SUTTER, JOHN FREDERICK, b Oak Harbor, Ohio, June 7, 43; m 65; c 2. GEOLOGY, GEOCHRONOLOGY. Educ: Capital Univ, BS, 65; Rice Univ, MA, 68, PhD(geol), 70. Prof Exp: Nat Res Coun resident res assoc, Manned Spacecraft Ctr, NASA, 69-70; asst prof earth & space sci, State Univ NY Stony Brook, 70-71; ASST PROF GEOL & MINERAL, OHIO STATE UNIV, 71- Mem: Geol Soc Am; Geochem Soc. Res: Potassium-argon geochronology of selected areas; Precambrian structural history of eastern United States. Mailing Add: Dept of Geol & Mineral Ohio State Univ Columbus OH 43210

SUTTER, JOHN RITTER, b Edwardsville, Ill, May 4, 30; m 58; c 3. PHYSICAL CHEMISTRY. Educ: Wash Univ, St Louis, AB, 51; Tulane Univ, MS, 56, PhD(chem), 59. Prof Exp: Asst chem, Tulane Univ, 54-59; jr chemist & AEC grant, Wash State Univ, 59-60; asst prof chem, La Polytech Inst, 60-62; ASST PROF CHEM, HOWARD UNIV, 62- Mem: AAAS; Am Chem Soc; Am Phys Soc. Res: Kinetics; fast reactions; thermodynamics; calorimetry. Mailing Add: Dept of Chem Howard Univ Washington DC 20059

SUTTER, MORLEY CARMAN, b Redvers, Sask, May 18, 33; m 57; c 3. PHARMACOLOGY. Educ: Univ Man, BSc & MD, 57, PhD(pharmacol), 63. Prof Exp: Pvt pract, Souris, Man, 57-58; asst resident med, Winnipeg Gen Hosp, 58-59; demonstr pharmacol, Univ Man, 59-63; Imp Chem Industs fel, Cambridge Univ, 63-65; asst prof, Univ Toronto, 65-66; from asst prof to assoc prof, 66-71, PROF PHARMACOL & CHMN DEPT, UNIV BC, 71- Concurrent Pos: Wellcome Found travel award, 63; supvr, Downing Col, Cambridge Univ, 63-65; Med Res Coun Can scholar, 66-71. Mem: AAAS; Pharmacol Soc Can; Can Soc Clin Invest; Brit Pharmacol Soc; Am Soc Pharmacol & Exp Therapeut. Res: Effects of adrenergic blocking agents in shock; mechanism of cardiac arrythmias induced by cyclopropane-epinephrine; pharmacology of veins. Mailing Add: Dept of Pharmacol Univ of BC Vancouver BC Can

SUTTER, PHILIP HENRY, b Mineola, NY, Dec 8, 30; m 55; c 4. PHYSICS. Educ: Yale Univ, BS, 52, MS, 54, PhD(physics), 59. Prof Exp: Res engr res labs, Westinghouse Elec Corp, 58-63, sr res engr, 63-64; asst prof physics, 64-67, ASSOC PROF PHYSICS, FRANKLIN & MARSHALL COL, 67- Concurrent Pos: NZ sr res fel physics & eng lab, Dept Sci & Indust Res, Wellington, NZ, 69-70. Mem: AAAS; Am Phys Soc. Res: Solid state physics; transport properties; ionic crystals; energy conversion. Mailing Add: Dept of Physics Franklin & Marshall Col Lancaster PA 17604

SUTTER, RICHARD P, b Birmingham, Ala, Mar 22, 37; m 64; c 3. BIOCHEMISTRY, MOLECULAR BIOLOGY. Educ: St Joseph's Col, Ind, BA, 59; Ohio State Univ, MSc, 61; Tufts Univ, PhD(biochem), 66. Prof Exp: Instr biochem, Univ Ill, Chicago, 66-67; from asst prof to assoc prof biol, 67-74, PROF BIOL, WVA UNIV, 74- Concurrent Pos: Adj biochemist, Presby St Luke's Hosp, Chicago, 66-67; vis assoc, Calif Inst Technol, 73-74. Mem: AAAS; Am Chem Soc; Am Soc Microbiol. Res: Sexual reproduction in fungi; trisporic acid synthesis; cellular and metabolic control mechanisms; development. Mailing Add: Dept of Biol WVa Univ Morgantown WV 26506

SUTTER, VERA LA VERNE, b Los Angeles, Calif, Apr 2, 24. MEDICAL MICROBIOLOGY, INFECTIOUS DISEASES. Educ: Univ Calif, Los Angeles, AB, 46, MA, 47, PhD(microbiol), 50; Am Bd Med Microbiol, dipl. Prof Exp: Asst bact, Univ Calif, Los Angeles, 45-50; chief bacteriologist, Clin Lab, Vet Admin Gen Hosp, Los Angeles, 50-62; asst res microbiologist, Sch Dent, Univ Calif, San Francisco, 62-67; DIR ANAEROBIC BACT RES LAB, WADSWORTH VET ADMIN HOSP CTR, 67- Concurrent Pos: Adj assoc prof, Dept Med, Sch Med, Univ Calif, Los Angeles, 73-; mem, Am Bd Med Microbiol, 75-; lectr & lab dir bact & parasitol, Mt St Mary's Col, Calif, 52-62. Mem: Am Soc Microbiol; fel Am Pub Health Asn; Am Acad Microbiol; Infectious Dis Soc Am. Res: Clinical bacteriology; epidemiology of hospital acquired infections; anaerobic bacteria in pathologic and pathophysiologic processes; normal intestinal microflora. Mailing Add: Anaerobic Bact Res Lab Wadsworth Vet Admin Hosp Ctr Los Angeles CA 90073

SUTTERLIN, ARNOLD M, b Boston, Mass, Aug 8, 39; m 65. ANIMAL PHYSIOLOGY. Educ: State Teachers Col Bridgewater, BS, 62; Univ Mass, PhD(zool), 66. Prof Exp: NIH fel, 66-68; MEM RES STAFF, BIOL STA, FISHERIES RES BD CAN, 68- Res: Physiology of fishes. Mailing Add: Biol Sta Fisheries Res Bd of Can St Andrews NB Can

SUTTIE, JOHN WESTON, b La Crosse, Wis, Aug 25, 34; m 55; c 2. BIOCHEMISTRY. Educ: Univ Wis, BSc, 57, MS, 58, PhD, 60. Prof Exp: Fel biochem, Nat Inst Med Res, Eng, 60-61; from asst prof to assoc prof, 61-69, PROF BIOCHEM, UNIV WIS-MADISON, 69- Concurrent Pos: Mem comns atmospheric fluorides & fluorosis, Nat Res Coun. Honors & Awards: Mead Johns Award, Am Inst Nutrit, 74. Mem: AAAS; Am Soc Exp Biol & Med; Am Soc Biol Chem; Am Inst Nutrit; Air Pollution Control Asn. Res: Toxicity of inorganic fluorides; protein synthesis in mammalian systems; mechanisms of action of vitamin K; chemistry of prothrombin. Mailing Add: Dept of Biochem Univ of Wis Madison WI 53706

SUTTKUS, ROYAL DALLAS, b Fremont, Ohio, May 11, 20; m 47; c 3. ICHTHYOLOGY, FISH BIOLOGY. Educ: Mich State Col, BS, 43; Cornell Univ, MS, 47, PhD(zool), 51. Prof Exp: Asst zool, Cornell Univ, 47-50; from asst prof to assoc prof, 50-60, PROF ZOOL, TULANE UNIV, 60- Mem: AAAS; Am Soc Ichthyologists & Herpetologists; Am Fisheries Soc; Soc Syst Zool; Soc Study Evolution. Res: Systematics of fresh and salt water fishes; zoogeography; growth and seasonal distribution; environmental biology; water quality and water pollution. Mailing Add: Dept of Biol Tulane Univ New Orleans LA 70118

SUTTLE, ANDREW DILLARD, JR, b West Point, Miss, Aug 12, 26. RADIOCHEMISTRY, NUCLEAR PHYSICS. Educ: Miss State Univ, BS, 44; Univ Chicago, PhD(chem), 52. Prof Exp: Vpres res & grad studies & dir, Miss Res Comn, Miss State Univ, 60-62; vpres res & prof chem, Tex A&M Univ, 62-71; PROF NUCLEAR BIOPHYSICS & RADIO BIOCHEM & SPEC ASST TO DIR, MARINE BIOMED INST, UNIV TEX MED BR, 71- Concurrent Pos: Sr scientist, Humble Oil & Refining Co, 52-62; spec asst to dir res & eng, Dept Defense, Washington, DC, 62-64, mem exec comt sci bd res & eng, 67-; mem, Atomic Indust Forum, 55. Mem: Am Chem Soc; Am Phys Soc; Am Nuclear Soc; Inst Elec & Electronics Eng. Res: Radiation chemistry; nuclear and thermal energy; petroleum industry. Mailing Add: Marine Biomed Inst Suite 831 200 University Blvd Galveston TX 77550

SUTTLE, JIMMIE RAY, b Forest City, NC, Dec 26, 32; m 51; c 3. PHYSICS, ELECTRICAL ENGINEERING. Educ: Presby Col, SC, BS, 58; Duke Univ, MAT, 60, MA, 65; NC State Univ, PhD(elec eng), 72. Prof Exp: Instr math, Presby Col, SC, 60-61; phys scientist, Info Processing Off, 61-65 & Res-Technol Div, 65-72, assoc dir, Electronics Div, 72-74, actg dir, 74-75, DIR, ELECTRONICS DIV, US ARMY RES OFF, 75- Mem: Inst Elec & Electronics Engrs. Res: Computer architecture; biomathematics; switching theory; electron paramagnetic resonance spectroscopy of organic solids. Mailing Add: US Army Res Off PO Box 12211 Research Triangle Park NC 27709

SUTTLES, WAYNE PRESCOTT, b Seattle, Wash, Apr 24, 18; m 41; c 7. ANTHROPOLOGY. Educ: Univ Wash, PhD(anthrop), 51. Prof Exp: From asst prof to assoc prof anthrop, Univ BC, 51-63; from assoc prof to prof, Univ Nev, 63-66; PROF ANTHROP, PORTLAND STATE UNIV, 66- Concurrent Pos: Am Coun Learned Socs fel, 61. Mem: Am Anthrop Asn; Am Ethnol Soc. Res: Cultural anthropology; relation of language to culture; culture change. Mailing Add: Dept of Anthrop Portland State Univ Portland OR 97207

SUTTNER, LEE JOSEPH, b Hilbert, Wis, June 3, 39; m 65; c 2. GEOLOGY. Educ: Univ Notre Dame, BS, 61; Univ Wis, MS, 63, PhD(geol), 66. Prof Exp: Asst prof geol, 66-71, ASSOC PROF GEOL, IND UNIV, BLOOMINGTON, 71- Mem: Nat Asn Geol Teachers; Soc Econ Paleontologists & Mineralogists; Geol Soc Am. Res: Stratigraphy and sedimentary petrology. Mailing Add: Dept of Geol Ind Univ Bloomington IN 47401

SUTTON, BLAINE MOTE, b Ft Recovery, Ohio, Jan 23, 21; m 46. MEDICINAL CHEMISTRY. Educ: Purdue Univ, BS, 42, MS, 48, PhD(pharmaceut chem), 50. Prof Exp: Group leader med chem, Smith Kline & French Labs, Philadelphia, 50-68, sr investr med chem, 68-71, asst dir med chem, 71-75, ASSOC DIR MED CHEM, SMITH KLINE LABS, 75- Concurrent Pos: Adj prof sch pharm, Temple Univ, 66. Mem: AAAS; Am Chem Soc; Am Pharmaceut Asn; NY Acad Sci; Acad Pharmaceut Sci. Res: Synthetic medicinal chemistry; sym pathomimetic amines, sedatives, antibiotics, antirheumatics and hypocholesteremics. Mailing Add: 2435 Byberry Rd Hatboro PA 19040

SUTTON, CHARLES SAMUEL, b Lima, Peru, July 15, 13; US citizen; m 46. MATHEMATICS. Educ: Mass Inst Technol, BS, 35, MS, 37. Prof Exp: Instr math, Tufts Col, 39-40; from asst prof to assoc prof, 40-69, PROF MATH, THE CITADEL, 69- Mem: Am Math Soc; Math Asn Am. Res: Analysis; iteration; functional equations. Mailing Add: Dept of Math The Citadel Charleston SC 29409

SUTTON, CONSTANCE RITA, b Minneapolis, Minn, Jan 29, 26; m 53; c 1. ANTHROPOLOGY. Educ: Univ Chicago, PhB, 46, MA, 54; Columbia Univ, PhD(anthrop), 69. Prof Exp: Lectr anthrop, Queens Col, NY, 60-61; from lectr to assoc prof, 61-73, chairwoman dept anthrop, 71-73, PROF ANTHROP, COL ARTS & SCI, NY UNIV, 73- Concurrent Pos: Wenner-Gren Found fel & Res Inst Study Man fel, Barbados, BWI, 69; NIH biomed support grant, NY Univ, 72-73, univ arts & sci res grant, 73. Mem: Fel Am Anthrop Asn; fel Royal Anthrop Inst Gt Brit & Ireland; fel Soc Appl Anthrop; Am Ethnol Soc; Inst Caribbean Studies. Res: Caribbean anthropology; social movements; migration and ethnicity; women and knowledge and power. Mailing Add: Dept of Anthrop NY Univ 25 Waverly Pl New York NY 10003

SUTTON, DALE DINKINS, b Ironton, Mo, Dec 16, 36; m 63; c 2. BOTANY. Educ: Univ Mo-Columbia, BS, 61; Northern Ariz Univ, MS, 68; Univ Nebr, Lincoln, PhD(bot), 71. Prof Exp: Teacher high schs, Mo, 61-67; teaching asst bot, Northern Ariz Univ, 67-68; asst, Univ Nebr, 68-71; PROF BIOL & CHMN DIV SCI & MATH, WSHORE COMMUNITY COL, SCOTTVILLE, MICH, 71- Mem: Am Bot Soc. Res: Anatomy and morphology of the embryo of the genus Querucs; grass leaf anatomy. Mailing Add: Div of Sci & Math WShore Community Col Scottville MI 49454

SUTTON, DALLAS ALBERT, b Grand Junction, Colo, Sept 12, 11; m 35; c 1. BIOLOGY. Educ: Univ Colo, AB, 39, PhD, 53; Northwestern Univ, MS, 40. Prof Exp: Prin pub sch, Colo, 34-45; instr biol, Mesa Col, 45-51; assoc prof biol & sci educ, Eastern Mont Col Educ, 54-57; PROF BIOL & SCI EDUC, CALIF STATE UNIV, CHICO, 57- Concurrent Pos: NSF grant. Mem: Am Soc Mammal. Res: Mammalogy; chipmunks of Colorado; chromosomes of the chipmunks, Genus eutamias. Mailing Add: Dept of Biol Calif State Univ Chico CA 95926

SUTTON, DAVID GEORGE, b San Francisco, Calif, Apr 17, 44; m 63; c 1. LASERS. Educ: Univ Calif, Berkeley, BS, 66; Mass Inst Technol, PhD(chem physics), 70. Prof Exp: Fel phys chem, Dept Chem, Mass Inst Technol, 70-71; fel phys chem, Dept Chem, Univ Toronto, 71-72; TECH STAFF LASERS, THE AEROSPACE CORP, 72- Mem: Am Phys Soc. Res: Experimental research in gas phase molecular exitation in gas phase molecular and energy transfer and electrically and chemically excited gas lasers. Mailing Add: The Aerospace Corp El Segundo CA 90009

SUTTON, DEREK, b Eng, July 15, 37; m 58; c 2. INORGANIC CHEMISTRY. Educ: Univ Nottingham, BSc, 58, PhD(chem), 63. Prof Exp: Asst lectr chem, Univ Nottingham, 62-64, lectr, 64-67; asst prof, 67-69, ASSOC PROF CHEM, SIMON FRASER UNIV, 69- Mem: The Chem Soc. Res: Synthesis and characterization of nitrato-complexes; study of arylazo and other complexes of transition metals related to the interaction of nitrogen with transition metals. Mailing Add: Dept of Chem Simon Fraser Univ Burnaby BC Can

SUTTON, DONALD DUNSMORE, b Oakland, Calif, June 8, 27; c 3. MICROBIOLOGY. Educ: Univ Calif, AB, 51, MA, 54, PhD(microbiol), 57. Prof Exp: Sr lab technician bact, Univ Calif, 50-53, asst, 53-55; asst prof & USPHS fel microbiol sch med, Ind Univ, 57-59; Waksman-Merck fel inst microbiol, Rutgers Univ, 59-60; assoc prof microbiol, 60-63, chmn dept biol, 65-70, PROF MICROBIOL, CALIF STATE UNIV, FULLERTON, 63- Concurrent Pos: Vis prof, Inst Appl Microbiol, Tokyo Univ, 70-71. Mem: Am Soc Microbiol; Am Inst Biol Sci; Mycol Soc Am; Brit Soc of Gen Microbiol. Res: Microbial physiology; metabolism of plant disease bacteria; mechanisms of growth inhibitors and antifungal agents; physiological basis of morphogenesis in fungi; location of enzymes in fungi; microbiology of food fermentations. Mailing Add: Dept of Biol Calif State Univ Fullerton CA 92634

SUTTON, EMMET ALBERT, b Toledo, Ohio, May 7, 35; m 58; c 2. ATOMIC PHYSICS. Educ: Cornell Univ, BEngPhys, 58, PhD(aeronaut eng), 61. Prof Exp: Asst prof physics, Hamilton Col, 61-62; asst prof aeronaut & astronaut, Purdue Univ, 62-65; prin res scientist, Avco Everett Res Lab, 65-70; V PRES, AERODYNE RES, INC, 70- Honors & Awards: Am Inst Aeronaut & Astronaut Award, 62. Mem: Am Inst Aeronaut & Astronaut. Res: Experimental chemical kinetics; hypersonic wake chemistry; rocket plume radiation. Mailing Add: Aerodyne Res Inc Tech-Ops Bldg South Ave Burlington MA 01803

SUTTON, GEORGE HARRY, b Chester, NJ, Mar 4, 27; m 47; c 4. SEISMOLOGY, GEOPHYSICS. Educ: Muhlenberg Col, BS, 50; Columbia Univ, MA, 53, PhD(geol), 57. Prof Exp: From res asst to res assoc, Lamont Geol Observ, Columbia Univ, 50-60, mem acad staff, 60-66, from asst prof to assoc prof geol, Univ, 60-66; prof & staff mem geosci, 66-71, PROF GEOPHYS, UNIV HAWAII, 71-, ASSOC DIR & GEOPHYSICIST, HAWAII INST GEOPHYS, 71- Concurrent Pos: Chief seismologist, Inst Sci Res Cent Africa, 55-56; prin investr & co-investr for Ranger, Surveyor & Apollo lunar seismog exps, NASA, 59-73, mem geophys working group planetology subcomt, 64-73, Surveyor sci eval adv team, 65-73; mem comt planetary surfaces & interiors, Space Sci Bd, Nat Acad Sci, 63-; mem ad hoc panel earthquake prediction, Off Sci & Technol, 64-65; mem staff, Hawaii Inst Geophys, 66-71. Mem: AAAS; Soc Explor Geophysicists; Geol Soc Am; Seismol Soc Am; Am Geophys Union. Res: Earthquake seismology; seismic wave propagation and source conditions; geophysical exploration of oceanic crustal and upper-mantle structure; lunar and planetary interiors; geophysical instrumentation for ocean-bottom and lunar use. Mailing Add: Hawaii Inst of Geophys Univ of Hawaii Honolulu HI 96822

SUTTON, GEORGE MIKSCH, b Lincoln, Nebr, May 16, 98. ORNITHOLOGY. Educ: Bethany Col, WVa, BS & ScD, 23; Cornell Univ, PhD, 32. Prof Exp: Asst cur birds, Carnegie Mus, 19-24; instr ornith, Univ Pittsburgh, 21-24; state ornithologist & chief bur res & info, State Bd Game Comnrs, Pa, 25-29; cur birds, Cornell Univ, 32-45; RES PROF ZOOL, CUR BIRDS & ORNITHOLOGIST, BIOL SURV, UNIV OKLA, 52- Concurrent Pos: Fel, Cranbrook Inst Sci; ornith ed, Am Col Dict & Encycl Arctica; hon trustee, Oglebay Inst. Mem Carnegie Mus ornith expeds, Labrador, 20, Hudson Bay, 23, 26, 31, Cape Sable region, Fla, 24, Southern Labrador, 28, Southampton Island, Hudson Bay, 29-30, Sask, 32, Rio Grande Valley, 33, BC, 34; Cornell exped, southern states, 35, Okla, 36-37, Mex, 38-39, Ariz, 40; Cornell & Carnegie Mus exped, Rio Grande Valley, 35; Cornell-Carleton exped, Mex, 41; independent exped, Mex, 47, 49, 50; Baffin Island, 53; Hudson Bay, 56; Iceland, 58; Victoria Island, 62; Jenny Lind Island, 66; Bathurst & Ellesmere Islands, 69. Mem: Cooper Ornith Soc; Wilson Ornith Soc (vpres, 25-29, pres, 41, 45-47); Am Geog Soc; Wildlife Soc; fel Am Ornith Union. Res: Life histories, plumages, distribution and taxonomy of birds; bird illustrations. Mailing Add: Dept of Zool Univ of Okla Norman OK 73069

SUTTON, HARRY ELDON, b Cameron, Tex, Mar 5, 27; m 62; c 2. HUMAN GENETICS. Educ: Univ Tex, BS, 48, MA, 49, PhD(biochem), 53. Prof Exp: Res scientist, Univ Tex, 48-52; asst biologist, Univ Mich, 52-56, instr human genetics, 56-57, asst prof, 57-60; assoc prof zool, 60-64, chmn dept, 70-73, assoc dean grad sch, 67-70 & 73-75, PROF ZOOL, UNIV TEX, AUSTIN, 64- Concurrent Pos: Mem comt personnel res, Am Cancer Soc, 61-64; mem genetics study sect, NIH, 63-67; mem adv coun, Nat Inst Environ Health Sci, 68-72, mem sci adv comt, 72-76; mem adv comt, Atomic Bom Casualty Comn; mem comt epidemiol & vet follow-up studies, Nat Acad Sci-Nat Res Coun; ed, Am J Human Genetics, 64-69. Mem: AAAS; Genetics Soc Am; Am Chem Soc; Am Soc Human Genetics (ed, Am J Human Genetics, 64-69); Am Soc Biol Chem. Res: Genetic control of protein structure; inherited variations in human metabolism: population genetics. Mailing Add: Dept of Zool Univ of Tex Austin TX 78712

SUTTON, JOHN CLIFFORD, b Halstead, Eng, Oct 10, 41; m 64; c 2. PLANT PATHOLOGY. Educ: Univ Nottingham, BSc, 65; Univ Wis, PhD(plant path), 69. Prof Exp: Asst prof plant path, 69-75, ASSOC PROF PLANT PATH, UNIV GUELPH, 75- Concurrent Pos: Nat Res Coun Can grant, 70-76. Mem: Can Phytopath Soc; Am Phytopath Soc. Res: Vesicular-arbuscular mycorrhizae; biology and ecology of soil borne plant pathogens; extension plant pathology. Mailing Add: Dept of Environ Biol Univ of Guelph Guelph ON Can

SUTTON, JOHN CURTIS, b Weiser, Idaho, May 13, 42; m 65; c 4. ORGANIC CHEMISTRY, ANALYTICAL CHEMISTRY. Educ: Univ Idaho, BS, 64, PhD(chem), 72. Prof Exp: ASSOC PROF CHEM, LEWIS-CLARK STATE COL, 70- Res: Utilization of tree bark for the sorption of heavy metal ions from aqueous solutions. Mailing Add: Dept of Chem Lewis-Clark State Col Lewiston ID 83501

SUTTON, JOHN WILLIAM, physical chemistry, see 12th edition

SUTTON, LOUISE NIXON, b Hertford, NC, Nov 4, 25; div. MATHEMATICS. Educ: Agr & Tech Col NC, BS, 46; NY Univ, MA, 51, PhD(math educ), 62. Prof Exp: Instr high sch, NC, 46-47; instr math, Agr & Tech Col NC, 47-50, asst prof, 51-57; asst prof, Del State Col, 57-62; assoc prof, 62-63, PROF MATH, ELIZABETH STATE UNIV, 63-, HEAD DEPT PHYS SCI & MATH, 62- Concurrent Pos: Asst dean women, Agr & Tech Col, 47-48; mem adv comt cert math & sci, Del State Bd Educ, 59-61, mem adv comt math, 61-62; mem bd dir, Perquimans County Indust Develop Corp, 67-72; mem gen adv comt, NC State Bd Soc Serv, 69-71; mem bd dir, Div Higher Educ, NC Asn Educr, 69-72; dir, NSF Inst, 71 & 72. Mem: Math Asn Am; Nat Educ Asn; Nat Asn Univ Women; Nat Coun Teachers Math. Res: Mathematics ecucation; concept learning in trigonometry; analytical geometry at college level. Mailing Add: 204 Parkview Dr Elizabeth City NC 27909

SUTTON, PAUL, b Hopkinsville, Ky, Sept 11, 29; m 54; c 2. SOIL FERTILITY. Educ: Univ Ky, BS, 51, MS, 57; Iowa State Univ, PhD(agron, soil fertil), 62. Prof Exp: Asst agronomist, Univ Ky, 54-55; asst horticulturist, Univ Fla, 61-67; assoc prof soils, Ohio Agr Res & Develop Ctr, 67-73, PROF AGRON, LAB ENVIRON SCI, OHIO STATE UNIV, 73- Mem: AAAS; Am Soc Agron. Res: Plant physiology. Mailing Add: Dept of Agron Lab Environ Sci Ohio State Univ Columbus OH 43210

SUTTON, PAUL MCCULLOUGH, b Ohio, Dec 3, 21; m 46; c 2. PHYSICS. Educ: Harvard Univ, BS, 43; Columbia Univ, MA, 48, PhD(physics), 53. Prof Exp: Asst physics, Columbia Univ, 50-52; res assoc physicist, Corning Glass Works, 52-54, supvr ultrasonics res, 54-56, supvr fundamental physics group, 56-59; sr staff scientist aeronutronic div, Ford Motor Co, 59-62, mgr appl physics dept, Philco Corp, 62-66, mgr physics lab, Philco-Ford Corp, 66-68, mgr physics & chem lab, 68-72, mgr res lab, 72-75, MGR RES LAB, AERONUTRONIC-FORD CORP, 75- Concurrent Pos: Lectr optics, Univ Calif, Irvine, 66. Mem: AAAS; fel Am Phys Soc; Am Ceramic Soc; Optical Soc Am. Res: Solid state physics; elastic constants; Debye temperature from elastic constants; photoelasticity; acoustic propagation; glass physics; dielectric properties of glasses; hypervelocity impact; space charge in glass; atmospheric turbulence; lasers. Mailing Add: Res Labs Aeronutronic-Ford Corp Ford Rd Newport Beach CA 92663

SUTTON, PAUL PORTER, b Baltimore, Md, Dec 19, 10; m 38. CHEMISTRY, PHYSICS. Educ: Johns Hopkins Univ, PhD(chem), 34. Prof Exp: Prof sci, Paducah Jr Col, 35; instr phys chem, Univ Ark, 26; prof chem, Bluefield Col, 36-37; from instr to assoc prof phys chem, 37-48, PROF PHYS CHEM, NC STATE UNIV, 48- Mem: Am Math Soc; Asn Symbolic Logic. Res: Measurement of electron affinities; theory of metals; evaporation rates of droplets. Mailing Add: Dept of Chem NC State Univ Raleigh NC 27607

SUTTON, ROBERT GEORGE, b Rochester, NY, June 17, 25; m 46; c 2. GEOLOGY. Educ: Univ Rochester, AB, 48, MS, 50; Johns Hopkins Univ, PhD, 56. Prof Exp: Instr geol, Alfred Univ, 50-52; jr instr, Johns Hopkins Univ, 52-54; from asst prof to assoc prof, 54-66, PROF GEOL, UNIV ROCHESTER, 66- Mem: AAAS; Geol Soc Am; Soc Econ Paleont & Mineral. Res: Paleozoic stratigraphy; sedimentology; sedimentary petrology. Mailing Add: Dept of Geol Sci Univ of Rochester Rochester NY 14627

SUTTON, ROGER BEATTY, b Lloydminster, Sask, Sept 14, 16; m 46; c 2. EXPERIMENTAL HIGH ENERGY PHYSICS. Educ: Univ Sask, BA, 38, MA, 39; Princeton Univ, PhD(physics), 43. Prof Exp: Res physicist, Off Sci Res & Develop, Princeton Univ, 42-43; physicist, Manhattan Dist, Los Alamos Sci Lab, 43-46; from asst prof to assoc prof physics, 46-56, PROF PHYSICS, CARNEGIE-MELLON UNIV, 56- Mem: Am Phys Soc; Am Asn Physics Teachers; Am Asn Physicists Med. Res: Molecular spectroscopy; low energy nuclear physics; radioactive isotopes; neutron physics; cyclotrons; study of x-rays from exotic atoms including antiprotonic, kaonic and sigma minus atoms. Mailing Add: Dept of Physics Carnegie-Mellon Univ Pittsburgh PA 15213

SUTTON, ROSCOE MURRAY DAVIDSON, b Toronto, Ont, Mar 22, 22. PHYTOPATHOLOGY, VIROLOGY. Educ: Univ Toronto, BA, 48, MA, 58, PhD(plant path), 60. Prof Exp: Head technician bact med sch, Queen's Univ, Ont, 46-47; asst plant pathologist sci serv, Can Dept Agr, 48-54, from assoc plant pathologist to plant pathologist res br, 54-67, res scientist, Cell Biol Res Inst, Ont, 67-71; RES COUN OFFICER, ENVIRON SECRETARIAT, DIV BIOL SCI, NAT RES COUN CAN, 71- Concurrent Pos: Secy subcomt biol, Assoc Comt Sci Criteria on Environ Qual. Honors & Awards: Wintercorbyn Award, 58. Mem: Am Phytopath Soc; Can Phytopath Soc; Can Soc Microbiol; Can Pub Health Asn. Res: Research, development and production of biological weapons; bacterial phytopathology, especially development of methods for the detection and identification of bacterial pathogens; studies of bacteriophages for phage typing, ecological and epidemiological studies; survey of pollution in Canada. Mailing Add: Environ Secretariat Nat Res Coun of Can 100 Sussex Dr Ottawa ON Can

SUTTON, RUSSELL PAUL, b Mo, July 31, 29; m 53; c 5. CHEMISTRY. Educ: Univ Mo-Columbia, BS, 51; State Univ Iowa, MS, 53, PhD(chem), 55. Prof Exp: Res chemist, E I du Pont de Nemours & Co, Inc, 55-58; from asst prof to assoc prof, 58-70, PROF CHEM, KNOX COL, ILL, 70- Mem: Am Chem Soc. Res: The chemistry of chalcones and flavylium compounds; gas chromatography. Mailing Add: Dept of Chem Knox Col Galesburg IL 61401

SUTTON, TURNER BOND, b Windsor, NC, Oct 24, 45. PLANT PATHOLOGY. Educ: Univ NC, BA, 68; NC State Univ, MS, 71, PhD(plant path), 73. Prof Exp: Res assoc plant path, Mich State Univ, 73-74; RES ASSOC PLANT PATH, NC STATE UNIV, 74- Mem: Am Phytopath Soc. Res: Apple diseases; epidemiology and control; pest management. Mailing Add: Dept of Plant Path NC State Univ PO Box 5397 Raleigh NC 27607

SUTTON, WILLIAM WALLACE, physiology, toxicology, see 12th edition

SUTTON, WILLIAM WALLACE, b Monticello, Miss, Dec 15, 30; m 54; c 6. PROTOZOOLOGY, CELL BIOLOGY. Educ: Dillard Univ, BA, 53; Howard Univ, MS, 59, PhD(zool), 65. Prof Exp: Med technician, DC Gen Hosp, 55-59; from instr to assoc prof biol, 59-69, actg chmn div natural sci, 69-70, PROF BIOL, DILLARD UNIV, 69-, CHMN DIV NATURAL SCI, 70- Concurrent Pos: Consult, NIH, 72-74 & 16 Inst Health Sci Consortium of NC & Va, 74-; assoc & regional liaison officer, Danforth Found, 75- Mem: Soc Protozoologists; Sigma Xi; Nat Inst Sci. Res: Radiation cell biology; responses of peritrichs to ionizing radiations; chemical analysis of the cyst wall and nutrition of peritrichs; isolation of nucleic acids from peritrichs. Mailing Add: Div of Natural Sci Dillard Univ New Orleans LA 70122

SUTULA, CHESTER LOUIS, b Erie, Pa, Feb 15, 33; m 55; c 8. PHYSICAL CHEMISTRY, CLINICAL CHEMISTRY. Educ: Col Holy Cross, BS, 54; Iowa State Univ, MS, 58, PhD(phys chem), 59. Prof Exp: Teaching asst, Iowa State Univ, 55-57; res scientist, Ames Lab, Iowa, 57-59; sr res scientist, Marathon Oil Co, Colo, 59-67; sr res scientist, 67-69, DIR, AMES RES LAB, AMES CO DIV, MILES LABS, INC, 69- Mem: AAAS; Am Chem Soc. Res: Surface chemistry; calorimetry; wetting properties of complex porous materials; structure of colloidal fluids; microbiology; immunoassay; instrumentation. Mailing Add: Ames Res Lab Miles Inc 1127 Myrtle St Elkhart IN 46514

SUYDAM, FREDERICK HENRY, b Lancaster, Pa, July 30, 23; m 44; c 3. ORGANIC CHEMISTRY. Educ: Franklin & Marshall Col, BS, 46; Northwestern Univ, PhD(chem), 50. Prof Exp: Instr chem, Franklin & Marshall Col, 46-47; asst, Northwestern Univ, 47-49; instr anat, Med Sch, Johns Hopkins Univ, 50-52; from asst prof to assoc prof, 52-62, PROF CHEM, FRANKLIN & MARSHALL COL, 62-, CHMN DEPT, 58- Mem: Am Chem Soc. Res: Peptide synthesis; reactions of amino acids; infrared absorption. Mailing Add: Dept of Chem Franklin & Marshall Col Lancaster PA 17604

SUZUKI, AKIO, b Tokyo, Japan, Apr 5, 37; m 64; c 1. PLANT BREEDING, EXPERIMENTAL STATISTICS. Educ: Hokkaido Univ, BS, 61; NC State Univ, MS, 66, PhD(genetics), 68. Prof Exp: Res assoc genetics & plant breeding, Nat Inst Genetics, Mishima, Japan, 61-63; PLANT BREEDER, GREAT WESTERN AGR RES CTR, 68- Mem: Am Soc Agron; Crop Sci Soc Am; Am Statist Asn; Biomet Soc Am. Res: Quantitative and population genetics; sugarbeet breeding and genetics; computerized variety testing. Mailing Add: Agr Res Ctr Great Western Sugar Co Longmont CO 80501

SUZUKI, DAVID TAKAYOSHI, b Vancouver, BC, Mar 24, 36; m 58; c 3. GENETICS. Educ: Amherst Col, BA, 58; Univ Chicago, PhD(zool), 61. Hon Degrees: LLD, Univ PEI, 74. Prof Exp: Res assoc genetics, Biol Div, Oak Ridge Nat Lab, 61-62; asst prof, Univ Alta, 62-63; from asst prof to assoc prof, 63-69, PROF ZOOL, UNIV BC, 69- Concurrent Pos: Res grants, Nat Res Coun Can, 62-, AEC, 64-69 & Nat Cancer Inst Can, 69- Mem: Genetics Soc Am; Genetics Soc Can; Can Soc Cell Biol (pres, 69-70). Res: Regulation of development and behavior; genetic organization of chromosomes; developmental and behavioral genetics. Mailing Add: Dept of Zool Univ of BC Vancouver BC Can

SUZUKI, GEORGE b Acampo, Calif, Mar 5, 22; m 45. MATHEMATICS. Educ: Univ Denver, BS, 44; Univ Minn, MA, 47, PhD(statist), 51. Prof Exp: Lectr & instr statist, Univ Minn, 46-51; actg chief struct anal br, Inter-indust Res Off, Dep Chief of Staff & Comptroller, US Dept Air Force, 51-53; statist specialist, Appl Math Lab, David Taylor Model Basin, US Dept Navy, 53-57, dir mgt sci staff, Navy Mgt Off, 57-64; dep chief tech anal div, Inst Appl Technol, 64-73, actg chief, 73-75, ASSOC DIR ADVAN PLANNING, INST APPL TECHNOL, NAT BUR STANDARDS, 75- Mem: Sigma Xi; Inst Math Statist; Am Statist Asn; Inst Mgt Sci. Res: Development of mathematical methods for solving management problems. Mailing Add: Rm B120 Technol Bldg Nat Bur of Standards Washington DC 20234

SUZUKI, HOWARD KAZURO, b Ketchikan, Alaska, Apr 3, 27; m 52; c 4. ANATOMY. Educ: Marquette Univ, BS, 49, MS, 51; Tulane Univ, PhD(anat), 55. Prof Exp: Asst zool & bot, Marquette Univ, 48-51; asst zool & anat, Tulane Univ, 51-55; instr anat, Sch Med, Yale Univ, 55-58; from asst prof to prof, Sch Med, Univ Ark, 58-70; assoc dean col health related professions, 70-71, actg dean, 71-72, prof anat, Col Med, 70-73, PROF NEUROSCI, COL MED, UNIV FLA, 72-, DEAN COL HEALTH RELATED PROFESSIONS, 72- Concurrent Pos: Mem gen res support prog adv comt, NIH; mem adv comt, Off Acad Affairs, US Vet Admin. Mem: Fel AAAS; Am Asn Anatomists; Asn Am Med Cols; Soc Exp Biol & Med; Am Soc Allied Health Professions. Res: Endocrine relations to bone; phagocytosis and reticuloendothelial system; neonatal human anatomy; comparative bone metabolism. Mailing Add: Col Health Related Professions Univ of Fla Gainesville FL 32610

SUZUKI, ISAMU, b Tokyo, Japan, Aug 4, 30; m 62; c 3. MICROBIOLOGY, BIOCHEMISTRY. Educ: Univ Tokyo, BSc, 53; Iowa State Univ, PhD(bact physiol), 58. Prof Exp: Fel microbiol, Western Reserve Univ, 58-60; instr, Univ Tokyo, 60-62; Nat Res Coun Can fel, 62-64, from asst prof to assoc prof, 64-69, PROF MICROBIOL, UNIV MAN, 69-, HEAD DEPT, 72- Mem: AAAS; Am Soc Microbiol; Can Soc Microbiol; Can Biochem Soc; NY Acad Sci. Res: Mechanism of the oxidation of inorganic sulfur compounds by Thiobacilli and ammonia by Nitrosomonas; physiology of autotrophic bacteria; carbon dioxide fixation enzymes; mechanism of enzyme reactions; kinetics; acting of thiamine pyrophosphate. Mailing Add: Dept of Microbiol Univ of Man Winnipeg MB Can

SUZUKI, KINUKO, b Hyogo, Japan, Nov 10, 33; m 60; c 1. PATHOLOGY, NEUROPATHOLOGY. Educ: MD, Osaka City Univ, 59. Hon Degrees: MA, Univ Pa, 71. Prof Exp: Asst prof path, Albert Einstein Col Med, 68; from asst prof to assoc prof, Sch Med, Univ Pa, 69-72; ASSOC PROF PATH, ALBERT EINSTEIN COL MED, 72- Mem: Am Asn Neuropath; Soc Neurosci. Res: Study of pathogenesis of developmental disorder of the central nervous system. Mailing Add: Dept of Path Albert Einstein Col of Med Bronx NY 10461

SUZUKI, KUNIHIKO, b Tokyo, Japan, Feb 5, 32; m 60; c 1. NEUROCHEMISTRY, BIOCHEMISTRY. Educ: Univ Tokyo, BA, 55, MD, 59. Prof Exp: Resident clin neurol, Albert Einstein Col Med, 60-62, instr neurol, 64, asst prof, 65-68; assoc prof, Sch Med, Univ Pa, 69-71, prof neurol & pediat, 71-72; PROF NEUROL & NEUROSCI, ALBERT EINSTEIN COL MED, 72- Concurrent Pos: Mem neurol B study sect, 71-75; mem adv bd, Nat Tay-Sachs & Allied Dis Asn; dep chief ed, J Neurochem, 75- Mem: Int Soc Neurochem; Soc Neurosci; Am Asn Neuropath; Am Soc Biol Chemists; Am Soc Neurochem. Res: Biochemistry of brain lipids, particularly gangliosides; biochemical and enzymatic studies of inherited metabolic disorders of the nervous system. Mailing Add: Dept of Neurol Albert Einstein Col of Med Bronx NY 10461

SUZUKI, MICHIO, b Taipei, Formosa, Feb 23, 27; m 59; c 2. PLANT PHYSIOLOGY, BIOCHEMISTRY. Educ: Tohuku Univ, Japan, BS, 52, PhD(agr chem), 62. Prof Exp: Asst plant physiol & biochem, Tohuku Univ, Japan, 52-66; RES SCIENTIST, RES BR, CAN DEPT AGR, 66- Concurrent Pos: Nat Res Coun Can res fel, 63-65. Mem: Can Soc Plant Physiologists; Am Soc Plant Physiologists; Agr Inst Can; Can Soc Agron; Crop Sci Soc Am. Res: Winter survival of perennial crops; vegetative regrowth of forage crops; metabolism of fructosan in grasses; plant nutrition; response of host plants to nematode infestation; evaluation of feed quality. Mailing Add: Res Sta Can Dept of Agr Charlottetown PE Can

SUZUKI, MINORU, b Tokyo, Japan, June 25, 28; m 62; c 2. PATHOLOGY, NEUROPATHOLOGY. Educ: Keio Univ, Japan, DrMed, 52, DrMedSci, 60; Am Bd Path, dipl & cert path anat, 59, cert clin path, 64, cert neuropath, 70. Prof Exp: Intern, Int St Luke's Hosp, Tokyo, 52-53; jr asst resident anat path, Mallory Inst Path, Boston City Hosp, Mass, 53-54; asst resident surg path, Hosp Univ Pa, 55-56, resident clin path, 56-58; resident exp path, Barnes Hosp, St Louis, Mo, 58-59; fel, Sch Med, Wash Univ, 59-60; pathologist, Atomic Bomb Casualty Comn, Japan, 60-62; chief path sect, Nat Cancer Ctr, Tokyo, 62-63; from asst prof to assoc prof, 63-72, PROF PATH, BAYLOR COL MED, 72- Concurrent Pos: Consult, Vet Admin Hosp, Tex, 66-; from assoc attend to attend physician, Ben Taub Gen Hosp, Houston, 67- Mem: Am Soc Exp Path; Am Soc Clin Path; Am Heart Asn; Am Asn Pathologists & Bacteriologists; Am Asn Neuropathologists. Res: Cardiovascular and cerebrovascular pathology, especially pathogenesis of atherosclerosis; human and experimental neuropathology. Mailing Add: Dept of Path Baylor Col of Med Houston TX 77025

SUZUKI, NOBORU, b Tokyo, Japan, Mar 18, 31; m 63. MATHEMATICS. Educ: Tohoku Univ, Japan, PhB, 53, PhM, 55, PhD(math), 63. Prof Exp: Asst & instr math, Tohoku Univ, Japan, 55-61; asst prof, Kanazawa Univ, Japan, 61-66; assoc prof, 66-71, PROF MATH, UNIV CALIF, IRVINE, 71- Mem: Am Math Soc; Math Asn Am; Math Soc Japan. Res: Functional analysis; theory of Von Neumann algebras; theory of operators on Hilbert space. Mailing Add: Dept of Math Univ of Calif Irvine CA 92664

SUZUKI, SHIGETO, b San Francisco, Calif, Feb 25, 25; m 53; c 1. ORGANIC CHEMISTRY. Educ: Univ Calif, Berkeley, BS, 55; Univ Southern Calif, PhD(chem), 59. Prof Exp: Sloan Found res fel org chem, Univ Calif, Berkeley, 59-60; SR RES ASSOC, CHEVRON RES CO, 60- Mem: Am Chem Soc; The Chem Soc. Res: Organic reaction mechanism, especially carbonium and carbanion rearrangements; organo-sulfur and organo-halogen chemistry. Mailing Add: Chevron Res Co 576 Standard Ave Richmond CA 94802

SUZUKI, TSUNEO, b Nagoya, Japan, Nov 23, 31; m 70; c 3. MICROBIOLOGY, BIOCHEMISTRY. Educ: Univ Tokyo, BS, 54, MD, 57; Hokkaido Univ, PhD(biochem), 69. Prof Exp: Japan Fel Asn fel & Fulbright travel grant, Univ Tokyo, 63; fel, Univ Wis, 63-66; fel, Univ Lausanne, 66-67; Ont Cancer Inst fel, Univ Toronto, 67-69; res assoc immunochem, Univ Wis, 69-70; ASST PROF IMMUNOCHEM, SCH MED, UNIV KANS MED CTR, KANSAS CITY, 70- Mem: Am Asn Immunol; Can Soc Immunol. Res: Immunochemistry of antibodies; structure-function relationships of antibodies and protein-protein interaction. Mailing Add: Dept of Microbiol Univ of Kans Sch of Med Kansas City KS 66103

SVACHA, ANNA JOHNSON, b Asheville, NC, Nov 27, 28; c 3. NUTRITION, BIOCHEMISTRY. Educ: Va Polytech Inst & State Univ, BS, 50; Univ Ariz, MS, 69, PhD(biochem, nutrit), 71. Prof Exp: Indust chemist, Hercules Powder Co, 51-53; physicist, Taylor Model Basin, US Navy, 53; high sch teacher, Tenn, 53-54; res asst anal chem, Tex Agr Exp Sta, 56-58; res asst org chem, Chas Pfizer & Co, Inc, 58-59; res asst nutrit, Univ Ariz, 67-68; asst poultry scientist, Ariz Agr Exp Sta, 71-72; ASST PROF NUTRIT, AUBURN UNIV, 72- Mem: AAAS; Poultry Sci Asn. Res: Appetite regulation with respect to protein and amino acid metabolism. Mailing Add: Dept of Nutrit & Foods Auburn Univ Auburn AL 36830

SVADLENAK, RUDOLF ELDO, inorganic chemistry, see 12th edition

SVANES, TORGNY, b Norway. ALGEBRA. Educ: Oslo Univ, MA, 65; MAss Inst Technol, PhD(math), 72. Prof Exp: Instr math, Oslo Univ, 66-69, NY State Univ Stony Brook, 72-73; asst prof Aarhus Univ, 73-75; ASST PROF MATH, PURDUE UNIV, 75- Mem: Am Math Soc. Res: Study of Schubert subvarieties of homogeneous spaces. Mailing Add: Dept of Math Purdue Univ West Lafayette IN 74907

SVEC, HARRY JOHN, b Cleveland, Ohio, June 24, 18; m 43; c 9. PHYSICAL CHEMISTRY, ANALYTICAL CHEMISTRY. Educ: John Carroll Univ, BS, 41; Iowa State Univ, PhD(phys chem), 49. Prof Exp: Asst chem, 41-43, jr chemist, Manhattan Proj, 43-46, res assoc, Inst Atomic Res, 46-50, asst prof chem, Univ & assoc chemist, Inst, 50-55, assoc prof & chemist, 55-60, PROF CHEM, IOWA STATE UNIV & SR CHEMIST, INST ATOMIC RES, 60- Concurrent Pos: Lectr, NATO Advan Study Inst Mass Spectros, 64; ed, Int J Mass Spectrometry Ion Physics, Am Soc Mass Spectrometry, 68- Mem: Am Chem Soc; Am Soc Testing & Mat; Am Soc Mass Spectrometry (pres, 74-76); Geochem Soc; The Chem Soc. Res: Metallurgy of rare metals; mass spectroscopy; mass spectrometry in physical, inorganic and analytical chemistry; corrosion mechanisms; determination of ultra trace levels of organic pollutants in water and air. Mailing Add: Dept of Chem Iowa State Univ Ames IA 50010

SVEC, LEROY VERNON, b Columbus, Nebr, Feb 27, 42; m 64; c 3. AGRONOMY, PLANT PHYSIOLOGY. Educ: Univ Nebr-Lincoln, BSc, 64; Purdue Univ, Lafayette, MSc, 68, PhD(agron), 70. Prof Exp: ASST PROF PLANT SCI, UNIV DEL, 69- Concurrent Pos: Mem, Coun Agr Sci & Technol. Mem: Am Soc Agron; Am Soc Plant Physiologists. Res: Plant metabolism and inorganic nutrition; plant stress resistance. Mailing Add: Dept of Plant Sci Univ of Del Col of Agr Sci Newark DE 19711

SVED, STEPHEN, biochemistry, see 12th edition

SVEDA, MICHAEL, b West Ashford, Conn, Feb 3, 12; m 36; c 2. CHEMISTRY. Educ: Univ Toledo, BS, 34; Univ Ill, PhD(chem, math), 39. Prof Exp: Asst chem, Univ Toledo, 31-34, teaching fel, 34-35; teaching asst, Univ Ill, 35-37, Eli Lilly res fel, 37-39; res chemist, E I du Pont de Nemours & Co, Inc, 39-44, res mgr, 44-47, new prod sales supvr, 47-51, prod mgr, 51-53, spec asst to mgt, 53-54; mgt consult, 55-60; dir acad proj, NSF, 61-62; corp assoc dir res, FMC Corp, New York, 62-64; RES & MGT COUN TO ACAD, INDUST & GOVT, 65- Concurrent Pos: Mem adv comt creativity in scientists & engrs, Rensselaer Polytech Inst, 65-68; res consult, NSF, 70. Mem: Am Chem Soc; AAAS. Res: Discovered cyclamate sweeteners; first application of Boolean algebra and theory of sets to people problems, and devised 3-dimensional models showing relationships; devised better way to take off human fat in obesity; interdisciplinary organizations broadly, including mathematical treatment for the first time. Mailing Add: 228 W Lane Revonah Woods Stamford CT 06905

SVEJDA, FELICITAS JULIA, b Vienna, Austria, Nov 8, 20; nat Can. ORNAMENTAL HORTICULTURE, PLANT BREEDING. Educ: State Univ Agr & Forestry, Austria, MSc, 46, PhD, 48. Prof Exp: Res asst rural econ, State Univ Agr & Forestry, Austria, 47-51; asst plant breeder, Swedish Seed Asn, 52-53; RES OFFICER, CAN DEPT AGR, 53- Mem: Am Soc Hort Sci; Genetics Soc Can; Can Soc Hort Sci; Agr Inst Can; Can Bot Asn. Res: Population biology; plant physiology; hybridization of ornamental plants. Mailing Add: Can Dept of Agr Ottawa ON Can

SVENDSEN, GERALD EUGENE, b Ashland, Wis, June 18, 40; m 61; c 2. ECOLOGY, ETHOLOGY. Educ: Univ Wis-River Falls, BS, 62; Univ Kans, MA, 64, PhD(behav ecol), 73. Prof Exp: Biologist, Fish-Pesticide Res Lab, US Fish & Wildlife Serv, 64-66 & Fish Control Lab, 66-68; from instr to asst prof biol, Viterbo Col, 66-70; ASST PROF ZOOL, OHIO UNIV, 73- Mem: Animal Behav Soc; Am Soc Mammalogists; Ecol Soc Am; Wildlife Soc; Am Soc Evolutionists. Res: Behavioral ecology; ethology of mammals; social systems analysis and evolution; spatial organization and distribution of terrestrial vertebrates; population biology of terrestrial vertebrates; vertebrate communication systems. Mailing Add: Dept of Zool & Microbiol Ohio Univ Athens OH 45701

SVENDSEN, KENDALL LORRAINE, b Greenville, Mich, June 24, 19; m m 43; c 3. GEOPHYSICS. Educ: Univ Mich, BS, 43. Prof Exp: Geophysicist, US Coast & Geod Surv, 46-70, chief geomagnetism div, 70-71, chief geomagnetic data div, Environ Data Serv, 71-72, CHIEF SOLID EARTH & MARINE GEOPHYS DATA SERV DIV, ENVIRON DATA SERV, NAT OCEANIC & ATMOSPHERIC ADMIN, BOULDER, 72- Concurrent Pos: Am Geophys Union liaison rep, Comn Geophys, Pan Am Inst Geog & Hist, 72-, alt US mem, 73-; co-chmn working group magnetic observ & mem working group collection & dissemination of data, Int Asn Geomagnetism & Aeronomy, 73- Mem: AAAS; Am Geophys Union; Am Cong Surv & Mapping; Am Polar Soc; Arctic Inst NAm. Res: Collection, processing and dissemination of geophysical data; geomagnetism. Mailing Add: 7350 Mt Meeker Rd Longmont CO 80501

SVENNE, JURIS PETERIS, b Riga, Latvia, Feb 14, 39; Can citizen; m 63; c 3. PHYSICS. Educ: Univ Toronto, BASc, 62; Mass Inst Technol, PhD(physics), 65. Prof Exp: Res assoc nuclear physics, Mass Inst Technol, 65-66; Nat Res Coun Can fels, Niels Bohr Inst, Copenhagen, Denmark, 66-68; res assoc, Inst Nuclear Physics, D'Orsay, France, 68-69; asst prof, 68-73, ASSOC PROF PHYSICS, UNIV MAN, 73- Concurrent Pos: Vis prof nuclear physics, Univ Oxford, 76. Mem: Am Phys Soc; Can Asn Physicists. Res: Theory of nuclear structure; Hartree-Fock theory; theory of nuclear reactions with 3-particle final states. Mailing Add: Dept of Physics Univ of Man Winnipeg MB Can

SVENSSON, ERIC CARL, b Hampstead, NB, Aug 13, 40; m 65; c 2. SOLID STATE PHYSICS. Educ: Univ NB, Fredericton, BSc, 62; McMaster Univ, PhD(physics), 67. Prof Exp: Asst res officer, 66-70, ASSOC RES OFFICER PHYSICS, ATOMIC ENERGY CAN, LTD, 71- Concurrent Pos: Guest scientist, Aktiebolaget Atomenergi, Studsvik, Sweden, 72-73. Mem: Am Phys Soc; Can Asn Physicists. Res: Neutron scattering; lattice dynamics; magnetic excitations; effects of impurities on excitation spectra; excitations in liquid helium. Mailing Add: Atomic Energy of Can Ltd Chalk River ON Can

SVETICH, GEORGE WILLIAM, physical chemistry, crystallography, see 12th edition

SVETLIK, JOSEPH FRANK, b Weimar, Tex, July 23, 18; m 42; c 5. RUBBER CHEMISTRY. Educ: Tex Tech Col, BS, 42. Prof Exp: Res chemist petrol, Tex Co, 42; res chemist, 42, sr sect chief rubber res, 49-55, mgr tech serv, Rubber Chem Div, 55-68, tech asst to div dir, 68-71, mgr tire technol br, Chem Dept, 71-74, DIR CARBON BLACK DEVELOP, PHILLIPS PETROL CO, 74- Mem: Am Chem Soc (treas, 52); Am Soc Testing & Mat; Am Inst Chem Engrs. Res: Grease manufacture; rubber compounding; correlation of manufacturing variables with physical properties of rubber; rubber testing devices; development of solution rubbers; compounding techniques for improved present and future automobile tires; carbon black development and process improvement. Mailing Add: Chem Dept Rubber Chem Div Phillips Petrol Co 1501 Commerce Dr Stow OH 44224

SVIHLA, RUTH DOWELL, zoology, botany, deceased

SVILOKOS, NIKOLA, b Zagreb, Yugoslavia, Oct 25, 31; nat US. MATHEMATICS, SYSTEM ANALYSIS. Educ: Loyola Col, Can, BSc, 53; Harvard Univ, AM, 59. Prof Exp: Mathematician-engr, Canadair Ltd, Gen Dynamics Corp, 56-58; res asst math, McGill Univ, 58-59; res mathematician, Dow Chem Co, 59-60; opers res analyst, Bankers Trust Co, 61-63; appl mathematician, Allied Chem Corp, 63-65; SYSTS ANALYST, NEW YORK STOCK EXCHANGE & PRES, EVE MATH, INC, 65- Mem: Am Math Soc; Math Asn Am; NY Acad Sci. Res: Application of computer systems in industry and finance; research in algebra, abstract analysis, tensor geometry and number theory. Mailing Add: 1060 Park Ave New York NY 10028

SVIRBELY, WILLIAM J, physical chemistry, see 12th edition

SVOBODA, GLENN RICHARD, b Racine, Wis, Nov 18, 30; m 57; c 3. POLYMER CHEMISTRY. Educ: Univ Wis, BS, 52, MS, 53, PhD(pharmaceut chem), 58. Prof Exp: Instr anal chem, Univ Wis, 58-59; res chemist, 59-61, mgr res lab, 62-64, dir res, 64-67, VPRES RES & DEVELOP, FREEMAN CHEM CORP, PORT WASHINGTON, 67- Concurrent Pos: Asst prof, Ore State Col, 58-59. Mem: AAAS; Am Chem Soc. Res: Natural products; organo-analytical techniques, especially electrochemistry and optical methods; polymer analysis by physical organic techniques; polymer and monomer synthesis; coatings; unsaturated polyester and urethane specialties; electrochemical and radiochemical syntheses. Mailing Add: 1525 Beechwood Lane Grafton WI 53024

SVOBODA, GORDON H, b Racine, Wis, Oct 29, 22; m 45; c 3. PHARMACOGNOSY. Educ: Univ Wis, BS, 44, PhD(pharmaceut chem), 49. Prof Exp: Actg instr pharmaceut chem, Univ Wis, 47-49; asst prof pharm, Univ Kans, 49-50; PHYTOCHEMIST & RES ASSOC, ELI LILLY & CO, 50- Concurrent Pos: Am Asn Cols Pharm-NSF vis scientist, 63-72; vis res prof, Univ Pittsburgh, 64-; mem biol & related res facilities vis comt, Bd Overseers, Harvard Univ. Honors & Awards: Am Pharmaceut Asn Award, 63; Ebert Prize, Am Soc Pharmacog, 67. Mem: Am Pharmaceut Asn; Am Soc Pharmacog (pres, 63-64); Int Pharmaceut Fedn; fel Acad Pharmaceut Sci. Mailing Add: 5302 Fallwood Dr Apt 105 Indianapolis IN 46220

SVOBODA, JAMES ARVID, b Great Falls, Mont, June 28, 34; m 60; c 4. INSECT PHYSIOLOGY. Educ: Col Great Falls, BS, 58; Mont State Univ, PhD(entom), 64. Prof Exp: Resident res assoc insect physiol, Pioneering Res Lab, 64-65, SR INSECT PHYSIOLOGIST, INSECT PHYSIOL LAB, AGR RES CTR, USDA, 65- Mem: AAAS; Entom Soc Am. Res: Metabolism of lipids in insects, specifically in sterols and their relationships to growth and metamorphosis; insect hormonal control mechanisms. Mailing Add: Insect Physiol Lab Agr Res Ctr Entom Bldg C USDA Beltsville MD 20705

SVOBODA, RUDY GEORGE, b Berwyn, Ill, Aug 15, 41; m 64; c 2. MATHEMATICS. Educ: Northern Ill Univ, BS, 66; Ohio Univ, MS, 67; Purdue Univ, PhD(math), 71.

Prof Exp: ASST PROF MATH, IND UNIV-PURDUE UNIV, FT WAYNE, 70- Mem: Am Math Soc; Math Asn Am; Soc Indust & Appl Math; Sigma Xi. Res: The study of differential-difference equations in various normed vector spaces to determine the completeness of the Eigen solutions. Mailing Add: Dept of Math Ind Univ-Purdue Univ 2101 Coliseum Blvd Ft Wayne IN 46805

SVOKOS, STEVE GEORGE, b Wierton, WVa, June 22, 34; m 60; c 3. BIOLOGICAL CHEMISTRY. Educ: Brooklyn Col, BS, 56; State Univ NY, MS, 62, PhD(bio-org chem), 64. Prof Exp: Chemist, Lederle Labs, Am Cyanamid Co, 56-60, res chemist, 65-69; regulatory liaison, Ayerst Labs Div, Am Home Prod Corp, NY, 69-72; DIR REGULATORY AFFAIRS, KNOLL PHARMACEUT CO, WHIPPANY, 72- Mem: Am Chem Soc; fel The Chem Soc. Res: Pharmaceutical administration; medicinal chemistry. Mailing Add: 59 First Ave Westwood NJ 07675

SWACKHAMER, FARRIS SAPHAR, b Cranford, NJ, May 31, 14; m 37; c 2. ORGANIC CHEMISTRY. Educ: Rutgers Univ, BS, 36; Polytech Inst Brooklyn, MS, 40. Prof Exp: Chemist, Am Cyanamid Co, 36-37, group leader, 37-41, tech rep, 45-48; sr technologist, Shell Chem Co, 48-50, asst dept mgr sales develop, 50, dept mgr, 50-51, mgr resins & plastics dept, 51-57, mgr sales develop, 57, dir tech serv labs, 57-63; assoc prof chem & chmn dept, 63-69, dir instnl res & asst to pres, 69-73, PROF CHEM, UNION COL, NJ, 73- Mem: AAAS; Am Chem Soc; Commercial Develop Asn (past pres); fel Am Inst Chemists. Res: Thermosetting polymers; market research; laboratory administration; nutrients in estuarine environment. Mailing Add: 10 Herning Ave Cranford NJ 07016

SWADER, FRED NICHOLAS, b Belle Vernon, Pa, Oct 9, 34; m 56; c 2. SOIL SCIENCE. Educ: Cornell Univ, BS, 61, MS, 63, PhD(agron, soil sci), 68. Prof Exp: Experimentalist, 61-63, res assoc agr eng, 63-67, asst prof, 67-74, ASSOC PROF SOIL SCI, CORNELL UNIV, 74- Concurrent Pos: Chmn, Cornell Agr Waste Mgt Conf, 70 & 71. Mem: Soil Conserv Soc Am; Am Soc Agr Engrs; Soil Sci Soc Am. Res: Plant-soil-water relationships, as influenced by the physical properties of various soils; soil management for recycling agricultural by-products. Mailing Add: Dept of Agron Emerson Hall Cornell Univ Ithaca NY 14853

SWADER, JEFF AUSTIN, JR, b Caruthersville, Mo, Jan 3, 38. PLANT PHYSIOLOGY. Educ: Univ Mo-Columbia, BS, 62, MS, 64; Univ Calif, Davis, PhD(plant physiol), 69. Prof Exp: Asst prof, 70-75, ASSOC PROF PLANT PHYSIOL, UNIV CALIF, RIVERSIDE, 75- Mem: Am Soc Plant Physiologists. Res: Subcellular organelles; enzyme isolation and reactions; herbicides; photosynthesis. Mailing Add: Dept of Plant Path & Physiol Va Polytech Inst & State Univ Blacksburg VA 24061

SWAFFORD, WILLIAM BRYSON, b Monterey, Tenn, Aug 23, 12; m 43; c 3. PHARMACY. Educ: Univ Tenn, BSPh, 48; Univ Miss, MS, 55; Memphis State Univ, MA, 58; Southern Univ, LLB, 62. Prof Exp: From instr to asst prof pharm, 48-59, asst prof pharm & pharm admin, 59-61, assoc prof pharmaceut admin & head dept, 61-63, PROF PHARM & CHMN DEPT, COL PHARM, UNIV TENN, MEMPHIS, 63-, ASST DEAN COL PHARM, 71- Mem: Am Pharmaceut Asn. Res: Drug and cosmetic formulas; legal aspects of pharmaceutical administration. Mailing Add: Col of Pharm Univ of Tenn Memphis TN 38103

SWAILES, GEORGE EDWARD, b Winnipeg, Man, July 20, 25; m 50; c 3. ENTOMOLOGY. Educ: Univ Man, BSA, 46; Colo Agr & Mech Col, MSc, 50; Iowa State Col, PhD(entom), 56. Prof Exp: ENTOMOLOGIST, RES STA, CAN DEPT AGR, 48- Res: cutworms damaging vegetable and field crops. Mailing Add: Can Agr Res Sta Lethbridge AB Can

SWAIMAN, KENNETH F, b St Paul, Minn, Nov 19, 31; m 73; c 4. NEUROCHEMISTRY, PEDIATRIC NEUROLOGY. Educ: Univ Minn, BA, 52, BS, 53, MD, 55; Am Bd Pediat, dipl; Am Bd Psychiat & Neurol, dipl. Prof Exp: Fel pediat, 56-57, Nat Inst Neurol Dis & Stroke spec fel pediat neurol, 60-63, from asst prof to assoc prof, 63-69, PROF PEDIAT & NEUROL, MED SCH, UNIV MINN, MINNEAPOLIS, 69-, DIR DIV PEDIAT NEUROL, 68- Mem: Am Neurol Asn; Am Acad Neurol; Am Acad Pediat; Child Neurol Soc (pres, 72-73); Soc Pediat Res. Res: Neurochemical changes in developing brain; energy and amino acid metabolism of immature brain. Mailing Add: Div of Pediat Neurol Univ of Minn Med Sch Minneapolis MN 55455

SWAIN, ANSEL PARRISH, b Valdosta, Ga, Sept 9, 09; m 41; c 2. BIOCHEMISTRY, ORGANIC CHEMISTRY. Educ: Emory Univ, BS, 32, MS, 34; Univ Ill, PhD(biochem), 39. Prof Exp: Biochemist, T D Spies Clin, Hillman Hosp, 39-41; chemist, Eastern Regional Res Lab, USDA, 41-43 & McNeil Labs, Inc, 43-58; ORG CHEMIST, EASTERN REGIONAL RES LAB, USDA, 58- Mem: Am Chem Soc. Res: Amino acids in nutrition of the white rat; protein metabolism in human malnutrition; milk proteins; synthesis of new drugs; tobacco, especially isolation of organic compounds; potato glyco alkaloids. Mailing Add: Eastern Regional Res Lab USDA 600 E Mermaid Lane Philadelphia PA 19118

SWAIN, CHARLES GARDNER, b Quincy, Mass, May 26, 17; m 45; c 2. CHEMISTRY. Educ: Harvard Univ, AB, 40, AM, 41, PhD(org chem), 44. Prof Exp: Nat Res Coun fel, Calif Inst Technol, 45-46; Am Chem Soc fel, 46-47, from instr to assoc prof, 47-58, PROF CHEM, MASS INST TECHNOL, 58- Concurrent Pos: Guggenheim fel, Univ London, 54-55. Honors & Awards: Petrol Chem Award, Am Chem Soc, 57. Mem: Am Chem Soc; Am Acad Arts & Sci. Res: Mechanism of organic reactions; effect of structure of reactants on structure of transition states; quantitative correlations between structure and reactivity; acid, base and polyfunctional catalysis; tracers and isotope effects. Mailing Add: Dept of Chem Mass Inst of Technol Cambridge MA 02139

SWAIN, EDWIN E, JR, chemistry, see 12th edition

SWAIN, ELISABETH RAMSAY, b Philadelphia, Pa, Feb 7, 17. ZOOLOGY. Educ: Wilson Col, BS, 38; Univ Pa, MA, 42, PhD, 53. Prof Exp: Instr physics, Wilson Col, 43-46, instr biol, 46-49; asst instr zool, Univ Pa, 51-54; from asst prof to assoc prof, 54-66, PROF BIOL, UNIV HARTFORD, 66-, CHMN DEPT, 54- Mem: AAAS. Res: Embryology. Mailing Add: Dept of Biol Univ of Hartford West Hartford CT 06117

SWAIN, FREDERICK MORRILL, JR, b Kansas City, Mo, Mar 17, 16; m 38; c 3. GEOLOGY. Educ: Univ Kans, AB, 38, PhD(stratig, paleont), 43; Pa State Col, MS, 39. Prof Exp: Geologist, Phillips Petrol Co, La, 41-43; asst prof mineral econ, Pa State Col, 43-46; from asst prof to assoc prof geol, 46-54, assoc chem dept geol & geophys, 59-61, PROF GEOL & GEOPHYS, UNIV MINN, MINNEAPOLIS, 54- Concurrent Pos: Assoc geologist, US Geol Surv, 44-46, geologist, 48-51, 61-; consult, Carter Oil Co, 51-53 & Pa RR, 54-57; part-time prof, Univ Del, 69- Honors & Awards: Award, Am Asn Petrol Geol, 49. Mem: Fel Geol Soc Am; Soc Econ Paleont & Mineral; Paleont Soc; Am Soc Limnol & Oceanog; Soc Econ Geol. Res: Stratigraphy; micropaleontology; organic geochemistry. Mailing Add: Dept of Geol & Geophys Univ of Minn Minneapolis MN 55455

SWAIN, HARRY SHELDON, b Prince Rupert, BC, July 26, 42; m 65. URBAN GEOGRAPHY, RESEARCH MANAGEMENT. Educ: Univ BC, BA, 64; Univ Minn, MA, 67, PhD(geog), 70. Prof Exp: Lectr geog, Univ Toronto & Scarborough Col, 68-70; asst prof, Univ BC, 70-72; sr res officer urban geog & transp econ, Ministry State Urban Affairs, Ottawa, 71, dir urban res, 71-73; proj leader, Int Inst Appl Syst Anal, Vienna, 74-75; POLICY ANALYST, TREAS BD SECRETARIAT, OTTAWA, CAN, 76- Concurrent Pos: Can Coun fel, St Catharine's Col, Cambridge Univ, 69-70; mem, Can Comt for Int Inst Appl Systs Anal, 76-; co-chmn, Intergovernmental Waterfront Comt, 76- Mem: AAAS; Asn Am Geog; Can Asn Geog; Regional Sci Asn; Am Geog Soc. Res: Urban policy analysis; aspects of urban simulation modelling; historical geography of western cities. Mailing Add: 1 Leona Dr Willowdale ON Can

SWAIN, HENRY HUNTINGTON, b Champaign, Ill, July 11, 23; m 48; c 2. PHARMACOLOGY. Educ: Univ Ill, AB, 43, BS, 49, MS & MD, 51. Prof Exp: Instr pharmacol, Univ Cincinnati, 52-54; from instr to assoc prof, 54-67, PROF PHARMACOL, MED SCH, UNIV MICH, ANN ARBOR, 67- Mem: Am Soc Pharmacol & Exp Therapeut. Res: Cardiovascular pharmacology, especially cardiac arrhythmias. Mailing Add: Dept of Pharmacol Univ of Mich Med Sch Ann Arbor MI 48104

SWAIN, HOWARD ALDRED, JR, b New York, NY, Mar 3, 28; m 51; c 3. PHYSICAL CHEMISTRY. Educ: Grove City Col, BS, 51; Univ Pa, PhD, 61. Prof Exp: High sch instr chem, NJ, 54-56; lab technician, Rohm and Haas Co, 56-57; chemist, Socony-Mobile Res & Develop, 57-58; asst chem, Univ Pa, 58-60; from asst prof to assoc prof, 60-70, PROF CHEM, WILKES COL, 70- Concurrent Pos: Oak Ridge Assoc Univs res partic, Savannah River Lab, SC, 67-68; consult, Vet Admin Hosp, Wilkes-Barre, Pa, 71; lectr, Col Miseracordia, 72; res partic water purification proj, Environ Protection Agency. Mem: Am Chem Soc. Res: Thermodynamics; radiochemistry. Mailing Add: Dept of Chem Wilkes Col Wilkes-Barre PA 18703

SWAIN, RICHARD RUSSELL, biochemistry, clinical chemistry, see 12th edition

SWAIN, WAYLAND ROGER, b Boone, Iowa, Jan 13, 38; m 60; c 1. PUBLIC HEALTH, ENVIRONMENTAL BIOLOGY. Educ: Ottawa Univ, BA, 60; Univ Minn, Minneapolis, MS, 65, PhD(environ biol), 69. Prof Exp: Water microbiologist, Univ Minn, Minneapolis, 62-64, instr environ microbiol, Sch Pub Health, 65-66; dir div res, Miller Hosp Res Complex, 69-71, asst prof prev med & actg chmn dept, 71-75, ASSOC PROF PREV MED & CHMN DEPT, SCH MED, UNIV MINN, DULUTH, 75-, DIR, LAKE SUPERIOR BASIN STUDIES CTR, 74- Concurrent Pos: Mem, Regional Med Prog Med Libr & Info Network Comt, 69- Mem: Am Pub Health Asn; Royal Soc Health; Int Asn Gt Lakes Res. Res: Limnology; aquatic biology, especially planktonic relationships; comparative physiology; delivery of health care; health care planning; psychosocial aspects of disease. Mailing Add: Dept of Prev Med Univ of Minn Sch of Med Duluth MN 55812

SWAISGOOD, HAROLD EVERETT, b Ashland, Ohio, Jan 19, 36; m 56; c 2. PROTEIN BIOCHEMISTRY. Educ: Ohio State Univ, BS, 58; Mich State Univ, PhD(chem), 63. Prof Exp: NIH fel, 63-64; from asst prof to assoc prof, 64-72, PROF FOOD CHEM, NC STATE UNIV, 72- Mem: AAAS; Am Chem Soc; Inst Food Technologists; Am Dairy Sci Asn; Am Soc Biol Chemists. Res: Physical-chemical characterization of proteins; studies of protein interactions and the relationship to biological activity; methods of preparation and characterization of enzymes covalently bound to surfaces. Mailing Add: Dept of Food Sci NC State Univ Raleigh NC 27607

SWAKON, EDWARD ANTONE, b North Dighton, Mass, Apr 9, 25; m 52; c 2. ORGANIC CHEMISTRY. Educ: Brown Univ, ScB, 47; Carnegie Inst Technol, MS, 50, DSc(org chem), 51. Prof Exp: Chemist, Standard Oil Co, Ind, 51-56 & Food Mach & Chem Corp, 56-57; CHEMIST, AMOCO RES CTR, 57- Mem: Am Chem Soc. Res: Organic chemistry in Ullman reaction; synthetic greases; dimethyl hydrazine; carbon monoxide; carbonyl sulfide; condensation polymers; petroleum pitches; petroleum oils and additives; cellulose preparations and modifications; agricultural research, herbicides and pesticides. Mailing Add: Amoco Res Ctr PO Box 400 Naperville IL 60540

SWALEN, JEROME DOUGLAS, b Minneapolis, Minn, Mar 4, 28; m 52; c 2. CHEMICAL PHYSICS. Educ: Univ Minn, BS, 50; Harvard Univ, AM, 54, PhD(chem physics), 56. Prof Exp: Fel, Div Pure Physics, Nat Res Coun Can, 56-57; physicist, Shell Develop Co, 57-62; mgr physics dept, 62-63, lab mgr, 63-67, mgr molecular physics dept, 67-73, RES STAFF MEM, IBM RES LAB, 73- Concurrent Pos: Vis prof, Phys Chem Inst, Univ Zurich, 72-73. Mem: Fel AAAS; fel Am Phys Soc; Am Chem Soc. Res: Laser spectroscopy of thin organic films and monolayers. Mailing Add: IBM Res Lab San Jose CA 95193

SWALHEIM, DONALD ARTHUR, b Cottage Grove, Wis, Oct 19, 13; m 44; c 1. CHEMISTRY. Educ: St Olaf Col, BS, 36; Univ Wis, MS, 38, PhD(inorg chem), 41. Prof Exp: Instr chem, Exten Div, Univ Wis, 38-40; res chemist, 41-46, res chemist & group leader, 46-52, res chemist & supvr, 52-72, RES ASSOC, E I DU PONT DE NEMOURS & CO, INC, 72- Honors & Awards: Past Pres Award, Am Electroplaters' Soc, 75. Res: Electroplating; electropolishing; continuous electrotinning of strip steel at high speeds; vinyl processes and applications for vinyl polymers; degreasing and phosphatizing of metals. Mailing Add: Indust Chem Dept Chestnut Run Lab E I du Pont de Nemours & Co Inc Wilmington DE 19898

SWALLOW, EARL CONNOR, b Montgomery County, Ohio, Dec 27, 41. ELEMENTARY PARTICLE PHYSICS, EXPERIMENTAL PHYSICS. Educ: Earlham Col, BA, 63; Washington Univ, MA, 65, PhD(physics), 70. Prof Exp: Resident student assoc elem particle physics, Argonne Nat Lab, 64 & 65, res staff assoc, 65-69, res asst for Washington Univ, 69-70; res assoc, 70-74, SR RES ASSOC ELEM PARTICLE PHYSICS, ENRICO FERMI INST, UNIV CHICAGO, 74- Concurrent Pos: Teaching asst, Washington Univ, 65. Mem: Am Phys Soc; Sigma Xi. Res: Fundamental interactions of elementary particles, especially weak interactions; experimental foundations of physical theory; relationship of experimental foundations to public policy. Mailing Add: Enrico Fermi Inst Univ of Chicago 5630 S Ellis Ave Chicago IL 60637

SWALLOW, RICHARD LOUIS, b Berwyn, Ill, June 16, 39; m 64; c 1. ZOOLOGY, BIOLOGY. Educ: Univ Ill, Urbana, BS, 63; Univ Mo-Columbia, MA, 66, PhD(zool), 68. Prof Exp: USPHS fel, Sch Med, Case Western Reserve Univ, 68-69; asst prof biol, Univ Houston, 69-73; ASSOC PROF BIOL, COKER COL, 73- Mem: Am Soc Zoologists. Res: Comparative physiology including control of metabolism by hormones in fish. Mailing Add: Coker Col Hartsville SC 29550

SWALLOW, RONALD JOSEPH, biophysics, see 12th edition

SWALLOW, WILLIAM HUTCHINSON, b Norwalk, Conn, Oct 21, 41. BIOSTATISTICS. Educ: Harvard Univ, AB, 64; Cornell Univ, MS, 68, PhD(biomet), 74. Prof Exp: ASST PROF STATIST, COOK COL, RUTGERS UNIV, NEW

BRUNSWICK, 73- Mem: Am Statist Asn; Sigma Xi. Res: Linear models; estimation of variance components; research directed at improving the teaching of statistics. Mailing Add: Dept of Statist & Comput Sci Cook Col Rutgers Univ New Brunswick NJ 08903

SWAMER, FREDERIC WURL, b Shawano, Wis, May 16, 18; m 46; c 3. ORGANIC CHEMISTRY. Educ: Lawrence Col, BA, 40; Univ Wis, MS, 42; Duke Univ, PhD(chem), 49. Prof Exp: Chemist, Appleton Water Purification Plant, Wis, 40; res chemist, Electrochem Dept, E I du Pont de Nemours & Co, 41-45; res assoc, Duke Univ, 49-50; res chemist, 50-64, RES ASSOC, ORG CHEM DEPT, EXP STA, E I DU PONT DE NEMOURS & CO, INC, WILMINGTON, DEL, 64- Mem: AAAS; Am Chem Soc. Res: Claisen condensation; physical properties of polymers; acetylene organoalkali compounds; fluorocarbons; heterogeneous catalysis. Mailing Add: Folly Hill Rd RD 4 West Chester PA 19380

SWAMINATHAN, SRINIVASA, b Madras, India, Aug 24, 26; m 52; c 1. MATHEMATICS. Educ: Presidency Col, Madras, India, BA, 47, MA, 48; Univ Madras, MSc, 50, PhD(math), 57. Prof Exp: Govt of France fel, Inst Henri Poincare, Paris, 57-58; lectr math, Univ Madras, 59-64; asst prof, Indian Inst Technol, Kanpur, 64-66; vis assoc prof, Univ Ill, Chicago Circle, 66-68; ASSOC PROF MATH, DALHOUSIE UNIV, 68- Concurrent Pos: Auth-mem comt for reorgn of curricula in math, Nat Coun Educ Res & Training, Govt of India, 64-67. Mem: Am Math Soc; Can Math Cong; Indian Math Soc. Res: Functional analysis; topology; geometry of Banach spaces; operator theory; paracompact spaces; fixed point theorems in analysis and topology. Mailing Add: 911 Greenwood Ave Halifax NS Can

SWAMY, VIJAY CHINNASWAMY, b Bombay, India, Oct 2, 38; m 72; c 1. PHARMACOLOGY. Educ: Bombay Univ, BSc, 59; Nagpur Univ, BPharm, 62; Ohio State Univ, MS, 64, PhD(pharmacol), 67. Prof Exp: Res asst pharmacol, Ohio State Univ, 64-67; res assoc, 67-69, ASST PROF PHARMACOL, STATE UNIV NY BUFFALO, 70- Mem: AAAS. Res: Smooth muscle pharmacology; hypertension; adrenergic mechanisms. Mailing Add: Dept of Biochem Pharmacol State Univ NY Buffalo NY 14214

SWAN, ALGERNON GORDON, b Andrews, NC, Jan 25, 23; m 47; c 2. PHYSIOLOGY, BIOPHYSICS. Educ: Univ NC, BA, 48, PhD(physiol), 60. Prof Exp: Chief biophys br, Aerospace Med Res Labs, Wright-Patterson AFB, US Air Force, Ohio, 60-62, dir life support res, Aerospace Med Div, Brooks AFB, Tex, 62-65, dir res, 65-68, dir test & eng, Air Force Spec Weapons Ctr, Kirtland AFB, NMex, 58-59, vcomdr & tech dir, 69-70, comdr & tech dir, 70-72; SR RES, BECTON, DICKINSON & CO, 72- Concurrent Pos: Mem, Comts Hearing, Bioacoust & Biomech, Nat Acad Sci-Nat Res Coun, 63-, Comt Nutrit, 64-; mem, Biosci Comt, NASA, 64-, Comt Biotechnol & Human Res, 65- & Comt Cardiopulmonary Res, Adv Group Aeronaut Res & Develop. Mem: Aerospace Med Asn. Res: Exercise and stress physiology; weapons effects; osmotic regulation and electrolyte flux in isolated tissues; human tolerance to aerospace stresses; nuclear environment simulation; instrumentation development; flight testing; qualification of instrumentation for space flight. Mailing Add: Becton Dickinson Res Ctr Research Triangle Park NC 27612

SWAN, DEAN GEORGE, b Wheatland, Wyo, Sept 16, 23; m 48; c 3. AGRONOMY. Educ: Univ Wyo, BS, 52, MS, 54; Univ Ill, PhD, 64. Prof Exp: Instr, Chadron High Sch, 52-53; instr weed res, Pendleton Exp Sta, Ore State Univ, 55-65; exten weed specialist, Univ Ariz, 65-66; EXTEN WEED SCIENTIST & ASSOC AGRONOMIST, WASH STATE UNIV, 66- Concurrent Pos: Sabbatical, Weed Res Orgn, Oxford, Eng, 72-73. Mem: Weed Sci Soc Am. Res: Weed control in crops, especially in winter wheat, peas and alfalfa. Mailing Add: Dept of Agron & Soils Wash State Univ Pullman WA 99163

SWAN, EMERY FREDERICK, b Northfield, NH, May 10, 16; m 39; c 3. INVERTEBRATE ZOOLOGY. Educ: Bates Col, BS, 38; Univ Calif, PhD(zool), 42. Prof Exp: Asst zool, Univ Calif, 38-42; lab technician & res asst, Am Brass Co, Conn, 42-46; asst prof natural sci, State Univ NY Teachers Col, New Paltz, 46-48; asst prof oceanog & resident scientist, Friday Harbor Labs, Univ Wash, 48-52; assoc prof, 52-64, PROF ZOOL, UNIV NH, 64- Concurrent Pos: Assoc, Harvard Univ, 53-57 & 65-71. Mem: AAAS; Ecol Soc Am; Am Soc Limnol & Oceanog; Soc Study Evolution; Am Soc Zoologists. Res: Biology of serpulimorph polychaetes and bivalve mollusks; growth, variation and taxonomy of the sea urchins of genus Strongylocentrotus. Mailing Add: Dept of Zool Spaulding Bldg Univ of NH Durham NH 03824

SWAN, FREDERICK ROBBINS, JR, b Hartford, Conn, Aug 14, 37; m 62; c 2. ECOLOGY. Educ: Middlebury Col, BA, 59; Univ Wis, MS, 61; Cornell Univ, PhD(conserv natural resources), 66. Prof Exp: Assoc prof, 66-74, actg chmn, Sch Natural Sci, 70-72, PROF BIOL, WEST LIBERTY STATE COL, 74- Concurrent Pos: Assoc, Dept Natural Resources, Cornell Univ, 74-75. Mem: Wildlife Soc; Ecol Soc Am. Res: Effects of fire on plant communities; measurement of light in forests. Mailing Add: Dept of Biol West Liberty State Col West Liberty WV 26074

SWAN, HAROLD JAMES CHARLES, b Sligo, Ireland, June 1, 22; US citizen; m 46; c 7. PHYSIOLOGY, CARDIOVASCULAR DISEASES. Educ: Univ London, MB, BS, 45, PhD(physiol), 51. Prof Exp: Res assoc, Mayo Clin, 51-53, Minn Heart Asn res fel, 53-54, consult cardiovasc dis, 55-65; DIR CARDIOL, CEDARS-SINAI MED CTR, LOS ANGELES, 65-; PROF MED, UNIV CALIF, LOS ANGELES, 66- Concurrent Pos: Assoc prof, Mayo Grad Sch, Univ Minn, 57-65; consult, Nat Heart Inst, 60-66; mem, Intersoc Comn Heart Dis Resources, 69- Honors & Awards: Walter Dixon Award, Brit Med Asn, 50. Mem: Am Physiol Soc; fel Am Col Physicians; fel Am Col Cardiol (pres, 73); Asn Univ Cardiol. Res: Ventricular function; myocardial hypertrophy; coronary arterial disease and myocardial ischemia and infarction. Mailing Add: 4833 Fountain Ave Los Angeles CA 90029

SWAN, HENRY STEWART DRUMMOND, b Pietermaritzburg, SAfrica, Feb 22, 18; Can citizen; m 48; c 2. FORESTRY. Educ: Cambridge Univ, BA, 39, MA, 41, PhD, 70; Oxford Univ, MA, 50. Prof Exp: Head tech sect, Long Ashton Res Sta, Bristol Univ, 50-52; head silvicult sect, Woodlands Res Div, Pulp & Paper Res Inst Can, 52-71, forestry liaison officer, 71-74; COORDR FOREST MGT GROUP, WOODLANDS SECT, CAN PULP & PAPER ASN, 74- Mem: Soc Am Foresters; Can Inst Forestry; Can Soc Plant Physiologists. Res: Tree physiology; mineral nutrition of tree species, particularly Canadian pulpwood species. Mailing Add: Can Pulp & Paper Asn 2300 Sun Life Bldg Montreal PQ Can

SWAN, JAMES BYRON, b Bloomington, Ill, Dec 9, 33; m 62. SOIL SCIENCE. Educ: Univ Ill, BS, 55, MS, 59; Univ Wis, PhD(soil physics), 64. Prof Exp: EXTEN SPECIALIST & PROF SOIL SCI, UNIV MINN, ST PAUL, 64- Mem: Soil Conserv Soc Am; Am Soc Agron. Res: Evapotranspiration measurement; amounts and timing of irrigation for irrigated row crops; effect of soil strength, soil temperature and soil water on plant growth. Mailing Add: Dept of Soil Sci Univ of Minn St Paul MN 55101

SWAN, KENNETH CARL, b Kansas City, Mo, Jan 1, 12; m 42; c 3. OPHTHALMOLOGY, PHARMACOLOGY. Educ: Univ Ore, BA, 33, MD, 36; Am Bd Ophthal, dipl, 40. Prof Exp: Assoc ophthal, Univ Iowa, 41-42, asst prof, 42-44; assoc prof, 44-45, PROF OPHTHAL & HEAD DEPT, MED SCH, UNIV ORE HEALTH SCI CTR, 45- Concurrent Pos: Proctor lectr, Univ Calif, 46; chmn bd, Am Bd Ophthal, 61; chmn sensory dis study sect, NIH, 61-63; mem adv coun, Nat Eye Inst, 69-71; consult, Nat Inst Neurol Dis & Blindness. Honors & Awards: Proctor Medal, Asn Res Vision & Ophthal, 53; Distinguished Serv Award, Univ Ore, 63, Med Sch Meritorious Achievement Award, 68. Mem: Am Ophthal Soc; Asn Res Vision & Ophthal; AMA; Am Acad Ophthal & Otolaryngol. Res: Ocular physiology, pharmacology and therapeutics; anomalies of binocular vision; tumors of the eyes; ocular manifestations of vascular diseases; surgical anatomy and pathology. Mailing Add: Dept of Ophthal Univ of Ore Med Sch Portland OR 97201

SWAN, KENNETH G, b White Plains, NY, Oct 2, 34; m 65; c 3. SURGERY. Educ: Harvard Univ, AB, 56; Cornell Univ, MD, 60. Prof Exp: Resident gen surg, New York Hosp-Cornell Med Ctr, 60-65; fel physiol, Gastrointestinal Res Lab, Vet Admin Ctr, Los Angeles, 65-66; resident thoracic surg, New York Hosp-Cornell Med Ctr, 66-68; dep dir div surg, Walter Reed Army Inst Res, 71-72, dir div surg, 72-73; DIR & ASSOC PROF DIV GEN & VASCULAR SURG, NJ MED SCH, 73- Mem: Am Physiol Soc; Am Gastroenterol Asn; Am Col Surgeons; Soc Univ Surgeons; Soc Thoracic Surgeons. Res: Splanchnic circulation; shock; vascular surgery and trauma. Mailing Add: Dept of Surg NJ Med Sch 100 Bergen St Newark NJ 07103

SWAN, LAWRENCE WESLEY, b Bengal, India, Mar 9, 22; m 46; c 3. BIOLOGY. Educ: Univ Wis, PhB, 42; Stanford Univ, MA, 47, PhD(biol), 52. Prof Exp: Res officer, Climatic Res Lab, Lawrence, Mass, 43-46; instr biol, Stanford Univ, 47-48 & Univ Santa Clara, 51-53; from instr to assoc prof, 54-72, PROF BIOL, SAN FRANCISCO STATE UNIV, 72- Concurrent Pos: Mem, Am Himalayan Expeds, Nepal, 54, 60-61, biol surv, Mt Orizaba, Mex, 64 & 65, biol world tour, 66, field studies, EAfrica, 69 & Galapagos Islands, 70. Mem: Ecol Soc Am; Am Soc Ichthyol & Herpet; Royal Geog Soc; Int Asn Quaternary Res; Am Inst Biol Sci. Res: High altitude ecology and the Aeolian zone; zoogeography of Asia; vertebrate evolution; science education in elementary schools. Mailing Add: Dept of Biol San Francisco State Univ San Francisco CA 94132

SWAN, PATRICIA B, b Hickory, NC, Oct 21, 37; m 62; c 2. NUTRITION, BIOCHEMISTRY. Educ: Univ NC, Greensboro, BS, 59; Univ Wis, MS, 61, PhD(biochem, nutrit), 64. Prof Exp: Res fel biochem, 64-65, from asst prof to assoc prof, 65-73, PROF NUTRIT, UNIV MINN, ST PAUL, 73- Mem: Am Inst Nutrit; Brit Nutrit Soc. Res: Amino acid metabolism; protein biosynthesis; vitamin metabolism. Mailing Add: Dept of Food Sci & Nutrit Univ of Minn St Paul MN 55108

SWAN, PAUL REESE, physics, see 12th edition

SWAN, PETER HOWARD, b Melbourne, Australia, Mar 25, 28. THEORETICAL PHYSICS, ATOMIC PHYSICS. Educ: Univ Melbourne, BSc, 48, MSc, 51; Univ London, PhD(math, nuclear physics), 53. Prof Exp: Hon res asst physics, Univ Col, Univ London, 51-54; Lyle res fel, Univ Melbourne, 54, lectr, 55-59, sr lectr, 60-64, reader, 64-65; RES PROF PHYSICS, LAVAL UNIV, 65- Concurrent Pos: Vis assoc prof, Rice Univ, 63-65. Mem: Am Phys Soc. Res: Collision theory; low energy nuclear physics; atomic and molecular physics; chemical physics. Mailing Add: Dept of Physics Laval Univ Quebec PQ Can

SWAN, RICHARD GORDON, b New York, NY, Dec 21, 33; m 63; c 2. MATHEMATICS. Educ: Princeton Univ, AB, 54, PhD(math), 57. Prof Exp: NSF res fel, Oxford Univ, 57-58; from instr to assoc prof, 58-65, PROF MATH, UNIV CHICAGO, 65- Concurrent Pos: Sloan fel, 60-65. Honors & Awards: Cole Prize, Am Math Soc, 70. Mem: Nat Acad Sci; AAAS; Math Asn Am; NY Acad Sci; Am Math Soc. Res: Algebraic K-theory; homological algebra. Mailing Add: Dept of Math Univ of Chicago Chicago IL 60637

SWAN, ROY CRAIG, JR, b New York, NY, June 7, 20; m 49; c 3. ANATOMY. Educ: Cornell Univ, AB, 41, MD, 47. Prof Exp: Intern med, New York Hosp, 47-48, asst resident, 48-49, resident endocrinol & metab, 49-50; asst med, Peter Bent Brigham Hosp, 50-52; from instr to assoc prof physiol, 52-59, prof anat, 59-70, JOSEPH C HINSEY PROF ANAT, MED COL, CORNELL UNIV, 70-, CHMN DEPT, 59- Concurrent Pos: Life Ins Med Res Fund fel, Harvard Med Sch, 50-52; Markle scholar, 54-59; res assoc, Cambridge Univ, 55-56; mem health res coun, City New York; consult, USPHS, 60-65 & Off Sci & Technol, 63-64; sect ed, Biol Abstr; mem & chmn anat test comt, Nat Bd Med Examr. Mem: Am Physiol Soc; Am Soc Clin Invest; Am Asn Anat. Res: Ion transport; muscle function and structure; fine structure of excitable cells. Mailing Add: Dept of Anat Cornell Univ Med Col New York NY 10021

SWANBORG, ROBERT HARRY, b Brooklyn, NY, Aug 27, 38; m 66; c 2. IMMUNOLOGY, IMMUNOCHEMISTRY. Educ: Wagner Col, BS, 60; Long Island Univ, MS, 62; State Univ NY Buffalo, PhD(immunol), 65. Prof Exp: NIH trainee immunochem, State Univ NY Buffalo, 65-66; from instr to assoc prof microbiol, 66-73, ASSOC PROF IMMUNOL & MICROBIOL, MED SCH, WAYNE STATE UNIV, 73- Concurrent Pos: Vis investr immunol, Wenner-Gren Inst, Sweden, 75-76. Mem: AAAS; Am Asn Immunol; Am Soc Exp Path; Am Soc Microbiol. Res: Immunochemical aspects of the immune response; mechanisms of self-tolerance and autoimmunity. Mailing Add: Dept Immunol & Microbiol Wayne State Univ Med Sch Detroit MI 48201

SWANEY, MILLER WOODSON, chemistry, see 12th edition

SWANHOLM, CARL E, b Boise, Idaho, June 2, 28; m 50; c 2. ORGANIC POLYMER CHEMISTRY. Educ: Univ Hawaii, BA, 53, PhD(org chem), 59. Prof Exp: NIH res fel, Univ Hawaii, 59-60; fel, Stanford Univ, 60-61; res chemist, Shell Develop Corp, 61-65; proj mgr develop, Boise Cascade Corp, 65-67, mgr prod develop dept, 67-70, mgr plastic bldg prod, 70-71; vpres opers, 71-74, PRES, BIO-DEGRADABLE PLASTICS, INC, 74-; VPRES, AM WESTERN CORP, 74- Mem: Am Chem Soc; The Chem Soc. Res: Polymer degradations and mechanisms. Mailing Add: 2021 E Alameda Dr Tempe AZ 85282

SWANK, GEORGE, JR, plant pathology, see 12th edition

SWANK, HOWARD WIGTON, b Butler, Ohio, Jan 30, 10; m 38; c 1. ANALYTICAL CHEMISTRY. Educ: Mt Union Col, BS, 32; Purdue Univ, MS, 34, PhD(chem), 37. Hon Degrees: DSc, Mt Union Col, 66; DSc, Purdue Univ, 62. Prof Exp: Asst, Purdue Univ, 32-37; res chemist, 37-43, res supvr, 43-47, tech supt, 47-50, mfg supt, 50, res dir, 50-51 & 54-55, tech mgr, 51-52, prod mgr, 52-54, mfg tech dir, 55-59, gen dir, 59-63, asst gen mgr, Textile Fibers Dept, 63-70, gen mgr, 70-72, V PRES & GEN MGR, TEXTILE FIBERS DEPT, E I DU PONT DE NEMOURS & CO, 72- Mem: Am Chem Soc. Res: Synthetic fibers. Mailing Add: Du Pont Textile Fibers Dept Tenth & Market Sts Wilmington DE 19898

SWANK, RICHARD TILGHMAN, b Drums, Pa, Feb 1, 42; m 66; c 2. BIOCHEMISTRY. Educ: Pa State Univ, BS, 64; Univ Wis-Madison, MS, 67, PhD(biochem), 69. Prof Exp: NIH fel, Lab Molecular Biol, Univ Wis-Madison, 69-70; res assoc mammalian biochem genetics, 70-72, SR CANCER RES SCIENTIST, ROSWELL PARK MEM INST, 72- Mem: AAAS; Am Chem Soc; Am Inst Biol Sci. Res: Genetic regulation of enzyme synthesis and degradation in mammals; biochemical mechanisms of enzyme subcellular localization in mammals; physical and chemical characterization of enzymes. Mailing Add: S-4407 Parker Rd Hamburg NY 14075

SWANK, ROBERT KESSLER, physics, see 12th edition

SWANK, ROLLAND LAVERNE, b Holland, Mich, Dec 31, 42; m 69. MATHEMATICS. Educ: Hope Col, BA, 65; Mich State Univ, MS, 66, PhD(math), 69. Prof Exp: Asst prof math, Allegheny Col, 69-74; PROGRAMMER, POWER & POWER, 74- Mem: Am Math Soc; Math Asn Am. Res: Topology; geometry. Mailing Add: Power & Power Meadville PA 16335

SWANK, ROY LAVER, b Camas, Wash, Mar 5, 09; m 37; c 3. NEUROLOGY. Educ: Univ Wash, BS, 30; Northwestern Univ, MD & PhD(anat), 35. Prof Exp: Asst anat, Med Sch, Northwestern Univ, 30-34; intern, Passavant Mem Hosp, Chicago, 34-35; house officer, Peter Bent Brigham Hosp, 36-41, jr assoc med, 41-42, assoc, 46-48; asst prof neurol, McGill Univ, 48-54; PROF NEUROL, SCH MED, UNIV ORE, PORTLAND, 54- Concurrent Pos: Fel, Harvard Med Sch, 37; Commonwealth Fund fel, Sweden & Montreal Neurol Inst, McGill Univ, 39-41; mem attend staff, Cushing Vet Admin Hosp, 46-48; lectr, Montreal Neurol Inst, McGill Univ, 48. Mem: Am Physiol Soc; Am Asn Anatomists; Am Neurol Asn; Can Neurol Asn. Res: Pyrimidal tracts; tissue staining; histochemical staining; vitamin deficiencies; electrophysiology; physiology of breathing. epidemiology of multiple sclerosis; fat metabolism and relationship to viscosity of blood; platelet adhesiveness and aggregation in surgical shock. Mailing Add: Dept of Neurol Univ of Ore Med Sch Portland OR 97201

SWANK, THOMAS FRANCIS, b Philadelphia, Pa, Nov 3, 37; m 63; c 2. COLLOID CHEMISTRY, PHOTOGRAPHIC CHEMISTRY. Educ: Villanova Univ, BS, 59; Univ Va, PhD(heterogeneous catalysis), 64. Prof Exp: Res chemist, Cabot Corp, 63-69; mgr, Ferro Fluidics Corp, 70-71; scientist, 71-76, SR SCIENTIST, POLAROID CORP, WALTHAM, 76- Mem: Am Chem Soc; Electron Micros Soc Am. Res: Heterogeneous catalysis; thin films; x-ray diffraction and spectroscopy; electron microscopy; structure of oxides; inorganic pigments; solid state physics; magnetic fluids; photographic science. Mailing Add: 21 Hitchin'post Rd Chelmsford MA 01824

SWANK, WENDELL GEORGE, biology, see 12th edition

SWANN, CHARLES PAUL, b Minneapolis, Minn, Dec 4, 18; m 51; c 3. NUCLEAR PHYSICS. Educ: Harvard Univ, BS, 41, MS, 43; Temple Univ, PhD(physics), 56. Prof Exp: Mech engr, Steam Div, Westinghouse Elec Corp, 43-46; NUCLEAR PHYSICIST, BARTOL RES FOUND, FRANKLIN INST, 46- Mem: Fel Am Phys Soc. Res: Nuclear structure studies; nuclear resonance fluorescence. Mailing Add: Bartol Res Found Franklin Inst Swarthmore PA 19081

SWANN, DALE WILLIAM, b Billings, Mont, Mar 11, 29; m 60; c 1. APPLIED MATHEMATICS, OPERATIONS RESEARCH. Educ: Yale Univ, BS, 51; Stanford Univ, PhD(math), 60. Prof Exp: Instr math, Stanford, 57-60; NATO fel sci, Cambridge Univ, 60-61; MEM TECH STAFF, BELL LABS, 61- Res: Mathematical economics; applied probability; quality theory and practice; reliability theory; ordinary differential equations; integral equations; asymptotic methods. Mailing Add: Qual Assurance Ctr Bell Labs Holmdel NJ 07733

SWANN, GORDON ALFRED, b Palisade, Colo, Sept 21, 31; m 53; c 4. GEOLOGY, ASTROGEOLOGY. Educ: Univ Colo, Boulder, BA, 58, PhD(geol), 62. Prof Exp: Geologist, 63-73, STAFF GEOLOGIST FOR TELEGEOL, US GEOL SURV, 73- Honors & Awards: Medal for Exceptional Sci Achievement, NASA, 71; Prof Excellence Award, Am Inst Prof Geologists, 72; Cert of Spec Commendation, Geol Soc Am, 73. Mailing Add: US Geol Surv 601 E Cedar Ave Flagstaff AZ 86001

SWANN, HENRY EDGAR, JR, physiology, toxicology, see 12th edition

SWANN, HOWARD STORY GRAY, b Chicago, Ill, Aug 4, 36. MATHEMATICS. Educ: Harvard Univ, AB, 58; Univ Chicago, MS, 59; Univ Calif, Berkeley, PhD(appl math), 68. Prof Exp: Asst math, Univ Chicago, 59-61; lectr, Univ Nigeria, 61-63; asst & instr, Univ Calif, Berkeley, 64-68; asst prof, Antioch Col, 68-70; ASSOC PROF MATH, SAN JOSE STATE UNIV, 70- Mem: Am Math Soc. Res: Functional analysis; differential equations; game theory; automata theory. Mailing Add: Dept of Math San Jose State Univ San Jose CA 95192

SWANN, SHERLOCK, JR, b Baltimore, Md, Sept 30, 00. ELECTROCHEMISTRY. Educ: Princeton Univ, BS, 22; Johns Hopkins Univ, PhD(org chem), 26. Prof Exp: Chemist, Columbia Gas Co, 26-27; asst, 27-29, from res assoc to res prof, 29-69, EMER PROF CHEM ENG, UNIV ILL, URBANA, 69- Concurrent Pos: Mem comt electrochem, Nat Res Coun, 38. Mem: AAAS; Am Chem Soc; hon mem Electrochem Soc (vpres, 41-43 & 56-58, pres, 58-59). Res: Organic electrochemistry. Mailing Add: 13 Roger Adams Lab Univ of Ill Urbana IL 61801

SWANN, WILLIAM B, b Hammonton, NJ, Feb 26, 23; m 49; c 3. ANALYTICAL CHEMISTRY. Educ: St Joseph's Col, Pa, BS, 43; Univ Del, MS, 53; Univ Pa, PhD, 63. Prof Exp: Res anal chemist, Socony Mobil Oil Co, 47-60; res anal chemist, Am Viscose Div, 60-65, HEAD ANAL GROUP, CHEM GROUP MGT, FMC CORP, 65- Mem: Am Chem Soc; Am Soc Testing & Mat. Res: Electroanalytical techniques; chromatography. Mailing Add: FMC Corp Chem Group Marcus Hook PA 19061

SWANSON, ARNOLD ARTHUR, b Rawlins, Wyo, Mar 11, 23; m 50; c 4. BIOCHEMISTRY, OPHTHALMOLOGY. Educ: Duke Univ, BA, 46; Trinity Univ, Tex, MA, 59; Tex A&M Univ, PhD(biochem), 61. Prof Exp: Res chemist, Med Sch, Temple Univ, 46-48; biochemist ophthal, US Air Force Sch Aviation Med, 50-59; sr chemist, USPHS, 61-63; chief res lab, Vet Admin Hosp, McKinney, Tex, 63-65 & Vet Admin Ctr, 65-68; ASSOC PROF BIOCHEM, MED UNIV SC, 68- Concurrent Pos: Dir, Swanson Biochem Labs, Inc, 52-58; consult, Southwestern Prods, Inc, 57- & Scott & White Hosp, 65-; adj prof, Baylor Univ, 66- Mem: Fel AAAS; Am Chem Soc; fel Am Inst Chem; Soc Exp Biol & Med; Asn Res Vision & Ophthal. Res: Proteolysis in normal and senile cataract lens; senile changes and mineral metabolism. Mailing Add: Dept of Biochem Med Univ of SC Charleston SC 29401

SWANSON, ARTHUR MARTIN, b Rockford, Ill, Jan 30, 11; m 38; c 2. BIOCHEMISTRY, FOOD SCIENCE. Educ: Univ Wis, BS, 35, MS, 36, PhD, 38. Prof Exp: Asst, Univ Wis, 35-38, res fel, 38-39; instr, NMex Agr & Mech Col, 39-40; prof chem, Westmont Col, 40-41; chemist, Borden Co, 41-46; from asst prof to assoc prof, 46-55, PROF FOOD SCI, UNIV WIS-MADISON, 55- Honors & Awards: Borden Award dairy mfg, 65. Mem: AAAS; Am Chem Soc; Am Asn Cereal Chem; Inst Food Technol; Am Dairy Sci Asn. Res: Chemistry of milk proteins; experimental baking; particle aggregation and high temperature- short time sterilization techniques. Mailing Add: 211 Babcock Hall Univ of Wis Madison WI 53706

SWANSON, AUGUST GEORGE, b Kearney, Nebr, Aug 25, 25; m 47; c 6. NEUROLOGY. Educ: Harvard Med Sch, MD, 49; Westminster Col, Mo, AB, 51. Prof Exp: Resident med, Sch Med, Univ Wash, 53-55, resident neurol, 55-57; asst resident, Boston City Hosp, 58; instr neurol, Sch Med, Univ Wash, 58-59, asst prof pediat & med, 59-63; vis res fel physiol, Oxford Univ, 63-64; assoc prof med, 64-70, PROF MED, SCH MED, UNIV WASH, 70- Concurrent Pos: Fel, Univ Wash, 55-57, assoc dean, Sch Med, 67-68, assoc dean acad affairs, 68-71; Markle scholar, 59; dir dep acad affairs, Asn Am Med Cols, 71- Mem: Inst of Med of Nat Acad Sci; Am Acad Neurol; Am Neurol Asn; Am Fedn Clin Res. Res: Facilitation of the development of medical education and biomedical research. Mailing Add: Dept of Acad Affairs Asn Am Med Col DuPont Circle NW Washington DC 20036

SWANSON, BARRY GRANT, b Green Lake, Wis, Apr 16, 44; m 70; c 2. FOOD SCIENCE. Educ: Univ Wis-Madison, BS, 66, MS, 70, PhD(food sci), 72. Prof Exp: Asst prof food sci, Univ Idaho, 72-73; ASST PROF FOOD SCI, WASH STATE UNIV, 73- Concurrent Pos: Travel award Spain, Inst Food Technologists, 74. Mem: Inst Food Technologists; Am Soc Hort Sci; Int Asn Milk Food Environ Sanitarians; Am Chem Soc; Potato Asn Am. Res: Analytical and toxicological studies of mycotoxins; nutrient retention in processed food products; resolution of undesirable flavor compounds in dry beans; food mycology. Mailing Add: Dept of Food Sci & Technol 375 Clark Hall Wash State Univ Pullman WA 99163

SWANSON, BASIL IAN, b Minn, Feb 13, 44; m 64; c 2. INORGANIC CHEMISTRY. Educ: Colo Sch Mines, BA, 66; Northwestern Univ, Evanston, PhD(chem), 70. Prof Exp: Fel, Los Alamos Sci Lab, Univ Calif, 70-71; res corp grant, NY Univ, 71-72, asst prof chem, 71-73; ASST PROF CHEM, UNIV TEX, AUSTIN, 73- Concurrent Pos: Vis staff mem, Los Alamos Sci Lab, 70- Mem: Am Chem Soc. Res: Study of structure and bonding in organic systems using crystallographic and vibrational spectroscopic techniques; study of structural phase changes in crystalline solids; photo-induced solid state reactions. Mailing Add: Dept of Chem Univ of Tex Austin TX 78712

SWANSON, CARL LOYAL WILLIAM, soils, deceased

SWANSON, CARL PONTIUS, b Rockport, Mass, June 24, 11; m 41; c 2. CYTOGENETICS. Educ: Mass State Col, BS, 37; Harvard Univ, MA, 39, PhD(biol), 41. Prof Exp: Sheldon traveling fel from Harvard Univ, Univ Mo, 41; asst prof bot, Mich State Col, 41-43; assoc biologist, NIH, 46; from assoc prof to prof bot, Johns Hopkins Univ, 46-56, William D Gill prof biol, 56-71, assoc dean undergrad studies, 66-71; assoc dir, Inst for Man & his Environ, 71-75, PROF BOT, UNIV MASS, AMHERST, 71- Concurrent Pos: Agt, USDA, 39; contract investr, Spec Proj Div, US Army, 46-49; pres, Int Photobiol Comt, 64-68. Mem: AAAS; Genetics Soc Am. Res: Cytogenetics of plants involving use of ionizing and photochemical radiations. Mailing Add: Dept of Bot Univ of Mass Amherst MA 01002

SWANSON, CARROLL ARTHUR, b Burlington, Iowa, Sept 6, 15; m 41; c 2. PLANT PHYSIOLOGY. Educ: Augustana Col, AB, 37; Ohio State Univ, MS, 38, PhD(plant physiol), 42. Prof Exp: From asst to asst prof bot, 38-48, res assoc, Manhattan Proj, Res Found, 44-46, assoc prof bot & plant path, 48-56, chmn dept, 67-69, assoc dean col biol sci, 69-70, PROF BOT & PLANT PATH, OHIO STATE UNIV, 56- Concurrent Pos: Asst gen foreman, Procter & Gamble Defense Corp, Miss, 43-44; prog dir, NSF, 59-60, consult, 60-66. Mem: AAAS; Am Soc Plant Physiologists; Bot Soc Am; Can Soc Plant Physiologists; Scandinavian Soc Plant Physiologists. Res: Translocation in phloem. Mailing Add: Dept of Bot Ohio State Univ Columbus OH 43210

SWANSON, CHARLES ANDREW, b Bellingham, Wash, July 11, 29; m 57; c 2. MATHEMATICS. Educ: Univ BC, BA, 51, MA, 53; Calif Inst Technol, PhD, 57. Prof Exp: From instr to assoc prof, 57-65, PROF MATH, UNIV BC, 65- Concurrent Pos: Assoc ed, Can J Math, 71- Mem: Am Math Soc; Math Asn Am; Can Math Cong. Res: Differential equations. Mailing Add: Dept of Math Univ of BC Vancouver BC Can

SWANSON, CHARLES RICHARD, b Minneapolis, Minn, Sept 11, 24; m 45; c 3. PLANT PHYSIOLOGY. Educ: NDak State Univ, BS, 48, MS, 49; Iowa State Univ, PhD(plant physiol), 53. Prof Exp: Plant physiologist, USDA, Md, 50-56, SDak, 56-60; assoc prof bot, NDak State Univ, 60-62; plant physiologist, 62-63, leader pesticide invests & metab in plants, 63-67, plant physiologist, Crops Res Div, Tex, 67-68, Miss, 68-70, lab chief, Southern Weed Sci Lab, 70-75, ASST TO DEP ADMINR, SOUTHERN REGION, AGR RES SERV, USDA, 75- Mem: AAAS; Weed Sci Soc Am (vpres, 73, pres-elect, 74, pres, 75). Res: Metabolic fate and mode of action of herbicides in plants. Mailing Add: Southern Region Agr Res Serv USDA 701 Loyola Ave Box 53326 New Orleans LA 70153

SWANSON, CLARENCE A E, b Lincoln, Nebr, Apr 2, 18; m 40; c 2. MATHEMATICS. Educ: Colo State Col, BA, 40, MA, 48. Prof Exp: Teacher & coach, High Sch, Colo, 40-41, Jr High Sch, 41-42 & High Sch, 42-45; teacher, Lamar High Sch & Lamar Jr Col, 45-46; teacher, Lamar Jr Col, 46-49, teacher math & acad dean, 49-54; teacher, 55-63, PROF MATH & HEAD DEPT, UNIV SOUTHERN COLO, 63- Mem: Math Asn Am; Am Math Soc. Res: Current curriculum developments in college and secondary mathematics. Mailing Add: Dept of Math Univ of Southern Colo Pueblo CO 81001

SWANSON, CURTIS JAMES, b Chicago, Ill, Dec 8, 41; m 65; c 2. COMPARATIVE PHYSIOLOGY, BIOCHEMISTRY. Educ: N Park Col, BA & BS, 64; Northern Ill Univ, MS, 66; Univ Ill, Urbana-Champaign, PhD(zool, physiol), 70. Prof Exp: Lectr biol, Univ Ill, 70; asst prof, 70-74, ASSOC PROF BIOL, WAYNE STATE UNIV, 74- Concurrent Pos: Grants, NSF, Wayne State Univ, 70-75 & NIH, 71-74; Riker res fel, Bermuda Biol Sta, 75. Mem: AAAS; Am Inst Biol Sci; Am Soc Zoologists; Am Physiol Soc. Res: Electron microscopy of muscle tissue; innervation and developmental neuromuscular physiology; control systems in development; protein biochemistry; theoretical and applied biomechanics; comparative ultrastructure of muscle. Mailing Add: Dept of Biol Wayne State Univ Detroit MI 48202

SWANSON, DAVID G, JR, b Chicago, Ill, Jan 14, 41. NUCLEAR CHEMISTRY, PHYSICAL CHEMISTRY. Educ: Northwestern Univ, BS, 64; Purdue Univ, PhD(nuclear & phys chem), 69. Prof Exp: Nuclear chemist, Sandia Corp, NMex, 69-73; NUCLEAR CHEMIST, AEROSPACE CORP, 73- Mem: Am Chem Soc; Am Phys Soc; Am Nuclear Soc. Res: Response of materials to radiation; nuclear reactions; radiation transport phenomena; high temperature physical chemistry; nuclear reactor safety. Mailing Add: Aerospace Corp El Segundo CA 90245

SWANSON, DAVID WENDELL, b Ft Dodge, Iowa, Aug 28, 30; m 53; c 3.

PSYCHIATRY. Educ: Augustana Col, Ill, BA, 52; Univ Ill, MD, 56. Prof Exp: Intern, Ill Cent Hosp, 56-57; resident psychiat, Ill State Psychiat Inst, 59-62, asst serv chief, 62-63; assoc prof & asst chmn dept, Stritch Sch Med, Loyola Univ Chicago, 63-70; assoc prof psychiat, Mayo Grad Sch Med & consult, Sect Psychiat, Mayo Clin, 70-74, PROF PSYCHIAT, MAYO MED SCH, UNIV MINN & HEAD SECT PSYCHIAT, MAYO CLIN, 74- Mem: AAAS; Am Col Psychiat; AMA; Am Psychiat Asn. Res: Paranoid and psychosomatic disorders. Mailing Add: Sect of Psychiat Mayo Clin Rochester MN 55901

SWANSON, DON R, b Los Angeles, Calif, Oct 10, 24; div; c 3. INFORMATION SCIENCE. Educ: Calif Inst Technol, BS, 45; Rice Univ, MA, 47; Univ Calif, Berkeley, PhD(physics), 52. Prof Exp: Res physicist, Radiation Lab, Univ Calif, 50-52; mem tech staff, Hughes Res & Develop Labs, 52-55; dept mgr comput appln, Thompson-Ramo-Wooldridge, Inc, 55-63; dean, 63-72, PROF LIBR SCI, LIB SCH, UNIV CHICAGO, SCI, LIBR SCH, UNIV CHICAGO, 63- Concurrent Pos: Mem sci info coun, NSF, 59-63; mem vis comt libr, Mass Inst Technol, 64-72; trustee, Nat Opinion Res Ctr, 64-73; mem adv comt, Libr Cong, 64-, toxicol info panel, President's Sci Adv Comt, 64-65, comt sci & tech commun, Nat Acad Sci, 66-70 & adv comt, Encyclop Britannica, 66- Res: Library science and education; computer systems analysis and applications; information processing; computer programming; indexing and retrieval of information. Mailing Add: Grad Libr Sch Univ of Chicago Chicago IL 60637

SWANSON, DONALD ALAN, b Tacoma, Wash, July 25, 38. GEOLOGY. Educ: Wash State Univ, BS, 60; Johns Hopkins Univ, PhD(geol), 64. Prof Exp: NATO fel, Ger, Italy & Canary Islands, 64-65; GEOLOGIST, US GEOL SURV, 65- Mem: AAAS; Geol Soc Am. Res: Petrology of volcanic rocks, especially from northwest United States; flowage mechanisms of ash flows; volcanology. Mailing Add: US Geol Surv 345 Middlefield Rd Menlo Park CA 94025

SWANSON, DONALD CHARLES, b Canon City, Colo, Sept 22, 26; m 50; c 2. PETROLEUM GEOLOGY, SEDIMENTOLOGY. Educ: Colo State Univ, BS, 50; Univ Tulsa, BS, 55. Prof Exp: Geol & geophys tax engr, Carter Oil Co, Okla, 51-56, jr geologist, Kans, 56, geologist, Ark, 56-57 & Okla, 57-60; geologist, Humble Oil Co, 60-62, sr geologist, Tex, 62-63 & Humble Res Ctr, 63-64, staff geologist, Humble Oil Co, Okla, 64-67; sr res geologist, Esso Prod Res Co, 67, sr res specialist, 67-74; RES ASSOC, EXXON PROD RES CO, 74- Honors & Awards: Levorsen Award, Am Asn Petrol Geologists, 68. Mem: Fel Geol Soc Am; Am Asn Petrol Geologists. Res: Clastic facies; determination of ancient sedimentary environments; paleogeography; methodology of environmental facies analyses; methodology of exploration; computer application to petroleum geology. Mailing Add: 13611 Kingsride Houston TX 77024

SWANSON, DONALD G, b Los Angeles, Calif, June 11, 35; m 60; c 3. PLASMA PHYSICS. Educ: Northwest Christian Col, BTh, 58; Univ Ore, BS, 58; Calif Inst Technol, MS, 61, PhD(physics), 63. Prof Exp: Fel, Calif Inst Technol, 63-64; asst prof, 64-68, ASSOC PROF ELEC ENG, UNIV TEX, AUSTIN, 68- Concurrent Pos: Consult, Advan Kinetics, Inc, Calif, 63- Mem: Am Phys Soc. Res: Compressional hydromagnetic waves; plasma-filled waveguide; ion cyclotron waves. Mailing Add: Dept of Elec Eng Univ of Tex Austin TX 78712

SWANSON, DONALD LEROY, b Montrose, SDak, Mar 24, 23; m 48; c 3. ANALYTICAL CHEMISTRY, PHYSICAL CHEMISTRY. Educ: SDak State Univ, BS, 47; Univ Wis, PhD(chem), 51. Prof Exp: Lab asst chem, Agr Exp Sta, SDak, 46-47; asst, Univ Wis, 47-51; res chemist, 51-58, group leader, 58-61, SECT MGR, AM CYANAMID CO, 62- Mem: Am Chem Soc. Res: Physical and mechanical properties of polymers; polymerization kinetics; copolymerization; radiation chemistry; analysis. Mailing Add: Sci Serv Dept Am Cyanamid Co Stamford CT 06904

SWANSON, DWIGHT WESLEY, b Harcourt, Iowa, Mar 15, 22; m 46; c 4. METEOROLOGY, DATA PROCESSING. Educ: Cornell Univ, BA, 46; Iowa State Univ, MS, 47. Prof Exp: Meteorologist, US Weather Bur, 47-51; meteorologist, 51-68, soil scientist, Soil Surv Interpretations, 68-70, data processing intern, Washington Data Processing Ctr, Statist Reporting Serv, 67-68, HEAD SOIL DATA STORAGE & RETRIEVAL UNIT, SOIL SURV, SOIL CONSERV SERV, USDA, WASHINGTON, DC, 70- Mem: Asn Comput Mach. Res: Effects of climate and weather on crops and soils; application of automatic data processing techniques in soil survey. Mailing Add: 15100 Donna Dr Silver Spring MD 20904

SWANSON, EARL HERBERT, JR, anthropology, deceased

SWANSON, ERIC WALLACE, b Knox, Ind, June 14, 18; m 41; c 3. DAIRY SCIENCE. Educ: Purdue Univ, BS, 39; Univ Mo, AM, 40, PhD(nutrit), 43. Prof Exp: Asst, Univ Mo, 39-40, asst instr dairy husb, 40-43, instr, 43-44 & 46-47; assoc prof dairying & assoc dairy husbandman, 47-56, PROF DAIRYING & ANIMAL SCIENTIST, AGR EXP STA, UNIV TENN, KNOXVILLE, 56- Mem: Am Inst Nutrit; Am Soc Animal Sci; Am Dairy Sci Asn. Res: Dairy cattle nutrition; reproductive physiology and diseases; milk secretion physiology; mastitis; nutritive value of proteins; thyroid function and iodine metabolism in cattle; effects of growth rates on lactation. Mailing Add: Dept of Animal Sci Univ of Tenn PO Box 1071 Knoxville TN 37901

SWANSON, ERNEST ALLEN, JR, b Miami, Fla, Apr 9, 36; m 67. ANATOMY, HISTOLOGY. Educ: Emory Univ, BA, 58, PhD(anat), 64. Prof Exp: Instr anat, Emory Univ, 64-65; instr, Univ Va, 65-67; asst prof, 67-72, ASSOC PROF ANAT, SCH DENT, TEMPLE UNIV, 72- Mem: Am Asn Anatomists. Res: Changes in the dental pulp associated with cholesterol induced arteriosclerosis. Mailing Add: Dept of Anatomic Sci Temple Univ Sch of Dent Philadelphia PA 19140

SWANSON, GUSTAV ADOLPH, b Mamre, Minn, Feb 13, 10; m 36; c 3. WILDLIFE ECOLOGY. Educ: Univ Minn, BS, 30, MS, 32, PhD(zool), 37. Prof Exp: Asst zool, Univ Minn, 30-34; biologist, State Dept Conserv, Minn, 35-36; asst prof game mgt, Univ Maine, 36-37; asst prof econ zool, Univ Minn, 37-41; assoc regional inspector, US Fish & Wildlife Serv, 41-42, chief sect coop wildlife res units, 44-46, chief div wildlife res, 46-48; assoc prof econ zool, Univ Minn, 42-44; prof conserv & head dept, Cornell Univ, 48-66; prof fishery & wildlife biol & head dept, 66-75, EMER PROF WILDLIFE BIOL, COLO STATE UNIV, 75- Concurrent Pos: Ed, J Wildlife Soc, 49-53; Am Scandinavian Found fel, Denmark, 54-55, Fulbright fel, 61-62; consult waterfowl res, Nature Conserv, Eng, Scotland & Northern Ireland, 55 & 60; dir, Cornell Biol Field Sta, 55-66, exec dir lab ornith, 58-61; fel, Rochester Mus, 56; consult, State Joint Legis Comt Rev Conserv Law, NY, 56-65 & natural resources, 56-66; Fulbright fel, NSW, Australia, 68. Honors & Awards: Aldo Leopold Mem Medalist, 73. Mem: Fel AAAS; hon mem Wildlife Soc (vpres, 45, pres, 54); Am Soc Mammalogy; Wilson Ornith Soc (treas, 38-42); Am Inst Biol Sci. Res: Wildlife management; conservation of natural resources; ornithology. Mailing Add: Dept of Fishery & Wildlife Biol Colo State Univ Ft Collins CO 80523

SWANSON, HARLEY DAMON, microbial genetics, microbial physiology, see 12th edition

SWANSON, HAROLD DUEKER, b Wichita, Kans, Mar 5, 30; m 55; c 3. CELL BIOLOGY. Educ: Friends Univ, BA, 53; Univ Kans, MA, 55; Univ Tenn, PhD(zool physiol), 60. Prof Exp: Asst zool, Univ Kans, 53-55 & Univ Tenn, 56-58; from asst prof to assoc prof, 60-74, PROF BIOL, DRAKE UNIV, 74- Res: Nucleocytoplasmic interaction; subcellular component isolation; regeneration of cirrhotic livers; cell physiology. Mailing Add: Dept of Biol Drake Univ Des Moines IA 50311

SWANSON, JACK LEE, b Aurora, Nebr, Oct 22, 34; m 56; c 3. PHYSICAL CHEMISTRY, BIOCHEMISTRY. Educ: Kearney State Coi, BS, 56; Univ Nebr, MS, 59, PhD(chem), 67. Prof Exp: Assoc prof chem, Kearney State Col, 58-71; DEAN SCH SCI & TECHNOL, CHADRON STATE COL, 71- Mem: AAAS; Am Chem Soc; Sigma Xi. Res: Infrared and ultraviolet spectroscopy; magneto-optical rotary dispersion; circular dichroism spectroscopy; medicinal chemistry. Mailing Add: Sch of Sci & Technol Chadron State Col Chadron NE 69337

SWANSON, JAMES A, b Aurora, Nebr, Oct 25, 35; m 57; c 2. PHYSICAL CHEMISTRY. Educ: Kearney State Col, BA, 57; Univ Nebr, MS, 59, PhD(chem), 62. Prof Exp: Part-time lab asst, Univ Nebr, 57-62; PROF CHEM, KEARNEY STATE COL, 62- Mem: Am Chem Soc. Res: Solution thermochemistry; thermodynamics. Mailing Add: Dept of Chem Kearney State Col Kearney NE 68847

SWANSON, JOHN LEE, b Hastings, Nebr, Aug 16, 36; m 68; c 5. EXPERIMENTAL PATHOLOGY, MICROBIOLOGY. Educ: Univ Nebr, BS, 59, MS, 61, MD, 62. Prof Exp: Pathologist res, Armed Forces Inst Path, 65-67; pathologist res, Walter Reed Army Inst Res, 67-68; asst prof microbiol, Col Physicians & Surgeons, Columbia Univ, 68-69; from asst prof to assoc prof microbiol, Mt Sinai Sch Med, 69-72; assoc prof path & microbiol, 72-75, PROF PATH & MICROBIOL, COL MED, UNIV UTAH, 75- Honors & Awards: Career Investr, Health Res Coun, City New York, 71. Mem: Am Soc Microbiol; Infectious Dis Soc Am. Res: Virulence factors of gonococci and pathogenesis of gonorrhea. Mailing Add: Dept of Path Med Ctr Univ of Utah Salt Lake City UT 84132

SWANSON, JOHN MELVIN, b Ishpeming, Mich, July 26, 15; m 41; c 3. TEXTILE CHEMISTRY. Educ: Mich Col Mining & Technol, BS, 37; Univ Wis, PhD(phys chem), 40. Prof Exp: Res chemist, 40-46, res supvr, 46-54, res mgr, 54-58, LAB DIR, E I DU PONT DE NEMOURS & CO, INC, 58- Mem: Am Chem Soc. Res: Molecular structure, physical properties and end use applications of synthetic yarns; solution properties of high polymers. Mailing Add: 1503 Veale Rd Westwood Manor Wilmington DE 19810

SWANSON, JOHN ROBERT, b Ft Collins, Colo, June 24, 39; m 62; c 1. BIOCHEMISTRY. Educ: Colo State Univ, BS, 61; Wash State Univ, PhD(biochem), 65; Am Bd Clin Chem, dipl. Prof Exp: NIH res fel biochem, Duke Univ, 65-67; clin chem training fel, Pepper Lab, Hosp Univ Pa, 67-69; asst prof, 69-74, PROF CLIN PATH, MED SCH, UNIV ORE, PORTLAND, 74- Mem: Am Asn Clin Chem; Am Chem Soc. Res: Mechanism of action of myosin; comparative enzymology of phospholucomutase; analytical methods for urinary protein, plasma renin and serum triglycerides. Mailing Add: Dept of Clin Path Univ of Ore Med Sch Portland OR 97201

SWANSON, JOHN WILLIAM, b Sioux City, Iowa, Oct 12, 17; m 41; c 3. PHYSICAL CHEMISTRY. Educ: Morningside Col, BA, 40. Hon Degrees: DSc, Morningside Col, 72. Prof Exp: Asst chem, Iowa State Col, 40-41; tech asst, 41-43, tech assoc, 44-45, res asst, 46-47, res assoc, 48-55, group leader surface & colloid chem, 53-55, group leader phys chem, 56-61, sr res assoc, 56-69, chmn phys chem dept, 62-69, DIR DIV NATURAL MAT & SYSTS, INST PAPER CHEM, 69- Concurrent Pos: Lectr, Lawrence Univ, 45-46; consult to numerous paper co, 50- Honors & Awards: Res & Develop Div Award, Tech Asn Pulp & Paper Indust, 74. Mem: AAAS; Am Chem Soc; fel Tech Asn Pulp & Paper Indust. Res: Surface and colloid chemistry of papermaking; polymer sorption at interfaces; surface area and bonding of cellulose fibers; paper sizing, coating; coagulation and retention of resins in aqueous systems; pollution abatement. Mailing Add: Div of Natural Mat & Systs Inst of Paper Chem PO Box 1039 Appleton WI 54911

SWANSON, LAWRENCE RAY, b Omaha, Nebr, Nov 4, 36; m 62; c 2. PHYSICS. Educ: Iowa State Univ, BS, 59; Fuller Theol Sem, BD, 63; Calif State Univ, Los Angeles, MS, 66; Univ Calif, Irvine, PhD(physics), 70. Prof Exp: Asst prof physics, Pasadena Col, 70-74; MEM FAC, DEPT PHYSICS, AZUSA PAC COL, 74- Mem: Am Asn Physics Teachers. Res: Theoretical solid state physics. Mailing Add: Dept of Physics Azusa Pac Col Hwy 66 at Citrus Azusa CA 91702

SWANSON, LEONARD WILLIAM, b Mt Vernon, Ohio, May 15, 13; m 39; c 1. MATHEMATICS. Educ: Kenyon Col, BS, 35; Univ Minn, MA, 40, PhD(math), 47. Prof Exp: High sch teacher, Minn, 35-37; instr math, Kenyon Col, 37-38; asst, Univ Minn, 38-41, instr math & mech, Sch Eng, 41-42; from asst prof to prof math & head dept, Coe Col, 42-50; dir appl sci dept, Midwest Area, Int Bus Mach Corp, 50-55; mgr opers res, Arthur Andersen & Co, 55-64; PROF QUANT METHODS & MANAGERIAL ECON, GRAD SCH MGT, NORTHWESTERN UNIV, EVANSTON, 64- Concurrent Pos: Sr assoc engr & math consult, Collins Radio Co, Iowa, 48-50; consult opers res, 65- Mem: Am Math Soc; Math Asn Am; Opers Res Soc Am; Soc Indust & Appl Math. Res: Theory of orthogonal polynomials; exterior ballistics; application of high speed computers; operations research; linear programming; probability and statistics; inventory theory; non-linear programming. Mailing Add: 517 N Merrill Park Ridge IL 60068

SWANSON, LLOYD VERNON, b Isanti, Minn, Oct 16, 38; m 66; c 2. REPRODUCTIVE ENDOCRINOLOGY. Educ: Univ Minn, St Paul, BS, 60, MS, 67; Mich State Univ, PhD(physiol), 70. Prof Exp: ASST PROF DAIRY PHYSIOL, ORE STATE UNIV, 71- Mem: AAAS; Am Dairy Sci Asn; Am Soc Animal Sci; Soc Study Reproduction. Res: Reproductive physiology of mammalian species, both male and female, with special interest in the endocrine control of ovulation and of spermatogenesis. Mailing Add: Dept of Animal Sci Ore State Univ Corvallis OR 97331

SWANSON, LYNN ALLEN, b Minneapolis, Minn, July 28, 42; m 67. ANALYTICAL CHEMISTRY. Educ: Univ Minn, Minneapolis, BChem, 64; Univ Iowa, MS, 68, PhD(anal chem), 70. Prof Exp: RES CHEMIST ANAL CHEM, COMMERCIAL SOLVENTS CORP, 69- Mem: AAAS; Am Chem Soc. Res: Trace analysis of pharmaceuticals, drugs and other additives in animal tissues and body fluids; general chromatography; spectrophotometry. Mailing Add: Res & Develop Commercial Solvents Corp Terre Haute IN 47808

SWANSON, LYNWOOD WALTER, b Turlock, Calif, Oct 7, 34; m 55; c 2. PHYSICAL CHEMISTRY. Educ: Univ of Pac, BSc, 56; Univ Calif, PhD(chem), 60. Prof Exp: Asst chemist, Univ Calif, Berkeley, 56-59; res assoc, Inst Study Metals, Univ Chicago, 59-61; sr scientist, Linfield Res Inst, 61-63; dir basic res, Field Emission Corp, 63-69; prof chem & dean fac, Linfield Col, 69-73; PROF APPL PHYSICS, ORE GRAD CTR, 73- Mem: Fel Am Phys Soc; fel Am Inst Chemists.

Res: Photochemistry; surface adsorption; field electron and ion microscopy; electron physics. Mailing Add: Ore Grad Ctr Beaverton OR 97005

SWANSON, MAX LYNN, b Hancock, Mich, Aug 5, 31, Can citizen; m 59; c 4. EXPERIMENTAL SOLID STATE PHYSICS. Educ: Univ BC, BA, 53, MSc, 54, PhD(metal physics), 58. Prof Exp: Res metallurgist, Metals Res Lab, Carnegie Inst Technol, 58-60; RES OFFICER METAL PHYSICS, CHALK RIVER NUCLEAR LABS, ATOMIC ENERGY CAN LTD, 60- Concurrent Pos: Guest scientist, Inst Physics, Max Planck Inst Metal Res, 65-66; vis prof, Univ Utah, 71-72. Mem: Can Asn Physicists. Res: Defect solid state physics, including irradiation damage in metals and semiconductor defect photoconductivity, plastic deformation of metals, ion channeling. Mailing Add: Atomic Energy of Can Ltd Chalk River ON Can

SWANSON, MILO HARLAND, poultry husbandry, see 12th edition

SWANSON, PAUL N, b San Mateo, Calif, June 29, 36; m 59; c 3. RADIO ASTRONOMY. Educ: Calif State Polytech Col, BS, 62; Pa State Univ, PhD(physics), 68. Prof Exp: Asst prof radio astron, Pa State Univ, University Park, 69-75; MEM STAFF, JET PROPULSION LAB, CALIF INST TECHNOL, 75- Mem: Am Astron Soc. Res: Millimeter wavelength radio astronomy and solar physics. Mailing Add: Jet Propulsion Lab Sect 823 Calif Inst of Technol Pasadena CA 91109

SWANSON, PHILLIP D, b Seattle, Wash, Oct 1, 32; m 57; c 5. NEUROLOGY, BIOCHEMISTRY. Educ: Yale Univ, BS, 54; Johns Hopkins Univ, MD, 58; Univ London, PhD(biochem), 64. Prof Exp: Fel neurol med, Sch Med, Johns Hopkins Univ, 59-62; Nat Inst Neurol Dis & Stroke spec fel, Univ London, 62-64; from asst prof to assoc prof, 64-73, PROF NEUROL, SCH MED, UNIV WASH, 73-, HEAD DIV, 67- Mem: Am Neurol Asn; Am Soc Clin Invest; Brit Biochem Soc. Res: Neurochemistry; cation transport and energy utilization in cerebral tissues; enzymes of importance in cation transport. Mailing Add: Div of Neurol Univ of Wash Sch of Med Seattle WA 98195

SWANSON, ROBERT ALLAN, b Chicago, Ill, Dec 16, 28; m 57. ELEMENTARY PARTICLE PHYSICS. Educ: Ill Inst Technol, BS, 51; Univ Chicago, MS, 53, PhD(physics), 58. Prof Exp: Res assoc physics, Univ Chicago, 58-59; asst prof, Princeton Univ, 59-60; from asst prof to assoc prof, 60-70, PROF PHYSICS, UNIV CALIF, SAN DIEGO, 70- Concurrent Pos: Vis assoc prof, Univ Chicago, 68-69; NSF fel, Univ Calif, 72- Mem: Am Phys Soc; Am Asn Physics Teachers. Res: Muonic atoms; experimental kaon physics. Mailing Add: Dept of Physics Univ of Calif at San Diego La Jolla CA 92093

SWANSON, ROBERT E, b Duluth, Minn, Dec 19, 24; m 47; c 2. MEDICAL PHYSIOLOGY. Educ: Univ Minn, BA, 49, PhD(physiol), 53. Prof Exp: Asst physiol, Univ Minn, 50-52, instr, 52-55; asst physiologist, Brookhaven Nat Lab, 55-58; asst prof physiol, Univ Minn, 58-61; assoc prof, 61-73, PROF PHYSIOL, MED SCH, UNIV ORE HEALTH SCI CTR, 73- Res: Renal, water and electrolyte balance. Mailing Add: Dept Physiol Sch Med Univ Ore Health Sci Ctr Portland OR 97201

SWANSON, ROBERT EARL, physics, see 12th edition

SWANSON, ROBERT HAROLD, b Los Angeles, Calif, Feb 15, 33, m 55; c 2. FOREST HYDROLOGY, FOREST PHYSIOLOGY. Educ: Colo State Univ, BSc, 59, MSc, 66. Prof Exp: Res forester hydrol, Rocky Mountain Forest & Range Exp Sta, US Forest Serv, 59-68; PROJ LEADER FOREST HYDROL, NORTHERN FOREST RES CTR, 68- Concurrent Pos: Res coordr, Alta Watershed Res Prog, 68-; res fel, Ministry of Works, NZ, 74-75. Mem: Can Inst Foresters; Sigma Xi. Res: Physiological bases for tree improvement, plant-water relation's forest arrangements streamflow interractions' watershed management simulation and evaluation techniques. Mailing Add: 5320-122nd St Edmonton AB Can

SWANSON, ROBERT JAMES, b St Petersburg, Fla, Nov 13, 45; m 67; c 2. ENDOCRINOLOGY. Educ: Wheaton Col, Ill, BS, 67; Fla State Univ, MS, 71, PhD(biol), 76. Prof Exp: Teacher gen sci, Madison High Sch, Fla, 67-68; instr anat & kinesiology, Fla State Univ, 69-70; ASST PROF ANAT, PHYSIOL & ENDOCRINOL, OLD DOMINION UNIV, 75- Mem: AAAS. Res: Female reproductive physiology, especially factors involved in ovulation, such as hormones, smooth muscle activity, nerve involvement and blood flow. Mailing Add: Dept of Biol Sci Old Dominion Univ Norfolk VA 23508

SWANSON, ROBERT LAWRENCE, b Baltimore, Md, Oct 11, 38; m 63; c 2. PHYSICAL OCEANOGRAPHY, CIVIL ENGINEERING. Educ: Lehigh Univ, BS, 60; Ore State Univ, MS, 65, PhD, 71. Prof Exp: With US Coast & Geodetic Surv, 60-66, commanding officer US Coast & Geodetic Surv Ship Marmer circulatory estuarine surv, 66-67, chief, Oceanog Div, Nat Ocean Surv, 69-72, PROJ MGR, NY BLIGHT PROJ, MANNED ENVIRON SYSTS ASSESSMENT, ENVIRON RES LABS, NAT OCEANIC & ATMOSPHERIC ADMIN, 72- Concurrent Pos: Prof asst, Col Gen Studies, George Washington Univ, 70-73. Honors & Awards: Karo Award, Soc Am Military Engrs, 73; Silver Medal, US Dept Com, 73; Prog Admin & Mgt Award, Nat Oceanic & Atmospheric Admin, 75. Mem: Am Soc Civil Engrs; Am Geophys Union; AAAS; Am Soc Photogramm; Marine Technol Soc. Res: Developing interrelationships and understanding between component parts of the coastal marine ecosystem; studying the impact of ocean dumping on marine ecosystem; specific interests in tides, tidal currents, tidal datums, marine boundaries. Mailing Add: MESA/NOAA Old Biol Bldg 004 State Univ of NY Stony Brook NY 11794

SWANSON, ROBERT NELS, b Ashland, Wis, Feb 4, 32; m 57; c 4. MICROMETEOROLOGY. Educ: Wis State Col, River Falls, BS, 53; Univ Mich, MS, 58. Prof Exp: Meteorologist, White Sands Missile Range, 58-61; staff scientist, GCA Corp, Utah, 61-72; METEOROLOGIST, PAC GAS & ELEC CO, SAN FRANCISCO, 72- Mem: Am Meteorol Soc; Royal Meteorol Soc; Air Pollution Control Asn. Res: Turbulence and diffusion as it applies to air pollution problems. Mailing Add: 1216 Babel Lane Concord CA 94521

SWANSON, ROGER GLENN, b Chicago, Ill, June 7, 23; m 57; c 2. GEOLOGY. Educ: Augustana Col, Ill, BA, 48; Univ Iowa, MS, 50. Prof Exp: Mem staff, Shell Oil Co, 50-59, area stratigr, 59-68; STAFF GEOLOGIST, SHELL DEVELOP CO, 68- Mem: Am Asn Petrol Geologists. Res: Stratigraphy; sedimentation. Mailing Add: Shell Develop Co Box 481 Houston TX 77001

SWANSON, RONALD FREDERICK, b Chicago, Ill, May 31, 40; m 62; c 2. BIOCHEMISTRY, DEVELOPMENTAL BIOLOGY. Educ: Univ Chicago, BS, 63, PhD(biochem), 67. Prof Exp: Res assoc biochem, Univ Chicago, 67-68; USPHS fel, Carnegie Inst Dept Embryol, 68-70; ASST PROF BIOL, UNIV VA, 70- Mem: Am Chem Soc. Res: Mechanism of protein synthesis by mitochondria from animal tissues; functional relationship between mitochondrial and nuclear genes in mitochondrial biogenesis. Mailing Add: Dept of Biol Univ of Va Charlottesville VA 22901

SWANSON, ROWENA WEISS, b Brooklyn, NY, Aug 3, 28. INFORMATION SCIENCE. Educ: Cath Univ, BSChemEng, 49; George Washington Univ, JD, 53. Prof Exp: Mem staff, Libr of Cong, 50-57; patent res specialist, Off Res & Develop, US Patent Off, 57-61; res adminr, Air Force Off Sci Res, 61-70; PROF LIBR & INFO SCI, UNIV DENVER, 70- Concurrent Pos: Consult, Orgn Am States, 76- Honors & Awards: Best Paper Award, Am Soc Info Sci, 75. Mem: AAAS; Asn Comput Mach; Inst Elec & Electronics Engrs; Am Soc Info Sci; Urban & Regional Info Systs Asn. Res: System analysis; organization theory and management applied to libraries and information systems; cost analysis and cost-benefit analysis; curriculum development for information professionals; information storage, retrieval and transfer. Mailing Add: 2923 S Steele St Denver CO 80210

SWANSON, SAMUEL EDWARD, b Woodland, Calif, Aug 1, 46; m 68. GEOCHEMISTRY, PETROLOGY. Educ: Univ Calif, Davis, BS, 68, MS, 70; Stanford Univ, PhD(geol), 74. Prof Exp: Field asst geol, US Geol Surv, 67; res asst geol, Univ Calif, Davis, 70; ASST PROF EARTH SCI, UNIV NC, CHARLOTTE, 74- Mem: Mineral Soc Am; Soc Environ Geochem & Health; Sigma Xi. Res: Application of geochemical techniques to the study of igneous and metamorphic rocks. Mailing Add: Dept of Geog & Earth Sci Univ of NC Charlotte NC 28223

SWANSON, SIGURD ARTHUR, b Ashland, Wis, June 18, 30; m 58; c 3. PHYSICAL CHEMISTRY. Educ: Wis State Univ, River Falls, BS, 52; Univ Iowa, MS, 55, PhD(phys chem), 63. Prof Exp: Asst prof chem, Concordia Col, Minn, 57-60; sr chemist, Minn Mining & Mfg Co, 63-67; res chemist, Kennametal, Inc, 67-75. Mem: Am Chem Soc; Am Vacuum Soc. Res: Magnetic and electric properties of metals and ceramics; powdered metallurgy; high temperature reactions of refractory metals. Mailing Add: Rte 1 Box 78 Mason WI 54856

SWANSON, TERRY B, physical chemistry, see 12th edition

SWANSON, VERN BERNARD, b Alta, Iowa, Feb 8, 25; m 47; c 3. ANIMAL BREEDING. Educ: NMex Agr & Mech Col, BS, 48, MS, 56; Iowa State Univ, PhD(animal breeding), 65. Prof Exp: Instr agr, NMex State Dept Voc Educ, 48-51; animal husbandman, US Dept Interior, 51-53 & USDA, 53-54; instr animal husb, NMex State Univ, 54-56; asst prof, 56-66, ASSOC PROF ANIMAL SCI, COLO STATE UNIV, 66- Mem: Am Soc Animal Sci. Res: Sheep production; breeding; wool technology. Mailing Add: Dept of Animal Sci Colo State Univ Ft Collins CO 80521

SWANSON, VIRGINIA LEE, b Sioux City, Iowa, June 15, 22; m 67. PATHOLOGY. Educ: Univ Southern Calif, BA, 47; Yale Univ, MD, 52; Am Bd Path, dipl, 58. Prof Exp: Intern path, Sch Med, Yale Univ, 52-53; USPHS res fel, Path-Anat Inst, Univ Copenhagen, 53-55; instr path, Sch Med, Yale Univ, 55-59; hosp pathologist, US Army Med Command, Tokyo, Japan, 59-60; hosp pathologist, Australian Pub Health Serv, Port Moresby Gen Hosp, Territories Papua & New Guinea, 60-61; res pathologist, US Army Med Command, Tokyo, 61-62; res pathologist & chief path div, US Army Trop Res Med Lab, San Juan, PR, 62-65; res pathologist pathologist, Armed Forces Inst Path & Walter Reed Army Inst Res, Washington, DC, 65-66; assoc prof, 66-71, PROF PATH, SCH MED, UNIV SOUTHERN CALIF, 71- Concurrent Pos: Asst resident, Grace-New Haven Community Hosp, 55-56, chief resident, 56-57, asst pathologist, 57-59; assoc pathologist, Children's Hosp, Los Angeles, 66-71 & 73-; prof, Sch Med, Univ Calif, San Diego, 71-73; chief lab serv, Vet Admin Hosp, San Diego, 71-73. Res: Pathology of the gastrointestinal tract; malabsorption; malnutrition; immunopathology. Mailing Add: Children's Hosp 4650 Sunset Blvd Los Angeles CA 90027

SWANSON-EARTLY, HEIDI H, b Copenhagen, Denmark, Apr 12, 29; nat Can; m 53; c 2. ENDOCRINOLOGY. Educ: McGill Univ, BSc, 48, MSc, 51, PhD(anat), 53. Prof Exp: Demonstr histol, McGill Univ, 48-52; demonstr, Univ Toronto, 52-53, res assoc path, 57-59; demonstr & lectr histol & embryol, Queen's Univ, Ont, 53-54; res assoc physiol, Univ Wash, 54-56, res assoc med, 56-57; sci officer endocrinol, Univ Hosp, State Univ Leiden, 61-64; SR LECTR ANAT, UNIV BIRMINGHAM, 64- Mem: Brit Soc Endocrinol; Asn Study Animal Behav; Brit Soc Study Fertil; Europ Soc Comp Endocrinol; U K Brain Res Asn. Res: Hypothalamic control of sexual function; influence of hormonal environment at birth on adult behavior; mechanisms for control of population density. Mailing Add: Dept of Anat Univ of Birmingham Birmingham England

SWANTON, MARGARET CATHERINE, b Washington, DC, Sept 26, 20; m 71. PATHOLOGY. Educ: Univ NC, AB, 43; Johns Hopkins Univ, MD, 46. Prof Exp: Intern, Univ Iowa Hosps, 46-47; fel, 47-48, from instr to assoc prof, 48-69, PROF PATH, SCH MED, UNIV NC, CHAPEL HILL, 69- Concurrent Pos: Attend pathologist, NC Mem Hosp, 52-; consult, Watts Hosp, 53- Mem: Am Soc Clin Path; Col Am Path; Am Soc Cytol; Int Acad Path; Asn Am Med Cols. Res: Cytopathology; hemophilic arthropathy. Mailing Add: Dept of Path Univ of NC Sch of Med Chapel Hill NC 27514

SWANZEY, EUGENE HARRY, b Lewistown, Mont, Nov 17, 21; m 52; c 4. MEDICINE. Educ: Univ Wash, BS, 43; McGill Univ, MD, CM, 51. Prof Exp: Pvt pract, 53-55; asst dir clin res, 56-58, DIR OFF GOVT CONTROLS, LEDERLE LABS, AM CYANAMID CO, 58- Res: Clinical research and toxicology of drugs; research administration and management. Mailing Add: Lederle Labs Am Cyanamid Co Pearl River NY 10965

SWARD, EDWARD LAWRENCE, JR, b Chicago, Ill, Aug 21, 33; m 57; c 2. PHYSICAL CHEMISTRY. Educ: Augustana Col, Ill, BA, 55; Univ Buffalo, PhD(chem), 61. Prof Exp: Res chemist, Mylar Res & Develop Lab, E I du Pont de Nemours & Co, Inc, 60-64, Du Pont de Nemours, Luxembourg, SA, 64-67 & Del, 67-69; mgr mkt develop, Celanese Res Co, 69-72; MGR LONG RANGE PLANNING, EL PASO PROD CO, 72- Concurrent Pos: Adj prof, Univ Tex, Permian Basin, 75- Mem: Commercial Develop Asn; Am Chem Soc. Res: Physical chemistry of polymers; kinetics; process development. Mailing Add: 4322 Springbrook Odessa TX 79762

SWARINGEN, ROY ARCHIBALD, JR, b Winston-Salem, NC, Feb 1, 42; m 69. ORGANIC CHEMISTRY. Educ: Univ NC, Chapel Hill, AB, 64; Univ Ill, Urbana, MS, 66, PhD(org chem), 69. Prof Exp: Res chemist org chem, R J Reynolds Tobacco Co, 69-70; sr develop chemist, 70-74, SECT HEAD DEVELOP RES, BURROUGHS WELLCOME CO, 74- Mem: Am Chem Soc; The Chem Soc. Res: Development research in pharmaceutical chemistry; synthetic organic chemistry; heterocyclic compounds. Mailing Add: Chem Develop Burroughs Wellcome Co Research Triangle Park NC 27709

SWARM, RICHARD LEE, b St Louis, Mo, June 9, 27; m 50; c 2. PATHOLOGY. Educ: Wash Univ, BA, 49, BS & MD, 50; Am Bd Path, dipl. Prof Exp: Intern, Barnes Hosp, St Louis, Mo, 50-51; instr & resident path, Wash Univ & Barnes Hosp, 51-54; pathologist, Nat Cancer Inst, 55-65; assoc prof path, Col Med, Univ Cincinnati, 65-68; DIR DEPT EXP PATH & TOXICOL, RES DIV, HOFFMANN-LA ROCHE, INC, 68- Concurrent Pos: Assoc clin prof path, Columbia Univ, 70- Mem: AAAS; Am Asn Cancer Res; Am Asn Pathologists & Bateriologists; fel Am Soc Clin Path; Col Am Pathologists. Res: Histopathology and toxicology in man and laboratory animals;

morphology of neoplasms and carcinogenesis; radiation injury; transplantation of tissues and tumors; ultrastructure of neoplastic cells. Mailing Add: Dept of Exp Path & Toxicol Res Dir Hoffman-La Roche Inc Nutley NJ 07110

SWART, WILLIAM LEE, b Brethren, Mich, July 13, 30; m 62; c 2. MATHEMATICS. Educ: Cent Mich Univ, BS, 58, MA, 62; Univ Mich, Ann Arbor, EdD(math educ), 69. Prof Exp: Teacher, Mesick Consol Schs, Mich, 58-61 & Livonia Pub Schs, 61-63; instr math, Eastern Mich Univ, 63-65; consult math educ, Genesee Intermediate Sch Dist, Mich, 65-67; from asst prof to assoc prof, 67-74, PROF MATH, CENT MICH UNIV, 74- Res: Learning of elementary mathematics; action research in public schools. Mailing Add: Dept of Math Cent Mich Univ Mt Pleasant MI 48858

SWARTOUT, JOHN ARTHUR, b Madison, Wis, Mar 17, 16; m 43; c 2. RESEARCH ADMINISTRATION, REACTOR ENGINEERING. Educ: Univ Buffalo, BA, 37; Northwestern Univ, PhD(phys chem), 40. Prof Exp: Res chemist, E I du Pont de Nemours & Co, 40-45; res chemist, Oak Ridge Nat Lab, 45-48, dir chem div, 49-50, asst res dir, 50-52; dir homogeneous reactor proj, 51-55, dep res dir, 52-55, dep dir, 55-64; asst gen mgr for reactors, AEC, 64-65; dir technol, 66-68, vpres, 68-74, CONSULT, UNION CARBIDE CORP, 74- Concurrent Pos: Res assoc, Metall Lab, Univ Chicago, 43, Clinton Labs, Tenn, 43-44 & Hanford Eng Works, Wash, 44-45. Mem: AAAS; Am Chem Soc; Am Nuclear Soc; Am Inst Chemists; Indust Res Inst. Res: High polymers; radiochemistry; chemical processes for isolation of fissionable elements; preparation of radioisotopes; development of nuclear reactors; development of nuclear power; environmental systems. Mailing Add: 32 W Beach Lagoon Rd Hilton Head Island SC 29928

SWARTS, DONALD EUGENE, organic chemistry, see 12th edition

SWARTS, ELWYN LOWELL, b Hornell, NY, Feb 26, 29; m 54; c 3. PHYSICAL CHEMISTRY. Educ: Hamilton Col, NY, AB, 49; Brown Univ, PhD, 54. Prof Exp: Mem fac, Alfred Univ, 53-56, res chemist, Knolls Atomic Lab, Gen Elec Co, NY, 56-57, res chemist, Glass Technol Lab, Ohio, 57-59; STAFF SCIENTIST, PPG INDUSTS, INC, 59- Concurrent Pos: Mem, Int Comn Glass. Mem: Am Chem Soc; Am Ceramic Soc. Res: Properties of glass; melting reactions. Mailing Add: Glass Res Ctr PPG Industs Inc Box 11472 Pittsburgh PA 15238

SWARTWOUT, JOSEPH RODOLPH, b Pascagoula, Miss, June 17, 25; m 48; c 5. OBSTETRICS & GYNECOLOGY. Educ: Tulane Univ, MD, 51. Prof Exp: Assoc & fel obstet & gynec, Harvard Med Sch, 53-55; instr, Sch Med, Tulane Univ, 55-60; asst prof, Sch Med, Univ Pittsburgh, 60-61; assoc prof, Sch Med, Emory Univ, 61-66; ASSOC PROF OBSTET & GYNEC, PRITZKER SCH MED, UNIV CHICAGO, 67- Concurrent Pos: Nat Found fel, 53; res assoc, Boston Lying-In Hosp, 53-55; Am Heart Asn fel, 57; prog coordr biomed ctr pop res, Biol Sci Div, Univ Chicago. Mem: Am Col Obstet & Gynec; Soc Gynec Invest; Am Oil Chemists Soc; Am Pub Health Asn. Res: Human reproduction; population; lipid metabolism and methodology; nutrition. Mailing Add: Dept of Obstet & Gynec Pritzker Sch Med Univ Chicago Chicago IL 60637

SWARTZ, BLAIR KINCH, b Detroit, Mich, Nov 5, 32; m 55; c 1. NUMERICAL ANALYSIS. Educ: Antioch Col, BS, 55; Mass Inst Technol, MS, 58; NY Univ, PhD(math), 70. Prof Exp: Asst biol, Sch Med & Dent, Univ Rochester, 51-52; asst chem, Detroit Edison Co, 52-53; asst physics, Antioch Col, 53-54; high sch teacher, 54-55; asst math, Mass Inst Technol, 55-58; res asst, 58, MEM STAFF, LOS ALAMOS SCI LAB, 59- Concurrent Pos: Asst, Am Optical Co, 55-56; lectr, State Univ NY Teachers Col New Paltz, 58 & Univ NMex, 59. Mem: Am Math Soc; Math Asn Am; Soc Indust & Appl Math. Res: Approximation theory. Mailing Add: 172 Paseo Penasco Los Alamos NM 87544

SWARTZ, CHARLES DANA, b Baltimore, Md, July 24, 15; m 49; c 3. PHYSICS. Educ: Johns Hopkins Univ, AB, 38, PhD(physics), 43. Prof Exp: Physicist, Manhattan Proj, SAM Labs, Columbia Univ, 42-46; assoc, Lab Nuclear Studies, Cornell Univ, 46-48; from instr to asst prof physics, Johns Hopkins Univ, 48-56; assoc prof, 56-62, PROF PHYSICS, UNION COL, NY, 62- Concurrent Pos: Fulbright lectr, Univ Ankara, 61-62; vis prof physics, Rensselaer Polytech Inst, 69-70. Mem: Am Phys Soc; Am Asn Physics Teachers. Res: Neutron physics; energy levels of light nuclei; science education; low-temperature physics. Mailing Add: Dept of Physics Union Col Schenectady NY 12308

SWARTZ, CHARLES J, pharmacy, pharmacology, see 12th edition

SWARTZ, CLIFFORD EDWARD, b Niagara Falls, NY, Feb 21, 25; m 46; c 6. EXPERIMENTAL HIGH ENERGY PHYSICS. Educ: Univ Rochester, AB, 45, MS, 46, PhD(physics), 51. Prof Exp: Assoc physicist, Brookhaven Nat Lab, 51-62; assoc prof, 57-67, PROF PHYSICS, STATE UNIV NY STONY BROOK, 67- Concurrent Pos: Ed, Physics Teacher. Mem: Am Phys Soc; Am Asn Physics Teachers. Res: Particle physics; high energy accelerators for nuclear physics research; science curriculum revision, kindergarten through college. Mailing Add: Dept of Physics State Univ of NY Stony Brook NY 11790

SWARTZ, DONALD PERCY, b Preston, Ont, Sept, 12, 21; US citizen; m 44; c 2. OBSTETRICS & GYNECOLOGY. Educ: Univ Western Ont, BA & MD, 51, MSc, 53. Prof Exp: Nat Res Coun Can grant, Univ Western Ont, 52-53; Am Cancer Soc fel, Johns Hopkins Hosp, 56-57, instr obstet & gynec, 57-58; lectr physiol, Univ Western Ont, 58-62; clin prof, Columbia Univ, 62-72, prof obstet & gynec, 72; PROF OBSTET & GYNEC & CHMN DEPT, ALBANY MED COL, 72-; OBSTETRICIAN-GYNECOLOGIST-IN-CHIEF, ALBANY MED CTR, 72- Concurrent Pos: Markle scholar, Univ Western Ont, 58-62; consult obstet & gynec, St Peter's Hosp, 72- Mem: Am Col Obstet & Gynec; Soc Study Reprod; Am Fertility Soc. Res: Fertility control; hormonal contraception; new approaches to pregnancy termination; gynecologic endocrinology. Mailing Add: Dept of Obstet & Gynec Albany Med Ctr Albany NY 12208

SWARTZ, FRANK JOSEPH, b Pittsburgh, Pa, Mar 22, 27; m 46; c 2. ANATOMY. Educ: Western Reserve Univ, BS, 49, MS, 51, PhD(zool), 55. Prof Exp: Asst biol, Western Reserve Univ, 49-52, Nat Cancer Inst fel, 55-56; from asst prof to assoc prof, 56-70, PROF ANAT, SCH MED, UNIV LOUISVILLE, 70- Concurrent Pos: Lectr & USPHS spec fel, Dept Anat, Harvard Med Sch, 69-70. Mem: Am Soc Anat; Am Soc Zoologists. Res: Human anatomy; cellular differentiation and genetic significance of polyploids in mammalian tissues. Mailing Add: Dept of Anat Univ of Louisville Sch of Med Louisville KY 40202

SWARTZ, GEORGE ALLAN, b Scranton, Pa, Dec 9, 30; m 54; c 3. PHYSICS. Educ: Mass Inst Technol, BS, 52; Univ Pa, MS, 54, PhD(physics), 58. Prof Exp: MEM TECH STAFF, DAVID SARNOFF RES CTR, RCA CORP, 58- Mem: Am Phys Soc. Res: Solid state microwave devices, particularly impact avalanche and transit time microwave sources. Mailing Add: David Sarnoff Res Ctr RCA Corp Princeton NJ 08540

SWARTZ, GORDON ELMER, b Buffalo, NY, May 12, 17; m 41; c 2. ZOOLOGY, EMBRYOLOGY. Educ: Univ Buffalo, BA, 39, MA, 41; NY Univ, PhD(biol), 46. Prof Exp: From asst to assoc prof, 39-62, PROF BIOL, STATE UNIV NY BUFFALO, 62- Mem: Fel AAAS; Am Micros Soc; Am Soc Zoologists; Am Asn Anat; NY Acad Sci. Res: Organogenesis; vertebrate experimental embryology; transplantation. Mailing Add: Dept of Biol State Univ of NY Buffalo NY 14214

SWARTZ, GWYNNE BURBANCK, physics, see 12th edition

SWARTZ, HAROLD M, b Chicago, Ill, June 22, 35; m 56; c 2. RADIOBIOLOGY, BIOPHYSICS. Educ: Univ Ill, BS & MD, 59; Univ NC, MS, 62; Georgetown Univ, PhD(biochem), 69. Prof Exp: Fel nuclear med, Walter Reed Army Inst Res, Med Corps, US Army, 62-64, res med officer, 64-68, chief dept biophys, 68-70, chief dept biol chem, 70; assoc prof radiol & biochem, 70-74, PROF RADIOL & BIOCHEM, MED COL WIS, 74-, DIR RADIATION BIOL & BIOPHYS LAB, 70- Mem: AAAS; Radiation Res Soc; Soc Nuclear Med; NY Acad Sci. Res: Free radicals and paramagnetic metal ions in biological systems; oxygen toxicity; radiation biology applied to radiation therapy; carcinogenesis. Mailing Add: Div of Radiol Med Col of Wis Milwaukee WI 53226

SWARTZ, HARRY, b Detroit, Mich, June 21, 11; m 42; c 1. ALLERGY. Educ: Univ Mich, AB, 30, MD, 33. Prof Exp: Clin asst allergy, Med Sch & Clins, NY Univ, 37-40 & Flower & Fifth Ave Hosp, New York, 40-42; asst chief allergy clin, Harlem Hosp, 46-48; prof med & chief allergy dept, NY Polyclin Med Sch & Hosp, 57-72; pres, Health Field Validation Corp, 67-72. Concurrent Pos: Chief allergy dept, Tilton Gen Hosp, Ft Dix, NJ, 42-46; clin asst, Inst Allergy, Roosevelt Hosp, 46-72; indust consult allergy; consult nutrit prod to pharmaceut 77 food indust, consult therapeut cosmetics to cosmetic indust & consult to publ indust; ed, Health Series, Med & Health Reporter, Issues in Current Med Pract & Med Opinion & Rev; sci dir, Ediciones, PLM, SA, 72-; ed-in-chief, Investigacion Medica Int, 73- Mem: AMA; fel Am Col Allergists; fel Am Acad Allergy; Am Geriat Soc; fel Royal Soc Health. Res: Clinical allergy; high protein vegetable source material as a partial answer to world hunger. Mailing Add: Apdo 752 Cuernavaca Morelos Mexico

SWARTZ, HARRY SIP, b Wichita, Kans, July 29, 25; m 47; c 4. PHARMACY ADMINISTRATION, PHARMACY. Educ: Albany Col Pharm, Union Univ, NY, BS, 51; Univ Colo, MS, 54; Univ Iowa, PhD(pharm, pharmaceut admin), 59. Prof Exp: Lab asst pharmaceut chem, Univ Colo, 52-54; instr pharm & pharmaceut admin, Creighton Univ, 54-55; instr pharm & pharmacist, Univ Iowa, 55-59; from asst prof to assoc prof, 59-67, PROF PHARM & PHARMACEUT ADMIN, FERRIS STATE COL, 67- Concurrent Pos: Consult community, hosp & mfg pharm & extended care. Mem: Am Pharmaceut Asn; Am Col Apothecaries. Res: Product development; hospital pharmacy and manufacturing; cosmetic pharmaceuticals; orthopedic and surgical garments. Mailing Add: Sch of Pharm Ferris State Col Big Rapids MI 49307

SWARTZ, HOWARD, b Toronto, Ont, Apr 3, 27; nat US; m 53; c 4. RADIOBIOLOGY. Educ: Univ Man, BSc, 52; Purdue Univ, MSc, 56, PhD(bionucleonics), 60. Prof Exp: Lectr, Sch Pharm, Univ Man, 52-57; from asst prof to assoc prof, 59-67, PROF PHARM, BUTLER UNIV, 67- Mem: Fel AAAS; Am Nuclear Soc; Am Chem Soc; Am Pharmaceut Asn; NY Acad Sci. Res: Bionucleonics; application of radioactive isotopes to biological research, particularly to biochemical and physiological problems; radiopharmaceuticals. Mailing Add: Dept of Bionucleonics Butler Univ Off 306 Indianapolis IN 46207

SWARTZ, JACOB, b Poznan, Poland, June 5, 21; US citizen; m 45; c 4. PSYCHIATRY, PSYCHOANALYSIS. Educ: Boston Univ, BS, 42, MD, 46. Prof Exp: Intern med, Boston City Hosp, 46-47; assoc psychiat, Med Col Ga, 48; resident psychiat, Boston Univ Hosp, 49-52, from instr to assoc prof, 50-69, PROF PSYCHIAT, SCH MED, BOSTON UNIV, 69-, ASSOC DEAN, 71- Concurrent Pos: Mem, Boston Psychoanal Soc & Inst, 57-, chmn bd trustees, 66-70, instr, 72-; dir psychiat clin, Boston Univ Hosp, 60-71; training & supv psychoanalyst, Boston Psychoanal Soc & Inst, 75- Mem: Am Psychiat Asn; Am Psychoanal Asn; Int Psychoanal Asn. Mailing Add: 80 E Concord St Boston MA 02118

SWARTZ, JEAN GIBSON, biochemistry, physiological chemistry, see 12th edition

SWARTZ, JOHN CROUCHER, b Syracuse, NY, Oct 25, 24; m 56; c 3. PHYSICS, MATERIALS SCIENCE. Educ: Yale Univ, BS, 46; Syracuse Univ, MS, 49, PhD, 52. Prof Exp: Res physicist, Consol Vacuum Corp, 52-55; sr scientist, E C Bain Lab Fundamental Res, US Steel Corp, 55-71 & Tyco Labs, Inc, Mass, 72-75; SR SCIENTIST, MOBIL TYCO SOLAR ENERGY CORP, 75- Mem: Am Phys Soc; Metall Soc; Am Asn Crystal Growth. Res: Crystal growth; metallurgy; defects in solids; experimental physics. Mailing Add: Mobil Tyco Solar Energy Corp 16 Hickory Dr Waltham MA 02154

SWARTZ, KARL D, b East Cleveland, Ohio, Sept 1, 38; c 2. SOLID STATE PHYSICS. Educ: Case Inst Technol, BS, 61; Univ Ill, MS, 63, PhD(physics), 66. Prof Exp: Res asst physics, Univ Ill, 62-66; asst prof, Worcester Polytech Inst, 66-70; ASSOC PROF PHYSICS, BALDWIN-WALLACE COL, 70- Res: Ultrasonics; anharmonicity and defect studies in solids; elasticity; mechanics. Mailing Add: Dept of Physics Baldwin-Wallace Col Berea OH 44017

SWARTZ, LESLIE GERARD, b Chicago, Ill, Aug 16, 30; m 58; c 4. PARASITOLOGY. Educ: Univ Ill, BS, 53, MS, 54, PhD(zool), 58. Prof Exp: Assoc prof, 58-70, PROF ZOOL, UNIV ALASKA, FAIRBANKS, 70- Concurrent Pos: Sr res assoc, Rice Univ, 66-67. Mem: AAAS; Am Soc Parasitol. Res: Helminth parasitology, especially ecology; avian, freshwater and general ecology. Mailing Add: Dept of Biol Sci Univ of Alaska Fairbanks AK 99701

SWARTZ, MARC JEROME, b Omaha, Nebr, Oct 31, 31; c 3. ANTHROPOLOGY. Educ: Wash Univ, AB, 52, MA, 53; Harvard Univ, PhD(anthrop), 58. Prof Exp: From instr to asst prof anthrop, Univ Mass, 57-59; asst prof, Univ Chicago, 59-64; assoc prof, Mich State Univ, 64-67, prof anthrop & African studies, 67-69; chmn dept anthrop, 71-74, PROF ANTHROP, UNIV CALIF, SAN DIEGO, 69- Concurrent Pos: NIMH grant, 62-63; Wenner-Gren Found Anthrop Res fel, 66; Soc Sci Res Coun grant, 67; vis prof anthrop, Cornell Univ, 67-68. Mem: Fel African Studies Asn; Am Anthrop Asn; fel Royal Anthrop Inst; Am Ethnol Soc; African Studies Asn. Res: Processual analysis of social and cultural systems, particularly politics and kinship in East Africa at the local level; development of processual theory; motivational aspects of society and culture. Mailing Add: Dept of Anthrop Univ of Calif at San Diego La Jolla CA 92037

SWARTZ, MARJORIE LOUISE, b Indianapolis, Ind, Feb 1, 24. INORGANIC CHEMISTRY. Educ: Butler Univ, BS, 46; Ind Univ, MS, 59. Prof Exp: Res assoc, 46-53, from instr to assoc prof, 53-69, PROF DENT MAT, SCH DENT, IND UNIV, INDIANAPOLIS, 69- Honors & Awards: Souder Award, Int Asn Dent Res, 68. Mem: AAAS; fel Am Col Dent; Am Dent Asn; Int Asn Dent Mat (pres, 65); hon mem Am Asn Women Dentists. Res: Physical and chemical properties of dental cements, resins and amalgams; effect of restorative materials on physical and chemical properties of tooth structure. Mailing Add: Ind Univ Sch of Dent 1121 W Michigan Indianapolis IN 46202

SWARTZ, ROBERT DAVID, b New York, NY, May 17, 37; m 71. URBAN GEOGRAPHY. Educ: Columbia Univ, AB, 59, BSIE, 60, MA, 62; Northwestern Univ, PhD(urban geog, Latin Am), 67. Prof Exp: Int researcher, Sears, Roebuck & Co, 63-67; ASSOC PROF GEOG, WAYNE STATE UNIV, 67- Concurrent Pos: Consult, City of Detroit, 71-74 & City of Warren, Mich, 72. Mem: Asn Am Geogr. Res: Retail location; Latin America. Mailing Add: Dept of Geog Rm 225 State Hall Wayne State Univ Detroit MI 48202

SWARTZ, THOMAS DAVID, organic chemistry, polymer chemistry, deceased

SWARTZ, WALTER H, b Marcellus, Mich, Feb 19, 22; m 43; c 2. DENTISTRY. Educ: Univ Mich, DDS, 45, MS, 47. Prof Exp: Inst dent, 45-51, from asst prof to assoc prof dent prosthesis, 51-61, PROF DENT, SCH DENT, UNIV MICH, ANN ARBOR, 61- Res: Oral Physiology related to lower denture stability; tissue fit of various denture bases; denture construction. Mailing Add: 1120 Chestnut Rd Ann Arbor MI 48104

SWARTZ, WILLIAM JOHN, b Portage, Wis, Aug 9, 20; m 48; c 2. MATHEMATICS. Educ: Mont State Univ, BS, 44; Mass Inst Technol, SM, 49; Iowa State Univ, PhD(math), 55. Prof Exp: Instr math, Mass Inst Technol, 47-48, Mont State Univ, 49-51 & Iowa State Univ, 51-55; from asst prof to assoc prof, 55-62, PROF MATH, MONT STATE UNIV, 62- Mem: Am Math Soc. Res: Differential equations. Mailing Add: Dept of Math Mont State Univ Bozeman MT 59715

SWARTZBERG, LEON, JR, anthropology, see 12th edition

SWARTZENDRUBER, DALE, b Parnell, Iowa, July 6, 25; m 49; c 4. SOIL PHYSICS. Educ: Iowa State Univ, BS, 50, MS, 52, PhD(soil physics), 54. Prof Exp: Asst soil physics, Iowa State Univ, 50-53; instr agr, Goshen Col, 53-54; asst soil scientist, Univ Calif, Los Angeles, 55-56; assoc prof, 56-63, PROF SOIL PHYSICS, PURDUE UNIV, WEST LAFAYETTE, 63- Concurrent Pos: Vis prof, Hebrew Univ Jerusalem & Volcani Inst, Rehovot, Israel, 71; vis scholar, Cambridge Univ, 71. Honors & Awards: Soil Sci Award, Soil Sci Soc Am, 75. Mem: AAAS; Soil Sci Soc Am; fel Am Soc Agron; Am Geophys Union; Int Soc Soil Sci. Res: Physics of soil and water, including water movement through saturated and unsaturated soils and porous media; soil, air, temperature, and structure; soil-water-plant relationships; hydrology and water resources. Mailing Add: Dept of Agron Purdue Univ West Lafayette IN 47906

SWARTZENDRUBER, DONALD CLAIR, b Kalona, Iowa, June 21, 30; m 55; c 2. ZOOLOGY. Educ: Univ Iowa, BA, 55, MS, 58, PhD(zool), 62. Prof Exp: Res asst biol, Oak Ridge Nat Lab, 59-60, res assoc, 62-63; biologist, 63-65, res assoc microbiol, Univ Mich, Ann Arbor, 65-68; SR SCIENTIST, MED DIV, OAK RIDGE ASSOC UNIVS, 68- Mem: AAAS; Electron Micros Soc Am; Am Asn Anatomists; Am Soc Cell Biol; Reticuloendothelial Soc. Res: Electron microscopic studies of lymphatic tissues; virus-cell relationships; elemental analysis of normal and malignant cells and cellular components. Mailing Add: Med Div Oak Ridge Assoc Univs PO Box 117 Oak Ridge TN 37830

SWARTZENDRUBER, LYDON JAMES, b Wellman, Iowa, Aug 8, 33; m 49. PHYSICS. Educ: Iowa State Univ, BS, 57; Univ Md, PhD(physics), 68. Prof Exp: PHYSICIST, NAT BUR STAND, 60- Mem: AAAS; Am Phys Soc; Am Soc Testing & Mat; Am Inst Mining, Metall & Petrol Engrs. Res: Solid state physics; semiconductors; magnetism; metallurgy. Mailing Add: Nat Bur of Stand Washington DC 20234

SWARTZENTRUBER, PAUL EDWIN, b Lagrange, Ind, Apr 23, 31; m 55; c 3. ORGANIC CHEMISTRY. Educ: Goshen Col, BA, 53; Univ Minn, Minneapolis, MS, 55; Univ Mo, PhD(org chem), 61. Prof Exp: Res chemist, Nat Cancer Inst, 56-59; asst ed org indexing dept, 61-63, assoc ed, 63-64, head org indexing dept, 64-71, mgr phys & inorg indexing dept, 71-72, mgr chem substance handling dept, 72-73, MGR CHEM TECHNOL, CHEM ABSTR SERV, 73- Mem: AAAS; Am Chem Soc. Res: Chemical nomenclature; computer storage and retrieval of chemical information; abstracting and indexing of chemical literature. Mailing Add: Chem Abstr Serv Columbus OH 43210

SWARTZWELDER, JOHN CLYDE, b Lynn, Mass, Apr 1, 11; m 64; c 2. MEDICAL PARASITOLOGY, TROPICAL PUBLIC HEALTH. Educ: Univ Mass, BS, 33; Tulane Univ, MS, 34, PhD(med protozool), 37; Am Bd Med Microbiol, dipl, 64. Prof Exp: Asst med parasitol, Sch Med, Tulane Univ, 33-37; from instr to prof med parasitol, Sch Med, 37-75, head dept trop med & med parasitol, 60-75, educ dir, Interam Training Prog Trop Med, 59-69, assoc dir, Int Ctr Med Res & Training, 61-69, dir, Interam Training Prog Trop Med & Int Ctr Med Res & Training, 69-75, EMER PROF MED PARASITOL, LA STATE UNIV MED CTR, NEW ORLEANS, 75- Concurrent Pos: Scientist, Charity Hosp, New Orleans, 38-; consult, Vet Admin Hosp, 49- Mem: Am Soc Trop Med & Hyg (vpres); Am Soc Parasitol; hon mem Mex Soc Parasitol. Res: Amebiasis; Chagas' disease; anthelminthics; research training in tropical medicine; medical education. Mailing Add: Dept Trop Med & Med Parasitol La State Univ Med Ctr New Orleans LA 70112

SWARZENSKI, WOLFGANG V, geology, see 12th edition

SWATEK, FRANK EDWARD, b Oklahoma City, Okla, June 4, 29; m 51; c 5. MICROBIOLOGY, MYCOLOGY. Educ: San Diego State Col, BS, 51; Univ Calif, Los Angeles, MA, 55, PhD(microbiol), 56. Prof Exp: From instr to assoc prof, 56-63, PROF MICROBIOL, CALIF STATE UNIV, LONG BEACH, 63-, CHMN DEPT, 60- Concurrent Pos: Consult, Dept Allergy & Dermat, Long Beach Vet Admin Hosp, 60-, Douglas Aircraft Co, Inc, 61- & Hyland Lab, 68-; lectr, Sch Med, Univ Southern Calif, 62- Honors & Awards: Carski Award, Am Soc Microbiol, 74. Mem: Am Soc Microbiol; Am Pub Health Asn; NY Acad Sci; fel Royal Soc Health; Sigma Xi. Res: Ecology and experimental pathology of deep mycoses, especially Coccidioides, Cryptococcus and Dermatophytes; industrial work on fungus deterioration of man-made products. Mailing Add: Dept of Microbiol Calif State Univ Long Beach CA 90805

SWAUGER, JAMES LEE, b West Newton, Pa, Nov 1, 13; m 43; c 3. ANTHROPOLOGY. Educ: Univ Pittsburgh, BS, 41, MLitt, 47. Hon Degrees: DSc, Waynesburg Col, 57. Prof Exp: Lab asst, Sect Archaeol & Ethnol, 35-41, asst preparator vert paleont, 41-42, custodian, Sect Archaeol & Ethnol, 46-48, cur, Sect Man, 49-63, asst dir, 55-64, ASSOC DIR, CARNEGIE MUS, 64- Concurrent Pos: Lectr, Univ Pittsburgh, 48-50, adj res prof anthrop, 71-; lectr, Duquesne Univ, 50-51; consult & lectr, Human Rels Inst, Pa State Univ, 51; dir, Ashdod Excavation Proj, 64-; mem, Pa Comn for Achievement Human Potential, 67- Mem: AAAS; Am Anthrop Asn; Am Asn Mus; Soc Am Archaeol; Int Coun Mus. Res: American Indian rock art; Philistine problems and megalith studies in Palestine and Arabia. Mailing Add: Carnegie Mus of Natural Hist Pittsburgh PA 15213

SWEARINGEN, JOHN JACOB, physiology, see 12th edition

SWEARINGIN, MARVIN LAVERNE, b Hamburg, Ill, Jan 23, 31; m 50; c 3. AGRONOMY. Educ: Univ Mo-Columbia, BS, 56, MS, 57; Ore State Univ, PhD(agron), 62. Prof Exp: Asst crops teaching, Univ Mo, 56-57; instr farm crops, Ore State Univ, 57-61; from asst prof to assoc prof, 61-75, PROF AGRON, PURDUE UNIV, WEST LAFAYETTE, 75- Concurrent Pos: Soybean res consult, Purdue Univ-Brazil Proj, Brazil, 67-69 & AID, Brazil, 71- Mem: Am Soc Agron; Crop Sci Soc Am. Res: Crop management systems for corn, soybeans and small grains. Mailing Add: Dept of Agron Purdue Univ West Lafayette IN 47906

SWEAT, CALVIN WAYNE, statistics, see 12th edition

SWEAT, FLOYD WALTER, b Salt Lake City, Utah, July 21, 41; m 65; c 2. BIOCHEMISTRY. Educ: Univ Utah, BS, 64, PhD(org chem), 68. Prof Exp: NIH fel, Harvard Univ, 68-70; ASST PROF BIOCHEM, UNIV UTAH, 70- Mem: AAAS; Am Chem Soc; The Chem Soc. Res: Enzyme purification and characterization; organic reaction mechanisms. Mailing Add: Dept of Biol Chem Univ of Utah Med Ctr Salt Lake City UT 84112

SWEAT, MAX LEROY, b Park City, Utah, July 8, 14; m 51; c 4. BIOCHEMISTRY. Educ: Univ Utah, BS, 38, PhD(biochem), 49; Utah State Univ, MS, 40. Prof Exp: Asst chemist, State Dept Agr, Utah, 40-41; chemist, State Dept Health, 41-45; asst biochem, Univ Utah, 45-49; biochemist, Agr Res Admin, USDA, 49; biochemist endocrinol sect, NIH, 49-51; asst res prof pharmacol, Univ Utah, 51-52; asst prof physiol, Med Sch, Western Reserve Univ, 52-56; from asst res prof to res prof obstet & gynec, 56-74, RES PROF ANAT, COL MED, UNIV UTAH, 74- Mem: AAAS; Endocrine Soc; Am Soc Biol Chemists. Res: Interrelationships of hormones and coenzymes and intracellular sites of steroid hormone synthesis; biosynthesis, metabolism and mechanism of action of steroid hormones. Mailing Add: Dept of Anat Univ of Utah Col of Med Salt Lake City UT 84132

SWEAT, ROBERT LEE, b Lamar, Colo, June 8, 31; m 53; c 2. VETERINARY MEDICINE, VIROLOGY. Educ: Colo State Univ, BS, 54, DVM, 56; Univ Nebr-Lincoln, MS, 62, PhD(med sci), 66. Prof Exp: Instr vet sci, Univ Nebr-Lincoln, 58-66, assoc prof, 66; vet virologist, Norden Labs, Inc, Nebr, 67; assoc res prof vet sci, Univ Idaho, 68-70- VET VIROLOGIST, FT DODGE LABS INC, 70- Concurrent Pos: Vet rep, Nebr State Bd Health, 66-67. Mem: AAAS; Am Vet Med Asn; US Animal Health Asn; Am Pub Health Asn; Conf Res Workers Animal Dis. Res: Veterinary science with emphasis on viral diseases and zoonosis. Mailing Add: PO Box 1082 Ft Dodge IA 50501

SWEDBERG, KENNETH C, b Brainerd, Minn, Apr 14, 30; m 58; c 2. PLANT ECOLOGY. Educ: St Cloud State Col, BS, 52; Univ Minn, MS, 56; Ore State Univ, PhD(plant ecol), 61. Prof Exp: Instr biol, Moorhead State Col, 56-58; asst prof, Wis State Univ-Stevens Point, 60-62; assoc prof, 62-69, PROF BIOL, EASTERN WASH STATE COL, 69- Mem: AAAS; Ecol Soc Am; Weed Sci Soc Am; Am Inst Biol Sci; Sigma Xi. Res: Plant synecology and autecology; experimental ecology of annual plants. Mailing Add: Dept of Biol Eastern Wash State Col Cheney WA 99004

SWEDES, JEAN SUSANNE, b Santa Monica, Calif, Apr 6, 46. MOLECULAR BIOLOGY. Educ: Occidental Col, BA, 67; Univ Calif, Los Angeles, PhD(molecular biol), 73. Prof Exp: Researcher biochem, Univ Calif, Los Angeles, 74-75; NIH FEL, UNIV CALIF, IRVINE, 75- Mem: Am Soc Microbiol. Res: Roles of adenine nucleotides in metabolic regulation; regulation of macromolecular biosynthesis especially regulation due to changes in energy supply; biochemical changes occurring in cells during nutrient starvations. Mailing Add: Dept of Molecular Biol & Biochem Univ of Calif Irvine CA 92717

SWEDISH, FRANK, b Scranton, Pa, Feb 17, 16; m 44; c 2. ORGANIC CHEMISTRY. Educ: Kent State Univ, BS, 38; Ohio State Univ, MS, 39. Prof Exp: Res chemist, B F Goodrich Co, 39-45 & Marathon Corp, 45-50; res chemist, 50-58, lab supvr, 58-64, RES SPECIALIST, MINN MINING & MFG CO, 64- Mem: Am Chem Soc; Tech Asn Pulp & Paper Indust. Res: Accelerators of vulcanization and antioxidants in rubber chemistry; pulp and paper chemistry; pressure sensitive tapes. Mailing Add: 3M Co Res Labs 2501 Hudson Rd 3M Ctr St Paul MN 55119

SWEDLUND, ALAN CHARLES, b Sacramento, Calif, Jan 21, 43; m 66; c 1. BIOLOGICAL ANTHROPOLOGY. Educ: Univ Colo, Ba, 66, MA, 68, PhD(anthrop), 70. Prof Exp: Asst prof anthrop, Prescott Col, 70-74; VIS ASST PROF ANTHROP, UNIV MASS, AMHERST, 74- Mem: AAAS; Am Asn Phys Anthrop; Am Soc Human Genetics; Am Eugenics Soc. Res: Demographic and human population genetics; paleoanthropology and paleodemography; osteology; demographic factors influencing non-random gene dispersion. Mailing Add: Dept of Anthrop Univ of Mass Amherst MA 01002

SWEEDLER, MOSS EISENBERG, b Brooklyn, NY, Apr 29, 42. METHEMATICS. Educ: Mass Inst Technol, BS, 63, PhD(math), 65. Prof Exp: Instr math, Mass Inst Technol, 65-67; from asst prof to assoc prof, 67-74, PROF MATH, CORNELL UNIV, 74- Mem: Am Math Soc. Res: Algebra. Mailing Add: Dept of Math White Hall Cornell Univ Ithaca NY 14850

SWEELEY, CHARLES CRAWFORD, b Williamsport, Pa, Apr 15, 30; m 50; c 2. BIOCHEMISTRY, ORGANIC CHEMISTRY. Educ: Univ Pa, BS, 52; Univ Ill, PhD(chem), 55. Prof Exp: Chemist, Nat Heart Inst, 55-60; asst res prof biochem, Univ Pittsburgh, 60-63, from assoc prof to prof, 63-68; PROF BIOCHEM, MICH STATE UNIV, 68- Concurrent Pos: Mem comt probs lipid anal, Nat Heart Inst, 58-59; consult, LKB Instruments, 65-71, Med Chem Study Sect, USPHS, 67-71 & Upjohn Co, 68-72; Guggenheim fel, Royal Vet Col, Stockholm, Sweden, 71. Mem: Am Chem Soc; Am Soc Biol Chem; Brit Biochem Soc. Res: Chemistry and metabolism of sphingolipids; sphingolipidoses; biochemistry of lysosomal hydrolases; analytical biochemistry; computer applications in gas chromatography and mass spectrometry; biochemistry of complex lipids and hormones of invertebrates. Mailing Add: Dept of Biochem Mich State Univ East Lansing MI 48823

SWEENEY, BEATRICE MARCY, b Boston, Mass, Aug 11, 14; m; c 4. BIOLOGICAL RHYTHMS. Educ: Smith Col, AB, 36; Radcliffe Col, PhD(biol), 42. Prof Exp: Lab asst endocrinol, Mayo Clin, 42; fel, Mayo Found, Univ Minn, 42-43; jr res biologist, Scripps Inst, Calif, 47-55, asst res biologist, 55-60, assoc res biologist, 60-61; res staff biologist, Yale Univ, 61-62, lectr biol, 62-67; lectr, 67-69, assoc prof, 69-71, PROF BIOL, UNIV CALIF, SANTA BARBARA, 71- Concurrent Pos: Consult, Monroe Labs, 59; mem, Nat Comt Photobiol, Nat Res Coun, 72-75. Mem: Fel AAAS; Am Soc Plant Physiol (secy-treas, 75-76); Soc Gen Physiol; Phycol Soc Am; Am Soc Cell Biol. Res: Nutrition; photosynthesis; bioluminescence; diurnal rhythms in marine dinoflagellates. Mailing Add: Dept of Biol Sci Univ of Calif Santa Barbara CA 93106

SWEENEY, DARYL CHARLES, b Oakland, Calif, Jan 21, 36; m 59; c 1. INVERTEBRATE PHYSIOLOGY, NEUROCHEMISTRY. Educ: Univ Calif, Berkeley, AB, 58; Harvard Univ, AM, 59, PhD(biol), 63. Prof Exp: Instr biol, Yale Univ, 63-65; from asst prof to assoc prof zool, 65-73, PROF ZOOL, UNIV ILL, URBANA-CHAMPAIGN, 73- Concurrent Pos: Consult, Nat Inst Neurol Dis & Stoke, 66-69. Mem: Am Soc Zool. Res: Neurochemistry and the behavior of invertebrate physiology. Mailing Add: Dept of Zool Univ of Ill at Urbana-Champaign Urbana IL 61801

SWEENEY, DAVID MICHAEL CURRIE, physics, see 12th edition

SWEENEY, GEORGE DOUGLAS, b Durban, SAfrica, Dec 21, 34; m 60; c 3. PHARMACOLOGY. Educ: Univ Cape Town, MB, ChB, 58, PhD(biochem), 63. Prof Exp: Sr lectr physiol, Univ Cape Town, 64-68; Ont fel, Col Physicians & Surgeons, Columbia Univ, 68-69; ASSOC PROF MED, McMASTER UNIV, 69- Mem: NY Acad Sci; Pharmacol Soc Can; Can Physiol Soc. Res: Hemoprotein synthesis and regulation; porphyrin metabolism; drug-mediated enzyme induction and liver injury. Mailing Add: Dept of Med McMaster Univ Hamilton ON Can

SWEENEY, HAROLD A, b Bayonne, NJ, Sept 7, 19; m 43; c 5. ANALYTICAL CHEMISTRY. Educ: Columbia Univ, AB, 41, MA, 42. Prof Exp: Chemist, Am Cyanamid Co, Conn, 46-53, supvr cent anal lab, La, 53-55; supvr wet chem lab, 55-62, MGR CHEM & INSTRUMENTAL LAB GROUP, KOPPERS CO, 62- Mem: Am Chem Soc. Res: Analysis of organic materials, paints, dyes, plastics, tar products, fuels and other petroleum products; catalysts; agricultural chemicals; surfactants; wafer. Mailing Add: Koppers Co Inc 440 College Park Dr Monroeville PA 15146

SWEENEY, MARTIN JOSEPH, biochemistry, organic chemistry, see 12th edition

SWEENEY, MARY ANN, b Hagerstown, Md, Sept 25, 45. PLASMA PHYSICS, ASTROPHYSICS. Educ: Mt Holyoke Col, BA, 67; Columbia Univ, MPhil, 73, PhD(astron), 74. Prof Exp: Instr astron, Fairleigh Dickinson Univ & William Paterson Col NJ, 72-73; fel plasma physics & fusion res, 74-76, STAFF MEM, SANDIA LABS, 76- Mem: Am Astron Soc; Am Phys Soc. Res: Target design calculations for electron and ion-beam fusion using hydrocodes; evolutionary models of degenerate dwarf stars and observational interpretation of these stars; equation of state formulations in high pressure regimes. Mailing Add: Sandia Lab Plasma Theory Div PO Box 5800 Albuquerque NM 87115

SWEENEY, MICHAEL ANTHONY, b Los Angeles, Calif, Dec 5, 31; m 54; c 3. PHYSICAL CHEMISTRY. Educ: Loyola Univ, Calif, BS, 53; Univ Calif, Berkeley, MS, 55, PhD(chem), 62. Prof Exp: Res chemist, Chevron Res Co, 61-66; asst prof, 66-72, ASSOC PROF CHEM, UNIV SANTA CLARA, 72- Concurrent Pos: Instr, Univ Exten, Univ Calif, Berkeley, 65-66; consult, USPHS, 68-71; vis res chemist, UK Atomic Energy Res Estab, Harwell, 72-73. Mem: AAAS; Am Chem Soc. Res: Radiation chemistry; chemical evolution. Mailing Add: Dept of Chem Univ of Santa Clara Santa Clara CA 95053

SWEENEY, MICHAEL JOSEPH, b Philadelphia, Pa, Jan 31, 39; m 62; c 3. IMMUNOLOGY, IMMUNOCHEMISTRY. Educ: Philadelphia Col Pharm & Sci, BSc, 66, BSc, 67; Temple Univ, PhD(microbiol, immunol), 71. Prof Exp: Fel immunol, Sch Med, Temple Univ, 71-72; ASST PROF IMMUNOL, FLA TECHNOL UNIV, 71- Mem: Am Soc Microbiol; Sigma Xi. Res: Studies of T and B cell populations in peripheral blood of immunosuppressed patients; immunologic, genetic and biochemical associations to the tumorous head phenotype in Drosophila melanogaster. Mailing Add: Dept of Biol Sci Fla Technol Univ Orlando FL 32816

SWEENEY, RICHARD F, b New York, NY, Aug 7, 21; m 56; c 2. ORGANIC CHEMISTRY. Educ: Queens Col, BS, 42, MS, NY Univ, 48, PhD(chem), 56. Prof Exp: Instr chem, NY Univ, 55-56; res chemist, Chem Div, 56-59, sr res chemist, 59-64, sr scientist, 64-69, tech supvr org fluorine chem, Indust Chem Div, 69-72, SUPVR SPEC CHEM DIV, BUFFALO RES LAB, ALLIED CHEM CORP, 72- Mem: Am Chem Soc; Am Assoc Textile Chem Colorists; Soc Petroleum Engrs. Res: Organic fluorine chemistry; fluoropolymers; coatings; surface active agents; interfacial phenomena relating to crude oil recovery. Mailing Add: 631 Stolle Rd Elma NY 14059

SWEENEY, ROBERT ANDERSON, b Freeport, NY, Oct 11, 40; m 63; c 2. PHYCOLOGY, LIMNOLOGY. Educ: State Univ Col Albany, BS, 62; Ohio State Univ, MS, 64, PhD(natural resources), 66. Prof Exp: From asst prof to assoc prof, 66-68, PROF BIOL, STATE UNIV NY COL BUFFALO, 68-, DIR, GREAT LAKES LAB, 67- Concurrent Pos: Consult, NY State Dept Environ Conserv, 70-; consult, US Army Corps Engrs, 71-; Dept Housing & Urban Develop grant, State Univ NY Buffalo, 71-72, US Army Corps Engrs grant, 71-, Environ Protection Agency grant, 72- Mem: AAAS; Int Asn Great Lakes Res; Phycol Soc Am; Am Inst Biol Sci; Am Soc Limnol & Oceanog. Res: Evaluation and solution of water pollution and eutrophication problems in the Great Lakes and their tributaries; interaction of pesticides and algae. Mailing Add: Great Lakes Lab State Univ of NY Col Buffalo NY 14222

SWEENEY, ROBERT MILTON, b Blue Island, Ill, Apr 27, 41; m 68; c 1. DEVELOPMENTAL BIOLOGY, ZOOLOGY. Educ: St Joseph's Col, Ind, BS, 63; Univ Ill, Urbana, MS, 65, PhD(zool), 68. Prof Exp: Instr zool, Univ Ill, Urbana, 68-69; asst prof biol, Elmhurst Col, 69-73; OPERS MGR, MACMILLAN SCI CO , TURTOX, 74- Mem: AAAS; Am Soc Zool; Soc Develop Biol; Am Inst Biol Sci. Res: Embryonic origin of ribs and ventrolateral muscles in Avian embryos; effects of nitrogen mustard on rib and muscle development; effect of tantalum foil blocks on somite migration; wing development. Mailing Add: 119 S Summit Ave Villa Park IL 60181

SWEENEY, THOMAS RICHARD, b Albany, NY, Sept 21, 14; m 41; c 2. MEDICINAL CHEMISTRY. Educ: Univ Md, BS, 37, PhD(org chem), 45. Prof Exp: Chemist, Briggs Filtration Co, DC, 40-41; chemist, NIH, Md, 41-47; res chemist, Univ Md, 47-50, US Naval Res Lab, 50-59; res chemist, 59-64, chief dept org chem, 64-69, DEP DIR DIV MED CHEM, WALTER REED ARMY INST RES, 69- Mem: Am Chem Soc. Res: Antiradiation and antiparasific agents. Mailing Add: Div of Medicinal Chem Walter Reed Army Inst Res washington DC 20012

SWEENEY, WILLIAM ALAN, b Can, Sept 12, 26; nat US; m 53; c 3. PETROLEUM CHEMISTRY. Educ: Univ BC, BASc, 49; Univ Wash, Seattle, PhD, 54. Prof Exp: Chemist, Can Industs, Ltd, 49-50; asst to vpres, Chevron Res Co, 75-76; SR RES ASSOC, CHEVRON RES CO, STANDARD OIL CO, CALIF, 54- Mem: Am Chem Soc. Res: Detergents; petrochemical processing; condensation polymers; biodegradability; desulfurization. Mailing Add: Chevron Res Co 576 Standard Ave Richmond CA 94802

SWEENEY, WILLIAM JOHN, b Oak Park, Ill, July 15, 40; m 64. MATHEMATICS. Educ: Univ Notre Dame, AB, 62; Stanford Univ, MS, 64, PhD(math), 66. Prof Exp: Instr math, Stanford Univ, 66-67; asst prof, Princeton Univ, 67-71; ASSOC PROF MATH, RUTGERS UNIV, NEW BRUNSWICK, 71- Mem: Math Asn Am; Am Math Soc. Res: Over-determined systems of linear partial differential equations. Mailing Add: Dept of Math Rutgers Univ New Brunswick NJ 08903

SWEENEY, WILLIAM MICHAEL, b Ashley, Pa, May 22, 21; m 51; c 3. INTERNAL MEDICINE. Educ: Univ Notre Dame, BS, 43; St Louis Univ, MD, 47. Prof Exp: Intern & resident internal med, Hosps, St Louis Univ, 47-51, Neilson fel biochem, Sch Med, 50-51, instr, 51-53; assoc dir clin pharmacol, 55-65, DIR MED RES, LEDERLE LABS, DIV AM CYANAMIDE CO, 65-; DIR RADIOISOTOPE LAB, BERGEN PINES COUNTY HOSP, PARAMUS, NJ, 61- Mem: AMA; Am Fedn Clin Res. Res: Medical research administration. Mailing Add: Lederle Labs Div Am Cyanamid Co Pearl River NY 10965

SWEENEY, WILLIAM MORTIMER, b Brooklyn, NY, Aug 4, 23; m 52; c 3. PETROLEUM CHEMISTRY. Educ: St John's Univ, BS, 43; Fordham Univ, MS, 47; Univ Colo, PhD(org chem), 52. Prof Exp: Chemist, Transformer Div, Gen Elec Co, Mass, 47-49; SR RES CHEMIST, BEACON RES LABS, TEXACO, INC, 52- Mem: Sigma Xi. Res: Organometallics, principally ferrocene; ester based synthetic luoricants for jet engines; fluorine chemistry; additives for gasoline and diesel fuels; pour depressants fuel oils; polymers. Mailing Add: Beacon Res Labs Texaco Inc Beacon NY 12508

SWEENEY, WILLIAM VICTOR, b Cleveland, Ohio, Jan 31, 47; m 68; c 1. BIOPHYSICS. Educ: Knox Col, BA, 68; Univ Iowa, MS, 70, PhD(chem), 73. Prof Exp: NIH fel biochem, Univ Calif, Berkeley, 73-75; ASST PROF CHEM, HUNTER COL, CITY UNIV NEW YORK, 75- Res: Physical properties of iron-sulfur proteins; nuclear magnetic resonance; electron paramagnetic resonance. Mailing Add: Dept of Chem Hunter Col 695 Park Ave New York NY 10021

SWEENY, ARTHUR, JR, b New York, NY, Jan 6, 07; m 32; c 2. ORGANIC CHEMISTRY. Educ: Harvard Univ, AB, 28; Columbia Univ, MA, 32. Prof Exp: Tutor chem, Hunter Col, 30-32, from instr to prof, 32-68; prof org chem, 68-74, EMER PROF ORG CHEM, LEHMAN COL, 74- Concurrent Pos: NY Div, Am Cancer Soc grant, 60- Mem: AAAS; Am Chem Soc. Res: Synthesis of phenanthrene derivatives and antitumor agents. Mailing Add: Dept of Org Chem Herbert H Lehman Col Bronx NY 10468

SWEENY, DANIEL MICHAEL, b Rockville Center, NY, Sept 25, 30; m 60; c 5. INORGANIC CHEMISTRY. Educ: Col Holy Cross, BSc, 52; Univ Notre Dame, PhD(chem), 55. Prof Exp: Res chemist, E I du Pont de Nemours & Co, 55-57; PROF INORG CHEM, BELLARMINE COL, KY, 57- Mem: Am Chem Soc; Mineral Soc Am; Soc Appl Spectros. Res: Physical properties and synthesis of coordination compounds; infrared spectroscopy; structural inorganic chemistry; interpretive spectroscopy. Mailing Add: Dept of Chem Bellarmine Col Louisville KY 40205

SWEENY, HALE CATERSON, b Anderson, SC, Mar 31, 25; m 48; c 3. MATHEMATICAL STATISTICS. Educ: Clemson Col, BME, 49; Va Polytech Inst, MS, 52, PhD, 56. Prof Exp: Design engr, Hunt Mach Works, 49-50; indust engr, Easfman Kodak Co, 51-52; instr indust eng, Va Polytech Inst, 52-53, asst prof statist, 53-56; res statistician, Atlantic Ref Co, 56-59; consult, 59-60; sr res statisfician, Res Triangle Inst, 60-64, mgr spec res, 64-72; HEAD BIOSTATIST DEPT, BURROUGHS WELLCOME CO, 72- Mem: Am Soc Mech Eng; Am Statist Asn; Inst Math Statist; Biomet Soc. Res: Development of statistical methodology application to production; chemical and clinical research; design of experiments; design of medical and veterinary clinical trials. Mailing Add: 3500 Cambridge Pd Durham NC 27707

SWEENY, JAMES GILBERT, b Philadelphia, Pa, Jan 18, 44. NATURAL PRODUCTS CHEMISTRY. Educ: Eckerd Col, BS, 65; Yale Univ, PhD(org chem), 69. Prof Exp: Fel chem, Yale Univ, 69-71, R Russell Agr Res Ctr, 71-72, Univ Glasgow, 72-73 & Univ Va, 73-74; RES SCIENTIST, COCA-COLA CO, 74- Mem: Am Chem Soc; The Chem Soc; Phytochem Soc NAm. Res: Synthesis of natural colorants and anthocyanins; synthesis and biosynthesis of monoterpenes and other volatile flavor compounds. Mailing Add: Corp Res Coca-Cola Co Atlanta GA 30301

SWEENY. KEITH HOLCOMB, b Tacoma, wash, Feb 6, 20; m 45; c 3. PHYSICAL CHEMISTRY. Educ: Univ Wash, Seattle, BS, 46; Ore State Univ, PhD(phys chem), 50. Prof Exp: Oceanog chemist, US Fish & Wild Life Serv, 41; res chemist, Kalunite, Inc, 42-45; instr chem, Univ Utah, 45-46; asst prin chemist, Aerojet-Gen Corp Div, 50-63, staff scientist, Space-Gen Corp Div, 63-69, indust wastes specialist, Envirogenics Co Div, Gen Tire & Rubber Co, El Monte, 69-75; MGR, WASTEWATER TREATMENT RES, ENVIROGENICS SYSTS CO, 75- Concurrent Pos: Consult rheol, Coast Proseal Div, Essex Chem, 72- Mem: Am Chem Soc; Soc Rheol; Sigma Xi. Res: Rheology of suspensions; fine particle technology; polymer chemistry and physics; aerosol physics; physical chemistry of rocket propellants; chemistry of aluminum; chemistry of pesticides; industrial waste treatment. Mailing Add: 2413 E Evergreen Ave West Covina CA 91791

SWEENY, WILFRED, organic chemistry, see 12th edition

SWEET, ARTHUR THOMAS, JR, b Salisbury, NC, Jan 19, 20; m 43; c 2. POLYMER CHEMISTRY. Educ: Univ NC, BS, 41; Ohio State Univ, PhD(chem). 48. Prof Exp: Staff chemist, Uranium Isotope Prod Dept, Tenn Eastman Corp, 44-46; res chemist, Nylon Res Div, 48-54, Dacron Res Div, 54-57 & Textile Fibers Patent Div, 57-74, PATENT ASSOC, TEXTILE FIBERS PATENT LIAISON DIV, E I DU PONT DE NEMOURS & CO, 74- Mem: Am Chem Soc. Res: Constitution of Grignard reagent; chemical characteristics of synthetic fibers; patent management. Mailing Add: Textile Fibers Patent Div E I du Pont de Nemours & Co Wilmington DE 19898

SWEET, BENJAMIN HERSH, b Boston, Mass, Dec 14, 24; m 56; c 3. VIROLOGY, IMMUNOLOGY. Educ: Tulane Univ, BS, 43; Boston Univ, MA, 48, PhD(med sci), 51. Prof Exp: Res assoc virol, Res Found, Children's Hosp, Cincinnati, 51-54; asst prof microbiol, Sch Med, Univ Md, 54-59; sr investr virol, Merck Sharp & Dohme Res Labs, 59-64; dir & mgr res & develop, Flow Labs, Inc, Md, 64-66; assoc dir life sci div, Gulf South Res Inst, 66-75; MGR, QUAL ASSURANCE BIOLABS, CUTTER LABS, 75- Mem: Am Soc Microbiol; Am Soc Trop Med & Hyg; Tissue Cult Asn; Soc Exp Biol & Med. Res: Arthropod borne, respiratory, oncogenic, latent and vaccine development viruses; viral immunology; ecology and zoonoses; immunology-adjuvants; water pollution; cell biology; quality assurance. Mailing Add: Cutter Labs Fourth & Parker St Berkeley CA 94710

SWEET, CHARLES EDWARD, b Elgin, Tex, Dec 27, 33; m 55; c 4. BACTERIOLOGY, MYCOLOGY. Educ: Univ Tex, BA, 55, MA, 63; Univ NC, MPH, 67, DPH(parasitol), 69. Prof Exp: Jr bacteriologist, Br Lab, NMex State Health Dept, 60-61; bacteriologist, 62, bacteriologist. NMex State Health Dept, 63, dir, 63-66, spec proj dir dept, 69-70, asst dir, Lab Servs, Tex State Health Dept, 70-73, DIR LAB, TEX STATE DEPT HEALTH, 70- Res: Serological means of identifying

Candida species and other yeastlike fungi. Mailing Add: Lab Sect Tex State Health Dept 1100 W 49th St Austin TX 78756

SWEET, CHARLES SAMUEL, b Cambridge, Mass, Apr 6, 42; m 68; c 2. PHARMACOLOGY. Educ: Northeastern Univ, BS, 66, MS, 68; Univ Iowa, PhD(pharmacol), 71. Prof Exp: Res asst pharmacol, Warner-Lambert Res Inst, NJ, 66-68; fel, Col Pharm, Northeastern Univ, 66-68; fel, Col Med, Univ Iowa, 68-71; fel res, Cleveland Clin Educ Found, 71-72; res fel, 72-75, SR RES FEL, MERCK INST THERAPEUT RES, 75- Mem: Coun Thrombosis, Am Heart Asn; Am Soc Pharmacol & Exp Therapeut. Res: Renin-angiotensin system in pathogenesis of experimental hypertension; participation of central nervous system in development and maintenance; action of antihypertensive drugs, particularly as they apply to known causes of hypertension. Mailing Add: Dept of Pharmacol Merck Inst for Therapeut Res West Point PA 19486

SWEET, DAVID PAUL, b Dixon, Mar 24, 48; m 68. ANALYTICAL CHEMISTRY. Educ: Bradley Univ, BA, 70; Univ Colo, PhD(anal chem), 74. Prof Exp: ANAL CHEMIST, DIV SYNTEX, ARAPAHOE CHEM INC, 74- Mem: Am Chem Soc; Am Soc Mass Spectrometry. Res: Chromatographic separations and trace analysis, especially using combined vapor phase chromatography-mass spectrometry and liquid chromatography-mass spectrometry.

SWEET, FREDERICK, b New York, NY, May 15, 38; m 62; c 4. BIOCHEMISTRY, ORGANIC CHEMISTRY. Educ: Brooklyn Col, BS, 60; Univ Alta, PhD(org chem), 68. Prof Exp: Substitute instr chem, Brooklyn Col, 60-62; instr, Bronx Community Col, 62-64; NIH res fel nucleoside chem, Sloan-Kettering Inst Cancer Res, 68-70; res assoc reproductive biochem, Univ Kans Med Ctr, Kansas City, 70-71; ASST PROF REPRODUCTIVE BIOCHEM, SCH MED, WASH UNIV, 71- Concurrent Pos: Lectr, Bronx Community Col, 68-70; vis asst prof biol sci, Southern Ill Univ, Edwardsville, 73-75; NIH res career develop award, 75. Mem: AAAS; Am Chem Soc; Chem Inst Can. Res: Reproductive biochemistry, mechanism of steroid action and metabolism; synthesis of affinity-labeling steroids; synthesis of nucleosides and nucleoside analogs; synthesis and nuclear magnetic resonance study of carbohydrates and carbohydrate analogs. Mailing Add: Dept of Obstet & Gynec Wash Univ Sch of Med St Louis MO 63110

SWEET, GEORGE H, b Texhoma, Okla, Feb 4, 34; m 55; c 3. IMMUNOLOGY. Educ: Wichita State Univ, BS, 60; Univ Kans, MA, 62, PhD(immunol), 65. Prof Exp: Immunologist, Armed Forces Inst Path, 65-66; from asst prof to assoc prof, 66-71, PROF BIOL, WICHITA STATE UNIV, 72- Mem: AAAS. Res: Cell biology; fungal serology; cellular immunology. Mailing Add: Dept of Biol Wichita State Univ Wichita KS 67208

SWEET, GERTRUDE EVANS, b Erie, Pa, Aug 27, 06; m 55. ZOOLOGY. Educ: Mt Holyoke Col, AB, 28; Univ Chicago, PhD(zool), 36. Prof Exp: Sci asst biol, Am Mus Natural Hist, 28-31; asst, Univ Chicago, 32-36; instr zool, Mt Holyoke Col, 36-38; biologist, Beloit Col, 38-44; asst prof, Wells Col, 44-47; prof & head dept, Lake Erie Col, 47-55; instr, 57-65, assoc prof, 65-71. EMER ASSOC PROF, BELOIT COL, 71- Mem: Am Soc Zool; Soc Study Evolution. Res: Aggregations studies; oxygen consumption of salamanders. Mailing Add: 811 Chapin St Beloit WI 53511

SWEET, HAVEN C, b Boston, Mass, Mar 1, 42; m 63; c 2. PLANT PHYSIOLOGY. Educ: Tufts Univ, BS, 63; Syracuse Univ, PhD(plant physiol), 67. Prof Exp: Res fel photobiol, Brookhaven Nat Labs, 67-68; sr res analyst bot, Brown & Root-Northrop, Tex, 68-69, supvr, 69-71; ASST PROF BIOL, FLA TECHNOL UNIV, 71- Mem: AAAS; Am Soc Plant Physiol; Bot Soc Am; Am Inst Biol Sci; Linnean Soc London. Res: Developing and utilizing methods for germ free plant research; effects of light and ethylene on plant growth; development of computer-assessment of plant taxonomic and ecological information. Mailing Add: Dept of Biol Fla Technol Univ Orlando FL 32816

SWEET, HERBERT C, b Syracuse, NY, Jan 14, 11; m 40; c 2. INTERNAL MEDICINE. Educ: Univ Mich, MD, 37, AM, 38. Prof Exp: Instr bact, Univ Mich, 32-38; from asst prof med to assoc prof internal med, 52-62, prof clin med, 62-70, CLIN PROF INTERNAL MED, SCH MED, ST LOUIS UNIV, 70-, DIR PULMONARY LAB, 50- Mem: Am Col Chest Physicians; Am Col Physicians. Res: Pulmonary function and emphysema. Mailing Add: 3 St James Ct St Louis MO 63119

SWEET, HERMAN ROYDEN, b Attleboro, Mass, Nov 3, 09; m 31; c 2. BOTANY, MICROBIOLOGY. Educ: Bowdoin Col, AB, 31; Harvard Univ, AB, 34, PhD(mycol), 40. Prof Exp: Asst, Harvard Univ, 36-37; instr, 37-42, from asst prof to assoc prof, 42-54, prof, 54-75, EMER PROF BIOL, TUFTS UNIV, 75- Concurrent Pos: Res assoc, Orchid Herbarium of Oakes Ames, Harvard Univ, 65- Mem: AAAS; Am Soc Microbiol; Brit Soc Gen Microbiol; Am Ornith Union; fel Linnean Soc London. Res: Orchidology. Mailing Add: Dept of Biol Tufts Univ Medford MA 02155

SWEET, LEONARD, b Akron, Ohio, Aug 28, 25; m 46; c 2. STATISTICS. Educ: Univ Akron, BAEd, 49; Kent State Univ, MEd, 54; Case Western Reserve Univ, PhD(statist), 70. Prof Exp: Teacher pub schs, Ohio, 49-57, supvr, 57-59; from asst prof to assoc prof, 59-69, PROF MATH, UNIV AKRON, 69- Concurrent Pos: Consult, Akron Pub Schs, Ohio, 62-65 & Addressograph Multigraph Corp, 71- Mem: Am Statist Asn; Math Asn Am. Res: Experimental design; symmetrical complementation designs. Mailing Add: Dept of Math Univ of Akron Akron OH 44325

SWEET, LOUISE ELIZABETH, b Ypsilanti, Mich, Oct 1, 16. CULTURAL ANTHROPOLOGY. Educ: Eastern Mich Univ, AB, 37; Univ Mich, MA, 39, PhD, 57. Prof Exp: Vis asst prof anthrop, Univ Kans, 57-58; Soc Sci Res Coun res grant, 58-59; res fel anthrop, Univ Mich, 59-60; assoc prof, Indiana State Col, Pa, 60-63; assoc prof, State Univ NY Binghamton, 63-71, chmn dept, 66-68; head dept, 71-74, PROF ANTHROP, UNIV MAN, 70- Concurrent Pos: Soc Sci Res Coun grant, 64-65; vis assoc prof, Univ Man, 68-69. Mem: AAAS; Am Anthrop Asn; Ethnol Soc; Am Ethnohist Soc; fel Royal Anthrop Inst Gt Brit & Ireland. Res: Ethnology; cultural ecology; culture change of the Levant and the Persian Gulf; comparative ethnohistory of Near Eastern enclaves. Mailing Add: Dept of Anthrop Univ of Man Winnipeg MB Can

SWEET, MELVIN MILLARD, b South Gate, Calif. NUMBER THEORY. Educ: Calif State Univ, Los Angeles, BA, 64, MA, 65; Univ Md, PhD(math), 72. Prof Exp: Mathematician, Nat Security Agency, 66-70; vis asst prof math, Univ Md, Baltimore County, 72-74; PERMENENT RES STAFF MEM MATH, COMMUN RES DIV, INST DEFENSE ANAL, 75- Mem: Am Math Soc; Math Asn Am. Res: Diophantine approximations. Mailing Add: Inst for Defense Anal Thanet Rd Princeton NJ 08540

SWEET, MERRILL HENRY, II, b Chicago Heights, Ill, Sept 5, 35; m 58; c 4. BIOLOGY. Educ: Univ Conn, BS, 58, PhD(entom), 63. Prof Exp: Res asst entom, Univ Conn, 62-63; asst prof, 63-66, ASSOC PROF BIOL, TEX A&M UNIV, 66- Mem: Ecol Soc Am; Assoc Trop Biol; Soc Study Evolution; Soc Syst Zool. Res: Systematics; ecology; behavior and life cycles of arthropods, especially hemipterous insects. Mailing Add: Dept of Biol Tex A&M Univ College Station TX 77843

SWEET, NORMAN JOSEPH, b Boston, Mass, Apr 15, 13; m 47; c 2. INTERNAL MEDICINE. Educ: Univ Calif, BS, 34, MD, 38; Am Bd Internal Med, dipl, 45. Prof Exp: Asst chief, 47-58, chief, 58-65, chief gold med serv, 60-62, chief blue med serv, 62-65, coordr emergency dept, 66-70, SR CONSULT MED, PSYCHIAT SERV, UNIV CALIF MED SERV, SAN FRANCISCO GEN HOSP, 70-, ASSOC PROF, SCH MED, UNIV CALIF, SAN FRANCISCO, 56- Concurrent Pos: Attend staff physician, Univ Calif Hosp, 45- & Vet Admin Hosp, San Francisco, 54-; consult to surgeon gen, Letterman Gen Hosp, US Army, 48-55, Far East Command, 53; hon vis physician, Nat Heart Hosp, London, Eng, 61-62; Commonwealth Fund fel, 61-62; consult, Mt Zion Hosp, 62-; vchmn dept med, Univ Calif, San Francisco, 57-65, head jr med student teaching, Sch Med & San Francisco Gen Hosp, 70- Honors & Awards: Hon fel, Royal Col Physicians, 42-46. Mem: Am Heart Asn; Am Fedn Clin Res. Res: Cardiac arrhythmias; teaching methodology; pharmacology of quinidine; organic disease in psychotic patients; cardiac effect antipsychotic drugs. Mailing Add: Univ of Calif Med Serv San Francisco Gen Hosp San Francisco CA 94110

SWEET, RICHARD CLARK, b Tarrytown, NY, Nov 28, 21; m 48; c 3. ANALYTICAL CHEMISTRY, PHYSICAL CHEMISTRY. Educ: Wesleyan Col, BA, 44, MA, 48; Rutgers Univ, PhD(chem), 52. Prof Exp: Chemist, 52-60, SUPVR METALL SYSTS APPLNS, PHILIPS LABS, N AM PHILIPS CO, BRIARCLIFF MANOR, 60- Mem: Am Chem Soc; Am Vacuum Soc; Sigma Xi; AAAS. Res: Spectroscopy; trace levels; ion exchange; polarography; water analysis; analytical methods; electronic components; vacuum techniques; ceramic-metal seals; cryogenic components design and fabrication; metals processing and joining techniques; welding and brazing. Mailing Add: 309 N Washington St NTarrytown Tarrytown NY 10591

SWEET, RICHARD FRANKLYN, b Tuscaloosa, Ala, Mar 3, 38; m 61; c 2. PHYSICS. Educ: Spring Hill Col, BS, 59; Univ Tex, Austin, MA, 61, PhD(physics), 64. Prof Exp: Actg asst prof, Univ Va, 64-65; from asst prof to assoc prof, Univ SAla, 66-73; ASSOC PROF PHYSICS, UNIV TEX, SAN ANTONIO, 73- Mem: Am Phys Soc; Am Asn Physics Teachers. Res: Theoretical nuclear physics, specifically, rearrangement collisions. Mailing Add: 210 E Skyview San Antonio TX 78228

SWEET, ROBERT DEAN, b Fairview, Ohio, Apr 6, 15; m 36; c 2. VEGETABLE CROPS. Educ: Ohio Univ, BS, 36; Cornell Univ, MS, 38, PhD(veg crops), 41. Prof Exp: Asst, 36-40, exten instr, 40-43, asst exten prof, 43-47, assoc prof, 47-49, PROF VEG CROPS, NY STATE COL AGR & LIFE SCI, CORNELL UNIV, 49- Mem: Am Soc Hort Sci; Weed Sci Soc Am. Res: Biological and chemical weed control. Mailing Add: Dept of Veg Crops NY State Col of Agr & Life Sci Ithaca NY 14850

SWEET, ROBERT MAHLON, b Omaha, Nebr, Sept 21, 43; m 66; c 3. MOLECULAR BIOLOGY. Educ: Calif Inst Technol, BS, 65; Univ Wis-Madison, PhD(phys chem), 70. Prof Exp: Lectr chem, Univ Wis-Madison, 70; fel molecular biol, Med Res Coun Lab Molecular Biol, Cambridge, Eng, 70-73; ASST PROF CHEM, UNIV CALIF, LOS ANGELES, 73- Concurrent Pos: Damon Runyon Mem Fund fel, 70-72; Europ Molecular Biol Orgn fel, 72. Mem: AAAS; Am Crystallog Asn; Sigma Xi. Res: Structure and function of enzymes, determined by x-ray diffraction techniques. Mailing Add: Dept of Chem Univ of Calif Los Angeles CA 90024

SWEET, ROGER GEORGE, b Chadwicks, NY, Nov 19, 17; m 40; c 4. PHYSICAL CHEMISTRY, INORGANIC CHEMISTRY. Educ: Colgate Univ, AB, 39; Johns Hopkins Univ, PhD(phys & inorg chem), 43. Prof Exp: Chemist, Johns Hopkins Univ, 41-43; proj leader, Linde Air Prod Co, NY, 43-48; res chemist, Catalyst Res Corp, Md, 48-51; dir opers, Spec Prod Div, Kennecott Copper Corp, NY, 51-71, Tex, 71-74; GEN MGR, QUE METAL POWDERS LTD, 74- Concurrent Pos: Sloan fel, Mass Inst Technol, 58-59. Mem: Am Chem Soc; Electrochem Soc; Am Phytopath Soc; fel Am Inst Chemists; Am Inst Mining, Metall & Petrol Engrs. Res: Metal powders, fungicides, tungsten, vanadium, uranium, selenium, silicon and copper chemicals; tungsten, titanium and zirconium metals. Mailing Add: Que Metal Powders Ltd Box 570 Sorel PQ Can

SWEET, ROGER SPENCER, organic chemistry, see 12th edition

SWEET, RONALD LANCELOT, b Bristol, Eng, Feb 6, 23; nat US; m 47. ORGANIC CHEMISTRY. Educ: Rutgers Univ, BSc, 44, MSc, 48, PhD, 55. Prof Exp: Fel petrol, Mellon Inst, 51-55; CHEMIST, PIGMENTS DEPT, E I DU PONT DE NEMOURS & CO, 55- Concurrent Pos: Co-adj, Rutgers Univ, 63-74. Mem: Am Chem Soc. Res: Pigments. Mailing Add: 632 Fairfield Circle Westfield NJ 07090

SWEET, THOMAS RICHARD, b Jamaica, NY, Sept 27, 21; m 48; c 2. CHEMISTRY. Educ: City Col New York, BS, 43; Ohio State Univ, PhD(chem), 49. Prof Exp: Asst, Manhattan Proj, War Res Div, Columbia Univ, 43-45 & Carbide & Carbon Chem Corp, 45-46; from asst prof to assoc prof, 49-65, PROF CHEM, OHIO STATE UNIV, 65- Mem: AAAS; Am Chem Soc. Res: Organic reagents, solvent extraction and trace metal analysis. Mailing Add: Dept of Chem Ohio State Univ 140 W 18th Ave Columbus OH 43210

SWEET, WALTER CLARENCE, b Denver, Colo, Oct 17, 27; m 57. GEOLOGY, INVERTEBRATE PALEONTOLOGY. Educ: Colo Col, BS, 50; Univ Iowa, MS, 52, PhD(geol), 54. Prof Exp: Instr, 54-57, from asst prof to assoc prof, 57-66, PROF GEOL, OHIO STATE UNIV, 66- Concurrent Pos: Fulbright res fel, Univ Oslo, 56-57; vis prof, Univ Colo, 58 & Univ Lund, Sweden, 66. Concurrent Pos: Consult, H A Brassert Co, Bolivia, 55. Mem: Fel Geol Soc Am; Paleont Soc; Soc Econ Paleont & Mineral; Am Asn Petrol Geol; Norweg Geol Soc. Res: Nautiloid cephalopods and conodonts; Ordovician and Permotriassic stratigraphy and paleontology. Mailing Add: Dept of Geol & Mineral Ohio State Univ Columbus OH 43210

SWEET, WILLIAM HERBERT, b Kerriston, Wash, Feb 13, 10; m 37; c 3. NEUROSURGERY. Educ: Univ Wash, SB, 30; Oxford Univ, BSc, 34; Harvard Univ, MD, 36. Hon Degrees: DSc, Oxford Univ, 57. Prof Exp: Instr neurosurg, Billings Hosp, Chicago, 39-40; Commonwealth Fund fel, Harvard Med Sch, 40-41; actg chief neurosurg serv, Birmingham United Hosp, 41-45; from instr to asst prof, 45-54, assoc clin prof, 54-58, assoc prof, 58-65, PROF SURG, HARVARD MED SCH, 65-; CHIEF NEUROSURG SERV, MASS GEN HOSP, 61- Concurrent Pos: Regional consult, Brit Emergency Med Serv, 41-45; asst, Mass Gen Hosp, 45-47, asst neurosurgeon, 47-48, assoc vis neurosurgeon, 48-58, vis neurosurgeon, 58-; lectr, Med Sch, Tufts Col, 47-51; neurosurgeon in chief, New Eng Ctr Hosp, 49-51; mem subcomt neurosurg, Nat Res Coun, 49-52 & mem subcomt neurol & neurosurg, 52-59; trustee, Assoc Univs, Inc, 58-; mem sci & technol adv comt, NASA, 64-70; mem neurol sci res training A comt, Nat Inst Neurol Dis & Stroke; honored guest, Cong Neurol Surgeons, 75. Honors & Awards: His Majesty's Medal for Serv in Cause of Freedom, 45. Mem: Am Acad Arts & Sci; Am Acad Neurol Surg; Am Neurol Asn; Am Surg Asn; Soc Neurol Surg. Res: Central nervous system; research in cerebrospinal and intracerebral fluid; brain tumors; mechanisms of pain and its neurosurgical control; abnormal behavior related to organic brain disease; irreversible coma; ethics of experimentation. Mailing Add: Mass Gen Hosp Neurosurg Serv Fruit St Boston MA 02114

SWEETING, LINDA MARIE, b Toronto, Ont, Dec 11, 41. ORGANIC CHEMISTRY. Educ: Univ Toronto, BSc, 64, MA, 65; Univ Calif, Los Angeles, PhD(org chem), 69. Prof Exp: Asst prof, Occidental Col, 69-70; asst prof, 70-75, ASSOC PROF, TOWSON STATE COL, 75- Mem: Am Chem Soc; AAAS; Asn Women in Sci Res: Application of nuclear magnetic resonance spectroscopy to organic chemistry; stereochemistry; weak organic bases. Mailing Add: Dept of Chem Towson State Col Baltimore MD 21204

SWEETING, ORVILLE JOHN, b Rochester, NY, Oct 26, 13; m 42; c 3. ORGANIC CHEMISTRY. Educ: Cornell Univ, AB, 35, PhD(chem), 42. Prof Exp: Instr high schs, NY, 35-38; chemist, Cornell Univ, 42-44; Ind Univ, 44-46; asst prof, Univ Colo, 46-49, assoc prof & dir isotopes lab, 49-52; res supvr, Olin Industs, Inc, 52-54, assoc dir, Film Res & Develop Dept, Olin Mathieson Chem Corp, 54-64; lectr chem & asst dir off teacher ed, Yale Univ, 64-69; vpres & provost, 69-71; PROF CHEM, QUINNIPIAC COL, 69-, DIR RES, 71-, CHMN DEPT CHEM, 71- Concurrent Pos: Consult, Conn State Dept Ed & US Off Ed. Mem: Am Chem Soc. Res: Organic synthesis; physical organic and polymer chemistry; syntheses with radioisotopes; polymer packaging films. Mailing Add: 108 Everit St New Haven CT 06511

SWEETMAN, BRIAN JACK, b Palmerston North, NZ, May 4, 36; m 62; c 3. ORGANIC CHEMISTRY, PHARMACOLOGY. Educ: Univ NZ, BSc, 58, MSc, 59; Univ Otago, NZ, PhD(org chem), 62. Prof Exp: Res officer div protein chem, Commonwealth Sci & Indust Res Orgn, Melbourne, Australia, 63-66; res assoc chem, 66-68, ASST PROF PHARMACOL IN RESIDENCE, VANDERBILT UNIV, 69- Mem: Am Chem Soc; Am Soc Mass Spectroscopy; Sigma Xi. Res: Mass spectrometry; analytical pharmacology; prostaglandins; vapor-phase analysis; medicinal and organosulfur chemistry; anti-radiation and anti-arthritic drugs; protein chemistry of keratin; natural products. Mailing Add: Dept of Pharmacol Vanderbilt Univ Nashville TN 37232

SWEETMAN, LAWRENCE, b La Junta, Colo, Feb 17, 42; m 70. BIOCHEMISTRY, PEDIATRICS. Educ: Univ Colo, BA, 64; Univ Miami, PhD(biochem), 69. Prof Exp: Res assoc biochem, Sloan-Kettering Inst Cancer Res, 68-72, instr, Sloan-Kettering Div, Grad Sch Med Sci, Cornell Univ, 69-72; ASST PROF PEDIAT, UNIV CALIF, SAN DIEGO, 72- Mem: Am Chem Soc. Res: Metabolism of inherited diseases in children; Lesch-Nyhan syndrome and organic acidurias. Mailing Add: Dept of Pediat Univ of Calif at San Diego La Jolla CA 92037

SWEETON, FREDERICK HUMPHREY, b Brattleboro, Vt, Oct 13, 16; m 53; c 3. PHYSICAL CHEMISTRY. Educ: Univ Conn, BS, 38; Yale Univ, PhD(phys chem), 41. Prof Exp: Chemist, Tenn Valley Authority, 41-43; CHEMIST, OAK RIDGE NAT LAB, 46- Mem: Am Chem Soc; Soil Sci Soc Am. Res: High temperature aqueous electrochemistry; soil chemistry; low temperature calorimetry; surface chemistry of slurries; chemistry of pressurized-water reactors. Mailing Add: 334 Louisiana Ave Oak Ridge TN 37830

SWEETSER, PHILIP BLISS, b Morrisville, Vt, Dec 26, 24; m 53; c 2. ANALYTICAL CHEMISTRY, PLANT BIOCHEMISTRY. Educ: Univ Vt, BS, 50; Princeton Univ, PhD(chem), 53. Prof Exp: RES SCIENTIST, CENT RES DEPT, E I DU PONT DE NEMOURS & CO, 53- Mem: AAAS; Am Chem Soc. Res: Metabolism and role of agrochemicals as plant growth regulators; separations. Mailing Add: 1006 Crestover Rd Wilmington DE 19803

SWELL, LEON, b New York, NY, July 26, 27; m; c 3. BIOCHEMISTRY. Educ: City Col New York, BS, 48; George Washington Univ, MS, 49, PhD(biochem), 52. Prof Exp: Lab asst, George Washington Univ, 49-51; chief biochemist, Vet Admin Ctr, Martinsburg, WVa, 51-64, chief lipid res lab, Vet Admin Hosp, Richmond, Va, 64-70; RES PROF BIOCHEM & MED, VA COMMONWEALTH UNIV, 70- Concurrent Pos: Assoc prof lectr, George Washington Univ, 59-; assoc res prof, Med Col Va, 64- Mem: AAAS; Am Soc Biol Chemists; Soc Exp Biol & Med; Am Inst Nutrit. Res: Cholesterol, lipid and electrolyte metabolism; enzymes. Mailing Add: Dept of Biochem Va Commonwealth Univ Box 606 Richmond VA 29219

SWENBERG, CHARLES EDWARD, b Meriden, Conn, Mar 11, 42. BIOPHYSICS. Educ: Univ Conn, BA, 62, MS, 63; Univ Rochester, PhD(physics), 68. Prof Exp: Res assoc physics, Univ Ill, Urbana, 67-69; assoc res scientist, Radiation & Solid State Lab, 69-70, asst prof, 70-73. SR SCIENTIST, RADIOL & SOLID STATE LAB, NY UNIV, 74- Concurrent Pos: Vis fel, Dept Physics, Univ Manchester, Eng, 75-; NIMH Post Doctoral Fel, St Elizabeth Hosp. Res: Primary processes in photobiology; optical properties of solids and liquids; primary processes in photosynthesis; exutable membranes. Mailing Add: NIMH-WAW Bldg St Elizabeth Hosp Washington DC 20032

SWENBERG, JAMES ARTHUR, b Northfield, Minn, Jan 15, 42; m 63; c 2. VETERINARY PATHOLOGY. Educ: Univ Minn, DVM, 66; Ohio State Univ, MS, 68, PhD(vet path), 70. Prof Exp: NIH trainee path, Ohio State Univ, 66-70, res assoc, 70, asst prof, 70-72, assoc prof, 72; RES SCIENTIST PATH, UPJOHN CO, 72- Concurrent Pos: Consult, Battelle Mem Inst, 71-72. Mem: Am Asn Cancer Res; AAAS; Am Asn Neuropathologists; Am Col Vet Pathologists. Res: Cancer research, including chemical carcinogenesis, neurooncogenesis and chemotherapy, and short-term tests for carcinogens; DNA damage/mutagensis; improved toxicology and data handling methods. Mailing Add: Path & Toxicol Res Unit Upjohn Co Kalamazoo MI 49001

SWENDSEID, MARIAN EDNA, b Petersburg, NDak, Aug 2, 18. BIOCHEMISTRY. Educ: Univ NDak, BA, 38, MA, 39; Univ Minn, PhD(physiol chem), 41. Prof Exp: Asst nutrit, Univ Ill, 42; res biochemist, Simpson Mem Inst, Univ Mich, 42-43; sr res chemist, Parke Davis & Co, Mich, 45-48; res biochemist, Simpson Mem Inst, Univ Mich, 48-52; assoc prof, 53-72, PROF NUTRIT, UNIV CALIF, LOS ANGELES, 72- Mem: Am Chem Soc; Am Soc Biol Chem. Res: Vitamin research; biochemical aspects of hematology; the use of carbon 13 in the study of intermediary metabolism; amino acids in nutrition. Mailing Add: Dept of Nutrit & Biol Chem Univ of Calif Sch of Pub Health Los Angeles CA 90024

SWENERTON, HELENE ROUPEN. b Norfolk, Va, Jan 13, 25; m 43; c 3. NUTRITION. Educ: Univ Calif, Davis, BS, 63, MS, 65, PhD(nutrit), 70. Prof Exp: Res nutritionist, 70-72, EXTEN NUTRITIONIST, UNIV CALIF, DAVIS, 72- Mem: AAAS; Teratology Soc; Am Dietetics Assoc. Res: Role of dietary zinc in mammalian growth and development; effects of maternal dietary deficiencies on fetal development. Mailing Add: Dept of Nutrit Univ of Calif Davis CA 95616

SWENSEN, ALBERT DONALD, b Provo, Utah, May 28, 15; m 37; c 6. BIOCHEMISTRY. Educ: Brigham Young Univ, AB, 37, MA, 38; La State Univ, PhD(biol chem), 41. Prof Exp: Biochemist, US Naval Res Lab, 41-47; from asst prof to assoc prof chem, 47-57, chmn dept, 60-63, PROF CHEM, BRIGHAM YOUNG UNIV, 57- Concurrent Pos: Fel, Univ Minn, 56; vis scientist, Charles F Kettering Res Lab, Yellow Springs, Ohio, 63-64. Mem: Am Chem Soc. Res: Enzymes; intermediary metabolism. Mailing Add: Dept of Chem Brigham Young Univ Provo UT 84602

SWENSEN, ALF WALDEMAR, physical chemistry, deceased

SWENSON, CHARLES ALLYN, b Clinton, Minn, Aug 31, 33; m 60; c 3. PHYSICAL CHEMISTRY. Educ: Gustavus Adolphus Col, BS, 55; Univ Iowa, PhD(chem), 59. Prof Exp: Asst, Univ Iowa, 55-56; asst prof chem, Wartburg Col, 58-60; res assoc, 60-62, from asst prof to assoc prof, 62-72, PROF BIOCHEM, UNIV IOWA, 72- Mem: Am Chem Soc; Biophys Soc. Res: Physical biochemistry; model studies for biopolymer interactions; spectroscopic and thermodynamic approaches; energy transduction in muscle contraction. Mailing Add: Dept of Biochemistry Univ of Iowa Iowa City IA 52242

SWENSON, CHRISTINA N, b Haynes, NDak, Dec 12, 12; m 41; c 4. PHYSICS. Educ: Gustavus Adolphus Col, BA, 40; Univ Minn, MA, 53. Prof Exp: Teacher high sch, Minn, 40-42 & Minnehaha Acad. Minneapolis. 46-60; from asst prof to assoc prof, 60-75, PROF PHYSICS, BETHANY COL, KANS, 75- Concurrent Pos: Res participation grant. Lab Atmospheric & Space Physics, Univ Colo, 6S; res, Kans State Univ, 69-70. Mem: Optical Soc Am; Am Asn Physics Teachers. Res: Basic physics; atomic and nuclear physics; beam foil spectroscopy. Mailing Add: Dept of Physics Bethany Col Lindsborg KS 67456

SWENSON, CLAYTON ALBERT, b Hopkins, Minn, Nov 11, 23; m 50; c 3. EXPERIMENTAL SOLID STATE PHYSICS. Educ: Harvard Univ, BS, 44; Oxford Univ, DPhil(physics), 49. Prof Exp: Instr physics, Harvard Univ, 49-52; res physicist, Div Indust Coop, Mass Inst Technol, 52-55; from asst prof to assoc prof, 55-60, PROF PHYSICS, IOWA STATE UNIV, 60-, CHMN DEPT, 75- Mem: Am Phys Soc. Res: Low temperatures; high pressures; low temperature thermodynamics; thermometry; equations of state of solids. Mailing Add: Dept of Physics Iowa State Univ Ames IA 50011

SWENSON, DONALD ADOLPH, b Camden, Ala, May 9, 32; m 55; c 3. PHYSICS, MATHEMATICS. Educ: Univ Ala, BS, 53; Univ Minn, MS, 56, PhD(physics), 58. Prof Exp: Physicist, Midwestern Univs Res Asn, 58-60 & 61-64; Ford Found fel, European Orgn Nuclear Res, Switz, 60-61; PHYSICIST, LOS ALAMOS SCI LAB, UNIV CALIF, 65- Mem: Am Phys Soc. Res: Particle accelerator, design and development; particle dynamics in circular and linear accelerators; proton beam measurement and diagnostics; development and use of computer control systems for particle accelerators. Mailing Add: Meson Phys Div Los Alamos Sci Lab Los Alamos NM 87544

SWENSON, EDWARD W, b Omaha, Nebr, Oct 24, 26; m 53; c 4. MEDICINE. Educ: Univ Minn, BA, 47; Univ Nebr, MD, 50. Prof Exp: Intern, Univ Chicago Clins, 50-51, Nat Tuberc Asn fel, Univ, 51-52; res fel, Pulmonary Function Lab, Munic Chest Hosp, Goteborg, Sweden, 54-56; chief, Pulmonary Function Lab, Munic Chest Hosp & Univ Gothenburg Lung Clin, 56-57; resident internal med, Med Ctr, Univ Calif, San Francisco, 57-58, San Francisco Heart Asn sr res fel, Cardiovasc Res Inst, 58-59; instr med & assoc staff mem, 59-60; asst prof, Sch Med, Univ Miami, 61-64; assoc prof med & chief pulmonary dis div, Col Med, Univ Fla, 64-75; MEM STAFF, SYNTEX RES DIV, 76- Concurrent Pos: Clin investr & chief cardiopulmonary lab, Vet Admin Hosp, Coral Gables, Fla, 61-64. Mem: Am Physiol Soc; Am Thoracic Soc; Am Heart Asn; Am Col Chest Physicians; Am Col Physicians. Res: Pulmonary medicine; cardiopulmonary pathophysiology. Mailing Add: Syntex Res Div Stanford Indust Park Palo Alto CA 94304

SWENSON, FRANK ALBERT, b Davenport, Iowa, Feb 3, 12; m 43; c 2. Educ: Augustana Col, AB, 36; Univ Iowa, MS, 40, PhD(geol), 42. Prof Exp: Surv man & geologist, US Army Corps Engrs, 37-38; asst geol, Univ Iowa, 38-42; jr geologist, Ground Water Br, 42, from asst geologist to geologist, Mil Geol Br, 42-46. ground water invests, 46-63, res geologist, US Geol Surv, 63-74; CONSULT GEOLOGIST-HYDROLOGIST, 75- Concurrent Pos: Consult, US Dept Justice. Mem: Fel Geol Soc Am; Int Geol Cong; Int Asn Hydrogeol. Res: Geology and ground water investigations in Montana and Wyoming; military intelligence; research in limestone hydrology; geochemistry. Mailing Add: 11615 W 31st Place Lakewood CO 80215

SWENSON, GARY RUSSELL, b Grantsburg, Wis, June 17, 41; m 67; c 2. ATMOSPHERIC PHYSICS. Educ: Wis State Univ-Superior, BS, 63; Univ Mich, Ann Arbor, MS, 68, PhD(atmospheric sci), 75. Prof Exp: SPACE SCIENTIST ATMOSPHERIC SCI, MARSHALL SPACE FLIGHT CTR, NASA, 68- Mem: Am Geophys Union; Sigma Xi. Res: Experimental research using remote sensing techniques, upper atmospheric phenomena, including aurora. Mailing Add: NASA Space Sci Lab Mail Code ES44 Marshall Space Flight Center AL 35812

SWENSON, GENE HOLSTROM, b Ft Dodge, Iowa, Dec 23, 31. VETERINARY PHARMACOLOGY. Educ: Iowa State Univ, DVM, 58; Univ Mo, MS, 70. Prof Exp: Vet pract, 58-68; res asst vet microbiol, Univ Mo, 68-70; field develop specialist, Merck & Co, 70-71; tech serv vet, Am Cyanamid Co, 72; RES VET PHARMACOL, UPJOHN CO, 72- Mem: Am Vet Med Asn; Am Asn Bovine Practr; US Animal Health Asn; Am Soc Vet Physiologists & Pharmacologists. Res: Etiology and pathogenesis of bovine mastitis in conjunction with the pharmacology and pharmacokinetics of drugs for therapy and prophylaxis of the disease. Mailing Add: Bldg 190 Upjohn Co Kalamazoo MI 49001

SWENSON, HENRY MAURICE, b Brooklyn, NY, Aug 13, 16; m 41; c 4. DENTISTRY. Educ: Univ Ill, BS, 41, DDS, 42; Am Bd Periodont, dipl, 51. Prof Exp: From instr to assoc prof, 45-62, PROF PERIODONT, SCH DENT, IND UNIV, INDIANAPOLIS, 62-, DIR CLIN, 56- Concurrent Pos: Consult, Vet Admin, 58- & US Dept Army, 59- Mem: Am Dent Asn; fel Am Col Dentists; Am Acad Periodont; Int Asn Dent Res. Res: Treatment and management of periodontal involvement. Mailing Add: 1121 W Michigan Indianapolis IN 46202

SWENSON, HERBERT ALFRED, b Portland, Ore, Feb 20, 11; m 43; c 2. HYDROLOGY. Educ: Ore State Univ, BS, 35. Prof Exp: Chemist, Crown Zellerbach Corp, Ore, 35-36 & US Army Corps Engrs, 36-37; chemist, US Geol Surv, Washington, DC, 38-61, asst chief, Qual of Water Br, 61-65, res hydrologist, 65-67; WATER RES SCIENTIST, OFF WATER RESOURCES RES, DEPT INTERIOR, 67- Mem: AAAS; Am Soc Civil Eng; Am Chem Soc; Am Water Works Asn. Res: International studies of water resources; water pollution; quality of natural waters; fluvial sediments; water use and reuse. Mailing Add:

SWENSON, HUGO NATHANAEL, b New Richland, Minn, Mar 11, 04; m 56. PHYSICS. Educ: Carleton Univ, AB, 25; Univ Ill, MS, 27, PhD(spectros), 30. Prof Exp: Asst physics, Univ Ill, 25-29; head dept, Earlham Col, 29-30; Am-Scand Found fel, Bohr's Inst, Denmark, 30-31; instr physics, Barnard Col, Columbia Univ, 31-37; instr physics, 37-41, from asst prof to assoc prof, 41-57, PROF PHYSICS, QUEENS

COL, NY, 57- Mem: AAAS; Am Phys Soc; Am Asn Physics Teachers. Res: Electronics. Mailing Add: 252-45 Brattle Ave Little Neck NY 11362

SWENSON, JACK SPENCER, b Seattle, Wash, Aug 5, 30; m 55; c 3. ORGANIC CHEMISTRY. Educ: Univ Wash, BS, 52; Univ Minn, Minneapolis, PhD(org chem), 56. Prof Exp: Res chemist, M W Kellogg Co, Pullman, Inc, 56-57 & Minn Mining & Mfg Co, 57-61; from asst prof to assoc prof chem, Grinnell Col, 61-70, chmn dept, 67-70; assoc prof, 70-74, PROF CHEM, NORTHERN ARIZ UNIV, 74-, CHMN DEPT, 70- Concurrent Pos: Consult, Minn Mining & Mfg Co, 61-65. Mem: Am Chem Soc; Am Sci Affil. Res: Adhesives; carbenes; latent curing reagents; Lossen rearrangements. Mailing Add: Dept of Chem Northern Ariz Univ Flagstaff AZ 86001

SWENSON, KNUD GEORGE, b Brookings, SDak, July 1, 23; m 46; c 2. ENTOMOLOGY. Educ: SDak State Col, BS, 48; Univ Calif, PhD, 51. Prof Exp: Asst entom, Iowa State Col, 48-49; asst prof, Exp Sta, Cornell Univ, 51-53; from assoc prof entom & assoc entomologist to prof entom & entomologist, 60-74, chmn dept entom, 71-73, EMER PROF ENTOM, ORE STATE UNIV, 74- Concurrent Pos: Agent, Agr Res Serv, USDA, 55-67; Guggenheim fel, Div Entom, Commonwealth Sci & Indust Res Orgn, Australia, 60-61. Mem: AAAS; Entom Soc Am; Am Phytopath Soc. Res: Insect transmission of palnt viruses; agricultural entomology; biology of Homoptera. Mailing Add: Dept of Entom Ore State Univ Corvallis OR 97331

SWENSON, LEONARD WAYNE, b Twin Falls, Idaho, June 11, 31; m 50; c 3. PHYSICS. Educ: Mass Inst Technol, BS, 54, PhD(physics), 60. Prof Exp: Instr physics, Northeastern Univ, 57-58 & Tufts Univ, 58-59; res assoc, Mass Inst Technol, 60-62; Bartol fel nuclear struct res group, Bartol Res Found, 62-64, res staff mem, 64-66; dir space radiation effects lab, Va Assoc Res Ctr, 67-68; ASSOC PROF PHYSICS, ORE STATE UNIV, 68- Concurrent Pos: Consult, Joseph Kaye & Co, 57-59. Mem: Am Phys Soc; Am Sci Affil. Res: Nuclear structure and reaction; high energy nuclear physics. Mailing Add: Dept of Physics Ore State Univ Corvallis OR 97331

SWENSON, MELVIN JOHN, b Concordia, Kans, Jan 14, 17; m 47; c 3. VETERINARY PHYSIOLOGY. Educ: Kans State Col, DVM, 43; Iowa State Col, MS, 47, PhD(path), 50. Prof Exp: Instr vet sci, La State Univ, 43; asst prof path, Iowa State Col, 49-50; assoc prof physiol, Kans State Col, 50-56; prof, Colo State Univ, 56-57; prof & head dept, 57-73, PROF PHYSIOL, IOWA STATE UNIV, 73- Mem: AAAS; Am Physiol Soc; Am Soc Vet Physiol & Pharmacol; Soc Exp Biol & Med; Am Vet Med Asn. Res: Need for trace minerals in animals; effect of antibiotics on growth and hematology; nutrient requirements of animals; histophysiology of nutritional deficiencies; anemias of farm animals. Mailing Add: Dept of Vet Anat Pharmacol & Physiol Iowa State Univ Col Vet Med Ames IA 50010

SWENSON, ORVAR, b Halsingborg, Sweden, Feb 7, 09; nat US; m 41; c 3. SURGERY. Educ: William Jewell Col, AB, 33; Harvard Med Sch, MD, 37. Prof Exp: Intern surg, Ohio State Univ Hosp, 37-38; house officer path, Children's Hosp, Boston, 38-39, house officer surg, 39-41; Cabot fel, Harvard Med Sch, 41-44, asst surg, 42-44, instr, 44-47, assoc, 47-50; assoc prof, Sch Med, Tufts Univ, 50-54, clin prof pediat surg, 54-57, prof, 57-60; prof surg, Med Sch, Northwestern Univ, Chicago, 60-73; PROF SURG, SCH MED, UNIV MIAMI, 73- Concurrent Pos: Surg house officer, Peter Bent Brigham Hosp, 39-41, asst res surgeon, 41-42, jr assoc, 44, res surgeon, 44-45; jr attend surgeon, Children's Hosp, Boston, 45, assoc vis surgeon, 45-47, surgeon, 47-50; vis surgeon, New Eng Peabody Home Crippled Children, 46-50, mem assoc staff, 50; lectr, Simmons Col, 48-50; sr surgeon, New Eng Ctr Hosp, 50-60; surgeon in chief, Boston Floating Hosp Infants & Children, 50-60 & Children's Mem Hosp, Chicago, 60-72. Mem: Assoc Asn Thoracic Surg; Am Urol Asn; Am Surg Asn; fel Am Col Surg; fel Am Acad Pediat. Res: Genito-urinary system. Mailing Add: PO Box 875 Biscayne Annex Miami FL 33152

SWENSON, PAUL ARTHUR, b St Paul, Minn, Feb 5, 20; m 42; c 2. CELL PHYSIOLOGY. Educ: Hamline Univ, BS, 47; Stanford Univ, PhD(biol), 52. Prof Exp: Instr physiol, Univ Mass, 50-S4, from asst prof to assoc prof, 54-66; RADIATION BIOPHYSICIST, BIOL DIV, OAK RIDGE NAT LAB, 66- Concurrent Pos: Vis assoc physiologist, Brookhaven Nat Lab, 56-57; USPHS spec fel, Oak Ridge Nat Lab, 62-64. Mem: AAAS; Biophys Soc; Am Soc Microbiol; Am Soc Photobiol. Res: Effects of ultraviolet and ionizing radiations on metabolic control in bacteria. Mailing Add: Biol Div Oak Ridge Nat Lab Oak Ridge TN 37830

SWENSON, RICHARD WALTNER, b New York, NY, May 17, 23; m 49; c 3. PHOTOGRAPHIC CHEMISTRY. Educ: Clark Col, Miss, AB, 48, AM, 49; Brown Univ, PhD(phys chem), 53. Prof Exp: Res chemist, 53-63, RES ASSOC, PHOTO PROD DEPT, E I DU PONT DE NEMOURS & CO, 63- Mem: Am Chem Soc; Soc Photog Sci & Eng(pres. 67-71). Res: Photographic chemistry, specifically emulsion chemistry. Mailing Add: 65 Edison Ave Tintou Falls NJ 07724

SWENSON, ROYAL JAY, b Pocatello, Idaho, July 9, 33; m 53; c 7. CROP SCIENCE, SOIL SCIENCE. Educ: Utah State Univ, BS, 58, MS, 60. Prof Exp: Res asst, Utah State Univ, 56-60; soil scientist, Bur Indian Affairs, Dept Interior, 60-64; dir res, Empire State Sugar Co, 64-66; asst prof, Kansas State Univ, 66-67; AGRONOMIST, CHEVRON CHEM CO, 67- Res: Field, greenhouse and laboratory research with crop and soil problems to determine more effective overall production programs for greater efficiency and higher economic returns to growers and industry. Mailing Add: Chevron Chem Co First Bank & Trust Suite 400 Richardson TX 75080

SWENTON, JOHN STEPHEN, b Kansas City, Kans, Dec 8, 40. PHYSICAL ORGANIC CHEMISTRY. Educ: Univ Kans, BA, 62; Univ Wis, PhD(chem), 65. Prof Exp: Nat Acad Sci-Nat Res Coun fel, Harvard Univ, 65-66; asst prof chem, 67-71, ASSOC PROF CHEM, OHIO STATE UNIV, 71- Concurrent Pos: Res Corp grant, 67-69; Eli Lilly res grant, 68-70. Mem: Am Chem Soc; The Chem Soc. Res: Mechanistic and exploratory organic photochemistry; effect of differing geometry for gound and excited states on photophysical and photochemical processes. Mailing Add: Dept of Chem Ohio State Univ 88 W 18th Ave Columbus OH 43210

SWERCZEK, THOMAS WALTER, b Cedar Rapids, Nebr, May 10, 39; m 64; c 3. VETERINARY PATHOLOGY. Educ: Kans State Univ, BS, 62, DVM, 64; Univ Conn, MS, 66, PhD(path), 69. Prof Exp: ASST PROF VET PATH, COL AGR, UNIV KY, 69- Mem: Am Vet Med Asn; Conf Res Workers Animal Dis. Res: Comparative pathology; pathogenesis of infectious diseases of horses. Mailing Add: Dept of Vet Sci Univ of Ky Col of Agr Lexington KY 40506

SWERDLOW, HERBERT, b New York, NY, Apr 8, 24; m 51; c 2. DENTISTRY. Educ: Brooklyn Col, BA, 47; NY Univ, DDS, 51; Univ Wash, MSc, 61. Prof Exp: Asst, Nat Inst Dent Res, 48-50, dent intern, USPHS Hosps, Baltimore, 51-52, mem dent staff, Galveston, Tex, 52-53, mem clin staff, Dent Dept, Clin Ctr, NIH, 53-59, resident prosthodontics, USPHS Hosp, Seattle, 59-61, clin assoc, Dent Dept, NIH, 61-64, CHIEF DENT SERV BR, NAT INST DENT RES, 64- Mem: Am Dent Asn; Am Prosthodontic Soc; fel Am Col Dent. Res: Pulpal reactions to dental procedures; prosthodontics. Mailing Add: 15300 SW 81st Ave Miami FL 33157

SWERDLOW, MARTIN A, b Chicago, Ill, July 7, 23; m 45; c 2. MEDICINE, PATHOLOGY. Educ: Univ Ill, BS, 45, MD, 47; Am Bd Path, dipl, 52. Prof Exp: Resident path, Michael Reese Hosp, Chicago, 48-50 & 51-52; pathologist, Menorah Med Ctr, 54-57; from asst prof to assoc prof path, Univ Ill Col Med, 57-60, clin assoc prof, 60-66, prof, 66-72, assoc dean, Abraham Lincoln Sch Med, 70-72; prof path, Sch Med, Univ Mo-Kansas City, 73-74; CHMN DEPT PATH, MICHAEL REESE HOSP & MED CTR, 74- Concurrent Pos: Pathologist, Englewood Hosp, Chicago; consult, Vet Admin Hosp, Hines & Cook County Hosp, Chicago; chmn dept path, Kansas City Gen Hosp, 73-74. Mem: Am Soc Clin Path; Col Am Pathologists; Am Acad Dermat; Asn Am Med Cols; NY Acad Sci. Res: Histopathology of skin disorders; experimental skin tumors. Mailing Add: Dept of Path Michael Reese Hosp & Med Ctr Chicago IL 60616

SWERLICK, ISADORE, b Philadelphia, Pa, Jan 23, 21; m 51; c 2. ORGANIC CHEMISTRY. Educ: Temple Univ, BA, 43; Duke Univ, PhD(chem), 50. Prof Exp: Asst, Duke Univ, 46-49; res chemist, 50-56, res supvr, 56-70, STAFF SCIENTIST, FILM DEPT, E I DU PONT DE NEMOURS & CO, 70- Mem: AAAS; Am Chem Soc. Res: Organics and polymers. Mailing Add: 3355 Cedar Grove Rd Richmond VA 23235

SWERLING, PETER, b New York, NY, Mar 4, 29; m 58; c 3. MATHEMATICS. Educ: Univ Calif, Los Angeles, PhD(math), 55. Prof Exp: Mem sr staff, Electronics Dept, Rand Corp, 52-56 & 57-61; res asst prof, Control Systs Labs, Univ Ill, 56-57; head adv systs off, Conductron Corp, 61-64; CONSULT, 64-; PRES, TECHNOL SERV CORP, 66- Concurrent Pos: Prof elec engr, Southern Calif, 65- Mem: Am Math Soc; Soc Indust & Appl Math; Inst Elec & Electronics Engrs. Res: Signal processing; random noise theory and application. Mailing Add: Technol Serv Corp 225 Santa Monica Blvd Santa Monica CA 90401

SWERN, DANIEL, b New York, NY, Jan 21, 16; m 38; c 2. ORGANIC CHEMISTRY. Educ: City Col New York, BS, 35; Columbia Univ, MA, 36; Univ Md, PhD(org chem), 40. Prof Exp: With USDA, 37-63; PROF CHEM & SR RES INVESTR, FELS RES INST, TEMPLE UNIV, 63- Concurrent Pos: Hon Res Fel, Univ London, 70-71; Flemming, Scott, Bond, Mattiello, Spencer, USDA & Bailey awards. Mem: AAAS; Am Asn Cancer Res: Am Chem Soc; Am Oil Chem Soc. Res: Cancer inhibition and carcinogenesis; chemistry of fats; organic peroxides; small rings; free radicals and olefins; polymerization; ylids; pseudohalogens. Mailing Add: Fels Res Inst Dept of Chem Temple Univ Philadelphia PA 19122

SWETHARANYAM, LALITHA, b Trivandrum, India; m 71. MATHEMATICAL ANALYSIS. Educ: Annamalai Univ, Madras, PhD(math), 66. Prof Exp: Lectr math & head dept, LVD Col, Raichur, India, 56-61; res fel, Annamalai Univ, Madras, 61-65, lectr, 65-69; ASSOC PROF, McNEESE STATE UNIV, 69- Mem: Am Math Soc; Indian Math Soc; Math Asn Am; Assoc Women in Math. Res: Functional analysis; point set topology. Mailing Add: Dept of Math Sci McNeese State Univ Lake Charles LA 70601

SWETITS, JOHN JOSEPH, b Passaic, NJ, Oct 1, 42; m 66; c 3. MATHEMATICAL ANALYSIS. Educ: Fordham Univ, BS, 64; Lehigh Univ, MS, 67, PhD(math), 68. Prof Exp: Instr math, Lafayette Col, 67-68, asst prof, 68-70; ASSOC PROF MATH, OLD DOM UNIV, 70- Mem: Am Math Soc; Math Asn Am; Soc Indust & Appl Math. Res: Summability theory; approximation theory. Mailing Add: Dept of Math Old Dominion Univ Norfolk VA 23508

SWETNICK, MARTIN JAY, physics, see 12th edition

SWETS, DON EUGENE, b Grand Rapids, Mich, Oct 7, 30; m 56; c 3. SOLID STATE PHYSICS, CHEMICAL PHYSICS. Educ: Univ Mich, BSE(physics) & BSE(math), 53, MS, 55. Prof Exp: SR RES PHYSICIST, RES LAB, GEN MOTORS CORP, 55- Res: Measurement of diffusion coefficients; hydrogen in steel; helium, neon, and hydrogen fused in quartz; growth of single crystals, especially tetragonal germanium dioxide; ribbon shaped germanium; hexamethylenetetramine IV-VI and III-V compounds. Mailing Add: Physics Dept Gen Motors Res Labs 12 Mile & Mound Rds Warren MI 48090

SWETT, JOHN EMERY, b San Francisco, Calif, Mar 19, 32; m 56; c 4. ANATOMY, NEUROPHYSIOLOGY. Educ: Univ Wash, AB, 56; Univ Calif, Los Angeles, PhD(anat), 60. Prof Exp: NSF fel, Univ Pisa, 60-61, NIH fel, 61-62; neurophysiologist, Good Samaritan Hosp, Portland, Ore, 62-66; assoc prof anat, State Univ NY Upstate Med Ctr, 66-67; ASSOC PROF ANAT, MED SCH, UNIV COLO, DENVER, 67- Concurrent Pos: NIH res grants, 64- Mem: Am Asn Anatomists; Am Physiol Soc; Soc Neurosci. Res: Anatomical and functional connections of sensory nerve fibers from muscle stretch receptors; the role of stretch receptor regulation of motor activity in posture and locomotion. Mailing Add: Dept of Anat Univ of Colo Med Sch Denver CO 80220

SWETT, KEENE, b Wilton, Maine, Nov 6, 32; m 54; c 2. GEOLOGY. Educ: Tufts Univ, BS, 55; Univ Colo, MS, 61; Univ Edinburgh, PhD(geol), 65. Prof Exp: Asst lectr geol, Univ Edinburgh, 63-66; asst prof, 66-70, ASSOC PROF GEOL, UNIV IOWA, 70- Concurrent Pos: NSF fel, 67- Mem: Geol Soc Am; Am Asn Petrol Geol; Soc Econ Paleont & Mineral; Int Asn Sedimentol. Res: Petrological studies of sediments and sedimentary rocks with especial regard to the post-depositional alterations and patterns of diagenesis; Cambro-Ordovician shelf sediments of western Newfoundland, northwest Scotland and central eastern Greenland; a case for continental plate separation. Mailing Add: Dept of Geol Univ of Iowa Iowa City IA 52240

SWEZ, JOHN ADAM, b Cleveland, Ohio, Nov 18, 41; m 65; c 1. BIOPHYSICS. Educ: Pa State Univ, BS, 63, MS, 65, PhD(biophys), 67. Prof Exp: Asst prof physics, 67-71, ASSOC PROF PHYSICS & DIR RADIATION LAB, IND STATE UNIV, TERRE HAUTE, 71- Mem: AAAS; Radiation Res Soc; Biphys Soc. Res: Degradation studies of DNA in Escherichia coli after ionizing radiation, physical characterization and effect of bacteriophage infection; injection of nucleic acid of bacteriophage T1 into its host Escherichia coli. Mailing Add: Dept of Physics Ind State Univ Terre Haute IN 47809

SWEZEY, ROBERT LEONARD, b Pasadena, Calif, Apr 30, 25; m 49; c 3. INTERNAL MEDICINE, PHYSICAL MEDICINE & REHABILITATION. Educ: Ohio State Univ, MD, 48; Am Bd Internal Med, dipl, 60; Am Bd Phys Med & Rehab, dipl, 69; Am Bd Rheumatology, dipl, 74. Prof Exp: Intern, Los Angeles County Gen Hosp, Calif, 48-49 & Am Hosp, Paris, 49-50; resident internal med, Wadsworth Gen Med & Surg Vet Admin Hosp, 51-54, fel rheumatol, Wadsworth Gen Med & Surg Vet Admin Hosp & Univ Calif, Los Angeles, 54-55, clin asst med, Univ Calif, Los Angeles, 55-57, clin instr, 57-64, asst prof, 64-65, career fel phys med & rehab, 65 & 66-67; acad career fel, Univ Minn, 65-66; assoc prof internal med, phys med & rehab, Sch Med, Univ Southern Calif, 67-73, prof med, phys med & rehab, 73-74; PROF MED & DIR DIV REHAB MED, SCH MED, UNIV CALIF, LOS ANGELES, 74- Mem: AMA; Am Acad Phys Med & Rehab; Am Rheumatism Asn;

fel Am Col Physicians; Asn Acad Physiatrists. Res: Rheumatology; clinical aspects of rheumatic diseases; mechanisms of arthritic deformities and their therapeusis. Mailing Add: Dept of Med Univ of Calif Sch Med Los Angeles CA 90024

SWEZEY, WILLIAM WEEKLEY, b Chippewa Lake, Ohio, Aug 10, 08; m 31; c 1. BIOLOGY. Educ: Mt Union Col, BS, 30; Johns Hopkins Univ, ScD(protozool), 33. Prof Exp: Asst protozool, Johns Hopkins Univ, 34-35; prof biol, York Col, 35-36; prof biol, Defiance Col, 36-46, dir pub rels, 43-45, asst to pres, 45-46; prof zool, 46-57, prof biol & dean col, 57-74, vpres acad affairs, 72-74, EMER PROF BIOL & EMER DEAN COL, GROVE CITY COL, 74- Concurrent Pos: Lectr, Ft Wayne Exten, Ind Univ, 39-46. Mem: Am Soc Microbiol; Am Soc Parasitol. Res: Parasitic infusoria; malaria. Mailing Add: Crawford Hall Grove City Col Grove City PA 16127

SWIATEK, KENNETH ROBERT, b Chicago, Ill, Dec 30, 35; m 60; c 4. NEUROSCIENCES. Educ: NCent Col, BS, 58; Univ Ill Med Ctr, PhD(biol sci), 65. Prof Exp: Res assoc biol chem, Dept Pediat, Univ Ill Col Med, 65-68; res scientist, 68-70, admin res sci, 70-75, DIR RES, ILL INST DEVELOP DISABILITIES, 75- Concurrent Pos: Grant carbohydrate metab newborn animal, NIH, 71-74. Mem: Am Chem Soc; Sigma Xi; AAAS; Soc Develop Biol; Am Asn Ment Deficiency. Res: Study of growth of nervous system with special emphasis on development of carbohydrate, ketone and amino acid metabolism in fetal and newborn brain tissue as affected by pain-relieving drugs of labor and delivery. Mailing Add: Ill Inst Develop Disabilities 1640 W Roosevelt Rd Chicago IL 60608

SWICK, KENNETH EUGENE. b Silver Lake, Ind, Jan 20, 36; m 57. MATHEMATICS. Educ: Anderson Col, BS, 57; Univ Southern Calif, MA, 64; Univ Iowa, PhD(math), 67. Prof Exp: Mathematician, Hughes Aircraft Corp, Calif, 62-64; asst prof, Cornell Col, 66-67 & Occidental Col, 67-70; ASST PROF MATH, QUEENS COL, NY, 70- Mem: Math Asn Am; Am Math Soc; Soc Indust & Appl Math. Res: Stability theory for nonlinear differential equations. Mailing Add: Dept of Math Queens Col New York NY 11367

SWICK, ROBERT WINFIELD, b Jackson, Mich, July 6, 25; m 47; c 4. BIOCHEMISTRY. Educ: Beloit Col, BS, 47; Univ Wis, MS, 49, PhD(biochem), 51. Prof Exp: Assoc biochemist, Div Biol & Med, Argonne Nat Lab, 51-69; PROF NUTRIT SCI, UNIV WIS-MADISON, 69- Mem: Am Soc Cell Biol; Am Inst Nutrit; Am Soc Biol Chem. Res: Control and kientics of protein metabolism; regulation of enzyme levels and metabolism. Mailing Add: Dept of Nutrit Sci Univ of Wis Madison WI 53706

SWICKLIK, LEONARD JOSEPH, b Nanticoke, Pa, Jan 26, 28; m 52; c 3. ORGANIC CHEMISTRY. Educ: Wilkes Col, BS, 49; Univ Pittsburgh, PhD(chem', 54. Prof Exp: Res chemist, E I du Pont de Nemours & Co, Va, 54; res chemist, 56-64, sr res chemist & group leader, 64-68, tech assoc, 68-70, head, Process Improv Lab, 71-73, SUPVR CHEM PROCESS, DISTILLATION PROD DIV, EASTMAN KODAK CO, 73- Mem: AAAS; Am Chem Soc. Res: Food applications of emulsifiers; fat and oil chemistry as applicable to edible products; food and animal feed applications of vitamins. Mailing Add: 92 Northwick Dr Rochester NY 14617

SWIDER, WILLIAM, JR, b Brooklyn, NY, Jan 5, 34; m 59; c 5. AERONOMY. Educ: Lehigh Univ, BS, 55, MS, 57; Pa State Univ, PhD(physics), 63. Prof Exp: Assoc physicist, Int Bus Mach Corp, S7-60; PHYSICIST, AIR FORCE CAMBRIDGE RES LABS, 64- Concurrent Pos: Res fel, Nat Ctr Space Res, Brussels, Belgium, 63-64; mem comn III, US Nat comt, Int Sci Radio Union, 66- Mem: Am Geophys Union; Sigma Xi. Res: Ionospheric physics; airglow. Mailing Add: Aeronomy Lab Air Force Cambridge Res Labs Laurence G Hanscom AFB Bedford MA 01731

SWIDLER, RONALD, organic chemistry, see 12th edition

SWIERSTRA, ERNEST EMKE, b Netherlands, Aug, 14, 30; Can citizen; m 62; c 2. ANIMAL PHYSIOLOGY. Educ: Univ Groningen, dipl, 51; Univ BC, BSA, 56, MSA, 58; Cornell Univ, PhD(physiol), 62. Prof Exp: Asst animal breeding, Cornell Univ, 58-62; SR RES SCIENTIST, RES STA, CAN DEPT AGR, 62- Mem: Am Soc Animal Sci; Can Soc Animal Sci; Brit Soc Study Fertil; Soc Study Reproduction. Res: Reproductive physiology. Mailing Add: Res Sta Can Dept of Agr Brandon MB Can

SWIFT, ABBOTT MONTAGUE, b New York, NY, Oct 21, 21; m 42; c 3. ORGANIC CHEMISTRY. Educ: Yale Univ, BS, 47, PhD(org chem', 50. Prof Exp: Chemist, 50-52, chemist in charge, Pre-Pilot Plant, 52-53, group leader, 53-55, supvr spec projs, Market Develop Dept, 55-61, chmn chem dept, Explosives & Mining, 61-63, asst mgr, Mkt Res Dept, Commercial Develop Div, 63-67, asst mgr, Mkt Develop Dept, 67-69, mgr tech admin, 69-70, tech mgr, Pigments Div, Color Pigments Dept, 71-74, MGR PLANNING & MKT RES, ORG CHEMICALS DIV, AM CYANAMID CO, 74- Mem: Am Chem Soc. Res: Chemical soil stabilization; corporate market research and research guidance; fuel cells; metal-air batteries; photochromic products; colored pigments. Mailing Add: Am Cyanamid Co Bound Brook NJ 08805

SWIFT, ARTHUR REYNDERS, b Worcester, Mass, July 25, 38; m 61; c 3. PHYSICS. Educ: Swarthmore Col, BA, 60; Univ Pa, PhD(physics), 64. Prof Exp: NATO fel physics, Cambridge Univ, 64-65; res assoc, Univ Wis, 65-67; asst prof, 67-70, ASSOC PROF PHYSICS, UNIV MASS, AMHERST, 70- Mem: Am Phys Soc. Res: Elementary particle theory. Mailing Add: Dept of Physics Univ of Mass Amherst MA 01002

SWIFT, BRINTON L. b Denver, Colo, June 12, 26; m 56; c 1. VETERINARY MEDICINE. Educ: Colo State Univ, DVM, 51. Prof Exp: Private practice, Buffalo, Wyo, 51-64; from asst prof to assoc prof, 64-69, PROF VET MED, UNIV WYO, 69- Mem: Am Vet Med Asn; Sigma Xi; Soc Theriogenology. Res: Bovine fetal disease. Mailing Add: Div Microbiol & Vet Med Univ of Wyo Laramie WY 82070

SWIFT, CAMM CHURCHILL, b Oakland, Calif, Sept 29, 40. ICHTHYOLOGY. Educ: Univ Calif, Berkeley, AB, 63; Univ Mich, Ann Arbor, MS, 65; Fla State Univ, PhD(biol), 70. Prof Exp: ASSOC CUR FISHES, NATURAL HIST MUS LOS ANGELES COUNTY, 70- Concurrent Pos: Adj asst prof, Dept Biol Sci, Univ Southern Calif, 72-; bd gov, Am Soc Ichthyol & Herpet, 73- Mem: Am Soc Ichthyol & Herpet; Soc Vert Paleont; Soc Study Evolution; Soc Syst Zool; Am Fisheries Soc. Res: Systematics and evolution of Recent and fossil, freshwater and marine shore fishes of North America. Mailing Add: Sect of Fishes Natural Hist Mus Los Angeles Co 900 Exposition Blvd Los Angeles CA 90007

SWIFT, CHARLES JAMES, b Eaton, Ohio, May 3, 18; m 46; c 2. COMPUTER SCIENCES. Educ: Haverford Col, BS, 40; Purdue Univ, MS, 42; Univ Pa, PhD(physics), 50. Prof Exp: Asst physicist, Sharples Res Corp, 46-47, asst prof, Am Univ, 50-51; physicist, Nat Bur Standards, 51-54; sr res engr, Gen Dynamics/Convair 54-59; SR SCIENTIST, COMPUT SCI CORP, EL SEGUNDO, 59- Mem: Am Phys Soc; Am Asn Physics Teachers; Asn Comput Mach. Res: Geomagnetic field; programming for computing machines. Mailing Add: 3520 S Cloverdale Ave Los Angeles CA 90016

SWIFT, DANIEL W, b Worcester, Mass, Mar 6, 35; m 61; c 2. PLASMA PHYSICS, SPACE PHYSICS. Educ: Haverford Col, BA, 57; Mass Inst Technol, MS, 59. Prof Exp: Mem staff, Lincoln Lab, Mass Inst Technol, 58-59; sr scientist, Res & Adv Develop Div, Avco Corp, 59-63; from asst prof to assoc prof, 63-72, PROF GEOPHYS, GEOPHYS INST, UNIV ALASKA, 72- Mem: AAAS; Am Geophys Union. Res: Theoretical studies of the earth's magnetosphere and auroral phenomena. Mailing Add: Geophys Inst Univ of Alaska Fairbanks AK 99701

SWIFT, DAVID LESLIE, b Chicago, Ill, Aug 7, 35; m 59; c 3. ENVIRONMENTAL MEDICINE, PHYSIOLOGY. Educ: Purdue Univ, BS, 57; Mass Inst Technol, SM, 59; Johns Hopkins Univ, PhD(chem eng), 63. Prof Exp: Chem engr, Argonne Nat Lab, Ill, 63-65; USPHS air pollution spec fel, London Sch Hyg, 65-66; asst prof, 66-70, ASSOC PROF ENVIRON MED, JOHNS HOPKINS UNIV, 70- Mem: Air Pollution Control Asn; Am Conf Govt Indust Hyg. Res: Physiology of respiratory tract; fate of inhaled particles and gases; fluid mechanics and transport in biological systems; air pollution transport. Mailing Add: Dept of Environ Med Johns Hopkins Univ Baltimore MD 21205

SWIFT, DONALD J P, b Dobbs Ferry, NY, July 26, 35; m 61; c 2. SEDIMENTOLOGY, OCEANOGRAPHY. Educ: Dartmouth Col, AB, 57; Johns Hopkins Univ, MA, 61; Univ NC, PhD(geol), 64. Prof Exp: Asst prof sedimentology, Dalhousie Univ, 63-66; assoc scientist marine geol, P R Nuclear Ctr, 66-67; assoc prof geol, Duke Univ, 67-68; Slover assoc prof, Inst Oceanog, Old Dom Univ, 68-71; RES OCEANOGR, ATLANTIC OCEANOG & METEOROL LABS, NAT OCEANIC & ATMOSPHERIC ADMIN, 71- Concurrent Pos: Res grant, Geol Surv Can, 64-65; res assoc, NS Mus, 65-66; assoc ed, Maritime Sediments, 65-66; res grant, Nat Res Coun Can & Defense Res Bd Can, 65-67, Coastal Eng Res Ctr, 69-71, US Geol Surv, 69-71 & NSF, 69-72; mem, Univ Senate. Old Dom Univ, 69-71, chmn, Univ Res Comt. 69-71; res grant, NASA, 70-71; proj leader, Continental Margin Sedimentation Proj, 72-; consult, Oceanog Panel, NSF, 74-; Res Assoc, Smithsoniam Inst, 74- Mem: AAAS; Soc Econ Paleont & Mineral; Geol Soc Am; Int Asn Sedimentol; Am Geophys Union. Res: Continental margin hydraulics; sediment transport. Mailing Add: Atlantic Oceanog & Meteorological Labs Nat Oceanic & Atmospheric Admin Miami FL 33149

SWIFT, DOROTHY GARRISON, b Flint, Mich, Aug 1, 39; m 61. BIOLOGICAL OCEANOGRAPHY. Educ: Swarthmore Col, BA, 61; Johns Hopkins Univ, MA, 67, PhD(oceanog), 73. Prof Exp: Guest student investr, Biol Dept, Woods Hole Oceanog Inst, 68-69; guest investr, Grad Sch Oceanog, 70-73, RES ASSOC, DEPT CHEM, UNIV RI, 74- Mem: Phycol Soc Am; Am Soc Limnol & Oceanog; Int Phycol Soc; Sigma Xi. Res: Biochemistry, nutrition and ecology of marine phytoplankton. Mailing Add: Dept of Chem Univ of RI Kingston RI 02881

SWIFT, ELIJAH, V, b Boston, Apr 12, 38; m 61. BIOLOGICAL OCEANOGRAPHY. Educ: Swarthmore Col, BA, 60; Johns Hopkins Univ, MA, 64, PhD(oceanog), 67. Prof Exp: NSF trainee, Woods Hole Oceanog Inst, 68-69; asst prof, 69-74, ASSOC PROF OCEANOG, UNIV RI, 74- Concurrent Pos: NSF grants, 71-76. Mem: Phycol Soc Am; Int Phycol Soc; Marine Biol Asn UK; Brit Phycol Soc; Am Soc Limnol & Oceanog. Res: Morphology, taxonomy, ecology and physiology of phytoplankton, particularly marine species. Mailing Add: Grad Sch of Oceanog Dept of Bot Univ of RI Kingston RI 02881

SWIFT, ERNEST HAYWOOD, o Chase City, Va, July 2, 97; m 22; c 1. ANALYTICAL CHEMISTRY. Educ: Univ Va, BS, 18; Calif Inst Technol, MS, 20, PhD(chem), 24; Randolph-Macon Col, LLD, 60. Prof Exp: Instr anal chem, 20-28, from asst prof to assoc prof, 28-43, prof, 43-67, chmn div chem & chem eng, 58-63, chmn faculty & faculty bd, 63-65, EMER PROF ANAL CHEM, CALIF INST TECHNOL, 67- Concurrent Pos: Guggenheim fel, 57-58; mem chem adv panel, NSF; adv bd, Ctr Adv Studies, Univ Va, 65- Honors & Awards: Fisher Award, Am Chem Soc, 55, Tolman Award, 62. Mem: Fel Am Inst Chem; Am Chem Soc; Am Acad Arts & Sci. Res: Analytical methods; half-cell potentials; distribution phenomena; coulometric methods; mechanism precipitation sulfides by thioacetamide. Mailing Add: Dept of Chem Calif Inst of Technol Pasadena CA 91125

SWIFT, FRED CALVIN, b Middleport, NY, Oct 16, 26; m 50; c 3. ENTOMOLOGY, ECOLOGY. Educ: Mich State Univ, BS, 50; Iowa State Col, MS, 52; Rutgers Univ, PhD, 58. Prof Exp: Entomologist, Niagara Chem Div, Food Mach & Chem Corp, 52-55; asst prof entom, Clemson Col, 58-59; assoc prof, 61-72, RES PROF ENTOM, RUTGERS UNIV, NEW BRUNSWICK, 72- Mem: Entom Soc Am; Ecol Soc Am. Res: Integrated control of fruit insect pests; ecology of the Phytoseiidae. Mailing Add: Dept of Entom & Econ Zool Col of Agr Env Sci Rutgers Univ New Brunswick NJ 08903

SWIFT, GEORGE HERBERT, JR, b Minot, NDak, July 1, 26; m 58; c 3. MATHEMATICS, COMPUTER SCIENCE. Educ: Univ Ore, BS, 49, MS, 51; Univ Wash, PhD, 54. Prof Exp: Instr math, Duke Univ, 54-56; appl sci rep, IBM Corp, Wash, 56-57, Colo, 57-58, asst engr prod develop lab, NY, 58-60, staff systs planner, 60-61, mgr large comput plans, Fed Systs Div, 61-64, northwest sci rep, 64-66; adj assoc prof info sci, Wash State Univ, 66-68; planner, IBM Corp, Fla, 68-72, mem staff, Develop Comput Systs, Gen Systs Div, 72-74, MEM STAFF, IBM CORP, MINN, 74- Mem: Am Math Soc; Math Asn Am; Asn Comput Mach. Res: Development of computer systems; irregular measures on topological spaces; irregular and n-valued irregular borel measures; matrix inversion by partition; information retrieval. Mailing Add: IBM Corp Dept 275 Bldg 040-3 Rochester MN 55901

SWIFT, GRAHAM, b Chesterfield, Eng, Apr 16, 39; m 61; c 2. ORGANIC POLYMER CHEMISTRY. Educ: Univ London, BSc, 61, PhD(org chem', 64. Prof Exp: NIH fel, Fels Res Inst, Philadelphia, 64-66; sci off, Imperial Chem Industs, Eng, 66-68; SR CHEMIST, ROHM AND HAAS CO, 68- Mem: The Chem Soc; Royal Inst Chem; Am Chem Soc. Res: Heterocyclic chemistry; synthesis and reactions of special aziridines, oxiranes and thiiranes; fatty acid chemistry; organic polymer coatings, especially synthesis and evaluation of novel coating compositions. Mailing Add: Rohm and Haas Co Spring House PA 19477

SWIFT, HAROLD EUGENE, b Butler, Pa, Mar 27, 36; m 60; c 3. PHYSICAL INORGANIC CHEMISTRY. Educ: Allegheny Col, BS, 58; Pittsburgh Univ, PhD(chem), 62. Prof Exp: Chemist, 62-63, res chemist, 62-63, sr res chemist, 63-65, SECT SUPVR, CATALYSIS SECT, GULF RES & DEVELOP CO, PITTSBURGH, 66- Mem: Am Chem Soc. Res: Heterogeneous and homogeneous catalysis for chemical, refining and pollution applications; catalyst preparation, characterization, evaluation and development. Mailing Add: 1410 Woodhill Dr Gibsonia PA 15044

SWIFT, HEWSON HOYT, b Auburn, NY, Nov 8, 20; m 42; c 2. CYTOLOGY. Educ: Swarthmore Col, BA, 42; Univ Iowa, MS, 45; Columbia Univ, PhD(zool), 50. Prof Exp: Instr zool, 49-51, from asst prof to assoc prof, 51-59, prof, 59-71, DISTINGUISHED SERV PROF BIOL, UNIV CHICAGO, 71-, CHMN DEPT, 72- Concurrent Pos: Mem, Cell Biol Study Sect, NIH, 58-62, Develop Biol Adv Panel, NSF, 62-65 & Etiology Cancer Adv Panel, Am Cancer Soc, 56-70; vis prof, Harvard

Univ, 70-71. Mem: Nat Acad Sci; Am Soc Cell Biol(pres, 64); Histochem Soc(pres, 73); Genetics Soc Am; Int Soc Cell Biol. Res: Cell biology; cytochemistry. Mailing Add: Whitman Lab Univ of Chicago Chicago IL 60637

SWIFT, IRVIN HENRY, physics, mathematics, see 12th edition

SWIFT, JACK BERNARD, b Ft Smith, Ark, Jan 3, 42; m 71. THEORETICAL SOLID STATE PHYSICS. Educ: Univ Ark, Fayetteville, BS, 63; Univ Ill, Urbana, MS, 65, PhD(physics), 68. Prof Exp: NSF fel, Max Planck Inst Physics & Astrophys, Munich, 68-69; res fel, Harvard Univ, 69-71; vis, Bell Tel Lab, 74; asst prof physics, 71-75; ASSOC PROF PHYSICS, UNIV TEX, AUSTIN, 75- Concurrent Pos: Res Fel, A P Sloan Found, 73-75. Res: Critical phenomena; light scattering properties of liquids; hydrodynamics. Mailing Add: Dept of Physics Univ of Tex Austin TX 78712

SWIFT, JONATHAN DEAN, b Portland, Ore, Nov 12, 18; m 47, 68. MATHEMATICS. Educ: Univ Calif, BA, 39; Calif Inst Technol, PhD(abstract algebra), 47. Prof Exp: From instr to assoc prof, 47-62, PROF MATH, UNIV CALIF, LOS ANGELES, 62- Mem: AAAS; Am Math Soc; Math Asn Am. Res: Galois fields; number theory; combinatorial analysis; finite geometries. Mailing Add: Dept of Math Univ of Calif Los Angeles CA 90024

SWIFT, LLOYD HARRISON, b Crete, Nebr, Sept 12, 20. PLANT MORPHOLOGY. Educ: Univ Nebr, AB, 41, MS, 60, PhD(bot), 62; Western Reserve Univ, MA, 42. Prof Exp: Lexicographer, World Pub Co, Ohio, 42; instr Eng, Ill Inst Technol, 44 & Univ Mo, 44-45; asst prof bot, Univ Alaska, 62-63; prof & chmn dept Univ Nebr, Ataturk Univ, Turkey, 63-65; RES ASSOC BOT, UNIV NEBR, 65- Mem: Am Inst Biol Sci. Res: Botanical bibliography and the classification and indexing of information important in botany; etymology and history of botanical terminology and nomenclature; phytography in plant morphology, taxonomy and physiology. Mailing Add: 2210 Sewell St Lincoln NE 68502

SWIFT, LLOYD WESLEY, JR, b San Francisco, Calif, July 11, 32; m 55; c 4. MICROMETEOROLOGY, FOREST HYDROLOGY. Educ: State Univ NY Col Forestry, Syracuse, BS, 54; NC State Univ, MS, 60; Duke Univ, DF, 72. Prof Exp: Jr forester, 54-55, res forester, 57-71, RES METEOROLOGIST, FOREST SERV, USDA, 72- Mem: Am Geophys Union; Am Meteorol Soc. Res: Energy balance of forested and logged mountain slopes; precipitation measurement and distribution over steep slopes; air circulation patterns in mountains; effect of slope aspect and inclination on forest microenvironment. Mailing Add: Coweeta Hydrol Lab Box 601 Franklin NC 28734

SWIFT, LYLE JAMES, b Eaton Co, Mich, Sept 17, 06; m 38; c 4. AGRICULTURAL CHEMISTRY. Educ: Mich State Col, BS, 35, MS, 36; Purdue Univ, PhD(agr biochem), 44. Prof Exp: Asst biochem, Mich State Col, 35-36; asst biochem, Purdue Univ, 36-37, asst chemist, 37-44, chemist, Fruit & Veg Prods Lab, USDA, 44-75; RETIRED. Mem: Am Chem Soc; Asn Off Anal Chem. Res: Isolation and identification of natural compounds; peppermint oil analysis; analytical methods and constituents of citrus juice lipids; origin of off-flavors in processed citrus commodities; bitter constituents of orange peel; agricultural biochemistry. Mailing Add: 1597 17th St NW Winter Haven FL 33880

SWIFT, ROBINSON MARDEN, b Wolfeboro, NH, May 6, 18; m 44; c 3. PHYSICAL CHEMISTRY. Educ: Univ NH, BS, 40; Northwestern Univ, MS, 48; Syracuse Univ, PhD(chem), 56. Prof Exp: Chemist, Bird & Son, Inc, Mass, 40-44; instr chem, Thiel Col, 47-53; from asst prof to assoc prof, 56-72, PROF CHEM, ST ANSELM'S COL, 72-; CHIEF CHEMIST, EDISON ELECTRONICS DIV, McGRAW-EDISON, 57- Mem: Am Chem Soc. Res: Thermodynamics; epoxy resins. Mailing Add: Dept of Chem St Anselm's Col Manchester NH 03102

SWIFT, TERRENCE JAMES, b Dubuque, Iowa, June 29, 37; m 65; c 2. PHYSICAL CHEMISTRY, BIOCHEMISTRY. Educ: Loras Col, BS, 59; Univ Calif, Berkeley, PhD(chem), 62. Prof Exp: NSF fel, Max Planck Inst Phys Chem, 62-63; asst prof chem, 63-67, ASSOC PROF CHEM, CASE WESTERN RESERVE UNIV, 67- Mem: Am Soc Biol Chem. Res: Magnetic resonance spectroscopy as applied to biological systems and processes. Mailing Add: Dept of Chem Case Western Reserve Univ Cleveland OH 44106

SWIFT, WILLIAM CLEMENT, b Lexington, Ky, Mar 17, 28; m 50; c 8. MATHEMATICS. Educ: Univ Ky, BS, 50, PhD(math), 55. Prof Exp: Instr math, Cornell Univ, 55-56; mem tech staff, Bell Tel Labs, Inc, 56-58; asst prof math, Rutgers Univ, 58-63; assoc prof, 63-69, PROF MATH, WABASH COL, 69- Mem: Am Math Soc; Math Asn Am. Res: Complex variables; conformal mapping; Taylor series; summability. Mailing Add: 116 N Grace Ave Crawfordsville IN 47933

SWIGART, JOHN IRVIN, b Metcalf, Ill, Jan 31, 04; m 28, 38. EXPERIMENTAL PHYSICS. Educ: Ill Wesleyan Univ, BS, 29; Ind Univ, AM, 30, PhD(physics), 38. Prof Exp: Lab asst, Ill Wesleyan Univ, 27-29; asst, Ind Univ, 29-30; asst prof, Bethany Col, WVa, 30-31; instr, 31-39, from asst prof to assoc prof. 39-45, PROF PHYSICS, UNIV UTAH, 46- Concurrent Pos: Consult, Follansbee Bros Co, WVa, 30-31, Peterson Eng Co, 43-48 & Lockheed Aircraft Corp, 61- Res: Acoustics and electricity; precision apparatus for measuring the velocity of sound in solid rods; frequency measurement in alternating current circuits; aeronautical physics; aerial navigation; physical properties of the upper atmosphere; high altitude research rockets. Mailing Add: 385 Third Ave Salt Lake City UT 84103

SWIGART, RICHARD HANAWALT, b Lewistown, Pa, July 7, 25; m 51; c 3. NEUROANATOMY. Educ: Univ NC, BA, 47; Univ Minn, PhD(anat), 53. Prof Exp: Asst anat, Univ Minn, 48-50, instr, 50-52, res assoc histochem, 52-53; from asst prof to assoc prof, 53-67, asst dean student affairs, 69-72, PROF ANAT, SCH MED, UNIV LOUISVILLE, 67-, ACTG DEAN, 72- Mem: Am Asn Anat; Biol Stain Comn; Histochem Soc; Soc Exp Biol & Med. Res: Chronic hypoxia, effect on cardiovascular and erythropoietic systems; carbohydrate metabolism in cardiac and skeletal muscle; effect of age on adaptive responses; neurological mutant mice. Mailing Add: Off of the Dean Health Sci Ctr Univ of Louisville Sch of Med Louisville KY 40202

SWIGER, ELIZABETH DAVIS, b Morgantown, WVa, June 27, 26; m 48; c 2. PHYSICAL CHEMISTRY. Educ: Univ WVa, BS, 48, MS, 52, PhD, 65. Prof Exp: Instr math, 54-56, instr chem, 56-60, from asst prof to assoc prof, 60-66, PROF CHEM, FAIRMONT STATE COL, 66- Mem: Am Chem Soc. Res: Nuclear quadrupole resonance spectroscopy; polarography; coordination compounds. Mailing Add: 1599 Hillcrest Rd Fairmont WV 26554

SWIGER, LOUIS ANDRE, b Waverly, Ohio, Sept 16, 32; m 53; c 3. ANIMAL GENETICS. Educ: Ohio State Univ, BSc, 54; Iowa State Univ, MSc, 57, PhD(animal breeding), 60. Prof Exp: Animal geneticist, USDA, 59-62; assoc prof animal sci & exp sta statist, Univ Nebr, 62-65; assoc prof, 65-70, PROF ANIMAL SCI, OHIO STATE UNIV, 70- Mem: Am Soc Animal Sci; Biomet Soc. Res: Population genetics and application to domestic animals. Mailing Add: Dept of Animal Sci Ohio State Univ Columbus OH 43210

SWIHART, JAMES CALVIN, b Elkhart, Ind, Feb 8, 27; m 47; c 2. THEORETICAL PHYSICS. Educ: Purdue Univ, BSChE, 49, MS, 51, PhD(physics), 55. Prof Exp: Danish govt fel, Inst Theoret Physics, Copenhagen Univ, 54-55; physicist, Argonne Nat Lab, 55-56; physicist, Res Ctr, Int Bus Mach Corp, 56-66; assoc prof, 66-67, assoc dean grad sch, 71-74, PROF PHYSICS, IND UNIV, BLOOMINGTON, 67- Mem: AAAS; fel Am Phys Soc; Biophys Soc. Res: Theory of solid state physics; superconductivity; many-body problem; biophysics. Mailing Add: Dept of Physics Ind Univ Bloomington IN 47401

SWIHART, THOMAS LEE, b Elkhart, Ind, July 29, 29; m 51; c 3. ASTROPHYSICS. Educ: Ind Univ, AB, 51, AM, 52; Univ Chicago, PhD(astrophys), 55. Prof Exp: Assoc prof physics & astron, Univ Miss, 55-57; mem staff, Los Alamos Sci Lab, 57-62; asst prof astrophys, Univ Ill, 62-63; assoc prof astron, 63-69, PROF ASTRON, UNIV ARIZ, 69-, ASTRONR, STEWARD OBSERV, 74- Concurrent Pos: Fulbright-Hays lectr, Aegean Univ, Turkey, 69-70. Mem: AAAS; Int Astron Union; Am Astron Soc. Res: Theoretical astrophysics; radiation transfer; polarization of radio sources; atmospheric structure of stars. Mailing Add: Steward Observ Univ of Ariz Tucson AZ 85721

SWIM, RICHARD TAYLOR, solid state physics, ocean engineering, see 12th edition

SWIMMER, ALVIN, mathematics, see 12th edition

SWINDALE, LESLIE D, b Wellington, NZ, Mar 16, 28; m 55; c 3. SOIL SCIENCE, RESEARCH ADMINISTRATION. Educ: Univ Victoria, NZ, 48, MSc, 50; Univ Wis, PhD(soil sci), 55. Prof Exp: Phys chemist, NZ Soil Bur, 49-57, sr phys chemist, 57-60; dir, NZ Pottery & Ceramics Res Asn, 60-63; chmn, Dept Agron & Soil Sci, 65-68, PROF SOIL SCI, UNIV HAWAII, MANOA, 63- Concurrent Pos: Fel, Univ Wis, 55-56; chief soil resources, Conservation and Development Serv, Land & Water Develop Div, Food & Agr Orgn, UN, Rome, 68-70. Mem: Soil Sci Soc Am; NZ Inst Chem; Royal Soc NZ; Int Soc Soil Sci; Am Soc Pub Admin. Res: Formation, transformation and properties of minerals in soils and clays. Genesis of soils, their characterization and uses. Mailing Add: Hawaii Agr Exp Sta Univ of Hawaii at Manoa Honolulu HI 96822

SWINDELL, ROBERT THOMAS, b Greenfield, Tenn, Feb 22, 38; m 61; c 1. ORGANIC CHEMISTRY. Educ: Memphis State Univ, BS, 61; Univ SC, PhD(org chem), 65. Prof Exp: NIH fel org photochem, Iowa State Univ, 65-66; ASST PROF CHEM, TENN TECHNOL UNIV, 66- Mem: Am Chem Soc; The Chem Soc. Res: Organic reaction mechanisms; organic photochemistry. Mailing Add: Dept of Chem Tenn Technol Univ Box 5055 Cookeville TN 38501

SWINDELL, WILLIAM, b Eng, June 3, 38; m 63; c 2. OPTICS. Educ: Univ Sheffield, BSc, 59, PhD(physics), 64. Prof Exp: Lectr physics, Melbourne Univ, 64-68; from asst prof to assoc prof, 68-75, PROF, OPTICAL SCI CTR, UNIV ARIZ, 75- Honors & Awards: Public Serv Group Achievement Award, NASA, 74. Mem: Optical Soc Am; fel Brit Inst Physics; assoc mem Australian Inst Physics. Res: Pioneer Jupiter image processing; polarized light; optical instrumentation; medical optics. Mailing Add: Optical Sci Ctr Univ of Ariz Tucson AZ 85721

SWINDELLS, FRANK EVANS, b Washington, DC, Feb 23, 99; m 28; c 2. PHYSICAL CHEMISTRY. Educ: Carnegie Inst Technol, BS, 21; Univ Pa, MS, 23, PhD(phys chem), 25. Prof Exp: Asst chemist, Nat Bur Stand, 21-22; res chemist, Patterson Screen Co, 25-35, res dir, 35-43, res dir, Patterson Screen Div, E I du Pont de Nemours & Co, 43-46; res assoc, Photoprods Dept, 46-48; engr mgr, Allen B Du Mont Labs, 58-60; consult, 60-63; sr scientist & br supt, Melpar, Inc, Va, 63-70; VPRES, ARTECH CORP, FALLS CHURCH, 70- Concurrent Pos: Mem, War Metall Bd, 44. Mem: AAAS; Am Chem Soc; Optical Soc Am; Electrochem Soc. Res: Determination of gases in metals; thermodynamics of solutions; preparation and properties of luminescent substances; manufacture of fluorescent screens; x-ray sensitometry; cathode ray tubes; photoconductors; thin film devices; electrochemical cells. Mailing Add: 859 N Jacksonville St Arlington VA 22205

SWINDLER, DARIS RAY, b Morgantown, WVa, Aug 13, 25; m 53; c 6. ANTHROPOLOGY. Educ: Univ WVa, AB, 50; Univ Pa, MA, 52, PhD(anthrop), 59. Prof Exp: Instr anat, Med Col, Cornell Univ, 56-57, Sch Med, Univ WVa, 57-59 & Med Col, Univ SC, 59-64; assoc prof anat & anthrop, Mich State Univ, 64-68; PROF ANTHROP, AFFILIATE REGIONAL PRIMATE RES CTR & UNIV WASH, 68-, ADJ CURATOR PRIMATE ANAT, THOMAS BURKE MEM WASH STATE MUS, 71- Mem: Am Asn Phys Anthrop; Am Asn Anat. Res: Evolution; primatology; physical anthropology; growth and development of primates; evolution of teeth. Mailing Add: Rm 5 Thomas Burke Mem Wash State Mus Univ of Wash Seattle WA 98195

SWINEBROAD, JEFF, b Nashville, Tenn, Mar 22, 26; m 53; c 2. ZOOLOGY. Educ: Ohio State Univ, BA, 49, MA, 50, PhD, 56. Prof Exp: Asst ornith, Univ Colo, 46-47; tech asst zool, Ohio Coop Wildlife Res Unit, 48; asst, Ohio State Univ, 49-50. asst instr, 51, instr conserv, Conserv Lab, 51-53 & Nat Audubon Soc, 54-56; zoologist, Rutgers Univ, 55-57, from asst prof to assoc prof, 57-66, prof biol, 66-68, chmn dept biol sci. 60-68. asst res specialist, Col Agr, 59-68; pop ecologist, USAEC, 68-72, chief, Ecol Sci Br, 72-74; DEP ASSOC DIR RES & DEVELOP PROGS, US ENERGY RES & DEVELOP ADMIN, 74- Mem: AAAS; Animal Behav Soc; Ecol Soc Am; Wilson Ornith Soc(secy, 67-); Cooper Ornith Union. Res: Avian anatomy; ecology and migration of birds; animal population behavior. Mailing Add: Div of Biomed and Environ Sci USAEC Washington DC 20545

SWINEFORD, ADA, b Chicago, Ill, July 12, 17. GEOLOGY. Educ: Univ Chicago, SB, 40, SM, 42; Pa State Univ, PhD(mineral), 54. Prof Exp: Geologist, State Geol Surv, Kans, 42-49; geologist & head petrog div, 49-66; from asst prof to assoc prof, 58-66; assoc prof, 66-70, PROF GEOL, WESTERN WASH STATE COL, 70- Concurrent Pos: Ed, Clays & Clay Minerals, Nat Clay Conf, 54 & 56-62. Mem: Fel AAAS; fel Geol Soc Am; fel Mineral Soc Am; Soc Econ Paleont & Mineral; Brit Mineral Soc. Res: Sedimentary petrography; clay mineralogy. Mailing Add: Dept of Geol Western Wash State Col Bellingham WA 98255

SWINEHART, BRUCE ARDEN, b Greentown, Ohio, Aug 28, 29; m 55; c 2. ANALYTICAL CHEMISTRY. Educ: Oberlin Col, AB, 51; Purdue Univ, MS, 53, PhD(anal chem), 55. Prof Exp: Res chemist, Mallinckrodt Chem Works, 55-60; res chemist, Wyandotte Chem Corp, Mich, 60-61, sr res chemist, 61-64, sect head, 64-67; RES CHEMIST, CORNING GLASS WORKS, 67- Mem: Am Chem Soc. Res: Uranium chemistry; ore analysis; gas chromatography; glass analysis; titrimetry. Mailing Add: Corning Glass Works Sullivan Park Corning NY 14830

SWINEHART, CARL FRANCIS, b Bainbridge, Ohio, Aug 26, 07; m 34; c 2. INORGANIC CHEMISTRY. Educ: Ohio Wesleyan Col, AB, 29; Western Reserve

Univ, PhD(inorg chem), 33. Prof Exp: Asst, Western Reserve Univ, 29-32; res chemist, Harshaw Chem Co, 32-52, assoc dir res, 52-60, dir tech develop, Inorg Prod, 60-62 & Crystal-Solid State, 62-67, proj dir crystal growth develop, Harshaw Div, Kewanee Oil Co, Ohio, 67-72; CONSULT, 72- Mem: Am Chem Soc. Res: Fluoride gases; manufacture of fluorides; synthetic crystal production. Mailing Add: 4102 Silsby Rd University Heights Cleveland OH 44118

SWINEHART, DONALD FOUGHT, b Strasburg, Ohio, Dec 30, 17; m 42; c 3. CHEMICAL KINETICS. Educ: Capital Univ, BS, 39; Ohio State Univ, MSc, 41, PhD(phys chem), 43. Prof Exp: Res chemist, Eastman Kodak Co, 43-44; res chemist & physicist, Manhattan Eng Dist, Los Alamos, NMex, 44-46; from asst prof to assoc prof, 46-61. PROF CHEM, UNIV ORE, 61- Concurrent Pos: Consult, AEC, 46-50. Mem: Am Chem Soc. Res: Polarographic analysis; adsorption to colloids; thermodynamics of electrolytic solutions; mass spectrometry; kinetics of unimolecular reactions; thermal conductivity of gas mixtures. Mailing Add: Dept of Chem Univ of Ore Eugene OR 97403

SWINEHART, JAMES HERBERT, b Los Angeles, Calif, Nov 22, 36; m 63; c 3. INORGANIC CHEMISTRY, BIOINORGANIC CHEMISTRY. Educ: Pomona Col, BA, 58; Univ Chicago, PhD(chem), 62. Prof Exp: NSF fel phys inorg chem, Max Planck Inst Phys Chem, Univ Göttingen, Ger, 62-63; from asst prof to assoc prof, 68-72, PROF CHEM, UNIV CALIF, DAVIS, 72- Concurrent Pos: Fel, John Simon Guggenheim Found, 69-70. Mem: Am Chem Soc. Res: Mechanisms of inorganic reactions; transition metals in the marine environment. Mailing Add: Dept of Chem Univ of Calif Davis CA 95616

SWINEHART, JAMES STEPHEN, b Cleveland, Ohio, July 27, 29; m 63; c 1. ORGANIC CHEMISTRY, SPECTROCHEMISTRY. Educ: Western Reserve Univ, BS, 50; Univ Cincinnati, MS, 51; New York Univ, PhD(chem), 59. Prof Exp: Asst, Western Reserve Univ, 49-50; res chemist, Merck & Co, 51-53; from asst prof org chem to assoc prof, Wagner Col, 57-61; assoc prof, Am Univ, 61-65; anal chemist, Atlantic Res Corp, 65-67; sr spectroscopist, Perkin Elmer Corp, 67-69 & Digilab, 69-70; PROF CHEM & CHMN DEPT, STATE UNIV NY COL CORTLAND, 70- Mem: Am Chem Soc. Res: Organic synthesis; natural products; instrumentation; infrared and nuclear magnetic resonance spectroscopy; information retrieval; cigarette smoke; gas chromatography; technical writing. Mailing Add: RD 3 11 Gwen Lane Cortland NY 13045

SWINEHART, PHILIP ROSS, b Los Alamos, NMex, May 20, 45. SOLID STATE ELECTRONICS. Educ: Ore State Univ, BS, 67; Ohio State Univ, MS, 68, PhD(elec eng), 74. Prof Exp: Res assoc, Electro Sci Lab, Ohio State Univ, 68-69; res engr, Ohio Semitronics Inc, 72-73; RES SCIENTIST LOW TEMP SENSORS & RADIATION DETECTION, LAKE SHORE CRYOTRONICS INC, 75- Mem: Inst Elec & Electronics Engrs. Res: Low temperature sensors and instrumentation. Mailing Add: Lake Shore Cryotronics Inc 4949 Freeway Dr E Columbus OH 43229

SWINFORD, KENNETH ROBERTS, b Trader's Point, Ind, July 8, 16; m 38; c 2. FORESTRY. Educ: Purdue Univ, BS, 37; Univ Fla, MSF, 48; Univ Mich, PhD, 60. Prof Exp: Exten ranger, State Forest Serv, Fla, 40-41; timber cruiser, Brooks-Scanlon Corp, 46; from asst prof to prof, 46-75, asst to dir forestry, 71-75, EMER PROF FORESTRY, UNIV FLA, 75- Concurrent Pos: Consult forester, F & W Forestry Serv Inc, Gainesville, Fla, 75-; teacher, Sch Forestry, Univ Fla, 75-76. Mem: Soc Am Foresters; Am Forestry Asn. Res: Management of forest lands, especially pine plantations; landscape forestry; outdoor recreational use of forests and wild lands. Mailing Add: F & W Forestry Serv Inc PO Box 13321 Gainesville FL 32604

SWINGLE, DONALD MORGAN, b Washington, DC, Sept 1, 22; m 43; c 3. APPLIED PHYSICS, SYSTEMS ENGINEERING. Educ: Wilson Teachers Col, BS, 43; NY Univ, MS, 47; Harvard Univ, AM, 48, MEngSci, 49, PhD(eng sci, appl physics), 50; George Washington Univ, MBA, 62; Indust Col Armed Forces, dipl, 62. Prof Exp: Engr, Signal Corps Eng Labs, US Dept Army, 46-47; res asst, Eng Res Lab, Harvard Univ, 49-50; physicist, Signal Corps Eng Labs, US Dept Army, 50-60, chief weather electronic res group, 50-53, chief meteorol techniques sect, 54-57, chief br, 58-60, physicist, sr res scientist & dep dir meteorol div, Electronics Labs, 61-63, res physicist, sr res physicist & chief meteorol res team A, US Army Electronics Command, 64-65, res physicist & sr scientist atmospheric sci lab, 65-66, res physicist & chief techniques & explor develop tech area, 66-71, sr scientist eng & explor develop tech area, 71-74, SR SCIENTIST, SPEC SENSORS TECH AREA, COMBAT SURVEILLANCE & TARGET ACQUISITION LAB, ARMY ELECTRONICS COMMAND, US DEPT ARMY, 74- Concurrent Pos: US mem, Comn Instruments & Methods Observ, World Meteorol Orgn, 53-59, US deleg, 53 & 57; adv, US Mil Acad, 59; reviewer res proposals, NSF, 60-; chmn, Nat Task Group Mesometeorol, Interdept Comt Atmospheric Sci, 63-74; Army rep, DOD Forum Environ Sci, 64-65; chmn, Nat Meso-Micrometeorol Res Facil Surv Group, Nat Ctr Atmospheric Res, 64; mem Army Res Coun, 64-65; consult, Livermore Res Lab, Univ Calif, 65. Mem: Am Meteorol Soc; sr mem Inst Elec & Electronics Engrs; Nat Soc Prof Eng; assoc fel Am Inst Aeronaut & Astronaut; fel NY Acad Sci. Res: Radar meteorology; atmospheric propagation, electromagnetic, acoustic waves; radioactive fallout prediction; meteorological techniques, applied meteorology, meteorological system engineering; atmospheric modification, management, mesometeorology; indirect sensory techniques; nuclear surveillance; research and development management. Mailing Add: 414 Prospect Ave Neptune NJ 07753

SWINGLE, HOMER DALE, b Hixson, Tenn, Nov 5, 16; m 42, 62; c 1. HORTICULTURE. Educ: Univ Tenn, BS, 39; Ohio State Univ, MS, 48; La State Univ, PhD(hort), 66. Prof Exp: Teacher high sch, 39-46; exten specialist hort, Agr Exten Serv, 46-47, from asst prof to assoc prof, 48-67, PROF, UNIV TENN, KNOXVILLE, 67- Concurrent Pos: Consult plant & soil water rels, Oak Ridge Nat Labs, 71-75. Mem: Am Soc Hort Sci. Res: Evaluation of vegetable varieties; chemical weed control in horticulture crops; mechanization of harvest. Mailing Add: Dept of Plant & Soil Sci Univ of Tenn Knoxville TN 37901

SWINGLE, HOMER SCOTT, fisheries, deceased

SWINGLE, KARL F, b Richland Center, Wis, Feb 16, 35; div; c 4. PHARMACOLOGY. Educ: Univ Wis, BA, 58; Univ Minn, PhD(pharmacol), 68. Prof Exp: Asst bacteriologist, Sioux City Dept Health, Iowa, 58-59; med technologist, Vet Admin Hosp, Minneapolis, 61-64; pharmacologist, 68-73, SUPVR PHARMACOL, RIKER LABS, MINN MINING & MFG CO, 73- Mem: Am Soc Pharmacol & Exp Therapeut; NY Acad Sci; Soc Exp Biol & Med; Am Chem Soc. Res: Anti-inflammatory and immunosuppressive drugs. Mailing Add: Riker Labs Minn Mining & Mfg Co PO Box 3221 St Paul MN 55101

SWINGLE, KARL FREDERICK, b Bozeman, Mont, Jan 7, 15; m 40; c 5. RADIOBIOLOGY. Educ: Mont State Col, BS, 37; Univ Wis, PhD(biochem), 42. Prof Exp: Res chemist, Inst Path, Western Pa Hosp, 42-43; asst res chemist, Univ Wyo, 43-45; assoc chemist, Mont State Col, 45-47, prof vet biochem, 57-61; supvry chemist, US Naval Radiol Defense Lab, 61-69; res chemist, 69-70, PHYSICIST, VET ADMIN HOSP, 70- Concurrent Pos: Assoc res radiobiologist, Univ Calif, Irvine, 70- Res: Radiation biochemistry; nucleic acid metabolism. Mailing Add: Vet Admin Hosp 5901 E Seventh St Long Beach CA 90801

SWINGLE, ROBERT SHELTON, II, b Canton, Ohio, July 17, 44; m 68. ANALYTICAL CHEMISTRY. Educ: Muskingum Col, BS, 66; Purdue Univ, MS & PhD(anal chem), 71. Prof Exp: CHEMIST, E I DU PONT DE NEMOURS & CO, INC, 71- Mem: AAAS; Am Chem Soc; Sigma Xi. Res: High speed liquid chromatography; gas and mass chromatography; x-ray photoelectron spectroscopy. Mailing Add: Exp Sta 228/222 E I du Pont de Nemours & Co Wilmington DE 19898

SWINGLEY, CHARLES STEPHEN, b Dallas, Tex, Nov 11, 43; m 64; c 2. PHYSICAL CHEMISTRY. Educ: Rochester Inst Technol, BS, 65; Wayne State Univ, PhD(chem), 70. Prof Exp: SR CHEMIST RES LABS, EASTMAN KODAK CO, 70- Mem: Soc Photog Sci & Eng. Res: Surface, colloid, polymer and photographic chemistry. Mailing Add: Res Labs Eastman Kodak Co 343 State St Rochester NY 14650

SWINK, LAURENCE N, b Enid, Okla, Oct 24, 34; m 59; c 1. CRYSTALLOGRAPHY. Educ: Univ Wichita, BA, 57; Iowa State Col, MSc, 59; Brown Univ, PhD(chem), 69. Prof Exp: Flight test technician, Cessna Aircraft Co, Kans, 56-57; mat engr, Chance-Vought Aircraft Co, Tex, 57; mat engr, Douglas Aircraft Co, Calif, 59-60; nuclear res officer, McClellan AFB, Calif, 60-63; mem tech staff crystallog, 66-75, MEM TECH STAFF, ELECTRO-OPTICS DIV, TEX INSTRUMENTS, INC, DALLAS, 75- Mem: Am Crystallog Asn. Res: Crystal structure determination by x-ray diffraction methods; crystal perfection study by x-ray and electron diffraction techniques; electron microprobe analysis; electron microscopy; auger spectroscopy, x-ray fluorescence. Mailing Add: 1313 Chickasaw Richardson TX 75080

SWINNER, PAUL ADOLPH, physical organic chemistry, see 12th edition

SWINNERTON, JOHN W, b Tiffin, Ohio, Jan 6, 31; m 53; c 3. CHEMICAL OCEANOGRAPHY. Educ: Mt Union Col, BS, 54; Pa State Univ, PhD(phys chem), 59. Prof Exp: RES CHEMIST, OCEAN SCI DIV, US NAVAL RES LAB, WASHINGTON, DC, 58- Mem: AAAS; Am Chem Soc; Am Soc Limnol & Oceanog. Res: Chemical oceanography; application of gas chromatography to oceanography. Mailing Add: 6620 Beddo St Alexandria VA 22306

SWINNEY, CHAUNCEY MELVIN, b Riverside, Calif, Sept 3, 18; m 42; c 3. ECONOMIC GEOLOGY, PETROLOGY. Educ: Pomona Col, BA, 40; Stanford Univ, PhD(geol), 49. Prof Exp: Tester, Union Oil Co, Calif, 42; geologist, US Geol Surv, 42-45; instr, Stanford Univ, 47-49, asst prof mineral sci, 49-56; supvr prod res div, Richfield Oil Corp, 59-66; MGR ENERGY RESOURCES EXP & DEVELOP, SOUTHERN CALIF EDISON CO, 66- Concurrent Pos: Geologist, US Geol Surv, 46-53; vpres, Mono Power Co, Southern Calif Edison Co, 73- Mem: Geol Soc Am; Am Asn Petrol Geol; Am Inst Mining Metall & Petrol Eng. Res: Energy resources exploration, development and production. Mailing Add: Southern Calif Edison Co 2244 Walnut Grove Ave Rosemead CA 91770

SWINNEY, HARRY LEONARD, b Opelousas, La, Apr 10, 39; m 67. FLUID PHYSICS. Educ: Southwestern Univ, Memphis, BS, 61; Johns Hopkins Univ, PhD(physics), 68. Prof Exp: Res assoc physics, Johns Hopkins Univ, 68-70, vis asst prof, 70-71; asst prof physics, NY Univ, 71-73; ASSOC PROF PHYSICS, CITY COL, CITY UNIV NEW YORK, 73- Mem: Am Phys Soc; Am Asn Physics Teachers. Res: Laser light scattering spectroscopic techniques are being used to investigate critical phenomena in fluids, turbulence, molecular collision processes in gases and the properties of biological macromolecules. Mailing Add: Dept of Physics City Col City Univ New York NY 10031

SWINSON, DEREK BERTRAM, b Belfast, N Ireland, Nov 5, 38; m 65. PHYSICS. Educ: Queen's Univ, Belfast, 60; Univ Alta, Calgary, MS, 61, PhD(physics), 65. Prof Exp: Asst prof, 65-71, ASSOC PROF PHYSICS, UNIV N MEX, 71- Mem: Brit Inst Physics. Res: Cosmic radiation, extensive air showers and related high energy interactions; mu-mesons underground and variations of their activity with solar cycle; sidereal cosmic ray anisotropies; consultant in accident reconstruction. Mailing Add: Dept of Physics Univ of NMex 800 Yale NE Albuquerque NM 87131

SWINTON, DAVID CHARLES, b St Charles, Ill, May 4, 43; m 70. CELL BIOLOGY. Educ: Brandeis Univ, AB, 65; Stanford Univ, PhD(biol), 72. Prof Exp: Res fel, 72-74, RES ASSOC BIOPHYS, UNIV CHICAGO, 74- Mem: Am Soc Cell Biol. Res: Information content of organelle DNA; expression of organelle DNA during mitotic cell cycle and during meiosis. Mailing Add: Dept of Biophys & Theoret Biol 920 E 58th St Chicago IL 60637

SWINTOSKY, JOSEPH VINCENT, b Kewaunee, Wis, Dec 14, 21; m 53; c 10. PHARMACY. Educ: Univ Wis, BS, 42, PhD(pharm), 48. Prof Exp: Instr pharm, Univ Wis, 47-49, asst prof, 49-53; sr scientist, Tablet Unit, Smith Kline & French Labs, 53, unit head, 53-57, group leader, Pharmaceut Res Sect, 57-59, prin res pharmacist, 59-67; DEAN COL PHARM, UNIV KY, 67- Concurrent Pos: Vis prof, Temple Univ, 58-67. Honors & Awards: Ebert Award, Am Pharmaceut Asn, 58; Res Achievement Award, Am Pharmaceut Asn Found, 64; Indust Pharmaceut Technol Award, Am Pharmaceut Asn, 75. Mem: AAAS; Am Chem Soc; Am Pharmaceut Asn; fel Acad Pharmaceut Sci (pres, 67-68); Polish Pharmaceut Soc. Res: Absorption, distribution and excretion of drugs; drug stability and stabilization; sustained-action medication; drug formulation; particle size measurement of powders. Mailing Add: Col of Pharm Univ of Ky Washington & Gladstone St Lexington KY 40506

SWINYARD, CHESTER ALLAN, b Logan, Utah, Oct 21, 06; m 29; c 2. ANATOMY. Educ: Utah State Univ, BS, 28, MS, 29; Univ Minn, PhD(neuroanat), 34; State Univ NY, MD, 51. Prof Exp: Instr anat, Tufts Col, 36-37; asst prof, Med Col SC, 37-40; from assoc prof to prof, Col Med, Univ Utah, 40-58; assoc dir, Children's Div, Inst Rehab Med, NY Univ, 58-60, dir, 60-67, dir pediat rehab res, 67-75; VIS EMER PROF SURG, STANFORD UNIV, 76- Concurrent Pos: Vis prof, State Univ NY, 49-50; fel, Med Ctr, NY Univ, 53-54. Mem: Am Asn Anatomists; Am Cong Rehab Med; Am Acad Cerebral Palsy; Am Acad Neurol. Res: Chronic muscular and neuromuscular disease; electromyography; rehabilitation. Mailing Add: Dept of Surg Stanford Univ Stanford CA 94305

SWINYARD, EWART AINSLIE, b Logan, Utah, Jan 3, 09; m 34; c 2. PHARMACOLOGY. Educ: Utah State Univ, BS, 32; Idaho State Col, BS, 36; Univ Minn, MS, 41; Univ Utah, PhD(pharmacol), 47. Prof Exp: From instr to asst prof pharm, Idaho State Col, 36-45, prof pharmacol, 45-47; PROF PHARMACOL & DIR PHARMACEUT RES, COL PHARM, UNIV UTAH, 47-, PROF PHARMACOL, COL MED, 67-, DEAN COL PHARM, 70- Concurrent Pos: Am Col Apothecaries fac fel; lectr, Col Med, Univ Utah, 45-67, chmn dept biopharmaceut sci, Univ, 65-71, distinguished res prof, 68-69; Rennebohm lectr, Univ Wis, 60; Kaufman lectr, Ohio State Univ, 63. Honors & Awards: Res Achievement Award, Am Pharmaceut Asn

Found. 65. Mem: Am Soc Pharmacol & Exp Therapeut; Am Pharmaceut Asn; NY Acad Sci. Res: Arsenical chemotherapy; body water and electrolyte distribution; experimental therapy of convulsive disorders; assay of anticonvulsant drugs; relationship between chemical structure and pharmacological activity of anticonvulsant drugs. Mailing Add: Col of Pharm Univ of Utah Salt Lake City UT 84112

SWINZOW, GEORGE K, b Kharkov, Russia, Feb 4, 15; nat US; m 40; c 1. PHYSICS. Educ: Univ Kharkov, Russia, BSc & AM, 40; Boston Univ, PhD, 57. Prof Exp: Eng geologist, Arctic Res Insts, Russia, 38-43; mining geologist, Berg & Heuttenverke, Ltd, Ger, 43-45; consult geologist, 45-52; longshoreman, Am Sugar Co, Mass, 52-55; asst prof geol, Grad Sch, Boston Univ, 55-58; RES GEOLOGIST, COLD REGIONS RES & ENG LAB, US ARMY CORPS ENGRS, 58- Mem: Geol Soc Am; Soc Cryobiol; AAAS; Sigma Xi. Res: Freezing as a physical and geological process; cryopedology; permafrost; physics of ice; problems of nucleation as a trigger of phase change; relation of ice physics to metallurgy. Mailing Add: Cold Regions Res & Eng Lab US Army Corps of Engrs Box 282 Hanover NH 03755

SWISCHUK, LEONARD EDWARD, b Bellevue, Alta, June 14, 37; m 60; c 4. RADIOLOGY. Educ: Univ Alta, BS & MD, 60. Prof Exp: Asst prof pediat, Med Ctr, Univ Okla, 66-68; assoc prof radiol & pediat, 68-70; assoc prof, 70-73, PROF RADIOL & PEDIAT, UNIV TEX MED BR GALVESTON, 73- Mem: AMA; Am Col Radiol; Am Acad Pediat; Radiol Soc NAm. Mailing Add: Dept of Radiol & Pediat Univ of Tex Med Br Galveston TX 77550

SWISHER, ELY MARTIN, b Bozeman, Mont, Sept 29, 15; m 40; c 2. ENTOMOLOGY. Educ: Willamette Univ, AB, 37; Ore State Col, MS, 41; Ohio State Univ, PhD(entom), 43. Prof Exp: Asst zool, Ohio State Univ, 40-43; midwest mgr, 43-53, mgr develop sect agr & sanit chem, 53-73, MGR GOVT REGULATORY RELS, ROHM AND HAAS CO, 73- Mem: Entom Soc Am; Weed Sci Soc Am. Res: Insecticides; fungicides; field evaluations; development of agricultural pesticide chemicals. Mailing Add: Govt & Reg Rels Rohm and Haas Co Independence Mall W Philadelphia PA 19105

SWISHER, HORTON EDWARD, b San Diego, Calif, Mar 1, 09; m 47; c 1. FOOD CHEMISTRY. Educ: Pomona Col, BA, 33; Claremont Cols, MA, 35. Prof Exp: Jr analyst, Am Potash & Chem Corp, 35-37; chemist & plant supvr, Armour & Co, Ill, 37-45; res chemist, 45-59, chief chemist, 59-63, asst mgr, Res & Develop Div, 63-67, Dir Res & Develop, Ont Prod Sect Lab, Sunkist Growers, Inc, 67-74; CONSULT BY-PRODS, CITRUS INDUST, 74- Concurrent Pos: Instr eve col, Texas Christian Univ, 43-45. Honors & Awards: Glycerine res award, 61; Indust Achievement Award, Inst Food Technol, 62, Food Man Year Award, 71. Mem: AAAS; Inst Food Technol; Am Chem Soc; fel Am Inst Chem. Res: Chemistry of citrus by-products, especially citrus oils, peel products, juices and beverages; dehydrated foods. Mailing Add: 595 West 25th St Upland CA 91786

SWISHER, JOSEPH VINCENT, b Kansas City, Mo, Jan 12, 32; m 60; c 3. ORGANIC CHEMISTRY. Educ: Cent Methodist Col, AB, 56; Univ Mo, PhD(org chem), 60. Prof Exp: Fel chem, Purdue Univ, 60-61; asst prof, 61-69, ASSOC PROF CHEM, UNIV DETROIT, 69- Mem: Am Chem Soc; Am Inst Chemists. Res: Oranosilicon chemistry; stereochemistry of addition reactions; reactions of metals with polyhalides. Mailing Add: Dept of Chem Univ of Detroit Detroit MI 48221

SWISHER, ROBERT DONALD, b Denver, Colo, Nov 16, 10; m 35; c 3. ORGANIC CHEMISTRY, ENVIRONMENTAL CHEMISTRY. Educ: Univ Mich, PhD(pharmaceut chem), 34. Prof Exp: Mem staff, 34-45, group leader, 45-73, SR ENVIRONMENTAL ADV, RES DEPT, MONSANTO CO, 73- Mem: Fel AAAS; Am Chem Soc; Am Oil Chem Soc; Soc Indust Microbiol; Water Pollution Control Fedn. Res: Sulfonation and sulfonic acid derivatives; detergents; biodegradation and environmental acceptability of surfactants and other materials. Mailing Add: 1894 Charmwood Ct Kirkwood MO 63122

SWISHER, SCOTT NEIL, b Le Center, Minn, July 30, 18; m 45; c 2. INTERNAL MEDICINE, HEMATOLOGY. Educ: Univ Minn, BS, 43, MD, 44; Am Bd Internal Med, dipl, 52. Prof Exp: Asst resident & fel med, Sch Med & Dent, Univ Rochester, 47-48; fel, Med Sch, Univ Minn, 48-49; fel med & hemat, Sch Med & Dent, Univ Rochester, 49-51, from instr to prof med, Univ, 51-67; PROF MED & CHMN DEPT, COL HUMAN MED, MICH STATE UNIV, 67- Concurrent Pos: From asst resident physician to chief resident physician, Ancker Hosp, St Paul, Minn, 48-49; from asst physician to sr assoc physician & head hemat unit, Strong Mem Hosp, Rochester, NY, 53-67; mem comt blood & related probs, Nat Res Coun, 54-, chmn, 60. Mem: AAAS; Am Soc Clin Invest; Asn Am Physicians; Am Col Physicians; Am Fedn Clin Res. Res: Mechanisms of destruction of erythrocytes by isoantibodies; human hemolytic disorders. Mailing Add: Dept Med B220 Life Sci I Col Human Med Mich State Univ East Lansing MI 48824

SWISLOCKI, NORBERT IRA, b Warsaw, Poland, Jan 11, 36; US citizen; m 63; c 2. BIOCHEMISTRY, ENDOCRINOLOGY. Educ: Univ Calif, Los Angeles, BS, 56, MA, 60, PhD(zool, endocrinol), 64. Prof Exp: USPHS fel biochem, Brandeis Univ, 64-66; from instr to asst prof, 67-73, ASSOC PROF BIOCHEM, GRAD SCH MED SCI, CORNELL UNIV, 73-, CHMN BIOCHEM UNIT, 75-, ASSOC, SLOAN-KETTERING INST CANCER RES, 66-, ASSOC MEM, 72- Mem: Assoc Am Physiol Soc; Am Soc Biol Chemists; Endocrine Soc. Res: Mechanisms of hormone action; membrane function. Mailing Add: Sloan-Kettering Inst Cancer Res 1275 York Ave New York NY 10021

SWISS, JACK, b Columbus, Nebr, Jan 30, 12; m 45; c 4. ORGANIC CHEMISTRY. Educ: Univ Nebr, BSc, 33, MSc, 34; Iowa State Univ, PhD(org chem), 39. Prof Exp: Fel Iowa State Univ, 39-40; res engr, Res Lab, 40-45, mgr chem develop sect, Insulation Dept, 45-52, mgr, Insulation Dept, Res Labs, 52-60, Insulation & Chem Dept, Mat Labs, 60-62 & Insulation & Chem Tech Dept, Res Labs, 62-66, dir, Chem Res, 66-74, DIV MGR, CHEM SCI, WESTINGHOUSE ELECTRIC CORP, 74- Concurrent Pos: Lectr, Pa State Col, 42-44 & Pittsburgh Univ, 44. Mem: Fel AAAS; fel Inst Elec & Electronics Eng; Am Chem Soc. Res: Organic arsenic and organosilicon compounds; organolithium; organomagnesium; high frequency and high voltage insulation; derivatives of biphenyl and dibenzofuran; surface coatings. Mailing Add: Westinghouse Res Labs Churchill Boro Pittsburgh PA 15235

SWISSLER, THOMAS JAMES, b Haddonfield, NJ, Dec 8, 41; m 69; c 1. ENVIRONMENTAL PHYSICS, MATHEMATICAL ANALYSIS. Educ: St Joseph's Col, Pa, BS, 64; State Univ NY Buffalo, PhD(physics), 71. Prof Exp: Res asst comput sci, Theol Biol Ctr, State Univ NY Buffalo, 70-74, res fel, 73-74; STAFF SCIENTIST PHYSICS, SYSTS & APPL SCI CO, RIVERDALE, MD, 74- Mem: AAAS; Am Meteorol Soc. Res: Stratospheric aerosols and ozone gases, their physical properties and distribution using remote sensing measurements from satellite. Mailing Add: NASA Langley Res Ctr MS 475 Hampton VA 23665

SWITENDICK, ALFRED CARL, b Batavia, NY, Oct 8, 31; m 63; c 2. SOLID STATE PHYSICS. Educ: Mass Inst Technol, SB, 53, PhD(solid state physics), 63; Univ Ill,

MS, 54. Prof Exp: Asst physics, Univ Ill, 53-54; asst physics, Mass Inst Technol, 56-62, res staff mem, 62-64; staff mem physics org solids, 64-68, staff mem solid state theory, 68-70, SUPVR, SOLID STATE THEORY, SANDIA LAB, 70- Mem: AAAS; Am Phys Soc. Res: Electronic energy bands of transition metals and transition metal compounds; intermetallic compounds; hydrogen in metals. Mailing Add: Div 5151 Sandia Labs Albuquerque NM 87115

SWITKES, EUGENE, b Newport News, Va, Dec 22, 43; m 69. QUANTUM CHEMISTRY. Educ: Oberlin Col, BA, 65; Harvard Univ, MS & PhD(theoret chem), 70. Prof Exp: NSF fels, Univ Edinburgh, 70-71 & Cambridge Univ, 71; ASST PROF CHEM, UNIV CALIF, SANTA CRUZ, 71- Concurrent Pos: Fel & responder, Neurosci Res Prog, Boulder Intensive Study Session, 72. Res: Theory of the electronic structure of molecules, quantum mechanics; information processing in the visual system; visual accommodation and perception. Mailing Add: Div of Natural Sci II Univ of Calif Santa Cruz CA 95060

SWITZER, BOYD RAY, b Harrisonburg, VA, Oct 3, 43; m 67; c 1. NUTRITION, BIOCHEMISTRY. Educ: Bridgewater Col, BA, 65; Univ NC, Chapel Hill, PhD(biochem), 71. Prof Exp: NIH fel, Univ Southern Calif, 71-72; ASST PROF NUTRIT, SCH PUB HEALTH, UNIV NC, CHAPEL HILL, 72-, ASST PROF BIOCHEM & NUTRIT, SCH MED, 74- Mem: AAAS; Am Chem Soc. Res: Hormonal regulation of protein metabolism as influenced by dietary nutrition; metabolic disorders of connective tissue. Mailing Add: Dept of Nutrit Univ of NC Sch of Pub Health Chapel Hill NC 27514

SWITZER, CLAYTON MACFIE, b London, Ont, July 17, 29; m 51; c 3. AGRICULTURE, WEED SCIENCE. Educ: Ont Agr Col, BSA, 51, MSA, 53; Iowa State Univ, PhD, 55. Prof Exp: Assoc prof bot, 55-65, assoc dean, 70-72, DEAN, ONT AGR COL, 72-, PROF BOT & HEAD DEPT, 65- Mem: Am Soc Plant Physiol; Weed Sci Soc Am; Can Soc Plant Physiol; Can Soc Hort Sci; Can Bot Soc. Res: Physiology of herbicide action; weed control; growth regulation of turfgrass. Mailing Add: Dept of Environ Biol Ont Agr Col Univ of Guelph Guelph ON Can

SWITZER, GEORGE LESTER, b Chester, WVa, Nov 5, 24; m 56; c 1. FOREST ECOLOGY, FOREST SOILS. Educ: Univ WVa, BS, 49; Yale Univ, MF, 50; State Univ NY Col Forestry, Syracuse, PhD(soils, physiol), 62. Prof Exp: From asst forester to assoc forester, Miss Agr Exp Sta, 50-62, from asst prof to assoc prof, 54-64, PROF SILVICULT, MISS STATE UNIV, 64-, FORESTER, MISS AGR EXP STA, 64- Mem: AAAS; Soc Am Foresters; Soil Sci Soc Am; Am Soc Agron; Ecol Soc Am. Res: Nutrient cycles in forest ecosystems; patterns of variation in forest tree species. Mailing Add: Dept of Forestry Miss State Univ Sch Forest Res Mississippi State MS 39762

SWITZER, GEORGE S, b Petaluma, Calif, June 11, 15; m 40; c 3. MINERALOGY. Educ: Univ Calif, AB, 37; Harvard Univ, PhD(mineral), 42. Prof Exp: Instr mineral, Yale Univ, 40-45; chief crystal engr, Majestic Radio Co, Ill, 45-46; dir res, Geomological Inst Am, Calif, 46-47; geologist, US Geol Surv, Washington, DC, 47-48; assoc cur mineral, US Nat Mus, 48-57, cur, 57-63, chmn, Dept Mineral Sci, 63-68; CUR MINERAL SCI, SMITHSONIAN INST, 68- Mem: Fel Mineral Soc Am (secy, 60-66); Geol Soc Am; Geochem Soc. Res: Morphological and x-ray crystallography; geochemistry; general systematic mineralogy. Mailing Add: Dept of Mineral Sci Smithsonian Inst Washington DC 20560

SWITZER, PAUL, b St Boniface, Can, Mar 4, 39; m 63; c 1. MATHEMATICAL STATISTICS, GEOLOGY. Educ: Univ Man, BA, 61; Harvard Univ, PhD(statist), 65. Prof Exp: ASSOC PROF STATIST & GEOL, STANFORD UNIV, 65- Mailing Add: Sequoia Hall Stanford Univ Stanford CA 94305

SWITZER, ROBERT LEE, b Clinton, Iowa, Aug 26, 40; m 65; c 2. BIOCHEMISTRY. Educ: Univ Ill, Urbana, BS, 61; Univ Calif, Berkeley, PhD(biochem), 66. Prof Exp: Fel biochem, Nat Heart Inst, 66-68; asst prof, 68-73, ASSOC PROF, UNIV ILL, URBANA, 73- Concurrent Pos: Fel, John Simon Guggenheim Mem Found, 75- Mem: Am Soc Biol Chem; Am Chem Soc; Am Soc Microbiol. Res: Microbial physiology and enzymology, particularly regulation of branched biosynthetic pathways, mechanisms of regulatory enzymes and regulation of enzymes during bacterial endospore formation. Mailing Add: Dept of Biochem Univ of Ill Urbana IL 61801

SWITZER, WILLIAM PAUL, b Dodge City, Kans, Apr 9, 27; m 51; c 2. ANIMAL PATHOLOGY, MICROBIOLOGY. Educ: Agr & Mech Col, Tex, DVM, 48; Iowa State Col, MS, 51, PhD, 54. Prof Exp: Asst diagnostician, Iowa Vet Diag Lab, Iowa State Univ, 48-52, from asst prof to prof vet hyg, Univ & Vet Med Res Inst, 52-74, PROF VET MICROBIOL & PREV MED, VET MED RES INST & ASSOC DEAN COL VET MED, IOWA STATE UNIV, 74- Mem: Am Soc Microbiol; Am Vet Med Asn. Res: Swine enteric and respiratory diseases; tissue culture; myoplasma. Mailing Add: Vet Med Res Inst Iowa State Univ Ames IA 50010

SWOBODA, ALLEN RAY, b Victoria, Tex, Oct 2, 38; m 59; c 2. SOIL CHEMISTRY, AGRONOMY. Educ: Tex A&M Univ, BS, 61, PhD(soil chem), 67; Va Polytech Inst, MS, 64. Prof Exp: Res assoc pesticide anal, 65-67; asst prof soil chem, 67-71, ASSOC PROF SOIL CHEM, TEX A&M UNIV, 71- Mem: Am Soc Agron; Soil Sci Soc Am; Int Soc Soil Sci; Clay Minerals Soc. Res: Thermodynamics of ion exchange; movement of nitrate and other ions through soils, disposal of waste in soils and the movement of polutants through soils. Mailing Add: Dept of Soil & Crop Sci Tex A&M Univ College Station TX 77843

SWOBODA, THOMAS JAMES, b Milwaukee, Wis, Feb 25, 21; m 47; c 2. PHYSICAL CHEMISTRY. Educ: Univ Ill, PhD(chem), 50. Prof Exp: RES CHEMIST, CENT RES DEPT, E I DU PONT DE NEMOURS & CO, INC, WILMINGTON, 50- Mem: Am Chem Soc. Res: Solid state chemistry; finely divided solids Mailing Add: Yeatman Station Rd RD 3 Newark DE 19711

SWOFFORD, HAROLD S, JR, b Spokane, Wash, July 24, 36; m 58; c 3. CHEMISTRY. Educ: Western Wash State Col, BA, 58; Univ Ill, Urbana, MS, 60, PhD(anal chem), 62. Prof Exp: Teaching asst chem, Univ Ill, Urbana, 58-60; asst prof, 62-66, ASSOC PROF CHEM, UNIV MINN, MINNEAPOLIS, 66- Mem: Am Chem Soc. Res: High temperature electrochemistry; fused salts. Mailing Add: Dept of Chem Univ of Minn Inst of Technol Minneapolis MN 55455

SWOOPE, CHARLES C, b Jersey City, NJ, July 7, 34; m 55; c 2. PROSTHODONTICS. Educ: Univ Md, DDS, 59; Univ Wash, MSD, 64. Prof Exp: Asst chief dent serv, USPHS Hosp, New Orleans, 64-67; dir grad & res prosthodontics, 67-71, assoc prof, 71-73, PROF PROSTHODONTICS, SCH DENT, UNIV WASH, 73- Concurrent Pos: Asst prof, Loyola Univ, 64-67; consult, USPHS Hosp, Seattle, Wash, 67-68. Mem: Am Prosthodont Soc; Am Col Prosthodont; Am Acad Denture Prosthetics; Am Equilibration Soc; Am Cleft Palate Asn. Res: Resilient lining materials; emotional evaluation of denture patients; bone changes; speech problems; force transmission to teeth. Mailing Add: Univ of Wash Sch Dent Seattle WA 98195

SWOPE, FRED C, b Lexington, Va, Mar 25, 35; m 64. FOOD SCIENCE, BIOCHEMISTRY. Educ: Univ Md, BS, 61; Mich State Univ, PhD(food sci), 68. Prof Exp: Asst prof, 68-74, ASSOC PROF BIOL, VA MIL INST, 74- Mem: Am Chem Soc. Res: Lipoproteins; structural studies on membranes. Mailing Add: Dept of Biol Va Mil Inst Lexington VA 24450

SWOPE, HENRIETTA HILL, b St Louis, Mo, Oct 26, 02. ASTRONOMY. Educ: Columbia Univ, AB, 25; Radcliffe Col, AM, 28. Prof Exp: Asst, Harvard Observ, 28-42; mem staff, Radiation Lab, Mass Inst Technol, 42-43; mathematician, Hydrographic Off, US Dept Navy, 43-47; assoc astron, Barnard Col, Columbia Univ, 47-52; asst, 52-62, RES FEL, MT WILSON & PALOMAR OBSERVS, 62- Honors & Awards: Cannon Prize, Am Astron Soc, 68; Barnard Distinguished Alumna Award, 75. Mem: Am Astron Soc; Royal Astron Soc. Res: Photometry; variable stars. Mailing Add: 135 S Holliston Ave Pasadena CA 91106

SWORD, CHRISTOPHER PATRICK, b San Fernando, Calif, Sept 9, 28; m 59; c 4. MICROBIOLOGY. Educ: Loyola Univ, Calif, BS, 51; Univ Calif, Los Angeles, PhD(microbiol), 59. Prof Exp: Fel & res assoc microbiol, Univ Kans, 58-59, from asst prof to prof, 59-70; PROF LIFE SCI & CHMN DEPT, IND STATE UNIV, TERRE HAUTE, 70- Concurrent Pos: President's fel, Soc Am Bact, 60. Mem: AAAS; Am Soc Microbiol; Reticuloendothelial Soc; Am Acad Microbiol; NY Acad Sci. Res: Biochemical and immunological mechanisms of pathogenesis; bacterial virulence; host-parasite interactions in Listeria monocytogenes infection; ultrastructure of bacteria and infected cells. Mailing Add: Dept of Life Sci Ind State Univ Terre Haute IN 47809

SWORDS, RUTH RILEY, b Itasca, Tex, Nov 1, 16; m 40; c 2. DENTISTRY, DENTAL HYGIENE. Educ: ETex State Teachers Col, BA, 38; Tex Wesleyan Col, BS, 61; Baylor Univ, DDS, 61. Prof Exp: Teacher elem sch, Tex, 38-42; eng draftsman, Gen Dynamics Corp, Ft Worth, 42-48; PROF DENT HYG & DIR, CARUTH SCH DENT HYG, COL DENT, BAYLOR UNIV, 62- Res: Clinical dental hygiene. Mailing Add: Caruth Sch of Dent Hyg Baylor Univ 800 Hall St Dallas TX 75226

SWORSKI, THOMAS JOHN, b Pittsburgh, Pa, Sept 8, 20; m 44; c 2. RADIATION CHEMISTRY. Educ: Duquesne Univ, BS, 42; Notre Dame Univ, PhD(phys chem), 51. Prof Exp: Chemist, Oak Ridge Nat Lab, 51-57 & Nuclear Res Ctr, Union Carbide Corp, 57-63; CHEMIST, OAK RIDGE NAT LAB, 64- Honors & Awards: Centennial of Sci Award, Notre Dame Univ, 65. Mem: Am Chem Soc; Radiation Res Soc. Res: Photochemistry; chemical kinetics. Mailing Add: Chem Div Oak Ridge Nat Lab Oak Ridge TN 37830

SWOYER, VINCENT HARRY, b Philadelphia, Pa, Mar 30, 32; m 60; c 3. ACADEMIC ADMINISTRATION, COMPUTER SCIENCE. Educ: Tufts Univ, BS, 54; Univ Rochester, MA, 60; Harvard Univ, EdD, 66. Prof Exp: Asst prof naval sci, Univ Rochester, 57-59, from asst dir to dir, Comput Ctr, 59-69; staff assoc, NSF, 69-70; DIR COMPUT CTR, UNIV ROCHESTER, RIVER CAMPUS, 70- Concurrent Pos: Lectr, Grad Sch Mgt, Univ Rochester, 63-; consult, NSF, 70- & Southern Regional Educ Bd, 71- Mem: Asn Comput Mach. Res: Monte Carlo studies, especially in area of multiple regression analysis; computer methods for statistical research applications. Mailing Add: Comput Ctr Univ of Rochester River Campus Sta Rochester NY 14627

SWYER, PAUL ROBERT, b London, Eng, May 21, 21; Can citizen; m 47; c 2. PEDIATRICS, NEONATOLOGY. Educ: Cambridge Univ, BA, 40, MB, BChir & MA, 43; FRCP, dipl child health, 48; FRCP(C). Prof Exp: Registr, Middlesex Hosp, 47-48; asst resident chest dis & med officer, Brompton Hosp for Dis of Chest, 48-49; registr, SWarwickshire Hosp, 49-50; asst med registr, Hosp Sick Children, 50-52; registr, Royal Hosp, Wolverhampton, 52-53; fel cardiol, 53-54, res assoc pediat, 54-60, asst scientist, 61-66, SR INVESTR RES INST & CHIEF DIV PERINATOLOGY, HOSP SICK CHILDREN, TORONTO, 67- Concurrent Pos: Can Dept Nat Health & Welfare grants, 54-67; sr staff physician, Hosp Sick Children, 64-; from asst prof to assoc prof, Univ Toronto, 65-75, prof, 75-; Med Res Coun Can grants, 66-74. Mem: Am Pediat Soc; Am Acad Pediat; Soc Pediat Res; Can Pediat Soc; Brit Med Asn. Res: Investigation of mechanics of breathing; energy metabolism, blood flow and pressures and methods of treatment of pulmonary disorders in newly born infants. Mailing Add: Hosp for Sick Children 555 University Ave Toronto ON Can

SY, JOSE, b Sorsogon, Philippines, Dec 10, 44. BIOCHEMISTRY. Educ: Adamson Univ, Manila, BS, 64; Duke Univ, PhD(biochem), 70. Prof Exp: Res Assoc, 70-74, ASST PROF, ROCKEFELLER UNIV, 74- Mem: AAAS; Am Chem Soc. Res: Nucleic acids and protein synthesis. Mailing Add: Box 227 Rockefeller Univ New York NY 10021

SYBERS, HARLEY D, b Tony, Wis, June 18, 33; m 58; c 2. PATHOLOGY, PHYSIOLOGY. Educ: Univ Wis-Madison, BS, 56, MS & MD, 63, PhD(physiol), 69; Am Bd Path, dipl & cert anat path & clin path, 68. Prof Exp: Resident path, Univ Wis-Madison, 64-68; asst prof, Univ Calif, San Diego, 69-75; ASSOC PROF PATH, BAYLOR COL MED, 75- Concurrent Pos: USPHS contract, Univ Calif, San Diego, 69-74; Nat Heart & Lung Inst grant, 74-77. Mem: Am Heart Asn; Am Soc Clin Path; Am Physiol Soc; Am Asn Path & Bact; Int Acad Path. Res: Cardiac pathophysiology. Mailing Add: Dept of Path Baylor Col of Med Houston TX 77025

SYBERT, JAMES RAY, b Greenville, Tex, Dec 25, 34; m 54; c 3. SOLID STATE PHYSICS, LOW TEMPERATURE PHYSICS. Educ: NTex State Univ, BA, 55, MA, 56; La State Univ, PhD(physics), 61. Prof Exp: Instr, 56-58, from asst prof to assoc prof, 61-67, PROF PHYSICS, N TEX STATE UNIV, 67-, CHMN DEPT, 69- Concurrent Pos: Adj prof, Southwest Ctr Advan Studies, Dallas, 67-70. Mem: Am Phys Soc; Am Asn Physics Teachers. Res: Electron transport in metals and semiconductors; size effect in metals; superconductivity; thermoelectricity and thermal conductivity in semiconductors; high magnetic field effects and magnetic field induced quantum oscillations in metals. Mailing Add: Dept of Physics NTex State Univ Denton TX 76203

SYBESMA, CHRISTIAAN, b Bandung, Indonesia, Aug 31, 28; m 57; c 3. BIOPHYSICS. Educ: Delft Univ Technol, MS, 56, PhD, 61. Prof Exp: Res assoc biol, Brookhaven Nat Lab, 62-63; sr sci officer, Biophys Lab, Univ Leiden, 63-64; res assoc bot & biophys, Univ Ill, Urbana, 66-67, from asst prof to assoc prof, 67-72; PROF BIOPHYS, FREE UNIV BRUSSELS, 72- Concurrent Pos: Res grants, CRR physiol & biophys & NSF, 67-69. Mem: AAAS; Biophys Soc; Belgian Biophys Soc; Belgian Biochem Soc. Res: Photobiology, particularly photosynthesis; bioenergetics. Mailing Add: Biophysics Lab Fac of Sci Free Univ of Brussels 1050 Brussels Belgium

SYDISKIS, ROBERT JOSEPH, b Bridgeport, Conn, Sept 19, 36; m 61; c 2. VIROLOGY. Educ: Univ Bridgeport, BA, 61; Northwestern Univ, PhD(microbiol), 65. Prof Exp: Fel, Univ Chicago, 65-67; asst prof virol, Sch Med, Univ Pittsburgh, 67-71; ASSOC PROF MICROBIOL, SCH DENT, UNIV MD, BALTIMORE CITY, 71- Mem: Am Soc Microbiol. Res: Structure of herpes virus, isolation and identification of structural and non-structural components synthesized in the cell; assembly of precursor components in vivo and in vitro. Mailing Add: Dept of Microbiol Univ of Md Sch of Dent Baltimore MD 21201

SYDNOR, KATHERINE LEE, b WVa, Jan 6, 11. MEDICINE. Educ: Univ Cincinnati, BS, 41, MS, 44, MD, 49. Prof Exp: Res assoc pharmacol, Col Med, Univ Utah, 51; instr physiol, Sch Med, Western Reserve Univ, 52-54; USPHS fel physiol, Ben May Lab Cancer Res, 55-56, asst prof, Ben May Lab Cancer Res & Univ Chicago, 56-59, assoc prof, 60-61; Eleanor Roosevelt fel cancer res, London, 61-62; assoc prof, 62-73, PROF MED, COL MED, UNIV KY, 73- Mem: AAAS; Endocrine Soc; Am Physiol Soc; Soc Exp Biol & Med. Mailing Add: Dept of Med Univ of Ky Med Ctr Lexington KY 40506

SYDNOR, THOMAS DAVIS, b Richmond, Va, Jan 27, 40; m 62; c 2. ORNAMENTAL HORTICULTURE. Educ: Va Polytech Inst & State Univ, BS, 62; NC State Univ, PhD(plant physiol), 72. Prof Exp: Landscape foreman, Southside Nurseries Inc, 62-63, vpres 65-69; ASST PROF ORNAMENTAL HORT, OHIO STATE UNIV, 72- Mem: Sigma Xi; Am Soc Hort Sci; Int Soc Arboricult. Res: The effects of growth regulatory chemicals and environmental conditions on growth and development of woody ornamentals during production, marketing and in landscape situations.

SYDOR, MICHAEL, b Prusseniv, Ukraine, Dec 25, 36; US citizen; m 62; c 3. ENVIRONMENTAL PHYSICS. Educ: Univ BC, BASc, 59; Univ NMex, PhD(physics), 64. Prof Exp: From asst prof physics to assoc prof, 68-74, PROF PHYSICS, UNIV MINN, DULUTH, 74- Mem: Int Asn Great Lakes Res; Soc Photo-Optical Engrs. Res: Light scattering from suspended solids; remote sensing of turbidity in lakes; studies of turbidity transport in Lake Superior; numerical modeling of transport processes in Lake Superior and Duluth Harbor; ice and snow studies through optical remote sensing; heat budget and ice growth in Duluth-Superior Harbor and Lake Superior. Mailing Add: Dept of Physics Univ of Minn Duluth MN 55812

SYDORIAK, STEPHEN GEORGE, b Passaic, NJ, Jan 6, 18; m 45; c 6. PHYSICS. Educ: Univ Buffalo, BA, 40; Yale Univ, PhD(physics), 48. Prof Exp: Mem staff, Radiation Lab, Mass Inst Technol, 41-45; MEM STAFF, LOS ALAMOS SCI LAB, UNIV CALIF, 48- Mem: Am Phys Soc. Res: Radar signal threshold studies; magnetic susceptibilities at low temperatures; liquid helium-three properties; 1962 helium-three vapor pressure scale of temperatures; hydrodynamic theory of channel boiling, solid hydrogen laser targets. Mailing Add: Los Alamos Sci Lab PO Box 1663 Los Alamos NM 87545

SYED, IBRAHIM BIJLI, b Bellary, India, Mar 16, 39; US citizen; m 64; c 2. MEDICAL PHYSICS, RADIOLOGICAL HEALTH. Educ: Univ Mysore, BSc, 60, MSc, 62; Johns Hopkins Univ, DSc(radiol sci), 72; Am Bd Radiol, dipl, 75. Prof Exp: Physicist, Victoria Hosp, Bangalore, India, 64-67; chief physicist, Halifax Infirmary, Can, 67-69; CHIEF PHYSICIST, MED PHYSICS DIV, BAYSTATE MED CTR, WESSON MEM UNIT, 73-; ASST CLIN PROF, SCH MED, UNIV CONN, 75- Concurrent Pos: Consult physicist, Bowring & Lady Curzon Hosp, 64, Bangalore Nursing Home, 64-67, Wing Mem Hosp, 73 & Mercy Hosp, 73-; vis prof, Springfield Tech Community Col, 73-; adj prof radiol, Holyoke Community Col, 73- Mem: Fel Inst Physics; fel Royal Soc Health; Am Asn Physicists in Med; Am Phys Soc; Soc Nuclear Med. Res: Radiopharmaceutical dosimetry; medical health physics; therapeutic radiological physics; effects of caffeine in man; estimation of absorbed dose to embryo and fetus from radiological and nuclear medicine procedures. Mailing Add: Med Physics Div Baystate Med Ctr Wesson Mem Unit 140 High St Springfield MA 01105

SYEKLOCHA, DELFA, b Vancouver, BC, Sept 12, 33. MEDICAL BACTERIOLOGY, IMMUNOLOGY. Educ: Univ BC, BA, 54; McGill Univ, MSc, 62, PhD(microbiol), 64. Prof Exp: NIH fel, Ont Cancer Inst, 64-65; ASST PROF MICROBIOL, UNIV BC, 65- Mem: Am Soc Microbiol; Can Soc Microbiol; Can Soc Immunol. Res: Immunizing potential of attenuated mutants of Pseudomonas aeruginosa strains and characterization of soluble antigens derived from the wild type; biological activities and characterization of a lethal exotoxin produced by Paeruginosa. Mailing Add: Dept of Microbiol Univ of BC Vancouver BC Can

SYKES, ALAN O'NEIL, b St Regis Falls, NY, May 19, 25; m 51; c 2. ACOUSTICS. Educ: Cornell Univ, AB, 48; Cath Univ Am, PhD, 68. Prof Exp: Physicist, David Taylor Model Basin, US Navy Dept, 48-52, electronic scientist, 52-54, physicist, 54-58, physicist & consult to head acoust div, 58-60, physicist & head, Noise Transmission & Radiation Sect, 60-61, physicist & head, Noise Res & Develop Br, 61-64, physicist & head, Struct Acoust Br, 64-65, PHYSICIST, SENSOR SYSTS PROG, OFF NAVAL RES, 65- Concurrent Pos: Mem hydroballistic adv comt, Navy Bur Ord, 56-60; mem struct impedence panel, Noise Adv Comt, Navy Bur Ships, 56-64, consult, Soc Automotive Engrs, 57- Mem: Acoust Soc Am. Res: Underwater acoustics with emphasis on signal processing and propagation; noise reduction, vibration theory and electroacoustic transducer development. Mailing Add: Off of Naval Res Code 222 800 N Quincy St Arlington VA 22217

SYKES, BRIAN DOUGLAS, b Montreal, Que, Aug 30, 43; m 68. PHYSICAL CHEMISTRY. Educ: Univ Alta, BSc, 65; Stanford Univ, PhD(chem), 69. Prof Exp: From asst prof to assoc prof chem, Harvard Univ, 69-75; ASSOC PROF CHEM, UNIV ALTA, 75- Concurrent Pos: NIH grant, Harvard Univ, 69-, Alfred P Sloan fel, 71-73. Mem: Am Chem Soc; Am Phys Soc. Res: Application of nuclear magnetic resonance to problems in biochemistry. Mailing Add: Dept of Biochem Univ of Alta Edmonton AB Can

SYKES, DONALD JOSEPH, b Buffalo, NY, Mar 16, 36; m 60; c 2. PHOTOGRAPHIC CHEMISTRY, STATISTICAL ANALYSIS. Educ: Rochester Inst Technol, BS, 58. Prof Exp: Dir appl res, Cormac Chem Corp, 59-63; res mgr, 63-69, asst dir res, 69-75, DIR RES, P A HUNT CHEM CORP, 75- Mem: Am Chem Soc; Soc Photog Scientists & Engrs. Mailing Add: 8 Sunset Lane Upper Saddle River NJ 07458

SYKES, DWANE JAY, b Mt Pleasant, Utah, July 4, 38; m 58; c 5. ENVIRONMENTAL SCIENCES, RANGE ECOLOGY. Educ: Utah State Univ, BS, 60; Iowa State Univ, PhD(plant physiol), 63. Prof Exp: Res asst plant ecol, Great Basin Res Sta, US Forest Serv, 58-59, res forester, Coweeta Hydrol Lab, 60, range conservationist, Intermountain Forest & Range Exp Sta, 61; asst prof agron, Purdue Univ & mem Brazilian Tech Asst Prog, Rural Univ, Brazil, 64-67; assoc prof land resources, Univ Alaska, Fairbanks, 67-74, head dept land resources & agr sci, 68-73; DIR, EYRING RES INST, 74- Concurrent Pos: Ecol & land resources consult, 65-; vpres plant prod resources, Twin Am Agr & Indust Developers, Inc, Kans, 67-; vpres, Aleutian Livestock Corp, 69-70; resident consult, Grumman Ecosysts Corp, 70-73; resident consult, Dames & Moore, 70-72; owner-oper, Frontier Int Consult Serv, 72-; pres & bd chmn, Frontier Int Land Corp, Fairbanks, 74-; mem adv comn, Alaska Bd Fish & Game, 69-73; vpres, Wilderness Assocs, Provo, 75-; adj prof bot & range sci, Brigham Young Univ, 75- Mem: Soil Sci Soc Am; Soc Range Mgt; Soil Conserv Soc Am; Asn Trop Biol; Int Soc Soil Sci. Res: Soil-plant moisture relations; agriculture and basic aspects of land management; international agriculture. Mailing Add: Eyring Res Inst 1455 W 820 North Provo UT 84601

SYKES, JAMES ENOCH, b Richmond, Va, Apr 12, 23; m 47; c 2. FISHERIES. Educ: Randolph-Macon Col, 48; Univ Va, MS, 49. Prof Exp: Lab instr biol, Univ Va, 48-49; fishery res biologist, US Fish & Wildlife Serv, 49-58, chief striped bass invests, 58-62, biol lab dir, Bur Com Fisheries, 62-71; DIR DIV FISHERIES, NAT MARINE FISHERIES SERV, 71- Mem: Am Fisheries Soc; Inst Fishery Res Biologists (pres, 75 & 76). Res: Coastal pelagic and offshore sport fisheries; assessment of marine resources; biology of Gulf and Atlantic menhaden. Mailing Add: Atlantic Estuarine Fisheries Ctr Beaufort NC 28516

SYKES, JOHN A, b Pinner, Eng, Dec 7, 18; US citizen; m 43; c 3. VIROLOGY, CANCER. Educ: LRFPS(G), 44; LRCP(E), 44; LRCS(E), 44; Univ London, cert pub health, 48; Univ Manchester, dipl bact, 49. Prof Exp: Bacteriologist, Med Res Coun Gt Brit, 48, assoc dir & asst bacteriologist, Pub Health Lab Serv, Suffolk, Eng, 49-54; med officer virus & zoonoses, Lab of Hyg, Dept Nat Health & Welfare Can, Ont, 54-56; from asst prof to assoc prof biol, Univ Tex Postgrad Sch Med, 58-66; DIR RES, CANCER RES DEPT, SOUTHERN CALIF CANCER CTR, CALIF HOSP MED CTR, 66- Concurrent Pos: Clin asst prof microbiol, Baylor Col Med, 56-66; assoc biologist, Sect Biol & Electron Micros, Univ Tex M D Anderson Hosp & Tumor Inst Houston, 62-64, assoc virologist, 64-66; assoc prof virol, Univ Tex Grad Sch Biomed Sci Houston, 64-66. Mem: Am Soc Microbiol; Tissue Cult Asn; Soc Cryobiol; Am Asn Cancer Res; fel Royal Soc Trop Med & Hyg. Res: Tissue culture; ultrastructural research. Mailing Add: Southern Calif Cancer Ctr 1414 S Hope St Los Angeles CA 90015

SYKES, LYNN RAY, b Pittsburgh, Pa, Apr 16, 37. GEOPHYSICS. Educ: Mass Inst Technol, BS & MS, 60; Columbia Univ, PhD(geol), 65. Prof Exp: Res scientist, Lamont Geol Observ, Columbia Univ, 62-65, res geophysicist, Inst Earth Sci, Environ Sci Serv Admin at Lamont Geol Observ, 65-68; assoc prof geol, 68-71, PROF GEOL, LAMONT-DOHERTY GEOL OBSERV, COLUMBIA UNIV, 71-, HEAD SEISMOL GROUP, 68- Concurrent Pos: Mem comts seismol & earthquake prediction, Nat Res Coun-Nat Acad Sci, 73-75. Honors & Awards: Macelwane Award, Am Geophys Union, 69, Bucher Medal, 75. Mem: Seismol Soc Am; fel Geol Soc Am; Soc Explor Geophys; fel Am Geophys Union; fel Royal Astron Soc. Res: Seismicity of the earth; world tectonics; mechanism of earthquakes; earth structure from surface waves; discrimination of underground explosions from earthquakes; earthquake prediction; tectonics. Mailing Add: Lamont-Doherty Geol Observ Columbia Univ Palisades NY 10964

SYKES, MARGUERITE PRINCE, b New York, NY, Dec 21, 13; m 48. CANCER. Educ: NY Univ, MD, 48. Prof Exp: Intern, Bellevue Hosp, 48-49, asst resident 4th div, 49-50; clin asst chemother dept, Mem Ctr Cancer & Allied Dis, Mem Hosp, 53-54; res assoc med, Sloan-Kettering Div, 54-57, ASST PROF MED, MED COL, CORNELL UNIV, 57- Concurrent Pos: Fel Sloan-Kettering Inst & spec fel med, Mem Hosp, 50-51, Damon Runyon fel, Mem Hosp, 51-53, spec fel, 52-57; assoc, Sloan-Kettering Inst, 59-60, assoc mem, 60-; asst attend physician, Mem Hosp, 57- Mem: James Ewing Soc; Am Soc Hemat; Harvey Soc; AMA; Am Asn Cancer Res. Res: Chemotherapeutic drugs in cancer research. Mailing Add: 45 Sutton Place S New York NY 10022

SYKES, PAUL JAY, JR, b Hummelstown, Pa, Aug 31, 18; m 48; c 1. PHYSICS, NUCLEAR ENGINEERING. Educ: Univ BC, BA, 48; Univ Calif, MA, 51. Prof Exp: Chief proj off nuclear eng test facil, Wright-Patterson Air Force Base, Ohio, 52-56, chief reactor hazards br, Spec Weapons Ctr, Kirtland Air Force Base, 57-58, proj officer, Anal Div, 58-60, proj officer, Physics Div, Air Force Weapons Lab, 60-64; ASST PROF PHYSICS, UNIV BC, 64- Mem: Am Asn Physics Teachers; Royal Astron Soc Can. Res: Design and engineering of nuclear research reactor facilities; reactor hazards and safeguards; nuclear weapon systems analysis. Mailing Add: 5616 Westport Pl West Vancouver BC Can

SYKORA, OSCAR P, b Nachod, Czech, June 22, 29; Can citizen; m 65. DENTISTRY. Educ: Sir George William Univ, BA, 54; Univ Montreal, MA, 55, PhD(slavic hist), 59; McGill Univ, DDS, 59. Prof Exp: From asst prof to assoc prof prosthodontics, McGill Univ, 61-71; ASSOC PROF PROSTHODONTICS, DALHOUSIE UNIV, 71- Mem: Am Prosthodontic Soc; Can Dent Asn; Can Inst Int Affairs; Can Acad Prosthodontics. Res: Fixed and removable prosthetic dentistry. Mailing Add: Fac of Dent Dalhousie Univ Halifax NS Can

SYLVESTER, EDWARD SANFORD, b New York, NY, Feb 29, 20; m 42; c 2. ENTOMOLOGY. Educ: Colo State Univ, BS, 43; Univ Calif, PhD(entom), 47. Prof Exp: Lab asst bot, Colo State Univ, 40, asst entom, 41-43; prin lab asst entom, Univ Calif, Berkeley, 45-47, instr entom univ & jr entomologist exp sta, 47-49, asst prof & asst entomologist, 49-55, lectr & assoc entomologist, 55-61, PROF ENTOM & ENTOMOLOGIST EXP STA, UNIV CALIF, BERKELEY, 61- Mem: Entom Soc Am; Am Phytopath Soc. Res: Insect vector-virus relationships; biostatistics; Aphidiae. Mailing Add: Div of Entom & Parasitol Univ of Calif Oxford Res Unit 366 Ocean View Ave Kensington CA 94707

SYLVIA, AVIS LATHAM, b Westerly, RI, Nov 16, 38. CELL PHYSIOLOGY. Educ: Univ NC, Greensboro, AB, 60; Univ Conn, MS, 66; Univ NC, Chapel Hill, PhD(physiol), 73. Prof Exp: Res assoc cell physiol, Sch Med, Univ NC, 73-74; fel, Ctr Aging & Human Develop, 74-76, RES ASSOC, DEPT PHYSIOL & PHARMACOL, DUKE UNIV MED CTR, 76- Mem: AAAS; Geront Soc. Res: Cellular oxidative metabolism, bioenergetics and redox phenomena; physiologiacl and biochemical aspects of development and aging in the mammalian central nervous system. Mailing Add: Dept of Physiol & Pharmacol Duke Univ Med Ctr Box 3709 Durham NC 27710

SYLWESTER, DAVID LUTHER, b Roseburg, Ore, Mar 17, 36; m 62; c 2. MATHEMATICAL STATISTICS, BIOSTATISTICS. Educ: Univ Ore, AB, 58; Ind Univ, AM, 60; Stanford Univ, PhD(statist), 66. Prof Exp: Asst math, Ind Univ, 58-60; res physicist, Arctic Inst NAm, 60-62; asst statist, Stanford Univ, 62-65; asst prof community med, Col Med, 65-69, asst prof math, 65-70, ASSOC PROF MATH, UNIV VT, 70- Concurrent Pos: Consult, Med Sch, Stanford Univ, 62-65 & Col Med, Univ Vt, 65- Mem: Inst Math Statist; Am Statist Asn; Biomet Soc. Res: Stochastic processes. Mailing Add: Dept of Math Univ of Vt Burlington VT 05401

SYLWESTER, ERHARDT PAUL, b Gaylord, Minn, 1906; m 36; c 3. BOTANY. Educ: St Olaf Col, BA, 30; Iowa State Univ, MS, 31, PhD(bot), 46. Prof Exp: Agent, USDA, Iowa, 30-31, seed analyst, 31-33; forest plant pathologist, Univ Iowa, 33-35; from actg head to head seed lab, 46-52, BOTANIST & PLANT PATHOLOGIST, AGR EXTEN SERV, IOWA STATE UNIV, 35-, PROF BOT, UNIV, 49- Concurrent Pos: Pres, Nat Asn Regional Weed Control Confs, 52-53. Honors & Awards: Super Serv Award, USDA; Meritorious Serv Award Agr, Wallace's Farmer. Mem: Hon mem Weed Sci Soc Am. Res: Herbicides; chemical and cultural weed and brush control; biology of horse nettle. Mailing Add: Dept of Bot & Plant Path 105 Bessey Hall Iowa State Univ Ames IA 50010

SYMBAS, PANAGIOTIS N, b Greece, Aug 15, 25; US citizen; m 65; c 2. THORACIC SURGERY, CARDIOVASCULAR SURGERY. Educ: Univ Salonika, MD, 54. Prof Exp: Intern surg, Vanderbilt Univ, 56-57, resident surgeon, 60-61, instr surg, 61-62; fel cardiovasc surg, St Louis Univ, 62-63; assoc thoracic surg, 64-65, from asst prof to assoc prof, 66-73, PROF SURG, THORACIC CARDIOVASC SURG DIV, SCH MED, EMORY UNIV, 73-, DIR SURG RES LAB, 70-; DIR THORACIC & CARDIOVASC SURG, GRADY MEM HOSP, ATLANTA, 68- Mem: Int Cardiovasc Soc; fel Am Col Chest Physicians; sr mem Am Fedn Clin Res; fel Am Col Cardiol; Soc Thoracic Surg. Mailing Add: Surg Res Lab Emory Univ Sch of Med Atlanta GA 30303

SYMCHOWICZ, SAMSON, b Krakow, Poland, Mar 20, 23; nat US; m 53; c 3. DRUG METABOLISM, BIOCHEMICAL PHARMACOLOGY. Educ: Chem Tech Col Eng, Czech, Chem Eng, 50; Polytech Inst Brooklyn, MS, 56; Rutgers Univ, PhD(physiol, biochem), 60. Prof Exp: Asst biochem, Allan Mem Inst, McGill Univ, 51-53; asst hormone res, Col Med, State Univ NY Downstate Med Ctr, 53-56; res biochemist, 56-66, sect leader, 66-70, head dept biochem, 70-73, ASSOC DIR BIOL RES, SCHERING CORP, 73- Mem: Am Chem Soc; Am Soc Pharmacol & Exp Therapeut; Am Inst Chemists; NY Acad Sci. Res: Biogenic amines. Mailing Add: Schering Corp Bloomfield NJ 07003

SYMINGTON, JANEY, b St Louis, Mo, June 29, 28; m 49; c 4. VIROLOGY, PLANT PHYSIOLOGY. Educ: Vassar Col, AB, 50; Radcliffe Col, PhD(biol), 59. Prof Exp: Res assoc & asst prof bot, 58-71, res asst prof biol, 71-73, RES ASSOC MICROBIOL, WASHINGTON UNIV, 73- Mem: Am Soc Microbiol; Sigma Xi; Am Soc Plant Physiol; Am Inst Biol Sci; Am Soc Cell Biol. Res: Virus structure and replication; interaction of viruses with antibodies and cell surfaces. Mailing Add: Dept of Microbiol & Immunol Washington Univ Box 8093 St Louis MO 63110

SYMKO, OREST GEORGE, b Ukraine, Jan 24, 39; Can citizen; m 62; c 2. PHYSICS. Educ: Univ Ottawa, BSc, 61, MSc, 62; Oxford Univ, DPhil(physics), 67. Prof Exp: Res officer physics, Clarendon Lab, Oxford Univ, 67-68; asst res physicist, Univ Calif, San Diego, 68-70; asst prof, 70-74, ASSOC PROF PHYSICS, UNIV UTAH, 74- Concurrent Pos: Res Corp grant, Univ Utah, 71; NSF grant, 72. Mem: Am Phys Soc. Res: Low temperature physics; magnetism; superconductivity; dilute alloys; nuclear magnetic resonance. Mailing Add: Dept of Physics Univ of Utah Salt Lake City UT 84112

SYMMES, DAVID, b New York, NY, Sept 4, 29; m 57; c 3. NEUROPHYSIOLOGY, PSYCHOLOGY. Educ: Harvard Univ, AB, 52; Univ Chicago, PhD(biopsychol), 55. Prof Exp: USPHS fel, Yale Univ, 57-59, from instr to asst prof physiol, Sch Med, 59-67; CHIEF SECT ON BRAIN & BEHAVIOR, BEHAV BIOL BR, NAT INST CHILD HEALTH & HUMAN DEVELOP, 67- Mem: Am Psychol Asn; Am Physiol Soc; Soc Neurosci. Res: Neuropsychology. Mailing Add: Behav Biol Br Nat Inst of Child Health & Human Develop Bethesda MD 20014

SYMON, KEITH RANDOLPH, b Ft Wayne, Ind, Mar 25, 20; m 43; c 4. MATHEMATICAL PHYSICS. Educ: Harvard Univ, SB, 42, AM, 43, PhD(theoret physics), 48. Prof Exp: From instr to assoc prof physics, Wayne Univ, 47-55; from asst prof to assoc prof, 55-57, PROF PHYSICS, UNIV WIS-MADISON, 57- Concurrent Pos: Head advan res group, Midwestern Univs Res Asn, 50-57, head theoret sect, 55-57, tech dir, 57-60. Mem: Fel Am Phys Soc; Am Asn Physics Teachers. Res: Plasma physics; numerical simulation of plasmas; orbit theory; design of high energy accelerators; theory of energy loss fluctuations of fast particles. Mailing Add: Dept of Physics Univ of Wis Madison WI 53706

SYMONS, DAVID THORBURN ARTHUR, b Toronto, Ont, July 24, 37; m 64; c 3. GEOPHYSICS, GEOLOGY. Educ: Univ Toronto, BASc, 60, PhD(econ geol), 65; Harvard Univ, MA, 61. Prof Exp: Nat Res Coun Can overseas fel paleomagnetism, Univ Newcastle, Eng, 65-66; res scientist rock magnetism, Geol Surv Can, 66-70; ASSOC PROF GEOPHYS & GEOTECTONICS, UNIV WINDSOR, 70-, HEAD DEPT GEOL, 73- Mem: Can Inst Mining & Metall; Am Geophys Union. Res: Application of paleomagnetic methods to geotectonic and geochronologic problems in the Canadian Cordillera and Shield and to the history of the earth's geomagnetic field. Mailing Add: Dept of Geol Univ of Windsor Windsor ON Can

SYMONS, PHILIP CHARLES, b Taunton, Somerset, Eng, June 4, 39; m 60; c 2. ELECTROCHEMISTRY. Educ: Univ Bristol, BSc, 60, PhD(electrochem), 63. Prof Exp: Sr chemist, Proctor & Gamble Ltd, Eng, 63-67 & Hooker Chem Corp, 68-69; res dir batteries, Udylite Co, Occidental Petrol Corp, 69-73; CHIEF SCIENTIST, ENERGY DEVELOP ASSOCS, 73- Mem: Am Chem Soc; Electrochem Soc; The Chem Soc. Res: High energy density batteries; electrochemical thermodynamics and kinetics; thermodynamics of phase transitions; electrochemical engineering. Mailing Add: Energy Develop Assocs 1100 W Whitcomb Ave Madison Heights MI 48071

SYMONS, PHILIP EDWARD KYRLE, animal behavior, see 12th edition

SYMPSON, ROBERT F, b Ft Madison, Iowa, June 21, 27; m 53; c 4. ANALYTICAL CHEMISTRY, ELECTROCHEMISTRY. Educ: Monmouth Col, BS, 50; Univ Ill, MS, 52, PhD, 54. Prof Exp: From asst prof to assoc prof, 54-65, PROF CHEM, OHIO UNIV, 65- Mem: Am Chem Soc; Electrochem Soc. Res: Polarography and amperometric titrations applied to analytical chemistry. Mailing Add: Dept of Chem Ohio Univ Athens OH 45701

SYNEK, MIROSLAV (MIKE), b Prague, Czech, Sept 18, 30; US citizen; m 65; c 2. ATOMIC PHYSICS, PHYSICAL CHEMISTRY. Educ: Charles Univ, Prague, MS, 56; Univ Chicago, PhD(physics), 63. Prof Exp: Technician chem, Inst Indust Med, Prague, Czech, 50-51; asst physics, Czech Acad Sci, 56-58; res asst, Univ Chicago, 58-62; asst prof physics, DePaul Univ, 62-65, assoc prof physics, 65-67, assoc prof chem, 66-67; prof physics, Tex Christian Univ, 67-71; regional sci adv & res scientist, Dept Physics, Univ Tex, Austin, 71-73, lectr, exten lectr & res scientist, Depts Physics, Chem, Astron & Math, 73-75; FAC MEM DIV EARTH & PHYS SCI, COL SCI & MATH, UNIV TEX, SAN ANTONIO, 75- Concurrent Pos: Consult, US Army, 58 & Physics Div, Argonne Nat Lab, 66-73; prin investr, Mat Lab, Wright-Patterson AFB, Ohio, 64-71; investr, Robert A Welch Found res grant, 69-71; referee manuscripts, Phys Rev; occasional lectr. Mem: Fel AAAS; Fel Am Phys Soc; Am Asn Physics Teachers. Res: Educational and computational physics; popularization of science; laser-active ions of rare earths; materials science; atomic wave functions; statistical mechanics; molecular dynamics. Mailing Add: Div of Earth & Phys Sci Univ of Tex Col of Sci & Math San Antonio TX 78285

SYNEK, ROSEMARIE WAHL, microbiology, biochemistry, see 12th edition

SYNER, FRANK N, b Springfield, Mass, Apr 25, 24; m 50; c 1. BIOCHEMISTRY. Educ: Univ Mass, BS, 51; Wayne State Univ, PhD(biochem), 60. Prof Exp:

Biochemist, Hawthorn Ctr, 60-62 & Lafayette Clin, 62-67; asst prof, 67-70, ASSOC PROF OBSTET & GYNEC, SCH MED, WAYNE STATE UNIV, 70- Res: Biochemical genetics and biochemistry of development; biochemistry of reproduction. Mailing Add: Dept of Obstet & Gynec Wayne State Univ Sch of Med Detroit MI 48207

SYNOVITZ, ROBERT J, b Milwaukee, Wis, Feb 3, 31; m 57; c 4. HEALTH SCIENCE. Educ: Wis State Univ, La Crosse, BS, 53; Ind Univ, Bloomington, MS, 56, HSD, 59. Prof Exp: Instr pub schs, Mo, 56-58; asst prof health sci, Eastern Ky Univ, 59-62; assoc prof physiol & health sci, Ball State Univ, 62-68; PROF HEALTH SCI & CHMN DEPT, WESTERN ILL UNIV, 68- Concurrent Pos: Teaching & res grant, Ball State Univ, 66-67. Mem: Am Sch Health Asn; Soc Pub Health Educ. Mailing Add: Dept of Health Sci Western Ill Univ Macomb IL 61455

SYNOWIEC, JOHN A, b Chicago, Ill, Sept 18, 37; m 67; c 2. MATHEMATICAL ANALYSIS. Educ: DePaul Univ, BS, 59, MS, 61, DMS, 62; Ill Inst Technol, PhD(math), 64. Prof Exp: Instr math, DePaul Univ, 61-62; from instr to asst prof, Ill Inst Technol, 63-67; asst prof, 67-74, ASSOC PROF MATH, IND UNIV NORTHWEST, 67- Mem: Am Math Soc; Math Asn Am. Res: Complex analysis; generalized functions; differential geometry. Mailing Add: Dept of Math Ind Univ Northwest Gary IN 46408

SYPERT, GEORGE WALTER, b Marlin, Tex, Sept 25, 41; m 71; c 1. NEUROSURGERY, NEUROPHYSIOLOGY. Educ: Univ Wash, BA, 63, MD, 67. Prof Exp: Resident neurosurg, Univ Wash, 68-73, instr, 73-74; ASST PROF NEUROSURG & NEUROSCI & ATTEND PHYSICIAN, SHANDS TEACHING HOSP, UNIV FLA & CHIEF NEUROSURG, GAINESVILLE VET ADMIN HOSP, 74- Concurrent Pos: Asst chief neurosurg, Fitsimons Gen Hosp, US Army, 68-70. Mem: Soc Neurosci; Cong Neurol Surgeons; Res Soc Neurol Surgeons; AAAS; Med Res Soc. Res: Neurophysiological and ionic mechanisms involved in synaptic transmission; neuronal repetitive firing mechanisms; pathophysiology of epilepsy; immunology of gliomas. Mailing Add: Neurol Surg JHMHC Box 265 Univ of Fla Gainesville FL 32610

SYPHAX, BURKE, b Washington, DC, Dec 18, 10; m 39; c 3. SURGERY. Educ: Howard Univ, BS, 32, MD, 36; Am Bd Surg, dipl, 43. Prof Exp: From asst to assoc prof, 40-58, chmn dept, 57-70, PROF SURG, SCH MED, HOWARD UNIV, 58-, CHIEF DIV GEN SURG, FREEDMEN'S HOSP, 51- Concurrent Pos: Fel, Rockefeller-Strong Mem Hosp, 41-42; consult, Vet Admin Hosp, DC, 56- Mem: AMA; Am Col Surg. Res: Gastroenterology; oncology; acute abdomen. Mailing Add: Rm 211 Annex 2 Freedmen's Hosp Sixth & Bryant St NW Washington DC 20001

SYPHERD, PAUL STARR, b Akron, Ohio, Nov 16, 36; m 54; c 4. MICROBIOLOGY. Educ: Ariz State Univ, BS, 59; Univ Ariz, MS, 60; Yale Univ, PhD(microbiol), 63. Prof Exp: NIH res fel biol, Univ Calif, San Diego, 62-64; from asst prof to assoc prof microbiol, Univ Ill, Urbana, 64-70; assoc prof, 70-72, PROF MICROBIOL, COL MED, UNIV CALIF, IRVINE, 72-, CHMN MICROBIOL, 74- Concurrent Pos: USPHS fel, Univ Calif, San Diego, 62-64. Mem: Am Soc Microbiologists; Am Soc Biol Chemists. Res: Structure and synthesis of ribosomes; regulation of nucleic acid synthesis; molecular basis of morphogenesis. Mailing Add: Dept of Microbiol Univ of Calif Col of Med Irvine CA 92664

SYSKI, RYSZARD, b Plock, Poland, Apr 8, 24; m 50; c 6. MATHEMATICS. Educ: Polish Univ Col, London, Dipl Ing, 50; Imp Col, Dipl, 51, BSc, 54, PhD(math), 60. Prof Exp: Sci officer, ATE Co, Eng, 51-52, sr sci officer, Hivac Ltd, 52-60; res assoc mgt sci & lectr math, 61-62, assoc prof, 62-66, chmn probability & statist div, 66-70, PROF MATH, UNIV MD, COLLEGE PARK, 66- Concurrent Pos: Chmn bibliog comt, Int Teletraffic Cong, 55-61, mem organizing comt, 67. Mem: Am Math Soc; fel Royal Statist Soc. Res: Congestion, queueing and probability theories; stochastic processes. Mailing Add: Dept of Math Univ of Md College Park MD 20742

SYTY, AUGUSTA, b Harbin, China. CHEMISTRY. Educ: Univ Tenn, Knoxville, BS, 64, PhD(anal chem), 68. Prof Exp: PROF ANAL CHEM, INDIANA UNIV PA, 68- Mem: Am Chem Soc. Res: Flame emission; atomic absorption; atomic fluorescence methods of analysis of nonmetals and other elements. Mailing Add: Dept of Chem Indiana Univ of Pa Indiana PA 15701

SZABO, ALEXANDER, b Copper Cliff, Ont, Mar 13, 31; m 57; c 4. SOLID STATE PHYSICS. Educ: Queen's Univ, BS, 53; McGill Univ, MS, 55; Tohoku Univ, Japan, DEng(physics), 70. Prof Exp: Res officer electron beams, 55-59, res officer microwave masers, 59-61, RES OFFICER LASERS & SOLID STATE PHYSICS, NAT RES COUN CAN, 61- Res: Study of coherence lifetimes and optical line widths of impurity ions in solids using fluorescence line narrowing and photon echo techniques. Mailing Add: Nat Res Coun Bldg M-50 Ottawa ON Can

SZABO, ARLENE SLOGOFF, b Philadelphia, Pa, Feb 19, 45; m 65; c 2. CELL BIOLOGY. Educ: Douglass Col, AB, 66, Rutgers Univ, MS, 68, PhD(cell biol), 71. Prof Exp: Res fel enzymol, Med Sch, Univ Southern Calif, 71-72; res asst plant physiol, Univ Calif, Los Angeles, 72-74; ASST PROF BIOL, LOYOLA MARYMOUNT UNIV, 74- Mem: Am Soc Cell Biol; AAAS; Asn Women in Sci; Sigma Xi. Res: Development and regulation of the peroxisome in yeast, as Saccharomyces cerevisiae. Mailing Add: Dept of Biol Loyola Marymount Univ Los Angeles CA 90045

SZABO, ARTHUR GUSTAV, b Toronto, Ont, Nov 19, 39; m 62; c 3. ORGANIC CHEMISTRY. Educ: Queen's Univ, Ont, BSc, 61; Univ Toronto, MA, 63, PhD(chem), 65. Prof Exp: Nat Res Coun Can fel chem, Southampton, 65-67; RES OFFICER, BIOL DIV, NAT RES COUN CAN, 67- Mem: Chem Inst Can. Res: Mechanistic organic photochemistry; nucleic acid photochemistry using conventional techniques as well as flash photolysis; biomolecule interactions using spectroscopic techniques; luminescence; absorption. Mailing Add: Div of Biol Sci Nat Res Coun Can 100 Sussex Dr Ottawa ON Can

SZABO, ATTILA, b Budapest, Hungary, Sept 6, 47; Can citizen; m 75. THEORETICAL BIOPHYSICAL CHEMISTRY. Educ: McGill Univ, BSc, 68; Harvard Univ, PhD(chem physics), 73. Prof Exp: Fel, Inst Physicochem Biol, Paris, 72-73; fel, Med Res Coun Lab Molecular Biol, Cambridge, Eng, 73-74; ASST PROF CHEM, IND UNIV, BLOOMINGTON, 74- Mem: Biophys Soc. Res: Structure-function relations in allosteric systems; effect of protein conformation on the spectroscopic properties of heme proteins. Mailing Add: Dept of Chem Ind Univ Bloomington IN 47401

SZABO, BARNEY JULIUS, b Debrecen, Hungary, Apr 11, 29; US citizen; m 56. GEOCHEMISTRY. Educ: Univ Miami, BS, 61, MS, 66. Prof Exp: Res chemist, Inst Marine Sci, Univ Miami, 57-66; RES CHEMIST, BR ISOTOPE GEOL, US GEOL SURV, 66- Res: Stable and radioactive trace elements analyses; uranium disequilibria dating; radio and isotopic chemistry of uranium, thorium, protactinium and artificial radioactive nuclides. Mailing Add: US Geol Surv Br Isotope Geol Denver Fed Ctr Denver CO 80225

SZABO, C KAROLY, b Mocs, Hungary, May 27, 23; US citizen; m 59; c 1. ORGANIC CHEMISTRY. Educ: Budapest Tech Univ, MS, 45, PhD(org chem), 56. Prof Exp: Res chemist, Arzola Ltd, Hungary, 46-47, res group leader, 48-49; res assoc, Chinoin Corp, 49-50; res scientist, Nat Inst Plant Protection, 50-52, head dept org chem, 52-56; sr res chemist, Stauffer Chem Co, 57-62, sr res assoc, 62-65; mgr org chem sect, Syracuse Univ Res Corp, 66-71; first officer indust develop, 70-75, SR OFFICER, UN INDUST DEVELOP ORGN, 75- Concurrent Pos: Mem nat comt plant protection, Nat Acad Sci, Hungary, 53-56; NIH grant, 67-69; indust consult. Mem: Am Chem Soc. Res: Research and development of agricultural chemicals, particularly chemistry of phosphorus insecticides, carbamate insecticides and herbicides, systemic fungicides and organic sulfur compounds. Mailing Add: UN Indust Develop Orgn Lerchenfelder Strasse 1 Vienna Austria

SZABO, GABOR, b Mar 20, 41; Can citizen. PHYSIOLOGY, BIOPHYSICS. Educ: Univ Montreal, BS, 64, MS, 66; Univ Chicago, PhD(physiol), 69. Prof Exp: Asst prof physiol, Med Ctr, Univ Calif, Los Angeles, 71-75; ASSOC PROF PHYSIOL, UNIV TEX MED BR GALVESTON, 75- Mem: AAAS; Biophys Soc. Res: Permeability in model and cell membranes. Mailing Add: Dept of Physiol & Biophys Univ of Tex Med Br Galveston TX 77550

SZABO, GEORGE, biology, see 12th edition

SZABO, KALMAN TIBOR, b Abda, Hungary, July 29, 21; US citizen; m 44; c 2. TERATOLOGY. Educ: Univ Budapest, BSc, 47; Rutgers Univ, MS, 62; Univ Vienna, MSc, 71, DSc(teratology), 73. Prof Exp: Res scientist genetics, Res Inst, Acad Sci, Fertod, Hungary, 53-56; res assoc reproductive physiol, Rutgers Univ, 60-62; toxicologist, 62-66, sr scientist & unit head teratology, 67-69, sr investr & group leader reproductive toxicol & teratology, 69-70, asst dir toxicol, 71-72, ASSOC DIR TOXICOL, SMITH KLINE & FRENCH LABS, 72- Concurrent Pos: Adv & partic, Adv Comt Protocols on Reproductive Studies, Safety Eval Food Additives & Pesticide Residues, Food & Drug Admin, 67; res assoc prof pediat, Med Col, Thomas Jefferson Univ, 75- Mem: Teratology Soc; Europ Teratology Soc; Soc Toxicol; Soc Study Reproduction; Am Genetic Asn. Res: Spontaneous and induced anomalies of the central nervous system; role of maternal nutritional deprivation in embryogenesis; evaluation of various factors affecting teratogenic response; drug toxicology. Mailing Add: 215 Morris Rd Ambler PA 19022

SZABO, STEVE STANLEY, b Westons Mills, NJ, Feb 10, 27; m 55; c 2. AGRONOMY. Educ: Rutgers Univ, BS, 52; Kans State Col, MS, 54; Univ Wis, PhD(agron), 58. Prof Exp: Asst prof agron, NMex State Univ, 57-61; plant physiologist, US Food & Drug Admin, 61-63 & Army Biol Labs, Ft Detrick, Md, 63-65; PLANT PHYSIOLOGIST, BOYCE THOMPSON INST PLANT RES, 65- Mem: Weed Sci Soc Am. Res: Chemical weed control; plant physiology. Mailing Add: 20 Milrose Lane Monsey NY 10952

SZABUNIEWICZ, MICHAEL, b Poland, Oct 11, 09. VETERINARY MEDICINE. Educ: Acad Vet Med, DVM, 34, DVSc(physiol), 37. Prof Exp: Dir, Exp Farm Kasese, Belgian Congo, 46-50; dir, Regional Diag Vet Lab & Exten Serv, 50-60; area vet, Agr Res Serv, USDA, 61-62; from asst prof to assoc prof vet physiol, 62-71, PROF VET PHYSIOL, TEX A&M UNIV, 71- Mem: AAAS; Am Vet Med Asn; Am Soc Vet Physiol & Pharmacol; Am Asn Lab Animal Sci. Res: Physiology; pharmacology; radiation research. Mailing Add: Dept of Vet Physiol & Pharm Tex A&M Univ College Station TX 77843

SZAKACS, ANTAL, pathology, see 12th edition

SZAKAL, ANDRAS KALMAN, b Szekesfehervar, Hungary, Sept 26, 36; US citizen; m 61; c 2. IMMUNOBIOLOGY, CANCER. Educ: Univ Colo, Boulder, BA, 61, MS, 63; Univ Tenn, PhD(immunobiol), 72. Prof Exp: Res asst histochem, Univ Colo, Boulder, 60-61; res assoc electron micros, Univ Mich, Ann Arbor, 65-66; staff biologist, Oak Ridge Nat Lab, 66-69, consult electron microscopist, 69-70, Oak Ridge Assoc Univs fel, 70-72, res biologist, Div Biol, 72-74; PRIN SCIENTIST LUNG IMMUNOBIOL, LIFE SCI DIV, MELOY LABS, INC, 74- Mem: AAAS. Res: Lung immunobiology, especially identification and isolation of lung cell types through cell-type specific antigens. Mailing Add: Meloy Labs Inc 6715 Electronic Dr Springfield VA 22151

SZAL, ROGER ANDREW, b Detroit, Mich, Jan 30, 42; m 72; c 1. INVERTEBRATE ZOOLOGY. Educ: Univ Notre Dame, BSc, 64; Stanford Univ, PhD(biol), 70. Prof Exp: ASST PROF BIOL, SOUTHEASTERN MASS UNIV, 70- Mem: Am Soc Zoologists. Res: Structure and function of invertebrate sense organs; invertebrate behavior. Mailing Add: Dept of Biol Southeastern Mass Univ North Dartmouth MA 02747

SZALAY, FREDERICK S, vertebrate paleontology, physical anthropology, see 12th edition

SZALAY, JEANNE, b Jan 17, 38; US citizen; c 2. CELL BIOLOGY. Educ: Wash Sq Col Arts & Sci, BA, 59; Columbia Univ, PhD(biol), 66. Prof Exp: Fel cell biol & cytol, Col Physicians & Surgeons, Columbia Univ, 66-67; fel, Albert Einstein Col Med, 67-69, res asst cell biol & cytol, 69-70, instr, Dept Anat, 70-72; ASST PROF BIOL, QUEENS COL, NY, 72- Mem: Asn Res Vision & Ophthalmol. Res: The fine-structure and function of blood vessels, with emphasis on the examination of cellular mechanisms involved in the regulation of vascular permeability and vascular smooth muscle activity. Mailing Add: Dept of Biol Queens Col Kissena Blvd Flushing NY 11367

SZANISZLO, PAUL JOSEPH, b Medina, Ohio, June 9, 39; m 60; c 2. MICROBIOLOGY, MYCOLOGY. Educ: Ohio Wesleyan Univ, BA, 61; Univ NC, MA, 64, PhD(bot), 67. Prof Exp: USPHS fel, Harvard Univ, 67-68; asst prof, 68-73, ASSOC PROF MICROBIOL, UNIV TEX, AUSTIN, 73- Mem: AAAS; Am Soc Microbiol; Mycol Soc Am. Res: Growth, development and differentiation in fungi with emphasis on cell wall chemistry and cell ultrastructure. Mailing Add: Dept of Microbiol Exp Sci Bldg Univ of Tex Austin TX 78712

SZANTO, JOSEPH, b Marcali, Hungary, Nov 4, 31; m 57; c 4. VETERINARY PARASITOLOGY. Educ: Col Vet Med, Budapest, Dr Vet, 55; Univ Ill, DVM, 61. Prof Exp: Vet practice, Hungary, 55-56; res asst vet parasitol, Col Vet Med, Univ Ill, 58-61; asst prof vet path, Univ Ky, 61-62; asst parasitologist, Chemagro Corp, Mo, 62-63; sr res parasitologist, Ciba Pharmaceut Co, 63-69; sr res parasitologist, 69-74, DIR VET CLIN RES & DEVELOP, E R SQUIBB & SONS, INC, 74- Mem: Am Vet Med Asn; Am Asn Vet Parasitol; Am Soc Parasitol; World Asn Adv Vet Parasitol. Res: Veterinary clinical research and drug development. Mailing Add: E R Squibb & Sons Inc Three Bridges NJ 08887

SZAP, PETER CHARLES, b New York, NY, Aug 20, 29; m 57; c 3. ANALYTICAL CHEMISTRY. Educ: Queens Col, NY, BS, 51; Fordham Univ, MS, 53. Prof Exp: Res chemist, Lever Bros Res & Develop Co, NJ, 53-59; anal res chemist, Toms River Chem Corp, 59-67, SECT LEADER ANAL QUAL CONTROL, CIBA-GEIGY CORP, 67- Concurrent Pos: Chem Abstractor, 51- Mem: Am Chem Soc; Sigma Xi; Soc Appl Spectros. Res: Analytical methods development in detergents, dyestuffs, resins and raw materials; analysis and applications of optical brighteners, especially infrared, ultraviolet and visible absorption spectroscopy; chromatography; liquid chromatography techniques. Mailing Add: 333 Killarney Dr Toms River NJ 08753

SZARA, STEPHEN ISTVAN, b Pestujhely, Hungary, Mar 21, 23; nat US; m 59; c 1. PSYCHOPHARMACOLOGY. Educ: Pazmany Peter Univ, Hungary, DSc(chem), 50; Med Univ Budapest, MD, 51. Prof Exp: Sci asst, Microbiol Inst, Univ Budapest, 49-50, asst prof, Biochem Inst, 50-53, chief, Biochem Lab, State Ment Inst, 53-56; vis scientist, Psychiat Clin Berlin, 57; vis scientist, Nat Inst Ment Health, 57-60, chief sect psychopharmacol, Lab Clin Psychopharmacol, Spec Ment Health Res, 60-71, CHIEF CLIN STUDIES SECT, CTR STUDIES NARCOTIC & DRUG ABUSE, NAT INST MENT HEALTH, 71- Concurrent Pos: Assoc clin prof, George Washington Univ. Mem: AAAS; Int Col Neuropsychopharmacol; fel Am Col Neuropsychopharmacol; Am Soc Pharmacol & Exp Therapeut. Res: Metabolism of psychotropic drugs, especially tryptamine derivatives; correlation between metabolism and psychotropic activity of drugs. Mailing Add: Ctr Studies Narcotic & Drug Abuse Nat Inst Ment Health Rockville MD 20852

SZAREK, STANLEY RICHARD, b Visalia, Calif, Nov 14, 47; m 74. PHYSIOLOGICAL ECOLOGY. Educ: Calif State Univ, Pomona, BS, 69; Univ Calif, Riverside, PhD(biol), 74. Prof Exp: ASST PROF BOT, ARIZ STATE UNIV, 74- Concurrent Pos: Consult NSF grant, Ore State Univ, 75- Mem: Am Soc Plant Physiologists; Sigma Xi. Res: Physiological and biophysical ecology of desert plants, emphasizing photosynthetic carbon metabolism and water relations. Mailing Add: Dept of Bot & Microbiol Ariz State Univ Tempe AZ 85282

SZAREK, WALTER ANTHONY, b St Catharines, Ont, Apr 19, 38. ORGANIC CHEMISTRY. Educ: McMaster Univ, BSc, 60, MSc, 62; Queen's Univ, Ont, PhD(org chem), 64. Prof Exp: Res fel chem, Ohio State Univ, 64-65; asst prof biochem, Rutgers Univ, 65-67; asst prof org chem, 67-71, ASSOC PROF ORG CHEM, QUEEN'S UNIV, ONT, 71- Mem: Am Chem Soc; Chem Inst Can; The Chem Soc; NY Acad Sci. Res: Structure and synthesis of carbohydrates and carbohydrate-containing antibiotics; biochemical aspects; conformational and mechanistic studies of carbohydrate reactions; photochemistry of carbohydrates; heterocyclic conformational analysis. Mailing Add: Dept of Chem Queen's Univ Kingston ON Can

SZARKA, LASZLO JOSEPH, b Szekesfehervar, Hungary, Sept 6, 35; stateless; m 63; c 2. BIOENGINEERING. Educ: Univ Med Sci, Hungary, MS, 61, PhD(phys chem), 67. Prof Exp: Asst res investr phys chem, Chinoin Pharmaceut Plant, Hungary, 61-63; sect head bioeng, Res Inst Pharmaceut Chem, Hungary, 64-72; head chemist, Pharmaceut Ctr, Hungary, 72-73; res & develop biochem, Blackman Labs Inc, 74; RES INVESTR BIOENG, SQUIBB INST MED RES, 74- Mem: Am Pharmaceut Asn. Res: Antibioticum and enzyme fermentation technology; batch, fed batch and continuous cultivation; transport phenomena, process control and reactor designing; optimization and scale-up problems; molecule biotransformations, steroids and hydrocarbons. Mailing Add: Squibb Inst for Med Res Georges Rd New Brunswick NJ 08903

SZASZ, GEORGE J, physical chemistry, see 12th edition

SZASZ, STEPHEN E, b Bucharest, Romania, Sept 15, 10; nat US; m 38. PHYSICAL CHEMISTRY. Educ: Univ Bucharest, MS, 31, MS, 32, PhD(chem), 39. Prof Exp: High sch teacher, Romania, 35-40; chemist exploitation res, Astra Romania Oil Co, 41-44; high sch teacher, Ger, 45-47; mem staff voc training, French Mil Govt, Ger, 47-51; asst proj chemist, Sinclair Res Labs, Inc, 52-58, sr proj engr, 58-60, from res assoc to sr res assoc, 60-66; CHIEF RESERVOIR ENGR, CALIF STATE LANDS DIV, 66- Mem: Am Chem Soc; Soc Petrol Eng; NY Acad Sci. Res: Petroleum production and exploration. Mailing Add: 5133 Vista Long Beach CA 90803

SZASZ, THOMAS STEPHEN, b Budapest, Hungary, Apr 15, 20; US citizen; div; c 2. PSYCHIATRY. Educ: Univ Cincinnati, AB, 41, MD, 44; Am Bd Psychiat & Neurol, dipl psychiat, 51. Hon Degrees: DSc, Allegheny Col, 75. Prof Exp: Intern med, Boston City Hosp, 44-45; resident psychiat, Clinics, Univ Chicago, 46-48; mem staff, Chicago Inst Psychoanal, 51-56; PROF PSYCHIAT, STATE UNIV NY UPSTATE MED CTR, 56- Concurrent Pos: Vis prof psychiat, Sch Med, Marquette Univ, 68; consult, Comt Ment Hyg, NY State Bar Asn, Judicial Conf Comt, Judicial Conf DC Circuit, Comt Laws Pertaining Ment Disorders & Res Adv Panel, Inst Study Drug Addiction; adv ed, J Forensic Psychol; mem bd consult, Psychoanal Rev, mem adv comt, Living Libraries, Inc; mem selection adv bd, Tort & Med Yearbk, Bobbs-Merrill Co. Honors & Awards: Holmes-Munsterberg Award Forensic Psychol, Int Acad Forensic Psychol, 69; Humanist of Year, Am Humanist Asn, 73; Am Inst Pub Serv Distinguished Serv Award, 74. Mem: Fel Royal Soc Health. Res: Epistemology of the behavioral sciences; history of psychiatry; psychiatry and law. Mailing Add: Dept of Psychiat State Univ NY Upstate Med Ctr Syracuse NY 13210

SZCZARBA, ROBERT HENRY, b Dearborn, Mich, Nov 27, 32; m 55; c 2. TOPOLOGY. Educ: Univ Mich, BS, 55; Univ Chicago, MS, 56, PhD(math), 60. Prof Exp: Off Naval Res res assoc math, 60-61, from asst prof to assoc prof, 61-74, PROF MATH, YALE UNIV, 74- Concurrent Pos: NSF fel, Inst Advan Study, 64-65. Mem: Am Math Soc. Res: Algebraic and differential topology and geometry of differentiable manifolds. Mailing Add: Dept of Math Yale Univ New Haven CT 06520

SZCZAWINSKI, ADAM FRANCISZEK, b Lwow, Poland, Oct 21, 13; nat Can; m 46; c 1. BOTANY. Educ: Univ Lwow, Poland, MPh, 37; Univ BC, PhD(bot), 53. Prof Exp: Instr, Univ Lwow, Poland, 37-39; asst, Univ BC, 50-52, lectr, 52-55; CUR BOT, PROV MUS, BC, 55- Mem: Ecol Soc Am; Mycol Soc Am; Am Soc Plant Taxon; Can Bot Asn; Int Asn Plant Taxon. Res: Ecology of Corticolous and Lignicolous plant communities of Pacific Northwest; forests of British Columbia; pollen survey of Victoria, British Columbia. Mailing Add: British Columbia Prov Mus Victoria BC Can

SZCZECH, GEORGE MARION, US citizen. VETERINARY PATHOLOGY, TERATOLOGY. Educ: Univ Minn, BS, 64, DVM, 66; Purdue Univ, PhD(vet path), 74. Prof Exp: Resident vet med, Colo State Univ, 66-67; vet, Sea Girt Animal Hosp, 67-68; sr scientist toxicol, Res Ctr, Mead Johnson & Co, 68-70; res fel vet path, Purdue Univ, 70-74; RES SCIENTIST PATH, UPJOHN CO, 74- Mem: Am Vet Med Asn; Am Col Vet Path; Teratology Soc; Sigma Xi. Res: Toxicologic pathology; pathology of mycotoxic diseases; clinical pathology and teratology. Mailing Add: Path-Toxicol Res Unit Upjohn Co Kalamazoo MI 49001

SZCZESNIAK, ALINA SURMACKA, food science, see 12th edition

SZCZESNIAK, RAYMOND ALBIN, b Buffalo, NY, Nov 28, 40; m 63; c 4. MEDICINAL CHEMISTRY, NUCLEAR MEDICINE. Educ: Fordham Univ, BS, 63; Univ Mich, Ann Arbor, MS, 65, PhD(med chem), 68. Prof Exp: Sr res scientist, E R Squibb & Sons, Inc, New Brunswick, 68-69, res chemist, 69-72, sr res chemist radiopharmaceut, 72-73; RADIOIMMUNOASSAY SPECIALIST, NUCLEAR-MED LABS, INC, 73- Mem: Soc Nuclear Med; Am Chem Soc. Res: Radioimmunoassay; radiomedicinal synthesis. Mailing Add: 4042 Mendenhall Dr Dallas TX 75234

SZE, HEVEN, b The Hague, Netherlands, Oct 22, 47; Chinese citizen; m 74. PLANT PHYSIOLOGY. Educ: Nat Taiwan Univ, BS, 68; Univ Calif, Davis, MS, 70; Purdue Univ, PhD(plant physiol), 75. Prof Exp: RES FEL BIOPHYS, HARVARD MED SCH, 75- Mem: Am Soc Plant Physiologists. Res: Mechanism of solute transport across membranes. Mailing Add: Biophys Lab Harvard Med Sch 25 Shattuck St Boston MA 02115

SZE, PAUL YI LING, b Shanghai, China. BIOCHEMISTRY, NEUROBIOLOGY. Educ: Nat Taiwan Univ, BS, 60; Duquesne Univ, MS, 62; Univ Wis, MS, 64; Univ Chicago, PhD(biochem), 69. Prof Exp: Asst prof, 69-72, ASSOC PROF BIOBEHAV SCI, UNIV CONN, 72- Mem: AAAS; Am Soc Neurochem. Res: Biochemistry of neurotransmitters; biochemical aspects of alcoholism; neurobiological effects of environmental stimulation. Mailing Add: U-154 Dept of Biobehav Sci Univ of Conn Storrs CT 06268

SZEBEHELY, VICTOR, b Budapest, Hungary, Aug 10, 21; nat US; m 70; c 1. ASTRONOMY. Educ: Budapest Tech Univ, DSc, 45. Prof Exp: Asst prof appl math, Budapest Tech Univ, 43-47; res asst, Pa State Univ, 47-48; assoc prof appl mech, Va Polytech Inst, 48-51; head ship dynamics br, David Taylor Model Basin, US Dept Navy, 51-57; mgr space mech, Missiles & Space Div, Gen Elec Co, 57-62; assoc prof astron, Yale Univ, 62-68; PROF AEROSPACE ENG & ENG MECH, UNIV TEX, AUSTIN, 68- Concurrent Pos: Lectr, McGill Univ, 48 & Univ Toronto, 49; prof lectr, Univ Md & George Washington Univ, 51-57; consult, NASA, Gen Elec Co, Int Tel & Tel Co, Bell Tel Labs & Inst Defense Anal. Honors & Awards: Sci Res Soc Award, Sigma Xi, 51; Officer, Order of Orange Nassau, Netherlands, 57. Mem: Am Inst Aeronaut & Astronaut; Am Soc Mech Engrs; Int Astron Union; Am Astron Soc. Res: Celestial mechanics; problem of three and n bodies; applied mechanics; analytical dynamics; continuum mechanics; applied mathematics; matrix and tensor analysis. Mailing Add! Dept of Aerospace Eng Univ of Tex Austin TX 78712

SZEBENYI, EMIL, b Budapest, Hungary, June 9, 20; US citizen; m 44; c 3. COMPARATIVE ANATOMY, EMBRYOLOGY. Educ: Budapest Tech Univ, dipl, 42, Doctoratus (animal genetics & husb), 43. Prof Exp: Asst prof animal genetics, Univ Agr Sci Hungary, 52-53, Hungarian Acad Sci fel & adj, 53-56; from asst prof to assoc prof, 62-73, PROF COMP ANAT, EMBRYOL & EVOLUTION, FAIRLEIGH DICKINSON UNIV, 73-, CHMN DEPT BIOL SCI, 71- Mem: AAAS; NY Acad Sci. Res: Anatomy. Mailing Add: Dept of Biol Sci Fairleigh Dickinson Univ Rutherford NJ 07070

SZEGHO, CONSTANTIN STEPHEN, b Nagybocsko, Hungary, Mar 15, 05; nat US; m 51; c 2. PHYSICS. Educ: Inst Technol, Munich, EE, 28; Inst Technol, Aachen, DSc(eng), 31. Prof Exp: Head cathode ray tube res dept, Baird TV, Ltd, London, 32-42; dir res, Rauland Corp, 42-51, vpres in chg res, 51-69; CONSULT, ZENITH RADIO CORP, 69- Mem: Am Phys Soc; fel Inst Elec & Electronics Eng. Res: Cathode-ray tubes, photoemissive and solid state devices. Mailing Add: Apt 3014 2800 Lake Shore Dr Chicago IL 60657

SZEGO, CLARA MARIAN, b Budapest, Hungary, Mar 23, 16; US citizen; m 43. CELL BIOLOGY. Educ: Hunter Col, BA, 37; Univ Minn, MS, 39, PhD(physiol chem), 42. Prof Exp: Asst physiol chem, Sch Med, Univ Minn, 40-42, asst physiol, 42, instr, 42-44; fel cancer, Minn Med Found, 44; assoc chemist, Off Sci Res & Develop, Nat Bur Stand, 44-45; res assoc, Worcester Found Exp Biol, 45-47; res instr physiol chem, Yale Univ, 47-48; asst clin prof biophys, 48-49, from asst prof to prof, 49-72, PROF BIOL, UNIV CALIF, LOS ANGELES, 72-, ASSOC MEM, MOLECULAR BIOL INST, 66- Concurrent Pos: Guggenheim fel, 56. Honors & Awards: Ciba Award, Endocrine Soc, 53; Gregory Pincus Mem Medallion, Worcester Found Exp Biol, 74. Mem: Fel AAAS; Am Physiol Soc; Am Soc Zoologists; Endocrine Soc; Am Soc Cell Biol. Res: Molecular mechanisms of endocrine regulation. Mailing Add: Dept of Biol Univ of Calif Los Angeles CA 90024

SZEKELY, ANDREW GEZA, b Temesvar, Rumania, Apr 15, 25; US citizen; m 51; c 6. PHYSICAL CHEMISTRY. Educ: Eötvös Lorand Univ, Budapest, dipl chem, 50. Prof Exp: Asst electrochem, Physico-Chem Inst, Eötvös Lorand Univ, Budapest, 49-51; res chemist, High Pressure Res Inst, Budapest, 51-56, dept head petrol refining, 56-57; res chemist, Tonawanda Res Lab, Linde Div, 58-62, sr res chemist, 62-66, develop assoc process metall, Newark Develop Labs, NJ, 67-70, SR DEVELOP ASSOC, GAS PROD DEVELOP, LINDE LABS, TARRYTOWN TECH CTR, UNION CARBIDE CORP, 70- Mem: AAAS; Am Inst Mining, Metall & Petrol Eng; Am Inst Chem. Res: Process metallurgy; thermodynamics and kinetics. Mailing Add: Linde Labs Union Carbide Corp Tarrytown NY 10591

SZEKELY, IVAN J, b Budapest, Hungary, Aug 13, 19; nat US; m. RESOURCE MANAGEMENT. Educ: Pazmany Peter Univ, Hungary, PhD, 43, MS, 44. Prof Exp: Asst prof chem, Bowling Green State Univ, 47-48; prof pharm, Univ Calif, 48-51; dir res & prod develop, Barnes-Hind Pharmaceut, Inc, 51-57, from exec vpres to pres, 57-69; PRES, PHARMACEUT RES INT, 71- Concurrent Pos: Pharmaceut mgt consult. Mem: Am Chem Soc; Am Pharmaceut Asn; NY Acad Sci. Res: Formulation of pharmaceutical dosage forms; ophthalmic solutions; physical-chemical properties of contact lens materials, especially contact lens solutions. Mailing Add: 13643 Wildcrest Dr Los Altos Hills CA 94022

SZEKELY, JOSEPH GEORGE, b Cleveland, Ohio, May 7, 40; m 67; c 2. BIOPHYSICS. Educ: Case Western Reserve Univ, BS, 62; State Univ NY Buffalo, PhD(physics), 67. Prof Exp: USPHS fel, Univ Tex M D Anderson Hosp & Tumor Inst, 67-68; ASSOC RES OFFICER, ATOMIC ENERGY CAN LTD, 68- Mem: Biophys Soc. Res: Membrane biophysics; computer models of cell growth; electron microscopy. Mailing Add: 30 Lansdowne Ave Pinawa MB Can

SZEKERES, GABOR LASZLO, organic chemistry, medicinal chemistry, see 12th edition

SZENT-GYORGYI, ALBERT, b Budapest, Hungary, Sept 16, 93; nat US; m 42; c 1. BIOCHEMISTRY. Educ: Univ Budapest, MD, 17; Cambridge Univ, PhD(chem), 27. Hon Degrees: Dr, Univ Lausanne, Univ Padua, Univ Paris, Oxford Univ, Univ Bordeaux, Oberlin Col & Brown Univ. Prof Exp: Prof med chem, Univ Szeged, 32-44; prof biochem, Univ Budapest, 45-47; DIR RES, INST MUSCLE RES, MARINE BIOL LAB, WOODS HOLE, MASS, 47- Concurrent Pos: Vis prof, Univ Liege, 35 & Harvard Univ, 36; Withering lectr, Univ Birmingham, Eng, 38; mem standing comt, Exec Comt, Bd Trustees, Marine Biol Lab, Woods Hole. Honors & Awards: Nobel Prize in Med, 37; Cameron Prize, 46; Lasker Award, Am Heart Asn, 54. Mem: Nat

Acad Sci; NY Acad Med; hon mem, NY Acad Sci; Royal Soc Edinburgh; Nat Acad Lincei. Res: Bioenergetics; biological oxidation; chemical physiology of muscle; submolecular biology. Mailing Add: PO Box 187 Woods Hole MA 02543

SZENT-GYORGYI, ANDREW GABRIEL, b Budapest, Hungary, May 16, 24; nat US; m 47; c 3. BIOCHEMISTRY. Educ: Univ Budapest, MD, 47; Dartmouth Col, MA, 63. Prof Exp: Instr, Univ Budapest, 46-47; fel, Neurophysiol Inst, Copenhagen, 48; mem inst muscle res, Marine Biol Lab, Woods Hole, 48-62, instr physiol, 53-58; prof biophys, Dartmouth Med Sch, 62-66; PROF BIOL, BRANDEIS UNIV, 66-, CHMN DEPT, 75- Concurrent Pos: Am Heart Asn estab investr, 55-62; USPHS res career award, 62-66; head physiol course, Marine Biol Lab, Woods Hole, 67-71, trustee, 70-76. Mem: Am Physiol Soc; fel Am Acad Arts & Sci; Muscular Dystrophy Asn; Biophys Soc (pres, 74-75); Soc Gen Physiol (pres, 71). Res: Chemistry and physiology of muscle contraction; structure of fibrous proteins. Mailing Add: Dept of Biol Brandeis Univ Waltham MA 02154

SZENTIVANYI, ANDOR, b Miskoic, Hungary, May 4, 26; US citizen; m 48; c 2. PHARMACOLOGY, ALLERGY. Educ: Debrecen Univ Med, MD, 50. Prof Exp: Asst prof med, Univ Med Sch Budapest, 53-56; Rockefeller fel med, Univ Chicago, 57-59; USPHS fel allergy & immunol, Univ Colo, Denver, 59-60, asst prof med, 61-65, from asst prof to assoc prof Pharmacol & microbiol, 63-67; prof med & chmn dept microbiol, Sch Med, Creighton Univ, 67-70; PROF PHARMACOL & CHMN DEPT, COL MED, UNIV S FLA, TAMPA, 70- Mem: Am Asn Immunol; Am Acad Allergy; Am Soc Pharmacol & Exp Therapeut; Am Soc Clin Pharmacol; Am Col Clin Pharmacol. Res: Pharmacological aspects of immune and hypersensitivity mechanisms. Mailing Add: 11603 Carrollwood Dr Tampa FL 33618

SZEPESI, BELA, b Ozd, Hungary, Nov 19, 38; US citizen; m 61; c 3. NUTRITION, MOLECULAR BIOLOGY. Educ: Albion Col, BA, 61; Colo State Univ, MS, 64; Univ Calif, Davis, PhD(comp biochem), 68. Prof Exp: RES CHEMIST, NUTRIT INST, AGR RES SERV, USDA, 69- Mem: Am Inst Nutrit; Brit Biochem Soc; Brit Nutrit Soc; Soc Exp Biol & Med. Res: Mechanism of overweight; control of gene expression in rat liver; disaccharide effect on enzyme induction; starvation-refeeding; hormone effect on rat liver enzymes. Mailing Add: Rm 313 Bldg 307 Nutrit Inst Agr Res Serv USDA Beltsville MD 20705

SZEPESI, ZOLTAN PAUL JOHN, b Sarosfa, Hungary, May 13, 12; nat US; m 42; c 1. ELECTRONICS. Educ: Univ Szeged, BSc, 32, MSc, 34; Univ Budapest, PhD(physics), 37. Prof Exp: Mem staff, Inst Theoret Physics, Hungary, 34-36; physicist, Tungsram Res Lab, 36-45; engr, Hungarian Tel & Tel Inst, 45-46; physicist, Pulvari Lab, 46-47; res assoc, Grenoble, 47-51; sr physicist, Can Marconi Co, Que, 51-58; fel engr, Electronic Tube Div, Westinghouse Elec Corp, NY, 58-72, FEL SCIENTIST, WESTINGHOUSE RES LAB, PA, 72- Mem: Can Asn Physicists; Fr Phys Soc. Res: Compton effect of gamma rays; noise effects in electron tubes; wave guides; slot antennas; high frequency oscillators; photoconductors; solid state display panels. Mailing Add: Westinghouse Res Lab Beulah Rd Pittsburgh PA 15235

SZEPSENWOL, JOSEL, b Radoszkosice, Poland, Feb 7, 05; nat US. ANATOMY. Educ: Univ Geneva, MD, 36. Prof Exp: Asst anat, Univ Geneva, 29-31, asst prof, 31-37; chief tissue cult sect, Univ Buenos Aires, 37-43; Knight fel, Yale Univ, 43-45; from instr surg to assoc prof anat, Emory Univ, 47-54; vis prof, Sch Med, Univ NC, 54-55; assoc prof, 55-60, PROF ANAT, SCH MED, UNIV PR, SAN JUAN, 60- Mem: Soc Exp Biol & Med; Am Physiol Soc; Am Asn Anatomists; Am Asn Cancer Res. Res: Experimental embryology; tissue culture; cancer. Mailing Add: Dept of Anat Univ of PR Sch of Med San Juan PR 00936

SZEPTYCKI, PAWEL, b Lwow, Poland, Feb 10, 35; m 61; c 3. MATHEMATICS. Educ: Univ Warsaw, MSc, 56; Univ SAfrica, PhD(math), 61. Prof Exp: Res officer math, SAfrican Coun Sci & Indust Res, 59-61; from asst prof to assoc prof, 61-67, PROF MATH, UNIV KANS, 67- Mem: Am Math Soc. Res: Mathematical analysis. Mailing Add: Dept of Math Univ of Kans Lawrence KS 66044

SZER, WLODZIMIERZ, b Warsaw, Poland, June 3, 24; m 48; c 2. BIOCHEMISTRY. Educ: Univ Lodz, MS, 50; Polish Acad Sci, PhD(biochem), 59. Prof Exp: Instr org chem, Univ Lodz, 51-53; res assoc, Inst Antibiotics, Poland, 54-56; asst prof biochem, Inst Biochem & Biophys, Polish Acad Sci, 59-62, dozent biochem, 63-67; PROF BIOCHEM, SCH MED, NY UNIV, 68- Concurrent Pos: Jane Coffin Childs Mem Fund fel, Sch Med, NY Univ, 63-64. Honors & Awards: J K Parnas Award, Polish Biochem Soc. Mem: Am Soc Biol Chemists; Harvey Soc. Res: Structure of polynucleotides; molecular mechanisms in protein synthesis. Mailing Add: Dept of Biochem NY Univ Med Ctr New York NY 10016

SZERB, JOHN CONRAD, b Budapest, Hungary, Feb 24, 26; nat US; m 57; c 2. PHYSIOLOGY. Educ: Univ Munich, MD, 50. Prof Exp: Lectr, 51-52, from asst prof to assoc prof, 52-63, PROF PHARMACOL, DALHOUSIE UNIV, 63-, CHMN DEPT PHYSIOL & BIOPHYS, 65- Concurrent Pos: Res fel, Pasteur Inst, Paris, 50-51; Nuffield fel, Cambridge Univ, 60-61, vis scientist, 70-71; mem, Med Res Coun Can, 65-70. Mem: Am Soc Pharmacol & Exp Therapeut; Pharmacol Soc Can; Can Physiol Soc; Brit Physiol Soc. Res: Release of transmitters and electrical activity in the central nervous system. Mailing Add: Dept Physiol & Biophys Dalhousie Univ Halifax NS Can

SZERENYI, PETER, b Budapest, Hungary, Feb 8, 43; US citizen; m 66. PHYSICAL CHEMISTRY. Educ: Ohio State Univ,PhD(electron spin resonance), 70. Prof Exp: Res assoc, Univ NC, 70-72; ASST PROF CHEM, TUFTS UNIV, 72- Mem: Am Chem Soc. Res: Triplet state properties of aromatic and of hydrogen bonded molecules; photophysical and photochemical processes of the triplet state. Mailing Add: Dept of Chem Tufts Univ Medford MA 02155

SZETO, GEORGE, b Hong Kong, Aug 10, 38; m 68. MATHEMATICS. Educ: United Col Hong Kong, BSc, 64; Purdue Univ, Lafayette, MA, 66, PhD(math), 68. Prof Exp: Asst prof, 68-71, ASSOC PROF MATH, BRADLEY UNIV, 71- Concurrent Pos: NSF res grant, 72-73. Mem: Am Math Soc. Res: Separable algebras; near rings. Mailing Add: Dept of Math Bradley Univ Peoria IL 61606

SZIKLAI, OSCAR, b Repashuta, Hungary, Oct 30, 24; Can citizen; m 49; c 4. BIOLOGY, FOREST GENETICS. Educ: Sopron Univ, Hungary, BSF, 46; Univ BC, MF, 61, PhD(forest genetics), 64. Prof Exp: Instr silvicult, Sopron Univ, Hungary, 46-47; asst forester, Hungarian Forest Serv, 47-49; res officer, Forest Res Inst, Budapest, 49-51; asst prof silvicult, Sopron Univ, Hungary, 51-56; asst prof, Sopron Forestry Sch, Univ BC, 57-59, lectr forestry, 59-61, instr, 61-64, from asst prof to assoc prof forest genetics, Univ, 64-71, PROF FOREST GENETICS, UNIV BC, 71- Concurrent Pos: Nat Res Coun Can res grants forest genetics, 62- Mem: Can Inst Forestry. Res: Forest biology; selection and hybridication of Salix and Populus genera and Pseudotsuga genus; variation and inheritance studies in western Canadian conifers. Mailing Add: Fac of Forestry Univ of BC Vancouver BC Can

SZILAGYI, PAUL JULIUS, organic chemistry, see 12th edition

SZINAI, STEPHEN SLOMO, b Gyongyos, Hungary, Nov 4, 22; m 48; c 1. CHEMISTRY, PHARMACOLOGY. Educ: Univ Budapest, MSc, 49; Hebrew Univ Jeruslaem, PhD(org chem), 56. Prof Exp: Res chemist, Govt Res Labs, Israel, 49-56; res fel, Weizmann Inst Sci, 56-57; head drug res labs, Reckitt & Sons Ltd, Eng, 57-64; head chem res, Lilly Res Ctr, Eng, 64-71; vis prof biopharm, 71-72, assoc prof med, 72-74, PROF PHARMACEUT CHEM, J HILLIS MILLER HEALTH CTR, UNIV FLA, 72-, PROF RADIOL, 74- Mem: Am Chem Soc; The Chem Soc; fel Royal Inst Chem; Brit Soc Drug Res. Res: Molecular pharmacology; transport of drugs in the body and effect of this on structure-activity relationships; rational design of biologically active molecules. Mailing Add: Health Ctr Univ of Fla Gainesville FL 32610

SZKOLNIK, MICHAEL, b Clifton, NJ, Aug 23, 20; m 44; c 5. PLANT PATHOLOGY. Educ: Rutgers Univ, BS, 43, PhD(plant path), 49. Prof Exp: Plant pathologist, Exp Plantations, Inc, Merck & Co, Guatemala, 49-51; from asst prof to assoc prof plant path, 51-61, PROF PLANT PATH, AGR EXP STA, NY STATE COL AGR & LIFE SCI, CORNELL UNIV, 61- Mem: Am Phytopath Soc. Res: Fruit diseases and control; fungicides; antibiotics; chemotherapy; influence of fungicidal sprays on plant growth, fruit yield and quality; behavior of stored fruit. Mailing Add: Dept of Plant Path Agr Exp Sta NY State Col of Agr & Life Sci Geneva NY 14456

SZMANT, HERMAN HARRY, b Kalisz, Poland, May 18, 18; nat US; m 41; c 2. ORGANIC CHEMISTRY. Educ: Ohio State Univ, BA, 40; Purdue Univ, PhD(chem), 44. Prof Exp: Res chemist, Monsanto Chem Co, 44-46; assoc prof chem, Duquesne Univ, 46-50, prof, 51-56; head dept chem & ctr chem res, Oriente, Cuba, 56-60; prof chem, Univ PR, San Juan & head phys sci div, PR Nuclear Ctr, 61-68; CHMN DEPT CHEM, UNIV DETROIT, 68- Concurrent Pos: Chem adv, US AID Mission, Dominican Repub, 68. Mem: AAAS; Am Chem Soc; NY Acad Sci; PR Acad Arts & Sci. Res: Physical organic chemistry; sulfur compounds; solvent effects. Mailing Add: Dept of Chem Univ of Detroit Detroit MI 48221

SZMUC, EUGENE JOSEPH, b Cleveland, Ohio, June 5, 27; m 63. GEOLOGY. Educ: Western Reserve Univ, BS, 51; Ohio State Univ, MS, 53, PhD, 57. Prof Exp: Asst geol, Univ Tenn, 52-53; asst prof, 57-71, ASSOC PROF GEOL, KENT STATE UNIV, 71- Mem: Geol Soc Am; Paleont Soc. Res: Stratigraphy of Mississippian rocks; paleontology of Mississippian invertebrate fossils. Mailing Add: Dept of Geol Kent State Univ Kent OH 44242

SZMUSZKOVICZ, JACOB, organic chemistry, see 12th edition

SZOLLOSI, DANIEL GABRIEL, b Mako, Hungary, July 21, 29; US citizen; m 53. EMBRYOLOGY, CELL BIOLOGY. Educ: Univ Santa Clara, BA, 56; Univ Wis, MS, 58, PhD(zool), 61. Prof Exp: Res asst, Univ Wis, 56-61; NIH fel, Marshall Lab Physiol of Reprod, Cambridge Univ, 61-62; from instr to asst prof biol struct, 62-68, ASSOC PROF BIOL STRUCT, UNIV WASH, 68- Concurrent Pos: NIH res grant, 64-67; sabbatical leave, Animal Res Sta, Cambridge, Eng, 70-71. Mem: Am Soc Cell Biol; Brit Soc Study Fertil; Am Asn Anat; Soc Develop Biol. Res: Ultrastructural study of gametogenesis in mammals and some invertebrates; cytodifferentiation during pre-implantation stages of mammalian embryos. Mailing Add: Dept of Biol Struct Univ of Wash Seattle WA 98195

SZONYI, GEZA, b Budapest, Hungary, Feb 7, 19; nat US; m 45; c 2. COMPUTER SCIENCE, INFORMATION SCIENCE. Educ: Univ Zurich, PhD, 45. Prof Exp: Res chemist, W Stark AG, Switz, 45-47; info scientist & head co, Chemolit, 47-51; res chemist, Can Industs, Ltd, 51-53 & Barrett Div, Allied Chem & Dye Corp, 53-57; lit scientist, Socony Mobile Oil Co, Inc, 57-58; info res chemist, Atlas Chem Industs, Inc, 58-64; chief lit chemist, Ciba Corp, NJ, 64-69; SUPVR SYSTS & STATIST, POLAROID CORP, 69- Concurrent Pos: Adv, Mass Manpower Comn. Mem: Am Chem Soc; Sigma Xi. Res: Chemical information research; statistics; scientific computer programming; computerized information retrieval systems; paper chemistry and technology; catalysis; physical and pharmaceutical chemistry; improved information handling methods; simulation. Mailing Add: Polaroid Corp 750 Main St Cambridge MA 02139

SZTANKAY, ZOLTAN GEZA, b Cleveland, Ohio, Apr 17, 37; m 70. PHYSICS. Educ: Valparaiso Univ, BS, 59; Univ Wis, MS, 61, PhD(physics), 65. Prof Exp: PHYSICIST, HARRY DIAMOND LABS, 65- Mem: Inst Elec & Electronics Engrs; Am Phys Soc. Res: Analysis of electrooptical, laser and infrared sensing systems; the propagation and back-scatter of laser beams in the atmosphere; the interaction of laser irradiation with a variety of solids. Mailing Add: Harry Diamond Labs Br 930 2800 Powder Mill Rd Adelphi MD 20783

SZU, SHOUSUN CHEN, b Chunking, China, Sept 25, 45; m 72; c 1. MOLECULAR BIOPHYSICS. Educ: Taiwan Prov Cheng-Kung Univ, BS, 66; Mich Tech Univ, MS, 67; Univ Calif, Davis, PhD(physics), 74. Prof Exp: Res assoc polymer hydrodynamics, Dept of Chem, Univ NC, 73-75; VIS FEL BIOPHYS, NAT CANCER INST, NIH, 75- Mem: AAAS; Biophys Soc. Res: Macromolecular conformations, including polypeptide and polynucleotide kinetics and protein foldings; macromolecular hydrodynamics. Mailing Add: Lab of Theoret Biol Nat Cancer Inst NIH Bethesda MD 20014

SZUCHET, SARA, b Poland. PHYSICAL BIOCHEMISTRY. Educ: Univ Buenos Aires, MSc, 56; Cambridge Univ, PhD(phys chem), 63. Prof Exp: Res asst, Lister Inst Prev Med, Univ London, 57-58; res assoc chem, Princeton Univ, 63-65; res assoc biol, 66-68, ASST PROF BIOPHYS SCI, STATE UNIV NY BUFFALO, 68- Mem: Brit Biophys Soc; Biophys Soc; Am Chem Soc; AAAS. Res: Structure of proteins and its relation to the function of the molecule; protein-protein interaction; protein-small ions interaction; metalloproteins; ultracentrifugation. Mailing Add: Dept of Biophys Sci State Univ NY Buffalo NY 14214

SZUHAJ, BERNARD F, b Lilly, Pa, Nov 27, 42; m 64; c 3. BIOCHEMISTRY, LIPID CHEMISTRY. Educ: Pa State Univ, BS, 64, MS, 66, PhD(biochem), 69. Prof Exp: Asst biochem, Pa State Univ, 64-68; scientist, Res Dept, 68-73, DIR FATS & OILS RES, FOOD & CHEM RES, CENT SOYA CO, INC, 73- Mem: AAAS; Am Oil Chem Soc; Am Chem Soc; Inst Food Technol; Sci Res Soc NAm. Res: Basic lipid research; research and development of fats and oil products; analytical biochemistry. Mailing Add: Cent Soya Co Food & Chem Res 1825 N Laramie Ave Chicago IL 60639

SZUMSKI, ALFRED JOHN, b South River, NJ, Jan 12, 26; m 50; c 1. PHYSIOLOGY, NEUROPHYSIOLOGY. Educ: Col William & Mary, BS, 51; Med Col Va, cert, 51, MS, 56, PhD(physiol), 64. Prof Exp: From instr to asst prof physiol, 63-66, ASST PROF PHYS THER, MED COL VA, 56-, ASSOC PROF PHYSIOL, 66- Concurrent Pos: Dept Health, Educ & Welfare res fel voc rehab admin, Nobel Inst Neurophysiol, Karolinska Inst, Sweden, 64-65 & res career award, 67; Humboldt Found fel & sr US scientist, Neurol Inst, Munich Tech Univ, 73-74. Mem: Am Phys Ther Asn; NY Acad Sci; Am Physiol Soc; Soc Neurosci. Res: Peripheral receptor and central influences on motor function; intracellular studies of motoneurones participating in natural reflexes; electromyography. Mailing Add: Dept Physiol Box 144 Med Col of Va Richmond VA 23298

SZUMSKI, STEPHEN ALOYSIUS, b DuPont, Pa, Dec 26, 19; m 46; c 10. MEDICAL ADMINISTRATION. Educ: Univ Ariz, BS, 47, MS, 49; Pa State Univ, PhD(bact), 51. Prof Exp: Microbiologist & group leader, 51-65, ASSOC DIR MED ADV DEPT, LEDERLE LABS, AM CYANAMID CO, 65- Mem: Am Soc Microbiologists. Res: Process development; microbial production of antibiotics; vitamins; enzymes; vaccines; reagins. Mailing Add: Lederle Labs Am Cyanamid Co Pearl River NY 10965

SZUREK, STANISLAUS (ANDREW), b Chicago, Ill, July 1, 07. PSYCHIATRY, PSYCHOANALYSIS. Educ: Univ Chicago, BS, 27, MA, 29; Rush Med Col, MD, 32; Am Bd Psychiat & Neurol, dipl child psychiat, 39. Prof Exp: Intern, Michael Reese Hosp, Chicago, 31-33; resident psychiat, Chestnut Lodge Sanitarium, Rockville, Md, 34; asst psychiat, Col Med Ill, 34-36, assoc, 36-39, asst prof, 39-43; assoc attend psychiatrist, Cook County Psychopath Hosp, 43; from assoc prof to prof psychiat, 47-74, EMER PROF PSYCHIAT, UNIV CALIF, SAN FRANCISCO, 74-, DIR CHILDREN'S SERV, LANGLEY PORTER NEUROPSYCHIAT INST, 46- Concurrent Pos: Chief of staff, Inst Juv Res, Chicago, 39-43; attend physician, Children's Serv, Ill Neuropsychiat Inst, 42-43; consult, Vet Admin Ment Hyg Clin, San Francisco, 47-51; expert consult, Off Surgeon Gen, Letterman Gen Hosp, 49-; chmn comn ther, Conf Inpatient Treatment Children, 55-56; assoc examr comt child psychiat, Am Bd Psychiat & Neurol, 56-60; prin investr, NIMH Training Grants Child Psychiat & Ment Retardation, Langley Porter Neuropsychiat Inst, Univ Calif, San Francisco, 58- Honors & Awards: Outstanding Civilian Serv Medal, Dept Army, 69; Royer Award, Univ Calif, San Francisco, 72. Mem: Am Psychiat Asn; Am Orthopsychiat Asn (vpres, 51, pres, 58); fel Am Acad Child Psychiat; fel Am Med Asn. Res: Clinical research in functional disorders of childhood, especially the psychotic disorders of preadolescent children. Mailing Add: 422 Judah St San Francisco CA 94122

SZÜSZ, PETER, b Novisad, Yugoslavia, Nov 11, 24; div. MATHEMATICS. Educ: Eötvös Lorand Univ, Budapest, PhD(math), 51; Hungarian Acad Sci, DMS, 62. Prof Exp: Res fel math, Math Inst, Hungarian Acad Sci, 50-65; vis prof, Memorial Univ, 65; vis assoc prof, Pa State Univ, 65-66; PROF MATH, STATE UNIV NY STONY BROOK, 66- Mem: Am Math Soc. Mailing Add: 284 Hallock Rd Stony Brook NY 11790

SZUTKA, ANTON, b Wediz, Ukraine, Apr 18, 20; nat US; m 57; c 1. ANALYTICAL CHEMISTRY. Educ: Univ Pa, MS, 55, PhD(chem), 59. Prof Exp: Chemist, Allied Chem & Dye Corp, 51-53; res assoc, Univ Pa, 54; res assoc, Hahnemann Med Col, 54-59, asst prof, 59-61; assoc prof chem, 61-64, PROF CHEM, UNIV DETROIT, 64- Mem: AAAS; Am Chem Soc; Radiation Res Soc. Res: Effects of radiation on porphines; radiation chemistry; photochemistry; exobiology. Mailing Add: Dept of Chem Univ of Detroit Detroit MI 48221

SZWARC, MICHAEL, b Bedzin, Poland, June 9, 09; m 33; c 3. PHYSICAL CHEMISTRY. Educ: Warsaw Polytech Inst, ChE, 32; Hebrew Univ, Israel, PhD(org chem), 42; Univ Manchester, PhD(phys chem), 47, DSc, 49. Prof Exp: Asst, Hebrew Univ, Israel, 34-42; lectr, Univ Manchester, 47-52; prof phys & polymer chem, 52-56, res prof, 56-64, DISTINGUISHED PROF PHYS & POLYMER CHEM, STATE UNIV NY COL ENVIRON SCI & FORESTRY, 64-, DIR POLYMER RES CTR, 67- Concurrent Pos: Baker lectr, Cornell Univ, 72. Mem: Am Chem Soc; fel Royal Soc. Res: Chemical kinetics; bond dissociation energies; reactivities of radicals; polymerization reactions; living polymers; reactivities of ions and ion-pairs; electron-transfer processes in aprotic solvents. Mailing Add: State Univ of NY Col of Environ Sci & Forestry Syracuse NY 13210

SZWED, JOHN FRANCIS, b Eutaw, Ala, Dec 24, 36; m 60; c 1. ANTHROPOLOGY. Educ: Marietta Col, BS, 58; Ohio State Univ, BSc, 59, MA, 60, PhD(anthrop), 65. Hon Degrees: MA, Univ Pa, 73. Prof Exp: Instr anthrop, Univ Cincinnati, 64-65; asst prof, Lehigh Univ, 65-67; asst prof, Temple Univ, 67-69; ASSOC PROF FOLKLORE, UNIV PA, 69- Concurrent Pos: Res fel, Inst Social & Econ Studies, Mem Univ Nfld, 62-64; NIMH proj grant, Ctr Urban Ethnog, Univ Pa, 69-; vis prof anthrop, New Sch Social Res, 69-70 & 75; mem comt eval govt agency res, Nat Acad Sci, 72-; Am Philos Soc grant, 73; vis prof, Swarthmore Col, 75; Rockefeller Found humanities fel, 75-76. Mem: Am Anthrop Asn; Am Folklore Soc. Res: Creole literatures; ethnicity; urban anthropology. Mailing Add: Ctr for Urban Ethnog Univ of Pa Philadelphia PA 19174

SZYBALSKI, ELIZABETH HUNTER, b Philadelphia, Pa, June 22, 27; m 55; c 2. MICROBIOLOGY. Educ: Duke Univ, BS, 48; Univ Pa, MS, 50, PhD(microbiol), 52. Prof Exp: Res assoc microbiol, Univ Pa, 52-54; microbiologist, Warner-Lambert Labs, NJ, 55-57; tech ed, Squibb Inst Med Res, NJ, 58-60; SR SCIENTIST, McARDLE LAB, UNIV WIS-MADISON, 60- Res: Molecular genetics of bacteriophages and mammalian cells; electron-micrographic mapping of nucleic acids. Mailing Add: McArdle Lab Univ of Wis Madison WI 53706

SZYBALSKI, WACLAW, b Lwow, Poland, Sept 9, 21; nat US; m 55; c 2. BIOCHEMISTRY, ONCOLOGY. Educ: Lwow Polytech Inst, ChEng, 44; Gliwice Polytech Inst, MChEng, 45; Gdansk Polytech Inst, DSc(bact chem), 49. Prof Exp: Dir chem, Agr Res Sta, Konskie, Poland, 44-45; asst prof indust microbiol, Gdansk Polytech Inst, 45-49; mycologist antibiotics, Wyeth Inc, West Chester, Pa, 50-51; staff mem microbiol genetics, Cold Spring Harbor Lab, NY, 51-54; assoc prof, Inst Microbiol, Rutgers Univ, 54-60; PROF ONCOL, McARDLE LAB CANCER RES, MED SCH, UNIV WIS-MADISON, 60- Concurrent Pos: Vis prof, Inst Technol, Copenhagen, Denmark, 47-50; dir regional lab, Bur Standards, Gdansk, Poland, 48-49; chmn, Gordon Conf Nucleic Acids; mem adv panel recombinant DNA molecules, NIH. Honors & Awards: K-A-Forster Award, Ger Acad Sci, 71. Mem: AAAS; Am Soc Microbiol; Am Asn Biol Chemists; hon mem Polish Asn Microbiol; hon mem Polish Med Alliance. Res: Molecular biology; molecular genetics; control of transcription and DNA replication; genetic and physical mapping and sequencing. Mailing Add: McArdle Lab Univ of Wis Med Sch Madison WI 53706

SZYDLIK, PAUL PETER, b Duryea, Pa, Aug 1, 33; m 62; c 3. PHYSICS. Educ: Univ Scranton, BS, 54; Univ Pittsburgh, MS, 57; Cath Univ Am, PhD(physics), 64. Prof Exp: Res asst nuclear physics, Radiation Lab, Univ Pittsburgh, 55-56; nuclear engr, Knolls Atomic Power Lab, Gen Elec Co, 56-60; res asst cosmic ray physics, Cath Univ Am, 60-61; res assoc theoret nuclear physics, Univ Calif, Davis, 64-66; appl physicist, Knolls Atomic Power Lab, Gen Elec Co, 66-68; assoc prof, 68-72, PROF PHYSICS, STATE UNIV NY COL PLATTSBURGH, 72- Mem: Am Phys Soc; Am Nuclear Soc. Res: Neutron cross sections; nuclear structure theory; cosmic ray physics. Mailing Add: Dept of Physics State Univ of NY Col Plattsburgh NY 12901

SZYMANSKI, CHESTER DOMINIC, b Bayonne, NJ, June 2, 30; m 56; c 3. BIOCHEMISTRY. Educ: The Citadel, BS, 51; Univ Miami, MS, 58; State Univ NY Col Forestry, Syracuse Univ, PhD(biochem), 62; Fairleigh Dickinson Univ, MBA, 74. Prof Exp: From chemist to sr chemist, 61-65, res assoc, 65-67, tech mgr, 67-74, ASSOC DIR APPL RES, NAT STARCH & CHEM CORP, 74- Mem: Am Chem Soc; The Chem Soc; Am Asn Cereal Chem. Res: Enzymic and chemical modification of starches; application of starches and synthetic polymers in industrial areas. Mailing Add: Nat Starch & Chem Corp 1700 W Front St Plainfield NJ 07063

SZYMANSKI, EDWARD STANLEY, b Philadelphia, Pa, Mar 24, 47. ENZYMOLOGY. Educ: St Josephs Col, Pa, BS, 69; Georgetown Univ, PhD(biochem), 74. Prof Exp: Res assoc, Dept Biol Sci, Purdue Univ, 74-75; DAIRY RES INC FEL, USDA, 75- Mem: Am Chem Soc; Sigma Xi. Res: Ligand binding to proteins; steroid reductases; enzyme kinetics; protein purification by biospecific affinity chromatography; function of milk proteins and specificity of milk enzymes. Mailing Add: Eastern Regional Res Ctr USDA 600 E Mermaid Lane Wyndmoor PA 19118

SZYMANSKI, HERMAN A, b Toledo, Ohio, Sept 24, 24; m 48; c 7. PHYSICAL CHEMISTRY. Educ: Univ Toledo, BE, 48; Univ Notre Dame, PhD(chem), 52. Prof Exp: Chemist, Am Can Co, 51-52; chmn dept natural sci, Loyola Univ, 52-54; res assoc, Argonne Nat Lab, 54; from instr to asst prof chem, Canisius Col, 54-56, assoc prof & chmn dept, 56-69; dean, 69-71, PRES, ALLIANCE COL, 72- Concurrent Pos: Dean, Cecil Community Col, 71-72; consult, various indust orgns. Mem: Am Chem Soc; Soc Appl Spectros (pres). Res: Spectroscopy; structural chemistry; radiation damage in solids; instrumentation; chemistry education. Mailing Add: Alliance Col Cambridge Springs PA 16403

SZYRYNSKI, VICTOR, b Oct 10, 13; nat Can; m 47; c 2. PSYCHIATRY. Educ: Univ Warsaw, MD, 38; Univ Ottawa, PhD(psychol), 49; FRCP(C), cert neurol, 52 & psychiat, 53; FRCP. Prof Exp: Demonstr neurol & psychiat, Univ Wilno, Poland, 39, sr resident & lectr, 39-40; lectr psychol, Polish Inst, Beirut, 46; consult, Guid Ctr, Univ Ottawa, 48-60, lectr psychiat, 48-49, assoc prof, 49-56, prof psychophysiol, 56-60, prof psychother, 58-60; consult, State Dept Health & dir, State Psychiat Clin, NDak, 60-61; assoc prof neurol & prof psychiat & chmn dept, Univ NDak, 61-64; PROF PSYCHIAT & PSYCHOTHER, UNIV OTTAWA, 64- Concurrent Pos: Attend neurologist, Ottawa Civic Hosp, 51-60, asst electroencephalographer, 54-60, consult psychiatrist, 57-60; neurologist & psychiatrist, Royal Can Air Force Hosp, Rockcliffe, 51-56, sr consult, 56-60; neurologist, Dept Vet Affairs, 51-60; consult neurologist & psychiatrist, State Hosp Jamestown, 60-64; chief dept psychiat, Ottawa Gen Hosp, 64-; dir, Ctr Pastoral Psychiat, St Paul Univ, Ont, 67- Honors & Awards: Gold Medal, Am Acad Psychosom Med, 59; Officer, Order of Polonia Restituta, Cross of Merit; Knight, Order of Holy Sepulchre. Mem: Fel Am Acad Psychosom Med (pres, 65-66); fel Am Col Physicians; fel Am Psychiat Asn; fel Am Acad Neurol; Sigma Xi. Res: Psychotherapy; community psychiatry; child psychiatry; clinical neurology. Mailing Add: Dept of Psychiat Univ of Ottawa Fac of Med Ottawa ON Can

T

TAAFFE, EDWARD JAMES, b Chicago, Ill, Dec 11, 21; m 48; c 8. GEOGRAPHY. Educ: NY Univ, BS, 44; Univ Ill, BS, 44; Univ Chicago, MS, 49, PhD(geog), 52. Prof Exp: Asst prof econ geog, Sch Com, Loyola Univ, Ill, 51-58; assoc prof geog, Northwestern Univ, 58-63; PROF GEOG & CHMN DEPT, OHIO STATE UNIV, 63- Concurrent Pos: Mem, Comn Col Geog, 64-68, comt geog, Earth Sci Div, Nat Acad Sci, 66-70, behav & soc sci surv comt, Nat Acad Sci-Soc Sci Res Coun, 67-70, mem bd dirs, Soc Sci Res Coun, 71- Mem: Asn Am Geogr (vpres, 70-71, pres, 71-72); Regional Sci Asn; Am Geog Soc. Res: Transportation and urban geography. Mailing Add: Dept of Geog Ohio State Univ Col Arts & Sci Columbus OH 43210

TAAFFE, ROBERT NORMAN, b Chicago, Ill, Sept 20, 29. ECONOMIC GEOGRAPHY. Educ: Ind Univ, AB & MA, 52; Univ Chicago, PhD(geog), 59. Prof Exp: Instr geog, Loyola Univ, Ill, 55-58; asst prof, Ind Univ, 59-63; from assoc prof to prof, Univ Wis-Madison, 63-70, chmn dept, 68-70; PROF GEOG & CHMN DEPT, IND UNIV, BLOOMINGTON, 70- Concurrent Pos: Mem joint Slavic comt, Am Coun Learned Socs & Soc Sci Res Coun, 65-68; mem comt on future, Inter-Univ Comt Travel Grants, 67-68. Mem: Asn Am Geog; Regional Sci Asn; Am Asn Advan Slavic Studies. Res: Economic geography of the Union of Soviet Socialist Republics; transportation geography; economic geography of Eastern Europe. Mailing Add: Dept of Geog Ind Univ Bloomington IN 47401

TAAGEPERA, MARE, b Narva, Estonia, May 16, 38; US citizen; m 61; c 3. PHYSICAL ORGANIC CHEMISTRY. Educ: Univ Del, BS, 60, MS, 63; Univ Pa, PhD(chem), 70. Prof Exp: Chemist, E I du Pont de Nemours & Co, Inc, 62-64; FEL CHEM, UNIV CALIF, IRVINE, 71- Mem: Am Chem Soc; Sigma Xi. Res: Mechanisms of organic reactions; rate and equilibrium studies; mass and ion cyclotron resonance spectroscopy. Mailing Add: Dept of Chem Univ of Calif Irvine CA 92664

TABACHNICK, IRVING I A, b New York, NY, July 20, 24; m 51; c 2. PHARMACOLOGY. Educ: Harvard Univ, AB, 48; Yale Univ, PhD(pharmacol), 53. Prof Exp: Pharmacologist & statistician, Baxter Labs, 53-55; pharmacologist, 55-58, sr pharmacologist, 58-60, sect head, 60-61, head dept biochem pharm, 61-64, head dept physiol & biochem, 64-68, assoc dir biol res div, 68-70, dir, 70-72, SR DIR BIOL RES & DEVELOP, SCHERING CORP, 72- Mem: AAAS; Am Soc Pharmacol & Exp Therapeut; Soc Exp Biol & Med; NY Acad Sci. Res: Histamine; diabetes; insulin; catecholamines; adrenergic receptors; anti-hormones. Mailing Add: Schering Corp 86 Orange St Bloomfield NJ 07003

TABACHNICK, JOSEPH, b New York, NY, May 14, 19; m 61; c 2. COMPARATIVE BIOCHEMISTRY. Educ: Univ Calif, Berkeley, BS, 42, MS, 47, PhD(comp biochem), 50. Prof Exp: Res asst food technol, Univ Calif, Berkeley, 46-51, lab instr, 47-49; res chemist, Turner Hall Corp, NJ, 51-52; res biochemist, 52-59, SR RES BIOCHEMIST & ASSOC MEM, DIV OF LABS & HEAD, LAB EXP DERMAT, ALBERT EINSTEIN MED CTR, 59- Concurrent Pos: Adj assoc prof, NY Med Col, 73- Mem: AAAS; Soc Invest Dermat; Am Soc Exp Path; Am Chem Soc; Am Soc Cell Biol. Res: Radiation biology and biochemistry of the skin; cytokinetics of repair in beta-irradiated skin; formation and metabolism of pyroglutamic acid and free amino acid in skin; suppression of radiation fibrosis. Mailing Add: Lab of Exp Dermat Albert Einstein Med Ctr Philadelphia PA 19141

TABACHNICK, MILTON, b New York, NY, June 25, 22; m 52. BIOCHEMISTRY. Educ: Univ Calif, AB, 47, MA, 49, PhD(biochem), 53. Prof Exp: Asst, Univ Calif, 49-52; Am Cancer Soc res fel biochem, Sch Med, Duke Univ, 52-53; instr, Inst Indust Med, Post-Grad Med Sch, NY Univ, 53-55; vis investr, Div Nutrit & Physiol, Pub Health Res Inst of City New York, 55-57; res assoc chem, Mt Sinai Hosp, 57-59; res assoc, NY State Psychiat Inst, 59-60; from asst prof to assoc prof, 61-69, PROF BIOCHEM, NEW YORK MED COL, 69-, DEAN GRAD SCH BASIC MED SCI, 71- Mem: Am Soc Biol Chem; Am Chem Soc; Endocrine Soc; Am Thyroid Asn; Brit Biochem Soc. Res: Protein chemistry; transport and mechanism of action of thyroid hormone. Mailing Add: Dept of Biochem New York Med Col Valhalla NY 10595

TABAKIN, BURTON SAMUEL, b Philadelphia, Pa, July 6, 21; div; c 5. MEDICINE. Educ: Univ Pa, AB, 43, MD, 47; Am Bd Internal Med, dipl, 55. Prof Exp: Fel physiol, Univ Vt & Trudeau Found, 51-52; from instr to assoc prof, 54-67, PROF MED, COL MED, UNIV VT, 67-, ACTG CHMN DEPT, 74-, DIR CARDIOPULMONARY LAB, 54-, DIR CARDIOL UNIT, 72- Concurrent Pos: Attend physician, Mary Fletcher Hosp & DeGoesbriand Mem Hosp, Burlington, Vt, 60-66; fel coun clin cardiol, Am Heart Asn. Mem: Fel Am Col Physicians; Am Heart Asn; Am Col Chest Physicians; Am Fedn Clin Res; fel Am Col Cardiol. Res: Clinical cardiopulmonary and exercise physiology. Mailing Add: Med Ctr Hosp of Vt Burlington VT 05401

TABAKIN, FRANK, b Newark, NJ, Sept 20, 35; m 63; c 2. THEORETICAL PHYSICS. Educ: Queens Col, NY, BS, 56; Mass Inst Technol, PhD(physics), 63. Prof Exp: Res assoc physics, Columbia Univ, 63-65; asst prof, 65-69, ASSOC PROF PHYSICS, UNIV PITTSBURGH, 69- Concurrent Pos: Res visitor, Oxford Univ, 70. Mem: Am Phys Soc. Res: Nuclear forces and matter; three-body problems; properties of nuclei; meson physics. Mailing Add: Dept of Physics Univ of Pittsburgh Pittsburgh PA 15213

TABATA, SUSUMU, b Steveston, BC, Dec 9, 25; m 59; c 3. PHYSICAL OCEANOGRAPHY. Educ: Univ BC, BA, 50, MA, 54; Univ Tokyo, DSc, 65. Prof Exp: Phys oceanogr, Pac Oceanog Group, Fisheries Res Bd Can, 52-70; res phys oceanogr, Offshore Oceanog Group, 71-75, RES SCIENTIST, INST OCEAN SCI, 75- Concurrent Pos: Asst scientist, Pac Oceanog Group, Fisheries Res Bd Can, 52-58, assoc scientist, 59-65, sr scientist, 66-70. Mem: Am Soc Limnol & Oceanog; Am Meteorol Soc; Am Geophys Union; fel Royal Meteorol Soc; Oceanog Soc Japan. Res: Circulation of inshore and offshore waters; processes affecting water properties; variability in the oceans; large-scale air-sea interactions. Mailing Add: Inst of Ocean Sci Patricia Bay 1230 Government St Victoria BC Can

TABATABAIAN, ALI MOHAMMAD, b Ghom, Iran, Dec 7, 33; m 61; c 3. MATHEMATICS. Educ: Univ Tehran, BA, 55; Univ Calif, Berkeley, MA, 59, PhD(math), 68. Prof Exp: Instr math, Univ Tehran, 55-57; asst prof, 66-72, ASSOC PROF MATH, CALIF STATE UNIV, SAN FRANCISCO, 72- Mem: Math Asn Am. Res: Probability; functional analysis. Mailing Add: Dept of Math Calif State Univ San Francisco CA 94132

TABB, DAVID LEO, b Louisville, Ky, Feb 8, 46; m 66; c 2. POLYMER SCIENCE. Educ: Univ Louisville, BChE, 69; Case Western Reserve Univ, MS, 72, PhD(polymer sci), 74. Prof Exp: Res eng chem eng, 69-70, RES CHEMIST POLYMER SCI, ELASTOMER CHEM DEPT, E I DU PONT DE NEMOURS & CO, INC, 74- Mem: Am Chem Soc. Res: Characterization of polymer structure and correlation to physical properties; thermoplastic elastomers; polymer blends; polymer processing technology; fourier transform infrared spectroscopy. Mailing Add: Elastomer Chem Dept/Exp Sta E I du Pont de Nemours & Co Wilmington DE 19898

TABBERT, ROBERT L, b Ripon, Wis, Sept 6, 28; m 52; c 2. GEOLOGY. Educ: Univ Wis, BS, 52, MSc, 54. Prof Exp: Micropaleontologist, Magnolia Petrol Co, 54-59; micropaleontologist, Socony Mobil's Field Res Lab, 59-62; palynologist, 62-67, supvr palynology group, 67-73, dir struct & stratig res, 73-75, SR RES ASSOC, GEOL SCI GROUP, ATLANTIC RICHFIELD CO, 75- Mem: Soc Econ Paleont & Mineral; Am Asn Petrol Geologists; Am Asn Stratig Palynologists; Geol Soc Am. Res: Biostratigraphy of Cretaceous-Tertiary sediments of Alaska; regional correlations in Mesozoic and Tertiary sediments of Arctic; structural geology, stratigraphy and petroleum exploration of Arctic. Mailing Add: PO Box 2819 Dallas TX 75080

TABBUTT, FREDERICK, physical chemistry, see 12th edition

TABELING, RAYMOND WILLIAM, analytical chemistry, see 12th edition

TABENKIN, BENJAMIN, b Peoria, Ill, Aug 20, 13; m 38; c 1. CHEMISTRY. Educ: Univ Ill, BS, 35; George Washington Univ, MA, 40. Prof Exp: Jr sci aide, Bur Plant Indust, USDA, 35-37, jr chemist, Bur Chem & Soils, 37-39, asst chemist, 39-40, assoc chemist, 40-41; sr res chemist & head gen microbiol, 41-62, HEAD MICROBIOL PILOT PLANT, HOFFMAN-LA ROCHE INC, 62- Mem: AAAS; Am Chem Soc; Am Soc Microbiologists; Soc Indust Microbiol; NY Acad Sci. Res: Biochemical activities of microorganisms; production of antibiotics; vitamins and enzymes; fungicides; antiseptics. Mailing Add: Hoffmann-La Roche Inc Nutley NJ 07110

TABER, BEN Z, b Providence, RI, Mar 17, 27; m; c 3. OBSTETRICS & GYNECOLOGY. Educ: Brown Univ, AB, 48; Harvard Univ, MD, 52; Am Bd Obstet & Gynec, dipl, 64. Prof Exp: Intern, Cook County Hosp, Chicago, 52-53; resident obstet & gynec, WSuburban Hosp, Oak Park, 54-55; resident, Cook County Hosp, Chicago, 55-57; obstetrician & gynecologist, pvt pract, El Paso, Tex, 58-63; med officer new drugs, Food & Drug Admin, 63-64; from assoc med dir to med dir, 65-73, vpres, 69-73, VPRES MED AFFAIRS, SYNTEX LABS, INC, 73- Concurrent Pos: Clin instr, Sch Med, Stanford Univ, 66-72, clin asst prof, 72- Mem: Fel Am Col Obstet & Gynec; AMA; Am Fertil Soc; Pan-Am Med Asn. Mailing Add: Syntex Labs Inc 3401 Hillview Ave Palo Alto CA 94304

TABER, CHARLES ALEC, b Texanna, Okla, Dec 10, 37; m 70. ICHTHYOLOGY. Educ: Northeastern State Col, BS, 61; Univ Okla, PhD(zool), 69. Prof Exp: ASST PROF BIOL, SOUTHWEST MO STATE UNIV, 69- Mem: Am Soc Ichthyologists & Herpetologists; Am Fisheries Soc. Res: Natural history and ecology of fishes. Mailing Add: Dept of Life Sci Southwest Mo State Univ Springfield MO 65806

TABER, DAVID, b New York, NY, July 14, 22; m 48; c 3. BIOMEDICAL ENGINEERING, ORGANIC CHEMISTRY. Educ: NY Univ, AB, 48; Polytech Inst Brooklyn, PhD(chem), 53. Prof Exp: Jr chemist, Air Reduction Co, Inc, 48-49; res assoc indust med, Postgrad Med Sch, NY Univ, 53-55; chemist, Gen Aniline & Film Corp, 55-58; sr chemist, Koppers Co, Inc, 58-63; sect head, Org Chem, Armour Grocery Prod Co, 63-65, res mgr new chem, Household Prod Res & Develop Dept, 65-68, mgr biol & med progs, Armour-Dial, Inc, 68-69, asst res dir, 69-75; VPRES & DIR TECH SERV, HOLLISTER, INC, 75- Mem: Am Chem Soc; Am Acad Dermat; Soc Cosmetic Chemists. Res: Organic synthesis, germicides, microbiology; product safety; regulatory affairs; clinical testing; medical and hospital devices; quality assurance. Mailing Add: Hollister Inc 3721 W Morse Ave Lincolnwood IL 60645

TABER, ELSIE, b Columbia, SC, May 3, 15. EMBRYOLOGY, REPRODUCTIVE PHYSIOLOGY. Educ: Univ SC, BS, 35; Stanford Univ, MA, 36; Univ Chicago, PhD(zool), 47. Prof Exp: Teacher biol, Greenwood High Sch, SC, 36-38; instr, Lander Col, 38-41; instr, Univ Chicago, 44-48, asst dean students, Div Biol Sci, 47-48; from asst prof to assoc prof, 48-65, PROF ANAT, MED UNIV SC, 65- Mem: AAAS; Am Soc Zool; Am Asn Anatomists; Am Inst Biol Sci; Soc Study Reproduction. Res: Developmental biology; endocrinology of reproductive systems. Mailing Add: Dept of Anat Med Univ of SC Charleston SC 29401

TABER, HARRY WARREN, b Longview, Wash, Oct 30, 35; c 3. MICROBIOLOGY, MOLECULAR BIOLOGY. Educ: Reed Col, BA, 57; Univ Rochester, PhD(biochem), 63. Prof Exp: AEC fel, Univ Rochester, 62-64; USPHS fel, Nat Inst Neurol Dis & Stroke, Bethesda, Md, 64-66; USPHS spec fel, Ctr Molecular Genetics, Gif-sur-Yvette, France, 66-67; asst prof, 68-73, ASSOC PROF MICROBIOL, SCH MED, UNIV ROCHESTER, 73- Concurrent Pos: NIH res career develop award, 74. Mem: Am Soc Microbiol. Res: Genetic regulation of membrane structure and function; bacterial sporulation; microbial redox enzyme systems. Mailing Add: Dept of Microbiol Univ Rochester Sch of Med Rochester NY 14642

TABER, HENRY GLENN, plant physiology, nutrition, see 12th edition

TABER, JOSEPH JOHN, b Adena, Ohio, Feb 6, 20; m 47; c 4. PHYSICAL CHEMISTRY, PETROLEUM ENGINEERING. Educ: Muskingum Col, BS, 42; Univ Pittsburgh, PhD, 55. Prof Exp: Asst prof naval sci, Ohio State Univ, 46; instr chem, Washington & Jefferson Col, 46-50; sr proj chemist, Gulf Res & Develop Co, 54-64; from asst prof to assoc prof chem, 64-72, adj assoc prof petrol eng, 67-72, PROF CHEM, GRAD FAC, UNIV PITTSBURGH, 72-, ADMIN OFFICER DEPT, 64-, PROF PETROL ENG, 72- Mem: Am Chem Soc; Am Inst Mining, Metall & Petrol Eng. Res: Surface chemistry; liquid-liquid and liquid-solid interfaces; effect of interfacial energies on capillarity and fluid flow in porous media; new methods of petroleum recovery. Mailing Add: 1703 President Dr Glenshaw PA 15116

TABER, RICHARD DOUGLAS, b San Francisco, Calif, Nov 22, 20; m 46; c 3. WILDLIFE ECOLOGY. Educ: Univ Calif, AB, 42, PhD, 51; Univ Wis, MS, 49. Prof Exp: Wildlife researcher & asst specialist, Univ Calif, 51-55, actg asst prof zool, 55-56; actg asst prof forestry, Univ Mont, 56-57, from assoc prof to prof, 58-68, assoc dir, Mont Forest & Conserv Exp Sta, 64-68; PROF FOREST ZOOL, UNIV WASH, 68- Concurrent Pos: US specialist forest-wildlife rels, Ger & Czech, 60 & Poland, 60 & 64; Fulbright res scholar, West Pakistan, 63-64; Guggenheim fel, 64; mem comn threatened deer, Int Union Conserv Nature, 73- Mem: Wildlife Soc; Ecol Soc Am; Am Soc Mammal. Res: Biology and conservation of free-living birds and mammals; wildlife and human culture; ungulate biology and effects on ecosystem. Mailing Add: Col of Forest Resources Univ of Wash Seattle WA 98195

TABER, RICHARD LAWRENCE, b Pontiac, Mich, Nov 9, 35; m 57; c 2. BIO-ORGANIC CHEMISTRY. Educ: Colo State Col, AB, 58; Univ NMex, PhD(org chem), 63. Prof Exp: From asst prof to assoc prof, 63-75, PROF CHEM, COLO COL, 75- Honors & Awards: Meritorious Serv Award, Am Chem Soc, 69. Mem: Am Chem Soc; Sigma Xi. Res: Enzymology of dihydroorotate dehydrogenase; organic liquid scintillation solutes. Mailing Add: Chem Dept Colo Col Colorado Springs CO 80903

TABER, ROBERT IRVING, b Perth Amboy, NJ, June 28, 36; m 60; c 3. PSYCHOPHARMACOLOGY, NEUROPHARMACOLOGY. Educ: Rutgers Univ, BS, 58; Med Col Va, PhD(pharmacol), 63. Prof Exp: Pharmacologist, 62-66, sr pharmacologist, 66-71, mgr pharmacol, 71-74, DIR BIOL RES, SCHERING CORP, BLOOMFIELD, 74- Mem: AAAS; assoc Am Col Neuropsychopharmacol; Am Soc Pharmacol & Exp Therapeut; Acad Pharmaceut Sci; Am Pharmaceut Asn. Res: Analgesics; drug effects on learning and memory. Mailing Add: 8 Sherbrook Dr Berkeley Heights NJ 07922

TABER, ROBERT WILLIAM, b Marietta, Ohio, Oct 4, 21; m 52; c 2. OCEANOGRAPHY, MARINE GEOLOGY. Educ: Marietta Col, BS, 48; Univ Mo, MA, 51. Prof Exp: Party chief oceanogr, US Navy Hydrographic Off, 51-56; head underwater sound sect, Naval Oceanog Off, 56-57, head systs anal group, 57-61; head geosci br, 61-62, adv develop progs, 62-68, CHIEF PROD CONTROL BR, NAT OCEANOG DATA CTR, NASA, 68- Mem: Marine Technol Soc. Res: Physical oceanography. Mailing Add: Nat Oceanog Data Ctr NASA Rockville MD 20852

TABER, STEPHEN, III, b Columbia, SC, Apr 17, 24; m 45; c 5. APICULTURE. Educ: Univ Wis, BS, 49. Prof Exp: Apiculturist, USDA, La, 50-65, APICULTURIST, USDA, ARIZ, 65- Mem: AAAS; Entom Soc Am; Am Genetic Asn; Bee Res Asn. Res: Bee behavior and genetics. Mailing Add: Bee Res Lab 2000 E Allen Rd Tucson AZ 85719

TABER, WILLARD ALLEN, b Marshalltown, Iowa, Feb 18, 25; m 50; c 2. MICROBIOLOGY. Educ: Univ Iowa, AB, 49, MS, 51; Rutgers Univ, PhD(microbiol), 54. Prof Exp: Asst, Univ Iowa, 49-51; fel, Rutgers Univ, 54-55; asst res officer, Prairie Regional Lab, Nat Res Coun Can, 55-61, sr res officer, 61-63; PROF BIOL, TEX A&M UNIV, 63- Concurrent Pos: NIH consult, 67-70. Mem: Am Soc Microbiologists; Soc Indust Microbiol. Res: Antifungal antibiotics; ecology of soil fungi; morphogenesis of fungi; metabolism of nonsugar carbon sources; mycology; secondary metabolism; transport in fungi, streptomyces. Mailing Add: Dept of Biol Tex A&M Univ College Station TX 77843

TABERSHAW, IRVING R, b New York, NY, June 14, 08; m 58; c 1. PREVENTIVE MEDICINE. Educ: Univ Pa, BS, 28; State Univ NY, MD, 33; Am Bd Prev Med, dipl. Prof Exp: Physician, Div Occup Hyg, Mass State Dept Labor, 42-45; dir div indust hyg, Ala State Dept Pub Health, 45-46; dir eastern med div, Liberty Mutual Ins Co, 46-51; med dir, Purolator Prod, NJ, 51-57; dir div indust hyg, NY State Dept Labor, 53-57; sci dir, Joseph E Seagram & Sons, Inc, 57-62; prof occup med & head off environ health & safety, 62-72, EMER PROF OCCUP MED, UNIV CALIF, BERKELEY, 72- Concurrent Pos: Assoc prof, Sch Pub Health, Columbia Univ, 46-62; lectr, NY Univ & Upstate Med Ctr, 46-56; assoc, NY Med Col, 53-57; consult, Nat Adv Comt Air Pollution, USPHS, 56-60, health & safety div, NY Opers Off, AEC, 47-64 & Bur Occup Health, Calif State Dept Pub Health, 66-; ed, J Occup Med, 67-; pres, Tabershaw-Cooper Assocs, 72-75. Mem: Am Med Asn; Am Pub Health Asn; Am Occup Med Asn; Am Acad Occup Med (pres, 57-58); Am Col Prev Med. Res: Occupational and environmental medicine. Mailing Add: 10230 Democracy Lane Potomac MD 20854

TABET, GEORGES ELIAS, b Cairo, Egypt, May 29, 13; nat US; m 43; c 5. CHEMISTRY. Educ: Am Univ Cairo, BSc, 33; Lehigh Univ, BSc & MSc, 40; Cornell Univ, PhD(org chem), 45. Prof Exp: Asst chem, Cornell Univ, 40-43; res chemist, Tenn Eastman Corp, NY, 44-45; res chemist, Ammonia Dept, E I du Pont de Nemours & Co, 45-53; staff chemist, Res & Develop, S C Johnson & Son Co, 53-62, sales mgr, 62-64, graphic arts indust mgr, Chem Div, 64-67; consult, Edward D Evans & Assocs, 67-68; RES MGR, PRINTING DEVELOP INC, 68- Mem: Am Chem Soc. Res: Carcinogenics; vinyl monomers; anthraquinone dyes; plasticizers for polyvinyl chloride; polymer intermediates; catalysis and high pressure reactions; fluorocarbons; synthetic organics; nylon and dacron intermediates; waxes; silicones, resins; coatings and inks; photoresists; printing plates; organic chemistry; research and development and marketing management. Mailing Add: 1116 Main St Racine WI 53403

TABIBIAN, RICHARD, b Detroit, Mich, June 1, 29; m 55; c 3. POLYMER CHEMISTRY. Educ: Wayne State Univ, BS, 51, MS, 54, PhD(chem), 56. Prof Exp: CHEMIST, E I DU PONT DE NEMOURS & CO, INC, 55- Mem: Am Chem Soc.

Res: Colloid chemistry. Mailing Add: Elastomers Dept Exp Sta E I du Pont de Nemours & Co Inc Wilmington DE 19898

TABIKH, ALI ABU, soil science, see 12th edition

TABISZ, GEORGE CONRAD, b New York, NY, Aug 28, 39; Can citizen; m 66; c 2. MOLECULAR PHYSICS. Educ: Univ Toronto, BASc, 61, MA, 63, PhD(physics), 68. Prof Exp: Nat Res Coun can fel, High Pressure Lab, Nat Ctr Sci Res, France, 68-70; asst prof physics, 70-75, ASSOC PROF PHYSICS, UNIV MANITOBA, 75- Mem: Optical Soc Am; Can Asn Physicists; Fr Phys Soc. Res: Molecular interactions in gases and liquids; visible and infrared absorption spectroscopy; laser raman scattering. Mailing Add: Dept of Physics Univ of Manitoba Winnipeg MB Can

TABLER, KENNETH AMBROSE, b Martinsburg, WVa, July 18, 26; m 50; c 4. MATHEMATICAL STATISTICS, ANALYTICAL STATISTICS. Educ: WVa Univ, BS, 48, MS, 49; Univ Ill, PhD(pop genetics, statist), 54. Prof Exp: Dairy husbandman, USDA, 49-56, anal statistician, Biomet Serv, 56-67; MATH STATISTICIAN, NAT CTR EDUC STATIST, US OFF EDUC, 67- Mem: Biomet Soc; Am Statist Asn; Asn Comput Machinery. Res: Statistical methods; data analysis; scientific computing and research in education. Mailing Add: Nat Ctr for Educ Statist 400 Maryland Ave SW Washington DC 20202

TABLER, RONALD DWIGHT, b Denver, Colo, May 18, 37; m 64; c 2. HYDROLOGY, WATERSHED MANAGEMENT. Educ: Colo State Univ, BS, 59, PhD(watershed mgt), 65. Prof Exp: RES HYDROLOGIST, ROCKY MOUNTAIN FOREST & RANGE EXP STA, US FOREST SERV, 59-, PROJ LEADER WATERSHED MGT, 64- Concurrent Pos: Lectr, Univ Wyo; consult, Wyo State Hwy Dept. Mem: AAAS; Soc Am Foresters; Am Geophys Union; Soil Conserv Soc; Am Water Resources Asn. Res: Hydrologic effects of vegetation conversion; physics of snow transport by wind; design of snow fence systems for highway protection; management of snow in windswept areas to increase water yields. Mailing Add: US Forest Serv Univ Sta Box 3313 Laramie WY 82070

TABOR, CELIA WHITE, b Boston, Mass, Nov 15, 18; m 46; c 4. PHARMACOLOGY. Educ: Radcliffe Col, BA, 40; Columbia Univ, MD, 43. Prof Exp: Intern med, Mass Gen Hosp, Boston, 44-45; asst resident, Univ Hosp, Vanderbilt Univ, 45-46; res assoc pharmacol, George Washington Univ, 47-49; MEM STAFF, LAB BIOCHEM PHARMACOL, NAT INST ARTHRITIS, METAB & DIGESTIVE DIS, 52- Mem: AAAS; Am Soc Biol Chemists; Am Soc Pharmacol & Exp Therapeut. Res: Biochemistry; enzymatic and metabolic studies of the polyamines. Mailing Add: Bldg 4 Rm 112 Nat Inst Arthritis Metab & Digestive Dis Bethesda MD 20014

TABOR, ELBERT CECIL, b Menefee Co, Ky, Sept 14, 06; m 31; c 2. BIOCHEMISTRY. Educ: Ky Wesleyan Col, AB, 29; Kans State Col, MS, 30. Prof Exp: Instr chem, Mich State Col, 30-40, asst prof biochem, 40-46; biochemist, Nutrit Sect, USPHS, 46-50, chemist, Nat Ctr Air Pollution Control, 50-68, asst chief div air qual, Nat Air Pollution Control Admin, 68-71; tech adv, Div Atmospheric Surveillance, Environ Protection Agency, 71-75; RETIRED. Mem: AAAS; Am Chem Soc; Air Pollution Control Asn. Res: Air monitoring methodology; air quality data; hazardous pollutants; air quality criteria. Mailing Add: 509 Emerson Dr Raleigh NC 27609

TABOR, HERBERT, b New York, NY, Nov 28, 18; m 46; c 4. PHARMACOLOGY, BIOCHEMISTRY. Educ: Harvard Univ, AB, 37, MD, 41. Prof Exp: Biochem researcher, Harvard Med Sch, 41-42; intern med, New Haven Hosp, Conn, 42; USPHS, 43-, CHIEF LAB BIOCHEM PHARMACOL, NAT INST ARTHRITIS, METAB & DIGESTIVE DIS, 61- Concurrent Pos: Field ed, J Pharmacol & Exp Therapeut, 60-68; assoc ed, J Biol Chem, 68-71, ed, 71- Mem: Am Soc Pharmacol & Exp Therapeut; Am Soc Biol Chem; Am Chem Soc; Am Acad Arts & Sci. Res: Biochemistry of amino acids and amines. Mailing Add: Bldg 4 Rm 110 Nat Inst Arthritis Metab & Digestive Dis Bethesda MD 20014

TABOR, ROBERT GEORGE, organic chemistry, medicinal chemistry, see 12th edition

TABOR, ROWLAND WHITNEY, b Denver, Colo, June 10, 32; m 57; c 2. GEOLOGY. Educ: Stanford Univ, BS, 54; Univ Wash, MS, 58, PhD(geol), 61. Prof Exp: Teaching asst geol, Univ Wash, 57-61; GEOLOGIST, US GEOL SURV, 61- Mem: Geol Soc Am. Res: Field mapping; igneous and metamorphic petrology. Mailing Add: U S Geol Surv 345 Middlefield Rd Menlo Park CA 94025

TABOR, THEODORE EMMETT, b Great Falls, Mont, Dec 28, 40; m 60; c 3. ORGANIC CHEMISTRY. Educ: Univ Mont, BA, 62; Kans State Univ, PhD(org chem), 67. Prof Exp: Res chemist, Dow Chem Co, 67-72, res specialist, Appl Res & Develop, Halogens Res Lab, 72-75, DIR ACAD EDUC, DOW CHEM CO, 75- Mem: Am Chem Soc; Sigma Xi; Am Asn Textile Chemists & Colorists. Res: Mechanisms of epoxide rearrangements; solvent processing of textiles; fire retardant chemicals for textiles and plastics. Mailing Add: Acad Educ Dow Chem Co 2030 Bldg Midland MI 48640

TABOR, WILLIAM JOSEPH, b Rockfall, Conn, Mar 22, 31; m 59; c 2. MAGNETISM. Educ: Rensselaer Polytech Inst, BS, 53; Harvard Univ, AM, 54, PhD(chem physics), 57. Prof Exp: MEM TECH STAFF, BELL LABS, INC, 59- Res: Microwave spectroscopy of gases and paramagnetic solids; quantum electronics; ferromagnetic solids; magneto-optics; magnetic bubble device technology. Mailing Add: Bell Labs Inc Murray Hill NJ 07974

TABORSKY, GEORGE, b Budapest, Hungary, Feb 12, 28; nat US; m 53; c 2. BIOCHEMISTRY. Educ: Brown Univ, BS, 51; Yale Univ, PhD(biochem), 56. Prof Exp: Am Cancer Soc fel, Carlsberg Lab, Denmark, 56-57; from instr to assoc prof biochem, Yale Univ, 57-70, dir grad studies in biochem, 64-67; assoc prof biochem, 70-72, PROF BIOCHEM, UNIV CALIF, SANTA BARBARA, 72-, CHMN DEPT, 73- Mem: Am Chem Soc; Biophys Soc; Am Soc Biol Chemists. Res: Chemistry and metabolism of phosphoproteins; bioenergetics; mechanism of enzyme action; protein chemistry. Mailing Add: Dept of Biol Sci Univ of Calif Santa Barbara CA 93106

TACHIBANA, DORA K, b Cipertino, Calif, Dec 19, 34. IMMUNOLOGY, MEDICAL MICROBIOLOGY. Educ: San Jose State Col, BA, 56; Stanford Univ, MA, 62, PhD(immunol), 64. Prof Exp: Pub health microbiologist trainee, Calif State Dept Pub Health, 56-57; clin lab technician trainee, San Jose Hosp, Calif, 57; pub health microbiologist, Santa Clara County Pub Health, 57-59; NIH fel immunol, Czech Acad Sci, 64-65; lab instr & res asst, Stanford Univ, 65-66; asst prof microbiol & immunol, San Francisco State Col, 66-69; res assoc med, Sch Med, Stanford Univ, 70; ASSOC PROF IMMUNOL & MICROBIOL, CALIF COL PODIATRIC MED, 70-; CLIN LAB DIR, CALIF PODIATRY HOSP, SAN FRANCISCO, 70- Concurrent Pos: Fac res award, San Francisco State Col, 68-69. Mem: Am Soc Microbiol; Am Asn Immunol; Am Soc Med Technol. Res: Microbial flora of the foot. Mailing Add: Dept of Basic Sci Calif Col of Podiat Med San Francisco CA 94115

TACHIBANA, HIDEO, b Los Altos, Calif, June 30, 25; m 64; c 2. PLANT PATHOLOGY. Educ: Univ Calif, Davis, BS, 57; Wash State Univ, PhD(plant path), 63. Prof Exp: RES PLANT PATHOLOGIST, USDA, 63- Mem: Am Phytopath Soc. Res: Soybean diseases; brown stem rot; breeding and screening for disease resistance. Mailing Add: 4024 Quebec Ames IA 50010

TACHMINDJI, ALEXANDER JOHN, b Athens, Greece, Feb 16, 28; US citizen; m 65. HYDRODYNAMICS. Educ: Durham Univ, BSc, 49, BSc(hons), 50; Mass Inst Technol, SM, 51. Prof Exp: With Swann, Hunter & W Richardson, Eng, 45-46; res assoc naval archit, Mass Inst Technol, 50-51; mem res staff, Ship Div, David W Taylor Model Basin, US Navy, 51-54, head res & propeller br, 54-59; head tactical warfare group, weapons syst eval div, Inst Defense Anal, 59-64, asst dir res & eng support div, 64-67, dep dir sci technol div, 67-69, dir systs eval div, 69-72; dir tactical technol off, Defense Advan Res Proj Agency, Dept Defense, 72-73, dep dir, Defense Advan Res Proj Agency, 73-75; CHIEF SCIENTIST, MITRE CORP, 75- Concurrent Pos: Consult, Am Bur Shipping, 56-59, Anti-Submarine Warfare Comt, Defense Sci Bd, 65-70 & Naval Surface Warfare Panel, 75-; ed, J Defense Res, Series B, 69- Mem: AAAS; Soc Naval Architects & Marine Engrs; assoc fel Am Inst Aeronaut & Astronaut; Opers Res Soc Am; fel Royal Inst Naval Architects. Res: Cavitation; super cavitation; potential theory; ship vibration and noise; systems analyses; hydroelasticity. Mailing Add: Mitre Corp Westgate Res Park McLean VA 22101

TACK, PETER ISAAC, b Marion, NY, Apr 15, 11; m 37; c 2. FISH BIOLOGY. Educ: Cornell Univ, BS, 34, PhD(agr), 39. Prof Exp: Asst biologist, State Conserv Dept, NY, 39-40, biologist, 40; instr zool, 40-43, asst prof & res asst, 43-46, assoc prof & res assoc, 46-50, prof fisheries & wildlife, 50-70, chmn dept, 50-69, PROF FISHERIES, WILDLIFE & ZOOL, MICH STATE UNIV, 70- Concurrent Pos: Mem, Governor's Botulism Control Comn, 64-65. Mem: Am Soc Limnol & Oceanog; Am Fisheries Soc; Am Soc Ichthyologists & Herpetologists. Res: Pond fish culture; population fluctuations of whitefish in Northern Lake Michigan and effects of pumped storage generating facility on Lake Michigan ecology. Mailing Add: Dept of Fisheries & Wildlife Mich State Univ East Lansing MI 48823

TACKER, MARTHA McCLELLAND, b Mineral Wells, Tex, Jan 16, 43; m 67; c 2. BIOCHEMISTRY, SCIENCE WRITING. Educ: Baylor Col Med, PhD(biochem), 69. Prof Exp: Res assoc biochem, Baylor Col Med, 69-70; res asst gastroenterol, Mayo Grad Sch Med, Univ Minn & Mayo Found, 70-71; instr physiol, Baylor Col Med, 71-74; SCI WRITING & EDITING, 74- Mem: AAAS; Sigma Xi; Am Med Writers Asn; Coun Biol Ed. Mailing Add: 2901 Wilshire Ave West Lafayette IN 47906

TACKER, WILLIS ARNOLD, JR, b Tyler, Tex, May 24, 42; m 67; c 2. CARDIOVASCULAR PHYSIOLOGY, MEDICAL EDUCATION. Educ: Baylor Univ, BS, 64, MD & PhD(physiol), 70. Prof Exp: Intern med, Mayo Clin, 70-71; mem fac physiol, Baylor Col Med, 71-74; MEM FAC BIOMED ENG, PURDUE UNIV, WEST LAFAYETTE, 74- Mem: AMA; Am Physiol Soc. Res: Life-threatening arrhythmia therapy; new teaching techniques and devices; diagnostic and therapeutic devices development. Mailing Add: Biomed Eng Ctr Eng Bldg Rm 34 Purdue Univ West Lafayette IN 47906

TACKETT, JAMES EDWIN, JR, b Los Angeles, Calif, Oct 8, 37; m 62; c 3. ANALYTICAL CHEMISTRY. Educ: Occidental Col, BA, 60; Univ Calif, Riverside, PhD(chem), 64. Prof Exp: Chemist, Union Carbide Corp, WVa, 64-65; SR RES CHEMIST, DENVER RES CTR, MARATHON OIL CO, 65- Mem: Am Chem Soc; Sigma Xi. Res: Applied spectroscopy; thermal analysis; analytical separations. Mailing Add: Marathon Oil Co Denver Res Ctr PO Box 269 Littleton CO 80120

TACKETT, JESSE LEE, b Dublin, Tex, Sept 20, 35; m 55; c 3. SOIL PHYSICS. Educ: Tex A&M Univ, BS, 57; Auburn Univ, MS, 61, PhD(soil physics), 63. Prof Exp: Instr agron, Tarleton State Col, 57-59; asst, Tex A&M Univ, 59-60; res soil scientist, Soil & Water Conserv Res Div, Agr Res Serv, USDA, 63-65; assoc prof agr, 65-70, PROF AGR & DEAN SCH AGR & BUS, TARLETON STATE COL, 70- Mem: Am Soc Agron; Soil Sci Soc Am. Res: Soil aeration and strength; plant growth relations. Mailing Add: Sch of Agr & Bus Tarleton State Univ Stephenville TX 76402

TACKETT, STANFORD L, b Virgie, Ky, Sept 5, 30; m 51; c 3. ANALYTICAL CHEMISTRY. Educ: Ohio State Univ, BS, 57, PhD(chem), 62. Prof Exp: Instr chem, Ohio State Univ, 61-62; asst prof, Ariz State Univ, 62-66; assoc prof, 66-69, PROF CHEM, INDIANA UNIV PA, 69-, CHMN DEPT, 74- Concurrent Pos: Consult, Off Res Anal, Holloman AFB, 63; NIH res grant, 64-66. Mem: AAAS; Am Chem Soc; Meteoritical Soc. Res: Analytical chemistry and electro-analytical techniques; chemistry of meteorites; chemistry of cyanocobalamin; pollution of streams by acid coal mine waste. Mailing Add: Dept of Chem Indiana Univ of Pa Indiana PA 15701

TACKLE, DAVID, b Calcutta, India, Aug 21, 21; US citizen; m 48; c 1. SILVICULTURE. Educ: Univ Calif, BS, 46, MS, 52. Prof Exp: Asst, Calif Forest & Range Exp Sta, US Forest Serv, 39-41, inspector, US Bur Entom & Plant Quarantine, 42, dist asst, Calif, 45-46, silviculturist, Forest Mgt Res, Calif Forest & Range Exp Sta, 46-51, res forester & proj leader, Intermountain Forest & Range Exp Sta, 51-64, supvry res forester & proj leader, Pac Northwest Forest & Range Exp Sta, 65-68, asst dir, 68-71, CHIEF CONIFER ECOL & MGT RES, US FOREST SERV, 71- Concurrent Pos: Consult, Mont State Univ, 59-64; consult travel grant, EAfrican Agr & Forestry Res Orgn, 67; mem forest mgt res adv comt, Ore State Univ, 68-70. Mem: Soc Am Foresters. Res: Forest management and research administration; management and utilization of lodgepole pine; forest wildlife relations. Mailing Add: US Forest Serv USDA Washington DC 20250

TADANIER, JOHN SOLOMAN, organic chemistry, see 12th edition

TADE, WILLIAM HOWARD, b Hillsboro, Iowa, Sept 6, 23; m 46; c 4. PATHOLOGY, PHYSIOLOGY. Educ: Univ Iowa, BA, 50, DDS, 54, MS, 60, PhD(physiol), 61. Prof Exp: From asst prof to assoc prof, 61-71, PROF ORAL PATH, COL DENT, UNIV IOWA, 71- Mem: AAAS; Am Acad Oral Path; Am Dent Asn; Int Asn Dent Res. Res: Histochemistry of dental pulp; keratinization of mucosa. Mailing Add: Dept of Oral Path Univ of Iowa Col of Dent Iowa City IA 52242

TADROS, MAHER EBEID, b Egypt, July 25, 43; m 72; c 1. SURFACE CHEMISTRY. Educ: Univ Assiut, BSc, 63; Ain Shams Univ, Cairo, MSc, 66; Clarkson Col Technol, MS, 71, PhD(inorg chem), 72. Prof Exp: Instr chem, Ain Shams Univ, Cairo, 63-68; res assoc surface chem, Univ Southern Calif, 72-74; RES SCIENTIST CHEM, MARTIN MARIETTA LAB, 74- Mem: Am Chem Soc. Res: Chemistry of the hydration reactions of portland cement constituents; kinetics of crystallization processes; surface phenomena of low energy surfaces. Mailing Add: Martin Marietta Lab 1450 S Rolling Rd Baltimore MD 21227

TAEBEL, WILBERT AUGUST, b Chicago, Ill, Apr 18, 07; m 31; c 2. CHEMISTRY. Educ: Elmhurst Col, BS, 35; Univ Ill, MS, 36, PhD(inorg chem), 38. Prof Exp: From asst to instr chem, Univ Ill, 38-41; chemist, Westinghouse Elec & Mfg Co, 41-46; dir chem prod dept, Eimer & Amend, 46-51; supt spec molybdenum prod dept, Lamp

Div, Westinghouse Elec Corp, NJ, 51-56, mgr parts eng, 56-61; chief chem div, Res Ctr, Curtiss-Wright Corp, 61-64; mgr res dept, Nat Beryllia Corp, 64-66; assoc prof inorg chem, 66-70, PROF INORG CHEM, BLOOMSBURG STATE COL, 70- Mem: Am Chem Soc; Am Soc Testing & Mat; Electrochem Soc; fel Am Inst Chemists; AAAS. Res: Rare earth; fluorescent materials; powder metallurgy; inorganic compounds. Mailing Add: Bloomsburg State Col Bloomsburg PA 17815

TAFFE, WILLIAM JOHN, b Albany, NY, Feb 3, 43; m 65; c 2. ATMOSPHERIC PHYSICS. Educ: Le Moyne Col, NY, BS, 64; Univ Chicago, SM, 67, PhD(geophys), 68. Prof Exp: Res physicist, Air Force Cambridge Res Labs, 68-69; asst prof physics, Colby Col, 69-71; asst prof natural sci, 71-75, ASSOC PROF NATURAL SCI, PLYMOUTH STATE COL, 75- Mem: Am Geophys Union; Am Asn Physics Teachers. Res: Atmospheric electricity; geophysical fluid dynamics; teaching of physics. Mailing Add: Dept of Natural Sci Plymouth State Col Plymouth NH 03264

TAFT, BRUCE A, b San Francisco, Calif, Jan 29, 30; m 61; c 1. PHYSICAL OCEANOGRAPHY. Educ: Stanford Univ, BS, 51; Univ Calif, San Diego, MS, 61, PhD(oceanog), 65. Prof Exp: Statistician, US Fish & Wildlife Serv, Calif, 55-58; res oceanogr, Scripps Inst, Univ Calif, 58-59; res asst oceanog, Johns Hopkins Univ, 59-60; res oceanogr, Scripps Inst Oceanog, Univ Calif, 60-65, asst res oceanogr, 65-68, asst prof oceanog, 68-73; RES ASSOC PROF OCEANOG, UNIV WASH, 73- Concurrent Pos: NSF grant, US-Japan Coop Sci Prog, 66- Mem: AAAS; Am Geophys Union; Oceanog Soc Japan. Res: Description of large scale oceanic circulation; velocity structure and distribution of properties in ocean currents. Mailing Add: Dept of Oceanog Univ of Wash Seattle WA 98195

TAFT, CLARENCE EGBERT, b Romeo, Mich, Nov 13, 06; m 30; c 2. BOTANY. Educ: Mich State Norm Col, AB, 29; Univ Okla, MS, 31; Ohio State Univ, PhD(bot), 34. Prof Exp: Asst bot, Univ Okla, 29-31; asst bot, 31-33, chief lab asst, 33-35, from instr to assoc prof, 35-54, PROF BOT, OHIO STATE UNIV, 54- Concurrent Pos: Exchange prof, Cornell Univ, 37; consult city algologist, Columbus, 42- Mem: Assoc Bot Soc Am; assoc Am Micros Soc (vpres, 47 & 48, pres, 49); Int Phycol Soc; Phycol Soc Am. Res: Freshwater algae; biological control of microorganisms in industrial and drinking water. Mailing Add: Dept of Bot Ohio State Univ Col Biol Sci Columbus OH 43210

TAFT, DAVID DAKIN, b Cleveland, Ohio, Mar 27, 38; m 61; c 3. POLYMER CHEMISTRY, ORGANIC CHEMISTRY. Educ: Kenyon Col, AB, 60; Mich State Univ, PhD(org chem), 63. Prof Exp: Sr res chemist, Archer Daniels Midland Co, 64-67; group leader, Resins, Ashland Chem Co, 67-70, mgr polymer chem, Ashland Oil, Inc, 70-72; vpres & gen mgr, Oxyplast, Inc, 74; dir com develop, 72-73, asst to pres, 72-74, DIR RES & DEVELOP, GEN MILLS CHEM, INC, 74- Mem: Am Chem Soc; Tech Asn Pulp & Paper Indust; Fedn Socs Paint Technol; Asn Finishing Processes (vpres). Res: Acrylic, polyester, epoxy, urethane coating and adhesive systems; water soluble coating and adhesive polymers; functional monomers; electrophotographic systems; powder coatings; non-yellowing isocyanates; nylon polymers; hydrophilic polymers. Mailing Add: Gen Mills Chem Inc 4620 W 77th St Minneapolis MN 55435

TAFT, DOROTHY LARUE, botany, see 12th edition

TAFT, EARL J, b New York, NY, Aug 27, 31; m 59; c 2. ALGEBRA. Educ: Amherst Col, BA, 52; Yale Univ, MA, 53, PhD(math), 56. Prof Exp: Instr math, Columbia Univ, 56-59; from asst prof to assoc prof, 59-66, PROF MATH, RUTGERS UNIV, 66- Concurrent Pos: NSF res grants, 63-67, 70-71 & 73; exec ed, Commun in Algebra, 74- Mem: Am Math Soc; Math Asn Am. Res: Nonassociative algebras; Hopf algebras; rings; groups. Mailing Add: Dept of Math Rutgers Univ New Brunswick NJ 08903

TAFT, EDGAR BRECK, b New Haven, Conn, Nov 16, 16; m 43. PATHOLOGY. Educ: Yale Univ, MD, 42; Univ Kans, MS, 50. Prof Exp: Nat Res Coun fel, Univ Kans, 48-49; Runyon clin res fel, Univ Kans & Univ Stockholm, 49-50; from asst pathologist to assoc pathologist, 52-60, PATHOLOGIST, MASS GEN HOSP, 61-; ASSOC CLIN PROF PATH, HARVARD MED SCH, 63- Concurrent Pos: From instr to assoc prof path, Harvard Med Sch, 47-63. Mem: AAAS; Histochem Soc; Soc Develop Biol; AMA; Am Asn Path & Bact; Am Asn Cancer Res; NY Acad Sci. Res: Liver disease; methods of histochemistry and cytochemistry. Mailing Add: Dept of Path Mass Gen Hosp Boston MA 02114

TAFT, HORACE DWIGHT, b Cincinnati, Ohio, Apr 2, 25; m 52; c 2. PHYSICS. Educ: Yale Univ, BA, 50; Univ Chicago, MS, 53, PhD, 55. Prof Exp: From instr to assoc prof physics, 56-65, dean col, 69-74, PROF PHYSICS, YALE UNIV, 65- Res: High energy physics. Mailing Add: Dept of Physics Yale Univ New Haven CT 06520

TAFT, JAY LESLIE, b Rockville Centre, NY, Mar 19, 44. BIOLOGICAL OCEANOGRAPHY. Educ: Lafayette Col, BA, 67; Johns Hopkins Univ, MA, 73, PhD(biol oceanog), 74. Prof Exp: Res assoc, 74, ASSOC RES SCIENTIST BIOL OCEANOG, CHESAPEAKE BAY INST, 74- Mem: AAAS; Am Soc Limnol & Oceanog; Phycol Soc Am; Estuarine Res Fedn. Res: Production and utilization of dissolved organic matter in estuaries and coastal ocean; nutrient cycling in estuaries. Mailing Add: Chesapeake Bay Inst Johns Hopkins Univ Baltimore MD 21218

TAFT, KINGSLEY ARTER, JR, b Cleveland, Ohio, Nov 17, 30; m 55; c 3. FORESTRY, GENETICS. Educ: Amherst Col, AB, 53; Univ Mich, BS, 57; NC State Univ, MS, 62, PhD(forestry, genetics), 66. Prof Exp: Forester, Nebo Oil Co, La, 57-59; res asst forestry, NC State, 59-63; forest geneticist, Div Forestry, 63-74, CHIEF FOREST & WILDLIFE RESOURCES BR, FISHERIES & WILDLIFE DEVELOP, TENN VALLEY AUTHORITY, 74- Mem: Soc Am Foresters. Res: Hardwood tree improvement and genetics. Mailing Add: Forest & Wildlife Resources Br Tenn Valley Authority Norris TN 37828

TAFT, ROBERT WHEATON, JR, b Lawrence, Kans, Dec 17, 22; m 44; c 3. PHYSICAL CHEMISTRY, ORGANIC CHEMISTRY. Educ: Univ Kans, BS, 44, MS, 46; Ohio State Univ, PhD(chem), 49. Prof Exp: Lab asst chem, Univ Kans, 44-46; asst, Ohio State Univ, 46-49; res assoc, Columbia Univ, 49-50; from asst prof to prof, Pa State Univ, 50-65; PROF CHEM, UNIV CALIF, IRVINE, 65- Concurrent Pos: Sloan fel, 55-57; Guggenheim fel, Harvard Univ, 58; consult, Sun Oil Co, 58- Mem: Am Chem Soc. Res: Kinetics; effect of molecular structure on reactivity; mechanisms of organic reactions; rate and equilibrium studies; fluorine nuclear magnetic resonance shielding. Mailing Add: Dept of Chem Univ of Calif Irvine CA 92650

TAFT, WILLIAM H, b San Mateo, Calif, Sept 21, 31; m 52; c 9. MARINE GEOLOGY. Educ: Stanford Univ, BS, 57, PhD(geol), 62; Univ SDak, MA, 58. Prof Exp: Res asst marine geol, Stanford Univ, 61-63; from asst prof to assoc prof geol, 63-75, dir sponsored res, 65-71, dir res inst, 66-71, PROF GEOL, UNIV S FLA, 75-, DIR RES, 71-, DIR GRAD STUDIES, 75- Mem: AAAS; Geol Soc Am; Mineral Soc Am. Res: Diagenesis of modern marine carbonate sediments and recrystallization of mestable carbonate minerals. Mailing Add: 4202 Fowler Ave Tampa FL 33620

TAFURI, JOHN FRANCIS, b St Barbara, Italy, Aug 4, 24; nat US; m 58; c 3. ENTOMOLOGY. Educ: Fordham Univ, BS, 44, MS, 48, PhD, 51. Prof Exp: Lab instr comp anat, Fordham Univ, 47-48, lab instr entom, 48, instr comp anat, Sch Educ, 50-51; from instr to assoc prof biol, 51-68, PROF BIOL, XAVIER UNIV, OHIO, 68- Concurrent Pos: Asst, Dept Animal Behavior, Am Mus Natural Hist, NY, 48-51. Mem: AAAS; Sigma Xi. Res: Electrical changes in tissues; social behavior in insects. Mailing Add: Dept of Biol Xavier Univ Cincinnati OH 45207

TAGER, MORRIS, b Riga, Latvia, Dec 25, 09; nat US. MICROBIOLOGY. Educ: Yale Univ, PhB, 31, MD, 36. Prof Exp: Med house officer, Beth Israel Hosp, Mass, 36-37; asst med, Sch Med, Yale Univ, 38-39, clin asst, 39-40, instr immunol & Am Col Physicians fel, 40-42, asst prof bact, 42-48; assoc prof microbiol & asst prof med, Sch Med, Western Reserve Univ, 48-51; PROF MICROBIOL & CHMN DEPT, DIV BASIC HEALTH SCI, EMORY UNIV, 51- Concurrent Pos: Asst res, New Haven Hosp, Conn, 38-39 & W W Winchester Hosp, 39-40; consult, Vet Admin, Atlanta, 52-; mem microbiol panel, US Off Naval Res, 59-65; mem bact & mycol study sect, USPHS, 62-64; mem bd regents, Nat Libr Med, 64-68; mem comn epidemiol surv, Armed Forces Epidemiol Bd, 69-73. Mem: AAAS; Am Soc Microbiologists; Am Asn Immunologists; Soc Exp Biol & Med; fel Am Acad Microbiol. Res: Bacteriology; staphylococci; mycology; blood clotting. Mailing Add: Dept of Microbiol Emory Univ 501 Woodruff Bldg Atlanta GA 30322

TAGGART, GEORGE BRUCE, b Philadelphia, Pa, Apr 8, 42. THEORETICAL SOLID STATE PHYSICS. Educ: Col William & Mary, BS, 64; Temple Univ, PhD(physics), 71. Prof Exp: Instr physics, Drexel Univ, 70; vis asst prof, Temple Univ, 70-71; ASST PROF PHYSICS, VA COMMONWEALTH UNIV, 71- Concurrent Pos: Fac res grantee, Va Commonwealth Univ, 73; consult, Temple Univ, 74, res assoc, 75; res assoc, Oak Ridge Nat Lab, 74; referee, Phys Rev & Phys Rev Lett, 75- Mem: AAAS; Am Phys Soc; Am Asn Physics Teachers. Res: Statistical mechanics of disordered multicomponent alloys and disordered and amorphous magnetic alloys; emphasis is on behavior of the critical temperature as a function of concentration and interaction type. Mailing Add: Dept of Physics Va Commonwealth Univ Richmond VA 23284

TAGGART, JAMES, seismology, geology, see 12th edition

TAGGART, JOHN VICTOR, b Brigham, Utah, Aug 29, 16; m 59. PHYSIOLOGY, MEDICINE. Educ: Univ Southern Calif, MD, 41. Prof Exp: Intern, Los Angeles County Hosp, Calif, 40-41; asst resident med, Clins, Univ Chicago, 41-42; res resident, Goldwater Mem Hosp, New York, 42-43; from asst & instr to assoc prof, 46-58, PROF MED, COL PHYSICIANS & SURGEONS, COLUMBIA UNIV, 58-, DALTON PROF PHYSIOL & CHMN DEPT, 62- Concurrent Pos: Nat Res Coun Welch fel internal med, 46-52; consult, Comn on Growth, Nat Res Coun, 53-54; Nathanson mem lectr, Univ Southern Calif, 55; consult, USPHS, 55-60; mem bd dirs, Russell Sage Inst Path, 55-, pres, 59-; career investr, Am Heart Asn, 58-62; mem exec comt, Health Res Coun, City of New York, 63-70, vchmn, 67-70; mem res career award comt, NIH, 66-69, chmn, 67-69. Mem: Am Physiol Soc; Am Soc Clin Invest; Harvey Soc (pres, 69); Asn Am Physicians. Res: Metabolic aspects of renal transport mechanisms; biochemistry; enzymology. Mailing Add: Dept of Physiol Columbia Univ New York NY 10032

TAGGART, MILLARD SEALS, JR, organic chemistry, see 12th edition

TAGGART, WILLIAM PAUL, b New York, NY, Apr 14, 41. POLYMER CHEMISTRY. Educ: Univ Rochester, BS, 62; Univ Mass, MS, 68, PhD(polymer sci & eng), 73. Prof Exp: Res chemist polymers, 62-74, RES SPECIALIST POLYMERS, MONSANTO CO, 74- Mem: Am Chem Soc. Res: The relationship between structural and morphological features of rubber modified glassy polymers and their performance properties; morphology of ion containing polymers using small angle x-ray scattering. Mailing Add: Dept of Res Monsanto Co 730 Worcester St Indian Orchard Springfield MA 01033

TAGGART, WILLIAM VROOM, biochemistry, see 12th edition

TAGUCHI, YOSHINORI, b Honjyo, Japan, Sept 25, 33; Can citizen; m 63; c 4. UROLOGY, IMMUNOLOGY. Educ: McGill Univ, BSc, 55, MD & CM, 59, PhD(exp med), 70; FRCS(C), 64. Prof Exp: Urologist, 66-72, DIR LABS, DEPT UROL, ROYAL VICTORIA HOSP, 71-; ASST PROF UROL, McGILL UNIV, 69- Mem: Can Urol Asn; Am Urol Asn; Can Soc Immunol; Transplantation Soc. Res: Cellular immunology; organ transplantation. Mailing Add: 1310 Greene Ave Westmount PQ Can

TAGUE, JEAN M, statistics, see 12th edition

TAHAN, THEODORE WAHBA, b Alexandria, Egypt, Jan 22, 36; m 62; c 1. RADIOTHERAPY. Educ: Alexandria Univ, MBBCh, 61, DMR&E, 64, MD & PhD(radiother), 69. Prof Exp: Resident radiother, Alexandria Univ Hosps, 62-64, clin demonstr, Fac Med, 64-69, lectr, 69-71; asst prof, 72-75, ASSOC PROF RADIOTHER, FAC MED, SHERBROOKE UNIV, 75- Concurrent Pos: Consult radiother, Hotel Dieu Hosp, St Vincent Hosp & Sherbrooke Hosp, Sherbrooke, 73. Mem: Can Radiol Asn; Can Oncol Soc; Egyptian Cancer Soc; Egyptian Radiol Soc; Radiol Soc NAm. Res: Elemental diet in irradiated patients; bronchial carcinoma, carcinoma of the bilharzial bladder, tongue and nasopharynx. Mailing Add: Radiother Dept Cent Hosp Univ of Sherbrooke Sherbrooke PQ Can

TAHIR-KHELI, RAZA ALI, b Hazara, West Pakistan, May 1, 36; m 62; c 1. THEORETICAL MAGNETISM. Educ: Oxford Univ, BA, 58, DPhil(physics), 62. Prof Exp: Res assoc physics, Univ Pa, 62-64; sr scientific officer, Pakistan Atomic Energy Comn, 64-66; from asst prof to assoc prof, 66-71, PROF PHYSICS, TEMPLE UNIV, 71- Concurrent Pos: Assoc, Exp Sta, E I du Pont de Nemours & Co, Del, 67- Mem: Am Phys Soc. Res: Solid state physics; many body physics; magnetism; order-disorder phenomena; random magnetism. Mailing Add: Dept of Physics Temple Univ Col Lib Arts Philadelphia PA 19122

TAHK, FREDERICK CHRISTOPHER, organic chemistry, see 12th edition

TAHMISIAN, THEODORE NEWTON, b Caesaria, Turkey, June 22, 09; US citizen; m 36; c 1. PHYSIOLOGY. Educ: Fresno State Col, AB, 35; Univ Iowa, MS, 39, PhD(zool), 42. Prof Exp: Asst biol, Fresno State Col, 34-37; tech asst zool, Univ Iowa, 38-42, res assoc, 42-43; res assoc med, Univ Chicago, 46; CHMN ELECTRON MICROS COMT & SR BIOLOGIST, ARGONNE NAT LAB, 46- Mem: Fel AAAS; Am Soc Zoologists; Am Physiol Soc; Soc Gen Physiol; Soc Exp Biol & Med. Res: Electron microscopical studies of radiation, induction and differentiation of cellular organelles; ultrastructural correlations in normal and cancer cells and viral invasions. Mailing Add: D202 Argonne Nat Lab 9700 S Cass Ave Argonne IL 60439

TAI, CLEMENT LEO, b Shanghai, China, Oct 19, 15; US citizen; m 46; c 2. APPLIED MECHANICS, APPLIED MATHEMATICS. Educ: Nat Cent Univ, China, BS, 41; Univ Colo, MS, 48; Polytech Inst Brooklyn, PhD(appl mech), 61. Prof Exp: Asst engr, Pub Health Admin, Repub China, 41-44, assoc engr, Ministry Commun, 44-46; sr designer struct eng, Devenco, Inc, 48-50; engr, Hardesty & Hanover, 50; struct engr, Chem Construct Corp, 50-61; res specialist struct dynamics, 61-63, sr tech specialist fluid, Space & space & struct dynamics, 63-66, proj engr, 66-67, MEM TECH STAFF, SPACE DIV, NAM ROCKWELL CORP, 67- Mem: Am Inst Aeronaut & Astronaut; Am Soc Mech Eng; Chinese Inst Eng. Res: Transient dynamics of orbiting space stations; cable dynamics; longitudinal oscillations of large liquid launch vehicles; oscillations of fluid in elastic containers; wave propagations in propellent lines; rotation of changing bodies. Mailing Add: NAm Rockwell Corp Space Div 12214 Lakewood Blvd Downey CA 90241

TAI, DOUGLAS L, b Hong Kong, Nov 6, 40; m 70. MEDICAL PHYSICS. Educ: Chinese Univ Hong Kong, BSc, 64; Cornell Univ, PhD(chem), 69. Prof Exp: Res assoc phys chem, Cornell Univ, 69-71; fac res assoc solid state chem, Ariz State Univ, 71-72, asst prof, 72-73; instr, Dept Chem, 73-74, INSTR, DEPT PHYSICS & ASTRON, UNIV KY, 75- Concurrent Pos: Trainee, Dept Radiation Med, Albert B. Chandler Med Ctr, Univ Ky, 75- Mem: Am Crystallog Asn; Am Chem Soc. Res: Teletherapy and intracavitary dosimetry. Mailing Add: Univ of Ky Dept Radiation Med Chandler Med Ctr Lexington KY 40506

TAI, HAN, b Yang Chow, China, Mar 20, 24; m 61; c 2. ANALYTICAL CHEMISTRY. Educ: Nanking Univ, BS, 49; Emory Univ, MS, 55, PhD(chem), 58. Prof Exp: Asst geol, Nat Taiwan Univ, Formosa, 51-54; res chemist, A E Staley Mfg Co, 58-66, head instrumental anal lab, 66-67, group leader anal labs, 67-70, SUPVR PESTICIDES MONITORING LAB, ENVIRON PROTECTION AGENCY, 70- Mem: Am Chem Soc; Soc Appl Spectros. Res: Chemical analysis of rocks and minerals, carbohydrates, polymers, pesticides; infrared spectroscopy; chromatography; environmental monitoring on pesticide residues. Mailing Add: Pesticides Monitoring Lab Environ Protection Agency Bay St Louis MS 39520

TAI, JULIA CHOW, b Shanghai, China. CHEMISTRY. Educ: Nat Taiwan Univ, BS, 57; Univ Okla, MS, 59; Univ Ill, Urbana, PhD(chem), 63. Prof Exp: Fels, Wayne State Univ, 63-68; asst prof chem, 69-73, ASSOC PROF CHEM, UNIV MICH, DEARBORN, 73- Mem: Am Chem Soc. Res: Quantum mechanical calculations of electronic spectra of molecules; structure of molecules. Mailing Add: Dept of Chem Univ of Mich Dearborn MI 48128

TAI, PETER YAO-PO, b Chutung, Taiwan, July 6, 37; US citizen; m 64; c 2. PLANT BREEDING. Educ: Nat Taiwan Univ, BS, 61; Tex A&M Univ, MS, 66; Okla State Univ, PhD(crop sci), 72. Prof Exp: Res assoc, 72-75, INSTR PLANT BREEDING, AGR EXP STA, UNIV GA, 75- Mem: Am Soc Agron; Crop Sci Soc Am; Am Peanut Res & Educ Asn; AAAS. Res: Inheritance of biochemical constituents in Arachis, phenogenetic studies of morphologic mutants and breeding for reproductive efficiency and pest resistance in peanuts. Mailing Add: Dept of Agron Coastal Plain Sta Univ of Ga Tifton GA 31794

TAI, WILLIAM, b Youngchow, Kiangsu, China, Mar 9, 34; m 63; c 4. CYTOGENETICS. Educ: Nat Chung Hsing Univ, BSc, 56; Utah State Univ, MSc, 64; Univ Utah, PhD(genetics), 67. Prof Exp: NIH fel biol sci, Stanford Univ, 67-69; asst prof bot & plant path, 69-72, ASSOC PROF BOT & PLANT PATH, MICH STATE UNIV, 72- Concurrent Pos: Vis prof, Nat Chung-Hsing Univ, 75-76. Mem: AAAS; Am Inst Biol Sci; Am Soc Cell Biol; Am Genetic Asn; Genetics Soc Am. Res: Plant cytogenetics; cytotaxonomy; plant breeding. Mailing Add: Dept of Bot & Plant Path Mich State Univ East Lansing MI 48824

TAIBLESON, MITCHELL H, b Oak Park, Ill, Dec 31, 29; m 49; c 3. MATHEMATICS. Educ: Univ Chicago, SM, 60, PhD(math), 62. Prof Exp: From asst prof to assoc prof math, 62-69, chmn dept, 70-73, res prof, Dept Psychiat, Med Sch, 73-75, PROF MATH, WASHINGTON UNIV, 69- Concurrent Pos: Mem, Inst Advan Study, 66-67. Mem: Am Math Soc; Math Asn Am. Res: Several dimensional harmonic analysis on real and local fields; Lipschitz and potential spaces; special functions; mathematical models in medical and behavioral science. Mailing Add: Dept of Math Washington Univ St Louis MO 63130

TAICHMAN, NORTON STANLEY, b Toronto, Ont, May 27, 36; m 58; c 5. PATHOLOGY, IMMUNOLOGY. Educ: Univ Toronto, DDS, 61, PhD(immunopath), 67; Harvard Univ, dipl periodont, 64. Prof Exp: Assoc dent, Fac Dent, Univ Toronto, 65-68, from lectr to asst prof path, 67-71; PROF PATH & CHMN DEPT, SCH DENT MED, UNIV PA, 72- Concurrent Pos: Assoc prof dent, Fac Dent, Univ Toronto, 68-72, assoc prof path, 71-72, mem, Inst Immunol, 70-72. Mem: Am Soc Exp Path; Can Soc Immunol; Can Fedn Biol Socs; Int Asn Dent Res; Can Dent Asn. Res: Inflammation, immunopathology, periodontal disease; immune deficiency syndromes; polymorphonuclear-leukocytes; platelets; lysosomes. Mailing Add: Dept of Path Univ of Pa Sch of Dent Med Philadelphia PA 19174

TAIMUTY, SAMUEL ISAAC, b West Newton, Pa, Dec 20, 17; m 53; c 2. PHYSICS. Educ: Carnegie Inst Technol, BS, 40; Univ Southern Calif, PhD(physics), 51. Prof Exp: Asst res physicist, Am Soc Heat & Ventilating Eng Lab, 40-42; physicist, Philadelphia Naval Shipyard, 42-44 & Long Beach Naval Shipyard, 44-46; sr physicist, US Naval Radiol Defense Lab, 50-52 & Stanford Res Inst, 52-72; SR PHYSICIST, LOCKHEED MISSILES & SPACE CO, 72- Mem: AAAS; Am Phys Soc; Sigma Xi; Am Asn Physicists in Med. Res: Heat transmission; magnetism; nuclear and radiation physics; radiation effects in solids; radiation dosimetry; industrial applications of radiation, ferroelectricity, thin films and organic dielectrics. Mailing Add: 2069 Edgewood Dr Palo Alto CA 94303

TAINITER, MELVIN, b Brooklyn, NY, Apr 24, 36; m 59; c 3. APPLIED MATHEMATICS. Educ: Brooklyn Col, BS, 58; NY Univ, MS, 61; Johns Hopkins Univ, PhD(statist), 65. Prof Exp: Mathematician, Thomas J Watson Res Ctr, Int Bus Mach Corp, 59-63; res assoc appl math, Carlyle Barton Lab, Johns Hopkins Univ, 63-65; mathematician, Brookhaven Nat Lab, 65-72; ASSOC PROF MATH, CITY COL NEW YORK, 72- Mem: Math Asn Am; Soc Indust & Appl Math; Am Math Soc. Res: Algebraic combinatorial theory; sequential decision theory. Mailing Add: Dept of Math City Col of New York New York NY 10031

TAINTER, FRANKLIN HUGH, b Winona, Minn, Apr 13, 41; m 66; c 1. FOREST PATHOLOGY. Educ: Univ Mont, BSF, 64; Univ Minn, St Paul, MS, 68, PhD(plant path), 70. Prof Exp: ASSOC PROF FOREST PATH, UNIV ARK, FAYETTEVILLE, 70- Mem: Am Phytopath Soc; Soc Am Foresters. Res: Ecology of forest tree diseases; American mistletoes; deterioration of wood products in service. Mailing Add: Dept of Plant Path Univ of Ark Fayetteville AR 72701

TAIT, COLUMBUS DOWNING, JR, b Valdosta, Ga, Sept 3, 23; m 56; c 2. MEDICINE, PSYCHOANALYSIS. Educ: Univ Va, BA, 43, MD, 47; Columbia Univ, cert psychoanal med, 57. Prof Exp: Intern med & path, Bellevue Hosp, New York, 47-48; resident psychiat, Compton Sanitarium, Calif, 48-49; AEC fel med sci, Duke Univ & Yale Univ, 49-50; resident, Rockland State Hosp, Orangeburg, NY, 50-51; pvt pract psychiat & psychoanal, 53-64; assoc prof, 64-67, PROF PSYCHIAT, SCH MED, EMORY UNIV, 67- Concurrent Pos: Consult & res psychiatrist, Youth Coun, Washington, DC, 54-62; instr psychoanal clin training & res, Col Physicians & Surgeons, Columbia Univ, 57-64; consult & res psychiatrist, Mobilization Youth, New York, 62-64; dir ment health res, Ga Ment Health Inst, 65-71; training psychoanalyst, Training Prog, Columbia Univ-Emory Univ, 65-; consult, Juv Courts, New York, 53-54 & Atlanta, Ga, 66-72; chief community ment health res, Ga Ment Health Inst, 71-73. Mem: AAAS; Am Psychoanal Asn; Int Psychoanal Asn; fel Am Psychiat Asn; AMA. Res: Study of the interaction of social and individual factors in juvenile delinquency and its prognosis and treatment; psychiatric program evaluation research and follow-up studies; psychoanalytic technique, especially in treating delinquents. Mailing Add: Dept of Psychiat Emory Univ Atlanta GA 30322

TAIT, JAMES SIMPSON, b Charlottetown, PEI, Feb 25, 30; m 58; c 2. FISH BIOLOGY. Educ: Dalhousie Univ, BSc, 50, MSc, 52; Univ Toronto, PhD(zool), 59. Prof Exp: Res scientist, Res Br, Ont Dept Lands & Forests, 58-64; asst prof biol, 64-70, ASSOC PROF BIOL, YORK UNIV, 70- Concurrent Pos: Consult, Res Br, Ont Dept Lands & Forests, 64-70. Mem: Am Fisheries Soc; Can Soc Zool. Res: Physiology of fish, particularly temperature and depth relations and swimbladder function; selective breeding of salmonid hybrids; effects of pesticides on fish. Mailing Add: Dept of Biol York Univ Downsview ON Can

TAIT, KEVIN S, b New York, NY, Nov 24, 33; m 59; c 4. APPLIED MATHEMATICS. Educ: Princeton Univ, AB, 55; Harvard Univ, PhD(appl math), 65. Prof Exp: Res staff mem, Sperry Rand Res Ctr, 64-67; assoc prof math, Boston Univ, 67-74; MEM STAFF, AERODYNE RES, 74- Mem: Soc Indust & Appl Math. Res: Optimal control theory; mathematical programming; numerical analysis. Mailing Add: Aerodyne Res South Ave Burlington MA 01803

TAIT, ROBERT JAMES, oceanography, see 12th edition

TAIT, ROBERT MALCOLM, b Newcastle on Tyne, Eng, Mar 28, 40; m 66. AGRICULTURE, ANIMAL HUSBANDRY. Educ: Univ Durham, BSc, 63; Univ Newcastle, Eng, PhD(animal nutrit), 66. Prof Exp: ASSOC PROF ANIMAL SCI, FAC AGR, UNIV BC, 66- Concurrent Pos: Res grants, Nat Res Coun Can, 75-76. Mem: Can Soc Animal Sci; Brit Soc Animal Prod; Nutrit Soc Can. Res: Intensive sheep production, especially nutrient requirements. Mailing Add: Dept of Animal Sci Univ of BC Fac of Agr Sci Vancouver BC Can

TAIT, WILLIAM CHARLES, b Waterloo, Iowa, Feb 9, 32; m 54; c 4. THEORETICAL SOLID STATE PHYSICS. Educ: Wabash Col, BA, 54; Cornell Univ, MA, 58; Purdue Univ, PhD, 62. Prof Exp: Teaching asst, Cornell Univ, 54-58; physicist, Res Lab, Bendix Corp, 58; instr physics, Wabash Col, 58-61; sr physicist, Cent Res, 62-66, res specialist, 67-68, supvr, 69-70, mgr cent res, 71-72, MGR DUPLICATING PROD, MINN MINING & MFG CO, 72- Mem: Am Phys Soc. Res: Solid state and quantum field theory; semiconductor lasers; photoconductors; electrophotography. Mailing Add: 14452 N 57th St Stillwater MN 55082

TAIZ, LINCOLN, b Philadelphia, Pa, Nov 5, 42; m 63; c 1. PLANT PHYSIOLOGY. Educ: Univ Utah, BS, 67; Univ Calif, Berkeley, PhD(bot), 71. Prof Exp: Actg asst prof bot, Univ Calif, Berkeley, 72-73; ASST PROF BIOL, UNIV CALIF, SANTA CRUZ, 73- Concurrent Pos: NSF grant, 75. Mem: Am Soc Plant Physiologists; Bot Soc Am. Res: The role of cell wall metabolism in plant growth and secretion; DNA metabolism in nondividing plant cells. Mailing Add: Thinmann Labs Div of Nat Sci Univ of Calif Santa Cruz CA 95064

TAJIMA, YUJI ALEXIS, b Sendai, Japan, Dec 29, 17; nat US; m 53. PLASTICS CHEMISTRY. Educ: Calif Inst Technol, BS, 40, MS, 41. Prof Exp: Res chemist, Velsicol Corp, 43-47; dir develop & prod labs, Julius Hyman & Co, 47-52, vpres & tech dir, Hyman Labs, Inc, 52-54; res scientist, NY Univ, 54-57; systs analyst, Olin Mathieson Chem Corp, 57-58; res scientist, NY Univ, 59-65; sr chemist, Aerojet-Gen Corp, 65-66; supvr advan propellant tech, Lockheed Propulsion Co, 66-71, sect chief chem, 71-75, SR RES SPECIALIST, LOCKHEED CALIF CO, 75- Mem: AAAS; Am Chem Soc; Am Inst Chem Engrs. Res: Kinetic theory; combustion; composites. Mailing Add: 1990 Fox Ridge Dr Pasadena CA 91107

TAKACS, GERALD ALAN, physical chemistry, see 12th edition

TAKAGI, SHUNSUKE, b Fukuoka, Japan, Mar 22, 19; m 43; c 3. MECHANICS. Educ: Univ Tokyo, BAgE, 43; Hokkaido Univ, DSc(physics), 56. Prof Exp: Asst prof math & mech, Tokyo Univ Agr & Technol, 47-60; contract scientist, 60-66, PHYS RES SCIENTIST, US ARMY COLD REGIONS RES & ENG LAB, 66- Concurrent Pos: Army Res Off-Durham grant, 71-73. Honors & Awards: Agr Eng Soc Japan Award, 59. Mem: Am Soc Civil Eng; Japan Soc Civil Eng. Res: Theoretical study of soil mechanics and soil physics; mathematical analysis of ice mechanics. Mailing Add: 17 Dresden Rd Hanover NH 03755

TAKAHASHI, AKIO, b Tokyo, Japan, Apr 15, 31; m 64; c 4. POLYMER CHEMISTRY. Educ: Tokyo Col Sci, BS, 57; Tokyo Inst Technol, MS, 60, PhD(polymer chem), 63. Prof Exp: Sr chemist, Mitsui Chem Indust, Inc, 63-65; sr chemist, Gaylord Assocs, Inc, NJ, 65-67, sect mgr, 67-70; res assoc polymer chem, 70-73, sr res assoc, 73-75, SCIENTIST & DISCIPLINE/PROG LEADER, HOOKER CHEM & PLASTICS CO, 75- Mem: Am Chem Soc; Japanese Soc Polymer Sci. Res: Polymer-organic chemistry; polymerization by free radical, Ziegler-Natta and ionic catalysts; polymerization through charge-transfer complexes; polybutadiene; polyvinyl chloride; polyethers; graft copolymers. Mailing Add: 62 Meadowview Lane Williamsville NY 14221

TAKAHASHI, ELLEN SHIZUKO, b Berkeley, Calif. PHYSIOLOGICAL OPTICS, OPTOMETRY. Educ: Univ Calif, Berkeley, BS, 52, MOpt, 53, PhD(physiol optics), 68. Prof Exp: Optometrist in pvt pract, 53-56; optometrist orthoptics, Stanford Univ Hosps, 56-62; clin instr & actg asst prof optom, Univ Calif, 62-67; NIH fel visual neurophysiol, Australian Nat Univ, 68-70; asst prof, 70-72, ASSOC PROF PHYSIOL OPTICS, UNIV ALA, BIRMINGHAM, 72-, DIR GRAD STUDIES, SCH OPTOM, 74- Concurrent Pos: Mem, Nat Adv Coun Health Prof Educ, 72-75. Mem: Am Acad Optom; Asn Res Vision & Ophthal; Soc Neurosci. Res: Visual neurophysiology; sensory and oculomotor disorders of binocular vision. Mailing Add: Sch of Optom Med Ctr Univ of Ala Birmingham AL 35294

TAKAHASHI, FRANCOIS IWAO, b Hakodate, Japan, Dec 12, 25; Can citizen; div; c 2. MICROBIAL GENETICS. Educ: Hakodate Fisheries Col, BA, 45; Kyushu Univ, MAS, 49; Univ Montreal, PhD(bact), 55. Prof Exp: Fel microbiol, Nat Res Coun Can, 56-58; res officer, Microbiol Res Inst, Can Dept Agr, Ottawa, 58-60, head genetics & taxon sect, 60-63; assoc prof biol, 64-67, PROF BIOL, McMASTER UNIV, 67- Mem: Am Soc Microbiol; Can Soc Microbiol; Can Soc Cell Biol. Res: Molecular biology of bacterial sporulation; transducing bacteriophage. Mailing Add: Dept of Biol McMaster Univ Hamilton ON Can

TAKAHASHI, HIRONORI, b Tokyo, Japan, June 5, 42. PLASMA PHYSICS. Educ: Keio Univ, Japan, BEng, 65; Mass Inst Technol, MS, 67, DSc(aeronaut & astronaut), 70. Prof Exp: Fel plasma physics, Univ Stuttgart, 71-72, guest lectr, 72-73; res assoc, 73-75, RES STAFF MEM, PLASMA PHYSICS LAB, PRINCETON UNIV, 75- Res: Plasma physics in connection with controlled thermonuclear fusion research; plasma waves, wave heating, tokamaks. Mailing Add: Plasma Physics Lab Princeton Univ Princeton NJ 08540

TAKAHASHI, MARK T, b Holtville, Calif, Feb 26, 36; m 61; c 2. PHYSICAL BIOCHEMISTRY. Educ: Oberlin Col, BA, 58; Univ Wis-Madison, PhD(phys chem), 63. Prof Exp: Res biochemist, Battelle Mem Inst, 64-67; ASST PROF PHYSIOL, RUTGERS MED SCH, COL MED & DENT NJ, 70- Concurrent Pos: NIH fel, Univ Mass, Amherst, 67-70. Mem: AAAS; Am Chem Soc. Res: Physical enzymology; membranes. Mailing Add: Dept of Physiol Rutgers Med Sch Piscataway NJ 08854

TAKAHASHI, SHUICHI, b Shiroishi, Japan, June 17, 28; m 55; c 1. MATHEMATICS. Educ: Tohoku Univ, Japan, PhD, 59. Prof Exp: From asst prof to assoc prof, 63-70, PROF MATH, UNIV MONTREAL, 70- Mem: Am Math Soc; Can Math Cong; Math Soc Japan. Res: Number theory. Mailing Add: Dept of Math Univ of Montreal PO Box 6128 Montreal PQ Can

TAKAHASHI, TARO, b Tokyo, Japan, Nov 15, 30; US citizen; m 66; c 2. GEOCHEMISTRY, GEOPHYSICS. Educ: Univ Tokyo, BEng, 53; Columbia Univ, PhD(geol), 57. Prof Exp: Res scientist, 57-59, RES ASSOC, LAMONT GEOL OBSERV, COLUMBIA UNIV, 59-; DISTINGUISHED PROF GEOL, QUEENS COL, NY, 71- Concurrent Pos: Lectr, Queens Col, NY, 58; res chemist, Scripps Inst Oceanog, Univ Calif, 59; asst prof, State Univ NY Col Ceramics, Alfred, 59-62, vis prof, 63; assoc prof, Univ Rochester, 62-70, prof, 70; vis asst prof, Columbia Univ, 66; vis assoc prof, Calif Inst Technol, 70. Mem: Am Geophys Union; Geol Soc Am; Am Phys Soc. Res: Thermodynamic and physical properties of metal oxides under extremely high pressures and temperatures; geochemistry of carbon dioxid- in the ocean atmosphere system; thermodynamics of metal sulfides in supercritical water vapor. Mailing Add: 187 E Hudson Ave Englewood NJ 07631

TAKAHASHI, WILLIAM NOBURU, b Oakland, Calif, Aug 14, 05; m 34; c 1. PLANT PATHOLOGY. Educ: Univ Calif, BS, 28, PhD(plant path), 32. Prof Exp: Asst plant path, Univ, 30-38, instr, 38-47, from asst prof to prof, 48-73, assoc, Exp Sta, 30-38, jr plant pathologist, 38-47, asst plant pathologist, 47, assoc plant pathologist, 52-58, plant pathologist, 58-73, EMER PROF PLANT PATH, UNIV CALIF, BERKELEY & EMER PLANT PATHOLOGIST, EXP STA, 73- Concurrent Pos: Dir agr res, War Relocation Authority, Ariz, 42-43; vis doctor, Cornell Univ, 43-44; Guggenheim Mem fel, Univ Rochester, 44-45. Mem: AAAS; Am Phytopath Soc; Bot Soc Am. Res: Fundamental research with plant viruses; electrophoresis; electromicrography; physiology. Mailing Add: Dept of Plant Path Univ of Calif Berkeley CA 94720

TAKAHASHI, YASUSHI, b Osaka, Japan, Dec 12, 24; m 59; c 2. THEORETICAL PHYSICS. Educ: Nagoya Univ, BSc, 51, DSc(physics), 54. Prof Exp: Nat Res Coun Can fel, 54-55; res assoc physics, Iowa State Univ, 55-57; res scholar, Dublin Inst Advan Studies, 57, from asst prof to prof, 57-68; PROF PHYSICS, UNIV ALTA, 68-, DIR THEORET PHYSICS INST, 69- Mem: Am Phys Soc; Can Asn Physicists; Royal Irish Acad; Ital Phys Soc. Res: Quantization of relativistic fields; quantum electrodynamics and field theory. Mailing Add: Dept of Physics Univ of Alta Edmonton AB Can

TAKANO, MASAHARU, b Tainan, Taiwan, Jan 20, 35; nat US; m 65; c 3. PHYSICAL CHEMISTRY, APPLIED PHYSICS. Educ: Hokkaido Univ, BSc, 57; Univ Tokyo, MSc, 59, DrSc(rheol), 63. Prof Exp: Nat Res Coun fel, McGill Univ, 63-65, fel, 65-67; res specialist, Corp Res Dept, 67-75, RES SPECIALIST & TECH TRANSLATOR, FOOD & FINE CHEM DIV, MONSANTO INDUST CHEM CO, 75- Mem: Soc Rheol; Am Chem Soc; AAAS; Am Phys Soc; Am Inst Chemists. Res: Rheology and physical chemistry of polymers and disperse systems; instrumentation and development of new techniques. Mailing Add: Monsanto Indust Chem Co 800 N Lindbergh Blvd St Louis MO 63166

TAKASHIMA, HERBERT T, organic chemistry, see 12th edition

TAKASHIMA, SHIRO, b Japan, May 12, 23; m 53; c 2. PHYSICAL BIOCHEMISTRY, NEUROPHYSIOLOGY. Educ: Univ Tokyo, BS, 47, PhD(biochem), 55. Prof Exp: Res fel phys chem, Univ Minn, 55-57; res assoc biomed eng, Univ Pa, 57-59; from assoc prof to prof protein res, Osaka Univ, 59-62; vis scientist, Walter Reed Med Ctr, 62-63; res assoc biomed eng, 63-64, asst prof, 64-70, ASSOC PROF BIOMED ENG, UNIV PA, 70- Mem: Biophys Soc; Am Chem Soc; NY Acad Sci; Bioeng Soc. Res: Dielectric relaxation of desoxyribonucleic acid; synthetic polynucleotides; polyamino acids and proteins; theory of dielectric relaxation; quantum chemistry of hydrogen bonds; electrical properties of excitable membranes. Mailing Add: Dept of Bioeng D2 Univ of Pa Philadelphia PA 19174

TAKASUGI, MITSUO, b Tacoma, Wash, Jan 28, 28; m 54; c 4. CANCER, IMMUNOLOGY. Educ: Univ Calif, Los Angeles, BA, 52, PhD(immunogenetics), 68; Univ Ore, MS, 62. Prof Exp: Sci teacher, Los Angeles City Schs, 53-64; fel, 53-64; Dept Tumor Biol, Karolinska Inst, Stockholm, Sweden, 68-69; RESEARCHER CANCER, DEPT SURG, UNIV CALIFORNIA, LOS ANGELES, 69- Mem: AAAS; Am Asn Cancer Res; Am Asn Immunol; Transplantation Soc. Res: Investigations into the cellular and humoral immune response to cancer. Mailing Add: Dept of Surg Univ of Calif Los Angeles CA 90024

TAKATS, STEPHEN TIBOR, b West Englewood, NJ, May 24, 30; m 60; c 3. CYTOLOGY, GENETICS. Educ: Cornell Univ, BS, 52; Univ Wis, MS, 54, PhD(genetics), 58. Prof Exp: Asst genetics, Univ Wis, 52-55; res collabr & USPHS res fel, Biol Dept, Brookhaven Nat Lab, 57-60; USPHS res fel biochem, Univ Glasgow, 60-61, Med Res Coun grant, 61; from asst prof to assoc prof biol, 61-69, chmn dept, 69-75, PROF BIOL, TEMPLE UNIV, 69- Mem: Bot Soc Am; Am Soc Plant Physiol; Genetics Soc Am; Am Soc Cell Biol. Res: Biochemical cytology; control of DNA synthesis in plant development. Mailing Add: Dept of Biol Temple Univ Philadelphia PA 19122

TAKAYAMA, KUNI, b Wapato, Wash, Feb 28, 32; m 59; c 2. BIOCHEMISTRY, MICROBIOLOGY. Educ: Ore State Univ, BS, 56; Univ Idaho, MS, 61, PhD(biochem), 64. Prof Exp: NIH grant, Inst Enzyme Res, Univ Wis-Madison, 64-65; proj assoc biochem of mycobact, Vet Admin Hosp, Madison, Wis, 65-67, res chemist, Tuberc Res Lab, 67-71, actg chief chemist, 68-69, chief chemist, 71-74, CHIEF RES CHEMIST, TUBERC RES LAB, VET ADMIN HOSP, MADISON, WIS, 74- Concurrent Pos: Proj assoc, Inst Enzyme Res, Univ Wis-Madison, 65-67, asst prof, 67-; NSF grants, Vet Admin Hosp & Univ Wis-Madison, 69-72, NIH grant, 73- Mem: AAAS; Am Soc Microbiol; Am Soc Biol Chem; Am Soc Cell Biol; Am Inst Biol Sci. Res: Biochemistry of mycobacteria; biosynthesis of mannophospholipids; synthesis of lipids; mode of action of isoniazid. Mailing Add: Tuberc Res Lab Vet Admin Hosp 2500 Overlook Terrace Madison WI 53705

TAKEDA, YASUHIKO, b Nagano, Japan, Mar 16, 27; m 57; c 4. PHYSIOLOGY, CLINICAL PATHOLOGY. Educ: Shinshu Univ, 46-48; Chiba Univ, MD, 52; Am Bd Path, dipl, 70. Prof Exp: From instr to asst prof, 63-69, ASSOC PROF MED, MED CTR, UNIV COLO, DENVER, 69- Concurrent Pos: Res fel, Div Lab Med, Med Ctr, Univ Colo, 58-60; Nat Res Coun Can res fel, McGill Univ, 60-63; Colo Heart Asn sr res fel med, Med Ctr, Univ Colo, Denver, 63-64, Am Heart Asn advan res fel, 64-66; NIH career develop award, 67-72; mem coun thrombosis, Am Heart Asn. Mem: AAAS; Am Physiol Soc; Am Soc Clin Path; Int Soc Thrombosis & Hemorrhagic Dis. Res: Regulation of plasma protein metabolism in health and disease; dynamics of thrombus formation and dissolution. Mailing Add: Dept of Med Univ of Colo Med Ctr Denver CO 80220

TAKEHARA, KENNETH N, biochemistry, toxicology, see 12th edition

TAKEKOSHI, TOHRU, b Taihoku, Taiwan, July 19, 35; Japanese citizen. POLYMER CHEMISTRY, ORGANIC CHEMISTRY. Educ: Kyoto Univ, BS, 58, MS, 60; Polytech Inst Brooklyn, PhD(polymer chem), 66. Prof Exp: STAFF CHEMIST, CORP RES & DEVELOP, GEN ELEC CO, 65- Mem: Am Chem Soc. Res: Organic synthesis and reaction mechanisms; polymer synthesis; fiber and plastics engineering; composites. Mailing Add: Gen Elec Corp Res & Develop Ctr PO Box 8 Schenectady NY 12302

TAKEMORI, AKIRA EDDIE, b Stockton, Calif, Dec 9, 29; m 58; c 2. PHARMACOLOGY. Educ: Univ Calif, AB, 51, MS, 53; Univ Wis, PhD(pharmacol), 58. Prof Exp: Res asst pharmacol, Univ Calif, 51-53; res asst, Univ Wis, 55-57, Am Cancer Soc fel, Enzyme Inst, 58-59; instr pharmacol, State Univ NY Upstate Med Ctr, 59-61, asst prof, 61-63; from asst prof to assoc prof, 63-69, PROF PHARMACOL, HEALTH SCI CTR, UNIV MINN, MINNEAPOLIS, 69- Concurrent Pos: Mem pharmacol A study sect, NIH, 71-74. Honors & Awards: Vis Scientist Award, Japan Soc Promotion Sci, 71; Alan Gregg Fellow Med Educ, China Med Bd NY, 71. Mem: AAAS; Am Soc Pharmacol & Exp Therapeut; Soc Exp Biol & Med; Pharmacol Soc Japan. Res: Mechanism of action of narcotic analgesics and narcotic antagonists; transport of drugs to the central nervous system; drug metabolism. Mailing Add: Dept Pharmacol 105 Millard Hall Univ of Minn Minneapolis MN 55455

TAKEMOTO, JAMES HIDEO, chemistry, see 12th edition

TAKEMOTO, KENNETH KANAME, b Hawaii, Sept 26, 20; m 51; c 2. VIROLOGY. Educ: George Washington Univ, BS, 49, MS, 50, PhD(- PhD(virol), 53. Prof Exp: USPHS res fel virol, Nat Microbiol Inst, 53-54; VIROLOGIST, NAT INST ALLERGY & INFECTIOUS DIS, 54- Mem: Am Soc Microbiol; Am Asn Immunologists; Soc Exp Biol & Med. Res: Anti-viral substances; tumor viruses; cell biology. Mailing Add: Nat Inst Allergy & Infect Dis Bethesda MD 20014

TAKEMURA, KAZ HORACE, b San Juan Bautista, Calif, Nov 2, 21; m 59; c 7. ORGANIC CHEMISTRY. Educ: Univ Calif, Los Angeles, BS, 47; Univ Ill, MS, 48, PhD(chem), 50. Prof Exp: Res fel chem, Ohio State Univ, 50-52; res chemist, Univ Calif, Berkeley, 52-53; chemist southern regional res lab, USDA, La, 53-56; asst prof chem, Loyola Univ, La, 56-58 & Univ Tulsa, 58-59; chemist, Sahyun Labs, 59-60; from asst prof to assoc prof chem, 60-66, chmn dept, 70-73, PROF CHEM, CREIGHTON UNIV, 66- Mem: AAAS; Am Chem Soc. Res: Organic synthesis and mechanisms. Mailing Add: Dept of Chem Creighton Univ Omaha NE 68178

TAKEO, MAKOTO, b Yamagata-shi, Japan, Apr 6, 20; m 47; c 3. PHYSICS. Educ: Tohuku Univ, Japan, BS, 43; Univ Ore, MS, 51, PhD(physics), 53. Prof Exp: Asst prof physics, Defense Acad, Japan, 53-56 & Portland State Col, 56-59; asst prof math, Univ Calgary, 59-62; assoc prof physics, 62-64, PROF PHYSICS, PORTLAND STATE UNIV, 64- Concurrent Pos: Consult, Sandia Corp, 57-60; vis prof, Univ Ore, 68-69. Mem: Am Phys Soc; Phys Soc Japan. Res: Theory on pressure broadening of spectral lines for neutral atoms; gas dynamics in ballistic compressors. Mailing Add: Dept of Physics Portland State Univ Portland OR 97207

TAKESAKI, MASAMICHI, b Sendai, Japan, July 18, 33; m 59; c 1. MATHEMATICS. Educ: Tohoku Univ, Japan, MS, 58, DSc(math), 65. Prof Exp: Res asst math, Tokyo Inst Technol, 58-63; assoc prof, Tohoku Univ, Japan, 63-70; PROF MATH, UNIV CALIF, LOS ANGELES, 70- Concurrent Pos: Fel, Sakkokai Found, 65-68; vis assoc prof, Univ Pa, 68-69 & Univ Calif, Los Angeles, 69-70; vis prof, Univ Aix-Marseille, 73-74 & Univ Bielefeld, 75-76. Mem: Am Math Soc; Math Soc Japan; Math Soc France. Res: Functional analysis; operator algebras; mathematical physics. Mailing Add: Dept of Math Univ of Calif Los Angeles CA 90024

TAKESHITA, TSUNEICHI, b Tokyo, Japan, Sept 13, 26; m 56; c 3. PHYSICAL ORGANIC CHEMISTRY. Educ: Waseda Univ, Japan, BSEng, 50; Univ Del, PhD(org chem), 62. Prof Exp: Chemist, Cent Res Inst, Japan Monopoly Corp, Tokyo, 50-64; asst prof catalysis, Res Inst Catalysis, Hokkaido Univ, 64-66; CHEMIST, ELASTOMER CHEM DEPT, E I DU PONT DE NEMOURS & CO, WILMINGTON, 66- Mem: Am Chem Soc; Chem Soc Japan; Sigma Xi. Res: Catalytic studies in organic chemistry; polymer chemistry. Mailing Add: 4 Alton Rd Polly Drummond Hill Newark DE 19711

TAKESUE, EDWARD I, b Honolulu, Hawaii, Dec 31, 27; m 57; c 4. PHARMACOLOGY, MEDICAL RESEARCH. Educ: Philadelphia Col Pharm, BS, 52; Purdue Univ, MS, 54, PhD(pharmacol), 55. Prof Exp: Pharmacologist, Columbus Pharmacol Co, 55-58, Lederle Labs, 58-62 & Sandoz Pharmaceut, NJ, 62-75; ASST DIR MED DEPT, PURDUE FREDERICK CO, 75- Mem: AAAS; Am Soc Pharmacol & Therapeut; Microcirc Soc; Am Rheumatism Asn; NY Acad Sci. Res: Pharmacology of antibiotics; lipemia clearing agents; antibiotic adjuvants; anti-inflammatory drugs. Mailing Add: Med Dept Purdue Frederick Co 50 Washington St Norwalk CT 06856

TAKETA, FUMITO, b Waimea Kauai, Hawaii, Mar 10, 26. BIOCHEMISTRY. Educ: Washington Univ, AB, 50; Univ Wis, PhD(biochem), 57. Prof Exp: Proj asst biochem, Univ Wis, 56-57, proj assoc, 57-58; from instr to assoc prof, Marquette Univ, 58-70; assoc prof, 70-72, PROF BIOCHEM, MED COL WIS, 72- Concurrent Pos: Fogarty sr int fel & vis prof biochem, King's Col, Univ London, 75-76. Mem: AAAS; Am Chem Soc; Am Soc Biol Chemists. Res: Protein and sulfhydryl chemistry; development of red blood cell; biosynthesis, structure and function of hemoglobin; metabolic and molecular control mechanisms. Mailing Add: Dept of Biochem Med Col of Wis Milwaukee WI 53233

TAKETA, SHOJIRO TOM, physiology, radiobiology, see 12th edition

TAKETOMO, YASUHIKO, b Tokyo, Japan; US citizen; m; c 2. PSYCHIATRY. Educ: First Col, Tokyo, BA, 42; Osaka Univ, MD, 45, DMedSc, 49; Columbia Univ, cert psychoanal med, 59. Prof Exp: Spec res fel, Ministry Educ, Japanese Govt, Med Sch, Osaka Univ, 45-47, asst biochem & neuropsychiat, 47-50; asst resident neuropsychiat, Albany Med Col, 50-51; resident/asst res psychiatrist, Worcester State Hosp, 51-52; res psychiatrist/sr res scientist, Res Facil, Rockland State Hosp, 52-64; res assoc neurol & psychiat, St Vincent's Hosp & Med Ctr, 64-67; asst prof psychiat, New York Med Col, 68-69; asst prof, 70-73, ASSOC PROF PSYCHIAT, ALBERT EINSTEIN COL MED, 73-; CHIEF, COGNITIVE DEVELOP SERV, BRONX PSYCHIAT CTR, 70- Concurrent Pos: Garioa scholar, US State Dept, Albany Med Col & Worcester State Hosp, 50-52; from res asst to res assoc, Col Physicians & Surgeons, Columbia Univ, 52-64; lectr, Med Sch, Osaka Univ, 54; vis scientist, NIH, 56; assoc, New York Sch Psychiat, 62-67; New York City Health Res Coun career scientist award, St Vincent's Hosp & New York Med Col, 64-69; mem acad fac, State Conn Dept Ment Health, 65-67; symp assoc, Ctr Res Math, Morphol & Psychol, 67-73; consult, Asn for Help of Retarded Children, 73- Mem: Am Psychiat Asn; fel Am Acad Psychoanal; Asn Psychoanal Med. Res: Psychopathology and treatment of psychiatric disorders of mental retardation; communicational behavior; cognitive and semiotic psychiatry; phenomenology of adaptational and existential crises; methodology of psychiatric research, especially the issue of temporality. Mailing Add: 1400 Waters Pl New York NY 10461

TAKEUTI, GAISI, b Isikawa, Japan, Jan 25, 26; m 47; c 2. MATHEMATICAL LOGIC. Educ: Univ Tokyo, PhD(math logic), 56. Prof Exp: Instr math, Univ Tokyo, 49-50; from asst prof to prof, Tokyo Univ Educ, 50-66; PROF MATH, UNIV ILL, URBANA, 66- Concurrent Pos: Mem, Inst Advan Study, 59-60 & 66-68. Mem: Am Math Soc; Asn Symbolic Logic; Math Soc Japan; Japan Asn Philos Sci. Res: Proof-theory; set-theory. Mailing Add: Dept of Math 273 Altgeld Hall Univ of Ill Urbana IL 61801

TAKIMOTO, HIDEYO HENRY, organic chemistry, see 12th edition

TAKKEN, EDWARD HAROLD, physics, see 12th edition

TAKMAN, BERTIL HERBERT, b Stockholm, Sweden, Aug 15, 21; m 43; c 5. CHEMISTRY. Educ: Univ Stockholm, PhD(org chem), 63. Prof Exp: INTERNAL CONSULT & HEAD CHEM SECT, ASTRA PHARMACEUT PROD, INC, 58- Concurrent Pos: Guest prof med chem, Northeastern Univ, 73-75. Mem: Am Chem Soc. Res: Medicinal and organic chemistry. Mailing Add: Res Labs Astra Pharmaceut Prod Inc Worcester MA 01606

TAKRURI, HARUN, b Hebron, Palestine, Aug 15, 42; US citizen. PHARMACY. Educ: Am Univ Beirut, BSc, 63; Univ Ill, MS, 66, PhD(pharm), 69. Prof Exp: Asst prof pharm, Col Pharm, Univ Ill Med Ctr, 69-70; SR PHARMACEUT CHEMIST, ELI LILLY & CO, 70- Mem: AAAS; Sigma Xi. Res: Colloid and polydisperse systems; mechanisms of drug release and absorption. Mailing Add: 4414 Fall Creekway N Indianapolis IN 46205

TAKVORIAN, KENNETH BEDROSE, b Philadelphia, Pa, Aug 24, 43; m 66; c 2. POLYMER CHEMISTRY, TEXTILE CHEMISTRY. Educ: Philadelphia Col Textiles & Sci, BS, 65; Clemson Univ, MS, 67, PhD(chem), 69. Prof Exp: RES CHEMIST, TEXTILE RES LAB, E I DU PONT DE NEMOURS & CO, INC, 69- Mem: Am Chem Soc; Am Asn Textile Chemists & Colorists. Res: All areas of textile chemistry; dyeing, finishing, and textile technology; polymer synthesis and morphology; metal chelates and organic synthesis and mechanisms. Mailing Add: Textile Res Lab E I du Pont de Nemours & Co Wilmington DE 19898

TALACKO, JOSEPH VENCESLAS, mathematical statistics, deceased

TALALAY, PAUL, b Berlin, Ger, Mar 31, 23; nat US; m 53; c 4. MOLECULAR PHARMACOLOGY. Educ: Mass Inst Technol, SB, 44; Yale Univ, MD, 48. Hon Degrees: DSc, Acadia Univ, 74. Prof Exp: Intern & asst resident, Surg Serv, Mass Gen Hosp, 48-50; asst prof, Ben May Lab Cancer Res, Univ Chicago, 50-57, from assoc prof to prof, Lab & Dept Biochem, Univ, 57-63; John Jacob Abel prof pharmacol & exp therapeut & dir dept pharmacol, 63-75, JOHN JACOB ABEL DISTINGUISHED SERV PROF PHARMACOL & EXP THERAPEUT, SCH MED, JOHNS HOPKINS UNIV, 75- Concurrent Pos: Am Cancer Soc scholar, 54-58; Charles Hayden Found prof, 58-63; mem pharm B study sect, NIH, 63-67; mem, Nat Adv Cancer Coun, 67-71; ed-in-chief, Molecular Pharm, 68-71; mem bd sci consults, Sloan-Kettering Inst, 71-; mem bd sci adv, Jane Coffin Childs Mem Fund for Med Res, 71-; Guggenheim Mem fel, 73-74. Honors & Awards: Theobold Smith Award, AAAS, 57. Mem: Fel AAAS; fel Am Acad Arts & Sci; Am Soc Biol Chemists; Am Soc Pharmacol & Exp Therapeut; Am Soc Clin Invest. Res: Molecular pharmacology; biochemistry; metabolism and mechanism of action of steroid hormones; amino acids; conformation and chemoteherapeutic activity. Mailing Add: Dept of Pharmacol Johns Hopkins Univ Sch of Med Baltimore MD 21205

TALAMO, BARBARA LISANN, b Washington, DC, May 30, 39; m 58; c 3. NEUROBIOLOGY, BIOCHEMISTRY. Educ: Radcliffe Col, AB, 60; Harvard Univ, PhD(biochem), 72. Prof Exp: Tutor biochem sci, Harvard Med Sch, 71-74; ASST PROF NEUROL & PHYSIOL CHEM, MED SCH, JOHNS HOPKINS UNIV, 74- Concurrent Pos: NSF fel neurobiol, Harvard Med Sch, 72-74. Mem: Soc Neurosci. Res: Neurotransmitter localization, physiology and metabolism; mechanisms of secretion. Mailing Add: Dept of Neurol Johns Hopkins Univ Sch of Med Baltimore MD 21205

TALATY, ERACH R, b Nagpur, India, Oct 20, 26; nat US; m 60. ORGANIC CHEMISTRY, ELECTROCHEMISTRY. Educ: Univ Nagpur, BSc, 48, MSc, 49, PhD(electrochem), 54; Ohio State Univ, PhD(org chem), 57. Prof Exp: Lectr chem, Col Sci, Univ Nagpur, 48-54; asst, Ohio State Univ, 56-57; sr res chemist, Columbia-Southern Chem Corp, 57-61; sr res chemist, Bridesburg Labs, Rohm and Haas Co, Pa, 61; fel, Harvard Univ, 61-62; assoc prof, Univ SDak, 62-64; res assoc, Iowa State Univ, 64-66; asst prof chem, La State Univ, New Orleans, 66-69; RES PROF CHEM, WICHITA STATE UNIV, 69- Mem: Am Chem Soc; The Chem Soc; Indian Chem Soc. Res: Isomerization of azobenzenes; electrodeposition of metals; phosgene chemistry; reactions of chloroformates and carbonates; addition and condensation polymers; acid chlorides; electron spin resonance; steroids; natural products; small-ring compounds; theoretical studies. Mailing Add: Dept of Chem Wichita State Univ Wichita KS 67208

TALBERT, GEORGE BRAYTON, b Ripon, Wis, July 21, 21; m 47; c 2. ENDOCRINOLOGY, ANATOMY. Educ: Univ NDak, BS, 41; Univ Wis, MA, 42, PhD(zool), 50. Prof Exp: From res assoc to assoc prof, 50-68, PROF ANAT, COL MED, STATE UNIV NY DOWNSTATE MED CTR, 68- Concurrent Pos: USPHS spec fel anat, Univ Birmingham, 63-64; USPHS res grant, Nat Inst Child Health & Human Develop, 66-76. Mem: Endocrine Soc; Am Asn Anatomists; Brit Soc Fertil; Geront Soc; Soc Study Reproduction. Res: Pituitary-gonadal relationship; sexual maturation; longevity; aging of reproductive system. Mailing Add: Dept of Anat State Univ of NY Downstate Med Ctr Brooklyn NY 11203

TALBERT, JAMES LEWIS, b Cassville, Mo, Sept 26, 31; m 58; c 2. PEDIATRIC SURGERY, THORACIC SURGERY. Educ: Vanderbilt Univ, BA, 53, MD, 57. Prof Exp: Intern surg, Johns Hopkins Hosp, 56-57, resident, 59-60 & 62-64, resident pediat surg, 64-65, instr surg, Sch Med, Johns Hopkins Univ, 65-66, asst prof surg & pediat surg, 66-67; assoc prof, 67-70, PROF SURG & PEDIAT, COL MED, UNIV FLA, 70- Concurrent Pos: Sr asst surgeon, Nat Heart Inst, 60-62; Garrett scholar pediat surg, Sch Med, Johns Hopkins Univ, 65-66; consult surg, Univ Hosp Jacksonville & Vet Admin Hosp, Gainesville, 72- Mem: Am Surg Asn; Soc Pediat Res; Soc Univ Surgeons; Am Pediat Surg Asn. Res: Congenital anomalies; cancer in childhood; metabolic responses to surgical stress in infants and children. Mailing Add: Dept of Surg Univ of Fla Col of Med Gainesville FL 32601

TALBERT, LUTHER M, b Abingdon, Va, Dec 30, 26; m 49; c 3. OBSTETRICS & GYNECOLOGY. Educ: Emory & Henry Col, BA, 49; Univ Va, MD, 53; Am Bd Obstet & Gynec, dipl. Prof Exp: From instr to assoc prof, 58-69, PROF OBSTET & GYNEC, SCH MED, UNIV NC, CHAPEL HILL, 69- Mem: Endocrine Soc; Soc Gynec Invest; Am Asn Obstet & Gynec; Am Fertil Soc; Am Col Obstet & Gynec. Res: Reproductive endocrinology and infertility. Mailing Add: N Lake Shore Ct Chapel Hill NC 27514

TALBERT, PRESTON TIDBALL, b Washington, DC, Feb 17, 25; m 56. ORGANIC CHEMISTRY. Educ: Howard Univ, BS, 50, MS, 52; Washington Univ, PhD(chem), 55. Prof Exp: Asst, Washington Univ, 51-52; res assoc, Univ Wash, 55-56, res instr, 56-59, NIH fel, 57-59; from asst prof to assoc prof bio-org chem, 59-70, PROF BIO-ORG CHEM, HOWARD UNIV, 70- Mem: Fel AAAS; fel Am Inst Chemists; Am Chem Soc; NY Acad Sci. Res: Mechanism of function, synthesis and degradation of biologically important compounds, especially nucleic, nucleosides, proteins and vitamins; enzyme function and mechanism of action. Mailing Add: Dept of Chem Howard Univ Washington DC 20059

TALBERT, RONALD EDWARD, b Toulon, Ill, May 20, 36; m 55; c 3. AGRONOMY, WEED SCIENCE. Educ: Univ Mo, BS, 58, MS, 60, PhD(field crops), 63. Prof Exp: Instr field crops, Univ Mo, 60-63; from asst prof to assoc prof agron, 63-73, PROF AGRON, UNIV ARK, FAYETTEVILLE, 73- Mem: AAAS; Weed Sci Soc Am; Am Soc Agron. Res: Use of herbicides in crops; behavior of herbicides in soils; physiological selectivity and action of herbicides. Mailing Add: Weed Sci & Physiol Lab Univ of Ark Rte 6 Box 83 Fayetteville AR 72701

TALBERT, WILLARD LINDLEY, JR, b Casper, Wyo, Mar 8, 32; m 52; c 4. NUCLEAR PHYSICS. Educ: Univ Colo, BA, 54; Iowa State Univ, PhD(physics), 60. Prof Exp: Res physicist, Ohio Oil Co, 59-62; from asst prof to assoc prof physics, 62-72, PROF PHYSICS, IOWA STATE UNIV, 72- Concurrent Pos: Prog dir nuclear sci, Ames Lab, US Energy Res & Develop Admin, 74- Mem: Fel Am Phys Soc. Res: Experimental nuclear spectroscopy, especially shortlived isotopes using on-line isotope separator. Mailing Add: Dept of Physics Iowa State Univ Ames IA 50010

TALBOT, ASHLEY FREDERICK, b New York, NY, Dec 8, 22; m 49; c 1. GEOGRAPHY. Educ: Univ Mich, BA, 48. Prof Exp: Asst ed, C S Hammond & Co, 49-59, sr ed, 59-66, EXEC ED, HAMMOND INC, MAPLEWOOD, NJ, 66- Res: Political, cultural, population, historical and regional geography; anthropology, history and demography, as related to political and historical ways. Mailing Add: 130 Prospect St East Orange NJ 07017

TALBOT, BERNARD, b Ville Montmorency, Que, Mar 15, 28. REHABILITATION MEDICINE. Educ: Laval Univ, BA, 49, MD, 58; FRCPS(C), 62. Prof Exp: Physiatrist, Rehab Inst Montreal, 62-68; PROF REHAB MED & CHMN SUB-DEPT REHAB MED, UNIV OTTAWA, 71-; PHYSIATRIST-IN-CHIEF, DEPT REHAB MED, ROYAL OTTAWA HOSP, 71- Concurrent Pos: Dir rehab progs, Hamilton Health Asn, 68 & Sherbrooke Regional Rehab Progs, 70; consult, Ottawa Gen Hosp, Perley Hosp & Nat Defence Med Ctr, 71 & Ottawa Civic Hosp, 72. Mem: Can Asn Phys Med & Rehab; Am Cong Rehab Med. Res: Regional rehabilitation program. Mailing Add: Dept of Rehab Med Royal Ottawa Hosp Ottawa ON Can

TALBOT, GERALD BYRON, b Wetaskiwin, Alta, Dec 12, 12; US citizen; m 46; c 2. FISH BIOLOGY. Educ: Univ Wash, BS, 44, MS, 48. Prof Exp: Fishery biologist sockeye salmon invests, Int Pac Salmon Fisheries Comn, BC, Can, 41-50; fishery biologist sockeye salmon invests, US Bur Com Fisheries, NC, 50-62, dir biol lab, 52-62, dir, Tiburon Marine Lab, 62-70, Washington, DC, 70-72, chief, Southeastern Reservoir Invests, 72-75, CONSULT, SOUTHEASTERN RESERVOIR INVESTS, BUR SPORT FISHERIES & WILDLIFE, 75- Honors & Awards: Award, Bur Com Fisheries, Dept Interior, 60; Spec Achievement Award, Bur Sport Fisheries & Wildlife, 70. Mem: AAAS; Am Fisheries Soc; Inst Fisheries Res Biol. Res: Anadromous, marine and freshwater fishes. Mailing Add: SE Resvr Invests Bur Sport Fisheries & Wildlife PO Box 429 Clemson SC 29631

TALBOT, JAMES LAWRENCE, b Epsom, Eng, Sept 6, 32; m 57; c 3. GEOLOGY. Educ: Cambridge Univ, BA, 54; Univ Calif, Berkeley, MA, 57; Univ Adelaide, PhD(geol), 63. Prof Exp: Lectr struct geol, Univ Adelaide, 58-63, sr lectr, 63-67; assoc prof geol, Lakehead Univ, 67-70; prof geol & chmn dept, 70-75, ACTG ACAD VPRES, UNIV MONT, 75- Concurrent Pos: Alexander von Humboldt Found fel, Univ Bonn, 63-64. Mem: Mineral Soc Am; Am Geophys Union; Australian Geol Soc; Geol Soc Am. Res: Structural analysis of basementcover complexes; analysis of strain in metamorphic rocks; studies on rock cleavage and mylonites. Mailing Add: Univ of Mont Missoula MT 59801

TALBOT, JOHN MAYO, b Sebastopol, Calif, May 8, 13; m 46; c 3. AEROSPACE MEDICINE, RADIOBIOLOGY. Educ: Univ Ore, AB, 35, MD, 38. Prof Exp: Med officer, US Army & US Air Force, 39-73; dir sci info, George Washington Univ Med Ctr, 73-74; CONSULT BIOMED RES, LIFE SCI RES OFF, FEDN AM SOCS EXP BIOL, 74- Concurrent Pos: Consult aerospace med, NASA, 73-; med consult, Environ Protection Agency, 73-75. Honors & Awards: Theodore C Lyster Award, Aerospace Med Asn, 67. Mem: Aerospace Med Asn; Int Acad Astronaut; Am Col Prev Med; Pan Am Med Asn. Res: Aerospace medicine and biology, radiobiology, toxicology of food chemicals; fatigue, crew performance, radiation protection, and life support systems in aerospace medicine. Mailing Add: Lands End Annapolis Rds Annapolis MD 21403

TALBOT, LEE MERRIAM, b New Beford, Mass, Aug 2, 30; m 59; c 1. ECOLOGY. Educ: Univ Calif, AB, 53, MA & PhD(geog range & vert ecol), 63. Prof Exp: Field biologist, Arctic Res Lab, Alaska, 51; staff ecologist, UNESCO-Int Union Conserv, Belg, 54-56; ecologist & dir EAfrican wildlife & wild land res proj, Nat Acad Sci-Nat Res Coun, Rockefeller Found, NY Zool Soc & Govt Kenya, 59-63; dir SE Asia proj, UNESCO-Int Union Conserv, 64-65; field rep int affairs ecol & conserv & res ecologist, Smithsonian Inst, 66-70, sci coordr conserv sect, Int Biol Prog, 66-70; sr scientist, President's Coun on Environ Qual, 70-75, ASST TO CHMN INT & SCI

AFFAIRS, PRESIDENT'S COUN ENVIRON QUAL, 75- Concurrent Pos: UNESCO lectr, Southeast Asia, 55, consult, 65-; leader, African wildlife exp, Wildlife Mgt Inst & Am Comt Int Wildlife Protection, 56; Taussig traveling fel, Univ Calif, 58-59; Pop Ref Bur consult, Govt Kenya & Tanganyika, 59-63, Hong Kong, 63, Philippines, 64, Indonesia & Thailand, 64-65 & Malaya, Sabah & Sarawak, 65; wildlife adv, UN Spec Fund & EAfrican Agr & Forestry Res Orgn, 63-64; assoc ecol, US Nat Zool Park, 66-; spec adv, Mus Natural Hist, Smithsonian Inst, 67-, res assoc, 74-; overseas consult, Fauna Preservation Soc, 67-; consult, UN Spec Fund, 63, Pac Sci Bd, Nat Acad Sci-Nat Res Coun, 64-65, Peace Corps, 66, Int Comn Nat Parks, 66-, Nat Park Serv, 68 & AID, 69; chmn, Am Comt Int Conserv, 74-; vpres, Int Union Conserv Nature & Natural Resources, 75- Honors & Awards: Wildlife Soc Award, 63; Albert Schweitzer Medal, Animal Welfare Inst, 75. Mem: Fel AAAS; Am Soc Mammal; Soc Range Mgt; Asn Am Geog; Wildlife Soc. Res: International conservation; wildlife, especially ecology and management; tropical land use and savannah ecology; conservation of renewable natural resources; methodology of ecological research and survey; environmental impact analysis; endangered species. Mailing Add: 6656 Chilton Ct McLean VA 22101

TALBOT, MARY, b Columbus, Ohio, Nov 30, 03. ECOLOGY. Educ: Denison Univ, BS, 25; Ohio State Univ, MA, 27; Univ Chicago, PhD(ecol), 34. Prof Exp: Asst zool, Ohio State Univ, 25-27; asst prof biol, Univ Omaha, 27-28; instr, Stephens Col, 28-30; asst, Univ Chicago, 31-34; instr, Mundelein Col, 35-36; prof & head dept, Lindenwood Col, 36-68, EMER PROF BIOL, THE LINDENWOOD COLS, 68- Res: Distribution, ecology, flights and populations of ants. Mailing Add: The Lindenwood Cols St Charles MO 63301

TALBOT, NATHAN BILL, b Boston, Mass, Nov 25, 09; m 34; c 2. PEDIATRICS. Educ: Harvard Univ, AB, 32, MD, 36. Prof Exp: Intern, Children's Hosp, 36-38; from asst to assoc prof, 39-62, CHARLES WILDER PROF PEDIAT & HEAD DEPT, HARVARD MED SCH, 62-; CHIEF, CHILDREN'S SERV, MASS GEN HOSP, 62- Concurrent Pos: Asst resident, Children's Hosp, 39-40, asst physician, 41-42; res fel pediat, Harvard Med Sch, 40-41; from asst physician to physician, Mass Gen Hosp, 42-62; consult, Children's Hosp Med Ctr, Boston Lying-In Hosp & Cambridge City Hosp. Honors & Awards: Mead Johnson Award & Borden Award, Am Acad Pediat. Mem: Fel Am Acad Arts & Sci; fel Am Soc Clin Invest; fel Am Pediat Soc; fel Soc Pediat Res; fel Endocrine Soc. Res: Interplay of physical, biologic, social, psychologic and behavioral factors on human development, health and disease. Mailing Add: Dept of Pediat Harvard Med Sch Boston MA 02115

TALBOT, PRUDENCE, b Mass, June 9, 44; m 68. REPRODUCTIVE PHYSIOLOGY, CELL BIOLOGY. Educ: Wilson Col, BA, 66; Wellesley Col, MA, 68; Univ Houston, PhD(cell biol), 72. Prof Exp: RES ASSOC MAMMALIAN FERTIL, UNIV HOUSTON, 72- Mem: Am Soc Cell Biol; Am Soc Zoologists; Soc Study Reprod; Sigma Xi. Res: Morphology, physiology and biochemistry of mammalian fertilization and the mechanism of mammalian ovulation. Mailing Add: Dept of Biol Univ of Houston Houston TX 77004

TALBOT, RAYMOND JAMES, JR, b Portsmouth, Va, Sept 17, 41; m 68. THEORETICAL ASTROPHYSICS. Educ: Mass Inst Technol, SB, 63, PhD(physics), 69. Prof Exp: Res assoc, 69-71, ASST PROF SPACE SCI, RICE UNIV, 71- Mem: Am Astron Soc; Sigma Xi. Res: Evolution of stars and galaxies; nucleosynthesis; pulsations of stars. Mailing Add: Dept of Space Physics & Astron Rice Univ Houston TX 77001

TALBOT, RICHARD B, physiology, see 12th edition

TALBOT, TIMOTHY RALPH, JR, b Berkeley, Calif, July 14, 16; m 43; c 5. RESEARCH ADMINISTRATION. Educ: Univ Pa, AB, 37, MD, 41. Prof Exp: Instr med, Sch Med, Boston Univ, 46-48; instr, Med Col, Cornell Univ, 48-51; assoc, 51-54, asst prof, 54-66, ASSOC PROF MED, SCH MED, UNIV PA, 66-; DIR, INST CANCER RES, 57-, PRES, 72-; PRES, FOX CHASE CANCER CTR, 74- Concurrent Pos: Asst, Sloan-Kettering Inst Cancer Res & asst attend physician, Mem Hosp, New York, 48-51; ward physician & mem hemat sect, Hosp Univ Pa, 51-; Nat Cancer Inst fel, Chester Beatty Res Inst, London, Eng, 56-57; dir, Fox Chase Ctr Cancer & Med Sci, 71-74. Mem: AAAS; Asn Am Cancer Insts; Am Fedn Clin Res; Am Asn Cancer Res; Am Clin & Climat Asn. Res: Hematology and cancer. Mailing Add: Fox Chase Cancer Ctr 7701 Burholme Ave Philadelphia PA 19111

TALBOT, WALTER RICHARD, b Pittsburgh, Pa, Dec 9, 09; m 36; c 2. MATHEMATICS. Educ: Univ Pittsburgh, AB, 31, MA, 33, PhD(math), 34. Prof Exp: From asst prof to prof math, Lincoln Univ, Mo, 34-63, head dept, 40-63, dean men, 39-44, registr, 46-48, actg dean instr, 55-57; PROF MATH & CHMN DEPT, MORGAN STATE COL, 63- Concurrent Pos: Consult curriculum resources group, Inst Serv to Educ, Mass, 64- Mem: Am Math Soc; Math Asn Am. Res: Mathematical and numerical analysis; computer science. Mailing Add: Dept of Math Morgan State Col Baltimore MD 21212

TALBOT, WILLIAM HENRY, b Holliston, Mass, Feb 10, 35; div. NEUROPHYSIOLOGY. Educ: Brown Univ, AB, 57; Rockefeller Univ, PhD(physiol), 64. Prof Exp: Fel, 64-66, from instr to asst prof, 66-73, ASSOC PROF PHYSIOL, SCH MED, JOHNS HOPKINS UNIV, 73- Mem: Am Physiol Soc; Sigma Xi. Res: Physiology of the central nervous system; use of digital computers in neurophysiological research. Mailing Add: Dept of Physiol Johns Hopkins Univ Sch of Med Baltimore MD 21205

TALBOTT, FRANCIS LEO, b Lancaster, Ohio, July 17, 03; m 63. NUCLEAR PHYSICS. Educ: St John's Univ, Ohio, AB, 24; Cath Univ, MA, 26, PhD(physics), 28. Prof Exp: From instr to prof physics, 26-69, EMER PROF PHYSICS, CATH UNIV AM, 69-; PROF PHYSICS, CHMN DEPT & RES COORDR, ALLENTOWN COL, 70- Concurrent Pos: Physicist, Johns Hopkins Univ, 43-46, proj supvr, 45-46; partic, Oak Ridge Nat Lab, 47-; coun mem, Oak Ridge Assoc Univs, 64-69. Mem: Am Phys Soc; Optical Soc Am; Am Asn Physics Teachers. Res: Proximity fuzes; photo neutrons; reactions of light nuclei; reactors; fundamental particles. Mailing Add: Dept of Physics Allentown Col Center Valley PA 18034

TALBOTT, RICHARD EVANS, b Oakland, Calif, May 11, 39; m 62; c 2. NEUROPHYSIOLOGY. Educ: Univ Calif, Berkeley, AB, 61; Univ Wash, PhD(physiol), 66. Prof Exp: Res physiologist, US Naval Radiol Defense Lab, Calif, 66-69; res assoc, 69-73, ASST PROF NEUROPHYSIOL, MED SCH, UNIV ORE, 73- Prof Exp: USPHS res career develop award, Med Sch, Univ Ore, 72- Mem: Am Physiol Soc; Int Brain Res Orgn; Soc Neurosci. Res: Neurophysiology of postural control; mode of central nervous system processing of information pertinent for behavior. Mailing Add: Dept of Physiol Univ of Ore Med Sch Portland OR 97201

TALBOTT, RICHARD LLOYD, b Chicago, Ill, July 15, 35; m 58; c 3. ORGANIC CHEMISTRY. Educ: DePauw Univ, BA, 57; Univ Ill, PhD(org chem), 60. Prof Exp: NSF fel org chem, Mass Inst Technol, 60-61; sr chemist, Cent Res Dept, 61-66, sr chemist, Indust Tape Div, 66-68, supvr adhesives res, 68-74, RES MGR, PACKAGING SYSTS DIV, 3M CO, 74- Mem: Am Chem Soc. Res: Pressure-sensitive adhesives chemistry; chemistry of fluorinated oxidants and fluorinated peroxides. Mailing Add: 3M Co 230-IS 3M Center St Paul MN 55101

TALBOTT, TED DELWYN, b Sudan, Tex, Oct 18, 29; m 55; c 2. ANALYTICAL CHEMISTRY. Educ: NTex State Col, BS, 51, MS, 55. Prof Exp: Indust chemist, E I du Pont de Nemours & Co, 55-57; anal chemist, Res Ctr, US Rubber Co, 57-64; SR RES CHEMIST, ANAL RES SECT, CHEMAGRO CORP, MOBAY CHEM CORP, 64- Mem: Am Chem Soc. Res: Methods development for various pesticides which include organic phosphorus compounds, carbonates and nitrogen hetercyclics. Mailing Add: Anal Res Sect Chemagro Corp Box 4913 Hawthorne Rd Kansas City MO 64120

TALBURT, WILLIAM FREDERICK, organic chemistry, see 12th edition

TALESNIK, JAIME, b Santiago, Chile, May 18, 16; m 41; c 2. PHARMACOLOGY, PHYSIOLOGY. Educ: Univ Chile, MD, 41. Prof Exp: Second chief instr physiol, Sch Med, Univ Chile, 41-49, asst prof, 50-58, assoc prof physiopath, 58-61, interim prof physiopath & dir dept exp med, 61-63; vis prof, 67-69, PROF PHARMACOL, FAC MED, UNIV TORONTO, 69- Concurrent Pos: Rockefeller Found grant, Banting & Best Dept Med Res, Univ Toronto, 46-47; Brit Coun grant, Nat Inst Med Res, London, Eng, 52. Mem: Chilean Biol Soc; Brit Physiol Soc; Am Soc Pharmacol & Exp Therapeut; Pharmacol Soc Can; NY Acad Sci. Res: Physiopharmacology of the coronary circulation; central modulation of cardiovascular reflexes. Mailing Add: Dept of Pharmacol Univ of Toronto Fac of Med Toronto ON Can

TALHAM, ROBERT J, b Cohoes, NY, May 27, 29; m 56; c 4. APPLIED MATHEMATICS, ACOUSTICS. Educ: State Univ NY Albany, BA, 55, MS, 56; Rensselaer Polytech Inst, PhD(appl math), 60. Prof Exp: Nat Acad Sci-Nat Res Coun res fel, Naval Res Lab, DC, 60-61; mem tech staff, Bell Tel Labs, NJ, 61-64; MGR UNDERSEA DEFENSE SYSTS ENG, HEAVY MIL ELECTRONICS DEPT, GEN ELEC CO, 64- Mem: Am Math Soc; Soc Indust & Appl Math; Acoust Soc Am. Res: Sonar systems; underwater acoustics; sound propagation in non-homogeneous medium; acoustic array design and development; signal processing. Mailing Add: General Elec Co HMES FRPI-A5 Farrell Rd Plant Syracuse NY 13201

TALIAFERRO, CHARLES M, b Leon, Okla, Mar 1, 40; m 60; c 2. PLANT BREEDING, PLANT GENETICS. Educ: Okla State Univ, BS, 62; Tex A&M Univ, MS, 65, PhD(plant breeding & genetics), 66. Prof Exp: Res agronomist, Agr Res Serv, USDA, 65-68; asst prof, 68-72, ASSOC PROF FORAGE BREEDING & GENETICS, OKLA STATE UNIV, 72- Mem: Am Soc Agron; Am Genetic Asn. Res: Basic genetic and breeding studies involving forage crops. Mailing Add: Dept of Agron Okla State Univ Stillwater OK 74075

TALIAFERRO, WILLIAM HAY, microbiology, immunology, deceased

TALLAN, HARRIS H, b New York, NY, July 9, 24; c 3. BIOCHEMISTRY. Educ: NY Univ, BA, 47; Yale Univ, PhD(biochem), 50. Prof Exp: Asst, Rockefeller Inst, 50-53, asst prof biochem, 53-59; biochemist, Res Labs, Geigy Chem Corp, 59-68; ASSOC RES SCIENTIST, DEPT PEDIAT RES, NY STATE INST RES IN MENT RETARDATION, 68- Concurrent Pos: NIH spec fel, Oxford Univ, 57-58. Mem: Am Soc Biol Chemists. Res: Enzymology; amino acid metabolism; inborn errors of metabolism; neurochemistry. Mailing Add: Dept of Pediat Res NY State Inst Res in Ment Retardation Staten Island NY 10314

TALLAN, IRWIN, b New York, NY, June 26, 27; m 59. GENETICS. Educ: Rutgers Univ, BA, 49; Ind Univ, PhD(genetics), 57. Prof Exp: Technician to H J Muller, Ind Univ, 50, asst to T M Sonneborn, 50-53; lectr zool, 56-58, from asst prof to assoc prof, 58-74, PROF ZOOL, UNIV TORONTO, 74- Mem: Am Soc Zool; Brit Soc Gen Microbiol. Res: Genetics; nucleo-cytoplasmic interactions in protozoans; infectivity of kappa and other plasmids. Mailing Add: Dept of Zool Univ of Toronto Toronto ON Can

TALLARIDA, RONALD JOSEPH, b Philadelphia, Pa, May 26, 37; m 58; c 3. BIOMATHEMATICS, PHARMACOLOGY. Educ: Drexel Inst, BS, 59, MS, 63; Temple Univ, PhD(pharmacol), 67. Prof Exp: Coop student, Philco Corp-Drexel Inst, 55-59; jr engr, Philco Corp, 59-60; from instr to asst prof math, Drexel Inst, 60-67; asst prof, 67-71, ASSOC PROF PHARMACOL, TEMPLE UNIV, 71- Concurrent Pos: Lectr, Philadelphia Col Pharm, 60, PMC Col, 61-62 & cardiovascular training grant prog, Med Sch, Temple Univ, 63-64; consult, Drexel Inst, 67- Mem: AAAS; Math Asn Am; Am Soc Pharmacol & Exp Therapeut. Res: Mathematical models for application to biology and medicine; drug receptor theory; pharmacology of vascular smooth muscle; pharmacology of morphine; drug induced disease. Mailing Add: Dept of Pharmacol Temple Univ Sch of Med Philadelphia PA 19122

TALLEDO, OSCAR EDUARDO, b Sullana, Peru, Aug 1, 29; US citizen; c 3. OBSTETRICS & GYNECOLOGY. Educ: San Marcos Univ, Lima, BS, 48, MD, 55; Am Bd Obstet & Gynec, spec cert div gynec oncol, 75. Prof Exp: Intern, San Marcos Univ, Lima, 54-55; intern, Crawford W Long Hosp, Emory Univ, 56-57, resident obstet & gynec, 57-58; resident, 58-60, fel, 60-61, from instr to assoc prof, 61-71, PROF OBSTET & GYNEC, MED COL GA, 71-, CHIEF GYNEC SERV, 74- Concurrent Pos: Nat Heart Inst grant obstet & gynec, Med Col Ga, 65, NIH grant, 68, dir obstet/gynec residency training prog, 74-; consult, Cent State Hosp, Macon City Hosp, Greenville Mem Hosp & Mem Med Ctr. Mem: Fel Am Col Obstet & Gynec; Soc Gynec Invest; AMA; Am Fertil Soc. Res: Physiology of pregnancy; vascular reactivity in pregnancy; fetal electrocardiography; uterine contractility studies; amniotic fluid. Mailing Add: Dept of Obstet & Gynec Med Col of Ga Augusta GA 30902

TALLENT, WILLIAM HUGH, b Akron, Ohio, May 28, 28; m 52; c 3. ORGANIC CHEMISTRY, BIOCHEMISTRY. Educ: Univ Tenn, BS, 49, MS, 50; Univ Ill, PhD(biochem), 53. Prof Exp: Asst, Univ Tenn, 49-50 & Univ Ill, 50-53; asst scientist, Nat Heart Inst, 53-57; res chemist, G D Searle & Co, 57-64; invests leader, Northern Regional Res Lab, 64-69, chief indust crops res, 69-75, DIR NORTHERN REGIONAL RES CTR, AGR RES SERV, USDA, 75- Concurrent Pos: Assoc ed, J Am Oil Chemists Soc, 70- Mem: Am Chem Soc; Am Oil Chemists Soc; Soc Econ Bot. Res: Application of chromatographic and spectroscopic methods to analysis, isolation and structure determination of terpenes, plant lipids, natural insecticides; plant enzymes; useful derivatives and synthetic modifications of natural products; research management. Mailing Add: Northern Regional Res Ctr Agr Res Serv USDA Peoria IL 61604

TALLER, ROBERT ARTHUR, b Cleveland, Ohio, Nov 15, 39; m 66; c 3. PHYSICAL ORGANIC CHEMISTRY. Educ: Bowling Green State Univ, BS, 61; Mich State Univ, MS, 63, PhD(phys org chem), 66. Prof Exp: Chemist, Union Carbide Corp, Tarrytown, 66-74, PROJ SCIENTIST, UNION CARBIDE, WVA, 74- Mem: AAAS; Am Chem Soc. Res: Conformational analysis; structure reactivity correlations; solubility parameters; structural studies by nuclear magnetic resonance; structure property relationships in polyurethane coatings. Mailing Add: Dept of Polyurethane Chem Union Carbide Corp PO Box 8361 South Charleston WV 25303

TALLEY, CHARLES PETER, b New York, NY, Aug 15, 41; m 68; c 3. ANALYTICAL CHEMISTRY. Educ: St Peter's Col, NJ, BS, 63; Polytech Inst New York, PhD(phys & org chem), 74. Prof Exp: Sr res chemist, Merck Sharp & Dohme Res Labs, Div Merck & Co, 68-72; res chemist, Anal Chem Div, Nat Bur Standards, 73; SR GROUP LEADER ANAL RES, CALGON CORP, SUBSID MERCK & CO, INC, 73- Mem: Am Chem Soc; AAAS. Res: Gas and high pressure liquid chromatography, especially as applied to the analysis of trace organics in biological, environmental and polymeric matrices. Mailing Add: Calgon Corp PO Box 1346 Pittsburgh PA 15230

TALLEY, CLAUDE PARKS, b Richmond, Va, June 13, 31; m 50; c 3. PHYSICAL CHEMISTRY. Educ: Univ Va, BS, 52; Univ Richmond, MS, 60. Prof Exp: Chemist, E I du Pont de Nemours & Co, 52-55; assoc scientist chem res, Texaco Exp, Inc, 55-56, sr scientist, 56-62, group leader, 62-64, proj supvr, 64-66, mkt supvr, 66-68; res assoc bus planning, Fibers Div, Allied Chem Corp, 69-71, dir indust prod, 71-73; INDUST CONSULT, 73- Concurrent Pos: Abstractor comt fire res, Nat Acad Sci-Nat Res Coun, 58-63. Mem: AAAS; Am Chem Soc. Res: High temperature chemistry; combustion kinetics; thermodynamics; chemical vapor plating; floating zone refining; preparation and properties of elemental boron; single crystal boron; boron filaments; composite structural materials; textile fibers; automotive air cushion systems; pollution control. Mailing Add: 3442 Northview Pl Richmond VA 23225

TALLEY, EUGENE ALTON, b Glenn Allen, Va, June 5, 11; m 44; c 2. AGRICULTURAL CHEMISTRY. Educ: Col William & Mary, BCh, 36; Univ Richmond, MS, 38; Ohio State Univ, PhD(org chem), 42. Prof Exp: Asst chemist carbohydrate div, Eastern Regional Res Lab, Bur Agr & Indust Chem, 42-44, chemist, 44-53, SR CHEMIST, PLANT PROD LAB, EASTERN REGIONAL RES LAB, AGR RES SERV, USDA, 53- Mem: Am Chem Soc; Am Potato Asn. Res: Synthesis of oligosaccharides; preparation of starch esters and ethers; nitrogen compounds in plants; glycoalkaloids. Mailing Add: Eastern Regional Res Lab USDA Agr Res Serv 600 E Mermaid Lane Philadelphia PA 19118

TALLEY, LAWERENCE HORACE, b Cincinnati, Ohio, Aug 16, 27. ACADEMIC ADMINISTRATION, SCIENCE EDUCATION. Educ: Ohio Univ, BS, 52, MS, 56; WVa Univ, EdD(sci educ), 72, CAS(higher educ admin), 74. Prof Exp: Res chemist, Mound Lab, 48-52; res scientist, Nat Lead Co, 56-58; sr res scientist, Parker Co, 58-60, dept dir, 60-62, bus mgr, 62-68; prof chem, 68-72, DIR SCH NATURAL SCI, WEST LIBERTY STATE COL, 72- Concurrent Pos: Assoc prof, Milton Col, 58-62. Mem: AAAS. Res: Educational psychology; general physical chemistry molecular structure; behavioral modification—science curriculum and methodology of teaching-learning. Mailing Add: 40 Heiskell Ave Wheeling WV 26003

TALLEY, ROBERT LEE, physics, engineering, see 12th edition

TALLEY, ROBERT MORRELL, b Erwin, Tenn, Mar 13, 24; m 48; c 2. PHYSICS. Educ: Univ SC, BS, 45; Univ Tenn, MS, 48, PhD(physics), 50. Prof Exp: Chief infrared br, US Naval Ord Lab, 51-57, chief solid state div, 57-58; VPRES & MGR LABS, SANTA BARBARA RES CTR, 58- Mem: Am Phys Soc; Optical Soc Am. Res: Infrared spectroscopy; intermetallic semiconductors; energy bands in solids; photodetectors; military infrared systems. Mailing Add: 417 Foxen Dr Santa Barbara CA 93105

TALLEY, SPURGEON MORRIS, b Atkins, Ark, May 6, 18; m 48; c 1. ANIMAL NUTRITION. Educ: Agr, Mech & Norm Col, Ark, BSA, 47; Kas State Univ, MS, 53, PhD(nutrit), 66. Prof Exp: Asst prof poultry sci & prod mgr, 54-66, ASSOC PROF ANIMAL NUTRIT, LINCOLN UNIV, MO, 66- Mem: Poultry Sci Asn; Am Soc Animal Sci. Res: Monogastric animals; nutrition of poultry and swine; plant proteins as sources of protein for the avian species; level of dietary protein and phase feeding on esophagoulcerogenesis of market swine; metabolizable energy requirements of market-type swine. Mailing Add: Dept of Agr Lincoln Univ Jefferson City MO 65101

TALLEY, THURMAN LAMAR, b Portales, NMex, July 26, 37; m 62; c 2. PHYSICS. Educ: Eastern NMex Univ, BS, 59, MS, 60; Fla State Univ, PhD(physics), 68. Prof Exp: Instr eng sci, Fla State Univ, 64-65; MEM STAFF, LOS ALAMOS SCI LAB, UNIV CALIF, 66- Concurrent Pos: Chmn, Joint AEC-Dept Defense Working Group Safeguard Sprint Nuclear Vulnerability & Effects, 68-71. Mem: Am Phys Soc. Res: Nuclear reaction theory; nuclear weapons design; nuclear weapons effects; computer simulation of complex physical phenomena. Mailing Add: Los Alamos Sci Lab PO Box 1663 Los Alamos NM 87544

TALLITSCH, ROBERT BOYDE, b Oak Park, Ill, June 3, 50; m 71. PHYSIOLOGY. Educ: NCent Col, BA, 71; Univ Wis-Madison, MS, 72, PhD(physiol), 75. Prof Exp: Res fel, Wis Heart Asn, 74-75; ASST PROF BIOL, AUGUSTANA COL, 75- Mem: Assoc Am Physiol Soc. Res: Cellular action of d-aldosterone upon sodium transport in single muscle cells. Mailing Add: Dept of Biol Augustana Col Rock Island IL 61201

TALLMAN, DENNIS EARL, b Bellefontaine, Ohio, Apr 23, 42; m 63; c 2. ANALYTICAL CHEMISTRY, BIOPHYSICAL CHEMISTRY. Educ: Ohio State Univ, BSc, 64, PhD(anal chem), 68. Prof Exp: NIH fel chem, Cornell Univ, 68-70; asst prof anal chem, 70-73, ASSOC PROF ANAL CHEM, NDAK STATE UNIV, 73- Concurrent Pos: Res Corp grant, NDak State Univ, 71-, Off Water Resources res grants, 71-74 & 75-77; NIH res grant, 74-76. Mem: AAAS; Am Chem Soc. Res: Stopped flow-temperature jump relaxation spectrometry; trace metal analysis; mechanisms of metal ion-activated enzymes; mechanisms of complex formation; analytical biochemistry; intermolecular and intramolecular hydrogen bonding. Mailing Add: Dept of Chem NDak State Univ Fargo ND 58102

TALLMAN, RALPH COLTON, b Cedar Rapids, Iowa, Oct 6, 05; m 29; c 1. ORGANIC CHEMISTRY. Educ: Cornell Col, Iowa, AB, 27; Cornell Univ, PhD(org chem), 31. Prof Exp: Instr org chem, Cornell Univ, 30-35; dir res, Schieffelin & Co, 35-43; proj leader cent res lab, Allied Chem & Dye Corp, 43-50; mgr planning & surv dept, Lion Oil Co Div, Monsanto Co, 50-51, dir res, 51-61, dir hydrocarbons div, 61-65, assoc dir res, Hydrocarbons & Polymers Div, 65-70; CONSULT, ENG DYNAMICS INT, 72- Mem: Am Chem Soc. Res: Drugs and pharmaceuticals; organic and industrial chemicals; administrative research. Mailing Add: Eng Dynamics Int 8420 Delmar Blvd St Louis MO 63124

TALLMAN, RICHARD LOUIS, b Wheeling, WVa, Apr 24, 31; m 56; c 3. CHEMISTRY, CORROSION. Educ: Kenyon Col, AB, 53; Univ Wis, PhD(phys chem), 60. Prof Exp: Sr chemist, Res Labs, Westinghouse Elec Corp, 59-73; SR RES CHEMIST, GEN MOTORS RES LABS, 73- Mem: Am Chem Soc; Nat Asn Corrosion Engrs. Res: High temperature metal combustion; x-ray diffraction; aluminum corrosion; microscopic, grammetryelectrochemical studies. Mailing Add: Dept of Phys Chem GM Res Labs Gen Motors Tech Ctr Warren MI 48090

TALMAGE, DAVID WILSON, b Kwangju, Korea, Sept 15, 19; US citizen; m 44; c 5. MEDICINE. Educ: Davidson Col, BS, 41; Washington Univ, MD, 44. Prof Exp: USPHS res fel, Washington Univ, 50-51; asst res prof path, Sch Med, Univ Pittsburgh, 51-52; from asst prof to assoc prof med, Sch Med, Univ Chicago, 52-59; from assoc dean to dean fac, 66-71, PROF MED & MICROBIOL, SCH MED, UNIV COLO, DENVER, 59- Concurrent Pos: Markle scholar med sci, 55-60; consult, Vet Admin Hosp, 59-; ed, J Am Acad Allergy, 63-67; dir, Webb-Waring Lung Inst, 73- Mem: Nat Acad Sci; AAAS; Am Soc Clin Invest; Am Asn Immunologists; Am Acad Allergy (pres, 65). Res: Immunology; allergy. Mailing Add: Dept of Microbiol Univ of Colo Med Ctr Denver CO 80220

TALMAGE, ROY VAN NESTE, b Moppo, Korea, Feb 9, 17; US citizen; m 42; c 3. PHYSIOLOGY. Educ: Maryville Col, AB, 38; Univ Richmond, MA, 40; Harvard Univ, PhD(endocrinol), 47. Prof Exp: Instr biol, Univ Richmond, 40-41; asst, Harvard Univ, 41-42 & 46-47; from instr to prof, Rice Univ, 47-70, chmn dept, 56-64, master, Wiess Col, 57-70; DIR ORTHOP RES & PROF SURG & PHARMACOL, SCH MED, UNIV NC, CHAPEL HILL, 70- Concurrent Pos: NIH res fel, State Univ Leiden, 64; mem, NIH Study Sects, 64-68 & 70-; gen chmn & co-chmn parathyroid confs, Houston, 60, Leiden, 64, Montreal, 67, Chapel Hill, 71 & Oxford, Eng, 74; staff biochemist, AEC, 69-70, mem nat adv dent res coun, Nat Inst Dent Res, 74-77. Mem: AAAS; Orthop Res Soc; Am Soc Zoologists; Soc Exp Biol & Med; Am Physiol Soc; Endocrine Soc. Res: Calcium regulating hormones and ion transport processes in bone. Mailing Add: Orthop Res Labs Univ of NC Chapel Hill NC 27514

TALMAN, JAMES DAVIS, b Toronto, Ont, July 24, 31; m 57; c 4. THEORETICAL PHYSICS. Educ: Univ Western Ont, BA, 53, MSc, 54; Princeton Univ, PhD, 59. Prof Exp: Instr physics, Princeton Univ, 57-59; asst prof, Am Univ Beirut, 59-60; from asst prof to prof math, 60-67, PROF APPL MATH, UNIV WESTERN ONT, 67- Concurrent Pos: Res asst, Univ Calif, Davis, 63-64. Mem: Am Phys Soc; Can Asn Physicists. Res: Quantal many-body problem, special functions. Mailing Add: Dept of Appl Math Univ of Western Ont London ON Can

TALMAN, RICHARD MICHAEL, b Toronto, Ont, Sept 24, 34; m 57; c 4. PHYSICS. Educ: Univ Western Ont, BA, 56, MA, 57; Calif Inst Technol, PhD(physics), 63. Prof Exp: From asst prof to assoc prof, 62-71, PROF PHYSICS, CORNELL UNIV, 71- Mem: Am Phys Soc. Res: Elementary and experimental particle physics; automatic scanning of spark pictures. Mailing Add: Dept of Physics Cornell Univ Ithaca NY 14850

TALNER, NORMAN STANLEY, b Mt Vernon, NY, Sept 28, 25; m 50; c 3. PEDIATRICS, CARDIOLOGY. Educ: Univ Mich, Ann Arbor, BS, 45; Yale Univ, MD, 49. Hon Degrees: MA, Yale Univ, 69. Prof Exp: Intern & resident pediat, Kings County Hosp, State Univ NY, 49-51; resident, Univ Hosp, Univ Mich, 51-52, instr, Med Sch, 54-56, Mich Heart Asn fel pediat cardiol, Hosp, 56-58, asst prof pediat, Univ, 58-60; from asst prof to assoc prof, 60-69, PROF PEDIAT, SCH MED, YALE UNIV, 69- Concurrent Pos: Attend physician, Yale-New Haven Hosp, 60-; USPHS career develop award, 62-72; examr, Sub-Bd Pediat Cardiol, Am Bd Pediat, 69-74; prog chmn, Am Heart Asn, 69-72; consult, Vet Admin, 72- Mem: Soc Pediat Res (mem secy, 69-72); Am Pediat Soc; Am Col Cardiol (asst secy, 72-74); cor mem Asn Europ Pediat Cardiol. Res: Cardiopulmonary physiology in infants and children. Mailing Add: Dept of Pediat Yale Univ Sch of Med New Haven CT 06510

TALSO, PETER JACOB, b Ishpeming, Mich, Sept 22, 21; m 43; c 4. MEDICINE. Educ: Wayne State Univ, BA, 43, MD, 45; Am Bd Internal Med, 54. Prof Exp: From instr to asst prof med, Univ Chicago, 50-52; asst prof, 53-58, PROF MED, STRITCH SCH MED, LOYOLA UNIV CHICAGO, 58-, CHMN DEPT, 63-; MED DIR, LITTLE COMPANY OF MARY HOSP, 71- Concurrent Pos: Consult, Hines Vet Admin Hosp, Maywood, 59-; attend physician, Cook County Hosp, 59-66, consult, 66- Mem: Fel Am Col Cardiol; fel Am Col Physicians. Res: Electrolyte and water metabolism; hypertension renal disease. Mailing Add: Little Company of Mary Hosp 2800 W 95th St Evergreen Park IL 60642

TALWANI, MANIK, b India, Aug 22, 33; m 58; c 3. GEOPHYSICS. Educ: Univ Delhi, BSc, 51, MSc, 53; Columbia Univ, PhD(geol), 60. Prof Exp: MEM STAFF, LAMONT-DOHERTY GEOL OBSERV, COLUMBIA UNIV, 57-, DIR, 73-, PROF GEOL, UNIV, 70- Concurrent Pos: Mem ocean affairs bd & chmn exec comt, Joint Oceanog Inst Deep Earth Sampling. Honors & Awards: Indian Geophys Union First Krishnan Medal, 65; James B Macelwane Award, Am Geophys Union, 67. Mem: Am Soc Explor Geophys; fel Am Geophys Union; Seismol Soc Am; fel Geol Soc Am; fel Royal Astron Soc. Res: Marine geophysics; oceanography, geodesy. Mailing Add: Lamont-Doherty Geol Observ Palisades NY 10964

TAM, ANDREW CHING, b Canton, China, Oct 13, 44; m 70; c 1. ATOMIC PHYSICS, MOLECULAR PHYSICS. Educ: Univ Hong Kong, BSc, 68, MSc, 70; Columbia Univ, PhD(physics), 75. Prof Exp: Fac fel physics, 70-72, preceptor, 71-72, res asst physics, 72-74, RES ASSOC, COLUMBIA RADIATION LAB, COLUMBIA UNIV, 75- Mem: Am Phys Soc; Sigma Xi. Res: Atomic and molecular spectroscopy; optical pumping; excimer lasers; laser interaction with atomic and molecular systems; laser-induced isotope separation. Mailing Add: Box 31 Pupin Columbia Radiat Lab 538 W120th St New York NY 10027

TAM, CHRISTOPHER K W, Chinese citizen. ACOUSTICS, APPLIED MATHEMATICS. Educ: McGill Univ, BEng, 62; Calif Inst Technol, MSc, 63, PhD(appl mech), 66. Prof Exp: Res fel, Calif Inst Technol, 66-67; asst prof, Mass Inst Technol, 67-71; ASSOC PROF MATH, FLA STATE UNIV, 71- Mem: Acoust Soc Am; Am Inst Aeronaut & Astronaut; Am Phys Soc; Am Geophys Union. Res: Physics of noise generation and propagation in aeroacoustics, including jet noise, airframe noise and duct acoustics; establishment of the importance of flow instabilities and large scale disturbances in noise generation. Mailing Add: Dept of Math Fla State Univ Tallahassee FL 32306

TAM, KWOK KUEN, b Hong Kong, Oct 30, 38; m 64; c 2. APPLIED MATHEMATICS. Educ: Univ Toronto, BASc, 62, MA, 63, PhD(appl math), 65. Prof Exp: Asst prof appl math, 65-69, ASSOC PROF APPL MATH, McGILL UNIV, 70- Concurrent Pos: Res fel, Harvard Univ, 71-72. Mem: Can Math Cong. Res: Fluid mechanics; construction of approximate solutions to some nonlinear boundary value problems. Mailing Add: Dept of Math McGill Univ Montreal PQ Can

TAM, KWOK-WAI, b Hong Kong, Mar 16, 38; US citizen; m 68; c 3. MATHEMATICAL ANALYSIS, OPERATIONS RESEARCH. Educ: Univ Wash, BS, 60, PhD(math), 67. Prof Exp: Teaching asst math, Univ Wash, 61-66; asst prof, 66-75, ASSOC PROF MATH, PORTLAND STATE UNIV, 75- Mem: Am Math Soc. Res: Mathematical programming. Mailing Add: Dept of Math Portland State Univ Portland OR 97201

TAMAKI, GEORGE, b Los Angeles, Calif, Mar 20, 31; m 56; c 2. ENTOMOLOGY. Educ: Univ Calif, Berkeley, BS, 60, PhD(entom), 65. Prof Exp: ENTOMOLOGIST, ENTOM RES DIV, USDA, 65- Mem: Entom Soc Am. Res: Insect population dynamics; insect ecology and biological control. Mailing Add: 3706 Nob Hill Blvd Yakima WA 98902

TAMAOKI, TAIKI, b Miki, Hyogo-Ken, Japan, Dec 3, 28; m 61; c 4. BIOCHEMISTRY, PLANT PATHOLOGY. Educ: Univ Tokyo, BSc, 51; Purdue Univ, MS, 58; Univ Wis, PhD(plant path), 60. Prof Exp: Fel oncol, McArdle Lab, Univ Wis, 61-64; asst prof biochem, 64-68, ASSOC PROF BIOCHEM, CANCER RES UNIT, UNIV ALTA, 68- Mem: AAAS; Am Chem Soc; Am Soc Biol Chem; Am Asn Cancer Res; Can Biochem Soc. Res: Regulation of protein and RNA synthesis in mammalian cells. Mailing Add: Cancer Res Unit Univ of Alta Edmonton AB Can

TAMAR, HENRY, b Vienna, Austria, Sept 15, 29; nat US; m 55; c 3. PROTOZOOLOGY; PHYSIOLOGY. Educ: NY Univ, AB, 49, MS, 51; Fla State Univ, PhD(physiol), 57. Prof Exp: Researcher, Lebanon Hosp, New York, 51; asst physiol, Fla State Univ, 51-55; asst prof biol, Am Int Col, 55-57; prof & head dept, Pembroke State Col, 57-62; ASSOC PROF ZOOL, IND STATE UNIV, TERRE HAUTE, 62- Concurrent Pos: Vis prof, Stephen F Austin State Col, 59, NC State Col, 61-62 & Marine Biol Labs, Woods Hole, 65. Mem: Am Micro Soc; Soc Protozool. Res: Principles of sensory physiology; locomotion, responses and structure of ciliates; jumping ciliates. Mailing Add: Dept of Life Sci Ind State Univ Terre Haute IN 47809

TAMARI, DOV, b Fulda, Ger, Apr 29, 11; m 48; c 3. MATHEMATICS. Educ: Hebrew Univ, Israel, MSc, 39; Univ Paris, Dr es Sc(math), 51. Prof Exp: Res fel math, Nat Ctr Sci Res, Paris, France, 49-53; sr lectr, Israel Inst Technol, 53-55, assoc prof, 55-59; prof, Univ Rochester, 59-60; mem, Inst Advan Study, 60-61; Orgn Am States vis prof, Univ Brazil, 61-62; res assoc, Univ Utrecht, 62; prof, Univ Caen, 62-63; chmn dept math, 64-67, PROF MATH, STATE UNIV NY BUFFALO, 63- Concurrent Pos: Vis prof, Hebrew Univ, Israel, 53-59; mem, Inst Advan Study, 67-68; vis prof, Israel Inst Technol, 73. Mem: Am Math Soc; Math Asn Am; Asn Symbolic Logic. Res: Algebra; semi-group, group and ring theory; embedding and word problems; topological semi-groups, groups and fields; mathematical logic; binary relations; partial algebras; combinatorial analysis. Mailing Add: Dept of Math State Univ NY 4246 Ridge Lea Rd Buffalo NY 14226

TAMARIBUCHI, KAY, chemistry, see 12th edition

TAMARIN, ARNOLD, b Chicago, Ill, Mar 27, 23; m 45; c 2. ORAL BIOLOGY, HISTOLOGY. Educ: Univ Ill, BS, 49, DDS, 51; Univ Wash, MSD, 60. Prof Exp: Assoc prof, 66-69, PROF ORAL BIOL, UNIV WASH, 69-, ADJ PROF, DEPT BIOL STRUCT, 74- Mem: Electron Micros Soc Am; Am Soc Cell Biol. Res: Comparative odontology; cell kinematics in the exocrine secretory process; exocrine collagen secretion in mytilus; ultrastructural morphology. Mailing Add: Dept of Oral Biol Univ of Wash Seattle WA 98195

TAMARIN, ROBERT HARVEY, b Brooklyn, NY, Dec 14, 42; m 68. GENETICS, ECOLOGY. Educ: Brooklyn Col, BS, 63; Ind Univ, PhD(zool), 68. Prof Exp: Comt Instnl Coop traveling scholar, Univ Wis, 67-68; USPHS fel genetics, Univ Hawaii, 68-70; Ford Found fel, Princeton Univ, 70-71; ASST PROF BIOL, BOSTON UNIV, 71- Concurrent Pos: NIH res grant. Mem: AAAS; Am Inst Biol Scientists; Am Soc Mammal; Am Soc Naturalists; Ecol Soc Am. Res: Population biology, including genetics, demography, reproductive physiology, behavior and general ecology of insular and mainland field mice to understand population regulation; tropical and radiation studies of field mice. Mailing Add: Dept of Biol Boston Univ Boston MA 02215

TAMARKIN, PAUL, b Leningrad, Russia, Apr 18, 22; nat US; m 43; c 3. PHYSICS, RESEARCH MANAGEMENT. Educ: Brown Univ, ScB, 42, PhD(physics), 49. Prof Exp: Asst chem, Univ Calif, 42-43; asst physics, Brown Univ, 43-44 & 47-49, asst prof, 52-56; physicist, David W Taylor Model Basin, US Dept Navy, 44-47; instr physics, Pa State Col, 49-50, asst prof, 50-52; physicist & group leader, Rand Corp, 56-65; dir, Adv Sensors Off, Adv Res Proj Agency, Off Secy Defense, DC, 65-67; asst to vpres, Rand Corp, 67-71, assoc head phys sci dept, 71-74; dir phys sci, Opers Anal Div, Gen Res Corp, 74-75; STAFF MEM, RIVERSIDE RES INST, 75- Concurrent Pos: Consult, Nat Comn Wiretapping & Electronic Surveillance, 74- Mem: Am Phys Soc; Am Geophys Union. Res: Shock waves, waves of finite amplitude in air; ultrasonics; underwater sound; solid state, low temperature, ionospheric physics; military systems research; management of government funded research; research and development assessment; system development management. Mailing Add: Riverside Res Inst 1701 N Ft Meyer Dr Arlington VA 22209

TAMASHIRO, MINORU, b Hilo, Hawaii, Sept 16, 24; m 52. INSECT PATHOLOGY. Educ: Univ Hawaii, BS, 51, MS, 54; Univ Calif, PhD, 60. Prof Exp: Asst entom, 51-54, jr entomologist, 54-55, from asst prof entom & asst entomologist to assoc prof entom & assoc entomologist, 57-73, PROF ENTOM & ENTOMOLOGIST, UNIV HAWAII, 73- Concurrent Pos: WHO consult, 63; NIH fel, 64-65. Mem: Entom Soc Am. Res: Microbial control; effect of pathogens and insecticides on biological control; termites, biology, ecology, control. Mailing Add: Dept of Entom Univ of Hawaii Honolulu HI 96822

TAMBASCO, DANIEL JOSEPH, b Amsterdam, NY, Mar 10, 36; m 66; c 2. THEORETICAL PHYSICS. Educ: Union Col, BS, 58; Univ Iowa, PhD(physics), 65. Prof Exp: Asst prof, 65-69, ASSOC PROF PHYSICS, MERRIMACK COL, 69- Mem: Am Phys Soc. Res: Field theory theory; statistical mechanics. Mailing Add: Dept of Physics Merrimack Col North Andover MA 01845

TAMBORSKI, CHRIST, b Buffalo, NY, Nov 12, 26; m 49; c 7. ORGANIC CHEMISTRY, FLUORINE CHEMISTRY. Educ: Univ Buffalo, BA, 49, PhD(org chem), 53. Prof Exp: Fel, Univ Buffalo, 53; SR SCIENTIST, AIR FORCE MAT LAB, WRIGHT-PATTERSON AFB, 55- Concurrent Pos: Consult, Childrens Hosp Res Found, 74- Honors & Awards: Jacobowitz Award, 52. Mem: Am Chem Soc; Sigma Xi. Res: High temperature stable fluids and elastomers for advanced aerospace applications; synthesis of organometallic compounds, heterocyclic compounds, organoaliphatic and aromatic fluorine compounds, anti-oxidants. Mailing Add: Air Force Mat Lab Wright-Patterson AFB Dayton OH 45433

TAMBURIN, HENRY JOHN, b Passaic, NJ, July 24, 44; m 68; c 2. INDUSTRIAL ORGANIC CHEMISTRY. Educ: Seton Hall Univ, BS, 66; Univ Md, College Park, PhD(org chem), 71. Prof Exp: From teaching asst to instr org chem, Univ Md, College Park, 66-72; res & develop chemist, Toms River Chem Corp, 72-75, SR DEVELOP CHEMIST, CIBA-GEIGY CORP, 75- Mem: Am Chem Soc. Res: The development of new and useful dyes for fibers and the introduction of modern technology in the dyestuff manufacturing process. Mailing Add: Bldg 216 Ciba-Geigy Corp PO Box 71 Toms River NJ 08753

TAMBURINO, JOHN, b Pittsburgh, Pa, Apr 24, 41. MATHEMATICAL LOGIC. Educ: Univ Pittsburgh, BS, 62, MA, 70, PhD(math), 72. Prof Exp: Teaching asst math, Univ Pittsburgh, 62-73; MATH CONSULT, CARDIAC OUTPUT, MONTEFIORE HOSP, 75- Mem: Sigma Xi. Mailing Add: 504 Sheridan Ave Pittsburgh PA 15206

TAMBURINO, LOUIS ANTHONY, b Pittsburgh, Pa, May 9, 36; m 58; c 1. MATHEMATICAL PHYSICS. Educ: Carnegie Inst Technol, BS, 57; Univ Pittsburgh, PhD(physics), 62. Prof Exp: Res assoc, Syracuse Univ, 63-64; res physicist, Aerospace Res Labs, 64-72, MATH PHYSICIST, AIR FORCE AVIONICS LAB, US AIR FORCE, 72- Mem: Am Phys Soc. Res: General relativity; gravitational radiation; plasma physics; tactical decision processes; inertial navigation; pattern recognition; optics; picture processing; cybernetics. Mailing Add: Systs Avionics Div Air Force Avionics Lab Bldg 620 Wright-Patterson AFB OH 45433

TAMBURRINI, VINCENT, JR, organic chemistry, see 12th edition

TAMBURRO, CARLO HORACE, b Caserta, Italy, Jan 20, 36; US citizen; m 71; c 1. INTERNAL MEDICINE, HEPATOLOGY. Educ: Georgetown Univ, BS, 58; Seton Hall Univ, MD, 62. Prof Exp: Intern med, Jersey City Med Ctr, NJ, 62-63, resident, 63-64; asst, Sch Med, Tufts Univ, 64-65; from asst to assoc prof med, NJ Med Sch, 69-74; ASSOC PROF MED, SCH MED, ASSOC IN ONCOL, CANCER CTR, CHIEF DIGESTIVE DIS & NUTRIT SECT & DIR, VINYL CHLORIDE PROJ, UNIV LOUISVILLE, 74- Concurrent Pos: Resident med, New Eng Ctr Hosp, Boston, 64-65; NIH fel hepatic dis, 65-68. Mem: Int Asn Study Liver; Am Asn Study Liver Dis; Am Soc Human Genetics; fel Am Col Nutrit; Am Fedn Clin Res. Res: Liver regeneration and metabolism; immunology and liver injury; vitamin metabolism and deficiency; alcohol metabolism and withdrawal syndromes; hepatitis, viral and drug addiction; liver disease and nutrition; industrial carcinogenesis, vinyl chloride; hepatic cancer. Mailing Add: Div of Hepatic Metab & Nutrit NJ Med Sch Newark NJ 07103

TAMBURRO, KATHLEEN O'CONNELL, b New York, NY, Oct 30, 42; m 71; c 3. PROTOZOOLOGY. Educ: Marymount Manhattan Col, BA, 64; Fordham Univ, MS, 65, PhD(biol, protozool), 68. Prof Exp: From res asst to res assoc biochem & physiol of protozoa, Haskins Labs, 65-74; ADMIN ASST & GRANT COORDR-WRITER, DEPT DIGESTIVE DIS & NUTRIT, SCH MED, UNIV LOUISVILLE, 76- Mem: AAAS; Soc Protozool; Am Soc Trop Med & Hyg; Am Soc Microbiologists. Res: Protozoa as pharmacological tools; chemotherapy of trypanosomatid parasites; biochemistry and physiology of Trypanosomatidae. Mailing Add: 512 Brandon Rd Louisville KY 40207

TAMERS, MURRY ALLEN, b Pittsburgh, Pa, June 21, 33; m 62; c 3. PHYSICAL CHEMISTRY, ENVIRONMENTAL CHEMISTRY. Educ: Princeton Univ, AB, 55; Yale Univ, MS, 57, PhD(phys chem), 58; Univ Paris, DSc(nuclear chem), 62. Prof Exp: Instr chem, Univ Tex, 58-60; mem staff, Saclay Nuclear Res Ctr, France, 60-62; head radiocarbon dating lab, Sci Res Inst, Caracas, Venezuela, 63-72; ASSOC PROF PHYS CHEM, NOVA UNIV, 73- Concurrent Pos: Head radiocarbon dating lab, Univ Tex, 62-65; guest prof, Univ Bonn, 66-67. Res: Nuclear chemistry and physics; age determinations with naturally occurring radioactive isotopes; ground water dating; radiochemistry techniques; liquid scintillation counting. Mailing Add: Nova Univ 3301 College Ave Ft Lauderdale FL 33314

TAMHANE, AJIT CHINTAMAN, b Bhiwandi, India, Nov 12, 46; m 75. APPLIED STATISTICS, MATHEMATICAL STATISTICS. Educ: Indian Inst Technol, Bombay, BTech(Hon), 68; Cornell Univ, MS, 73, PhD(statist), 75. Prof Exp: Jr engr design, Larsen & Toubro Ltd, Bombay, 68-70; ASST PROF INDUST ENG & MGT SCI, NORTHWESTERN UNIV, 75- Mem: Inst Math Statist; Am Statist Asn. Res: Statistical multiple decision procedures; multivariate analysis; statistical applications in engineering, biology, economics and social sciences. Mailing Add: Dept of Indust Eng Technol Inst Northwestern Univ Evanston IL 60201

TAMIMI, HAMDI AHMAD, b Hebron, Palestine, Aug 20, 24; US citizen; m 54; c 5. MICROBIOLOGY, INFECTIOUS DISEASES. Educ: Sterling Col, BS, 50; Univ Colo, Denver, MS, 52, PhD(microbiol), 57. Prof Exp: From res assoc to instr microbiol, Washington Univ, 57-60; asst prof, Col Physicians & Surgeons, 60-65, PROF MICROBIOL & CHMN DEPT, SCH DENT, UNIV OF THE PAC, 65- Concurrent Pos: Prin investr, USPHS grants, Washington Univ, 57-60, Col Physicians & Surgeons & Univ of the Pac, 60-69; lectr, Exten Div, Univ Calif, Berkeley, 63-; hosp consult, Mary's Help Hosp, Daly City, 65-71. Mem: NY Acad Sci; Am Soc Microbiol. Res: Mechanisms by which microorganisms cause disease; study of staphylococci, enteric organisms and indigenous microbiota as to growth requirements, toxins, enzymes and other characteristics which enable them to cause disease. Mailing Add: 113 Yolo St Corte Madera CA 94925

TAMIMI, YUSUF NIMR, b Nablus, Jordan, Nov 15, 31; m 63; c 2. SOIL CHEMISTRY. Educ: Purdue Univ, BS, 57; NMex State Univ, MS, 60; Univ Hawaii, PhD(soil chem), 64. Prof Exp: Asst agronomist, Univ, 63-70, ASSOC SOIL SCIENTIST, AGR EXP STA, UNIV HAWAII, 70- Mem: Am Soc Agron; Mineral Soc Am; Int Soc Soil Sci. Res: Chemistry of soil phosphorous; field crops, pasture fertilization and forest tree nutrition; forest soils. Mailing Add: Agr Exp Sta Univ of Hawaii Hilo HI 96720

TAMIR, HADASSAH, b Haifa, Israel, Oct 5, 30; US citizen; m 49; c 2. NEUROSCIENCE, BIOCHEMISTRY. Educ: Hebrew Univ, Jerusalem, MSc, 55; Israel Inst Technol, DSc(chem), 59. Prof Exp: Mem res staff, Princeton Univ, 65-67; res assoc, Med Sch, Columbia Univ, 67-71; SR RES SCIENTIST, DIV NEUROSCI, PSYCHIAT INST OF STATE OF NY, 71- Concurrent Pos: Res fel biochem, Pub Health Res Inst, City of New York, 59-63; res fel biochem & bact, Med Sch, NY Univ, 63-65. Res: Effects of drugs on release and uptake of biogenic amines in nerve endings and cell-free systems. Mailing Add: Psychiat Inst of State of NY 722 W 168th St New York NY 10032

TAMIR, THEODOR, b Bucharest, Roumania, Sept 17, 27; m 49; c 2. ELECTROPHYSICS, ELECTRICAL ENGINEERING. Educ: Israel Inst Technol, BS, 53, Dipl Ing, 54, MS, 58; Polytech Inst Brooklyn, PhD(electrophys), 62. Prof Exp: Res engr, Sci Dept, Ministry Defense, Israel, 53-56; instr elec eng, Israel Inst Technol, 56-58; res assoc, 58-62, from asst prof to assoc prof, 62-69, PROF ELECTROPHYS, POLYTECH INST NEW YORK, 69-, HEAD, DEPT ELEC ENG & ELECTROPHYS, 74- Concurrent Pos: Consult indust & govt labs; co-ed, Appl Phys, 72-; adv ed, Optics Commun, 75- Mem: Inst Elec & Electronics Engrs; Brit Inst Elec Eng; Int Union Radio Sci; Optical Soc Am. Res: Electromagnetic wave propagation in non-uniform media and periodic structures; radiation and diffraction phenomena; properties of configurations supporting surface, leaky, lateral and other wave types; elastic and optical waves. Mailing Add: Dept of Elec Eng & Electrophys Polytech Inst of New York Brooklyn NY 11201

TAMM, IGOR, b Tapa, Estonia, Apr 27, 22; nat US; m 53; c 3. VIROLOGY, MEDICINE. Educ: Yale Univ, MD, 47. Prof Exp: Intern med, Grace-New Haven Community Hosp Univ Serv, 47-48, asst resident, 48-49; from asst to assoc prof, 49-64, PROF MED, ROCKEFELLER UNIV, 64-, PROF VIROL, 73-, SR PHYSICIAN,

HOSP, 64- Concurrent Pos: Asst, Sch Med, Yale Univ, 47-49; from asst physician to physician, Rockefeller Univ Hosp, 49-64; assoc mem comn acute respiratory dis, Armed Forces Epidemiol Bd, 61-; mem virol & rickettsiology study sect, NIH, 64-68; mem bd sci consults, Sloan-Kettering Inst Cancer Res, 66-; centennial lectr, Univ Ill, 68; mem study panel allergy & infectious dis, Health Res Coun City of New York, 68-; mem adv comt, Am Cancer Soc, 69- Honors & Awards: Alfred Benzon Prize, 67. Mem: AAAS; Harvey Soc; Am Soc Clin Invest; Soc Exp Biol & Med; Am Soc Microbiol. Res: Viral replication and virus-induced alterations in cells; biosynthesis of nucleic acids and proteins; interaction of viruses with mucoproteins. Mailing Add: Dept of Virol Rockefeller Univ New York NY 10021

TAMMEN, JAMES F, b Sacramento, Calif, Feb 27, 25; m 47; c 2. PLANT PATHOLOGY. Educ: Univ Calif, BS, 49, PhD(plant path), 54. Prof Exp: Jr plant pathologist, Calif State Bur Plant Path, 49-50; asst plant path, Univ Calif, 50-53, jr plant pathologist, 53-54; chief pathologist, Fla State Plant Bd, 54-56; from asst prof to assoc prof plant path, 56-63, head dept, 63-74, PROF PLANT PATH, PA STATE UNIV, UNIVERSITY PARK, 63- Mem: Am Phytopath Soc. Res: Diseases of floricultural crops; epidemiology. Mailing Add: Dept of Plant Path Pa State Univ University Park PA 16802

TAMOR, STEPHEN, b New York, NY, Nov 29, 25; m 49; c 4. THEORETICAL PHYSICS. Educ: City Col New York, BS, 44; Univ Rochester, PhD, 50. Prof Exp: Physicist, Oak Ridge Nat Lab, 50-52; physicist, Radiation Lab, Univ Calif, 52-55; physicist, Res Lab, Gen Elec Co, NY, 55-66, Space Sci Lab, Pa, 66-71; PHYSICIST, SCI APPLN, INC, 71- Concurrent Pos: Guggenheim fel, 63-64. Mem: Am Phys Soc. Res: Meson theory; nuclear and plasma physics; reactor theory. Mailing Add: Sci Appln Inc PO Box 2351 La Jolla CA 92037

TAMORRIA, CHRISTOPHER RICHARD, b Washington, DC, June 20, 32; m 61; c 1. MEDICINAL CHEMISTRY, COMMUNICATION SCIENCE. Educ: Georgetown Univ, BS, 54, MS, 58; Univ Md, PhD(med chem), 61. Prof Exp: Asst org chem, Georgetown Univ, 54-55; chemist, Food & Drug Admin, 55-56; asst org chem, Georgetown Univ, 56-57; asst inorg chem, Univ Md, 57-58; org chemist, Pharmaceut Prod Develop Sect, Lederle Labs, Am Cyanamid Co, 60-68, mgr regulatory agencies & info processing, Med Res Div, Cyanamid Int, 68-70; sr tech assoc, US Pharmacopeia, Md, 70-73; DIR SCI COMMUN, PURDUE FREDERICK CO & AFFIL, 73- Honors & Awards: Gold Medal, Am Inst Chemists, 54. Mem: Am Chem Soc; Am Pharmaceut Asn; fel Am Found Pharmaceut Educ; Am Inst Chemists; Drug Info Asn. Res: Partial synthesis of steroids; correlation of structure and biological activity, especially in the synthesis of new steroid homologs and tetracycline antibiotics. Mailing Add: 27 Maymont Lane Trumbull CT 06611

TAMPAS, JOHN PETER, b Burlington, Vt, May 18, 29; m 62; c 4. RADIOLOGY. Educ: Univ Vt, BS, 51, MD, 54. Prof Exp: Teaching fel pediat radiol, Children's Hosp of Los Angeles, Univ Southern Calif, 60-61; NIH res fel cardiovasc radiol, Nat Heart Inst, 61-62; from asst prof to assoc prof, 62-69, PROF RADIOL & CHMN DEPT, UNIV VT, 70- Concurrent Pos: James Picker Found scholar radiol res, Univ Vt, 62-65; from asst attend radiologist to attend radiologist, Mary Fletcher Hosp & DeGoesbriand Mem Hosp, 62-; physician-in-residence, Vet Admin Hosp, 72- Mem: AMA; fel Am Col Radiol; Soc Pediat Radiol; Am Roentgen Ray Soc; Radiol Soc NAm. Res: Basic and clinical problems in radiology; pediatric and cardiovascular radiology. Mailing Add: Dept Radiol Mary Fletcher Unit Med Ctr Hosp of Vt Burlington VT 05401

TAMRES, MILTON, b Warsaw, Poland, Mar 12, 22; US citizen; m 60; c 2. PHYSICAL INORGANIC CHEMISTRY. Educ: Brooklyn Col, BA, 43; Northwestern Univ, PhD(phys chem), 49. Prof Exp: Anal chemist, Celanese Corp Am, Md, 43-44; asst, Northwestern Univ, 44-47; from instr to asst prof chem, Univ Ill, 48-53; from asst prof to assoc prof, 53-63, PROF CHEM, UNIV MICH, ANN ARBOR, 63- Concurrent Pos: Guggenheim fel, 59-60; mem, Adv Coun Col Chem, 62-66; Am Chem Soc-Petrol Res Fund int fel, 66-67; vis scholar, Univ Tokyo, 74. Mem: AAAS; Am Chem Soc; fel Am Inst Chemists. Res: Electron donor-acceptor interactions; basicities of cyclic compounds. Mailing Add: Dept of Chem Univ of Mich Ann Arbor MI 48104

TAMS, WILLIAM P, b Richmond, Va, June 14, 16; m 37; c 2. INDUSTRIAL ORGANIC CHEMISTRY. Educ: Princeton Univ, AB, 38. Prof Exp: Anal chemist, Org Chem Dept, E I du Pont de Nemours & Co, Inc, 39-40, prod supvr synthetics, 40-45, prod supvr synthetics, NJ, 46-50, prod supvr develop, 50-53, chemist, Tex, 54-55, develop supvr, NJ, 55-62, res & develop div head, Jackson Lab, 63-74; RETIRED. Mem: Am Chem Soc; Am Inst Chem Engrs. Res: Aromatic intermediates for dyestuff, agriculture, rubber chemicals and textile fibers. Mailing Add: 5610 Kirkwood Hwy Wilmington DE 19808

TAMSITT, JAMES RAY, b Big Spring, Tex, Nov 22, 28. ZOOLOGY. Educ: Univ Tex, BA, 51, MA, 53, PhD(vert ecol), 58. Prof Exp: Lectr zool, Univ Man, 57-58; instr biol, ETex State Univ, 58-59; prof zool, Univ of the Andes, Colombia, 59-65; NIH fel med zool, Sch Med, Univ PR, San Juan, 65-67; assoc cur, Dept Mammal, 67-73, ASSOC PROF ZOOL, UNIV TORONTO, 69-, CUR, ROYAL ONT MUS, 73- Concurrent Pos: Vis prof biol, Pontificia Univ Javeriana, Colombia, 75; sr Fulbright-Hays fel, 75. Mem: Fel AAAS; Am Soc Mammalogists; Soc Study Evolution. Res: Ecology, natural history, taxonomy and ectoparasites of Neotropical bats. Mailing Add: Royal Ont Mus 100 Queen's Park Toronto ON Can

TAMSKY, MORGAN JEROME, b St Louis, Mo, July 26, 42; m 66. PHYSICAL CHEMISTRY. Educ: Washington Univ, BA, 64; Univ Kans, PhD(chem), 70. Prof Exp: Sr chemist, 69-74, RES SPECIALIST, 3M CO, 74- Mem: Am Chem Soc. Res: Surface phenomenon. Mailing Add: 1548 Duluth St St Paul MN 55106

TAMSMA, ARJEN, dairying, see 12th edition

TAMURA, TSUNEO, b Hawaii, Nov 15, 25; m 51; c 3. SOILS. Educ: Univ Hawaii, BS, 48; Univ Wis, MS, 51, PhD(soils), 52. Prof Exp: From asst soil scientist to assoc soil scientist, Conn Agr Exp Sta, 52-57; chemist, Westinghouse Elec Corp, 57; CHEMIST, OAK RIDGE NAT LAB, 57- Mem: AAAS; Am Soc Agron; Am Chem Soc; Sigma Xi; Mineral Soc Am. Res: Soil chemistry and genesis; soil clay mineralogy; radioactive waste disposal; health physics; toxic metals in environment. Mailing Add: 897 W Outer Dr Oak Ridge TN 37830

TAN, AGNES WEN HUI, biochemistry, see 12th edition

TAN, AH-TI CHU, b Amoy, China, Sept 24, 35; Can citizen. CHEMISTRY, BIOCHEMISTRY. Educ: Mapua Inst Technol, BSChem, 57; Adamson Univ, Manila, BSChE, 58; McGill Univ, MSc, 62, PhD(chem kinetics), 66. Prof Exp: Lectr phys chem, Adamson Univ, Manila, 58-60; res chemist, Bathurst Paper Co, 65-66; assoc biochem, Col Med, Univ Vt, 66-68, vis asst prof, 68-69; asst prof ophthal, 69-72, ASST PROF ANESTHESIA, FAC MED, McGILL UNIV, 72- Concurrent Pos: NIH grant biochem, Col Med, Univ Vt, 66-68; Que Med Res Coun grant ophthal, Fac Med, McGill Univ, 69-72; prof assoc anaesthesia res, McGill Univ, 72- Mem: Am Chem Soc; Chem Inst Can; Can Biochem Soc; Soc Neurosci; Int Soc Neurochem. Res: Physicochemical studies of proteins; brain cell membranes; neurotransmitters; molecular mechanism of synaptic transmission; neuroendocrinology. Mailing Add: Apt 817 1500 Stanley St Montreal PQ Can

TAN, BIAN DJOEN, b Kediri, Java, Indonesia, May 25, 35. PARASITOLOGY, RADIATION BIOLOGY. Educ: Bandung Tech Inst, BSc, 60, MSc, 62; Univ Tenn, PhD(zool), 65. Prof Exp: Res assoc dept entom & zool, Univ Tenn, Knoxville, 65-68, asst prof, 67-68; res assoc, Dept Environ Health, 70-71, res asst, Dept Pharmacol, 71-72, RES ASSOC & LECTR ZOOL, UNIV TORONTO, 72- Concurrent Pos: Instr indust microbiol, Centenial Col, 74-75. Mem: Am Soc Parasitol; Can Pub Health Asn. Res: Host parasite relationship of Hymenolepis microstoma and the effects of ionizing radiation on the system; physiology of lifting and back injury; cardiovascular system and prostaglandins, cyclic adenosine monophosphate, adenyl cyclase; Hymenolepis microstoma and the host gastrointestinal function. Mailing Add: Dept of Zool Univ of Toronto Toronto ON Can

TAN, BOEN HIE, analytical biochemistry, see 12th edition

TAN, CELINE G L, b Singapore, Nov 22, 43; m 66. BIOCHEMISTRY. Educ: Univ Singapore, BSc, 66; Univ Man, MSc, 68, PhD(biochem), 70. Prof Exp: NIH fel smooth muscle chem, Carnegie-Mellon Univ, 70-71; NIH fel Protein Struct, Yale Univ, 72-74; MEM STAFF, DIV MED BIOCHEM, UNIV CALGARY, 74- Res: Protein structure function; smooth muscle biochemistry; electron microscopy. Mailing Add: Div of Med Biochem Univ of Calgary Calgary AB Can

TAN, CHARLES HUA-MIN, b Peiping, China, Apr 4, 30; m 59; c 2. INTERNAL MEDICINE, BIOCHEMISTRY. Educ: Nat Taiwan Univ, BM, 56; Tulane Univ, PhD(biochem), 67. Prof Exp: Resident & instr internal med, Med Sch, Tulane Univ, 62-67; ASST PROF BIOCHEM & INTERNAL MED, SCH MED, ST LOUIS UNIV, 67- Concurrent Pos: Assoc investr, Touro Res Inst, La, 66-67. Mem: AAAS; Am Fedn Clin Res. Res: Metabolism of vitamin B-12 binding substances in mammalian system. Mailing Add: 11717 Arboroak Dr St Louis MO 63126

TAN, CHARLOTTE, b Kiang-Si, China, Apr 19, 23; US citizen; m 59; c 1. CANCER. Educ: Hsiang-Ya Med Col, China, MD, 47; Am Bd Pediat, dipl, 54. Prof Exp: Resident internal med, Nanking Cent Hosp, China, 47-48; intern & resident, St Barnabas Hosp, Newark, NJ, 48-49; res resident hemat & pediat resident, Children's Hosp Philadelphia, 50-51; res fel chemother, 52-54, res assoc, 55-57, asst, 57-62, assoc, 60-62, ASSOC MEM, SLOAN-KETTERING INST, 62-; ASSOC PROF PEDIAT, MED COL, CORNELL UNIV, 70- Concurrent Pos: Spec fel med, Memorial Ctr, New York, 52-55, spec fel pediat, 55-57, clin asst, Pediat Serv, 57-58; instr med, Sloan-Kettering Div, Grad Sch Med Sci, Med Col, Cornell Univ, 54-55, instr med, Med Col, 55-57, instr pediat, 58 & 62-70; clin asst pediatrician, James Ewing Hosp, 57-58, asst vis pediatrician, 58-62, assoc vis pediatrician, 62-; from asst attend pediatrician to assoc attend pediatrician, Memorial Hosp, 58-70, attend pediatrician, 70-, assoc chmn chemother, pediat, 74-; vis prof, Nat Taiwan Univ Med Col, 66-67. Mem: Am Acad Pediat; Am Asn Cancer Res; Am Fedn Clin Res; AMA; Am Soc Clin Oncol. Mailing Add: Sloan-Kettering Cancer Ctr 44 E 68th St New York NY 10021

TAN, ENG M, b Malaysia, Aug 26, 26; US citizen; m 62; c 2. IMMUNOLOGY. Educ: Johns Hopkins Univ, AB, 52, MD, 56. Prof Exp: Asst prof med, Washington Univ, 65-67; assoc mem, Dept Exp Path, 67-70, HEAD DIV ALLERGY & IMMUNOL, SCRIPPS CLIN & RES FOUND, 70- Res: Autoimmune diseases. Mailing Add: Scripps Clin & Res Found 476 Prospect St La Jolla CA 92037

TAN, FRANCIS C, b Manila, Philippines, Sept 21, 39; m 71. CHEMICAL OCEANOGRAPHY. Educ: Cheng Kung Univ, Taiwan, BSc, 61; McGill Univ, MSc, 65; Pa State Univ, PhD(geochem), 69. Prof Exp: NSF fel, Pa State Univ, 69-70; geochemist, Minn Geol Surv, Univ Minn, 70-71; RES SCIENTIST, ATLANTIC OCEANOG LAB, BEDFORD INST OCEANOG, CAN DEPT ENVIRON, 72- Mem: Geochem Soc; fel Geol Asn Can; Am Geophys Union. Res: Stable isotope oceanography, marine geochemistry. Mailing Add: Atlantc Oceanog Lab Bedford Inst Can Dept of Environ Dartmouth NS Can

TAN, HENRY HARRY, b Sukabumi, Indonesia, Dec 15, 24; US citizen; m 59; c 5. ORGANIC CHEMISTRY. Educ: Hope Col, AB, 55; Univ Mich, Ann Arbor, MS, 58, PhD(chem), 62. Prof Exp: Res chemist, 62-64, tech field rep, Appl Chem & Mkt, 64-68, res chemist, 68-73, MEM TECH SERV & DEVELOP STAFF, E I DU PONT DE NEMOURS & CO, INC, 73- Mem: Am Chem Soc. Res: Organic syntheses; applied chemistry; political, social and economic affairs of Asia. Mailing Add: 103 Colorado Ave Shipley Heights Wilmington DE 19803

TAN, HENRY S I, b Bandung, Indonesia, Mar 26, 32; US citizen. PHARMACEUTICAL CHEMISTRY. Educ: Univ Indonesia, BSPharm, 54, MSPharm, 56; Univ Ky, PhD(pharmaceut sci), 71. Prof Exp: Instr & assoc prof, Bandung Inst Technol, 57-66; ASST PROF PHARM, UNIV CINCINNATI, 71- Mem: Am Pharmaceut Asn; Acad Pharmaceut Sci; Am Asn Col Pharm. Res: Developmental analysis procedures drugs. Mailing Add: Univ of Cincinnati Cincinnati OH 45221

TAN, JAMES CHIEN-HUA, b Nanchang, China, Oct 8, 35; m 65; c 2. GENETICS, STATISTICS. Educ: Chung-Shing Univ, Taiwan, BS, 57; Mont State Univ, MS, 61; NC State Univ, PhD(genetics), 68. Prof Exp: Asst prof biol, Slippery Rock State Col, 65-66; asst prof, 66-70, ASSOC PROF BIOL, VALPARAISO UNIV, 70- Mem: Am Inst Biol Scientists; Genetics Soc Am. Res: Cytogenetics and statistical biology. Mailing Add: Dept of Biol Valparaiso Univ Valparaiso IN 46383

TAN, JULIA S, b Taipei, Taiwan; US citizen. POLYMER SCIENCE, PHYSICAL CHEMISTRY. Educ: Nat Taiwan Univ, BA, 61; Wesleyan Univ, MA, 63; Yale Univ, PhD(chem), 66. Prof Exp: Asst prof chem, Wesleyan Univ, 66-69; res assoc biophys, Univ Rochester, 69-70; RES CHEMIST, RES LABS, EASTMAN KODAK CO, 70- Mem: Am Chem Soc. Res: Solution properties of polyelectrolytes. Mailing Add: B-81 Res Labs Eastman Kodak Co Rochester NY 14650

TAN, KIM H, b Djakarta, Indonesia, Mar 24, 26; US citizen; m 57; c 1. TROPICAL AGRICULTURE. Educ: Univ Indonesia, MSc, 55, PhD(soil sci), 58. Prof Exp: Assoc prof soil sci fac agr, Univ Indonesia, 58-64, prof fac agr & agr acad, 64-67, head dept fac agr, 65-67; technician soil anal nitrogen lab, Agr Res Serv, USDA, Colo, 67-68; asst prof soil sci, Dept Agron, 68-73, ASSOC PROF AGRON, UNIV GA, 73- Concurrent Pos: Rockefeller Found grant/fel, NC State Univ, 60-61 & Cornell Univ, 61; mem comt VIII, Southern Regional Coop Soil Surv, Soil Conserv Serv, USDA, 72- Mem: Clay Mineral Soc Am; Am Soc Agron; Soil Sci Soc Am; Int Soc Soil Sci. Res: Pedology; genesis and characterization of soils and organic matter in soils; effect of organic matter on soil properties and plant growth; chemistry and mineralogy of soils. Mailing Add: Dept of Agron Univ of Ga Athens GA 30602

TAN, KOK-CHIANG, b Singapore, May 31, 37; m 62; c 2. GEOGRAPHY. Educ: Nanyang Univ, BA, 60; Univ Western Ont, MA, 62; Univ London, PhD(geog), 66. Prof Exp: Asst prof geog, 66-70, ASSOC PROF GEOG, UNIV GUELPH, 70- Concurrent Pos: Vis assoc prof, Nanyang Univ, 72-73. Mem: Asn Am Geog; Can Asn Geog. Res: The impact of geography on decision-making and the spatial consequence of political and economic policies in newly emerging states; urbanization of the Third World. Mailing Add: Dept of Geog Univ of Guelph Guelph ON Can

TAN, KOK-KEONG, b Shanghai, China, June 1, 43; m 69; c 3. MATHEMATICS. Educ: Nanyang Univ, BSc, 66; Univ BC, PhD(Math), 70. Prof Exp: Teacher high sch, Malaysia, 66; ASST PROF MATH, DALHOUSIE UNIV, 70- Mem: Am Math Soc; Math Asn Am; Can Math Cong. Res: Functional analysis and topology, in particular, fixed point theorems. Mailing Add: Dept of Math Dalhousie Univ Halifax NS Can

TAN, LIAT, b Semarang, Java, Apr 1, 29; m 63; c 2. ORGANIC BIOCHEMISTRY. Educ: Univ Amsterdam, BSc, 53; Univ Münster, MSc, 55; Univ Freiburg, Dr rer nat, 58. Prof Exp: Res chemist, Steroid Res Lab, Leo Pharmaceut Prod, Denmark, 59-60; sr res chemist, Union Chimique Belge SA, 60-62; res fel org chem, Laval Univ, 62-63; examr steroid chem, Can Patent Off, 63-66; Welch res fel & instr biochem & nutrit, Med Br, Univ Tex, 66-67; asst prof biochem, 67-70, ASSOC PROF BIOCHEM, UNIV SHERBROOKE, 70- Mem: AAAS; Am Chem Soc; Chem Soc Belg; NY Acad Sci; Can Biochem Soc. Res: Synthesis and biochemistry of steroids; steroidases, specificity and mechanism of action; microbiological transformations; chemical endocrinology; physiologically active natural products; mechanism of biological oxidations at inactive sites in steroids. Mailing Add: Dept of Biochem Hosp Ctr Univ of Sherbrooke Sherbrooke PQ Can

TAN, MENG HEE, b Kuala Pilah, Malaysia, Mar 30, 42; Can citizen; m 70; c 3. INTERNAL MEDICINE, MEDICAL RESEARCH. Educ: Dalhousie Univ, BSc, 65, MD, 69; FRCP(C), 75. Prof Exp: Lectr med, 74-75, ASST PROF MED, DALHOUSIE UNIV, 75- Concurrent Pos: Res fel, Harvard Med Sch, 71-73; res fel, Med Coun Can, 71-74, centennial fel, 74-75; res fel, Cardiovasc Res Inst, San Francisco, 73-75. Mem: Am Col Physicians; Am Diabetes Asn; Am Fedn Clin Res; Royal Col Physicians & Surgeons Can. Res: Roles of lipoprotein lipase and apoproteins in triglyceride metabolism in diabetes mellitus. Mailing Add: 5849 University Ave Halifax NS Can

TAN, PETER CHING-YAO, b Canton, China, Nov 1, 23; Can citizen; m. MATHEMATICAL STATISTICS, CIVIL ENGINEERING. Educ: Sun Yat-Sen Univ, BSc, 43; Luther Col, BA, 57; Univ Sask, MA, 59; Univ Toronto, PhD(math), 68. Prof Exp: Instr math, Univ Sask, 59-60; lectr, Univ Toronto, 62-65; spec lectr, Univ Sask, Regina Campus, 66-68; vis asst prof statist, Stanford Univ, 68-69; asst prof math, 69-73, ASSOC PROF MATH, CARLETON UNIV, 73- Concurrent Pos: Vis lectr, Univ Guyana, 75-76. Mem: Can Soc Hist Philos Math; Can Math Cong; Math Asn Am; Inst Math Statist; Am Statist Asn. Res: Statistical inference; mathematical and statistical methods in civil engineering. Mailing Add: Dept of Math Carleton Univ Colonel By Drive Ottawa ON Can

TAN, WAI-YUAN, b China, Aug 14, 34; m 64; c 1. BIOSTATISTICS. Educ: Taiwan Prov Col Agr, BA, 55; Nat Taiwan Univ, MS, 59; Univ Wis, MS(math) & MS(statist), 63, PhD(statist), 64. Prof Exp: Asst res fel biostatist, Inst Bot, Acad Sinica, Taiwan, 59-61, assoc res fel, 64-67 & res fel, 67-68; assoc prof statist, Nat Taiwan Univ, 65-67; assoc prof biostatist, Biol Res Ctr, Taiwan, 65-67, prof, 67-68; asst prof statist, Univ Wis-Madison, 68-72; assoc prof statist, Wash State Univ, 73-75; PROF MATH, MEMPHIS STATE UNIV, 75- Concurrent Pos: Vis assoc prof, Tsing Hua Univ, Taiwan, 65-67; statist adv, Joint Inst Indust Res, Taiwan, 65-67. Mem: Am Statist Asn; Biomet Soc; Inst Math Statist; Royal Statist Soc. Res: Genetics; multivariate analysis, mathematical genetics and quantitative genetics; growth curve analysis. Mailing Add: Dept of Math Memphis State Univ Memphis TN 38152

TAN, YEN T, b Hong Kong, Feb 12, 40. SOLID STATE PHYSICS, SURFACE PHYSICS. Educ: Columbia Univ, BS, 62; Yale Univ, PhD(chem), 66. Prof Exp: RES ASSOC, RES LABS, EASTMAN KODAK CO, 66- Concurrent Pos: Adj prof, Rochester Inst Technol, 74-75. Mem: Am Chem Soc; Am Inst Mining, Metall & Petrol Eng; Am Vacuum Soc. Res: Surface properties of solids; transport phenomena; thermodynamics. Mailing Add: Res Labs B-81 Eastman Kodak Co Rochester NY 14650

TAN, YIN HWEE, b Singapore, Sept 2, 42; m; c 1. VIROLOGY, GENETICS. Educ: Univ Singapore, BSc Hons, 65; Univ Man, PhD(biochem), 69. Prof Exp: Med Res Coun Can fel biochem & virol, Univ Pittsburgh, 69-71; Med Res Coun Can fel somatic cell genetics, Yale Univ, 72-74; asst prof pediat, Johns Hopkins Univ, 74-75; ASSOC PROF, DIV MED BIOCHEM, UNIV CALGARY, 75- Concurrent Pos: Sect chief molecular genetics, Lab Cellular Comp Physiol, Nat Inst Aging, 74-75. Mem: Am Soc Microbiol; Can Biochem Soc. Res: Regulation of chromosome 21-directed anti-viral genes; molecular genetics of host cell-virus interaction; molecular genetics of interferon; biochemical genetics; genetics of the anti-tumor effect of human interferon. Mailing Add: Div of Med Biochem Univ of Calgary Fac of Med Calgary AB Can

TAN, ZOILO CHENG-HO, b Bulan, Philippines, Oct 18, 40; m 67; c 2. PHOTOGRAPHIC CHEMISTRY. Educ: Cheng Kung Univ, Taiwan, BS, 63; Univ Ark, Fayetteville, MS, 66; Mass Inst Technol, PhD(nuclear chem), 69. Prof Exp: Res asst chem, Univ Ark, Fayetteville, 63-65 & Mass Inst Technol, 65-69; SR RES CHEMIST, RES LABS, EASTMAN KODAK CO, 69- Res: Solvent extraction from molten salts; electron exchange reaction in mixed solvents; solubility and complex formation of silver with halides and thiol compounds; photographic research. Mailing Add: 171 Chimney Hill Rd Rochester NY 14612

TANABE, ALVIN MASAAKI, entomology, insect pathology, see 12th edition

TANABE, MASATO, b Stockton, Calif, Jan 18, 25; m 55; c 3. ORGANIC CHEMISTRY. Educ: Univ Calif, BS, 47, PhD(chem), 51. Prof Exp: Chemist, US Naval Radiation Defense Lab, 47-48; res chemist, Riker Labs, Inc, 51-57; sr org chemist, 57-72, DIR DEPT PHARMACEUT CHEM, STANFORD RES INST, MENLO PARK, 72- Concurrent Pos: Fulbright res fel, Japan, 54-55. Mem: Am Chem Soc. Res: Medicinal chemistry; steroids; alkaloids; natural products; biosynthesis. Mailing Add: 972 Moreno Ave Palo Alto CA 94303

TANABE, TSUNEO Y, b Blackfoot, Idaho, Nov 17, 18; m 66; c 2. PHYSIOLOGY. Educ: Iowa State Col, BS, 42; Cornell Univ, MS, 45; Univ Wis, PhD(physiol of reprod), 48. Prof Exp: Asst prof, 49-50, 52-68, ASSOC PROF DAIRY PHYSIOL, PA STATE UNIV, UNIVERSITY PARK, 68- Concurrent Pos: Gosney fel, Calif Inst Technol, 51. Mem: AAAS; Am Soc Animal Sci; Am Dairy Sci Asn. Res: Embryology; dairy physiology; endocrinology. Mailing Add: Dept of Dairy Sci Pa State Univ University Park PA 16802

TANADA, TAKUMA, b Honolulu, Hawaii, Oct 30, 19; m 47; c 2. PLANT PHYSIOLOGY. Educ: Univ Hawaii, BS, 42, MS, 44; Univ Ill, PhD(bot), 50. Prof Exp: Asst soil chemist, Agr Exp Sta, Univ Hawaii, 42-44; sci consult natural resources sect, Supreme Comdr Allied Powers, US Army, Tokyo, 46-47; plant physiologist, Agr Res Serv, USDA, 50-57; agron res adv, Int Coop Admin, Ceylon, 57-60; RES PLANT PHYSIOLOGIST, AGR RES SERV, USDA, 60- Mem: Am Soc Plant Physiol; Bot Soc Am; Japanese Soc Plant Physiol. Res: Radiation biology; photosynthesis; mineral nutrition. Mailing Add: 12920 Moray Rd Silver Spring MD 20906

TANADA, YOSHINORI, b Puuloa, Oahu, Hawaii, June 8, 17; m 49; c 2. INSECT PATHOLOGY. Educ: Univ Hawaii, BS, 40, MS, 45; Univ Calif, PhD(entom), 53. Prof Exp: Asst zool, Univ Hawaii, 43-45, jr entomologist exp sta, 45-53, asst entomologist, 53-56, asst prof zool & entom, 54-56, assoc prof & assoc entomologist, 56; asst insect pathologist, Lab Insect Path, 56-59, assoc insect pathologist, 59-64, lectr, 61-65, chmn div invert path, 64-65, INSECT PATHOLOGIST, EXP STA, UNIV CALIF, BERKELEY, 64-, PROF ENTOM, UNIV, 65- Concurrent Pos: Consult, US Army, Okinawa, 50, SPac Comn, Pac Sci Bd, Nat Res Coun, 59, UN Develop Prog, Western Samoa, 71 & Food & Agr Orgn, UN, 71; Fulbright res scholar, Japan, 62-63. Mem: Fel AAAS; Entom Soc Am; Soc Protozool; Soc Invert Path; Am Inst Biol Scientists. Res: General insect pathology; epizootiology of diseases of insects; resistance of insects to diseases; microbial and biological control; economic entomology. Mailing Add: Div of Entom & Parasitol 333 Hilgard Hall Univ of Calif Berkeley CA 94720

TANAKA, FRED SHIGERU, b Shoshone, Idaho, Aug 1, 37; m 66; c 2. PESTICIDE CHEMISTRY. Educ: Ore State Univ, BS, 59, PhD(org chem), 66. Prof Exp: Chief chemist, Wash State Dept Health, 66-67; RES CHEMIST USDA, 67- Mem: Am Chem Soc. Res: Organic microsynthesis of radiochemically labeled compounds; photochemical degradation of pesticides; identification of biological metabolites; organic and biological reaction mechanisms. Mailing Add: Metab & Radiation Res Lab State Univ Sta Fargo ND 58102

TANAKA, JOHN, b San Diego, Calif, June 18, 24; m 59; c 2. INORGANIC CHEMISTRY. Educ: Univ Calif, Los Angeles, BA, 51; Iowa State Univ, PhD, 56. Prof Exp: From asst prof to assoc prof chem, SDak State Univ, 56-63; NASA fel, Univ Pittsburgh, 63-65; from asst prof to assoc prof chem, 65-75, PROF CHEM, UNIV CONN, 75-, DIR HONORS PROG, 71- Mem: Am Chem Soc; The Chem Soc. Res: Synthesis and properties of ternary hydrides; reactions of boron hydrides; vacuum line syntheses. Mailing Add: Dept of Chem Univ of Conn Storrs CT 06268

TANAKA, KATSUMI, b San Francisco, Calif, Mar 1, 25; m 53; c 1. PHYSICS. Educ: Univ Calif, AB, 49, PhD(physics), 52. Prof Exp: Assoc physicist, Argonne Nat Lab, 52-64; PROF PHYSICS, OHIO STATE UNIV, 64- Concurrent Pos: Vis prof, Univ Naples, 60-61. Mem: Fel Am Phys Soc. Res: Elementary particle physics. Mailing Add: Dept of Physics Ohio State Univ Columbus OH 43210

TANAKA, KOUICHI ROBERT, b Fresno, Calif, Dec 15, 26; m 65; c 3. MEDICINE, HEMATOLOGY. Educ: Wayne State Univ, BS, 49, MD, 52. Prof Exp: Intern, Los Angeles County Gen Hosp, 52-53; resident path, Detroit Receiving Hosp, Mich, 53-54, resident med, 54-57; instr med & jr res hematologist, 57-59, asst prof med, Sch Med & asst res hematologist, Med Ctr, 59-61, assoc prof med, Sch Med, 61-68, PROF MED, SCH MED, UNIV CALIF, LOS ANGELES, 68-, ATTEND PHYSICIAN, MED CTR, 58-; CHIEF DIV HEMAT, HARBOR GEN HOSP, 61-, ASSOC CHMN DEPT MED, 71- Concurrent Pos: Consult, US Naval Hosp, Long Beach. Mem: AAAS; fel Am Col Physicians; Asn Am Physicians; Am Soc Hemat; Am Soc Clin Invest; Am Fedn Clin Res. Res: Internal medicine; red cell metabolism. Mailing Add: Dept of Med Harbor Gen Hosp Torrance CA 90509

TANAKA, RYO, b Matsumoto-shi, Japan, Dec 1, 28; m 59. NEUROCHEMISTRY, BIOCHEMISTRY. Educ: Univ Tokyo, MD, 54, PhD(biochem), 59. Prof Exp: Fulbright exchange scholar, 59; res assoc biochem, Med Ctr, Univ Ala, 59-60; res assoc neurochem, Med Ctr, Univ Ill, 60-63; res assoc biochem, Col Med, Univ Western Ont, 63-65; assoc prof biochem, Sch Pharmaceut Sci, Showa Univ, Japan, 65-69; asst prof, 66-70, ASSOC PROF NEUROCHEM, CTR BRAIN RES, MED CTR, UNIV ROCHESTER, 70- Mem: Am Soc Biol Chem; Am Soc Neurochem; Int Soc Neurochem; Japanese Biochem Soc. Res: Lipid-protein interaction; synapse biochemistry; biochemistry and enzymology of the membrane. Mailing Add: Ctr for Brain Res Univ of Rochester Med Ctr Rochester NY 14642

TANAKA, TOYOICHI, b Nagaoka, Japan, Jan 4, 46; m 70; c 2. BIOPHYSICS. Educ: Univ Tokyo, BS, 68, MA, 70, DSc(physics), 72. Prof Exp: Fel biophys, 72-75, ASST PROF PHYSICS, MASS INST TECHNOL, 75- Concurrent Pos: Res assoc med physics, Retina Found, 73- Mem: Phys Soc Japan; Biophys Soc Japan. Res: Laser scattering spectroscopy; critical phenomena of macromolecular solutions with applications to cataract and atherosclerosis; blood circulation; taxis of microorganisms; physics of gels. Mailing Add: Rm 13-2009 Mass Inst of Technol 77 Massachusetts Ave Cambridge MA 02139

TANAKA, YASUOMI, b Tokyo, Japan, Dec 5, 39; m 74; c 1. FOREST ECOLOGY. Educ: Tokyo Univ Educ, BS, 62; Duke Univ, MF, 67, PhD(forest ecol), 70. Prof Exp: Silviculturist, Agr Farm Monte D'Este, Brazil, 62-63; res fel forestry, Univ Sao Paulo, Brazil, 64-65; instr, Univ Parana, Brazil, 65-66; res assoc ecol, Ecol Sci Div, Oak Ridge Nat Lab, 70-71; FOREST NURSERY ECOLOGIST, WEYERHAEUSER FORESTRY RES CTR, 71- Mem: Sigma Xi; Japanese Forestry Soc; Soc Am Foresters. Res: Physiological and ecological aspect of seedling production; seedling dormancy, growth, nutrition in the greenhouse and openbed nursery; seed technology. Mailing Add: 505 N Pearl St Weyerhaeuser Forestry Res Ctr Centralia WA 98531

TANANBAUM, HARVEY DALE, b Buffalo, NY, July 17, 42; m 64; c 2. ASTROPHYSICS. Educ: Yale Univ, BA, 64; Mass Inst Technol, PhD(physics), 68. Prof Exp: Staff scientist, Am Sci & Eng, Inc, 68-73; ASTROPHYSICIST, SMITHSONIAN INST ASTROPHYS OBSERV, 73- Concurrent Pos: Assoc astron, Harvard Col Observ, Harvard Univ, 73- Mem: AAAS; Am Astron Soc. Res: X-ray astronomy, especially discrete cosmic x-ray sources with satellite payloads. Mailing Add: Ctr for Astrophys 60 Garden St Cambridge MA 02138

TANCZOS, FRANK I, b Northampton, Pa, Jan 2, 21. CHEMICAL PHYSICS. Educ: Moravian Col, BS, 39; Cath Univ, PhD(physics), 56. Prof Exp: Res chemist cent labs, Lehigh Portland Cement Co, Pa, 39-43; phys chemist, Bur Ord, 46-59, tech dir supporting res, Bur Naval Weapons, 59-66, TECH DIR RES & TECHNOL, AIR SYSTS COMMAND, DEPT NAVY, 66- Concurrent Pos: Mem, NASA res adv comts, Chem Energy Processes, 59-60, mem chem energy systs, 60-61 & air-breathing propulsion systs, 62-; lectr grad sch eng, Cath Univ, 60-61. Honors & Awards: Superior Civilian Serv Award, Dept Navy, 65. Mem: Am Chem Soc; Am Phys Soc; assoc fel Am Inst Aeronaut & Astronaut. Res: New cement composition chemistry; molecular vibrational relaxation theory; energy conversion; propellant chemistry and thermodynamics; rocket and air-breathing jet propulsion principles; hypersonic air-breathing propulsion principles. Mailing Add: 1500 Massachusetts Ave NW Washington DC 20005

TANDBERG-HANSSEN, EINAR ANDREAS, solar physics, astrophysics, see 12th edition

TANDLER, BERNARD, b Brooklyn, NY, Feb 18, 33; m 55; c 2. CYTOLOGY, ELECTRON MICROSCOPY. Educ: Brooklyn Col, BS, 55; Columbia Univ, AM, 57; Cornell Univ, PhD(cytol), 61. Prof Exp: Res fel, Sloan-Kettering Inst, 61-62; instr anat, Sch Med, NY Univ, 62-63; assoc biol, Sloan-Kettering Inst Cancer Res, Cornell Univ, 63-67, asst prof, Grad Sch, Sloan-Kettering Div, 66-67; assoc prof, 67-72, PROF ORAL BIOL & MED, SCH DENT, CASE WESTERN RESERVE UNIV, 72-, ASSOC PROF ANAT, SCH MED, 67- Concurrent Pos: Lectr, Brooklyn Col, 61-63; vis prof anat, Univ Copenhagen, 73; vis assoc prof, Sch Med, Stanford Univ, 75. Mem: Am Soc Cell Biol; Am Asn Anatomists; Electron Micros Soc Am; Int Asn Dent Res. Res: Mitochondrial biogenesis; ultrastructure of normal and neoplastic salivary glands; hepatic ultrastructure. Mailing Add: Dept Oral Biol & Med Sch of Dent Case Western Reserve Univ Cleveland OH 44106

TANEJA, VIDAY SAGAR, b India, Sept 7, 31; m 62; c 2. MATHEMATICAL STATISTICS. Educ: Panjab Univ India, BA, 50, MA, 52; Univ Minn, MA, 63; Univ Conn, PhD(statist), 66. Prof Exp: Lectr math, Doaba Col, 53-59; instr, Univ Minn, Morris, 64-65; asst prof math statist, NMex State Univ, 66-70; assoc prof, 70-74, PROF MATH STATIST, WESTERN ILL UNIV, 74- Mem: Inst Math Statist; Opers Res Soc Am; Am Statist Asn; Sigma Xi. Res: Statistical methodology, statistical analysis, time series and operations research. Mailing Add: Dept of Math Western Ill Univ Macomb IL 61455

TANENBAUM, BASIL SAMUEL, b Providence, RI, Dec 1, 34; m 56; c 3. IONOSPHERIC PHYSICS. Educ: Brown Univ, BS, 56; Yale Univ, MS, 57, PhD(physics), 60. Prof Exp: Staff physicist res div, Raytheon Co, 60-63; prof eng, Case Western Reserve Univ, 63-75; DEAN FAC, HARVEY MUDD COL, 75- Concurrent Pos: Vis scientist, Arecibo Observ, 67-68; Sigma Xi Res Award, 69; sci adv comt, Nat Astron & Ionospheric Ctr, 72- Mem: Am Phys Soc; AAAS; Inst Elec & Electronic Eng. Res: Sound propagation; theory of turbulence; ionospheric physics; waves in plasmas; radiation in a plasma; kinetic theory of gas mixtures and plasmas; shock wave theory; energy conversion; solar energy. Mailing Add: Dean of Fac Harvey Mudd Col Claremont CA 91711

TANENBAUM, STUART WILLIAM, b New York, NY, July 15, 24; m 62; c 2. BIOCHEMISTRY, MICROBIOLOGY. Educ: City Col New York, BS, 44; Columbia Univ, PhD, 51. Prof Exp: Instr chem, City Col New York, 48; Am Cancer Soc fel, Stanford Univ, 51-52; lectr bact, Univ Calif, 52; res assoc biol, Stanford Univ, 52-53; from res assoc to prof microbiol, Col Physicians & Surgeons, Columbia Univ, 53-73; DEAN SCH BIOL CHEM & ECOL, STATE UNIV NY COL ENVIRON SCI & FORESTRY, 73-; ADJ PROF MICROBIOL, STATE UNIV NY UPSTATE MED CTR, 73- Concurrent Pos: Mem panel molecular biol, NSF, 62-63 & 72-73, prog dir molecular biol sect, 71-72; State Univ NY fac exchange scholar, 74-; trustee, Forestry Fedn, Syracuse, NY, 74- Mem: Am Soc Biol Chemists; Brit Biochem Soc; Am Soc Microbiol; Am Chem Soc. Res: Fungal metabolism; antibiotic biosynthesis; bacterial physiology; immunochemistry; cell and molecular biology. Mailing Add: Sch of Biol Chem & Ecol SUNY Col Environ Sci & Forestry Syracuse NY 13210

TANFORD, CHARLES, b Halle, Ger, Dec 29, 21; nat US; div; c 3. BIOCHEMISTRY. Educ: NY Univ, BA, 43; Princeton Univ, MA, 44, PhD(phys chem), 47. Prof Exp: Asst, Princeton Univ, 43-44; chemist, Tenn Eastman Corp, 44-45; asst, Princeton Univ, 45-46; Lalor fel phys chem, Harvard Med Sch, 47-49; from asst prof to prof, Univ Iowa, 49-60; prof, 60-71, JAMES B DUKE PROF PHYS BIOCHEM, MED CTR, DUKE UNIV, 71- Concurrent Pos: Guggenheim fel, Yale Univ, 56-57; consult, USPHS, 59-63; USPHS res career award, 62; vis prof, Harvard Univ, 66. Mem: Nat Acad Sci; Am Acad Arts & Sci; AAAS; Am Chem Soc; Biophys Soc. Res: Physical chemistry of proteins and related substances; structure of membranes and lipoproteins; structure and function of antibodies. Mailing Add: Dept of Biochem Duke Univ Med Ctr Durham NC 27710

TANG, ALFRED SHO-YU, b Shanghai, China, Sept 9, 34; US citizen; m 69; c 2. ALGEBRA. Educ: Univ Hong Kong, BSc, 56; Univ SC, MS, 60; Univ Calif, Berkeley, PhD(math), 69. Prof Exp: ASSOC PROF MATH, SAN FRANCISCO STATE UNIV, 66- Mem: Am Math Soc; Math Asn Am. Mailing Add: Def Math San Francisco State Univ San Francisco CA 94132

TANG, ANDREW H, b Canton, China, Feb 10, 36; US citizen; m 64; c 2. PHARMACOLOGY. Educ: Howard Col, BS, 60; Purdue Univ, MS, 62, PhD(pharmacol), 64. Prof Exp: Res assoc, 64-70, SR RES SCIENTIST PHARMACOL, UPJOHN CO, 70- Mem: AAAS; Am Soc Pharmacol & Exp Therapeut. Res: Pharmacology of the central nervous system; spinal cord physiology; behavioral pharmacology. Mailing Add: Upjohn Co Kalamazoo MI 49001

TANG, CHUNG LIANG, b Shanghai, China, May 14, 34; US citizen; m 58; c 3. PHYSICS. Educ: Univ Wash, BS, 55; Calif Inst Technol, MS, 56; Harvard Univ, PhD(appl physics), 60. Prof Exp: Res staff mem, Raytheon Co, 60-61, sr res scientist, 61-63, prin res scientist, 63-64; assoc prof elec eng, 64-68, PROF ELEC ENG, CORNELL UNIV, 68- Concurrent Pos: Consult res div, Raytheon Co, 64-72; assoc ed, J Quantum Electronics, Inst Elec & Electronics Eng, 69- Mem: Fel Am Phys Soc. Res: Quantum electronics; electromagnetic theory. Mailing Add: Dept of Elec Eng Cornell Univ Ithaca NY 14850

TANG, CHUNG-MUH, b Tungkang, Taiwan, Oct 20, 36; m 65; c 2. METEOROLOGY. Educ: Nat Taiwan Univ, BS, 59; Univ Calif, Los Angeles, MA, 65, CPhil, 69, PhD(meteorol), 70. Prof Exp: Res asst, Taiwan Rain Stimulation Res Inst, 61-62; NSF grant & asst res meteorologist, Univ Calif, Los Angeles, 70; Defense Dept grant & res staff meteorologist, Yale Univ, 70-75; ASST PROF PHYSICS & ATMOSPHERIC SCI, DREXEL UNIV, 75- Mem: AAAS; Am Meteorol Soc. Res: Theoretical studies of large scale atmospheric motions. Mailing Add: Dept Physics & Atmospheric Sci Drexel Univ Philadelphia PA 19104

TANG, CHUNG-SHIH, b China, Jan 8, 38; m 65; c 2. PLANT CHEMISTRY. Educ: Taiwan Univ, BS, 60, MS, 62; Univ Calif, Davis, PhD(agr chem), 67. Prof Exp: Res chemist, Univ Calif, Davis, 67-68; asst prof agr biochem, 68-73, ASSOC PROF AGR BIOCHEM, UNIV HAWAII, 73- Mem: AAAS; Am Chem Soc; Am Soc Plant Physiol. Res: Volatile compounds of tropical fruit; naturally occurring isothiocyanates. Mailing Add: Dept Agr Biochem Univ of Hawaii 1825 Edmondson Rd Honolulu HI 96822

TANG, JAMES JUH-LING, b Tientsin, China, Mar 8, 37; US citizen; m 65; c 2. APPLIED MECHANICS, HEAT TRANSFER. Educ: Nat Taiwan Univ, BS, 59; Univ Mo-Rolla, MS, 63; Yale Univ, PhD(appl mech), 70. Prof Exp: Engr, Weiskopf & Pickworth, 63-65; sr res engr, 70-75, RES ASSOC, AM CAN CO, 75- Mem: Am Soc Mech Engrs; Soc Rheology; Am Inst Physics. Res: Elastic-plastic material behavior; heat transfer of industrial processes; computerized process control. Mailing Add: 30 Windsor Dr Princeton Junction NJ 08550

TANG, JORDAN J N, b Foochow, China, Mar 23, 31; m 58; c 2. BIOCHEMISTRY. Educ: Taiwan Prov Col, BS, 53; Okla State Univ, MS, 57; Univ Okla, PhD(biochem), 61. Prof Exp: Res asst biochem, Okla State Univ, 55-57; res asst, 57-58, biochemist, 61-63, assoc biochem, 63-65, assoc prof, 65-69, head neurosci sect & actg head, Found, 70-71, HEAD LAB PROTEIN STUDIES, OKLA MED RES FOUND, 71-; PROF BIOCHEM, SCH MED, UNIV OKLA, 71- Concurrent Pos: Res assoc biochem, Sch Med, Univ Okla, 62-63, asst prof & asst head dept, 63-67, assoc prof, 67-70; vis scientist, Lab Molecular Biol, Cambridge, Eng, 65-66. Mem: AAAS; Am Chem Soc; Am Soc Biol Chemists. Res: Structure and function of proteins; structure of gastric proteolytic enzymes. Mailing Add: Okla Med Res Found 825 NE 13th St Oklahoma City OK 73104

TANG, KWONG-TIN, b Feb 24, 36; US citizen. PHYSICS, PHYSICAL CHEMISTRY. Educ: Univ Wash, BS, 58, MA, 59; Columbia Univ, PhD(physics), 65. Prof Exp: Res assoc chem, Columbia Univ, 65-66; physicist, Collins Radio Co, 66-67; from asst prof to assoc prof, 67-72, PROF PHYSICS & CHMN DEPT, PAC LUTHERAN UNIV, 72- Concurrent Pos: Consult, Boeing Co, 70 & 72; vis lectr, Univ Wash, 71; Res Corp grant, Pac Lutheran Univ, 71-74. Mem: Am Phys Soc; Am Asn Physics Teachers. Res: Atomic and molecular collision; scattering theory; reaction rates; intermolecular forces; optical dispersion; lattice vibration. Mailing Add: Dept of Physics Pac Lutheran Univ Tacoma WA 98447

TANG, LUN HAN, b Kwangtung, China, Nov 21, 29; m 63; c 2. NUCLEAR SCIENCE, ENGINEERING. Educ: Univ Taiwan, BS, 53, MS, 59; Univ Mich, MS, 62, PhD(nuclear sci), 65. Prof Exp: Instr mech eng, Univ Taiwan, 53-57; res asst nuclear eng, Univ Mich, 62-64; asst prof, 64-68, ASSOC PROF PHYSICS, UNIV DETROIT, 68- Mem: Am Phys Soc. Res: Perturbed gamma-gamma angular correlations; nuclear bound state problems. Mailing Add: Dept of Physics Univ of Detroit Detroit MI 48221

TANG, PEI CHIN, b Hupei, China, Sept 14, 14; nat US; m; c 6. NEUROPHYSIOLOGY, NEUROANATOMY. Educ: Tsing Hua Univ, China, BS, 42; Univ Wash, PhD(physiol), 53. Prof Exp: Instr physiol, Med Sch, Peking Univ, 45-48; instr pharmacol, Univ Wash, 52-55; asst prof anat, Univ Tex Southwestern Med Sch Dallas, 55-56; neurophysiologist, Air Force Sch Aviation Med, 56-60; chief neurophysiol br, Civil Aeromed Res Inst, Fed Aviation Agency, Okla, 60-66; chief physiol sci div, Res Dept, Naval Aerospace Med Inst, 66-70; ASSOC PROF PHYSIOL, CHICAGO MED SCH, 70- Mem: Am Physiol Soc; Am Asn Anat; Aerospace Med Asn; Soc Exp Biol & Med. Res: Central nervous control of respiration, micturition and vasomotor activity. Mailing Add: Dept of Physiol Chicago Med Sch Chicago IL 60612

TANG, ROBERT CHENG-WEI, horticulture, see 12th edition

TANG, RUEN CHIU, b Kiangsu, China, Oct 31, 34; m 60; c 3. FOREST PRODUCTS. Educ: Nat Chung-Hsin Univ, Taiwan, BS, 57; NC State Univ, PhD(wood sci), 68. Prof Exp: Teacher, Kung Hua Sch Elec Technol, 56-57; wood technologist, Taiwan Forest Bur, 59-63; res asst, US Naval Res, NC State Univ, 63-66; teaching asst wood mech, Univ Wash, 66-67; State of Ky & US Air Force res assoc, Inst Theoret & Appl Mech, 68, State of Ky res assoc wood mech, Univ, 69, asst prof wood sci, 70-73, ASSOC PROF WOOD SCI, UNIV KY, 74- Mem: Soc Wood Sci & Technol; Soc Exp Stress Anal; Forest Prod Res Soc; Soc Am Foresters. Res: Anisotropic elasticity; composite materials; fiber mechanics; noise control. Mailing Add: Dept of Forestry Univ of Ky Lexington KY 40506

TANG, TERRY CHU, b Kwei-chou, China; US citizen. MICROBIOLOGY. Educ: Mt Marty Col, BA, 62; Ill Inst Technol, MS, 67, PhD(microbiol), 70. Prof Exp: ASST PROF MICROBIOL, VA COMMONWEALTH UNIV, 73- Mem: Am Soc Microbiol. Res: Physiology and metabolic regulation of pathogenic spirochaetes. Mailing Add: Dept of Microbiol Box 847 MCV Sta Va Commonwealth Univ Richmond VA 23298

TANG, VICTOR KUANG-TAO, b Peiping, China, Mar 13, 29; US citizen; m 63. STATISTICS, MATHEMATICS. Educ: Nat Taiwan Univ, BA, 56; Univ Wash, MA, 63; Iowa State Univ, PhD(statist), 71. Prof Exp: ASSOC PROF MATH, HUMBOLDT STATE UNIV, 63- Mem: Inst Math Statist; Am Statist Asn. Res: Applied and mathematical statistics. Mailing Add: Dept of Math Humboldt State Univ Arcata CA 95521

TANG, WEN, b Nanking, China, Sept 19, 21; m 48; c 5. DYNAMIC METEOROLOGY. Educ: Nat Cent Univ, China, BS, 45; NY Univ, MS, 58, PhD(meteorol), 60. Prof Exp: Teacher, Rotary High Sch, 45-46; weather officer meteorol, Chinese Govt, 46-56; from asst scientist to assoc scientist, NY Univ, 56-60; staff scientist, GCA Corp, 60-68; asst prof meteorol, 68-70, ASSOC PROF METEOROL, LOWELL TECHNOL INST, 70- Mem: Am Meteorol Soc; Am Geophys Union. Res: Synoptic meteorology; atmospheric circulations of various scales. Mailing Add: Dept of Meteorol Lowell Technol Inst Lowell MA 01854

TANG, YAU-CHIEN, b China, Aug 7, 28; nat US; m 60; c 2. PHYSICS. Educ: Univ Ill, PhD(physics), 58. Prof Exp: Res assoc, Fla State Univ, 58-62 & Brookhaven Nat Lab, 62-64; assoc prof, 64-74, PROF PHYSICS, UNIV MINN, MINNEAPOLIS, 74- Mem: Am Phys Soc. Res: Low energy nuclear physics. Mailing Add: Dept of Physics & Astron Univ of Minn Minneapolis MN 55455

TANG, YI-NOO, b Hunan, China, Feb 28, 38; m 64; c 2. PHYSICAL CHEMISTRY, RADIOCHEMISTRY. Educ: Chung Chi Col, Hong Kong, BA, 59; Univ Kans, PhD(chem), 64. Prof Exp: Fel, Univ Kans, 64-65; fel, Univ Calif, Irvine, 65-66, instr chem, 66-67; asst prof, 67-69, ASSOC PROF CHEM, TEX A&M UNIV, 69- Mem: Am Chem Soc; Am Inst Physics. Res: Hot atom chemistry; unimolecular reactions; photochemistry; carbene chemistry; gas chromatography; silicon chemistry. Mailing Add: Dept of Chem Tex A&M Univ College Station TX 77843

TANGHERLINI, FRANK R, b Boston, Mass, Mar 14, 24; m 60; c 4. PHYSICS. Educ: Harvard Univ, BS, 48; Univ Chicago, MS, 52; Stanford Univ, PhD(physics), 59. Prof Exp: NSF fel physics, Niels Bohr Inst, Copenhagen, Denmark, 58-59 & Sch Theoret & Nuclear Physics, Naples, Italy, 59-60; res assoc, Univ NC, Chapel Hill, 60-61; asst prof, Duke Univ, 61-64; assoc prof, George Washington Univ, 64-66; sci assoc space res, Ion Lab, Tech Univ Denmark & Danish Space Res Inst, 66-67; ASSOC PROF PHYSICS, COL HOLY CROSS, 67- Concurrent Pos: Sr res engr, Gen Dynamics/Convair, 52-55; lectr, Univ NC, Chapel Hill, 61; vis scientist, Int Ctr Theoret Physics, 73-74. Mem: AAAS; Am Phys Soc; Ital Phys Soc. Res: Mathematical biology; cybernetics; high energy physics; foundations of special relativity; general relativity; dimensionality of space; electron theory; ionosphere and space physics; theoretical physics; pionmuon model; exotic atoms, relativity and quantum optics. Mailing Add: Dept of Physics Col of the Holy Cross Worcester MA 01610

TANGORA, MARTIN CHARLES, b New York, NY, June 21, 36. TOPOLOGY. Educ: Calif Inst Technol, BS, 57; Northwestern Univ, MS, 58, PhD(math), 66. Prof Exp:

Instr math, Northwestern Univ, 66-67 & Univ Chicago, 67-69; temp lectr, Univ Manchester, 69-70; asst prof, 70-72, ASSOC PROF MATH, UNIV ILL, CHICAGO CIRCLE, 72- Concurrent Pos: Sr vis fel, Univ Oxford, 73-74. Mem: Am Math Soc; Math Asn Am. Res: Algebraic topology; homotopy theory; cohomological methods. Mailing Add: Dept of Math Box 4348 Univ of Ill Chicago Circle Chicago IL 60680

TANI, SMIO, b Tokyo, Japan, Feb 24, 25; m 58; c 3. THEORETICAL PHYSICS. Educ: Univ Tokyo, BS, 46, ScD, 55. Prof Exp: Res assoc, Kyoto Univ, 51-54, Tokyo Univ Educ, 54-57, Case Western Reserve Univ, 57-59 & Wash Univ, 59-60; from res scientist to sr res scientist physics, NY Univ, 60-65; assoc prof, 65-68, PROF PHYSICS, MARQUETTE UNIV, 68- Mem: Am Phys Soc. Res: Theory of scattering; canonical transformation in classical and quantum mechanics; atomic physics; elementary particle physics. Mailing Add: Dept of Physics Marquette Univ Milwaukee WI 53233

TANIKELLA, MURTY SUNDARA SITARAMA, b Amalapuram, India, Dec 5, 38; m 67. PHYSICAL CHEMISTRY, POLYMER CHEMISTRY. Educ: Osmania Univ, India, BSc, 57, MSc, 59; Princeton Univ, MA, 64; Univ Pittsburgh, PhD(phys, physico-org chem), 67. Prof Exp: Lectr chem, Osmania Univ, India, 59-62; fel with Prof K S Pitzer, Rice Univ, 67-68; fel, Nat Res Coun Can, 68-69; Nat Res Coun Can fel, Univ Calgary, 69-70; res chemist, Carothers Res Lab, 70-74, SR RES CHEMIST, CAROTHERS RES LAB, E I DU PONT DE NEMOURS & CO, INC, 74- Concurrent Pos: Fulbright travel grant, 62. Mem: Am Chem Soc. Res: Thermodynamics of hydrogen bonding; spectroscopy; structure of water; problems in textile fibers chemistry. Mailing Add: Carothers Res Lab E I du Pont de Nemours & Co Inc Wilmington DE 19898

TANIMOTO, DOUGLAS HIDETOSHI, physics, quantum optics, see 12th edition

TANIMOTO, TAFFEE TADASHI, b Kobe, Japan, Dec 15, 17; nat US; m 46; c 4. MATHEMATICS, GEOMETRY. Educ: Univ Calif, Los Angeles, AB, 42; Univ Chicago, MS, 46; Univ Pittsburgh, PhD(math), 50. Prof Exp: Instr math, Ill Inst Technol, 46-49; instr math, Allegheny Col, 49-51, asst prof, 51-54; mathematician, Int Bus Mach Corp, 54-61; head pattern recognition lab, Melpar, Inc, 61-63; staff mathematician, Honeywell, Inc, 63-65; PROF MATH & CHMN DEPT, UNIV MASS, BOSTON, 65-, DIR GRAD PROGS, 75- Mem: Am Math Soc; Math Asn Am. Res: Geometry and analysis. Mailing Add: Dept of Math Univ of Mass Harbor Campus Boston MA 02125

TANIS, ELLIOT ALAN, b Grand Rapids, Mich, Apr 23, 34; m; c 3. MATHEMATICS, STATISTICS. Educ: Cent Col, Iowa, BA, 56; Univ Iowa, MS, 60, PhD(math), 63. Prof Exp: Lectr math, Univ Iowa, 63; asst prof math statist, Univ Nebr, 63-65; assoc prof math, 65-71, PROF MATH & CHMN DEPT, HOPE COL, 71- Mem: Am Math Soc; Math Asn Am; Inst Math Statist; Am Statist Asn. Res: Mathematical statistics; distribution theory; use of the computer in statistics. Mailing Add: Dept of Math Hope Col Holland MI 49423

TANIS, JAMES IRAN, b Zeeland, Mich, Oct 8, 34; m 70; c 1. EXPLORATION GEOPHYSICS. Educ: Mich Technol Univ, BS, 57, MS, 58; Univ Utah, PhD(geophys), 63. Prof Exp: Geophysicist, Shell Oil Co, 62-70; GEOPHYSICIST, CONTINENTAL OIL CO, 70- Mem: Am Geophys Union; Soc Explor Geophysicists. Res: Crustal studies. Mailing Add: Continental Oil Co Box 1267 Ponca City OK 74601

TANJI, KENNETH K, b Honolulu, Hawaii, Jan 13, 32; m 52; c 3. WATER POLLUTION. Educ: Univ Hawaii, BA, 54; Univ Calif, Davis, MS, 61. Prof Exp: Lab technician, 58-67, assoc specialist water & soil chem, 67-70, SPECIALIST WATER & SOIL CHEM, UNIV CALIF, DAVIS, 71-, LECTR WATER SCI, 71- Mem: AAAS; Am Soc Agron; fel Am Inst Chemists; Am Chem Soc; Int Soc Soil Sci. Res: Computer modeling; aquatic chemistry; water quality and pollution. Mailing Add: 537 Reed Dr Davis CA 95616

TANK, RONALD W, b Milwaukee, Wis, June 14, 29; m 55; c 2. GEOLOGY. Educ: Univ Wis, BS, 51, MS, 55; Univ Copenhagen, dipl geol, 60; Ind Univ, PhD(geol), 62. Prof Exp: Explor geologist, Stand Oil Co Calif, 55-59; asst prof geol & cur, 62-64, ASSOC PROF GEOL, LAWRENCE UNIV, 64- Mem: Soc Econ Paleont & Mineral; Nat Asn Geol Teachers. Res: Clay mineralogy; economic geology; environmental geology. Mailing Add: Dept of Geol Lawrence Univ Appleton WI 54911

TANKERSLEY, ROBERT WALKER, JR, b Watsonville, Calif, June 18, 27; m 51; c 3. VIROLOGY, BACTERIOLOGY. Educ: Stanford Univ, AB, 52, MA, 54, PhD(med microbiol), 56. Prof Exp: Instr virol, Med Sch, Univ Minn, 56-58, instr bact & virol, 58-60; from asst prof to assoc prof microbiol, Med Col Va, 60-68; DIR MICROBIOL RES, A H ROBINS PHARMACEUT CO, 68- Concurrent Pos: USPHS fel, Med Sch, Univ Minn, 56-58. Mem: Am Soc Microbiol. Res: Antiviral chemotherapy; cell-virus relationships. Mailing Add: 1211 Sherwood Ave Richmond VA 23220

TANNAHILL, MARY MARGARET, b Weatherford, Tex, Apr 30, 44. POLYMER CHEMISTRY. Educ: Tex Tech Univ, BS, 66; Mich State Univ, PhD(phys chem), 73. Prof Exp: Asst chem, Mich State Univ, 66-72; trainee physiol, Univ Tex Med Br, Galveston, 72-73; res chemist, Union Carbide Corp, 74-75; TECH SPECIALIST POLYMER CHEM, GULF OIL CHEM CO, 75- Concurrent Pos: Consult, Shriners Burns Inst, Galveston, Tex, 74- Mem: Am Chem Soc. Res: Product development for polypropylene; physiology of burns. Mailing Add: Gulf Oil Chem Co Chemical Row Orange TX 77630

TANNENBAUM, CARL MARTIN, b New York, NY, Apr 1, 40. BIOCHEMISTRY. Educ: City Col New York, BS, 60; Univ Ariz, MS, 68; Univ Nebr, PhD(chem), 74. Prof Exp: Chemist, Ciba Pharmaceut Co, 61-64; fel physiol, Med Sch, Yale Univ, 72-75; ASST PROF BIOCHEM, SWISS FED INST TECHNOL, 75- Mem: Am Chem Soc. Res: Epithelial membrane transport of sugars and amino acids; characterization of the factors involved in transport. Mailing Add: Biochem Lab Swiss Fed Inst Technol Universitatstr 16 CH-8006 Zurich Switzerland

TANNENBAUM, HAROLD E, b New York, NY, Dec 31, 14; m 37; c 2. SCIENCE EDUCATION. Educ: Columbia Univ, MA, 37, EdD, 50. Prof Exp: Teacher, Park Sch, Ohio, 37-42; head sci dept, Elisabeth Irwin High Sch, NY, 44-52; prof sci educ, State Univ NY Teachers Col, New Paltz, 52-61; chmn div curric & instruct, Grad Sch Educ, Yeshiva Univ, 61-64; PROF SCI EDUC, HUNTER COL, 64-, CHMN DEPT CURRIC & TEACHING, 68- Concurrent Pos: Sci Manpower Comt fel, Columbia Univ, 58-59; consult, State Educ Depts, NH, Va, NDak & NY. Mem: AAAS; Nat Sci Teachers Asn. Mailing Add: Dept of Curric & Teaching Hunter Col 695 Park Ave New York NY 10021

TANNENBAUM, HARVEY, b New York, NY, June 26, 23; m 46; c 3. PHYSICAL CHEMISTRY. Educ: NY Univ, BS, 48. Prof Exp: Mem staff, 49-65, CHIEF REMOTE SENSING, DEFENSE SYST DIV, DIR DEVELOP & ENG, EDGEWOOD ARSENAL, 65- Concurrent Pos: Partic, NATO Experts Panel Laser Remote Sensing of Atmosphere, 75-; mem, Joint Army Navy NASA Air Force Comt Propulsion Hazards, 75- Mem: Optical Soc Am; Sigma Xi. Res: Infrared physics; trace gas detection; pollution monitoring instrumentation; remote sensing instrumentation; electro-optical systems; spectroscopy. Mailing Add: Dir of Develop & Eng Edgewood Arsenal Aberdeen Proving Ground MD 21010

TANNENBAUM, IRVING ROBERT b Spring Lake, NJ, Feb 24, 26; m 51; c 4. PHYSICAL CHEMISTRY. Educ: Va Polytech Inst, BS, 46, MS, 48; Univ Ill, PhD(phys chem), 51. Prof Exp: Instr math, Va Polytech Inst, 46-47; asst phys chem, Univ Ill, 47-50; mem staff chem res, Los Alamos Sci Lab, Univ Calif, 51-56; sr phys chemist, Atomics Int Div, NAm Aviation, Inc, 56-61; scientist, Electro-Optical Systs, Inc, 61-63; sr scientist, Heliodyne Corp, 63-65; PRES & TECH DIR CHEMATICS RES CORP, RESEDA, 65- Concurrent Pos: Adj prof, Univ Calif, Los Angeles, 58- Mem: Am Chem Soc. Res: Theory of liquid mixtures; inorganic and plutonium chemistry; hydrides; x-ray crystallography; chemical kinetics; re-entry physics. Mailing Add: 8354 Etiwanda Ave Northridge CA 91324

TANNENBAUM, MICHAEL J, b Bronx, NY, Mar 10, 39. HIGH ENERGY PHYSICS. Educ: Columbia Univ, AB, 59, MA, 60, PhD(physics), 65. Prof Exp: Vis scientist, Europ Orgn Nuclear Res, 65-66; from asst prof to assoc prof physics, Harvard Univ, 66-71; ASSOC PROF PHYSICS, ROCKEFELLER UNIV, 71- Concurrent Pos: Ernest Kempton Adams traveling fel from Columbia Univ, 65-66; NSF fel, 66; Alfred P Sloan Found fel, 67-69; mem prog adv comt, Fermi Nat Lab, 72-75. Mem: Am Phys Soc. Res: Muon elastic and inelastic scattering; muon g-2; muon tridents; photoproduction with a tagged beam; single leptons, lepton pairs and other new phenomena in proton-proton interactions. Mailing Add: Dept of Exp Physics Rockefeller Univ New York NY 10021

TANNENBAUM, ROBERT S, b Cleveland, Ohio, Apr 17, 42; m 63; c 2. COMPUTER SCIENCE. Educ: Haverford Col, BA, 62; Columbia Univ, MA, 65, EdD(sci educ), 68. Prof Exp: Comput asst, Standard Oil Co, NJ , 60-61; teacher high schs, Ill & NY, 62-65; instr sci educ, Teachers Col, Columbia Univ, 65-67, asst & assoc, Comput Ctr, 66-68; ASSOC PROF MED COMPUT SCI & CHMN DEPT, SCH HEALTH SCI, HUNTER COL, CITY UNIV NEW YORK, 68- Concurrent Pos: Proj evaluator, Environ Health Sci Masters Deg, Hunter Col, 71-; consult, Consortium Occup Ther Educ, 74- Mem: Asn Develop Computerized Instrnl Systs; Soc Advan Med Systs; Asn Comput Mach; AAAS; Nat Sci Teachers Asn. Res: Applications of computers in health science education and in medicine and related health sciences. Mailing Add: Med Comput Sci Sch Health Sci Hunter Col Rm 213 118 E 107th St New York NY 10029

TANNENBAUM, STANLEY, b New York, NY, Mar 1, 25; m 47; c 4. INORGANIC CHEMISTRY. Educ: City Col New York, BS, 46; Ohio State Univ, PhD(chem), 49. Prof Exp: Res chemist, Nat Adv Comt Aeronaut, 50-53; res chemist reaction motors div, Thiokol Chem Corp, 53-69; prog mgr, 69-74, MGR TECH SERV, RONSON METALS CORP, 74- Mem: Am Chem Soc. Res: Synthesis of silicon and boron containing chemicals; physical and thermochemical properties of materials; alteration of properties of rocket propellants; determination of fire and explosive hazards of chemicals and chemical processes. Mailing Add: 19 Raynor Rd Morristown NJ 07960

TANNENBAUM, STEVEN ROBERT, b Brooklyn, NY, Feb 23, 37; m 59; c 2. FOOD CHEMISTRY. Educ: Mass Inst Technol, BS, 58, PhD(food sci), 62. Prof Exp: Res assoc food technol, 62-63, from instr to assoc prof food sci, 63-74, PROF FOOD CHEM, MASS INST TECHNOL, 74- Concurrent Pos: Consult, UN Develop Prog & Protein Adv Group; mem panel food supply, White House Conf Food, Nutrit & Health; mem comt food stand & fortification policy, Nat Acad Sci-Nat Res Coun; vis prof, Hebrew Univ Jerusalem, 73-74. Honors & Awards: Samuel Cate Prescott Award, Inst Food Technologists, 70. Mem: Am Chem Soc; Inst Food Technologists; Am Inst Nutrit; Brit Nutrit Soc; fel Brit Inst Food Sci & Technol. . Res: Chemistry of nitrates, nitrites and N-nitroso compounds; chemical carcinogens; enzymes; protein technology. Mailing Add: Dept Nutrit & Food Sci Rm 16-209 Mass Inst Technol Cambridge MA 02139

TANNENWALD, LUDWIG MAX, b Frankfurt, Ger, Feb 21, 28; nat US; m 63; c 2. OPTICAL PHYSICS. Educ: Univ Chicago, BS, 46, MS, 49, PhD(physics), 51. Prof Exp: Physicist, US Naval Ord Test Sta, 51-55, assoc res scientist, 55-56, res scientist, 56-58, group leader, 58-59, staff scientist, 59-72, STAFF SCIENTIST, LOCKHEED MISSILE & SPACE CO, 72- Concurrent Pos: Res physicist, Univ Calif, Berkeley, 71-72. Mem: Am Phys Soc. Res: Quantum mechanical calculations of collision processes and statistical mechanics. Mailing Add: Lockheed Missile & Space Co 3251 Hanover Palo Alto CA 94304

TANNENWALD, PETER ERNEST, b Kiel, Ger, May 30, 26; nat US. PHYSICS. Educ: Univ Calif, AB, 47, PhD(physics), 52. Prof Exp: Asst physics, Univ Calif, 47-51, Radiation Lab, 50-52; physicist, 52-59, asst group leader, 59-62, group leader, 62-63, asst head, 63-65, assoc head, 65-74, SR SCIENTIST, SOLID STATE DIV, LINCOLN LAB, MASS INST TECHNOL, 74- Mem: Fel Am Phys Soc. Res: Solid state physics and quantum electronics; microwave resonance in ferrites; spin wave resonance in magnetic films; masers; millimeter waves; microwave ultrasonics; lasers; Raman and Brillouin spectroscopy; far infrared quantum electronics and submillimeter wave technology. Mailing Add: Solid State Div Lincoln Lab Mass Inst of Technol Lexington MA 02173

TANNER, ALAN ROGER, b Port Lavaca, Tex, Jan 2, 41; m 69; c 1. INDUSTRIAL ORGANIC CHEMISTRY, ELECTROCHEMISTRY. Educ: Univ Tex, Austin, BS, 64, PhD(org chem), 69. Prof Exp: Appln res chemist, Jefferson Chem Co, 69-71; RES & DEVELOP CHEMIST, SOUTHWESTERN ANAL CHEM, INC, 71- Concurrent Pos: Instr org chem, St Edward's Univ, 74-75. Mem: Am Chem Soc. Res: Conventional and electrochemical synthetic approaches to new and existing marketable products. Mailing Add: Southwestern Anal Chem Inc 821 E Woodward Austin TX 78704

TANNER, ALLAN BAIN, b New York, NY, May 27, 30; m 65; c 2. GEOPHYSICS, GEOCHEMISTRY. Educ: Mass Inst Technol, SB, 52. Prof Exp: Geophysicist theoret geophys br, US Geol Surv, DC, 54-69, GEOPHYSICIST ISOTOPE BR, US GEOL SURV, 69- Concurrent Pos: Comt mem, Nat Coun Radiation Protection & Measurements, 73-74. Mem: Geochem Soc; Soc Explor Geophys; Am Geophys Union. Res: Behavior of radon isotopes in natural environments; nuclear geophysics and geochemistry; x-ray fluorescence; isotope geology; gamma-ray spectrometry; in situ neutron activation analysis; health physics. Mailing Add: US Geol Surv 990 Nat Ctr Reston VA 22092

TANNER, CHAMP BEAN, b Idaho Falls, Idaho, Nov 16, 20; m 41; c 5. MICROCLIMATOLOGY, SOIL PHYSICS. Educ: Brigham Young Univ, BS, 42; Univ Wis, PhD(soils), 50. Prof Exp: From instr to assoc prof soils, 50-61, PROF METEOROL & SOIL SCI, UNIV WIS-MADISON, 61- Mem: AAAS; Soil Sci Soc Am; Am Soc Agron; Am Geophys Union; Ecol Soc Am. Res: Evapotranspiration;

plant environment; soil moisture and structure. Mailing Add: Dept of Soil Sci Univ of Wis Madison WI 53706

TANNER, CHARLES E, b Preston, Cuba, Sept 10, 32; Can citizen; m 57; c 2. IMMUNOLOGY, PARASITOLOGY. Educ: Purdue Univ, BS, 53; McGill Univ, MS, 56, PhD, 57. Prof Exp: Teaching fel bact & immunol, McGill Univ, 55-57; Nat Res Coun Can overseas fel, 57-58; from asst prof to assoc prof parasitol, 58-71, PROF PARASITOL, INST PARASITOL, MACDONALD COL, McGILL UNIV, 71-, ASSOC MEM MICROBIOL, FAC AGR, UNIV, 73- Concurrent Pos: Mem, Int Comn Trichinellosis, 73- Mem: AAAS; Am Soc Parasitologists; Can Soc Microbiol; Can Soc Immunol; Can Soc Zool. Res: Immunology of host-parasite relations; immunochemistry of parasite antigens; serological diagnosis of parasite infections. Mailing Add: Inst of Parasitol Macdonald Col McGill Univ Ste Anne de Bellevue PQ Can

TANNER, CLARA LEE, b Biscoe, NC, May 28, 05; m 36; c 1. CULTURAL ANTHROPOLOGY, ARCHAEOLOGY. Educ: Univ Ariz, AB, 27, MA, 28. Prof Exp: From instr to assoc prof anthrop, 28-68, PROF ANTHROP, UNIV ARIZ, 68- Concurrent Pos: Vis prof anthrop, Univ Denver, 48. Mem: Fel Am Anthrop Asn; Soc Am Archaeol; Sigma Xi. Res: Recent and contemporary Southwest Indian crafts and easel art; prehistoric Southwest Indian art; detailed analysis of Apache and all Southwestern Indian baskets. Mailing Add: Dept of Anthrop Univ of Ariz Tucson AZ 85721

TANNER, DAVID, b Brooklyn, NY, Aug 7, 28; m 53; c 4. POLYMER CHEMISTRY. Educ: NY Univ, BA, 50; Brooklyn Polytech Inst, PhD(polymer chem), 54. Prof Exp: Res asst polymer chem, Univ Ill, 53-54; res chemist exp sta, Nylon Res Div, 54-59, res assoc, 59-62, supvr res, 62-63, sr supvr technol, Nylon Tech Div, 63-65, res mgr, Orlon-Lycra Res Div, 65-68, lab dir, Benger Lab, 68-69, tech mgr, Dacron Div, 70-72, TECH MGR, ORLON-ACETATE-LYCRA DIV, E I DU PONT DE NEMOURS & CO, INC, 72- Mem: Am Chem Soc. Res: Organic synthesis; radiation chemistry; fiber technology including polyamides, acrylics, elastomers and high temperature fibers. Mailing Add: Textile Fibers Dept Nemours Bldg E I du Pont de Nemours & Co Inc Wilmington DE 19898

TANNER, DAVID JOHN, b Redfield, SDak, May 7, 42; m 61; c 2. LOW TEMPERATURE PHYSICS. Educ: Univ Chicago, SB, 60, SM, 61, PhD, 66. Prof Exp: Asst prof physics, Wayne State Univ, 65-70, lead programmer, 70-72; PHYSICIST-PROGRAMMER, KMS FUSION, INC, 72- Mem: Am Phys Soc. Res: Ions in liquid helium; laser-plasma interaction. Mailing Add: KMS Fusion Inc PO Box 1567 3941 Research Park Dr Ann Arbor MI 48106

TANNER, DENNIS DAVID, b Montreal, Que, Mar 6, 30; US citizen; m 60; c 2. ORGANIC CHEMISTRY. Educ: Univ Calif, Los Angeles, BSc, 53; Stanford Univ, MSc, 57; Univ Colo, PhD(chem), 61. Prof Exp: Asst, Stanford Univ, 56-57 & Univ Colo, 57-60; res fel chem, Columbia Univ, 61-63; annual asst prof, 63-65, from asst prof to assoc prof, 65-75, PROF CHEM, UNIV ALTA, 75- Mem: Am Chem Soc; Chem Inst Can. Res: Mechanisms of free radical reactions; free radical and ionic rearrangement mechanisms; organic photochemistry. Mailing Add: Dept of Chem Univ of Alta Edmonton AB Can

TANNER, EARL C, b Providence, RI, Nov 16, 19; m 57; c 2. PHYSICS. Educ: Brown Univ, AB, 41, AM, 47, ScM, 59; Harvard Univ, PhD(hist), 51. Prof Exp: Asst to dir, Plasma Physics Lab, 58-64, ASST DIR, PLASMA PHYSICS LAB, PRINCETON UNIV, 64- Mem: Am Phys Soc. Res: Plasma physics. Mailing Add: Plasma Physics Lab Princeton Univ PO Box 451 Princeton NJ 08540

TANNER, FRED WILBUR, JR, bacteriology, see 12th edition

TANNER, GARY DALE, b Cherokee, Okla, Oct 27, 42; m 65; c 3. ENTOMOLOGY, BIOLOGY. Educ: Bob Jones Univ, BS, 64; Clemson Univ, MS, 66; Va Polytech Inst & State Univ, PhD(entom), 71. Prof Exp: Res asst entom, Clemson Univ, 64-66; asst chief eval sect, Aedes Aegypti Eradication Prog, Nat Ctr Dis Control, Dept Health, Educ & Welfare, Ga, 66-68; ASST PROF BIOL, GRACE COL, 71- Mem: AAAS; Entom Soc Am; Am Mosquito Control Asn. Res: Host preferences and feeding activities of Culicoides biting gnats. Mailing Add: Dept of Biol Grace Col Winona Lake IN 46590

TANNER, GEORGE ALBERT, b Vienna, Austria, Aug 2, 38; US citizen; m 62; c 2. PHYSIOLOGY. Educ: Cornell Univ, AB, 59; Harvard Univ, PhD(physiol), 64. Prof Exp: Nat Heart Inst res trainee physiol, Med Col, Cornell Univ, 64-67; asst prof, 67-72, ASSOC PROF PHYSIOL, SCH MED, IND UNIV, INDIANAPOLIS, 72- Concurrent Pos: Vis prof, Heidelberg Univ, 74-75. Mem: Int Soc Nephrology; Am Soc Nephrology; Am Physiol Soc. Res: Renal function. Mailing Add: Dept of Physiol Ind Univ Med Ctr Indianapolis IN 46202

TANNER, HOWARD ALLEN, fisheries, see 12th edition

TANNER, JACK WILLIAM, agronomy, see 12th edition

TANNER, JAMES GORDON, b St Thomas, Ont, June 11, 31; m 56; c 3. GEOPHYSICS. Educ: Univ Western Ont, BSc, 54, MSc, 58; Univ Durham, PhD, 69. Prof Exp: Res scientist, Dom Observ, 54-72, CHIEF GRAVITY & GEODYNAMICS DIV, EARTH PHYSICS BR, SCI & TECHNOL SECTOR, DEPT ENERGY, MINES & RESOURCES, CAN GOVT, 72- Concurrent Pos: Demonstr, Univ Western Ont, 57. Mem: Am Geophys Union; Geol Asn Can. Res: Solid earth geophysics; structural studies of earth's crust; numerical analysis; potential theory. Mailing Add: Earth Phys Br Sci & Technol Sect Dept of Energy Mines & Resources Ottawa ON Can

TANNER, JAMES MERVIL, b Jesup, Ga, Dec 29, 34; m 56; c 3. THEORETICAL PHYSICS. Educ: Ga Inst Technol, BS, 56, MS, 61, PhD(physics), 64. Prof Exp: Assoc scientist, Westinghouse Elec Corp, 56-58; instr physics, Ga Inst Technol, 58-62, asst prof, 64-65; assoc prof, Univ NC, Charlotte, 65-67; ASSOC PROF PHYSICS, GA INST TECHNOL, 67- Mem: Am Phys Soc; Am Asn Physics Teachers. Res: Mathematical physics. Mailing Add: Dept of Physics Ga Inst of Technol Atlanta GA 30332

TANNER, JAMES TAYLOR, b Homer, NY, Mar 6, 14; m 41; c 3. ANIMAL ECOLOGY. Educ: Cornell Univ, BS, 35, MS, 36, PhD(ornith), 40. Prof Exp: Asst prof biol, Tenn State Teachers Col, Johnson City, 40-42 & ETenn State Col, 46; from asst prof to assoc prof zool, 47-63, dir grad prog ecol, 69-74, PROF ZOOL, UNIV TENN, KNOXVILLE, 63- Mem: AAAS; Ecol Soc Am; Am Ornith Union. Res: Ecological and population studies of animals. Mailing Add: Univ of Tenn 408 Tenth St Knoxville TN 37916

TANNER, JAMES THOMAS, b Franklin, Ky, Apr 23, 39; m 64. RADIOCHEMISTRY. Educ: Eastern Ky State Col, BS, 61; Univ Ky, PhD(radiochem), 66. Prof Exp: Res chemist, Carnegie-Mellon Univ, 66-68, lectr chem, 68-69; RES CHEMIST, FOOD & DRUG ADMIN, 69- Mem: Fel AAAS; fel Meteoritical Soc; Am Chem Soc. Res: Neutron activation analysis; trace elements in meteorites; trace element distribution in foods, drugs and consumer products. Mailing Add: HFF-145 Food & Drug Admin Washington DC 20204

TANNER, JOHN EYER, JR, b Cleveland, Ohio, Apr 30, 30; m 66; c 3. PHYSICAL CHEMISTRY. Educ: Oberlin Col, AB, 51; Ind Univ, MS, 54; Univ Wis, PhD(phys chem), 66. Prof Exp: Res asst phys chem, Am Found Biol Res, Wis, 59-62; res fel, Pa State Univ, 66-68; res asst, Max Planck Inst Med Res, Ger, 68-69; fel, Sci Res Staff, Ford Motor Co, Mich, 69-71; RES CHEMIST, NAVAL WEAPONS SUPPORT CTR, 71- Mem: Am Chem Soc; Biophys Soc; Am Phys Soc. Res: Nuclear magnetic resonance; emission and absorption spectroscopy; thermodynamics. Mailing Add: Appl Sci Dept Naval Weapons Support Ctr Crane IN 47522

TANNER, JOSEPH JARRATT, b Port Arthur, Tex, Aug 5, 20; m 47; c 2. PETROLEUM GEOLOGY. Educ: Rutgers Univ, BS, 42; Princeton Univ, AM & PhD(geol), 49. Prof Exp: Photogeologist, 49-53, asst dir explor projs, 53-56, sr foreign geologist, 56-59, mgr, Australian Opers, Brisbane, 59-69, mgr, Far East, Singapore, 69-72, SR CONSULT SCIENTIST, PHILLIPS PETROL CO, BARTLESVILLE, OKLA, 72- Mem: Geol Soc Am; Am Asn Petrol Geol. Res: Petroleum exploration. Mailing Add: Phillips Petroleum Co 634 FPB Bartlesville OK 74003

TANNER, JURATE E, b Kaunas, Lithuania, Apr 21, 26; Can citizen; m 57; c 2. MICROBIOLOGY. Educ: McGill Univ, BSc, 52, MSc, 54, PhD(immunol), 56. Prof Exp: Res officer bact, Animal Dis Res Inst, Hull, Que, 56-57; fel, Pasteur Inst Paris, 57-58; lectr, 63-67, ASST PROF BACT, JOHN ABBOTT COL, MACDONALD COL, McGILL UNIV, 67- Mem: Can Soc Microbiol. Res: Bacteriology. Mailing Add: Dept of Biol John Abbott Col Ste Anne de Bellevue PQ Can

TANNER, LLOYD GEORGE, b Cozad, Nebr, Oct 3, 18; c 2. GEOLOGY. Educ: Univ Nebr, BS, 51, MS, 56. Prof Exp: Field supvr, Works Progress Admin, 40-41, asst cur vert paleont, State Mus, 51-56, instr vert paleont, 56-70, ASSOC CUR VERT PALEONT, STATE MUS, 56-, COORDR SYST COLLECTION, UNIV NEBR, LINCOLN, 69-, ASST PROF GEOL, UNIV, 70- Concurrent Pos: Co-prin investr fauna & stratig,sequence, Big Bone Lick, Ky, 62-; mem, Yale Peabody Mus Egyptian Expeds, 65-67. Res: Vertebrate paleontology; stratigraphy; Pleistocene geology. Mailing Add: Rm W-436 Nebraska Hall Univ of Nebr Lincoln NE 68508

TANNER, NOALL STEVAN, b Ogden, Utah, Sept 21, 34; m 55; c 2. PHARMACOGNOSY, PHARMACOLOGY. Educ: Univ Utah, BS, 56, PhD(pharmacog), 66. Prof Exp: Asst prof pharmacog, Col Pharm, Butler Univ, 63-66; ASSOC PROF PHARMACOL, COL PHARM, NDAK STATE UNIV, 67- Res: Biopharmaceutical sciences; environmental study of Veratrum californicum; relation of neurotransmitter to behavior in rats; study and screening of natural products. Mailing Add: Dept of Pharmacol NDak State Univ Fargo ND 58102

TANNER, RAYMOND LEWIS, b Memphis, Tenn, Dec 11, 31, m 68; c 3. RADIOLOGICAL PHYSICS, HEALTH PHYSICS. Educ: Memphis State Univ, BS, 53; Vanderbilt Univ, MS, 55; Univ Calif, Los Angeles, PhD(med physics), 67. Prof Exp: Asst prof physics, Memphis State Univ, 55-62; vis physicist, Harbor Gen Hosp, Torrance, Calif, 63-64; assoc prof, 67-70, PROF MED PHYSICS, CTR HEALTH SCI, UNIV TENN, MEMPHIS, 70- Concurrent Pos: Consult self-radiation protection, 55- Mem: Am Asn Physicists in Med (pres, 73-74); Health Physics Soc; Am Col Radiol; Radiol Soc NAm; Sigma Xi. Res: Radiation dosimetry. Mailing Add: Ctr for Health Sci Univ of Tenn Memphis TN 38163

TANNER, ROGER LEE, analytical chemistry, photochemistry, see 12th edition

TANNER, WARD DEAN, JR, b Jacksonville, Fla, Dec 6, 18. WILDLIFE MANAGEMENT, ECOLOGY. Educ: Univ Minn, BS, 41; Pa State Univ, MS, 48; Iowa State Univ, PhD(zool), 53. Prof Exp: Asst wildlife mgt, Pa State Univ, 41, 46-48; refuge mgr, Bombay Hook Nat Wildlife Refuge, US Fish & Wildlife Serv, 48-50; from instr to assoc prof biol, 53-74, PROF BIOL, GUSTAVUS ADOLPHUS COL, 74- Mem: Wildlife Soc. Res: Ecology of wildlife and ruffed grouse; botany; forestry; fisheries management. Mailing Add: Dept of Biol Gustavus Adolphus Col St Peter MN 56082

TANNER, WILLIAM FRANCIS, JR, b Milledgeville, Ga, Feb 4, 17; m 38; c 3. GEOLOGY. Educ: Baylor Univ, BA, 37; Tex Tech Col, MA, 39; Univ Okla, PhD(geol), 53. Prof Exp: Asst geol, Baylor Univ, 35-37; oil ed, Amarillo Times, 39-41 & 45-46; asst prof geol & journalism, Okla Baptist Univ, 46-51; spec instr, Univ Okla, 51-54; vis lectr geol, 54-56, from assoc prof to prof, 56-74, REGENTS PROF GEOL, FLA STATE UNIV, 74- Concurrent Pos: Geologist, Shell Oil Co, 54; ed, Coastal Res Notes, 62-; NSF vis scientist, 65-66. Mem: Fel AAAS; fel Geol Soc Am; Soc Econ Paleontologists & Mineralogists; Am Soc Photogram; Seismol Soc Am. Res: Sedimentology; stratigraphy; geomorphology; hydrodynamics; beach and shore processes; structural, areal, field and subsurface geology; rheology and deformation of materials; circular statistics; paleogeography; paleoclimatology; petroleum exploration and resources; other energy sources. Mailing Add: Dept of Geol Fla State Univ Tallahassee FL 32306

TANNER, WILMER W, b Fairview, Utah, Dec 17, 09; m 35; c 3. ZOOLOGY. Educ: Brigham Young Univ, AB, 36, MA, 37; Univ Kans, PhD(vert zool), 49. Prof Exp: Teacher high sch, Utah, 37-46; instr zool, Univ Kans, 47-48; assoc prof vert zool, 49-62, PROF VERT ZOOL, BRIGHAM YOUNG UNIV, 62- Concurrent Pos: Consult, AEC Proj, 59-; ed, Herpetologica, 60-, secy-treas, Acad Conf, 66; res fel, Brigham Young Univ, 67-74. Mem: AAAS; Soc Syst Zool (vpres, 69, pres, 70); Am Soc Ichthyol & Herpet. Res: Herpetology, including taxonomy, anatomy, geographical distribution and natural history. Mailing Add: Dept of Zool Brigham Young Univ Provo UT 84601

TANQUARY, ALBERT CHARLES, b Columbus, Kans, Mar 9, 29; m 49; c 5. POLYMER SCIENCE. Educ: Kans State Col, Pittsburg, BS, 50; Okla State Univ, MS, 52, PhD(chem), 54. Prof Exp: Sr chemist cent res dept, Minn Mining & Mfg Co, 54-55, group supvr, 55-58, res mgr fibers dept, 59-61, mgr, 62-63; group leader res & develop div, Union Camp Corp, 63-65; HEAD POLYMER DIV, SOUTHERN RES INST, 66- Mem: Am Chem Soc; Am Asn Textile Chemists & Colorists; Fiber Soc; Soc Plastics Engrs; Tech Asn Pulp & Paper Indust. Res: Polymer and textile chemistry; mechanical properties of polymers, fibers and films; fiber spinning; biomaterials research; biomedical engineering. Mailing Add: Polymer Div Southern Res Inst 2000 Ninth Ave S Birmingham AL 35205

TANSEY, MICHAEL RICHARD, b Oakland, Calif, Mar 27, 43; m 63; c 2. MYCOLOGY. Educ: Univ Calif, Berkeley, BA, 65, PhD(bot), 70. Prof Exp: NSF fel & res assoc microbiol, 70-71, ASST PROF PLANT SCI, IND UNIV, BLOOMINGTON, 71- Mem: AAAS; Am Soc Microbiol; Bot Soc Am; Mycol Soc

Am; Brit Mycol Soc. Res: Biology of thermophilic fungi; cellulose biodegradation; heated habitats. Mailing Add: Dept of Plant Sci Ind Univ Bloomington IN 47401

TANSEY, ROBERT PAUL, SR, b Newark, NJ, Apr 27, 14; m 41; c 4. PHARMACY. Educ: Rutgers Univ, BS, 38, MS, 50. Prof Exp: Control pharmacist, Res & Develop Labs, Burroughs Wellcome Co, Inc, 40-43; asst dept head prod & control, E R Squibb & Sons, 43-45; head, Pharmaceut Res & Develop Lab, Maltbie Labs, 45-50; assoc dept head, Merck & Co, 50-53; sect leader, Schering Co, Inc, 53-58; res coordr, Strong Cobb Arner Co, Inc, 58-62; regist pharmacist, Saywell Pharm, Ohio, 62-63; TECH DIR & VPRES, VET LABS, INC, 63- Mem: Animal Health Inst; Am Pharmaceut Asn. Res: Pharmaceutical development; formulation and methods analysis; production processing techniques and control methods; pharmaceutical plant and equipment design; development of special techniques for control and sustained release medicinal forms; plant management. Mailing Add: 11141 Glen Arbor Rd Kansas City MO 64114

TANSY, MARTIN F, b Wilkes Barre, Pa, Mar 8, 37; m 64; c 3. PHYSIOLOGY. Educ: Wilkes Col, BA, 59; Jefferson Med Col, MS, 61, PhD(physiol), 64. Prof Exp: Res asst physiol, Jefferson Med Col, 59-61, fel, 62-64; from asst prof to assoc prof, 64-72, PROF PHYSIOL, SCHS DENT, PHARM & ALLIED HEALTH PROF, TEMPLE UNIV, 72-, CHMN DEPT, 64- Mem: Fel Am Col Nutrit; Soc Exp Biol & Med; Am Physiol Soc; Am Pharmaceut Asn; Am Fedn Clin Res. Res: Gastrointestinal and radiation physiology. Mailing Add: Dept of Physiol & Biophys Temple Univ Dent Sch Philadelphia PA 19140

TANTRAPORN, WIROJANA, b Chonburi, Thailand, Apr 17, 31; m 53; c 1. EXPERIMENTAL SOLID STATE PHYSICS. Educ: Univ Denver, BS, 52; Univ Mich, MS, 53, PhD(physics), 58. Prof Exp: Lectr physics, Univ Mich, 58-59; physicist electronics lab, 59-62, PHYSICIST, RES & DEVELOP CTR, GEN ELEC CO, 62- Mem: AAAS. Res: Thin films; insulators; solid state microwave devices; electron devices; computer simulations. Mailing Add: Gen Elec Res & Develop Ctr Schenectady NY 12301

TANTRAVAHI, RAMANA V, b Samalkot, India, Dec 1, 35; m 62; c 2. HUMAN GENETICS, CYTOGENETICS. Educ: Andhra Univ, BS(Hons), 56, MS, 57; Harvard Univ, PhD(biol & plant genetics), 67. Prof Exp: Lectr bot, Andhra Univ, 58-62; res fel biol, Harvard Univ, 67-69; Govt India pool officer plant genetics, Agr Univ, Bangalore, 69-70; res assoc plant genetics, Suburban Exp Sta, Univ Mass, 70-73; RES ASSOC HUMAN GENETICS, COLUMBIA UNIV MED CTR, 73- Res: Mammalian cytogenetics and mouse cytogenetics; gene mapping by cytogenetic and somatic cell genetic techniques; evolutionary aspects of animal and plant communities. Mailing Add: Dept of Human Genetics & Develop Columbia Univ Med Ctr New York NY 10032

TANTTILA, WALTER H, b Sax, Minn, Nov 21, 22; m 51; c 3. PHYSICS. Educ: Univ Minn, BChE, 48, MA, 50; Univ Wash, PhD(physics), 55. Prof Exp: Instr physics, Minn State Teachers Col, Winona, 50-51; res physicist, Minneapolis-Honeywell Regulator Co, 51-52; asst physics, Univ Wash, 52-55; asst prof, Mich State Univ, 55-58; from asst prof to assoc prof, 58-72, PROF PHYSICS, UNIV COLO, BOULDER, 72- Mem: Am Phys Soc. Res: Nuclear and electron magnetic resonance; low temperature phenomena. Mailing Add: Dept of Physics Univ of Colo Boulder CO 80302

TANZ, RALPH, b New York, NY, Oct 10, 25; m 52; c 3. PHARMACOLOGY, PHYSIOLOGY. Educ: Univ Rochester, BA, 48; Harvard Med Sch, 49-52; Univ Colo, PhD(pharmacol), 58. Prof Exp: Asst pharmacol, Sch Med, Univ Colo, 54-57; instr, Med Units, Univ Tenn, 57-59; sr instr, Sch Med, Western Reserve Univ, 59-62; asst prof pharmacol, New York Med Col, 62-63; head cardiovasc sect, Dept Pharmacol, Geigy Res Labs, NY, 63-69; ASSOC PROF PHARMACOL, SCH MED, UNIV ORE HEALTH SCI CTR, 69- Concurrent Pos: NIH career develop award, 61; chmn sect, Gordon Res Conf, 66; Fogarty sr int fel, Heart Res Labs, Dept Med, Univ Cape Town Med Sch, SAfrica, 76-77. Mem: AAAS; Am Soc Pharmacol & Exp Therapeut; Cardiac Muscle Soc (pres, 64); Am Heart Asn; Soc Exp Biol & Med. Res: Isolated cardiac tissue; effect of cardiac glycosides and catecholamines on cardiac muscle; antiarrhythmics; antihypertensives. Mailing Add: Dept of Pharmacol Sch of Med Univ of Ore Health Sci Ctr Portland OR 97201

TANZER, CHARLES, b New York, NY, Dec 4, 12; m 37. SCIENCE EDUCATION, MEDICAL MICROBIOLOGY. Educ: Long Island Univ, BS, 33; NY Univ, MS, 36, PhD(bact), 41. Prof Exp: Asst bact, Col Med, NY Univ, 34-38; instr, Bronx High Sch Sci, 38-43 & Dewitt Clinton High Sch, 45-49; chmn sci dept, Seward Park High Sch, 49-57; prin jr high schs, New York, 57-65; prof & coordr sci, Teacher Educ Prog, 65-73, EMER PROF TEACHER EDUC PROG, HUNTER COL, 73- Concurrent Pos: Adj assoc prof, Long Island Univ, 51-65; Ford Found fel, 57. Honors & Awards: Meritorious Serv Award, Am Cancer Soc, 73. Mem: AAAS; Am Soc Microbiol; Am Cancer Soc; Nat Asn Biol Teachers; Nat Sci Teachers Asn. Res: Serology; parasitology; intestinal parasites. Mailing Add: 600 W 218th St New York NY 10034

TANZER, JASON MICHAEL, b Boston, Mass, Feb 25, 38; m 62; c 2. DENTISTRY, MICROBIAL PHYSIOLOGY. Educ: Bates Col, BS, 59; Tufts Univ, DMD, 63; Georgetown Univ, PhD(physiol), 69. Prof Exp: Comn officer student training & extern prog fel, NIH, 62; prin investr, Nat Inst Dent Res, 66-70; assoc prof, 70-73, actg head dept, 74-75, PROF GEN DENT, HEALTH CTR, UNIV CONN, HARTFORD, 73- Concurrent Pos: Consult, Adv Group Dent Drugs, Food & Drug Admin, 71-72; Vet Admin clin investr grant microbial physiol, Vet Admin Hosp, Newington, Conn, 70-73. Mem: Int Asn Dent Res; Am Soc Microbiol. Res: Sugar and phosphate metabolism of oral bacteria; bacterial genetics; microbiology of dental plaque; antibacterial agents to prevent tooth decay and periodontal disease; immunization against tooth decay; preventive dentistry. Mailing Add: 64 Harvest Lane West Hartford CT 06117

TANZER, MARVIN LAWRENCE, b New York, NY, Jan 26, 35; m 54; c 4. BIOCHEMISTRY. Educ: Mass Inst Technol, SB, 55; NY Univ, MD, 59. Prof Exp: Intern med, Johns Hopkins Hosp, Baltimore, 59-60, asst resident, 60-61; res fel, Mass Gen Hosp-Harvard Med Sch, 61-62 & 64-65, asst biologist, 65-68, assoc, 67-68; from asst prof to assoc prof, 68-75, PROF BIOCHEM, HEALTH CTR, UNIV CONN, 75- Concurrent Pos: Arthritis Found fel, Mass Gen Hosp, Boston, 61-62, Am Heart Asn fel, 64-66; Am Heart Asn estab investr, Mass Gen Hosp, Boston, 66-68 & Med Sch, Univ Conn, 68-71; investr, Marine Biol Lab, Woods Hole, 66-71; tutor biochem sci, Harvard Univ, 67-68; Josiah Macy, Jr Found fac scholar award, 74-75; vis prof dermat, Univ Liege, Belg, 74-75. Mem: Sigma Xi; Am Heart Asn; Am Chem Soc; Am Soc Biol Chemists. Res: Properties, function and synthesis of connective tissue components. Mailing Add: Dept of Biochem Univ of Conn Health Ctr Farmington CT 06032

TANZER, RADFORD CHAPPLE, b Little Falls, NY, Sept 16, 05; m 43. MEDICINE. Educ: Dartmouth Col, BS, 25; Harvard Med Sch, MD, 29. Prof Exp: Instr anat, 39-42, from instr to asst prof surg, 42-49, from asst prof surg to prof plastic surg, 49-71, EMER PROF PLASTIC SURG, DARTMOUTH MED SCH, 71- Concurrent Pos: Consult, Vet Admin. Mem: Am Soc Plastic & Reconstruct Surg; Am Soc Surg of the Hand; Am Asn Plastic Surg (pres, 71-72). Res: Surgery of the hand; ear reconstruction. Mailing Add: 8 N Balch St Hanover NH 03755

TANZI, FAUSTO, b Leghorn, Italy, Mar 24, 17; nat US. INTERNAL MEDICINE. Educ: Ill Inst Technol, BS, 46; Univ Chicago, MD, 50. Prof Exp: Intern, Clins, Univ Chicago, 50-51, resident med, 51-54, instr & chief resident, 54-55, secy dept, 55-61, asst prof, Sch Med, 55-60, assoc prof, 60-67; ASSOC PROF MED, SCH MED, UNIV SOUTHERN CALIF, 67- Concurrent Pos: Coman fel, 51-52; asst med dir, Barlow Hosp, Los Angeles, 67-70, med dir, 70- Mem: AAAS; Am Thoracic Soc; Am Fedn Clin Res; Am Heart Asn; Asn Am Med Cols. Res: Clinical research and treatment of chronic respiratory diseases. Mailing Add: Barlow Hosp 2000 Stadium Way Los Angeles CA 90026

TAO, KAR-LING JAMES, b Canton, China, Aug 20, 41; m 69; c 2. PLANT PHYSIOLOGY. Educ: New Asia Col, BS, 64; Tuskegee Inst, MSc, 68; Univ Wis, PhD(hort & bot), 71. Prof Exp: Sec educ teacher sci, Tsung Tsin Mid Sch, 65-66; assoc plant biochem, Cornell Univ, 71-72; RES ASSOC PLANT PHYSIOL, NY STATE AGR EXP STA, 73- Mem: Am Soc Plant Physiol; AAAS. Res: Plant hormones, seed dormancy and seed germination; protein and RNA synthesis in plants; seed deteiolation and seed vigor. Mailing Add: Dept of Seed & Vegetable Sci NY State Agr Exp Sta Geneva NY 14456

TAO, ROBERT CHI-MEI, b Szechuen, China, Mar 12, 43; m 70; c 1. NUTRITION. Educ: Chung-Hsiung Univ, Taiwan, BS, 65; Laval Univ, MS, 69; Univ Mo, PhD(nutrit), 73. Prof Exp: Fel nutrit, Univ Mo, 74; MGR BIOCHEM, CONTECH LABS, PET INC, 74- Mem: Nutrit Soc; Inst Food Technologists. Res: Study nutritional values of manufactured foods and implement known nutrition knowledge in new product development. Mailing Add: Pet Inc Greenville IL 62246

TAO, SHU-JEN, b Soochow, China, Oct 7, 28; m 58; c 3. NUCLEAR SCIENCE, PHYSICAL CHEMISTRY. Educ: Amoy Univ, BSc, 49; Univ NSW, MEngSc, 61, PhD(nuclear chem), 64. Prof Exp: Scientist, Taiwan Rain Stimulation Res Inst, 51-59; res fel nuclear & radiation chem, Australian AEC, 60-61; SR STAFF SCIENTIST, NEW ENG INST MED RES, 65- Mem: AAAS; Am Phys Soc; Am Chem Soc; NY Acad Sci. Res: Positron physics; positronium chemistry; fast timing electronic instruments; applied statistics. Mailing Add: 12 Woodchuck Lane Wilton CT 06897

TAPAS, JOHN C, biochemistry, organic chemistry, see 12th edition

TAPE, GERALD FREDERICK, b Ann Arbor, Mich, May 29, 15; m 39; c 3. SCIENCE ADMINISTRATION. Educ: Eastern Mich Univ, AB, 35; Univ Mich, MS, 36, PhD(physics), 40. Hon Degrees: DSc, Eastern Mich Univ, 64. Prof Exp: Asst physics, Eastern Mich Univ, 33-35 & Univ Mich, 36-39; instr, Cornell Univ, 39-42; staff mem radiation lab, Mass Inst Technol, 42-46; from asst prof to assoc prof physics, Univ Ill, 46-50; asst to dir, Brookhaven Nat Lab, 50-51, dep dir, 51-62; from vpres to pres, Assoc Univs, Inc, 62-63; comnr, US AEC, 63-69; PRES, ASSOC UNIVS, INC, 69- Concurrent Pos: Mem, President's Sci Adv Comn, 69-73; mem high energy adv panel, US AEC, 69-74, sr tech adv Geneva IV, 71; mem, Defense Sci Bd, 70-74, chmn, 70-73; mem bd dirs, Atomic Indust Forum, 70-73; bd trustees, Sci Serv, 70-; mem sci adv comt, Int Atomic Energy Agency, 72-74, US rep, 73-; consult, Nat Security Coun, 74- Honors & Awards: Cert Appreciation, Army-Navy, 47 & Dept of State, 69; Meritorious Civilian Serv Medal, Secy of Defense, 69; Distinguished Pub Serv Medal, Dept of Defense, 73. Mem: Fel AAAS; fel Am Phys Soc; fel Am Nuclear Soc; Am Astron Soc. Res: Nuclear physics; particle physics; accelerator development; reactor development; applications of atomic energy; radioastronomy. Mailing Add: Assoc Univs Inc 1717 Massachusetts Ave NW Washington DC 20036

TAPE, NORMAN WILSON, food science, see 12th edition

TAPER, CHARLES DANIEL, b PEI, Aug 18, 11. HORTICULTURE. Educ: Univ Man, PhD(bot), 53. Prof Exp: CHMN DEPT HORT, McGILL UNIV, 71- Mem: Can Soc Plant Physiologists; Agr Inst Can; Can Hort Soc. Res: Malus tissue culture studies. Mailing Add: Box 286 Macdonald College PQ Can

TAPIA, FERNANDO, b Panama, Apr 8, 22; m 47; c 3. PSYCHIATRY. Educ: Univ Iowa, BA, 44, MD, 47; Am Bd Psychiat & Neurol, dipl psychiat, 60, dipl child psychiat, 66. Prof Exp: Intern, Santo Thomas Hosp, Panama, 48; dir, Boquette Sanit Unit-Panama, 48-54; resident psychiat, Barnes Hosp, Wash Univ, 54-57; asst dir out-patient clin, Malcolm Bliss Ment Health Ctr, 57-58; chief psychiatrist, Child Guid Clin, St Louis County Health Ctr, 58-59, dir ment health div, 59-61; from asst prof to prof psychiat, Sch Med, Univ Mo-Columbia, 61-72, chief sect child psychiat, 61-72; PROF PSYCHIAT & BEHAV SCI, COL MED, UNIV OKLA, 71-, CHIEF MENT HEALTH SERV, UNIV HOSP & CLINS, 72- Concurrent Pos: Instr, Wash Univ, 57-61; consult, St Louis County Juv Court, 57-61 & Convent of the Good Shepherd, 57-61; dir children's serv, Mid-Mo Ment Health Ctr, 66-72. Mem: Fel Am Psychiat Asn. Mailing Add: Univ of Okla Health Sci Ctr PO Box 26901 Oklahoma City OK 73190

TAPIA, RICHARD, b Santa Monica, Calif, Mar 25, 39; m 60; c 2. NUMERICAL ANALYSIS. Educ: Univ Calif, Los Angeles, BA, 61, MA, 66, PhD(math), 67. Prof Exp: Mathematician, Todd Shipyards, Calif, 61-63 & Int Bus Mach Corp, 63-65; actg asst prof math, Univ Calif, Los Angeles, 67-68; vis asst prof, US Army Math Res Ctr, Univ Wis-Madison, 68-70; ASSOC PROF MATH SCI, RICE UNIV, 70- Mem: Am Math Soc. Mailing Add: Dept of Math Sci Rice Univ Houston TX 77001

TAPIA, SANTIAGO, b Santiago, Chile, Nov 13, 39. ASTROPHYSICS. Educ: Univ Chile, Lic, 65; Univ Ariz, PhD(astron), 75. Prof Exp: Lab asst crystallog, Univ Chile, 61-62, teaching asst physics, 63-65, res asst astron, 66-69; RES ASSOC ASTRON, UNIV ARIZ, 75- Concurrent Pos: Lectr astron, Tech Sch Aeronaut, Chilean Air Force, 69. Mem: Am Astron Soc. Res: Observational study of optical properties of BL Lacertae objects and their relation to quasi-stellar objects; observational study of linear polarization in stars and nebulae; optical radiometry of faint astronomical objects. Mailing Add: Lunar & Planetary Lab Univ of Ariz Tucson AZ 85721

TAPLEY, DONALD FRASER, b Woodstock, NB, May 19, 27; nat US; m 57; c 3. INTERNAL MEDICINE. Educ: Acadia Univ, BSc, 48; Univ Chicago, MD, 52. Prof Exp: Intern & asst resident, Presby Hosp, New York, 52-54; fel physiol chem, Johns Hopkins Univ, 54-56; from asst prof to assoc prof, 56-72, actg dean fac med, 73-74, PROF MED, COL PHYSICIANS & SURGEONS, COLUMBIA UNIV, 72-, ASSOC DEAN FAC AFFAIRS, 70-, DEAN FAC MED, 74- Concurrent Pos: Fel physiol chem, Oxford Univ, 56-57; from asst attend physician to attend, Presby Hosp, 57- Mem: Am Soc Clin Invest; Am Thyroid Asn; Endocrine Soc; Harvey Soc. Res: Intermediary metabolism; thyroid physiology. Mailing Add: Off of the Dean Columbia Univ New York NY 10032

TAPLEY, NORAH DUVERNET, b St John, NB, July 25, 21; nat US. MEDICINE. Educ: Mt Holyoke Col, BA, 45; Columbia Univ, MD, 49. Prof Exp: Intern, Hartford

Hosp, 49-50, jr asst med resident, 50-51; resident radiol, Columbia-Presby Med Ctr, 51-54; instr, Columbia Univ, 54-58, asst prof, 58-62; assoc radiotherapist, 63-68, assoc prof, 63-71, RADIOTHERAPIST, UNIV TEX M D ANDERSON HOSP & TUMOR INST HOUSTON, 68-, PROF RADIOTHER, GEN FAC, 71- Concurrent Pos: From asst radiologist to assoc attend, Columbia-Presby Med Ctr, 54-62; dir sect radiation ther, Presby Hosp, 58-62; Nat Cancer Inst spec fel radiobiol, Sch Med, Stanford Univ, 60-61; spec asst to assoc dir grants & training, Nat Cancer Inst, 61-63. Mem: Fel Am Col Radiol; Am Radium Soc; AMA; Pan Am Med Asn; Soc Therapeut Radiol. Res: Clinical applications of the electron beam; normal tissue and tumor response; electron beam dosimetry. Mailing Add: Dept of Radiother M D Anderson Hosp & Tumor Inst Houston TX 77030

TAPLIN, GEORGE VORCE, b Rochester, NY, Jan 2, 10; m 32; c 2. MEDICINE, RADIOLOGY. Educ: Union Univ, NY, BA, 32; Univ Rochester, MD, 36. Prof Exp: Intern med, Sch Med, Univ Rochester, 36-37, asst resident, 37-38, instr, 39-47; ASSOC CLIN PROF, SCH MED, UNIV CALIF, LOS ANGELES, 47-, PROF RADIOL, 58-, RES PHYSICIAN AEC PROF, LAB NUCLEAR MED & RADIATION BIOL, 53-, ASSOC DIR LAB, 69- Concurrent Pos: Fel med, Sch Med, Univ Rochester, 38-39; res grants, Lederle Labs, G D Searle & Co, Chem Corps, US Army, Abbott Labs, USPHS & E R Squibb & Sons; consult, Bur Pneumonia Control, State Dept Health, NY, 39-42; asst physician, Strong Mem & Rochester Munic Hosps, 39-47; jr attend, Highland & Genessee Hosps, Rochester, NY, 40-41, sr attend, 46-47; attend consult, Birmingham Vet Admin Hosp, Van Nuys, 47-51; assoc resident physician, Lab Nuclear Med & Radiation Biol, Sch Med, Univ Calif, Los Angeles, 47-53, asst dir lab, 68-69; consult, USPHS, 47- & Food & Drug Admin, 68-71; sr physician & dir blood bank, Vet Admin Ctr, Los Angeles, 51-56; vis physician, Los Angeles County Harbor Gen Hosp, Torrance, 56-; mem subcomt inhalation hazards, Comt Biol Effects Atomic Radiation, Nat Acad Sci-Nat Res Coun, 56-; US del, Int Conf Radioisotopes in Sci Res, UNESCO, Paris, 57; US del & tech adv, Int Conf Peaceful Uses Atomic Energy, Geneva, 58; res collabr, Med Dept, Brookhaven Nat Lab, 58-; panel mem radiopharmaceut, US Pharmacopoeia. Mem: Radiation Res Soc; Reticuloendothelial Soc; Soc Nuclear Med (pres, 69-70); Soc Exp Biol & Med; The Chem Soc. Res: Diseases of the lungs; aerosol therapy; inhalation hazards from radioactive particles; chemical methods of radiation dosimetry; new radiotracer methods for investigation of diseases of the liver, kidneys and reticuloendothelial system. Mailing Add: Lab Nuclear Med & Radiation Biol Univ of Calif Los Angeles CA 90024

TAPP, CHARLES MILLARD, b Memphis, Tenn, Nov 9, 36; m 55; c 1. PHYSICS. Educ: Union Univ, BA, 58; Memphis State Univ, BS, 60; Univ Va, MS, 62, PhD(physics), 64. Prof Exp: Staff mem radiation calibration, Nat Bur Standards, 60; tech staff mem radiation damage, 64-66, div supvr vacuum tube physics & technol, 66-69, dept mgr vacuum tube devices, 69-71, DEPT MGR MICROELECTRONIC COMPONENTS, SANDIA LABS, 71- Concurrent Pos: Assoc ed transactions parts, hybrids & packaging, Inst Elec & Electronics Engrs, 74- Mem: Inst Elec & Electronics Engrs; Am Phys Soc. Res: Solid state surface effects; microelectronic thin and thick film processes; vacuum tube design, development; radiation effects in devices; neutron sources. Mailing Add: Orgn 2430 Sandia Labs Albuquerque NM 87115

TAPP, WILLIAM JOUETTE, b Quincy, Ill, July 26, 18; m 46; c 1. ORGANIC CHEMISTRY. Educ: Univ Ill, BS, 39; Cornell Univ, PhD(org chem), 43. Prof Exp: Asst chem, Cornell Univ, 40-41 & 41-43, Nat Defense Res Comt fel, 40-41; res & develop chemist, 43-46, proj leader, 46-54, staff asst, 55-56, patent adminr, 57-66, asst dir pharmaceut tech, 66-67, MGR ADMIN, UNION CARBIDE CORP, 67- Mem: Am Chem Soc. Res: Organic nitrogen and sulfur compounds; synthetic lubricants; industrial organic synthesis. Mailing Add: Corp Res Dept Tarrytown Tech Ctr Tarrytown NY 10591

TAPPA, DONALD W, b New York, NY, Jan 25, 34; m 58; c 1. ECOLOGY. Educ: Brooklyn Col, BS, 57; Williams Col, MA, 59; Yale Univ, PhD(biol), 64. Prof Exp: Asst prof biol, Lycoming Col, 64-65; from asst prof to assoc prof biol, Wilkes Col, 65-74, dean acad affairs, 72-74; UNDERGRAD DEAN, COL ST ROSE, 74- Concurrent Pos: Adj prof, Temple Univ, 67-73. Res: Limnology; plankton dynamics. Mailing Add: Col of Saint Rose Albany NY 12203

TAPPAN, DONALD VESTER, b Orestes, Ind, May 5, 25; m 49; c 5. BIOCHEMISTRY. Educ: Purdue Univ, BS, 49; Univ Wis, MS, 51, PhD(biochem), 53. Prof Exp: Res assoc, Andean Inst Biol, Lima, 53-55; proj assoc oncol, McArdle Mem Lab, Univ Wis, 55-56; from asst prof to assoc prof biochem, Univ Maine, 56-59; chief biochem sect, 69-70, HEAD BIOCHEM BR, SUBMARINE MED RES LAB, US NAVAL SUBMARINE BASE, 71- Mem: AAAS; Am Chem Soc; fel Am Inst Chemists; NY Acad Sci. Res: Metabolic influences of environmental factors; endocrine control of metabolism; hyperbaric biochemistry; nutrition. Mailing Add: Biochem Br Submarine Med Res Lab US Naval Submarine Base Groton CT 06340

TAPPE, JOHN, b Cincinnati, Ohio, Mar 22, 32; m 58; c 1. GEOLOGY, GEOCHEMISTRY. Educ: Univ Cincinnati, BA, 58, PhD(igneous petrol), 66. Prof Exp: Dir drilling opers, Earth Sci Labs, Cincinnati, 61-62; asst prof geol & petrol, 66-70, ASST PROF GEOL, TEMPLE UNIV, 70- Mem: AAAS; Geol Soc Am; Am Asn Petrol Geologists; Am Geophys Union. Res: Igneous petrology; trace element distribution; distribution of stress fields during rock deformation. Mailing Add: Dept of Geol Temple Univ Philadelphia PA 19122

TAPPEINER, JOHN CUMMINGS, II, b Los Angeles, Calif, Dec 15, 34; m 65; c 2. FOREST ECOLOGY, SILVICULTURE. Educ: Univ of Calif, Berkeley, BS, 57, MS, 61, PhD(forestry), 66. Prof Exp: Forester, US Forest Serv, 59-63; res asst forest ecol, Univ Calif, Berkeley, 63-66; Ford Found teaching fel forestry, Agr Univ Minas Gerais, 66-67; from asst prof to assoc prof, Forest Res Ctr, Univ Minn, St Paul, 68-73; REGIONAL SILVICULTURIST, US FOREST SERV, 73- Mem: Soc Am Foresters; Ecol Soc Am. Res: Natural regeneration of Sierra Nevada Douglas fir and ponderosa pine; ecology of hazel and understory vegetation; biomass and nutrient content of shrubs and herbs; effect of mechanized harvesting of forest soils. Mailing Add: US Forest Serv 630 Sansome St San Francisco CA 94111

TAPPEL, ALOYS LOUIS, b St Louis, Mo, Nov 21, 26; m 51; c 6. BIOCHEMISTRY. Educ: Iowa State Univ, BS, 48; Univ Minn, PhD(biochem), 51. Prof Exp: From instr to assoc prof, 51-61, PROF FOOD SCI & BIOCHEMIST, UNIV CALIF DAVIS, 61- Concurrent Pos: Guggenheim fel, 65-66. Honors & Awards: Borden Award, Am Inst Nutrit, 73. Mem: Am Chem Soc; Am Oil Chem Soc; Am Soc Biol Chem; Am Inst Nutrit. Res: Oxidant molecular damage and biological protection systems. Mailing Add: Dept of Food Sci & Technol Univ of Calif Davis CA 95616

TAPPEN, NEIL CAMPBELL, b Jacksonville, Fla, Feb 26, 20; m 52; c 2. PHYSICAL ANTHROPOLOGY, PRIMATOLOGY. Educ: Univ Fla, AB, 41; Univ Chicago, MA, 49, PhD(anthrop), 52. Prof Exp: Res assoc human biol, Univ Mich, 51-52; from assoc anthrop to instr, Univ Pa, 52-54; from instr anat to asst prof, Emory Univ, 54-59; assoc prof phys anthrop, Tulane Univ, 59-65; prof anthrop, 65-69, EARNEST A HOOTON PROF ANTHROP, UNIV WIS, MILWAUKEE, 69- Concurrent Pos: NIH grants, Emory Univ, 55-59, Tulane Univ, 60-65 & Univ Wis, Milwaukee, 65-71; Fulbright sr res scholar, Makerere Col, Uganda, 56-57; NSF grant, Univ Wis, Milwaukee, 71-74. Mem: AAAS; Am Anthrop Asn; Am Asn Phys Anthropologists; Am Anat Asn; Int Primatological Soc. Res: Organization of bone; non-human primates and their relationship to human evolution; structure of bone in fossil hominids; problems of human evolution. Mailing Add: Dept of Anthrop Univ of Wis Milwaukee WI 53201

TAPPER, DANIEL NAPTALI, b Philadelphia, Pa, Dec 5, 29; m 59. NEUROPHYSIOLOGY, RADIATION BIOLOGY. Educ: Rutgers Univ, BS, 51; Univ Pa, VMD, 55; Cornell Univ, PhD(physiol), 59. Prof Exp: Asst physiol, 55-59, res assoc radiation biol, 59-61, from asst prof to assoc prof, 61-69, PROF RADIATION BIOL, CORNELL UNIV, 69- Concurrent Pos: NIH spec fel, Stockholm, 65-66. Mem: AAAS; Am Physiol Soc. Res: Behavior; receptor physiology; neurophysiological and behavioral effects of ionizing radiation. Mailing Add: Dept of Biol Sci Cornell Univ Ithaca NY 14850

TAPPERT, FREDERICK DRACH, b Philadelphia, Pa, Apr 21, 40. MATHEMATICAL PHYSICS. Educ: Pa State Univ, BS, 62; Princeton Univ, PhD(physics), 67. Prof Exp: Mem tech staff, Bell Labs, 67-73; RES SCIENTIST, COURANT INST MATH SCI, NY UNIV, 74- Concurrent Pos: Vis staff mem, Los Alamos Sci Lab, 74-; consult, Stanford Res Inst, 74 & Sci Applns, Inc, 74- Mem: Am Phys Soc; AAAS; Soc Indust & Appl Math; Am Math Soc; Asn Comput Mach. Res: Theory and numerical simulation of wave propagation effects in plasmas, gases, liquids and solids. Mailing Add: Courant Inst of Math Sci NY Univ 251 Mercer St New York NY 10012

TAPPHORN, RALPH M, b Grinnell, Kans, July 26, 44; m 69; c 1. NUCLEAR PHYSICS. Educ: Ft Hays Kans State Col, BS, 66; Kans State Univ, PhD(physics), 70. Prof Exp: Res assoc nuclear physics, Ballistics Res Lab, Aberdeen Proving Ground, Md, 70-72; SR SCIENTIST NUCLEAR PHYSICS, SCHLUMBERGER-DOLL RES CTR, 72- Mem: Am Phys Soc; Sigma Xi; Inst Elec & Electronics Engrs. Res: Geophysical exploration with gamma-ray spectroscopy; nuclear detectors and instrumentation; radiation damage investigations. Mailing Add: Schlumberger-Doll Res Ctr PO Box 307 Ridgefield CT 06877

TAPPMEYER, WILBUR PAUL, b Owensville, Mo, May 19, 22; m 47; c 5. INORGANIC CHEMISTRY. Educ: Southeast Mo State Col, AB, 45; Univ Mo, BS, 47, PhD(inorg chem), 61. Prof Exp: Teacher high sch, Mo, 44-45; asst chem, Mo Sch Mines, 45-46; prof, Southwest Baptist Col, 47-60; from asst prof to assoc prof, 60-66, PROF CHEM, UNIV MO, ROLLA, 66- Mem: Am Chem Soc.. Res: Solid phase extraction of metal chelates; dimeric and polymeric properties of certain metal acetates and other alkonates. Mailing Add: Dept of Chem Univ of Mo Rolla MO 65401

TAPSCOTT, ROBERT EDWIN, b Terre Haute, Ind, June 10, 38; m 67; c 1. CHEMISTRY. Educ: Univ Colo, BS, 64; Univ Ill, Urbana, PhD, 68. Prof Exp: Asst prof, 68-72, ASSOC PROF CHEM, UNIV N MEX, 72- Concurrent Pos: Petrol Res Fund grant, 69-71. Mem: AAAS; The Chem Soc; Am Chem Soc. Res: Stereoselective effects in transition metal complexes; geometry and structure of coordination compounds. Mailing Add: Dept of Chem Univ of NMex Albuquerque NM 87131

TARAGIN, MORTON FRANK, b Washington, DC, Feb 1, 44; m 68; c 2. SOLID STATE PHYSICS. Educ: George Washington Univ, BS, 65, MPh, 69, PhD(physics), 70. Prof Exp: ASST PROF PHYSICS, GEORGE WASHINGTON UNIV, 70- Mem: Am Phys Soc. Res: Mössbauer effect; spectroscopy studies of glass structure. Mailing Add: Dept of Physics George Washington Univ Washington DC 20052

TARAIL, ROBERT, b New York, NY, Aug 1, 19; c 3. MEDICINE, COMMUNITY HEALTH. Educ: Univ Minn, AB, 39, MD, 43. Prof Exp: Intern, Michael Reese Hosp, Chicago, 43-44; intern internal med, Univ Minn Hosps, 46; asst res, Minneapolis Vet Hosp, 46-47; res fel med, Metab Div, Yale Univ, 47-49; instr biochem, Sch Med, Boston Univ, 49-50; res fel res med, Univ Pittsburgh, 50-52, instr res med, 52-53; head metab sect, Roswell Park Mem Inst, 53-63; staff physician, Vet Admin Hosp, Martinez, Calif, 63-71 & San Francisco, 71-72; MED DIR, METHADONE PROG, S SAN MATEO COUNTY, 73- Concurrent Pos: Asst prof, Sch Med, Univ Buffalo, 55-63; physician, San Mateo County Jail, Calif, 73- Mem: Soc Exp Biol & Med; Am Fedn Clin Res; fel Am Col Physicians. Res: Metabolic disorders; potassium studies; hypoglycemic action of tris hydroxymethylaminomethane. Mailing Add: 500 Arguello St Redwood City CA 94063

TARANIK, JAMES VLADIMIR, b Los Angeles, Calif, Apr 23, 40; m 71; c 2. EXPLORATION GEOLOGY, PHOTOGEOLOGY. Educ: Stanford Univ, BS, 64; Colo Sch Mines, PhD(geol), 74. Prof Exp: Chief remote sensing, Iowa Geol Surv, 71-74; PRIN REMOTE SENSING SCIENTIST, EARTH RESOURCES OBSERV SYST DATA CTR, US GEOL SURV, 75- Concurrent Pos: Adj prof geol, Univ Iowa, 71-; vis prof civil eng, Iowa State Univ, 72-74; consult, Earth Resources Technol Satellite Follow on Eval Panel Geol, Goddard Space Flight Ctr, NASA, Synchronous Observ Satellite Prog Eval, Geol Applns, Active microwave Syst Eval Workshop Earth-Land Panel, Geol Landuse Water, Johnson Space Ctr, 74; mem adv coun, Kuwait Inst Sci Res J Sci & Technol, 76- Mem: Geol Soc Am; Am Asn Petrol Geologists; Am Inst Aeronaut & Astronaut; Am Inst Prof Geologists; Soc Mining Engrs. Res: Development of applications of remote sensing technology to mineral and mineral fuel exploration; assessment of environmental and engineering geologic aspects of mineral resource development; engineering geology and geohydrology of civil works site selection. Mailing Add: Applns Br EROS Data Ctr US Geol Surv Sioux Falls SD 57198

TARANTA, ANGELO, b Rome, Italy, Apr 18, 27; m 57. INTERNAL MEDICINE. Educ: Univ Rome, MD, 49. Prof Exp: Instr microbiol, 55-58, adj asst prof, 58-60, asst prof med, 60-65, ASSOC PROF MED, SCH MED, NY UNIV, 65- Concurrent Pos: Res assoc, Irvington House Inst, 55-60, res dir, 60-63, assoc dir, 65-70; dir med, Cabrini Health Care Ctr, 73- Res: Rheumatic diseases; immunology. Mailing Add: Dept of Med NY Univ Sch of Med New York NY 10016

TARANTINO, LAURA MARY, b Exeter, Pa, Feb 6, 47. BIOCHEMISTRY. Educ: Col Misericordia, BS, 68; Cornell Univ, PhD(biochem), 75. Prof Exp: ASSOC RES SCIENTIST MED, COL PHYSICIANS & SURGEONS, COLUMBIA UNIV & ROOSEVELT HOSP, 75- Mem: AAAS; Am Chem Soc. Res: Enzyme regulation in the central nervous system; role and control of hexosemonophosphate shunt in cells; metabolic effects of insulin; diabetes. Mailing Add: Roosevelt Hosp 428 W 59th St New York NY 10019

TARAPCHAK, STEPHEN J, b Staten Island, NY, Mar 20, 42; m 65. LIMNOLOGY, PHYCOLOGY. Educ: Clarion State Col, BS, 64; Ohio Univ, MS, 66; Univ Minn, PhD(ecol), 73. Prof Exp: Res fel, Limnol Res Ctr, Univ Minn, Minneapolis, 73-74; RES SCIENTIST BIOL OCEANOG, GT LAKES ENVIRON RES LAB, NAT

OCEANIC & ATMOSPHERIC ADMIN, 74- Concurrent Pos: Consult, Environ Statements Syst Div, Argonne Nat Lab, 73-74. Mem: Am Phycol Soc; Am Soc Limnol & Oceanog; Int Limnol Soc. Res: Phytoplankton ecology. Mailing Add: Gt Lakes Environ Res Lab 2300 Washtenaw Ave Ann Arbor MI 48104

TARAS, MICHAEL ANDREW, b Olyphant, Pa, Sept 6, 21; m 48; c 4. FOREST PRODUCTS, WOOD TECHNOLOGY. Educ: Pa State Univ, BS, 42, MF, 48; PhD(wood NC State Univ, PhD(wood technol), 65. Prof Exp: Forest prod technologist, Forest Prod Lab, 48-54, FOREST PROD TECHNOLOGIST, SOUTHEASTERN FOREST EXP STA, FORESTRY SCI LAB, US FOREST SERV, 54- Mem: Forest Prod Res Soc; Soc Wood Sci & Technol; Int Asn Wood Anat. Res: Forestry; wood anatomy related to wood identification, quality, seasoning and moisture movement through wood; in log and tree classification systems and wood weight-volume relationships. Mailing Add: Southeastern Forest Exp Sta Forestry Sci Lab US Forest Serv Carlton St Athens GA 30602

TARAS, PAUL, b Tunisia, May 12, 41; m 63; c 2. NUCLEAR PHYSICS. Educ: Univ Toronto, BScEng Phys, 62, MSc, 63, PhD(nuclear physics), 65. Prof Exp: Asst prof, 65-70, ASSOC PROF PHYSICS, UNIV MONTREAL, 70- Mem: Am Phys Soc; Can Asn Physicists. Res: Nuclear spectroscopy via heavy ion induced reactions. Mailing Add: Dept of Physics Univ of Montreal Montreal PQ Can

TARASZKA, ANTHONY JOHN, b Wallington, NJ, Feb 19, 35; m 60. ANALYTICAL CHEMISTRY, PHARMACEUTICAL CHEMISTRY. Educ: Rutgers Univ, BS, 56, MS, 58; Univ Wis, PhD(pharmaceut chem), 62. Prof Exp: Res assoc anal res & develop, 62-63, head dept, 63-66, mgr control res & develop, 66-70, dir control, 70-74, VPRES CONTROL, UPJOHN CO, 74- Mem: Am Chem Soc; Am Pharmaceut Asn. Res: Trace component analysis; separation techniques; reaction mechanisms. Mailing Add: 410 Marylynn Lane Kalamazoo MI 49002

TARASZKA, MILDRED J, b Wild Rose, Wis, Mar 27, 35; m 60. PHYSICAL CHEMISTRY. Educ: Wis State Univ-Stevens Point, BS, 56; Univ Wis, PhD(phys chem), 62. Prof Exp: RES ASSOC, UPJOHN CO, 62- Res: Solution kinetics and pharmacokinetics. Mailing Add: Upjohn Co Kalamazoo MI 49001

TARBELL, DEAN STANLEY, b Hancock, NH, Oct 19, 13; m 42; c 3. ORGANIC CHEMISTRY. Educ: Harvard Univ, AB, 34, MA, 35, PhD(org chem), 37. Prof Exp: Asst, Radcliffe Col, 36-37; fel, Univ Ill, 37-38; from instr to prof org chem, Univ Rochester, 38-60, Houghton prof chem, 60-66, chmn dept, 64-66; distinguished prof, 67-75, HARVIE BRANSCOM DISTINGUISHED PROF CHEM, VANDERBILT UNIV, 75- Concurrent Pos: Guggenheim fels, Oxford Univ, 46-47 & Stanford Univ, 61-62; Fuson lectr, Univ Nev, 72; mem, NIH study sects; consult, Walter Reed Army Inst Res, 72- Honors & Awards: C H Herty Medal, Am Chem Soc, 73. Mem: Nat Acad Sci; Am Chem Soc; Am Acad Arts & Sci; The Chem Soc; Hist Sci Soc. Res: Reaction phenolic ethers; organic sulfur compounds; structure and synthesis of natural products; structural and theoretical organic chemistry; structure of antibiotics; history of organic chemistry in the United States. Mailing Add: Dept of Chem Vanderbilt Univ Nashville TN 37235

TARBLE, RICHARD DOUGLAS, b Kewanee, Ill, Aug 27, 19; m 43; c 2. HYDROLOGY. Educ: Univ Nev, BS, 49; Agr & Mech Col, Tex, MS, 57. Prof Exp: Meteorol aide, US Weather Bur, 46-49, meteorologist snow res, 50-52, hydrol res, 52-55, radar hydrol res, 55-57, radar hydrologist prog develop, 57-60, asst chief river servs, 60-63, hydrologist in charge, Fed-State River Forecast Ctr, Calif, 63-71; UN-WORLD METEOROL ORGN EXPERT HYDROMETEOROL, CONSULT, MALAYSIAN METEOROL SERV, KUALA LUMPUR, 71- Concurrent Pos: From instr to asst prof, Univ Nev, 46-50. Mem: Am Meteorol Soc; Am Geophys Union. Res: Application of meteorology to hydrology; effect of snow melt on stream flow; quantitative precipitation forecasting; use of radar in flood prediction; development of river and flood forecasting procedures; snow pillow and precipitation gage research and development. Mailing Add: 4a Jalan 9/3 Petaling Jaya Selangor Malaysia

TARBY, THEODORE JOHN, b Auburn, NY, May 9, 41; m 64; c 1. ANATOMY, NEUROPHYSIOLOGY. Educ: Calif Inst Technol, BS, 64; Univ Calif, Los Angeles, PhD(anat), 68. Prof Exp: ASST PROF ANAT, MED CTR, UNIV COLO, DENVER, 68- Concurrent Pos: Milhelm Found Cancer Res grant, Med Ctr, Univ Colo, Denver, 71- Mem: AAAS; Am Asn Anat. Res: Cerebral tissue impedance and extracellular space; blood-brain barrier; olfactory function in normal salmon; glial physiology. Mailing Add: Dept of Anat Univ of Colo Med Ctr Denver CO 80220

TARCZY, E K, b Budapest, Hungary, Sept 28, 31; Can citizen; m 55; c 2. ORGANIC CHEMISTRY, BIOLOGICAL CHEMISTRY. Educ: Eötvös Lorand Univ, Budapest, dipl chem, 54. Prof Exp: Res chemist, G Richter Pharmaceut Co, Budapest, 54-55, Hungarian Pharma Co, 55-56 & Procter & Gamble Co Can Ltd, 57-61; res assoc biochem res & prod, 61-73, asst dir, Insulin Div, 73-74, DIR INSULIN DIV, CONNAUGHT LABS LTD, 74- Mem: Chem Inst Can; Can Diabetic Asn. Res: Preparative biochemistry in protein hormones; insulin; peptide synthesis. Mailing Add: Connaught Labs Ltd 1755 Steeles Ave W Willowdale ON Can

TAREN, JAMES A, b Toledo, Ohio, Nov 10, 24. NEUROSURGERY. Educ: Univ Toledo, BS, 48; Univ Mich, MD, 52; Am Bd Neurol Surg, dipl, 60. Prof Exp: Intern surg, Univ Hosp, Univ Mich, 52-53, resident, 53-54; teaching fel neurosurg, Harvard Med Sch, 54-55; resident, 55-57, from instr to assoc prof, 57-69, PROF NEUROSURG, MED SCH, UNIV MICH, ANN ARBOR, 69-, DIR NEUROBEHAV SCI PROG, 75- Concurrent Pos: Res fel, Boston Children's Hosp, 54-55; asst surg, Peter Bent Brigham Hosp, Boston, 54-55; actg chief neurosurg, Vet Admin Hosp, Ann Arbor, 58-73; chief neurosurg, Wayne County Gen Hosp, 58-72. Mem: AAAS; Asn Am Med Cols; Cong Neurol Surg; NY Acad Sci. Res: Central nervous system; stereotoxic neurosurgery. Mailing Add: Dept of Surg Univ of Mich Ann Arbor MI 48106

TARG, RUSSELL, b Chicago, Ill, Apr 11, 34; m 58; c 3. PHYSICS. Educ: Queens Col, NY, BS, 54. Prof Exp: Res asst phsyics, Columbia Univ, 54-56; engr, Sperry Gyroscope Co, 56-59; res assoc plasmas, Polytech Inst Brooklyn, 59; physicist, TRG, Inc, 59-62; eng specialist, Sylvania Elec Co, 62-72; SR RES PHYSICIST, BIOENG LAB, STANFORD RES INST, 72- Mem: Am Phys Soc; Optical Soc Am; Am Inst Elec & Electroncis Engrs. Res: Electron beam-plasma interactions; gas laser research; laser detection; modulation and frequency control; research in extra sensory perception. Mailing Add: Stanford Res Inst K 1027 Menlo Park CA 94025

TARGOWSKI, STANISLAW P, b Nagorzyce, Poland, Nov 13, 40; m 70. IMMUNOLOGY, VETERINARY MICROBIOLOGY. Educ: Univ Warsaw, DVM, 63; Univ Wis, MS, 69, PhD(immunol), 72. Prof Exp: Resident vet, Kroplewo Stud Farm, 63-66; res asst vet sci, Pa State Univ, 66-67; res asst vet microbiol, Univ Wis, 67-72; res assoc, Purdue Univ, 72-74; ASST PROF MICROBIOL, STATE UNIV NY BUFFALO, 74- Mem: Am Soc Microbiol; Am Vet Med Asn; Am Col Vet Microbiologists. Res: Bacteriology, virology with particular emphasis on immunity in neoplastic and infectious diseases, diagnostic microbiology and clinical immunology. Mailing Add: Dept of Microbiol State Univ of NY Buffalo NY 14214

TARJAN, ARMEN CHARLES, b Chambridge, Mass, Dec 10, 20; m 45; c 2. PLANT NEMATOLOGY. Educ: Rutgers Univ, BS, 47; Univ Md, MS, 49, PhD(plant path), 51. Prof Exp: Asst nematologist, USDA, Md, 50-51; asst res prof plant path, Univ RI, 51-55; PROF NEMATOL, AGR RES & EDUC CTR, UNIV FLA, 55- Mem: Soc Nematol. Res: Chemical control of plant nematodes; taxonomic studies of nematodes; biological control of nematodes; nematophagus fungi. Mailing Add: Agr Res & Educ Ctr Univ of Fla Lake Alfred FL 33850

TARJAN, GEORGE, b Zsolna, Hungary, June 18, 12; US citizen; m 41; c 2. PSYCHIATRY. Educ: Pazmany Peter Univ, Hungary, MD, 35. Prof Exp: Resident physician, Mercy Hosp, Janesville, Wis, 40-41; asst physician, Utah State Hosp, Provo, 41-43; dir clin serv, Peoria State Hosp, Ill, 46-47; dir clin serv, Pac State Hosp, Pomona, Calif, 47-49, supt & med dir, 49-65; from asst clin prof to clin prof, 53-65, clin prof, Sch Pub Health, 61-65, PROF PSYCHIAT, SCHS MED & PUB HEALTH, UNIV CALIF, LOS ANGELES, 65-, PROG DIR MENT RETARDATION & CHILD PSYCHIAT, NEUROPSYCHIAT INST, 65- Concurrent Pos: Mem, President's Panel Ment Retardation, 61-62 & President's Comt, 66-71, consult, 71- Honors & Awards: Int Award Distinguished Leadership in Field Ment Retardation, Kennedy Found, 71; Am Asn Ment Deficiency Leadership Award, 70; President's Comt Ment Retardation Award Merit, 72; Am Psychiat Asn Distinguished Serv Award, 73. Mem: Am Psychiat Asn; Am Asn Ment Deficiency (pres, 58-59); AMA; Am Acad Child Psychiat; Group Advan Psychiat (pres, 71-73). Res: Child psychiatry; mental retardation. Mailing Add: Neuropsychiat Inst 760 Westwood Plaza Los Angeles CA 90024

TARKOW, HAROLD, b Milwaukee, Wis, Dec 9, 12; m 43; c 2. PHYSICAL CHEMISTRY. Educ: Univ Wis, BS, 34, PhD(phys chem), 39. Prof Exp: Res chemist, Portland Cement Asn, Chicago, 40-42; inst chem, Bradley Col, 42-43; from chemist to supvry chemist, 43-68, chief div wood chem res, 68-72, asst dir, Chem Utilization & Protection Res, 72-75, ASST DIR, FIBER & CHEM RES, FOREST PROD LAB, US FOREST SERV, 75- Mem: Am Chem Soc. Res: Colloid chemical phenomena; absorption; swelling; permeability; thermodynamics of solution; wood, pulp and paper; carbohydrates; synthetic resins and plastics. Mailing Add: Forest Prod Lab US Forest Serv Madison WI 53705

TARLETON, GADSON JACK, JR, b Sumter, SC, Apr 29, 20; m 49; c 2. RADIOLOGY. Educ: Morris Col, AB, 39; Meharry Med Col, MD, 44. Prof Exp: Resident radiol & orthop, Hubbard Hosp & Meharry Med Col, 44, resident radiol, 45-48; assoc prof, 49-52, PROF RADIOL, MEHARRY MED COL, 52-, CHMN DEPT & DIR TUMOR CLIN, 49- Concurrent Pos: Fel radiother, Bellevue Hosp, New York, 48-49; vis scholar, Columbia Presby Hosp, New York, 49. Mem: Radiol Soc NAm; AMA; Nat Med Asn; Am Col Radiol. Mailing Add: Dept of Radiol Meharry Med Col Nashville TN 37208

TARLETON, RAYMOND JOSEPH, b Nov 19, 25; US citizen; m 50. BIOCHEMISTRY. Educ: Univ Minn, BA, 48, MA, 52. Prof Exp: Control chemist, Seeger Refrigerator Co, Whirlpool Corp, 47-48; managing ed, 50-56, bus mgr, 56-58, exec secy, 58-65, EXEC V PRES, AM ASN CEREAL CHEMISTS, 65- Concurrent Pos: Consult, AID, 66-67; Mem: Am Phytopath Soc (vpres, 67-); Am Soc Brewing Chemists (vpres, 72-); League Int Food Educ (secy, 69, pres, 71; Am Chem Soc; AAAS. Res: Nutrition, specifically human with emphasis on B12; techniques of scientific writing, editing and publishing, including production, sales and distribution of books and journals; administration. Mailing Add: Am Asn of Cereal Chemists 3340 Pilot Knobb Rd St Paul MN 55121

TARLOV, ALVIN R, b South Norwalk, Conn, July 11, 29; m 56; c 5. INTERNAL MEDICINE, BIOCHEMISTRY. Educ: Dartmouth Col, BA, 51; Univ Chicago, MD, 56. Prof Exp: Intern, Philadelphia Gen Hosp, 56-57; resident med, Univ Chicago, 57-58, res assoc, 58-62; res assoc biochem, Harvard Med Sch, 62-64; from asst prof to assoc prof, 64-70, PROF MED & CHMN DEPT, UNIV CHICAGO, 70- Concurrent Pos: USPHS res career develop award, 62-; Markle scholar, 66- Mem: Am Fedn Clin Res. Res: Metabolism of red blood cells; inherited disorders of red cell metabolism; biochemistry of red cell and mitochondrial membranes; membrane lipids. Mailing Add: Dept of Internal Med Univ of Chicago Chicago IL 60637

TARNEY, ROBERT EDWARD, b Hammond, Ind, Jan 8, 31; m 66; c 3. ORGANIC CHEMISTRY. Educ: Purdue Univ, BS, 52; Univ Wis, PhD, 58. Prof Exp: RES CHEMIST, E I DU PONT DE NEMOURS & CO, INC, 57- Res: Synthesis of monomers for and polymers of elastomeric materials. Mailing Add: RD 2 Box 350 Hockessin DE 19707

TARNOWSKI, GEORGE SERGE, b St Petersburg, Russia, Aug 29, 11; US citizen; m 42; c 2. CANCER. Educ: Univ Belgrade, MD, 33. Prof Exp: Med officer, UN Relief & Rehab Admin, Ger, 45-47; res assoc antibiotics, Sharp & Dohme Inc, 49-51; ASSOC MEM CANCER RES, SLOAN KETTERING INST CANCER RES, 51- Mem: Am Asn Cancer Res. Res: Experimental cancer therapy; immunobiology of tumors. Mailing Add: Sloan Kettering Inst 145 Boston Post Rd Rye NY 10580

TARPLEY, ANDERSON RAY, JR, b New Orleans, La, Sept 19, 44; m 69. PHYSICAL CHEMISTRY, SPECTROSCOPY. Educ: Ga Inst Technol, BS, 66; Emory Univ, MS, 70, PhD(phys chem), 71; Ga State Univ, MBA, 72. Prof Exp: Res chemist, Eastman Kodak Co, NY, 66-67; instr res, Emory Univ, 71-72; SR CHEMIST ANAL SERV, TENN EASTMAN CO, EASTMAN KODAK CO, 72- Concurrent Pos: NIH fel med chem, Emory Univ, 71. Mem: AAAS; Am Chem Soc; Am Inst Chem; Am Mgt Asn. Res: Nuclear magnetic resonance spectroscopy; technical management; uses of computers in science; molecular orbital calculations; mass spectrometry. Mailing Add: Anal Serv Lab Org Chem Div Tenn Eastman Co Kingsport TN 37660

TARPLEY, JERALD DAN, b Lubbock, Tex, July 13, 42; m 65; c 1. ATMOSPHERIC PHYSICS. Educ: Tex Technol Col, BS, 64; Univ Colo, PhD(astrogeophys), 69. Prof Exp: Advan Study Prog fel, Nat Ctr Atmospheric Res, 69-70, physicist, Environ Res Labs, 70-73, PHYSICIST, NAT ENVIRON SATELLITE SERV, NAT OCEANIC & ATMOSPHERIC ADMIN, 73- Mem: Am Geophys Union; Am Meteorol Soc. Res: Physics of upper atmosphere and ionosphere; geomagnetism; remote sensing of the environment. Mailing Add: Nat Environ Satellite Serv Nat Oceanic & Atmospheric Admin Washington DC 20233

TARPLEY, WALLACE ARMELL, b Norwood, Ga, Feb 13, 34; m 59; c 2. INSECT ECOLOGY. Educ: Univ Ga, BSEd, 54, PhD(zool), 67; Clemson Univ, MS, 56. Prof Exp: Asst prof entom, Va Polytech Inst & State Univ, 60-64; ASSOC PROF BIOL, E TENN STATE UNIV, 64- Mem: AAAS; Entom Soc Am; Ecol Soc Am. Res: Ecological terminology, specifically the preparation of an ecological glossary; history of ecological terms; ecology of fresh water insects. Mailing Add: Box 2602 Dept of Biol E Tenn State Univ Johnson City TN 37601

TARPLEY, WILLIAM BEVERLY, JR, physical chemistry, organic chemistry, see 12th edition

TARR, BETTY R, b Milwaukee, Wis, Oct 10, 13; m 42. CHEMISTRY. Educ: Toledo Univ, BS, 35, MS, 37; Univ Ill, PhD(chem), 41. Prof Exp: Tech asst, Patent Dept, Phillips Petrol Co, Okla, 41-54, mgr tech info sect, Res & Develop Dept, 54-55; res librn, Technicolor Corp, 55-65; sr res chemist, Bell & Howell Res Labs, Calif, 65-68; asst prof, 68-71, ASSOC PROF CHEM, LOS ANGELES SOUTHWEST COL, 71- Mem: AAAS; Am Chem Soc. Mailing Add: Los Angeles Southwest Col Los Angeles CA 90047

TARR, CHARLES EDWIN, b Johnstown, Pa, Jan 14, 40; m 64. PHYSICS. Educ: Univ NC, Chapel Hill, BS, 61, PhD(physics), 66. Prof Exp: Asst physics, Univ NC, Chapel Hill, 62-66, res assoc, summer 66 & Univ Pittsburgh, 66-68; asst prof, 68-74, ASSOC PROF PHYSICS, UNIV MAINE, ORONO, 74- Mem: Am Phys Soc. Res: Nuclear magnetic resonance; electron paramagnetic resonance; instrumentation for nuclear magnetic resonance. Mailing Add: Dept of Physics Univ of Maine Orono ME 04473

TARR, DONALD ARTHUR, b Norfolk, Nebr, Aug 1, 32; m 57; c 2. INORGANIC CHEMISTRY. Educ: Doane Col, AB, 54; Yale Univ, MS, 56, PhD(chem), 59. Prof Exp: From instr to asst prof chem, Col Wooster, 58-65; asst prof, 65-67, ASSOC PROF CHEM, ST OLAF COL, 67- Concurrent Pos: Danforth teaching fel, 58-; res assoc, Univ Colo, 64-65; vis res fel, Univ Kent, 72-73. Mem: AAAS; Am Chem Soc. Res: Structure and stability of metal complexes; reactions of coordinated ligands. Mailing Add: Dept of Chem St Olaf Col Northfield MN 55057

TARR, HUGH LEWIS AUBREY, b Clevedon, Eng, Nov 17, 05; m 36; c 3. FISHERIES. Educ: Univ BC, BSA, 26, MSA, 28; McGill Univ, PhD(bact), 31; Cambridge Univ, PhD(biochem), 34. Prof Exp: Asst, Univ BC, 26-28; asst indust & cellulose chem, McGill Univ, 29; res bacteriologist, Rothamsted Exp Sta, 34-38; from res bacteriologist to dir technol sta, Pac Environ Inst, Fisheries Res Bd Can, 38-70; RETIRED. Honors & Awards: Gold Medal, Prof Inst Pub Servs Can. Mem: Fel Inst Food Technol; Prof Inst Pub Servs Can; fel Royal Soc Can; Brit Biochem Soc. Res: Agricultural bacteriology and biochemistry of fish; bacterial enzymes; biochemical methylation of furine bases in echinoderms. Mailing Add: Pac Environ Inst 4160 Marine Dr West Vancouver BC Can

TARRANT, PAUL, b Birmingham, Ala, Nov 1, 14; m 37, 72; c 3. CHEMISTRY. Educ: Howard Col, BS, 36; Purdue Univ, MS, 38; Duke Univ, PhD(chem), 44. Prof Exp: Instr, Ala Bd Educ, 38-40; chemist, Shell Develop Co, Calif, 40-41; asst, Duke Univ, 41-44; res chemist, Am Cyanamid Co, Conn, 44-46; from instr to assoc prof, 46-57, PROF CHEM, UNIV FLA, 57- Concurrent Pos: Chief investr, Off Naval Res Proj, 47-50 & 53-56; dir, Off Qm Gen Res Proj, 51-67, US Air Force Res Proj, 54-66 & 72-74, NSF, 68-71, NASA, 69-71 & Mass Inst Technol, 69-71; dir res, PCR, 53-66, vpres, 66-68; consult, Redstone Arsenal, US Dept Army & Naval Ord Lab, Calif; adv, Cotton Chem Lab, USDA; ed, Fluorine Chem Rev. Mem: Am Chem Soc. Res: Preparations and reactions of fluorine containing organic compounds and inert polymers. Mailing Add: Dept of Chem Univ of Fla Gainesville FL 32601

TARSEY, ALEXANDRE ROLF, b Memphis, Tenn, Jan 11, 23; m 57. APPLIED CHEMISTRY. Educ: La State Univ, Baton Rouge, BS, 46; Univ Mich, MSE, 47; Tulane Univ, PhD(inorg chem), 51. Prof Exp: Fabrication chemist, S Puerto Rico Sugar Co, 47-48; fel, Inst Phys Chem, Paris, 51-52; sr res engr, Titanium Metals Co Am, 52-58; specialist, Electrodialysis Proj, Arid Zone Res Inst, Beersheba, Israel, 58-59; sr chemist, Elmore Facil, Brush Beryllium Co, Ohio, 59-62; specialist, Electrochem & Surface Chem Unit, Rocketdyne Div, NAm Rockwell Corp, 62-66, sr specialist, Autonetics Div, 66-69; mgr testing & anal, Xerox Med Diag Opers, Calif, 69-70; chief pollution control sect, Sanit Br, US Army Eng Command, Europe, 70-74; PRES, AM SOLAR ENERGY CORP, 74- Concurrent Pos: Lectr phys chem, Univ Nev, Las Vegas, 55-56; mem working group on liquid propellant test methods, Interagency Chem Rocket Propulsion Group, 64-66; mem, Nev State Energy Resources Adv Bd, 74- Mem: Am Chem Soc; NY Acad Sci. Res: Design of experiments, tests and analyses; water pollution and purification; ion exchange, electrodialysis and electrochemistry; surface chemistry; rheology; industrial wastes; nonferrous extractive metallurgy. Mailing Add: Am Solar Energy Corp Suite 20 2960 Westwood Las Vegas NV 89109

TARSHIS, IRVIN BARRY, b Portland, Ore, May 12, 14; m 44; c 2. PARASITOLOGY, ZOOLOGY. Educ: Ore State Col, BA, 38; Univ Calif, Berkeley, PhD(parasitol), 53. Prof Exp: Res scientist, State Dept Fish & Game, Calif, 48-51; agr res scientist, USDA, Ga, 53-54; sect dep chief & med entomologist, Ft Detrick, Md, 54-57; asst prof entom, Univ Calif, Los Angeles, 57-63; RES PARASITOLOGIST, PATUXENT WILDLIFE RES CTR, 64- Concurrent Pos: Scripps res fel, Zool Soc San Diego, 60-61; consult, Israel, 60-61; del, Int Cong Entom, Montreal, Can, 56, Vienna, Austria, 60, Canberra, Australia, 72; del Int Cong Trop Med & Malaria, Rio de Janeiro, Brazil, 63. Honors & Awards: Gorgas Mem Award, 34; Animal Care Panel Res Award, 63. Mem: Fel AAAS; Sigma Xi; Am Soc Parasitol; Entom Soc Am; Am Soc Trop Med & Hyg. Res: Biology and epidemiology of arthropods affecting man and animal; host-parasite relationships; transmission of arthropod-borne diseases to wildlife; biology of Simuliidae; membrane feeding haematophagous arthropods; prevention and control of arthropods infesting animals and structures. Mailing Add: 17219 Emerson Dr Silver Spring MD 20904

TARSKI, ALFRED, b Warsaw, Poland, Jan 14, 20; nat US; m 29; c 2. MATHEMATICS. Educ: Univ Warsaw, PhD(math), 23. Prof Exp: Instr logic, Pedagog Inst, Fical, Warsaw, 22-25; docent & adj prof math & logic, Univ Warsaw, 25-39; res assoc math, Harvard Univ, 39-41; Guggenheim fel, 41-42 & 55-56; mem, Inst Advan Study, 42; lectr, 42-45, from assoc prof to prof, 45-68, res prof, Inst Basic Res in Sci, 58-60, EMER PROF MATH, UNIV CALIF, BERKELEY, 68- Concurrent Pos: Prof, Zeromski's Lycee, Warsaw, 25-39; Rockefeller Found fel, 35; vis prof, City Col New York, 40; Sherman mem lectr, Univ Col, Univ London, 50, 66; lectr, Sorbonne, 55; vis prof, US Dept State Smith-Mundt grant, Nat Univ Mex, 57; vis Flint prof philos, Univ Calif, Los Angeles, 67; pres, Int Union Hist & Philos Sci, 57, chmn, US Nat Comt, 62-63. Honors & Awards: Alfred Jurzykowski Found Award, 66. Mem: Nat Acad Sci; Am Math Soc; Asn Symbolic Logic (pres, 44-46); Royal Neth Acad Sci; hon mem Neth Math Soc. Res: Mathematical logic and metamathematics; set theory; measure theory; general algebra. Mailing Add: 462 Michigan Ave Berkeley CA 94707

TARTAKOFF, JOSEPH, surgery, deceased

TARTAR, VANCE, b Corvallis, Ore, Sept 15, 11; m 50; c 3. EXPERIMENTAL MORPHOLOGY. Educ: Univ Wash, Seattle, BS, 33, MS, 34; Yale Univ, PhD(zool), 38. Prof Exp: Seessel fel, Yale Univ, 39; instr zool, Univ Vt, 39-42 & Yale Univ, 42; asst aquatic biologist, US Fish & Wildlife Serv, 42; biologist, Wash State Dept Fisheries, 43-50; res assoc prof, 51-61, RES PROF ZOOL, UNIV WASH, 61- Mem: Am Soc Zool; Soc Protozool; Soc Develop Biol. Res: Oyster and cell biology; experimental morphogenesis in ciliates; protozoology; cell biology; the biology of Stentor. Mailing Add: Univ of Wash Field Lab Nahcotta WA 98637

TARTER, CURTIS BRUCE, b Louisville, Ky, Sept 26, 39; m 64; c 1. ASTROPHYSICS, THEORETICAL PHYSICS. Educ: Mass Inst Technol, SB, 61; Cornell Univ, PhD(astrophys), 67. Prof Exp: Sr scientist, Aeronutronic Div, Philco-Ford Corp, 67; physicist, 67-69, group leader, 69-73, DEP DIV LEADER, THEORET PHYSICS, LAWRENCE LIVERMORE LAB, 73- Concurrent Pos: Lectr, Dept Appl Sci, Univ Calif, Davis, 71- Mem: Am Phys Soc; Am Astron Soc. Res: Theoretical description of the properties of matter at high temperatures and densities; theoretical astrophysics, particularly x-ray astronomy and the dynamical phases of stellar evolution. Mailing Add: Lawrence Livermore Lab PO Box 808 L-71 Livermore CA 94550

TARTER, DONALD CAIN, b Somerset, Ky, July 22, 36; m 60; c 3. ZOOLOGY. Educ: Georgetown Col, Ky, BS, 58; Miami Univ, Ohio, MAT, 62; Univ Louisville, PhD(zool), 68. Prof Exp: Teacher chem biol, Bradford High Sch, Ohio, 58-60 & Tipp City High Sch, Ohio, 60-64; teacher biol, Ky Southern Col, 64-68; TEACHER ZOOL, MARSHALL UNIV, 68- Mem: Am Fisheries Soc; Am Entom Soc; Sigma Xi; Am Soc Ichthyologists & Herpetologists. Res: Taxonomy and ecology of fishes and aquatic insects. Mailing Add: Dept of Biol Sci Marshall Univ Huntington WV 25705

TARTER, MICHAEL E, b New York, NY, Dec 20, 38; m 61; c 2. BIOSTATISTICS. Educ: Univ Calif, Los Angeles, AB, 59, MA, 61, PhD(biostatist), 64. Prof Exp: From asst prof to assoc prof biostatist, Univ Mich, Ann Arbor, 64-67; ASSOC PROF BIOSTATIST, DEPT MED & MATH, UNIV CALIF, IRVINE, 68- & DIV MEASUREMENT SCI, UNIV CALIF, BERKELEY, 70- Concurrent Pos: Consult, Upjohn Drug Co, Med Diag Corp, Regional Med Asn, Pac Med Ctr & Presch & Adolescent Proj. Mem: Fel Am Statist Asn; Asn Comput Mach. Res: Graphical biometry; computational aspects of statistical procedures; biostatistical consultation training; nonparametric density estimation; programmed and computer assisted instruction; sorting theory. Mailing Add: Biostatist Univ Rm 32 Earl Warren Hall Univ of Calif Berkeley CA 94720

TARTOF, KENNETH D, b Detroit, Mich, Dec 30, 41; m 67. GENETICS. Educ: Univ Mich, BS, 63, PhD(genetics), 68. Prof Exp: NIH res fel, 68-70, res assoc, 70-71, ASST MEM, INST CANCER RES, 71- Mem: AAAS; Genetics Soc Am. Res: Structure and function of genes; genetic control of gene redundancy; regulation of DNA and RNA metabolism. Mailing Add: Inst for Cancer Res 7701 Burholme Ave Philadelphia PA 19111

TARVER, FRED RUSSELL, JR, b Knoxville, Tenn, Mar 7, 25; m 50; c 3. FOOD TECHNOLOGY, BACTERIOLOGY. Educ: Univ Tenn, BSA, 50, MA, 54; Univ Ga, PhD(food technol, bact), 63. Prof Exp: Mem staff, Security Mills, Inc, Tenn, 50-54; instr poultry, Univ Tenn, 54-56; asst prof, Univ Fla, 56-60, 62-63; exten assoc prof, 63-75, EXTEN PROF FOOD SCI, NC STATE UNIV, 75. Mem: Inst Food Technol; Poultry Sci Asn; World Poultry Sci Asn. Res: Product development and marketing of new poultry and egg products; sanitation in processing plants, with emphasis on ecology; 4-H club activities connected with poultry and egg products. Mailing Add: Dept Food Sci 129-D Schaub Hall NC State Univ Raleigh NC 27607

TARVER, HAROLD, b Wigan, Eng, June 7, 08; nat US; m; c 2. CHEMISTRY. Educ: Univ Alta, BS, 32, MSc, 35; Univ Calif, PhD(biochem), 39. Prof Exp: Asst, Univ Alta, 32-35; asst, Sch Med, Univ Calif, Berkeley, 36-39, fel, 39-41; from instr to prof biochem, 41-75, EMER PROF BIOCHEM, SCH MED, UNIV CALIF, SAN FRANCISCO, 75- Mem: AAAS; Am Soc Biol Chem; Soc Exp Biol & Med. Res: Metabolism of sulfur and protein-isotopic studies. Mailing Add: Dept of Biochem & Biophys Univ of Calif Med Ctr San Francisco CA 94143

TARVER, MAE-GOODWIN, b Selma, Ala, Aug 9, 16. STATISTICS. Educ: Univ Ala, BS, 39, MS, 40. Prof Exp: Res engr, 41-48, proj engr statist, 48-54, RES STATISTICIAN & INTERNAL QUAL CONTROL CONSULT, RES & DEVELOP DEPT, CONTINENTAL CAN CO, INC, 54- Concurrent Pos: Instr, Ill Inst Technol, 57-62, adj assoc prof, 63- Honors & Awards: Lisy Award, Am Soc Qual Control, 61. Mem: Am Statist Asn; fel Am Soc Qual Control; Biomet Soc; Inst Food Technol. Res: Multivariate statistical analysis of industrial research data; statistical and information theory appraisal of sensory tests; quality engineering applied to container fabrication and packaging of food products. Mailing Add: Continental Can Co Inc 1200 W 76th St Chicago IL 60620

TARVIN, DONALD, b Roberts, Ill, Apr 9, 05; m 37; c 2. CHEMISTRY. Educ: Ill State Univ, BE, 28; Univ Ill, MS, 30, PhD(chem), 33. Prof Exp: Chemist, State Water Surv, Ill, 33-35; assoc chemist water & silt lab, Tenn Valley Authority, 35-39; chemist, Gen Chem Div, Allied Chem Co, 39-44; chief chemist, Calco Chem Div, Am Cyanamid Co, 44-53; chemist & assoc, Floyd G Browne Assocs, Ltd, Consult Engrs, 53-72; RETIRED. Concurrent Pos: Instr, Univ Ill, 33-34. Mem: Am Chem Soc; Am Water Works Asn; Water Pollution Control Fedn; fel Am Inst Chem. Res: Water, sewage and industrial waste investigations and treatment. Mailing Add: 663 King Ave Marion OH 43302

TARWATER, JAN DALTON, b Ft Worth, Tex, Sept 30, 37; m 58; c 3. ALGEBRA. Educ: Tex Tech Col, BS, 59; Univ NMex, MA, 61, PhD(math), 65. Prof Exp: Asst prof math, Western Mich Univ, 65-67 & NTex State Univ, 67-68; from asst prof to assoc prof, 68-73, assoc chmn dept, 72-73, PROF MATH & CHMN DEPT, TEX TECH UNIV, 73- Mem: Math Asn Am; Am Math Soc; Soc Indust & Appl Math. Res: Algebra, especially Abelian groups and homological algebra; graph theory; history of mathematics. Mailing Add: Dept of Math Tex Tech Univ Lubbock TX 79409

TARWATER, OLIVER REED, b Chattanooga, Tenn, Mar 12, 44; m 66. SYNTHETIC ORGANIC CHEMISTRY, POLYMER CHEMISTRY. Educ: Maryville Col, BS, 66; Purdue Univ, Lafayette, MS, 69, PhD(med chem), 70. Prof Exp: Res chemist, Personal Care Div, Gillette Co, Chicago, 70-74; SR CHEMIST, SOUTHERN RES INST, BIRMINGHAM, 74- Mem: Am Chem Soc. Res: Synthetic organic and synthetic polymer chemistry related to biomaterials. Mailing Add: Southern Res Inst 2000 Ninth Ave South Birmingham AL 35205

TARZWELL, CLARENCE MATTHEW, b Deckerville, Mich, Sept 29, 07; m 38; c 3. WATER POLLUTION, POLLUTION BIOLOGY. Educ: Univ Mich, AB, 30, MS, 32, PhD(aquatic biol, fisheries mgt), 36. Hon Degrees: ScD, Baldwin-Wallace Col, 67. Prof Exp: Stream improv supvr, Mich State Dept Conserv, 33-34; asst aquatic biologist, US Bur Fisheries, 34; asst range mgr, US Forest Serv, NMex, 35, 36-38; from asst aquatic biologist to assoc aquatic biologist, Tenn Valley Authority, Ala, 38-43; asst & sr sanitarian, USPHS, Ga, 43-46, sr biologist, Tech Develop Div, 46-48, chief aquatic biol sect, Environ Health Ctr, Cincinnati, Ohio, 48-53, sci dir, Robert A Taft Sanit Eng Ctr, 53-65; dir, Nat Marine Water Qual Lab, Environ Protection Agency, RI, 65-72, sr res adv res & monitoring, 72-75; ENVIRON CONSULT, 75- Concurrent Pos: Mem aquatic life adv comt, Ohio River Valley Sanit Compact, 52-68; chmn & mem comt on stand bioassay methods, Toxicity Comt, Water Pollution

Control Fedn, 56-75; mem adv comt control stream temperatures, Pa Sanit Water Bd, 59-62; mem adv comt water qual criteria, Calif State Water Pollution Control Bd, 61-63; mem subcomt res, comt pest control & wildlife rels, Nat Acad Sci-Nat Res Coun, 61-63; US deleg int meetings sci res water pollution, Orgn Econ Coop & Develop, Paris, France, 61-62, chmn comt on toxic mat in aquatic environ, 62-67; mem expert adv panel environ health, WHO, 62-65; actg dir nat water qual lab, Fed Water Pollution Control Admin, Duluth, Minn, 64-67; adj prof, Col Resource Develop, Univ RI, 65-; chmn nat tech adv comt water qual req fish, other aquatic life & wildlife, Secy Interior, 67-68; mem comt power plant siting, Nat Acad Eng-Nat Acad Sci, 71-72; mem nat temperature comt, Environ Protection Agency, 71-72; mem, NAm Game Policy Comt, 71-73. Honors & Awards: Conservationist of the Yr Award, State of Ohio, 61; Prof Conservationist Award, Am Motors Corp, 62; Aldo Leopold Medal, Wildlife Soc, 63; USPHS Meritorious Serv Medal, 64; Distinguished Career Award, Environ Protection Agency, 73 & Commendable Serv Medal, 74; Award of Excellence, Am Fisheries Soc, 74. Mem: Hon mem Am Fisheries Soc; hon mem Wildlife Soc; Am Soc Ichthyol & Herpet; Am Soc Limnol & Oceanog; Atlantic Estuarine Res Soc. Res: Water pollution control; determination of water quality criteria and standards for aquatic life; standard bioassay methods; environmental improvement and management; toxicity of wastes to aquatic life; aquatic biology and ecology. Mailing Add: Box 159 Old Post Rd Wakefield RI 02879

TASAKI, ICHIJI, b Fukushim-Ken , Japan, Oct 21, 10; nat US; m; c 2. NEUROPHYSIOLOGY. Educ: Keio Univ, Japan, MD, 38. Hon Degrees: DS, Uppsala Univ, 72. Prof Exp: Privat-docent physiol, Keio Univ, 38-42, privat-docent physics & prof physiol, Med Col, 42-51; prof physiol, Nihon Univ, Tokyo, 51; res assoc, Cent Inst Deaf, St Louis, Mo, 51-53; chief spec senses sect, Lab Neurophysiol, Nat Inst Neurol Dis & Blindness, 53-61; mem staff, 61-66, CHIEF LAB NEUROBIOL, NIMH, 66- Concurrent Pos: Mem, Marine Biol Lab, Woods Hole, Mass. Mem: Am Acad Neurol; Physiol Soc Japan. Res: Nerve and sense organs; electrophysiology. Mailing Add: Lab of Neurobiol Rm ID-02 Bldg 36 NIMH Bethesda MD 20014

TASCH, AL FELIX, JR, b Corpus Christi, Tex, May 12, 41; m 63; c 2. SOLID STATE PHYSICS. Educ: Univ Tex, Austin, BS, 63; Univ Ill, Urbana, MS, 65, PhD(physics), 69. Prof Exp: RES SCIENTIST, SOLID STATE PHYSICS, CENT RES LABS, TEX INSTRUMENTS, INC, 69- Mem: Am Phys Soc. Res: Solid state device physics and processing technology; recombination-generation and trapping in semiconductors; metal-oxide-silicon field effect transistors; microwave transistors; charge-coupled devices. Mailing Add: MS 134 Cent Res Labs Tex Instruments Inc PO Box 5936 Dallas TX 75222

TASCHDJIAN, CLAIRE LOUISE, b Berlin, Ger, Jan 7, 14; nat US; m 44; c 1. MEDICINE, MYCOLOGY. Educ: Catholic Univ, Peking, China, BSc, 45; Wagner Col, MS, 60. Prof Exp: Asst, Cenozoic res lab, Peking Union Med Col, China, 40-41; teacher, Peking Am High Sch, China, 45-48; mycologist, Sect Dermat, Dept Med, Univ Chicago, 49-52; clin asst dermat & syphilol, Postgrad Med Sch, NY Univ, 52-56; asst prof biol, St Francis Col, NY, 64-69; RES ASSOC PEDIAT, MAIMONIDES MED CTR, 56-; ASSOC PROF BIOL, ST FRANCIS COL, NY, 69- Concurrent Pos: Lectr, State Univ NY Downstate Med Ctr, 63-; vis prof, Dept Path, Univ Ulm, 70-71. Honors & Awards: Outstanding Clin Paper, Maimonides Med Ctr, 72. Mem: Med Mycol Soc Am; Am Soc Microbiol; NY Acad Sci; Int Soc Human & Animal Mycol. Res: Immunologic, immunochemical and histochemical factors in candidiasis; epidemiology, serology, pathogenesis and serodiagnosis of systemic candidiasis; physiology of Candida albicans in vivo and in vitro; science education; evolution. Mailing Add: 109-50 117th St S Ozone Park Sta Jamaica NY 11420

TASCHDJIAN, EDGAR, b Vienna, Austria, Mar 8, 04; nat US; m 44; c 3. GENETICS. Educ: Vienna State Col Agr, DrIng(Agr), 28. Prof Exp: Chief sect plant breeding, State Agr Inst, Sao Paulo, Brazil, 28-30; correspondent, Nat Mus Natural Hist, France, 32-33; agr adv, J G White Eng Corp, NY, 33-34; prof biol, Fu Jen Univ, China, 36-46, chief sect plant prod, 46-47; assoc prof bot, Univ Loyola, Ill, 48-54; assoc prof, 54-69, chmn dept, 61-69, EMER PROF BIOL, ST FRANCIS COL, NY, 69- Concurrent Pos: Adv, Chinese Catholic Agr Res Asn, 45-47; vis prof, Haigazian Col, Lebanon, 69-70 & Dept Path, Univ Ulm, 70-71. Mem: AAAS; Genetics Soc Am; Am Soc Agron; Am Soc Plant Physiol; Am Genetic Asn. Res: Environmental science; blood coagulation; topological biostatistics; agricultural genetics; biophysics; membrane transport; motoric systems. Mailing Add: 109-50 117th St South Ozone Park NY 11420

TASCHEK, RICHARD FERDINAND, b Chicago, Ill, June 5, 15; m 42; c 4. PHYSICS. Educ: Univ Wis, BA, 36, PhD(physics), 41; Univ Fla, MS, 38. Prof Exp: Teaching asst, Univ Wis, 38-40; res physicist, Oldbury Electro-Chem Co, NY, 41-42; Nat Defense Res Comt physicist, Princeton Univ, 42-43; physicist, 43-62, div leader exp physics, 62-70, asst dir res, 71-72, ASSOC DIR RES, LOS ALAMOS SCI LAB, 72- Concurrent Pos: Mem nuclear cross sect adv group, AEC, 48-57, chmn, 53-57, mem tripartite nuclear cross sect comt, 56-61, chmn, 61; mem Euro-Am nuclear data comt, 57-72, chmn, 60-62; mem int nuclear data working group, 63-72; adv, Int Nuclear Data Comt, 63-73, ex-officio mem, 72-74; mem standing comt controlled thermonuclear reactions & ad hoc adv comt on Los Alamos meson proj; co-chmn, Vis Comt Lab Nuclear Sci, Mass Inst Technol, 67; mem adv comt neutron physics, Oak Ridge Nat Lab; mem Nuclear Physics Panel, Physics Surv Comt, Div Phys Sci, Nat Res Coun; mem Off Stand Ref Data Eval Panel, Nat Bur Stand, & Inst Basic Stand Eval Panel, 71-75; mem, Ctr Radiation Res, Nat Bur Stand Panel, 71-76, chmn, 72-; mem adv comt, Univ Alaska Geophys Inst, 73-76; mem, Numerical Data Adv Bd, US Nat Comt for CODATA, 73-, chmn, 74- Mem: Fel AAAS; fel Am Phys Soc; fel Am Nuclear Soc; Am Geophys Union. Res: Nuclear reactions and scattering; nuclear properties; neutron physics; accelerators and detectors; space physics; Vela satellite program; controlled thermonuclear reactions. Mailing Add: Mail Stop 102 Box 1663 Los Alamos Sci Lab Los Alamos NM 87545

TASCHENBERG, EMIL FREDERICK, b Cumberland, Md, June 4, 16; m 47; c 2. ENTOMOLOGY. Educ: Gettysburg Col, AB, 38; Cornell Univ, PhD(econ entom), 45. Prof Exp: Asst entom, Exp Sta, 39-45, from asst prof to assoc prof, 45-59, PROF ENTOM, VINEYARD LAB, AGR EXP STA, NY STATE COL AGR & LIFE SCI, CORNELL UNIV, 59- Mem: AAAS; Entom Soc Am. Res: Biology and control of grape and small fruit insects; insect toxicology; mass rearing methods for economic insects and several saturniid moths and insect sex pheromones. Mailing Add: Dept of Entom Agr Exp Sta Cornell Univ Geneva NY 14456

TASHIAN, RICHARD EARL, b Cranston, RI, Oct 7, 22; m 68; c 1. BIOCHEMICAL GENETICS. Educ: Univ RI, BS, 47; Purdue Univ, MS, 49, PhD(zool), 51. Prof Exp: Asst, Purdue Univ, 48-51; asst prof biol, Long Island Univ, 51-54, actg chmn dept, 54; sci assoc, Dept Trop Res, NY Zool Soc, 54-55; res assoc, Inst Study Human Variation, Columbia Univ, 56-57; from res assoc to assoc prof, 57-70, PROF HUMAN GENETICS, MED SCH, UNIV MICH, ANN ARBOR, 70- Concurrent Pos: Vis scientist, Dept Chem, Carlsberg Lab, Copenhagen, Denmark, 68-69. Mem: Fel AAAS; Am Soc Human Genetics; Am Soc Zoologists; Brit Biochem Soc. Res: Biochemical genetics and protein evolution; evolution of the structure and function relationships of carbonic anhydrase; comparative enzymology; isozymes; primate evolution. Mailing Add: Dept of Human Genetics Univ of Mich Med Sch Ann Arbor MI 48104

TASHIRO, HARUO, b Selma, Calif, Mar 24, 17; m 42; c 3. ECONOMIC ENTOMOLOGY. Educ: Wheaton Col, BS, 45; Cornell Univ, MS, 46, PhD(entom), 50. Prof Exp: Asst entom, Cornell Univ, 47-50; entomologist, USDA, 50-67; PROF ENTOM, AGR EXP STA, CORNELL UNIV, 67- Concurrent Pos: Assoc prof, Cornell Univ, 58-63; res assoc, Univ Calif, Riverside, 63-67. Mem: Entom Soc Am. Res: Biology; biological and chemical control of turf insects; insects; ornamental plants; nursery and permanent plantings; insect pathology. Mailing Add: Dept of Entom Agr Exp Sta Cornell Univ Geneva NY 14456

TASHIRO, JOHN T, pharmacognosy, natural products chemistry, see 12th edition

TASHJIAN, ARMEN H, JR, b Cleveland, Ohio, May 2, 32; m 55; c 3. ENDOCRINOLOGY, PHARMACOLOGY. Educ: Yale Univ, 50-53; Harvard Univ, MD, 57. Prof Exp: Intern med, Harvard Serv, Boston City Hosp, 57-58, asst resident, 58-59; clin res assoc, Metab Dis Br, Nat Inst Arthritis & Metab Dis, 59-61; Nat Found res fel, 61-63, from instr to assoc prof, 63-70, PROF PHARMACOL, SCH DENT MED & SCH MED, HARVARD UNIV, 70- Mem: AAAS; Am Soc Pharmacol & Exp Therapeut; Endocrine Soc; Am Soc Cell Biol; Tissue Cult Asn. Res: Purification, bioassay, immunoassay and immunochemistry, structure-function interrelationships and mechanism of action of protein and peptide hormones; cell biology; establishment, control of function and growth, and immunology of differentiated, clonal strains of animal cells in culture. Mailing Add: Lab of Pharmacol Harvard Sch of Dent Med Boston MA 02115

TASHJIAN, ROBERT JOHN, b Worcester, Mass, Feb 4, 30. VETERINARY MEDICINE. Educ: Clark Univ, AB, 51; Univ Pa, VMD, 56. Hon Degrees: DSc, Hartwick Col, 68. Prof Exp: Staff vet, Animal Med Ctr, New York, 56-58, head med, 58-62, chief of staff, 62-63, dir, 63-74; PRES, NEW ENG INST COMP MED, 74- Concurrent Pos: Res assoc, Vet Admin Hosp, Bronx, NY, 60-64; instr, Sch Vet Med, Univ Pa, 61-63; prin investr, Nat Heart Inst grant, 62-66, sponsor grant, 66-69; lectr animal sci & mem grad fac, Univ Maine, Orono, 67-, vis prof, 74-; affil prof, Colo State Univ, 67-; adj prof, Univ Mass, Amherst, 69-70; dir, Duke Farms, Somerville, NJ, 71-74; consult to bd trustees, Animal Med Ctr, New York, 74-; mem coun clin cardiol, Am Heart Asn. Mem: AAAS; Am Asn Lab Animal Sci; Am Asn Vet Clinicians; Am Vet Med Asn; Am Vet Radiol Soc. Res: Internal medicine, especially comparative cardiovascular disease. Mailing Add: New Eng Inst of Comp Med 306 Main St Worcester MA 01608

TASHLICK, IRVING, b New York, NY, July 5, 28; m 50. ORGANIC CHEMISTRY, POLYMER CHEMISTRY. Educ: City Col New York, BS, 49; Polytech Inst Brooklyn, MS, 53, PhD(org chem), 58. Prof Exp: Chemist, Plastics Div, Monsanto Co, 58-60, group leader, 60-62; mgr chem res, Res & Eng Div, Int Pipe & Ceramics Corp, 62-66; vpres, Wharton Industs, Inc, 68-72; PRES, ALVA-TECH INC, 73- Mem: Am Chem Soc; Soc Plastics Eng. Res: Long term properties of plastics, coatings and adhesives; polyurethanes. Mailing Add: 311 Newark Ave Bradley Beach NJ 07720

TASI, JAMES, b New York, NY, Dec 6, 33; m 60; c 3. MECHANICS. Educ: NY Univ, BCE, 55; Univ Ill, MS, 56; Columbia Univ, PhD(mech), 62. Prof Exp: Engr, Markin Co, Md, 57-58, assoc res scientist, Colo, 61-65; fel mech, Johns Hopkins Univ, 65-66; assoc prof, 66-72, PROF MECH, STATE UNIV NY STONY BROOK, 72- Mem: Am Soc Mech Engrs. Res: Acoustic vibrations of structures; thermoelastic dissipation in crystalline solids; wave propagation; stability; mechanical properties of solids. Mailing Add: Dept of Mech State Univ of NY Stony Brook NY 11790

TASKER, CLINTON WALDORF, b Syracuse, NY, Sept 14, 18; m 41; c 3. CHEMISTRY. Educ: Syracuse Univ, BSc, 41, MSc, 44; McGill Univ, PhD(cellulose chem), 47. Prof Exp: Res chemist, Sylvania Indust Corp, Va, 41-43; lab asst pulp & paper technol, State Univ NY Col Forestry, Syracuse Univ, 43-44; lab demonstr org & inorg chem, McGill Univ, 44-46; sr res chemist, Sylvania Div, Am Viscose Corp, 47-53, tech supt, 53-59, mgr supt, 59-61; dir tech res & develop, 61-65, vpres corp res & develop, 65-68, Paperboard Div, 68-72, V PRES RES & DEVELOP, PACKAGING CORP AM, TENNECO INC, 72- Mem: Am Inst Chemists; NY Acad Sci; Am Chem Soc; Tech Asn Pulp & Paper Indust. Res: Alkaline chemical pulping processes; chemistry and structure of cellulose ethers; synthesis of plasticizers for cellulose; tosyl and iodo derivatives of some hydroxyethyl ethers. Mailing Add: 1860 Surfside Dr Manistee MI 49660

TASKER, JOHN B, b Concord, NH, Aug 28, 33; m 61, c 3. VETERINARY MEDICINE, CLINICAL PATHOLOGY. Educ: Cornell Univ, DVM, 57, PhD(vet path), 63. Prof Exp: Am Vet Med Asn fel, Cornell Univ, 61-63; from asst prof to assoc prof vet med, Colo State Univ, 63-67; assoc prof vet path, 67-69, PROF CLIN PATH, NY STATE VET COL, CORNELL UNIV, 69- Mem: Am Vet Med Asn; Am Soc Vet Clin Path. Res: Veterinary clinical pathology, especially hematology and fluid balance disorders. Mailing Add: Dept of Clin Path NY State Vet Col Cornell Univ Ithaca NY 14850

TASKER, RONALD REGINALD, b Toronto, Ont, Dec 18, 27; m 55; c 4. NEUROSURGERY. Educ: Univ Toronto, BA, 48, MD, 52, MA, 54; FRCS(C), 59. Prof Exp: Asst physiol, Banting & Best Dept Med Res, 48-52, mem clin & res staff, 61-66, ASST PROF SURG, FAC MED, UNIV TORONTO, 66-; ASSOC PROF NEUROSURG, TORONTO GEN HOSP, 73- Concurrent Pos: Resident fel neurosurg, Fac Med, Univ Toronto, 58-59; McLaughlin traveling fel, Mass Gen Hosp, 59 & Univ Wis, 60; mem clin & res staff, Toronto Gen Hosp, 61-66, asst prof neurosurg, 66-73. Mem: Am Asn Neurol Surg; Can Neurosurg Soc; Can Med Asn; Am Acad Neurol Surg; Am Soc Res Stereoencephalotomy. Res: Hyperkinetic disorders; dyskinesias and stereotatic surgery; sensory physiology and pain. Mailing Add: Rm 121•Toronto Gen Hosp Toronto ON Can

TASLITZ, NORMAN, b New York, NY, Feb 12, 29; m 56; c 3. NEUROANATOMY. Educ: NY Univ, BS, 51; Univ Pa, cert phys ther, 52; Stanford Univ, PhD(anat), 63. Prof Exp: Physical therapist, Univ Wis Hosps, Madison, 52-54 & Wis Neurol Found, Madison, 54-58; asst prof anat, Sch Phys Ther, 63-71, from instr to sr instr, Dept Anat, 63-67, ASST PROF ANAT, DEPT ANAT, CASE WESTERN RESERVE UNIV, 67- Mem: Am Asn Anatomists; Sigma Xi; AAAS; Am Asn Univ Prof. Res: Development of biological models to demonstrate the hemodynamic, metabolic and electrophysiologic performance of intact central nervous system tissue under control conditions, at various subnormal temperature levels and following trauma or periods of circulatory arrest. Mailing Add: Dept of Anat Sch of Med Case Western Reserve Univ Cleveland OH 44106

TASMAN, WILLIAM S, US citizen; m; c 3. OPHTHALMOLOGY. Educ: Haverford Col, BA, 51; Temple Univ, MD, 55. Prof Exp: ASSOC RETINA SERV, WILLS EYE HOSP, 62- Concurrent Pos: Heed fel, 61-62; Retina Found fel, 62. Mem: Am Col

Physicians; Am Col Surg; Am Ophthal Soc. Res: Retinal diseases in children; retinal detachment surgery. Mailing Add: Wills Eye Hosp 187 E Evergreen Ave Philadelphia PA 19118

TASSAVA, ROY A, b Ironwood, Mich, July 5, 37; m 61; c 3. DEVELOPMENTAL BIOLOGY, COMPARATIVE ENDOCRINOLOGY. Educ: Northern Mich Univ, BS, 59; Brown Univ, MAT, 65; Mich State Univ, PhD(zool), 68. Prof Exp: Pub sch teacher, 59-64; NIH res fel zool, Mich State Univ, 68-69; asst prof, 69-73, ASSOC PROF ZOOL, OHIO STATE UNIV, 73- Concurrent Pos: Sigma Xi res award, 68. Mem: Am Inst Biol Sci; Am Soc Zoologists. Res: Role of hormones and nerves in amphibian limb regeneration; pituitary, thyroid and adrenal hormone physiology in amphibians. Mailing Add: Dept of Zool Ohio State Univ Columbus OH 43210

TASSONI, JOSEPH PAUL, b Port Washington, NY, Mar 10, 23; m 55; c 3. ENTOMOLOGY. Educ: Univ Utah, BA, 37; NY Univ, MS, 51, PhD(biol), 53. Prof Exp: From instr to asst prof anat, histol & neuroanat, New York Med Col, 53-56; from asst prof to assoc prof anat, 56-69, asst dean, 71-73, PROF ANAT, NJ MED SCH, COL MED & DENT NJ, 69-, DIR ADMIS, 71-, ASSOC DEAN, 73- Mem: Am Asn Anat. Res: Physiological entomology; protein changes in metamorphosis and starvation; diabetes; alloxan and other drugs; enzymology; muscular dystrophy; biochemistry and histochemistry of neuromuscular diseases. Mailing Add: NJ Med Sch Col of Med & Dent of NJ Newark NJ 07107

TASSOUL, JEAN-LOUIS, b Brussels, Belg, Nov 1, 38; m 66. ASTROPHYSICS. Educ: Free Univ Brussels, LSc, 61, DSc, 64. Prof Exp: Res fel, Nat Found Sci Res, Belg, 65-66; res assoc, Univ Chicago, 66-67 & Princeton Univ, 67-68; from asst prof to assoc prof, 68-75, PROF PHYSICS, UNIV MONTREAL, 75- Mem: Int Astron Union. Res: Rotating stars and nonlinear instabilities. Mailing Add: Dept of Physics Univ Montreal PO Box 6128 Montreal PQ Can

TATA, ROBERT JOSEPH, b New Britain, Conn, Mar 3, 35; m 58; c 3. GEOGRAPHY OF LATIN AMERICA, ECONOMIC GEOGRAPHY. Educ: Syracuse Univ, AB, 57, MA, 61, PhD(geog), 68. Prof Exp: Teaching asst geog, Syracuse Univ, 57-58; asst prof, 64-69, ASSOC PROF GEOG, FLA ATLANTIC UNIV, 69-, CHMN DEPT, 70- Concurrent Pos: Mem permanent consult fac, US Army Command & Gen Staff Col, 67- Mem: Asn Am Geogr; Am Geog Soc. Res: Development and revolutionary change of Latin America; geographic methodology. Mailing Add: Dept of Geog Fla Atlantic Univ Boca Raton FL 33432

TATARCZUK, JOSEPH RICHARD, b Portland, Maine, June 15, 36; m 64; c 4. NUCLEAR PHYSICS, COMPUTER ENGINEERING. Educ: Col of the Holy Cross, BS, 58; Rensselaer Polytech Inst, MS, 61, PhD(physics), 65. Prof Exp: Res asst nuclear physics, Rennselaer Polytech Inst, 59-65; res assoc, Nuclear Physics Div, Max Planck Inst Chem, 65-66; res assoc neutron physics, 66-70, ASST TO DIR LINAC OPERS & SUPPORT SERV, LINEAR ACCELERATOR LAB, RENNSELAER POLYTECH INST, 70- Concurrent Pos: Chief physicist, Nuclear Med Serv, Albany Vet Admin Hosp, 72-; mem adj fac nuclear struct, comput & nuclear instrumentation & nuclear med, Dept Nuclear Eng, Rennselaer Polytech Inst, 74- Mem: Am Phys Soc. Res: Neutron physics; gamma ray spectroscopy; nuclear, accelerator and computer instrumentation. Mailing Add: Linear Accelerator Lab Rennselaer Polytech Inst Troy NY 12181

TATE, BRYCE EUGENE, b Girard, Ill, Apr 15, 20; m 59; c 2. ORGANIC CHEMISTRY. Educ: Univ Ill, BS, 42; Univ Wis, MS, 44, PhD(chem), 50. Prof Exp: Res fel chem, Harvard Univ, 53-55; chemist, Chas Pfizer & Co, 55-59, proj leader, 59-61, group supvr, 61-68, MGR CHEM PROD RES, PFIZER INC, 68- Mem: Am Chem Soc; Am Inst Chemists; Soc Petrol Engrs. Res: Synthetic organic chemistry; mechanism of organic reactions; polymer chemistry. Mailing Add: Cent Res Pfizer Inc Eastern Point Rd Groton CT 06340

TATE, CHARLES FRANK, JR, b Tazewell, Va, July 30, 14; m 44; c 3. MEDICINE. Educ: Ky State Teachers Col, Morehead, BS, 39; Univ Va, MD, 49. Prof Exp: Clin asst prof, 55-57, from asst prof to assoc prof, 57-75, PROF MED, JACKSON MEM HOSP & SCH MED, UNIV MIAMI, 75-, CHIEF PULMONARY DIS SECT, HOSP, 60- Concurrent Pos: Attend physician, Vet Admin Hosp, Coral Gables, 57-58, consult, 58- Mem: Am Thoracic Soc; Am Heart Asn; Am Col Chest Physicians; Am Col Physicians; Am Soc Internal Med. Res: Pulmonary diseases and function; clinical effects of smoking. Mailing Add: 6800 SW 126th Terr Miami FL 33156

TATE, CHARLES LUTHER, b Springfield, Mass, Aug 2, 38; m 59; c 2. PATHOLOGY, VETERINARY MEDICINE. Educ: Univ Pa, VMD, 66. Prof Exp: Fel nutrit path, Dept Nutrit, Mass Inst Technol, 66-67; pathologist, Div Vet Med, Walter Reed Army Inst Res, DC, 67-69; from res fel path to sr res fel path, 69-74, DIR TOXICOL & PATH, MERCK, SHARPE & DOHME RES LABS, WEST POINT, 74- Concurrent Pos: Adj asst prof, Sch Vet Med, Univ Pa, 72- Mem: Am Vet Med Asn; Int Acad Path; Soc Pharmacol & Environ Path. Res: Developmental, nutritional and toxicologic pathology. Mailing Add: 119 Oberlin Terrace Lansdale PA 19446

TATE, DAVID PAUL, b Chicago, Ill, Dec 10, 31; m 53; c 3. ORGANIC CHEMISTRY. Educ: Hamline Univ, BS, 53; Purdue Univ, MS, 55, PhD, 58. Prof Exp: Sr chemist, Stand Oil Co, Ohio, 57-63; mgr polymerization, 63-71, ASST DIR RES, FIRESTONE TIRE & RUBBER CO, 71- Mem: Am Chem Soc. Res: Lithium amine reductions; phosphorous compounds; organometallic chemistry; elastomer synthesis. Mailing Add: Firestone Tire & Rubber Co Akron OH 44317

TATE, FRED ALONZO, b Rice Lake, Wis, Jan 8, 20; m 47; c 2. ORGANIC CHEMISTRY, INFORMATION SCIENCE. Educ: Ohio Univ, BS, 47; Harvard Univ, MA, 50, PhD(chem), 51. Prof Exp: Asst prof chem, Ohio Univ, 51-53; from asst ed to assoc ed, Chem Abstr Serv, 53-57; admin asst, Res Labs, Gen Motors Corp, 57-58; mgr sci info sect, Wyeth Labs, 58-61; asst dir, 61-70, assoc dir, 71-74, ASSOC DIR PLANNING & DEVELOP, CHEM ABSTR SERV, 74- Mem: Am Chem Soc; Am Soc Info Sci. Res: Data and information management systems; automated and manual information-handling systems; publishing systems for technical information; information utilization systems; business management systems associated with foregoing areas. Mailing Add: Chem Abstr Serv 2540 Olentangy River Rd Columbus OH 43210

TATE, HERMAN DOUGLAS, entomology, see 12th edition

TATE, JAMES LEROY, JR, b Mauston, Wis, Apr 10, 40; m 63; c 1. ORNITHOLOGY. Educ: Northern Ill Univ, BS, 62; Univ of the Pac, MS, 64; Univ Nebr, Lincoln, PhD(zool), 69. Prof Exp: Asst dir lab ornith, Cornell Univ, 69-73, assoc prof wildlife sci, 71-74; SR ENVIRON SCIENTIST, ATLANTIC RICHFIELD CO, 74- Concurrent Pos: Grants, NSF, 72 & Nat Aububon Soc, 73. Mem: Am Ornithologists' Union; Wilson Ornith Soc (secy, 71-); Cooper Ornith Soc; Brit Ornithologists' Union; Soc Vert Paleontologists. Res: Ecology of hole-nesting bird species; paleornithology; wildlife management. Mailing Add: PO Box 2043 Denver CO 80201

TATE, LAURENCE GRAY, b Cambridge, Eng, Feb 10, 45; US citizen. INSECT TOXICOLOGY, INSECT PHYSIOLOGY. Educ: Limestone Col, BS, 66; Univ SC, MS, 68, PhD(biol), 71. Prof Exp: USPHS res assoc insect toxicol, NC State Univ, 71-74; ASST PROF BIOL, UNIV S ALA, 74- Mem: Sigma Xi. Res: Metabolism of xenobiotics in tissues of marine crustaceans; carbohydrate metabolism in insects. Mailing Add: Dept of Biol Sci Univ of S Ala Mobile AL 36688

TATE, PARR ALLEN, b Montreal, Que, Apr 15, 25. RESEARCH ADMINISTRATION. Educ: McGill Univ, BS, 46, MS, 47; Mass Inst Technol, PhD(physics), 53. Prof Exp: Jr res officer optics, Nat Res Coun Can, 47-50, asst res officer elec measurements, 53-56; defence sci serv officer, Defence Res Bd Can, 56-72; dir gen technol, Dept Nat Defence, Can, 72-75; DIR NBC DEFENCE, DEFENCE RES ESTAB OTTAWA, 75- Res: Optics of aerial photography; microwave spectroscopy of alkali halide vapors; alternating current measurements; thermal radiation from explosions; theory of ionization chambers. Mailing Add: Defence Res Estab Ottawa ON Can

TATE, ROBERT FLEMMING, b Oakland, Calif, Dec 15, 21. MATHEMATICAL STATISTICS. Educ: Univ Calif, AB, 44, PhD(math statist), 52; Univ NC, MS, 49. Prof Exp: Lectr, Univ Calif, 51-53; from instr to assoc prof math, Univ Wash, Seattle, 53-65; assoc prof, 65-67, PROF MATH, UNIV ORE, 67- Mem: Am Math Soc; fel Inst Math Statist (assoc secy, 63-); Am Statist Asn. Res: Theory of correlation and of estimation. Mailing Add: Dept of Math Univ of Ore Eugene OR 97403

TATE, ROBERT LEE, III, b Victoria, Tex, Dec 1, 44; m 71; c 1. SOIL MICROBIOLOGY. Educ: Univ Ariz, BS, 66, MS, 67; Univ Wis-Madison, PhD(bact), 70. Prof Exp: Scholar bact, Univ Calif, Los Angeles, 70-72, res assoc, Dept Agron, Cornell Univ, 72-75; ASST PROF MICROBIOL, AGR RES & EDUC CTR, FLA, 75- Mem: Am Soc Microbiol. Res: Microbial interactions with organic soils; reactions involved with the subsidence of these soils and the products produced during this subsidence as they involve water quality. Mailing Add: Agr Res & Educ Ctr PO Drawer A Belle Glade FL 33430

TATELMAN, MAURICE, b Omaha, Nebr, Dec 6, 17; m 61; c 2. MEDICINE. Educ: Univ Nebr, AB, 40, MD, 42. Prof Exp: From asst prof to assoc prof, 50-61, PROF RADIOL, COL MED, WAYNE STATE UNIV, 61- Concurrent Pos: Chmn dept radiol, Sinai Hosp Detroit, 68-; consult, Wayne County Gen Hosp, Darborn Vet Admin Hosp, Detroit Mem Hosp & Highland Park Gen Hosp. Mem: Fel Am Col Radiol; Am Roentgen Ray Soc. Res: Clinical diagnostic radiology; neuroradiology. Mailing Add: 6767 W Outer Dr Detroit MI 48235

TATEOSIAN, LOUIS HAGOP, b Chelsea, Mass, Mar 23, 37; m 63; c 1. DENTAL MATERIALS. Educ: Case Western Reserve Univ, BS, 59; Univ Md, MS, 62. Prof Exp: Res chemist, Naval Res Lab, 62-65; CHIEF CHEMIST, DENTSPLY INT, INC, 65- Mem: Am Chem Soc. Res: Fluorinated epoxy resin synthesis; acrylic polymer preparation; adhesives and coatings. Mailing Add: Cent Res Lab Dentsply Int Inc 550 W College Ave York PA 17404

TATHAM, GEORGE, b Kendal, Eng, Jan 24, 07. GEOGRAPHY OF EUROPE, POLITICAL GEOGRAPHY. Educ: Univ Liverpool, BA, 29, MA, 32; Clark Univ, PhD, 34. Prof Exp: Lectr geog, Univ Col, Univ London, 30-32, Univ Birmingham, 34-35, Armstrong Col, 35-36 & Univ Col, 36-39; from asst prof to prof, Univ Toronto, 38-60; prof, York Univ, 60-61, dean fac, 61-68, dean students, 62-68, PROF GEOG, YORK UNIV & MASTER, McLAUGHLIN COL, 68- Mem: Asn Am Geogr; Can Inst Int Affairs. Mailing Add: McLaughlin Col York Univ 4700 Keele St Downsview ON Can

TATINI, SITA RAMAYYA, b Mortha, India, Oct 6, 35; m 54; c 3. MICROBIOLOGY, FOOD SCIENCE. Educ: Univ Madras, BVSc, 57; Univ Minn, MS, 66, PhD(food sci & indust), 69. Prof Exp: Vet asst surgeon, Andhra Animal Husb Dept, State of Andhra Pradesh, India, 58-61; res fel, 69, NIH-Food & Drug Admin res grant, 71-74, ASST PROF FOOD MICROBIOL, UNIV MINN, ST PAUL, 69- Mem: AAAS; Inst Food Technol; Am Soc Microbiol; Am Dairy Sci Asn; Int Asn Milk, Food & Environ Sanit. Res: Growth, survival and production of enterotoxins by staphylococci in food products; developing rapid methods for assessment of psychrophilic bacteria in milk. Mailing Add: Dept of Food Sci & Indust Inst of Agr Univ of Minn St Paul MN 55101

TATOR, BENJAMIN ALMON, b Poughkeepsie, NY, Aug 13, 14; m 51; c 5. GEOLOGY. Educ: Columbia Univ, BA, 39; La State Univ, MS, 41, PhD(geomorphol), 48. Prof Exp: Asst, Sch Geol, La State Univ, 39-41, instr geol & geog, 47-48, from asst prof to assoc prof, 49-53; dir photogeol res unit, Gulf Oil Corp, 53-60, staff geologist, Latin Am Gulf Oil Co, 61-71; V PRES, WEEKS-TATOR CONSULT GEOLOGISTS, INC, 71- Concurrent Pos: Petrol geologist adv, Res & Training Inst, UN Tech Assistance Prog, India, 64-65; mem, Nat Coun Geol Educ. Mem: Geol Soc Am; Am Soc Photogram; Soc Econ Paleontologists & Mineralogists; Am Geog Soc; Am Asn Petrol Geologists. Res: Geomorphology of the Colorado Rockies and desert landscapes; photogeology; exploration and research regional geology. Mailing Add: 7600 Red Rd Suite 201 South Miami FL 33143

TATOR, CHARLES HASKELL, b Toronto, Ont, Aug 24, 36; m 60; c 2. NEUROSURGERY. Educ: Univ Toronto, MD, 61, MA, 63, PhD(neuropath), 65. Prof Exp: From assoc to asst prof, 69-74, ASSOC PROF, DEPT SURG, UNIV TORONTO, 74-; SURGEON, DIV NEUROSURG, SUNNYBROOK MED CTR, 69- Res: Brain tumor research and spinal cord injury research. Mailing Add: Div of Neurosurg Sunnybrook Med Ctr 2075 Bayview Ave Toronto ON Can

TATRO, MAHLON CHARLES, b Enosburg Falls, Vt, Aug 29, 22; m 52; c 3. ENVIRONMENTAL SCIENCES. Educ: Univ Mo, BA, 49; Univ Mass, MS, 51, PhD(food tech), 52. Prof Exp: Control chemist, Linde Air Prod Co Div, Union Carbide & Carbon Chem Co, 48-49; anal chemist, State Bur Labs, Vt, 52-54; chief chemist, Loma Linda Food Co, Calif, 54-56; agr chemist, Agr & Natural Resources Div, Int Coop Admin, 56-58, agr prog officer, 58-63; res assoc prof, Natural Resources Inst, Univ Md & dir seafood processing lab, 63-74; RES ASSOC PROF, CTR ENVIRON & ESTUARINE STUDIES, UNIV MD, 74- Concurrent Pos: Lab dir, Edward W McCready Mem Hosp, 68- Res: Studies on home sewage disposal systems utilizing Evapo-transpiration beds including modifications of same. Mailing Add: Box 775 Univ of Md Ctr Environ & Estuarine Studies Cambridge MD 21613

TATRO, PETER RICHARD, b Winthrop, Mass, Jan 20, 36; m 57; c 3. PHYSICAL OCEANOGRAPHY, ACOUSTICS. Educ: Ga Inst Technol, BME, 57; Mass Inst Technol, PhD(oceanog), 66. Prof Exp: Res oceanogr, Fleet Numerical Weather Ctr, Calif, 66-69; spec asst for ocean sci, Off Naval Res, Washington, DC, 69-72, head, Acoust Environ Support Detachment, Arlington, 72-75, SPEC ASST OCEANOGR OF THE NAVY, OFF NAVAL RES, ALEXANDRIA, 75- Concurrent Pos: Consult,

Int Decade Ocean Explor, NSF, 72-75. Honors & Awards: Navy Achievement Medal; Navy Expeditionary Medal. Res: Application of advanced digital technology to the problem of predicting the acoustic characteristics of the oceans. Mailing Add: 2604 Ryegate Lane Alexandria VA 22308

TATSCHL, ANNEHARA KATHLEEN, b Portales, NMex, Apr 17, 42; m 65; c 1. BOTANY. Educ: Univ NMex, BS, 64, MS, 66; Univ Kans, PhD(bot), 70. Prof Exp: ASST CUR HERBARIUM & INSTR BOT, KANS STATE UNIV, 68- Res: Taxonomy of angiosperms. Mailing Add: Div of Biol Kans State Univ Manhattan KS 66502

TATSUMOTO, MITSUNOBU, b Junsho, Japan, Mar 19, 23; m 48; c 2. GEOCHEMISTRY, GEOCHRONOLOGY. Educ: Tokyo Bunrika Univ, BS, 48, DSc(inorg chem), 57. Hon Degrees: Dr, Univ Paris VII, 75. Prof Exp: Res asst chem, Tokyo Kyoiku Univ, 48-57, lectr, 57; res fel oceanog, Scripps Inst, Calif, 57-58; Welch Found fel, 59; res fel oceanog, Tex A&M Univ, 59; res fel geochem, Calif Inst Technol, 59-62; CHEMIST, US GEOL SURV, 62- Concurrent Pos: Vis prof, Inst Geophys, Univ Paris VI, 71 & 74. Mem: Geochem Soc; Am Geophys Union. Res: Isotope geochemistry; natural radioactivity; meteorite and lunar chronology; marine geochemistry. Mailing Add: Isotope Geol Br US Geol Surv Fed Ctr Denver CO 80225

TATTER, DOROTHY, b Chicago, Ill, Apr 11, 22; m 49; c 3. MEDICINE. Educ: Rosary Col, BS, 43; Univ Ill, MD, 47. Prof Exp: From instr to asst prof, 49-70, ASSOC PROF PATH, SCH MED, UNIV SOUTHERN CALIF, 70- Concurrent Pos: Resident path, Los Angeles County Gen Hosp, 49-52, head physician, Autopsy Dept Labs, 52- Mem: AMA. Res: Pathology. Mailing Add: Dept Labs & Path Univ of Southern Calif Med Ctr Los Angeles CA 90033

TATTERSALL, IAN MICHAEL, b Paignton, Devon, Eng, May 10, 45. PHYSICAL ANTHROPOLOGY, PRIMATOLOGY. Educ: Cambridge Univ, BA, 67, MA, 70; Yale Univ, MPhil, 70, PhD(geol), 71. Prof Exp: ASST CUR PHYS ANTHROP, AM MUS NATURAL HIST, 71- Concurrent Pos: Vis lectr, Grad Fac, New Sch Social Res, 71-72; adj asst prof, Lehman Col, City Univ New York, 71-74. Mem: Am Asn Phys Anthropologists; Soc Vert Paleont; Am Anthrop Asn; Soc Syst Zool; AAAS. Res: Evolution, functional anatomy, ecology and behavior of the primates, particularly of the Malagasy lemurs; primate systematics; evolutionary theory in relation to phylogenetic reconstruction and other systematic applications. Mailing Add: Dept Anthrop Am Mus Natural Hist Cent Park W at 79th St New York NY 10024

TATTRIE, NEIL HERBERT, biochemistry, see 12th edition

TATUM, EDWARD LAWRIE, microbiology, biochemistry, deceased

TATUM, HOWARD JAMES, b Philadelphia, Pa, May 22, 15; m 46; c 3. OBSTETRICS & GYNECOLOGY. Educ: Univ Wis, BA, 36, PhD(pharmacol), 41, MD, 43; Am Bd Obstet & Gynec, dipl, 52. Prof Exp: Asst pharmacol & toxicol, Med Sch, Univ Wis, 36-40, instr, 40-41; intern, Univ Hosp, Univ Calif, 43-44; resident obstet & gynec, Wis Gen Hosp, 46-49; from instr to assoc prof, Sch Med, La State Univ, 49-58; prof, Med Sch, Univ Ore, 58-66; ASSOC DIR, BIO-MED DIV, POP COUN, ROCKEFELLER UNIV, 66- Concurrent Pos: Vis prof, Sch Med, Univ Valle, Colombia, 57-58; vis prof, Univ Buenos Aires, 64-65; affil, Rockefeller Univ, 66-; clin prof, Med Col, Cornell Univ, 67- Honors & Awards: Am Asn Obstet & Gynec Award, 55. Mem: AAAS; Soc Gynec Invest; fel Am Asn Obstet & Gynec; fel Am Col Obstet & Gynec; fel Am Gynec Invest. Res: Pharmacology and toxicology of the barbituric acid derivatives and analeptics; genito-urinary surgery; physiology; biochemistry; cancer; biology of reproduction and contraception. Mailing Add: Bio-Med Div Pop Coun Rockefeller Univ 66th & York Ave New York NY 10021

TATUM, JAMES PATRICK, b Dallas, Tex, July 6, 38. PHYSICAL CHEMISTRY. Educ: Rice Univ, BA, 61; Fla State Univ, PhD(phys chem), 66. Prof Exp: Res assoc chem, Univ Ill, 66-68; ASST PROF CHEM, IND STATE UNIV, TERRE HAUTE, 68- Mem: Am Phys Soc. Res: Theoretical chemistry. Mailing Add: Dept of Chem Ind State Univ Terre Haute IN 47809

TATUM, LOYD ALLEN, plant breeding, see 12th edition

TATUM, WILLIAM EARL, b Ft Payne, Ala, Sept 13, 33; m 53; c 4. ORGANIC CHEMISTRY. Educ: Chattanooga Univ, BA, 55; Univ Tenn, PhD(chem), 58. Prof Exp: Res chemist, Exp Sta, 58-61, tech rep, Venture Develop Sect, 61-63, staff scientist, Yerkes Res Lab, 63-64, res supvr, Circleville Res Lab, Ohio, 64-67, develop supvr, Circleville Plant, 67-68, tech supt, Florence Plant, SC, 68-70, cellophane prod mgr, Del, 70-71, dir prod & tech div, 71-73, venture mgr, 73-74, DIR SPECIALTY MKT DIV, E I DU PONT DE NEMOURS & CO, INC, DEL, 75- Mem: Am Chem Soc; Am Phys Soc. Res: Polymer chemistry and engineering; synthetic organic chemistry. Mailing Add: Film Dept E I du Pont de Nemours & Co Inc Wilmington DE 19898

TAUB, AARON M, vertebrate morphology, pharmacology, see 12th edition

TAUB, ABRAHAM, b New York, NY, Sept 21, 01. PHARMACEUTICAL CHEMISTRY. Educ: Columbia Univ, BS, 22, AM, 27. Prof Exp: From instr to prof chem, 22-65, distinguished serv prof, 65-69, EMER DISTINGUISHED SERV PROF CHEM, COL PHARMACEUT SCI, COLUMBIA UNIV, 69- Concurrent Pos: Asst, Revision Comt, US Pharmacopeia, 20-30; consult chemist, 22-; asst, Nat Formulary, 37; consult, Nat Asn Pharmaceut Mfrs, 70- Honors & Awards: Rusby Award, 62; Man of Year Award, Nat Asn Pharmaceut Mfrs, 72. Mem: AAAS; Am Chem Soc; Am Pharmaceut Asn; fel Am Inst Chem; NY Acad Sci. Res: Quantitative color standards; deterioration of medicinals; development of analytical methods; stability of parenteral solutions; chromatography; radioisotope tracer techniques. Mailing Add: 1080 Fifth Ave New York NY 10028

TAUB, ABRAHAM HASKEL, b Chicago, Ill, Feb 1, 11; m 33; c 3. MATHEMATICS. Educ: Univ Chicago, BS, 31; Princeton Univ, PhD(math physics), 35. Prof Exp: Asst, Princeton Univ, 34-35 & Inst Adv Study, 35-36 & 40-41; from instr to prof math, Univ Wash, Seattle, 36-48; res prof appl math, Univ Ill, Urbana, 48-64, head digital comput lab, 61-64; dir comput ctr, 64-68, PROF MATH, UNIV CALIF, BERKELEY, 64- Concurrent Pos: Theoret physicist, Div 2, Nat Defense Res Comt, Princeton Univ, 42-45; mem, Guggenheim Post-Serv, 47-48 & Guggenheim fel, 53; mem, Appl Math Adv Coun, Nat Bur Stand, 49-54, chmn adv panel, Appl Math Div, 51-60; mem comt on training & res in math, Nat Res Coun, 52-54; mem rev comt, Appl Math Div, Argonne Nat Lab, 60-62. Mem: AAAS; Am Math Soc; fel Am Phys Soc; Math Asn Am; Soc Indust & Appl Math. Res: Relativity; quantum mechanics; interaction of shock waves; digital computers; numerical analysis. Mailing Add: Dept of Math Univ of Calif Berkeley CA 94720

TAUB, ARTHUR, b New York, NY, Jan 4, 32; m 63; c 1. NEUROPHYSIOLOGY, NEUROLOGY. Educ: Yeshiva Univ, BA, 52; Mass Inst Technol, SM, 53, PhD(neurophysiol), 64; Yale Univ, MD, 57; Am Bd Psychiat Neurol, dipl, 75. Prof Exp: NIH fel neurophysiol, Mass Inst Technol, 64-66, res assoc, 66-68; dir neurosurg res lab, 69-75, dir pain clin, 71-75, asst prof, 68-72, assoc prof neurophysiol & neurol, 72-75, PROF CLIN ANESTHESIOL & DIR SECT STUDY & TREATMENT OF PAIN, DEPT ANESTHESIOL, SCH MED, YALE UNIV, 75- Concurrent Pos: NIH res career develop award, Mass Inst Technol & Yale Univ, 66-73; resident neurol, Sch Med, Yale Univ, 69-72; assoc neurologist, Yale New Haven Hosp, 71-75, attend neurologist, 75-; Royal Soc Med Found traveling fel, UK, 72. Mem: Am Acad Neurol; Sigma Xi; Soc Neurosci; Int Asn Study Pain. Res: Pain; application of basic scientific approach to modalities for control; neuropharmacology; computer simulation; clinical control by medical and surgical means; anesthesiology. Mailing Add: Dept of Anesthesiol Yale Univ Sch of Med New Haven CT 06504

TAUB, DAVID, b New York, NY, Nov 13, 19; m 44; c 2. ORGANIC CHEMISTRY. Educ: City Col New York, BS, 40; Harvard Univ, AM, 46, PhD, 50. Prof Exp: Chemist, Manhattan Proj, Kellex Corp, 42-46; USPHS fel, Harvard Univ, 49-51; sr chemist, 51-65, sect head process res, 65-68, SR RES FEL, MERCK SHARP & DOHME RES LABS, 68- Mem: Am Chem Soc; The Chem Soc; Swiss Chem Soc. Res: Organic synthesis; natural products. Mailing Add: Merck Sharp & Dohme Res Labs Rahway NJ 07065

TAUB, EDWARD, b Brooklyn, NY, Oct 22, 31; m 59. NEUROPSYCHOLOGY. Educ: Brooklyn Col, BA, 53; Columbia Univ, MA, 59; NY Univ, PhD(psychol), 70. Prof Exp: Res asst neuropsychol, Columbia Univ, 56; res asst, Dept Exp Neurol, Jewish Chronic Dis Hosp, 57-60, res assoc, 60-68; CHIEF, BEHAV BIOL CTR, INST BEHAV RES, 68- Concurrent Pos: Asst prof, Sch Med, Johns Hopkins Univ, 70- Mem: Am Psychol Asn; Soc Neurosci; Biofeedback Res Soc; Psychonomic Soc; Am Physiol Soc. Res: The role of somatosensory feedback and spinal reflexes in movement and learning; fetal surgery; transcendental meditation as a therapy for alcoholism; biofeedback and self-regulation of human hand temperature. Mailing Add: Behav Biol Ctr Inst Behav Res 2429 Linden Lane Silver Spring MD 20910

TAUB, FRIEDA B, b Newark, NJ, Oct 11, 34; m 54; c 1. ECOLOGY, NUTRITION. Educ: Rutgers Univ, BA, 55, MA, 57, PhD(zool), 59. Prof Exp: Fisheries biologist, 59-61, instr food sci, 61-62, from res asst prof to res assoc prof, 62-71, PROF FISHERIES, COL FISHERIES, UNIV WASH, 71- Concurrent Pos: Res grants, 61-74. Mem: Fel AAAS; Ecol Soc Am; Am Soc Limnol & Oceanog; Am Soc Ichthyol & Herpet; Am Fisheries Soc. Res: Aquatic food chains; ecosystems. Mailing Add: Col of Fisheries Univ of Wash Seattle WA 98195

TAUB, IRWIN ALLEN, physical chemistry, inorganic chemistry, see 12th edition

TAUB, JOHN MARCUS, b Chicago, Ill, July 26, 47. PSYCHOPHYSIOLOGY. Educ: Univ Calif, Santa Cruz, AB, 69, MS & PhD(biopsychol), 72. Prof Exp: Res biophysicist, Univ Calif, Santa Cruz, 72-73; fel neurosci, Brain Res Inst, Univ Calif, Los Angeles, 73-75; DIR, SLEEP & DREAM LAB & ASST PROF PSYCHIAT & PSYCHOL, SCH MED, UNIV VA, 75- Concurrent Pos: NIH res fel, 73. Mem: AAAS; Asn Psychophysiol Study Sleep; Soc Neurosci. Res: Investigations on the behavioral and psychophysiological effects of acute and chronic variations in the length and time of sleep in young adults. Mailing Add: Sleep & Dream Lab Dept of Psychiat Sch of Med Univ of Va Charlottesville VA 22901

TAUB, ROBERT NORMAN, b Brooklyn, NY, Apr 21, 36; m 68; c 1. IMMUNOLOGY, HEMATOLOGY. Educ: Yeshiva Univ, AB, 57; Yale Univ, MD, 61; Univ London, PhD(biol), 69. Prof Exp: Intern med, New Eng Med Ctr Hosps, Boston, 61-62; intern path, Sch Med, Yale Univ, 62-63; clin & res fel hemat, New Eng Med Ctr Hosps, 63-65, asst resident med, 65-66; NIH res fel immunol, Nat Inst Med Res, London, 66-68; assoc hemat, 68-69, asst prof med, 69-72, ASSOC PROF MED, MT SINAI SCH MED, 72-, HEAD TRANSPLANTATION IMMUNOL LAB, MT SINAI HOSP, 70- Concurrent Pos: Res fel path, Sch Med, Yale Univ, 62-63; Leukemia Soc Am scholar award, 68-73; attend physician, Mt Sinai Hosp, 69-70; USPHS res career develop award, 75- Honors & Awards: Emil Conason Mem Res Award, Mt Sinai Sch Med, 71. Mem: Transplantation Soc; Am Soc Clin Invest; Am Soc Hemat; Am Asn Immunologists; Am Soc Exp Path. Res: Immunosuppression; transplantation immunology; function of thymus and other lymphoid organs; clinical immunology; immunology of leukemia. Mailing Add: Dept of Med Mt Sinai Sch of Med New York NY 11029

TAUB, STEPHEN ROBERT, b Jamaica, NY, Nov 30, 33; m 63; c 2. GENETICS. Educ: Rochester Univ, AB, 55; Univ Ind, PhD(zool), 60. Prof Exp: Instr biol, Harvard Univ, 60-63; asst prof, Princeton Univ, 63-69; from asst prof to assoc prof, Richmond Col, City Univ New York, 69-74; CHMN & PROF BIOL, GEORGE MASON UNIV, FAIRFAX, 74- Mem: AAAS; Genetics Soc Am; Am Soc Cell Biol. Res: Extra-chromosomal and developmental genetics. Mailing Add: Dept of Biol George Mason Univ Fairfax VA 22030

TAUBE, CLARENCE MARTIN, b Baroda, Mich, Jan 14, 12; m 55. FISHERIES. Educ: Kalamazoo Col, AB, 34; Mich State Col, MS, 42. Prof Exp: Mem staff, 46-51, asst fisheries biologist, 51-64, fisheries res biologist, Fisheries Div, Inst Fisheries Res, State Dept Natural Resources, Mich, 64-75; RETIRED. Mem: Am Fisheries Soc; Wildlife Soc. Res: Fisheries biology; relationships between trout and coho salmon. Mailing Add: 2078 Scio Rd Dexter MI 48130

TAUBE, HENRY, b Neudorf, Sask, Nov 30, 15; nat US; m 40, 52; m 4. INORGANIC CHEMISTRY. Educ: Univ Sask, BS, 35, MS, 37; Univ Calif, PhD(chem), 40. Prof Exp: Instr chem, Univ Calif, 40-41; from instr to asst prof, Cornell Univ, 41-46; res assoc, Nat Defense Res Comt, 44-45; from asst prof to prof chem, Univ Chicago, 46-61, chmn dept, 55-59; chmn dept, 72-74, PROF CHEM, STANFORD UNIV, 61- Concurrent Pos: Guggenheim fel, 49 & 55. Honors & Awards: Award, Am Chem Soc, 55, Howe Award, 60, Distinguished Serv Award, 67 & Nichols Medal & Willard Gibbs Medal, 71; Chandler Award, Columbia Univ, 64; Kirkwood Award, Yale Univ & Am Chem Soc, 66. Mem: Nat Acad Sci; Am Chem Soc; Am Acad Arts & Sci. Res: Chemistry of complex ions; new aquo ions, nitrogen as a ligand; mechanism of electron transfer; new unstable species by matrix isolation technique. Mailing Add: Dept of Chem Stanford Univ Stanford CA 94305

TAUBENECK, WILLIAM HARRIS, b Marshall, Ill, Aug 27, 33. PETROLOGY. Educ: Ore State Col, BS, 49, MS, 50; Columbia Univ, PhD(geol), 55. Prof Exp: From instr to assoc prof, 51-65, PROF GEOL, ORE STATE UNIV, 65- Concurrent Pos: NSF fel, 59-61; Guggenheim fel, 63-64. Mem: AAAS; Geol Soc Am; Mineral Soc Am. Res: Layering in igneous rocks; petrogenesis of granite rocks; Columbia River basalt; thermal metamorphism; rock forming minerals; general geology of northeastern Oregon and western Idaho; evolution of the Pacific Northwest. Mailing Add: Dept of Geol Ore State Univ Corvallis OR 97331

TAUBER, ARTHUR, b New York, NY, June 2, 28; m 56; c 3. CHEMISTRY, MAGNETISM. Educ: NY Univ, BA, 51, MA, 52; Polytech Inst Brooklyn, MS, 59, PhD, 72. Prof Exp: Phys chemist, Sig Corps Eng Labs, 52-56, res phys scientist,

Electronics Res & Develop Labs, 56-63, RES PHYS SCIENTIST, ELECTRONICS COMMAND, US ARMY, 63- Honors & Awards: Meritorious Civilian Serv Medal, US Dept Army, 63. Mem: Am Chem Soc; Mineral Soc Am; Am Crystallog Asn; Am Asn Crystal Growth. Res: Crystal and solid state chemistry; x-ray crystallography; solid state chemistry and physics of ferrites for microwave magnetic devices; rare-earth permanent magnet alloys; rare-earth alloys; hydrogen absorbers; storage of hydrogen. Mailing Add: US Army Electronics Command ATT AMSEL-TL-ESA Fort Monmouth NJ 07703

TAUBER, CATHERINE ANN, entomology, see 12th edition

TAUBER, GERALD ERICH, b Vienna, Austria, Oct 31, 22; nat US; m 56; c 3. THEORETICAL PHYSICS. Educ: Univ Toronto, BA, 46; Univ Minn, MA, 47, PhD(theoret physics), 51. Prof Exp: Asst physics, Univ Minn, 46-50; lectr, McMaster Univ, 50-52; fel, Nat Res Coun Can, 52-54; from asst prof to prof physics, Case Western Reserve Univ, 54-68; vis prof, 65-68, PROF PHYSICS, TEL AVIV UNIV, 68- Concurrent Pos: Vis prof, Nat Res Coun Can, 55; consult, Franklin Inst, 56-; vis prof, Israel Inst Technol, 59-60. Mem: Am Phys Soc; Can Asn Physicists; Israeli Phys Soc; Int Soc Gen Relativity & Gravitation. Res: Independent particle model in nuclear physics; atomic physics; cosmology; astrophysics; general relativity; cosmological models; field theory; transport phenomena; elementary particles in relativity; collective motion; relativistic astrophysics. Mailing Add: Dept of Phys & Astron Tel Aviv Univ Tel Aviv Israel

TAUBER, MAURICE JESSE, b Cracow, Can, Oct 21, 37; m 66; c 3. ENTOMOLOGY. Educ: Univ Man, BS, 58, MS, 59; Univ Calif, Berkeley, PhD, 66. Prof Exp: Asst mgr, Tauber, Good & Assocs, 57-59, gen mgr, 61-62; PROF ENTOM, CORNELL UNIV, 66- Mem: AAAS; Am Soc Zool; Animal Behav Soc; Entom Soc Am; Entom Soc Can. Res: Species specific behavior; Circadian rhythms; physiology of insect reproduction. Mailing Add: Dept of Ecol & Evolutionary Biol Cornell Univ Ithaca NY 14850

TAUBER, OSCAR EARNST, b Decatur, Ill, May 23, 08; m 40; c 2. PHYSIOLOGY. Educ: James Millikin Univ, BA, 30; Iowa State Col, MS, 32, PhD(gen physiol), 35. Prof Exp: Tech asst, State Natural Hist Surv, Ill, 30; instr, James Millikin Univ, 31-32; from instr to prof zool & entom, 46-69, in charge zool, 60-62, chmn dept, 62-73, DISTINGUISHED PROF SCI & HUMANITIES, IOWA STATE UNIV, 69- Concurrent Pos: Marine Biol Lab, Woods Hole, Mass, 34; vis prof, Univ Hawaii, 62. Mem: AAAS; Am Physiol Soc; Entom Soc Am; Soc Exp Biol & Med; Sigma Xi. Res: Chitin distribution; hemocytes and circulation in invertebrates; factors affecting mitosis in cell cultures; fat metabolism in insects; chronic poisonings in mammals; cytochemical effects of dietary deficiencies and food additives. Mailing Add: Dept of Zool Iowa State Univ Ames IA 50011

TAUBER, SELMO, b Shanghai, China, Aug 24, 20; nat US; m 50; c 1. APPLIED MATHEMATICS. Educ: Beirut Sch Eng, Lebanon, dipl, 43; Univ Lyons, France, Lic es sc, 47; Univ Vienna, Austria, DrPhil, 50. Prof Exp: Head sci dept, High Sch, Lebanon, 51-55; design engr, 56-57; assoc prof, 57-63, PROF MATH, PORTLAND STATE UNIV, 63- Concurrent Pos: From instr to asst prof, Univ Kans, 57-59. Mem: Am Math Soc; Soc Indust & Appl Math; Math Asn Am. Res: Engineering structural problems; finite differences; differential equations; combinatorial analysis; system analysis; mathematical models in air pollution. Mailing Add: Dept of Math Portland State Univ Portland OR 97207

TAUBER, STEPHEN JULIUS, information science, see 12th edition

TAUBLER, JAMES H, b Cokeville, Pa, Mar 30, 35; m 59; c 1. MICROBIOLOGY. Educ: St Vincent Col, BA, 57; Cath Univ, MS, 59, PhD(microbiol), 65. Prof Exp: Microbiologist, Philadelphia Gen Hosp, 59-69; asst prof, 69-71, PROF BIOL, ST VINCENT COL, 71- Concurrent Pos: Lectr, Holy Family Col, Pa, 66-; dir, Delmont Labs, Swarthmore, 66- Mem: AAAS; Am Soc Microbiol; Brit Soc Gen Microbiol. Res: Staphylococcal alpha toxin and its relationship to pathogenesis; delayed hypersensitivity and its effects in staphylococcal infections. Mailing Add: Dept of Biol St Vincent Col Latrobe PA 15650

TAUBMAN, MARTIN ARNOLD, b New York, NY, July 10, 40; m 65; c 1. IMMUNOLOGY, ORAL BIOLOGY. Educ: Brooklyn Col, BS, 61; Columbia Univ, DDS, 65; State Univ NY Buffalo, PhD(oral biol), 70. Prof Exp: ASSOC STAFF MEM & HEAD IMMUNOL DEPT, FORSYTH DENT CTR, 70- Concurrent Pos: Nat Inst Dent Res grant, Forsyth Dent Ctr, 72-75, NIH career develop award, 72-77. Mem: AAAS; Int Asn Dent Res. Res: Secretory immunoglobulins; effect of secretory antibodies on oral microorganisms. Mailing Add: Forsyth Dent Ctr 140 Fenway Boston MA 02115

TAUBMAN, ROBERT EDWARD, b New York, NY, Jan 12, 21; m 43; c 3. PSYCHIATRY, PSYCHOLOGY. Educ: City Col New York, BA, 41; Columbia Univ, MS, 42, PhD(psychol), 48; Univ Nebr, MD, 60. Prof Exp: Chief psychol serv, Hastings State Hosp, Nebr, 52-56; resident physician psychiat, Ore State Hosp, Salem, 61-64; assoc prof, 64-70, PROF PSYCHIAT, MED SCH, UNIV ORE, 70- Concurrent Pos: NIMH fel ment retardation, Letchworth Village, Thiells, NY, 62; consult, Ore Fairview Home, Div Voc Rehab, Marion County Juv Dept, 62-64; pvt pract, 64-; attend physician, Vet Admin Hosp, Portland, Ore, 64-; psychiat dir, Physicians Inst, Ore Acad Family Physicians, 65- Mem: Am Pyschiat Asn; AMA; Soc Teachers Family Med. Res: Aging; dying; death; self growth among physicians and other professionals; emotional components of life-threatening diseases; psychiatry for non-psychiatric physicians. Mailing Add: Dept of Psychiat Ore Univ Med Sch Portland OR 97201

TAUBMAN, SHELDON BAILEY, b Cleveland, Ohio, Oct 8, 36; m 60; c 2. IMMUNOLOGY, PATHOLOGY. Educ: Northwestern Univ, Evanston, BA, 58; Case Western Reserve Univ, PhD(biochem), 64, MD, 66. Prof Exp: ASST PROF PATH, HEALTH CTR, UNIV CONN, 68- Concurrent Pos: NIH fel, 64-66, fel path, NY Univ, 66-68 & grant, Health Ctr, Univ Conn. Mem: AAAS; Harvey Soc; Am Asn Path & Bact. Res: Biochemical mechanisms of inflammation. Mailing Add: Dept of Path Univ of Conn Health Ctr Farmington CT 06032

TAUC, JAN, b Pardubice, Czech, Apr 15, 22; m 47; c 2. SOLID STATE PHYSICS. Educ: Czech Inst Technol, IngDr, 49; Czech Acad Sci, DrSc(physics), 56; Charles Univ, Prague, RNDr, 56. Prof Exp: Scientist microwave res, Sci & Tech Res Inst, Tanvald & Prague, Czech, 49-52; dept head inst solid state physics, Czech Acad Sci, 53-69; prof exp physics, Charles Univ, Prague, 64-69, dir inst physics, 68-69; mem tech staff, Bell Labs, 69-70; PROF ENG & PHYSICS, BROWN UNIV, 70- Concurrent Pos: UNESCO res fel, Harvard Univ, 61-62; vis prof, Univ Paris, 69; consult, Bell Labs, 70- Honors & Awards: Nat Prize for Sci, Czech Govt, 55 & 69. Mem: Am Phys Soc; Europ Phys Soc. Res: Optical properties of crystalline and amorphous solids; electronic states; lattice vibrations. Mailing Add: Div of Eng Brown Univ Providence RI 02912

TAUCCI, ENES BARBARA, b Phillipsburg, NJ, Jan 1, 32. MATHEMATICS, BIOMETRY. Educ: Univ Akron, BA, 59; Iowa State Univ, MS, 62; Yale Univ, PhD(biomet), 66. Prof Exp: Asst math, Iowa State Univ, 61-62; asst prof, Univ Omaha, 66-67; ASSOC PROF MATH, UNIV AKRON, 67- Mem: AAAS; Math Asn Am; Am Math Soc. Res: Matrix and probability theory; methods to study aggregation of rare diseases in time and space. Mailing Add: PO Box 604 Phillipsburg NJ 08865

TAUER, KENNETH J, b Minn, Apr 5, 23; m 44; c 4. PHYSICAL CHEMISTRY, CHEMICAL PHYSICS. Educ: Univ Minn, PhD, 51. Prof Exp: Researcher, Gen Elec Co, Wash, 51-53; asst prof chem, Boston Col, 53-56; MEM RES STAFF, MAT RES AGENCY LAB, US ARMY MAT & MECH RES CTR, 57- Mem: Am Chem Soc; Am Phys Soc. Res: Electronics structure of transition metals and alloys; magnetism; electronic transport; magnetoresistance; Hall effect. Mailing Add: US Army Mat & Mech Res Amra Bldg 292 Watertown MA 02172

TAUER, RITA JEAN, mathematics, see 12th edition

TAUFEN, HARVEY JAMES, b Lewiston, Idaho, Mar 30, 19; m 42; c 5. ORGANIC CHEMISTRY. Educ: Gonzaga Univ, BS, 40; Ill Inst Technol, MS, 41, PhD(org chem), 44. Prof Exp: From asst chem to instr, Ill Inst Technol, 41-44; foreman chem div, Manhattan Proj, Tenn Eastman Corp, 44-45; res chemist, 45-50, supvr, 50-52, mgr synthetic res, 52-56, dir develop, Synthetic Dept, 56-59, dir develop & opers, Int Dept, 59-62, asst gen mgr, 62-63, gen mgr dept, 63-70, DIR, INT DEPT, HERCULES INC, 66-, VPRES, 70- Concurrent Pos: Chmn, Gordon Res Conf, 57. Mem: Am Chem Soc; Sigma Xi; Soc Indust Chem. Res: Determination of structure of organic molecules using Raman spectroscopy; new industrial chemical products from naval stores; detection of rotational isomers and structural configurations in organic molecules by means of the Raman effect; air oxidation of hydrocarbons. Mailing Add: Hercules Inc Wilmington DE 19899

TAULBEE, ORRIN EDISON, b Taulbee, Ky, Oct 18, 27; m 55; c 4. COMPUTER SCIENCE. Educ: Berea Col, BA, 50; Mich State Univ, MA, 51, PhD(math), 57. Prof Exp: Res mathematician, Univac Div, Sperry Rand Corp, 55-58; math specialist, Lockheed Aircraft Corp, 58-59; assoc prof math, Mich State Univ, Oakland, 59-61; mgr info sci, Goodyear Aerospace Corp, 61-66; dir mgt info systs & comput ctr, 66-70, PROF COMPUT SCI & CHMN DEPT, UNIV PITTSBURGH, 66- Concurrent Pos: Consult, US Air Force, 66-71, Pittsburgh Bd Educ, 67-72 & Pa Dept Educ, 72-; ed bd, Encycl Libr & Info Sci, 65- & J Info Systs, 73- Mem: Am Math Soc; Math Asn Am; Asn Comput Mach; Am Soc Info Sci; NY Acad Sci. Res: Combinatorial analysis; linear programming; applications of computers; pattern recognition; classification theory; factor and numerical analysis; instructional computing; computing resources management; management information systems; information processing systems. Mailing Add: Dept of Comput Sci Univ of Pittsburgh Pittsburgh PA 15260

TAUNTON-RIGBY, ALISON, b Barnsley, Eng, Apr 23, 44; m 66; c 2. CHEMISTRY, BIOCHEMISTRY. Educ: Bristol Univ, BSc, 65, PhD(chem), 68. Prof Exp: SR RES CHEMIST, COLLAB RES INC, 68-, HEAD, CHEM RES, PROD DIV, 70-, PROJ LEADER, 70- Mem: AAAS; Am Chem Soc. Res: Nucleic acid chemistry; immunology; psychoactive drugs; carcinogens; pteridines; enzymology. Mailing Add: Chem Res Lab Collab Res Inc 1365 Main St Waltham MA 02154

TAURINS, ALFRED, b Dauguli, Latvia, Aug 20, 04; Can citizen; m 30; c 2. ORGANIC CHEMISTRY. Educ: Univ Latvia, ChEng, 30, DrChem, 36. Prof Exp: Instr chem, Univ Latvia, Riga, 30-38, privat-docent, Org Chem, 38-39; prof chem, Latvian Agr Acad, Yelgava, 39-44; from assoc prof to prof, 49-72, EMER PROF ORG CHEM, McGILL UNIV, 72- Concurrent Pos: Morberg fel, Univ Latvia, 39. Mem: Am Chem Soc; Chem Inst Can; fel Royal Soc Can. Res: Synthesis and spectra of heterocyclic compounds; thiazoloisoquinolines, isothiazolopyridines, naphthyridines. Mailing Add: Dept of Chem McGill Univ Montreal PQ Can

TAUROG, ALVIN, b St Louis, Mo, Dec 5, 15; m 40; c 2. BIOCHEMISTRY, PHYSIOLOGY. Educ: Univ Calif, Los Angeles, BA, 37, MA, 39; Univ Calif, Berkeley, PhD(physiol), 43. Prof Exp: Asst chem, Univ Calif, Los Angeles, 37-38; asst physiol, Univ Calif, Berkeley, 42-43, chemist, Radiation Lab, 44-45, res assoc physiol, 46-59; assoc prof, 59-63, PROF PHARMACOL, UNIV TEX HEALTH SCI CTR DALLAS, 63- Mem: AAAS; Am Chem Soc; Am Soc Biol Chem; Endocrine Soc; Am Thyroid Asn. Res: Thyroid physiology and biochemistry; iodine metabolism. Mailing Add: Dept of Pharmacol Univ of Tex Health Sci Ctr Dallas TX 75235

TAUSS, KURT H, b Ger, Mar 30, 14; nat US; m 45; c 4. ORGANIC CHEMISTRY. Educ: Univ Zurich, PhD(chem), 42. Prof Exp: Res chemist, Eastman Kodak Labs, 47-50 & Nat Aniline Div, Allied Chem & Dye Co, 51-53; sect leader res, Courtaulds Inc, 53-67; prof chem eng, Univ SAla, 67-69; SR RES ASSOC, ERLING RIIS RES LAB, INT PAPER CO, 69- Mem: AAAS; Am Inst Chem Eng. Res: Cellulose chemistry; man-made fibers; dyes and pigments; color and constitution; environmental control; operational research; management theory. Mailing Add: Erling Riis Res Lab Int Paper Co PO Box 2328 Mobile AL 36601

TAUSSIG, ANDREW, b Budapest, Hungary, Dec 6, 29; Can citizen; m 59; c 3. MICROBIOLOGY, BIOCHEMISTRY. Educ: McGill Univ, BSc, 52, PhD(biochem), 55. Prof Exp: Res fel biochem, McGill Univ, 55-61; from asst bacteriologist to bacteriologist, Jewish Gen Hosp, 61-67; vis scientist, Dept Biochem, McGill Univ, 67-70; BACTERIOLOGIST, MICROBIOL LAB, 69-; DIR LABS, MT SINAI HOSP, ST AGATHE, QUE, 74- Concurrent Pos: Bacteriologist, Bellechasse Hosp, Montreal, 73-, Jewish Convalescent Hosp, Chomedey & Mt Sinai Hosp, St Agathe, Que. Mem: Am Soc Microbiol; Can Soc Microbiol. Res: Nucleic acids of bacteria and bacteriophage; induced enzyme synthesis in bacteria; diagnostic bacteriology. Mailing Add: Microbiol Lab 5845 Cote de Neiges Rd Montreal PQ Can

TAUSSIG, HELEN BROOKE, b Cambridge, Mass, May 24, 98. MEDICINE. Educ: Univ Calif, AB, 21; Johns Hopkins Univ, MD, 27. Hon Degrees: DSc, Boston Univ, 48, Goucher Col, 49, Univ NC, 50, Nat Univ Athens, 57, Harvard Univ, 59; LLD, Hood Col, 50; DS, Northwestern Univ, 51, Columbia Univ, 51, Woman's Med Col, Pa, 51, Middlebury Col, 52, Western Col, 53. Prof Exp: Archibald fel med, Johns Hopkins Hosp, 27-28, intern pediat, 28-30, from assoc prof to prof pediat, Johns Hopkins Univ, 46-63, EMER PROF PEDIAT, JOHNS HOPKINS UNIV, 63-, PHYSICIAN-IN-CHGE CARDIAC CLIN, HARRIET LANE HOME, JOHNS HOPKINS HOSP, 30- Concurrent Pos: Thomas River Mem res fel award, 63-69; mem bd, Int Cardiol Found, 67- Honors & Awards: Chevalier Legion d'Honneur, France, 47; Mead-Johnson Award, 48; Passano Award, 48; Lasker Award, 54; Feltrinelli Prize, Rome, Italy, 54; Gardner Award, 59; Gold Heart Award, Am Heart Asn, 63; Medal of Freedom, 64; Founders Award, Radcliffe Col, 66; John Phillips Mem Award, Am Col Physicians, 66; Carl Ludwig Medal of Honor, Ger, 67; William F Faulkes Award, Nat Rehab Asn, 71; Howland Award, Am Pediat Soc, 71. Mem: Nat Acad Sci; master Am Col Physicians; Am Pediat Soc; Soc Pediat Res; Am Heart Asn. Res: Congenital malformations of the heart; blue babies. Mailing Add: 1100 Hollins Lane Baltimore MD 21209

TAUSSIG, ROBERT TRIMBLE, b St Louis, Mo, Apr 26, 38; m 63; c 1. PLASMA PHYSICS. Educ: Harvard Univ, BA, 60; Columbia Univ, MA, 63, PhD(appl math, plasma physics), 65. Prof Exp: Asst math, Columbia Univ, 61-63 & plasma physics, 63-65, res assoc, 65-66; NSF res grant, Nagoya Univ, 66-67; instr & res fel, Harvard Univ, 67-68; from asst prof to assoc prof plasma physics, 69-74, ASSOC PROF ENG SCI, SCH ENG & APPL SCI, COLUMBIA UNIV, 74- Mem: Am Phys Soc. Res: Free surface phenomena in hydrodynamics; high temperature gas dynamics; nonlinear wave propagation; radiative transfer under laboratory and astrophysical conditions; computer analysis of engineering problems. Mailing Add: Plasma Res Lab Sch Eng & Appl Sci Columbia Univ New York NY 10027

TAUSSIG, STEVEN J, b Timisoara, Rumania, June 2, 14; US citizen; m 65; c 1. INDUSTRIAL MICROBIOLOGY. Educ: Univ Prague, Czech, Chem Eng, 37; Bucharest Polytech Inst, PhD(biochem), 58. Prof Exp: Chemist, Solventul SA, Timisoara, 37-48; asst prof, Polytech Inst Timisoara, 48-57; res fel, Agr Res Inst, Timisoara, 57-59; mgr plant & equip sales, Int Chem Corp, NY, 60-61; tech dir, Pac Labs, Inc, Honolulu, 61-62 & Pac Enzyme Prod, 62-63; res biochemist, Dole Co, Inc, 63-64, dir lab serv, 64-73; PRES, CHEM CONSULTS INT INC, HONOLULU, 73- Concurrent Pos: Tech consult, Pac Labs, Inc & Pac Biochem Co, 63- & Monsanto Co, Mo, 64-65. Mem: AAAS; Am Chem Soc; Soc Indust Microbiol. Res: Fermentations; bacterial and other hydrolytic enzymes; chemical equilibria in esterification reactions; cancer research; study of effects of Bromelash on cancer. Mailing Add: Apt 2406 469 Ena Rd Honolulu HI 96815

TAUSSKY, OLGA (MRS JOHN TODD), b Olomouc, Czech, Aug 30, 06; nat US; m 38. MATHEMATICS, NUMBER THEORY. Educ: Univ Vienna, PhD(math), 30; Cambridge Univ, MA, 37. Prof Exp: Asst, Univ Göttingen, 31-32; asst, Math Inst, Univ Vienna, 32-34; lectr, Univ London, 37-44; sci officer, Ministry Aircraft Prod, Eng, 43-46; res, Dept Sci & Indust Res, 46-47; math consult, Nat Bur Stand, 47-48 & 49-57; res assoc, 57-71, PROF MATH, CALIF INST TECHNOL, 71- Concurrent Pos: Mem, Inst Advan Study, 48; Fulbright vis prof, Univ Vienna, 65. Honors & Awards: Ford Prize, Math Asn Am. Mem: Am Math Soc; Math Asn Am; corresp foreign mem, Acoustics Acad Sci. Res: Algebra; algebraic number theory; matrix theory; numerical analysis. Mailing Add: Dept of Math Calif Inst Technol Pasadena CA 91125

TAUSTER, SAMUEL JACK, physical chemistry, see 12th edition

TAUTVYDAS, KESTUTIS JONAS, b Telsiai, Lithuania, Jan 1, 40; US citizen; m 62; c 4. BIOLOGY. Educ: Univ Md, BS, 63; Cornell Univ, MS, 65; Yale Univ, PhD(cell & molecular develop biol), 69. Prof Exp: Res assoc cell biol, Univ Colo, 69-71; ASST PROF PLANT PHYSIOL, MARQUETTE UNIV, 71- Mem: Am Soc Plant Physiol; Am Soc Cell Biol. Res: Mechanisms of plant hormone action; biochemistry of cell nuclei; control of plant growth and development. Mailing Add: Dept of Biol Marquette Univ Milwaukee WI 53233

TAUXE, WELBY NEWLON, b Knoxville, Tenn, May 24, 24; c 4. NUCLEAR MEDICINE, CLINICAL PATHOLOGY. Educ: Univ Tenn, Knoxville, BS, 44, MD, 50; Univ Minn, Minneapolis, MS, 58. Prof Exp: Mayo Found fel, Mayo Clin, 54-58, chief nuclear med dept, 58-72; PROF NUCLEAR MED & CLIN PATH, COL MED, UNIV ALA, BIRMINGHAM, 72- Concurrent Pos: Consult, AEC, 62-70; treas, Am Bd Nuclear Med, 71; consult, Am Nat Standards Inst, 72. Mem: Am Soc Clin Path; Soc Nuclear Med. Res: Copper kinetics in Wilson's disease; use of radioactive materials in diagnosis. Mailing Add: Dept of Nuclear Med Univ of Ala Birmingham AL 35299

TAVAKOLIAN, BAHRAM MEHDI, b Teheran, Iran, Sept 24, 45; US citizen; m 66; c 1. ETHNOLOGY. Educ: Univ Calif, Los Angeles, AB, 65, MA, 66, PhD(anthrop), 74. Prof Exp: Asst prof antrhop, Antioch Col, 70-71; instr, 71-74, ASST PROF ANTHROP, MT HOLYOKE COL, 74- Mem: Fel Am Anthrop Asn; Soc Appl Anthrop. Res: Role of ideology and world view in ecological adaptation, especially nomadic pastoralist societies of the Middle East; cultural factors in urban migration and adjustment. Mailing Add: Dept of Sociol & Anthrop Merrill House Mt Holyoke Col South Hadley MA 01075

TAVANO, DONALD C, b Newark, NY, Aug 26, 36; m 58; c 2. PUBLIC HEALTH EDUCATION, MEDICAL EDUCATION. Educ: State Univ NY Col Cortland, BS, 60; Univ Ill, MS, 61; Univ Mich, MPH, 63; Mich State Univ, PhD(educ), 71. Prof Exp: Instr health educ, State Univ NY Col Cortland, 61-62; dir health educ, Saginaw City & County Health Depts, 63-65; consult health educ, Mich Dept Pub Health, 65-66; asst prof, 66-72, ASSOC PROF COMMUNITY MED, MICH STATE UNIV, 72- Concurrent Pos: Vis lectr, Sch Pub Health, Univ Mich, 72; consult, Mich Dept Educ, 66-, Gov Off Health & Med Affairs & Nat Bd Examrs Osteop Physicians & Surgeons Inc, 75. Mem: Soc Pub Health Educ; Am Pub Health Asn; Asn Behav Sci Med Educ; Asn Teachers Prev Med. Res: Patient education; the impact of patient education on health care cost containment, patient compliance and treatment outcomes. Mailing Add: Dept of Community Med Mich State Univ East Lansing MI 48824

TAVARES, DONALD FRANCIS, b East Providence, RI, Apr 1, 31; m 66; c 2. ORGANIC CHEMISTRY. Educ: Brown Univ, AB, 53; Yale Univ, MSc, 55, PhD(org chem), 59. Prof Exp: Teaching asst chem, Yale Univ, 53-55 & 56-59; teacher, Prospect Sch, Conn, 55-56; res assoc, Univ Ill, Urbana, 59-60 & Duke Univ, 60-61; res assoc chem, Univ Alta, 61-62; asst prof, 62-69, asst dean arts & sci, 67-70, ASSOC PROF CHEM, UNIV CALGARY, 69- Honors & Awards: Schering Found Prize, Yale Univ, 59. Mem: AAAS; Am Chem Soc; Chem Inst Can; The Chem Soc. Res: Study of organic reaction mechanisms; carbanions; organic photochemistry; chemistry of epoxides; chemistry of organic sulfur compounds. Mailing Add: Dept of Chem Univ of Calgary Calgary AB Can

TAVARES, ISABELLE IRENE, b Merced, Calif, Oct 6, 21. MYCOLOGY, LICHENOLOGY. Educ: Univ Calif, PhD(bot), 59. Prof Exp: Sr lab technician protozool, 49-52, from herbarium botanist to sr herbarium botanist, 52-68, ASSOC SPECIALIST BOT, UNIV CALIF, BERKELEY, 68- Mem: Bot Soc Am; Mycol Soc Am; Int Asn Plant Taxon. Res: Laboulbeniales; Usnea. Mailing Add: Dept of Bot Univ of Calif Berkeley CA 94720

TAVASSOLI, MEHDI, b Tehran, Iran, Mar 30, 33; US citizen; m 66; c 3. HEMATOLOGY. Educ: Tehran Univ Sch Med, MD, 61. Prof Exp: Intern, Cambridge City Hosp, Mass, 61-62; resident, Cook County Hosp, Chicago, 63-64 & Carney Hosp, Boston, 64-66; hematologist, Tufts-New Eng Med Ctr, 66-72 & from instr to asst prof med, Sch Med, Tufts Univ, 69-72; HEMATOLOGIST, SCRIPPS CLIN & RES FOUND, 72- Concurrent Pos: Charlton fel, Tufts Univ, 68; res fel, The Med Found, Boston, 70; vis investr anat, Med Sch, Johns Hopkins Univ, 70-71; consult hemat, Vet Admin Hosp & Univ Calif, San Diego, 73- Honors & Awards: John Larkin Award, The Guild of St Luke, 66. Mem: Am Soc Hemat; Am Col Physicians; Int Soc Exp Hemat. Res: Structural basis of hemopoiesis and the function of microenvironmental factor in immunohemopoiesis. Mailing Add: Dept of Hemat Scripps Clin & Res Found La Jolla CA 92037

TAVEL, MORTON, b Brooklyn, NY, June 14, 39; m 69; c 1. THEORETICAL PHYSICS. Educ: City Col New York, BS, 60; Stevens Inst Technol, MS, 62; Yeshiva Univ, PhD(physics), 64. Prof Exp: Res assoc, Brookhaven Nat Lab, 64-66, asst scientist, 66-67; asst prof, 67-70, ASSOC PROF PHYSICS, VASSAR COL, 70- Mem: AAAS; NucAm Nuclear Soc; Am Phys Soc. Res: Quantum field theory; plasma physics; transport theory. Mailing Add: Dept of Physics Vassar Col Poughkeepsie NY 12601

TAVEN, RONALD EARL, horticulture, see 12th edition

TAVERAS, JUAN M, b Dom Repub, Sept 27, 19; nat US; m 47; c 3. RADIOLOGY. Educ: Norm Sch Santiago, Dom Repub, BS, 37; Univ Santo Domingo, MD, 43; Univ Pa, MD, 49. Prof Exp: Instr radiol, Col Physicians & Surgeons, Columbia Univ, 50-52, from asst prof to prof, 52-65; prof radiol & chmn dept, Sch Med, Wash Univ, radiologist-in-chief, Barnes & Allied Hosps, Univ Med Ctr & dir, Mallinckrodt Inst Radiol, 65-71; RADIOLOGIST-IN-CHIEF, MASS GEN HOSP, BOSTON, 71- Concurrent Pos: Dir radiol, Neurol Inst, New York, 52-65; from asst to attend radiologist, Presby Hosp, New York, 50-65; mem neurol study sect, Nat Inst Neurol Dis & Stroke, 64-68; consult, US Marine Hosp, Staten Island, NY, Bronx Vet Admin Hosp, St Barnabas Hosp Chronic Dis, New York, Morristown Mem Hosp, NJ, St Louis City Hosp & Jewish Hosp, St Louis, 65-71. Mem: AMA; Am Neurol Asn; fel Am Col Radiol; Am Roentgen Ray Soc; Am Soc Neuroradiol (pres, 62-64). Res: Radiologic aspect of neurological science, especially cerebral angiography and cerebral vascular disease. Mailing Add: Mass Gen Hosp Boston MA 02114

TAVES, DONALD R, b Aberdeen, Idaho, July 22, 26; m 51; c 4. MEDICINE, RADIOBIOLOGY. Educ: Washington Univ, BS, 49, MD, 53; Univ Calif, MPH, 57; Univ Rochester, PhD(radiation biol), 63. Prof Exp: Health officer, Shasta County Health Dept, 57-60; USPHS fel, 60-63; asst prof radiation biol & biophys, 63-67, ASSOC PROF PHARMACOL, SCH MED, UNIV ROCHESTER, 67- Mem: Soc Toxicol; Am Soc Pharmacol & Exp Therapeut. Res: Pharmacology and toxicology of fluoride. Mailing Add: Dept of Pharmacol Univ of Rochester Sch of Med Rochester NY 14642

TAVES, MILTON ARTHUR, b Aberdeen, Idaho, Aug 14, 25; m 45; c 2. ORGANIC CHEMISTRY. Educ: Univ Utah, BS, 45; Mass Inst Technol, PhD(org chem), 48. Prof Exp: Lab asst chem, Univ Utah, 43-45; asst, Mass Inst Technol, 45-46; res chemist, Res Ctr, 48-54, res supvr, 54-60 & 62-64, tech asst to dir res, 60-62, RES MGR, SYNTHETIC RES DIV, RES CTR, HERCULES, INC, 64- Mem: AAAS; Am Chem Soc. Res: Heterocyclics; hydrogen peroxide; hydroperoxide chemistry; auto-oxidation and catalytic oxidation of organic compounds; terpenes; condensation polymers; resins; plasticizers; process development; agricultural chemicals. Mailing Add: Hercules Res Ctr Wilmington DE 19899

TAVILL, ANTHONY SYDNEY, b Manchester, Eng, July 15, 36; m 59; c 3. GASTROENTEROLOGY. Educ: Univ Manchester, MB & ChB, 60, MD, 70; Royal Col Physicians London, MRCP, 63. Prof Exp: Med Res Coun travelling fel med, Albert Einstein Col Med, 66-68; lectr, Royal Free Hosp, Sch Med, Univ London, 68-71; sr clin scientist, Div Clin Invest, Med Res Coun Clin Res Ctr, Eng, 71-72, consult gastroenterol & liver dis, 72-75; HEAD GASTROENTEROL & PROF MED, CASE WESTERN RESERVE UNIV, CLEVELAND METROP GEN HOSP, 75- Mem: Med Res Soc Gt Brit; Brit Soc Gastroenterol; Europ Soc Clin Invest; Am Asn Study Liver Dis; Int Asn Study Liver. Res: Control mechanisms in hepatic protein metabolism in gastrointestinal and renal disease, homeostasis of plasma transport proteins and metabolic interrelationships between iron transport and storage proteins of the liver. Mailing Add: Dept of Gastroenterol Cleveland Metrop Gen Hosp Cleveland OH 44109

TAVOLGA, MARGARET CORDSEN, b New York, NY, Mar 24, 21; m 46. ENDOCRINOLOGY. Educ: NY Univ, BA, 42, MS, 44, PhD(endocrinol), 48. Prof Exp: Am Cancer Soc grant, Am Mus Natural Hist, 48-50; from asst prof to assoc prof, 57-70, PROF BIOL, FAIRLEIGH DICKINSON UNIV, 70- Mem: AAAS; Am Soc Zool. Res: Adrenal cortex physiology; sex hormone physiology in fishes; effects of hormones on pigment in fishes; sexual and social behavior in porpoises. Mailing Add: Dept of Biol Sci Fairleigh Dickinson Univ Teaneck NJ 07666

TAVOLGA, WILLIAM NICOLAI, b New York, NY, Feb, 9, 22; m 46. ANIMAL BEHAVIOR. Educ: City Col New York, BS, 43; NY Univ, MS, 46, PhD(embryol), 49. Prof Exp: Teaching fel, 43, tutor biol, 44-49, from instr to assoc prof, 49-66, PROF BIOL, CITY COL NEW YORK, 67-; RES ASSOC, AM MUS NATURAL HIST, 53-; RES ASSOC, MOTE MARINE LAB, 66- Concurrent Pos: USPHS res fel, 53-55; Guggenheim fel, 67-68. Mem: AAAS; Am Soc Ichthyol & Herpet; Am Soc Zool; Animal Behav Soc. Res: Behavior and bio-acoustics of fishes; animal communication; fish parasitology, embryology and endocrinology. Mailing Add: Dept of Biol City Col of New York New York NY 10031

TAVORMINA, PETER ALBERT, biochemistry, deceased

TAX, SOL, b Chicago, Ill, Oct 30, 07; m 33; c 2. CULTURAL ANTHROPOLOGY, APPLIED ANTHROPOLOGY. Educ: Univ Wis, PhB, 31; Univ Chicago, PhD(anthrop), 35. Hon Degrees: DrHumaneLett, Univ Wis, Milwaukee, 69; DrLaws, Wilmington Col, Ohio & DSSc, Univ Valle de Guatemala, 74; DSc, Beloit Col, Wis, 75. Prof Exp: Investr & ethnologist, Carnegie Inst, 34-46; from res assoc anthrop to assoc prof, 40-48, chmn dept, 55-58, PROF ANTHROP, UNIV CHICAGO, 48- Concurrent Pos: Ed, Am Anthropologist, 52-55, Current Anthrop, 59-74, Viking Fund Publs, 60-69; mem, US Nat Comn UNESCO, 60-66; coordr, Am Indian Conf, Chicago, 61; res assoc, Wenner-Gren Found Anthrop Res, 61-; dir, Carnegie Corp Cross-cult Educ Proj, 62-67; partic, NSF Re-Study Guatemalan Indians, 63-67; mem bd dirs, The Bridge, Ctr Advan Intercult Studies, 65-70 & Coun Study Mankind, 65-; consult, US Off Educ, 65-; anthrop adv to secy, Smithsonian Inst, 66-, dir, Ctr Study Man, 70-; pres, Int Union Anthrop & Ethnol Sci & of IXth Int Cong, 68-73; gen ed, World Anthrop, 73- Honors & Awards: Viking Fund Medal Anthrop, 61; Spec Plaque, Diputacion Provincial de Santander, Spain, 73. Mem: Fel AAAS; Fel Am Anthrop Asn (pres, 58-59); fel Am Folklore Soc; hon fel Royal Anthrop Inst Gt Brit & Ireland; hon fel Ethnol Soc Hungary. Res: North American Indians; middle American ethnology; action anthropology; North and Middle America (Mexico, Guatemala) Indian cultural and social anthropology. Mailing Add: Apt 307 1700 E 56th St Chicago IL 60637

TAYAMA, HARRY K, b Los Angeles, Calif, May 26, 35; m 61; c 2. HORTICULTURE, FLORICULTURE. Educ: Univ Ill, BS, 58, MS, 59; Ohio State Univ, PhD(hort), 63. Prof Exp: Asst prof hort, Pa State Univ, 63-64; from asst prof to assoc prof, 64-71, asst prof hort, Ohio Agr Res & Develop Ctr, 66-67, PROF HORT, OHIO AGR RES & DEVELOP CTR, OHIO STATE UNIV, 71- Honors & Awards: Kenneth Post Award, 66. Mem: Am Soc Hort Sci. Res: Ecological factors affecting growth and flowering of florist crops; plant growth regulators. Mailing Add: Dept of Hort Ohio State Univ 2001 Fyffe Ct Columbus OH 43210

TAYBACK, MATTHEW, b Tarrytown, NY, June 30, 19; m 45; c 3. PUBLIC HEALTH. Educ: Harvard Univ, AB, 39; Columbia Univ, AM, 40; Johns Hopkins Univ, ScD(biostatist), 53. Prof Exp: Res assoc, NY State Psychiat Inst, 40-42; res statistician, Md State Health Dept, 46-48, dir bur biostatist, Baltimore, 48-53, dir statist sect, 53, asst comnr health, 57-63, dep comnr health, 63-69, asst secy health & ment hyg & sci affairs, 69-73; from asst prof to assoc prof, 52-65, PROF HYG & PUB HEALTH, SCH MED, UNIV MD, BALTIMORE CITY, 65-; STATE DIR ON AGING, 74- Concurrent Pos: Lectr, Johns Hopkins Univ, 51-; vis prof, Univ Philippines, 57; consult, US Army, 57-, WHO, 60- & USAID, 61-; vchmn, State Comn on Aging, Md, 61-75; chmn, State Adv Bd Price Comn, 72-74; chmn, State Emp Ret Rev Bd, 74- Mem: Am Pub Health Asn. Res: Epidemiology; demography; health services administration. Mailing Add: State Dir on Aging State Off Bldg 301 W Preston St Baltimore MD 21201

TAYBI, HOOSHANG, b Malayer, Iran, Oct 22, 19; nat US; m 55; c 2. RADIOLOGY. Educ: Univ Teheran, MD, 44; NY Univ, MSc, 54; Am Bd Pediat, dipl, 54; Am Bd Radiol, dipl, 59. Prof Exp: Intern, St Vincent's Hosp, Staten Island, NY, 49-50; resident pediat, Children's Hosp, Akron, Ohio, 52-54; fel pediat roentgenol, Children's Hosp, Cincinnati, 54-56; resident gen radiol, King's County Hosp, Brooklyn, NY, 58-59; from asst prof to assoc prof radiol, Sch Med, Univ Okla, 60-62; assoc prof radiol & chief sect pediat roentgenol, Med Ctr, Ind Univ, 62-64, prof, Sch Med, 64; dir div pediat roentgenol, Children's Hosp & Adult Med Ctr, San Francisco, 64-66; DIR DEPT RADIOL, CHILDREN'S HOSP MED CTR NORTHERN CALIF, 67- Concurrent Pos: Assoc clin prof radiol, Sch Med, Univ Calif, San Francisco, 64-70, clin prof, 70-; consult, Letterman Gen Hosp & Children's Hosp San Francisco. Honors & Awards: Cert Merit, Am Roentgen Ray Soc, 55. Mem: Am Col Radiol; Am Roentgen Ray Soc; Soc Pediat Radiol (pres, 74); AMA. Res: Pediatric radiology. Mailing Add: Children's Hosp Med Ctr 51st & Grove Sts Oakland CA 94609

TAYLOR, ALAN NEIL, b Franklin, NY, Sept 10, 34; m 55; c 2. PHYSIOLOGY, BIOCHEMISTRY. Educ: Ohio State Univ, BS, 57; Cornell Univ, MS, 60, PhD(phys biol), 69. Prof Exp: Res assoc mineral metab, NY State Vet Col, Cornell Univ, 60-66, NIH traineeship, 66-69, sr res assoc membrane transport, 69-75; ASSOC PROF, BAYLOR COL DENT, DALLAS, 76- Mem: AAAS; Am Inst Nutrit; Soc Exp Biol & Med. Res: Mineral metabolism; membrane transport; vitamin D metabolism. Mailing Add: Dept of Microanat Baylor Col Dent 800 Hall St Dallas TX 75226

TAYLOR, ALAN WORMALD, physical chemistry, colloid chemistry, see 12th edition

TAYLOR, ALBERT CECIL, b Hopei, China, Aug 15, 05; US citizen; m 35; c 2. BIOLOGY. Educ: Taylor Univ, BA, 30; Univ Ky, MA, 34; Univ Chicago, PhD(zool), 42. Prof Exp: Teacher biol acad, Chicago Evangel Inst, 30-32; teacher sci, Asheland Jr High Sch, Louisville, 34-35; teacher zool, NC State Teachers Col, Asheville, 35-39; asst zool, Univ Chicago, 42-43; res assoc, 43-45; asst prof anat, Col Dent, NY Univ, 45-49, assoc prof, Col Dent & Grad Sch, 49-55; assoc, Rockefeller Inst, 55-58, assoc prof develop biol, 58-65; prof anat, Univ Tex Dent Br Houston & Grad Sch Biomed Sci & mem dent sci inst, Houston, 65-74; MEM FAC PHYSIOL & BIOPHYS, UNIV ILL, URBANA, 74- Mem: Soc Exp Biol & Med; Am Asn Anatomists; Am Soc Cell Biol; Soc Develop Biol; Soc Cryobiol. Res: Development nervous system; freezing and freeze-drying of tissues; cell interactions in development; cell contact and adhesion; lysis of tissue collagen. Mailing Add: Dept of Physiol & Biophys Univ of Ill 524 Burrill Hall Urbana IL 61801

TAYLOR, ALBERT EDWARD, b Kilgore, Idaho, Mar 9, 08; m 31; c 3. ANALYTICAL CHEMISTRY. Educ: Univ Kans, BA, 30, MA, 34; Univ Mich, PhD(chem), 45. Prof Exp: From instr to prof, 30-73, dir AEC res proj, 51-61, dir grad studies, 57-67, head div phys sci, 61-65, asst dean col liberal arts, 65-67, dean grad sch, 67-73, EMER PROF CHEM, IDAHO STATE UNIV, 73- Mem: Fel AAAS; Am Chem Soc. Res: Use of exchange resins; analytical chemistry of thorium, uranium and strontium; photodecomposition; spectrophotometry; radiochemistry. Mailing Add: 805 S 19th Ave Pocatello ID 83201

TAYLOR, ALBERT LEE, b Florence, Colo, Nov 30, 01; m 28; c 1. NEMATOLOGY. Educ: George Washington Univ, BS, 34. Prof Exp: Nematologist, USDA, 35-46; technologist, Shell Chem Corp, 46-49; nematologist, USDA, 49-56, leader, Nematol Invests, 56-64; consult nematologist, Food & Agr Orgn, UN, Cyprus, 64-65 & Thailand, 65; nematologist, USDA, Fla, 66-67 & Food & Agr Orgn, Fiji, 67-68, nematologist & proj mgr, Thailand, 68-72; CHIEF NEMATOLOGIST, DIV PLANT INDUST, FLA DEPT AGR & CONSUMER SERV, 72- Res: Plant nematology; chemical control of plant nematodes. Mailing Add: Div of Plant Indust PO Box 1269 Gainesville FL 32601

TAYLOR, ALBERT WILLIAM, physiology, physical education, see 12th edition

TAYLOR, ALBERTO SIDNEY, plant morphology, genetics, see 12th edition

TAYLOR, ALFRED HENRY, JR, b Norfolk, Va, Apr 28, 07; m 41; c 2. PHYSICAL CHEMISTRY. Educ: Univ NH, BS, 30, MS, 33; Columbia Univ, PhD(phys chem), 41. Prof Exp: Instr anal chem, Univ NH, 30-33; asst chem, Columbia Univ, 34-37, col pharm, 37-39; res chemist, Exp Res Labs, Burroughs Wellcome & Co, Inc, 39-42, Manhattan Proj, SAM Labs, Columbia Univ, 42-45 & Carbide & Carbon Chem Co, NY, 45-46; res chemist, 46-53, mgr anal div, 53-58, asst dir indust gases & cryogenics res div, 58-66, tech dir, Rare & Specialty Gases Dept, Airco Indust Gas Div, Air Reduction Co, Inc, 66-72; CONSULT, 72- Concurrent Pos: Vol exec, P T Aneka Gas Industri, Jakarta, Indonesia, 73-74. Mem: Am Chem Soc. Res: Reaction kinetics; physical measurements; instrumental analysis; absorption spectroscopy; rare gases. Mailing Add: 329 S Northfield Rd Millington NJ 07946

TAYLOR, ALLEN LEVORE, physics, see 12th edition

TAYLOR, ALTON ROBERT, b Bolton Landing, NY, Mar 11, 07; m 30; c 3. MICROBIOLOGY. Educ: Wesleyan Univ, BS, 30; Princeton Univ, PhD(endocrine biol), 39. Prof Exp: Asst biol, Princeton Univ, 30-34, res assoc & instr endocrine physiol, 34-39; res assoc exp surg, Sch Med, Duke Univ, 39-47; res virologist, Parke, Davis & Co, 47-55, lab dir virus res, 55-69; PROF MICROBIOL, SCH MED, WAYNE STATE UNIV, 69- Mem: AAAS; Electron Micros Soc Am; Am Asn Immunol; NY Acad Sci. Res: Electron microscopy; instrumentation and methods, teaching methods and research.

TAYLOR, ANDREW RONALD ARGO, b Ottawa, Ont, July 6, 21; m 47; c 4. BOTANY. Educ: Univ Toronto, BA, 43, PhD(bot), 55. Prof Exp: Asst bot, Univ Toronto, 43-46; from asst prof to assoc prof biol, 46-62, PROF BOT, UNIV NB, FREDERICTON, 62-, ACTG DEAN SCI, 74- Concurrent Pos: Asst biologist, Fisheries Res Bd Can, 48-60; hon lectr, Univ St Andrews, 60-61. Mem: Bot Soc Am; Phycol Soc Am; Can Bot Asn; Can Soc Plant Physiol (vpres, 64-65); Brit Phycol Soc. Res: Developmental morphology, ecology and taxonomy of north Atlantic marine algae. Mailing Add: Dept of Biol Univ of NB Fredericton NB Can

TAYLOR, ANGUS ELLIS, b Craig, Colo, Oct 13, 11; m 36; c 3. MATHEMATICS. Educ: Harvard Univ, SB, 33; Calif Inst Technol, PhD(math), 36. Prof Exp: Instr math, Calif Inst Technol, 36-37; Nat Res Coun fel, Princeton Univ, 37-38; from instr to prof math, 38-65, vpres acad affairs, Univ Calif, Los Angeles, 65-75, UNIV PROVOST, UNIV CALIF (SYSTEMWIDE ADMIN), 75- Concurrent Pos: Fulbright res fel, Univ Mainz, 55. Mem: Am Math Soc; Math Asn Am. Res: Theory of functions; linear operators and spectral theory; history of mathematics. Mailing Add: Off of Pres Univ Hall Univ of Calif Berkeley CA 94720

TAYLOR, ANNA NEWMAN, b Vienna, Austria, Oct 28, 33; US citizen; m 59. PHYSIOLOGY, ANATOMY. Educ: Western Reserve Univ, AB, 55; PhD(physiol), 61. Prof Exp: Am Heart Asn res fel physiol, Western Reserve Univ, 61-63, instr, 62-63; Am Heart Asn adv res fel, Lab Neurophysiol, Henri-Rousselle Hosp, Paris, France, 63-64; asst prof physiol, Western Reserve Univ, 64-65; asst prof anat, Dept Anat & Psychiat, Col Med, Baylor Univ, 65-67; asst res anatomist, Univ Calif, Los Angeles, 67-68, from asst prof to assoc prof anat, 68-75; ASSOC PROF ANAT, DEPT PHYSIOL, CORNELL MED COL, NEW YORK, 75- Concurrent Pos: USPHS fel, 61; res specialist, Houston State Psychiat Inst, 65-67; NIMH res scientist develop award, 72- Mem: Fel AAAS; Am Physiol Soc; Endocrine Soc; Am Asn Anat; Soc Neurosci. Res: Neuroendocrinology; central nervous system control of pituitary-adrenal function; actions of hormones on central nervous system. Mailing Add: Dept of Physiol Cornell Med Col New York NY 10021

TAYLOR, ARDELL NICHOLS, b Terral, Okla, Jan 19, 17; m 43; c 2. PHYSIOLOGY. Educ: Tex Tech Col, BS, 39; Univ Tex, MA, 41, PhD(zool sci), 43. Hon Degrees: DSc, Lincoln Col, 71. Prof Exp: Tutor zool, Univ Tex, 39-42, from instr to asst prof physiol, Sch Med, 43-46; from asst prof to assoc prof, Sch Med, Univ Okla, 46-51, prof, chmn dept & assoc dean, 51-60; assoc secy, Coun Med Educ & dir, Dept Allied Med Prof & Serv, AMA, 60-67; dean sch related health sci, Chicago Med Sch, 67-69, PRES, UNIV HEALTH SCI-CHICAGO MED SCH, 69- Mem: Soc Exp Biol & Med; Am Physiol Soc. Res: Nucleic acid metabolism on ova; nerve conduction and facilitation; hypertension; experimental vascular physiology; dynamics of circulation; medical education. Mailing Add: Off of the Pres Univ of Health Sci-Chicago Med Sch Chicago IL 60612

TAYLOR, AUBREY BRYANT, b Palmyra, Mo, Dec 11, 03; m 37; c 1. ZOOLOGY. Educ: Brown Univ, PhB, 27; Univ Iowa, MS, 29, PhD(zool), 35. Prof Exp: From asst zool to instr, Univ Iowa, 27-30, asst 30-35, Rockefeller res assoc, 35-36; from instr to prof, 36-72, EMER PROF PHYSIOL, UNIV ILL, URBANA, 72- Concurrent Pos: USPHS spec fel, 65-66. Mem: AAAS; Am Soc Zool; Am Physiol Soc; Am Soc Cell Biol. Res: Cellular physiology; bioelectric currents; parasites of acrididae; electrocardiograms of insects; cytophysiology; absorption of fat; electron microscopy; ultrastructure changes associated with cellular aging; electron microscope studies of the junctional area between longitudinal and circular muscles of the cat's small intestine. Mailing Add: Dept of Physiol & Biophys Univ of Ill Urbana IL 61801

TAYLOR, AUBREY ELMO, b El Paso, Tex, June 4, 33; m 54; c 3. PHYSIOLOGY, BIOPHYSICS. Educ: Tex Christian Univ, AB, 60; Univ Miss, PhD(physiol, biophys), 64. Prof Exp: Res asst learning theory, Bell Helicopter Co, Tex, 59-60; prof math, Exten Ctr, 60-65, from asst prof to assoc prof, Med Ctr, 64-71, PROF PHYSIOL & BIOPHYS, MED CTR, UNIV MISS, 71- Mem: AAAS; Am Math Soc; Soc Indust & Appl Math; Biophys Soc; Am Physiol Soc. Res: Irreversible thermodynamics and membrane biophysics to mammalian physiology, especially in fields of cardio-pulmonary and intestinal dynamics. Mailing Add: Dept of Physiol & Biophys Univ of Miss Med Ctr Jackson MS 39216

TAYLOR, AUSTIN LAURENCE, b Vancouver, BC, Jan 23, 32; US citizen; m 56; c 2. MICROBIOLOGY. Educ: Western Md Col, BS, 54; Univ Calif, Berkeley, PhD(bact), 61. Prof Exp: Res assoc bact genetics, Brookhaven Nat Lab, 61-62; res microbiologist, NIH, 62-65; asst prof, 65-70, ASSOC PROF MICROBIOL, SCH MED, UNIV COLO, DENVER, 70- Concurrent Pos: USPHS res grant, 66- Mem: AAAS; Genetics Soc Am; Am Soc Microbiol. Res: Radiation biology; bacterial and bacteriophage genetics. Mailing Add: Dept of Microbiol Med Ctr Univ of Colo Denver CO 80220

TAYLOR, BARRIE FREDERICK, b Nottingham, Eng, Nov 21, 39; m 65. MARINE MICROBIOLOGY. Educ: Univ Leeds, BSc, 62, PhD(biochem), 65. Prof Exp: NSF grant bact, Rutgers Univ, 65-67; res assoc microbiol, Univ Tex, Austin, 67-69; asst prof biol oceanog, 70-74, ASSOC PROF MARINE SCI, BIOL & LIVING RESOURCES DIV, SCH MARINE & ATMOSPHERIC SCI, UNIV MIAMI, 74- Concurrent Pos: NSF res grants, 70-72 & 73-75; NIH grant, 74-75. Mem: AAAS; Am Soc Microbiol; Am Soc Limnol & Oceanog. Res: Microbial biochemistry; autotrophic and lithotrophic micro-organisms; aromatic degradation by microbes. Mailing Add: Dept of Functional Biol Marine Inst Univ of Miami Miami FL 33149

TAYLOR, BARRY EDWARD, b Potsdam, NY, July 7, 47; m 68; c 1. SOLID STATE CHEMISTRY. Educ: State Univ NY Col Fredonia, BS, 69; Brown Univ, PhD(chem), 74. Prof Exp: Teaching asst chem, State Univ NY Col Fredonia, 66-67, res asst, 68-69; teaching assoc chem, Brown Univ, 69-73; chemist, 73-74, RES CHEMIST, EXP STA, E I DU PONT DE NEMOURS & CO, 74- Mem: Am Chem Soc; AAAS; Electrochem Soc. Res: Preparative solid state chemistry; crystallographic and physical properties of oxides, halides and sulfides; chemistry of alkali metal compounds; study of ionic conductivity in solids. Mailing Add: Exp Sta Bldg356/345 E I du Pont de Nemours & Co Wilmington DE 19898

TAYLOR, BARRY NORMAN, b Philadelphia, Pa, Mar 27, 36; m 58; c 3. SOLID STATE PHYSICS, LOW TEMPERATURE PHYSICS. Educ: Temple Univ, AB, 57; Univ Pa, MS, 60, PhD(physics), 63. Prof Exp: From instr to asst prof physics, Univ Pa, 63-66; physicist, RCA Labs, NJ, 66-70; chief, Absolute Elec Measurements Sect, 70-74, CHIEF, ELEC DIV, NAT BUR STAND, 74- Concurrent Pos: Consult, Philco Corp, 64-65; instr math, Rider Col, 69-70; mem, Nat Acad Sci-Nat Res Coun-Nat Acad Eng Adv Comt Fundamental Constants. Honors & Awards: RCA Outstanding Achievement Award in Sci, 69; John Price Wetherill Medal, Franklin Inst, Philadelphia & Silver Medal Award, Dept Com, Washington, 75. Mem: Fel Am Phys Soc; sr mem Inst Elec & Electronics Eng. Res: Precision measurements and fundamental constants; absolute electrical measurements; superconductivity; electron tunneling; Josephson effects. Mailing Add: Rm B258 Bldg 220 Nat Bur Stand Washington DC 20234

TAYLOR, BETTY, b Waukon, Iowa, Aug 11, 14. FOODS, NUTRITION. Educ: Iowa State Univ, BS, 36; Columbia Univ, MA, 39; Mich State Univ, PhD(foods & nutrit), 59. Prof Exp: Dietitian, Post-Grad Hosp, 38-43; asst prof foods & nutrit, Univ Southwestern La, 46-47; from asst prof to assoc prof, 47-59, actg chmn dept home econ, 73-74, PROF FOODS & NUTRIT, WESTERN MICH UNIV, 59- Mem: Am Dietetic Asn; Inst Food Technol; Am Home Econ Asn; Soc Nutrit Educ. Res: Experimental foods; nutrition education; dietetics. Mailing Add: Dept of Home Econ Western Mich Univ Kalamazoo MI 49008

TAYLOR, BILLY G, b Booneville, Miss, Aug 11, 24; m 49; c 2. ONCOLOGY,

SURGERY. Educ: Harvard Med Sch, MD, 48. Prof Exp: From intern to resident surg, Grady Mem Hosp, Atlanta, Ga, 48-50; resident, Hosp, Emory Univ, 50-51; resident, Mem Ctr Cancer & Allied Dis, New York, 53-57; instr surg, 57-61, assoc prof clin surg, 64-72, assoc prof surg, 72-74, PROF SURG, LA STATE UNIV MED CTR, NEW ORLEANS, 74- Concurrent Pos: Spec fel head & neck surg with Dr Hayes Martin, Mem Ctr Cancer & Allied Dis, New York, 57; chief surg, Vet Admin Hosp, New Orleans, 64-72, sr physician, 72-; consult, USPHS Hosp, New Orleans, 64-; active staff, Touro Infirmary, New Orleans, 64-; sr vis surgeon, Charity Hosp, New Orleans, 64- Mem: James Ewing Soc; Soc Head & Neck Surg; AMA; fel Am Col Surg; Am Thoracic Soc. Res: Oncologic and head and neck surgery; physical and chemical properties of human cadaver blood; clincial studies of ameloblastoma of the mandible and carotid body tumors; clinical studies on carcinoma of the male breast and carcinoma on the larynx. Mailing Add: Vet Admin Hosp 1601 Perdido St New Orleans LA 70140

TAYLOR, BOYD EUGENE, b Atlanta, Ga, July 31, 01; wid. GENETICS. Educ: Emory Univ, BS, 23; Univ Madrid, MA, 25; Sorbonne, MS, 26. Prof Exp: Sci writer, Hearst Newspapers, 19-40; INDEPENDENT PRESS, 40-; DIR, GENETICS RES LIBR, 50- Concurrent Pos: Instr gen sci, High Sch, Ga, 22-27; prof mil sci & tactics, 22-27; chmn hist house mus corp, 40-; mem, Human Genetics Coun, 50- & Pop Ref Bur, 60-; sustaining mem, Coun Basic Educ, 60-; chmn, Mt Rushmore Mem Comn. Mem: AAAS; Am Genetic Asn; Genetics Soc Am; Pop Asn Am; fel Royal Soc Health. Res: Demography; electronic language laboratory; direct method language instruction; ethnic population survey of Atlanta; inventor of the Taylorscope, a stereo viewer for aerial photography. Mailing Add: 327 St Paul Ave SE Atlanta GA 30312

TAYLOR, BRUCE CAHILL, b Cleveland, Ohio, June 5, 42; div; c 2. BIOMEDICAL ENGINEERING, INSTRUMENTATION. Educ: Hiram Col, BA, 64; Kent State Univ, MA, 66, PhD(physiol), 71. Prof Exp: Assoc dir res, Vascular Res Lab, Akron City Hosp, 71-75; SR RES SCIENTIST, ABCOR, INC, 75- Concurrent Pos: Biomed consult, Midgard Electronics, 75- Mem: Am Soc Artifical Internal Organs; Asn Advan Med Instrumentation; Biomed Eng Soc. Res: Design and development of new types of medical instrumentation and new techniques for biomaterials development. Mailing Add: Abcor Inc 341 Vassar St Cambridge MA 02139

TAYLOR, CARL ERNEST, b Mussoorie, India, July 26, 16; m 43; c 3. PREVENTIVE MEDICINE, EPIDEMIOLOGY. Educ: Muskingum Col, BS, 37; Harvard Univ, MD, 41, MPH, 51, DrPH, 53; FRCP(C), 45. Hon Degrees: DSc, Muskingum Col, 62. Prof Exp: Med officer, Gorgas Hosp, 41-44; chief med serv, USPHS Marine Hosp, Pittsburgh, 44-46; hosp supt, Fatehgarh, India, 47-50; instr, Sch Pub Health, Harvard Univ, 51-53; prof prev med, Christian Med Col, India, 53-56; assoc prof prev med & pub health & dir prog for teachers, Sch Pub Health, Harvard Univ, 56-61; PROF & CHMN DEPT INT HEALTH, SCH HYG & PUB HEALTH, JOHNS HOPKINS UNIV, 61- Mem: Am Soc Trop Med & Hyg; Asn Teachers Prev Med; Am Pub Health Asn; Royal Soc Trop Med & Hyg. Res: International health; health planning in developing countries; population dynamics; medical education; epidemiology of leprosy and nutrition and infections; integration of health and family planning. Mailing Add: Dept Int Health Sch of Hyg & Pub Health Johns Hopkins Univ Baltimore MD 21205

TAYLOR, CARLTON FULTON, plant pathology, see 12th edition

TAYLOR, CHARLES ARTHUR, JR, b Ithaca, NY, July 28, 13; m 37, 69; c 2. PLANT TAXONOMY. Educ: Cornell Univ, BS, 35, MS, 39. Prof Exp: Assoc prof, 49-67, PROF BOT, S DAK STATE UNIV, 67-, CUR, HERBARIUM, 72- Mem: AAAS; Am Soc Plant Taxon; Int Asn Plant Taxon. Res: Flora of South Dakota, monographic studies in Poaceae. Mailing Add: Dept of Bot & Biol SDak State Univ Brookings SD 57006

TAYLOR, CHARLES BRUCE, b Hecla, SDak, Feb 6, 15; m 38; c 2. PATHOLOGY. Educ: Univ Minn, BS, 38, MB, 40, MD, 41. Prof Exp: Intern, Univ Hosp, Univ Minn, 40-41, fel physiol, 41-42; res assoc & asst prof path, Col Med, Univ Ill, 46-51, assoc prof, 54-59; assoc prof, Univ NC, 51-54; prof, Med Sch, Northwestern Univ, 69-70; prof, Med Ctr, Univ Ala, Birmingham, 70-72; PROF PATH, ALBANY MED COL, 72-; DIR RES, VET ADMIN HOSP, ALBANY, 72- Concurrent Pos: Res assoc, Presby Hosp, Chicago, 46-51, attend pathologist, 54-59; chmn dept path, Evanston Hosp, Ill, 59-70; mem coun arteriosclerosis, Am Heart Asn. Mem: Am Physiol Soc; Am Soc Exp Path; Am Heart Asn; Am Asn Path & Bact; AMA. Res: Arteriosclerosis and its pathogenesis; human metabolism of cholesterol; degenerative pulmonary diseases; pathogenesis of gall stones. Mailing Add: Vet Admin Hosp Albany NY 12208

TAYLOR, CHARLES ELLETT, b Chicago, Ill, Sept 9, 45; m 69. POPULATION GENETICS. Educ: Univ Calif, Berkeley, AB, 68; State Univ NY, Stony Brook, PhD(ecol & evolution), 73. Prof Exp: ASST PROF BIOL, UNIV CALIF, RIVERSIDE, 74- Mem: Genetics Soc Am; Soc Study Evolution; Am Ecol Soc. Res: Population genetics and ecology. Mailing Add: Dept of Biol Univ of Calif Riverside CA 92502

TAYLOR, CHARLES EMORY, b White Plains, NY, Mar 2, 40; m 70. NUCLEAR MAGNETIC RESONANCE. Educ: Williams Col, BA, 61, MA, 63; Mich State Univ, PhD(physics), 67. Prof Exp: ASST PROF PHYSICS, ANTIOCH COL, 67- Mem: Am Phys Soc; Am Asn Physics Teachers. Res: Nuclear spin lattice relaxation in antiferromagnetics; holography; holaesthetics. Mailing Add: Dept of Physics Antioch Col Yellow Springs OH 45387

TAYLOR, CHARLES JOEL, b Portland, Ore, Apr 12, 19; m 52; c 3. PHYSICS. Educ: Univ Ill, BS, 40, MS, 48, PhD(physics), 51. Prof Exp: Asst nuclear physics, Univ Ill, 50-51; physicist, NAm Aviation, Inc, 51-52; PHYSICIST, LAWRENCE LIVERMORE LAB, UNIV CALIF, 52- Mem: Am Phys Soc. Res: Scintillation counters; neutron physics. Mailing Add: Lawrence Livermore Lab Univ of Calif Livermore CA 94550

TAYLOR, CHARLES LUTHER, b Wilkes-Barre, Pa, Sept 4, 21; m 49; c 2. PHYSICAL METEOROLOGY. Educ: Pa State Univ, BS, 42, MS, 47. Prof Exp: Instr meteorol, Pa State Univ, 43-44 & 46-47; instr physics, Wilkes Col, 47-51; ASSOC PROF METEOROL, NAVAL POSTGRAD SCH, 54- Mem: Am Meteorol Soc; Am Geophys Union. Res: Atmospheric optics; radiation; sound and atmospheric electricity. Mailing Add: Dept of Meteorol Naval Postgrad Sch Monterey CA 93940

TAYLOR, CHARLES PATRICK STIRLING, b Toronto, Ont, May 11, 30; m 55; c 4. BIOPHYSICS. Educ: Univ BC, BA, 52; Oxford Univ, BA, 54, MA, 57; Univ Pa, PhD(biophys), 60. Prof Exp: Childs Mem Fund fel biophys, Cambridge Univ, 60-61; asst prof physics, Univ BC, 61-67; assoc prof, 68-74, PROF BIOPHYS, UNIV WESTERN ONT, 74- Mem: Sigma Xi; Biophys Soc; Can Soc Cell Biol; Am Soc Photobiol. Res: Electron paramagnetic resonance and optical spectroscopy of cytochromes and other hemoproteins; cellular respiration and oxidase action spectra; muscle contraction. Mailing Add: Dept of Biophys Univ of Western Ont London ON Can

TAYLOR, CHARLES RICHARD, b Phoenix, Ariz, Sept 8, 39. COMPARATIVE PHYSIOLOGY, ENVIRONMENTAL PHYSIOLOGY. Educ: Occidental Col, BA, 60; Harvard Univ, MA, 62, PhD(biol), 63. Prof Exp: Res fel mammal, Mus Comp Zool, Harvard Univ, 64-67; res assoc zool, Duke Univ, 68-70; assoc prof, 70-74, PROF BIOL, HARVARD UNIV, 74-, ALEXANDER AGASSIZ PROF ZOOL, 74-, DIR CONCORD FIELD STA, 70- Concurrent Pos: Hon lectr, Univ Col, Nairobi, Kenya & attached res officer, EAfrican Vet Res Orgn, 64-67; res assoc, Mus Comp Zool, Harvard Univ, 68- Mem: AAAS; Am Soc Mammal; Am Physiol Soc; Am Soc Zool; NY Acad Sci. Res: Water metabolism; temperature regulation; respiratory and exercise physiology; physiology of wild and domestic ruminants. Mailing Add: Dept of Biol Harvard Univ Cambridge MA 02138

TAYLOR, CHARLES WILLIAM, b Duluth, Minn, Sept 26, 30; m 53; c 4. ORGANIC CHEMISTRY. Educ: Univ Minn, BA, 52; Univ Wis, PhD(chem), 57. Prof Exp: Sr chemist, 57-72, RES SPECIALIST, MED PROD DIV, CENT RES DEPT, 3M CO, 72- Mem: Am Chem Soc. Res: Fluorocarbon chemistry; biomedical materials; thermosetting acrylics; pressure sensitive adhesives. Mailing Add: Cent Res Dept 3M Co 3M Ctr Bldg 236-3 St Paul MN 55101

TAYLOR, CHARLOTTE CLARKE, b St Louis, Mo, May 25, 25; m 65; c 2. ENVIRONMENTAL CHEMISTRY. Educ: Tufts Univ, BS, 44; Univ Calif, MA, 50, PhD(chem), 56. Prof Exp: Lectr sci, Los Angeles City Cols, 56-61; lectr chem, Mirman Sch Gifted Children, 63-64; asst prof, Mt St Mary's Col, 65-68; lectr chem, Exten Div, Univ Calif, Los Angeles, 58-61, vol res assoc med biochem, Med Ctr, 71-75; FREE LANCE WRITER ENVIRON TOXICOL, 75- Concurrent Pos: Res assoc, Univ Calif, Los Angeles, 58-63; consult environ toxicol, 72- Mem: Environ Defense Fund; Soil & Health Found. Res: The toxicology of pesticides and environmental pollutants; carcinogenic, mutagenic and other adverse effects on human health of millions of chemicals in our food, air and water. Mailing Add: 6982 Wildlife Rd Malibu CA 90265

TAYLOR, CLAYBORNE D, b Kokomo, Miss, July 15, 38; m 63; c 3. ELECTROMAGNETICS. Educ: Miss State Univ, BS, 61; NMex State Univ, MS, 64, PhD(physics), 65. Prof Exp: Staff mem, Sandia Corp, 65-67; from asst prof to assoc prof physics, Miss State Univ, 67-71; prof elec eng, Univ Miss, 71-72; prof elec eng, 72-74, PROF ELEC ENG & PHYSICS, MISS STATE UNIV, 74- Concurrent Pos: Assoc mem US Nat Comt, Int Union Radio Sci. Mem: Am Phys Soc; Inst Elec & Electronics Engrs. Res: Field and antenna theories; electromagnetic boundary value problems. Mailing Add: Dept of Physics Box 5167 Miss State Univ Mississippi State MS 39762

TAYLOR, CONSTANCE ELAINE SOUTHERN, b Washington, DC, Nov 14, 37; m 59; c 3. ECOLOGY, SYSTEMATIC BOTANY. Educ: Univ Okla, BS, 59, MS, 61, PhD(plant ecol & syst bot), 75. Prof Exp: Teacher pub schs, Okla, 63-64; instr, 70-74, ASST PROF BIOL, SOUTHEASTERN OKLA STATE UNIV, 74- Mem: Am Soc Plant Taxonomists; Nat Wildlife Fedn. Res: The genus Solidago, or goldenrods, Oklahoma vascular plants and chemotaxonomy of bacteria. Mailing Add: Dept of Biol Southeastern Okla State Univ Durant OK 74701

TAYLOR, D DAX, b Chicago, Ill, Oct 10, 37; m 61; c 3. ANATOMIC PATHOLOGY, CLINICAL PATHOLOGY. Educ: Amherst Col, AB, 59; Univ Mo-Columbia, MD, 63; Am Bd Path, dipl & cert anat path & clin path, 68. Prof Exp: Resident path, Sch Med, Univ Mo-Columbia, 63-68, instr, 68-69, asst prof path & asst dean sch med, 69-72; ASSOC PROF PATH & ASSOC DEAN MED EDUC, SCH MED, SOUTHERN ILL UNIV, SPRINGFIELD, 72- Mem: Asn Am Med Cols; Am Soc Clin Path. Res: Medical education, student and curriculum evaluation; platelet patho-physiology. Mailing Add: Southern Ill Univ Sch of Med PO Box 3926 Springfield IL 62708

TAYLOR, DALE FREDERICK, b Evansville, Ind, June 16, 44; m 67; c 2. ELECTROCHEMISTRY. Educ: Univ Toronto, BSc, 66, MSc, 68, PhD(chem), 70. Prof Exp: Staff scientist battery res, 70-73, mgr personnel admin, 74, STAFF SCIENTIST, MAT SCI & ENG SECTOR, GEN ELEC CORP RES & DEVELOP, 75- Mem: Electrochem Soc. Res: Growth kinetics and structural properties of anodic oxide films on aluminum; corrosion of boiling water reactor structural materials. Mailing Add: Gen Elec Res & Develop Ctr PO Box 8 Schenectady NY 12301

TAYLOR , DALE L, b Studley, Kans, Sept 30, 36; m 58; c 3. ECOLOGY. Educ: Kans State Univ, BS, 58, MS, 60; Univ Wyo, PhD(ecol), 69. Prof Exp: Instr zool, Kans State Univ, 60-61; instr biol, Augustana Col, 61-64; instr zool, Univ Wyo, 64-68; asst prof, Millikin Univ, 68-70; ASSOC PROF & HEAD BIOL, STERLING COL, 70- Concurrent Pos: Collabr, Nat Park Serv, 70-74, res biologist, 74-75. Mem: Ecol Soc Am; Am Inst Biol Sci; Am Soc Mammal. Res: Biotic succession following forest fires in Yellowstone and Grand Teton National Parks. Mailing Add: Dept of Biol Sterling Col Sterling KS 67579

TAYLOR, DAVID COBB, b Portland, Maine, June 7, 39; m 71. ANALYTICAL CHEMISTRY. Educ: Bowdoin Col, AB, 61; Wesleyan Univ, MA, 63; Univ Conn, PhD(chem), 71. Prof Exp: Asst prof, 68-72, ASSOC PROF CHEM, SLIPPERY ROCK STATE COL, 72- Mem: AAAS; Am Chem Soc; fel Am Inst Chemists; Nat Sci Teachers Asn. Res: Electroanalytical chemistry and multicomponent systems. Mailing Add: Dept of Chem Slippery Rock State Col Slippery Rock PA 16057

TAYLOR, DAVID LAWRENCE, b Trenton, NJ, Feb 22, 36; m 58; c 4. PAPER CHEMISTRY, CHEMICAL ENGINEERING. Educ: Princeton Univ, BSChE, 57; Lawrence Univ, MS, 59, PhD(polymer phys chem), 62. Prof Exp: ST PROJ MGR, MONSANTO CO, 62- Mem: Tech Asn Pulp & Paper Indust; Am Chem Soc. Res: Paper chemical additives and coatings; water soluble polymers; polyelectrolytes. Mailing Add: Monsanto Co 800 N Lindbergh Blvd St Louis MO 63166

TAYLOR, DAVID LEROY, chemistry, see 12th edition

TAYLOR, DAVID RUXTON FRASER, b Scotland; Can citizen; c 3 GEOGRAPHY. Educ: Univ Edinburgh, MA, 60, PhD(geog), 66; Harvard Univ, Cert Comput Mapping, 67. Prof Exp: Educ officer, Govt Kenya, 61-66; from asst prof to assoc prof, 66-75, PROF GEOG, CARLETON UNIV, 75- Concurrent Pos: Brit Coun lectr, 66; Int Develop Res Ctr grant, Univ Nairobi & Carleton Univ, Muranga Prov, Kenya, 72-73. Res: Geography of Africa; growth centre concept and its application to rural development in Kenya; application and development of computer mapping techniques to both Canadian and African statistics and other areas. Mailing Add: Dept of Geog Carleton Univ Ottawa ON Can

TAYLOR, DAVID WARD, b Chesterfield, Eng, Aug 18, 38; m 65; c 2. THEORETICAL PHYSICS, SOLID STATE PHYSICS. Educ: Oxford Univ, BA, 61, MA, 65, PhD(physics), 65. Prof Exp: Mem tech staff, Bell Tel Labs, NJ, 64-67; asst prof, 67-72, ASSOC PROF THEORET SOLID STATE PHYSICS, McMASTER

UNIV, 72- Mem: Can Phys Soc; Am Phys Soc. Res: Dynamics of disordered crystals; calculations of phonons and phonon dependent properties of metals and alloys. Mailing Add: Dept of Physics McMaster Univ Hamilton ON Can

TAYLOR, DEE C, b Coalville, Utah, Dec 28, 22; m 41; c 2. ANTHROPOLOGY, ARCHAEOLOGY. Educ: Univ Utah, BA, 51, MA, 53; Univ Mich, PhD(anthrop), 57. Prof Exp: From asst prof to assoc prof, 57-67, PROF ANTHROP, UNIV MONT, 67- Concurrent Pos: Horace H Rackham fel, Univ Mich, 57; anthrop consult, Proj Head Start, 65. Mem: Fel Am Anthrop Asn; Soc Am Archaeol; Am Ethnol Soc. Res: North American prehistoric archaeology and historic site archaeology; anthropological theory; race and minorities. Mailing Add: Dept of Anthrop Univ of Mont Missoula MT 59801

TAYLOR, DENNIS RILEY, organic chemistry, see 12th edition

TAYLOR, DERMOT BROWNRIGG, b Ireland, Mar 30, 15; US citizen; m 45, 65; c 1. PHARMACOLOGY. Educ: Trinity Col, Dublin, MD, 37, MB, BCh & BAO, 38. Prof Exp: Asst physiol, Trinity Col, Dublin, 38-39; lectr, King's Col, Univ London, 39-45, lectr pharmacol, 45-50; assoc prof, Univ Calif, San Francisco, 50-53; chmn dept, 53-68, PROF PHARMACOL, UNIV CALIF, LOS ANGELES, 53- Concurrent Pos: Univ London traveling fel, Yale Univ, 48. Mem: Am Soc Pharmacol & Exp Therapeut; Brit Physiol Soc; The Chem Soc; Brit Biochem Soc; Brit Pharmacol Soc. Res: Mode of action of neuromuscular blocking agents. Mailing Add: 6982 Wildlife Malibu CA 90265

TAYLOR, DIANE, b Covina, Calif. TROPICAL MEDICINE. Educ: Univ Hawaii, BA, 68, MS, 70, PhD(zool), 75. Prof Exp: Instr biol, Univ Hawaii, 70-73, FEL TROP MED, SCH MED, UNIV HAWAII, 75- Res: Cell mediated immune studies of parasitic infections with special emphasis on malaria and trematodes. Mailing Add: Leahi Hosp 3675 Kilauea Ave Dept of Trop Med & Med Microbiol Honolulu HI 96816

TAYLOR, DONALD CURTIS, b London, Ky, June 16, 39; m 67; c 1. MATHEMATICS. Educ: Univ Ky, BS, 61, MS, 64, PhD(math), 67. Prof Exp: Elec eng, Westinghouse Elec Corp, 61-62; assoc prof math, Univ Mo-Columbia, 67-73; ASSOC PROF MATH, MONT STATE UNIV, 73- Concurrent Pos: Fel, La State Univ, Baton Rouge, 69-70. Mem: Am Math Soc. Res: Functional analysis. Mailing Add: Dept of Math Mont State Univ Bozeman MT 59715

TAYLOR, DONALD FRANCIS, b Woodstock, Ill, Jan 24, 03; m 29; c 1. INORGANIC CHEMISTRY. Educ: Univ Ill, BS, 25. Prof Exp: Chief chemist, 25-59, tech dir metal prods div, 59-66, sr scientist, 66-68, CONSULT, FRANSTEEL METALL CORP, 68- Mem: Am Chem Soc; Electrochem Soc; Am Soc Metals. Res: Chemistry and metallurgy of refractory metals and refractory metal carbides; electrodeposition of metals. Mailing Add: 928 Rogers Ct Waukegan IL 60085

TAYLOR, DONALD FULTON, b Charles Town, WVa, July 10, 32; m 52; c 5. PATHOBIOLOGY, BIOCHEMISTRY. Educ: Shepherd Col, BA, 57; Johns Hopkins Univ, DSc(pathobiol), 70. Prof Exp: Chmn biol dept high sch, Pa, 57-65; PROF SCI, CHEYNEY STATE COL, 70-, PROG DIR BIOMED RES, 71-, COORDR, HEALTH RELATED PROFESSIONS, 70-72 &, 74- Concurrent Pos: Dir & counsr, Cheyney Div, Cheyney-Hahnemann Med Prog, 70; vis prof, Hahnemann Med Col & Hosp, 71-; consult new access routes to med careers, Am Fedn Negro Affairs, 71. Mem: AAAS; Am Pub Health Asn. Res: Amino acid utilization of normal and rous sarcoma virus infected chick embryo fibroblast cultures. Mailing Add: Health Related Professions Box 135 Cheyney State Col Cheyney PA 19319

TAYLOR, DONALD JAMES, b Dayton, Ohio, Mar 6, 33; m 55; c 2. ASTRONOMY. Educ: Calif Inst Technol, BS, 55, MS, 58; Univ Wis, PhD(astron), 63. Prof Exp: Proj assoc space astron lab, Univ Wis, 63-65; asst prof astron, Univ Ariz, 65-71; ASSOC PROF PHYSICS, UNIV NEBR, LINCOLN, 71- Mem: Int Astron Union; Am Astron Soc; Astron Soc Pac. Res: Planetary astronomy; astronomical instrumentation; pulsars; applications of area scanning to astronomical problems. Mailing Add: Behlen Lab of Physics Univ of Nebr Lincoln NE 68508

TAYLOR, DONALD RUDOLPH, JR, b Philadelphia, Pa, Nov 17, 26. ELECTRONICS, BIOMEDICAL ENGINEERING. Educ: Drexel Inst Technol, BS in EE, 50. Prof Exp: Proj engr, Philco Corp, 49-57, res specialist, 59-67; sr res scientist, Krusen Ctr Res & Eng, 67-72, DIR ENG SERV, REHAB ENG CTR, MOSS REHAB HOSP, 72- Concurrent Pos: Adj prof, Sch Med, Temple Univ, 67- Res: Prosthetics and orthotics; rehabilitation engineering; circuit and patent research; analog and digital circuits and systems; pattern recognition; electro-optical systems; myoelectric signal processing and control; cybernetic systems. Mailing Add: 322 W Willowgrove Ave Philadelphia PA 19118

TAYLOR, DONALD STINSON, b Albany, Ore, June 1, 14; m 38. CHEMISTRY. Educ: Linfield Col, BA, 35; Calif Inst Technol, PhD(chem), 38. Prof Exp: Res chemist, Shell Develop Co, Calif, 38-41; res chemist, E I du Pont de Nemours & Co, 41-44, lab supvr, 44-45; res chemist, Pac Coast Borax Co, 45-50, res engr, 50-54, dir indust res, 55-56, US Borax & Chem Corp, 56-57, vpres res, 57, vpres & gen mgr, US Borax Res Corp, 58-59, vpres tech, 59-69, vpres financial, 69-75, DIR, US BORAX & CHEM CORP, 69-, SPEC CONSULT, 75- Mem: Am Chem Soc; Am Inst Chem. Res: Processing of inorganic chemicals; chemistry of boron compounds. Mailing Add: 800 San Rafael Terr Pasadena CA 91105

TAYLOR, DOROTHY JANE, b Waco, Tex. BIOLOGY. Educ: Rice Univ, BA, 43; Iowa State Univ, MS, 47; George Washington Univ, PhD(biol), 57. Prof Exp: Tech asst, Synthetic Rubber Lab, Humble Oil Co, 43-45; lab instr zool, biol & physiol, Iowa State Univ, 45-47; parasitologist, Lab Trop Dis, NIH, 47-58, biologist, 58-75, CHIEF, BREAST CANCER PROG COORD BR, DIV CANCER BIOL & DIAG & EXEC SECY, BREAST CANCER TASK FORCE, NAT CANCER INST, NIH, 75- Mem: Fel AAAS; Am Asn Cancer Res; Sigma Xi. Res: Experimental biology of breast cancer; malaria and amebiasis; in vitro cultivation; experimental chemotherapy; nutritional aspects; endocrine tumors; host-tumor biology and therapy. Mailing Add: Breast Cancer Coord Br NCI Landow Bldg NIH Bethesda MD 20014

TAYLOR, DOUGLAS HIRAM, b Doddsville, Miss, Dec 15, 39; m 61. ETHOLOGY, ECOLOGY. Educ: Univ Dayton, BS, 66, MS, 68; Miss State Univ, PhD(zool), 70. Prof Exp: NSF fel, Univ Notre Dame, 70-71; asst prof, 71-75, ASSOC PROF ZOOL, MIAMI UNIV, 75- Mem: AAAS; Am Soc Zoologists; Animal Behav Soc; Ecol Soc Am. Res: Animal behavior, ecology and orientation; agonistic behavior; behavioral aspects of ecology of vertebrates. Mailing Add: Dept of Zool Miami Univ Oxford OH 45056

TAYLOR, DUANE FRANCIS, b Iowa City, Iowa, Sept 30, 25; m 50; c 7. DENTAL MATERIALS. Educ: Univ Mich, BSE, 49, MSE, 50; Georgetown Univ, PhD(biochem), 61. Prof Exp: Head dept dent mat, Sch Dent, Wash Univ, St Louis, 50-54; phys metallurgist, Dent Sect, Nat Bur Stand, 54-61; dir mat res, CMP Industs, 61-63; prof dent, Sch Dent, 63-74, PROF OPER DENT-DENT SCI, DENT RES CTR, UNIV NC, CHAPEL HILL, 74- Concurrent Pos: Consult, US Army, 67- & NIH, 69- Mem: AAAS; Am Soc Metals; Int Asn Dent Res. Res: Dental amalgams; cobalt-chromium alloys; denture base materials; polymerization mechanisms; properties of multiphase solids; materials for implant prosthesis. Mailing Add: Dent Res Ctr Univ of NC Chapel Hill NC 27514

TAYLOR, DWIGHT WILLARD, b Pasadena, Calif, Jan 18, 32. INVERTEBRATE ZOOLOGY. Educ: Univ Mich, BS, 53; Univ Calif, MA, 54, PhD, 57. Prof Exp: Geologist, US Geol Surv, DC, 55-67; assoc prof zool, Ariz State Univ, 67-69; res assoc malacol, San Diego Mus Natural Hist, 69-74; PROF ZOOL, PAC MARINE STA, UNIV PAC, 74- Mem: AAAS; Soc Syst Zool; Soc Study Evolution; Am Malacol Union. Res: Freshwater mollusks, ecology, taxonomy and distribution; biogeography. Mailing Add: Pac Marine Sta Dillon Beach CA 94929

TAYLOR, EDWARD ALAN, b Richmond, Va, Apr 23, 43. ORGANIC CHEMISTRY. Educ: Univ NC, Chapel Hill, BS, 65; Ga Inst Technol, PhD(org chem), 70. Prof Exp: SR DEVELOP CHEMIST, CIBA-GEIGY CHEM CORP, 70- Mem: Am Chem Soc. Res: Heterocyclic chemistry; light stabilizers; sequestering agents; triazine synthesis; catalytic hydrogenation; nuclear magnetic resonance spectroscopy. Mailing Add: Develop Dept CIBA-Geigy Chem Corp McIntosh AL 36553

TAYLOR, EDWARD CHRISTOPHER, theoretical physics, see 12th edition

TAYLOR, EDWARD CURTIS, b Springfield, Mass, Aug 3, 23; m 46; c 2. ORGANIC CHEMISTRY. Educ: Cornell Univ, AB, 46, PhD(org chem), 49. Hon Degrees: DSc, Hamilton Col, 69. Prof Exp: Merck fel, Zurich Tech Univ, 49-50; Du Pont fel, Univ Ill, 50-51, instr chem, 51-53, asst prof org chem, 53-54; from asst prof to assoc prof, 54-64, PROF CHEM, PRINCETON UNIV, 64-, A BARTON HEPBURN PROF ORG CHEM, 66-, CHMN DEPT CHEM, 74- Concurrent Pos: Consult, Procter & Gamble Co, 52-; NSF sr fac fel, Harvard Univ, 59; Fulbright scholar, 60; vis prof inst org chem, Stuttgart Tech Univ, 60; vis lectr, Weizmann Inst, 60; mem, Chem Adv Comt, Air Force Off Sci Res, 62-; consult, Eastman Kodak Co, 65-; distinguished vis prof, Univ Buffalo, 69; ed org chem, Wiley Intersci, Inc, 69-; vis prof, Univ E Anglia, 69 & 72; consult, Eli Lilly & Co, 70- & Tenn Eastman Co, 71-; H J Backer lectr, Univ Groningen, 71; ed, Advances in Org Chem; co-ed, Gen Heterocyclic Chem & Chem of Heterocyclic Compounds; mem ed adv bd, J Org Chem, Synthetic Commun, Heterocycles & Chem Substruct Index, 73- Honors & Awards: Res Awards, Smith Kline & French Found, 55, Hoffmann-La Roche Found, 64 & 65, S B Penick Found, 69, 70, 71 & 72 & Ciba Pharmaceut Co, 71; Creative Work Award, Am Chem Soc, 74. Mem: AAAS; Am Chem Soc; fel NY Acad Sci; The Chem Soc; Ger Chem Soc. Res: Organic synthesis; heterocyclic chemistry, particularly pyrimidines, purines and pteridines; organothallium chemistry; natural products; photochemistry. Mailing Add: Frick Chem Lab Princeton Univ Princeton NJ 08540

TAYLOR, EDWARD DONALD, b Clifton, Tex, Sept 30, 40; m 65; c 1. PHYSICAL CHEMISTRY. Educ: Univ Tex, Austin, BS, 63; Tex Tech Univ, PhD(chem), 67. Prof Exp: Asst prof & fel chem, Fla State Univ, 67-68; PROF CHEM, ODESSA COL, 68- Concurrent Pos: Fel, Tex A&M Univ, 71. Honors & Awards: Eastman Fel, Eastman Kodak Co, 68. Mem: Am Chem Soc. Res: Investigations of thermoluminescence via electron paramagnetic resonance to understand the mechanism of the process. Mailing Add: Odessa Col Box 3752 Odessa TX 79760

TAYLOR, EDWARD GODFREY, b Pembroke Dock, Wales, Aug 8, 10; nat US; m 39; c 3. PHYSICAL CHEMISTRY. Educ: Univ Wales, BSc, 32, PhD(phys chem), 35; Brown Univ, ScM, 38. Prof Exp: Asst, Univ Col, Wales, 34-36; from lectr to asst prof chem, Queen's Univ, Can, 38-45; chief chemist & dir res, Sheffield Tube Corp, 45-46; from asst prof to prof, 46-70, EBENEZER FITCH EMER PROF CHEM, WILLIAMS COL, 70- Mem: Am Chem Soc; assoc Royal Inst Chem. Res: Conductance studies; electrolytes. Mailing Add: Erinsville ON Can

TAYLOR, EDWARD MORGAN, b Rapid City, SDak, Dec 27, 33; m 58; c 2. GEOLOGY, PETROLOGY. Educ: Ore State Univ, BS, 57, MS, 60; Wash State Univ, PhD(geol), 67. Prof Exp: Instr geol, Ore State Univ, 62-63 & Wash State Univ, 64-65; asst prof, 66-71, ASSOC PROF GEOL, ORE STATE UNIV, 71- Mem: Geol Soc Am; Mineral Soc Am. Res: Volcanic petrology of Cascade Range of California, Oregon and Washington. Mailing Add: Dept of Geol Ore State Univ Corvallis OR 97331

TAYLOR, EDWARD STEWART, b Hecla, SDak, Aug 20, 11; m 40; c 3. OBSTETRICS & GYNECOLOGY. Educ: Univ Iowa, BA, 33, MD, 36; Am Bd Obstet & Gynec, dipl, 46. Prof Exp: PROF OBSTET & GYNEC & HEAD DEPT, MED CTR, UNIV COLO, DENVER, 47- Concurrent Pos: Dir, Am Bd Obstet & Gynec, 60-69; consult, Surgeon Gen, US Dept Air Force; ed, Obstet & Gynec Surv J; coun mem, Nat Inst Child Health & Human Develop. Mem: Am Gynec Soc; Am Asn Obstet & Gynec; Am Col Surg; Am Col Obstet & Gynec. Res: Cancer of the cervix; physiology of pregnancy. Mailing Add: Dept of Obstet & Gynec Univ of Colo Med Ctr Denver CO 80220

TAYLOR, EDWARD WYLLYS, b Richmond, Va, Mar 1, 27; m 50; c 3. CHEMISTRY, PSYCHOLOGY. Educ: Hampden-Sydney Col, BS, 49. Prof Exp: Res chemist, Va-Carolina Chem Corp, 49-55, asst plant supt, 55-58, asst plant mgr, 58-59, proj mgr, 59-61; exec vpres, Gulf Design Corp, 61-72; pres, Conserv, Inc, 72-74, PRES, PINE-LAKE CHEM, INC, 74- Mem: NY Acad Sci; Am Inst Chem; Am Inst Chem Eng; Am Chem Soc. Res: Fluorine recovery and utilization; phosphate processing; uranium recovery; fertilizer manufacturing; air pollution control; plant design, engineering and construction. Mailing Add: Conserv Inc Number 1 Lone Palm Pl Lakeland FL 33801

TAYLOR, EDWIN FLORIMAN, b Oberlin, Ohio, June 22, 31; m 55; c 3. PHYSICS. Educ: Oberlin Col, AB, 53; Harvard Univ, MA, 54, PhD(physics), 57. Prof Exp: Asst prof physics, Wesleyan Univ, 56-64; vis assoc prof, 64-66, SR RES SCIENTIST, EDUC RES CTR, MASS INST TECHNOL, 66- Mem: Am Phys Soc; Am Asn Physics Teachers. Res: Solid state physics; educational writing and research in mechanics, special relativity and quantum physics; computer-assisted learning. Mailing Add: 86 Oxford Rd Newton Centre MA 02159

TAYLOR, EDWIN WILLIAM, b Toronto, Ont, June 8, 29; m 56; c 3. BIOPHYSICS. Educ: Univ Toronto, BA, 52; McMaster Univ, MSc, 55; Univ Chicago, PhD(biophys), 57. Prof Exp: Instr physics, Ont Agr Col, 52-53; res assoc biol, Mass Inst Technol, 57-59; from instr to prof biophys, 59-74, PROF BIOPHYS & THEORET BIOL, UNIV CHICAGO, 74- Mem: Biophys Soc. Res: Mechanochemical systems; muscle; flagella; protoplasmic streaming; mechanism of cell division; physical protein chemistry. Mailing Add: Dept of Biophys Univ of Chicago 5640 Ellis Ave Chicago IL 60637

TAYLOR, ELIZABETH DANIELS, organic chemistry, biochemistry, see 12th edition

TAYLOR, ELLA RICHARDS, b Cumberland, Md, Apr 2, 25; m 52; c 6.

ANALYTICAL CHEMISTRY. Educ: Wheaton Col, Ill, BS, 47; Univ Ill, PhD(chem), 52. Prof Exp: Chemist, Celanese Corp Am, 47-48; asst, Univ Ill, 48-52; teacher high sch, Ark, 52-53; ANAL CHEMIST, TAYLOR SEED FARMS, INC, 53- Mem: Sigma Xi. Res: Preservation of meat; determination of calcium; separation of red from white rice. Mailing Add: Taylor Seed Farms Inc Hickory Ridge AR 72347

TAYLOR, ELLISON HALL, b Kalamazoo, Mich, Sept 6, 13; m 38; c 2. PHYSICAL CHEMISTRY. Educ: Cornell Univ, BChem, 35; Princeton Univ, MA, 37, PhD(phys chem), 38. Prof Exp: Instr chem, Univ Utah, 38-40; instr chem eng, Cornell Univ, 40-42; res chemist, Div War Res, Columbia Univ, 42-45; res chemist, Clinton Labs, Tenn, 45-48, asst dir chem div, 46-48; actg dir chem div, 48, prob leader, 48-49, from assoc dir to dir chem div, 49-51, asst dir res, 51-54, dir chem div, 54-75, SR RES STAFF MEM, OAK RIDGE NAT LAB, 75- Concurrent Pos: Vis prof, Cornell Univ, 65. Mem: AAAS; Am Chem Soc; Am Phys Soc. Res: Heterogeneous catalysis; chemical problems related to isotope separations; radiation chemistry; chemical kinetics; molecular beams in chemistry. Mailing Add: Chem Div Oak Ridge Nat Lab PO Box X Oak Ridge TN 37830

TAYLOR, ELMORE HECTOR, b Melfort, Sask, Sept 17, 31; m 58; c 2. PHARMACOGNOSY, PHYTOCHEMISTRY. Educ: Univ Sask, BSP, 56, MS, 57; Purdue Univ, PhD(pharmacog), 60. Prof Exp: Res asst, Sask Res Coun, 56-57; asst, Purdue Univ, 57-60; asst prof pharmacog & actg chmn dept, Col Pharm, Univ Tenn, Memphis, 60-62, assoc prof & chmn dept, 62-68; vis prof, Univ London, 69-70; PROF PHARMACOG, UNIV ILL MED CTR, 71-, ASST DEAN CURRICULAR AFFAIRS, 72- Mem: Am Asn Cols Pharm; Am Soc Pharmacog (vpres, 67-68, pres, 68-69); Acad Pharmaceut Sci. Res: Microbiological transformations; biosynthesis; microbial chemistry; phytochemistry; enzymatic synthesis of natural drug products. Mailing Add: Dept Pharmacog & Pharmacol Col of Pharm Univ Ill Med Ctr PO Box 6998 Chicago IL 60680

TAYLOR, EUGENE ALFRED, b Fitchburg, Mass, Sept 15, 26; m 51; c 1. GEODESY. Educ: Univ Mass, BS, 50; Ohio State Univ, MS, 68. Prof Exp: Mem hydrographic & oceanog oper, US Coast & Geod Surv, 50-52 & 58-61, mem tech staff, Geod Div, 53-58, in-chg data acquisition phase, Satellite Triangulation Div, 61-64, chief satellite triangulation div, Environ Sci Serv Admin, US Coast & Geod Surv, 64-69, commanding officer, Ship Pathfinder, 69-70, chief opers & requirement div, 71; ASSOC DIR OFF FLEET OPERS, NAT OCEAN SURV, NAT OCEANIC & ATMOSPHERIC ADMIN, 72- Honors & Awards: Gold Medal, Dept Com, 65. Mem: Am Geophys Union; Am Soc Photogram. Res: Satellite triangulation; astronomy. Mailing Add: C7 Off Fleet Oper Nat Ocean Surv Nat Oceanic & Atmospheric Admin Rockville MD 20853

TAYLOR, EUGENE EMERSON, b Moscow, Idaho, Mar 13, 21; m 44; c 3. PSYCHIATRY, PUBLIC HEALTH. Educ: Univ Idaho, BS, 42; Washington Univ, MD, 45; Univ NC, MPH, 50. Prof Exp: Instr, Sch Pub Health, Univ NC, 51-54; dir commun & chronic dis, Louisville-Jefferson County Bd Health, Ky, 54-56; psychiat resident, NC Mem Hosp, 58-61; asst prof psychiat, Med Sch, Univ Ore, 61-64; psychiatrist, Multnomah County Health Div, Portland, 64-65; dir community serv sect, Ore Ment Health Div, 65-69; STAFF PSYCHIATRIST, MULTNOMAH COUNTY MENT HEALTH CLIN, 69- Res: Community psychiatry. Mailing Add: Albina Multiserv Ctr 5022 N Vancouver Ave Portland OR 97217

TAYLOR, FLETCHER BRANDON, JR, b Aug 24, 29; m 54; c 4. INTERNAL MEDICINE. Educ: Stanford Univ, BS, 52; Univ Calif, San Francisco, MD, 56. Hon Degrees: MS, Univ Pa, 71. Prof Exp: Intern, Southern Pac Hosp, San Francisco, 56-57, resident surg, 57-58; res assoc protein chem, London Hosp, Eng, 58-59; mem res staff thrombosis, Cardiovasc Res Inst, San Francisco, 59-65; from asst prof to assoc prof med, Hosp Univ Pa, Philadelphia, 65-74, co-chmn div allergy & immunol, 68-74; PROF PATH & MED & DIR DIV EXP PATH & MED, HEALTH SCI CTR, UNIV OKLA, 74- Concurrent Pos: Resident med, Univ Calif Hosp, San Francisco, 59-62; consult hemat, NASA Manned Space Flight Ctr, 68-; mem thrombosis coun, Am Heart Asn, 71- Honors & Awards: Cochems Prize Cardiovasc Res, 68; Louis Pasteur Lectr Award, Univ Paris, 69. Mem: Int Soc Thrombosis & Haemostasis; Am Asn Immunol; Am Soc Clin Invest; Am Physiol Soc; Am Fedn Clin Res. Res: Thrombosis and protein chemistry. Mailing Add: Div of Exp Path & Med Univ of Okla Health Sci Ctr Oklahoma City OK 73190

TAYLOR, FLOYD HECKMAN, b North Versailles, Pa, May 6, 26; m 55; c 2. MATHEMATICS. Educ: Bucknell Univ, BS, 49; Univ Pittsburgh, MS, S3, ScD, 64. Prof Exp: Asst proj engr math, Sperry Gyroscope Co, NY, 53-54; instr, Hood Col, 54-56; mathematician, Atlantic Div, Aerojet Gen Corp Div, Gen Tire & Rubber Co, 56-57 & US Army Biol Labs, 57-64; asst prof biostatist, Grad Sch Pub Health & asst prof prev med, Sch Med, 64-70, assoc prof community med, 70-75, RES PROF COMMUNITY MED, SCH MED, UNIV PITTSBURGH, 75- Concurrent Pos: Instr, Frederick Community Col, 57-61. Mem: Math Asn Am; Am Statist Soc; Biomet Soc. Res: Numerical analysis; applications of mathematics to digital computers; mathematical models as applied to biology; biostatistics. Mailing Add: Sch of Med Univ of Pittsburgh Pittsburgh PA 15213

TAYLOR, FLOYD LEO, mathematics, see 12th edition

TAYLOR, FRANCIS B, b New York, NY, June 15, 25. MATHEMATICS. Educ: Manhattan Col, AB, 44; Columbia Univ, AM, 47, PhD(math educ), 59. Prof Exp: From instr to assoc prof, 47-65, head dept, 64-72, PROF MATH, MANHATTAN COL, 65- Concurrent Pos: Lectr, Col Mt St Vincent, 53-56; NSF fac fel, 57-58; lectr, Sch Gen Studies, Hunter Col, 59-60. Mem: Math Asn Am; Am Statist Asn; Inst Math Statist; Sigma Xi. Res: Mathematical statistics. Mailing Add: Dept of Math Manhattan Col Bronx NY 10471

TAYLOR, FRANK JOHN RUPERT (MAX), b Cairo, Egypt, July 17, 39; m 63; c 2. MARINE BIOLOGY. Educ: Univ Cape Town, BSc, 59, Hons, 60, PhD(marine bot), 65. Prof Exp: Res asst marine phytoplankton, Inst Oceanog, Univ Cape Town, 60-64; from asst prof to assoc prof, 65-75, PROF BIOL & OCEANOG, INST OCEANOG, UNIV BC, 75- Concurrent Pos: Can-France Exchange fel, France, 72; vis scientist, Phuket Marine Biol Ctr, Thailand, 73. Mem: Am Soc Limnol & Oceanog; Marine Biol Asn UK. Res: Taxonomy and distributional ecology of unicellular marine organisms, principally diatoms and dinoflagellates; undergraduate biology; marine phytoplankton ecology. Mailing Add: Inst of Oceanog Univ of BC Vancouver BC Can

TAYLOR, FRED HERBERT, b Groton, Mass, Apr 27, 10; m 40, 71; c 3. BOTANY. Educ: Mass State Col, BS, 33; Harvard Univ, AM, 34, PhD(bot), 39. Prof Exp: Asst bot, Radcliffe Col, 35; asst biol, Harvard Univ, 36-39; from instr to asst prof bot, Clemson Agr Col, 39-43; from asst prof to assoc prof, 43-55, plant morphologist, Exp Sta, USDA, 43-62, PROF BOT, UNIV VT, 55- Concurrent Pos: Res fel biol, Harvard Univ, 63-64. Mem: AAAS; Bot Soc Am; Int Asn Wood Anat. Res: Anatomy of vascular plants; anatomy and physiology of the sugar maple; economic botany. Mailing Add: Dept of Bot Univ of Vt Burlington VT 05401

TAYLOR, FRED M, b Chanute, Kans, Aug 21, 19; m 42; c 5. PEDIATRICS. Educ: Stanford Univ, AB, 41, MD, 44. Prof Exp: From instr to prof pediat, Baylor Col Med, 48-69; prof, Univ Tex Med Sch San Antonio, 69-72, from asst dean to assoc dean acad develop, 69-72; PROF PEDIAT, BAYLOR COL MED, 72-, DIR, OFF CONTINUING EDUC, 72- Mem: AAAS; Am Acad Pediat; Soc Res Child Develop. Res: Infant nutrition and feeding; developmental behavior of infants, children and adolescents, including behavioral, learning and intellectual handicaps. Mailing Add: Off of Continuing Educ Baylor Col of Med Houston TX 77025

TAYLOR, FRED WILLIAM, b Springcreek, WVa, Jan 17, 32; m 54; c 4. WOOD SCIENCE, WOOD TECHNOLOGY. Educ: Va Polytech Inst, BS, 53; NC State Col, MWT, 54, PhD(wood sci & technol), 65. Prof Exp: Asst prof forestry, Univ Vt, 54-55; asst timber buyer, J B Belcher Lumber Co, 55-56; wood technologist, Pulaski Veneer & Furniture Co, 56-59; res asst, NC State Univ, 59-62; wood utilization exten specialist, Univ Mo, 63-65; PROF WOOD SCI & TECHNOL & ASST DIR, MISS FORESTRY PROD UTILIZATION LAB, MISS STATE UNIV, 65- Concurrent Pos: Coun Sci & Indust Res grant, SAfrica, 71-72. Mem: Forest Prod Res Soc; Soc Wood Sci & Technol; Int Asn Wood Anatomists; Inst Wood Sci. Res: Natural variation in the anatomical structure of angiosperm xylem, especially genetic implications of property variations. Mailing Add: Miss Forest Prod Lab PO Drawer FP Mississippi State MS 39762

TAYLOR, FREDERICK HENRY CARLYLE, zoology, see 12th edition

TAYLOR, G JEFFREY, b Port Jefferson, NY, June 27, 44; m 65; c 3. GEOCHEMISTRY, PETROLOGY. Educ: Colgate Univ, AB, 66; Rice Univ, MA, 68, PhD(geol), 70. Prof Exp: Smithsonian Res Found res fel, lunar mineral & petrol, Smithsonian Astrophys Observ, 70-72, res assoc, 72-73; asst prof, Wash Univ, 73-76; SR RES ASSOC, INST METEORITICS, UNIV NMEX, 76- Concurrent Pos: Assoc, Harvard Univ, 70-; vis scientist, Lunar Sci Inst, 74-76. Honors & Awards: Nininger Meteorite Prize, Ctr Meteorite Studies, Ariz State Univ, 69. Mem: Am Geo- phys Union; Geochem Soc; Meteoritical Soc. Res: Petrologic and chemical nature of the moon, meteorite and earth, with emphasis on their origins and thermal histories. Mailing Add: Dept of Geol Univ of NMex Albuquerque NM 87131

TAYLOR, GARY N, b Plainfield, NJ, Oct 19, 42; m 64; c 3. ORGANIC CHEMISTRY. Educ: Princeton Univ, BA, 64; Yale Univ, MS, 66, PhD(org chem), 68. Prof Exp: Air Force Off Sci Res fel, Calif Inst Technol, 68-69; MEM TECH STAFF, BELL LABS, 69- Mem: Am Chem Soc. Res: Synthetic organic chemistry; photochemistry; polymer chemistry; radiation chemistry; lithography. Mailing Add: Bell Labs Murray Hill NJ 07974

TAYLOR, GEORGE ALLAN, b Hinton, Okla, June 18, 37; m 60; c 2. PLANT BREEDING, GENETICS. Educ: Utah State Univ, MS, 65; Iowa State Univ, PhD(plant breeding), 67. Prof Exp: Assoc secy, Crop Qual Coun, 67-69; Mont Wheat Res & Mkt Comt grant, 69-76, ASSOC PROF AGRON & GENETICS, MONT STATE UNIV, 69- Mem: Am Soc Agron; Crop Sci Soc Am; Am Inst Biol Sci. Res: Plant breeding, genetics and plant physiology as related to improvement of plant species. Mailing Add: Dept of Plant & Soil Sci Mont State Univ Bozeman MT 59715

TAYLOR, GEORGE FREDERIC, b Denver, Colo, Mar 6, 08; m. COMPUTER SCIENCE, METEOROLOGY. Educ: Calif Inst Technol, BS, 29, MS, 31, PhD(geol), 33. Prof Exp: Geologist, Union Oil Co Calif, 29-30; jr geologist, US Geol Surv, 30-31; meteorologist, Western Air Express, 33-35; chief meteorologist, Am Airlines, 35-37 & Western Air Express Corp, 37-39; instr, Air Corps Tech Sch, US Army, 39-42; dispatcher, Trans World Airlines, Inc, 46-48, author, 53-54; staff scientist & mgr flight data processing, Lockheed Missiles & Space Co, 55-61, mgr info anal ctr, 61-65, resident mgr, Houston Info Systs, 65, asst gen mgr Houston Aerospace Systs Div, Lockheed Electronics Co, 66, MGR COMPUT CTR, LOCKHEED MISSILES & SPACE CO, SUNNYVALE, 67- Concurrent Pos: Asst, Calif Inst Technol, 29-33; spec adv, US AEC. Mem: AAAS; Am Meteorol Soc; assoc fel Am Inst Aeronaut & Astronaut; Am Geophys Union. Res: Data processing and analysis. Mailing Add: 40 Hillbrook Dr Portola Valley CA 94025

TAYLOR, GEORGE RUSSELL, b Pittsburgh, Pa, Aug 31, 23; m 48; c 1. PHYSICAL CHEMISTRY. Educ: Carnegie Inst Technol, DSc(chem), 51. Prof Exp: Group leader, Goodyear Synthetic Rubber Plant, Tex, 48-49; fel, Mellon Inst, 49-50, sr fel, 51-55; adv scientist, supvry engr & mgr chem, Atomic Power Dept, 55-66, MGR CHEM, ADV REACTORS DIV, WESTINGHOUSE ELEC CORP, 67- Mem: AAAS; Am Chem Soc; Am Nuclear Soc; Am Soc Testing & Mat. Res: Quantum chemistry; elastomers; birefringence; crystallization, viscoelasticity and strength; colloid chemistry; high temperature water chemistry; liquid sodium chemistry. Mailing Add: Advan Reactors Div Westinghouse Elec Corp Madison PA 15663

TAYLOR, GEORGE STANLEY, b Jackson, NC, Nov 29, 20; m 47; c 3. SOIL PHYSICS. Educ: NC State Col, BS, 43, MS, 48; Iowa State Univ, PhD(soil physics), 50. Prof Exp: Asst, NC State Col, 47-48 & Iowa State Univ, 49-50; from asst prof to assoc prof, 51-61, PROF AGRON, OHIO STATE UNIV, 61- Concurrent Pos: Vis assoc prof, Univ Calif, 58-59; consult, US Agency Int Develop, Punjab, India, 68 & 70. Mem: Am Soc Agron; Soil Sci Soc Am. Res: Water flow in porous media; land drainage; two-dimensional modeling of heat and water flow in porous media by numerical techniques; disposal wastes in soil. Mailing Add: Dept of Agron Ohio State Univ 1883 Neil Ave Columbus OH 43210

TAYLOR, GEORGE THOMAS, b Asheboro, NC, July 18, 35. DEVELOPMENTAL BIOLOGY, ZOOLOGY. Educ: Guilford Col, AB, 57; Univ NC, Chapel Hill, MA, 64; Univ Mass, Amherst, PhD(zool), 70. Prof Exp: Instr zool, Atlantic Christian Col, 62-65; asst prof human anat, Col Osteop Med & Surg, 69-73; ASST PROF PHYSIOL, SOUTHERN ILL UNIV, 73- Mem: AAAS; Am Soc Cell Biol; Am Soc Zoologists; Soc Develop Biol; Am Ornithologists Union. Res: Cytochemical and cytological aspects of oocyte differentiation and early development in marine invertebrates; special interest in changes in subcellular morphology and function during embryonic cytodifferentiation. Mailing Add: Dept of Physiol Life Sci II Southern Ill Univ Carbondale IL 62901

TAYLOR, GERALD C, b Oregon, Mo, Sept 10, 19; m 50; c 4. BACTERIOLOGY, VIROLOGY. Educ: Univ Kans, BA, 49, MA, 51, PhD(bact), 55. Prof Exp: Lab instr bact & virol, Univ Kans, 50-52, asst rickettsiae, 52-55; asst, Lab Br, Commun Dis Ctr, USPHS, 55-57, lab supvr & unit chief, Tissue Cult Unit, 57-71, LAB SUPVR & UNIT CHIEF, LAB BR, CTR DIS CONTROL, USPHS, 71- Honors & Awards: Commendation Medal, USPHS, 72. Mem: AAAS; Am Soc Microbiol; Sigma Xi. Res: Human and animal virology; development of tissue culture in the field of virology. Mailing Add: Tissue Cult Unit Ctr Dis Control 1600 Clifton Rd Atlanta GA 30333

TAYLOR, GERALD REED, JR, b Bloxom, Va, Apr 17, 37; m 60; c 2. PHYSICS. Educ: Va Polytech Inst & State Univ, BS, 59, MS, 61; Univ Va, PhD(physics), 67. Prof Exp: Assoc physicist, Texaco Exp Inc, 60-62; res physicist solid state physics, Linde Div, Union Carbide Corp, 67-69; ASSOC PROF PHYSICS, MADISON COL,

VA, 69- Honors & Awards: J Shelton Horsley Res Award, Va Acad Sci, 62. Mem: Am Phys Soc. Res: Low Temperature solid state physics; ferromagnetism; amorphous materials; magneto-thermal conductivity; resistivity, other transport properties; physics education and instruction; application of computers to physics instruction. Mailing Add: Dept of Physics Madison Col Harrisonburg VA 22801

TAYLOR, GOERGE ALBERT, horticulture, plant physiology, see 12th edition

TAYLOR, GORDON, b Ladner, BC, May 9, 23; m 47; c 2. GEOGRAPHY. Educ: Univ BC, BA, 49, MA, 50. Prof Exp: Asst geog, Univ Minn, 50-52; res officer recreation planning, Nat Parks Br, 54-68; asst chief planning, Nat Parks Serv, 68-72; chief res & develop div, Travel Indust Br, 72-74; ASST DIR RES & DEVELOP, CAN GOVT OFF TOURISM, 74- Concurrent Pos: Chmn res comt, Fed Prov Parks Conf, 64-; sessional lectr, Dept Geog, Carleton Univ, Ottawa, 74. Mem: Travel Res Asn (pres, 76); Can Asn Geographers. Res: Recreation geography; population geography, regional planning; economic, social and environmental impacts of leisure land use. Mailing Add: 12 Mulvagh Ave Ottawa ON Can

TAYLOR, GORDON STEVENS, b Danbury, Conn, Nov 12, 21; m 46; c 1. PLANT PATHOLOGY. Educ: Univ Conn, BS, 47; Iowa State Col, MS, 49, PhD(corn stalk rot), 52. Prof Exp: Asst corn leaf blight & stalk rot, Iowa State Col, 48-52; asst agr scientist, 52-53, assoc agr scientist, Valley Lab, 53-55, agr scientist, 55-60, CHIEF AGR SCIENTIST, VALLEY LAB, CONN AGR EXP STA, 60- Mem: AAAS; Am Phytopath Soc; Am Inst Biol Sci; Air Pollution Control Asn. Res: Helminthosporium leaf blight of corn; stalk rot of corn in relation to the corn borer; verticillium wilt of potatoes; tobacco diseases; nematodes; effects of air pollution on plants. Mailing Add: Conn Agr Exp Sta Valley Lab PO Box 248 Windsor CT 06095

TAYLOR, HAROLD ALLISON, JR, b Richmond, Va, Oct 18, 42; m 65; c 2. HUMAN GENETICS, BIOCHEMISTRY. Educ: Univ Tenn, BS, 65, MS, 67, PhD(zool), 71. Prof Exp: Fel pediat, Sch Med, Johns Hopkins Univ, 71-73, instr, 73-75; asst dir, Genetics Lab, John F Kennedy Inst, Baltimore, 73-75; DIR LABS, GREENWOOD GENETIC CTR, 75- Mem: AAAS; Am Soc Human Genetics. Res: Lysosomal storage diseases; inborn errors of metabolism; genetics of mental retardation. Mailing Add: Greenwood Genetics Ctr Genet Lab 1020 Spring St at Ellenberg Greenwood SC 29646

TAYLOR, HAROLD ERNEST, b St Johns, Nfld, May 6, 13; m 36; c 2. PATHOLOGY. Educ: Dalhousie Univ, MD, 36; FRCP(C), 54; FRCP(E), 56. Hon Degrees: LLD, Dalhousie Univ, 68. Prof Exp: Asst prof path, Dalhousie Univ, 40-42 & 45-47; dir path. Shaughnessy Hosp, Vancouver, 47-49; from assoc prof to prof & head dept, Univ BC, 51-68; dir awards prog, Med Res Coun Can, 68-75; PROF PATH & HEAD DEPT, UNIV OTTAWA, 75- Concurrent Pos: Assoc dir path, Vancouver Gen Hosp, 49-54, exec dir path, 54-68; consult, BC Cancer Inst, 54-68; mem, Med Res Coun Can, 60-66. Mem: Am Asn Path & Bact; Am Soc Exp Path; Int Acad Path; Can Med Asn. Res: Transplantation immunology; anti-lymphocyte serum and cellular immunity. Mailing Add: Dept of Path Univ of Ottawa Fac of Med Ottawa ON Can

TAYLOR, HAROLD EVANS, b Philadelphia, Pa, Sept 13, 39; m 64; c 3. PLASMA PHYSICS, SPACE PHYSICS. Educ: Haverford Col, BA, 61; Mass Inst Technol, MS, 62; Univ Iowa, PhD(physics), 66. Prof Exp: Res assoc physics, Univ Iowa, 66; Nat Acad Sci assoc, Goddard Spaceflight Ctr, NASA, 66-68; res assoc plasma physics, Princeton Univ, 68-71; ASST PROF ASTROPHYS, STOCKTON STATE COL, 71- Concurrent Pos: Consult, Los Alamos Sci Labs, 67. Mem: AAAS; Fedn Am Sci; Int Solar Energy Soc; Am Phys Soc; Am Geophys Union. Res: Space plasma physics, including physics of the magnetosphere; physics of the interplanetary medium; cosmic rays; solar and wind energy devices. Mailing Add: Fac of Natural Sci & Math Stockton State Col Pomona NJ 08240

TAYLOR, HAROLD LELAND, b Cambridge, Kans, May 3, 20; m 43; c 1. BIOCHEMISTRY. Educ: Southwestern Col, Kans, AB, 42; Univ Kans, PhD(biochem), 55. Hon Degrees: DSc, Southwestern Col, Kans, 73. Prof Exp: Res assoc, Harvard Univ, 42-46; biochemist, State Health Dept, Mich, 46-51 & 53-56; mgr immunochem dept, Pitman-Moore Div, Dow Chem Co, 56-63, tech asst to dir biol labs, 63-64, res chemist, Dow Res Labs, Zionsville, 64-72, clin monitor, Med Dept, 72-74, CLIN INVESTR, DOW CHEM CO, 74- Mem: Am Chem Soc; Sigma Xi; NY Acad Sci; Am Soc Clin Pharmacol & Therapeut. Res: Fractionation of human blood plasma; physiological effects of antithyroid drugs in rats; isolation of immune fraction from hyperimmune canine plasma; control of biologicals production; drug metabolism; serum cholesterol lowering drugs. Mailing Add: Dow Chem Co PO Box 68511 Indianapolis IN 46268

TAYLOR, HAROLD MELLON, organic chemistry, agricultural chemistry, see 12th edition

TAYLOR, HAROLD NATHANIEL, b Baltimore, Md, May 18, 21; m 42; c 3. CHEMISTRY. Educ: Johns Hopkins Univ, BE, 42, MS, 45; Cornell Univ, PhD(phys chem), 49. Prof Exp: Jr chem engr, Tenn Valley Authority, Wilson Dam, Ala, 42-43; jr instr chem eng, Johns Hopkins Univ, 43-44, asst, Off Rubber Reserve, 44, & Manhattan Proj, SAM Labs, Columbia Univ, 44-45; res chemist, Manhattan Proj, Carbide & Carbon Chem Co, 45-46 & Ammonia Dept, Exp Sta, E I du Pont de Nemours & Co, 49-53; PRES, HAGERSTOWN LEATHER GOODS CO, 53- Concurrent Pos: Dir, Cent Chem Co, 70- & Antietam Bank Co, 74- Mem: AAAS; Am Chem Soc. Res: Polymer chemistry; polymerization; plastic applications. Mailing Add: 1877 Fountain Head Rd Hagerstown MD 21740

TAYLOR, HARRY WILLIAM, b Saskatchewan, Can, Sept 28, 25; m 49; c 2. NUCLEAR PHYSICS. Educ: Univ Man, BSc, 51, MSc, 52, PhD(physics), 54. Prof Exp: Lectr physics, Univ Man, 52-53; fel, Cosmic Ray Sect, Nat Res Coun Can, 54-55; asst prof physics, Queen's Univ, Ont, 55-61; assoc prof, Univ Alta, 61-65; assoc prof, 65-71, PROF PHYSICS, UNIV TORONTO, 71- Mem: Fel AAAS; Fel Am Phys Soc; fel Brit Inst Physics; fel Inst Nuclear Eng. Res: Gamma-ray spectroscopy; angular correlations of successive gamma-rays. Mailing Add: Dept of Physics Univ of Toronto Toronto ON Can

TAYLOR, HENRY LONGSTREET, b St Paul, Minn, Feb 2, 12; m 41; c 3. PHYSIOLOGY. Educ: Harvard Univ, AB, 35; Univ Minn, PhD(physiol), 42. Prof Exp: Asst, 40-42, res assoc, 42-44, from asst prof to assoc prof, 44-56, actg dir lab physiol hyg, 51-52, PROF PHYSIOL HYG, UNIV MINN, MINNEAPOLIS, 56- Concurrent Pos: Mem coun arteriosclerosis, Am Heart Asn. Mem: AAAS; Am Physiol Soc; Soc Exp Biol & Med; Geront Soc; Am Heart Asn. Res: Exercise; nutrition; semi-starvation; cardiovascular function in man; epidemiology of heart disease. Mailing Add: Gate 27 Stadium Lab Physiol Hyg Sch of Pub Health Univ of Minn Minneapolis MN 55455

TAYLOR, HERBERT BRADLEY, b Pensacola, Fla, July 31, 29; m 52; c 2. SURGICAL PATHOLOGY. Educ: George Washington Univ, BS, 50, MD, 54. Prof Exp: Sr pathologist, Armed Forces Inst Path, 58-70; PROF PATH, MED CTR, ST LOUIS UNIV, 70-, CHMN DIV GEN PATH, 73-, CHIEF SURG PATH, HOSP, 70- Concurrent Pos: Consult, Obstet & Gynec Adv Comt, Food & Drug Admin, 68-71, Med Lab 5th US Army, 71 & Nat Cancer Inst, 71-; mem epidemiol subcomt, Nat Breast Cancer Task Force, 71- Honors & Awards: Commendation Award, Vet Admin, 67. Mem: Fel Am Soc Clin Path; assoc fel Am Col Obstet & Gynec; Int Acad Path. Res: Breast diseases; gynecologic pathology; oncology. Mailing Add: Dept of Path St Louis Univ Med Ctr St Louis MO 63104

TAYLOR, HERBERT CECIL, b Houston, Tex, Nov 29, 24; m 48; c 2. ANTHROPOLOGY. Educ: Univ Tex, BA & MA, 49; Univ Chicago, PhD(anthrop), 51. Prof Exp: Asst, Univ Chicago, 49-51; from asst prof to assoc prof, 51-62, chmn dept sociol & anthrop, 60-65, PROF ANTHROP, WESTERN WASH STATE COL, 62-, DEAN RES & GRANTS, 65- Concurrent Pos: Vis asst prof, Univ BC, 54. Mem: Fel Am Anthrop Asn. Res: Social change; Southwestern archaeology; Northwest Coast ethno-history. Mailing Add: Dept of Anthrop-Sociol Western Wash State Col Bellingham WA 98225

TAYLOR, HOWARD CANNING, JR, b New York, NY, Feb 17, 00; m 24; c 3. OBSTETRICS & GYNECOLOGY. Educ: Yale Univ, PhB, 20; Columbia Univ, MD, 24. Hon Degrees: DSc, NY Univ, 55. Prof Exp: From asst to prof gynec & obstet, Col Med, NY Univ, 35-39; prof gynec, Sch Med, Univ Pa, 39-40; from asst to prof gynec & obstet, Col Med, NY Univ, 40-46; prof obstet & gynec & chmn dept, 46-65, dir, Int Inst Study Human Reprod, 65-70, EMER PROF OBSTET & GYNEC, COL PHYSICIANS & SURGEONS, COLUMBIA UNIV, 65-; SR CONSULT, POP COUN NEW YORK, 70- Concurrent Pos: From asst to attend gynecologist, Roosevelt Hosp, New York, 27-47; from asst to attend surgeon, Mem Hosp, 27-39 & 40-46; assoc & attend obstetrician & gynecologist, Bellevue Hosp, 35-39 & 40-46; dir, Sloane Hosp Women, 46-65; ed-in-chief, Am J Obstet & Gynec, Am Gynec Soc, 59-69. Mem: Fel Am Gynec Soc (secy, 41-46, pres, 57-58); Harvey Soc; Am Cancer Soc (pres, 54); fel Am Pub Health Asn; fel Am Asn Obstet & Gynec (pres, 66-67). Res: Cancer of female reproductive organs; physiology of pregnancy. Mailing Add: Box 328 Southport CT 06430

TAYLOR, HOWARD EDWARD, b Ft Worth, Tex, Feb 1, 22; m 44; c 3. MATHEMATICS. Educ: Rice Inst, BA, 42, MA, 48, PhD(math), 50; Calif Inst Technol, MS, 43. Prof Exp: Instr meteorol, Calif Inst Technol, 43-44; asst math, Rice Inst, 46-50; from instr to assoc prof, Fla State Univ, 50-69; CALLAWAY PROF MATH, WGA COL, 69- Concurrent Pos: Vis assoc prof, Univ Chicago, 56-57; assoc chmn dept math, Fla State Univ, 64-69; mem comt math exam, Col Entrance Exam Bd, 70-, chmn, 75-; mem, Nat Coun Teachers Math. Mem: Am Math Soc; Am Meteorol Soc; Math Asn Am. Res: Functions of a complex variable; analysis. Mailing Add: Dept of Math WGa Col Carrollton GA 30117

TAYLOR, HOWARD LAWRENCE, b Kansas City, Mo, May 23, 38; c 2. APPLIED MATHEMATICS, RESEARCH ADMINISTRATION. Educ: Austin Col, AB, 59; Univ Kans, MA, 62, PhD(math), 68. Prof Exp: Lab asst physics, Austin Col, 57-59; asst math, Univ Kans, 59-67; mathematician, 60-70, MGR RES, SUN OIL CO, 70- Mem: Math Asn Am; Soc Indust & Appl Math; Soc Petrol Engrs; Soc Explor Geophysicists. Res: Development of mathematical methods to analyze geophysical data and solve reservoir engineering problems; simulation models and inverse problems. Mailing Add: Sun Oil Co 503 N Cent Expressway Richardson TX 75080

TAYLOR, HOWARD MELVIN, b Pride, Tex, Jan 20, 24; m 48; c 2. SOIL SCIENCE. Educ: Tex Tech Col, BS, 49; Univ Calif, PhD, 57. Prof Exp: Soil scientist, Soil Conserv Serv, USDA, 49-54; res asst irrig, Univ Calif, 55-57, instr, 57; SOIL SCIENTIST, AGR RES SERV, USDA, 57- Concurrent Pos: FAO Andre Mayer fel, Australia, 65-66. Mem: Fel AAAS; fel Am Soc Agron; Soil Sci Soc Am; Am Soc Agr Eng; Soil Conserv Soc Am. Res: Soil physics, particularly soil-plant water relations and effects of soil compaction on plant growth. Mailing Add: Dept of Agron Iowa State Univ Ames IA 50010

TAYLOR, HOWARD MILTON, III, b Baltimore, Md, May 9, 37; m 64. APPLIED MATHEMATICS, STATISTICS. Educ: Cornell Univ, BME, 60, MIndustEng, 61; Stanford Univ, PhD(math, statist), 65. Prof Exp: Res assoc & lectr appl probability, Stanford Univ, 64-65; from asst prof to assoc prof appl probability, 65-74, ASSOC PROF OPER RES, CORNELL UNIV, 74-, GRAD FAC REP, 74- Concurrent Pos: NSF fel & vis asst prof, Univ Calif, Berkeley, 68-69; on leave at Math Inst, Oxford Univ, 72-73. Mem: Inst Math Statist; Am Statist Asn; Inst Mgt Sci; Opers Res Soc Am. Res: Applied probability; Markov sequential decision processes; optimal stopping problems; quality control. Mailing Add: Dept of Indust Eng & Oper Res Col of Eng Cornell Univ Ithaca NY 14850

TAYLOR, HOWARD S, b New York, NY, Sept 17, 35; m 59; c 2. CHEMISTRY, PHYSICS. Educ: Columbia Univ, BA, 56; Univ Calif, Berkeley, PhD(chem), 59. Prof Exp: NSF fel chem, Free Univ Brussels, 59-61; from asst prof to prof, 61-74, Humboldt prof, 74-75, PROF CHEM, UNIV SOUTHERN CALIF, 75- Concurrent Pos: Consult, Jet Propulsion Lab, Calif Inst Technol, 60- & NAm Aviation, Inc, 65-66; guest prof, Univ Freiburg, 67; staff scientist, Los Alamos Nat Lab, 73. Mem: Am Chem Soc; Am Phys Soc; The Chem Soc. Mailing Add: Dept of Chem Univ of Southern Calif Univ Park Los Angeles CA 90007

TAYLOR, HUGH ALAN, b New York, NY, Apr 26, 38; m 65; c 2. ANALYTICAL CHEMISTRY. Educ: NY Univ, AB, 59; Princeton Univ, MA, 62, PhD(anal chem), 63. Prof Exp: Res specialist, Chemstrand Res Ctr, 63-73; res group leader, 73-75, SR RES SPECIALIST, MONSANTO TRIANGLE PARK DEVELOP CTR, 75- Mem: Am Chem Soc. Res: Photochemistry; organic functional group analysis; ultraviolet and visible spectroscopy; physical testing of surface coatings and adhesives; preparation and testing of energy absorbing foams; gas and liquid chromatography. Mailing Add: Monsanto Triangle Park Devel Ctr Box 12274 Research Triangle Park NC 27709

TAYLOR, HUGH P, JR, b Holbrook, Ariz, Dec 27, 32; m 59. PETROLOGY, GEOCHEMISTRY. Educ: Calif Inst Technol, BS, 54, PhD(geochem), 59; Harvard Univ, AM, 55. Prof Exp: Asst prof geol, Calif Inst Technol, 59-61; asst prof geochem, Pa State Univ, 61-62; from asst prof to assoc prof, 62-69, PROF GEOL, CALIF INST TECHNOL, 69- Mem: AAAS; Geol Soc Am; Am Geophys Union; Mineral Soc Am. Res: Oxygen, hydrogen, carbon and silicon isotopic compositions of igneous and metamorphic minerals and rocks, meteorites and the moon; ore deposits and hydrothermal alteration; ultramafic rocks of southeast Alaska. Mailing Add: Div of Geol & Planetary Sci Calif Inst of Technol Pasadena CA 91109

TAYLOR, IAIN EDGAR PARK, b Chester, Eng, Aug 18, 38; m 67. PLANT PHYSIOLOGY, PLANT BIOCHEMISTRY. Educ: Univ Liverpool, BSc, 61, PhD(bot), 64. Prof Exp: Teacher, Blundell's Sch, Tiverton, Eng, 64-66; res assoc bot, Univ Tex, Austin, 67-68, vis asst prof, 68; asst prof, 68-71, ASSOC PROF BOT, UNIV BC, 71- Mem: Am Soc Plant Physiol; Can Bot Asn; Can Soc Plant Physiol; Brit Biochem Soc; Brit Inst Biol. Res: Plant and fungal cell walls and cell wall associated proteins; studies on nitrogen metabolism during germination of mosses. Mailing Add: Dept of Bot Univ of BC Vancouver BC Can

TAYLOR, ISAAC MONTROSE, b Morgantown, NC, June 15, 21; m 46; c 5. MEDICINE. Educ: Univ NC, AB, 42; Harvard Univ, MD, 45. Prof Exp: Intern, Mass Gen Hosp, 45-46, resident physician, 47; asst med adv, Harvard Med Sch, 48; chief med res, Mass Gen Hosp, 51; from asst prof to assoc prof, 52-64, dean sch med, 64-71, PROF MED, SCH MED, UNIV NC, CHAPEL HILL, 64- Concurrent Pos: Nat Res Coun fel, Harvard Med Sch, 48-50; Markle scholar, 54-61; manpower consult, Tristate Regional Med Prog, 71-72, assoc dir manpower, 72- Mem: Soc Exp Biol & Med; Am Chem Soc; AMA; Am Fedn Clin Res. Res: Metabolism of electrolytes. Mailing Add: Univ of NC Sch of Med Chapel Hill NC 27514

TAYLOR, JACK, b Tiger, Ga, Feb 23, 22; m 43. PLANT PATHOLOGY. Educ: Univ Ga, BSA, 50, MSA, 51; NC State Col, PhD(plant path), 57. Prof Exp: Asst plant pathologist, Exp Sta, Univ Ga, 50-55; asst, NC State Col, 55-57; from asst plant pathologist to plant pathologist, Exp Sta, 57-68, PROF PLANT PATH, UNIV GA, 68- Mem: Am Phytopath Soc; Mycol Soc Am. Res: Fungus and bacterial diseases of plants. Mailing Add: Dept of Plant Path & Plant Genet Univ of Ga Athens GA 30601

TAYLOR, JACK CROSSMAN, b Washington, DC, Nov 9, 24; m 52; c 2. ANIMAL HUSBANDRY. Educ: Va Polytech Inst, BS, 50, MS, 57. Prof Exp: Instr, Agr Exp Sta, Univ Va, 51-57; asst prof animal husb, Miss State Univ, 57-63; res geneticist, Swine Res Br, Animal Husb Res Div, Agr Res Serv, USDA, 63-67; SUPVR ANIMAL HUSB, FOOD & DRUG ADMIN, 67- Mem: Am Soc Animal Sci. Res: Evaluation of data for animal safety and efficacy of new animal drugs; development of standards and regulations for animal feed ingredients derived from animal, human & industrial wastes. Mailing Add: Bur of Vet Med Food & Drug Adm 5600 Fishers Ln Rockville MD 20852

TAYLOR, JACK DEAN, b Banff, Alta, Jan 13, 29; m 52; c 4. PHARMACOLOGY. Educ: Univ Alta, BS, 50, MSc, 52; Univ Sask, PhD(physiol, pharmacol), 57. Prof Exp: Lectr biochem, Univ Alta, 52-54, sessional instr pharmacol, 57-58, instr, 58-59, lectr, 59-60, from asst prof to assoc prof, 60-68, hon assoc prof, 68-72; dir sci opers, S Hanson & Assocs, 68-75; DIR SPEC HEALTH PROJS, GOVT OF ALTA, 75- Concurrent Pos: Biochemist, Misericordia Hosp, Edmonton, 52-54, asst dir dept lab med, 57-63, consult biochemist & pharmacologist, 63- Mem: Can Soc Clin Chemists (past pres); Can Soc Forensic Sci (past pres). Res: Distribution, excretion, metabolism of drugs; pathological chemistry. Mailing Add: 6703 Hardisty Dr Edmonton AB Can

TAYLOR, JACK ELDON, b Emporia, Kans, Jan 16, 26; m 48; c 2. PHYSICS. Educ: Univ Wis, PhB, 46, MS, 49, PhD(physics), 51. Prof Exp: Asst physics, Univ Wis, 47-51; res assoc, Gen Elec Co, 51-61; sr res staff mem, Gen Dynamics/Electronics, 61-71; PRIN ENGR, STROMBERG CARLSON CORP, 71- Mem: Am Phys Soc; Inst Elec & Electronics Eng; Optical Soc Am; Sigma Xi. Res: Ultra high vacuum techniques; gas lasers; optical communication systems; atmospheric optical transmission; solid state devices; LSI logic arrays; switching matrices; material and manufacturing problems. Mailing Add: 31 Old Pond Rd Rochester NY 14625

TAYLOR, JACK HOWARD, b Memphis, Tenn, July 7, 22; m 44; c 4. PHYSICS. Educ: Southwestern at Memphis, BS, 44; Johns Hopkins Univ, PhD(physics), 52. Prof Exp: Physicist, US Naval Res Lab, 44-46; instr physics, Southwestern at Memphis, 46-47; asst, Radiation Lab, Johns Hopkins Univ, 48-50; physicist, Exp Sta, E I du Pont de Nemours & Co, 52-53; asst prof physics, Univ of the South, 53-54; consult infrared, US Naval Res Lab, 54-56; assoc prof, 56-60, PROF PHYSICS, SOUTHWESTERN AT MEMPHIS, 60-, HEAD DEPT, 56-, PRIN INVESTR INFRARED, AIR FORCE CAMBRIDGE RES CTR CONTRACT, 59-, DIR LAB ATMOSPHERIC & OPTICAL PHYSICS, 64- Concurrent Pos: Consult, Electro-Optics Group, Pan Am World Airways, Inc & Patrick Air Force Base, Fla, 63- Mem: Fel AAAS; Am Phys Soc; fel Optical Soc Am; Am Asn Physics Teachers. Res: Physics and military applications of infrared; atmospheric physics and transmission in infrared; time-dependent infrared phenomena and infrared techniques. Mailing Add: Dept of Physics Southwestern at Memphis Memphis TN 38112

TAYLOR, JACK NEEL, b Jacksonville, Fla, Aug 15, 18; div; c 2. UROLOGY. Educ: Ohio State Univ, AB, 40; Harvard Univ, MD, 43; Am Bd Urol, dipl, 54. Prof Exp: Asst prof genito-urinary surg, 51-66, ASSOC PROF UROL, COL MED, OHIO STATE UNIV, 66- Mem: AMA; Am Urol Asn; fel Am Col Surgeons. Res: Clinical urology. Mailing Add: 1275 Olentagy River Rd Columbus OH 43212

TAYLOR, JACKSON JOHNSON, b Winnabow, NC, Nov 20, 18; m 51; c 3. PHYSICS. Educ: Univ Richmond, BS, 42; Cornell Univ, MS, 48. Prof Exp: From instr to asst prof, 48-56, chmn div sci, 57-72, ASSOC PROF PHYSICS, UNIV RICHMOND, 56-, CHMN DEPT, 51-54, 55-58 & 63- Concurrent Pos: Assoc prof, Sch Pharm, Med Col Va, 51-61. Mem: Am Phys Soc; Am Asn Physics Teachers. Res: Evaporation of chlorine atoms from silver chloride crystals; teaching undergraduate physics, especially lecture demonstrations and curriculum development. Mailing Add: Dept of Physics Univ of Richmond Richmond VA 23173

TAYLOR, JAMES A, JR, b Woonsocket, RI, 1939; m 60; c 3. BIOCHEMISTRY, PHARMACOLOGY. Educ: Providence Col, BS, 60; Purdue Univ, MS, 63, PhD(biochem), 66. Prof Exp: Resident res assoc, Div Biol & Med, Argonne Nat Lab, 65-67; SR RES SCIENTIST, MED RES LABS, PFIZER INC, 67- Mem: AAAS. Res: Drug metabolism and pharmacokinetics. Mailing Add: Cent Res Labs Pfizer Inc Groton CT 06340

TAYLOR, JAMES ADDISON, b Sidney, Ohio, Nov 24, 32; m 60; c 2. GEOGRAPHY. Educ: Kent State Univ, BA, 54; Univ Ill, MS, 56, PhD(geog), 62. Prof Exp: Instr geog, Bowling Green State Univ, 60-63; asst prof geog & head dept geog geol, Bradley Univ, 63-66, assoc prof geog & head dept geog & earth sci, 66-69, from asst to assoc dean, Col Lib Arts & Sci, 66-69; DEAN, COL LIB ARTS, ALFRED UNIV, 69- Mem: Asn Am Geogr. Res: Physical and economic geography, especially Anglo-America; industrial development and location; university academic administration. Mailing Add: Off of the Dean Col of Lib Arts Alfred Univ Alfred NY 14802

TAYLOR, JAMES EARL, b Beverly, Ohio, Sept 7, 16; m 45; c 3. RESEARCH ADMINISTRATION. Educ: Western Reserve Univ, AB, 38, MS, 40; Univ Pa, PhD(physics), 43. Prof Exp: Asst physics, Western Reserve Univ, 38-40 & Univ Pa, 40-42; physicist, Norden Labs Corp, 43-44, head res & develop sect, 44-50; from supvr electronics group & dir exp lab to lab dir, M Ten Bosch, Inc, 50-63; mem tech planning dept, 63-66, mgr res, Tech Planning Off, 66-69 & commun & educ, 69-70, TECH ASST, XEROX CORP, 70- Mem: Am Phys Soc; Am Ord Asn. Res: Ultrasonic studies; mass spectroscopy; isotope separation; ordnance research and development; research and development management; education. Mailing Add: 114 Xerox Corp Phillips Rd Webster NY 14580

TAYLOR, JAMES HERBERT, b Corsicana, Tex, Jan 14, 16; m 46; c 3. BIOLOGY. Educ: Southeastern State Col, BS, 39; Univ Okla, MS, 41; Univ Va, PhD(biol), 44. Prof Exp: Teacher high sch, Okla, 39-40; asst prof plant sci, Univ Okla, 46-47; assoc prof bot, Univ Tenn, 47-51; from asst prof to assoc prof, Columbia Univ, 51-58, prof cell biol, 58-64; PROF BIOL SCI, FLA STATE UNIV, 64-, ASSOC DIR, INST MOLECULAR BIOPHYS, 72- Concurrent Pos: Guggenheim Found fel, Calif Inst Technol, Pasadena, 59-60. Mem: Am Soc Cell Biol (pres, 70-71); Genetics Soc Am; Biophys Soc. Res: Culture of excised anthers; autoradiographic studies of nucleic acid and protein synthesis at the cellular level; chromosome duplication and structure; mechanisms of DNA replication in chromosomes; molecular organization of chromosomes. Mailing Add: Inst of Molecular Biophys Fla State Univ Tallahassee FL 32306

TAYLOR, JAMES KENNETH, b Fall River, Mass, July 28, 29; m 53; c 2. ECOLOGY. Educ: State Teachers Col Bridgewater, BScEd, 51; Columbia Univ, MA, 55. Prof Exp: Teacher pub schs, Mass, 51-56; from instr to asst prof, 56-64, ASSOC PROF BIOL, WESTFIELD STATE COL, 64-, CHMN DEPT BIOL, 74- Mem: AAAS; Nat Asn Biol Teachers; Ecol Soc Am; Nat Sci Teachers Asn; Am Forestry Asn. Res: Ecological basis of conservation; dissemination of ecological principles and their applications to solution of local problems. Mailing Add: Dept of Biol Sci Westfield State Col Westfield MA 01085

TAYLOR, JAMES LEE, b Berkey, Ohio, Jan 6, 31; m 58; c 2. HORTICULTURE. Educ: Ohio State Univ, BS, 53; Mich State Univ, MS, 57; Univ Ill, PhD(bot), 60. Prof Exp: PROF HORT & EXTEN SPECIALIST, MICH STATE UNIV, 60- Concurrent Pos: Chmn bd trustees, Nat Jr Hort Found, Inc. Mem: Am Hort Soc; Am Soc Hort Sci; Nat Jr Hort Asn. Res: Evaluation of vegetable varieties; nut tree culture and physiology; evaluation of Carpathian walnuts. Mailing Add: Dept of Hort Mich State Univ East Lansing MI 48824

TAYLOR, JAMES LESTER, b Knoxville, Tenn, Sept 26, 04; m 33; c 2. ORGANIC CHEMISTRY. Educ: Ohio Wesleyan Univ, AB, 29; Syracuse Univ, MS, 32; Univ NC, PhD(chem), 44. Prof Exp: Instr chem, 36-41, res dir flax, Exp Sta, 41-44, Ga Tech Alumni Found fel, 44-45, assoc prof textile chem, 45-47, prof, 47-58, dir, 58-74, EMER PROF TEXTILE ENG & EMER DIR, A FRENCH TEXTILE SCH, 74- Concurrent Pos: Mem, Nat Coun Textile Educ. Mem: Am Chem Soc; Am Soc Eng Educ; Am Asn Textile Chem & Colorists. Res: Chemical treatments of textile fibers and fabrics; cellulose chemistry; applications of various chemicals to warp sizing cotton; fiber blends; permanent press fabrics; fiber physics. Mailing Add: A French Textile Sch Ga Inst Technol Atlanta GA 30332

TAYLOR, JAMES ROBERT, b Dubuque, Iowa, Apr 3, 23; m 44; c 3. ANALYTICAL CHEMISTRY. Educ: Univ Dubuque, BS, 47. Prof Exp: Chemist, 48-52, res chemist, 52-65, supvr routine chem, 65-69, res chemist, 69-73, SECT HEAD ANAL SERV, SWIFT & CO, 73- Mem: Am Oil Chemists Soc; Am Chem Soc; Am Soc Testing & Mat. Res: Development of new analytical methods for food products and improving existing methods; thin layer chromatography, gas chromatography and atomic absorption spectroscopy. Mailing Add: Swift & Co 1919 Swift Dr Oak Brook IL 60521

TAYLOR, JAMES WELCH, b Newton, Miss, Sept 17, 35; m 57; c 2. ANALYTICAL CHEMISTRY. Educ: Vanderbilt Univ, BA, 56; Ga Inst Technol, MS, 58; Univ Ill, Urbana, PhD(chem), 64. Prof Exp: Develop chemist, Mobil Oil Co, 58-61; asst prof chem, Tulane Univ, La, 64-66; from asst prof to assoc prof, 66-73, PROF CHEM, UNIV WIS-MADISON, 73- Mem: Am Chem Soc; The Chem Soc. Res: Photoionization mass spectrometry; photoelectron spectroscopy; isotope kinetics; rates and mechanisms of reactions; analytical instrumentation. Mailing Add: Dept of Chem Univ of Wis 1101 University Ave Madison WI 53706

TAYLOR, JAMES WILLIAM, b Mansfield, La, Mar 17, 21; m 47; c 3. GEOGRAPHY. Educ: Northwestern State Col, La, BA, 46; La State Univ, MA, 48, PhD, 56. Prof Exp: Instr geog, Ind Univ, 49-50; from instr to assoc prof, 50-60, PROF GEOG, CALIF STATE UNIV, SAN DIEGO, 60- Mem: Asn Am Geog. Res: Human; historical. Mailing Add: Dept of Geog Calif State Univ San Diego CA 92182

TAYLOR, JAMES WOODALL, b Sumner Co, Tenn, Oct 25, 14; m 44; c 2. PHYSICAL GEOGRAPHY. Educ: Austin Peay State Col, BS, 47; Syracuse Univ, MA, 50; Ind Univ, PhD(geog), 55. Prof Exp: Instr geog, George Peabody Col, 50-51; from instr to asst prof, Memphis State Col, 53-56; from asst prof to prof, Wis State Univ, Eau Claire, 56-69; PROF GEOG, WESTERN KY UNIV, 69- Concurrent Pos: Consult, Audio-Visual Ctr, Ind Univ, 61-62; Wis State Univ grant, Univ Chicago, 65. Mem: AAAS; Asn Am Geog; Am Geog Soc; Nat Coun Geog Educ. Res: Climatology; geomorphology; regional geography of North America; geography of agriculture. Mailing Add: Dept of Geog Western Ky Univ Bowling Green KY 42101

TAYLOR, JAY EUGENE, b Stayton, Ore, Feb 2, 18; m 48; c 4. CHEMISTRY. Educ: Ore State Col, BA, 40; Univ Wis, MS, 43; Purdue Univ, PhD, 47. Prof Exp: Instr chem, Univ Wis, 46-48; asst prof org chem, Miami Univ, 48-52; fel phys chem, Ohio State Univ, 52-54; asst prof, Univ Nebr, 54-60; assoc prof phys-org chem, 60-63, PROF CHEM, KENT STATE UNIV, 63- Mem: Am Chem Soc. Res: Mechanisms of oxidation and decomposition reactions; flow techniques applied to kinetic studies of both liquid and gaseous systems; homogeneous versus heterogeneous gas-phase pyrolysis studies using a wall-less reactor. Mailing Add: Dept of Chem Kent State Univ Kent OH 44242

TAYLOR, JEAN MARIE, b Protection, NY, Nov 21, 32. TOXICOLOGY, PHARMACOLOGY. Educ: Keuka Col, BA, 54; Univ Rochester, MS, 56, PhD(pharm), 59. Prof Exp: Res assoc pharm, Atomic Energy Proj, Univ Rochester, 54-59; pharmacologist, 59-71, ACTG CHIEF, CHRONIC TOXICOL BR, DIV TOXICOL, US FOOD & DRUG ADMIN, 71- Mem: Soc Toxicol. Res: Toxicity of flavoring agents; hepatotoxins. Mailing Add: Div of Toxicol Food & Drug Admin Washington DC 20204

TAYLOR, JERRY DUNCAN, b Plumerville, Ark, June 5, 38; m 64; c 3. MATHEMATICS. Educ: State Col Ark, BA, 60; Univ Ark, MS, 64; Fla State Univ, PhD(math educ), 69. Prof Exp: ASSOC PROF MATH, CAMPBELL COL, NC, 62- Mem: Math Asn Am. Res: New methods of extending the field of rational numbers to the field of real numbers. Mailing Add: Dept of Math Campbell Col Buies Creek NC 27506

TAYLOR, JOCELYN MARY, b Portland, Ore, May 30, 31; m 72. ZOOLOGY. Educ: Smith Col, BA, 52; Univ Calif, Berkeley, MA, 53, PhD(zool), 59. Prof Exp: Asst zool, Conn Col, 53-54; Fulbright grantee, Australia, 54-55; assoc zool, Univ Calif, 59; from instr to asst prof, Wellesley Col, 59-65; assoc prof, 65-74, PROF ZOOL, UNIV BC, 74- Concurrent Pos: Sigma Xi grant, 61-62; Lalor Found grant, 62-63; NSF grants, 63-71 & Australia, 63-64 & 65; Nat Res Coun Can res grants, 66-77, travel grant, Div Animal Physiol, Commonwealth Sci & Indust Res Orgn, Australia, 71-72. Mem: AAAS; Am Soc Mammalogists; Soc Study Reprod; Am Soc Zool; Australian Soc Mammal. Res: Reproductive biology of mammals; evolution of Australasian murid rodents; marsupial placentation. Mailing Add: Dept of Zool Univ of BC Vancouver BC Can

TAYLOR, JOE SAMUEL, animal nutrition, animal physiology, see 12th edition

TAYLOR, JOHN CHRISTOPHER, b Chelmsford, Eng, Jan 17, 36; Can citizen; m 69. MATHEMATICS. Educ: Acadia Univ, BSc, 55; Queen's Univ, Ont, MA, 57; McMaster Univ, PhD(math), 60. Prof Exp: J F Ritt instr math, Columbia Univ, 60-63; from asst prof to assoc prof, 63-74, PROF MATH, MCGILL UNIV, 74- Mem: Am Math Soc; Can Math Cong. Res: Analysis; axiomatic and probabilistic potential theory. Mailing Add: Dept of Math McGill Univ Montreal ON Can

TAYLOR, JOHN DALLAS, b Hamilton, Ind, Jan 21, 40; m 64; c 1. GEOLOGY. Educ: Purdue Univ, Lafayette, BS, 62; Univ Ark, Fayetteville, MS, 64; Univ Iowa, PhD(geol), 72. Prof Exp: Prod geologist petrol, Humble Oil & Refining Co, Tex, 64-66; ASST PROF GEOL, UNIV ALA, 71- Mem: Geol Soc Am; Paleont Soc. Res: Mississippian and Pennsylvanian ammonoids; Mississippian conodonts; Mississippian stratigraphy. Mailing Add: Dept of Geol & Geog Univ of Ala University AL 35486

TAYLOR, JOHN DIRK, b Mecca, Calif, Mar 31, 39; m 63; c 2. DEVELOPMENTAL BIOLOGY. Educ: Univ Ariz, BS, 62, PhD(biol), 67. Prof Exp: NSF, US-Japan Coop Sci Prog fel biol, Keio Univ, Japan, 67-68; from asst prof to assoc prof, 68-75, PROF BIOL, WAYNE STATE UNIV, 75-, ASSOC PROF COMP MED, SCH MED, 72-, CHMN DEPT BIOL, 74- Concurrent Pos: Asian Found guest lectr, Univs in Seoul, Korea, 68. Mem: AAAS; Am Soc Cell Biol; Am Soc Zool; Soc Develop Biol. Res: Biochemical and ultrastructural aspects of developmental processes with particular emphasis on cell differentiation and cell death in embryonic development and cell renewal in adult organisms. Mailing Add: Dept of Biol Wayne State Univ Detroit MI 48207

TAYLOR, JOHN FULLER, b Jamestown, NY, June 10, 12; m 35; c 2. BIOCHEMISTRY. Educ: Cornell Univ, AB, 33; Johns Hopkins Univ, PhD(physiol chem), 37; Am Bd Clin Chem, dipl. Prof Exp: Nat Res Coun fel med sci, Harvard Med Sch, 37-39, teaching fel biol chem, 39-41; biochemist, Lederle Labs, Inc, 41-43; asst prof biol chem, Sch Med, Washington Univ, 43-52; chmn dept, 52-72, PROF BIOCHEM, SCH MED, UNIV LOUISVILLE, 52- Concurrent Pos: Vis prof, Univ Oslo, 55; Commonwealth Fund fel, Cambridge, Eng & Rome, Italy, 61-62. Mem: Am Chem Soc; Am Soc Biol Chemists; Soc Exp Biol & Med; Biophys Soc; Brit Biochem Soc. Res: Hemoglobin; enzyme proteins; quinones; hemochromogens; oxidation-reduction potentials of biological systems; sulfhydryl groups of proteins; physical chemistry of proteins and polysaccharides; ultracentrifugation. Mailing Add: Dept of Biochem PO Box 1055 Univ of Louisville Health Sci Ctr Louisville KY 40201

TAYLOR, JOHN GARDINER VEITCH, b Toronto, Ont, Sept 22, 26; m 55. PHYSICS. Educ: McMaster Univ, BSc, 50; Univ Sask, MSc, 52. Prof Exp: Res physicist photonuclear reactions, Univ Sask, 53; res physicist atomic mass measurements, McMaster Univ, 54-55; RES PHYSICIST RADIOACTIVITY STAND, ATOMIC ENERGY CAN, LTD, 56- Mem: Am Phys Soc; Can Asn Physicists. Res: Nuclear physics; radioactivity. Mailing Add: Physics Div Atomic Energy of Can Ltd Chalk River ON Can

TAYLOR, JOHN H, b Louisville, Ky, Mar 3, 11; m; c 1. ORGANIC CHEMISTRY. Educ: Lincoln Univ, BA, 30; Univ Pa, MS, 34; Univ Del, PhD(org chem), 53. Prof Exp: Asst chem, Howard Univ, 30-31; actg head sci dept, Howard High Sch, 32-61; res chemist, Eastern Lab, E I du Pont de Nemours & Co, Inc, 61 & Rapavno Develop Lab, 62; head sci & math dept, 63-71, PROF NATURAL SCI & MATH, CHEYNEY STATE COL, 74-, DIR, DIV NATURAL SCI & MATH, 71- Concurrent Pos: Res chemist, Jackson Lab, E I du Pont de Nemours & Co, 55 & Miscellaneous Intermediates Lab, 56-58; res chemist, Cent Res & Polymer Lab, Exp Sta, 60, Atlas Chem Industs, Inc, 57 & Hercules Powder Co, 59; consult, Apex Beauty Supplies, NJ, 48-53 & Gloplant Chem Co, NY, 53-57. Mem: AAAS; Am Chem Soc; Nat Sci Teachers Asn; fel Am Inst Chemists. Res: Shielding of the exocyclic and cyclic double bonds in the bicyclo-terpenes. Mailing Add: Div of Natural Sci & Math Cheyney State Col Cheyney PA 19319

TAYLOR, JOHN HALL, b Greenfield, Mass, Feb 18, 22; m 48; c 4. VISION. Educ: Wesleyan Univ, AB, 47; Univ Mich, MA, 50, PhD(psychol), 52. Prof Exp: Asst psychol, Univ Mich, 48-52, res assoc, 52-54; asst res psychologist, Scripps Inst Oceanog, 54-60, assoc res psychologist, 60-65, res psychologist, Visibility Lab, 65-72 & Inst Pure & Appl Phys Sci, 72-74, RES PSYCHOLOGIST, UNIV CALIF SAN DIEGO, 75- Concurrent Pos: Mem exec coun, Comt Vision, Armed Forces-Nat Res Coun, 71-72, staff sci adv, 72-75; consult, Int Comn Illum, 72- Mem: AAAS; Optical Soc Am; Asn Res Vision & Ophthal; Int Acad Astronaut. Res: Psychophysiology of vision sensory psychology; visual problems of space exploration; vision underwater; visual sensitivity in humans; photic responses of marine forms. Mailing Add: Dept of Psychol (C-009) Univ of Calif San Diego La Jolla CA 92093

TAYLOR, JOHN JACOB, b Dayton, Ohio, June 26, 28. MICROBIOLOGY. Educ: Heidelberg Col, BS, 50; Univ Ohio, MS, 52; Ohio State Univ, PhD, 57. Prof Exp: Instr bot, Univ Ohio, 50-52; res specialist toxicol, Nat Cash Register Co, 52-54; asst, Res Found, Ohio State Univ, 56-57; from asst prof to assoc prof, 57-69, PROF MICROBIOL, UNIV MONT, 69- Mem: Mycol Soc Am; Med Mycol Soc of the Americas; Can Soc Microbiol. Res: Morphology and physiology of fungi; host parasite relationships; antibiotic resistance in pathogenic fungi. Mailing Add: Dept of Microbiol Univ of Mont Missoula MT 59801

TAYLOR, JOHN JOSEPH, b Hackensack, NJ, Feb 27, 22; m 43; c 2. MATHEMATICS. Educ: St John's Univ, NY, AB, 42; Univ Notre Dame, MS, 47. Hon Degrees: DS, St John's Univ, NY, 74. Prof Exp: Appl mathematician, Bendix Aviation Corp, 46-47; scientist, Physics Dept, Kellex Corp, NY, 47-50; scientist, Bettis Atomic Power Lab, Westinghouse Elec Corp, 50-52, mgr shielding physics, 52-55 & surface ship proj, Physics Dept, 55-58, mgr reactor develop, 58-65 & mat develop, 65-67, mgr eng, Atomic Power Div, 67, eng mgr, PWR Plant Div, 67-70, gen mgr, Breeder Reactor Div, 70-74; VPRES, ADVAN NUCLEAR SYSTS, 74- Concurrent Pos: Consult, US Govt Acct Off, 74-; mem subcomt sci & pub policy, Nat Res Coun, 75- Honors & Awards: Westinghouse Award of Merit, 57. Mem: Nat Acad Eng; AAAS; Am Phys Soc; Am Nuclear Soc. Res: Nuclear reactors; shielding and reactors for atomic submarines; digital computer techniques in reactor design; nuclear fuel development. Mailing Add: Advan Nuclear Systs PO Box 355 Pittsburgh PA 15230

TAYLOR, JOHN JOSEPH, b Westhampton Beach, NY, May 15, 25; m 46; c 2. ANATOMY. Educ: Hofstra Col, BA, 53; Cornell Univ, MS, 56; Univ Buffalo, PhD(anat), 59. Prof Exp: Asst to Dr G N Papanicolaou, Med Col, Cornell Univ, 50-53, asst anat, 54-56; from asst to instr, Sch Med, Univ Buffalo, 56-60; from asst prof to assoc prof, Sch Med, Univ NDak, 60-66; ASSOC PROF ANAT, SCH MED, ST LOUIS UNIV, 66-, ASSOC PROF NEUROANAT IN NEUROL, 67- Concurrent Pos: Lectr, Drexel Inst, 62-64. Mem: Soc Neurosci; Am Asn Anatomists; Electron Micros Soc Am; Am Soc Cell Biol; Am Asn Hist Med. Res: Electron microscopy and histochemistry of connective tissue and central nervous system. Mailing Add: Dept of Anat St Louis Univ Sch of Med St Louis MO 63104

TAYLOR, JOHN KEENAN, b Mt Rainier, Md, Aug 14, 12; m 38; c 2. PHYSICAL CHEMISTRY, ANALYTICAL CHEMISTRY. Educ: George Washington Univ, BS, 34; Univ Md, MS, 38, PhD(chem), 41. Prof Exp: Sci aide, 29-36, chemist, 36-40, asst chemist, 40-42, assoc chemist, 42-44, chemist, 44-61, chief microchem anal sect, 61-73, CHIEF, AIR & WATER POLLUTION ANAL SECT, NAT BUR STANDARDS, 73- Concurrent Pos: Adj prof, Am Univ. Honors & Awards: Rosa Award Stand Activ, Nat Bur Standards, 74. Mem: Am Chem Soc; Electrochem Soc; Air Pollution Control Soc. Res: Optical properties and density of water; preparation of pure platinum metals; electrochemistry of solutions; standard electrode potentials; separation of isotopes; polarography; coulometry; microchemical and environmental analysis; development of analytical methods and standard reference materials for air and water pollution analysis. Mailing Add: B326 Chem Bldg Nat Bur of Standards Washington DC 20234

TAYLOR, JOHN LANGDON, JR, b Brooklyn, NY, Nov 3, 28; m 61. MEDICAL EDUCATION. Educ: Univ Chicago, PhB, 50; Univ Calif, Los Angeles, PhD(physiol), 57. Prof Exp: Grad res anatomist, Univ Calif, Los Angeles, 56-57, grad res physiologist, 57, asst res physiologist, 57-59; from instr to asst prof physiol, Col Med, Univ Ill, 59-65; assoc prof biol, Wayne State Univ, 65-68; assoc prof physiol & asst to dean col osteop med, Mich State Univ, 69-71, assoc prof med educ, Col Osteop Med & asst dean student affairs, Univ, 71-72, assoc prof acad affairs, Col Osteop Med, 72-75; PROF OSTEOP MED & ASSOC DEAN STUDENT AFFAIRS, COL OSTEOP MED, OHIO UNIV, 75- Mem: AAAS; Am Educ Res Asn; Nat Coun Measurement in Educ; Soc Gen Physiol; Biophys Soc. Res: Neuro-regulatory mechanism; biological rhythms; test scoring methods; career choices in the health professions. Mailing Add: Col of Osteop Med Ohio Univ Athens OH 45701

TAYLOR, JOHN LIPPINCOTT, invertebrate zoology, marine ecology, see 12th edition

TAYLOR, JOHN MITCHELL, biochemistry, see 12th edition

TAYLOR, JOHN ROBERT, b London, Eng, Feb 2, 39; m 62; c 2. THEORETICAL PHYSICS. Educ: Cambridge Univ, BA, 60; Univ Calif, Berkeley, PhD(physics), 63. Prof Exp: NATO fel physics, Cambridge Univ, 62-64; instr, Princeton Univ, 64-66; from asst prof to assoc prof, 66-72, PROF PHYSICS, UNIV COLO, BOULDER, 72- Mem: Am Phys Soc. Res: Quantum theory; quantum theory of scattering. Mailing Add: 3290 Heidelberg Dr Boulder CO 80303

TAYLOR, JOHN RONALD ERIC, poultry pathology, see 12th edition

TAYLOR, JOSEPH HOOTON, JR, b Philadelphia, Pa, Mar 29, 41; m 63; c 2. ASTRONOMY. Educ: Haverford Col, BA, 63; Harvard Univ, PhD(astron), 68. Prof Exp: Lectr astron & res assoc, Harvard Col Observ, Harvard Univ, 68-69; asst prof, 69-72, ASSOC PROF ASTRON, UNIV MASS, 72- Concurrent Pos: Grant, Res Corp, Univ Mass, 68-69, NSF, 69-72 & NASA, 71-72; consult, Mass Gen Hosp, 71- Mem: Am Astron Soc; Int Union Radio Sci; Int Astron Union. Res: Radio astronomy; pulsars; design and development of radio telescopes and information processing systems. Mailing Add: Dept of Physics & Astron Univ of Mass Amherst MA 01002

TAYLOR, JOSEPH LAWRENCE, b Apr 7, 41; US citizen; m 59; c 2. MATHEMATICAL ANALYSIS. Educ: La State Univ, BS, 63, PhD(math), 64. Prof Exp: Instr math, Harvard Univ, 64-65; from asst prof to assoc prof, 65-71, PROF MATH, UNIV UTAH, 71- Concurrent Pos: Sloan fel. Honors & Awards: Steele Prize, Am Math Soc, 75. Mailing Add: Dept of Math Univ of Utah Salt Lake City UT 84112

TAYLOR, JULIUS DAVID, b Erie, Pa, Dec 18, 13; m 38; c 1. BIOCHEMISTRY. Educ: Univ Pittsburgh, BS, 36; Univ Rochester, PhD(biochem), 40. Prof Exp: Instr biochem, Sch Med & Dent, Univ Rochester, 40-41; res org chemist, Distillation Prod, Inc, NY, 41-43; res biochemist, Eaton Labs, Inc, 43-47; chem pharmacologist, 48-60, from asst dept head to dept head pharmacol, 60-64, drug eval specialist, Dept Drug Regist, 65-69 & Div Regulatory Affairs, 69-70, DATA SPECIALIST, DEPT EXP BIOMET, ABBOTT LABS, 70- Concurrent Pos: Lectr, Med Sch, Northwestern Univ, Chicago, 59-62; prof, Sch Med, Univ Chicago, 62- Mem: AAAS; Am Chem Soc; Soc Exp Biol & Med; Am Soc Pharmacol & Exp Therapeut; Drug Info Asn. Res: Drug enzymology, metabolism and kinetics; pharmacology; toxicology; bionics; simulation. Mailing Add: Dept Exp Biomet Abbott Labs North Chicago IL 60064

TAYLOR, JULIUS HENRY, b NJ, Feb 15, 14; m 37; c 2. SOLID STATE PHYSICS. Educ: Lincoln Univ, Pa, AB, 38; Univ Pa, MA, 41, PhD(physics), 51. Prof Exp: Instr, WVa State Col, 45-46; res physicist, Univ Pa, 47-49; PROF PHYSICS & HEAD DEPT, MORGAN STATE COL, 50- Mem: Am Phys Soc; Am Asn Physics Teachers. Res: Pressure dependence of the electrical properties of semiconductors. Mailing Add: Dept of Physics Morgan State Col Baltimore MD 21239

TAYLOR, KATHLEEN C, b Cambridge, Mass, Mar 16, 42. PHYSICAL CHEMISTRY. Educ: Rutgers Univ, New Brunswick, AB, 64; Northwestern Univ, Evanston, PhD(phys chem), 68. Prof Exp: Fel, Univ Edinburgh, 68-70; assoc sr res chemist, 70-74, SR RES CHEMIST, RES LABS, GEN MOTORS CORP, 74- Mem: Am Chem Soc; Catalysis Soc; The Chem Soc. Res: Surface chemistry; heterogeneous catalysis; catalytic control of automobile exhaust emissions. Mailing Add: Phys Chem Dept Gen Motors Res Labs 12 Mile & Mound Roads Warren MI 48090

TAYLOR, KEITH MAR, b Thatcher, Ariz, Nov 15, 22; m 46; c 3. ORGANIC CHEMISTRY. Educ: Univ Ariz, BS, 45, MS, 47; Purdue Univ, PhD(chem), 50. Prof Exp: Instr chem, Univ Ariz, 45-46; asst instr, Purdue Univ, 47-48; res sect leader, 52-53, RES GROUP LEADER, RES CTR, MONSANTO CO, 53- Res: Natural products, composition of tree fractions; fluorinated organic nitrogen compounds; petrochemicals. Mailing Add: 349 Elmcrest Dr Ballwin MO 63011

TAYLOR, KENNETH BOIVIN, b Columbus, Ohio, Aug 7, 35; m 58; c 2. BIOCHEMISTRY, ENZYMOLOGY. Educ: Oberlin Col, AB, 57; Case Western Reserve Univ, MD, 61; Mass Inst Technol, PhD, 67. Prof Exp: Res assoc, Mass Inst Technol, 64-67, asst prof biol, 67-70; ASSOC PROF BIOCHEM, UNIV ALA, BIRMINGHAM, 70- Res: Enzyme mechanism specifically applied to oxygenases of the central nervous system and their relationships with behavioral states. Mailing Add: Dept of Biochem Univ of Ala Birmingham AL 35294

TAYLOR, KENNETH GRANT, b Paterson, NJ, May 12, 36; m 61; c 3. ORGANIC CHEMISTRY. Educ: Calvin Col, AB, 57; Wayne State Univ, PhD(org chem), 63. Prof Exp: Res assoc, Mass Inst Technol, 63-64; sr res assoc chem, Wayne State Univ, 64-66; from asst prof to assoc prof, 66-73, PROF CHEM, UNIV LOUISVILLE, 73- Concurrent Pos: Assoc prof, Univ Nancy I, France, 74-75. Mem: Am Chem Soc;

AAAS. Res: Synthesis of strained carbocyclic and heterocyclic compounds; chemistry of azoxy compounds; chemical carcinogenesis; cycloaddition reactions. Mailing Add: Dept of Chem Univ of Louisville Louisville KY 40208

TAYLOR, KENNETH MONROE, b Bethany, Ill, June 5, 16; m 44; c 2. GENETICS. Educ: Univ Calif, Los Angeles, AB, 40, MA, 42, PhD(zool), 48. Prof Exp: Asst zool, Univ Calif, Los Angeles, 40-41 & 46; instr, Univ Kans, 48-49; from asst prof to prof zool, 49-74, chmn dept, 54-57, PROF BIOL, CALIF STATE UNIV, SAN DIEGO, 74- Concurrent Pos: USPHS spec fel, Inst Cancer Res, Philadelphia, 66-67. Mem: AAAS; Am Soc Human Genetics. Res: Human genetics; cytogenetics. Mailing Add: Dept of Biol Calif State Univ 5402 College Ave San Diego CA 92115

TAYLOR, KENNETH ORLEN, b Porters, Tex, Sept 2, 39; m 65; c 3. COMPUTER SCIENCES. Educ: Univ Houston, BS, 67, MS, 74. Prof Exp: Drilling res technologist, 67-69, tech programmer, 69-70, syst engr, 70-73, develop eng supvr, 73-74, eng supvr, 74-75, PROJ MGR RES & DEVELOP, BAROID DIV, NL INDUSTS, 75- Mem: Soc Petrol Engrs. Res: Electrical, mechanical, and computer application directed to the oil well drilling environment. Mailing Add: Baroid Div NL Industs PO Box 1675 Houston TX 77001

TAYLOR, KIRMAN, b Yorkshire, Eng, Sept 30, 20; nat US; m 41; c 2. ANALYTICAL CHEMISTRY, CLINICAL CHEMISTRY. Educ: Queen's Col, NY, BS, 41; Polytech Inst Brooklyn, MS, 43. Prof Exp: Instr, Polytech Inst Brooklyn, 46; res chemist, Celotex Corp, 47-48; res chemist & group leader, Westvaco Chem Div, Food Machinery & Chem Corp, 48-52; res assoc, George Washington Univ, 52-54; group leader, Diamond Alkali Co, 54-65, proj mgr, 65-67, ASSOC DIR RES DEPT, RES CTR, DIAMOND SHAMROCK CORP, 68- Concurrent Pos: Lectr, Wagner Col, 49-52 & Lake Erie Col, 60-67. Mem: Am Chem Soc; NY Acad Sci; AAAS. Res: Building materials technology; inorganic phosphates; coordination compounds; inorganic polymers; plastic and metal coatings; nuclear fuels; water chemistry; electroless and electrolytic plating; hydrometallurgy; management of analytical environmental chemical facilities. Mailing Add: 11612 Basswood Rd Chardon OH 44024

TAYLOR, LARRY THOMAS, coordination chemistry, bioinorganic chemistry, see 12th edition

TAYLOR, LAURISTON SALE, b Brooklyn, NY, June 1, 02; m 25; c 2. MEDICAL PHYSICS. Educ: Cornell Univ, AB, 26. Hon Degrees: DSc, Univ Pa, 60 & St Procopius Col, 65. Prof Exp: Asst, Heckscher Found, Cornell Univ, 24-27; from asst physicist to assoc physicist, Nat Bur Stand, 27-35, sr physicist & chief x-ray sect, 35-41, chief proving ground group, 42-43, x-ray sect, 42-43, 46-49, asst chief atomic physics div, 47-51, chief radiol physics lab, 49-51, atomic radiation physics div, 51-60, radiation physics div, 60-62, assoc dir, 62-64; spec asst to pres, Nat Acad Sci, 65-69, exec dir adv comt emergency planning, 65-71; PRES, NAT COUN RADIATION PROTECTION & MEASUREMENTS, 64- Concurrent Pos: Mem, Int Comn Radiol Protection, 28-69, secy, 37-50, emer mem, 69-; mem, Int Comn Radiol Units & Measurements 28-69, secy, 34-50, chmn, 53-69, hon chmn & emer mem, 69-; chmn, Nat Coun Radiation Protection & Measurements, 29-64, chmn subcomt regulation of radiation exposure, 53-57, subcomt permissible exposure under emergency conditions, 55-59; dir Pan-Am Cancer Union, 39; mem sect E, div A, Nat Defense Res Comt, Off Sci Res & Develop, 42; nuclear comt Z-54, safety code, indust use of radiation, Am Nat Stand Comt, 42-46, nuclear stand bd, 56-; consult, Air Force Opers Anal Div, Dept Defense, 47-52, mem, guided missiles countermeasures panel, Res & Develop Bd, 47-52, adv comt radiol defense, 48-51, comt weapons systs eval with Nat Acad Sci, 53-55, consult, Weapons Systs Eval Group, Joint Chiefs of Staff, 54-65, mem interagency comt biomed weapons effects tests, 57, consult, Inst Defense Anal, 57-67, mem ad hoc sci adv comt, Armed Forces Radiobiol Res Facil & chmn panel radiol instruments, 62-65, mem nuclear weapons effects res med adv group, 65-71, chmn rev comt, Armed Forces Radiobiol Res Inst, 66; chief biophys br, AEC, 48-49; mem subcomt radiobiol, Comt Nuclear Sci, Nat Acad Sci-Nat Res Coun, 49-54, consult comt med & surg, 54, comt radiol, 54, mem adv comt civil defense, 54-65, chmn, 57-65; sr consult, Civil Serv Bd Expert Exam, Civil Serv Comn, 52-58, chmn, 62-65; mem comt to investigate fed civil defense admin, Joint House & Senate Appropriation Comt, 52; US deleg, Atoms for Peace Conf, 55 & 58; mem interagency comt int rels, Bus Defense Serv Admin, 55-58, interdept comt radiation protection of food, 56-65; mem expert adv panel radiation, WHO, 56-72; comt occup safety & health, Int Labor Orgn, 56-59, expert adv panel radiation protection, 57; WHO deleg, Int Conf Radioisotopes, UNESCO, 57; US deleg, Int Stand Orgn, Geneva, 57, mem tech comts 12 & 85, 58-60; consult, Div Radiol Health & mem adv comt radiation, Dept Health, Educ & Welfare, 58-62; consult, Adv Comt Isotope & Radiation Develop, AEC, 58-63, mem rev comt radiation & human spermatogenesis, Seattle, 65, radium toxicity, Washington, DC, 66 & health physics div, Oak Ridge, 71; consult, Adv Comt Health Physics, Oak Ridge Nat Lab, 58-61; mem working group, Fed Radiation Coun, 59-64, rep for Nat Coun Radiation Protection & Measurements & Dept Com; mem comt ionizing radiation, Int Bur Weights & Measures, 59-65; mem panel, Develop Basic Radiation Safety Stand, Int Atomic Energy Agency, 60-64; rev comt, Div Biol & Med Res, Argonne Nat Lab, 60-70, chmn div radiol physics, 63-66; mem rev comt, Naval Radiol Defense Lab, 66; personnel selection panel, Naval Res Lab, 66; prog adv comt, Off Emergency Planning, Exec Off of the President, 66-; mem, President's Coun Recreation & Natural Beauty, 66-69; pres, Second Int Cong Med Physics, 70. Honors & Awards: Except Serv Gold Medal, Dept Com, 49; Sylvanus Thompson Medal, Brit Inst Radiol, 50; Lester Medal Lectr, Soc Nondestructive Test, 54; Gold Medal, Radiol Soc NAm, 55; Janeway Medal, Am Radium Soc, 56; Edward Bennett Rosa Award, Nat Bur Stand, 65; Gold Medal, Am Col Radiol, 65; Distinguished Serv Award, Exec Off of the President, 68 & Citation, 73; Distinguished Serv Award, Int Comn Radiation Units & Measurements, 69; Citation, US AEC, 69; Distinguished Serv Award, Japanese Soc Radiol, 69; Gold Medal, Royal Swed Acad Sci, 73; Gold Medal, XIII Int Cong Radiol, Madrid, 73; Distinguished Achievement Award, Health Physics Soc, 74. Mem: Am Asn Physicists in Med; assoc fel Am Col Radiol; fel Am Phys Soc; Am Roentgen Ray Soc; Health Physics Soc (pres, 58-59). Res: X-ray absorption; dosage; standards, protection; high voltage measurements; ionization of liquids; proximity fuzes; variable time oscillators. Mailing Add: 7407 Denton Rd Bethesda MD 20014

TAYLOR, LAWRENCE AUGUST, b Paterson, NJ, Sept 14, 38; m 60; c 2. GEOCHEMISTRY, MINERALOGY. Educ: Ind Univ, Bloomington, BS, 60, MS, 63; Lehigh Univ, PhD(geochem), 68. Prof Exp: Instr geol, Univ Del, 63-64; fel geochem, Geophys Lab, Carnegie Inst Washington, 68-70 & Max Planck Inst Nuclear Physics, 70-71; asst prof geosci, Purdue Univ, West Lafayette, 71-74; MEM FAC, DEPT GEOL, UNIV TENN, KNOXVILLE, 74- Concurrent Pos: Res grants, Geochem Sect, NSF, 72; prin investr, NASA Manned Spacecraft Ctr, 72-; co-investr, Max Planck Inst Nuclear Physics & NASA. Mem: Mineral Soc Am; Meteoritical Soc. Res: Experimental geochemistry and mineralogy into the stability relations of sulfide and oxide compounds at low and high pressures and application of these data to natural rocks and minerals; geochemistry of lunar rocks, meteorites and ore deposits. Mailing Add: Dept of Geol Univ of Tenn Knoxville TN 37916

TAYLOR, LAWRENCE DOW, b Boston, Mass, Oct 6, 32; m 55; c 2. GLACIOLOGY, GEOMORPHOLOGY. Educ: Dartmouth Col, BA, 54, MA, 58; Ohio State Univ, PhD(geol), 62. Prof Exp: Geologist, US Geol Surv, Greenland, 54-55; res assoc, Northwest Greenland Glaciol, Dartmouth Col & Air Force Cambridge Res Labs, 57-58; res assoc, Southeast Alaska Glaciol, Inst Polar Studies, Ohio State Univ, 59-62; chief glaciologist, SPole Traverse, US Antarctic Res Prog, NSF, 62-63; asst prof geol, Col Wooster, 63-64; asst prof, 64-70, ASSOC PROF GEOL, ALBION COL, 70-, CHMN DEPT, 64- Concurrent Pos: NSF res grants, 60-61 & 62-63, field inst grant, Can Rockies, 68; Kellogg Found res & teaching grant, 71-72. Mem: AAAS; fel Geol Soc Am; Am Polar Soc; Nat Asn Geol Teachers; Arctic Inst NAm. Res: Glacial geology; structure of lake ice, Greenland; structure and flow of Alaskan glaciers; snow stratigraphy, Antarctica; microparticles in Antarctic ice. Mailing Add: Dept of Geol Sci Albion Col Albion MI 49224

TAYLOR, LEIGHTON ROBERT, JR, b Glendale, Calif, Nov 17, 40; m 63; c 2. ICHTHYOLOGY. Educ: Occidental Col, BA, 62; Univ Hawaii, MS, 65; Scripps Inst Oceanog, Univ Calif, San Diego, PhD(marine biol), 72. Prof Exp: Res asst, Scripps Inst Oceanog, 69-71, mus cur, Scripps Aquarium-Mus, La Jolla, Calif, 71-72; asst leader res, Hawaii Coop Fishery Res Unit, 72-75; DIR, WAIKIKI AQUARIUM, UNIV HAWAII, 75- Concurrent Pos: Mem grad fac, Dept Zool, Univ Hawaii, 72-; mem ed bd, Pac Sci, 73- Mem: Am Soc Ichthyologists & Herpetologists; Am Fisheries Soc; Am Soc Zoologists; Sigma Xi. Res: Taxonomy and ecology of tropical marine fishes and sharks. Mailing Add: Waikiki Aquarium 2777 Kalakaua Ave Honolulu HI 96815

TAYLOR, LEWIS WALTER, b East Troy, Wis, Oct 23, 00; m 23; c 1. POULTRY HUSBANDRY. Educ: Univ Wis, BS, 22, PhD(genetics, poultry husb), 31; Kans State Col, MS, 25. Prof Exp: Instr poultry husb, Mass Col, 22-23; asst, Kans State Col, 23-24 & Univ Ky, 27-28; assoc, 28-31, from asst prof to prof, 31-68, from actg head to head poultry div, 33-51, EMER PROF POULTRY HUSB, UNIV CALIF, BERKELEY, 68- Mem: AAAS; Soc Study Evolution; Cooper Ornith Soc; fel Poultry Sci Asn (pres, 40); Teratology Soc. Res: Poultry genetics and incubation; embryonic mortality; hereditary resistance to disease. Mailing Add: Dept of Poultry Husb Univ of Calif Berkeley CA 94720

TAYLOR, LINCOLN HOMER, b Wolsey, SDak, Oct 26, 20; m 46; c 5. AGRONOMY. Educ: SDak State Col, BS, 42; Iowa State Univ, MS, 49, PhD, 51. Prof Exp: From asst prof to assoc prof agron, Univ Maine, 51-55; PROF AGRON, VA POLYTECH INST & STATE UNIV, 55- Mem: Crop Sci Soc Am; Am Soc Agron. Res: Forage crop and turfgrass breeding; genetics. Mailing Add: Dept of Agron 7 Smyth Hall Va Polytech Inst & State Univ Blacksburg VA 24061

TAYLOR, LLOYD DAVID, b Boston, Mass, Jan 11, 33; m 57; c 3. CHEMISTRY. Educ: Boston Col, BS, 54; Mass Inst Technol, PhD(org chem), 58. Prof Exp: Res assoc, Mass Inst Technol, 57-58; scientist, 58-65, res group leader, 65-68, MGR POLYMER RES LAB, POLAROID CORP, CAMBRIDGE, 68- Mem: AAAS; Am Chem Soc; Am Inst Chem; Soc Photog Sci & Eng. Res: Polymer chemistry; photographic chemistry; syntheses of novel monomers and polymers; plastics; solubility and diffusional phenomena; critical temperature behaviour and mixing. Mailing Add: One Maureen Rd Lexington MA 02173

TAYLOR, LYLE HERMAN, b Paton, Iowa, Oct 23, 36; m 59; c 3. LASERS. Educ: Iowa State Univ, BS, 58; NMex State Univ, MS, 61; Univ Kans, PhD(physics), 68. Prof Exp: Asst physicist, White Sands Missile Range, 58-61; assoc physicist, Midwest Res Inst, 61-64, tech consult, 64-67; SR SCIENTIST, WESTINGHOUSE RES LABS, 67- Mem: Am Phys Soc. Res: Phonon-phonon interactions; physical adsorption; cleavage energies; F-center wave functions; spin-spin interactions; rare earth fluorescence; crystal field theory; laser pumps; holographic strain analysis; gas laser computer simulation. Mailing Add: Gas Laser Res Westinghouse Res Labs Pittsburgh PA 15235

TAYLOR, LYNN JOHNSTON, b Akron, Ohio, June 10, 36; m 61; c 2. ORGANIC CHEMISTRY, POLYMER CHEMISTRY. Educ: Harvard Univ, AB, 58; Mass Inst Technol, PhD(org chem), 63. Prof Exp: Sr res chemist, Agr Div, Monsanto Co, 63-65; res scientist, Okemos Res Lab, 66-70, sr res scientist, 70-74, CHIEF PLASTICS MATERIALS RES, OWENS-ILL, INC, 74- Mem: Am Chem Soc; Am Inst Chem; The Chem Soc. Res: Kinetics and mechanisms of organic chemical reactions; polymer synthesis, modification and degradation; thermal analysis of polymers. Mailing Add: Owens-Ill Tech Ctr PO Box 1035 Toledo OH 43666

TAYLOR, MARIE CLARK, b Sharpsburg, Pa, Feb 16, 11; m 48; c 1. BOTANY. Educ: Howard Univ, BS, 33, MS, 35; Fordham Univ, PhD(bot), 41. Prof Exp: Teacher pub schs, DC, 38-44; from asst prof to assoc prof, 45-65, PROF EOT, HOWARD UNIV, 65-, HEAD DEPT, 47- Concurrent Pos: Fund Advan Educ fel, 52-53; dir, Inst Radiation Biol, US AEC, 58-66. Mem: AAAS; Bot Soc Am; Nat Asn Biol Teachers; Nat Inst Sci. Res: Morphogenesis in angiosperms; influence of environmental factors on tissue development. Mailing Add: 6415 13th St NW Washington DC 20012

TAYLOR, MARY LOWELL BRANSON, b Coeur d'Alene, Idaho, Nov 24, 32; m 55; c 2. MICROBIAL PHYSIOLOGY. Educ: Univ Idaho, BS, 54; Univ Ill, PhD(bact), 59. Prof Exp: Asst, Univ Ill, 54-57; res assoc, Emory Univ, 57-59; USPHS fel, Oak Ridge Nat Lab, 59-61; RES ASSOC BIOL, PORTLAND STATE UNIV, 61- Mem: Am Soc Microbiol; Am Soc Plant Physiol. Res: Bacterial physiology; carbohydrate synthesis and dissimilation; alcohol oxidation; natural products oxidation; enzyme biosynthesis. Mailing Add: Dept of Biol Portland State Univ PO Box 751 Portland OR 97207

TAYLOR, MARY MARSHALL, b Swarthmore, Pa, Jan 9, 29; m 50, 71; c 3. MICROBIOLOGY. Educ: Bryn Mawr Col, BA, 50, PhD(biol), 64. Prof Exp: Sr scientist, Sci Info Dept, Smith Kline & French Labs, 64-68; instr microbiol, Med Col Pa, 68-69; teacher, Baldwin Sch, Pa, 69-71; asst prof microbiol & biol, 72-75, DIR ACAD SERV & ASSOC PROF BIOL SCI, MONTGOMERY COUNTY COMMUNITY COL, 75- Mem: AAAS. Res: Gastrointestinal effects of bacterial endotoxins; normal flora of intestine and oral cavity. Mailing Add: Dept of Sci Montgomery County Community Col 340 DeKalb Pike Blue Bell PA 19422

TAYLOR, MERLIN GENE, b Zanesville, Ohio, May 11, 36; m 63; c 2. PHYSICS. Educ: Muskingum Col, BS, 58; Brown Univ, MSc, 65, PhD(physics), 67. Prof Exp: Res asst, High Energy Physics Lab, Brown Univ, 58-66; asst prof physics, Wilkes Col, 66-67 & Am Univ Cairo, 67-69; asst prof, 69-71, ASSOC PROF PHYSICS, BLOOMSBURG STATE COL, 71- Mem: Am Phys Soc; Am Asn Physics Teachers. Res: Use of computers in physics teaching; nuclear physics; activation analysis. Mailing Add: Dept of Physics Bloomsburg State Col Bloomsburg PA 17815

TAYLOR, MERREL ARTHUR, b Macy, Ind, July 6, 13; m 43; c 2. CONSERVATION, ENVIRONMENTAL BIOLOGY. Educ: Asst prof zool, San Diego State Col, 46-60; instr biol, San Diego City Col, 60-61; asst prof zool, San Diego State Col, 61-66; PROF BIOL, GROSSMONT COL, 66- Concurrent Pos: Biologist, US Naval Electronics Lab, 51-53. Mem: Soil Conserv Soc Am; Soc Am

Foresters. Res: Natural history of vertebrates; marine biology; deep scattering; fish; conservation of renewable resources. Mailing Add: Dept of Biol Grossmont Col 8800 Grossmont College Blvd El Cajon CA 92020

TAYLOR, MICHAEL DEE, b New York, NY, Dec 17, 40; m 70. MATHEMATICS. Educ: Univ Fla, BA, 63; Fla State Univ, MS, 65, PhD(math), 69. Prof Exp: Asst prof, 68-72, ASSOC PROF MATH, FLA TECHNOL UNIV, 72- Mem: Math Asn Am; Am Math Soc. Res: Topology of 3-manifolds and probabilistic metric spaces. Mailing Add: Dept of Math Sci Fla Technol Univ PO Box 25000 Orlando FL 32816

TAYLOR, MICHAEL LEE, b Rockville, Ind, May 27, 41; m 65; c 2. ANALYTICAL CHEMISTRY, MEDICINAL CHEMISTRY. Educ: Purdue Univ, Lafayette, BS, 63, MS, 65, PhD(med chem), 67. Prof Exp: Nat Res Coun resident res assoc, Chem Lab, Aerospace Res Labs, Wright-Patterson AFB, Ohio, 70-71, res scientist anal mass spectros, 71-75; RES ASSOC PROF CHEM, WRIGHT STATE UNIV, 75- Concurrent Pos: Mem bd dirs, Spectronics, Inc, 74- Res: Development of ultratrace analytical methods for determination of chlorocarbons and metals in biological and environmental samples. Mailing Add: Dept of Chem Wright State Univ Dayton OH 45431

TAYLOR, MILTON WILLIAM, b Glasgow, Scotland, Dec 10, 31; US citizen; m 57; c 2. MICROBIOLOGY, GENETICS. Educ: Cornell Univ, BS, 61; Stanford Univ, PhD(biol), 66. Prof Exp: NIH fel virol, Univ Calif, Irvine, 66-67; asst prof, 67-70, ASSOC PROF MICROBIOL, IND UNIV, BLOOMINGTON, 70- Mem: AAAS; Am Soc Microbiol; Am Asn Cancer Res; Tissue Cult Asn; Genetics Soc Am. Res: Cancer research; viral oncolysis; controlling mechanisms in animal virus infection in tissue culture; somatic cell genetics. Mailing Add: Dept of Microbiol Ind Univ Bloomington IN 47401

TAYLOR, MODDIE DANIEL, b Nymph, Ala, Mar 3, 12; m 37; c 1. INORGANIC CHEMISTRY. Educ: Lincoln Univ, Mo, BS, 35; Univ Chicago, SM, 38, PhD(chem), 41. Prof Exp: Instr chem, Lincoln Univ, Mo, 35-41; assoc chemist, Manhattan Engrs, Ill, 43-45; prof chem, Lincoln Univ, Mo, 45-48; assoc prof, 48-59, PROF CHEM, HOWARD UNIV, 59-, CHMN DEPT, 69- Concurrent Pos: Ford Found fel, 52-53; Welch vis scholar, Prairie View Agr & Mech Col, 60-61; NSF fac fel, 61-62. Honors & Awards: Mfg Chem Asn Medal, 60. Mem: AAAS; Am Chem Soc; Nat Inst Sci. Res: Synthesis, spectra and structure of normal and coordination compounds of the transition and inner transition elements; TGA, DGA and DSC investigation of these compounds. Mailing Add: Dept of Chem Howard Univ 525 College St Washington DC 20011

TAYLOR, MORRIS CHAPMAN, b Fulton, Ky, May 28, 39; m 60; c 1. NUCLEAR PHYSICS, MEDICAL PHYSICS. Educ: Univ Tenn, BS, 62; Univ Calif, Los Angeles, MS, 64; Rice Univ, MA, 66, PhD(physics), 68. Prof Exp: Lab technician, Oak Ridge Nat Lab, 61-62; mem tech staff, Hughes Aircraft Co, 62-64; asst prof physics, St Louis Univ, 68-69; mem staff, Columbia Sci Res Inst, Houston, 69-71; AEC & State of Tex joint sr fel, Rice Univ & Univ Tex M D Anderson Hosp & Tumor Inst, 71; CHIEF SCIENTIST, COLUMBIA SCI INDUSTS, 72- Concurrent Pos: Prof lectr, St Louis Univ, 69- Mem: Instrument Soc Am; Am Phys Soc. Res: Experimental nuclear physics; radiological physics; nondestructive elemental analysis. Mailing Add: Appl Res Div Columbia Sci Industs PO Box 9908 Austin TX 78766

TAYLOR, MORRIS D, b Mitchell, Ind, Apr 14, 34; m 56; c 3. SCIENCE EDUCATION, PHYSICAL CHEMISTRY. Educ: Purdue Univ, BS, 56, PhD(sci educ), 66. Prof Exp: Instr physics & educ, Purdue Univ, 60-63; from asst prof to prof chem & educ, 63-70, PROF CHEM, EASTERN KY UNIV, 70- Mem: AAAS; Nat Sci Teachers Asn. Mailing Add: Dept of Chem Eastern Ky Univ Richmond KY 40475

TAYLOR, MURRAY EAST, b Casselton, NDak, Apr 14, 15; m 50; c 1. ANALYTICAL CHEMISTRY. Educ: Univ Wash, BS, 48, MS, 52, PhD(chem), 60. Prof Exp: Microchemist, Univ Wash, 48-52, res instr chemother, 52-58; res chemist, Boeing Airplane Co, 58-63; assoc staff scientist, Missile Div, Chrysler Corp, 63-64; SPECIALIST ENGR, BOEING CO, SEATTLE, 64- Mem: AAAS; Am Chem Soc; NY Acad Sci. Res: Instrumental analysis; microchemistry; thermochemistry; radiation effects on materials. Mailing Add:

TAYLOR, NEIL WELLING, biophysics, see 12th edition

TAYLOR, NORMAN BURWELL GEORGE, b Ramsgate Eng, Aug 13, 17; Can citizen; div; c 5. POPULATION STUDIES. Educ: Univ Toronto, MD, 42; Univ Western Ont, PhD(med res), 49. Prof Exp: Asst prof med res, Univ Western Ont, 50-54; head physiol group, Defence Res Estab Toronto, Defence Res Bd, 54-57, supt physiol wing, 57-65, actg chief supt, Estab, 65-66, dir physiol wing, 67-68; sci coordr environ physiol lab, Univ Calif, Berkeley, 68-70; lectr concentration in pop dynamics, Univ Wis-Green Bay, 70-71, chmn dept, 70-73, prof, 71-75; RETIRED. Concurrent Pos: Nat Res Coun Can med fel, McGill Univ & Univ Western Ont, 47-49; Nuffield traveling fel med, Oxford Univ, 49-50; sr fel, Univ Western Ont, 50-54, hon lectr, 54-64; course mem, Nat Defence Col, Can, 66-67. Mem: AAAS. Res: Teaching and administration of population studies. Mailing Add: 2207 Russell St Berkeley CA 94705

TAYLOR, NORMAN FLETCHER, b Newcastle Upon Tyne, Eng, Mar 4, 28; m 48; c 2. BIOCHEMISTRY. Educ: Univ Oxford, BA, 53, MA, 56, DPhil(biochem), 56; FRIC. Prof Exp: Exchange vis scientist, Dept Pharmacol, Univ Calif, Los Angeles, 57-59; Sci Res Coun fel, Dept Biochem, Univ Oxford, 59-60; sr lectr chem, Bristol Col Sci & Technol, Eng, 62-65; reader & head biochem, Univ Bath, Eng, 65-73; PROF BIOCHEM, UNIV WINDSOR, 73- Concurrent Pos: Vis prof, Dept Chem, Temple Univ, 67; fel, Canterbury Col, Ont, 74- Mem: Am Soc Biol Chemists; Can Biochem Soc; Brit Biochem Soc; Brit Chem Soc. Res: Microbial and mammalian metabolism of synthetic fluorinated carbohydrates and related compounds; mechanism of transport across biological membranes; insect biochemistry. Mailing Add: Dept of Chem Univ of Windsor Windsor ON Can

TAYLOR, NORMAN LINN, b Augusta, Ky, July 18, 26; m 51; c 5. AGRONOMY. Educ: Univ Ky, BS, 49, MS, 51; Cornell Univ, PhD(plant breeding), 53. Prof Exp: Asst agronomist, 53-56, assoc prof, 56-66, ASSOC AGRONOMIST, UNIV KY, 56-, PROF AGRON, 66- Honors & Awards: Cert of Merit, Am Forage & Grasslands Coun, 72. Mem: AAAS; Crop Sci Soc Am; Am Soc Agron; Am Genetic Asn; Genetics Soc Can. Res: Forage crops genetics and breeding; interspecific hybridization in the genus Trifolium. Mailing Add: Dept of Agron Univ of Ky Lexington KY 40506

TAYLOR, OLIVER CLIFTON, b Hallet, Okla, Nov 29, 18; m 40; c 3. HORTICULTURE. Educ: Okla Agr & Mech Col, MS, 51; Mich State Col, PhD, 53. Prof Exp: Asst county agent, Okla Exten Serv, 47-49; instr agr, Northeast Okla Agr & Mech Col, 50-51; horticulturist, 53, HORTICULTURIST, STATEWIDE AIR POLLUTION RES CTR, UNIV CALIF, RIVERSIDE, 53-, LECTR PLANT SCI, 74- Mem: AAAS; Air Pollution Control Asn; Am Soc Hort Sci. Res: Citriculture; biological effects of air pollutants and plant physiological responses to air pollutants. Mailing Add: Statewide Air Pollution Res Ctr Univ of Calif Riverside CA 92502

TAYLOR, PALMER WILLIAM, b Stevens Point, Wis, Oct 3, 38; m 65; c 1. PHARMACOLOGY. Educ: Univ Wis-Madison, BS, 60, PhD(pharm), 64. Prof Exp: Res assoc pharmacol, NIH, Bethesda, Md, 64-68; NIH vis fel, Molecular Pharmacol Res Unit, Cambridge Univ, 68-70; asst prof, 70-74, ASSOC PROF PHARMACOL, SCH MED, UNIV CALIF, SAN DIEGO, 74- Concurrent Pos: NIH grant pharmacol, Sch Med, Univ Calif, San Diego, 70- Mem: Am Soc Biol Chemists; Am Soc Pharmacol & Exp Therapeut. Res: Drug-protein interactions; kinetic and physical-chemical studies. Mailing Add: Div of Pharmacol Sch of Med Univ of Calif at San Diego La Jolla CA 92014

TAYLOR, PATRICK TIMOTHY, b Mt Vernon, NY, Mar 20, 38. MARINE GEOPHYSICS. Educ: Mich State Univ, BS, 60; Pa State Univ, MS, 62; Stanford Univ, PhD(geophys), 65. Prof Exp: Res asst marine geophys, Lamont-Doherty Geol Observ, Columbia Univ, 65-66; GEOPHYSICIST, US NAVAL OCEANOG OFF, 66- Honors & Awards: Kaminiski Award, Sci Res Soc Am, 70. Mem: Am Geophys Union; Geol Soc Am; Royal Astron Soc; Soc Explor Geophys; Seismol Soc Am. Res: Interpretation of marine geophysical data, such as gravity, magnetics, seismic and heat flow. Mailing Add: Code 6120 US Naval Oceanog Off Washington DC 20373

TAYLOR, PAUL DUANE, b Warren, Ohio, July 8, 40; m 65; c 4. CHEMISTRY. Educ: Ind Inst Technol, BS, 62; Long Island Univ, MS, 64; Univ Cincinnati, PhD(inorg chem), 69. Prof Exp: Teaching asst, Long Island Univ, 62-64; res chemist, Res & Eng Develop Dept, M W Kellogg Co, 64-66; teaching asst, Univ Cincinnati, 66-68; from res chemist to sr res chemist, 69-74, RES SUPVR, CELANESE RES CO, 74- Mem: Am Chem Soc. Res: Organometallics as they apply to homogeneous and heterogeneous catalysis. Mailing Add: Celanese Res Co PO Box 1000 Summit NJ 07901

TAYLOR, PAUL JOHN, b Chicago, Ill, Jan 30, 39; m 60; c 1. ANALYTICAL CHEMISTRY. Educ: Northern Ill Univ, BS, 64, PhD(anal chem), 71. Prof Exp: US AEC fel, Purdue Univ, Lafayette, 70-72; ASST PROF CHEM, WRIGHT STATE UNIV, 72- Mem: Am Chem Soc. Res: Coordination compounds and their analytical applications, gas chromatography; computers and their applications to analytical chemistry; liquid crystals and mass spectroscopy. Mailing Add: Dept of Chem Wright State Univ Dayton OH 45431

TAYLOR, PAUL M, b Baltimore, Md, June 26, 27; m 55; c 4. PEDIATRICS, PHYSIOLOGY. Educ: Johns Hopkins Univ, AB, 47, MD, 51. Prof Exp: Res fel, 54-56, from instr to assoc prof, 56-71, PROF PEDIAT, OBSTET & GYNEC, SCH MED, UNIV PITTSBURGH, 71- Concurrent Pos: USPHS fel, Nuffield Inst Med Res, Oxford Univ, 59-60; dir div neonatology & chief dept pediat, Magee-Women's Hosp, 65-; vis prof, Inst Path, Univ Geneva, 71-72. Mem: Soc Pediat Res; Am Physiol Soc. Res: Physiology of the newborn infant; role of lung edema in neonatal lung problems. Mailing Add: Dept of Pediat Magee-Women's Hosp Pittsburgh PA 15213

TAYLOR, PAUL PEAK, b Childress, Tex, May 11, 21; m 45; c 2. PEDODONTICS. Educ: Baylor Univ, DDS, 44; Univ Mich, MS, 51. Prof Exp: PROF GRAD PEDODONTICS & CHMN DEPT, COL DENT, BAYLOR UNIV, 60-; DIR TRAINING & ASST CHIEF DENT SERV, CHILDREN'S MED CTR, 60-, DIR DENT, 66- Concurrent Pos: Consult, Scottish Rite Hosp for Crippled Children, Dallas, Tex, 60- & Denton State Sch, 64-; proctor, Am Bd Pedodont, 66-67. Mem: Fel Am Col Dent; Am Acad Pedodont; Am Dent Asn; Am Soc Dent for Children. Res: Physiological responses of pulp tissues and the testing of patient responses to dental stimuli. Mailing Add: Dept of Pedodont Baylor Univ Col of Dent Dallas TX 75226

TAYLOR, PETER ANTHONY, b Liverpool, Eng, June 9, 32; m 68; c 2. PHYSICAL CHEMISTRY, TEXTILE ENGINEERING. Educ: Liverpool Col Technol, Eng, ARIC, 56; Univ Manchester, PhD(chem), 63. Prof Exp: Jr asst analyst, Liverpool City Pub Health Dept, 49-52; analyst, Distillers Co (Biochem), Ltd, 52-56, res chemist, 58-60 & Fibers Div, Allied Chem Corp, 63-66; PROJ MGR, PHILLIPS FIBERS CORP, 66- Mem: Royal Inst Chem; Am Chem Soc. Res: Polymerization kinetics, free radical and condensation; man-made fiber rheology, processing and dyeing; polymer pigmentation; textile and fiber finishing; spin finish development. Mailing Add: Res Dept Phillips Fibers Corp PO Box 66 Greenville SC 29602

TAYLOR, PETER BERKLEY, b Yonkers, NY, Dec 1, 33; m 58; c 3. MARINE ECOLOGY. Educ: Cornell Univ, BS, 55; Univ Calif, Los Angeles, MS, 60; Univ Calif, San Diego, PhD(marine biol), 64. Prof Exp: NSF res fel, 63-64; asst prof oceanog, Univ Wash, 64-71; MEM FAC, EVERGREEN STATE COL, 71- Mem: AAAS; Ecol Soc Am; Am Soc Limnol & Oceanog; World Maricult Soc. Res: Ecology of marine benthic organisms; estuarine ecology; venomous marine animals. Mailing Add: Evergreen State Col Olympia WA 98505

TAYLOR, PHILIP LIDDON, b London, Eng, Oct 17, 37; m 66. SOLID STATE PHYSICS. Educ: Univ London, BSc, 59; Cambridge Univ, PhD(physics), 62. Prof Exp: Magnavox res fel mat sci, 62-64, from asst prof to assoc prof, 64-74, PROF PHYSICS, CASE WESTERN RESERVE UNIV, 74- Res: Theoretical solid state physics. Mailing Add: Dept of Physics Case Western Res Univ Univ Circle Cleveland OH 44106

TAYLOR, PHILIP SEYFANG, b Minneapolis, Minn, Dec 16, 21; m 42. ENTOMOLOGY. Educ: Univ Minn, BS, 48, MS, 52. Prof Exp: Lab technician, Univ Minn, 47-48, asst, 48-52; cur zool & ed supt, 52-58, DIR, SCI MUS MINN, ST PAUL INST, 58- Concurrent Pos: Dir, Urban Lab, Inc, 75- Mem: Am Inst Biol Sci; Int Coun Mus; Am Asn Mus; Nat Asn Biol Teachers. Res: Biology and ecology of insects; indentification of zoological remains associated with archaeological sites; science education. Mailing Add: Sci Mus of Minn 30 ETenth St St Paul MN 55101

TAYLOR, R PERRY, b Birmingham, Eng, Jan 24, 29; m 52; c 2. PHYSICAL CHEMISTRY. Educ: Univ Birmingham, BSc, 50, PhD(chem), 53. Prof Exp: Res assoc, Northwestern Univ, 53-54; res fel, Univ Calif, Calif, Los Angeles, 54-55; sect leader phys chem, Richfield Oil Corp, 56-62; res specialist, NAm Aviation, Inc, 62-63; proj mgr, Martin Marietta Corp, 63-65; dir appl res, Columbia Cellulose Co, Ltd, BC, 65-67; dir new prod develop, Mead Corp, 67-74; STAFF MEM, INT PAPER CO, NY, 74- Mem: Am Chem Soc. Res: Polymerization kinetics; photochemistry; radiochemistry; aerothermodynamics; high temperature structures and materials; pulp and paper technology; digitally controlled printing and publishing systems. Mailing Add: International Paper Co Tuxedo Park NY 10987

TAYLOR, RALPH E, b Lykens, Ohio, Sept 25, 05; m 34; c 1. GEOLOGY. Educ: Univ Mich, AB, 27, MS, 28; La State Univ, PhD(geol), 38. Prof Exp: Field geologist, Ark Natural Gas Co, La, Tex, Kans & Nebr, 29-32 & USDA, 33-34; asst, La State Univ, 34-37; explor geologist, Freeport Sulphur Co, US, Can, Cuba, Mex & Africa, 37-49; sr staff geologist, Humble Oil & Refining Co, 49-67; CONSULT GEOLOGIST, 67-

Concurrent Pos: Dir & mgr explor, Phelan Sulphur Co, 68-71; vpres & dir, Frontier Petrol Co, Inc & Frontier Sulphur Co, Inc, NY, 68-71; vpres, Transworld Minerals Australia, Australia & SEAsia, 70-72; consult geologist, var co & individuals in Australia & NZ, 73. Mem: Geol Soc Am; Soc Econ Geol; Mineral Soc Am; Am Inst Mining, Metall & Petrol Eng; Australian Geol Soc. Res: Geology of cap rock and salt of salt domes; limestones; economic geology of petroleum, sulphur, salt, manganese, nickel, iron, potash, phosphates and uranium; North and South America, Africa, Europe, Australia and the Middle and Far East. Mailing Add: PO Drawer A La Porte TX 77571

TAYLOR, RAYMOND DEAN, b Okemah, Okla, Aug 18, 28; m 61; c 2. LOW TEMPERATURE PHYSICS. Educ: Kans State Col, Pittsburg, BS, 50; Rice Inst, PhD(phys chem), 54. Prof Exp: Asst, Rice Inst, 51-52; mem staff, 54-73, ASSOC GROUP LEADER, LOS ALAMOS SCI LAB, 73- Mem: Am Chem Soc; Am Phys Soc. Res: Low temperature calorimetry; cryogenics; transport and state properties of liquid helium; Mössbauer effect; superconductivity; magnetism; direct current superconducting power transmission lines. Mailing Add: Los Alamos Sci Lab Box 1663 MS764 Los Alamos NM 87545

TAYLOR, RAYMOND JOHN, b Ada, Okla, Jan 20, 30; m 59; c 3. PLANT TAXONOMY. Educ: ECent State Col, BSEd, 54; Univ Okla, MNS, 61, PhD(plant ecol), 67. Prof Exp: Teacher high sch, Okla, 54-55 & Okla City Pub Sch Syst, 55-61; asst prof biol, Southeastern State Col, 61-63; asst bot, Univ Okla, 63-65; assoc prof, 65-74, PROF BIOL, SOUTHEASTERN STATE UNIV, 74- Concurrent Pos: Grants, Southeastern State Col Res Found, 68 & NIH, 72. Mem: Am Soc Plant Taxonomists. Res: Dauphine Island, Alabama, Alaska, Costa Rica and Oklahoma plants; aquatic microphytes. Mailing Add: Dept of Biol Southeastern State Univ Durant OK 74701

TAYLOR, RAYMOND L, b Providence, RI, July 3, 30; m 55; c 5. LASERS, CHEMICAL PHYSICS. Educ: Brown Univ, ScB, 55; Calif Inst Technol, PhD(chem), 60. Prof Exp: Asst, Calif Inst Technol, 55-59; prin res scientist, Avco Everett Res Lab, Inc, 59-73; PRIN SCIENTIST & MEM BD DIRS, PHYS SCI INC, 73- Concurrent Pos: Chmn atomic physics res comt, Avco Everett Res Lab, Inc, 70-73; mem, Dept Transp Comt Stratospheric Chem, 73-75. Honors & Awards: Silver Combustion Medal, Combustion Inst, 68. Mem: Am Chem Soc; Am Phys Soc; Combustion Inst; Sigma Xi. Res: Radiation and energy transfer processes in gases; optical experiments and instrumentation; molecular gas laser device research and development; laser applications. Mailing Add: Phys Sci Inc 18 Lakeside Off Park Wakefield MA 01880

TAYLOR, RAYMOND LEECH, b New York, NY, Sept 8, 01; m 25; c 2. ZOOLOGY, BOTANY. Educ: Cornell Univ, BS, 24; Harvard Univ, MS, 27, ScD(forest entom), 29. Prof Exp: Asst entomologist, State Forest Serv, Maine, 29-30; instr entom, State Univ, NY Col Forestry, Syracuse, 30-31; from asst prof to assoc prof biol, Col William & Mary, 31-46; from assoc prof to prof & res head, Sampson Col, 46-49; from asst admin secy to assoc admin secy, AAAS, 49-67, mem bd biol abstracts, 53-59; consult, 67-75; RETIRED. Concurrent Pos: In-chg nature study, Mt Desert Island Biol Lab, 31, dir, Dorr Sta, 32; mem staff, Mt Lake Biol Sta, Univ Va, 34, 37 & 56; mem staff, Univ NC, 38-39, Keenan vis prof, 39; collabr, Am Coun Educ, Ill, 40; mem staff, Audubon Nature Ctr, Conn, 46; registr, Int Cong Parasitol, 71. Mem: AAAS; Am Inst Biol Sci. Res: Insect parasites of white pine weevil; bionomics of birch leaf-mining sawfly, growth and physiology of insect larvae; parasitic Hymenoptera of North America; history of ornamental plants in America; arrangements for scientific meetings. Mailing Add: 1820 NJohnson St Arlington VA 22207

TAYLOR, RHODA E, b Hartford City, Ind, Feb 20, 36; m 57; c 2. REPRODUCTIVE ENDOCRINOLOGY. Educ: Asbury Col, BA, 57; Purdue Univ, MS, 63, PhD(physiol), 65. Prof Exp: Asst prof biol, Ind Univ, Kokomo, 65-67; ASSOC PROF BIOL, SLIPPERY ROCK STATE COL, 67- Mem: AAAS; Am Soc Zool; Sigma Xi. Res: Maternal behavior; reproductive physiology and behavior. Mailing Add: Dept of Biol Slippery Rock State Col Slippery Rock PA 16057

TAYLOR, RICHARD MELVIN, b Salt Lake City, Utah, Aug 19, 29; m 55; c 4. AGRONOMY, PLANT PHYSIOLOGY. Educ: Utah State Univ, BS, 58, MS, 59; Iowa State Univ, PhD(agron, plant physiol), 64. Prof Exp: Instr agron, Iowa State Univ, 59-64; ASST AGRONOMIST, TEX A&M UNIV, 64- Mem: Am Soc Agron; Crop Sci Soc Am. Res: Effects of soluble salts and temperature upon the germination and emergence of seeds and fruiting patterns of cotton; production and management; root physiology and native plant domestication. Mailing Add: Tex Agr Res Ctr 10601 NLoop Rd El Paso TX 79927

TAYLOR, RICHARD MORELAND, b Owensboro, Ky, Oct 10, 87; m 26; c 3. PUBLIC HEALTH. Educ: Univ Mich, MD, 10; Johns Hopkins Univ, DrPH, 26. Prof Exp: Asst bact, Sch Med, NY Univ, 10-11; instr, NY Post-Grad Med Sch, 11-12, instr path, 12, from asst prof to prof, 13-19; med dir, Am Red Cross Comn, Poland, 20-21; mem staff, Int Health Div, Rockefeller Found, 22-53; dir, Dept Virol, US Naval Res Unit No 3, Egypt, 54-56; lectr epidemiol & prev med, Med Sch, Yale Univ, 56-60; lectr epidemiol, Sch Pub Health, Univ Calif, Berkeley, 60-70; RETIRED. Honors & Awards: Achievement Award in Arbovirology, Am Comt Arbovirology, 66. Mem: AAAS; Am Soc Trop Med & Hyg; Am Pub Health Asn. Res: Virology; epidemiology. Mailing Add: 1332 Rockledge Lane Apt 4 Walnut Creek CA 94595

TAYLOR, RICHARD RAY, b Prairieburg, Iowa, Nov 21, 22; m 50; c 5. MEDICINE. Educ: Univ Chicago, BS, 44, MD, 46; Am Bd Internal Med, dipl, 55. Prof Exp: Intern, Highland Alameda County Hosp, Oakland, Calif, 46-47; Med Corps, US Army, 47-, ward officer gen med, Army-Navy Gen Hosp, 47-51, resident internal med, Letterman Army Hosp, San Francisco, 51-52, resident cardiol, 52-53, internist, Hemorrhagic Fever Ctr, Korea, 54, resident pulmonary dis, Fitzsimons Army Hosp, Denver, 55, chief non-tuberc chest sect, 56-57, commanding officer, US Army Med Res & Develop Unit, 57-58, dep commanding officer, US Army Med Res & Nutrit Lab, 58-59, chief biophys & astronaut res br, US Army Med Res & Develop Command, Off Surgeon Gen, 59, chief res div, 59-62, prog planning officer, 62-63, dep comdr, 63-64, staff surgeon, Joint US Mil Adv Group, Thailand & surgeon, Mil Assistance Command, Thailand, 64-65, chief biol & med sci div, Off Dir Defense Res & Eng, Off Secy Defense, 66-69, command surgeon, US Mil Assistance Command, Viet Nam, 69-70, commanding gen, US Army Med Res & Develop Command, Washington, DC, 70-73, DEP SURGEON GEN & SURGEON GEN, US ARMY, 73- Concurrent Pos: Asst clin prof, Sch Med, Univ Colo, 57-59. Mem: AMA; fel Am Col Physicians; fel Am Col Chest Physicians; Am Thoracic Soc; NY Acad Sci. Res: Pulmonary and infectious diseases. Mailing Add: Surgeon Gen Dept of the Army Washington DC 20310

TAYLOR, RICHARD WILTON, physical chemistry, inorganic chemistry, see 12th edition

TAYLOR, ROBERT BARTLEY, b Pendleton, Ore, Oct 15, 26; m 49; c 3. CULTURAL ANTHROPOLOGY. Educ: Wheaton Col, BS, 49; Univ Ore, MS, 51, PhD(anthrop), 60. Prof Exp: Instr anthrop, Wheaton Col, 51-54; from temp instr to asst prof anthrop & sociol, 57-67, ASSOC PROF ANTHROP, KANS STATE UNIV, 67- Concurrent Pos: Vis asst prof sociol, Univ NMex, 59-60. Mem: Fel Am Anthrop Asn; Soc Appl Anthrop. Res: Cultural change processes; Mesoamerican ethnology. Mailing Add: Dept of Sociol & Anthrop Kans State Univ Manhattan KS 66506

TAYLOR, ROBERT BRUCE, biochemistry, clinical chemistry, see 12th edition

TAYLOR, ROBERT BURNS, JR, b Downingtown, Pa, Oct 13, 20; m 45, 72; c 1. ORGANIC CHEMISTRY, INFORMATION SCIENCE. Educ: Swarthmore Col, AB, 41; Ohio State Univ, MSc, 42; Pa State Col, PhD(org chem), 45. Prof Exp: Instr chem, Pa State Col, 45-46; res chemist, 46-53, res supvr, 53-56, sr patent chemist, 56-60, from supvr to sr supvr textile fibers patent div, 60-71, MGR CENT PATENT INDEX, E I DU PONT DE NEMOURS & CO, 71- Mem: Am Chem Soc; Sigma Xi. Res: Synthetic antimalarials; synthetic fibers; patents. Mailing Add: Cent Patent Index 3211 Centre Rd E I du Pont de Nemours & Co Inc Wilmington DE 19898

TAYLOR, ROBERT CHARLES, physical chemistry, inorganic chemistry, see 12th edition

TAYLOR, ROBERT CLEMENT, b Mankato, Minn, Dec 2, 35; m 57; c 6. PHYSIOLOGY, ZOOLOGY. Educ: Mankato State Col, BS, 57; Univ SDak, MS, 61; Univ Ariz, PhD(physiol), 66. Prof Exp: ASST PROF ZOOL, UNIV GA, 66- Mem: AAAS; Am Soc Zoologists. Res: Comparative physiology; neurophysiology. Mailing Add: Dept of Zool Univ of Ga Athens GA 30601

TAYLOR, ROBERT COOPER, b Colorado Springs, Colo, May 5, 17; m 42; c 2. PHYSICAL CHEMISTRY. Educ: Kalamazoo Col, AB, 41; Brown Univ, PhD(phys chem), 47. Prof Exp: Asst chem, Brown Univ, 41-42, res chemist, Manhattan Proj, 42-46, instr phys chem, 47-49; from instr to assoc prof, 49-62, actg chmn, 66, PROF CHEM, UNIV MICH, ANN ARBOR, 62-, ASSOC CHMN, 67- Concurrent Pos: Consult, W J Barrow Res Lab, 74- Mem: AAAS; Am Chem Soc; Am Phys Soc. Res: Molecular spectroscopy and structure; boron hydride derivatives; Lewis complexes; liquid ammonia solutions; hydrogen bonded substances; uranium chemistry. Mailing Add: Dept of Chem Univ of Mich Ann Arbor MI 48104

TAYLOR, ROBERT CRAIG, b Franklin, Pa, Jan 26, 39; m 66; c 1. INORGANIC CHEMISTRY. Educ: Col Wooster, BA, 60; Princeton Univ, MA, 62, PhD(chem), 64. Prof Exp: NATO fel, Imp Col, Univ London, 64-65; asst prof chem, Univ Ga, 65-72; ASSOC PROF CHEM, OAKLAND UNIV, 72- Mem: Am Chem Soc; The Chem Soc; Sigma Xi. Res: Transition metal chemistry; organophosphorous chemistry; lanthanide shift reagents; transition metal catalyzed stereospecific polymerizations of diolefins; transition metal complexes as antitumor agents; role of molybdenum in enzymes. Mailing Add: Dept of Chem Oakland Univ Rochester MI 48063

TAYLOR, ROBERT E, b Milwaukee, Wis, Dec 22, 23; m 53; c 2. SOIL CONSERVATION. Educ: Univ Wis, BS, 48. Prof Exp: Soil conservationist, Soil Conserv Serv, 48-53 & Soil & Water Conserv Res Div, 53-57, soil scientist, 57-66, asst to br chief, 66-72, ASST AREA DIR, AGR RES SERV, USDA, 72- Mem: AAAS; Am Soc Agron; Soil Sci Soc Am; Crop Sci Soc Am; Soil Conserv Soc Am. Res: Crop and soil management practices, pasture renovation and seedbed preparation methods, strip cropping, interseeding forage crops in row crops and mulching systems for runoff and erosion control. Mailing Add: Agr Res Serv UMC 48 Utah State Univ Logan UT 84322

TAYLOR, ROBERT E, b Havelock, Nebr, July 24, 20; m 47; c 3. PHYSIOLOGY, BIOPHYSICS. Educ: Univ Ill, BS, 42; Univ Rochester, PhD(physiol), 50. Prof Exp: Mem staff, Radiation Lab, Off Sci Res & Develop, Mass Inst Technol, 42-45; Merck fel physiol, Univ Chicago, 50-51, Nat Heart Inst fel, 51-52; asst prof neurophysiol, Col Med, Univ Ill, 52-53; NSF fel, Physiol Lab, Cambridge Univ, 53-54, Nat Inst Neurol Dis & Blindness fel, 54-55 & Univ Col, Univ London, 55-56; PHYSIOLOGIST, LAB BIOPHYS, NAT INST NEUROL & COMMUN DISORDERS & STROKE, 56- Mem: Am Physiol Soc; Biophys Soc; NY Acad Sci; Soc Gen Physiol; hon mem Chilean Biol Soc. Res: Properties of natural excitable membranes; muscle contraction activation. Mailing Add: Bldg 36 Rm 2A31 Nat Inst Neurol & Commun Disorders & Stroke Bethesda MD 20014

TAYLOR, ROBERT EDWARD, organic chemistry, see 12th edition

TAYLOR, ROBERT EMERALD, JR, b Polk City, Fla, Aug 21, 30; m 51; c 4. PHYSIOLOGY. Educ: Southern Col Pharm, BS, 58; Univ Fla, MS, 61, PhD(physiol), 63. Prof Exp: NIH fel pharmacol, Univ Vt, 63-65; from asst prof to prof physiol & biophys, Sch Med, Univ Ala, Birmingham, 65-73, coord correlated basic med sci educ, 71-73, assoc dir off undergrad med educ, 72-73; PROF PHYSIOL & BIOPHYS & ASSOC DEAN COL BASIC MED SCI, UNIV TENN CTR HEALTH SCI, MEMPHIS, 73- Mem: Biophys Soc; Am Soc Zool; Endocrine Soc; Am Physiol Soc. Res: Experimental endocrinology; control of water and electrolyte balance; active ion transport; developmental physiology. Mailing Add: Col of Basic Med Sci Univ of Tenn Ctr for Health Sci Memphis TN 38163

TAYLOR, ROBERT FRANKLIN, b Cambridge, Ill, Feb 7, 15; m 41; c 4. RESOURCE MANAGEMENT, RESEARCH ADMINISTRATION. Educ: Beloit Col, BS, 37; Univ Wis, PhD(chem), 41. Prof Exp: Res chemist, Com Solvents Corp, Ind, 41-44, tech serv rep, 44-47; lab supvr, Pittsburgh Plate Glass Co, Wis, 48-50; proj mgr plastic pipe proj, Minn Mining & Mfg Co, 50-57; regional mgr, Arthur D Little, Inc, Ill, 57-72; consult, 72-74; DIR, MGT CTR, COL ST THOMAS, 74- Mem: AAAS; Am Mkt Asn; Am Chem Soc. Res: Marketing chemical products; new product invention, development and market introduction; new business management; organizational development; dynamics of operating groups; problems of service industries; management and organizational training; evaluation of the effectiveness of training programs and techniques. Mailing Add: Mgt Ctr Col of St Thomas St Paul MN 55105

TAYLOR, ROBERT GAY, b Cleveland, Ohio, July 8, 40. MICROBIOLOGY, ENVIRONMENTAL SCIENCES. Educ: Wittenberg Univ, BS, 63; John Carroll Univ, MS, 66; Tex A&M Univ, PhD(environ studies), 69. Prof Exp: Sr bacteriologist, Cleveland Dept Pub Health, 63-65; asst prof, 69-73, ASSOC PROF MICROBIOL, EASTERN N MEX UNIV, 73- Concurrent Pos: Dept Interior, Water Resources Res Inst grant, Eastern NMex Univ, 71-72; vis scientist, Lawrence Berkeley Lab, Univ Calif, 72; consult, Dept Civil Eng, Univ Tex, El Paso, 72- Honors & Awards: Outstanding Res Award, Nat Air Pollution Control Asn, 67. Mem: AAAS; Am Chem Soc. Res: Biomedical biochemical mechanisms; environmental microbiology with respect to water treatment and contamination. Mailing Add: PO Box 2296 Portales NM 88130

TAYLOR, ROBERT JAMES, statistics, operations research, see 12th edition

TAYLOR, ROBERT JOE, b Pomona, Calif, May 1, 45; m 67; c 2. POPULATION ECOLOGY. Educ: Stanford Univ, AB, 67; Univ Calif, Santa Barbara, MS, 70,

PhD(biol), 72. Prof Exp: Res assoc ecol, Princeton Univ, 71-72; ASST PROF ECOL, UNIV MINN, ST PAUL, 72- Mem: Ecol Soc Am; Brit Ecol Soc; Soc Pop Ecol. Res: The influence of predatory behavior upon population in space and time; patterns of species diversity on peninsulas. Mailing Add: Dept of Ecol & Behav Biol 310 Biol Ctr Univ of Minn St Paul MN 55108

TAYLOR, ROBERT JOSEPH, b Salt Lake City, Utah, Dec 10, 41; m 67; c 3. ENGINEERING PHYSICS. Educ: Univ Utah, BA, 67; Cornell Univ, PhD(appl physics), 71. Prof Exp: Scientist metall, Res Ctr, Kennecott Copper Co, 67; scientist acoust, Interand Corp, 71-72; PHYSICIST, JOHNS HOPKINS UNIV APPL PHYSICS LAB, 72- Mem: Am Phys Soc. Res: Electromagnetic wave propagation through the ionosphere for global dissemination of submicrosecond time from satellites; systems studies for determining the lowest cost energy conversion systems. Mailing Add: Johns Hopkins Univ Appl Physics Lab Johns Hopkins Rd Laurel MD 20810

TAYLOR, ROBERT LAWRENCE, mathematics, see 12th edition

TAYLOR, ROBERT LEE, b Palmer, Nebr, July 17, 25; m 45. MEDICAL MYCOLOGY. Educ: Nebr Wesleyan Univ, AB, 47; Univ Nebr, MA, 50; Duke Univ, PhD(microbiol), 54; Am Bd Microbiol, dipl. Prof Exp: Asst chief bact & chief mycol sect, Lab Serv, Fitzsimons Army Hosp, Med Serv Corp, US Army, Colo, 50-55, chief bact sect, Med Res & Develop Unit, 55-58 & mycol sect, Middle Am Res Unit, Ancon, CZ, 58-61, mycologist, Walter Reed Army Inst Res, 61-63, chief mycol sect, 63-66, chief dept bact & mycol, SEATO Med Res Lab, Bangkok, Thailand, 66-67; PROF MICROBIOL, UNIV TEX HEALTH SCI CTR, SAN ANTONIO, 68-, DEP CHMN DEPT, 72- Mem: Fel Am Acad Microbiol; Am Soc Microbiol; Mycol Soc Am; Am Thoracic Soc; Med Mycol Soc Am. Res: Medical microbiology. Mailing Add: Dept of Microbiol Univ of Tex Health Sci Ctr San Antonio TX 78284

TAYLOR, ROBERT LEE, b Tenn, July 23, 43; m 68. MATHEMATICS, STATISTICS. Educ: Univ Tenn, Knoxville, BS, 66; Fla State Univ, MS, 69, PhD(statist), 70. Prof Exp: Teaching asst math, Univ Tenn, Knoxville, 66-67; asst prof, 71-74, ASSOC PROF MATH, UNIV SC, 74- Mem: Math Asn Am; Inst Math Statist. Res: Probability; probabilistic functional analysis and statistical applications. Mailing Add: Dept of Math & Comput Sci Univ of SC Columbia SC 29208

TAYLOR, ROBERT MACKAY, b Vancouver, BC, Sept 9, 13; m 41; c 2. INTERNAL MEDICINE. Educ: Univ Toronto, BA, 36, MD, 39. Prof Exp: Dir med div, Atomic Energy Can Ltd, Ont, 50-55; EXEC DIR, NAT CANCER INST CAN & EXEC VPRES, CAN CANCER SOC, 55- Concurrent Pos: Asst prof med, Univ Toronto, 55-; consult, Ont Cancer Inst, 58-; secy-gen, Int Union Against Cancer, 66-74, vpres, 74- Mem: Am Asn Cancer Res; Can Soc Clin Invest; Can Med Asn. Res: Nutrition; clinical uses of radioisotopes in cancer research. Mailing Add: Can Cancer Soc Nat Cancer Ins Can 77 Bloor St W Toronto ON Can

TAYLOR, ROBERT MARTIN, b Los Angeles, Calif, Aug 8, 09; m 41; c 3. GEOGRAPHY. Educ: Univ Wash, BA, 30, MA, 31, PhD(geog), 56. Prof Exp: Foreign serv officer, US Dept State, 31-53; teaching asst geog, Univ Wash, 53-56; asst prof, Univ Toronto, 56-57; assoc prof int trade, resources & transp & geog, Univ Tex, Austin, 57-66; prof geog & chmn dept, Tex Christian Univ, 66-74; RETIRED. Mem: Asn Am Geog. Res: Economic geography. Mailing Add: 12001 Maplerock Ct San Antonio TX 78230

TAYLOR, ROBERT MORGAN, b Orange, NJ, May 13, 41; m 65; c 2. ANALYTICAL CHEMISTRY, ELECTROCHEMISTRY. Educ: Williams Col, BA, 63; Pa State Univ, PhD(chem), 68; Drexel Univ, MBA, 74. Prof Exp: From scientist to sr scientist, 68-70, PRIN SCIENTIST, TECH CTR, LEEDS & NORTHRUP CO, 70- Mem: Am Chem Soc; Electrochem Soc. Res: Potentiometric and voltammetric analysis; high temperature electrochemistry; molten and solid electrolytes; thermometric analysis; chemical instrumentation. Mailing Add: Leeds & Northrup Co Dickerson Rd North Wales PA 19454

TAYLOR, ROBERT THOMAS, b Harrison, Ark, Sept 14, 36; m 58; c 2. BIOCHEMISTRY. Educ: Univ Calif, Los Angeles, BA, 59; Univ Calif, Berkeley, PhD(biochem), 64. Prof Exp: USPHS res fel, Nat Heart Inst, 64-66, staff res fel, 66-68; RES BIOCHEMIST, BIOMED DIV, LAWRENCE LIVERMORE LAB, 68- Mem: Am Soc Biol Chem; AAAS. Res: Enzymology, one-carbon enzymes, B-12 methyltransferases, folate coenzyme metabolism; heavy metal methylation; alkyl-metal cellular toxicity. Mailing Add: Biomed Div Lawrence Livermore Lab Livermore CA 94550

TAYLOR, ROBERT TIECHE, b San Diego, Calif, June 29, 32; m 64; c 2. MEDICAL ENTOMOLOGY. Educ: Okla State Univ, BS, 54, MS, 57, PhD(entom), 60. Prof Exp: Asst prev med officer, US Army, Ft Stewart, Ga, 54-56; malaria specialist, Pan Am Health Orgn, 60-61; spec asst wood preserv & entom, Are Pub Works Off, Chesapeake, US Navy, 61-63; malaria specialist, Malaria Eradication Prog, Port-au-Prince, Haiti, 63-65; sr scientist, 65-72, SCIENTIST DIR ENTOM, BUR TROP DIS, CTR DIS CONTROL, USPHS, 72- Mem: Am Soc Trop Med & Hyg; Royal Soc Trop Med & Hyg; Entom Soc Am; Am Mosquito Control Asn. Res: Conducting and supervising investigations on chemical control of mosquitoes, triatomidae and simuliidae. Mailing Add: Bur of Trop Dis Ctr for Dis Control Atlanta GA 30333

TAYLOR, ROBERT WILLIAM, b Dallas, Tex, Feb 10, 32; m 55; c 3. COMPUTER SCIENCE. Educ: Univ Tex, Austin, BA, 57, MA, 64. Prof Exp: Res scientist psychoacoustics, Defense Res Lab, Univ Tex, Austin, 55-59; teacher math, Howey Acad, Fla, 59-60; systs engr systs design, Martin Co, Fla, 60-61; sr res scientist man-mach systs res, ACF Electronics, Md, 61-62; res mgr electronics & control, Off Advan Res & Technol, NASA Hq, Washington, DC, 62-65; res dir comput sci, Advan Res Projs Agency, Off Secy Defense, 65-69; res dir, Info Res Lab, Univ Utah, 69-70; PRIN SCIENTIST & ASSOC MGR, COMPUT SCI LAB, XEROX PALO ALTO RES CTR, 70- Concurrent Pos: Mem comts vision & bioacoust, Nat Res Coun, 62-65; mem comput sci & eng bd, Nat Acad Sci, 67-69; mem electronic data processing adv bd, Dept Defense, 68-69; mem comput sci adv bd, Stanford Univ, 71-, lectr, 75- Honors & Awards: Cert Appreciation, Advan Res Projs Agency, Off Secy Defense, 69. Mem: Asn Comput Mach; Inst Elec & Electronics Engrs. Res: Interactive information processing and communications systems; central nervous system; computer graphics; artificial intelligence; research and development management. Mailing Add: Comput Sci Lab Xerox Palo Alto Res Ctr 3333 Coyote Hill Rd Palo Alto CA 94304

TAYLOR, RONALD, b Victor, Idaho, Oct 16, 32; m 55; c 2. BOTANY, GENETICS. Educ: Idaho State Univ, BS, 56; Univ Wyo, MS, 60; Wash State Univ, PhD(bot, genetics), 64. Prof Exp: Asst prof bot, 64-68, assoc prof biol, 68-72, PROF BIOL, WESTERN WASH STATE COL, 72- Concurrent Pos: Environ consult, Northwest Pollution Authority, 74- Mem: AAAS; Bot Soc Am; Genetics Soc Am; Am Soc Plant Taxon; Sigma Xi. Res: Cytotaxonomy, chemotaxonomy and evolution of selected higher plant taxa; biosystematics and ecology of Picea. Mailing Add: Dept of Biol Western Wash State Col Bellingham WA 98225

TAYLOR, RONALD CHARLES, b Port Huron, Mich, Nov 28, 32; m 64; c 1. METEOROLOGY. Educ: Univ Calif, Los Angeles, BA, 59; Univ Hawaii, PhD(meteorol), 68. Prof Exp: Asst prof meteorol, St Louis Univ, 68-69, Univ Hawaii, 69-74 & State Univ NY, Brockport, 75; ASSOC PROF METEOROL, INST FLUID DYNAMICS, UNIV MD, COLLEGE PARK, 75- Concurrent Pos: Res contract, US Navy Weather Res Facility, Norfolk, Va, 69- Mem: AAAS; Am Meteorol Soc; Am Geophys Union; Meteorol Soc Japan. Res: Tropical meteorology air-sea interaction; polar meteorology, Antarctic, synoptic and physical. Mailing Add: Inst of Fluid Dynam & Appl Math Meteorol Prog Univ Md College Park MD 20742

TAYLOR, RONALD LEE, b Oakland, Calif, Jan 25, 38; div; c 2. FORENSIC SCIENCE. Educ: San Jose State Univ, AB, 60; Univ Minn, MS, 62, PhD(entom), 64. Prof Exp: Asst res pathobiologist, Univ Calif, Irvine, 64-69; from asst prof to assoc prof biol, Calif State Col, San Bernardino, 69-73; DIR FORENSIC LABS, LOS ANGELES COUNTY DEPT CHIEF MED EXAMR-CORONER, 73- Concurrent Pos: Assoc clin prof path, Sch Med, Univ Southern Calif, 73- Mem: AAAS; Am Inst Biol Sci; Am Inst Forensic Sic (vpres, 75-); Am Acad Forensic Sci; Electron Micros Soc Am. Res: Insects as human food; arthropod integument; application of scanning electron microscopy and energy dispersive x-ray analysis to problems of forensic science; invertebrate pathology. Mailing Add: 4026 Via Padova Claremont CA 91711

TAYLOR, ROSCOE L, b Wolsey, SDak, Dec 1, 23; m 52; c 8. AGRONOMY. Educ: SDak State Col, BS, 48; Iowa State Col, MS, 50. Prof Exp: Agronomist, 51-56, head agron dept, 56-67, RES AGRONOMIST, AGR EXP STA, UNIV ALASKA, 67- Mem: Crop Sci Soc Am; Am Soc Agron; Biomet Soc. Res: Cereal and forage crop breeding; genetics and speciation in naturally occuring grass and legume plant populations. Mailing Add: Inst of Agr Sci Box AE Palmer AK 99645

TAYLOR, ROY LEWIS, b Olds, Alta, Apr 12, 32. PLANT TAXONOMY. Educ: Sir George Williams Univ, BSc, 57; Univ Calif, Berkeley, PhD(bot), 62. Prof Exp: Teaching bot, Univ Calif, Berkeley, 58-60, assoc, 61-62; res officer, Plant Res Inst, Can Dept Agr, 62-65, chief taxon sect, 65-68; DIR BOT GARDEN, UNIV BC, 68- Concurrent Pos: Can rep on subcomt Pac plant genetics, Pac Sci Asn, 63-66; mem gov bd, Biol Coun Can, 66-69, secy, 69-72, from vpres to pres, 72-74. Mem: Int Orgn Biosyst; Am Soc Plant Taxon; fel Linnean Soc London; Can Bot Asn (secy, 65-66, vpres, 66-67, pres, 67-68); Am Asn Bot Gardens & Arboretums (vpres, 73-75, pres, 75-77). Res: Systematic botany of North American Saxifragaceae; systematics and cytotaxonomy of the vascular plants of the Queen Charlotte Islands. Mailing Add: Bot Garden Univ of BC Vancouver BC Can

TAYLOR, ROYAL ERVIN, JR, b Los Angeles, Calif, Jan 15, 38; m 69; c 2. ANTHROPOLOGY. Educ: Pac Union Col, AB, 60; Univ Calif, Los Angeles, MA, 65 & 67, PhD(anthrop), 70. Prof Exp: Asst prof anthrop, Calif State Univ, Northridge, 67-69; asst prof, 69-76, ASSOC PROF ANTHROP, UNIV CALIF, RIVERSIDE, 76- Concurrent Pos: NSF fel, 72-73. Mem: Am Anthrop Asn; Soc Am Archaeol; Am Quaternary Asn. Res: Archaeometry, the application of physical science technologies in archaeology specifically the development and application of dating methods, in particular radiocarbon and fluorine diffusion methods. Mailing Add: Radiocarbon Lab Dept of Anthrop Univ of Calif Riverside CA 90650

TAYLOR, RUSSELL JAMES, JR, b Rockville, Conn, Mar 8, 35. BIOCHEMISTRY. Educ: Bates Col, BS, 57; Ohio State Univ, MSc, 62, PhD(biochem), 64. Prof Exp: Res chemist, Parke, Davis & Co, Mich, 57-59; asst biochem, Ohio State Univ, 59-64; res biochemist, Lederle Labs, Am Cyanamid Co, NY, 64-69; sr res biochemist, 69-70, GROUP LEADER BIOCHEM, MCNEIL LABS, INC, 70- Mem: AAAS; Am Chem Soc; NY Acad Sci. Res: Intermediary metabolism and enzymology of amino acids, catecholamines, serotonin and histamine; coenzymes; enzyme inhibitors. Mailing Add: McNeil Labs Inc Ft Washington PA 19034

TAYLOR, SAMUEL EDWIN, b Tuskegee, Ala, Oct 19, 41; m 61; c 2. PHARMACOLOGY. Educ: Univ Ala, Tuscaloosa, BS, 63; Univ Ala, Birmingham, PhD(pharmacol), 71. Prof Exp: NIH trainee pharmacol, Univ Tenn Med Units, 71-72; ASST PROF PHARMACOL, SCH DENT, UNIV ORE HEALTH SCI CTR, 72- Res: Autonomic/cardiovascular pharmacology; effects of autonomic agents on bronchial and vascular smooth muscle; differentiation of autonomic receptors; cardiovascular shock. Mailing Add: Dept of Pharmacol Sch of Dent Univ Ore Health Sci Ctr Portland OR 97201

TAYLOR, SAMUEL G, III, b Elmhurst, Ill, Sept 2, 04; m 38; c 3. ONCOLOGY. Educ: Yale Univ, BA, 27; Univ Chicago, MD, 32. Prof Exp: Dir steroid tumor clin, Univ Ill, 47-71; dir cancer ctr planning, Rush Presby-St Luke's Med Ctr, 72-75, PROF MED, RUSH MED COL, 71-, ASSOC DIR RUSH CANCER CTR, 75- Concurrent Pos: Assoc attend physician, Presby Hosp, 48-60; from asst prof to prof, Univ Ill Col Med, 48-72; rep dept med, Tumor Coun, Univ Ill, 48-72; mem consult staff, Lake Forest Hosp, 54-61; from attend physician to sr attend physician, Presby-St Luke's Hosp, Chicago, 61-, head sect oncol, 61-71, consult, Sect Oncol, 71-; consult, Cancer Control Prog, USPHS, 59-63. Mem: Endocrine Soc; Am Asn Cancer Res; Am Col Physicians; Am Radium Soc; Am Soc Clin Oncol. Res: Systematic therapy for cancer. Mailing Add: Dept of Med Rush Med Col Chicago IL 60612

TAYLOR, SNOWDEN, b New York, NY, June 25, 24; m 49; c 6. HIGH ENERGY PHYSICS, PARTICLE PHYSICS. Educ: Stevens Inst Technol, ME, 50; Columbia Univ, AM, 57, PhD(physics), 59. Prof Exp: Asst & lectr physics, Columbia Univ, 52-58; from instr to assoc prof, 58-72, PROF PHYSICS, STEVENS INST TECHNOL, 72- Honors & Awards: Ottens Award, 63. Mem: Am Phys Soc. Res: Strong interaction physics, primarily using bubble chamber techniques. Mailing Add: Dept of Physics Stevens Inst of Technol Hoboken NJ 07030

TAYLOR, STEPHEN KEITH, b Los Angeles, Calif, Mar 28, 44; m 69; c 1. ORGANIC CHEMISTRY. Educ: Pasadena Col, BA, 69; Univ Nev, Reno, PhD(org chem), 74. Prof Exp: RES CHEMIST, E I DU PONT DE NEMOURS & CO, INC, 73- Mem: Am Chem Soc; Sigma Xi. Res: Reactions of epoxides; photographic chemistry. Mailing Add: Photo Prod Dept Exp Sta E352/129 E I du Pont de Nemours & Co Wilmington DE 19898

TAYLOR, STERLING ELWYNN, b Salt Lake City, Utah, Mar 12, 42; m 64; c 2. AGRICULTURAL METEOROLOGY, ECOLOGY. Educ: Utah State Univ, BS, 66; Washington Univ, PhD(biol), 71. Prof Exp: Atmospheric physicist, US Army Atmospheric Sci Lab, 71-73; AGR METEOROLOGIST, US DEPT COM, NAT WEATHER SERV, 73- Mem: Ecol Soc Am; Am Inst Biol Scientists; Sigma Xi. Res: Influence of climate on plant and animal distribution and biological adaptation to environment; applications of remote sensing technology to environmental and biological measurements. Mailing Add: Environ Studies Serv Ctr Auburn Univ Auburn AL 36830

TAYLOR, STEVE L, b Portland, Ore, July 19, 46; m 73. FOOD SCIENCE, TOXICOLOGY. Educ: Ore State Univ, BS, 68, MS, 69; Univ Calif, Davis, PhD, 73. Prof Exp: Res assoc, Dept Food Sci, Univ Calif, Davis, 73-74, fel toxicol, 74-76; RES CHEMIST FOOD TOXICOL, DEPT NUTRIT, LETTERMAN ARMY INST RES, 76- Concurrent Pos: Fel, Nat Inst Environ Health Sci, 74-76. Mem: Inst Food Technologists; Am Chem Soc. Res: Toxicological evaluation of food chemicals and products including improved analytical methods, evaluation of relative hazard, analysis of toxin levels in foods, and establishment of toxicological guidelines for military subsistence. Mailing Add: Food Hyg Div Dept of Nutrit Letterman Army Inst of Res Presidio San Francisco CA 94129

TAYLOR, STUART ROBERT, b Brooklyn, NY, July 15, 37; m 63; c 3. PHYSIOLOGY, BIOPHYSICS. Educ: Cornell Univ, BA, 58; Columbia Univ, MA, 61; NY Univ, PhD(physiol), 66. Prof Exp: Lab asst zool, Columbia Univ, 59-60, lectr, 60-61; res asst physiol, Inst Muscle Dis, Inc, 62-67; Dept Health, Educ & Welfare rehab res fel, Univ Col, Univ London, 67-69; from instr to asst prof pharmacol, State Univ NY Downstate Med Ctr, 69-71; consult pharmacol, physiol & biophys, Mayo Grad Sch Med, 71-75, ASSOC PROF PHARMACOL, PHYSIOL & BIOPHYS, MAYO MED SCH, UNIV MINN, 75- Concurrent Pos: Mem res allocations comt, Minn Heart Asn, 72-74, mem bd dirs, 73-75; mem physiol study sect, Div Res Grants, NIH, 73-77; estab investr, Am Heart Asn, 74-79. Mem: Am Physiol Soc; Biophys Soc; Brit Biophys Soc; assoc Brit Physiol Soc; Soc Gen Physiol. Res: Physiology of stimulus-response coupling in contractile cells. Mailing Add: Mayo Grad Sch of Med Mayo Found Rochester NY 55901

TAYLOR, SUSAN SEROTA, b Racine, Wis, June 20, 42; m 65; c 1. BIOCHEMISTRY. Educ: Univ Wis-Madison, BA, 64; Johns Hopkins Univ, PhD(biochem), 68. Prof Exp: NIH fel, Med Res Coun Lab Molecular Biol, Cambridge Univ, 68-70; NIH fel, 71-72, ASST PROF CHEM, UNIV CALIF, SAN DIEGO, 72- Concurrent Pos: NIH career develop award & res grant, Univ Calif, San Diego, 72- Mem: Am Chem Soc. Res: Protein chemistry; amino acid sequencing. Mailing Add: 4080 Basic Sci Bldg Univ of Calif San Diego La Jolla CA 92093

TAYLOR, TERRY MAC, b Bertram, Tex, June 23, 43; m 65; c 1. RUMINANT NUTRITION. Educ: Southwest Tex State Univ, BS, 66; Tex A&M Univ, PhD(animal nutrit), 73. Prof Exp: SR RES SCIENTIST ANIMAL NUTRIT, NORWICH PHARMACAL CO, DIV MORTON-NORWICH PROD, INC, 73- Mem: Am Soc Animal Sci. Res: Determination of efficacy of new drugs for improvement of production efficiency of cattle and sheep, generally involving studies of ruminal nitrogen and energy metabolism. Mailing Add: Norwich Pharmacal Co PO Box 191 Norwich NY 13815

TAYLOR, THEODORE BREWSTER, b Mexico City, Mex, July 11, 25; nat US; m 48; c 5. APPLIED PHYSICS. Educ: Calif Inst Technol, BS, 45; Cornell Univ, PhD(theoret physics), 54. Prof Exp: Physicist, Radiation Lab, Univ Calif, 46-49, theoret physicist, Los Alamos Sci Lab, 49-56; nuclear physicist, High Energy Fluid Dynamics Dept & chmn, Gen Atomic Div, Gen Dynamics Corp, 56-64; dep dir, Defense Atomic Support Agency, 64-66; chmn bd, Int Res & Tech Corp, 67-76; PROF AEROSPACE & MECH SCI, PRINCETON UNIV, 76- Concurrent Pos: Consult, Govt & Indust, 56-; lectr, San Diego State Col, 57. Honors & Awards: Lawrence Mem Award, AEC, 65; Meritorious Civilian Serv Medal, Secy Defense, 66. Mem: AAAS; Am Phys Soc; Solar Energy Soc. Res: International control and development of nuclear energy; nuclear explosives and effects of nuclear explosions; space propulsion; pollution control; technology assessment; solar energy systems; controlled environment agriculture. Mailing Add: Dept Aerospace & Mech Sci Eng Quad Princeton Univ Princeton NJ 08540

TAYLOR, THOMAS IVAN, chemistry, deceased

TAYLOR, THOMAS NEWTON, b Cedar Rapids, Iowa, June 21, 44. SURFACE PHYSICS. Educ: Iowa State Univ, BS, 66; Brown Univ, MS & PhD(physics), 73. Prof Exp: Fel surface physics, Lawrence Berkeley Lab & Lawrence Livermore Lab, 73-75; STAFF PHYSICIST, LOS ALAMOS SCI LAB, 75- Mem: Am Phys Soc; Am Vacuum Soc. Res: Surface properties of actinide elements and their compounds; emphasis on catalysis and corrosion; low energy electron diffraction, Auger electron spectroscopy and allied techniques. Mailing Add: Los Alamos Sci Lab CMB-8 MS-734 PO Box 1663 Los Alamos NM 87545

TAYLOR, THOMAS NORWOOD, b Lakewood, Ohio, June 14, 37; m 59; c 5. PALEOBOTANY. Educ: Miami Univ, AB, 60; Univ Ill, Urbana, PhD(bot), 64. Prof Exp: NSF res fel, Yale Univ, 64-65; from asst prof to prof biol sci, Univ Ill, Chicago Circle, 65-72, dir scanning electron microscope lab, 67-72; prof bot, Ohio Univ, 72-74; PROF BOT & CHMN DEPT, OHIO STATE UNIV, 74- Concurrent Pos: NSF res grants, 64-; NSF res grants paleobot, 65-74; res assoc, Geol Dept, Field Mus Natural Hist, Chicago, 67-; Ill Acad Sci res grants, 70-72. Mem: Bot Soc Am; Brit Paleont Asn; Int Orgn Paleobot; Int Asn Plant Taxon. Res: Structure and evolution of Paleozoic vascular plants; electron microscopy of fossil pollen and spores; morphology of extant vascular plants. Mailing Add: Dept of Bot Ohio State Univ 1735 Neil Ave Columbus OH 43210

TAYLOR, THOMAS TALLOTT, b Montpelier, Ind, Apr 18, 21; m 58; c 2. PHYSICS. Educ: Purdue Univ, BS, 42; Calif Inst Technol, MS, 53, PhD(physics), 58. Prof Exp: Engr, Gen Elec Co, 42-46; res physicist, Hughes Aircraft Co, 46-54; asst prof physics, Univ Calif, Riverside, 58-63; from asst prof to assoc prof, 63-71, PROF PHYSICS, LOYOLA UNIV LOS ANGELES, 71-, CHMN DEPT, 67- Concurrent Pos: Instr, Eng Exten, Univ Calif, Los Angeles, 47-50 & 57-58. Mem: AAAS; Am Phys Soc; fel Inst Elec & Electronics Eng; Am Asn Physics Teachers. Res: Electromagnetic theory; antennas; lattice sums in crystals. Mailing Add: 6622 W87th St Los Angeles CA 90045

TAYLOR, TIMOTHY H, b Sawyer, Ky, July 4, 18; m 45; c 2. AGRONOMY. Educ: Univ Ky, BS, 48, MS, 50; Pa State Univ, PhD(agron), 55. Prof Exp: Asst agronomist, Va Agr Exp Sta, 49-55; assoc agronomist, Agr Exp Sta, 55-60, assoc prof, 60-66, PROF AGRON, UNIV KY, 67- Concurrent Pos: Consult, 59-; vis scientist, Am Soc Agron; vis biologist, Am Inst Biol Sci. Honors & Awards: Merit Cert, Am Forage & Grassland Coun; Centennial Medal, US Weather Serv. Mem: Fel Am Soc Agron; Crop Sci Soc Am; Brit Grassland Soc; Am Forage & Grassland Coun; Sigma Xi. Res: Ecology of humid temperate grasslands; forage crop ecology and physiology; ecology of cultivated grasslands. Mailing Add: Dept of Agron Univ of Ky Lexington KY 40506

TAYLOR, W J RUSSELL, b Winnipeg, Man, May 6, 30; m 63; c 2. CLINICAL PHARMACOLOGY. Educ: Univ Man, BA, 51, MD & BSc, 56; Univ Toronto, MA, 59, PhD, 69. Prof Exp: Intern, Winnipeg Gen Hosp, Man, 55-56; teacher, Inst Aviation Med, Toronto, Ont, 56-58, Sch Aviation Med, 56-61 & Dept Pharmacol & Therapeut, 58-65, in chg undergrad teaching of pharmacol, Fac Dent, 62-65; res dir spec treatment unit, Newark City Hosp, NJ, 65-67, course coordr & maj lectr, Educ Comt, 65-67; ASSOC PROF MED, HAHNEMANN MED COL, 67-, DIR, HAHNEMANN MED CLIN, 68-; DIR CLIN PHARMACOL-TOXICOL CTR, PHILADELPHIA GEN HOSP, 67- Concurrent Pos: Fel, Sunnybrook Vet Hosp, Toronto, 58-60; mem attend staff, Ajax & Pickering Gen Hosp, Ont, 56-58; mem attend & teaching staff, Toronto Western Hosp, 58-66; res asst, Toronto Gen Hosp, 58-66; dir clin pharmacol group, NIH, 62-65; ed, J Appl Therapeut, 62-; clin pharmacologist, Hoffmann-La Roche, Inc, NJ, 65-67; clin toxicologist, Essex County Med Examr Off, 66-; ed, Int J Clin Pharmacol, Therapeut & Toxicol, 67-; mem staff, Hahnemann Hosp, Philadelphia, 67-; sr attend physician, Philadelphia Gen Hosp, 68-, consult clin toxicologist, 71-; dir, Anvil Res Assocs Ltd, 69- Mem: Aerospace Med Asn; fel Am Soc Clin Pharmacol & Therapeut; Am Fedn Clin Res; NY Acad Sci; Int Soc Clin Pharmacol (vpres, 71-). Res: Internal medicine; clinical toxicology; health care research. Mailing Add: Clin Pharmacol-Toxicol Ctr Philadelphia Gen Hosp Philadelphia PA 19104

TAYLOR, WALTER ALLEN, chemistry, see 12th edition

TAYLOR, WALTER FULLER, b Boston, Mass, Dec 5, 40; m 67. MATHEMATICS. Educ: Swarthmore Col, AB, 62; Harvard Univ, MA, 63, PhD(math), 68. Prof Exp: Asst prof, 67-72, NSF grants, 69-74, ASSOC PROF MATH, UNIV COLO, 72- Concurrent Pos: Fulbright Found sr res fel, Univ New SWales, Australia, 75. Mem: Am Math Soc. Res: Model theory, universal algebra and topology. Mailing Add: Dept of Math Univ of Colo Boulder CO 80309

TAYLOR, WALTER HERMAN, JR, b Laurens, SC, July 5, 31; m 55; c 2. MICROBIAL PHYSIOLOGY. Educ: Duke Univ, MA, 54; Univ Ill, PhD(bact), 59. Prof Exp: Asst bact, Univ Ill, 54-57 & Emory Univ, 57-59; res assoc biol div, Oak Ridge Nat Lab, 59, USPHS fel, 59-61; from asst prof to assoc prof, 61-69, PROF BIOL, PORTLAND STATE UNIV, 69- Mem: AAAS; Am Soc Microbiol; Brit Soc Gen Microbiol; Am Chem Soc. Res: Bacterial physiology; carbohydrate metabolism; protein biosynthesis; pyrimidine metabolism in microorganisms. Mailing Add: Dept of Biol Portland State Univ PO Box 751 Portland OR 97207

TAYLOR, WALTER KINGSLEY, b Calhoun, Ky, Nov 12, 39; m 68. ORNITHOLOGY, VERTEBRATE ECOLOGY. Educ: Murray State Univ, BS, 62; La Tech Univ, MS, 64; Ariz State Univ, PhD(zool), 67. Prof Exp: Asst, La Tech Univ, 62-64 & Ariz State Univ, 64-67; asst prof biol, 69-71, ASSOC PROF BIOL, FLA TECHNOL UNIV, 71- Mem: Am Ornith Union; Wilson Ornith Soc; Cooper Ornith Soc; Am Zool Soc. Res: Breeding biology; migratory biology; population ecology of birds. Mailing Add: Dept of Biol Fla Technol Univ Orlando FL 32816

TAYLOR, WALTER ROWLAND, b Baltimore, Md, Dec 31, 18; m 49; c 2. OCEANOGRAPHY. Educ: Wash Col, Md, BS, 40; Univ Wis, MS, 47, PhD(biochem), 49. Prof Exp: Asst chem, Ga Inst Technol, 40-41; asst biochem, Univ Wis, 46-49; res assoc chem, Univ Ill, 49-51; instr physiol chem, Sch Med, 51-56, res assoc, Chesapeake Bay Inst, 56-74, from asst prof to assoc prof oceanog, 58-68, assoc prof earth & planetary sci, 68-74, PRIN RES SCIENTIST, CHESAPEAKE BAY INST, JOHNS HOPKINS UNIV, 74-, ASST DIR, INST, 75- Mem: AAAS; Am Soc Limnol & Oceanog; Phycol Soc Am; Ecol Soc Am; Int Phycol Soc. Res: Primary production in marine and estuarine environments; nutrition, physiology and ecology of marine organisms. Mailing Add: Chesapeake Bay Inst Johns Hopkins Univ Baltimore MD 21218

TAYLOR, WALTER WILLARD, b Chicago, Ill, Oct 17, 13; m 37, 62, 70; c 3. ANTHROPOLOGY, ARCHAEOLOGY. Educ: Yale Univ, AB, 35; Univ NMex, 37; Harvard Univ, PhD(anthrop), 43. Prof Exp: Instr sci, Ariz State Teachers Col, 37-38; asst instr anthrop, Harvard Univ, 40-42; instr, Univ Tex, 42-43; Rockefeller Found fel, NMex, 46; vis asst prof, Univ Wash, 49; Guggenheim fel, NMex, 50-51; vis prof, Univ Wash, 53; prof, Nat Sch Anthrop & Hist, Mex, 55-58; prof, 58-74, chmn dept, 58-63, EMER PROF ANTHROP, SOUTHERN ILL UNIV, CARBONDALE, 74- Concurrent Pos: Vis prof, Technol Inst Merida, Washington Univ & St Michael's Col, NMex; lectr man & cult, Int Seminars, Friends Serv Comt, 48-53; Fulbright res award, US Dept of State, 62-63; NSF grants, Mex, 64, Spain, 67 & NMex, 71-73; Wenner-Gren Found grant, NMex, 69-70; consult, Black Mesa Proj, Southern Ill Univ, 74- Honors & Awards: Kaplan Res Award, Sigma Xi, 73. Mem: Fel AAAS; fel Am Anthrop Asn; Soc Am Archaeol; Sigma Xi; Mex Anthrop Soc. Res: Archaeological theory and method; cultural theory; archaeology of the Greater Southwest; neolithic and beaker ware cultures of Western Europe, especially Spain. Mailing Add: PO Box 5334 Santa Fe NM 87501

TAYLOR, WARREN EGBERT, b Colorado Springs, Colo, Nov 15, 20; m 47; c 3. METROLOGY. Educ: Kalamazoo Col, BA, 47; Ohio State Univ, PhD(physics), 52. Prof Exp: Mem tech staff, 52-70, PROJ LEADER, VACUUM METROLOGY GROUP, SANDIA LABS, 70- Mem: AAAS; Am Phys Soc; Am Vacuum Soc. Res: Vacuum technology; nuclear radiation measurements; health physics. Mailing Add: Div 9531 Sandia Labs Albuquerque NM 87115

TAYLOR, WELTON IVAN, b Birmingham, Ala, Nov 12, 19; m 45; c 2. MICROBIOLOGY. Educ: Univ Ill, Urbana, AB, 41, MS, 47, PhD(bact), 48. Prof Exp: From instr to asst prof bact, Univ Ill Col Med, 48-54; res bacteriologist, Swift & Co, 54-59; supvr clin microbiol, Children's Mem Hosp, 59-64; bacteriologist in chief, West Suburban Hosp, Oak Park, Ill, 64-69; ASSOC PROF MICROBIOL, UNIV ILL MED CTR, 69- Concurrent Pos: Consult, Chicago Park Dist, 49 & Inst Gas Technol, Ill Inst Technol, 60-; Nat Inst Allergy & Infectious Dis spec res fel, Inst Pasteur, France & Cent Pub Health Lab, Eng, 61-62; consult microbiologist, Northwest Community Hosp, Arlington Heights, Ill, 63-70, Jackson Park & Englewood Hosps, Chicago, 64-, Armour & Co, 66-68, Resurrection Hosp, Chicago, 67-, Grant Hosp, Chicago, 69-75, St Mary of Nazareth Hosp Ctr, Chicago, 73- & Swedish Covenant Hosp, Chicago, 74-; Nat Commun Dis Ctr res grant, 66-68; Dept Army res contract, 71-73. Mem: Am Soc Microbiol. Res: Detection of Vibrio parahemolyticus in routine stool analysis; methods for detection of Salmonella and Shigella with minimal laboratory facilities; rapid identification procedures for non-enteric pathogens. Mailing Add: 7621 S Prairie Ave Chicago IL 60619

TAYLOR, WENDELL HERTIG, b Uniontown, Pa, June 15, 05. CHEMISTRY. Educ: Princeton Univ, BS, 26, AM, 30, PhD(chem), 33. Prof Exp: Res chemist, Exp Sta, E I du Pont de Nemours & Co, 26-29; asst chem, Princeton Univ, 29-31 & 32-33, from instr to asst prof, 33-43; chmn dept sci, 43-70, EMER CHMN DEPT SCI, LAWRENCEVILLE SCH, 70- Concurrent Pos: Mem chem bond approach comt, NSF, 59- Mem: Am Chem Soc. Res: Organic sulfur compounds; phenolaldehyde condensations; hydramides; history of science. Mailing Add: 122 Patton Ave Princeton NJ 08540

TAYLOR, WILBUR SPENCER, b Junction City, Kans, Apr 4, 28; m 52; c 3. INDUSTRIAL MICROBIOLOGY. Educ: Kans State Univ, BS, 49, MS, 50. Prof Exp: Microbiologist, 50-55, group leader res, 55-71, group supvr, 71-73, MGR BIO-PAINT & PLASTICS, R T VANDERBILT CO, INC, 73- Mem: Am Soc Testing & Mat; Am Asn Textile Chemists & Colorists; Soc Indust Microbiol. Res: Research and development of industrial biocides. Mailing Add: R T Vanderbilt Co Inc 33 Winfield St Norwalk CT 06855

TAYLOR, WILLIAM C, inorganic chemistry, see 12th edition

TAYLOR, WILLIAM CLYNE, b Aberdeen, Scotland, Mar 26, 24; Can citizen; m 54; c 3. PEDIATRICS. Educ: Aberdeen Univ, MB & ChB, 45; Univ London, DCH, 50; FRCP(C), 58. Prof Exp: PROF PEDIAT, UNIV ALTA, 57- Concurrent Pos: Mead Johnson res fel, Univ Alta, 58-59; Schering traveling fel, Africa, Australia & NZ, 66-67; Brit Commonwealth grant, UK, 66. Mem: Can Soc Clin Invest; Am Acad Pediat; Can Pediat Soc. Res: Evaluation of undergraduate and postgraduate medical education; delivery of health care in the Northwest Territories of Canada. Mailing Add: Clin Sci Bldg Univ of Alta Edmonton AB Can

TAYLOR, WILLIAM DANIEL, b Cardiff, Gt Brit, May 25, 34. BIOPHYSICS. Educ: Univ Manchester, BSc, 56, PhD(phys chem), 59. Prof Exp: Fel physics, Pa State Univ, 59-61; res fel chem, Univ Manchester, 61-63; from asst prof to assoc prof biophys, 63-71, PROF BIOPHYS & HEAD DEPT, PA STATE UNIV, UNIVERSITY PARK, 71- Concurrent Pos: Mem ad hoc comt theoret biol, NASA, 64-; consult, HRB Singer, Inc, Pa, 66-70. Mem: AAAS; Am Chem Soc; Biophys Soc; Fedn Am Sci; The Chem Soc. Res: Physical chemistry of polymers and biopolymers; replication of nucleic acids; radiation biology. Mailing Add: Dept of Biophys Pa State Univ University Park PA 16802

TAYLOR, WILLIAM EDWIN, b Worland, Wyo, Nov 15, 21; m 43; c 1. MATERIALS SCIENCE. Educ: Purdue Univ, BS, 42, PhD(metall eng), 50. Prof Exp: Instr metall eng, Purdue Univ, 46-48, asst solid state physics, 48-50; metallurgist, Oak Ridge Nat Lab, 50-52; res physicist & mgr mat dept, Semiconductor Prod Div, Motorola, Inc, Ariz, 52-60; prof metall, Mich State Univ, 60-62; mem tech staff, Semiconductor Prod Div, Motorola, Inc, 62-69; vpres, Finch & Taylor, 69-75; SR MATERIALS ENGR, SPECTROLAB, INC, 75- Concurrent Pos: Lectr, Univ Tenn, 51-52 & Univ Ariz, 56-57; consult, Motorola, Inc, 60-61, Control Data Corp, 60-62 & Oak Ridge Nat Lab, 61-62. Mem: AAAS; Am Soc Metals; Am Inst Mining, Metall & Petrol Eng. Res: Semiconductor metallurgy and devices. Mailing Add: 12500 Gladstone Sylmar CA 91342

TAYLOR, WILLIAM EWART, b Toronto, Ont, Nov 21, 27; m 52; c 3. ANTHROPOLOGY, ARCHAEOLOGY. Educ: Univ Toronto, BA, 51; Univ Ill, AM, 52; Univ Mich, PhD(anthrop), 65. Hon Degrees: PhD, Univ Calgary. Prof Exp: Res asst, Univ Mich, 54-56; arctic archaeologist, 56-63, div chief archaeol, 63-67, DIR NAT MUS MAN, NAT MUS CAN, 67- Concurrent Pos: Vchmn bd gov, Arctic Inst NAm; Can mem permanent coun, Int Union Prehist & Prohist Sci; secy Can comt & mem permanent coun, Int Union Anthrop & Ethnol Sci; mem adv comt, Comn Can Studies; mem adv comt, Inst Social & Econ Res, Memorial Univ. Mem: Fel AAAS; fel Royal Anthrop Inst; fel Am Anthrop Asn; fel Sigma Xi. Res: Archaeology of Arctic North America. Mailing Add: Nat Mus Man Nat Mus Can Ottawa ON Can

TAYLOR, WILLIAM F, b Cincinnati, Ohio, Oct 14, 21; m 43; c 3. BIOSTATISTICS. Educ: Univ Calif, PhD(math statist), 51. Prof Exp: Instr biostatist, Univ Calif, 50-51; chief dept biometrics, Sch Aviation Med, US Air Force, 51-58; assoc prof biostatist, Sch Pub Health, Univ Calif, Berkeley, 58-63, prof, 63-67; HEAD MED RES STATIST SECT, MAYO CLIN, 67-, MEM FAC DEPT MED STATIST & EPIDEMIOL, 74- Concurrent Pos: Consult, Am Joint Comn Cancer Staging, 71-; mem, US Nat Comn Vital & Health Statist, 72- Mem: Biomet Soc; fel Am Pub Health Asn; fel Am Statist Asn. Res: Design of clinical trials; sequential analysis in medicine; reference value problems. Mailing Add: Med Res Statist Mayo Clin Rochester MN 55901

TAYLOR, WILLIAM H, II, b Philadelphia, Pa, Dec 17, 38; m 57; c 2. SOLID STATE PHYSICS. Educ: Johns Hopkins Univ, BES, 60; Princeton Univ, MSE, 61, AM, 61, PhD(solid state sci), 64. Prof Exp: Staff mem, Redstone Arsenal Res Div, Rohm and Haas Co, Ala, 63-64; chief, Solid State Br, Explosives Lab, Picatinny Arsenal, 66-68; vpres, Data Sci Ventures, Inc, 68-72; vpres, White, Weld & Co, Inc, 72-73; VPRES, CROCKER CAPITAL CORP, 73- Mem: Am Phys Soc. Res: Investment banking, venture capital financing of technology companies; point defects in solids; properties of solid explosives; soft contact lens materials. Mailing Add: Crocker Capital Corp Rm 600 111 Sutter St San Francisco CA 94104

TAYLOR, WILLIAM HENRY, JR, chemistry, see 12th edition

TAYLOR, WILLIAM IRVING, b NZ, July 23, 23; m 52; c 3. ORGANIC CHEMISTRY. Educ: Univ Auckland, PhD(chem), 48, DSc, 68. Prof Exp: Nat Res Coun scholar, Switz, 48-49; Nat Res Coun Can fel, 50; Imp Chem Industs fel, Cambridge Univ, 51-52; assoc prof chem, Univ NB, 52-55; chemist, Ciba Pharmaceut Co, 55-62, dir natural prod, 63-67, dir biochem res, 67-68; dir fragrance res, 68-71, VPRES RES & DEVELOP, INT FLAVORS & FRAGRANCES, INC, 71-, DIR CHEM SYNTHESIS & DEVELOP, 72- Concurrent Pos: Guest investr, Rockefeller Univ, 66-67. Mem: Am Chem Soc; Am Soc Pharmacog; NY Acad Sci; fel The Chem Soc. Res: Synthesis and structural elucidation of natural products, especially in field of flavor and aroma chemicals. Mailing Add: Int Flavors & Fragrances Inc 1515 Hwy 36 Union Beach NJ 07735

TAYLOR, WILLIAM JAPE, b Bonneville, Miss, Sept 5, 24; m 48; c 4. INTERNAL MEDICINE, CARDIOLOGY. Educ: Yale Univ, BS, 44; Harvard Med Sch, MD, 47. Prof Exp: Intern, Second Harvard Med Serv, Boston City Hosp, 47-48; asst resident med, Duke Univ, 48-50, instr, Sch Med & Hosp, 54-55; instr, Sch Med, Univ Pittsburgh, 55-58; chief, Div Cardiol, 58-74, from asst prof to assoc prof, 58-64, prof med, 64-74, DISTINGUISHED SERV PROF MED, COL MED, UNIV FLA, 74- Concurrent Pos: Fel coun clin cardiol, Am Heart Asn; Am Col Physicians res fel med, Duke Univ, 50-51, univ res fel, Sch Med & Hosp, 51-52, USPHS res fel, 54-55; vis prof med, Fac Health Sci, Univ Ife, Ile-Ife, Nigeria, 74-75. Mem: Am Fedn Clin Res; Asn Univ Cardiol; fel Am Col Cardiol; fel Am Col Physicians. Mailing Add: Dept of Med Univ of Fla Col of Med Gainesville FL 32601

TAYLOR, WILLIAM JOHN, JR, chemistry, see 12th edition

TAYLOR, WILLIAM JOHNSON, b Chengtu, China, Dec 3, 16; US citizen; m 49; c 4. PHYSICAL CHEMISTRY, THEORETICAL CHEMISTRY. Educ: Denison Univ, AB, 37; Ohio State Univ, PhD(phys chem), 42. Prof Exp: Instr chem, Univ Calif, 41-42; res assoc, Thermochem Sect, Nat Bur Standards, 42-47; spectros & low-temperature res, Cryogenic Lab, 47-50, from asst prof to assoc prof chem, 50-61, PROF CHEM, OHIO STATE UNIV, 61- Mem: Am Chem Soc; Am Phys Soc. Res: Statistical mechanics and thermodynamics; theory of long-chain molecules; molecular vibrations; dipole moments; self-consistent field and localized molecular orbitals; configuration interaction method and correlation energy; foundations of quantum mechanics. Mailing Add: Dept of Chem Ohio State Univ Columbus OH 43210

TAYLOR, WILLIAM L, b Corsicana, Tex, Oct 17, 26; m 50; c 3. METEOROLOGY, ELECTROMAGNETISM. Educ: Okla State Univ, BS, 50. Prof Exp: Physicist, Nat Bur Standards, Washington, DC, 50-51, Alaska, 51-52 & Colo, 52-65; physicist, Inst Telecommun Sci & Aeronomy, Environ Sci Serv Admin, 65-70, PHYSICIST, ENVIRON RES LABS, NAT OCEANIC & ATMOSPHERIC ADMIN, US DEPT COM, 70- Mem: AAAS; Am Geophys Union; Sigma Xi; Int Union Radio Sci. Res: Radio propagation between the earth and ionosphere in the lower frequency bands, using lightning discharges as the source; measurement of radio noise from severe thunderstorms and tornadoes. Mailing Add: Environ Res Lab NOAA US Dept Com Boulder CO 80302

TAYLOR, WILLIAM LEROY, physics, physical chemistry, see 12th editlon

TAYLOR, WILLIAM LLOYD, inorganic chemistry, see 12th edition

TAYLOR, WILLIAM RALPH, b Spring Hill, Kans, Aug 29, 19; wid; c 6. ICHTHYOLOGY. Educ: Univ Kans, AB, 47; Univ Mich, MS, 51, PhD(zool), 55. Prof Exp: Asst mus natural hist, Univ Kans, 38-41; asst, Div Fishes, Mus Zool, Univ Mich, 47-51; biologist, Inst Fisheries Res, State Dept Conserv, Mich, 52-53; aquatic biologist, State Wildlife & Fisheries Comn, La, 54-56; ASSOC CURATOR FISHES, NAT MUS NATURAL HIST, 56- Mem: Soc Syst Zool (treas, 70-72); Am Fisheries Soc; Am Soc Ichthyologists & Herpetologists; Am Soc Limnol & Oceanog. Res: Taxonomy and distribution of fishes; fishery biology; vertebrate zoology. Mailing Add: Div Fishes Nat Mus Natural Hist Washington DC 20560

TAYLOR, WILLIAM RANDOLPH, b Philadelphia, Pa, Dec 21, 95; m 26; c 2. BOTANY. Educ: Univ Pa, BS, 16, MS, 17, PhD(bot), 20. Prof Exp: Asst bot, Univ Pa, 15-17, from instr to asst prof, 18-27, prof, 27-30; prof, 30-66, Russel lectr, 64, EMER PROF BOT, UNIV MICH, ANN ARBOR, 66-, CURATOR ALGAE, 30- Prof Exp: Res assoc, Acad Natural Sci Philadelphia, 25-30 & Farlow Herbarium, Harvard Univ, 62-; mem corp, Woods Hole Marine Biol Lab, 25-, trustee, 39-, chmn libr comt, 47-53; mem, Hancock Exped, Galapagos Islands, SAm & Cent Am, 33-34 & Cent Am & Caribbean, 39; mem corp, Bermuda Biol Sta, 35-, trustee, 53-; mem adv comn, Charles Darwin Biol Sta, Galapagos, Ecuador; sr biologist, Oceanog Sect, Bikini Exped, Crosroads, 46; mem expeds to Jamaica, 56, Lesser Antilles, 66, 67 & 68 & Bermuda. Honors & Awards: Retzius Medal, Univ Lund, 48. Mem: AAAS; Am Soc Naturalists (secy, 43-46); Bot Soc Am (vpres, 56); Phycol Soc Am (pres, 48); fel Am Acad Arts & Sci. Res: Fresh water algae; North Atlantic marine algae, marine algae of the tropical Americas, Falkland Islands and Magellan Strait, Galapagos Islands, Pacific and Indian Oceans. Mailing Add: Dept of Bot Univ of Mich Ann Arbor MI 48104

TAYLOR, WILLIAM WEST, b Northampton Co, NC, Dec 4, 23; m 53; c 4. PHARMACY. Educ: Univ NC, BS, 47, PhD(pharm), 62. Prof Exp: Intern hosp pharm, Duke Hosp, Durham, NC, 46-47, staff pharmacist, 48; chief pharmacist, Strong Mem Hosp, Rochester, NY, 48; instr hosp pharm, 52-62, ASST PROF HOSP PHARM, UNIV NC, CHAPEL HILL, 62- Concurrent Pos: Chief pharmacist, NC Mem Hosp, 52-68, assoc dir, Div Pharm Serv, 68-74, special formulations pharmacist, 75- Mem: Am Pharmaceut Asn; Am Soc Hosp Pharmacists; Am Asn Cols Pharm. Res: Hospital pharmacy; drug control; special compounding and dosage preparation; purification and formulation of dyes for clinical purposes. Mailing Add: Rte 2 Whitfield Rd Chapel Hill NC 27514

TAYLORSON, RAYMOND BRIERLY, b Providence, RI, June 10, 32; m 56; c 2. PLANT PHYSIOLOGY, HORTICULTURE. Educ: Univ RI, BS, 54; Univ Wis, PhD(hort, bot), 58. Prof Exp: Plant physiologist, Ga Coastal Plain Exp Sta, 65, PLANT PHYSIOLOGIST, PLANT INDUST STA, PLANT SCI RES DIV, AGR RES SERV, USDA, 66- Mem: Am Soc Plant Physiol; Weed Sci Soc Am; Am Soc Hort Sci. Res: Natural and synthetic growth regulators; weed biology; weed seed germination; weed control. Mailing Add: Weed Invests Plant Sci Res Div Agr Res Serv Plant Indust Sta Beltsville MD 20705

TAYMOORIAN, FATEMEH MOEZIE, organic chemistry, physical chemistry, see 12th edition

TAYMOR, MELVIN LESTER, b Brockton, Mass, Feb 10, 19; m 42; c 3. OBSTETRICS & GYNECOLOGY. Educ: Johns Hopkins Univ, AB, 40; Tufts Univ, MD, 43; Am Bd Surg, dipl, 54. Prof Exp: From instr to asst prof, 56-69, ASSOC PROF GYNEC, HARVARD MED SCH, 69-; CHIEF GYNEC, PETER BENT BRIGHAM HOSP, BOSTON, 70- Concurrent Pos: USPHS fel, 50-51; NIH fels, 60-61; mem, Pop Coun, 58-60. Mem: Am Fertil Soc (treas, 61-66); Endocrine Soc; Soc Gynec Invest; Am Col Obstet & Gynec; Am Gynec Soc. Res: Gonadotropins in reproductive physiology. Mailing Add: Peter Bent Brigham Hosp 721 Huntington Ave Boston MA 02115

TAYSOM, ELVIN DAVID, b Rockland, Idaho, Aug 5, 17; m 39; c 3. ANIMAL HUSBANDRY. Educ: Univ Idaho, BS, 40; Utah State Univ, MS, 50; Wash State Univ, PhD, 61. Prof Exp: Asst animal sci, Utah State Univ, 49-50; asst, Wash State Univ, 50-53, sheep specialist, 53; from asst prof to assoc prof, 53-74, PROF ANIMAL SCI, ARIZ STATE UNIV, 74- Mem: Am Soc Animal Sci. Res: Mineral metabolism and its relation to the formation to urinary calculi and hormone functions in the body. Mailing Add: Dept of Animal Husb Ariz State Univ Tempe AZ 85281

TAYYEB, A, b Sitapur, India, Mar 20, 22; Can citizen; m 48; c 3. GEOGRAPHY. Educ: Muslim Univ, Aligarh, India, BA, 43, MA, 44; Univ Toronto, MA, 60. Prof Exp: Asst geologist, Geol Surv India, 44-46, surv officer, 46-48; lectr geog, Govt Col Rawalpindi, WPakistan, 48-50; from lectr to asst prof, 52-67, ASSOC PROF GEOG, UNIV TORONTO, 67- Concurrent Pos: Consult, Govt Prov Ont, 52-60; Humanities & Soc Res Coun Can grant, 60-; Can Coun grant, Iraq, 68- Mem: Am Geog Soc; Can Asn Geog; Can Inst Int Affairs. Res: Planning; conservation; political geography. Mailing Add: Dept of Geog Univ of Toronto Toronto ON Can

TAZUMA, JAMES JUNKICHI, b Seattle, Wash, July 17, 24; m 54; c 2. CHEMISTRY. Educ: Univ Wash, BSc, 48, PhD(org chem), 53. Prof Exp: Parke Davis Co fel, Wayne State Univ, 52-53; res chemist, Henry Ford Hosp, 53-54 & Food Mach & Chem Corp, NJ, 55-58; sr res chemist, 58-65, SECT HEAD PETROCHEM PROCESS RES, GOODYEAR TIRE & RUBBER CO, 65- Mem: Am Chem Soc. Res: Petroleum chemistry; catalysis; reaction mechanism; process development; rubber chemicals development. Mailing Add: Goodyear Tire & Rubber Co 142 Goodyear Blvd Akron OH 44316

TCHEN, TCHE TSING, b Peiping, China, Oct 1, 24; nat US; m 60. BIOCHEMISTRY. Educ: Aurora Univ, China, ChemE, 48; Univ Chicago, PhD(biochem), 54. Prof Exp: Asst, Univ Chicago, 51-54, res assoc, 54-55; res fel, Harvard Univ, 55-58; assoc prof biochem, 58-61, PROF BIOCHEM, WAYNE STATE UNIV, 61- Concurrent Pos: Am Cancer Soc scholar, 56-58; mem, Physiol Chem Study Sect, USPHS, 60-64; ed, Arch Biochem & Biophys, 68-72 & J Biol Chem, Am Soc Biol Chem, 68-72. Mem: Am Chem Soc; Am Soc Biol Chem. Res: Cellular differentiation; mechanism of cell death; action of ACTH and MSH; pigment cells. Mailing Add: 435 Dept of Chem Wayne State Univ Detroit MI 48202

TCHERTKOFF, VICTOR, b Lausanne, Switz, Aug 7, 19; US citizen; m 42; c 2. PATHOLOGY. Educ: City Col New York, BS, 40; New York Med Col, MD, 43; Am Bd Path, dipl anat, 61. Prof Exp: Asst pathologist, Metrop Hosp, New York, 57-60; assoc prof, 61-67, PROF PATH, NEW YORK MED COL, 67-; DIR PATH LABS, METROP HOSP, 67- Concurrent Pos: Pathologist-in-chg, Metrop Hosp, 61-66. Mem: Col Am Path; Am Soc Clin Path; Int Acad Path; Pan-Am Med Asn; Am Soc Contemporary Med & Surg. Mailing Add: Off of Dir of Labs Metrop Hosp 1901 First Ave New York NY 10029

TCHEUREKDJIAN, NOUBAR, b Beirut, Lebanon, Jan 4, 37; m 71. PHYSICAL CHEMISTRY. Educ: Ill Inst Technol, BS, 58; Lehigh Univ, MS, 60, PhD(phys chem), 63. Prof Exp: SR RES CHEMIST, S C JOHNSON & SON, INC, 63- Mem: Am Chem Soc. Res: Colloid and surface chemistry; physical chemistry; aerosol technology. Mailing Add: Phys Res S C Johnson & Son Inc Racine WI 53403

TCHIR, MORRIS FREDERICK, b Edmonton, Alta, Jan 17, 45; m 67; c 2. CHEMISTRY. Educ: Univ Alta, BSc, 65; Univ Western Ont, PhD(org chem), 69. Prof Exp: Nat Res Coun Can fel chem, Royal Inst Gt Brit, London, 69-71; ASST PROF ORG CHEM, UNIV WATERLOO, 71- Concurrent Pos: Ramsay fel, Univ Col, Univ London, 69-71; Nat Res Coun Can res grants, 71- Mem: Am Soc Photobiol. Res: Photochemistry of aromatic carbonyl compounds; chemical and spectroscopic characterization of orth-quinodimethanes. Mailing Add: Dept of Chem Univ of Waterloo Waterloo ON Can

TCHOLAKIAN, ROBERT KEVORK, b Jerusalem, Palestine, Apr 26, 38; US citizen; c 1. BIOCHEMISTRY, REPRODUCTIVE BIOLOGY. Educ: Berea Col, BS, 58; Fla State Univ, MS, 63; Med Col Ga, PhD(physiol, biochem), 67. Prof Exp: NIH fels, Steroid Biochem Inst, Univ Utah, 67-68 & Univ Southern Calif, 68-69; instr reprod, Univ Southern Calif, 69-70; asst prof endocrinol, M D Anderson Hosp & Tumor Inst, 70-71; ASST PROF REPROD MED & BIOL, UNIV TEX MED SCH HOUSTON, 71- Mem: AAAS; Am Soc Zool; Soc Study Reprod; Am Soc Archology. Res: Comparative endocrinology, primarily sexual differentiation; reproductive biology and steroid biochemistry as related to action of hormones. Mailing Add: Dept of Reprod Med & Biol Univ of Tex Med Sch Houston TX 77025

TEABEAUT, JAMES ROBERT, II, b Fayetteville, NC, Aug 27, 24. PATHOLOGY. Educ: Duke Univ, MD, 47; Am Bd Path, dipl, 53. Prof Exp: Intern path, Duke Hosp, Durham, NC, 48, intern internal med, 49; Rockefeller res fel legal med, Harvard Med Sch, 49-51; chief div forensic path, Armed Forces Inst Path, 51-54; asst prof path, Sch Med, Univ Tenn, 54-59; coroner, Shelby County, Tenn, 55-59; assoc prof, 59-69, PROF PATH, MED COL GA, 69- Concurrent Pos: Lederle med fac award, 54-55; Med examr, State of Ga, 59-; consult, Vet Admin Hosp, Augusta, Ga, 59- & US Army Hosp, Ft Gordon, Ga, 60-, lectr, Mil Police Sch, 63-, hon mem staff, 66- Mem: Col Am Path; Am Soc Clin Path; AMA; Int Acad Path; Int Acad Forensic Sci. Res: Forensic pathology; pathology of human cardiovascular disease. Mailing Add: Dept of Path Med Col of Ga Augusta GA 30902

TEACH, EUGENE GORDON, b Hayward, Calif, Oct 27, 26; m 54; c 1. ORGANIC CHEMISTRY. Educ: St Mary's Col, Calif, BS, 51; Univ Notre Dame, PhD(chem), 53. Prof Exp: Res assoc, Univ Calif, Los Angeles, 53-54, res chemist, USDA, 54-57; RES ASSOC, STAUFFER CHEM CO, 57-, SUPVR, 74- Mem: AAAS; Am Chem Soc; The Chem Soc. Res: Actylene-allene chemistry; high temperature polymers; fluorine chemistry; agricultural chemistry. Mailing Add: 1200 S 47th St Richmond CA 94804

TEACH, WILLIAM CHARLES, b St Louis, Mo, Feb 5, 18; m 43; c 4. CHEMISTRY. Educ: Grinnell Col, BA, 40; State Col Wash, MS, 41. Concurrent Pos: Chemist, Koppers Co, Inc, 42-45; fel, Mellon Inst, 45-52; sr chemist, Res Dept, Koppers Co, Inc, 52-53, asst chief chemist, Chem Div, 53-54; mgr styrene polymer develop, 55-57, mgr impact polystyrene res, 58-61; sr scientist, 61-63, mgr consumer prod res, 63-67, mgr tech serv, 67-69, MGR CONSUMER PROD RES, US BORAX RES CORP, 69- Mem: Am Chem Soc; Am Asn Textile Chemists & Colorists; fel Am Inst Chem; Am Oil Chemists Soc. Res: Thermoplastics; vinyl polymerization; latex coatings; physical testing of plastics; detergents; cleaners. Mailing Add: US Borax Res Corp Box 4111 Anaheim CA 92803

TEAFORD, MARGARET ELAINE, b Union Star, Mo, Feb 2, 28. BIOCHEMISTRY, CLINICAL CHEMISTRY. Educ: Northwest Mo State Col, BS, 50; Univ Mo-Columbia, MS, 59, PhD(biochem), 64; Am Bd Clin Chemists, dipl. Prof Exp: Med technologist, Methodist Hosp, St Joseph, Mo, 51-52, chief med technol, 52-56; NIH traineeship, Univ Wash, 64-66; TECH DIR LAB MED, ALLEN MED LABS, LTD, 66- Mem: AAAS; Am Asn Clin Chemists; Asn Clin Sci. Res: Methodology in clinical chemistry and establishing normal values for the various biochemical parameters in the human. Mailing Add: Allen Med Labs Ltd 2821 N Ballas Rd St Louis MO 63131

TEAGUE, CLAUDE EDWARD, JR, b Sanford, NC, Sept 9, 24; m 68; c 3. CHEMISTRY. Educ: Univ NC, AB, 47, PhD(chem), 50. Prof Exp: Res chemist, Am Viscose Corp, 50-51; res chemist, 52-60, mgr chem res, 60-70, ASST DIR RES, R J REYNOLDS TOBACCO CO, 70- Mem: Am Chem Soc. Res: Synthetic organic chemistry; polymers and synthetic fibers; tobacco chemistry. Mailing Add: R J Reynolds Tobacco Co Fourth & Main Sts Winston-Salem NC 27102

TEAGUE, DAVID BOYCE, b Franklin, NC, May 17, 37; m 64; c 2. APPLIED MATHEMATICS. Educ: NC State Col, BSEE, 59, MS, 61; NC State Univ, PhD(appl math), 65. Prof Exp: Instr math, NC State Univ, 64-65; asst prof, Univ NC, Charlotte, 65-68; ASSOC PROF MATH, WESTERN CAROLINA UNIV, 68- Res: Elasticity; mathematical theory of elasticity; mixed boundary value problems in elasticity. Mailing Add: Dept of Math Western Carolina Univ Cullowhee NC 28723

TEAGUE, DWIGHT MAXWELL, b Shepherd, Mich, July 30, 16; m 39, 65; c 5. INDUSTRIAL CHEMISTRY, AUTOMOTIVE ENGINEERING. Educ: Wayne Univ, BS, 38, MS, 41; Univ Mich, PhD(biochem), 50. Prof Exp: Lab asst, Wayne Univ, 35-38; res chemist, Children's Fund, Mich, 38-41; asst, Med Sch, Univ Mich, 41-43; supvr chem lab, Manhattan Proj, 43-45, asst dept head chem res, 45-56, managing engr chem res, 56-69, coordr emissions res, 69-72, CHIEF RES SCIENTIST, CHRYSLER CORP, 72- Concurrent Pos: Instr, Eve Sch, Chrysler Inst Eng, 45-50, head dept chem, 50-60; res assoc & lectr, Wayne State Univ, 48-; lectr, Detroit Inst Technol, 60-; lectr, Univ Detroit, 70-, mem, Eng Adv Coun, 73-; chmn panel plant damage from air pollution, Coord Res Coun, 71- Mem: AAAS; Am Chem Soc; Electron Micros Soc Am (secy, 57-58, pres, 61); Soc Automotive Engr; Sigma Xi. Res: Electron microscopy, especially of metals, metal surfaces, paints and paint weathering; instrumental analysis; absorption spectroscopy; physical and biological chemistry; friction and lubrication; adhesives; paint; polymers; friction materials; fuel cells and batteries; automotive exhaust emissions; catalytic materials for automotive emissions. Mailing Add: Chrysler Corp Dept 9010 PO Box 1118 Detroit MI 48231

TEAGUE, HAROLD JUNIOR, b Fayetteville, NC, Nov 5, 41; m 70. ORGANIC CHEMISTRY. Educ: Methodist Col, NC, BS, 64; NC State Univ, MS, 67, PhD(org chem), 70. Prof Exp: ASSOC PROF CHEM, PEMBROKE STATE UNIV, 70- Mem: Am Chem Soc. Res: Mechanistic rearrangement studies of sulfur containing ring compounds. Mailing Add: Dept of Chem Pembroke State Univ Pembroke NC 28372

TEAGUE, HOWARD STANLEY, b Rockville, Ind, Jan 16, 22; m 43; c 4. NUTRITION. Educ: Univ Nebr, BS, 48, MS, 49; Univ Minn, PhD, 52. Prof Exp: From asst prof to prof animal sci, Ohio Agr Res & Develop Ctr, 52-72; MEM STAFF, ANIMAL SCI RES DIV, US MEAT ANIMAL RES CTR, 72- Mem: Am Soc Animal Sci; Am Inst Nutrit; Soc Study Reproduction. Res: Nutrition relative to growth and reproductive processes in domestic animals. Mailing Add: Animal Sci Res Div US Meat Animal Res Cr PO Box 166 Clay Center NE 68933

TEAGUE, KEFTON HARDING, b Siler City, NC, Sept 30, 20; m 43; c 2. ECONOMIC GEOLOGY. Educ: NC State Col, BS, 41. Prof Exp: JR assoc geologist, Tenn Valley Authority, 42-51; geologist, US Geol Surv, 51-55; geologist, 55-62, CHIEF DIV GEOLOGIST, INT MINERALS & CHEM CORP, 62- Mem: Geol Soc Am; Soc Econ Geol; Clay Minerals Soc. Res: Areal and structural geology of Georgia; mica pegmatites; sillimanite; kyanite; talc; feldspar; barite; bentonite; quartz. Mailing Add: Int Min & Chem Co IMC Plaza Libertyville IL 60048

TEAGUE, MARION WARFIELD, b Arkadelphia, Ark, July 6, 41; m 62; c 3. PHYSICAL INORGANIC CHEMISTRY. Educ: Ouachita Baptist Col, BS, 63; Purdue Univ, MS, 68, PhD(chem), 71. Prof Exp: Res chemist, Aberdeen Res & Develop Ctr, 68-70; ASSOC PROF CHEM, HENDRIX COL, 70- Mem: AAAS; Am Chem Soc. Res: Non-coplanar aromatic systems; electron transfer mechanism. Mailing Add: Dept of Chem Hendrix Col Conway AR 72032

TEAGUE, PERRY OWEN, b Marshall, Tex, July 13, 36; m 64; c 2. IMMUNOLOGY. Educ: NTex State Univ, BA, 58, MA, 61; Univ Okla, PhD(immunol), 66. Prof Exp: Res assoc immunol, Univ Southern Calif, 66; res fel, Univ Minn, 66-68; asst prof, 68-73, ASSOC PROF PATH, MED SCH, UNIV FLA, 73- Concurrent Pos: NIH fel pediat, Univ Minn, 67-68. Mem: Am Asn Immunol; Soc Exp Biol & Med; assoc Am Soc Clin Path. Res: Age-associated and early decline of thymus dependent lymphocyte function; diagnostic clinical immunology. Mailing Add: Dept of Path Univ of Fla Col of Med Gainesville FL 32610

TEAGUE, PEYTON CLARK, b Montgomery, Ala, June 26, 15; m 37; c 1. ORGANIC CHEMISTRY. Educ: Auburn Univ, BS, 36; Pa State Univ, MS, 37; Univ Tex, PhD(org chem, biochem), 42. Prof Exp: Res chemist, Am Agr Chem Co, NJ, 37-39; instr chem, Auburn Univ, 41-42; res chemist, US Naval Res Lab, DC, 42-45; asst prof chem, Univ Ga, 45-48 & Univ Ky, 48-50; assoc prof chem, 50-56, assoc dean grad sch, 66-68, PROF CHEM, UNIV SC, 56- Concurrent Pos: Vis prof, Univ Col, Dublin, 63-64. Mem: Am Chem Soc; Phytochem Soc NAm (pres, 69-70). Res: Chemistry and stereochemistry of flavonoids; nitrogen heterocycles; organolithium compounds; aldehyde condensations. Mailing Add: Dept of Chem Univ of SC Columbia SC 29208

TEAGUE, ROBERT STERLING, b Montgomery, Ala, July 13, 13; m 35, 61; c 5. PHARMACOLOGY, ENDOCRINOLOGY. Educ: Univ Ala, AB, 33; Northwestern Univ, MS, 36; Univ Chicago, MD, 37, PhD(pharmacol), 39. Prof Exp: Instr pharmacol, Tulane Univ, 39-43, asst prof, 43; assoc prof physiol & pharmacol, 43-48, PROF PHARMACOL & CHMN DEPT, SCHS MED & DENT, UNIV ALA, BIRMINGHAM, 48- Concurrent Pos: Ledyard fel, New York Hosp, 41-42. Mem: Soc Exp Biol & Med; Am Soc Pharmacol & Exp Therapeut; Endocrine Soc; Am Soc Zool; Soc Toxicol. Res: Synthetic estrogen metabolism and toxicology; pituitary hormones, especially melanocyte stimulating hormones; animal color change; adrenergic mechanisms; alcohol and acetaldehyde toxicology. Mailing Add: Univ of Ala Med Ctr Birmingham AL 35294

TEAGUE, TOMMY KAY, b Crossett, Ark, July 11, 43; m 65; c 1. ALGEBRA. Educ: Hendrix Col, AB, 65; Univ Kans, MA, 67; Mich State Univ, PhD(math), 71. Prof Exp: Asst prof math, Gustavus Adolphus Col, 71; ASST PROF MATH, HENDRIX COL, 71-, COORDR COMPUT SYST & SERV, 75- Mem: Am Math Soc; Math Asn Am. Res: Varieties of groups; embeddings of groups; computer science. Mailing Add: Dept of Math Hendrix Col Conway AR 72032

TEAL, GORDON KIDD, b Dallas, Tex, Jan 10, 07; m 31; c 3. PHYSICAL CHEMISTRY. Educ: Baylor Univ, AB, 27; Brown Univ, ScM, 28, PhD(chem), 31. Hon Degrees: LLD, Baylor Univ, 69; ScD, Brown Univ, 69. Prof Exp: Res chemist, Bell Tel Labs, Inc, 30-53; asst vpres & dir mat & components res, Tex Instruments, Inc, 53-55, asst vpres & dir res, 55-61, asst vpres res & eng, 61-65, int tech dir, 63-65; dir, Inst Mat Res, Nat Bur Standards, 65-67; asst vpres in charge tech develop, Equip Group, Tex Instruments, Inc, 67-68, vpres corp develop & chief scientist, 68-72; CONSULT, 72- Concurrent Pos: Res assoc, Columbia Univ, 32-35; consult, Dept Defense, NASA, Nat Acad Sci-Nat Acad Eng & Nat Bur Standards, 56-64 & 70-; mem panel selenium, Nat Acad Sci-Nat Res Coun, 56, mem panel semiconductors, 57, mem mat adv bd, 60-64, mem ad hoc comt mat & processes for electronic devices, 70-71; mem mat panel, Adv Group Electronic Parts, Off Asst Secy Defense, 56-59; dir at large, Inst Radio Eng, 59 & 62; mem adv panel to Inst Appl Technol, Nat Bur Standards, 69-75, mem comt electronic technol issues study, 72-74, chmn panel to evaluate electronic technol div, 72-75; mem, US-India Nat Acad Sci Workshop Indust Res Mgt, 70; chmn US Nat Acad Sci deleg to Ceylon, Indust Res Mgt Workshop, 70; mem aeronaut & space eng bd, Nat Acad Eng, 70-72; mem adv panel to Nat Acad Sci-Nat Acad Eng-Nat Res Coun. Coun; chmn bd comt comput in educ, Brown Univ, 71-75, mem & emer trustee; trustee, Baylor Univ & Baylor Univ Med Ctr, Dallas; contribr, Comt Surv Mat Sci & Eng, Nat Acad Sci, 72; mem US Nat Acad Sci deleg to Joint Repub China-US Workshop Indust Innovation & Prod Develop, 75. Honors & Awards: Inventor of the Year Award, Patent, Trademark & Copyright Res Inst, George Washington Univ, 66; Golden Plant Award, Am Acad Achievement, 67; Cert Appreciation & Honor Scroll, Nat Bur Standards & US Dept Com, 67; Medal of Honor, Inst Elec & Electronics Eng, 68; Creative Invention Award, Am Chem Soc, 70. Mem: Nat Acad Engr; fel AAAS (vpres & chmn indust sci sect, 69); fel Inst Elec & Electronics Engr; fel Am Inst Chem; Am Chem Soc. Res: Raman spectra; photoelectric and secondary emission phenomena; television; pyrolytically deposited films; microwave attenuator materials; silicon carbide varistors; germanium and silicon single crystals; transistors; co-developer of the junction transistor. Mailing Add: 5222 Park Ln Dallas TX 75220

TEAL, JOHN JEROME, JR, b New York, NY, Feb 7, 21; m 50; c 4. HUMAN ECOLOGY. Educ: Harvard Univ, BS, 44; Yale Univ, MA, 46. Prof Exp: Res assoc arctic geog, McGill Univ, 51-52; assoc prof anthrop, Univ Vt, 58-59; PROF HUMAN ECOL, UNIV ALASKA, FAIRBANKS, 64-; PRES ANIMAL DOMESTICATION, INST NORTHERN AGR RES, 54- Concurrent Pos: Sr fel, Arctic Inst NAm, 51-52; hon fel, Am Scand Found, 50-52, Lithgow Osborne traveling lectr, 66-67. Mem: AAAS; Am Soc Mammal; Am Anthrop Asn; Am Geog Soc; Am Forestry Asn. Res: Domestication of musk oxen; human ecology in arctic regions. Mailing Add: Box 291 College AK 99701

TEAL, JOHN MOLINE, b Omaha, Nebr, Nov 9, 29; m 50; c 2. MARINE BIOLOGY, ECOLOGY. Educ: Harvard Univ, AB, 51, MA, 52, PhD, 55. Prof Exp: Asst prof zool, Marine Inst, Univ Ga, 55-59; asst zool & oceanog, Inst Oceanog, Dalhousie Univ, 59-61; assoc scientist, 61-71, SR SCIENTIST, WOODS HOLE OCEANOG INST, 71- Mem: AAAS; Am Soc Limnol & Oceanog; Ecol Soc Am; Am Ornith Union. Res: Ecological energetics; effects of hydrostatic pressure; hydrocarbon pollution in oceans; salt marsh ecology. Mailing Add: Woods Hole Oceanog Inst Woods Hole MA 02543

TEANEY, DALE T, b Monrovia, Calif, May 19, 33; m; c 1. SOLID STATE PHYSICS, BIOPHYSICS. Educ: Pomona Col, BA, 55; Univ Calif, Berkeley, PhD(physics), 60. Prof Exp: Res assoc physics, Atomic Energy Res Estab, Eng, 60-62; RES STAFF MEM, WATSON RES CTR, IBM CORP, 62- Concurrent Pos: Staff scientist, Nat Acad Sci Phys Surv Comt, 70-71. Mem: AAAS; fel Am Phys Soc. Res: Magnetic resonance; calorimetry; liquid crystals; lipid bilayers. Mailing Add: IBM Watson Res Ctr PO Box 218 Yorktown Heights NY 10598

TEARE, FREDERICK WILSON, b Lacombe, Alta, June 9, 25; m 50; c 4. PHARMACEUTICAL CHEMISTRY, RADIOPHARMACY. Educ: Univ Alta, BSc, 49, MSc, 51; Univ NC, PhD(pharmaceut chem), 55. Prof Exp: Res pharmacist, Chas Pfizer Co, NY, 55-57; asst prof pharmaceut chem, 57-64, ASSOC PROF PHARMACEUT CHEM, UNIV TORONTO, 64- Concurrent Pos: Sabbatical, Drugs Metab Labs, Ciba Ltd, Switz. Mem: Can Pharmaceut Asn; Soc Nuclear Med. Res: Pharmacology; pharmaceutics; physiochemical analysis and separation of components of medicinal interest; quantitative analysis of drugs and metabolites in products and in biological media; drug metabolism; radiopharmaceuticals; determination of, trace levels of toxic metals and their effect on drug-metabolizing enzymes. Mailing Add: Fac of Pharm Univ of Toronto Toronto ON Can

TEARE, IWAN DALE, b Moscow, Idaho, July 24, 31; m 52; c 4. CROP PHYSIOLOGY. Educ: Univ Idaho, BS, 53; Wash State Univ, MS, 59; Purdue Univ, PhD(crop physiol, ecol), 63. Prof Exp: County agent, Idaho, 56-57; instr agron, Purdue Univ, 61-63; asst prof, Wash State Univ, 63-69; ASSOC PROF AGRON, KANS STATE UNIV, 69- Mem: Crop Sci Soc Am; Am Soc Agron; Am Soc Plant Physiologists; Sigma Xi. Res: Crop ecology; modeling the plant canopy in relation to the evapotranspiration process; assessing the interrelationships in the soil-plant-air contiuum; effect of phytoclimate on physio- logical processes of plants in relation to predicting water stress, irrigation scheduling and maximization of crop yield. Mailing Add: Evapotranspiration Lab Kans State Univ Manhattan KS 66502

TEAS, DONALD CLAUDE, b San Antonio, Tex, 1927; m 56; c 3. PHYSIOLOGICAL PSYCHOLOGY. Educ: Univ Tex, BA, 50, MA, 51, PhD(psychol), 54. Prof Exp: Res scientist, Defense Res Lab & instr, Univ Tex, 54-56; asst prof psychol, Drake Univ, 56-58; res assoc, Cent Inst Deaf, St Louis, Mo, 58-61; fel, Res Lab Electronics, Mass Inst Technol, 61-62; asst res prof otolaryngol, Sch Med, Univ Pittsburgh, 62-66, assoc res prof, 67; assoc prof psychol, speech & biomed, 67-71, PROF PSYCHOL, SPEECH & BIOMED, INST ADVAN STUDY COMMUN PROCESSES, UNIV FLA, 71- Concurrent Pos: Nat Inst Neurol Dis & Blindness spec fel, 58-62, career res develop award, 62-67. Mem: AAAS; Acoust Soc Am; Psychonomic Soc. Res: Physiological acoustics; cochlear processes; extracellular recording acoustic-evoked neural activity. Mailing Add: Dept of Speech Univ of Fla Inst Advan Study Commun Process Gainesville FL 32601

TEAS, HOWARD JONES, b Rolla, Mo, Sept 4, 20; m 42; c 4. GENETICS. Educ: La State Univ, AB, 42; Stanford Univ, MA, 46; Calif Inst Technol, PhD(genetics), 47. Prof Exp: Asst genetics, Carnegie Inst, 42-43; biologist, Oak Ridge Nat Lab, 47-48; res fel, Calif Inst Technol, 48-49, sr res fel, 50-53; plant physiologist, USDA, 53-56; assoc prof biochem, Univ Fla, 56-60; head agr bio-sci div, Nuclear Ctr, Univ PR, 60-62; prog dir, NSF, 62-64; chmn div biol sci, Univ Ga, 64-67; PROF BIOL, UNIV MIAMI, 67- Concurrent Pos: Mem bd dirs, Orgn Trop Studies, 67-72. Mem: AAAS; Ecol Soc Am; Am Soc Plant Physiol; Am Soc Biol Chem; Radiation Res Soc. Res: Plant physiology; tropical biology; physiological ecology. Mailing Add: 6700 SW 130th Terr Miami FL 33156

TEASDALE, JOHN G, b Utah, June 11, 13; m 42; c 3. PHYSICS. Educ: Univ Calif, Los Angeles, AB, 36, PhD(physics), 50. Prof Exp: Physicist, US Navy Radio & Sound Lab, 41 & Manhattan Proj, Radiation Lab, Univ Calif, 42-45; res fel, Calif Inst Technol, 50-52, sr res fel, 52-56; from asst prof to assoc prof physics, 56-62, PROF PHYSICS, SAN DIEGO STATE UNIV, 62- Concurrent Pos: Consult, Convair Div, Gen Dynamics Corp, 60. Mem: Am Phys Soc; Am Asn Physics Teachers; Sigma Xi. Res: Nuclear physics. Mailing Add: Dept of Physics San Diego State Univ San Diego CA 92115

TEASDALE, WILLIAM BROOKS, b Brownsville, Pa, July 19, 39; m 63; c 2. ORGANIC CHEMISTRY. Educ: Geneva Col, BS, 61. Prof Exp: Prod chemist, 61-62, develop chemist, 62-73, SR DEVELOP CHEMIST, EASTMAN KODAK CO, 73- Mem: Am Chem Soc; Soc Photog Scientists & Engrs. Res: Research and development of organic chemical processes to be used in the production of photographic chemicals. Mailing Add: Bldg 303 Kodak Park Rochester NY 14650

TEASDALL, ROBERT DOUGLAS, b London, Ont, Dec 9, 20; nat US; m 48; c 2. NEUROLOGY. Educ: Univ Western Ont, MD, 46, PhD, 50; Am Bd Psychiat & Neurol, dipl, 57. Prof Exp: Asst res neurol, Baltimore City Hosps, 52-53; asst, Johns Hopkins Hosp, 53-54, from instr to assoc prof neurol med, Sch Med, Johns Hopkins Univ, 54-73; CHIEF DIV NEUROL, HENRY FORD HOSP, 74-; CLIN PROF NEUROL, UNIV MICH, ANN ARBOR, 75- Mem: Am Physiol Soc; AMA; Am Neurol Asn; Am Acad Neurol. Res: Neurophysiology. Mailing Add: Div of Neurol Henry Ford Hosp Detroit MI 48202

TEATE, JAMES LAMAR, b Moultrie, Ga, Mar 4, 32; m 53; c 3. FOREST ECOLOGY. Educ: Univ Ga, BS, 54, MF, 56; NC State Univ, PhD(forestry, ecol), 67. Prof Exp: Info & educ forester, Fla Forest Serv, 56-58; instr forestry & res asst, Auburn Univ, 58-60; instr & asst forester, Miss State Univ, 60-62; assoc prof forestry, Wis State Univ-Stevens Point, 65-67; ASSOC PROF FOREST RECREATION, OKLA STATE UNIV, 67-; RES SPECIALIST, OKLA AGR EXP STA, 67- Concurrent Pos: Proj consult statewide comprehensive outdoor recreation plant, Okla Indust Develop & Parks Dept, 69-70. Mem: Soc Am Foresters; Am Forestry Asn. Res: Environmental aspects of forest management; habitat requirements of wildlife species, especially deer; forest recreation planning, development, site management. Mailing Add: Dept of Forestry Okla State Univ Stillwater OK 74074

TEATER, ROBERT WOODSON, b Ky, Feb 27, 27; m 52; c 4. AGRONOMY. Educ: Univ Ky, BS, 51; Ohio State Univ, MS, 55, PhD(agron), 57. Prof Exp: Asst prof agron, Ohio State Univ & Agr Exp Sta, 57; exec asst to dir, Ohio Dept Natural Resources, 61-63, asst dir, 63-69; ASSOC DEAN, COL AGR & HOME ECON, OHIO STATE UNIV, 69-, DIR, SCH NATURAL RESOURCES, 71-, CHMN & PROF, DEPT NATURAL RESOURCES & PROF AGRON, 73- Mem: Am Soc Agron; Soil Conserv Soc Am. Res: Soil fertility; plant nutrition; conservation; natural resources; environmental science. Mailing Add: Sch Nat Resources 113 Agr Admin Ohio State Univ 2120 Fyffe Rd Columbus OH 43210

TEBBE, ROBERT FREDERICK, chemistry, see 12th edition

TEBEAU, CARL PRESTON, chemistry, see 12th edition

TE BEEST, DAVID ORIEN, b Baldwin, Wis, Nov 9, 46; m 73; c 2. PLANT PATHOLOGY. Educ: Univ Wis-Stevens Point, BS, 68, Univ Wis-Madison, MS, 71, PhD(plant path), 74. Prof Exp: Res asst plant path, Univ Wis-Madison, 68-74; RES ASSOC PLANT PATH, UNIV ARK, FAYETTEVILLE, 75- Mem: Am Phytopath Soc; Sigma Xi. Res: Biological control of weeds with plant pathogens; pesticide physiology of fungi; physiology of plant disease; biological control of plant diseases. Mailing Add: Dept of Plant Path Univ of Ark Fayetteville AR 72701

TEBO, EDITH JANSSEN, b East Orange, NJ, Aug 16, 23; m 50; c 3. PHYSICS, ASTROPHYSICS. Educ: Vassar Col, AB, 44; Univ Va, PhD(astrophys), 49. Prof Exp: Instr astron, Vassar Col, 44-45; asst, Yerkes Observ, Univ Chicago, 45-47; instr, Univ Va, 47-49; res assoc, Harvard Col Observ, 49-50; chief visibility group, Atmospheric Physics Sect, US Army Signal Res & Develop Lab, 52-56, asst chief, Wave Propagation Sect, Phys Sci Div, 56-58, res physicist, Explor Res Div, 58-61; leader laser tech team, Electro-Optics Tech Area, 61-63, LEADER LASER COMPONENTS TEAM, LASER TECH AREA, COMBAT SURVEILLANCE & TARGET ACQUISITION LAB, US ARMY ELECTRONICS COMMAND, FT MONMOUTH, 63- Concurrent Pos: Mem comn radio astron, Int Sci Radio Union; chmn working group reconnaissance & ranging, Laser Adv Group, US Army Materiel Command. Mem: Am Astron Soc; Optical Soc Am. Res: Classification of stellar spectra; stellar motions; electromagnetic theory; quantum theory; atmospheric physics; radio astronomy; electro-optics; laser propagation, reconnaissance and ranging devices and safety. Mailing Add: Box 12M RD 1 Eatontown NJ 07724

TEBO, HEYL GREMMER, b Atlanta, Ga, Oct 17, 16; m 40. ANATOMY. Educ: Oglethorpe Univ, AB, 37, MA, 39; Emory Univ, DDS, 47. Prof Exp: Instr anat, Oglethorpe Univ, 38-39; teaching fel oral surg & anat, Univ Tex Dent Br Houston, 47-48, instr anat & surg, 48-50; asst prof diag & radiol, Sch Dent, Univ Ala, Birmingham, 50-52; asst chief dent serv, Vet Admin Hosp, Houston, Tex, 52-61; PROF ANAT, UNIV TEX DENT BR HOUSTON, 62-, PROF ANAT, MED SCH, 74- Concurrent Pos: Clin assoc prof, Univ Tex Dent Br Houston, 52-61; consult, Vet Admin Hosp, Houston, Tex, 62- Mem: Fel AAAS; Am Asn Anat; Int Asn Dent Res; Am Acad Dent Radiol; Am Asn Phys Anthrop. Res: Osteology of head; radiographic anatomy; personality characteristics of patients; oral pathology related to radiography of head. Mailing Add: Dept of Anat Univ of Tex Dent Br PO Box 20068 Houston TX 77025

TEBROCK, HARRY, preventive medicine, see 12th edition

TECHO, ROBERT, b New York, NY, Jan 1, 31; m 55; c 3. INFORMATION SCIENCE. Educ: Ga Inst Technol, BChE, 53, MS, 58, PhD(chem eng), 61. Prof Exp: Sr res engr, Eng Exp Sta, Ga Inst Technol 59-65; assoc prof info systs, Ga State Univ, 69-74; CONSULT, CHEM ENG & COMPUT ANAL, 65-; PROF INFO SYSTS, GA STATE UNIV, 74- Mem: Am Chem Soc; Am Inst Chem Eng; Asn Comput Mach; AAAS. Res: Computer applications of engineering problems, including systems analysis for pipeline operations, hydraulic transients, and power analysis; computer science, including teleprocessing information systems and data communications design. Mailing Add: Ga State Univ University Plaza Atlanta GA 30303

TECOTZKY, MELVIN, b Chicago, Ill, Feb 17, 24; m 56; c 2. INORGANIC CHEMISTRY. Educ: Univ Ill, BS, 48, PhD(chem), 53. Prof Exp: Res asst inorg chem, Univ Ill, 51-53, fel, 53-54; res chemist, Diversey Corp, 54-56; sr chemist, W R Grace & Co, 56-59; res chemist & proj leader, FMC Corp, 59-61; staff scientist, Missiles & Space Div, Lockheed Aircraft Corp, Calif, 61-68; DIR RES, CHEM PROD DIV, US RADIUM CORP, HACKETTSTOWN, 68- Mem: Am Chem Soc; Am Phys Soc; Electrochem Soc; The Chem Soc. Res: Rare earths; solid state chemistry; luminescence; chelates; thorium; uranium; sulfides; hydrazine; phosphates; transition elements; electronic materials; magnetic materials; non-aqueous solvents. Mailing Add: 27 N Linden Ln Mendham NJ 07945

TEDESCHI, CESARE GEORGE V, pathology, deceased

TEDESCHI, DAVID HENRY, b Newark, NJ, Feb 20, 30. PHARMACOLOGY. Educ: Rutgers Univ, BSc, 52; Univ Utah, PhD, 55. Prof Exp: Asst pharmacol, Univ Utah, 52-54; assoc dir pharmacol, Smith Kline & French Labs, Pa, 55-68; dir pharmacol, Geigy Pharmaceut, NY, 68-70, dir pharmacol & dep dir biol res, Ciba-Geigy Corp, 70-72; DIR CENT NERV SYST DIS THER, RES SECT, LEDERLE LABS DIV, AM CYANAMID CO, 72- Honors & Awards: Am Pharmaceut Asn Found Award in Pharmacodynamics, 69. Mem: Am Col Neuropsychopharmacol; Am Soc Pharmacol & Exp Therapeut; Soc Exp Biol & Med; Int Col Neuropsychopharmacol; Int Soc Biochem Pharmacol. Res: Neuropsychopharmacology; site and mechanism of action of drugs on central nervous system. Mailing Add: Lederle Labs Div Am Cyanamid Co Pearl River NY 10965

TEDESCHI, HENRY, b Novara, Italy, Feb 3, 30; nat US; m 57; c 3. PHYSIOLOGY. Educ: Univ Pittsburgh, BS, 50; Univ Chicago, PhD(physiol), 55. Prof Exp: Res assoc & instr, Univ Chicago, 55-57, asst prof, 57-60; from asst prof to assoc prof physiol, Univ Ill Col Med, 60-65; PROF BIOL, STATE UNIV NY ALBANY, 65- Concurrent Pos: NIH spec res fel, Oxford Univ, 71-72. Mem: Biophys Soc; Am Soc Cell Biol; Am Physiol Soc; Soc Gen Physiol; Am Soc Biol Chem. Res: Cell physiology; structural and functional organization of the cell; intracellular membranes. Mailing Add: Dept of Biol Sci State Univ NY Albany NY 12222

TEDESCHI, RALPH EARL, b Newark, NJ, Nov 20, 27; m 51; c 2. PHARMACOLOGY. Educ: Rutgers Univ, BS, 51; Med Col Va, PhD(pharmacol), 54. Prof Exp: Resident pharmacol, Oxford Univ, 54-55; res assoc, Div Metab Res, Jefferson Med Col, 55-56; assoc dir pharmacol, Smith Kline & French Labs, 56-69; head dept pharmacol, Wm S Merrell Co, Ohio, 69-71; head dept, 71-72, dir clin pharmacol, 72-73, dir develop, 73-74, TECH ASST TO DIR PHARMACEUT RES & DEVELOP, HUMAN HEALTH RES & DEVELOP LABS, DOW CHEM CO, ZIONSVILLE, 74- Mem: Am Soc Pharmacol & Exp Therapeut; Acad Pharmaceut Sci; Am Pharmaceut Asn; Soc Exp Biol & Med. Res: Neuropharmacology; pharmacology of the autonomic nervous system; cardiovascular pharmacology. Mailing Add: 7850 Holly Creek Lane Indianapolis IN 46240

TEDESCHI, ROBERT JAMES, b Woodside, NY, July 25, 21; m 52; c 3. ORGANIC CHEMISTRY. Educ: Cornell Univ, AB, 44, MS, 45, PhD(org chem), 47. Prof Exp: Microanal chemist, Wyeth Inst, Pa, 46; res chemist, Calco Chem Div, Am Cyanamid Co, NJ, 47-53; sect head, Cent Res Labs, Air Reduction Co, 53-59, proj leader, 59-64, supvr plastics div, Airco Chem & Plastics Div, 64-69, supvr org chem, 69-71, dir res & develop, Acetylenic Chem Div, 71-74, ASSOC DIR RES, ACETYLENIC CHEM

DIV, AIR PROD & CHEM CO, MIDDLESEX, 74- Mem: Am Chem Soc; fel Am Inst Chem; NY Acad Sci. Res: Acetylene chemistry; high pressure synthesis; catalytic reactions; organic chemicals development; industrial applications for acetylenic chemicals; organo metallics; chelates and complexes; corrosion inhibitors; perfumery intermediates; acetylenic surfactants; specialty monomers and polymers. Mailing Add: RD 2 Box 143 Whitehouse Station NJ 08889

TEDESCO, THOMAS ALBERT, b York, Pa, Dec 5, 35. HUMAN GENETICS, BIOCHEMICAL GENETICS. Educ: Franklin & Marshall Col, BS, 60; Univ Pa, PhD(biol), 69. Prof Exp: Res technician pediat, Hosp, 60-61, res asst, 61-62, res assoc genetics, 62-65, from instr to asst prof pediat, Sch Med, 65-72, ASST PROF PEDIAT & MED GENETICS, SCH MED, UNIV PA, 72- Mem: AAAS; Am Soc Human Genetics. Res: Inborn errors in metabolism; enzyme deficiency disease. Mailing Add: Dept of Genetics Univ of Pa Sch of Med Philadelphia PA 19104

TEDFORD, MYRON DUNCAN, b The Dalles, Ore, Jan 23, 20; m 54; c 3. ANATOMY, HUMAN GENETICS. Educ: Univ Ore, BS, 49, MS, 50. Prof Exp: From instr to assoc prof, 50-66, PROF HUMAN ANAT, DENT SCH, UNIV ORE, 66- Mem: AAAS; Soc Human Genetics; hon mem Acad Gen Dent; Int Asn Dent Res; Sigma Xi; Am Asn Univ Prof. Res: Human anatomy; maintenance of pregnancy in ovariectomized golden hamsters with desoxycorticosterone acetate; experimental pseudohermaphroditism in guinea pigs; histochemical localization of specific and nonspecific cholinesterases in decalcified human teeth. Mailing Add: Dept of Anat Univ of Ore Dent Sch 611 SW Campus Dr Portland OR 97201

TEDFORD, RICHARD HALL, b Los Angeles, Calif, Apr 25, 29; m 54. VERTEBRATE PALEONTOLOGY, STRATIGRAPHY. Educ: Univ Calif, Los Angeles, BS, 51; Univ Calif, Berkeley, PhD(paleont), 60. Prof Exp: Instr geol, Univ Calif, Riverside, 59-60, lectr, 60-61, from asst prof to assoc prof, 61-66; assoc curator vert paleont, 66-69, CURATOR VERT PALEONT, AM MUS NATURAL HIST, 69- Mem: Soc Vert Paleont; Paleont Soc; Am Soc Mammal. Res: Phylogeny, geographic distribution and paleoecology of Carnivora, Marsupials and other mammals; stratigraphy and chronology of Cenozoic rocks. Mailing Add: Am Mus Natural Hist Central Park W at 79th St New York NY 10024

TEDROW, JOHN CHARLES FREMONT, b Rockwood, Pa, Apr 21, 17; m 43; c 2. SOILS. Educ: Pa State Univ, BS, 39; Mich State Univ, MS, 40; Rutgers Univ, PhD, 50. Prof Exp: Jr soil surveyor, Soil Conserv Serv, USDA, 41, soil scientist, 46-47; from instr to assoc prof, 48-66, PROF SOILS, RUTGERS UNIV, NEW BRUNSWICK, 66- Concurrent Pos: Sr pedologist, Arctic Soil Invests, 53; prin investr, Arctic Inst NAm, 55-67; Antarctic pedologic investr, NSF, 61-63; ed in chief, Soil Sci. Mem: Soil Sci Soc Am; Am Polar Soc; Am Arbit Asn; Am Geophys Union; fel Am Soc Agron. Res: Soil morphology; genesis and survey; soils of the Arctic and Alpine regions. Mailing Add: Dept of Soils Lipman Hall Rutgers Univ New Brunswick NJ 08903

TEDROW, PAUL MULLER, b Ware, Mass, Apr 5, 40. LOW TEMPERATURE PHYSICS. Educ: Mass Inst Technol, SB, 61; Cornell Univ, PhD(physics), 66. Prof Exp: Fel physics, Cornell Univ, 66-67; MEM STAFF, FRANCIS BITTER NAT MAGNET LAB, MASS INST TECHNOL, 67- Mem: Am Phys Soc. Res: Superconductivity; ferromagnetism. Mailing Add: Francis Bitter Nat Magnet Lab Mass Inst of Technol Cambridge MA 02139

TEEGARDEN, DAVID MORRISON, b Dayton, Ohio, Jan 10, 41; m 66; c 3. ORGANIC CHEMISTRY. Educ: Ohio Wesleyan Univ, AB, 63; Univ Mich, MS, 65, PhD(org chem), 72. Prof Exp: Asst prof chem, Univ Wis-Platteville, 69-73; ASST PROF CHEM, ST JOHN FISHER COL, 73- Mem: Am Chem Soc. Res: Stereochemistry and mechanisms of reactions of norbornyl compounds. Mailing Add: Dept of Chem St John Fisher Col Rochester NY 14618

TEEGARDEN, KENNETH JAMES, b Chicago, Ill, May 13, 28; m 59. PHYSICS. Educ: Univ Chicago, AB, 47, BS, 50; Univ Ill, MS, 51, PhD, 54. Prof Exp: Res assoc, 54-58, asst prof, 58-59, sr res assoc, 60-61, assoc prof, 61-66, PROF OPTICS, INST OPTICS, UNIV ROCHESTER, 66- Concurrent Pos: Alfred P Sloan Found fel, Univ Rochester, 59-63. Mem: Fel Am Phys Soc; Optical Soc Am. Res: Electronic properties of ionic solids and the solid rare gases. Mailing Add: Inst of Optics Univ of Rochester Rochester NY 14627

TEEGUARDEN, DENNIS EARL, b Gary, Ind, Aug 21, 31; m 54; c 2. FORESTRY ECONOMICS. Educ: Mich Tech Univ, BS, 53; Univ Calif, Berkeley, MF, 58, PhD(agr econ), 64. Prof Exp: Res asst, Pac Southwest Forest & Range Exp Sta, US Forest Serv, Berkeley, Calif, 57; asst specialist, Agr Exp Sta, 58-63, actg asst prof, Sch Forestry, 63, from asst prof to assoc prof, 64-73, PROF FORESTRY, SCH FORESTRY, UNIV CALIF, BERKELEY, 73- Mem: Soc Am Foresters; Am Econ Asn. Res: Application of operations research techniques to problems of resource allocation in public and private forestry enterprises. Mailing Add: Univ of Calif Sch of Forestry Berkeley CA 94720

TEEKELL, ROGER ALTON, b Elmer, La, Mar 3, 30; m 53; c 4. METABOLISM, BIOCHEMISTRY. Educ: La State Univ, BS, 51, MS, 55, PhD(nutrit, biochem), 58. Prof Exp: Asst, La State Univ, 54-58; res scientist, Agr Res Lab, Univ Tenn-AEC, 58-61; res chemist, Dow Chem Co, Tex, 61-63; from assoc prof to prof physiol, 63-74, PROF POULTRY SCI, LA STATE UNIV, BATON ROUGE, 74- Mem: AAAS; Poultry Sci Asn; World Poultry Sci Asn. Res: Intermediary metabolism of amino acids and lipids using labeled compounds. Mailing Add: Dept of Poultry Sci La State Univ Baton Rouge LA 70803

TEERI, ARTHUR EINO, b Dover, NH, July 29, 16; m 41; c 2. BIOCHEMISTRY. Educ: Univ NH, BS, 37, MS, 40; Rutgers Univ, PhD(biochem), 43. Prof Exp: Asst physiol chem, Univ NH, 38-40; res fel biochem, Res Labs, US Dept Interior, 40-41; asst prof, 43-53, PROF BIOCHEM, UNIV NH, 53- Mem: Fel AAAS; Am Chem Soc; Am Inst Nutrit; NY Acad Sci. Res: Physiological chemistry; animal and human nutrition; vitamins; clinical methods. Mailing Add: PO Box 146 Durham NH 03824

TEERI, JAMES ARTHUR, b Exeter, NH, Feb 28, 44; m 67. ECOLOGY, POLAR BIOLOGY. Educ: Univ NH, BS, MS, 68; Duke Univ, PhD(bot), 72. Prof Exp: ASST PROF BIOL, UNIV CHICAGO, 72- Concurrent Pos: Assoc ed, Paleobiol, 74-; mem, Comn Optical Radiation Measurement, 74 Mem: AAAS; Sigma Xi; Ecol Soc Am; Am Inst Biol Sci; Arctic Inst NAm. Res: Evolution of plant growth responses to environmental fluctuation. Mailing Add: Barnes Lab Univ of Chicago 5630 S Ingleside Ave Chicago IL 60637

TEERLINK, WILFORD JOHN, organic chemistry, see 12th edition

TEETER, HOWARD MAPLE, b Peoria, Ill, Oct 14, 15; m 42; c 3. ORGANIC CHEMISTRY. Educ: Bradley Univ, BS, 37; Univ Ill, MS, 38, PhD(org chem), 40. Prof Exp: Instr phys & org chem, Bradley Univ, 40-42; res chemist, Northern Regional Res Lab, 42-45, head oil prod invests, Oilseed Crops Lab, 45-60, asst dir

northern mkt & nutrit res div, Agr Res Serv, 60-72, ASST TO DEP ADMINR N CENT REGION, AGR RES SERV, USDA, 72- Concurrent Pos: Lectr, Bradley Univ, 46-48. Honors & Awards: Bond Award, Am Oil Chemists Soc, 60; Superior Serv Awards, USDA, 61 & 63. Mem: AAAS; Am Chem Soc; Am Oil Chemists Soc. Res: Organic syntheses involving higher unsaturated fatty acids; polymerization; polyesters; polyamides; synthetic resins; chemicals and intermediates from fats and oils. Mailing Add: USDA Agr Res Serv 2000 W Pioneer Pkwy Peoria IL 61614

TEETER, JAMES WALLIS, b Hamilton, Ont, Mar 14, 37; m 60; c 3. GEOLOGY, PALEONTOLOGY. Educ: McMaster Univ, BSc, 60, MSc, 62; Rice Univ, PhD(paleont), 66. Prof Exp: Asst prof, 65-69, ASSOC PROF GEOL, UNIV AKRON, 69- Concurrent Pos: Faculty res grant, Univ Akron, 69, 73 & 74. Mem: AAAS; Sigma Xi; Geol Soc Am; Am Asn Petrol Geol; Soc Econ Paleont & Mineral. Res: Paleoecological study of Pleistocene Ostracoda from the Fairlawn Mastodon Site, Ohio; Key Largo limestone facies; Ordovician Nautiloid touchmarks; Ostracoda and enviroments of Caloosahatchee Formation; living Pelecypod behavior; marine Ostracoda dispersal. Mailing Add: Dept of Geol Univ of Akron Akron OH 44325

TEETER, RICHARD MALCOLM, b Berkeley, Calif, Feb 24, 26; m 49, 68; c 2. ORGANIC CHEMISTRY. Educ: Univ Calif, BS, 49; Univ Wash, PhD(chem), 54. Prof Exp: Sr res chemist, 54-68, SR RES ASSOC, CHEVRON RES CO, 68- Mem: Am Chem Soc; Am Soc Mass Spectrometry. Res: Analytical mass spectrometry; preparation of derivatives to aid analysis; reaction mechanisms in mass spectrometry; application of computers to mass spectrometry. Mailing Add: Chevron Res Co 576 Standard Ave Richmond CA 94802

TEETERS, WILBER OTIS, b Toledo, Ohio, Aug 14, 08; m 30; c 3. CHEMISTRY. Educ: Univ Iowa, BS, 30; Univ Ill, MS, 33, PhD(org chem), 35. Prof Exp: Anal & res chemist, Jackson Lab, E I du Pont de Nemours & Co, 30-31; res chemist, Rohm & Haas Co, Pa, 35-36; res chemist, Ammonia Dept, El du Pont de Nemours & Co, 36-40; supvr org res, Barrett Div, Allied Chem Corp 40-46; prin res chem supvr, Sun Chem Corp, 46-49; assoc dir petrol & chem res labs, M W Kellogg Co Div, Pullman, Inc, 49-57; assoc dir res & asst to vpres in charge res, Johnson & Johnson, Inc, 57-73; RETIRED. Concurrent Pos: Chmn bd, NJ Coun Res & Develop, 63-64; Indust mkt res consult. Mem: AAAS; Am Chem Soc; fel Am Inst Chem. Res: Organic synthesis; plastics; resins; organic pigments and medicinals; protein chemistry; textile chemicals. Mailing Add: 369 Windsor Rd River Edge NJ 07661

TEETERS, WILBUR OLDROYD, b Dover, Ohio, May 31, 08; m 31; c 3. PHYSICAL CHEMISTRY. Educ: Butler Univ, BS, 29; NY Univ, MS, 31, PhD(phys chem), 35. Prof Exp: Gen mgr, Hoke, Inc, New York, 35-37; engr, Air Reduction Sales Co, 37-38; PRES, HOKE, INC, 39- Mem: AAAS; Am Chem Soc; fel Am Inst Chem. Res: Sensitization of cuprous oxide barrier layer photoelectric cells. Mailing Add: Hoke Inc 1 Tenakill Park Cresskill NJ 07626

TEETERS, WILLIAM DALE, b Englewood, NJ, May 4, 42; m 61; c 2. THEORETICAL HIGH ENERGY PHYSICS, THEORETICAL PHYSICS. Educ: Univ Iowa, BA, 64, MS, 66, PhD(physics), 68. Prof Exp: Asst prof, 68-73, ASSOC PROF PHYSICS, CHICAGO STATE UNIV, 73- Concurrent Pos: Staff mem, Physics Div, Argonne Nat Lab, 75-76. Mem: Am Phys Soc. Res: Elementary particle theory with emphasis on weak interactions; nuclear theory; relativistic treatment of bound states and applications to quark model. Mailing Add: Dept of Phys Sci Chicago State Univ Chicago IL 60628

TEFFT, MELVIN, b Dec 15, 32; US citizen. RADIOTHERAPY. Educ: Harvard Univ, AB, 54; Boston Univ, MD, 58; Am Bd Radiol, dipl, 63. Prof Exp: Intern med, Boston City Hosp, 58-59; resident radiol, Mass Mem Hosp, 59-62; asst, Harvard Med Sch, 62-65, instr, 66-67, clin assoc, 67-69, asst prof, 69-70, from asst prof to assoc prof radiation ther, 70-73; prof radiol, Cornell Univ, 73-75; PROF RADIATION MED, BROWN UNIV, 75-; RADIOTHERAPIST, DEPT RADIATION ONCOL & ASSOC MEM, DEPT PEDIAT, RI HOSP, 75- Concurrent Pos: Asst radiol, Children's Hosp Med Ctr, 62-64, secy, Radiation Safety Comt, 62-67, chmn, 67-70, radiotherapist, Med Ctr & Tumor Ther Div, 64-70, chief, Div Radiother & Nuclear Med, 67-69, radiotherapist-in-chief, Dept Radiation Ther, 69-70, mem, Staff Exec Comt, 69-70, mem subcomt clin invest, Med Ctr & subcomt med adv bd, Clin Res Ctr, 70; consult radiol, Lemuel Shattuck Hosp, 62-68, Mass Gen Hosp, 64-71 & Boston Lying-In Hosp, 66-74; consult radiother, Lemuel Shattuck Hosp, 66-70; consult, Dept Radiother, Tufts Med Sch at Lemuel Shattuck Hosp, 69-70; mem, Children's Cancer Chemother Group A, 69-; consult, Dept Radiation Ther, Mem Hosp, New York, 72-74; asst radiol, Sch Med, Boston Univ, 62-68; NIH clin fel radiation ther, Mass Gen Hosp, 63-64; assoc radiol, Peter Bent Brigham Hosp, 66-70; mem, Hepatoma Protocol Writing Comt, Children's Cancer Study Group A, 69-, Rhabdomyosarcoma Protocol Comt, 70- & Criteria, Data & Statist Comt, 73-; assoc prof therapeut radiol, Med Sch, Tufts Univ, 69-70; assoc attend radiotherapist, Mem Hosp, 70-71, attend radiotherapist, 73-75, dir med educ, Dept Radiation Ther, 73-75; assoc mem, Sloan-Kettering Inst, 70-71; assoc radiotherapist, Mass Gen Hosp, 71-73; attend radiologist, New York Hosp, 73-75. Mem: Fel Am Col Radiol; Am Soc Therapeut Radiologists; Am Radium Soc; Int Soc Pediat Oncol; AMA. Res: Pediatric oncology, combined radiation therapy and chemotherapy to enhance local control and disease-free survival; evaluation of normal tissue sensitivity by combined modalities of treatment. Mailing Add: Dept of Radiation Oncol RI Hosp Providence RI 02902

TEFFT, STANTON KNIGHT, b Chicago, Ill, Jan 25, 30; m 62; c 2. CULTURAL ANTHROPOLOGY, ETHNOLOGY. Educ: Mich State Univ, BA, 52; Univ Wis, MS, 55; Univ Minn, PhD(anthrop), 60. Prof Exp: Asst prof anthrop, Westminster Col, 61-64; ASSOC PROF ANTHROP, WAKE FOREST UNIV, 64- Concurrent Pos: NIH grant, Westminster Col, 62-63 & Sigma Xi fel, 62-63; Soc Psychol Studies Social Issues grant, Wake Forest Univ, 70 & Dept Health, Educ & Welfare grant, 71-72. Mem: AAAS; fel Am Anthrop Asn; Am Sociol Asn; Am Soc Ethnohist; Am Acad Polit & Soc Sci. Res: Cross-cultural study of primitive peacemaking and warfare regulation; values and cultural change; cultural adaptation among the Plains Indians; ethnohistory of the Wind River Shoshone primitive war. Mailing Add: Dept of Sociol & Anthrop Wake Forest Univ Box 7805 Winston-Salem NC 27109

TEFFT, WAYNE EARL, solid state physics, see 12th edition

TEGENKAMP, THOMAS RICHARD, b Dayton, Ohio, Oct 23, 29; m 54; c 3. BIOLOGY, GENETICS. Educ: Ohio State Univ, BS, 52, MS, 54, PhD(bot, genetics), 61. Prof Exp: Asst prof, 62-72, ASSOC PROF BIOL, OTTERBEIN COL, 72- Res: Biomagnetic research on meiosis and genetic transmission. Mailing Add: Dept of Biol Otterbein Col Westerville OH 43081

TEGGINS, JOHN E, b Wallasey, Eng, Jan 6, 37; m 60; c 3. INORGANIC CHEMISTRY. Educ: Univ Sheffield, BSc, 58; Boston Univ, AM, 60, PhD(chem), 65. Prof Exp: Res chemist, Courtaulds Can, Ltd, 60-62; res assoc radiochem, Iowa State Univ, 65-66; from asst prof to assoc prof chem, 66-75, PROF CHEM, AUBURN UNIV, MONTGOMERY, 75- Mem: Am Chem Soc; sr mem Chem Inst Can. Res:

Determination of thermodynamic and kinetic data for coordination complexes in aqueous solution. Mailing Add: 4425 Shamrock Ln Montgomery AL 36106

TEICHER, HARRY, b Middle Village, NY, Jan 11, 27; m 51; c 3. COLLOID CHEMISTRY. Educ: Queens Col, NY, BS, 48; Syracuse Univ, MS, 50, PhD(chem), 53. Prof Exp: Asst chem, Syracuse Univ, 49-53; res chemist, 53-56, RES GROUP LEADER, RES & DEVELOP DEPT, MONSANTO CO, 56- Mem: AAAS; Am Chem Soc; Sigma Xi; Am Asn Textile Chemists & Colorists. Res: Silica. Mailing Add: Monsanto Co Res & Develop Dept 800 N Lindbergh Blvd St Louis MO 63166

TEICHER, HENRY, b Jersey City, NJ, July 9, 22. MATHEMATICAL STATISTICS. Educ: Univ Iowa, BA, 46; Columbia Univ, MA, 47, PhD(math statist), 50. Prof Exp: Asst prof math, Univ Del, 50-51; from asst prof to prof math statist, Purdue Univ, 51-67; vis prof, Columbia Univ, 67-68; PROF MATH STATIST, RUTGERS UNIV, NEW BRUNSWICK, 68- Concurrent Pos: Vis asst prof, Stanford Univ, 55-56; vis assoc prof & mem inst math sci, NY Univ, 60-61. Mem: Am Math Soc; fel Inst Math Statist. Res: Probability and mathematical statistics, especially stopping rules, limit distributions and mixtures of distributions. Mailing Add: Statist Ctr Rutgers Univ New Brunswick NJ 08903

TEICHER, JOSEPH D, b New York, NY, Aug 1, 17; m 42; c 2. PSYCHIATRY. Educ: City Col New York, BS, 33; Columbia Univ, MA, 34; NY Univ, MD, 40; NY Psychoanal Inst, cert, 51. Prof Exp: Consult, Pub Schs, Bronxville, NY, 46-52; assoc clin prof psychiat, 53-60, PROF CHILD PSYCHIAT, SCH MED, UNIV SOUTHERN CALIF, 60- Concurrent Pos: Dir child guid clin, St Luke's Hosp, New York, 47-52; dir, Child Guid Clin Los Angeles, 52-; chief psychiat, Children's Hosp, 53-58; mem fac, Southern Calif Psychoanal Inst, 54- Mem: Fel Am Psychiat Asn; Am Psychoanal Asn. Res: Normal and atypical child development; adolescent therapy techniques; studies in attempted suicide in adolescence; alcoholism and drug abuse in adolescence. Mailing Add: Dept of Psychiat Univ of Southern Calif Sch Med Los Angeles CA 90033

TEICHERT, CURT, b Koenigsberg, Prussia, May 9, 05; m 28. GEOLOGY, PALEONTOLOGY. Educ: Univ Koenigsberg, PhD(geol), 28; Univ Western Australia, DSc, 44. Prof Exp: Asst, Freiburg Univ, 27-29; Rockefeller fel geol, 30; res fel, Univ Copenhagen, 33-37; res lectr, Univ Western Australia, 37-46; asst chief govt geologist, Victoria Mines Dept, 46-47; sr lectr, Univ Melbourne, 47-53; prof geol, NMex Inst Mining & Technol, 53; geologist, US Geol Surv, 54-64, chief petrol geol lab, 54-58, staff geologist, 58-61, geol adv, US AID, Pakistan, 61-64; REGENTS DISTINGUISHED PROF GEOL, UNIV KANS, 64- Concurrent Pos: Fulbright traveling scholar, 51; guest prof, Univs Göttingen, Bonn & Freiburg, 58; consult, Caltex, 40-41, Stand Vacuum, 48-50, Australian Bur Mining Resources, 48-51 & Shell Oil Co, 53-54; mem, Danish Exped, Greenland, 31-32; US coordr, Cent Treaty Orgn stratig working group, 63- Honors & Awards: Syme Prize, Australia, 43. Mem: Geol Soc Am; Paleont Soc (pres, 71-72); Soc Econ Paleont & Mineral; Australian Geol Soc (secy, 51-53); Am Asn Petrol Geol. Res: Paleozoic stratigraphy and paleontology; ancient and modern coral reefs; fossil cephalopods; sedimentation; paleoecology; stratigraphy of southwestern Asia. Mailing Add: Dept of Geol Univ of Kans Lawrence KS 66044

TEICHMAN, ROBERT, b Berlin, Ger, June 22, 23; nat US; m 62; c 2. BIOSTATISTICS. Educ: Univ Conn, BS, 51, MS, 53; NC State Univ, PhD(animal sci), 67. Prof Exp: Asst animal nutrit, Univ Conn, 50-58, asst exp statist, NC State Univ, 58-60, asst statistician, 60-61; mgr res statist dept, Ralson Purina Co, 61-67; assoc prof exp statist, NC State Univ, 67-69; MGR BIOSTATIST, ICI UNITED STATES INC, 69- Mem: Am Statist Asn; Biomet Soc; Am Soc Pharmacol & Exp Therapeut. Res: Design and analysis of biological experiments; research in statistical procedures. Mailing Add: Planning & Develop Dept ICI United States Inc Wilmington DE 19897

TEICHMANN, THEODOR, b Königsberg, Ger, Sept 16, 23; nat US; m 53; c 2. THEORETICAL PHYSICS. Educ: Univ Cape Town, BSc, 43, MSc, 45; Princeton Univ, Am, 47, PhD(physics), 49. Prof Exp: Jr lectr elec eng, Univ Cape Town, 44-46; res assoc physics, Princeton, 50-52; res physicist, Res & Develop Labs, Hughes Aircraft Co, 52-55; mgr systs anal & simulation, Missiles & Space Div, Lockheed Aircraft Corp, 55-56, mgr nuclear physics, 56-57, sci asst to dir res, 57-60; consult scientist satellite systs, 60; mem res & develop staff, Spec Nuclear Effects Lab, Gen Atomic Div, Gen Dynamics Corp, 60-68; prin scientist, KMS Technol Ctr, Calif, 68-72, PRIN SCIENTIST, KMS FUSION INC, ANN ARBOR, 72- Mem: AAAS; Am Phys Soc; Am Math Soc; Am Nuclear Soc; Soc Indust & Appl Math; sr mem Am Astronaut Soc. Res: Operations research; continuum mechanics; energy conversion; systems analysis; application of inertially confined fusion to energy production on both electric and gas sectors. Mailing Add: 11901 Trailwood Rd Plymouth MI 48170

TEICHNER, ROBERT W, b Newark, NJ, Oct 14, 12; m 45; c 4. CHEMISTRY, SOLID STATE ELECTRONICS. Educ: Polytech Inst Brooklyn, BChE, 34; Harvard Univ, MA, 36. Prof Exp: Chief chemist, Duplicator Supplies Div, Remington Rand, Inc, 36-50 & Polychrome Corp, 50-52; sr scientist, Lab Advan Res, Sperry Rand Corp, 53-55; chief chemist, Mergenthaler Linotype Co, 55-58; sr scientist, Shockley Transistor Co Div, Clevite Corp, 58-62; SR SCIENTIST, OPTO ELECTRONICS DIV, HEWLETT-PACKARD, 62- Mem: Am Chem Soc. Res: Microminiaturization of solid state devices; packaging; interconnections; products and processes for the graphic arts. Mailing Add: 3537 Murdoch Dr Palo Alto CA 94306

TEICHNER, VICTOR JEROME, b New York, NY, Oct 22, 26; m 55; c 1. PSYCHIATRY, PSYCHOANALYSIS. Educ: Temple Univ, MD, 49; Columbia Univ, cert psychoanal med, 63. Prof Exp: Sr psychiatrist, Bellevue Psychiat Hosp, New York, 54-59; DIR DEPT PSYCHIAT, METROP HOSP CTR, 72-; PROF CLIN PSYCHIAT, NY MED COL, 73- Concurrent Pos: Ed, Bull Asn Psychoanal Div, 71-75, consult ed, 75-; asst ed, J Am Acad Psychoanal, 72- Mem: Fel Am Acad Psychoanal; fel Am Psychiat Asn; Am Psychoanal Asn; Am Soc Adolescent Psychiat. Res: Psychotherapy; confidentiality of health records. Mailing Add: 145 E 84th St New York NY 10028

TEICHROEW, DANIEL, b Can, Jan 5, 25; nat US; m 50; c 1. MATHEMATICS. Educ: Univ Toronto, BA, 48, MA, 49; Univ NC, PhD(statist), 53. Prof Exp: Res assoc, Univ NC, 51-52; mathematician, Nat Bur Stand, DC & Inst Numerical Anal, Univ Calif, Los Angeles, 52-55; sr electronics appln specialist, Nat Cash Register Co, 55, spec rep prod develop, 55-56, head bus systs anal, 56-57; from assoc prof to prof mgt, Grad Sch Bus, Stanford Univ, 57-64; prof orgn sci & head div, Case Western Reserve Univ, 64-68; PROF INDUST ENG & CHMN DEPT, UNIV MICH, ANN ARBOR, 68- Concurrent Pos: Lectr, Sch Bus Admin, Univ Southern Calif, 56-57; ed sci & bus appln sect, Communications, Asn Comput Mach, 63- Mem: Asn Comput Mach; Inst Mgt Sci (vpres, 67-); Opers Res Soc Am; Inst Math Statist; Am Math Soc. Res: Development and application of scientific techniques to organizational problems, particularly operations research, management science and computer techniques. Mailing Add: Dept of Indust Eng 231 W Eng Bldg Univ of Mich Ann Arbor MI 48104

TEIGER, MARTIN, b New York, Dec 30, 36; m 64; c 1. PHYSICS, ASTRONOMY. Educ: Columbia Univ, AB, 58, MA, 60, PhD(physics), 65. Prof Exp: Lectr physics, City Col New York, 61-65, instr, 65-66; from asst prof to assoc prof, 66-75, PROF PHYSICS, LONG ISLAND UNIV, 75-, CHMN DEPT, 69- Mem: AAAS; Am Phys Soc; Am Pub Health Asn. Res: Planetary atmospheres and surface environments; radiative transfer theory; numerical methods for computers; environmental management. Mailing Add: Dept of Physics Long Island Univ Brooklyn NY 11201

TEIPEL, JOHN WILLIAM, b Covington, Ky, Feb 17, 43; m 66; c 2. BIOCHEMISTRY. Educ: Rockhurst Col, BA, 64; Duke Univ, PhD(biochem), 68. Prof Exp: Am Cancer Soc fel biochem, Univ Calif, Berkeley, 68-70; asst prof chem & biochem, Univ Ill, Urbana, 70-72; scientist, 72-74, sr scientist, 74-75, PRINCIPAL SCIENTIST, ORTHO RES FOUND, 75- Res: Physical biochemistry of proteins and enzymes; immunochemistry. Mailing Add: Biochem Dept Ortho Res Found Raritan NJ 08869

TEITEL, SIDNEY, b Detroit, Mich, May 7, 17; m 38; c 3. MEDICINAL CHEMISTRY. Educ: Brooklyn Col, BS, 36, MS, 41; Tohoku Univ, Japan, PhD, 72. Prof Exp: Chemist, Paragon Oil Co, NY, 36-37; asst, Felton Chem Co, 37-42; res chemist, Franco-Am Chem Works, NJ, 42-43; res fel, 43-68, RES GROUP CHIEF, CHEM RES DEPT, HOFFMANN-LA ROCHE, INC, 68- Concurrent Pos: Instr, Bergen Jr Col, 49-51. Mem: Am Chem Soc; Sigma Xi. Res: Synthesis of pharmaceuticals and aromatic chemicals; semi-synthetic penicillins; antibacterial and entiemetic agents; synthesis of alkaloids. Mailing Add: Chem Res Dept Hoffmann-La Roche Inc Nutley NJ 07110

TEITELBAUM, CHARLES LEONARD, b Brooklyn, NY, June 14, 25; m 50. ANALYTICAL CHEMISTRY. Educ: Brooklyn Col, BA, 45; Purdue Univ, MS, 48, PhD(org chem), 51. Prof Exp: Res chemist, Heyden Chem Corp, 50-53; prin chemist, Battelle Mem Inst, 53-58; chemist, Coty, Inc Div, Chas Pfizer & Co, 58-65 & Florasynth, Inc, 65-66; RES SPECIALIST, GEN FOODS CORP, 66- Mem: Am Chem Soc. Res: Analysis of natural products relating to odor and flavor. Mailing Add: 225 W 11th St New York NY 10014

TEITELBAUM, HARRY ALLEN, b Brooklyn, NY, Oct 7, 07; m 47; c 2. NEUROLOGY, PSYCHIATRY. Educ: Univ Md, BS, 29, MD, 35, PhD(anat, neurol), 36; Am Bd Psychiat & Neurol, dipl. Prof Exp: Instr anat, Sch Med, Univ Md, 36-37; intern, Montefiore Hosp, New York, 37-38; intern psychiat, Bellevue Hosp, 38-39, resident neurol, 39-40; asst, Sch Med, Univ Md, Baltimore City, 40-47, assoc, 48-51, asst prof, 51-65; asst prof psychiat, 62-72, EMER ASST PROF PSYCHIAT, SCH MED, JOHNS HOPKINS UNIV, 72-; ASSOC PROF NEUROL, UNIV MD, BALTIMORE CITY, 65- Concurrent Pos: Head neuropsychiat div, Sinai Hosp, 48-65, head div neurol, 65-; instr, Pavlovian Lab, Phipps Psychiat Clin, Johns Hopkins Hosp, 52-69; asst ed, Pavlovian 66. Mem: AAAS; Pavlovian Soc NAm (secy-treas, 65-); Am Asn Anat; Am Psychiat Asn; AMA. Res: Anatomy of autonomic nervous systems; innervation endocrine glands and viscera; physiology of respiratory reflexes; aphasia; autonomic drug action on urinary bladder; muscle fasciculations and conditioned reflexes; clinical neurology and psychiatry. Mailing Add: 200 W Cold Spring Ln Baltimore MD 21210

TEITELL, LEONARD, microbiology, see 12th edition

TEITLER, SIDNEY, b New York, NY, July 1, 30. PHYSICS. Educ: Long Island Univ, BS, 51; Univ Ill, MS, 53; Syracuse Univ, PhD(physics), 57. Prof Exp: PHYSICIST, US NAVAL RES LAB, 57- Mailing Add: Code 4105 US Naval Res Lab Washington DC 20375

TEJA, JAGDISH SINGH, b Jhingran, India, Oct 2, 37; m 63; c 2. PSYCHIATRY. Educ: Panjab Univ, MD, 60; Mysore Univ, DPM, 64; All India Inst Med Sci, MD(psychiat), 67. Prof Exp: Tutor psychiat, All India Inst Med Sci, 65-67, lectr, 67-68; asst prof, Postgrad Inst Med Educ & Res, 68-70; asst prof, 71, ASSOC PROF PSYCHIAT & DIR RES, SCH MED, UNIV VA, 72- Concurrent Pos: Consult, Vet Admin Hosp, Salem, Va, 73- Mem: Am Psychiat Asn; Sigma Xi; Royal Col Psychiat, Eng; Indian Psychiat Soc. Res: Transcultural psychiatry, especially culture specific syndromes and relationship of family patterns to mental illness; psychopharmacology and biochemical psychiatry, especially lithium; depression in youth. Mailing Add: Dept of Psychiat Sch of Med Univ of Va Box 203 Charlottesville VA 22901

TEJADA, CARLOS, b Guatemala City, Guatemala, Dec 14, 25; m 51; c 6. PATHOLOGY, NUTRITION. Educ: San Carlos Univ Guatemala, MD, 49, BS, 71. Prof Exp: Kellog Found fel, Mass Gen Hosp, 51-53; prof path & dir dept, Sch Med, San Carlos Univ Guatemala, 54-64; chief, Div Clin Path, 55-64, chief, Div Educ, 64-74, DIR, INST NUTRIT CENT AM & PANAMA, 75- Concurrent Pos: Asst chief, Dept Path, Gen Hosp, Guatemala, 54-58; NIH grant, Inst Nutrit Cent Am & Panama, 55-69; chief lab path, Roosevelt Hosp, Guatemala, 58-64; WHO consult, 60; consult, Inst Int Med, Med Ctr, La State Univ, 67; Josiah Macy Jr Found grant, 67-; Williams Waterm Res Corp grant, 71- Mem: Int Acad Path; Am Soc Clin Path; Cent Am Asn Path; Latin Am Asn Anat Path; Guatemala Col Med. Res: Nutritional pathology; atherosclerosis. Mailing Add: Inst Nutrit of Cent Am & Panama PO Box 1188 Guatemala City Guatemala

TEKEL, RALPH, b New York, NY, May 27, 20; m 60; c 2. ORGANIC CHEMISTRY. Educ: Polytech Inst Brooklyn, BS, 41; Purdue Univ, MS, 47, PhD(chem), 49. Prof Exp: Asst tech dir, Vitamins, Inc, Ill, 48-49; res assoc, Carter Prod, NJ, 49-51; mgr pilot plant, Nat Drug Co, 51-60; asst to mgr chem div, Wyeth Labs, 60-63; dir org res, Betz Lab, 63-65; from asst prof to assoc prof org chem, 65-74, ASSOC PROF CHEM, LA SALLE COL, 74- Concurrent Pos: Lectr, Holy Family Col, Pa, 66-67; consult, Am Electronic Labs, 66- Mem: AAAS; fel Am Inst Chem; Am Chem Soc. Res: Medicinals; biochemicals; halogen chemicals; pilot plant development; continuous thin layer chromatography. Mailing Add: Dept of Chem La Salle Col Philadelphia PA 19141

TEKELI, SAIT, b Samsun, Turkey, June 14, 32; m 59; c 1. VETERINARY MEDICINE, PATHOLOGY. Educ: Univ Ankara, DVM, 54, PhD(path), 58; Univ Wis-Madison, MS, 62, PhD(avian leukosis), 64. Prof Exp: Dist vet, Dept Agr, Samsun, Turkey, 54-55; res asst animal path, Univ Ankara, 55-60; res asst vet sci, Univ Wis-Madison, 60-64, res asst bovine leukosis, 66-67; res pathologist, Norwich Pharmacal Corp, 67-69; PATHOLOGIST, ABBOTT LABS, 69- Mem: Soc Toxicol; Int Acad Path; Soc Pharmacol & Environ Path. Res: Drug toxicity; drug-induced lesions as well as chemical carcinogens. Mailing Add: Dept of Path Abbott Labs North Chicago IL 60064

TELANG, VASANT G, b Kumta, India, July 18, 35. MEDICINAL CHEMISTRY. Educ: Univ Bombay, BS, 56, MS, 64; Univ RI, PhD(pharmaceut chem), 68. Prof Exp: Lab instr pharmaceut chem, Univ Bombay, 59-62, asst prof pharm, 62-63; teaching asst pharmaceut chem, Col Pharm, Univ RI, 63-68; NIH res specialist, Col Pharm, Univ Minn, Minneapolis, 68-74; MEM STAFF, COL PHARM, HOWARD UNIV, 74- Concurrent Pos: Consult, Suneeta Labs, India, 70- Mem: Indian Pharmaceut Asn;

Am Chem Soc. Res: Mechanism of action of narcotic analgesics and their antagonists. Mailing Add: Dept of Biomed Chem Howard Univ Col of Pharm Washington DC 20059

TELEFUS, DAVID, organic chemistry, see 12th edition

TELEGDI, VALENTINE LOUIS, b Budapest, Hungary, Jan 11, 22; nat US; m 50. PHYSICS. Educ: Univ Lausanne, MSc, 46; Swiss Fed Inst Technol, PhD(physics), 50. Prof Exp: Asst physics, Swiss Fed Inst Technol, 47-50; from instr to prof, 50-71, ENRICO FERMI DISTINGUISHED SERV PROF PHYSICS, UNIV CHICAGO, 71- Concurrent Pos: Vis mem, H H Wills Lab, Univ Bristol, 48; lectr, Northwestern Univ, 53-54; vis res fel, Calif Inst Technol, 53; Ford fel & NSF vis scientist, Europ Orgn Nuclear Res, Geneva, 59; Loeb vis prof, Harvard Univ, 66; univ lectr, NY Univ, 67. Mem: Am Phys Soc. Res: Nuclear emulsion technique; experiment and theory of interaction of nuclei with photons; Compton effect of proton; symmetry properties of weak interactions; muon decay and absorption; decay of free neutron; magnetic properties of the muon; hypernuclei; long-lived strange particles. Mailing Add: Dept of Physics Univ of Chicago 5640 Ellis Ave Chicago IL 60637

TELEKI, GEZA, b Budapest, Hungary, Nov 27, 11; m 55. GEOLOGY. Educ: Univ Vienna, PhD, 37. Prof Exp: Geologist, Hungarian Geol Surv, 36-40; prof econ geol & geog, Royal Francis Joseph Univ, Hungary, 40-44; prof econ geog, Tech & Eng Univ, Budapest, 44-48; sr researcher geog, Univ Va, 49-50, sr researcher & assoc prof foreign affairs, 50-55; sr researcher, Arctic Inst NAm, 56-58; assoc prof, 58-59, PROF GEOL, GEORGE WASHINGTON UNIV, 59-, CHMN DEPT, 58- Concurrent Pos: Minister Pub Educ, Hungarian Govt, 44-45. Mem: Geol Soc Am; Am Soc Photogram; Asn Am Geog. Res: Photointerpretation, especially in geomorphology; arctic sea-ice research; environmental geology. Mailing Add: Dept of Geol George Washington Univ Washington DC 20006

TELEKI, PAUL GEZA, physical oceanography, marine geology, see 12th edition

TELES, MORRIS, geometry, see 12th edition

TELFAIR, DAVID, b Sabina, Ohio, Aug 12, 12; m 39; c 4. PHYSICS. Educ: Earlham Col, AB, 36; Haverford Col, MS, 37; Pa State Col, PhD(physics), 41. Prof Exp: Res physicist, Monsanto Chem Co, Mass, 41-46; from asst prof to assoc prof, 46-61, PROF PHYSICS, EARLHAM COL, 61- Mem: Am Phys Soc; Am Asn Physics Teachers. Res: Ultrasonic measurement in gases; physical test methods for plastics; molecular versus physical properties of plastics; natural radioactivity of soils. Mailing Add: 548 S Round Barn Rd Richmond IN 47374

TELFER, NANCY, b San Francisco, Calif, Apr 15, 30. MEDICINE, NUCLEAR MEDICINE. Educ: Stanford Univ, AB, 51; Woman's Med Col Pa, MD, 56; Am Bd Internal Med, dipl, 63; Am Bd Nuclear Med, dipl, 72. Prof Exp: Intern, Los Angeles County-Univ Southern Calif Med Ctr, 56-57, resident internal med, 57-60; instr med, Ctr Health Sci, Univ Calif, Los Angeles, 60-61, asst prof, 62-67; asst prof, 67-69, ASSOC PROF RADIOL & MED, LOS ANGELES COUNTY-UNIV SOUTHERN CALIF MED CTR, 69- Concurrent Pos: Los Angeles County Heart Asn res fel, Isotope Lab Med, Cantonal Hosp, Geneva, Switz, 61-62; Kate Meade Hurd fel, Woman's Med Col Pa, 61-62. Mem: AAAS; Am Fedn Clin Res; fel Am Col Physicians; Soc Nuclear Med; fel Am Col Nuclear Physicians. Res: Body electrolyte composition using radioactive tracers and the dilution principle. Mailing Add: LAC-USC Med Ctr 1200 N State St Box 772 Los Angeles CA 90033

TELFER, WILLIAM HARRISON, b Seattle, Wash, June 21, 24; m 50; c 2. REPRODUCTIVE BIOLOGY, DEVELOPMENTAL BIOLOGY. Educ: Reed Col, BA, 48; Harvard Univ, MS, 49, PhD(biol), 52. Prof Exp: Jr fel, Harvard Soc Fels, 52-54; from asst prof to prof biol, 54-73, chmn grad group, 60-70, PROF ZOOL & CHMN DEPT BIOL, UNIV PA, 73- Concurrent Pos: Guggenheim fel, Stanford Univ, 60-61; NSF sr fel, Univ Miami, 68-69; staff mem, NIH training prog in fertilization & gamete physiology, Marine Biol Lab, Woods Hole, Mass, 71- Mem: Am Soc Zool; Soc Devleop Biol. Res: Physiology and developmental aspects of egg formation in insects, crustaceans and other invertebrates. Mailing Add: Dept of Biol Univ of Pa Philadelphia PA 19104

TELFORD, HORACE SPOONER, entomology, aquatic biology, see 12th edition

TELFORD, IRA ROCKWOOD, b Idaho Falls, Idaho, May 6, 07; m 33; c 4. ANATOMY. Educ: Univ Utah, AB, 31, AM, 33; George Washington Univ, PhD(anat), 42. Prof Exp: Sch teacher, Idaho, 33-37; instr anat, Sch Med, George Washington Univ, 41-43, from asst prof to assoc prof, 43-47; prof & chmn dept, Sch Dent, Univ Tex, 47-53; prof & chmn dept, Sch Med, George Washington Univ, 53-72; PROF ANAT, SCHS MED & DENT, GEORGETOWN UNIV, 72- Mem: Soc Exp Biol & Med; Am Asn Anat; Am Acad Neurol. Res: Histology, muscle and nerve studies in vitamin E deficiency; vitamins; cancer in dietary deficiencies; muscular dystrophy. Mailing Add: Dept of Anat Georgetown Univ Schs Med & Dent Washington DC 20007

TELFORD, JAMES WARDROP, b Merbein, Australia, Aug 16, 27; m 54; c 3. ATMOSPHERIC PHYSICS, COMPUTER SCIENCE. Educ: Univ Melbourne, BSc, 50, DSc(atmospheric convection), 70; Univ Sydney, dipl numerical anal automatic comput, 62. Prof Exp: Sr res scientist, Radiophys Div, Commonwealth Sci & Indust Res Orgn, 50-65; vis scientist, Dept Cloud Physics, Imp Col, Univ London, 65-66; sr res scientist, Radiophys Div, Commonwealth Sci & Indust Res Orgn, 66-67; DEP DIR LAB ATMOSPHERIC PHYSICS & RES PROF ATMOSPHERIC MOTION, DESERT RES INST, UNIV NEV SYST, RENO, 67- Concurrent Pos: Dept Defense res contract, Lab Atmospheric Physics, Desert Res Inst, Reno, 67-72, NSF res grants, 70-74, lectr, Univ Nev, Reno, 71- Mem: Am Meteorol Soc; fel Royal Meteorol Soc. Res: Experimental and theoretical work on coalescence mechanisms in clouds; theory of stochastic coalescence in warm clouds; theory of clear air convection; airborne air motion measuring system. Mailing Add: Lab of Atmospheric Physics Desert Res Inst Sage Bldg Stead Campus Univ Nev Syst Reno NV 89507

TELFORD, RUTH JANE, b Stanford, NY, Sept 8, 20. ANESTHESIOLOGY. Educ: Fla State Univ, BS, 42; Emory Univ, MS, 43; Baylor Univ, MD, 50; Am Bd Anesthesiol, dipl, 58. Prof Exp: Res chemist, Dept Nutrit, Univ Ala, 43-44; asst pharmacol, Baylor Col Med, 44-46; intern, Rochester Gen Hosp, 50-51, resident anesthesia, 52-54; assoc pharmacol, Med Sch, Univ Louisville, 51-52; from instr to assoc prof, 54-69, PROF ANESTHESIOL, BAYLOR COL MED, 69- Concurrent Pos: Assoc, Jefferson Davis Hosp, 54-, mem attend staff; consult, Tex Children's Hosp, Houston, 59-, St Luke's Hosp, Vet Admin Hosp & Tex Inst Rehab, 60-; mem attend staff, Ben Taub Gen Hosp & Methodist Hosp. Mem: Am Soc Anesthesiol; AMA. Mailing Add: Dept of Anesthesiol Baylor Col of Med Houston TX 77025

TELFORD, SAM ROUNTREE, JR, b Winter Haven, Fla, Aug 25, 32; m 57; c 3. EPIZOOTIOLOGY. Educ: Univ Va, BA, 55; Univ Fla, MS, 61; Univ Calif, Los Angeles, PhD(zool), 64. Prof Exp: Lectr zool, Univ Calif, Los Angeles, 64-65; Nat Inst Allergy & Infectious Dis res fel parasitol, Inst Infectious Dis, Univ Tokyo, 65-67; mem staff, Gorgas Mem Lab, CZ, 67-70; int assoc cur, 70, asst cur, 70-73, FIELD RES ASSOC, FLA STATE MUS, UNIV FLA, 73-, ASST PROF BIOL SCI & ZOOL, UNIV, 70- Concurrent Pos: Med Zoologist, WHO, Geneva, Switz, 73, Chagas Dis Vector Res Unit, Acarigua, Venezuela, 73-75 & Vertebrate Pest Control Ctr, Karachi, Pakistan, 75- Mem: Am Soc Ichthyologists & Herpetologists; Am Soc Parasitol; Soc Protozool; Soc Study Amphibians & Reptiles; World Fedn Parasitologists. Res: Herpetology; parasitology; ecology; population dynamics of reptilian host-parasite associations; lower vertebrate parasitology; ecology and systematics of reptiles and amphibians; saurian malaria; zoonotic disease. Mailing Add: Fla State Mus Univ of Fla Gainesville FL 32601

TELFORD, WILLIAM MURRAY, b Ottawa, Ont, Aug 7, 17; m 42; c 2. PHYSICS, GEOPHYSICS. Educ: McGill Univ, BSc, 39, MSc, 41, PhD(physics), 49. Prof Exp: Engr, Res Enterprises, Ont, 41-45; from asst prof to assoc prof geophys, 60-75, RES ASSOC PHYSICS, McGILL UNIV, 45-, PROF GEOPHYS, 75- Res: Radar; high energy accelerators; applied geophysics. Mailing Add: Dept of Mining Eng & Appl Geophys McGill Univ Montreal PQ Can

TELKES, MARIA, b Budapest, Hungary, Dec 12, 00; nat US. ENERGY CONVERSION. Educ: Univ Budapest, BA, 20, PhD(phys chem), 24. Hon Degrees: DSc, St Joseph Col, Conn, 57. Prof Exp: Instr physics, Univ Budapest, 23-24; biophysicist, Cleveland Clin Found, 26-37; engr, Res Dept, Westinghouse Elec & Mfg Co, 37-39; res assoc, Mass Inst Technol, 39-53; proj dir solar energy prog, Res Div, NY Univ, 53-58; res dir, Solar Energy Lab, Curtiss-Wright Corp, 58-60; dir res, Cryo-Therm, Inc, Pa, 60-64; mgr thermodyn lab, Melpar Inc, Westinghouse Air Brake Co, 64-69; sr res specialist, Nat Ctr Energy Mgt & Power, Univ Pa, 69-72; SR SCIENTIST, INST ENERGY CONVERSION, UNIV DEL, 72- Concurrent Pos: Adj prof, Inst Energy Conversion, Univ Del, 72- Mem: Am Chem Soc; Solar Energy Soc; Am Soc Heating, Refrigerating & Air-Conditioning Engr. Res: Solar-thermal storage materials used in solar heated and cooled buildings; thermoelectric generators; semiconductors; phase-change thermal control of terrestrial and space applications. Mailing Add: Inst Energy Conversion Univ of Del Newark DE 19711

TELL, BENJAMIN, b Philadelphia, Pa, Dec 11, 36; m 66; c 2. PHYSICS. Educ: Columbia Univ, BA, 58; Univ Mich, Ann Arbor, MS, 60, PhD(physics), 63. Prof Exp: MEM TECH STAFF, BELL TEL LABS, 63- Mem: Am Phys Soc. Res: Optical and electrical properties of semiconductors. Mailing Add: Bell Tel Labs Holmdel NJ 07733

TELLE, JOHN MARTIN, b Akron, Ohio, Nov 3, 47; m 68; c 3. LASERS. Educ: Univ Colo, BS, 69; Cornell Univ, MS, 72, PhD(physics), 75. Prof Exp: RES SCIENTIST LASER PHYSICS, LOS ALAMOS SCI LAB, 75- Mem: Am Phys Soc. Res: Electronic to vibrational energy transfer processes and lasers employing these processes. Mailing Add: Los Alamos Sci Lab Los Alamos NM 87545

TELLER, DANIEL MYRON, b Nashville, Tenn, Feb 10, 30; m 58; c 2. ORGANIC CHEMISTRY. Educ: Northwestern Univ, BS, 52; Loyola Univ, Ill, MS, 54; Mich State Univ, PhD(org chem), 59. Prof Exp: Res chemist, Fabrics & Finishes Dept, E I du Pont de Nemours & Co, 59-60; RES SCIENTIST, WYETH LABS, INC, PHILADELPHIA, 60- Mem: Am Chem Soc. Res: Thiophene derivatives; steroid synthesis; medicinal chemistry. Mailing Add: 824 Devon State Rd Devon PA 19333

TELLER, DAVID CHAMBERS, b Wilkes-Barre, Pa, July 25, 38; m 60; c 1. PHYSICAL BIOCHEMISTRY. Educ: Swarthmore Col, BA, 60; Univ Calif, Berkeley, PhD(biochem), 65. Prof Exp: Asst prof, 65-70, ASSOC PROF BIOCHEM, UNIV WASH, 70- Concurrent Pos: Consult, Spinco Div, Beckman Instruments, 66. Mem: Am Soc Biol Chem. Res: Physical chemistry and equilibria of proteins; non-covalent association. Mailing Add: Dept of Biochem Univ of Wash Seattle WA 98105

TELLER, DAVID NORTON, b New York, NY, Oct 1, 36; m 59. NEUROCHEMISTRY, PSYCHOPHARMACOLOGY. Educ: Brooklyn Col, BS, 57; NY Univ, MS, 60, PhD(cytochem), 64. Prof Exp: Biologist, Fine Organics, Inc, 56-57; res asst hemat & nutrit, New York Med Col, 57-59; sr res scientist, NY State Ment Hyg, Manhattan State Hosp, NY State Res Inst, 59-66, assoc res scientist, 66-76; ASSOC PROF, DEPT PSYCHIAT & BEHAV SCI, MED SCH, UNIV LOUISVILLE, 76- Concurrent Pos: Lectr, Dept Psychiat, New York Med Col, 66-67 & Grad Div, Fairleigh Dickinson Univ, 68-71. Mem: Am Chem Soc; Am Soc Neurochem; Am Soc Pharmacol & Exp Therapeut; Am Soc Test & Mat; Int Col Neuropsychopharmacol. Res: Drug binding and transport; subcellular particle preparation; molecular pharmacology. Mailing Add: 9406 Doral Ct Number 6 Louisville KY 40220

TELLER, DAVIDA YOUNG, b Yonkers, NY, July 25, 38; m 60; c 2. VISION, PSYCHOLOGY. Educ: Swarthmore Col, BA, 60; Univ Calif, Berkeley, PhD(psychol), 65. Prof Exp: Res asst prof psychol, 65-67, actg asst prof, 67-68, from asst prof to assoc prof psychol & physiol, 68-74, PROF PSYCHOL & PHYSIOL, UNIV WASH, 74- Concurrent Pos: Nat Inst Neurol Dis & Blindness res grant, 68-71; Nat Eye Inst res grant, 71-74; mem comt vision, Nat Res Coun, 71-76; mem vision res & training comt, Nat Eye Inst, 72-; affil, Regional Primate Res Ctr, 73- & Child Develop & Ment Retardation Ctr, 75-; NSF res grant, 75. Mem: AAAS; Optical Soc Am; Asn Res Vision & Ophthal. Res: Psychophysical studies of vision, especially of spatial interactions in vision; development of vision in human and monkey infants. Mailing Add: Dept of Psychol Univ of Wash Seattle WA 98195

TELLER, EDWARD, b Budapest, Hungary, Jan 15, 08; nat US; m 34; c 2. PHYSICS. Educ: Univ Leipzig, PhD, 30. Hon Degrees: Many from various cols & univ in US, 54-64. Prof Exp: Res assoc, Univ Leipzig, 29-31 & Univ Göttingen, 31-33; Rockefeller fel, Copenhagen, 34; lectr, Univ London, 34-35; prof physics, George Washington Univ, 35-41 & Columbia Univ, 41-42; physicist, Manhattan Eng Dist, Univ Chicago, 42-43 & Los Alamos Sci Lab, 43-46; prof physics, Univ Chicago, 46-52; prof, 53-60, univ prof, 60-75, EMER PROF PHYSICS, UNIV CALIF, BERKELEY, 75-; SR RES FEL, HOOVER INST WAR, PEACE & REVOLUTION, STANFORD UNIV, 75- Concurrent Pos: Asst dir, Los Alamos Sci Lab, 49-52; consult, Lawrence Radiation Lab, Livermore, 52-53, dir, 58-60, assoc dir, Lawrence Livermore Lab, Univ Calif, 54-72, assoc dir-at-large, 72-; mem sci adv bd, US Air Force; gen adv comt, USAEC; dir, Thermo Electron Corp; mem, President's Foreign Intel Adv Bd; consult, Comn Critical Choices of Americans, 74- Honors & Awards: Priestley Mem Award, Dickinson Col, 57; Einstein Award, 59; Gen Donovan Mem Award, 59; Award, Midwest Res Inst, 60; Living Hist Award, Res Inst Am, 60; Golden Plate Award, 61; White & Fermi Awards, 62; Robins Award Am, 63; Harvey Prize, The Technion, Haifa, Israel, 75. Mem: Nat Acad Sci; fel Am Nuclear Soc; fel Am Phys Soc; Am Ord Asn; Am Acad Arts & Sci. Res: Chemical, molecular and nuclear physics; quantum theory. Mailing Add: Hoover Inst Stanford Univ Stanford CA 94305

TELLER, JAMES TOBIAS, b Evanston, Ill, Aug 1, 40; m 63; c 2. GEOLOGY. Educ: Univ Cincinnati, BS, 62, PhD(geol), 70; Ohio State Univ, MS, 64. Prof Exp: Field geologist, Inst Polar Studies, 64-65; petrol geologist, Atlantic Richfield Co, 65-67; asst prof, 70-73, ASSOC PROF GEOL, UNIV MAN, 73- Concurrent Pos: Nat Res Coun Can grants, Univ Man, 70-75, Geol Surv Can grant, 71-72; geol consult, Underwood McLellan & Assoc, 75. Mem: Geol Soc Am; Soc Econ Paleont & Mineral; fel Geol Asn Can. Res: Glacial sedimentation; glacial and lacustrine sedimentation and stratigraphy of southern Manitoba. Mailing Add: Dept of Earth Sci Univ of Man Winnipeg MB Can

TELLER, JOHN ROGER, b Cincinnati, Ohio, June 30, 32; m 60; c 2. ALGEBRA. Educ: Univ Cincinnagi, BS, 55, MA, 59; Tulane Univ, PhD(math), 64. Prof Exp: Asst prof math, Univ NH, 64-65; ASSOC PROF MATH, GEORGETOWN UNIV, 65- Mem: Am Math Soc; Math Asn Am. Res: Mathematical research in partially ordered groups. Mailing Add: Dept of Math Georgetown Univ Washington DC 20057

TELLER, LEO, b Ceske Budejovice, Czech, July 20, 31; m 57; c 4. WATERSHED MANAGEMENT, ENVIRONMENTAL SCIENCE. Educ: Univ Melbourne, BScF, 58; Yale Univ, MF, 60; Univ Wash, PhD(watershed mgt), 63. Prof Exp: Asst forester, Forests Comn Victoria, Australia, 54-63, forest hydrologist, 63-66; forestry officer watershed mgt, UN Food & Agr Orgn, 66-69; ASSOC PROF WATERSHED MGT, COLO STATE UNIV, 69- Concurrent Pos: Coordr San Juan ecol proj, US Bur Reclamation, 70- Mem: Am Geophys Union; Am Water Resources Asn. Res: Water quality; weather modification ecology. Mailing Add: Dept of Watershed Sci Colo State Univ Ft Collins CO 80521

TELLER, MORRIS N, b New York, NY, Jan 18, 15; m 42; c 3. CANCER. Educ: Brooklyn Col, BS, 40; Univ Minn, MS, 46, PhD(plant path), 48. Prof Exp: Asst plant path, Univ Minn, 44-48; assoc res microbiologist, Parke, Davis & Co, 48-52, head microbiol sect, 52-54; asst, Div Steroid Biol, 54-56, assoc & head rat host studies sect, Div Human Tumor Exp Chemother, 54-60, head cancerigenesis & chemother sect, 60-64, SECT HEAD, DIV EXP CHEMOTHER, SLOAN-KETTERING INST CANCER RES, 63-, ASSOC MEM, 60-; ASST PROF BIOL, SLOAN-KETTERING DIV, MED COL, CORNELL UNIV, 58- Mem: AAAS; Am Asn Cancer Res; Am Asn Immunol; Geront Soc. Res: Experimental chemotherapy; aging, immunity and cancer; carcinogenesis. Mailing Add: Walker Lab Sloan-Kettering Inst Rye NY 10580

TELLER, MORTON HERMAN, b Newburg, NY, May 29, 21; m 43; c 2. PHYSICS. Educ: Univ Fla, BSE, 43, MS, 49. Prof Exp: Instr, 42-48, curator, 49-67, asst prof, 67-74, EMER ASST PROF PHYSICS, UNIV FLA, 74- Mem: Am Phys Soc. Mailing Add: Rte 2 Box 435 Melrose FL 32666

TELLINGHUISEN, JOEL BARTON, b Cedar Falls, Iowa, May 27, 43; m 72. CHEMISTRY, PHYSICS. Educ: Cornell Univ, AB, 65; Univ Calif, Berkeley, PhD(chem), 69. Prof Exp: Res assoc chem, Univ Canterbury, 69-71; res assoc physics, Univ Chicago, 71-73; Nat Res Coun res assoc, Nat Oceanic & Atmospheric Admin, Boulder, Co, 73-75; ASST PROF CHEM, VANDERBILT UNIV, 75- Res: Molecular and atomic physics; optical spectroscopy. Mailing Add: Dept of Chem Vanderbilt Univ Nashville TN 37235

TELSER, ALVIN GILBERT, b Chicago, Ill, May 11, 39; m 67; c 2. BIOCHEMISTRY, CELL BIOLOGY. Educ: Univ Chicago, BS, 61, PhD(biochem), 68. Prof Exp: Helen Hay Whitney Found fel develop biol, Brandeis Univ, 68-70; Helen Hay Whitney Found fel cell biol, Yale Univ, 70-71; ASST PROF ANAT & CELL BIOL, MED SCH, NORTHWESTERN UNIV, CHICAGO, 71- Mem: AAAS; Soc Develop Biol; Am Soc Cell Biol; NY Acad Sci; Sigma Xi. Res: Biochemical aspects of differentiation and development in eukaryotic systems with emphasis on understanding regulatory mechanisms at a molecular level. Mailing Add: Dept of Anat Northwestern Univ Med Sch Chicago IL 60611

TEMIN, HOWARD MARTIN, b Philadelphia, Pa, Dec 10, 34; m 62; c 2. ONCOLOGY, VIROLOGY. Educ: Swarthmore Col, BA, 55; Calif Inst Technol, PhD(biol), 59. Hon Degrees: Dr, Swarthmore Col, 72; Dr, New York Med Col, 72. Prof Exp: Asst prof, 60-64, assoc prof, 64-69, PROF ONCOL, UNIV WIS-MADISON, 69-, WIS ALUMNI RES FOUND PROF CANCER RES, 71-, AM CANCER SOC PROF VIRAL ONCOL & CELL BIOL, 74- Concurrent Pos: Mem virol & rickettsiol study sect, NIH, 71-75; assoc ed, J Cell Physiol, 66- & Cancer Res, 71-74. Honors & Awards: Warren Triennial Prize, Mass Gen Hosp, 71 (shared); Pap Award, Pap Inst Miami, 72; Bertner Award, Univ Tex M D Anderson Hosp & Tumor Inst Houston, 72; US Steel Award, Nat Acad Sci, 72; Award Enzyme Chem, Am Chem Soc, 73; Award Distinguished Achievement, Mod Med, 73; Griffuel Prize, Asn Develop Res Cancer, Villejuif, France, 73; Dyer Lectr Award, NIH, 74; Clowes Lectr Award, Am Asn Cancer Res, 74; Int Award, Gaindner Found, Toronto, 74 (shared); Albert Lasker Award Basic Med Sci, 74; Nobel Prize in Physiol or Med, 75 (shared). Mem: Nat Acad Sci; fel Am Acad Arts & Sci. Res: Replication of and mechanism of neoplastic transformation by RNA tumor viruses; RNA-directed DNA synthesis and protoviruses; control of multiplication of cultured cells. Mailing Add: McArdle Lab Univ of Wis Madison WI 53706

TEMIN, RAYLA GREENBERG, b New York, NY, May 4, 36; m 62; c 2. GENETICS. Educ: Brooklyn Col, BS, 56; Univ Wis, MS, 58, PhD(genetics), 63. Prof Exp: Proj assoc, 63-72, ASST SCIENTIST MED GENETICS, UNIV WIS-MADISON, 72- Res: Population genetics of Drosophila melanogaster; heterozygous effects of mutations. Mailing Add: Dept of Med Genetics Univ of Wis Madison WI 53706

TEMIN, SAMUEL CANTOR, b Washington, DC, Nov 4, 19; m 47; c 2. POLYMER CHEMISTRY. Educ: Wilson Teachers Col, BS, 39; Univ Md, MS, 43, PhD(org chem), 49. Prof Exp: Res chemist, Army Chem Ctr, Md, 42-44; asst gen org chem & biochem, Univ Md, 46-48; res chemist, Indust Rayon Corp, Ohio, 49-53, res supvr, 53-58; mgr polymer chem group, Explor Sect, Koppers Co, Inc, 58-65; asst dir, Fabric Res Labs, Inc, Mass, 65-72; SECT HEAD POLYMER CHEM, LEXINGTON LAB, KENDALL CO, COLGATE-PALMOLIVE CO, LEXINGTON, 72- Concurrent Pos: Lectr, Pa State Univ, 63-65 & Northeastern Univ, 66- Mem: Am Chem Soc; Fiber Soc; Int Asn Dent Res; Soc Plastics Engr. Res: Polymer synthesis and structural relationships, adhesives, dental materials, monomer synthesis.

TEMKIN, AARON, b Morristown, NJ, Aug 15, 29; m 58; c 2. ATOMIC PHYSICS. Educ: Rutgers Univ, BS, 51; Mass Inst Technol, PhD(physics), 56. Prof Exp: Fulbright fel, Ger, 56-57; physicist, US Naval Res Lab, 57-58; physicist, Nat Bur Standards, 58-60; PHYSICIST THEORET STUDIES GROUP, GODDARD SPACE FLIGHT CTR, NASA, 60- Honors & Awards: Exceptional Performance Award, Goddard Space Flight Ctr, NASA, 71. Mem: Fel Am Phys Soc. Res: Scattering of electrons from atoms, polarized orbitals, nonadiabatic theory; oxygen, hydrogen; threshold law for electron-atom impact ionization; symmetric Euler angle decomposition of three body problem; calculation of autoionization states; scattering of electrons from diatomic molecules, rotational and vibrational excitation. Mailing Add: Code 602 NASA-Goddard Space Flight Ctr Greenbelt MD 20771

TEMKIN, RICHARD JOEL, b Boston, Mass, Jan 8, 45; c 2. LASERS, PLASMA PHYSICS. Educ: Harvard Col, BA, 66; Mass Inst Technol, PhD(physics), 71. Prof Exp: Res fel physics, Harvard Univ, 71-74; STAFF MEM PHYSICS, FRANCIS BITTER NAT MAGNET LAB, MASS INST TECHNOL, 74 Concurrent Pos: IBM Corp fel, 72-74. Mem: Am Phys Soc. Res: Submillimeter lasers, both theory and experiment; laser breakdown and heating of gases; optical and submillimeter diagnostics of plasmas. Mailing Add: Francis Bitter Nat Magnet Lab Mass Inst Technol 170 Albany St Cambridge MA 02139

TEMMER, GEORGES MAXIME, b Vienna, Austria, Apr 10, 22; nat US; m 43. NUCLEAR PHYSICS. Educ: Queens Col, NY, BS, 43; Univ Calif, MA, 44, PhD(physics), 49. Prof Exp: Asst physics, Univ Calif, 43-44 & 46-49; res assoc, Univ Rochester, 49-51; physicist, Nat Bur Standards, 51-53; mem staff terrestrial magnetism, Carnegie Inst, 53-63; PROF PHYSICS & DIR, NUCLEAR PHYSICS LAB, RUTGERS UNIV, NEW BRUNSWICK, 63- Concurrent Pos: Guest investr, Cryogenic Sect, Nat Bur Standards, 53-55; Guggenheim Mem fel, Paris & Copenhagen, 56-57; vis prof, Univ Md, 59; prof, Fla State Univ, 60-63; Rutgers Res Coun fac fel, 68-69 & 75. Honors & Awards: Lindback Found Award for Excellence in Res, Rutgers Univ, 73. Mem: Fel Am Phys Soc. Res: Nuclear reaction mechanisms; very short lifetimes; scattering; angular correlation; low temperature nuclear alignment; gamma-ray spectroscopy; Coulomb excitation; polarized particle sources. Mailing Add: Dept of Physics Rutgers Univ New Brunswick NJ 08903

TEMPELIS, CONSTANTINE H, b Superior, Wis, Aug 27, 27; m 55; c 2. IMMUNOLOGY. Educ: Wis State Col, Superior, BS, 50; Univ Wis, MS, 53, PhD(med microbiol), 55. Prof Exp: Proj assoc immunol, Univ Wis, 55-57; instr microbiol, Sch Med, Univ WVa, 57-58; from asst res immunologist to assoc res immunologist, 58-66, lectr, 60-66, assoc prof-in-residence immunol, 67-70, assoc prof, 70-72, PROF IMMUNOL, SCH PUB HEALTH, UNIV CALIF, BERKELEY, 72- Concurrent Pos: NIH career develop award, 65-70. Mem: AAAS; Am Asn Immunol; NY Acad Sci; Sigma Xi. Res: Production of immunologic unresponsiveness in chickens to soluble protein antigens; immunologic approaches to study of feeding habits of arthropod vectors. Mailing Add: Sch of Pub Health Univ of Calif Berkeley CA 94720

TEMPERLEY, JUDITH KANTACK, b Meriden, Conn, Feb 12, 36; m 56. NUCLEAR PHYSICS. Educ: Univ Rochester, BS, 57; Univ Ore, MS, 59, PhD(physics), 65. Prof Exp: RES PHYSICIST, US ARMY NUCLEAR DEFENSE LAB & US ARMY BALLISTIC RES LABS, 65- Mem: Am Phys Soc; Sigma Xi. Res: Application of nuclear physics techniques to problems in materials science and radiation phenomenology. Mailing Add: US Army Ballistic Res Labs Aberdeen Proving Ground MD 21005

TEMPEST, BRUCE DEAN, b Catasauqua, Pa, Nov 3, 35; m 59; c 3. INFECTIOUS DISEASES. Educ: Lafayette Col, AB, 57; Univ Pa, MD, 61. Prof Exp: Resident med, Philadelphia Gen Hosp, 61-65; fel allergy & immunol, Univ Pa Hosp, 65-67; chief internal med, Dept Health, Educ & Welfare, USPHS, Tuba City, 67-70, Gallup Indian Med Ctr, 70-71, CLIN DIR, SURVEILLANCE PROJ, GALLUP INDIAN MED CTR, 71- Concurrent Pos: Clin asst prof, Dept Family & Community Med, Sch Med, Univ NMex, 73- Mem: Fel Am Col Physicians. Res: Epidemiology of pneumonia, especially pneumococcal pneumonia and the study of clinical manifestations; efficacy of vaccines in pneumonia prevention. Mailing Add: 1603 Monterey Dr Gallup NM 87301

TEMPLE, AUSTIN LIMIEL, (JR), b Leesville, La, Nov 3, 40; m 62; c 2. APPLIED MATHEMATICS. Educ: Centenary Col La, BS, 62; La State Univ, Baton Rouge, MA, 64; George Peabody Col, PhD(math), 71. Prof Exp: Instr math, Northwestern State Univ, 64-65, asst prof, 67-69; instr, Vanderbilt Univ, 69-70; asst prof, 70-75, ASSOC PROF MATH, NORTHWESTERN STATE UNIV, 75- Mem: Math Asn Am. Res: Mathematics education, in-service curriculum for elementary school teachers; develop and standardize tests for college freshmen math courses; develop computer statistical library. Mailing Add: Dept of Math Northwestern State Univ Natchitoches LA 71457

TEMPLE, CARROLL GLENN, b Hickory, NC, Mar 7, 32: m 56; c 2. ORGANIC CHEMISTRY, MEDICINAL CHEMISTRY. Educ: Lenoir-Rhyne Col, BS, 54; Birmingham-Southern Col, MS, 58; Univ NC, PhD(org chem), 62. Prof Exp: Assoc chemist, 55-59, res chemist, 60-64, SR CHEMIST, SOUTHERN RES INST, 64- Mem: Am Chem Soc. Res: Synthesis of potential antimalarian and anticancer drugs. Mailing Add: Southern Res Inst 2000 Ninth Ave S Birmingham AL 35205

TEMPLE, DAVIS LITTLETON, JR, b Tupelo, Miss, June 10, 43; m 66. MEDICINAL CHEMISTRY. Educ: Univ Miss, BS, 66, PhD(med chem), 69. Prof Exp: Res assoc, La State Univ, New Orleans, 69-70; SR INVESTR CHEM RES, MEAD JOHNSON & CO, DIV BRISTOL-MYERS CO, 70- Mem: AAAS; Am Chem Soc. Res: Design and synthesis of drugs for the respiratory area, including antiallergy and mucolytic agents as well as beta-adrenergic agonist and hetercyclic bronchodilators. Mailing Add: Chem Res Mead Johnson & Co Evansville IN 47721

TEMPLE, KENNETH LOREN, b St Paul, Minn, Mar 22, 18; m 43; c 3. GEOENVIRONMENTAL SCIENCE. Educ: Middlebury Col, AB, 40; Univ Wis, MS, 42; Rutgers Univ, PhD(microbiol), 48. Prof Exp: Chemist, US Naval Res Lab, DC, 42-45; instr bact, Univ RI, 48; assoc res specialist, Eng Exp Sta, WVa Univ, 48-53; microbiologist, Tex Co, 53-55; assoc prof microbiol, Agr Exp Sta, Mont State Univ, 55-61; sr res specialist, Commonwealth Sci & Indust Res Orgn, Australia, 61-63; PROF MICROBIOL, MONT STATE UNIV, 64- Mem: Soc Gen Microbiol; Am Chem Soc; Am Soc Microbiol; Am Acad Microbiol. Res: Autotrophic bacteria; microbiology of thermal waters; coal mines; geomicrobiology. Mailing Add: Dept Microbiol Mont State Univ Bozeman MT 59715

TEMPLE, ROBERT DWIGHT, b Des Moines, Iowa, July 1, 41; m 65; c 4. ORGANIC CHEMISTRY. Educ: Clemson Col, BS, 62; Fla State Univ, PhD(chem), 66. Prof Exp: Chemist, E I du Pont de Nemours & Co, 62; RES CHEMIST, PROCTER & GAMBLE CO, 66-, SECT HEAD, 72- Mem: AAAS; Am Chem Soc; The Chem Soc. Res: Physical organic chemistry. Mailing Add: Procter & Gamble Co Miami Valley Labs Cincinnati OH 54239

TEMPLE, ROBERT S, animal breeding, genetics, see 12th edition

TEMPLE, STANLEY, b New York, NY, Aug 17, 30; m 57; c 2. ORGANIC CHEMISTRY. Educ: NY Univ, AB, 52, PhD(org chem), 58. Prof Exp: Res fel cancer steroids, Med Sch, Univ Va, 58-60; res chemist, Plant Tech Sect, Org Chem Dept, 60, res chemist, Process Dept, 60-61, res chemist, Res & Develop Div, Org Chem Dept, 61-74, SR RES CHEMIST, JACKSON LABS, CHAMBERS WORKS, E I DU PONT DE MOURS & CO, INC, 74- Mem: AAAS; Sigma Xi; Am Chem Soc; The Chem Soc; Am Oil Chemists Soc. Res: Fluorochemicals; fluorinated polymers; surface active agents; bio-organic chemistry; cosmetics. Mailing Add: Jackson Lab Chambers Works E I du Pont de Nemours & Co Inc Wilmington DE 19899

TEMPLE, WADE JETT, b Petersburg, Va, Feb 8, 33. MOLECULAR PHYSICS. Educ: Randolph-Macon Col, BS, 54; Univ WVa, MS, 60, PhD(physics), 64. Prof Exp: Instr physics, Randolph-Macon Col, 55-58; from instr to asst, Univ WVa, 60-64; assoc prof, 64-71, PROF PHYSICS, RANDOLPH-MACON COL, 71- Mem: Am Asn Physics Teachers; Am Phys Soc. Res: Electron spin resonance; microwave spectroscopy of free radicals. Mailing Add: Dept of Physics Randolph-Macon Col Ashland VA 23005

TEMPLE, WILLIAM BENSON, b Sallis, Miss, Feb 2, 13; m 38; c 4. APPLIED MATHEMATICS. Educ: La Col, BS, 35; La State Univ, MA, 37; Univ Tex, PhD, 54. Prof Exp: Actg head dept, La Col, 37-38; instr & asst prof, Agr & Mech Col, Tex, 38-48; assoc prof, 48-54, PROF MATH, LA TECH UNIV, 54-, HEAD DEPT, 59- Mem: Am Math Soc; Math Asn Am. Res: Analysis. Mailing Add: Dept of Math La Tech Univ Ruston LA 71270

TEMPLEMAN, GARETH J, b Little Falls, NY, Apr 21, 37; m 70; c 1. PHYSICAL CHEMISTRY, ANALYTICAL CHEMISTRY. Educ: Ohio Wesleyan Univ, BA, 60; State Univ NY Buffalo, PhD(chem), 70. Prof Exp: Res fel phys org chem, State Univ NY Buffalo, 69-70; scientist, 70-71, GROUP LEADER INSTRUMENTATION, CORP RES, PILLSBURY CO, 71- Mem: Am Chem Soc; Am Oil Chemists Soc. Res: Nuclear magnetic resonance; mass spectrometry; electrochemistry; flavors; perception; food chemistry; microwave heating; fats and oils; thermal chemistry. Mailing Add: Pillsbury Co Corp Res 311 Second St Minneapolis MN 55414

TEMPLEMAN, WILFRED, b Bonavista, Nfld, Feb 22, 08; m 37; c 4. MARINE BIOLOGY, FISHERIES. Educ: Dalhousie Univ, BSc, 30; Univ Toronto, MA, 31, PhD(marine biol), 33. Prof Exp: Sci asst zool, Biol Bd Can, 30-34; lectr zool, McGill Univ, 34-36; from assoc prof to prof biol, Mem Univ Nfld, 36-44, head dept, 36-43; dir, Nfld Govt Lab, 44-49; dir, St John's Biol Sta, Fisheries Res Bd Can, 49-72; J L PATON PROF MARINE BIOL & FISHERIES, MEM UNIV NFLD, 72- Concurrent Pos: Vis prof, Mem Univ Nfld, 57-72; fisheries adv, Govt Nfld, 73-75. Mem: Am Fishers Soc; Am Soc Ichthyologists & Herpetologists; Can Soc Zoologists; fel Royal Soc Can; Marine Biol Asn UK. Res: Life history of lobster, capelin, dogfish, cod, haddock and redfish; distribution, systematics and life history of fish species in western North Atlantic. Mailing Add: Queen's Col Mem Univ Nfld St John's NF Can

TEMPLER, DAVID ALLEN, b Chicago, Ill, July 23, 42; m 65. POLYMER CHEMISTRY. Educ: Northwestern Univ, Evanston, BS, 64; Ind Univ, Bloomington, PhD(org chem), 68. Prof Exp: Sr chemist, 68-745, LAB HEAD, LATIN AM OPER, TECH SERV LAB, ROHM AND HAAS, 75- Mem: AAAS; Am Chem Soc; The Chem Soc. Res: Polymer synthesis and characterization with regard to applications in the area of organic coatings; medicinal chemistry; leather chemistry; textile chemistry; ion exchange. Mailing Add: Rohm and Haas Latin Am Oper 12906 SW 89th Ct Miami FL 33156

TEMPLETON, ARCH W, b Madison, Wis, Mar 30, 32; m 55; c 4. MEDICINE, RADIOLOGY. Educ: Univ Omaha, BA, 54; Univ Nebr, MD, 57. Prof Exp: Asst prof med, Wash Univ, 63-64; assoc prof, Univ Mo, 64-68; PROF RADIOL & CHMN DEPT, MED CTR, UNIV KANS, 68- Mem: Radiol Soc NAm; AMA; Asn Univ Radiol. Res: Computer research in medicine; vascular radiology. Mailing Add: Dept of Radiol Univ of Kans Med Ctr Kansas City MO 66103

TEMPLETON, CHARLES CLARK, b Houston, Tex, Oct 4, 21; m 44; c 4. PHYSICAL CHEMISTRY. Educ: La Polytech Inst, BS, 42; Univ Wis, MS, 47, PhD(chem), 48. Prof Exp: Jr res chemist, Shell Oil Co, 42; res chemist, Univ Wis, 46-48; instr, Univ Mich, 48-50; STAFF RES CHEMIST, BELLAIRE RES CTR, SHELL DEVELOP CO, 50- Mem: Am Chem Soc; Am Inst Mining, Metall & Petrol Eng. Res: Solvent extraction; phase equilibria; electrochemistry; multiphase fluid flow; petroleum production. Mailing Add: Shell Develop Co PO Box 481 Houston TX 77001

TEMPLETON, DAVID HENRY, b Houston, Tex, Mar 2, 20; m 48; c 2. PHYSICAL CHEMISTRY. Educ: La Polytech Inst, BS, 41; Univ Tex, MA, 43; Univ Calif, PhD(chem), 47. Prof Exp: Instr chem, Univ Tex, 42-44; res chemist, Metall Lab, Univ Chicago, 44-46; res chemist, Radiation Lab, 46-47, from instr to assoc prof, Univ, 47-58, dean col chem, 70-75, PROF CHEM, UNIV CALIF, BERKELEY, 58- Concurrent Pos: Guggenheim Mem fel, 53. Mem: AAAS; Am Chem Soc; Am Crystallog Asn. Res: Properties of radioactive isotopes; nuclear reactions; structures of crystals. Mailing Add: Dept of Chem Univ of Calif Berkeley CA 94720

TEMPLETON, FREDERIC EASTLAND, b Portland, Ore, May 11, 05; m 36, 75; c 2. RADIOLOGY. Educ: Washington Univ, BS, 27; Univ Ore, MD, 31. Prof Exp: Univ Chicago fel radiol, Univ Stockholm, 33-34; from instr to assoc prof roentgenol, Univ Chicago, 35-43; head dept radiol, Cleveland Clin, 43-45; prof & exec officer, 47-53, clin prof, 53-68, prof, 68-75, EMER PROF RADIOL, MED SCH, UNIV WASH, 75- Mem: Am Gastroenterol Asn; Am Roentgen Ray Soc (1st vpres, 54); fel AMA; Radiol Soc NAm; hon mem Mex Soc Radiol & Phys Ther. Res: Radiologic gastroenterology. Mailing Add: Univ Hosp Univ of Wash Med Sch Seattle WA 98195

TEMPLETON, GEORGE EARL, b Little Rock, Ark, June 27, 31; m 58; c 4. PLANT PATHOLOGY. Educ: Univ Ark, BS, 53, MS, 54; Univ Wis, PhD(plant path), 58. Prof Exp: Asst, Univ Wis, 56-58; from asst prof to assoc prof, 58-67, PROF PLANT PATH, UNIV ARK, FAYETTEVILLE, 67- Mem: Am Phytopath Soc; Mycol Soc Am; Am Soc Plant Physiol. Res: Physiology of parasitism; diseases of rice; biological control of weeds with plant palliogen. Mailing Add: Dept of Plant Path Univ of Ark Fayetteville AR 72701

TEMPLETON, GORDON HUFFINE, b Edowah, Tenn, July 17, 40; m 64; c 1. PHYSIOLOGY. Educ: Univ Tenn, Knoxville, BS, 63; Southern Methodist Univ, MS, 66; Univ Tex Southwestern Med Sch Dallas, PhD(biophys), 70. Prof Exp: Instrumentation engr, Gen Dynamics Corp, Tex, 63-66; instr, 70-71, ASST PROF PHYSIOL, UNIV TEX SOUTHWESTERN MED SCH DALLAS, 71- Concurrent Pos: Nat Heart & Lung Inst spec fel, Univ Tex Southwestern Med Sch Dallas, 70-72; adj prof, Southern Methodist Univ, 70-; Nat Heart & Lung Inst res grant, 75-; Ischemic Scor grant, 75- Mem: Am Heart Asn; Inst Elec & Electronics Engrs; Am Fedn Clin Res; Am Physiol Soc. Res: Cardiac muscle mechanics; measurement of the mechanical properties of both in vivo and in vitro preparations of heart muscle. Mailing Add: Dept of Physiol Univ of Tex Southwestern Med Sch Dallas TX 75235

TEMPLETON, IAN M, b Rugby, Eng, July 31, 29; m 56; c 2. METAL PHYSICS, LOW TEMPERATURE PHYSICS. Educ: Oxford Univ, MA, 50, DPhil(physics), 53. Prof Exp: Fel physics, Nat Res Coun Can, 53-54; mem staff, Res Lab, Assoc Elec Industs, Rug-yRugby, 55-57; asst res officer,57-60, assoc res officer, 60-64, sr res officer, 64-71, PRIN RES OFFICER PURE PHYSICS, NAT RES COUN CAN, 71- Concurrent Pos: Secy, Nat Comt, Int Union Pure & Appl Physics, mem very low temp comn, 72-; comn mem, Int Inst Refrigeration, 72- Mem: Fel Brit Inst Physics; fel Royal Soc Can; Am Phys Soc; Can Asn Physicists. Res: Noise in semiconductors; superconductive devices; thermoelectricity; Fermi surfaces. Mailing Add: Nat Res Coun Physics Div Sussex Dr Ottawa ON Can

TEMPLETON, JAMES R, comparative physiology, deceased

TEMPLETON, JOE WAYNE, b Loraine, Tex, July 18, 41; m 62; c 1. IMMUNOLOGY, GENETICS. Educ: Abilene Christian Col, BS, 64; Ore State Univ, PhD(genetics), 68. Prof Exp: Res fel genetics, Ore State Univ, 65-68; asst prof med genetics, Med Sch, Univ Ore, 68-75; ASST PROF MICROBIOL & IMMUNOL, BAYLOR COL MED, HOUSTON, 75-; ASSOC PROF VET MED & SURG, COL VET MED, TEX A&M UNIV, 75- Mem: Genetics Soc Am; Am Genetic Asn. Res: Immunogenetics of organ transplantation, especially in the dog; general canine genetics. Mailing Add: Inst of Comp Med Baylor Col of Med Houston TX 77025

TEMPLETON, JOHN CHARLES, b Buffalo, NY, June 7, 43; m 67. INORGANIC CHEMISTRY. Educ: Col Wooster, BA, 65; Wesleyan Univ, MA, 67; Univ Colo, Boulder, PhD(inorg chem), 70. Prof Exp: ASST PROF CHEM, WHITMAN COL, 70- Concurrent Pos: NSF grant, Whitman Col, 71-73. Mem: Am Chem Soc; Sigma Xi (vpres, 72-73, pres, 73-74). Res: Studies of transition-metal complex ions in concentrated acid solutions, correlations with acidity functions; reaction kinetics and mechanisms; ultraviolet-visible spectroscopy. Mailing Add: Dept of Chem Whitman Col Walla Walla WA 99362

TEMPLETON, JOHN Y, III, b Portsmouth, Va, July 1, 17; m 43; c 4. SURGERY. Educ: Davidson Col, BS, 37; Jefferson Med Col, MD, 41. Prof Exp: Clin prof surg, Jefferson Med Col, 57-64; prof, Univ Pa, 65-67; Samuel D Gross prof & head dept, 67-70, PROF SURG, JEFFERSON MED COL, 70- Concurrent Pos: Am Cancer Soc clin fel, Jefferson Hosp, 50-51, Runyon fel, 51-52. Mem: Am Surg Asn; Soc Thoracic Surg; Am Col Surg; Soc Vascular Surg; Int Soc Surg. Res: General, cardiac and gastrointestinal surgery. Mailing Add: 130 S Ninth St Philadelphia PA 19107

TEMPLETON, MCCORMICK b Cincinnati, Ohio, May 12, 23; m 54; c 2. ANATOMY. Educ: Columbia Univ, AB, 48; Univ Kans, PhD(anat), 58. Prof Exp: Asst zool, Columbia Univ, 48-49, asst oncol & path, Med Ctr, Univ Kans, 49-51, instr anat, 51-54, asst neuroembryol, 54-57; instr anat, Med Sch, Northwestern Univ, 57-65; asst prof, 65-69, ASSOC PROF ANAT, SCH DENT, UNIV SOUTHERN CALIF, 69-, ACTG CHMN DEPT, 71- Mem: AAAS; Am Asn Anat; Biol Stain Comn; Am Soc Cell Biol. Res: Medical anatomy; histochemistry and microchemistry of nervous system of embryo and adult vertebrates; electrophoresis of esterases in blood and digestive tract. Mailing Add: Dept of Anat Univ of Southern Calif Los Angeles CA 90007

TEMPLETON, WILLIAM CHELCY, JR, b Ackerman, Miss, June 14, 18; m 46; c 1. AGRONOMY. Educ: Miss State Univ, BS, 38; Univ Ill, MS, 39; Purdue Univ, PhD, 60. Prof Exp: Instr, 39-42, from asst prof to assoc prof, 46-60, PROF AGRON, UNIV KY, 60- Mem: Am Soc Agron. Res: Forage crop production and utilization; physiology and ecology of pasture plants; techniques of measuring grassland productivity; forage plant-grazing animal complex. Mailing Add: Dept of Agron Univ of Ky Lexington KY 40506

TEMPLETON, WILLIAM LEES, b London, Eng, Apr 15, 26; m 52; c 3. RADIATION ECOLOGY, MARINE ECOLOGY. Educ: Univ St Andrews, BSc, 50, Hons, 51. Prof Exp: Sr biologist, UK Atomic Energy Authority, Windscale, Eng, 51-65; sr res scientist radioecol, 65-68, mgr aquatic ecol, 68-69, ASSOC DEPT MGR ECOSYSTS, PAC NORTHWEST LABS, BATTELLE MEM INST, 69- Concurrent Pos: Consult, Int Atomic Energy Agency, 60-; mem panel radioactivity in marine environ, Nat Acad Sci-Nat Res Coun, 68-72, mem panel energy & environ, 75-; affil prof, Col Fisheries, Univ Wash, 74- Mem: Fel AAAS; Marine Biol Asn UK; Am Soc Limnol & Oceanog; Brit Ecol Soc; Brit Freshwater Biol Asn. Res: Waste management practice as related to radioecology and limnology of fresh and marine waters; effects of low level chronic pollution by heavy metal, radiation and oil. Mailing Add: Pac Northwest Labs Ecosyst Dept Battelle Mem Inst PO Box 999 Richland WA 99352

TEN BRINK, NORMAN WAYNE, b Shelby, Mich, May 17, 43; m 67; c 1. QUATERNARY GEOLOGY. Educ: Univ Mich, Ann Arbor, BS, 66; Franklin & Marshall Col, MS, 68; Univ Wash, PhD(geol), 71. Prof Exp: Contract geologist, Geol Surv Greenland, 69-71; res fel geol, Inst Polar Studies & Dept Geol, Ohio State Univ, 71-72, asst dir, Inst Polar Studies, 72-73; ASST PROF GEOL, COL ARTS & SCI, GRAND VALLEY STATE COLS, 73- Concurrent Pos: Field leader NSF grant, Univ Wash, 70-71; prin investr NSF grants, Ohio State Univ, 72-74 & 73- Mem: AAAS; Am Quaternary Asn; Arctic Inst NAm; Geol Asn Can; Geol Soc Am. Res: Glacial geology and geomorphology of Arctic, Antarctic and alpine areas. Mailing Add: Dept of Geol Col Arts & Sci Grand Valley State Cols Allendale MI 49401

TEN BROECK, WALTER TRYON LIVINGSTON, JR, chemistry, see 12th edition

TENBROEK, BERNARD JOHN, b Grand Rapids, Mich, Mar 29, 24; m 48; c 4. ZOOLOGY. Educ: Calvin Col, BA, 49; Univ Colo, MA, 55, PhD(zool), 60. Prof Exp: From instr to assoc prof, 55-66, PROF BIOL, CALVIN COL, 66-, CHMN DEPT, 61- Mem: AAAS; Am Soc Zool. Res: Studies on thyroid and pituitary function in neotenic forms of Ambystoma tigrinum. Mailing Add: Dept of Biol Calvin Col Grand Rapids MI 49506

TEN CATE, ARNOLD RICHARD, b Accrington, Eng, Oct 21, 33; m 56; c 3. ANATOMY, DENTISTRY. Educ: Univ London, BDS, 60, BSc, 55, PhD(anat), 58. Prof Exp: Sr lectr dent sci, Royal Col Surgeons Eng, 61-63; sr lectr anat in dent, Guy's Hosp Med Sch, Univ London, 63-68; prof anat & dent, 68-71, PROF BIOL SCI & CHMN DIV, FAC DENT, UNIV TORONTO, 71- Honors & Awards: Colyer Prize, Royal Soc Med, 62; Milo Hellman Award, Am Asn Orthod, 75. Mem: Anat Soc Gt Brit & Ireland. Res: Dental histology; development of periodontium and connective tissue remodeling. Mailing Add: Div of Biol Sci Fac of Dent Univ of Toronto 124 Edward St Toronto ON Can

TENCZA, THOMAS MICHAEL, b Wallington, NJ, July 8, 32; m 59. ORGANIC CHEMISTRY. Educ: Columbia Univ, AB, 54; Seton Hall Univ, MS, 64, PhD(chem), 66; Fairleigh Dickenson Univ, MBA, 71. Prof Exp: Res chemist, S B Penick Co, 57-60; DIR RES & DEVELOP COORD, BRISTOL-MYERS CO, HILLSIDE, 60- Mem: Am Chem Soc; Am Pharmaceut Asn; Am Mgt Asn; Soc Cosmetic Chemists. Res: Research and development management; product development; rearrangement reactions of small ring compounds; botanical drugs. Mailing Add: Hillside NJ

TENCZAR, FRANCIS J, b Chicago, Ill, Sept 18, 22; m 47; c 4. PATHOLOGY. Educ: Univ Notre Dame, BS, 45; Univ Ill, MD, 46; Am Bd Path, cert anat path, 53, cert clin path, 55. Prof Exp: Resident path, Cincinnati Gen Hosp, Ohio, 49-51; resident, Chicago Wesley Mem Hosp, 51-53; asst prof path, Univ Cincinnati, 53-54; instr, Med Sch, Northwestern Univ, 54-56, asst prof, 56-61; med dir, Minneapolis War Mem Blood Bank, 61-63; asst prof path, Northwestern Univ, 63-64; assoc prof path, Univ Ill, Chicago Circle, 64-73, med dir, Res & Educ Hosps Blood Bank, 64-73; CONSULT, BLOOD BANK, NORTHWESTERN MEM HOSP, 73- Concurrent Pos: Attend pathologist, Daniel Drake Mem Hosp, Cincinnati, Ohio, 53-54; assoc pathologist, Chicago Wesley Mem Hosp, 54-61; consult, Vet Res Hosp, 57-61; pathologist, St Francis Hosp, Blue Island, Ill, 63-64; consult, Westside Vet Admin Hosp, 64-73. Mem: Am Soc Clin Path; Col Am Path; AMA. Res: Clinical and anatomical pathology; immunohematology; blood banking. Mailing Add: Blood Bank Northwestern Mem Hosp Chicago IL 60611

TENDAM, DONALD JAN, b Hamilton, Ohio, May 28, 16; m 39; c 3. NUCLEAR PHYSICS. Educ: Miami Univ, AB, 40; Purdue Univ, MS, 42, PhD(physics), 49. Prof Exp: Asst, 40-42, from instr to assoc prof, 42-60, PROF PHYSICS, PURDUE UNIV, WEST LAFAYETTE, 60-, ASSOC HEAD DEPT, 66- Mem: AAAS; Am Phys Soc; Am Asn Physics Teachers. Res: Particle accelerators; radioactive tracers; nuclear reactions; deuteron-bombarded semiconductors. Mailing Add: Dept of Physics Purdue Univ West Lafayette IN 47906

TENDLER, MOSES DAVID, b New York, NY, Aug 7, 26; m 48; c 7. MICROBIOLOGY. Educ: NY Univ, BA, 47, MA, 51; Columbia Univ, PhD, 57. Prof Exp: From instr to assoc prof, 52-63, asst dean, 56-59, PROF BIOL, YESHIVA UNIV, 63- Concurrent Pos: Consult, Eli Lilly & Co, 59-61 & Hoffmann-La Roche Inc, 63-; res dir, Thermobios Pharmaceut Corp, 63-; mem res adv coun, NY Cancer Res Inst, 65-71. Mem: AAAS; Am Soc Microbiol; Torrey Bot Club; Am Soc Clin Pharmacol & Therapeut; NY Acad Sci. Res: Nutrition of thermophilic actinomycetes; antibiotic and antitumor agents proliferated by thermophilic organisms; physiological problems of thermophily. Mailing Add: Dept of Biol Yeshiva Univ 500 W 185th St New York NY 10033

TEN EICK, ROBERT EDWIN, b Portchester, NY, Oct 14, 37; m 62; c 2. ELECTROPHYSIOLOGY, PHARMACOLOGY. Educ: Columbia Univ, BS, 63, PhD(pharmacol), 68. Prof Exp: Guest investr cardiac electrophysiol, Rockefeller Univ, 68; asst prof, 68-74, ASSOC PROF PHARMACOL, MED SCH, NORTHWESTERN UNIV, CHICAGO, 74- Concurrent Pos: NIH trainee cardiac electrophysiol, Rockefeller Univ, 68; Chicago Heart Asn res grant; vis prof, Physiol Inst, Univ Saarland, 74-75; NIH res career develop award, 75-80. Mem: Am Physiol Soc; Am Heart Asn; Am Soc Pharmacol & Exp Therapeut. Res: Electrophysiological basis of cardiac dysrhythmias associated with heart disease; mode of action of pharmacological and physiological antiarrhythmic interventions; electrophysiology of the heart; myocardial membrane currents and their relation to cardiac electrical activity and cardiac contraction. Mailing Add: Dept of Pharmacol Northwestern Univ Med Sch Chicago IL 60611

TENEN, STANLEY S, experimental psychology, see 12th edition

TENENBAUM, JOEL, b Brooklyn, NY, Dec 17, 40; m 67. DYNAMIC METEOROLOGY. Educ: Calif Inst Technol, BS, 62; Harvard Univ, AM, 63, PhD(physics), 68. Prof Exp: Res assoc physics, Stanford Linear Accelerator Ctr, 68-71; asst prof, 71-76, ASSOC PROF PHYSICS, STATE UNIV NY COL PURCHASE, 76- Concurrent Pos: Consult, Inst Space Studies, NASA Goddard Space Flight Ctr, 72- Mem: AAAS; Am Phys Soc; Am Asn Physics Teachers. Res: Modeling of large scale processes in dynamic meteorology; elementary particle physics. Mailing Add: Div of Natural Sci State Univ NY Col Purchase NY 10577

TENENBAUM, LEON EDWARD, b New York, NY, July 12, 15; m 42; c 2. CHEMISTRY. Educ: City Col New York, BS, 36; Columbia Univ, PhD(chem), 40; New York Law Sch, LLB, 62. Prof Exp: Res chemist, Int Vitamin Co, NY, 40-41; Upjohn fel, Upjohn Co, Mich, 41-43, res chemist, 43-46; sr res chemist, Nepera Chem Co, 46-51, patient liaison, 50-51, head dept biochem, 51-57; patent coordr, Vick Chem Co, 57-59; patent agent, Langner, Parry, Card & Langner, 59-61; dir org chem dept, Purdue, Frederick Corp, 61-62; PATENT COUNSEL, US VITAMIN & PHARMACEUT CORP, REVLON INC, 62- Concurrent Pos: Lectr, City Col New York, 53-57. Mem: AAAS; Am Chem Soc; NY Acad Sci. Res: Organic chemistry; enzymes; vitamin synthesis; antibiotics; hormones; medicinals; pyridine chemistry; patent law. Mailing Add: Revlon Inc Legal Dept 767 Fifth Ave New York NY 10022

TENENBAUM, SAUL, b New York, NY, Nov 3, 17; m; c 3. MICROBIOLOGY. Educ: Wash State Univ, BS, 43; Long Island Univ, MS, 64. Prof Exp: Jr seafood inspector, US Food & Drug Admin, 42-44; asst chemist, US Maritime Comn, 44-45; sr biochemist, Standard Brands, 45-46; bacteriologist, Atlantic Yeast Co, 46-48; chief bacteriologist, Premo Pharmaceut Lab, 48-57; group leader, 57-65, mgr, 65-67, ASST DIR MICROBIOL, REVLON RES CTR, 67- Concurrent Pos: Lectr, Fairleigh-Dickinson Univ, 51-56; adj assoc prof pharmaceut sci, Sch Pharm, St John Univ, 73- Mem: Soc Indust Microbiol; Am Soc Microbiol; Soc Cosmetic Chemistry; NY Acad Sci. Res: Development and evaluation of biostatic and biocidal agents; pseudomonads; preservation; microbial content; skin microbiology; immunology; hypersensitive state and agents; analytical immunological technics; mutagenicity, topical and ocular infection. Mailing Add: Revlon Res Ctr 945 Zerega Ave Bronx NY 10473

TENENHOUSE, ALAN M, b Montreal, Que, Aug 8, 35; m 61; c 2. BIOCHEMISTRY, ENDOCRINOLOGY. Educ: McGill Univ, BSc, 55, PhD(biochem), 59, MD & CM, 62. Prof Exp: ASST PROF PHARMACOL, McGILL UNIV, 68- Concurrent Pos: Fel biochem, Univ Wis, 63-65; NIH fel, 64-66; fel biochem, Univ Pa, 65-68; Univ Pa plan scholar, 66-68. Mem: Endocrine Soc; Can Biochem Soc; Can Pharmacol Soc. Res: Mechanism of hormone action with emphasis on role of 3' 5' AMP and calcium in hormone action; biochemistry and physiology of parathyroid hormone and calcitonin. Mailing Add: Dept of Pharmacol & Therapeut McGill Univ McIntyre Med Ctr Montreal PQ Can

TENER, GORDON MALCOLM, b Vancouver, BC, Nov 24, 27. BIOCHEMISTRY. Educ: Univ BC, BA, 49; Univ Wis, MS, 51, PhD(biochem), 53. Prof Exp: Rockefeller fel biochem, Inst Phys-Chem Biol, Paris, France, 53-54; res scientist, BC Res Coun, Vancouver, 54-60; from asst prof to assoc prof, 60-67, PROF BIOCHEM, UNIV BC, 67- Concurrent Pos: Merck Sharp & Dohme Award, Chem Inst Can, 64. Mem: Am Chem Soc; Am Soc Biol Chem; Can Biochem Soc; The Chem Soc. Res: Nucleotide and oligonucleotide synthesis and isolation; purification and properties of transfer ribonucleic acids; gene lccalization in Drosophila. Mailing Add: Dept of Biochem Univ of BC Vancouver BC Can

TEN EYCK, DAVID RODERICK, b Portland, Ore, Oct 24, 29. MEDICINE, PUBLIC HEALTH. Educ: Occidental Col, AB, 51; Univ Southern Calif, MD, 55; Johns Hopkins Univ, MPH, 67, DPH, 73. Prof Exp: Med officer, Med Corps, US Navy, 55-69, dep dir, US Naval Med Res Unit 3, Addis Ababa, Ethiopia, 69-71, cmndg officer, US Naval Med Res Unit 1, Univ Calif, Berkeley, 71-75, cmndg officer, Naval Health Res Ctr, San Diego, 75; PVT PRACT, 75- Mem: AMA; Am Pub Health Asn. Res: Tropical medicine and epidemiology; serologic diagnosis of parasitic diseases; diagnosis and prevention of coccidioidomycosis. Mailing Add: 17412 Hart St Van Nuys CA 91406

TENG, CHING SUNG, b Amoy, Fukien, Nov 20, 37; US citizen; m 64; c 2. REPRODUCTIVE BIOLOGY, BIOCHEMISTRY. Educ: Tunghai Univ, Taiwan, BS, 60; Univ Tex, Austin, MS, 64, PhD(biochem), 67. Prof Exp: Res assoc biochem, Univ Tex, Austin, 67-69; guest investr, Rockefeller Univ, 69-71; asst prof develop biol, State Univ NY Stony Brook, 71-73; ASSOC PROF CELL BIOL, BAYLOR COL MED, 73- Concurrent Pos: NIH res fel, Cancer Inst, 69-70, NIH spec res fel, 70-71, NIH grant award, 73- Mem: Am Soc Cell Biol. Res: Steroid hormone-controlled sex organ differentiation. Mailing Add: Dept of Cell Biol Baylor Col Med Houston TX 77025

TENG, CHRISTINA WEI-TIEN TU, b Kuming, Yunnan, China, July 23, 42; m 64; c 2. CELL BIOLOGY, BIOCHEMISTRY. Educ: Tunghai Univ, Taiwan, BS, 63; Univ Tex, Austin, PhD(biol), 69. Prof Exp: Guest investr cell biol, Rockefeller Univ, 69-71; sr res assoc med, Brookhaven Nat Lab, 71-73; ASST PROF CELL BIOL, BAYLOR COL MED, 73- Mem: Am Soc Cell Biol. Res: Hormone controlled sex organ differentiation; study of the regulatory mechanism for gene activation in mammalian system. Mailing Add: Dept of Cell Biol Baylor Col of Med Houston TX 77025

TENG, EVELYN LEE, b Chungking, China, Feb 8, 38; m 63; c 2. PSYCHOBIOLOGY. Educ: Nat Taiwan Univ, BS, 59; Stanford Univ, MA, 60, PhD(psychol), 63. Prof Exp: Res fel psychobiol, Calif Inst Technol, 63-69, sr res fel, 69-72; asst prof neurol, 72-75, ASSOC PROF NEUROL, SCH MED, UNIV SOUTHERN CALIF, 75- Concurrent Pos: Vis assoc biol, Calif Inst Technol, 72- Mem: Am Psychol Asn. Res: Functional relationship between brain and behavior, especially higher cognitive functions in man. Mailing Add: 1474 Rose Villa St Pasadena CA 91106

TENG, JAMES, b Hong Kong, Dec 4, 29; US citizen; m 57; c 2. CHEMISTRY. Educ: Tri-State Col, BS, S2; Case Western Reserve Univ, MS, 61, PhD(org chem), 67. Prof Exp: Chem engr, Radio Receptor Co, 52-53; process engr, Nylonge Corp, 53-56, res supvr, 56-61, tech supvr, 61-66; fel, Purdue Univ, 66-67; res proj mgr, 68-75, RES MGR ADVAN PROD, ANHEUSER BUSCH, INC, 75- Mem: AAAS; Am Inst Chem; NY Acad Sci; Am Chem Soc. Res: Carbohydrate chemistry; regenerated cellulose; cellulose derivatives; starch derivatives. Mailing Add: Cent Res Dept Anheuser Busch Inc St Louis MO 63118

TENG, JON IE, b Kienow, China, Oct 19, 30; m 58; c 2. STEROID CHEMISTRY, NATURAL PRODUCTS CHEMISTRY. Educ: Nat Taiwan Univ, BS, 55; SDak State Univ, MS, 62; Univ Fla, PhD(agr biochem), 65. Prof Exp: Agr scientist, Taiwan Sugar Corp, Inc, 56-60; res assoc nitrogen metab in hort plants, Univ Ill, Urbana, 65-66; res chemist, Am Crystal Sugar Co, 66-68; NIH fel steroid biochem, Med Sch, Univ Minn, 68-69, res assoc pharmacol, 69-70; res biochemist, 70-75, RES SCIENTIST, UNIV TEX MED BR GALVESTON, 75- Mem: Am Chem Soc; Inst Am Chemists. Res: Steroid biosynthesis and metabolism; drug metabolism; plant nutrition and biochemistry; steroid biochemistry; cholesterol metabolism in mammalian liver, kidney, brain and aortal tissues. Mailing Add: Dept of Biochem Hendrix Bldg Univ of Tex Med Br Galveston TX 77550

TENG, LEE CHANG-LI, b Peiping, China, Sept 5, 26; nat US; m 61; c 1. THEORETICAL PHYSICS. Educ: Fu Jen Univ, China, BS, 46; Univ Chicago, MS, 48, PhD(physics), 51. Prof Exp: Cyclotron asst, Univ Chicago, 49-51; lectr physics, Univ Minn, 51-52, asst prof, 52-53; assoc prof, Univ Wichita, 53-55; assoc physicist, Particle Accelerator Div, Argonne Nat Lab, 55-61, head theory group, 56-62, sr physicist, 61-67, dir, 62-67; HEAD ACCELERATOR THEORY SECT, NAT ACCELERATOR LAB, 67- Mem: Fel Am Phys Soc. Res: High energy accelerators and instrumentation; high energy and nuclear physics; quantum field theory. Mailing Add: Nat Accelerator Lab PO Box 500 Batavia IL 60510

TENG, LINA CHEN, b Fukien, China, Dec 8, 39; US citizen; m 65. DRUG METABOLISM. Educ: Nat Taiwan Univ, BS, 63; Utah State Univ, PhD(org chem), 67. Prof Exp: SR RES CHEMIST, A H ROBINS CO, INC, 73- Mem: Am Chem Soc; Sigma Xi; AAAS. Res: Studies of the metabolism, mainly isolation and identification of the metabolites, of the existing or research drugs in animals and human beings. Mailing Add: Res Labs A H Robins Co Inc 1211 Sherwood Ave Richmond VA 23220

TENG, NELSON N H, b Peking, China. CELL BIOLOGY, BIOPHYSICS. Educ: Univ Ill, Urbana, BS, 65; Univ Pa, MS, 66; Univ Calif, Berkeley, PhD(biophys), 72. Prof Exp: Res assoc cell biol, Mass Inst Technol, 72-75; RES ASSOC CELL BIOL, DEPT PHARMACOL, SCH MED, UNIV MIAMI, 75- Concurrent Pos: Hereditary Dis Found fel, 76; Muscular Dystrophy Asn res fel. Mem: Biophys Soc; AAAS. Res: Membrane surface phenomena with emphasis in hereditary disease. Mailing Add: Dept of Pharmacol Univ of Miami Sch of Med Miami FL 33136

TENG, TA-LIANG, b China, July 3, 37; m 63; c 2. GEOPHYSICS. Educ: Univ Taiwan, BS, 59; Calif Inst Technol, PhD(geophys, appl math), 66. Prof Exp: Res fel geophys, Seismol Lab, Calif Inst Technol, 66-67; from asst prof to assoc prof geophys, 67-74, ASSOC PROF GEOL SCI, UNIV SOUTHERN CALIF, 74- Mem: AAAS; Am Geophys Union; Seismol Soc Am. Res: Elastic wave propagations; observational and theoretical seismology; elastic and anelastic properties of the earth and planetary interiors. Mailing Add: Dept of Geol Sci Univ of Southern Calif Los Angeles CA 90007

TENGERDY, ROBERT PAUL, b Budapest, Hungary, Dec 17, 30; US citizen; m 53; c 2. IMMUNOCHEMISTRY, MICROBIOLOGY. Educ: Budapest Tech Univ, Dipl Chem Eng, 53; St John's Univ, NY, PhD(microbial biochem), 61. Prof Exp: Asst prof biochem eng, Budapest Tech Univ, 53-56; res biochemist, Chas Pfizer & Co, NY, 57-61; asst prof chem & microbiol, 61-64, assoc prof biochem & microbiol, 64-71, PROF MICROBIOL, COLO STATE UNIV, 71- Concurrent Pos: Europ Molecular Biol Orgn fel, Pasteur Inst Paris, 68; Humboldt fel, Max Planck Inst, Univ Göettingen, 68. Mem: AAAS; Am Asn Immunol; Am Soc Microbiol. Res: Immunochemistry of proteins and polyribonucleotides; stress immunochemistry; nutritional aspects of immunology; waste conversion by microbes. Mailing Add: Dept of Microbiol Colo State Univ Ft Collins CO 80523

TENNANT, BUD C, b Burbank, Calif, Nov 10, 33; m 63; c 3. VETERINARY MEDICINE, GASTROENTEROLOGY. Educ: Univ Calif, BS , 57, DVM, 59. Prof Exp: Intern, Sch of Vet Med, Univ Calif, 59, from asst prof to assoc prof, 62-72; res assoc, Dept Surg, Albert Einstein Col Med, 62; res fel, Gastrointestinal Unit, Mass Gen Hosp, 6869; PROF COMP GASTROENTEROL, NY STATE COL VET MED, CORNELL UNIV, 72- Mem: Am Col Vet Internal Med; Am Gastroenterol Asn; Am Inst Nutrit; Am Vet Med Asn; Soc Exp Biol in Med. Res: Diseases of the gastrointestinal tract and liver of domestic animals; pathogenesis of neonatal enteric infections; influence of microorganisms on intestinal function. Mailing Add: NY State Col of Vet Med Cornell Univ Ithaca NY 14853

TENNANT, CHARLES BEARD, b Syracuse, NY, July 21, 21; m 51; c 2. INORGANIC CHEMISTRY. Educ: Hamilton Col, BS, 43; Cornell Univ, PhD(chem), 50. Prof Exp: Lab asst, Cornell Univ, 46-48; investr, 50-59, supvr, 59-64, asst chief, 64-66, MGR APPLNS RES DIV, RES DEPT, NJ ZINC CO, 66- Mem: Am Chem Soc; Am Soc Metals; Am Inst Mining, Metall & Petrol Eng. Res: Research and applications of white pigment materials and of nonferrous metals, zinc oxide, titanium dioxide, zinc metal and metal powders. Mailing Add: 366 Edgemont Ave Residence Park Palmerton PA 18071

TENNANT, JUDITH R, b Sioux City, Iowa, Apr 3, 19. VIROLOGY. Educ: Ohio Wesleyan Univ, BA, 40; Univ Minn, Minneapolis, MS, 57, PhD(virol, cancer biol), 61. Prof Exp: Fel, Jackson Lab, 61-63; res assoc, Sloan-Kettering Inst Cancer Res, 64-67, assoc, 67-72; ASSOC EXEC ED, J CELL BIOL, ROCKEFELLER UNIV, 73- Mem: Am Asn Immunol; Transplantation Soc; Coun Biol Ed; Am Asn Cancer Res; NY Acad Sci. Res: Cancer biology, etiology, natural resistance, genetic factors and immuno-therapy in leukemogenesis. Mailing Add: Rockefeller Univ 1230 York Ave New York NY 10021

TENNANT, RAYMOND WALLACE, b West Frankfort, Ill, Sept 19, 37; m 60; c 4. MICROBIOLOGY, VIROLOGY. Educ: Univ Notre Dame, MS, 61; Georgetown Univ, PhD(microbiol), 63. Prof Exp: Virologist, Dept Virus Res, Microbiol Assocs, Inc, Md, 61-65; USPHS fel replication of DNA viruses, Albert Einstein Med Ctr, 65-66; VIROLOGIST, BIOL DIV, OAK RIDGE NAT LAB, 66-, PROF, GRAD SCH BIOMED SCI, 68- Mem: AAAS; Am Asn Cancer Res; Am Soc Microbiol. Res: Epizoology of murine viruses; replication of DNA containing viruses; murine leukemia; RNA tumor virus cell biology. Mailing Add: 106 Windham Rd Oak Ridge TN 37830

TENNANT, WILLIAM EMERSON, b Washington, DC, Oct 8, 45; m 68; c 2. SOLID STATE PHYSICS. Educ: Harvard Univ, AB, 67; Univ Calif, Berkeley, PhD(solid state physics), 74. Prof Exp: MEM TECH STAFF PHYSICS, ROCKWELL INT SCI CTR, 73- Res: Collective excitations in solids; semiconductor devices; optical nondestructive evaluation methods; low frequency excitations of organic compounds; radiation damage; crystal alloys and defects; solar energy collection. Mailing Add: Rockwell Int Sci Ctr 1049 Camino Dos Rios Thousand Oaks CA 91360

TENNENHOUSE, GERALD JAY, ceramics, inorganic chemistry, see 12th edition

TENNENT, DAVID MADDUX, b Bryn Mawr, Pa, Oct 2, 14; m 45; c 4. BIOCHEMISTRY. Educ: Yale Univ, AB, 36, PhD(org chem), 40. Prof Exp: Asst appl physiol, Yale Univ, 40-42; biochemist, Merck Inst Therapeut Res, 42-60; asst dir res, Hess & Clark Div, Richardson-Merrell Inc, 60-63, dir res & develop, 63-69, vpres & dir res & develop, 69-75, CONSULT VET AFFAIRS, RHODIA INC, 75- Concurrent Pos: Fel, Coun Arteriosclerosis, Am Heart Asn. Mem: Fel AAAS; Am Chem Soc; Am Soc Biol Chem; Soc Exp Biol & Med. Res: Pharmacology of drugs; bacterial pyrogens; cholesterol metabolism and experimental atherosclerosis; medications to improve performance and health of production animals; FDA applications. Mailing Add: 74 Morgan Ave Ashland OH 44805

TENNENT, HOWARD GORDON, b Quebec, Que, Feb 29, 16; US citizen; m 48; c 4. ORGANOMETALLIC CHEMISTRY. Educ: Rensselaer Polytech Inst, BS, 37, MS, 39; Univ Wis, PhD(phys chem), 42. Prof Exp: Res chemist, 42-47, mgr res div cellulose prod, 47-51, exp sta, 52, cent res div, 53-57, res assoc, 58-66, SR RES ASSOC, HERCULES, INC, 66- Mem: Am Chem Soc; Sci Res Soc Am. Res: Chemistry and applications of organometallic compounds in catalysis. Mailing Add: Res Ctr Hercules Inc Wilmington DE 19899

TENNESSEN, KENNETH JOSEPH, b Ladysmith, Wis, June 10, 46; m 67; c 2. FRESH WATER ECOLOGY. Educ: Univ Wis, BS, 68; Univ Fla, MS, 73, PhD(entom), 75. Prof Exp: BIOLOGIST THERMAL EFFECTS AQUATIC INSECTS, TENN VALLEY AUTHORITY, 75- Mem: Entom Soc Am; NAm Benthological Soc; Int Soc Odontol. Res: Investigation of effects of heated discharge from electric generating plants on the survival, growth, reproductive capacity and distribution of dominant aquatic insects in the Tennessee River Valley. Mailing Add: Aquatic Biol Sect Tenn Valley Authority Muscle Shoals AL 35660

TENNEY, LINWOOD POWERS, organic chemistry, see 12th edition

TENNEY, ROBERT IMBODEN, b Decatur, Ill, Nov 15, 10; m 35; c 3. MICROBIOLOGY. Educ: Univ Ill, BS, 32. Prof Exp: Master brewer, Rockford Brewing Co, 32-34; chief chemist, Best Malt Prod Co, 34-35; chemist & instr, Wahl-Henius Inst, 35-38, tech dir, 38-42, pres, 42-58; pres, Sassy Int, 56-66; VPRES & SECY, OZONE, INC, 71- Concurrent Pos: Tech dir, Fleischmann Malting Co, 58-61, vpres, 61-64, exec vpres, 64-66; consult to food & fermentation industs. Mem: AAAS; Am Soc Qual Control; Am Soc Brewing Chemists (pres, 67-68); Am Chem Soc; Inst Food Technol. Res: Brewing; production of foods through fermentations or enzymes; sterilization by ozonation; ozone treatment of waste materials. Mailing Add: PO Box 123 Winnetka IL 60093

TENNEY, STEPHEN MARSH, b Bloomington, Ill, Oct 22, 22; m 47; c 3. PHYSIOLOGY. Educ: Dartmouth Col, AB, 43; Cornell Univ, MD, 46. Prof Exp: Asst prof physiol, Dartmouth Med Sch, 51-54; asst prof physiol & med, Sch Med & Dent, Univ Rochester, 54-56, assoc prof, 56; PROF PHYSIOL & CHMN DEPT, DARTMOUTH MED SCH, 56- Concurrent Pos: Markle scholar, 54-59. Mem: Am Physiol Soc; Am Acad Arts & Sci; Nat Inst Med; Am Soc Clin Invest. Res: Physiology of circulation and respiration. Mailing Add: Dept of Physiol Dartmouth Med Sch Hanover NH 03755

TENNEY, WILTON R, b Buckhannon, WVa, July 2, 28; m 51. PLANT PATHOLOGY. Educ: WVa Wesleyan Col, BS, 50; Univ WVa, MS, 52, PhD(plant path), 55. Prof Exp: Plant scientist, Chem Res & Develop Labs, Army Chem Ctr, Md, 57; from asst prof to assoc prof, 57-71, PROF BIOL, UNIV RICHMOND, 71- Res: Physiology of fungi; host-parasite relations in fungus diseases of plants; toxicity of freshwater bryozoans. Mailing Add: Dept of Biol Univ of Richmond Richmond VA 23173

TENNILLE, AUBREY W, b Baker Co, Ga, Feb 4, 29; m 53; c 1. SOIL FERTILITY, SOIL MICROBIOLOGY. Educ: Univ Ga, BSA, 50; Okla State Univ, MSA, 55; Univ Fla, PhD(soils), 59. Prof Exp: Lab asst soil microbiol, Univ Fla, 58-59; asst county agent, Coop Exten, Univ Ga, 59-60, exten specialist, 60-62; from asst prof to assoc prof, 62-66, PROF AGRON, ARK STATE UNIV, 66- Mem: Am Soc Agron; Soil Sci Soc Am; Soil Conserv Soc Am. Res: Fertility research on zinc and manganese of rice soils of Arkansas and soil salt problems of eastern Arkansas. Mailing Add: Dept of Agron Ark State Univ State University AR 72467

TENNILLE, NEWTON BRIDGEWATER, b South Richmond, Va, Mar 15, 08; m 35; c 3. VETERINARY RADIOLOGY. Educ: Ohio State Univ, DVM, 33. Prof Exp: Vet, WToledo Animal Hosp, 36-48; assoc prof vet med & surg, 49-53, prof vet med & surg & head radiol div, 53-73, EMER PROF RADIOL, COL VET MED, OKLA STATE UNIV, 73- Mem: Am Vet Radiol Soc (pres, 57-58); Am Vet Med Asn; Soc Nuclear Med; Am Animal Hosp Asn. Res: Improved methods for diagnosis and therapy; use of radioisotopes in animals. Mailing Add: 603 Arapaho Dr Stillwater OK 74074

TENNISON, ROBERT L, b Cleburne, Tex, July 12, 33; m 58; c 3. MATHEMATICS. Educ: Howard Payne Col, BA, 54, MEd, 57; Okla State Univ, MS, 59, PhD(math), 64. Prof Exp: Asst prof math, ECent State Univ, 59-61, assoc prof, 64-65; asst prof, 65-67, ASSOC PROF MATH, UNIV TEX, ARLINGTON, 67- Mem: Math Asn Am. Res: Convexity; linear geometry. Mailing Add: Dept of Math Univ of Tex Arlington TX 76010

TENNISSEN, ANTHONY CORNELIUS, b Akron, Ohio, Feb 14, 20; m 52; c 2. GEOLOGY. Educ: Univ Tulsa, BS, 50; Syracuse Univ, MS, 52; Univ Mo-Rolla, PhD(geol), 63. Prof Exp: Geologist, US AEC, 52-54; head geologist, Ideal Cement Co, 54-57; geologist, Mo Geol Surv, 59-63; from asst prof to assoc prof, 63-74, PROF GEOL, LAMAR UNIV, 74- Mem: Mineral Soc Am; Am Crystallog Asn. Res: Characterization of minerals and inorganic compounds by x-ray analysis; crystal-structure analysis of minerals and other compounds. Mailing Add: Dept of Geol Lamar Univ Beaumont TX 77710

TENNYSON, MARILYN ELIZABETH, b Pittsburgh, Pa, June 17, 48. SEDIMENTARY PETROLOGY, TECTONICS. Educ: Middlebury Col, AB, 70; Univ Wash, MS, 72, PhD(geol), 74. Prof Exp: ASST PROF GEOL, WHITTIER COL, 74- Mem: Geol Soc Am; Am Geophys Union. Res: Cordilleran tectonics, particularly Mesozoic history of the Pacific Northwest and California from structural, stratigraphic and sedimentary petrological standpoints; Colorado River sedimentation. Mailing Add: Dept of Geol Whittier Col Whittier CA 90608

TENNYSON, RICHARD HARVEY, b Minneapolis, Minn, Oct 11, 21; m 45; c 9. ORGANIC CHEMISTRY. Educ: St Mary's Col, Minn, BS, 46; Univ Ill, PhD(org chem), 52. Prof Exp: Res chemist, Corn Prod Co, 50-55, anal res supvr, 56-61; sect leader, Qual Control Dept, 61-68, DIR NUTRIT QUAL CONTROL, MEAD JOHNSON & CO, 68- Mem: Am Chem Soc; Inst Food Technol. Res: Quality control administration; analytical methods development; physical, chemical, biological and microbiological control; nutritionals and pharmaceuticals. Mailing Add: 1365 Mesker Park Dr Evansville IN 47712

TENORE, KENNETH ROBERT, b Boston, Mass, Mar 22, 43. BIOLOGICAL OCEANOGRAPHY. Educ: St Anselm Col, AB, 65; NC State Univ, MS, 67, PhD(zool), 70. Prof Exp: Investr biol oceanog, Woods Hole Oceanog Inst, 70-72, asst scientist, 72-75; ASST PROF BIOL OCEANOG, SKIDAWAY INST OCEANOG, 75- Mem: Am Soc Limnol & Oceanog; Estuarine Res Fedn; Ecol Soc Am; AAAS. Res: Bioenergetics of detrital food chains in marine benthic communities. Mailing Add: Skidaway Inst of Oceanog Savannah GA 31401

TENSMEYER, LOWELL GEORGE, b Pocatello, Idaho, Feb 21, 28; m 54; c 4. PHYSICAL CHEMISTRY. Educ: Univ Utah, BA, 52, PhD(combustion), 57. Prof Exp: Asst prof chem, Ohio Univ, 56-57 & Utah State Univ, 57-59; Petrol Inst fel ceramics, Pa State Univ, 59; res scientist, Linde Div, Union Carbide Corp, 60-63; SR PHYS CHEMIST, ELI LILLY & CO, 63- Mem: Am Chem Soc; Am Phys Soc; Coblentz Soc. Res: Molecular spectroscopy; adsorption; crystal growth and purification; lasers; photochemistry and photobiology. Mailing Add: Eli Lilly & Co 740 S Alabama St Indianapolis IN 46206

TENZER, RUDOLF KURT, b Jena, Ger, Oct 9, 20; nat US; m 47; c 4. PHYSICS. Educ: Univ Frankfort, Dipl & Dr rer nat, 50. Prof Exp: Scientist radiation temperature measurements, Hartmann & Braun Co, Ger, 48-53; scientist magnetics; Ind Gen Corp, 53-69, mgr res, 66-69; mgr res, 69-74, MGR RES & MFG ENG, IGC DIV, ELECTRONIC MEMORIES & MAGNETICS CORP, 74- Mem: Inst Elec & Electronics Eng; Am Phys Soc. Res: Permanent magnets; magnetization process; domain theory; ferrites; high temperature properties; temperature measurements by radiation; color pyrometers. Mailing Add: 1075 Brookdale Dr Martinsville NJ 08836

TEODORO, ROSARIO REYES, b Philippines, Sept 24, 13; nat US; m 32; c 2. BACTERIOLOGY. Educ: Univ Mich, PhD(bact), 45. Prof Exp: Bacteriologist, Western Condensing Co, 45-51; res bacteriologist, Henry Ford Hosp, 52-53; from asst prof to assoc prof bact, 53-74, ASSOC PROF BIOL, WAYNE STATE UNIV, 74- Concurrent Pos: NSF fac fel, Oxford Univ, 59. Mem: AAAS; Am Soc Microbiol. Res: Growth factors and metabolism of butyl alcohol organisms; beta-propiolactone and other virucidal agents; riboflavin production by microorganisms. Mailing Add: Dept of Biol Wayne State Univ Detroit MI 48202

TE PASKE EVERETT RUSSELL, b Sheldon, Iowa, Sept 15, 30; m 51; c 4. BIOLOGY, ANIMAL BEHAVIOR. Educ: Westmar Col, AB, 51; State Col Iowa, MA, 57; Okla State Univ, PhD(zool), 63. Prof Exp: Teacher pub schs, Iowa, 52-61; from asst prof to assoc prof, 63-71, PROF BIOL UNIV NORTHERN IOWA, 71- Mem: AAAS; Animal Behav Soc; Am Inst Biol Sci; Poultry Sci Asn. Res: Morphology and taxonomy of Chiroptera; breeding behavior in the Japanese quail; social behavior in chickens. Mailing Add: Dept of Biol Univ of Northern Iowa Cedar Falls IA 50613

TEPE, JOHN BERNARD, analytical chemistry, agricultural chemistry, see 12th edition

TEPFER, SANFORD SAMUEL, b Brooklyn, NY, Mar 24, 18; m 42; c 4. PLANT MORPHOLOGY. Educ: City Col New York, BS, 38; Cornell Univ, MS, 39; Univ Calif, PhD(bot), 50. Prof Exp: Asst bot, Univ Calif, 47-50; instr, Univ Ariz, 50-53, res assoc agr, 53-54; instr biol, Ore Col Educ, 54-55; from asst prof to assoc prof, 55-67, co-chmn dept, 68-71, PROF BIOL, UNIV ORE, 67-, HEAD DEPT, 72- Concurrent Pos: NSF sci fac fel, 65; Fulbright lectr & vis prof, Univ Paris, 71-72. Mem: AAAS; Bot Soc Am; Soc Develop Biol; Int Soc Plant Morphol. Res: Developmental studies of shoot apex and flowers; culture of floral buds; floral morphogenesis. Mailing Add: Dept of Biol Univ of Ore Eugene OR 97403

TEPHLY, THOMAS R, b Norwich, Conn, Feb 1, 36; m 60; c 2. PHARMACOLOGY. Educ: Univ Conn, BS, 57; Univ Wis, PhD(pharmacol), 62; Univ Minn, MD, 65. Prof Exp: Instr pharmacol, Univ Wis, 62; from asst prof to assoc prof, Univ Mich, Ann Arbor, 65-71; PROF PHARMACOL & DIR TOXICOL CTR, UNIV IOWA, 71- Concurrent Pos: Am Cancer Soc res scholar, 62-65. Honors & Awards: John J Abel Award, Am Soc Pharmacol & Exp Therapeut, 71. Mem: Am Soc Pharmacol & Exp Therapeut; Am Soc Biol Chem; Soc Toxicol. Res: Biochemical pharmacology and toxicology; drug metabolism; methanol and ethanol metabolism; heme biosynthesis. Mailing Add: Dept of Pharmacol Univ of Iowa Iowa City IA 52240

TEPLEY, LEE RAYMOND, physics, see 12th edition

TEPLEY, NORMAN, b Denver, Colo, Dec 14, 35; m 68; c 2. SOLID STATE PHYSICS, MEDICAL PHYSICS. Educ: Mass Inst Technol, SB, 57, PhD(physics), 63. Prof Exp: Asst prof physics, Wayne State Univ, 63-69; ASSOC PROF PHYSICS, OAKLAND UNIV, 69- Concurrent Pos: Vis prof, Dept Physics, Univ Lancaster, 70. Mem: AAAS; Am Phys Soc. Res: Physics of metals; ultrasonic studies of Fermi surfaces; electronic structures of metals; properties of superconductors; magnetic fields arising from living systems. Mailing Add: Dept of Physics Oakland Univ Rochester MI 48063

TEPLICK, JOSEPH GEORGE, b Philadelphia, Pa, Sept 29, 11; m 37; c 3. MEDICINE, RADIOLOGY. Educ: Univ Pa, AB, 31, MS, 32, MD, 36, MSc, 42. Prof

Exp: Assoc radiol, Jefferson Med Col, 43-48; chief & dir radiol, Kensington Hosp, 49-63; clin assoc prof radiol, 63-69, clin assoc prof diag radiol, 69-71, PROF RADIOL, HAHNEMANN MED COL, 71-, DIR DIV GEN DIAG, 74- Concurrent Pos: Dir, Curtis X-ray Dept, Jefferson Med Col, 45-48; chief radiol, Albert Einstein Med Ctr, 50-53; vis radiologist, Philadelphia Gen Hosp, 60-; assoc, Sch Med, Univ Pa, 60-; mem staff, Hahnemann Hosp, 63- Mem: Fel Am Col Radiol; Radiol Soc NAm; Roentgen Ray Soc; Am Thoracic Soc; NY Acad Sci. Res: Hapato-splenography; intravenous and parenteral radiopaque emulsions. Mailing Add: Dept of Diag Radiol Hahnemann Med Col Philadelphia PA 19102

TEPLITZ, VIGDOR LOUIS, b Cambridge, Mass, Feb 5, 37; m 61. HIGH ENERGY PHYSICS. Educ: Mass Inst Technol, SB, 58; Univ Md, PhD(physics), 62. Prof Exp: Physicist, Lawrence Radiation Lab, Univ Calif, Berkeley, 62-64; NATO fel physics, Europ Orgn Nuclear Res, 64-65; asst prof, Mass Inst Technol, 65-69, assoc prof, 69-73; PROF & HEAD DEPT PHYSICS, VA POLYTECH INST & STATE UNIV, 73- Concurrent Pos: Coun mem, Fedn Am Scientists, 72-76. Mem: Fel Am Phys Soc; Fedn Am Sci. Res: Elementary particle theory; S-matrix theory; weak interaction models; phenomenology and data analysis. Mailing Add: Dept of Physics Va Polytech Inst & State Univ Blacksburg VA 24061

TEPLY, LESTER JOSEPH, b Muscoda, Wis, Apr 22, 20; m 50; c 3. BIOCHEMISTRY. Educ: Univ Wis, BA, 40, MS, 44, PhD(biochem), 45. Prof Exp: Asst biochem, Univ Wis, 40-45; tech secy, Food Composition Comt, Nutrit Biochemist Coord, Nat Res Coun, 45; biochemist, USPHS, 45-46; res biochemist, Columbia Univ, 46-48; res biochemist, Enzyme Inst, Univ Wis, 48-51, asst dir labs, Wis Alumni Res Found, 51-55, dir lab projs, 55-60; SR NUTRITIONIST, UNICEF, NY, 60- Mem: AAAS; Am Chem Soc; Am Soc Biol Chem; Am Pub Health Asn; Am Inst Nutrit. Res: Nutrition; vitamins; enzymes; animal nutrition; microbiological nutrition and metabolism; B-complex vitamins; food technology. Mailing Add: United Nations Children's Fund United Nations Bldg New York NY 10017

TEPLY, MARK LAWRENCE, b Lincoln, Nebr, Jan 11, 42; m 68; c 1. ALGEBRA. Educ: Univ Nebr, BA, 63, MA, 65, PhD(math), 68. Prof Exp: Asst prof, 68-73, ASSOC PROF MATH, UNIV FLA, 73- Mem: Am Math Soc; Math Asn Am. Res: Noncommutative rings and their modules; torsion theories; filters of ideals; homological algebra; direct sum decompositions of modules. Mailing Add: Dept of Math Univ of Fla Gainesville FL 32601

TEPPER, BYRON SEYMOUR, b New Bedford, Mass, Apr 12, 30; m 55; c 2. MICROBIOLOGY. Educ: Northeastern Univ, BS, 51; Univ Mass, MS, 53; Univ Wis, PhD(microbiol), 57. Prof Exp: Res assoc biochem, Univ Ill Col Med, 57-59; asst prof, 60-68, ASSOC PROF PATHOBIOL, SCH HYG, JOHNS HOPKINS UNIV, 68-, BIOHAZARDS SAFETY OFFICER, JOHNS HOPKINS MED INSTS, 74- Concurrent Pos: Assoc biochemist, Leonard Wood Mem Leprosy Res Lab, Baltimore, 59-65, microbiologist, 65-74. Mem: AAAS; Am Soc Microbiol; Soc Gen Microbiol; Int leprosy Asn; Am Acad Microbiol. Res: Host dependent microorganisms; human and murine leprosy; mycobacterial physiology; photobiology. Mailing Add: Safety Dept Hampton House Johns Hopkins Med Insts Baltimore MD 21205

TEPPER, HERBERT BERNARD, b Brooklyn, NY, Dec 25, 31; m 59; c 2. PLANT ANATOMY. Educ: State Univ NY Col Forestry, Syracuse Univ, BS, 53, MS, 58; Univ Calif, Davis, PhD(bot), 62. Prof Exp: Res asst forest bot, State Univ NY Col Forestry, Syracuse Univ, 56-58; res forester, US Forest Serv, 58-59; res asst, Univ Calif, Davis, 59-62; from instr to assoc prof, 62-67, PROF FOREST BOT, STATE UNIV NY COL ENVIRON SCI & FORESTRY, SYRACUSE, 67-, CHMN DEPT FOREST BOT & PATH, 65- Mem: AAAS; Am Soc Plant Physiol; Bot Soc Am; Int Soc Plant Morphol. Res: Morphogenesis in the shoot apex; seed germination; bud and cambial reactivation. Mailing Add: Dept of Forest Bot & Path State Univ of NY Col of Environ Sci & Forestry Syracuse NY 13210

TEPPER, LLOYD BARTON, b Los Angeles, Calif, Dec 21, 31; m 57; c 2. TOXICOLOGY, OCCUPATIONAL HEALTH. Educ: Dartmouth Col, AB, 54; Harvard Univ, MD, 57, MIH, 60, ScD(occup med), 62; Am Bd Prev Med, dipl, 64. Prof Exp: Fel, Mass Gen Hosp, 58-60; fel, Mass Inst Technol, 59-61; physician, Eastman Kodak Co, 61-62 & AEC, 62-65; assoc dir occup med & inst environ health, Kettering Lab, Univ Cincinnati, 65-72, assoc prof environ health, Univ, 65-71, prof, 71-72, assoc prof med, 69-72; ASSOC COMNR SCI, FOOD & DRUG ADMIN, 72- Mem: Am Acad Occup Med. Res: Industrial and environmental toxicology, especially as related to toxicology of beryllium, lead and other industrial metals; environmental and medical standards. Mailing Add: Food & Drug Admin 5600 Fishers Lane Rockville MD 20852

TEPPER, MORRIS, b Palestine, Mar 1, 16; nat US; m; c 2. METEOROLOGY, SCIENCE ADMINISTRATION. Educ: Brooklyn Col, BA, 36, MA, 38; Johns Hopkins Univ, PhD(fluid mech), 52. Prof Exp: Qualifications analyst & chief, Phys Sci Unit, US Civil Serv Comn, 39-43; chief, Severe Local Storms Res Unit, US Weather Bur, 46-59; dir meteorol prog, NASA, 59-65, dep dir space applns progs & dir meteorol, 66-69, DEP DIR EARTH OBSERVS PROGS & DIR METEOROL, NASA, 69- Concurrent Pos: Mem staff, USDA Grad Sch, 52-; mem, US Nat Comt Int Hydrol Decade & chmn work group remote sensing in hydrol, Nat Acad Sci, 71-75, liaison rep, US Comt Global Atmospheric Res Prog, mem, Comt Int Environ Progs, US Interagency Comts; chmn, Working Group 6, Comt Space Res, Int Coun Sci Unions; mem, Int Comn Space Sci Bd. Honors & Awards: Meisinger Award, Am Meteorol Soc, 50; Except Serv Medal, NASA, 66; Gold Medal, Nat Ctr Space Studies, France, 72. Mem: Fel Am Meteorol Soc; assoc fel Am Inst Aeronaut & Astronaut. Res: Satellite meteorology; mesometeorology; severe local storms; space applications; earth observation satellites; remote sensing; global atmospheric research. Mailing Add: 107 Bluff Terr Silver Spring MD 20902

TEPPERMAN, HELEN MURPHY, b Hartford, Conn, Jan 9, 17; m 43; c 3. PHYSIOLOGICAL CHEMISTRY. Educ: Mt Holyoke Col, BA, 38; Yale Univ, PhD(physiol chem), 42. Prof Exp: Asst, Mem Hosp, NY, 42-43 & Yale Univ, 43-44; pharmacologist, Med Res Lab, Edgewood Arsenal, Md, 44-45; from instr to assoc prof, 46-72, PROF PHARMACOL, STATE UNIV NY UPSTATE MED CTR, 72- Mem: Am Physiol Soc; Endocrine Soc. Res: Endocrinology and metabolism; parathyroid hormones. Mailing Add: Dept of Pharmacol 766 Irving Ave State Univ NY Upstate Med Ctr Syracuse NY 13210

TEPPERMAN, JAY, b Newark, NJ, Mar 23, 14; m 43; c 3. MEDICINE. Educ: Univ Pa, AB, 33; Columbia Univ, MD, 38. Prof Exp: Intern, Bassett Hosp, NY, 38-40; hon fel, Sch Med, Yale Univ, 40-41, Coxe fel, 41-42, asst, Aeromed Unit, Dept Physiol, 42-44; assoc prof pharmacol, 46-53, PROF EXP MED, COL MED, STATE UNIV NY UPSTATE MED CTR, 53- Concurrent Pos: Mem metab study sect, NIH & physiol comt, Nat Bd Med Examr, 63-67, mem pharmacol comt, 72-75; consult, Vet Admin, DC, 64-67 & Food & Drug Admin, 69- Mem: Soc Exp Biol & Med; Am Physiol Soc; Endocrine Soc; Am Soc Pharmacol & Exp Therapeut; Am Soc Biol Chem. Res: Endocrinology and metabolism. Mailing Add: Dept of Pharmacol State Univ of NY Col of Med Syracuse NY 13210

TEPPERT, WILLIAM ALLAN, SR, b Oshkosh, Nebr, Oct 10, 15; m 39; c 2. PHARMACOLOGY. Educ: Univ Ill, BS, 43 & 48, MS, 47; Univ Iowa, PhD(zool), 58. Prof Exp: Asst mammalian physiol, Univ Ill, 42-43 & 46-48; from instr to asst prof biol, 48-57, from asst prof to assoc prof pharmacol, 49-64, PROF PHARMACOL, COL PHARM, DRAKE UNIV, 64- Mem: Am Pharmaceut Asn. Res: Cell physiology and pharmacology; cellular neuropharmacology. Mailing Add: Col of Pharm Drake Univ Des Moines IA 50310

TEPPING, BENJAMIN JOSEPH, b Philadelphia, Pa, Jan 29, 13; m 40; c 2. MATHEMATICAL STATISTICS, APPLIED STATISTICS. Educ: Ohio State Univ, BA, 33, MA, 35, PhD(math), 39. Prof Exp: Math statistician, US Bur Census, 40-55; mathematician, Nat Analysts, Inc, 55-60; chief statist adv group, Surv & Res Corp, Korea, 60-63; chief, Res Ctr Measurement Methods, US Bur Census, 63-73; STATIST CONSULT, 73- Concurrent Pos: Lectr, USDA Grad Sch, 41-52, Am Univ, 43 & Univ Mich, 48-53; adj assoc prof, Univ Pa, 56-60; assoc ed, J Am Statist Asn, 64-66; mem vis lectr prog, Inst Math Statist, 69-72. Honors & Awards: Meritorious Serv Award, Dept Com, 50; Cult Medal, Repub Korea, 63. Mem: Fel AAAS; Fel Am Statist Asn; Am Math Soc; Inst Math Statist; Inter-Am Statist Inst. Res: Sampling theory and methods; measurement problems in censuses and surveys; problems of matching lists. Mailing Add: 401 Apple Grove Rd Silver Spring MD 20904

TERADA, KAZUJI, b Honolulu, Hawaii, Jan 4, 27. INORGANIC CHEMISTRY. Educ: Univ Hawaii, BA, 52, MS, 54; Univ Utah, PhD, 61. Prof Exp: CHEMIST, RES & DEVELOP, ROCKY FLATS DIV, ROCKWELL INT, GOLDEN, 60- Mem: Am Chem Soc. Mailing Add: 161 Linden Ave Boulder CO 80302

TERANDO, M LORETTA, b Spring Valley, Ill, Dec 31, 29. PSYCHOPHYSIOLOGY, TRANSACTIONAL ANALYSIS. Educ: St Ambrose Col, AB, 56; St Louis Univ, PhD(protein chem), 66; Bradley Univ, MA, 74. Prof Exp: Fel biol free radicals, Wash Univ, 66-67, res assoc, 67-68; asst prof biochem & microbiol, Marycrest Col, 68-74; intern, Dist 150, Peoria, Ill, 74-75; SCH PHYSIOLOGIST, PEKIN HIGH SCH, 75- Mem: Am Soc Zool; Sigma Xi; Am Personnel & Guid Asn. Res: Biological rhythms and influence of environment; clinical psychology; addiction; drug effects on intracranial stimulation; psychotherapy and behavior disorders in school environment. Mailing Add: 1213 West Thrush Peoria IL 61604

TERANGO, LARRY, b Clarksburg, WVa, Nov 30, 25; m 51; c 2. SPEECH PATHOLOGY, AUDIOLOGY. Educ: Kent State Univ, BA, 50, MA, 54; Case Western Reserve Univ, PhD(speech path & audiol), 66. Prof Exp: Clinician, Painesville City Schs, 52-59; instr speech path & audiol, Kent State Univ, 61-62; asst prof, San Jose State Col, 62-63; instr speech path & audiol, Kent State Univ, 63-64; instr speech & dir speech & hearing clin, Ohio State Univ, 64-66; assoc prof speech & dir speech & hearing clin, Univ Wyo, 66-68; prof, chmn dept spec educ & dir speech & hearing clin, ETenn State Univ, 68-74; PROF HEALTH SCI & DIR SPEECH & HEARING CTR, WESTERN CAROLINA UNIV, 74- Concurrent Pos: Vpres, Wyo Cleft Palate Eval Team, 66-68. Mem: Am Speech & Hearing Asn; Coun Except Children; Nat Educ Asn; Am Cleft Palate Asn. Res: Vocal characteristics in the male voice; language dysfunction; multidisciplinary approach to study of neurological disturbances. Mailing Add: Speech & Hearing Ctr PO Box A-W Cullowhee NC 28723

TERANISHI, ROY, b Stockton, Calif, Aug 1, 22; m 44; c 1. ORGANIC CHEMISTRY. Educ: Univ Calif, BS, 50; Ore State Col, PhD, 54. Prof Exp: Instr chem, Portland State Col, 53-54; RES CHEMIST, USDA, 54- Mem: Am Chem Soc. Res: Gas chromatography; flavor chemistry. Mailing Add: 89 Kingston Rd Kensington CA 94707

TERASAKI, PAUL ICHIRO, b Los Angeles, Calif, Sept 10, 29; m 56; c 4. IMMUNOLOGY. Educ: Univ Calif, Los Angeles, BA, 50, MA, 52, PhD(zool), 56. Prof Exp: Res asst zool, 52-54, res asst, Atomic Energy Proj, 54-55, res zoologist, Dept Surg, 55-56, jr res zoologist, 56-57, asst res zoologist, 58-61, assoc res zoologist, 61-62, assoc prof surg, 62-66, PROF SURG, CTR HEALTH SCI, UNIV CALIF, LOS ANGELES, 66- Concurrent Pos: Res fel zool with Prof P B Medawar, Univ Col, Univ London, 57-58; USPHS career develop award, 63-; mem transplantation & immunol adv comt, NIH, 67-70; mem nomenclature comt leukocyte antigens, WHO. Honors & Awards: Modern Med Award Distinguished Achievement. Mem: Am Soc Cell Biol; Am Asn Immunol; Soc Cryobiol; Am Soc Immunol; Int Transplantation Soc. Res: Transplantation immunology; homotransplantation; leucocyte typing. Mailing Add: Dept of Surg Univ of Calif Ctr for Health Sci Los Angeles CA 90024

TERASMAE, JAAN, b Estonia, May 28, 26; nat Can; m 54. GEOLOGY, PALYNOLOGY. Educ: Univ Uppsala, Fil Kand, 51; McMaster Univ, PhD, 55. Prof Exp: Asst, Palynological Lab Stockholm, 50-51; geologist, Geol Surv Can, 55-68; PROF GEOL, BROCK UNIV, 68-, CHMN DEPT GEOL SCI, 69- Concurrent Pos: Mem assoc comt, Quaternary Res & Comt Muskeg Res, Nat Res Coun Can. Mem: AAAS; Int Glaciol Soc; Int Limnol Soc; Int Peat Soc; Soc Environ Geochem & Health. Res: Pleistocene chronology, geology and stratigraphy; paleobotany of Pleistocene deposits. Mailing Add: Dept of Geol Sci Brock Univ St Catherines ON Can

TERBORGH, JOHN J, b Washington, DC, Apr 16, 36. POPULATION BIOLOGY, PLANT PHYSIOLOGY. Educ: Harvard Univ, AB, 58, AM, 60, PhD(biol), 63. Prof Exp: Staff scientist, Tyco Labs, Inc, 63-65; asst prof bot, Univ Md, 65-71; ASSOC PROF BIOL, PRINCETON UNIV, 71- Concurrent Pos: Res grants, Am Philos Soc, Am Mus Natural Hist & Nat Geog Soc, 64-67; NSF res grants, 68- Mem: AAAS; Am Soc Naturalists; Soc Study Evolution; Ecol Soc Am. Res: Tropical ecology; population biology of birds. Mailing Add: Dept of Biol Princeton Univ Princeton NJ 08540

TERDIMAN, JOSEPH FRANKLIN, b New York, NY, Feb 14, 40; m 65; c 2. BIOMEDICAL ENGINEERING. Educ: Cornell Univ, BEngPhys, 61; NY Univ, MD, 65; Univ Ill Med Ctr, PhD(physiol), 72. Prof Exp: Res scientist pop exposure studies, Nat Ctr Radiol Health, 67-69; MED INFO SCIENTIST, KAISER FOUND RES INST, 69- Concurrent Pos: Lectr, Univ Ill, Chicago Circle, 65-67 & Sch Optom, Univ Calif, Berkeley, 69- Mem: AAAS; Soc Advan Med Systs; Biomed Eng Soc. Res: Development of medical information systems for hospital automation; patient monitoring, diagnosis and therapy; integration of computers and engineering methods with classical neurophysiological techniques; neurophysiology. Mailing Add: Kaiser Found Res Inst 3779 Piedmont Ave Oakland CA 94611

TEREPKA, ANTHONY RAYMOND, b Newark, NJ, Aug 5, 25; m 47; c 5. INTERNAL MEDICINE. Educ: Univ Rochester, MD, 51. Prof Exp: Intern med, Strong Mem Hosp, Rochester, NY, 51-52, asst resident, 53-54; instr med, 55-58, asst prof radiation biol & med, 59-65, ASSOC PROF RADIATION BIOL & BIOPHYS & ASST PROF MED, SCH MED & DENT, UNIV ROCHESTER, 66- Concurrent Pos: USPHS sr res fel, Karolinska Inst, Sweden, 61-62. Mem: Endocrine Soc; Am Fedn Clin Res. Res: Endocrinology; calcium metabolism; radiation biology. Mailing Add: Univ of Rochester Med Ctr Rochester NY 14642

TERESA, GEORGE WASHINGTON, b Osceola, Ark, Nov 23, 23; m 54. BACTERIOLOGY. Educ: Ark Agr & Mech Col, BS, 52; Univ Ark, MS, 55; Kans State Univ, PhD(microbiol), 59. Prof Exp: Asst prof bact, Auburn Univ, 59-61 & Univ RI, 61-62; assoc prof biol, Wichita State Univ, 62-68; ASST PROF BACT & BIOCHEM, UNIV IDAHO, 68- Mem: AAAS; Am Soc Microbiol. Res: Immunology and pathogenic bacteriology; immunology and pathogenic mechanism of Gram anaerobes. Mailing Add: Dept of Bacteriol & Biochem Univ of Idaho Moscow ID 83843

TERESHKOVICH, GEORGE, b New York, NY, Mar 18, 30; m 55; c 1. HORTICULTURE. Educ: La Polytech Inst, BS, 52; Univ Ga, MS, 57; La State Univ, PhD(hort, agron), 63. Prof Exp: Res asst, Univ Ga, 52-54, asst prof hort, 56-60; res assoc, La State Univ, 60-63; asst prof, Univ Ga, 63-68; assoc prof hort, 68-74, PROF & ACTG CHMN DEPT PARK ADMIN, LANDSCAPE ARCHIT & HORT, TEX TECH UNIV, 75- Mem: Am Soc Hort Sci. Res: Cultural and adaptability studies with vegetable and ornamental crops. Mailing Add: Dept of Park Admin Tex Tech Univ Lubbock TX 79409

TERESI, JOSEPH DOMINIC, b San Jose, Calif, Aug 18, 15; m 47; c 6. BIOCHEMISTRY, HEALTH PHYSICS. Educ: San Jose State Col, AB, 38; Univ Wis, PhD(biochem), 43. Prof Exp: Res assoc, Manhattan Proj, Chicago, 43, Clinton Labs, Tenn, 44, & Univ Chicago, 45; res assoc, Stanford Univ, 47-51, actg instr, 48-50, actg asst prof, 50-51; chemist, US Naval Radiol Defense Lab, 51-69; sr biochemist, Stanford Res Inst, 69-71; sr scientist, San Francisco Bay Marine Res Ctr, 71-73; res assoc div nuclear med, Stanford Univ, 73-74; SR ENG, FAST BREEDER REACTOR DEPT, GEN ELEC CO, 74- Mem: Fel AAAS; Am Chem Soc; Health Physics Soc; Am Nuclear Soc. Res: Analysis of radionuclides in biological materials; radiation effects; aerospace nuclear safety; radiation protection; radiation ecology and internal emitters; bay sediment analysis; radiological assessment; nuclear reactor safety analysis. Mailing Add: 1395 Villa Dr Los Altos CA 94022

TERESINE (LEWIS), MARY, b St Louis, Mo, July 6, 08. MATHEMATICS. Educ: Fontbonne Col, AB, 38; Cath Univ, MA, 44, PhD(math), 47. Prof Exp: Instr high sch, Colo, 38-43; asst math, Cath Univ, 43-44 & 46-47; from instr to prof, 47-74, dean women, 51-66, EMER PROF MATH, FONTBONNE COL, 74- Mem: Am Math Soc; Math Asn Am. Res: Number theory; theory of congruences; construction and application of magic rectangles. Mailing Add: Dept of Math Fontbonne Col 6800 Wydown Blvd St Louis MO 63105

TERHAAR, CLARENCE JAMES, b Cottonwood, Idaho, Apr 29, 26; m 57; c 4. TOXICOLOGY. Educ: Univ Idaho, BS, 53, MS, 54; Kans State Col, PhD(parasitol), 57. Prof Exp: Asst prof entom, Kans State Col, 57-58; TOXICOLOGIST, EASTMAN KODAK CO, 58- Mem: Entom Soc Am; Am Soc Parasitol; Am Micros Soc; Soc Toxicol; Am Indust Hyg Asn. Res: Invertebrate and mammalian toxicology. Mailing Add: Health & Safety Lab Eastman Kodak Co Rochester NY 14650

TER HAAR, GARY L, b Zeeland, Mich, May 2, 36; m 56; c 2. INORGANIC CHEMISTRY. Educ: Hope Col, BA, 58; Univ Mich, MS, 60, PhD(chem), 62. Prof Exp: RES ASSOC, ETHYL CORP, 62- Mem: Am Chem Soc. Res: Inorganic coordination chemistry; environmental research. Mailing Add: Ethyl Corp Res Labs Detroit MI 48220

TERHUNE, ROBERT W, physics, see 12th edition

TERJUNG, WERNER HEINRICH, b Müllheim-Ruhr, WGer, Feb 27, 31; US citizen; m 54; c 2. GEOGRAPHY. Educ: Calif State Col, Long Beach, BA, 62, MA, 63; Univ Calif, Los Angeles, PhD(geog), 66. Prof Exp: ASSOC PROF GEOG, UNIV CALIF, LOS ANGELES, 66- Concurrent Pos: NSF res grant, 66-67; res grant, Univ Calif, Los Angeles, 67-68; US Dept Interior & Univ Calif Water Resources Ctr res grants, 71-72. Mem: Int Soc Biometeorol; Asn Am Geog; Am Geog Soc; African Studies Asn. Res: Physiological climatology; energy balance climatology; microclimatology; biogeography. Mailing Add: Dept of Geog Univ of Calif Los Angeles CA 90024

TERKLA, LOUIS GABRIEL, b Anaconda, Mont, Mar 24, 25; m 49; c 2. DENTISTRY. Educ: Univ Ore, DMD, 52. Prof Exp: From instr to assoc prof, 52-61, asst to dean dent sch, 60-66, asst dean acad affairs, 66-67, PROF DENT, DENT SCH, UNIV ORE, 61-, DEAN DENT SCH, 67- Mem: AAAS; Am Dent Asn; Int Asn Dent Res; fel Am Col Dent (pres, 73-74); Am Asn Dent Schs (pres, 75-76). Res: Clinical behavior of dental materials; cavity preparation design; testing to predict technical performance of dental students. Mailing Add: Univ of Ore Dent Sch 611 SW Campus Dr Portland OR 97201

TERMAN, CHARLES RICHARD, b Mansfield, Ohio, Sept 8, 29; m 51; c 2. ANIMAL ECOLOGY, ANIMAL BEHAVIOR. Educ: Albion Col, AB, 52; Mich State Univ, MS, 54, PhD(behav pop dynamics), 59. Prof Exp: Assoc prof biol, Taylor Univ, 61-63, actg dir res, 62-63; from asst prof to assoc prof, 63-69, PROF BIOL, COL WILLIAM & MARY, 69- Concurrent Pos: Nat Inst Ment Health fel, Sch Hyg & Pub Health, Johns Hopkins Univ, 59 & Penrose Res Lab, 59-61; NATO sr sci fel, Eng & exchange scientist, US & Polish Nat Acads Sci, 74. Honors & Awards: NIH Career Develop Award, 70- Mem: Fel AAAS; Animal Behav Soc; Am Sci Affil; Am Soc Mammal; Am Soc Nat. Res: Population dynamics; socio-biological factors influencing the growth and physiology of populations; reproductive physiology; behavioral ecology. Mailing Add: Lab of Endocrinol & Pop Ecol Dept Biol Col of William & Mary Williamsburg VA 23185

TERMAN, GILBERT LEROY, b Columbia City, Ind, Oct 18, 13; m 41; c 3. AGRONOMY. Educ: Kans State Col, BS, 38; Univ Wis, PhD(soils), 41. Prof Exp: Asst agron, Univ Ky, 42-46; agronomist, Univ Maine, 46-52; agriculturist, Int Coop Admin, Lebanon, 52-54; AGRONOMIST, TENN VALLEY AUTHORITY, 54- Concurrent Pos: Vis prof, Univ Nebr, Lincoln, 67-68. Mem: Fel Am Soc Agron; Soil Sci Soc Am. Res: Soil chemistry, fertilizer experiments; soils and crops; experimental design. Mailing Add: Tenn Valley Authority Muscle Shoals AL 35660

TERMAN, MAX R, b Mansfield, Ohio, Apr 15, 45; m 68. ECOLOGY, ETHOLOGY. Educ: Spring Arbor Col, BA, 67; Mich State Univ, MS, 69, PhD(zool), 73. Prof Exp: PROF BIOL, TABOR COL, 69- Mem: Ecol Soc Am; Am Soc Mammalogists; Animal Behav Soc; Am Inst Biol Sci; Sigma Xi. Res: Interspecific competition between rodents; rodent population dynamics; behavior of rodents. Mailing Add: 612 S Lincoln Hillsboro KS 67063

TERMINE, JOHN DAVID, b Brooklyn, NY, Sept 25, 38; m 61; c 4. BIOCHEMISTRY. Educ: St John's Univ, NY, BS, 60; Univ Md, MS, 63; Cornell Univ, PhD(biochem), 66. Prof Exp: Teaching asst chem, Univ Md, 60-63; asst res scientist, Hosp Spec Surg, New York, 63-66; from instr to asst prof biochem, Med Col, Cornell Univ, 66-70; spec res fel, 70-73, RES BIOCHEMIST, MOLECULAR STRUCT SECT, LAB BIOL STRUCT, NAT INST DENT RES, NIH, 73- Mem: AAAS; Int Asn Dent Res; Am Chem Soc; Biophys Soc; NY Acad Sci. Res: Calcification biochemistry; physical biochemistry; protein structural biochemistry; spectroscopy. Mailing Add: Lab of Biol Struct Nat Inst of Dent Res NIH Bethesda MD 20014

TERMINIELLO, LOUIS, b New York, NY, Sept 22, 30; m 53; c 5. ENZYMOLOGY. Educ: Fordham Univ, BS, 52, MS, 54, PhD, 57. Prof Exp: Proj leader, Nat Dairy Res Labs, Nat Dairy Prods Corp, 57-58; res assoc, 58-62, res biologist, 62-69, GROUP LEADER, STERLING-WINTHROP RES INST, 69- Res: Isolation and characterization of proteins and enzymes; mechanism of enzyme action; chemical modification of enzymes and the relationship of structure to activity; therapeutic application of enzymes; fibrinolysis; fibrinolytic enzymes. Mailing Add: Sterling-Winthrop Res Inst Rennselaer NY 12144

TERNAY, ANDREW LOUIS, JR, b New York, NY, Aug 29, 39; m 61; c 2. ORGANIC CHEMISTRY. Educ: City Col New York, BS, 59; NY Univ, MS, 62, PhD(chem), 63. Prof Exp: NSF fel & res assoc chem, Univ Ill, 63-64; instr, Case Western Reserve Univ, 64-65, asst prof, 65-69; ASSOC PROF CHEM, UNIV TEX, ARLINGTON, 70- Concurrent Pos: Grants, Nat Cancer Inst, Dept Chem, Case Western Reserve Univ, 66-69 & Welch Found, Univ Tex, 72-; consult, Arbrook, Inc, 71- Mem: AAAS; Am Chem Soc; The Chem Soc. Res: Drug design; molecular spectroscopy; application of stereochemistry to synthesis of new drugs, especially psychoactive materials. Mailing Add: Dept of Chem Univ of Tex Arlington TX 76010

TERNBERG, JESSIE L, b Corning, Calif, May 28, 24. SURGERY. Educ: Grinnell Col, AB, 46; Univ Tex, PhD(biochem), 50; Washington Univ, MD, 53. Prof Exp: Instr surg, 59-65, assoc prof, 65-71, PROF PEDIAT SURG, SCH MED, WASHINGTON UNIV, 71- Res: Free radicals in biological systems as studies by electron spin resonance spectrometer. Mailing Add: Dept of Surg Washington Univ 4960 Audubon Ave St Louis MO 63110

TERNER, CHARLES, b Lublin, Poland, Apr 30, 16; nat US; m 45; c 3. BIOCHEMISTRY. Educ: Univ London, BSc, 44, DSc(biochem), 69; Univ Sheffield, PhD(biochem), 49. Prof Exp: Mem staff, Med Res Coun Unit for Res Cell Metab, Dept Biochem, Univ Sheffield, 47-50; sr sci officer, Dept Physiol, Nat Inst Res in Dairying, Eng, 50-55; mem staff, Worcester Found Exp Biol, 55-59; PROF BIOL, BOSTON UNIV, 59- Concurrent Pos: Vis scientist, Pop Coun, Rockefeller Univ, 72. Mem: AAAS; Am Soc Biol Chem; Am Chem Soc; Soc Exp Biol & Med; Soc Study Reproduction. Res: Biochemistry of male reproductive tissues; embryonic development of fish. Mailing Add: Biol Sci Ctr 2 Cummington St Boston Univ Boston MA 02215

TERNES, JOSEPH WAYNE, b Milwaukee, Wis, Aug 6, 37. PSYCHOPHARMACOLOGY. Educ: Univ Calif, Riverside, BA, 62; San Diego State Univ, MS, 65; Fla State Univ, PhD(psychol), 69. Prof Exp: NIH fel behav biol & ecol, Univ Minn, 69-70, res assoc behav biol, 70-71; vis scientist radiobiol, Univ PR, 71-72, vis prof psychol, 71-72, asst prof, 72-75; STAFF SCIENTIST, MONELL CHEM SENSES CTR, UNIV PA, 75- Mem: Am Psychol Asn; Psychonomic Soc; Sigma Xi; Animal Behav Soc; Int Soc Biol Rhythms. Res: Effects of opiate dependence upon olfaction and taste; conditioned taste preference and aversions in psychopharmacology. Mailing Add: Monell Chem Senses Ctr Univ Pa 3500 Market St Philadelphia PA 19104

TERPKO, STEPHEN PAUL, inorganic chemistry, organic chemistry, see 12th edition

TER-POGOSSIAN, MICHEL MATHEW, b Berlin, Ger, Apr 21, 25; nat US. MEDICAL PHYSICS. Educ: Univ Paris, BA, 42; Washington Univ, MS, 48, PhD(nuclear physics), 50. Prof Exp: From instr to prof radiation sci, 50-71, PROF RADIOL, SCH MED, WASHINGTON UNIV, 71- Mem: Fel Am Phys Soc; Am Nuclear Soc; Am Radium Soc; Radiation Res Soc; hon fel Am Col Radiol. Res: Medical applications of short-lived isotopes; gamma ray spectroscopy; scintillation counters; radiation dosimetry; radiobiology; lasers in biology; reconstructive tomography in radiologic imaging; positron emission imaging. Mailing Add: Div of Radiation Sci Washington Univ Sch of Med St Louis MO 63110

TERRACIO, LOUIS, b Beaver Falls, Pa, Feb 28, 48; m 70; c 1. HISTOLOGY. Educ: Geneva Col, BS, 70; Univ Minn, PhD(anat), 75. Prof Exp: SR INSTR ANAT, HAHNEMANN MED COL & HOSP, 75- Mem: Soc Cryobiol; Electron Micros Soc Am; AAAS. Res: Myocardial revascularization; cardiovascular histodynamics; freeze-drying of biological tissues for electron microscopy. Mailing Add: Dept of Anat Hahnemann Med Col & Hosp 230 N Broad St Philadelphia PA 19102

TERRANOVA, ANDREW CHARLES, b Cleveland, Ohio, Aug 29, 35. ENTOMOLOGY, TOXICOLOGY. Educ: Ohio State Univ, BSc, 60, MSc, 61, PhD(entom), 65. Prof Exp: INSECT PHYSIOLOGIST, METAB & RADIATION RES LAB, AGR RES SERV, USDA, 65- Mem: AAAS; Entom Soc Am; Am Chem Soc. Res: Metabolism and mode of action of insect chemosterilants; biochemistry and physiology of insect reproduction. Mailing Add: Metab & Radiation Res Lab Agr Res Serv USDA Fargo ND 58102

TERRANT, SELDON W, b Cleveland, Ohio, Nov 17, 18; m 40; c 1. CHEMISTRY, INFORMATION SCIENCE. Educ: Ohio Univ, BS, 40; Case Inst Technol, MS, 48, PhD(chem), 50. Prof Exp: Anal chemist, Carbide & Carbon Chem Corp, 40-42; instr chem, Case Inst Technol, 49-52; res chemist, E I du Pont de Nemours & Co, NC, 52-57; tech dir, Stand Prod Co, Ohio, 57-59; asst ed, Chem Abstr Serv, 59-60, assoc ed, 60-62; sr scientist, Southern Res Inst, Ala, 62-63; assoc ed, Chem Abstr Serv, 63-64, managing ed spec publ & serv, 64-68, sr staff adv, 68-72; HEAD RES & DEVELOP, BKS & JOURS DIV, AM CHEM SOC, 72- Mem: Am Chem Soc; Am Soc Info Sci; Spec Libr Asn; Soc Tech Commun. Res: Polarography; stereoisomerism of polymers; textile fibers; rubber processing; chemical literature. Mailing Add: Bks & Jours Div Am Chem Soc 1155 16th St NW Washington DC 20036

TERRAS, AUDREY ANNE, b Wash, DC, Sept 10, 42; m 65. NUMBER THEORY. Educ: Univ Md, College Park, BS, 64; Yale Univ, MA, 66, PhD(math), 70. Prof Exp: Instr math, Univ Ill, Urbana, 68-70; asst prof, Univ PR, Mayagüez, 70-71 & Brooklyn Col, 71-72; ASST PROF MATH, UNIV CALIF, SAN DIEGO, 72- Concurrent Pos: NSF grant, 74- Mem: Am Math Soc; Math Asn Am; Am Women in Math; Soc Indust & Appl Math; AAAS. Res: Zeta functions; automorphic forms of matrix argument; harmonic analysis on homogeneous spaces. Mailing Add: Dept of Math Univ of Calif at San Diego La Jolla CA 92093

TERREAULT, BERNARD J E J, b Montreal, Que, Mar 29, 40; m 68; c 2. NUCLEAR SCIENCE. Educ: Univ Montreal, BSc, 60, MSc, 62; Univ Ill, Urbana-Champaign, PhD(physics), 68. Prof Exp: Fel, Lab High Energy Physics, Polytech Sch, Paris, 69-70; res assoc high energy physics, Ohio Univ, 70-71, asst prof, 71-72; ASSOC PROF NUCLEAR SCI, ENERGY CTR, NAT INST SCI RES, UNIV QUE, 72- Mem: Am Phys Soc; AAAS; Can Asn Physicists. Res: Radiation effects in nuclear materials of interest in controlled thermonuclear fusion, mainly low-energy light-ion bombardment and 14-MEV neutron bombardment. Mailing Add: Energy Ctr Nat Inst for Sci Res Univ of Que CP 1020 Varennes PQ Can

TERRELL, EDWARD EVERETT, b Wilmington, Ohio, Oct 6, 23; m 50; c 3. PLANT TAXONOMY. Educ: Wilmington Col, AB, 47; Cornell Univ, MS, 49; Univ Wis, PhD, 52. Prof Exp: Muellhaupt scholar bot, Ohio State Univ, 52-53; assoc prof biol & head dept sci, Pembroke State Col, 54-56; assoc prof biol, Guilford Col, 56-60; BOTANIST, AGR RES SERV, USDA, 60- Mem: AAAS; Bot Soc Am; Am Soc Plant Taxon; Torrey Bot Club; Int Asn Plant Taxon. Res: Plant taxonomy and ecology; taxonomy of Houstonia; taxonomy of grasses. Mailing Add: Agr Res Ctr West Beltsville MD 20705

TERRELL, GLEN EDWARD, b Humble, Tex, Nov 17, 39; m 59; c 2. NUCLEAR PHYSICS. Educ: Univ Tex, Austin, BS, 62, MA, 64, PhD(physics), 66. Prof Exp: Asst prof, 66-74, ASSOC PROF PHYSICS, UNIV TEX, ARLINGTON, 74- Mem: Am Phys Soc. Res: Computer-assisted instruction; low energy nuclear physics; polarization of protons elastically scattered by several nuclei; gamma ray directional correlation. Mailing Add: Dept of Physics Univ of Tex Arlington TX 76010

TERRELL, JAMES, (JR), b Houston, Tex, Aug 15, 23; m 45; c 3. PHYSICS. Educ: Rice Univ, BA, 44, MA, 47, PhD(physics), 50. Prof Exp: Res asst physics, Rice Univ, 50; asst prof, Case Western Reserve Univ, 50-51; MEM STAFF, LOS ALAMOS SCI LAB, UNIV CALIF, 51- Mem: Fel AAAS; fel Am Phys Soc; Am Astron Soc. Res: Astrophysics; fission; relativity; nuclear physics; lasers. Mailing Add: Los Alamos Sci Lab Box 1663 Univ of Calif Los Alamos NM 87545

TERRELL, JOHN HART, b Magnolia, Ark, Feb 3, 34; m 59; c 5. APPLIED PHYSICS. Educ: Univ Ark, BS, 56; Brandeis Univ, MA, 61, PhD(physics), 64. Prof Exp: Res asst physics, Harvard Univ, 60-61; staff physicist, Nat Magnet Lab, Mass Inst Technol, 61-62; res fel physics, Brandeis Univ, 64-65; sr physicist, Mithras Div, Sanders Assocs, Mass, 65-69; consult appl physics, Keystone Comput Assocs, Inc, 69-71; staff scientist, Commun Systs Div, GTE Sylvania, Inc, Needham, 71-72; vpres eng, Thermo Magnetics, Inc, Woburn, Mass, 72-73; CONSULT, 73- Mem: Inst Elec & Electronics Engrs. Res: Electromagnetic pulse and internal electromagnetic pulse effects following a nuclear event; non-destructive testing applications. Mailing Add: PO Box 134 Lincoln Center MA 01773

TERRELL, ROSS CLARK, b Oneonta, NY, Sept 22, 25; m 56. ORGANIC CHEMISTRY. Educ: Hartwick Col, BS, 50; Columbia Univ, PhD(org chem), 55. Prof Exp: Res chemist, Shulton Inc, 55-59; sr chemist, Air Reduction Co, Inc, 59-67, supvr chem res, Ohio Med Prods, 67-75, MGR CHEM PROD DEVELOP, OHIO MED PRODS, DIV OF AIRCO INC, 75- Concurrent Pos: Hon vis res fel, Univ Birmingham, 66-67. Res: Organic synthesis; terpenes; surfactants; anesthetics; fluorine chemistry; pharmaceuticals. Mailing Add: Ohio Med Prods 1930 Losantiville Ave Cincinnati OH 45237

TERRELL, ROY PAUL, b Adolphus, Ky, Aug 28, 14; m 45. GEOGRAPHY. Educ: Western Ky Univ, BS, 37; George Peabody Col Teachers, MA, 41; Clark Univ, PhD(geog), 49. Prof Exp: Asst prof geog, Univ Fla, 46-47; assoc prof geog & econ, Auburn Univ, 47-50; prof & head dept geog & geol, Western Ky Univ, 60-67, E Jeffries prof geog, 67-69; PROF GEOG, MID TENN STATE UNIV, 69- Concurrent Pos: State coordr, Ky, Nat Coun Geog Educ, 38- Mem: Asn Am Geog; Nat Coun Geog Educ. Res: Climates and population potential for tropical highlands; geography of the South, especially manufacturing and agriculture. Mailing Add: Dept of Geog-Earth Sci Mid Tenn State Univ Tenn Blvd Murfreesboro TN 37130

TERRELL, THOMAS R, mathematics, statistics, see 12th edition

TERRES, GERONIMO, b Santa Barbara, Calif, July 1, 25; m 47; c 5. IMMUNOBIOLOGY. Educ: Univ Calif, BA, 50; Stanford Univ, MA, 51; Calif Inst Technol, PhD(biol), 56. Prof Exp: Asst, Calif Inst Technol, 52-55; assoc scientist microbiol, Brookhaven Nat Lab, 55-60; asst prof human physiol, Sch Med, Stanford Univ, 60-69; ASSOC PROF PHYSIOL, SCH MED, TUFTS UNIV, 69- Concurrent Pos: NIH sr res fel, 60-62; USPHS res career develop award, 62-69; vis asst prof, Harvard Med Sch, 67-68. Mem: AAAS; Am Asn Immunol; Soc Exp Biol & Med; Am Physiol Soc; Radiation Res Soc. Res: Immune degradation; acquired immune tolerance in mice; initiation and control of the immune response. Mailing Add: Dept of Physiol Tufts Univ Sch of Med Boston MA 02111

TERRES, JOHN KENNETH, b Philadelphia, Pa, Dec 17, 05; m 50. ENVIRONMENTAL BIOLOGY. Prof Exp: Field biologist, Soil Conserv Serv, USDA, 35-42; pub info specialist, Am Chem Soc, 45-46; consult field biologist, 46-48; actg ed, Audubon Mag, Nat Audubon Soc, 48-49, managing ed, 49-55, ed, 55-60; GEN ED NATURE BOOKS, J B LIPPINCOTT CO, 59-; WRITER, 60- Concurrent Pos: Consult ed, Doubleday & Co, Inc, 66-68; mem lab ornith, Cornell Univ, 71. Honors & Awards: John Burroughs Medal, John Burroughs Mem Asn, Inc, 71. Mem: Wildlife Soc: assoc Am Soc Mammal; assoc Wilson Ornith Soc; assoc Cooper Ornith Soc; Wilderness Soc. Res: Behavior and life histories of birds, mammals and insects; plant taxonomy, distribution and ecology; animal ecology. Mailing Add: 345 E 57th St New York NY 10022

TERRIERE, LEON C, b Stone, Idaho, June 17, 20; m 44; c 3. BIOCHEMISTRY. Educ: Univ Idaho, BS, 43; Ore State Col, PhD(chem), 50. Prof Exp: From asst prof to assoc prof, Agr Exp Sta, 50-61, PROF BIOCHEM, ORE STATE UNIV, 61-, PROF INSECT TOXICOL, AGR CHEM & ENTOM, 70- Res: Chemistry of insecticides; insect toxicology. Mailing Add: Dept of Entom Ore State Univ Cor 1063 Corvallis OR 97331

TERRIERE, ROBERT T, b Seattle, Wash, July 17, 26; m 58; c 2. GEOLOGY. Educ: Calif Inst Technol, BS, 49; Pa State Col, MS, 51; Univ Tex, PhD(geol), 60. Prof Exp: Geologist, US Geol Surv, 51-58; res geologist, 58-70, RES ASSOC, CITIES SERV OIL CO, 70- Mem: Fel Geol Soc Am; Am Asn Petrol Geol. Res: Physical stratigraphy; sedimentary petrography; petroleum geology. Mailing Add: 3819 S Troost Tulsa OK 74105

TERRILL, CLAIR ELMAN, b Rippey, Iowa, Oct 27, 10; m 32; c 2. ANIMAL BREEDING. Educ: Iowa State Col, BS, 32; Univ Mo, PhD, 36. Prof Exp: Asst animal husb, Univ Mo, 32-36; asst animal husbandman, Exp Sta, Univ Ga, 36; asst animal husbandman, Sheep Exp Sta, Bur Animal Indust, Idaho, 36-37, assoc animal husbandman, Western Sheep Breeding Lab & Sheep Exp Sta, Agr Res Serv, 53-55, chief sheep & fur animal res br, Animal Husb Res Div, 55-72, NAT PROG STAFF SCIENTIST FOR SHEEP & OTHER ANIMALS, AGR RES SERV, USDA, 72- Concurrent Pos: Dir, Am Forage & Grass Land Coun, 63-65; mem, World Asn Animal Prod Coun, 63- Honors & Awards: Achievement Award, Ital Exp Inst & Ital Soc Advan Zootech; Distinguished Achievement Award, Sheep Indust Develop. Mem: AAAS; Genetics Soc Am; hon fel Am Soc Animal Sci (secy-treas, 60-62, vpres, 63, pres, 64); Am Meat Sci Asn; Am Genetic Asn (vpres, 69, pres, 70). Res: Animal genetics; reproductive physiology; sheep, goat and fur animal breeding and production. Mailing Add: Nat Prog Staff Agr Res Ctr-West Agr Res Serv USDA Beltsville MD 20705

TERRILL, THOMAS ROBERT, b Charlottesville, Va, June 7, 33; m 51; c 5. PLANT BREEDING, GENETICS. Educ: Pa State Univ, BS, 55, MS, 61; NC State Univ, PhD(crop sci), 65. Prof Exp: Instr agron, Pa State Univ, 55-61; res asst crop sci, NC State Univ, 61-64, from instr to asst prof, 64-66; ASST PROF AGRON, SOUTHERN PIEDMONT RES & CONTINUING EDUC CTR, VA POLYTECH INST & STATE UNIV, 66- Concurrent Pos: VChmn agron sect, Tobacco Workers Conf, 67, chmn, 68-69. Mem: Am Soc Agron; Am Genetic Asn. Res: Agronomy; micronutrient and macronutrient utilization by tobacco plants; topping, spacing, water utilization and other cultural variables with tobacco; genetic systems controlling alkaloids, quality and growth characteristics of tobacco. Mailing Add: Dept of Agron Va Polytech Inst & State Univ Blacksburg VA 24061

TERRIS, MILTON, b New York, NY, Apr 22, 15; m 41, 71; c 2. EPIDEMIOLOGY. Educ: Columbia Col, AB, 35; NY Univ, MD, 39; Johns Hopkins Univ, MPH, 44. Prof Exp: Intern, Harlem Hosp, New York, 39-41 & Bellevue Psychiat Hosp, New York, 41-42; apprentice epidemiologist, State Dept Health, NY, 42-43, asst dist health officer, 44-46; med assoc subcomt on med care, Am Pub Health Asn, 46-48, staff dir, 48-51; assoc prev med & pub health, Sch Med, Univ Buffalo, 52-54, from asst prof to assoc prof, 54-58; prof epidemiol, Sch Med, Tulane Univ, 58-60; head chronic dis unit, Div Epidemiol, Pub Health Res Inst New York, 60-64; PROF PREV MED, NEW YORK MED COL, 64-, CHMN DEPT COMMUNITY & PREV MED, 68- Concurrent Pos: Asst dean post-grad educ, Univ Buffalo, 51-58. Mem: Am Pub Health Asn (pres, 66-67); Am Epidemiol Soc; Asn Teachers Prev Med (pres, 61-62); Int Epidemiol Asn. Res: Epidemiology of cancer; cirrhosis of liver; prematurity; heart disease. Mailing Add: Dept of Community & Prev Med NY Med Col Fifth Ave & 106th St New York NY 10029

TERROUX, FERDINAND RICHARD, b Montreal, Que, 02; m 30; c 2. EXPERIMENTAL PHYSICS. Educ: Loyola Univ, BA, 21; McGill Univ, BSc, 25, MSc, 26; Cambridge Univ, PhD, 31. Prof Exp: Sr demonstr physics, 31-34, lectr, 35-41, from asst prof to assoc prof, 42-75, assoc dean fac grad studies, 63-75, CUR, RUTHERFORD MUS, McGILL UNIV, 66- Res: Electrical discharge in gases; applications of the expansion chamber to the study of fast beta-rays; nuclear physics. Mailing Add: Rutherford Mus Rm 108 PO Box 6070 Macdonald Physics Bldg McGill Univ Montreal PQ Can

TERRY, DANIEL HETFIELD, b North Plainfield, NJ, June 18, 12; m 41; c 2. ORGANIC CHEMISTRY. Educ: Randolph-Macon Col, BS, 36; Univ Va, PhD(org chem), 40. Prof Exp: Res chemist, Jackson Lab, E I du Pont de Nemours & Co, 40-45; res & process develop chemist, Gen Aniline & Film Corp, 45-49, tech serv mgr, Antara Chem Div, 49-52; dir res, Bon Ami Co, 52-56, vpres, 56-57; asst dir res & develop, Boyle-Midway, Inc Div, Am Home Prod Corp, 57-58, dir, 58-65, tech dir, Home Prod Int, Ltd, 65-66, Boyle-Midway Int, Inc, 66-75; RETIRED. Mem: Fel Am Chem Soc; Am Oil Chem Soc; Am Mgt Asn; fel Am Asn Textile Chem & Colorists; Am Inst Chem. Res: Consumer household products; cosmetics and pharmaceuticals; cleansers; insecticides; deodorants; polishes; detergents; sanitizers; toothpaste; aspirin; decongestants; hair sprays. Mailing Add: 9122 Lakewood Dr Seminole FL 33542

TERRY, DAVID LEE, b Burkley, Ky, Mar 22, 36; m 62; c 3. SOIL SCIENCE. Educ: Univ Ky, BS, 58, MS, 61; NC State Univ, PhD(soil sci), 68. Prof Exp: Agronomist, Univ Ky, 59-60; instr, 62-66, ASSOC PROF SOIL SCI, NC STATE UNIV, 71- Mem: Am Soc Agron. Res: Soil fertility, including water and ion movement in very sandy soils. Mailing Add: Dept of Soil Sci NC State Univ Raleigh NC 27606

TERRY, EDGAR RAYMOND, mathematics, philosophy, see 12th edition

TERRY, GLENN A, inorganic chemistry, see 12th edition

TERRY, HERBERT, b New York, NY, Jan 30, 22; m 57; c 2. OPERATIONS RESEARCH. Educ: City Col New York, BS, 42; Polytech Inst Brooklyn, BChE, 46. Prof Exp: Group leader, Foster D Snell, Inc, 42-46; tech serv mgr, Shawinigan Resins Corp, 47-63; tech mgr coatings div, Hooker Chem Corp, 63-67; res dir polymer appln, Foster D Snell, Inc, Subsidiary Booz, Allen & Hamilton, Inc, 67-69, res dir, 69-71, vpres, 71-74; PRES, DECISIONEX, INC, 74- Mem: Am Chem Soc; Opers Res Soc Am; Inst Mgt Sci. Res: Applications of decision theory and operations research to computer programming for financial analysis. Mailing Add: Decisionex Inc 21 Charles St Westport CT 06880

TERRY, JOHN CHRISTOPHER, analytical chemistry, inorganic chemistry, see 12th edition

TERRY, LUTHER LEONIDAS, b Red Level, Ala, Sept 15, 11; m 40; c 3. MEDICINE. Educ: Birmingham-Southern Col, BS, 31; Tulane Univ, MD, 35; Am Bd Internal Med, dipl, 43. Hon Degrees: DSc, Birmingham-Southern Col, 61, Jefferson Med Col, 62, Tulane Univ, 64, Union Col, 64, Univ RI, 64, Rose Polytech Inst, 65, McGill Univ, 66 & Univ Ala, 66; Dr Med Sci, Woman's Med Col Pa, 64; LLD, Univ Alaska, 64, Calif Col Med, 65 & Marquette Univ, 68. Prof Exp: Intern, Hillman Hosp, Birmingham, Ala, 35-36; asst resident med, Univ Hosp, Cleveland, Ohio, 36-37; resident City Hosp, 37-38, intern path & asst admitting officer, 38-39; instr med, Washington Univ, 39-40; from instr to asst prof med, prev med & pub health, Univ Tex, 40-42; mem med serv staff, USPHS Hosp, Baltimore, Md, 42-43, chief med serv, 43-53, asst dir, Nat Heart Inst, 58-61, Surgeon Gen, USPHS, 61-65; vpres med affairs, 65-71, PROF MED & COMMUNITY MED, UNIV PA, 65- Concurrent Pos: Assoc prof, Univ Tex, 42-46; instr, Sch Med, Johns Hopkins Univ, 44-53, asst prof, 53-61; mem med div, Strategic Bombing Surv, Japan, 45-46; staff mem, Subcomt Invest Malmedy Atrocities, Senate Comt Mil Affairs, 49; chief cardiovasc clin, USPHS Hosp, Baltimore, Md, 50-53, chief gen med & exp therapeut, Nat Heart Inst, 50-58; mem cardiovasc study sect, NIH, 50-55, chmn med bd, Clin Ctr, 53-55, mem, 53-58, dir residency training prog, Nat Heart Inst, 53-61, chmn cardiovasc res training comt, 57-61, mem comt civilian health requirements, USPHS, 56-58, mem adv comt nutrit, Indian Health Serv, 57-61; chmn, Nat Interagency Coun Smoking & Health, 67-; chmn adv comt vet med res & educ, Nat Acad Sci, 69-71; spec consult, Am Cancer Soc, 72-75; pub trustee, Nutrit Found; trustee, Inst Med Res, Camden, NJ; dir, Med Alert Found, chmn bd, 75; dir, Elwyn Inst; mem adv bd, Leonard Wood Mem & Nat Health & Safety Comt; chmn, State Adv Comt Health for Appalachia, Pa; mem, Nat Bd Med Examr; mem expert comt cardiovasc dis, WHO, 74-; res fel pneumonia, Washington Univ, 39-40. Honors & Awards: Bruce Award, Am Col Physicians, 65; Hilleboe Prize lect, NY State Annual Health Conf, 65; Distinguished Serv Med, USPHS, 65. Mem: Hon fel Am Col Chest Physicians; hon mem Am Hosp Asn; hon fel Am Col Dent; fel & hon master Am Col Physicians; hon fel Royal Soc Health. Res: Experimental therapeutics; cardiovascular diseases. Mailing Add: Col Hall Univ of Pa Philadelphia PA 19104

TERRY, MILTON EVERETT, b Windham, Conn, June 14, 16; m 37; c 3. MATHEMATICAL STATISTICS. Educ: Acadia Univ, BSc, 37; Univ NC, PhD(math statist), 51. Prof Exp: Instr math & French, Cascadilla, 37-39; asst prof math, Blue Ridge Col, 39-41; instr, Randolph-Macon Woman's Col, 46; assoc prof statist, Va Polytech Inst & State Univ, 49-52; statistician, 52-60, head dept statist, 60-63, asst dir traffic studies, 63-64, dir opers res & comput, 64-65, dir comput projs res ctr, 65-71, HEAD STATIST COMPUT RES, BELL TEL LABS, INC, 71- Concurrent Pos: Adj prof, Rutgers Univ, 55-62; vis prof, Inst Statist Studies & Res, Univ Cairo, 69-71; mem adv coun, Qm Food & Container Inst; mem subcomt adv bd, Nat Res Coun. Mem: Fel Am Soc Qual Control; fel Am Statist Asn; Inst Math Statist. Res: Use of electronic computer systems; design of management information and analytic systems. Mailing Add: 381 Creek Bed Rd Mountainside NJ 07092

TERRY, NORMAN, b Maidstone, Eng, Sept 5, 39; m 68; c 2. PLANT PHYSIOLOGY. Educ: Southampton Univ, BSc, 61; Nottingham Univ, MSc, 63, PhD(plant physiol), 66. Prof Exp: Res fel plant physiol, Div Biosci, Nat Res Coun Can, 66-68; asst specialist plant physiol, Dept Soils & Plant Nutrit, 68-72, ASST PROF ENVIRON PLANT PHYSIOL, UNIV CALIF, BERKELEY, 72- Mem: Am Soc Plant Physiologists; AAAS; Crop Sci Soc Am; Brit Soc Exp Biol. Res: Environmental and internal factors involved in the regulation of photosynthesizing and growing plant systems with particular emphasis on the effects of trace elements at deficient or toxic levels. Mailing Add: Dept of Soils & Plant Nutrit Univ of Calif Berkeley CA 94720

TERRY, ONA JOY, b Houston, Tex, Nov 13, 21; m 46; c 1. ORGANIC CHEMISTRY. Educ: Univ Houston, BS, 42, MS, 45. Prof Exp: Asst prof, 58, ASSOC PROF CHEM, TARLETON STATE UNIV, 58- Mem: Am Chem Soc; Am Asn Univ Profs. Res: Organic analysis and preparation. Mailing Add: Dept of Phys Sci Tarleton State Univ Tarleton Station TX 76402

TERRY, ORVILLE WHITFIELD, b Orient, NY, Apr 12, 14; m 50. MARINE ECOLOGY. Educ: Cornell Univ, BS, 35, MS, 37; State Univ NY Stony Brook, PhD(biol sci), 70. Prof Exp: Res asst agr, Cornell Univ, 46; teacher high schs, NY, 61-63; res assoc marine sci, 70-76, ADJ ASSOC PROF MARINE SCI, MARINE SCI RES CTR, STATE UNIV NY STONY BROOK, 76- Concurrent Pos: Sci ed, NY Sea Grant Inst, State Univ NY-Cornell Univ, 72- Mem: AAAS; World Maricult Soc; Am Fisheries Soc; Phycol Soc Am; Am Soc Agr Engrs. Res: Physiology, ecology and culture of seaweeds; mariculture of shellfish; management of tidal wetlands. Mailing Add: Marine Sci Res Ctr State Univ NY Stony Brook NY 11794

TERRY, PAUL H, b Fall River, Mass, June 22, 28. ORGANIC CHEMISTRY. Educ: Southeastern Mass Univ, BS, 51; Univ Mass, MS, 59, PhD(org chem), 63. Prof Exp: Chemist, Dept Geront, Wash Univ, 52-53; RES CHEMIST, INSECT CHEMOSTERILANTS LAB, AGR ENVIRON QUAL INST, AGR RES SERV, USDA, 63- Mem: AAAS; Am Chem Soc; Am Inst Chem. Res: Synthesis and mode of action of insect chemosterilants; phosphorus chemistry, especially phosphoramides. Mailing Add: Insect Chemosterilants Lab Agr Environ Qual Inst Agr Res Ctr Beltsville MD 20705

TERRY, RAYMOND DOUGLAS, b Southampton, NY, Apr 19, 45. MATHEMATICS. Educ: State Univ NY, Stony Brook, BS, 66; Mich State Univ, MS, 68, PhD(math), 72. Prof Exp: Teaching asst math, Mich State Univ, 66-72; instr math, Ga Inst Technol, 72-74; ASST PROF MATH, CALIF POLYTECH STATE UNIV, SAN LUIS OBISPO, 74- Mem: Am Math Soc; Math Asn Am; Soc Indust & Appl Math. Res: Higher-order delay and functional differential equations. Mailing Add: Dept of Math Calif Polytech State Univ San Luis Obispo CA 93407

TERRY, RICHARD BENJAMIN, embryology, cytology, see 12th edition

TERRY, RICHARD D, b Salt Lake City, Utah, Jan 29, 24; m 48; c 1. OCEANOGRAPHY, SPACE SCIENCES. Educ: Univ Southern Calif, AB, 50, MS, 56, PhD, 65. Prof Exp: Res asst oceanog, Allan Hancock Found, Univ Southern Calif, 50-55, res assoc, 55-60; res specialist, Autonetics Div, NAm Aviation Inc, 60-64, dir ocean eng, Gen Off, 64-65, spec asst oceanology, Ocean Systs Oper, 66-68; pres, Seaonics Int Inc, 68-70; EXEC DIR, RICHARD TERRY & ASSOCS/ENVIRON SCI & SERV, 70- Concurrent Pos: Consult oceanog, marine geology, 55-, Dep Asst Secy Defense, 64 & US Dept Navy, 64-68; mem, Community Regional & Natural Resources Develop Group Comt, US Chamber Commerce, 66-67; prog dir, Nat Oceanog Govt & Indust Report, 64-65; chmn bd, Red Sea Enterprises Co, Ltd, 70- Mem: Am Geophys Union; Geol Soc Am. Res: Marine geology; submarine topography; sediment; physical and chemical oceanography; ocean engineering; deep submergence technology; environmental sciences; basic and applied research in environmental sciences. Mailing Add: 3903 Calle Abril San Clemente CA 92672

TERRY, ROBERT DAVIS, b Hartford, Conn, Jan 13, 24; m 52; c 1. NEUROPATHOLOGY. Educ: Williams Col, Mass, BA, 46; Union Univ, NY, MD, 50. Prof Exp: Asst neuropathologist, Montefiore Hosp, New York, 55-59; PROF PATH, ALBERT EINSTEIN COL MED, 59-, CHMN DEPT, 70- Mem: Electron Micros Soc Am; Am Asn Path & Bact; Am Neurol Asn; Am Asn Neuropath (pres, 69-70); Am Acad Neurol. Res: Electron microscope studies of pathology of the nervous system; biologic studies of aging brain. Mailing Add: Dept of Path Albert Einstein Col of Med Bronx NY 10461

TERRY, ROBERT JAMES, b Crockett, Tex, May 1, 22; m 48; c 1. ZOOLOGY, DEVELOPMENTAL BIOLOGY. Educ: Tex Southern Univ, BS, 46; Univ Atlanta, MS, 49; Univ Iowa, PhD(zool, bact), 54. Prof Exp: Instr biol, 48-49, asst prof biol & bact, 50-52, assoc prof biol & actg head dept, 53-54, dean col arts & sci, 69-71, PROF BIOL & HEAD DEPT, TEX SOUTHERN UNIV, 54-, DEAN FACULTIES, 71-, VPRES ACAD AFFAIRS, 74- Concurrent Pos: Consult, State Teachers Asn Tex, 55; vis scientist, Tex Acad Sci, 60-66; sci consult, Govt India, 65; NSF consult, 67; mem gen res support prog adv comt, NIH, 72-76. Mem: Am Soc Zool; Nat Inst Sci. Res: Neurobiology, particularly regeneration of central nervous tissue in amphibia and in mice; influence of thyroxine on development of specific tissues in amphibians. Mailing Add: Sch of Arts & Sci Tex Southern Univ Houston TX 77004

TERRY, ROBERT LEE, b Mt Holly, NJ, Aug 7, 18; m 43; c 2. CELL PHYSIOLOGY. Educ: Earlham Col, AB, 39; Univ Pa, PhD(zool), 48. Prof Exp: Instr biol, Philadelphia Col Pharm, 46-47; asst instr zool, Univ Pa, 47-48; asst prof biol, Union Col, NY, 48-51; asst prof zool, Iowa State Col, 51-52; assoc prof, 52-67, PROF BIOL, COLBY COL, 67- Mem: AAAS; Am Soc Zool. Res: Granular components of frog eggs; ionic requirements of bacteria; surface properties of myxomycete plasmodia. Mailing Add: Dept of Biol Colby Col Waterville ME 04901

TERRY, ROGER, b Waterville, NY, May 8, 17; m 42; c 2. SURGICAL PATHOLOGY. Educ: Colgate Univ, AB, 39; Univ Rochester, MD, 44. Prof Exp: Intern path, Med Ctr, Univ Rochester, 44-45, instr, Univ, 45-51, from asst prof to prof, 51-69; PROF PATH, MED CTR, UNIV SOUTHERN CALIF, 69-, HEAD SURG PATHOLOGIST, LOS ANGELES COUNTY-UNIV SOUTHERN CALIF MED CTR, 69- Concurrent Pos: Actg pathologist, Park Ave Hosp, Rochester, NY, 45-47; resident, Med Ctr, Univ Rochester, 45-51; actg pathologist, Genesee Hosp, Rochester, 47-50 & Highland Hosp, 49-51; co-exec dir, Calif Tumor Tissue Registry, Los Angeles, 69. Mem: Am Soc Exp Path; Am Soc Clin Path; Am Asn Path & Bact; Am Geriat Soc; Int Acad Path. Res: Metabolic bone diseases. Mailing Add: LAC-USC Med Ctr Box 39 1200 N State St Los Angeles CA 90033

TERRY, SAMUEL MATTHEW, b Nashville, Tenn, Jan 23, 15; m 43; c 2. CHEMISTRY. Educ: Vanderbilt Univ, BA, 43. Prof Exp: Jr chemist, Winthrop Chem Co, NY, 43-44; chemist, Publicker Industs, Inc, 44-46; res engr, Battelle Mem Inst, 46-51; res fel, Mellon Inst, 51-54; res chemist, Pittsburgh Plate Glass Co, 54-56; mgr prod develop, Reynolds Chem Prods Co, 56-66; vpres & tech dir, Atlantis Chem Corp, 66-67; TECH DIR, HOOVER CHEM PROD DIV, HOOVER BALL & BEARING CO, 67- Mem: Am Chem Soc. Res: Physical, organic and polymer chemistry. Mailing Add: 1560 Marian Ave Ann Arbor MI 48103

TERRY, STUART LEE, b Chicago, Ill, Apr 8, 42; c 2. POLYMER CHEMISTRY. Educ: Cornell Univ, BChemE, 65, PhD(chem eng), 69; Rensselaer Polytech Inst, MS, 74. Prof Exp: NSF trainee polymer chem, Cornell Univ, 64-68 & asst, 65-68; sr res chemist, 68-75, GROUP LEADER POLYMER SYNTHESIS, MONSANTO CO, 75- Res: Identification and preparation of polymeric materials with morphologies required for industrial and consumer end use performance. Mailing Add: Monsanto Co 730 Worcester St Indian Orchard MA 01151

TERRY, THOMAS MILTON, b Knoxville, Tenn, Apr 2, 39; m 64; c 2. BIOPHYSICS, MICROBIOLOGY. Educ: Yale Univ, BA, 61, MS, 63, PhD(molecular biophys), 67. Prof Exp: USPHS fel biophys, Univ Geneva, 67; asst prof microbiol, Albert Einstein Col Med, 68-69; ASST PROF MICROBIOL, UNIV CONN, 69- Mem: AAAS; Am Soc Microbiol. Res: Biological membrane structure and function; molecular biology of mycoplasma; ultrastructure of bacteria; bacteriophage development. Mailing Add: Dept of Microbiol U-44 Univ of Conn Storrs CT 06268

TERRY, WILLIAM DAVID, b New York, NY, Oct 22, 33; m 66; c 4. IMMUNOLOGY. Educ: Cornell Univ, BA, 54; State Univ NY Downstate Med Ctr, MD, 58. Prof Exp: Intern, Jewish Hosp Brooklyn, NY, 58-59, asst resident, 59-61; NIH trainee, Sch Med, Univ Calif, San Francisco, 61-62; res assoc, Immunol Sect, Gen Labs & Clins, 62-64, sr investr, 64-71, chief br, 71-73, ASSOC DIR IMMUNOL, DIV CANCER BIOL & DIAG, NAT CANCER INST, 73- Mem: Am Asn Immunol; Am Fedn Clin Res; Am Soc Clin Invest; Am Asn Cancer Res. Res: Nature of the immune response, particularly as it relates to the recognition of and reaction against tumors by the tumor bearing host. Mailing Add: Div of Cancer Biol & Diag Nat Cancer Inst Bethesda MD 20014

TERSHAK, DANIEL R, b Wilkes-Barre, Pa, Nov 19, 36; m 67; c 2. VIROLOGY, MICROBIOLOGY. Educ: King's Col, Pa, BS, 58; Yale Univ, PhD(virol), 62. Prof Exp: Instr microbiol, Yale Univ, 62-63, fel, 63-64; asst prof, 64-69, ASSOC PROF MICROBIOL, PA STATE UNIV, 69- Concurrent Pos: USPHS grant, 65- Mem: Am Soc Microbiol. Res: Nucleic acid and protein synthesis; biochemical genetics. Mailing Add: Dept Microbiol S-101 Frear Bldg Pa State Univ University Park PA 16802

TERSHAKOVEC, GEORGE ANDREW, b Lviv, Ukraine, May 6, 14; nat US; m 54; c 2. BIOCHEMISTRY. Educ: Lviv Univ, MD, 39. Prof Exp: Sr asst biochem, Sch Med, Lviv Univ, 39-41; biochemist & biologist pharmaceut mfg, Laokoon, Lviv, 41-44; asst physiol chem, Sch Med, Univ Vienna, 45-49; instr path, Med Res Unit, 50-52, from instr to assoc prof, 52-60, PROF BIOCHEM, SCH MED, UNIV MIAMI, 60- Mem: AAAS; Am Chem Soc; Sigma Xi; Shevchenko Sci Soc. Res: Carbohydrate metabolism in muscle and liver; enzymatic synthesis of adenosine-5'-phosphate and adenosine triphosphate from adenosine; dynamics of inflammation and repair; clinical biochemistry; gerontology. Mailing Add: Dept of Biochem Sch of Med Univ of Miami PO Box 520875 Miami FL 33152

TERSS, ROBERT H, b East St Louis, Ill, Sept 13, 25; m 55. ORGANIC CHEMISTRY. Educ: Wash Univ, AB, 49; Univ Kans, PhD(chem), 53. Prof Exp: Asst instr chem, Univ Kans, 49-51; res chemist, Dept Org Chem, 53-57, res supvr dyes, 57-58, head div dyes, 59-63, patents & intel, 63-65, photochem, 65-70, supt dyes & chem qual control, 70-71, supt dye mfg, 71-73, DIV HEAD DYES PROCESS, E I DU PONT DE NEMOURS & CO, INC, 73- Mem: Am Chem Soc; Am Asn Textile Chem & Colorists. Res: Dyes; heterocycles; photochemistry; information handling systems; photographic materials. Mailing Add: 708 Ambleside Dr Wilmington DE 19808

TERTZAKIAN, GERARD, b Cairo, Egypt, Dec 14, 37; Can citizen; m 59; c 2. ORGANIC CHEMISTRY. Educ: Univ Sheffield, BSc, 58; Univ Sask, MSc, 60, PhD(org chem), 62. Prof Exp: Res fel mycol chem, Nat Res Coun Can, 62-64; res assoc org chem, San Jose State Col, 64-65, asst prof chem, 65-66; chemist, R & L Molecular Res, Ltd, Alta, 66-68; mgr specialty chem & tech sales, Raylo Chem Ltd, 68-71; PRES TEROCHEM LABS LTD, 71- Mem: Sr mem Chem Inst Can; The Chem Soc. Res: Synthetic organic chemistry. Mailing Add: Terochem Labs Ltd PO Box 8188 Sta F Edmonton AB Can

TERWEDOW, HENRY ALBERT, JR, b Hoboken, NJ, July 22, 46; m 67; c 3. MEDICAL ENTOMOLOGY. Educ: Univ Notre Dame, BS, 68, PhD(biol), 74; Montclair State Col, MA, 69; Am Registry Prof Entomologists, cert, 75. Prof Exp: Res assoc entom, Sch Med, Univ Md, 73-75; FEL ENTOM, UNIV CALIF, BERKELEY, 75- Mem: Entom Soc Am; Am Soc Trop Med & Hyg; Am Soc Parasitologists; Genetics Soc Am; Am Mosquito Control Asn. Res: Vectorial capacity and genetic control of mosquito species involved in the transmission of filariae and arboviruses. Mailing Add: Div of Entom & Parasitol 201 Wellman Hall Univ Calif Berkeley CA 94720

TERWILLIGER, CHARLES, JR, b Laramie, Wyo, July 16, 18; m 41; c 3. PLANT ECOLOGY, RANGE SCIENCE. Educ: Colo State Univ, BS, 41, MF, 47; Univ Wyo, PhD(plant sci), 60. Prof Exp: From instr to assoc prof, 50-67, actg head dept, 63-66, PROF RANGE SCI, COLO STATE UNIV, 67- Concurrent Pos: Co-ed current lit sect, J Soc Range Mgt, 61-65. Mem: AAAS; Soc Range Mgt; Soc Am Foresters; Am Inst Biol Sci. Res: Ecology of the sagebrush plant community; range soils and soil fertilization. Mailing Add: Dept of Range Sci Colo State Univ Ft Collins CO 80521

TERWILLIGER, DON WILLIAM, b Klamath Falls, Ore, Mar 27, 42; m 70. SOLID STATE PHYSICS. Educ: Calif Inst Technol, BS, 64; Univ Ore, MA, 66, PhD(physics), 70. Prof Exp: ASST PROF PHYSICS, MIDDLEBURY COL, 70- Mem: AAAS; Am Phys Soc; Am Asn Physics Teachers. Res: Low temperature physics; electron scattering from imperfections in metals; Fermi surface studies. Mailing Add: Dept of Physics Middlebury Col Middlebury VT 05753

TERWILLIGER, KENT MELVILLE, b San Jose, Calif, June 17, 24; m 51; c 4. HIGH ENERGY PHYSICS. Educ: Calif Inst Technol, BS, 49; Univ Calif, PhD(physics), 52. Prof Exp: From instr to assoc prof, 52-65, PROF PHYSICS, UNIV MICH, ANN ARBOR, 65- Concurrent Pos: John Simon Guggenheim fel, 64-65; mem high energy physics adv panel, AEC, 68-71; vis scientist, Europ Orgn Nuclear Res, 72. Mem: Am Phys Soc. Res: Elementary particle scattering experiments; high energy accelerators. Mailing Add: Dept of Physics Univ of Mich Ann Arbor MI 48104

TERZAGHI, BETTY ERICKSON, molecular biology, genetics, see 12th edition

TERZAGHI, ERIC, molecular biology, see 12th edition

TERZAGHI, MARGARET, b Boston, Mass, May 7, 41. CANCER. Educ: Boston Univ, AB, 64, MS, 69; Harvard Univ, MS, 70, DSc(radiation biol, physiol), 74. Prof Exp: Res asst cancer res, Harvard Univ, 70-75, res assoc, 75; RES ASSOC CANCER RES, OAK RIDGE NAT LAB, 75- Res: An examination of possible in vitro models of in vivo carcinogenesis induced by chemical carcinogens and/or radiation. Mailing Add: Biol Div Carcinogenesis Group Oak Ridge Nat Lab PO Box Y Oak Ridge TN 37830

TERZAGHI, RUTH DOGGETT, b Chicago, Ill, Oct 14, 03; m 30; c 2. GEOLOGY. Educ: Univ Chicago, BS, 24, MS, 25; Harvard Univ, PhD(geol), 30. Prof Exp: Instr geol, Goucher Col, 25-26; instr, Wellesley Col, 26-28; researcher, 30-43; geol consult, 43-75; RETIRED. Concurrent Pos: Lectr eng geol, Harvard Univ, 57-61, res fel, 63-70. Mem: Fel Geol Soc Am; hon mem Asn Eng Geol. Res: Concrete deterioration due to defective aggregate or environmental factors; petrology of igneous rocks; ground water and engineering geology. Mailing Add: 3 Robinson Circle Winchester MA 01890

TERZIAN, LEVON ARAM, parasitology, deceased

TERZIAN, YERVANT, b Feb 9, 39; m 66; c 2. ASTRONOMY. Educ: Am Univ Cairo, BSc, 60; Univ Ind, MA, 63, PhD(astron), 65. Prof Exp: Res assoc radio astron, Cornell Univ Ctr Radiophysics & Space Res & Arecibo Ionospheric Observ, 65-67, asst prof, 67-72, ASSOC PROF ASTRON, CORNELL UNIV, 72-, ASST DIR, CTR RADIOPHYSICS & SPACE RES, 68-, GRAD FAC REP, 74- Mem: Int Union Radio Sci; Int Astron Union; Am Astron Soc. Res: Radio astronomical studies of interstellar matter; radio properties of extragalactic nebulae and other radio sources; radio emission from planetary nebulae; pulsars. Mailing Add: Ctr Radiophysics & Space Res Space Sci Bldg Cornell Univ Ithaca NY 14850

TERZUOLI, ANDREW JOSEPH, b Brooklyn, NY, Oct 5, 14; m 42; c 4. MATHEMATICS. Educ: Brooklyn Col, BA, 36; NY Univ, MS, 48. Prof Exp: PROF MATH, POLYTECH INST BROOKLYN, 46- Concurrent Pos: Consult statist, Syska & Hennessey, NY, 74. Mem: AAAS; Am Math Soc; Am Meteorol Soc; Math Asn Am; Inst Math Statist. Res: Probability; mathematical statistics. Mailing Add: Dept Math Polytech Inst Brooklyn 333 Jay St Brooklyn NY 11201

TERZUOLO, CARLO A, b Acqui, Italy, Sept 2, 25; nat US; m 54; c 1. PHYSIOLOGY. Educ: Univ Turin, MD, 49. Prof Exp: Asst, Univ Turin, 48-49; Ital Res Coun fel, 50-51; asst prof, Free Univ Brussels, 51-53; asst, Univ Calif, Los Angeles, 54-56, res assoc, 57-59; PROF PHYSIOL, UNIV MINN, MINNEAPOLIS, 59-, DIR LAB NEUROPHYSIOL, 71- Concurrent Pos: Multiple Sclerosis Soc fel, 56-57; Fulbright res fel, Univ Pisa, 67-68. Mem: Am Physiol Soc; Biophys Soc; Soc Neurosci. Res: Nerve cell and receptor physiology; dynamic characteristics of neuronal systems controlling movements; vestibular, cerebellar and segmental reflex mechanisms; models of brain functions. Mailing Add: Lab of Neurophysiol Univ of Minn Minneapolis MN 55455

TESAR, CHARLES, b New York, NY, Dec 31, 06; m 36. BIOCHEMISTRY. Educ: Columbia Univ, AB, 32, MA, 40, PhD(biochem), 46. Prof Exp: Res assoc, Univ Chicago, 45-47; asst prof physiol chem, 47-72, assoc prof urol, 63-72, dir, Brady Res Lab, Hosp, 47-73, EMER ASST PROF PHSYIOL HOPKINS CHEM & EMER ASSOC PROF UROL, SCH MED, JOHNS HOPKINS UNIV, 72- Concurrent Pos: Fulbright scholar, Inst Jules Bordet, Brussels, 59-60. Mem: AAAS; Am Chem Soc. Res: Metabolism of amino acids; partial synthesis of steroids; hormone inhibition. Mailing Add: Brady Res Lab Johns Hopkins Hosp Monument St & Rutledge Ave Baltimore MD 21205

TESAR, JOSEPH THOMAS, b Vrutky, Czech, Mar 7, 28; US citizen; m 64; c 1. INTERNAL MEDICINE, IMMUNOLOGY. Educ: Charles Univ, Prague, MD, 51. Prof Exp: USPHS trainee & Laidlaw fel rheumatol, 65-68, assoc internal med, 69-73, ASST PROF MED, SCH MED, NORTHWESTERN UNIV, CHICAGO, 73-; STAFF RHEUMATOLOGIST, VET ADMIN RES HOSP, CHICAGO, 71- Concurrent Pos: Attend physician, Northwestern Mem Hosp, Chicago. Mem: Am Fedn Clin Res; Am Rheumatism Asn. Res: Biological properties of immune complexes; immunological diseases; rheumatoid arthritis. Mailing Add: Dept of Med Northwestern Univ Sch of Med Chicago IL 60611

TESAR, MILO, b Nebr, Apr 7, 20; m 44; c 4. AGRONOMY. Educ: Univ Nebr, BS, 41; Univ Wis, MS, 47, PhD(agron), 49. Prof Exp: From asst prof to assoc prof, 49-58, actg chmn dept, 64-66, PROF CROP SCI, MICH STATE UNIV, 58- Concurrent Pos: NATO fel, Grassland Res Inst, Eng, 59-60; consult, Univ Ryukus, 67; mem, Int Grassland Cong, Australia, 70, Russia, 74. Mem: Am Soc Agron. Res: Forage physiology and digestibility; legume seeding establishment; pasture renovation; recycling waste water through forage. Mailing Add: 509 Kedzie Dr East Lansing MI 48823

TESCHAN, PAUL E, b Milwaukee, Wis, Dec 15, 23; m 48; c 2. MEDICINE. Educ: Univ Minn, BS, 46, MD & MS, 48; Am Bd Internal Med, dipl, 55. Prof Exp: Intern, Res & Educ Hosp, Univ Ill, 48-49, resident internal med, Presby Hosp, Chicago, 49-50, ward officer, Metab Ward, Dept Hepatic & Metab Dis, Walter Reed Army Inst Res, 50-53, resident internal med, Barnes Hosp, St Louis, Mo, 53-54, chief renal br, Surg Res Unit, Brooke Army Med Ctr, Ft Sam Houston, Tex, 54-60, asst commandant, Walter Reed Army Inst Res, Walter Reed Army Med Ctr, 60-63, dep dir div basic surg res, 64-65, dep dir div surg, 65-66, chief dept metab & dir div med, 66-69; ASSOC PROF MED & UROL, VANDERBILT UNIV, 69- Concurrent Pos: Med Corps, US Army, 48-69; fel cardiorenal dis, Peter Bent Brigham Hosp, Boston, 50; chief renal insufficiency ctr, Korea, 52-53; consult, Surgeon Gen, US Army, 60-69; chief dept surg physiol, Walter Reed Army Inst Res, Walter Reed Army Med Ctr, 61-66, dep dir div basic surg res, 62-63, chief renal-metab serv, Walter Reed Gen Hosp, 66-69; chief US Army med res team, Vietnam, 63-64; dir, Tenn Mid-South Regional Med Prog, 69-72. Mem: Am Fedn Clin Res; fel Am Col Physicians; Soc Artificial Internal Organs; Am Physiol Soc; Int Soc Nephrol. Res: Pathogenesis and prevention of acute renal failure; prophylactic dialysis; uremia. Mailing Add: Dept of Med & Urol Vanderbilt Univ Sch of Med Nashville TN 37232

TESCHE, FREDERICK RUTLEDGE, b San Jose, Calif, Aug 11, 21; m 42; c 5. PHYSICS. Educ: Univ Calif, BA, 43; Univ Calif, Los Angeles, MA, 49, PhD(physics), 51. Prof Exp: Engr, Consol Vultee Aircraft, 46-51; mem staff, Los Alamos Sci Lab, 51-57, group leader, 57-65, assoc div leader, 65-68; dep dir mil appln, AEC, 68-71, sci rep, London, 71-72, spec asst to dir, Div Controlled Thermonuclear Res, 72-73, SCI ADV TO MGR, US ENERGY RES & DEVELOP ADMIN, SAN FRANCISCO OPERS OFF, 73- Concurrent Pos: Consult prof, Univ NMex, 57-61. Mem: Sr mem Inst Elec & Electronics Engrs. Res: Electromagnetism; flash radiography; accelerator development. Mailing Add: 3728 Mosswood Dr Lafayette CA 94549

TESH, ROBERT BRADFIELD, b Wilmington, Del, Jan 22, 36; m 60; c 2. EPIDEMIOLOGY. Educ: Franklin & Marshall Col, BS, 57; Jefferson Med Col, MD, 61; Tulane Univ, MS, 67. Prof Exp: Intern, San Francisco Gen Hosp, Calif, 61-62; resident pediat, Gorgas Hosp, 62-63; physician, USPHS, Peace Corps, Recife, Brazil, 63-65; NIH fel infectious dis, Depts Pediat & Epidemiol, Sch Med, Tulane Univ, 65-67; epidemiologist, Mid Am Res Unit, 67-72, EPIDEMIOLOGIST, PAC RES SECT, NIH, 72- Mem: AAAS; Am Soc Trop Med & Hyg; Am Soc Microbiol; Royal Soc Trop Med & Hyg. Res: Entomology; microbiology; public health; virology. Mailing Add: NIH Pac Res Sect Box 1680 Honolulu HI 96806

TESK, JOHN A, b Chicago, Ill, Oct 19, 34; m DENTAL RESEARCH. Educ: Northwestern Univ, BS, 57, MS, 60, PhD(mat sci), 63. Prof Exp: Asst prof metall, Univ Ill, Chicago, 64-68; asst metallurgist, Metall Div, Argonne Nat Lab, 68-70; asst mgr res & develop, 70-71, DIR RES & DEVELOP, DENT DIV, HOW-MEDICA, INC, CHICAGO, 71- Concurrent Pos: Consult, Argonne Nat Lab, 64-67. Honors & Awards: Grainger Award, Univ Ill, Chicago, 64. Mem: Int Asn Dent Res; Am Soc Prev Dent; Am Inst Mining & Metall Engrs; Am Soc Metals; Am Soc Eng Educ. Res: Radiation damage in metals at low temperature; point defects in metals; dental and medical materials and devices; biomaterials. Mailing Add: 25 Lake Dr Plainfield IL 60544

TESKE, RICHARD GLENN, b Cleveland, Ohio, Aug 16, 30. ASTRONOMY. Educ: Bowling Green State Univ, BS, 52; Ohio State Univ, MA, 56; Harvard Univ, PhD, 61. Prof Exp: From instr to asst prof, 61-6/, ASSOC PROF ASTRON, McMATH-HULBERT OBSERV, UNIV MICH, ANN ARBOR, 67- Mem: AAAS; Int Astron Union; Am Astron Soc. Res: Astronomical and solar spectroscopy; solar physics; solar x-radiation. Mailing Add: Dept of Astron Univ of Mich Ann Arbor MI 48104

TESKE, RICHARD H, b Christiansburg, Va, July 22, 39; m 61; c 2. VETERINARY MEDICINE, TOXICOLOGY. Educ: Va Polytech Inst, BA, 62; Univ Ga, DVM, 65; Univ Fla, MS, 66. Prof Exp: Asst prof vet sci, Univ Fla, 67; dir toxicol, Hill Top Res, Inc, Ohio, 67-70; VET MED OFFICER, DIV VET MED RES, BUR VET MED, FOOD & DRUG ADMIN, 70- Res: Comparative pharmacology and toxicology. Mailing Add: Div of Vet Med Res Bur of Vet Med Food & Drug Admin 5600 Fishers Lane Rockville MD 20852

TESMER, IRVING HOWARD, b Buffalo, NY, May 31, 26; m 64; c 2. STRATIGRAPHY, PALEONTOLOGY. Educ: Univ Buffalo, BA, 46, MA, 48; Syracuse Univ, PhD(geol), 54. Prof Exp: From instr to asst prof geol, Univ NH, 50-55; instr, Rutgers Univ, 55-57; from asst prof to assoc prof, 57-63, chmn dept, 66-69, PROF GEOL, STATE UNIV NY COL BUFFALO, 63- Concurrent Pos: Mem, Paleont Res Inst. Mem: Fel AAAS; Geol Soc Am; Am Asn Petrol Geol; Paleont Soc. Res: Devonian stratigraphy and paleontology; geology of western New York. Mailing Add: Dept of Geosci 1300 Elmwood Ave State Univ of NY Col at Buffalo Buffalo NY 14222

TESMER, JOSEPH RANSDELL, b Lafayette, Ind, Sept 9, 39; m 62; c 2. EXPERIMENTAL NUCLEAR PHYSICS. Educ: Purdue Univ, BS, 62; Univ Wash, PhD(nuclear physics), 71. Prof Exp: Engr, Boeing Co, 62-64; res assoc nuclear physics, Purdue Univ, 71-73; asst scientist physics, Univ Wis, 73-75; STAFF MEM NUCLEAR PHYSICS, LOS ALAMOS SCI LAB, 75- Mem: Am Phys Soc; AAAS; Am Asn Physicists in Med. Res: Stripping of high energy negative hydrogen beams, and negative hydrogen beam production. Mailing Add: Los Alamos Sci Lab Group P-11 MS 808 PO Box 1663 Los Alamos NM 87545

TESORO, GIULIANA C, b Venice, Italy, June 1, 21; nat US; m 43; c 2. ORGANIC POLYMER CHEMISTRY. Educ: Yale Univ, PhD(org chem), 43. Prof Exp: Res chemist, Calco Chem Co, NJ, 34-44; res chemist, Onyx Oil & Chem Co, 44-46, head org synthesis dept, 46-55, asst dir res, 55-57, assoc dir, 57-58; asst dir org res, Cent Res Lab, J P Stevens & Co, Inc, 58-68; sr scientist, Textile Res Inst, NJ, 68-69; sr scientist, 69-71, DIR CHEM RES, BURLINGTON INDUSTS, INC, 71- Concurrent Pos: Vis prof, Mass Inst Technol, 72-; mem comt on fire safety aspects of polymeric materials, Nat Acad Sci, 72- Mem: Am Chem Soc; Am Asn Textile Chem & Colorists; Fiber Soc (pres, 74); Am Inst Chemists; Textile Inst Gt Brit. Res: Synthesis of pharmaceuticals; textile chemicals; germicides; polymers; chemical modification of fibers; synthesis and rearrangement of glycols in the hydrogenated naphthalene series; polymer flammability and flame retardants. Mailing Add: 278 Clinton Ave Dobbs Ferry NY 10522

TESS, ROY WILLIAM HENRY, b Chicago, Ill, Apr 25, 15; m 44; c 2. POLYMER CHEMISTRY. Educ: Univ Ill, BS, 39; Univ Minn, PhD(org chem), 44. Prof Exp: Lab asst, Underwriters Labs, Inc, Ill, 36-37; asst chem, Univ Minn, 39-44; chemist, Shell Develop Co, 44-59, res supvr, 59-62 & 64-67, Royal Dutch Shell Plastics Lab, Holland, 62-63, tech planning supvr, Coatings, Shell Chem Co, NY, 67-70, Tex, 70-73, TECH SUPVR SOLVENTS BUS CTR, SHELL CHEM CO, 73- Concurrent Pos: Trustee, Paint Res Inst. Mem: Fedn Soc Paint Technol (pres, 73-); Soc Cosmetic Chem; Am Chem Soc; fel Am Inst Chem; Am Oil Chem Soc. Res: Epoxy, allyl, alkyd, polyester and hydrocarbon resins; surface coatings; varnishes and drying oils; polyols; polar and hydrocarbon solvents; high polymer latices; atmospheric chemistry. Mailing Add: Shell Chem Co One Shell Plaza Houston TX 77002

TESSEL, RICHARD EARL, b Cincinnati, Ohio, Sept 6, 44. NEUROPHARMACOLOGY, PSYCHOPHARMACOLOGY. Educ: Univ Calif, Los Angeles, BA, 66; Univ Ill, Chicago Circle, MA, 69; Univ Mich, PhD(pharmacol), 74. Prof Exp: Nat Inst Drug Abuse fel pharmacol, Sch Med, Univ Colo, 74-75; ASST PROF PHARMACOL, SCH PHARM, UNIV KANS, 75- Res: Role of biogenic amine release in brain in modulating the reinforcing, locomotor-stimulant, stereotypic and operant-schedule effects of amphetamines and its congeners. Mailing Add: Dept of Pharmacol & Toxicol Sch of Pharm Univ of Kans Lawrence KS 66045

TESSER, HERBERT, b Jersey City, NJ, Mar 25, 39; m 61; c 2. PHYSICS. Educ: Polytech Inst Brooklyn, BS, 60; Stevens Inst Technol, MS, 63, PhD(physics), 68. Prof Exp: Metrol engr, Kearfott Corp, 60-62; res asst physics, Stevens Inst Technol, 64-67; ASSOC PROF PHYSICS, PRATT INST, 67- Concurrent Pos: Res consult, NRA, Inc, 71-72. Mem: Am Phys Soc. Res: Electrodynamics; statistical mechanics; relativity. Mailing Add: Dept of Physics Pratt Inst Brooklyn NY 11205

TESSIERI, JOHN EDWARD, b Vineland, NJ, Sept 3, 20; m 43; c 3. RESEARCH ADMINISTRATION. Educ: Pa State Univ, BS, 42, MS, 47; Stanford Univ, PhD(org chem), 50. Prof Exp: Chemist & asst to asst dir res, Texaco, Inc, NY, 49, group leader, 55, asst supvr lubricants res, 55, Tex, 55-56, supvr chem res, 56-57, asst dir res, 57-60, dir fuels & chem res, 60-62, vpres, Texaco Exp Inc, Va, 62-63, exec vpres, 63-65, pres, 65-66, mgr sci planning, Texaco Inc, 67-68, asst to pres, 68-69, staff coordr strategic planning group exec off, 69-70, gen mgr strategic planning, 70-71, VPRES RES & TECH DEPT, TEXACO INC, 71- Mem: Am Chem Soc; Indust Res Inst; AAAS; Sci Res Soc Am; Dirs Indust Res. Res: Product and process development, including petrochemicals, fuels and lubricants; exploration and production research; coal beneficiation, gasification and liquefaction. Mailing Add: Texaco Inc PO Box 509 Beacon NY 12508

TESSLER, ARTHUR NED, b New York, NY, Feb 21, 27; m 53; c 4. UROLOGY. Educ: NY Univ, AB, 48, MD, 52; Am Bd Urol, dipl. Prof Exp: Investr, USPHS grant, 59-62, PROF CLIN UROL, SCH MED, NY UNIV, 72- Concurrent Pos: Consult, Vet Admin Hosp, NY, 70- Honors & Awards: Carl Hartman Award, Am Fertil Soc, 63. Mem: AMA; Am Fertil Soc; Am Col Surg. Mailing Add: 566 First Ave New York NY 10016

TESSLER, GEORGE, b Brooklyn, NY, Mar 7, 36; m 69. PHYSICS. Educ: Brooklyn Col, BS, 57; Univ Pa, MS, 59, PhD(physics), 64. Prof Exp: SR SCIENTIST, BETTIS ATOMIC POWER LAB, WESTINGHOUSE ELEC CORP, 63- Mem: Am Phys Soc; AAAS. Res: Acquisition of neutron cross section data for use in reactor design. Mailing Add: Bettis Atomic Power Lab Westinghouse Elec Corp PO Box 79 West Mifflin PA 15122

TESSLER, MARTIN MELVYN, b Brooklyn, NY, Sept 12, 37; m 62; c 1. ORGANIC CHEMISTRY. Educ: Brooklyn Col, BS, 58; Univ Kans, PhD(chem), 62. Prof Exp: Chemist, Enjay Chem Intermediates Lab, Esso Res & Eng Co, 65-68; proj supvr, 68-72, RES ASSOC NAT STARCH & CHEM CORP, PLAINFIELD, 72- Mem: AAAS; Am Chem Soc; The Chem Soc. Res: Starch chemistry.

TESSMAN, ETHEL STOLZENBERG, molecular biology, see 12th edition

TESSMAN, IRWIN, b New York, NY, Nov 24, 29; m 49; c 1. BIOPHYSICS, MOLECULAR BIOLOGY. Educ: Cornell Univ, AB, 50; Yale Univ, MS, 51, PhD(physics), 54. Prof Exp: NSF fel, Cornell Univ, 54-55, Am Cancer Soc fel, 55-57; fel, Mass Inst Technol, 57-58, res assoc biol, 58-59; assoc prof biophys, 59-62, PROF BIOL, PURDUE UNIV, 62- Concurrent Pos: NSF sr fel, Harvard Med Sch, 67; prof molecular biol, Univ Calif, 69-72. Honors & Awards: Gravity Res Found Prize, 53; Sigma Xi Res Award, Purdue Univ, 66. Mem: AAAS; Genetics Soc Am; Am Biophys Soc. Res: Radiocarbon dating; ionization by charged particles; molecular genetics; reproduction of bacterial viruses; molecular studies of replication, mutation, recombination and function of genetic material. Mailing Add: Dept of Biol Sci Purdue Univ Lafayette IN 47907

TESSMAN, JACK ROBERT, b New York, NY, May 5, 19; m 63; c 2. PHYSICS. Educ: City Col New York, BS, 38; Univ Calif, Berkeley, MA, 50, PhD(physics), 52. Prof Exp: Tutorial fel, City Col New York, 39; physicist, Mat Div, Wright-Patterson Air Force Base, 39-46; asst, Univ Calif, Berkeley, 46-50; asst prof physics, Pa State Univ, 52-56; from asst prof to assoc prof, 56-72, vis lectr, 55-56, PROF PHYSICS, TUFTS UNIV, 72- Concurrent Pos: Vis assoc prof, Univ Calif, 61; lectr, Mass Inst Technol, 64-65, staff mem sci teaching ctr, 62-65; NSF Sci Fac fel, 62-63. Mem: AAAS; Am Phys Soc; Am Asn Physics Teachers. Res: Solid state physics; magnetic phenomena; electromagnetism; science education. Mailing Add: Dept of Physics Tufts Univ Medford MA 02155

TESSMER, CARL FREDERICK, b North Braddock, Pa, May 28, 12; m 39; c 2. PATHOLOGY. Educ: Univ Pittsburgh, BS, 33, MD, 35; Am Bd Path, dipl, 41. Prof Exp: Resident path, Presby Hosp, Pittsburgh, 36-37; fel, Mayo Clin, 37-38; resident pathologist, Queen's Hosp, Honolulu, Hawaii, 39-40; chief lab, Tripler Gen Hosp, Honolulu, Med Corps, US Army, 42-45, chief radiologic safety, Bikini, 46, pathologist, US Naval Med Res Inst, 46-48, dir atomic bomb casualty comn, Nat Res Coun, 48-51, commanding officer, Army Med Res Lab, Ft Knox, Ky, 51-54, chief basic sci div & radiation path br, Armed Forces Inst Path, Walter Reed Army Med Ctr, DC, 54-60, commanding officer, 406th Med Gen Lab, Japan, 60-62, pathologist, Walter Reed Army Inst Res, 62-63; chief exp path sect, Univ Tex M D Anderson Hosp & Tumor Inst Houston, 63-71, prof path, 63-73; CHIEF LAB SERV, VET ADMIN CTR, 73- Concurrent Pos: Hektoen lect, 60; Armed Forces Inst Path Centennial lect, 62; mem grad fac, Univ Tex Grad Sch Biomed Sci, 63-73, path coordr, Univ Tex Med Sch Houston, 71-73; consult, Walter Reed Army Med Ctr, 64-; mem, USPHS Adv Comt, Collab Radiol Health Animal Res Lab, 65-70; mem, subcomt 34, Nat Comt Radiation Protection, 70-; consult, Radiation Bioeffects & Epidemiol Adv Comn, Food & Drug Admin, Dept Health, Educ & Welfare, 72- Mem: Fel Am Soc Clin Path; Soc Nuclear Med; Am Asn Path & Bact; fel Col Am Path; Int Acad Path. Res: Morphologic and experimental radiation pathology; trace elements; copper metabolism. Mailing Add: Lab Serv Vet Admin Ctr Temple TX 76501

TEST, CHARLES EDWARD, b Indianapolis, Ind, Jan 10, 16; m 38; c 4. MEDICINE. Educ: Princeton Univ, AB, 37; Univ Chicago, MD, 41. Prof Exp: Instr med, Univ Chicago, 49-51; from asst prof to assoc prof, 53-63, PROF MED, SCH MED, IND UNIV, INDIANAPOLIS, 63- Mem: Endocrine Soc; Am Diabetes Asn; fel Am Col Physicians. Res: Endocrinology; metabolism. Mailing Add: Dept of Med Ind Univ Sch of Med Indianapolis IN 46207

TEST, FREDERICK HAROLD, b Rolla, Mo, Apr 16, 12; m 40. ZOOLOGY. Educ: Purdue Univ, BS, 34; Cornell Univ, AM, 35; Univ Calif, PhD(zool), 40. Prof Exp: Asst zool, Univ Calif, 35-38 & 39-40, asst, Mus Vert Zool, 38-39; from instr to prof zool, 40-74, PROF ECOL & EVOLUTIONARY BIOL-BIOL SCI, UNIV MICH, ANN ARBOR, 74-, FAC COUNSR, 55- Concurrent Pos: Vis investr, Rancho Grande Biol Sta, Venezuela, 51, 56 & 60; vis prof, Cent Univ Venezuela, 60, hon prof, 68; Univ Mich Fac Res Fund grants, 42, 45, 48, 51 & 58; NSF grant, 56. Mem: AAAS; Am Soc Zool; Ecol Soc Am (treas, 50-54); Soc Study Evolution; Am Soc Mammal. Res: Ecology of vertebrates; speciation; carotenoid pigmentation; population studies; ecology of Venezuelan cloud forest; animal behavior. Mailing Add: 1204 Henry St Ann Arbor MI 48104

TESTA, ANTHONY CARMINE, b New York, NY, Nov 19, 33; m 62. PHYSICAL CHEMISTRY, PHOTOCHEMISTRY. Educ: City Col New York, BS, 55; Columbia Univ, MA, 58, PhD(chem), 61. Prof Exp: Res chemist, Cent Res Div, Lever Bros Co, 61-63; from asst prof to assoc prof, 63-71, PROF CHEM, ST JOHN'S UNIV, NY, 71- Concurrent Pos: Res leave, Max-Planck Inst Spectros, Univ Göttingen, 70-71. Mem: NY Acad Sci; Am Chem Soc. Res: Photochemistry and flash photolysis of molecules in solution; luminescence spectroscopy; fluorescence and phosphorescence. Mailing Add: Dept of Chem St John's Univ Jamaica NY 11432

TESTA, RAYMOND THOMAS, b New Haven, Conn, Dec 21, 37; m 62; c 3. MICROBIOLOGY, BIOCHEMISTRY. Educ: Providence Col, BS, 59; Syracuse Univ, MS, 64, PhD(microbiol), 66. Prof Exp: Res asst microbiol, Syracuse Univ, 64-66; scientist, 66-68, sr scientist, 68-72, prin microbiologist, 72-74, MGR, ANTIBIOTICS SCREENING & FERMENTATION DEPT, SCHERING CORP, 74- Mem: AAAS; Am Soc Microbiol; Soc Indust Microbiol; NY Acad Sci. Res: Antibiotics; factors affecting the production and biosynthesis of antibiotic; spore formation. Mailing Add: Microbiol Div Schering Corp 60 Orange St Bloomfield NJ 07003

TESTARDI, LOUIS RICHARD, b Philadelphia, Pa, Sept 23, 30; m 57; c 4. SOLID STATE PHYSICS. Educ: Univ Calif, Berkeley, AB, 56; Univ Pa, MS, 60, PhD(physics), 63. Prof Exp: Res physicist, Elec Storage Battery Co, 57-58 & Franklin Inst Labs, 58-62; res asst, Univ Pa, 63; RES PHYSICIST, BELL TEL LABS, 63- Mem: Am Phys Soc. Res: Transport, optical, magnetic and ultrasonic properties of solids; low temperature physics; superconductivity. Mailing Add: Bell Tel Labs Murray Hill NJ 07971

TESTER, ALBERT LEWIS, biology, deceased

TESTER, ALLEN CRAWFORD, b New Haven, Mo, Oct 30, 97; m 20; c 1. GEOLOGY. Educ: Univ Kans, AB & AM, 21; Univ Wis, PhD(geol), 29. Prof Exp: Instr geol, Univ Kans, 21-22 & Univ Wis, 24-25; assoc, Univ Iowa, 25-29, from asst prof to assoc prof, 29-37; geologist in charge dist, Socony-Vacuum Co, Columbia, SAm, 38-40; prof, 40-42, 46-66, EMER PROF GEOL, UNIV IOWA, 67- Concurrent Pos: Asst state geologist, State Geol Surv, Iowa, 33-38; vpres, Explor & Dir, Vermilion Cliffs Mining Corp, 53-58. Honors & Awards: Ben Parker Medal, Am Inst Prof Geol, 72. Mem: AAAS; fel Geol Soc Am; Soc Econ Paleont & Mineral; Am Asn Petrol Geol; Am Inst Prof Geol (vpres, 63-65, pres, 67). Res: Sedimentology; stratigraphy; ground water and mineral resources. Mailing Add: 2315 Rochester Ave Apt 111 Iowa City IA 52240

TESTER, CECIL FRED, b Boone, NC, May 23, 38; m 67; c 2. BIOCHEMISTRY, ENVIRONMENTAL MANAGEMENT. Educ: Appalachian State Univ, BS, 60; Univ Ga, PhD(biochem), 67. Prof Exp: Teacher city schs, NC, 60-61; res asst biochem, Univ Ga, 64-65, teaching asst, 65-66; AEC fel, Purdue Univ, Lafayette, 67-68; res chemist & USDA grant, NC State Univ, 68-75; RES CHEMIST, BELTSVILLE AGR RES CTR, WEST, AGR RES SERV, USDA, 75- Mem: Am Soc Agron. Res: Biological waste management and plant interactions. Mailing Add: Rm 337 Bldg 007 BARC West Beltsville MD 20705

TESTER, JOHN ROBERT, b New Ulm, Minn, Nov 18, 29; m 60; c 2. ECOLOGY. Educ: Univ Minn, BS, 51; Colo State Univ, MS, 53; Univ Minn, PhD(wildlife ecol), 60. Prof Exp: Res asst wildlife biol, Colo Game & Fish Dept, 51-53; game biologist, Minn Div Game & Fish, 54-56; asst scientist ecol, Mus Natural Hist, 56-60, from instr to assoc prof, 60-70, PROF ECOL & BEHAV BIOL, UNIV MINN, ST PAUL, 70-, HEAD DEPT, 73- Concurrent Pos: NSF fel, Aschoff Div, Max Planck Inst Physiol of Behav & Aberdeen Univ, 69-70; mem behav sci training comt, NIH. Mem: Fel AAAS; Ecol Soc Am; Am Soc Mammal; Wildlife Soc. Res: Population ecology; biotelemetry; wildlife management; animals behavior. Mailing Add: Dept of Ecol & Behav Biol Univ of Minn St Paul MN 55108

TESTERMAN, JACK DUANE, b Marietta, Okla, Dec 13, 33; m 53; c 3. STATISTICS, COMPUTER SCIENCE. Educ: Okla State Univ, BA, 55, MS, 57; Univ Tex, PhD, 72. Prof Exp: Res statistician, Jersey Prod Res Co, 57-63 & Phillips Petrol Co, 63; assoc prof, 63-69, registr, 65-70, PROF MATH & STATIST, UNIV SOUTHWESTERN LA, 69-, DIR INSTNL RES, 70-, VPRES UNIV RELATIONS, 73- Concurrent Pos: Chmn inst studies & opers anal comt, Am Asn Col Registrars & Admissions Officers, 74-; chmn ad hoc comt ways & means, Am Statist Asn, 75- Mem: Am Statist Asn; Math Asn Am; Asn Comput Mach; Asn Instnl Res; Data Process Mgt Asn. Res: Application of statistics; data analysis and use of computers. Mailing Add: Univ of Southwestern La PO Box 2331 Lafayette LA 70501

TESTERMAN, JOHN KENDRICK, b Galveston, Tex, Feb 25, 45; m 66; c 1. COMPARATIVE PHYSIOLOGY. Educ: Loma Linda Univ, BA, 67; Univ Calif, Irvine, PhD(biol), 71. Prof Exp: ASST PROF BIOL, LOMA LINDA UNIV, LA SIERRA CAMPUS, 71- Mem: AAAS; Am Inst Biol Sci; Am Soc Zoologists. Res: Organic compounds in sea water and their significance to marine organisms; comparative physiology of marine invertebrates; effects of pollutants on marine life. Mailing Add: Dept of Biol Loma Linda Univ La Sierra Campus Riverside CA 92505

TETENBAUM, MARVIN, b Brooklyn, NY, June 27, 21; m 54; c 3. PHYSICAL CHEMISTRY. Educ: NY Univ, BChE, 42; Polytech Inst Brooklyn, MChE, 47, PhD(chem), 54. Prof Exp: Res asst, Metall Lab, Univ Chicago, 42-43; res engr sam labs, Columbia Univ, 43-44; chem engr, Ballistics Res Lab, Ord Dept, US Dept Army, 47-48; radio chemist, US Naval Res Lab, 48-56; sr engr, Aircraft Nuclear Propulsion Dept, Gen Elec Co, Ohio, 56-59; CHEMIST, ARGONNE NAT LAB, 59- Concurrent Pos: Mem staff, Atomic Energy Res Estab, Harwell, Eng, 66-67. Mem: AAAS; Am Chem Soc; Sci Res Soc Am. Res: High temperature chemistry. Mailing Add: Chem Eng Div Argonne Nat Lab 9700 S Cass Ave Argonne IL 60439

TETENBAUM, MARVIN THEODORE, b Brooklyn, NY, Mar 13, 32; m 57; c 2. ORGANIC CHEMISTRY. Educ: Brooklyn Col, BS, 54; Duke Univ, AM, 55, PhD(chem), 57. Prof Exp: Org res chemist, A E Stanley Mfg Co, Ill, 57-61; ORG RES CHEMIST, CENT RES LAB, ALLIED CHEM CORP, 61- Concurrent Pos: Eve instr, Farleigh Dickinson Univ. Res: Synthetic organic chemistry; condensations and cyclizations using alkali amides; intermediates and resins from amino acids; fungicides based on rhodanines; aerobic unsaturated polyesters; monomer and polymer synthesis; emulsion polymerizations; nylon and heterocyclic chemistry. Mailing Add: Cent Res Lab Allied Chem Corp Morristown NJ 07961

TETENBAUM, SIDNEY JOSEPH, b Bronx, NY, Aug 30, 22; m 53; c 2. ELECTROMAGNETISM. Educ: City Col New York, BS, 43; NY Univ, MEE, 48; Yale Univ, PhD(physics), 52. Prof Exp: Sci investr, Electron Tubes Panel, Res & Develop Bd, Columbia Univ, 46-48; asst physics, Yale Univ, 49-52; res physicist, Varian Assocs, 52-54; eng specialist, Sylvania Elec Prod Inc, 54-63; staff scientist, Lockheed Res Lab, 63-74, HEAD ELECTROMAGNETICS GROUP, COMMUN SCI LAB, LOCKHEED RES LAB, 74- Res: Plasma physics; gaseous electronics; microwave devices; electromagnetic wave propagation. Mailing Add: 1650 Edmonton Ave Sunnyvale CA 94087

TETER, MICHAEL PHILLIP, physics, applied mathematics, see 12th edition

TETERIS, NICHOLAS JOHN, b Martins Ferry, Ohio, Jan 14, 29; m 61; c 2. OBSTETRICS & GYNECOLOGY. Educ: Washington & Jefferson Col, BA, 50; Ohio State Univ, MD, 54, MSc, 61; Am Bd Obstet & Gynec, dipl, 65. Prof Exp: Asst prof, 65-67, assoc prof & asst dean col med, 67-70, PROF OBSTET & GYNEC, COL MED, OHIO STATE UNIV, 70- Concurrent Pos: Cancer fel obstet & gynec, Col Med, Ohio State Univ, 62-64; consult, US Air Force Hosps, Wright-Patterson & Lockborne Air Force Bases, 62; asst dir, Ohio State Univ Hosps, 62- Mem: AMA; Am Col Obstet & Gynec; Am Col Surg. Res: Fetology; gynecologic cancer; obstetrical emergencies. Mailing Add: Dept of Obstet & Gynec Ohio State Univ Hosps Columbus OH 43210

TETERUCK, WALTER R, b Toronto, Ont, Apr 14, 34; m 58; c 1. PROSTHODONTICS, DENTAL MATERIALS. Educ: McGill Univ, BSc, 57, DDS, 61; Ind Univ, MSD, 63. Prof Exp: From instr to asst prof fixed prosthodontics, Sch Dent, Univ Ky, 63-66, asst prof restorative dent, 66-67; ASSOC PROF FIXED PROSTHODONTICS & HEAD DIV, FAC DENT, UNIV WESTERN ONT, 67- Mem: Assoc Am Dent Asn; Can Dent Asn; Int Asn Dent Res. Res: Fit of dental casting alloys employing a variety of different investing materials; accuracy of dental cast mounting procedures; problems and solutions of dental occlusion; hydrocolloidgypsum incompatibility. Mailing Add: Div of Fixed Prosthodontics Univ of Western Ont Fac of Dent London ON Can

TETRAULT, ROBERT CLOSE, b Walhalla, NDak, Nov 25, 33; m 58; c 4. ENTOMOLOGY. Educ: NDak State Univ, BS, 58, MS, 63; Univ Wis, PhD(entom), 67. Prof Exp: Asst prof, 65-71, ASSOC PROF ENTOM, PA STATE UNIV, UNIVERSITY PARK, 71- Mem: Entom Soc Am. Res: Taxonomy of Coleoptera, especially the family Helodidae. Mailing Add: Dept of Entom 106 Patterson Bldg Pa State Univ University Park PA 16802

TETREAULT, FLORENCE G, b Detroit, Mich. STATISTICS. Educ: Univ Detroit, BS; Univ Mich, MA; Iowa State Univ, PhD(statist), 65. Prof Exp: Instr statist, Iowa State Univ, 60-64; assoc prof, 64-72, PROF MATH, UNIV DETROIT, 72- Mailing Add: Dept of Math Univ of Detroit Detroit MI 48154

TETTENHORST, RODNEY TAMPA, b St Louis, Mo, Feb 1, 34; m 60; c 1. MINERALOGY. Educ: Wash Univ, BS, 55, MA, 57; Univ Ill, Urbana, PhD(mineral), 60. Prof Exp: From asst prof to assoc prof, 60-75, PROF MINERAL, OHIO STATE UNIV, 75- Mem: Clay Minerals Soc; Mineral Soc Am; Mineral Soc Gt Brit & Ireland. Res: Clay mineralogy. Mailing Add: Dept of Mineral & Geol Ohio State Univ Columbus OH 43210

TEUBER, HANS-LUKAS, b Berlin, Ger, Aug 7, 16; nat US; m 41; c 2. PSYCHOPHYSIOLOGY. Educ: Harvard Univ, PhD, 47. Hon Degrees: Dr, Univ Lyon & Univ Geneva, 75. Prof Exp: Asst psychol, Harvard Univ, 41-42; from res assoc to prof psychiat & neurol, Col Med, NY Univ, 47-61; PROF PSYCHOL, CHMN DEPT & DIR PSYCHOPHYSIOL LAB, MASS INST TECHNOL, 61- Concurrent Pos: Dir psychophysiol lab, NY Univ-Bellevue Med Ctr, 47-61; mem biosci subcomt, NASA; consult head injury sect, NIH, 67-; area consult, Vet Admin; Eastman prof, Oxford Univ, 71-72; Killian lectr, Mass Inst Technol, 76-77. Honors & Awards: Karl Spencer Lashley Award, 66. Mem: Nat Acad Sci; Int Brain Res Orgn; Am Neurol Asn; Am Acad Neurol; Asn Res Nerv & Ment Dis. Res: Somatosensory changes after penetrating brain wounds in man; visual field defects after penetrating missile wounds of the brain. Mailing Add: Dept of Psychol Bldg E10-012 Mass Inst of Technol Cambridge MA 02139

TEUFEL, HUGO, JR, mathematics, see 12th edition

TEUFER, GUNTER, inorganic chemistry, see 12th edition

TEUKOLSKY, SAUL ARNO, b Johannesburg, SAfrica, Aug 2, 47; m 71; c 1. THEORETICAL ASTROPHYSICS. Educ: Univ Witwatersrand, BSc Hons(physics) & BSc Hons(appl math), 70; Calif Inst Technol, PhD(physics), 73. Prof Exp: Res assoc physics, Calif Inst Technol, 73-74; ASST PROF PHYSICS, CORNELL UNIV, 74- Concurrent Pos: Alfred P Sloan Found res fel, 75- Mem: Am Phys Soc; Am Astron Soc. Res: General relativity and relativistic astrophysics. Mailing Add: Lab of Nuclear Studies Cornell Univ Ithaca NY 14853

TEUMAC, FRED N, b Little Ferry, NJ, Feb 2, 31; m 57; c 3. CHEMISTRY, SCIENCE ADMINISTRATION. Educ: Rutgers Univ, BS, 52; Univ Fla, MS, 58, PhD(chem), 61. Prof Exp: Asst to chief chemist, Burry Biscuit Co, 54-56; res chemist, Dow Chem Co, 61-62, sr res chemist, 62-65; sr process engr, Fiber Industs, 65-66; vpres & tech dir, Wica Chem Co, NC, 66-68; group mgr res, Plastic & Indust Prod Div, 68-72, DEVELOP MGR COATED FABRICS, UNIROYAL INC, 72- Mem: Am Chem Soc. Res: Fluorine chemistry; practical applications for fluorocarbons, aziridine derivatives, composite structures, solution chemistry with particular interest in chemical cleaning and corrosion; rubber lattices; shoe-making materials; polyurethane coatings, exotic polyurethane coatings. Mailing Add: 1446 Riding Mall South Bend IN 46614

TEUSCHER, GEORGE WILLIAM, b Chicago, Ill, Jan 11, 08; m 34; c 2. DENTISTRY. Educ: Northwestern Univ, DDS, 29, MSD, 36, MA, 40, PhD(educ), 42. Hon Degrees: ScD, NY Univ, 65. Prof Exp: From instr to assoc prof, 31-45, dean, Dent Sch, 53-72, PROF PEDODONTICS, DENT SCH, NORTHWESTERN UNIV, 45- Concurrent Pos: Regent, Nat Libr Med; ed, J Am Soc Dent for Children, 68- Mem: Am Soc Dent for Children; Am Dent Asn; Am Acad Pedodontics; fel Am Col Dent; Int Asn Dent Res. Res: Dental caries; reactions of dental pulp in children; sodium fluoride; prevention in clinical dentistry for children; principles of dental education. Mailing Add: Northwestern Univ Dent Sch 311 E Chicago Ave Chicago IL 60611

TEVEBAUGH, ARTHUR DAVID, b Knox Co, Ind, Nov 25, 17; m 43; c 2. PHYSICAL CHEMISTRY. Educ: Purdue Univ, BS, 40; Iowa State Univ, PhD(phys chem), 47. Prof Exp: Asst chem, Iowa State Univ, 40-42; res chemist, Manhattan Proj, 42-47; res chemist, Knolls Atomic Power Lab, Gen Elec Co, 47-55; supvr reactor chem unit, 50-54, actg mgr chem & chem eng sect, 54-55, sr res chemist, Res & Develop Lab, 55-63; sr chemist & sect mgr chem eng div, 63-69, assoc dir div, 69-72, dir lab prog planning off, 72-73, DIR COAL PROGS, ARGONNE NAT LAB, 73- Mem: AAAS; Am Chem Soc; Am Nuclear Soc; Am Inst Chem Eng. Res: Analytical, radio and soil chemistry; production and handling of fluorine; chemical problems related to development and operation of nuclear reactors and reactor fuel reprocessing; polymer research; thermoelectric materials; fuel cells and batteries; electrochemistry; reactor safety. Mailing Add: 540 W Bonnie Brae Rd Hinsdale IL 60521

TEVETHIA, MARY JUDITH (ROBINSON), b Ft Wayne, Ind, Feb 25, 39; m 65; c 2. MOLECULAR BIOLOGY, GENETICS. Educ: Mich State Univ, BS, 60, MS, 62, PhD(microbiol), 64. Prof Exp: Fel microbiol, Emory Univ, 64-65; fel biol, Univ Tex, M D Anderson Hosp & Tumor Inst, Houston, 65-72, asst biologist & asst prof biol, 72-73; ASST PROF PATH, SCH MED, TUFTS UNIV, 73- Concurrent Pos: NIH fels, 64-65, 65- Mem: Am Soc Microbiol. Res: Genetic studies on simian papova virus SVAO. Mailing Add: Dept of Path Tufts Univ Sch of Med Boston MA 02111

TEVETHIA, SATVIR S, b Buland Shahr, India, Aug 5, 36; m 65; c 1. VIROLOGY, IMMUNOLOGY. Educ: Agra Univ, BSc, 54, BVSc, 58; Mich State Univ, MS, 62, PhD(microbiol), 64. Prof Exp: Vet asst surg, Indian Govt, 58-59; res asst microbiol, Mich State Univ, 60-64; from asst prof to assoc prof virol, Baylor Col Med, 71-73; ASSOC PROF PATH, SCH MED, TUFTS UNIV, 73- Concurrent Pos: Fel virol, Baylor Col Med, 64-66; Nat Cancer Inst res career develop award, 67-71. Mem: AAAS; Am Soc Microbiol; Transplantation Soc; Am Soc Cell Biol; Am Asn Cancer Res. Res: Tumor viruses and immunology. Mailing Add: Dept of Path Tufts Univ Sch of Med Boston MA 02111

TEVIOTDALE, BETH LUISE, b Long Beach, Calif, July 17, 40; div. PHYTOPATHOLOGY. Educ: Pomona Col, BA, 62; Univ Calif, Davis, MS, 70, PhD(plant path), 74. Prof Exp: Lab technician bot, Calif Inst Technol, 62-63; lab technician immunol, Univ Calif, Los Angeles, 65-68; EXTEN SPECIALIST PLANT PATH, SAN JOAQUIN VALLEY RES & EXTEN CTR, UNIV CALIF, 75- Mem: Sigma Xi. Res: Vegetable and tree crops, with major emphasis on disease-free potatoes for seed and deep bark canker of walnuts. Mailing Add: San Joaquin Valley Res & Exten Ctr Univ Calif 9240 S Riverbend Rd Parlier CA 93648

TEW, JOHN GARN, b Mapleton, Utah, Oct 26, 40; m 65; c 4. IMMUNOLOGY, MICROBIOLOGY. Educ: Brigham Young Univ, BS, 66, MS, 67, PhD(microbiol), 70. Prof Exp: NIH fel, Case Western Reserve Univ, 70-72; ASST PROF MICROBIOL, VA COMMONWEALTH UNIV, 72- Mem: Am Soc Microbiol. Res: Role of persisting antigen in the induction maintenance and regulation of the humoral immune response; role of beta-lysin in innate immunity. Mailing Add: Dept of Microbiol Med Col Va Sta PO Box 847 Richmond VA 23298

TEW, RICHARD WILCOX, b Chicago, Ill, May 10, 27; m 50; c 3. MICROBIOLOGY. Educ: Harvard Univ, BA, 49, MA, 51; Univ Wis, PhD(bot, bact), 59. Prof Exp: Res chemist, Prod Res Dept, Oscar Mayer & Co, 53-56; asst prof microbiol, Miss State Univ, 59-60; scientist, Aerojet-Gen Corp, 60-65; biol scientist, US Naval Ord Test Sta, China Lake, 65-68; consult biologist, US Army Advan Materiel Concepts Agency, Washington, DC, 68-70; ASSOC PROF BIOL, UNIV NEV, LAS VEGAS, 70- Mem: Am Soc Microbiol. Res: Microbial ecology; aquatic microbiology; aerobiology. Mailing Add: Dept of Biol Univ of Nev Las Vegas NV 89101

TEW, RONALD KAY, soil science, see 12th edition

TEWARI, PARAM HANS, b Gorakhpur, India, Jan 31, 29; m; c 2. SURFACE CHEMISTRY, COLLOID CHEMISTRY. Educ: Lucknow Univ, India, BS, 50, MS, 52, PhD(chem), 59. Prof Exp: Asst prof chem, Lucknow Univ, India, 58-59; asst prof & reader, Gorakhpur Univ, India, 59-62; res assoc, Univ Md, College Park, 62-65, Univ Southern Calif, 66 & Univ Alta, 67-68; RES OFFICER, ATOMIC ENERGY CAN, WHITESHELL NUCLEAR RES ESTAB, 69- Mem: Am Chem Soc; Chem Inst Can. Res: High temperature chemistry of metal oxides-water interfaces, surface and colloid chemistry of filtration, particle deposition, and corrosion of metal oxides and sulfides. Mailing Add: Atomic Energy of Can Whiteshell Nuclear Res Estab Pinawa MB Can

TEWARSON, REGINALD P, b Pauri, India, Nov 17, 30; m 60; c 2. APPLIED MATHEMATICS, COMPUTER SCIENCE. Educ: Univ Lucknow, BSc, 50; Agra Univ, MSc, 52; Boston Univ, PhD(appl math), 61. Prof Exp: Lectr physics, Messmore Col, India, 50-51; lectr math, Univ Lucknow, 52-57; sr mathematician, Honeywell Inc, 60-64; from asst prof to assoc prof, 64-69, PROF APPL MATH, STATE UNIV NY STONY BROOK, 69- Concurrent Pos: Vis prof, Oxford Univ, 70-71. Mem: Am Math Soc; Soc Indust & Appl Math; Asn Comput Mach. Res: Sparse matrices; linear algebra; numerical analysis. Mailing Add: Dept of Appl Math & Statist State Univ of NY Stony Brook NY 11794

TEWELES, SIDNEY, meteorology, see 12th edition

TEWES, HOWARD ALLAN, b Los Angeles, Calif, May 1, 24; m 53; c 3. NUCLEAR CHEMISTRY. Educ: Univ Calif, Los Angeles, BS, 48, MS, 50, PhD(chem), 52. Prof Exp: Asst chem, Univ Calif, Los Angeles, 49-52; chemist, Calif Res & Develop Co, 52-53; CHEMIST, LAWRENCE LIVERMORE LAB, UNIV CALIF, 53- Mem: AAAS; Am Chem Soc; Am Phys Soc. Res: Proton induced reactions of thorium; spallation reactions; high energy neutron induced nuclear reactions; industrial applications of nuclear explosives; analysis of environmental impacts of advanced energy resource recovery technologies. Mailing Add: Lawrence Livermore Lab PO Box 808 Livermore CA 94550

TEWHEY, JOHN DAVID, b Lewiston, Maine, Feb 14, 43; m 65; c 3. PETROLOGY. Educ: Colby Col, BA, 65; Univ SC, MS, 68; Brown Univ, PhD(geol), 75. Prof Exp: STAFF GEOLOGIST, LAWRENCE LIVERMORE LAB, UNIV CALIF, 74- Concurrent Pos: Consult geologist, Div Geol, SC State Develop Bd, 73- Mem: Geol Soc Am; Am Geophys Union; Soc Mining Engrs. Res: Evaluation of the geochemical controls of mineral equilibria in contact metamorphic environments; regional tectonics of the Southern Appalachians. Mailing Add: Lawrence Livermore Lab Univ of Calif Livermore CA 94550

TEWKSBURY, CHARLES ISAAC, b Portsmouth, NH, Feb 26, 25; m 49; c 3. RUBBER CHEMISTRY. Educ: Univ NH, BS, 48, MS, 49; Princeton Univ, PhD(chem), 53. Prof Exp: Lab asst, Univ NH, 48-49; asst, Princeton Univ, 49-53; phys chemist, Nat Res Corp, 53-54, proj mgr, 54-57, sr chemist, 57-59; res chemist, Monsanto Chem Co, 59; group leader, Cabot Corp, 59-61; asst assoc dir new prod res, 61-63; RES DIR, ODELL CO, 63-; TREAS, FAY SPECIALTIES, INC, 71-, PRES, 74- Mem: Am Chem Soc. Res: Gas kinetics; hydrocarbon oxidation; heterogeneous catalysis; metals; high temperature phenomena; radiation chemistry; polymerization; polymer characterization; elastomers; adhesives. Mailing Add: Odell Co 60 Acton St PO Box 201 Watertown MA 02172

TEWKSBURY, DUANE ALLAN, b Osceola, Wis, Oct 4, 36; m 58; c 2. BIOCHEMISTRY. Educ: St Olaf Col, BA, 58; Univ Wis, MS, 60, PhD(biochem), 64. Prof Exp: RES BIOCHEMIST, MARSHFIELD CLIN FOUND FOR MED RES & EDUC, 64- Concurrent Pos: Wis Heart Asn grant, 67- Mem: Am Chem Soc; NY Acad Sci; Am Fedn Clin Res. Res: Biochemical studies of peptide hormone systems; isolation and characterization of the protein components of the renin-angiotension system. Mailing Add: Marshfield Med Found 510 N St Joseph Ave Marshfield WI 54449

TEWKSBURY, JAMES G, physical chemistry, see 12th edition

TEWS, JEAN KRING, b Ogdensburg, NY, May 21, 28; m 56; c 1. BIOCHEMISTRY. Educ: St Lawrence Univ, BS, 49; Univ Wis, MS, 52, PhD(biochem), 54. Prof Exp: Asst biochem, Univ Wis, 50-54; biochemist, Galesburg State Res Hosp, 54-55; proj assoc physiol, 55-62, res assoc, 63-66, RES ASSOC BIOCHEM, UNIV WIS-MADISON, 67- Mem: Am Inst Nutrit; Am Soc Neurochem. Res: Enzyme activities and nutrition; neurochemistry; factors influencing chemical components of brain; amino acids; amino acid transport. Mailing Add: Dept of Biochem Univ of Wis Madison WI 53706

TEWS, LEONARD L, b Rush Lake, Wis, May 28, 34; m 60; c 3. BOTANY, MYCOLOGY. Educ: Wis State Univ, Oshkosh, BS, 56; Ind Univ, MA, 58; Univ Wis, PhD(bot, mycol), 65. Prof Exp: Teaching asst bot, Ind Univ, 56-58; teacher high sch, 58-61; res asst mycol, Univ Wis, 61-63; from instr to assoc prof, 64-75, PROF BIOL, UNIV WIS-OSHKOSH, 75- Concurrent Pos: Water Resources res grant, Water Resources Ctr, 69-70. Mem: Mycol Soc Am; Brit Mycol Soc; Int Asn Gt Lakes Res. Res: Effects of soil fumigants and fungicides on microfungi of a marsh; microfungi of lakes and cattail marshes; lignin decomposition. Mailing Add: Dept of Biol Univ Wis Oshkosh WI 54901

TEXON, MEYER, b New York, NY, Apr 23, 09; m 41; c 2. MEDICINE. Educ: Harvard Univ, AB, 30; NY Univ, MD, 34; Am Bd Internal Med & Am Bd

Cardiovasc Dis, dipl, 44. Prof Exp: Asst prof, 57-73, ASSOC PROF FORENSIC MED, POST-GRAD MED SCH, NY UNIV, 73- Concurrent Pos: Asst med examr, City of New York, 57-; assoc physician, French Hosp & New York Infirmary; attend physician, Manhattan Eye, Ear & Throat Hosp; consult cardiovasc dis, Bur Hearings & Appeals, Social Security Agency, Dept Health, Educ & Welfare; fel coun clin cardiol & coun atherosclerosis, Am Heart Asn. Honors & Awards: Hektoen Silver Medal, AMA, 58. Mem: AMA; Am Heart Asn; Am Col Physicians; fel Am Col Cardiol; NY Acad Med. Res: Cardiovascular disease; internal medicine; atherosclerosis; hemodynamics; heart disease and industry; role of vascular dynamics in the development of atherosclerosis. Mailing Add: 3 E 68th St New York NY 10021

TEXTER, ELMER CLINTON, JR, b Detroit, Mich, June 12, 23; m 49; c 3. MEDICAL EDUCATION, GASTROENTEROLOGY. Educ: Mich State Univ, BA, 43; Wayne State Univ, MD, 46. Prof Exp: Asst resident, Goldwater Mem Hosp, NY Univ, 50-51; instr, Sch Med, Duke Univ, 51-53; assoc, Med Sch, Northwestern Univ, Chicago, 53-56, asst chief gastroenterol clins, 53-55, dir training prog in gastroenterol, 55-63, from asst prof to assoc prof med, 58-68; chmn dept clin physiol, Olen B Culbertson Res Ctr, Scott & White Clin, Temple, Tex, 68-72; PROF MED PHYSIOL & BIOPHYS, SCH MED & ASST DEAN SCH HEALTH RELATED PROF, MED CTR, UNIV ARK, LITTLE ROCK, 72-, ASSOC CHIEF OF STAFF FOR EDUC & ACTG CHIEF GASTROENTEROL, VET ADMIN HOSP, 72- Concurrent Pos: Res fel med, Col Med, Cornell Univ, 48-50; attend physician, Passavant Mem Hosp, 53-68; ed, Am J Digestive Dis, 56-68; chief gastroenterol sect, Vet Admin Res Hosp, 59-63; consult, US Naval Hosp, Great Lakes, 63-68; attend physician, Cook County Hosp, Ill, 66-68; mem sr staff, Scott & White Mem Hosp, Temple Univ, 68-72; consult, Santa Fe Hosp, Vet Admin Ctr, Temple, 68-72; William Beaumont Hosp, Dept Army, El Paso, 68-, consult to Surgeon Gen, US Army, 70; lectr, Univ Fla, 69; adj prof physiol, Univ Tex Southwestern Med Sch Dallas, 69-72; fac coordr allied health training prog, Temple Jr Col, 69-72; attend physician, Univ Hosp, Univ Ark, Little Rock, 72- Mem: Am Physiol Soc; Am Gastroenterol Asn; fel Am Col Physicians; Am Fedn Clin Res; Am Med Writers' Asn (pres-elect, 72-). Res: Gastrointestinal physiology and pathophysiology; health care delivery systems. Mailing Add: Vet Admin Hosp Univ of Ark Med Ctr Little Rock AR 72206

TEXTOR, ROBERT BAYARD, b Cloquet, Minn, Mar 13, 23; m 67. CULTURAL ANTHROPOLOGY. Educ: Univ Mich, BA, 45; Cornell Univ, PhD, 60. Prof Exp: Civilian officer, Info & Educ, Mil Govt Allied Occup, Japan, 46-48; cultural anthrop consult, US Int Coop Admin, Thailand, 57-58; res fel & res assoc anthrop & Southeast Asia studies prog, Yale Univ, 59-61; res fel statist, Harvard Univ, 62-64; assoc prof, 64-68, PROF EDUC & ANTHROP, STANFORD UNIV, 68- Concurrent Pos: Consult, Peace Corps, 61-62. Mem: Fel Am Anthrop Asn; Am Educ Res Asn; Asn Asian Studies; Soc Appl Anthrop; Siam Soc. Res: Religion, magic and divination; educational and political anthropology; community development; cross-national and cross-cultural analysis; statistical and formal methods of ethnography; anthropology of Japan, Thailand and Southeast Asia. Mailing Add: Dept of Anthrop Stanford Univ Stanford CA 94305

TEXTOR, ROBIN EDWARD, b Detroit, Mich, Mar 19, 43; m 65; c 2. MATHEMATICS. Educ: Tenn Polytech Inst, BS, 64; Univ Tenn, Knoxville, MS, 68; Drexel Univ, PhD(math), 72. Prof Exp: Comput appln programmer, Comput Technol Ctr, Union Carbide Corp, Tenn, 64-69; asst prof math, Univ SC, 71-72; COMPUT APPLN ANALYST, OAK RIDGE NAT LAB, UNION CARBIDE CORP, 72-, SECT HEAD, 73- Mem: Math Asn Am; Am Math Soc; Soc Indust & Appl Math. Res: Singular hyperbolic partial differential equations; numerical solution of fluid flow problems and partial differential equations. Mailing Add: Comput Sci Div Union Carbide Corp PO Box P Oak Ridge TN 31830

TEXTORIS, DANIEL ANDREW, b Cleveland, Ohio, Jan 19, 36; m 59; c 3. GEOLOGY. Educ: Case Western Reserve Univ, BA, 58; Ohio State Univ, MS, 60; Univ Ill, PhD(geol), 63. Prof Exp: Asst prof geol, Univ Ill, 63-65; from asst prof to assoc prof, 65-73, asst dean res admin, 68-74, PROF GEOL, UNIV NC, CHAPEL HILL, 73-, ASSOC DEAN RES ADMIN, 74- Concurrent Pos: Geologist, Diamond Alkali Co, 57-60; consult, Southern Ill-Pa Coal, 61-65; coord NSF sci develop prog, 67-74; consult, Carbonate Dredging, 68-70. Mem: AAAS; Geol Soc Am; Soc Econ Paleont & Mineral; Am Asn Petrol Geol; Nat Coun Univ Res Adminrs. Res: Sedimentary geology; carbonate petrography; diagenesis of sediments; petrology and geochemistry of volcanic tuff; interpretation of ancient carbonate environments; paleoecology. Mailing Add: Dept of Geol Univ of NC Chapel Hill NC 27514

TEYLER, TIMOTHY JAMES, b Portland, Ore, Nov 25, 42; m 66; c 1. NEUROSCIENCES. Educ: Ore State Univ, BS, 64; Univ Ore, MS, 68, PhD(psychol), 69. Prof Exp: Asst prof psychol, Univ Southern Calif, 68-69; lectr psychobiol, Univ Calif, Irvine, 69-72, assoc res psychobiologist, 73-74; lectr, 74-75, ASSOC PROF PSYCHOL, HARVARD UNIV, 75- Concurrent Pos: NIMH fel, Univ Calif, Irvine, 70-72; NSF sr fel & Fulbright scholar, Inst Neurophysiol, Univ Oslo, 73-74. Mem: Soc Neurosci; Psychonomic Soc. Res: Neurobiological correlates of behavioral plasticity; neurolinguistics; magnetoencephalography. Mailing Add: Dept of Psychol & Social Rels Harvard Univ Cambridge MA 02138

THACHER, HENRY CLARKE, JR, b New York, NY, Aug 8, 18; m 42; c 5. COMPUTER SCIENCE, NUMERICAL ANALYSIS. Educ: Yale Univ, AB, 40; Harvard Univ, MA, 42; Yale Univ, PhD(phys chem), 49. Prof Exp: Instr chem, Yale Univ, 46-48; asst prof, Ind Univ, 49-54; task scientist, Aeronaut Res Lab, Wright Air Develop Ctr, US Air Force, Ohio, 54-58; assoc chemist, Argonne Nat Lab, 58-66; prof comput sci, Univ Notre Dame, 66-71; PROF COMPUT SCI, UNIV KY, 71- Concurrent Pos: Consult, Argonne Nat Lab, 66- Mem: AAAS. Res: Numerical approximation and computer programming; computation and approximation of special functions. Mailing Add: Dept of Comput Sci Univ of Ky Lexington KY 40506

THACHER, PHILIP DURYEA, b Palo Alto, Calif, Jan 13, 37; m 63; c 2. SOLID STATE PHYSICS. Educ: Calif Inst Technol, BS, 58; Cornell Univ, PhD(physics), 65. Prof Exp: STAFF MEM, SANDIA LABS, 65- Mem: AAAS; Am Phys Soc; Inst Elec & Electronics Eng. Res: Optical effects in ferroelectric ceramics and crystals; laser energy; radiometry. Mailing Add: Div 9532 Sandia Labs Albuquerque NM 87115

THACHET, THOMAS, biochemistry, see 12th edition

THACKER, EDWARD JESSE, b Concord, NH, Aug 29, 15; m 40; c 3. NUTRITION. Educ: Univ Mass, BS, 37; Pa State Univ, MS, 42; Cornell Univ, PhD(animal nutrit), 54. Prof Exp: Asst & instr animal nutrit, Pa State Univ, 37-43; biochemist plant soil & nutrit lab, 46-59, agr adminstr agr res serv, 59-64, DIR & INVESTS LEADER, METAB & RADIATION RES LAB, USDA, 64- Mem: Fel AAAS; Am Chem Soc; Am Soc Animal Sci; Am Inst Nutrit. Res: Energy, mineral, pesticide metabolism; interrelationships of plant, soil and animal nutrition. Mailing Add: Metab & Radiation Res Lab USDA State Univ Sta Fargo ND 58102

THACKER, JOHN CHARLES, b Clinton, Okla, Oct 29, 43; m 68; c 2. STATISTICAL ANALYSIS. Educ: Cornell Univ, BS, 66; Brown Univ, PhD(appl math), 74. Prof Exp: MEM TECH STAFF STATIST, AEROSPACE CORP, 74- Mem: Inst Math Statist; Am Statist Asn; Soc Indust & Appl Math. Res: Statistical inference on stochastic processes; time series analysis; stochastic point processes; application of statistical techniques to air and water pollution problems. Mailing Add: Aerospace Corp PO Box 92957 Los Angeles CA 90009

THACKER, RAYMOND, b Ashton-U-Lyne, UK, May 9, 32. ELECTROCHEMISTRY, PHYSICAL CHEMISTRY. Educ: Univ Manchester, BSc, 52, MSc, 53, PhD(phys chem), 55. Prof Exp: Sci officer, UK Atomic Energy Auth Indust Group, 55-58; res assoc electrode kinetics, Univ Pa, 58-60; SR RES CHEMIST, RES LABS, GEN MOTORS CORP, 60- Mem: Am Chem Soc; Electrochem Soc; assoc Royal Inst Chem. Res: Thermodynamic properties of nonelectrolyte solutions; electrochemistry of surfaces; fuel cell electrode processes; batteries. Mailing Add: Gen Motors Corp Res Labs 12 Mile & Mound Rds Warren MI 48090

THACKER, WILLIAM CARLISLE, b Albany, Ga, May 11, 44. FLUID DYNAMICS. Educ: Ga Inst Technol, BS, 65; Univ Ill, MS, 67, PhD(physics), 71. Prof Exp: Fel meteorol, Univ Chicago & Nat Ctr Atmospheric Res, 71-72; resident res assoc oceanog, Nat Res Coun-Nat Oceanic & Atmospheric Admin, 72-74; RES PHYSICIST, SEA AIR INTERACTION LAB, NAT OCEANIC & ATMOSPHERIC ADMIN, DEPT COM, 74- Mem: Am Geophys Union. Res: Oceanic and atmospheric circulation and turbulent mixing. Mailing Add: NOAA/AOML Sea Air Interaction Lab 15 Rickenbacker Causeway Miami FL 33149

THACKERAY, ERNEST RUSSEL, b Can, Oct 6, 06; m 35; c 1. PHYSICS. Educ: Univ Sask, BA, 27, MA, 29; Univ Wis, PhD(physics), 49. Prof Exp: From instr to assoc prof physics, Regina Col, 29-49; assoc prof, 49-50, prof & head dept, 51-66, distinguished prof, 66-74, DISTINGUISHED EMER PROF PHYSICS, UNIV AKRON, 74- Mem: Am Phys Soc; Optical Soc Am. Res: Spectroscopy; absorption spectrum of atomic sodium. Mailing Add: Dept of Physics Univ of Akron Akron OH 44304

THADDEUS, PATRICK, b June 6, 32; US citizen; m 63; c 2. PHYSICS, ASTROPHYSICS. Educ: Univ Del, BSc, 53; Oxford Univ, MA, 55; Columbia Univ, PhD(physics), 60. Prof Exp: Res physicist, Radiation Lab, Columbia Univ, 60-61; Nat Acad Sci fel astrophys, 61-64, RES PHYSICIST, GODDARD INST, SPACE STUDIES, 64- Concurrent Pos: Adj asst prof, Columbia Univ, 64-66, adj prof, 71-; adj assoc prof, NY Univ, 63-; vis assoc prof, State Univ NY Stony Brook, 66- Honors & Awards: Outstanding Achievement Medal, NASA, 70. Mem: Am Phys Soc; Am Astron Soc. Res: Radio and optical astronomy; interstellar molecules; microwave spectroscopy; masers. Mailing Add: Inst Space Studies 2880 Broadway New York NY 10025

THAELER, CHARLES SCHROPP, JR, b Philadelphia, Pa, Jan 9, 32; m 57; c 3. ZOOLOGY. Educ: Earlham Col, AB, 54; Univ Calif, Berkeley, MA, 60, PhD(zool), 64. Prof Exp: Actg instr zool, Univ Calif, Berkeley, 63-64, actg asst cur, Mus Vert Zool, 63-64; asst prof zool, South Bend Campus, Ind Univ, 64-66; from asst prof to assoc prof, 66-74, PROF BIOL, NMEX STATE UNIV, 74- Concurrent Pos: NSF res grants, 68-70 & 72-74. Mem: Fel AAAS; Am Soc Mammal; Soc Study Evolution; Ecol Soc Am; Soc Syst Zool. Res: Mammalian systematics and cytogenetics; evolution and ecology, especially taxonomy; evolution, cytogenetics and distribution of geomyids; taxonomy of microtine rodents. Mailing Add: Dept of Biol NMex State Univ Las Cruces NM 88001

THAEMERT, JONA CARL, b Sylvan Grove, Kans, May 25, 22; m 47; c 2. ANATOMY. Educ: Univ Denver, BS, 50, MA, 52; Univ Colo, PhD(anat), 59. Prof Exp: Instr zool, Univ Denver, 50-52; instr biol, Ft Lewis Col, 52-55; asst prof anat, Sch Med, Univ Mo, 59-61; asst prof, Univ Colo, 61-66; head sect electron micros, Congenital Heart Dis Res & Training Ctr, Hektoen Inst Med Res, 66-74; ASSOC PROF ANAT, CHICAGO MED SCH, 74- Concurrent Pos: Assoc prof anat, Sch Med, Northwestern Univ, Chicago, 66-71. Mem: Sigma Xi; Am Asn Anat; Am Soc Cell Biol; Electron Micros Soc Am. Res: Cytology of muscle and nerve; electron microscopy. Mailing Add: Dept of Anat Chicago Med Sch Chicago IL 60612

THAKKAR, ARVIND LAVJI, b Karachi, WPakistan, Apr 19, 39; m 73. PHYSICAL PHARMACY. Educ: Univ Bombay, BPharm, 61; Columbia Univ, MS, 64; Univ Wash, PhD(phys pharm), 67. Prof Exp: Teaching asst col pharm, Columbia Univ, 61-63; col pharm, Univ Wash, 64-66; sr pharmaceut chemist, 67-75, RESEARCH SCIENTIST, RES LABS, ELI LILLY & CO, 75- Mem: Am Pharmaceut Asn; Acad Pharmaceut Sci; NY Acad Sci. Res: Surface activity of drugs; micellar solubilization; complex formation; polymorphism; solubility. Mailing Add: Lilly Res Labs Indianapolis IN 46206

THAL, ALAN PHILIP, b Cape Town, SAfrica, July 15, 25; nat US; m 49; c 2. SURGERY, EXPERIMENTAL MEDICINE. Educ: Univ Cape Town, MB & ChB, 49; Univ Minn, PhD, 56; Am Bd Surg, dipl, 57; Am Bd Thoracic Surg, dipl, 61. Prof Exp: Intern surg, Johns Hopkins Hosp, 51-52; res asst & resident, Univ Hosps, Univ Minn, 52-56, from asst prof to assoc prof, 56-61; prof & chmn dept, Med Col, Wayne State Univ, 61-66; PROF SURG, SCH MED, UNIV KANS, 66- Concurrent Pos: Estab investr, Am Heart Asn, 56-61. Mem: Am Soc Exp Path; Soc Univ Surg; Am Heart Asn; fel Am Col Surg; Am Surg Asn. Res: Biologic effects of bacterial toxins; gastrointestinal physiology, particularly the pancreas; coronary sclerosis; shock. Mailing Add: Dept of Surg Univ of Kans Med Ctr Kansas City KS 66103

THALACKER, VICTOR PAUL, b Muscatine, Iowa, Apr 21, 41; m 70; c 2. ORGANIC CHEMISTRY. Educ: Wis State Univ-Stevens Point, BS, 63; Univ Ariz, PhD(org chem), 68. Prof Exp: SR CHEMIST ORG CHEM, 3M CO, 67- Mem: Am Chem Soc. Res: Insect attractants; natural products; pheromones; polymer processing; water based adhesives; plant culture. Mailing Add: 3M Co Bldg 209-BC 3M Ctr St Paul MN 55119

THALE, THOMAS RICHARD, b Indianapolis, Ind, June 4, 15; m 41; c 5. PSYCHIATRY. Educ: Loyola Univ Chicago, BSM, 38, MD, 39. Prof Exp: Resident psychiat, Manteno State Hosp, Ill, 39 & Norwich State Hosp, Conn, 39-43; instr psychol, Univ Conn, 42-45; instr psychiat & pub health, Yale Univ, 43-45; instr psychiat, Wash Univ, 45-50; from instr to assoc prof, 50-69, PROF PSYCHIAT, MED SCH & ASSOC PROF SOCIAL WORK, SCH SOCIAL SERV, ST LOUIS UNIV, 70- Concurrent Pos: Med adv, Bur Hearings & Appeals, Social Security Admin, 67-; consult, Family & Childrens Soc Greater St Louis. Mem: AMA; fel Am Psychiat Asn. Res: Psychological evaluation chemotherapy. Mailing Add: Dept Psychiat St Louis Univ Med Sch 1221 S Grand Blvd St Louis MO 63104

THALER, ALVIN ISAAC, b New York, NY, Aug 26, 38; m 59; c 2. ALGEBRA. Educ: Columbia Univ, AB, 59; Johns Hopkins Univ, PhD(math), 66. Prof Exp: Assoc math appl physics lab, Johns Hopkins Univ, 62-64; instr, Col Notre Dame, Md, 64-66; asst prof, Univ Md, 66-71; PROG DIR ALGEBRA, NSF, 71- Mem: Am Math Soc;

Math Asn Am. Res: Algebraic number theory; algebraic geometry. Mailing Add: Math Sci Sect NSF Washington DC 20550

THALER, JON JACOB, b Richland, Wash, Feb 3, 47; m 68; c 1. EXPERIMENTAL HIGH ENERGY PHYSICS. Educ: Columbia Univ, BA, 67, MA, 69, PhD(physics), 72. Prof Exp: Instr, 71-75, ASST PROF PHYSICS, PRINCETON UNIV, 75- Mem: Am Phys Soc. Res: Investigation of high energy processes as tests of the Quark-Parton model, scaling, and the possible existence of new quantum numbers. Mailing Add: Dept of Physics Princeton Univ Princeton NJ 08540

THALER, M MICHAEL, b Brzezany, Poland, Sept 29, 34; Can citizen; m 66; c 2. PEDIATRICS, DEVELOPMENTAL BIOLOGY. Educ: Univ Toronto, MD, 58. Prof Exp: Intern, Mt Zion Hosp, San Francisco, 58-59; jr pediat resident, Children's Hosp, Detroit, 59-60; sr resident, Boston City Hosp, Mass, 60-61; asst resident path, Hosp for Sick Children, Toronto, 61-62; res fel pediat path, Univ Toronto, 62; vis resident, Hosp St Antoine, Paris, 63; res fel path chem, Hosp for Sick Children, Toronto, 63-65; res fel develop biochem, Harvard Med Sch, 65-67; from instr to asst prof, 67-72, ASSOC PROF PEDIAT, SCH MED, UNIV CALIF, SAN FRANCISCO, 71- Concurrent Pos: Vis scientist, Wash Univ, 64; Josiah Macy Jr Found fac scholar, 74-75; vis scientist, Weizmann Inst Sci, Israel. Mem: Soc Pediat Res; Am Soc Clin Invest; Am Soc Biol Chemists; Am Asn Study Liver Dis; Int Asn Study Liver. Res: Liver disease of newborn infants; perinatal and developmental aspects of hepatic metabolism; bilirubin metabolism; pediatric agstrointestinal function and pathology. Mailing Add: Dept of Pediat Univ of Calif Med Sch San Francisco CA 94122

THALER, OTTO FELIX, b Vienna, Austria, June 17, 23; US citizen; m 47; c 3. PSYCHIATRY, PSYCHOANALYSIS. Educ: Univ Rochester, MD, 49; State Univ NY Downstate Med Ctr, cert psychoanal, 66. Prof Exp: PROF PSYCHIAT, SCH MED, UNIV ROCHESTER, 55- Mem: Am Psychiat Asn. Mailing Add: Dept of Psychiat Univ of Rochester Sch of Med Rochester NY 14642

THALER, RAPHAEL MORTON, b Brooklyn, NY, May 19, 25; m 52; c 4. THEORETICAL PHYSICS. Educ: NY Univ, AB, 47; Brown Univ, ScM, 49, PhD(physics), 50. Prof Exp: Res assoc theoret physics, Yale Univ, 50-52; mem staff, Los Alamos Sci Lab, 53-60; assoc prof, 60-64, PROF PHYSICS, CASE WESTERN RESERVE UNIV, 64-, VCHMN DEPT, 67- Concurrent Pos: Res assoc, Mass Inst Technol, 57-58; consult, Argonne Nat Lab & Lewis Res Lab, NASA. Mem: Fel Am Phys Soc; Ital Phys Soc. Res: Nuclear and elementary particle physics. Mailing Add: Dept of Physics Case Western Reserve Univ University Circle Cleveland OH 44106

THALER, WARREN ALAN, b New York, NY, Jan 7, 34; m 56; c 1. ORGANIC CHEMISTRY. Educ: City Col New York, BS, 56; Columbia Univ, MA, 58, PhD(chem), 61. Prof Exp: RES ASSOC CORP RES LAB, EXXON RES & ENG CO, 60- Mem: Am Chem Soc. Res: Free radical chemistry; sulfur chemistry; additive substitution reactions; stereochemistry of free radical reactions; cationic polymerization; elastomer chemistry. Mailing Add: Corp Res Lab Exxon Res & Eng Co Linden NJ 07036

THALER, WILLIAM JOHN, b Baltimore, Md, Dec 4, 25; m 51; c 2. PHYSICS. Educ: Loyola Col, BS, 47; Cath Univ, MS, 49, PhD(physics), 51. Prof Exp: Physicist, Baird Assocs, Inc, 47; instr physics, Cath Univ, 47-48, asst, 48-51; PHYSICIST, US OFF NAVAL RES, 51-; PROF PHYSICS, GEORGETOWN UNIV, 60- Mem: Am Phys Soc; Acoust Soc Am. Res: Ultrasonic studies of relaxation phenomena in gases; propagation of shock waves in liquids, solids and gases; effects of ultrasonics on biological media; laser research. Mailing Add: 5532 Summit St Centreville VA 22020

THALGOTT, FRED WILLIAM, b Zeigler, Ill, July 27, 15; m 51; c 6. MATHEMATICS. Educ: Univ Southern Ill, BEd, 37; Univ Ill, MS, 41, BS, 47. Prof Exp: Teacher high sch, Ill, 37-41; assoc mech engr, Oak Ridge Nat Lab, 47-48; assoc mech engr, 48-52, head nuclear eng sect reactor eng div, 52-55, assoc dir Idaho div, 56-58, dep dir reactor physics div, 68-70, SR PHYSICIST, ARGONNE NAT LAB, 55-, DEP DIR APPL PHYSICS DIV, 70- Mem: Am Math Soc; Am Phys Soc; fel Am Nuclear Soc. Res: Reactor theory and design; kinetic behavior of nuclear reactors. Mailing Add: Argonne Nat Lab PO Box 2528 Idaho Falls ID 83401

THALL, PETER FRANCIS, b Stillwater, Okla, Aug 5, 49. MATHEMATICAL STATISTICS. Educ: Mich State Univ, BS, 71; Fla State Univ, MS, 73, PhD(statist & probability), 75. Prof Exp: Biomet trainee statist, Fla State Univ, 71-75; ASST PROF STATIST, UNIV TEX, DALLAS, 75- Mem: Am Statist Asn. Res: Robust statistical inference; probabilistic information theory; stochastic point processes and dependence structure; clustering models. Mailing Add: Grad Prog Math Sci Univ Tex at Dallas Box 688 Richardson TX 75080

THALMANN, HANS ERNST, paleontology, deceased

THALMANN, ROBERT H, b San Antonio, Tex, Nov 14, 39. ANATOMY, PSYCHOLOGY. Educ: Univ Tex, BA, 61; Univ Mich, MA, 64, PhD(psychol), 67. Prof Exp: USPHS fel, Emory Univ, 67-69; ASST PROF ANAT, BAYLOR COL MED, 69- Mem: AAAS. Res: Neural basis of appetitive behavior, such as feeding, drinking and self stimulation. Mailing Add: Dept of Cell Biol Baylor Col of Med Houston TX 77025

THAMES, HOWARD DAVIS, JR, b Monroe, La, Aug 3, 41; m 66; c 2. BIOMATHEMATICS. Educ: Rice Univ, BA, 63, PhD(chem), 70. Prof Exp: Proj investr, 70-71, ASST PROF BIOMATH, UNIV TEX M D ANDERSON HOSP & TUMOR INST, HOUSTON, 71- Mem: Am Asn Physicists in Med. Res: Kinetics and interactions of reaction systems inside mitochondria, or affecting it externally, and how they determine its functions. Mailing Add: Dept of Biomath M D Anderson Hosp & Tumor Inst Houston TX 77025

THAMES, JOHN LONG, b Richmond, Va, Sept 29, 24; m 48; c 4. WATERSHED MANAGEMENT. Educ: Univ Fla, BSF, 50; Univ Miss, MS, 59; Univ Ariz, PhD, 66. Prof Exp: Res forester, US Forest Serv, 50-67; assoc prof watershed hydrol, 67-69, PROF WATERSHED HYDROL, UNIV ARIZ, 69-, PROG CHMN WATERSHED HYDROL, 75- Concurrent Pos: Coordr, Int Biol Prog; consult, Argonne Nat Lab, Am Smelting & Refining Co & Shelly Loy. Mem: Soil Sci Soc Am; Soc Am Foresters; Sigma Xi; Am Geophys Union. Res: Plant-soil-water relations; hydrologic modeling; decision analyses; hydrology of surface mined lands. Mailing Add: Sch of Renewable Natural Resources Univ of Ariz Tucson AZ 85721

THAMES, SHELBY FRELAND, b Hattiesburg, Miss, Aug 10, 36; m 54; c 3. POLYMER CHEMISTRY, ORGANIC CHEMISTRY. Educ: Univ Southern Miss, BS, 59, MS, 61; Univ Tenn, PhD(org chem), 64. Prof Exp: From instr to assoc prof chem, 60-70, dean col sci, 71-74, PROF POLYMER SCI, UNIV SOUTHERN MISS, 70-, DEAN COL SCI & TECHNOL, 74-; DIR SOUTHERN INST SURFACE COATINGS, 68- Concurrent Pos: Res grants, Walter Reed Inst Res, 64-68, Diamond-Shamrock Corp, 66-68, Inst Copper Res Asn, NASA, Masonite Corp & Stand Paint & Varnish Co; Petrol Res Fund fel award, 68-69. Mem: Am Chem Soc; Am Inst Chem. Res: Organosilicon, synthetic organic and organometallic chemistry. Mailing Add: Box 5165 Southern Sta Univ of Southern Miss Hattiesburg MS 39401

THAMES, WALTER HENDRIX, JR, b Richmond, Va, July 29, 18; m 43; c 2. NEMATOLOGY. Educ: Univ Fla, BSA, 47, MS, 48, PhD, 59. Prof Exp: Asst entomologist, Everglades Exp Sta, Univ Fla, 48-55, asst soil microbiol, 57-59; assoc prof, 59-68, PROF PLANT PATH & PHYSIOL, TEX A&M UNIV, 68- Mem: Am Phytopath Soc; Soc Syst Zool; Soc Nematol. Res: Biology and control of nematodes; plant nematology. Mailing Add: Dept of Plant Sci Tex A&M Univ College Station TX 77843

THAMM, RICHARD C, JR, b Danville, Va, Jan 6, 30; m 67; c 2. ORGANIC CHEMISTRY. Educ: Wash State Univ, BS, 53; Univ Ill, PhD(org chem), 56. Prof Exp: CHEMIST ELASTOMER CHEM DEPT, E I DU PONT DE NEMOURS & CO, INC, 56- Mem: Am Chem Soc. Res: Organic chemistry as related to synthetic elastomers. Mailing Add: 1211 Bruce Rd Carrcroft Wilmington DE 19803

THAMPI, NAGENDRAN SANKARANARAYANAN, b Trivandrum, India, Feb 9, 31; m 70; c 1. BIOCHEMISTRY, PHARMACOLOGY. Educ: Univ Col, Trivandrum, India, BSc, 50; Birla Inst Technol, India, BPharm, 53; Banaras Hindu Univ, MPharm, 56; Tulane Univ La, PhD(biochem), 68. Prof Exp: Lectr med chem, Maharaja's Col, India, 53-54; asst prof, Sagar Univ, 56-59; asst prof, Banaras Hindu Univ, 59-62; res assoc biochem & Drexel Univ, 68-70; clin chemist, Grad Hosp, Univ Pa, 70-71; DIR CLIN LABS & BIOCHEM RES, NORRISTOWN STATE HOSP, 71- Concurrent Pos: Assoc path, Sch Med, Univ Pa, 70-72; adj prof biochem, Drexel Univ, Philadelphia, 75. Mem: Am Chem Soc; Soc Neurosci; The Chem Soc; NY Acad Sci; Am Soc Clin Chem. Res: Drug design, drug metabolism, steroids, vasoactive peptides, clinical enzymology, cyclic nucleotides, neurochemistry. Mailing Add: Res Labs Norristown State Hosp Norristown PA 19401

THANASSI, JOHN WALTER, b St Louis, Mo, Oct 2, 37; m 64; c 3. BIOCHEMISTRY. Educ: Lafayette Col, BA, 59; Yale Univ, PhD(biochem), 63. Prof Exp: Fel chem, Cornell Univ, 63-64; fel, Univ Calif, Santa Barbara, 64-65; staff fel, Lab Chem, NIH, 65-67; asst prof, 67-72, ASSOC PROF BIOCHEM, COL MED, UNIV VT, 72- Concurrent Pos: USPHS res grants, 68-71 & 72- Mem: AAAS; Am Chem Soc; Am Soc Biol Chem. Res: Enzyme mechanisms. Mailing Add: Dept of Biochem Univ of Vt Col of Med Burlington VT 05401

THANIGASALAM, KANDIAH, b Jaffna, Ceylon, Oct 26, 39. NUMBER THEORY. Educ: Univ Ceylon, BS, 60; Univ London, MS, 64; Pa State Univ, PhD(math), 70. Prof Exp: Asst lectr math, Univ Ceylon, 60-62; lectr math, Lanchester Col Technol, Eng, 65-67; sr lectr math, Constantine Col Technol, Eng, 67-68; asst prof math, Fordham Univ, 70-71; ASST PROF MATH, PA STATE UNIV, BEAVER, 71- Res: Analytic theory of numbers, with special interest in Waring's problem and its generalizations. Mailing Add: Dept of Math Beaver Campus Pa State Univ Monaca PA 15061

THANOS, ANDREW, b Canton, Ohio, June 12, 22; m 43; c 2. MICROBIOLOGY, PLANT PATHOLOGY. Educ: Kent State Univ, AB, 47; Univ WVa, MS, 48; Mich State Univ, PhD(plant path, mycol, bact), 52. Prof Exp: Teaching asst bot mycol & plant path, Mich State Univ, 48-50, instr bot & plant path, 50-51; sr res microbiologist, Eli Lilly & Co, 52-61; head microbiol, NAm Aviation, Inc, 62-67; dir microbiol, RPC Corp, 67-73; pres, Lume Corp, 73-74; OWNER, POLYCHROME SPEC, 74- Concurrent Pos: Lectr, Butler Univ, 58-60. Mem: AAAS; Am Soc Microbiol; Soc Indust Microbiol; Mycol Soc Am. Res: Development of rapid biochemical detection systems utilizing microorganisms as sensors; microbiological deterioration of materials; microbiological desalination of sea water; fungal physiology; marine microbiology; rapid detection of air-borne pathogens. Mailing Add: Polychrome Spec 15214 Grevillea Ave Lawndale CA 90260

THAPAR, MANGAT RAI, b Khanna, India, Oct 4, 39; m 68; c 1. SEISMOLOGY, GEOPHYSICS. Educ: Indian Sch Mines, Dhanbad, BSc, 61, MSc & AISM, 62; Univ Western Ont, PhD(geophys), 68. Prof Exp: Sr sci officer seismol, Coun Sci & Indust Res, New Delhi, 62-64; teaching fel seismol & geophys, Univ Western Ont, 68-71; NSF res fel seismol, Univ Pittsburgh, 71-72; chief res engr, Seismograph Serv Corp, 72; analyst seismol, Phillips Petrol Co, 72-74; RES GEOPHYSICIST, CITIES SERV OIL CO, 74- Concurrent Pos: Lectr, Univ Pittsburgh, 71-72. Mem: Am Geophys Union; Soc Exp Geophys; Indian Soc Earthquake Technol. Res: Applied geophysics; experimental, theoretical, model earthquake and lunar seismology; digital data processing techniques; geophysical exploration techniques; seismic interpretation research; seismic wave propagation; 2-dimensional digital seismic modeling. Mailing Add: Cities Serv Oil Co Explor Prod Res PO Box 50408 Tulsa OK 74150

THARIN, JAMES COTTER, b West Palm Beach, Fla, Mar 22, 31; m 55; c 2. GEOLOGY. Educ: St Joseph's Col, Ind, BS, 54; Univ Ill, Urbana, MS, 58, PhD(geol), 60. Prof Exp: Instr phys sci, Univ Ill, Urbana, 59-61; petrol geologist, Chevron Oil Co, 61-63; asst prof geol, Wesleyan Univ, 63-67; assoc prof & head dept, 67-74, PROF GEOL & CHMN DEPT, HOPE COL, 74- Mem: AAAS; Geol Soc Am; Soc Econ Geol; Soc Econ Paleont & Mineral. Res: Sedimentation; Pleistocene geology; glacial geology and clay mineralogy; textural studies of glacial drift in northwestern Pennsylvania, Calgary area, Alberta and Connecticut Valley; petroleum exploration in Mississippi Delta. Mailing Add: Dept of Geol Hope Col Holland MI 49423

THARP, A G, b Franklinton, Ky, Jan 6, 27. INORGANIC CHEMISTRY. Educ: Univ Ky, BS, 51; Purdue Univ, PhD, 57. Prof Exp: Asst, Purdue Univ, 51-54; asst res engr, Univ Calif, 54-55, res engr, 59; res chemist, Oak Ridge Nat Lab, 56-59; from asst prof to prof inorg chem, 59-74, PROF CHEM, CALIF STATE UNIV, LONG BEACH, 74- Mem: Am Chem Soc. Res: Structural investigations of high melting silicides, germanides and carbides; thermodynamic properties of high melting inorganic compounds. Mailing Add: Dept of Chem Calif State Univ 6101 E Seventh St Long Beach CA 90801

THARP, GERALD D, b Wahoo, Nebr, Aug 9, 32; m 57; c 4. VERTEBRATE PHYSIOLOGY. Educ: Univ Nebr, BS, 58, MS, 61; Univ Iowa, PhD(physiol), 65. Prof Exp: Instr physiol, Univ Iowa, 64-65; asst prof, Wis State Univ, Oshkosh, 65-67; asst prof, 67-72, ASSOC PROF PHYSIOL, UNIV NEBR, LINCOLN, 72- Mem: Am Physiol Soc; Am Col Sports Med. Res: Exercise and stress physiology especially the effects of training. Mailing Add: Sch of Life Sci Univ of Nebr Lincoln NE 68588

THARP, VERNON LANCE, b Hemlock, Ohio, Mar 13, 17; m 40; c 6. VETERINARY MEDICINE. Educ: Ohio State Univ, DVM, 40. Prof Exp: Field vet, US Bur Animal Indust, 40-42; instr vet surg & clin, 42-47, from asst prof to assoc prof vet med, 47-55, dir vet clin, 47-71, chmn dept vet clin sci, 60-71, dir equine res ctr, 65-71, dir food animal res ctr, 70-71, PROF VET MED, COL VET MED, OHIO STATE UNIV, 55-; ASSOC DEAN, 72- Concurrent Pos: Harness Tracks Am, NY Racing Asn & New York Jockey Club grants equine res col vet med, Ohio State Univ, 65-72. Mem: Am Vet Soc Study Breeding Soundness; Am Asn Bovine Practitioners; Am Vet Med Asn; Am Asn Vet Clinicians (secy-treas, 62). Res: Funding, administration and performance of research in environmental health, nutrition and diseases of domestic and laboratory animals. Mailing Add: Col of Vet Med Ohio State Univ 1900 Coffey Rd Columbus OH 43210

THATCHER, EVERETT WHITING, b Jefferson, Ohio, Jan 24, 04; m 28; c 2. PHYSICS. Educ: Oberlin Col, AB, 26, AM, 27; Univ Mich, PhD(physics), 31. Prof Exp: Asst physics, Purdue Univ, 26-27; instr, Univ Nebr, 27-29; instr, Univ Mich, 29-31; from asst prof to assoc prof, Union Univ, NY, 31-46, instr aerodyn, meteorol & radio civilian pilot training prog, 39-43, coordr, 40-43, instr electronics eng defense training, 40-41; head res dept, US Naval Electronics Lab, 46-53, head spec res div, 53-65; CHMN SPACE SCI & MEM EXEC COUN & ADV COMT, COMN EDUC RESOURCES, 65- Concurrent Pos: Dep tech dir joint task force I, Washington, DC & Bikini, 46. Mem: AAAS; fel Am Phys Soc; Inst Elec & Electronics Eng; Am Geophys Union. Res: Propagation of radio waves; statistical fluctuations in electron currents under space charge; thermal agitation of electric charge in conductors; multiple space-charge; properties of monomolecular films. Mailing Add: 3808 Liggett Dr San Diego CA 92106

THATCHER, JAMES W, b Schenectady, NY, Mar 25, 36; m 58; c 3. COMPUTER SCIENCE, MATHEMATICS. Educ: Pomona Col, BA, 58; Univ Mich, MA, 61, PhD(commun sci), 64. Prof Exp: Assoc engr, Convair Astronaut, 58-59; staff programmer comput ctr, Univ Mich, 59-60, res assoc automata logic of comput group, 62-63; STAFF MEM AUTOMATA, THOMAS J WATSON RES CTR, IBM CORP, 63- Concurrent Pos: Lectr, NY Univ, 65; asst prof, Fisk Univ, 65-66; sr vis fel, Sch Artificial Intel, Edinburgh Univ, 72-73. Mem: Am Math Soc; Asn Symbolic Logic; Asn Comput Mach. Res: Programming theory; theoretical computer science; mathematical semantics. Mailing Add: Thomas J Watson Res Ctr IBM Corp Box 218 Yorktown Heights NY 10598

THATCHER, ROBERT CLIFFORD, b Boonville, NY, Jan 11, 29; m 49; c 4. FOREST ENTOMOLOGY. Educ: State Univ NY, BS, 53, MS, 54; Auburn Univ, PhD, 71. Prof Exp: Biol aide, Southern Forest Exp Sta, La, 54, entomologist, 54-68, asst br chief forest insects, 68-73, proj leader, Southern Forest Exp Sta, 73-74, PROG MGR, OFF OF SECY, USDA SOUTHERN PINE BEETLE PROG, US FOREST SERV, 74- Concurrent Pos: Instr, Stephen F Austin State Univ, 57 & 63. Mem: Soc Am Foresters; Entom Soc Am. Res: Forest ecology; interdisciplinary research and development activities relating to southern pine beetle pest management. Mailing Add: USDA Southern Pine Beetle Prog 2500 Shreveport Hwy Pineville LA 71360

THATCHER, THEODORE OSSIP, b Ogden, Utah, May 3, 10; m 38; c 3. ENTOMOLOGY. Educ: Utah State Univ, BS, 32, MS, 35; Univ Calif, PhD(entom), 48. Prof Exp: Field aide, Forest Insect Lab, Bur Entom, USDA, Idaho, 36; foreman, Insect Control Crew, Minidoka Nat Forest, 36; ranger & naturalist, Nat Park Serv, Utah & Nev, 37-41; asst, Univ Calif, 46-48; from asst prof to prof entom, Colo State Univ, 48-73; RETIRED. Concurrent Pos: With US Opers Mission, Int Coop Admin, Pakistan, 61-63. Mem: Entom Soc Am; Entom Soc Can; Coleopterists Soc. Res: Taxonomy and biology of adults and larvae of Scolytidae. Mailing Add: 644 S 500 E Logan UT 84321

THATCHER, VERNON EVERETT, b Talent, Ore, Feb 4, 29. ZOOLOGY, PARASITOLOGY. Educ: Ore State Univ, BA, 52, MA, 54; La State Univ, PhD, 61. Prof Exp: Marine biologist, Res Found, Tex A&M Univ, 57; asst zool, La State Univ, 58-60; res assoc prev med & pub health, Med Br, Univ Tex, 60-62; staff ecologist, Gorgas Mem Lab, Panama, 62-67; asst prof parasitol, Sch Med, Tulane Univ, 67-70; assoc prof biol, Univ Valle, Colombia, 70-73; assoc res, Found Higher Educ, Cali, 73-74; ASSOC PROF BIOL, UNIV VALLE, COLOMBIA, 74- Mem: Am Soc Parasitol; Am Micros Soc; Cooper Ornith Soc. Res: Life cycles and phylogeny of helminths; parasites of freshwater fishes; mariculture. Mailing Add: Dept of Biol Div of Sci Univ del Valle Apdo Aereo 15042 Cali Colombia

THATCHER, WALTER EUGENE, b Evanston, Ill, Jan 22, 27; m 49; c 11. ANALYTICAL CHEMISTRY. Educ: Northwestern Univ, BS, 50; Univ Ill, PhD(chem), 55. Prof Exp: RES SPECIALIST, CENT RES LABS, 3M CO, 54- Mem: Am Chem Soc; Am Crystallog Asn. Res: Analytical chemistry in x-ray diffraction and fluorescence. Mailing Add: Cent Res Labs 3M Co 201-1E PO Box 33221 St Paul MN 55133

THATCHER, WAYNE RAYMOND, b Montreal, Que, May 23, 42. SEISMOLOGY. Educ: McGill Univ, BSc, 64; Calif Inst Technol, MS, 67, PhD(geophys), 72. Prof Exp: Geophysicist, Nat Ctr Earthquake Res, 71-75, GEOPHYSICIST, OFF EARTHQUAKE STUDIES, US GEOL SURV, 75- Concurrent Pos: Vis prof, Stanford Univ, 72- Mem: Am Geophys Union; Seismol Soc Am. Res: Aspects of earthquake source mechanism; seismic surface wave propagation; crustal structure; microearthquakes; crustal deformation and earthquake prediction. Mailing Add: Off of Earthquake Studies US Geol Surv 345 Middlefield Rd Menlo Park CA 94025

THATCHER, WILLIAM WATTERS, b Baltimore, Md, Jan 12, 42; m 62; c 3. REPRODUCTIVE PHYSIOLOGY, REPRODUCTION ENDOCRINOLOGY. Educ: Univ Md, BS, 63, MS, 65; Mich State Univ, PhD(dairy sci), 68. Prof Exp: NIH fel, Mich State Univ, 68-69; asst prof, 69-74, ASSOC PROF REPRODUCTIVE PHYSIOL & ENDOCRINOL, UNIV FLA, 74- Mem: Am Dairy Sci Asn; Soc Study Reproduction; Am Soc Animal Sci. Mailing Add: Dept of Dairy Sci Univ of Fla Gainesville FL 32611

THAU, MARCUS, b Tarnov, Austria, Feb 10, 97; m 30; c 1. CHEMISTRY, ENGINEERING. Educ: Univ Vienna, PhD(chem), 22. Prof Exp: Res chemist, Williamburgh Chem Co & Calco Chem Co, NY, 24-26; dir res, Lacquer & Chem Corp, 26-27, tech dir, 27-48; tech dir, Metric Lacquer Mfg Co, NJ, 48-50; tech dir, pres & owner, Universal Coatings Inc, 50-72; RES CONSULT, 72- Honors & Awards: Paint Pioneer Cert, New York Paint, Varnish & Lacquer Asn, 52. Mem: Am Chem Soc. Res: Dyes; synthetic resins; protective and decorative coatings; insulation; aircraft coatings to increase speed in flight; dielectric strenth; infrared baking; printed circuits. Mailing Add: 405 E 72nd St New York NY 10021

THAU, ROSEMARIE B ZISCHKA, b Vienna, Austria, Mar 15, 36; m 70. BIOCHEMISTRY, ENDOCRINOLOGY. Educ: Univ Vienna, BS, 54, PhD(chem), 63. Prof Exp: Res assoc exp surg med ctr, Duke Univ, 63-65; instr pediat, State Univ NY Downstate Med Ctr, 65-70, asst prof, 70-72; STAFF SCIENTIST, POP COUN, ROCKEFELLER UNIV, 72- Mem: AAAS. Res: Virology; tobacco mosaic virus; avian tumor virus; chemical causes of mental retardation; amino acid metabolism; lipid metabolism; hormone metabolism in reproductive physiology. Mailing Add: Pop Coun Rockefeller Univ York Ave & 66th St New York NY 10021

THAW, RICHARD FRANKLIN, b Denver, Colo, Nov 13, 20; m 42; c 1. BOTANY, NATURAL HISTORY. Educ: Ore State Col, BS, 43, MEd, 47, MS, 53, EdD(gen sci, sci educ), 58. Prof Exp: Teacher high sch, Ore, 47-58; sci coordr, San Diego County Schs, Calif, 58-59; assoc prof biol & natural sci, 59-68, PROF BIOL & NATURAL SCI, SAN JOSE STATE UNIV, 68- Concurrent Pos: Crown-Zellerbach fel, 54. Honors & Awards: Achievement Award, Nat Sci Teachers Asn, 56. Mem: Nat Sci Teachers Asn; Nat Audubon Soc. Res: Plant taxonomy; science teaching methods. Mailing Add: Dept of Natural Sci San Jose State Univ San Jose CA 95192

THAXTON, GEORGE DONALD, b Richmond, Va, Feb 28, 31; m 54; c 4. NUCLEAR PHYSICS, SOLID STATE PHYSICS. Educ: Richmond Univ, BS, 59; Univ NC, PhD(physics), 65. Prof Exp: Res assoc physics, Fla State Univ, 64-66; ASST PROF PHYSICS, AUBURN UNIV, 66- Mem: Am Phys Soc; Am Asn Physics Teachers. Res: Theory of direct nuclear reactions; theory of band structure of solids. Mailing Add: Dept of Physics Auburn Univ Auburn AL 36830

THAXTON, JAMES PAUL, b Longview, Miss, Sept 6, 41; m 65; c 3. AVIAN PHYSIOLOGY. Educ: Miss State Univ, BS, 64, MS, 66; Univ Ga, PhD(physiol), 71. Prof Exp: Instr biol, Northeast La Univ, 66-67; asst prof poultry sci, 71-73, ASSOC PROF POULTRY SCI, NC STATE UNIV, 73- Concurrent Pos: Consult, Nat Inst Environ Health Sci, 74- Honors & Awards: Poultry Sci Assoc Res Award, Poultry Sci Asn, 74. Mem: AAAS; Poultry Sci Asn; World Poultry Sci Asn; Sigma Xi. Res: Effects of environmental parameters, such as temperatures, toxins and heavy metals, on the immunological responsiveness of the avian species. Mailing Add: Dept of Poultry Sci PO Box 5703 NC State Univ Raleigh NC 27607

THAYER, CHARLES WALTER, b Springfield, Vt, May 18, 44; m 67. INVERTEBRATE PALEONTOLOGY, PALEOECOLOGY. Educ: Dartmouth Col, BA, 66; Yale Univ, MPhil, 69, PhD(geol), 72. Prof Exp: ASST PROF GEOL, UNIV PA, 71- Mem: AAAS; Brit Palaeont Asn; Paleont Soc; Int Palaeont Union; Nat Speleol Soc. Res: Adaptation and functional morphology of invertebrates, especially brachiopods and bivalves; paleozoic communities. Mailing Add: Dept of Geol Univ of Pa Philadelphia PA 19174

THAYER, DONALD WAYNE, b Kansas City, Mo, Jan 15, 37; m 69. MICROBIOLOGY, BIOCHEMISTRY. Educ: Kans State Univ, BS, 62, MS, 63; Colo State Univ, PhD(microbiol), 66. Prof Exp: NSF resident res assoc, Naval Med Res Inst, 66-67, res chemist, 66-69; asst prof biol, 69-72, ASSOC PROF BIOL, TEX TECH UNIV, 72- Concurrent Pos: Lectr, Soc Sigma Xi-Sci Res Soc Am regional lect exchange prog, 71. Mem: AAAS; Am Soc Microbiol; Am Chem Soc; Inst Food Technol. Res: Single-cell protein and carbohydrate metabolism; production of single-cell protein from cellulose waste products and the physiology of the associated cellulolytic bacteria. Mailing Add: Dept of Biol Sci Tex Tech Univ Lubbock TX 79409

THAYER, GORDON WALLACE, b Weymouth, Mass, Feb 28, 40; m 63; c 2. ECOLOGY. Educ: Gettysburg Col, BA, 62; Oberlin Col, MA, 64; NC State Univ, PhD(zool), 69. Prof Exp: FISHERY BIOLOGIST, ATLANTIC ESTUARINE FISHERIES CTR, NAT MARINE FISHERIES SERV, NAT OCEANIC & ATMOSPHERIC ADMIN, 68- Mem: Am Soc Limnol & Oceanog; Ecol Soc Am; Estuarine Res Fedn; Sigma Xi. Res: Energetics of eelgrass communities; dynamics of zooplankton populations; influence of detritus in invertebrate and vertebrate food webs; phytoplankton ecology and nutrient limiting factors. Mailing Add: Atlantic Estuarine Fisheries Ctr Beaufort NC 28516

THAYER, HELEN IDA, organic chemistry, see 12th edition

THAYER, JOHN STEARNS, b Glen Ridge, NJ, Apr 1, 38. INORGANIC CHEMISTRY, ORGANOMETALLIC CHEMISTRY. Educ: Cornell Univ, BA, 60; Univ Wis, PhD(chem), 64. Prof Exp: Asst prof chem, Ill Inst Technol, 64-66; ASST PROF CHEM, UNIV CINCINNATI, 66- Concurrent Pos: Frederick Gardner Cottrell grant, 67-73. Mem: AAAS; Am Chem Soc. Res: Organometallic compounds of silicon, germanium, tellurium and platinum; biological aspects of organometallic chemistry; history of organometallics. Mailing Add: Dept of Chem Univ of Cincinnati Cincinnati OH 45221

THAYER, KEITH EVANS, b Lime Springs, Iowa, Feb 5, 28; m 53; c 4. DENTISTRY. Educ: Cornell Col, BA, 51; Univ Iowa, DDS, 55, MS, 56. Prof Exp: From instr to assoc prof, 56-63, PROF FIXED PROSTHODONTICS, COL DENT, UNIV IOWA, 63-, HEAD DEPT, 60- Concurrent Pos: Attend dentist, Vet Admin Hosp, Iowa City; vis Fulbright prof, Univ Singapore, 68-69; consult, Am Dent Asn Vietnam Educ Proj, 72; consult, Vet Admin Hosp. Mem: Int Dent Fedn; Int Asn Dent Res; Am Dent Asn. Res: Rubber base and silicone impression materials; gingival retraction agents and their effect on oral tissues; occlusion. Mailing Add: Univ of Iowa Col of Dent Iowa City IA 52240

THAYER, LEWIS A, b Mercur, Utah, Dec 25, 03; m 29; c 4. ANALYTICAL CHEMISTRY. Educ: Wash State Univ, BS, 25, MS, 26; Stanford Univ, PhD(microbiol), 35. Prof Exp: Instr bot & chem, Sacramento Jr Col, 32-33; from instr to prof bot, microbiol & chem, Centenary Col, 33-46; prof chem, 46-74, dean fac & acad vpres, 58-69, EMER PROF CHEM & DEAN FAC, LEWIS & CLARK COL, 74- Mem: Am Chem Soc. Res: Biogenetic origin of petroleum; plant growth hormones. Mailing Add: Dept of Chem Lewis & Clark Col 0615 SW Palatine Hill Rd Portland OR 97219

THAYER, PAUL ARTHUR, b New York, NY, Apr 30, 40; m 66; c 1. MARINE GEOLOGY, SEDIMENTOLOGY. Educ: Rutgers Univ, BA, 61; Univ NC, PhD(geol), 67. Prof Exp: Develop geologist, Calif Co Div, Chevron Oil Co, 67-68; asst prof geol, Tex A&I Univ, 68-70; ASST PROF MARINE SCI RES, UNIV NC, WILMINGTON, 70- Concurrent Pos: Tex A&I Univ fac res grants, 68-69 & 69-70; Soc Sigma Xi grant, 69; petrol geologist, BP Alaska Explor Inc, 75-; instrnl sci equipment prog award, NSF, 72-74 & 75-77; res grant, AEC, 74. Mem: AAAS; Geol Soc Am; Int Asn Sedimentol; Soc Econ Paleont & Mineral; Am Asn Petrol Geol. Res: Petrology of clastic sedimentary rocks; reconstruction of depositional environments within ancient sedimentary rocks; Triassic nonmarine stratigraphy; provenance, dispersal and origin of modern and ancient sediments; sedimentology of modern carbonate sediments. Mailing Add: Dept of Marine Sci Res Univ of NC Wilmington NC 28401

THAYER, PAUL LOYD, b Centralia, WVa, Feb 25, 28; m 53; c 3. PLANT PATHOLOGY. Educ: Marietta Col, BS, 52; Ohio State Univ, MS, 55, PhD, 58. Prof Exp: Asst plant pathologist, Everglades Exp Sta, Univ Fla, 58-65; plant pathologist, 65-72, NORTHEASTERN REGIONAL PLANT SCI RES MGR, ELI LILLY & CO, 72- Mem: Am Phytopath Soc. Res: Bacterial and fungus diseases of vegetable crops; fungicide and nematocide evaluation. Mailing Add: Plant Sci Field Res Greenfield Labs Eli Lilly & Co PO Box 708 Greenfield IN 46140

THAYER, PHILIP CRABB, solid state physics, see 12th edition

THAYER, PHILIP STANDISH, b Pelham, Mass, Oct 1, 23; m 46; c 3. MICROBIOLOGY. Educ: Amherst Col, BA, 48, MA, 49; Calif Inst Technol, PhD(biochem), 52. Prof Exp: Merck fel bact, Univ Calif, 51-53; instr chem, Univ Calif, Los Angeles, 53-55; mem staff biol, 55-72, VPRES & HEAD LIFE SCI DIV,

ARTHUR D LITTLE, INC, 72- Res: Tissue culture; chemotherapy and radiation biology. Mailing Add: Life Sci Div Arthur D Little Inc 30 Memorial Dr Cambridge MA 02140

THAYER, RICHARD P, statistics, see 12th edition

THAYER, ROLLIN HAROLD, b St Francis Mission, SDak, Dec 30, 16; m 44; c 1. POULTRY NUTRITION. Educ: Okla State Univ, BS, 40; Univ Nebr, MS, 42; Wash State Univ, PhD, 55. Prof Exp: From asst prof to prof poultry sci, 43-74, PROF ANIMAL SCI & NUTRIT, OKLA STATE UNIV, 74- Mem: AAAS; Am Poultry Sci Asn; Am Inst Nutrit; World Poultry Sci Asn. Res: Estrogens in poultry fattening; layer breeder hen requirements; nutritive requirements for breeder turkeys. Mailing Add: Dept of Animal Sci & Indust Okla State Univ Stillwater OK 74074

THAYER, SCOTT DWIGHT, b Gibbstown, NJ, Apr 1, 22; m 45; c 4. AIR POLLUTION, MICROMETEOROLOGY. Educ: Middlebury Col, AB, 43; Mass Inst Technol, MS, 48. Prof Exp: Phys scientist, Chem Corps Lab, US Army, 48-57, oper res analyst, Army Oper Res Group, 57-64; sr scientist, Travelers Res Ctr, 64-67; VPRES & SR SCIENTIST, GEOMET INC, 67- Honors & Awards: Spec Act or Serv Award, Dept Army, 61. Mem: Am Meteorol Soc; Oper Res Soc Am; Air Pollution Control Asn. Res: Development and validation of simulation models for prediction of air pollutant concentrations on various geographic scales. Mailing Add: Geomet Inc 15 Firstfield Rd Gaithersburg MD 20760

THAYER, THOMAS PRENCE, b Scarsdale, NY, May 22, 07; m 31; c 2. GEOLOGY. Educ: Univ Ore, BA, 29; Northwestern Univ, MA, 31; Calif Inst Technol, PhD(phys geol), 34. Prof Exp: Instr geol, Univ Nev, 34-35; jr geologist, US Eng Dept, 35-37; asst geologist, 37-42, assoc geologist, 42-70, GEOLOGIST, US GEOL SURV, 70- Honors & Awards: Distinguished Serv Award, US Dept Interior, 73. Mem: AAAS; fel Geol Soc Am; Mineral Soc Am; fel Soc Econ Geol; fel Am Geophys Union. Res: Geology of chromite deposits; ultramafic and related rocks; geology of northeastern Oregon; Tertiary volcanic rocks of Oregon. Mailing Add: Eastern Mineral Resources Br US Geol Surv Stop 954 Reston VA 22092

THAYER, WALTER RAYMOND, JR, b Providence, RI, Apr 16, 29; m 55; c 3. MEDICINE, GASTROENTEROLOGY. Educ: Providence Col, BS, 50; Tufts Univ, MD, 54. Hon Degrees: MA, Brown Univ, 66. Prof Exp: Resident gastroenterol, Sch Med, Yale Univ, 61-62, instr med, 60-62, asst prof, 62-66; assoc prof, 66-70, PROF MED, BROWN UNIV, 70-; CHIEF GASTROENTEROL, RI HOSP, 66- Concurrent Pos: Fel clin gastroenterol, Sch Med, Yale Univ, 59-60; NSF fel, Wenner-Gren Inst, Stockholm, 71-72. Mem: Am Soc Clin Invest; Am Gastroenterol Asn. Res: Immunology in gastrointestinal diseases; gastric secretion. Mailing Add: Dept of Med Brown Univ Providence RI 02912

THAYNE, WILLIAM V, b Binghamton, NY, July 23, 41; m 63; c 3. ANIMAL BREEDING, STATISTICS. Educ: Cornell Univ, BS, 63; Univ Ill, MS, 67, PhD(dairy sci), 71. Prof Exp: Res asst dairy sci, Univ Ill, 63-67; instr animal sci, 67-70, asst prof statist & comput sci, 73-75, ASST PROF ANIMAL SCI, WVA UNIV, 70-, ASSOC PROF STATIST & COMPUT SCI, 75- Mem: Am Dairy Sci Asn; Biomet Soc; Am Genetic Asn. Res: Dairy cattle genetics. Mailing Add: 1016 AS WVA Univ Morgantown WV 26505

THEARD, LESLIE (PETER) M, physical chemistry, see 12th edition

THEARD, LOWELL PAUL, physical chemistry, chemical physics, see 12th edition

THEDFORD, ROOSEVELT, b Greene Co, Ala, Apr 16, 37; m 60; c 5. ORGANIC BIOCHEMISTRY, MOLECULAR BIOLOGY. Educ: Clark Col, BS, 59; Univ Buffalo, MA, 62; State Univ NY Buffalo, PhD(biochem), 73. Prof Exp: Cancer res scientist, Roswell Park Mem Inst, 61-69 & 72-74; ASSOC PROF CHEM, CLARK COL, 74- Mem: Am Chem Soc; Sigma Xi. Res: Study of the physicochemical and biological properties of alkylated synthetic homopoly ribonucleotides and determination of the functions of strategically located modified nucleosides as found in transfer RNA. Mailing Add: Dept of Chem Clark Col 240 Chestnut St SW Atlanta GA 30314

THEDFORD, WILLIAM ANDREW, b Okla City, Okla, Nov 15, 39; m 62; c 2. TOPOLOGY, ALGEBRA. Educ: Okla State Univ, BS, 62, MS, 64; NMex State Univ, PhD(math), 70. Prof Exp: ASST PROF MATH SCI, VA COMMONWEALTH UNIV, 70- Mem: Am Math Soc; Math Asn Am. Res: The development of mathematical models for qualitative analysis of physical problems. Mailing Add: Phys Sci Bldg Dept of Math Sci Va Commonwealth Univ Richmond VA 23284

THEIL, ELIZABETH, b Jamaica, NY, Mar 29, 36; m 57; c 2. BIOCHEMISTRY, DEVELOPMENTAL BIOLOGY. Educ: Cornell Univ, BS, 57; Columbia Univ, PhD(biochem), 62. Prof Exp: Res asst genetics of microorganisms, State Univ NY Downstate Med Ctr, 57-58; res assoc chem, Fla State Univ, 64-66; res assoc animal sci & biochem, 67-69, instr biochem, 69-71, asst prof, 71-75, ASSOC PROF BIOCHEM, NC STATE UNIV, 75- Concurrent Pos: United Med Health Serv, NC, grant dept biochem, NC State Univ, 70-71, Res Corp grant, 70-71 & 72-; Nat Inst Arthritis & Metab Dis grant, 71-74; NSF res grant, 75-; res grant, NC Agr Exp Sta, 74- Mem: Sigma Xi; Am Chem Soc; Am Soc Biol Chem. Res: Biochemistry of development; regulation of protein synthesis; iron metabolism and iron storage. Mailing Add: Dept of Biochem 334 Polk Hall NC State Univ Raleigh NC 27607

THEIL, GEORGE B, b Brooklyn, NY, Aug 5, 25; m 48; c 3. INTERNAL MEDICINE. Educ: Marquette Univ, MD, 48. Prof Exp: Fel renal dis, Univ Calif, San Francisco, 58-60; from asst prof to assoc prof internal med, Univ Iowa, 64-70; PROF INTERNAL MED, MED COL WIS, 70-, VCHMN DIV MED, 70- Concurrent Pos: Consult, Vet Admin Hosp, 66- Mem: Am Physiol Soc; Am Fedn Clin Res; Soc Gen Physiol. Res: Renal physiology and diseases. Mailing Add: Med Serv Vet Admin Ctr Hosp Wood WI 53193

THEIL, MICHAEL HERBERT, b Brooklyn, NY, Nov 2, 33; m 57; c 2. POLYMER CHEMISTRY. Educ: Cornell Univ, AB, 54; Polytech Inst Brooklyn, PhD(chem), 63. Prof Exp: Sr res chemist, Res Ctr of Tex, US Chem Co, 62-64; res assoc chem, Fla State Univ, 64-66; ASST PROF TEXTILE CHEM, NC STATE UNIV, 66- Mem: Am Chem Soc; Japanese Soc Polymer Sci; Fiber Soc. Res: Phase transitions of polymers; polymerization mechanisms; copolymer statistics. Mailing Add: Dept of Textile Chem NC State Univ Raleigh NC 27607

THEILE, FRED CHARLES, b West Orange, NJ, Mar 6, 22; m 46; c 2. ORGANIC CHEMISTRY, MATHEMATICS. Educ: Rutgers Univ, BScChem, 43. Prof Exp: Res chemist, US Naval Res Lab, 43-45, chemist, Norda Essential Oil & Chem Co, 45-47; plant mgr aromatic chem, Orbis Prod Co, 47-51; prod mgr, Shulton Inc, 51-56, perfumery adminr, 56-58, res adminr, NJ, 58-67, exec asst to pres, Jacqueline Cochran Div, NY, 67-70, mkt dir, Carven Parfums, 68-70; exec vpres, Lautier Fils Inc, 70-74, GEN MGR, LAUTIER FIL DIV & VPRES, RHODIA INC, 74- Mem: AAAS; Am Chem Soc; Soc Cosmetic Chem. Res: Perfumery; microbiology. Mailing Add: Lautier Fils Inc 300 Webro Rd Parsippany NJ 07054

THEILEN, ERNEST OTTO, b Columbus, Nebr, June 4, 23. INTERNAL MEDICINE. Educ: Univ Nebr, BA, 45, MD, 47; Am Bd Internal Med, dipl, 55; Am Bd Cardiovasc Dis, dipl, 65. Prof Exp: Instr med, 51-52, assoc, 52, from asst prof to assoc prof internal med, 55-63, PROF INTERNAL MED, COL MED, UNIV IOWA, 63- Concurrent Pos: Fel coun clin cardiol & coun circulation, Am Heart Asn. Mem: AMA; Am Fedn Clin Res; Am Col Cardiol; Am Physiol Soc; Am Col Physicians. Res: Cardiovascular physiology; electrocardiography; clinical cardiology. Mailing Add: Dept of Med Univ Hosps Iowa City IA 52242

THEILEN, GORDON H, b Montevideo, Minn, May 29, 28; m 53; c 3. VETERINARY MEDICINE. Educ: Univ Calif, BS, 53, DVM, 55. Prof Exp: Pvt pract, Ore, 55-56; lectr vet med, 56-57, from instr to asst prof, 57-62, from asst prof to prof clin sci, 62-74, PROF SURG, SCH VET MED, UNIV CALIF, DAVIS, 74- Concurrent Pos: Spec fel, Leukemia Prog, Nat Cancer Inst, 64-65; NY Cancer Res Inst fel tumor immunol, with Chester Beatty, Univ London, 72-73; mem sci & rev comt & bd dirs, Leukemia Soc Am, 71-76. Mem: Am Vet Med Asn; Am Asn Cancer Res; Am Asn Vet Clin. Res: Leukemia-sarcoma and myeloproliferative disease complex and subjects dealing with clinical oncology, particularly tumor biology. Mailing Add: Dept of Surg Sch of Vet Med Univ of Calif Davis CA 95616

THEILHEIMER, FEODOR, b Gunzenhausen, Ger, June 18, 09; nat US; m 48; c 1. MATHEMATICS. Educ: Berlin Univ, PhD(math), 36. Prof Exp: From instr to asst prof math, Trinity Col, 42-48; mathematician naval ord lab, 48-53, MATHEMATICIAN NAVAL SHIP RES & DEVELOP CTR, US DEPT NAVY, BETHESDA, MD, 53- Mem: Am Math Soc; Math Asn Am; Soc Indust & Appl Math; Asn Comput Mach. Res: Numerical analysis; fluid dynamics. Mailing Add: 2608 Spencer Rd Chevy Chase MD 20015

THEILHEIMER, WILLIAM, b Augsburg, Ger, Oct 11, 14; nat US; m 56; c 1. ORGANIC CHEMISTRY. Educ: Basel Univ, PhD(org chem), 40. Prof Exp: Asst to Prof Erlenmeyer, Basel Univ, 40-47; consult, 48-59, lit chemist sci info dept, 59-63, RESIDENT CONSULT, HOFFMANN-LA ROCHE, INC, 64- Concurrent Pos: Ed, Synthetic Methods Org Chem, 44- Mem: Am Chem Soc. Res: Synthetic methods. Mailing Add: 318 Hillside Ave Nutley NJ 07110

THEILING, LOUIS FOSTER, JR, organic chemistry, see 12th edition

THEIMER, EDGAR E, b Newark, NJ, June 29, 15; m 58; c 1. ANALYTICAL CHEMISTRY, PHARMACEUTICAL CHEMISTRY. Educ: Polytech Inst Brooklyn, BChE, 36; NY Univ, MS, 39. Prof Exp: Chief chemist, Metrop Labs, Inc, 40-58; chief chemist, Pharmich Div, Mich Chem Corp, 58-61; staff chemist, Int Flavors & Fragrances, Inc, 61-64; chief anal chemist, Smith, Miller & Patch, Inc, 64-72; CHIEF ANAL CHEMIST, COOPER LABS, CEDAR KNOLLS, 72- Mem: Am Chem Soc; Parenteral Drug Asn. Res: Analytical methods development. Mailing Add: 9 Meade Court Somerset NJ 08873

THEIMER, ERNST THEODORE, b Newark, NJ, Sept 23, 10; m 49; c 2. CHEMISTRY. Educ: Univ Cincinnati, AB, 30, MA, 31; NY Univ, PhD(org chem), 45. Prof Exp: Res chemist, Van Ameringen-Haebler, Inc, 31-57, dir res, 57-58, Van Ameringen-Haebler Div, Int Flavors & Fragrances, Inc, 59-60, dir res & develop, 61-68, chief sci off, 68-71, vpres, 63-71, CONSULT, INT FLAVORS & FRAGRANCES, INC, 71- Mem: Am Chem Soc; Asn Res Dirs. Res: Organic synthesis; terpene chemistry; instrumental analysis. Mailing Add: Int Flavors & Fragrances Inc 1515 Highway 36 Union Beach NJ 07735

THEIMER, OTTO, b Vienna, Austria, Feb 22, 18; m 45; c 2. PHYSICS. Educ: Univ Vienna, BS, 39; Munich Tech Univ, MS, 43, Dr rer nat (chem physics), 45. Prof Exp: Asst physics, Graz Tech Univ, 46-52, privat docent, 49-52; asst prof, Univ BC, 52-55; assoc prof, Univ Okla, 55-59; RES PROF RES CTR, NMEX STATE UNIV, 59- Concurrent Pos: Fel, Edinburgh Univ, 51; vis staff mem, Los Alamos Sci Labs, 65-73. Mem: Fel AAAS; fel Am Phys Soc; Am Asn Physics Teachers. Res: Molecular spectroscopy; light scattering; physical adsorption of gases; quantum chemistry; statistical mechanis; plasma and solid state physics; atomic physics; chemical physics; electrodynamics; electrohydrodynamics; plasma physics; quantum electrodynamics; atmospheric physics. Mailing Add: NMex State Univ Res Ctr Las Cruces NM 88003

THEIMER, ROSE, physics, see 12th edition

THEINE, ALICE, b Menomonee, Wis, Feb 23, 38. ORGANIC CHEMISTRY. Educ: Alverno Côl, BA, 59; Marquette Univ, MS, 65; La State Univ, PhD(chem), 72. Prof Exp: Teacher sci & math, St Boniface High Sch, Iowa, 59-63 & St Joseph High Sch, Wis, 63-66; instr, 66-70, ASST PROF CHEM, ALVERNO COL, 72- Mem: Am Chem Soc. Res: Development of improved methods of competence-based instruction in undergraduate chemistry courses; investigation of cationic intermediates in oxidative decarboxylation reactions by product analysis studies. Mailing Add: 3401 S 39th St Milwaukee WI 53215

THEINER, MICHA, b Rehovoth, Israel, May 6, 36; m 63; c 2. BIOCHEMISTRY. Educ: Israel Inst Technol, BSc, 58; Univ Pittsburgh, PhD(biochem), 63. Prof Exp: Res assoc food safety, Mass Inst Technol, 63-66; asst prof biochem, Carnegie-Mellon Univ, 66-69; asst prof, 69-75, ASSOC PROF BIOCHEM, SCH DENT MED, UNIV PITTSBURGH, 75- Mem: Am Chem Soc; Inst Food Technol; Int Asn Dent Res; NY Acad Sci; Am Soc Prev Dent. Res: Dental biochemistry; gas chromatography; clinical chemistry; clinical nutrition and counseling; educational techniques. Mailing Add: Dept of Biochem 528 Salk Hall Univ of Pittsburgh Sch Dent Med Pittsburgh PA 15261

THEIS, CHARLES VERNON, b Newport, Ky, Mar 27, 00; m 27; c 1. GEOLOGY. Educ: Univ Cincinnati, CE, 22, PhD(geol), 29. Prof Exp: Instr geol, Univ Cincinnati, 23-29; jr geologist, US Corps Engrs, 29-30; from asst geol to res hydrologist, US Geol Surv, 30-75; RETIRED. Concurrent Pos: Geologist, Cincinnati Resource Surv, 24-25. Honors & Awards: Distinguished Serv Award, US Dept Interior, 58. Mem: AAAS; Geol Soc Am; Am Geophys Union. Res: Ground water hydrology; mine drainage problems; fundamental principles of ground water movement; waste disposal hydrology. Mailing Add: US Geol Surv PO Box 4369 Albuquerque NM 87106

THEIS, GAIL ANN, b New York, NY, Mar 7, 33; m 61; c 2. IMMUNOLOGY. Educ: Cornell Univ, BA, 54; NY Univ, MS, 57; Univ Pa, PhD(biol), 61. Prof Exp: Instr path, Sch Med, NY Univ, 70-72; ASST PROF PATH, NEW YORK MED COL, 72- Concurrent Pos: Arthritis Found fel, NY Univ, 69-72; Leukemia Soc Am spec fel, New York Med Col, 72-74; USPHS res career develop award, 74- Mem: Am Asn Immunol. Mailing Add: Dept of Path New York Med Col Valhalla NY 10595

THEIS, JEROLD HOWARD, b Richmond, Calif, July 29, 38; m 67; c 1. MEDICAL MICROBIOLOGY, PARASITOLOGY. Educ: Univ Calif, Berkeley, AB, 60; Univ

Calif, Davis, DVM, 64, PhD(comp path), 72. Prof Exp: Asst res vet, George Williams Hooper Found, Univ Calif, San Francisco Med Ctr & Repub Singapore, 64-67; ASST PROF MED MICROBIOL, SCH MED, UNIV CALIF, DAVIS, 70- Concurrent Pos: USPHS fel, Univ Calif, Davis, 67-69; consult, Sacramento Med Ctr, Calif, 71- & Primate Res Ctr, Davis, 72- Honors & Awards: Grand Prize, Int Med Film Festival, Brussels, Belg, 72. Mem: Am Vet Med Asn. Res: Mechanisms of transmission of arthropod borne disease agents at the host-arthropod interface. Mailing Add: Dept of Med Microbiol Univ of Calif Sch of Med Davis CA 95616

THEIS, RICHARD JAMES, b Cincinnati, Ohio, Nov 30, 37; m 61; c 3. ORGANIC POLYMER CHEMISTRY. Educ: Xavier Univ, BS, 60, MS, 62; Univ Cincinnati, PhD(chem), 66. Prof Exp: Res chemist, 66-72, STAFF SCIENTIST FILM DEPT, E I DU PONT DE NEMOURS & CO, INC, 73- Mem: Am Chem Soc. Res: Barrier coatings for films; films for packaging uses; adherable films; filled films. Mailing Add: E I du Pont de Nemours & Co Inc PO Box 89 Circleville OH 43113

THEISEN, CYNTHIA THERES, b Dearborn, Mich. ORGANIC CHEMISTRY. Educ: Siena Heights Col, BS, 60; Purdue Univ, Lafayette, MS, 63; St John's Univ, PhD(org chem), 67. Prof Exp: Assoc ed org chem, 67-75, SR ASSOC ED ORG CHEM, CHEM ABSTR SERV, 75- Mem: Am Chem Soc. Mailing Add: Dept 51 Chem Abstr Serv Columbus OH 43210

THEISEN, WILFRED ROBERT, b Sept 5, 29; US citizen. HISTORY OF SCIENCE, PHYSICS. Educ: St John's Univ, BA, 52; Univ Colo, MS, 63; Univ Wis, PhD(hist of sci), 72. Prof Exp: ASST PROF PHYSICS & HIST OF SCI, ST JOHN'S UNIV, MINN, 55- Res: Medieval optical manuscripts of Euclid. Mailing Add: St John's Univ Collegeville MN 56321

THEISS, JEFFREY CHARLES, b Stamford, Conn, Aug 29, 46; m 68; c 1. PHARMACOLOGY. Educ: Univ RI, BS, 68; Brown Univ, PhD(med sci), 73. Prof Exp: Res assoc biochem pharmacol, Roger Williams Gen Hosp & Brown Univ, 71-73; res fel biochem pharmacol, Univ Calif, San Diego & L C Strong Res Found, 73-75; ASST RES SCI PHARMACOL & TOXICOL, UNIV CALIF, SAN DIEGO, 75- Mem: Sigma Xi; AAAS. Res: Mechanisms by which various nucleotides inhibit growth of mammary and lung tumors in mice; screening of chemicals for carcinogenic potency by the A mouse lung tumor bioassay. Mailing Add: Dept of Community Med Univ of Calif at San Diego La Jolla CA 92037

THEISSEN, DONALD RAYMOND, organic chemistry, see 12th edition

THEKAEKARA, MATTHEW POTHEN, b Changanacherry, India, Mar 21, 14; nat US. PHYSICS. Educ: Madras Univ, AB, 37, MS, 39; Johns Hopkins Univ, PhD(physics), 56. Prof Exp: Asst prof physics, St Aloysius' Col, Madras Univ, 39; asst prof, St Joseph's Col, 41-44; assoc prof & chmn dept, Loyola Col, 48-52; asst prof physics & astron, Georgetown Univ, 57-60, assoc prof physics, 60-64, actg chmn dept, 60-62; consult, 62-64, PHYSICIST, GODDARD SPACE FLIGHT CTR, NASA, 64- Concurrent Pos: Vis lectr, Pace Col, 65-70. Honors & Awards: NASA Exceptional Performance Award, 70; Space Environ Award, Inst Environ Sci, 71. Mem: Optical Soc Am; Inst Environ Sci; Am Asn Physics Teachers; Solar Energy Soc; Am Soc Testing & Mat. Res: Solar physics; solar constant and solar spectral irradiance; solar irradiance on ground; solar energy utilization; space simulation; quantum mechanism; space optics; absolute radiometry; interferometry; philosophy of science. Mailing Add: Code 912 NASA Goddard Space Flight Ctr Greenbelt MD 20771

THEKKEKANDAM, JOSEPH THOMAS, b Kerala, India, May 14, 38; m 70. ORGANIC CHEMISTRY. Educ: Univ Kerala, BSc, 59; Univ India, MS, 62; Univ Detroit, PhD(chem), 70. Prof Exp: Asst lectr chem, MS, Univ India, 62-66; teaching asst, Univ Detroit, 66-70; Coun Tobacco Res fel, A&T State Univ, NC, 70-71; CHEMIST-IN-CHARGE RES & DEVELOP, KAY CHEM CO, 71- Concurrent Pos: Fulbright travel grant, Inst Int Educ, India, 66. Mem: Am Chem Soc. Res: Photochemistry of ammonium azides and amines; pesticide metabolism; synthetic detergents. Mailing Add: Res & Develop Kay Chem Co Greensboro NC 27409

THELEN, CHARLES JOHN, b Cedar Rapids, Nebr, Mar 2, 21; m 47; c 1. ORGANIC CHEMISTRY, AEROSPACE TECHNOLOGY. Educ: Univ Iowa, BS, 42, PhD(org chem), 49. Prof Exp: Chemist, B F Goodrich Co, 42; asst, Univ Iowa, 47-49; fel, Univ Calif, Los Angeles, 49-50; res chemist, US Naval Ord Test Sta, 50-58, head explosives & pyrotech div, 58-60, head propellants div, 60-68, MISSILE PROPULSION TECHNOL ADMINR, NAVAL WEAPONS CTR, 69- Concurrent Pos: Navy mem exec comt, Joint Army, Navy, NASA, Air Force Interagency Propulsion Comt, 74-; Michelson Lab fel mgt, Naval Weapons Ctr, 72. Mem: Am Chem Soc; Sigma Xi. Res: Missile propulsion technology and high polymers. Mailing Add: Naval Weapons Ctr Code 4505 China Lake CA 93555

THELEN, EDMUND, b Berkeley, Calif, May 8, 13; m 65; c 2. MATERIALS SCIENCE, INDUSTRIAL CHEMISTRY. Educ: Univ Calif, Berkeley, BS, 34. Prof Exp: Asst chemist, Certain-teed Prod Corp, 34-36; res chemist, OC Field Gasoline Corp, 36-41; asst to exec engr Eclipse-Pioneer Div, Bendix Aviation Corp, 46-47; sr res engr, 47-51, mgr colloids & polymers lab, 51-74, co-dir dent mat sci ctr, 69-74, DIR PHYS & LIFE SCI DEPT, FRANKLIN INST RES LABS, 74- Concurrent Pos: Instr, Hahnemann Med Col, 64-74. Honors & Awards: Spec Recognition Award, Am Soc Landscape Arch, 74. Mem: Am Chem Soc; Am Soc Testing & Mat; Soc Rheol; Sigma Xi; Solar Energy Soc. Res: Nonmetallic materials; materials and processes to promote health, safety, energy conservation and environment; porous pavements; dental materials; solar applications. Mailing Add: Franklin Inst Res Labs 20th & Race St Philadelphia PA 19103

THELEN, THOMAS HARVEY, b Albany, Minn, Aug 11, 41; m 64; c 2. GENETICS. Educ: St John's Univ, Minn, BS, 64; Univ Minn, PhD(genetics), 69. Prof Exp: Cytogeneticist, Minn Dept Health, 69-70; asst prof, 70-74, ASSOC PROF BIOL, CENT WASH STATE COL, 70- Res: Human genetics and cytogenetics and related areas. Mailing Add: Dept of Biol Cent Wash State Col Ellensburg WA 98926

THELIN, JACK HORSTMANN, b Kearny, NJ, Aug 15, 12; m 39; c 2. CHEMISTRY. Educ: Maryville Col, BA, 38; Univ Tenn, MS, 39; Rutgers Univ, PhD(phys chem), 43. Prof Exp: Res chemist, Calco Chem Co Div, 36-41, develop chemist, 41-45, asst chief develop chemist dye intermediates div, 45-46, chief develop chemist, 46-53, RES CHEMIST RES DIV, AM CYANAMID CO, 53- Mem: Am Soc Testing & Mat; Am Chem Soc. Res: Sulfonation; nitration and alkylation; system ammonium sulfamate; sodium nitrate; ammonium nitrate; polymer physical testing equipment; polymer tribology. Mailing Add: 126 E Spring St Somerville NJ 08876

THELMAN, JOHN PATRICK, b Richmond Hill, NY, Dec 25, 42; m 65; c 3. CELLULOSE CHEMISTRY. Educ: State Univ NY Stony Brook, BS, 64; State Univ NY Buffalo, PhD(org chem), 69. Prof Exp: Res chemist, 68-71, res group leader acetate sect, 71-74, RES SUPVR ACETATE SECT, EASTERN RES DIV, ITT RAYONNIER INC, WHIPPANY, 74- Mem: Am Chem Soc; Am Inst Chem. Res: Polysaccharides; natural products; end-use product development. Mailing Add: 4 Sunnyside Dr Kenvil NJ 07847

THENEN, SHIRLEY WARNOCK, b San Mateo, Calif; m 62; c 2. NUTRITION, BIOCHEMISTRY. Educ: Univ Calif, Berkeley, AB, 57, PhD(nutrit), 70. Prof Exp: Res fel hemat & med, Harvard Med Sch & Mass Gen Hosp, 70-72; ASST PROF NUTRIT, SCH PUB HEALTH, HARVARD UNIV, 72- Concurrent Pos: NIH res career develop award, 75-80. Mem: AAAS; Asn Women Sci. Res: The significance of folic acid and vitamin B-12 metabolism in the hemopoietic system and during pregnancy; experimental obesity, both genetic and acquired. Mailing Add: Dept of Nutrit Harvard Univ Sch of Pub Health Boston MA 02115

THEOBALD, CLEMENT WALTER, b Lincoln, Nebr, Mar 24, 18; m 36; c 3. ORGANIC CHEMISTRY. Educ: Univ Nebr, AB, 39, MA, 40; Univ Ill, PhD(org chem), 43. Prof Exp: Res chemist chem dept, 43-51, res supvr, 51-53, res mgr film dept, 53-55, lab dir, Yerkes Labs, NY, 55-57, asst dir res div fabrics & finishes dept, 57, dir, 57-69, dir electrochem dept, 69-72, dir res div indust chem dept, 72-73, VCHMN & EXEC DIR COMT EDUC AID, E I DU PONT DE NEMOURS & CO, INC, 73- Mem: AAAS; Am Chem Soc. Res: Synthetic resins and plastics; paints, varnishes and enamels; fine chemicals; rubber; polymer chemistry; applications for polymeric materials. Mailing Add: 213 W 14th St Wilmington DE 19801

THEOBALD, J KARL, b Prescott, Ariz, Feb 18, 21; m 42, 67. PHYSICS. Educ: Univ Calif, PhD(physics), 52. Prof Exp: PHYSICIST, LOS ALAMOS SCI LAB, 52- Mem: Am Geophys Union. Res: Gaseous electronics; atmospheric physics. Mailing Add: Los Alamos Sci Lab PO Box 1663 Los Alamos NM 87544

THEOBALD, WILLIAM L, b New York, NY, Feb 12, 36. SYSTEMATIC BOTANY, HORTICULTURE. Educ: Rutgers Univ, BS, 58, MS, 59; Univ Calif, Los Angeles, PhD(bot), 63. Prof Exp: Lectr biol, Univ Calif, Santa Barbara, 63-65; NSF fel, Jodrell Lab, Royal Bot Gardens, Eng, 65-66; fel, Evolutionary Biol, Harvard Univ, 66-67; asst prof biol, Occidental Col, 67-71; assoc prof bot, Univ Hawaii, 71-75; DIR, PAC TROP BOT GARDEN, LAWAI, HAWAII, 75- Concurrent Pos: Res assoc bot, Univ Hawaii, 75- Mem: Bot Soc Am; Am Soc Plant Taxon; fel Linnean Soc London; Int Asn Plant Taxon; Am Asn Bot Gardens & Arboreta. Res: Flora of Hawaii, Pacific Islands and Ceylon; systematics of Gesneriaceae, Acanthaceae, Araliaceae, Bignoniaceae, Umbelliferae; trichome anatomy and classification in angiosperms; comparative anatomical studies of Umbelliferae. Mailing Add: Pac Trop Bot Garden PO Box 340 Lawai Kanai HI 96765

THEODORATUS, ROBERT JAMES, b Bellingham, Wash, June 24, 28; m 62; c 3. CULTURAL ANTHROPOLOGY, ETHNOLOGY. Educ: Wash State Col, BA, 50; Univ Wash, MA, 53, PhD(anthrop), 61. Prof Exp: Res analyst anthrop, Human Rels Area Files, 61-62; asst prof, Sacramento State Col, 62-66; ASSOC PROF ANTHROP, COLO STATE UNIV, 66- Mem: AAAS; fel Am Anthrop Asn; fel Royal Anthrop Inst Gt Brit & Ireland; Am Ethnol Soc; Polynesian Soc. Res: Pacific Northwest Indian ethnology; bibliographic resources in ethnology; European folk ethnology; religions of tribal and peasant peoples; Siberian ethnology. Mailing Add: Dept of Anthrop Colo State Univ Fort Collins CO 80523

THEODORE, JOSEPH M, JR, b Fall River, Mass, Apr 29, 31; m 55; c 5. PHARMACY. Educ: New Eng Col Pharm, BS, 55; Univ Wis, MS, 58; Mass Col Pharm, PhD(pharm), 65. Prof Exp: Instr pharm, New Eng Col Pharm, 58-61; from instr to asst prof, Northeastern Univ, 62-66; assoc prof, 66-74, PROF PHARM, OHIO NORTHERN UNIV, 74- Mem: Am Pharmaceut Asn; Am Col Apothecaries; Am Soc Hosp Pharmacists. Res: Spectrofluorometric analysis of drugs; solid state reactions occurring in certain tablet formulations. Mailing Add: Dept of Pharm Ohio Northern Univ Ada OH 45810

THEODORE, TED GEORGE, b Los Angeles, Calif, Aug 19, 37; m 61; c 2. ECONOMIC GEOLOGY. Educ: Univ Calif, Los Angeles, AB, 61, PhD(geol), 67. Prof Exp: GEOLOGIST, US GEOL SURV, 67- Mem: AAAS; Geol Soc Am; Soc Econ Geol. Res: Genesis of porphyrytype copper deposits; geochemistry of base-metal ore deposits; fabrics of metamorphic terranes. Mailing Add: US Geol Surv 345 Middlefield Rd Menlo Park CA 94025

THEODORE, THEODORE SPIROS, b Braddock, Pa, Nov 6, 33; m 57; c 3. MICROBIAL PHYSIOLOGY. Educ: Univ Pitteburgh, BSc, 55, MSc, 57, PhD(bact), 62. Prof Exp: Asst bact, Univ Pittsburgh, 57-62; staff fel, 62-65, RES MICROBIOLOGIST, NIH, 65- Mem: AAAS; Am Soc Microbiol. Res: Bacterial physiology and nutrition; intermediary and mineral metabolism; biochemical genetics. Mailing Add: NIH Bldg 5 Rm 237 Bethesda MD 20014

THEODORIDES, VASSILIOS JOHN, b Konstantia, Greece, Feb 20, 31; US citizen; m 58; c 3. VETERINARY PARASITOLOGY. Educ: Univ Thessaloniki, DVM, 56; Boston Univ, MA, 60, PhD(parasitol), 63. Prof Exp: Asst to prof clins vet sch, Univ Thessaloniki, 56-57; teaching fel microbiol, Boston Univ, 58-62; lectr micros anat, 62-63; res parasitologist, Charles Pfizer & Co, Inc, 63-65; sr microbiologist parasitologist, Smith Kline & French Labs, 65-66, group leader microbiol, 66-68, group leader animal health dept, 68, assoc dir res chemother, 68-73, MGR PARASITOL, SMITH KLINE CORP, 73- Concurrent Pos: Adj assoc prof sch vet med, Univ Pa, 72- Mem: Am Soc Microbiol; Am Vet Med Asn; Am Soc Parasitol; Am Soc Trop Med & Hyg; NY Acad Sci. Res: Morphology, physiology and electron microscopy of Trichomonas; development of chemotherapeutic agents for the control of gastrointestinal nematodes of domestic animals. Mailing Add: 1621 Margo Lane West Chester PA 19380

THEOKRITOFF, GEORGE, b Eng, Apr 7, 24; Can citizen. PALEONTOLOGY. Educ: Univ London, BSc, 45, MSc, 48, PhD(geol), 61. Prof Exp: Instr geol, Mt Holyoke Col, 54-56 & Bucknell Univ, 56-60; asst prof geol, Univ NH, 60-61; assoc prof geol, St Lawrence Univ, 64-67; ASSOC PROF GEOL, RUTGERS UNIV, NEWARK, 67- Mem: Geol Soc Am; Geol Soc London; Paleont Soc. Res: Cambrian paleontology and stratigraphy, including morphology, taxonomy and evolution of Cambrian trilobites; ecology of Cambrian organisms; biogeography and biostratigraphy of North Atlantic region. Mailing Add: Dept of Geol Rutgers Univ Newark NJ 07102

THEOLOGIDES, ATHANASIOS, b Ptolemais, Greece, Feb 5, 31; US citizen; m 65; c 2. INTERNAL MEDICINE, ONCOLOGY. Educ: Aristoteles Univ, MD, 55; Univ Minn, Minneapolis, PhD(med & biochem), 67. Prof Exp: From instr to assoc prof, 65-74, Nat Cancer Inst grant, Med Ctr, 69-75, PROF MED, MED SCH, UNIV MINN, MINNEAPOLIS, 74- Mem: Am Asn Cancer Res; Am Fedn Clin Res; Am Soc Clin Oncol; Soc Exp Biol & Med; Am Soc Hemat. Res: Medical oncology; tumor-host metabolic interrelationships. Mailing Add: Dept of Med Univ of Minn Med Ctr Minneapolis MN 55455

THEOPHANIDES, THEOPHILE, inorganic chemistry, spectroscopy, see 12th edition

THERIAULT, FREDERIC RUSSELL, information science, see 12th edition

THERIOT, EDWARD DENNIS, JR, b Baton Rouge, La, Mar 19, 38; m 60; c 2. PHYSICS. Educ: Duke Univ, BS, 60; Yale Univ, MS, 61, PhD(physics), 67. Prof Exp: NATO vis scientist fel physics, Europ Orgn Nuclear Res, Geneva, Switz, 67-68; res fel, Los Alamos Sci Lab, 68-69; PHYSICIST, FERMI NAT ACCELERATOR LAB, 69- Mem: Am Phys Soc. Res: Elementary particle physics, particularly relating to weak and electromagnetic interactions; hyperon decays; particle production; positronium. Mailing Add: Fermi Nat Accelerator Lab PO Box 500 Batavia IL 60510

THERIOT, LEROY JAMES, b Port Arthur, Tex, Apr 11, 35; m 58; c 3. INORGANIC CHEMISTRY. Educ: Southwestern La Univ, BS, 57; Tulane Univ, PhD(chem), 62. Prof Exp: Chemist, Ethyl Corp, 62-63; fel, Harvard Univ, 63-64 & Univ Tex, 64-65; asst prof, 65-70, ASSOC PROF CHEM, N TEX STATE UNIV, 70- Mem: Am Chem Soc. Res: Preparation and electronic structure of metal complexes. Mailing Add: Dept of Chem NTex State Univ Denton TX 76203

THERRIAULT, DONALD G, b Claremont, NH, June 14, 27; m 52; c 4. BIOCHEMISTRY. Educ: Univ Ottawa, BSc, 50; Univ NH, MS, 52; Univ Louisville, PhD, 60. Prof Exp: Asst biochem, Ind Univ, 52-54; res biochemist, US Army Med Res Lab, 54-62, res biochemist, US Army Res Inst Environ Med, 62-66, chief biochem & pharmacol div, 66-72; HEALTH SCIENTIST ADMINR, THROMBOSIS & HEMORRHAGIC DIS BR, NAT HEART & LUNG INST, 72- Mem: AAAS; Am Soc Biol Chem; fel Am Inst Chem; fel NY Acad Sci; Am Oil Chem Soc. Res: Application of physicochemical methods to the problem of interaction of calcium ions and phospholipids with purified plasma proteins involved in blood coagulation. Mailing Add: Thrombosis & Hemorrhagic Dis Br Nat Heart & Lung Inst Bethesda MD 20014

THERRIEN, CHESTER DALE, b Coos Bay, Ore, June 18, 36; m 60; c 2. MYCOLOGY, CYTOCHEMISTRY. Educ: St Ambrose Col, BA, 62; Univ Tex, PhD(bot), 66. Prof Exp: Asst prof biol, 65-70, ASSOC PROF BIOL, PA STATE UNIV, 70- Mem: Bot Soc Am; Genetics Soc Am. Res: Nucleic acids and nucleoproteins is the plasmodial slime molds. Mailing Add: Dept of Biol 208 Life Sci Bldg I Pa State Univ University Park PA 16802

THEUER, RICHARD CHARLES, b Hoboken, NJ, June 15, 39; m 62; c 3. NUTRITION. Educ: St Peter's Col, NJ, BS, 60; Univ Wis, Madison, MS, 62, PhD(biochem), 65; Ind State Univ, MBA, 73. Prof Exp: Res asst biochem, Univ Wis, 60-62 & 64-65; sr scientist, 65-67, group leader, 67-68, sect leader, 68-70, DIR DEPT NUTRIT RES, MEAD JOHNSON RES CTR, 70- Mem: AAAS; Am Chem Soc; Am Inst Nutrit. Res: Mineral metabolism; infant nutrition and formulas; special clinical formulas. Mailing Add: Dept of Nutrit Res Mead Johnson Res Ctr Evansville IN 47721

THEUER, WILLIAM JOHN, b New York, NY, Nov 27, 35; m 61; c 3. ORGANIC CHEMISTRY. Educ: Queens Col, BS, 57; Univ Del, PhD(chem), 65. Prof Exp: Sr scientist, Sandoz Pharmaceut, 65-67; res chemist, Celanese Res Co, 67-68, sr res chemist, Celanese Fibers Co, 68-71, group leader, Celanese Fibers Mkt Co, 71-74, TECH MGR, CELANESE COATINGS & SPEC CO, 74- Mem: Am Chem Soc; fel Am Inst Chem; Am Asn Textile Chemists & Colorists; Int Disposable & Nonwoven Asn. Res: Heterocyclic chemistry; photochemistry; cellulose and fiber chemistry; spinning research; pharmaceutical chemistry; industrial uses of fibers; textile polymer chemistry. Mailing Add: 4033 Silverbell Dr Charlotte NC 28211

THEURER, CLARK BRENT, b Logan, Utah, Oct 17, 34; m 56; c 4. ANIMAL NUTRITION. Educ: Utah State Univ, BS, 56; Iowa State Univ, MS, 60, PhD(animal nutrit), 62. Prof Exp: Asst prof animal sci, Va Polytech Inst, 62-64; assoc prof, 64-71, PROF ANIMAL SCI, UNIV ARIZ, 71-, ANIMAL SCIENTIST, AGR EXP STA, 74- Mem: Am Soc Animal Sci; Am Dairy Sci Asn; Soc Range Mgt. Res: Regulation of voluntary feed intake in ruminants; starch, protein and volatile fatty acid metabolism in ruminants. Mailing Add: Dept of Animal Sci Univ of Ariz Tucson AZ 85721

THEURER, JESSOP CLAIR, b Logan, Utah, Sept 4, 28; m 53; c 4. PLANT GENETICS, PLANT BREEDING. Educ: Utah State Univ, BS, 53, MS, 57; Univ Minn, PhD(plant genetics), 62. Prof Exp: Res fel oats radiation, Univ Minn, 61-62; res agronomist, 62-63, GENETICIST, CROPS RES DIV, AGR RES SERV, USDA, 63- Concurrent Pos: Int farm youth exchange student, Lebanon & Syria, 52; assoc ed, Crop Sci Soc Am, 70-72. Mem: Am Soc Agron; Am Soc Sugar Beet Technol. Res: Agronomy; cytology; plant pathology; breeding, genetics and disease resistance of sugar beets. Mailing Add: Crops Res Lab Utah State Univ UM63 Logan UT 84322

THEUSCH, COLLEEN JOAN, b Milwaukee, Wis, Dec 18, 32. NUMBER THEORY, OPERATIONS RESEARCH. Educ: Dominican Col, Wis, BEd, 61; Univ Detroit, MA, 66; Mich State Univ, PhD(math), 71. Prof Exp: Instr math, Col Racine, 70-71; MATHEMATICIAN, RES & DEVELOP DEPT, RICHMAN BROS CO, 71- Mem: Am Math Soc. Res: Development of grading and marker making system and numerically controlled cutting via laser cutters in men's clothing manufacturing. Mailing Add: 14567 Madison Ave Apt 608 Lakewood OH 44107

THEUWS, JACQUES ANTOINE, b ommel, Belg, Nov 15, 14. CULTURAL ANTHROPOLOGY. Hon Degrees: PhD, Cath Univ Louvain. Prof Exp: Asst lectr anthrop, Nat Univ Zaire, Kinshasa Campus, 62-65; prof, Univ Congo, 65-69; PROF CULT ANTHROP, UNIV WINDSOR, 70- Mem: Can Sociol & Anthrop Asn. Res: Phenomenology; Luba culture. Mailing Add: Dept of Anthrop Univ of Windsor Windsor ON Can

THEWS, ROBERT LEROY, b Fairmont, Minn, June 13, 39; c 2. THEORETICAL PHYSICS. Educ: Mass Inst Technol, SB, 62, PhD(physics), 66. Prof Exp: Physicist, Lawrence Berkeley Lab, Univ Calif, 66-68; res assoc & asst prof physics, Univ Rochester, 68-70; asst prof, 70-75, ASSOC PROF PHYSICS, UNIV ARIZ, 75- Mem: Am Phys Soc. Res: Theoretical high energy elementary particle physics. Mailing Add: Dept of Physics Univ of Ariz Tucson AZ 85721

THEYS, JOHN C, US citizen. ASTROPHYSICS. Educ: NC State Univ, BS, 64; Columbia Univ, PhD(astron), 73. Prof Exp: Fel astron, Kitt Peak Nat Observ, 73-74; ASST PROF ASTRON, STATE UNIV NY STONY BROOK, 74- Mem: Am Astron Soc; Astron Soc Pac. Res: Interacting galaxies; dynamics of the interstellar medium; stellar systems. Mailing Add: Dept of Earth & Space Sci State Univ NY Stony Brook NY 11794

THIBAULT, NEWMAN WILLIAM, b Newmarket, NH, May 26, 10; m 38; c 3. MINERALOGY. Educ: Dartmouth Col, AB, 32; Syracuse Univ, AM, 34; Univ Mich, PhD(mineral), 43. Prof Exp: Asst geol, Syracuse Univ, 32-34; asst, Va Polytech Inst & State Univ, 34-35; lab asst mineral, Univ Mich, 36; res crystallographer, 36-52, chief phys res secr, Res & Develop Dept, 52-53, asst dir, 53-62, assoc dir, 62-63, sir res & develop, Abrasive Div, 63-67, Grinding Wheel Div, 67-69, VPRES TECH, GRINDING WHEEL DIV, NORTON CO, 69- Mem: Fel Geol Soc Am; fel Mineral Soc Am. Res: Application of mineralogical techniques to problens in the abrasive industry; vitrified, organic and metal bonded grinding wheels and associated products. Mailing Add: Grinding Wheel Div Norton Co 1 New Bond St Worcester MA 01606

THIBAULT, ROGER EDWARD, b Salem, Mass, June 28, 47; m 70. ECOLOGY. Educ: Univ Wis, BS, 69; Univ Conn, PhD(ecol), 74. Prof Exp: NDEA fel ecol, Univ Conn, 71-74; instr zool, Iowa State Univ, 74-75; ASST PROF BIOL, BOWLING GREEN STATE UNIV, 75- Mem: Am Soc Ichthyol & Herpet; Ecol Soc Am; Soc Study Evolution; Sigma Xi; AAAS. Res: Aquatic ecology; aquatic entomology and ichthyology; evolution of unisexual vertebrates. Mailing Add: Dept of Biol Sci Bowling Green State Univ Bowling Green OH 43402

THIBAULT, THOMAS DELOR, b Claremont, NH, Aug 14, 42; m 69; c 3. ORGANIC CHEMISTRY. Educ: Providence Col, BS, 64; Mass Inst Technol, PhD(org chem), 69. Prof Exp: Sr org chemist, 69-74, RES SCIENTIST, ELI LILLY & CO, 74- Mem: Am Chem Soc. Res: Synthesis of biologically active structures; reaction mechanisms; new synthetic reactions. Mailing Add: Eli Lilly & Co Box 708 Greenfield IN 46140

THIBEAULT, JACK CLAUDE, b Lowell, Mass, June 23, 46; m 66; c 1. INORGANIC CHEMISTRY. Educ: Lowell Technol Inst, BS, 67; Calif Inst Technol, PhD(inorg chem), 72. Prof Exp: Fel theoret chem, Cornell Univ, 72-74; RES CHEMIST, ROHM AND HAAS CO, 74- Res: Organometallic catalysis; process research; theoretical chemistry. Mailing Add: Rohm and Haas Co 5000 Richmond St Philadelphia PA 19137

THIBERT, ROGER JOSEPH, b Tecumseh, Ont, Aug 29, 29; m 54; c 2. CHEMISTRY. Educ: Univ Western Ont, BA, 51; Univ Detroit, MS, 54; Wayne State Univ, PhD(biochem), 58. Prof Exp: Lectr chem, 53-56, from asst prof to assoc prof, 57-67, assoc dean arts & sci, 64-70, PROF CHEM, UNIV WINDSOR, 67- Concurrent Pos: Instr sch nursing, Grace Hosp, 54-; res assoc sch med, Wayne State Univ, 71-72, prof path, 72-; assoc dir clin chem lab, Detroit Gen Hosp, Mich, 71- Mem: Fel AAAS; Am Chem Soc; fel Chem Inst Can; Am Soc Biol Chem; Am Asn Clin Chem. Res: Clinical biochemistry. Mailing Add: Dept of Chem Univ of Windsor Windsor ON Can

THIBODEAU, GARY ARTHUR, b Sioux City, Iowa, Sept 26, 38; m 64; c 2. PHYSIOLOGY, PHARMACOLOGY. Educ: Creighton Univ, BS, 62; SDak State Univ, MS, 67, MS, 70, PhD(physiol), 71. Prof Exp: Mem prof serv staff, Baxter Labs, Inc, 63-65; ASST PROF ENTOM-ZOOL, SDAK STATE UNIV, 65- Mem: AAAS; Am Inst Biol Sci; Am Pub Health Asn. Res: Animal physiology; pharmacology of hypolipedemic agents; pathological dyslipemias; thyroid physiology; vascular morphology. Mailing Add: Dept of Entom-Zool SDak State Univ Brookings SD 57006

THICKSTUN, WILLIAM RUSSELL, JR, b Washington, DC, Oct 14, 22; m 54. MATHEMATICS. Educ: Univ Md, BSc, 47, MA, 49, PhD, 52. Prof Exp: Asst math, Univ Md, 48-52, instr, 52-53; mathematician, US Naval Ord Lab, Md, 53-68; ASSOC PROF MATH, CLARKSON COL TECHNOL, 68- Mem: Am Math Soc; Soc Indust & Appl Math; Am Inst Aeronaut & Astronaut. Res: Applied mathematics; fluid dynamics. Mailing Add: Dept of Math Clarkson Col of Technol Potsdam NY 13676

THIE, JOSEPH ANTHONY, b Indianapolis, Ind, Dec 15, 27. REACTOR PHYSICS. Educ: Univ Notre Dame, BS, 47, PhD(physics), 51. Prof Exp: Instr physics, Univ Dayton, 47-48; AEC fel, Cornell Univ, 51-52; assoc physicist, Argonne Nat Lab, 53-60; partner & consult, McLain Rodger Assocs, 60-61; INDEPENDENT NUCLEAR REACTOR CONSULT, 61- Mem: Am Nuclear Soc. Res: Reactor safety; reactor random fluctuation phenomena; heavy water exponential experiments. Mailing Add: PO Box 517 Barrington IL 60010

THIEBAUX, HELEN JEAN, b Washington, DC, Aug 17, 35; m 62; c 5. APPLIED STATISTICS. Educ: Reed Col, BA, 57; Stanford Univ, PhD(statist), 64. Prof Exp: Res asst econ & indust, Ivan Block & Assoc, Econ & Indust Consult, 58; statist analyst, Med Sch, Univ Ore, summer 60; asst pub health analyst, Calif Dept Pub Health, summer 61; asst prof statist, Univ Conn, 64-65; asst prof, Univ Mass, Amherst, 65-66 & 67-71; lectr, Univ Colo, Boulder, 72-73; consult, Nat Ctr Atmospheric Res, 72-74; ASSOC PROF MATH, DALHOUSIE UNIV, 75- Mem: Am Statist Asn; Can Meteorol Soc. Res: Models and testing procedures for multivariate systems with diverse information sources; applications to meteorology. Mailing Add: Dept of Math Dalhousie Univ Halifax NS Can

THIEBAUX, MARTIAL LEON, JR, theoretical physics, atmospheric physics, see 12th edition

THIEBERGER, PETER, b Vienna, Austria, Sept 19, 35; m 63; c 1. NUCLEAR PHYSICS. Educ: Balseiro Inst Physics, Argentina, MS, 59; Univ Stockholm, Fil lic, 61, Fil Dr(physics), 62. Prof Exp: Physicist, Bariloche Atomic Ctr, Argentine AEC, 62-65; res asst, 65-67, from asst physicist to assoc physicist, 67-70, physicist, 70-74, head, Tandem Van De Graaff Facil Opers, 71-75, SR PHYSICIST, BROOKHAVEN NAT LAB, 74-, GROUP LEADER TANDEM VAN DE GRAAF FACIL OPERS & DEVELOP GROUP, 75- Concurrent Pos: Asst prof, Balseiro Inst Physics, 62-63, prof, 63-65; vis scientist, Res Inst Physics, Stockholm, Sweden, 68-69; consult, Tennelec Instrument Co, 67-71. Mem: Am Phys Soc. Res: Nuclear structure; measurements of half-lives and g-factors of nuclear states; development of nuclear instruments and methods; nuclear reactions with heavy ions; accelerator development and operation. Mailing Add: Physics Dept Brookhaven Nat Lab Upton NY 11973

THIEDE, FREDERICK C, physiology, see 12th edition

THIEDE, HENRY A, b Rochester, NY, Oct 2, 26; m 51; c 2. OBSTETRICS & GYNECOLOGY. Educ: Univ Buffalo, MD, 49. Prof Exp: Intern surg, Buffalo Gen Hosp, 49-50; asst resident obstet & gynec, Genesee & Strong Mem Hosps, Rochester, 52-54; resident, Genesee Hosp, 54-56; from instr to assoc prof, Sch Med, Univ Rochester, 57-66; asst dean, 70-73, PROF OBSTET & GYNEC & CHMN DEPT, SCH MED, UNIV MISS, 67-, ASSOC DEAN SCH MED, 73- Mem: Am Col Obstet & Gynec; Soc Gynec Invest; Am Gynec Soc; Soc Human Genetics. Res: Biology of reproduction and reproduction wastage. Mailing Add: Univ of Miss Med Ctr 2500 N State St Jackson MS 39216

THIEDE, JÖRN, b Berlin, Ger, Apr 14, 41; m 70; c 2. GEOLOGICAL OCEANOGRAPHY. Educ: Kiel Univ, Diplom, 67, Dr rer nat(sci), 71. Prof Exp: Amanuensis geol, Aarhus Univ, 67-70; res asst geol, Kiel Univ, 70-71; amanuensis geol, Aarhus Univ, 71-72, lectr geol, 72; univ lectr micropaleont, Bergen Univ, 73, sr lectr, 73-74; ASST PROF GEOL OCEANOG, ORE STATE UNIV, 74- Concurrent Pos: Mem exec comt, Climate Long-Range Invests, Mapping & Predictions, NSF-Int Decade Ocean Explor, 75-; mem, Joint Oceanog Insts Deep Earth Sampling Panel Passive Continental Margins, 75-; core cur, Sch Oceanog, Ore State Univ, 75- Mem: Geol Soc Am; Soc Econ Paleontologists & Mineralogists; Ger Geol Union; Ger Paleont Soc; Danish Geol Soc. Res: Paleo-oceanography and paleoclimatology;

ecology and paleo-ecology of planktonic protozoans, of mero-planktonic and nektonic mollusks; sedimentology of biogenous deep-sea sediments; history of the oceans. Mailing Add: Sch of Oceanog Ore State Univ Corvallis OR 97331

THIEL, THOMAS J, b Upper Sandusky, Ohio, Dec 31, 28; m 55; c 4. SOIL PHYSICS, HYDROLOGY. Educ: Ohio State Univ, BS, 56, MS, 59. Prof Exp: Soil scientist, 57-63, soil scientist & asst adminr, 63-72, PROG ANALYST, PROG PLANNING & REV, SOIL & WATER CONSERV RES DIV, AGR RES SERV, USDA, 72- Mem: Am Soc Agron; Soil Sci Soc Am; Soil Conserv Soc. Res: Analog, numerical, model and field studies on the drainage of agricultural lands, specifically on drainage of heavy-clay lake-bed soils, sloping fragipan soils and peat soils under artesian pressure; computerized information retrieval. Mailing Add: USDA Agr Res Serv NCent Region 2000 W Pioneer Pkwy Peoria IL 61614

THIELE, ELIZABETH HENRIETTE, b Portland, Ore, Apr 26, 20. BIOCHEMISTRY. Educ: Univ Wash, BS, 42; Univ Pa, MS, 43, PhD(bact), 51. Prof Exp: Technician, Merck Sharp & Dohme, 43-44, RES ASSOC, MERCK INST THERAPEUT RES, 44- Mem: Am Soc Microbiol; Reticuloendothelial Soc. Res: Enzymology; immunology. Mailing Add: Merck Inst for Therapeut Res Rahway NJ 07065

THIELE, VICTORIA FLORENCE, b New York, NY, Oct 22, 33. NUTRITION. Educ: Hunter Col, BA, 55, MS, 57; Univ Maine, Orono, PhD(biochem), 65. Prof Exp: Instr chem, Univ Maine, Orono, 60-61; PROF NUTRIT, SYRACUSE UNIV, 64- Mem: Am Inst Nutrit; Soc Nutrit Educ; Am Dietetic Asn; Am Chem Soc; Nutrit Today Soc. Res: Vitamin B6; nutritional status surveys; nutrition education. Mailing Add: Dept of Human Nutrit 200 Slocum Hall Syracuse Univ Syracuse NY 13210

THIELEN, LAWRENCE EUGENE, b Chicago, Ill, Sept 2, 21; m 50; c 2. ORGANIC CHEMISTRY. Educ: Loyola Univ, Ill, BS, 42. Prof Exp: Chemist, Kankakee Ord Works, E I du Pont de Nemours & Co, 42; res chemist, Pure Oil Co, 43 & 46, G D Searle & Co, 46-56, Nalco Chem Co, 56-58 & Abbott Labs, 58-62; GROUP LEADER POLYMER & ORG CHEM, R R DONNELLEY & SONS CO, 62- Mem: Am Chem Soc. Res: Pharmaceuticals; synthesis of plastics and pharmaceuticals; inks; coatings; printing process technology. Mailing Add: R R Donnelley & Sons Co R&D Bldg 3-5 330 E 22nd St Chicago IL 60616

THIELGES, BART A, b Chicago, Ill, June 16, 38; m 60; c 3. FOREST GENETICS. Educ: Southern Ill Univ, BS, 63; Yale Univ, MF, 64, MPhil, 67, PhD(forest genetics), 67. Prof Exp: Res asst plant anat, Southern Ill Univ, 62-63; res asst forest genetics, Yale Univ, 63-67; asst prof, Ohio Agr Res & Develop Ctr, 67-71; ASSOC PROF, SCH FORESTRY & WILDLIFE MGT, LA STATE UNIV, BATON ROUGE, 71- Mem: Bot Soc Am; Soc Am Foresters. Res: Breeding of forest trees; natural variation studies; experimental taxonomy; biochemical systematics and genetics. Mailing Add: Sch Forestry & Wildlife Mgt La State Univ Baton Rouge LA 70803

THIELKING, DAVID H, b Amsterdam, NY, Apr 30, 21; m 53; c 2. PHYSICS. Educ: St Lawrence Univ, BS, 42; Univ Buffalo, EdM, 48, EdD(educ sci), 57. Prof Exp: Jr engr, Gen Tel & Electronics Corp, 42-46; fel physics, Univ Buffalo, 46-48; from asst prof to assoc prof, 48-63, PROF PHYSICS, STATE UNIV NY COL BUFFALO, 63- Mem: Am Asn Physics Teachers. Mailing Add: Dept of Physics State Univ of NY Col Buffalo NY 14222

THIELKING, WILLIAM HENRY, organic chemistry, see 12th edition

THIELMAN, LEROY OSWALD, b Blackduck, Minn, Feb 26, 42; m 70. LASERS. Educ: St Cloud State Col, BA, 69; Univ Idaho, PhD(physics), 74. Prof Exp: ASSOC SCIENTIST PHYSICS, LOCKHEED MISSILES & SPACE CO INC, 74- Res: The study of atmospheric aerosol scattering of 10 micrometer radiation using carbon dioxide laser heterodyne detection. Mailing Add: 596 La Conner Dr 6 Sunnyvale CA 94087

THIELMANN, VERNON JAMES, b Hastings, Minn, June 4, 37; m 63; c 2. ANALYTICAL CHEMISTRY. Educ: Northern State Col, BS, 63; Univ SDak, MNS, 68; Baylor Univ, PhD(chem), 74. Prof Exp: Teacher pub schs, SDak & Iowa, 63-69; asst prof chem, Morningside Col, 69-72; ASST PROF CHEM, SOUTHWEST MO STATE UNIV, 74- Mem: Am Chem Soc; Sigma Xi. Res: Preparation of complex cation exchanged montmorillonite clays and subsequent study of their structure, thermal stability, surface area and effectiveness for use as gas chromatographic packing materials. Mailing Add: Dept of Chem Southwest Mo State Univ Springfield MO 65802

THIEME, FREDERICK PATTON, b Seattle, Wash, Feb 20, 14; m 48; c 5. ANTHROPOLOGY. Educ: Univ Wash, AB, 36; Columbia Univ, PhD, 50. Hon Degrees: LLB, Colo State Univ, 70. Prof Exp: Social Sci Res Coun res assoc, Puerto Rico, 48-49; asst prof anthrop, Univ Mich, 49-55, assoc prof, 55-58, chmn dept, 57-58; provost, Univ Wash, 58-63, vpres, 63-69; pres, , 69-74, PROF ANTHROP, UNIV COLO, BOULDER, 69- Concurrent Pos: Mem, Nat Sci Bd, 64-, Nat Comn Accrediting, 70-74 & Nat Bd Grad Educ, 71-74 & 72-75; comnr, Educ Comn of the States, 69-75; mem, Nat Adv Dent Res Coun & White House Panel on Privacy & Behav Res, Off Sci & Technol. Mem: Fel AAAS; fel Am Anthrop Asn; Am Asn Phys Anthrop. Res: Physical anthropology; human genetics and growth. Mailing Add: Dept of Anthrop Univ of Colo Boulder CO 80302

THIEME, MELVIN T, b Decatur, Ind, Oct 27, 25; m 49; c 3. PHYSICS, MATHEMATICS. Educ: Purdue Univ, BS, 49, MS, 51, PhD(nuclear physics), 55. Prof Exp: Staff mem, Los Alamos Sci Lab, 55-63 & Lawrence Radiation Lab, 63-64, STAFF MEM, LOS ALAMOS SCI LAB, 64- Mem: Am Phys Soc. Res: Beta-ray spectroscopy; nuclear weapon design and development with specialization in shock hydrodynamics. Mailing Add: 79 Mesa Verde Dr Los Alamos NM 87544

THIEN, LEONARD BENEDICT, botany, evolution, see 12th edition

THIEN, STEPHEN JOHN, b Clarence, Iowa, Apr 11, 44; m 66; c 2. AGRONOMY. Educ: Iowa State Univ, BS, 66; Purdue Univ, Lafayette, MS, 68, PhD(agron), 71. Prof Exp: Asst agron, Purdue Univ, 66-70; ASST PROF AGRON, KANS STATE UNIV, 70- Mem: Am Soc Agron; Soil Sci Soc Am; Am Soc Plant Physiol. Res: Soil fertility; plant nutrition and physiology; soil management. Mailing Add: Dept of Agron Waters Hall Kans State Univ Manhattan KS 66506

THIERET, JOHN WILLIAM, b Chicago, Ill, Aug 1, 26; m 50; c 5. BOTANY. Educ: Utah State Univ, BS, 50, MS, 51; Univ Chicago, PhD(bot), 53. Prof Exp: Cur econ bot, Field Mus Natural Hist, 53-62; from assoc prof to prof biol, Univ Southwestern La, 62-73, Edwin Lewis Stephens prof sci, 72-73; PROF BOT & CHMN DEPT BIOL SCI, NORTHERN KY UNIV, 73- Concurrent Pos: Adv, Encycl Britannica, 59- Mem: Soc Econ Bot. Res: Flora of central and southeastern United States; taxonomy of Scrophulariaceae and Gramineae; seed and fruit classification and morphology; economic botany. Mailing Add: Dept of Biol Sci Northern Ky Univ Highland Heights KY 41076

THIERRIN, GABRIEL, b Surpierre, Switz, Dec 22, 21; m 51; c 2. MATHEMATICS, COMPUTER SCIENCE. Educ: Univ Fribourg, DSc, 51; Univ Paris, DSc, 54. Prof Exp: Sci assoc, Nat Ctr Sci Res, France, 52-54 & Nat Found Sci Res, Switz, 54-57; prof math, Inst Higher Studies, Tunisia, 57-58 & Univ Montreal, 58-70; PROF MATH, UNIV WESTERN ONT, 70- Mem: Asn Comput Mach; Am Math Soc; Can Math Soc; Can Info Processing Soc. Res: Algebra; theories of rings and semi-groups; systems theory; theory of automata and languages. Mailing Add: Dept of Math Univ of Western Ont London ON Can

THIERS, HARRY DELBERT, b Ft McKavett, Tex, Jan 22, 19; m 53; c 1. MYCOLOGY. Educ: Schreiner Inst, AB, 38; Univ Tex, BA, 41, MA, 47; Univ Mich, PhD, 55. Prof Exp: Asst, Univ Tex, 39-41, tutor bot, 45-47, instr bot & mycol, Tex A&M Univ, 47-50, asst biol, 50-55, assoc prof, 55-59; assoc prof, 59-63, PROF BIOL, SAN FRANCISCO STATE UNIV, 63- Concurrent Pos: Assoc plant pathologist, USDA, 48. Mem: Bot Soc Am; Mycol Soc Am; Int Asn Plant Taxon; Brit Mycol Soc; Am Bryol & Lichenological Soc. Res: Taxonomy of fleshy fungi of California and West Coast of North America. Mailing Add: Dept of Biol San Francisco State Univ San Francisco CA 94132

THIERS, RALPH EDWARD, analytical chemistry, see 12th edition

THIERSCH, JOHANNES BERNHARD, b Freiberg, Ger, June 3, 10; nat US; m; c 3. PATHOLOGY, CANCER. Educ: Univ Univ Freiburg, DrMedHabil, 35; Univ Berne, MD, 35; Univ Adelaide, MD, 38; Wash Univ, MD, 51; Am Bd Path, dipl, 50. Prof Exp: Jr res asst path, Univ Freiburg, 34-35; prof path & head dept, Nat Tung Chi Univ, China, 36-38; clin pathologist & head dept, Inst Med & Vet Sci, Australia, 38-47; asst prof path, Med Sch, Univ Wash, 50-52, assoc prof path & coordr cancer res, 52-54, assoc res prof path, 54-55, assoc prof path & coordr cancer res, 52-54, assoc res prof path, 54-55, assoc res prof pharmacol, 55-63, clin prof pharmacol, 63-70; PRES & DIR, INST BIOL RES, 55-, PRES & DIR, CYTODIAG LAB, 56- Concurrent Pos: Span Nat res fel nerv & brain dis, Nat Cancer Inst, Spain, 35-36; Brit Anti-Cancer Campaign grant, Univ Adelaide, 38-47; fel cancer res, Mem Hosp, Mem Ctr Cancer & Allied Dis, New York, 47-50; USPHS cancer res fel, Sloan-Kettering Inst, New York, 48-49; State of Wash res grant, 52; USPHS res grant, 54-63; Population Coun, Inc grant, 55-60; Neale res pathologist, Univ Adelaide, 38-47; dir lab, Seattle Gen Hosp, 51-71; sr consult, US Naval Hosp, Bremerton, 52-; lectr, Postgrad Med Sch, Univ London, 53, Max Planck Inst, Heidelberg, 53 & Univs, Union SAfrica & Uganda, 54; dir lab, Ballard Gen Hosp, 54-; lectr, Int Conf Planned Parenthood Asn, Japan, 55, India, 59, Karolinska Inst, Sweden, 56 & Molten Inst, Cambridge Univ, 60; pathologist, Edmunds Mem Hosp, 62- Mem: Soc Exp Biol; Asn Path & Bact; fel Col Am Path. Res: Effect of folic acid antagonists, antimetabolites and related compounds on mammalian reproduction; cancer, especially leukemia; nitrogen mustards and substitute ethylenimines; control of reproduction; anatomical distribution of residual carcinoma of the cervix. Mailing Add: Cytodiag Lab 5334 Tallman NW Seattle WA 98107

THIES, CURT, physical chemistry, see 12th edition

THIES, HERMAN RODERICK, chemistry, see 12th edition

THIES, RICHARD WILLIAM, b Detroit, Mich, Sept 16, 41; m 64; c 1. ORGANIC CHEMISTRY. Educ: Univ Mich, BS, 63; Univ Wis, PhD(org chem), 67. Prof Exp: NIH fel org chem, Univ Calif, Los Angeles, 67-68; asst prof, 68-75, ASSOC PROF ORG CHEM, ORE STATE UNIV, 75- Mem: Am Chem Soc; The Chem Soc. Res: Medium sized ring chemistry; carbonium ion chemistry; thermal rearrangements; synthesis of hormone analogs. Mailing Add: Dept of Chem Ore State Univ Corvallis OR 97331

THIES, ROGER E, b Bronxville, NY, June 30, 33; div; m 71; c 2. PSYCHOSOMATIC MEDICINE. Educ: Bates Col, BS, 55; Harvard Univ, AM, 57; Rockefeller Univ, PhD(physiol), 61; Univ Okla, MA, 75. Prof Exp: Guest investr neurophysiol & NIH fel, Rockefeller Univ, 60-61; from instr to asst prof physiol, Sch Med, Wash Univ, 61-65; lectr, Makerere Univ Col, Uganda, 65-67; ASSOC PROF PHYSIOL & BIOPHYS, MED CTR, UNIV OKLA, 67-, ASSOC PROF BIOL PSYCHOL, HEALTH SCI CTR, 72- Concurrent Pos: Hon res asst & NIH spec fel, Univ Col, Univ London, 64-65. Mem: Am Physiol Soc; Am Orthopsychiat Asn; Guild Struct Integration. Res: Mammalian neuromuscular transmission; affective education and innovative teaching; evaluation of team building activities for health professionals; synaptic transmission; direct manipulation of jaw muscles for relief of tension; ways of health professionals caring for themselves. Mailing Add: 616 Miller Ave Norman OK 73069

THIESFELD, VIRGIL ARTHUR, b Glencoe, Minn, Oct 26, 37; m 59; c 2. BOTANY, PLANT PHYSIOLOGY. Educ: Luther Col, Iowa, BA, 59, Univ SDak, Vermillion, MA, 63; Univ Okla, PhD(bot), 65. Prof Exp: Teacher high sch, Iowa, 59-60 & Minn, 60-62; res asst bot, Univ Okla, 63-65; asst prof, 65-68, ASSOC PROF BIOL & CHMN DEPT, UNIV WIS-STEVENS POINT, 68- Concurrent Pos: Res grant, Univ Wis-Stevens Point, 68-69. Mem: Am Soc Plant Physiol. Res: Plant growth regulators. Mailing Add: Dept of Biol Univ of Wis Stevens Point WI 54481

THIESSEN, GEORGE JACOB, b Russia, May 7, 13; nat US; m 46; c 3. PHYSICS. Educ: Univ Sask, BS, 35, MSc, 37; Columbia Univ, PhD(physics), 41. Prof Exp: Radon technician, Sask Cancer Comn, 36-37; res officer, 41-52, sr res officer, 53-55, PRON RES OFFICER, NAT RES COUN CAN, 55- Concurrent Pos: Sci investr, Dept Munitions & Supply, 45; spec lectr, McGill Univ, 48-56 & Univ Ottawa, 56-61. Mem: Fel Acoust Soc Am; fel Royal Soc Can. Res: Applied physics; acoustics. Mailing Add: Div of Physics Nat Res Coun Ottawa ON Can

THIESSEN, HENRY ARCHER, b Teaneck, NJ, Nov 8, 40; m 62; c 1. PHYSICS. Educ: Calif Inst Technol, BS, 61, MS, 62, PhD(physics), 67. Prof Exp: STAFF MEM MEDIUM ENERGY PHYSICS, LOS ALAMOS SCI LAB, 66- Mem: Am Phys Soc. Res: Experimental medium energy physics. Mailing Add: Los Alamos Sci Lab PO Box 1663 Los Alamos NM 87544

THIESSEN, REINHARDT, JR, b Kiel, Wis, Oct 20, 13; m 38; c 2. BIOCHEMISTRY. Educ: Univ Pittsburgh, BS, 34. Prof Exp: Asst, Univ Pittsburgh, 35-37; head chemist res & control lab, Repub Yeast Corp, NJ, 37-41; proj leader cent labs, 42-46, head biol sect, 46-48, head nutrit sect, 48-55, from asst lab dir to lab dir, 55-62, sr res specialist, Tech Ctr, 62-69, res assoc & area mgr nutrit sci, 69-72, CORP RES MGR NUTRIT SCI, TECH CTR, GEN FOODS CORP, 72- Mem: AAAS; Am Pub Health Asn; Am Chem Soc; Inst Food Technol; Am Inst Nutrit. Res: Nutrition; toxicology; bacteriology; protein nutrition; carbohydrate nutrition; cacao chemistry; tracers in nutrition and toxicology; food chemistry; dental caries. Mailing Add: 586 Gilbert Ave Pearl River NY 10965

THIESSEN, WILLIAM ERNEST, b Kansas City, Mo, Sept 17, 34; m 60; c 1. ORGANIC CHEMISTRY. Educ: Univ Calif, Berkeley, BS, 56, PhD(chem), 60. Prof Exp: Instr chem, Univ Wash, 60-62; asst prof, Univ Calif, Davis, 62-68; NIH spec fel, Chem Div, Oak Ridge Nat Lab, 68-70, RES CHEMIST, CHEM DIV, OAK RIDGE

NAT LAB, 70- Mem: Am Chem Soc; Am Crystallog Asn. Res: Structure elucidation of complex natural products and accurate molecular geometry of organic compounds by x-ray and neutron diffraction. Mailing Add: Chem Div Oak Ridge Nat Lab Oak Ridge TN 37830

THIGPEN, CHARLES, statistics, see 12th edition

THILENIUS, OTTO G, b Bad Soden, Ger, July 7, 29; US citizen; m 56; c 2. PEDIATRIC CARDIOLOGY, PHYSIOLOGY. Educ: Univ Frankfurt, MD, 53; Univ Chicago, PhD(physiol), 62. Prof Exp: Resident, 57-59, instr, 61-62, asst prof pediat & physiol, 64-69, assoc prof pediat, 69-72, PROF PEDIAT, UNIV CHICAGO, 72- Concurrent Pos: USPHS fel physiol, Univ Chicago, 59-61; fel cardiol, Harvard Univ, 62-64. Mem: Am Heart Asn; Am Physiol Soc; Am Acad Pediat; Am Col Cardiol; Soc Pediat Res. Mailing Add: Dept of Pediat Univ of Chicago Chicago IL 60637

THILL, BRUCE PALMER, organic chemistry, polymer chemistry, see 12th edition

THILO, EDWARD RUDOLF, b Philadelphia, Pa, Jan 18, 16; m 41; c 3. PHYSICS. Educ: Temple Univ, BS, 37, MA, 39. Prof Exp: Physicist, Beacon Res Labs, Tex Co, NY, 40-41; physicist, 41-57, DIR PHYSICS RES LAB, PITMAN DUNN LABS, FRANKFORD ARSENAL, ARMAMENTS COMMAND, US DEPT ARMY, 57- Mem: AAAS; Am Phys Soc; Sigma Xi. Res: High speed radiography; x-ray diffraction; ballistic measurement and instrumentation. Mailing Add: Physics Res Lab Frankford Arsenal Philadelphia PA 19137

THIMANN, KENNETH VIVIAN, b Ashford, Eng, Aug 5, 04; nat US; m 29; c 3. PLANT PHYSIOLOGY. Educ: Univ London, BSc, 24, PhD(biochem), 28. Hon Degrees: AM, Harvard Univ, 60; PhD, Univ Basel, 60; Dr, Univ Clermont-Ferrand, 61. Prof Exp: Instr bact, Univ London, 26-28, Beit fel, 29-30; instr biochem, Calif Inst Technol, 30-35; lectr bot, 35-36, asst prof plant physiol, 36-39, tutor, 36-42, assoc prof, 39-46, dir biol labs, 46-50, prof, 46-62, Higgins prof biol, 62-65, EMER HIGGINS PROF BIOL, HARVARD UNIV, 65-; PROF BIOL, UNIV CALIF, SANTA CRUZ, 65- Concurrent Pos: Guggenheim fel, 50-51 & 58; exchange prof, France, 54-55; provost, Crown Col, Univ Calif, Santa Cruz, 65-72; pres, Int Bot Cong, Seattle, 69 & Int Plant Growth Substance Cong, Tokyo, 73; mem exec comt, Assembly Life Sci, Nat Res Coun, 73-; vis prof, Univ Mass, Amherst, 74. Honors & Awards: Hales Prize, Am Soc Plant Physiol, 36. Mem: Fel Nat Acad Sci; Am Soc Naturalists (pres, 54); Bot Soc Am (pres, 60); Am Soc Plant Physiol (pres, 50); Soc Gen Physiol (pres, 49). Res: Physiology of bacteria, protozoa and fungi; growth, auxins and correlation in plants; general plant biochemistry. Mailing Add: Thomann Labs Univ of Calif Santa Cruz CA 95064

THIO, ALAN POO-AN, b Jatinegara, Indonesia, Jan 17, 31; US citizen; m 57; c 3. MEDICINAL CHEMISTRY, PESTICIDE CHEMISTRY. Educ: Univ Indonesia, BS, 54, MS, 57; Univ Ky, PhD(org chem), 61. Prof Exp: Assoc prog orgc chem, Bandug Inst Technol, 60-67; assoc med chem, Col Pharm, 6770, SR CHEM ANALYST, DIV REGULATORY SERV, UNIV KY, 70- Concurrent Pos: Asbtractor, Chem Abstr Serv, 66- Mem: Am Chem Soc; Asn Offs Anal Chemists. Res: Development of analytical procedures for the quantitative determination of pesticide residues in feeds, fertilizers and soil. Mailing Add: Div of Regulatory Serv Univ of Ky Lexington KY 40506

THIRGOOD, JACK VINCENT, b Northumberland, Eng, Apr 5, 24; m 49; c 3. FORESTRY. Educ: Univ Wales, BSc, 50; Univ BC, MF, 61; Ore State Univ, MF, 65; State Univ NY Col Forestry, Syracuse, PhD, 71. Prof Exp: With, US Forestry Comn, 40-44, silviculture asst, Res Br, 50-54; silviculturist, Cyprus Forest Serv, 54-56; adv & dir Nat Forest Res Inst Iraq, Food & Agr Orgn, UN, 56-57; forestry consult, US, 57-58; forester, BC Forest Serv, 58; res forester, Celgar Ltd, BC, 59; asst prof silviculture, State Univ NY Col Forestry, Syracuse, 60-62; prof silviculture & forest bot & head dept, Univ Liberia, Found Mutual Asst in Africa & UK Ministry Overseas Develop, 62-64; forestry consult & area dir, Tilhill Forestry & Adv Ltd, Eng, 64-67; ASSOC PROF SILVICULTURE & FOREST POLICY, UNIV BC, 68- Concurrent Pos: Consult indust land reclamation, 58- Mem: Int Soc Trop Foresters; corp mem Soc Am Foresters; Brit Ecol Soc; Sigma Xi; Can Inst Forestry. Res: Silviculture, forest management, history and policy; temperate, tropical and arid zone forestry; industrial land reclamation. Mailing Add: Fac of Forestry Univ of BC Vancouver BC Can

THIRTLE, JOHN ROBSON, b English, WVa, Oct 3, 14; m 42; c 3. ORGANIC CHEMISTRY. Educ: Univ Rochester, BS, 40; Iowa State Col, PhD(org chem), 43. Prof Exp: Sr res assoc, Nat Defense Res Comt, Manhattan Proj, Iowa State Col, 42-44; SR RES ASSOC, EASTMAN KODAK CO, 44- Mem: Am Chem Soc; Soc Photog Sci & Eng. Res: Dibenzofuran analgesics; heterocyclic systems; uranium chemistry; organometallic compounds; phenylenediamines; highly active and nontoxic color developing agents; color photography. Mailing Add: Color Photog Div Res Labs Eastman Kodak Co Rochester NY 14650

THIRUGNANAM, MUTHUVELU, b India, July 20, 40; m 74. PLANT NEMATOLOGY, ENTOMOLOGY. Educ: Annamalai Univ, India, BS, 62, MS, 65; Rutgers Univ, PhD(nematol), 73. Prof Exp: Instr entom, Annamalai Univ, India, 62-65, lectr, 66-69; asst prof, Rutgers Univ, 73-75; ENTOMOLOGIST, ROHM AND HAAS CO RES LABS, 75- Mem: Entom Soc Am; Soc Nematologists; AAAS. Res: Development of insecticides and nematicides; culture procedure for insects and nematodes; studies on the structure-activity relationship in insecticidal compounds. Mailing Add: Rohm and Haas Res Labs Norristown & McKean Rds Spring House PA 19477

THIRUVATHUKAL, JOHN VARKEY, b Shertallay, India, Aug 4, 39; m 71. GEOPHYSICS, OCEANOGRAPHY. Educ: St Louis Univ, BS, 61; Mich State Univ, MS, 63; Ore State Univ, PhD(geophys), 68. Prof Exp: Res asst geophys, Ore State Univ, 63-67; from instr to asst prof geol, DePauw Univ, 67-70; ASST PROF GEOL, MONTCLAIR STATE COL, 70- Concurrent Pos: Consult, Off Earth Sci, Nat Acad Sci, 69-; adj prof, Fairleigh Dickinson Univ. Mem: Am Geophys Union; Soc Explor Geophys. Mailing Add: Dept of Phys-Geosci Montclair State Col Upper Montclair NJ 07043

THIRUVATHUKAL, KURIAKOSE V, b Shertallay, Kerala, India, May 1, 25; m 65; c 3. ZOOLOGY, MORPHOLOGY. Educ: Univ Kerala, BSc, 53; Boston Col, MS, 56; St Louis Univ, PhD(biol, histol), 60. Prof Exp: Instr zool, anat, histol & animal tech, Duquesne Univ, 59-60; asst prof zool & histol, Aquinas Col, 60-62; asst prof biol, Gonzaga Univ, 62-65 & Canisius Col, 65-68; chmn dept, 68-71, PROF BIOL, LEWIS UNIV, 68- Mem: Am Micros Soc; Soc Syst Zool; Indian Soc Animal Morphol & Physiol; Am Soc Zoologists; Am Asn Univ Profs (secy, 75-). Res: Histology and morphology of reptiles; vertebrate zoology; coelacanth morphology; research in reptilia and coelacanth. Mailing Add:

THIRUVENGADA, SESHAN, b Madras, India, Oct 22, 39; m 70; c 1. POLYMER CHEMISTRY, PHYSICAL CHEMISTRY. Educ: St Joseph's Col, Madras, BSc, 57; Purdue Univ, Lafayette, PhD(phys chem), 69. Prof Exp: Jr sci asst, Anal Div, Atomic Energy Estab, Bombay, India, 58-59; sci off radiochem spectros, Plutonium Proj, Atomic Res Ctr, Trombay, 60-63; res asst molecular spectros, Dept Chem, Purdue Univ, Lafayette, 63-67; RES CHEMIST, CORP RES CTR, UNIROYAL, INC, 68- Concurrent Pos: Adj lectr, Dept Chem, Hunter Col, 71-72. Mem: Sigma Xi; Am Chem Soc. Res: Physical chemistry of polymers; polymer fractionation; gel permeation chromatography; surface characterization in relation to rubber adhesion; vulcanization chemistry; vibrational spectra of large molecules; thermodynamics. Mailing Add: Corp Res Ctr UniroyalInc Middlebury CT 06749

THOA, NGUYEN BICH, b US citizen. PHARMACOLOGY, PHYSIOLOGY. Educ: Sorbonne, BA, 53; Saigon Univ, MS, 58; Howard Univ, MS, 63, PhD(pharmacol), 65. Prof Exp: Instr pharmacol, Sch Med & Dent, Georgetown Univ, 65-68; vis scientist, Lab Clin Sci, NIMH, 68-72, CIVIL SERV FEL, NINDA, 72- Concurrent Pos: AMA Educ & Res Found grant, Med Sch, Georgetown Univ, 67-70; NIMH grant, Lab Clin Sci, NIMH, 74- Mem: Am Soc Pharmacol & Exp Therapeut. Res: Correlation between transmitters physiology and biochemistry; transmitters biosynthesis, release and metabolism; serotonergic and adrenergic pathways in the central nervous system; effects of behavior on biogenic pathways in the central nervous system; effects of behavior on biogenic amines status in the periphery and in the central nervous system. Mailing Add: NIH Lab of Clin Sci Rm 25-46 Bethesda MD 20014

THOBURN, JAMES MILLS, analytical chemistry, see 12th edition

THODE, HENRY GEORGE, b Dundurn, Sask, Sept 10, 10; m 35; c 3. PHYSICAL CHEMISTRY. Educ: Univ Saskatchewan, BSc, 30, MSc, 32; Univ Chicago, PhD(chem), 34. Hon Degrees: DSc, Univ Toronto, 55 & McMaster Univ, 73; LLD, Univ Saskatchewan, 58. Prof Exp: Instr chem, Pa Col Women, 35-36; asst, Columbia Univ, 36-38; mem staff, US Rubber Co, 39; from asst prof to assoc prof, 39-44, dir res, 47-61, head dept, 48-52, prin, Hamilton Col, 49-63, from vpres to pres univ, 57-72, PROF PHYS CHEM, McMASTER UNIV, 44- Concurrent Pos: Res chemist, Nat Res Coun Can, 43-46, mem, 55-61; mem, Order of Brit Empire, 46, ed adv bd, J Inorg & Nuclear Chem, 54-, Defence Res Bd Can, 55-61 & bd gov, Ont Res Found; mem bd dirs, Atomic Energy Can Ltd, Steel Co Can Ltd & Fidelity Mortgage & Savings; mem comn atomic weights, Inorg Chem Div, Int Union Pure & Appl Chem, 63-, ed adv bd, Earth &Planetary Sci Letters, 65- & Companion, Order of Can, 67; mem, Can Nat Comt, Int Union Pure & Appl Chem, 75- Honors & Awards: Medal, Chem Inst Can, 57; Tory Medal, Royal Soc Can, 59. Mem: Am Chem Soc; fel Royal Soc Can (pres, 59-60); hon fel Chem Inst Can (pres, 51); fel Royal Soc. Res: Equilibrium and kinetic isotope effects; significance of isotope abundance variations in nature; mass spectrometry; mass and charge distributions of fission products; isotope abundances in terrestrial, meteoritic and lunar materials. Mailing Add: Nuclear Res Bldg McMaster Univ 1280 Main St Hamilton ON Can

THOE, ROBERT STEVEN, b Pensacola, Fla, Aug 19, 45; m 68; c 1. ATOMIC PHYSICS. Educ: Baylor Univ, BS, 68; Univ Conn, MS, 70, PhD(physics), 73. Prof Exp: ASST RES PROF PHYSICS, UNIV TENN, 73- Concurrent Pos: Consult, Union Carbide Corp, Oak Ridge Nat Lab, 75- Mem: Am Phys Soc. Res: The study of atomic collision phenomena, primarily by the detection and measurement of the non-characteristic radiations emitted during violent ion-atom collisions. Mailing Add: Bldg 5500 Box X Oak Ridge Nat Lab Oak Ridge TN 37830

THOENNES, LAWRENCE ANTHONY, b Kansas City, Mo, Feb 9, 09; m 43; c 6. PHYSIOLOGY, PHARMACOLOGY. Educ: DePaul Univ, AB, 30, MS, 32; St Louis Univ, PhD(physiol), 40. Prof Exp: From instr to asst prof biol, St Mary's Col, 33-42; res assoc, Kraft Foods Co, 42-44; head biol res, Kendall Co Div, Colgate-Palmolive Co, 44-73; RETIRED. Concurrent Pos: Consult, Sherwood Med Indust, 71-74. Mem: Am Pharmaceut Asn; Am Soc Testing & Mat. Res: Low temperature physiology; skin physiology; suture absorption; wound healing; mammalian toxicology; toxicity of polymeric materials. Mailing Add: 7447 W Rosedale Chicago IL 60631

THOM, KARLHEINZ, physics, see 12th edition

THOMA, GEORGE EDWARD, b Dayton, Ohio, Aug 9, 22; m 49; c 5. INTERNAL MEDICINE, NUCLEAR MEDICINE. Educ: Univ Dayton, BS, 43; St Louis Univ, MD, 47. Prof Exp: From instr to asst prof internal med, 51-61, asst to vpres, 62-67, asst vpres med ctr, 67-73, ASSOC PROF INTERNAL MED, ST LOUIS UNIV, 61-, SECT HEAD, SECT NUCLEAR MED, 59-, VPRES MED CTR, 73- Concurrent Pos: Dir radioisotopes lab, Med Ctr, St Louis Univ, 49-51 & 54-; consult, Health Physics Div, Oak Ridge Nat Lab, 54-, Lockheed Aircraft Corp, 54-, US Army, 56-, Med Div, Oak Ridge Inst Nuclear Studies, 58-, Div Radiol Health, USPHS, 60-, Div Compliance, US AEC, 62- & Am Pub Health Asn, 62-; ed, J Nuclear Med, 59- Mem: Soc Nuclear Med; Radiation Res Soc; Health Physics Soc; Am Thyroid Asn; Am Soc Internal Med. Res: Radiobiology; thyroid function; clinical application of radioisotopes. Mailing Add: Radioisotopes Lab St Louis Univ Med Ctr St Louis MO 63104

THOMA, GEORGE WILLIAM, b Easton, Pa, Jan 6, 21; m 44; c 2. PATHOLOGY. Educ: Lafayette Col, AB, 42; Univ Pa, MD, 45. Prof Exp: Asst prof path & legal med, Med Col Va, 52-57; clin assoc prof path, Med Br, 57-61, PROF PATH, UNIV TEX DENT BR & GRAD SCH BIOMED SCI HOUSTON, 61- Concurrent Pos: Asst chief med examr, State of Va, 51-57; pathologist, Galveston County Hosp & St Mary's Infirmary, 57- & Univ Tex M D Anderson Hosp & Tumor Inst, 61- Mem: AMA; Am Asn Path & Bact; Col Am Path; Int Acad Path. Res: Forensic and oral pathology. Mailing Add: Dept of Path Univ of Tex Dent Br Houston TX 77025

THOMA, JOHN ANTHONY, b Springfield, Ill, Dec 6, 32; m 58; c 4. BIOCHEMISTRY. Educ: Bradley Univ, AB, 54; Iowa State Univ, PhD(biochem), 58. Prof Exp: Res assoc enzymol, Brookhaven Nat Lab, 58-60; asst prof chem, Ind Univ, 60-65; assoc prof, 66-70, vchmn dept, 73-75, PROF CHEM, UNIV ARK, FAYETTEVILLE, 70- Concurrent Pos: Mem ed adv bd, J Chromatographic Sci, 69-; vis fel, Dent Res Inst, Sydney, Australia, 74; consult, Cargill, 75. Mem: AAAS; Am Chem Soc; Am Asn Biol Chemists; Fed Am Scientists. Res: Mechanism of enzyme action, enzyme subsite mapping; carbohydrate chemistry; molecular diseases. Mailing Add: Dept of Chem Univ of Ark Fayetteville AR 72701

THOMA, RICHARD WILLIAM, b Milwaukee, Wis, Dec 7, 21; m 52; c 4. BIOCHEMISTRY. Educ: Univ Wis, BSc, 47, MSc, 49, PhD(biochem), 51. Prof Exp: Res assoc sect microbiol, Squibb Inst Med Res, 51-61, res supvr microbiol develop, 62-68, ASST DIR BIOL PROCESS DEVELOP, E R SQUIBB & SONS, 68- Mem: AAAS; Am Acad Microbiol; Am Chem Soc; Am Soc Microbiol; NY Acad Sci. Res: Fermentation process research and development. Mailing Add: 1831 Mountain Top Rd Bridgewater NJ 08807

THOMA, ROY E, b San Antonio, Tex, May 12, 22; m 53; c 2. INORGANIC CHEMISTRY, PHYSICAL CHEMISTRY. Educ: Univ Tex, MA, 48. Prof Exp: Assoc prof, Sam Houston State Col, 48-51; asst prof, Tex Tech Col, 51-52; chemist, 52-60, proj chemist molten salt reactor progs, 67-71, TASK GROUP DIR, OAK RIDGE NAT LAB, 71- Concurrent Pos: Mem bd dirs, Environ Systs Corp, 73- Mem: Am

Chem Soc; Sigma Xi; fel Am Ceramic Soc; Am Nuclear Soc; fel AAAS. Res: Physical chemistry of inorganic fused salts, particularly determinations of phase equilibria and interrelationships of crystal structures in condensed systems of these materials; nuclear power stations, especially floating nuclear power plants. Mailing Add: Oak Ridge Nat Lab Oak Ridge TN 37830

THOMAN, CHARLES J, b Wilkes-Barre, Pa, Nov 4, 28. ORGANIC CHEMISTRY. Educ: Spring Hill Col, BS, 53; Fordham Univ, MS, 56; Woodstock Col, STB, 59, STM, 60; Univ Mass, Amherst, PhD(org chem), 66. Prof Exp: Instr chem, Univ Scranton, 53-55; res asst with L A Carpino, Univ Mass, 66; asst prof, 66-69, ASSOC PROF CHEM, UNIV SCRANTON, 69- Mem: AAAS; fel Am Inst Chem; Am Chem Soc; Am Asn Jesuit Sci. Res: N-nitrosoketimines; chemistry of sydnones; nucleic acid antimetabolites. Mailing Add: Dept of Chem Univ of Scranton Scranton PA 18510

THOMAN, RICHARD SAMUEL, b Lamar, Colo, May 10, 19; m 42; c 2. ECONOMIC GEOGRAPHY, URBAN GEOGRAPHY. Educ: Univ Colo, BA, 41, MA, 48; Univ Chicago, PhD(geog), 53. Prof Exp: Instr geog, Gorham State Col, 45-48; instr econ geog, Univ Mo, 48-51; res assoc, Univ Chicago, 53-55; assoc prof & head dept, Univ Omaha, 55-57; pvt res & writing, 57-61; from assoc prof to prof geog, Queen's Univ, Ont, 61-71; pvt res & writing, 71-72; PROF ECON GEOG & REGIONAL PLANNING, CALIF STATE UNIV, HAYWARD, 72- Concurrent Pos: Off Naval Res fel, Univ Chicago, 53-55; Govt Can Dept Labour fel, Queen's Univ, Ont, 65-66, Dept Indust fel, 66; mem comn col geog, Asn Am Geogr & NSF, 65-67; dir regional develop, Dept Treas & Econ, Govt Ont, 67-71; mem comn regional aspects econ develop, Int Geog Union, 67-; mem, Can Coun Urban & Regional Res; chmn bd trustees, Land Econ Found, 73-; NSF travel grant, Mysore, India, 74. Mem: Asn Am Geogrs; Am Geog Soc; Can Asn Geog; Can Asn Am Studies (vpres, 65-67). Res: Economic geography resulting in policy measures, regionally expressed, with particular reference to the United States and Canada; changing location of manufacturing industries and establishments in United States. Mailing Add: Dept of Geog Calif State Univ Hayward CA 94542

THOMAS, ALAN, b Evansburg, Pa, Jan 1, 23; m 44; c 3. FOOD SCIENCE. Educ: Pa State Univ, BS, 49; Univ Minn, MS, 50, PhD(dairy technol), 54. Prof Exp: Qual control technician, Am Cyanamid Co, 46-47; asst, Univ Minn, 49-50, instr dairy technol, 51-54; instr chem & dairy technol, Temple Univ, 50-51; res assoc, Bowman Dairy Co, Ill, 54-56; res proj mgr, M&M's Candies, Food Mfrs, Inc, NJ, 56-60, prod develop mgr, 61-64; vpres res & develop, Mars Candies, Ill, 64-67, VPRES RES & DEVELOP, M&M/MARS, 67- Honors & Awards: Res Award, Nat Confectioners Asn US, 71. Mem: AAAS; Am Dairy Sci Asn; Am Asn Cereal Chem; Inst Food Technol. Res: Food science and technology; applied research in confectioneries. Mailing Add: M&M/Mars High St Hackettstown NJ 07840

THOMAS, ALEXANDER, b New York, NY, Jan 11, 14; m 38; c 4. PSYCHIATRY. Educ: City Col New York, BS, 32; NY Univ, MD, 36; Am Bd Psychiat & Neurol, dipl, 48. Prof Exp: From instr to assoc prof, 48-66, PROF PSYCHIAT, SCH MED, NY UNIV, 66-; DIR PSYCHIAT DIV, BELLEVUE HOSP, NEW YORK, 68- Concurrent Pos: Assoc attend psychiatrist, Bellevue & Univ Hosps, 58-68, attend psychiatrist, 68- Mem: Fel Am Psychiat Asn. Res: Longitudinal study of child development; psychosomatic medicine. Mailing Add: Dept of Psychiat NY Univ Sch of Med New York NY 10016

THOMAS, ALEXANDER EDWARD, III, b Chicago, Ill, May 3, 30; m 56; c 3. ORGANIC CHEMISTRY, ANALYTICAL CHEMISTRY. Educ: Univ Ill, BS, 55; DePaul Univ, MS, 61. Prof Exp: Res chemist, Cent Org Res Lab, Glidden Co, 55-58, sect head anal chem, Durkee Foods Group, 58-66, mgr chem res dept, 66-68, mgr chem res, Dwight P Joyce Res Ctr, 68-71, MGR APPLNS RES, DWIGHT P JOYCE RES CTR, GLIDDEN-DURKEE DIV, SCM CORP, 71- Mem: Am Chem Soc; Am Oil Chem Soc; Soc Appl Spectros. Res: Analytical chemistry of glycerides, surfactants and protective coatings; chromatographic and instrumental methods; synthesis of organic azides. Mailing Add: 16335 Ramona Dr Middleburg Heights OH 44130

THOMAS, ALFORD MITCHELL, b Bunnlevel, NC, July 24, 42; m 70. MEDICINAL CHEMISTRY. Educ: Campbell Col, NC, BA, 64; Univ NC, Chapel Hill, PhD(org chem), 69. Prof Exp: NIH fel, Univ Va, 69-71, res assoc org chem, 71-72; SR CHEMIST, ABBOTT LABS, 72- Mem: Am Chem Soc; The Chem Soc; AAAS. Res: Synthesis of biological peptides. Mailing Add: Dept 482 Res Div Abbott Labs North Chicago IL 60064

THOMAS, ANN P, b Pottanat Vempally, India, Dec, 28, 34; US citizen; m 61; c 2. ORGANIC CHEMISTRY. Educ: Univ Madras, BS, 55; Univ Kerala, MS, 58; George Washington Univ, PhD(chem), 70. Prof Exp: Chemist, Health Serv Div, USV Pharmaceut Corp, 72-74; res assoc chem, Col Pharm, Howard Univ, 74-75; RES ASSOC SOLAR STUDIES, GODDARD SPACE FLIGHT CTR, NASA, 75- Concurrent Pos: Lectr chem, George Washington Univ, 70-71; instr chem, Prince Georges Col, Md, 72-73. Mem: Am Chem Soc; Am Asn Clin Chemists. Res: Radiative and photochemical properties of the upper atmosphere, especially total and spectral irradiance of the sun and pollution monitoring. Mailing Add: Code 912 Rm 395 Bldg 22 Goddard Space Flight Ctr Greenbelt MD 20771

THOMAS, ARTHUR L, b New York, NY, July 24, 28. CHEMISTRY. Educ: Columbia Col, AB, 51; Princeton Univ, PhD, 56. Prof Exp: Engr photo prod, E I du Pont de Nemours & Co, Inc, 55-58, res supvr, 58-59; chem engr, Standard Ultramarine & Color Co, 60-65 & MHD, Inc & Plasmachem Inc, 65-68; from instr to asst prof chem, Calif State Polytech Col, San Luis Obispo, 69-73; vis asst prof immunochem, Columbia Univ, 73; SCI ED, RONALD PRESS CO, 74- Mem: AAAS; Am Chem Soc. Res: Pigments; high temperature reactions. Mailing Add: 29 Claremont Ave New York NY 10027

THOMAS, ARTHUR NORMAN, b Los Angeles, Calif, Jan 27, 31; m 50; c 5. SURGERY, THORACIC SURGERY. Educ: Stanford Univ, BS, 53; Univ BC, MD, 57. Prof Exp: Intern surg, San Francisco Gen Hosp, 57-58, resident, 58-59; resident, Univ Hosp, 60-62, clin instr, Sch Med, 63-68, asst prof, 68-73, ASSOC PROF SURG, SCH MED, UNIV CALIF, SAN FRANCISCO, 73-; CHIEF THORACIC SURG, SAN FRANCISCO GEN HOSP, 70- Concurrent Pos: NIH res fel, Univ Hosp, Univ Calif, San Francisco, 59-60, NIH fel, Cardiovasc Res Inst, 63-65; asst chief surg, Vet Admin Hosp, San Francisco, 66-70, attend physician thoracic surg, 70- Mem: Soc Thoracic Surg; Samson Thoracic Surg Soc; Am Asn Thoracic Surg. Res: Thoracic surgery; cardiopulmonary research. Mailing Add: Dept Surg San Francisco Gen Hosp 1001 Potrero Ave San Francisco CA 94110

THOMAS, AUBREY STEPHEN, JR, b Wolfeboro, NH, Nov 4, 33; m 56. PLANT PHYSIOLOGY. Educ: Keene State Col, BEd, 62; Univ NH, MS, 64, PhD(bot, plant physiol), 67. Prof Exp: From asst prof to assoc prof, 67-74, ASSOC PROF BIOL & CHMN DEPT, MERRIMACK COL, 74- Mem: AAAS; Am Soc Plant Physiol; Bot Soc Am; Nat Asn Biol Teachers; Am Inst Biol Sci. Res: Ecological plant physiology; effect of light, radiation, electrical fields and other environmental factors upon the growth and development of the plant and its physiology. Mailing Add: Dept of Biol Merrimack Col North Andover MA 01845

THOMAS, BARBARA SMITH, b Palo Alto, Calif, Oct 2, 42; m 73; c 1. TOPOLOGY. Educ: Reed Col, BA, 64; Carnegie-Mellon Univ, MS, 70, PhD(math), 73. Prof Exp: Mellon fel, Univ Pittsburgh, 72-73; ASST PROF MATH, MEMPHIS STATE UNIV, 73- Mem: Am Math Soc; Math Asn Am; Am Asn Univ Profs. Res: Topological structure of free topological groups and free products of topological groups; epimorphisms in the category of Hausdorff topological groups; related problems of categorical topology. Mailing Add: Dept of Math Sci Memphis State Univ Memphis TN 38152

THOMAS, BARRY, b Eng, Dec 31, 41; US citizen. ZOOLOGY, ENVIRONMENTAL EDUCATION. Educ: Calif State Univ, Fullerton, BA, 67, MA, 68; Univ BC, PhD(zool), 71. Prof Exp: Lectr ecol, Univ Calgary, 71-72; asst prof, 72-75, PROF ENVIRON EDUC, CALIF STATE UNIV, FULLERTON, 75- Concurrent Pos: Pres, BioReCon, 71-; dir, Tucker Wildlife Sanctuary, 72- Mem: Am Soc Mammal; Audubon Soc. Res: Karyotaxonomy of island rodents; urbanization effects on wild animal populations; biological impact statements. Mailing Add: Dept of Sci Educ Calif State Univ Fullerton CA 92634

THOMAS, BARRY HOLLAND, b Lancaster, Eng, June 1, 39; m 66; c 2. BIOCHEMISTRY, PHARMACOLOGY. Educ: Univ Liverpool, BSc, 62, PhD(pharmacol), 65. Prof Exp: Asst lectr pharmacol, Univ Liverpool, 65-67, lectr, 67-69; RES SCIENTIST, HEALTH PROTECTION BR, DEPT NAT HEALTH & WELFARE CAN, 69- Mem: Brit Biochem Soc; Brit Pharmacol Soc; Pharmacol Soc Can. Res: Role of drug metabolism in the toxicity of drugs; toxicity of drug interactions. Mailing Add: Kars ON Can

THOMAS, BENJAMIN EARL, b Boise, Idaho, July 30, 13; m 40; c 4. GEOGRAPHY. Educ: Univ Idaho, BS, 34, MS, 35; Ohio State Univ, MA, 40; Harvard Univ, PhD(geog), 47. Prof Exp: Teacher soc sci, Soda Springs High Sch, Idaho, 35-39; Ford Found foreign field fel, EAfrica, 57-58; vis prof, Univ EAfrica, Kampala, Uganda, 59-60; actg dir, African Studies Ctr, Univ Calif, Los Angeles, 63-64, assoc dir, 64-65, actg dir, 65-66, chmn dept, 71-74, PROF GEOG, UNIV CALIF, LOS ANGELES, 47- Honors & Awards: Bronze Star Medal. Mem: Asn Am Geog; African Studies Asn (pres, 68-69). Res: Geography of Africa and the Middle East; history of geographic thought. Mailing Add: Dept of Geog Univ of Calif Los Angeles CA 90024

THOMAS, BENJAMIN WILLIAM, b Jackson, Miss, Mar 11, 08; m 44; c 4. PHYSICS. Educ: Miss Col, BA, 31; Pa State Univ, MS, 34, PhD(physics), 38. Prof Exp: Sr res specialist, Humble Oil & Ref Co, 38-56; dir instrumentation physics res, Tex Butadiene & Chem Corp, 56-61; pres & owner, Tirco, Inc, 61-71; PROF TECHNOL, SAN JACINTO COL, 71- Mem: Am Chem Soc; fel Instrument Soc Am. Res: Engineering and research in industrial process control instrumentation, including computerized control hardware and onstream analyzers. Mailing Add: Dept of Instrumentation San Jacinto Col 8060 Spencer Hwy Pasadena TX 77505

THOMAS, BERT O, b Lead, SDak, May 15, 26; m 49; c 2. ZOOLOGY. Educ: Colo State Univ, BS, 50, MS, 52; Univ Minn, PhD, 59. Prof Exp: Biologist, Hydrol Surv, US Fish & Wildlife Serv, 50; asst zool, Univ Minn, 52-57; biologist, State Dept Health, Minn, 57-59; from asst prof to prof zool & chmn dept biol sci, 59-74, PROF BIOL, UNIV NORTHERN COLO, 74- Concurrent Pos: NSF lectr, Univ Colo, 59-61; tech adv, Continental Mach, Inc, 55-57; consult, Wilkie Found, 57-59. Mem: AAAS; Am Soc Limnol & Oceanog. Res: Plankton communities; ichthyology; radio ecology; aquatic biology. Mailing Add: Dept of Biol Sci Univ of Northern Colo Greeley CO 80631

THOMAS, BERWYN BRAINERD, b Iowa City, Iowa, Apr 6, 19; m 50; c 3. CHEMISTRY. Educ: Univ Ariz, BS, 40; Lawrence Univ, PhD(paper chem), 44. Prof Exp: Res chemist, 47-60, group leader, 60-72, TECH INFO SPECIALIST, OLYMPIC RES DIV, ITT RAYONIER INC, 72- Mem: Am Chem Soc; Tech Asn Pulp & Paper Indust. Res: Preparation and use of cellulose in chemical conversion processes and paper manufacture; methods of evaluation; information retrieval. Mailing Add: Rte 4 Box 30 Shelton WA 98584

THOMAS, BILLY SEAY, b Tenn, Dec 31, 26; m 53; c 1. THEORETICAL PHYSICS. Educ: Wayne State Univ, BS, 53; Vanderbilt Univ, MS, 55, PhD(physics), 59. Prof Exp: Nat Res Coun fel, Argonne Nat Lab, 58-59; asst prof physics, Vanderbilt Univ, 59-60; ASST PROF PHYSICS, UNIV FLA, 60- Mem: Am Phys Soc. Res: Atomic and molecular scattering; elementary particle physics. Mailing Add: Dept of Physics Univ of Fla Gainesville FL 32601

THOMAS, BLAKEMORE EWING, b Okmulgee, Okla, Nov 6, 17; m 45; c 3. ECONOMIC GEOLOGY. Educ: Univ Calif, AB, 40; Calif Inst Technol, MS, 43, PhD(geol), 49. Prof Exp: Instr meteorol & geol, Calif Inst Technol, 42-46; consult geologist, 46-49; asst prof econ geol, Univ Kans, 49-51; consult petrol & mining geol, William Ross Cabeen & Assoc, 51-56; assoc prof, 56-63, chmn dept, 63-68, PROF GEOL, SAN DIEGO STATE UNIV, 63- Concurrent Pos: Geol consult, US Dept Justice, US Dept Army-Corps Engrs, 60-; dir, Comet Petrol Corp, Wichita Kans, 68-72 & New Camp Minerals, Inc, Wichita, Kans, 70- Mem: Geol Soc Am; Am Asn Petrol Geol; Am Inst Mining, Metall & Petrol Eng; Sigma Xi. Res: Areal and structural geology; exploration, petroleum and mining. Mailing Add: Dept Geol Sci San Diego State Univ San Diego CA 92182

THOMAS, BRANTLEY DENMARK, JR, chemistry, see 12th edition

THOMAS, BRUCE ROBERT, b Guthrie Center, Iowa, Jan 1, 38; m 60; c 3. EXPERIMENTAL PHYSICS. Educ: Grinnell Col, BA, 60; Cornell Univ, PhD(theoret physics), 65. Prof Exp: Asst prof physics, Grinnell Col, 65-67; asst prof, 67-70, ASSOC PROF PHYSICS, CARLETON COL, 70- Mem: Am Phys Soc; Am Asn Physics Teachers. Mailing Add: Dept of Physics Carleton Col Northfield MN 55057

THOMAS, BYRON HENRY, b Oakland, Calif, Oct 9, 97; m 22; c 3. NUTRITION, BIOCHEMISTRY. Educ: Univ Calif, BS, 22; Univ Wis, MS, 24, PhD(animal nutrit), 29. Prof Exp: Instr animal husb, Univ Calif, 22-23; asst, Univ Wis, 24-25; dir nutrit, Walker-Gordon Lab Co, NJ, 29-31; prof animal husb & head animal nutrit & chem, Exp Sta, 31-49, PROF BIOCHEM, IOWA STATE UNIV, 49- Mem: Am Chem Soc; Am Soc Animal Sci; Soc Exp Biol & Med; Am Dairy Sci Asn; Am Inst Nutrit. Res: Irregularities in nutrition in the production of congenital abnormalities in mammals, and their effect on the nutrition, embryology, and histology of affected fetuses. Mailing Add: Dept of Biochem & Biophys Iowa State Univ Ames IA 50010

THOMAS, CARL OWENS, physical chemistry, see 12th edition

THOMAS, CARMEN CHRISTINE, b Ger, Apr 15, 08; US citizen. DERMATOLOGY. Prof Exp: Univ Del, AB, 29; Woman's Med Col Pa, MD, 32;

Univ Pa, ScD(dermat, syphil), 40. Prof Exp: Intern, Philadelphia Gen Hosp, 32-34, asst chief res physician, 34-35; from instr to asst prof dermat & syphil, Grad Sch Med, Univ Pa, 36-67; instr, 36-39, dir dept oncol, 52-66, prof, 41-69, EMER PROF DERMAT & SYPHIL, WOMAN'S MED COL PA, 69- Concurrent Pos: Pvt pract; consult, Philadelphia Gen Hosp, 44-, Vet Admin Hosp, 53-, Mem Hosp, Roxborough, Elwyn Inst & Devereaux Sch. Mem: Fel Am Acad Dermat. Res: Tuberculin sensitivity in dermatoses; immunologic aspects of sarcoidosis; cutaneous streptomycin sensitivity; neoplasia skin. Mailing Add: 1930 Chestnut St Philadelphia PA 19103

THOMAS, CAROLYN EYSTER, b Toledo, Ohio, July 14, 28; m 53. NEUROANATOMY, HISTOLOGY. Educ: Univ Toledo, BS, 48; Northwestern Univ, Chicago, MS, 51, PhD(anat), 53. Prof Exp: Instr anat, Med Sch, Northwestern Univ, Chicago, 53-58, asst prof, 58-66; asst prof, 66-68, ASSOC PROF ANAT, CHICAGO MED SCH-UNIV HEALTH SCI, 68- Mem: Am Asn Anat; Am Soc Zool. Res: Electrophysiological studies of cat cerebellum; spinal cord structure; muscular architecture of the heart. Mailing Add: Dept of Anat Chicago Med Sch-Univ Health Sci Chicago IL 60612

THOMAS, CAROLYN MARGARET, b Brownville, Maine, Feb 25, 41. ANALYTICAL CHEMISTRY. Educ: Univ Maine, BA, 63; Northeastern Univ, PhD(anal chem), 68. Prof Exp: Teaching fel anal chem, Northeastern Univ, 63-67, lectr, Lincoln Col, 67-68; res chemist, US Army Natick Labs, 67-68; asst prof, 68-73, ASSOC PROF ANAL CHEM, WAYNESBURG COL, 73- Mem: Am Chem Soc; Sigma Xi. Res: Molten salts; electrochemistry; stability of foods; atomic absorption and emission spectroscopy applied to water pollution. Mailing Add: Dept of Chem Waynesburg Col Waynesburg PA 15370

THOMAS, CHARLES ALLEN, b Scott Co, Ky, Feb 15, 00; m 26; c 4. CHEMISTRY. Educ: Transylvania Univ, BA, 20; Mass Inst Technol, Ms, 24; Transylvania Univ, DSc(org chem), 33; Hobart Col, LLD, 50; Univ Mo-Rolla, DEng, 65; Simpson Col, DSc, 67. Hon Degrees: DSc, Wash Univ, 47. Prof Exp: Res chemist, Gen Motors Res Corp, 23-24 & Ethyl Gasoline Corp, 24-25; res chemist, Thomas & Hochwalt Labs, Inc, 26-28, pres, 28-36; cent res dir, Monsanto Co, Ohio, 36-45, vpres & tech dir, Mo, 45-46, exec vpres, 47-50, pres, 51-60, chmn bd, 60-65, chmn finance comt, 65-68; RETIRED. Concurrent Pos: Cur, Transylvania Univ; mem corp, Mass Inst Technol; chmn bd trustees, Wash Univ. Honors & Awards: US Medal for Merit, 46; Medal, Indust Res Inst, 47; Gold Medal, Am Inst Chemists, 48; Perkin Medal, Soc Chem Indust, 53; Priestley Medal, Am Chem Soc, 55. Mem: Nat Acad Sci; Soc Chem Indust; Electrochem Soc; Am Chem Soc (pres, 58); Am Inst Chemists. Res: Development of tetraethyl lead; bromine from sea water; effect of alkali metals on combustion; synthetic resins; synthetic styrene and rubber; rocket propellants; plutonium. Mailing Add: 7701 Forsyth Blvd St Louis MO 63105

THOMAS, CHARLES ALLEN, JR, b Dayton, Ohio, July 7, 27; m 51; c 2. BIOPHYSICAL CHEMISTRY. Educ: Princeton Univ, AB, 50; Harvard Univ, PhD(phys chem), 54. Prof Exp: Res scientist, Eli Lilly & Co, Ind, 54-55; Nat Res Coun fel physics, Univ Mich, 55-56 , instr, 56-57; from asst prof to prof biophys, Johns Hopkins Univ, 57-67; PROF BIOL CHEM, HARVARD MED SCH, 67- Concurrent Pos: NSF sr fel, Weizmann Inst, 65. Mem: Am Acad Arts & Sci. Res: Molecular anatomy of viral and bacterial chromosomes; genetic recombination, organization and function of higher chromosomes. Mailing Add: Dept of Biol Chem Harvard Med Sch Boston MA 02115

THOMAS, CHARLES CARLISLE, JR, b Rochester, NY, Aug 18, 25; m 46; c 4. ACADEMIC ADMINISTRATION, NUCLEONICS. Educ: Univ Iowa, BS, 47; Univ Rochester, MS, 50; Am Inst Chemists, cert; Inst Nuclear Mat Mgt, cert. Prof Exp: Nuclear chemist, US Bur Mines, Okla, 50-51; sr engr, Bausch & Lomb Optical Co, NY, 51-52; tech engr, Aircraft Nuclear Propulsion Dept, Gen Elec Co, Ohio, 52-53; prin chemist, Battelle Mem Inst, 53-55; fel engr, Westinghouse Elec Corp, Pa, 55-60; proj leader radiation chem, Quantum, Inc, 60-62; res mgr, Western NY Nuclear Res Ctr, Inc, 62-72; actg dir, 72-74, DIR NUCLEAR SCI & TECHNOL FAC, STATE UNIV NY BUFFALO, 74- Concurrent Pos: Int Atomic Energy Agency vis prof, Tsing Hua, China, 64-65; adj assoc prof eng sci, aerospace & nuclear eng, State Univ NY Buffalo, 73- Mem: AAAS; Am Chem Soc; Am Nuclear Soc; fel Am Inst Chemists; Radiation Res Soc. Res: Radiation chemistry and biology; neutron activation analysis; dosimetry of high intensity radiation; reactor technology; environmental analysis; radiation effects on materials. Mailing Add: 136 Berryman Dr Buffalo NY 14226

THOMAS, CHARLES DANSER, b Weston, WVa, Nov 5, 08; m 35; c 2. PHYSICS. Educ: WVa Univ, AB & MS, 30; Univ Chicago, PhD(physics), 37. Prof Exp: From instr to prof, 32-74, chmn dept, 54-69, EMER PROF PHYSICS, WVA UNIV, 74- Mem: Am Phys Soc. Res: Field theory; electromagnetic waves; nuclear spectroscopy. Mailing Add: Dept of Physics WVa Univ Morgantown WV 26506

THOMAS, CHARLES GOMER, b Columbia, Mo, Oct 9, 41; m 63; c 2. MATHEMATICS, COMPUTER SCIENCES. Educ: Pomona Col, BA, 62; Cambridge Univ, BA, 64; Univ Ill, Urbana, PhD(math), 68. Prof Exp: Mem, Inst Advan Study, Princeton Univ, 68-69; asst prof math, Univ Wash, 69-75; RES ENGR, FLOW RES, INC, 75- Mem: Math Asn Am. Mailing Add: 2519 E Calhoun St Seattle WA 98112

THOMAS, CHARLES HILL, b Dexter, Ga, Jan 31, 22; m 45; c 1. POULTRY HUSBANDRY, GENETICS. Educ: Univ Ga, BSA, 52, MSA, 53; NC State Univ, PhD(genetics), 56. Prof Exp: Asst prof poultry husb, Miss State Univ & asst poultry husbandman, Agr Exp Sta, 56-58, assoc prof & assoc poultry husbandman, 58-66, PROF POULTRY SCI, MISS STATE UNIV & POULTRY GENETICIST, AGR EXP STA, 66- Mem: Am Poultry Sci Asn; Am Genetic Asn. Res: Inheritance of resistance to insecticides in Drosophila melanogaster. Mailing Add: Dept of Poultry Sci Miss State Univ Box 298 Mississippi State MS 39762

THOMAS, CHARLES I, b Warren, Ohio, Jan 1, 08; m 31; c 2. OPHTHALMOLOGY. Educ: Wesleyan Univ, BS, 30; Western Reserve Univ, MD, 35; Am Bd Ophthal, dipl, 43. Prof Exp: From demonstr to assoc clin prof, 40-60, PROF OPHTHAL, UNIV HOSPS, MED SCH, CASE WESTERN RESERVE UNIV, 60- Mem: AAAS; AMA; Asn Res Vision & Ophthal; Pan-Am Med Asn; Am Col Surg. Res: Uptake of radioactive isotopes in disease of the eye; disease of the cornea. Mailing Add: Dept of Ophthal Case Western Reserve Univ Cleveland OH 44106

THOMAS, CHARLES S, b Leominster, Mass, Oct 30, 29; m 53; c 2. MAMMALOGY, SCIENCE EDUCATION. Educ: Mass State Teachers Col Fitchburg, BS, 54; Cornell Univ, MS, 57, PhD(sci educ, mammal, ornith), 65. Prof Exp: Assoc prof biol, 59-74, PROF BIOL SCI, STATE UNIV NY COL BROCKPORT, 74- Mem: Nat Asn Res Sci Teaching. Res: Development of audio-video-tutorial program of instruction in principles of biology utilizing audio tapes and television modules. Mailing Add: Dept of Sci State Univ of NY Col Brockport NY 14420

THOMAS, CHARLES W, paleontology, geophysics, see 12th edition

THOMAS, CLARENCE DELMAR, b Bozeman, Mont, Oct 14, 00; m 24; c 2. ATOMIC PHYSICS. Educ: Northeastern Mo State Col, BS, 27; Univ Mo, AM, 31, PhD(physics), 38. Prof Exp: Asst physics, Univ Mo, 29-30; from instr to asst prof, Mo Sch Mines, 30-42; prof & head dept, Univ Tulsa, 46-66, prof, 66-67; prof physics, Oral Roberts Univ, 67-73; RETIRED. Concurrent Pos: Instr, US Naval Acad, 42-46. Mem: Am Phys Soc; Am Soc Eng Educ; Am Asn Physics Teachers; Sigma Xi. Res: Oxyluminescence efficiencies; diffraction of x-rays by liquids. Mailing Add: 1536 S College Ave Tulsa OK 74104

THOMAS, CLARENCE HENRY, physical chemistry, see 12th edition

THOMAS, CLAUDEWELL SIDNEY, b New York, NY, Oct 5, 32; m 68; c 3. PSYCHIATRY, PUBLIC HEALTH. Educ: Columbia Univ, BA, 52; State Univ NY Downstate Med Ctr, MD, 56; Am Bd Psychiat, dipl, 62; Yale Univ, MPH, 64. Prof Exp: Chief emergency treatment serv, Ment Health Ctr, New Haven, Conn, 65-67; educ dir psychiat emergency serv, Yale Univ, 67-68, dir social & community psychiat training, 68-73; dir div ment health prog, NIMH, 73-74; CHMN DEPT PSYCHIAT, COL MED NJ, 74- Concurrent Pos: Consult, Compass Club, New Haven, 63-65; consult psychiatrist, Div Alcoholism, State of Conn, 63-65; vol consult, Caribbean Fed Ment Health, New York, 64; dir Hill-West Haven div & chief unit III, Conn Ment Health Ctr, 67-68; mem ad hoc comt minority admin, Yale Univ, 68-70; soc sci mem, Nat Ctr Health Serv Res & Develop, 69-70; consult, Wash Sch Psychiat, A K Rice Inst Wash; prof sociol, Rutgers Univ; attend psychiatrist, Harrison Martland Hosp, Newark; consult, Cancer Clin Belle Mead, St Joseph's Hosp, Patterson. Mem: Fel Am Psychiat Asn; fel Am Pub Health Asn; fel Royal Soc Health; fel NY Acad Med; NY Acad Sci. Res: Application of theory and concepts in the areas of social and community psychiatry to further the understanding of mental health needs of individuals and groups. Mailing Add: Dept of Psychiat Col of Med of NJ Newark NJ 07103

THOMAS, CLAYTON JAMES, b St Joseph, Mo, Oct 29, 20; m 42; c 5. MATHEMATICS, OPERATIONS RESEARCH. Educ: Univ Chicago, BS, 42, MS, 47. Prof Exp: Lectr, Univ Chicago, 46-47, sr mathematician & group leader, Inst Air Weapons Res, 47-55; from opers analyst to chief res group , Opers Anal Off, 55-71, ASST FOR OPERS RES, ASST CHIEF OF STAFF, STUDIES & ANAL, HQ, US DEPT AIR FORCE, 71- Concurrent Pos: Instr, Roosevelt Univ, 48-50. Honors & Awards: Lanchester Prize, 57. Mem: AAAS; Am Math Soc; Asn Comput Mach. Res: Theory of games; probability; mathematical models of combat; simulation and personnel models; econometrics; meteorology. Mailing Add: 413 River Bend Rd Great Falls VA 22066

THOMAS, CLAYTON LAY, b Metropolis, Ill, Dec 23, 21; m 50; c 4. MEDICINE. Educ: Univ Ky, BS, 44; Med Col Va, MD, 46; Harvard Univ, MPH, 58. Prof Exp: Intern med, Montreal Gen Hosp, 46-47; instr, US Naval Sch Aviation Med, 53-54; instr, Col Med, Univ Utah, 54-56; med dir, 58-70, VPRES MED AFFAIRS, TAMPAX INC, 69- Concurrent Pos: Clin asst, Harvard Med Serv, Boston City Hosp, 56-57; res fel path, Mallory Inst Path, 56-57; consult, Flight Safety Found, 57-58 & Parachutes, Inc, Mass, 60; consult, Dept Pop Sci, Sch Pub Health, Harvard Univ, fel epidemiol, 57-59; mem med & training serv comt, US Olympic Comt, 66-73; pres, Balloon Sch Mass, Inc, 70-; Fed Aviation Admin pilot exam-lighter-than-air-free balloon, 72-; pres, Pop Res Found, 74-; mem med dept vis comt, Mass Inst Technol, 75- Mem: Am Fertil Soc; Am Sch Health Asn; Am Col Health Asn; Aerospace Med Asn; NY Acad Sci. Res: Medical ecology; physiology of menstruation; bacteriuria; medical aspects of sport parachuting and hot air ballooning; aerospace medicine; epidemiology; medical lexicography; sports medicine; health and sex education; physiology of reproduction. Mailing Add: Festiniog Farm Dingley Dell Palmer MA 01069

THOMAS, CLINTON EDWARD, b Erie, Pa, Nov 13, 22; m 46; c 2. PHYSICS, ASTRONOMY. Educ: Univ Mich, BS, 48, MS, 49 & 53. Prof Exp: Instr high sch, Mich, 49-59; ASSOC PROF PHYSICS & ASTRON, EASTERN MICH UNIV, 59- Concurrent Pos: Sci adv, Nat Teacher Educ Ctr, Ministry Educ, Mogidiscio, Somalia Repub, Africa, 67-69. Mem: AAAS; Am Asn Physics Teachers; Am Astron Soc. Res: Determination of orbital elements; computer generation of star charts. Mailing Add: Dept of Physics & Astron Eastern Mich Univ Ypsilanti MI 48197

THOMAS, COLIN GORDON, JR, b Iowa City, Iowa, July 25, 18; m 46; c 4. SURGERY. Educ: Univ Chicago, BS, 40, MD, 43. Prof Exp: Assoc surg, Col Med, Univ Iowa, 50-51, asst prof, 51-52; from asst prof to assoc prof, 52-61, PROF SURG, SCH MED, UNIV NC, CHAPEL HILL, 61-, CHMN DEPT, 66- Mem: Soc Univ Surg; Am Asn Cancer Res; Am Thyroid Asn; Am Surg Asn; Soc Surg Alimentary Tract. Res: Thyroid cancer; gastrointestinal disorders. Mailing Add: Dept of Surg NC Mem Hosp Chapel Hill NC 27514

THOMAS, DAN ANDERSON, b Ooltewah, Tenn, Oct 1, 22; m 44; c 2. PHYSICS. Educ: Univ Chattanooga, BS, 45; Vanderbilt Univ, PhD(physics), 52. Prof Exp: Asst prof physics, Univ of the South, 49-51; res physicist, US Naval Ord Lab, 51-52; from assoc prof to prof physics, Rollins Col, 52-63; PROF PHYSICS & DEAN OF FACULTIES, JACKSONVILLE UNIV, 63-, VPRES, 67- Concurrent Pos: Consult, US Naval Underwater Sound Reference Lab, 53-63. Mem: Fel AAAS; Am Phys Soc; Acoustical Soc Am; Am Asn Physics Teachers. Res: Beta ray spectroscopy; underwater acoustics; wave motion in solids; vibration of plates. Mailing Add: Off of Dean Faculties Jacksonville Univ Jacksonville FL 32211

THOMAS, DANIEL WALTER, organic chemistry, see 12th edition

THOMAS, DAVID BARTLETT, medicine, epidemiology, see 12th edition

THOMAS, DAVID DALE, b Lansing, Mich, Sept 18, 49; m 75. MOLECULAR BIOPHYSICS. Educ: Stanford Univ, BS, 71, PhD(biophys), 76. Prof Exp: Res asst physics, High Energy Physics Lab, Stanford Univ, 70; res asst, Dept Genetics, Stanford Univ, 71, student biophys, Dept Biol Sci, 71-75; FEL BIOPHYS, DEPT MUSCLE RES, BOSTON BIOMED RES INST, 76- Concurrent Pos: NSF fel, 71; fel, Muscular Dystrophy Asns Am, Inc, 76-77. Mem: AAAS; Fedn Am Scientists; Biophys Soc. Res: Molecular dynamics in muscle contraction as studied by spectroscopic probe methods; electron paramagnetic resonance studies of spin-labeled muscle proteins, such as actin and myosin. Mailing Add: Boston Biomed Res Inst Dept Muscle Res 20 Staniford St Boston MA 02114

THOMAS, DAVID GILBERT, b London, Eng, Aug 4, 28; US citizen; m 57; c 2. PHYSICAL INORGANIC CHEMISTRY. Educ: Oxford Univ, BA, 49, MA & DPhil(chem), 52. Prof Exp: Head semiconductor electronics res lab, 62-68, dir electron device process & battery lab, 68-69, EXEC DIR ELECTRONIC DEVICE, PROCESS & MAT DIV, BELL LABS, 69- Honors & Awards: Oliver E Buckley Solid State Physics Award, Am Phys Soc, 69. Mem: Fel Am Phys Soc; AAAS; Electrochem Soc. Res: Optical and electrical properties of semiconductors. Mailing Add: Bell Labs Murray Hill NJ 07974

THOMAS, DAVID HURST, b Oakland, Oakland , Calif, May 27, 45; m 69. ANTHROPOLOGY, ARCHAEOLOGY. Educ: Univ Calif, Davis, BA, 67, MA, 68, PhD(anthrop), 71. Prof Exp: Asst prof anthrop & chmn dept, City Col New York, 71-72; ASST CUR, NAM ARCHAEOL, AM MUS NATURAL HIST, 72- Concurrent Pos: Assoc ed, J Human Ecology, 72-74; adj asst prof, City Col NY, 74- Mem: Am Anthrop Asn; Soc Am Archaeol; NY Acad Sci; Soc Syst Zool. Res: Prehistoric popultaion ecology of American desert west; statistics and computer applications to archaeology; numerical taxonomy; Great Basin ethnology; faunal analysis of archaeological sites; functional analysis of lithic artifacts. Mailing Add: Dept of Anthrop Am Mus Natural Hist Central Park West at 79th St New York NY 10024

THOMAS, DAVID JOHN, b Portland, Ore, Sept 11, 45. ETHNOLOGY, ECONOMIC ANTHROPOLOGY. Educ: Stanford Univ, BS, 67; Univ Mich, Ann Arbor, MA, 70, PhD(anthrop), 73. Prof Exp: Field technician anthrop, Dept Human Genetics, Univ Mich, Ann Arbor, 70-71; ASST PROF ANTHROP, VANDERBILT UNIV, 74- Res: Primitive trade systems; South American Indians; indigenous peoples of Melanesia; shamanism and social control; egalitarian societies. Mailing Add: Dept of Sociol & Anthrop Box 1811 Sta B Vanderbilt Univ Nashville TN 37235

THOMAS, DAVID PHILLIP, b Wasco, Ore, July 7, 18; m 43; c 3. FORESTRY. Educ: Univ Wash, BSF, 41, MF, 47. Prof Exp: Res assoc, Eng Exp Sta, Univ Wash, 46, assoc forestry, Col Forestry, 46-47; instr wood technol, State Univ NY Col Forestry, Syracuse, 47-50; from asst prof to assoc prof wood sci & technol, 50-66, spec asst to provost, 64-72, dir inst forest products, 66-72, assoc dean col, 73-75, PROF FOREST RESOURCES, COL FOREST RESOURCES, UNIV WASH, 66-, CHMN FOREST MGT & SOC SCI DIV, 75- Concurrent Pos: Mem, King Co Environ Develop Comn, 73- & State of Wash Forest Pract Adv Comt, 74- Mem: AAAS; Soc Am Foresters; Forest Prod Res Soc; Soc Wood Sci & Technol; Am Forestry Asn. Res: Economics and technology of utilizing forest crops; forest practices and their impact on environmental quality. Mailing Add: Col Forest Resources AR-10 Univ of Wash Seattle WA 98195

THOMAS, DEMPSEY LEE, b Macclenny, Fla, Feb 2, 33; m 58; c 2. BIOLOGICAL SCIENCES. Educ: Univ Fla, BSA, 55, MAg, 56; Univ Tex, Austin, PhD(bot), 68. Prof Exp: Instr high schs, Fla, 56-64;asst prof, Concordia Lutheran Col, 65-67; asst ASSOC PROF BIOL, UNIV NEW ORLEANS, 68- Mem: AAAS; Am Soc Plant Physiol; Phycol Soc Am; Bot Soc Am; Am Inst Biol Sci. Res: Biochemical taxonomy and physiology of algae; physiological chemistry of legumes; immunology of lower plants; effects of different forms of mercury on blue-green and green algae. Mailing Add: Dept of Biol Sci Univ New Orleans Lakefront New Orleans LA 70122

THOMAS, DON WYLIE, b Spanish Fork, Utah, Aug 8, 23; m 52; c 6. VETERINARY SCIENCE, ANIMAL SCIENCE. Educ: Utah State Univ, BS, 49; Iowa State Univ, DVM, 53. Prof Exp: Vet pathologist, Calif State Dept Agr, 53-54; PROF VET & ANIMAL SCI & EXTEN VET, UTAH STATE UNIV, 54- Concurrent Pos: State del nat plans conf, USDA, 56, 58 & 62; vchmn, Utah Herd Health & Mastitis Comt, 58- Mem: Am Asn Exten Vet; Am Asn Vet Nutritionists. Res: Prevention of diseases in cattle, sheep, horses and poultry. Mailing Add: Dept of Animal Sci Utah State Univ Logan UT 84322

THOMAS, DONALD C, b Cincinnati, Ohio, Sept 26, 35; m 57; c 3. VIROLOGY, MICROBIOLOGY. Educ: Xavier Univ, BS, 57; Univ Cincinnati, MS, 59; St Louis Univ, PhD(microbiol), 68. Prof Exp: Res assoc bact, Univ Cincinnati, 59-61; instr biol, Thomas More Col, 61-63; instr med microbiol, 68-69, ASST PROF MICROBIOL & PEDIAT, OHIO STATE UNIV, 69-, ASSOC DIR PROG DEVELOP ASSISTANCE DIV, RES FOUND, 71- Mem: AAAS; Fedn Am Sci; Am Soc Microbiol. Res: Biochemistry of virus replication; macromolecular synthesis and characterization of virus particles; oncogenic and common cold viruses. Mailing Add: Prog Develop Assistance Div Ohio State Univ Res Found Columbus OH 43212

THOMAS, DONALD HENRY, b Detroit, Mich, Feb 25, 37; m 62; c 3. APPLIED MATHEMATICS, APPLIED STATISTICS. Educ: Wayne State Univ, BS, 59, MA, 64, PhD(math), 70. Prof Exp: Res mathematician, Gen Motors Res Labs, 60-68; asst prof math, Univ Detroit, 68-69; RES MATHEMATICIAN, GEN MOTORS RES LABS, 69- Mem: Soc Indust & Appl Math; Inst Math Statist. Res: Numerical analysis, approximation theory and statistics with special emphasis on spline approximations in one and several variables to problems of computer-aided design and automation. Mailing Add: Math Dept Gen Motors Res Labs Warren MI 48090

THOMAS, DUDLEY WATSON, b Los Angeles, Calif, Apr 14, 20; m 46; c 2. BIOCHEMISTRY. Educ: Univ Calif, AB, 42; Calif Inst Technol, MS, 47, PhD(org chem), 51. Prof Exp: Chemist, US Rubber Co, 42-43; biochemist, Rohm & Haas Co, 51-60; res fel chem, Calif Inst Technol, 60-62; asst prof biochem, Sch Dent, Univ Southern Calif, 62-67; ASST PROF ZOOL, CALIF STATE UNIV, LOS ANGELES, 67-, CHMN DEPT BIOL, 71- Mem: AAAS; Am Chem Soc; NY Acad Sci. Res: Enzymes; structural-biological activity relationships; connective tissue metabolism. Mailing Add: 1460 Linda Ridge Rd Pasadena CA 91103

THOMAS, E LLEWELLYN, b Salisbury, Eng, Dec 15, 17; m 47. MEDICINE, ELECTRICAL ENGINEERING. Educ: Univ London, BSc, 51; McGill Univ, MD & CM, 55. Prof Exp: Jr engr, Brit Broadcasting Co, 37-39; asst controller telecommun, Govt Malaya, 45-51; res assoc, Med Col, Cornell Univ, 56-58; sci officer, Defence Res Med Labs, Toronto, 58-60; PROF PHARMACOL, UNIV TORONTO, 61-, ASSOC DEAN MED, 74- Concurrent Pos: Prof psychol, Univ Waterloo, 62-64. Mem: Nat Soc Prof Eng; Inst Elec & Electronics Eng; fel Royal Soc Arts; Can Med Asn; fel Royal Soc Can. Mailing Add: Inst of Biomed Eng Univ of Toronto Toronto ON Can

THOMAS, EDWARD CARL, b Braceville, Ill, Oct 15, 10; m 39; c 2. PHYSICS. Educ: Carleton Col, BA, 33. Prof Exp: Jr metallurgist, Carnegie Ill Steel Corp, 33-34; prod operator, Inland Steel Corp, 34-42; res engr, Kellog Switchboard & Supply Co, 42-46; res physicist, Great Lakes Carbon Corp, 46-56, group leader res & develop, 56-62, sr process engr, 62-64, asst tech supt prod, 64-65, tech supt, 65-66, tech serv adminr, 66-75; RETIRED. Mem: Electrochem Soc; Am Phys Soc. Res: Carbon technology; petrographic methods of analysis using photomicroscopy. Mailing Add: 125 Jordan Rd Williamsville NY 14221

THOMAS, EDWARD DONNALL, b Mart, Tex, Mar 15, 20; m 42; c 3. INTERNAL MEDICINE, ONCOLOGY. Educ: Univ Tex, BA, 41, MA, 43; Harvard Med Sch, MD, 46; Am Bd Internal Med, dipl, 53. Prof Exp: From intern to sr asst resident med, Peter Bent Brigham Hosp, Boston, 46-52, chief med res, 52-53, hematologist, 53; res assoc, Cancer Res Found, Children's Med Ctr, 53-55; hematologist & asst physician, Mary Imogene Bassett Hosp, 55-56; assoc clin prof med, Col Physicians & Surgeons, Columbia Univ, 56-63; PROF MED, SCH MED, UNIV WASH, 63- Concurrent Pos: Fel med, Mass Inst Technol, 50-51; instr, Harvard Med Sch, 53; physician-in-chief, Mary Imogene Bassett Hosp, 56-63; clin prof, Albany Med Col, 58-63. Mem: Am Soc Clin Invest; Asn Am Physicians; NY Acad Sci; Int Soc Hemat. Res: Marrow biochemistry and transplantation; irradiation effects; hematology. Mailing Add: Div of Oncol Univ of Wash Sch of Med Seattle WA 98195

THOMAS, EDWARD SANDUSKY, JR, b Kansas City, Mo, Jan 11, 38; m 60. MATHEMATICS. Educ: Whittier Col, BA, 59; Univ Wash, MS, 61; Univ Calif, Riverside, PhD(math), 64. Prof Exp: Asst prof math, Univ Mich, Ann Arbor, 65-69; ASSOC PROF MATH, STATE UNIV NY ALBANY, 69- Mem: Am Math Soc. Res: Topology, dynamical systems. Mailing Add: Dept of Math State Univ of NY Albany NY 12203

THOMAS, EDWARD WILFRID, b Croydon, Eng, May 9, 40. PHYSICS. Educ: Univ London, BSc, 61, PhD(physics), 64. Prof Exp: Asst res physicist, 64-65, from asst prof to assoc prof, 65-73, PROF PHYSICS, GA INST TECHNOL, 73- Concurrent Pos: Consult, Oak Ridge Nat Lab, 65-75. Mem: Am Phys Soc; Optical Soc Am; Brit Inst Physics. Res: Collisions between atomic, ionic and molecular systems, particularly on the formation of excited states; development of photon and particle detection techniques. Mailing Add: Sch of Physics Ga Inst of Technol Atlanta GA 30332

THOMAS, ELIZABETH FLOWERS, textiles, see 12th edition

THOMAS, ELIZABETH WADSWORTH, b Washington, DC, May 23, 44; m 70; c 1. ANALYTICAL CHEMISTRY. Educ: ECarolina Univ, AB, 66; Univ Va, PhD(phys & org chem), 70. Prof Exp: Res assoc pharmacol, Med Sch, Univ Va, 70-72; lectr, Univ Wis, Parkside & Barat Col, 73-74; ANAL CHEMIST, ABBOTT LABS, 75- Mem: Am Chem Soc; Sigma Xi. Res: Development of analytical methodology for the analysis of new pharmaceutical products and new consumer products; solution of unusual complaints about current marketed products. Mailing Add: 3925 Dorchester Ave Gurnee IL 60031

THOMAS, ELMER LAWRENCE, b Springfield, Ohio, Jan 18, 16; m 41; c 3. FOOD SCIENCE. Educ: Ohio State Univ, BSc, 41; Univ Minn, MSc, 43, PhD, 50. Prof Exp: Res asst, 41-45, instr dairy prod, 46-50, asst prof dairy technol, 51-54, assoc prof dairy indust, 54-60, PROF FOOD SCI, UNIV MINN, ST PAUL, 61- Mem: Am Dairy Sci Asn; Inst Food Technol. Res: Physical, chemical and engineering aspects of dairy products processing; sensory testing of foods. Mailing Add: Dept of Food Sci & Nutrition Univ of Minn St Paul MN 55108

THOMAS, ESTES CENTENNIAL, III, b Plaquemine, La, Dec 13, 40; m 64; c 3. PHYSICAL CHEMISTRY. Educ: La State Univ, Baton Rouge, BS, 62; Stanford Univ, PhD(phys chem), 66. Prof Exp: Fel, Princeton Univ, 66-67; chemist, Shell Develop Co, 67-72, sr engr, 72-75, STAFF ENGR, SHELL OIL CO, 75- Mem: Am Chem Soc; Am Inst Mining, Metall & Petrol Eng; Soc Prof Well Log Analysts. Res: Physical characteristics and transport of energy through interstices of subterranean earthen formations, particularly those containing hydrocarbons. Mailing Add: Coastal Div Shell Oil Co Box 60123 New Orleans LA 70160

THOMAS, FORREST DEAN, II, b Provo, Utah, Dec 27, 30; m 52; c 2. INORGANIC CHEMISTRY. Educ: Brigham Young Univ, BS, 55; Pa State Univ, PhD(chem), 59. Prof Exp: From asst prof to assoc prof, 59-73, PROF CHEM, UNIV MONT, 73- Mem: Am Chem Soc. Res: Organic and inorganic synthesis; preparation of coordination compounds. Mailing Add: Dept of Chem Univ of Mont Missoula MT 59801

THOMAS, FRANK BANCROFT, b Camden, Del, June 14, 22; m 60; c 2. HORTICULTURE, FOOD TECHNOLOGY. Educ: Univ Del, BS, 48; Pa State Univ, MS, 49, PhD(hort), 55. Prof Exp: From instr to asst prof hort, Pa State Univ, 49-58; food processing exten specialist, 58-61, exten assoc prof, 61-66, EXTEN PROF FOOD SCI, NC STATE UNIV, 66- Concurrent Pos: Vis fel food technol, Mass Inst Technol, 54-55; consult cryogenic foods, 63-64; mem NC Gov Comn Com Fisheries, 64-65, mem Com & Sport Fisheries Adv Comt, 75-76; sabbatical, Dept Food Sci & Technol, Univ Hawaii, 68; partic, Food & Agr Orgn Conf Fish Qual & Inspection, Halifax, Can, 69 & Conf Fishery Prod Technol, Japan, 73; prog leader Food Sci Seafood Adv Serv, NC Sea Grant Prog, 70- Mem: Inst Food Technologists. Res: Post-harvest physiology of fruits and vegetables; food processing; chemical and microbiological changes in seafoods; flavor and color evaluation; extension and applied research on seafood utilization. Mailing Add: Dept of Food Sci NC State Univ Raleigh NC 27607

THOMAS, FRANK HARRY, b Alamo, Ga, Oct 16, 32; m 52; c 3. SOILS, INORGANIC CHEMISTRY. Educ: Univ Ga, BSA, 54, MS, 56, PhD(soil phosphorus), 59. Prof Exp: Asst chemist, Everglades Exp Sta, 59-66; assoc chemist, 66-68, prof chem & chmn div sci-math, 68-75, asst acad dean, 73-75, ACAD DEAN, ABRAHAM BALDWIN AGR COL, 75- Mem: Am Chem Soc; AAAS; Am Soc Allied Health Professions. Mailing Add: Off of Acad Dean Abraham Baldwin Agr Col Tifton GA 31794

THOMAS, FRANK HENRY, b Chicago, Ill, Jan 25, 32; m 58; c 3. RESOURCE GEOGRAPHY, URBAN GEOGRAPHY. Educ: Univ Ill, BA, 55; Northwestern Univ, MSc, 56, PhD(geog), 60. Prof Exp: Prof geog, Southern Ill Univ, Carbondale, 59-73, chmn dept, 66-72; PROF GEOG & CHMN DEPT, GA STATE UNIV, 73- Concurrent Pos: Vis prof geog, Univ Liverpool, 62-63; chmn soc sci comt, Univ Coun Water Resources, 71-73; staff specialist, US Water Resources Coun, 75-76. Mem: Asn Am Geog; Inst Brit Geog; Am Water Resources Asn; Regional Sci Asn. Res: Regional growth centers; metropolitan resource management. Mailing Add: Dept of Geog Ga State Univ Atlanta GA 30303

THOMAS, GARLAND LEON, b Topeka, Kans, Aug 29, 20; m 48; c 3. PHYSICS. Educ: Drury Col, BS, 42; Univ Mo, AM, 48, PhD(physics), 54. Prof Exp: Fel engr, atomic power dept, Westinghouse Elec Corp, 53-59; assoc prof physics, Drury Col, 59-66 & Fla Inst Technol, 66-71; MEM STAFF, FLA PLANNING DEPT, BREVARD CO, 71- Mem: Biophys Soc; Am Phys Soc; Am Asn Physics Teachers; Am Nuclear Soc. Res: Remote sensing; nuclear reactor physics; ultrasonic cavitation. Mailing Add: 200 E Southgate Blvd Melbourne FL 32901

THOMAS, GARTH JOHNSON, b Pittsburgh, Kans, Sept 8, 16; m 45; c 2. NEUROPSYCHOLOGY. Educ: Kans State Teachers Col Pittsburg, AB, 38; Univ Kans, MA, 40; Harvard Univ, AM, 43, PhD(exp psychol), 48. Prof Exp: Asst instr, Univ Kans, 38-41; tutor, Harvard Univ, 41-43 & 47-48; from instr to asst prof psychol, Univ Chicago, 48-54; res assoc & assoc prof, Neuropsychiat Inst, Col Med, Univ Ill, 54-57, res prof, Biophys Res Lab, Dept Elec Eng & Dept Physiol & Biophys, 57-66; PROF PHYSIOL PSYCHOL, CTR BRAIN RES, UNIV ROCHESTER, 66-, DIR CTR, 70- Concurrent Pos: Dept Defense NIMH & NSF res grants, Univ Chicago, Univ Ill & Univ Rochester, 51-; mem psychol sci fel rev panel, NIMH, 64-69 & psychobiol rev panel NSF, 68-71; consult ed, McGraw-Hill Encycl Sci & Technol, 64-, Contemp Psychol, 64-73 & Sinauer Assocs Press, 68-; assoc ed, J Comp & Physiol Psychol, 69-73, ed, 74- Mem: AAAS; Am Physiol Soc; Psychonomic Soc; Soc Exp Psychol; Soc Neurosci. Res: Brain function; studies of behavioral effects of central nervous system lesions; psychophysiology of sensation. Mailing Add: Ctr for Brain Res Box 605 Med Ctr Univ of Rochester Rochester NY 14642

THOMAS, GARY E, b Lookout, WVa, Oct 25, 34; m 61; c 2. ATMOSPHERIC PHYSICS. Educ: NMex State Univ, BS, 57; Univ Pittsburgh, PhD(physics), 63. Prof Exp: Res assoc, Aeronomy Serv, Nat Ctr Sci Res, France, 62-63; mem tech staff space physics lab, Aerospace Corp, 65-67; assoc prof astro-geophys, 67-74, PROF ASTRO-GEOPHYS, UNIV COLO, BOULDER, 74- Mem: AAAS; Am Geophys Union. Res: Theoretical study of upper atmosphere; radiative transfer; application of spectroscopic remote sensing data to study of atmospheric structure; measurement techniques and theoretical study of interplanetary medium; investigation of lunar atmosphere experiment. Mailing Add: Lab for Atmos & Space Physics Univ of Colo Boulder CO 80302

THOMAS, GEORGE BRINTON, JR, b Boise, Idaho, Jan 11, 14; m 36; c 3. MATHEMATICS. Educ: State Col Wash, AB, 34, AM, 36; Cornell Univ, PhD(math), 40. Prof Exp: Instr math, Cornell Univ, 37-40; from instr to assoc prof, 44-60, asst elec eng, 43-45, exec officer dept math, 50-59, PROF MATH, MASS INST TECHNOL, 60- Mem: AAAS; Am Math Soc; Am Soc Eng Educ; Math Asn Am (1st vpres, 58-59); Inst Math Statist. Res: Probability; theory of numbers. Mailing Add: Dept of Math Room 2-361 Mass Inst of Technol Cambridge MA 02139

THOMAS, GEORGE E, b Bloomington, Ill, July 18, 21; m 44; c 2. EXPERIMENTAL NUCLEAR PHYSICS. Educ: Ill Wesleyan Univ, BS, 43. Prof Exp: Jr physicist, 43-66, SCI ASSOC EXP NUCLEAR PHYSICS, ARGONNE NAT LAB, 66- Mem: Am Phys Soc; Int Nuclear Target Develop Soc. Res: Research, development and production of thin film targets used particularly at tandem and dynamatron accelerators, and nuclear physics experiments associated with these targets. Mailing Add: Argonne Nat Lab Physics Div 9700 S Cass Ave Argonne IL 60439

THOMAS, GEORGE HOWARD, b Minerva, Ohio, Apr 27, 36; m 60; c 3. PEDIATRICS. Educ: Western Md Col, AB, 59; Univ Md, PhD, 63. Prof Exp: Instr biochem, Sch Med, Univ Md, 63; asst pediat, Sch Med, 65-67, instr, 67-68, ASST PROF PEDIAT & MED, SCH MED & DIR GENETICS LAB, JOHN F KENNEDY INST, JOHNS HOPKINS UNIV, 68- Mem: AAAS; Am Soc Human Genetics; Soc Pediat Res. Res: Human genetics. Mailing Add: J F Kennedy Inst Genetics Lab 707 N Broadway Baltimore MD 21205

THOMAS, GEORGE JOSEPH, JR, physical chemistry, biophysics, see 12th edition

THOMAS, GEORGE RICHARD, b Bethlehem, Pa, Feb 1, 20; m 55; c 4. ORGANIC CHEMISTRY. Educ: Bowdoin Col, BS, 41; Northwestern Univ, PhD(chem), 48; Harvard Univ, adv mgt prog, 62. Prof Exp: Res fel, Univ Ill, 48, Harvard Univ, 49 & Boston Univ, 49-54. Prof Exp: US Army, 54-, chief dyestuffs res sect, Qm Res & Eng Command, Natick Labs, 54-56, chief, Chem & Plastics Div, 56-62, assoc dir & dir res, Clothing & Org Mat Lab, 62-68, CHIEF, ORG MAT LAB, ARMY MAT & MECH RES CTR, US ARMY, 68- Concurrent Pos: Pres, Thomason Chem, Inc, 54-60; vis prof, Boston Univ, 64-65. Mem: Fel Am Inst Chemists; AAAS; Am Chem Soc; Res Eng Soc Am; Am Soc Testing & Mat. Res: Military applications of polymers as fibers, films, foams, elastomers and rigid and reinforced plastics; materials research; polymer research; operations and management research; business administration. Mailing Add: Org Mat Lab Army Mat & Mech Res Ctr Watertown MA 02172

THOMAS, GERALD ANDREW, b Birmingham, Ala, Oct 8, 11; m 40; c 4. RADIOCHEMISTRY. Educ: Birmingham-Southern Col, BS & MS, 32; Univ Fla, PhD(chem), 52. Prof Exp: Teacher pub schs, Ala, 32-39; asst, Johns Hopkins Univ, 39-40; res chemist, Niagara Alkali Co, 40-45; from instr to assoc prof chem, Univ Fla, 46-57; chmn div sci, math & eng, 57-63, PROF CHEM, SAN FRANCISCO STATE UNIV, 57- Concurrent Pos: NSF fel, 54-56; lectr & scientist, US AEC Latin-Am Prog, Atoms in Action, 65-68; consult, Oak Ridge Assoc Univs, 63- Mem: Am Chem Soc; Am Nuclear Soc. Res: Physical properties of organic compounds; nucleonics; radioisotopes. Mailing Add: Dept of Chem San Francisco State Univ San Francisco CA 94132

THOMAS, GERALD WAYLETT, b Small, Idaho, July 3, 19; m 53; c 3. RANGE ECOLOGY, AGRONOMY. Educ: Univ Idaho, BS, 41; Tex A&M Univ, MS, 51, PhD(range mgt), 54. Prof Exp: Range & work unit conservationist, US Soil Conserv Serv, Idaho, 46-50; asst, Tex A&M Univ, 51, assoc prof range & forestry, 51-56; res coordr, Tex Agr Exp Sta, 56-58; dean agr, Tex Tech Univ, 58-70; PRES, NMEX STATE UNIV, 70- Concurrent Pos: Consult, Agr Res Serv, USDA, 59. Honors & Awards: Distinguished Flying Cross; Air Medal. Mem: Fel AAAS; Soc Range Mgt; Soil Conserv Soc Am. Res: Response of vegetation to grazing management; wildlife-livestock relationships; range improvement; brush control; relation of rainfall to vegetation and ranching risk range pitting and reseeding; relation of vegetation to soils; progress and change in American agriculture; ecology of food production. Mailing Add: Off of the Pres NMex State Univ Las Cruces NM 88003

THOMAS, GORDON ALBERT, b Kingston, Pa, June 8, 43; m 66; c 1. EXPERIMENTAL SOLID STATE PHYSICS. Educ: Brown Univ, ScB, 65; Univ Rochester, PhD(physics), 71. Prof Exp: Fel physics, Univ Rochester, 71-72; MEM TECH STAFF PHYSICS, BELL LABS, 72- Mem: Am Phys Soc. Res: Studies of transport properties and spectroscopy that probe the metal-insulator transition in liquids and solids. Mailing Add: Bell Labs Murray Hill NJ 07974

THOMAS, GRANT WORTHINGTON, b Washington, DC, Feb 23, 31; m 51; c 5. SOIL CHEMISTRY. Educ: Brigham Young Univ, BS, 53; NC State Univ, MS, 56, PhD(soils), 58. Prof Exp: Asst prof agron, Va Polytech Inst, 58-60, assoc prof, 60-64; prof soils, Tex A&M Univ, 64-68; PROF AGRON, UNIV KY, 68- Concurrent Pos: Vis prof, Univ Calif, Riverside, 62-63. Mem: Am Soc Agron. Res: Reactions and movement of solutes in soils. Mailing Add: Dept of Agron Univ of Ky Lexington KY 40506

THOMAS, GUSTAVE DANIEL, b Bay St Louis, Miss, Aug 4, 40; m 60; c 3. ENTOMOLOGY. Educ: Miss State Univ, 62, MS, 64; Univ Mo-Columbia, PhD(entom), 67. Prof Exp: RES ENTOMOLOGIST, BIOL CONTROL INSECTS RES LAB, N CENT REGION, AGR RES SERV, USDA, 67- Concurrent Pos: Res assoc, Univ Mo, 67- Mem: Entom Soc Am; Int Orgn Biol Control. Res: Population dynamics; economic thresholds; relationships of parasites and predators to their hosts. Mailing Add: Biol Contr Insects Res Lab Agr Res Serv USDA Box A Columbia MO 65201

THOMAS, HAROLD ALBERT, b Newport, Ore, Aug 12, 11; m 35; c 3. PHYSICS. Educ: Ore State Col, BS, 36, MS, 37; Agr & Mech Col, Tex, PhD, 47. Prof Exp: Instr, Purdue Univ, 37-42; asst prof, Agr & Mech Col, Tex, 42-44, assoc prof elec eng & supvr, Mass Spectrometer Lab, 46-47; res engr, Res Labs, Westinghouse Elec Corp, Pa, 44-45; physicist, Atomic Physics Div, Nat Bur Stand, 47-51, chief, Radio Stand Div, 54-56; chief, Fuze Div, US Naval Ord Lab, 51-54; head electronics div, Gen Atomic Div, Gen Dynamics Corp, 56-69; MGR NUCLEAR INSTRUMENTATION & CONTROL DEPT, ELECTRONIC SYSTS DIV, GEN ATOMIC CO, 69- Concurrent Pos: US del, Int Electrotech Comn, 45. Mem: Inst Elec & Electronics Engrs. Res: Electronic ordnance devices; reactor instrumentation. Mailing Add: Nucl Instrum & Contr Dept Gen Atomic Co Electronic Systs Div Box 81608 San Diego CA 92138

THOMAS, HAROLD LEE, b Westphalia, Kans, July 2, 34; m 56; c 3. MATHEMATICS, STATISTICS. Educ: Kans State Col Pittsburg, BS, 58, MS, 59; Okla State Univ, PhD(statist), 64. Prof Exp: Instr math, Kans State Col Pittsburg, 59-60; asst, Okla State Univ, 60-64; from asst prof to assoc prof, 64-69, PROF MATH, KANS STATE COL PITTSBURG, 69- Mem: Math Asn Am; Am Statist Asn. Res: Factorial experiments in experimental designs. Mailing Add: Dept of Math Kans State Col Pittsburg KS 66762

THOMAS, HAROLD TODD, b Seattle, Wash, Feb 3, 42; c 1. PHOTOCHEMISTRY. Educ: Calif Inst Technol, BS, 64; Wesleyan Univ, MA, 66; Princeton Univ, PhD(chem), 70. Prof Exp: Fel chem, Bell Tel Labs, 69-70; SR RES CHEMIST, EASTMAN KODAK CO, 70- Mem: Am Chem Soc. Res: Photochemistry of ordered systems; spectroscopy and photochemistry of adsorbed molecules. Mailing Add: Res Labs Eastman Kodak Co Rochester NY 14650

THOMAS, HENRY CARRISON, b Cheraw, SC, Jan 30, 10; m 41; c 3. PHYSICAL CHEMISTRY. Educ: Univ NC, BS, 31, MS, 32; Yale Univ, PhD, 35; Cath Univ, Louvain, Belg, DSc, 74. Prof Exp: Res chemist, Explosives Dept, E I du Pont de Nemours & Co, 35-37; from instr to assoc prof chem, Yale Univ, 37-57; prof, 57-75, EMER PROF CHEM, UNIV NC, CHAPEL HILL, 75- Mem: Am Chem Soc. Res: Ion exchange; chromatography; chemical kinetics; diffusion. Mailing Add: 304 Venable Hall Dept Chem Univ of NC Chapel Hill NC 27514

THOMAS, HENRY COFFMAN, b Sacramento, Ky, Dec 29, 18; m 44; c 1. PHYSICS. Educ: Western Ky State Col, BS, 43; Vanderbilt Univ, MS & PhD, 49. Prof Exp: From asst to assoc prof physics, Miss State Col, 49-55; assoc prof & head dept, Bradley Univ, 55-58; chmn dept, 58-74, PROF PHYSICS, TEX TECH UNIV, 58-, ASSOC DEAN ARTS & SCI, 74- Mem: Am Phys Soc. Res: Low energy nuclear physics. Mailing Add: Dept of Physics Tex Tech Univ Col of Arts & Sci Lubbock TX 79409

THOMAS, HERBERT REX, b Riverside, Calif, Apr 13, 13; m 40; c 2. PLANT PATHOLOGY. Educ: Univ Calif, BS, 35, PhD(plant path), 41. Prof Exp: Asst, Univ Calif, 37; agent plant path, Div Fruits & Veg Crops & Dis, Bur Plant Indust, 37-41, from asst pathologist to assoc pathologist, 41-43, bur plant indust, Soils & Agr Eng, 46-47, from pathologist to sr pathologist, 47-54, exp sta adminr hort & veg crops, Off Exp Stas, 54-56, asst dir, Crops Res Div, Agr Res Serv, 56-65, dir, 65-70, dep adminr plant sci & entom, 70-72, DEP ADMINR WESTERN REGION, AGR RES SERV, USDA, 72- Mem: AAAS; Am Phytopath Soc. Res: Administration of regional program of agricultural research in plant, entomological, livestock, veterinary, soil, water, air, marketing, nutrition and engineering sciences. Mailing Add: Agr Res Serv USDA 2840 Telegraph Ave Berkeley CA 94705

THOMAS, HERIBERTO VICTOR, b Panama City, Repub Panama, Mar 17, 17; US citizen; m 50; c 1. PREVENTIVE MEDICINE. Educ: Univ Southern Calif, AB, 50, MS, 60; Univ Calif, Los Angeles, MPH, 64; Univ Calif, Berkeley, PhD, 68. Prof Exp: Res fel pharmacol & biochem, Univ Southern Calif, 53-56, res assoc, Sch Med, 57-61; res assoc physiol & biochem, St Joseph Hosp, Burbank, 61-64; res chemist, 64-66, res specialist, 66-72, coordr, Sickle Cell Anemia Prog, 72-74, CHIEF GENETIC DIS, CALIF STATE DEPT HEALTH, 74-; CONTROL UNIT LECTR, HEALTH & MED SCI, UNIV CALIF, BERKELEY, 73- Concurrent Pos: Reviewer & consult sci, AAAS & NSF, 72- & Nat Heart & Lung Inst, NIH, Rev Br, 75- Honors & Awards: Macgee Award, Am Oil Chem Soc, 63. Mem: AAAS; Am Chem Soc; NY Acad Sci; Sigma Xi. Res: Physiological chemistry of air pollutants and their health effects; lipid metabolism; structural changes in lung tissue as a consequence of adverse ambient conditions; prenatal diagnosis of disabling genetic diseases; genetic counseling as a method of prevention. Mailing Add: Calif State Dept Health 2151 Berkeley Way Berkeley CA 94704

THOMAS, HERMAN HOIT, b Raleigh, NC, Dec 26, 31; m 57; c 2. GEOCHEMISTRY. Educ: Lincoln Univ, Pa, BA, 58; Univ Pa, PhD(geochem), 73. Prof Exp: Chemist, US Geol Surv, 58-64, Fairchild-Hiller Corp, 64-65 & Melpar Corp, 65-66; SPACE SCIENTIST GEOCHEM, GODDARD SPACE FLIGHT CTR, NASA, 66- Mem: Am Geophys Union. Res: Composition of the Earth; earthquake prediction and monitoring; composition and age of the solar system. Mailing Add: Code 922 Goddard Space Flight Ctr NASA Greenbelt MD 20771

THOMAS, HOLLIS ALLEN, b Providence, RI, Nov 30, 31; m 55; c 3. ENTOMOLOGY. Educ: Univ RI, BSc, 53; Rutgers Univ, MSc, 55, PhD(entom), 57. Prof Exp: Asst prof entom, Tex Col Arts & Indust, 57-59; entomologist, State Forest Serv, Maine, 59-65; ASSOC ENTOMOLOGIST, US FOREST SERV, 65- Mem: Entom Soc Am; Entom Soc Can. Res: Insect biology; beneficial insects; Acarina. Mailing Add: Forestry Sci Lab USDA PO Box 12254 Research Triangle Park NC 27709

THOMAS, HOWARD MAJOR, b Elwood, Nebr, Feb 14, 18; m 43; c 6. PHYSICAL CHEMISTRY. Educ: Nebr State Teachers Col, 42; Univ Iowa, PhD(chem), 49. Prof Exp: Asst prof chem & head dept, St Ambrose Col, 49-52; from assoc prof to prof, Univ SDak, 53-58; PROF CHEM & CHMN DEPT, UNIV WIS-SUPERIOR, 58- Concurrent Pos: Vis prof, Univ Wis, 64-65; Fulbright lectr phys chem, Univ Col Cape Coast, Ghana, 68-69. Mem: AAAS; Am Chem Soc. Res: Chemical kinetics; azeotropic solution; audio-visual aids for chemistry teaching. Mailing Add: Dept of Chem Univ of Wis Superior WI 54880

THOMAS, HUBERT LANDON, organic chemistry, see 12th edition

THOMAS, HUGO F, environmental geology, natural resources, see 12th edition

THOMAS, JACK WARD, b Ft Worth, Tex, Sept 7, 34; m 57; c 2. WILDLIFE BIOLOGY. Educ: Tex A&M Univ, BS, 57; WVa Univ, MS, 69; Univ Mass, PhD, 73. Prof Exp: Biologist, Tex Game & Fish Comn, 57-62; res biologist, Tex Parks & Wildlife Dept, 62-67; asst prof wildlife mgt, WVa Univ & wildlife res biologist, Forestry Sci Lab, Northeastern Forest Exp Sta, US Forest Serv, 67-71, proj dir environ forestry res, Pinchot Inst Environ Forestry, 71-73, PROJ LEADER, RANGE & WILDLIFE HABITAT RES, PAC NORTHWEST FOREST EXP STA, US FOREST SERV, 73- Mem: AAAS; Wildlife Soc; Wilson Ornith Soc; Am Ornith Union; Am Soc Mammal. Res: Mobility and home range management of deer and turkeys; population dynamic of deer; disease impact on deer and antelope populations; wildlife habitat research; sociobioeconomic implications of game habitat manipulation; habitat requirements for wildlife in urbanizing areas; relationships of wild and domestic ungulates to forested ranges; non-consumptive utilization of wildlife. Mailing Add: US Forest Serv Range & Wildlife Habitat Lab La Grande OR 97850

THOMAS, JAMES ARTHUR, b International Falls, Minn, Apr 22, 38; m 61; c 2. BIOCHEMISTRY. Educ: St Olaf Col, BA, 60; Univ Wis, MS, 63, PhD(biochem), 66. Prof Exp: USPHS fel biochem, Univ Minn, 67-69; asst prof, 69-75, ASSOC PROF BIOCHEM, IOWA STATE UNIV, 75- Mem: AAAS; Am Chem Soc; Am Soc Biol Chemists. Res: Enzymology of glycogen metabolism and protein phosphorylation-dephosphorylation. Mailing Add: Dept of Biochem & Biophys Iowa State Univ Ames IA 50010

THOMAS, JAMES BLAKE, anatomy, embryology, see 12th edition

THOMAS, JAMES E, b Marshall, Mo, May 17, 26; m 52; c 2. PHYSICS. Educ: Mo Valley Col, BS, 50; Univ Mo-Rolla, MS, 55; Univ Mo-Columbia, PhD(physics), 63. Prof Exp: Teacher high sch, Mo, 50-52; instr math, Univ Mo-Rolla, 52-55; assoc prof physics, Mo Valley Col, 55-61; ASSOC PROF PHYSICS, KANS STATE COL PITTSBURG, 63- Mem: Am Phys Soc; Am Asn Physics Teachers. Res: Small angle x-ray diffraction; particle size and structure determination, solid state; radiation damage in perfect crystals. Mailing Add: Dept of Physics Kans State Col Pittsburg KS 66762

THOMAS, JAMES H, b Cardston, Alta, Jan 9, 36; m 59; c 4. PLANT BREEDING, GENETICS. Educ: Utah State Univ, BSc, 61, MS, 63; Univ Alta, PhD(genetics), 66. Prof Exp: Res asst agron, Utah State Univ, 60-63; teacher sci, Taber Sch Div, Alta, 63-64; res asst genetics, Univ Alta, 64-66; res officer, Can Dept Agr, 66-67 & Rudy Patrick Seed Co, 67-69; seed specialist & adv, Utah State Univ-AID, Bolivia, 69-72, ASST PROF PLANT BREEDING, UTAH STATE UNIV, 72- Concurrent Pos: Dryland agr adv, Coun US Univs for Soil & Water Develop in Arid & Semi-Arid Areas, Iran, 74- Mem: Am Soc Agron; Crop Sci Soc Am; Am Inst Biol Sci. Res: Plant breeding in forages, grass and alfalfa. Mailing Add: Dept of Plant Sci UMC 63 Utah State Univ Logan UT 84322

THOMAS, JAMES POSTLES, b Washington, DC, July 28, 39; m 65; c 3. ZOOLOGY, ECOLOGY. Educ: Univ NC, Chapel Hill, AB, 62; Univ Ga, MS, 66, PhD(zool), 70. Prof Exp: Ga Power Co fel aquatic ecol, Univ Ga, 70; res assoc biol oceanog, Univ Wash, 70-72, NSF grant, 71-72; FISHERIES BIOLOGIST & INVESTS CHIEF BIOL OCEANOG, NAT MARINE FISHERIES SERV, 72- Mem: AAAS; Am Soc Limnol & Oceanog; Ecol Soc Am. Res: Energy flow and cycling of carbon in marine and estuarine ecosystems; primary productivity and the release, assimilation and oxidation of phytoplankton-derived dissolved organic matter in marine waters; seabed oxygen consumption and primary productivity. Mailing Add: US Dept Com Sandy Hook Lab Nat Marine Fisheries Serv Highlands NJ 07732

THOMAS, JAMES WILLIAM, b Ironwood, Mich, Nov 11, 41; m 67; c 1. MATHEMATICS. Educ: Mich Technol Univ, BS, 63; Univ Ariz, MS, 65, PhD(math), 67. Prof Exp: Asst prof math, Univ Wyo, 67-72; ASSOC PROF MATH, COLO STATE UNIV, 72- Concurrent Pos: NSF sci develop grant, Univ Ariz, 70-71. Mem: Am Math Soc. Res: Applied mathematics; hydrodynamics; nonlinear functional analysis; applications of nonlinear functional analysis. Mailing Add: Dept of Math Colo State Univ Ft Collins CO 80521

THOMAS, JEROME FRANCIS, b Chicago, Ill, Jan 8, 22; m; c 3. AIR POLLUTION, WATER POLLUTION. Educ: DePaul Univ, BS, 43; Univ Calif, PhD(chem), 50. Prof Exp: Res chemist, 50-55, assoc prof, 55-72, PROF SANIT ENG, UNIV CALIF, BERKELEY, 72-, CHMN, DIV HYDRAUL & SANIT ENG, 73- Concurrent Pos: Consult, AEC-Energy Res Develop Admin, 69-; adj prof, Environ Protection Agency, 70-; consult, Nat Acad Sci, 73- Mem: Am Chem Soc; Am Soc Eng Educ; Soc Appl Spectros; Am Water Works Asn. Res: Sanitary chemistry; chemical aspects applied to air, water pollution and quality control. Mailing Add: Dept Hydraul & Sanit Eng 659 Davis Hall Univ Calif Berkeley CA 94720

THOMAS, JESS WILLIAM, b Bellaire, Ohio, Nov 17, 17; m 48; c 2. APPLIED PHYSICS. Educ: Carnegie Inst Technol, BS, 39; Notre Dame Univ, MS, 62. Prof Exp: Chem engr, Army Chem Corps, 40-54; chem engr, Oak Ridge Nat Lab, Union Carbide & Carbon Chem Corp, 54-56; chem engr, Whirlpool Corp, 56-63; AEROSOL PHYSICIST, HEALTH & SAFETY LAB, US ENERGY RES & DEVELOP ADMIN, 63- Mem: Am Indust Hyg Asn; Health Physics Soc. Res: Generation and measurement of size, density, charge and diffusion rate of aerosols; efficiency of air filters and carbon beds; measurement of radon, thoron, and daughters. Mailing Add: 550 Fairmount Ave Chatham NJ 07928

THOMAS, JOAB LANGSTON, b Holt, Ala, Feb 14, 33; m 54. BOTANY. Educ: Harvard Univ, AB, 55, AM, 57, PhD, 59. Prof Exp: Cytotaxonomist, Arnold Arboretum, Harvard Univ, 59-61; from asst prof to prof biol, Univ Ala, 61-76, asst dean, Col Arts & Sci, 65; CHANCELLOR, NC STATE UNIV, RALEIGH, 76- Concurrent Pos: Dir, Herbarium, Univ Ala, 61-76, dir arboretum, 64-76. Mem: Am Soc Plant Taxon; Bot Soc Am; Int Asn Plant Taxon. Res: Systematics and cytogenetics of higher plants. Mailing Add: Off of the Chancellor NC State Univ Raleigh NC 27607

THOMAS, JOHN, b Rome, Ga, May 13, 18; m 44; c 2. MEDICINE. Educ: Morris Brown Col, BA, 40; Meharry Med Col, MD, 44. Prof Exp: From instr to prof med, 49-70, PROF INTERNAL MED, MEHARRY MED COL, 70-, DIR CARDIOVASC DIS CLIN RES CTR, 61- Concurrent Pos: Fel med & cardiol, Meharry Med Col, 45-46, Nat Heart Inst trainee, 49-52; res collabr, Brookhaven Nat Lab, 61-62. Mem: Fel Am Col Physicians; Nat Med Asn. Res: Cardiovascular disease; hypertension and myocardial infarction. Mailing Add: Meharry Med Col Nashville TN 37208

THOMAS, JOHN A, b La Crosse, Wis, Apr 6, 33; m 57; c 2. PHARMACOLOGY. Educ: Univ Wis-La Crosse, BS, 56; Univ Iowa, MA, 58, PhD(physiol), 61. Prof Exp: Instr physiol, Univ Iowa, 60-61; asst prof pharmacol, Sch Med, Univ Va, 61-64; assoc prof, Sch Med, Creighton Univ, 64-67; asst dean, 73-75, PROF PHARMACOL, SCH MED, WVA UNIV, 67-, ASSOC DEAN, 75- Mem: Am Soc Pharmacol & Exp Therapeut; Endocrine Soc; Soc Toxicol. Res: Endocrine pharmacology, mechanism of action of androgens; prostate gland neoplasms; pesticides and reproduction. Mailing Add: Dept of Pharmacol WVa Univ Med Ctr Morgantown WV 26505

THOMAS, JOHN ALVA, b Berwyn, Ill, May 9, 40; m 65; c 2. BIOCHEMISTRY. Educ: DePauw Univ, AB, 62; Univ Ill, Urbana, PhD(biochem), 68. Prof Exp: NIH fel, Univ Pa, 68-70; ASST PROF BIOCHEM, SCH MED, UNIV SDAK, 70- Mem: AAAS; Biophys Soc; Am Chem Soc. Res: Mechanism of action peroxidases and catalases; biological halogenation reactions; bioenergetics and oxidative phosphorylation; enzyme kinetics. Mailing Add: Dept of Biochem Univ of SDak Sch of Med Vermillion SD 57069

THOMAS, JOHN CUNNINGHAM, organic chemistry, see 12th edition

THOMAS, JOHN EUGENE, b Youngstown, Ohio, July 5, 14; m 42; c 3. PHYTOPATHOLOGY. Educ: Ohio State Univ, BSc, 41; Univ Wis, PhD(plant path), 47. Prof Exp: From instr to asst prof plant path, Univ Wis, 45-50; assoc prof, 50-57, PROF PLANT PATH, OKLA STATE UNIV, 57-, HEAD DEPT, 67- Mem: Am Inst Biol Sci; Am Phytopath Soc; Mycol Soc Am. Res: Diseases of trees. Mailing Add: Dept of Plant Path Okla State Univ Stillwater OK 74074

THOMAS, JOHN HUNTER, b Beuthen, Ger, Mar 26, 28; US citizen; m 53 & 66. PLANT SYSTEMATICS. Educ: Calif Inst Technol, BS, 49; Stanford Univ, AM, 49, PhD, 59. Prof Exp: Asst, Stanford Univ, 49-51 & 53-55, curatorial asst, 55-56; instr, Occidental Col, 56-58; from asst cur to cur, 58-72, DIR DUDLEY HERBARIUM, STANFORD UNIV, 72-, ASSOC PROF BIOLSCI, UNIV, 69- Concurrent Pos: Cur, Dept Bot, Calif Acad Sci, 69- Mem: Am Soc Plant Taxon; Bot Soc Am; Soc Study Evolution; Am Fern Soc. Res: Flora of central and lower California and Alaska; management of systematic collections; information storage and retrieval in systematic collections; botanical history. Mailing Add: 838 Cedro Way Stanford CA 94305

THOMAS, JOHN JENKS, b Boston, Mass, Dec 15, 36; m 65. GEOLOGY. Educ: Williams Col, BA, 61; Northwestern Univ, MS, 65; Univ Kans, PhD(geol), 68. Prof Exp: Asst prof, 68-75, ASSOC PROF GEOL, SKIDMORE COL, 75- Mem: AAAS; Geol Soc Am; Mineral Soc Am; Am Geophys Union; Nat Asn Geol Teachers. Res: Structural geology; tectonics; metamorphic petrology. Mailing Add: Dept of Geol Skidmore Col Saratoga Springs NY 12866

THOMAS, JOHN KERRY, b Llanelly, South Wales, Gt Brit, May 16, 34; m 59; c 3. PHYSICAL CHEMISTRY. Educ: Univ Manchester, BSc, 54, PhD(chem), 57, DSc, 69. Prof Exp: Nat Res Coun Can fel chem, 57-58; sci off, Atomic Energy Res Estab, Eng, 58-60; res assoc, Argonne Nat Lab, 60-62, assoc chemist, 62-70; PROF CHEM, UNIV NOTRE DAME, 70- Mem: Radiation Res Soc; Am Chem Soc; The Chem Soc. Res: Polymer and radiation chemistry; photochemistry. Mailing Add: Dept of Chem Univ of Notre Dame Notre Dame IN 46556

THOMAS, JOHN M, b Wilmar, Calif, Sept 17, 36; m 57; c 3. BIOMETRICS, ECOLOGY. Educ: Calif State Polytech Col, BS, 58; Wash State Univ, MS, 60; Univ Ariz, PhD(biochem, nutrit), 65. Prof Exp: Instr avian physiol, Calif State Polytech Col, 60-61; res assoc biochem, Univ Ariz, 61-65; sr res scientist, Pac Northwest Labs, Battelle Mem Inst, 61-71; gen ecologist, Div Biol & Med, Energy Res & Develop Admin, 71-72; RES ASSOC, PAC NORTHWEST LABS, BATTELLE MEM INST, 72- Concurrent Pos: Mem task group, Int Comt Radiation Protection, 74- Mem: AAAS; Am Statist Asn; Biomet Soc. Res: Biology and ecology; methods development for field surveys and the prediction of effects of various insults on humans based on laboratory and field animal data. Mailing Add: 124 Oakmont Ct Richland WA 99352

THOMAS, JOHN MARTIN, b Omaha, Nebr, Oct 17, 10; m 36; c 4. PEDIATRICS. Educ: Grinnell Col, AB, 32; Yale Univ, MD, 37. Prof Exp: From instr to asst prof pediat, Col Med, Creighton Univ, 40-49; from asst prof to assoc prof, 49-58, PROF PEDIAT, UNIV NEBR MED CTR, OMAHA, 58-, ASST PROF REHAB, 66- Mem: AMA; Am Acad Pediat. Mailing Add: Swanson Prof Ctr 8601 W Dodge Rd Omaha NE 68114

THOMAS, JOHN PAUL, plant breeding, genetics, see 12th edition

THOMAS, JOHN PAUL, b Kokomo, Ind, June 29, 33; m 61; c 2. PHYSICAL INORGANIC CHEMISTRY. Educ: Purdue Univ, BS, 54, PhD(phys chem), 62. Prof Exp: Res assoc phys chem, Univ Minn, 62-63; asst prof phys inorg chem, Ill Inst Technol, 63-69; assoc prof, 69-72, actg chmn, Div Sci & Math, 72-73, PROF CHEM, SOUTHWEST MINN STATE UNIV, 72- Mem: Am Chem Soc; The Chem Soc. Res: Spectroscopic studies of chemical bonding; infrared, visible-ultraviolet and electron paramagnetic resonance spectroscopy. Mailing Add: Chem Prog Southwest Minn State Univ Marshall MN 56258

THOMAS, JOHN PELHAM, b Ashby, Ala, Apr 18, 22; m 45; c 6. MATHEMATICS. Educ: Auburn Univ, BS, 46; Univ Va, MAT, 61; Univ SC, PhD(math), 65. Prof Exp: Asst county agt, Agr Exten Serv, Auburn, 46-53; farmer, 53-54; jr & high sch teacher, Ala, 55-60; asst prof math, Univ NC, 64-67; PROF MATH & HEAD DEPT, WESTERN CAROLINA UNIV, 67- Mem: Math Asn Am. Res: Maximal topological spaces; separation axioms; properties preserved under strengthening and weakening of topologies. Mailing Add: Dept of Math Western Carolina Univ Cullowhee NC 28723

THOMAS, JOHN RICHARD, b Anchorage, Ky, Aug 26, 21; m 44; c 2. PHYSICAL CHEMISTRY. Educ: Univ Calif, BS, 43, PhD(phys chem), 47. Prof Exp: Asst Nat Defense Res Comt, Univ Calif, 43-44, Manhattan Dist, 44-47; res assoc, US AEC Contract, Gen Elec Co, 47-48; res chemist, Calif Res Corp, 48-49; asst chief chem br, US AEC, 49-51; sr res scientist, Calif Res Corp, 51-67, mgr res & develop, Ortho Div, Chevron Chem Co, 67-68, asst secy, Stand Oil Co Calif, 68-70, PRES, CHEVRON RES CO, 70- Res: Free radicals; oxidation kinetics; electron spin resonance. Mailing Add: Chevron Res Co Richmond CA 94802

THOMAS, JOHN WILLIAM, b Spanish Fork, Utah, Mar 25, 18; m 45; c 4. ANIMAL NUTRITION, DAIRY SCIENCE. Educ: Utah State Univ, BS, 40; Cornell Univ, PhD(nutrit), 46. Prof Exp: Res assoc, Nat Defense Res Comt, Northwestern Univ, 42-45 & Carnegie Inst Technol, 45; nutritionist & biochemist, Bur Dairy Indust, USDA, 46-53 & dairy husb res br, 53-60; PROF DAIRY SCI, MICH STATE UNIV, 60- Honors & Awards: Am Feed Mfrs Asn Award, 53; USDA Superior Serv Award, 59; Borden Award, Am Dairy Sci Asn, 74. Mem: Fel AAAS; Am Chem Soc; Am Soc Animal Sci; Am Dairy Sci Asn; Am Inst Nutrit. Res: Mineral and vitamin requirements and functions in feeding of dairy cattle; forage evaluation and preservation; thyroid active stimulants for cattle; rumen functions; calf nutrition. Mailing Add: Dept of Dairy Sci Mich State Univ East Lansing MI 48824

THOMAS, JOHNNY RAY, b Albuquerque, NMex, Aug 7, 41; m 61; c 4. PLANT BREEDING, AGRONOMY. Educ: NMex State Univ, BS, 63; Ore State Univ, MS, 65, PhD(plant breeding), 66. Prof Exp: Plant breeder, Farmers Forage Res Corp, 66-70; vpres & res dir, Rudy-Patrick Co, 70-72, RES DIR, N AM PLANT BREEDERS, 72- Mem: Am Soc Agron. Res: Plant breeding in forage and cereal crops. Mailing Add: N Am Plant Breeders Rte 2 Brookston IN 47923

THOMAS, JON CHARLES, b Ft Wayne, Ind, Apr 23, 48. ASTRONOMY. Educ: Ind Univ, Bloomington, BS, 70; Univ Mich, Ann Arbor, MS, 72, PhD(astron), 75. Prof Exp: MOREHEAD FEL & LECTR ASTRON, UNIV NC, CHAPEL HILL, 75- Mem: Am Astron Soc. Res: Empirical mass determination for O stars; spatial distribution of highly-reddened, late-type supergiants in the galaxy. Mailing Add: Dept of Physics & Astron Univ of NC Chapel Hill NC 27514

THOMAS, JOSEPH CALVIN, b Churubusco, Ind, May 2, 33; m 55; c 1. SCIENCE EDUCATION, ORGANIC CHEMISTRY. Educ: Asbury Col, AB, 54; Univ Ky, MA, 55, EdD(sci educ, chem), 61. Prof Exp: Instr chem, Asbury Col, 54-59 & Jessamine County High Sch, 59-60; asst prof sci, 61-65, chmn dept chem, 63-74, PROF CHEM, UNIV N ALA, 65- Concurrent Pos: Sci consult local pub sch systs, 63- Mem: AAAS; Am Chem Soc; Nat Sci Teachers Asn. Res: Preparation and continued education of secondary school science teachers, particularly their preparation in the physical sciences. Mailing Add: Dept of Chem Univ of N Ala Florence AL 35630

THOMAS, JOSEPH CHARLES, b Mt Union, Pa, Oct 16, 45; m 65; c 2. MATHEMATICS. Educ: Shippensburg State Col, BS, 66; Pa State Univ, MA, 68; Kent State Univ, PhD(math), 75. Prof Exp: Instr, 68-70, ASST PROF MATH, LOCK HAVEN STATE COL, 72- Res: Torsion theories generated by ideals; global dimension of associative rings with identity. Mailing Add: Dept Math & Comput Sci Lock Haven State Col Lock Haven PA 17745

THOMAS, JOSEPH FRANCIS, JR, b Chicago, Ill, Feb 29, 40; m 67. SOLID STATE PHYSICS, MATERIALS SCIENCE. Educ: Cornell Univ, BEP, 63; Univ Ill, Urbana, MS, 65, PhD(physics), 68. Prof Exp: Asst prof physics, Univ Va, 67-72; asst prof, 72-75, ASSOC PROF PHYSICS, WRIGHT STATE UNIV, 75- Mem: AAAS; Am Phys Soc; Metall Soc; Am Inst Metall Engrs. Res: Physics of metals and alloys; cohesion and structure; dislocations and internal friction; plastic deformation. Mailing Add: Dept of Physics Wright State Univ Dayton OH 45431

THOMAS, JOSEPH JAMES, b Columbia, Pa, Sept 10, 09; m 32, 51; c 7. BIOCHEMISTRY. Educ: Pa State Univ, BS, 30, MS, 32, PhD(biochem), 35. Prof Exp: Asst res, NY Exp Sta, Geneva, 30; instr agr biochem, Pa State Univ, 31-36; biochemist, Rohm and Haas, 36-41; from asst dir to dir res, S D Warren Co, 42-68, vpres res, 68-72; tech consult, Edward C Jordan Co, Inc, 73-76; TECH PULP, PAPER & CHEM CONSULT, 73- Mem: Am Chem Soc; Tech Asn Pulp & Paper Indust. Res: Synthetic resins; functional uses of pulp and paper. Mailing Add: 16234 N 111th Ave Sun City AZ 85351

THOMAS, JOSEPHUS, JR, b Linton, Ind, Nov 28, 27; m 50; c 3. ANALYTICAL CHEMISTRY. Educ: Ind State Univ, BS, 53; Univ Ill, PhD(anal chem), 57. Prof Exp: Asst anal chem, Univ Ill, 53-56; res chemist, E I du Pont de Nemours & Co, Inc, 57-62; PHYS CHEMIST, ILL STATE GEOL SURV, 62- Mem: Am Chem Soc. Res: Surfaces, sintering, x-ray and electron diffraction; instrumental methods of analysis; thermal analysis; phase transformations in solids. Mailing Add: Ill State Geol Surv 305 Natural Resources Bldg Urbana IL 61801

THOMAS, JULIAN EDWARD, SR, b Yazoo City, Miss, Aug 1, 37; m 56; c 3. MICROBIAL PHYSIOLOGY. Educ: Fisk Univ, AB, 59; Atlanta Univ, MS, 67, PhD(biol), 71; Southern Univ, MST, 68. Prof Exp: Teacher pub schs, Ga, 59-65; instr biol & chem, SC State Col, 67-69; fel microgenetics, Argonne Nat Lab, 71-73; ASSOC PROF BIOL, TUSKEGEE INST, 73- Concurrent Pos: Consult, Argonne Ctr Educ Affairs, 75-76. Mem: Sigma Xi; Fedn Am Scientists; AAAS; Am Soc Microbiol. Res: Involvement of transfer RNA in the regulation of enzyme synthesis by repression, and derepression, control. Mailing Add: Dept of Biol Tuskegee Inst Tuskegee Institute AL 36088

THOMAS, KEITH SKELTON, b Tallahassee, Fla, Aug 29, 36; m 63; c 3. PHYSICS. Educ: Tulane Univ, BS, 58; Johns Hopkins Univ, PhD(physics), 64. Prof Exp: Instr physics, Johns Hopkins Univ, 63-64; STAFF MEM, LOS ALAMOS SCI LAB, UNIV CALIF, 64- Res: Crystal spectroscopy of rare earth chlorides; high density and high temperature plasmas with emphasis on diagnostic techniques. Mailing Add: Los Alamos Sci Lab Group CTR-7 Los Alamos NM 87544

THOMAS, KOTTARATHIL MATHEW, nuclear physics, see 12th edition

THOMAS, KURIAN K, b Kottayam, India, May 5, 34; m 65; c 2. PHYSIOLOGY, BIOCHEMISTRY. Educ: Univ Kerala, BSc, 54, MSc, 56; Univ Fla, PhD(zool), 64. Prof Exp: Lectr zool, C M S Col, Univ Kerala, 56-61; res assoc biol, Northwestern Univ, 65-68; asst prof, 68-72, ASSOC PROF BIOL, ST MARY'S UNIV, NS, 72- Mem: AAAS; Am Soc Zoologists. Res: Lipids and lipoprotein metabolism in insects. Mailing Add: Dept of Biol St Mary's Univ Halifax, NS Can

THOMAS, LARRY EMERSON, b Indianapolis, Ind, Dec 27, 43. APPLIED MATHEMATICS. Educ: Rose Polytech Inst, BS, 66; Rensselaer Polytech Inst, MS, 68, PhD(math), 70. Prof Exp: Asst prof, 70-72, ASSOC PROF MATH, ST PETER'S COL, NJ, 72- Mem: Math Asn Am; Soc Indust & Appl Math. Res: Differential equations. Mailing Add: Dept of Math St Peter's Col Jersey City, NJ 07306

THOMAS, LAZARUS DANIEL, b Toledo, Ohio, Oct 21, 25; m 50; c 5. PHYSICAL CHEMISTRY. Educ: Univ Mich, BS, 48, MS, 49. Prof Exp: Teaching fel, Univ Mich, 49-50; RES SUPVR, LIBBEY-OWENS-FORD CO, 51- Mem: Electrochem Soc; Am Electroplaters Soc; Am Chem Soc; Am Ceramic Soc. Res: Films on glass; semiconductors; surface chemistry of glass; electrochemistry. Mailing Add: Libbey-Owens-Ford Co 1701 E Broadway Toledo OH 43605

THOMAS, LEO ALVON, b Gifford, Idaho, Mar 19, 22; c 2. PARASITOLOGY, MEDICAL MICROBIOLOGY. Educ: Univ Idaho, BS, 49; Univ Mich, MS, 50; Tulane Univ, PhD(parasitol, med microbiol), 55. Prof Exp: With virus labs, Rockefeller Found, NY, 55-57; MED BACTERIOLOGIST, ROCKY MOUNTAIN LAB, USPHS, 57- Res: Ecology and classification of arthropod-borne viruses. Mailing Add: USPHS Rocky Mountain Lab Hamilton MT 59840

THOMAS, LEWIS, b Flushing, NY, Nov 25, 13; m 43; c 3. INTERNAL MEDICINE, PATHOLOGY. Educ: Princeton Univ, BS, 33; Harvard Univ, MD, 37. Hon Degrees: MA, Yale Univ, 69; ScD, Univ Rochester, 74. Prof Exp: Intern, Boston City Hosp, 37-39; intern, Neurol Inst, New York, 39-41; Tilney Mem fel, Thorndike Lab, Boston City Hosp, 41-42; vis investr, Rockefeller Inst, 42-46; asst prof pediat, Sch Med, Johns Hopkins Univ, 46-48; from assoc prof to prof med, Sch Med, Tulane Univ, 48-50; prof pediat & med & dir pediat res labs, Heart Hosp, Univ Minn, 50-54; prof path & chmn dept, Sch Med, NY Univ, 54-58, prof med & chmn dept, 58-66, dean, 66-69; prof path & chmn dept, Sch Med Yale Univ, 69-72, dean sch med, 72-73; PRES & CHIEF EXEC OFFICER, MEM SLOAN-KETTERING CANCER CTR, 73-; PROF MED & PATH, MED SCH, CORNELL UNIV, 73-, CO-DIR, GRAD SCH MED SCI, 74- Concurrent Pos: Consult, Surgeon Gen, US Dept Army, 52; mem path study sect, Path Training Comt, NIH, 54-58, Nat Adv Health Coun, 58-62 & Nat Adv Child Health & Human Develop Coun, 63-67; mem comn streptococcal dis, Armed Forces Epidemiol Bd, NIH, 54 & 58; consult, Manhattan Vet Admin Hosp, 54-69; mem, Bd Health, New York, 56-69; dir third & fourth med divs, Bellevue Hosp, 58-66, pres med bd, 63-66, dir med, Univ Hosp, 58-66; mem bd sci consult, Sloan-Kettering Inst Cancer Res, 66-72; mem, President's Sci Adv Comt, 67-70; consult, Surgeon Gen, USPHS; mem bd dirs, Pub Health Res Inst, New York; mem med adv comt, Ctr Biomed Educ, City Univ New York, 72-; prof biol, SKI Div, Grad Sch Med Sci, Cornell Univ, 73-; mem comt educ, Div Biol & Med Sci, Brown Univ, 74-; mem sci adv comt, Inst Cancer Res, Fox Chase, 74-; mem comt health professions of bd trustees, Sch Med, Univ Pittsburgh, 74-; mem med adv comt, Irvington House Inst, New York, 74-; adj prof, Rockefeller Univ, 75-; trustee, Rockefeller Univ, 75-, Draper Lab, 75- & John Simon Guggenheim Mem Found, 75-; mem bd dirs, Josiah Macy Jr Found, 75- Honors & Awards: Distinguished Achievement Award, Modern Med, 75; Nat Book Award for Arts & Letters, 75. Mem: Nat Acad Sci; Am Acad Arts & Sci; fel NY Acad Sci; Practitioners Soc; Am Soc Clin Oncol. Res: Infectious disease; hypersensitivity; pathogenicity of mycoplasmas. Mailing Add: Mem Sloan-Kettering Cancer Ctr 1275 York Ave New York NY 10021

THOMAS, LEWIS EDWARD, b Lima, Ohio, May 18, 13; m 40; c 4. ORGANIC CHEMISTRY. Educ: Ohio Northern Univ, BS, 35; Purdue Univ, MS, 37. Prof Exp: Asst chem, Purdue Univ, 35-39; from instr to asst prof, Va Mil Inst, 40-45; develop engr, 45-50, tech serv & lab supvr, 50-70, asst mgr lab, 70-73, MGR LAB, SUN OIL CO, 73- Concurrent Pos: Vis scientist, Ohio Acad Sci, NSF vchmn bd trustees, Univ Toledo & Toledo-Lucas County Libr Syst. Mem: Nat Soc Prof Engrs; Am Chem Soc; Am Inst Chem Engrs. Res: Chlorination of aliphatic hydrocarbons; selective solvents for olefin and diolefin purification; pyrolysis of chlorinated aliphatic hydrocarbons; esterification of alcohol ethers. Mailing Add: 4148 Deepwood Lane Toledo OH 43614

THOMAS, LEWIS JONES, JR, b Philadelphia, Pa, Dec 13, 30; m 55; c 2. ANESTHESIOLOGY. Educ: Haverford Col, BS, 53; Washington Univ, MD, 57; Am Bd Anesthesiol, dipl, 63. Prof Exp: Intern med, Bronx Munic Hosp, NY, 57-58; USPHS res fel, Sch Med, Washington Univ, 59-60; resident anesthesiol, Barnes Hosp, St Louis, Mo, 60-62; staff anesthesiologist, Clin Ctr, NIH, 62-64; asst prof anesthesiol, 64-74, asst prof physiol & biophys, 70-74, assoc dir biomed comput lab, 72-75, ASSOC PROF ANESTHESIOL, PHYSIOL & BIOPHYS, SCH MED, WASHINGTON UNIV, 74-, DIR BIOMED COMPUT LAB, 75- Concurrent Pos: USPHS res career develop award, 66-71. Honors & Awards: Borden Award, Asn Am Med Cols, 57. Mem: AAAS; Am Physiol Soc; AMA. Res: Respiratory physiology; biomedical computer applications; anesthesia. Mailing Add: Biomed Comput Lab Washington Univ Sch of Med St Louis MO 63110

THOMAS, LLEWELLYN HILLETH, b London, Eng, Oct 21, 03; nat US; m 33; c 3. PHYSICS. Educ: Cambridge Univ, BA, 24, PhD(theoret physics), 27, MA, 28, DSc, 65. Prof Exp: From asst prof to prof physics, Ohio State Univ, 29-43; physicist & ballistician, Ballistic Res Lab, Aberdeen Proving Ground, Md, 43-45; prof physics, Ohio State Univ, 45-46; mem sr staff, Watson Sci Comput Lab, Columbia Univ, 46-48, prof physics, Univ, 50-68, EMER PROF PHYSICS, COLUMBIA UNIV, 68-; PROF PHYSICS, NC STATE UNIV, 68- Mem: Nat Acad Sci; AAAS; fel Am Phys Soc; Royal Astron Soc. Res: Theoretical astrophysics; atomic physics; relativity theory; nuclear, atomic and molecular structure; field theory; computational methods. Mailing Add: Dept of Physics NC State Univ Raleigh NC 27607

THOMAS, LLOYD BREWSTER, b LeMars, Iowa, June 2, 09; m 35; c 6. CHEMISTRY. Educ: Univ Mo, AB, 30; Univ Minn, PhD(phys chem), 35. Prof Exp: Asst chem, Univ Minn, 31-35; from instr to assoc prof, 35-50, PROF CHEM, UNIV MO-COLUMBIA, 50- Concurrent Pos: Res worker, Nat Defense Res Comt, 42-44; proj dir, Off Ord Res, 51- Honors & Awards: Res Award, Sigma Xi, 74. Res: Photochemistry; mercury sensitized reactions; heat conduction in rarefied gases; electron activated reactions; thermal accommodation coefficients and adsorption; surface chemistry; reaction kinetics. Mailing Add: Dept of Chem Univ of Mo Columbia MO 65201

THOMAS, LLYWELLYN MURRAY, b Detroit, Mich, Sept 23, 22; m 47; c 6. NEUROSURGERY. Educ: Wayne State Univ, BA, 49, MD, 52. Prof Exp: Assoc prof, 65-68, asst chmn dept, 65-70, PROF NEUROSURG & CHMN DEPT, SCH MED, WAYNE STATE UNIV, 70-, ASSOC DEAN HOSP AFFAIRS, 72- Concurrent Pos: Sr attend, Detroit Gen Hosp, 65- & Grace Hosp, Detroit, 70-; consult, Harper Hosp, 71- & Children's Hosp Mich, 71- Mem: Am Asn Neurol Surg; Am Col Surg; AMA; Cong Neurol Surg. Res: Head injury. Mailing Add: Dept of Neurosurg Wayne State Univ Sch of Med Detroit MI 48202

THOMAS, LOUIS BARTON, b Medicine Lodge, Kans, June 8, 19; m 44; c 3. PATHOLOGY. Educ: Col Idaho, AB, 40; Univ Chicago, MD, 45; Am Bd Path, dipl, 52. Prof Exp: Resident path, Univ Minn, 48-51; spec fel neuropath, Mayo Clin, 51-52; resident, Mem Ctr Cancer & Allied Dis, New York, 52-53; head surg path & post-mortem serv, Clin Ctr, NIH, 53-69; CHIEF LAB PATH, NAT CANCER INST, 69- Concurrent Pos: Clin prof, Schs Med & Dent, Georgetown Univ, 72- Mem: Am Asn Path & Bact; fel Col Am Path; Am Asn Cancer Res; Am Soc Exp Path; Int Acad Path. Res: Diagnostic and research pathology, particularly cancer; leukemia and malignant lympomas. Mailing Add: Lab of Path Rm 2A-29 Clin Ctr Nat Cancer Inst Bethesda MD 20014

THOMAS, LOWELL PHILLIP, b Miami, Fla, Dec 2, 33; m 56; c 3. BIOLOGICAL OCEANOGRAPHY. Educ: Univ Miami, BS, 55, MS, 59, PhD(marine biol), 65. Prof Exp: Res aide marine biol, 58-59, from instr to asst prof, 60-70, ASSOC PROF MARINE BIOL, ROSENTIEL INST, MARINE & ATMOSPHERIC SCI, UNIV MIAMI, 70- Concurrent Pos: Vis lectr, Fla State Univ, 63; mem staff, Orgn Trop Studies, Costa Rica, 67, Bermuda, Biol Sta, 68 & Fairleigh-Dickinson Field Sta, St Croix, 69 & 70. Mem: Marine Biol Asn UK; Sigma Xi. Res: Systematics and developmental anatomy of echinoderms; ecology and behavior of marine animals. Mailing Add: Rosenstiel Inst Marine & Atmos Sci Univ of Miami Miami FL 33149

THOMAS, LYELL JAY, JR, b Madison, Wis, Apr 17, 25; m 48; c 2. PHARMACOLOGY. Educ: Oberlin Col, AB, 48; Univ Pa, PhD(zool), 53. Prof Exp: Instr pharmacol, Woman's Med Col, Pa, 52-55; asst prof biol, 55-60, assoc prof, 60-62, ASSOC PROF PHARMACOL, UNIV SOUTHERN CALIF, 62- Mem: Am Physiol Soc; Soc Gen Physiol; Cardiac Muscle Soc. Res: Cellular physiology and pharmacology of heart muscle; excitation contraction coupling in heart muscle; mechanism of insulin secretion. Mailing Add: Dept of Pharmacol Univ of Southern Calif Med Sch Los Angeles CA 90033

THOMAS, MARTHA JANE BERGIN, b Boston, Mass, Mar 13, 26; m 55; c 4. ANALYTICAL CHEMISTRY, PHYSICAL CHEMISTRY. Educ: Radcliffe Col, AB, 45; Boston Univ, AM, 50, PhD(chem), 52. Prof Exp: Sr engr in chg chem lab, 45-59, group leader lamp mat eng labs, Lighting Prod Div, 59-66, sect head chem & phosphor lab, Sylvania Lighting Ctr, 66-72, MGR TECH ASSISTANCE LABS, GTE SYLVANIA LIGHTING PROD GROUP, SYLVANIA ELEC PROD, INC, GEN TEL & ELECTRONICS CORP, DANVERS, 72- Concurrent Pos: Instr eve div, Boston Univ, 52-70; adj prof chem, Univ RI, 74- Honors & Awards: Nat Achievement Award, Soc Women Engrs, 65; Golden Plate, Am Acad Achievement, 66. Mem: Am Chem Soc; Electrochem Soc; fel Am Inst Chemists. Res: Phosphors; photoconductors; ion exchange membranes; complex ions; instrumental analysis. Mailing Add: 18 Cabot St Winchester MA 01890

THOMAS, MARTIN LEWIS HALL, b Feb 9, 35; Can citizen; m 56; c 2. BIOLOGY. Educ: Univ Durham, BSc, 56; Univ Toronto, MSA, 62; Dalhousie Univ, PhD, 70. Prof Exp: Assoc scientist, Biol Sta, Fisheries Res Bd Can, Ont, 56-62, scientist, Biol Sub-Sta, 62-70; asst prof, 70-74, ASSOC PROF BIOL, UNIV NB, ST JOHN, 70- Concurrent Pos: Mem, Int Oceanog Found, 55- Mem: Nat Shellfisheries Asn; Marine Biol Asn UK; Brit Ecol Soc. Res: Ecology of larval lampreys; estuarine ecology; marine benthic ecology; ecology of freshwater bivalve molluscs; marine pollution. Mailing Add: Dept of Biol Univ of NB St John NB Can

THOMAS, MARY BETH, b Sewanee, Tenn, Mar 2, 41. DEVELOPMENTAL BIOLOGY, CELL BIOLOGY. Educ: Agnes Scott Col, BA, 63; Univ NC, Chapel Hill, MA, 70, PhD(zool), 71. Prof Exp: Vis asst prof biol, 71-72, ASST PROF BIOL, WAKE FOREST UNIV, 72- Mem: AAAS; Am Soc Cell Biol. Res: Biological motility; structure and function of muscles and microtubules; spermiogenesis in flatworms; invertebrate embryology. Mailing Add: Dept of Biol Wake Forest Univ Winston-Salem NC 27109

THOMAS, MCCALIP JOSEPH, b Yazoo City, Miss, Jan 1, 14; m 59; c 2. CHEMISTRY. Educ: Miss State Col, BS, 36; Vanderbilt Univ, MS, 38, PhD(org chem), 41. Prof Exp: Asst chemist, Exp Sta, Miss State Col, 36-37; Glidden-Upjohn-Abbott fel, Northwestern Univ, 41-42; from res chemist to sr res chemist, A E Staley Mfg Co, 42-48, asst to mgr, Mkt Develop Dept, 48-52, asst mgr, 52-61; mem staff, Applns Res Dept, Nat Cash Register Co, 61-67, sci liaison & mem staff tech support, 67-68; exec vpres, Hill Top Res, Inc, 68-73, DIR MKT, HILL TOP TESTING SERV, INC DIV, AM BIOMED CORP, 73- Concurrent Pos: Mem indust adv comt soup & gravy bases, Qm Food & Container Inst, 54 & task group, Res & Develop, 55. Mem: Am Chem Soc; Am Oil Chem Soc; Am Pharmaceut Asn; Inst Food Technol; fel Am Inst Chemists. Res: Research, development and testing in the biological, toxicological, chemical, medical and microbiological fields. Mailing Add: Hill Top Testing Serv Inc Div Am Biomed Corp Miamisville OH 45147

THOMAS, MICHAEL DAVID, b Merthyr Tydfil, Wales, Jan 2, 42. GEOPHYSICS. Educ: Univ Wales, BS, 64, PhD(geol), 68. Prof Exp: Fel magnetic interpretation, Geol Surv Can, 68-69; geophysicist, Survair Ltd, 69-71; RES SCIENTIST GRAVITY INTERPRETATION, EARTH PHYSICS BR, DEPT ENERGY, MINES & RESOURCES, CAN, 72- Mem: Geol Asn Can; Can Geophys Union. Res: Geological interpretation of gravity anomalies within the Canadian Precambrian shield with emphasis on the anomalies along the structural province boundaries and over anorthositic-gabbroic intrusions and complexes. Mailing Add: Gravity & Geodyn Div Dept of Energy Mines & Resources Ottawa ON Can

THOMAS, MIRIAM MASON HIGGINS, b Chicago, Ill, June 22, 20; m 47; c 1. NUTRITION. Educ: Bennett Col, NC, BS, 40; Univ Chicago, MS, 42. Prof Exp: Res assoc food chem, Div Biol Sci, Univ Chicago, 42-45; RES ASSOC NUTRIT, FOOD LAB, NUTRIT DIV, US ARMY NATICK LABS, 45- Concurrent Pos: Vis fac lectr, Dept Nutrit & Food Sci, Mass Inst Technol, 74-76; Dept Defense Sec Army fel, 75. Mem: AAAS; Soc Nutrit Educ; Coblentz Soc; Inst Food Technol; Sigma Xi. Res: Chemical aspects of protein and amino acid metabolism; effects of processing and storage on the nutritive quality of military rations and vitamin fortification of ration components. Mailing Add: Food Lab Nutrit Div US Army Natick Labs Natick MA 01760

THOMAS, MITCHELL HOWARD, JR, plant pathology, see 12th edition

THOMAS, MONTCALM TOM, b Brooklyn, Conn, Feb 5, 36; m 62; c 1. PHYSICS. Educ: Univ Conn, BA, 57, MS, 59; Brown Univ, PhD(physics), 66. Prof Exp: Mem tech staff, Bell Tel Labs, NJ, 65-68; asst prof physics, Wash State Univ, 68-74; MEM STAFF, PAC NORTHWEST LABS, BATTELLE MEM INST, 74- Mem: Am Phys Soc; Am Vacuum Soc. Res: Solid state, atomic and molecular physics; surface structure and kinetics of solids; thin films in solid state physics; photoelectric phenomena; low energy electron diffraction; high vacuum techniques. Mailing Add: Pac Northwest Labs Battelle Mem Inst PO Box 999 Richland WA 99352

THOMAS, MORGAN D, b Pontlliw, S Wales, Jan 5, 25; m 49; c 4. ECONOMIC GEOGRAPHY. Educ: Queen's Univ, Belfast, PhD(geog), 54. Prof Exp: Instr geog & conserv, Univ Mich, 55-56, lectr conserv, 56-57; assoc prof geog, Mont State Univ, 57-59, chmn dept, 58-59; assoc prof, 59-65, assoc dean, Grad Sch, 69-75, PROF GEOG, UNIV WASH, 65-, ACTG DEAN, GRAD SCH, 75- Concurrent Pos: Exchange fel, Univ Mich, 54-55; ed, Regional Sci Asn, 62-; Ford Found study res grant, 67-68; NSF grant growth poles & regional develop, 71-76. Mem: Asn Am Geog; Am Econ Asn; Regional Sci Asn; Int Regional Sci Asn (pres, 75-76). Res: Economic geography, especially manufacturing; regional economic growth and industrial development. Mailing Add: Dept of Geog Univ of Wash Seattle WA 98105

THOMAS, MORLEY KEITH, b Middlesex Co, Ont, Aug 19, 18; m 42; c 2. METEOROLOGY, CLIMATOLOGY. Educ: Univ Western Ont, BA, 41; Univ Toronto, MA, 49. Prof Exp: Meteorologist, Atmospheric Environ Serv, Can, 45-51 & div bldg res, Nat Res Coun Can, 51-53; supt climat opers, 53-72, DIR METEOROL APPLNS BR, ATMOSPHERIC ENVIRON SERV, 72- Concurrent Pos: Assoc comt snow & ice mech, Nat Res Coun Can, 59-65 & subcomt meteorol & atmospheric sci, 67-70; mem working group on climatic atlases, World Meteorol Orgn, 60-65, chmn, 65-69; mem, Nat Adv Comt Geog Res, 65-70. Mem: Am Meteorol Soc; Royal Meteorol Soc (treas, Can Br, 50-51, secy, 64-66, vpres, 66-67); Can Meteorol Soc (vpres, 67-68, pres, 68-70); Can Asn Geog. Res: Atlases; urban climates; climatic change; climatological services; meteorological applications. Mailing Add: Meteorol Appls Br Atmos Environ Serv 4905 Dufferin St Downsview ON Can

THOMAS, NORMAN DWIGHT, b Kansas City, Mo, Feb 21, 26; m 54; c 2. ANTHROPOLOGY, ETHNOLOGY. Educ: Univ NMex, BA, 51, MA, 56; Univ Calif, Berkeley, PhD(anthrop), 67. Prof Exp: Cur anthrop, San Diego Mus Man, 58-62; asst prof, Univ Nebr, Lincoln, 66-68; asst prof, Northern Ariz Univ, 68-73; asst prof, 73-75, ASSOC PROF ANTHROP, TEX A&M UNIV, 75- Concurrent Pos: Wenner-Gren Found res grant, Tuxtla Gutierrez, Chiapas, Mex, 70-71. Mem: Am Anthrop Asn; Mex Anthrop Soc. Res: Ethnology of contemporary Middle America; social structure and ceremonialization of Middle American Indian and peasant communities; urbanization of traditional societies. Mailing Add: Dept of Sociol & Anthrop Tex A&M Univ College Station TX 77843

THOMAS, NORMAN RANDALL, b Caerphilly, Wales, Dec 22, 32; m 54; c 5. DENTISTRY, PHYSIOLOGY. Educ: Bristol Univ, BDS, 57, BSc, 60, PhD(dent), 65. Prof Exp: Med Res Coun sci asst path res, Royal Col Surgeons, Eng, 60-62; lectr dent med, Bristol Univ, 62-66, lectr anat, 66-68; PROF DENT & HONS PROF MED, UNIV ALTA, 68- Mem: Can Dent Asn; Can Asn Anat; Anat Soc Gt Brit & Ireland; Int Asn Dent Res. Res: Collagen formation and maturation in tooth eruption; neurophysiology of orofacial complex. Mailing Add: 3050A Dent-Pharm Ctr Univ of Alta Edmonton AB Can

THOMAS, NORMAN WILLIAM, organic chemistry, see 12th edition

THOMAS, OSCAR OTTO, b Aline, Okla, Feb 16, 19; m 49; c 3. ANIMAL NUTRITION. Educ: Okla State Univ, BS, 41; Wash State Univ, MS, 48; Okla State Univ, PhD(animal nutrit), 52. Prof Exp: From asst prof to assoc prof, 51-59, PROF ANIMAL NUTRIT & ANIMAL NUTRITIONIST, EXP STA, MONT STATE UNIV, 59- Concurrent Pos: Mem subcomt prenatal & postnatal mortality bovines & comt animal health, Nat Res Coun. Mem: Fel AAAS; Am Soc Animal Sci. Res: Nutritive requirements of range livestock; energy, protein and phosphorus; utilization of barley in cattle fattening; growth stimulants. Mailing Add: Dept of Animal & Range Sci Mont State Univ Bozeman MT 59715

THOMAS, OWEN PESTELL, b Middelburg, SAfrica, Apr 28, 33; m 67; c 1. POULTRY NUTRITION. Educ: Univ Natal, BSc, 54, MSc, 62; Univ Md, College Park, PhD(poultry nutrit), 66. Prof Exp: Nutritionist, United Oil & Cake Mills Ltd, 55-61; asst, Univ Md, College Park, 64-66, res assoc, 66-68; nutritionist, United Oil & Cake Mills Ltd, 68-70; asst prof, 70-72, ASSOC PROF POULTRY SCI, UNIV MD, COLLEGE PARK, 72-, CHMN DEPT, 71- Mem: Poultry Sci Asn. Res: Protein and amino acid requirements of broilers; body composition and pigmentation of broilers. Mailing Add: Dept of Poultry Sci Univ of Md College Park MD 20742

THOMAS, PATRICIA ZEIS, b Buffalo, NY, Aug 10, 27. BIOCHEMISTRY, ENDOCRINOLOGY. Educ: Univ Buffalo, BA, 49, MD, 52; Am Bd Pediat, dipl, 57. Prof Exp: Intern pediat, Buffalo Children's Hosp, NY, 52-53, resident, 53-54; resident, Univ Iowa, 54-55; fel pediat rheumatol, 57-58; res assoc biochem, Sch Med, Univ Pittsburgh, 59-60; Staff scientist, Worcester Found Exp Biol, 61-70; guest investr & fel training prog exec ed, Rockefeller Univ, 70-71; with med div, Ciba-Geigy Corp, 72-73; MANAGING ED PROC, NAT ACAD SCI, 73- Concurrent Pos: Fel pediat endocrinol, Philadelphia Children's Hosp, Pa, 55-56 & Univ NC, 56-57; pediat consult, State Serv Crippled Children, State Univ Iowa, 58; fel steroid biochem, Clark Univ & Worcester Found Exp Biol, 58-59. Res: Steroid biochemistry; peripheral metabolism of steroids. Mailing Add: Nat Acad Sci 2101 Constitution Ave Washington DC 20418

THOMAS, PAUL CLARENCE, b Watsonville, Calif, Nov 26, 28; m 56; c 1. PLANT BREEDING, VEGETABLE CROPS. Educ: Col Agr, Univ Calif, BS, 50. Prof Exp: Veg breeder, W Atlee Burpee Co, 50-58; DIR RES, PETOSEED CO, INC, 58- Mem: Am Soc Hort Sci. Res: Development of processing tomatoes, fresh market tomatoes, Open Pollinated and F-1 hybrids, bedding plant hybrid vegetables; quality studies for processing and fresh market. Mailing Add: Petoseed Res Ctr Rte 4 Box 1255 Woodland CA 95695

THOMAS, PAUL DAVID, b Bellwood, Pa, Mar 8, 26; m 51; c 4. ORGANIC CHEMISTRY. Educ: Rutgers Univ, BA, 49; Univ Ill, PhD(org chem), 54. Prof Exp: Chemist, Fries Bros Chem Mfg Co, 49-51; res chemist, Tidewater Assoc Oil Co, 51; res chemist, 54-71, res supvr org chem res & develop, 71-73, PATENT CHEMIST, CHAS PFIZER & CO, INC, 73- Mem: Am Chem Soc; Inst Food Technol. Res: Food additives and flavors. Mailing Add: 271 Plant St Groton CT 06340

THOMAS, PAUL EMERY, b Phoenix, Ariz, Feb 15, 27; m 58; c 2. TOPOLOGY. Educ: Oberlin Col, BA, 50; Oxford Univ, BA, 52; Princeton Univ, PhD(math), 55. Prof Exp: Res instr, Columbia Univ, 55-56; from asst prof to assoc prof, 56-63, PROF MATH, UNIV CALIF, BERKELEY, 63- Concurrent Pos: NSF fel, 58-59; Guggenheim Mem Found fel, 61; prof, Miller Inst, 66-67; ed, Proc, Am Math Soc, 68-71; mem div math sci, Nat Res Coun, 70-72. Mem: Am Math Soc. Res: Algebraic topology; homotopy theory; differentiable manifolds. Mailing Add: Dept of Math Univ of Calif Berkeley CA 94720

THOMAS, PAUL MILTON, b Sligo, Pa, Dec 1, 29; m 51; c 1. IMMUNOBIOLOGY, ICHTHYOLOGY. Educ: Allegheny Col, BS, 58; Univ Mich, MA, 59, MS, 62, PhD(sci admin, ichthyol), 64. Prof Exp: Instr biol, Houghton Col, 59-62; teacher high sch, Mich, 62-64; from asst prof to assoc prof biol, Pasadena Col, 64-67; res fel, Calif Inst Technol, 67-68; PROF BIOL, EDINBORO STATE COL, 68-, CHMN DEPT, 69- Concurrent Pos: Res fel radiation biol, Cornell Univ, 66-67. Mem: AAAS; Am Fisheries Soc. Res: Fish anesthetics; effects of industrial pollution of fish; radiation effects on elasmobranch antibody response; sexual dimorphism in fish; artificial fish shelters; administrative effects on science education. Mailing Add: Dept of Biol Edinboro State Col Edinboro PA 16444

THOMAS, PERCY LEROY, b Rimbey, Alta, July 4, 40; m 65; c 2. GENETICS, MYCOLOGY. Educ: Univ Alta, BSc, 63, MSc, 65; Australian Nat Univ, PhD(genetics), 70. Prof Exp: RES SCIENTIST MICROBIAL GENETICS, RES BR, CAN DEPT AGR, 65- Mem: Genetics Soc Can. Res: Genetics of virulence of cereal diseases. Mailing Add: Res Sta Can Dept Agr 25Dafoe Rd Winnipeg MB Can

THOMAS, PRENTICE MARQUET, JR, b Talladega, Ala, Oct 25, 44; m 67. ARCHAEOLOGY, CULTURAL ANTHROPOLOGY. Educ: Univ SC, BA, 66; Tulane Univ, PhD(anthrop), 72. Prof Exp: Asst prof anthrop, Univ Tenn, Knoxville, 70-74; MEM FAC, UNIV OF THE AMERICAS, 74- Mem: Am Anthrop Asn; Soc Am Archaeol. Res: Archaeological settlement pattern survey at the ancient Maya city of Becan in Campeche, Mexico. Mailing Add: Univ of the Americas PO Box 507 Dept B-1 Puebla Mexico

THOMAS, RALPH EDWARD, b Dayton, Ohio, Mar 14, 24; m 48, 65; c 4. OPERATIONS RESEARCH, STATISTICS. Educ: Ohio Wesleyan Univ, BA, 48; Ohio State Univ, MA, 50; PhD(opers Case Western Reserve Univ, PhD(opers res), 66. Prof Exp: Mathematician, Battelle Mem Inst, 52-55 & NAm Aviation, Inc, 55-57; math statistician, 57-63, chief appl statist, 63-66, CHIEF STATIST & MATH MODELING, BATTELLE MEM INST, 67- Concurrent Pos: Consult statist, 55-57; lectr, Ohio State Univ, 58-59 & Case Western Reserve Univ, 65-66. Mem: Am Math Soc; Math Asn Am; Inst Math Statist. Res: Development of statistical and mathematical models for applications to physical and human systems. Mailing Add: Battelle Mem Inst 505 King Ave Columbus OH 43201

THOMAS, RALPH HAROLD, b Reading, Eng, Nov 27, 32; m 58; c 3. HEALTH PHYSICS. Educ: Univ London, BSc, 55, PhD(nuclear physics), 59; Am Bd Health Physics, cert, 69. Prof Exp: Res physicist, Assoc Elec Industs, UK, 58-59; prin sci officer, Rutherford High Energy Lab, Sci Res Coun, UK, 59-68; sr health physicist, Stanford Univ, 68-70; HEAD DEPT HEALTH PHYSICS, LAWRENCE BERKELEY LAB, UNIV CALIF, 70- Concurrent Pos: Lectr, Reading Col Technol, 55-63; vis scientist, Lawrence Berkeley Lab, Univ Calif, 63-65, Europ Orgn Nuclear Res, Geneva, Switz, 66 & Brookhaven Nat Lab, 70; mem working group, Int Comn Radiol Protection, 66-70; chmn adv panel accelerator radiation safety, US Atomic Energy Comn, 69-72; mem, Comt High Energy & Space Disimetry, Int Comn Radiation Units, 73- Mem: Health Physics Soc; fel Brit Inst Physics. Res: Accelerator radiation problems; high energy dosimetry; cosmic-ray produced neutrons; heavy ion radiobiology. Mailing Add: Dept of Health Physics Lawrence Berkeley Lab Univ Calif Berkeley CA 94720

THOMAS, RICHARD CARROLL, animal science, animal breeding, see 12th edition

THOMAS, RICHARD DEAN, microbiology, biochemistry, see 12th edition

THOMAS, RICHARD DEAN, b Payson, Utah, Feb 14, 47; m 70; c 2. TOXICOLOGY, MEDICINAL CHEMISTRY. Educ: Utah State Univ, BS, 71; Colo State Univ, PhD(chem), 74. Prof Exp: Sr metab chemist agr chem, Biochem Dept, Agr Div, Ciba-Geigy Corp, 74-76; TOXICOLOGIST & CRITERIA DOC MGR,

CTR OCCUP & ENVIRON SAFETY & HEALTH, STANFORD RES INST, 76- Mem: Am Chem Soc; AAAS; Am Inst Chemists; Am Soc Appl Spectros. Res: Investigation into the toxicology, metabolism and environmental impact of chemicals; physiological impairment and potential for cancer and disease production related to chemical exposure; setting tolerances as they relate to health and regulation of chemicals. Mailing Add: Stanford Res Inst 1611 N Kent St Rosslyn Plaza Arlington VA 22209

THOMAS, RICHARD GARLAND, b Houston, Tex, June 23, 23; m 71. PHYSICS. Educ: Hampton Inst, BS, 43; Columbia Univ, MA, 50; Univ Calif, Berkeley, PhD(physics), 59. Prof Exp: Sr scientist physics, Gen Elec Co, 59-63 & Lawrence Livermore Lab, 63-68; PROF PHYSICS & HEAD DEPT, PRAIRIE VIEW AGR & MECH COL, 68- Mem: AAAS; Am Phys Soc; Am Asn Physics Teachers. Res: Low energy nuclear physics; x-ray spectroscopy. Mailing Add: Dept of Physics Prairie View Agr & Mech Col Prairie View TX 77445

THOMAS, RICHARD JOSEPH, b Wilkes-Barre, Pa, Nov 29, 28; m 51; c 2. WOOD TECHNOLOGY. Educ: Pa State Univ, BS, 54; NC State Univ, MWT, 55; Duke Univ, DF, 67. Prof Exp: Tech rep, Nat Casein NJ, 55-57, sales mgr, 57; from asst prof to assoc prof wood & paper sci, 57-71, PROF WOOD & PAPER SCI, SCH FOREST RESOURCES, NC STATE UNIV, 71- Mem: Int Asn Wood Anat; Soc Wood Sci & Technol. Res: Study of wood ultrastructure, particularly relationships of ultrastructure to physical properties and function within the plant; investigations of differentiation of cell wall and cell wall markings. Mailing Add: Dept of Wood & Paper Sci NC State Univ PO Box 5126 Raleigh NC 27607

THOMAS, RICHARD NELSON, b Omaha, Nebr, Mar 3, 21; m 45; c 1. ASTROPHYSICS Educ: Harvard Univ, BS, 42, PhD(astron), 48. Prof Exp: Ballistician, Ballistic Res Lab, Aberdeen Proving Ground, Md, 42-45; Jewett fel, Inst Advan Study, 48-49; assoc prof astron, Univ Utah, 48-53; vis lectr, Harvard Univ, 52-53, lectr observ, 53-57; consult astrophys to dir, Boulder Labs, Nat Bur Stand, 57-62; fel, Joint Inst Lab Astrophys, 62-74, ADJOINT PROF ASTROPHYS, UNIV COLO, BOULDER, 62- Concurrent Pos: Mem bd dirs, Annual Rev, Inc; vis prof, Col de France, Paris, 73-75 & Univ Paris, 61, 75-76. Mem: Int Astron Union; Am Astron Soc; Am Phys Soc; Royal Astron Soc. Res: Stellar atmospheres; solar physics; astroballistics; non-equilibrium thermodynamics. Mailing Add: Dept of Astrogeophys Univ of Colo Boulder CO 80302

THOMAS, RICHARD SANBORN, b Madison, Wis, June 14, 27; m 57; c 2. BIOPHYSICS. Educ: Oberlin Col, BA, 49; Univ Calif, Berkeley, PhD(biophys), 55. Prof Exp: Am Cancer Soc fel cancer res & NSF fel cytochem, Carlsberg Lab, Denmark, 55-57; asst res biophysicist, Virus Lab, Univ Calif, Berkeley, 58-60; RES PHYSICIST, WESTERN REGIONAL RES LAB, USDA, 60- Concurrent Pos: USPHS spec fel, Electron Micros Lab, Swiss Fed Inst Technol, 67-68; consult microscopy & biosci appln plasma chem, Tegal Corp, Richmond, Calif, 72- Mem: AAAS; Electron Micros Soc Am; Am Soc Cell Biol; Biophys Soc; Microbeam Anal Soc. Res: Biological ultrastructure and fine cytochemistry; development of techniques for electron microscopic cytochemistry, especially by plasma etching; intracellular mineral deposits, bacterial spores, keratin, microfibrillar proteins, virus particles; plant tissues. Mailing Add: Western Regional Res Lab USDA Albany CA 94710

THOMAS, ROBERT, b Atlanta, Ga, Aug 27, 34; m 69. CRYSTALLOGRAPHY, PHYSICAL CHEMISTRY. Educ: Boston Univ, AB, 55, PhD(phys chem), 65. Prof Exp: NIH res assoc chem, Univ Colo, 64-66; AEC res assoc, 66-68, ASSOC SCIENTIST CHEM, BROOKHAVEN NAT LAB, 68- Mem: Am Crystallog Asn; Am Chem Soc. Res: Crystal structure determination by x-ray development of computer controlled, parallel, single crystal x-ray data collection systems. Mailing Add: Dept of Chem Brookhaven Nat Lab Upton NY 11973

THOMAS, ROBERT E, b Salineville, Ohio, Feb 17, 36; m 62; c 2. PHYSIOLOGY, BIOCHEMISTRY. Educ: Kent State Univ, BS, 61, MA, 63, PhD(biol sci), 66. Prof Exp: Instr biol sci, Kent State Univ, 63-64; from asst prof to assoc prof, 66-74, PROF BIOL SCI, CALIF STATE UNIV, CHICO, 74- Mem: AAAS; Am Inst Biol Sci; Sigma Xi. Res: Sublethal effects of water soluble oil fractions on salmon respiration. Mailing Add: Dept of Biol Sci Calif State Univ Chico CA 95926

THOMAS, ROBERT EUGENE, b Iowa, Oct 15, 19; m 43; c 3. PSYCHIATRY. Educ: Univ Southern Calif, AB, 42, MD, 51; Johns Hopkins Univ, MPH, 55. Prof Exp: Intern, Santa Fe Coastlines Hosp, Los Angeles, Calif, 50-51; psychiat resident, Vet Admin Hosp, Perry Point & Baltimore, Md, 51-53; chief div ment health, Wash County Dept Health, Hagerstown, 53; from instr to asst prof pub health admin, Sch Hyg, Johns Hopkins Univ, 54-58; regional ment health adminr, 58-68, REGIONAL MENT HEALTH DIR, DIV LOCAL PROGS, CALIF DEPT MENT HYG, 68- Concurrent Pos: Chief div ment health, State Dept Health, Md, 53-55; mem, Gov Comn Ment Health, 54; lectr, Sch Pub Health & asst clin prof, Dept Psychiat, Sch Med, Univ Calif, Los Angeles, 60-70; psychiat resident ment health admin in pub health, Sch Hyg, Johns Hopkins Univ; assoc clin prof, Dept Psychiat, Loma Linda Sch Med; dir, Hemet Valley Community Ment Health Ctr, 69- Mem: Fel Am Psychiat Asn; AMA; fel Am Pub Health Asn; fel Am Orthopsychiat Asn. Res: Mental health administration in public health. Mailing Add: 1116 E Lathan Ave Hemet CA 92343

THOMAS, ROBERT GLENN, b Watertown, NY, Oct 9, 26; m 49; c 3. RADIOBIOLOGY, BIOPHYSICS. Educ: St Lawrence Univ, BS, 49; Univ Rochester, PhD(radiation biol), 55. Prof Exp: From instr to asst prof radiation biol, Univ Rochester, 55-61; from sect head to dept head radiobiol, Lovelace Found Med Educ & Res, 61-74; GROUP LEADER MAMMALIAN BIOL, LOS ALAMOS SCI LAB, 74- Concurrent Pos: Mem task group, Biol Effects Radiation on Lung Comt 1, Int Comn Radiol Protection, 68-; mem sci comts 30 & 31, Nat Coun Radiation Protection & Measurements. Mem: Am Radiation Res Soc; Health Physics Soc; Reticuloendothelial Soc; Am Indust Hyg Asn. Res: Toxicity of inhaled radioactive materials; application of experimental results to practical hazards evaluation in nuclear industry; toxicity of inhaled fossil fuel products. Mailing Add: 103 La Vista Dr Los Alamos NM 87544

THOMAS, ROBERT JAMES, organic chemistry, see 12th edition

THOMAS, ROBERT JAY, b Harvey, Ill, Mar 30, 30; m 53; c 2. COMPUTER SCIENCES. Educ: Oberlin Col, BA, 52; Ind Univ, MS, 54; Univ Ill, MS, 58, PhD(math), 64. Prof Exp: Dir recreational ther, Cent State Ment Hosp, Indianapolis, Ind, 54-55; adv, 3-2 combined eng prog, 62-72, dir comput ctr, 63-66, from instr to assoc prof math, 58-71, PROF MATH, DePAUW UNIV, 71- Concurrent Pos: Pace res appointment, Argonne Nat Lab, 67, comput consult, 67-71. Mem: AAAS; Asn Comput Mach; Am Asn Sex Educrs & Counrs. Res: Pattern recognition; determination of cell motility by computer; human sexuality. Mailing Add: Dept of Math DePauw Univ Greencastle IN 46135

THOMAS, ROBERT JOSEPH, b Lowell, Mass, July 13, 12; m 42; c 4. CHEMISTRY. Educ: Lowell Textile Inst, BTC, 34; Univ Notre Dame, MS, 37, PhD(org chem), 39. Prof Exp: Textile chemist, Apponaug Co, RI, 34-36; res chemist, Tech Lab, 39-42, Jackson Lab, 42-43, Manhattan Proj, Chambers Works, 43-44, tech lab, 44-50, supvr dyeing develop div, 50-65, INTERDEPT LIAISON, TECH LAB, E I DU PONT DE NEMOURS & CO, INC, 65- Concurrent Pos: Mem bd dirs, Lowell Tech Inst Res Found. Mem: Am Chem Soc; Am Asn Textile Chem & Colorists. Res: Dye application to textile fibers; textile chemistry; interdepartmental relations. Mailing Add: 10 Sack Ave Penns Grove NJ 08069

THOMAS, ROBERT JOSEPH, b Chicago, Ill, Feb 16, 40; m 69; c 3. CELL BIOLOGY, EPIDEMIOLOGY. Educ: Loyola Univ Chicago, BS, 62; Iowa State Univ, BS, 64, PhD(cell biol), 66; Johns Hopkins Univ, MPH, 76. Prof Exp: Instr cell biol, State Univ NY Albany, 66-67, res assoc & lectr, 67-68; instr radiol, State Univ NY Upstate Med Ctr, 68-69; asst prof anat, Sch Med, Creighton Univ, 69-75, asst prof prev med & pub health, 72-75. Mem: Am Soc Cell Biol; Electron Micros Soc Am. Res: Drug toxicity during early development; cancer epidemiology; psoriasis. Mailing Add: 2111 S 133rd Ave Omaha NE 68144

THOMAS, ROBERT L, b Dover-Foxcroft, Maine, Oct 10, 38; m 62; c 1. SOLID STATE PHYSICS. Educ: Bowdoin Col, AB, 60; Brown Univ, PhD(physics), 65. Prof Exp: Res asst physics, Brown Univ, 60-65; res assoc, 65-66, asst prof, 66-70, ASSOC PROF PHYSICS, WAYNE STATE UNIV, 70- Mem: Am Phys Soc. Res: Ultrasonics. Mailing Add: Dept of Physics Wayne State Univ Detroit MI 48202

THOMAS, ROBERT MALCOLM, b Toowoomba, Australia, June 23, 15; US citizen; m 38, 63; c 3. ORGANIC CHEMISTRY, MATHEMATICAL ANALYSIS. Educ: Univ Queensland, AB, 37; Univ Otago, MSc, 40; Univ Nottingham, PhD(org & phys chem), 48. Prof Exp: Sr chemist, Dhrangadnra Chem Works, India, 48-55; org chem consult, Govt India, 55-57; lectr org & phys chem, Univ Western Australia, 57-62; org chemist, Western Fine Chems, Inc, Calif, 62-65, sr scientist, 65-69; head dept natural prod, Stone Chem Corp, 69-71; consult org chem, 71-75; ORG CHEMIST & SR MATHEMATICIAN, FURDEN & ASSOCS, SCOTTSDALE, AZ, 75- Concurrent Pos: Lectr chem, Indian Inst Technol, Kanpur, 56-57. Mem: AAAS; The Chem Soc. Res: Organophosphorus and steroid chemistry; natural products; ultraviolet and fluorescence spectroscopy of proteins. Mailing Add: 2133 W Monroe Phoenix AZ 85009

THOMAS, ROBERT NELSON, b Oakmont, Pa, July 17, 26; m 55; c 2. GEOGRAPHY, POPULATION STUDIES. Educ: Ind Univ, Educ: Indiana Univ, Pa, BS, 50; Univ Pittsburgh, MA, 56; Pa State Univ, PhD(geog), 68. Prof Exp: Instr geog, Oakmont Jr High Sch, 50-55, Hampton Twp Sr High Sch, 56-60; asst prof, Indiana Univ, Pa, 62-65, assoc prof, 65-69; assoc prof, 69-75, ASST DIR, LATIN AM STUDIES CTR, MICH STATE UNIV, 74-, PROF GEOG, 75- Concurrent Pos: Urban planning adv, US AID, Guatemala, 65-66. Mem: Asn Am Geogrs; Pop Asn Am; Conf Latin Am Geogrs (pres, 72-73); Latin Am Studies Asn. Res: Internal migration systems of Latin America. Mailing Add: Dept of Geog Mich State Univ East Lansing MI 48824

THOMAS, ROBERT OSBORNE, plant physiology, see 12th edition

THOMAS, ROBERT P, b Miami, Fla, June 27, 27, m; c 6. OPHTHALMOLOGY. Educ: Univ NC, AB, 46, MD, 54. Prof Exp: Instr ophthal, Bowman Gray Sch Med, 60-61; instr, 61-62, asst prof, 62-66, PROF OPHTHAL & CHMN DEPT, MED COL GA, 66- Concurrent Pos: Consult, Ft Gordon, Ga, Vet Admin Hosp, Augusta & Reidsville State Prison, 61-67. Mem: Am Acad Ophthal & Otolaryngol. Mailing Add: Dept of Ophthal Med Col of Ga Augusta GA 30902

THOMAS, ROBERT SPENCER DAVID, b Toronto, Ont, July 29, 41; m 65. MATHEMATICS. Educ: Univ Toronto, BSc, 64; Univ Waterloo, MA, 65; Univ Southampton, PhD(math), 68. Prof Exp: Lectr math, Univ Waterloo, 65-66 & Univ Zambia, 68-70; asst prof comput sci, 70-72, ASSOC PROF COMPUT SCI, UNIV MAN, 72- Mem: Can Math Cong; Inst Math & Appln. Res: Knots; braids; computation with braids; application of mathematics. Mailing Add: Dept of Comput Sci Univ of Man Winnipeg MB Can

THOMAS, ROGER JERRY, b Detroit, Mich, July 3, 42; m 66; c 1. SOLAR PHYSICS. Educ: Univ Mich, Ann Arbor, BS, 64, MS, 66, PhD(astron), 70. Prof Exp: Nat Acad Sci-Nat Res Coun resident res assoc solar physics, 70-71, ASTROPHYSICIST, GODDARD SPACE FLIGHT CTR, NASA, 71- Concurrent Pos: Proj scientist orbiting solar observ satellite prog, Goddard Space Flight Ctr, NASA, 75- Mem: Int Astron Union; Am Astron Soc; Astron Soc Pac. Res: Solar x-ray and extreme ultraviolet astronomy; solar activity; solar flares; solar corona. Mailing Add: Code 682 Goddard Space Flight Ctr NASA Greenbelt MD 20771

THOMAS, RONALD EMERSON, b Ont, Can, Apr 19, 30; m 62; c 3. MATHEMATICAL STATISTICS, OPERATIONS RESEARCH. Educ: Queen's Univ, Ont, BA, 52, MA, 58; Univ NC, PhD(math statist), 62. Prof Exp: Actuarial asst, Excelsior Life Inst Co, 52-57; mem tech staff, 62-68, SUPVR APPL PROBABILITY GROUP, BELL LABS, 68- Concurrent Pos: Adj prof, Fairleigh Dickinson Univ, 64-65. Mem: Opers Res Soc Am; Am Statist Asn. Res: Mathematical studies of probability; statistical methodology; graph theory and network design. Mailing Add: Opers Anal Ctr Bell Labs Holmdel NJ 07733

THOMAS, RONALD LESLIE, b Edmonton, Alta, Can, June 29, 35; m 59; c 3. AGRONOMY. Educ: Univ Alta, BSc, 57, MSc, 59; Ohio State Univ, PhD(soils), 63. Prof Exp: From asst prof to assoc prof, 63-71, PROF SOIL SCI, UNIV GUELPH, 71- Mem: Am Soc Agron; Agr Inst Can; Can Soc Soil Sci. Res: Soil organic matter chemistry, the reactions, nature and importance of organic matter and its decomposition. Mailing Add: Dept of Land Resource Sci Univ of Guelph Guelph ON Can

THOMAS, RONALD S, underwater acoustics, electrical engineering, see 12th edition

THOMAS, ROY, b Prince Rupert, BC, Aug 8, 14; m 41; c 4. PHYSICS. Educ: Univ Alta, BSc, 38; Univ Calif, PhD(physics), 42. Prof Exp: Instr physics, Washington Univ, 42-46; asst prof, St Louis Univ, 46-48; assoc prof, 48-53, PROF PHYSICS, UNIV NMEX, 53- Mem: Am Phys Soc. Res: Theoretical problems in atomic physics. Mailing Add: Dept of Physics Univ of NMex Albuquerque NM 87131

THOMAS, ROY DALE, b Sevier Co, Tenn, Nov 12, 36; m 59; c 3. PLANT TAXONOMY. Educ: Carson-Newman Col, BS, 58; Southeastern Baptist Theol Sem, BD, 62; Univ Tenn, PhD(plant taxon), 66. Prof Exp: ASSOC PROF BIOL, NORTHEASTERN LA UNIV, 66-, CUR HERBARIUM, 74- Mem: Am Soc Plant Taxon; Bot Soc Am; Int Soc Plant Taxon; Soc Econ Bot; Am Fern Soc. Res: Vegetation and flora of Chilhowee Mountain in east Tennessee; flora of northeast Louisiana; Ophioglossaceae of the Gulf South. Mailing Add: Dept of Biol Northeast La Univ Monroe LA 71201

THOMAS, ROY ORLANDO, b Oneida, Tenn, Dec 15, 21; m 45; c 1. ANIMAL NUTRITION, DAIRY HUSBANDRY. Educ: Berea Col, BS, 46; Univ Tenn, MS, 52; Mich State Univ, PhD(animal nutrit), 64. Prof Exp: Teacher, Lewis County, Ky Bd Educ, 46 & Scott County, Tenn Bd Educ, 46-51; cow tester exten serv, Univ Tenn, 52; fieldman, Nashville Milk Producers, Inc, 52-53; asst dairy husbandman, Univ Tenn, 53-61; res asst animal nutrit, Mich State Univ, 61-64; asst prof, 64-72, ASSOC PROF ANIMAL NUTRIT, W VA UNIV, 72- Mem: AAAS; Am Dairy Sci Asn; Am Soc Animal Sci. Res: Evaluation of feed materials, methods of feeding and the effect of these materials and methods on production and well-being of animals. Mailing Add: Div of Animal & Vet Sci WVa Univ Morgantown WV 26506

THOMAS, RUTH BEATRICE, b Ringgold, La. BOTANY. Educ: Northwestern State Col, La, BS, 40; George Peabody Col, MA, 44; Vanderbilt Univ, PhD(biol), 51. Prof Exp: Pub sch teacher, La, 40-44; instr, Sullins Col, 44-46; instr, George Peabody Col, 46-48; asst prof biol, Millikin Univ, 51-54; prof, Eastern NMex Univ, 54-63; assoc prof, 64-70, PROF BIOL, SAM HOUSTON STATE UNIV, 70- Mem: Fel AAAS; Bot Soc Am; Nat Asn Biol Teachers. Res: Gymnosperm gametophyte development; descriptive morphology. Mailing Add: Dept of Biol Sam Houston State Univ Huntsville TX 77340

THOMAS, SARAH NELL, b Gainesville, Ga. PHYSIOLOGY, RADIATION BIOLOGY. Educ: Brenau Col, BA, 48; Univ Denver, MS, 57; Tex Woman's Univ, PhD(radiation biol), 70. Prof Exp: Teacher & head dept sci, Pub Schs, Ga, 48-60; assoc prof biol & chmn dept, Brenau Col, 60-67; instr, Tex Woman's Univ, 70; ASSOC PROF BIOL, LANGSTON UNIV, 70- Concurrent Pos: NSF traineeship, 67-69. Mem: AAAS; Am Inst Biol Sci. Res: Gonad development in male rats irradated the first day of postnatal life. Mailing Add: PO Box 902 Guthrie OK 73044

THOMAS, T ROY, chemical physics, see 12th edition

THOMAS, TELFER LAWSON, b Montreal, Que, June 1, 32; m 56; c 2. ORGANIC CHEMISTRY. Educ: McGill Univ, BS, 53, PhD(org chem), 57. Prof Exp: Res chemist, Imp Oil Ltd, 57-59 & Gen Aniline & Film Co, 59-62; SR RES CHEMIST PHARMACEUT DIV, PENNWALT CORP, 62- Mem: Am Chem Soc. Res: Synthesis of new compounds for discovery of useful drugs. Mailing Add: 17 Candle Wood Dr Pittsford NY 14534

THOMAS, THOMAS DARRAH, b Glen Ridge, NJ, Apr 8, 32; m 56; c 4. NUCLEAR CHEMISTRY, PHYSICAL CHEMISTRY. Educ: Haverford Col, BS, 54; Univ Calif, PhD(chem), 57. Prof Exp: From instr to asst prof chem, Univ Calif, 57-59; vis assoc chemist, Brookhaven Nat Lab, 59-60, assoc chemist, 60-61; from asst prof to assoc prof chem, Princeton Univ, 61-71; PROF CHEM, ORE STATE UNIV, 71- Concurrent Pos: Consult, Los Alamos Sci Lab, 65; Guggenheim fel, Univ Calif, Berkeley, 69. Mem: AAAS; Am Chem Soc; Am Phys Soc. Res: Medium energy nuclear reactions; x-ray photoelectron spectroscopy. Mailing Add: Dept of Chem Ore State Univ Corvallis OR 97331

THOMAS, THURLO BATES, biology, deceased

THOMAS, TIMOTHY FARRAGUT, b Cleveland, Ohio, June 15, 38; m 63. PHYSICAL CHEMISTRY. Educ: Oberlin Col, AB, 60; Univ Ore, PhD(chem kinetics), 64. Prof Exp: Res assoc chem, Brandeis Univ, 64-66; asst prof, 66-73, ASSOC PROF CHEM, UNIV MO-KANSAS CITY, 73- Concurrent Pos: Nat Res Coun sr res assoc, Air Force Cambridge Res Lab, 75- Mem: Am Chem Soc; Am Phys Soc; Am Soc Mass Spectrom; The Chem Soc; Sigma Xi. Res: Unimolecular reaction kinetics; photochemistry of gases; fluorescence lifetimes and quantum yields; mass spectrometry and electron spin resonance; photodissociation spectra of gaseous ions. Mailing Add: Dept of Chem Univ of Mo-Kansas City Kansas City MO 64110

THOMAS, TRACY YERKES, b Alton, Ill, Jan 8, 99; m 28; c 1. MATHEMATICS. Educ: Rice Inst, AB 21; Princeton Univ, AM, 22, PhD(math), 23. Prof Exp: Nat Res Coun fel physics, Univ Chicago, 23-24; Nat Res Coun fel math, Univ Zurich, 24-25 & Harvard Univ & Princeton Univ, 25-26; from asst prof to assoc prof math, Princeton Univ, 26-38; prof, Univ Calif, Los Angeles, 38-44, fac res lectr, 43; prof & chmn dept, 44-54, head grad inst appl math, 50-54, dir grad inst math & mech, 54-56, distinguished serv prof math, 56-69, EMER PROF MATH, IND UNIV, BLOOMINGTON, 69- Concurrent Pos: Consult appl math staff, US Naval Res Lab, 51-68 & Rand Corp, Santa Monica, Calif, 60-69; vis prof, Dept Math, Univ Calif, San Diego, 62-63 & Sch Eng, Univ Calif, Los Angeles, 65-66, 67-68 & 69-70; ed, J Math & Mech. Mem: Nat Acad Sci; AAAS; Am Math Soc (vpres, 40-42); Math Asn Am; Soc Eng Sci. Res: Tensor analysis and differential geometry; theory of relativity; supersonic flow and shock wave theory; plasticity theory; cosmology; gas dynamics; fracture; extended theory of conditions for discontinuities over moving surfaces. Mailing Add: Boelter Hall Rm 5732 Sch of Eng Univ of Calif Los Angeles CA 90024

THOMAS, TUDOR LLOYD, b Utica, NY, Apr 23, 21; m 44; c 4. PHYSICAL CHEMISTRY. Educ: Univ Mich, BS, 43, MS, 46, PhD(chem), 49. Prof Exp: Res chemist, 49-56, develop supvr, 56-58, from asst mgr to mgr molecular sieve develop, 58-63, mgr molecular sieve prod, 63-67, GEN MGR MOLECULAR SIEVE DEPT, LINDE DIV, UNION CARBIDE CORP, 67- Mem: AAAS; Am Chem Soc; Am Mgt Asn. Res: Adsorption; catalysis. Mailing Add: 111 Hawthorne Pl Briarcliff Manor NY 10510

THOMAS, VERA, b Prague, Czech, May 2, 28; US citizen; m 67. CHEMISTRY, TOXICOLOGY. Educ: Charles Univ, Prague, MS, 52; Czech Acad Sci, PhD(chem), 62. Prof Exp: Res assoc indust toxicol, Inst Indust Hyg & Occup Dis, Prague, Czech, 49-67; ASST PROF DRUG METAB, SCH MED, UNIV MIAMI, 67- Concurrent Pos: Secy comt maximum allowable concentration toxic compounds, Czech Ministry Health, 62-67; consult, WHO, Chile & Venezuela, 67. Mem: NY Acad Sci; Am Conf Govt Indust Hygienists. Res; Uptake, distribution, metabolism and excretion of drugs, especially of volatile compounds; pesticides distribution. Mailing Add: 145 SE 25 Rd Miami FL 33129

THOMAS, VIRGINIA LEE, b Traer, Iowa, Sept 22, 16. MICROSCOPY. Prof Exp: Spectrographer & x-ray diffractionist, Rock Island Arsenal, 42-45; in charge electron micros res labs, US Rubber Co, 45-54; chief spectrographer, Driver-Harris Co, 54-56; group leader, Cent Res Labs, Interchem Corp, 56-62; res scientist, Res Div, Am Radiator-Stand Sanit Corp, NJ, 62-68; SUPVR ELECTRON MICROS, DEPT MICROBIOL, RUTGERS MED SCH, COL MED & DENT NJ, 68- Mem: AAAS; Electron Micros Soc Am. Res: Use of electron microscopy in the graphic arts; microscopy of inks, pigment dispersions, fibers, plastics, ceramics, glasses and enamels; whiteware; microscopy of viruses; tissue culture; antibody-antigen reactions. Mailing Add: Rutgers Med Sch, Col Med & Dent NJ Dept Microbiol Piscataway NJ 08854

THOMAS, VIRGINIA LYNN, b Graham, Tex, Mar 29, 43. MEDICAL MICROBIOLOGY. Educ: Baylor Univ, BSc, 65, MSc, 67; Univ Tex Med Sch, San Antonio, PhD(microbiol), 73. Prof Exp: Teaching fel biol, Baylor Univ, 65-66, res asst & technician microbiol, Col Dent & Med Ctr, Dallas, 66-68; sr res & teaching asst microbiol, Univ Tex Med Sch, San Antonio, 68-70, teaching fel, 70-73; instr, 73-75, ASST PROF MICROBIOL, UNIV TEX HEALTH SCI CTR, SAN ANTONIO, 75- Mem: Am Soc Microbiol; Sigma Xi; AAAS. Res: Host-parasite relationships in bacterial diseases; immunologic aspects of urinary tract infection; immunofluorescence procedure for localizing the site of urinary tract infection. Mailing Add: Dept of Microbiol Univ of Tex Health Sci Ctr San Antonio TX 78284

THOMAS, WALTER DILL, JR, b St Louis, Mo, July 3, 18; m 39; c 2. PHYTOPATHOLOGY. Educ: Colo State Univ, BS, 39; Univ Minn, MS, 43, PhD(phytopath), 47. Prof Exp: Instr bot, Colo State Univ, 39-41; asst phytopath, Univ Minn, 41-44, 46; from asst prof plant path & asst plant pathologist to prof & Plant pathologist, Agr Exp Sta, 46-54; dir res, Arboriculture Serv & Supply Co, Colo, 54-55; lead res biologist, Ortho Div, Chevron Chem Co, 55-66, tech asst to mgr res & develop, 66-67, forestry specialist, 67-70; pres, Forest & Environ Protection Serv, 70-71; vpres Natural Resouces Mgt Corp & mem bd dir, Environ Home & Garden Serv, 72-74; PRES, FOREST-AGR ENVIRON PROTECTION SERV, 74- Mem: Am Phytopath Soc; Soc Am Foresters; Asn Consult Foresters; Am Soc Consult Arborists. Res: Disease of potatoes, beans, onions and ornamental plants; forest diseases; agricultural pesticides; mycorrhizae; air pollution damage to plants; remote sensing of forest diseases and insects. Mailing Add: Forest-Agr Environ Protection Serv PO Box 350 Lafayette CA 94549

THOMAS, WALTER IVAN, b Elwood, Nebr, Mar 28, 19; m 41; c 1. AGRONOMY, GENETICS. Educ: Iowa State Univ, BS, 49, MS, 53, PhD(plant breeding genetics), 55. Prof Exp: Res asst agron, Iowa State Univ, 49-50, 53-55, asst prof, 55-59; assoc prof, 59-63, head dept, 64-69, PROF AGRON, PA STATE UNIV, UNIVERSITY PARK, 63-, ASSOC DEAN RES, COL AGR & ASSOC DIR AGR EXP STA, 69- Mem: Fel AAAS; Am Soc Agron; Sigma Xi. Res: Plant breeding, genetics and pathology. Mailing Add: 229 Agr Admin Bldg Pa State Univ University Park PA 16802

THOMAS, WALTER MORELAND, b Boston, Mass, July 8, 16; m 43; c 3. CHEMISTRY. Educ: Northeastern Univ, BS, 39; Polytech Inst Brooklyn, MS, 45, PhD(chem), 50. Prof Exp: Lab asst, Hunt-Spiller Mfg Co, Mass, 35-39; chemist, Bennett, Inc, 39-40; chemist, 40-54, group leader, 54-57, SECT MGR, AM CYANAMID CO, 57- Mem: Am Chem Soc. Res: Emulsions for paper; melamine resins; vinyl type polymers; wet strength paper; acrylic fibers. Mailing Add: 13 Hillcrest Ave Darien CT 06820

THOMAS, WALTER WILLIAM, b Vohwinkel, Ger, Jan 4, 13; nat US; m 50; c 3. ORGANIC CHEMISTRY. Educ: Univ Pa, BA, 34; Univ Marburg, PhD(chem), 39. Prof Exp: Anal chemist, E I du Pont de Nemours & Co, 34-35, res chemist, 39-41, process develop chemist, 41-49; RES CHEMIST, HERCULES INC RES CTR, 49- Mem: Am Chem Soc; AAAS; Am Asn Textile Chemists & Colorists. Res: Pharmacology; organic syntheses; pharmaceuticals; insecticides; polymers and dyestuffs; textile fibers; photopolymerization processes. Mailing Add: 101 Dickinson Lane Wilmington DE 19807

THOMAS, WILBUR ADDISON, b Louisville, Miss, June 26, 22; m 49; c 2. PATHOLOGY. Educ: Univ Miss, BA, 41; Univ Tenn, MD, 46; Am Bd Path, dipl. Prof Exp: Intern, Baptist Hosp, Memphis, Tenn, 46-47, asst resident path, 49-50; asst resident & resident, Mass Gen Hosp, Boston, 50-52; instr, Harvard Med Sch, 52-53; from instr to assoc prof, Sch Med, Wash Univ, 53-59; CYRUS STRONG MERRILL PROF PATH & CHMN DEPT, ALBANY MED COL, 59- Mem: Am Soc Exp Path; Am Asn Path & Bact; AMA; Col Am Path. Res: Arteriosclerosis. Mailing Add: Dept of Path Albany Med Col Albany NY 12208

THOMAS, WILLIAM ANDREW, b Berea, Ky, July 23, 36; m 57; c 2. GEOLOGY. Educ: Univ Ky, BS, 56, MS, 57; Va Polytech Inst, PhD(geol), 60. Prof Exp: Geologist, Calif Co, 59-63; from assoc prof to prof geol, Birmingham-Southern Col, 63-70, chmn dept, 67-70; assoc prof, Queens Col, NY, 70-72, chmn dept, 71-72; PROF GEOL & CHMN DEPT, GA STATE UNIV, 72- Mem: Geol Soc Am; Am Asn Petrol Geol; Soc Econ Paleontologists & Mineralogists. Res: Stratigraphy and structure of contemporaneously deformed sediments; stratigraphic and structural continuity of Appalachian and Ouachita mountains; stratigraphy of Gulf Coastal Plain; Appalachian structure and stratigraphy; Mississippian stratigraphy. Mailing Add: Dept of Geol Ga State Univ Atlanta GA 30303

THOMAS, WILLIAM ARTHUR, b Pittsburgh, Pa, Jan 20, 39; m 61; c 1. ECOLOGY. Educ: Purdue Univ, BS, 60; Univ Minn, MS, 64, PhD, 67; Univ Tenn, JD, 72. Prof Exp: Ecologist, Oak Ridge Nat Lab, 67-73; asst prof law, Univ Tenn, 72-73; RES ATTY, AM BAR FOUND, 73- Mem: AAAS; Am Bar Asn; Int Acad Law & Sci; Int Coun Environ Law. Res: Incorporation of science into the legal system; indicators of environmental quality; technology assessment. Mailing Add: Am Bar Found 1155 E 60th St Chicago IL 60637

THOMAS, WILLIAM BENJAMIN, b Monroe, Wis, Mar 14, 06; m 32; c 1. ORGANIC CHEMISTRY. Educ: Univ Ill, BS, 28; Univ Wis, PhD(org chem), 32. Prof Exp: Asst chem, Univ Wis, 28-34; from instr to asst prof, Bates Col, 34-43; res chemist, Maine Mills Lab, 43-45; from asst prof to assoc prof anal chem, 45-55, prof chem, 55-74, dean fac, 71-72, EMER PROF CHEM, BATES COL, 74- Concurrent Pos: Consult, Bates Mfg Co, 45-57. Mem: Am Chem Soc. Res: Acetoacetic ester condensation; chemical modification of cotton; resin treatment of cotton. Mailing Add: Dept of Chem Bates Col Lewiston ME 04240

THOMAS, WILLIAM CLARK, JR, b Bartow, Ga, Apr 7, 19; m 46. INTERNAL MEDICINE, ENDOCRINOLOGY. Educ: Univ Fla, BS, 40; Cornell Univ, MD, 43. Prof Exp: Intern med, New York Hosp, 44, asst resident med, 46-49; pvt pract, 49-54; NIH fels, Johns Hopkins Univ, 54-57; asst prof med & chief div postgrad educ, 57-60, chief endocrine div, 57-70, assoc prof med, 60-63, dir clin res ctr, 62-68, PROF MED, COL MED, UNIV FLA, 63- Concurrent Pos: Chief med serv, Vet Admin Hosp, Gainesville, Fla, 68-73, assoc chief of staff for res, 73-; hon res fel, Univ Manchester, 69-70. Mem: AAAS; Am Clin & Climat Asn; Am Diabetes Asn; Endocrine Soc; Am Fedn Clin Res. Res: Mineral metabolism; clinical research; factors affecting renal calculus formation. Mailing Add: Med Serv Vet Admin Hosp Gainesville FL 32601

THOMAS, WILLIAM CLAUDE, heat transfer, see 12th edition

THOMAS, WILLIAM G, b Ipswich, SDak, July 11, 17; m 40; c 2. PHYSICAL CHEMISTRY. Educ: SDak State Col, BS, 39; Univ Mich, MS, 48; Mich State Univ, PhD(chem), 54. Prof Exp: Instr chem, SDak State Col, 46; asst prof & mem bd examrs, Mich State Univ, 49-53; from asst prof to prof chem, Cent Mich Univ, 53-59; PROF CHEM, EASTERN NMEX UNIV, 59- Mem: Fel AAAS; Am Chem Soc. Res: Structure and properties of matter; x-ray crystallography, primarily binary crystalline fluorides; electrochemistry, primarily conductivity of fluoride solutions. Mailing Add: Dept of Chem Eastern NMex Univ Portales NM 88130

THOMAS, WILLIAM GRADY, b Charlotte, NC, Mar 21, 34; m 55; c 3. AUDIOLOGY. Educ: Appalachian State Univ, BS, 57; Washington Univ, MA, 61; Univ Fla, PhD(auditory physiol), 68. Prof Exp: From instr to asst prof, 61-70, ASSOC PROF AUDIOL, MED SCH, UNIV NC, CHAPEL HILL, 70-, DIR HEARING & SPEECH, 61-, DIR AUDITORY RES LAB, 68- Concurrent Pos: Fac res grant, Univ NC, Chapel Hill, 68-69, Off Naval Res grants, 69-76, NIH grant, 72-75; consult, Exp Diving Unit, Dept Navy, 69- & Nat Inst Environ Health Sci, 72-; Rockefeller Found grant, 75-76. Honors & Awards: Cert of Recognition, Am Speech & Hearing Asn, 69. Mem: AAAS; Am Speech & Hearing Asn; Acoust Soc Am. Res: Auditory physiology and psychoacoustics, particularly the effects of drugs and environmental conditions on the ear; basic electrophysiology of the auditory system. Mailing Add: Dept of Surg Univ of NC Sch of Med Chapel Hill NC 27514

THOMAS, WILLIAM HEWITT, b Riverside, Calif, Dec 25, 26; m 56; c 2. BIOLOGICAL OCEANOGRAPHY. Educ: Pomona Col, BA, 49; Univ Md, MS, 52, PhD, 54. Prof Exp: Lab asst, Regional Salinity Lab, USDA, 46-47, plant physiol, 48; lab asst plant physiol, Citrus Exp Sta, Univ Calif, 49-50 & Univ Md, 51-54; jr res biologist, 54-56, asst res biologist, 56-64, ASSOC RES BIOLOGIST, SCRIPPS INST OCEANOG, UNIV CALIF, SAN DIEGO, 64- Mem: AAAS; Am Soc Plant Physiol; Am Soc Limnol & Oceanog; Phycol Soc Am. Res: Mineral nutrition and nitrogen metabolism of algae; primary production in the ocean; cultural requirements of marine phytoplankton; marine pollution. Mailing Add: Scripps Inst of Oceanog Univ of Calif at San Diego La Jolla CA 92093

THOMAS, WILLIAM LEROY, b Long Beach, Calif, Mar 18, 20; m 42, 64; c 7. CULTURAL GEOGRAPHY, GEOGRAPHY OF SOUTHEAST ASIA. Educ: Univ Calif, Los Angeles, AB, 42, MA, 48; Yale Univ PhD(geog, Southeast Asian Studies), 55. Prof Exp: Instr geol & geog, Rutgers Univ, 47-50; asst dir res, Wenner-Gren Found Anthrop Res, 50-57; from asst prof to assoc prof geog, Univ Calif, Riverside, 57-63; prof anthrop & geog & chmn dept, 63-71, chmn dept geog, 71-74, PROF GEOG & SOUTHEAST ASIA STUDIES, CALIF STATE UNIV, HAYWARD, 71- Concurrent Pos: Asst res, Southeast Asia Studies Prog, Yale Univ, 49-50; consult, Far East Prog Div, Econ Coop Admin, 50-51; prin investr, NSF fel, Wenner-Gren Found, 55-56; lectr geog, Yale Univ, 55-56; consult, Nat Acad Sci-Nat Res Coun, 57; US Navy geog res contract, Pac Missile Range, Univ Calif, Riverside & Berkeley, 59-61; organizer, Sect Geog, Tenth Pac Sci Conf, Nat Acad Sci-Univ Hawaii, 59-61; Off Naval Res res fel, Prov Ilocos Norte, Philippines, 61-62; vis prof anthrop & geog, La State Univ, Baton Rouge, 66; vis prof geog, Univ Wis-Madison, 66; vis prof, Univ Toronto, 68 & 69; res assoc, Inst Philippine Cult, Ateneo de Manila Univ, 70; Nat Acad Sci air photo interpretation study, Calif State Univ, Hayward, 72-73; consult comt effects herbicides in Vietnam, Div Biol & Agr, Nat Acad Sci, 72-73; Fulbright sr scholar, Univ Western Australia, 74. Honors & Awards: Citation Meritorious Contrib to Geog, Asn Am Geog, 61. Mem: Asn Am Geog; Am Geog Soc; Pac Sci Asn; Asn Asian Studies. Res: Man's role and impact on changing the face of the earth, particularly California, the Pacific Islands, and island and mainland Southeast Asia; history of geographic thought. Mailing Add: Dept of Geog Calif State Univ Hayward CA 94542

THOMAS, WILLIAM ROBB, b Toronto, Kans, Dec 17, 26; m 54; c 1. FOOD SCIENCE, DAIRY BACTERIOLOGY. Educ: Okla State Univ, BSc, 50; Ohio State Univ, MSc, 52; Iowa State Univ, PhD(dairy bact) 61. Prof Exp: Asst dairy tech, Ohio State Univ, 50-52; mem sanit stand staff, Evaporated Milk Asn, Ill, 52-53; asst prof dairy mfg, Univ Wyo, 53-59; res asst, Iowa State Univ, 59-61; assoc prof dairy mfg, Univ Wyo, 61-65; exten food technologist, Univ Calif, Davis, 65-68; ASSOC DEAN & DIR RESIDENT INSTR, COL AGR SCI, COLO STATE UNIV, 69- Concurrent Pos: Appointee, Int Sci & Educ Coun, 74- Mem: Nat Asn Col & Teachers Agr (pres, 75-76); Am Dairy Sci Asn; Inst Food Technol; Int Asn Milk, Food & Environ Sanit. Res: Dairy technology; thermoduric bacteria; lipolytic enzymes of milk; consumer and market analysis of dairy products; dairy plant operation analysis. Mailing Add: Col of Agr Sci Colo State Univ Ft Collins CO 80523

THOMAS, WILLIAM STEPHEN, biology, see 12th edition

THOMAS, WINFRED, b Geneva, Tex, June 6, 20; m 43; c 1. AGRONOMY. Educ: Prairie View State Col, BS, 43; Cornell Univ, MS, 47; Ohio State Univ, PhD(agron), 54. Prof Exp: Instr agron, Ala Agr & Mech Col, 47-51; asst, Ohio State Univ, 51-53; AGRONOMIST, ALA A&M UNIV, 53-, DEAN SCH AGR & ENVIRON SCI, 73- Mem: Am Soc Agron; Soil Sci Soc Am. Res: Effect of foliar applied fertilizers on growth and composition of corn; plant population-nitrogen relationships of corn; nitrogen-sulfur-protein relationships of corn. Mailing Add: Dept of Agron Ala A&M Univ Box 202 Normal AL 35762

THOMAS A KEMPIS, MARY, b Manley, Iowa, May 15, 96. MATHEMATICS. Educ: Col St Theresa, AB, 20; Univ Mich, AM, 26, PhD(math), 36. Prof Exp: Instr high sch, Minn, 20-22, prin, 24-26; instr high sch, Ohio, 22-24; head dept physics, Col St Teresa, 26-39, head dept math, 39-49; prin, St Mary High Sch, 49-50 & St Joseph Cent High Sch, 50-53; instr math & sci, Cotter Sr High Sch, 53-54; head dept math, Col St Teresa, Minn, 54-69, prof, 69-74; RETIRED. Mem: Math Asn Am. Res: Linear and quadratic equations; history of mathematics. Mailing Add: Dept of Math Col of St Teresa Winona MN 55987

THOMASIAN, ARAM JOHN, b Boston, Mass, Aug 12, 24; m 53; c 3. ELECTRICAL ENGINEERING, STATISTICS. Educ: Brown Univ, BSc, 49; Harvard Univ, MA, 51; Univ Calif, PhD(math statist), 56. Prof Exp: PROF ELEC ENG & STATIST, UNIV CALIF, BERKELEY, 56- Res: Information theory; probability; electroencephatography. Mailing Add: Dept of Elec Eng Univ of Calif Berkeley CA 94720

THOMASON, BERENICE MILLER, b Birmingham, Ala, Mar 10, 24; m 44; c 1. MICROBIOLOGY. Educ: Ga State Col, BS, 60. Prof Exp: Med technologist, Sta Hosp, Ft Benning, Ga, 43-45 & Thayer Gen Hosp, Nashville, Tenn, 45; pub health technologist, Muscogee County Health Dept, Columbus, Ga, 48-51; bacteriologist, Commun Dis Ctr, Atlanta, Ga, 51-53 & Third Army Labs, Ft McPherson, Ga, 53-54; bacteriologist, 54-63, RES MICROBIOLOGIST, CTR DIS CONTROL, USPHS, 63- Mem: Am Soc Microbiol; Sigma Xi. Res: Development and application of fluorescent antibody technic for rapid detection of pathogenic bacteria. Mailing Add: Anal Bacteriol Br Ctr for Dis Control Atlanta GA 30333

THOMASON, IVAN J, b Burney, Calif, June 27, 25; m 50; c 5. PLANT NEMATOLOGY, PLANT PATHOLOGY. Educ: Univ Calif, BS, 50; Univ Wis, MS, 52, PhD(plant path), 54. Prof Exp: Res asst plant path, Univ Wis, 50-54; jr nematologist, Citrus Res Ctr, Agr Exp Sta, 54-56, from asst nematologist to assoc nematologist, 54-67, chmn dept nematol, 63-70, prof nematol, 67-73, PROF NEMATOL & PLANT PATH, UNIV CALIF, RIVERSIDE, 73-, NEMATOLOGIST, CITRUS RES CTR, AGR EXP STA, 67- Concurrent Pos: Mem subcomt nematodes, Agr Bd, Nat Acad Sci-Nat Res Coun, 64-67; mem Univ Calif-AID Pest Mgt Study Team, Southeast Asia, 71; mem agr pest control adv comt, Calif State Dept Agr, 72- & Pest Control Advisors Comt, Dir Adv Comt on APCA, Calif Dept Food & Agr, 74-76. Mem: Am Phytopath Soc; Soc Nematol (pres, 75-76); Soc Europ Nematol. Res: Biology and control of nematodes attacking sugarbeets and dry beans; efficacy and mode of action of nematicides. Mailing Add: 4686 Holyoke Pl Riverside CA 92507

THOMASON, STEVEN KARL, b Salem, Ore, June 2, 40; m 60; c 2. MATHEMATICAL LOGIC. Educ: Univ Ore, BS, 62; Cornell Univ, PhD, 66. Prof Exp: Asst prof, 66-70, ASSOC PROF MATH, SIMON FRASER UNIV, 70- Concurrent Pos: Vis asst prof, Univ Calif, Berkeley, 68-69. Mem: Am Math Soc; Can Math Cong; Math Asn Am; Asn Symbolic Logic. Res: Recursion theory and modal logic. Mailing Add: Dept of Math Simon Fraser Univ Burnaby BC Can

THOMASON, WILLIAM HUGH, b Hampton, Ark, Apr 4, 45; m 69; c 1. PHYSICAL CHEMISTRY, CORROSION. Educ: Hendrix Col, BA, 67; La State Univ, Baton Rouge, PhD(phys chem), 75. Prof Exp: RES SCIENTIST PHYS CHEM & CORROSION RES & DEVELOP, CONTINENTAL OIL CO, 75- Mem: Am Chem Soc; Optical Soc Am; Soc Petrol Engrs; Nat Asn Corrosion Engrs. Res: Corrosion problems in oil production, particularly those caused by hydrogen sulfide; water treating and scale problems. Mailing Add: 2509 Woodthrush Ponca City OK 74601

THOMASSEN, KEITH I, b Harvey, Ill, Nov 22, 36; m 57; c 2. PLASMA PHYSICS. Educ: Chico State Col, BS, 58; Stanford Univ, MS, 60, PhD(elec eng), 63. Prof Exp: Res assoc plasma physics, Stanford Univ, 62-63, NATO fel, 63-64, res assoc, 64-65, lectr, 64-68, res physicist, 65-68; from asst prof to assoc prof elec eng, Mass Inst Technol, 68-73; asst ctr div leader, 73-74, ASSOC CTR DIV LEADER, LOS ALAMOS SCI LAB, 74- Concurrent Pos: Consult, Lincoln Labs, Lexington, Mass, 68-74. Mem: Am Phys Soc; Am Nuclear Soc. Res: Fusion reactor design; component development; plasma research, energy storage and transfer. Mailing Add: Los Alamos Sci Lab Los Alamos NM 87544

THOMASSEN, PAUL R, JR, b Richmond, Utah, June 28, 16; m 43; c 3. BACTERIOLOGY. Educ: Stanford Univ, AB, 44, AM, 46. Prof Exp: Asst bact, Stanford Univ, 44-46; from instr to assoc prof, 46-64, coordr basic & clin sci, 60, PROF ORAL DIAG & CHMN DEPT, SCH DENT, UNIV OF THE PAC, 64- Mem: AAAS; Am Dent Asn; Int Asn Dent Res. Res: Electron microscopy of bacteria and bacteriophage; phosphatase inhibitors of dental caries; bacteriology of root canals. Mailing Add: 2155 Webster St San Francisco CA 94115

THOMASSEN, ROBERT WILLIAM, b Newman Grove, Nebr, June 2, 31; m 58; c 3. VETERINARY PATHOLOGY. Educ: Colo State Univ, BS, 55, DVM, 56. Prof Exp: Instr anat, Colo State Univ, 56-57; asst chief path sect, US Army Med Res & Nutrit Lab, 57-60; chief path sect, US Army Trop Res Med Lab, 60-63; pathologist, Armed Forces Inst Path, 63-66 & US Army Med Unit, Ft Detrick, Md, 66-67; CHIEF, COLLAB RADIOL HEALTH LAB, COLO STATE UNIV, 67- Concurrent Pos: Affil fac, Colo State Univ, 70- Mem: Am Col Vet Pathologists; Int Acad Path; Am Vet Med Asn. Res: Investigation of long-term effects of single low doses of gamma radiation in the beagle. Mailing Add: Collab Radiol Health Lab Colo State Univ Foothills Campus Ft Collins CO 80523

THOMASSON, CLAUDE LARRY, b Blue Grass, Va, Mar 6, 32; m 57; c 3. PHARMACY. Educ: Univ Cincinnati, BS, 54; Univ Fla, PhD(pharm), 57. Prof Exp: Assoc prof pharm, Southern Col Pharm, Mercer, 57-61, prof & chmn dept, 61-64; assoc prof, WVa Univ, 64-66; ASSOC PROF PHARM, AUBURN UNIV, 66- Mem: Am Pharmaceut Asn; Acad Pharmaceut Sci; AAAS. Res: Dispensing and clinical pharmacy. Mailing Add: Sch of Pharm Auburn Univ Auburn AL 36830

THOMASSON, MAURICE RAY, b Columbia, Mo, Sept 3, 30; m 56; c 3. GEOLOGY. Educ: Univ Mo, BA, 53, MA, 54; Univ Wis, PhD(geol), 59. Prof Exp: Geologist, Shell Oil Co, La, 59-68, mgr geol dept, Shell Develop Co, Tex, 68-70, div explor mgr, Shell Oil Co, La, 70-72, mgr forecasting, planning & econ, 72-74, MEM STAFF, SHELL INT PETROL CO, LTD, LONDON, 74- Mem: Geol Soc Am; Soc Econ Paleont & Mineral; Am Asn Petrol Geol. Res: Paleogeography and sedimentation of western North America; paleocurrent and basin studies; general stratigraphy and stratigraphic paleontology. Mailing Add: Shell Int Petrol Co Ltd Shell Centre London SE 1 England

THOME, FREDERICK A, b Union, NJ, Oct 27, 34; m 61; c 1. ANALYTICAL CHEMISTRY. Educ: Upsala Col, BS, 57; Ohio Univ, MS, 59. Prof Exp: Sr chemist, Stand Oil Co, Ohio, 61-64; RES CHEMIST, R J REYNOLDS TOBACCO CO, 64- Mem: Am Chem Soc; Sigma Xi. Res: Gas chromatography-mass spectrometry coupling; mass spectrometry of natural products. Mailing Add: R J Reynolds Tobacco Co Res Dept 115 Chestnut St SE Winston-Salem NC 27102

THOME, GEORGE DURST, b Detroit, Mich, Feb 15, 36; m 60; c 2. RADIOPHYSICS, GEOPHYSICS. Educ: Antioch Col, BS, 59; Cornell Univ, MS, 62, PhD(elec eng), 66. Prof Exp: Engr radio physics, Raytheon Corp, 59-61; res assoc ionospheric physics, Cornell Univ, 66-68; SCIENTIST, RADIO PHYSICS, RAYTHEON CORP, 68- Concurrent Pos: Mem comn III, Union Radio Sci. Mem: Am Geophys Union. Res: Man-made and natural ionospheric disturbances and radar techniques for studying them; traveling ionospheric disturbances; auroral irregularities; chemical releases; antenna arrays; computer graphics. Mailing Add: Raytheon Corp Equip Div Wayland MA 01778

THOMEIER, SIEGFRIED, b Aussig, Czech, Dec 19, 37; Ger citizen. MATHEMATICS, TOPOLOGY. Educ: Univ Frankfurt, Dipl Math, 62, Dr Phil Nat(math), 65. Prof Exp: Sci asst math, Frankfurt Univ, 63-65; assoc prof, Math Inst, Aarhus, Denmark, 65-68; PROF MATH, MEM UNIV NFLD, 68- Concurrent Pos: Nat Res Coun Can res grant, Mem Univ Nfld, 69-; vis prof math, Univ Konstanz, Ger, 75-76. Mem: AAAS; Am Math Soc; London Math Soc; Ger Math Soc; Math Soc Belg. Res: Homotopy theory; homotopy groups of special topological spaces; homology theory; homological and categorical algebra; number theory and combinatorial mathematics; applications of topology. Mailing Add: Dept of Math Mem Univ of Nfld St John's NF Can

THOMERSON, JAMIE E, b Ft McKavett, Tex, May 7, 35; m 57; c 2. ICHTHYOLOGY, ZOOLOGY. Educ: Univ Tex, BS, 57; Tex Tech Col, MS, 61; Tulane Univ, PhD(zool), 65. Prof Exp: High sch teacher, 58-59; asst prof zool, 65-69, ASSOC PROF ZOOL, SOUTHERN ILL UNIV, EDWARDSVILLE, 69- Mem: Am Soc Ichthyologists & Herpetologists; Am Fisheries Soc; Asn Trop Biol; Soc Syst Zool. Res: Fish systematics, ecology, behavior and genetics. Mailing Add: Dept of Biol Sci Southern Ill Univ Edwardsville IL 62025

THOMISON, JOEL DOUGLAS, b Pendleton, Ore, Oct 19, 45; m 69. MATHEMATICS. Educ: Yale Univ, BS, 68; Univ Md, College Park, MA, 71, PhD(math), 74. Prof Exp: Mathematician, Naval Res Lab, 68-73; ASST PROF MATH, ITHACA COL, 75- Mem: Am Math Soc; Math Asn Am. Res: Investigate methods of completing locally convex function spaces in such a way that the completion is a function space with original domain. Mailing Add: Ithaca Col Ithaca NY 14850

THOMMARSON, RONALD L, physical chemistry, see 12th edition

THOMMES, GLEN ANTHONY, physical chemistry, see 12th edition

THOMMES, ROBERT CHARLES, b Chicago, Ill, Aug 31, 28. ZOOLOGY. Educ: De Paul Univ, BS, 50, MS, 52; Northwestern Univ, PhD, 56. Prof Exp: From instr to assoc prof, 56-67, chmn dept, 68-70, PROF BIOL, DE PAUL UNIV, 67- Mem: AAAS; Soc Zool; Soc Develop Biol; Soc Exp Biol & Med. Res: Developmental endocrinology. Mailing Add: Dept of Biol Sci De Paul Univ Chicago IL 60614

THOMPKINS, LEON, b Augusta, Ga, Nov 4, 36; m 62; c 2. PHARMACEUTICAL CHEMISTRY. Educ: Morehouse Col, BS, 61; Univ Calif, San Francisco, PhD(pharmaceut chem), 68. Prof Exp: Chemist, Hyman Labs, Fundamental Res Co, Inc, 62-63; SR PHARMACEUT CHEMIST, ELI LILLY & CO, 68- Mem: Am Pharmaceut Asn; Am Chem Soc; Sigma Xi; AAAS; NY Acad Sci. Res: Development and bioavailability of human drug dosage forms. Mailing Add: Lilly Res Lab Dept MC747 307 E McCarty St Indianapolis IN 46206

THOMPSON, ALAN BRUCE, b Newcastle-upon-Tyne, Eng, May 1, 47; m 69. PETROLOGY. Educ: Manchester Univ, BSc, 68, PhD(geol), 71. Prof Exp: Lectr geol, Manchester Univ, 70-73; ASST PROF GEOL, HARVARD UNIV, 73- Concurrent Pos: Res fel geol, Harvard Univ, 72. Mem: Fel Mineral Soc Am; Am Geophys Union; Int Geochem Soc. Res: Kinetics and rates of geological processes; petrological constraints on mass and heat transfer in the Earth's crust and upper mantle; application of thermodynamics to petrology; fluids in the Earth's crust, their origin and motion during metamorphism. Mailing Add: Dept of Geol Sci Harvard Univ 20 Oxford St Cambridge MA 02138

THOMPSON, ALAN MORLEY, b Omaha, Nebr, Sept 2, 25; m 50; c 2. PHYSIOLOGY. Educ: Iowa State Univ, BS, 49; Univ Minn, PhD(physiol), 56. Prof Exp: Asst physiol, Univ Minn, 52-54, teaching intern, Ford Found, 54-55, asst, Univ, 55, instr, 55-56; res fel physiol, 56-57; instr, Univ Tex Southwestern Med Sch Dallas, 58-59; asst prof zool, Iowa State Univ, 59-62; assoc prof, 62-66, PROF PHYSIOL, UNIV KANS MED CTR, KANSAS CITY, 66-, ASSOC DEAN GRAD SCH, 70- Mem: Am Physiol Soc; Soc Neurosci. Res: Transport of materials; cerebral blood flow and permeability; tissue carbon dioxide, electrolytes and acid-base balance. Mailing Add: Dept of Physiol Univ of Kans Med Ctr Kansas City KS 66103

THOMPSON, ALBERT JOHNSON, JR, b Graham, NC, Nov 15, 08; m 33; c 2. CHEMISTRY. Educ: Davidson Col, BS, 30; NC State Univ, BS, 31. Prof Exp: Control chemist, Chem Control Div, Merck & Co, Inc, 34-37, control chemist & supvr, 37-46, chief label & regist sect label & govt dept, 46-48, mgr govt liaison dept, 48-51, qual testing dept, 51-54, stand & specif dept, 54-56 & stands & anal res, 56-65; CHEMIST, BUR DRUGS, FOOD & DRUG ADMIN, 66- Mem: Am Chem Soc; Am Pharmaceut Asn; Acad Pharmaceut Sci. Res: Review and evaluation of manufacturing controls; technical aspects of labeling. Mailing Add: 5919 Sherborn Lane Springfield VA 22152

THOMPSON, ALLAN M, b Ithaca, NY, May 22, 40; m 66. PETROLOGY, SEDIMENTOLOGY. Educ: Carleton Col, BA, 62; Brown Univ, ScM, 64, PhD(stratig), 68. Prof Exp: Asst prof, 67-72, ASSOC PROF GEOL & GEOCHEM, UNIV DEL, 72- Concurrent Pos: Assoc geologist, Del Geol Surv, 72- Mem: Geol Soc Am; Am Asn Petrol Geol; Soc Econ Paleont & Mineral. Res: Sedimentology, plutonic petrology and structure; stratigraphy, structure and petrology of Appalachian orogen. Mailing Add: Dept of Geol Univ of Del Newark DE 19711

THOMPSON, ALONZO CRAWFORD, b Tifton, Ga, June 4, 28; m 55; c 3. ORGANIC CHEMISTRY, PHARMACEUTICAL CHEMISTRY. Educ: Berry Col, AB, 53; Univ Miss, MS, 55, PhD(chem), 62. Prof Exp: Prof chem, Miss Delta Jr Col, 55-56 & Southern State Col, 56-58; RES CHEMIST, BOLL WEEVIL RES LAB, USDA, 62- Concurrent Pos: Adj asst prof biochem, Miss State Univ, 74- Mem: Am Chem Soc; Am Inst Chemists. Res: Pharmaceutical synthesis; natural products; insects and plant stimulants. Mailing Add: Boll Weevil Res Lab USDA Box 5367 Mississippi State MS 39762

THOMPSON, ANSON ELLIS, b Eugene, Ore, Apr 9, 24; m 45; c 4. PLANT BREEDING, PLANT GENETICS. Educ: Ore State Univ, BS, 48; Cornell Univ, PhD(plant breeding), 52. Prof Exp: Lab instr bot, Ore State Univ, 47-48; asst plant breeding, Cornell Univ, 48-51; from instr to assoc prof veg crops, Univ Ill, Urbana, 51-63, prof plant genetics, 63-71, asst dir agr exp sta, 67-69; prof hort & head dept hort & landscape archit, 71-75, PROF HORT & HORTICULTURIST, DEPT PLANT SCI, UNIV ARIZ, 75- Concurrent Pos: Univ Ky-Int Coop Admin vis prof, Univ Indonesia, 58-60; res admin consult, Univ Ill-AID Col Contract Team, Uttar Pradesh Agr Univ, India, 67 & party chief & adv, 67-69. Honors & Awards: Woodbury Award, Am Soc Hort Sci, 65, Asgrow Award, 66. Mem: Fel AAAS; fel Am Soc Hort Sci; Am Hort Soc; Int Soc Hort Sci; Sigma Xi. Res: Vegetable breeding and genetics; inheritance and breeding methods for disease resistance and quality constituents; domestication of xerophytic gourd species; breeding; selection; native plant materials for landscaping in arid lands. Mailing Add: Dept of Plant Sci Univ Ariz Tucson AZ 85721

THOMPSON, ANTHONY C, b Middlesbrough, Eng, Apr 4, 37; m 63; c 3. MATHEMATICS. Educ: Univ London, BSc, 59; Univ Newcastle, PhD(math), 63. Prof Exp: Instr math, Yale Univ, 62-63; lectr pure math, Univ Col Swansea, Wales, 63-66; asst prof math, 66-69, ASSOC PROF MATH, DALHOUSIE UNIV, 69- Concurrent Pos: Nat Res Coun res grant, 67-72. Mem: Am Math Soc; Can Math Cong; London Math Soc. Res: Functional analysis. Mailing Add: Dept of Math Dalhousie Univ Halifax NS Can

THOMPSON, ANTHONY RICHARD, b Hull, Yorkshire, Eng, Apr 7, 31; m 63; c 1. RADIO ASTRONOMY. Educ: Univ Manchester, BSc, 52, PhD, 56. Prof Exp: Electronic engr, Elec & Musical Instruments Ltd, Eng, 55-57; res assoc, Harvard Col Observ, 57-61, res fel, 61-62; radio astronr, Stanford Univ, 62-70, sr res assoc, 70-72; VLA proj engr, 73-74, DEP MGR VLA PROJ, NAT RADIO ASTRON OBSERV, 75- Concurrent Pos: Vis sr res fel, Calif Inst Technol, 66-71, vis assoc, 71-72. Mem: Int Astron Union; Int Union Radio Sci; Am Astron Soc. Res: Structure of cosmic radio sources, solar radio astronomy; design of radio telescopes. Mailing Add: Nat Radio Astron Observ PO Box O Socorro NM 87801

THOMPSON, ARTHUR CARSTEN, b Detroit, Mich, Sept 21, 19; m 50; c 3. PHYSICAL CHEMISTRY. Educ: Wayne State Univ, BS, 46, MS, 49; Wash State Univ, 56. Prof Exp: From instr to asst prof phys chem, Univ Idaho, 49-57; sr res chemist, Nalco Chem Co, 57-62; ASSOC PROF CHEM, MARIETTA COL, 62- Mem: Am Chem Soc; Sigma Xi. Res: Physical chemistry of polymer solutions; collodal clays; silicates; coagulation; acidic properties of bentonite. Mailing Add: Dept of Chem Marietta Col Marietta OH 45750

THOMPSON, ARTHUR HOWARD, b Duluth, Minn, June 15, 18; m 45; c 4. POMOLOGY. Educ: Univ Minn, BS, 41; Univ Md, PhD, 45. Prof Exp: Asst pomologist, USDA, Wash, 45-48, assoc horticulturist, 48-50, horticulturist, WVa, 50-52; PROF POMOL, UNIV MD, COLLEGE PARK, 52- Mem: Am Soc Hort Sci. Res: Chemical thinning of apples and peaches; chemical control of preharvest drop; fruit tree nutrition. Mailing Add: Dept of Hort Univ of Md College Park MD 20742

THOMPSON, ARTHUR HOWARD, b Salt Lake City, Utah, Mar 3, 42; m 65. SOLID STATE PHYSICS. Educ: Ohio State Univ, BSc & MSc, 66; Stanford Univ, PhD(physics), 70. Prof Exp: Fel, Stanford Univ, 70; group leader physics, Syva Co, Calif, 70-71; SR RES PHYSICIST, EXXON RES & ENG CO, 71- Mem: Am Phys Soc. Res: Super conductivity; semiconductors; transport and magnetic properties. Mailing Add: Exxon Res & Eng Co PO Box 45 Linden NJ 07036

THOMPSON, AYLMER HENRY, b Ill, Sept 11, 22; m 41; c 3. METEOROLOGY. Educ: Univ Calif, Los Angeles, MA, 48, PhD(meteorol), 60. Prof Exp: Lectr meteorol, Univ Calif, Los Angeles, 48-52; asst prof, Univ Utah, 52-60; PROF METEOROL, TEX A&M UNIV, 60- Concurrent Pos: Sci adv Found Glacier & Environ Res, Wash. Mem: Am Meteorol Soc; Am Geophys Union; foreign mem Royal Meteorol Soc; Int Glaciol Soc; Sigma Xi. Res: Synoptic meteorology of sub-tropics; satellite meteorology; inversions; meteorology of glaciated regions. Mailing Add: Dept of Meteorol Tex A&M Univ College Station TX 77843

THOMPSON, BOBBY BLACKBURN, b Lumber City, Ga, May 15, 33; m 59; c 1. MEDICINAL CHEMISTRY, ORGANIC CHEMISTRY. Educ: Berry Col, BA, 55; Univ Miss, MS, 56, PhD(org med chem), 63. Prof Exp: Asst prof, 60-70, ASSOC PROF MED CHEM, SCH PHARM, UNIV GA, 70- Mem: Am Chem Soc; Am Pharmaceut Asn. Res: Synthesis of organic and heterocyclic compounds of potential medicinal value; structural elucidation by chemical and instrumental means. Mailing Add: Sch of Pharm Univ of Ga Athens GA 30601

THOMPSON, BONNIE CECIL, b Baird, Tex, Dec 18, 35; m 59; c 1. SOLID STATE PHYSICS. Educ: North Tex State Univ, BA, 57, MA, 58; Univ Tex, PhD(physics), 65. Prof Exp: Instr physics, North Tex State Univ, 58-60; teaching asst, Univ Tex, Austin 60-61, res scientist, 61-65; asst prof, 65-74, ASSOC PROF PHYSICS, UNIV TEX, ARLINGTON, 74- Mem: Am Phys Soc. Res: Nuclear spin-lattice relaxation processes in solids; nuclear magnetic resonance of biological molecules. Mailing Add: Dept of Physics Univ of Tex Arlington TX 76019

THOMPSON, BRIAN J, b Glossop, Eng, June 10, 32; m 56; c 2. OPTICS. Educ: Univ Manchester, BScTech, 55, PhD(physics), 59. Prof Exp: Demonstr physics fac tech, Univ Manchester, 55-56, asst lectr, 57-59; lectr, Leeds Univ, 59-62; sr physicist, Tech Opers, Inc, 63-65, mgr phys optics dept, 65-66, dir optics dept, 66-67, mgr tech opers west & tech dir, Beckman & Whitley Div, 67-68; PROF OPTICS & DIR INST OPTICS, UNIV ROCHESTER, 68-, DEAN COL ENG & APPL SCI, 74- Concurrent Pos: Adj prof, Northeastern Univ, 66-67. Mem: Am Phys Soc; fel Optical Soc Am; fel Brit Inst Physics & Phys Soc; Soc Photo-Optical Instrumentation Engrs (pres, 74-75, 75-76). Res: Diffraction; interference; partial coherence; holography; application to particle sizing; optical data processing; x-ray diffraction; soft x-ray spectroscopy; optical properties of materials. Mailing Add: Col of Eng & Appl Sci Univ of Rochester Rochester NY 14627

THOMPSON, BUFORD DALE, b Lake Wales, Fla, Oct 22, 22; m 44; c 3. HORTICULTURE, VEGETABLE CROPS Educ: 0Univ Fla, BSA, 48, MSA, 49, PhD(hort), 54. Prof Exp: Asst prof veg crops & asst horticulturist, 49-60, assoc prof & assoc horticulturist, 60-66, PROF VEG CROPS & HORTICULTURIST, AGR EXP STA, UNIV FLA, 66- Honors & Awards: Vaughan Award, Am Soc Hort Sci, 62. Mem: Am Soc Hort Sci; Am Soc Plant Physiol. Res. Biological and chemical changes involved in the post harvest handling, transportation and storage of horticultural crops. Mailing Add: 725 NW 40th Terr Gainesville FL 32601

THOMPSON, CARL EUGENE, b Lucinda, Pa, June 14, 41; m 64; c 4. ANIMAL BREEDING, ANIMAL GENETICS. Educ: Pa State Univ, University Park, BS, 63, MS, 68; Va Polytech Inst & State Univ, PhD(animal breeding & genetics), 71. Prof Exp: Asst county agr agent, Agr Exten Serv, Pa State Univ, 63-66; asst prof animal sci, Ft Hays Kans State Col, 71-73; ASST PROF BEEF CATTLE BREEDING, CLEMSON UNIV, 74- Mem: Am Soc Animal Sci; Am Genetics Asn; Sigma Xi. Res: Beef cattle breeding and genetics; cross-breeding; genetic-environmental interaction; reproductive physiology. Mailing Add: Dept of Animal Sci Clemson Univ Clemson SC 29631

THOMPSON, CAROL, physics, see 12th edition

THOMPSON, CHARLES CALVIN, b Los Angeles, Calif, May 4, 35; m 60; c 2. ORAL PATHOLOGY. Educ: St Martin's Col, BA, 57; Univ Ore, DMD, 62; Emory Univ, MSD, 68. Prof Exp: Pvt dent pract, Ore, 64-66; resident path, Dent Sch, Emory Univ, 66-68; asst prof oral diag & med, Dent Sch, Univ Calif, Los Angeles, 68-69; ASSOC PROF PATH, DENT SCH & ASST PROF DENT, MED SCH, UNIV ORE, 69- Concurrent Pos: Nat Inst Dent Res fel, Emory Univ, 66-68; attend consult, Wadsworth Vet Admin Hosp, Los Angeles, 68-69; oral pathologist & clin consult, 71-; instr, Mt Hood Community Col, 72-73. Mem: Am Acad Oral Path; Am Soc Forensic Odontol; Int Soc Forensic Odontol-Stomatol. Res: Oncology, chemical, physical carcinogenesis; teratology, induction of developmental defects and explanation; bone, especially effects of dimethyl sulfoxide on bone. Mailing Add: Dept of Path Sch of Dent Univ of Ore Health Sci Ctr Portland OR 97201

THOMPSON, CHARLES DENISON, b Niagara Falls, NY, May 4, 40. CORROSION. Educ: Oberlin Col, BA, 62; Am Univ, MS, 68, PhD(chem), 71. Prof Exp: Chemist, US Army Environ Hyg Agency, 63-65; chemist biochem, Aldridge Assocs, 69; Welch fel, Rice Univ, 71-72; CORROSION ENGR, KNOLLS ATOMIC ENERGY LAB, GEN ELEC CO, 73- Mem: Electrochem Soc. Res: High temperature materials corrosion and electrochemistry. Mailing Add: Knolls Atomic Energy Lab G2-154 Schenectady NY 12301

THOMPSON, CHARLES FREDERICK, b Dayton, Ohio, Oct 1, 43; m 67. POPULATION ECOLOGY, ORNITHOLOGY. Educ: Ind Univ, BA, 67, MA, 70, PhD(zool), 71. Prof Exp: Res fel ecol, Univ Ga, 71-72, asst prof zool, 72-73; teaching fel zool, Miami Univ, 73-75; ASST PROF BIOL, STATE UNIV NY COL GENESEO, 75- Concurrent Pos: Vis asst prof zool, Ind Univ, 74. Mem: Ecol Soc Am; Brit Ecol Soc; Am Ornithologists Union; Brit Ornithologists Union; Neth Ornithologists Union. Res: Regulation and dynamics of bird populations; avian breeding adaptations; structure and evolution of avian social organization. Mailing Add: Dept of Biol State Univ NY Col Geneseo NY 14454

THOMPSON, CHARLES RAYMOND, b Ft Thomas, Ky, Feb 16, 16; m 39; c 2.

PHARMACOLOGY. Educ: Univ Cincinnati, BA, 37, MA, 40. Prof Exp: Asst physiol, Univ Cincinnati, 38-40; pharmacologist, Wm S Merrell Co, 40-43, pharmacologist & head toxicol & endocrinol, 43-58; dir pharmacol res, 58-71, dir exp ther, 71-73, dir qual assurance tech serv, 73-75, DIR CORP COMPLIANCE, CUTTER LABS, INC, 76- Mem: AAAS; Soc Exp Biol & Med; Endocrine Soc; Soc Study Reprod; NY Acad Sci. Res: Histology; sex hormones; endocrine pharmacology; toxicology; prosthetics research. Mailing Add: Qual Assurance Cutter Labs Inc Fourth & Parker St Berkeley CA 94710

THOMPSON, CHESTER RAY, b Storrs, Utah, May 27, 15; m 40; c 4. BIOCHEMISTRY. Educ: Utah State Univ, BS, 38; Univ Wis, MS, 41, PhD(biochem), 43. Prof Exp: Chemist, Rocky Mountain Packing Corp, Utah, 36-39; biochemist, Univ Wis, 39-43, Forest Prods Lab, US Forest Serv, 43 & Purdue Univ, 44-45; plant biochemist, Univ Chicago, 45-49; head forage invest, Field Crops Lab, Western Utilization Res & Develop Div, USDA, 49-60; proj leader, 60-67, RES BIOCHEMIST, STATEWIDE AIR POLLUTION RES CTR, AGR AIR RES PROG, UNIV CALIF, RIVERSIDE, 67- Mem: AAAS; Am Soc Plant Physiol; Am Chem Soc. Res: Chemical stabilization of carotene in alfalfa; occurrence of anti-oxidants in natural products; carotenoids in green plants; saponins and estrogens in forages; effects of air pollution on plants and human beings. Mailing Add: Statewide Air Pollution Res Ctr Univ of Calif Riverside CA 92502

THOMPSON, CLARENCE GARRISON, b Corvallis, Ore, Nov 3, 18. ENTOMOLOGY. Educ: Ore State Col, BS, 40; Univ Calif, MS, 47, PhD, 50. Prof Exp: Jr insect pathologist, Univ Calif, 50-51, asst insect pathologist, 51-53; insect pathologist, 53-70, ENTOMOLOGIST, US FOREST SERV & PROF ENTOM, AGR EXP STA, ORE STATE UNIV, USDA, 70- Res: Insect pathology. Mailing Add: US Forest Serv Ore State Univ Corvallis OR 97331

THOMPSON, CLARENCE HENRY, JR, b Perry, Kans, May 4, 18; m 42; c 3. VETERINARY MEDICINE. Educ: Kans State Col, DVM, 41, MS, 47. Prof Exp: Asst vet private vet hosp, 41; jr vet tuberc eradication, Bur Animal Indust, 41-46, vet pathologist path div, 47-54, adminstr state exp sta div, Agr Res Serv, 54-63, asst to dir animal disease & parasite res div, 63-71, STAFF ASST TO ADMINR, AGR RES SERV, USDA, 71- Mem: Am Vet Med Asn. Res: Animal diseases, especially virus diseases of poultry, in fields of epizoology, bacteriology, pathology and immunology. Mailing Add: 6203 87th Ave Hyattsville MD 20784

THOMPSON, CLIFFORD FRANCIS, b Washington, DC, Aug 30, 32; m 56; c 5. ORGANIC POLYMER CHEMISTRY. Educ: Univ Md, BS, 55; Univ Ill, PhD(org chem), 58. Prof Exp: Res chemist, Dow Chem Co, 58-63, proj leader polymer chem, 63, mkt analyst, 64, mgr mkt res, 64-65, mgr res & develop, Dow Smith Inc, 65-67, tech dir chem, plastics & metals, Dow Chem Europe SA, 67-71, DIR RES & DEVELOP, WESTERN DIV, DOW CHEM USA, 71- Mem: Am Chem Soc Plastics Indust. Res: Exploratory research; process development; product development; manufacturing technical support for organic chemicals; inorganics; plastics; resins; membranes; medical devices; electro chemicals. Mailing Add: Walnut Creek Res Ctr Dow Chem USA 2800 Mitchell Dr Walnut Creek CA 94598

THOMPSON, CLIFTON C, b Franklin, Tenn, Aug 16, 39; m 59; c 1. PHYSICAL CHEMISTRY. Educ: Middle Tenn State Univ, BS, 61; Univ Miss, PhD(phys & inorg chem), 64. Prof Exp: Res assoc, Univ Tex, 64-65; asst prof spectrochem, Rutgers Univ, 65; asst prof chem, Marshall Univ, 65-66; assoc prof, Mid Tenn State Univ, 66-68; from asst prof to assoc prof, Memphis State Univ, 68-74; PROF CHEM, SOUTHWEST MO STATE UNIV, 74-, DEAN, SCH SCI & TECHNOL, 74- Concurrent Pos: Lectr, Kanawha Valley grad ctr, WVa Univ, 66; NSF acad year exten grant, 66-68. Mem: Am Chem Soc; The Chem Soc. Res: Spectral, thermodynamic and kinetic studies of molecular complexes; quantum chemistry; computer applications to physical systems. Mailing Add: Dept of Chem Southwest Mo State Univ Springfield MO 65802

THOMPSON, CRAYTON BEVILLE, b Paris, Tex, Dec 28, 20; m 55; c 1. ORGANIC CHEMISTRY. Educ: Univ Tex, BS, 42; Univ Ill, MS, 47, PhD(chem), 49. Prof Exp: Chem engr, Freeport Sulphur Co, 42-46; lab asst chem, Univ Ill, 47-49; RES & DEVELOP CHEMIST, EASTMAN KODAK CO, 49- Mem: Am Chem Soc. Res: Synthesis of amino acids; antihalation backings for photographic films; abrasion resistant and anitstatic applications for plastics; adhesion. Mailing Add: Bldg 7 Kodak Park Eastman Kodak Co Rochester NY 14650

THOMPSON, DANIEL JAMES, b Terre Haute, Ind, Feb 10, 42; m 65; c 2. TERATOLOGY, TOXICOLOGY. Educ: Ind State Univ, BS, 64, MA, 66. Prof Exp: SR RES TOXICOLOGIST, DOW CHEM CO, 66- Mem: Teratology Soc; Environ Mutagen Soc; Soc Toxicol. Res: Reproductive physiology; perinatal toxicology. Mailing Add: Path-Toxicol Dept Dow Chem Co PO Box 68511 Indianapolis IN 46268

THOMPSON, DANIEL QUALE, b Madison, Wis, Oct 3, 18; m 53; c 4. WILDLIFE ECOLOGY, CONSERVATION. Educ: Univ Wis-Madison. BS, 42, MS, 50; Univ Mo, PhD(field zool), 55. Prof Exp: Instr field zool, Univ Mo, 50-54; from asst prof to assoc prof biol, Ripon Col, 55-62; WILDLIFE RES BIOLOGIST, US BUR SPORT FISHERIES & WILDLIFE & LEADER NY COOP WILDLIFE RES UNIT, CORNELL UNIV, 62- Concurrent Pos: Mem grad fel panel, NSF, 71-72; ed, J Wildlife Mgt, 73-74. Mem: Am Soc Mammal; Wildlife Soc; Wilson Ornith Soc; Ecol Soc Am. Res: Wildlife conservation; ecology of terrestrial vertebrates. Mailing Add: Dept of Natural Resources Fernow Hall Cornell Univ Ithaca NY 14850

THOMPSON, DAVID BRUCE, mathematics, computer science, see 12th edition

THOMPSON, DAVID DUVALL, b Ithaca, NY, June 1, 22; m 45; c 4. INTERNAL MEDICINE. Educ: Cornell Univ, BA, 43, MD, 46. Prof Exp: Intern & resident med, New York Hosp, 46-50; res fel physiol, Med Col, Cornell Univ, 50-51; instr physiol, Med Col, Cornell Univ, 51-53; resident physician, NIH, 53-55; asst prof physiol, 55-57, assoc prof med, 57-64, PROF MED, MED COL, CORNELL UNIV, 64-; ATTEND PHYSICIAN, DEPT MED, NEW YORK HOSP, 64-, DIR, 67- Concurrent Pos: Clin instr, Sch Med, George Washington Univ, 54-55; resident physician, NIH, 55-56; Lederle award, 55-57; from asst attend to actg physician-in-chief, Dept Med, New York Hosp, 57-67; asst vis physician, 2nd Med Div, Bellevue Hosp, 57-66, assoc vis physician, 66-; attend physician, Vet Admin Hosp, 58-; mem med adv bd, Kidney Found NY, 62-; actg chmn dept med, Cornell Univ, 65-67. Mem: Am Soc Clin Invest; Am Physiol Soc; Harvey Soc; Am Fedn Clin Res; Am Col Physicians. Res: Renal and electrolyte physiology. Mailing Add: New York Hosp 525 E 68th St New York NY 10021

THOMPSON, DAVID J, b Danville, Ind, Apr 17, 34; m 68; c 2. GENETICS, PLANT BREEDING. Educ: Univ Idaho, BS, 54, MS, 56; Cornell Univ, PhD(plant breeding), 60. Prof Exp: Co-geneticist, 60-63, RES DIR GENETICS & PLANT BREEDING, FERRY-MORSE SEED CO, 63- Mem: Int Soc Hort Sci; Am Soc Hort Sci. Res: Genetics, cytology and physiology of male-sterility and self-incompatibility in plants. Mailing Add: Ferry-Morse Seed Co San Juan Bautista CA 95045

THOMPSON, DAVID JOHN, b Cincinnati, Ohio, Jan 11, 45; m 72. ASTROPHYSICS. Educ: Johns Hopkins Univ, BA, 67; Univ Md, PhD(physics), 73. Prof Exp: Res assoc physics, Univ Md, 73; PHYSICIST, GODDARD SPACE FLIGHT CTR, NASA, 73- Mem: Am Phys Soc. Res: Gamma ray astronomy and its relationship to cosmic ray physics and other aspects of astrophysics; high energy gamma ray detectors, particularly spark chambers. Mailing Add: Code 662 Goddard Space Flight Ctr NASA Greenbelt MD 20771

THOMPSON, DAVID WALLACE, b Chicago, Ill, Jan 27, 42; m 63; c 2. INORGANIC CHEMISTRY, ORGANOMETALLIC CHEMISTRY. Educ: Wheaton Col, BS, 63; Northwestern Univ, Evanston, PhD(chem), 68. Prof Exp: Asst prof, 67-70, ASSOC PROF CHEM, COL WILLIAM & MARY, 70- Mem: Am Chem Soc. Res: Coordination chemistry of group IV elements; use of transition elements to catalyze organic reactions. Mailing Add: Dept of Chem Col of William & Mary Williamsburg VA 23185

THOMPSON, DON DEAN, physical chemistry, see 12th edition

THOMPSON, DONALD ENRIQUE, b Chicago, Ill, Jan 2, 31; m 57; c 2. ANTHROPOLOGY, ARCHAEOLOGY. Educ: Harvard Univ, AB, 53, MA, 61, PhD(anthrop), 62. Prof Exp: Field res asst archaeol, Field Mus Natural Hist, NMex, 50; field res asst, Carnegie Inst Washington, Mex, 53; field res asst, Field Mus Natural Hist, Peru, 56; Fulbright fel, San Marcos Univ, Lima, 59-60; instr anthrop, Univ Wis, 61-62; from asst prof to assoc prof, 62-72, PROF ANTHROP, UNIV WIS-MADISON, 72-, CHMN DEPT, 73- Concurrent Pos: NSF fels, Peru, 64-65, 66-67 & 70-73. Mem: AAAS; Am Anthrop Asn; Soc Am Archaeol; Archaeol Inst Am; Am Ethnol Soc. Res: Archaeology, ethnology and ethnohistory of Latin America, especially the Central Andean area and to a lesser extent Mesoamerica; late Prehispanic villages in the Eastern Andes. Mailing Add: Dept of Anthrop Social Sci Bldg Univ of Wis-Madison Madison WI 53706

THOMPSON, DONALD F, organic chemistry, see 12th edition

THOMPSON, DONALD LEO, b Keota, Okla, Dec 31, 43; m 65; c 3. PHYSICAL CHEMISTRY. Educ: Northeastern State Col, BS, 65; Univ Ark, Fayetteville, PhD(phys chem), 70. Prof Exp: Res assoc theoret chem, Univ Calif, Irvine, 70-71; MEM STAFF THEORET CHEM, LOS ALAMOS SCI LAB, 71- Concurrent Pos: Vis assoc prof physics, Univ Miss, 75-76. Mem: Am Chem Soc., Res: Theoretical molecular dynamics; reaction kinetics and intermolecular energy transfer. Mailing Add: Los Alamos Sci Lab PO Box 1663 Los Alamos NM 87544

THOMPSON, DONALD LEROY, b Highland Park, Mich, Nov 15, 32; m 62; c 1. HEALTH PHYSICS. Educ: City Col New York, BS, 61; Long Island Univ, MS, 66; St Johns Univ, PhD(phys chem), 72. Prof Exp: Sr scientist med physics, Radiation Physics Lab, State Univ NY Downstate Med Ctr, 62-65, asst dir, 65-69, co-dir, 69-72; actg dep dir, Div Radioactive Mat & Nuclear Med, 74-75, health physicist, 72-74, BR CHIEF HEALTH PHYSICS, RADIOACTIVE MAT BR, BUR RADIOL HEALTH, 75- Concurrent Pos: Instr radiol, State Univ NY, 65-69, asst prof radiol sci, 69-72; instr, Found Advan Educ Sci, 73- Mem: Am Asn Physicists in Med; Health Physics Soc; Am Pub Health Asn. Res: Radiation exposures related to medical and consumer products. Mailing Add: Bur of Radiol Health 5600 Fishers Lane Rockville MD 20852

THOMPSON, DONALD LORAINE, b SDak, Feb 10, 21; m 49; c 1. AGRONOMY. Educ: SDak State Col, BS, 47, MS, 49; Iowa State Col, PhD(corn breeding), 53. Prof Exp: Asst small grain breeding, SDak State Col, 47-49; corn breeding, Iowa State Col, 49-52; PROF CROP SCI & RES AGRONOMIST, USDA, NC STATE UNIV, 52- Mem: Am Soc Agron. Res: Practical and theoretical aspects of corn breeding relating to quantitative genetics; disease resistance; relationships among economic traits; maximum production; forage evaluation. Mailing Add: Dept of Crop Sci & Genetics NC State Univ PO Box 5126 Raleigh NC 27607

THOMPSON, DONALD OSCAR, b Clear Lake, Iowa, Feb 27, 27; m 46; c 3. SOLID STATE PHYSICS. Educ: Univ Iowa, BA, 49, MS, 50, PhD, 53. Prof Exp: Physicist, Union Carbide Nuclear Co, 56-64; ASSOC DIR, N AM ROCKWELL SCI CTR, 64- Mem: Fel Am Phys Soc; Am Inst Mining, Metall & Petrol Eng. Res: Radiation damage in metals, particularly interaction of radiation-produced defects and dislocations; anharmonic and nonlinear effects in solids; materials research in support of nondestructive testing; nondestructive testing apparatus. Mailing Add: NAm Rockwell Sci Ctr 1049 Camino Dos Rios Thousand Oaks CA 91360

THOMPSON, DONOVAN JEROME, b Stoughton, Wis, Jan 30, 19; m 42; c 2. BIOSTATISTICS. Educ: St Olaf Col, BA, 41; Univ Minn, MA, 47; Iowa State Col, PhD(statist), 52. Prof Exp: Mem staff dept math, Univ Minn, 46-47 & Statist Lab, Iowa State Col, 47-53; prof biostatist, Grad Sch Pub Health, Univ Pittsburgh, 53-66; PROF BIOSTATIST, SCH PUB HEALTH & COMMUNITY MED, UNIV WASH, 66-, CHMN DEPT, 73- Mem: AAAS; Biomet Soc; Am Pub Health Asn; Am Statist Asn; Inst Math Statist. Res: Statistical theory and methodology. Mailing Add: Sch of Pub Health & Community Med Univ of Wash Seattle WA 98195

THOMPSON, DOUGLAS STUART, b Richmond, Calif, Dec 5, 39; m 70. PHYSICAL CHEMISTRY. Educ: Univ Calif, Berkeley, BS, 61; Mass Inst Technol, PhD(phys chem), 65. Prof Exp: NIH trainee, Univ Calif, San Diego, 66-67; RES PHYS CHEMIST, E I DU PONT DE NEMOURS & CO, INC, 68- Mem: AAAS; Am Phys Soc. Res: Characterization of macromolecules; inelastic light scattering; electron spin resonance; nuclear magnetic resonance; dye binding; trace gas analysis; ion molecule reactions. Mailing Add: Eng Physics Lab Du Pont Exp Sta E I du Pont de Nemours & Co Inc Wilmington DE 19898

THOMPSON, EDWARD IVINS BRADBRIDGE, b Burlington, Iowa, Dec 20, 33; m 57; c 2. MOLECULAR BIOLOGY, CELL BIOLOGY. Educ: Rice Inst, BA, 55; Harvard Med Sch, MD, 60. Prof Exp: Intern & resident med, Presby Hosp, Col Physicians & Surgeons, Columbia Univ, 60-62; res assoc neurochem, Lab Clin Sci, NIMH, 62-64, res scientist molecular biol, Lab Molecular Biol, Nat Inst Arthritis & Metab Dis, 64-69, sr res scientist molecular & cell biol, 69-73, HEAD SECT BIOCHEM GENE EXPRESSION, LAB BIOCHEM, NAT CANCER INST, 73- Mem: Am Chem Soc; Tissue Cult Asn; Am Soc Cell Biol; Am Soc Biol Chemists; Am Asn Cancer Res. Res: Endocrinology; regulation of gene expression in eukaryotic cells; mechanism of steroid hormone action; effects of steroids in malignant cells. Mailing Add: Lab of Biochem Bldg 37 Rm 4C09 Nat Cancer Inst Bethesda MD 20014

THOMPSON, EDWARD VALENTINE, b Sharon, Conn, Feb 6, 35; m 56; c 1. PHYSICAL CHEMISTRY. Educ: Cornell Univ, AB, 56; Polytech Inst Brooklyn, PhD(phys chem), 62. Prof Exp: Chemist, Am Cyanamid Co, 56-57, res chemist, 61-66; ASSOC PROF CHEM ENG, UNIV MAINE, ORONO, 66- Mem: Am Chem

Soc. Res: Polymer chemistry and physics; theory of viscoelasticity; thermodynamics. Mailing Add: Dept of Chem Eng Univ of Maine Orono ME 04473

THOMPSON, EMMANUEL BANDELE, b Zarla, Nigeria, Mar 15, 28. PHARMACOLOGY. Educ: Rockhurst Col, BS, 55; Univ Mo-Kansas City, BS, 59; Univ Nebr, Lincoln, MS, 63; Univ Wash, PhD(pharmacol), 66. Prof Exp: Hosp pharmacist, Univ Kans Med Ctr, 59-60; retail pharmacist, Cundiff Drug Store, 61; sr res pharmacologist, Baxter Labs Inc, Ill, 63-66; asst prof, 69-73, ASSOC PROF PHARMACOL, COL PHARM, UNIV ILL MED CTR, 73- Concurrent Pos: Univ Ill Grad Col grant, 69-70 & Exten, 70-71; USPHS grant, 72-74; prin res investr & consult, West Side Vet Admin Hosp, Chicago, 71- Mem: NY Acad Sci; Am Asn Cols Pharm; Am Pharmaceut Asn. Res: Cardiovascular pharmacology. Mailing Add: Dept of Pharmacol Univ of Ill Med Ctr Chicago IL 60612

THOMPSON, EMMETT FRANK, b El Reno, Okla, Nov 6, 36; m 61; c 3. FOREST ECONOMICS. Educ: Okla State Univ, BS, 58; NC State Univ, MS, 60; Ore State Univ, PhD(forest econ), 66. Prof Exp: From asst prof to prof forestry, Va Polytech Inst, 62-73; PROF FORESTRY & HEAD DEPT, MISS STATE UNIV, 73- Mem: Soc Am Foresters; Forest Products Res Soc. Res: Economics of forest resource management. Mailing Add: Dept of Forestry PO Drawer FD Miss State Univ Mississippi State MS 39762

THOMPSON, ERIC DOUGLAS, b Buffalo, NY, Mar 24, 34; m 60; c 3. SOLID STATE PHYSICS. Educ: Mass Inst Technol, SB & SM, 56, PhD(physics), 60. Prof Exp: NSF res fel, 62-63; from asst prof to assoc prof, 63-69, PROF ENG, CASE WESTERN RESERVE UNIV, 69- Concurrent Pos: Sr res assoc, Jet Propulsion Lab, 72-73. Mem: Am Phys Soc; Inst Elec & Electronics Eng. Res: Theory of magnetism; collective modes in ferromagnetic metals; magnetic neutron scattering; domain wall dynamics; solid state microwave active devices; Josephson Junction devices; thermal atomic scattering. Mailing Add: Elec Eng & Appl Physics Case Western Reserve Univ Cleveland OH 44106

THOMPSON, ERIC FONTELLE, JR, b Montgomery, Ala, Sept 29, 32; m 60; c 4. ZOOLOGY. Educ: Huntingdon Col, AB, 54; Univ Ga, MS, 56, PhD(zool), 59. Prof Exp: Instr, 59-61, asst prof, 61-75, ASSOC PROF BIOL, UNIV SC, 75- Concurrent Pos: Consult, Campbell Soup Co; SC Elec & Gas. Mem: Ecol Soc Am; Am Soc Ichthyologists & Herpetologists; Am Soc Zoologists. Res: Animal behavior; physiology of behavior; biological effects of magnetic fields; ecology. Mailing Add: Dept of Biol Univ of SC Columbia SC 29208

THOMPSON, ERNEST FREEMAN, oceanography, see 12th edition

THOMPSON, EVAN M, b Payson, Utah, Aug 7, 33; m 59; c 5. PHYSICAL ORGANIC CHEMISTRY. Educ: Brigham Young Univ, BA, 60, PhD(org chem), 65. Prof Exp: Charles F Kettering & Great Lakes Cols Asn teaching fel chem, Antioch Col, 64-65; from asst prof to assoc prof, 65-74, PROF CHEM, CALIF STATE COL, STANISLAUS, 74-, DEAN SCH NATURAL SCI, 70- Mem: AAAS; Am Chem Soc. Res: Organic reaction mechanisms; kinetics; chemical education. Mailing Add: Sch Natural Sci Calif State Col Stanislaus Turlock CA 95380

THOMPSON, FAY MORGEN, b St Paul, Minn, Dec 13, 35; m 55; c 2. OCCUPATIONAL HEALTH, ENVIRONMENTAL CHEMISTRY. Educ: Univ Minn, BA, 63, PhD(org chem), 70. Prof Exp: Instr chem, Macalester Col, 67-68; INSTR OCCUP HEALTH, SCH PUB HEALTH, UNIV MINN, 70- Mem: Am Indust Hyg Asn; Am Chem Soc; Sigma Xi. Res: Collection and analysis of selected air contaminants; high purity water systems; hazardous waste disposal. Mailing Add: 229 Kennard St St Paul MN 55106

THOMPSON, FRED G, b Cleveland, Ohio, Nov 13, 34; m 57; c 1. MALACOLOGY. Educ: Univ Mich, BS, 58; Wayne State Univ, MA, 61; Univ Miami, PhD(zool), 64. Prof Exp: Res scientist, Univ Miami, 64-66; interim assoc cur, 66-71, ASST CUR MALACOL, FLA STATE MUS, UNIV FLA, 71- Concurrent Pos: NIH res grant systs Amnicolidae, 64-67. Res: Parasitology. Mailing Add: Fla State Mus Univ of Fla Gainesville FL 32601

THOMPSON, FREDERIC CHRISTIAN, b Boston, Mass, Apr 24, 44; m 72. ENTOMOLOGY, EVOLUTIONARY BIOLOGY. Educ: Univ Mass, Amherst, BS, 66, PhD(entom), 69. Prof Exp: Med entomologist, Pac Prog, Smithsonian Inst, Washington, DC, 66-67; NY State Coun Arts curatorial fel & res fel, Am Mus Natural Hist, NY City, 72-74; RES ENTOMOLOGIST, SYST ENTOM LAB, IIBIII, AGR RES SERV, USDA, 74- Mem: AAAS; Entom Soc Am; Am Entom Soc; Soc Syst Zool. Res: Systematics and zoogeography of syrphid flies of the world; taxonomy of medically important flies. Mailing Add: System Entom Lab IIBIII Agr Res Serv USDA c/oUS Nat Mus Washington DC 20560

THOMPSON, FREDERICK NIMROD, JR, b Newport News, Va, Dec 9, 39; m 62; c 2. REPRODUCTIVE ENDOCRINOLOGY. Educ: Wake Forest Univ, BS, 61; Univ Ga, DVM, 65; Iowa State Univ, PhD(physiol), 73. Prof Exp: Instr physiol, Iowa State Univ, 67-73; ASST PROF PHYSIOL, UNIV GA, 73- Mem: Soc Study Reprod; Sigma Xi; Am Vet Med Asn. Res: Hormonal control of partarition and cardiovascular effects of adrenal corticosteroids. Mailing Add: Dept of Physiol & Pharmacol Univ of Ga Athens GA 30601

THOMPSON, GARY GENE, b Beach, NDak, Oct 18, 40; m 70. GEOLOGY, PALYNOLOGY. Educ: Univ NDak, BS, 62; Mich State Univ, PhD(geol), 69. Prof Exp: Geologist, Shell Develop Co, 68-69 & Shell Oil Co, 69-70; ASST PROF GEOL, SALEM STATE COL, 71- Mem: AAAS; Soc Econ Paleontologists & Mineralogists; Am Asn Stratig Palynologists. Res: Cretaceous, tertiary and quaternary palynomorph biostratigraphy and paleoecology. Mailing Add: Dept of Earth Sci Salem State Col Salem MA 01970

THOMPSON, GARY HAUGHTON, b Long Beach, Calif, Mar 25, 35; m 64; c 3. PHYSICAL CHEMISTRY, INORGANIC CHEMISTRY. Educ: Univ Colo, Boulder, BS, 60; Univ Utah, PhD(chem), 69. Prof Exp: Engr, Hercules Powder Co, 60-63; RES CHEMIST, SAVANNAH RIVER LAB, E I DU PONT DE NEMOURS & CO, INC, 69- Mem: Am Chem Soc. Res: Chromatography, including gas, liquid and ion exchange; radioiodine sorption, gas-solid and gas-liquid systems; radioactive waste management and solvent extraction processes. Mailing Add: Savannah River Lab E I du Pont de Nemours & Co Inc Aiken SC 29801

THOMPSON, GARY LYNN, b Olustee, Okla, Sept 24, 38; m 61; c 2. ECONOMIC GEOGRAPHY, GEOGRAPHY OF THE SOVIET UNION. Educ: Univ Okla, BA, 60, MA, 62; Mich State Univ, PhD(geog), 68. Prof Exp: Personnel technician, Okla State Personnel Bd, 60-62; instr geog, Mich State Univ, 65-66; asst prof, Univ Okla, 67-70; asst prof, Univ Miami, 70-71; ASSOC PROF GEOG, UNIV OKLA, 72- Concurrent Pos: Chmn geog sect, Okla Acad Sci, 68-69; vis asst prof geog, Western Wash State Col, 69; assoc ed, J Geog, 70-71. Mem: Asn Am Geogr; Am Asn Advan Slavic Studies; Regional Sci Asn. Res: Regional economic specialization. Mailing Add: Dept of Geog Univ of Okla Norman OK 73069

THOMPSON, GEOFFREY, b Stockton-on-Tees, Durham, Eng, Oct 18, 35; m 61; c 3. GEOCHEMISTRY, OCEANOGRAPHY. Educ: Univ Manchester, BSc, 61, PhD(geochem), 65. Prof Exp: Res chemist, Imp Chem Industs, UK, 58-59; geologist, Transvaal Gold Mines, SAfrica, 60; asst scientist, 65-70, ASSOC SCIENTIST, WOODS HOLE OCEANOG INST, 70- Concurrent Pos: Res assoc dept mineral sci, Smithsonian Inst, 70-; assoc ed, Geochimica et Cosmochmica Acta, 73- & J Marine Res, 74- Mem: AAAS; Geochem Soc; Am Geophys Union; Soc Appl Spectros. Res: Geochemistry of oceanic crust; origin, evolution, composition, geochemistry of ocean sediments and marine organisms. Mailing Add: Dept Chem Woods Hole Oceanog Inst Woods Hole MA 07543

THOMPSON, GEORGE ALBERT, b Swissvale, Pa, June 5, 19; m 44; c 3. GEOPHYSICS, GEOLOGY. Educ: Pa State Col, BS, 41; Mass Inst Technol, MS, 42; Stanford Univ, PhD(geol), 49. Prof Exp: Actg instr, 47-48, lectr, 48-49, from asst prof to assoc prof, 49-60, PROF GEOPHYS, STANFORD UNIV, 60-, CHMN DEPT, 67- Concurrent Pos: Geologist & geophysicist, US Geol Surv, 42-; NSF fel, 56-57; Guggenheim fel, 63-64; G K Gilbert award seismic geol, 64-; mem, Geodynamics Comt, Nat Res Coun, 75-78. Mem: AAAS; Soc Explor Geophys; Soc Econ Geol; Geol Soc Am; Am Geophys Union. Res: Structure and geophysics of Basin Range Province; crust-mantle structure; geophysics of ultramafic rocks; geology of quicksilver deposits. Mailing Add: Dept of Geophys Stanford Univ, Stanford CA 94305

THOMPSON, GEORGE REX, b Oakley, Idaho, July 24, 43; m 63; c 1. TOXICOLOGY, PHARMACOLOGY. Educ: Ore State Univ, BS, 65, PhD(toxicol & pharmacol), 69. Prof Exp: Res asst toxicol & pharmacol, Ore State Univ, 66-69; researcher toxicol, Mason Res Inst, Mass, 69-72; supvr toxicol, Biomed Res Lab, ICI Am Inc, 72-73; HEAD SECT GEN TOXICOL, ABBOTT LABS, 73- Res: Toxicity of marihuana or THC, cyclamate/cyclohexylamine, new drugs, anticancer compounds and pesticides; delineation of normal physiological parameters via the utilization of toxic materials. Mailing Add: Abbott Labs Abbott Park D-468 North Chicago IL 60064

THOMPSON, GEORGE RICHARD, b Ann Arbor, Mich, Apr 2, 30; m 57; c 3. INTERNAL MEDICINE, RHEUMATOLOGY. Educ: Univ Mich, Ann Arbor, BS, 50, MD, 54. Prof Exp: Intern, Ohio State Univ Hosp, 54-55; resident, Hosp, 55-58, from instr to asst prof, 62-69, ASSOC PROF INTERNAL MED, MED SCH, UNIV MICH, ANN ARBOR, 69-; DIR RHEUMATOL SECT, WAYNE COUNTY GEN HOSP, ELOISE, 63- Concurrent Pos: USPHS fel rheumatol, Rackham Arthritis Res Unit, Univ Mich Hosp, Ann Arbor, 60-62, assoc physician, 63- Mem: Am Rheumatism Asn; fel Am Col Physicians; Am Fedn Clin Res. Res: Arthritis; mucopolysaccharide metabolism; gout and urate excretion; rubella-associated arthritis. Mailing Add: Dept of Med Wayne County Gen Hosp Eloise MI 48132

THOMPSON, GERALD LEE, b Swea City, Iowa, Mar 16, 45; m 67; c 2. ORGANIC CHEMISTRY. Educ: Iowa State Univ, BS, 68; Ohio State Univ, PhD(chem), 72. Prof Exp: NIH fel, Harvard Univ, 72-74; SR ORG CHEMIST, ELI LILLY & CO, 74- Mem: Am Chem Soc. Res: Antitumor drug design; alkaloid synthesis. Mailing Add: Lilly Res Labs Eli Lilly & Co Indianapolis IN 46206

THOMPSON, GERALD LUTHER, b Rolfe, Iowa, Nov 25, 23; m 54; c 3. APPLIED MATHEMATICS. Educ: Iowa State Col, BS, 44; Mass Inst Technol, SM, 48; Univ Mich, PhD(math), 53. Prof Exp: Instr math, Princeton Univ, 51-53; asst prof, Dartmouth Col, 53-58; prof, Ohio Wesleyan, 58-59; PROF MATH & INDUST ADMIN, GRAD SCH INDUST ADMIN, CARNEGIE-MELLON UNIV, 59- Concurrent Pos: Consult, Princeton Univ, Int Bus Mach Corp, Sandia Corp, Beth Steel Corp & McKinsey & Co; Inst Mgt Sci rep, Math Div, Nat Res Coun, 71-73; mem, Sealift Readiness Comt, Nat Acad Sci, 74-75. Mem: Am Math Soc; Soc Indust & Appl Math; Economet Soc; Math Asn Am; Opers Res Soc Am. Res: Applications of mathematics to managerial and behavioral sciences; mathematical economics; control theory; graph theory and combinatorial problems; game theory. Mailing Add: Grad Sch of Indust Admin Carnegie-Mellon Univ Schenley Park Pittsburgh PA 15213

THOMPSON, GORDON WILLIAM, b Vancouver, BC. EPIDEMIOLOGY. Educ: Univ Alta, DDS, 65; Univ Toronto, MScD, 67, PhD(epidemiol & biostatist), 71. Prof Exp: ASSOC DEAN & ASSOC PROF DENT, UNIV TORONTO, 69- Mem: Biomet Soc; Int Asn Dent Res. Res: Biostatistical and computer applications in the fields of dental science and growth and development of humans with particular emphasis on the craniofacial complex and stomatognathic system. Mailing Add: Fac of Dent Univ of Toronto 124 Edward St Toronto ON Can

THOMPSON, GRANT, b Ogden, Utah, Feb 26, 27; m 49; c 5. ORGANIC CHEMISTRY. Educ: Univ Utah, BA, 50, PhD(chem), 53. Prof Exp: Res chemist, E I du Pont de Nemours & Co, 53-58; proj chemist, 58-59, sr chemist, 59-60, supvr new propellants sect, 60-62, mgr propellant develop dept, Wasatch Div, 63-75, MGR, RES & DEVELOP LABS, WASATCH DIV, THIOKOL CORP, 75- Mem: Am Chem Soc. Res: Mechanism of propellant cure; curing agents and catalysts for hydrocarbon propellants; mechanism of hydrocarbon propellant aging; high energy oxidizers and propellants. Mailing Add: Res & Develop Labs Wasatch Div Thiokol Corp PO Box 524 Brigham City UT 84302

THOMPSON, GRANVILLE BERRY, animal nutrition, see 12th edition

THOMPSON, GUY A, JR, b Rosedale, Miss, May 31, 31; m 60; c 3. BIOCHEMISTRY. Educ: Miss State Univ, BS, 53; Calif Inst Technol, PhD(biochem), 59. Prof Exp: NSF res fel chem, Univ Manchester, 59-60; res assoc biochem, Univ Wash, 60-62, from instr to asst prof, 62-67; assoc prof, 67-74, PROF BOT, UNIV TEX, AUSTIN, 74- Concurrent Pos: NIH res career develop award, 63-67. Mem: AAAS; Am Soc Biol Chem; Am Oil Chem Soc; Am Chem Soc. Res: Lipid metabolism; biochemistry of membranes. Mailing Add: Dept of Bot Univ of Tex Austin TX 78712

THOMPSON, HAROLD G, b New York, NY, May 31, 33; m 59; c 4. ORGANIC CHEMISTRY. Educ: Wagner Col, BS, 54; Syracuse Univ, MS, 56; Univ Md, PhD(org chem), 62. Prof Exp: Res chemist, 61-66, chief chemist, WVa, 66-70, chief chemist, Rubber Chem Dept, NJ, 70-71, mgr tech serv, 72-73, mgr pharmaceut mfg, 73-75, MKT MGR ELASTOMERS & POLYMER ADDITIVES DEPT, AM CYANAMID CO, BOUND BROOK, NJ, 75- Mem: Am Chem Soc. Res: Oxidation of mercaptans to disulfides using manganese dioxide; development of dyes, intermediates, optical brighteners, rubber processing chemicals and specialty elastomers. Mailing Add: 514 Lyme Rock Somerville NJ 08876

THOMPSON, HARTWELL GREENE, JR, b Hartford, Conn, Aug 30, 24; m 55; c 4. NEUROLOGY. Educ: Yale Univ, BA, 46; Cornell Univ, MD, 50. Prof Exp: From asst to assoc neurol, Col Physicians & Surgeons, Columbia Univ, 57-59; from asst prof to assoc prof, Univ Wis, 59-64; prof & chmn dept, Sch Med, WVa Univ, 64-69; prof

& assoc dean student affairs, Sch Med, Univ Pa, 69-73; PROF NEUROL & DEAN CHARLESTON DIV, W VA UNIV MED CTR, 73- Concurrent Pos: Consult, Vet Admin Hosp, Clarksburg, WVa, 64- Mem: AMA; Am Acad Neurol. Res: Neurology training programs and undergraduate education in neurology; multiple sclerosis; neurological complications of neoplastic diseases. Mailing Add: PO Box 2867 Charleston WV 25330

THOMPSON, HARVEY E, b Valders, Wis, Oct 30, 20; m 53; c 2. AGRONOMY. Educ: Univ Wis, BS, 47, MS, 48, PhD(agron, econ entom), 51. Prof Exp: ENTEN AGRONOMIST, IOWA STATE UNIV, 50- Mem: Am Soc Agron. Res: Forage and grain crop production. Mailing Add: 117 Agron Bldg Dept of Agron Iowa State Univ Ames IA 50010

THOMPSON, HAZEN SPENCER, b Frelighsburg, Que, Sept 10, 28; m 56; c 2. PLANT PATHOLOGY. Educ: McGill Univ, BSc, 50; Univ Toronto, MA, 52, PhD(plant path), 55. Prof Exp: Demonstr bot, Univ Toronto, 50-53; demonstr res off, Plant Res Inst, Res Br, Can Dept Agr, 51-65, fungicide liaison officer, Sci Info Sect, 65-73; PESTICIDE REV BIOLOGIST, ENVIRON PROTECTION SERV, ENVIRON CAN, 73- Mem: Can Phytopath Soc. Res: Chemical control of plant diseases; fungicides and nematocides; effects of pesticides on organisms in the environment. Mailing Add: Environ Protection Serv Environ Can Ottawa ON Can

THOMPSON, HENRY JOSEPH, b Mamaroneck, NY, Sept 5, 21; m 47; c 3. BIOLOGY. Educ: Whittier Col, AB, 47; Stanford Univ, MA, 48, PhD, 52. Prof Exp: Instr biol, Whittier Col, 48-49; actg instr, Stanford Univ, 51-52; instr bot, 52-54, from asst prof to assoc prof, 54-66, PROF BOT, UNIV CALIF, LOS ANGELES, 66- Mem: Am Soc Plant Taxon; Soc Study Evolution. Res: Systematics and evolution. Mailing Add: Dept of Biol Univ of Calif Los Angeles CA 90024

THOMPSON, HENRY THERON, b Ephraim, Utah, Jan 19, 11; m 45; c 4. ORGANIC CHEMISTRY. Educ: Univ Utah, AB, 32; Harvard Univ, PhD(org chem), 37. Prof Exp: Asst, Harvard Univ, 37-39; res chemist, E I du Pont de Nemours & Co, 39-42; res chemist, Gen Aniline & Film Corp, 42-51, asst to dir res, 51-56; analyst opers res lab, Gen Elec Co, 56-61; tech asst, Indust Chem Div, Allied Chem Corp, 61-70; BUDGET ANALYST, MONTCLAIR STATE COL, 70- Mem: AAAS; Am Chem Soc. Res: Oxidation-reduction potentials of quinone anils. Mailing Add: 16 Deer Run Circle Chatham NJ 07928

THOMPSON, HERBERT BRADFORD, b Detroit, Mich, Apr 22, 27; m 49. STRUCTURAL CHEMISTRY. Educ: Olivet Col, BS, 48; Oberlin Col, AM, 50; Mich State Col, PhD(chem), 53. Prof Exp: Res instr chem, Mich State Univ, 53-55; from asst prof to assoc prot, Gustavus Adolphus Col, 55-63; res assoc, Inst Atomic Res, Iowa State Univ, 63-65; res assoc, Univ Mich, 65-67; chmn dept chem, 68-69, 74-75, PROF CHEM, UNIV TOLEDO, 67- Mem: Am Chem Soc; Am Phys Soc. Res: Molecular structure and geometry; conformational analysis; data acquisition and computer applications in chemistry; electron diffraction; dipole moments. Mailing Add: Dept of Chem Univ of Toledo Toledo OH 43606

THOMPSON, HERBERT ELMER, biochemistry, see 12th edition

THOMPSON, HERBERT STANLEY, b China, June 12, 32; nat US; m 53; c 5. OPHTHALMOLOGY, NEUROLOGY. Educ: Univ Minn, BA, 53, MD, 61; Univ Iowa, MS, 66. Prof Exp: Instr, 67, asst prof, 67-71, ASSOC PROF OPHTHAL, UNIV IOWA, 71- Concurrent Pos: Nat Inst Neurol Dis & Blindness spec fel clin neuro-ophthal, Univ Calif, San Francisco, 66-67; Nat Inst Neurol Dis & Blindness res career develop award, 68; hon clin asst neuro-ophthal, Nat Hosps Nerv Dis, Queen Sq, London, 72-73. Mem: Asn Res Vision & Ophthal; Ophthal Soc UK; Fr Soc Ophthal; Am Acad Ophthal & Otolaryngol. Res: Neuro-ophthalmology, especially of the autonomic nervous system. Mailing Add: Dept of Ophthal Univ of Iowa Hosps Iowa City IA 52240

THOMPSON, HERMAN O, b Earl Park, Ind, Jan 21, 11; m 38; c 2. PHARMACY. Educ: Univ NC, BS, 37; Purdue Univ, MS, 40, PhD, 44. Prof Exp: Sales & med rep, Eli Lilly & Co, Ind, 38-39; asst pharmaceut bot, Purdue Univ, 40-42; asst prof pharm, Univ Ga, 44-45; assoc prof mfg pharm, Univ Ill, 45-46; assoc prof pharm, 46-51, PROF PHARM, SCH PHARM, UNIV NC, CHAPEL HILL, 51- Mem: Am Pharmaceut Asn; Acad Pharmaceut Sci; NY Acad Sci. Res: Red squill; assay method for capsaicin in capsicum; enteric coatings; flavored pharmaceutical vehicles; tablet lubrication and tabletting; tablet lubrication, binders, particle size and hopper flow; drug stability. Mailing Add: Dept of Pharm Univ of NC Sch of Pharm Chapel Hill NC 27514

THOMPSON, HOWARD E, b Bristolville, Ohio, Jan 23, 07. ORGANIC CHEMISTRY, MATHEMATICS. Educ: Hiram Col, AB, 29; Western Reserve Univ, MA, 31, PhD(phys chem), 34. Prof Exp: Res asst chem, Western Reserve Univ, 29-34; res chemist, Harshaw Chem Co, 34-36; independent consult chemist, Cleveland, Ohio, 36-42; res chemist, Civil Serv, Ft Detrick, 46-51; res chemist, 51-71, ORG CHEMIST, DUGWAY PROVING GROUND, US DEPT ARMY, 71- Mem: Am Chem Soc; Math Asn Am. Res: Synthesis of organic chemical compounds. Mailing Add: PO Box 455 Dugway UT 84022

THOMPSON, HOWARD K, JR, b Boston, Mass, May 19, 28; m 64; c 3. COMPUTER SCIENCE, BIOMATHEMATICS. Educ: Yale Univ, BA, 49; Columbia Univ, MD, 53. Prof Exp: Intern internal med, 1st Med Div, Bellevue Hosp, NY, 53-54, jr asst res, 55; clin fel, Mary I Bassett Hosp, Cooperstown, 54; cardiovasc res fel, Duke Hosp, Durham, NC, 58, sr asst res internal med, 58-59, cardiovasc res fel, 59-60, chief res, 60-61; biophys fel, Mass Inst Technol, 61-62; assoc med, Duke Univ, 62-65, asst prof biophys, 65-69, asst prof biomath, 66-69, assoc physiol, 63-69; PROF MED, BAYLOR COL MED, 71- Concurrent Pos: Assoc, Am Bd Internal Med, 63- Mem: Fel Am Col Cardiol; NY Acad Sci. Res: Indicator dilution method of blood flow estimation; estimation of regional cerebral blood flow by radio-xenon inhalation; biostatistics; computers in medicine. Mailing Add: F405 Methodist Hosp Houston TX 77025

THOMPSON, HUGH ERWIN, b Newport, RI, Aug 4, 17; m 46; c 4. ENTOMOLOGY. Educ: Univ RI, BS, 47; Cornell Univ, PhD(entom), 54. Prof Exp: Asst state entomologist, State Dept Agr & Conserv, RI, 47-48; entomologist, State Dept Agr, Pa, 53-56; asst prof entom, 56-63, ASSOC PROF ENTOM, KANS STATE UNIV, 63- Mem: Entom Soc Am; Soc Am Foresters. Res: Biology and control of insects attacking shade trees and ornamental plants; insect transmission of tree diseases. Mailing Add: Dept of Entom Kans State Univ Manhattan KS 66506

THOMPSON, HUGH WALTER, b New York, NY, Dec 7, 36; m 64. ORGANIC CHEMISTRY. Educ: Cornell Univ, AB, 58; Mass Inst Technol, PhD(org chem), 63. Prof Exp: NIH res fel, Columbia Univ, 62-64; from asst prof to assoc prof chem, 64-72, PROF CHEM, RUTGERS UNIV, NEWARK, 72- Mem: Am Chem Soc. Res: Mechanisms and stereochemical courses of organic reactions; compounds of unusual symmetry and stereochemistry; development of new synthetic methods. Mailing Add: Dept of Chem Rutgers Univ Newark NJ 07102

THOMPSON, IDA, b USA, Jan 21, 38. PALEOBIOLOGY. Educ: Univ Chicago, BA, 65, MS, 68, PhD(geol), 72. Prof Exp: Asst prof geol, Northern Ill Univ, 72-73; ASST PROF GEOL, PRINCETON UNIV, 73- Mem: Int Soc Chronobiol; Soc Vert Paleont; Anglo-Am Snail Watching Soc. Res: Biological rhythms of shell growth; annelid ecology and taxonomy; marine ecosystems and biogeography. Mailing Add: Dept of Geol & Geophys Sci Princeton Univ Princeton NJ 08540

THOMPSON, JACK COATS, b Donner Summit, Calif, Oct 5, 09; m 34. METEOROLOGY. Educ: Univ Calif, Los Angeles, AB, 48. Prof Exp: Aerologist, Sacramento City Airport Comn, Calif, 28-29, meteorol observer, 36-38, meteorologist, 38-47, dist forecaster, 47-53, meteorologist in charge, 53-56; res meteorologist, Washington, DC, 56-60, dir tech plans systs anal & design, 60-65; PROF METEOROL, SAN JOSE STATE UNIV, 65- Concurrent Pos: Consult, World Meteorol Orgn, Geneva, Switz, 65- Honors & Awards: Silver medal, US Dept Commerce, 55. Mem: Am Meteorol Soc; Am Geophys Union. Res: Basic experience in meteorological applications for aviation, agriculture and other operations; weather prediction; meteorological economics; application of decision theory to meteorological problems. Mailing Add: Dept of Meteorol San Jose State Univ San Jose CA 95192

THOMPSON, JAMES ARTHUR, b Sturgeon Bay, Wis, Aug 15, 31; m 55; c 3. ENVIRONMENTAL MANAGEMENT. Educ: St Olaf Col, BA, 55; Iowa State Univ, MS, 58. Prof Exp: Anal chemist, Ames Lab, AEC, 55-59; res chemist anal div, Alcoa Res Lab, 59-61, sr chemist, Warrick Opers, Ind, 61-70, chief chemist, Wenatchee Works, 70-75, NORTHWEST ENVIRON MGR, ALUMINUM CO AM, 75- Mem: AAAS; Am Chem Soc; Air Pollution Control Asn; Sigma Xi; Am Indust Hyg Asn. Res: Instrumental analysis; industrial hygiene; air and water pollution. Mailing Add: Aluminum Co of Am PO Box 221 Wenatchee WA 98801

THOMPSON, JAMES BURLEIGH, JR, b Calais, Maine, Nov 20, 21; m 57. PETROLOGY, GEOCHEMISTRY. Educ: Dartmouth Col, AB, 42; Mass Inst Technol, PhD, 50. Prof Exp: Instr geol, Dartmouth Col, 42; asst, Mass Inst Technol, 46-47, instr, 47-49; instr petrol, 49-50, asst prof petrog, 50-55, assoc prof mineral, 55-60, PROF MINERAL, HARVARD UNIV, 60- Concurrent Pos: Ford Found fel, 52-53; Guggenheim fel, 63. Honors & Awards: A L Day Medal, Geol Soc Am, 64. Mem: Nat Acad Sci; AAAS; fel Geol Soc Am; fel Mineral Soc Am; fel Am Acad Arts & Sci. Res: Metamorphic petrology; geology of New England. Mailing Add: Dept of Geol Sci Harvard Univ Cambridge MA 02138

THOMPSON, JAMES CHARLES, b San Antonio, Tex, Aug 16, 28; m 67; c 5. PHYSIOLOGY, SURGERY. Educ: Agr & Mech Col Tex, BS, 48; Univ Tex, MD, 51, MA, 52; Am Bd Surg, dipl. Prof Exp: Intern, Univ Tex Med Br Galveston, 51-52; asst resident surg, Hosp, Univ Pa, 52-54 & 56-58, chief resident, 58-59; from asst surgeon to assoc surgeon, Pa Hosp, 59-63; head physician, Harbor Gen Hosp, 63-67, chief surg, 67-70; PROF SURG & CHMN DEPT, UNIV TEX MED BR GALVESTON, 70-, CHIEF SURG, HOSP, 70- Concurrent Pos: Fel, Harrison Dept Surg Res, Sch Med, Univ Pa, 52-54 & 56-57, Albert & Mary Lasker fel, 57-59; John A Hartford Found grants, 60-; NIH grants, 60-; asst instr surg, Sch Med, Univ Pa, 53-54 & 56-58, from instr to assoc instr, 58-61, asst prof, 61-63; from assoc prof to prof, Sch Med, Univ Calif, Los Angeles, 63-70. Mem: AAAS; Am Surg Asn; Am Asn Surg of Trauma; Am Col Surgeons; Am Fedn Clin Res. Res: Gastric physiology; general surgery; organ transplantation; humoral control of gastric secretion; histamine metabolism; secretion in isolated tissue; radioimmunoassay and metabolism of gastrointestinal hormones, gastrin, cholecystokinin and secretin; molecular heterogeneity of gastrointestinal hormones. Mailing Add: Dept of Surg Univ of Tex Med Br Galveston TX 77550

THOMPSON, JAMES CHARLTON, b Leeds, UK, Jan 4, 41; m 65; c 2. INORGANIC CHEMISTRY. Educ: Cambridge Univ, BA, 62, PhD(chem), 65. Prof Exp: Fel, Rice Univ, 65-67; asst prof chem, 67-72, ASSOC PROF CHEM, UNIV TORONTO, 72-, ASSOC CHMN DEPT, 74- Mem: The Chem Soc; Chem Inst Can; Am Chem Soc. Res: Studies on the synthesis, structures and properties of silicon compounds, particularly those with fluorine or hydrogen bound to silicon. Mailing Add: Dept of Chem Univ of Toronto Toronto ON Can

THOMPSON, JAMES CHILTON, b Ft Worth, Tex, June 14, 30; m 55; c 3. PHYSICS. Educ: Tex Christian Univ, BA, 52; Rice Inst, MA, 54, PhD(physics), 56. Prof Exp: From asst prof to assoc prof, 56-67, PROF PHYSICS, UNIV TEX, AUSTIN, 67- Mem: Am Phys Soc. Res: Transport coefficients in solid and liquid metals; metal-ammonia solutions; metal-nonmetal transition; amorphous semiconductors. Mailing Add: Dept of Physics Univ of Tex Austin TX 78712

THOMPSON, JAMES EDWIN, b Maryville, Mo, Feb 2, 36; m 65; c 2. ORGANIC CHEMISTRY. Educ: Cent Methodist Col, AB, 56; Univ Mo, PhD(chem cyclopropanes), 61. Prof Exp: CHEMIST, PROCTER & GAMBLE CO, 61- Mem: Am Chem Soc. Res: Electronic effects in cyclopropanes; synthetic lipid and phospholipid, organo-phosphorus and organo-sulfur chemistry; radiochemical synthesis. Mailing Add: Procter & Gamble Co Miami Valley Labs Box 175 Cincinnati OH 45239

THOMPSON, JAMES LOWRY, b Syracuse, NY, Oct 5, 40; m 63; c 2. APPLIED MATHEMATICS, ENGINEERING SCIENCE. Educ: Brown Univ, AB, 62; Johns Hopkins Univ, PhD(mech), 68. Prof Exp: Asst prof math & eng sci, State Univ NY Buffalo, 68-74; RES MECH ENGR, US ARMY TANK AUTOMOTIVE COMMAND, 74- Concurrent Pos: NSF res grant, State Univ NY Buffalo, 71-73. Mem: AAAS; Am Math Soc; Soc Natural Philos; Soc Indust & Appl Math; Int Soc Terrain-Vehicle Systs. Res: Analysis and optimization of complex systems; continuum mechanics. Mailing Add: 1448 Anita Ave Grosse Pointe Woods MI 48236

THOMPSON, JAMES MARION, b Findlay, Ohio, July 26, 26; m 53; c 3. PLANT BREEDING, PLANT GENETICS. Educ: Ohio State Univ, BSc, 50, MS, 54, PhD(agron), 63. Prof Exp: Fel corn breeding, Agron Dept, Ohio State Univ, 54-56; corn breeder, Steckley Hybrid Corn Co, 57-62; geneticist, Blairsville, Ga, 63-70, GENETICIST, AGR RES SERV, USDA, BYRON, GA, 70- Mem: Am Genetic Asn; Am Pomol Soc. Res: Apple breeding projects designed to develop new varieties adapted in the Southern Coastal Plain and in the Southern Appalachian Mountains; plum breeding project; genetic research in pears, apples and plums. Mailing Add: SE Fruit & Tree Nut Res Sta Agr Res Serv USDA PO Box 87 Byron GA 31008

THOMPSON, JAMES NEAL, JR, b Lubbock, Tex, May 24, 46. GENETICS. Educ: Univ Okla, BS, 68, BA, 68; Univ Cambridge, PhD(genetics), 73. Prof Exp: Fel genetics, Univ Cambridge, 73-75; ASST PROF ZOOL, UNIV OKLA, 75- Concurrent Pos: Marshall scholar, Univ Cambridge, 70-73. Mem: Genetics Soc Am; Genetical Soc Gt Brit; Am Soc Mammalogists; Soc Study Human Biol. Res: Development and genetics of quantitative characters; genetic determination of patterns. Mailing Add: Dept of Zool Univ of Okla Norman OK 73069

THOMPSON, JAMES OLIVER, b North Battleford, Sask, Apr 30, 14; nat US; m 39; c 3. PHYSICAL CHEMISTRY. Educ: Univ London, BSc, 40; Univ Toronto, MA, 42; Univ Wis, PhD(phys chem), 48. Prof Exp: Chemist, Brit Drug Houses Ltd, London, 37-40; demonstr chem, Univ Toronto, 40-42; chemist, Imp Oil, Ltd & Polymer Corp, Ltd, Ont, Can, 42-45; asst chem, Univ Wis. 45-48; proj engr chem, Chrysler Corp, 48-49; res assoc chem, Inst Paper Chem, 49-53; mgr mat & process eng, Supplies Div, Int Bus Mach Corp, 53-66, sr chemist, Info Records Div, 66-68, SR CHEMIST, GEN PRODS DIV, IBM CORP, 68- Res: Molecular size distribution and degradation of polymeric systems; papers; inks; printing; detergent cleaning systems. Mailing Add: 308 Harding Ave Los Gatos CA 95030

THOMPSON, JAMES ROBERT, b Memphis, Tenn, June 18, 38; m 67. MATHEMATICS, STATISTICS. Educ: Vanderbilt Univ, BE, 60; Princeton Univ, MA, 63, PhD(math), 65. Prof Exp: Asst prof, Vanderbilt Univ, 64-67; asst prof math, Ind Univ, Bloomington, 67-70; ASSOC PROF MATH, RICE UNIV, 70- Concurrent Pos: Vpres opers res, Robert M Thrall & Assocs, Inc, 70-; adj assoc prof, Univ Tex MD Anderson Hosp & Tumor Inst, 72- Mem: Am Math Soc; Inst Math Statist; Am Statist Asn. Res: Biomathematics; estimation theory; time series. Mailing Add: Dept of Math Sci Rice Univ Houston TX 77001

THOMPSON, JAMES SCOTT, b Saskatoon, Sask, July 31, 19; m 44; c 2. ANATOMY, GENETICS. Educ: Univ Sask, BA, 40, MA, 41; Univ Toronto, MD, 45. Prof Exp: Res assoc med, Univ Toronto, 46-48; lectr & asst prof anat, Univ Western Ont, 48-50; assoc prof & prof, Univ Alta, 50-62; PROF ANAT, UNIV TORONTO, 66- CHMN DEPT, 66- Concurrent Pos: Exec secy & asst dean, Univ Alta, 53-62; vis investr, Jackson Lab, 62-63. Mem: AAAS; Am Asn Anatomists; Genetics Soc Am; Can Asn Anatomists (vpres, 64-65 & 73-75, pres, 75-77); Genetics Soc Can. Res: Genetic factors in cardiovascular disease; matters related to medical education and teaching of genetics. Mailing Add: Dept of Anat Univ of Toronto Toronto ON Can

THOMPSON, JAMES TIPTON, b Murray, Ky, Apr 23, 41; m 63; c 2. ANIMAL SCIENCE, ANIMAL NUTRITION. Educ: Murray State Univ, BS, 63; Univ Ky, MS, 64, PhD(animal sci), 66. Prof Exp: Res asst animal sci, Univ Ky, 63-66; asst prof agr, 66-70, ASSOC PROF ANIMAL SCI, ILL STATE UNIV, 70- Mem: Am Soc Animal Sci. Res: Nitrogen metabolism in ruminants; urea utilization of ruminant microorganisms. Mailing Add: Dept of Agr Ill State Univ Normal IL 61761

THOMPSON, JERRY NELSON, b Cincinnati, Ohio, Apr 2, 39; m 65; c 2. GENETICS, BIOCHEMISTRY. Educ: Univ Cincinnati, BS, 64; Ind Univ, PhD(med genetics), 70. Prof Exp: Res asst teratology, Cincinnati Children's Hosp Res Found, 61-64; ASST PROF BIOCHEM & PEDIAT, MED CTR, UNIV ALA, BIRMINGHAM, 72- Concurrent Pos: USPHS fel, Univ Chicago, 70-72; Nat Found March Dimes Basil O'Connor starter res grant, Med Ctr, Univ Ala, Birmingham, 74-76. Mem: Am Soc Human Genetics; Tissue Cult Asn. Res: Biochemical and genetic studies of cell cultures of various genetic lysosomal storage diseases. Mailing Add: Lab of Med Genetics Univ of Ala Med Ctr Birmingham AL 35294

THOMPSON, JESSE CLAY, JR, b Hot Springs, Va, Sept 17, 26; m 50; c 3. SYSTEMATICS. Educ: Hampden-Sydney Col, BS, 49; Univ Va, PhD, 56. Prof Exp: From asst prof to assoc prof biol, Hollins Col, 55-63; prof & chmn dept, Hampden-Sydney Col, 63-67; prof, Queens Col, NC, 67-69; PROF & CHMN DEPT, ROANOKE COL, 69- Concurrent Pos: Mem, Va Inst Marine Sci, 61, Int Indian Ocean Exped, 63, Mt Lake Biol Sta, 65 & Eniwetok Marine Biol Lab, 66; res partic, Palmer Sta, Antarctica, 68-69. Mem: AAAS; Am Inst Biol Sci; Soc Protozool. Res: Morphology; systematics and geographical distribution of hymenostome ciliated protozoa. Mailing Add: Dept of Biol Roanoke Col Salem VA 24153

THOMPSON, JESSE ELDON, b Laredo, Tex, Apr 7, 19; m 44; c 4. SURGERY. Educ: Univ Tex, BA, 39; Harvard Univ, MD, 43. Prof Exp: Instr surg, Boston Univ, 51-54; from asst prof to assoc prof, 54-68, CLIN PROF SURG, UNIV TEX HEALTH SCI CTR DALLAS, 68-; CO-CHIEF GEN SURG & CHIEF CONSULT PERIPHERAL VASCULAR SURG, BAYLOR UNIV MED CTR, DALLAS, 68- Concurrent Pos: Rhodes scholar physiol & Fulbright fel, Oxford Univ, 49-50. Mem: Int Soc Surg; Soc Vascular Surg; Am Surg Asn; Am Col Surgeons; AMA. Res: Vascular surgery; clinical investigation of hypertension, gastric physiology, peripheral vascular diseases and strokes; surgical management of vascular diseases. Mailing Add: 3600 Gaston Ave Dallas TX 75246

THOMPSON, JOHN, b Talara, Peru, Apr 21, 24; US citizen; m 50; c 1. HISTORICAL GEOGRAPHY, GEOGRAPHY OF LATIN AMERICA. Educ: Stanford Univ, BA, 47; Univ Calif, Berkeley, MA, 51; Stanford Univ, PhD(geog), 58. Prof Exp: From instr to asst prof geog, Stanford Univ, 53-64; assoc prof, 64-66, head dept, 66-75, PROF GEOG, UNIV ILL, URBANA, 66- Concurrent Pos: Fulbright award, Univ Chile, 57-58, Univ Brazil, 60 & Asia Found-Pakistan Ministry Educ, 63; head, Latin Am Studies Unit, Lang Develop Br, Off Educ, Dept Health, Educ & Welfare, Washington, DC, 62-64; dir, Ctr Latin Am Studies, Univ Ill, 64-66; mem adv screening comt, Coun Int Exchange Scholars, 74- & Nat Endowment for Humanities, 75- Mem: Asn Am Geog; Latin Am Studies Asn. Res: Urban food supply problems in Latin America; agricultural and settlement geography; history of natural sciences in Latin America. Mailing Add: Dept Geog 220 Davenport Hall Univ of Ill Urbana IL 61801

THOMPSON, JOHN BROCKWAY, physical chemistry, chemical engineering, see 12th edition

THOMPSON, JOHN C, JR, b Thomas, WVa, Oct 4, 30; m 54; c 3. PHYSICAL BIOLOGY. Educ: Va Polytech Inst, BS, 51, MS, 58; Cornell Univ, PhD(agr econ), 62. Prof Exp: Res assoc phys biol, 61-65, asst prof environ radiation biol, 65-68, ASSOC PROF ENVIRON RADIATION BIOL, CORNELL UNIV, 68- Mem: Health Physics Soc. Res: Radioactive contamination of the food chain, sampling techniques, controlled human studies, radionuclide deposition and cycling, world wide evaluation of fallout; biological costs of energy production; comparative environmental analyses and energy options. Mailing Add: Dept of Phys Biol Cornell Univ Ithaca NY 14853

THOMPSON, JOHN DANIEL, b Plains, Ga, July 10, 27; m 53; c 2. OBSTETRICS & GYNECOLOGY. Educ: Emory Univ, BS, 48, MD, 51. Prof Exp: Instr gynec, Sch Med, Johns Hopkins Univ, 54-56; instr obstet & gynec, Sch Med, La State Univ, 56-57, asst prof, 57-59; assoc prof, 59-70, PROF OBSTET & GYNEC & CHMN DEPT, SCH MED, EMORY UNIV, 70- Concurrent Pos: Markle scholar med sci, 57. Res: Gynecologic oncology; ovarian transplantation. Mailing Add: Dept of Obstet & Gynec Emory Univ Sch of Med Atlanta GA 30322

THOMPSON, JOHN DARRELL, b Mitchell, SDak, Sept 13, 33; m 57; c 4. PHYSICS, MOLECULAR BIOPHYSICS. Educ: Augustana Col, SDak, BA, 55; Iowa State Univ, MS, 62; Univ Wis, PhD(biophys), 67. Prof Exp: From instr to asst prof, 57-68, ASSOC PROF PHYSICS, AUGUSTANA COL, S DAK, 68- Mem: AAAS; Am Asn

Physics Teachers. Res: Structure and function of Escherichia coli ribosomes; hormonal control of protein synthesis in the chick embryo; interfacing computers to laboratory equipment. Mailing Add: Dept of Physics Augustana Col Sioux Falls SD 57102

THOMPSON, JOHN EVELEIGH, b Toronto, Ont, May 30, 41; m 65. BIOCHEMISTRY, CELL BIOLOGY. Educ: Univ Toronto, BSA, 63; Univ Alta, PhD(plant biochem), 66. Prof Exp: Fel med biochem, Univ Birmingham, 66-67; asst prof biol, 68-72, ASSOC PROF BIOL, UNIV WATERLOO, 72- Mem: Can Soc Plant Physiol; Am Soc Plant Physiol. Res: Effects of cell differentiation on membrane structure and function in plant and animal systems; mode of membrane biosynthesis; role of membranes in cell-cell interaction. Mailing Add: Dept of Biol Univ of Waterloo Waterloo On Can

THOMPSON, JOHN FANNING, b Ithaca, NY, May 24, 19; m 43; c 5. PLANT BIOCHEMISTRY. Educ: Oberlin Col, AB, 40; Cornell Univ, PhD(biochem), 44. Prof Exp: Instr biochem, Cornell Univ, 44-45; res assoc bot, Univ Chicago, 46-47; NIH fel, Univ Rochester, 47-49, res assoc, 49-50; from instr to asst prof bot, 50-55, ASSOC PROF BOT, CORNELL UNIV, 55-; PLANT PHYSIOLOGIST, PLANT, SOIL & NUTRIT LAB, USDA, 52- Concurrent Pos: NSF sr fel, 59-60. Mem: Am Soc Plant Physiol; Am Chem Soc; Am Soc Biol Chem. Res: Nitrogen and sulfur metabolism and mineral nutrition of plants; chromatographic techniques; control mechanisms. Mailing Add: US Plant Soil & Nutrit Lab USDA Tower Rd Ithaca NY 14850

THOMPSON, JOHN HAROLD, JR, b Amsden, Ohio, Jan 2, 21; m 43; c 3. PARASITOLOGY. Educ: Heidelberg Col, AB, 43; Ohio State Univ, MS, 48; Univ Minn, PhD, 52. Prof Exp: From instr path to asst prof clin path, 52-70, ASSOC PROF CLIN PATH, MAYO MED SCH, UNIV MINN, 70- Concurrent Pos: Consult, Mayo Clin, 52- Mem: AAAS; Am Soc Parasitol; Wildlife Dis Asn; Am Fedn Clin Res. Res: Blood coagulation. Mailing Add: Dept of Clin Path Mayo Med Sch Univ of Minn Rochester MN 55455

THOMPSON, JOHN LESLIE, b New Castle, Pa, July 11, 17; m 42; c 2. ENVIRONMENTAL SCIENCES. Educ: Slippery Rock State Teachers Col, BS, 40; Univ Wis, MS, 48, PhD(geog), 56. Prof Exp: Teacher gen sci, Sharpsville High Sch, Pa, 40-41; from asst prof to assoc prof geog, 49-61, PROF GEOG, MIAMI UNIV, 61- Mem: Asn Am Geog; Am Geog Soc; Nat Coun Geog Educ; Conserv Educ Asn. Res: Social aspects of environmental problems. Mailing Add: Dept of Geog Miami Univ Oxford OH 45056

THOMPSON, JOHN R, b Beltrami, Minn, Oct 6, 18; m 47; c 4. AGRONOMY. Educ: Univ Minn, BS, 48, MS, 52; Iowa State Univ, PhD(crop prod), 64. Prof Exp: From instr to assoc prof agron, Univ Minn, 52-67; SUPT, HAWAII AGR EXP STA, UNIV HAWAII, 67-, AGRONOMIST, 74- Mem: Am Soc Agron; Corp Sci Soc Am. Res: Seed and crop production; plant breeding. Mailing Add: Hawaii Agr Exp Sta Univ of Hawaii 461 W Lanikaula Hilo HI 96720

THOMPSON, JOHN RICHARD, invertebrate zoology, see 12th edition

THOMPSON, JOHN S, b Lincoln, Nebr, Oct 29, 28; m 54, 72; c 5. INTERNAL MEDICINE, IMMUNOLOGY. Educ: Univ Calif, Berkeley, BA, 49; Univ Chicago, MD, 53. Prof Exp: Intern, Univ Chicago Hosps, 53-54; jr asst resident, Presby Hosp, New York, 54-55; sr asst resident med, Univ Chicago Hosps, 57-58, resident, 58-59; from instr to assoc prof, Sch Med, Univ Chicago, 59-69; vchmn vet affairs, 69-71, PROF MED, UNIV IOWA, 69-, V CHMN DEPT, 71- Concurrent Pos: Nat Cancer Inst res fel, 58-60; Lederle med fac award, 66-68; chief med & chief sect allergy & clin immunol, Vet Admin Hosp, Iowa City. Mem: Fel Am Col Physicians; fel Am Acad Allergy; Transplantation Soc; Am Asn Immunologists. Res: HLA- and B-cell typing and genetics; natural immunosuppressive factors in fetal preservation. Mailing Add: Dept of Med Univ of Iowa Hosps Iowa City IA 52240

THOMPSON, JOHN STARK, physiological chemistry, see 12th edition

THOMPSON, JOSEPH KYLE, b Columbus, Ohio, Oct 2, 20; m 56; c 3. PHYSICAL INORGANIC CHEMISTRY. Educ: Sterling Col, BA, 42; Univ Kans, MA, 49, PhD, 50. Hon Degrees: DSc, Sterling Col, 67. Prof Exp: Chemist, 42-46, CHEMIST, US NAVAL RES LAB, 50- Mem: Am Chem Soc; Sigma Xi; Am Inst Chemists. Res: Kinetics and mechanisms of adsorption and filtration, particularly air cleaning devices; oxides of alkali and alkaline earth metals; nuclear magnetic resonance. Mailing Add: US Naval Res Lab Code 6180 Washington DC 20390

THOMPSON, JOSEPH LIPPARD, b Newport News, Va, May 12, 32; m 55; c 3. RADIOCHEMISTRY, PHYSICAL CHEMISTRY. Educ: Va Polytech Inst, BS, 54; Pa State Univ, MS, 59, PhD(chem), 63. Prof Exp: Nat Res Coun res asst, Nat Bur Stand, Washington, DC, 63-64; asst prof chem, 64-75, ASSOC PROF CHEM, IDAHO STATE UNIV, 75- Mem: Am Chem Soc. Res: Chemical effects of nuclear transformations; Mössbauer effect; atmospheric monitoring. Mailing Add: Dept of Chem Idaho State Univ Pocatello ID 83201

THOMPSON, JULIA ANN, b Little Rock, Ark, Mar 13, 43. ELEMENTARY PARTICLE PHYSICS, HIGH ENERGY PHYSICS. Educ: Cornell Col, BA, 64; Yale Univ, MS, 66, PhD(physics), 69. Prof Exp: Res assoc physics, Brookhaven Nat Lab, 69-71; res assoc & assoc instr, Univ Utah, 71-72; ASST PROF PHYSICS, UNIV PITTSBURGH, 72- Mem: Am Phys Soc. Res: Hyperon interactions and decays, particularly hyperon resonances and the decay of cascade or sigma to lambda, electron, neutrino. Mailing Add: Dept of Physics Univ of Pittsburgh Pittsburgh PA 15260

THOMPSON, KENNETH, b Leeds, Eng, Nov 26, 23; nat US; m 50, 67; c 3. GEOGRAPHY. Educ: Univ Cambridge, BA, 50; Northwestern Univ, MS, 53, PhD(geog), 55. Prof Exp: From instr to assoc prof, 55-70, chmn dept, 63-70, PROF GEOG, UNIV CALIF, DAVIS, 70- Mem: Asn Am Geogr; Am Geog Soc. Res: Historical geography. Mailing Add: Dept of Geog Univ of Calif Davis CA 95616

THOMPSON, KENNETH DAVID, b Wimbeldon, NDak, Apr 13, 40; m 65; c 2. MICROBIOLOGY, IMMUNOLOGY. Educ: Univ NDak, BS, 63; MS, 67, PhD(microbiol), 70. Prof Exp: NIH fel, 70-72, instr, 72-73, ASST PROF MICROBIOL & IMMUNOL, HEALTH SCI CTR, TEMPLE UNIV, 73- Mem: Am Soc Microbiol; Reticuloendothelial Soc. Res: Tumor immunology and clinical immunology. Mailing Add: Dept of Microbiol & Immunol Temple Univ Sch of Med Philadelphia PA 19122

THOMPSON, KENNETH ROY, b Berkeley, Calif, Dec 12, 41; m 67; c 1. PHYSICAL CHEMISTRY, ANALYTICAL CHEMISTRY. Educ: DePauw Univ, BA, 63; Case Inst Technol, PhD(chem), 68. Prof Exp: Res assoc chem, Univ Fla, 68-69; RES CHEMIST, ENG PHYSICS LAB, EXP STA, E I DU PONT DE NEMOURS & CO, INC, 69- Mem: Am Chem Soc; Am Phys Soc. Res: High temperature chemistry; chemical vapor transport reactions; process analyzers. Mailing Add: Eng Physics Lab Exp Sta E I du Pont de Nemours & Co Wilmington DE 19898

THOMPSON, LAFAYETTE, JR, agronomy, weed science, see 12th edition

THOMPSON, LANCELOT CHURCHILL ADALBERT, b Jamaica, West Indies, Mar 3, 25; US citizen; m 52; c 2. INORGANIC CHEMISTRY. Educ: Morgan State Col, BS, 52; Wayne State Univ, PhD(inorg chem), 56. Prof Exp: Instr chem, Wolmers Boys Sch, Jamaica, 55-56; Int Nickel Co fel, Pa State Univ, 57; from asst prof to assoc prof inorg chem, 58-66, asst dean col arts & sci, 64-66, PROF CHEM & DEAN STUDENT SERV, UNIV TOLEDO, 66-, VPRES STUDENT AFFAIRS, 68- Concurrent Pos: Consult, Owens-Ill Glass Co, Ohio, 62-64. Mem: AAAS; Am Chem Soc. Res: Determination of structure of coordination compounds; coordination polymers; solubility of hydrous oxides. Mailing Add: Dept of Chem Univ of Toledo Toledo OH 43606

THOMPSON, LARRY CLARK, b Hoquiam, Wash, June 13, 35; m 55; c 2. INORGANIC CHEMISTRY. Educ: Willamette Univ, BS, 57; Univ Ill, MS, 59, PhD(inorg chem), 60. Prof Exp: From asst prof to assoc prof, 60-68, PROF CHEM, UNIV MINN, DULUTH, 68-, HEAD DEPT, 72- Concurrent Pos: Vis prof, Univ Sao Paulo, 69. Mem: Am Chem Soc. Res: Coordination chemistry of the rare earth elements; high coordination numbers; ligands with unusual steric requirements. Mailing Add: Dept of Chem Univ of Minn Duluth MN 55812

THOMPSON, LARRY FLACK, b Union City, Tenn, Aug 31, 44; m 64; c 2. POLYMER CHEMISTRY. Educ: Tenn Technol Univ, BS, 66, MS, 68; Univ Mo-Rolla, PhD(chem), 71. Prof Exp: MEM TECH STAFF CHEM & THIN FILMS, BELL LABS, 70- Concurrent Pos: Guest prof, Rutgers Univ, 72. Mem: Am Chem Soc. Res: Electron beam polymer resist studies for microfabrication of integrated electronics; thin polymer films for use in microelectronic fabrication. Mailing Add: 6F225 Bell Labs 600 Mountain Ave Murray Hill NJ 07974

THOMPSON, LAURA, b Honolulu, Hawaii, Jan 23, 05; m 63. CULTURAL ANTHROPOLOGY, APPLIED ANTHROPOLOGY. Educ: Mills Col, BA, 27; Univ Calif, Berkeley, PhD(anthrop), 33. Prof Exp: Asst ethnologist, Bishop Mus, Honolulu, 29-34; Inst Pac Rels grant, Honolulu, 34; social scientist, Naval Govt Guam, 38-40; social scientist, Community Surv Educ, Hawaii, 40-41; coordr Indian educ res, US Off Indian Affairs, Univ Chicago, 41-47; res consult, Inst Ethnic Affairs, Wash, 46-54; prof anthrop, City Col New York 54-56; vis prof, Univ NC, 57-58; vis prof, NC State Col, 59-60; prof, Univ Southern Ill, 61-62; prof, San Francisco State Col, 62-63; CONSULT ANTHROPOLOGIST, 63- Concurrent Pos: Bernice P Bishop fel, Yale Univ, Fiji Islands, 33-34; consult, US Naval Govt Guam, 38-39 & US Off Indian Affairs, 42-44; adv, Policy Bd, US Nat Indian Inst, Wash, 48; Viking Fund grant & Wenner-Gren fel, NY & Iceland, 48 & 51; Rockefeller Found grants, 51 & 52; distinguished vis prof anthrop, Pa State Univ, 61; consult, Hutterite Socialization Proj, Pa State Univ, 62-65 & Centennial Joint Sch Syst, Pa, 64-66. Mem: AAAS; fel Am Anthrop Asn; NY Acad Sci. Res: Comparative interdisciplinary research in small communities, Fiji, Guam, Hawaii, Hopi, Navaho, Sioux and Papago Indians, Iceland, Lower Saxons of West Germany; human ecology; demography; ethnopsychology; ecosystem approach toward population control; unified anthropology. Mailing Add: 1530 Palisades Ave Apt 14H Ft Lee NJ 07024

THOMPSON, LAWRENCE HADLEY, b Tyler, Tex, July 22, 41; m 68; c 2. CELL BIOLOGY. Educ: Univ Tex, Austin, BS, 63, MS, 67, PhD(biophys), 69. Prof Exp: Fel cell biol, Ont Cancer Inst, 69-71, staff physicist, 71-73; SR SCIENTIST CELL BIOL, LAWRENCE LIVERMORE LAB, UNIV CALIF, 73- Mem: Am Soc Cell Biol. Res: Study of mechanisms of somatic cell mutation and development of somatic cell genetics through the isolation, characterization and applications of conditional lethal mutants of cultured mammalian cells. Mailing Add: Biomed Div L-523 Livermore Lab Univ of Calif PO Box 808 Livermore CA 94550

THOMPSON, LEIF HARRY, b Chadron, Nebr, Dec 6, 43; c 1. REPRODUCTIVE PHYSIOLOGY. Educ: Univ Nebr, BS, 67; NC State Univ, MS, 70, PhD(animal sci), 72. Prof Exp: ASST PROF REPRODUCTIVE PHYSIOL, TEX TECH UNIV, 72- Concurrent Pos: Gen livestock prod consult, Area cattle & sheep ranchers, 72- Mem: Am Soc Animal Sci; Soc Study Fertil; Soc Study Reprod. Res: Influence of environment, nutrition, development, hormonal therapy and selection on reproductive efficiency in swine, beef cattle and sheep and hormonal regulation of growth of feedlot animals. Mailing Add: Dept of Animal Sci Tex Tech Univ Lubbock TX 79409

THOMPSON, LEONARD GARNETT, JR, b Deer Lodge, Mont, Mar 24, 03; m 47; c 2. SOILS. Educ: Mont State Col, BS, 28; Iowa State Col, MS, 29, PhD(soil bact), 31. Prof Exp: Teacher agr & physics, Tenn State Teachers Col, Johnson City, 31-32; farm mgr, 32-34, 43-44; instr bact, Univ Tenn, 34-35; prof agron, Panhandle Agr & Mech Col, 35; asst soil surveyor, Soil Conserv Serv, USDA, 35-40; asst soil surveyor grazing serv, US Dept Interior, 40-42; soil chemist, Univ Fla, 45-46, North Fla Exp Sta, 46-62, agr exp sta, 62-74; RETIRED. Res: Soil management; bacteriology, fertility and survey. Mailing Add: Agr Exp Sta Gainesville FL 32601

THOMPSON, LEWIS CHISHOLM, b Brechenridge, Tex, Jan 18, 26; m 55; c 2. NUCLEAR PHYSICS. Educ: Rice Univ, BA, 50, MA, 52, PhD(physics), 54. Prof Exp: Physicist, Naval Res Lab, 54-56; sr nuclear engr, Gen Dynamics/Convair, 56-59; asst prof physics, Univ Ga, 59-63; assoc prof, La Sierra Col, 65-70; PROF PHYSICS, LOMA LINDA UNIV, 70- Mem: Am Phys Soc. Res: Energy levels of light nuclei; nuclear instruments; nuclear shielding; low energy particle accelerators. Mailing Add: 5513 Wentworth Dr Riverside CA 92505

THOMPSON, LLOYD G D, geophysics, geodesy, see 12th edition

THOMPSON, LOREN EDWARD, JR, b Salem, WVa, May 8, 37; m 58; c 3. GEOPHYSICS. Educ: Marietta Col, BS, 60; Ohio Univ, MS, 63. Prof Exp: Geophys interpreter explor, Phillips Petrol Co, 63-67; ADVAN GEOPHYSICIST EXPLOR RES, DENVER RES CTR, MARATHON OIL CO, 67- Mem: Soc Explor Geophysicists; Soc Prof Well Log Analysts. Res: Acoustical, electrical, nuclear and mechanical borehole geophysical devices, including electronics, data acquisition and interpretation. Mailing Add: Marathon Oil Co Denver Res Ctr PO Box 269 Littleton CO 80120

THOMPSON, LOUIS MILTON, b Throckmorton, Tex, May 15, 14; m 37; c 5. SOILS. Educ: Agr & Mech Col, Tex, BS, 35; Iowa State Col, MS, 47, PhD, 50. Prof Exp: Sr foreman laborers, Soil Conserv Serv, USDA, Tex, 35-36, jr soil surveyor, Tex & La, 39-40; instr agron, Agr & Mech Col, Tex, 36-39, 40-42; asst prof soils, 47-50, prof in charge farm oper curriculum, 50-58, ASSOC DEAN AGR, COL AGR, IOWA STATE UNIV, 58- Concurrent Pos: Vis prof, Univ Ill, 58. Mem: Am Soc Agron; Soil Sci Soc Am; AAAS. Res: Developing statistical models to determine influence of weather variables on production of corn, soybeans and wheat. Mailing Add: Col of Agr Iowa State Univ Ames IA 50010

THOMPSON, LUCIEN ORRIN, b Eau Claire, Wis, Sept 10, 16; m 40; c 6. PETROLEUM GEOLOGY. Educ: Univ Minn, BA, 38, MA, 39. Prof Exp: Dist geologist, Stand Oil Co, 46-48, Slick-Moorman Oil Co, 48-49, Heyser & Heard, 50-51; div geologist, Am-Maracaibo & Felmont Oil Corp, 53-58, chief geologist, Felmont Petrol Corp, 58-60; consult geologist, Tex, 60-68; chief geologist, Int Oils Explor Netherlands, Australia, 68-69; consult geologist, 69-74; CHIEF GEOLOGIST, PETROL EXPLOR & OPERATING CORP, 74- Concurrent Pos: Chief geologist, Explor & Develop Inc, Tulsa, Okla, 67-68. Mailing Add: 1501 W Pecan Ave Midland TX 79701

THOMPSON, LYELL, b Rock Island, Ill, May 10, 24; m 46; c 5. SOILS. Educ: Okla State Univ, BS, 48; Ohio State Univ, PhD(soils), 52. Prof Exp: Asst prof, Ohio State Univ, 51-53; soil scientist, Noble Found, Okla, 53-58; assoc prof agron, 58-69, PROF AGRON, UNIV ARK, FAYETTEVILLE, 69- Mem: Am Soc Agron; Soil Sci Soc Am. Effect of soil fertility, trace element availability and soil acidity upon crop production; increase of food production. Mailing Add: Dept of Agron Univ of Ark Fayetteville AR 72701

THOMPSON, MAJOR CURT, b Cullman, Ala, May 25, 37; m 62; c 2. INORGANIC CHEMISTRY. Educ: Birmingham-Southern Col, BS, 59; Ohio State Univ, MS, 61, PhD(inorg chem), 63. Prof Exp: CHEMIST, SAVANNAH RIVER LAB, ATOMIC ENERGY DIV, E I DU PONT DE NEMOURS & CO, INC, 63- Mem: Am Chem Soc. Res: Synthesis of binary compounds of actinide elements which are stable at high temperature; complexes of the actinides and lanthanides; solvent extraction. Mailing Add: Savannah River Lab E I du Pont de Nemours & Co Inc Aiken SC 29801

THOMPSON, MALCOLM J, b Baldwin, La, Feb 15, 27; m 53; c 3. ORGANIC CHEMISTRY. Educ: Xavier Univ, La, BS, 50, MS, 52. Prof Exp: Instr chem, Xavier Univ, La, 52-54; chemist, US Bur Mines, 54-55; org chemist, NIH, 55-60; res org chemist, Chem Warfare Labs, Army Chem Ctr, Md, 60-62; RES CHEMIST, INSECT PHYSIOL LAB, USDA, 62- Mem: AAAS; Am Chem Soc. Res: Chemistry of steroids, sapogenins and natural products; synthesis and structural elucidations; insect hormones, isolation and structural elucidation of insect molting hormones; feeding stimulants; synthesis of compounds with gonadotropic and juvenile hormone activity. Mailing Add: Inst Physiol Lab Agr Res Ctr Entom Bldg C Beltsville MD 20705

THOMPSON, MARCUS LUTHER, b Liberty, Miss, July 12, 06; m; c 2. GEOLOGY, PALEONTOLOGY. Educ: Miss State Col, BS, 30, Univ Iowa, MS, 33, PhD(paleont), 34. Prof Exp: Instr, Miss State Col, 30-31; res assoc, Univ Iowa, 34-36; paleontologist, Shell Oil Co, 37-38; geologist & paleontologist, Phillips Petrol Co, 38-39; geologist, NMex Bur Mines, 39-42; from asst prof to assoc prof geol, Univ Kans, 43-46, prof & chmn dept, 54-57; from assoc prof to prof, Univ Wis, 46-54; prin geologist, State Geol Surv, Ill, 57-67, prin res geologist, 67-74; RETIRED. Concurrent Pos: Geologist, State Geol Surv, Kans, 42-46; vis Fulbright prof, Kyushu Univ, 60; del, Int Geol Cong Mex, 56; ed, J Paleont, Soc Econ Paleont & Mineral, 57-62. Mem: Paleont Soc; fel Geol Soc Am; Soc Econ Paleont & Mineral; Am Asn Petrol Geol. Res: Upper Paleozoic micropaleontology and stratigraphy. Mailing Add: 303 W Vermont Urbana IL 61801

THOMPSON, MARGARET A WILSON, b Northwich, Eng, Jan 7, 20; Can citizen; m 44; c 2. GENETICS. Educ: Univ Sask, BA, 43; Univ Toronto, PhD(human genetics), 48. Prof Exp: Lectr zool, Univ Toronto, 47-48 & Univ Western Ont, 48-50; lectr, Univ Alta, 50-59, asst prof human genetics, 59-62; vis investr, Jackson Lab, 62-63; res assoc pediat & lectr zool, 63-64, asst prof pediat & zool, 64, assoc prof zool, 65-70, assoc prof med cell biol, 69-72, assoc prof med genetics, 72-73, ASSOC PROF PEDIAT, UNIV TORONTO, 66-, PROF MED GENETICS, 73- Concurrent Pos: Muscular Dystrophy Asn Can res fel, 62-63; sr staff geneticist, Hosp for Sick Children, Toronto, 63-; mem bd dirs, Am Soc Human Genetics, 75-78. Mem: Am Soc Human Genetics; Genetics Soc Am; Genetics Soc Can (pres, 72-73). Res: Human genetics. Mailing Add: Dept of Genetics Hosp for Sick Children Toronto ON Can

THOMPSON, MARTIN LEROY, b Kindred, NDak, Jan 8, 35; m 63; c 3. INORGANIC CHEMISTRY. Educ: Concordia Col, Moorhead, Minn, BA, 56; Ind Univ, PhD(inorg chem), 64. Prof Exp: From instr to asst prof, 62-69, ASSOC PROF CHEM, LAKE FOREST COL, 69- Mem: Am Chem Soc. Res: Inorganic chemistry of volatile silicon and boron compounds. Mailing Add: Dept of Chem Johnson Sci Bldg Lake Forest Col Lake Forest IL 60045

THOMPSON, MARVIN P, b Troy, NY, June 22, 33; m 53; c 3. BIOCHEMISTRY. Educ: Kans State Univ, BS, 56, MS, 57; Mich State Univ, PhD(food sci), 60. Prof Exp: Biochemist, 60-71, CHIEF BIOCHEMIST, MILK PROPERTIES LAB, EASTERN REGIONAL RES LAB, USDA, 71- Concurrent Pos: Prof, Pa State Univ, 65- Honors & Awards: Borden Award, 70; Arthur S Flemming Award, 71; Superior Serv Award, USDA, 71. Mem: Am Chem Soc; Am Dairy Sci Asn; Am Soc Biol Chem. Res: Isolation and properties of milk proteins; genetic polymorphism of milk proteins; structure of casein micelles; accelerated curing of cheese. Mailing Add: USDA Eastern Regional Res Lab Milk Properties Lab Philadelphia PA 19118

THOMPSON, MARVIN PETE, JR, b Mackville, Ky, Sept 28, 41; m 62; c 3. WILDLIFE ECOLOGY, MAMMALOGY. Educ: Univ Ky, BS, 63; Kans State Univ, MS, 67; Southern Ill Univ, Carbondale, PhD(zool), 71. Prof Exp: Asst prof, 68-73, ASSOC PROF BIOL, EASTERN KY UNIV, 73-, FAC RES GRANT, 69- Mem: Wildlife Soc; Am Soc Mammal; Nat Wildlife Fedn. Res: Woodchuck ecology and physiology; ecology of pest mammals. Mailing Add: Dept of Biol Sci Eastern Ky Univ Richmond KY 40475

THOMPSON, MARY E, b Minneapolis, Minn, Dec 21, 28. PHYSICAL INORGANIC CHEMISTRY. Educ: Col St Catherine, BA, 53; Univ Minn, MS, 58; Univ Calif, Berkeley, PhD(chem), 64. Prof Exp: Instr sci & math, Derham Hall High Sch, 53-57, 58-59; res asst chem, Lawrence Radiation Lab, Calif, Berkeley, 61-64; lab instr, 53-56, asst prof, 64-69, ASSOC PROF CHEM & CHMN DEPT, COL ST CATHERINE, 69- Mem: AAAS; Am Chem Soc; fel Am Inst Chem; The Chem Soc. Res: Hydrolytic polymerization in aqueous solutions; kinetics; magnetic susceptibility of solutions of transition metal polymers. Mailing Add: Dept of Chem Col of St Catherine 2004 Randolph Ave St Paul MN 55105

THOMPSON, MARY ELEANOR, b Cleveland, Ohio, Nov 5, 26. GEOCHEMISTRY. Educ: Boston Univ, BA, 48; Harvard Univ, MA, 63, PhD(geol), 64. Prof Exp: Mineralogist, US Geol Surv, 48-57; electrode chemist, EPSCO, Inc, Mass, 62-63; res assoc geochem, Dept Geol, Univ SC & electrode chem, Dept Geol & Sch Med, Stanford Univ, 64-67; RES SCIENTIST & MGR CHEM LIMNOL, CAN CTR INLAND WATERS, 67- Concurrent Pos: Co-recipient, NSF grant, Univ SC, 65-66; res assoc, Dept Geol, McMaster Univ, 68-69. Mem: Int Asn Sci Hydrol; Geochem Soc; Am Chem Soc; Am Soc Limnol & Oceanog; Int Asn Gt Lakes Res. Res: Low temperature aqueous geochemistry; specific-ion electrodes; chemical limnology. Mailing Add: Can Ctr for Inland Waters Box 5050 Burlington ON Can

THOMPSON, MAXINE MARIE, b Bloomington, Ill, Nov 3, 26; m 53; c 2. GENETICS, HORTICULTURE. Educ: Univ Calif, BS, 48, MS, 51, PhD(genetics), 60. Prof Exp: Jr specialist viticulture, Univ Calif, Davis, 62-63; asst prof biol, Wis State Univ, Oshkosh, 63-64; res assoc hort, 64-67, asst prof bot, 66-68, asst prof hort, 68-71, ASSOC PROF HORT, ORE STATE UNIV, 71- Mem: AAAS; Bot Soc Am;

Am Soc Hort Sci. Res: Cytological and botanical studies related to horticultural problems, especially horticultural breeding. Mailing Add: Dept of Hort Ore State Univ Corvallis OR 97331

THOMPSON, MAYNARD, b Michigan City, Ind, Sept 8, 36; m 55; c 2. MATHEMATICS. Educ: DePauw Univ, AB, 58; Univ Wis, MS, 59, PhD(math), 62. Prof Exp: Lectr, 62-64, from asst prof to assoc prof, 64-74, PROF MATH, IND UNIV, BLOOMINGTON, 73-, CHMN DEPT, 74- Concurrent Pos: Res assoc, Univ Md, 70-71. Mem: Am Math Soc; Math Asn Am; Soc Indust & Appl Math. Res: Approximation theory; complex analysis; mathematical biology. Mailing Add: Dept of Math Ind Univ Bloomington IN 47401

THOMPSON, MICHAEL BRUCE, b Kansas City, Mo, Aug 25, 39; m 67; c 1. EMBRYOLOGY. Educ: Baker Univ, BS, 63; Kans State Univ, MS, 67, PhD(biol), 69. Prof Exp: Head dept biol, 70-74, ASST PROF BIOL, MINOT STATE COL, 69-, CHMN DIV SCI & MATH, 74- Mem: Soc Study Reproduction; Am Soc Zool. Res: Developmental placentation. Mailing Add: Div of Sci Minot State Col Minot ND 58701

THOMPSON, MILTON AVERY, b Salem, Ore, July 5, 29; m 57; c 3. ENVIRONMENTAL MANAGEMENT. Educ: San Jose State Col, BA, 51; Ore State Univ, MS, 53, PhD(phys chem), 57. Prof Exp: From chemist to sr chemist, Dow Chem Co, 57-61, res supvr, 62-65, sr res mgr, 66-68, dir chem res & develop, 69-70, mgr environ sci, 70-74; MGR ENVIRON SCI & WASTE CONTROL, ROCKWELL INT, 75- Mem: Am Chem Soc. Res: Plutonium chemistry; plutonium processing, revovery and corrosion; nonaqueous plutonium chemistry. Mailing Add: 931 Gapter Rd Boulder CO 80303

THOMPSON, MILTON D, b Haxtun, Colo, Aug 25, 09; m 38. ZOOLOGY. Educ: Univ Minn, BS, 35; Lincoln Col, Ill, DH, 70. Prof Exp: Dir, Minneapolis Sci Mus, 36-51; asst dir, 51-62, DIR, ILL STATE MUS, 63-, DICKSON MOUNDS MUS, 74- Mem: Asn Sci Mus Dirs (secy-treas, 61-64, 67-); Am Asn Mus; Wilson Ornith Soc. Res: Ornithology; herpetology. Mailing Add: Ill State Mus Spring & Edwards St Springfield IL 62706

THOMPSON, NEAL PHILIP, b Brooklyn, NY, July 18, 36; m 58; c 5. PLANT PHYSIOLOGY, PLANT ANATOMY. Educ: Wheaton Col, Ill, BS, 57; Miami Univ, Ohio, MA, 62; Princeton Univ, PhD(biol), 65. Prof Exp: Asst prof plant physiol & asst plant physiologist, 65-72, ASSOC PROF PLANT PHYSIOL & ASSOC PLANT PHYSIOLOGIST, UNIV FLA, 72- Mem: Am Chem Soc. Res: Developmental structure of higher plants; translocation of materials, exogenously applied or endogenous, in higher plants, their effects on anatomical structure and their metabolism; pesticides in the environment, particularly as related to birds and fish. Mailing Add: Pesticide Res Lab Univ of Fla Gainesville FL 32611

THOMPSON, NOEL PAGE, b San Francisco, Calif, Oct 22, 29; m 54; c 2. BIOMEDICAL ENGINEERING. Educ: Stanford Univ, BA, 51, MS, 61; Univ Calif, Los Angeles, MD, 55. Prof Exp: Intern med, Univ Hosps, Univ Wis, 55-56; chief bioeng & physiol div, Palo Alto Med Res Found, 58-73; CHIEF MED INSTRUMENTATION LAB, PALO ALTO MED CLIN, 64- Concurrent Pos: Consult assoc prof, Stanford Univ, 61- & Univ Santa Clara, 62-68. Mem: AMA; Am Inst Ultrasonics in Med; Am Acad Family Physicians; Inst Elec & Electronics Engrs. Res: Theoretical and applied biomedical engineering; mathematics and electronic instruments as applied to research and in the practice of medicine. Mailing Add: Palo Alto Med Clin 300 Homer Ave Palo Alto CA 94301

THOMPSON, NORMAN ROBERT, b Strathroy, Ont, Sept 13, 15; nat US; m 43; c 1. PLANT BREEDING, GENETICS. Educ: Ont Agr Col, BSA, 39; Univ Toronto, MSA, 49; Mich State Univ, PhD, 52. Prof Exp: Asst potato prod, Ont Agr Col, 39-42; res officer, Can Dept Agr, 46-53; from asst prof to assoc prof plant breeding, 53-65, PROF PLANT BREEDING, MICH STATE UNIV, 65- Mem: Potato Asn Am. Res: Potato varieties with higher nutritional values from species hybrids; inheritance studies of proteins, especially the amino acid methionine; rest period; processing qualities. Mailing Add: Dept of Crop & Soil Sci Mich State Univ East Lansing MI 48823

THOMPSON, NORMAN STROM, b Ft William, Ont, Nov 10, 23; m 51; c 4. ORGANIC CHEMISTRY. Educ: Univ Man, BSc, 50, MSc, 52; McGill Univ, PhD(wood chem), 54. Prof Exp: Res chemist, Rayonier, Inc, Wash, 53-60; res assoc, 60-69, PROF CHEM & SR RES ASSOC, INST PAPER CHEM, 69- Mem: AAAS; Am Chem Soc; Am Tech Asn Pulp & Paper Indust; Can Pulp & Paper Asn. Res: Location and composition of the carbohydrate constituents of wood and their behavior during pulping. Mailing Add: Inst of Paper Chem Appleton WI 54911

THOMPSON, OWEN EDWARD, b St Louis, Mo, Nov 20, 39; m 72; c 1. METEOROLOGY, ATMOSPHERIC PHYSICS. Educ: Univ Mo-Columbia, BS, 61, MS, 63, PhD(atmospheric sci), 66. Prof Exp: Instr physics & math, Stephens Col, 64-66; instr atmospheric sci, Univ Mo-Columbia, 66-68; asst prof, 68-72, ASSOC PROF METEOROL, UNIV MD, COLLEGE PARK, 72- Concurrent Pos: Wallace Eckert vis scientist, IBM Thomas J Watson Res Ctr, Yorktown Heights, NY, 75-76. Mem: Am Meteorol Soc; Am Geophys Union. Res: Dynamical and physical meteorology; atmospheric waves and oscillations; micrometeorology and boundary layer studies; forest environment; satellite meteorology; meteorological instrumentation; educational film making. Mailing Add: Meteorol Prog Univ of Md College Park MD 20742

THOMPSON, PATRICK HALEY, b Galveston, Tex, Feb 20, 33; m 57; c 3. ENTOMOLOGY. Educ: Auburn Univ, BS, 55; Univ Wis, MS, 62, PhD(entom), 64. Prof Exp: Res asst & proj assoc med entom, Univ Wis, 60-64; asst prof, Univ Md, 64-65 & Rutgers Univ, 65-67; RES ENTOMOLOGIST, AGR RES SERV, US DEPT AGR, 67- Res: Behavior, ecology and taxonomy of Tabanidae. Mailing Add: Agr Res Serv US Dept of Agr College Station TX 77840

THOMPSON, PAUL EVERETT, parasitology, see 12th edition

THOMPSON, PAUL O, b Stoughton, Wis, Feb 12, 21; m 54; c 2. PSYCHOACOUSTICS, BIOACOUSTICS. Educ: St Olaf Col, BA, 43; Univ Southern Calif, MA, 50. Prof Exp: Res psyhcologist, US Navy Electronics Lab, 48-67; RES PSYCHOLOGIST, NAVAL UNDERSEA CTR, 67- Mem: AAAS; Acoust Soc Am. Res: Speech intelligibility, intensity and pitch sensation and perception; thresholds. Mailing Add: Naval Undersea Ctr Code 4013 San Diego CA 92132

THOMPSON, PAUL WOODARD, b Manchester, NH, May 21, 09; m 36; c 2. CHEMISTRY. Educ: Univ Ill, BS, 30, MS, 32. Prof Exp: Chemist, State Water Surv, Ill, 30-32; res chemist, Sherwin Williams Paint & Varnish Co, Ill, 35-39, Acme White Lead & Color Works, Mich, 39-42 & Ethyl Corp, 42-71; RES ASSOC ECOL, CRANBROOK INST SCI, 56-, FEL, 66- Honors & Awards: Oakleaf Award, The Nature Conservancy, 75. Mem: AAAS; Am Chem Soc. Res: Oxidation reactions of tetraethyl lead; stability of halogen compounds and fuels; ecology and flora of Michigan; petroleum chemistry; conservation of natural areas; ecological survey, Sleeping Bear Dunes National Lakeshore, Huron Mountains, Michigan, and Michigan prairies. Mailing Add: Cranbrook Inst of Sci Bloomfield Hills MI 48013

THOMPSON, PETER ERVIN, b Urbana, Ill, Mar 20, 31; m 60; c 2. GENETICS. Educ: Purdue Univ, BS, 54, MS, 56; Univ Tex, PhD(genetics), 59. Prof Exp: NIH fel zool, Univ Calif, Berkeley, 59-60; res assoc biol, Oak Ridge Nat Lab, 60-61; from asst prof to assoc prof genetics, Iowa State Univ, 61-68; prof, 68-72, HEAD DEPT ZOOL, UNIV GA, 72- Concurrent Pos: Vis lectr, Univ Wis, 63, 64 & 66. Mem: Genetics Soc Am; Am Soc Nat. Res: Invertebrate and primate genetics; genetic control of protein synthesis; developmental regulation of gene activities; hemoglobin structures and evolution. Mailing Add: Dept of Zool 722 Biol Sci Bldg Univ of Ga Athens GA 30601

THOMPSON, PETER TRUEMAN, b Palmerton, Pa, Oct 15, 29; m 53; c 4. PHYSICAL CHEMISTRY. Educ: Johns Hopkins Univ, AB, 51; Univ Pittsburgh, PhD(phys chem), 57. Prof Exp: Res asst, Univ Pittsburgh, 51-56, res assoc & instr, 56-58; from instr to assoc prof, 58-73, PROF CHEM, SWARTHMORE COL, 73- Concurrent Pos: NSF sci fac fel, Cambridge Univ, 65-66; vis adj prof, Univ Del, 73-74. Mem: Am Chem Soc; Sigma Xi. Res: Physical chemistry of electolyte solutions both aqueous and non-aqueous. Mailing Add: Dept of Chem Swarthmore Col Swarthmore PA 19081

THOMPSON, PHEBE KIRSTEN, b Glace Bay, NS, Sept 5, 97; nat US; m 23; c 4. ENDOCRINOLOGY, GERIATRICS. Educ: Dalhousie Univ, MD & CM, 23. Prof Exp: Asst biochem, Sch Pub Health, Harvard Univ, 24-26; asst & res fel med, Metab Lab, Mass Gen Hosp, Boston, 26-29; res endocrinol, Cent Free Dispensary & Rush Med Col, 30-46; med ed & writing, 46-53; ED, J AM GERIAT SOC, 54- Concurrent Pos: Managing ed, J Clin Endocrinol & Metab, Endocrine Soc, 54-61, consult ed, J Clin Endocrinol & Metab & Endocrinol, 61-64. Honors & Awards: Thewlis Award, Am Geriat Soc, 66; Cert of Appreciation, Am Thyroid Asn, 66. Mem: AAAS; Endocrine Soc; fel Am Geriat Soc; fel Geront Soc; Am Genetic Asn. Res: Medical writing and editing. Mailing Add: 2337 N Commonwealth Ave Chicago IL 60614

THOMPSON, PHILIP DUNCAN, b Rossville, Ind, Apr 6, 22; m 44; c 5. METEOROLOGY. Educ: Univ Chicago, SB, 43; Mass Inst Technol, ScD(meteorol), 53. Prof Exp: US Air Force, 42-, proj officer meteorol, Univ Calif, Los Angeles, 45-46, meteorol group, Inst Advan Study, 46-48, chief atmospheric anal lab, Air Force Cambridge Res Ctr, 48-51, dir, Joint Geophys Res Directorate & Air Weather Serv Prediction Proj, 53-54, chief res & develop sect, Joint Numerical Prediction Unit, 54-58, ASSOC DIR, NAT CTR ATMOSPHERIC RES, US AIR FORCE, 60- Concurrent Pos: Lectr, Mass Inst Technol, 53; exchange lectr, Inst Meteorol, Univ Stockholm; lectr, Univ Colo, 64-; mem subcomt meteorol probs, Nat Adv Comt Aeronaut, 49-51; mem comt atmospheric sci, Nat Acad Sci-Nat Res Coun, 61- Honors & Awards: Legion of Merit, 57; Meisinger Award, Am Meteorol Soc, 60. Mem: Sigma Xi; Royal Meteorol Soc. Res: Mathematical and physical basis of weather prediction; theory of large scale disturbances in atmospheric and oceanic currents; theory of turbulence. Mailing Add: Nat Ctr Atmospheric Res Lab for Atmospheric Sci Boulder CO 80302

THOMPSON, PHILLIP EUGENE, b York, Pa, Nov 14, 46; m 73. SOLID STATE PHYSICS. Educ: Lebanon Valley Col, BS, 68; Univ Del, PhD(physics), 75. Prof Exp: Asst engr, York Div, Borg-Warner Corp, 69-70; ASST PROF PHYSICS, LEBANAN VALLEY COL, 74- Mem: Am Phys Soc; AAAS. Res: Radiation damage in solid state materials, particularly alkali halide crystals bombarded with charged particles in the energy interval of million electron volts per atomic mass unit; damage production. Mailing Add: Dept of Physics Lebanon Valley Col Annville PA 17003

THOMPSON, PHILLIP GERHARD, b Eagle Grove, Iowa, Jan 28, 30; m 55; c 2. INORGANIC CHEMISTRY. Educ: St Olaf Col, BA, 54; Cornell Univ, PhD(inorg chem), 59. Prof Exp: Fulbright scholar & Ramsay fel, Cambridge Univ, 59; sr chemist, Cent Res Labs, 3M Co, 59-64, sr res chemist, Contract Res Lab, Cent Res, 64-68, sr res scientist, Magnetic Prod Div, 68-70; TECH DIR & CONSULT, THOMPSON ASSOCS, 70- Concurrent Pos: Mem, Metrop Coun Comprehensive Health Planning Bd, Metrop Coun, 68-70; 3M Indust lectr, 3M Co, 68-70; vis Ramsay res fel, Dept Physics & Chem, St Olaf Col, 70-71; mem & secy, St Paul Environ Qual Bd, 70-72; comprehensive environ health fel, Dept Environ Health, Sch Pub Health, Univ Minn, 72 & 73; comnr, St Paul Water Bd, 72-, vpres, 74-; consult, Environ Health & Safety, Univ Minn, 74, vis lectr, 74- Mem: Air Pollution Control Asn; AAAS; Am Chem Soc; Am Indust Hyg Asn; Am Water Works Asn. Res: Fluorine, oxygen, nitrogen, boron and hydride chemistry; environmental chemistry; air and respirable mass sampling; water quality; trace contaminants; transformation of environmental pollutants; new synthetic techniques; unusual oxygen fluorine compounds; propellants; high performance sealants; ferrites; Mossbauer spectroscopy. Mailing Add: 229 Kennard St St Paul MN 55106

THOMPSON, QUENTIN ELWYN, b Woodstock, Ill, Oct 20, 24; m 49; c 4. INDUSTRIAL ORGANIC CHEMISTRY. Educ: Bradley Univ, BS, 48; Univ Wis, PhD(chem), 51. Prof Exp: Res chemist, 51-59, scientist, 59-67, ADV SCIENTIST, MONSANTO CO, 67- Mem: Am Chem Soc; Sigma Xi. Res: Organic synthesis, analysis and structural identification; organic chemistry of sulfur, phosphorus and ozone; traction fluids and lubricants; dielectric fluids. Mailing Add: Monsanto Co Res Dept 800 N Lindbergh Blvd St Louis MO 63166

THOMPSON, RALPH J, b Greenville, Tex, Apr 11, 30; m 59; c 2. PHYSICAL CHEMISTRY. Educ: ETex State Col, BS & MS, 54; Univ Tex, Austin, PhD(chem), 63. Prof Exp: Asst prof chem, Univ Tex, Arlington, 55-59; fel, Ind Univ, 63-65; assoc prof, 65-70, PROF CHEM, EASTERN KY UNIV, 70- Mem: Am Chem Soc. Res: Nuclear magnetic resonance of boron; nucleic acid research as related to brain function. Mailing Add: Dept of Chem Eastern Ky Univ Richmond KY 40475

THOMPSON, RALPH J, JR, b Los Angeles, Calif, Jan 27, 28; m 48; c 3. SURGERY. Educ: La Sierra Col, BS, 50; Loma Linda Univ, MD, 51; Am Bd Surg, dipl, 61. Prof Exp: Instr, 59-61, asst prof, 61-64, ASSOC PROF SURG, SCH MED, LOMA LINDA UNIV, 64- Concurrent Pos: Fel cancer surg, Mem Sloan-Kettering Cancer Ctr, 60-61. Mem: James Ewing Soc; Am Soc Clin Oncol. Res: Cancer surgery. Mailing Add: Dept of Surg Loma Linda Univ Med Ctr Loma Linda CA 92354

THOMPSON, RAYMOND HARRIS, b Portland, Maine, May 10, 24; m 48; c 2. ANTHROPOLOGY, ARCHAEOLOGY. Educ: Tufts Univ, BS, 47; Harvard Univ, AM, 50, PhD(anthrop), 55. Prof Exp: Asst dir, Peabody Mus Upper Gila Archaeol Exped, Harvard Univ, 49; asst prof anthrop & cur, Mus Anthrop, Univ Ky, 52-56; asst dir, Archaeol Field Sch, 56-60, from asst prof to assoc prof anthrop, 56-64, dir archaeol field sch, 61-65, PROF ANTHROP & HEAD DEPT & DIR ARIZ STATE MUS, UNIV ARIZ, 64- Concurrent Pos: Ed, Soc Am Archaeol J, 58-62; mem adv panel, Prog Anthrop, NSF, 63-65; mem Nat Acad Sci-Nat Res Coun panel, NSF grad fels, 64-66; mem USPHS res nursing in patient care rev comt, 67-; mem, Nat Acad Sci Comn Educ in Agr & Natural Resources Comt Soc Sci, 68; chmn, Comt for the

Recovery of Archaeol Remains, 73- Mem: Fel AAAS; fel Am Anthrop Asn; Soc Am Archaeol (pres elect, 75-76, pres, 76-77); Soc Ethnohist; Asn Mus. Res: Archaeology of New World; specialization in southwestern United States and Mesoamerica; archaeological theory; educational, administrative and publication policy in anthropology. Mailing Add: Dept of Anthrop Univ of Ariz Tucson AZ 85721

THOMPSON, RAYMOND K, b Vermillion, SDak, June 16, 16. NEUROSURGERY. Educ: Univ Md, BS, 37, MD, 41. Prof Exp: Asst neurosurg, 47-50, instr & dir res, 50-51, assoc, 51-53, asst prof, 53-62, assoc prof neurol surg, 62-73, CLIN PROF NEUROL SURG, SCH MED, UNIV MD, BALTIMORE CITY, 73-, NEUROSURGEON, UNIV HOSP, 47- Concurrent Pos: Consult, Provident Hosp, 47-58; mem courtesy staff, Church Home & Hosp, 47-62, assoc staff mem, 62-; attend neurosurgeon, St Agnes Hosp, 47-72, chief dept neurosurg, 59-71, hon staff mem, 72-; attend neurosurgeon, Md Gen Hosp, 47-, chief dept neurosurg, 59-70; attend neurosurgeon, SBaltimore Gen, Bon Secours & Lutheran Hosps, 47- & Franklin Square Hosp, 48-; neurosurgeon, Mercy Hosp, 47-; from asst vis neurosurgeon to vis neurosurgeon, Baltimore City Hosps, 48-54, consult, 54-67; vis neurosurgeon, St Joseph's Hosp, 48-52, chief dept neurosurg, 52-71, assoc staff mem, 72-; consult, USPHS Hosp, 48-73; assoc staff mem neurosurg, Hosp for Women of Md, 61-65 & Greater Baltimore Med Ctr, 65- Mem: Fel Am Col Surgeons; Neurosurg Soc Am (vpres, 58, secy, 59-62, pres, 64); Am Asn Neurol Surgeons; AMA; Cong Neurol Surgeons (pres, 58). Res: Neurophysiologic research relative to the brain vasculature and stress patterns of neurone conduction. Mailing Add: Dept of Neurosurg Univ of Md Sch of Med Baltimore MD 21201

THOMPSON, RAYMOND MELVIN, b Judith Gap, Mont, May 19, 18; m 41; c 2. GEOLOGY. Educ: Mont Sch Mines, BS, 41; Univ Wyo, MA, 50. Prof Exp: Archeol foreman, Nat Youth Admin, Mont, 36; supvr archeol projs & in chg excavations & state lab, Works Prog Admin, 38-39; seismic helper, Magnolia Petrol Co, Tex, 41, jr geologist, 41-42; jr geologist, US Geol Surv, Ark, 42-43, jr geologist & party chief, Ga, 43, asst geologist, Wyo, 43-44, assoc geologist, 46-47, geologist, 47-50; div geologist, BA Oil Co, 50-51; mgr, Rocky Mt Div, Hiawatha Oil Co, 51-53; consult geologist & engr, 53-68; exec vpres, Inlet Oil Corp, 69-73, sr vpres explor, 71-73, dir & vpres, Inlet North Sea, Ltd, 71-73; PRES, TMT Corp, 73-; PRES, VIPONT MINING CORP, 76- Concurrent Pos: Assoc geologist & party chief arctic explor, Naval Petrol Reserve, US Geol Surv, 47; mem, Panel Deep Ocean Mining, Nat Acad Sci, 73-75. Mem: Fel Geol Soc London; Am Inst Mining, Metall & Petrol Engrs; fel Geol Soc Am; Am Asn Petrol Geologists; assoc Arctic Inst NAm. Res: Stratigraphy and structure of the Rocky Mountain area; oil and gas possibilities and stratigraphy and structure of arctic Alaska, North Africa and Europe; bauxite; high alumina clays; uranium; minerals. Mailing Add: 6901 S Yosemite No 208 Englewood CO 80110

THOMPSON, RICHARD BAXTER, b Fresno, Calif, June 1, 26; m 50; c 3. FISH BIOLOGY Educ: San Jose State Col, BA, 50; Univ Wash, PhD(fisheries), 66. Prof Exp: Fishery res biologist, Fisheries Res Inst, Univ Wash, 50-54; FISHERY RES BIOLOGIST, NORTHWEST FISHERIES CTR, NAT MARINE FISHERIES SERV, NAT OCEANIC & ATMOSPHERIC ADMIN, 54- Mem: Am Fisheries Soc; Am Inst Fishery Biol. Res: Fish ethology; orientation and navigation of anadromous fishes; sensory perception of fishes; marine game fishery resources and utilization. Mailing Add: 3244 NE 104th Seattle WA 98125

THOMPSON, RICHARD BRUCE, b Fargo, NDak, Oct 12, 39; m 61; c 2. MATHEMATICS. Educ: Univ Northern Iowa, BA, 61; Univ Wis, MS, 63, PhD(topology), 67. Prof Exp: Asst prof, 67-70, ASSOC PROF MATH, UNIV ARIZ, 70- Concurrent Pos: NSF res grants, 68-71. Mem: Am Math Soc; Math Asn Am. Res: Topological fixed point theory, particularly semicomplexes, quasi-complexes and local and global fixed point indices. Mailing Add: Dept of Math Univ of Ariz Tucson AZ 85721

THOMPSON, RICHARD CLAUDE, b Kansas City, Mo, Mar 12, 39. INORGANIC CHEMISTRY. Educ: Univ Chicago, BS, 61; Univ Md, PhD(chem), 65. Prof Exp: Resident res assoc, Chem Div, Argonne Nat Lab, 65-66; asst prof chem, Ill Inst Technol, 66-67; asst prof, 67-70, ASSOC PROF CHEM, UNIV MO-COLUMBIA, 70- Concurrent Pos: Consult, Argonne Nat Lab, 66- Mem: Am Chem Soc. Res: Kinetics and mechanisms of inorganic reactions. Mailing Add: Dept of Chem Univ of Mo Columbia MO 65201

THOMPSON, RICHARD FREDERICK, b Portland, Ore, Sept 6, 30; m 60; c 3. PHYSIOLOGICAL PSYCHOLOGY. Educ: Reed Col, BA, 52; Univ Wis, MS, 53, PhD(physiol psychol), 56. Prof Exp: NIH fel physiol, Univ Wis, 56-59; from asst prof to prof med psychol, Med Sch, Univ Ore, 59-67; prof, Univ Calif, Irvine, 67-73; prof psychol, Harvard Univ, 73-75; PROF PSYCHOL, UNIV CALIF, IRVINE, 75- Concurrent Pos: Nat Inst Ment Health res career award, 62-67, 67-73; mem adv panel psychobiol, NSF, 67-70; mem res scientist rev comt, Nat Inst Ment Health, 69-74; mem comt biol bases soc behav, Social Sci Res Coun, 72; mem US Nat Comt for the Int Brain Res Orgn, 75-78. Honors & Awards: Commonwealth Award, 66; Distinguished Sci Contrib Award, Am Psychol Asn, 75. Mem: Am Physiol Soc; Soc Neurosci; Am Psychol Asn; fel AAAS; Int Neuropsychol Soc. Res: Neurophysiology; cerebral cortex and behavior; neural basis of learning. Mailing Add: Dept of Psychobiol Univ of Calif Irvine CA 92664

THOMPSON, RICHARD JOHN, b Chapman Ranch, Tex, Aug 9, 27; m 52; c 2. ANALYTICAL CHEMISTRY, AIR POLLUTION. Educ: Univ Tex, BS, 52, MA, 56, PhD(inorg chem), 59. Prof Exp: Asst prof chem, Lamar State Col, 57-58 & North Tex State Univ, 59-62; from asst prof to assoc prof, Tex Tech Col, 62-68; chief metals & adv anal unit, Air Qual & Emission Data Prog & supvry res chemist, Nat Air Pollution Control Admin, 68-69, chief lab serv br, Div Air Qual & Emissions Data, Bur Criteria & Standards, 69-71; chief, Air Qual Anal Lab Br, Div Atmospheric Surveillance, 71-73, chief, Qual Assurance & Environ Monitoring Lab, 73-75, CHIEF, ANAL CHEM BR, ENVIRON MONITORING SUPPORT LAB, ENV RES CTR, ENVIRON PROTECTION AGENCY, 75- Concurrent Pos: Res grants, Res Corp, 60-, Welch Found, 61-69 & NSF, 64-68, 74-75; adj prof, NC State Univ, 74; consult, Lawrence Livermore Lab, 70-72 & NC State Univ, 75- Mem: Fel AAAS; Am Chem Soc; Sigma Xi. Res: Inorganic and analytical chemistry of rhenium; inorganic syntheses; non-aqueous solvent chemistry; development of methods for atmospheric pollutants, including trace elements and precipitation components. Mailing Add: Anal Chem Br MS-78 Environ Protection Agency Research Triangle Park NC 27711

THOMPSON, RICHARD MICHAEL, b Thief River Falls, Minn, May 30, 45; m 67; c 1. BIOCHEMISTRY, ANALYTICAL CHEMISTRY. Educ: Univ Minn, BChem, 67; Univ Wis, PhD(biochem), 71. Prof Exp: ASST PROF PEDIAT, SCH MED, IND UNIV, INDIANAPOLIS, 73- Concurrent Pos: NIH trainee, Baylor Col Med, 71-72, Nat Heart Inst res associateship, 72-73. Mem: AAAS; Am Chem Soc; Am Soc Mass Spectrometry. Res: Biochemical genetics; gas phase analytical techniques; drug metabolism; identification of natural products. Mailing Add: Dept of Pediat Ind Univ Sch of Med Indianapolis IN 46202

THOMPSON, RICHARD SCOTT, b Lubbock, Tex, May 24, 39; m 67; c 1. PHYSICS. Educ: Calif Inst Technol, BS, 61; Harvard Univ, AM, 62, PhD(physics), 66. Prof Exp: NSF fel, Ctr Nuclear Res, France, 66; mem, Inter-Acad Exchange Prog, Inst Theoret Physics, Moscow, 66-67; vis foreign scientist grant, Ctr Nuclear Res, France, 67-68; asst physicist, Brookhaven Nat Lab, 68-70; asst prof physics, 70-72, ASSOC PROF PHYSICS, UNIV SOUTHERN CALIF, 72- Mem: Am Phys Soc. Res: Superconductivity. Mailing Add: Dept of Physics Univ of Southern Calif Los Angeles CA 90007

THOMPSON, ROBERT, b Chicago, Ill, June 21, 27; m 58; c 1. NEUROPSYCHOLOGY. Educ: Univ Tampa, BS, 50; Fla State Univ, MS, 52; Univ Tex, PhD(psychol), 55. Prof Exp: Asst, Yerkes Labs Primate Biol, Univ Fla, 51-52; asst prof psychol, La State Univ, 54-57; sr res psychologist, Southeast La Hosp, New Orleans, 57-59; assoc prof psychol, Peabody Col, 59-60; assoc prof psychiat, Med Ctr, Univ Calif, Los Angeles & res Psychologist, Neuropsychiat Inst, 60-63; PROF PSYCHOL, LA STATE UNIV, 63- Mem: AAAS; fel Am Psychol Asn; Am Asn Anat. Res: Neuropsychology of memory and learning. Mailing Add: Dept of Psychol La State Univ Baton Rouge LA 70803

THOMPSON, ROBERT CHARLES, b Winnipeg, Man, Apr 21, 31; m 60. MATHEMATICS. Educ: Univ BC, BA, 54, MA, 56; Calif Inst Technol, PhD(math), 60. Prof Exp: Defense sci officer, Defence Res Bd, Can, 56-57; from instr to asst prof math, Univ BC, 60-64; from asst prof to assoc prof, 64-69, PROF MATH, UNIV CALIF, SANTA BARBARA, 69- Concurrent Pos: Ed, J Linear & Multilinear Algebra. Mem: Am Math Soc; Math Asn Am; Soc Indust & Appl Math. Res: Algebra, especially linear algebra and number theory. Mailing Add: Dept of Math Univ of Calif Santa Barbara CA 93106

THOMPSON, ROBERT DEANE, b Grand Rapids, Mich, Oct 8, 09; m 32; c 3. INSTRUMENTATION. Educ: Univ Mich, BSE(chem) & BSE(math), 31, PhD(phys chem), 37. Prof Exp: Asst chem, Univ Mich, 36-37; Rackham fel, Nat Bur Standards, 37-38, res chemist & tech adv, 38-43; glass prod engr, Commercial & Glass Prod Div, Taylor Instrument Co, 43-54, chief develop engr, 54-60, coordr indust develop, 60-72, mgr long range planning, 72-74; RETIRED. Concurrent Pos: Mem adv comt commercial standards, Nat Bur Standards. Honors & Awards: Award of Merit, Am Soc Test & Mat, 65. Mem: Am Chem Soc; fel Am Soc Test & Mat; Instrument Soc Am. Res: Precision electrical conductivity measurements involving redetermination of standard reference values; precision thermometry; comparison of thermodynamic and international temperature scales; manufacture of thermometers, barometers, hydrometers, compasses, medical and industrial process control instruments. Mailing Add: 66 Stone Island Lane Penfield NY 14526

THOMPSON, ROBERT EDWARD, b Waubay, SDak, June 2, 15; m 51. PHARMACOLOGY. Educ: Univ Md, BS, 38, MS, 40, PhD(pharmacol), 43. Prof Exp: Pharmacist, Morgan & Millard, Md, 36-42; asst pharmacol, Sch Pharm, Univ Md, 38-43; head pharmacol res, Armour Pharmaceut Co, 43-52; res assoc, Marvin R Thompson, Inc, 52-54; head pharmaceut develop, Armour Pharmaceut Co, 54-62; dir pharmaceut res & develop, 62-70, sr dir pharmaceut sci, 70-75, SR ADV PHARMACEUT SCI, SCHERING CORP, 75- Concurrent Pos: Lectr, Sch Med, Univ Ill, 47-50. Mem: Endocrine Soc; Soc Exp Biol & Med; NY Acad Sci. Res: Quantitative pharmacology; pharmaceutical chemistry drug product research and development. Mailing Add: Schering Corp Bloomfield NJ 07003

THOMPSON, ROBERT GENE, b Hiddenite, NC, Dec 23, 31; m 53; c 2. PHYSICAL ORGANIC CHEMISTRY. Educ: Univ NC, BS, 52; Univ Tenn, MS, 54, PhD(chem), 56. Prof Exp: Res chemist, 56-60, sr res chemist, 60-61, supvr res, 61-63, tech, 63-64, sr supvr, 64-66, tech supt, 66-71, RES MGR, E I DU PONT DE NEMOURS & CO, INC, 71- Mem: Am Chem Soc. Res: Vinyl and condensation polymers; synthetic textile fibers. Mailing Add: Textile Fibers Dept Chestnut Run Wilmington DE 19801

THOMPSON, ROBERT GRAY, chemical engineering, research administration, see 12th edition

THOMPSON, ROBERT HARRY, b Columbus, Ohio, May 2, 24; m 47. MATHEMATICS, COMPUTER SCIENCE. Educ: Sterling Col, BS, 45, DSc, 69; Univ Kans, MA, 51. Prof Exp: Prof math, Sterling Col, 47-67; ASSOC PROF MATH, WASHBURN UNIV TOPEKA, 67- Mem: Math Asn Am. Res: General mathematics. Mailing Add: 5530 W 25th Terr Topeka KS 66614

THOMPSON, ROBERT JAMES, b Dayton, Ohio, Sept 21, 30; m 56; c 2. MATHEMATICS. Educ: Ohio State Univ, BSc, 52, MSc, 54, PhD(math), 58. Prof Exp: Asst math, Ohio State Univ, 52-53, from asst instr to instr, 53-58; mem staff, 58-65, supvr appl math div II, 65-72, supvr numerical anal div, 72-75, SUPVR APPL MATH DIV, SANDIA LAB, 75- Mem: Am Math Soc; Math Asn Am; Soc Indust & Appl Math. Res: Applied mathematics; numerical analysis. Mailing Add: Dept 5120 Sandia Lab Albuquerque NM 87115

THOMPSON, ROBERT JOHN, JR, b San Francisco, Calif, Nov 10, 17; m 45; c 3. PHYSICAL CHEMISTRY. Educ: Univ Calif, Los Angeles, BS, 40; Univ Rochester, PhD(phys chem), 46. Prof Exp: Control chemist, Eastman Kodak Co, 37-41; res assoc, George Washington Univ, 43-46; sr res engr, M W Kellogg Co Div, Pullman, Inc, 46-53 & Bendix Aviation Corp, 53-54; vpres & dir res div, Rocketdyne Div, NAm Aviation, Inc, Calif, 54-71, vpres & gen mgr, Rocketdyne Solid Rocket Div, 71-72, sr staff scientist, Rocketdyne Div, NAm Rockwell Corp, 72-73; STAFF ASST TO DIR, APPL PHYSICS LAB, JOHNS HOPKINS UNIV, 74- Concurrent Pos: Mem subcomt rocket engines, NASA, 51-54 & subcomt aircraft fuels, 58, mem res adv comt energy processes, 59- Mem: AAAS; Am Chem Soc; Am Inst Aeronaut & Astronaut; Am Inst Chem Engrs. Res: Guided missiles; rocket and jet propulsion; propellants and fuels; combustion; heat transfer and fluid flow; chemical processes, thermodynamics and kinetics; radiation and spectra; space science; energy processes, systems and applications. Mailing Add: 12912 Ruxton Rd Silver Spring MD 20904

THOMPSON, ROBERT KRUGER, b Jeffersonville, Ohio, Jan 15, 22; m 43; c 3. ENTOMOLOGY. Educ: Ohio State Univ, BSc, 47, MSc, 48, PhD(entom), 50. Prof Exp: Field aide, Bur Entom & Plant Quarantine, US Dept Agr, 46-47; field res entomologist, 50-51, field res supvr, 51-65, SUPVR BIOL RES LABS, ORTHO DIV, CHEVRON CHEM CO, 66- Mem: Entom Soc Am. Res: Chemical control of insects, weeds and plant diseases. Mailing Add: Ortho Div Chevron Chem Co 940 Hensley Richmond CA 94801

THOMPSON, ROBERT MILTON, organic chemistry, see 12th edition

THOMPSON, ROBERT NATHAN, industrial hygiene, see 12th edition

THOMPSON, ROBERT POOLE, b Winnipeg, Man, Feb 8, 23; m 53; c 5. DEVELOPMENTAL BIOLOGY. Educ: Univ Western Ont, BA, 49, MSc, 53; Univ Toronto, PhD(zool), 63. Prof Exp: Res officer entom, Can Dept Agr, 53-56; from instr to assoc prof biol, St Francis Xavier Univ, 56-67; assoc prof, 67-72, PROF BIOL, STATE UNIV NY COL BROCKPORT, 72- Concurrent Pos: Nat Res Coun

Can grants, 63-68; State Univ NY Res Found fac res fel, 68-69. Mem: AAAS; Soc Develop Biol; Soc Study Reproduction; Am Soc Zoologists. Res: Homologous inhibition in the development of pattern in the embryo; time of eruption of the third molar tooth. Mailing Add: Dept of Biol Sci State Univ of NY Col Brockport NY 14420

THOMPSON, ROBERT QUINCY, biochemistry, see 12th edition

THOMPSON, ROBERT RICHARD, b Springfield, Mo, Mar 30, 31; m 55; c 2. ORGANIC GEOCHEMISTRY. Educ: Drury Col, BS, 53; Wash Univ, St Louis, 55-56, PhD(org chem), 57. Prof Exp: Sr res engr, Pan Am Petrol Corp, Standard Oil Co, Ind, 57-61, tech group supvr, 61-65, staff res scientist, 65-71; res group supvr, 71-75, RES SECT MGR, AMOCO PROD CO, 75- Mem: Am Chem Soc; Geochem Soc. Res: Organic geochemistry; origin of oil; geochemical prospecting. Mailing Add: Res Dept Amoco Prod Co PO Box 591 Tulsa OK 74102

THOMPSON, ROBERT WALDER, physics, see 12th edition

THOMPSON, RODGER IRWIN, b Texarkana, Tex, Aug 9, 44; m 67; c 2. ASTROPHYSICS. Educ: Mass Inst Technol, SB, 66, PhD(physics), 70. Prof Exp: Asst prof optical sci, 70-71, asst prof astron, 71-74, ASSOC PROF ASTRON, STEWARD OBSERV, UNIV ARIZ, 74- Mem: Am Phys Soc; Am Astron Soc. Res: Theoretical astrophysics including molecular physics, stellar evolution and nucleosynthesis; observational infrared and visible spectroscopy with Fourier transform spectrometers. Mailing Add: Steward Observ Univ of Ariz Tucson AZ 85721

THOMPSON, RONALD EARL, organic chemistry, physical chemistry, see 12th edition

THOMPSON, RONALD HALSEY, b Brooklyn, NY, Apr 29, 26; m 51; c 2. INSTRUMENTATION, MATERIALS SCIENCE. Educ: Adelphi Col, BA, 50; Columbia Univ, MA, 51; Univ Pa, PhD, 59. Prof Exp: Technician, Cornell Univ, 51-53 & Univ Pa, 53-55; physiologist, Nat Insts Health, 55-75, Sci Dir, 68-75; TEACHER, NORTHERN VA COMMUNITY COL, 74- Mem: AAAS; Inst Elec & Electronics Eng. Mailing Add: 3200 Shoreview Rd Triangle VA 22172

THOMPSON, RONALD HOBART, b Memphis, Tenn, Feb 21, 35; m 60; c 4. NUCLEAR CHEMISTRY. Educ: La Tech Univ, BS, 61, MS, 68; Univ Ark, PhD(chem), 72. Prof Exp: Chemist, Western Elec Corp, 68-70; ASST PROF CHEM, LA TECH UNIV, 72- Mem: Am Chem Soc; AAAS. Res: Cosmology; concentration of trace elements on the earth; solar wind composition; medical applications of radioisotopes. Mailing Add: Dept of Chem La Tech Univ Ruston LA 71270

THOMPSON, RORY, b Seattle, Wash, May 10, 42. PHYSICAL OCEANOGRAPHY. Educ: San Diego State Col, AB, 62, MS, 64; Mass Inst Technol, PhD(meteorol), 68. Prof Exp: Res asst math statist, San Diego State Col, 62-64; fel phys oceanog, Woods Hole Oceanog Inst, 68-69; asst prof atmospheric sci, Ore State Univ, 69-70; asst scientist, 70-72, ASSOC SCIENTIST PHYS OCEANOG, WOODS HOLE OCEANOG INST, 72- Concurrent Pos: Vis lectr, Univ Western Australia, 75. Mem: Am Meteorol Soc; Am Geophys Union; Royal Meteorol Soc. Res: Geophysical fluid dynamics; computational mathematics; simulation. Mailing Add: C304 Woods Hole Oceanog Inst Woods Hole MA 02543

THOMPSON, ROSEMARY ANN, b San Diego, Calif, May 15, 45; m 67; c 1. MARINE BIOLOGY. Educ: Univ Mo-Columbia, BA, 67; Univ Calif, San Diego, PhD(marine biol), 72. Prof Exp: Res assoc marine biol, Univ Southern Calif, 72-73; MARINE BIOLOGIST, HENNINGSON, DURHAM & RICHARDSON, 74- Concurrent Pos: Consult environ scientist, EG&G Co, 74; consult, Hinningson, Durham & Richardson, 74. Res: Marine and aquatic biology of fish as related to pollution and modification of the environment as well as mariculture. Mailing Add: Henningson Durham & Richardson Ecosci 804 Anacapa St Santa Barbara CA 93101

THOMPSON, ROY CHARLES, JR, b Kansas City, Mo, June 19, 20; div; c 4. RADIATION BIOLOGY, BIOCHEMISTRY. Educ: Univ Tex, BA, 40, MA, 42, PhD(biochem), 44. Prof Exp: Tutor, Univ Tex, 40-41, instr, 41-43, res assoc biochem, 43-44, asst prof chem, 47-50; res chemist, Manhattan Dist, US Army Engrs Plutonium Proj, Metall Lab, Univ Chicago, 44-46; res chemist, Radiation Lab, Univ Calif, 46-47; res chemist, Gen Elec Co, Washington, 50-65; STAFF SCIENTIST, BIOL DEPT, PAC NORTHWEST LAB, BATTELLE MEM INST, 65- Concurrent Pos: Mem comt int exposure, Int Comn Radiol Protection. Mem: AAAS; Radiation Res Soc; Health Physics Soc. Res: Radiochemical study of biochemical processes; evaluation of hazards from internally deposited radioisotopes; especially plutonium and other transuranium elements. Mailing Add: Biol Dept Pac Northwest Lab Battelle Mem Inst Richland WA 99352

THOMPSON, ROY DANIEL, dairy science, animal physiology, see 12th edition

THOMPSON, ROY LLOYD, b Minn, Apr 29, 27; m 54; c 3. AGRONOMY. Educ: Univ Minn, BS, 51, MS, 59; Pa State Univ, PhD(agron), 67. Prof Exp: Field supvr, Minn Crop Improv Asn, 49-51; agronomist, Univ Minn, Morris, 56-57 & Rockefeller Found, 67-72; EXTEN AGRONOMIST, UNIV MINN, ST PAUL, 72- Mem. Am Soc Agron; Crop Sci Soc Am. Res: Applied crop physiology in the management of field crops for the development and improvement of crop production systems. Mailing Add: Dept of Agron Univ of Minn St Paul MN 55101

THOMPSON, RUFUS HENNEY, b St Louis, Mo, Dec 16, 08; m 38; c 1. MORPHOLOGY. Educ: Univ Kans, AB, 36, AM, 37; Stanford Univ, PhD(biol), 41. Prof Exp: Asst instr, Univ Kans, 36-38, instr bot, 41-42; asst, Stanford Univ, 38-41; hydrograph investr, State Dept Res & Educ, Md, 46-47; from asst prof to assoc prof bot, 47-59, PROF BOT, UNIV KANS, 59- Concurrent Pos: Fulbright scholar, Auckland, 54-55; Guggenheim fel, 57-58; prin investr, US Antarctic Res Prog, 60-61. Mem: Bot Soc Am; Phycol Soc Am; Am Bryol & Lichenological Soc. Res: Morphology and taxonomy of the Algae and Hepaticae; pure culture of Chrysophyceae and Dinophyceae; algal life cycles. Mailing Add: Dept of Bot Univ of Kans Lawrence KS 66045

THOMPSON, RUTHERFORD BOSTON, JR, polymer chemistry, see 12th edition

THOMPSON, SAMUEL, III, b Dallas, Tex, Aug 12, 32. PETROLEUM GEOLOGY, STRATIGRAPHY. Educ: Southern Methodist Univ, BS, 53; Univ NMex, MS, 55. Prof Exp: Petrol geologist, Exxon Corp, 54-74; PETROL GEOLOGIST, NMEX BUR MINES & MINERAL RESOURCES, 74- Mem: Am Asn Petrol Geologists; Soc Econ Paleontologists & Mineralogists; Int Asn Sedimentologists. Res: Regional evaluation of the potential for petroleum exploration in southwestern New Mexico; system for analysis of sedimentary units; physico-stratigraphy, chronostratigraphy and eustatic geochronology. Mailing Add: NMex Bur Mines & Mineral Resources Socorro NM 87801

THOMPSON, SAMUEL ALCOTT, b Lynchburg, Va, June 20, 98; m 32; c 2. SURGERY. Educ: Wake Forest Col, BS, 18; Jefferson Med Col, MD, 20; Bd Thoracic Surg, dipl, 48. Hon Degrees: DSc, Wake Forest Col, 60. Prof Exp: Assoc prof surg, 37-60, PROF CLIN SURG, NEW YORK MED COL, 60- Concurrent Pos: Attend surgeon, Flower & Fifth Ave Hosp, New York, 35-; attend thoracic surgeon, Metrop Hosp, 37-68, Rikers Island Hosp, 39- & St Clare's Hosp, 40-71; consult, St Joseph's Hosp, 41-, Paterson Gen Hosp, NJ, 48-, Passaic Gen Hosp, NJ, 57-, St Clare's Hosp, NY Infirmary & Metrop Hosp. Mem: Fel Am Asn Thoracic Surg; fel AMA; fel Am Col Surgeons; fel Am Col Chest Physicians; fel Am Col Cardiol. Res: Bronchial catheterization; resuscitation; surgical treatment of angina pectoris; thoracic surgery. Mailing Add: 888 Park Ave New York NY 10021

THOMPSON, SAMUEL LEE, b Hopkinsville, Ky, Oct 24, 41; m 59; c 2. THEORETICAL PHYSICS. Educ: Murray State Univ, BS, 62; Univ Ky, PhD(physics), 66. Prof Exp: TECH STAFF MEM, SANDIA CORP, 66- Mem: Am Phys Soc. Res: Equation of state; hydrodynamics; radiation transport; molecular relaxation. Mailing Add: Div 5162 Sandia Corp PO Box 5800 Albuquerque NM 87115

THOMPSON, SAMUEL STANLEY, JR, b Monroe, La, Nov 25, 36; m 63; c 3. PHYTOPATHOLOGY. Educ: La Polytech Inst, BS, 59; Purdue Univ, MS, 61, PhD(plant path), 65. Prof Exp: EXTEN PLANT PATHOLOGIST, RURAL DEVELOP CTR, 62- Mem: Am Phytopath Soc. Res: Extension service educational work on tobacco, soybean and peanut diseases. Mailing Add: Rural Develop Ctr PO Box 48 Tifton GA 31794

THOMPSON, SAMUEL WESLEY, II, b Moberley, Mo, July 25, 25; m 46; c 3. VETERINARY PATHOLOGY, HISTOCHEMISTRY. Educ: Iowa State Univ, DVM, 46, MS, 56. Prof Exp: Post vet, Army Posts in US & Guam, 46-51; lab officer virol, Walter Reed Army Inst Res, DC, 51-52, Fourth Army Area Med Lab, Ft Sam Houston, Tex, 52-53; prof med sci & tactics, Vet Reserve Officers Training Corps, Iowa State Univ, 53-56, lab officer path, Armed Forces Inst Path, 56-57, chief path div, Army Med Res & Nutrit Lab, Univ Colo, 57-63; staff pathologist, Middle Am Res Unit, CZ, Panama, 63-64; command officer med res, Army Med Res Unit, Panama, 64-65; chief vet path, Walter Reed Army Inst Res, 65-66, sr pathologist, Armed Forces Inst Path, 66-67; mgr path, Ciba Pharmaceut Co, 67-71, MGR PATH, CIBA-GEIGY PHARMACEUT DIV, 71- Concurrent Pos: Affiliate prof, Col Vet Med, Colo State Univ, 63-66; lectr, Georgetown Univ, 66-67; nat proj officer, Toxicity Testing Group of Surgeon Gen Intravenous Alimentation Task Force, US Army Med Res & Develop Command, 61-63; consult, chief of Vet Corps, US Army, 66-67; hon trustee, Mem Adv Bd & Progs Dir, Charles Louis Davis DVM Found Advan Vet Path; consult vet toxicol, Surgeon Gen US Army, 72- Mem: Am Vet Med Asn; Am Col Vet Path; Am Col Vet Toxicol; Am Inst Nutrit; Histochem Soc. Res: Comparative histochemistry; pathology of laboratory animals; spontaneous and induced lesions and application of histochemistry to the study of these diseases. Mailing Add: Path Dept Subdiv of Toxicol Ciba-Geigy Pharmaceut Div 556 Morris Ave Summit NJ 07901

THOMPSON, SANFORD P, b Flushing, NY, Dec 10, 21; m 44, 71; c 2. THEORETICAL PHYSICS. Educ: Union Col, BS, 42; Yale Univ, MS, 43, PhD(physics), 50. Prof Exp: Physicist, US Naval Res Lab, 43-63; PROF PHYSICS, RANDOLPH-MACON COL, 63- Mem: Acoust Soc Am. Res: Electromechanical reciprocity; mechanical shock and vibration; mathematics of optimization. Mailing Add: PO Box 547 Ashland VA 23005

THOMPSON, STANLEY GERALD, b Los Angeles, Calif, Mar 9, 12; m 38; c 2. CHEMISTRY. Educ: Univ Calif, Los Angeles, AB, 34; Univ Calif, Berkeley, PhD(chem), 48. Prof Exp: Chemist, Stand Oil Co Calif, 34-42, Metall Lab, Univ Chicago, 42-44 & 45-46 & Hanford Eng Works, E I du Pont de Nemours & Co, Inc, 44-45; CHEMIST, LAWRENCE BERKELEY LAB, UNIV CALIF, 46- Concurrent Pos: Guggenheim fel, Sweden, 55 & Denmark, 65. Honors & Awards: Award, Am Chem Soc, 65. Mem: Assoc Am Chem Soc; assoc Am Phys Soc. Res: Nuclear chemistry on transuranium elements; co-discoverer of berkelium, californium, einsteinium, fermium and mendelevium; nuclear fission process; research on fission and heavy ion reactions; developed concept of chemical process for first industrial scale separation and isolation of plutonium. Mailing Add: 85 Fairlawn Dr Berkeley CA 94708

THOMPSON, STEVEN RISLEY, b Hermiston, Ore, Dec 3, 38; m 68. ZOOLOGY, GENETICS. Educ: Portland State Col, BS, 61; Ore State Univ, MS, 64, PhD(zool), 66. Prof Exp: Res assoc genetics, Univ Notre Dame, 66-68; ASST PROF GENETICS, ITHACA COL, 68- Mem: Genetics Soc Am; Am Soc Zoologists. Res: Developmental genetics of Drosophila Melanogaster. Mailing Add: Dept of Biol Ithaca Col Ithaca NY 14850

THOMPSON, THOMAS ARTHUR, biochemistry, see 12th edition

THOMPSON, THOMAS EATON, b San Mateo, Calif, Aug 10, 38; div; c 2. SOLID STATE PHYSICS. Educ: Univ Calif, Berkeley, AB, 60; Univ Pa, MS, 62, PhD(physics), 69. Prof Exp: ASST PROF SOLID STATE ELECTRONICS, UNIV PA, 70- Mem: Am Phys Soc. Res: Electronic structure of metals and semiconductors; graphite intercalation compounds; magnetic quantum effects in solids; ultrasonics; surface acoustic waves. Mailing Add: Moore Sch of Elec Eng Univ of Pa Philadelphia PA 19174

THOMPSON, THOMAS EDWARD, b Cincinnati, Ohio, Mar 15, 26; m 53; c 4. BIOCHEMISTRY. Educ: Kalamazoo Col, BA, 49; Harvard Univ, PhD(biochem), 55. Prof Exp: From asst prof to assoc prof physiol chem, Sch Med, Johns Hopkins Univ, 58-66; PROF BIOCHEM & CHMN DEPT, SCH MED, UNIV VA, 66- Concurrent Pos: NIH res fel biochem, Harvard Univ, 55-57; Swed-Am exchange fel, Am Cancer Soc, LKB-Produkter Fabrikasaktiebolog, Sweden, 57-58; hon res fel, Birmingham, Eng, 58. Mem: Am Chem Soc; Biophys Soc; Soc Develop Biol; Am Soc Biol Chemists; The Chem Soc. Res: Physical chemistry of proteins; lipid protein interactions; biological membrane structure. Mailing Add: Dept of Biochem Univ of Va Sch of Med Charlottesville VA 22901

THOMPSON, THOMAS LEO, b Gering, Nebr, Dec 8, 22; m 46; c 3. BACTERIAL PHYSIOLOGY. Educ: Univ Nebr, AB, 48, MS, 50; Univ Tex, PhD, 53. Prof Exp: From asst prof to assoc prof, 52-64, PROF MICROBIOL, UNIV NEBR, LINCOLN, 64-, ACTG CHMN DEPT, 70- Mem: Am Soc Microbiol. Res: Bacterial physiology, mainly resistance to antibiotics; bacterial genetics in relation to thermophily and the transformation phenomenon. Mailing Add: Dept of Microbiol Univ of Nebr Sch of Life Sci Lincoln NE 68508

THOMPSON, THOMAS LUMAN, b Boulder, Colo, Dec 25, 27; m 56; c 6. GEOLOGY. Educ: Univ Colo, BA, 50; Stanford Univ, PhD(geol), 62. Prof Exp: Geologist, Phillips Petrol Corp, 50-51; geologist, Pan Am Petrol Corp, 69-72, STAFF RES SCIENTIST, AMOCO PROD CO, 72- Mem: Geol Soc Am; Am Asn Petrol

Geol; Am Geophys Union. Res: World tectonics; structure of continental margins; deep water petroleum and mineral exploration. Mailing Add: Res Dept Amoco Prod Co PO Box 591 Tulsa OK 74102

THOMPSON, THOMAS LUTHER, b Houston, Tex, Feb 28, 38. GEOLOGY. Educ: Univ Kans, BS, 60, MS, 62; Univ Iowa, PhD(geol), 65. Prof Exp: GEOLOGIST, MO GEOL SURV & WATER RESOURCES, 65-, CHIEF AREAL GEOL & STRATIG, 71- Mem: Geol Soc Am; Paleont Soc; Soc Econ Paleont & Mineral; Int Paleont Union; Pander Soc. Res: Stratigraphy; biostratigraphy; micropaleontology; correlation of Paleozoic strata through the use of paleontology. Mailing Add: Mo Geol Surv & Water Resources PO Box 250 Rolla MO 65401

THOMPSON, THOMAS WILLIAM, marine biology, botany, see 12th edition

THOMPSON, THOMAS WILLIAM, b Canton, Ohio, May 25, 36; m 66; c 1. SPACE PHYSICS. Educ: Case Inst Technol, BS, 58; Yale Univ, ME, 59; Cornell Univ, PhD(elec eng), 66. Prof Exp: Engr, Sylvania Elec Prod, 59-61; res asst radar astron, Arecibo Observ, Cornell Univ, 63-64, 66-69; MEM TECH STAFF SPACE PHYSICS, JET PROPULSION LAB, CALIF INST TECHNOL, 69- Mem: Int Union Radio Sci. Res: Radar astronomy; mapping of lunar radar echoes; resolution of the delay-Doppler ambiguity. Mailing Add: 3043 Cloudcrest Rd La Crescenta CA 91214

THOMPSON, TOMMY BURT, b Tucumcari, NMex, Apr 3, 38; m 58; c 4. ECONOMIC GEOLOGY, PETROLOGY. Educ: Univ NMex, BS, 61, MS, 63, PhD(geol), 66. Prof Exp: From asst prof to assoc prof geol, Okla State Univ, 66-73; ASSOC PROF GEOL, COLO STATE UNIV, 73- Mem: Geol Soc Am; Am Inst Mining, Metall & Petrol Engrs; Soc Econ Geologists. Res: Conceptual models for exploration of metallic resources; mineral resources of central Colorado; igneous petrology; hydrothermal alteration of igneous rocks; exploration for uranium in sedimentary and igneous rocks. Mailing Add: Dept of Earth Resources Colo State Univ Ft Collins CO 80523

THOMPSON, VENAN E, neuropsychology, neurophysiology, see 12th edition

THOMPSON, VICTOR CARL, b Ozawkie, Kans, May 6, 20; m 45. APICULTURE, ENTOMOLOGY. Educ: Kans State Univ, BS, 44; Iowa State Univ, MS, 55. Prof Exp: Queen breeder, Puett Co, Ga, 44; res asst apicult, Fruit & Truck Br Exp Sta, Ark, 45-51; res assoc, Iowa State Univ, 51-62; RES ASSOC, OHIO STATE UNIV, 62- Mem: Entom Soc Am; Am Beekeeping Fedn; Bee Res Asn. Res: Biology of the honey bee; disease resistance; genetics of behavior in the honey bee. Mailing Add: Dept of Entom Ohio State Univ 1735 Neil Ave Columbus OH 43210

THOMPSON, VINTON NEWBOLD, b Mt Holly, NJ, July 24, 47; m 75. INDUSTRIAL HYGIENE. Educ: Harvard Univ, AB, 69; Univ Chicago, PhD(genetics), 74. Prof Exp: Indust hygienist, Ill Dept Labor, 75; INDUST HYGIENIST, OCCUP SAFETY & HEALTH ADMIN, US DEPT LABOR, 75- Mailing Add: Occup Safety & Health Admin US Dept of Labor 230 S Dearborn Chicago IL 60604

THOMPSON, WALLACE WATROUS, organic chemistry, see 12th edition

THOMPSON, WALTER LEE, organic chemistry, see 12th edition

THOMPSON, WARREN CHARLES, b Santa Monica, Calif, May 22, 22; m 48; c 4. OCEANOGRAPHY. Educ: Univ Calif, Los Angeles, BA, 43; Univ Calif, San Diego, MS, 48; Agr & Mech Col Tex, PhD, 53. Prof Exp: Asst, Scripps Inst, Univ Calif, San Diego, 46-47, 47-48; proj dir, Res Found, Agr & Mech Col, Tex, 50-52; assoc prof aerol & oceanog, 53-60, PROF OCEANOG, NAVAL POSTGRAD SCH, 60- Concurrent Pos: Assoc petrol engr, Humble Oil & Ref Co, La, 47; sci liaison officer, London Br, US Off Naval Res, 60-61; vpres, Oceanog Serv Inc, Santa Barbara, 65-66; consult, 70- Mem: AAAS; Geol Soc Am; Soc Econ Paleont & Mineral; Am Meteorol Soc; Am Asn Petrol Geol. Res: Submarine geology relating to the continental shelves and shorelines; ocean waves; shallow water processes. Mailing Add: Dept of Oceanog Naval Postgrad Sch Monterey CA 93940

THOMPSON, WARREN ELWIN, b Joliet, Ill, June 15, 30; m 62; c 2. PHYSICAL CHEMISTRY. Educ: Univ Wis, BS, 51; Harvard Univ, AM, 53, PhD(chem), 56. Prof Exp: Fulbright scholar, Kamerlingh Onnes Lab, Univ Leiden, 55-57; instr & asst prof chem, Univ Calif, Berkeley, 57-59; asst prof, Case Western Reserve Univ, 59-65; PROG MGR, OFF INT PROGS, NAT SCI FOUND, 65- Mem: Am Chem Soc; AAAS. Res: Molecular spectroscopy; photochemistry; chemical studies related to astronomy. Mailing Add: Off of Int Progs Nat Sci Found 1800 G St NW Washington DC 20550

THOMPSON, WARREN SLATER, b Utica, Miss, Aug 19, 29; m 53; c 4. WOOD SCIENCE & TECHNOLOGY. Educ: Auburn Univ, BS, 51, MS, 55; NC State Univ, PhD, 60. Prof Exp: Asst forester, Miss State Univ, 53-54; wood technologist, Masonite Corp, 57-59; asst & assoc prof forestry, La State Univ, 59-64; DIR, FOREST PROD LAB, MISS STATE UNIV, 64-, PROF WOOD SCI & TECHNOL & HEAD DEPT, 74- Mem: Forest Prod Res Soc; Soc Wood Sci & Technol. Res: Wood pathology and preservation. Mailing Add: Forest Prod Lab Miss State Univ Mississippi State MS 39762

THOMPSON, WAYNE HOWARD, animal pathology, see 12th edition

THOMPSON, WILFRED ROLAND, JR, b Indianola, Miss, Sept 30, 33; m 55; c 2. AGRONOMY. Educ: Miss State Univ, BS, 55, MS, 60, PhD(agron), 66. Prof Exp: Asst county agent, Miss Agr Exten Serv, 57-58; asst agronomist, Miss Agr Exp Sta, 60-67; asst prof turf mgt, Miss State Univ, 62-67; AGRONOMIST, POTASH INST NAM, 67- Mem: Am Soc Agron. Res: Soil fertility; crop and soil management; research oriented educational organization. Mailing Add: 810 Howard Rd Starkville MS 39759

THOMPSON, WILLIAM, JR, b Hyannis, Mass, Dec 4, 36; m 59; c 4. ACOUSTICS. Educ: Mass Inst Technol, BS, 58; Northeastern Univ, MS, 63; Pa State Univ, PhD(eng acoust), 71. Prof Exp: Jr engr, Raytheon Co, 58-60; sr engr, Cambridge Acoust Assocs, Inc, 60-66; res asst transducer studies, 66-72, ASST PROF ACOUST MECH, APPL RES LAB, PA STATE UNIV, 72- Mem: Acoust Soc Am; Inst Elec & Electronics Eng. Res: Electroacoustic transducer design, construction and calibration; acoustic radiation and scattering; underwater acoustics. Mailing Add: 601 Glenn Rd State College PA 16801

THOMPSON, WILLIAM A, b Moorestown, NJ, Oct 25, 36; m 61; c 1. SOLID STATE PHYSICS. Educ: Drexel Inst Tech, BS, 59; Univ Pittsburgh, PhD(physics), 64. Prof Exp: RES STAFF MEM PHYSICS, THOMAS J WATSON RES CTR, IBM CORP, 64- Mem: Am Phys Soc; Fedn Am Sci. Res: Electron tunneling properties of superconductors and magnetic semiconductors. Mailing Add: Thomas J Watson Res Ctr IBM Corp PO Box 218 Yorktown Heights NY 10598

THOMPSON, WILLIAM ALFRED, JR, statistics, see 12th edition

THOMPSON, WILLIAM BALDWIN, b Meriden, Conn, Mar 28, 35; m 60; c 3. GEOPHYSICS. Educ: Mass Inst Technol, BS & MS, 58, PhD(geophys), 63. Prof Exp: Mem tech staff, Bellcomm, Inc, 63-72; HEAD DEPT, BELL TEL LABS, 72- Mem: Am Geophys Union. Res: Electromagnetic cavity resonance phenomena in the earth's atmosphere; physical properties of lunar and planetary surfaces and atmospheres; scientific mission planning for Apollo lunar exploration and planetary missions; economic analyses of telephone network costs and investment; business economics. Mailing Add: Bell Tel Labs 600 Mountain Ave Murray Hill NJ 07974

THOMPSON, WILLIAM BELL, b Belfast, Northern Ireland, Feb 27, 22; m 52, 72; c 2. PLASMA PHYSICS. Educ: Univ BC, BA, 45, MA, 47; Univ Toronto, PhD(math), 50; Oxford Univ, MA, 62. Prof Exp: Sr fel, UK Atomic Energy Authority, Harwell, Eng, 50-53, sr scientist plasma theory, 53-60, dep chief scientist, Culham Lab, 60-62; prof, Oxford Univ, 63-65; PROF PHYSICS, UNIV CALIF, SAN DIEGO, 65- Concurrent Pos: Vis researcher, Plasma Physics Lab, Princeton Univ, 60; vis prof, Univ Calif, San Diego, 60-61 & Univ Colo, 72; ed, Advances in Plasma Physics. Mem: Fel Am Phys Soc; Can Asn Physicists; Am Math Soc; Can Cong Math; fel Royal Astron Soc. Res: Theoretical physics; controlled thermonuclear fusion. Mailing Add: Dept of Physics Univ of Calif at San Diego La Jolla CA 92037

THOMPSON, WILLIAM BENBOW, JR, b Detroit, Mich, July 26, 23; m 47, 58; c 3. OBSTETRICS & GYNECOLOGY. Educ: Univ Southern Calif, AB, 47, MD, 51; Am Bd Obstet & Gynec, dipl. Prof Exp: Intern, Harbor Gen Hosp, Los Angeles, 51-52; resident obstet & gynec, Galliinger Munic Hosp, Washington, DC, 52-53; from resident to sr resident, George Washington Univ Hosp, 53-55; asst, La State Univ, 55-56; clin instr, Univ Calif, Los Angeles, 56-62, asst clin prof, 62-64; assoc prof, Calif Col Med, 64-66; asst prof, 66-73, assoc dean med student serv, 69-73, ASSOC PROF OBSTET & GYNEC & ACTG CHMN DEPT, UNIV CALIF, IRVINE-CALIF COL MED, 73- Concurrent Pos: Dir obstet & gynec, Orange County Med Ctr, Orange, 67- Mem: Fel Am Col Obstetricians & Gynecologists; fel Am Col Surgeons; AMA. Res: Techniques in tubal sterilization. Mailing Add: Calif Col of Med Univ of Calif Irvine CA 92664

THOMPSON, WILLIAM D, b Montgomery, Ala, Apr 21, 36; m 63. NEUROPHYSIOLOGY. Educ: Auburn Univ, DVM, 62; Okla State Univ, PhD(physiol), 67. Prof Exp: Instr physiol, Okla State Univ, 65-67; res assoc, New York Med Col, 67-68; res assoc, Lab Perinatal Physiol, San Juan, PR, 68-69, STAFF ASSOC, LAB NEUROL CONTROL, NAT INST NEUROL DIS & STROKE, 69- Concurrent Pos: NIH trainee, Okla State Univ, 63-65. Mem: Am Soc Vet Physiol & Pharmacol. Res: Suprasegmental somatic motor control and spinal cord physiology. Mailing Add: Lab of Neural Control Nat Inst Neurol Dis & Stroke Bethesda MD 20014

THOMPSON, WILLIAM DONALD, b Kansas City, Mo, June 27, 21; m 42; c 4. PSYCHOLOGY, PHYSIOLOGY. Educ: Univ Kans, BA, 47, MA, 48, PhD(psychol, zool), 54. Prof Exp: Instr psychol, Univ Kans, 49-54; res assoc psychophysiol, Fels Res Inst, Antioch Col, 54-57, asst prof psychol, Col, 55-57; assoc prof, 57-62, PROF PSYCHOL, BAYLOR UNIV, 62-, DIR GRAD STUDIES, 62-, CHMN DEPT PSYCHOL, 69- Concurrent Pos: Sr consult, Vet Admin Hosp, Waco, Tex. Mem: Am Psychol Asn; Soc Psychophysiol Res. Res: Psychophysiology; instrumentation; behavioral biophysics. Mailing Add: Dept of Psychol Baylor Univ Waco TX 76703

THOMPSON, WILLIAM E, physical chemistry, see 12th edition

THOMPSON, WILLIAM LAY, b Austin, Tex, Feb 16, 30; m 58; c 3. VERTEBRATE ZOOLOGY. Educ: Univ Tex, BA, 51, MA, 52; Univ Calif, Berkeley, PhD(zool), 59. Prof Exp: From asst prof to assoc prof, 59-71, PROF BIOL, WAYNE STATE UNIV, 71- Mem: AAAS; Am Soc Zool; Animal Behav Soc; Am Ornith Union; Wilson Ornith Soc. Res: Animal behavior, especially communication and habitat selection in birds. Mailing Add: Dept of Biol Wayne State Univ Detroit MI 48202

THOMPSON, WILLIAM M, b Dallas, Tex, Nov 27, 34; m 68; c 1. COMPUTER SCIENCES. Educ: Univ Tex, BA, 59; Univ Tex Southwestern Med Sch Dallas, PhD(biophys), 68. Prof Exp: Instr biophys, Univ Tex Southwestern Med Sch Dallas, 68-69, asst prof, 69-70; ASST PROF PREV MED & COMMUNITY HEALTH & DIR RES COMPUT CTR, UNIV TEX MED BR GALVESTON, 70- Mem: Am Asn Anat; Inst Elec & Electronics Eng. Mailing Add: Dept of Prev Med & Commun Health Univ of Tex Med Br Galveston TX 77550

THOMPSON, WILLIAM OXLEY, II, b Richmond, Va, Apr 25, 41; m 63; c 2. MATHEMATICAL STATISTICS, APPLIED STATISTICS. Educ: Univ Va, BA, 63; Va Polytech Inst & State Univ, PhD(statist), 68. Prof Exp: Teacher high sch, Va, 63-64; ASST PROF STATIST, UNIV KY, 67-; MATH STATISTICIAN, ADDICTION RES CTR, NIMH, 72- Concurrent Pos: Consult, Clin Res Ctr, NIMH, 68-71 & Addiction Res Ctr, 70-72. Mem: Am Statist Asn; Biomet Soc. Res: Experimental designs and analysis for estimating linear and nonlinear models; estimation of variance components; development of statistical methodology in fields of application; biostatistics. Mailing Add: NIMH Addiction Res Ctr PO Box 2000 Lexington KY 40507

THOMPSON, WILLIAM RAE, b Brooklyn, NY, July 29, 96; m 42. MATHEMATICAL STATISTICS. Educ: Columbia Univ, AB, 23; Yale Univ, PhD(math), 30. Prof Exp: Asst, Lab Biophys Res, Mem Hosp, New York, 22-24; path, Sch Med, Yale Univ, 24-36; sr biochemist, Div Labs & Res, State Dept Health, Albany, NY, 36-65, sr res scientist biomet, 65-66, vol worker, 66-; RETIRED. Honors & Awards: Gov Alfred E Smith Award, Am Soc Pub Admin, 52. Mem: AAAS; Am Math Soc; Am Soc Biol Chem; Soc Indust & Appl Math; fel Am Statist Asn. Res: Sequential analysis and objective steering of experimental course; enzymes; irradiation; physico-chemical methods; microvolumetric calibration apparatus; number theory; statistical methodology; analysis without assumed distribution forms; bioassay; electronic computer programming. Mailing Add: 1 Darroch Rd Delmar NY 12054

THOMPSON, WILLIAM RAYMOND, b Brooklyn, NY, Dec 25, 42; m 66; c 2. BIOLOGICAL CHEMISTRY. Educ: Harvard Univ, AB, 64; Cornell Univ, PhD(chem), 72. Prof Exp: NIH fel, Cornell Univ, 72-73; ASST PROF CHEM, UNIV TEX, DALLAS, 74- Mem: AAAS; Am Chem Soc. Res: Purification of membrane proteins; investigation of protein-lipid interactions. Mailing Add: Dept of Chem Univ Tex at Dallas PO Box 688 Richardson TX 75080

THOMPSON, WILLIAM TALIAFERRO, JR, b Petersburg, Va, May 26, 13; m 41; c 3. MEDICINE. Educ: Davidson Col, AB, 34; Med Col Va, MD, 38; Am Bd Internal Med, dipl, 46. Hon Degrees: ScD, Davidson Col, 75. Prof Exp: From instr to assoc prof, 46-59, chmn dept, 59-73, PROF MED, MED COL VA, 59- Concurrent Pos: Chief med serv, McGuire Vet Admin Hosp, 55-59, consult, 59-; trustee, Mary Baldwin Col, 59-64, Union Theol Sem, 60-70 & Davidson Col, 65-73; mem med adv bd, Nemours Found, Del, 60-, chmn bd mgrs, 65- Mem: Am Psychosom Soc; Am

Fedn Clin Res; fel Am Col Physicians; Am Clin & Climat Asn; NY Acad Sci. Res: Internal medicine, especially pulmonary pathophysiology. Mailing Add: 4602 Sulgrave Rd Richmond VA 23219

THOMPSON, WILMER LEIGH, JR, b Shreveport, La, June 25, 38; m 57; c 1. CLINICAL PHARMACOLOGY, MEDICINE. Educ: Col Charleston, BS, 58; Med Univ SC, MS, 60, PhD(pharmacol), 63; Johns Hopkins Univ, MD, 65; Am Bd Internal Med, dipl, 71. Prof Exp: Intern, Osler Med Serv, Johns Hopkins Hosp, 65-66, resident, 66-67 & 69-70; asst prof med & pharmacol, Sch Med, Johns Hopkins Univ, 70-74; ASSOC PROF MED & PHARMACOL, SCH MED, CASE WESTERN RESERVE UNIV, 74- Concurrent Pos: Fels med, Sch Med, Johns Hopkins Univ, 65-67 & 69-70; staff assoc, Nat Cancer Inst, 67-69; asst physician & dir med intensive care unit, Johns Hopkins Hosp, 70-74; assoc physician & dir clin pharmacol prog, Univ Hosps Cleveland, 74-; adj prof, Sch Libr Sci, Case Western Reserve Univ, 74-; adj assoc prof, Dept Pharmacol, Col Med, Ohio State Univ, 75- Mem: Am Soc Pharmacol & Exp Therapeut; Am Col Physicians; Am Fedn Clin Res. Res: Critical care medicine; human pharmacokinetics; drug interactions; clinical toxicology; cardiovascular pharmacology; clinical drug trials. Mailing Add: Wearn 344 Univ Hosps Cleveland OH 44106

THOMPSON, WYNELLE DOGGETT, b Birmingham, Ala, May 25, 14; m 38; c 4. BIOCHEMISTRY, ORGANIC CHEMISTRY. Educ: Birmingham-Southern Col, BS, 34, MS, 35; Univ Ala, MS, 56, PhD(biochem), 60. Prof Exp: Instr chem, Birmingham-Southern Col, 35-36; high sch instr gen sci, Ala, 36-37; jr chemist, Bur Home Econ, USDA, Washington, DC, 37-38; high sch instr gen sci, Ala, 40-41; instr chem, Birmingham-Southern Col, 41-44; instr, Exten Ctr, Univ Ala, 50-52 & 53-55; from asst prof to assoc prof, 55-67, PROF CHEM, BIRMINGHAM-SOUTHERN COL, 67- Mem: Am Chem Soc; Am Inst Chemists. Res: Protozoa growth and culture; enzymes, structure of and assays techniques; allosteric effects; fluorescence produced. Mailing Add: Dept of Chem Birmingham-Southern Col Birmingham AL 35204

THOMS, RICHARD EDWIN, b Olympia, Wash, June 5, 35; m 67. PALEONTOLOGY. Educ: Univ Wash, Seattle, BS, 57, MS, 59; Univ Calif, PhD(paleont), 65. Prof Exp: Teaching asst peleont, Univ Calif, 60-64; asst prof, 64-70, ASSOC PROF GEOL, PORTLAND STATE UNIV, 70- Mem: Paleont Soc. Res: West coast marine Tertiary biostratigraphy; ichnology. Mailing Add: Dept of Earth Sci Portland State Univ Portland OR 97207

THOMSEN, DONALD LAURENCE, JR, mathematics, see 12th edition

THOMSEN, HARRY LUDWIG, b Boise, Idaho, June 14, 11; m 35; c 3. GEOLOGY. Educ: Oberlin Col, AB, 32, MA, 34. Prof Exp: Seismic party chief, Shell Oil Co, Calif, 35-38, seismologist, 38-41, div geophysicist, Calif & Rocky Mt area, 41-48, dist geologist, Colo, 48-51, area geologist, Okla, 51-53; spec assignment, Bataafse Petrol Maatschappij NV, Holland, 53-54; div explor mgr, Shell Oil Co, Mont, 54-60, mgr explor econ, NY, 60-66, sr staff geologist, Shell Develop Co, Tex, 66-69, spec asst to vpres explor, Shell Oil Co, 69-70; GEOL CONSULT, 70-; GEOPHYSICIST, US GEOL SURV, 74- Mem: Am Asn Petrol Geol; fel Geol Soc Am; Soc Explor Geophys. Res: Exploration for oil and gas; development and application of methods for evaluating petroleum exploration opportunities; estimation of undiscovered oil and gas resources. Mailing Add: 3097 S Steele St Denver CO 80210

THOMSEN, JOHN STEARNS, b Baltimore, Md, June 10, 21; m 52; c 4. ATOMIC PHYSICS. Educ: Johns Hopkins Univ, BE, 43, PhD(physics), 52. Prof Exp: Elec engr, Gen Elec Co, 43-45; asst prof physics, Univ Md, 50-51; res staff asst, Radiation Lab, Johns Hopkins Univ, 51-52, res scientist, 52-53, 54-55; asst prof, Stevens Inst Technol, 53-54; asst prof mech eng, 55-61, res scientist, 62-70, FEL BY COURTESY PHYSICS, JOHNS HOPKINS UNIV, 70- Concurrent Pos: Mem comt fundamental constants, Nat Res Coun, 61-72, chmn, 69-71. Mem: Fel Am Phys Soc; Am Asn Physics Teachers. Res: Thermodynamics; irreversible processes; nonlinear electrical circuits; heat conduction with temperature dependent properties; statistical evaluation of atomic constants; x-ray wave lengths and precision experiments. Mailing Add: Dept of Physics Johns Hopkins Univ Baltimore MD 21218

THOMSEN, WARREN JESSEN, b Iowa, Mar 5, 22; m 43; c 3. MATHEMATICS. Educ: Iowa State Col, BS, 42; Univ Iowa, MS, 47, PhD(math), 52. Prof Exp: Instr math, Wis State Col, Whitewater, 52-53; res mathematician, Appl Physics Lab, Johns Hopkins Univ, 53-56; assoc prof math, Western Ill Univ, 56-57; prof & head dept, Mankato State Col, 57-65; head dept, 65-73, PROF MATH, MOORHEAD STATE COL, 65- Mem: Meteoritical Soc; Math Asn Am. Res: Meteors; resistance of air to meteorites. Mailing Add: Dept of Math Moorhead State Col Moorhead MN 56560

THOMSON, ALAN, b Passaic, NJ, July 1, 28; m 53; c 6. GEOLOGY. Educ: WVa Univ, BS, 52, MS, 54; Rutgers Univ, PhD(geol), 57. Prof Exp: Asst geol, WVa Univ, 52-54; petrogr, Shell Oil Co, 57-65, sr geologist, 65, res geologist, 65-67, res assoc, 67-72, staff res geologist, 72-73, STAFF GEOLOGIST, SHELL OIL CO, 73- Concurrent Pos: Instr, Odessa Col, 58-62. Mem: Fel Geol Soc Am; Soc Econ Paleont & Mineral; Int Asn Sedimentol. Res: Petrology of sedimentary rocks; determination of depositional environments. Mailing Add: Shell Oil Co Box 60775 New Orleans LA 70160

THOMSON, ALEXANDER, b Scotland, June 8, 21. MEDICINE. Educ: Univ BC, BA, 47; McGill Univ, MD & CM, 51. Prof Exp: DIR MED ADV DEPT, LEDERLE LABS, AM CYANAMID CO, 59- Mem: Asn Am Med Cols; Can Med Asn; Col Gen Pract Can. Res: Medical research in pharmaceuticals. Mailing Add: Med Adv Dept Lederle Labs Am Cyanamid Co Pearl River NY 10965

THOMSON, ANDREW, meteorology, deceased

THOMSON, ANDREW, b Gary, Ind, Sept 2, 25; m 50; c 4. INTERNAL MEDICINE, IMMUNOLOGY. Educ: Ind Univ, BS, 48, MD, 51. Prof Exp: Intern, Univ Clins, Univ Chicago, 51-52, resident, 52-55, asst prof med, 56-63; asst prof, 63-67, ASSOC PROF MED, UNIV ILL COL MED, 67- Res: Immune mechanisms in the human as they apply to clinical medicine. Mailing Add: 1725 W Harrison Chicago IL 60612

THOMSON, ASHLEY EDWIN, b Regina, Sask, June 6, 21; m 47; c 7. MEDICINE, PHARMACOLOGY. Educ: Univ Sask, BA, 43; Univ Man, MD, 45, MSc, 48; FRCPS(C). Prof Exp: PROF MED, PHARMACOL & THERAPEUT, UNIV MAN, 65-; DIR RENAL UNIT, HEALTH SCI CTR, 67- Mem: Can Soc Nephrology; Am Soc Nephrology; Int Soc Nephrology; Am Physiol Soc; Can Soc Clin Invest. Res: Hemodialysis and transplantation. Mailing Add: Renal Unit Health Sci Ctr 700 William Ave Winnipeg MB Can

THOMSON, BETTY FLANDERS, b Cleveland, Ohio, May 10, 13. BOTANY. Educ: Mt Holyoke Col, AB, 35, AM, 38; Columbia Univ, PhD(bot), 42. Prof Exp: Asst bot, Mt Holyoke Col, 36-38 & Barnard Col, Columbia Univ, 38-41; instr, Univ Vt, 41-43; from instr to assoc prof, 43-61, PROF BOT, CONN COL, 61-, CHMN DEPT, 70- Concurrent Pos: NSF fac fel, 58-59. Mem: AAAS; Bot Soc Am; Soc Develop Biol; Ecol Soc Am; Torrey Bot Club. Res: Developmental botany; plant growth; anatomy; ecology. Mailing Add: Dept of Bot Conn Col New London CT 06320

THOMSON, CECIL LYMAN, b Dundas, Ont, Aug 13, 13; m 40; c 2. VEGETABLE CROPS. Educ: Univ Toronto, BSA, 37; Univ Minn, MS, 45. Prof Exp: Asst veg crops, Hort Exp Sta, Ont Agr Col, 37-40, asst, Col, 40-45, asst res prof, 44-46; PROF PLANT & SOIL SCI, UNIV MASS, AMHERST, 46- Mem: Am Soc Hort Sci. Res: Weed control, culture of vegetable crops. Mailing Add: Dept of Plant & Soil Sci Bowditch Hall Univ of Mass Amherst MA 01002

THOMSON, DALE S, b Cleveland, Ohio, Feb 12, 34; m 56; c 3. DEVELOPMENTAL BIOLOGY. Educ: Cedarville Col, AB, 56; Ohio State Univ, MS, 62, PhD(zool), 65. Prof Exp: Teacher, Miami Christian Sch, 56-57; from asst prof to assoc prof biol, Cedarville Col, 57-67; ASSOC PROF BIOL, MALONE COL, 67-, CHMN DIV SCI & MATH, 69- Mem: Am Sci Affil. Res: Developmental biology, especially cellular ultrastructure with respect to gland development in the chick; immune response in mice. Mailing Add: 1144 Shelley St NE North Canton OH 44721

THOMSON, DAVID JAMES, b Victoria, BC, June 25, 44. ACOUSTICS. Educ: Univ Victoria, BS, 66, PhD(geophys), 73. Prof Exp: Fel physics, Univ Victoria, BC, 72-74; DEFENCE SCI SERV OFFICER, DEFENCE RES ESTAB PAC, 74- Res: Geomagnetism; numerical modelling of currents induced in non-uniform conductors; numerical modelling of underwater sound propagation in horizontally stratified and range dependent environments. Mailing Add: DREP/DND Forces Mail Off Victoria BC Can

THOMSON, DENNIS WALTER, b New York, NY, Mar 14, 41; m 65. METEOROLOGY. Educ: Univ Wis-Madison, BS, 63, MS, 64, PhD(meteorol), 68. Prof Exp: Ger Acad Exchange Serv fel, Univ Hamburg, 68-69; vis asst prof meteorol, Univ Wis-Madison, 69-70; asst prof, 70-72, ASSOC PROF METEOROL, COL EARTH & MINERAL SCI, PA STATE UNIV, UNIVERSITY PARK, 72- Concurrent Pos: Vis scientist, Nat Ctr Atmospheric Res, Boulder, Colo, 72-; chmn res aviation adv panel, NCAR, 73-75. Mem: Am Meteorol Soc. Res: Physical meteorology; indirect atmospheric sounding; meteorological measurements and instrumentation systems; inadvertant weather modification. Mailing Add: Dept of Meteorol Pa State Univ 506 Deike Bldg University Park PA 16802

THOMSON, DONALD A, b Detroit, Mich, Apr 9, 32; m 57; c 4. MARINE ECOLOGY, ICHTHYOLOGY. Educ: Univ Mich, BS, 55, MS, 57; Univ Hawaii, PhD(zool), 63. Prof Exp: Asst prof, 63-68, ASSOC PROF ZOOL, UNIV ARIZ, 68-, CUR FISHES, 66-, CHMN MARINE SCI PROG, ECOL & EVOLUTIONARY BIOL DEPT, 73- Concurrent Pos: Coordr, Ariz-Sonora Marine Sci Prog, 64-66; prin investr, Off Naval Res, 65-69; chief scientist, R/V Te Vega, Stanford Exped 16, 67; assoc investr, Off Saline Water, 68. Mem: AAAS; Am Soc Ichthyol & Herpet; Soc Syst Zool; Ecol Soc Am. Res: Community ecology, species diversity and stability of marine shore fishes in the Gulf of California. Mailing Add: Dept of Ecol & Evolutionary Biol Univ of Ariz Tucson AZ 85721

THOMSON, DUNCAN MACLAREN, b Chicago, Ill, Oct 24, 09; m 44; c 3. PHYSIOLOGY. Educ: Univ Chicago, SB, 32; Northwestern Univ, MS, 41; Univ Calif, PhD(physiol), 53. Prof Exp: Asst prof biol, Westmont Col, 45-47; asst, Univ Calif, 47-50; instr zool, Univ Ariz, 50; instr, Univ Southern Calif, 51-52; assoc prof biol, Whitworth Col, 52-57; asst prof, Eastern Wash State Col, 57-61, assoc prof, 61-64; chmn sci dept, Modesto Jr Col, Calif, 64-66, instr biol, 66-67; ASST PROF BIOL, DES MOINES COL OSTEOP MED & SURG, 67- Mem: AAAS. Res: Androgen-estrogen effects on rats; effects of age and low phosphorus rickets on calcium 45 metabolism in rats; factors in strontium 90 in rats; cardiovascular-respiratory response in exercise. Mailing Add: Dept of Clin Sci Des Moines Col Osteop Med & Surg Des Moines IA 50312

THOMSON, GEORGE WILLIS, b Seward, Ill, July 10, 21; m 45; c 3. FORESTRY. Educ: Iowa State Univ, BS, 43, MS, 47, PhD(silvicult, soils), 56. Prof Exp: Instr gen forestry, 47-52, asst prof forest mgt, 52-56, assoc prof, 56-60, actg head forestry dept, 67 & 75, PROF MENSURATION & PHOTOGRAM, IOWA STATE UNIV, 60-, CHMN DEPT, 75- Mem: Soc Am Foresters; Soc Range Mgt; Am Soc Photogram. Res: Forest management employing aerial photogrammetry; forest regulation; range management; case studies in land management; remote sensing from LANDSAT-1 with emphasis on forest type delineation. Mailing Add: Dept of Forestry Iowa State Univ Ames IA 50011

THOMSON, GERALD EDMUND, b New York, NY, June 6, 32; m 58; c 2. INTERNAL MEDICINE, NEPHROLOGY. Educ: Queens Col, BS, 55; Howard Univ, MD, 59. Prof Exp: Clin dir dialysis unit, Kings County Hosp, State Univ NY Downstate Med Ctr, 65-67; assoc dir med, Coney Island Hosp, Brooklyn, 67-70; chief div nephrology, 70-71, DIR MED, HARLEM HOSP CTR, 71-; PROF, COLUMBIA UNIV, 72- Concurrent Pos: NY Heart Asn fel renal dis, State Univ NY Downstate Med Ctr, 65-65; mem med adv bd, NY Kidney Found, 71-; mem, Health Res Coun City of New York, 72-75; mem, Health Res Coun State of NY, 75-; mem hypertension info comt & mem educ adv comt, NIH, 73; mem bd dirs, NY Heart Asn, 73- Mem: Am Soc Nephrology. Res: Hypertension. Mailing Add: Harlem Hosp Ctr 506 Lenox Ave New York NY 10037

THOMSON, GORDON MERLE, b Madison, Wis, May 3, 41; m 63; c 2. BIOSTATISTICS. Educ: Cornell Univ, BS, 63; Iowa State Univ, MS, 66, PhD(animal breeding statist), 68. Prof Exp: Assoc statist & comput sci, Iowa State Univ, 68-71; STATISTICIAN, RALSTON PURINA CO, 71-, MGR BIOL SERV, RES 900, 75- Mem: Am Statist Asn; Am Dairy Sci Asn; Am Soc Animal Sci; Sigma Xi. Res: Application of statistics to biological research data; conduct of biological experiments in the areas of texicology and efficacy of feed additives. Mailing Add: Ralston Purina Co Checkerboard Sq Plaza St Louis MO 63188

THOMSON, JAMES ALEX L, b Vancouver, BC, Aug 4, 28. ENGINEERING PHYSICS. Educ: Univ BC, BASc, 52; Calif Inst Technol, PhD(eng sci), 58. Prof Exp: Sr staff scientist, Space Sci Lab, Gen Dynamics-Convair, 58-67; mem tech staff, Sci & Technol Div, Inst Defense Anal, 67-68; prof eng sci, Res Inst Eng Sci & co-dir, Wayne State Univ, 68-72; VPRES ENG PHYSICS, PHYS DYNAMICS, INC, 69- Mem: Am Geophys Soc; Am Phys Soc. Res: Fluid dynamics; ionospheric physics; radiative transfer; oceanography. Mailing Add: Physical Dynamics Inc PO Box 1069 Berkeley CA 94701

THOMSON, JAMES EDGAR, geology, see 12th edition

THOMSON, JAMES EMSLIE, food technology, see 12th edition

THOMSON, JEFFREY JOHN, b Tientsin, China, Oct 16, 41; US citizen; m 65; c 2. PLASMA PHYSICS. Educ: Stanford Univ, BS, 64, MS, 65; Univ Calif, Irvine, MA, 70, PhD(physics), 73. Prof Exp: Sr res engr rocket systs, Autonetics/NAm Aviation, 66-70; PHYSICIST LASER FUSION, LAWRENCE LIVERMORE LAB, 73- Mem:

Am Phys Soc. Res: Interaction of intense radiation with plasma, including heating and transport properties; plasma turbulence, especially renormalized operator theory. Mailing Add: Lawrence Livermore Lab (L-545) Box 808 Livermore CA 94550

THOMSON, JOHN DAVID, b Onawa, Iowa, Apr 11, 12; m 42; c 3. PHYSIOLOGY. Educ: Morningside Col, AB, 33; Univ Chicago, MS, 36; Univ Iowa, PhD, 40. Prof Exp: Asst zool, 37-39, from asst to asst prof physiol & biophys, 40-54, ASSOC PROF PHYSIOL & BIOPHYS, COL MED, UNIV IOWA, 54- Mem: AAAS; Am Physiol Soc; Soc Exp Biol & Med. Res: Physiology of muscle; neurobiology. Mailing Add: 5-472 Basic Sci Bldg Univ of Iowa Iowa City IA 52240

THOMSON, JOHN FERGUSON, b Garrett, Ind, Apr 18, 20; m 43, 73; c 2. PHARMACOLOGY, BIOCHEMISTRY. Educ: Univ Chicago, SB, 41, SM, 42, PhD(pharmacol), 47. Prof Exp: Res assoc anat, Univ Chicago, 43-45, from res assoc to asst prof pharmacol, 46-51, mem staff, Toxicity Lab, 43-51; assoc pharmacologist, 51-62, sr biologist, 62-69, actg dir, 69-70 & 74-75, ASSOC DIR, DIV BIOL & MED RES, ARGONNE NAT LAB, 70-74 & 75- Concurrent Pos: Adj prof, Northern Ill Univ & Univ Ill, Chicago Circle, 71- Mem: AAAS; Soc Exp Biol & Med; Sigma Xi; Am Physiol Soc; Radiation Res Soc. Res: Radiation biology; cytoenzymology; isotope toxicity. Mailing Add: Div of Biol & Med Res Argonne Nat Lab Argonne IL 60439

THOMSON, JOHN OLIVER, b Cleveland, Ohio, Feb 2, 30; m 55; c 3. PHYSICS. Educ: Williams Col, AB, 51; Univ Ill, MS, 53, PhD, 56. Prof Exp: Fulbright grant, Univ Rome, 56-57 & Univ Padua, 57-58; from asst prof to assoc prof physics, 58-72, PROF PHYSICS, UNIV TENN, KNOXVILLE, 72- Concurrent Pos: Consult, Physics Div, Oak Ridge Nat Lab, 58-; assoc prof, Memphis State Univ, 66-68. Mem: Am Phys Soc. Res: Solid state and low temperature physics. Mailing Add: Dept of Physics & Astronomy Univ of Tenn Knoxville TN 37916

THOMSON, JOHN WALTER, b Scotland, July 9, 13; nat US; m 37; c 5. BOTANY. Educ: Columbia Univ, AB, 35; Univ Wis, MA, 37, PhD, 39. Prof Exp: Dir staff, Sch Nature League, Am Mus Natural Hist, 39-41; tutor biol, Brooklyn Col, 41; instr, Wis State Teachers Col, Superior, 42-44; from asst prof to assoc prof bot, 44-62, chmn dept bot & zool, Exten, 48-67, PROF BOT, UNIV WIS-MADISON, 62- Concurrent Pos: Exchange prof, Helsinki Univ, 65-66. Mem: Am Bryol & Lichenological Soc (vpres, Am Bryol Soc, 56-57, pres, 58-59); Am Soc Plant Taxon; Torrey Bot Club (secy, 40); Int Soc Plant Taxon. Res: Lichens; taxonomy and ecology of North American lichens, especially Arctic lichens; monographs, Peltigera, Physcia, Cladonia, Baemyces. Mailing Add: 236 Birge Hall Univ of Wis Dept of Bot Madison WI 53706

THOMSON, JUNIUS RICHARD, b Rutherfordton, NC, Feb 24, 27; m 48; c 2. BIOLOGY. Educ: Emory Univ, BS, 48, MS, 49. Prof Exp: Asst bacteriologist, Commun Dis Ctr, US Pub Health Serv, Ga, 49-50; instr biol, Univ Chattanooga, 50-51; asst scientist, Oak Ridge Nat Lab, 51-52; res biologist, Southern Res Inst, Ala, 52-64; sr biologist, Midwest Res Inst, Mo, 64-66; res biologist, R J Reynolds Tobacco Co, 66-68; prof assoc, Adv Ctr Toxicol, Nat Acad Sci, 68-69; dir, Sci Ctr Pinellas Co, Fla, 69-70; admin analyst, Div Med Servs, Tenn Valley Auth, 70-71; CONSULT TOXICOL & BIOMED SCI, 71- Mem: Fel AAAS; Am Soc Zool; Soc Prof Biol; Sigma Xi; Am Pub Health Asn. Res: Experimental therapeutics and toxicology; mechanism of action of antimetabolites and drug resistance phenomena; combination chemotherapy; nutritional deficiencies; literature analysis; biomedical instrumentation. Mailing Add: 1670 E Clifton Rd NE Atlanta GA 30307

THOMSON, KEITH PATRICK BOWMER, b Belfast, Northern Ireland, Mar 6, 39; Can citizen; m 64; c 2. PHYSICS. Educ: Queen's Univ, Belfast, BSc, 62; Univ Toronto, MA, 63, PhD(physics), 70. Prof Exp: Meteorologist, Can Dept Transport, 62-63; scientist physics, Radio Corp Am, Quebec, 63-65; Nat Res Coun Can fel agr physics, Macdonald Col, McGill Univ, 70; res scientist inland waters, 70-75, HEAD APPLN DEVELOP REMOTE SENSING, CAN CTR, 75- Honors & Awards: Cert, Am Water Res Asn, 73. Mem: Can Remote Sensing Soc (secy-treas, 75-76). Res: Development of remote sensing applications; water resources; limnology; atmospheric physics; underwater optics. Mailing Add: Can Ctr for Remote Sensing 2464 Sheffield Rd Ottawa ON Can

THOMSON, KEITH STEWART, b Heanor, Eng, July 29, 38; m 63. ZOOLOGY, PALEONTOLOGY. Educ: Univ Birmingham, BSc, 60; Harvard Univ, AM, 61, PhD(biol), 63. Prof Exp: NATO sci fel & temporary lectr zool, Univ Col, Univ London, 63-65; asst prof biol, 65-70, ASSOC CUR VERT ZOOL, PEABODY MUS NATURAL HIST & ASSOC PROF BIOL, YALE UNIV, 70- Mem: Am Soc Zool; Soc Vert Paleont; fel Zool Soc London. Res: Vertebrate biology, especially phylogeny; paleontology and functional biology; origin of adaptations and of major groups. Mailing Add: Div of Vert Zool Peabody Mus of Natural Hist Yale Univ New Haven CT 06520

THOMSON, KENNETH CLAIR, b Gunnison, Utah, Mar 5, 40; m 61; c 3. ECONOMIC GEOLOGY, PETROGRAPHY. Educ: Univ Utah, BS, 63, PhD, 70. Prof Exp: Illustrator-geologist, Utah Geol Surv, 59-68; asst prof geol, 68-71, ASSOC PROF GEOL, SOUTHWEST MO STATE UNIV, 71- Mem: Nat Speleol Soc; Mineral Soc Am. Res: Geochemistry as applied to mineral prospecting; speleology. Mailing Add: Dept of Geol Southwest Mo State Univ Springfield MO 65802

THOMSON, KER CLIVE, b Toronto, Ont, Mar 2, 28; US citizen; m 55; c 2. GEOPHYSICS, SEISMOLOGY. Educ: Univ BC, BA, 52; Colo Sch Mines, DSc(geophys), 65. Prof Exp: Seismologist, Seismograph Serv Corp, Okla, 52-54; seismologist, Standard Oil Co Calif, 54-55, seismic party chief, 55-58; instr physics, Colo Sch Mines, 58-61; res physicist, 61-65, br chief seismol, 65-75, DIR TERRESTRIAL SCI LAB, AIR FORCE CAMBRIDGE RES LABS, 75- Concurrent Pos: Lectr, Gordon Col, 66-69 & Boston Col, 72. Mem: Seismol Soc Am; Am Geophys Union; Soc Explor Geophys; Europ Asn Explor Geophys; Am Sci Affil. Res: Earthquake focal mechanism; theoretical seismology; model seismology; wave propagation in absorptive media; seismic radiation in tectonically stressed media; nuclear test detection; structural vibration; terrestrial gravity. Mailing Add: Air Force Cambridge Res Labs LW Hanscom Field Bedford MA 01730

THOMSON, MICHAEL GEORGE ROBERT, b Portsmouth, Eng, Mar 7, 41; m 74. ELECTRON OPTICS. Educ: Univ Cambridge, BA, 62, PhD(physics), 67. Prof Exp: Res assoc physics, Univ Chicago, 67-70; mem res staff physics, Res Lab Electronics, Mass Inst Technol, 70-73; MEM TECH STAFF, BELL LABS, 74- Mem: Electron Micros Soc Am; Am Phys Soc. Res: Electron optics for electron beam lithography; theory of image formation in the conventional and the scanning transmission electron microscopes; aberration theory for quadrupole and other unconventional electron lenses. Mailing Add: Bell Labs Murray Hill NJ 07974

THOMSON, REGINALD GEORGE, b Woodstock, Ont, Apr 7, 34; m 57; c 3. VETERINARY PATHOLOGY. Educ: Univ Toronto, DVM, 59, MVSc, 63; Cornell Univ, PhD(vet path), 65. Prof Exp: Assoc prof, 65-67, PROF PATH, ONT VET COL, UNIV GUELPH, 67- Concurrent Pos: Res grants, Can Dept Agr, 66-72, Nat Res Coun Can, 68-72. Mem: Am Col Vet Path; Am Vet Med Asn; Can Vet Med Asn; Int Acad Path. Res: Bone osteophyte formation; respiratory diseases in cattle. Mailing Add: Dept of Path Univ of Guelph Ont Vet Col Guelph ON Can

THOMSON, RICHARD EDWARD, b Comox, BC, Apr 14, 44. PHYSICAL OCEANOGRAPHY. Educ: Univ BC, BS, 67, PhD(physics & oceanog), 71. Prof Exp: RES SCIENTIST, INST OCEAN SCI, ENVIRON CAN, 71- Mem: Can Meteorol Soc; Can Geog Soc. Res: Wave propagation in random media; energetics of planetary waves and internal gravity waves; vorticity mixing and redistribution in the ocean. Mailing Add: Inst of Ocean Sci Environ Cam 512 Fed Bldg Victoria BC Can

THOMSON, ROBERT RITTER, physics, see 12th edition

THOMSON, TOM RADFORD, b Hachiman, Japan, Nov 7, 18; nat US; m 46; c 5. PHYSICAL ORGANIC CHEMISTRY. Educ: Univ Calif, BS, 39; Kans State Univ, MS, 40, PhD(chem), 45. Prof Exp: Asst chem, Kans State Univ, 39-40, instr, 42, chemist, 42-47; plant chemist, Cascade Frozen Foods, Wash, 41; from assoc prof to prof chem, Adams State Col, 47-61, head dept, 47-61; assoc prof, 61-67, PROF CHEM, ARIZ STATE UNIV, 67- Concurrent Pos: Sigma Xi res grant, 54; res assoc, Univ Calif, 54; vis scholar, Utah, 59; Res Corp grant, 59-60; resident author, Addison-Wesley Publ Co, Calif, 67-68. Mem: Am Chem Soc. Res: Carbohydrates; starch chemistry; chemical education. Mailing Add: 4820 N Granite Reef Scottsdale AZ 85281

THOMSON, WILLIAM ALEXANDER BROWN, b London, Eng, Dec 21, 28; m 56; c 4. BIOCHEMISTRY, ANALYTICAL CHEMISTRY. Educ: St Andrews Univ, BSc, 51, dipl ed, 52; Univ Wis, MS, 58, PhD(biochem), 61. Prof Exp: Chemist, Can Packers Ltd, 53-54; from asst scientist to assoc scientist, Fisheries Res Bd Can, 54-63; sr chemist, Am Potato Co, Idaho, 63-64; asst prof food sci, NC State Univ, 64-66; sr chemist, Marine Colloids, Inc, Maine, 66-68, mgr tech opers, 68-70; SECT HEAD, PROD DEVELOP DEPT, ROSS LABS, 70- Mem: Am Chem Soc; Inst Food Technol; Brit Biochem Soc; The Chem Soc. Res: Chemistry of polysaccharides from marine algae; biochemistry of fishery products; fluid nutritional products; dietary products for inborn errors of metabolism. Mailing Add: Prod Develop Dept Ross Labs 625 Cleveland Ave Columbus OH 43216

THOMSON, WILLIAM WALTER, b Chico, Calif, Oct 11, 30; c 2. BOTANY, CYTOLOGY. Educ: Sacramento State Col, BA, 53, MA, 60; Univ Calif, Davis, PhD(bot), 63. Prof Exp: Asst res botanist, Air Pollution Res Ctr, 63-64, from asst prof to assoc prof, 64-74, PROF BIOL, UNIV CALIF, RIVERSIDE, 74- Mem: AAAS; Bot Soc Am; Electron Micros Soc Am. Res: Ultrastructure of chloroplasts and plant membranes as related to development, physiology and stress conditions. Mailing Add: Dept of Biol Univ of Calif Riverside CA 92502

THONNARD, NORBERT, b Berlin, Ger, Jan 22, 43; US citizen; m 64; c 3. ASTROPHYSICS, ATOMIC PHYSICS. Educ: Fla State Univ, BA, 64; Univ Ky, MS, 69, PhD(physics), 71. Prof Exp: Physicist, US Army Engr Res & Develop Labs, 64-66; consult astrophys, Battelle Pac Northwest Labs, Battelle Mem Inst, 72; fel, 70-72, STAFF MEM, DEPT TERRESTRIAL MAGNETISM, CARNEGIE INST, 72- Mem: Am Phys Soc; Am Astron Soc; AAAS. Res: Extragalactic 21 centimeter hydrogen line studies, observational cosmology, microwave emission from interstellar molecules; development of instrumentation for radio and optical astronomy; studies of x-ray emission from atomic collisions. Mailing Add: Carnegie Inst 5241 Broad Branch Rd NW Washington DC 20015

THOR, DANIEL EINAR, b Davenport, Iowa, Sept 4, 38; m 71; c 1. IMMUNOLOGY, MICROBIOLOGY. Educ: Univ Ill Med Ctr, MD, 63, PhD(immunol, microbiol), 68. Prof Exp: Technologist, Blood Bank, Presby-St Luke's Hosp, Chicago, 60-63, intern surg, 63-64, resident & instr, 64-65; res investr immunol, NIH, 68-71; ASSOC PROF IMMUNOL, MICROBIOL & PATH, UNIV TEX HEALTH SCI CTR SAN ANTONIO, 71- Concurrent Pos: Asst prof, Univ Ill Med Ctr & lectr & asst prof, Postgrad Prog, NIH, 68-71; reviewer, Depts Med & Surg, Vet Admin Merit Rev Bd in Immunol, 72-; mem comt immunodiag, Nat Cancer Inst. Honors & Awards: Borden Award, Univ Ill Med Ctr, 63. Mem: AAAS; Fedn Am Socs Exp Biol; Am Asn Immunologists; Am Soc Microbiol; Reticuloendothelial Soc. Res: Cellular immunology and delayed type hypersensitivity; mechanisms of chemical mediator production, especially migration inhibition factor; chemical carcinogenesis and tumor immunity; RNA mechanisms of immunity; immune reconstruction with transfer factor. Mailing Add: Dept of Microbiol Univ of Tex Health Sci Ctr San Antonio TX 78284

THOR, EYVIND, b Oslo, Norway, Nov 24, 28; US citizen; m 56; c 3. FOREST GENETICS, FOREST ECOLOGY. Educ: Univ Wash, Seattle, BS, 54, MS, 56; NC State Univ, PhD(forestry), 61. Prof Exp: Asst prof forestry res, 59-65, assoc prof forestry, 65-71, PROF FORESTRY, UNIV TENN, 71- Concurrent Pos: Elwood L Demmon res award, 61; US Forest Serv res grant, 66-68. Mem: Soc Am Foresters; Ecol Soc Am. Res: Forest ecology; breeding of trees for timber production (pines and hardwoods), Christmas trees (pine and spruce) and disease resistance (American chestnut). Mailing Add: Forestry Dept Univ of Tenn Knoxville TN 37916

THORBECKE, GEERTRUIDA JEANETTE, b Neth, Aug 2, 29; m 57; c 3. IMMUNOLOGY, EXPERIMENTAL PATHOLOGY. Educ: State Univ Groningen, MD, 50. Prof Exp: Asst histol, State Univ Groningen, 48-54; asst path, State Univ Leiden, 56-57; from res assoc to assoc prof, 57-70, PROF PATH, SCH MED, NY UNIV, 70- Concurrent Pos: Foreign Opers Mission to Neth scholar, Lobund Inst, Ind, 54-56; USPHS res grants, 59-, USPHS res career develop award, 61-71; career scientist award, Health Res Coun, City of New York, 71-72. Mem: Am Soc Exp Path; Soc Exp Biol & Med; Am Asn Immunologists; Transplantation Soc; Brit Soc Immunol. Res: Antibody formation; serum proteins; lymphoid tissues; immunological tolerance; tumor immunity. Mailing Add: Dept of Path NY Univ Sch of Med New York NY 10016

THOREN, CONRAD JOSEPH, b Chicago, Ill, Mar 10, 15; m 52. GEOGRAPHY, CARTOGRAPHY. Educ: Univ Chicago, AB, 37, SM, 41. Prof Exp: Geog attache, US Dept State, England & India, 46-55, consul, Mozambique, Africa, 56-57, geog attache, Denmark, 58-59; statistician, Outdoor Recreation Resources Rev Comn, Off of the Pres, 60-61; GEOGR, BUR CENSUS, US DEPT COM, 62- Mem: Asn Am Geog; Am Geog Soc; Royal Geog Soc; Inst Brit Geog. Res: General geography, especially cartography and map information; population geography; recreation and transportation geography and tourism; United States, Western Europe, Asian sub-continent and Africa south of the Sahara. Mailing Add: 3900 Watson Pl NW Apt B4D Washington DC 20016

THORESEN, ASA CLIFFORD, b Blenheim, NZ, Sept 9, 30; nat US; m 52; c 2. ORNITHOLOGY. Educ: Emmanuel Missionary Col, BA, 54; Walla Walla Col, MA, 58; Ore State Col, PhD, 60. Prof Exp: Vis prof marine inverts, Biol Sta, Walla Walla Col, 60, 70; from asst prof to assoc prof, 60-67, PROF BIOL & CHMN DEPT, ANDREWS UNIV, 67- Concurrent Pos: Leader biol expeds Peru, 64-65, 68 & SPac & Australia, 72; NSF grant, NZ, 66-67. Mem: AAAS; Cooper Ornith Soc; Am Ornith Union. Res: Life history and behaviorial studies of oceanic birds, particularly of the family Alcidae; ultrastructural studies of avian tissues. Mailing Add: Box 147 Andrews Berrien Springs MI 49104

THORESEN, TIMOTHY HANS HALE, history of anthropology, social anthropology, see 12th edition

THORHAUG, ANITRA L, b Chicago, Ill, June 1, 40. BIOPHYSICS, ALGOLOGY. Educ: Univ Miami, BS, 63, MS, 65, PhD(marine sci), 69. Prof Exp: Environ Sci Serv Admin assoc biophys, Atlantic Labs, Miami, Fla, 68-69; res scientist, Algology Div Fish & Appl Estuarine Ecol, 69-71, RES SCIENTIST MICROBIOL, SCH MED, UNIV MIAMI, 71- Concurrent Pos: Fed Water Qual Admin grant, Univ Miami, 69-70, NSF inst grant, 70-71, Nat Oceanic & Atmospheric sea grant & NSF grant biophysics, 70-71, AEC grant, 70; fel, Weizmann Inst Sci, 71-; vis scientist, Donner Lab, Univ Calif, Berkeley, 71. Mem: AAAS; Bot Soc Am; Phycol Soc Am; Am Soc Zool; Brit Phycol Soc. Res: Membrane transport biophysics; membrane transport and physiology of giant algal cells; near-shore macro-plant ecology; physiology of tropical macro-flora-temperature, salinity, light, sediment. Mailing Add: Rosenstiel Sch Marine Sci Univ of Miami 10 Rickenbacker Miami FL 33149

THORINGTON, RICHARD WAINWRIGHT, JR, b Philadelphia, Pa, Dec 24, 37; m 67; c 1. BIOLOGY. Educ: Princeton Univ, BA, 59; Harvard Univ, MA, 63, PhD(biol), 64. Prof Exp: Primatologist, New Eng Regional Primate Res Ctr, 64-69, assoc mammal, Mus Comp Zool, 64-69; CUR PRIMATES, BIOL PROG, SMITHSONIAN INST, 69- Mem: AAAS; Am Soc Mammal; Soc Study Evolution; Int Primatol Soc. Res: Form and function of mammals; thermoregulation and thermal effects on development; primate ecology and taxonomy. Mailing Add: Primate Res Prog Smithsonian Inst Washington DC 20560

THORLAND, RODNEY HAROLD, b Lake Mills, Iowa, Feb 16, 41. SOLID STATE PHYSICS. Educ: Luther Col, BA, 64; Emory Univ, MSc, 69, PhD(physics), 71. Prof Exp: Vol, US Peace Corps, 64-67; asst prof physics, Kennesaw Jr Col, 71-72; res assoc physics, Emory Univ, 72-73; ASST PROF PHYSICS, VOL STATE COMMUNITY COL, 73- Mem: Am Phys Soc; Am Asn Physics Teachers; assoc Sigma Xi. Res: Nuclear magnetic resonance, electron paramagnetic resonance and far infrared spectroscopy. Mailing Add: Vol State Community Col Nashville Pike Gallatin TN 37066

THORMAN, CHARLES HADLEY, b Albany, Calif, June 14, 36; m 57; c 2. GEOLOGY. Educ: Univ Redlands, BS, 58; Univ Wash, MS, 60, PhD(geol), 62. Prof Exp: Geologist, Humble Oil & Ref Co, 62-65 & Olympic Col, Wash, 65-68; vis asst prof geol, Univ Ore, 68-71; GEOLOGIST, OFF INT GEOL, US GEOL SURV, 71- Mem: Am Asn Petrol Geol; Geol Soc Am. Res: Pretertiary tectonics of the eastern basin and range; structure of western Liberia. Mailing Add: US Geol Surv Fed Ctr Denver CO 80225

THORMAR, HALLDOR, b Iceland, Mar 9, 29; m 62; c 3. VIROLOGY. Educ: Copenhagen Univ, PhD(cell physiol), 56, DrPhil(virol), 66. Prof Exp: Res scientist, State Serum Inst, Copenhagen, Denmark, 60-62; from res scientist to assoc res scientist, Inst Exp Path, Keldur, Iceland, 62-67; CHIEF RES SCIENTIST VIROL, INST RES MENT RETARDATION, 67- Concurrent Pos: Investr virol, Sci Res Inst, Caracas, Venezuela, 65-66. Mem: Am Soc Microbiol. Res: Cell physiology; effect of temperature and temperature changes on cell growth and division; slow virus infections, particularly visna in sheep; isolation and study of visna virus and comparison of visna virus to other viruses; pathogenesis of slow virus infections. Mailing Add: Inst for Res in Ment Retardation 1050 Forest Hill Rd Staten Island NY 10314

THORN, FRANK b New York, NY, Apr 17, 40; m 59; c 4. PSYCHOPHYSIOLOGY. Educ: Rensselaer Polytech Inst, BS, 61; Univ Rochester, PhD(psychol), 67. Prof Exp: Teaching asst psychol, Univ Rochester, 61-62, USPHS trainee physiol psychol, 62-64, USPHS trainee visual sci, Ctr Visual Sci, 64-66, USPHS trainee physiol, Ctr Brain Res, 67, instr comp psychol, Univ, 65; USPHS fel brain res, Brain Res Inst, Univ Calif, Los Angeles, 67-69; ASSOC PROF PHYSIOL, COL OPTOM, PAC UNIV, 69- Concurrent Pos: Bd mem, Ore Zool Res Ctr, 72-; fel, Med Ctr, Univ Rochester, 75- Mem: AAAS; Asn Res Vision & Ophthal. Res: Electrophysiology of visual system; neural mechanisms of visual detection and pattern recognition; electrodiagnostics; visual development. Mailing Add: Col of Optom Pac Univ Forest Grove OR 97116

THORN, GEORGE DENIS, b London, Eng, Feb 16, 21; m 44; c 3. AGRICULTURAL CHEMISTRY. Educ: Univ Alta, BSc, 43; Queen's Univ, Ont, MA, 44; McGill Univ, PhD(chem), 47. Prof Exp: Asst prof org chem & biochem, Mt Allison Univ, 47-51; CHEMIST, PESTICIDE RES INST, CAN DEPT AGR, 51- Concurrent Pos: Hon lectr, Univ Western Ont, 58. Mem: Am Phytopath Soc; fel Chem Inst Can. Res: Heterocyclic compounds; chemistry and fungicidal action of dithiocarbamates; relationship of chemical structure to biological activity; chemistry, mode of action and metabolism of systemic fungicides. Mailing Add: Pesticide Res Inst Can Dept of Agr Univ PO London ON Can

THORN, JOHN ANTHONY, biochemistry, see 12th edition

THORN, JOHN PAUL, b Clarksburg, WVa, Oct 22, 21; m 42; c 4. RESEARCH ADMINISTRATION. Educ: Rutgers Univ, BS, 48. Prof Exp: Proj engr process develop, Govt Rubber Labs, Ohio, 44-46; res chemist, Chem Res Div, Esso Res & Eng Co, 48-54, from group head to div dir, Enjay Labs, 54-66, mgr, Enjay Polymer Labs, 66-67, mgr, Baytown Res & Develop Div, 67-73, mgr plastics technol, Esso Chem Co, 69-73; MGR NEW VENTURES, EXXON CHEM CO USA, 73- Mem: Am Chem Soc; Soc Plastics Eng. Res: Research and process development on petroleum and petrochemical products, including plastics, synthetic rubber, lower alcohols and oxo products; product applications research. Mailing Add: 5016 Ashwood Dr Baytown TX 77520

THORN, RICHARD MARK, b New Castle, Pa, Mar 8, 47; m 69. IMMUNOBIOLOGY. Educ: Univ Calif, San Diego, BA, 69; Univ Pa, PhD(molecular biol), 74. Prof Exp: Fel immunol, Sch Med, Johns Hopkins Univ, 74-76; SCIENTIST II, FREDERICK CANCER RES CTR, 76- Res: T cell mediated cytolysis; mechanism of target cell destruction; generation of precursor cells and their induction; antigenic specificity of induction and cytolysis and its role in the control of infectious disease and cancer. Mailing Add: 2 North O'Neill Labs 5601 Loch Raven Blvd Baltimore MD 21239

THORN, ROBERT JEROME, physical chemistry, see 12th edition

THORN, ROBERT NICOL, b Coeur d'Alene, Idaho, Aug 31, 24; m 62; c 4. PHYSICS. Educ: Harvard Univ, PhD(physics), 53. Prof Exp: DIV LEADER, THEORET DESIGN DIV, LOS ALAMOS SCI LAB, 53- Concurrent Pos: Mem sci adv group, Space Systs Div, US Air Force, 62-63, sci adv bd, 62-, nuclear panel, 64-; mem sci adv group, Defense Nuclear Agency & Defense Intel Agency. Honors & Awards: E O Lawrence Award, Atomic Energy Comn, 67. Mem: AAAS; Am Phys Soc. Res: Classical theoretical physics; quantum and nuclear physics; weapons systems analysis and design. Mailing Add: 981 Barranca Rd Los Alamos NM 87544

THORNBER, JAMES PHILIP, b Hebden Bridge, Eng, Dec 22, 34; m 60; c 2. PLANT BIOCHEMISTRY. Educ: Cambridge Univ, BA, 58, MA, 61, PhD(biochem), 62. Prof Exp: Sci off plant biochem, Twyford Labs Ltd, Arthur Guinness, Son & Co, Ltd, 61-67; res assoc biol, Brookhaven Nat Lab, 67-69, asst scientist, 69-70; asst prof bot, 70-72, assoc prof biol, 72-75, PROF BIOL, UNIV CALIF, LOS ANGELES, 75- Concurrent Pos: NSF grant, Univ Calif, Los Angeles, 71- Mem: Am Soc Biol Chem; Am Soc Photobiol; Am Soc Plant Physiol. Res: Photosynthesis; organization of chlorophyll in plants and bacteria; chlorophyll-protein complexes; photochemical reaction centers; membrane composition, structure and biogenesis. Mailing Add: Dept of Biol Univ of Calif Los Angeles CA 90024

THORNBER, KARVEL KUHN, b Portland, Ore, Apr 7, 41; m 67; c 2. SOLID STATE PHYSICS. Educ: Calif Inst Technol, BS, 63, MS, 64, PhD(elec eng), 66. Prof Exp: Res fel elec eng, Stanford Univ, 66-68; res fel physics, Univ Bristol, 68-69; MEM TECH STAFF, BELL LABS, 69- Res: Quantum mechanics and statistical mechanics of dissipation systems, transport properties and noise theory. Mailing Add: Bell Labs 600 Mountain Ave Murray Hill NJ 07974

THORNBERRY, HALBERT HOUSTON, b Corydon, Ky, Dec 28, 02; m 46; c 1. PLANT PATHOLOGY. Educ: Univ Ky, BS, 25, MS, 26; Univ Minn, PhD(plant path), 34. Prof Exp: Asst plant path, Univ Minn, 26-28 & Univ Ill, 28-31; fel, Rockefeller Inst Technol, 31-35; jr plant pathologist, Citrus Exp Sta, Univ Calif, 35-36; asst pathologist, Univ Ky, 36-37; pathologist, Bur Plant Indust, US Dept Agr, 37-38; from asst prof to prof, 38-71, EMER PROF PLANT PATH, UNIV ILL, URBANA, 71-; CONSULT PLANT HEALTH & MGT, 72- Concurrent Pos: Res award, Soc Am Florists, 68. Mem: AAAS; Am Phytopath Soc; Am Soc Microbiol; Am Chem Soc. Res: Phytovirology; chemopathology; antibiotics; bacterial diseases. Mailing Add: 1602 S Hillcrest St PO Box 128 Urbana IL 61801

THORNBURG, DAVID DEVOE, b Chicago, Ill, Apr 25, 43; m 66; c 1. PHYSICAL METALLURGY. Educ: Northwestern Univ, BS, 67; Univ Ill, Urbana, MS, 69, PhD(metall), 71. Prof Exp: MEM RES STAFF DEVICE PHYSICS, XEROX PALO ALTO RES CTR, 71- Mem: Inst Elec & Electronics Engrs; Am Phys Soc; Am Vacuum Soc; Am Inst Metall Engrs. Res: Physics of amorphous semiconductor devices and thermodynamic and kinetic properties of non-crystalline materials. Mailing Add: Xerox Palo Alto Res Ctr 3333 Coyote Hill Rd Palo Alto CA 94304

THORNBURG, WAYNE, biophysics, deceased

THORNBURG, WENDELL LEWIS, chemistry, see 12th edition

THORNBURGH, DALE A, b Tiffin, Ohio, Dec 1, 31; m 61; c 3. FOREST ECOLOGY. Educ: Univ Wash, BS, 59, PhD(forestry), 69; Univ Calif, Berkeley, MS, 62. Prof Exp: Lectr silviculture, Univ Wash, 63-64; from asst prof to assoc prof forest ecol, 64-74, PROF FOREST ECOL, HUMBOLDT STATE UNIV, 74- Mem: Ecol Soc Am; Soc Am Foresters. Res: Carying capacity of subalpine meadows in wilderness areas; development of forest habitat types and successional models. Mailing Add: Dept of Forestry Humboldt State Univ Arcata CA 95521

THORNBURY, JOHN R, b Cleveland, Ohio, Mar 16, 29; m 55; c 2. RADIOLOGY. Educ: Miami Univ, AB, 50; Ohio State Univ, MD, 55. Prof Exp: From instr to asst prof radiol, Univ Colo, 62-63; asst prof, Univ Iowa, 63-66 & Univ Wash, 66-68; assoc prof, 68-71, PROF RADIOL, UNIV MICH, ANN ARBOR, 71- Mem: Asn Univ Radiologists; Radiol Soc NAm; Am Roentgen Ray Soc. Res: Urological radiology; radiographic contrast materials, methods of enhancing effects; applications of computer technology to diagnostic radiology; visual perception and information processing in diagnostic radiology. Mailing Add: Dept of Radiol S4432 Univ Hosp Univ of Mich Ann Arbor MI 48104

THORNDIKE, ALAN MOULTON, b Montrose, NY, June 27, 18; m 42; c 5. PHYSICS. Educ: Wesleyan Univ, BA, 39; Columbia Univ, AM, 40; Harvard Univ, PhD(chem physics), 47. Prof Exp: Asst physicist, Div War Res, Univ Calif, 41-42; res assoc, Div War Res, Columbia Univ, 42-43; field serv consult, Off Sci Res & Develop, 43-45; asst scientist, Brookhaven Nat Lab, 47-52, from assoc scientist to scientist, 52-58; vis prof, Johns Hopkins Univ, 58-59; scientist, 59-65, sr physicist, 65-70, assoc chmn physics dept, 70-73, SR PHYSICIST, BROOKHAVEN NAT LAB, 73- Concurrent Pos: Consult, Off Technol Assessment, 74- Mem: Am Phys Soc. Res: Primary cosmic radiation; meson physics; elementary particles; electronic data processing; high energy interactions; computer simulation. Mailing Add: Physics Dept Brookhaven Nat Lab Upton NY 11973

THORNDIKE, EDWARD HARMON, b Pasadena, Calif, Aug 2, 34; m 55; c 3. ELEMENTARY PARTICLE PHYSICS. Educ: Wesleyan Univ, AB, 56; Stanford Univ, MS, 57; Harvard Univ, PhD(physics), 60. Prof Exp: Res fel physics, Harvard Univ, 60-61; from asst prof to assoc prof, 61-72, PROF PHYSICS, UNIV ROCHESTER, 72- Concurrent Pos: NSF sr fel, Univ Geneva & Europ Orgn Nuclear Res, 70. Mem: Am Phys Soc. Res: Nucleon-nucleon interactions; few nucleon problems; high energy photoproduction processes. Mailing Add: Dept of Physics Univ of Rochester Rochester NY 14627

THORNDIKE, EDWARD MOULTON, b New York, NY, Sept 25, 05; m 30; c 3. PHYSICS. Educ: Wesleyan Univ, BS, 26; Columbia Univ, MA, 27; Calif Inst Technol, PhD(physics), 30. Prof Exp: Fel physics, Calif Inst Technol, 30-31; instr, Polytech Inst Brooklyn, 31-38; from instr to asst prof, Queen's Col, NY, 38-43; assoc prof, Univ Southern Calif, 43-44; from asst prof to prof, 44-70, EMER PROF PHYSICS, QUEEN'S COL, NY, 70- Concurrent Pos: Physicist, Woods Hole Oceanog Inst, 42-43, consult, 44-45; consult, Columbia Univ, 44, 45-48, physicist, 45, res assoc, 50- Mem: AAAS. Res: Optics; oceanography. Mailing Add: Lamont-Doherty Geol Observ Palisades NY 10964

THORNE, CHARLES JOSEPH, b Pleasant Grove, Utah, May 28, 15; m 42; c 2. APPLIED MATHEMATICS. Educ: Brigham Young Univ, AB, 36; Iowa State Col, MS, 38, PhD(math physics), 41. Prof Exp: Asst math, Iowa State Col, 38-41; instr, Univ Mich, 41-43; asst prof, La State Univ, 43-44; develop engr, Curtiss-Wright Corp, 44-45; from asst prof to prof math, Univ Utah, 45-55; res scientist, Res Dept, US Naval Ord Test Sta, 55-60, sr res scientist & head math div, 60-61; supvry mathematician, Pac Missile Range, 61-65, sr opers res analyst, Naval Missile Ctr, 65-75, HEAD ASSESSMENT DIV, PAC MISSILE TEST CTR, US DEPT NAVY, 75- Concurrent Pos: Assoc prof, Univ Calif, Los Angeles, 48-49, lectr, 56-68; sr investr, US Navy Projs, 49-51; dir & prin investr, US Army Ord Projs, 51-55. Mem: Am Math Soc; Am Soc Mech Eng; Math Asn Am; Soc Indust & Appl Math. Res: Differential equations; boundary value problems; elasticity; analysis; operations research; numerical analysis. Mailing Add: 1447 Sunrise Ct Camarillo CA 93010

THORNE, CURTIS BLAINE, b Pine Grove, WVa, May 13, 21; m 59; c 1.

MICROBIAL GENETICS. Educ: WVa Wesleyan Col, BS, 43; Univ Wis, MS, 44, PhD(biochem), 48. Prof Exp: Biochemist, US Army Biol Labs, 48-61; prof bact genetics, Ore State Univ, 61-63; biochemist, US Army Biol Labs, 63-66; PROF BACT GENETICS, UNIV MASS, AMHERST, 66- Concurrent Pos: Waksman hon lectr, 59. Mem: AAAS; Am Soc Microbiol; Am Soc Biol Chem; Brit Soc Gen Microbiol; Genetics Soc Am. Res: Bacterial genetics; transformation and transduction; genetics of Bacillus species. Mailing Add: 45 Western Lane Amherst MA 01002

THORNE, DAVID WYNNE, b Perry, Utah, Dec 19, 08; m 37; c 5. SOIL FERTILITY. Educ: Utah State Univ, BS, 33; Iowa State Univ, MS, 34, PhD(soils), 36. Hon Degrees: DSc, Utah State Univ, 75. Prof Exp: Asst soils, Iowa State Univ, 36, actg res asst prof, 36-37; assoc prof, Agr & Mech Col Tex, 37-39; assoc prof, 39-47, prof agron & head dept, 47-55, dir, Agr Exp Sta & Univ Res, 55-65, vpres res, 65-73, dir agr exp sta, 73-74, EMER PROF SOIL SCI, UTAH STATE UNIV, 74-; AGR CONSULT, 74- Concurrent Pos: Chief soils & fertilizer res br, Tenn Valley Auth, 53-54; int consult land, water & agr develop, mem, Bd Agr & Renewable Resources, Nat Res Coun, 71-; mem bd govs, Int Crops Res Inst for Semi-Arid Tropics, India, 71-; agr res, United Nations Develop Prog, Ethiopis, S Yeman, & watershed mgt, India, 74, 75. Mem: Fel AAAS (vpres & chmn sect O, 60-61); fel Am Soc Agron; Soil Sci Soc Am (pres, 55-56); fel Soil Conserv Soc Am; Am Inst Biol Sci. Res: Enzyme systems of Rhizobium; microbial population of soil; fertility problems of high-lime soils; management of irrigated soils; research administration. Mailing Add: Utah State Univ Logan UT 84321

THORNE, JAMES MEYERS, b Logan, Utah, June 3, 37; m 60; c 2. PHYSICAL CHEMISTRY. Educ: Utah State Univ, BS, 61; Univ Calif, Berkeley, PhD(chem), 66. Prof Exp: ASSOC PROF CHEM, BRIGHAM YOUNG UNIV, 66- Concurrent Pos: Vis staff, Laser Div, Los Alamos Sci Lab, Univ Calif, 72- Mem: Am Chem Soc. Res: Applications of lasers to nuclear fusion, nonlinear optics; magneto and electrooptics. Mailing Add: 1119 E 2620 North Provo UT 84601

THORNE, JOHN CARL, b Ft Dodge, Iowa, Feb 24, 43; m 70; c 1. PLANT BREEDING, GENETICS. Educ: Augustana Col, Ill, BA, 65; Iowa State Univ, MS, 67, PhD(plant breeding), 69. Prof Exp: Res assoc soybean breeding, Iowa State Univ, 67-69; PLANT BREEDER, NORTHRUP, KING & CO, 69- Mem: Am Soc Agron; Am Soybean Asn. Res: Plant breeding and genetics related to soybean variety development. Mailing Add: Northrup King & Co PO Box 49 Washington IA 52353

THORNE, KIP STEPHEN, b Logan, Utah, June 1, 40; m 60; c 2. ASTROPHYSICS, THEORETICAL PHYSICS. Educ: Calif Inst Technol, BS, 62; Princeton Univ, AM, 63, PhD(theoret physics), 65. Prof Exp: NSF vis fel physics, Princeton Univ, 65-66; Alfred P Sloan res fel, 66-, assoc prof, 67-70, PROF, CALIF INST TECHNOL, 70- Concurrent Pos: Lectr, Enrico Fermi Int Sch Physics, Varenna, Italy, 65, 69; Fulbright lectr, Sch Theoret Physics, Les Houches, France, 66; Guggenheim fel, Inst Astrophys, Paris, France, 67; vis assoc prof, Univ Chicago, 68; vis assoc prof, Moscow State Univ, 69, vis prof, 75; adj prof, Univ Utah, 71-; mem Int Comt Gen Relativity and Gravitation, 71-. Honors & Awards: Sci Writing Award, Am Inst Physics-US Steel Corp, 69. Mem: AAAS; fel Am Acad Arts & Sci; fel Am Phys Soc; Am Astron Soc; Int Astron Union. Res: Theoretical astrophysics; gravitation theory. Mailing Add: 106-38 Calif Inst of Technol Pasadena CA 91125

THORNE, MARLOWE DRIGGS, b Perry, Utah, Nov 4, 18; m 41; c 4. AGRONOMY. Educ: Utah State Agr Col, BS, 40; Iowa State Col, MS, 41; Cornell Univ, PhD, 48. Prof Exp: Soil physicist & head dept agron, Pineapple Res Inst, Univ Hawaii, 47-54; soil scientist & irrig work proj leader, Eastern Soil & Water Mgt Sect, Soil & Water Conserv Res Br, Agr Res Serv, US Dept Agr, 55-56; prof agron & head dept, Okla State Univ, 56-63; head dept, 63-70, PROF AGRON, UNIV ILL, URBANA, 63- Concurrent Pos: Water technol adv, GB Pant Univ, Pantnagar, India, 70-72. Mem: Soil Sci Soc Am; Am Soc Agron; Soil Conserv Soc Am. Res: Irrigation; mulching; tillage. Mailing Add: Dept of Agron Univ of Ill Urbana IL 61801

THORNE, MELVYN CHARLES, b San Francisco, Calif, Dec 27, 32; m 58; c 2. EPIDEMIOLOGY. Educ: Univ Calif, AB, 56; Harvard Univ, MD, 60; Johns Hopkins Univ, MPH, 68. Prof Exp: Epidemiologist, Field Epidemiol Res Sta, Nat Heart Inst, 61-63 & 65-66; Peace Corps physician, Morocco, 63-65; resident internal med, Mary Imogene Bassett Hosp, 66-67; rep, Pop Coun, Tunisia, 68-72; ASST PROF INT HEALTH, SCH HYG & PUB HEALTH, JOHNS HOPKINS UNIV, 72- Concurrent Pos: Tech adv, Urban Life-Pop Educ Inst, 73-76; health syst adv, Overseas Develop Coun Seminars, Priv Vol Orgns, 74-76; consult, Health Educ, Porter Novelli & Assocs. Mem: Am Pub Health Asn; Pop Asn Am. Res: Simple, effective methods to introduce health services into populations currently deprived of them, and to introduce population education into school systems with focus on using information in decisions. Mailing Add: Dept of Int Health Johns Hopkins Univ Baltimore MD 21205

THORNE, OAKLEIGH II, b New York, NY, Oct 12, 28; m 53; c 3. BIOLOGY, ECOLOGY. Educ: Yale Univ, BS, 51, MS, 53; Univ Colo, PhD(zool), 58. Prof Exp: PRES, THORNE FILMS, INC, 54-, CHMN, THORNE ECOL INST, 55- Mem: Fel AAAS; Ecol Soc Am; Wildlife Soc; Am Soc Mammalogists; Am Ornith Union. Res: Biological photography; populations studies; animal behavior; bird banding. Mailing Add: Thorne Ecol Inst 934 Pearl Boulder CO 80302

THORNE, RICHARD MANSERGH, b Birmingham, Eng, July 25, 42; m 63; c 1. SPACE PHYSICS, PLASMA PHYSICS. Educ: Univ Birmingham, BSc, 63; Mass Inst Technol, PhD(physics), 68. Prof Exp: Asst prof, 68-71, ASSOC PROF METEOROL, UNIV CALIF, LOS ANGELES, 71- Concurrent Pos: NSF grants, 71-75; mem nat comt, Int Union Radio Sci, 71-; consult, Jet Propulsion Lab, Calif Inst Technol. Mem: Am Geophys Union; Int Union Radio Sci. Res: Structure and stability of radiation belts; magnetosphere-ionosphere interactions; wave propagation in anistotropic media. Mailing Add: Dept of Meteorol Univ of Calif Los Angeles CA 90024

THORNE, ROBERT FOLGER, b Spring Lake, NJ, July 13, 20; m 47; c 1. SYSTEMATICS, BIOGEOGRAPHY. Educ: Dartmouth Col, AB, 41; Cornell Univ, MS, 42, PhD(bot), 49. Prof Exp: Asst bot, Cornell Univ, 45-46, instr, 48-49; from asst prof to prof, Univ Iowa, 49-62; PROF BOT, CLAREMONT GRAD SCH, 62-; CUR & TAXONOMIST, RANCHO SANTA ANA BOT GARDEN, 62- Concurrent Pos: Fulbright res scholar, Univ Queensland, 59-60; NSF sr fel, 60; chmn adv coun, Flora NAm Proj. Mem: Bot Soc Am; Am Soc Plant Taxon (secy, 57-58, pres, 68); Int Soc Plant Morphol; Int Asn Plant Taxon; Fr Soc Biogeog. Res: Phylogeny and geography of flowering plants; floristics; marine phanerogams. Mailing Add: Rancho Santa Ana Bot Garden Claremont CA 91711

THORNER, MELVIN WILFRED, b New York, NY, July 19, 07; m 76; c 1. NEUROSCIENCES. Educ: Univ Pa, BS, 28, MS, 37, DSc, 40; Jefferson Univ, MD, 32. Prof Exp: Mem fac neurol, Univ Pa, 36-61; prof biomed, Drexel Univ, 61-63; assoc dir res neuropharmacol, Schering Corp, 65-70; dir, Res Neurol, Traverse City State Hosp, 70-72; CHIEF NEUROL, DOWNEY VET ADMIN HOSP, 72-; ASSOC PROF NEUROL, NORTHWESTERN UNIV, 72- Mem: Am Acad Neurol; Am Neurol Asn; Am Psychiat Asn; AAAS. Res: Chemical imprinting of mammalian nervous systems; the application of holographic techniques to the electroencephalogram. Mailing Add: 331 E Blodgett Ave Lake Bluff IL 60044

THORNER, ROBERT M, biostatistics, public health, see 12th edition

THORNEYCROFT, IAN HALL, physiology, endocrinology, see 12th edition

THORNGATE, JOHN HILL, b Eau Claire, Wis, Dec 23, 35; m 56; c 3. PHYSICS. Educ: Ripon Col, BA, 57; Vanderbilt Univ, MS, 61, PhD, 73. Prof Exp: Inspector health physics, Oak Ridge Opers Off, US Atomic Energy Comn, 59; res group leader radiation dosimetry, Health Physics Div, 64-75, asst sect chief, 73-75, HEALTH PHYSICIST, OAK RIDGE NAT LAB, 60- Mem: Health Physics Soc; Am Phys Soc; Sigma Xi. Res: Radiation dosimetry and spectrometry. Mailing Add: PO Box X Oak Ridge Nat Lab Oak Ridge TN 37830

THORNTON, CHARLES DE WANE, b Brownsville, Pa, Mar 3, 15; m 35; c 5. CHEMISTRY, PHYSICS. Educ: Univ Pittsburgh, BS, 39. Hon Degrees: DSc, Ind Tech Col, 57. Prof Exp: Asst, Mellon Inst Indust Res, 39-40; from res chemist to res supvr, Tenn Eastman Co Div & Res Labs, Eastman Kodak Co, 40-48; from tech asst to chief off opers anal & planning, US Atomic Energy Comn, 48-56; asst to pres atomic energy & dir res & develop, Farnsworth Electronics Co, 56-58; vpres phys sci, Components & Instrumentation Labs, Int Tel & Tel Corp, 58-60, US group tech dir, 60-67; exec vpres & dir Cleveland Container Corp, 67-70; dir nuclear mat safeguards div, US Atomic Energy Comn, 70-72, spec asst energy policy, 72-75, SPEC ASST DIR, US ENERGY RES & DEVELOP ADMIN, 75- Concurrent Pos: Chmn nationwide fissionable standards samples comt, US Atomic Energy Comn, 49-55, consult, 56-70, chmn adv comt standard reference mat & methods of measurements, 56-; comn nuclear energy, Int Tel & Tel Corp, 57-58; mem comt nuclear instrumentation & controls, Atomic Indust Forum, 57-58. Mem: Fel AAAS; Inst Nuclear Mat Mgt; Am Chem Soc; Opers Res Soc Am; Am Nuclear Soc. Res: Analytical instrumentation; spectroscopy; uranium and transuranics engineering and economics of nuclear energy; systems analysis materials accountability systems; cost-benefit analysis; energy converters; special purpose electronic devices; automation. Mailing Add: 404 Chesapeake Dr Great Falls VA 22066

THORNTON, CHARLES PERKINS, b Indianapolis, Ind, Jan 1, 27; m 54; c 2. PETROLOGY. Educ: Univ Va, AB, 49; Yale Univ, MS, 50, PhD(geol), 53. Prof Exp: Field geologist, State Geol Surv, Va, 50-52; from instr to asst prof petrog, Pa State Univ, 52-61; asst prof geol, Bucknell Univ, 61-63; assoc prof, 63-69, PROF GEOL, PA STATE UNIV, UNIVERSITY PARK, 69- Mem: Geol Soc Am. Res: Geology of central Shenandoah Valley, Virginia; petrography and petrology of volcanic rocks; volcanology. Mailing Add: Dept of Geosci 201 Dieke Bldg Pa State Univ University Park PA 16802

THORNTON, CHARLES STEAD, biology, deceased

THORNTON, CLARENCE GOULD, physical chemistry, see 12th edition

THORNTON, DANIEL MCCARTY, b Richmond, Va, Jan 30, 18; m 46; c 4. TEXTILES. Educ: Univ Richmond, BS, 38; Univ Pa, MS, 40. Prof Exp: Asst instr chem, Drexel Inst Technol, 39-40; asst chemist, Wortendyke Mfg Co, Va, 40-41; res chemist, 41-43, technologist, 43-48, group leader, 48-49, field res supvr, 49, mgr orlon sales develop, 49-51, orlon customer serv, 51-52, staple fibers customer serv, 52-54, nylon tech serv, 54-60, mgr mkt, Intimate Apparel Indust, 60-64, MKT RES MGR, TEXTILE FIBERS, E I DU PONT DE NEMOURS & CO, INC, 64- Concurrent Pos: Vchmn textile sect, NY Bd Trade; mem, Indust Sector Adv Comt on Textiles & Apparel, Dept of Com, 74- Mem: Am Mkt Asn. Res: Markets and uses for fibers; consumer and trade motivational studies; psychology and sociology; textiles and clothing; marketing research. Mailing Add: Textile Fivers Div E I du Pont de Nemours & Co Inc Wilmington DE 19898

THORNTON, EDWARD RALPH, b Syracuse, NY, July 19, 35; m 69; c 1. ORGANIC CHEMISTRY, BIO-ORGANIC CHEMISTRY. Educ: Syracuse Univ, BA, 57; Mass Inst Technol, PhD(org chem), 59. Hon Degrees: MA, Univ Pa, 71. Prof Exp: NIH fel, Mass Inst Technol, 59-60 & Harvard Univ, 60-61; from asst prof to assoc prof, 61-69, PROF CHEM, UNIV PA, 69- Mem: Fedn Am Sci; Am Chem Soc; The Chem Soc. Res: Structure and mechanism in organic and biological chemistry; glycolipid chemistry and membrane structure. Mailing Add: Dept of Chem D5 Univ of Pa Philadelphia PA 19174

THORNTON, ELIZABETH K, b Brooklyn, NY, June 4, 40. PHYSICAL ORGANIC CHEMISTRY. Educ: Mt Holyoke Col, AB, 61; Univ Pa, PhD(org chem), 66. Prof Exp: Teaching asst chem, Univ Pa, 61-62, NIH fel org chem, 63-66; NATO fel, Swiss Fed Inst Technol, 66-68; asst prof, 68-75, ASSOC PROF CHEM, WIDENER COL, 75- Mem: AAAS; Fedn Am Sci; Asn Women Sci. Res: Kinetic isotope effects and reaction mechanisms; organic biochemistry. Mailing Add: Dept of Chem Widener Col Chester PA 19013

THORNTON, ERLY J, b Ala, Sept 20, 08; m 36. POULTRY HUSBANDRY. Educ: Tuskegee Inst, BS, 33; Univ Mass, MS, 44. Prof Exp: Asst poultryman, Tuskegee Inst, 32-33; poultryman, Tenn State Univ, 33-35; poultryman & head dept poultry husb, Va State Col, 45-46; asst, Univ Mass, 46-47; head dept, 47-66, PROF POULTRY HUSB, TENN STATE UNIV, 47- Mem: Poultry Sci Asn. Res: Possible genetic factors and mortality in relation to the sex ratio of chickens at hatching. Mailing Add: Dept of Poultry Husbandry Tenn State Univ Nashville TN 37203

THORNTON, GEORGE DANIEL, b Elberton, Ga, Aug 10, 10; m 39. SOILS. Educ: Univ Ga, BS, 36, MS, 38; Iowa State Col, PhD(soil fertility), 47. Prof Exp: County agr agent, Ga Exten Serv, 36; asst soil surveyor, Ga State Col, 36, instr agron, 37-40; asst agronomist, Exp Sta, Univ Ga, 40-41; asst prof soils & asst soil microbiologist, Col Agr, Univ Fla, 41-45; asst, Iowa State Col, 45-47; assoc prof soils & assoc soil microbiology, 47-51, prof soils, 51-71, soil microbiologist, 51-56, asst dean col, 56-71, EMER PROF SOILS, COL AGR, UNIV FLA, 71- Mem: Fel Am Soc Agron; Soil Sci Soc Am. Res: Soil microbiology. Mailing Add: PO Box 833 Venice FL 33595

THORNTON, GEORGE FRED, b Newton, Mass, Mar 8, 33; m 63; c 2. INTERNAL MEDICINE, INFECTIOUS DISEASES. Educ: Harvard Univ, AB, 55; Boston Univ, MD, 59. Prof Exp: Instr clin med, Sch Med, Yale Univ, 64-65; instr med, Johns Hopkins Univ, 65-67; asst prof, 67-72, ASSOC PROF CLIN MED, SCH MED, YALE UNIV, 72-; DIR MED SERV, WATERBURY HOSP, 72-; ASSOC CLIN PROF MED, UNIV CONN, 75- Concurrent Pos: Fel allergy & infectious dis, Johns Hopkins Univ, 62-64. Mem: AMA; Am Fedn Clin Res; Infectious Dis Soc Am. Res: Clinical epidemiology. Mailing Add: Med Serv Waterbury Hosp Waterbury CT 06720

THORNTON, JOHN ALEXANDER, b Olympia, Wash, Jan 3, 33; m 61. SOLID STATE PHYSICS, PLASMA PHYSICS. Educ: Univ Wash, Seattle, BS, 57, MS, 59; Northwestern Univ, PhD(plasma physics), 63. Prof Exp: Asst engr, Appl Physics Lab, Univ Wash, 57-58; dir diag res, Space Sci Labs, Litton Industs, Inc, Calif, 63-68; dir,

68-73, VPRES RES & DEVELOP, TELIC CORP, SANTA MONICA, 73- Mem: Am Phys Soc; Am Vacuum Soc; Am Inst Aeronaut & Astronaut; Am Soc Mech Eng. Res: Plasma discharges; thin film deposition techniques, including sputtering, evaporation and chemical vapor deposition; plasma and radiation chemistry; physics and metallurgy of metallic and non-metallic coatings; vacuum technology; laser interactions with surfaces. Mailing Add: 1280 Barrington Apt 22 Los Angeles CA 90025

THORNTON, JOHN WILLIAM, b Shawnee, Okla, Apr 21, 36; m 57; c 3. ZOOLOGY, CYTOLOGY. Educ: Okla State Univ, BS, 58; Univ Wash, PhD(zool), 64. Prof Exp: From asst prof to assoc prof, 60-74, PROF ZOOL, OKLA STATE UNIV, 74- Concurrent Pos: USPHS res grant, 65-68; staff biologist, Comn Undergrad Educ Biol Sci, 70-71; mem adv comt, Purdue Minicourse Proj. Mem: AAAS; Am Soc Zoologists; Am Inst Biol Sci. Res: Cellular ultrastructure; cell and tissue culture; undergraduate curricular improvement; investigative laboratories. Mailing Add: Dept of Zool Okla State Univ Stillwater OK 74074

THORNTON, KENT W, b Ames, Iowa, Apr 29, 44; m 66; c 1. AQUATIC ECOLOGY, SYSTEMS SCIENCE. Educ: Univ Iowa, BA, 67, MS, 69; Okla State Univ, PhD(ecol), 72. Prof Exp: Teaching asst zool, Univ Iowa, 68-69; lectr environ systs theory, Okla State Univ, 72; asst prof biol, Bowling Green State Univ, 73-74; SYSTS ECOLOGIST, WATERWAYS EXP STA, US ARMY ENGRS, 74- Concurrent Pos: NSF fel, Ctr Systs Sci, Okla State Univ, 72-73; mem methods ecosyst anal, Nat Comn Water Qual, 74; actg br chief, Waterways Exp Sta, US Army Engrs, 75. Mem: AAAS; Am Inst Biol Sci; Ecol Soc Am; NAm Benthological Soc; Int Soc Limnol. Res: Systems theoretical approach to the conceptualization, analysis and application of mathematical ecosystem models for watershed-reservoir planning and management; sampling theory approach to dynamic systems. Mailing Add: 111 Signal Hill Dr Vicksburg MS 39180

THORNTON, MELVIN CHANDLER, b Sioux City, Iowa, July 2, 35; m 58; c 4. MATHEMATICS. Educ: Univ Nebr, BS, 57; Univ Ill, MS, 61, PhD(math), 65. Prof Exp: Asst prof math, Univ Wis, 65-69; asst prof, 69-73, ASSOC PROF MATH, UNIV NEBR, LINCOLN, 73- Mem: Am Math Soc; Math Asn Am. Res: General topology. Mailing Add: Dept of Math Univ of Nebr Lincoln NE 68588

THORNTON, MELVIN LEROY, b Billings, Mont, Nov 7, 28; m 52; c 2. BOTANY. Educ: Univ Denver, BA, 52; Tufts Univ, MA, 58; Univ Mont, PhD(bot), 69. Prof Exp: NSF fel, Birkbeck Col, Univ London, 69-70; ASST PROF BOT, UNIV MONT, 71- Mem: AAAS; Brit Mycol Soc; Mycol Soc Am. Res: Dispersal of fungi; ecology of zoosporic fungi. Mailing Add: Dept of Bot Univ of Mont Missoula MT 59801

THORNTON, PAUL A, b Campbell Co, Ky, June 29, 25; m 45; c 2. PHYSIOLOGY, NUTRITION. Educ: Univ Ky, BS, 49, MS, 53; Mich State Univ, PhD(nutrit), 56. Prof Exp: Asst & assoc prof nutrit, Colo State Univ, 56-62, assoc prof physiol, 63-64; vis prof, 62-63, ASSOC PROF PHYSIOL, UNIV KY, 64-; RES PHYSIOLOGIST, VET ADMIN HOSP, LEXINGTON, 64- Mem: Soc Exp Biol & Med; Am Inst Nutrit; Am Physiol Soc; Geront Soc. Res: Skeletal physiology and the influence of age on bone tissue change; endocrinological and other environmental factors which affect bone. Mailing Add: Vet Admin Hosp 15A Leestown Rd Lexington KY 40507

THORNTON, ROBERT LYSTER, b Wootton, Eng, Nov 29, 08; nat US; m 38; c 3. PHYSICS. Educ: McGill Univ, BSc, 30, PhD(spectros), 33. Prof Exp: Demonstr, McGill Univ, 30-31; Moyse traveling scholar from McGill Univ, Univ Calif, 33-34, res assoc, Radiation Lab, 34-36, instr physics, Univ Mich, 36-38; res assoc, Radiation Lab, Univ Calif, 38-39; assoc prof, Wash Univ, 40-45; res physicist, Manhattan Dist Proj, 42-43, assoc dir, Lawrence Radiation Lab, 58-72, prof, 45-72, EMER PROF PHYSICS, UNIV CALIF, BERKELEY, 72- Concurrent Pos: Asst dir, Process Improv Div, Tenn Eastman Corp, 43-45. Mem: Fel Am Phys Soc. Res: Nuclear physics; accelerator design and construction. Mailing Add: 522 Cragmont Ave Berkeley CA 94708

THORNTON, ROBERT MELVIN, b Auburn, Calif, Nov 14, 37; m 57, 71; c 2. BIOLOGY, PLANT PHYSIOLOGY. Educ: Calif Inst Technol, BS, 59; Harvard Univ, MA, 61, PhD(biol), 66. Prof Exp: Sr scientist, Biol/Eng, Appl Sci Corp, Calif, 61-63; instr, Ojai Valley Sch, 63-64; instr biol, Univ Calif, Santa Cruz, 66-67, asst prof, 67-68, asst prof bot, 68-74, ASSOC PROF BOT, UNIV CALIF, DAVIS, 74- Concurrent Pos: NSF res grant, Univ Calif, Davis, 69-72. Mem: AAAS; Am Soc Plant Physiol; Bot Soc Am. Res: Regulatory mechanisms with particular reference to photoresponse mechanisms in the development of fungi. Mailing Add: Dept of Bot Univ of Calif 218 Robbins Hall Davis CA 95616

THORNTON, ROGER LEA, b Wilmington, Del, Mar 9, 35; m 58; c 3. ORGANIC CHEMISTRY. Educ: Univ Del, BS, 57; Mass Inst Technol, PhD(org chem), 61. Prof Exp: Res chemist, Del, 61-65, Va, 65-67, SR RES CHEMIST, E I DU PONT DE NEMOURS & CO, INC, DEL, 67- Res: Vapor-phase catalytic reactions; thermally stable condenstaion polymers; emulsion polymerization. Mailing Add: Org Chem Dept E I du Pont de Nemours & Co Inc Wilmington DE 19898

THORNTON, ROY FRED, b Upper Darby, Pa, Feb 27, 41; m 66; c 2. CHEMICAL ENGINEERING, ELECTROCHEMISTRY. Educ: Johns Hopkins Univ, BS, 63, PhD(chem eng), 67. Prof Exp: Chem engr, Battery Bus Dept, Gen Elec Co, 67-69, STAFF MEM, ELECTROCHEM, GEN ELEC CO CORP RES & DEVELOP, 69- Mem: Am Inst Chem Eng; Electrochem Soc. Res: Development of electrochemical energy storage systems. Mailing Add: Box 8 Corp Res & Develop Gen Elec Co Schenectady NY 12305

THORNTON, STEPHEN THOMAS, b Kingsport, Tenn, Oct 2, 41; m 61; c 2. EXPERIMENTAL NUCLEAR PHYSICS. Educ: Univ Tenn, Knoxville, BS, 63, MS, 64, PhD(physics), 67. Prof Exp: US Atomic Energy Comn fel, Univ Wis-Madison, 67-68; asst prof, 68-72, ASSOC PROF PHYSICS, UNIV VA, 72- Concurrent Pos: Consult, Physics Div, Oak Ridge Nat Lab, 72-; Fulbright-Hays sr fel, Max Planck Inst, Heidelberg, 73-74. Mem: AAAS; Am Phys Soc. Res: Neutron polarization; nuclear structure studies from experimental heavy ion nuclear physics. Mailing Add: Dept of Physics Univ of Va Charlottesville VA 22903

THORNTON, WILLIAM ANDRUS, JR, b Buffalo, NY, June 16, 23; m 44; c 4. PHYSICS. Educ: Univ Buffalo, BA, 48; Yale Univ, MS, 49, PhD(physics), 51. Prof Exp: Res assoc labs, Gen Elec Co, 51-56; sr res engr, 56-59, fel res engr, 59-65, mgr phosphor res, 65-67, RES ENG CONSULT, WESTINGHOUSE ELEC CORP, BLOOMFIELD, 67- Mem: Am Phys Soc; Optical Soc Am; Electrochem Soc. Res: Light and color. Mailing Add: Westinghouse Elec Corp Lamp Div Res Lab Bloomfield NJ 07003

THORNTON, WILLIAM EDGAR, b Faison, NC, Apr 14, 29; m; c 2. MEDICINE, ASTRONAUTICS. Educ: Univ NC, BS, 52, MD, 63. Prof Exp: Chief engr, Electronics Div, Del Mar Eng Labs, Calif, 55-59; intern, Wilford Hall Hosp, Lackland AFB, US Air Force, Tex, 64, assigned to Aerospace Med Div, Brooks AFB, 65-67; SCIENTIST-ASTRONAUT, JOHNSON SPACE CTR, NASA, 67- Concurrent Pos: Officer-in-chg instrumentation lab, Flight Test Air Proving Ground, consult to Air Proving Ground Command. Honors & Awards: Exceptional Serv Medal, NASA, 72; Crew Mem on Skylab Med Exp Altitude Test. Res: Physics; medical experiments baseline data and evaluation of equipment, operations and procedures. Mailing Add: NASA Johnson Space Ctr Houston TX 77058

THORNTON, WILLIAM NORMAN, JR, b Courtland, Va, July 26, 12; m 46; c 3. OBSTETRICS & GYNECOLOGY. Educ: Univ Va, BS, 33, MD, 36. Prof Exp: From asst prof to assoc prof, 46-50, PROF OBSTET & GYNEC & CHMN DEPT, SCH MED, UNIV VA, 50- Concurrent Pos: Dir, Am Bd Obstet & Gynec, 63-69, chmn, 70-73. Mem: Am Gynec Soc; Am Asn Obstetricians & Gynecologists (treas, 66, pres, 70); fel Am Col Surgeons; fel Am Col Obstetricians & Gynecologists. Res: Uterine contractility and placental transfer. Mailing Add: Dept of Obstet & Gynec Univ of Va Hosp Charlottesville VA 22901

THORP, EDWARD O, b Chicago, Ill, Aug 14, 32; m 56; c 3. MATHEMATICS. Educ: Univ Calif, Los Angeles, BA, 53, MA, 55, PhD(math), 58. Prof Exp: Instr math, Univ Calif, Los Angeles, 58-59; C L E Moore instr, Mass Inst Technol, 59-61; from asst prof to assoc prof, NMex State Univ, 61-65; assoc prof, 65-67, PROF MATH, UNIV CALIF, IRVINE, 67-; PRES, OAKLEY SUTTON MGT CORP, 72-, CHMN, OAKLEY SUTTON SECURITIES CORP, 72- Concurrent Pos: Res grants, NSF, 62-64 & US Air Force Off Sci Res, 64-74. Mem: Am Math Soc; Am Stat Asn; fel Inst Math Stat; Am Finance Asn; Am Econ Asn. Res: Functional analysis; probability theory; game theory; statistics; mathematical finance; numerical solution Stefan problems; parabolic differential equations. Mailing Add: Dept of Math Univ of Calif Irvine CA 92664

THORP, ELDON MARION, b Okla, Jan 9, 07. SOIL CONSERVATION. Educ: Tex Tech Col, AB, 27; Univ Calif, MS, 31, PhD(oceanog), 34. Prof Exp: Res assoc, Scripps Inst, Univ Calif, 34; prof geol & geog & head dept, Baylor Univ, 34-39; asst geologist, 39-72, GEOLOGIST, SOIL CONSERV SERV, US DEPT AGR, 72- Mem: AAAS; fel Geol Soc Am; assoc Soc Econ Paleont & Mineral. Res: Geology of ocean bottoms; micropaleontology; soil erosion; modern stream sediments. Mailing Add: Eng & Water Shed Planning Unit US Dept of Agr Rm 345 Fed Bldg Lincoln NE 68508

THORP, FRANK KEDZIE, b Denver, Colo, Apr 29, 36; m 65; c 1. BIOCHEMISTRY, PEDIATRICS. Educ: Mich State Univ, BA, 55; Univ Chicago, MD, 60, PhD(biochem), 62. Prof Exp: Intern pediat, Univ Chicago, 61-62; resident, Children's Hosp Med Ctr, Boston, 62-63; instr, 65-66, asst prof, 66-72, ASSOC PROF PEDIAT, UNIV CHICAGO, 72- Concurrent Pos: NIH res fel biochem, Children's Hosp Med Ctr, Boston, 63-65; Joseph P Kennedy, Jr scholar, 66-; Am Acad Pediat grant, 69-; dir ment develop clin, Joseph P Kennedy, Jr Ment Retardation Res Ctr, 71- Mem: Am Acad Pediat. Res: Teaching and clinical work in pediatrics; research in metabolic diseases of children; embryogenesis of connective tissue and control mechanisms of mucopolysaccharide metabolism. Mailing Add: Dept of Pediat Univ of Chicago Chicago IL 60637

THORP, NORMAN, physical chemistry, chemical engineering, see 12th edition

THORP, ROBBIN WALKER, b Benton Harbor, Mich, Aug 26, 33; m 54, 67; c 3. INSECT TAXONOMY, ECOLOGY. Educ: Univ Mich, BS, 55, MS, 57; Univ Calif, Berkeley, PhD(entom), 64. Prof Exp: Jr specialist, Univ Calif, Berkeley, 62-63, asst specialist, 63-64, asst res entomologist, 64; asst apiculturist, 64-72, ASSOC PROF ENTOM & ASSOC APICULTURIST, UNIV CALIF, DAVIS, 72- Mem: AAAS; Ecol Soc Am; Soc Syst Zool; Entom Soc Am; Soc Study Evolution. Res: Pollination ecology, especially bee and flower relationships; ecology and systematics of bees and ecology of their biotic enemies; coevolution and coadaptation of pollinating insects and entomophilous angiosperms. Mailing Add: Dept of Entom Univ of Calif Davis CA 95616

THORP, WILLIAM T S, animal pathology, poultry pathology, see 12th edition

THORPE, BERT DUANE, b Spanish Fork, Utah, Sept 21, 29; m 55; c 6. IMMUNOLOGY, WILDLIFE DISEASES. Educ: Univ Utah, BS(chem) & BS(bact), 58, PhD(microbiol), 63. Prof Exp: Asst chem, Univ Utah, 55-57, asst bact, 57, res bacteriologist, Epizool Lab, 58-61, res instr ecol & epizool, 61-63, from asst res prof to assoc res prof, 63-68, dir epizool lab, 61-68, clin lectr microbiol, 67-68; PROF MICROBIOL, UNIV NORTHERN COLO, 68- Concurrent Pos: Lectr, Brigham Young Univ, 65-66; consult, Dept Defense, 67-; vpres nat comt, Int Northwestern Conf Dis Man. Mem: Am Chem Soc; Am Asn Immunol; Am Soc Microbiol; Am Soc Trop Med & Hyg; Soc Exp Biol & Med. Res: Host mechanisms of resistance to infectious diseases; new methods of detection and isolation of microorganisms; zoonoses; animal infections and human diseases; natural and acquired immunity. Mailing Add: Dept of Biol Sci Univ of Northern Colo Greeley CO 80639

THORPE, JOHN ALDEN, b Lewiston, Maine, Feb 29, 36; m 59; c 2. GEOMETRY, COSMOLOGY. Educ: Mass Inst Technol, SB, 58; Columbia Univ, AM, 59, PhD(math), 63. Prof Exp: Instr math, Columbia Univ, 63; C L E Moore instr, Mass Inst Technol, 63-65; asst prof, Haverford Col, 65-68; ASSOC PROF, STATE UNIV NY STONY BROOK, 68- Concurrent Pos: Mem, Inst Adv Study, 67-68. Mem: Am Math Soc; Math Asn Am. Res: Differential geometry; general relativity. Mailing Add: Dept of Math State Univ of NY Stony Brook NY 11794

THORPE, MARTHA CAMPBELL, b Tullahoma, Tenn, Apr 28, 22; m 43; c 2. PHYSICAL ORGANIC CHEMISTRY. Educ: Vanderbilt Univ, BA, 44; Samford Univ, MA, 68. Prof Exp: Anal chemist, E I du Pont de Nemours & Co, Inc, 44-45; SR CHEMIST MOLECULAR SPECTROS, SOUTHERN RES INST, 61- Mem: Am Chem Soc; Int Soc Magnetic Resonance. Res: H-1 and C-13 nuclear magnetic resonance spectroscopy of organic compounds. Mailing Add: Southern Res Inst 2000 Ninth Ave S Birmingham AL 35205

THORPE, MICHAEL FIELDING, b Bromley, Eng, Mar 12, 44. THEORETICAL PHYSICS. Educ: Univ Manchester, BSc, 65; Oxford Univ, DPhil(physics), 68. Prof Exp: Res assoc physics, Brookhaven Nat Lab, 68-70; asst prof, 70-74, ASSOC PROF PHYSICS, YALE UNIV, 74- Concurrent Pos: Yale jr fac fel & guest scientist, Max Planck Inst Solids, Stuttgart, 72-73. Mem: Am Phys Soc; Brit Inst Physics. Res: Theoretical solid state physics, including low temperature excitations, magnetism and amorphous solids. Mailing Add: Becton Ctr Yale Univ 15 Prospect St New Haven CT 06520

THORPE, NEAL OWEN, b Wausau, Wis, Sept 8, 38; m 60; c 3. BIOCHEMISTRY. Educ: Augsburg Col, BA, 60; Univ Wis, Madison, PhD(physiol chem), 64. Prof Exp: USPHS fel, 65-66; Am Heart Asn adv res fel, 66-67; ASSOC PROF BIOL, AUGSBURG COL, 67- Concurrent Pos: Res Corp grant & Am Heart Asn grant, 69-71; regional dir Grants, Res Corp, 73-74. Res: Immunochemistry and the structure of sigma virus in Drosophila. Mailing Add: Dept of Biol Augsburg Col 21st Ave S at 8th St Minneapolis MN 55404

THORPE, RALPH IRVING, b Halls Harbour, NS, Feb 29, 36; m 68; c 2. ECONOMIC GEOLOGY. Educ: Acadia Univ, BSc, 58; Queen's Univ, Ont, MSc, 63; Univ Wis-Madison, PhD(econ geol), 67. Prof Exp: RES SCIENTIST MINERAL DEPOSITS, GEOL SURV CAN, 65- Mem: Mineral Soc Am; Mineral Asn Can; Can Inst Mining & Metall; Geol Asn Can. Res: Genesis of metalliferous ore deposits; lead isotope interpretations; ore mineralogy. Mailing Add: Mineral Deposits Sect Geol Surv Can Ottawa ON Can

THORPE, TREVOR ALLEYNE, b Barbados, WI, Oct 18, 36; m 63; c 2. PLANT PHYSIOLOGY. Educ: Allahabad Agr Inst, BScAgr, 61; Univ Calif, Riverside, MS, 64, PhD(plant sci & physiol), 68. Prof Exp: Nat Res Coun fel & res plant physiologist, Fruit & Vegetable Chem Lab, US Dept Agr, Calif, 68-69; asst prof bot, 69-73, ASSOC PROF BOT, UNIV CALGARY, 73-, ASST DEAN FAC ARTS & SCI, 74- Concurrent Pos: Chmn, Int Asn Plant Tissue Cult, 74-78. Mem: Can Soc Plant Physiol; Am Soc Plant Physiol; Japanese Soc Plant Physiol; Scand Soc Plant Physiol; Tissue Cult Asn. Res: Experimental plant morphogenesis; cytology, physiology and biochemistry of organ formation in tissue culture systems; plant propagation by tissue culture methods. Mailing Add: Dept of Biol Univ of Calgary Calgary AB Can

THORSEN, ARTHUR C, b Portland, Ore, July 27, 34. EXPERIMENTAL SOLID STATE PHYSICS. Educ: Reed Col, BA, 56; Rice Inst, MA, 58, PhD(physics), 60. Prof Exp: Res physicist, Atomics Int Div, NAm Rockwell Corp, 60-63, mem tech staff, Sci Ctr, 63-67, mem tech staff, Autonetics Div, 67-70, res & technol div, Anaheim, Ca, 70-73 & Rockwell Int Sci Ctr, 73-74, PROG MGR IR&D, ROCKWELL INT SCI CTR, 74- Mem: Am Phys Soc. Res: Low temperature solid state physics; electronic structure of metals; transport properties of semiconductors; thin films; superconductivity. Mailing Add: Rockwell Int Sci Ctr 1049 Camino dos Rios Thousand Oaks CA 91360

THORSEN, JAN, b Toronto, Ont, Oct 18, 32; m 60; c 2. VETERINARY VIROLOGY. Educ: Ont Vet Col, DVM, 55; Univ Toronto, dipl, 62, PhD(virol), 65. Prof Exp: Vet officer, Her Majesty's Overseas Civil Serv, 55-59; vet, First St Vet Clin, Alta, 59-61; res assoc virol, Univ Toronto, 62-65; sr virologist, Pitman-Moore Div, Dow Chem Co, 65-67; ASSOC PROF VIROL, ONT VET COL, UNIV GUELPH, 67- Concurrent Pos: Asst dir vet serv, Kenya, 71-73. Mem: Am Soc Microbiol; Can Soc Microbiol; Can Vet Med Asn; Royal Col Vet Surg. Res: Studies of veterinary viruses from the point of view of etiological studies; virus classification and vaccine development. Mailing Add: Dept of Vet Microbiol Univ of Guelph Guelph ON Can

THORSETT, GRANT OREL, b Shelton, Wash, Jan 25, 40; m 63; c 3. MOLECULAR BIOLOGY. Educ: Wash State Univ, BS, 62; Yale Univ, MS, 65, PhD(molecular biophys), 69. Prof Exp: Asst prof biol, 67-72, ASSOC PROF BIOL, WILLAMETTE UNIV, 72- Concurrent Pos: Instr, Proj Newgate, Ore State Penitentiary, 71-72; NSF res partic, Willamette Univ, 71-73. Mem: AAAS. Res: Bacterial transformation; biochemical systematics; bacterial biochemistry; use of computers in undergraduate curricula. Mailing Add: Dept of Biol Willamette Univ Salem OR 97301

THORSON, JOHN WELLS, b Detroit, Mich, Feb 25, 33; m 64. NEUROPHYSIOLOGY, BIOPHYSICS. Educ: Rensselaer Polytech Inst, BS, 55, MS, 58; Univ Calif, Los Angeles, PhD(zool), 65. Prof Exp: Physicist, Gen Elec Co, 55-60; NIH trainee biophys, Univ Calif, Los Angeles, 60-65; NSF fel physiol, Max Planck Inst Biol, 65-66 & Oxford Univ, 66-67; asst prof neurosci, Univ Calif, San Diego, 67-68, res scientist, 68-69; vis lectr zool, Oxford Univ, 69-70; res fel, Max Planck Inst Physiol of Behav, 70-72; AFFIL ZOOL, OXFORD UNIV, 72- Concurrent Pos: Mass Inst Technol neurosci res prog fel, Univ Colo, 66; prin investr, Air Force Off Sci Res grants, 67-69 & 69-70; consult, Max Planck Inst, 75- Mem: AAAS; Sigma Xi; Antiquarium Horological Soc. Res: Experimental and theoretical analysis of the dynamics of biological systems; visual movement perception; macromolecular basis of muscle contraction; sensory adaptation; methodology in teaching applied mathematics; theory of pendulum-clock escapements. Mailing Add: The Old Marlborough Arms Combe Oxford England

THORSON, RALPH EDWARD, b Chatfield, Minn, June 25, 23; m 52; c 3. MEDICAL PARASITOLOGY, VETERINARY PARASITOLOGY. Educ: Univ Notre Dame, BS, 48, MS, 49; Johns Hopkins Univ, ScD(hyg), 52. Prof Exp: Instr parasitol, Sch Hyg & Pub Health, Johns Hopkins Univ, 52-53; assoc prof, Sch Vet Med, Ala Polytech Inst, 53-57, prof, 58-59; group leader parasitic chemother, Res Div, Am Cyanamid Co, 57-58; prof biol & head dept & Lobund Labs, Univ Notre Dame, 59-64; prof parasitol & chmn dept trop health, Am Univ Beirut, 64-66; PROF BIOL, UNIV NOTRE DAME, 66- Concurrent Pos: Mem awards comt, Sigma Xi, 67- Mem: Fel AAAS; Am Soc Parasitol; Am Soc Zool; Am Soc Trop Med & Hyg; fel Am Acad Microbiol. Res: Immunology of parasitic infections, especially helminths; physiology of parasitic helminths. Mailing Add: Dept of Biol Univ of Notre Dame Notre Dame IN 46556

THORSON, THOMAS BERTEL, b Rowe, Ill, Jan 12, 17; m 41; c 2. ZOOLOGY. Educ: St Olaf Col, BA, 38; Univ Wash, Seattle, MS, 41, PhD(zool), 52. Prof Exp: Teacher high sch, Mont, 38-39 & Wash, 42-43; instr zool, Yakima Jr Col, Wash, 46-48, Univ Nebr, 48-50 & San Francisco State Col, 52-54; asst prof, SDak State Univ, 54-56; from asst prof to assoc prof, 56-61, chmn dept, 67-71, PROF ZOOL, UNIV NEBR, LINCOLN, 61-, VDIR SCH LIFE SCI, 75- Concurrent Pos: NSF-NIH & Off Naval Res grants, Field Expeds Cent Am, SAm & Nigeria, 60- Mem: Fel AAAS; Am Soc Zool; Ecol Soc Am; Am Soc Ichthyol & Herpet; Am Fisheries Soc. Res: Water economy of amphibians in relation to terrestrialism; ecological and phylogenetic significance of body water partitioning in vertebrates; osmoregulation of elasmobranchs; fresh water elasmobranch biology. Mailing Add: Dept of Zool Univ of Nebr Lincoln NE 68508

THORSON, WALTER ROLLIER, b Tulsa, Okla, Sept 3, 32. THEORETICAL CHEMISTRY. Educ: Calif Inst Technol, BS, 53, PhD(chem), 57. Prof Exp: NSF fel chem, Harvard Univ, 56-57; instr, Tufts Univ, 57-58; asst prof phys chem, Mass Inst Technol, 58-64, assoc prof chem, 64-68; PROF CHEM, UNIV ALTA, 68- Mem: Am Phys Soc; Can Asn Physicists. Res: Theory of atomic collisions; electronic structure of molecules and solids; quantum mechanics. Mailing Add: Dept of Chem Univ of Alta Edmonton AB Can

THORSTEINSON, ASGEIR JONAS, b Winnipeg, Man, Sept 2, 17; m 43; c 4. ENTOMOLOGY. Educ: Univ Man, BSA, 41; Univ London, PhD(entom), 46. Prof Exp: Insect physiologist, Forest Insect Sci Serv, Dominion Dept Agr, Can, 47-48; from asst prof to assoc prof entom, 48-59, actg chmn dept, 54-59, PROF ENTOM & HEAD DEPT, UNIV MAN, 59- Mem: AAAS; Entom Soc Am; Entom Soc Can. Res: Insect behavior. Mailing Add: Dept of Entom Univ of Man Winnipeg MB Can

THORSTENSEN, THOMAS CLAYTON, b Milwaukee, Wis, Nov 29, 19; m; c 3. CHEMISTRY. Educ: Univ Minn, BS, 42; Lehigh Univ, MS, 47, PhD, 49. Prof Exp: Chemist, S B Foot Tanning Co, 42-44; asst, Lehigh Univ, 46-49, res assoc, 49-51; res chemist, J S Young Co, 51-55; proj dir, Res Found, Lowell Technol Inst, 55-59; OWNER-DIR, THORSTENSEN LAB, 59- Concurrent Pos: Vis prof, Lowell Technol Inst, 60-; consult aid to underdeveloped nations, UN & Dept State. Mem: Am Chem Soc; Am Leather Chem Asn (pres, 74). Res: Mineral tannages chromium; iron; aluminum and zirconium; theory of mineral tannages; synthetic tanning agents. Mailing Add: Thorstensen Lab 66 Littleton Rd Westford MA 01886

THORSTENSON, DONALD CARL, b Chicago, Ill, Jan 4, 41; m 62; c 1. GEOCHEMISTRY. Educ: Monmouth Col, BA, 62; Univ Ill, Urbana, MS, 64; Northwestern Univ, PhD(geol), 69. Prof Exp: ASST PROF GEOL, SOUTHERN METHODIST UNIV, 69- Mem: AAAS. Res: Low temperature aqueous geochemistry. Mailing Add: Dept of Geol Sci Southern Methodist Univ Dallas TX 75222

THORUP, JAMES TAT, b Salt Lake City, Utah, Dec 20, 30; m 58; c 5. AGRONOMY, SOIL FERTILITY. Educ: Brigham Young Univ, BA, 55; NC State Col, MS, 57; Univ Calif, Davis, PhD(soils, plant nutrit), 66. Prof Exp: Teacher high schs, Calif, 61-66; AGRONOMIST, CHEVRON CHEM CO, 66- Concurrent Pos: Teacher eve div, Santa Monica City Col, 62-63 & Orange Coast Col, 63-66. Mem: Am Soc Agron; Soil Sci Soc Am. Res: Factors affecting plant growth in sodic soils; pH effect on plant growth and water uptake. Mailing Add: Chevron Chem Co PO Box 5458 Fresno CA 93755

THORUP, OSCAR ANDREAS, JR, b Washington, DC, Mar 12, 22; m 44; c 1. MEDICINE. Educ: Univ Va, BA, 44, MD, 46; Am Bd Internal Med, dipl, 55. Prof Exp: Asst resident neurosurg, Hosp, Univ Va, 49-50, from asst resident to resident internal med, 50-52, asst to dean sch med, 53-67, dir teacher's preventorium, 53-54, from instr to assoc prof internal med, 53-66; prof & head dept, Col Med, Univ Ariz, 67-74; PROF INTERNAL MED, DIR CONTINUING EDUC & ASSOC DEAN, SCH MED, UNIV VA, 74- Concurrent Pos: Fel internal med, Hosp, Univ Va, 50; res fel, Univ NC, 52-53. Mem: AMA; Am Fedn Clin Res; Am Col Physicians. Res: Hematology, particularly red blood cell enzymes and proteins. Mailing Add: Univ of Va Sch of Med Charlottesville VA 22901

THORUP, RICHARD M, b Salt Lake City, Utah, Dec 20, 30; m 57; c 3. AGRONOMY, SOIL FERTILITY. Educ: Brigham Young Univ, BA, 55; NC State Univ, MS, 57; Univ Calif, Davis, PhD(soil sci, plant nutrit), 62. Prof Exp: Agronomist, 60-61, field agronomist, 61-67, regional agronomist, 67-75, NAT MGR AGRON, ORTHO DIV, CHEVRON CHEM CO, 75- Mem: Am Soc Agron; Soil Sci Soc Am; Crop Sci Soc Am. Res: Chemistry of phosphates in the soil, including solubility and interrelationships with soil moisture; maximum fertility studies with field and tree crops; micronutrients; effect of fertilizers on environment. Mailing Add: Chevron Chem Co 200 Bush St San Francisco CA 94120

THOUEZ, JEAN-PIERRE MARY, b Poitiers, France, Jan 27, 42; m 69; c 2. URBAN GEOGRAPHY, ECONOMIC GEOGRAPHY. Educ: Advan Sch Com, Poitiers, BA, 64; Univ Grenoble, MA, 66, PhD, 68. Prof Exp: Economist, Thomas Connell Explor Co, 69-70; asst prof, 70-75, ASSOC PROF GEOG, UNIV SHERBROOKE, 75- Concurrent Pos: Consult, Urban Renewal Comn City Sherbrooke, 71, Res Ctr Regional & Urban Planning, 71 & Res Comn, Can Inst Int Affairs, 72; Can Coun Arts grant, Univ Sherbrooke, 70-73; vis res, Tufts Univ & fel World Peace Found, 74-76. Mem: Asn Am Geog; Fr Geog Asn; Can Asn Geog. Res: Simulation game for teaching and research purposes; medical geography and migration decision process. Mailing Add: Dept of Geog Univ of Sherbrooke Sherbrooke PQ Can

THOURSON, THOMAS LAWRENCE, b Chicago, Ill, Dec 30, 25; m 47; c 3. PHYSICS. Educ: Ill Inst Technol, BS, 50; Northwestern Univ, MS, 52, PhD, 64. Prof Exp: Physicist, Int Harvester Co, 52-54, Stand Oil Co, Ind, 54-56 & Borg Warner Corp, 56-65; PHYSICIST, XEROX CORP, 65- Mem: Soc Photog Sci & Eng. Res: Xerography; physics of xerographic processes; xeroradiography; image processing. Mailing Add: Xerox Corp Pasadena CA 91107

THRAILKILL, JOHN VERNON, b San Diego, Calif, Aug 31, 30; m 52. GEOLOGY, GEOCHEMISTRY. Educ: Univ Colo, AB, 53, MS, 55; Princeton Univ, PhD(geol), 65. Prof Exp: Geologist, Continental Oil Co, 55-61; asst prof, 65-69, ASSOC PROF GEOL, UNIV KY, 69-, CHMN DEPT, 74- Mem: AAAS; Nat Speleol Soc; Geol Soc Am; Geochem Soc; Am Geophys Union. Res: Low temperature and solution geochemistry; hydrogeology of limestone terrains. Mailing Add: Dept of Geol Univ of Ky Lexington KY 40506

THRALL, ROBERT MCDOWELL, b Toledo, Ill, Sept 23, 14; m 36; c 3. MATHEMATICS. Educ: Ill Col, AB, 35; Univ Ill, AM, 35, PhD(math), 37. Hon Degrees: ScD, Ill Col, 60. Prof Exp: Instr math, Univ Mich, 37-40; mem, Inst Advan Study, 40-42; from asst prof to prof math, Univ Mich, Ann Arbor, 42-69; PROF MATH SCI & CHMN DEPT, RICE UNIV, 69- Concurrent Pos: Res mathematician appl math group, Nat Defense Res Comt, Columbia Univ, 44-45; mem staff radiation lab & sect chief & ed-in-chief, Mars, Mass Inst Technol, 44-46; prof opers anal, Univ Mich, Ann Arbor, 56-69, head opers res dept, 57-60, res mathematician inst sci & technol, 60-69; consult, Rand Corp, Weapon Syst Eval Group, US Dept Defense & Math Steering Comt, Army Res Off, 58-; ed-in-chief, Mgt Sci, 61-69; adj prof, Dept Comput Sci & Inst Rehab & Res, Baylor Col Med, 71- & Univ Tex Sch Pub Health, Houston, 72-; pres, Robert M Thrall & Assoc, Inc; vis prof quant methods, Univ Houston & sr scientist NSF Industs Studies, 74-75. Mem: AAAS; Am Math Soc; Soc Indust & Appl Math; Opers Res Soc Am; Math Asn Am; Inst Mgt Sci (pres, 69-70). Res: Representations of groups; rings and lie rings; operations research linear and nonlinear programming and game theory; theory of application of mathematical models. Mailing Add: Dept of Math Sci Rice Univ PO Box 1892 Houston TX 77001

THRASHER, DONALD MILLER, b Bloomington, Ind, Jan 8, 29; m 53; c 3. ANIMAL SCIENCE, ANIMAL HUSBANDRY. Educ: Purdue Univ, BS, 51, MS, 55, PhD(animal nutrit), 57. Prof Exp: Asst, Purdue Univ, 53-57; from asst prof to assoc prof, 57-60, PROF ANIMAL SCI, LA STATE UNIV, BATON ROUGE, 65- Mem: Am Soc Animal Sci; Sigma Xi. Res: Large animal nutrition, especially swine nutrition and production; energy and amino acid sources for pigs; feeding management. Mailing Add: Dept of Animal Sci Louisiana State Univ Baton Rouge LA 70803

THRASHER, GEORGE W, b Bloomington, Ind, July 8, 31; m 53; c 1. ANIMAL NUTRITION. Educ: Purdue Univ, BS, 52, MS, 54, PhD, 58. Prof Exp: Asst animal nutrit, Chas Pfizer & Co, 54; voc agr teacher, Morgan County Schs, Ind, 54-56; asst animal nutrit, Purdue Univ, 56-58, exten swine specialist, 58-59; res animal nutritionist, Com Solvents Corp, 59-64; ASST DIR, ANIMAL HEALTH PROD RES, PFIZER INC, 64- Mem: Am Soc Animal Sci. Res: Antibiotics; hormones; anthelmintics; antimicrobials; chemotherapeutics; minerals; vitamins. Mailing Add: Animal Health Res Pfizer Inc Terre Haute IN 47808

THRASHER, JACK DWAYNE, cell biology, see 12th edition

THREEFOOT, SAM ABRAHAM, b Meridian, Miss, Apr 10, 21; m 54; c 3. CARDIOVASCULAR DISEASES. Educ: Tulane Univ, BS, 43, MD, 45; Am Bd Internal Med, dipl, 53. Prof Exp: Intern, Michael Reese Hosp, Chicago, 45-47; from instr to prof med, Tulane Univ, 48-70; PROF MED & ASST DEAN, MED COL GA

& CHIEF OF STAFF, FOREST HILLS DIV, VET ADMIN HOSP, AUGUSTA, 70- Concurrent Pos: Fel med, Sch Med, Tulane Univ, 47-49; from asst vis physician to sr vis physician, Charity Hosp La, New Orleans, 47-69, consult, 69-70; consult, Lallie Kemp Charity Hosp, Independence, 51-53; clin asst, Touro Infirmary, 53-56, dir res & med studies, 53-63, jr staff mem, 56-60, sr assoc, 60-63, sr dept med, 63-70, dir res, Touro Res Inst, 53-70; mem exec comt, Coun on Circulation, Am Heart Asn, 68-, from vchmn to chmn, 71-75, chmn credentials comt, 72-73; mem bd consult, Int Soc Lymphology, 70-; mem bd dirs, Am Heart Asn, 66-70 & 72-75, mem exec comt, 69-70 & 73-75. Mem: Soc Nuclear Med; Am Fedn Clin Res; fel Am Col Physicians; Am Heart Asn; fel Am Col Cardiol. Res: Electrolyte turnover in congestive heart failure; anatomy and physiology of lymphatics as a transport system and their role in pathogenesis of disease. Mailing Add: Forest Hills Div Vet Admin Hosp Augusta GA 30904

THREET, RICHARD LOWELL, b Browns, Ill, Nov 17, 24; m 46; c 4. GEOLOGY. Educ: Univ Ill, BS & AB, 47, AM, 49; Univ Wah, Seattle, PhD(geol), 52. Prof Exp: From instr to asst prof geol, Univ Nebr, 51-57; asst prof, Univ Utah, 57-61; from asst prof to assoc prof, Calif State Univ, San Diego, 61-68, PROF GEOL, SAN DIEGO STATE UNIV, 68- Concurrent Pos: Vis prof, Ohio State Univ, 53, 63, 66-60, 72-73, Col Southern Utah, 54-55 & Univ Ill, 57; chmn dept geol, San Diego State Univ, 72-73. Mem: Am Inst Prof Geol; Geol Soc Am; Am Soc Photogram; Nat Asn Geol Teachers. Res: Colorado plateau geology; geomorphology; structural geology; photogeology. Mailing Add: Dept of Geol Sci San Diego State Univ San Diego CA 92182

THREINEN, DAVID TRONVIG, b Kenosha, Wis, July 21, 34; m 56; c 3. GEOLOGY. Educ: Beloit Col, BS, 56; Northwestern Univ, MS, 61. Prof Exp: Sr geologist, Pan-Am Petrol Corp, 56-69; dist geologist, Rocky Mountain Dist, J M Huber Corp, Denver, 69-74, GEOLOGIST, J M HUBER CORP, HOUSTON, 74- Mem: Am Asn Petrol Geologists. Res: Gulf Coast Pleistocene stratigraphy. Mailing Add: J M Huber Corp 2000 West Loop S Houston TX 77027

THRELFALL, WILLIAM, b Preston, Eng, Oct 14, 39; m 65; c 2. PARASITOLOGY, ORNITHOLOGY. Educ: Univ Wales, BSc, 62, PhD(agr zool), 65. Prof Exp: Assoc prof, 65-75, PROF BIOL, MEM UNIV NFLD, 75- Mem: Sci fel Zool Soc London; Am Ornith Union; Wildlife Disease Asn; Can Soc Zool; Brit Trust Ornith; Brit Ornith Union. Res: Ecological and geographical aspects of parasitology; helminthology; breeding biology and migratory movements of marine birds. Mailing Add: Dept of Biol Mem Univ of Nfld St John's NF Can

THRELKELD, STEPHEN FRANCIS H, b Watford, Eng, Dec 27, 24; m 52; c 2. BIOLOGY, GENETICS. Educ: Univ Alta, BSc, 57, MSc, 58; St Catharine's Col, Cambridge, PhD(bot), 61. Prof Exp: From asst prof to assoc prof genetics, 61-71, chmn res unit biochem, biophys & molecular biol, 64-68, assoc chmn dept biol, 66-68, PROF GENETICS, McMASTER UNIV, 71- Mem: Am Soc Naturalists; Genetics Soc Am; Genetics Soc Can; Chem Inst Can; Can Soc Cell Biol. Res: Neurospora; recombination. Mailing Add: Dept of Biol McMaster Univ Hamilton ON Can

THRIFT, FREDERICK AARON, b St George, Ga, Oct 6, 40; m 67; c 2. ANIMAL BREEDING. Educ: Univ Fla, BSA, 62; Univ Ga, MS, 65; Okla State Univ, PhD(animal breeding), 68. Prof Exp: ASSOC PROF ANIMAL SCI, UNIV KY, 67- Mem: Biomet Soc; Am Soc Animal Sci. Res: Beef cattle and sheep breeding research. Mailing Add: Dept of Animal Sci Univ of Ky Lexington KY 40506

THROCKMORTON, JAMES RODNEY, b St John, Wash, Sept 4, 36; m 58; c 3. PESTICIDE CHEMISTRY. Educ: Univ Idaho, BS, 58, MS, 60; Univ Minn, PhD(org chem), 64. Prof Exp: Sr chemist, Imaging Res Lab, 64-66, Contract Res Lab, 66-71, RES SPECIALIST, 3M CO, 74- Mem: Am Chem Soc. Res: Organic fluorochemicals; imaging technology. Mailing Add: 555 Aberdeen Curve St Paul MN 55119

THROCKMORTON, LYNN HIRAM, b Loup City, Nebr, Dec 20, 27. ZOOLOGY. Educ: Univ Nebr, BS, 49, MS, 56; Univ Tex, PhD(zool), 59. Prof Exp: Instr zool, Univ Nebr, 56; spec instr, Univ Tex, 59-60; vis asst prof, Univ Calif, 60-61; res assoc, 61-62; from instr to assoc prof, 62-71, PROF BIOL, UNIV CHICAGO, 71- Mem: Genetic Soc Am; Entom Soc Am; Am Soc Naturalists; Am Soc Zoologists; Soc Syst Zool. Res: Taxonomy; phylogeny and biogeography of Drosophila and other Drosophilids; biochemical evolution and speciation of Drosophila; utilization and evaluation of computer methods in taxonomy. Mailing Add: Dept of Biol Univ of Chicago Chicago IL 60637

THROCKMORTON, MORFORD CHURCH, b Waynesburg, Pa, July 28, 19. PHYSICAL CHEMISTRY. Educ: Grove City Col, BS, 40; Western Reserve Univ, MS, 41, PhD(phys chem), 44. Prof Exp: Lab asst gen chem & qual anal, Grove City Col, 38-40; chemist, Cleveland Clin Res Found, Ohio, 40-41, Res Labs, Stand Oil Co, 42 & Texaco, Inc, 43-54, group leader, 54-59; sr res chemist, Firestone Tire & Rubber Co, 60-63; sr res chemist, 64-68, RES SCIENTIST, GOODYEAR TIRE & RUBBER CO, 68- Mem: AAAS; Am Chem Soc. Res: Catalysis; synthetic fuels; development of synthetic rubber; stereospecific polymerization; petrochemicals. Mailing Add: 967 Newport Rd Akron OH 44303

THROCKMORTON, PETER E, b St Paul, Minn, Jan 20, 27; m 48; c 3. ORGANIC CHEMISTRY. Educ: Univ Minn, BCHE, 48, MS, 55; Kans State Univ, PhD, 60. Prof Exp: Asst res engr, Tainton Co, 48-49; res engr, Glenn L Martin Aircraft Co, 49-52; chemist, Gen Mills Co, 52-56; asst, Kans State Univ, 56-59; assoc chemist, Midwest Res Inst, 59-65; sr res chemist, Ashland Oil, Inc, 65-73, SR RES CHEMIST, ASHLAND CHEM CO, 73- Mem: Am Chem Soc. Res: Heterocyclic and organometallic compounds in organic chemistry; synthesis; process research; surfactants from cornstarch;sulfur derivatives. Mailing Add: Ashland Chem Co PO Box 2219 Columbus OH 43216

THRON, WOLFGANG JOSEPH, b Ribnitz, Ger, Aug 17, 18; US citizen; m 53; c 5. MATHEMATICS. Educ: Princeton, AB, 39; Rice Inst, MA, 42, PhD(math), 43. Prof Exp: Instr math, Harvard Univ, 43-44; from instr to assoc prof, Wash Univ, St Loiuis, 46-54; assoc prof, 54-57, PROF MATH, UNIV COLO, BOULDER, 57- Concurrent Pos: Vis prof, Free Univ Berlin, 51, Philippines, 66-67, Univ Erlangen, 70-71 & Punjab Univ, India, 74-75; res grant, Air Force Off Sci Res, Ger, 57-58; vis prof & Fulbright lectr, India, 62-63. Mem: Am Math Soc. Res: Complex variables, analysis of convergence and truncation errors of infinite processes in particular continued fractions; general topology, lattice of topologies, proximity contiguity and nearness spaces, extensions of spaces. Mailing Add: Dept of Math Univ of Colo Boulder CO 80302

THRONEBERRY, GLYN OGLE, b Rule, Tex, Nov 1, 27; m 48; c 1. PLANT PHYSIOLOGY. Educ: NMex State Univ, BS, 50; Iowa State Col, MS, 52, PhD(plant physiol), 53. Prof Exp: Plant physiologist, Kans State Col, USDA, 54-55; asst prof biol, 55-57, from asst prof to assoc prof, 57-67, PROF BOT & ENTOM, NMEX STATE UNIV, 67- Mem: AAAS; Am Soc Plant Physiol; Am Phytopath Soc. Res: Plant host-pathogen relationships; intermediary metabolism; fungus physiology; plant biochemistry. Mailing Add: Dept of Bot & Entom NMex State Univ Las Cruces NM 88003

THROOP, LEWIS JOHN, b Detroit, Mich, June 18, 29; m 54; c 1. ANALYTICAL CHEMISTRY. Educ: Wayne State Univ, BS, 54, MS, 56, PhD(anal chem), 57. Prof Exp: Chemist, Syntex, SA, Mex, 57-59; group leader anal chem, Mead Johnson & Co, 59-64; dept head, 64-71, ASST DIR ANAL CHEM, INST ORG CHEM, SYNTEX CORP, 71- Mem: Am Chem Soc; Am Pharmaceut Asn. Res: Electrochemistry; organic structure characterization; laboratory automation. Mailing Add: Syntex Corp 3401 Hillview Ave Palo Alto CA 94304

THROW, FRANCIS EDWARD, b Ottumwa, Iowa, Oct 4, 12; m 38; c 3. RESEARCH ADMINISTRATION. Educ: Park Col, BA, 33; Univ Mich, MS, 36, PhD(physics), 40. Prof Exp: Instr physics, Milwaukee State Teachers Col, 38; instr physics & math, Polytech Inst PR, 40-41; instr physics, Altoona Undergrad Ctr, 41-42; graound sch instr math, physics & theory of flight, US Navy Pre-Flight Sch, Iowa, 42-44; prof physics & head dept, Cornell Col, 44-52; chmn dept physics, Wabash Col, 52-56; asst dir physics div, 56-73, ASS DIR RADIOL & ENVORON RES DIV, ARGONNE NAT LAB, 73- Mem: Am Phys Soc; Am Asn Physics Teachers. Res: Technical editing; analytical mechanics; discharges in gases. Mailing Add: Radiol & Environ Res Div Argonne Nat Lab 9700 S Cass Ave Argonne IL 60439

THROWER, PETER ALBERT, b Norfolk, Eng, Jan 9, 38; m 60. PHYSICS, MATERIALS SCIENCE. Educ: Cambridge Univ, BA, 60, MA, 63; PhD(physics), 69. Prof Exp: Sci officer, Atomic Energy Res Estab, Eng, 60-65, sr sci officer, 65-69; ASSOC PROF MAT SCI, PA STATE UNIV, UNIVERSITY PARK, 69- Concurrent Pos: Ed, Chem & Physics of Carbon. Mem: Am Soc Metals. Res: Structure and properties of carbon and graphite; irradiation damage to graphite; electron microscopy; mineral microstructures. Mailing Add: 302 Mineral Sci Bldg Pa State Univ University Park PA 16802

THRUPP, LAURI DAVID, b Sask, Nov 30, 30; US citizen; m 52; c 4. INFECTIOUS DISEASES, MICROBIOLOGY. Educ: Stanford Univ, AB, 51; Univ Wash, MD, 55. Prof Exp: Asst chief & chief polio surveillance, Epidemiol Br, Nat Commun Dis Ctr, 56-58; resident physician, Boston City Hosp & Thorndike Serv, 58-60, jr asst physician, Harvard Serv, 61-63; asst prof med & med microbiol, Sch Med, Univ Southern Calif, 63-66, asst prof med, 66-68; ASSOC PROF MED & HEAD DIV INFECTIOUS DIS, UNIV CALIF, IRVINE, 68- Concurrent Pos: Life Ins Med Res Found res fel med & bact, Boston City Hosp & Harvard Med Sch, 60-63; asst chief commun dis, Los Angeles County Gen Hosp, 63-64, attend physician, 63-, med microbiologist, 64-65; consult, Los Angeles County Health Dept, 64- & Calif State Health Dept, 66-; chief infectious dis serv, Orange County Med Ctr, 68- Mem: Am Fedn Clin Res; Am Soc Microbiol; Am Pub Health Asn; Infectious Dis Soc Am; NY Acad Sci. Res: Clinical and experimental pyelonephritis, pathogenesis and immune response, role of bacterial L-forms; gram-negative hospital-acquired infections; meningitis, clinical and immunological: bacteriology. Mailing Add: Infectious Dis Serv Univ of Calif Irvine Orange CA 92668

THUAN, TRINH XUAN, b Hanoi, Vietnam, Aug 20, 48. ASTROPHYSICS. Educ: Calif Inst Technol, BS, 70; Princeton Univ, PhD(astrophysics), 74. Prof Exp: RES FEL ASTROPHYS, CALIF INST TECHNOL, 74- Mem: Am Astron Soc. Res: Study of the formation, clustering and evolution of galaxies and of cosmological questions both theoretical and observational; problems of interstellar matter. Mailing Add: Dept of Astron 105-24 Calif Inst of Technol Pasadena CA 91125

THUILLIER, RICHARD HOWARD, b New York, NY, Apr 3, 36; m 66; c 3. METEOROLOGY, AIR POLLUTION. Educ: Fordham Univ, BS, 59; NY Univ, MS, 63. Prof Exp: Instr meteorol, State Univ NY Maritime Col, 63-66; dir res, Weather Engrs Panama, Inc, 66-68; from air pollution meteorologist to sr air pollution meteorologist, 68-73, CHIEF RES & PLANNING, BAY AREA AIR POLLUTION CONTROL DIST, 73- Mem: Am Meteorol Soc. Res: Development and application of mathematical models of air pollutant transport and dispersion processes for use in assessing the air quality impact of land use and transportation decisions. Mailing Add: Bay Area Air Pollut Control Dist 939 Ellis St San Francisco CA 94109

THULINE, HORACE CROCKETT, b Berar, India, Aug 18, 22; US citizen; m 41; c 9. MEDICINE. Educ: Seattle Pac Col, BS, 49; Univ Wash, MD, 53. Prof Exp: Asst lab dir clin path, Children's Orthop Hosp, 54-56; biochemist, Ment Health Res Inst, State of Wash, 56-58; clin pathologist, Mary Bridge Children's Hosp, 58-59; res consult, Rainier Sch, State of Wash, 59-60, lab dir clin path, 59-75; SUPVR BIRTH DEFECTS STUDY & COUNSELING PROG, DSHS, STATE OF WASH, 75- Concurrent Pos: From clin instr to clin assoc prof pediat, Sch Med, Univ Wash, 54-; consult, Vet Admin Hosp, American Lake, 60-69. Mem: AAAS; Am Asn Mental Deficiency; Am Soc Human Genetics; Down's Syndrome Cong. Res: Genetics and cytogenetics of handicapping conditions; animal models of human diseases. Mailing Add: Birth Def Study & Counselng Prog 1704 NE 150th St Seattle WA 98155

THUM, ALAN BRADLEY, b Washington, DC, May 30, 43; m 66; c 1. MARINE ECOLOGY. Educ: Univ Redlands, BS, 65; Univ Pac, MS, 67; Ore State Univ, PhD(marine ecol), 71. Prof Exp: Lectr invert zool, Univ Cape Town, 71-75; SR SCIENTIST & ENVIRON CONSULT, LOCKHEED MARINE BIOL LAB, LOCKHEED AIRCRAFT SERV, 75- Concurrent Pos: Consult, Bur Land Mgt, Univ Southern Calif, 76- Mem: Ecol Soc Am; Am Soc Limnol & Oceanog; Int Asn Meiobenthologists; Royal Soc SAfrica. Res: Ecology of interstitial meiofauna; ecology and systematics of turbellaria; reproductive ecology of marine benthic invertebrates. Mailing Add: Lockheed Marine Biol Lab Lockheed Aircraft Serv 6350 Yarrow Dr Suite A Carlsbad CA 91008

THUMM, BYRON ASHLEY, b Malden, WVa, Jan 2, 23; m 56. ANALYTICAL CHEMISTRY. Educ: Morris Harvey Col, BS, 45; Duke Univ, PhD(chem), 51. Prof Exp: Res chemist, Am Viscose Div, FMC Corp, 51-63; ASSOC PROF CHEM, STATE UNIV NY COL, FREDONIA, 63- Mem: Am Chem Soc. Res: Solution kinetics and equilibrium; rayon spinning process; water analysis; formaldehyde complexes. Mailing Add: Dept of Chem State Univ of NY Col Fredonia NY 14063

THUN, RUDOLF EDUARD, b Berlin, Ger, Jan 30, 21; nat US; m 44; c 2. SOLID STATE PHYSICS. Educ: Univ Frankfurt, dipl, 54, PhD(physics), 55. Prof Exp: Physicist, Ger Gold & Silver Separation Plant, 51-55 & US Army Engrs Res & Develop Labs, Ft Belvoir, Va, 55-59; var sci & managerial positions, Int Bus Mach Corp, 59-67; mgr microelectronics, Missile Systs Div, 67-70, mgr electronic prod design lab, 70-75, MGR ADVAN ELECTRONICS LAB, MISSILE SYSTS DIV, RAYTHEON CO, BEDFORD, 75- Honors & Awards: Corps Engrs Award, US Army, 59; 1975 Contributions Award, Inst Elec & Electronics Engrs, 75. Mem: Fel Am Phys Soc; Optical Soc Am; fel Inst Elec & Electronics Eng. Res: Microelectronics; physics of thin films; electron diffraction and microscopy; x-ray diffraction; physical and electron optics; solid state devices and integrated circuits. Mailing Add: 228 Heald Rd Carlisle MA 01741

THUN, WAYNE EARL, inorganic chemistry, see 12th edition

THURBER, DAVID LAWRENCE, b Oneonta, NY, Dec 29, 34; m 64; c 1. GEOCHEMISTRY. Educ: Union Col, NY, BS, 56; Columbia Univ, MA, 58, PhD(geol), 63. Prof Exp: Res asst geochem, Lamont Geol Observ, Columbia Univ, 56-63, res scientist, 63-64, res assoc, 64-66; assoc prof geol, 66-70, PROF EARTH & ENVIRON SCI, QUEENS COL, NY, 70- Concurrent Pos: Am Geophys Union vis lectureship, 64; lectr, Queens Col, NY, 65-66; vis res assoc, Lamont Geol Observ, 66-70, vis sr res assoc, Lamont-Doherty Geol Observ, 70- Mem: AAAS; Am Geophys Union; Geochem Soc. Res: General geochemistry; stable and radioisotope geochemistry; geochronology; hydrochemistry. Mailing Add: Dept of Earth & Environ Sci Queens Col Flushing NY 11367

THURBER, GEORGE A, b Liscomb, Iowa, Apr 21, 07; m 30; c 2. MEDICAL ENTOMOLOGY. Educ: Iowa State Univ, BS, 31, MS, 32. Prof Exp: Flight instr, US Army Air Force, 42-45; instr biol & chem, Iowa Pub Sch, 46-58; from asst prof & exec asst to dean sch med to assoc prof, 58-70, PROF MED ENTOM, SCH MED, LA STATE UNIV MED CTR, NEW ORLEANS, 70-, EXEC ASST TO CHANCELLOR, MED CTR, 64- Concurrent Pos: Co-dir training prog trop med, La State Univ, 58-; consult, USPHS Hosp, Carville, La, 65- & New Orleans PsychoAnal Inst, 66-; mem, Nat Coun Int Health. Mem: Entom Soc Am; Am Soc Trop Med & Hyg; Am Soc Parasitologists; Asn Am Med Cols. Res: Relative toxicity of insecticides; vectors of Chagas' disease; education in tropical medicine. Mailing Add: Off of Chancellor La State Univ Med Ctr New Orleans LA 70112

THURBER, JAMES KENT, b Utica, NY, Oct 29, 33. APPLIED MATHEMATICS. Educ: Brooklyn Col, BS, 55; NY Univ, PhD, 61. Prof Exp: Asst appl math, NY Univ, 57-61; asst prof math, Adelphi Univ, 61-64; assoc math, Brookhaven Nat Lab, 64-66, mathematician, 66-69; PROF MATH, PURDUE UNIV, LAFAYETTE, 69- Mem: Am Math Soc; Am Nuclear Soc; Math Asn Am; Soc Indust & Appl Math; NY Acad Sci. Res: Neutron transport; kinetic theory of gases; asymptotic analysis; mathematical programming; applications of nonstandard analysis. Mailing Add: Div of Math Sci Purdue Univ Lafayette IN 47907

THURBER, ROBERT EUGENE, b Bayshore, NY, Oct 11, 32; m 53; c 4. PHYSIOLOGY, BIOPHYSICS. Educ: Col of the Holy Cross, BS, 54; Adelphi Univ, MS, 61; Univ Kans, PhD(physiol), 65. Prof Exp: Res assoc radiation biol, Brookhaven Nat Lab, Assoc Univs Inc, 56-61; from instr to assoc prof physiol, Med Col Va, Va Commonwealth Univ, 64-69; assoc prof, Jefferson Med Col, Thomas Jefferson Univ, 69-70; PROF PHYSIOL & CHMN DEPT, SCH MED, E CAROLINA UNIV, 70- Concurrent Pos: Consult, US Vet Admin, Va State Bd Med Examr & NASA, Va, 66-69, US Naval Hosp, Portsmouth, 68-69 & Psychol Consult Inc, 68-70; mem bd dirs, NC Heart Asn, 72- Mem: AAAS; Am Physiol Soc; NY Acad Sci; Sigma Xi. Res: Nonequilibrium transfer and distribution of electrolytes; renal transport; radiation biology and carbohydrate metabolism. Mailing Add: Dept of Physiol ECarolina Univ Sch of Med Greenville NC 27834

THURBER, WALTER ARTHUR, b East Worcester, NY, Nov 27, 08; m 34; c 2. ORNITHOLOGY. Educ: Union Col NY, BS, 33; NY State Col Teachers, MS, 38; Cornell Univ, PhD(nature study), 41. Prof Exp: Teacher NY schs, 26-29, 33-38; asst ed, Cornell Univ, 38-39; instr physics & phys sci, State Univ NY Teachers Col, Cortland, 40-43; asst prof physics & phys & earth sci, 43-48, prof sci, 48-58; vis prof sci educ, Syracuse Univ, 58-61, adj prof, 61-72; TEXTBOOK WRITER, 61- Concurrent Pos: Consult, NY State Educ Dept, 43-55, NY pub schs, 47-52, Orgn Cent Am States, 66-, Ministerio de Educacion, El Salvador, 72 & Direccion General de Recursas Naturales, El Salvador, 74-; field collabr, Lab Ornith, Cornell Univ, 70-; vis prof, Nat Univ El Salvador, 71- Mem: Fel AAAS; NY Acad Sci; Am Ornith Union; Wilson Ornith Soc; Cooper Ornith Soc. Res: Elementary and secondary science education; organization of syllabuses and textbooks. Mailing Add: 101 Sycamore St Apt 14 Liverpool NY 13088

THURBER, WILLIAM SAMUELS, b Ann Arbor, Mich, Mar 6, 22; m 43; c 4. ORGANIC CHEMISTRY. Educ: Mich State Univ, BS, 46, MS, 48. Prof Exp: Asst dir styrene polymerization lab, Dow Chem Co, 54-57, dir, Strosackers Res & Develop Group, 57-61, plant supt, 61-64, tech dir, Saginaw Bay Res Dept, 64-68, admin asst to div dir res, 68-70, sect mgr process develop & eng, 70-76, MGR EQUAL EMPLOY OPPORTUNITY, MICH DIV, DOW CHEM USA, 76- Mem: Am Chem Soc. Res: Styrene polymers; antioxidants; polyglycols; surface active agents. Mailing Add: EEO Mich Div Dow Chemicals USA Midland MI 48640

THURBERG, FREDERICK PETER, b Weymouth, Mass, Aug 31, 42; m 64; c 2. PHYSIOLOGY, MARINE BIOLOGY. Educ: Univ Mass, BA, 64, MEd, 66; Univ NH, MS, 69, PhD(zool, physiol), 72. Prof Exp: Teacher pub schs, Mass, 64-67; PHYSIOLOGIST, NAT MARINE FISHERIES SERV, NAT OCEANIC & ATMOSPHERIC ADMIN, 71- Mem: Estuarine Res Fedn; Nat Shellfisheries Asn. Res: Physiological ecology; effects of pollutants on marine organisms; invertebrate physiology; marine biotoxins; red tides. Mailing Add: Mid Atlantic Coastal Fisheries Nat Marine Fisheries Serv Milford CT 06460

THURESON-KLEIN, ASA KRISTINA, b Sveg, Sweden, May 31, 34; m 61; c 2. BIOLOGY, PHARMACOLOGY. Educ: Univ Stockholm, MA, 58, PhD(biol), 68. Prof Exp: Instr bot, Univ Stockholm, 58-64; res assoc, 65-68, asst prof, 69-74, ASSOC PROF PHARMACOL, MED CTR, UNIV MISS, 74- Concurrent Pos: Pharmaceut Mfrs Asn Found fel pharmacol & morphol, Med Ctr, Univ Miss, 71-73. Mem: Am Soc Pharmacol & Exp Therapeut; Electron Micros Soc Am; Soc Neurosci; Sigma Xi. Res: Effects of pharmacological agents on uptake, storage and release of neurotransmitter from catecholamine storage vesicles, using combined biochemical and morphological methods; adrenergic innervation of brown adipose tissue; mechanisms of transmitter-release from nerve-endings at veins and arteries; morphological changes in blood-vessels after subarachnoid hemorrhage and vasospasm. Mailing Add: Dept of Pharmacol Univ of Miss Med Ctr Jackson MS 39216

THURLOW, JOHN FRANK, b Poland, Maine, Oct 16, 21; m 49, 53. BIOCHEMISTRY. Educ: Bates Col, BS, 43; Calif Inst Technol, PhD(biochem), 48. Prof Exp: Res scientist, E I du Pont de Nemours & Co, 48-50; res assoc chem biol coord ctr, Nat Res Coun, 50-51, head biochemist, 51-52; res chemist & mkt analyst, S D Warren Co, 52-55; PRES, GORHAM INT INC, 55- Mem: Am Chem Soc; Tech Asn Pulp & Paper Indust. Res: Industrial chemistry, especially technical and market development and foreign licensing for the paper industry and its suppliers; worldwide consulting. Mailing Add: 209 Mosher Rd South Windham ME 04082

THURMAIER, ROLAND JOSEPH, b Chicago, Ill, June 25, 28; m 55; c 4. ORGANIC CHEMISTRY, POLYMER CHEMISTRY. Educ: Bradley Univ, BS, 50; Univ Iowa, MS, 58, PhD(org chem), 60. Prof Exp: Plant chemist, Corn Prod Ref Corp, 51-55; res chemist, E I du Pont de Nemours & Co, 60-66; ASST PROF ORG CHEM, UNIV WIS-STEVENS POINT, 66- Concurrent Pos: Mem, Stevens Point Transit Comn, 70; mem study comt mass transit, 72. Mem: AAAS; Am Chem Soc. Res: Polymers; stabilizers for polyurethanes. Mailing Add: Dept of Chem Univ of Wis Stevens Point WI 54481

THURMAN, DUANE EDWARD, b Wichita, Kans, Nov 8, 41; m 64; c 2. ENTOMOLOGY, AGRICULTURAL CHEMISTRY. Educ: Mo Sch Mines, BS, 63; Univ Kans, PhD(org chem), 67. Prof Exp: Proj scientist agr chem, 69-74, SEVIN PROD DEVELOP MGR, AGR PROD DIV, UNION CARBIDE CORP, 74- Mem: Entom Soc Am; Weed Sci Soc Am; Am Soybean Asn. Res: Product development programs for SEVIN carbonyl insecticide; organic chemistry, pesticides, herbicides and insecticides. Mailing Add: Union Carbide Corp Agr Prod Div PO Box 1906 Salinas CA 93901

THURMAN, GEORGE RAYMOND, b Fayette, Mo, Oct 2, 12; m 40; c 2. MATHEMATICS, PHYSICS. Educ: Cent Methodist Col, AB, 34; Univ Mo, AM, 36, PhD(math), 39. Prof Exp: From asst to instr math, Univ Mo, 36-40; asst prof, The Citadel, 40-42; from assoc prof to prof, Tusculum Col, 42-47; tech engr, Carbide & Carbon Chem Co, Tenn, 44-46; physicist, 47-55, mgr defense res div, 55-58, dir eng lab, Calif, 58-60, SR RES ASSOC, RES LAB, FIRESTONE TIRE & RUBBER CO, 60- Res: Electronics; high polymer physics; sound and vibration; physics of rubber. Mailing Add: Res Lab Firestone Tire & Rubber Co Akron OH 44317

THURMAN, LLOY DUANE, b Oconto, Nebr, Sept 3, 33; m 57; c 3. PLANT ECOLOGY. Educ: Univ Nebr, BS, 59, MS, 61; Univ Calif, Berkeley, 66. Prof Exp: Asst prof biol, Southern Calif Col, 65-67; from asst prof to assoc prof biol, 67-72, chmn dept natural sci, 69-73, PROF BIOL, ORAL ROBERTS UNIV, 72- Mem: AAAS; Bot Soc Am; Ecol Soc Am; Soc Study Evolution; Am Inst Biol Sci. Res: Ecology and systematics of Orthocarpus, a genus of chlorophyllous root parasites; curricula and teaching methods in biology; human disturbance of ecosystems. Mailing Add: Dept of Natural Sci Oral Roberts Univ 7777 S Lewis Tulsa OK 74105

THURMAN, MELBURN D, b Redford, Mo, Oct 31, 41; m 65; c 2. ANTHROPOLOGY. Educ: Univ Chicago, AB, 65; Univ Calif, Los Angeles, MA, 68; Univ Calif, Santa Barbara, PhD(anthrop), 73. Prof Exp: Asst prof anthrop, Univ Md, College Park, 70-74; vis asst prof, Purdue Univ, 74-75; ASST PROF ANTHROP, PRINCETON UNIV, 75- Concurrent Pos: Archaeol consult, Purdue Univ Prog Geosci Applns Archaeol & Battleground Hist Corp, 74-76. Mem: Fel Am Anthrop Asn; AAAS; Soc Am Archaeol; Soc Hist Archaeol. Res: Cultural evolution and cultural ecology; ethnohistory of the Plains, especially development of nomadic adaptations, and the Eastern Woodlands, especially Delaware Indians and the Algonkian prophetic tradition; eastern United States archaeology. Mailing Add: 200 Green Annex Dept of Anthrop Princeton Univ Princeton NJ 08540.

THURMAN, RICHARD GARY, b Wichita, Kans, Mar 1, 40; m 70. CHEMISTRY. Educ: NMex State Univ, BS, 62, MS, 65; Univ Ariz, PhD(chem), 71. Prof Exp: Asst prof chem, Univ Ariz, 70-71; asst prof, 71-74, ASSOC PROF CHEM, UNIV NEBR-OMAHA, 74- Mem: Am Chem Soc. Res: Liquid and gas chromatography—study of the separation processes; computer-controlled chemical instrumentation; use of computers in chemical education. Mailing Add: Dept of Chem Univ of Nebr Omaha NE 68101

THURMAN, RONALD GLENN, b Carbondale, Ill, Nov 25, 41. BIOCHEMISTRY, PHARMACOLOGY. Educ: St Louis Col Pharm, BS, 63; Univ Ill, PhD(pharmacol), 67. Prof Exp: ASST PROF BIOPHYS & PHYS BIOCHEM, JOHNSON RES FOUND, UNIV PA, 71- Prof Exp: Fel, Johnson Res Found, Univ Pa, 67-69; NATO fel, Inst Physiol Chem, Munich, Ger, 69-70, Alexander von Humboldt fel, 70-71; NIMH career develop award, 71. Mem: AAAS; Am Pharmaceut Soc. Res: Drug and alcohol metabolism. Mailing Add: Johnson Res Found Univ of Pa Philadelphia PA 19104

THURMAN, WILLIAM GENTRY, b Jacksonville, Fla, July 1, 28; m 49; c 3. PEDIATRICS, ONCOLOGY. Educ: Univ NC, BS, 49; McGill Univ, MD & CM, 54. Prof Exp: Asst prof pediat, Tulane Univ, 60-61; assoc prof, Emory Univ, 61-62; prof, Cornell Univ, 62-64; prof & chmn dept, Sch Med, Univ Va, 64-73, dir, Ctr Delivery Health Care, 69-73; dean sch med, Tulane Univ, 73-75; PROF PEDIAT & PROVOST, UNIV OKLA HEALTH SCI CTR, 75- Concurrent Pos: Fel hemat & oncol, 58-60; Markle scholar acad med, 59-64; consult, US Air Force, 59-, Comn on Cancer, 63- & Comn on Pediat Hemat, 65-; chmn dept pediat, Mem Sloan-Kettering Cancer Ctr, 62-64; prog consult, Nat Found, 64-; prog consult pediat, VI, 66-; mem, Nat Rev Comt Regional Med Prog. Mem: Soc Pediat Res; Am Soc Hemat; Am Soc Human Genetics; Am Asn Cancer Res; Am Pediat Soc. Res: Immunologic abnormalities associated with malignancy in children; clinical management of children with malignancy; evaluation of various drugs; methods and models for delivery of health care. Mailing Add: 633 NE 14th St Oklahoma City OK 73190

THURMAN-SWARTZWELDER, ERNESTINE H, b Atkins, Ark, Mar 7, 20; m 64; c 2. MEDICAL ENTOMOLOGY, RESEARCH ADMINISTRATION. Educ: Col Ozarks, BS, 44; Univ Md, PhD(bionomics of mosquitoes), 58; Am Registry Cert Entomologists. Prof Exp: Asst, Tulane Univ, 44; entomologist, Commun Dis Ctr, USPHS, Fla, 45-48; entomologist in-chg ident unit, Bur Vector Control, Calif, 48-51, entomologist, Nat Microbiol Inst, Md, 51, training adv malaria control, Div Int Health, US Opers Mission, Thailand, 51-53, entomologist & spec consult, Washington, DC, 53-54, exec secy, Trop Med & Parasitol Study Sect, Div Res Grants, NIH, 54-64, res adminr collab studies, Nat Heart Inst, 64-67; asst prof, 67-70, ASSOC PROF PATH, LA STATE UNIV MED CTR, NEW ORLEANS, 70- Mem: Fel AAAS; fel Am Pub Health Asn; Entom Soc Am; Am Soc Trop Med & Hyg; Am Mosquito Control Asn (rec secy, 61-64). Res: Control of vectors and vector-borne diseases; taxonomy of mosquitoes; tropical medicine; science administration. Mailing Add: 30 Versailles Blvd New Orleans LA 70125

THURMON, JOHN C, b Redford, Mo, Mar 4, 30; m 56; c 1. VETERINARY ANESTHESIOLOGY. Educ: Univ Mo, BS, 60, DVM, 62, MS, 67. Prof Exp: From instr to asst prof vet med, 62-70, ASSOC PROF VET ANESTHESIOL & HEAD DIV, COL VET MED, UNIV ILL, URBANA, 71-, ASSOC PROF PHYSIOL & PHARMACOL, 75-, ASSOC PROF BIOENG, COL ENG, 72- Concurrent Pos: Nat Heart Inst fel, Baylor Col Med, 69; vis prof anesthesiol, Col Med, Univ Ill, Chicago, 70; consult, Bristol Labs, Syracuse, NY, 70-74, Affil Labs, Whitehall, Ill, 72-74, Bay Vet Corp, Shawnee Mission, Kans, 73-, Norden Labs, Lincoln, Nebr, 74- & Dept Med Sci, Southern Ill Univ, 75- Mem: Am Soc Vet Anesthesiol (pres elect, 73, pres, 74 & 76); Am Soc Anesthesiologists; Int Anesthesia Res Soc; Am Col Vet Anesthesiologists (chmn, 75). Res: General anesthesia and its effects on homeostatic mechanisms of domestic and wild animals; development and design of equipment for use in veterinary anesthesia. Mailing Add: Dept of Vet Clin Med Univ of Ill Col of Vet Med Urbana IL 61801

THURMON, THEODORE FRANCIS, b Baton Rouge, La, Oct 20, 37; m 61; c 2. MEDICAL GENETICS. Educ: La State Univ, Baton Rouge, BS, 60; La State Univ, New Orleans, MD, 62. Prof Exp: Intern, Naval Hosp, Pensacola, Fla, 62-63, resident, Philadelphia, Pa, 63-65, pediatrician, cytogeneticist & cardiologist, St Albans, NY, 65-

68; ASSOC PROF PEDIAT, DIR HERITABLE DIS CTR, MED SCH, LA STATE UNIV, NEW ORLEANS, 69- Concurrent Pos: USPHS grant, Johns Hopkins Hosp, Baltimore, 68-69; consult, La State Dept Hosps, 70- Res: Genetic regulation. Mailing Add: Sch of Med La State Univ New Orleans LA 70112

THURMOND, CARL DRYER, chemistry, see 12th edition

THURMOND, JOHN TYDINGS, b Dallas, Tex, Oct 22, 41; m 69; c 2. VERTEBRATE PALEONTOLOGY, ENVIRONMENTAL GEOLOGY. Educ: St Louis Univ, BS, 63; Southern Methodist Univ, MS, 67, PhD(geol), 69. Prof Exp: Res asst to pres, Inst Study Earth & Man, Southern Methodist Univ, 70; ASST PROF GEOL, BIRMINGHAM-SOUTHERN COL, 70- Concurrent Pos: Vis prof fac chem sci & pharm, San Carlos Univ, Guatemala, 71; geologist, Harbert Construct Co, Ala, 71-72; asst cordr coop univ upper-div prog, Univ Ala, Gadsden, 72. Mem: AAAS; Paleont Soc; Soc Vert Paleont. Res: Paleoecology, functional morphology, taxonomy of Mesozoic/Cenozoic fishes; Pleistocene paleoecology; Cretaceous marine reptiles; environmental geology; data retrieval in paleontology. Mailing Add: Box A-30 Birmingham-Southern Col Birmingham AL 35204

THURMOND, WILLIAM, b Lodi, Calif, Apr 11, 26; m 49; c 2. DEVELOPMENTAL PHYSIOLOGY. Educ: Univ Calif, Berkeley, AB, 48, MA, 50, PhD(zool), 57. Prof Exp: Instr zool, San Mateo Community Col, Calif, 49-50; from instr to assoc prof, 51-64, PROF ZOOL, CALIF POLYTECH STATE UNIV, SAN LUIS OBISPO, 64- Concurrent Pos: Vis prof, Univ Frankfort, 69-70. Mem: Fel AAAS; Am Soc Zool; Am Inst Biol Sci. Res: Development of hypothalamic and pituitary control of the adrenal cortex and pigment affector systems in amphibia. Mailing Add: Dept of Biol Sci Calif Polytech State Univ San Luis Obispo CA 93401

THURNAU, DONALD HARLAN, physics, see 12th edition

THURNAUER, PETER G, theoretical physics, see 12th edition

THURNER, JOSEPH JOHN, b Middletown, NY, Oct 26, 20; m 48; c 2. INORGANIC CHEMISTRY. Educ: Hartwick Col, BS, 49; Harvard Univ, MA, 51. Prof Exp: From instr to assoc prof chem, 51-69, chmn dept chem, 67-70, PROF CHEM, COLGATE UNIV, 69-, DIR DIV NATURAL SCI & MATH, 70- Concurrent Pos: Consult, Indium Corp Am, 52-62. Mem: AAAS; The Chem Soc; Am Chem Soc; Electrochem Soc. Res: Organometallic compounds of germanium; organometallic and inorganic compounds; alloys of indium; synthesis and analysis of indium-bearing substances. Mailing Add: Dept of Chem Colgate Univ Hamilton NY 13346

THUROW, GORDON RAY, b Aurora, Ill, Feb 13, 29; m 55; c 5. ZOOLOGY, ANATOMY. Educ: Univ Chicago, PhB, 48, BS, 50, MS, 51; Ind Univ, PhD, 55. Prof Exp: Asst zool & ornith, Ind Univ, 52 & 54, tech adv ed film, 55; assoc prof natural sci, Newberry Col, 57-59; fel anat, Med Col SC, 59-61; asst prof, Univ Kans, 61-66; ASSOC PROF BIOL, WESTERN ILL UNIV, 66- Mem: Am Soc Ichthyol & Herpet; Soc Study Amphibians & Reptiles; Soc Study Evolution; Ecol Soc Am; Am Asn Anat. Res: Morphological, physiological and behavioral adaptations of vertebrates; herpetology; ecology. Mailing Add: Dept of Biol Western Ill Univ Macomb IL 61455

THURSTON, CLAUDE ELMORE, analytical chemistry, see 12th edition

THURSTON, EARLE LAURENCE, b New York, NY, Jan 17, 43; m 62; c 3. BOTANY, CELL BIOLOGY. Educ: State Univ NY, Geneseo, BS, 64; Iowa State Univ, MS, 67, PhD(bot), 69. Prof Exp: NSF fel cell res inst, Univ Tex, Austin, 69-70; ASSOC PROF CELL BIOL & COORDR ELECTRON MICROS CTR, TEX A&M UNIV, 70- Mem: Bot Soc Am; Electron Micros Soc; Am Soc Cell Biol. Res: Developmental morphology and ultrastructure; electron microprobe analysis. Mailing Add: Dept of Biol Electron Micros Ctr Tex A&M Univ College Station TX 77843

THURSTON, GEORGE BUTTE, b Austin, Tex, Oct 8, 24; m 47; c 2. BIOMEDICAL ENGINEERING, BIOPHYSICS. Educ: Univ Tex, BS, 44, MA, 48, PhD(physics), 52. Prof Exp: Res scientist, Defense Res Lab, Tex, 49-52; asst prof physics, Univ Wyo, 52-53 & Univ Ark, 53-54; from assoc prof to prof, Okla State Univ, 54-68; PROF MECH ENG & BIOMED ENG, UNIV TEX, AUSTIN, 68- Concurrent Pos: Consult, US Naval Ord Test Sta, 54-55; res physicist, Univ Mich, 58-59; NSF fel, Macromolecules Res Ctr, Strasbourg, 63-64; US sr scientist, Alexander von Humboldt Found, WGer, 75. Mem: Fel Am Phys Soc; fel Acoust Soc Am; Soc Rheol. Res: Acoustics; rheology; polymer science; macromolecules; optical and electrical properties of solutions. Mailing Add: Dept of Mech Eng Univ of Tex Austin TX 78712

THURSTON, HERBERT DAVID, b Sioux Falls, SDak, Mar 24, 27; m 51; c 3. PLANT PATHOLOGY. Educ: Univ Minn, MS, 53, PhD(plant path), 58. Prof Exp: Asst plant path, Univ Minn, 50-53, instr, 53-54; asst plant pathologist, Rockefeller Found, Colombia, 54-56; instr plant path, Univ Minn, 56-57; from assoc plant pathologist to plant pathologist, Rockefeller Found, Colombia, 58-67; PROF PLANT PATH, CORNELL UNIV, 67- Mem: Am Phytopath Soc; Potato Asn Am. Res: Diseases of potatoes; tropical plant pathology; nature of resistance to fungus diseases. Mailing Add: Dept of Plant Path Plant Sci Bldg Cornell Univ Ithaca NY 14850

THURSTON, HUGH ANSFRID, b UK, 22; m 62. ALGEBRA. Educ: Univ Cambridge, PhD(math), 48. Prof Exp: Lectr math, Bristol Univ, 46-57; asst prof, 58-59, ASSOC PROF MATH, UNIV BC, 60- Mem: Can Math Cong. Res: Abstract algebra; theory of congruences. Mailing Add: Dept of Math Univ of BC Vancouver BC Can

THURSTON, JOHN ROBERT, b Maumee, Ohio, May 6, 26; m 53; c 2. BACTERIOLOGY, IMMUNOLOGY. Educ: Ohio State Univ, BSc, 49, MSc, 51, PhD(bact), 55. Prof Exp: Res microbiologist, Ohio Tuberc Hosp, Columbus, 55-57; microbiologist, Trudeau Found, Inc, NY, 57-61; RES MICROBIOLOGIST, NAT ANIMAL DIS LAB, USDA, 61- Mem: Am Soc Microbiol. Res: Serologic investigations of mycobacteria, serologic diagnosis in tuberculosis; intradermal tuberculin testing of cattle; glycoprotein changes in tuberculous cattle; serology of nocardiosis and aspergillosis. Mailing Add: Nat Animal Dis Lab PO Box 70 Ames IA 50010

THURSTON, RICHARD, b Orono, Maine, Feb 3, 22; m 48; c 2. ENTOMOLOGY. Educ: Univ Conn, BS, 47; Rutgers Univ, PhD(entom), 52. Prof Exp: Asst entom, Univ Maine, 47-48; from asst entomologist to assoc entomologist, 52-59, from asst prof to assoc prof, 57-66, PROF ENTOM, UNIV KY, 66- Concurrent Pos: Adv entom, Ky Team, Thailand, 73-75. Mem: Entom Soc Am. Res: Biology, ecology and control of tobacco insects; resistance to insects in Nicotiana tabacum, Mailing Add: Dept of Entom Univ of Ky Lexington KY 40506

THURSTON, ROBERT NORTON, b Kilbourne, Ohio, Dec 31, 24; m 49; c 4. PHYISCS. Educ: Ill Inst Technol, BS, 45; Ohio State Univ, MS, 48, PhD(phyiscs), 52. Prof Exp: Teacher high sch, Ohio, 46-47; instr aeronaut eng, Ohio State Univ, 49-50; MEM TECH STAFF, BELL TEL LABS, MURRAY HILL, 52- Mem: Am Phys Soc; Acoust Soc Am; Inst Elec & Electronics Eng. Res: Mechanics; crystal physics; communications science. Mailing Add: Squire Terrace RD 1 Colts Neck NJ 07722

THURSTON, WILLIAM, b New York, NY, Nov 1, 15; m 37; c 1. GEOLOGY. Educ: Columbia Univ, AB, 38, AM, 43, PhD, 52. Prof Exp: Asst geologist, Cuban-Am Manganese Corp, Cuba, 38-39; asst geol, Brown Univ, 39-40; trustees asst, Columbia Univ, 40-42; geologist, US Geol Surv, 42-51, mining co, Mex, 51-52 & Nicaro Nickel Co, Cuba, 52-54; assoc prof, Sch Eng, La Polytech Inst, 54-55; exec secy div earth sci, Nat Acad Sci-Nat Res Coun, 55-59; staff geologist, US Geol Surv, 59-67; asst to sci adv, US Dept Interior, 67-70, spec asst to dir, US Geol Surv, 70-76; STAFF OFFICER, NAT ACAD SCI-NAT RES COUN, 76- Concurrent Pos: Mem & nat del, Int Geol Cong, 64, 68 & 72. Mem: Fel AAAS, Geol Soc Am & Soc Econ Geol. Res: Economic geology; geology of pegmatites and fluorspar. Mailing Add: 4200 Massachusetts Ave NW Washington DC 20016

THURSTONE, FREDERICK LOUIS, b Chicago, Ill, Feb 12, 32; m 54; c 3. BIOMEDICAL ENGINEERING, ELECTRICAL ENGINEERING. Educ: Univ NC, BS, 53; NC State Univ, MS, 57, PhD(elec eng), 61. Prof Exp: Instr elec eng, NC State Univ, 56-61, asst prof, 61-62; asst prof biomed eng, Bowman Gray Sch Med, 62-66, assoc prof, 66-67, dir dept, 62-67; assoc prof, 67-70, PROF BIOMED ENG, DUKE UNIV, 70- Mem: Inst Elec & Electronics Eng; Acoust Soc Am; Instrument Soc Am; Am Inst Ultrasonics in Med. Res: Fetal electrocardiography; diagnostic ultrasound; ultrasound scanning and imaging systems; ultrasound holography and visual reconstruction. Mailing Add: Biomed Eng Dept Duke Univ Durham NC 27706

THURTELL, GEORGE WILLIAM, agricultural meteorology, soil physics, see 12th edition

THUSS, WILLIAM GETZ, JR, b Birmingham, Ala, Nov 18, 24; m 49; c 3. INDUSTRIAL MEDICINE, Educ: Univ Md, MD, 48; Univ Cincinnati, ScD(indust med), 56. Prof Exp: Intern, New Haven & Grace-New Haven Community Hosps, Yale Univ, 48-49; res path, Sch Med, Baylor Univ & Vet Admin Hosp, 49-50; ASSOC CLIN PROF PUB HEALTH & EPIDEMIOL & CHIEF OCCUP MED SECT, SCH MED, UNIV ALA, BIRMINGHAM, 56- Concurrent Pos: Aviation med examr, Fed Aviation Agency, 61-; pvt pract; consult. Mem: Indust Med Asn; Am Indust Hyg Asn; Asn Teachers Prev Med; Am Med Asn; Am Acad Occup Med. Res: Occupational medicine. Mailing Add: Thuss Clin 2124 Fourth Ave S Birmingham AL 35233

THUT, CARL CLARENCE, physical organic chemistry, see 12th edition

THUT, PAUL DOUGLAS, b Exeter, NH, Feb 1, 43; m 66; c 3. PSYCHOPHARMACOLOGY. Educ: Hamilton Col, AB, 65; Univ RI, MS, 68; Dartmouth Col, PhD(pharmacol), 71. Prof Exp: Asst prof pharmacol, Col Med, Univ Ariz, 70-74; ASSOC PROF PHARM, SCH DENT, UNIV MD, 74- Concurrent Pos: Pharmaceut Mfrs Found grant, 72-73; consult, Vet Admin Hosp, Tucson, 72- Mem: Soc Neurosci. Res: Psychomotor effects of l-dehydroxphenylalanine; effects of RNA in learning and alcohol metabolism on behavioral performance. Mailing Add: Dept of Pharmacol Univ of Md Sch of Dent Baltimore MD 21201

THWAITE, ROBERT DAVID, b Larchmont, NY, Oct 2, 26; m 50; c 2. GEOLOGY, PHYSICAL SCIENCE. Educ: Dartmouth Col, BA, 49. Prof Exp: Res asst, Carnegie Inst Geophys Lab, 49-52; engr, Am Instrument Co, 52-54; res assoc fel, Portland Cement Asn, 54-60; ENGR RES DEPT BETHLEHEM STEEL CORP, 60- Mem: AAAS. Res: High temperature phase equilibrium studies of anhydrous silicate rock-forming systems; mineralogical and petrographic analysis of ferrous and non-ferrous raw materials. Mailing Add: Homer Res Lab Bethlehem Steel Corp Bethlehem PA 18016

THWAITES, THOMAS TURVILLE, b Madison, Wis, Aug 21, 31; m 53; c 2. NUCLEAR PHYSICS. Educ: Univ Wis, BS, 53; Univ Rochester, MA, 56, PhD(physics), 59. Prof Exp: Physicist, Stromberg Carlson Div, Gen Dynamics Corp, 55-56; asst prof physics, 59-66, ASSOC PROF PHYSICS, PA STATE UNIV, UNIVERSITY PARK, 66- Mem: Am Phys Soc. Res: Scattering of 200 million electron volts polarized protons; radioactive decay schemes; radiative capture of charged particles; scattering of charged particles. Mailing Add: Dept of Physics Pa State Univ University Park PA 16802

THWAITES, WILLIAM MUELLER, b Madison, Wis, July 10, 33; m 55; c 3. GENETICS. Educ: Univ Wis, BS, 55; Univ Mich, MS, 62, PhD(genetics), 65. Prof Exp: Asst prof, 65-70, ASSOC PROF BIOL, SAN DIEGO STATE UNIV, 70- Concurrent Pos: Prin investr, NSF res grant, 66-68; fel, Battelle Mem Inst, Richland, Washington, 72-73; vis prof, Instituto de Investigaciones Biomedicas, Mexico City, 73-74. Mem: Genetics Soc Am. Res: General, microbial and biochemical genetics; regulation of enzyme synthesis in fungi. Mailing Add: Dept of Biol San Diego State Univ San Diego CA 92182

THWEATT, JOHN G, b Norton, Va, Nov 21, 32; m 54; c 2. ORGANIC CHEMISTRY. Educ: Ga Inst Technol, BS, 54, PhD(org chem), 61. Prof Exp: Instr chem, Ga Inst Technol, 57-59; res chemist, 60-63, sr res chemist, res labs, 63-75, SR CHEMIST DEVELOP DEPT, ORG CHEM DIV, TENN EASTMAN CO, 75- Mem: Am Chem Soc. Res: Chemistry of enamines and other electron rich olefins; process development, especially aliphatic syntheses; process improvement. Mailing Add: Bldg 231 Tenn Eastman Co Kingsport TN 37662

THWING, HENRY WARREN, b Orange, Va, Mar 16, 30; m 58; c 2. ALGEBRA. Educ: Yale Univ, BS, 51; Univ Va, MA, 55; Fla State Univ, PhD(math), 69. Prof Exp: Master, Woodberry Forest Sch, 55-63; asst prof math, 63-69, ASSOC PROF MATH, STETSON UNIV, 69- Res: Integral derivations of P-ADIC fields. Mailing Add: 1108 E University Ave De Land FL 32720

THYAGARAJAN, B S, b Tiruvarur, India, July 14, 29; m 56; c 3. ORGANIC CHEMISTRY. Educ: Loyola Col, Madras, India, MA, 51; Presidency Col, Madras, MSc, 53, PhD(chem), 56. Prof Exp: Fel, Northwestern Univ, 56-58; fel, Univ Wis, 58-59; reader org chem, Madras Univ, 60-68; prof, Univ Idaho, 69-74; PROF CHEM & DIR EARTH & PHYS SCI DIV, UNIV TEX, SAN ANTONIO, 74- Honors & Awards: Intrasci Res Award, 66. Mem: AAAS; NY Acad Sci; The Chem Soc; Pharmaceut Soc Japan. Res: Heterocyclic chemistry; aromaticity; molecular migrations. Mailing Add: Earth & Phys Sci Div Univ of Tex San Antonio TX 78284

THYER, NORMAN HAROLD, b Gloucester, Eng, Sept 3, 29; m 65; c 3. METEOROLOGY, MATHEMATICS. Educ: Univ Birmingham, BSc, 57; Univ Wash, Seattle, PhD(meteorol), 62. Prof Exp: Meteorol asst, Air Ministry, UK, 49-50 & Falkland Islands Dependencies Surv, 50-53; tech asst aeronaut eng, Gloster Aircraft Co, 54; micrometeorologist, Univ Wash, Seattle, 60, meteorol training expert, World Meteorol Orgn, 62-63; asst prof meteorol & physics, Univ BC, 63-66; res assoc meteorol, McGill Univ, 66-68; asst prof, 68-71, ASSOC PROF MATH, NOTRE DAME UNIV, NELSON, 71- Mem: Royal Meteorol Soc; Can Meteorol Soc. Res:

Studies of local winds near valleys and convectional storms. Mailing Add: R R 2 Nelson BC Can

THYGESEN, KENNETH HELMER, b Cambridge, NY, June 30, 37; m 55; c 2. EXPERIMENTAL SOLID STATE PHYSICS. Educ: Washington & Lee Univ, 58; Clarkson Col Technol, MS, 60, PhD(physics), 67. Prof Exp: Instr physics, Clarkson Col Technol, 59-65; ASSOC PROF PHYSICS, STATE UNIV NY COL POTSDAM, 67- Mem: Am Asn Physics Teachers. Res: Solid state physics-imperfections in metal crystals; plastic deformation, electric, magnetic and thermal properties of metals and alloys. Mailing Add: Dept of Physics State Univ of NY Col Potsdam NY 13676

THYGESON, PHILLIPS, b St Paul, Minn, Mar 28, 03; m 25; c 2. OPHTHALMOLOGY. Educ: Stanford Univ, AB, 25, MD, 28; Univ Colo, OphD, 30, MS, 33; Am Bd Ophthal, dipl, 31. Hon Degrees: LLD, Univ Calif, San Francisco, 71. Prof Exp: Intern, Colo Gen Hosp, 27-28, assoc vis ophthalmologist, 28-31; asst prof ophthal, Col Med, Univ Iowa, 31-36; asst prof, Col Physicians & Surgeons, Columbia Univ, 36-39, prof & exec off dept, 39-42; from assoc clin prof to prof, 47-70, EMER PROF OPHTHAL, SCH MED, UNIV CALIF, SAN FRANCISCO, 70-, DIR PROCTOR FOUND RES OPHTHAL, 59- Concurrent Pos: Nat Res Coun fel med, Colo Gen Hosp, 31-32; trustee, Proctor Found Res Ophthal, 47-; consult, WHO, 53-, Div Indian Health, USPHS, 54-, Commun Dis Ctr, 56- & Food & Drug Admin, 73-; mem comn sensory dis study sect, NIH, 54-60. Honors & Awards: Medal, AMA, 37; Howe Medal, Am Ophthal Soc, 49; Proctor Medal, Asn Res Vision & Ophthal, 50; Chibret Gold Medal, Int Orgn Against Trachoma, 66. Mem: Am Soc Microbiol; Am Ophthal Soc; fel AMA; Asn Res Vision & Ophthal; fel Am Acad Ophthal & Otolaryngol. Res: Microbiology and external disease of the eye. Mailing Add: Proctor Found Univ of Calif 95 Kirkham St San Francisco CA 94122

THYR, BILLY DALE, b Kansas City, Kans, June 1, 32; m 58; c 3. PLANT PATHOLOGY, MYCOLOGY. Educ: Univ Ottawa, Kans, BA, 59; Wash State Univ, PhD(plant path), 64. Prof Exp: RES PLANT PATHOLOGIST, AGR RES SERV, USDA, 63- Mem: AAAS; Am Phytopath Soc; Mycol Soc Am. Res: Bacterial diseases of tomato and other vegetables; control of bacterial diseases through host resistance; alfalfa diseases; interaction of organisms infecting alfalfa. Mailing Add: USDA Agr Res Serv WR PO Box 8858 Univ Station Reno NV 89507

THYSEN, BENJAMIN, b Bronx, NY, July 27, 32; m 62; c 1. OBSTETRICS & GYNECOLOGY, BIOCHEMISTRY. Educ: City Col New York, BS, 54; Univ Mo, MS, 63; St Louis Univ, PhD(biochem), 67. Prof Exp: Res asst obstet & gynec & biochem, Albert Einstein Col Med, 59-61; asst biochem, Sch Med, Univ Mo, 61-63; instr, St Louis Univ, 67-68; sr res scientist, Technicon Corp, 68-69, group leader, 69-70; ASST PROF OBSTET, GYNEC, BIOCHEM & LAB MED & CHIEF ENDOCRINE LABS, ALBERT EINSTEIN COL MED, 71- Mem: AAAS; Endocrine Soc; Am Chem Soc; Am Asn Clin Chemists; Soc Study Reprod. Res: Estrogen metabolism; laboratory medicine; mechanism of action of the steroid hormones; methodology; endocrinology. Mailing Add: Dept of Obstet & Gynec Albert Einstein Col of Med Bronx NY 10461

TIAO, GEORGE CHING-HWUAN, b London, Eng, Nov 8, 33; Chinese citizen; m 58; c 4. MATHEMATICAL STATISTICS, ECONOMIC STATISTICS. Educ: Nat Taiwan Univ, BA, 55; NY Univ, MBA, 58; Univ Wis-Madison, PhD(econ), 62. Prof Exp: Asst prof statist & bus, Univ Wis-Madison, 62-65; vis assoc prof statist, Harvard Bus Sch, 65-66; assoc prof, 66-68, PROF STATIST & BUS, UNIV WIS-MADISON, 68- Concurrent Pos: Vis lectr, Post Col Prof Educ, Carnegie Mellon Univ, 66-; statist consult to var indust co, 66-; vis prof, Univ Essex, 70-71; chmn dept statist, Univ Wis-Madison, 73-75. Mem: Fel Royal Statist Soc; fel Am Statist Asn; fel Inst Math Statist; AAAS; Air Pollution Control Asn. Res: Bayesian methods in statistics; time series analysis; statistical analysis of air pollution data; economic and business forecasting. Mailing Add: Dept of Statist Univ of Wis Madison WI 53706

TIBBETTS, FRED ERNEST, III, organic chemistry, organometallic chemistry, see 12th edition

TIBBETTS, GARY GEORGE, b Omaha, Nebr, Oct 12, 39; m 64. PHYSICS. Educ: Calif Inst Technol, BS, 61; Univ Ill, Urbana, MS, 63, PhD(physics), 67. Prof Exp: Ger Res Asn grant, vis scientist, Munich Tech Univ, 67-69; SR RES PHYSICIST, GEN MOTORS RES LABS, 69- Mem: Am Vacuum Soc. Res: Surface physics; plasma-surface interactions; adsorption of gases on surfaces; electronic and chemical properties of surfaces. Mailing Add: Physics Dept 12 Mile & Mound Rd Gen Motors Res Labs Warren MI 48090

TIBBETTS, MERRICK SAWYER, b Keene, NH, Dec 30, 25; m 50; c 4. ORGANIC CHEMISTRY. Educ: Univ NH, BS, 48, MS, 51; Stevens Inst Technol, PhD(org chem), 66. Prof Exp: Plant chemist, Rubber Div, Eberhard Faber Pencil Co, 51-53; sr chemist, Bendix Corp, 53-62; proj leader synthetic org chem, Int Flavors & Fragrances, 66-69; sr res chemist, Florasynth, Inc, 69-70; SR RES SCIENTIST, PEPSICO INC, LONG ISLAND CITY, NY, 70- Mem: Am Chem Soc; Am Soc Test & Mat; Inst Food Technol; Sigma Xi. Res: Conformational analysis; organic synthesis; subjective-objective correlation; carbohydrate chemistry. Mailing Add: Pepsico Inc 4600 5th St Long Island City NY 11101

TIBBITTS, FORREST DONALD, b Tacoma, Wash, Jan 23, 29; m 51; c 1. EMBRYOLOGY. Educ: East Wash Col, BA, 51; Ore State Col, MA, 55, PhD, 58. Prof Exp: Instr sci, Ore Col Educ, 57-59; asst prof zool, 59-69, PROF BIOL, UNIV NEV, RENO, 69-, PROF ANAT, MED SCH, 71- Mem: AAAS; Am Asn Anat; Am Soc Zool. Res: Reproductive biology; placentation. Mailing Add: Dept of Biol Univ of Nev Reno NV 89507

TIBBITTS, THEODORE WILLIAM, b Melrose, Wis, Apr 10, 29; wid; c 2. HORTICULTURE, ENVIRONMENTAL PHYSIOLOGY. Educ: Univ Wis, BS, 50, MS, 52, PhD(hort, agron), 53. Prof Exp: From asst prof to assoc prof, 55-71, PROF HORT, UNIV WIS-MADISON, 71- Concurrent Pos: Res engr space biol, NAm Aviation, Inc, Calif, 65-66; bot adv, Manned Space Craft Ctr, Tex, 68-69; mem rev panel, Skylab Biol Exps, NASA, 70-71; res leave, Lab Plant Physiol Ond, Netherlands, 74. Honors & Awards: Marion Meadows Award, Am Soc Hort Sci, 68. Mem: AAAS; Am Soc Hort Sci; Am Soc Plant Physiol; Int Soc Hort Sci; Am Inst Biol Sci. Res: Environmental physiology of vegetable crops; physiological breakdowns under favorable environments; air pollution; geophysical environment of plants; standardization in plant growth chambers. Mailing Add: Dept of Hort Univ of Wis Madison WI 53706

TIBBLES, JOHN JAMES, b Toronto, Ont, Mar 16, 24; m 51; c 2. FISHERIES. Educ: Ont Agr Col, BSA, 51; Univ Wis, MS, PhD(fisheries), 56. Prof Exp: Assoc scientist, Fisheries Res Bd Can, 56-65, DIR SEA LAMPREY CONTROL CTR, FISHERIES SERV, CAN DEPT ENVIRON, 66- Mem: Am Fisheries Soc. Res: Sea lamprey control. Mailing Add: Sea Lamprey Control Ctr Can Dept Agr Ship Canal PO Sault Ste Marie ON Can

TIBBS, JOHN FRANCISCO, b Pacific Grove, Calif, Oct 12, 38; m 66. PROTOZOOLOGY. Educ: Fresno State Col, BA, 60; Univ Southern Calif, MS, 64, PhD(biol), 68. Prof Exp: Asst prof zool, 68-70, DIR BIOL STA, UNIV MONT, 70-, ASSOC PROF ZOOL, 73- Mem: AAAS; Soc Protozool. Res: Ecology of Arctic and Antarctic protozoans and marine invertebrates; taxonomy of the Phaeodarina; ecology of the Sarcodina. Mailing Add: Univ of Mont Biol Sta Bigfork MT 59911

TIBBY, RICHARD BITNER, b Salt Lake City, Utah, Mar 12, 11; m 50; c 1. OCEANOGRAPHY, MARINE BIOLOGY. Educ: Univ Calif, Berkeley, AB, 32; Univ Calif, Los Angeles, PhD(oceanog), 43; Univ Southern Calif, MA, 34. Prof Exp: High sch teacher, Calif, 33-35; fisheries biologist, Calif State Fish & Game Comn, 35-37; asst zool, Univ Calif, Los Angeles, 37-38, asst zool, Scripps Inst, 38-41, oceanogr, Div War Res, Navy Radio & Sound Lab, Point Loma, 41-43; consult oceanog, 45-47; asst prof zool, 47-48, prof basic sci, 48-61, PROF BIOL, UNIV SOUTHERN CALIF, 61-, DIR, CATALINA MARINE SCI CTR, 68- Concurrent Pos: Oceanogr, US Fish & Wildlife Serv, 38-41; consult, marine industs, 45-; oceanogr, Catalina Marine Sci Found, 48-65, asst dir, Catalina Marine Sci Ctr, 65-68; mem, Calif State Comn Marine & Coastal Resources, 69- Mem: Am Geophys Union; Marine Technol Soc; Am Soc Oceanog. Res: Oceanic circulation; eddy diffusion; light absorption; military oceanography; oceanic waste disposal and pollution; academic and industrial program development and management in oceanography and other marine sciences. Mailing Add: Catalina Marine Sci Ctr Univ of Southern Calif Los Angeles CA 90007

TICE, DAVID ANTHONY, b Brooklyn, NY, Dec 31, 29; m 52; c 3. THORACIC SURGERY, CARDIOVASCULAR SURGERY. Educ: Columbia Univ, BA, 51; NY Univ, MD, 55; Am Bd Surg, dipl, 61; Bd Thoracic Surg, dipl, 65. Prof Exp: Intern, Third Surg Div, Bellevue Hosp, 55-56, asst resident, 56-59; resident, Bellevue Hosp & asst surg, NY Univ, 59-60; res asst, Univ, 60-62, from instr to asst prof, 60-67, from asst attend to assoc attend, Univ Hosp, 60-68, assoc prof, Sch Med, 67-73, PROF SURG, SCH MED, NY UNIV, 73- Concurrent Pos: NY Heart Asn fel, Sch Med, NY Univ, 60-62; asst attend surg, Methodist Hosp Brooklyn, 60-63; from asst vis physician to assoc vis physician, Bellevue Hosp, 60-66, vis physician, 66-; consult cardiovasc surg, NY State Dept Health, 62 & Lutheran Med Ctr, 70; asst med examr, City of New York, 62-73; attend cardiovasc surg, New York Vet Admin Hosp, 63-66, attend thoracic surg, 64-66, chief surg serv, 67-74; attend surg, NY Univ Hosp, 68-; mem, Vet Admin Res & Educ Coun, Washington, DC, 72-73. Mem: Am Asn Thoracic Surg; fel Am Col Cardiol; fel Am Col Surgeons; Soc Univ Surgeons; Transplantation Soc. Res: Thoracic and cardiovascular physiology and disease. Mailing Add: Dept of Surg NY Univ Sch of Med New York NY 10016

TICE, LINWOOD FRANKLIN, b Salem, NJ, Feb 17, 09; m 29; c 2. PHARMACEUTICAL CHEMISTRY. Educ: Philadelphia Col Pharm, BS, 33, MSc, 35; St Louis Col Pharm, DSc, 54. Prof Exp: Res fel, Wm R Warner & Co, 31-35; res fel, Edible Mfrs Res Soc, 35-38; asst prof, 38-40, dean, 59-75, dir sch pharm, 40-71, asst dean, 41-56, assoc dean, 56-59, PROF PHARM, PHILADELPHIA COL PHARM, 40-, EMER DEAN, 75- Concurrent Pos: Res fel, Sharp & Dohme, 38-40; ed, Am J Pharm, 40-; tech ed, El Farmaceutico, 41-59 & Pharm Int, 47-59; mem revision comt, US Pharmacopoeia, 40-60, bd trustees, 60-70, 70-75; dir, Am Found Pharmaceut Ed, 54-59; mem Am Coun Pharmaceut Ed, 60-66; mem, Philadelphia Med-Pharmaceut Sci, 63-. Honors & Awards: Remington Honor Medal, Am Pharmaceut Asn, 71. Mem: AAAS (vpres, 71-72); Am Chem Soc; Am Pharmaceut Asn (pres elect, 65-66, pres, 66-67); Asn Cols Pharm (pres, 55-56); fel Am Inst Chem. Res: Pharmaceuticals; surfactants; proteins; emulsions. Mailing Add: Philadelphia Col of Pharm & Sci 43rd St & Kingsessing Ave Philadelphia PA 19104

TICE, RUSSELL L, b Parkersburg, WVa, Dec 5, 32; m 56; c 4. PHYSICAL CHEMISTRY. Educ: Marshall Col, BS, 60; Univ Calif, Los Angeles, PhD(chem), 65. Prof Exp: Res asst chem, Univ Calif, Los Angeles, 62-65; from asst prof to assoc prof, 65-75, PROF CHEM, CALIF STATE POLYTECH COL, 75- Concurrent Pos: NSF fel, Ind Univ, 66. Mem: Am Chem Soc; Am Phys Soc. Res: Investigation of electron-atom and electron-molecule collision cross sections by cyclotron resonance; electron spin resonance of radicals. Mailing Add: Dept of Chem Calif State Polytech Col San Luis Obispo CA 93401

TICHAUER, ERWIN RUDOLPH, b Berlin, Ger, Apr 27, 18; m 46. BIOMEDICAL ENGINEERING, OCCUPATIONAL HEALTH. Educ: Technische Hochschule, dipl, 38; Albertus Univ, ScD(phys sci), 40. Prof Exp: Dep dir area team 1065, UNRRA, 46-47; engr in-chg trainiing & res, FAMIC Ltda, Chile, 47-50; works mgr design & develop, PMS Pty Ltd, Australia, 50-53; specialist lectr eng, Univ Queensland, 53-56; expert productivity, UN Tech Assistance Admin, 56-59; sr lectr indust eng, Univ New SWales, 60-64; prof, Tex Tech Univ, 64-67; res prof, 67-68, PROF BIOMECH & DIR DIV, CTR SAFETY & INST REHAB MED, MED CTR, NY UNIV, 68- Concurrent Pos: Distinguished vis prof, San Marcos Univ, Lima, 58; hon consult, Royal S Sidney Hosp, 60-64; consult, UNICEF, 61, Major Corp, 64- & Waterbury Hosp, Conn, 69-; vis assoc prof, Tex Tech Univ, 63; mem, Australian Coun Rehab of Disabled, 64-; guest lectr, USPHS, 67; chmn subcomt biomech, Comt Z-94, Am Nat Stand Inst, 68-; Am Soc Mech Engrs rep, US Nat Comt Eng, Med & Biol, Nat Acad Eng-Nat Res Coun, 69-72; mem bd trustees comt, NJ Inst Technol, 69-; chmn biomed eng, NY Acad Med, 71-72; mem comt prosthetics res in Vet Admin, Nat Res Coun, 75- Honors & Awards: Gilbreth Medal, Soc Advan Mgt; Golden Plate Award, Acad Achievement; Outstanding Achievement Award & Distinguished Res Award, Am Inst Indust Engrs; Metrop Life Award, Nat Safety Coun. Mem: AAAS; fel Royal Soc Health; Am Soc Mech Engrs; Am Inst Indust Engrs; Am Soc Eng Educ. Res: Occupational biomechanics, pharmacokinesiology and rehabilitation medicine; anatomy; medical education; medical thermography; preventive occupational medicine, safety and traumatology; functional anatomy and physiology applied to design of tasks, tools and equipment for both healthy and disabled workers; work stress on women. Mailing Add: Inst of Rehab Med NY Univ Med Ctr 400 E 34th St New York NY 10016

TICHENOR, ROBERT LAUREN, b Ft Atkinson, Wis, Sept 1, 18; m 43; c 4. PHYSICAL CHEMISTRY. Educ: Mont State Univ, BS, 39; Harvard Univ, AM, 41, PhD(chem), 42. Prof Exp: Res chemist, Eastman Kodak Co, 42-44, Manhattan Proj, Tenn, 44-45; group leader, Thos A Edison, Inc, 45-51; RES CHEMIST, E I DU PONT DE NEMOURS & CO, INC, 51- Concurrent Pos: Rhodes scholar, Oxford Univ, 46-47. Mem: AAAS; Electrochem Soc. Res: Cellulose esters; electrochemistry of nickel; technology of synthetic fibers. Mailing Add: 437 Walnut Ave Waynesboro VA 22980

TICHO, HAROLD KLEIN, b Brno, Czech, Dec 21, 21; nat US; m. EXPERIMENTAL HIGH ENERGY PHYSICS. Educ: Univ Chicago, PhD(physics), 49. Prof Exp: Asst, Univ Chicago, 42-48; from asst prof to assoc prof, 48-58, chmn dept, 67-71, PROF PHYSICS, UNIV CALIF, LOS ANGELES, 58-, DEAN DIV PHYS SCI, 74- Concurrent Pos: Guggenheim fel, 66-67, 73-74. Mem: Am Phys Soc. Res: High energy nuclear physics; elementary particles. Mailing Add: Dept of Physics Univ of Calif Los Angeles CA 90024

TICKLE, ROBERT SIMPSON, b Norfolk, Va, July 31, 30; m 55; c 1. NUCLEAR

PHYSICS. Educ: US Mil Acad, BS, 52; Univ Va, MS, 58, PhD(nuclear physics), 60. Prof Exp: From instr to assoc prof, 60-68, PROF PHYSICS, UNIV MICH, ANN ARBOR, 68- Mem: Am Phys Soc; Am Asn Physics Teachers. Res: Measurement of photonuclear cross sections; accelerator design and development; study of nuclear structure with charged particle experiments. Mailing Add: Dept of Physics Univ of Mich Ann Arbor MI 48104

TICKNER, ALFRED WILLIAM, b Moose Jaw, Sask, Feb 3, 22; m 48; c 2. PHYSICAL CHEMISTRY, RESEARCH ADMINISTRATION. Educ: Univ Sask, BE, 44, MSc, 47; Univ Toronto, PhD(chem), 49. Prof Exp: Res fel mass spectrometry, Nat Res Coun Can, 49-51; res fel collision processes, Univ Col, London, 51-52; asst res officer, Mass Spectrometry, 52-54, assoc res officer, 55-62, sr res officer, 63-64, chief personnel serv, 64-70, actg secy, Coun, 71-72, CHIEF PERSONNEL ADV, NAT RES COUN CAN, 73- Mem: Chem Inst Can. Res: Reactions in electric discharges; mass spectrometry as applied to chemical analysis; metal vapors; administration. Mailing Add: Nat Res Coun Can Ottawa ON Can

TICKNOR, LELAND BRUCE, b Centralia, Wash, May 9, 22; m 52. PHYSICAL CHEMISTRY. Educ: Univ Wash, BS, 44; Mass Inst Technol, PhD(phys chem), 50. Prof Exp: Mem staff, Dept Metall, Mass Inst Technol, 50-52; instr chem, Swarthmore Col, 52-54; phys chemist & sect leader, Res & Develop Dept, Viscose Div, FMC Corp, 54-65; sr res chemist, Appl Res Lab, US Steel Corp, 65-68; SR RES CHEMIST, DOW BADISCHE CO, 68-, GROUP LEADER RES DEPT, 75- Mem: Fiber Soc; Am Chem Soc. Res: Thermodynamics of solutions; retained energies of cold working in metals; cellulose chemistry; cellulose fibers; metal coatings; acrylic fibers. Mailing Add: 107 Indian Springs Rd Williamsburg VA 23185

TICKNOR, ROBERT LEWIS, b Portland, Ore, Oct 26, 26; m 50; c 3. ORNAMENTAL HORTICULTURE. Educ: Ore State Univ, BS, 50; Mich State Univ, MS, 51, PhD(pomol), 53. Prof Exp: Asst hort, Mich State Univ, 50-53; asst prof nursery culture, Univ Mass, 53-57, assoc prof, 57-59; assoc prof hort, 59-72, PROF HORT, ORE STATE UNIV, 72- Mem: Am Soc Hort Sci; Am Asn Bot Gardens & Arboretums; Int Plant Propagators Soc. Res: Chemical weed control; plant nutrition, propagation and materials. Mailing Add: North Willamette Exp Sta Rte 2 Box 254 Aurora OR 97002

TIDBALL, CHARLES STANLEY, b Geneva, Switz, Apr 15, 28; US citizen; m 52. COMPUTER SCIENCE, MEDICAL EDUCATION. Educ: Wesleyan Univ, BA, 50; Univ Rochester, MS, 52; Univ Wis, PhD(physiol), 55; Univ Chicago, MD, 58. Prof Exp: Asst physiol, Univ Wis, 52-55; res asst surg, Univ Chicago, 55-56, asst physiol, 56-58, res asst, 57; intern, Madison Gen Hosp, Wis, 58-59; physician, Mendota State Hosp, 59; asst res prof, 59-63, from assoc prof from actg chmn to chmn dept, 63-71, to prof, 63-65, HENRY D FRY PROF PHYSIOL, MED CTR, GEORGE WASHINGTON UNIV, 65-, RES PROF MED, 72-, DIR COMPUT ASSISTED EDUC, 73- Concurrent Pos: USPHS fel, 60-61, USPHS res career develop award, 61-63. Mem: AAAS; Asn Am Med Cols; Am Physiol Soc; Digital Equip Comput Users Soc; Asn Develop Comput-Based Instruct Systs. Res: Mini-computer time-sharing systems; computer assisted education; medical computer utilization; graduate and medical education; physiology. Mailing Add: Ross Hall Rm 457 George Washington Univ Med Ctr Washington DC 20037

TIDBALL, MARY ELIZABETH, b Anderson, Ind, Oct 15, 29; m 52. MEDICAL PHYSIOLOGY, INSTITUTIONAL RESEARCH. Educ: Mt Holyoke Col, BA, 51; Univ Wis, MS, 55, PhD(physiol), 59; Wilson Col, Dr Sci, 73; Trinity Col, Dr Sci, 74. Prof Exp: Asst physiol, Univ Wis, 52-55, 58-59; asst histochem, Univ Chicago, 55-56, physiol, 56-58; USPHS fel, Nat Heart Inst, 59-61; staff pharmacologist, Hazelton Labs, Inc, 61-62; asst res prof pharmacol, 62-64, assoc res prof physiol, 64-70, res prof, 70-71, PROF PHYSIOL, GEORGE WASH UNIV, 71- Concurrent Pos: Consult, Hazelton Labs, Inc. 62-63; assoc sci coordr, Food & Drug Admin, Sci Assocs Training Prog, 66-67; consult, Food & Drug Admin, 66-68; trustee, Mt Holyoke Col, 68-73, vchmn, 72-73; vchmn, Cedar Crest Conf Undergrad Educ Women, 69-70; trustee educ comt, Am Youth Found, 71-72; trustee, Hood Col, 72-; nat adv coun on NIH training progs & fels, Nat Acad Sci, 72-; exec secy, Nat Acad Sci-Nat Res Coun Comn on Human Resources in Comt on the Educ & Employment of Women in Sci & Eng, 74-75; consult, inst res, Wellesley Col, 74-75 & Woodrow Wilson Nat Fel Found, 74-75; chmn task force on women, Am Physiol Soc, 73-; chmn review panels, NSF, 74-; chmn educ comt, Bd Trustees Hood Col, 74- Mem: AAAS; Am Physiol Soc; Am Asn Higher Educ. Res: coupling; autacoids and neuroendocrinology; education of women; institutional research; science literacy. Mailing Add: 4100 Cathedral Ave NW Washington DC 20016

TIDBALL, RONALD RICHARD, b Greeley, Colo, July 13, 30; m 53, 72; c 3. SOIL SCIENCE. Educ: Colo State Univ, BS, 52; Univ Wash, MS, 57; Univ Calif, Berkeley, PhD(soil sci), 65. Prof Exp: Res Asst forestry, Univ Wash, 55-S7 & Weyerhauser Co, 57; res asst soil sci, Univ Calif, 58-64; SOIL SCIENTIST, ENVIRON GEOCHEM, US GEOL SURV, 65- Mem: Am Quaternary Asn. Res: Study of distribution of chemical elements in natural materials of earth's crust and natural landscapes. Mailing Add: US Geol Surv Box 25046 Fed Ctr Denver CO 80225

TIDD, CHARLES WHARTON, b Round Mountain, Tex, Feb 15, 03; m 25; c 1. PSYCHIATRY. Educ: Northwestern Univ, BS, 30, MS, 31, MD, 33; Am Bd Psychiat & Neurol, dipl, 47. Prof Exp: Intern, US Naval Hosp, San Diego, Calif, 32-33; res physician neurol & psychiat, Menninger Clin, Topeka Kans, 33-34, staff psychiatrist, 36-38; res physician neurol & psychiat, Compton Sanitarium, Calif, 34-36; pvt practice, 38-42, 46-53; prof psychiat & head dept, 53-68, EMER PROF PSYCHIAT, SCH MED, UNIV CALIF, LOS ANGELES, 68- Concurrent Pos: Training analyst, Los Angeles, Inst Psychoanal, 46-; consult, Vet Admin Hosp, Brentwood, 51-, Metrop State Hosp, 53-, Camarillo State Hosp, 56- & Dept Pub Health, Calif, 56-; teacher, Mt Zion Hosp & Med Ctr, Univ Calif, San Francisco, 69- Mem: Am Psychiat Asn; Am Psychoanal Asn; AMA; Asn Am Med Cols. Res: Psychoanalysis; ethology; human emotional development. Mailing Add: The Sequoias 1400 Geary Blvd Apt 7-P San Francisco CA 94109

TIDD, JOSEPH SHEPARD, b Taunton, Mass, Apr 8, 07; m 36; c 3. BOTANY. Educ: Dartmouth Col, BS, 28, AM, 31; Univ Mich, PhD(bot), 35. Prof Exp: Teacher high sch, Mass, 28-29; instr bot, Dartmouth Col, 29-31, 35-41, asst prof, 41-43; asst, Herbarium, Univ Mich, 31-33; asst bot garden, 33-34; clerk, Los Angeles Seed Co, Calif, 43-44; pathologist emergency plant dis prev proj, USDA, 44-45; PLANT PATHOLOGIST, ASGROW SEED CO, INC, THE UPJOHN CO, 45- Mem: Am Phytopath Soc. Res: Physiologic specialization in barley powdery mildew; genetics of disease resistance in plants; breeding for disease resistance in vegetable crops. Mailing Add: Asgrow Seed Co The Upjohn Co Kalamazoo MI 49001

TIDD, ROBERT FREDERICK, b Buffalo, NY, June 22, 24; m 47; c 7. MATHEMATICS. Educ: Univ Buffalo, MA, 51, PhD(math), 59. Prof Exp: From asst prof to prof math & head dept, Canisius Col, 47-66; PROF MATH & CHMN DEPT, NDAK STATE UNIV, 66-, DIR STUDENT ACAD AFFAIRS, 73- Mem: Am Math Soc; Math Asn Am; NY Acad Sci. Res: Complex variable. Mailing Add: Dept of Math NDak State Univ Fargo ND 58102

TIDD, WILBUR METELLUS, zoology, entomology, deceased

TIDEMAN, PHILIP LUNDSTEN, b Cokato, Minn, May 24, 26; m 53; c 5. RESOURCE GEOGRAPHY, AGRICULTURAL GEOGRAPHY. Educ: Univ Minn, BA, 49; St Cloud State Univ, BS, 51; Univ Nebr, MA, 53, PhD(geog), 67. Prof Exp: Teacher geog, Brainerd Pub Schs, Minn, 52-54; teacher, Great Falls Pub Schs, Mont, 54-56; asst prof, St Cloud State Univ, 57-66; assoc prof, Univ Wyo, 66-70; PROF GEOG & CHMN DEPT, ST CLOUD STATE UNIV, 70- Concurrent Pos: Off Water Resources Res, Dept Interior fel, Minn, 70-72; Asn Am Geogr fel, Lower Miss Valley, 72; consult, Upper Miss River Basin Comn & Coronet Instructional Media, 75. Mem: Asn Am Geogr; Nat Coun Geog Educ; Conserv Educ Asn. Res: Agricultural land use changes; water resource management. Mailing Add: Dept of Geog St Cloud State Univ St Cloud MN 56301

TIDESWELL, NORMAN WILSON, inorganic chemistry, see 12th edition

TIDRICK, ROBERT THOMPSON, b Doleib Hill, Sudan, Aug 4, 09; US citizen; m; c 4. SURGERY. Educ: Tarkio Col, AB, 32; Wash Univ, MD, 36; Am Bd Surg, dipl, 43. Prof Exp: Intern, Univ Hosp, Univ Iowa, 36-37; intern, Am Hosp, Assiut, Egypt, 37-38; asst resident surg, Univ Hosp, Univ Iowa, 38-39, resident, 39-44, asst pathologist, 39-40, instr orthop path, 40-42, asst & instr surg, 42-44, from assoc to prof, Col Med, 44-69, head dept, 51-69; PROF SURG, MED COL OHIO, 69- Concurrent Pos: Mem nat sci adv bd, Nat Found Ileitis & Colitis. Honors & Awards: Burdick Award, 40. Mem: Soc Exp Biol & Med; Am Geriat Soc; Soc Univ Surgeons; AMA; Am Surg Asn. Res: Diseases of the colon; cancer; traumatology. Mailing Add: Dept of Surg Med Col of Ohio Toledo OH 43614

TIDRIDGE, WILLIAM ALBERT, inorganic chemistry, see 12th edition

TIDWELL, JAMES THOMAS, molecular biology, biochemistry, see 12th edition

TIDWELL, THOMAS TINSLEY, b Atlanta, Ga, Feb 20, 39; m 71. ORGANIC CHEMISTRY. Educ: Ga Inst Technol, BS, 60; Harvard Univ, AM, 63, PhD(reaction mechanisms), 64. Prof Exp: NIH fel, Univ Calif, San Diego, 64-65; asst prof chem, Univ SC, 65-72; ASSOC PROF CHEM, SCARBOROUGH COL, UNIV TORONTO, 72- Concurrent Pos: NIH fel, Univ EAnglia, 66-67. Mem: Am Chem Soc; The Chem Soc. Res: Steric strain; reactions of peroxides; reactivity of enolates; phosphate ester hydrolysis. Mailing Add: Dept of Chem Scarborough Col U of Toronto West Hill ON Can

TIDWELL, TROY HASKELL, JR, b New Boston, Tex, June 8, 37; m 57; c 3. ELECTROCHEMISTRY. Educ: NTex State Univ, BS, 59, MS, 60; La State Univ, PhD(chem), 64. Prof Exp: Asst prof chem, NTex State Univ, 63-68; sr electrochemist, Columbus Labs, Battelle Mem Inst, 68-71; CHIEF TECH BR, ENFORCEMENT DIV, REGION II, US ENVIRON PROTECTION AGENCY, 71- Concurrent Pos: Consult, Am Foods Inc & Acme Brick, 66; vis asst chemist, Brookhaven Nat Lab, 66-67. Mem: Electrochem Soc. Res: Molten salt diffusion; polarography and double layer studies. Mailing Add: Enforcement Div US EPA 26 Federal Plaza New York NY 10007

TIDWELL, WILLIAM LEE, b Greenville, SC, Jan 14, 26; m 46; c 1. MICROBIOLOGY. Educ: Univ SC, BS, 45; Univ Hawaii, MS, 48; Univ Calif, Los Angeles, PhD(microbiol), 51. Prof Exp: Asst bact, Univ Hawaii, 46-48 & Univ Calif, Los Angeles, 48-51; asst prof biol, Agr & Mech Col, Tex, 51-55, asst res bact, Eng Exp Sta, 52-55; asst prof biol, 55-58, assoc prof bact, 58-62, PROF MICROBIOL, CALIF STATE UNIV, SAN JOSE, 62- Concurrent Pos: Consult, State Col Affairs, Calif State Employees Asn, 66-68. Mem: AAAS; Am Soc Microbiol; NY Acad Sci. Res: General bacteriology; water and sewage bacteriology. Mailing Add: Dept of Biol Sci Calif State Univ San Jose CA 95192

TIECKE, RICHARD WILLIAM, b Muscatine, Iowa, Apr 5, 17. PATHOLOGY. Educ: Univ Iowa, BS, 40, DDS, 42, MS, 47; Am Bd Path, dipl. Prof Exp: Resident path, Univ Chicago, 47-49; assoc prof oral path, Georgetown Univ, 52-54; assoc prof path, Sch Dent, Northwestern Univ, 55-62; ASST EXEC DIR SCI AFFAIRS, AM DENT ASN, 71- Concurrent Pos: Consult, US Naval Hosp, Great Lakes, 55-70, Vet Admin Res Hosp, Chicago, 56-70, Vet Admin West Side Hosp & Pub Health Hosp, 56-71; prof, Col Dent, Univ Ill, Chicago, 62-; consult, Nat Cancer Inst; surgeon gen, Div Chronic Dis, USPHS; pres, Am Bd Oral Path, 67-71; prof path, Schs Dent & Med, Northwestern Univ, 74- Mem: Int Acad Oral Path; Am Dent Asn; fel Am Col Dent; Am Acad Oral Path (pres, 57); Int Asn Dent Res. Res: Cancer; pathologic physiology of oral disease. Mailing Add: 179 E Lake Shore Dr Chicago IL 60611

TIECKELMANN, HOWARD, b Chicago, Ill, Oct 29, 16; m 42; c 5. BIO-ORGANIC CHEMISTRY. Educ: Carthage Col, BA, 42; Univ Buffalo, PhD(org chem), 48. Prof Exp: Chemist, Armour Res Found, Ill, 46; from instr to prof, 46-61, chmn dept, 70-74, DISTINGUISHED TEACHING PROF CHEM, STATE UNIV NY BUFFALO, 75- Mem: AAAS; The Chem Soc; Am Inst Chemists; Am Chem Soc. Res: Heterocyclic compounds; pyridines; pyrimidines, alkylations and rearrangements; natural products. Mailing Add: Dept of Chem State Univ of NY Buffalo NY 14214

TIEDCKE, CARL HEINRICH WILHELM, chemistry, see 12th edition

TIEDEMAN, GEORGE TRENT, b Ignacio, Colo, Aug 6, 35; m 61; c 3. QRGANIC CHEMISTRY, POLYMER CHEMISTRY. Educ: Univ Chicago, AB, 54, MS, 57; Univ Colo, PhD(org chem), 63. Prof Exp: Chemist, Pasadena Found Med Res, 57-58 & Atomics Int Div, NAm Aviation, Inc, Calif, 63-68; CHEMIST, WEYERHAEUSER CO, 68- Mem: Sigma Xi; Am Chem Soc. Res: Stereochemistry of ionic and free radical reactions; electron spin resonance studies of reactions and kinetics of free atoms; kinetic and product studies of pyrolytic and radiolytic oxidation reactions; resins and polymers; mammalian pheromones; animal attractants and repellants; plant growth biochemicals. Mailing Add: Weyerhaeuser Co 3400 13th Ave SW Seattle WA 98134

TIEDEMAN, JOHN ALBERT, b Schenectady, NY, Mar 8, 04; m 32; c 2. PHYSICS. Educ: Union Col, NY, BS, 26, MS, 28; Univ Va, PhD(physics), 31. Prof Exp: Asst gen eng lab, Gen Elec Co, 26-27, engr, 28-29; from asst prof to assoc prof physics, Womans Col, Univ NC, 31-41; dir educ, Ansco Div, Gen Aniline & Film Corp, 45-48; physicist, Appl Physics Lab, Johns Hopkins Univ, 48-68; opers analyst, Kettelle Assocs, Va, 68-72; LECTR PHYSICS, UNIV NC, WILMINGTON, 72- Concurrent Pos: Vis prof, US Naval Postgrad Sch, 66-67. Mem: Am Phys Soc; Am Asn Physics Teachers. Res: Color photography; operations analysis; mechanism of the electric spark. Mailing Add: 312 Stradleigh Rd Wilmington NC 28401

TIEDEMANN, ALBERT WILLIAM, JR, b Baltimore, Md, Nov 7, 24; m 53; c 4. ANALYTICAL CHEMISTRY, SCIENCE ADMINISTRATION. Educ: Loyola Col,

Md, BS, 47; NY Univ, MS, 49; Georgetown Univ, PhD(chem), 58. Prof Exp: Instr chem, Mt St Agnes Col, 50-55; sr res chemist, Emerson Drug Co, 55-56, chief chemist, Emerson Drug Co Div, Warner-Lambert Pharmaceut Co, 56-60; supvr anal group, Allegany Ballistic Lab, Hercules, Inc, 61-68, supt tech serv, Radford Army Ammunition Plant, 68-72; DIR CONSOLIDATED LABS, COMMONWEALTH VA, 72- Mem: Am Soc Qual Control; fel Am Inst Chemists; Am Mgt Asn. Res: Development of analytical methods. Mailing Add: Div of Consol Labs One N 14th St Richmond VA 23219

TIEDEMANN, CLIFFORD E, b Oak Park, Ill, June 30, 38; m 60; c 1. GEOGRAPHY. Educ: Univ Ill, AB, 60; Univ Okla, MA, 63; Mich State Univ, PhD(geog), 66. Prof Exp: Asst prof geog, Univ Calif, Los Angeles, 65-67; asst prof, 67-70, ASSOC PROF GEOG & HEAD DEPT, UNIV ILL, CHICAGO CIRCLE, 70- Mem: Asn Am Geogrs; Regional Sci Asn. Res: Analysis of geographic patterns. Mailing Add: Dept of Geog Univ of Ill at Chicago Circle Chicago IL 60680

TIEDJE, JAMES MICHAEL, b Newton, Iowa, Feb 9, 42; m 65; c 3. MICROBIAL ECOLOGY, SOIL MICROBIOLOGY. Educ: Iowa State Univ, BS, 64; Cornell Univ, MS, 66, PhD(soil microbiol), 68. Prof Exp: Asst prof, 68-73, ASSOC PROF MICROBIAL ECOL, MICH STATE UNIV, 73- Concurrent Pos: Eli Lilly career develop grant, 74; vis assoc prof, Univ Ga, 74-75; ed, Appl Microbiol, 74-; consult, NSF, 74- Mem: Am Soc Microbiol; Am Soc Agron; Soil Sci Soc Am; Am Inst Biol Sci; AAAS. Res: Microbial metabolism of pesticides, chelants and other organic chemicals; microbial activities in eutrophic lakes; exobiology; nitrogen cycle. Mailing Add: Dept of Microbiol & Pub Health Mich State Univ East Lansing MI 48824

TIEDJENS, VICTOR ALPHONS, b Brillion, Wis, June 13, 95; m 23; c 2. HORTICULTURE. Educ: Univ Wis, BS, 21, MS, 22; Rutgers Univ, PhD(plant physiol), 32. Prof Exp: Asst agron, Univ Wis, 19-23; asst res prof veg gardening, Mass Col, 23-30; res specialist hort, Rutgers Univ, 30-32, assoc prof veg crops, 34-45; dir res, Yoder Bros, Ohio, 32-34; dir, Va Truck Exp Sta, 45-51; dir res, Na-Churs Plant Food Co & Berlou Mfg Co, 51-55; VPRES & DIR RES, GROWERS CHEM CORP, 55- Honors & Awards: Thomas Roland Award, Mass Hort Soc, 53. Mem: AAAS; Am Chem Soc. Res: Plant nutrition; nitrogen assimilation in plants; plant fertilizer relationships; liquid fertilizers; soilless culture of plants. Mailing Add: Growers Chem Corp PO Box 1750 Milan OH 44846

TIEDKE, KENNETH EARLE RIORDAN, b Neshkoro, Wis, Mar 3, 11; m 46; c 2. APPLIED ANTHROPOLOGY, ETHNOLOGY. Educ: Univ Wis, BA, 38, MA, 39. Prof Exp: Psychologist, Vet Advisement Bur, Brooklyn Col, 45-46; prof anthrop, Mich State Univ, 48-57; chief community develop adv, US Opers Mission, Iraq, Nepal & Malawi, 57-66; dir div soc sci, 70-72, PROF ANTHROP, SOUTHAMPTON COL, LONG ISLAND UNIV, 66-, ASSOC DEAN COL, 72- Concurrent Pos: Soc Sci Res Coun grant, 46-48; consult, US Indian Serv, 48-54; Rockefeller Found fel, Mich State Univ, 50-51; consult, Gov Mich, 50-54; dir res social change, Inter-Am Inst Agr Sci, 53-55. Mem: Fel AAAS; Am Geog Soc; Am Anthrop Asn; Am Ethnol Soc; Soc Appl Anthrop. Res: American Indian; culture change; technology; man's adaptation to environment; community analysis. Mailing Add: Southampton Col Long Island Univ Southampton NY 11968

TIEFEL, RALPH MAURICE, b Brazil, Ind, Sept 3, 28; m 66. BOTANY. Educ: Cent Mo State Col, BS, 53; Univ Mo, MA, 55, PhD(bot), 57. Prof Exp: Assoc prof, 57-59, PROF BIOL, CARTHAGE COL, 60- Mem: Soc Econ Bot; Torrey Bot Club. Res: Soil and plant relationships; meristems; history of science. Mailing Add: Dept of Biol Carthage Col Kenosha WI 53140

TIEFENTHAL, HARLAN E, b Kalamazoo, Mich, Aug 23, 22; m 42; c 4. ORGANIC CHEMISTRY. Educ: Kalamazoo Col, BA, 44; Univ Ill, MS, 48; Mich State Col, PhD(chem), 50; Univ Chicago, MBA, 65. Prof Exp: Fel chem, Mellon Inst, 50-54, group leader, Koppers Co, Inc, 54-58, mgr fine chem group, 58-62; asst res dir, Armour Indust Chem Co, 62-65, dir spec projs, 66-68; OWNER, TIEFENTHAL ASSOCS, 68- Concurrent Pos: Exec dir, Assoc Cols Chicago Area, 69-; lectr, George Williams Col, 72- Mem: Am Chem Soc. Res: Grignard reactions; dehydrogenation reactions; dehydrocyclizations; polymerization; alkylation; hydrogenation; organic synthesis; amination; aliphatic amine chemistry. Mailing Add: 4544 Grand Ave Western Springs IL 60558

TIEH, THOMAS TA-PIN, b Peking, China, May 2, 34; US citizen; m 62; c 4. MINERALOGY. Educ: Univ Ill, Urbana, BS, 58; Stanford Univ, MS, 60, PhD(geol), 65. Prof Exp: Geologist, Bear Creek Mining Co, 60; res assoc geol, Univ Hawaii, 62-63; asst cur mining & petrol, Stanford Univ, 65-66; asst prof, 66-71, ASSOC PROF GEOL, TEX A&M UNIV, 71- Concurrent Pos: Welch Found grant, Tex A&M Univ, 69-72. Mem: Brit Mineral Soc. Res: Petrology and geochemistry. Mailing Add: Dept of Geol Tex A&M Univ College Station TX 77843

TIEMAN, CHARLES HENRY, JR, b Los Angeles, Calif, Sept 3, 26; div; c 2. ORGANIC CHEMISTRY. Educ: Univ Calif, Los Angeles, BS, 49; Univ Colo, PhD(chem), 53. Prof Exp: Asst, Univ Colo, 50-53; res & develop chemist, Aerojet-Gen Corp, Gen Tire & Rubber Co, 53-58; RES CHEMIST, SHELL DEVELOP CO, 58- Mem: Am Chem Soc. Res: Organic syntheses and mechanisms. Mailing Add: Biol Sci Res Ctr Shell Develop Co PO Box 4248 Modesto CA 95352

TIEMAN, SUZANNAH BLISS, b Washington, DC, Oct 10, 43; m 69. NEUROSCIENCES, VISION. Educ: Cornell Univ, AB, 65; Stanford Univ, PhD(physiol), 74. Prof Exp: NAT EYE INST FEL, DEPT ANAT, UNIV CALIF MED CTR, SAN FRANCISCO, 74- Mem: AAAS; Soc Neurosci. Res: The anatomical and physiological bases of visual perception, and its development, particularly its susceptibility to restricted environments. Mailing Add: Dept of Anat Univ of Calif Med Ctr San Francisco CA 94143

TIEMANN, JEROME J, b Yonkers, NY, Feb 21, 32; m 57. SOLID STATE PHYSICS. Educ: Mass Inst Technol, ScB, 53; Stanford Univ, PhD(physics), 60. Prof Exp: Asst, Stanford Univ, 53-55 & 56-57; PHYSICIST, RES & DEVELOP CTR, GEN ELEC CO, 57- Concurrent Pos: Consult, Radiation Lab, Univ Calif, 55-57. Mem: Inst Elec & Electronics Engrs. Res: Quantum mechanics; electronics; solid state electronic device phenomena; electronic circuit and system design. Mailing Add: Corp Res & Develop Ctr Gen Elec Co 1 River Rd Schenectady NY 12301

TIEMEIER, OTTO WILLIAM, zoology, see 12th edition

TIEMSTRA, PETER J, b Chicago, Ill, Oct 7, 23; m 46; c 3. FOOD CHEMISTRY. Educ: Northwestern Univ, BS, 44; Univ Chicago, BS, 47, MS, 48. Prof Exp: Anal chemist, Swift & Co, 48-49, anal res chemist, 49-53, from asst head chemist to head chemist, 53-63, head res & develop, Stabilizer Div, 63-65; res chemist, Derby Foods, Inc, Ill, 65-66, tech dir, 66-69; dir res & qual assurance, 69-70, DIR SPEC SERV, SWIFT GROCERY PROD CO, CHICAGO, 70- Mem: Am Chem Soc; Am Oil Chem Soc; Am Soc Qual Control; Inst Food Technologists; Am Asn Candy Technologists. Res: Quality control; product research and development. Mailing Add: 6543 Pontiac Dr LaGrange IL 60525

TIEN, HSIN TI, b Peking, China, Feb 1, 28; US citizen; m 53; c 4. BIOPHYSICS. Educ: Univ Nebr, BS, 53; Temple Univ, MA, 60, PhD(chem), 63. Prof Exp: Proj engr, Allied Chem Corp, 56-57; med scientist, Eastern Pa Psychiat Inst, 57-63; assoc prof chem, Northeastern Univ, 63-66; PROF BIOPHYS, MICH STATE UNIV, 66- Concurrent Pos: Grants, Res Corp, 64-65, NIH, 64- & Off Saline Water, US Dept Interior, 68-71. Mem: AAAS; Am Chem Soc; Biophys Soc. Res: Physical chemical investigations of membranes, particularly molecular lipid membranes; photosynthesis and vision; specific electrodes; ion-exchange equilibria; bilayer lipid membranes as models of biological membranes. Mailing Add: Dept of Biophys Mich State Univ East Lansing MI 48824

TIEN, REX YUAN, b Hupei, China, Aug 10, 35; m; c 3. ORGANIC CHEMISTRY. Educ: Chung Hsing Univ, Taiwan, BS, 58; Univ RI, PhD(chem), 68. Prof Exp: NSF fel chem, State Univ NY Albany, 67-68; SR CHEMIST, AM HOECHST CORP, 68- Mem: Am Chem Soc. Res: Applied chemistry, dyes and pigments; photochemistry; organometallic chemistry, kinetics and instrumentation. Mailing Add: 129 Quidnick St Coventry RI 02816

TIEN, WEICHEN, b Anhwei, China, July 12, 38; US citizen; m 66; c 2. MICROBIAL GENETICS, INDUSTRIAL MICROBIOLOGY. Educ: Nat Taiwan Univ, BS, 60; Univ Ky, MS, 65, PhD(microbiol), 68. Prof Exp: Sr microbiologist, S B Penick & Co, CPC Int, 68-70; sr scientist antibiotics, Wyeth Labs, Inc, 70-71, supvr, 71-74; TEAM LEADER FERMENTATION, PFIZER INC, 74- Mem: Am Soc Microbiol; Sigma Xi. Res: Bacterial and fungal genetics and cultural development for new antibiotics. Mailing Add: Pfizer Inc Groton CT 06340

TIERKEL, ERNEST SHALOM, b Philadelphia, Pa, July 2, 17; m 58; c 2. VETERINARY PUBLIC HEALTH, EPIDEMIOLOGY. Educ: Univ Pa, AB, 38, VMD, 42; Columbia Univ, MPH, 46; Am Bd Vet Pub Health, dipl, 54. Prof Exp: Vet pathologist, Bur Animal Indust, USDA, 42-45; officer-in-chg rabies res unit, Virus Lab, Commun Dis Ctr, USPHS, 46-49, asst chief vet pub health prog, 50-52, dir nat rabies control prog, 54-64; Commun Dis Ctr consult vet pub health, Va State Health Dept, 53-54; dep dir health serv, AID, 64-66; Commun Dis Ctr-AID consult zoonoses & vet epidemiol, Nat Inst Commun Dis, New Delhi, India, 66-68; asst surgeon gen & dir off sci, US Dept Health, Educ & Welfare, 68-73; CHIEF, BUR DIS CONTROL, DEL DEPT HEALTH, 73- Concurrent Pos: Mem, Expert Panel Rabies, WHO, 50-, Expert Comt, Rome, 53, Paris, 56 & 61, Geneva, 59 & 65, Expert Panel Zoonoses, 64-, Expert Comt, Geneva, 66; vis lectr, Univ Pa, 53 & 60-66, Harvard Univ, 53, Univ Ga, 53-54 & 60-64, Columbia Univ, 54-55, Emory Univ, 59 & 60-63 & Johns Hopkins Univ, 64-65. Mem: AAAS; Sigma Xi; Asn Mil Surg US; fel Am Pub Health Asn; Conf Pub Health Vets (pres, 57-58). Res: Rabies; public health zoonoses; environmental and international health. Mailing Add: 189 S Fairfield Dr Dover DE 19901

TIERNAN, ROBERT JOSEPH, b Boston, Mass, Dec 14, 35; m 59; c 4. PHYSICS, CERAMICS. Educ: Boston Col, AB, 57, MS, 59; Mass Inst Technol, PhD(ceramics), 69. Prof Exp: Res asst accelerators, Grad Sch, Boston Col, 57-59; solid state physicist, Naval Res Lab, 59-62; atomic physicist, Nat Bur Stand, 63; develop engr, Sylvania, 63-64; AEC res asst, Grad Sch, Mass Inst Technol, 64-69; SR RES SCIENTIST, RAYTHEON CO, WALTHAM, 69- Concurrent Pos: Res assoc, Argonne Nat Lab, 75. Res: Solid state physics; electronic, optical and magnetic properties of ceramics. Mailing Add: 224 North St Stoneham MA 02180

TIERNAN, THOMAS ORVILLE, b Chattanooga, Tenn, July 22, 36; m 61; c 1. CHEMICAL PHYSICS, ANALYTICAL CHEMISTRY. Educ: Univ Windsor, BSc, 58; Carnegie Inst Technol, MS, 60, PhD(chem), 66. Prof Exp: Ohio State Univ Res Found res chemist, Wright-Patterson AFB, 60-61, res chemist, Off Aeronaut Space Res, Aerospace Labs, 61-67, group leader high energy chem kinetics, 67-75; PROF CHEM, WAYNE STATE UNIV, 75- Mem: AAAS; Am Chem Soc; Am Phys Soc; Am Soc Testing & Mat; fel Am Inst Chemists. Res: Mass spectrometry; gas phase kinetics; ion and electron impact-phenomena; plasma characterization and diagnostics; lasers; gaseous electronics; analytical methods development; environmental monitoring, materials characterization. Mailing Add: 121 Woodfield Pl Dayton OH 45459

TIERNEY, DONALD FRANK, b Butte, Mont, May 24, 31; m 54; c 2. PHYSIOLOGY, MEDICINE. Educ: Univ Calif, Berkeley, BA, 53; Univ Calif, San Francisco, MD, 56. Prof Exp: Intern, Philadelphia Gen Hosp, 57; asst resident, Univ Pa, 58; asst resident, Univ Calif, San Francisco, 59, asst prof physiol, 65-68; assoc prof, 68-75, PROF MED, UNIV CALIF, LOS ANGELES, 75- Concurrent Pos: USPHS fels, Univ Calif, San Francisco, 59-65; mem pulmonary dis adv comt, Nat Heart & Lung Inst, 73-76. Mem: Am Physiol Soc; Am Thoracic Soc; Am Soc Clin Invest. Res: Pulmonary physiology, biochemistry and metabolism. Mailing Add: Ctr for Health Sci Univ of Calif Los Angeles CA 90024

TIERNEY, JOHN A, b New Britain, Conn, Nov 17, 17; m 41; c 2. MATHEMATICS. Educ: Cent Conn State Col, BE, 39; Columbia Univ, MA, 42; Univ Md, PhD(math), 51. Prof Exp: Instr math, Univ Norwich, 43-45 & Vanderbilt Univ, 45-46; from asst prof to assoc prof, 46-62, PROF MATH, US NAVAL ACAD, 62- Concurrent Pos: Math sci adminr, US Army Res Off, 61-62. Mem: Am Math Soc; Math Asn Am. Res: Physical sciences. Mailing Add: 1204 Poplar Ave Annapolis MD 21401

TIERNEY, MYLES, b New York, NY, Sept 3, 37; m 61; c 2. ALGEBRA, MATHEMATICAL LOGIC. Educ: Brown Univ, BA, 59; Columbia Univ, PhD(math), 65. Prof Exp: Asst prof math, Rice Univ, 65-66; guest prof, Res Inst Math, Swiss Fed Inst Technol, 66-68; assoc prof, 68-72, PROF MATH, RUTGERS UNIV, 72- Concurrent Pos: Assoc prof, Univ Paris, 74-75. Mem: Am Math Soc. Res: Category theory; sheaf theory; set theory; logic; algebraic topology; homological algebra. Mailing Add: Dept of Math Rutgers Univ New Brunswick NJ 08903

TIERS, GEORGE VAN DYKE, b Chicago, Ill, Mar 23, 27; m 50; c 1. PHYSICAL CHEMISTRY, ORGANIC CHEMISTRY. Educ: Univ Chicago, SB, 46, SM, 50, PhD(chem), 56. Prof Exp: Asst pharmacol, Univ Chicago, 45-46, chemist, Ord Res Proj, 48-49; res assoc, 51-65, CORP SCIENTIST, CENT RES DEPT, 3M CO, 65- Honors & Awards: Carbide Award, Am Chem Soc, 59. Mem: Am Chem Soc. Res: Nuclear magnetic resonance spectroscopy; fluorine, organic, dye, polymer and physical-organic chemistry; duplicating and imaging technology. Mailing Add: 3M Ctr St Paul MN 55133

TIERSON, WILLIAM CORNELIUS, b Newark, NY, Dec 19, 25; m 47; c 7. FOREST ECOLOGY, WILDLIFE ECOLOGY. Educ: State Univ NY Col Forestry, Syracuse, BSF, 49, MF, 67. Prof Exp: Forester, NY State Col Environ Sci & Forestry, Syracuse, 51-68, forest mgr, 68-72, DIR WILDLIFE RES, ADIRONDACK ECOL CTR, 72- Mem: Soc Am Foresters; Wildlife Soc. Res: Northern hardwood silviculture and ecology; forest game management; wildlife population dynamics; forest fertilization and growth. Mailing Add: Adirondack Ecol Ctr NY State Col Environ Sci & For Newcomb NY 12852

TIERSTEN, HARRY FRANK, b Brooklyn, NY, Jan 4, 30; m 53; c 2. APPLIED MECHANICS. Educ: Columbia Univ, BS, 52, MS, 56, PhD(appl mech), 61. Prof Exp: Stress analyst, Grumman Aircraft Eng Corp, 52-53; struct designer, J G White Eng Corp, 53-56; instr civil eng, City Col New York, 56-60; res asst appl mech, Columbia Univ, 60-61; mem tech staff, Bell Tel Labs, 61-68; PROF MECH, RENSSELAER POLYTECH INST, 68- Mem: Am Phys Soc; Acoust Soc Am; Am Soc Mech Eng; Soc Natural Philos; Inst Elec & Electronics Engr. Res: Elasticity; couple stress elasticity; electromagnetism; electrostriction; piezoelectricity; magnetism; magnetoelasticity; waves; vibrations. Mailing Add: Dept of Mech Rensselaer Polytech Inst Troy NY 12181

TIERSTEN, MARTIN STUART, b Aug 7, 31; US citizen; m 53; c 2. PHYSICS. Educ: Queens Col, NY, BA, 53; Columbia Univ, AM, 58, PhD(theoret solid state physics), 62. Prof Exp: Tutor physics, 57-62, instr, 62-63, asst prof, 63-70, ASSOC PROF PHYSICS, CITY COL NEW YORK, 70- Mem: Am Phys Soc; Am Asn Physics Teachers. Res: Theoretical physics. Mailing Add: Dept of Physics City Col of New York New York NY 10031

TIESZEN, LARRY L, b Marion, SDak, Mar 2, 40; m 59; c 2. PLANT PHYSIOLOGY. Educ: Augustana Col, SDak, BA, 61; Univ Colo, PhD(bot), 65. Prof Exp: Kettering fel biol, Albion Col, 65-66; asst prof, Univ Minn, Duluth, 66; from asst prof to assoc prof, 66-75, PROF BIOL, AUGUSTANA COL, S DAK, 75- Concurrent Pos: Res grants, Sigma Xi, 65-67, Arctic Inst NAm, 66-68 & NSF, 70-76; NSF vis prof, Univ Nairobi, Kenya, 75. Mem: AAAS; Am Soc Plant Physiol; Ecol Soc Am; Arctic Inst NAm; Can Soc Plant Physiol. Res: Photosynthesis and pigments in arctic and alpine grasses; environmental influence of photosynthesis; growth under extreme conditions; physiological adaptation; United States tundra biome program; photosynthesis and water stress in finger millet and other tropical grasses. Mailing Add: Dept of Biol Augustana Col Sioux Falls SD 57102

TIETJEN, JAMES JOSEPH, b New York, NY, Mar 29, 33; m 58; c 2. PHYSICAL CHEMISTRY. Educ: Iona Col, BS, 56; Pa State Univ, MS, 58, PhD(chem), 63. Prof Exp: Mem tech staff mat sci, 63-69, group head, 69-70, DIR MAT RES, RCA LABS, RCA CORP, 70- Concurrent Pos: Mem, Solid State Sci Adv Panel, Nat Acad Sci, 73- Honors & Awards: David Sarnoff Outstanding Achievement Awards, 67 & 70. Mem: Am Chem Soc; Am Inst Mining Metall & Petrol Engrs; Electrochem Soc. Res: Displays, semiconductor materials and devices; insulators; metallic systems; optical phenomena; electron optics; negative electron affinity effects; luminescent materials. Mailing Add: RCA Labs 201 Washington Rd Princeton NJ 08540

TIETJEN, JOHN H, b Jamaica, NY, June 19, 40; m 68; c 1. BIOLOGICAL OCEANOGRAPHY, INVERTEBRATE ZOOLOGY. Educ: City Col New York, BS, 61; Univ RI, PhD(oceanog), 66. Prof Exp: From asst prof to assoc prof, 66-75, PROF BIOL, CITY COL NEW YORK, 75- Concurrent Pos: Consult, Millstone Point Co, Conn & Northeast Utilities Co, 68-; res grants, NSF, 68- & Nat Oceanic & Atmospheric Admin, 73-75. Mem: AAAS; Am Soc Zoologists; Am Soc Limnol & Oceanog; Marine Biol Asn UK. Res: Estuarine ecology; physiological ecology of meiofauna; ecology and distribution of deep sea meiofauna; bionomics of marine nematodes. Mailing Add: Dept of Biol City Col of New York New York NY 10031

TIETJEN, WILLIAM LEIGHTON, b Americus, Ga, Jan 3, 37; m 68; c 1. ENTOMOLOGY, ZOOLOGY. Educ: Univ Ga, BS, 58; Univ Tenn, PhD(radiation biol), 67. Prof Exp: Res asst marine biol, Marine Inst, Univ Ga, 60; asst prof, 67-71, ASSOC PROF BIOL, GA SOUTHWESTERN COL, 71- Mem: AAAS; Am Inst Biol Sci; Am Soc Limnol & Oceanog; Ecol Soc Am; Entom Soc Am. Res: Arthropod metabolism and energy flow; environmental effects on physiology of arthropods. Mailing Add: Dept of Biol Ga Southwestern Col Americus GA 31709

TIETZ, NORBERT W, b Stettin, Ger, Nov 13, 26; US citizen; m 59; c 4. CLINICAL CHEMISTRY. Educ: Stuttgart Tech Univ, PhD(natural sci), 50. Prof Exp: Res fel biochem, Univ Munich, 51-54; res fel clin chem, Rockford Mem Hosp, 54-55 & Univ Chicago, 55-56; head div biochem, Reid Mem Hosp, Richmond, Ind, 56-59; assoc path, Chicago Med Sch-Univ Health Sci, 59-64, from asst prof to assoc prof clin path, 64-69, prof clin chem, 69-76; DIR CLIN CHEM, UNIV KY MED CTR & PROF PATH, COL MED, 76- Concurrent Pos: Dir clin chem, Mt Sinai Hosp Med Ctr, 59-76; consult, Dept Health, State of Ill, 67-; consult, Vet Admin Hosp, Hines, Ill, 74-76 & Vet Admin Hosp, Lexington, Ky, 76- Honors & Awards: Chicago Clin Chem Award, 71; Award Outstanding Effort in Educ & Training, Am Asn Clin Chem, 76. Mem: Fel AAAS; Am Asn Clin Chem; Am Chem Soc; Am Soc Clin Path; fel Am Inst Chemists. Res: Methodology related to clinical chemistry; lipid metabolism related to atherosclerosis; enzyme chemistry. Mailing Add: Dept of Path Univ of Ky Med Ctr Lexington KY 40506

TIETZ, RAYMOND FRANK, organic chemistry, see 12th edition

TIETZ, WILLIAM JOHN, JR, b Chicago, Ill, Mar 6, 27; m 53; c 3. NEUROPHYSIOLOGY. Educ: Swarthmore Col, BA, 50; Univ Wis, MS, 52; Colo State Univ, DVM, 57; Purdue Univ, PhD(physiol), 61. Prof Exp: Instr vet sci, Purdue Univ, 57-59, from instr to assoc prof physiol, 59-64; sect leader, Collab Radiol Health Lab, 64-67, assoc prof physiol & biophys, Univ, 64-67, chmn dept, 67-70, vpres student-univ rels, relations, 70-71, PROF PHYSIOL & BIOPHYS, COLO STATE UNIV, 67-, DEAN COL VET MED & BIOMED SCI, 71- Mem: AAAS; Am Physiol Soc; Am Vet Med Asn; Conf Res Workers Animal Dis; Am Soc Vet Physiol & Pharmacol. Res: Radiation biology; veterinary neurophysiology and neurosurgery; effects of ionizing radiation on early embryogenesis. Mailing Add: Col of Vet Med & Biomed Sci Colo State Univ Ft Collins CO 80521

TIETZE, CHRISTOPHER, demography, statistics, see 12th edition

TIETZE, FRANK, b Manila, PI, Aug 19, 24; nat US; m 54; c 4. BIOCHEMISTRY. Educ: Trinity Col, Conn, BS, 45; Northwestern Univ, MS, 47. 47, PhD(biochem), 49. Prof Exp: USPHS fel, Duke Univ, 49-50; USPHS fel, Univ Wash, 50-51, instr biochem, 51-52; instr, Univ Pa, 52-56; RES BIOCHEMIST, NAT INST ARTHRITIS & METAB DIS, 56- Mem: AAAS; Am Soc Biol Chem. Res: Mechanisms of disulfide bond reduction. Mailing Add: Nat Inst Arthritis & Metab Dis Bethesda MD 20014

TIFFANY, BURRIS DWIGHT, b Kansas City, Mo, Dec 6, 20; m 54; c 3. ORGANIC CHEMISTRY. Educ: Univ Kansas City, BA, 42; Univ Minn, MS, 46; Univ Wis, PhD(org chem), 49. Prof Exp: Asst, Univ Minn, 42-44; asst antimalarials, Off Sci Res & Develop, 44-45 & Abbott Labs, 45-47; Du Pont fel, Mass Inst Technol, 49-50; instr, Univ Ky, 50-51; sr res chemist, 51-71, RES ASSOC, UPJOHN CO, 71- Concurrent Pos: Vis scholar, Univ Calif, Los Angeles, 66-67. Mem: Am Chem Soc. Res: Organic synthesis; antithrombotics; large scale preparation of medicinal compounds. Mailing Add: Dept of Chem Res Prep Upjohn Co Kalamazoo MI 49002

TIFFANY, LOIS HATTERY, b Collins, Iowa, Mar 8, 24; m 45; c 3. PLANT PATHOLOGY, MYCOLOGY. Educ: Iowa State Col, BS, 45, MS, 47, PhD, 50. Prof Exp: From instr to assoc prof, 50-65, PROF BOT & PLANT PATH, IOWA STATE UNIV, 65- Mem: Am Phytopath Soc; Mycol Soc Am. Res: Forage crop and grass diseases; ascomycetes. Mailing Add: Dept of Bot & Plant Path Iowa State Univ Ames IA 50010

TIFFANY, LOYD WAYNE, medical microbiology, see 12th edition

TIFFANY, MARY LOIS, b Pittsburgh, Pa, Aug 22, 19; m 42; c 5. BIOPHYSICS. Educ: Univ Pittsburgh, BS, 38; Univ Ill, MS, 40; Univ Mich, MA, 46, PhD(biophys), 71. Prof Exp: Res asst biol, Univ Pittsburgh, 34-38; teaching asst physiol, Univ Mich, 39-40; res asst biophys, Mass Inst Technol, 40-41; res physicist, War Dept, Camp Evans, 42 & Raytheon Res Labs, 43; RES ASSOC BIOPHYS, UNIV MICH, ANN ARBOR, 60- Concurrent Pos: Fels, NIH, 72-73 & Univ Mich, 75-76. Mem: Am Chem Soc; Biophys Soc. Res: Biological macromolecules; x-ray diffraction; conformation in solution; electron microscopy; blood platelet studies. Mailing Add: 1828 Vinewood Blvd Ann Arbor MI 48104

TIFFANY, OTHO LYLE, b Flint, Mich, Nov 26, 19; m 42; c 5. PHYSICS. Educ: Univ Mich, BS, 43, MS, 46, PhD(physics), 50. Prof Exp: Mem staff, Radiation Lab, Mass Inst Technol, 43-45; res engr, Willow Run Res Ctr, Mich, 49-58; chief scientist, Aerospace Systs Div, 58-70, DIR SPACE & EARTH SCI, BENDIX CORP, 70- Mem: Am Phys Soc; Am Geophys Union; Inst Elec & Electronics Engrs. Res: Space sciences, geophysics, oceanography, environmental research. Mailing Add: 1828 Vinewood Blvd Ann Arbor MI 48104

TIFFANY, SHARON WESTON, b Houston, Tex, Dec 15, 43; m 72. ANTHROPOLOGY. Educ: Univ Calif, Los Angeles, AB, 66, MA, 68, PhD(anthrop), 72. Prof Exp: ASST PROF ANTHROP, UNIV WIS-WHITEWATER, 72- Mem: Am Anthrop Asn; Asn Social Anthrop Oceania. Res: Pacific Island communities; land tenure; social change; legal anthropology; non-unilineal descent systems. Mailing Add: Dept of Anthrop Univ of Wis Whitewater WI 53190

TIFFANY, WALTER WARREN, b Milwaukee, Wis, Oct 2, 43; m 72. ANTHROPOLOGY. Educ: Stanford Univ, BA, 65; Univ Calif, Los Angeles, MA(psychol) & MA(anthrop), 67, PhD(anthrop), 71. Prof Exp: ASST PROF ANTHROP, UNIV WIS-WHITEWATER, 71- Res: Polynesia; political organization; ambilineal descent systems; social structural change; warfare. Mailing Add: Dept of Sociol & Anthrop Univ of Wis Whitewater WI 53190

TIFFIN, DONALD LLOYD, b London, Ont, Nov 16, 31; m 58; c 3. MARINE GEOPHYSICS, GEOLOGY. Educ: Univ BC, BASc, 65, PhD(geophys), 69. Prof Exp: Res scientist geosci, Marine Sci Br, 69-71, RES SCIENTIST GEOSCI, CAN DEPT ENERGY, MINES & RESOURCES, GEOL SURV CAN, 71- Concurrent Pos: Pres, Geo-Marine Serv Ltd, 68- Mem: Am Geophys Union. Res: Geophysics and geology of continental margins. Mailing Add: Geol Surv Can 505 100 W Pendar St Vancouver BC Can

TIFFNEY, WESLEY NEWELL, b Ilion, NY, Apr 7, 09; m 36; c 2. MYCOLOGY, MARINE BIOLOGY. Educ: Bates Col, BS, 33; Harvard Univ, MA, 35, PhD(mycol), 36. Prof Exp: Prof biol, Am Int Col, 36-46; prof, 46-74, EMER PROF BIOL, BOSTON UNIV, 74-, CUR HERBARIUM, 70-; BOT CUR, GRAY MUS, MARINE BIOL LAB, WOODS HOLE, 74- Concurrent Pos: Chmn sci, Boston Univ, 46-64, adv & lectr, Nursing Training, 50-57, coordr freshman biol prog, 57-66, marine prog, 68-72; adv & res dir, Optical Res Labs, 48-57; educ adv, Mass State Prog Community Cols, 59-60; consult, Asn Gen & Lib Educ, 60; cur nonvascular plants, NEng Bot Club, 65-71; vis investr, Systs-Ecol Prog, Marine Biol Lab, Woods Hole, 68-72. Mem: Bot Soc Am; Mycol Soc Amn. 17 Res: Ecology and taxonomy of nonvascular plants especially marine organisms of the salt marsh and dune areas; soil and aquatic fungi of salt marshes and dunes of sea margin and offshore islands. Mailing Add: 226 Edge Hill Rd Sharon MA 02067

TIFFT, WILLIAM GRANT, b Derby, Conn, Apr 5, 32; m 65; c 6. ASTRONOMY. Educ: Harvard Univ, AB, 54; Calif Inst Technol, PhD, 58. Prof Exp: Hon res fel astron, Australian Nat Univ, 58-60; res assoc astron & physics, Vanderbilt Univ, 60-61; astronr, Lowell Observ, UnivxAriz,x61-64Ariz, 61-64; assoc prof, 64-73, PROF ASTRON, UNIV ARIZ, 73- Concurrent Pos: NSF fel, 58-60. Mem: Am Astron Soc; Royal Astron Soc; Int Astron Union. Res: Optical stellar astronomy; galactic structure and extragalactic problems; interpretation of the red-shift in galaxies and clusters of galaxies. Mailing Add: Steward Observ Univ of Ariz Tucson AZ 85721

TIGCHELAAR, EDWARD CLARENCE, b Hamilton, Ont, Feb 10, 39; m 62; c 2. GENETICS, PLANT BREEDING. Educ: Univ Guelph, BScA, 62; Purdue Univ, MSc, 64, PhD(genetics, plant breeding), 66. Prof Exp: Asst prof hort, Univ Guelph, 66-67; asst prof, 67-73, ASSOC PROF HORT, PURDUE UNIV, WEST LAFAYETTE, 73- Concurrent Pos: Asst prof, Int Prog Agr, AID, Brazil, 67-69. Mem: Genetics Soc Can. Res: Genetics of anthocyanin formation and photocontrol of anthocyanin synthesis; vegetable genetics and breeding. Mailing Add: Dept of Hort Purdue Univ West Lafayette IN 47906

TIGCHELAAR, PETER VERNON, b Chicago, Ill, May 15, 41; m 63; c 2. PHYSIOLOGY. Educ: Calvin Col, AB, 63; Univ Ill, Urbana, MS, 66, PhD(physiol), 69. Prof Exp: NIH fel endocrinol, Univ Ill, 70-71; asst prof physiol, Sch Med, Ind Univ-Purdue Univ, Indianapolis, 71-75; ASSOC PROF BIOL, CALVIN COL, 75- Concurrent Pos: NIH fel endocrinol, 69-70. Mem: AAAS; Endocrine Soc; Am Physiol Soc. Res: Mammalian reproductive physiology; synthesis, control and effects of mammalian gonadotrophic hormones. Mailing Add: Dept of Biol Calvin Col Grand Rapids MI 49506

TIGER, LIONEL, b Montreal, Que, Feb 5, 37; m ; c 1. BIOLOGICAL ANTHROPOLOGY, SOCIAL STRUCTURE. Educ: McGill Univ, BA, 57, MA, 59; London Sch Econ, PhD(polit sociol), 63. Prof Exp: Asst prof sociol, Univ BC, 63-68; assoc prof anthrop, Livingston Col, 68-70, assoc prof, Fac Grad Studies & dir grad progs, Univ, 70-72, PROF ANTHROP, GRAD SCH, RUTGERS UNIV, 72- Concurrent Pos: Can Coun spec award soc sci & Nat Res Coun Assoc Comt Exp Psychol, 66-67; Can Coun-Killam bequest & Guggenheim fel, 68; co-sr investr human aggress, Guggenheim Found, 72-75, consult & res dir, 72- Mem: Am Sociol Asn; Can Sociol & Anthrop Asn; Asn Study Animal Behav; Royal Anthrop Inst Gt Brit & Ireland; Am Anthrop Asn. Res: Human evolution; nature and expression of sex differences; male groups; theoretical implications of biosociological research. Mailing Add: Guggenheim Found Res Off 17 W Ninth St New York NY 10011

TIGERTT, WILLIAM DAVID, b Wilmer, Tex, May 22, 15; m 38; c 2. PATHOLOGY. Educ: Baylor Univ, MD, 37, AB, 38; Am Bd Path, dipl, 42. Prof Exp: Intern, Baylor Hosp, 37-38, instr path, Baylor Col Med, 38-40; pathologist, Brooke Gen Hosp, Med Corps, US Army, 40-43, commanding officer, 26th Army Med Lab, Southwest Pac, 44-46 & 406th Med Gen Lab, Tokyo, 46-49, asst commandant, Army Med Serv Grad

Sch, 49-54, chief spec opers br, Walter Reed Army Inst Res, 54-56, commanding officer, Army Med Unit, Ft Detrick, Md, 56-61, med officer, Field Teams, Walter Reed Army Inst Res, 61-63, dir & commandant, 63-68, commanding gen, Madigan Army Hosp, 72; assoc prof med, 56-68, PROF EXP MED, SCH MED, UNIV MD, BALTIMORE CITY, 69-, PROF PATH, 71- Concurrent Pos: Fel path, Baylor Col Med, 38-40; lab consult, Far East Command, US Army, 46-49, consult, Surgeon Gen, 60-68 & 73-; consult, WHO, 61-62 & 69-71; dir comn malaria, Armed Forces Epidemiol Bd, 67-69; chmn, US Army Med Res & Develop Command Adv Panel, 73- Mem: Am Asn Immunologists; AMA; fel Am Col Physicians. Res: Infectious diseases; immunology. Mailing Add: Dept of Path Univ of Md Sch of Med Baltimore MD 21201

TIGGES, JOHANNES, b Rietberg, Ger, July 7, 31; m 59; c 2. NEUROANATOMY. Educ: Univ Münster, PhD(zool), 61. Prof Exp: Res assoc neuroanat, Max Planck Inst Brain Res, 61-62; res assoc vision in primates, Yerkes Primate Labs, Fla, 62-63; res assoc neuroanat & physiol, Max Planck Inst Brain Res, 63-65; asst prof, 67-75, ASSOC PROF ANAT, EMORY UNIV, 75-, NEUROANATOMIST, YERKES PRIMATE RES CTR, 66- Res: Light and electron microscopy of primate visual system. Mailing Add: Yerkes Primate Res Ctr Emory Univ Atlanta GA 30322

TIGNER, MAURY, b Middletown, NY, Apr 22, 37; m 60; c 2. HIGH ENERGY PHYSICS. Educ: Rensselaer Polytech Inst, BS, 58; Cornell Univ, PhD(physics), 63. Prof Exp: Res assoc, 63-68, SR RES ASSOC PHYSICS, CORNELL UNIV, 68- Mem: Am Phys Soc; Am Vacuum Soc. Res: Design and development of new particle accelerators and improvement of existing designs. Mailing Add: Lab of Nuclear Studies Cornell Univ Ithaca NY 14850

TIHEN, JOSEPH ANTON, b Harper, Kans, Nov 20, 18; m 40; c 7. ZOOLOGY. Educ: Univ Kans, AB, 40; Univ Rochester, PhD(zool), 45. Prof Exp: Asst instr zool, Univ Kans, 40-41; asst instr, Univ Rochester, 41-44, res assoc mouse genetics unit, Manhattan Proj, 44-46; asst prof zool, Tulane Univ, 46-47; asst prof, Univ Fla, 50-57, res assoc, AEC Proj, Sch Med, 57-58; asst prof zool, Univ Ill, 58-61; asst prof, 61-65, assoc prof biol, 65-70, PROF BIOL, UNIV NOTRE DAME, 70- Mem: Am Soc Zool; Soc Study Amphibians & Reptiles; Soc Syst Zool; Soc Study Evolution; Am Soc Ichthyologists & Herpetologists. Res: Systematics, Cenozoic paleontology and phylogeny of reptiles and amphibians. Mailing Add: Dept of Biol Univ of Notre Dame Notre Dame IN 46556

TIKSON, MICHAEL, b Campbell, Ohio, Nov 22, 24; m 57; c 4. MATHEMATICS. Educ: Youngstown Univ, BS, 48; Lehigh Univ, MA, 49; Mass Inst Technol, MS, 56. Prof Exp: Instr math, Lehigh Univ, 49-50; guest worker, Nat Bur Standards, DC, 51-52; sr mathematician, Wright Air Develop Div, US Air Force Res & Develop Command, 52-56, chief anal sect, Digital Comput Br, 56-58, br chief, 58-60; consult & head digital comput ctr, 60-66, assoc mgr, Systs & Electronics Dept, Columbus Labs, 66-70, mgr comput systs & applns, 70-74, MGR, COMPUT & INFO SYSTS DEPT, COLUMBUS LABS, BATTELLE MEM INST, 74- Mem: Simulation Coun; Asn Comput Mach. Res: Management of research and operations in computer and information systems. Mailing Add: Battelle Mem Inst Columbus Labs 505 King Ave Columbus OH 43201

TIKTOPOULOS, GEORGE S, b Salonica, Greece, Oct 15, 31. THEORETICAL HIGH ENERGY PHYSICS. Educ: Univ Chicago, MS, 61, PhD(physics), 63. Prof Exp: Res assoc physics, Princeton Univ, 63-65, asst prof, 65-72; ASSOC PROF PHYSICS, UNIV CALIF, LOS ANGELES, 72- Mailing Add: Dept of Physics Univ of Calif Los Angeles CA 90024

TILDEN, JAMES WILSON, b Philo, Calif, Dec 31, 04; m 43; c 3. ENTOMOLOGY. Educ: San Jose State Col, AB, 42; Stanford Univ, MA, 47, PhD(biol), 48. Prof Exp: From asst prof to prof, 48-73, EMER PROF ENTOM, SAN JOSE UNIV, 73- Mem: Entom Soc Am; Lepidop Soc; Soc Syst Zool. Res: Insect ecology; habits of insects, especially food relationships; taxonomy and distribution of butterflies, especially Lycaenidae and Hesperiidae; insect morphology; distribution and life history of Coleoptera. Mailing Add: Dept of Biol Sci San Jose Univ San Jose CA 95112

TILDON, J TYSON, b Baltimore, Md, Aug 7, 31; m 55; c 2. BIOCHEMISTRY. Educ: Morgan State Col, BS, 54; Johns Hopkins Univ, PhD(biochem), 65. Prof Exp: Res asst chem, Sinai Hosp, Baltimore, Md, 54-59; asst prof, Goucher Col, 67-68; res asst prof biochem & pediat, 68-71, assoc prof, 71-74, PROF PEDIAT, SCH MED, UNIV MD, BALTIMORE CITY, 74-, ASSOC PROF BIOCHEM, 72-, DIR PEDIAT RES, 70- Concurrent Pos: Helen Hay Whitney fel biochem, Brandeis Univ, 65-67; lectr, Antioch Col, Baltimore Campus, 72- Mem: AAAS; Am Soc Biol Chemists; Am Soc Neurochem; Am Chem Soc; Tissue Cult Asn. Res: Developmental biochemistry and control processes. Mailing Add: Dept of Pediat Univ of Md Sch of Med Baltimore MD 21201

TILFORD, SHELBY GRANT, molecular physics, atomic physics, see 12th edition

TILL, HARRIS RAYMOND, JR, polymer chemistry, organic chemistry, see 12th edition

TILL, JAMES EDGAR, b Lloydminster, Sask, Aug 25, 31; m 59; c 3. BIOPHYSICS, CELL BIOLOGY. Educ: Univ Sask, BA, 52, MA, 54; Yale Univ, PhD(biophys), 57. Prof Exp: Biophysicist, 57-69, HEAD BIORES DIV, ONT CANCER INST, 69-; PROF BIOPHYS, UNIV TORONTO, 65- Concurrent Pos: Res fel microbiol, Connaught Med Res Labs, 56-57. Honors & Awards: Gairdner Found Award, 69. Mem: Biophys Soc; Can Soc Cell Biol; Soc Exp Hemat; Am Asn Cancer Res; fel Royal Soc Can. Res: Cellular differentiation. Mailing Add: Ont Cancer Inst Biores Div 500 Sherbourne St Toronto ON Can

TILL, MICHAEL JOHN, b Independence, Iowa, July 30, 34; m 67. PEDODONTICS. Educ: Univ Iowa, DDS, 61, MS, 63; Univ Pittsburgh, MEd & PhD(higher educ), 70. Prof Exp: Instr pedodontics, Univ Iowa, 61-63; pedodontist, Eastman Inst, Stockholm, Sweden, 63-64; asst prof, Royal Dent Col, Denmark, 64-66; asst prof, Univ Pittsburgh, 66-70; PROF PEDODONTICS, UNIV MINN, MINNEAPOLIS, 70- Mem: Am Dent Asn; Int Asn Dent Res; Am Educ Res Asn; Am Soc Dent for Children; Am Acad Pedodont. Res: Dental educational research. Mailing Add: Univ of Minn Sch of Dent Minneapolis MN 55455

TILLAY, ELDRID WAYNE, b Yerington, Nev, Feb 26, 25; m 53; c 2. INORGANIC CHEMISTRY. Educ: Pac Union Col, BA, 50; Stanford Univ, MS, 52; La State Univ, PhD(inorg chem), 67. Prof Exp: Res asst chem, Stanford Univ, 52-57; instr, Sacramento City Col, 57-60; from asst prof to assoc prof, 60-72, PROF CHEM, PAC UNION COL, 72-, HEAD DEPT, 74- Mem: Am Chem Soc. Res: Organometallic chemistry and bio-inorganic chemistry. Mailing Add: Dept of Chem Pacific Union College Angwin CA 94508

TILLER, CALVIN OMAH, b Richmond, Va, June 22, 25; m 52; c 2. PHYSICS. Educ: Col William & Mary, BS, 48; Syracuse Univ, MS, 50. Prof Exp: Qual engr, Eastman Kodak Co, 50-51; supvr, Optical Eng Dept, Otis Elevator Co, 51-55; physicist, Titmus Optical Co, 55-56; sr res physicist, Va Inst Sci Res, 56-68; RES SCIENTIST, RES & DEVELOP CTR, PHILIP MORRIS, INC, 68- Honors & Awards: J Sheldon Horsley Award, Va Acad Sci, 60; IR100 Award, Indust Res, Inc, 72. Mem: Electron Micros Soc Am; Am Phys Soc; Am Vacuum Soc. Res: Solid state physics of thin metallic films; electron microscopy and diffraction; physical optics; thermal analysis of tobacco. Mailing Add: Phys Res Dept Philip Morris Res & Dev Ctr Richmond VA 23224

TILLER, RALPH EARL, b Birmingham, Ala, Nov 16, 25; m 49; c 3. PEDIATRICS. Educ: Birmingham-Southern Col, BS, 47; Tulane Univ, MD, 51; Am Bd Pediat, dipl, 56. Prof Exp: Intern, Fitzsimmons Army Hosp, Denver, 51-52; resident pediat, Tulane Univ, 53-54, chief resident, 54-55; pvt pract, Columbus, Ga, 55-67; assoc prof, 67-71, PROF PEDIAT, SCH MED, UNIV ALA, BIRMINGHAM, 71- Concurrent Pos: Consult, Martin Army Hosp, US Army Med Corps, Ft Benning, Ga, 59-67; dir, State Crippled Children's Seizure Clin, Columbus, 60-67; chief pediat serv, Med Ctr, Columbus, 62-67; consult, Muscogee Health Dept, 62-67; dir outpatient clin, Children's Hosp, Birmingham, 67-69, dir inpatient teaching serv, Cystic Fibrosis Care Teaching & Res Ctr & Pediat Chest Dis Clin, 69- Mem: Am Acad Pediat; Am Thoracic Soc. Res: Tuberculosis. Mailing Add: Children's Hosp 1601 Sixth Ave S Birmingham AL 35233

TILLER, RICHARD EDWARD, b Washington, DC, Nov 11, 18; m 48. ENVIRONMENTAL HEALTH. Educ: Univ Md, BS, 41, MS, 42, PhD(zool), 47. Prof Exp: Biologist, State Dept Res & Educ, Md, 43-47 & 51-56; instr zool, Univ Md, 47-48; res biologist, US Fish & Wildlife Serv, 48-51; analyst, Opers Res Off, Johns Hopkins Univ, 56-61 & Res Anal Corp, 61-68; assoc prof estuarine ecol, Charles County Community Col, 71-72; NATURAL RESOURCES/COMPREHENSIVE HEALTH PLANNER, TRI-COUNTY COUN SOUTHERN MD, 72- Res: Aquatic ecology, water pollution—land use relationships, compilation and analysis of data for implementation of regional community and environmental health programs. Mailing Add: Tri-Co Coun Southern Md PO Box 301 Waldorf MD 20601

TILLERY, BILL W, b Muskogee, Okla, Sept 15, 38; m 59; c 2. SCIENCE EDUCATION. Educ: Northeastern Okla State Univ, BS, 60; Univ Northern Colo, MA, 65, EdD(sci educ), 67. Prof Exp: Teacher pub schs, Okla & Colo, 60-64; res assoc sci, Univ Northern Colo, 66-67; asst prof sci educ, Fla State Univ, 67-69; assoc prof & dir sci ctr, Univ Wyo, 69-73; assoc prof, 73-75, PROF SCI EDUC, ARIZ STATE UNIV, 75- Mem: Nat Sci Teachers Asn; Asn Educ Teachers Sci; Nat Asn Res Sci Teaching. Res: The use of science materials for motivation and intellectual development of undergraduate non-science majors. Mailing Add: Dept of Physics Ariz State Univ Tempe AZ 85281

TILLES, HARRY, b Buffalo, NY, Mar 22, 23; m 48; c 4. ORGANIC CHEMISTRY. Educ: Univ Buffalo, BA, 48; Univ Calif, PhD(org chem), 51. Prof Exp: Res chemist org synthesis, Nat Aniline Div, Allied Chem & Dye Corp, 51-53; res chemist, 53-59, group leader indust chem group, 59-62, prof officer, Nat Cancer Inst contract, 62-64, sr res chemist, 64-68, res assoc, Western Res Ctr, 68-74, SR RES ASSOC, RICHMOND RES CTR, STAUFFER CHEM CO, 74- Mem: Am Chem Soc. Res: Synthesis of organic chemicals for agricultural screening; optimization of chemical synthesis processes. Mailing Add: Stauffer Chem Co Res Ctr 1200 S 47th St Richmond CA 94803

TILLETT, STEPHEN SZLATENYI, systematic botany, see 12th edition

TILLETT, WILLIAM SMITH, medicine, deceased

TILLEY, AUBRA EVERETT, b Marshall, Ark, Feb 8, 18; m 40; c 2. GEOPHYSICS, ELECTRICAL ENGINEERING. Educ: Univ Okla, BSEE, 46. Prof Exp: Elec engr, Phillips Petrol Co, 40-41 & 46-51; supvr auto pilot develop, NAm Aviation, Inc, 51-53; supvr geophysics develop, Calif Res Corp, 53-62; chief engr, Chevron Res Corp, Stand Oil Co Calif, 62-66, mgr systs & eng, Chevron Oil Co, 66-68; PRES, INPUT/OUTPUT, INC, 68- Mem: Soc Explor Geophys; Inst Elec & Electronics Engrs; Audio Eng Soc; Am Soc Oceanog. Res: Geophysical instrumentation; geophysical data processing systems; data displays; applied geophysics; analog computers; digital computing systems. Mailing Add: Input/Output Inc 8009 Harwin Dr Houston TX 77036

TILLEY, DAVID RONALD, b Fuquay Springs, NC, Mar 10, 30; m 65; c 1. NUCLEAR PHYSICS. Educ: Univ NC, BS, 52; Vanderbilt Univ, MS, 54; Johns Hopkins Univ, PhD(nuclear physics), 58. Prof Exp: Jr instr physics, Johns Hopkins Univ, 53-58; res assoc nuclear physics, Duke Univ, 58-61, asst prof, 61-66; assoc prof, 66-72, PROF PHYSICS, NC STATE UNIV, 72- Concurrent Pos: Staff physicist, Triangle Univs Nuclear Lab, 66- Mem: Am Phys Soc. Res: Gamma ray spectroscopy; nuclear reactions. Mailing Add: Dept of Physics NC State Univ Raleigh NC 27607

TILLEY, DONALD E, b Flushing, NY, July 6, 25; Can citizen; m 48; c 3. PHYSICS. Educ: McGill Univ, BSc, 48, PhD(physics), 51. Prof Exp: Res assoc physics, Radiation Lab, McGill Univ, 51-52; from asst prof to assoc prof, 52-57, head dept physics, 61-71, PROF PHYSICS, ROYAL MIL COL, QUE, 57-, DEAN SCI & ENG, 69- Mem: Am Phys Soc; Can Asn Physicists. Res: Physics of dielectrics; nuclear reactions; radioactive isotopes. Mailing Add: 24 St Denis St St Jean PQ Can

TILLEY, GEORGE LEVIS, b Doylestown, Pa, July 10, 38. PHYSICAL CHEMISTRY. Educ: Calif State Polytech Col, San Luis Obisop, BS, 61; Purdue Univ, PhD(chem), 67. Prof Exp: RES CHEMIST, UNION OIL CO, CALIF, 67- Mem: Am Chem Soc. Res: Industrial catalysis. Mailing Add: Union Oil Res Ctr PO Box 76 Brea CA 92621

TILLEY, JAMES NOEL, organic chemistry, see 12th edition

TILLEY, JOHN LEONARD, b New York, NY, June 4, 28; m 51. MATHEMATICS. Educ: Univ Pa, BS, 50; Univ Fla, MEd, 54, PhD(math), 61. Prof Exp: Teacher high sch, Fla, 54-56 & St Petersburg Jr Col, 56-58; instr math, Univ Fla, 60-61; from asst prof to assoc prof, Clemson Univ, 61-64; assoc prof, 64-69, actg head dept math, 71-72, PROF MATH & DIR S D LEE HONS PROG, MISS STATE UNIV, 69- Concurrent Pos: Mem, Nat Collegiate Hons Coun. Mem: Nat Coun Teachers Math; Math Asn Am; Soc Indust & Appl Math. Res: Classical methods of applied mathematics. Mailing Add: Dept of Math Miss State Univ Mississippi State MS 39762

TILLEY, STEPHEN GEORGE, b Lima, Ohio, July 21, 43; m 65; c 1. POPULATION BIOLOGY, HERPETOLOGY. Educ: Ohio State Univ, BS, 65; Univ Mich, Ann Arbor, MS, 67, PhD(zool), 70. Prof Exp: ASST PROF BIOL SCI, SMITH COL, 70-; RES ASSOC, DEPT VERT ZOOL, SMITHSONIAN INST, 74- Honors & Awards: Stoye Award, Am Soc Ichthyol & Herpet, 70. Mem: Soc Study Evolution; Am Soc Ichthyol & Herpet; Soc Study Amphibians & Reptiles; AAAS; Ecol Soc Am. Res: Population biology and evolution of amphibians, especially desmognathine salamanders; genetic structures of populations. Mailing Add: Dept of Biol Sci Smith Col Northampton MA 01060

TILLING, ROBERT INGERSOLL, b Shanghai, China, Nov 26, 35; US citizen; m 62; c 2. GEOLOGY. Educ: Pomona Col, BA, 58; Yale Univ, MS, 60, PhD(geol), 63. Prof Exp: Geologist, US Geol Surv, Va, 62-72, Hawaiian Volcano Observ, 72-75, SCIENTIST-IN-CHG, HAWAIIAN VOLCANO OBSERV, US GEOL SURV, 75- Mem: Geol Soc Am; Mineral Soc Am; Geochem Soc. Res: Igenous petrology and volcanology. Mailing Add: Hawaiian Volcano Observ Hawaii Volcanoes Nat Park HI 96718

TILLITSON, EDWARD WALTER, b Charlevoix Co, Mich, Jan 13, 03; m 30; c 3. DENTAL MATERIALS, CHEMICAL ENGINEERING. Educ: Univ Mich, BS, 29, MS, 32. Prof Exp: Res & develop chem engr, Whiting Swenson Co, 29-31; chief chemist, Iodent Chem Co, 32-35; sr res chemist, Parke Davis & Co, 35-42; head develop labs, Gelatin Prod Div, R P Scherer Corp, 42-46; assoc prof chem eng, Wayne State Univ, 46-63; res assoc, Sch Dent, Univ Mich, 63-73; CONSULT ENGR, 73- Concurrent Pos: Consult, 46- Mem: Am Chem Soc; Int Asn Dent Res. Res: Materials science; polymers; casualty investigation; custom research instrumentation; friction and wear; stress analysis. Mailing Add: 255 Ridgemont Rd Grosse Pointe Farms MI 48236

TILLMAN, ALLEN DOUGLAS, b Rayville, La, June 9, 16; m 45; c 5. ANIMAL HUSBANDRY. Educ: Southwestern La Univ, BS, 40; La State Univ, MS, 42; Pa State Univ, PhD, 52. Prof Exp: Asst, La State Univ, 40-42; instr animal nutrit, Pa State Univ, 46-48; asst prof, La State Univ, 48-52; from assoc prof to prof animal husb, Okla State Univ, 52-73; VIS PROF ANIMAL HUSB, UNIV GADJAH MADA & FIELD OFFICER, ROCKEFELLER FOUND, INDONESIA, 73- Concurrent Pos: Gen Educ Bd fel, 51-52; res partic, Oak Ridge Inst Nuclear Studies, 56-57; consult, AEC, 57-, gen chmn, Radioisotope Conf, 59; consult, USDA exchange team, USSR, 59; consult, Estab Labs for Food & Agr Orgn, Arg, 61 & nutrit study, Libya, 62; Fulbright lectr, Univ Col, Dublin, 62-63; Nat Feed Ingredients travel fel, Europe, 66; mem, Comt Animal Nutrit, Nat Res Coun-Nat Acad Sci, 67-73; Ford Found head animal prod div, EAfrica Agr & Res Orgn, Nairobi, Kenya, 69-71. Honors & Awards: Am Soc Animal Sci Award, 59; Tyler Award, 67. Mem: AAAS; Am Soc Animal Sci; Poultry Sci Asn; Am Inst Nutrit. Res: Metabolism of protein; energy and minerals; vitamin and mineral deficiencies; use of radioisotopes in mineral studies. Mailing Add: Rockefeller Found PO Box 63 Yogyakarta Indonesia

TILLMAN, LARRY JAUBERT, b Bay St Louis, Miss, Aug 20, 48; m 70; c 1. HISTOLOGY. Educ: Univ Miss, BA, 70, MS, 72, PhD(histol, electron micros), 74. Prof Exp: CHIEF ELECTRON MICROS, DEPT PATH, BROOKE ARMY MED CTR, 74- Concurrent Pos: Clin appointee, Dept Anat, Micros Anat Sect, Univ Tex Health Sci Ctr & Med Sch, San Antonio, 75- Mem: Soc Armed Forces Med Lab Scientists; Electron Micros Soc Am; AAAS. Res: Correlation of the fine structure of the cells comprising the uriniferous tubules of Gallus domesticus to their function; use of transmission and scanning electron microscopy in the diagnosis of renal and tumor disease. Mailing Add: Electron Micros Sect Anat Path Brooke Army Med Ctr Ft Sam Houston TX 78234

TILLMAN, RICHARD MILTON, b Muskogee, Okla, Sept 7, 28; m 48; c 2. ORGANIC CHEMISTRY. Educ: Southern Methodist Univ, BS, 52, MS, 53. Prof Exp: From asst res chemist to sr res chemist, 53-61, res group leader, 61-63, tech asst, 63-64, supvry res scientist, 64-67, mgr, Plant Foods Res Div, 67-72, assoc mgr petrol prod div, Res & Develop, 72-75, MGR RES SERV DIV, RES & DEVELOP, CONTINENTAL OIL CO, 75- Mem: Sr mem Am Chem Soc; Soc Automotive Engrs. Res: Hydrocarbon fuels; inorganic colloids; petroleum-based specialties; plant foods; phosphate rock; phosphoric acid; sulfur; inorganic fluorides; nitrogen compounds; agricultural chemicals and specialties; environmental chemistry. Mailing Add: 2400 Wildwood Ponca City OK 74601

TILLMAN, STEPHEN JOEL, b Springfield, Mass, Mar 31, 43; m 65; c 2. MATHEMATICS. Educ: Brown Univ, ScB, 65, PhD(math), 70. Prof Exp: Instr math, Brown Univ, 69-70; asst prof, 70-75, ASSOC PROF MATH, WILKES COL, 75- Mem: Am Math Soc; Math Asn Am; Opers Res Soc Am. Res: Teaching and developing additional courses in operations research and related areas. Mailing Add: Dept of Math Wilkes Col Wilkes-Barre PA 18703

TILLMANNS, EMMA-JUNE H, b New York, NY, June 4, 19. INFORMATION SCIENCE. Educ: Hunter Col, AB, 41; NY Univ, MS, 47, PhD(org chem), 54. Prof Exp: Chemist, Burroughs Wellcome Co, Inc, 41-53; sr chemist, Redstone Res Div, Rohm & Haas Co, 53-56; asst ed, Org Indexing Dept, Chem Abstr, 56-60, assoc ed, 60; res chemist, Info Sect, Chem Res Dept, Atlas Chem Indust, Inc, 60-67, sr res chemist, 67-72, res info sect, ICI Am, Inc, 72-74, SUPVR BIOMED INFO, RES INFO SERV, ICI UNITED STATES, INC, 74- Mem: Am Chem Soc; Drug Info Asn; Am Soc Info Sci. Res: Olefin-nitrile reaction; organic nomenclature. Mailing Add: PO Box 1725 Wilmington DE 19899

TILLOTSON, DONALD BEARSE, b Brooklyn, NY, July 5, 14; m 40; c 2. MATHEMATICS. Educ: Eastern Nazarene Col, AB, 36; Boston Univ, MA, 37; Univ Kans, PhD(math educ), 62. Prof Exp: Instr, Eastern Nazarene Acad, 37-41; instr, Spring Arbor Jr Col, 41-43; instr math & physics, Northwest Nazarene Col, 43-45, head math dept, 45-59; asst, Univ Kans, 60-61; asst prof, Baker Univ, 61-62; HEAD MATH DEPT, NORTHWEST NAZARENE COL, 62- Mem: Math Asn Am. Mailing Add: Dept of Math Northwest Nazarene Col Nampa ID 83651

TILLOTSON, JAMES E, b Cambridge, Mass, Feb 9, 29; m 56; c 2. FOOD SCIENCE. Educ: Harvard Univ, AB, 53; Boston Univ, MA, 56; Mass Inst Technol, PhD(food sci), 64; Univ Del, MBA, 69. Prof Exp: Teacher, Manter Hall Sch, 57-63; res asst nutrit & food sci, Mass Inst Technol, 61-63; res chemist indust & biochem dept, E I du Pont de Nemours & Co, 64-65, develop dept, 65-66, tech rep indust & biochem dept, 66-69; DIR RES & DEVELOP, OCEAN SPRAY CRANBERRIES, INC, 69- Mem: Fel Am Inst Chem; Am Chem Soc; prof mem Inst Food Technol; Soc Nutrit Educ. Res: Commercial development of food products and agricultural chemicals; research management; technical forecasting; government regulation of agribusiness; sweeteners. Mailing Add: 240 Forest Ave Cohasset MA 02025

TILLOTSON, JAMES GLEN, b Brandon, Man, July 20, 23; m 48; c 3. PHYSICS. Educ: Univ Man, BSc, 45; Univ Western Ont, MSc, 47. Prof Exp: Asst prof physics, Univ NB, 47-53; dir appl physics sect, Can Armament Res & Develop Estab, 53-55; PROF PHYSICS, ACADIA UNIV, 55- Concurrent Pos: Consult, NB Dept Health, 51-52; Defence Res Bd Can grant, 57-59. Mem: Inst Elec & Electronics Engrs; Can Asn Physicists. Res: Non-linear vibrations; acoustic radiation. Mailing Add: Dept of Physics Acadia Univ Wolfville NS Can

TILLSON, ALBERT HOLMES, b Washington, DC, May 27, 10; m 45; c 3. HISTOLOGY, CRYSTALLOGRAPHY. Educ: Col Wooster, BA, 34; Univ Md, MS, 35, PhD(plant morphol), 38. Prof Exp: Asst bot, Univ Md, 34-37; jr morphologist, USDA, 37-38, jr botanist, Div Plant Explor & Introd, Bur Plant Indust, 38-42; microanalyst, Food & Drug Admin, Fed Security Agency, 46-53, Dept Health, Educ & Welfare, 53-68; MICROBIOLOGIST, SPEC TESTING & RES LAB, DRUG ENFORCEMENT ADMIN, US DEPT JUSTICE, 68- Mem: Asn Off Anal Chem. Res: Microscopy and optical crystallography of drugs; microscopic identification of drug tablets and capsules. Mailing Add: DEA Spec Testing Lab 7704 Old Springhouse Rd McLean VA 22101

TILLSON, DAVID STANLEY, b Albany, NY, Mar 5, 28; m 54; c 3. ANTHROPOLOGY. Educ: Bates Col, AB, 49; Syracuse Univ, MA, 55, DSS, 57. Prof Exp: Prog asst community develop, Int Coop Admin, 57-61; asst prof anthrop & sociol, Ohio Univ, 61-62; PROF ANTHROP, STATE UNIV NY COL BROCKPORT, 62- Mem: Fel Am Anthrop Asn. Res: Religion and community and national development, especially economic development. Mailing Add: Dept of Anthrop State Univ of NY Col Brockport NY 14420

TILLSON, HENRY CHARLES, b Philadelphia, Pa, Sept 16, 23; m 50; c 2. RUBBER CHEMISTRY. Educ: Mass Inst Technol, SB, 44; Pa State Univ, MS, 48, PhD(org chem), 51. Prof Exp: Res chemist, Res Ctr, 50-52, tech rep, 52-54, RES CHEMIST, RES CTR, HERCULES INC, 54- Mem: AAAS; Am Chem Soc. Res: Organic nitrogen compounds-nitramines; emulsion polymerization vinyl and condensation polymers, especially protective coatings; rubber compounding and crosslinking. Mailing Add: Hercules Inc Research Center Wilmington DE 19899

TILLY, LAURENCE JOHN, biology, aquatic ecology, see 12th edition

TILMANN, JEAN PAUL, b Beal City, Mich, Apr 13, 28. GEOGRAPHY. Educ: Cent Mich Univ, BA, 57, MA, 63; Univ Mich, PhD(geog), 69. Prof Exp: Jr high sch teacher, Blessed Sacrament Sch, Mich, 52-58; prin, St Basil Sch, 58-62; chmn dept geog, 65-74, ASST PROF GEOG, AQUINAS COL, 65-, DIR ENVIRON STUDIES OFF, RESOURCES & PROGS, 72- Concurrent Pos: Consult, W K Kellogg Found, 71, W K Kellogg Found grant, 72-74; mem comt studies, Aquinas Col, 71-73, mem libr comt, 72-; mem global ministry comt, Grand Rapids Dominican Sisters, 72- Mem: Asn Am Geogr; Am Geog Soc. Res: History and philosophy of geographical thought; building models for environmental studies; role of the college community-neighborhood development; environmental education. Mailing Add: Dept of Geog Aquinas Col Grand Rapids MI 49506

TILSON, BRET RANSOM, b Yuba City, Calif, May 19, 37. MATHEMATICS. Educ: Mass Inst Technol, BS, 60; Univ Calif, Berkeley, PhD(math), 69. Prof Exp: Asst prof math, Columbia Univ, 69-74; ASST PROF MATH, QUEEN'S COL, NY, 74- Res: Decomposition and complexity of finite semigroups; automata theory. Mailing Add: Dept of Math Queen's Col Flushing NY 11367

TILTON, BERNARD ELLSWORTH, b Hanford, Calif, Oct 3, 23; m 46; c 3. PHARMACOLOGY. Educ: Pac Union Col, BA, 46; Loma Linda Univ, MD, 48; Univ Calif, Los Angeles, MS, 56, PhD, 60. Prof Exp: Asst prof, 53-61, ASSOC PROF PHARMACOL, SCH MED, LOMA LINDA UNIV, 61- Concurrent Pos: Fel clin pharmacol, Univ Mich, 65-66. Mem: AAAS; Am Soc Clin Pharmacol & Therapeut. Res: Mode of action for neuromuscular blocking agents; clinical cardiovascular pharmacology. Mailing Add: Barton & Anderson Loma Linda CA 92354

TILTON, GEORGE ROBERT, b Danville, Ill, June 3, 23; m 48; c 4. GEOCHEMISTRY. Educ: Univ Ill, BS, 47; Univ Chicago, PhD(chem), 51. Prof Exp: Asst, Univ Chicago, 47-51; mem staff, Dept Terrestrial Magnetism, Carnegie Inst, 51-56, phys chemist, Geophys Lab, 56-65; PROF GEOCHEM, UNIV CALIF, SANTA BARBARA, 65- Concurrent Pos: Assoc ed, Geochimica Cosmochimica Acta, 74- Mem: AAAS; Geochem Soc; fel Am Geophys Union; fel Geol Soc Am; Meteoritical Soc. Res: Distribution of uranium, thorium and lead; isotopic composition of lead in terrestrial and meteoritic materials; geologic age of minerals. Mailing Add: Dept of Geol Sci Univ of Calif Santa Barbara CA 93106

TILTON, JAMES EARL, b Decatur, Ill, Aug 1, 38; m 60; c 3. ANIMAL PHYSIOLOGY. Educ: Ill State Norm Univ, BS, 61; Okla State Univ, MS, 64, PhD(animal breeding), 66. Prof Exp: Asst prof, 65-68, ASSOC PROF ANIMAL SCI, NDAK STATE UNIV, 68- Mem: Am Soc Animal Sci. Res: Embryonic mortality in adrenalectomized ewes; estrous control in the ovine. Mailing Add: Dept of Animal Sci NDak State Univ Fargo ND 58102

TILTON, RICHARD C, b Quincy, Mass, Mar 24, 36; m 58; c 3. MICROBIOLOGY. Educ: Tufts Univ, BS, 58; Univ Mass, MS, 63, PhD(microbiol), 65. Prof Exp: Instr biol & Off Naval Res fel, Boston Univ, 65-66; asst prof microbiol, 66-68, asst prof lab med, 68-72, chief clin microbiol, 68-72, ASSOC PROF LAB MED, SCH MED, UNIV CONN, 72- Concurrent Pos: Univ Res Coun res fel, 66-67; dir microbiol div, McCook-Univ Hosp, 72- Mem: AAAS; Am Soc Microbiol. Res: Metabolism of metals and pyruvate by fecal streptococci; sulfur-oxidizing bacteria; clinical microbiology. Mailing Add: Dept of Lab Med Univ of Conn Sch of Med Hartford CT 06112

TIMASHEFF, SERGE NICHOLAS, b Paris, France, Apr 7, 26; nat US; m 53; c 1. PHYSICAL BIOCHEMISTRY. Educ: Fordham Univ, BS, 46, MS, 47, PhD(chem), 51. Prof Exp: Instr chem, Fordham Univ, 47-49, lectr, 49-50; res fel, Calif Inst Technol, 51 & Yale Univ, 51-55; prin phys chemist, Eastern Regional Res Lab, USDA, Pa, 55-66, head pioneering res lab, Mass, 66-73; PROF BIOCHEM, BRANDEIS UNIV, 66- Concurrent Pos: NSF sr res fel, Macromolecule Res Ctr, France, 59-60; adj prof, Drexel Inst Technol, 63-64; vis prof, Univ Ariz, 66; mem, Fordham Univ Coun, 68-; mem, Biophys-Phys Biochem Study Sect, NIH, 68-72; Guggenheim fel, Inst Molecular Biol, Paris, 72-73; vis prof, Univ Paris, 72-73; co-ed, Biol Macromolecules; exec ed, Archives of Biochem & Biophys. Honors & Awards: Am Chem Soc Award, 63 & 66, Arthur H Flemming Award, 64; Distinguished Serv Award, USDA, 65; Sci Achievement Award, Fordham Univ, 67. Mem: AAAS; Am Chem Soc; Am Soc Biol Chemists; Biophys Soc. Res: Structure and interactions of proteins and nucleic acids; physical methods of high polymer studies; solution thermodynamics of macromolecules. Mailing Add: Dept of Biochem Brandeis Univ Waltham MA 02154

TIMBERLAKE, JOSEPH WILLIAM, b Kansas City, Kans, Sept 5, 40; m 67. CLINICAL BIOCHEMISTRY. Educ: Univ Mo-Kansas City, BA, 63, MS, 69; Univ Kans Med Ctr, Kansas City, PhD(biochem), 74. Prof Exp: Technologist clin chem, Res Hosp & Med Ctr, 63-65; develop chemist, Univ Kans Med Ctr, Kansas City, 66-69; lab dir, Statlabs of Kans, Inc, 73-75; CLIN BIOCHEMIST, KANSAS CITY GEN HOSP-UNIV MO MED SCH, KANSAS CITY, 75- Mem: Am Asn Clin Chemists. Res: Developmental clinical biochemistry. Mailing Add: Dept of Path Kansas City Gen Hosp & Med Ctr 24th & Cherry St Kansas City MO 64108

TIMBERLAKE, WILLIAM EDWARD, b Washington, DC, May 2, 48; m 69; c 2. MYCOLOGY, DEVELOPMENTAL BIOLOGY. Educ: State Univ NY Col Forestry, BS, 70; State Univ NY Col Environ Sci & Forestry, MS, 72, PhD(biol), 74. Prof Exp: Assoc, Univ Geneva, 74; ASST PROF BIOL, WAYNE STATE UNIV, 74- Mem: AAAS; Am Soc Microbiol; Mycol Soc Am; Sigma Xi. Res: Molecular mechanisms of steroid hormone action in the water mold Achlya. Mailing Add: Dept of Biol Wayne State Univ Detroit MI 48202

TIMBLIN, LLOYD O, JR, b Denver, Colo, June 25, 27; m 50; c 1. APPLIED PHYSICS, PHYSICAL SCIENCE. Educ: Univ Colo, BS, 50; Univ Denver, MS, 67. Prof Exp: Physicist, 50-58, head Spec Invests Lab Sect, 58-63, chief Chem Eng Br, 63-70, CHIEF APPL SCI BR, DIV GEN RES, US BUR RECLAMATION, 70- Concurrent Pos: Mem Colo adv coun, Sem Environ Arts & Sci; chmn US team, US/USSR Joint Study Plastic Films & Soil Stabilizers, 75- Honors & Awards: Superior Performance Awards, US Dept Interior, 60, Sigma Xi, 60, 61 & Sigma Pi Sigma, 63. Mem: Am Phys Soc; Am Water Works Asn; Nat Asn Corrosion Engrs; Am Soc Testing & Mat; Ecol Soc Am. Res: Engineering physics; corrosion engineering; corrosion mechanisms; cathodic protection; protective coatings; radioisotopes applications; water quality and pollution control; applied ecology; materials analysis and development; water treatment and desalting; weed control. Mailing Add: Div Gen Res US Bur Reclamation Fed Off Bldg 1961 Stout Denver CO 80202

TIMELL, TORE ERIK, b Stockholm, Sweden, Mar 31, 21; nat US; m 47; c 3. ORGANIC CHEMISTRY. Educ: Royal Inst Technol, Sweden, ChemE, 46, lic, 48, DrTech(cellulose chem), 50. Prof Exp: Chief asst, Royal Inst Technol, Sweden, 46-50; res assoc chem, McGill Univ, 50-51; res assoc, State Univ NY Col Forestry, Syracuse, 51-52; Hibbert Mem fel, McGill Univ, 52, res assoc, 53-62; PROF FOREST CHEM, STATE UNIV NY COL ENVIRON SCI & FORESTRY, SYRACUSE, 62- Concurrent Pos: Chemist, Pulp & Paper Res Inst Can, 53-59, res group leader, 60-62. Mem: Am Chem Soc; Soc Wood Sci & Technol; Int Acad Wood Sci; Int Asn Wood Anat; Swed Chem Soc. Res: Chemistry of wood and bark; ultrastructure, cytology and physiology of wood and bark; reaction wood. Mailing Add: State Univ of NY Col of Environ Sci & Forestry Syracuse NY 13210

TIMIAN, ROLAND GUSTAV, b Langdon, NDak, Mar 5, 20; m 49; c 5. PLANT VIROLOGY. Educ: NDak Agr Col, BS, 49, MS, 50; Iowa State Col, PhD(plant path), 53. Prof Exp: Asst bot, NDak Agr Col, 47-49, fed agent pathologist, 49-50; path res asst, Iowa State Col, 50-53; PATHOLOGIST, AGR RES SERV, USDA, NDAK STATE UNIV, 53- Concurrent Pos: Tech adv barley, NCent Region, Agr Res Serv, USDA, 73- Mem: Am Phytopath Soc; Sigma Xi. Res: Cereal virus diseases; diseases in barley, especially virus diseases. Mailing Add: Dept of Plant Path NDak State Univ Fargo ND 58102

TIMIN, MITCHELL E, ecology, computer science, see 12th edition

TIMIRAS, PAOLA SILVESTRI, b Rome, Italy, July 21, 23; nat US; m 46; c 2. DEVELOPMENTAL PHYSIOLOGY, NEUROENDOCRINOLOGY. Educ: Univ Rome, MD, 47; Univ Montreal, PhD(exp med, surg), 52. Prof Exp: Asst prof exp med & surg, Univ Montreal, 50-51, asst prof physiol, 51-53; asst prof pharmacol, Univ Utah, 54-55; asst physiologist, 55-58, from asst prof to assoc prof, 58-67, PROF PHYSIOL, UNIV CALIF, BERKELEY, 67- Mem: AAAS; Endocrine Soc; Am Soc Pharmacol & Exp Therapeut; Am Physiol Soc; Geront Soc. Res: Endocrinology; environmental physiology; aging. Mailing Add: Dept of Physiol & Anat Univ of Calif Berkeley CA 94720

TIMKEN, RICHARD LUCIEN, zoology, ecology, deceased

TIMM, EUGENE ALVIN, b Neenah, Wis, Oct 3, 25; m 51; c 3. MICROBIOLOGY. Educ: Univ Wis, BS, 51, MS, 53, PhD(bact, virol), 55. Prof Exp: Asst virol, Univ Wis, 51-55; from assoc res virologist to sr res virologist, 55-63, dir biol qual control, 63-65, dir biol & pharmaceut control, 65-68, ASST DIR DIV QUAL CONTROL & GOVT REGULATIONS, PARKE, DAVIS & CO, 68- Concurrent Pos: Mem, Rev Comt, US Pharmacopoeia, 70- Mem: AAAS; Am Soc Microbiol; Pharmaceut Mfrs Asn; NY Acad Sci. Res: Viral immunology; virus inactivation and vaccines; serological tests; factors affecting infant immune response. Mailing Add: Parke Davis & Co PO Box 118 Detroit MI 48232

TIMM, GERALD WAYNE, b Brandon, Minn, Dec 9, 40. BIOMEDICAL ENGINEERING, ELECTRICAL ENGINEERING. Educ: Univ Minn, Minneapolis, BEE, 63, MS, 65, PhD(elec eng), 67. Prof Exp: ASSOC PROF NEUROL, MED SCH, UNIV MINN, MINNEAPOLIS, 67- Concurrent Pos: Res fel elec eng, Univ Minn, Minneapolis, 67, NIH trainee, 70-72, NIH grant, 74-; mem grad fac, Univ Minn, 71-; consult, Baylor Col Med, 72-, Am Med Systs, Inc, 73- & Purdue Univ, 74- Mem: AAAS; Inst Elec & Electronics Eng; sr mem Instrument Soc Am; NY Acad Sci. Res: Lower urinary tract and reproductive systems function; investigations of instrumentation systems to diagnose and treat impaired genito-urinary and gastrointestinal function. Mailing Add: Dept of Neurol Univ of Minn Med Sch Minneapolis MN 55455

TIMM, HERMAN, b Kenosha, Wis, Aug 26, 26; m 49; c 7. AGRONOMY. Educ: Cornell Univ, BS, 51; Mich State Univ, MS, 53; Pa State Univ, PhD(agron), 56. Prof Exp: SPECIALIST, UNIV CALIF, DAVIS, 56- Mem: Soil Sci Soc Am; Am Soc Hort Sci. Res: Ecology and physiology of the potato; changes in cultural practices for high yields and better quality control of crops; agricultural recycling of waste products; effect on soil-water-plant relationships. Mailing Add: Dept of Veg Crops Univ of Calif Davis CA 95616

TIMMA, DONALD LEE, b Lebanon, Kans, Sept 1, 22; m 45; c 1. CHEMISTRY. Educ: Kans State Col, BS, 44; Ohio Univ, PhD(chem), 49. Prof Exp: Spectrographer, Tenn Eastman Co, 44-45; asst chem, Ohio Univ, 46-48; spectrographer, Monsanto Chem Co, 45-46, res physicist, Mound Lab, 49-50, res group leader, 50-51, res sect chief, 51-57; dir anal chem & methods develop, 57-67, exec dir control lab, 67, VPRES QUAL CONTROL, MEAD JOHNSON & CO, 67- Mem: Am Chem Soc. Res: Analytical instrumentation; emission and absorption spectroscopy. Mailing Add: Qual Control Mead Johnson & Co Evansville IN 47712

TIMME, ROBERT WILLIAM, b Victoria, Tex, July 22, 40; m 62; c 2. APPLIED PHYSICS, Educ: Tex A&M Univ, BS, 62; Rice Univ, MA, 69, PhD(physics), 70. Prof Exp: Res physicist, Ames Res Lab, NASA, 62-65, aerospace engr, Manned Spacecraft Ctr, 65-66; res assoc solid state physics, Rice Univ, 66-70; prin engr electromagnetics, Lockheed Electronics Co, Inc, 70-71; RES PHYSICIST, NAVAL RES LAB, 71- Concurrent Pos: Mem, Transducer Mat Comt, Naval Sea Systs Command, 73-; mem oceanology adv group, Naval Res Lab, 75-; prof oceanology, Fla Inst Technol, 75- Honors & Awards: NASA Achievement Awards, Manned Spacecraft Ctr, 66 & Hq, 73; Res Publ Award, Naval Res Lab, 76. Mem: Am Phys Soc; Acoust Soc Am. Res: Characterization of the effects of stress, temperature and time on the piezoelectric, magnetostricture, elastic and acoustic properties of materials applicable to sonar systems. Mailing Add: Naval Res Labs PO Box 8337 Orlando FL 32806

TIMMER, KATHLEEN MAE, b Ellsworth, Mich, July 21, 42. ALGEBRA. Educ: Calvin Col, BS, 64; Purdue Univ, MS, 66; Colo State Univ, PhD(math), 72. Prof Exp: Instr math, Calvin Col, 66-68; ASST PROF MATH, JACKSONVILLE UNIV, 72- Mem: Am Math Soc. Mailing Add: Box 205 Jacksonville Univ Jacksonville FL 32211

TIMMER, LAVERN WAYNE, b West Olive, Mich, Aug 16, 41; m 63; c 2. PLANT PATHOLOGY. Educ: Mich State Univ, BS, 63; Univ Calif, Riverside, PhD(plant path), 69. Prof Exp: Purdue Univ-Ford Found fel plant path, Latin Am, 66-68; asst prof, 70-74, ASSOC PROF PLANT PATH, CITRUS CTR, TEX A&I UNIV, 74- Mem: Am Phytopath Soc; Int Soc Citricult. Res: Postharvest decays and soil-borne and virus diseases of citrus. Mailing Add: Tex A&I Univ Citrus Ctr Weslaco TX 98596

TIMMER, RICHARD F, anatomy, see 12th edition

TIMMERMANN, DAN, JR, b New Braunfels, Tex, Oct 8, 33; m 55; c 2. BOTANY, CROP BREEDING. Educ: Tex A&M Univ, BS, 55, PhD(bot), 67; Ohio State Univ, MA, 62. Prof Exp: Teacher high sch, Tex, 58-61 & 62-63; asst prof, 67-70, ASSOC PROF BOT, ARK STATE UNIV, 70- Concurrent Pos: Consult plant breeding, Plant Res Div, Bryco, Inc, 74- Mem: Bot Soc Am. Res: Electron microscopy of plant cells; effects of gaseous air pollutants on the ultrastructure of specific cell organelles; cotton and soybean breeding. Mailing Add: Box 787 State University AR 72467

TIMMONS, FRANCIS LEONARD, weed science, see 12th edition

TIMMONS, RICHARD B, b Sherbrooke, Que, June 23, 38; m 63; c 3. PHYSICAL CHEMISTRY. Educ: St Francis Xavier Univ, BS, 58; Cath Univ Am, PhD(chem), 62. Prof Exp: Fel chem kinetics, Brookhaven Nat Lab, 62-64; asst prof chem, Boston Col, 64-65; asst prof, 65-68, ASSOC PROF CHEM, CATH UNIV AM, 68- Mem: Am Chem Soc. Res: Chemical kinetics; photochemistry, particularly in the vacuum ultraviolet region; kinetic isotope effects on reaction rates; chemistry of air pollution. Mailing Add: Dept of Chem Cath Univ of Am Washington DC 20017

TIMNICK, ANDREW, b Kremianka, Russia, Dec 29, 18; nat US; m 43; c 2. ANALYTICAL CHEMISTRY. Educ: Wartburg Col, BA, 40; Univ Iowa, MS, 42, PhD(phys chem), 47. Prof Exp: Asst, Univ Iowa, 40-42 & 46-47; control & develop chemist, Polymer Corp, Ont, 43-45; res chemist, Imp Oil, Ltd, 45-46; asst prof chem, WVa Univ, 47-49; from asst prof to assoc prof, 49-65, dir labs, 64-69, PROF CHEM, MICH STATE UNIV, 65- Concurrent Pos: Sr chemist, Oak Ridge Nat Lab, 52, consult, 53-62; vis prof, Univ Newcastle, 68. Mem: AAAS; Am Chem Soc. Res: Spectrophotometric, spectrofluorometric and electrometric methods of analysis. Mailing Add: Dept of Chem Mich State Univ East Lansing MI 48824

TIMOFEEFF, NICOLAY P, b Montreal, Que, Jan 12, 30; m 62; c 2. ENVIRONMENTAL SCIENCE, GEOGRAPHY. Educ: McGill Univ, BSc, 54; Columbia Univ, MA, 63, PhD(geog), 67. Prof Exp: Lectr geol & geog, Brooklyn Col, 57-58; lectr earth sci, Adelphi Suffolk Col, 63-64; lectr geog & geol, Hunter Col, 60-63; lectr earth sci, Barnard Col, Columbia Univ, 64-66; lectr phys geog, Columbia Univ, 64-66; assoc prof geog, 66-74, PROF GEOG, STATE UNIV NY BINGHAMTON, 74-, CHMN DEPT, 71- Concurrent Pos: State Univ NY organized res grants, 68-69 & 71-72; US mem nat comt, Int Geog Union spec travel award for excellence, 68. Mem: Am Asn Geogr; Int Union Geod & Geophys. Res: Application of multivariate analysis in evaluating the resources and physical geography of regions; quantitative methods in geography. Mailing Add: Dept of Geog State Univ of NY Binghamton NY 13901

TIMON, WILLIAM EDWARD, JR, b Natchitoches, La, Jan 13, 24; m 46, 56; c 6. MATHEMATICAL STATISTICS. Educ: Northwestern State Col La, BS, 50; Tulane Univ, MS, 51; Okla State Univ, PhD, 62. Prof Exp: Chief comput, Western Geophys Co, 45-46; instr, Tulane Univ, 50-51; mathematician, Esso Stand Oil Co, 51-53; instr math, La State Univ, 53-54; asst prof, Northwestern State Col La, 54-56 & Southwestern La Inst, 56-57; from asst prof to prof, Northwestern State Col La, 57-65, head dept, 62-65; prof, Parsons Col, 65-74, chmn dept, 67-74; PROF MATH, GLASSBORO STATE COL, 74- Mem: Math Asn Am; Nat Coun Teachers Math. Res: Analysis of the slipped-block design. Mailing Add: Dept of Math Glassboro State Col Glassboro NJ 08028

TIMONY, PETER EDWARD, b Orange, NJ, Dec 30, 43; m 66; c 3. ORGANIC CHEMISTRY. Educ: Fairleigh Dickinson Univ, BA, 67; Univ Notre Dame, PhD(org chem), 72. Prof Exp: Res chemist, 71-75, SR RES CHEMIST, STAUFFER CHEM CO, 75- Mem: Am Chem Soc; Am Soc Lubrication Engrs. Res: Mechanistic organoboron chemistry; synthetic lubricants. Mailing Add: Eastern Res Ctr Stauffer Chem Co Dobbs Ferry NY 10522

TIMOTHY, DAVID HARRY, b Pittsburgh, Pa, June 9, 28; m 53; c 3. PLANT GENETICS, EVOLUTION. Educ: Pa State Univ, BS, 52, MS, 55; Univ Minn, PhD(plant genetics), 56. Prof Exp: Asst geneticist, Rockefeller Found, 56-58, assoc geneticist, 58-61; assoc prof, 61-66, PROF CROP SCI, NC STATE UNIV, 66- Concurrent Pos: Consult, Latin Am Sci Bd, Nat Acad Sci, 64-65; mem, Adv Comt, Orgn Trop Studies, 68-70; mem, Exec Comt, Southern Pasture & Forage Crop Improvement Conf, 68-72, chmn, 71; mem, Nat Cert Grass Variety Rev Bd, 68-74 & Nat Found Seed Proj Planning Conf, 64-68. Mem: Fel AAAS; Asn Trop Biol; Am Soc Agron; Crop Sci Soc Am; Am Inst Biol Sci. Res: Origin, race inter-relationships and evolution of maize; corn and forage grass breeding; germ plasm resources; evaluation and improvement methods in the Gramineae; cytotaxonomy; evolution in domesticated grasses and their wild relatives. Mailing Add: Dept of Crop Sci NC State Univ Raleigh NC 27607

TIMOTHY, JOHN GETHYN, b Ripley, Eng, Sept 23, 42. SPACE PHYSICS. Educ: Univ London, BS, 63, PhD(space physics), 67. Prof Exp: Res asst space physics, Mullard Space Sci Lab, Univ Col London, 67-71; physicist, 71-72, SR PHYSICIST, HARVARD COL OBSERV, 73- Mem: Optical Soc Am; Am Geophys Union; Am Astron Soc. Res: Space astronomy; instrumentation for photometric measurements at extreme ultraviolet and photoelectric detection systems, imaging and nonimaging for soft x-ray wavelengths; photoelectric systems, imaging and nonimaging, for use at visible, ultraviolet and soft x-ray wavelengths. Mailing Add: Ctr for Astrophys Harvard Col Observ 60 Garden St Cambridge MA 02138

TIMOURIAN, HECTOR, b Mex, Aug 24, 33; US citizen; m 58; c 2. DEVELOPMENTAL BIOLOGY. Educ: Univ Calif, Los Angeles, BA, 55, PhD(zool), 60. Prof Exp: Commonwealth Sci & Indust Res Orgn res fel immunol, Queensland Univ, 60-61; res zoologist, Univ Calif, Los Angeles, 61-62; NIH res fel biol, Calif Inst Technol, 62-64; asst prof biol, San Fernando Valley State Col, 64-65; BIOLOGIST, LAWRENCE LIVERMORE LAB, UNIV CALIF, 65- Mem: AAAS; Am Soc Zoologists. Res: Effects of metal ions on embryonic development, sperm activity and fertilization. Mailing Add: Lawrence Livermore Lab Univ of Calif Livermore CA 94550

TIMOURIAN, JAMES GREGORY, b New York, NY, May 5, 41. MATHEMATICS. Educ: Syracuse Univ, PhD(math), 67. Prof Exp: Asst prof math, Univ Tenn, 67-69; asst prof, 69-70, ASSOC PROF MATH, UNIV ALTA, 70- Res: Differential topology; singularities of maps on manifolds. Mailing Add: Dept of Math Univ of Alta Edmonton AB Can

TIMUSK, THOMAS, b Estonia, June 3, 33; Can citizen; m 57; c 2. PHYSICS. Educ: Univ Toronto, BA, 57; Cornell Univ, PhD(physics), 61. Prof Exp: Res assoc physics, Cornell Univ, 61-62; asst, Univ Frankfurt, 62-64; res asst prof, Univ Ill, Urbana, 64-65; from asst prof to assoc prof, 65-73, PROF PHYSICS, McMASTER UNIV, 74- Concurrent Pos: Sloan fel, 66-68. Mem: Am Phys Soc; Can Asn Physicist. Res: Solid state physics; localized vibrations; far infrared spectroscopy. Mailing Add: Dept of Physics McMaster Univ Hamilton ON Can

TINCHER, WAYNE COLEMAN, b Frankfort, Ky, Jan 15, 35; m 57; c 3. PHYSICAL CHEMISTRY. Educ: David Lipscomb Col, BA, 56; Vanderbilt Univ, PhD(chem), 60. Prof Exp: Res chemist, Chemstrand Res Ctr, Monsanto Co, 60-65, group leader spectros, 65-71; ASSOC PROF TEXTILE CHEM, A FRENCH TEXTILE SCH, GA INST TECHNOL, 71- Mem: AAAS; Am Chem Soc; Am Asn Textile Chemists & Colorists; Sigma Xi. Res: Nuclear magnetic resonance spectra and structure of polymers; mechanics of polymer degradation; fiber and fabric flammability; textile process water pollution control. Mailing Add: A French Textile Sch Ga Inst of Technol Atlanta GA 30332

TINDALL, CHARLES GORDON, JR, b Trenton, NJ, Sept 3, 42; m 64; c 2. FORENSIC SCIENCE. Educ: Col Wooster, BA, 64; Ohio State Univ, MS, 67, PhD(org chem), 70. Prof Exp: Fel med chem, Nucleic Acid Res Inst, Int Chem & Nuclear Corp, 70-72; TECH SUPVR FORENSIC CHEM, NJ STATE POLICE SOUTH REGIONAL LAB, 72- Concurrent Pos: Adj prof, Stockton State Col, 74- & Ocean County Col, 75- Mem: Am Chem Soc; The Chem Soc; Forensic Sci Soc. Res: Detection of accelerants in arson investigation; characterization of hair; toxicology and detection of drugs and poisons in blood and urine. Mailing Add: NJ State Police SRegional Lab Rte 30 PO Box 126 Hammonton NJ 08037

TINDALL, GEORGE TAYLOR, b Magee, Miss, Mar 13, 28; m 47; c 4. NEUROSURGERY. Educ: Univ Miss, AB, 48; Johns Hopkins Univ, MD, 52. Prof Exp: Intern gen surg, Johns Hopkins Univ, 52-53; resident neurosurg, Med Ctr, Duke Univ, 55-61, from asst prof to assoc prof, 61-68; chief neurosurg serv, Vet Admin Hosp, 61-68; prof neurosurg & chief div, Univ Tex Med Br Galveston, 68-73; PROF NEUROSURG, SCH MED, EMORY UNIV, 73- Mem: Cong Neurol Surgeons; Soc Neurol Surgeons; Soc Univ Neurosurgeons (pres, 65). Res: Hypophysectomy and neuroendocrinology; measurement of the cerebral circulation; physiologic changes induced by increases in intracranial pressure, hemorrhage and effect of various pharmacologic agents; cranial aneurysms. Mailing Add: Div of Neurosurg Emory Univ Clin Atlanta GA 30322

TINDELL, RALPH S, b Tampa, Fla, Jan 16, 42; m 63; c 2. MATHEMATICS. Educ: Univ SFla, BA, 63; Fla State Univ, MS, 65, PhD(math), 67. Prof Exp: Res assoc math, Univ Ga, 66; vis mem & grantee, Inst Advan Study, 66-67; asst prof, Univ Ga, 67-70; ASSOC PROF MATH, STEVENS INST TECHNOL, 70- Concurrent Pos: Res assoc, Inst Advan Study, 69-70. Mem: Am Math Soc; Asn Mems Inst Advan Study. Res: Geometric topology; graph theory. Mailing Add: Dept of Math Stevens Inst of Technol Hoboken NJ 07030

TINER, JACK DALTON, zoology, see 12th edition

TING, ANNSHENG CHIEN, b Taipei, Rep China. NUMERICAL ANALYSIS. Educ: Nat Taiwan Univ, BA, 69; Cornell Univ, MS, 73, PhD(comput sci), 74. Prof Exp: Asst prof comput sci, Va Polytech Inst & State Univ, 74-75; ASST PROF COMPUT SCI, PA STATE UNIV, 75- Mem: Asn Comput Mach. Res: Develop algorithms for constrained and unconstrained optimization. Mailing Add: Dept of Comput Sci Pa State Univ University Park PA 16802

TING, ER YI, b Shantung, China, June 3, 19; US citizen; m 58; c 3. PHYSIOLOGY, INTERNAL MEDICINE. Educ: Nat Defense Med Col, China, MD, 48. Prof Exp: Resident internal med, Bronx Munic Hosp Ctr, NY, 54-56; instr med, Albert Einstein Col Med, 59-63; ASST PROF MED, NEW YORK MED COL & DIR PULMONARY FUNCTION RES LAB, METROP HOSP CTR, 63- Concurrent Pos: Fel respiratory physiol, Univ Buffalo, 56-57; fel med, State Univ NY Downstate Med Ctr, 57-59; fel med & cardiorespiratory physiol, Albert Einstein Col Med-Bronx Munic Hosp Ctr, 59-63; mem, Int Union Physiol Sci, 65- Mem: AAAS; Am Physiol Soc; Am Col Chest Physicians; Am Thoracic Soc; Am Fedn Clin Res. Res: Cardiorespiratory physiology, especially respiratory mechanics; cardiopulmonary diseases. Mailing Add: Pulmonary Function Res Lab Metrop Hosp 1901 First Ave New York NY 10029

TING, FRANCIS TA-CHUAN, b Tsingtao, China, Apr 26, 34; US citizen; m 66. GEOLOGY. Educ: Nat Taiwan Univ, BS, 57; Univ Minn, MS, 62; Pa State Univ, PhD, 67. Prof Exp: Asst prof geol, Macalester Col, 68-69; res assoc coal petrol, Pa State Univ, 69-70; from asst prof to assoc prof geol, Univ NDak, 70-74; ASSOC PROF GEOL, WVA UNIV, 74- Concurrent Pos: Fel, Pa State Univ, 67-68 & Univ Minn, 68-69; NATO sr fel, 73. Mem: Sigma Xi; Geol Soc Am; Bot Soc Am; Soc Econ Paleont & Mineral; Geochem Soc. Res: Coal petrology and chemistry; paleobotany. Mailing Add: Dept of Geol & Geog WVa Univ Morgantown WV 26506

TING, IRWIN PETER, b San Francisco, Calif, Jan 13, 34; m 52; c 1. PLANT PHYSIOLOGY, METABOLISM. Educ: Univ Nev, BS, 60, MS, 61; Iowa State Univ, PhD(plant physiol, biochem), 64. Prof Exp: NSF fel, 64-65; plant physiologist, 65-66, from asst prof to assoc prof plant physiol & metab, 66-72, PROF BIOL, UNIV CALIF, RIVERSIDE, 72- Concurrent Pos: Vis fel, Australian Nat Univ, 72; exchange prof, Univ Paris, 74. Mem: AAAS; Bot Soc Am; Am Soc Plant Physiologists. Res: Gas transfer between plant surfaces and environment; carbon dioxide metabolism; plant isoenzymes. Mailing Add: Dept of Biol Univ of Calif Riverside CA 92502

TING, LU, b China, Apr 18, 25; nat US; m 51; c 3. APPLIED MATHEMATICS. Educ: Chiao Tung Univ, China, BS, 46; Mass Inst Technol, SM, 48; Harvard Univ, MS, 49; NY Univ, ScD(aero eng), 51. Prof Exp: Res assoc aerodyn, NY Univ, 51-52; spec design engr, Foster Wheeler Corp, 52-55; res prof aerodyn, Polytech Inst Brooklyn, 55-64; prof aeronaut & astronaut, 64-68, PROF MATH, NY UNIV, 68- Concurrent Pos: Consult, Noise Reduction & Control Br, NASA Langley Res Ctr, 74- Mem: Soc Indust & Appl Math; Am Inst Aeronaut & Astronaut; Am Phys Soc. Res: Shock deflections; boundary layer theory; supersonic wing-body interference; space mechanics; nonlinear wave propagations; perturbation methods; aeroacoustics. Mailing Add: Dept of Math NY Univ Washington Sq New York NY 10012

TING, ROBERT CHIN-YAO, b Shanghai, China, Nov 15, 29; US citizen; m 62; c 3. MICROBIOLOGY. Educ: Amherst Col, AB, 53, MA, 56; Univ Ill, PhD(microbiol), 60. Prof Exp: Res fel biol, Calif Inst Technol, 60-62; vis scientist, Nat Cancer Inst, 62-64, microbiologist, 64-69; sci dir, Bionetics Res Lab, Litton Industs, 69-73; VPRES, BIOTECH RES LAB, 73- Concurrent Pos: Vis prof, Col Med, Nat Taiwan Univ, 66. Mem: AAAS; Tissue Cult Asn; Am Asn Cancer Res. Res: Bacteriophage; bacteria genetics; tumor virus; tissue culture; tumor immunology in experimental animals. Mailing Add: Biotech Res Lab 12061 Twinbrook Pkwy Rockville MD 20852

TING, SAMUEL C C, b Ann Arbor, Mich, Jan 27, 36; m 60; c 2. PARTICLE PHYSICS. Educ: Univ Mich, BSE(physics) & BSE(math), 59, MS, 60, PhD(physics), 62. Prof Exp: Ford Found fel, Europ Coun Nuclear Res, Switz, 63-64; instr physics, Columbia Univ, 64-65, asst prof, 65-67; assoc prof, 67-69, PROF PHYSICS, MASS INST TECHNOL, 69- Concurrent Pos: Ground leader, Deutches Electronen Synchrotronen, Hamburg, Ger, 66; assoc ed, Nuclear Physics B, 70. Mem: Fel Am Phys Soc; fel AAAS; fel Am Acad Arts & Sci. Res: Experimental particle physics; quantum electrodynamics; interactions of photons with matter. Mailing Add: Lab for Nuclear Sci Mass Inst of Technol Cambridge MA 02139

TING, SHIH-FAN, b Changteh, China, Sept 27, 17; US citizen; m 47; c 1. CHEMISTRY. Educ: Univ Chekiang, BS, 41; Univ Ala, MS, 57, PhD(chem), 60. Prof Exp: Asst prof chem, Fisk Univ, 60-65; fel, Duquesne Univ, 65-66; assoc prof, 66-69, PROF CHEM, MILLERSVILLE STATE COL, 69- Concurrent Pos: NSF grant, 63-65. Mem: Am Chem Soc. Res: High-frequency titration; rhenium chemistry; coordination compounds; nuclear magnetic resonance studies of hydrogen bonding. Mailing Add: Dept of Chem Millersville State Col Millersville PA 17551

TING, SIK VUNG, b Shanghai, China, Mar 3, 18; nat US; m 46; c 3. HORTICULTURE. Educ: Mich State Univ, BS, 41; Ohio State Univ, MS, 43, PhD(hort), 52. Prof Exp: Asst, Ohio State Univ, 41-43 & 49-52, Agr Exp Sta, 43-45; assoc prof hort, Nanking Univ, 47-49; asst horticulturist, 52-60, assoc biochemist, 60-68, RES BIOCHEMIST, CITRUS EXP STA, FLA STATE CITRUS COMN & PROF BIOCHEM, FLA STATE UNIV, 68- Mem: AAAS; Am Soc Hort Sci; Am Chem Soc; Inst Food Technologists; NY Acad Sci. Res: Biochemistry of horticultural plants, especially chemical components of citrus fruit. Mailing Add: Citrus Exp Sta Univ of Fla PO Box 1088 Lake Alfred FL 33850

TING, TSUAN WU, b Anking, Anhwei, China, Oct 10, 22; nat US; m 57; c 2. MATHEMATICS. Educ: Nat Cent Univ, China, BS, 47; Univ RI, MS, 56; Ind Univ, MS, 59, PhD(math), 60. Prof Exp: Technician, China 60th Arsenal, 47-49, assoc engr, 49-53; res asst eng, Univ RI, 54-56; math, Ind Univ, 56-59; sr mathematician, Gen Motors Res Labs, 60-61; asst prof mech, Univ Tex, 61-63; vis mem, Courant Inst Math Sci, NY Univ, 63-64; assoc prof math, NC State Univ, 64-66; PROF MATH, UNIV ILL, URBANA, 66- Mem: Soc Indust & Appl Math; Tensor Soc; Am Math Soc; Soc Natural Philos. Res: Theory of partial differential equations; mathematical physics; principles of continuum mechanics and differential geometry. Mailing Add: Dept of Math Univ of Ill Urbana IL 61801

TING, WILLIAM SU, b Hong Kong, Oct 4, 14; US citizen; m 58; c 2. GEOLOGY, PALYNOLOGY. Educ: Yenching Univ, China, BSc, 32; Glasgow Univ, PhD(geog), 37. Prof Exp: Prof geomorphol, Nat Cent Univ, China, 37-48; educ officer, Northcote Training Col, Hong Kong, 49-56; res geologist, Univ Calif, Los Angeles, 59-69; PROF GEOG, CALIF STATE UNIV, LOS ANGELES, 69- Concurrent Pos: Leader sci exped, Sinkiang, Acad Sinica, 43-44; prof, Sun Yat-Sen Univ, 44-45. Res: Geomorphology, ethology and ancient history of China; paleobotany. Mailing Add: 727 N Beverly Glen Blvd Los Angeles CA 90024

TING, YU-CHEN, b Honan, China, Oct 3, 20; m 60; c 2. PLANT CYTOLOGY, PLANT GENETICS. Educ: Nat Honan Univ, China, BS, 44; Cornell Univ, MSA, 52; La State Univ, PhD(hort genetics), 54. Prof Exp: Asst bot, Nat Honan Univ, 44-47; res fel, Harvard Univ, 54-62; from asst prof to assoc prof biol, 62-67, PROF BIOL, BOSTON COL, 67- Mem: AAAS; Genetics Soc Am; Am Genetic Asn; Bot Soc Am; Am Soc Hort Sci. Res: Cytology and genetics of Ipomoea batatas and related species; flower induction and site of synthesis of pigments in sweet potato plants; cytogenetics of maize and its relatives. Mailing Add: Dept of Biol Boston Col Chestnut Hill MA 02167

TINGA, JACOB HINNES, b Wilmington, NC, Mar 9, 20; m 57; c 2. HORTICULTURE, PLANT PHYSIOLOGY. Educ: NC State Col, BS, 42; Cornell Univ, MS, 52, PhD, 56. Prof Exp: Assoc prof hort, Va Polytech Inst, 55-68; PROF HORT, UNIV GA, 68- Mem: Am Soc Hort Sci. Res: Effect of environment on growth of horticultural crops; ornamental and greenhouse crops; winter protection; container production; propagation of large cuttings of woody plants; nursery economics. Mailing Add: Dept of Hort Univ of Ga Athens GA 30601

TING-BEALL, HIE PING, b Sibu, Malaysia, Dec 15, 40; US citizen; m 70. CELL PHYSIOLOGY. Educ: Greensboro Col, BS, 63; Tulane Univ, MS, 65, PhD(physiol), 67. Prof Exp: NIH fel biophys, Mich State Univ, 67-69, res assoc biochem, 70-72; NIH trainee phys biochem, Johnson Res Found, Univ Pa, 69-70; res assoc physiol & pharm, 72-74, ASST MED RES PROF PHYSIOL & PHARM & ANAT, DUKE UNIV MED CTR, 75- Concurrent Pos: K E Osserman fel, Myasthenia Gravis Found, 73-74; fac res award, Duke Univ Med Ctr, 76. Mem: Biophys Soc; Sigma Xi. Res: Structure and function of cell membranes; biophysics of bimolecular lipid membranes. Mailing Add: Dept of Anat Duke Univ Med Ctr Durham NC 27710

TINGEY, DAVID THOMAS, b Salt Lake City, Utah, Jan 30, 41; m 68; c 1. PLANT PHYSIOLOGY, AIR POLLUTION. Educ: Univ Utah, BA, 66, MA, 68; NC State Univ, PhD(plant physiol), 72. Prof Exp: Plant physiologist, Environ Protection Agency, NC State Univ, 68-73, PLANT PHYSIOLOGIST, ENVIRON PROTECTION AGENCY, NAT ECOL RES LAB, 73- Concurrent Pos: Asst prof bot, Ore State Univ, 74- Mem: Am Soc Plant Physiol; Scand Soc Plant Physiol; Am Soc Agron. Res: Studying the effects of atmospheric pollutants on plant physiology. Mailing Add: Environ Protection Agency Nat Ecol Res Lab Corvallis OR 97330

TINGEY, FRED HOLLIS, mathematical statistics, see 12th edition

TINGEY, GARTH LEROY, b Woodruff, Utah, Apr 14, 32; m 53; c 7. PHYSICAL CHEMISTRY. Educ: Brigham Young Univ, BS, 54, MS, 59; Pa State Univ, PhD(phys chem), 63. Prof Exp: Sr scientist, Gen Elec Corp, 63-65; sr res scientist, 65-66, unit mgr mat, 66-70, RES ASSOC, PAC NORTHWEST LABS, BATTELLE MEM INST, 70- Mem: Am Chem Soc. Res: Radiation chemical studies of gaseous mixtures; inert gas sensitized radiolysis reactions; chemical kinetics; chemical studies of carbon and graphite; high temperature gas cooled nuclear reactor technology. Mailing Add: Pac NW Lab Battelle Mem Inst Box 999 Richland WA 99352

TINGEY, WARD M, b Brigham City, Utah, Apr 9, 44; m 68; c 1. ENTOMOLOGY. Educ: Brigham Young Univ, BS, 66, MS, 68; Univ Ariz, PhD(entom), 72. Prof Exp: Asst res entomologist, Univ Calif, Davis, 72-74; ASST PROF ENTOM, CORNELL UNIV, 74- Mem: Entom Soc Am; AAAS. Res: Genetic resistance of crop plants to insect pests; behavioral and developmental responses of insects to resistant host plants; host plant growth and development responses to insect injury; pest management. Mailing Add: Dept of Entom Comstock Hall Cornell Univ Ithaca NY 14853

TINGLE, WILLIAM HERBERT, b Parnassus, Pa, Aug 31, 17; m 45; c 2. INSTRUMENTATION, SPECTROSCOPY. Educ: Univ Pittsburgh, BS, 49. Prof Exp: Instr eng physics, Univ Pittsburgh, 49-50; spectroscopist anal chem, 51-62, sect head anal chem, 63-72, SCI ASSOC EQUIP DEVELOP, ALCOA LABS, 73- Concurrent

Pos: Chmn subcomt, Am Soc Testing & Mat, 62-73; adv, USA Adv Group, Int Stand Orgn Comt 79, 70-76; secretariat, Int Stand Orgn, 70-76. Mem: Sigma Xi; Optical Soc Am; Soc Appl Spectros; Am Chem Soc. Res: Evaluating principles of measurement and control in chemical and metallurgical processes, infrared measurement of temperature, and high gradient magnetic separation. Mailing Add: Alcoa Tech Ctr Alcoa Center PA 15069

TINGLEY, ARNOLD JACKSON, b Point de Bute, June 9, 20; m 46; c 2. MATHEMATICS. Educ: Mt Alison Univ, BA, 49; Univ Minn, PhD(math), 52. Prof Exp: Instr math, Univ Nebr, 52-53; from asst prof to assoc prof, 53-62, head dept, 66-73, PROF MATH, DALHOUSIE UNIV, 62-, REGISTR, 73- Mem: Math Asn Am; Can Math Cong. Res: Analysis. Mailing Add: Dept of Math Dalhousie Univ Halifax NS Can

TINGSTAD, JAMES EDWARD, b Virginia, Minn, Aug 28, 28; m 50; c 3. PHARMACY. Educ: Univ Wis, BS, 53, MS, 55, PhD(pharm), 57. Prof Exp: Res assoc, Upjohn Co, 57-63, head prod develop, 63-66, mgr pharm res, 66-68; prof pharm, Sch Pharm, Univ Calif, San Francisco, 68-70; head pharmaceut res & develop, Endo Labs, 71-73; MGR PHARM DEVELOP, RIKER LABS INC, SUBSID 3M CO, 73- Mem: Am Pharmaceut Asn. Res: New product development; preformulation; stability; biopharmaceutics. Mailing Add: 218-3 3M Ctr Riker Labs Inc St Paul MN 55101

TINKER, DAVID OWEN, b Toronto, Ont, Jan 25, 40; m 62; c 3. BIOCHEMISTRY, PHYSICAL CHEMISTRY. Educ: Univ Toronto, BSc, 61; Univ Wash, PhD(biochem), 65. Prof Exp: Nat Res Coun Can fel, Univ London, 65-66; asst prof, 66-71, ASSOC PROF BIOCHEM, UNIV TORONTO, 71- Concurrent Pos: Assoc ed, Can J Biochem, 74- Mem: Am Chem Soc; Can Biochem Soc. Res: Structure, occurrence, metabolism and physical chemical properties of complex lipids from biological membranes. Mailing Add: Dept of Biochem Fac of Med Univ of Toronto Toronto ON Can

TINKER, HAROLD BURNHAM, inorganic chemistry, see 12th edition

TINKER, JOHN FRANK, b Wis, Mar 25, 22; m 48. CHEMISTRY, INFORMATION SCIENCE. Educ: Univ Va, BS, 43; Harvard Univ, PhD(chem), 51. Prof Exp: Instr chem, Harvard Univ, 50-52; INFO SCIENTIST, EASTMAN KODAK CO, 52- Mem: Am Chem Soc. Res: Chemical information. Mailing Add: Kodak Park B-83 Rochester NY 14650

TINKER, ROBERT FREDERICK, b Dec 11, 41; US citizen; m 65; c 2. PHYSICS. Educ: Swarthmore Col, BA, 63; Stanford Univ, MS, 64; Mass Inst Technol, PhD(physics), 70. Prof Exp: Instr math & physics, Stillman Col, 64-66; instr, Wellesley Col, 68-69; staff physicist, Comn Col Physics, 70-71; ASST PROF PHYSICS, AMHERST COL, 71- Concurrent Pos: Consult, Tech Educ Res Ctr, Cambridge, Mass, 71- Res: Elementary excitations in superfluid helium at low temperatures; science teaching methods and equipment, especially for slow learners and disadvantaged students; application of new technology to teaching. Mailing Add: Dept of Physics Amherst Col Amherst MA 01002

TINKER, SPENCER WILKIE, b Anamoose, NDak, Jan 29, 09; m 38; c 1. ICHTHYOLOGY. Educ: Univ Wash, BS, 31; Univ Hawaii, MS, 34. Prof Exp: Instr zool, 33-34, ed, 35-55, from asst prof to prof, 55-72, dir, Waikiki Aquarium, 40-72, EMER PROF ZOOL, UNIV HAWAII, 72- Res: Indo-Pacific fish and molluscs; whales. Mailing Add: 1121 Hunakai St Honolulu Hi 96816

TINKHAM, MICHAEL, b Green Lake Co, Wis, Feb 23, 28; m 61; c 2. SOLID STATE PHYSICS. Educ: Ripon Col, AB, 51; Mass Inst Technol, MS, 51, PhD(physics), 54. Prof Exp: NSF fel, Clarendon Lab, Oxford Univ, 54-55; res physicist, Univ Calif, Berkeley, 55-57, lectr, 56-57, from asst prof to prof physics, 57-66; GORDON McKAY PROF APPL PHYSICS & PROF PHYSICS, HARVARD UNIV, 66-, CHMN DEPT PHYSICS, 75- Concurrent Pos: Guggenheim fel, 63-64; NSF sr fel, Cavendish Lab, Cambridge Univ, 71-72. Honors & Awards: Buckley Prize, Am Phys Soc, 74. Mem: Nat Acad Sci; fel Am Acad Arts & Sci; fel Am Phys Soc; fel AAAS. Res: Superconductivity; magnetism; far infrared. Mailing Add: Dept of Physics Harvard Univ Cambridge MA 02138

TINKLE, DONALD WARD, b Dallas, Tex, Dec 3, 30; m 51; c 4. VERTEBRATE ZOOLOGY. Educ: Southern Methodist Univ, BS, 52; Tulane Univ, MS, 55, PhD, 56. Prof Exp: Asst prof biol, WTex State Col, 56-57; res assoc, Tex Tech Col, 57, from asst prof to prof, 57-65; PROF ZOOL, UNIV MICH, ANN ARBOR, 65-, DIR MUS ZOOL, 75- Concurrent Pos: Grants, Am Philos Soc, 58, NSF, 58-, NIH, 60-62 & AEC, 60-67; Maytag Chair vert ecol, Ariz State Univ, 72; consult, Adv Panel, NSF, 74-76. Mem: Fel AAAS; Am Soc Ichthyol & Herpet; Ecol Soc Am; Am Soc Naturalists. Res: Evolution of life history characteristics; general population biology and systematics of reptiles. Mailing Add: Univ of Mich Mus of Zool Room 1080 Ann Arbor MI 48104

TINKLEPAUGH, ROBERT LEIF, physical chemistry, see 12th edition

TINKLIN, GWENDOLYN L, b Corning, Kans, May 8, 10. FOODS. Educ: Kans State Univ, BS, 40, MS, 44. Prof Exp: Pub sch teacher, Kans, 30-43; asst home econ, Kans State Univ, 43, asst home economist, Exp Sta, 43-45, instr foods & nutrit, Univ, 45-47; instr, Iowa State Univ, 47-48; from asst prof to prof, 49-74, actg head dept, 53-55 & 63-66, EMER PROF FOODS & NUTRIT, KANS STATE UNIV, 74- Concurrent Pos: Kans State Univ-AID adv home sci, Andhra Pradesh Agr Univ, India, 66-67. Mem: Am Home Econ Asn; Inst Food Technologists. Res: Functional properties of dried and frozen egg products; organoleptic and chemical investigations of frozen fruits and vegetables and baked products. Mailing Add: Dept of Foods & Nutrit Kans State Univ Justin Hall Manhattan KS 66506

TINLINE, ROBERT DAVIES, b Moose Jaw, Sask, Aug 4, 25; m 48; c 5. PLANT PATHOLOGY. Educ: Univ Sask, BA, 48; Univ Wis, MS, 52, PhD(plant path), 54. Prof Exp: Tech officer I, 48-49, tech officer II, 49-51, agr res officer, 51-67, RES SCIENTIST, AGR CAN, 67- Mem: Mycol Soc Am; Am Phytopath Soc; Can Phytopath Soc; Agr Inst Can. Res: Root and leaf diseases of cereals; variability and genetics of plant pathogenic fungi. Mailing Add: Agr Can Res Sta 107 Science Crescent Saskatoon SK Can

TINNEY, FRANCIS JOHN, b Brooklyn, NY, July 31, 38. MEDICINAL CHEMISTRY. Educ: St John's Univ, NY, BS, 59, MS, 61; Univ Md, PhD(aza steroids), 65. Prof Exp: Ortho Res Found fel chem, Univ Md, 65-66; assoc res chemist, 66-69, res chemist, 69-70, SR RES CHEMIST, PARKE, DAVIS & CO, 70- Mem: Am Chem Soc; Am Pharmaceut Asn. Res: Organic medicinal chemistry; steroids; heterocyclic steroids; heterocyclics; organic nitrogen containing compounds; peptides. Mailing Add: Parke Davis & Co 2800 Plymouth Rd Ann Arbor MI 48106

TINNIN, ROBERT OWEN, b Santa Barbara, Calif, Sept 6, 43; m 65. PLANT ECOLOGY. Educ: Univ Calif, Santa Barbara, BA, 65, PhD(ecol), 69. Prof Exp: Asst prof, 69-74, ASSOC PROF ECOL, PORTLAND STATE UNIV, 74- Mem: Ecol Soc Am. Res: Plant interactions, particularly between dwarf mistletoes and their hosts. Mailing Add: Dept of Biol Portland State Univ Portland OR 97207

TINOCO, IGNACIO, JR, b El Paso, Tex, Nov 22, 30; m 51; c 1. PHYSICAL CHEMISTRY. Educ: Univ NMex, BS, 51, DSc, 72; Univ Wis, PhD(chem), 54. Prof Exp: Res fel chem, Yale Univ, 54-56; from instr to assoc prof, 56-66, PROF CHEM, UNIV CALIF, BERKELEY, 66- Concurrent Pos: Guggenheim fel, 64. Honors & Awards: Calif Sect Award, Am Chem Soc, 65. Mem: Am Chem Soc; Am Phys Soc; Biophys Soc. Res: Biophysical chemistry. Mailing Add: Dept of Chem Univ of Calif Berkeley CA 94720

TINOCO, JOAN W H, nutrition, see 12th edition

TINSLEY, BEATRICE MURIEL, b Chester, Eng, Jan 27, 41; NZ citizen; m 61; c 2. THEORETICAL ASTROPHYSICS, COSMOLOGY. Educ: Univ Canterbury, BSc, 61, MSc, 63; Univ Tex, Austin, PhD(astron), 67. Prof Exp: Fel, Univ Tex, Austin, 67-68, vis scientist physics, Univ Tex, Dallas, 69-73, asst prof astron, 73-74; ASSOC PROF ASTRON, YALE UNIV, 75- Concurrent Pos: Sloan fel, 75-77. Honors & Awards: Annie Jump Cannon Prize, Am Astron Soc & Am Asn Univ Women, 74. Mem: Am Astron Soc; Royal Astron Soc UK; Int Astron Union. Res: Theoretical studies of the evolution of galaxies, especially with application to chemical evolution and cosmology. Mailing Add: Yale Univ Observ Box 2023 Yale Sta New Haven CT 06520

TINSLEY, BRIAN ALFRED, b Wellington, NZ, Apr 23, 37; m 61; c 2. SPACE PHYSICS. Educ: Univ Canterbury, BSc, 58, MSc, 61, PhD(physics), 63. Prof Exp: Res assoc atmospheric & space sci, 63-65, res scientist, 65-67, asst prof, 67-70, ASSOC PROF PHYSICS, UNIV TEX, DALLAS, 70- Mem: Am Geophys Union. Res: Radio and optical observations of airglow, natural and artificial aurorae; design of grille spectrometers for measurement of airglow, thermal and astronomical emissions; theoretical studies of terrestrial hydrogen and photoelectron emissions. Mailing Add: Dept of Physics Univ of Tex PO Box 688 Richardson TX 75080

TINSLEY, IAN JAMES, b Sydney, Australia, Sept 23, 29; m 55; c 2. BIOCHEMISTRY. Educ: Univ Sydney, BSc, 50; Ore State Univ, MS, 55, PhD(food sci), 58. Prof Exp: Res officer, Commonwealth Sci & Indust Res Orgn, Australia, 50-53; from asst prof to assoc prof, 57-70. PROF BIOCHEM, ORE STATE UNIV, 70- Honors & Awards: Florasynth Award, Inst Food Technologists, 55. Mem: Am Inst Nutrit; Am Oil Chem Soc. Res: Lipid metabolism; essential fatty acid nutrition; biochemical effects of pesticide ingestion; interactions of pesticides with lipids. Mailing Add: Dept of Agr Chem Ore State Univ Corvallis OR 97331

TINSLEY, SAMUEL WEAVER, b Hopkinsville, Ky, July 15, 23; m 45; c 3. ORGANIC CHEMISTRY. Educ: Western Ky State Col, BS, 44; Northwestern Univ, PhD(org chem), 50. Prof Exp: Asst prof chem, Tex Tech Col, 49-50; res chemist, 50-60, asst dir org res, 60-64, assoc dir res & develop, 64-67, mgr new chem, 67-71, mgr corp res, 71-74, DIR CORP RES, UNION CARBIDE CHEM CORP, 74- Mem: Am Chem Soc. Res: Aromatic synthesis; peracids; epoxides. Mailing Add: Tarrytown Tech Ctr Union Carbide Chem Co Tarrytown NY 10591

TINSMAN, JAMES HERBERT, JR, b Philadelphia, Pa, Apr 22, 30; m 56; c 4. PHYSICAL ANTHROPOLOGY. Educ: Univ Pa, AB, 56, AM, 60; Univ Colo, MA, 66, PhD(anthrop), 71. Prof Exp: Instr philos & econ, 59-60, from asst to assoc prof, 60-71, PROF ANTHROP, KUTZTOWN STATE COL, 71-, CHMN DEPT SOC SCI, 74- Concurrent Pos: Lectr philos, Muhlenberg Col, 60-65 & anthrop, Univ Colo, 69-71; co-ed, Newsletter Pa Anthropologists. Mem: Fel Am Anthrop Asn; fel Soc Appl Anthrop; fel Am Soc Human Genetics; Am Asn Phys Anthrop; Soc Am Archaeol. Res: Contemporary human variation, anthropometric, anthroscopic and serological, as relates to human genetics, population structure and, ultimately, human evolution. Mailing Add: Dept of Anthrop Kutztown State Col Kutztown PA 19530

TINT, HOWARD, b Philadelphia, Pa, Jan 22, 17; m 41; c 2. BIOCHEMISTRY. Educ: Univ Pa, AB, 37, PhD(mycol, plant path), 43. Prof Exp: Leader scouting crews, Bur Entom & Plant Quarantine, USDA, Washington, DC, 39-40, biol tech aide, Bur Plant Indust, 41, physiologist, Bur Agr & Indust Chem, 43-45; biochemist, Wyeth, Inc, 45-50, sr res biochemist, 50-56, supvr biologics lab, Wyeth Labs, 56-60, dir, Prod Develop Div, 60-63, DIR, BIOL & CHEM DEVELOP DIV, WYETH LABS, 63- Concurrent Pos: Microbiologist, Off Sci Res & Develop, Johnson Found, Univ Pa, 45. Mem: Am Chem Soc; Sigma Xi; NY Acad Sci. Res: Microbiology; physiology; virology; tissue culture; cancer immunology; chemical development and pilot-plant; fine-chemicals production. Mailing Add: Wyeth Labs Box 8299 Philadelphia PA 19101

TINTI, DINO S, b San Bernardino, Calif, Feb 20, 41; m 62; c 3. PHYSICAL CHEMISTRY. Educ: Univ Calif, Riverside, BA, 62; Calif Inst Technol, PhD(chem), 68. Prof Exp: Res chemist, Univ Calif, Los Angeles, 67-70; asst prof, 70-74, ASSOC PROF CHEM, UNIV CALIF, DAVIS, 74- Concurrent Pos: Sloan fel, 74- Mem: Am Phys Soc. Res: Optical and magnetic resonance spectroscopy of the triplet state; interactions among excited electronic states; matrix isolation spectroscopy. Mailing Add: Dept of Chem Univ of Calif Davis CA 95616

TINTNER, GERHARD, b Nürnberg, Ger, Sept 29, 07; nat US; m 41; c 1. MATHEMATICS. Educ: Univ Vienna, PhD(econ, statist), 29. Prof Exp: Res assoc, Austrian Inst Trade Cycle Res, 29-36; res assoc, Cowles Found & lectr economet, Colo Col, 36-37; asst prof econ, math & statist, Iowa State Univ, 37-39, assoc prof, 39-46, prof, 46-62; prof, Univ Pittsburgh, 62; distinguished prof econ & math, 63-74, EMER DISTINGUISHED PROF ECON & MATH, UNIV SOUTHERN CALIF, 74- Concurrent Pos: Rockefeller fels, Harvard Univ & Columbia Univ, 34-36 & Univ Cambridge & Univ Paris; lectr, Grad Sch & statistician, USDA, 44; res assoc, Univ Cambridge, 48-49; vis prof, Univ Vienna, 56-57 & Lisbon Tech, 59; econometrician, UN, India, 50, Ecuador, 62, Food & Agr Orgn, India, 65. Mem: Fel Economet Soc; fel Am Statist Asn; fel Inst Math Statist. Res: Econometrics; time series analysis; stochastic linear programming: empirical models of economic development. Mailing Add: Col of Letts Arts & Sci Univ of Southern Calif Dept of Econ Los Angeles CA 90007

TINUS, RICHARD WILLARD, b Orange, NJ, Mar 26, 36; m 58; c 2. PLANT PHYSIOLOGY. Educ: Wesleyan Univ, BA, 58; Duke Univ, MF, 60; Univ Calif, Berkeley, PhD(plant physiol), 65. Prof Exp: Plant physiologist, Agr Res Serv, 65-68, PRIN PLANT PHYSIOLOGIST, US FOREST SERV, USDA, 68- Honors & Awards: Superior Serv Award, USDA, 75. Mem: Am Soc Plant Physiologists; Soc Cryobiol; Soc Am Foresters. Res: Development of greenhouse container systems for tree seedling production; vegetative propagation of pine; cold and drought resistance of trees. Mailing Add: US Forest Serv First & Brander St Bottineau ND 58318

TIN-WA, MAUNG, b Rangoon, Burma, May 12, 40. PHARMACOGNOSY,

NATURAL PRODUCTS CHEMISTRY. Educ: Univ Rangoon, BSc, 61; Ohio State Univ, MSc, 65; Univ Pittsburgh, PhD(pharmacog), 69. Prof Exp: Teaching asst chem, Univ Rangoon, 61-62; res asst pharmacog, Ohio State Univ, 65; teaching asst, Univ Pittsburgh, 69; NSF fel plant sci, Univ Calif, Riverside, 70-71; NIH res assoc natural anticancer agents, Univ Ill Med Ctr, 71-72, Nat Cancer Inst contract sr investr, 72-74, asst prof pharmacog, Med Ctr, 73-75; RES DIR, RES & CONSULT ASSOCS, 75- Concurrent Pos: Assoc ed, Pharmacog Titles, 72-74. Mem: Am Soc Pharmacog; Phytochem Soc NAm; Bot Soc Am; Am Pharmaceut Asn. Res: Isolation and structure elucidation of biologically active natural products; marine natural products; natural toxins. Mailing Add: Res & Consult Assocs 1100 Quail St Suite 100 Newport Beach CA 92660

TIPHANE, MARCEL, b Montreal, Que, June 2, 17; m 45; c 5. GEOLOGY. Educ: Univ Montreal, BA, 39; Laval Univ, BASc, 44; McGill Univ, MSc, 47. Prof Exp: From asst prof to assoc prof, 44-69, PROF GEOL, UNIV MONTREAL, 69- Concurrent Pos: Consult, Calumet Uranium Mines, Ltd, 54-56; NSF Inst Oceanog grant, 59. Mem: Fel Geol Asn Can; Can Inst Mining & Metall. Res: Petrography; marine sedimentation. Mailing Add: Dept of Geol Univ of Montreal PO Box 6128 Montreal PQ Can

TIPLER, PAUL A, b Antigo, Wis, Apr 12, 33; m 58; c 2. NUCLEAR PHYSICS. Educ: Purdue Univ, BS, 55; Univ Ill, PhD(physics), 62. Prof Exp: Asst prof physics, Wesleyan Univ, 61-62; asst prof, 62-69, ASSOC PROF PHYSICS, OAKLAND UNIV, 69- Mem: Am Phys Soc; Am Asn Physics Teachers. Res: Low energy nuclear physics. Mailing Add: Dept of Physics Oakland Univ Rochester MI 48063

TIPPER, DONALD JOHN, b Birmingham, Eng, July 21, 35; m 65; c 3. MICROBIOLOGY, MOLECULAR BIOLOGY. Educ: Univ Birmingham, BSc, 56, PhD(chem), 59. Prof Exp: Res assoc immunochem, Dept Surg Res, St Luke's Hosp, 59-60; chemist, Guinness's Brewery, Dublin, Ireland, 60-62; res asst pharmacol, Wash Univ, 62-64; res asst, Univ Wis-Madison, 64-65, from asst prof to assoc prof, 65-71; PROF MICROBIOL & CHMN DEPT, MED SCH, UNIV MASS, 71- Concurrent Pos: USPHS career develop award, 68-71. Mem: Am Soc Microbiol; Brit Soc Gen Microbiol; Am Soc Biol Chemists. Res: Structure and biosynthesis of bacterial cell walls; mode of action of penicillins; biosynthesis of bacillus spore cortex and coats; yeast RNA polymerases. Mailing Add: Dept of Microbiol Univ of Mass Med Sch Worcester MA 01605

TIPPER, RONALD CHARLES, biological oceanography, see 12th edition

TIPPING, RICHARD H, b Abington, Pa, Aug 31, 39. MOLECULAR PHYSICS. Educ: Pa State Univ, BSc, 63, MSc, 66, PhD(physics), 69. Prof Exp: Asst prof, 69-74, ASSOC PROF PHYSICS, MEM UNIV NFLD, 74- Mem: Am Phys Soc; Can Asn Physicists. Res: Molecular spectral line shapes; intensities and spectroscopic constants of diatomic molecules. Mailing Add: Dept of Physics Mem Univ of Nfld St John's NF Can

TIPPINS, HAMLIN HANNIBAL, entomology, see 12th edition

TIPPINS, HARRY H, JR, b Claxton, Ga, Jan 8, 33; m 53; c 3. SOLID STATE PHYSICS, OPTICAL PHYSICS. Educ: Univ Fla, BSEE, 55; Univ Ill, MS, 59, PhD(physics), 62. Prof Exp: Mem tech staff, Bell Tel Labs, 55-57 & res tech staff, Aerospace Corp, 61-73; CONSULT AUTONETICS DIV, ROCKWELL INT & JET PROPULSION LAB, CALIF INST TECHNOL, 73- Mem: Am Phys Soc. Res: Optical, magnetic and electrical properties of solids; transport phenomena; photoconductivity; magnetoresistance; band structure; optical and magnetic resonance spectroscopy; crystal field theory; color centers; solar cells; auger spectroscopy; cryogenics; radiation effects; fiber optics; proximity sensors. Mailing Add: 5007 W 133rd St Hawthorne CA 90250

TIPPLES, KEITH H, b Cambridge, Eng, Feb 4, 36; m 62; c 3. CEREAL CHEMISTRY. Educ: Univ Birmingham, BSc, 59, PhD(appl biochem), 62. Prof Exp: Nat Res Coun Can res fel, 63-64, RES SCIENTIST, GRAIN RES LAB, CAN GRAIN COMN, 64- Mem: Am Asn Cereal Chem. Res: Wheat dormancy and alpha emylase activity; factors affecting the bread baking process, related particularly to new no-bulk-fermentation-time methods. Mailing Add: Grain Res Lab Can Grain Comn 1404-303 Main St Winnipeg MB Can

TIPPO, OSWALD, b Milo, Maine, Nov 27, 11; m 34; c 2. BOTANY. Educ: Univ Mass, BS, 32; Harvard Univ, AM, 33, PhD(bot), 37. Hon Degrees: DSc, Univ Mass, 54. Prof Exp: Asst biol, Radcliffe Col, 35-37; instr bot, Univ Ill, 37-39, assoc, 39-41, from asst prof to prof, 41-55, actg head dept, 47-48, chmn, 48-55, dean grad col, 53-55; Eaton prof & chmn dept, dir Marsh Bot Garden & dir bot labs, Yale Univ, 55-60; provost, Univ Colo, 60-63; exec dean arts & sci, NY Univ, 63-64; provost, 64-70, chancellor, 70-71, COMMONWEALTH PROF BOT & HIGHER EDUC, UNIV MASS, AMHERST, 71- Concurrent Pos: Assoc biologist, Wood Sect, Testing Lab, US Navy Yard, Philadelphia, 43-44, biologist, 44-45; ed in chief, Am J Bot, 51-53; trustee, Biol Abstr, 57-63, pres, 63-64. Mem: AAAS (vpres, 58); Bot Soc Am (vpres, 54, pres, 55); fel Am Acad Arts & Sci. Res: Plant anatomy and phylogeny; phylogeny of angiosperms. Mailing Add: Dept of Bot Univ of Mass Amherst MA 01002

TIPS, ROBERT LEONARD, b San Antonio, Tex, Feb 13, 27; m 50; c 4. PEDIATRICS, GENETICS. Educ: Univ Tex, BA, 49, MA, 50, MD, 56; Univ Notre Dame, PhD(genetics), 52; Am Bd Pediat, dipl, 61. Prof Exp: Asst anat, Univ Tex Med Br Galveston, 52-53, asst bact, 53-54, asst biol, 54, asst prof bact & parasitol, 54-58, intern, 57, res assoc human genetics, 58-59, resident pediat, 59; from instr to asst prof, Med Sch, Univ Ore, 59-63; assoc prof genetics, 63-65, CLIN ASST PROF PEDIAT, BAYLOR COL MED, 63- Concurrent Pos: Asst prof, Univ Houston, 53; chmn div genetics & assoc sect scientist, Ore Primate Res Ctr, 59-63; chmn div genetics, Houston State Psychiat Inst, Tex Med Ctr, 63-65; dir, Med Genetics Ctr, Pasadena, 65-70; chmn dept pediat, Southmore Hosp, 65-70; assoc, Women's & Childrens' Clin, 65-68; med dir, Ctr Biolab; consult perinatal res coun, Nat Inst Neurol Dis & Blindness. Mem: AAAS; Am Soc Human Genetics; fel Am Asn Ment Deficiency; AMA; fel Am Acad Pediat. Res: Primate and immunological genetics. Mailing Add: Rte 4 Box 129A Center TX 75935

TIPSWORD, RAY FENTON, b Beecher City, Ill, Sept 9, 31; m 52; c 2. PHYSICS. Educ: Eastern Ill Univ, BSEd, 53; Southern Ill Univ, MS, 57; Univ Ala, PhD(physics), 62. Prof Exp: Instr physics, Mo Sch Mines, 57-59; fel, Univ Ala, 62-63; asst prof, 63-68, ASSOC PROF PHYSICS, VA POLYTECH INST & STATE UNIV, 68- Res: Nuclear magnetic resonance and low temperature physics. Mailing Add: Tom's Creek Rd Blacksburg VA 24060

TIPTON, ANN BAUGH, b Freeport, Tex, Sept 4, 38; m 63. PHYSICAL CHEMISTRY, FUEL SCIENCE. Educ: Southwestern Univ, Tex, BS, 60; Univ Tex, Austin, MA, 63, PhD(phys chem), 66. Prof Exp: Asst prof chem, Southwestern Univ, Tex, 64-67; Welch fel, Univ Tex, Austin, 66-69; tech specialist, Lockheed Propulsion Co, 69-75; SR RES CHEMIST, OCCIDENTAL RES CORP, 75- Honors & Awards: Citation Merit Chem, Southwestern Univ, Tex, 72. Mem: AAAS; Am Chem Soc; Am Phys Soc; Soc Appl Spectros. Res: Microwave spectroscopy and molecular structure; aging studies of solid propellants and composites; instrument automation, computer programming; synthetic fuels; coal conversion, desulfurization, combustion kinetics and liquefaction. Mailing Add: Occidental Res Corp 1855 Carrion Rd La Verne CA 91750

TIPTON, CARL LEE, b Collins, Iowa, July 26, 31; m 57; c 3. BIOCHEMISTRY. Educ: Univ Nebr, BS, 54, MS, 57; Univ Ill, PhD, 61. Prof Exp: Assoc & instr, 61-62, asst prof, 62-67, ASSOC PROF BIOCHEM, IOWA STATE UNIV, 67- Concurrent Pos: NIH sr fel, Univ Calif, Davis, 69-70. Mem: AAAS; Am Chem Soc; Phytochem Soc NAm; Am Soc Plant Physiologists. Res: Biochemistry of complex lipids; biochemistry of pest resistance in plants. Mailing Add: Dept of Biochem & Biophys Iowa State Univ Ames IA 50010

TIPTON, CHARLES M, b Evanston, Ill, Nov 29, 27; m 53; c 4. PHYSIOLOGY. Educ: Springfield Col, BS, 52; Univ Ill, MS, 53, PhD(physiol), 62. Prof Exp: High sch teacher, Ill, 53-55; teaching asst health educ, Univ Ill, 55-57, instr, 57-58, asst physiol, 58-61; asst prof, Springfield Col, 61-63; from asst prof to assoc prof, 63-73, PROF PHYSIOL, UNIV IOWA, 73- Concurrent Pos: Chmn study sect appl physiol & bioeng, NIH, 73-76. Mem: Am Physiol Soc; fel Am Heart Asn; Am Col Sports Med (pres-elect, 73-74). Res: Exercise physiology, including bradycardia of training, mechanisms of cardiac hypertrophy, ligamentous strength, endocrines and training; exercise testing; pharmacological differences and training. Mailing Add: Exercise Physiol Lab Univ of Iowa Iowa City IA 52240

TIPTON, GEORGE MURTHA, b Guthrie, Okla, Dec 31, 10. ANALYTICAL CHEMISTRY. Educ: St Mary's Col, Kans, AB, 31; St Louis Univ, PhD(chem), 49. Prof Exp: High sch instr, Colo, 39-40; instr chem, Regis Col, 49-51, head dept, 49-57, from asst prof to assoc prof, 51-61; assoc prof, 61-69, actg head dept, 65-66, PROF CHEM, ST LOUIS UNIV, 69-, DIR PREMED STUDIES, 68- Mem: AAAS; Am Chem Soc. Res: Dithizone as analytical reagent; absorption spectroscopy; solvent extraction. Mailing Add: Dept of Chem St Louis Univ St Louis MO 63103

TIPTON, HENRY C, b Oakville, Tenn, Mar 4, 15; m 43; c 4. NUTRITION, BIOCHEMISTRY. Educ: Miss State Univ, BS, 39, MS, 63, PhD(animal nutrit), 65. Prof Exp: ASST PROF BIOL SCI, UNIV S FLA, 65- Res: Animal nutrition; relative biological value of the isomers and analogue of methionine in poultry diets. Mailing Add: Dept of Biol Sci Univ of S Fla Tampa FL 33620

TIPTON, ISABEL HANSON, b Monroe, Ga, June 17, 09; m 34; c 4. PHYSICS. Educ: Univ Ga, BS, 29, MS, 30; Duke Univ, PhD(physics), 34. Prof Exp: Instr physics & chem, Cox Col, 34; tutor math, Ohio State Univ, 35-37; instr physics, Univ Ala, 43-44, res assoc physiol, 44-46; from instr to prof, 48-72, EMER PROF PHYSICS, UNIV TENN, KNOXVILLE, 72- Concurrent Pos: Consult, Oak Ridge Nat Lab, 50-; chmn, Bd Dirs, Stewart Labs, Inc, Tenn, 68-72. Mem: Am Phys Soc; Soc Appl Spectros; Health Physics Soc. Res: Raman effect in gases; spectrographic analysis of human tissue, diets and excreta. Mailing Add: 113 W 23rd St Long Beach Southport NC 28461

TIPTON, KENNETH WARREN, b Belleville, Ill, Nov 14, 32; m 57; c 2. PLANT GENETICS, AGRONOMY. Educ: La State Univ, BS, 55, MS, 59; Miss State Univ, PhD(agron, genetics), 69. Prof Exp: Teaching asst, 58-59, from asst prof to assoc prof, 59-74, PROF AGRON, AGR EXP STA, LA STATE UNIV, BATON ROUGE, 74- Mem: Am Soc Agron; Crop Sci Soc Am. Res: Varietal evaluation and adaptation of grain sorghum; varietal evaluation, breeding and genetics of small grains. Mailing Add: 210 AG Ctr Dept of Agron La State Univ Baton Rouge LA 70803

TIPTON, MERLIN J, b Watertown, SDak, Mar 23, 30; m 54; c 3. GEOLOGY. Educ: Univ SDak, BA, 55, MA, 58. Prof Exp: Geologist, 56-62, asst state geologist, 62-68, ASSOC STATE GEOLOGIST, SDAK GEOL SURV, 68- Mem: Fel Geol Soc Am. Res: Pleistocene geology and hydrogeology. Mailing Add: State Geol Surv Sci Ctr Univ SDak Vermillion SD 57069

TIPTON, SAMUEL RIDLEY, b Sylvester, Ga, Oct 6, 07; m 34; c 4. PHYSIOLOGY. Educ: Mercer Univ, AB, 28; Duke Univ, PhD(zool, physiol), 33. Prof Exp: Fel physiol, Sch Med, Univ Rochester, 33-34, instr, 34-35; instr, Ohio State Univ, 35-41; asst prof, Col Med, Wayne State Univ, 41-43; assoc prof, Sch Med, Univ Ala, 43-47; prof, 47-72, head dept, 67-71, EMER PROF ZOOL, UNIV TENN, KNOXVILLE, 72- Concurrent Pos: Consult, Oak Ridge Nat Lab, 47-70. Mem: AAAS; Am Physiol Soc; Am Soc Zoologists; Soc Gen Physiol; Wilson Ornith Soc. Res: Cell and tissue respiration; ionic equilibria in nerve; potassium chloride contraction in muscle; endocrine and dietary effects on tissue metabolism; protein synthesis and respiratory enzyme activity; x-ray and polysome formation. Mailing Add: 113 W 23rd St Long Beach Southport NC 28461

TIPTON, VERNON JOHN, b Springville, Utah, July 12, 20; m 43; c 5. PARASITOLOGY, MEDICAL ENTOMOLOGY. Educ: Brigham Young Univ, BS, 48, MS, 49; Univ Calif, PhD, 59. Prof Exp: Entomologist, Walter Reed Army Inst Res, Med Serv Corps, US Army, 49-52, Calif, 52-54, commanding officer, 37th Prev Med Co, Korea, 54-55, chief entom sect, 5th Army Med Lab, St Louis, Mo, 56-59, chief environ health br, Off Surgeon, Ft Amador, CZ, 57-62, chief entom br dept prev med, Med Field Serv Sch, Ft Sam Houston, Tex, 62-66, chief dept entom, 406th Med Lab, 66-68; assoc prof zool, 68-71, PROF ZOOL, BRIGHAM YOUNG UNIV, 71-, DIR CTR HEALTH ENVIRON STUDIES, 75- Res: Siphonaptera; mesostigmatid mites; systematics; bionomics. Mailing Add: Ctr Health & Environ Studies 785 WIDB Brigham Young Univ Provo UT 84601

TIRMAN, ALVIN, b Brooklyn, NY, Nov 27, 31. PHYSICAL CHEMISTRY, MATHEMATICS. Educ: Hofstra Col, AB, 53; Bowling Green State Univ, MA, 65; Carnegie-Mellon Univ, PhD(chem), 70. Prof Exp: Asst prof math, 69-75, ASSOC PROF MATH, KINGSPORT UNIV CTR, E TENN STATE UNIV, 75- Mem: Math Asn Am; Am Chem Soc. Res: Electrode kinetics. Mailing Add: Dept of Math Kingsport Univ Ctr E Tenn State Univ Kingsport TN 37662

TIRPAK, MICHAEL ROBERT, polymer chemistry, see 12th edition

TIRTHA, RANJIT, b Jullundur City, India, Jan 3, 28; m 53; c 3. GEOGRAPHY. Educ: Govt Col, Ludhiana, India, MA, 48; Univ NC, PhD(geog), 61. Prof Exp: Lectr geog, Govt Col, Mandi, India, 50-58; sr lectr, Govt Col, Solan, India, 61-62; lectr human geog, Univ Delhi, 62-66; asst prof geog, Wis State Univ, Oshkosh, 66-67; assoc prof, Northern Ill Univ, 67-70; ASSOC PROF GEOG & GEOL, EASTERN MICH UNIV, 70- Mem: Asn Am Geog; Am Geog Soc; Asn Asian Studies. Res: Population geography; city growth in India. Mailing Add: Dept of Geog & Geol Eastern Mich Univ Ypsilanti MI 48197

TISCHENDORF, JOHN ALLEN, b Lincoln City, Ind, July 22, 29; m 59; c 3. APPLIED STATISTICS. Educ: Evansville Col, AB, 50; Purdue Univ, MS, 52,

PhD(math statist), 55. Prof Exp: Mem tech staff, 57-59, supvr reliability & statist studies, Allentown, 59-64, SUPVR STATIST APPLN, BELL TEL LABS, HOLMDEL, 64- Concurrent Pos: Vis lectr, Rutgers Univ, 66 & 67 & Stanford Univ, 68-69. Mem: Am Statist Asn; Am Soc Qual Control. Res: Statistics applied to management problems; data analysis; statistical consulting; mathematical modeling; reliability; sampling; order statistics. Mailing Add: Laurelwood RD 1 Colt's Neck NJ 07722

TISCHER, ROBERT GEORGE, b Duluth, Minn, May 31, 12; m 41; c 4. MICROBIOLOGY, FOOD TECHNOLOGY. Educ: La State Univ, BS, 39, MS, 41; Univ Mass, PhD(food technol), 44. Prof Exp: Food technologist, Owens-Ill Glass Co, Calif, 44-46; dir qual control, John Inglis Frozen Foods, 46-47; head food processing lab, Iowa State Col, 47-53; dir food labs, Qm Food & Container Inst, 53-57; PROF MICROBIOL & HEAD DEPT, MISS STATE UNIV, 57- Mem: Am Soc Microbiol; Am Chem Soc; Inst Food Technologists. Res: Bacteriology of paper mill wastes; efficiency of home canning; glass packaging; freezing; quality of canned corn and grape juice; mathematical treatment of meat processing and microbiological problems; research management; microbiology of waste; stabilization ponds; feeding systems for long space voyages. Mailing Add: Dept of Microbiol Miss State Univ PO Drawer MB Mississippi State MS 39762

TISCHER, THOMAS NORMAN, b Milwaukee, Wis, Apr 21, 34. PHOTOGRAPHIC CHEMISTRY. Educ: Marquette Univ, BS, 56, MS, 58; Univ Wis, PhD(anal chem), 61. Prof Exp: RES ASSOC, EASTMAN KODAK CO, 61- Mem: Am Chem Soc; Soc Photog Scientists & Engrs; Soc Motion Picture & TV Engrs. Res: Photographic film and process chemistry and research; thin-layer chromatography; analytical separations. Mailing Add: 115 Heritage Circle Rochester NY 14615

TISCHIO, JOHN PATRICK, b Newark, NJ, Mar 17, 42; m 66; c 1. BIOCHEMISTRY. Educ: Fairleigh Dickinson Univ, BS, 65; Univ Rochester, PhD(biochem), 71. Prof Exp: ASST PROF BIOCHEM, PHILADELPHIA COL PHARM & SCI, 70- Concurrent Pos: Lectr, Wagner Free Inst Sci, 71- Mem: AAAS; Am Chem Soc; Am Asn Cols Pharm; Sigma Xi. Res: Primary sequence determination of proteins; solid phase methodology; protein structure and function correlations with emphasis on membrane proteins and the recognition sites involved in antigen-antibody immuno reactions. Mailing Add: Philadelphia Col of Pharm & Sci 43rd St & Woodland Ave Philadelphia PA 19104

TISCHLER, HERBERT, b Detroit, Mich, Apr 28, 24; m 54; c 2. INVERTEBRATE PALEONTOLOGY. Educ: Wayne State Univ, BS, 50; Univ Calif, Berkeley, MA, 55; Univ Mich, PhD(geol), 61. Prof Exp: Instr geol, Wayne State Univ, 56-58; assoc prof earth sci, Northern Ill Univ, 58-65; PROF GEOL & CHMN DEPT, UNIV NH, 65- Concurrent Pos: NSF sci fac fel, Columbia Univ, 64-65. Mem: Fel Geol Soc Am; Paleont Soc; Am Asn Petrol Geol; Soc Econ Paleont & Mineral; Nat Asn Geol Teachers. Res: Paleoecology of marine invertebrates; ecology of benthic foraminifera; carbonate petrology; stratigraphy. Mailing Add: Dept of Earth Sci James Hall Univ of NH Durham NH 03824

TISDALE, EDWIN WILLIAM, b Argyle, Man, Mar 10, 10; nat US; m 36; c 2. RANGE SCIENCE. Educ: Univ Man, BSc, 30; Univ Minn, MS, 45, PhD(plant ecol), 48. Prof Exp: Asst agrostologist, Dom Range Exp Sta, Alta, 31-35, agrostologist range & forage crops res, BC, 35-37, officer-in-charge, Dom Range Exp Substa, BC, 37-41, officer-in-charge range res, Sask, 41-47; assoc prof range mgt, 47-53, prof & assoc dir forest, wildlife & range exp sta, 53-75, EMER PROF RANGE MGR, COL FORESTRY, WILDLIFE & RANGE SCI, UNIV IDAHO, 75- Concurrent Pos: NSF fel, Univ Calif, Berkeley, 58-59; consult, Dry-Lands Res Inst, Mid East & NAfrica, 65-66; consult grazing resources, US Peace Corps, 74-75. Mem: Fel AAAS; Ecol Soc Am; Soc Range Mgt (vpres, 56, pres, 57). Res: Ecology and grazing resources of the Pacific Northwest; ecology and management of sagebrush-grass and mountain shrub vegetation; ecotypic variation in range species. Mailing Add: 915 W C St Moscow ID 83843

TISDALE, SAMUEL LUTHER, soil chemistry, see 12th edition

TISHKOFF, GARSON HAROLD, b Rochester, NY, Aug 8, 23; m 59. MEDICINE, HEMATOLOGY. Educ: Univ Rochester, BS, 44, PhD(pharmacol), 51, MD, 53. Prof Exp: AEC assoc, Univ Rochester, 46-52; instr med, Med Sch, Tufts Univ, 56-57 & Harvard Med Sch, 57-60; from asst prof to assoc prof, Sch Med, Univ Calif, Los Angeles, 60-71; PROF MED, MICH STATE UNIV & DIR, LANSING REGIONAL BLOOD PROG, AM RED CROSS, 71- Concurrent Pos: Nat Acad Sci-Nat Res Coun fel, New Eng Ctr Hosp, Tufts Univ, 55-56; assoc, Beth Israel Hosp, Boston, 57-60. Mem: AAAS; Am Soc Hemat; Int Soc Hemat; Int Soc Thrombosis & Haemostasis; Int Soc Blood Transfusion. Res: Internal medicine; hemolytic anemia; blood coagulation; blood transfusion. Mailing Add: 1800 E Grand River Ave Lansing MI 48911

TISHLER, FREDERICK, b Boston, Mass, Aug 9, 35; m 57; c 2. ANALYTICAL CHEMISTRY. Educ: Mass Col Pharm, BS, 57; Univ Mich, MS, 59, PhD(pharmaceut chem), 61. Prof Exp: Sr chemist, Ciba Pharmaceut Co, 61-66, from asst mgr to mgr anal chem, 67-69, mgr, Geigy Pharm Co, 69-70, asst dir, 70-73, ASST TO EXEC DIR QUAL CONTROL, PHARMACEUT DIV, CIBA-GEIGY, 73- Mem: Am Pharmaceut Asn; Am Chem Soc; NY Acad Sci. Res: Fluorescent determination and mechanism of fluorescence of steroidal and nonsteroidal compounds; reaction kinetics; functional group analysis; drug analysis of biological material. Mailing Add: CIBA-GEIGY Pharmaceut Div Summit NJ 07901

TISHLER, MAX, b Boston, Mass, Oct 30, 06; m 34; c 2. ORGANIC CHEMISTRY. Educ: Tufts Col, BS, 28; Harvard Univ, AM, 33, PhD(org chem), 34. Hon Degrees: DSc, Tufts Col, 56, Univ Strathclyde, 69, Rider Col, 70, Fairfield Univ, 72 & Upsala Col, 72; DS, Bucknell Univ, 62 & Philadelphia Col Pharm, 66; DEng, Stevens Inst Technol, 66. Prof Exp: Asst, Tufts Col, 28; res assoc, Harvard Univ, 34-36, instr chem, 36-37; res chemist, Merck & Co, Inc, 37-41, sect head in chg process develop, 41-44, dir develop res, 44-53, process res & develop div, 53-54, vpres & exec dir sci activities, 54-56, pres, Merck Sharp & Dohme Res Labs, NJ, 57-70, mem bd dirs, Merck & Co, Inc, 62-70, sr vpres res & develop, 69-70; prof chem, 70-74, EMER PROF CHEM, WESLEYAN UNIV, 74-, UNIV PROF SCI, 72- Concurrent Pos: Estab Max Tishler lectr, Harvard & Max Tishler scholar, Tufts Univ, 51; ed, Org Syntheses, 60-61; life trustee, Tufts Univ & Union Jr Col; assoc trustee sci, Univ Pa, 62-66; Rennebohn lectr, Univ Wis, 63; Du Pont lectr, Dartmouth Col, 69; Welch Found lectr, Univ Tex, 69; Cecil C Brown lectr, Stevens Inst Technol, 71; Kauffman mem lectr, Ohio State Univ, 67; mem bd visitors, Sch Pub Health & Dept Chem, Harvard Univ; mem admin bd, Tufts-New Eng Med Ctr; sci bd gov, Weizmann Inst Sci. Honors & Awards: Indust Res Inst Medal, 61; Swedish Royal Acad Eng Sci Award, 64; Sturmer Lect Award, Philadelphia Col Pharm; Soc Chem Indust Medal, 63; Chem Pioneer Award, Am Inst Chemists, 68, Freedman Patent Award, 70; Priestley Medal, Am Chem Soc, 70. Mem: Nat Acad Sci; Am Chem Soc (pres, 72); fel Soc Chem Indust; fel Am Inst Chemists; Am Acad Arts & Sci. Res: Development of processes; synthesis or isolation of pharmacologically active compounds; synthesis of vitamins, drugs, alkaloids and amino acids; isolation alkaloids; antibiotics; steroid synthesis. Mailing Add: Dept of Chem Wesleyan Univ Middletown CT 06457

TISHLER, PETER VERVEER, b Boston, Mass, July 18, 37; m 60; c 2. MEDICAL GENETICS. Educ: Harvard Univ, AB, 59; Yale Univ, MD, 63. Prof Exp: Staff assoc, Nat Inst Arthritis & Metab Dis, 66-68; house officer internal med II & IV, Harvard Med Serv, 63-66, res fel med, Thorndike Mem Lab & Channing Lab, 68-69, STAFF MEM MED & GENETICS, CHANNING LAB, BOSTON CITY HOSP & HARVARD MED SCH, 69- Concurrent Pos: Asst physician, Dept Pediat, Boston City Hosp, 69-, assoc vis physician, Dept Med, 71-; assoc, Ctr Human Genetics, Harvard Med Sch, 71, asst prof med, 72. Mem: AAAS; NY Acad Sci; Am Soc Human Genetics; Am Fedn Clin Res. Res: Genetics of chronic disease; biochemistry of diseases of porphyrin metabolism; gene mapping. Mailing Add: Channing Lab Boston City Hosp Boston MA 02118

TISI, GENNARO MICHAEL, b New York, NY, Sept 26, 35; m 57; c 3. PULMONARY DISEASES, INTERNAL MEDICINE. Educ: Fordham Univ, BS, 56; Georgetown Univ, MD, 60. Prof Exp: From intern to resident med, Georgetown Univ Hosp, 60-65, pulmonary fel, 65-67; res fel, Cardiovasc Res Inst, Univ Calif, San Francisco, 67-68; asst prof, 68-73, ASSOC PROF MED, SCH MED, UNIV CALIF, SAN DIEGO, 73- Concurrent Pos: NIH fel, 65-68, spec fel, 68; pulmonary consult, US Naval Hosp, San Diego, 68- & US Naval Hosp, Camp Pendleton, 72-; chief, Pulmonary Sect, San Diego Vet Hosp, 72- Mem: Am Col Physicians; Am Col Chest Physicians; Am Physiol Soc; Am Fedn Clin Res. Res: Pulmonary mechanics. Mailing Add: San Diego Vet Admin Hosp La Jolla CA 92037

TISINGER, RICHARD MARTIN, JR, b Harrisonburg, Va, Oct 26, 29; m 55; c 3. DATA PROCESSING. Educ: Reed Col, BA, 51; Johns Hopkins Univ, PhD(nuclear physics), 63. Prof Exp: Staff mem weapons div, Los Alamos Sci Lab, Univ Calif, 52-56; jr instr, Johns Hopkins Univ, 57-61, res asst, 61-63; staff mem weapons div, 63-74, STAFF MEM LASER DIV, LOS ALAMOS SCI LAB, UNIV CALIF, 74- Res: Fast data acquisition in laser spectroscopy. Mailing Add: 3 Hopi Lane Los Alamos NM 87544

TISLOW, RICHARD FREDERICK, b Czernowitz, Austria, Mar 6, 06; nat US; m 46. PSYCHIATRY, PHARMACOLOGY. Educ: Univ Vienna, MD, 31; Univ Warsaw, MedSciD, 36. Prof Exp: Sr asst, Dept Gen & Exp Path, Univ Warsaw, 34-38; asst, Sch Med, Johns Hopkins Univ, 39; asst, Peter Bent Brigham Hosp, Harvard Med Sch, Boston, Mass, 40-43; dir biol res labs, Schering Corp, 43-53; chief pharmacologist, Res Div, Wyeth Labs, Inc, Pa, 53-71; DIR PSYCHIAT EMERGENCY SERV, WEST PHILADELPHIA COMMUNITY MENT HEALTH CONSORTIUM, 71- Concurrent Pos: Instr dept psychiat & pharmacol, Med Sch, Univ Pa, 71- Mem: Am Psychiat Asn; Am Soc Pharmacol & Exp Therapeut; Am Physiol Soc; Soc Exp Biol & Med; fel Am Col Neuropsychopharmacol. Res: Steroid hormones, adrenal function, extraction hypophyseal corticotropin; antithyroid drugs; development of chlorpheniramine, carphenazine, iprindol; psychosomatic medicine; psychophysiological signs and signals; ego strength parameters as outcome predictor in drug addiction. Mailing Add: W Philadelphia Community Ment Health Consortium Box 8076 Philadelphia PA 19101

TISONE, GARY C, b Boulder, Colo, Dec 24, 37; m 59; c 4. ATMOSPHERIC PHYSICS. Educ: Univ Colo, BS, 59, PhD(physics), 67. Prof Exp: Physicist, Vallecitos Atomic Lab, Gen Elec Co, 59-61; physicist, Nat Bur Standards, 61-62; res assoc atomic physics, Univ Colo, 66-67; STAFF MEM, SANDIA CORP, 67- Mem: Am Phys Soc. Res: Electron-negative ion collisions; photodetachment of negative ions; gaseous electronics; physics of the upper atmosphere; high power gas laser research. Mailing Add: 10212 Chapala Place NE Albuquerque NM 87111

TISONE, THOMAS C, b Boulder, Colo, June 9, 40; m 66; c 2. MATERIALS SCIENCE. Educ: Colo Sch Mines, BS, 62; Northwestern Univ, PhD(mat sci), 67. Prof Exp: Mem tech staff, Bell Tel Labs, 67-74; MGR, GOULD LABS, 74- Mem: Am Soc Metals; Am Vacuum Soc. Res: Materials and materials process development for electronics applications. Mailing Add: 326 Forest Lane Schaumburg IL 60192

TISUE, GEORGE THOMAS, b Carroll, Iowa, Nov 25, 40; m 61; c 2. ENVIRONMENTAL CHEMISTRY, ANALYTICAL CHEMISTRY. Educ: Beloit Col, BS, 61; Yale Univ, PhD(org chem), 66. Prof Exp: Res chemist tech ctr, Celanese Chem Co, Tex, 66 & 68; NIH fel plant biochem, Univ Freiburg, 66-67; from asst prof to assoc prof chem, Beloit Col, 68-74; CHEMIST, ARGONNE NAT LAB, 74- Mem: AAAS; Am Chem Soc. Res: Biogeochemical cycling and effects of heavy metal pollutants in the Great Lakes. Mailing Add: Radiol & Environ Res Div Argonne Nat Lab Argonne IL 60439

TISZA, LASZLO, b Budapest, Hungary, July 7, 07; nat US; div. PHYSICS. Educ: Univ Budapest, PhD(physics), 32. Prof Exp: Res assoc, Phys Tech Inst, Kharkov, 35-37; res assoc, Col France, 37-40; instr physics, 41-45, from asst prof to assoc prof, 45-60, prof, 60-73, EMER PROF PHYSICS & SR LECTR, MASS INST TECHNOL, 73- Concurrent Pos: Guggenheim fel, 62-63; vis prof, Univ Paris, 62-63. Mem: Fel Am Phys Soc; fel Am Acad Arts & Sci. Res: Theoretical physics; statistical thermodynamics; foundations of quantum mechanics. Mailing Add: Dept of Physics Mass Inst of Technol Cambridge MA 02139

TISZA, VERONICA ELIZABETH BENEDEK, b Szeged, Hungary, Aug 7, 12; nat US; m 38. MEDICINE. Educ: Eötvös Lorand Univ, Budapest, MD, 37; Am Bd Pediat, dipl, 51; Am Bd Psychiat & Neurol, dipl, 55. Prof Exp: Asst prof pediat, Sch Med, Tufts Univ, 52-61, asst prof psychiat, 52-59, assoc prof, 59-61; assoc prof child psychiat, Sch Med, Univ Pittsburgh, 61-68; vol psychiatrist, Hamstead Heath Nursery, Eng, 68-69; asst prof, 69-72, ASSOC PROF PSYCHIAT, HARVARD MED SCH, 72- Concurrent Pos: Dir psychiat serv children, New Eng Med Ctr, 52-61; staff psychiatrist, Judge Baker Guid Ctr, 56-61; lectr, Sch Social Work, Boston Univ, 59-61; assoc dir psychiat clin, Children's Hosp, Pittsburgh, Pa, 61-68; sr assoc psychiat & dir psychiat training for pediatricians, Children's Hosp Med Ctr, Boston, 69-; dir psychiat training, Children's Hosp Med Ctr-Judge Baker Guid Ctr, 71- Mem: Am Psychiat Asn; Am Orthopsychiat Asn; fel Am Acad Pediat; Am Acad Child Psychiat. Res: Emotional problems of hospitalized children; development problems of childrem with congenital malformations. Mailing Add: 300 Longwood Ave Boston MA 20115

TITCHENER, EDWARD BRADFORD, b Cambridge, Mass, July 15, 27; m 52; c 2. BIOCHEMISTRY. Educ: Univ Mich, BS, 51; Ohio State Univ, MS, 54, PhD(physiol chem), 56. Prof Exp: Asst biochem, Ohio State Univ, 53-56; from asst prof to assoc prof, 58-74, PROF BIOCHEM, UNIV ILL COL MED, 74- Concurrent Pos: NIH trainee, Enzyme Inst, Univ Wis, 56-58. Mem: AAAS; Am Chem Soc; Am Soc Biol Chemists; NY Acad Sci. Res: Transfer RNA. Mailing Add: Dept of Biochem Univ of Ill Col of Med Chicago IL 60612

TITCHENER, JAMES LAMPTON, b Binghamton, NY, Apr 9, 22; m; c 4. PSYCHIATRY, PSYCHOANALYSIS. Educ: Princeton Univ, AB, 46; Duke Univ, MD, 49; Chicago Psychoanal Inst, cert psychoanal, 64. Prof Exp: Intern psychiat,

Walter Reed Gen Hosp, 49-50; resident, Cincinnati Gen Hosp, 50-54; from instr to assoc prof, 54-68, PROF PSYCHIAT, MED CTR, UNIV CINCINNATI, 68-; MEM FAC PSYCHOANAL, CINCINNATI PSYCHOANAL INST, 74- Concurrent Pos: Career investr, USPHS, 55-60 & NIMH, 57-62; training & supv analyst, Cincinnati Psychoanal Inst, 74-; attend physician, Cincinnati Gen Hosp. Mem: Fel AAAS; fel Am Psychiat Asn; Am Psychosom Soc; Am Psychoanal Asn. Res: Study of the effects of physical trauma on personality functioning; family and marital dynamics and marital therapy; psychological trauma resulting from disasters such as the Buffalo Creek disaster. Mailing Add: Dept of Psychiat Univ of Cincinnati Col of Med Cincinnati OH 45267

TITELBAUM, SYDNEY, b Luck, Russia, Apr 24, 13; US citizen; m 39; c 1. ANIMAL PHYSIOLOGY, EVOLUTIONARY BIOLOGY. Educ: Univ Chicago, PhB, 33, PhD(physiol), 38, JD, 42. Prof Exp: Teaching asst physiol, Univ Chicago, 34-38; chief physiologist, Chicago Biol Res Lab, 39-59; lectr forensic med sch med, Loyola Univ Chicago, 46-64; PROF BIOL, CITY COLS CHICAGO, 60- Concurrent Pos: Consult-examr, NCent Asn Cols & Sec Schs, 66-; dean, Bogan Col, 67-68. Honors & Awards: Award, Asn Off Racing Chem, 60. Mem: AAAS; Am Inst Biol Sci; Asn Off Racing Chem (ed, 56-58, pres, 57). Res: Physiology of sleep. Mailing Add: Dept of Biol Loop Col 64 E Lake St Chicago IL 60601

TITIEV, MISCHA, b Kremenchug, Russia, Sept 11, 01; nat US; m 35; c 1. ETHNOLOGY. Educ: Harvard Univ, AB, 23, AM, 24, PhD(anthrop), 35. Prof Exp: Asst mus cur & jr archaeologist, Nat Park Serv, 35-36; instr anthrop, 36-39, from asst prof to assoc prof, 39-51, prof, 51-70, EMER PROF ANTHROP, UNIV MICH, ANN ARBOR, 70- Concurrent Pos: Fulbright award, Australia, 54. Mem: Fel Am Anthrop Asn. Res: Ethnology of the Hopi Indians; social organization of Japanese in Peru and Japan; ethnology of Araucanian Indians of Chile. Mailing Add: 910 Heather Way Ann Arbor MI 48104

TITKEMEYER, CHARLES WILLIAM, b Rising Sun, Ind, Jan 14, 19; m 47; c 2. ANATOMY, BACTERIOLOGY. Educ: Ohio State Univ, DVM, 49; Mich State Univ, MS, 51, PhD, 56. Prof Exp: From instr to prof anat, Mich State Univ, 49-69; PROF VET ANAT & HEAD DEPT, SCH VET MED, LA STATE UNIV, BATON ROUGE, 69- Concurrent Pos: Consult vet, Univ Ky Contract Team, Bogor, Indonesia, 60-62; prof & head dept vet sci, Univ Nigeria, 66-68. Mailing Add: 1148 Aurora Pl Baton Rouge LA 70806

TITLE, REUBEN SEYMOUR, physics, see 12th edition

TITLEY, SPENCER ROWE, b Denver, Colo, Sept 27, 28; m 51; c 3. GEOLOGY, GEOCHEMISTRY. Educ: Colo Sch Mines, GeolE, 51; Univ Ariz, PhD(geol), 58. Prof Exp: Jr geologist, NJ Zinc Co, 51, staff geologist, 53-55, res geologist, 58-60; instr geol, 57-58, from asst prof to assoc prof, 60-67, PROF GEOL, UNIV ARIZ, 67- Concurrent Pos: Geologist, US Geol Surv, 63- Mem: Soc Econ Geol; Geol Soc Am; Soc Explor Geophys; Am Inst Mining, Metall & Petrol Eng; Am Geophys Union. Res: Geology of the porphyry copper deposits; mineralogy of copper-lead-zinc deposits; geology and mineral deposits of Pacific Basin. Mailing Add: Dept of Geosci Univ of Ariz Tucson AZ 85721

TITMAN, PAUL WILSON, b Lowell, NC, Aug 30, 20. BOTANY. Educ: Belmont Abbey Col, BS, 39; Univ NC, AB, 41, MA, 49; Harvard Univ, PhD(biol), 52. Prof Exp: Instr bot, Univ NC, 48-49; asst prof, Univ Louisville, 52-53; instr, Univ Conn, 53-55; assoc prof, 55-61, PROF BOT, CHICAGO STATE UNIV, 61- Concurrent Pos: Res resident, Harvard Univ, 54; chmn coun faculties, Ill State Bd Cols & Univs; consult. Mem: Bot Soc Am; fel Royal Hort Soc. Res: Morphogenesis; systematic anatomy; microbiology; paleontology. Mailing Add: Dept of Biol Chicago State Univ Chicago IL 60628

TITONE, LUKE VICTOR, b Marsala, Italy, Oct 25, 11; nat US; m 39; c 4. PHYSICS. Educ: NY Univ, MS, 40. Prof Exp: Instr physics, NY Univ, 40-51; PROF PHYSICS, MANHATTAN COL, 51- Mem: Am Phys Soc. Res: Atomic and nuclear physics. Mailing Add: 676 N Broadway Yonkers NY 10701

TITTERTON, PAUL JAMES, b Copiague, NY, Feb 23, 40; m 63; c 3. ELECTRO OPTICS. Educ: Boston Col, BS, 61; Brandeis Univ, MS, 63, PhD(physics), 67. Prof Exp: SR ENG SPECIALIST, GTE SYLVANIA INC, 66- Mem: Optical Soc Am; Am Meteorol Soc. Res: Atmospheric effects on laser beams; precise optical ranging techniques; optimum optical communication methodology; sensitive optical receiver research. Mailing Add: GTE Sylvania Inc PO Box 188 Mountain View CA 94042

TITTIGER, FRANZ, b Bela Crkva, Yugoslavia, June 22, 29; Can citizen; m 59; c 2. MEAT SCIENCE, MICROBIOLOGY. Educ: Univ Munich, Vet, 54, Dr Med Vet, 55. Prof Exp: Pvt pract, 55-59; vet inspector, Contagious Dis Div, 59-63, pathologist, Animal Path Div, 64-67, DIR GUELPH AREA LAB, ANIMAL PATH DIV, CAN AGR, 67- Res: Meat hygiene, detection of noxious substances, food poisoning organisms, parasites, chemical residues in meat products; composition and adulteration of meat products. Mailing Add: Animal Path Lab Health Animal Br Can Agr 620 Gordon St Guelph ON Can

TITTLE, CHARLES WILLIAM, b Bonham, Tex, Nov 11, 17; m 43; c 7. PHYSICS. Educ: NTex State Univ, BS, 39, MS, 40; Mass Inst Technol, PhD(physics), 48. Prof Exp: Instr physics, NTex State Univ, 40-41; instr pre-radar, Southern Methodist Univ, 43; asst prof physics, NTex State Univ, 43-44; instr elec commun radar sch, Mass Inst Technol, 44-45; from asst prof to prof physics, NTex State Univ, 48-51; head nuclear physics sect, Gulf Res & Develop Co, 51-55; dir western div, Tracerlab, Inc, 55-56, assoc tech dir, 56-57; prof nuclear eng, 57-63, chmn dept mech eng, 61-65, chmn dept physics, 65-75, PROF PHYSICS & MECH ENG, SOUTHERN METHODIST UNIV, 63- Concurrent Pos: Consult nuclear shielding proj, Mass Inst Technol, 48-51; consult, S W Marshall, Jr, 47, Gulf Res & Develop Co, 50, Atlantic Refining Co, 57-58, Western Co, 58-, Well Reconnaissance, Inc, 62-64, Ling-Temco-Vought, 63-64, Mobil Oil Corp, 63- & Nuclear-Chicago Corp, 64-66. Mem: Am Phys Soc; Am Nuclear Soc; Am Asn Physics Teachers. Res: Neutron physics; nuclear well logging; applications of radioisotopes; boundary value problems; fundamental concepts of physics. Mailing Add: Dept of Physics Southern Methodist Univ Dallas TX 75275

TITTLER, IRVING ALBERT, b New York, NY, Aug 28, 08; m 35; c 2. ZOOLOGY. Educ: Bethany Col, WVa, BS, 29; Columbia Univ, MA, 31, PhD(zool), 35. Prof Exp: Asst biol, Bethany Col, WVa, 28-29; asst zool, Columbia Univ, 29-33; instr biol exten div, Brooklyn Col, 35; instr, Yeshiva Col, 35-37; instr, 37-46, from asst prof to assoc prof, 46-58, PROF BIOL, BROOKLYN COL, 58- Concurrent Pos: Am Cancer Soc fel, Hopkins Marine Sta, Stanford Univ, 47-48. Mem: Am Soc Zool; fel Am Cancer Soc; Soc Protozool; Soc Exp Biol & Med; NY Acad Sci. Res: Protozoology; carcinogens; microbiology; growth and metabolism of protozoa. Mailing Add: Dept of Biol Brooklyn Col Brooklyn NY 11210

TITTMAN, JAY, b Bayonne, NJ, Dec 28, 22; m 44; c 3. NUCLEAR SCIENCE. Educ: Drew Univ, AB, 43; Columbia Univ, AM, 48, PhD(physics), 51. Prof Exp: Nuclear physicist, 51-54, assoc res physicist, 54-55, res physicist, 55-66, head dept physics res, Schulmberger-Doll Res Ctr, 67-72, SR STAFF SCIENTIST, SCHLUMBERGER WELL SURV CORP, 66-, HEAD ENG PHYSICS DEPT, 72- Concurrent Pos: Adj prof phys sci, New Eng Inst, 71-72; mem subcomt nuclear logging, Am Petrol Inst, 73- Mem: AAAS; Am Phys Soc; Soc Petrol Engrs. Res: Slow neutron scattering; nuclear detection instruments; slowing down of neutrons; gamma ray scintillation spectroscopy; gamma ray diffusion; ion accelerators; fast neutron interactions; sonics; geophysical instruments and methods. Mailing Add: 12918 Hermitage Lane Houston TX 77024

TITTMANN, BERNHARD R, b Moshi, Tanzania, Sept 15, 35; US citizen; m 66; c 1. SOLID STATE PHYSICS, ACOUSTICS. Educ: George Washington Univ, BS, 57; Univ Calif, Los Angeles, PhD(solid state physics), 65. Prof Exp: Mem tech staff res & develop aerospace & systs group, Hughes Aircraft Co, 57-61; res asst solid state physics, Univ Calif, Los Angeles, 61-65, asst prof & fel, 65-66; MEM TECH STAFF, SCI CTR, ROCKWELL INT, 66- Mem: Am Phys Soc; Am Geophys Union. Res: Acoustic surface waves; acoustic properties of lunar rock; ultrasonic absorption of type I and type II superconductors; dislocation-electron interaction; superconductivity in high pressure polymorphs of semiconductors; ferromagnetic modes in epitaxial single crystal yttrium iron garnet; microwave conformal array antennas. Mailing Add: 248 Teasdale St Thousand Oaks CA 91360

TITUS, CHARLES JOSEPH, b Mt Clemens, Mich, June 23, 23; m 49; c 2. MATHEMATICS. Educ: Univ Detroit, BSc, 44; Brown Univ, ScM, 45; Syracuse Univ, PhD(math), 48. Prof Exp: Instr math, Syracuse Univ, 48; from instr to assoc prof, 49-62, PROF MATH, UNIV MICH, ANN ARBOR, 63- Concurrent Pos: Vis prof, Univ Calif, 58-59; mem, Inst Defense Anal, 60-61. Mem: Am Math Soc. Res: Complex variables and generalizations; transformation semigroups; qualitative theory of differential equations; communications; geometric analysis; differential topology. Mailing Add: Dept of Math Univ of Mich Ann Arbor MI 48104

TITUS, DONALD DEAN, b Worland, Wyo, Mar 22, 44; m 66; c 2. INORGANIC CHEMISTRY. Educ: Univ Wyo, BS, 66; Calif Inst Technol, PhD(chem), 71. Prof Exp: ASST PROF CHEM, TEMPLE UNIV, 71- Concurrent Pos: Am Chem Soc Petrol Res Fund grant, 71-74. Mem: Am Chem Soc; Am Crystallog Asn. Res: Transition-metal complexes, non-rigidity in hydrides, the trans-influence and selenium complexes; x-ray crystal structures. Mailing Add: Dept of Chem Temple Univ Philadelphia PA 19122

TITUS, DUDLEY SEYMOUR, b Ithaca, NY, Mar 18, 29; m 54. FOOD TECHNOLOGY. Educ: Cornell Univ, BS, 52; State Col Wash, MS, 54; Univ Ill, PhD(food tech), 57. Prof Exp: Asst food technol, State Col Wash, 52-54; food technologist, Merck & Co, Inc, 57-67; head food technol sect, Mallinckrodt Chem Works, 67-72, res & develop scientist, 72, MKT RES SPECIALIST, MALLINCKRODT, INC, 72- Mem: Am Asn Cereal Chem; Inst Food Technol. Res: Food microbiology and preservation; human nutrition; cereal products canned and frozen foods; dairy products; food additives. Mailing Add: Mallinckrodt Inc PO Box 5439 St Louis MO 63147

TITUS, ELWOOD OWEN, b Rochester, NY, Sept 20, 19; m 51; c 2. ORGANIC CHEMISTRY. Educ: Williams Col, BA, 41; Columbia Univ, PhD(chem), 47. Prof Exp: Asst antimalarials div war res, Columbia Univ, 43-46; res assoc antibiotics & metab prod, Squibb Inst Med Res, 46-50; CHEMIST CHEM PHARMACOL LAB, NAT HEART & LUNG INST, 50- Concurrent Pos: Mem, Nat Res Coun, 66- Mem: Am Chem Soc; Am Soc Biol Chem; Am Soc Pharmacol & Exp Therapeut; NY Acad Sci; Biophys Soc. Res: Organic synthesis; biochemistry; metabolism of biologically active compounds; application of counter-current distribution to metabolic studies on 4-amino-quinoline antimalarials; mechanism of action of steroids and catecholamines; lipid metabolism; biochemistry of cell membranes. Mailing Add: Sect on Molecular Pharmacol Nat Heart & Lung Inst Bethesda MD 20014

TITUS, FRANK BETHEL, JR, hydrogeology, see 12th edition

TITUS, JACK L, b South Bend, Ind, Dec 7, 26; m 49; c 5. PATHOLOGY. Educ: Univ Notre Dame, BS, 48; Wash Univ, MD, 52; Univ Minn, PhD(path), 62. Prof Exp: Physician, Rensselaer, Ind, 53-57; assoc prof path, Mayo Grad Sch Med, Univ Minn, 61-72, prof, Mayo Clin, 71-72; PROF PATH & CHMN DEPT, BAYLOR COL MED, 72- Concurrent Pos: Fel path, Mayo Grad Sch Med, Univ Minn, 57-61; chief path serv, Methodist Hosp; pathologist-in-chief, Harris County Hosp Dist. Honors & Awards: Billings Gold Medal, AMA, 68, Hoektoen Gold Medal, 69. Mem: AMA; Am Asn Pathologists & Bacteriologists; Am Soc Clin Path; Int Acad Path; Col Am Pathologists. Res: Cardiac conduction system; cardiac anomalies; valvular heart disease; ischemic heart disease; atherosclerosis. Mailing Add: Dept of Path Baylor Col of Med Houston TX 77025

TITUS, JOHN S, b Mich, Apr 19, 23; m 46; c 4. POMOLOGY. Educ: Mich State Col, BS, 46, MS, 47; Cornell Univ, PhD(pomol), 51. Prof Exp: Instr hort, Mich State Col, 46-48; asst, Cornell Univ, 49-51; instr pomol, 51-53, from asst prof to assoc prof, 53-66, PROF POMOL, DEPT HORT, UNIV ILL, URBANA, 66- Concurrent Pos: Res assoc, Univ Calif, Davis, 62-63; Fulbright-Hays lectr, Univ Col, Dublin, 71-72. Honors & Awards: Gourley Award, Am Soc Hort Sci, 74. Mem: Fel

TITUS, RICHARD LEE, b Dayton, Ohio, Aug 6, 34. ORGANIC CHEMISTRY. Educ: DePauw Univ, BA, 56; Mich State Univ, PhD(chem), 64. Prof Exp: Instr chem, Univ Toledo, 62-64, asst prof, 64-67; from asst prof to assoc prof, 67-73, PROF CHEM, UNIV NEV, LAS VEGAS, 73- Mem: AAAS; Am Chem Soc. Res: Synthesis of heterocyclic compounds. Mailing Add: Dept of Chem Univ of Nev Las Vegas NV 89154

TITUS, ROBERT CHARLES, b Paterson, NJ, Aug 9, 46. INVERTEBRATE PALEONTOLOGY. Educ: Rutgers Univ, BS, 68; Boston Univ, AM, 71, PhD(geol), 74. Prof Exp: Instr geol, Windham Col, 73-74; ASST PROF GEOL, HARTWICK COL, 74- Mem: Sigma Xi; Paleont Soc; Paleont Asn; Geol Soc Am; Soc Econ Paleontologists & Mineralogists. Res: Paleontology of Middle Ordovician fossil invertebrate benthic communities. Mailing Add: Dept of Geol Hartwick Col Oneonta NY 13820

TITUS, WALTER FRANKLIN, b Mamaroneck, NY, Nov 9, 25; m 52; c 4. PHYSICS. Educ: Amherst Col, BA, 48; Harvard Univ, PhD(physics), 54. Prof Exp: Physicist, Nat Bur Standards, 54-58; asst prof physics, Dartmouth Col, 58-65; ASSOC PROF PHYSICS, UNION COL, NY, 65- Mem: Am Phys Soc; Am Asn Physics Teachers. Res: Nuclear and high energy electron physics; particle detection. Mailing Add: Dept of Physics Union Col Schenectady NY 12308

TITUS, WILLIAM JAMES, b Oakland, Calif, Dec 13, 41; m 67. LOW TEMPERATURE PHYSICS, STATISTICAL MECHANICS. Educ: Univ Calif, Davis, BS, 63; Stanford Univ, MS, 65, PhD(physics), 68. Prof Exp: Res assoc physics,

Univ Minn, 68-70; ASST PROF PHYSICS, CARLETON COL, 70- Mem: Am Phys Soc. Mailing Add: Dept of Physics Carleton Col Northfield MN 55057

TIVIN, FRED, analytical chemistry, see 12th edition

TIXIER, MAURICE PIERRE, b Clermont, France, Feb 1, 13; nat US; m 39; c 2. GEOPHYSICS. Educ: Ecole des Arts et Metiers d'Erquelinnes, Belg, Eng, 32. Prof Exp: Field engr, Societe de Prospection Electrique, Paris, 34-35; dist engr, Tex, 35-39, area mgr, Colo, 41-49, chief petrol engr, Tex, 49-52, chief field develop engr, 52-57, mgr field develop, 57-66, dir field interpretation, 66-69, dir prod logging, 69-72, TECH ADV, SCHLUMBERGER WELL SURV CORP, 72- Honors & Awards: Gold Medal, Soc Prof Well Log Analysts, 70. Mem: Soc Explor Geophys; Am Asn Petrol Geol; Am Geophys Union; Am Inst Mining, Metall & Petrol Eng; Soc Petrol Eng. Res: Electrical logging; electrochemistry; well bore geophysics. Mailing Add: 2319 Bolsover Rd Houston TX 77005

TIZARD, IAN RODNEY, b Belfast, Northern Ireland, Oct 27, 42; m 69; c 2. IMMUNOLOGY. Educ: Univ Edinburgh, BVMS, 65, BSc, 66; Cambridge Univ, PhD(immunol), 69. Prof Exp: Med Res Coun Can fel, Univ Guelph, 69-71; vet res officer, Animal Dis Res Asn, 71-72; asst prof vet immunol, 72-74, ASSOC PROF VET IMMUNOL, UNIV GUELPH, 74- Mem: Can Soc Immunol; Brit Soc Immunol; Royal Col Vet Surg; Royal Soc Trop Med & Hygience. Res: Cellular immunity; cytophilic antibodies; immunity to Protozoa; toxoplasmosis; trypanosomiasis. Mailing Add: Dept of Vet Microbiol Univ of Guelph Guelph ON Can

TJALMA, RICHARD ARLEN, b Holland, Mich, Sept 2, 29; m 51; c 2. VETERINARY MEDICINE. Educ: Mich State Univ, BS, 50, DVM, 54; Harvard Univ, MS, 65; Am Bd Vet Pub Health, dipl, 65. Prof Exp: Asst prof infectious dis res sect col med, Univ Iowa, 56-60; chief epizootiology sect, Nat Cancer Inst, 61-68; epidemiologist, Mayo Clin, 68-69; asst to dir, Nat Inst Environ Health Sci, 69-73; ASST DIR, NAT CANCER INST, 73- Concurrent Pos: Lectr col vet med, Ohio State Univ, 55-56; consult, US AID, 62; consult col vet med, Mich State Univ, 62-64, from assoc prof to prof, 64-68; consult lab infectious dis res, Col Med, Univ Iowa, 62-; consult, WHO, 65. Mem: Fel Am Pub Health Asn; Am Vet Med Asn. Res: Infectious and noninfectious disease epidemiology; comparative medicine. Mailing Add: Nat Cancer Inst Bethesda MD 20014

TJIO, JOE HIN, b Java, Indonesia, Feb 11, 19; US citizen; m 48; c 1. CYTOGENETICS. Educ: Univ Colo, PhD(biophys, cytogenetics), 60. Hon Degrees: Dr, Univ Claude Bernard, France, 74. Prof Exp: Head cytogenetics, Estacion Exp de Aula Dei, 48-59; res biologist, 59-74, CHIEF CYTOGENETICS SECT LAB EXP PATH, NAT INST ARTHRITIS & METAB DIS, 74- Concurrent Pos: Res assoc genetics inst, Univ Lund, 59- Honors & Awards: Super Serv Award, Dept Health, Educ & Welfare, NIH, 74. Mem: Genetics Soc Am; Am Soc Human Genetics; Am Genetic Asn; Am Soc Nat. Res: Plant, animal and human cytogenetics; mammalian cytogenetics. Mailing Add: Nat Inst of Health Bldg 10 Room 9D-04 Bethesda MD 20014

TJIOE, SARAH ARCHAMBAULT, b Philadelphia, Pa, Oct 12, 44; m 67; c 1. PHARMACOLOGY. Educ: Univ Pa, BA, 66, PhD(pharmacol), 71. Prof Exp: Instr, 72-75, ASST PROF PHARMACOL, COL MED, OHIO STATE UNIV, 75- Concurrent Pos: NIH training grant pharmacol, Col Med, Ohio State Univ, 71-72, Pharmaceut Mfrs Asn Found fel pharmacol-morphol, 72-74. Mem: Soc Neurosci. Res: Effects of psychoactive agents on brain mitochondria. Mailing Add: Dept of Pharmacol Ohio State Univ Col of Med Columbus OH 43210

TJOSTEM, JOHN LEANDER, b Sisseton, SDak, June 6, 35; m 62; c 3. MICROBIOLOGY, PLANT PHYSIOLOGY. Educ: Concordia Col, Moorhead, Minn, BA, 59; NDak State Univ, MS, 62, PhD(bot), 68. Prof Exp: Instr biol, Luther Col, Iowa, 62-65; assoc prof, Concordia Col, Moorhead, Minn, 67-68; ASSOC PROF BIOL, LUTHER COL, IOWA, 68- Mem: AAAS; Am Soc Plant Physiol. Res: Metabolic pathways of certain carbohydrates in algae. Mailing Add: Dept of Biol Luther Col Decorah IA 62101

TKACHEFF, JOSEPH, JR, b Waterbury, Conn, Feb 13, 26; m 54; c 2. PHARMACY. Educ: RI Col Pharm, BSc, 46; Philadelphia Col Pharm, MSc, 47. Prof Exp: Pharmacist, E M Altman & Waterbury Drug Co, Conn, 45-46; control chemist, G F Harvey Co, 47-49, res & develop chemist, 49-50, plant chemist, 50-51, prod mgr, 51-58; SR RES PHARMACIST & GROUP LEADER PROD DEVELOP & RES, SOLID DOSAGE SECT, STERLING-WINTHROP RES INST, 58- Mem: Am Chem Soc; Am Pharmaceut Asn; Acad Pharmaceut Sci. Res: Pharmaceutical chemistry; physiological biochemistry. Mailing Add: 7245 Russell Rd Greenfield Center NY 12833

TKACHUK, RUSSELL, b Redwater, Alta, Dec 25, 30; m 57; c 3. ORGANIC CHEMISTRY. Educ: Univ BC, BA, 54, MSc, 56; Univ Sask, PhD(chem), 59. Prof Exp: RES SCIENTIST, GRAIN RES LAB, CAN GRAIN COMN, 59- Concurrent Pos: Adj prof, Univ Man; fel, St Vincent's Sch Med Res, Melbourne, Australia, 68-69. Mem: Am Asn Cereal Chem; Am Chem Soc; Chem Inst Can; Can Inst Chem; Sigma Xi. Res: Chemistry of amino acids, peptides, enzymes and proteins in wheat. Mailing Add: Grain Res Lab Can Grain Comn 303 Maine St Winnipeg MB Can

TOALSON, WILMONT, b Clark, Mo, Feb 13, 08; m 31. APPLIED MATHEMATICS. Educ: William Jewell Col, 29; Univ Kans, AM, 37. Prof Exp: Teacher high sch, Kans, 37-41; instr math, Pratt Jr Col, 41-43; from asst prof to assoc prof, 46-63, dept adv, 56-74, PROF MATH, FT HAYS KANS STATE COL, 63- Concurrent Pos: Consult, US Agency for Int Develop Prog, India, 66,67. Mem: Math Asn Am. Res: Laplace transform. Mailing Add: Dept of Math Ft Hays Kansas State Col Hays KS 67601

TOBA, HACHIRO HAROLD, b Puunene, Hawaii, Aug 24, 32; m 58; c 3. ENTOMOLOGY. Educ: Univ Hawaii, BS, 57, MS, 61; Purdue Univ, PhD(entom), 66. Prof Exp: RES ENTOMOLOGIST, AGR RES SERV, USDA, 65- Mem: AAAS; Entom Soc Am; Sigma Xi. Res: Development of new, safe and effective methods for control of soil insect pests of vegetables. Mailing Add: Yakima Agr Res Lab USDA-Agr Res Serv 3706 W Nob Hill Blvd Yakima WA 98902

TOBACH, ETHEL, b Miaskovka, Russia, Nov 7, 21; nat US; m 47. ANIMAL BEHAVIOR. Educ: Hunter Col, BA, 49; NY Univ, MA, 52, PhD(comp psychol, physiol psychol), 57. Hon Degrees: DSc, Long Island Univ, 75. Prof Exp: Res assoc, Payne Whitney Psychiat Clin, New York, 49-53 & Pub Health Res Inst NY, Inc, 53-56; res fel comp & physiol psychol, Am Mus Natural Hist, 57-61; asst prof comp physiol & exp psychol, Sch Med, NY Univ, 61-65; assoc cur, 64-69, CUR, DEPT ANIMAL BEHAV, AM MUS NATURAL HIST, 69- Concurrent Pos: NIMH career develop awards, 61-74. Mem: AAAS; Am Psychol Asn; Animal Behav Soc; Psychonomic Soc; NY Acad Sci. Res: Development and evolution of behavior; emotional behavior; social behavior; neurohormonal relationships; sensory processes; autonomic phenomena. Mailing Add: Dept Animal Behav Am Mus Nat Hist Cent Park West at 79th St New York NY 10024

TOBEN, HOWARD RAY, b Keokuk, Iowa, Apr 15, 41; m 65; c 1. IMMUNOBIOLOGY, ANATOMY. Educ: Univ Miami, BS, 63, MS, 67; Ohio State Univ, PhD(anat), 71. Prof Exp: ASST PROF CELL BIOL, UNIV TEX HEALTH SCI CTR DALLAS, 73- Concurrent Pos: NIH-AID fel develop immunobiol, Univ Ala, Birmingham, 71-73. Mem: Am Asn Anatomists. Res: Immunobiology of human lymphocytes in normal and disease states. Mailing Add: Dept of Cell Biol Univ of Tex Health Sci Ctr Dallas TX 75235

TOBERMAN, RALPH OWEN, b Harrisburg, Pa, Sept 23, 23; m 46; c 3. ANALYTICAL CHEMISTRY. Educ: Albright Col, BS, 49. Prof Exp: Control chemist, Pa Ohio Steel Corp, 49-51; res asst anal chem, Sharp & Dohme, Inc, 51-52, res assoc, 52-63, unit head anal res, 63-65, specifications assoc, 65-66, HEAD CONTAINER & PROD TESTING UNIT, MERCK SHARP & DOHME RES LABS, 66- Mem: Am Chem Soc; Drug Info Asn. Res: Analysis of pharmaceutical products; methods development; specifications for new products; package research and product testing. Mailing Add: Hickory Hill Dr RD 1 Norristown PA 19401

TOBEY, ARTHUR ROBERT, b Portland, Ore, Aug 4, 20; wid; c 3. APPLIED PHYSICS. Educ: Yale Univ, BS, 42, MS, 46, PhD(physics), 48. Prof Exp: Mem staff radiation lab, Mass Inst Technol, 42-45; asst cosmic ray proj, Off Naval Res, Yale Univ, 47-48; asst prof physics, State Col Wash, 48-50; supvr physics, Armour Res Found, 50-52; sr scientist, 52-53; supvr TV res, 53-56, group head video systs lab, 56-59, STAFF SCIENTIST ENG DIV, STANFORD RES INST, 59- Mem: Am Phys Soc. Res: Radar and radar-type systems; neutron component of cosmic rays; electromechanical and electro-optical devices; electronic instrumentation; communication theory; man-computer systems; satellite systems modelling. Mailing Add: Info Systs Group Stanford Res Inst Menlo Park CA 94025

TOBEY, FRANK LINDLEY, JR, b Coeur d'Alene, Idaho, Aug 28, 23; m 57; c 1. VISION, PHOTOMETRY. Educ: Univ Mich, BSCh, 47, MSCh, 48, MS, 50, PhD(physics), 62. Prof Exp: Res assoc physics, Univ Mich, 50-60; res physicist, Cornell Aeronaut Lab, 62-64; fel lab astrophys, Harvard Col Observ, 64-65; fel shocktube physics, McDonnell-Douglas Corp, 66-69; res instr spectros of vision, Med Sch, Wash Univ, St Louis, 70-74; ASST PROF SPECTROS OF VISION, MED SCH, UNIV FLA, 74- Concurrent Pos: Asst prof physics & astron, Southern Ill Univ, Edwardsville, 70. Mem: AAAS; Am Phys Soc; Asn Res Vision & Ophthal Sigma Xi. Res: Measurement of transition probabilities; atomic and molecular parameters; properties of space; plasma diagnostics; optical properties of the retina and retinal receptors; experimental determination of receptor waveguide properties; physics of vision. Mailing Add: Dept of Ophthal Box J284 JHMHC Univ of Fla Col of Med Gainesville FL 32610

TOBEY, ROBERT ALLEN, b Owosso, Mich, May 26, 37; m 60; c 2. CELL BIOLOGY, CANCER. Educ: Mich State Univ, BS, 59; Univ Ill, PhD, 63. Prof Exp: STAFF MEM, CELLULAR RADIOBIOL SECT, LOS ALAMOS SCI LAB, UNIV CALIF, 64- Mem: AAAS; Am Soc Cell Biol; Am Soc Biol Chem; Am Soc Microbiol; Am Asn Cancer Res. Res: Factors controlling traverse of the life cycle; sequential biochemical markers in the mammalian cell cycle; control of mammalian cell proliferation; effects of anticancer drugs and environmental pollutants on cell kinetics. Mailing Add: Cellular-Molecular Radiobiol Univ of Calif Los Alamos Sci Lab Los Alamos NM 87545

TOBEY, STEPHEN WINTER, b Chicago, Ill, Jan 9, 36; m 53; c 4. PHARMACEUTICAL CHEMISTRY. Educ: Ill Inst Technol, BS, 57; Univ Wis, MS, 59, PhD(inorg chem), 65. Prof Exp: Instr phys chem, WVa Wesleyan Col, 59-61; asst prof inorg chem, Purdue Univ, 64-65; res chemist, Eastern Res Lab, Dow Chem USA, 65-68, sr res chemist, 68-70, res dir, 70-74, dir chem lab, 73-75, MGR CHEM PROCESS DEVELOP, DOW LEPETIT RES & DEVELOP, 75- Concurrent Pos: Vis prof, Harvard Univ, 65. Mem: Am Chem Soc. Res: Process engineering development. Mailing Add: Dow Lepetit Res & Develop Midland MI 48640

TOBIA, ALFONSO JOSEPH, b Brooklyn, NY, June 19, 42. PHARMACOLOGY. Educ: St Louis Col Pharm, BS, 65; Purdue Univ, MS & PhD(pharmacol), 69. Prof Exp: Asst prof pharmacol, Univ Ga, 69-74; SR INVESTR CARDIOVASC PHARMACOL, SMITH KLINE & FRENCH LABS, 74- Concurrent Pos: Ga Heart Asn res grant, Univ Ga, 71-72, Nat Heart & Lung Inst res grant hypertension, 71-; mem coun high blood pressure & coun basic sci, Am Heart Asn. Mem: AAAS; Am Heart Asn; NY Acad Sci. Res: Reflex vasodilation and hypertension; cardiovascular and biochemical pharmacology. Mailing Add: Cardiovasc Pharmacol Smith Kline & French Labs Philadelphia PA 19101

TOBIAN, LOUIS, b Dallas, Tex, Jan 26, 20; m 51. INTERNAL MEDICINE. Educ: Univ Tex, BA, 40; Harvard Med Sch, MD, 44. Prof Exp: House officer med, Peter Bent Brigham Hosp, Boston, 44; asst resident, Univ Hosp, Univ Calif, 44-45 & Parkland Hosp, Tex, 45-46; asst prof med, Univ Tex Southwestern Med Sch Dallas, 54; assoc prof, 54-64, PROF MED, SCH MED, UNIV MINN, MINNEAPOLIS, 64- Concurrent Pos: Res fel med, Univ Tex Southwestern Med Sch Dallas, 46-51; res fel biochem, Harvard Med Sch, 51-54; estab investr, Am Heart Asn, 51-56, mem coun arteriosclerosis, 56-, chmn, Coun High Blood Pressure Res, 72; George Brown lectureship, 69; chmn task force hypertension, Nat Heart & Lung Inst, 72-73, mem adv comt hypertension res ctrs, 72-74; chmn comt hypertension & renal vascular dis, NIH Kidney Res Surv Group, 74-75. Mem: Am Clin & Climat Asn; Asn Am Physicians; Am Soc Clin Invest; Am Physiol Soc; fel NY Acad Sci. Res: Hypertension; renal circulation; sodium excretion. Mailing Add: Dept of Med Univ of Minn Hosp Minneapolis MN 55455

TOBIAS, CHARLES W, b Budapest, Hungary, Nov 2, 20; nat US; m 50; c 3. ELECTROCHEMISTRY, CHEMICAL ENGINEERING. Educ: Univ Tech Sci, Budapest, ChemEng, 42, PhD(phys chem), 46. Prof Exp: Res & develop engr, United Incandescent Lamp & Elec Co, Ltd, Hungary, 42-43, 46-47; instr chem eng, 47-48, lectr, 48-50, from asst prof to assoc prof, 50-60, chmn dept chem eng, 67-72, PROF CHEM ENG, UNIV CALIF, BERKELEY, 60-, PRIN INVESTR LAWRENCE RADIATION LAB, 54- Concurrent Pos: Instr phys chem, Univ Tech Sci, Budapest, 45-46; consult res prof, Miller Inst, 58-59. Honors & Awards: Acheson Medal & Prize, Electrochem Soc, 72. Mem: Fel AAAS; Am Chem Soc; Int Soc Electrochem (pres-elect, 75-77); Electrochem Soc (pres, 72); Am Inst Chem Eng. Res: Current distribution in electrolytic cells; mass transfer in electrode processes; electrolytic oxidation and reduction; electrodeposition; batteries; fuel cells. Mailing Add: Dept of Chem Eng Univ of Calif Berkeley CA 94720

TOBIAS, CORNELIUS ANTHONY, b Budapest, Hungary, May 28, 18; nat US; m 43; c 2. BIOPHYSICS. Educ: Univ Calif, Berkeley, MA, 40, PhD(nculear physics), 42. Prof Exp: Physicist, 42-45, from instr to assoc prof biophys, 45-55, vchmn dept physics in-chg of med physics, 60-67, chmn div med physics, 67-71, PROF MED PHYSICS, DONNER LAB, UNIV CALIF, BERKELEY, 55-, PROF ELEC ENG, 65-, CHMN GRAD GROUP BIOPHYS & MED PHYSICS, 69- Concurrent Pos: Fel med physics, Univ Calif, Berkeley, 45-47; Guggenheim fel, Karolinska Inst, Sweden, 56-57; vis prof, Harvard Univ, 60; mem subcomt, Nat Res Coun, mem comt radiol,

Nat Acad Sci-Nat Res Coun; mem radiation study sect, NIH, 60-63; pres radiation biophys comn, Int Union Pure & Appl Physics, 69-72, coun mem, Int Union Pure & Appl Biophys, 69-75. Honors & Awards: Lawrence Mem Award, 63; Annual Award, Am Nuclear Soc Aerospace Div, 72. Mem: Am Asn Physicists in Med; NY Acad Sci; Am Phys Soc; Radiation Res Soc (pres, 62-63); Biophys Soc. Res: Biological effects of radiation; cancer research; space medicine. Mailing Add: 363 Donner Lab Univ of Calif Berkeley CA 94720

TOBIAS, IRWIN, b New York, NY, Mar 31, 35; m 60; c 2. PHYSICAL CHEMISTRY. Educ: Brooklyn Col, BS, 56; Princeton Univ, MA, 58, PhD(chem), 60. Prof Exp: Res assoc, Inst Molecular Physics, Univ Md, 59-61; from asst prof to assoc prof, 61-71, PROF CHEM, RUTGERS UNIV, NEW BRUNSWICK, 71- Concurrent Pos: Consult, Gen Tel & Electronics Corp, 61-64; consult, Am Optical Co, 64- Res: Theoretical chemistry; lasers; interaction between light and matter. Mailing Add: Sch of Chem Rutgers Univ New Brunswick NJ 08903

TOBIAS, JERRY VERNON, b St Louis, Mo, Oct 14, 29; c 2. PSYCHOACOUSTICS. Educ: Univ Mo, AB, 50; Univ Iowa, MA, 54; Western Reserve Univ, PhD(audition), 59. Prof Exp: Asst, Univ Iowa, 53-54; instr speech path & audiol, Ball State Teachers Col, Ind, 54-56; res assoc audition, Western Reserve Univ, 56-59; res scientist psychophys, Defense Res Lab & vis asst prof psychol, Univ Tex, 59-61; assoc prof, 63-71, PROF PSYCHOL, UNIV OKLA, 71-; SUPVRY PSYCHOLOGIST, CIVIL AEROMEDICAL INST, FED AVIATION ADMIN, 61- Concurrent Pos: Mem comt hearing, bioacoust & biomech, Nat Acad Sci-Nat Res Coun. Mem: Fel Acoust Soc Am; Sigma Xi; fel Am Speech & Hearing Asn. Res: Physiological and psy- chological acoustics; audition; experimental phonetics; physiological psychology; psychophysics; sensation and perception; auditory time constants; binaural audition; noise hazards and control. Mailing Add: Civil Aeromedical Inst Fed Aviation Admin Aeronaut Ctr PO Box 25082 Oklahoma City OK 73125

TOBIAS, JOSEPH, b Olomouc, Moravia, Sept 13, 20; nat US; m 49; c 3. DAIRY SCIENCE, FOOD SCIENCE. Educ: Univ Ga, BS, 42; Univ Ill, PhD(dairy tech), 52. Prof Exp: Instr & first asst, 48-53, from asst prof to assoc prof, 53-64, PROF DAIRY TECH, UNIV ILL, URBANA, 64- Mem: Am Dairy Sci Asn. Res: Bacteriological evaluation of high-temperature short-time pasteurization of milk and other dairy products; ice cream technology; electrophoresis of milk proteins; flavor chemistry; micronutrients and microconstituents in milk. Mailing Add: 103 Dairy Manufactures Univ of Ill Urbana IL 61801

TOBIAS, PETER STEPHEN, biochemistry, see 12th edition

TOBIAS, ROBERT FRED, electrochemistry, see 12th edition

TOBIAS, RUSSELL STUART, b Columbus, Ohio, Sept 11, 30; m 56; c 3. BIOINORGANIC CHEMISTRY, ORGANOMETALLIC CHEMISTRY. Educ: Ohio State Univ, BSc, 52, PhD(chem), 56. Prof Exp: NSF fel, Royal Inst Technol, Sweden, 56-57; sr chemist, E I du Pont de Nemours & Co, NJ, 57-58; res assoc chem, Univ NC, 58-59; from asst prof to prof, Univ Minn, Minneapolis, 59-69; PROF CHEM, PURDUE UNIV, 69-, HEAD DIV INORG CHEM, CHEM DEPT, 73- Concurrent Pos: Consult, Mobil Oil Corp, 64-68; mem subcomt inorg nomenclature, Nat Acad Sci-Nat Res Coun, 64-; NSF sr fel, Munich Technol Univ & Imp Col, Univ London, 67-68; mem ed bd, J Coord Chem, 70-; vis scientist, NIH, 76. Mem: AAAS; Am Chem Soc; fel The Chem Soc. Res: Applications of Raman spectroscopy in the study of inorganic systems; interactions of metal ions with nucleotides and polynucleotides. Mailing Add: Dept of Chem Purdue Univ Lafayette IN 47907

TOBIASON, FREDERICK LEE, b Pe Ell, Wash, Sept 15, 36; m 61; c 3. PHYSICAL CHEMISTRY. Educ: Pac Lutheran Univ, BA, 58; Mich State Univ, PhD(phys chem), 63. Prof Exp: Res assoc nuclear magnetic resonance spectros, Emory Univ, 63-64; res chemist, Benger Lab, E I du Pont de Nemours & Co, Inc, 64-66; from asst prof to assoc prof, 66-73, PROF PHYS CHEM, PAC LUTHERAN UNIV, 73- Concurrent Pos: Consult, Reichhold Chem, Inc, 67-; chmn chem dept, Pac Lutheran Univ, 73-76 & regency prof, 75-76. Mem: AAAS; Am Chem Soc. Res: Molecular structure; microwave spectroscopy; nuclear magnetic resonance spectroscopy; electric dipole moments; molecular characterization of polymers; wood chemistry; adhesives; polymer chain configuration calculations. Mailing Add: Dept of Chem Pac Lutheran Univ Tacoma WA 98447

TOBIE, JOHN EDWIN, b Collison, Ill, Dec 26, 11; m 47. IMMUNOLOGY. Educ: Univ Ill, AB, 35; La State Univ, MS, 36; Tulane Univ, PhD(parasitol), 40. Prof Exp: Tech asst, NIH, 37-40; instr parasitol, Tulane Univ, 41-43; parasitologist, Lab Trop Dis, NIH, 43-57, biologist, Lab Immunol, 57-61, actg chief, 61-63; chief lab germfree animal res, Nat Inst Allergy & Infectious Dis, 63-69, chief lab microbial immunity, 69-70, asst sci dir lab & clin res, 70-72; CONSULT, NAT ACAD SCI, 74- Concurrent Pos: Fel, La State Univ, 34-37; fel trop med, Tulane Univ, 40-41. Honors & Awards: Dept Health, Educ & Welfare Superior Serv Award, 63. Mem: Fel AAAS; Am Soc Trop Med & Hyg; Am Asn Immunologists; NY Acad Sci. Res: Immunological response to parasites in germfree and conventional hosts; fluorescent antibody studies on antibody production and immunoglobulin synthesis in malaria and other protozoa diseases; cellular localization of antibodies by fluorescence. Mailing Add: 5902 Wilson Lane Bethesda MD 20034

TOBIESSEN, PETER LAWS, b Philadelphia, Pa, Mar 30, 40; m 68. PLANT ECOLOGY. Educ: Wesleyan Univ, BA, 63; Pa State Univ, University Park, MS, 66; Duke Univ, PhD(bot), 71. Prof Exp: ASST PROF BIOL, UNION COL, NY, 70- Mem: Am Soc Plant Physiol; Ecol Soc Am. Res: Physiological plant ecology. Mailing Add: Dept of Biol Union Col Schenectady NY 12308

TOBIN, ALLAN JOSHUA, b Manchester, NH, Aug 22, 42; m 68; c 1. DEVELOPMENTAL BIOLOGY, MOLECULAR BIOLOGY. Educ: Mass Inst Technol, BS, 63; Harvard Univ, PhD(biophys), 69. Prof Exp: USPHS fels, Weizman Inst Sci, 69-70 & Mass Inst Technol, 70-71; asst prof biol, Harvard Univ, 71-74; ASST PROF BIOL, UNIV CALIF, LOS ANGELES, 74- Res: Erythroid cell development. Mailing Add: Dept of Biol Univ of Calif Los Angeles CA 90024

TOBIN, CHARLES EMIL, b Charleston, SC, Oct 26, 11; m 42; c 4. ANATOMY, EMBRYOLOGY. Educ: Col Charleston, BS, 34; Yale Univ, PhD(zool), 38. Prof Exp: Asst zool, Col Charleston, 32-34; asst, Yale Univ, 34-37; from instr to prof anat, Sch Med & Dent, Univ Rochester, 38-68; prof human biol, Sch Dent & clin assoc prof anat & pediat, Sch Med, 68-73, EMER PROF HUMAN BIOL, UNIV COLO MED CTR, DENVER, 73- Concurrent Pos: Ed, Shearer's Manual Human Dissection. Mem: Am Asn Anat. Res: Gross anatomy; fascia; lungs; injection methods; adrenals; vitamin E; fetal function of the adrenal cortex in the rat; general function of the adrenal cortex as related to reproduction in the rat. Mailing Add: Univ of Colo Med Ctr Denver CO 80220

TOBIN, DANIEL F, b New York, NY, Aug 8, 08; m 35; c 1. DENTISTRY. Educ: NY Univ, DDS, 31. Prof Exp: Chief oper dent & asst dir pedodontics, Murry & Leonie Guggenheim Dent Clin, 40-44, dir pedodontics, 46-63; prof restorative dent & dean prof dent educ, 63-69, EMER DEAN DENT, COL DENT, COL MED & DENT, NJ, 69- Concurrent Pos: Dean's comt, East Orange Vet Admin Hosp, 63-; prof oper dent & chmn dept, NY Univ, 70-72. Mem: Am Dent Asn; Am Soc Dent for Children; Am Asn Hosp Dent (pres, 63-65). Res: Pedodontics; professional dental education, especially public health dentistry; institutional dental care and dentistry for children. Mailing Add: 360 First Ave New York NY 10010

TOBIN, ELAINE MUNSEY, b Louisville, Ky, Dec 23, 44; m 68; c 2. PLANT DEVELOPMENT. Educ: Oberlin Col, BA, 66; Harvard Univ, PhD(biol), 72. Prof Exp: Fel biol, Brandeis Univ, 73-75; ASST PROF BIOL, UNIV CALIF, LOS ANGELES, 75- Mem: Am Soc Plant Physiologists; AAAS. Res: Control of plant development by light. Mailing Add: Dept of Biol Univ of Calif Los Angeles CA 90024

TOBIN, JOHN ROBERT, JR, b Elgin, Ill, Dec 18, 17; m 42; c 1. INTERNAL MEDICINE, CARDIOLOGY. Educ: Univ Notre Dame, BS, 38; Univ Chicago, MD, 42; Univ Minn, MS, 50; Am Bd Internal Med, dipl, 52; Am Bd Cardiovasc Dis, dipl, 66. Prof Exp: Intern med & surg, Presby Hosp, Chicago, 42-43; resident path, Univ Chicago Clins, 46-47; asst staff, Mayo Clin, 50-51; staff, Rickwood Clin, Spokane, Wash, 51-55; dir adult cardiol, Cook County Hosp, Ill, 59-69; PROF MED, STRITCH SCH MED, LOYOLA UNIV CHICAGO, 62-, CHMN DEPT, 69- Concurrent Pos: Fel med, Mayo Found, 47-50, NIH spec fel physiol, 63-64; assoc & attend staff med, Cook County Hosp, 55-69; physician-in-chief, Loyola Hosp, 69- Mem: Fel Am Col Physicians; fel Am Col Cardiol. Res: Cardiovascular physiology and disease. Mailing Add: 2160 S First Ave Maywood IL 60153

TOBIN, LESLIE WARREN, agronomy, see 12th edition

TOBIN, MARVIN CHARLES, b St Louis, Mo, Jan 10, 23; m 47; c 3. PHYSICAL CHEMISTRY, PHYSICS. Educ: Wash Univ, BA, 47; Ind Univ, MA, 49; Univ Conn, PhD(chem), 52. Prof Exp: Sr chemist, Arthur D Little, Inc, 52-53; group leader spectros, Olin Mathieson Chem Corp, 53-57; res chemist, Am Cyanamid Co, 57-60, group leader polymer physics, 60-62, solid state physics, 62-64; sr staff scientist, Perkin-Elmer Corp, 64-70; prof physics, Univ Bridgeport, 70-75, lectr eve div, 59-68, adj prof, 68-70; adj prof, 70-75, MEM FAC, DEPT PHYS SCI, NEW ENG INST, 75- Mem: AAAS; NY Acad Sci; Am Phys Soc. Res: Polymer and solid state physics; molecular and laser spectroscopy. Mailing Add: Dept Phys Sci New England Inst Ridgefield CT 06877

TOBIN, RICHARD BRUCE, b Buffalo, NY, Mar 6, 25; m 47; c 5. PHYSIOLOGY. Educ: Union Univ, NY, BS, 45; Univ Rochester, MD, 49. Prof Exp: Instr physiol & med, Sch Med & Dent, Univ Rochester, 54-59, asst prof & asst physician, 59-63; assoc prof med, 66-68, assoc prof physiol, 67-68, PROF MED & BIOCHEM, UNIV NEBR MED CTR, OMAHA, 68-; STAFF PHYSICIAN, VET ADMIN HOSP, OMAHA, 73- Concurrent Pos: USPHS fel biochem, Univ Amsterdam, 63-65; assoc chief of staff, Vet Admin Hosp, Omaha, 66-73. Mem: AAAS; Am Fedn Clin Res; Am Diabetes Asn; Biophys Soc; Soc Exp Biol & Med. Res: Acid-base balance; cellular energetics; nutrition and regulation of intermediary metabolism. Mailing Add: Vet Admin Hosp 4101 Woolworth Ave Omaha NE 68105

TOBIN, SIDNEY MORRIS, b Toronto, Ont, Jan 18, 23; m 49; c 4. MEDICAL RESEARCH, OBSTETRICS & GYNECOLOGY. Educ: Univ Toronto, MD, 46; FRCS(C), 51. Prof Exp: Clin lectr, 61-71, ASST PROF OBSTET & GYNEC, UNIV TORONTO, 71- Concurrent Pos: Chmn, Res Adv Comt, Res Dept, Mt Sinai Hosp, Toronto, 70- Mem: Can Med Asn; Soc Obstetricians & Gynecologists Can; Royal Col Physicians & Surgeons Can; Can Oncol Soc; Can Fel Travel Soc. Res: Studies to elucidate the role of herpes simplex virus type II as a carcinogen or co-carcinogen in the etiology of squamous cell carcinoma of the cervix in humans. Mailing Add: 69 Dunveegan Rd Toronto ON Can

TOBIN, THOMAS VINCENT, b Plymouth, Pa, Apr 8, 26; m 47; c 1. BIOLOGY. Educ: King's Col, BS, 51; Boston Col, MS, 53. Prof Exp: Asst biol, Boston Col, 51-52; from instr to asst prof, 52-62, ASSOC PROF BIOL, KING'S COL, 62-, DIR EDUC TV, 66-, CHMN DIV NATURAL SCI, 71- Mem: AAAS. Res: Chemoreception and proprioception in the crayfish; cold acclimitization; influence of drugs on animal behavior. Mailing Add: Dept of Biol King's Col Wilkes-Barre PA 18711

TOBIS, JEROME SANFORD, b Syracuse, NY, July 23, 15; m 38; c 3. MEDICINE. Educ: City Col New York, BS, 36; Chicago Med Sch, MD, 43. Prof Exp: Consult phys med & rehab, Vet Admin Hosp, Bronx, NY, 48-53; prof phys med & rehab & dir, NY Med Col, 52-70; PROF PHYS MED & REHAB & CHMN DEPT, UNIV CALIF, IRVINE-CALIF COL MED, 70- Concurrent Pos: Baruch fel, Columbia Univ, 47; dir phys med & rehab, Metrop Hosp, 52-70 & Bird S Coler Hosp, 52-70; consult, City Hosp, 48 & Bur Handicapped Children, New York City Dept Health, 53-70; ed, Arch Phys Med & Rehab, 59-73; chief rehab med, Montefiore Hosp Med Ctr, 61-70; dir dept phys med & rehab, Orange County Med Ctr, 70- Mem: AAAS; AMA; Am Pub Health Asn; Am Cong Rehab Med; Am Acad Phys Med & Rehab. Res: Hemiplegia; cardiac rehabilitation; rehabilitation of handicapped children; geriatrics. Mailing Add: Dept of Phys Med & Rehab Univ of Calif Col of Med Irvine CA 92664

TOBISCH, OTHMAR TARDIN, b Berkeley, Calif, June 18, 32; m 64; c 1. STRUCTURAL GEOLOGY. Educ: Univ of Calif, Berkeley, BA, 58, MA, 60; Univ London, PhD(struct geol), 63. Prof Exp: Fulbright fel, Innsbruck, Austria, 63-64; res geologist, US Geol Surv, 64-69; asst prof, 69-72, ASSOC PROF EARTH SCI, UNIV CALIF, SANTA CRUZ, 72- Mem: Geol Soc Am; fel Geol Soc London, 66- Res: Polyphase deformation in orogenic belts; quantitative strain determination of deformed rocks; genesis of orogenic belts. Mailing Add: Div of Natural Sci Univ of Calif Santa Cruz CA 95060

TOBKES, MARTIN, b New York, NY, Feb 8, 28; m 52; c 2. ORGANIC CHEMISTRY. Educ: City Col New York, BS, 48; Polytech Inst Brooklyn, MS, 54, PhD(org chem), 63. Prof Exp: Res & develop chemist, Nopco Chem Co, 49-54 & Charles Bruning Co, 54-56; res org chemist, Ethicon, Inc, 56-58; res fel org chem, Polytech Inst Brooklyn, 59-63; develop chemist, 63-72, GROUP LEADER CHEM DEVELOP, LEDERLE LABS, AM CYANAMID CO, 72- Mem: Am Chem Soc. Res: Vitamin synthesis and stability; Perkin condensations; corrosion; light sensitive coatings; medicinal synthesis; polymers; antibiotics; steroids. Mailing Add: Pearl River NY

TOBLER, WALDO RUDOLPH, b Portland, Ore, Nov 16, 30; c 2. GEOGRAPHY, CARTOGRAPHY. Educ: Univ Wash, BA, 55, MA, 57, PhD(geog), 61. Prof Exp: PROF GEOG, UNIV MICH, ANN ARBOR, 61- Concurrent Pos: Dir several res projs, Univ Mich, Ann Arbor, 61-73; NSF res grants, 66-68 & 72-74; mem, Comt Adv Earth Oriented Applications of Earth Satellites, Nat Acad Sci, 66-70; US mem, Comn Geog Data Processing, Int Geog Union, 68-76. Honors & Awards: Meritorious Contribr Award, Asn Am Geogrs, 68. Mem: Urban & Regional Info Systs Asn; Regional Sci Asn; Psychomet Soc; fel Am Geog Soc; Am Soc Photogram. Res:

Mathematical analysis of geographic-cartographic problems. Mailing Add: Dept of Geog Univ of Mich Ann Arbor MI 48104

TOBOCMAN, WILLIAM, b Detroit, Mich, Mar 14, 26; m 50; c 2. THEORETICAL NUCLEAR PHYSICS. Educ: Mass Inst Technol, SB, 50, PhD(physics), 53. Prof Exp: Res assoc, Cornell univ, 53-54; mem, Inst Adv Study, 54-56; NSF fel, Univ Birmingham, 56-57; asst prof physics, Rice Univ, 57-60; assoc prof, 60-66, PROF PHYSICS, CASE WESTERN RESERVE UNIV, 66- Concurrent Pos: Sloan Found fel, 61-64; res fel, Weizmann Inst, 63-64. Mem: Am Phys Soc. Res: Theory of nuclear reactions; quantum mechanical many-body problem. Mailing Add: Dept of Physics Case Western Reserve Univ Cleveland OH 44106

TOBUREN, LARRY HOWARD, b Clay Center, Kans, July 9, 40; m 62; c 2. ATOMIC PHYSICS, MOLECULAR PHYSICS. Educ: Kans State Teachers Col, BA, 62; Vanderbilt Univ, PhD(physics), 68. Prof Exp: SR RES SCIENTIST, PAC NORTHWEST LABS, BATTELLE MEM INST, 67- Mem: Am Phys Soc. Res: Atomic and molecular collision processes; Auger electron studies; measurement of inner shell ionization cross sections and continuum electron distributions resulting from charged particle impact. Mailing Add: Pac Northwest Labs Battelle Mem Inst PO Box 999 Richland WA 99352

TOBY, SIDNEY, b London, Eng, May 30, 30; m 53; c 2. PHYSICAL CHEMISTRY. Educ: Univ London, BSc, 52; McGill Univ, PhD, 55. Prof Exp: Fel photochem, Nat Res Coun, Can, 55-57; from instr to assoc prof, 57-69, PROF CHEM, RUTGERS UNIV, NEW BRUNSWICK, 69- Mem: AAAS; Am Chem Soc. Res: Kinetics of gaseous reactions; photochemistry; chemiluminescence. Mailing Add: Sch of Chem Rutgers Univ New Brunswick NJ 08903

TOCCHINI, JOHN JOSEPH, b San Mateo, Calif, July 20, 12; m 40; c 1. DENTISTRY. Educ: Univ of the Pac, BS, 36, DDS, 37; Am Bd Pedodont, dipl. Prof Exp: From clin instr to assoc clin prof oper dent, 39-53, dean sch dent, 53-70, PROF PEDODONTICS, SCH OF DENT, UNIV OF THE PAC, 53- Mem: Am Soc Dent for Children; Am Dent Asn; fel Am Col Dent; Am Acad Pedodont. Res: Pedodontics. Mailing Add: 9250 Alcosta Blvd San Ramon CA 94583

TOCCI, PAUL M, b Brooklyn, NY, Nov 11, 33; m 68; c 2. BIOCHEMISTRY. Educ: Johns Hopkins Univ, BA, 55; Univ Md, PhD(biochem), 64. Prof Exp: ASST PROF PEDIAT, SCH MED, UNIV MIAMI, 67- Concurrent Pos: Dir, Biochem Genetics Lab, 64-; pres, TLC Corp, Miami; mem, Int Fedn Clin Chem. Mem: AAAS; Am Asn Clin Chemists; fel Am Inst Chemists; Am Soc Human Genetics; Am Fedn Clin Res. Res: Amino acid metabolism; inborn errors of metabolism. Mailing Add: Dept of Pediat Univ of Miami Sch of Med Miami FL 33152

TOCCO, DOMINICK JOSEPH, b New York, NY, Jan 25, 30; m 52; c 4. BIOCHEMISTRY, PHARMACOLOGY. Educ: St John's Univ, BS, 51, MS, 53; Georgetown Univ, PhD(chem), 60. Prof Exp: Biochemist, US Army Chem Ctr, Md, 53-55, Nat Heart Inst, 55-60, Merck Inst Therapeut Res, 60-66 & Shell Develop Co, 66-70; BIOCHEMIST, MERCK SHARP & DOHME, 70- Mem: Am Soc Pharmacol & Exp Therapeut. Res: Transport of drugs and natural substances across biological membranes; drug metabolism; pharmacodynamics; experimental enzyme kinetics. Mailing Add: Merck Sharp & Dohme West Point PA 19486

TOCHER, DON, b Hollister, Calif, May 19, 26; m 63; c 1. SEISMOLOGY. Educ: Univ Calif, Berkeley, BA, 45, MA, 52, PhD(seismol), 55. Prof Exp: Res fel comt exp geol & geophys, Harvard Univ, 55-56; from asst res seismologist to assoc res seismologist, Univ Calif, Berkeley, 56-64; dir earthquake mechanism lab, Nat Oceanic & Atmospheric Admin, 64-74; VPRES & CHIEF SEISMOLOGIST, WOODWARD-CLYDE CONSULTS, 74- Concurrent Pos: Res assoc, Seismographic Sta, Univ Calif, Berkeley, 64-; mem, Earthquake Eng Res Inst; mem subcomt recent crustal movements, Int Union Geod & Geophys, 63-69; US del, US-Japan Conf Earthquake Prediction Probs, 64, 66 & 68; mem comt earthquake prediction, President's Off Sci & Technol, 64-66, comt Alaska Earthquake, Nat Acad Sci-Nat Res Coun & chmn panel seismol, 64-72. Mem: Seismol Soc Am (vpres, 62-63, 72, secy, 65-71, pres, 73-74), fel Geol Soc Am; Am Geophys Union; fel Royal Astron Soc. Res: Physics of earth's interior; earthquake energy; crustal movements; earth strains; surface faulting and fault creep; micro-earthquakes; seismicity of Western United States; engineering seismology. Mailing Add: 2740 Derby St Berkeley CA 94705

TOCHER, RICHARD DANA, b Oakland, Calif, Oct 8, 35; m 61; c 3. PLANT PHYSIOLOGY. Educ: Stanford Univ, AB, 57; Univ Wash, Seattle, MS, 63, PhD(bot), 65. Prof Exp: Nat Res Coun fel marine bot, Atlantic Regional Lab, 65-66; asst prof, 66-70, ASSOC PROF BIOL, PORTLAND STATE UNIV, 70- Mem: Am Soc Plant Physiol. Res: Plant physiology and biochemistry, especially marine algae and parasitic angiosperms. Mailing Add: Dept of Biol Portland State Univ Portland OR 97207

TOCHER, STEWART ROSS, b Santa Rosa, Calif, May 7, 23; m 52; c 1. FORESTRY. Educ: Univ Calif, BS, 49, MF, 50. Prof Exp: Asst prof forestry, Utah State Univ, 52-56, assoc prof, 57-66; lectr, 66-70, Samuel T Dana Prof Outdoor Recreation, 70-75, PROF NATURAL RESOURCES, UNIV MICH-ANN ARBOR, 75- Mem: Soc Am Foresters. Res: Forest recreation planning and management; aerial photogrametric techniques in forest management. Mailing Add: 501 Huron View Dr Ann Arbor MI 48103

TOCKER, STANLEY, organic chemistry, polymer chemistry, see 12th edition

TOCUS, EDWARD C, b Youngstown, Ohio, Apr 22, 25; m 53; c 2. PHARMACOLOGY. Educ: Grinnell Col, AB, 50; Univ Chicago, MS, 56, PhD(pharmacol), 59. Prof Exp: Res asst radiol, Wash Univ, 51-53; res assoc med, Univ Chicago, 56-60; pharmacologist, Lederle Labs, Am Cyanamid Co, NY, 60-66; pharmacologist, Bur Med, 66-70, supvry pharmacologist, Div Neuropharmacol Drugs, 70-72, CHIEF OF DRUG ABUSE STAFF, BUR DRUGS, FOOD & DRUG ADMIN, 72- Concurrent Pos: Mem Am del, Int Atoms for Peace Conf, Geneva, 58; consult, WHO, Geneva, Switz, 75- Res: Development of new drugs and regulations controlling the national and international use of drugs; develop and implement programs to control illicit use of drugs. Mailing Add: Food & Drug Admin Bur of Drugs 5600 Fishers Lane Rockville MD 20852

TOCZEK, DONALD RICHARD, b LaPorte, Ind, Nov 21, 38; m 72; c 1. ENTOMOLOGY, BOTANY. Educ: Purdue Univ, BS, 61; NDak State Univ, MS, 63, PhD(entom), 67. Prof Exp: Res asst entom, NDak State Univ, 61-67; ASSOC PROF BIOL, HILLSDALE COL, 67- Mem: AAAS; Entom Soc Am; Ecol Soc Am; Entom Soc Can; Am Inst Biol Sci. Res: Basic botany; invertebrate zoology. Mailing Add: Dept of Biol Hillsdale Col Hillsdale MI 49242

TODARO, GEORGE JOSEPH, b New York, NY, July 1, 37; m 62; c 3. CANCER. Educ: Swarthmore Col, BA, 58; NY Univ, MD, 63. Prof Exp: Intern path, 62-63, fel, 64-65, asst prof path, NY Univ Sch Med, 65-67; from staff assoc to head molecular biol sect, Viral Carcinogenesis Br, 67-70, CHIEF, VIRAL LEUKEMIA & LYMPHOMA BR, NAT CANCER INST, NIH, 70- Concurrent Pos: Career develop award, USPHS, 67. Honors & Awards: Superior Serv Award, Dept Health Educ & Welfare, 71; Parke-Davis Award, Am Soc Exp Path, 75. Mem: Am Soc Microbiol; Am Asn Cancer Res; Soc Exp Biol Med. Res: Virus and genetic factors in cancer etiology. Mailing Add: Nat Cancer Inst Bldg 37 Rm 1B-22 9000 Rockville Pike Bethesda MD 20014

TODD, ARLIE C, parasitology, see 12th edition

TODD, CHARLES WYVIL, b Hutchinson, Minn, Feb 3, 18; m 53; c 3. BIOCHEMISTRY. Educ: Univ Minn, BA, 50; Univ Rochester, PhD(chem), 43. Prof Exp: Asst, Univ Rochester, 40-43; res chemist, Exp Sta, E I du Pont de Nemours & Co, Inc, 43-51, res supvr, 51-60; vis lectr, Biochem Div, Univ Ill, 63-66; sr res scientist, Dept Biol, 67-71, DIR DEPT IMMUNOL, CITY OF HOPE NAT MED CTR, 71- Concurrent Pos: Guggenheim fel, 61; NSF fel, Pasteur Inst, 62-63. Res: Immunology. Mailing Add: Dept of Immunol City of Hope Nat Med Ctr Duarte CA 91010

TODD, CLEMENT JAMESON, b Los Angeles, Calif, June 8, 21; m 50; c 3. METEOROLOGY. Educ: Univ Calif, Los Angeles, BS, 50; Calif Inst Technol, MS, 52. Prof Exp: Proj mgr, Weather Modification Co, 54-56; res meteorologist weather modification, Meteorol Res Inc, 56-65; res scientist weather modification, Navy Weather Res Facil, US Navy, 65-70; ASST CHIEF WEATHER MODIFICATION, DIV ATMOSPHERIC WATER RESOURCES MGT, BUR RECLAMATION, US DEPT INTERIOR, 70- Mem: Am Meteorol Soc; Royal Meteorol Soc; Weather Modification Asn (pres, 61 & 62). Res: Development of the technology for precipitation management of the Bureau of Reclamation's research effort in weather modification. Mailing Add: Atmospheric Water Resources Mgt Code 1200 Bur Reclamation Bldg 67 Denver CO 80225

TODD, DAVID, b New York, NY, Mar 29, 16; m 41 & 51; c 5. CHEMISTRY. Educ: Swarthmore Col, BA, 38; Harvard Univ, MA, 40, PhD(org chem), 42. Prof Exp: Lab asst, Harvard Univ, 40-42, chemist comt med res proj, 42-43; res chemist, Oceanog Inst, Woods Hole, 43-45 & Med Col, Cornell Univ, 45-46; asst prof chem, Amherst Col, 46-52; res assoc, Worcester Found Exp Biol, 52-56; res fel, Cambridge Univ, 56-57; from asst prof to assoc prof, 57-64, PROF CHEM, WORCESTER POLYTECH INST, 64- Concurrent Pos: Vis lectr, Clark Univ, 53-54; chemist, Off Sci Res & Develop, Harvard Univ, 43. Mem: Am Chem Soc. Res: Resin acids; hydrazones; penicillin analogs; steroids. Mailing Add: Dept of Chem Worcester Polytech Inst Worcester MA 01609

TODD, EDWARD LAWRENCE, entomology, see 12th edition

TODD, EDWARD PAYSON, b Newburyport, Mass, Jan 26, 20; m 50; c 3. ATMOSPHERIC PHYSICS, SCIENCE ADMINISTRATION. Educ: Mass Inst Technol, BS, 42; Univ Colo, PhD(physics), 54. Prof Exp: Res physicist, United Shoe Mach Corp, Mass, 46-49; supvr appl res, Pitney-Bowes, Inc, Conn, 54-57; mem res staff physics, Univ Colo, 57-59, tech dir upper air lab, 59-63; assoc prog dir atmospheric sci sect, 60-61, prog dir aeronomy, 63-65, actg sect head atmospheric sci sect, 63-64, spec asst to assoc dir res, 65-66, dep assoc dir res, 66-70, dep asst dir res, 70-75, DEP ASST DIR ASTRON, ATMOSPHERIC, EARTH & OCEAN SCI, NSF, 75- Concurrent Pos: Mem fed comt meteorol serv & supporting res, NSF, 70-, chmn interdept comt atmospheric sci, 73- Honors & Awards: Distinguished Serv Medal, NSF, 71. Mem: Am Geophys Union; Am Meteorol Soc. Mailing Add: 312 Van Buren St Falls Church VA 22046

TODD, EDWIN HARKNESS, b Loris, Sc, Apr 8, 18; m 44; c 1. PLANT PATHOLOGY. Educ: Univ Ga, BS, 49, MS, 50; La State Univ, PhD(plant path), 53. Prof Exp: Plant pathologist, USDA, Tex, 52-54, US Army Biol Labs, Md, 54 & USDA, Fla, 54-64; assoc dir res, 64-67, VPRES RES, US SUGAR CORP, 67- Res: Plant mycology and physiology; agronomy; diseases of sugar-cane, rice and St Augustine grass, especially control, isolation, culture and identification of disease pathogens; culture and varietal development of commercial sugarcanes. Mailing Add: Res Dept US Sugar Corp PO Drawer 1207 Clewiston FL 33440

TODD, EWEN CAMERON DAVID, b Glasgow, Scotland, Dec 25, 39; m 67; c 1. MICROBIOLOGY. Educ: Glasgow Univ, BSc, 63, PhD(bact taxon), 68. Prof Exp: Asst lectr bact, Glasgow Univ, 65-68; res scientist, 68-70, HEAD, METHODOLOGY SECT, FOOD MICROBIOL, HEALTH PROTECTION BR, DEPT NAT HEALTH & WELFARE, 71- Mem: Brit Soc Appl Bact; Brit Soc Gen Microbiol. Res: Development of methods for food microbiology; public health aspects of food; microbial taxonomy. Mailing Add: Microbiol Div Food Res Labs Dept Nat Health & Welfare Ottawa ON Can

TODD, FRANK ARNOLD, b Merrill, Iowa, Sept 11, 11; m 36; c 2. VETERINARY MEDICINE. Educ: Iowa State Col, DVM, 33; Yale Univ, MPH, 35; Am Bd Vet Pub Health, dipl. Prof Exp: Secy res & develop bd, Off Secy Defense, US Army, 49-51; consult vet serv, Fed Civil Defense Admin, 51-54; asst to adminstr, Agr Res Serv, USDA, 54-65; WASH REP, AM VET MED ASN, 65- Mem: Am Vet Med Asn; Am Pub Health Asn; NY Acad Sci. Res: Epidemiology and epizootiology of animal diseases; exotic animal diseases; biological and chemical warfare defense; atomic energy. Mailing Add: 145 S Aberdeen St Arlington VA 22204

TODD, GLEN CORY, b Crawfordsville, Ind, May 10, 31; m 54; c 4. PATHOLOGY. Educ: Ind Cent Col, BA, 54; Univ Pa, VMD, 58; Cornell Univ, PhD(path), 65; Am Col Vet Path, dipl. Prof Exp: Vet, Agr Res Serv, USDA, 58-62; teaching assoc path, Cornell Univ, 62-65; pathologist, Agr Res Serv, USDA, 65-66 & Vet Res Div, Food & Drug Admin, 66-67; SR PATHOLOGIST, LILLY RES LABS, ELI LILLY & CO, 67- Mem: Am Vet Med Asn; NY Acad Sci; Int Acad Path. Res: Physiopathology of vitamin E and selenium; nutritional hepatic necrosis; aflatoxicosis in animals; induced myopathies; developmental and neoplastic diseases. Mailing Add: 1410 Bittersweet Dr Greenfield IN 46140

TODD, GLENN WILLIAM, b Kansas City, Mo, Sept 30, 27; m 51; c 4. PLANT PHYSIOLOGY. Educ: Univ Mo, AB, 49, MA, 50, PhD, 52. Prof Exp: Res fel plant physiol, Calif Inst Technol, 52-53; jr biochemist, Citrus Exp Sta, Riverside, Calif, 53-54, asst biochemist, 54-58; from asst prof to assoc prof, 58-66, PROF BOT, OKLA STATE UNIV, 67-, DIR SCH BIOL SCI, 73- Concurrent Pos: Fulbright prof, Cairo Univ, 66-67; actg dir sch biol sci, Okla State Univ, 72-73. Mem: AAAS; Am Soc Plant Physiol; Crop Sci Soc Am. Res: Physiological responses of plants to environmental stress; drought, heat and cold resistance. Mailing Add: Sch of Biol Sci Okla State Univ Stillwater OK 74074

TODD, GORDON LIVINGSTON, b Princeton, WVa, Mar 17, 44; m 68; c 2. ANATOMY, ELECTRON MICROSCOPY. Educ: Kenyon Col, AB, 66; Med Col Ga, MS, 69, PhD(anat), 72. Prof Exp: ASST PROF ANAT, SCH MED, CREIGHTON UNIV, 72- Concurrent Pos: Ga Heart Asn grant, Med Col Ga, 71-72.

Mem: AAAS; Am Asn Anat. Res: Lymphatics; autonomic innervation. Mailing Add: Dept of Anat Sch of Med Creighton Univ Omaha NE 68178

TODD, HAROLD DAVID, b Mt Vernon, Ill, Nov, 19, 44; m 69. CHEMISTRY. Educ: Univ Ill, Urbana, BS, 66; Johns Hopkins Univ, PhD(chem), 71. Prof Exp: ASST PROF CHEM & DIR COMPUT CTR, WESLEYAN UNIV, 71- Mem: Am Phys Soc. Res: Theoretical chemistry with emphasis upon mathematical and computational problems. Mailing Add: Dept of Chem Wesleyan Univ Middletown CT 06457

TODD, HARRY FLYNN, JR, b Baton Rouge, La, Apr 9, 41. ANTHROPOLOGY, MEDICAL ANTHROPOLOGY. Educ: La State Univ, BS, 62; George Washington Univ, JD, 65; Univ Calif, Berkeley, MA, 70, PhD(anthrop), 72. Prof Exp: Fel, Ctr Study Law & Soc & actg asst prof anthrop, Univ Calif, Berkeley, 72-73; lectr & fel, Med Anthrop Prog; Dept Int Health, 74-76, ADJ ASST PROF ANTHROP, UNIV CALIF, SAN FRANCISCO, 76- Concurrent Pos: Co-ed, Med Anthrop Newsletter. Mem: Fel Am Anthrop Asn; fel Geront Soc; Law & Soc Asn; Soc Med Anthrop; fel Royal Anthrop Inst. Res: Legal problems, needs, behavior and perceptions of the elderly in an urban setting, with a focus on both formal and informal mechanisms and agencies used for dispute settlement and conflict management. Mailing Add: Med Anthrop Prog 1320 Third Ave Univ of Calif San Francisco CA 94143

TODD, HOLLIS N, b Glens Falls, NY, Jan 28, 14; m 36; c 1. PHOTOGRAPHY, PHYSICS. Educ: Cornell Univ, BA, 34, MEd, 35. Prof Exp: PROF PHYSICS, ROCHESTER INST TECHNOL, 46- Mem: Soc Photog Sci & Eng; Soc Motion Picture & TV Engrs. Res: Photographic physics, science and engineering. Mailing Add: Sch of Photographic Arts & Sci Rochester Inst of Technol Rochester NY 14623

TODD, JAMES HOPKINS, b Greenwich, Conn, May 24, 16; m 39; c 2. GEOLOGY. Educ: Dartmouth Col, AB, 38; Univ Minn, PhD(geol), 42. Prof Exp: Instr geol, Univ Minn, 39-42; geologist & geophysicist, Calif Co, La, 42-46, dist geologist, 47, geophys supt, 48-49, div exp supt, 50-54; mgr explor dept, Stand Oil Co, Tex, 54-55, vpres & dir, 55-59, consult, Standard Oil Co, Calif, 59-60; vpres explor, prod, land & legal, 61-70, VPRES & GEN MGR EXPLOR, WESTERN DIV, CHEVRON OIL CO, 70- Mem: Soc Explor Geophys; Am Asn Petrol Geol. Res: Pleistocene history of the upper Mississippi River; economic geology; petrology. Mailing Add: Chevron Oil Co PO Box 599 Denver CO 80201

TODD, JAMES WYATT, b Houston Co, Ala, Dec 16, 42; m 64; c 2. ENTOMOLOGY. Educ: Auburn Univ, BS, 66, MS, 68; Clemson Univ, PhD(entom), 73. Prof Exp: From field res asst to res asst entom, Auburn Univ, 64-68; ASST PROF ENTOM, UNIV GA, 68- Mem: Entom Soc Am. Res: Development of soybean insect pest management systems including host plant resistance; utilization of natural control agents; chemical and microbial pesticides; cultural control practices and economic injury thresholds. Mailing Add: Ga Coastal Plain Exp Sta PO Box 748 Tifton GA 31794

TODD, JAY, JR, b Seattle, Wash, May 31, 15; m 43; c 2. PHYSICS, Educ: Univ Wash, Seattle, 38, MSF, 46, PhD(physics), 52. Prof Exp: Asst, Douglas Fir Plywood Asn, 39-41; lab technician, Boeing Aircraft Co, 41-46; physicist, Appl Physics Lab, Univ Wash, Seattle, 52, Sandia Corp, 52-57 & Los Alamos Sci Lab, 57-75; RETIRED. Res: Cosmic rays; shock waves; hydrodynamics. Mailing Add: 376 Venado Los Alamos NM 87544

TODD, JERRY WILLIAM, b La Crosse, Wis, Jan 7, 30; m 56; c 1. ANALYTICAL CHEMISTRY. Educ: Wis State Univ, Platteville, BS, 51; Univ Wis, PhD(anal chem), 60. Prof Exp: Teacher high sch, Wis, 51-52; chemist, Liberty Powder Defense Corp, Olin Mathieson Chem Corp, 52-56; sr chemist, Cent Res Lab, 60-69, SR RES SPECIALIST, AGRICHEM ANAL LAB, 3M Co, 69- Mem: Am Chem Soc. Res: Gas and liquid chromatography, ur-vis-IR spectroscopy; titrimetry. Mailing Add: 3M Co 3M Ctr St Paul MN 55101

TODD, JOHN, b Carnacally, Ireland, May 16, 11; nat US; m 38. NUMERICAL ANALYSIS. Educ: Queen's Univ, Belfast, BSc, 31. Prof Exp: Lectr, Queen's Univ, Belfast, 33-37 & King's Col, London, 37-49; expert appl math, Nat Bur Stand, 47-48, chief comput lab, 49-54, numerical anal, 54-57; PROF MATH, CALIF INST TECHNOL, 57- Concurrent Pos: Scientist, Brit Admiralty, 39-46; Fulbright prof, Uni Univ Vienna, 65. Mem: Am Math Soc; Soc Indust & Appl Math; Math Asn Am; Asn Comput Mach. Res: Mathematical analysis; algebra. Mailing Add: Dept of Math 253-37 Calif Inst Technol Pasadena CA 91109

TODD, KENNETH S, JR, b Three Forks, Mont, Aug, 25, 36. VETERINARY PARASITOLOGY. Educ: Mont State Univ, BS, 62, MS, 64; Utah State Univ, PhD(zool), 67. Prof Exp: Asst zool, Utah State Univ, 64-67; asst prof, 67-71, ASSOC PROF VET PARASITOL, UNIV ILL, URBANA, 71- Concurrent Pos: Actg dir ctr human ecol, Univ Ill, Urbana, 72. Mem: Am Soc Parasitol; Soc Protozool; Wildlife Dis Asn; Soc Nematol; Am Micros Soc. Res: Parasites of wildlife and domestic animals; ecology of parasitism. Mailing Add: Col of Vet Med Univ of Ill Urbana, IL 61801

TODD, LEE JOHN, b Denver, Colo, Nov 8, 36; m 60; c 2. INORGANIC CHEMISTRY. Educ: Univ Notre Dame, BS, 58; Fla State Univ, MS, 60; Ind Univ, PhD(chem), 63. Prof Exp: Res assoc chem, Mass Inst Technol, 63-64; asst prof inorg chem, Univ Ill, Urbana, 64-68; assoc prof, 68-74, PROF CHEM, IND UNIV, BLOOMINGTON, 74- Mem: Am Chem Soc. Res: Inorganic and physical inorganic chemistry of boron and organometallic compounds. Mailing Add: Dept of Chem A669 Chem Bldg Indiana Univ Bloomington IN 47401

TODD, LEONARD, b Glasgow, Scotland, Feb 7, 40. THEORETICAL MECHANICS. Educ: Strathclyde Univ, BSc, 61; Cambridge Univ, PhD(appl math), 64. Prof Exp: C L E Moore instr appl math, Mass Inst Technol, 64-66; lectr, Strathclyde Univ, 66-69; ASSOC PROF APPL MATH, LAURENTIAN UNIV, 69- Concurrent Pos: Nat Res Coun Can res grants, 69- Res: Theoretical fluid mechanics; asymptotic expansions with emphasis on applications. Mailing Add: Dept of Math Laurentian Univ Sudbury ON Can

TODD, LESLIE JAMES, physical chemistry, see 12th edition

TODD, LESLIE JAY, chemistry, deceased

TODD, MARGARET EDNA, b New York, NY, Sept 14, 24; m 66. PHYSIOLOGY. Educ: Mich State Univ, BS, 46; Fordham Univ, MS, 55, PhD, 58. Prof Exp: Med technician, Woman's Hosp & State Dept Health, Detroit, 46; med technician, 47-53, res asst blood coagulation & vascular dis, 58-72, ASST PROF BIOCHEM, DEPT SURG, MED COL, CORNELL UNIV, 72-; SR SCIENTIST, TECHNICIAN INSTRUMENT CORP, 73- Concurrent Pos: Grant, Churchill Hosp, Oxford, Eng, 62 & Mayo Clin, 63. Mem: AAAS; Am Soc Clin Path; Soc Study Blood; NY Acad Sci; Int Soc Hemat. Res: Insect physiology; blood components; separation of amino acids by paper chromatography; human blood coagulation; electrophoresis of coagulation factors; blood coagulation and anticoagulation therapy in relation to heart disease, pregnancy and kidney disease; use of the nonhuman primate as an animal model to simulate man in physiological and pharmacological studies. Mailing Add: 20 Myrtle Ave Edgewater NJ 07020

TODD, MARY ELIZABETH, b Kingston, Ont; m 72. ANATOMY. Educ: Univ BC, BA, 57, MSc, 59; Univ Glasgow, PhD(zool), 62. Prof Exp: Lectr & head biol dept, United Col, Man, 62-63; from lectr to sr lectr anat, Univ Glasgow, 62-71; vis asst prof anat, Univ BC, 70-71; asst prof anat, Univ Western Ont, 72; sessional lectr oral biol & zool, 72-73, asst prof, 73-75, ASSOC PROF ANAT, UNIV BC, 75- Mem: Anat Soc Gt Brit; Can Asn Anatomists; Am Asn Anatomists; Sigma Xi. Res: Ultrastructure of vascular tissues; specifically developmental studies on components of the walls of vessels and investigation of cultured vascular smooth muscle cells. Mailing Add: Dept of Anat Univ of BC Vancouver BC Can

TODD, NEIL BOWMAN, b Cambridge, Mass, Jan 3, 36. EVOLUTIONARY BIOLOGY, GENETICS. Educ: Univ Mass, BS, 59; Harvard Univ, PhD(biol), 63. Prof Exp: Geneticist, Animal Res Ctr, Med Sch, Harvard Univ, 63-68; DIR & GENETICIST, CARNIVORE GENETICS RES CTR, 68- Concurrent Pos: Geneticist, Bio-Res, Inst, Mass, 68-69; res dir, Faunalabs, Inc, 68-72; adj prof dept biol, Boston Univ, 71-76. Res: Chromosomal mechanisms in the origin and evolution of mammals; population genetics and mutantallele frequencies in domestic cats. Mailing Add: Carnivore Genetics Res Ctr Newtonville MA 02160

TODD, PAUL WILSON, b Bangor, Maine, June 15, 36; m 57; c 4. BIOPHYSICS. Educ: Bowdoin Col, AB, 59; Mass Inst Technol, BS, 59; Univ Rochester, MS, 60; Univ Calif, Berkeley, PhD(biophys), 64. Prof Exp: Lectr med physics, Univ Calif, Berkeley, 64-66; asst prof, 66-72, ASSOC PROF BIOPHYS, PA STATE UNIV, 73- Concurrent Pos: Eleanor Roosevelt Int Cancer res fel, 67-68; mem biomed steering comt, Los Alamos Meson Physics Facility, 70-, chmn, 73-74 & vis staff mem, Los Alamos Sci Lab, 74; vis fel, Princeton Univ, 71-72; mem biol comt, Argonne Univ Asn, 73-75 & chmn, 75; consult, Oak Ridge Nat Lab, 75- Mem: AAAS;Biophys Soc; Am Soc Cell Biol; Tissue Cult Asn; Am Soc Photobiol; NY Acad Sci. Res: Radiation physics; cellular radiation biology; basic research related to radiation therapy; chemistry of cell surface; mammalian cell culture; cell electrophoresis; automated cytology. Mailing Add: 618 Life Sci Bldg PA State Univ University Park PA 16802

TODD, ROBERT EMERSON, b Hartford, Conn, Dec 10, 06; m 32; c 3. ZOOLOGY, EMBRYOLOGY. Educ: Bowdoin Col, BS, 29; Harvard Univ, MA, 35, PhD(zool), 38. Prof Exp: Asst zool, Harvard Univ, 35-38; instr biol, 38-43, from asst prof to prof, 43-72, EMER PROF, ZOOL, COLGATE UNIV, 72- Concurrent Pos: Chmn dept zool, Colgate Univ, 55-61 & biol, 64-67. Mem: Am Soc Zool; AAAS. Res: Experimental embryology; cellular components of centrifuged frog eggs. Mailing Add: 58 Payne St Hamilton NY 13346

TODD, SAMUEL SPAULDING, b Arcata, Calif, Feb 15, 07; m 38; c 3. PHYSICAL CHEMISTRY, THERMOCHEMISTRY. Educ: Stanford Univ, AB, 27, ChE, 29, PhD(chem), 36. Prof Exp: Teacher high sch, Calif, 29-31; lab asst chem, Stanford Univ, 33-35, asst, 35-36; prof chem & physics, Albany Col, 36-38; assoc prof math, Hastings Col, 38-39; assoc prof chem, NDak Agr Col, 39-40; instr, San Jose State Col, 40-42; asst chemist, Southern Regional Res Lab, USDA, 42-44; assoc chemist, Petrol Exp Sta, US Dept Interior, 44-47; phys chemist, Minerals Thermodyn Br, US Bur Mines, Calif, 47-55, phys chemist, Bartlesville Petrol Res Ctr, 55-66, PROJ LEADER, BARTLESVILLE ENERGY RES CTR, ENERGY RES & DEVELOP ADMIN, 66- Concurrent Pos: Asst, Stanford Univ, 41. Mem: AAAS; Am Chem Soc. Res: Thermodynamic properties of inorganic and organic substances from thermal measurements; vapor-flow calorimetry and heats of vaporization. Mailing Add: Bartlesville Energy Res Ctr ERDA PO Box 1398 Bartlesville OK 74003

TODD, TERRENCE PATRICK, b Brantford, Ont, Aug 16, 46; m 70; c 3. GEOPHYSICS. Educ: Univ Toronto, BS, 69; Mass Inst Technol, PhD(geophys), 73. Prof Exp: Res asst geophys, Dept Earth & Planetary Sci, Mass Inst Technol, 69-73; RES GEOPHYSICIST, SHELL DEVELOP CO, 73- Res: Elastic properties of rock and other materials. Mailing Add: Shell Develop Co PO Box 481 Houston TX 77001

TODD, WILBERT REMINGTON, b Milwaukee, Wis, July 5, 06; m 36; c 3. BIOCHEMISTRY. Educ: Univ Wis, BS, 28, MS, 31, PhD(agr chem), 33. Prof Exp: Food chemist, Jewett & Sherman, Wis, 28-29; chief chemist, Nutrit Res Lab, 33-35, from instr to prof, 35-72, EMER PROF BIOCHEM, MED SCH, UNIV ORE, 72- Concurrent Pos: Chemist, Ore Racing Comn, 46-59, racing comnr, 58- Mem: Am Soc Biol Chem; Am Inst Nutrit. Res: Trace elements in animal nutrition; vitamin assays; glycogen synthesis in rats; methods of determinations of carbohydrates; amino acids feeding and glycogen formation in rats. Mailing Add: Dept of Biochem Univ of Ore Med Sch Portland OR 97201

TODD, WILLIAM HAWORTH, organic chemistry, see 12th edition

TODD, WILLIAM MCCLINTOCK, b Colon, Panama, July 17, 25; US citizen; m 47; c 2. MEDICAL MICROBIOLOGY, VIROLOGY. Educ: Univ Ga, BS, 50; Vanderbilt Univ, MS, 55, PhD(microbiol), 57. Prof Exp: Asst prof microbiol, Sch Med, Univ Miss, 57-63; ASSOC PROF MICROBIOL, MED UNITS, UNIV TENN, MEMPHIS, 63-, PROF, CTR FOR HEALTH SCI, 65- Concurrent Pos: USPHS fel biochem & res assoc, Vanderbilt Univ, 58-60. Mem: Am Soc Microbiol; Tissue Cult Asn. Res: Host-virus relationships. Mailing Add: Dept of Microbiol Univ of Tenn Ctr for Health Sci Memphis TN 38163

TODHUNTER, ELIZABETH NEIGE, b Christchurch, NZ, July 6, 01; nat US. NUTRITION. Educ: Univ NZ, BS, 26, MS, 28; Columbia Univ, PhD(chem), 33. Prof Exp: From asst prof to assoc prof nutrit, State Col Wash, 34-41; from assoc prof to prof & head dept, Univ Ala, 41-43, dean sch home econ, 53-66; NUTRIT CONSULT, 66-; VIS PROF NUTRIT, SCH MED, VANDERBILT UNIV, 67- Concurrent Pos: Consult to Surgeon Gen, US Air Force, 65-67; vis lectr & sci writing, 66- Mem: AAAS; Am Chem Soc; Am Dietetic Asn (pres, 57-58); Am Inst Nutrit. Res: Vitamin C in foods and body fluids; vitamin A in foods; human food consumption; dietary studies; nutritional status measurements; nutrition of elderly; history of nutrition. Mailing Add: Div of Nutrit Vanderbilt Univ Sch Med Nashville TN 37232

TODSEN, THOMAS KAMP, b Pittsfield, Mass, Oct 21, 18; m 39; c 2. SCIENCE ADMINISTRATION, BOTANY. Educ: Univ Fla, BS, 39, MS, 42, Beaumont Mem Res fel & PhD(org chem), 50. Prof Exp: Asst & instr, Univ Fla, 47-50; instr, NMex Col Agr & Mech Arts, 50-51; chief chemist, 51-53, chief warheads engr, 53-58, sci adv off, 58-59, land combat systs eval, 59-66, dir test opers, 66-69, dir SSMPO, 69-72, TECH DIR ARMY MISSILE TEST & EVAL, WHITE SANDS MISSILE RANGE, 72- Mem: AAAS; Am Chem Soc; Sigma Xi. Res: Naturally occurring plant constituents; plant taxonomy; plant distribution. Mailing Add: 2000 Rose Lane Las Cruces NM 88001

TODY, WAYNE H, fishery biology, see 12th edition

TOENISKOETTER, RICHARD HENRY, b St Louis, Mo, Mar 21, 31; m 53; c 6. INDUSTRIAL CHEMISTRY. Educ: Univ St Louis, BS, 52, MS, 56, PhD(chem), 58. Prof Exp: Res chemist, Union Carbide Corp, 57-67; sr res chemist, ADM Chem Co Div, Ashland Oil Co, 67-68, group leader, Ashland Chem Co, 68-70, mgr inorg chem res & develop div, 70-73, MGR FOUNDRY RES, ASHLAND CHEM CO, 73- Mem: Am Chem Soc; Am Ceramic Soc; Sigma Xi. Res: Organic and inorganic polymer chemistry; coordination and fluorine compounds; boron hydrides; materials and foundry products research. Mailing Add: 6771 Masefield St Worthington OH 43085

TOENNIES, JAN PETER, b Philadelphia, Pa, May 3, 30; m 66; c 2. MOLECULAR PHYSICS. Educ: Brown Univ, PhD(chem), 57. Prof Exp: Asst, 57-67, docent, 67-71, HON PROF, INST PHYSICS, UNIV BONN, 71-; DIR, MAX PLANCK INST FLUID DYNAMICS, 69- Concurrent Pos: Guest docent, Gothenburg Univ, 66-75; apl prof, Univ Göttingen, 72- Mem: Am Phys Soc; Ger Phys Soc; Europ Phys Soc. Res: Molecular beam investigations of elastic, inelastic and reactive collisions; theory of inelastic scattering; chemical reactions in shock waves. Mailing Add: Max Planck Inst for Fluid Dynamics Böttingerstrasse 6-8 Göttingen West Germany

TOENNIESSEN, GARY HERBERT, b Lockport, NY, July 9, 44; m 67. ENVIRONMENTAL SCIENCES. Educ: State Univ NY Buffalo, BA, 66; Univ NC, MS, 68, PhD(environ sci), 71. Prof Exp: Prog assoc, 71-72, ASST DIR NATURAL & ENVIRON SCI, ROCKEFELLER FOUND, 72- Mem: Am Soc Microbiol; Water Pollution Control Fedn. Res: Structure and function of aquatic ecosystems in particular and environmental problems in general. Mailing Add: Rockefeller Found 1133 Ave of the Americas New York NY 10036

TOEPFER, ALAN JAMES, b Chicago, Ill, Oct 20, 41; m 64; c 2. PHYSICS. Educ: Marquette Univ, BS, 62; Univ Southern Calif, MS, 64, PhD(physics), 68. Prof Exp: Res asst plasma physics, Royal Inst Technol, Sweden, 64-65; staff mem, Aerospace Corp, 65-66; STAFF MEM PLASMA PHYSICS, SANDIA LABS, 68- Mem: AAAS; Am Phys Soc. Res: Plasma and nuclear physics. Mailing Add: Sandia Labs, Albuquerque NM 87115

TOEPFER, EDWARD WILLIAM, food chemistry, nutrition, see 12th edition

TOEPLITZ, BARBARA KEELER, b Danbury, Conn, Feb 14, 15; m 71. SPECTROSCOPY. Educ: Columbia Univ, BS, 39; Rutgers Univ, MS, 60. Prof Exp: Dietician, Methodist Hosp, Brooklyn, NY, 35-39, Four Winds, Katonah, NY, 39-41 & York Hosp, Pa, 41-43; spectrogr, Curtiss Propeller Div, Curtiss Wright Corp, 43-45; asst spectros, 46-47, res assoc, 57-60, SR RES CHEMIST, E R SQUIBB & SONS, INC, 60- Mem: Am Crystallog Asn; Soc Appl Spectros; Coblentz Soc. Res: X-ray crystallography; infrared spectroscopy. Mailing Add: E R Squibb & Sons Inc PO Box 4000 Princeton NJ 08540

TOETZ, DALE W, b Milwaukee, Wis, Sept 23, 37; m 65; c 2. FISH BIOLOGY, LIMNOLOGY. Educ: Univ Wis, BS, 59, MS, 61; Ind Univ, PhD(zool), 65. Prof Exp: Actg instr zool, Univ Wis, Milwaukee, 61-62; teaching asst, Ind Univ, 62-65; asst prof, 65-69, ASSOC PROF ZOOL, OKLA STATE UNIV, 69- Concurrent Pos: Water Resources Res Inst res grant, 66-; assoc res biologist, Scripps Inst Oceanog, 74. Mem: Ecol Soc Am; Am Soc Limnol & Oceanog; Int Soc Theoret & Appl Limnol. Res: Limnology of nitrogen; year class formation in fish; productivity of aquatic macrophytes. Mailing Add: Sch of Biol Sci Okla State Univ Stillwater OK 74074

TOEWS, CORNELIUS J, b Altona, Man, Mar 22, 37; m 61; c 3. ENDOCRINOLOGY, BIOCHEMISTRY. Educ: Univ Man, BSc & MD, 63; Queen's Univ, Ont, PhD(biochem), 67; FRCPS(C), 69. Prof Exp: ASST PROF BIOCHEM & MED, MED SCH, McMASTER UNIV, 71- Concurrent Pos: Med Res Coun Can Centennial res fel med, Joslin Res Lab, Harvard Univ, 68-71; jr assoc med, Peter Bent Brigham Hosp, Boston, 70-71; instr, Med Sch, Harvard Univ, 70-71. Mem: Can Soc Clin Invest; Can Biochem Soc; Am Diabetes Asn; Can Diabetes Asn. Res: Regulation of gluconeogenesis in the liver; regulation of glycolysis and intermediary metabolism in skeletal muscle. Mailing Add: Dept of Med McMaster Univ Med Sch Hamilton ON Can

TOEWS, DANIEL PETER, b Grande Prairie, Alta, Dec 18, 41; m 64; c 2. ANIMAL PHYSIOLOGY. Educ: Univ Alta, BSc, 63, MSc, 66; Univ BC, PhD(zool), 69. Prof Exp: Asst prof zool, Univ Alta, 69-71; ASST PROF BIOL, ACADIA UNIV, 71- Mem: AAAS; Can Soc Zool; Brit Soc Exp Biol. Res: Comparative respiration and circulation in fishes and amphibians. Mailing Add: Dept of Biol Acadia Univ Wolfville NS Can

TOEWS, KORNELIUS GERHARD, b Russia, Aug 25, 06; nat Can; m 33; c 5. MATHEMATICS. Educ: Univ Sask, BSc, 33, MSc, 34, BEd, 38. Prof Exp: Prin math & sci, Rosthern Jr Col, 38-51; prof math & phys, Luther Col, 52-57; from asst prof to prof, 57-74, EMER PROF MATH, UNIV SASK, REGINA, 74- Mem: Can Math Cong. Res: Grassmann theory in its relation to linear associative algebra. Mailing Add: Dept of Math Univ of Sask Regina SK Can

TOFE, ANDREW JOHN, b New York, NY, May 6, 40. NUCLEAR MEDICINE. Educ: Univ Dayton, BS, 63; Fla State Univ, PhD(nuclear chem), 69. Prof Exp: Staff scientist activation, Miami Valley Lab, 70-72; SCIENTIST NUCLEAR MED, PROCTER & GAMBLE CO, 72- Mem: Soc Nuclear Med; Am Chem Soc. Res: Utilization of radioisotopes in conjunction with organ specific chemical compounds for early detection of human disease or adnormalities. Mailing Add: Miami Valley Lab Procter & Gamble Co PO Box 39175 Cincinnati OH 45247

TOFFEL, GEORGE MATHIAS, b Greensburg, Pa, Jan 28, 11; m 38; c 4. CHEMISTRY. Educ: Vanderbilt Univ, BA, 35, MS, 36. Prof Exp: Instr chem, Vanderbilt Univ, 34-36; teacher high sch, Ga, 36-37; head sci dept, Marion Inst, 37-47; asst prof, 47-51, ASSOC PROF CHEM, UNIV ALA, TUSCALOOSA, 51- Concurrent Pos: Vis prof, Univ Hawaii, 64-65. Mem: Am Chem Soc; fel Am Inst Chem. Res: Electro-organic chemistry; free radicals; resistor research; magnetic alloys of manganese; reaction mechanism studies with carbon-14 fatty acid esters; ketonization of fatty acids. Mailing Add: Box H Dept of Chem Univ of Ala University AL 35486

TOFT, DAVID ORVILLE, biochemistry, endocrinology, see 12th edition

TOFT, ROBERT JENS, b Wis, Mar 2, 33; m 56; c 3. ACADEMIC ADMINISTRATION. Educ: Beloit Col, BA, 55; Rice Univ, MA, 57, PhD(biol), 60. Prof Exp: Asst biol, Rice Univ, 59-60, res asst, 60; from instr to asst prof, Bowdoin Col, 60-63; asst physiologist, Argonne Nat Lab, 63-64; from asst prof to assoc prof biol, 64-70, asst dean, 70; prog dir, NSF, 70-72; DEAN COL IV, GRAND VALLEY STATE COLS, 72-, PROG DEVELOP OFFICER, 75- Concurrent Pos: Consult div biol & med res, Argonne Nat Lab, 64-; assoc dir col teacher progs NSF, 68-69. Mem: AAAS; Am Inst Biol Sci; Am Asn Higher Educ. Res: Bone and thyroid metabolism; metabolism and toxicity of radionuclides in beagles; alternative instructional modes for undergraduate education; designed and started a totally self-paced modular college. Mailing Add: Grand Valley State Cols Allendale MI 49401

TOGASAKI, ROBERT K, b San Francisco, Calif, July 24, 32; m 59. PLANT PHYSIOLOGY, CELL BIOLOGY. Educ: Haverford Col, BA, 56; NIH fel & PhD(biochem), Cornell Univ, 64. Prof Exp: Res fel biol, Harvard Univ, 67, lectr, 67-68; asst prof, 68-73, ASSOC PROF PLANT SCI, IND UNIV, BLOOMINGTON, 73- Concurrent Pos: NIH fel, 65-67. Mem: Am Soc Plant Physiol; Am Soc Cell Biol; Genetic Soc Am; Soc Protozoologists. Res: Photosynthetic carbon metabolism and its regulation; biochemical and genetic analysis of photosynthetic mechanisms and its regulation in Chlamydomonas reinhardi, a model eukaryotic photosynthetic organism. Mailing Add: Dept of Plant Sci Ind Univ Bloomington IN 47401

TOGLIA, JOSEPH U, b Pescopagano, Italy, Apr 24, 27; US citizen; m; c 3. NEUROLOGY. Educ: Liceo Scientifico, Avellino, Italy, BS, 45; Univ Rome, MD, 51. Prof Exp: Staff neurologist, Baylor Col Med, 60-63; prof neurol & otorhinol, 66-74, PROF NEUROL & CHMN DEPT, SCH MED, TEMPLE UNIV, 74-; CHIEF NEUROL, PHILADELPHIA GEN HOSP, 66- Concurrent Pos: Attend physician, Temple Univ Hosp, 66-; consult, Vet Admin Hosp, 66- & NIH, 73- Mem: AMA; Am Acad Neurol; Am Acad Ophthal & Otolaryngol; Pan-Am Med Asn; Ital Med Asn. Res: Electronystagmography; clinical vestibular physiology. Mailing Add: Temple Univ Hosp 3401 N Broad St Philadelphia PA 19140

TOGO, YASUSHI, b Sapporo, Japan, May 9, 20; m 48; c 1. INFECTIOUS DISEASES, VIROLOGY. Educ: Univ Tokyo, MD, 45, DMedSci, 52. Prof Exp: Asst serol, Sch Med, Univ Tokyo, 45; microbiologist, US Army Tokyo Army Hosp, 45-50; asst med, Sch Med, Univ Tokyo, 50-55; asst res, 56, from instr to asst prof, 59-67, ASSOC PROF MED, SCH MED, UNIV MD, BALTIMORE CITY, 67- Concurrent Pos: Fel, Univ Md, Baltimore City, 56-59. Mem: Am Soc Microbiol; Infectious Dis Soc Am. Res: Clinical and laboratory studies of viral and other microbial vaccines and antiviral compounds. Mailing Add: Div of Infectious Dis Univ of Md Sch of Med Baltimore MD 21201

TOGURI, JAMES M, b Vancouver, BC, Sept 22, 30; m 57; c 5. CHEMISTRY. Educ: Univ Toronto, BASc, 55, MASc, 56, PhD(metall), 58. Prof Exp: Nat Res Coun Can fel metall, Imp Col Sci & Technol, Univ London, 58-59; fel inorg chem, Tech Univ Norway, 59-61; res assoc chem, Inst Metals, Univ Chicago, 61-62; group leader, Noranda Res Ctr, Que, 62-63, head dept, 63-66; assoc prof, 66-69, PROF METALL & MAT SCI, UNIV TORONTO, 69- Concurrent Pos: Royal Norweg Sci Coun fel, 60-61; grants, Nat Res Coun Can & Defence Res Bd Can, 66-; ed-in-chief, Can Metall Quart, 67- Mem: Am Inst Mining, Metall & Petrol Engrs; Can Inst Mining & Metall. Res: High temperature chemistry; thermodynamic properties; kinetics high temperature; fused salt chemistry. Mailing Add: Dept of Metall & Mat Sci Univ of Toronto Toronto ON Can

TOHVER, HANNO TIIT, b Tartu, Estonia, Dec 18, 35; Can citizen. PHYSICS. Educ: Queen's Univ, Ont, BS, 57, MS, 59; Purdue Univ, PhD, 68. Prof Exp: Asst, Purdue Univ, 60-67; asst prof, 68-71, ASSOC PROF PHYSICS, UNIV ALA, BIRMINGHAM, 71- Mailing Add: Dept of Physics Univ of Ala Birmingham AL 35294

TOIGO, ANGELO, b Chicago, Ill, Apr 4, 26; m 50; c 2. MEDICINE. Educ: Beloit Col, BS, 46; Northwestern Univ, MD, 50; Am Bd Internal Med, dipl, 58; Am Bd Pulmonary Dis, dipl, 62. Prof Exp: Resident internal med, Vet Admin Hosp, Hines, Ill, 51-54, staff physician, 56-58, sect chief med serv, 58-66, asst chief med serv, 66-70; DIR MED, CONEMAUGH VALLEY MEM HOSP, 70- Mem: Fel Am Col Physicians; Am Thoracic Soc. Res: Chronic respiratory disease, its epidemiology and treatment. Mailing Add: Conemaugh Valley Mem Hosp Johnstown PA 15905

TOIVOLA, PERTTI TOIVO KALEVI, b Pori, Finland, Aug 15, 46; US citizen; m 73. MEDICAL PHYSIOLOGY, NEUROENDOCRINOLOGY. Educ: Univ Wash, BA, 68, PhD(physiol & biophys), 72. Prof Exp: Res asst physiol, Regional Primate Res Ctr, Univ Wash, 68-69, sr fel endocrinol, Dept Med, 73; ASST SCIENTIST, REGIONAL PRIMATE RES CTR, UNIV WIS, 73-, ASST PROF PHYSIOL, SCH MED, 74- Mem: Int Soc Neuroendocrinol; Am Physiol Soc. Res: Central nervous system regulation of the endocrine system. Mailing Add: Dept of Physiol Sch of Med Univ of Wis Madison WI 53706

TOJI, LORRAINE HELLENGA, b Three Oaks, Mich, Oct 22, 38; m 65. BIOCHEMISTRY. Educ: Hope Col, AB, 60; Wayne State Univ, MS, 62; Univ Pa, PhD(biochem), 69. Prof Exp: Instr chem, Hope Col, 61-64; RES ASSOC BIOCHEM, INST MED RES, CAMDEN, 69- Mem: AAAS; Am Chem Soc. Res: Induction and characterization of somatic cell mutants in control; virus transformed cells. Mailing Add: RD 2 Box 96 Sewell NJ 08080

TOKAY, ELBERT, b Brooklyn, NY, May 27, 16; m 40; c 2. BIOLOGY. Educ: City Col New York, AB, 36; Univ Chicago, PhD(physiol), 41. Prof Exp: Asst physiol, Univ Chicago, 39-41; instr, 41-43, from asst prof to assoc prof, 47-58, PROF PHYSIOL, VASSAR COL, 58- Mem: AAAS. Res: Drugs and other factors affecting central nervous potentials and metabolism. Mailing Add: Dept of Biol Vassar Col Poughkeepsie NY 12601

TOKAY, F HARRY, b St Paul, Minn, July 30, 36. AUDIOLOGY, SPEECH & HEARING SCIENCES. Educ: St Cloud State Col, BS, 60; Mich State Univ, MA, 62, PhD(audiol), 66. Prof Exp: Asst prof audiol, Cent Mich Univ, 65-66 & Univ Mass, Amherst, 67-74; ASSOC PROF AUDIOL & CHMN COMMUN DISORDERS PROG, UNIV NH, 74- Mem: Am Speech & Hearing Asn; Acoust Soc Am. Res: Audiology with children. Mailing Add: Commun Disorders Prog Univ of NH Durham NH 03824

TÖKES, LASZLO GYULA, b Budapest, Hungary, July 7, 37; US citizen. ORGANIC CHEMISTRY. Educ: Univ Southern Calif, BA, 61; Stanford Univ, PhD(chem), 65. Prof Exp: Nat Ctr Sci Res, France fel, Univ Strasbourg, 65-66; fel, 66-67, DEPT HEAD SPECTROS, SYNTEX RES DIV, SYNTEX CORP, 67- Mem: Am Chem Soc; Am Soc Mass Spectrometry. Res: Mass spectrometric fragmentation mechanisms; isotope labeling studies; structure elucidations by using advanced spectroscopic methods; natural product chemistry with special interest in marine chemistry. Mailing Add: Inst of Org Chem Syntex Res Div Syntex Corp Palo Alto CA 94304

TOKES, ZOLTAN ANDRAS, b Budapest, Hungary, May 14, 40; m 72; c 1. BIOCHEMISTRY, DEVELOPMENTAL BIOLOGY. Educ: Univ Southern Calif, BSc, 64; Calif Inst Technol, PhD(biochem), 70. Prof Exp: Lectr biochem, Univ Malaya, 70-71; res immunol, Basel Inst Immunol, Hoffmann-LaRoche, Inc, 71-74; ASST PROF BIOCHEM, UNIV SOUTHERN CALIF & DIR CELL MEMBRANE LABS, LOS ANGELES COUNTY UNIV SOUTHERN CALIF CANCER CTR, 74- Mem: Tissue Cult Asn; Malaysian Biochem Soc. Res: Recognition of cell surface changes with differentiation; cell-cell interactions and their immunological relevance.

Mailing Add: Dept of Biochem Sch of Med Univ of Southern Calif 2025 Zonal Ave Los Angeles CA 90033

TOKITA, NOBORU, b Sapporo, Japan, Feb 20, 23; m 53. PHYSICS. Educ: Hokkaido Univ, BS, 47, DrSci, 56. Prof Exp: Res mem, Kobayashi Inst Physics, Japan, 47-52; asst prof physics, Waseda Univ, Japan, 52-57; res assoc polymer sci, Duke Univ, 57-60; sr res physicist, 60-68, RES ASSOC RES CTR, UNIROYAL INC, 68- Res: Polymer physics; vibration and sound; rheology of elastomer and plastics. Mailing Add: Oxford Mgt & Res Ctr Uniroyal Inc Middlebury CT 06749

TOKO, HARVEY VERNER, plant pathology, entomology, see 12th edition

TOKOLI, EMERY G, b Budapest, Hungary, June 6, 23, nat US; m 48; c 1. ORGANIC CHEMISTRY. Educ: Eötvös Lorand Univ, Budapest, MS, 47. Prof Exp: Res chemist, Chinoin Co Ltd, Hungary, 45-47; dir res, Fine Orgs, Inc, NJ, 48-60; sr res chemist, Minn Mining & Mfg, 60-64; res scientist, Union-Camp Paper Corp, 64-66; SCIENTIST, RES & ENG DIV, XEROX CORP, 66- Concurrent Pos: Instr, Fairleigh Dickinson, 58-59. Mem: Am Chem Soc; fel Am Inst Chem. Res: Nucleophilic substitution reactions; Friedel crafts alkylations and acylations; thermally stable polymers and intermediates; fluorocarbon and silicone chemistry; lignin chemistry; organic photoconductors; redox polymers. Mailing Add: Res & Engr Div Xerox Corp Webster NY 14580

TOKSOZ, MEHMET NAFI, b Antakya, Turkey, Apr 18, 34. GEOPHYSICS. Educ: Colo Sch Mines, GpE, 58; Calif Inst Technol, MS, 60, PhD(geophys, elec eng), 63. Prof Exp: Res fel geophys, Calif Inst Technol, 63-65; from asst prof to assoc prof, 65-71, PROF GEOPHYS, MASS INST TECHNOL, 71-, DIR, GEORGE WALLACE JR GEOPHYS OBSERV, DEPT EARTH & PLANETARY SCI, 75- Mem: AAAS; Am Geophys Union; Seismol Soc Am; Soc Explor Geophys. Res: Seismology and structure of planetary interiors. Mailing Add: Dept of Earth & Planetary Sci Mass Inst of Technol Cambridge MA 02139

TOKUDA, SEI, b Ewa, Hawaii, Aug 17, 30; m 55; c 3. IMMUNOLOGY, MICROBIOLOGY. Educ: Univ Hawaii, BS, 53; Univ Wash, PhD(microbiol), 60. Prof Exp: Trainee microbiol, Univ Wash, 59-60, res instr immunol, 62-63; asst prof immunol & microbiol, Univ Vt, 63-66; from asst prof to assoc prof, 66-74, PROF IMMUNOL, SCH MED, UNIV N MEX, 74- Concurrent Pos: Res fel immunochem, Calif Inst Technol, 60-62; NIH fel, 61-62, NIH res grant, 64-; Am Cancer Soc res grant, 64; USPHS career develop award, 68-73; vis investr, Jackson Lab, 65 & Scripps Clin & Res Found, La Jolla, Calif, 70-71. Mem: AAAS; Am Asn Immunologists; Am Soc Microbiol; Transplantation Soc. Res: Hypersensitivity reaction in mice; transplantation immunity; biological and serological activity of immunoglobulin and immunoglobulin subunits. Mailing Add: Dept of Microbiol Univ of NMex Sch of Med Albuquerque NM 87131

TOKUHATA, GEORGE K, b Matsue, Japan, Aug 25, 24; US citizen; m 49. EPIDEMIOLOGY, PUBLIC HEALTH. Educ: Keio Univ, Japan, BA, 50; Miami Univ, MA, 53; Univ Iowa, PhD(behav sci), 56; Johns Hopkins Univ, DPH(epidemiol), 62. Prof Exp: Res assoc ment health, Mich State Dept Ment Health, 56-59; spec asst div chief, Div Chronic Dis, USPHS, 59-62, prin epidemiologist, 62-63; assoc prof prev med, Col Med, Univ Tenn & chief epidemiol, St Jude Children's Res Hosp, 63-67; RES DIR, DIV RES & BIOSTATIST, PA STATE DEPT HEALTH, 67- Concurrent Pos: USPHS fel, Med Ctr, Johns Hopkins Univ, 59-60, NIH fel, 60-61; Mary Reynold Babcock Found res grant, 59-61; Food & Drug Admin res grant; US Consumer Prod Safety Comn grant, 73-74; Maternal & Child Health Serv grant, 74-; Nat Ctr Health Statist grant, 74-; prof epidemiol & biostatist, Grad Sch Pub Health, Univ Pittsburgh; assoc prof community med, Col Med, Temple Univ. Mem: Fel Am Pub Health Asn; fel Am Sociol Asn. Res: Genetics aspects of cancer and other chronic diseases in adults and children; epidemiology of chronic diseases; evaluation of public health programs research and development. Mailing Add: Bur of Health Res Pa State Dept Health PO Box 90 Harrisburg PA 17120

TOKUHIRO, TADASHI, b Yokohama, Japan, Feb 26, 30; m 56; c 2. PHYSICAL CHEMISTRY, CHEMICAL PHYSICS. Educ: Tokyo Col Sci, BS, 57; Tokyo Inst Technol, MS, 59, PhD(phys chem), 62. Prof Exp: Matsunago Sci Found grant & res assoc, Tokyo Inst Technol, 64-65; univ fel & res assoc, Ohio State Univ, 65-69; asst prof, 69-73, ASSOC PROF CHEM, UNIV DETROIT, 73- Mem: Am Chem Soc; Am Phys Soc; Int Soc Magnetic Resonance. Res: Magnetic resonance spectroscopy; liquid structure, quantum chemistry; vibrational relaxation. Mailing Add: Dept of Chem Univ of Detroit Detroit MI 48221

TOKUMARU, TADASU, b Matsuyama, Japan, Oct 7, 28; m 59; c 3. VIROLOGY. Educ: Okayama Univ, MD, 52; Univ Pa, MSc, 57. Prof Exp: Intern, Tokyo Med Col Hosp, Japan, 52-53; mem med staff, 59-60; res mem, Max Planck Inst Virus Res, Ger, 56-57; instr pediat, Univ Pa, 59-60; asst prof microbiol, Dartmouth Med Sch, 60-62; asst prof pediat, Sch Med, Univ Pa, 62-69; ASST PROF MICROBIOL, COL PHYSICIANS & SURGEONS, COLUMBIA UNIV, 69- Concurrent Pos: Res fel virol, Children's Hosp, Philadelphia, Pa, 54-56; res fel pharmacol, Sch Med, Univ Pa, 57-59. Mem: Am Asn Immunol. Res: Virus-host interaction of animal viruses. Mailing Add: Eye Inst 635 W 165th St New York NY 10032

TOKUNAGA, CHIYOKO, b Kure, Hiroshima, Japan, Dec 7, 14. GENETICS. Educ: Hiroshima Univ, BA, 39; Kyoto Univ, ScD(genetics), 51. Prof Exp: Prof biol, Kobe Col, 51-62; biologist, Lawrence Radiation Lab, Univ Calif, 62-71; ASSOC RES ZOOLOGIST, DEPT ZOOL, UNIV CALIF, BERKELEY, 71- &DEPT MOLECULAR BIOL, 74- Concurrent Pos: Fulbright exchange scholar genetics, 57-59, res assoc, 61-62. Honors & Awards: Annual Prize, Zool Soc Japan, 59. Mem: Genetics Soc Am. Res: Pattern formation in Drosophila. Mailing Add: Dept of Molecular Biol Univ of Calif Berkeley CA 94720

TOLAND, WILLIAM GRIDLEY, JR, b Springfield, Mass, Nov 29, 17; m 39; c 2. INDUSTRIAL CHEMISTRY. Educ: Antioch Col, BS, 40; Purdue Univ, MS, 41, PhD(org chem), 44. Prof Exp: Chem analyst, Eastman Kodak Co, NY, 37-39; asst, Antioch Col, 39-40 & Purdue Univ, 40-42; res chemist in charge explor process res lab, Calif Res Corp, 44-50, sr res chemist, 50-56, supvr res chemist, 56-57, sr res assoc, 57-59, res scientist, 59-61, actg mgr chem res div, 61-62, sect supvr, 62, spec assignment on staff of vpres, 62-63; mgr res & develop dept, Ortho Div, Chevron Chem Co, 63-67, vpres chem res dept, Chevron Res Co, Richmond, 67-68, VPRES RES & DEVELOP, CHEVRON CHEM CO, SAN FRANCISCO, 68- Concurrent Pos: Lectr, 57- Honors & Awards: Chem Pioneering Award, Am Inst Chem, 69. Mem: AAAS; Am Chem Soc. Res: Petrochemical processes; oxidation; halogenation; sulfur chemistry; pesticides; research administration. Mailing Add: 10 Madeleine Lane San Rafael CA 94901

TOLBERG, ADELAIDE BROKAW, b Orange, NJ, Jan 26, 26; m 56; c 2. PHYSIOLOGY, MICROBIOLOGY. Educ: Swarthmore Col, BA, 47; Rochester Univ, MS, 50; Stanford Univ, PhD(physiol), 51. Prof Exp: Asst physiol & biol, Stanford Univ, 50-51, NIH fel microbiol, Hopkins Marine Sta, 51-53; asst prof bact, Univ Calif, Riverside, 54-56, asst res physiologist, Berkeley, 56-71; CONSULT, 71- Mem: AAAS. Res: Water conservation in desert rodents; physiology of dehydration; effects of potassium deficiency; bacterial physiology; electrolyte exchange in yeast; red cell and chloroplast permeability. Mailing Add: 84 Kingston Rd Kensington, CA 94707

TOLBERG, RONALD STANLEY, physical chemistry, see 12th edition

TOLBERG, WESLEY E, physical chemistry, inorganic chemistry, see 12th edition

TOLBERT, BERT MILLS, b Twin Falls, Idaho, Jan 15, 21; m 59; c 4. BIOCHEMISTRY, NUTRITION. Educ: Univ Calif, Berkeley, BS, 42, PhD(chem), 45. Prof Exp: Teaching asst, Univ Calif, Berkeley, 42-44, res chemist, Lawrence Radiation Lab, 44-57; assoc prof, 57-61, PROF CHEM, UNIV COLO, 61- Concurrent Pos: USPHS fel, 52-53; Int Atomic Energy Agency vis prof, Univ Buenos Aires, 62-63; biophysicist, US AEC, Washington, DC, 67-68; vis staff, Los Alamos Sci Lab, 70-; consult, Surgeon Gen Off, US Army. Mem: Am Chem Soc; Am Soc Biol Chem; Am Inst Nutrit; Radiation Res Soc; Soc Exp Biol Med. Res: Metabolism and function of ascorbic acid; radiation chemistry of proteins; catabolism of labeled compounds to 14 carbon dioxide; application of C-14 and H-3 to biochemistry; instrumentation in radiochemistry; synthesis of labeled compounds; use of stable isotopes. Mailing Add: Dept of Chem Univ of Colo Boulder CO 80302

TOLBERT, CHARLES RAY, b Van, WVa, Nov 14, 36; m 67; c 3. ASTRONOMY. Educ: Univ Richmond, BS, 58; Vanderbilt Univ, MS, 60, PhD(physics, astron), 63. Prof Exp: Res assoc, Kapteyn Astron Lab, Netherlands, 63-67; res assoc ctr advan studies, 67-69, asst prof, 69-70, ASSOC PROF, UNIV VA, 70- Mem: Am Astron Soc; Int Astron Union; Int Union Radio Sci; AAAS. Res: Photoelectric photometry of variable stars and binary systems; 21 centimeter radio-astronomical studies of high galactic latitudes; reduction techniques. Mailing Add: Leander McCormick Observ Univ of Va Charlottesville VA 22903

TOLBERT, GENE EDWARD, b Concordia, Kans, Sept 22, 25; m 54; c 3. ECONOMIC GEOLOGY. Educ: Colo Col, BA, 49; Harvard Univ, MA, 57, PhD, 62. Prof Exp: Geologist, US Geol Surv, Alaska, 49-52, foreign br, Brazil, 52-55; geologist, Hanna Mining Co, 57-59; prof, Univ Sao Paulo, 59-63; geologist, US Geol Surv, Pakistan, 63-65; consult, UN, 66; geologist, US Steel Corp, Brazil, 66-70; managing dir, Terraserv Projetos Geologicos Ltd, Rio de Janeiro, 71-75; CONSULT GEOLOGIST, 75- Mem: Geol Soc Am; Am Soc Econ Geologists; Geochem Soc; Brazilian Geol Soc. Res: Economic geology; exploration geology; mineral deposits of Brazil. Mailing Add: 3663 N N Harrison St Arlington VA 22207

TOLBERT, MARGARET ELLEN MAYO, b Suffolk, Va, Nov 24, 43; m 72; c 1. BIOCHEMISTRY. Educ: Tuskegee Inst, BS, 67; Wayne State Univ, MS, 68; Brown Univ, PhD(biochem), 74. Prof Exp: Instr, Opportunities Industrialization Ctr, 71-72; instr math, 69-70, res technician biochem, 69, ASST PROF CHEM, TUSKEGEE INST, 73- Mem: Sigma Xi; Am Chem Soc; Orgn Black Scientists; AAAS. Res: Metabolic studies involving isolated rat hepatic cells. Mailing Add: Dept of Chem Tuskegee Inst Tuskegee Institute AL 36088

TOLBERT, NATHAN EDWARD, b Twin Falls, Idaho, May 19, 19; m 52; c 3. BIOCHEMISTRY. Educ: Univ Calif, BS, 41; Univ Wis, MS, 48, PhD(biochem), 50. Prof Exp: Asst chem dept viticult, Col Agr, Univ Calif, 41-43, biochemist, Radiation Lab, 50; res admin, US AEC, 50-52; sr biochemist, Oak Ridge Nat Lab, 52-58; PROF BIOCHEM, MICH STATE UNIV, 58- Mem: Am Chem Soc; Am Soc Biol Chem; Am Soc Plant Physiol. Res: Plant biochemistry and plant growth substances; glycolic acid metabolism, biosynthesis and function; photosynthesis and relation to plant growth; microbodies and peroxisomes. Mailing Add: Dept of Biochem Mich State Univ East Lansing MI 48823

TOLBERT, ROBERT JOHN, b Pelican Rapids, Minn, Apr 16, 28; m 53; c 2. PLANT ANATOMY. Educ: Moorhead State Univ, BS & BA, 55; Rutgers Univ, PhD(bot), 59. Prof Exp: From asst prof to assoc prof biol, Univ WVa, 59-63; assoc prof, 63-65, PROF BIOL, MOORHEAD STATE UNIV, 65- Mem: AAAS; Bot Soc Am. Res: Anatomical investigation of vegetative shoot apices; comparative studies of shoot apices in the order Malvales. Mailing Add: Dept of Biol Moorhead State Univ Moorhead MN 56560

TOLBERT, THOMAS WARREN, b Greenwood, SC, Dec 1, 45. PHYSICAL CHEMISTRY, MOLECULAR SPECTROSCOPY. Educ: Wofford Col, BS, 67; State Univ NY Binghamton, PhD(chem), 74. Prof Exp: Asst prof chem, Wofford Col, 72-74; proj scientist, 74-75, LAB SUPVR RES & DEVELOP, PARKE-DAVIS MED SURG DIV, 75- Mem: Am Chem Soc. Res: Interaction characteristics of reacting molecules. Mailing Add: 127 Shannon St Greenwood SC 29646

TOLBERT, TOMMY LYLE, b Webb City, Okla, Apr 15, 33; m 55; c 2. ORGANIC CHEMISTRY, POLYMER CHEMISTRY. Educ: Univ Okla, BS, 55, MS, 56, PhD(org chem), 58. Prof Exp: Res chemist, Chemstrand Co, 59-60, Chemstrand Res Ctr, Inc, 60-65, sr res chemist, 63-64, group leader explor polymer & fiber res, 64-65; proj mgr fiber reinforced composites res, Cent Res Dept, Monsanto Co, 65-69, proj develop mgr, Textiles Div, 69-70, res mgr corp res dept, 70-72, res sect mgr, Monsanto Indust Chem, 72-74, DIR TECH INDUST FIBERS, MONSANTO TEXTILES CO, 74- Concurrent Pos: Affil prof, Wash Univ, 68-72. Mem: AAAS; Am Chem Soc; Am Inst Chem Eng; Soc Plastics Eng; NY Acad Sci. Res: Aliphatic imine chemistry; vinyl monomers; fiber-reinforced plastics; rubber chemicals; nylon, aramide and polyheterocyclic fibers. Mailing Add: PO Box 12830 Pensacola FL 32575

TOLDERLUND, DOUGLAS STANLEY, b Newport, RI, Jan 14, 39; m 61; c 2. MARINE ECOLOGY. Educ: Brown Univ, BA, 60; Columbia Univ, PhD(marine geol), 69. Prof Exp: Sr ecologist, Raytheon Co, 69-70; ASSOC PROF MARINE SCI, US COAST GUARD ACAD, 70- Mem: Am Soc Limnol & Oceanog; Am Fisheries Soc; Int Oceanog Found. Res: Estuarine ecology, especially water quality and finfish studies. Mailing Add: Ocean Sci Sect US Coast Guard Acad New London CT 06320

TOLER, ROBERT WILLIAM, b Norphlet, Ark, Dec 15, 28; m 50; c 4. VIROLOGY, PLANT PATHOLOGY. Educ: Univ Ark, BS, 50, MS, 58; NC State Univ, PhD(plant virol), 62. Prof Exp: Res technician rice br exp sta, Univ Ark, 51-54, specialist & plant pathologist, Agr Mission, Panama, 55-57; res plant pathologist, coastal plain exp sta, Agr Res Serv, USDA, Univ Ga, 61-65; assoc prof, 69-74, PROF PLANT PATH, TEX A&M UNIV, 74-, CEREAL VIROLOGIST, 66- Concurrent Pos: Consult, Foy Pittman Rice Farms, Ark, 50-51, Campos Manola Arca, SA, Manziuillo, Cuba, 54 & Ford Found, Antonia Narro Col Agr, Coahuila, Mex, 66; dir plant protection lab, Remote Sensing Ctr, Tex A&M Univ, 71- Mem: Am Phytopath Soc. Res: Physiological effects and host response in plant pathology; cereal virology identification and transmission of viruses that cause cereal diseases. Mailing Add: Dept of Plant Sci Tex A&M Univ College Station TX 77843

TOLGYESI, EVA, b Budapest, Hungary. ORGANIC CHEMISTRY, POLYMER CHEMISTRY. Educ: Budapest Technol Univ, BSc, 53; Univ Leeds, PhD(textile chem), 59. Prof Exp: Asst prof chem, Budapest Technol Univ, 53-56; sr chemist, Harris Res Labs, 65-69, PROJ SUPVR POLYMER CHEM, GILLETTE RES INST, 69- Mem: Am Chem Soc; Soc Cosmetic Chemists; Am Asn Textile Chemists & Colorists. Res: Keratin chemistry; chemical modification of wool; moth-proofing; cationic surfactants; fluoropolymers. Mailing Add: Gillette Res Inst 1413 Research Blvd Rockville MD 20850

TOLGYESI, WILLIAM STEVEN, b Budapest, Hungary, Mar 29, 31; m 54; c 3. ORGANIC CHEMISTRY. Educ: Budapest Tech Univ, BSc, 53; Univ Leeds, PhD(keratin chem), 59. Prof Exp: Asst prof org chem technol, Budapest Tech Univ, 53-54; res chemist, Telecommun Res Inst, Hungary, 54-56, Res Inst, Hungarian Acad Sci, 56 & Dow Chem Can, Ltd, 59-65; PRIN SCIENTIST COSMETIC CHEM, GILLETTE RES INST, 65- Mem: Fiber Soc; Soc Cosmetic Chemists; Am Chem Soc. Res: Keratin fiber and cosmetic chemistry and physics. Mailing Add: Gillette Res Inst 1413 Research Blvd Rockville MD 20850

TOLIMIERI, RICHARD, b New York, NY, Nov 19, 41; m 68; c 2. PURE MATHEMATICS. Educ: City Col New York, BS, 63; Columbia Univ, PhD(math), 69. Prof Exp: Gibbs instr math, Yale Univ, 69-71; asst prof, Lehman Col, 71-72; mem, Inst Advan Studies, 72-73; asst prof, Lehman Col, 73-74; ASSOC PROF MATH, UNIV CONN, 74- Res: Analysis on non-Abelian groups and the application of such studies to special function and number theory. Mailing Add: Dept of Math Univ of Conn Storrs CT 06250

TOLIN, SUE ANN, b Montezuma, Ind, Nov 29, 38. PLANT VIROLOGY, PHYTOPATHOLOGY. Educ: Purdue Univ, BS, 60; Univ Nebr, MS, 62, PhD(bot), 65. Prof Exp: Res asst plant path, Univ Nebr, 60-65; res assoc bot & plant path, Purdue Univ, 65-66; asst prof, 66-72, ASSOC PROF PLANT PATH, VA POLYTECH INST & STATE UNIV, 72- Mem: Am Phytopath Soc; Am Soc Microbiol. Res: Identification, purification and characterization of plant pathogenic viruses; electron microscopy; mechanisms of resistance of plants to viruses. Mailing Add: Dept of Plant Path & Physiol Va Polytech Inst & State Univ Blacksburg VA 24061

TOLIVER, ADOLPHUS PRESTON, molecular biology, biochemistry, see 12th edition

TOLKMITH, HENRY, b Berlin, Ger, Feb 13, 10; nat US; m 37; c 2. ORGANIC CHEMISTRY, BIO-ORGANIC CHEMISTRY. Educ: Tech Univ Berlin, ScD(chem), 35. Prof Exp: Instr gen chem, Tech Univ Berlin, 36, asst prof inorg chem, 37; chemist, I G Farben, Ger, 38-39, div mgr org phosphorous compounds, 40-45, chem consult, Org Indust, 46-47; from res chemist to assoc scientist, 48-69, RES SCIENTIST, DOW CHEM CO, 69- Honors & Awards: Morrison Award, NY Acad Sci, 58. Mem: Fel AAAS; fel Am Inst Chem; fel The Chem Soc; NY Acad Sci. Res: Phosphorous chemistry; molecular structure and biological behavior; chemotherapeutics. Mailing Add: 305 W Carpenter Midland MI 48640

TOLL, JOHN SAMPSON, b Denver, Colo, Oct 25, 23; m 70; c 2. THEORETICAL PHYSICS. Educ: Yale Univ, BS, 44; Princeton Univ, AM, 48, PhD(physics), 52. Hon Degrees: DSc, Univ Md, 73 & Univ Wroclaw, 74. Prof Exp: Managing ed & actg chmn, Yale Sci Mag, 43-44; asst, Princeton Univ, 46-48; theoret physicist, Los Alamos Sci Lab, 50-51; staff mem & assoc dir, Proj Matterhorn, Forrestal Res Ctr, Princeton Univ, 51-53; prof physics & chmn dept physics & astron, Univ Md, 53-65; PROF PHYSICS & PRES, STATE UNIV NY STONY BROOK, 65- Concurrent Pos: Guggenheim Mem Found fel, Inst Theoret Physics, Univ Copenhagen & Univ Lund, 58-59; US deleg & head sci secretariat, Int Conf High Energy Physics, 60; mem-at-lg US nat comn, Int Union Pure & Appl Physics, 61-63; mem, Gov Adv Comt Atomic Energy, State of NY, 66-70; Nordita vis prof, Niels Bohr Inst Theoret Physics, Univ Copenhagen, 75-76. Mem: AAAS; Am Phys Soc; Am Asn Physics Teachers; Nat Sci Teachers Asn; Fedn Am Scientists (chmn, 61-62). Res: Elementary particle theory; scattering. Mailing Add: State Univ of NY Stony Brook NY 11790

TOLLE, JON WRIGHT, b Mattoon, Ill, June 26, 39; m 64. MATHEMATICS, OPERATIONS RESEARCH. Educ: DePauw Univ, BA, 61; Univ Minn, PhD(math), 66. Prof Exp: Res assoc, Argonne Nat Lab, 63; instr math, Univ Minn, 66-67; asst prof, 67-73, ASSOC PROF MATH & OPERS RES, UNIV NC, CHAPEL HILL, 73-, CHMN CURRIC OPERS RES, 74- Concurrent Pos: Vis assoc prof, Grad Sch Bus, Univ Chicago, 75. Mem: Opers Res Soc Am; Am Math Soc; Soc Indust & Appl Math; Math Asn Am. Res: Mathematical programming; optimization theory; difference equations. Mailing Add: Dept of Math Univ of NC Chapel Hill NC 27514

TOLLE, LAWRENCE DUDLEY, pharmaceutics, biochemistry, see 12th edition

TOLLEFSON, CHARLES IVAR, b Moose Jaw, Sask, Oct 2, 18; m 40; c 2. BIOCHEMISTRY. Educ: Univ Sask, BSA, 40, MSc, 47; Univ Minn, PhD(agr biochem), 50. Prof Exp: Asst radioactive ruthenium, Univ Sask, 45-47; asst lactose, Univ Minn, 47-50; res biochemist, Stine Lab, E I du Pont de Nemours & Co, 50-57; res sect leader nutrit, 57-72, MGR RES & DEVELOP, R T FRENCH CO, 72- Mem: AAAS; Am Chem Soc; Inst Food Technologists. Res: Selenium in grains; adsorption and solvent distribution of ruthenium; nutritional effects of lactose; new growth factors; nutrition of cage birds; food dehydration. Mailing Add: R T French Co 1 Mustard St Rochester NY 14609

TOLLEFSON, ERIC LARS, b Moose Jaw, Sask, Oct 15, 21; m 47; c 3. PHYSICAL CHEMISTRY, CHEMICAL ENGINEERING. Educ: Univ Sask, BA, 43, MA, 45; Univ Toronto, PhD(phys chem), 48. Prof Exp: Demonstr chem, Univ Sask, 41-43 & 45, asst, Directorate Chem Warfare, 43-44; jr res officer, Nat Res Coun Can, 45; lab asst, Univ Toronto, 45-47, lectr, 47-48; jr res officer, Nat Res Coun Can, 48-49, asst res officer, 50-51; chemist, Process Res Div, Stanolind Oil & Gas Co, Okla, 51-52, sr chemist, 53-56; head phys chem sect, Res Dept, Can Chem Co, Ltd, 56-66, supt chem develop dept, 65-66, tech mgr, Can Chem Co Div, Chem-Cell Ltd, Alta, 66-67; assoc prof, 67-70, actg head dept, 71, PROF CHEM ENG, UNIV CALGARY, 70-, HEAD DEPT, 72- Mem: Am Chem Soc; fel Chem Inst Can. Res: Kinetics; atomic hydrogen with acetylene; oxidation of ethylene; Fischer-Tropsch synthesis; alcohol dehydrogenation; liquid phase oxidation of hydrocarbons; biological oxidation; activated carbon adsorption of pollutants; reduction of nitrogen oxides in stack gases. Mailing Add: Dept of Chem Eng Univ of Calgary Calgary AB Can

TOLLEFSON, JEFFREY L, b Hampa, Idaho, July 30, 42; m 65; c 2. MATHEMATICS. Educ: Univ Idaho, BS, 65; Mich State Univ, MS, 66, PhD(math), 68. Prof Exp: NASA trainee, Mich State Univ, 65-68; asst prof math, Tulane Univ, 68-71 & Tex A&M Univ, 71-74; ASSOC PROF MATH, UNIV CONN, 74- Mem: Am Math Soc. Res: Topology of manifolds. Mailing Add: Dept of Math Univ of Conn Storrs CT 06268

TOLLEFSRUD, PHILIP BJØRN, b Fargo, NDak, June 11, 38; m 59; c 2. EXPERIMENTAL NUCLEAR PHYSICS. Educ: Univ NDak, BA, 64; Univ Wis, MS, 66, PhD(physics), 69. Prof Exp: STAFF MEM PHYSICS, SANDIA CORP, 69- Mem: Am Phys Soc; Sigma Xi. Res: Isotopic spin, fast burst reactors and simulation sciences. Mailing Add: 11408 Golden Gate Ave NE Albuquerque NM 87111

TOLLER, LOUIS, b Philadelphia, Pa, Dec 2, 17; m 43; c 3. PHYSICS. Educ: Temple Univ, BS, 38; Duke Univ, PhD(neutron cross sect), 55. Prof Exp: Asst prof physics, Univ Louisville, 54-55; prof, Madison Col, 55-59; PROF PHYSICS, ALMA COL, 59- Concurrent Pos: NSF sci fac fel, Stanford Univ, 65-66. Mem: Am Asn Physics Teachers; Math Asn Am; Am Phys Soc. Res: Neutron and low temperature physics. Mailing Add: Dept of Physics Alma Col Alma MI 48801

TOLLES, WALTER EDWIN, b Moline, Ill, Feb 1, 16; m 37; c 1. BIOPHYSICS, PHYSIOLOGY. Educ: Antioch Col, BS, 39; Univ Minn, MS, 41; State Univ NY Downstate Med Ctr, PhD(biophys & physiol), 69. Prof Exp: Asst, Kettering Found, Antioch Col, 37-39; physicist, Div War Res, Airborne Instruments Lab, Columbia Univ, 42-45; supvr, Airborne Instruments Lab, Inc, 45-54, head dept med & biol physics, 54-69; dir, Inst Oceanog & Marine Biol, 59-69; instr, 69-70, ASSOC PROF OBSTET & GYNEC, STATE UNIV NY DOWNSTATE MED CTR, 70- Concurrent Pos: Consult, NIH. Mem: AAAS; Biophys Soc; Am Soc Limnol & Oceanog; Am Phys Soc; Am Soc Cytol. Res: Magnetic techniques in undersea warfare; electronic countermeasures; high speed micro-scanning systems associated data handling system; physiological monitoring systems; clinical instrumentation; diagnostic computer methods. Mailing Add: Dept of Obstet & Gynec State Univ NY Downstate Med Ctr Brooklyn NY 11203

TOLLES, WILLIAM MARSHALL, b New Britain, Conn, June 30, 37; m 59; c 2. PHYSICAL CHEMISTRY. Educ: Univ Conn, BA, 58; Univ Calif, Berkeley, PhD(phys chem), 62. Prof Exp: Fel, Rice Univ, 61-62; from asst prof to assoc prof, 62-73, PROF CHEM, NAVAL POSTGRAD SCH, 73- Concurrent Pos: Consult, Naval Weapons Ctr, China Lake, summers, 66- Mem: Am Phys Soc; Am Chem Soc. Res: Microwave spectroscopy; rotational spectra of molecules; electron spin resonance; microwave properties of materials; non-linear molecular spectroscopy. Mailing Add: Code 61TL Dept of Physics & Chem Naval Postgrad Sch Monterey CA 93940

TOLLESTRUP, ALVIN V, b Los Angeles, Calif, Mar 22, 24; m 44; c 2. PHYSICS. Educ: Calif Inst Technol, PhD(physics), 50. Prof Exp: Res fel physics, 50-53, from asst prof to assoc prof, 53-62, PROF PHYSICS, CALIF INST TECHNOL, 62- Mem: Am Phys Soc. Res: Nuclear disintegration energy value determinations; interaction of 500 million electron volts; gamma rays with hydrogen and deuterium. Mailing Add: Dept of Physics Calif Inst of Technol Pasadena CA 91109

TOLLETT, JAMES TERRELL, nutrition, biochemistry, see 12th edition

TOLLIN, GORDON, b New York, NY, Dec 26, 30; m 55; c 3. BIOPHYSICAL CHEMISTRY. Educ: Brooklyn Col, BS, 52; Iowa State Univ, PhD, 56. Prof Exp: Res assoc chem, Fla State Univ, 56; chemist, Lawrence Radiation Lab, Univ Calif, 56-59, NSF fel, 56-57; from asst prof to assoc prof, 59-67, PROF CHEM, UNIV ARIZ, 67- Concurrent Pos: Sloan fel, 62-66. Mem: AAAS; Am Chem Soc; Biophys Soc; Am Soc Biol Chem. Res: Mechanism of enzyme action; photosynthesis; free radicals in biological energy conversion; biological oxidation-reduction; phototaxis in microorganisms. Mailing Add: Dept of Chem Univ of Ariz Tucson AZ 85721

TOLLMAN, JAMES PERRY, b Chadron, Nebr, Nov 6, 04; m 29; c 3. PATHOLOGY. Educ: Univ Nebr, BSc, 27, MD, 29; Am Bd Path, dipl, 37. Prof Exp: Intern & resident, Peter Bent Brigham Hosp, Boston, Mass, 29-31; from asst prof clin path to prof path, 31-74, chmn dept path & bact, 48-54, dean, 52-64, EMER PROF PATH & EMER DEAN, COL MED, UNIV NEBR, 74- Concurrent Pos: Vis prof, Univ Chiengmai, 64-65. Mem: Am Soc Clin Path; Am Asn Pathologists & Bacteriologists; fel AMA. Res: Effects of dusts on tissues; effect of toxic gases; tissue changes in endocrine disturbances. Mailing Add: 4441 E Sixth St Tucson AZ 85711

TOLMACH, LEONARD JOSEPH, b New York, NY, Apr 18, 23; m 45; c 3. CELL BIOLOGY, RADIOBIOLOGY. Educ: Univ Mich, BS, 43; Univ Chicago, PhD(chem), 51. Prof Exp: Jr chemist, Manhattan Proj, 44-46; from instr to asst prof biophys, Sch Med, Univ Colo, 51-58; assoc prof, 58-64, PROF RADIATION BIOL, SCH MED, WASHINGTON UNIV, 64-, PROF ANAT, 69- Concurrent Pos: Mem comt molecular biol, Washington Univ, 61-72, chmn, 64-66; NSF sr fel, 63-64; mem biophys sci training comt, Nat Inst Gen Med Sci, 66-68. Mem: Radiation Res Soc; Biophys Soc; Am Asn Cancer Res. Res: Effects of radiations and other toxic agents on cell proliferation. Mailing Add: Dept of Anat Washington Univ Sch of Med St Louis MO 63110

TOLMAN, CHADWICK ALMA, b Oct 11, 38; US citizen; m 62; c 3. PHYSICAL CHEMISTRY, INORGANIC CHEMISTRY. Educ: Mass Inst Technol, BS, 60; Univ Calif, Berkeley, PhD(phys chem), 64. Prof Exp: Fel & res assoc chem, Mass Inst Technol, 64-65; CHEMIST, EXP STA, E I DU PONT DE NEMOURS & CO, INC, 65- Mem: Am Chem Soc. Res: Mechanisms of homogeneous catalysis by transition metal complexes; kinetics and equilibria of organometallic reactions. Mailing Add: Exp Sta E I du Pont de Nemours & Co Inc Wilmington DE 19898

TOLMAN, EDWARD LAURIE, b Chelsea, Mass, Oct 9, 42; m 67; c 1. PHARMACOLOGY. Educ: Univ Mass, BA, 64, MA, 65; State Univ NY Upstate Med Ctr, PhD(pharmacol), 70. Prof Exp: NIH fel physiol, Milton S Hershey Med Ctr, Pa State Univ, 69-71, res assoc, 71-72; SR RES BIOLOGIST, LEDERLE LABS, 72- Mem: AAAS. Res: Disorders of carbohydrate and lipid metabolism; prostaglandins and inflammation. Mailing Add: Metab Dis Res Sect Lederle Lab Div Am Cyanamid Co Middletown Rd Pearl River NY 10965

TOLMAN, RICHARD LEE, organic chemistry, biochemistry, see 12th edition

TOLMSOFF, WALTER JOHN, b Independence, Ore, Dec 1, 34; m 53; c 4. PLANT PATHOLOGY, BIOCHEMISTRY. Educ: Ore State Univ, BS, 56, MS, 59; Univ Calif, Davis, PhD(plant path), 65. Prof Exp: Instr res, Ore State Univ, 57-58; lab technician, Univ Calif, 59-63; plant pathologist, Crop Protection Inst, Durham, NH, 63-66, asst dir, 66; res plant pathologist, Crops Res Div, 67-70, PLANT PATHOLOGIST, NAT COTTON PATH RES LAB, AGR RES SERV, USDA, 70- Mem: Bot Soc Am; fel Am Inst Chem; Am Phytopath Soc; Am Soc Plant Physiol. Res: Mitochondrial metabolism in plant pathogenic fungi; respiratory enzymes, pathways, sites of inhibition by fungicides; physiology, genetics and cytology of Verticillium; Verticillium diseases in plants. Mailing Add: Nat Cotton Path Res Lab PO Drawer JF College Station TX 77840

TOLNAI, SUSAN, b Budapest, Hungary, Nov 29, 28; Can citizen; m 50; c 2. CELL BIOLOGY. Educ: Eötvös Lorand Univ, Budapest, MD, 53. Prof Exp: Bacteriologist, Lab Hyg, Dept Nat Health & Welfare, Can, 57-58, biologist, 58-62; lectr, 62-63, from asst prof to assoc prof, 63-71, PROF HISTOL & EMBRYOL, FAC MED, UNIV OTTAWA, 71- Concurrent Pos: Nat Acad Sci Hungary fel, 53-56. Mem: Tissue Cult Asn; Can Soc Cell Biol; NY Acad Sci; Am Soc Cell Biol; Can Soc Immunol. Res:

Cytology of ascites cells; lysosomal enzymes in cultured cells. Mailing Add: Dept of Histol & Embryol Fac of Med Univ of Ottawa Ottawa ON Can

TOLSTEAD, WILLIAM LAWRENCE, b Howard Co, Iowa, Nov 25, 09. BOTANY. Educ: Luther Col, Iowa, BS, 33; Iowa State Univ, MS, 36; Univ Nebr, PhD(plant ecol), 42. Prof Exp: Biologist, Conserv & Surv Div, Univ Nebr, 35-42; assoc prof, 57-67, PROF BIOL, DAVIS & ELKINS COL, 67-, CHMN DEPT, 72- Mem: Ecol Soc Am; Am Hort Soc. Res: Plant breeding rhododendron. Mailing Add: Dept of Biol Davis & Elkins Col Elkins WV 26241

TOLSTED, ELMER BEAUMONT, b Philadelphia, Pa, Apr 28, 20. MATHEMATICS. Educ: Univ Chicago, BS, 40, MS, 41; Brown Univ, PhD(math), 46. Prof Exp: Instr math, Brown Univ, 42-47; from asst prof to assoc prof, 47-61, PROF MATH, POMONA COL, 61- Concurrent Pos: Fulbright exchange prof, Eng, 49-50; instr, Claremont Inst Music, 51- Honors & Awards: Ford Award, Math Asn Am, 65. Mem: Am Math Soc; Math Asn Am. Res: Subharmonic functions. Mailing Add: Dept of Math Pomona Col Claremont CA 91711

TOLSTOY, IVAN, b Baden-Baden, Ger, Mar 30, 23; nat US; m 47, 64; c 3. GEOPHYSICS. Educ: Univ Sorbonne, Lic es sc, 45; Columbia Univ, MA, 47, PhD(geophys), 50. Prof Exp: Mem sci staff, Lamont Geol Observ, Columbia Univ, 48-51; sr res engr, Stanolind Oil & Gas Co, 51-53; res scientist, Hudson Lab, Columbia Univ, 53-60, sr res assoc, 62-67, assoc dir, 64-67, prof ocean eng, 67-68; PROF GEOL & FLUID DYNAMICS, GEOPHYS FLUID DYNAMICS INST, FLA STATE UNIV, 68- Concurrent Pos: Chief scientist, MidAtlantic Ridge Exped, Colombia, 50; consult, Gen Elec Co & Carter Oil Co, 60-62; vis prof, Univ Leeds, 71-72. Mem: AAAS; Am Phys Soc; fel Acoust Soc Am; Seismol Soc Am; Soc Explor Geophys. Res: Theory of acoustic and elastic wave propagation; hydrodynamics; theoretical mechanics; seismology; submarine topology and geology. Mailing Add: Geophys Fluid Dynamics Inst Fla State Univ Tallahassee FL 32306

TOM, BALDWIN HENG, b San Francisco, Calif, Sept 19, 40. TRANSPLANTATION IMMUNOLOGY. Educ: Univ Calif, Berkeley, BA, 63; Univ Ariz, MS, 67, PhD(microbiol), 70. Prof Exp: Fel, Stanford Univ Sch Med, 70-72, res assoc immunol, 72-73; instr, 73-74, assoc, 74-75, ASST PROF SURG & PHYSIOL, NORTHWESTERN UNIV SCH MED, 75- Mem: NY Acad Sci; Soc Exp Biol & Med; Tissue Cult Asn; Transplantation Soc; Sigma Xi. Res: Dissection of the cellular and molecular bases for immune reactivities, especially as they relate to transplantation rejection and to tumor resistance. Mailing Add: Dept of Surg Northwestern Univ Sch of Med Chicago IL 60611

TOM, THEODORE BENTON, b Carterville, Ill, Sept 19, 18; m 41; c 4. ORGANIC CHEMISTRY. Educ: Southern Ill Univ, BS, 40; Ohio State Univ, MSc, 42, PhD(org chem), 43. Prof Exp: Res chemist, Standard Oil Co, Ind, 43-61; asst res dir, Am Oil Co, 62-65, dir fuels res, 65-68, mgr res, 68-69, vpres res & develop, 69-75; RETIRED. Mem: Am Chem Soc; Soc Automotive Engrs; Indust Res Inst; Am Petrol Inst. Res: Pure hydrocarbon synthesis; desulfurization petroleum additives; oil qualities and stability; mercaptan extraction from naphthas; motor oil formulations; automotive research; petroleum processing. Mailing Add: Am Oil Co PO Box 431 Whiting IN 46394

TOMAJA, DAVID LOUIS, b Bridggport, Conn, July 15, 46; m 69. ORGANOMETALLIC CHEMISTRY. Educ: Univ Conn, BA, 68; State Univ NY Albany, PhD(chem), 74. Prof Exp: Chemist, Gen Elec Res & Develop Ctr, 68-70; RES CHEMIST, PHILLIPS PETROL RES & DEVELOP CTR, 74- Mem: Am Chem Soc; The Chem Soc; Sigma Xi. Res: Synthesis of organometallic compounds; elucidation of their structure and bonding by spectroscopic techniques; application of TIN-119 moessbauer spectroscopy to chemical problems. Mailing Add: Phillips Petrol Co 231-RB1 Res & Develop Ctr Bartlesville OK 74004

TOMALIA, DONALD ANDREW, b Owosso, Mich, Sept 5, 38; m 59; c 4. PHYSICAL ORGANIC CHEMISTRY. Educ: Univ Mich, BS, 61; Bucknell Univ, MS, 62; Mich State Univ, PhD(phys org chem), 68. Prof Exp: RES MGR SPECIALTY PROD, CHEM PROD RES LAB, DOW CHEM CO, 62- Mem: Am Chem Soc. Res: Functional monomers and polymers usually incorporating heterocyclic moieties; unusual cross linking devices; onium type chemistry; water borne polymer systems. Mailing Add: Chem Prod Res Lab Dow Chem Co Midland MI 48640

TOMAN, FRANK R, b Ellsworth, Kans, June 6, 39; m 62; c 2. BIOCHEMISTRY. Educ: Kans State Univ, BS, 61, MS, 63, PhD(biochem), 67. Prof Exp: Asst prof, 66-69, ASSOC PROF BIOCHEM, WESTERN KY UNIV, 69- Concurrent Pos: Res assoc, Plant Res Lab, Mich State Univ-AEC, 73-74. Mem: Am Chem Soc. Res: Plant proteins and enzymes; plant growth retardant chemicals; microtubules. Mailing Add: Dept of Biol Western Ky Univ Bowling Green KY 42101

TOMAN, KAREL, b Pilsen, Czech, Mar 19, 24; m 48; c 1. CRYSTALLOGRAPHY. Educ: Prague Tech Univ, Ing Chem, 48, Dr Tech(chem), 51; Czech Acad Sci, DrSc(physics), 65. Prof Exp: Res officer, Inst Metals, Czech, 50-56; sr res officer, Inst Solid State Physics, 56-62; head lab crystallog, Inst Macromolecular Chem, 62-68; Sci Res Coun UK sr vis res fel, Univ Birmingham, 68-69; res fel, Inst Mat Sci, Univ Conn, 69-70; PROF CRYSTALLOG, WRIGHT STATE UNIV, 70- Concurrent Pos: Nicolet fel, McGill Univ, 66-67. Mem: Am Crystallog Asn; Mineral Soc Am. Res: Crystal structure and imperfections. Mailing Add: Dept of Geol Wright State Univ Dayton OH 45431

TOMAN, KURT, b Vienna, Austria, Aug 11, 21; US citizen; m 58; c 3. IONOSPHERIC PHYSICS. Educ: Vienna Tech Univ, MS, 49; Univ Ill, Urbana, PhD(elec eng), 52. Prof Exp: Lab engr, Lecher Inst, Reichenau, Austria, 43-44 & Ctr Tube Res, Tanvald, Czech, 44-45; asst electronics, Univ Ill, Urbana, 49-52, res assoc, 52; re fel, Harvard Univ, 52-55; physicist, 55-63, supvry res physicist & br chief ionospheric radio physics, 63-73, SR SCIENTIST, AIR FORCE CAMBRIDGE RES LAB, 73- Concurrent Pos: Mem nat comn II, Int Union Radio Sci, 61-, working group III-9, Radio Wave Propagation in Ionosphere, 74- Mem: AAAS; Inst Elec & Electronics Engrs; Am Geophys Union; Sigma Xi. Res: Dynamics of the ionosphere; radio wave propagation; spectral analysis of ionospheric waves; wave interference phenomena; group and phase path measurements; ionospheric Doppler variations. Mailing Add: Air Force Cambridge Res Lab LI Laurence G Hanscom Field Bedford MA 01731

TOMANEK, GERALD WAYNE, b Quinter, Kans, Sept 16, 21; m 45; c 3. BOTANY. Educ: Ft Hays Kans State Col, AB, 42, MS, 47; Univ Nebr, PhD, 51. Prof Exp: From asst prof to assoc prof, 47-67, PROF BIOL, FT HAYS KANS STATE COL, 67-, CHMN DIV NATURAL SCI & MATH, 56-, ACTG PRES COL, 75- Concurrent Pos: Consult, Int Coop Admin Arg, 61. Res: Grassland ecology. Mailing Add: Div of Natural Sci & Math Ft Hays Kans State Col Hays KS 67601

TOMARELLI, RUDOLPH MICHAEL, b Pittsburgh, Pa, Jan 10, 17; m 52; c 4. BIOCHEMISTRY, NUTRITION. Educ: Univ Pittsburgh, BS, 38, MS, 41; Western Reserve Univ, MS, 42, PhD(biochem), 43. Prof Exp: Asst, Mellon Inst, 39-41; res chemist, Wyeth Inst Appl Biochem, 43-47; instr nutrit, Univ Pa, 47-50; res chemist, Wyeth Labs, Inc, 50-55, sr investr, Wyeth Inst Med Res, 55-70, MGR NUTRIT DEPT, WYETH LABS, INC, 70- Concurrent Pos: Asst prof, St Joseph's Col, Pa, 48-50. Mem: Am Chem Soc; Am Soc Biol Chemists; Am Inst Nutrit. Res: Infant nutrition. Mailing Add: Nutrit Dept Wyeth Labs Inc Radnor PA 19087

TOMAS, FRANCISCO, b Montreal, Que, Dec 21, 30; m 60; c 2. PHYSICS, SCIENCE ADMINISTRATION. Educ: Sir George Williams Univ, BSc, 56. Prof Exp: Cur physics, Lab, Sir George Williams Univ, 56-65, dir, 65-73; PLANNING OFFICER, CONCORDIA UNIV, 73- Concurrent Pos: Consult, Int Youth Sci Week, 66-67. Mem: Can Asn Physicists. Res: Laboratory instruction and administration; programming; computerization of physics laboratory student testing. Mailing Add: Sci Planning Off Concordia Univ Montreal PQ Can

TOMASCH, WALTER J, b Cleveland, Ohio, July 26, 30; m 55; c 2. EXPERIMENTAL SOLID STATE PHYSICS. Educ: Case Western Reserve Univ, BS, 52, PhD(physics), 58; Rensselaer Polytech Inst, MS, 55. Prof Exp: Sr physicist, Atomics Int Div, NAm Aviation Corp, 58-68; PROF PHYSICS, UNIV NOTRE DAME, 68- Mem: Fel Am Phys Soc. Res: Superconductivity; physics of metals and alloys. Mailing Add: Dept of Physics Univ of Notre Dame Notre Dame IN 46556

TOMASCHKE, HARRY E, b Kendall, NY, Apr 25, 29; m 53; c 4. SURFACE PHYSICS. Educ: Mich State Univ, BS, 56; Univ Ill, MS, 58, PhD(physics), 64. Prof Exp: Res assoc, Coord Sci Lab, Univ Ill, Urbana, 58-64; ASSOC PROF PHYSICS, GREENVILLE COL, 64- Mem: Am Phys Soc. Res: Electrical breakdown in vacuum; absorption of gases under ultrahigh vacuum conditions. Mailing Add: Dept of Physics Greenville Col Greenville IL 62246

TOMASELLI, VINCENT PAUL, b Weehawken, NJ, May 3, 41; m 66; c 2. PHYSICS. Educ: Fairleigh Dickinson Univ, BS, 62, MS, 64; NY Univ, PhD(physics), 71. Prof Exp: Res physicist, Uniroyal Res Ctr, 64-66; from instr to asst prof, 66-74, ASSOC PROF PHYSICS, FAIRLEIGH DICKINSON UNIV, 74- Mem: Am Phys Soc. Res: Far-infrared spectroscopy with applications to biological materials; optical instrumentation. Mailing Add: Dept of Physics Fairleigh Dickinson Univ Teaneck NJ 07666

TOMASHEFSKI, JOSEPH FRANCIS, b Plymouth, Pa, Dec 30, 22; m 49; c 3. PHYSIOLOGY. Educ: Hahnemann Med Col, MD, 47; Am Bd Prev Med, dipl & cert aerospace med. Prof Exp: Asst biol, Temple Univ, 41-43; intern, Wilkes-Barre Gen Hosp, Pa, 47-48, resident med, 48-49; resident pulmonary dis, Jefferson Med Col & Hosp, 49-51; asst prof med & physiol, Col Med, Ohio State Univ, 53-66, assoc prof prev med & physiol, 66-72; DIR PULMONARY FUNCTION LAB & INHALATION THER & STAFF PHYSICIAN, DEPT PULMONARY DIS, CLEVELAND CLIN, 71-, HEAD DEPT, 73- Concurrent Pos: Dir res, Ohio Tuberc Hosp, 53-67; med res consult, Battelle Mem Inst, 59-67, med dir & med res adv, 67-17; dir pulmonary function labs, Univ Hosps, Ohio State Univ, 67; consult, US Air Force & Vet Admin; clin prof prev med, Col Med, Ohio State Univ, 72- Mem: Am Physiol Soc; Am Thoracic Soc; fel Am Col Chest Physicians; AMA; Aerospace Med Asn. Res: Pulmonary diseases; respiratory physiology; aviation medicine; pulmonary function testing; environmental medicine; biomedical engineering. Mailing Add: Cleveland Clin 9500 Euclid Ave Cleveland OH 44106

TOMASHEFSKY, PHILIP, b Brooklyn, NY, May 4, 24; m 48; c 2. EXPERIMENTAL PATHOLOGY. Educ: City Col New York, BS, 46, MS, 51; NY Univ, MS, 63, PhD(biol), 69. Prof Exp: Chemist, Funk Found, 48-65 & US Vitamin Corp, 65; biochemist, Dept Urol, 65-69, assoc, 69-71, ASST PROF PATH, COL PHYSICIANS & SURGEONS, COLUMBIA UNIV, 71- Mem: Fel AAAS; NY Acad Sci; Sigma Xi. Res: Neoplastic and hyperplastic growth of the kidney and prostate and other urological tissues. Mailing Add: Dept of Urol Col of Phys & Surg Columbia Univ New York NY 10032

TOMASI, GORDON ERNEST, b Denver, Colo, Dec 16, 30; m 54; c 3. BIOCHEMISTRY, ORGANIC CHEMISTRY. Educ: Colo State Col, BA, 57, MA, 58; Univ Louisville, PhD(biochem), 63. Prof Exp: From asst prof to assoc prof, 62-70, PROF CHEM, UNIV NORTHERN COLO, 70- Res: Bioenergetics; mechanisms of organic reaction and enzyme catalyzed reactions. Mailing Add: Dept of Chem Univ of Northern Colo Greeley CO 80631

TOMASI, THOMAS B, JR, b Barre, Vt, May 24, 27; m 48; c 3. IMMUNOLOGY, BIOCHEMISTRY. Educ: Dartmouth Col, AB, 50; Univ Vt, MD, 54; Rockefeller Univ, PhD, 65. Prof Exp: Instr med, Col Physicians & Surgeons, Columbia Univ, 57-58; asst physician, Rockefeller Univ, 58-60; asst prof med, Univ Vt, 60-61, actg chmn dept, 60-62, assoc prof & chmn dept, 62-65; prof, State Univ NY Buffalo, 65-74, dir, Div Immunol & Arthritis, 72-74; PROF IMMUNOL, MAYO MED SCH, 74- Concurrent Pos: Sr investr, Arthritis & Rheumatism Found, 60-65; dir, NIH training grant arthritis & metab dis; chief med, DeGoesbriand Hosp, 61-65; asst physician, Rockefeller Univ, 64-65; mem gen med study sect, NIH. Mem: Arthritis & Rheumatism Asn; Asn Am Physicians; Am Chem Soc; Am Fedn Clin Res; Am Soc Clin Invest. Res: Internal medicine; immunological diseases; immunochemistry. Mailing Add: Dept of Immunol Mayo Med Sch Rochester MN 55901

TOMASINO, CHARLES, organic chemistry, see 12th edition

TOMASULO, JOSEPH ANTHONY, anatomy, histology, see 12th edition

TOMASZ, ALEXANDER, b Budapest, Hungary, Dec 23, 30; US citizen; m 56; c 1. BIOCHEMISTRY, CELL BIOLOGY. Educ: Pazmany Peter Univ, Budapest, dipl, 53; Columbia Univ, PhD(biochem), 61. Prof Exp: Res assoc cytochem, Inst Genetics, Hungarian Nat Acad, 53-56; Am Cancer Soc fel & guest investr genetics, 61-63, asst prof genetics & biochem, 63-67, ASSOC PROF GENETICS & BIOCHEM, ROCKEFELLER UNIV, 67- Mem: AAAS; Am Soc Microbiol; Am Soc Cell Biol; Harvey Soc. Res: Biosynthesis and functioning of cell surface structures; cell to cell interactions; control of cell division; molecular genetics. Mailing Add: Lab of Genetics Rockefeller Univ New York NY 10021

TOMASZ, MARIA, b Szeged, Hungary, Oct 18, 32; US citizen; m 56, c 2. ORGANIC CHEMISTRY, BIOCHEMISTRY. Educ: Univ Eötvös Lorand, Budapest, dipl chem, 56; Columbia Univ, MA, 59, PhD(chem), 62. Prof Exp: Res assoc, Rockefeller Inst, 61-62, res assoc biochem, NY Univ, 62-64, instr, 64-66; asst prof, 66-71, ASSOC PROF CHEM, HUNTER COL, 71- Mem: Am Chem Soc. Res: Chemistry of nucleic acids; chemical basis of action of mutagens, carcinogens and antibiotics. Mailing Add: Dept of Chem Hunter Col New York NY 10021

TOMASZEWSKI, JOSEPH EDWARD, organic chemistry, pharmacodynamics, see 12th edition

TOMB, ANDREW SPENCER, systematic botany, palynology, see 12th edition

TOMBAUGH, LARRY WILLIAM, b Erie, Pa, Jan 28, 39; m 60; c 2. FOREST ECONOMICS. Educ: Pa State Univ, BS, 60; Colo State Univ, MS, 63; Univ Mich, PhD(resource econ), 68. Prof Exp: Economist, NCent Forest Exp Sta, 66-69; prin economist, Southeastern Forest Exp Sta, 69-71; prog mgr, 71-75, DIR, DIV ADVAN ENVIRON RES & TECHNOL, NAT SCI FOUND, 75-; PROF APPL SCI, GEORGE WASHINGTON UNIV, 75- Concurrent Pos: Lectr resource econ, Univ Mich, 66-69. Mem: Fel AAAS; Soc Am Foresters. Res: Economic incentives for environmental control and the economic and environmental consequences of new human settlements. Mailing Add: Nat Sci Found 1800 G St NW Washington DC 20550

TOMBER, MARVIN L, b South Bend, Ind, Aug 4, 25; m 48; c 2. ALGEBRA. Educ: Univ Notre Dame, BS, 46; Univ Pa, PhD(math), 52. Prof Exp: Instr math, Amherst Col, 52-55; from asst prof to assoc prof, 55-65, PROF MATH, MICH STATE UNIV, 65- Mem: Am Math Soc; Math Asn Am. Res: Non-associative algebras. Mailing Add: Dept of Math Mich State Univ East Lansing MI 48823

TOMBES, AVERETT SNEAD, b Easton, Md, Sept 13, 32; m 57; c 4. INVERTEBRATE PHYSIOLOGY. Educ: Univ Richmond, BS, 54; Va Polytech Inst, MS, 56; Rutgers Univ, PhD, 61. Prof Exp: Asst prof, Clemson Univ, 61-65; NIH res fel, Univ Va, 65-66; assoc prof, 66-74, PROF ZOOL, CLEMSON UNIV, 74- Concurrent Pos: Nat Ctr Sci Res France vis prof, Univ Lille, 71-72. Mem: AAAS; Am Soc Cell Biol; Inst Soc, Ethics & Life Sci; Am Physiol Soc; Am Soc Zoologists. Res: Physiology and endocrinology of invertebrates; ultrastructure of invertebrate neuroendocrine tissue; physiology of adult insect diapause. Mailing Add: Dept of Zool Clemson Univ Clemson SC 29631

TOMBLIN, FRED FITCH, b Ventura, Calif, Apr 19, 41; m 66; c 1. SPACE SCIENCE, NUCLEAR PHYSICS. Educ: Harvey Mudd Col, BS, 63; Univ Calif, Santa Barbara, 65, PhD(phyiscs), 67. Prof Exp: Mem tech staff space sci, Bellcomm, Inc, 67-72, MEMTECH STAFF, SYSTS ANAL, BELL TEL LABS, 72- Mem: Am Phys Soc. Res: Solar and stellar x-ray emission processes; atmospheric x-ray fluorescence; radiation belt phenomena. Mailing Add: Bell Tel Labs Murray Hill NJ 07974

TOMBOULIAN, PAUL, b Rochester, NY, Oct 19, 34; m 57; c 3. ORGANIC CHEMISTRY. 'Educ: Cornell Univ, AB, 53; Univ Ill, PhD(org chem), 56. Prof Exp: Res fel chem, Univ Minn, 56-59, instr, 57-58; from asst prof to assoc prof, 59-67, PROF CHEM, OAKLAND UNIV, 67-, CHMN DEPT, 62-, COORDR ENVIRON STUDIES, 70- Mem: Am Chem Soc. Res: Instrument design; instrumental analysis; water quality studies. Mailing Add: Dept of Chem Oakland Univ Rochester MI 48063

TOMBRELLO, THOMAS ANTHONY, JR, b Austin, Tex, Sept 20, 36; div; c 3. NUCLEAR PHYSICS. Educ: Rice Univ, BA, 58, MA, 60, PhD(physics), 61. Prof Exp: Res fel physics, Calif Inst Technol, 61-63; from instr to asst prof, Yale Univ, 63-64; res fel, 64-65, from asst prof to assoc prof, 65-71, PROF PHYSICS, CALIF INST TECHNOL, 71- Concurrent Pos: NSF fel, 61-62; mem prog adv comt, Los Alamos Meson Physics Facil, 70-; Alfred P Sloan Found fel, 71-73; consult, Los Alamos Sci Lab & NSF; assoc ed, Nuclear Physics, 72-; mem comt nuclear sci, Nat Res Coun, 74- Mem: Fel Am Phys Soc. Res: Nuclear structure; applications of nuclear physics; space physics; theoretical atomic physics. Mailing Add: Kellogg Radiation Lab 106-38 Calif Inst of Technol Pasadena CA 91125

TOMBROPOULOS, ELIAS GEORGE, b Alexandria, UAR, Aug 9, 25; US citizen; m 65; c 1. BIOCHEMISTRY, NUTRITION. Educ: Agr Col Athens, Greece, BS, 50; Iowa State Univ, MS, 54; Univ Calif, Davis, PhD(nutrit), 59. Prof Exp: Trainee under Dr I L Chaikoff, Dept Physiol, Univ Calif, Berkeley, 59-61; biological scientist, Gen Elec Co, 61-62, sr scientist, 62-64; sr scientist, 64-68, RES ASSOC BIOL DEPT, PAC NORTHWEST LAB, BATTELLE MEM INST, 68- Mem: Brit Biochem Soc. Res: Methods for removal of inhaled insoluble and soluble radioactive particles from the lungs; general biochemistry of the lung, especially biosynthesis of lung surfactants; lipid metabolism. Mailing Add: PO Box 533 Richland WA 99352

TOMCUFCIK, ANDREW STEPHEN, b Czech, Oct 26, 21; nat US; m 54; c 5. ORGANIC CHEMISTRY. Educ: Fenn Col, BS, 43; Western Reserve Univ, MS, 48; Yale Univ, PhD(org chem), 51. Prof Exp: Res chemist uranium refining, Mallinckrodt Chem Works, 46; instr chem, Fenn Col, 46-47; res chemist uranium refining, Mallinckrodt Chem Works, 49; res chemist, Am Cyanamid Co, 50-53, res chemist, Lederle Labs, 53-55, group leader, 55-74, DEPT HEAD, LEDERLE LABS, AM CYANAMID CO, 74- Mem: AAAS; Am Chem Soc; NY Acad Sci; The Chem Soc. Res: Chemotherapy of cancer; parasitic infections; heterocyclic chemistry; cardiovascular-renal diseases. Mailing Add: Lederle Labs Am Cyanamid Co Pearl River NY 10965

TOMECKO, JOSEPH WENCESLAUS, physical chemistry, organic chemistry, see 12th edition

TOMER, KENNETH BEAMER, b New Kensington, Pa, Mar 13, 44; m 66; c 1. MASS SPECTROMETRY, ORGANIC CHEMISTRY. Educ: Ohio State Univ, BS, 66; Univ Colo, PhD(chem), 70. Prof Exp: Fel photochem, H C Orsted Inst, Univ Copenhagen, 70-71; fel mass spectrometry, Dept Chem, Stanford Univ, 71-73; asst prof chem, Brooklyn Col, 73-75; ASST PROF PEDIAT, MED SCH, UNIV PA, 75- Mem: Am Chem Soc; The Chem Soc; Am Soc Mass Spectrometry. Res: Investigations of fragmentation mechanisms of organic compounds in a mass spectrometer; clinical and biochemical applications of mass spectrometry. Mailing Add: Childrens Hosp of Pa Metab Res 34th & Civic Ctr Blvd Philadelphia PA 19104

TOMES, DWIGHT TRAVIS, b Bowling Green, Ky, Sept 21, 46; m 66; c 3. PLANT BREEDING, PLANT GENETICS. Educ: Western Ky Univ, BS, 68; Univ Ky, PhD(crop sci), 75. Prof Exp: ASST PROF CROP SCI, UNIV GUELPH, 75- Mem: Am Genetics Asn; Genetics Soc Can; Am Soc Agron; Agr Inst Can. Res: Plant breeding and genetics of forage legumes with emphasis on seed yield and seedling vigor of Lotus corniculatus L; tissue culture of forage legumes including anther culture, and asexual propagation. Mailing Add: Dept of Crop Sci Ont Agr Col Univ of Guelph Guelph ON Can

TOMES, MARK LOUIS, b Ft Wayne, Ind, Nov 15, 17; m 44; c 2. GENETICS. Educ: Ind Univ, AB, 39; Agr & Mech Col, Tex, MS, 41; Purdue Univ, PhD(genetics), 52. Prof Exp: Chief plant res, Stokely-Van Camp, Inc, 46-48; asst geneticist, 48-53, ASSOC GENETICIST, EXP STA, PURDUE UNIV, WEST LAFAYETTE, 53-, PROF GENETICS, 59-, HEAD DEPT BOT & PLANT PATH, 69- Concurrent Pos: Vis prof, Pa State Univ, 66-67. Mem: AAAS; Am Soc Hort Sci; Genetics Soc Am; Am Genetic Asn. Res: Genetics and breeding of horticultural crops. Mailing Add: Dept of Bot & Plant Path Life Sci Bldg Purdue Univ West Lafayette IN 47907

TOMETSKO, ANDREW M, b Mt Pleasant, Pa, Feb 13, 38; m 65; c 3. BIOCHEMISTRY, ORGANIC CHEMISTRY. Educ: St Vincent Col, BS, 60; Univ Pittsburgh, PhD(biochem), 64. Prof Exp: Res assoc biochem, Brookhaven Nat Lab, 64-65, asst scientist, 65-66; ASST PROF BIOCHEM, SCH MED & DENT, UNIV ROCHESTER, 66- Mem: AAAS; Am Chem Soc. Res: Chemical synthesis of polypeptides; protein structure and function; separation techniques in protein chemistry; computer simulation; instrument design and construction. Mailing Add: Dept of Biochem Univ of Rochester Sch of Med & Dent Rochester NY 14620

TOMEZSKO, EDWARD STEPHEN JOHN, b Philadelphia, Pa, Apr 9, 35; m 62; c 3. PHYSICAL CHEMISTRY. Educ: Villanova Univ, BS, 57; Pa State Univ, MS, 61, PhD(phys chem), 62. Prof Exp: Resident res assoc phys chem, Nat Bur Standards, 62-64; sr res chemist, Arco Chem Co Div, Atlantic Richfield Co, 64-71; ASST PROF CHEM, PA STATE UNIV, DELAWARE COUNTY CAMPUS, 71- Mem: Am Chem Soc; Catalysis Soc. Res: Thermodynamics; homogeneous and heterogeneous catalysis; inorganic and organic synthesis. Mailing Add: Dept of Chem Pa State Univ Delaware County Campus Media PA 19063

TOMIC, ERNST ALOIS, b Vienna, Austria, Feb 1, 26; m 52; c 2. INORGANIC CHEMISTRY, PHYSICAL CHEMISTRY. Educ: Univ Vienna, PhD, 56. Prof Exp: Asst inorg geochem & anal chem, Second Chem Inst, Univ Vienna, 55-57; res chemist, Explosives Dept, 58-68, sr res chemist, 68-70, sr res chemist, Polymer Intermediates Dept, 70-74, STAFF RES CHEMIST, POLYMER INTERMEDIATES DEPT, EXP STA, E I DU PONT DE NEMOURS & CO, INC, 74- Res: Ion exchange; coordination chemistry; radiochemistry; inorganic synthesis; molten salts; electrochemistry; hydrometallurgy. Mailing Add: 1430 Emory Rd Wilmington DE 19803

TOMICH, CHARLES EDWARD, b Gallup, NMex, Oct 23, 37; m 59; c 3. DENTISTRY, ORAL PATHOLOGY. Educ: Loyola Univ, La, DDS, 61; Ind Univ, Indianapolis, MSD, 68; Am Bd Oral Path, dipl. Prof Exp: ASSOC PROF ORAL PATH, SCH DENT, IND UNIV, INDIANAPOLIS, 69- Concurrent Pos: USPHS training grant, Sch Dent, Ind Univ, Indianapolis, 66-69; consult, Nat Inst Dent Res, 71- Mem: Fel Am Acad Oral Path; Int Asn Dent Res; Int Acad Path. Res: In vivo hard tissue marking agents; salivary gland histochemistry; oral neoplasms. Mailing Add: Dept of Oral Path Sch of Dent Ind Univ 1121 W Michigan St Indianapolis IN 46202

TOMICH, PROSPER QUENTIN, b Orange Vale, Calif, Oct 11, 20; m 46; c 5. VERTEBRATE ZOOLOGY, ANIMAL ECOLOGY. Educ: Univ Calif, Berkeley, AB, 43; Univ Calif, Davis, PhD(zool), 59. Prof Exp: Lab asst plague res, Hooper Found, Univ Calif, 43-44, res zoologist, Hastings Natural Hist Reservation, 47-52; assoc zool, Univ Calif, Davis, 56-59; ANIMAL ECOLOGIST, STATE DEPT HEALTH, HAWAII, 59- Concurrent Pos: Naval med res unit adv, Egyptian Govt, 46-47; arbovirus res training grant, Pa State Univ, 66-67; mem island ecosysts proj, Int Biol Prog, Univ Hawaii, 69-75. Mem: AAAS; Ecol Soc Am; Am Soc Mammalogy; Wildlife Soc; Am Ornithologists Union. Res: Rodents, fleas, and plague; field ecology of mule deer, ground squirrel, mongoose, and of Arctic birds and mammals; leptospirosis in populations of small mammals; history and adaptation of mammals in Hawaiian Islands. Mailing Add: Res Unit Dept of Health Honokaa HI 96727

TOMIKEL, JOHN, b Cuddy, Pa, Apr 30, 28; m 49; c 2. EARTH SCIENCE. Educ: Clarion State Col, BS, 51; Univ Pittsburgh, MLitt, 56, PhD(higher educ), 70; Syracuse Univ, MS, 62. Prof Exp: Sci teacher, Fairview High Sch, 51-63; asst prof earth sci, Edinboro State Col, 63-65; PROF GEOG, CALIFORNIA STATE COL, PA, 65- Mem: Nat Asn Geol Teachers. Res: Earth science education. Mailing Add: Dept of Earth Sci California State Col of Pa California PA 15419

TOMIMATSU, TOSHIO, b Watsonville, Calif, June 8, 19; m 44; c 2. PHYSICAL CHEMISTRY. Educ: Univ Calif, BS, 41, MS, 56. Prof Exp: Assoc chemist, Western Utilization Res & Develop Div, 46-64, RES CHEMIST, WESTERN REGIONAL RES LAB, AGR RES SERV, USDA, 64- Mem: Am Chem Soc. Res: Light scattering; circular dichroism; Raman spectroscopy. Mailing Add: Western Regional Res Lab 800 Buchanan St Berkeley CA 94710

TOMISEK, ARTHUR JOHN, b Chicago, Ill, Oct 8, 20; m 47. BIOCHEMISTRY. Educ: Univ Ill, BS, 42; Ore State Univ, PhD(org chem), 48. Prof Exp: Asst, Univ Mo, 42-43 & Ore State Col, 44-47; chemist, Nutrit Res Labs, Univ Chicago, 47-48; res assoc, Univ Wis, 48-49; fel, Inst Enzyme Res, 49-51; biochemist, Southern Res Inst, Ala, 51-71; BIOCHEMIST, BIOCHEM DEPT, MICHAEL REESE HOSP, 72- Mem: Am Chem Soc; Am Soc Biol Chemists; Am Asn Cancer Res; Am Asn Clin Chemists. Res: Properties of nitrogen heterocycles; isolation and structure determination of antibiotics; enzymes; photosynthesis; formate metabolism. Mailing Add: Biochem Dept Michael Reese Hosp Chicago IL 60616

TOMIZAWA, HENRY HIDEO, b San Francisco, Calif, Feb 12, 26; m; c 5. BIOCHEMISTRY. Educ: Iowa State Col, BS, 49; Univ Ill, PhD(chem), 52. Prof Exp: From res assoc med to res asst prof & lectr biochem, Univ Wash, 52-59; sr res assoc, Fels Res Inst, Ohio, 59-65; res dir, Reference Lab, Calif, 65-67; clin chem consult, 67-68; chief clin chem lab, Los Angeles Vet Admin Hosp, 68-69; head toxicol-spec chem & qual control, Reference Lab, 69-70; TECH DIR, BIO-REAGENTS & DIAG, INC, 70- Mem: Am Chem Soc; Am Soc Biol Chemists; Am Asn Clin Chemists. Res: Research and development; clinical tests. Mailing Add: Bio-Reagents & Diag Inc 17392 Daimler St Irvine CA 92705

TOMIZUKA, CARL TATSUO, b Tokyo, Japan, May 24, 23; nat US; m 56; c 4. SOLID STATE PHYSICS. Educ: Univ Tokyo, BS, 45; Univ Ill, MS, 51, PhD(physics), 54. Prof Exp: Asst physics, Univ Ill, 51-54, res assoc, 54-55, res asst prof physics & elec eng, 55-56; asst prof physics, Inst Study Metals, Univ Chicago, 56-60; PROF PHYSICS, UNIV ARIZ, 60-, HEAD DEPT, 70- Honors & Awards: Creative Teaching Award, Univ Ariz Found, 73. Mem: Am Phys Soc; Phys Soc Japan. Res: Solid state diffusion; anelasticity; high pressure; magnetism. Mailing Add: Dept of Physics Univ of Ariz Tucson AZ 85721

TOMKIEWICZ, MICHA, b Warsaw, Poland, May 25, 39; Israeli citizen; c 1. PHYSICAL CHEMISTRY. Educ: Hebrew Univ, MS, 63, PhD(chem), 69. Prof Exp: Instr phys chem, Hebrew Univ, 67-69; fel, Univ Guelph, 69-71; Nat Inst Gen Med Sci fel biophys, Univ Calif, Berkeley, 71-72, Nat Inst Gen Med Sci spec fel, 72-73; FEL BIOPHYS, THOMAS J WATSON RES CTR, IBM CORP, 73- Res: Using biophysical and electrochemical approaches to photolyse water with visible radiation for the purpose of converting solar energy to useful chemical energy. Mailing Add: Watson Res Ctr IBM Corp PO Box 218 Yorktown Heights NY 10598

TOMKINS, DAVID FRANCIS, b Dubuque, Iowa, Oct 10, 47; m 66; c 2. ANALYTICAL CHEMISTRY. Educ: Univ Dubuque, BS, 69; Univ Iowa, PhD(anal chem), 72. Prof Exp: Sr chemist anal chem, Hoffmann-La Roche Inc, 72-74; SR CHEMIST ANAL CHEM, MONSANTO CO, 74- Res: Development of high pressure liquid chromatographic techniques and equipment in support of research projects and plant operations; development of new detectors and columns for high pressure liquid chromatography. Mailing Add: Corp Res Dept Monsanto Co 800 N Lindbergh Ave St Louis MO 63166

TOMKINS, FRANK SARGENT, b Petoskey, Mich, June 24, 15; m 42, 63; c 1. ATOMIC SPECTROSCOPY. Educ: Kalamazoo Col, BS, 37; Mich State Col, PhD(phys chem), 42. Prof Exp: Physicist, Buick Motor Div, Gen Motors Corp, Ill, 41-43; physicist, 43-46, SR SCIENTIST & GROUP LEADER, ARGONNE NAT LAB, 46- Concurrent Pos: Guggenheim fel, Nat Ctr Sci Res, France, 60-61; consult, Bendix Corp, 63-75; Argonne Nat Lab-Argonne Univs Asn distinguished appt, 75. Mem: AAAS; fel Optical Soc Am; assoc Am Phys Soc; assoc Fr Phys Soc. Res: Physical chemistry; optical spectroscopy. Mailing Add: Chem Div Argonne Nat Lab Argonne IL 60439

TOMKINS, GORDON MAYER, molecular biology, biochemistry, deceased

TOMKINS, JOHN PRESTON, b Ellenton, Pa, May 5, 18; m 46; c 2. HORTICULTURE. Educ: Pa State Univ, BS, 40, MS, 42; Cornell Univ, PhD(pomol), 51. Prof Exp: Res assoc pomol, Exp Sta, State Univ NY Col Agr, Cornell Univ, 46-49; asst prof & exten specialist, Mich State Col, 50-53; horticulturist, Welch Grape Juice Co, 53; from asst prof to assoc prof pomol, Exp Sta, State Univ NY Col Agr, Cornell Univ, 54-62; ASSOC PROF POMOL, CORNELL UNIV, 62- Res: Culture of strawberries, raspberries and grapes. Mailing Add: Dept of Pomol Cornell Univ Ithaca NY 14850

TOMLIN, DON C, b Meridian, Idaho, Aug 29, 32; m 58; c 1. ANIMAL NUTRITION. Educ: Calif State Polytech Col, BSc, 55; Univ Fla, MSc, 56, PhD(animal nutrit), 60. Prof Exp: Res asst, Univ Fla, 55-60; res asst animal nutrit, Madera Milling Co, Calif, 60; fel forage eval, Ohio Agr Exp Sta, 61-62; res officer, Exp Farm, Can Dept Agr, BC, 62-65; res assoc range livestock nutrit, Utah Agr Exp Sta, 65-67; animal scientist, US Sheep Exp Sta, USDA, Idaho, 67-70; asst prof animal sci, Univ Alaska, Fairbanks, 70-75; CONSULT, AGRO-NORTH CONSULTS, 75- Mem: AAAS; Am Soc Animal Sci; Soc Range Mgt. Res: Ruminant nutrition; evaluation and utilization of native and domestic forages. Mailing Add: Box 1094 Palmer AK 99645

TOMLINSON, BARRETT LYNN, physical chemistry, see 12th edition

TOMLINSON, EVERETT PARSONS, b Montclair, NJ, Sept 18, 14; m 37; c 3. NUCLEAR PHYSICS, HIGH ENERGY PHYSICS. Educ: Yale Univ, BS, 36; Calif Inst Technol, PhD(physics), 42. Prof Exp: Asst, Calif Inst Technol, 38-41; instr physics, Princeton Univ, 46-50, lectr, 50-52; lectr, Bryn Mawr Col, 52-53; res assoc, Princeton Univ, 53-56, mem proj res staff, 56-67; MEM FAC, CAPE COD COMMUNITY COL, 67- Mem: AAAS; Am Phys Soc; Am Asn Physics Teachers. Res: Beta ray spectroscopy; high energy accelerators; electricity and magnetism. Mailing Add: Fox Hill Rd Chatham MA 02633

TOMLINSON, GEORGE HERBERT, b Fullerton, La, May 2, 12; m; c 3. CHEMISTRY. Educ: Bishop's Univ, Can, BA, 31; McGill Univ, PhD(chem), 35. Prof Exp: Res assoc cellulose & indust chem, McGill Univ, 35-36; chief chemist, Howard Smith Chem Ltd, 36-40; res dir, Howard Smith Paper Mills Ltd, 41-60; res dir, 61-70, VPRES RES & ENVIRON TECHNOL, DOMTAR LTD, 70- Honors & Awards: Medal, Tech Asn Pulp & Paper Indust, 69. Mem: AAAS; Can Pulp & Paper Asn; Tech Asn Pulp & Paper Indust; Am Soc Testing & Mat; Am Chem Soc. Mailing Add: Domtar Ltd 395 Blvd deMaisonneuve W Montreal PQ Can

TOMLINSON, GERALDINE ANN, b Vancouver, BC, Feb 5, 31; m 57. MICROBIOLOGY, BIOCHEMISTRY. Educ: Univ BC, BSA, 57, PhD(agr microbiol), 64; Univ Calif, Berkeley, MA, 60. Prof Exp: Res assoc comp biol, Kaiser Res Ctr Comp Biol, 60-61; develop pharmacol, Dept Pediat, State Univ NY Buffalo, 64-65; asst prof biol, Rosary Hill Col, 65-66; res assoc agr biochem, Stauffer Agr Ctr, Calif, 66-67; asst prof, 67-72, ASSOC PROF BIOL, UNIV SANTA CLARA, 72- Concurrent Pos: Ames Res Ctr, NASA-Univ Santa Clara res grants, Ames Res Ctr, 68- Mem: AAAS; Am Soc Microbiol. Res: Microbial physiology and biochemistry; extremely halophilic bacteria. Mailing Add: Dept of Biol Univ of Santa Clara Santa Clara CA 95053

TOMLINSON, GLEN E, b Rushville, Ill, Nov 2, 28; c 4. MEDICINE. Educ: Univ Ill, BS, 53, MD, 56. Prof Exp: Family med pract, Thomsen Clin, 61-62; pvt pract, 62-71; PROF FAMILY PRACT & HEAD DEPT, ABRAHAM LINCOLN SCH MED, UNIV ILL COL MED, 71- Concurrent Pos: Attend physician, St Francis Hosp, 69-; attend physician family pract & dir dept, Cook County Hosp, 71-; lectr, Ill Acad Family Physicians, 72; mem, West Side Health Planning Task Force, 72- Mem: AMA; Soc Teachers Family Med. Res: Family practice. Mailing Add: 1825 W Harrison St Chicago IL 60612

TOMLINSON, GUS, b Dickson, Tenn, Apr 30, 33; m 52; c 2. ELECTRON MICROSCOPY, CELL PHYSIOLOGY. Educ: George Peabody Col, BS, 58; Vanderbilt Univ, PhD(molecular biol), 62. Prof Exp: From asst prof to assoc prof, 62-66, PROF BIOL & CHMN DEPT, GEORGE PEABODY COL, 66- Concurrent Pos: Nat Cancer Inst fel, Swiss Fed Inst Technol, 63-64, NIH res grant electron micros, 65-68; ed, J Tenn Acad Sci, 70- Mem: AAAS; Am Physiol Soc; Electron Micros Soc Am. Res: Correlation of ultrastructural and biochemical phenomena in differentiating soil amoeba by utilizing current techniques of electron microscopy, biochemistry, and freeze-etching procedures. Mailing Add: Dept of Biol George Peabody Col Nashville TN 37203

TOMLINSON, HARLEY, b Tunbridge, Vt, July 20, 32; m 58; c 2. PLANT PATHOLOGY. Educ: Univ Vt, BS, 59, MS, 61, PhD(bot), 65. Prof Exp: ASST PLANT PATHOLOGIST, CONN AGR EXP STA, 65- Mem: Am Phytopath Soc. Res: Chemistry of natural products; biochemical plant pathology; effects of air pollutants on plants. Mailing Add: Conn Agr Exp Sta Box 1106 New Haven CT 06504

TOMLINSON, HAZEL M, b Huntingdon Valley, Pa, May 15, 07. CHEMISTRY. Educ: Temple Univ, AB, 26, QAM, 28; Columbia Univ, PhD(chem), 39. Prof Exp: From instr to asst prof, 28-48, ASSOC PROF CHEM, TEMPLE UNIV, 48-, ASST DEAN COL LIB ARTS, 70- Mem: Am Chem Soc. Res: Physical and analytical chemistry; effect of dielectric on kinetics of ions in solution; redox titrations in nonaqueous media. Mailing Add: 76686 Williams Way Elkins Park Philadelphia PA 19117

TOMLINSON, JACK TRISH, b Bakersfield, Calif, Aug 22, 29; m 63; c 3. INVERTEBRATE ZOOLOGY. Educ: Univ Calif, AB, 50, MA, 52, PhD(zool), 56. Prof Exp: Instr biol, Oakland City Col, 54-57; from instr to assoc prof, 57-68, PROF BIOL, SAN FRANCISCO STATE UNIV, 68-, CHMN DEPT, 75- Mem: AAAS; Animal Behav Soc; Soc Syst Zool; Paleont Soc. Res: Invertebrate physiology; animal behavior. Mailing Add: Dept of Biol San Francisco State Univ San Francisco CA 94132

TOMLINSON, MICHAEL, b Leeds, Eng, Mar 30, 29; m 59; c 3. PHYSICAL CHEMISTRY. Educ: Univ Leeds, BSc, 49. Prof Exp: Asst exp officer radiation chem, Atomic Energy Res Estab, Harwell, Eng, 49-54; exp officer, Chalk River Nuclear Labs, Atomic Energy Can Ltd, 54-57; sr sci officer, Atomic Energy Res Estab, Harwell, Eng, 57-62; assoc res officer, Chalk River Nuclear Labs, 62-63, assoc res officer mat sci, Whiteshell Nuclear Res Estab, 63-71, HEAD RES CHEM BR, WHITESHELL NUCLEAR RES ESTAB, ATOMIC ENERGY CAN LTD, 71- Mem: Chem Inst Can. Res: Chemistry for nuclear power. Mailing Add: Whiteshell Nuclear Res Estab Pinawa MB Can

TOMLINSON, MICHAEL BANGS, b Miami, Fla, Oct 2, 37; m 63; c 1. MATHEMATICS. Educ: Reed Col, BA, 60; Univ Ore, MS, 62, PhD(math), 68. Prof Exp: Sci programmer, Lawrence Radiation Lab, Univ Calif, Berkeley, 63-65; asst prof math, Va Polytech Inst & State Univ, 68-69; asst prof, 69-70, ASSOC PROF MATH, UNIV MASS, BOSTON, 70- Mem: Am Math Soc. Res: Functional analysis; function algebras; Banach algebras. Mailing Add: Dept of Math Univ of Mass Boston MA 02116

TOMLINSON, NEIL, biochemistry, bacteriology, see 12th edition

TOMLINSON, PHILIP BARRY, b Leeds, Eng, Jan 17, 32; m 65; c 2. BOTANY. Educ: Univ Leeds, BSc, 53, PhD(bot), 55; Harvard Univ, AM, 71. Prof Exp: Fel bot, Univ Malaya, 55-56; lectr, Univ Col Ghana, 56-59 & Univ Leeds, 59-60; res scientist, Fairchild Trop Garden, Fla, 60-71; PROF BOT, HARVARD UNIV, 71- Concurrent Pos: Forest anatomist, Cabot Found, Harvard Univ, 65-71. Res: Morphology and anatomy of monocotyledons, especially palms; tropical botany. Mailing Add: Harvard Forest Petersham MA 01366

TOMLINSON, RAYMOND VALENTINE, b Smithers, BC, July 25, 27; m 57. BIOCHEMISTRY. Educ: Univ BC, BA, 54, MSc, 56; Univ Calif, Berkeley, PhD(biochem), 61. Prof Exp: Head technician, Children's Hosp, Vancouver, BC, 4953; res assoc, BC Neurol Inst, Vancouver, 56-57; fel, Univ BC, 61-64; asst res prof, State Univ NY Buffalo, 64-66; PRIN SCIENTIST, SYNTEX CORP RES DIV, 66- Mem: Am Pharmaceut Asn. Res: Intermediary metabolism of drugs. Mailing Add: Syntex Crop Res Div 3401 Hillview Ave Palo Alto CA 94304

TOMLINSON, RICHARD HOWDEN, b Montreal, Que, Aug 2, 23; m 49. PHYSICAL CHEMISTRY, INORGANIC CHEMISTRY. Educ: Bishop's Univ, Can, BSc, 43; McGill Univ, PhD(chem), 48. Prof Exp: Nat Res Coun Can fel, Cambridge Univ, 49; asst prof phys & inorg chem, 52-58, chmn dept, 67-74, PROF PHYS CHEM, McMASTER UNIV, 58- Mem: The Chem Soc. Res: Diffusion; mass spectrometry; polymerization radiochemistry. Mailing Add: Dept of Chem McMaster Univ Hamilton ON Can

TOMLINSON, WALTER JOHN, III, b Philadelphia, Pa, Apr 3, 38; m 61; c 1. PHYSICS. Educ: Mass Inst Technol, BS, 60, PhD(physics), 63. Prof Exp: Consult, Edgerton, Germeshausen & Grier, Inc, 60-63, sr scientist, 63; MEM TECH STAFF, BELL LABS, 65- Mem: Am Phys Soc; Optical Soc Am. Res: Integrated optics; photochemistry; gaseous optical masers, atomic and molecular; magnetic field effects in optical masers; isotope shifts in radioactive nuclei. Mailing Add: 22 Indian Creek Rd Holmdel NJ 07733

TOMLJANOVICH, NICHOLAS MATTHEW, b Susak, Yugoslavia, Mar 5, 39; US citizen; m 66; c 1. THEORETICAL PHYSICS, ELEMENTARY PARTICLE PHYSICS. Educ: City Col New York, BS, 61; Mass Inst Technol, PhD(physics), 66. Prof Exp: Physicist, US Weather Bur, 62; teaching asst physics, Mass Inst Technol, 62-64; res asst elem particle physics, Lab Nuclear Sci, 64-66; MEM TECH STAFF, MITRE CORP, 66- Concurrent Pos: Physicist, Nat Bur Standards, 63. Mem: Am Phys Soc. Res: Radar detection theory and electromagnetic propagation; holography; modern optics; scattering theory; plasma physics. Mailing Add: 13 Oxbow Lane Burlington MA 01803

TOMOMATSU, HIDEO, b Tokyo, Japan, June 8, 29. ORGANIC CHEMISTRY. Educ: Waseda Univ, Japan, BEn, 53; Univ of the Pac, MSc, 60; Ohio State Univ, PhD(org chem), 65. Prof Exp: Res chemist, Hodogaya Chem Co Ltd, Japan, 53-59 & Jefferson Chem Co, Inc, Tex, 65-71; PROJ LEADER, RES LABS, QUAKER OATS CO, 71- Honors & Awards: Awarded 14 US & 3 foreign patents. Mem: Am Chem Soc; The Chem Soc. Res: Synthesis of new elastomer of high polymer and new industrial chemicals. Mailing Add: Res Labs Quaker Oats Co 617 W Main St Barrington IL 60010

TOMONTO, JAMES R, b White Plains, NY, Apr 14, 32; m 56; c 5. NUCLEAR PHYSICS. Educ: Villanova Univ, BS, 54; Rensselaer Polytech Inst, MS, 59. Prof Exp: Engr, Airborne Instruments Lab, 57; anal physicist, Nuclear Power Eng Div, Alco Prod Inc, 58-59; exp physicist, Knolls Atomic Power Lab, 59-64; mgr nuclear eng dept, Gulf United Nuclear Fuels Corp, NY, 64-74; MGR NUCLEAR ANAL DEPT, FLA POWER & LIGHT CO, 74- Mem: Am Nuclear Soc. Res: Design and analysis of water moderated power and research reactors; development of analysis methods relating to use of plutonium as a fuel in thermal power and fast breeder reactors. Mailing Add: 14311 SW 74 Ct Miami FL 33158

TOMOZAWA, YUKIO, b Iyo-City, Japan, Sept 3, 29; nat US; m 57; c 2. THEORETICAL HIGH ENERGY PHYSICS. Educ: Univ Tokyo, BSc, 52, DSc(physics), 61. Prof Exp: Asst physics, Univ Tokyo, 56-57, Tokyo Univ Educ, 57-59, Cambridge Univ, 59-60 & Univ Col, Univ London, 60-61; res assoc, Inst Physics, Univ Pisa, 61-64; mem, Inst Advan Study, 64-66; from asst prof to assoc prof, 66-72, PROF PHYSICS, UNIV MICH, ANN ARBOR, 72- Mem: Am Phys Soc. Res: Symmetries in elementary particle physics; theory of weak interactions; axiomatic field theory; quantum field theory. Mailing Add: Randall Lab of Physics Univ of Mich Ann Arbor MI 48104

TOMPA, ALBERT S, b Trenton, NJ, Aug 26, 31; m 57; c 5. ANALYTICAL CHEMISTRY, PHYSICAL CHEMISTRY. Educ: St Joseph Col, BS, 54; Fordham Univ, MS, 57, PhD(anal chem), 60. Prof Exp: Lab instr anal chem, Fordham Univ, 54-59; RES CHEMIST, NAVAL ORD STA, 60- Mem: AAAS; Am Chem Soc. Res: Infrared, nuclear magnetic resonance and thermal analysis study of polymers; molecular structure of organic compounds; trace analysis of inorganic compounds; chemical degradation of propellants and recovery of ingredients; effect of additives on the thermal decomposition of polymers; kinetics of decomposition of liquid gas propellants. Mailing Add: Naval Ord Sta Indian Head MD 20640

TOMPA, FRANK WILLIAM, b New York, NY, Nov 5, 48; m 72; c 2. COMPUTER SCIENCES. Educ: Brown Univ, ScB, 70, ScM, 70; Univ Toronto, PhD(comput sci), 74. Prof Exp: Lectr comput sci, Univ Toronto, 74; ASST PROF COMPUT SCI, UNIV WATERLOO, 74- Mem: Asn Comput Mach. Res: Data structures design and specification; systems design for interactive computer graphics; programming languages design and specification. Mailing Add: Dept of Comput Sci Univ of Waterloo Waterloo ON Can

TOMPKIN, GERVAISE WILLIAM, physical chemistry, see 12th edition

TOMPKIN, ROBERT BRUCE, b Akron, Ohio, Apr 2, 37; m 61; c 3. FOOD MICROBIOLOGY. Educ: Ohio Univ, BSc, 59; Ohio State Univ, MSc, 61, PhD(microbiol), 63. Prof Exp: Res microbiologist, 64-65, head microbiol res div, 65-66, CHIEF MICROBIOLOGIST, RES & DEVELOP CTR, SWIFT & CO, 66- Concurrent Pos: Mem registry comt, Nat Registry Microbiologists, 74-77. Mem: Am Soc Microbiol; Am Meat Sci Asn; Inst Food Technologists; Int Asn Milk, Food & Environ Sanit; Am Acad Microbiol. Res: Prevention of food-borne diseases. Mailing Add: Res & Develop Ctr Swift & Co 1919 Swift Dr Oak Brook IL 60521

TOMPKINS, DANIEL REUBEN, b New York, NY, Oct 2, 31; m 64; c 2. PLANT PHYSIOLOGY, HORTICULTURE. Educ: Univ Md, BS, 59, MS, 62, PhD, 63. Prof Exp: Asst horticulturist, Western Wash Res & Exten Ctr, Wash State Univ, 62-68; from assoc prof to prof hort food sci, Univ Ark, Fayetteville, 69-75; HORTICULTURIST, COOPER STATE RES SERV, USDA, 75- Mem: AAAS; Am Soc Hort Sci; Am Soc Plant Physiologists; Sigma Xi. Res: Growth substances; physiology of horticultural plants and horticultural research administration. Mailing Add: Coop State Res Serv USDA Washington DC 20250

TOMPKINS, DONALD ROY, JR, b Calif. PHYSICS. Educ: Univ NDak, BS, 55; Univ Colo, MS, 58; Univ Ariz, PhD(physics), 64. Prof Exp: Asst prof physics, La State Univ, 64-67 & Univ Ga, 67-70; assoc prof, Univ Wyo, 70-75; PRES, TERRENE CORP, 75- Mem: Am Phys Soc. Res: Geophysics; atmospheric physics. Mailing Add: Terrene Corp 505 N Alamo St Refugio TX 78377

TOMPKINS, E CROSBY, b Milroy, Ind, May 23, 39; m 66; c 2. TOXICOLOGY, PHARMACOLOGY. Educ: Purdue Univ, BS, 61, MS, 64, PhD(physiol), 66. Prof Exp: Res asst pop genetics, Purdue Univ, 61-64, res asst physiol, 64-66; sr pharmacologist, Res Dept, R J Reynolds Tobacco Co, NC, 66-70; SR SCIENTIST, DEPT PATH & TOXICOL, MEAD JOHNSON & CO, 70- Mem: AAAS; NY Acad Sci; Drug Info Asn. Res: Reproductive physiology. Mailing Add: Dept of Path & Toxicol Mead Johnson & Co Evansville IN 47721

TOMPKINS, GARY ALVIN, b Denver, Colo, June 28, 38; m 69; c 1. SOIL SCIENCE, ENVIRONMENTAL CHEMISTRY. Educ: Colo State Univ, BS, 64; Univ Ariz, MS, 67; NC State Univ, PhD(soil sci), 73. Prof Exp: Res instr soil sci, NC State Univ, 71-73; asst res soil scientist, Univ Calif, Berkeley, 73-75; RES BIOLOGIST ENVIRON RES, LAWRENCE LIVERMORE LAB, UNIV CALIF, 75- Mem: Soc Environ Geochem & Health; Soil Sci Soc Am; Am Soc Plant Physiologists. Res: Heavy metal and trace element biological cycling through soil-water-plant systems; environmental assessment of new energy sources; transuranic element movement in food webs. Mailing Add: Biomed & Environ Res Div Lawrence Livermore Lab Livermore CA 94550

TOMPKINS, JOHN CARTER, b Los Angeles, Calif, Sept 27, 46. EXPERIMENTAL HIGH ENERGY PHYSICS. Educ: Univ Calif, Los Angeles, BS, 68, MS, 69, PhD(physics), 73. Prof Exp: Adj asst prof physics, Univ Calif, Los Angeles, 74; RES ASSOC HIGH ENERGY PHYSICS, FERMI NAT ACCELERATOR LAB, 74- Res: Measurement of the form factors of the pion and kaon in high energy elastic scattering experiments. Mailing Add: Fermilab PO Box 500 Batavia IL 60510

TOMPKINS, PAUL CARTER, b Walla Walla, Wash, Apr 11, 14; m 42; c 2. BIOCHEMISTRY. Educ: Whitman Col, AB, 35; Univ Calif, PhD(biochem), 41. Prof Exp: Asst biochem, Univ Calif, 40-41; from asst to res assoc chem, Stanford Univ, 41-43; res assoc, Metall Lab, Univ Chicago, 43-45; sr chemist, Clinton Lab, Tenn, 45-47; prin biochemist, Oak Ridge Nat Lab, 48-49; staff adv to sci dir, US Naval Radiol Defense Lab, 49-50, from assoc sci dir to sci dir, 50-60; chief, Res Br, Div Radiol Health, USPHS, 60-61; dep dir, Off Radiation Stand, US Atomic Energy Comn, 61-62; exec dir, Fed Radiation Coun, 62-70; actg dir, Criteria & Stand Div, Off Radiation Progs, Environ Protection Agency, 70-72, sr sci adv, Off Res & Develop, 72-76; RETIRED. Concurrent Pos: Consult, Joint Comt Atomic Energy, US Cong, 57, 59. Mem: Health Physics Soc; Am Chem Soc; Am Nuclear Soc; Radiation Res Soc. Res: Vitamin analysis; stability of protein solutions; separation; purification and use of radioisotopes, particularly high activity procedures; health physics; nuclear weapon effects. Mailing Add: 125 Newport Dr Oak Ridge TN 37830

TOMPKINS, ROBERT, b Rhinebeck, NY, June 13, 41; m 62; c 3. DEVELOPMENTAL GENETICS. Educ: Earlham Col, AB, 62; Ind Univ, PhD(zool), 68. Prof Exp: NIH fel microbiol, Univ Ill, 68-69; fel, Inst Cancer Res, Col Physicians & Surgeons, Columbia Univ, 69; ASST PROF BIOL, PRINCETON UNIV, 69- Res: Biochemistry of amphibian development. Mailing Add: Dept of Biol Princeton Univ Princeton NJ 08540

TOMPKINS, RONALD K, b Malta, Ohio, Oct 14, 34; m 56; c 3. SURGERY. Educ: Ohio Univ, BA, 56; Johns Hopkins Univ, MD, 66; Ohio State Univ, MS, 68. Prof Exp: NIH trainee surg & fel phys chem, Col Med, Ohio State Univ, 66-69, instr & fel phys chem, 66-69, instr surg, 68-69; asst prof, 69-73, ASSOC PROF SURG, SCH MED, UNIV CALIF LOS ANGELES, 73- Concurrent Pos: NIH res grants, Inst Arthritis & Metab Dis, Dept Health Educ & Welfare, 6871; res grants, John A Hartford Found, Inc, 70-77; consult, Sepulveda Vet Admin Hosp, 71-, Rand Corp Study Cholecystectomy, 76; mem, Prog Comt, Soc Univ Surgeons, 74-77, Asn Acad Surg, 75-77; hosp res, Am Col Surgeons Southern Calif chapter, 72-; mem, Long Range Planning Comt, Soc Surg Alimentary Tract, 74- Honors & Awards: Student Res Prize, Am Gastroenterol Asn, 71, Am Fedn Clin Res Western Sect, 72. Mem: Soc Clin Surg; Am Surg Asn; Soc Univ Surgeons; Soc Surg Alimentary Tract; Am Inst Nutrit. Res: Biochemical and nutritional research related to diseases of the gastrointestinal tract, especially hepatobiliary and pancreatic diseases. Mailing Add: Dept of Surg Sch of Med Ctr for Health Sci UCLA Los Angeles CA 90024

TOMPKINS, VICTOR NORMAN, b Milbrook, NY, May 30, 13; m 38; c 4 PATHOLOGY. Educ: Cornell Univ, AB, 34; Union Univ, NY, MD, 38; Am Bd Path, dipl, 46. Prof Exp: Resident path, New Eng Deaconess Hosp, 39-40, Albany Hosp, 40-41 & Pondville Hosp, 41-42; sr pathologist, Div Labs & Res, State Dept Health, NY, 47-49, asst dir in charge diag labs, 49-56, assoc dir, 56-58, dir, 58-68; PROF PATH, ALBANY MED COL, 60- Concurrent Pos: Assoc prof path, Albany Med Col, 53-60. Mem: AAAS; Am Soc Human Genetics; Am Soc Clin Path; Am Soc Exp Path; AMA. Res: Immunology. Mailing Add: Dept of Path Albany Med Col Albany NY 12201

TOMPSETT, RALPH RAYMOND, b Tidioute, Pa, Oct 8, 13; m 42; c 4. INTERNAL MEDICINE. Educ: Cornell Univ, AB, 34, MD, 39. Prof Exp: From instr to assoc prof med, Med Col, Cornell Univ, 46-57; PROF MED, UNIV TEX HEALTH SCI CTR, DALLAS, 57- Concurrent Pos: Attend physician, Med Ctr, Baylor Univ, 57- & Parkland Mem Hosp, 60- Mem: Am Soc Clin Invest; Asn Am Physicians; master Am Col Physicians; Am Fedn Clin Res. Res: Infectious diseases. Mailing Add: Baylor Univ Med Ctr Dallas TX 75246

TOMPSON, CLIFFORD WARE, b Mexico, Mo, Dec 12, 29; m 51; c 3. PHYSICS. Educ: Univ Mo, BS, 51, AM, 56, PhD(physics), 59. Prof Exp: Physicist, US Navy Electronics Lab, San Diego, 51-55; assoc prof, 59-72, PROF PHYSICS, UNIV MO-COLUMBIA, 72- Mem: Am Phys Soc; Am Asn Physics Teachers. Res: X-ray diffraction; neutron diffraction; structure of liquids; lattice vibrations. Mailing Add: Dept of Physics Univ of Mo-Columbia Columbia MO 65201

TOMPSON, ROBERT NORMAN, b Adrian, Mich, Jan 7, 20; m 47; c 1. MATHEMATICAL ANALYSIS; APPLIED MATHEMATICS. Educ: Adrian Col, ScB, 41; Univ Nev, MS, 49; Brown Univ, PhD(math), 53. Prof Exp: Res inspector ord mat, US War Dept, 42-43; from asst to instr math, Univ Nev, 46-49; asst prof, Fla State Univ, 53-54; mem tech staff, Bell Tel Labs, Inc, 54-55; asst prof, Fla State Univ, 55-56; mathematician & programmer, Int Bus Mach Corp, 56; assoc prof, 56-64, PROF MATH, UNIV NEV, RENO, 64- Mem: AAAS; Am Math Soc; Am Phys Soc. Res: Measure and integration theory; topology; systems theory; probability theory. Mailing Add: 997 Meadow St Reno NV 89502

TOMUSIAK, EDWARD LAWRENCE, b Edmonton, Alta, Mar 3, 38; m 61; c 1. THEORETICAL NUCLEAR PHYSICS. Educ: Univ Alta, BSc, 60, MSc, 61; McGill Univ, PhD(theoret physics), 64. Prof Exp: NATO overseas fel, Oxford Univ, 64-66; asst prof, 66-70, ASSOC PROF PHYSICS, UNIV SASK, 70- Mem: Can Asn Physicists. Res: Nuclear structure calculations using realistic two-nucleon potentials; nuclear models and electromagnetic interactions with nuclei. Mailing Add: Dept of Physics Univ of Sask Saskatoon SK Can

TON, BUI AN, b Hanoi, Vietnam, Jan 23, 37; m 63; c 2. MATHEMATICAL ANALYSIS. Educ: Saigon Univ, BSc, 59; Mass Inst Technol, PhD(math), 64. Prof Exp: Res staff mathematician, Yale Univ, 63-64; vis asst prof, Math Res Ctr, Univ Wis, 64-65; Nat Res Coun Can fel, 65-66; asst prof math, Univ Montreal, 66-67; assoc prof, 67-71, PROF MATH, UNIV BC, 71- Mem: Can Math Cong. Res: Partial differential equations. Mailing Add: Dept of Math Univ of BC Vancouver BC Can

TONASCIA, JAMES A, b Los Banos, Calif, Mar 2, 44; m 65. BIOSTATISTICS. Educ: Univ San Francisco, BS, 65; Johns Hopkins Univ, PhD(biostatist), 70. Prof Exp: ASST PROF BIOSTATIST, JOHNS HOPKINS UNIV, 70- Mem: Am Statist Asn; Biomet Soc; Inst Math Statist; Math Asn Am; Royal Statist Soc. Res: Biostatistical methods; epidemiology; statistical computing. Mailing Add: Sch of Hyg & Pub Health Johns Hopkins Univ Baltimore MD 21205

TONDEUR, PHILIPPE, b Zurich, Switz, Dec 7, 32; m 65. MATHEMATICS. Educ: Univ Zurich, PhD(math), 61. Prof Exp: Res fel math, Univ Paris, 61-63; lectr, Univ Zurich, 63-64; res fel, Harvard Univ, 64-65; lectr, Univ Calif, Berkeley, 65-66; assoc prof, Wesleyan Univ, 66-68; PROF MATH, UNIV ILL, URBANA, 68- Mem: Am Math Soc; Math Soc France; Swiss Math Soc. Res: Geometry and topology. Mailing Add: Dept of Math Univ of Ill Urbana IL 61801

TONDRA, RICHARD JOHN, b Canton, Ohio, Jan 23, 43; m 66; c 2. MATHEMATICS. Educ: Univ Notre Dame, BS, 65; Mich State Univ, MS, 66, PhD(topology, manifold theory), 68. Prof Exp: ASSOC PROF MATH, IOWA STATE UNIV, 68- Mem: Am Math Soc. Res: Topological and piecewise linear manifold theory. Mailing Add: Dept of Math Iowa State Univ Ames IA 50011

TONE, JAMES N, b Grinnell, Iowa, Feb 9, 33; m 54; c 2. ANIMAL PHYSIOLOGY. Educ: Coe Col, BA, 54; Drake Univ, MA, 61; Iowa State Univ, PhD(animal physiol), 63. Prof Exp: PROF PHYSIOL, ILL STATE UNIV, 63- Mem: AAAS. Res: Histophysiological effects of food additives on mammals. Mailing Add: Dept of Biol Sci Ill State Univ Normal IL 61761

TONELLI, ALAN EDWARD, b Chicago, Ill, Apr 14, 42; m 74; c 2. POLYMER PHYSICS. Educ: Univ Kans, BS, 64; Stanford Univ, PhD(polymer chem), 68. Prof Exp: MEM TECH STAFF POLYMER PHYSICS, BELL LABS, 68- Mem: Am Chem Soc; Am Phys Soc. Res: Study of the conformations and physical properties of synthetic and biological macromolecules. Mailing Add: Bell Labs 600 Mountain Ave Murray Hill NJ 07974

TONELLI, GEORGE, b Tenafly, NJ, Feb 20, 21; m 55; c 4. VETERINARY MEDICINE. Educ: Parma Univ, DVM, 48. Prof Exp: Biologist, Peters Serum Co, Kans, 50; res vet, Animal Indust Sect, 51-55, group leader & pharmacologist, Exp Therapeut Sect, 55-60, group leader endocrinol, Endocrine Res Dept, 60-68, group leader, Toxicol Dept, 68-73, MGR TOXICOL/PHARMACOL EVAL, INT DIV, LEDERLE LABS, AM CYANAMID CO, 73- Mem: Am Soc Pharmacol & Exp Therapeut; Endocrine Soc. Mailing Add: Lederle Labs Am Cyanamid Co Pearl River NY 10965

TONEY, FRANK MORGAN, b Nashville, Tenn, Dec 28, 25; m 54; c 2. PHYSICAL CHEMISTRY. Educ: Mid Tenn State Col, BS, 49; Polytech Inst Brooklyn, PhD(chem), 56. Prof Exp: SR RES CHEMIST, FIBERS DIV, FMC CORP, 56- Mem: Am Chem Soc. Res: Physical chemistry of polymer solutions; electrode reactions; kinetics of organic reactions. Mailing Add: Fibers Div FMC Corp Marcus Hook PA 19061

TONEY, FRED, JR, b Mooresboro, NC, Aug 12, 37; m 64; c 1. MATHEMATICS. Educ: NC State Univ, BS, 59, MS, 61, PhD(math), 68. Prof Exp: Asst math, NC State Univ, 59-61; asst prof, Wilmington Col, NC, 61-64; asst, NC State Univ, 64-67, instr, 67-68; PROF MATH & CHMN DEPT, UNIV NC, WILMINGTON, 68- Mem: Math Asn Am; Am Math Soc. Res: Abstract algebra. Mailing Add: Dept of Math Univ of NC Wilmington NC 28401

TONEY, JOE DAVID, b Rosston, Ark, Aug 12, 42; m 64; c 2. INORGANIC CHEMISTRY. Educ: Univ Ark, Pine Bluff, BS, 64; Univ Ill, Urbana, PhD(chem), 69. Prof Exp: TEACHER CHEM, CALIF STATE UNIV, FRESNO, 69- Concurrent Pos: Vis res assoc chem, Argonne Nat Lab, 72; lectr health manpower prog, Pacific Col, 73. Mem: Am Chem Soc. Res: Syntheses and structural characterization of transition metal chelate compounds involving amino acid ligands; visible, infrared and Raman studies of bonding in chelate complexes. Mailing Add: Calif State Univ Fresno CA 93740

TONEY, MARCELLUS E, JR, b Baltimore, Md, Dec 25, 20; m 45; c 1. CLINICAL MICROBIOLOGY. Educ: Va Union Univ, BS, 42; Meharry Med Col, MT, 46; Cath Univ Am, MS, 53, PhD(zool), 56. Prof Exp: Assoc prof zool, 57-71, PROF BIOL, VA UNION UNIV, 71- Concurrent Pos: Vis prof, Va State Col, 59-; chief premed adv, Va Union Univ, 63-, coordr biol, 74-; res grants, NSF & Nat Urban League; consult, Richmond Math & Sci Ctr, NSF & Friends Adoption Asn; US Dept Health, Educ & Welfare grant pesticide res, 72-77. Mem: AAAS; Am Inst Biol Sci; Am Asn Biol Teachers; NY Acad Sci; Sigma Xi. Res: Microbiology; endocrinology. Mailing Add: Dept of Zool Va Union Univ Richmond VA 23220

TONG, BOK YIN, b Shanghai, China, Mar 5, 34; m 69; c 1. SOLID STATE PHYSICS, BIOPHYSICS. Educ: Univ Hong Kong, BSc, 57; Univ Calif, Berkeley, MA & MLS, 59; Univ Calif, San Diego, PhD(solid state physics), 67. Prof Exp: Asst librn, Oriental

Collection, Univ Hong Kong, 59-61, asst lectr math, 61-65, lectr math, 65-67; asst prof, 67-69, ASSOC PROF PHYSICS, UNIV WESTERN ONT, 69- Mem: Am Phys Soc. Res: Theory of metals, surface physics and amorphous material; DNA molecules, muscle contraction and membrane activity. Mailing Add: Dept of Physics Univ of Western Ont London ON Can

TONG, JAMES LUN, b San Francisco, Calif, Aug 9, 12. MICROBIOLOGY. Educ: Univ Calif, AB, 36; Univ Mich, AM, 39; Univ Colo, PhD(bact), 48. Prof Exp: Asst, Univ Mich, 41-42; from instr to assoc prof bact, Univ Denver, 45-68; RES MICROBIOLOGIST, VET ADMIN HOSP, 62- Concurrent Pos: Instr, Sch Nursing, Denver Gen Hosp, 48-51, chief bacteriologist, 49-53; chief bacteriologist, Vet Admin Hosp, Denver, 53-60; vol fac, Sch Med, Univ Col, 72- & Sch Dent, 73- Mem: AAAS; Am Soc Microbiol; NY Acad Sci. Res: Immunology; clinical microbiology; mechanism of specific immunologic tolerance. Mailing Add: Vet Admin Hosp 1055 Clermont Denver CO 80220

TONG, JAMES YING-PEH, b Shanghai, China, Dec 8, 26; US citizen; m 51; c 3. PHYSICAL INORGANIC CHEMISTRY. Educ: Univ Calif, Berkeley, BS, 50, MS, 51; Univ Wis, PhD(chem), 54. Prof Exp: Res chemist, Le Roy Res Lab, Durex Plastics & Chem, Inc, 53-54; res assoc phys inorg chem, Univ Ill, 54-57; from asst prof to assoc prof, 57-68, PROF CHEM, OHIO UNIV, 68- Mem: Am Chem Soc. Res: Equilibrium and kinetics studies of homogeneous and heterogeneous reaction in solution and radiochemistry; chemistry and history of photography; water chemistry. Mailing Add: Dept of Chem Ohio Univ Athens OH 45701

TONG, LEE KARL JAN, b China, Jan 12, 13; nat US; m 44. c 2. PHYSICAL CHEMISTRY. Educ: Univ Calif, BS, 38, PhD(chem), 41. Prof Exp: Fel chem, Univ Calif, 41-44; chemist, 44-53, res assoc, 53-66, SR RES ASSOC, EASTMAN KODAK CO, 66- Mem: Am Chem Soc. Res: Reaction kinetics; photographic chemistry. Mailing Add: Eastman Kodak Res Labs 1669 Lake Ave Rochester NY 14650

TONG, LONG SUN, b China, Aug 20, 15; US citizen; m 39. HEAT TRANSFER. Educ: Chinese Nat Inst Technol, BS, 40; Univ Fla, MS, 53; Stanford Univ, PhD(mech eng), 56. Prof Exp: Asst prof mech eng, Ord Eng Col, Taiwan, 47-52; sr engr, Atomic Power Dept, Westinghouse Elec Corp, 56-59, supvr thermal & hydraul design, Atomic Power Dept, 59-62, adv engr, 63-65, mgr thermal & hydraul design & develop, 65-66, mgr thermal & hydraul eng, 66-70, consult engr, PWR Syst Div, 70-72, sr consult, 72-73; ASST DIR, DIV REACTOR SAFETY RES, NUCLEAR REGULATORY COMN, 73- Concurrent Pos: Lectr, Univ Pittsburgh, 57-60, adj prof, 65-72; lectr, Carnegie Inst Technol, 61-67. Honors & Awards: Westinghouse Order of Merit, 69; Mem Award, Heat Transfer Div, Am Soc Mech Eng, 73. Mem: Am Soc Mech Eng; fel Am Nuclear Soc; Am Soc Eng Educ. Res: Fluid flow; thermal and hydraulic design; analysis and development of pressurized water reactors; research in water reactor safety. Mailing Add: 9733 Lookout Pl Gaithersburg MD 20760

TONG, MARY POWDERLY, b New York, NY, May 24, 24; m 56; c 5. MATHEMATICS. Educ: St Joseph's Col, NY, BA, 50; Columbia Univ, MA, 51, PhD, 69. Prof Exp: Instr math, St Joseph's Col, NY, 51-54, City Col New York, 54 & Columbia Univ, 54-59; asst prof, Univ Conn, 60-66; from asst prof to assoc prof, Fairfield Univ, 66-70; PROF MATH, WILLIAM PATERSON COL NJ, 70- Concurrent Pos: Delta Epsilon Sigma res fel, 68; NSF fac fel, 59-60. Mem: Am Math Soc; Math Asn Am; NY Acad Sci. Res: Topology; foundations of mathematics; applications of mathematics. Mailing Add: 725 Cooper Ave Oradell NJ 07649

TONG, STEPHEN S C, analytical chemistry, physical chemistry, see 12th edition

TONG, WINTON, b Los Angeles, Calif, May 3, 27; m 51; c 3. PHYSIOLOGY, BIOCHEMISTRY. Educ: Univ Calif, Berkeley, BS, 47, PhD(thyroid function), 53. Prof Exp: Res physiologist, Univ Calif, Berkeley, 47-62; from asst prof to assoc prof, 62-71, dir grad studies, 70-72, PROF PHYSIOL, SCH MED, UNIV PITTSBURGH, 71- Concurrent Pos: USPHS res career develop award, 62- Honors & Awards: Van Meter Prize, Am Thyroid Asn, 64. Mem: AAAS; Am Soc Biol Chemists; Am Physiol Soc; Endocrine Soc. Res: Mechanisms in the biosynthesis of thyroid hormones and thyroglobulin; mechanism of action of thyrotropin; physiology of thyroid function; cultivation of thyroid cells in vitro. Mailing Add: Dept of Physiol Univ of Pittsburgh Sch of Med Pittsburgh PA 15261

TONG, YULAN CHANG, b Nanking, China, Oct 21, 35; m 65. ORGANIC CHEMISTRY. Educ: Cheng Kung Univ, Taiwan, BS, 56; Univ Ill, MS, 58, PhD(org chem), 61. Prof Exp: Res assoc, Univ Mich, 61-62; org res chemist, Edgar C Britton Res Lab, Mich, 62-65, res chemist, Res Lab, Western Div, 66-70, sr res chemist, 70-72, RES SPECIALIST, RES LAB, WESTERN DIV, DOW CHEM CO, 72- Mem: Am Chem Soc. Res: Heterocyclic chemistry. Mailing Add: Western Div Res Lab Dow Chem Co 2800 Mitchell Dr Walnut Creek CA 94598

TONG, YUNG LIANG, b Shantung, China, July 11, 35; m 65; c 2. STATISTICS, MATHEMATICS. Educ: Nat Taiwan Univ, BS, 58; Univ Minn, Minneapolis, MA, 63, PhD(statist), 67. Prof Exp: Asst prof statist, Univ Nebr, Lincoln, 67-69; vis asst prof, Univ Minn, 69-70; asst prof, 70-72, ASSOC PROF STATIST, UNIV NEBR-LINCOLN, 72- Mem: Am Statist Asn; Inst Math Statist. Res: Mathematical and applied statistics; multiple decision theory; multivariate analysis; sequential analysis. Mailing Add: Dept of Math Univ of Nebr Lincoln NE 68508

TONGREN, JOHN CORBIN, b Ridgway, Pa, Sept 22, 09; m 38; c 3. CHEMISTRY. Educ: Pa State Col, BS, 32; Lawrence Col, MS, 34, PhD(chem), 37. Prof Exp: Res chemist, 37-47, tech asst to paper mill supt, 47-53, res & develop chemist, 53-57, tech dir, Watervliet Paper Co Div, 57-62, sr scientist res & develop, 62-68, HEAD PROCESS RES SECT, HAMMERMILL PAPER CO, 68- Mem: Am Chem Soc; Tech Asn Pulp & Paper Indust. Res: Deterioration of color in bleached sulfite pulp; sulfite pulping process; pulp bleaching; analytical methods for chemical wood pulps; semichemical pulping process and recovery; paper coating; stock preparation; paper manufacturing. Mailing Add: 316 W 39th St Erie PA 16508

TONI, YOUSSEF TANIOUS, b Girga, Egypt, Apr 2, 29; m 61; c 1. GEOGRAPHY. Educ: Alexandria Univ, BA, 52; Durham Univ, PhD(geog), 56. Prof Exp: Prof geog, Damascus Univ, 57-60; sr lectr, Ain Shams Univ, Cairo, 60-66; assoc prof, Univ Jordan, 66-67; assoc prof, 67-71, PROF GEOG, LAURENTIAN UNIV, 71- Concurrent Pos: Mem, Permanent Coun Place-Names Arab World, 63-; mem, Nat Comn Atlas Arab World, 64- Honors & Awards: UNESCO Cert Microclimat for Ecol & Soil Sci, 60. Mem: Fel Egyptian Geog Soc; fel Royal Geog Soc; Asn Am Geogr. Res: Nomadism; population studies; geographical terminology; biogeography; Arab world; arid lands; Cyrenaica; the Levant; Egyptian deserts; land reclamation; regional development. Mailing Add: Dept of Geog Laurentian Univ Sudbury ON Can

TONIK, ELLIS J, b Philadelphia, Pa, Jan 9, 21; m 48; c 3. MEDICAL MICROBIOLOGY. Educ: Roanoke Col, BS, 50. Prof Exp: Med bacteriologist diag bact, Dept Pub Welfare, Ill, Chicago State Hosp, 50-51, supvry bacteriologist, East Moline State Hosp, 51-52 & Kankakee State Hosp, 52; med bacteriologist, Process Res & Pilot Plants Div, Chem Corps Res & Develop Labs, US Dept Army, 52-60, supvry bacteriologist, Tech Eval Div, 60, actg chief animal path sect, 60-61, chief exp animal sect, Appl Aerobiol Div, 61-71, sr investr, Microbiol Res Div, 71-72; SR MICROBIOLOGIST, FT HOWARD VET ADMIN HOSP, 72- Mem: Am Soc Microbiol; Sigma Xi. Res: Experimental respiratory diseases of laboratory animals; aerobiological research and technology; experimental and clinical pathology; immunology; chemotherapeutic agents; virulence of airborne particulates; laboratory diagnosis and assay methods; management of laboratory animals. Mailing Add: 526 Mary St Frederick MD 21701

TONKING, WILLIAM HARRY, b Newton, NJ, Apr 22, 27; m 64; c 2. MINING GEOLOGY. Educ: Princeton Univ, AB, 49, PhD(geol), 53. Prof Exp: Asst geol, Princeton Univ, 49-50 & 51-53; asst instr, Northwestern Univ, 50-51; geologist, Bear Creek Mining Co, 53-55 & Stand Oil Co, Tex, 55-62; dep mgr, Mohole Proj, 62-67, MGR SPEC PROJS, BROWN & ROOT INC, 67- Honors & Awards: Silver Medallist, Royal Soc Arts, 66. Mem: Geol Soc Am; Royal Soc Arts. Res: Petrology; volcanic rocks and ore deposits in the Southwest; petroleum geology; deep ocean engineering, geology and geophysics. Mailing Add: Brown & Root Inc PO Box 3 Houston TX 77001

TONKS, DAVID BAYARD, b Edmonton, Alta, Aug 31, 19; m 46; c 2. BIOCHEMISTRY, CLINICAL CHEMISTRY. Educ: Univ BC, BA, 41; McGill Univ, PhD(org chem), 49. Prof Exp: Sr chemist, Clin Labs, Lab Hyg, Dept Nat Health & Welfare, Can, 48-57; asst biochemist, Biochem Dept & Res Inst, Hosp Sick Children, Toronto, 57-62; from asst prof to assoc prof, 62-75, PROF MED, McGILL UNIV, 75-; DIR DIV CLIN CHEM, DEPT MED, MONTREAL GEN HOSP, 62- Concurrent Pos: Chem Inst Can fel, 62-; tech dir, Seaforth Clin Labs, Montreal, 64-; lab consult, Douglas Hosp, Verdun, 65, Reddy Mem Hosp, Montreal, 66-70 & Cybermedix Ltd, Montreal, 69-; mem bd dirs, Bio Res Labs, Point Claire, 65-72; Can nat rep, Int Comn Clin Chem, 66-70; secy sect clin chem, Int Union Pure & Appl Chem, 67-71, pres sect, 71-75, bur mem, 71-75, past-pres, 75-, assoc mem comn toxicol, 73-, mem Can nat comt, 74-; mem exec bd, Int Fedn Clin Chem, 67-75. Honors & Awards: Warner-Chilcott Award, Can Soc Lab Technologists, 67; Ames Award, Can Soc Clin Chem, 68; Ann Award, Que Corp Hosp Biochemists, 72. Mem: Am Asn Clin Chemists; Can Biochem Soc; fel Chem Inst Can; Can Soc Clin Chem (from secy to pres, 57-66, treas, 74-76); NY Acad Sci. Res: Development of synthetic antigens for serodiagnosis of syphilis; quality control and evaluation of laboratory precision in clinical chemistry laboratories; analytical methods in clinical chemistry. Mailing Add: Div of Clin Chem Montreal Gen Hosp Montreal PQ Can

TONKYN, RICHARD GEORGE, b Portland, Ore, Mar 26, 27; m 48; c 6. ORGANIC CHEMISTRY, POLYMER CHEMISTRY. Educ: Reed Col, BA, 48; Univ Ore, MA, 51; Univ Wash, PhD(org chem), 60. Prof Exp: Instr org chem, Univ Ore, 52; supvr res, Anal Labs, Titanium Metals Corp Am, Nev, 52-54; sr res anal chemist, Allegheny Ludlum Steel Corp, Pa, 54-55; res engr, Boeing Airplane Co, Wash, 55-59; NSF fel, Univ Col, Univ London, 60-61; chemist, Union Carbide Corp, NJ, 61-67; proj scientist, 67-69; sr res chemist, 69-70, group leader, 70-72, MGR ORG RES & PROCESS DEVELOP, BETZ LABS, INC, 72- Mem: Am Chem Soc; The Chem Soc. Res: Monomer and polymer synthesis; high pressure technology; polyelectrolyte synthesis; chemicals and processes for water treatment and water pollution control. Mailing Add: Betz Labs Inc Trevose PA 19047

TONN, ROBERT J, b Watertown, Wis, June 23, 27; m 61; c 2. MEDICAL ENTOMOLOGY. Educ: Colo State Univ, BS & MS, 49; Okla State Univ, PhD(entom), 59; Univ Okla, MPMPH, 64. Prof Exp: Asst prof biol, Tex Lutheran Col, 57-59; asst prof, Wis State Univ, 59-61; res assoc trop med & parasitol, Int Ctr Med Res & Training, Sch Med, La State Univ, 62-64; chief encephalitis field sta, Mass State Dept Pub Health, 64-66; instr trop pub health, Sch Pub Health, Harvard Univ, 65-66; proj leader, Aedes Res Unit, Bangkok, Thailand, 66-70, mem staff, Japanese Encephalitis Virus Res Unit, Seoul, Korea, 70, proj leader, EAfrican Aedes Res Unit, Dar Es Salaam, Tanzania, 70-72, PROJ LEADER, CHAGAS DIS VECTOR RES UNIT, ACARIGUA, VENEZUELA, WHO, 73- Concurrent Pos: Fel med zool, Sch Trop Med, Univ PR, 61; lectr, Sch Med, Tufts Univ, 65-66; consult, Mass State Dept Pub Health, 66. Mem: Am Soc Trop Med & Hyg; Am Soc Parasitologists; Am Mosquito Control Asn; Royal Soc Trop Med & Hyg. Res: Arbo viruses. Mailing Add: WHO-CDVRU Apartado 2030 Las Delicius Maracay-Acarigua Venezuela

TONNA, EDGAR ANTHONY, b Malta, May 10, 28; nat US; m 51; c 4. CELL PHYSIOLOGY, CELL CHEMISTRY. Educ: St John's Univ, NY, BS, 51; NY Univ, MS, 53, PhD(biol), 56. Prof Exp: Res collabr div exp path, Med Res Ctr, Brookhaven Nat Lab, 56-59, head histochem & cytochem res lab, 59-67; PROF HISTOL, GRAD SCH BASIC MED SCI, NY UNIV, 67-, DIR INST DENT RES, COL DENT, 71-, DIR LAB CELLULAR RES, 67- Concurrent Pos: Res biochemist, Hosp Spec Surg, New York, 53-56, head histochem & cytochem res lab, 56-59; adj assoc prof, Grad Sch, Long Island Univ, 56-62; consult radiobiol, Inst Dent Res, NY Univ, 64-67. Mem: Fel Geront Soc; Histochem Soc; Am Asn Anatomists; fel Royal Micros Soc; Int Asn Dent Res. Res: Cellular contribution to skeletal and dental development, growth, repair and disease during aging; autoradiographic, cytochemical and cytological studies using optical analytical and electron microscopic techniques to determine biochemical and cell morphological changes in skeletal and dental parameters; cell gerontology. Mailing Add: Inst for Dent Res NY Univ Col of Dent 339 E 25th St New York NY 10010

TONNDORF, JUERGEN, b Göttingen, Ger, Feb 1, 14; nat US; m 40. PHYSIOLOGY. Educ: Univ Kiel, MD, 38; Univ Heidelberg, PhD(otolaryngol), 45. Prof Exp: Privat-docent otolaryngol, Univ Heidelberg, 45-47; mem res & teaching staff, Sch Aviation Med, US Air Force, Randolph Field, Tex, 47-53; from asst res prof to res prof otolaryngol, Univ Hosps, Univ Iowa, 53-62; PROF OTOLARYNGOL, COL PHYSICIANS & SURGEONS, COLUMBIA UNIV, 62- Concurrent Pos: USPHS res career award, 63; mem study sect commun dis, NIH, 64-68, mem proj rev comt, Commun Dis Prog, 68-71; mem sci rev comt, Deafness Res Found, 64-68; mem acoust mgt bd, Am Nat Stand Inst, 72-; mem panel ear, nose & throat devices, Food & Drug Admin, 75- Honors & Awards: Von Eicken Award, 42; Ludwig Haymann Award, 70. Mem: Fel Acoust Soc Am; Am Otol Soc; Am Acad Ophthal & Otolaryngol; Asn Res Otolaryngol; hon mem Ger Otolaryngol Soc. Res: Auditory physiology, especially in reference to the ear. Mailing Add: Dept Otolaryngol Fowler Mem Lab Col Phys & Surg Columbia Univ New York NY 10032

TONNE, PHILIP CHARLES, b Chicago, Ill, Apr 2, 38; m 63; c 2. MATHEMATICS. Educ: Marquette Univ, BS, 60; Univ NC, MA, 63, PhD(math), 65. Prof Exp: Instr math, Univ NC, 65-66; asst prof, 66-71, ASSOC PROF MATH, EMORY UNIV, 71- Mem: Am Math Soc. Res: Classical analysis. Mailing Add: Dept of Math Emory Univ Atlanta GA 30322

TONNIS, JOHN A, b Scottsburg, Ind, Apr 18, 39; m 72. ORGANIC CHEMISTRY. Educ: Hanover Col, BA, 61; Ind Univ, MS, 64, PhD(org chem), 68. Prof Exp: Res chemist, Reilly Tar & Chem Co, 64-65; fel org chem, Ind Univ, 68; asst prof, 68-70,

ASSOC PROF CHEM, UNIV WIS-LA CROSSE, 70- Mem: AAAS; Am Chem Soc. Res: Preparation and antitumor evaluation of aldehydes and aldehyde derivatives; preparation and use of sulfonamides as organic chelating reagents; new synthetic methods in organic chemistry. Mailing Add: Dept of Chem Univ of Wis La Crosse WI 54601

TONZETICH, JOHN, b Nanaimo, BC, Oct 28, 41. GENETICS. Educ: Univ BC, BSc, 63; Duke Univ, MA, 67, PhD(zool), 72. Prof Exp: ASST PROF BIOL, BUCKNELL UNIV, 70- Mem: Genetics Soc Am; Am Soc Zoologists; AAAS; Sigma Xi. Res: Chromosomal inversions in Drosophila species; radiation induced aberrations. Mailing Add: Dept of Biol Bucknell Univ Lewisburg PA 17837

TOOGOOD, GERALD EDWARD, b Surrey, Eng, Apr 9, 38; m 64. INORGANIC CHEMISTRY. Educ: Univ Nottingham, BSc, 59, PhD(chem), 62. Prof Exp: Fel & res assoc chem, Argonne Nat Lab, 62-64; asst prof, 64-67, ASSOC PROF CHEM, UNIV WATERLOO, 67- Concurrent Pos: Vis prof, Univ Western Australia, 71. Mem: AAAS; Am Chem Soc; Can Inst Chemists; The Chem Soc. Res: Chemistry of lanthanide elements and compounds in non-aqueous media. Mailing Add: Dept of Chem Univ of Waterloo Waterloo ON Can

TOOGOOD, JOHN ALFRED, b Chancellor, Alta, Jan 7, 15; m 41; c 4. SOILS. Educ: Univ Alta, BSc, 41; Univ Minn, PhD, 48. Prof Exp: Teacher & prin schs, Can, 34-37, teacher high sch, 41-45; from asst prof to assoc prof, 48-58, PROF SOIL SCI & CHMN DEPT, UNIV ALTA, 59- Concurrent Pos: Registr, Alta Inst Agrologists, 52-63, pres, 67-68. Res: Soil conservation. Mailing Add: Dept of Soil Sci Univ of Alta Edmonton AB Can

TOOHEY, LOREN MILTON, b Antioch, Nebr, Mar 2, 21; m 42; c 4. GEOLOGY. Educ: Univ Nebr, BSc, 48, MSc, 50; Princeton Univ, PhD(geol), 53. Prof Exp: Field & res assoc, State Mus, Univ Nebr, 36-50; res assoc, Frick Lab, Am Mus Natural Hist, 50-56; geologist, Carter Oil Co, Ill, 56-60; GEOLOGIST, EXXON CO, USA, 60- Mem: Geol Soc Am; Soc Vert Paleont; Am Asn Petrol Geologists; Soc Econ Paleontologists & Mineralogists. Res: Cenozoic vertebrate paleontology and stratigraphy; geology of petroleum; sedimentology. Mailing Add: Exxon Co USA Midland TX 79701

TOOHIG, TIMOTHY E, b Lawrence, Mass, Feb 17, 28. EXPERIMENTAL HIGH ENERGY PHYSICS. Educ: Boston Col, BS, 51; Univ Rochester, MS, 53; Johns Hopkins Univ, PhD(physics), 62; Woodstock Col, STB, 64, STL, 65. Prof Exp: Asst dir admin res, Inst Natural Sci, Woodstock Col, 63-66; assoc physicist, Brookhaven Nat Lab, 67-70; assoc head neutrino lab sect, Nat Accelerator Lab, 70-74; ASST TO HEAD RES DIV, FERMI NAT ACCELERATOR LAB, 74- Mem: Am Phys Soc. Res: Elementary particle physics; investigation of the properties and interactions of elementary particles. Mailing Add: Fermi Nat Accelerator Lab PO Box 500 Batavia IL 60510

TOOKER, EDWIN WILSON, b Concord, Mass, May 9, 23; m 46; c 3. ECONOMIC GEOLOGY. Educ: Bates Col, BS, 47; Lehigh Univ, MS, 49; Univ Ill, PhD(geol), 52. Prof Exp: NSF fel, Univ Ill, 52-53; geologist, Base & Ferrous Metals Br & Pac Mineral Resources Br, 53-71, chief, Pac Mineral Resources Br, 71-72, CHIEF, OFF MINERAL RESOURCES, US GEOL SURV, 72- Mem: AAAS; Soc Mining Engrs; Soc Econ Geologists; Clay Minerals Soc; Geol Soc Am. Res: Geology of base and precious metal ore deposits; environment of ore deposition; wall rock alteration; base metal commodity resources. Mailing Add: Nat Ctr 913 Reston VA 22092

TOOKER, ELISABETH (JANE), b Brooklyn, NY, Aug 2, 27. ANTHROPOLOGY. Educ: Radcliffe Col, BA, 49, PhD(anthrop), 58; Univ Ariz, MA, 53. Prof Exp: Instr anthrop, Univ Buffalo, 57-60; teaching asst, Harvard Univ, 60-61; asst prof, Mt Holyoke Col, 61-65, NIH grant, 64-65; asst prof, 65-67, ASSOC PROF ANTHROP, TEMPLE UNIV, 67- Mem: Fel AAAS; fel Am Anthrop Asn. Res: North American Indians. Mailing Add: Dept of Anthrop Temple Univ Philadelphia PA 19122

TOOKEY, HARVEY LLEWELLYN, b Hooper, Nebr, Dec 2, 22; m 50; c 4. BIOCHEMISTRY. Educ: Univ Nebr, AB, 44, MS, 50; Purdue Univ, PhD(biochem), 55. Prof Exp: Sr control chemist, Norden Lab, Nebr, 46-48; PRIN CHEMIST, NORTHERN REGIONAL RES LAB, USDA, 55- Mem: Am Chem Soc. Res: Enzymes of lipid metabolism; proteinases, enzymes acting on glucosinolates; chemistry of natural products; alkaloids; non-infectious diseases of cattle. Mailing Add: Northern Regional Res Lab USDA Peoria IL 61604

TOOLAN, HELENE WALLACE, b Chicago, Ill, Feb 7, 12; m 30, 45; c 3. PATHOLOGY. Educ: Univ Chicago, BS, 29; Cornell Univ, PhD(path), 46. Prof Exp: Res asst path, Med Col, Cornell Univ, 46-50, from asst prof to assoc prof, Sloan-Kettering Div, 52-64; ASSOC SCIENTIST, SLOAN-KETTERING INST, 64-; DIR, PUTNAM MEM HOSP INST MED RES, 64-; ASSOC PROF EXP PATH, COL MED, UNIV VT, 64- Mem: AAAS; Am Asn Cancer Res; Am Asn Path & Bact; Am Soc Exp Path; Harvey Soc. Res: Heterologous transplantation of human tissues; immunology; viruses and cancer; congenital deformities. Mailing Add: Putnam Mem Hosp Inst for Med Res Monument Circle Bennington VT 05201

TOOLE, EBEN RICHARD, b Baraboo, Wis, July 1, 13; m 42; c 2. FOREST PATHOLOGY. Educ: Syracuse Univ, BS, 35; Duke Univ, AM, 38, PhD(forest path), 40. Prof Exp: Jr forester, US Forest Serv, 35-37; jr forester, Div Forest Path, Bur Plant Indust, USDA, 38-39, from jr forest pathologist to forest pathologist, 40-45, assoc pathologist, Bur Plant Industs, Soils & Agr Eng, 45-52; pathologist, US Forest Serv, Miss, 52-68; asst prof, 68-73, ASSOC PROF WOOD SCI & TECHNOL & ASSOC TECHNOLOGIST, FOREST PROD UTILIZATION LAB, MISS STATE UNIV, 73- Mem: Soc Am Foresters; Am Phytopath Soc. Res: Forest tree diseases. Mailing Add: Forest Prod Utilization Lab Miss State Univ PO Box FP Mississsippi State MS 39762

TOOLE, JAMES FRANCIS, b Atlanta, Ga, Mar 22, 25; m 52; c 4. MEDICINE. Educ: Princeton Univ, BA, 47; Cornell Univ, MD, 49; LaSalle Exten Univ, LLB, 63. Prof Exp: Intern med, Univ Pa, 49-50, resident neurol, 53-55, instr, 57-60, assoc, 60-61, prof, 62-67, WALTER C TEAGLE PROF NEUROL, BOWMAN GRAY SCH MED, 67- Concurrent Pos: Fulbright fel, Nat Hosp, London, 55-56; mem res comt, Am Heart Asn, 65, mem med ethics comt, 66, ed, Current Concepts Cerebrovasc Dis-Stroke, 69-72; chmn, Sixth & Seventh Princeton Conf Cerebrovasc Dis, 68 & 70; vis prof, Univ Calif, San Diego, 69-70; mem exam bd, Nat Bd Med Examr & Am Bd Psychol & Neurol. Mem: Asn Res Nerv & Ment Dis; fel Am Col Physicians; fel Am Acad Neurol; Am Neurol Asn; Am Fedn Clin Res. Res: Cerebral circulation and cerebrovascular diseases; physiology and pathology of the brain. Mailing Add: Dept of Neurol Bowman Gray Sch of Med Winston-Salem NC 27103

TOOLE, RICHARD CHARLES, inorganic chemistry, see 12th edition

TOOLEY, WILLIAM HENRY, b Berkeley, Calif, Nov 18, 25. PEDIATRICS. Educ: Univ Calif, MD, 49. Prof Exp: From clin instr to asst clin prof, 56-61, from asst prof to assoc prof, 61-72, chief newborn serv, hosp, 62-71, PROF PEDIAT, MED CTR, UNIV CALIF, SAN FRANCISCO, 72-, CHIEF, DIV PEDIAT PULMONARY DIS, UNIV HOSPS, 71-, SR STAFF MEM, CARDIOVASC RES INST, 63- Concurrent Pos: Pvt pract, Calif, 56-58. Mem: Am Acad Pediat; Soc Pediat Res; Am Pediat Soc. Res: Cardiopulmonary disease; neonatal medicine. Mailing Add: Dept of Pediat Univ of Calif San Francisco CA 94122

TOOM, PAUL MARVIN, b Pella, Iowa, Apr 1, 42; m 65; c 1. BIOCHEMISTRY. Educ: Cent Col, Iowa, BA, 64; Colo State Univ, PhD(biochem), 69. Prof Exp: Res asst biochem, Colo State Univ, 67-68, asst prof, 69-70; asst prof, 70-73, ASSOC PROF BIOCHEM, UNIV SOUTHERN MISS, 73- Mem: AAAS; Am Chem Soc; Int Soc Toxinology; Am Asn Clin Chemists; Sigma Xi. Res: Biochemistry of marine toxins; clinical biochemistry; insect biochemistry; analytical biochemistry. Mailing Add: Dept of Chem Univ of Southern Miss Box 337 Hattiesburg MS 39401

TOOME, VOLDEMAR, b Estonia, Sept 10, 24; m 52. PHYSICAL CHEMISTRY. Educ: Univ Bonn, dipl, 48 & 52, Dr rer nat, 54. Prof Exp: Sci asst, Inst Phys Chem, Univ Bonn, 54; dept head control div, Merck & Co, Ger, 54-57; sr phys res chemist, 57-68, res fel, 67-70, GROUP LEADER, HOFFMANN-LA ROCHE, INC, 70- Mem: Am Chem Soc; NY Acad Sci. Res: Ultraviolet and infrared spectroscopy; optical rotatory dispersion and circular dichroism; instrumental analysis; polarography; electrochemistry; dissociation constants of organic acids and bases; physical organic chemistry; microanalysis. Mailing Add: Dept of Phys Chem Hoffmann-La Roche Inc Nutley NJ 07110

TOOMEY, DONALD FRANCIS, b New York, NY, Apr 15, 27; m 52; c 4. GEOLOGY, PALEONTOLOGY. Educ: Univ NMex, BS, 51, MS, 53; Rice Univ, PhD(geol), 64. Prof Exp: Stratigr, Shell Oil Co, 53-56, geologist, Shell Develop Co, 56-64, staff res geologist, Res Ctr, Amoco Prod Co, 64-72; CHMN FAC EARTH SCI, UNIV TEX OF THE PERMIAN BASIN, 73- Concurrent Pos: Mem bd dirs, Cushman Found Foraminiferal Res, 62-, fel, 64-, vpres, 65-68, pres, 68-69. Mem: Paleont Soc; Soc Econ Paleontologists & Mineralogists; Am Asn Petrol Geologists. Res: Paleozoic algae and foraminifers; carbonate rock facies trends; development of paleoecologic criteria in relation or organic buildups through geologic time; recent environmental studies and paleoenvironmental implications. Mailing Add: Rte 1 Box 604 Odessa TX 79763

TOOMEY, JAMES MICHAEL, b Boston, Mass, Mar 2, 30; m 54; c 6. OTORHINOLARYNGOLOGY. Educ: Col of Holy Cross, Mass, BS, 51; Harvard Sch Dent Med, DMD, 55; Boston Univ, MD, 58. Prof Exp: Intern surg, Wash Univ-Barnes Hosp, Med Ctr, 58-59, asst resident surg, 61-62, asst resident otolaryngol, 62-64, chief resident, 64-65, instr, 64-65; sr instr, Med Ctr, Univ Rochester, 65-66, asst prof, 66-68; ASSOC PROF OTOLARYNGOL & HEAD DEPT, HEALTH CTR, UNIV CONN, 68- Concurrent Pos: NIH training grant, Sch Med, Wash Univ, 63-65; consult, Newington Vet Admin Hosp, Conn, 68-; lectr, Dept Speech, Univ Conn, 69-; consult, Hartford Hosp, 69-, St Francis Hosp, Hartford, 69- & Rocky Hill Vet Admin Hosp, Conn, 71- Mem: Am Laryngol, Rhinol & Otol Soc; Soc Univ Otolaryngol; Am Acad Facial Plastic & Reconstruct Surg; Am Soc Maxillofacial Surg. Res: Skin flap physiology; laryngeal physiology; experimental laryngeal surgery. Mailing Add: Dept of Otorhinolaryngol Univ of Conn Health Ctr Hartford CT 06112

TOOMEY, JOSEPH EDWARD, b Somerville, NJ, Aug 8, 43; m 67; c 1. SYNTHETIC ORGANIC CHEMISTRY, STRUCTURAL CHEMISTRY. Educ: Rider Col, BS, 70; Purdue Univ, PhD(org chem), 76. Prof Exp: RES CHEMIST, REILLY LAB, REILLY TAR & CHEM CO, 75- Mem: Am Chem Soc. Res: Synthesis, isolation and structural determination of pyridine chemicals; prediction of optical rotatory power, its relation to molecular structure and absolute configuration; unusual condensation reactions. Mailing Add: Reilly Lab Reilly Tar & Chem Co 1500 S Tibbs Ave Indianapolis IN 46241

TOOMEY, RICHARD E, biochemistry, see 12th edition

TOOMRE, ALAR, b Rakvere, Estonia, Feb 5, 37; US citizen; m 58; c 3. ASTRONOMY, APPLIED MATHEMATICS. Educ: Mass Inst Technol, BS(aeronaut eng) & BS(physics), 57; Univ Manchester, PhD(fluid mech), 60. Prof Exp: C L E Moore instr math, Mass Inst Technol, 60-62; fel astrophys, Inst Advan Study, 62-63; from asst prof to assoc prof, 63-70, PROF APPL MATH, MASS INST TECHNOL, 70- Concurrent Pos: Guggenheim fel astrophys, Calif Inst Technol, 69-70, Fairchild scholar, 75. Mem: AAAS; Am Acad Arts & Sci; Am Astron Soc. Res: Dynamical studies of galaxies; aerodynamics; rotating fluids. Mailing Add: Rm 2-371 Dept of Math Mass Inst of Technol Cambridge MA 02139

TOONE, ELAM COOKSEY, JR, b Richmond, Va, Nov 4, 08; m 36, 64; c 3. MEDICINE. Educ: Hampden-Sydney Col, BA, 29; Med Col Va, MD, 34; Am Bd Internal Med, dipl, 41. Hon Degrees: LLD, Hampden-Sydney Col, 73. Prof Exp: Asst, 36-37, instr, 38-40, assoc, 41-46, from asst prof to assoc prof, 46-60, chmn div connective tissue dis, Dept Med, 68-74, PROF MED, MED COL VA, 60-; CHMN RHEUMATOL SECT, MED SERV, McGUIRE VET ADMIN HOSP, 75- Concurrent Pos: Consult, McGuire Vet Admin Hosp, 58-75; mem med & sci comt, Arthritis & Rheumatism Found, 59-, chmn intercoun med comt, 58-59. Honors & Awards: Distinguished Serv Award, Va Chapter, Arthritis Found, 66. Mem: AMA; Am Clin & Climat Asn; Am Rheumatism Asn (vpres, 63-64); Am Soc Clin Invest. Res: Rheumatology, peripheral rheumatoid arthritis and rheumatoid spondylites; clinical manifestations and course; evaluation of certain therapeutic agents, including x-ray, gold salts, cortisone and phenylbutazone; toxic manifestations of cortisone and phenylbutazone. Mailing Add: Med Serv McGuire Vet Admin Hosp Richmond VA 23249

TOONEY, NANCY MARION, b Ilion, NY, Feb 19, 39. BIOCHEMISTRY, BIOPHYSICS. Educ: State Univ NY Albany, BS, 60, MS, 61; Brandeis Univ, PhD(biochem), 66. Prof Exp: Teaching intern biochem, Dept Chem & Biol, Hope Col, 66-67; fel biophys, Childrens Cancer Res Found & Sch Med, Harvard Univ, 67-73; ASST PROF BIOCHEM, POLYTECH INST NY, 73- Concurrent Pos: NIH fel, 67-70, res career develop award, Nat Heart & Lung Inst, 75-78. Mem: AAAS; Am Chem Soc; Biophys Soc. Res: Protein chemistry, electron microscopy and optical methods of analysis; structure and function of the blood clotting protein, fibrinogen; biological macromolecules. Mailing Add: Dept of Chem Polytech Inst NY 333 Jay St Brooklyn NY 11201

TOOP, EDGAR WESLEY, b Chilliwack, BC, Feb 26, 32; m 59; c 4. ORNAMENTAL HORTICULTURE. Educ: Univ BC, BSA, 55; Ohio State Univ, MSc, 57, PhD(plant path), 60. Prof Exp: Res officer, Can Dept Agr, 55-56; instr bot, Ohio State Univ, 61-62; asst prof, 62-69, ASSOC PROF HORT, UNIV ALTA, 69- Mem: Agr Inst Can; Can Soc Hort Sci. Res: Greenhouse flower crops; herbaceous ornamentals. Mailing Add: Dept of Plant Sci Univ of Alta Edmonton AB Can

TOOPER, EDWARD BENJAMIN, biochemistry, see 12th edition

TOOPS, EDWARD CHASSELL, b Columbus, Ohio, Aug 15, 26; m 51; c 4. REACTOR PHYSICS. Educ: Ohio Wesleyan Univ, BA, 47; Ind Univ, MS, 49, PhD(physics), 51. Prof Exp: Physicist, E I du Pont de Nemours & Co, 51-55, Combustion Eng, Inc, Conn, 55-70 & Gen Elec Co, 70-71; PHYSICIST, POWER GENERATION DIV, BABCOCK & WILCOX CO, 71- Mem: Am Nuclear Soc. Res: Particle and nuclear reactions; nuclear reactor physics; mathematical analysis; recycle of plutonium in reactor fuels. Mailing Add: 3135 Sedgewick Dr Lynchburg VA 24503

TOOPS, EMORY EARL, JR, physical chemistry, deceased

TOOTHILL, RICHARD B, b Philadelphia, Pa, July 28, 36; m 59; c 2. ORGANIC CHEMISTRY. Educ: Lehigh Univ, BS, 58; Mass Inst Technol, MS, 60; Univ Del, PhD(org chem), 64. Prof Exp: Tech serv rep paper chem, Hercules Powder Co, 60-61, chemist, 61-62; res chemist, 64-67, GROUP LEADER, BOUND BROOK LABS, AM CYANAMID CO, 67- Mem: Am Chem Soc. Res: Thiosemicarbazones; s-triazines; benzothiazoles; anthraquinone derivatives. Mailing Add: Org Chem Am Cyanamid Co Bound Brook NJ 08805

TOP, FRANKLIN HENRY, SR, b Grand Rapids, Mich, Feb 1, 03; m 32; c 2. PREVENTIVE MEDICINE, EPIDEMIOLOGY. Educ: Calvin Col, AB, 25; Univ Pa, MD, 28; Johns Hopkins Univ, MPH, 35; Am Bd Prev Med, dipl, 49. Prof Exp: Intern, Harper Hosp, Detroit, 28-29; sr intern, Women's Hosp, 29-30; resident, Commun Dis Div, Herman Kiefer Hosp, 30-31, resident physician & actg asst to med dir, 31-35; from asst prof to assoc prof prev med, Col Med, Wayne State Univ, 36-43, clin prof, 43-50, actg head dept, 49-50; prof epidemiol & pediat, Col Med, Univ Minn, 50-52; prof hyg & prev med & head dept, 52-66, dir, Inst Agr Med, 55-71, prof prev med & environ health & head dept, Col Med, 66-71, EMER PROF PREV MED & ENVIRON HEALTH & EMER DIR, INST AGR MED, 71- Concurrent Pos: Med dir, Commun Dis Div, Herman Kiefer Hosp, 35-37, dir, Hosp, 47-50; epidemiologist, City Dept Health, Detroit, 35-42; lectr, Sch Pub Health & Postgrad Sch, Univ Mich, 35-50; ed, Hist Am Epidemiol, 52; mem & pres, Iowa Bd Health, 55- Consult, Hemolytic Streptococcus Comn, US Secy, War, 41-47, Surgeon Gen, US Army, Ger, 46; Univ Hosp, Univ Minn, 50-52, Diphtheria Teaching Mission, US Dept State, Ger, 51 & Univ Hosp, Univ Iowa, 52-71. Mem health res study sect, NIH, 59-63; mem bd sci counr, Nat Inst Allergy & Infectious Dis, 60-64, chmn, 63-64. Honors & Awards: Grulee Award, Am Acad Pediat, 69. Mem: Am Epidemiol Soc (pres, 64); fel AMA; fel Am Pub Health Asn; fel Am Col Physicians; fel Am Acad Pediat. Res: Immunization; agricultural medicine. Mailing Add: Inst for Agr Med Univ of Iowa Oakdale Campus Oakdale IA 52319

TOPAZIAN, RICHARD G, b Greenwich, Conn, Feb 2, 30; m 58; c 4. ORAL SURGERY. Educ: Houghton Col, BA, 51; McGill Univ, DDS, 55; Univ Pa, cert oral surg, 57; Am Bd Oral Surg, dipl, 64. Prof Exp: Lectr dent & oral surg, Christian Med Col, Vellore, India, 59-61, reader, 61-63; from asst prof to assoc prof oral surg, Col Dent, Univ Ky, 63-67; prof & chmn dept, Sch Dent & Sch Med, Med Col Ga, 67-75; PROF ORAL & MAXILLOFACIAL SURG & HEAD DEPT, SCH DENT MED & PROF SURG, SCH MED, UNIV CONN, FARMINGTON, 75- Concurrent Pos: Consult, USPHS Hosp, Lexington, Ky, 64-67, US Army, Ft Jackson, SC, 67-75, Vet Admin Hosps, Augusta, Ga, 67-75 & Newington, Conn, 75- & Coun Dent Educ, Am Dent Asn; mem adv bd biomat res & adj prof, Clemson Univ; mem adv comt, Am Bd Oral Surg, 67-75. Mem: AAAS; Am Dent Asn; Am Soc Oral Surgeons; Int Asn Dent Res; fel Am Col Dent. Res: Dental education; teaching and research in oral surgery; bone pathology and diseases of the temporomandibular joint. Mailing Add: Oral & Max Surg Sch of Dent Med Univ of Conn Health Ctr Farmington CT 06032

TOPEL, DAVID G, b Lake Mills, Wis, Oct 24, 37; m. ANIMAL Educ: Univ Wis, BS, 60; Kans State Univ, MS, 62; Mich State Univ, PhD(food sci), 65. Prof Exp: Instr food sci, Mich State Univ, 64-65; asst prof animal sci, 65-67, assoc prof animal sci 18food technol, 68-73, PROF ANIMAL SCI & FOOD TECHNOL, IOWA STATE UNIV, 73- Concurrent Pos: USPHS fel, Univ Wis, 68; Fulbright grant, Denmark, 71-72. Mem: Am Soc Animal Sci; Inst Food Technol; Am Meat Sci Asn. Res: Influence of adrenal gland function, body composition and growth patterns on stress adaptation and muscle characteristics in the pig. Mailing Add: Dept of Animal Sci Iowa State Univ Ames IA 50010

TOPHAM, RICHARD WALTON, b Montgomery, WVa, May 22, 43; m 67; c 1. BIOCHEMISTRY. Educ: Hampden-Sydney Col, BS, 65; Cornell Univ, PhD(biochem), 70. Prof Exp: NIH fel, Fla State Univ, 69-71; asst prof, 71-75, ASSOC PROF CHEM, UNIV RICHMOND, 75- Concurrent Pos: Res scientist, Res Corp grant, 75-77. Mem: Am Chem Soc. Res: Role of copper-containing enzymes of blood serum in iron metabolism; characterization of enzymes and enzyme reactions of sterol biosynthesis. Mailing Add: Dept of Chem Univ of Richmond Richmond VA 23173

TOPHAM, WILLIAM SANFORD, b Cedar City, Utah, Sept 22, 37; m 60; c 2. BIOPHYSICS, BIOENGINEERING. Educ: Univ Utah, BS, 60, PhD(biophys, bioeng), 65. Prof Exp: Res engr, Cardiovasc Lab, Latter-Day Saints Hosp, Salt Lake City, Utah, 62-64; ASST PROF BIOPHYS & BIOENG, UNIV UTAH, 64-, MGR DEPT BIOPHYS, UTAH BIOMED TEST LAB, 70- Mem: Asn Advan Med Instrumentation. Res: Cardiovascular physiology; biomedical application of computers; instrumentation. Mailing Add: Dept of Biophysics Utah Biomed Test Lab Salt Lake City UT 84112

TOPLISS, JOHN G, b Mansfield, Eng, June 3, 30; nat US; m 58; c 2. MEDICINAL CHEMISTRY. Educ: Univ Nottingham, BSc, 51, PhD(chem), 54. Prof Exp: Res fel chem, Royal Inst Technol, Stockholm, Sweden, 54-56 & Columbia Univ, 56-57; from res chemist to sr res chemist, 57-66, sect leader, 66-68, from asst dir to assoc dir chem res, 68-73, DIR CHEM RES, SCHERING CORP, 73- Mem: Am Chem Soc; The Chem Soc; NY Acad Sci. Res: Synthesis and structure-activity relationships of drugs. Mailing Add: Schering Corp 60 Orange St Bloomfield NJ 07003

TOPOFF, HOWARD RONALD, b New York, NY, May 7, 41. BIOLOGY, ANIMAL BEHAVIOR. Educ: City Col New York, BS, 64, PhD(biol), 68. Prof Exp: Lectr biol, City Col New York, 67-68; res fel animal behav, 68-70, RES ASSOC ANIMAL BEHAV, AM MUS NATURAL HIST, 70-; ASST PROF PSYCHOL, HUNTER COL, 70- Mem: AAAS; Animal Behav Soc; NY Acad Sci. Res: Behavioral development in social insects; insect communication, behavior and physiology. Mailing Add: Am Mus of Natural Hist Central Park W at 79th St New York NY 10024

TOPOL, LEO ELI, b Boston, Mass, Apr 15, 26; m 48; c 2. PHYSICAL CHEMISTRY. Educ: Northeastern Univ, BS, 46; Univ Minn, Minneapolis, PhD(phys chem), 52. Prof Exp: Res chemist, Oak Ridge Nat Lab, 52-57; res specialist, Atomics Int Div, NAm Aviation, Inc, 57-63, mem tech staff, Sci Ctr, 64-69 & Atomics Int Div, 69-71, MEM TECH STAFF, SCI CTR, NAM ROCKWELL CORP, 71- Honors & Awards: Awarded 4 US Patents on Gas Pollution Monitors, 72-75. Mem: Am Chem Soc; Sigma Xi. Res: Electrolytes; new glass compositions, electrochemistry; thermodynamics and phase studies of fused salt systems and molten metal-metal salt solutions; electrochemistry of porous media; high-conducting solid electrolytes; solid electrochemical gas pollutant sensors. Mailing Add: 23435 Strathern St Canoga Park CA 91304

TOPOLESKI, LEONARD DANIEL, b Wilkes-Barre, Pa, Apr 11, 35; m 58; c 3. PLANT BREEDING, VEGETABLE CROPS. Educ: Pa State Univ, BS, 57, MS, 59; Purdue Univ, PhD(genetics, plant breeding), 62. Prof Exp: Asst prof, 62-68, ASSOC PROF VEG CROPS, NY STATE COL AGR & LIFE SCI, CORNELL UNIV, 68- Mem: Am Soc Hort Sci. Res: Genetics; vegetative hybridization; physiology of interspecific incompatability. Mailing Add: Dept of Veg Crops NY State Col of Agr & Life Sci Ithaca NY 14850

TOPOREK, MILTON, b New York, NY, Apr 18, 20; m 42; c 3. BIOCHEMISTRY. Educ: Brooklyn Col, BA, 40; George Washington Univ, MA, 48; Univ Rochester, PhD(biochem), 52. Prof Exp: Res assoc org chem, George Washington Univ, 48; res assoc biochem, Univ Rochester, 48-52; res chemist, Univ Mich, 52-57; from asst prof to assoc prof, 58-72, PROF BIOCHEM, JEFFERSON MED COL, THOMAS JEFFERSON UNIV, 72- Mem: AAAS; Am Chem Soc; Am Inst Nutrit; Am Inst Chemists; Am Soc Biol Chemists. Res: Control of plasma protein synthesis by liver, relationship to disease states; vitamin B-12, intrinsic factor relationships. Mailing Add: Dept of Biochem Jefferson Med Col Thomas Jefferson Univ Philadelphia PA 19107

TOPP, ALLAN CRICKINGTON, b Halifax, NS, Mar 17, 10; nat US; m 49; c 2. PHYSICAL CHEMISTRY. Educ: Dalhousie Univ, BSc, 37, MS, 39; McGill Univ, PhD(phys chem), 41. Prof Exp: Res chemist, Aluminum Co Can, 41-44; asst res chemist, Nat Res Coun Can, 44-46; asst prof, 46-50, ASSOC PROF CHEM, GA INST TECHNOL, 50- Mem: Am Chem Soc. Res: Chemistry of aluminate solutions; purification of beryllium; adsorption by charcoal; production of war gases; separation of rare earth elements; electrochemistry; separation of stable isotopes. Mailing Add: Dept of Chem Ga Inst of Technol Atlanta GA 30332

TOPP, STEPHEN V, b Longview, Tex, Oct 19, 37; m 57; c 2. REACTOR PHYSICS. Educ: Col William & Mary, BS, 59; Univ Va, MS, 60, PhD(physics), 62. Prof Exp: Res physicist, 62-67, sr physicist, 67-69, ASST CHIEF SUPVR, SAVANNAH RIVER LAB, E I DU PONT DE NEMOURS & CO, INC, 69- Concurrent Pos: Instr physics, Univ SC, 62- Mem: Am Phys Soc; Am Nuclear Soc; Opers Res Soc Am. Res: Neutron polarization measurements, reactor criticality measurements and lattice calculations for heavy water reactors; application of computer techniques to weapons production system modeling; decision modeling for nuclear waste management. Mailing Add: Savannah River Lab E I du Pont de Nemours & Co Inc Aiken SC 29801

TOPP, WILLIAM CARL, b Cleveland, Ohio, Feb 3, 48; div. CELL BIOLOGY. Educ: Oberlin Col, BA, 69; Princeton Univ, MA, 71, PhD(chem), 73. Prof Exp: Res assoc physics, Princeton Univ, 73, instr chem, 73-74, res assoc, 74; RES ASSOC BIOL, COLD SPRING HARBOR LAB QUANT BIOL, 74- Mem: Sigma Xi; Am Phys Soc. Res: Virus/cell interactions; cell growth control; cellular differentiation and teratocarcinomas. Mailing Add: Cold Spring Harbor Lab Cold Spring Harbor NY 11724

TOPP, WILLIAM ROBERT, b Milwaukee, Wis, May 27, 39. MATHEMATICS. Educ: St Louis Univ, BA, 63, MA, 64; Univ Wash, MS, 67, PhD(math), 68. Prof Exp: Instr math, Univ Seattle, 67-68; asst prof, Marquette Univ, 69-70; asst prof, 70-74, ASSOC PROF MATH, UNIV OF THE PAC, 74- Mem: Math Asn Am. Res: Rings; algebras. Mailing Add: Dept of Math Univ of the Pac Stockton CA 95204

TOPPEL, BERT JACK, b Chicago, Ill, July 2, 26; m 50; c 2. REACTOR PHYSICS. Educ: Ill Inst Technol, BS, 48, MS, 50, PhD(physics), 52. Prof Exp: Instr physics, Ill Inst Technol, 49-51; assoc physicist, Brookhaven Nat Lab, 52-56; ASSOC PHYSICIST, APPL PHYSICS DIV, ARGONNE NAT LAB, 56- Mem: Am Phys Soc; Am Nuclear Soc. Res: Nuclear reactions initiated by charged particles and neutrons; scintillation detector studies of gamma ray events; reactor critical facility experimentation; theoretical reactor physics calculations; reactor physics computer code development. Mailing Add: Appl Physics Div Argonne Nat Lab Argonne IL 60439

TOPPEN, DAVID LIVINGSTONE, inorganic chemistry, see 12th edition

TOPPER, YALE JEROME, b CHicago, Ill, Aug 11, 16; m 56; c 4. BIOCHEMISTRY. Educ: Northwestern Univ, BS, 42; Harvard Univ, MA, 43, PhD(chem), 47. Prof Exp: Assoc nutrit & physiol, Pub Health Res Inst, City of NY, Inc, 48-53; Am Heart Asn res fel, Biochem Res Lab, Mass Gen Hosp, 53-54; mem staff, 54-62, CHIEF SECT INTERMEDIARY METAB, NAT INST ARTHRITIS & METAB DIS, 62- Mem: AAAS; Am Soc Biol Chem. Res: Biochemistry of development and differentiation. Mailing Add: Lab of Biochem & Metab Nat Inst Arthritis & Metab Dis Bethesda MD 20014

TOPPING, JOSEPH JOHN, b Amsterdam, NY, Oct 9, 42; m 65; c 2. ANALYTICAL CHEMISTRY. Educ: Le Moyne Col, NY, BS, 65; Univ NH, MS, 67, PhD(chem), 69. Prof Exp: AEC fel anal chem, Ames Lab, Iowa State Univ, 69-70; ASST PROF CHEM, TOWSON STATE COL, 70- Mem: AAAS; Am Inst Chemists; Am Chem Soc. Res: Instrumental analytical chemistry; development of new methods of separation and analysis of trace metals; chromatography. Mailing Add: Dept of Chem Towson State Col Towson MD 21204

TOPPING, NORMAN HAWKINS, b Flat River, Mo, Jan 12, 08; m 30; c 2. INFECTIOUS DISEASES. Educ: Univ Southern Calif, AB, 33, MD, 36. Prof Exp: Intern, USPHS, Marine Hosps, San Francisco, Calif & Seattle, Wash, 36-37; mem staff med res viral & rickettsial dis, NIH, Md, 37-48, assoc dir, Insts, 48-52; vpres in-chg med affairs, Univ Pa, 52-58; pres, 58-70, CHANCELLOR, UNIV SOUTHERN CALIF, 70- Concurrent Pos: Chmn res comt & mem comn virus res & epidemiol, Nat Found, 58- Honors & Awards: Ashford Award, 43; Medal, US Typhus Comn, 45. Mem: AAAS; Am Epidemiol Soc; Soc Exp Biol & Med; Asn Am Physicians. Res: Virus and rickettsial diseases. Mailing Add: Off Chancello Univ Southern Cal Suite 1202 3810 Wilshire Blvd Los Angeles CA 90010

TORACK, RICHARD M, b Passaic, NJ, July 23, 27; m 53; c 4. PATHOLOGY. Educ: Seton Hall Univ, BS, 48; Georgetown Univ, MD, 52. Prof Exp: Asst pathologist, Montefiore Hosp, 58-59, asst neuropathologist, 59-61; asst prof path, New York Hosp-Cornell Med Ctr, 62-65, assoc prof, 65-68, assoc attend pathologist, 62-68; assoc prof, 68-70, PROF PATH & ANAT, WASH UNIV, 70- Concurrent Pos: Nat Cancer Inst fel path, Montefiore Hosp, 58-59; NIH res fel, Yale Med Sch, 61-62; consult, Mem Hosp, 64-68; assoc attend, Barnes Hosp, 68- Mem: AAAS; Am Asn Neuropath; Am Asn Path & Bact; Histochem Soc; Am Neurol Asn. Res: Electron histochemistry of disease of the nervous system. Mailing Add: Dept of Path Wash Univ St Louis MO 63110

TORALBALLA, GLORIA C, b Philippines, Jan 18, 15; nat US; m 46; c 1. CHEMISTRY, BIOCHEMISTRY. Educ: Univ Philippines, BS, 36, MS, 38; Univ

Mich, PhD(chem), 42. Prof Exp: Fel, Univ Mich, 43-44; res asst org chem, Columbia Univ, 44-46; instr anal & org chem, Marquette Univ, 49-52; res assoc biochem, Columbia Univ, 54-57, asst prof chem, Barnard Col, 58-63; asst prof, Hunter Col, 64-70; assoc prof, 70-75, PROF CHEM, LEHMAN COL, 75- Mem: Am Chem Soc. Res: Structure and mechanism of action of porcine pancreatic amylase; analytical studies of metallo-biochemicals. Mailing Add: Dept of Chem Lehman Col Bronx NY 10468

TORALBALLA, LEOPOLDO VASQUEZ, b Philippines, Nov 15, 10; m 46; c 1. MATHEMATICS. Educ: Univ Mich, PhD(math), 41. Prof Exp: Instr math, Univ Mich, 43-44; fel, Princeton Univ, 44-45; asst prof, Fordham Univ, 46-49; assoc prof, Marquette Univ, 49-53; from asst prof to prof, 53-74, EMER PROF MATH, NY UNIV, 74- Mem: Am Math Soc; Math Asn Am. Res: Topological algebra. Mailing Add: 1205 Tryon Ave Englewood NJ 07630

TORASKAR, JAYASHREE RAVALNATH, b May 21, 38; Indian citizen. X-RAY ASTRONOMY, PARTICLE PHYSICS. Educ: Bombay Univ, BSc, 59, MSc, 61; Columbia Univ, PhD(physics), 69. Prof Exp: Res assoc physics, Columbia Univ, 69-73 & Stanford Univ, 74-75; GUEST RES ASSOC PHYSICS, BROOKHAVEN NAT LAB, 75- Mem: Am Phys Soc. Res: Experimental x-ray astronomy. Mailing Add: 9 Warner St Hampton Bays NY 11946

TORBIT, CHARLES ALLEN, JR, b Fountain, Colo, Nov 8, 24; m 41; c 4. DEVELOPMENTAL BIOLOGY. Educ: Colo State Univ, BS, 62, PhD(cell biol), 72; Colo Col, MA, 66. Prof Exp: Fel physiol, Sch Med, Univ Kans, 72-74; embryologist, Codding Embryol Sci Inc, 74-75; CELL BIOLOGIST, STANFORD RES INST, 75- Mem: Sigma Xi; AAAS; Electron Micros Soc Am; Soc Study Reproduction; Int Embryo Transfer Soc. Res: Cell biology of ovulation, fertilization and early development of the mammalian egg and its hormonal control. Mailing Add: Stanford Res Inst 333 Ravenswood Ave Menlo Park CA 94025

TORCH, REUBEN, b Chicago, Ill, Dec 20, 26; m 49; c 3. PROTOZOOLOGY. Educ: Univ Ill, BS, 47, MS, 48, PhD(zool), 53. Prof Exp: Asst zool, Univ Ill, 47-53; from instr to assoc prof, Univ Vt, 53-65; from asst dean to actg dean col arts & sci, 66-73, PROF BIOL, OAKLAND UNIV, 65-, DEAN COL ARTS & SCI, 73- Mem: Soc Protozoologists; Am Micros Soc; Am Soc Zoologists; Am Soc Cell Biol. Res: Taxonomy of marine psammophilic ciliates; nucleic acid synthesis and regeneration in ciliates. Mailing Add: Col of Arts & Sci Oakland Univ Rochester MI 48063

TORCHIA, DENNIS ANTHONY, b Reading, Pa, June 15, 39; m 67; c 1. BIOPHYSICS. Educ: Univ Calif, Riverside, BA, 61; Yale Univ, MS, 64, PhD(physics), 67. Prof Exp: NIH fel, Med Sch, Harvard Univ, 67-69; mem tech staff polymer chem, Bell Labs, 69-71; physicist, Polymers Div, Nat Bur Standards, 71-74; BIOPHYSICIST, NAT INST DENT RES, 74- Mem: Am Chem Soc; Am Phys Soc; Biophys Soc. Res: Proton and carbon-13 magnetic resonance studies of the molecular conformaton and motion of cyclic peptides, polypeptides, proteins and polymers. Mailing Add: Rm 106 Bldg 30 Nat Insts Health Bethesda MD 20014

TORCHIANA, MARY LOUISE, b Philadelphia, Pa, July 22, 29. PHYSIOLOGY, PHARMACOLOGY. Educ: Immaculata Col, Pa, BA, 51; Temple Univ, MS, 60; Boston Univ, PhD(physiol), 64. Prof Exp: Res assoc pharmacol, 52-58 & 63-65, sr res pharmacologist, 65, res fel pharmacol, 65-72, DIR GASTROINTESTINAL RES, MERCK SHARP & DOHME RES LABS, 72- Mem: Am Soc Pharmacol & Exp Therapeut. Res: Catecholamine distribution; cardiovascular physiology and pharmacology; pharmacology of the gastrointestinal tract. Mailing Add: Merck Sharp & Dohme Res Labs Box 26-208 West Point PA 19486

TORCHINSKY, ALBERTO, b Buenos Aires, Arg, Mar 9, 44; m 69; c 1. MATHEMATICAL ANALYSIS. Educ: Univ Buenos Aires, Licenciado, 66; Univ Wis, Milwaukee, MS, 67; Univ Chicago, PhD(math), 71. Prof Exp: Asst prof, Cornell Univ, 71-75; ASST PROF MATH, IND UNIV, BLOOMINGTON, 75- Mem: Am Math Soc. Res: Problems related to singular integrals; Hp spaces and applications to differential equations. Mailing Add: Ind Univ Swain Hall E Bloomington IN 47401

TORDELLA, JOHN P, b Garrett, Ind, May 24, 19; m 43; c 9. CHEMISTRY. Educ: Loyola Univ, Ill, BS, 41; Univ Ill, MS, 42, PhD(chem anal), 44. Prof Exp: Asst, Univ Ill, 41-44; res assoc, Hanford Eng Works, Wash, 44-46, Ammonia Dept, 46-51, Polychem Dept, 51-60, Electrochem Dept, 60-70, RES ASSOC, PLASTICS DEPT, E I DU PONT DE NEMOURS & CO, INC, 70- Mem: AAAS; Am Chem Soc; Soc Rheol. Res: Rheology of molten polymers; adhesion; structure of polymer wax blends; high polymer physics. Mailing Add: E I du Pont de Nemours & Co Inc Wilmington DE 19898

TORDOFF, HARRISON BRUCE, b Mechanicville, NY, Feb 8, 23; m 46; c 3. ZOOLOGY. Educ: Cornell Univ, BS, 46; Univ Mich, MA, 49, PhD(zool), 52. Prof Exp: Cur birds, Sci Mus, Inst Jamaica, BWI, 46-47; asst prof zool, Univ & asst cur birds, Mus, Univ Kans, 50-57, assoc prof zool & assoc cur birds, 57; from asst prof to prof zool, Univ & cur birds, Mus Zool, Univ Mich, Ann Arbor, 57-70; PROF ECOL & BEHAV BIOL & DIR, BELL MUS NATURAL HIST, UNIV MINN, MINNEAPOLIS, 70- Concurrent Pos: Ed, Wilson Bull, 52-54. Mem: Soc Vert Paleont; Cooper Ornith Soc; Wilson Ornith Soc; fel Am Ornithologists Union. Res: Ornithology; systematics; paleontology; morphology; behavior; breeding biology. Mailing Add: Bell Mus of Natural Hist Univ of Minn Minneapolis MN 55455

TORDOFF, WALTER, III, b Newton, Mass, Jan 2, 43; m 65; c 2. POPULATION BIOLOGY. Educ: Univ Mass, BA, 65; Colo State Univ, MS, 67, PhD(zool), 71. Prof Exp: Asst prof, 70-75, ASSOC PROF BIOL SCI, CALIF STATE COL, STANISLAUS, 75- Mem: Am Soc Ichthyologists & Herpetologists; Soc Study Evolution; Soc Study Amphibians & Reptiles; Sigma Xi. Res: Ecology and genetics of chapparal and montane populations of reptiles and amphibians, particularly Hyla regilla. Mailing Add: Dept of Biol Sci Calif State Col Stanislaus Turlock CA 95380

TORELL, DONALD THEODORE, b Mont, Oct 19, 26; m 50; c 2. ANIMAL SCIENCE. Educ: Mont State Col, BS, 49; Univ Calif, MS, 50. Prof Exp: Assoc animal husb, Univ Calif, 49-50, res asst beef cattle invest, 50-51; instr, Ariz State Col, 51; LIVESTOCK SPECIALIST, HOPLAND FIELD STA, UNIV CALIF, 51- Concurrent Pos: Fulbright res sr scholar, Uganda, 61-62; specialist, Univ Chile-Univ Calif Coop Prog, 69-70. Mem: Am Soc Animal Sci; Soc Range Mgt. Res: Sheep nutrition, genetics, physiology and general sheep improvement. Mailing Add: Univ of Calif Hopland Field St 4070 University Rd Hopland CA 95449

TOREN, ERIC CLIFFORD, JR, b Chicago, Ill, Sept 16, 33; m 63; c 1. ANALYTICAL CHEMISTRY. Educ: Northwestern Univ, BS, 55; Univ Ill, MS, 60, PhD(chem), 61. Prof Exp: Instr chem, Duke Univ, 61-62, from asst prof to assoc prof, 62-70; ASSOC PROF MED & PATH, MED CTR, UNIV WIS-MADISON, 70- Concurrent Pos: Consult, Foote Mineral Corp, 64-65. Mem: Am Chem Soc; Am Asn Clin Chemists; Instrument Soc Am. Res: Kinetic methods of analysis; analytical instrumentation and automation laboratory computing. Mailing Add: Dept of Med Univ of Wis Med Ctr Madison WI 53706

TOREN, GEORGE ANTHONY, b Chicago, Ill, June 12, 24; m 49. ORGANIC CHEMISTRY. Educ: Hope Col, AB, 48; Purdue Univ, MS, 51, PhD(chem), 53. Prof Exp: PROD CONTROL SPECIALIST, MINN MINING & MFG CO, ST PAUL, 53- Mem: Am Chem Soc. Res: Boron and graphite advanced composites; pressure sensitive tapes. Mailing Add: 678 E Eldridge Ave Maplewood MN 55117

TOREN, PAUL EDWARD, b Lincoln, Nebr, July 18, 23; m 49; c 3. ANALYTICAL CHEMISTRY. Educ: Univ Nebr, AB, 47, MS, 48; Univ Minn, PhD(chem), 54. Prof Exp: Chemist, Phillips Petrol Co, 53-59; sr chemist, 59-67, RES SPECIALIST, CENT RES LABS, 3M CO, 67- Mem: Am Chem Soc. Res: Electroanalytical chemistry; analytical instrumentation. Mailing Add: Cent Res Labs 3M Co PO Box 33221 St Paul MN 55133

TORESON, WILFRED EARL, b Calif, Dec 25, 16; m 45; c 1. PATHOLOGY. Educ: McGill Univ, MD, 42, MSc, 48, PhD(path), 50; Am Bd Path, dipl & cert clin path, 53. Prof Exp: Lectr path, McGill Univ, 46-50, asst prof, 50; from instr to prof, Univ Calif, San Francisco & pathologist, Univ Hosp, 50-66; prof path, State Univ NY Downstate Med Ctr & dir labs, Univ Hosp, 66-70; PROF PATH, SCH MED, UNIV CALIF, DAVIS, 70- Concurrent Pos: Dir labs, South Pac Hosp, Calif, 52-58, consult, 58-66; consult, Letterman Army Hosp, 58-66; attend, Ft Miley Vet Admin Hosp, 60-66; dir anat path, Sacramento Med Ctr, 70- Mem: Am Asn Pathologists & Bacteriologists; AMA; Col Am Pathologists; Am Soc Clin Path; Int Acad Path. Res: Isolated myocarditis; cancer of the esophagus; experimental diabetes; automation and computers in clinical pathology. Mailing Add: 2315 Stockton Blvd Sacramento CA 95817

TORFASON, WILMER ESPLIN, b Sask, Can, Mar 3, 20; m 42; c 2. HORTICULTURE. Educ: Univ Man, BSA, 49; Univ Alta, MSc, 52; Univ Minn, PhD, 64. Prof Exp: VEG PHYSIOLOGIST, RES STA, CAN DEPT AGR, 49- Mem: Am Soc Hort Sci; Potato Asn Am; Can Soc Hort Sci. Res: Vegetable crops, especially physiological effects of environmental conditions and cultural practices. Mailing Add: Res Sta Can Dept of Agr Lethbridge AB Can

TORGERSON, DAVID FRANKLYN, b Winnipeg, Man, July 11, 42; m 66; c 2. MASS SPECTROMETRY. Educ: Univ Man, BSc, 65, MSc, 66; McMaster Univ, PhD(chem), 69. Prof Exp: Asst prof chem, Dept Chem, 69-70, res scientist, Cyclotron Inst, 70-74, SR SCIENTIST CHEM, CYCLOTRON INST, TEX A&M UNIV, 74- Concurrent Pos: Consult, NIH Workshop on Mass Spectrometry, 75. Mem: Am Phys Soc; AAAS; Am Soc Mass Spectros; Can Inst Chem; Am Chem Soc. Res: Plasma desorption mass spectrometry of involatile solids and thin films; nuclear chemistry; atomic masses; new instrumentation. Mailing Add: Cyclotron Inst Tex A&M Univ College Station TX 77843

TORGERSON, RONALD THOMAS, b Minneapolis, Minn, Sept 20, 36; m 63; c 2. HIGH ENERGY PHYSICS, THEORETICAL PHYSICS. Educ: Col St Thomas, BS, 58; Univ Chicago, MS, 62, PhD(physics), 65. Prof Exp: Instr physics, Univ Notre Dame, 65-68; asst prof, Ohio State Univ, 68-73; RES ASSOC PHYSICS, UNIV ALTA, 73- Mem: AAAS; Am Phys Soc. Res: Quantum field theory; high energy collisions; pi pi scattering; weak and electromagnetic interactions. Mailing Add: Dept of Physics Univ of Alta Edmonton AB Can

TORGESEN, JOHN LAU, physical chemistry, see 12th edition

TORGESON, DEWAYNE CLINTON, b Ambrose, NDak, Oct 1, 25; m 59; c 3. PLANT PATHOLOGY. Educ: Iowa State Univ, BS, 49; Ore State Univ, PhD(plant path), 53. Prof Exp: PLANT PATHOLOGIST, BOYCE THOMPSON INST PLANT RES, INC, 52-, PROG DIR BIOREGULANT CHEM, 63- Mem: AAAS; Am Inst Biol Sci; Am Phytopath Soc; Torrey Bot Club. Res: Fungicides; discovery and development of pesticides; fate of pesticides in the environment. Mailing Add: Boyce Thompson Inst Plant Res 1086 N Broadway Yonkers NY 10701

TORIBARA, TAFT YUTAKA, b Seattle, Wash, Apr 10, 17; m 48; c 2. BIOPHYSICS, CHEMISTRY. Educ: Univ Wash, BS, 38, MS, 39; Univ Mich, PhD(chem), 42. Prof Exp: Res chemist, Dept Eng Res, Univ Mich, 42-48; scientist chem, Atomic Energy Proj, 48, from asst prof to assoc prof, 50-63, PROF RADIOBIOL & BIOPHYS, MED SCH, UNIV ROCHESTER, 63- Concurrent Pos: Nat Inst Gen Med Sci spec res fel, Univ Tokyo, 60-61. Mem: AAAS; Am Chem Soc. Res: Binding of ions and small molecules to serum proteins; analytical chemistry of trace materials in biological systems. Mailing Add: Dept of Radiobiol & Biophys Univ of Rochester Med Ctr Rochester NY 14642

TORIO, JOYCE CLARKE, b Biddeford, Maine, Oct 1, 34; m 55. SCIENCE ADMINISTRATION, INFORMATION SCIENCE. Educ: Rutgers Univ, BS, 56, MS, 61, PhD(hort, soils), 65. Prof Exp: Lab technician soils, Rutgers Univ, 56-61, res assoc cranberry cult, 61-65; ed biochem, hort & soils, Chem Abstr Serv, Am Chem Soc, 65-69; staff officer, Agr Bd & Renewable Resources, Nat Acad Sci, 69-74; HEAD INFO SERV, INT RICE RES INST, 74- Mem: Am Soc Hort Sci; Am Chem Soc; Am Inst Biol Sci. Res: Pomology, mineral nutrition and plants; soil fertility and analysis; plant physiology and pathology; rice culture and associated multiple cropping systems research-information management. Mailing Add: Int Rice Res Inst PO Box 933 Manila Philippines

TORKELSON, ARNOLD, b Thompson, NDak, Oct 28, 22; m 44; c 4. ORGANOMETALLIC CHEMISTRY. Educ: Univ NDak, BSc, 46; Purdue Univ, MS, 48, PhD, 50. Prof Exp: Asst, Purdue Univ, 46-48; prod develop chemist, 50-58, mgr anal chem, 58-65, mgr fluid prod develop, 65-72, mgr specialities develop, 72-76, MGR FLUIDS RESINS & SPECIALTIES PROD DEVELOP, SILICONE PROD DEPT, GEN ELEC CO, 76- Mem: AAAS; Am Chem Soc. Res: Synthesis of organosilicon compounds; rate studies on the cleavage of silicon-carbon bond; product development and research on silicone fluids, resins and specialty products. Mailing Add: Gen Elec Co Specialty Div Silicone Prod Dept Waterford NY 12188

TORKELSON, THEODORE RUBEN, b St James, Minn, Nov 25, 26; m 52; c 5. TOXICOLOGY. Educ: Gustavus Adolphus Col, BA, 51; Univ Nebr, MA, 54; Univ Pittsburgh, ScD(hyg), 66. Prof Exp: Toxicologist, 53-61 & 62-74, OCCUP HEALTH ASSOC, DOW CHEM CO, 74- Mem: Soc Toxicol; Am Indust Hyg Asn. Res: Industrial and solvent toxicology; industrial hygiene; industrial toxicology and occupational health. Mailing Add: Dow Chem Co Midland MI 48640

TORLEY, ROBERT EDWARD, b Monmouth, Ill, Jan 28, 18; m 41; c 2. CHEMISTRY. Educ: Monmouth Col, BS, 39; Univ Iowa, MS, 41, PhD(chem), 42. Prof Exp: Chemist, Am Cyanamid Co, Conn, 43-51, chemist chem processing plant, Idaho, 51-53, gen supt, 53, asst plant mgr, Bridgeville Plant, Pa, 53-56, asst plant mgr, Res Div, NY, 56-58, contract mgr govt solid rocket propellant contract, Stamford Labs, 58-62, dir contract res dept, 62-63, dir phys dept, 63-71, dir sci serv dept, 71-

75; VPRES & DIR TECHNOL, T&E CTR, EVANS PROD CO, 74- Mem: Fel AAAS; Am Chem Soc; Am Phys Soc. Mailing Add: Evans Prod Co T&E Ctr 1115 SE Crystal Lake Dr Corvallis OR 97330

TORMEY, JOHN MCDIVIT, b Baltimore, Md, Oct 7, 34; m 59; c 2. PHYSIOLOGY, CELL BIOLOGY. Educ: Loyola Col, Md, BS, 56; Johns Hopkins Univ, MD, 61. Prof Exp: Fel ophthal, Johns Hopkins Univ, 61-62, instr, 62-63; res fel biol, Harvard Univ, 63-64; asst prof anat & ophthal, Johns Hopkins Univ, 64-66; staff assoc phys biol, Nat Inst Arthritis & Metab Dis, 66-68; asst prof, 68-70, ASSOC PROF PHYSIOL, UNIV CALIF, LOS ANGELES, 70- Concurrent Pos: Nat Inst Neurol Dis & Blindness fel, 61-63, spec fel, 63-66, res grants, 65-66 & 68- Mem: Am Physiol Soc; Am Soc Cell Biol; Am Asn Anat. Res: Relationship between structure and function of body tissues, especially epithelia; development of methods for localizing transport functions. Mailing Add: Dept of Physiol Univ of Calif Ctr for Health Sci Los Angeles CA 90024

TORNABENE, THOMAS GUY, b Cecil, Pa, May 6, 37; m 62; c 3. MICROBIOLOGY. Educ: St Edward's Univ, BS, 59; Univ Houston, MS, 62, PhD(biol chem), 67. Prof Exp: Instr biol, Univ Houston, 62-65; fel biochem, Nat Res Coun, Ottawa, Can, 67-68; asst prof, 68-73, ASSOC PROF MICROBIOL, COLO STATE UNIV, 73- Mem: Am Soc Microbiol; Am Oil Chem Soc. Res: Biogenesis and distribution of microbial hydrocarbons; microbial lipids and carbohydrates; metabolic pathways and mechanisms of synthesis of biochemical compounds. Mailing Add: Dept of Microbiol Colo State Univ Fort Collins CO 80521

TORNETTA, FRANK JOSEPH, biology, see 12th edition

TORNHEIM, LEONARD, b Chicago, Ill, Aug 21, 15; c 4. NUMERICAL ANALYSIS. Educ: Univ Chicago, SB, 35, SM, 36, PhD(math), 38. Prof Exp: Instr math, Chicago Pub Jr Cols, 38-40; instr math & statist, Antioch Col, 40-41; sect chief, Comput Sect, Ballistic Res Lab, Aberdeen Proving Ground, Md, 43-45; instr math, Princeton Univ, 46; from instr to asst prof, Univ Mich, 46-55; lectr, Univ Calif, 55-56; SR RES ASSOC, CHEVRON RES CO, 56- Mem: Soc Indust & Appl Math; Math Asn Am. Res: Numerical analysis; industrial mathematics. Mailing Add: PO Box 1627 Richmond CA 94802

TORNQVIST, ERIK GUSTAV MARKUS, b Lund, Sweden, Jan 13, 24; m 69; c 1. POLYMER CHEMISTRY, BIOCHEMISTRY. Educ: Royal Inst Technol, Sweden, MSc, 48; Univ Wis, MS, 53, PhD(biochem), 55. Prof Exp: First res asst, Div Food Chem, Royal Inst Technol, Sweden, 49-51; res asst, Dept Biochem, Univ Wis, 51-55; res chemist, Chem Res Div, Esso Res & Eng Co, 55-58, res assoc, 58-66, SR RES ASSOC, ENJAY POLYMER LABS, LINDEN, EXXON RES & ENG CO, 66- Mem: AAAS; Am Chem Soc; NY Acad Sci; Swedish Asn Eng & Archit. Res: Organometallic chemistry and catalysis; polymer chemistry, especially synthesis and mechanisms of polymerization; biotechnical production of protein, fat, vitamins and antibiotics. Mailing Add: 38 Mareu Dr Scotch Plains NJ 07076

TORO-GOYCO, EFRAIN, b Cabo Rojo, PR, Mar 14, 31; m 55; c 5. PHYSICAL CHEMISTRY, BIOCHEMISTRY. Educ: Univ PR, BS, 54, LLB, 64; Harvard Univ, MA, 56, PhD(biochem), 58. Prof Exp: Assoc, 58, from asst prof to assoc prof, 58-69, PROF BIOCHEM, SCH MED, UNIV PR, SAN JUAN, 69-, CHMN DEPT, 72- Concurrent Pos: Lederle Med Fac Award, 66-69; asst chief radioisotope serv, San Juan Vet Admin Hosp, 59-65, consult, 65- Mem: Am Chem Soc; affil AMA; Am Soc Biol Chemists. Res: Immunochemistry; enzymology; marine pharmacology; biochemistry of schistosomes. Mailing Add: Dept of Biochem Univ of PR Sch of Med San Juan PR 00905

TOROK, ANDREW, JR, b Hopewell, Va, Oct 30, 25; m 51; c 2. CHEMISTRY. Educ: Pa State Univ, BS, 49; Stevens Inst Technol, MS, 56. Prof Exp: Anal chemist, William P Warner, Inc, 49-51; res chemist, Venus Pen & Pencil Co, 51-53, chief chemist, 53-57, tech dir, 57-61; prod develop mgr, Ga Kaolin Co, 61-74; DIR RES & DEVELOP, FABER-CASTELL CORP, 74- Mem: Am Chem Soc; Am Ceramic Soc; Clay Minerals Soc; NY Acad Sci; Fine Particle Soc (treas, 70-73). Res: Clays and clay products, especially application in new fields. Mailing Add: 44 Long Ridge Rd Dover NJ 07801

TOROK, ELIZABETH ESTHER, biochemistry, see 12th edition

TOROK, NICHOLAS, b Budapest, Hungary, June 13, 09; US citizen; m 39. OTOLARYNGOLOGY. Educ: Eötvös Lorand Univ, Budapest, 34. Prof Exp: From instr to asst prof otolaryngol, Eötvös Lorand Univ, Budapest, 40-47; from instr to assoc prof, 50-68, PROF OTOLARYNGOL, UNIV ILL COL MED, 68- Concurrent Pos: Consult, Chicago Read Hosp, Ill State Psychiat Inst, Ill Hosp Sch & Michael Reese Hosp. Honors & Awards: Award, Am Acad Ophthal & Otolaryngol, 69; NASA Skylab Achievement Award. Mem: Am Acad Ophthal & Otolaryngol; Am Laryngol, Rhinol & Otol Soc; Am Neurotol Soc (pres, 73-74); Am Acad Cerebral Palsy; affil Royal Soc Med. Res: Otology; neuro-ortology; vestibular studies. Mailing Add: Eye & Ear Infirmary Univ of Ill Col of Med Chicago IL 60612

TOROP, WILLIAM, b New York, NY, Jan 12, 38; m 60; c 2. INORGANIC CHEMISTRY, SCIENCE EDUCATION. Educ: Univ Pa, AB, 59, MS, 61, EdD(sci educ), 68. Prof Exp: Prof employee chem, Upper Darby Sr High Sch, Pa, 60-68; asst prof chem & sci educ, St Joseph's Col (Pa), 68-71; PROF CHEM, WEST CHESTER STATE COL, 71- Concurrent Pos: Elem sci consult, Interboro Sch Dist, Pa, 69-70; elem sci consult, Marple Newtown Sch Dist, 70-72; dir, Del Valley Inst Sci Educ, 71- Mem: Am Chem Soc. Res: Use of written laboratory reports in high school chemistry; trivalent basic polyphosphates; evaluation of elementary science programs; computer managed and computer assisted instruction. Mailing Add: Dept of Chem West Chester State Col West Chester PA 19380

TOROSIAN, GEORGE, b Racine, Wis, Jan 1, 36; m 64; c 2. PHARMACY, PHARMACOLOGY. Educ: Univ Wis-Madison, BS, 62, MS, 64, PhD(pharm), 66. Prof Exp: Sr pharm chemist, Menley & James Labs Div, Smith Kline & French Labs, Inc, 66-69; ASSOC PROF PHARM, COL PHARM, UNIV FLA, 69- Mem: Am Pharmaceut Asn; Am Asn Cols Pharm; Acad Pharmaceut Sci. Res: Product development and design; biopharmaceutics; solution kinetics. Mailing Add: Col of Pharm Univ of Fla Gainesville FL 32610

TORO VIZCARRONDO, CARLOS E, applied statistics, analytical statistics, see 12th edition

TORP, BRUCE ALAN, b Duluth, Minn, Sept 5, 37; m 60; c 3. INORGANIC CHEMISTRY. Educ: Univ Minn, BA, 59; Iowa State Univ, MS, 62, PhD(inorg chem), 64. Prof Exp: Sr chemist, 64-68, supvr inorg chem res group, 68-71, lab mgr, Physics & Mat Res Lab, 71-74, DIR MAT & ELECTRONICS RES LAB, CENT RES LABS, 3M CO, 75- Mem: Am Chem Soc. Res: Coordination, transition metal and solid state chemistry; semiconductor research. Mailing Add: Cent Res Labs 3M Co PO Box 33221 St Paul MN 55101

TORRANCE, DANIEL J, b Peking, China, Nov 14, 21; US citizen; m 51; c 2. RADIOLOGY. Educ: Univ Wash, BSc, 44; Johns Hopkins Univ, MD, 49. Prof Exp: Intern med, Johns Hopkins Hosp, 49-50, fel path, 50-51, asst resident radiol, 51-53, from instr to assoc prof, Sch Med, Johns Hopkins Univ, 53-63; head div, Scripps Clin & Res Found, La Jolla, Calif, 63-66; assoc prof, Sch Med, Wash Univ, 66-68; prof, Univ Calif, Los Angeles, 68-72; CHIEF RADIOLOGIST, BAY HARBOR HOSP, CALIF, 72- Concurrent Pos: Consult, USPHS Hosp, Baltimore, Md, 54-; radiologist, Johns Hopkins Hosp, 55-63; assoc radiologist, Mallinckrodt Inst Radiol, Barnes Hosp, St Louis, 66-; chief dept radiol, Harbor Gen Hosp, Torrance, Calif; clin prof radiol, Univ Calif, Los Angeles, 72- Mem: Am Col Radiol. Res: Chest radiograph in connection with the pulmonary circulation; problems in the radiography of pulmonary atelectasis; radiographic manifestations of pulmonary edema. Mailing Add: Dept of Radiol Bay Harbor Hosp Harbor City CA 90810

TORRANCE, ELLEN MCCORMICK, b Cleveland, Ohio, Mar 23, 41. PURE MATHEMATICS, ACTUARIAL MATHEMATICS. Educ: Barnard Col, Columbia Univ, AB, 62; Stanford Univ, MS, 63; Univ Ill, Urbana, PhD(math), 68. Prof Exp: Instr math, Lamar Univ, 63-64; asst prof, Mt Holyoke Col, 68-71; assoc prof math & chmn dept, Sterling Col, 72-74; vis asst prof, Kans State Univ, 74-75; ACTUARIAL TRAINEE ASST TO THE PRES, COLOGNE LIFE REINSURANCE CO, 75- Mem: Am Math Soc; Math Asn Am. Res: Functional analysis; number theory; actuarial studies. Mailing Add: Cologne Life Reinsurance Co 1200 Bedford St Stamford CT 06905

TORRANCE, ESTHER MCCORMICK, b Ft Wayne, Ind, Aug 12, 09; m 34; c 2. MATHEMATICS. Educ: Columbia Univ, AB, 31; Cornell Univ, MA, 32; Brown Univ, PhD(math), 39. Prof Exp: Asst prof math, Fresno State Univ, 67-73; RETIRED. Mem: Am Math Soc. Res: Topology. Mailing Add: 200 Glenwood Circle Apt 436 Monterey CA 93940

TORRANCE, JERRY BADGLEY, JR, b San Diego, Calif, July 20, 41; m 64; c 2. PHYSICS. Educ: Stanford Univ, BS, 63; Univ Calif, Berkeley, MA, 66; Harvard Univ, PhD(appl physics), 69. Prof Exp: RES PHYSICIST, THOMAS J WATSON RES CTR, IBM CORP, 69- Mem: Am Phys Soc. Res: Solid state phyiscs, particularly magnetic semiconductors, optical and infrared properties, magnetic insulators. Mailing Add: Phys Sci Dept Box 218 Thomas J Watson Res Ctr IBM Corp Yorktown Heights NY 10598

TORRE, FRANK JOHN, b Newark, NJ, Oct 6, 44; m 68; c 1. PHYSICAL CHEMISTRY. Educ: Monmouth Col NJ, BS, 67; Rutgers Univ, PhD(phys chem), 71. Prof Exp: Res chem, Bell Tel Labs, 67-68; fel, Univ Rochester, 71-73; ASST PROF CHEM, SPRINGFIELD COL, 73- Mem: Am Chem Soc. Mailing Add: Dept of Chem Springfield Col Springfield MA 01109

TORRE-BUENO, JOSE ROLLIN, b Tucson, Ariz, Nov 20, 48; m 69. PHYSIOLOGY. Educ: State Univ NY Stony Brook, BS, 70; Rockefeller Univ, PhD(physiol), 75. Prof Exp: RES ASSOC PHYSIOL, DUKE UNIV, 75-, NIH FEL, 76- Mem: Animal Behav Soc. Res: Energetics and behavior during flight and migration in birds. Mailing Add: Dept of Zool Duke Univ Durham NC 27706

TORRENCE, ROBERT JAMES, b Pittsburgh, Pa, June 7, 37; m 59. THEORETICAL PHYSICS. Educ: Carnegie-Mellon Univ, BS, 59; Univ Pittsburgh, PhD(physics), 65. Prof Exp: Res assoc physics, Syracuse Univ, 65-67; adj prof, Ctr Advan Studies, Nat Polytech Inst, Mex, 67-68; asst prof, 68-70, ASSOC PROF MATH, UNIV CALGARY, 70-, CHMN DIV APPL MATH, 75- Res: General relativity with emphasis on gravitational radiation. Mailing Add: Dept of Math Univ of Calgary Calgary AB Can

TORRES, ANDREW MARION, b Albuquerque, NMex, Jan 20, 31; m 55; c 4. BOTANY. Educ: Univ Albuquerque, BS, 52; Univ NMex, MS, 58; Ind Univ, PhD(bot), 61. Prof Exp: Instr biol, Wis State Univ, Oshkosh, 60-61; asst prof bot & genetics, Univ Wis, Milwaukee, 61-64; assoc prof, 64-70, assoc dean grad sch, 69-72, PROF BOT & GENETICS, UNIV KANS, 70- Concurrent Pos: NSF grants, 61-; Ford Found sr adv, Univ Oriente, Venezuela, 66-67; chief of party, Aid to higher educ, Dominican Republic, 68. Mem: AAAS; Am Soc Plant Taxon; Bot Soc Am; Soc Study Evolution; Genetics Soc Am. Res: Cytogenetics; chemosystematics of Compositae; alcohol dehydrogenase isozymes of sunflowers; genetics; subunit structure; activities. Mailing Add: Dept of Bot Univ of Kans Lawrence KS 66044

TORRES, FERNANDO, b Paris, France, Nov 29, 24; m 55; c 1. NEUROLOGY, NEUROPHYSIOLOGY. Educ: Ger Col, Colombia, BA, 41; Nat Univ Colombia, MD, 48; Am Bd Psychiat & Neurol, dipl, 61. Prof Exp: Asst neurosurg, Inst Cancer, Buenos Aires, Arg, 49-50; resident neurol, Montefiore Hosp, New York, 53-55; from instr to assoc prof, 56-64, PROF NEUROL, UNIV MINN, MINNEAPOLIS, 64- Concurrent Pos: Fel, Johns Hopkins Hosp, 50-52, NIH res fel, 52-53; NIH spec fel, LaSalpetriere Hosp, Paris, 63-64; asst, Columbia Univ, 54-55; consult prof, Univ PR, 61- Mem: AAAS; fel Am Acad Neurol; Am Neurol Asn; Soc Neurosci; Am Electroencephalog Soc. Res: Electroencephalography; clinical neurophysiology; epilepsy; cerebrovascular physiology; developmental cerebral physiology. Mailing Add: Box 28 Dept of Neurol Univ of Minn Hosp Minneapolis MN 55455

TORRES-BLASINI, GLADYS, b PR; m; c 2. MICROBIOLOGY. Educ: Univ PR, BS, 48; Univ Mich, Ann Arbor, MS, 52, PhD(bact), 53; Duke Univ, cert mycol, 54. Prof Exp: Teaching asst bact, Univ Mich, 52; from assoc to assoc prof, 54-64, mem admis comt, 67-70, actg chmn, 70, mem admis comt, Sch Dent, 70, PROF MICROBIOL, SCH MED, UNIV PR, SAN JUAN, 64-, PRES ADMIS COMT, 71- Concurrent Pos: USPHS fel mycol, 54; Hoffmann-La Roche grant fungistatic drugs, 54, Trichophyton species, 56; Vet Admin Hosp grant, 56 & 57; Univ PR Sch Med & NIH grant, 58; NIH grant, 58, 60, 64 & 68-70; lectr, Hahnemann Med Sch, 62; Univ PR Med Sch Gen Res Funds grant, 64; mem, Study Sect Res, Vet Admin Hosp, San Juan, PR, 70. Mem: Am Soc Microbiol; PR Pub Health Asn; PR Soc Microbiol; Tissue Cult Asn. Res: Bacteriology. Mailing Add: Dept of Microbiol & Immunol Univ of PR Sch of Med San Juan PR 00931

TORRES-PEIMBERT, SILVIA, b Mexico City, Mex, June 26, 40; m 62; c 2. ASTROPHYSICS. Educ: Nat Univ Mex, BS, 64; Univ Calif, Berkeley, PhD(astron), 69. Prof Exp: From asst prof to assoc prof, 69-72, PROF ASTRON, INST ASTRON, NAT UNIV MEX, 72- Mem: Am Astron Soc; Int Astron Union; Acad Sci Invest Mex. Res: Physical conditions of interstellar matter, particularly the determination of abundances and energy input; observations and theoretical models of planetary nebulae and H I I regions. Mailing Add: Inst of Astron PO Box 70-264 Mexico 20 D F Mexico

TORRES-PINEDO, RAMON, b Burgos, Spain, Apr 3, 29; US citizen; m 57; c 4. PEDIATRICS, GASTROENTEROLOGY. Educ: Univ Granada, BS, 48; Univ Madrid, MD, 56. Prof Exp: Intern, San Juan City Hosp, PR, 58-59; resident pediat, San Juan City Hosp & Univ Hosp, 59-61; assoc, Univ Hosp, Univ PR, 63-65, from asst prof to prof, 65-75, asst dir pediat res, Clin Res Ctr, Sch Med, 63-75, prof physiol

& head dept, 66-75; PROF PEDIAT & CHIEF PEDIAT GASTROENTEROL, OKLA CHILDREN'S MEM HOSP, OKLAHOMA CITY, 75- Concurrent Pos: Fels pediat res, Michael Reese Hosp & Med Ctr, Univ Ill, 61-63; consult physician, San Juan City Hosp, 64-75. Mem: PR Med Asn; Pediat Res Soc Mex. Res: Pediatric research; electrolyte transport; intermediary metabolism; nutrition. Mailing Add: Dept of Pediat Gastroenterol Okla Children's Mem Hosp Oklahoma City OK 73126

TORRES-RODRIGUEZ, VICTOR M, b Coamo, PR, Feb 17, 26; US citizen; m 48; c 7. MEDICINE. Educ: Univ PR, Rio Piedras, BS, 47; Columbia Univ, MD, 51. Prof Exp: Dermatologist, US Army Hosp, Ft Jackson, SC, 57-59; from asst prof to assoc prof, 59-67, PROF DERMAT, SCH MED, UNIV PR, RIO PIEDRAS, 67-, CHIEF SECT, 65-, DIR DERMAT RESIDENCY PROG, AFFILIATED HOSPS, 66- Concurrent Pos: Fel dermat, Columbia-Presby Med Ctr, 54-57, Lederle Int fel dermatopath, 64-65; consult, Rodriguez Army Hosp, San Juan, 62-, various pvt hosps, 63- & Vet Admin Hosp, 70- Mem: Am Acad Dermat; Am Soc Dermatopath; Am Dermat Asn. Res: Pathology of certain tropical dermatoses, chiefly granulomas. Mailing Add: Div of Dermat Univ Hosp Caparra Heights Sta Rio Piedras PR 00935

TORREY, HENRY CUTLER, b Yonkers, NY, Apr 4, 11; m 37; c 2. MAGNETIC RESONANCE. Educ: Univ Vt, BS, 32; Columbia Univ, AM, 33, PhD(physics), 37. Hon Degrees: DSc, Univ Vt, 65. Prof Exp: Instr physics, Princeton Univ, 37; instr Pa State Col, 37-41, asst prof, 41-42; mem staff, Radiation Lab, Mass Inst Technol, 42-46; assoc prof physics, 46-50, chmn dept, 59-64, dean grad sch & dir res coun, 65-74, PROF PHYSICS, RUTGERS UNIV, NEW BRUNSWICK, 50- Concurrent Pos: Consult, Calif Res Corp, Standard Oil Co Calif, 52-; Guggenheim fel & Rutgers Univ fac fel, Univ Paris, 64-65. Mem: AAAS; fel Am Phys Soc; Am Asn Physics Teachers. Res: Molecular beams; viscosity of gases; nuclear magnetic moments; radio frequency spectroscopy; nuclear magnetic and paramagnetic resonance; free radicals; semiconductors; crystal rectifiers. Mailing Add: Dept of Physics Rutgers Univ New Brunswick NJ 08903

TORREY, JOHN GORDON, b Philadelphia, Pa, Feb 22, 21; m 49; c 5. PLANT PHYSIOLOGY. Educ: Williams Col, BA, 42; Harvard Univ, MA, 47, PhD, 50. Prof Exp: Harvard Univ traveling fel, Cambridge Univ, 48-49; from instr to assoc prof bot, Univ Calif, 49-60; dir, Cabot Found, 66-75, PROF BOT, HARVARD UNIV, 60- Concurrent Pos: Guggenheim fel, 65-66; hon sr res fel, Univ Glasgow, 73. Mem: AAAS; Am Acad Arts & Sci; Bot Soc Am; Am Soc Plant Physiol; Soc Develop Biol (pres, 63). Res: Physiology of root growth; physiology and biochemistry of tissue differentiation; root nodules in legumes and non-legumes. Mailing Add: Cabot Found Harvard Univ Petersham MA 01366

TORREY, RUBYE PRIGMORE, b Sweetwater, Tenn, Feb 18, 26; m 57; c 2. RADIATION CHEMISTRY, ANALYTICAL CHEMISTRY. Educ: Tenn State Univ, BS, 46, MS, 48; Syracuse Univ, PhD(chem), 68. Prof Exp: Res assoc & instr chem, 48-57, from asst prof to assoc prof, 57-72, PROF CHEM, TENN STATE UNIV, 72- Concurrent Pos: Asst lectr, Syracuse Univ, 63-68; US AEC res grant & res collabr, Brookhaven Nat Lab, 70- Mem: AAAS; Am Chem Soc. Res: Electro-analytical chemistry; gas phase reaction mechanisms using alpha radiolysis and high-pressure impact mass spectrometry; effects of various factors on polarographic diffusion coefficients using chronopotentiometric technique. Mailing Add: Dept of Chem Tenn State Univ Nashville TN 37203

TORREY, THEODORE WILLETT, b Woodbine, Iowa, Jan 14, 07; m 38. DEVELOPMENTAL ANATOMY. Educ: Univ Denver, AB, 27; Harvard Univ, AM, 29, PhD(zool), 32. Prof Exp: From instr to prof zool, 32-72, chmn dept, 48-66, EMER PROF ZOOL, IND UNIV, BLOOMINGTON, 72- Mem: AAAS; Am Soc Zool; Soc Exp Biol & Med. Res: Nervous system; degeneration of nerves and sense organs; embryological sense organs; embryology of urogenital system. Mailing Add: 421 Clover Lane Bloomington IN 47401

TORRIANI GORINI, ANNAMARIA, b Milan, Italy, Dec 19, 18; nat US; m 60; c 1. BACTERIAL PHYSIOLOGY. Educ: Univ Milan, PhD(natural sci), 42. Prof Exp: Res asst physiol & bact, Pasteur Inst, Univ Paris, 48-55; Fulbright fel microbiol, NY Univ, 55-58; res assoc, Biol labs, Harvard Univ, 58-59; from res assoc to asst prof, 59-69, ASSOC PROF BIOL, MASS INST TECHNOL, 69- Concurrent Pos: NIH res career award, 63- Mem: Am Soc Biol Chemists; Am Soc Microbiol. Res: Control of protein synthesis; bacterial genetics; bacterial spores germination. Mailing Add: Dept of Biol Mass Inst of Technol Cambridge MA 02139

TORRIE, BRUCE HAROLD, b Toronto, Ont, Mar 24, 37; m 60; c 2. SOLID STATE PHYSICS. Educ: Univ Toronto, BASc, 59; McMaster Univ, PhD(physics), 63. Prof Exp: Res fel, Atomic Energy Res Estab, Harwell, Eng, 62-65; asst prof, 65-70, ASSOC PROF PHYSICS, UNIV WATERLOO, 70- Mem: Can Asn Physicists. Res: Critical and magnetic scattering of neutron and laser beams. Mailing Add: Dept of Physics Univ of Waterloo Waterloo ON Can

TORTORELLO, ANTHONY JOSEPH, b Chicago, Ill, Sept 26, 45; m 71; c 1. SYNTHTIC ORGNAIC CHEMISTRY, ORGANIC POLYMER CHEMISTRY. Educ: St Joseph's Col, BS, 67; Loyola Univ, Chicago, MS, 70, PhD(chem), 75. Prof Exp: RES SCIENTIST CHEM, AM CAN CO, BARRINGTON, 74- Mem: Am Chem Soc; Sigma Xi; AAAS. Res: Monomer synthesis for photopolymerizable coatings; polymer synthesis for metal coatings; synthesis of photosensitive polymerization initiators. Mailing Add: 449 E Ct Elmhurst IL 60126

TORY, ELMER MELVIN, b Vermilion, Alta, Dec 10, 28; m 56; c 2. APPLIED MATHEMATICS, CHEMICAL ENGINEERING. Educ: Univ Alta, BSc, 52; Purdue Univ, PhD(chem eng), 61. Prof Exp: Res chemist, Aluminium Labs Ltd, 54-58; asst prof chem eng, McMaster Univ, 60-63; assoc chem engr, Brookhaven Nat Lab, 63-65; assoc prof, 65-73, PROF MATH, MT ALLISON UNIV, 73- Mem: Can Math Cong; Can Soc Chem Eng; Can Soc Mech Eng; Can Soc Hist Philos Math; AAAS. Res: Theoretical and experimental studies of settling of slurries; computer simulation of random packing of spheres. Mailing Add: Dept of Math Mt Allison Univ Sackville NB Can

TORZA, SERGIO, b Bergamo, Italy, May 26, 39; m 70; c 1. SURFACE CHEMISTRY, HYDRODYNAMICS. Educ: Polytech Inst Milan, BChE & MChE, 65; McGill Univ, PhD(phys chem), 70. Prof Exp: Res scientist, Union Camp Corp, 70-72; MEM TECH STAFF SURFACE CHEM, BELL LABS, 72- Mem: Am Phys Soc; Am Chem Soc; AAAS. Res: Newtonian, non-Newtonian and anisotropic fluids in motion; suspensions and emulsions; rheocapillarity. Mailing Add: 1A-348 Bell Labs Murray Hill NJ 07974

TOSCH, WILLIAM CONRAD, physical chemistry, see 12th edition

TOSH, FRED EUGENE, b Bemis, Tenn, Feb 13, 30; m 55; c 3. MEDICINE, EPIDEMIOLOGY. Educ: Univ Tenn, MD, 54; Univ Calif, MPH, 63. Prof Exp: Intern, Baptist Hosp, Memphis, Tenn, 54-55; med epidemiologist, Commun Dis Ctr, USPHS, 55-57; pvt pract, Tenn, 57-58; med epidemiologist, Kansas City Field Sta, Ctr Dis Control, USPHS, 58-64, chief pulmonary mycoses unit, 64-66, dep dir ecol invests prog, 67-73, DIR DIV QUAL & STAND, USPHS REGIONAL OFF, COLO, 73- Concurrent Pos: Resident, Mo State Sanitorium, 60-61; instr med, Univ Kans, 64-70, asst clin prof, 70-73. Mem: AMA; Am Thoracic Soc; Am Epidemiol Soc. Res: Epidemiology and research of the pulmonary mycoses. Mailing Add: USPHS 11037 Fed Off Bldg 19th & Stout Sts Denver CO 80202

TOSHACH, SHEILA, b Drumheller, Alta, June 2, 21. MEDICAL BACTERIOLOGY. Educ: Univ Alta, BSc, 43; Univ Toronto, MA, 50, dipl, 61. Prof Exp: Tech bact, 43-48, sr tech, 50-55, asst bacteriologist, 56-66, SR ASST BACTERIOLOGIST, PROV LAB PUB HEALTH, ALTA, 67-; ASSOC PROF BACT, UNIV ALTA, 67- Mem: Am Pub Health Asn; Can Soc Microbiol; Can Pub Health Asn. Res: Bacteriophage; brucellosis; Neisseria; fluorescent antibodies in bacteriological diagnosis. Mailing Add: Dept of Bact Univ of Alta Edmonton AB Can

TOSI, JOSEPH ANDREW, JR, b Worcester, Mass, July 1, 21; m 48; c 3. ECOLOGY. Educ: Mass State Col, BS, 43; Yale Univ, MF, 48; Clark Univ, PhD(geog), 59. Prof Exp: Forester & ecologist, Northern Zone, Inter-Am Inst Agr Sci, 51-52, Andean Zone, 52-63; resident staff geogr, Cent Am Field Prog, Assoc Cols Midwest, 64-67; LAND-USE ECOLOGIST & ADMINR, TROP SCI CTR, 67- Concurrent Pos: Consult, Forest Surv, Venezuela, 55; consult ecol surv, Colombia, 59-60; consult to comn on environ landscape planning, Int Union for Conserv Nature & Natural Resources, 73-; mem Inst Ecol. Honors & Awards: Order of Agr Merit, Grade of Comdr, Govt of Peru, 74. Mem: AAAS; Soc Am Foresters; Asn Am Geog; Sigma Xi. Res: Tropical ecology; bioclimatology; land utilization, especially tropical rural areas; biogeography; economic geography; life zone ecological theory; economic botany. Mailing Add: Trop Sci Ctr Apt 83870 San Jose Costa Rica

TOSI, OSCAR I, b Buenos Aires, Arg, June 17, 23; US citizen. AUDIOLOGY, ACOUSTICS. Educ: Univ Buenos Aires, ScD; Ohio State Univ, PhD, 65. Prof Exp: Assoc prof physics, Univ Buenos Aires, 51-62; res assoc voice commun, Ohio State Univ, 63-65; from asst prof to assoc prof, 65-70, PROF AUDIOL & ACOUST PHONETICS, MICH STATE UNIV, 70- Concurrent Pos: Dept Justice-Mich State Police grant, 68-71; expert witness on voice identification, Fed & State Courts, US & Can, 68-; vpres, Int Asn Voice Identification, Inc, 72; elected staff mem voice commun tech comt, Acoust Soc Am, 74-77. Mem: Acoust Soc Am; Am Asn Physics Teachers; Am Speech & Hearing Asn; Int Asn Logopedics & Phoniatrics; Int Col Exp Phonology. Res: Voice spectrography and identification; low levels of human acoustical energy; voice identification; articulatory pauses. Mailing Add: 247 Auditorium Mich State Univ East Lansing MI 48824

TOSKEY, BURNETT ROLAND, b Seattle, Wash, May 27, 29. MATHEMATICS. Educ: Univ Wash, BS, 52, MA, 58, PhD(algebra), 59. Prof Exp: From instr to assoc prof, 58-69, PROF MATH, SEATTLE UNIV, 69- Mem: Math Asn Am. Res: Abelian groups; ring theory; homological algebra; additive groups of rings. Mailing Add: Dept of Math Seattle Univ Seattle WA 98122

TOSONI, ANTHONY LOUIS, b Italy, Apr 15, 20; Can citizen; m 47; c 7. CHEMISTRY. Educ: Univ Toronto, BA, 42, MA, 44, PhD(antibiotics), 47. Prof Exp: Res mem, 44-72, asst dir, 72-73, DIR, PLASMA PROD DIV, CONNAUGHT LABS, 73- Mem: Am Chem Soc; Chem Inst Can. Res: Antibiotics; fermentation; enzymes; blood products. Mailing Add: Connaught Labs Ltd 1755 Steeles Ave W Willowdale ON Can

TOSSELL, WILLIAM ELWOOD, b Can, Jan 3, 26; m 47; c 3. AGRONOMY, CROP BREEDING. Educ: Univ Toronto, MSA, 48; Univ Wis, PhD(agron), 53. Prof Exp: Lectr field husb, Ont Agr Col, Univ Guelph, 48-50, from asst prof to prof, 50-61, head dept crop sci, univ, 61-66, assoc dean, 66-70, DEAN RES, UNIV GUELPH, 70- Mem: AAAS; Am Soc Agron; Crop Sci Soc Am; Can Soc Agron (pres, 67-68); Soc Res Admin. Res: Forage breeding; genetics; forage crop production. Mailing Add: 76 Callander Dr Guelph ON Can

TOSTESON, DANIEL CHARLES, b Milwaukee, Wis, Feb 5, 25; m 49, 69; c 6. PHYSIOLOGY, BIOPHYSICS. Educ: Harvard Univ, MD, 49. Prof Exp: Intern med, Col Physicians & Surgeons, Columbia Univ, 49-51; intern, Dept Med, Brookhaven Nat Labs, 51-53; mem, Lab Kidney & Electrolyte Metab, Nat Heart Inst, 53-55 & 57-58; from assoc prof to prof physiol, Sch Med, Washington Univ, 58-71; James B Duke Distinguished Prof physiol & pharmacol, Sch Med, Duke Univ, 71-75, chmn dept, 71-75; LOWELL T COGGESHALL PROF MED SCI, SCH MED, UNIV CHICAGO, 75-, DEAN, DIV BIOL SCI & PRITZKER SCH MED, 75- Concurrent Pos: NSF fel, Zoophysiol Univ Copenhagen, 55-56; NSF fel, Physiol Lab, Cambridge Univ, 56-57. Mem: AAAS; Nat Inst Med; Am Physiol Soc; Soc Gen Physiologists; Biophys Soc. Res: Membrane physiology. Mailing Add: Div of Biol Sci Univ of Chicago Sch of Med Chicago IL 60637

TOSTEVIN, JAMES EARLE, b Mandan, NDak, June 28, 38; m 65; c 1. PAPER CHEMISTRY. Educ: Carleton Col, BA, 60; Inst Paper Chem, MS, 62, PhD(paper chem), 66. Prof Exp: Group leader anal, Columbia Cellulose Co, Ltd, 66-69; RES CHEMIST, OLYMPIC RES DIV, ITT RAYONIER INC, 69- Mem: Am Chem Soc; Tech Asn Pulp & Paper Indust. Res: Research into application and properties of natural cellulose fibers. Mailing Add: 1043 Connection St Shelton WA 98584

TOTEL, GREGORY LEE, b Ottawa, Ill, Apr 27, 45; m 68. PHYSIOLOGY. Educ: Luther Col, BA, 67; Univ Ill, MS, 69, PhD(physiol), 70; St Louis Univ, MD, 75. Prof Exp: Asst physiol, Univ Ill, 67-69; INSTR PHYSIOL, ST LOUIS UNIV, 71- Mem: AMA. Res: Human environmental physiology; renal hypertension. Mailing Add: Dept of Physiol St Louis Univ Sch of Med St Louis MO 63103

TOTH, BELA, b Pecs, Hungary, Oct 26, 31; US citizen; m 63; c 4. PATHOLOGY, ONCOLOGY. Educ: Univ Vet Sci, Budapest, DVM, 56. Prof Exp: From res asst to res assoc oncol, Chicago Med Sch, 59-63, asst prof, 63-66; fel exp biol, Weizmann Inst, 66-67; assoc prof path, 68-72, PROF PATH, EPPLEY INST RES CANCER, COL MED, UNIV NEBR, OMAHA, 72- Concurrent Pos: USPHS trainee path, 61-63, res career develop award, 69; Eleanor Roosevelt int cancer res fel, 67-68. Mem: AAAS; Am Asn Cancer Res; Am Soc Exp Path; NY Acad Sci; Am Asn Path & Bact. Res: Experimental oncology; chemical carcinogenesis; leukemogenesis. Mailing Add: Eppley Inst for Res in Cancer Univ of Nebr Col of Med Omaha NE 68105

TOTH, JOZSEF, b Bekes, Hungary, June 22, 33; Can citizen; m 56; 56; c 2. GEOPHYSICS. Educ: Univ Utrecht, BSc, 58, MSc, 60, PhD(hydrogeol), 65. Prof Exp: From jr res officer to assoc res officer hydrogeol, 60-68, HEAD GROUND WATER DIV HYDROGEOL, RES COUN ALTA, 68- Concurrent Pos: Mem subcomt hydrol, assoc comt geod & geophys, Nat Res Coun Can, 63-68; lectr, Univ Alta, 66-71. Honors & Awards: O E Meinzer Award, Geol Soc Am, 65. Mem: Geol Soc Am. Res: Hydrogeology; theoretical and practical investigations of the factors controlling natural movement of groundwater as applied to problems of geology, regional water balance and local water supplies. Mailing Add: Dept of Geol Res Coun of Alta 87th Ave & 114th St Edmonton AB Can

TOTH, KENNETH STEPHEN, b Shanghai, China, Mar 17, 34; m 56; c 2. NUCLEAR PHYSICS. Educ: San Diego State Col, AB, 54; Univ Calif, PhD(chem), 58. Prof Exp: Asst chem, Univ California, 54-55, asst, Lawrence Radiation Lab, 55-58; Fulbright fel, Inst Theoret Physics, Denmark, 58-59; NUCLEAR CHEMIST, OAK RIDGE NAT LAB, 59- Concurrent Pos: Guggenheim fel, Niels Bohr Inst, Copenhagen, Denmark, 65-66; exchange physicist joint inst for nuclear res, Dubna, USSR, Nat Acad Sci, 75. Mem: Am Chem Soc; Am Phys Soc. Res: Nuclear properties of radioactive isotopes in rare earth region; low energy; heavy-ion nuclear reactions. Mailing Add: Physics Div Holifield Nat Lab Oak Ridge TN 37830

TOTH, LOUIS ANDREW, b South Bend, Ind, Aug 23, 09; m 38; c 2. PHYSIOLOGY. Educ: Univ Ky, AB, 31, MS, 32; Univ Rochester, PhD(physiol), 36. Prof Exp: Asst physiol, Univ Ky, 31-32; from instr to asst prof, Sch Med, Tulane Univ, 36-42; from asst prof to assoc prof, 46-53, PROF PHYSIOL, SCH DENT, LA STATE UNIV, NEW ORLEANS, 53-, DIR INTRAMURAL ATHLETICS, 71- Mem: Am Physiol Soc; Soc Exp Biol & Med. Res: Kidney function; anoxia; posture and circulation; ureteral peristalis. Mailing Add: Dept of Physiol La State Univ Med Ctr New Orleans LA 70119

TOTH, LOUIS MCKENNA, physical chemistry, see 12th edition

TOTH, ROBERT ALLEN, b Richmond, Ind, Aug 10, 39; m 59; c 3. PHYSICS. Educ: Earlham Col, AB, 62; Fla State Univ, MS, 66, PhD(physics), 69. Prof Exp: Physicist, Infrared Spectros, Nat Bur Standards, 62-66; instr Earlham Col, 66-67; assoc, Fla State Univ, 69-70; SR SCIENTIST, INFRARED SPECTROS & REMOTE SENSING, JET PROPULSION LAB, 70- Mem: Optical Soc Am. Res: Infrared spectroscopy; high resolution, its application to laboratory and theoretical data and to remote sensing of the atmosphere. Mailing Add: Planetary Atmospheres Sect Jet Propulsion Lab Pasadena CA 91103

TOTH, ROBERT S, b Detroit, Mich, Sept 4, 31; m 53; c 3. SOLID STATE PHYSICS. Educ: Wayne State Univ, AB, 54, MS, 55, PhD(physics), 60. Prof Exp: Res assoc physics, Wayne State Univ, 55-60; sr scientist, sci lab, Ford Motor Co, 60-69; VPRES, SENSORS, INC, 69- Mem: Am Phys Soc. Res: Metal oxide semi-conductors; crystal structure theory of alloy phases; magnetic structure of metals and alloys; thin film physics; epitaxy; thermoelectricity; infrared physics. Mailing Add: Sensors Inc 3908 Varsity Dr Ann Arbor MI 48104

TOTH, STEPHEN JOHN, b Elizabeth, NJ, Feb 19, 12; m 46; c 1. SOIL CHEMISTRY. Educ: Rutgers Univ, BS, 33, MS, 35, PhD(soil chem), 37. Prof Exp: Specialist forest soils, 37-39, instr agr chem, 39-42, from asst prof to assoc prof soils, 46-56, PROF SOILS, RUTGERS UNIV, NEW BRUNSWICK, 56-, ASSOC RES SPECIALIST, 47-, ASST SOIL CHEMIST, NJ AGR EXP STA, 39- Mem: Fel AAAS; Am Chem Soc; Soil Sci Soc Am; fel Am Inst Chem; fel Am Geog Soc. Res: Soil chemistry; colloids; nutrition; radioisotopes; fertilizers; water quality; bottom sediments; wildlife crops; composts. Mailing Add: Dept of Soils & Crops Rutgers Univ New Brunswick NJ 08903

TOTH, WILLIAM JAMES, b Carteret, NJ, Jan 20, 36; m 67; c 3. POLYMER CHEMISTRY. Educ: Rutgers Univ, New Brunswick, BA, 68; Princeton Univ, MS, 71, PhD(chem), 72. Prof Exp: SR RES CHEMIST, MOBIL CHEM CO, MOBIL OIL CORP, 63- Mem: Am Chem Soc. Res: Chemical mechanical, dielectric, rheological and physical properties of new polymers; physical chemistry of liquid crystals; characterization of monomeric and polymeric liquid crystals. Mailing Add: 7 Wendy Way Milltown NJ 08850

TOTO, PATRICK D, b Niles, Ohio, Jan 6, 21; m 45; c 3. ORAL PATHOLOGY. Educ: Kent State Univ, BS, 48; Ohio State Univ, DDS, 48, MS, 50; Am Bd Oral Path, dipl, 48. Prof Exp: Asst prof, 50-53 & 55-57, clin dir, 55-57, assoc prof, dir res & coordr grad studies, 57-76, PROF ORAL PATH & ORAL DIAG & CHMN DEPT ORAL PATH, SCH DENT, LOYOLA UNIV CHICAGO, 71- Concurrent Pos: Consult, Vet Admin Hosps, Hines, Ill, 53 & Chicago, 61- Mem: AAAS; Am Soc Clin Path; Am Soc Zoologists; NY Acad Sci; Am Dent Asn. Res: Ultramicroscopic study of plasma cell differentiation; generation cycle of age changes in the oral epithelium; induction of oral cancer; differentiation of tissue from undifferentiated connective tissue cells; globulin production in the oral mucosa; pathogenesis of periodontitis. Mailing Add: Dept of Oral Path Loyola Univ of Chicago Sch Dent Maywood IL 60153

TOTON, EDWARD THOMAS, b Philadelphia, Pa, Dec 6, 42; m 70. ASTROPHYSICS. Educ: St Joseph's Col (Pa), BS, 64; Univ Md, College Park, PhD(physics), 69. Prof Exp: Air Force Off Sci Res fel, Inst Advan Study, 69-70; NSF fel, Inst Theoret Physics, Univ Vienna, 70-71; res assoc & assoc instr astrophys, Univ Utah, 71-72; RES ASSOC PHYSICS, UNIV PA, 72- Concurrent Pos: Vis asst prof physics, St Joseph's Col (Pa), 74-75; consult, Naval Res Lab, Washington, DC, 75- Mem: Am Phys Soc; AAAS. Res: Astrophysical studies related to structure of neutron stars, nature of radiation from galaxies, nature of universe at moment of creation; research in combustion physics, including flame propagation, ignition, quenching and noise generation. Mailing Add: Dept of Physics Univ of Pa Philadelphia PA 19104

TOTTEN, DON EDWARD, b US citizen. CULTURAL GEOGRAPHY. Educ: Univ Chicago, MA, 50; Univ Heidelberg, PhD(soc geog, geol), 57. Prof Exp: Lectr geog, Europ Div, Univ Md, 50-55, asst dir, 55-65; PROF GEOG, CLARION STATE COL, 65- Concurrent Pos: Guest lectr, Univ Heidelberg, 60-65. Mem: Asn Am Geog; Nat Coun Geog Educ; Asn Can Studies US. Res: Environmental education. Mailing Add: Clarion State Col Clarion PA 16214

TOTTEN, JAMES EDWARD, b Saskatoon, Sask, Aug 9, 47; m 68; c 1. GEOMETRY. Educ: Univ Regina, BA, 67; Univ Waterloo, MMath, 69, PhD(geom), 74. Prof Exp: Nat Res Coun Can fel geom, Univ Math Inst, Tübingen, WGer, 74-76; ASST PROF MATH, ST MARY'S UNIV, NS, 76- Res: Linear spaces, a set of elements called points and distinguished subsets of points called lines, such that two points determine a unique line and all lines have at least two points. Mailing Add: Dept of Math St Mary's Univ Halifax NS Can

TOTTEN, ROBERT STOY, pathology, deceased

TOTTEN, ROGER EARL, genetics, biochemistry, see 12th edition

TOTTEN, STANLEY MARTIN, b Lodi, Ohio, July 15, 36; m 58; c 5. GEOLOGY. Educ: Col Wooster, BA, 58; Univ Ill, MS, 60, PhD(geol), 62. Prof Exp: From asst prof to assoc prof, 62-71, PROF GEOL, HANOVER COL, 71- Concurrent Pos: NSF fel, Univ Birmingham, 68-69. Mem: Fel Geol Soc Am; Soc Econ Paleont & Mineral; fel Geol Soc London; Glaciol Soc; Am Quaternary Asn. Res: Glacial geology; Pleistocene and Paleozoic stratigraphy; sedimentary petrology. Mailing Add: Dept of Geol Hanover Col Hanover IN 47243

TOTTER, JOHN RANDOLPH, b Saragosa, Tex, Jan 7, 14; m 38; c 3. BIOCHEMISTRY. Educ: Univ Wyo, AB, 34, AM, 35; Univ Iowa, PhD(biochem), 38. Prof Exp: Instr chem, Univ Wyo, 35-36; asst biochem, Univ Iowa, 36-38; asst Univ WVa, 38-39; asst, Sch Med, Univ Ark, 39-42, from asst prof to assoc prof, 42-52; biochemist, Oak Ridge Nat Lab, 52-56; biochemist, USAEC, 56-58; biochemist, Univ of the Repub, Uruguay, 58-60; prof chem & chmn div biol sci, Univ Ga, 60-62; assoc dir res, Div Biol & Med, USAEC, 63-67, dir, 67-72; assoc dir biomed & environ sci, 72-74, BIOCHEMIST, OAK RIDGE NAT LAB, 74- Concurrent Pos: Nutrit biochemist, Univ Alaska, 47; prof biochem, Univ Tenn, 75- Mem: Am Chem Soc; Soc Exp Biol & Med; Am Soc Biol Chem; Am Soc Nat; Soc Nuclear Med. Res: Amino acid and formate metabolism; synthesis and metabolism of pterins; radiation effects; luminescence. Mailing Add: 109 Wedgewood Dr Oak Ridge TN 37830

TOTTON, EZRA LESTER, b Sedalia, NC, Nov 5, 08; m 35. ORGANIC CHEMISTRY, BIOCHEMISTRY. Educ: Knoxville Col, BS, 35; Univ Iowa, MS, 43; Univ Wis, PhD(biochem), 48. Prof Exp: Chmn dept, 49-69, PROF CHEM, NC CENT UNIV, 49- Concurrent Pos: NSF fel, Stanford Univ, 59-60. Mem: Am Chem Soc; Nat Inst Sci; Am Soc Biol Chem. Res: Synthesis of phosphorylated sugars; acyloin condensation with unsaturated esters; chemistry of D-glucarid acid; synthesis of physiologically active compounds. Mailing Add: Dept of Chem NC Cent Univ Durham NC 27707

TOTUSEK, ROBERT, b Garber, Okla, Nov 3, 26; m 47; c 3. ANIMAL NUTRITION. Educ: Okla Agr & Mech Col, BS, 49; Purdue Univ, MS, 50, PhD(animal nutrit), 52. Prof Exp: Asst, Purdue Univ, 49-50, instr, 50-52; from asst prof to assoc prof, 52-60, PROF ANIMAL HUSB, OKLA STATE UNIV, 60- Mem: Am Soc Animal Sci. Res: Range cow nutrition and management. Mailing Add: Dept of Animal Sci & Indust Okla State Univ Stillwater OK 74074

TOU, JAMES CHIEH, b Su-yang, China, Apr 25, 36; US citizen; m 64; c 3. ANALYTICAL CHEMISTRY, PHYSICAL CHEMISTRY. Educ: Taiwan Norm Univ, BSc, 61; Univ Utah, PhD(chem), 66. Prof Exp: Teaching asst chem, Taiwan Norm Univ, 60-61; res asst, Univ Utah, 62-65; from res chemist to sr res chemist, Chem Physics Res Lab, 65-75, SR ANAL SPECIALIST CHEM, ANAL LAB, DOW CHEM CO, 75- Honors & Awards: V A Stenger Anal Sci Award, Dow Chem Co, 75. Mem: Am Chem Soc. Res: Organic mass spectrometry; chemical ionization; electron impact and field ionization; gas-chromatography-mass spectrometry; chemical property and analysis of bis-chloromethyl ether and chloromethyl methyl ether. Mailing Add: Anal Labs B-574 Mich Div Dow Chem Co Midland MI 48640

TOUBA, ALI R, b Tabriz, Iran, Apr 25, 25; m 57; c 4. FOOD TECHNOLOGY. Educ: Rutgers Univ, BSc, 51, MSc, 52; Univ Ill, PhD(food technol), 56. Prof Exp: Asst food microbiol, Univ Ill, 53-56; assoc technologist food res, Res Ctr, Gen Foods Corp, 56-63; proj mgr food res, Tronchemics Res, Inc, 63-65; res assoc explor food res, 65-70, HEAD EXPLOR FOOD RES, GEN MILLS, INC, 70- Concurrent Pos: Tech consult, Teheran, Iran, 60-63. Mem: Am Chem Soc; Inst Food Technologists; Am Asn Cereal Chemists; Am Soc Microbiol. Res: Food texture; fabricated foods; gums; space foods; freeze drying beverages; flavors; cereals and snacks; desserts; dehydrated products. Mailing Add: 4609 Island View Dr Mound MN 55364

TOUBASSI, ELIAS HANNA, b Jaffa, Israel, May 28, 43; US citizen; m 67; c 2. MATHEMATICS. Educ: Bethel Col (Kans), AB, 66; Lehigh Univ, MS, 69, PhD(math), 70. Prof Exp: Sr tech aide prog design, Bell Tel Labs, 66-67; res assoc, 70-71, asst prof, 70-75, ASSOC PROF MATH, UNIV ARIZ, 75- Mem: Am Math Soc; Math Asn Am. Res: Algebra, specifically infinite abelian groups. Mailing Add: Dept of Math Univ of Ariz Tucson AZ 85721

TOUCHBERRY, ROBERT WALTON, b Manning, SC, Oct 27, 21; m 48; c 4. ANIMAL BREEDING. Educ: Clemson Col, BS, 45; Iowa State Col, MS, 47, PhD(animal breeding, genetics), 48. Prof Exp: Asst dairy sci, Univ Ill, Urbana, 48-49, asst prof dairy cattle genetics, 49-55, assoc prof genetics in dairy sci, 55-59, prof, 59-70; PROF ANIMAL SCI & HEAD DEPT, UNIV MINN, ST PAUL, 70- Concurrent Pos: Fulbright res fel, Denmark, 56-57; geneticist, Div Biol & Med, US AEC, 67-68. Honors & Awards: Animal Breeding & Genetics Award, Am Soc Animal Sci, 71. Mem: AAAS; Am Soc Human Genetics; Genetics Soc Am; Am Diary Sci Asn; Am Genetic Asn. Res: Population genetics; quantitative genetics; effects of crossbreeding on the growth and milk production of dairy cattle; effects of x-irradiation on quantitative traits of mice and fruit flies; statistical studies of animal records. Mailing Add: Dept of Animal Sci Univ of Minn St Paul MN 55101

TOUCHBURN, SHERMAN PAUL, poultry nutrition, see 12th edition

TOUCHETTE, NORMAN WALTER, b East St Louis, Ill, Jan 2, 25; m 48; PLASTICS CHEMISTRY. Educ: Ill Col, AB, 48; Inst Textile Technol, MS, 50. Prof Exp: From chemist to chief chemist, Fulton Bag & Cotton Mills, Ga, 50-56; res chemist, 56-63, RES GROUP LEADER, MONSANTO CO, 63- Mem: Am Chem Soc; Soc Plastics Eng. Res: Plasticizer application and polymer modification. Mailing Add: Monsanto Co 800 N Lindbergh Blvd St Louis MO 63166

TOUCHSTONE, JOSEPH CARY, b Soochow, China, Nov 27, 21; US citizen; m 55; c 3. ORGANIC CHEMISTRY, BIOCHEMISTRY. Educ: Stephen F Austin State Teachers Col, BS, 43; Purdue Univ, MS, 46, PhD(biochem), 53. Prof Exp: Asst, Purdue Univ, 43-45; res assoc, Univ Tex, Southwestern Med Sch, 46-49; res assoc med, 52-56, assoc, Pepper Lab Clin Chem, Univ Hosp, 52-56, asst res prof obstet & gynec & res assoc, Harrison Dept Surg Res, Univ, 56-63, res assoc prof, 63-67, assoc prof res surg, 63-68, RES PROF OBSTET & GYNEC, SCH MED, UNIV PA, 67-, DIR STEROID LAB & PROF RES SURG, 68- Concurrent Pos: NIH res career award, 61-71; pres & co-founder, Chromatog Forum, 66-67, pres, 71-72. Mem: Am Chem Soc; Endocrine Soc; Am Soc Biol Chemists; Am Asn Clin Chemists; Am Acad Forensic Sci. Res: Steroid chemistry; organic synthesis; isolation and metabolism of steroid hormones; chromatography; phenol chemistry; adrenal physiology; gas chromatography of steroids; thin layer and liquid chromatography. Mailing Add: Univ of Pa Hosp Philadelphia PA 19104

TOUGER, JEROLD STEVEN, b Brooklyn, NY, Aug 6, 45; m 69; c 1. EXPERIMENTAL SOLID STATE PHYSICS. Educ: Cornell Univ, BA, 66; City Univ New York, PhD(physics), 74. Prof Exp: ASST PROF PHYSICS, CURRY COL, 74- Res: Thermoelectric power; transport properties in magnetic alloys; structural linguistics applied to scientific and mathematical discourses; teaching methods in physics and calculus. Mailing Add: Div of Sci & Math Curry Col Milton MA 02186

TOUGH, JAMES THOMAS, b Chicago, Ill, May 4, 38; m 60; c 2. LOW TEMPERATURE PHYSICS. Educ: Univ Ill, BS, 60; Univ Wash, PhD(liquid helium), 64. Prof Exp: Res assoc low temperature physics, 64-65, asst prof, 65-68, ASSOC PROF PHYSICS, OHIO STATE UNIV, 68- Mem: Am Phys Soc; Fedn Am Scientists. Res: Hydrodynamics of liquid helium II; properties of liquid helium three-helium four solutions. Mailing Add: 174 W 18th Ave Columbus OH 43210

TOULMIN, LYMAN DORGAN, JR, b Mobile, Ala, July 4, 04; m 45. PALEONTOLOGY, GEOLOGY. Educ: Univ Ala, AB, 26, MA, 34; Princeton Univ,

PhD(geol), 40. Prof Exp: Asst, Univ Ala, 25-27; teacher schs, Ala, 29-34; teacher, Univ Mil Sch, 34-35; asst, Princeton Univ, 35-38; instr geol, Agr & Mech Col, Tex, 38-41, asst prof, 41-42; staff geologist, State Geol Surv, Ala, 42-45; assoc prof geol, Birmingham-Southern Col, 45-48; prof, 48-75, EMER PROF GEOL, FLA STATE UNIV, 75- Concurrent Pos: Vis prof, La State Univ, 53-54. Mem: Paleont Soc; Geol Soc Am; Soc Econ Paleont & Mineral; Am Asn Petrol Geol. Res: Micropaleontology; paleontology and str.atigraphy of the Coastal Plain; geology and structure of the Jackson fault. Mailing Add: Dept of Geol Fla State Univ Tallahassee FL 32306

TOULMIN, PRIESTLEY, III, b Birmingham, Ala, June 5, 30; m 52; c 2. GEOLOGY. Educ: Harvard Univ, AB, 51, PhD(geol), 59; Univ Colo, MS, 53. Prof Exp: Geologist, 53-56, chief br exp geochem & mineral, 66-72, GEOLOGIST, US GEOL SURV, 58- Concurrent Pos: Lectr vis geol scientist prog, Am Geol Inst, 64; adj assoc prof, Columbia Univ, 66; scientist, NASA Proj Viking, 68-, team leader, inorg chem invest, 72-;ed J Translations, Am Geochem Soc, 65-68; assoc ed Am Mineralogist, J Mineral Soc Am, 74-76. Mem: AAAS; fel Mineral Soc Am; Geochem Soc; fel Geol Soc Am; Geochem Soc. Res: Igneous and sulfide petrology; phase equilibria and thermochemistry of ore minerals; mineralogy and geochemistry of Mars. Mailing Add: Geol Div US Geol Surv 959 Nat Ctr Reston VA 22092

TOUPIN, RICHARD A, b Miami, Fla, Aug 20, 26; m 50; c 3. MATHEMATICAL PHYSICS, CONTINUUM MECHANICS. Educ: Univ SC, BS, 46; Univ Hawaii, MS, 49; Syracuse Univ, PhD(physics), 61. Prof Exp: Instr physics, Univ Hawaii, 49-50; res asst theoret mech, US Naval Res Lab, 50-62; res asst appl math, res ctr, 62-74, DIR MATH SCI, IBM RES DIV, 74- Mem: Am Math Soc; Soc Natural Philos (secy, 65-67). Res: Elasticity and electromagnetic theories; relativity mechanics; dielectrics; differential geometry. Mailing Add: IBM Res Ctr Yorktown Heights NY 10598

TOUPS, POLLY ANTICICH, b New Orleans, La, May 15, 29; m 46; c 3. ANTHROPOLOGY, ARCHAEOLOGY. Educ: La State Univ, BA, 47; Tulane Univ La, PhD(anthrop), 70. Prof Exp: Asst prof anthrop, Fresno State Col, 65-66; asst prof, Southern Univ, New Orleans, 66-67; res asst, Idaho State Univ, 67-69; ASST PROF SOCIOL & ANTHROP, WESTERN KY UNIV, 69- Res: Plateau prehistory. Mailing Add: Dept of Sociol & Anthrop Western Ky Univ Bowling Green KY 42101

TOURGEE, RONALD ALAN, b Wakefield, RI, May 2, 38; c 3. MATHEMATICAL STATISTICS. Educ: Univ RI, BS, 60, MS, 62; Univ SFla, PhD(math), 75. Prof Exp: Teacher math, Keene State Col, 64-66 & Mt Holyoke Col, 66-68; TEACHER MATH, KEENE STATE COL, 68- Mem: Am Math Soc; Am Statist Asn. Res: Stochastic systems; mathematical statistics; applied probability. Mailing Add: Dept of Math Keene State Col Keene NH 03431

TOURIAN, ARA YERVANT, b Jerusalem, May 19, 33; US citizen; m 59; c 3. BIOCHEMICAL GENETICS. Educ: Am Univ Beirut, BS, 55; Iowa State Univ, MD, 58. Prof Exp: Intern med, Washington Hosp Ctr, DC, 58-59; resident neurol, NY Univ Med Ctr, 62-63, chief resident, 64-65; instr & fel biophys & neurol, Med Ctr, Univ Colo, 65-69; ASSOC PROF MED, MED CTR, DUKE UNIV, 69- Concurrent Pos: NIH res career develop award, Med Ctr, Duke Univ; vis scientist cell biol, Dept Zool, Cambridge Univ, 75-76. Mem: AAAS; Am Soc Neurochem; Am Acad Neurol; Cambridge Philos Soc. Res: Metabolic and genetic control mechanisms of phenylalanine hydorxylase; biochemical genetics utilizing tissue culture, cell and minisepregant chromosomal hybridization for the study of genetic complementation. Mailing Add: Duke Univ Med Ctr M3066 Durham NC 27710

TOURIGNY, GUY J, b Ponteix, Sask, Aug 22, 36; m 60; c 2. PHYSICAL ORGANIC CHEMISTRY. Educ: Univ Ottawa, BA, 55; Univ Alta, BSc, 62, PhD(chem), 67. Prof Exp: Res assoc, Univ Ill, Urbana, 67-68; asst prof, 68-74, ASSOC PROF CHEM, UNIV SASK, 74- Mem: Chem Inst Can. Res: Mechanism of reactions; solvolysis reactions; stereochemistry. Mailing Add: Dept of Chem & Chem Eng Univ of Sask Saskatoon SK Can

TOURIN, RICHARD HAROLD, b New York, NY, Dec 4, 22; m 48; c 2. ENERGY CONVERSION. Educ: City Col New York, BS, 47; NY Univ, MS, 48. Prof Exp: Res physicist, Control Instrument Div, Warner & Swasey Co, 48-51, chief physicist, 51-59, mgr res lab, 59-63, div mgr, 63-71; dir mkt, Klinger Sci Apparatus Corp, 71-73; DIR NEW PROG DEVELOP, NY STATE ENERGY RES & DEVELOP AUTHORITY, 73- Concurrent Pos: Adj instr, Cooper Union, 55-60; US mem joint comt, Int Flame Res Found, 66-68; mem energy res adv comt, Stevens Inst Technol, 75- Mem: Fel Optical Soc Am; Am Phys Soc; Combustion Inst. Res: Spectroscopic gas temperature measurement; optical physics; energy conversion; remote sensing. Mailing Add: NY State Energy R&D Authority 230 Park Ave New York NY 10017

TOURING, ROSCOE MANVILLE, b Winnipeg, Man, June 30, 24; m 52; c 4. GEOLOGY. Educ: Univ Man, BS, 45; Stanford Univ, MS, 51, PhD(geol), 59. Prof Exp: Geologist, Int Petrol Co, Ecuador, Colombia & Peru, 46-50; geologist, Humble Oil & Ref Co, Calif & Ore, 52-64, res assoc, Esso Prod Res Co, Tex, 64-66, explor mgr, Minerals Dept, 66-67, mgr stratig geol div, 67-69, mgr non-hydrocarbon minerals study group, Standard Oil Co (NJ), NY, 69-71, SR GEOL SCIENTIST, MINERALS DEPT, EXXON CO USA, 72- Mem: AAAs; Geol Soc Am; Am Asn Petrol Geol. Res: Adviser minerals exploration program. Mailing Add: Minerals Dept Exxon Co USA PO Box 2180 Houston TX 77001

TOURNEY, GARFIELD, b Quincy, Ill, Feb 6, 27; m 50; c 3. PSYCHIATRY. Educ: Univ Ill, BS, 46, MD, 48; State Univ Iowa, MS, 52. Prof Exp: Asst prof psychiat, Sch Med, Univ Miami, 54-55; from asst prof to prof, Sch Med, Wayne State Univ, 55-67; prof, Univ Iowa, 67-71; co-chmn dept, 71-73, PROF PSYCHIAT, SCH MED, WAYNE STATE UNIV, 71-, CHMN DEPT, 73- Concurrent Pos: Assoc examr, Am Bd Psychiat & Neurol, 67- Mem: Fel Am Psychiat Asn; AMA; Am Psychosom Soc; fel Am Col Psychiat. Res: Biochemical and clinical studies of schizophrenia and depressive illnesses. Mailing Add: Lafayette Clin 951 E Lafayette Ave Detroit MI 48207

TOURTELLOTTE, CHARLES DEE, b Kalamazoo, Mich, Aug 28, 31; m 55; c 4. INTERNAL MEDICINE, BIOCHEMISTRY. Educ: Johns Hopkins Univ, AB, 53; Temple Univ, MS & MD, 57; Am Bd Internal Med, dipl. Prof Exp: Intern med, Univ Mich, 57-58, resident & jr clin instr, 58-60; instr med & biochem, 63-65, from asst prof to assoc prof, 65-72, res asst prof biochem, 65-71, actg chief sect rheumatol, 66-67, PROF MED, SCH MED, TEMPLE UNIV, 72-, CHIEF SECT RHEUMATOL, SCH MED & UNIV HOSP, 67- Concurrent Pos: USPHS trainee rheumatol, Temple Univ, 60-61; Helen Hay Whitney Found fel biochem, Rockefeller Univ, 61-63; Arthritis Found fel, 63-66; mem, Gov Bd, Arthritis Found; consult, Vet Admin Hosp, Wilmington, Del, Episcopal Hosp, St Christopher's Hosp Children, Shriners Hosp Crippled Children & Children's Heart Hosp, Philadelphia. Mem: Fel Am Col Physicians; Am Rheumatism Asn; Am Fedn Clin Res. Res: Biochemistry and physiology of connective tissue; endochondral ossification; amino acid metabolism; histidine; heritable disorders of bone and connective tissues; rheumatic diseases; medical education. Mailing Add: Sect of Rheumatol Temple Univ Sch of Med Philadelphia PA 19140

TOURTELLOTTE, MARK ETON, b Worcester, Mass, Oct 25, 28; m 53; c 3. BIOCHEMISTRY. Educ: Dartmouth Col, BA, 50; Univ Conn, MS, 53, PhD(microbiol), 60. Prof Exp: From asst instr to instr bact, Univ Conn, 53-60; res assoc biophys, Yale Univ, 60-62; assoc prof, 63-67, PROF ANIMAL PATH, UNIV CONN, 67- Mem: AAAS; Am Soc Microbiol; Am Asn Avian Path; fel Am Inst Chem; NY Acad Sci. Res: Immunology; diagnostic bacteriology; lipids; chemistry and biosynthesis in mycoplasma; structure and function of biomembranes. Mailing Add: Dept of Pathbiol Univ of Conn Storrs CT 06268

TOURTELLOTTE, WALLACE WILLIAM, b Great Falls, Mont, Sept 13, 24; m 53; c 4. NEUROLOGY. Educ: Univ Chicago, PhB & BS, 45, PhD(biochem neuropharmacol), 48, MD, 51; Am Bd Psychiat & Neurol, dipl, 60. Prof Exp: Res assoc & instr pharmacol, Univ Chicago, 48-51; intern med, Sch Med, Univ Rochester, 51-52; resident neurol, Med Sch, Univ Mich, 54-57; from asst prof to prof, 57-71; PROF NEUROL & V CHMN DEPT, UNIV CALIF, LOS ANGELES, 71-; CHIEF NEUROL SERV, VET ADMIN WADSWORTH HOSP CTR, LOS ANGELES, 71- Concurrent Pos: Consult, Vet Admin Hosp, Ann Arbor, Mich, 58-71; chief neurol serv, Wayne County Gen Hosp, Detroit, 59-71; mem, Multiple Sclerosis Res Comt, Int Comn Correlation Neurol & Neurochem, World Fedn Neurol, 59-; vis assoc prof, Washington Univ, 63-64; asst examr, Am Bd Psychiat & Neurol, 64-; mem, Med Adv Bd, Nat Multiple Sclerosis Soc, 71-; exchange biomed investr, Vet Admin-Fr Nat Inst Health & Med Res, Paris, 72. Honors & Awards: Mitchell Award, Am Acad Neurol, 59- Mem: AAAS; Am Neurol Asn; Am Soc Pharmacol & Exp Therapeut; Asn Res Nerv & Ment Dis; Am Acad Neurol. Res: Organic neurology and neurochemical correlations in multiple sclerosis patients. Mailing Add: Neurol Serv Vet Admin Wadsworth Hosp Ctr Los Angeles CA 90073

TOURTELOT, HARRY ALLISON, b Lincoln, Nebr, June 15, 18; m 40, 65; c 6. GEOLOGY. Educ: Univ Nebr, AB, 40. Prof Exp: Proj technician, State Geol Surv, Ala, 40-42; GEOLOGIST, US GEOL SURV, 42- Mem: Geol Soc Am; Geochem Soc; Soc Econ Paleont & Mineral; Int Asn Sedimentol; Am Asn Petrol Geol. Res: Stratigraphy of continental tertiary rocks; geologic structure of Central Wyoming; geochemistry of sedimentary rocks; petrology of shale; environmental geochemistry. Mailing Add: US Geol Surv Fed Ctr Denver CO 80225

TOUSEY, RICHARD, b Somerville, Mass, May 18, 08; m 32; c 1. PHYSICS. Educ: Tufts Univ, AB, 28; Harvard Univ, AM, 29, PhD(physics), 33. Hon Degrees: ScD, Tufts Univ, 62. Prof Exp: Instr physics, Harvard Univ, 33-36, tutor, 34-36, Cutting fel, 35-36; res instr, Tufts Univ, 36-41; head instrument sect, 42-45, head micron waves br, 45-48, PHYSICIST, US NAVAL RES LAB, 41-, HEAD ROCKET SPECTROS BR, 58- Concurrent Pos: Darwin lectr, Royal Astron Soc, 63; Russell lectr, Am Astron Soc, 66; mem comt vision, Armed Forces-Nat Res Coun. Honors & Awards: Hulburt Award, 48; Medal, Photog Soc Am, 59; Ives Medal, Optical Soc Am, 60; Prix Ancel, Photog Soc France, 62; Draper Medal, Nat Acad Sci, 63; Distinguished Achievement Award, US Navy, 63; Eddington Medal, Royal Astron Soc, 64; Except Sci Achievement Medal, NASA, 74. Mem: Nat Acad Sci; fel Am Phys Soc; fel Optical Soc Am; Am Astron Soc (vpres,64-66); fel Am Acad Arts & Sci. Res: Optical properties of the atmospheres; spectroscopy from rockets; physiological optics; photographic photometry; vacuum ultraviolet. Mailing Add: Code 7140 US Naval Res Lab Washington DC 20375

TOUSIGNANT, WILLIAM FRANCIS, organic chemistry, see 12th edition

TOUSIGNAUT, DWIGHT R, b Ironwood, Mich, Dec 4, 33; m 64; c 3. PHARMACY. Educ: Univ Mich, BS, 59; Univ Calif, Pharm D, 61. Prof Exp: Residency hosp pharm, San Francisco Med Ctr, Univ Calif, 59-61; pharmacist, Queen Elizabeth Hosp & Royal Perth Hosp, Australia, 62-63; pharmacist, Stanford Med Ctr, 63-64; Fulbright prof hosp pharm, Cairo, 64-66; DIR DEPT PROF PRACT & ED INT, PHARMACEUT ABSTRACTS, AM SOC HOSP PHARMACISTS, 66- Concurrent Pos: Mem nomenclature adv comt, Nat Libr Med. Honors & Awards: Bristol Award, 59. Mem: Am Pharmaceut Asn; Am Soc Hosp Pharmacists; Drug Info Asn (vpres, 73-74). Res: Drug information processing; pharmacy education; drug absorption from implanted or injected routes of administration; griseofulvin solubility studies; plastic drug sorption studies. Mailing Add: Am Soc of Hosp Pharmacists 4630 Montgomery Ave Bethesda MD 20014

TOUSSIENG, POVL WINNING, b Nysted, Denmark, Sept 5, 18; US citizen. PSYCHIATRY. Educ: Copenhagen Univ, MD, 45. Prof Exp: Resident gen psychiat, Menninger Sch Psychiat, 50-53, John Harper Seeley fel child psychiat, Children's Div, Menninger Clin, 53-55, staff psychiatrist, 55-65; assoc prof child psychiat & pediat, 65-69, PROF CHILD PSYCHIAT, HEALTH SCI CTR, COL MED, UNIV OKLA, 69- Concurrent Pos: Consult, Kans Indust Sch Boys, 53-61; mem fac, Menninger Sch Psychiat, 53-65; consult, Kans Neurol Inst, 64-65, Minn Dept Ment Health, Minneapolis, 65-66 & Spec Subcomt Indian Educ, US Senate Comt Labor & Pub Welfare, 69; mem, Nat Drafting Comt Juv Studies Proj, 73-; mem bd, Psychiat Outpatients Ctr Am, 73- Mem: Fel Am Psychiat Asn; fel Am Orthopsychiat Asn; Soc Res Child Develop. Res: Childhood autism; coping devices of normal and disturbed children; various modalities of psychotherapy; adolescent experience in changing times; delinquency; adoption; delivery systems of help. Mailing Add: Dept of Psychiat Univ of Okla Col of Med Oklahoma City OK 73105

TOUSTER, OSCAR, b New York, NY, July 3, 21; m 44; c 1. MOLECULAR BIOLOGY, BIOCHEMISTRY. Educ: City Col New York, BS, 41; Oberlin Col, MA, 42; Univ Ill, Urbana, PhD(biochem), 47. Prof Exp: Chemist, Atlas Powder Co, 42-43; res biochemist, Abbott Labs, 44-45; from instr to assoc prof biochem, 47-58, PROF BIOCHEM, VANDERBILT UNIV, 58-, PROF MOLECULAR BIOL, 73-, CHMN DEPT MOLECULAR BIOL, 63- Concurrent Pos: Guggenheim fel, Oxford Univ, 57-58; H Hughes investr, Vanderbilt Univ & Oxford Univ, 57-60; consult, NIH, 61-70; mem, Subcomt Metab Intermediates, Nat Res Coun, 66-; mem, Bd Dirs, Oak Ridge Assoc Univs, 73-, vpres, 74-; mem, Sci Adv Bd, Eunice Kennedy Shriver Ctr Ment Retardation, Waltham, Mass, 74- Honors & Awards: Theobald Smith Award Med Sci, AAAS, 56. Mem: Fel AAAS; Am Soc Biol Chemists; Biochem Soc; Am Inst Biol Sci; Am Chem Soc. Res: Lysosome biochemistry; membrane enzymes; carbohydrate metabolism; beta-glucuronidase chemistry and action; liver glycosidases. Mailing Add: Dept of Molecular Biol Vanderbilt Univ Nashville TN 37235

TOVE, SAMUEL B, b Baltimore, Md, July 29, 21; m 45; c 3. BIOCHEMISTRY, NUTRITION. Educ: Cornell Univ, BS, 43; Univ Wis, MS, 48, PhD(biochem), 50. Prof Exp: Asst, Univ Wis, 46-50; from asst res prof to assoc res prof animal sci, 50-60, PROF BIOCHEM, NC STATE UNIV, 60-, HEAD DEPT, 75- Concurrent Pos: William Neal Reynolds prof biochem, NC State Univ, 75. Mem: AAAS; Am Chem Soc; Soc Exp Biol & Med; Am Inst Nutrit; Am Soc Biol Chemists. Res: Lipid and intermediary metabolism. Mailing Add: Dept of Biochem NC State Univ Raleigh NC 27607

TOVE, SHIRLEY RUTH, b New York, NY, Jan 31, 25; m 45; c 3. BACTERIOLOGY, BIOCHEMISTRY. Educ: Cornell Univ, BS, 45; Univ Wis, MS, 48, PhD(bact, biochem), 50. Prof Exp: Instr chem, NC State Univ, 50-51, bact, 51-52; vis teacher biol, NC Col Durham, 64-65; assoc prof, 65-72, chmn dept, 65-75, consult planning, Div Natural Sci & Math, 65, PROF BIOL, SHAW UNIV, 72- Mem: Am Soc Microbiol. Res: Biochemistry of nitrogen fixation. Mailing Add: Dept of Biol Shaw Univ Raleigh NC 27602

TOVELL, WALTER MASSEY, b Toronto, Ont, June 25, 16; m 72. GEOLOGY. Educ: Univ Toronto, BA, 40, PhD, 54; Calif Inst Technol, MS, 42. Prof Exp: Geologist, Calif Standard Co, 42-46; lectr geol, Univ Toronto, 49-50, assoc prof, Univ & Col Educ, 59-64; mus asst, 46-48, cur geol dept, 48-72, assoc dir, 71-73, dir pro tem, 72-73, DIR ROYAL ONT MUS, 73-; ASSOC PROF GEOL, UNIV TORONTO, 64-, ASSOC PROF GEOG, FAC EDUC, 66- Concurrent Pos: Asst prof geol, Univ Toronto, 62-64; mem & vchmn info & educ comt, Met Toronto & Region Conserv Authority, 68-74, chmn, 75- Mem: Fel Geol Asn Can (secy-treas, 60-62); Mus Dirs Asn Can. Res: Stratigraphy and Pleistocene geology; research on geology history of Great Lakes with special emphasis on Georgian Bay. Mailing Add: Royal Ont Mus 100 Queen's Park Toronto ON Can

TOVERUD, SVEIN UTHEIM, b Oslo, Norway, Dec 14, 29; m 54; c 3. PHARMACOLOGY, ENDOCRINOLOGY. Educ: Harvard Univ, DMD, 54; Norweg State Dent Sch, Cand Odont, 56; Univ Oslo, Dr Odont, 64. Prof Exp: Asst oper dent, Univ Oslo, 56, instr, 62-63, res assoc, 63-64, from asst prof to assoc prof, 65-70; ASSOC PROF PHARMACOL, SCH MED, UNIV NC, CHAPEL HILL, 69-, ASSOC PROF ORAL BIOL, SCH DENT, 69- Concurrent Pos: Res fel physiol & biochem, Univ Oslo, 56-62; USPHS int fel, Sch Dent Med, Harvard Univ, 64. Mem: AAAS; Int Asn Dent Res. Res: Hormonal regulation of calcium metabolism; influence of nutrition and drugs, especially vitamins and hormones on bones and teeth. Mailing Add: Dent Res Ctr Univ of NC Chapel Hill NC 27514

TOWBIN, EUGENE JONAS, b New York, NY, Sept 18, 18; m 49; c 4. INTERNAL MEDICINE, PHYSIOLOGY. Educ: NY Univ, BA, 41; Univ Colo, MS, 42; Univ Rochester, MD & PhD(physiol), 49. Prof Exp: Asst psychol, Univ Rochester, 42-44, asst physiol, 44-47, intern med, Duke Univ, 49-50, resident, 50-52, clin asst prof med, 55-56, from asst prof med to asst prof physiol, 56-65, from assoc prof med to assoc prof physiol, 62-69, PROF MED & PHYSIOL, SCH MED, UNIV ARK, LITTLE ROCK, 69-, ASSOC DEAN SCH MED, 68-; CHIEF OF STAFF, VET ADMIN HOSP, 68- Concurrent Pos: Fel cardiol, Duke Univ, 52; ward physician, Vet Admin Hosp, 55-58, exec secy & mem res comt, 56-58, asst dir prof serv for res & educ, 58-61, assoc chief of staff for res & educ, 61-72. Mem: Am Fedn Clin Res; Geront Soc; Am Col Physicians; Am Physiol Soc; Soc Exp Biol & Med. Res: Water and electrolyte metabolism; physiological regulation of thirst and hunger. Mailing Add: Vet Admin Hosp 300 E Roosevelt Rd Little Rock AR 72206

TOWE, ARNOLD LESTER, b Patterson, Calif, July 25, 27; wid. PHYSIOLOGY, BIOPHYSICS. Educ: Pac Lutheran Col, BA, 48; Univ Wash, PhD(psychol, physiol), 53. Prof Exp: Res assoc, 53-54, from instr to asst prof anat & physiol, 54-58, from asst prof to assoc prof physiol & biophys, 58-65, PROF PHYSIOL & BIOPHYS, SCH MED, UNIV WASH, 65- Concurrent Pos: Mem, NIH Study Sect, 66-70. Res: Neurophysiology, particularly analysis of sensory and motor systems, including gross potentials and single unit activity; cortical physiology. Mailing Add: Dept of Physiol & Biophys Univ of Wash Sch of Med Seattle WA 98195

TOWE, GEORGE COFFIN, b Passaic, NJ, Nov 28, 21; m 47; c 1. PHYSICS, SCIENCE EDUCATION. Educ: Hamilton Col (NY), BS, 43; Univ Mich, MS, 47, PhD(chem), 54. Prof Exp: Physicist, US Naval Ord Lab, 43-45; res assoc, Eng Res Inst, Univ Mich, 46-53; res engr, Sci Lab, Ford Motor Co, 53-55; from asst prof to assoc prof physics, Mont State Col, 55-61; prof physics, head dept & chmn div natural sci, Findlay Col, 61-62; assoc prof physics, 62-65, chmn dept, 65-72, PROF PHYSICS, ALFRED UNIV, 65-, CHMN DIV SPEC PROGS, 74- Concurrent Pos: Lectr, Univ Wyo, 59; vis scientist, Atomic Energy Res Estab, Eng, 67-68; consult, Oak Ridge Inst Nuclear Studies, 66- Mem: Fel Brit Inst Physics; Am Asn Physics Teachers. Res: Radioactivity; radiation; solid state diffusion; nuclear activation analysis. Mailing Add: Dept of Physics Box 832 Alfred Univ Alfred NY 14802

TOWE, KENNETH MCCARN, b Jacksonville, Fla, Jan 31, 35. GEOLOGY, ELECTRON MICROSCOPY. Educ: Duke Univ, AB, 56; Brown Univ, MSc, 59; Univ Ill, PhD(geol), 61. Prof Exp: Res assoc electron micros, Univ Ill, 61-62; Ford Found res fel geol, Calif Inst Technol, 62-64; GEOLOGIST-ELECTRON MICROSCOPIST, DEPT PALEOBIOL, SMITHSONIAN INST, 64- Concurrent Pos: Vis prof geol-Paleont Inst, Univ Tübingen, 73. Mem: AAAS; Geol Soc Am; Clay Minerals Soc; Mineral Soc Am. Res: Biomineralogy; clay mineralogy; application of electron microscopy to geology and paleontology. Mailing Add: Dept of Paleobiol Smithsonian Inst Washington DC 20560

TOWELL, DAVID GARRETT, b Fillmore, NY, May 30, 37; m 60; c 2. GEOCHEMISTRY. Educ: Pa State Univ, BS, 59; Mass Inst Technol, PhD(geochem), 63. Prof Exp: Res fel geochem, Calif Inst Technol, 63-64; asst prof, 64-68, ASSOC PROF GEOCHEM, IND UNIV BLOOMINGTON, 68- Mem: AAAS; Geochem Soc; Geol Soc Am; Mineral Soc Am. Res: General inorganic, rare-earth, trace element and isotope geochemistry; electron-probe microanalysis of minerals and rocks; chemical thermodynamics; phase equilibria; radiochemistry; radioactivation analysis; radiotracer studies. Mailing Add: Dept of Geol Ind Univ Bloomington IN 47401

TOWELL, EDWARD EMERSON, b Blakely, Ga, Aug 31, 13; m 37. CHEMISTRY. Educ: Col Charleston, BS, 34; Univ NC, PhD(org chem), 44. Prof Exp: Teacher high sch, SC, 36-41; lab asst org chem, Univ NC, 41-42; assoc prof, 43-50, dean col, 58-64 & 68-70, PROF CHEM, COL CHARLESTON, 50- Mem: Fel AAAS; Am Chem Soc. Mailing Add: Dept of Chem Charleston Col Charleston SC 29401

TOWER, DONALD BAYLEY, b Orange, NJ, Dec 11, 19; m 47; c 1. RESEARCH ADMINISTRATION. Educ: Harvard Univ, AB, 41, MD, 44; McGill Univ, MSc, 48, PhD(exp neurol), 51. Prof Exp: Intern surg, Univ Minn Hosps, 44-45; asst resident neurosurg, Montreal Neurol Inst, McGill Univ, 48-49, assoc neurochemist, 51-53, lectr exp neurol, Fac Med, Univ, 51-52, asst prof, 52-53; chief, Sect Clin Neurochem, 53-60, chief, Lab Neurochem, 61-73, DIR, NAT INST NEUROL & COMMUN DIS & STROKE, 73- Concurrent Pos: Res fel neurochem, Montreal Neurol Inst, McGill Univ, 47-51; Markle scholar med sci, 51-53; clin clerk, Nat Hosp, London, Eng, 51; assoc prof, Sch Med & consult, Georgetown Univ, 53-; mem, Neurol Study Sect, Div Res Grants, NIH, 54-61; mem, US Bd Civil Serv Exam, 61-67; chmn, Neurochem Deleg to USSR, US-USSR Exchange Prog Health & Med Sci, 69; mem, Basic Res Task Force, Adv Comt Epilepsies, USPHS, 69-73; chief ed, J Neurochem, 69-73; mem, Neurochem Panel, Int Brain Res Orgn, mem, Cent Coun, 74- Mem: Am Acad Neurol; Am Neurol Asn; Am Soc Biol Chemists; Int Soc Neurochem; Am Soc Neurochem (treas, 70-75). Res: Biochemistry of brain, nerve and muscle; epilepsy; metabolic studies of epileptic brain tissue; neural proteins; amino acids and electrolytes; history of neurochemistry. Mailing Add: Nat Inst Neurol & Commun Dis & Stroke Bethesda MD 20014

TOWERS, BARRY, b Toledo, Ohio, July 20, 38; m 63. FOREST PATHOLOGY, MYCOLOGY. Educ: Thiel Col, BA, 61; Duke Univ, MF, 61, DF(forest path), 65. Prof Exp: Res asst forest path, Duke Univ & Southern Forest Dis & Insect Res Coun, 62; fel phytotoxic air pollutants, Sch Pub Health, Univ NC, Chapel Hill, 65-68; FOREST PATHOLOGIST, DIV FOREST PEST MGT, PA DEPT ENVIRON RESOURCES, 68- Mem: AAAS; Am Phytopath Soc; Soc Am Foresters; Int Shade Tree Conf. Res: Diseases of forest trees and coniferous plantations, particularly root rot diseases; phytotoxicity of air pollutants. Mailing Add: Div of Forest Pest Mgt Pa Dept of Environ Resources Middletown PA 17057

TOWERS, BERNARD, b Preston, Eng, Aug 20, 22; m 71; c 3. PEDIATRICS, DEVELOPMENTAL ANATOMY. Educ: Univ Liverpool, MB, ChB, 47; Cambridge Univ, MA, 54; Royal Col Physicians, Licentiate, 47. Prof Exp: House surgeon, Royal Infirmary, Liverpool, 47; asst lectr anat, Bristol Univ, 49-50; lectr anat & histol, Univ Wales, 50-54; lectr anat, Cambridge Univ, 54-70, dir med studies, 64-70; PROF PEDIAT & ANAT, SCH MED, UNIV CALIF, LOS ANGELES, 71- Concurrent Pos: Fel med, Jesus Col, Cambridge Univ, 57-70; ed, Brit Abstr Med Sci, 54-56; chmn, Teilhard Ctr Future Man, London, 66-69; consult, Inst Human Values in Med, 71- Mem: Anat Soc Gt Brit & Ireland; Am Asn Anatomists; fel Royal Soc Med; Brit Soc Hist Med. Res: Fetal and neonatal lung; development of the heart and congenital anomalies; early detection of myocardial ischemia; primate evolution, especially human; medical history. Mailing Add: Dept of Pediat UCLA Ctr Health Sci Los Angeles CA 90024

TOWERS, GEORGE HUGH NEIL, b Bombay, India, Sept 28, 23; m 44; c 5. PLANT BIOCHEMISTRY. Educ: McGill Univ, BSc, 50, MSc, 51; Cornell Univ, PhD, 54. Prof Exp: From asst prof to assoc prof bot, McGill Univ, 53-62; sr res officer, Nat Res Coun Can, 62-64; head dept, 64-70, PROF BOT, UNIV BC, 70- Honors & Awards: Lalor Found Award, 55. Mem: Bot Soc Am; Can Soc Plant P sio Physiol (pres, 65-66); Linnean Soc London. Res: Plant metabolism. Mailing Add: Dept of Bot Univ of BC Vancouver BC Can

TOWERY, BEVERLY TODD, b Springfield, Ill, Dec 9, 15; m 42; c 3. MEDICINE. Educ: Western Ky State Col, BS, 36; Vanderbilt Univ, MD, 40. Prof Exp: Intern med, Univ Hosp, Vanderbilt Univ, 40-41; resident, Boston City Hosp, 41-42; resident, Mass Gen Hosp, 46; from instr to assoc prof, Med Sch, Vanderbilt Univ, 49-56; chmn dept med, 56-70, PROF MED, UNIV LOUISVILLE, 56-, CHIEF SECT ENDOCRINOL, 70- Concurrent Pos: Res fel, Thorndike Mem Lab, 47-48; Markle scholar med sci, 49-54. Res: Endocrinology; thyroid function; goiter; metabolic bone disease. Mailing Add: Dept of Med Univ of Louisville Sch of Med Louisville KY 40202

TOWLE, ALBERT, b Stockton, Calif, May 19, 25; m 46; c 3. INVERTEBRATE PHYSIOLOGY, SCIENCE EDUCATION. Educ: Col of Pac, AB, 46; San Jose State Col, MA, 53; Stanford Univ, PhD(biol), 62. Prof Exp: Chmn dept math & sci, James Lick High Sch, Calif, 50, chmn dept sci, 51, teacher biol, physiol & chem, 53-57 & 58-59, chmn dept sci, 61-64 & 65-66; asst biol, Stanford Univ, 57-58; lectr biol, 64-65, chmn dept marine biol, 72-75, PROF BIOL CALIF STATE UNIV, SAN FRANCISCO, 66- Concurrent Pos: Lectr, NSF partic, San Jose State Col, 60 & Purdue Univ, 65, partic, NSF Social Psychol Workshop & Conf for Dirs, DC, 68, dir, NSF-Nat Asn Sec Sch Prin Inst Sec Sch Adminr, 68; Nat Sch Teachers Asn Res Antarctica, 70; Smithsonian sponsored attendance and presentation of paper on Sipunculida, Yugoslavia, 70; res in the Galapagos, 74-75. Mem: AAAS; Am Inst Biol Sci; Nat Asn Biol Teachers; Nat Sci Teachers Asn. Res: Teaching of high school biology; behavior and ecology of terrestrial isopods; reproductive physiology of marine invertebrates; distribution of Sipunculida in the Galapagos Archepelago. Mailing Add: Dept Marine Biol Sch Natural Sci Calif State Univ San Francisco CA 94132

TOWLE, DAVID WALTER, b Concord, NH, May 26, 41; m 74. MOLECULAR BIOLOGY, MEDICAL ETHICS. Educ: Univ NH, BS, 65, MS, 67; Dartmouth Col, PhD(biol sci), 71. Prof Exp: Asst prof, 70-75, ASSOC PROF BIOL, UNIV RICHMOND, 75- Mem: AAAS; Am Soc Zoologists. Res: Biochemistry and physiology of osmoregulation in marine organisms; bioethics. Mailing Add: Dept of Biol Univ of Richmond Richmond VA 23173

TOWLE, LAIRD C, b Exeter, NH, Sept 13, 33; m 56; c 4. HIGH PRESSURE PHYSICS, SOLID STATE PHYSICS. Educ: Univ NH, BSc, 55, MSc, 58; Univ Va, PhD(physics), 62. Prof Exp: Res physicist, Avco Corp, Mass, 62-63; res physicist, Allis-Chalmers, Wis, 63-66; SECT HEAD HIGH PRESSURE PHYSICS, US NAVAL RES LAB, 66- Mem: Am Phys Soc. Res: Mechanical and electrical properties of solids under high pressure; equations of state of solids; solid-solid phase transitions; low temperature heat capacity of solids. Mailing Add: Cryogenics Branch Naval Res Lab Code 6434 Washington DC 20375

TOWLE, LOUIS WALLACE, b Frog Mountain, Ala, Nov 21, 08; m 31; c 2. ANALYTICAL CHEMISTRY. Educ: Univ Ariz, BS, 30, MS, 32. Prof Exp: Instr chem, Ariz State Col, 33; teacher high sch, Ariz, 33-36; res chemist, 37-44, tech serv supvr, 44-51, tech dir, 51-54, gen supt, 54-65, vpres, 69-71, GEN MGR, APACHE POWDER CO, 65-, PRES, 71- Mem: Am Chem Soc; fel Am Inst Chem; Am Inst Mining, Metall & Petrol Eng. Res: Preparation and uses of acetylene di-carboxylic acid; analysis of albumen and globulin blood proteins; nitroglycerin blasting explosives; manufacturing heavy chemicals; nitric and sulphuric acids; ammonium nitrate; anhydrous ammonia; ammonium nitrate blasting agents. Mailing Add: Apache Powder Co PO Box 700 Benson AZ 85602

TOWLE, MARGARET ASHLEY, b Atlanta, Ga, Jan 12, 03; m 30. ETHNOBOTANY. Educ: Oglethorpe Univ, BA, 24; Columbia Univ, PhD, 58. Prof Exp: Res fel ethnobot, Harvard Univ, 58-64, cur ethnobot collections, 64-73, hon cur, 73-76; RETIRED. Mem: Soc Am Archaeol; Soc Econ Bot. Res: Ethnobotany of Central Andes; evolution of agriculture. Mailing Add: Bot Mus Harvard Univ Cambridge MA 02138

TOWLE, PHILIP HAMILTON, b San Francisco, Calif, Aug 11, 18. ORGANIC CHEMISTRY. Educ: Stanford Univ, AB, 40; Mass Inst Technol, PhD(org chem), 48. Prof Exp: Res chemist labs, Standard Oil Co (Ind), 48-54, group leader, 54-61, res & develop, Amoco Chem Corp, 61-64, sect leader, 64-67, div dir org chem res, 67-71, DIR RES SERV, STANDARD OIL CO (IND), 71- Mem: AAAS; Am Chem Soc. Res: Petrochemicals; aromatic oxidations; information systems. Mailing Add: Res Serv Dept Res Ctr Standard Oil Co (Ind) PO Box 400 Naperville IL 60540

TOWLER, MARTIN LEE, b Hockley, Tex, Sept 18, 10; m 40; c 5. NEUROLOGY, PSYCHIATRY. Educ: Univ Tex, MD, 35; Am Bd Psychiat & Neurol, dipl, 42. Prof Exp: Intern, Med Br, Univ Tex, 36, resident neurol & psychiat, 39, instr, 39-41; PROF NEUROL & PSYCHIAT, UNIV TEX MED BR GALVESTON, 46- Concurrent Pos: Rockefeller Found fel, Sch Med, Univ Colo, 41-42; pvt pract; consult, Surg Gen, US Army, 49; consult, Lackland AFB Hosp, 54- Mem: Am EEG Soc; fel Am Psychiat Asn; AMA; Asn Am Med Cols; fel Am Acad Neurol. Res: Effect of drugs on the electroencephalograph pattern; clinical value and limitation of antidepressant drugs. Mailing Add: 200 University Blvd Galveston TX 77550

TOWLER, OSCAR ALWYNNE, JR, nuclear physics, see 12th edition

TOWNE, DUDLEY HERBERT, b Schenectady, NY, Nov 7, 24. THEORETICAL PHYSICS. Educ: Yale Univ, BS, 47; Harvard Univ, MA, 49, PhD(physics), 54. Prof Exp: From instr to assoc prof, 52-63, PROF PHYSICS, AMHERST COL, 63- Concurrent Pos: Staff mem, Rockefeller Found, 63-64. Mem: AAAS; Am Phys Soc. Res: Scattering of electromagnetic radiation; broadening of spectral lines; wave propagation in inhomogeneous media. Mailing Add: Dept of Physics Amherst Col Amherst MA 01002

TOWNE, JACK C, b New York, NY, Apr 23, 27; m 50; c 2. BIOCHEMISTRY, RADIOCHEMISTRY. Educ: Univ Calif, Los Angeles, BS, 50; Univ Wis, MS, 52, PhD(biochem), 55. Prof Exp: USPHS fel, 54-56; dir biochem lab, Inst Psychosom & Psychiat Res & Training, 56-58; prin scientist biochem, Vet Admin Hosp, Tucson, Ariz, 58-70; PROF CHEM, UNIV DALLAS, 70-, CHMN DEPT, 72- Concurrent Pos: Asst prof, Med Sch, Northwestern Univ, 58-65; holder & co-investr, NIH grants, 59-65; lectr, Univ of the Andes, Venezuela, 64; res assoc, Col Med, Univ Ariz, 65-70. Mem: AAAS; Am Chem Soc; Am Inst Chem; NY Acad Sci. Res: Enzymology; intermediary and amine metabolism; radiometric syntheses and analyses. Mailing Add: Dept of Chem Univ of Dallas Irving TX 75060

TOWNEND, ROBERT EDWARD, b Mobile, Ala, Feb 22, 22; m 55; c 2. PHYSICAL BIOCHEMISTRY. Educ: Temple Univ, BA, 52, MA, 55, PhD(phys chem), 62. Prof Exp: Chemist, 55-62, res chemist, 62-66, SR RES CHEMIST, EASTERN REGIONAL RES LAB, USDA, 66- Mem: Am Chem Soc; Am Soc Biol Chemists; AAAS. Res: Binding of ions to proteins; transport phenomena and association of proteins; optical methods applied to structure of biological macromolecules. Mailing Add: Dairy Lab USDA 600 E Mermaid Lane Philadelphia PA 19118

TOWNER, GEORGE RUTHERFORD, b New York, NY, Sept 15, 33. INFORMATION SCIENCE. Educ: Univ Calif, Berkeley, BA, 56, MA, 57. Prof Exp: Asst dir, Kaiser Found Res Inst, 58-60; res engr, Advan Instrument Corp 60-61; pres, Berkeley Instruments Corp, 62-67; CHMN BD, TOWNER SYSTS CO, 68- Concurrent Pos: Lectr, Conf on Automation, Univ Hawaii, 66. Mem: Instrument Soc Am. Res: Cybernetics; man-machine systems; meteorological data systems; information theory. Mailing Add: Towner Systs Co 14666 Doolittle Dr San Leandro CA 94577

TOWNER, HOWARD FROST, b Los Angeles, Calif, Aug 10, 43; m 65; c 2. BIOLOGY, ECOLOGY. Educ: Univ Calif, Riverside, AB, 65; Stanford Univ, PhD(biol), 70. Prof Exp: NIH fel, Univ Calif, Los Angeles, 70-71, asst res neurologist, Ctr Health Sci, 71-72; ASST PROF BIOL, LOYOLA MARYMOUNT UNIV, 72- Concurrent Pos: Consult, Wadsworth Hosp, US Vet Admin, 72- Mem: Am Soc Zool. Res: Ecology of desert organisms; cytogenetics and plant evolution. Mailing Add: Loyola Marymount Univ Los Angeles CA 90045

TOWNER, IAN STUART, b Hastings, Eng, May 24, 40; m 66; c 2. THEORETICAL NUCLEAR PHYSICS. Educ: Univ London, BSc, 62, PhD(nuclear physics), 66. Prof Exp: Res assoc, Nuclear Physics Lab, Oxford Univ, 65-70; RES OFFICER NUCLEAR PHYSICS, CHALK RIVER NUCLEAR LABS, ATOMIC ENERGY CAN LTD, 70- Mem: Brit Inst Physics. Res: Nuclear structure; models. Mailing Add: Physics Div Chalk River Nuclear Labs Chalk River ON Can

TOWNER, RICHARD HENRY, b Gunnison, Colo, Oct 7, 48. ANIMAL GENETICS. Educ: Colo State Univ, BS, 70; Univ Wis-Madison, MS, 73, PhD(genetics, meat & animal sci), 75. Prof Exp: GENETICIST, H&N INC, 75- Mem: Am Soc Animal Sci. Res: Improvement of existing strains and development of new strains of leghorn chickens by utilizing quantative genetic principles. Mailing Add: 15305 NE 40th St Redmond WA 98052

TOWNES, ALEXANDER SLOAN, b Birmingham, Ala, June 19, 29; m 51; c 5. INTERNAL MEDICINE, IMMUNOLOGY. Educ: Vanderbilt Univ, BA, 50, MD, 53. Prof Exp: From instr to assoc prof, Johns Hopkins Univ, 61-72; chief sect rheumatol, 72-75, PROF MED, COL MED, UNIV TENN, MEMPHIS, 72-; CHIEF MED SERV, MEMPHIS VET ADMIN HOSP, 75- Concurrent Pos: Fel med, Sch Med, Johns Hopkins Univ, 59-61; asst physician in chief, Baltimore City Hosps, 63-70. Mem: AAAS; Am Rheumatism Asn; Am Fedn Clin Res; fel Am Col Physicians; Am Asn Immunologists. Res: Clinical medicine and rheumatology; role complement and immune reactions in pathogenesis of rheumatic diseases; correlation of clinical findings with immunologic changes and effects of therapy. Mailing Add: Vet Admin Hosp 1030 Jefferson Ave Memphis TN 38104

TOWNES, CHARLES HARD, b Greenville, SC, July 28, 15; m 41; c 4. PHYSICS. Educ: Furman Univ, BA & BS, 35; Duke Univ, MA, 37; Calif Inst Technol, PhD(physics), 39. Hon Degrees: Twenty from US & foreign univs & cols, 60-69. Prof Exp: Asst physics, Calif Inst Technol, 37-39; mem tech staff, Bell Tel Labs, 39-47; from assoc prof to prof physics, Columbia Univ, 48-61, chmn dept, 52-55, exec dir radiation lab, 50-52; prof physics & provost, Mass Inst Technol, 61-66, inst prof physics, 66-67; UNIV PROF PHYSICS, UNIV CALIF, BERKELEY, 67- Concurrent Pos: Adams fel, 50; Guggenheim fel, 55-56; Fulbright lectr, Paris, 55-56 & Tokyo, 56; lectr, Enrico Fermi Int Sch Physics, 55 & 60, dir, 63; Scott lectr, Cambridge Univ, 63; centennial lectr, Univ Toronto, 67; vpres & dir res, Inst Defense Anal, 59-61; trustee, Salk Inst Biol Studies, 63-68, Carnegie Inst, 65- & Rand Corp, 65-70; mem bd dirs, Perkin-Elmer Corp, 66-; mem corp, Woods Hole Oceanog Inst, 69-, trustee, 71-74; mem sci adv bd, US Dept Air Force, 58-61; chmn sci & tech adv comt manned space flight, NASA, 64-69, mem space prog adv coun, 71-; mem, President's Sci Adv Comt, 66-70, vchmn, 67-69; chmn, President-Elect's Task Force on Space, 68, mem, President's Task Force Nat Sci Policy, 69; mem coun, Nat Acad Sci, 69-72, chmn space sci bd, 70-73; consult, Sci & Technol Bur, Arms Control & Disarmament Agency, 70-; chmn sci adv comt, Gen Motors Corp, 71-73. Honors & Awards: Nobel Prize in Physics, 64; Award, Res Corp, 58; Liebmann Mem Prize, Inst Radio Engrs, 58, Sarnoff Award, Inst Elec Engrs, 61 & Medal of Honor, Inst Elec & Electronics Engrs, 67; Except Serv Award, US Air Force, 59; Comstock Award, Nat Acad Sci, 59, Carty Medal, 62; Ballantine Medal, Franklin Inst, 59 & 62; Rumford Premium, Am Acad Arts & Sci, 61; Beckman Award, Instrument Soc Am, 61; Scott Award, City of Philadelphia, 63; Young Medal & Prize, Brit Inst Physics & Phys Soc, 63; Priestley Award, Dickinson Col, 66; C E K Mees Medal, Optical Soc Am, 68; Michelson-Morley Award, 70; Wilhem-Exner Award, 70; Medal of Honor, Univ Liege, 71. Mem: Nat Acad Sci; fel Am Phys Soc (pres, 67); hon mem Optical Soc Am; Am Philos Soc; Am Astron Soc. Res: Molecular and nuclear structure; radio and infrared astronomy; microwave spectroscopy; optics; quantum electronics. Mailing Add: Dept of Physics Univ of Calif Berkeley CA 94720

TOWNES, CHARLES HENRY, b Petersburg, Va, May 30, 15; m 40. MEDICINE, PHYSICS. Educ: Va State Col, BS, 35; Pa State Col, MS, 38, PhD(physics), 42; Howard Univ, MD, 47. Prof Exp: Prin, Lightfoot Training Sch, Va, 35-36; teacher physics, Va State Col, 36-37, instr chem & physics, 38-44; lectr, Howard Univ, 44-47; intern, Freedman's Hosp, Washington, DC, 47-48; physician, chief health serv & med dir, 48-68, COL PHYSICIAN & MED DIR, MEM HOSP, VA STATE COL, 68- Concurrent Pos: Resident physician, US Navy contract, Pa State Col, 40-41, Hamilton Watch Co contract, 41; med dir, Mercy Hosp. Mem: Nat Med Asn. Res: Physical chemistry; x-ray of age hardening in light alloys and orientation in metals; conductivity of binary mixtures of fatty acids, bones and teeth; vitamin assay and change; problems in student health. Mailing Add: Dept of Health Serv Va State Col Petersburg VA 23806

TOWNES, HENRY KEITH, JR, b Greenville, SC, Jan 20, 13; m 37; c 2. ENTOMOLOGY. Educ: Furman Univ, BS & BA, 33; Cornell Univ, PhD(entom), 37. Hon Degrees: LLD, Furman Univ, 75. Prof Exp: Instr zool, Syracuse Univ, 37-38; entom, Cornell Univ, 38-40; Nat Res Coun fel, Acad Natural Sci, Pa, 40-41; jr entomologist, Bur Entom & Plant Quarantine, USDA, 41, from assoc entomologist to entomologist, 41-49; from assoc prof to prof entom, NC State Col, 49-56; ADJ PROF, UNIV MICH, ANN ARBOR, 56-; DIR, AM ENTOM INST, 61- Concurrent Pos: Adv, Pl Govt, 52-54; mem econ surv, US Com Co, Micronesia, 46; mem exped, Wash & Ore, 40, Ariz, 47, Colo & Calif, 48, SAm, 65-66, SAfrica, 70-71, Alaska, 73, Calif & Ariz, 74 & Nfld, 75; vis prof, Mich State Univ, 72 & Carleton Univ, 75. Mem: Fel Entom Soc Am; cor mem Am Entom Soc; Entom Soc Can; cor mem Netherlands Entom Soc; cor mem Chilean Entom Soc. Res: Insect taxonomy, especially parasitic Hymenoptera. Mailing Add: Am Entom Inst 5950 Warren Rd Ann Arbor MI 48105

TOWNES, MARJORIE CHAPMAN, b Pawcatuck, Conn, Mar 28, 09; m 37; c 2. ENTOMOLOGY. Educ: Mt Holyoke Col, BA, 31; Cornell Univ, MA, 32, PhD(bot), 35. Prof Exp: Instr bot, Mt Holyoke Col, 35-37; mem entom expeds, Wash & Ore, 40, Ariz, 47, Colo & Calif, 48, Philippines, 52-54 & Japan, 54; RES ASSOC, AM ENTOM INST, 61- Concurrent Pos: Res assoc, Univ Mich, Ann Arbor, 64-72; mem entom exped, S Am, 65-66, S Africa, 70-71. Res: Insect taxonomy and ecology, especially Ichneumonidae. Mailing Add: Am Entom Inst 5950 Warren Rd Ann Arbor MI 48105

TOWNES, MARY MCLEAN, b Southern Pines, NC, July 12, 28; m 54; c 2. CELL PHYSIOLOGY. Educ: NC Col Durham, BS, 49, MSPH, 50; Univ Mich, MS, 53, PhD(cell physiol), 62. Prof Exp: From instr to assoc prof, 50-68, PROF BIOL, NC CENT UNIV, 68- Concurrent Pos: Consult biol improv prog, NSF & NC Acad Sci Prog High Sch Teachers Biol, 65-66; consult minority access to res careers, Nat Inst Gen Med Sci, NIH, 75-77. Mem: AAAS; Soc Gen Physiologists; NY Acad Sci; Sigma Xi; Am Soc Zoologists. Res: pH relations of contractility of glycerinated stalks of Vorticella convallaria; contractile properties of glycerinated stalks of Vorticella. Mailing Add: Dept of Biol NC Cent Univ Durham NC 27707

TOWNES, PHILIP LEONARD, b Salem, Mass, Feb 18, 27; m 56. GENETICS. Educ: Harvard Univ, AB, 48; Univ Rochester, PhD(zool), 53, MD, 59. Prof Exp: Asst biol, 48-51, from instr to assoc prof anat, 52-66, asst prof pediat, 65-69, PROF ANAT, PROF GENETICS, SCH MED & DENT, UNIV ROCHESTER, 66-, PROF PEDIAT, 69- Mem: AAAS; Soc Pediat Res; Am Pediat Soc; Am Soc Human Genetics. Res: Experimental embryology; cell movements; biochemical aspects of development; enzymes; proteins; metabolic inhibitors; physiology of development; human genetics and embryology. Mailing Add: Div of Genetics Univ Rochester Sch Med & Dent Rochester NY 14620

TOWNLEY, CHARLES WILLIAM, b East Liverpool, Ohio, Oct 27, 34; m 57; c 2. NUCLEAR CHEMISTRY, RESEARCH ADMINISTRATION. Educ: Ohio State Univ, BSc, 56, PhD(nuclear chem), 59. Prof Exp: Sr chemist, 59-62, fel, 62-65, chief chem physics res, 65-67, chief struct physics res, 67-70, mgr mat sci, 70-73, mgr info & commun systs, 73-74, MGR WILLIAM F CLAPP LABS, COLUMBUS LABS, BATTELLE MEM INST, 74- Mem: AAAS; fel Am Inst Chem; Am Chem Soc. Res: Radiochemical separations; reactor fuel development; radioisotope applications; decay scheme studies; activation analysis; fission product release and chemistry; research management. Mailing Add: PO Box 1637 Duxbury MA 02332

TOWNLEY, JOHN LEWIS, III, b Fergus Falls, Minn, Jan 11, 21; m 55; c 3. PETROLEUM. Educ: Univ Minn, BA, 48. Prof Exp: Petrol geologist, Magnolia Petrol Co Div, Mobil Oil Corp, 48-50, petrol geologist, 50-59, staff geologist, 59-61, CHIEF PROD GEOLOGIST, MOBIL OIL CAN, LTD, 61- Mem: Am Asn Petrol Geologists; Can Soc Petrol Geologists. Mailing Add: Mobil Oil Can Ltd Box 800 Calgary AB Can

TOWNLEY, JUDY ANN, b San Antonio, Tex, Sept 19, 46. INFORMATION SCIENCE. Educ: Univ Tex, Austin, BA, 68; Harvard Univ, SM, 69, PhD(appl math), 73. Prof Exp: LECTR & RES FEL APPL MATH, HARVARD UNIV, 73-, DIR INFO SCI PROG, 75-; MEM STAFF, MASS COMPUT ASSOCS, INC, 73- Mem: Asn Comput Mach; Inst Elec & Electronics Engrs. Mailing Add: Aiken Computation Lab 33 Oxford St Harvard Univ Cambridge MA 02138

TOWNLEY, ROBERT WILLIAM, b Lampasas, Tex, Apr 28, 07; m 29; c 2. CHEMISTRY. Educ: Austin Col, BA, 29; Univ Tex, MA, 35, PhD(phys chem), 38. Prof Exp: Anal chemist, First Tex Chem Mfg Co, 31-33; from asst to instr chem, Univ Tex, 35-37; bacteriologist, State Dept Health, Tex, 37-38, chemist, 38-39, chief chemist, 39-41; res chemist, Humble Oil & Ref Co, 41-42; indust hyg engr, USPHS, Md, 42-44; res chemist, Ciba Pharmaceut Prod, Inc, NJ, 44-50; assoc prof chem, Drew Univ, 50-54; head res dept, Personal Prod Corp, 54-57; dir Townley Res & Consult, 57-73; assoc prof Fairleigh Dickinson Univ, 58-59; CONSULT, 73- Mem: Am Chem Soc. Res: Foods; drugs; water; corrosion; air and water pollution; industrial hygiene; microbiology. Mailing Add: 470 Long Hill Rd Gillette NJ 07933

TOWNLEY-SMITH, THOMAS FREDERICK, b Scott, Sask, Aug 27, 42; m 63; c 3. PLANT BREEDING. Educ: Univ Sask, BSA, 64, MSc, 65; Univ Guelph, PhD(plant breeding), 69. Prof Exp: Res asst plant breeding, Univ Guelph, 65-68; RES SCIENTIST, WHEAT BREEDING, RES STA, CAN DEPT AGR, 68- Concurrent Pos: Sr wheat breeder, Plant Breeding Sta, Can Int Develop Agency, Njoro, Kenya, East Africa, 72-74. Mem: Genetics Soc Can; Can Soc Agron. Res: Breeding durum wheat; genetics and cytogenetics of wheat. Mailing Add: Can Dept of Agr Res Sta Box 1030 Swift Current SK Can

TOWNS, CLARENCE, JR, b Little Rock, Ark, July 22, 16; m 44; c 3. PATHOLOGY, HISTOLOGY. Educ: Cent YMCA Col, BS, 42; Univ Ill, DDS, 45, MS, 74; Am Bd Endodontic, dipl, 57. Prof Exp: Instr histol, 68-69, asst instr basic sci, 70-72, ASST PROF HISTOL, COL DENT, UNIV ILL, 74- Mem: Sigma Xi; Am Acad Forensic Dent; Am Soc Oral Med; Am Asn Endodontics; Am Dent Asn. Res: Exfoliative cytology of the oral mucosa in the male Negro nonsmoker and smoker; ultra structures study of oral mucosa; comparison of normal and hyperkerototic human oral mucosa. Mailing Add: 200 E 75th St Chicago IL 60619

TOWNS, DONALD LIONEL, b Sioux City, Iowa, Mar 8, 35; m 60; c 2. PHYSICAL ORGANIC CHEMISTRY, CHEMICAL ENGINEERING. Educ: Ga Inst Technol, BCheE, 57; Univ Wis, PhD(chem), 63. Prof Exp: Process res chemist, 62-72, sr process res chemist, Niagara Chem Div, 72-73, process eng group leader, Indust Chem Div, 73-74, tech mgr, 74-75, FURADAN PROD MGR, AGR CHEM DIV,

FMC CORP, 75- Mem: Am Chem Soc; The Chem Soc; Am Inst Chem Engrs. Res: Technical improvement, environmental protection and production of the insecticide Furadan. Mailing Add: Agr Chem Div FMC Corp 100 Niagara St Middleport NY 14105

TOWNS, ROBERT LEE ROY, b Bartlesville, Okla, Oct 27, 40; m 60; c 2. CHEMISTRY. Educ: Univ New Orleans, BS, 65; Univ Tex, Austin, PhD(phys chem), 69. Prof Exp: Vis asst prof chem, Univ New Orleans, 70-71; asst prof & petrol fund grant chem, Tex A&M Univ, 71-73; ASSOC PROF CHEM, CLEVELAND STATE UNIV, 73- Concurrent Pos: NATO grant, Advan Study Inst, Univ York, Eng, 71. Mem: Am Chem Soc; Am Crystallog Asn; Am Inst Physics; Am Asn Clin Chemists. Res: X-ray fluorescence; trace and ultratrace metal analysis in human tissues and body fluids; instrument design, development and automation; x-ray crystallography; molecular structure determination; structure/activity relationships in Treflan herbicides. Mailing Add: Dept of Chem Cleveland State Univ Cleveland OH 44115

TOWNSEND, ALDEN MILLER, b Tulsa, Okla, Mar 4, 42; m 68; c 2. PLANT GENETICS, TREE PHYSIOLOGY. Educ: Pa State Univ, University Park, BS, 64; Yale Univ, MF, 66; Mich State Univ, PhD(plant genetics & physiol), Mich State Univ, 69. Prof Exp: RES GENETICIST, SHADE TREE & ORNAMENTAL PLANTS LAB, AGR RES SERV, USDA, 70- Mem: Genetics Soc Am; Scand Soc Plant Physiol; Genetics Soc Am; Am Phytopath Soc; Int Soc Arboricult. Res: Breeding of urban trees for resistance to diseases, air pollutants, cold and soil stresses; physiological genetics of Acer rubrum and elm hybrids. Mailing Add: Shade Tree & Ornamental Plants Lab Agr Res Serv USDA Delaware OH 43015

TOWNSEND, ALEXANDRA A, b Harvey, Ill, Dec 2, 38. PHYSIOLOGY, OSTEOPATHY. Educ: Hope Col, AB, 62; Chicago Col Osteop, DO, 67. Prof Exp: Instr med terminology, Cent YMCA Community Col, 65; instr physiol & pharmacol, 66-67, res assoc physiol, 68-71, ASSOC PROF PHYSIOL & PHARMACOL, CHICAGO COL OSTEOP MED, 71- Concurrent Pos: Pvt practr, 68- Honors & Awards: Borden Award Osteop Med, 67. Mem: AAAS; Am Heart Asn; Am Osteop Asn; Am Physiol Soc. Res: Investigation of hepatic and renal circulation; case studies in osteopathic medicine. Mailing Add: Dept of Physiol & Pharmacol Chicago Col of Osteopath Med Chicago IL 60615

TOWNSEND, CHARLEY E, b Decatur Co, Kans, July 2, 29; m 59; c 1. PLANT BREEDING, GENETICS. Educ: Kans State Univ, BS, 50, MS, 51; Univ Wis, PhD(agron, plant path), 56. Prof Exp: RES GENETICIST, AGR RES SERV, USDA & COLO STATE UNIV, 56- Mem: Am Soc Agron; Crop Sci Soc Am. Res: Breeding legumes for western ranges and pastures. Mailing Add: Crop Res Lab Colo State Univ Ft Collins CO 80523

TOWNSEND, EDWIN C, b Vienna, WVa, July 7, 36; m 58; c 3. BIOMETRY. Educ: Univ WVa, BS, 58, MS, 64; Cornell Univ, PhD(biomet), 68. Prof Exp: Staff asst comput, 61-62, res assoc, 62-63, ASSOC PROF STATIST, W VA UNIV, 68- Mem: Biomet Soc; Am Statist Asn. Mailing Add: Dept of Statist WVa Univ Morgantown WV 26505

TOWNSEND, FRANK MARION, b Stamford, Tex, Oct 29, 14; m 51; c 2. PATHOLOGY. Educ: Tulane Univ, MD, 38. Prof Exp: US Air Force, 40-65, pathologist, Sch Med, Washington Univ, 45-47, instr clin path, Col Med, Univ Nebr, 47-48, assoc pathologist, Scott & White Clin, Temple Univ, 49, regional consult path, Vet Admin Hosp, Tex & La, 50, chief lab serv, Lackland AFB, 50-54, dep dir, Armed Forces Inst Path, 55-59, dir, 59-63, vcommander, Aerospace Med Div, Air Force Systs Command, Brooks AFB, Tex, 63-65; pathologist, Tex State Dept Health, 65-69; clin prof path, 69-72, PROF PATH & CHMN DEPT, UNIV TEX HEALTH SCI CTR, SAN ANTONIO, 72- Concurrent Pos: Assoc prof, Med Br, Univ Tex, 49-58, lectr, 58-63, assoc prof, Postgrad Sch, 53-54; consult, Surgeon Gen, 54-63 & NASA, 67-74; mem, Joint Comt Aviation Path, 56-63, chmn, 60-62; mem, Armed Forces Comt Bioastronaut, Nat Res Coun, 59-60, Nat Adv Cancer Coun, 59-63, Exp Adv Panel Cancer, WHO, 58- & Proj Mercury Recovery Team, NASA, 60-63. Honors & Awards: Moseley Award, Aerospace Med Asn, 62; Founders Medal, Asn Mil Surgeons US, 62. Mem: AAAS; fel Am Soc Clin Path; fel AMA; fel Am Col Physicians; fel Am Col Path. Res: Aerospace and respiratory disease pathology. Mailing Add: 10406 Mt Marcy Dr San Antonio TX 78213

TOWNSEND, GORDON FREDERICK, b Toronto, Ont, Mar 4, 15; m 40; c 2. APICULTURE. Educ: Ont Agr Col, BSA, 38; Univ Toronto, MSA, 42. Prof Exp: Asst med res, Banting Inst, Can, 38; from asst prof to assoc prof apicult, 39-49, head dept, 39-71, PROF APICULT, ONT AGR COL, UNIV GUELPH, 49- Concurrent Pos: Dir, Can-Kenya Apicult Prog, 71-74. Mem: Bee Res Asn. Res: Chemistry of royal jelly; processing and packing of honey; pollination of tree fruits. Mailing Add: Dept of Environ Biol Ont Agr Col Univ of Guelph Guelph ON Can

TOWNSEND, HOWARD GARFIELD, JR, b Rochester, NY, Sept 10, 38; m 64; c 2. ENTOMOLOGY. Educ: Cornell Univ, BS, 60; Va Polytech Inst, MS, 63; Pa State Univ, PhD(entom), 70. Prof Exp: Res asst entom, Va Polytech Inst, 60-62; experimentalist II, NY Agr Exp Sta, Geneva, 63-65; instr, Pa State Univ, 65-69; RES ENTOMOLOGST, STATE FRUIT EXP STA, SOUTHWEST MO STATE UNIV, 70- Mem: Entom Soc Am. Res: Insect and mite pests of pome and stone fruits, grapes and small fruits. Mailing Add: State Fruit Exp Sta Southwest Mo State Univ Mountain Grove MO 65711

TOWNSEND, JOAN B, b Dallas, Tex, July 9, 33; m 70; c 2. ANTHROPOLOGY, ARCHAEOLOGY. Educ: Univ Calif, Los Angeles, BA, 59, PhD(anthrop), 65. Prof Exp: Asst prof anthrop, Los Angeles State Col, 63; instr, Southern Ill Univ, Carbondale, 63-64; from instr to asst prof, 64-72, ASSOC PROF ANTHROP, UNIV MAN, 72- Concurrent Pos: Northern Studies grant, Univ Man, 66-, Nat Res Coun Can grant, 67-69, Can Coun grant, 70-, Res Bd grant, 72- Mem: Fel Am Anthrop Asn; fel Royal Anthrop Inst Gt Brit & Ireland; NY Acad Sci; fel Soc Appl Anthrop; Fel Arctic Inst NAm. Res: Ethnohistory and ethnography of Tanaina Athapaskan Indians and Eskimos of southwestern Alaska; Athapaskan Indians; culture change. Mailing Add: Dept of Anthrop Univ of Man Winnipeg MB Can

TOWNSEND, JOEL IVES, b Greenwood, SC, Aug 20, 20. GENETICS. Educ: Univ SC, BS, 41; Columbia Univ, PhD(zool), 52. Prof Exp: Lectr, Columbia Univ, 50-51; asst prof zool, Univ Tenn, 52-60; asst prof genetics, 60-62, ASSOC PROF HUMAN GENETICS, MED COL VA, VA COMMONWEALTH UNIV, 62- Concurrent Pos: Prof, Univ Rio Grande do Sul, Brazil, 54. Mem: AAAS; Soc Study Evolution; Genetics Soc Am; Am Eugenics Soc; Am Genetic Asn. Res: Population genetics; genetics of human isolates; genetics and cytology of marginal populations of Drosophila. Mailing Add: Prog Human Genetics Med Col of Va Va Commonwealth Univ Richmond VA 23219

TOWNSEND, JOHN FORD, b Kansas City, Mo, Jan 14, 36; m 59; c 3. PATHOLOGY. Educ: Univ Mo-Columbia, AB, 58, MD, 61; Am Bd Path, dipl, 67. Prof Exp: Intern, Med Br, Univ Tex, 61-62; resident, 62-66, from asst prof to assoc prof, 68-75, PROF PATH & V CHMN DEPT, SCH MED, UNIV MO-COLUMBIA, 75- Concurrent Pos: Chief path serv, Vet Admin Hosp, Columbia, Mo, 72-75. Res: Electrical energy transport through tissue; uterine peroxidase; diabetes using animal model Mystromys albicandatus. Mailing Add: Dept of Path Sch of Med Univ of Mo Columbia MO 65201

TOWNSEND, JOHN MARSHALL, b Amarillo, Tex, Sept 1, 41. MEDICAL ANTHROPOLOGY. Educ: Univ Calif, Berkeley, BA, 63; Univ Calif, Santa Barbara, MA, 67, PhD(anthrop), 72. Prof Exp: Asst prof anthrop, Univ Mont, 72-73; ASST PROF ANTHROP, SYRACUSE UNIV, 73- Mem: Am Anthrop Asn; fel Soc Appl Anthrop; Soc Med Anthrop. Res: Cross-cultural mental health; labeling theory; health care delivery and human fertility. Mailing Add: 500 University Pl Syracuse NY 13210

TOWNSEND, JOHN ROBERT, b Brooten, Minn, Oct 26, 25; m 48; c 2. PHYSICS. Educ: Cornell Univ, BS, 45, PhD(physics), 51. Prof Exp: Physicist, Hanford Atomic Prod Opers, Gen Elec Co, Wash, 51-54; from instr to assoc prof, 54-66, PROF PHYSICS, UNIV PITTSBURGH, 66- Mem: Am Phys Soc; Am Asn Physics Teachers. Res: Radiation effects in solids; defects in metals. Mailing Add: Dept of Physics Univ of Pittsburgh Pittsburgh PA 15213

TOWNSEND, JOHN WILLIAM, JR, b Washington, DC, Mar 19, 24; m 48; c 4. PHYSICS. Educ: Williams Col, BA, 47, MA, 49. Hon Degrees: DSc, 61. Prof Exp: Asst physics, Williams Col, 47-49; physicist, US Naval Res Lab, DC, 49-50, unit head, 50-52, sect head, 52-53, asst br head, 53-55, head rocket sonde br, 55-58; chief space sci div, NASA, 58-59, asst dir, Goddard Space Flight Ctr, 59-65, dep dir, 65-68; dep adminstr, Environ Sci Serv Admin, 68-70, ASSOC ADMINR, NAT OCEANIC & ATMOSPHERIC ADMIN, 70- Concurrent Pos: Mem comt aeronomy, Int Union Geod & Geophys; mem tech panel on rocketry, US Nat Comt, Int Geophys Year; exec secy, US Rocket & Satellite Res Panel, 58-; mem Int Acad Astronaut, Int Astronaut Fedn. Honors & Awards: Meritorious Civilian Serv Award, US Dept Navy, 57; Outstanding Leadership Medal, NASA, 62, Distinguished Serv Medal, 71; Arthur S Fleming Award, 63. Mem: Nat Acad Eng; Am Phys Soc; fel Am Meteorol Soc; Sigma Xi; Am Geophys Union. Res: Space science and space applications; aeronomy; upper atmosphere physics; composition of the upper atmosphere; mass spectrometry; scientific, meteorological and communications satellites; design and development of sounding rockets. Mailing Add: Nat Oceanic & Atmospheric Admin US Dept of Commerce Rockville MD 20852

TOWNSEND, JONATHAN, b Colo, July 17, 22; m 55; c 1. PHYSICS. Educ: Univ Denver, BS, 43; Washington Univ, MA, 48, PhD(physics), 51. Prof Exp: Engr, Gen Elec Co, 43-44; physicist, Carbide & Carbon Chem Co, 45-46; asst prof physics, 51-57, ASSOC PROF PHYSICS & BOT, WASHINGTON UNIV, 57- Mem: AAAS; Am Phys Soc. Res: Nuclear and paramagnetic resonance; free radicals in biology; photosynthesis; electronics. Mailing Add: Dept of Physics Washington Univ St Louis MO 63130

TOWNSEND, LEE HILL, b Leland, Miss, Jan 20, 03; m 40; c 1. ENTOMOLOGY. Educ: Univ Va, BS, 25; Univ Ill, MS, 32, PhD(entom), 35. Prof Exp: Instr high sch, Miss, 25-27; asst entom, Univ Ill, 27-30; asst entomologist, State Natural Hist Surv, Ill, 32-36; from instr to prof agr entom, Univ Ky, 36-70, chmn dept, 57-70, prof entom, 70-73; RETIRED. Mem: Emer mem Entom Soc Am. Res: Larvae and anatomy of Neuroptera; larvae of cylorrhaphous Diptera. Mailing Add: 731 Sunset Dr Lexington KY 40502

TOWNSEND, LEROY B, b Lubbock, Tex, Dec 20, 33; m 53; c 2. MEDICINAL CHEMISTRY. Educ: NMex Highlands Univ, BA, 55, MS, 57; Ariz State Univ, PhD(chem), 65. Prof Exp: Res assoc chem, Ariz State Univ, 60-65; res assoc, 65-67, asst res prof, 67-69; asst prof, 69-71, assoc prof med, chem, 71-75, ASST RES PROF CHEM, UNIV UTAH, 67-, PROF MED CHEM, 75- Concurrent Pos: Consult, Heterocyclic Chem Corp, 69- Mem: Am Chem Soc; The Chem Soc; Int Soc Heterocyclic Chem (treas, 76-77). Res: Nitrogen heterocycles, for example pyrrole, pyrimidine, imidazo(4,5-c)pyridine, pyrazole, pyrazolo(3,4-d)pyrimidine, purine, pyrrolo(2,3-d)pyrimidine, pyrazolo(4,3-d)pyrimidine imidazole and the nucleosides arabinofuranosides, ribopyranosides, 2'-deoxyribofuranosides and ribofuranosides of these systems with biological and chemotherapeutic interest as well as structure elucidation and chemical synthesis of certain antibiotics. Mailing Add: Dept of Chem Univ of Utah Salt Lake City UT 84112

TOWNSEND, PATRICIA KATHRYN, b Chicago, Ill, July 2, 41; m 62; c 1. CULTURAL ANTHROPOLOGY. Educ: Univ Mich, AB, 62, AM, 63, PhD(anthrop), 69. Prof Exp: Asst prof anthrop, Eastern Mich Univ, 68-73; fel ment retardation ctr, Univ Calif, Los Angeles, 74-75. Mem: Am Anthrop Asn. Res: Cultural ecology; cultural adaptation to tropical forest; social organization; New Guinea, Amazonia. Mailing Add: 193 Mt Vernon Rd Amherst NY 14226

TOWNSEND, RALPH N, b Normal, Ill, May 20, 31; m 58; c 2. MATHEMATICS. Educ: Ill Wesleyan Univ, BS, 53; Univ Ill, MS, 55, PhD(math), 58. Prof Exp: Asst math, Univ Ill, 54-58; asst prof math, San Jose State Col, 58-60; from asst prof to assoc prof, 60-71, asst dean, Col Arts & Sci, 69-75, PROF MATH, BOWLING GREEN STATE UNIV, 71-, ASSOC DEAN, COL ARTS & SCI, 75- Mem: Am Math Soc; Math Asn Am. Res: Analysis, including Schwartz distributions; complex analysis. Mailing Add: Dept of Math Bowling Green State Univ Bowling Green OH 43402

TOWNSEND, SAMUEL FRANKLIN, b Montague, Mich, Mar 22, 35; m 58; c 2. BIOLOGY, ANATOMY. Educ: Kalamazoo Col, AB, 57; Univ Mich, MS, 59, PhD(anat), 61. Prof Exp: From asst prof to assoc prof biol, Kalamazoo Col, 61-69, chmn dept, 66-69; ASSOC PROF ANAT, COL MED, UNIV CINCINNATI, 69- Concurrent Pos: Partic, NSF Res Participation Prog Col Teachers, 64-66. Res: Cellular differentiation in the adult rat; healing and control of experimental ulcers. Mailing Add: Dept of Anat Univ of Cincinnati Col of Med Cincinnati OH 45219

TOWNSEND, WILLIAM E, animal husbandry, bacteriology, see 12th edition

TOWNSHEND, JOHN LYNDEN, b Hamilton, Ont, Feb 6, 26; m 53; c 2. PLANT PATHOLOGY. Educ: Univ Western Ont, BSc, 51, MSc, 52; Imp Col, Univ London, dipl, 65. Prof Exp: Tech officer, Forest Path Unit, Forest Prod Lab, Dept Northern Affairs & Nat Resources, Can, 52; RES SCIENTIST, RES BR, CAN DEPT AGR, 52- Mem: Soc Nematol; Can Phytopath Soc; Soc Europ Nematol. Res: Ecology of plant parasitic nematodes; forage nematodes. Mailing Add: Res Sta Box 185 Can Dept Agr Vineland Station ON Can

TOWNSLEY, PHILIP MCNAIR, b Vancouver, BC, Nov 21, 25; m 52; c 3. INDUSTRIAL MICROBIOLOGY. Educ: Univ BC, BSA, 49; Univ Calif, Berkeley, MS, 50, PhD(comp biochem), 56. Prof Exp: Biochemist, Dept Agr, Govt Can, 56-61, group leader, Process & Prod Group, Fishery Res Bd, 61-63; group leader biochem, BC Res Coun, 63-67; PROF INDUST MICROBIOL, UNIV BC, 67- Concurrent Pos: Mem bd dirs, John Dunn Agencies Ltd & Pac Micro-Bio Cult Ltd. Mem: Inst Food

Technol; Can Inst Food Sci & Technol; Int Asn Plant Tissue Cult; Can Soc Microbiol. Res: Practical application of basic research. Mailing Add: Dept Food Sci Fac of Agr Sci Univ of BC Vancouver BC Can

TOWNSLEY, SIDNEY JOSEPH, b Colorado Springs, Colo, Aug 6, 24; m 50; c 5. RADIOBIOLOGY. Educ: Univ Calif, AB, 47; Univ Hawaii, MS, 50; Yale Univ, PhD(zool), 54. Prof Exp: Asst prof marine zool & asst, Marine Lab, 54-60, assoc prof marine biol, 60-66, prof, 66-70, PROF MARINE ZOOL, UNIV HAWAII, 70- Mem: Am Soc Zoologists; Am Soc Limnol & Oceanog; Ecol Soc Am. Res: Ecology of radioisotopes in marine organisms; systematics; stomatopod Crustacea and cephalopod mollusks; histochemistry and physiology of heavy metals. Mailing Add: Dept of Zool Univ of Hawaii Honolulu HI 96822

TOWNSLEY, WILLIAM W, JR, b Pensacola, Fla, Aug 17, 33; m 59; c 2. PLANT PATHOLOGY & PHYSIOLOGY. Educ: Wheaton Col (Ill), BS, 55; Univ Del, MS, 61; Univ Md, PhD(plant path), 65. Prof Exp: Assoc prof biol, West Chester State Col, 64-67; asst prof, Lake Forest Col, 67-69; exec vpres, Nat Liquid Fertilizer, 69-70; PRES, BOT CONSULT, INC, 70- Mailing Add: 2730 Wildwood Lane Deerfield IL 60015

TOWSE, DONALD FREDERICK, b Somerville, Mass, Dec 5, 24; m 45; c 6. GEOLOGY. Educ: Mass Inst Technol, BS, 48, PhD(geol), 51; Am Inst Prof Geologists, cert. Prof Exp: Geologist, Amerada Petrol Corp, 48-49, 50-51; asst prof geol, Univ NDak, 51-54; geologist, State Geol Surv, NDak, 51-54; consult geologist, 54-56; vis assoc prof, Univ Calif, Los Angeles, 56-57; proj geologist, Kaiser Aluminum & Chem Corp, 57-60, sr proj geologist, Kaiser Cement & Gypsum Corp, 60-71, sr geologist, Kaiser Explor & Mining Co, 71-73; GEOLOGIST, LAWRENCE LIVERMORE LAB, UNIV CALIF, 74- Honors & Awards: Pres Award, Am Asn Petrol Geologists, 52. Mem: Am Asn Petrol Geologists; Am Inst Mining, Metall & Petrol Engrs. Res: Stratigraphy and petroleum resources; carbonate oil reservoirs; cement raw materials; laterites; uranium ores; geothermal resources; geology and economics of geothermal energy deposits. Mailing Add: Lawrence Livermore Lab Box 3011 San Jose CA 95116

TOXEN, ARNOLD MARTIN, solid state physics, low temperature physics, see 12th edition

TOY, ARTHUR DOCK FON, b Canton, China, Sept 13, 13, nat US; m 42; c 3. INDUSTRIAL CHEMISTRY. Educ: Univ Ill, BS, 39, MS, 40, PhD(chem), 42. Prof Exp: Res chemist, Victor Chem Works Div, Stauffer Chem Co, 42-53, dir org res, 53-59, assoc dir res, 59-63, dir res, 63-65; vis scientist, Cambridge Univ, 65-66; sr scientist, 66-68, sr scientist & actg mgr, Chem Res Dept, 68-70, sr scientist & actg mgr, Specialties Dept, 70-72, chief scientist, 72-74, DIR EASTERN RES CTR, STAUFFER CHEM CO, 75- Mem: Am Chem Soc; NY Acad Sci; The Chem Soc; Soc Ger Chem. Res: Organic phosphorus compounds for plastic applications and insecticides; allyl aryl-phosphonate flame resistant plastic; economic process for synthesis of phosphorus insecticides; aquo ammono phosphoric acids; organic reaction mechanisms. Mailing Add: Eastern Res Ctr Stauffer Chem Co Dobbs Ferry NY 10522

TOY, GERALD ROBERT, organic chemistry, see 12th edition

TOY, MADELINE SHEN, b Shanghai, China, Nov 6, 28; US citizen; m 51; c 1. POLYMER & ORGANIC CHEMISTRY. Educ: Col St Teresa (Minn), BS, 49; Univ Wis, MS, 51; Ohio State Univ, MS, 57; Univ Pa, PhD(org chem), 59. Prof Exp: Proj chemist, Freelander Res & Develop Div, Dayco Corp, Calif, 59, mgr org lab, 60; asst prof, Calif State Univ, Northridge, 60-61; staff mem, Int Tel & Tel Corp, Fed Labs, 61-63; res scientist, astropower lab, Douglas Aircraft Corp, Newport Beach, 64-69, head polymer sci, Douglas Adv Res Labs, McDonnell Douglas Corp, 69-70; sr polymer chemist, Stanford Res Inst, Menlo Park, 71-75; SR SCIENTIST, SCI APPLN, INC, 75- Mem: AAAS; Am Chem Soc; fel Am Inst Chemists; NY Acad Sci; fel The Chem Soc. Res: Optically active polymers; fire retardant polyurethanes; high temperature plastics; thermoplastic films; high energy perfluorinated salts; surface polymerizations on metal substrates; fluoroelastomers; multifunctional fluoropolymers; perfluoropolymer-forming reactions; flame resistant surface treatments. Mailing Add: Sci Appln Inc 1257 Tasman Dr Sunnyvale CA 94086

TOY, STEPHEN THOMAS, b Danville, Pa, Oct 1, 39; m 73; c 1. IMMUNOLOGY, VIROLOGY. Educ: Susquehanna Univ, BA, 61; Univ Fla, PhD(microbiol), 66. Prof Exp: Sr lectr virol, Case Western Reserve Univ, 68-69, asst prof, 69-71; asst prof, Thomas Jefferson Univ, 71-74; RES IMMUNOLOGIST, CENT RES DEPT, EXP STA, E I DU PONT DE NEMOURS & CO, INC, 74- Concurrent Pos: Rosalie B Hite fel, Univ Tex, Austin, 66; trainee E I du Pont Exp Sta, Del, 66-68; clin chemist, St Luke's Hosp, Cleveland, 69-70. Mem: AAAS; Am Soc Microbiol. Res: Cell membrane research involving tumor-specific antigens on surface of cells and lymphocytic response to these antigens. Mailing Add: Exp Sta Cent Res Dept E I du Pont de Nemours & Co Wilmington DE 19898

TOY, TERRENCE JOSEPH, b Sidney, Ohio, Aug 10, 46; m 68. GEOMORPHOLOGY, GEOGRAPHY. Educ: State Univ NY Buffalo, BA, 69, MA, 70; Univ Denver, PhD(geog), 73. Prof Exp: ASST PROF GEOG, UNIV DENVER, 73- Concurrent Pos: Hydrol technician surface hydrol, US Geol Surv, 73 & 74, prin investr mine reclamation, 75-76. Mem: Sigma Xi; Nat Geog Soc. Res: Climatic appraisal of the rehabilitation potential of strippable coal lands in the Powder River basin, Wyoming and Montana. Mailing Add: Dept of Geog Univ of Denver Denver CO 80210

TOYAMA, THOMAS KAZUO, b Minot, NDak, Feb 25, 24. HORTICULTURE. Educ: Univ Minn, BS, 53, PhD(plant breeding), 61. Prof Exp: Res officer, Can Dept Agr, 61-63; horticulturist, 63-70, res horticulturist, 70-75, SUPVRY RES HORTICULTURIST, AGR RES SERV, USDA, 75- Concurrent Pos: Horticulturist, Wash State Univ, 63-70. Mem: AAAS; Am Soc Hort Sci; Am Pomol Soc. Res: Plant breeding; breeding, genetics and cytology of stone fruits. Mailing Add: Irrigated Agr Res & Exten Ctr Prosser WA 99350

TOZER, THOMAS NELSON, b San Diego, Calif, July 4, 36; m 60. PHARMACEUTICAL CHEMISTRY. Educ: Univ Calif, San Francisco, BS & PharmD, 59, PhD(pharmaceut chem), 63. Prof Exp: Lectr chem & pharmaceut chem, Univ Calif, San Francisco, 63; NIMH fel, 63-65; asst prof, 65-74, ASSOC PROF PHARM & PHARMACEUT CHEM, UNIV CALIF, SAN FRANCISCO, 74- Concurrent Pos: Pharmacist, 59-; fel, Lab Chem Pharmacol, Nat Heart Inst, 65; vis asst prof, Columbia Univ, 66; USPHS consult, 66- Mem: AAAS; Am Pharmaceut Asn; Acad Pharmaceut Sci; Am Chem Soc. Res: Biopharmaceutics; stability of free radicals formed from phenothiazine and related compounds and their electron-spin-resonance spectra; catecholamine and indole metabolism; biological transport systems, especially the brain; drug metabolism and pharmacokinetics. Mailing Add: Sch of Pharm Univ of Calif San Francisco CA 94122

TOZLOSKI, ALBERT HENRY, b Sunderland, Mass, Apr 20, 26; m 52. ECONOMIC ENTOMOLOGY, INVERTEBRATE ZOOLOGY. Educ: Univ Mass, BS, 50, MS, 52, PhD(entom, bot), 54. Prof Exp: Instr biol, Teachers Col, Conn, 55-59, asst prof biol, 59-65, chmn dept, 65-74, PROF BIOL, CENT CONN STATE COL, 74- Mem: AAAS; Ecol Soc Am; Entom Soc Am. Res: Ecology; invertebrate zoology. Mailing Add: Dept of Biol Cent Conn State Col New Britain CT 06050

TRABANT, EDWARD ARTHUR, b Los Angeles, Calif, Feb 28, 20; m 43; c 3. APPLIED MATHEMATICS. Prof Exp: Occidental Col, AB, 41; Calif Inst Technol, PhD(appl math), 47. Prof Exp: From instr to prof math & eng sci, Purdue Univ, 47-60; dean sch eng, State Univ NY Buffalo, 60-66; vpres acad affairs, Ga Inst Technol, 66-68; PRES, UNIV DEL & PROF ENG SCI, COL ENG, 68- Concurrent Pos: Consult, Allison Div, Gen Motors, Ind, 50-55, Argonne Nat Lab, Ill, 55-61; Carborundum Corp, NY, 64-68 & Army Sci Adv Panel, 66-71. Mem: Am Soc Eng Educ; Am Soc Mech Eng; Am Math Soc; Am Nuclear Soc. Mailing Add: Off of the Pres Univ of Del Newark DE 19711

TRABER, DANIEL LEE, b Victoria, Tex, Apr 28, 38; m 59. PHYSIOLOGY, PHARMACOLOGY. Educ: St Mary's Univ, Tex, BA, 59; Univ Tex, MA, 62, PhD(physiol), 65. Prof Exp: Asst physiol, Med Br, Univ Tex, 60-65; fel pharmacol, Col Med, Ohio State Univ, 65-66; asst prof physiol & res asst prof anesthesiol, 66-70, dir, Interdisciplinary Labs, 70-72, ASSOC PROF ANESTHESIOL & PHYSIOL, UNIV TEX MED BR GALVESTON, 70-, DIR, INTEGRATED FUNCTIONAL LAB, 72- Concurrent Pos: Chief, Div Anesthesia Res, Shriners Burn Inst, Galveston, 71- Mem: AAAS; Am Physiol Soc; Am Burn Asn; Soc Exp Biol & Med; Am Soc Pharmacol & Exp Therapeut. Res: Pharmacology of anesthetic agents; shock; autonomic pharmacology; endotoxemia. Mailing Add: Dept of Anesthesiol & Physiol Univ of Tex Med Br Galveston TX 77550

TRACE, JAMES CHALMERS, b Columbus, Ohio, Oct 22, 24; m 48; c 2. VETERINARY MEDICINE. Educ: Ohio State Univ, DVM, 49. Prof Exp: Vet inspector virus-serum control, US Bur Animal Indust, 49-51; asst supt biol farm, Biol Prod, 51-53, asst dir pharmaceut prod develop, 53-59, dir, 59-62, vpres res & develop, 62-70, EXEC V PRES RES & CONTROL, FT DODGE LABS DIV, AM HOME PROD CORP, 70- Mem: Am Soc Vet Physiol & Pharmacol; Am Vet Med Asn; Animal Health Asn. Res: Veterinary pharmacology; chemotherapy; immunology. Mailing Add: Ft Dodge Labs Div Am Home Prod Corp 800 5th St NW Ft Dodge IA 50501

TRACE, ROBERT DENNY, b Zanesville, Ohio, Oct 27, 17; m 50; c 2. GEOLOGY. Educ: Southern Methodist Univ, BS, 40; Univ Calif, Los Angeles, MA, 47. Prof Exp: GEOLOGIST, US GEOL SURV, 42- Concurrent Pos: Instr, Hopkinsville Community Col, Ky, 66-70. Mem: Fel Geol Soc Am; Soc Econ Geol. Res: Stratigraphic and structural geology of fluorspar-zinc-lead deposits. Mailing Add: 412 Eagle St Princeton KY 42445

TRACEY, JOSHUA IRVING, JR, b New Haven, Conn, May 2, 15; m 46; c 2. GEOLOGY. Educ: Yale Univ, BA, 37, MS, 43, PhD(geol), 50. Prof Exp: GEOLOGIST, US GEOL SURV, 42- Mem: AAAS; Soc Econ Paleont & Mineral; Geol Soc Am; Am Asn Petrol Geol; Am Geophys Union. Res: Bauxite; geology and ecology of coral reefs; geology of the Pacific Islands; tertiary stratigraphy of southwestern Wyoming. Mailing Add: Nat Ctr Stop 915 US Geol Surv Reston VA 22092

TRACEY, MARTIN LOUIS, JR, b Boston, Mass, Mar 3, 43; m 66; c 2. GENETICS. Educ: Providence Col, AB, 65; Brown Univ, PhD(biol), 71. Prof Exp: Fel genetics, Univ Calif, Davis, 71-73; dir genetics, Bodega Marine Lab, Calif, 73-74; ASST PROF BIOL, BROCK UNIV, 74- Mem: AAAS; Genetics Soc Am; Am Soc Naturalists; Can Genetics Soc; Soc Study Evolution. Res: Speciation and genetic differentiation; recombination; polymorphism; genetic organization; genetic control of behavior; genetics of human population; aquaculture. Mailing Add: Dept of Biol Sci Brock Univ St Catherines ON Can

TRACHEWSKY, DANIEL, b Montreal, Que, Nov 4, 40; m 67; c 2. ENDOCRINOLOGY, MOLECULAR BIOLOGY. Educ: McGill Univ, BSc, 61, MSc, 63, PhD(biochem), 66. Prof Exp: Res assoc biochem of reproduction, Biomed Div Pop Coun, Rockefeller Univ, 65-67; dir molecular biol, Montreal Clin Res Inst, 68-75; asst prof med, McGill Univ, 69-75; from asst prof to assoc prof med, Univ Montreal, 69-75; ASSOC PROF MED, BIOCHEM & MOLECULAR BIOL, HEALTH SCI CTR, UNIV OKLA, 75- Concurrent Pos: Grants, Can Med Res Coun, Que Med Res Coun & Que Heart Found, Montreal Clin Res Inst, 68-; off referee, reviewing grant applns, Can Med Res Coun, 69- Mem: NY Acad Sci; Endocrine Soc; Am Fedn Clin Res; Int Platform Asn; Soc Exp Biol & Med. Res: Mechanism of action of steroid and peptide hormones at the molecular level; pathophysiology of hypertension. Mailing Add: Health Sci Ctr Univ of Okla PO Box 26901 Oklahoma City OK 73190

TRACHT, MYRON EDWARD, b New York, NY, June 8, 28; m 56; c 3. PATHOLOGY. Educ: Princeton Univ, AB, 48, Univ Chicago, MS, 54, PhD(physiol) & MD, 55. Prof Exp: Pathologist, US Naval Med Res Inst, 56-58; asst prof, 61-69, ASSOC PROF PATH, COL PHYSICIANS & SURGEONS, COLUMBIA UNIV, 69-; DIR LABS, HOLY NAME HOSP, 65- Concurrent Pos: USPHS res fel path, Mt Sinai Hosp, 58-61; res assoc, Sch Med, Georgetown Univ, 56-58; asst attend pathologist, Presby Hosp, New York, 61-63; assoc pathologist, Beth Israel Hosp, 63-65. Res: Fat transport and metabolism; liver injury; endocrine influences on metabolism. Mailing Add: Dept of Path Holy Name Hosp 718 Teaneck Rd Teaneck NJ 07666

TRACHTENBERG, EDWARD NORMAN, b New York, NY, Dec 8, 27; m 54; c 3. ORGANIC CHEMISTRY. Educ: NY Univ, AB, 49; Harvard Univ, AM, 51, PhD(org chem), 53. Prof Exp: Instr chem, Columbia Univ, 53-58; from asst prof to assoc prof, 58-70, PROF CHEM, CLARK UNIV, 70- Concurrent Pos: NSF fel, Univ London, 67-68. Mem: Am Chem Soc; Am Inst Chem; The Chem Soc. Res: Mechanism of organic reactions; organic synthesis; selenium dioxide oxidation of organic compounds; 1, 3-dipolar cycloadditions. Mailing Add: Dept of Chem Clark Univ Worcester MA 01610

TRACHTENBERG, ISAAC, b New Orleans, La, Aug 20, 29; m 57; c 2. ELECTROCHEMISTRY. Educ: Rice Inst, BA, 50; La State Univ, MS, 52, PhD(chem), 57. Prof Exp: Assoc chemist, Am Oil Co, Tex, 57-59, chemist, 59-60, sr chemist & group leader, 60; mem tech staff, 60-63, br head basic electrochem, 63-66, chem kinetics, 66-68, systs anal process control, 69-70, br head environ monitoring, 71-72, br head sensors res & develop, 72-74, mgr qual & reliability assurance, 74-75, MGR PROCESS CONTROL, SEMICONDUCTOR GROUP, TEX INSTRUMENTS, 74- Concurrent Pos: VChmn & chmn, Gordon Res Con Electrochem, 68-69; ed j battery div, Electrochem Soc, 69-72. Mem: Am Chem Soc; Electrochem Soc. Res: Semiconductors; ion-selective electrochemical sensors; environmental sciences and monitoring; semiconductor device quality and reliability.

Mailing Add: Semiconductor Group PO Box 5012 Tex Instruments MS 17 Dallas TX 75222

TRACHTENBERG, MICHAEL CARL, b New York, NY, June 22, 41; m 64; c 4. NEUROANATOMY, NEUROPHYSIOLOGY. Educ: City Col New York, BA, 62; Univ Calif, Los Angeles, PhD(anat neurosci), 67. Prof Exp: Res biologist, Boston Vet Admin Hosp, 69-70; INSTR NEUROL, SCH MED, BOSTON UNIV, 69-; ASST NEUROPHYSIOLOGIST, McLEAN HOSP, 71- Concurrent Pos: Nat Inst Neurol Dis & Stroke res fel surg, Med Sch, Harvard Univ, 67-69; Eric Slack-Gyr Found vis researcher neuroanat & neurophysiol, Univ Zurich, 70-71; instr anat, Med Sch, Harvard Univ, 71-72. Mem: Soc Neurosci; Am Epilepsy Soc. Res: Correlation of structure, function and behavior in the visual system; supportive functions of neurological cells examined biophysically, physiologically and anatomically. Mailing Add: McLean Hosp Neuropsychol Lab PO Box 8 Belmont MA 02178

TRACHTMAN, MENDEL, b May 6, 29; US citizen; m 50; c 3. RADIATION CHEMISTRY, PHOTOCHEMISTRY. Educ: Temple Univ, AB, 51; Drexel Inst Technol, MS, 56; Univ Pa, PhD(chem), 61. Prof Exp: Res chemist, Frankford Arsenal, Philadelphia, Pa, 51-67; PROF CHEM, PHILADELPHIA COL TEXTILES & SCI, 67-, Concurrent Pos: Res Corp res grant, Philadelphia Col Textiles & Sci, 72-74. Mem: Am Chem Soc. Res: Determination of quenching cross sections of aromatic molecules; fading properties of various dye molecules. Mailing Add: Dept of Chem Philadelphia Col of Textiles & Sci Philadelphia PA 19144

TRACIE, CARL JOSEPH, b Sexsmith, Alta, May 27, 39; m 61; c 3. HISTORICAL GEOGRAPHY, GEOGRAPHY OF WESTERN CANADA. Educ: Univ Alta, BEd, 65, MA, 67, PhD(geog), 70. Prof Exp: ASST PROF GEOG, UNIV SASK, 70- Concurrent Pos: Can Coun res grant, Univ Sask, 71-73 & 74-75. Mem: Can Asn Geog; Am Asn Geog. Res: Process of agricultural settlement, particularly on the forest-grassland fringe; environmental perception and its influence on agricultural settlement; gold-rush settlement; ethnic group settlement in Western Canada. Mailing Add: Dept of Geog Univ of Sask Saskatoon SK Can

TRACTON, MARTIN STEVEN, b Brockton, Mass, Feb 9, 45; m 66; c 1. METEOROLOGY. Educ: Univ Mass, Amherst, BS, 66; Mass Inst Technol, MS, 69, PhD(meteorol), 72. Prof Exp: Asst prof meteorol, Naval Postgrad Sch, 72-75; RES METEOROLOGIST, NAT METEOROL CTR, NAT OCEANIC & ATMOSPHERIC ADMIN, 75- Mem: Am Meteorol Soc. Res: Synoptic-dynamic aspects of the role of cumulus convection in the development of extratropical cyclones; test and evaluation of numerical prediction models. Mailing Add: Nat Meteorol Ctr Nat Oceanic & Atmospheric Admin Washington DC 20231

TRACY, C RICHARD, b Glendale, Calif, May 24, 43; m 67; c 1. ECOLOGY. Educ: Calif State Univ, Northridge, BA, 66, MS, 68; Univ Wis, PhD(zool), 72. Prof Exp: Res assoc environ studies, Univ Wis, 72-73, lectr, 73-74, asst scientist, 73-74; ASST PROF ZOOL, COLO STATE UNIV, 74-, NATURAL RESOURCE ECOL LAB, 75- Concurrent Pos: Asst prof biol, Univ Mich Biol Sta, 74-; nat adv comt person, Univ Wis, Biotron, 74- Mem: AAAS; Am Inst Biol Sci; Ecol Soc Am; Sigma Xi; Am Soc Ichthyologists & Herpetologists. Res: Biophysical ecological techniques to study evolutionary ecological questions of adaptations to physical environments; dispersal; dispersion; habitat selection and space utilization in animals and plants. Mailing Add: Dept of Zool-Entom Colo State Univ Ft Collins CO 80523

TRACY, DAVID J, b Covington, Ky, Jan 22, 37. ORGANIC CHEMISTRY. Educ: Villa Madonna Col, AB, 59; Univ Ill, MS, 61, PhD(chem), 64. Prof Exp: Res specialist, 64-49, TECH ASSOC, GAF CORP, 69- Mem: Am Chem Soc. Mailing Add: 209 Comly Rd Apt M-24 Lincoln Park NJ 07035

TRACY, DERRICK SHANNON, b Mirzapur, India, July 1, 33; m 69. MATHEMATICAL STATISTICS. Educ: Univ Lucknow, BSc, 51, MSc, 53; Univ Mich, MS, 60, ScD(math), 63. Prof Exp: Lectr math, Ewing Col, Allahabad, 53-55; sr lectr math statist, Govt Col Bhopal, 56-59; res asst math, Univ Mich, 61-63; asst prof statist, Univ Conn, 63-65; assoc prof, 65-70, PROF MATH, UNIV WINDSOR, 70- Concurrent Pos: Nat Res Coun Can res grants, 66-; Defence Res Bd Can res grants, 68-; consult, Bell Tel Co Can & Ont Inst Studies Educ, 67-; consult, Walter Reed Army Res Inst, 68; vis prof, Univ Calif, Riverside, 72-73 & UnivWaterloo, 74; assoc ed Can J Statist, 73-; Can Coun travel grant to Poland, 75. Mem: Am Statist Asn; Inst Math Statist; Am Math Soc; Math Asn Am; Biomet Soc. Res: Products of generalized k-statistics; finite moment formulae; symmetric functions; matrix derivatives in multivariate analysis. Mailing Add: Dept of Math Univ of Windsor Windsor ON Can

TRACY, JAMES ESTEL, organic chemistry, see 12th edition

TRACY, JAMES FRUEH, b Isle of Pines, July 29, 16; US citizen; m 57; c 2. NUCLEAR PHYSICS. Educ: Univ Ill, BS, 40; Univ Calif, MS, 49, PhD(physics), 53. Prof Exp: Elec engr, Gen Elec Co, NY, 40-46; opers analyst, Broadview Res & Develop, Calif, 53; PHYSICIST, LAWRENCE LIVERMORE LAB, UNIV CALIF, 53- Mem: Am Phys Soc. Res: Physics design of nuclear weapons; application of nuclear explosives to industry and science. Mailing Add: 1262 Madison Ave Livermore CA 94550

TRACY, JOSEPH CHARLES, JR, b Wilkes Barre, Pa, Jan 15, 43; m 66; c 3. SOLID STATE PHYSICS, SURFACE PHYSICS. Educ: Rensselaer Polytech Inst, BEE, 64; Cornell Univ, PhD(appl physics), 68. Prof Exp: Fel surface physics, NAm Rockwell Sci Ctr, Calif, 68-69, mem tech staff, 69-70; mem tech staff, Bell Tel Labs, Murray Hill, 70-74; MEM STAFF, DEPT PHYSICS, GEN MOTORS RES LAB, 74- Mem: Am Phys Soc; Am Vacuum Soc. Res: Structural aspects of gas adsorption on metallic surfaces; instrumentation for surface studies of the solid state; electronic properties of semiconductor surfaces. Mailing Add: Dept of Physics Gen Motors Res Lab GM Tech Ctr Warren MI 48090

TRACY, JOSEPH WALTER, b Seattle, Wash, June 22, 24; m 50; c 5. INORGANIC CHEMISTRY, MEAT SCIENCE. Educ: Univ Wash, BS, 51, MS, 54, PhD(chem), 60. Prof Exp: Assoc res engr, Boeing Airplane Co, 54-58; prof chem, Northwest Nazarene Col, 60-70; CHEMIST CHG MEAT INSPECTION LAB, IDAHO STATE DEPT AGR, 70- Mem: Am Chem Soc; Asn Off Anal Chem. Res: Meat analysis; x-ray crystallography; crystal structures of simple inorganic compounds. Mailing Add: 823 Ninth Ave So Nampa ID 83651

TRACY, M JOANNA, b Newark, NJ, Oct 22, 00. EXPERIMENTAL MEDICINE. Educ: Fordham Univ, BS, 28; Catholic Univ Am, MS, 38; Inst Divi Thomae, PhD, 60. Prof Exp: Chmn, Nat Sci Div, 46-72, PROF BIOL & CHEM, CALDWELL COL FOR WOMEN, 41- Concurrent Pos: Researcher, Inst Divi Thomae. Mem: AAAS; Am Chem Soc; Am Asn Biol Teachers; Nat Sci Teachers Asn. Res: Testing efficacy of a beef brain extract on staphlococcus aureus infections in mice; testing extent of immunity from infection following injections of beef brain extract; immunity in C3H mice to transplantable lymphosarcoma 6C3HED following injections of viable tumor cells. Mailing Add: Natural Sci Div Caldwell Col for Women Caldwell NJ 07006

TRACY, PHILIP T, b Chester, Pa, Jan 14, 30; m 50; c 2. PLASMA PHYSICS, ELECTROMAGNETICS. Educ: San Diego State Col, BS, 58. Prof Exp: Physicist, Gen Dynamics Corp, 58-61; sr physicist, Geophys Corp Am, 61-62; res scientist, Kaman Nuclear, 62-69, sr scientist, KMS Technol Ctr, 69-73, RES SCIENTIST, KAMAN SCI CORP, 73- Mem: Am Phys Soc. Res: Electromagnetic theory. Mailing Add: Kaman Sci Corp PO Box 7463 Colorado Springs CO 80933

TRACY, RICHARD E, b Klamath Falls, Ore, Apr 30, 34; m 62; c 2. PATHOLOGY. Educ: Univ Chicago, BA, 55, MD & PhD(path), 61. Prof Exp: Intern, Presby Hosp, Denver, Colo, 61-62; res assoc path, Univ Chicago, 62-64, instr, 64-65; asst prof, Med Sch, Univ Ore, 65-67; asst prof, 67-73, ASSOC PROF PATH, LA STATE UNIV SCH MED, NEW ORLEANS, 73- Concurrent Pos: USPHS trainee, 62-65. Honors & Awards: Bausch & Lomb Medal, 61; Joseph A Capps Prize, 65. Mem: AAAS; Am Heart Asn; Am Soc Exp Path. Res: Arteriosclerotic and hypertensive cardiovascular and renal disease; biostatistics. Mailing Add: Dept of Path La State Univ Sch of Med New Orleans LA 70112

TRACY, WILLIAM E, b Memphis, Tenn, Aug 16, 34; m 56; c 3. PEDODONTICS. Educ: Univ Tenn, DDS, 61; Am Bd Pedodont, dipl, 71. Prof Exp: Res asst dent mat, Dent Sch, 60-61, instr pediat, Med Sch & instr dent, Dent Sch, 61-62, asst prof dent, 62-68, ASSOC PROF DENT, DENT SCH, UNIV ORE, 68-, INSTR PEDIAT & DENT MED, MED SCH, 67-, MEM STAFF, CHILD STUDY CLIN, 71- Concurrent Pos: Nat Inst Dent Res fel growth & develop, Dent Sch, Univ Ore, 61-62, spec fel, 65-; Nat Inst Child Health & Human Develop career develop award, 66-67, res grant, 66-71. Mem: AAAS; Am Acad Pedodont; Am Soc Dent Children; Am Dent Asn; Int Asn Dent Res. Res: Growth and development of children; dentofacial growth. Mailing Add: 12555 SW Third Beaverton OR 97005

TRAFICANTE, DANIEL DOMINICK, b Jersey City, NJ, Nov 20, 33; m 55; c 4. STRUCTURAL CHEMISTRY. Educ: Syracuse Univ, BS, 55; Mass Inst Technol, PhD(org chem), 62. Prof Exp: Instr gen chem, US Air Force Acad, 62-63, res assoc, org chem, Frank J Seiler Res Lab, 63-66, chief anal & prog br, 544th aerospace reconnaissance tech wing, 66-67; res assoc chem, Univ Calif, Davis, 67-68; lectr chem & dir undergrad labs, 68-71, DIR CHEM SPECTROMETRY LAB, MASS INST TECHNOL, 70- Concurrent Pos: Prof chem & head dept, Bellevue Col, 66-67; consult, Nuclear Advan Corp, 75- Mem: Am Chem Soc. Res: Nuclear magnetic resonance spectroscopy. Mailing Add: Dept of Chem Rm 18-023 Mass Inst of Technol Cambridge MA 02139

TRAGER, WILLIAM, b Newark, NJ, Mar 20, 10; m 35; c 3. PARASITOLOGY. Educ: Rutgers Univ, BS, 30; Harvard Univ, AM, 31; PhD(biol), 33. Hon Degrees: ScD, Rutgers Univ, 65. Prof Exp: Nat Res Coun fel med, 33-34; fel, Rockefeller Inst, 34-35, asst, 35-40, assoc 40-50, assoc mem, 50-59, assoc prof, 59-64, PROF PARASITOL, ROCKEFELLER UNIV, 64- Concurrent Pos: Ed J, Soc Protozool, 53-65; mem study sect parasitol & trop med, Nat Inst Allergy & Infectious Dis, 54-58 & 66-70; mem training grant comt, 61-64; mem malaria comn, Armed Forces Epidemiol Bd, 65-70; guest investr, W African Inst Trypanosomiasis Res, 58-59; vis prof, Fla State Univ, 62, Med Sch, Puerto Rico, 63 & Med Sch, Nat Univ Mex, 65; Guggenheim found fel, 73. Mem: Nat Acad Sci; Am Soc Parasitol (pres, 74); Soc Protozool (pres, 60-61); Am Soc Trop Med & Hyg (vpres, 64-65); NY Acad Sci. Res: Insect physiology; physiology of parasitisms; cultivation of intracellular parasites; malaria. Mailing Add: Rockefeller Univ York Ave & 66th St New York NY 10021

TRAGER, WILLIAM FRANK, b Winnipeg, Man, Oct 17, 37; m 60; c 3. PHARMACEUTICAL CHEMISTRY, ORGANIC CHEMISTRY. Educ: Univ San Francisco, BSc, 60; Univ Wash, PhD(pharmaceut chem), 65. Prof Exp: NIH fel pharm, Chelsea Col Sci & Technol, London, 65-67; fel, Univ Wash, 67; asst prof chem & pharmaceut chem, Col Pharm, Univ San Francisco, 67-72; ASSOC PROF PHARMACEUT CHEM, COL PHARM, UNIV WASH, 72- Mem: Am Chem Soc. Res: Drug metabolism studies and mass spectroscopy. Mailing Add: Col of Pharm Univ of Wash Seattle WA 98105

TRAGESER, MILTON B, b Orange, NJ, Apr 12, 28; m 54; c 3. PHYSICS. Educ: Mass Inst Technol, BS, 51. Prof Exp: Asst navig gyros, 51-52, aircraft guid & navig systs, 52-58, head space syst studies, 58-61, Apollo Study Prog, 61, dir Apollo Guid & Navig Prog, 61-66, ADV TECH DIR, INSTRUMENTATION LAB, MASS INST TECHNOL, 66- Concurrent Pos: Mem ad hoc comt, US Air Force Sci Adv Bd Space Technol, 66-67. Res: Gyroscope development; navigation and guidance system design; guidance phenomenology; gravity gradiometers. Mailing Add: DL 11-237B Draper Lab Mass Inst Technol 37 Cambridge Pkwy Cambridge MA 02139

TRAGGIS, DEMETRIUS G, b Winsted, Conn, Oct 28, 16; m 46; c 1. PEDIATRICS. Educ: Johns Hopkins Univ, AB, 38, MD, 42. Prof Exp: Intern & asst resident pediat, Johns Hopkins Hosp, 42-43 & 46-47; intern & asst resident pediat, Univ Hosp, Univ Mich, 47-48, chief resident, 48-49, instr, 49-50; from asst prof to assoc prof, Sch Med, Univ Miami, 56-67; SR PEDIATRICIAN, CHILDREN'S CANCER RES FOUND, 67- Concurrent Pos: Asst prof pediat, Harvard Med Sch; physician, Children's Hosp Med Ctr, Boston. Mem: Am Acad Pediat. Res: Prematurity and cancer in children. Mailing Add: Children's Cancer Res Found 35 Binney St Boston MA 02115

TRAHAN, DONALD HERBERT, b North Adams, Mass, Mar 14, 30; m 61; c 1. MATHEMATICS. Educ: Univ Vt, BS, 52; Univ Nebr, MA, 54; Univ Pittsburgh, PhD, 61. Prof Exp: Instr math, Univ Mass, 56-59; asst prof, Univ Pittsburgh, 61-65; asst prof & chmn dept, Chatham Col, 65-66; ASSOC PROF MATH, NAVAL POSTGRAD SCH, 66- Concurrent Pos: Hays-Fulbright lectr, Nat Univ Ireland, 63-64. Mem: Am Math Soc; Math Asn Am. Res: Complex variables; univalent function theory; real analysis. Mailing Add: 79 Via Ventura Monterey CA 93940

TRAHANOVSKY, WALTER SAMUEL, b Conemaugh, Pa, June 15, 38; m 67; c 3. ORGANIC CHEMISTRY. Educ: Franklin & Marshall Col, BS, 60; Mass Inst Technol, PhD(chem), 63. Prof Exp: NSF fel chem, Harvard Univ, 63-64; from instr to assoc prof, 64-74, PROF CHEM, IOWA STATE UNIV, 74- Concurrent Pos: A P Sloan fel, 70-72. Mem: AAAS; Am Chem Soc; The Chem Soc. Res: Physical-organic chemistry, including the study of oxidations and reductions of organic compounds; oxidative cleavages; cyclization reactions; free radicals; carbocations; arene tricarbonylchromium complexes; flash vacuum thermolysis; methylenecyclobutenones; tropolone derivatives. Mailing Add: Dept of Chem Iowa State Univ Ames IA 50011

TRAIL, CARROLL C, b Forney, Tex, Dec 25, 27; m 51; c 3. NUCLEAR PHYSICS. Educ: Agr & Mech Col Tex, BS, 49, MS, 51, PhD, 56. Prof Exp: Asst physics, Argonne Nat Lab, 56-60, assoc physicist, 60-64; assoc prof, 64-68, PROF PHYSICS, BROOKLYN COL, 68-, CHMN DEPT, 69- Concurrent Pos: Vis prof, Linear Accelerator Lab, Orsay, France, 70-71. Mem: Fel Am Phys Soc. Res: Low energy. Mailing Add: Dept of Physics Brooklyn Col Brooklyn NY 11210

TRAIL, STANLEY M, b Bristol, Conn, Mar 19, 30; m 66. STATISTICS. Educ: Bowling Green State Univ, BA, 51, BS, 54; Univ Conn, MA, 55, PhD(educ measurement), 61; Okla State Univ, PhD(statist), 67. Prof Exp: Pub sch teacher, Ohio, 54; asst prof math & psychol, RI Col, 57-62; staff asst math, Okla State Univ, 62-67; asst prof, 67-71, ASSOC PROF MATH, NORTHERN ILL UNIV, 71- Mem: Am Statist Asn; Math Asn Am. Res: Obtaining and combining estimators in experimental designs based on balanced and partially balanced incomplete block designs. Mailing Add: Dept of Math Sci Northern Ill Univ De Kalb IL 60115

TRAIN, CARL T, b Lindsborg, Kans, Jan 19, 39; m 60; c 3. PARASITOLOGY, INVERTEBRATE ZOOLOGY. Educ: Bethany Col (Kans), BS, 61; Kans State Univ, MS, 63, PhD(parasitol), 67. Prof Exp: Asst zool, Kans State Univ, 61-64, instr, 66-67; asst prof, 67-73, ASSOC PROF BIOL, SOUTHEAST MO STATE UNIV, 73- Mem: Am Soc Parasitol; Am Micros Soc. Res: General parasitology, biology and reproduction of nematodes; life cycles of parasites; anthelmintic studies and testing; application of statistical measurements to parasitological studies. Mailing Add: Dept of Biol Southeast Mo State Univ Cape Girardeau MO 63701

TRAINA, VINCENT MICHAEL, b Oceanside, NY, May 8, 43. TOXICOLOGY. Educ: Rutgers Univ, BA, 65, MS, 70, PhD(physiol), 73. Prof Exp: Res investr toxicol, Squibb Inst Med Res, 66-74; MGR TOXICOL, CIBA-GEIGY CORP, 74- Honors & Awards: Supvry Develop Prog Award, Squibb Inst, 73. Mem: Sigma Xi; Environ Mutagen Soc; Am Soc Zoologists; Am Inst Biol Sci. Res: Body fluid volumes and concentrations and electrolyte changes during periods of prolonged starvation; all phases of toxicology. Mailing Add: Ciba-Geigy Corp Morris Ave Summit NJ 07901

TRAINER, DANIEL OLNEY, b Chicago, Ill, July 13, 26; m 55; c 2. WILDLIFE DISEASES. Educ: Ripon Col, BS, 50; Univ Wis-Madison, MS, 55, PhD(bact), 61. Prof Exp: Res virologist, Fromm Labs, 55-56; pathologist, Wis Conserv Dept, 56-62; from asst prof to prof vet sci, Univ Wis-Madison, 62-71; DEAN, COL NATURAL RESOURCES, UNIV WIS-STEVENS POINT, 71- Honors & Awards: Distinguished Serv Award, Wildlife Dis Asn, 73. Mem: AAAS; Wildlife Dis Asn (vpres, 66-68, pres, 68-70); Wildlife Soc; Am Inst Biol Sci; Soc Am Foresters. Res: Ecology of disease, especially diseases of wild or natural populations. Mailing Add: Col of Natural Resources Univ of Wis Stevens Point WI 54481

TRAINER, JOHN EZRA, SR, b Allentown, Pa, Feb 8, 14; m 39; c 2. ORNITHOLOGY. Educ: Muhlenberg Col, BS, 35; MS, Cornell Univ, 38, PhD(ornith), 46. Prof Exp: Teacher biol, Tenn State Teachers Col, 38-39; from instr to prof, 39-65, SR PROF BIOL, MUHLENBERG COL, 65- Mem: Wilson Ornith Soc; Am Ornith Union. Res: Hearing ability of birds; auditory acuity of certain birds; respiration of birds; vertebrate morphology. Mailing Add: Dept of Biol Muhlenberg Col Allentown PA 18104

TRAINER, JOHN EZRA, JR, b Allentown, Pa, Aug 31, 43; m 67; c 3. HELMINTHOLOGY, BIOLOGICAL CHEMISTRY. Educ: Muhlenburg Col, BS, 65; Wake Forest Univ, MA, 67; Univ Okla, PhD(zool), 71. Prof Exp: Teaching asst, Wake Forest Univ, 65-67; teaching asst, Univ Okla, 67-69; ASST PROF BIOL, JACKSONVILLE UNIV, 71- Concurrent Pos: Teaching asst biol sta, Mich State Univ, 66 & Univ Okla, 67. Mem: Am Soc Parasitol; Am Soc Zool; Am Inst Biol Sci; AAAS. Res: Ecology, biochemistry and ultrastructure of the Pentastomida and other parasitic helminths. Mailing Add: Dept of Biol Jacksonville Univ Jacksonville FL 32211

TRAINER, JOSEPH B, b Ft Wayne, Ind, Nov 23, 12; m 40; c 2. MEDICAL PHYSIOLOGY. Educ: Univ Wash, BS, 39, MS, 40; Univ Ore, MD, 46. Prof Exp: Asst prof physiol, Med Sch, 49-55, assoc prof med, 55-70, dir student health serv, Med, Dent & Nursing Schs, 51-69, ASSOC PROF PHYSIOL, MED SCH, UNIV ORE, 55-, PROF MED, 70- Concurrent Pos: Chmn health serv dirs, Ore State Syst of Higher Educ & consult to chancellor, 60-67; mem, Governor's Comt Family Life, 65-66; regional dir, Am Social Health Asn, 65-67. Mem: Fel Am Asn Marriage & Family Counr. Res: Sex education of physicians, medical students and nursing students; premarital counseling. Mailing Add: Dept of Med Univ of Ore Sch of Med Portland OR 97201

TRAINOR, FRANCIS RICE, b Pawtucket, RI, Feb 11, 29; m 56. PHYCOLOGY. Educ: Providence Col, BS, 50; Vanderbilt Univ, MA, 53, PhD(biol), 57. Prof Exp: From instr to assoc prof, 57-67, PROF BOT, UNIV CONN, 67- Concurrent Pos: Consult, Elec Boat Div, Gen Dynamics Corp, Conn, 58-61; Fulbright res scholar, Stockholm, 70; Fulbright lectr, Greece & Yugoslavia, 71. Honors & Awards: Distinguished Fac Award, Univ Conn, 62; Darbaker Award, Bot Soc Am, 65. Mem: Bot Soc Am; Phycol Soc Am (vpres, 68, pres, 69); Brit Phycol Soc; Int Phycol Soc; Am Inst Biol Sci. Res: Sexual reproduction in unicellular algae; algal nutrition and morphogenesis; eutrophication. Mailing Add: Dept of Biol Sci Univ of Conn Storrs CT 06268

TRAINOR, LYNNE E H, b Chamberlain, Sask, Dec 4, 21; m 47; c 3. THEORETICAL PHYSICS. Educ: Univ Sask, BA, 46, MA, 47; Univ Minn, PhD(physics), 51. Prof Exp: Fel, Nat Res Coun Can, 51-52; asst prof physics, Queen's Univ, Ont, 52-55; vis prof, Univ BC, 55-56; from asst prof to prof, Univ Alta, 56-63; PROF PHYSICS, UNIV TORONTO, 63- Concurrent Pos: Chmn, North York Bd Educ, 70-72. Mem: Am Phys Soc; Can Asn Physicists (secy, 66-68, hon secy-treas, 68-70). Res: Nuclear structure theory; properties of nuclear matter; statistical mechanics of Bose-Einstein systems; properties of thin helium films; electron scattering; theoretical biology. Mailing Add: Dept of Physics Univ of Toronto Toronto ON Can

TRAISE, THORNTON, b Chicago, Ill, Nov 18, 28; m 50; c 3. ORGANIC CHEMISTRY. Educ: Blackburn Col, AB, 50; DePaul Univ, MS, 55. Prof Exp: Res chemist, Victor Chem Works, 50-56; res chemist, Stand Oil Co, Inc, 56-66, res proj supvr, Tuloma Gas Prod Co, Ind, 66-68, res proj supvr, Am Oil Co, Chicago, 68-74, MERCHANDISE MGR FERTILIZER, AMOCO OIL CO, CHICAGO, 74- Mem: AAAS; Am Chem Soc. Res: Lubrication; lubricating grease; polymerization; organophosphorus chemistry; fertilizers; plant nutrients; coordination compounds; nitrogen chemistry; agricultural chemicals. Mailing Add: 1205 Dorothy Dr Crete IL 60417

TRAISMAN, HOWARD SEVIN, b Chicago, Ill, Mar 18, 23; m 56; c 3. PEDIATRICS. Educ: Northwestern Univ, BS, 43, BM, 46, MD, 47. Prof Exp: Intern, Cook County Hosp, Chicago, 46-47; resident, Children's Mem Hosp, 49-51; from instr to assoc, 52-57, from asst prof to assoc prof, 62-73, PROF PEDIAT, MED SCH, NORTHWESTERN UNIV, CHICAGO, 73- Concurrent Pos: Attend pediatrician, Children's Mem Hosp, 52- & Northwestern Mem Hosp, 67- Mem: Am Acad Pediat; Am Pediat Soc; Endocrine Soc; Lawson Wilkins Pediat Endocrine Soc; Am Diabetes Asn. Res: Juvenile Diabetes Mellitus. Mailing Add: Dept of Pediat Northwestern Univ Med Sch Chicago IL 60611

TRAITOR, CHARLES EUGENE, b West Frankfort, Ill, Jan 28, 34; m 57; c 2. PHARMACOLOGY, TOXICOLOGY. Educ: St Louis Col Pharm, BS, 60; Purdue Univ, MS, 63, PhD(pharmacol), 65. Prof Exp: RES PHARMACOLOGIST & RES GROUP LEADER, TOXICOL EVAL DEPT, LEDERLE LABS DIV, AM CYANAMID CO, 65- Res: Toxicology of drugs on various organ systems; biochemical and other methods of detection of toxicological changes. Mailing Add: Toxicol Eval Dept Lederle Labs Pearl River NY 10965

TRAJMAR, SANDOR, b Bogacs, Hungary, Sept 7, 31; US citizen; m 57; c 1. MOLECULAR PHYSICS. Educ: Debrecen Univ, dipl, 55; Univ Calif, Berkeley, PhD(phys chem), 61. Prof Exp: Chemist, N Hungarian Chem Works, 55-57; chemist, Stauffer Chem Co, 57-58; teaching asst phys chem, Univ Calif, Berkeley, 58-59, res asst, Lawrence Radiation Lab, 59-61; SR SCIENTIST, JET PROPULSION LAB, CALIF INST TECHNOL, 61-, HEAD MOLECULAR SPECTROS GROUP, 70- Concurrent Pos: Res fel, Calif Inst Technol, 64-66, sr res fel, 69- Honors & Awards: NASA Medal Except Sci Achievement, 73. Mem: Am Chem Soc; Am Phys Soc. Res: High temperature chemistry; molecular spectroscopy; low-energy electron scattering; atomic physics. Mailing Add: 4800 Oak Grove Dr Pasadena CA 91103

TRAKATELLIS, ANTHONY C, b Thessaloniki, Greece, Sept 4, 31. BIOCHEMISTRY, MOLECULAR BIOLOGY. Educ: Nat Univ Athens, MD, 55, Dr(biochem), 58, MS, 61. Prof Exp: Res assoc biochem, Univ Pittsburgh, 61-64, asst prof, 64-65; assoc scientist, Brookhaven Nat Lab, 65-68; assoc prof, 68-71, PROF BIOCHEM, MT SINAI SCH MED, 71- Mem: AAAS; Am Chem Soc; Brit Biochem Soc; NY Acad Sci; Am Soc Biol Chem. Res: Nucleic acids and protein biosynthesis. Mailing Add: Dept of Biochem Mt Sinai Sch of Med New York NY 10029

TRALLI, NUNZIO, b New York, NY, Dec 5, 17; m 57. PHYSICS. Educ: City Col New York, BS, 39; NY Univ, MS, 44, PhD(physics), 51. Prof Exp: Physicist, US Testing Co, Inc, NJ, 40-43 & Powers Elec & Commun Co, NY, 43 & 44-45; res assoc, Tiffany Found, NY, 45; from asst prof to assoc prof, St John's Univ, NY, 45-51; scientist, Atomic Energy Div, H K Ferguson Co, Inc, NY, 51-52; scientist, Walter Kidde Nuclear Labs, Inc, 52-53, sr scientist, 53-55; sr scientist, Nuclear Develop Corp Am, 55-59, mgr theoret phys sect, 59-61; mgr, United Nuclear Corp, 61-62; chmn dept physics, 62-67 & 74-75, dir sci div, 67-70, PROF PHYSICS, C W POST COL, LONG ISLAND UNIV, 62- Concurrent Pos: Adj assoc prof, Hofstra Col, 53-62. Mem: AAAS; Am Phys Soc; Am Nuclear Soc; NY Acad Sci. Res: Nuclear physics; electromagnetic theory; mathematical methods of physics. Mailing Add: C W Post Col Dept of Physics Long Island Univ Greenvale NY 11548

TRAMA, FRANCESCO BIAGIO, b Philadelphia, Pa, Dec 13, 27; m 54. LIMNOLOGY. Educ: Temple Univ, AB, 48, MA, 50; Univ Mich, PhD(zool), 57. Prof Exp: Asst limnol, Acad Natural Sci Philadelphia, 50-53; asst prof zool, Chicago Teachers Col, 57-60; asst prof, 60-63, ASSOC PROF ZOOL, RUTGERS COL, RUTGERS UNIV, NEW BRUNSWICK, 63-, ASSOC DEAN, 73- Concurrent Pos: Res assoc, Great Lakes Res Inst, Univ Mich, 57-61. Mem: Fel AAAS; Am Soc Limnol & Oceanog; Ecol Soc Am. Res: Trophic dynamics and energy transfer in aquatic ecosystem; primary productivity in fresh water. Mailing Add: Dept of Zool Rutgers Col of Rutgers Univ New Brunswick NJ 08903

TRAMBARULO, RALPH, b East Longmeadow, Mass, Jan 24, 25; m 55; c 4. PHYSICS, PHYSICAL CHEMISTRY. Educ: Yale Univ, BS, 44, PhD(phys chem), 49. Prof Exp: Res assoc, Duke Univ, 49-52; asst prof physics, Pa State Col, 52-53 & Univ Del, 53-56; MEM TECH STAFF, BELL TEL LABS, 56- Mem: Inst Elec & Electronics Engrs. Res: Microwave physics and spectroscopy; microwave integrated circuits. Mailing Add: Bell Tel Labs Box 400 Holmdel NJ 07733

TRAMELL, PAUL RICHARD, b El Centro, Calif, Mar 10, 43; m 67; c 1. PHARMACOLOGY, BIOCHEMISTRY. Educ: Fresno State Col, BA, 65, MA, 67; Rice Univ, PhD(biochem), 70. Prof Exp: Dir instrumentation, Cent Calif Med Labs, 65-67; biochemist, Abbott Labs, 71-72; SR PHARMACOLOGIST, ALZA CORP, 72- Concurrent Pos: Nat Inst Dent Res fel, Rice Univ, 70-; NIH fel pharmacol, Med Sch, Stanford Univ, 70-71. Mem: AAAS; Am Asn Clin Chem. Res: Enzymology; drug metabolism; pharmacokinetics. Mailing Add: Dept of Pharmacol Alza Corp Palo Alto CA 94304

TRAMMEL, KENNETH, b Skipperville, Ala, Oct 30, 37; m 58; c 4. ENTOMOLOGY. Educ: Univ Fla, BS, 60, PhD(entom), 65. Prof Exp: Res assoc citrus pest control, Citrus Exp Sta, Univ Fla, 64-65, asst entomologist, 65-67; entomologist, CIBA Agrochem Co, Fla, 67-69; ASSOC PROF ENTOM, NY STATE AGR EXP STA, CORNELL UNIV, 69- Mem: Entom Soc Am. Res: Apple and pear pest management; field application of sex pheromones for monitoring and control of pests. Mailing Add: Dept of Entom NY State Agr Exp Sta Geneva NY 14456

TRAMMELL, GEORGE THOMAS, b Marshall, Tex, Feb 5, 23; m 45; c 4. THEORETICAL PHYSICS. Educ: Rice Inst, BA, 44; Cornell Univ, PhD, 50. Prof Exp: Physicist, Oak Ridge Nat Lab, 50-61; PROF PHYSICS, RICE UNIV, 61- Mem: Am Phys Soc. Res: Particle theory; solid state theory. Mailing Add: Dept of Physics Rice Univ Houston TX 77001

TRAMMELL, JACK HARMAN, JR, b Hamilton, Tex, Dec 16, 39; m 63; c 2. POULTRY NUTRITION. Educ: Tex A&M Univ, BS, 62, MS, 66, PhD(animal nutrit, biochem), 69. Prof Exp: Poultry nutritionist, Commercial Solvents Corp, 69-75; DIR TECH SERV, HERIDER FARMS INC, 75- Mem: Poultry Sci Asn; World Poultry Sci Asn; Am Soc Animal Sci. Mailing Add: Box 271 Center TX 75935

TRAMONDOZZI, JOHN EDMUND, b Malden, Mass, Aug 28, 42. ORGANIC CHEMISTRY. Educ: Boston Col, BS, 64, PhD(chem), 72. Prof Exp: Asst prof chem, 69-75, ASSOC PROF CHEM, CURRY COL, 75- Mem: Am Chem Soc; AAAS. Res: Reactions and syntheses of organic sulfur compounds, especially sufinates and sulfones; organic reactions in fused salt media. Mailing Add: Dept of Chem Sci Div Curry Col Milton MA 02186

TRAMPUS, ANTHONY, b Cleveland, Ohio, July 22, 27. MATHEMATICS. Educ: Case Inst Technol, BS, 51, PhD(math), 57; George Washington Univ, MS, 53. Prof Exp: Mathematician, Nat Bur Stand, DC, 51-53, Firestone Tire & Rubber Co, 53-56 & Gen Elec Co, 57-63; staff mathematician, Interstate Electronics Corp, Calif, 63-70; res mathematician, Univ Dayton Res Inst, 70-72; MATHEMATICIAN CONSULT, NAT SPACE TECHNOL LABS, SPERRY RAND CORP, 72- Mem: Am Math Soc; Soc Indust & Appl Math; Math Asn Am. Res: Function theory in linear algebra; mathematical analysis in science and engineering. Mailing Add: Apt 40 Louisville Garden Apts Bay St Louis MS 39520

TRAMS, EBERHARD GEORG, b Berlin, Ger, Jan 30, 26; nat US; m 50; c 3. BIOCHEMISTRY, PHARMACOLOGY. Educ: Andreas Gym, Berlin, BS, 46; George Washington Univ, PhD, 54. Prof Exp: Res assoc chemother, Cancer Clin, George Washington Univ, 51-55, asst prof pharmacol, Sch Med, 55-58; biochemist, Sect Lipid Chem, 58-60, actg chief, 60-64, CHIEF SECT PHYSIOL & METAB, LAB NEUROCHEM, NAT INST NEUROL & COMMUN DIS & STROKE, 64- Mem: Soc Exp Biol & Med; Am Soc Pharmacol & Exp Therapeut; Am Soc Biol Chemists;

Soc Gen Physiologists; Am Soc Neurochem. Res: Neurochemistry; structure and function of complex lipids and plasma membrane; marine biology; chemical pharmacology. Mailing Add: Lab of Neurochem Nat Inst of Neurol & Commun Dis Bethesda MD 20014

TRANK, JOHN W, b Minneapolis, Minn, July 24, 28; m 52; c 3. PHYSIOLOGY, ELECTRICAL ENGINEERING. Educ: Univ Minn, BEE, 51, MS, 56, PhD(physiol), 61. Prof Exp: From instr biophys to instr physiol, Univ Minn, 54-61; lectr, McGill Univ, 61-63, asst prof, 63-64; asst prof, 64-70, ASSOC PROF PHYSIOL, MED CTR, UNIV KANS, 70- Concurrent Pos: Dep dir univ surg clin, Montreal Gen Hosp, 61-64. Mem: AAAS; Am Physiol Soc; Inst Elec & Electronics Engrs; Biophys Soc. Res: Engineering analysis of cardiovascular control and instrumentation for biological research; bioengineering. Mailing Add: Dept of Physiol Univ of Kans Med Ctr Kansas City KS 66103

TRANKLE, ROBERT JOHN, b Rapid City, SDak, June 19, 22; m 46; c 3. BOTANY, MICROBIOLOGY. Educ: Univ SDak, AB, 49, MA, 51; Univ Iowa, PhD(bot), 63. Prof Exp: Instr bact, Univ SDak, 51-52; asst prof biol, Univ Omaha, 52-58; ASSOC PROF BIOL, FRANKLIN COL, 63- Mem: Bot Soc Am. Res: Comparative anatomy of the xylem elements of the genus Selaginella. Mailing Add: Dept of Biol Franklin Col Franklin IN 46131

TRAN-MANH, NGO, b Vietnam, Oct 9, 35; m 68; c 1. NUCLEAR MEDICINE, MEDICAL PHYSICS. Educ: Univ Saigon, PCB, 57, MD, 64; Univ Calif, Berkeley, PhD(med physics), 69. Prof Exp: Intern med, Choray Hosp, Univ Saigon, 62-63, res asst nuclear med, Med Sch, 63-65, consult, Choray Hosp, 64-65; ASST PROF NUCLEAR MED & RADIOBIOL, MED SCH, UNIV SHERBROOKE, 69- Concurrent Pos: US AEC grant, Donner Lab, Lawrence Berkeley Lab, Univ Calif, 65-69; grant, Univ Sherbrooke, 72. Mem: Am Fedn Clin Res; Soc Nuclear Med; Soc Radiation Res; Can Soc Biochem; Can Soc Clin Invest. Res: Quantitative studies in intermediary metabolism; developments of combined radioisotope and biochemical techniques for clinical diagnosis purposes. Mailing Add: Dept of Nuclear Med & Radiobiol Univ of Sherbrooke Med Ctr Sherbrooke PQ Can

TRANNER, FRANK, b Chelsea, Mass, May 22, 22; m 49; c 4. PHARMACY, CHEMISTRY. Educ: Mass Col Pharm, BS, 43. Prof Exp: Develop & control chemist, Hat Corp Am, 44-46; develop chemist, Remington Rand Inc, 47-54 & Rilling-Dermetics Inc, 54-56; asst chief chemist, Germaine Monteil Cosmetiques, 57-59; develop chemist & sect head, 60-69, res mgr, 69-72, DIR TECH SERV, CHESEBROUGH-POND'S INC, 72- Mem: Fel Am Inst Chemists; Am Chem Soc; Am Pharmaceut Asn; Soc Cosmetic Chem. Res: Research and development of cosmetic, toiletry and pharmaceutical products and processes. Mailing Add: 25 Skyview Dr Trumbull CT 06611

TRANQUADA, ROBERT ERNEST, b Los Angeles, Calif, Aug 27, 30; m 51; c 3. INTERNAL MEDICINE. Educ: Pomona Col, BA, 51; Stanford Univ, MD, 55. Prof Exp: Intern med, Med Ctr, Univ Calif, Los Angeles, 55-56, asst resident, 56-57; intermediate resident, Vet Admin Hosp, Los Angeles, 57-58; from asst prof to assoc prof med, Sch Med, Univ Southern Calif, 60-68, chmn dept community med & pub health, 66-69, prof, 68-75, assoc dean, Sch Med, 69-75; PROF MED & ASSOC DEAN EXTRAMURAL & POSTGRAD PROGS, SCH MED, UNIV CALIF, LOS ANGELES, 76- Concurrent Pos: USPHS fel metab dis, Med Ctr, Univ Calif, Los Angeles, 58-59; USPHS fel diabetes, Sch Med, Univ Southern Calif, 59-60; med dir, Los Angeles County-Univ Southern Calif Med Ctr, 69-74; regional dir, Cent Health Serv Region, Los Angeles County Dept Health Serv, 74-75. Mem: Am Diabetes Asn; Am Fedn Clin Res; Endocrine Soc; Am Pub Health Asn. Res: Community medicine; Diabetes Mellitus; oral hypoglycemic agents; lactic acidosis; spontaneous hypoglycemia; medical care research. Mailing Add: Sch of Med UCLA Health Sci Ctr Los Angeles CA 90024

TRANSTRUM, LLOYD G, b Idaho, Nov 11, 18; m 42; c 2. AGRICULTURE, ENVIRONMENTAL SCIENCES. Educ: Utah State Univ, BS, 49, MSc, 51. Prof Exp: Asst to pres, Sioux City Stockyards Corp, Iowa, 51-52; sr agr tech asst to vpres, Utah opers, 52-53, supvr agr res, Columbia-Geneva Div, 54-55, GEN SUPVR AGR DIV, US STEEL CORP, PROVO, 56- Concurrent Pos: Adv res comt air pollution effect on agr, Robert A Taft Eng Ctr, USPHS, Ohio, 58; mem indust adv comt air pollution res, Boyce Thompson Inst Plant Res & Univ Wis, 54-, comt chmn, 62-63; indust coordr, Fluoride Res Projs, Utah State Univ & Stanford Res Inst, 54-63; adv environ health comt, Salt Lake County Community Serv Coun, 66; mem, Joint City-County Air Pollution Adv Comt, 66-; mem air pollution comt, Utah County, 67-; adv comt air pollution effects agr, Nat Air Pollution Control Admin, 70; adv res comt air pollution effects agr, Agr Res Serv, USDA, 72-; vchmn, Utah Air Conserv Comt, 73- Mem: Fel AAAS; Am Soc Animal Sci; Asn Iron & Steel Eng; Air Pollution Control Asn. Res: Methods for monitoring community air pollution; effects of air pollution on livestock and agricultural crops; agricultural uses of coal chemical derivatives; fluoride, ozone, sulfur dioxide and nitrogen oxides. Mailing Add: 745 E 300 North St American Fork UT 84003

TRANSUE, LAURENCE FREDERICK, b Summerfield, Kans, Apr 2, 14; m 58. PHYSICAL CHEMISTRY. Educ: Tarkio Col, AB, 36; Univ Nebr, AM, 39, PhD(phys chem), 41. Prof Exp: Asst, Univ Nebr, 37-41; res chemist, 41-50, supvr, 50-52, SUPT & RES MGR PHOTO-PROD DEPT, E I DU PONT DE NEMOURS & CO, INC, 52- Mem: Am Chem Soc; Soc Photog Sci & Eng. Res: Surface films; chemistry and physics of photography. Mailing Add: 110 Wendover Rd Rochester NY 14610

TRANSUE, WILLIAM REAGLE, b Pen Argyl, Pa, Nov 30, 14; m 36; c 3. MATHEMATICAL ANALYSIS. Educ: Lafayette Col, BS, 35; Lehigh Univ, MA, 39, PhD(math), 41. Prof Exp: Asst, Inst Advan Study, 42-43; assoc physicist, Ord Dept, US Dept Army, 43-44, physicist, 44-45; assoc prof math, Kenyon Col, 45-48; asst, Inst Advan Study, 48-49; prof math, Kenyon Col, 49-66; PROF MATH, STATE UNIV NY BINGHAMTON, 66. Concurrent Pos: Fulbright scholar, Italy, 51-52; NSF fac fel, Paris, 60-61. Mem: Am Math Soc; Math Asn Am. Res: Functional analysis; theory of measure and integration. Mailing Add: Dept of Math State Univ of NY Binghamton NY 13901

TRAPANI, IGNATIUS LOUIS, b San Francisco, Calif, Nov 19, 25; m 52; c 2. PHYSIOLOGY, IMMUNOLOGY. Educ: Univ San Francisco, BS, 48, MS, 50; Stanford Univ, PhD(physiol), 56. Prof Exp: Asst physiol, Stanford Univ, 51-54, jr res assoc, 54-56; USPHS fel, Calif Inst Technol, 56-58, res fel immunochem, 56-60; asst chief dept exp immunol, Nat Jewish Hosp, 60-69, actg chief, 63-69; asst prof microbiol, Univ Colo Med Ctr, Denver, 65-69; PROF CHEM, COLO MOUNTAIN COL, 69-, CHMN DEPT SCI & MATH, 75- Mem: AAAS; Am Physiol Soc; Am Asn Immunol; Soc Exp Biol & Med; NY Acad Sci. Res: Physiological and physicochemical properties of plasma substitutes; interaction of macromolecules with erythrocytes; physiology and immunology of animals at high altitude and low temperatures; antigen-antibody complexes, immunophysiological parameters of antibody formation. Mailing Add: Dept of Chem Colo Mountain Col Glenwood Springs CO 81601

TRAPANI, ROBERT-JOHN, b New York, NY, Sept 8, 29; m 54; c 2. IMMUNOLOGY. Educ: NY Univ, AB, 49; Cath Univ Am, PhD, 60. Prof Exp: Med bacteriologist, Walter Reed Army Inst Res, 53-55; med bacteriologist, Nat Cancer Inst, 55-61; DIR DEPT IMMUNOL, MICROBIOL ASSOCS, INC, 61- Res: Natural resistance, immunity and influencing factors; Gram-negative endotoxins; human histocompatability; transplantation immunity. Mailing Add: Microbiol Assocs Inc 11810 Parklawn Dr Rockville MD 20852

TRAPASSO, LOUIS E, organic chemistry, see 12th edition

TRAPIDO, HAROLD, b Newark, NJ, Dec 10, 16; m 53; c 1. BIOLOGY. Educ: Cornell Univ, BS, 38, AM, 39, PhD(zool), 43. Prof Exp: Asst zool, Cornell Univ, 38-42; biologist, Gorgas Mem Lab, 46-56; dep dir, Virus Res Ctr, India, 56-62; mem staff, Rockefeller Found & vis prof, Univ Del Valle, Colombia, 64-70; PROF TROP MED & MED PARASITOL, SCH MED, LA STATE UNIV, NEW ORLEANS, 70- Concurrent Pos: Consult, USPHS, 47-49; consult, Div Med & Pub Health, Rockefeller Found, 50 & 52; mem, Expert Panel Malaria, WHO, 52-58, Expert Panel Virus Dis, 58-; mem, Sci Adv Bd, Indian Coun Med Res, 57-60. Mem: Fel AAAS; Am Soc Ichthyologists & Herpetologists; Am Soc Mammalogists; Am Soc Trop Med & Hyg; Soc Study Evolution. Res: Biology of arthropods and ecology of arthropod borne virus diseases. Mailing Add: Dept of Trop Med La State Univ Sch of Med New Orleans LA 70112

TRAPOLD, JOSEPH HUGH, b Wilkes-Barre, Pa, Mar 20,. 26; m 49; c 5. PHARMACOLOGY. Educ: Univ Scranton, BS, 47; Univ Tenn, MS, 50, PhD(pharmacol), 51. Prof Exp: Assoc pharmacologist, Ciba Pharmaceut Prod, Inc, 52-54; from instr to assoc prof pharmacol, Sch Med, La State Univ, 54-58; dir pharmacol labs, 58-63, dir pharmacol sect, 63-65, DIR BIOL RES, SANDOZ PHARMACEUT, 65-, VPRES, 68- Concurrent Pos: NIH grants, Nat Heart Inst, 55-58; consult, Vet Admin Hosp, New Orleans, La, 57-58. Mem: Am Soc Pharmacol & Exp Therapeut; Am Col Clin Pharmacol & Therapeut. Res: Adrenergic blockage; cardiovascular effects of vasopressin; effects of Reserpine; influence of ganglion-blockade upon total and regional cardiovascular hemodynamics. Mailing Add: Sandoz Pharmaceut Div Sandoz Inc East Hanover NJ 07936

TRAPP, ALLAN LAVERNE, b Stockbridge, Mich, July 20, 32; m 55; c 4. VETERINARY PATHOLOGY. Educ: Mich State Univ, BS, 54, DVM, 56; Iowa State Univ, PhD(vet path), 60. Prof Exp: Vet livestock investr, Animal Dis Eradication Br, USDA, 56-57; res assoc animal dis res, Iowa State Univ, 57-60; asst prof, Ohio Agr Exp Sta, Ohio State Univ, 60-65, assoc prof, Ohio Agr Res & Develop Ctr, 65-66; assoc prof animal dis res, 66-70, PROF ANIMAL DIS DIAG WORK & TEACHING, MICH STATE UNIV, 70- Concurrent Pos: Mem, Med Adv Coun, Detroit Zoo, 69- Mem: Wildlife Dis Asn; Am Vet Med Asn. Res: Respiratory diseases of cattle; gastrointestinal diseases of cattle and swine. Mailing Add: Vet Diag Lab Mich State Univ Dept of Path East Lansing MI 48823

TRAPP, CHARLES ANTHONY, b Chicago, Ill, July 9, 36; m 58; c 2. PHYSICAL CHEMISTRY. Educ: Loyola Univ, Ill, BS, 58; Univ Chicago, MS, 60, PhD(chem), 63. Prof Exp: NSF fel physics, Oxford Univ, 62-63; asst prof chem, Ill Inst Technol, 63-69; assoc prof, 69-74, ASSOC PROF CHEM, UNIV LOUISVILLE, 74- Concurrent Pos: Petrol Res Fund starter grant, 63-64, type A res grant, 65-68; NSF res grant, 64-67; consult, Argonne Nat Lab, 69-; Res Corp grant, 70-72. Mem: Am Phys Soc. Res: Magnetic properties of matter; electron spin resonance in transition metal compounds, organic free radicals and biologically important compounds; electrical conductivity studies of nonmetals. Mailing Add: Dept of Chem Univ of Louisville Louisville KY 40208

TRAPP, GENE ROBERT, b Hammond, Wis, June 16, 38; m 60. MAMMALOGY. Educ: Wash State Univ, BS, 60; Univ Alaska, MS, 62; Univ Wis, PhD(zool), 72. Prof Exp: Res asst, Coop Wildlife Res Unit, Univ Alaska, 60-62; teaching asst, Dept Biol, Univ NMex, 62-63; wildlife biologist, US Soil Conserv, Honesdale, Pa, 63-64 & Br River Basin Studies, US Fish & Wildlife Serv, Tex, 64-65; teaching asst, Dept Zool, Univ Wis-Madison, 65-70; collabr carnivore res, US Nat Park Serv, Zion Nat Park, 67-69; asst prof, 70-76, ASSOC PROF BIOL, CALIF STATE UNIV, SACRAMENTO, 76- Mem: Am Soc Mammalogists; Ecol Soc Am; Animal Behav Soc; Wildlife Soc; AAAS. Res: The behavioral ecology of mammals, especially carnivores. Mailing Add: Dept of Biol Sci Calif State Univ Sacramento CA 95819

TRAPP, GEORGE E, JR, b Pittsburgh, Pa, June 30, 44; m 68; c 2. APPLIED MATHEMATICS. Educ: Carnegie-Mellon Univ, BS, 66, MS, 67, PhD(math), 70. Prof Exp: ASSOC PROF COMPUT SCI, W VA UNIV & CONSULT, WESTINGHOUSE ELEC CORP, 70- Mem: Sigma Xi; Am Math Soc. Res: Algebraic analysis of electrical networks and numerical analysis. Mailing Add: Dept of Statist & Comput Sci WVa Univ Morgantown WV 26506

TRAPPE, JAMES MARTIN, b Spokane, Wash, Aug 16, 31; m 63; c 4. MYCOLOGY, FOREST PATHOLOGY. Educ: Univ Wash, BS, 53, PhD(forest bot), 62; State Univ NY, MS, 55. Prof Exp: Forester, Colville Nat Forest, 53-56, res forester, Pac Northwest Forest & Range Exp Sta, 56-65, PROJ LEADER & PRIN MYCOLOGIST, PAC NORTHWEST FOREST & RANGE EXP STA, US FOREST SERV, 65-; ASSOC PROF BOT, ORE STATE UNIV, 65- Concurrent Pos: NSF grants, 66-67, 68-69, 70-72 & 74; Am Philos Soc grants for mycol res, Univ Torino, Italy, 67-68, Nat Polytech Inst, Mex, 72; Japan Soc Prom Sci res fel, 75. Mem: Fel AAAS; Brit Mycol Soc; Mex Soc Mycol; Mycol Soc Am; Japanese Mycol Soc. Res: Taxonomy of fungi, especially hypogeous species, Mycorrhizae; biological control of root diseases. Mailing Add: Pac NW Forest & Range Exp Sta 3200 Jefferson Way Corvallis OR 97331

TRASK, NEWELL JEFFERSON, JR, b Boston, Mass, July 15, 30; m 59; c 3. GEOLOGY. Educ: Mass Inst Technol, BS, 52; Univ Colo, Boulder, MA, 56; Harvard Univ, PhD(geol), 65. Prof Exp: Geologist, Calif Co, 56-60; GEOLOGIST, US GEOL SURV, 64- Mem: AAAS; Am Geophys Union. Res: Astrogeology; structural geology; stratigraphy. Mailing Add: Nat Ctr US Geol Surv 12201 Sunrise Valle Dr Reston VA 22092

TRASLER, DAPHNE GAY, b Iquique, Chile, July 2, 26; Can citizen; m 51; c 2. GENETICS. Educ: McGill Univ, BSc, 48, MSc, 54, PhD, 58. Prof Exp: Demonstr genetics, McGill Univ, 47; chief asst plant genetics, Inst Cotton Genetics, Peru, 49-51; demonstr genetics, 52-53, res assoc develop genetics & teratol, 58-70, ASSOC PROF BIOL, McGILL UNIV, 70- Concurrent Pos: Grants, NSF, 59-62, Asn Aid Crippled Children, 65-66, NIH, 66-69 & Nat Res Coun Can, 70. Mem: Genetics Soc Can; Teratology Soc (pres, 72-73). Res: Study of the factors leading to Cleft palate and cleft lip in mice; mouse teratology. Mailing Add: McGill Univ Dept of Biol PO Box 6070 Montreal PQ Can

TRASON, WINONA B, invertebrate zoology, see 12th edition

TRAUB, ALAN CUTLER, b Hartford, Conn, Jan 20, 23; m 51; c 3. ELECTRO-

OPTICS. Educ: Trinity Col, Conn, BS, 47; Univ Cincinnati, MS, 49, PhD(physics), 52. Prof Exp: Res physicist, Am Optical Co, 52-56; res physicist, Fenwal, Inc, 56-61, chief res engr, 61-63; mem tech staff, Mitre Corp, Mass, 63-70; chief scientist, Foto-Mem, Inc, 70-71; consult, 71-73; prod develop engr, Identicon Corp, 73-74; ADVAN DEVELOP MGR, VANZETTI INFRARED & COMPUT SYSTS, INC, 74- Concurrent Pos: Res fel, Tufts Univ, 72-73. Honors & Awards: Soc Tech Writers & Publ Award of Excellence, 70. Mem: Optical Soc Am. Res: Spectrophotometry; colorimetry; thin optical films; optical and thermal sensors; visual perception; three-dimensional displays; fiber optics; aerospace electro-optical instrumentation; atmospheric optical propagation; optical communications systems; laser applications; optical memories. Mailing Add: 56 Donna Rd Framingham MA 01701

TRAUB, EUGENE FREDERICK, b Dubuque, Iowa, Nov 15, 94; m 22, 49; c 4. DERMATOLOGY. Educ: Univ Mich, BS, 16, MD, 18. Prof Exp: Intern & res physician, Univ Mich Hosp, 19-20; res physician, NY Skin & Cancer Hosp, 21-47; prof dermat, 47-55, EMER PROF DERMAT, NY MED COL, 55-; ED-IN-CHIEF, CUTIS, 65- Concurrent Pos: Vis prof, Univ Vt, 28-29; prof, NY Postgrad Med Sch, Columbia Univ, 41-47; clin prof, Temple Univ, 55-59; consult & med dir, Skin & Cancer Hosp, Philadelphia, 54-, Cambridge-Md Hosp, 55-, East Shore State Hosp, 58- & Easton Mem Hosp, 64-66. Honors & Awards: Atlantic Dermat Conf Award Merit, 70. Mem: Am Dermat Asn; Am Acad Dermat; fel NY Acad Med. Res: Carcinoma of the skin. Mailing Add: Cambridge MD 21613

TRAUB, FREDERICK BEDRICH, microbiology, deceased

TRAUB, JOSEPH FREDERICK, b WGer, June 24, 32; nat US; m 69; c 2. COMPUTER SCIENCE. Educ: City Col New York, BS, 54; Columbia Univ, PhD(appl math), 59. Prof Exp: Mem tech staff, Bell Tel Labs, Inc, NJ, 59-70; PROF COMPUT SCI & MATH & HEAD DEPT COMPUT SCI, CARNEGIE-MELLON UNIV, 71- Concurrent Pos: Vis assoc prof comput sci, Stanford Univ, 66; mem coun, Conf Bd Math Sci, 66-; consult, Int Math & Statist Libraries, Inc; ed jour, Asn Comput Mach, 70- & J Numerical Anal, Soc Indust & Appl Math, 70-; consult, Math Software Alliance Planning Proj, NSF, 74-, mem cent steering comt comput sci & eng res study, 74-; chmn, President's Adv Comt Comput Sci, Stanford Univ, 74-75; mem adv comt, Inst Defense Anal, 76- Mem: Fel AAAS; Asn Comput Mach; Soc Indust & Appl Math. Res: Computational complexity; parallel computation; algorithmic analysis; numerical mathematics; large scientific problems. Mailing Add: Dept of Comput Sci Carnegie-Mellon Univ Pittsburgh PA 15213

TRAUB, RICHARD KIMBERLEY, b Bessemer, Mich, Mar 13, 34; m 56; c 2. MEDICAL RESEARCH. Educ: Johns Hopkins Univ, BES, 56; Towson State Col, MA, 71; Univ Del, MS, 75. Prof Exp: Res engr, Plastics Dept, Du Pont Co, 59-62; chief of group, Defense Develop & Eng Lab, 62-72, Human Factors Eng, 72-74, PROG DIR PROPHYLAXIS & THER, BIOMED LAB, EDGEWOOD ARSENAL, MD, 74- Concurrent Pos: Mem, Pyrotech Comt, Am Ordnance Asn, 62-72; clin psychologist, Harford County, State Md, 71-; intel specialist, Surgeon Gen, US Army & mem, Acad Coun, Edgewood Arsenal, 73- Mem: Am Psychol Asn. Res: Therapy, prophylaxis and skin decontamination of anti-cholinesterase chemical warfare agents including basic mechanisms, searches, development and standardization. Mailing Add: 900 Country Club Rd Havre de Grace MD 21078

TRAUB, ROBERT, b New York, NY, Oct 26, 16; m 39; c 2. MEDICAL ENTOMOLOGY. Educ: City Col New York, BS, 38; Cornell Univ, MS, 39; Univ Ill, PhD(med entom), 47. Prof Exp: Asst entom, Univ Ill, 39-41; chief dept parasitol, Med Ctr, US Army, 47-55, commanding officer, Med Res Unit, Malaya, 55-59, commanding officer, Med Res & Develop Command, 59-62; PROF MICROBIOL, SCH MED, UNIV MD, 62- Concurrent Pos: Parasitologist, 4th Hoogstraal Exped, Mex, 41; field dir, Army Med Res Units, Malaya, NBorneo & Labrador; Comn Hemorrhagic Fever, Korea, 52; Univ Md Sch Med Res Units, Pakistan, Ethiopia, Thailand & New Guinea; hon assoc, Field Mus, Ill, Smithsonian Inst, DC & Bishop Mus, Honolulu; consult ectoparasite-borne dis, WHO, US Army & US Navy. Mem: AAAS; Entom Soc Am; fel Am Soc Parasitol; fel Am Soc Trop Med & Hyg; Soc Syst Zool. Res: Ecology and control of vectors and reservoirs of disease; systematics of Siphonaptera and trombiculid mites. Mailing Add: Dept of Microbiol Univ of Md Sch of Med Baltimore MD 21201

TRAUB, WESLEY ARTHUR, b Milwaukee, Wis, Sept 25, 40; m 63; c 1. ASTROPHYSICS, PLANETARY ATMOSPHERES. Educ: Univ Wis-Milwaukee, BS, 52; Univ Wis, MS, 64, PhD(physics), 68. Prof Exp: PHYSICIST, CTR ASTROPHYS, SMITHSONIAN & HARVARD COL OBSERVS, 68- Concurrent Pos: Res assoc eng & appl physics, Harvard Univ, 68-74. Mem: AAAS; Am Astron Soc; Am Phys Soc; Optical Soc Am. Res: Optical and far-infrared spectroscopy of planets and interstellar medium; instrumentation of high-resolution spectrometers. Mailing Add: Smithsonian & Harvard Col Observs 60 Garden St Cambridge MA 02138

TRAUGER, DAVID LEE, b Ft Dodge, Iowa, June 16, 42; m 64; c 2. ZOOLOGY, ECOLOGY. Educ: Iowa State Univ, BS, 64, MS, 67, PhD(animal ecol), 71. Prof Exp: Instr zool & entom, Iowa State Univ, 67-70, asst prof zool & entom & exec secy environ coun, 70-72; wildlife res biologist, 72-75, ASST DIR NORTHERN PRAIRIE WILDLIFE RES CTR, FISH & WILDLIFE SERV, US DEPT INTERIOR, 75- Concurrent Pos: Wildlife technician, Northern Prairie Wildlife Res Ctr, 66-70. Mem: Wildlife Soc; Arctic Inst NAm; Am Ornith Union; Cooper Ornith Union; Ecol Soc Am. Res: Breeding biology, population dynamics, habitat requirements of waterfowl, particularly diving ducks in prairie parklands and subarctic taiga. Mailing Add: Northern Prairie Wildlife Res Ctr PO Box 1747 Jamestown ND 58401

TRAUGER, DONALD BYRON, b Exeter, Nebr, June 29, 20; m 45; c 2. PHYSICS. Educ: Nebr Wesleyan Univ, AB, 42. Prof Exp: Physicist, Manhattan Proj, Columbia Univ, 42-44; engr, Union Carbide Corp, 44-54; head irradiation eng dept, 54-64, dir gas-cooled reactor prog, 64-70, ASSOC DIR REACTOR & ENG SCI, OAK RIDGE NAT LAB, 70- Mem: AAAS; fel Am Nuclear Soc; Am Phys Soc; Sigma Xi. Res: Reactor technology; gas cooled reactor fuels; nuclear irradiation tests of fuels and materials; liquid metals; behavior of gases; isotope separation. Mailing Add: Oak Ridge Nat Lab Oak Ridge TN 37830

TRAUGH, JOLINDA ANN, b Detroit, Mich. BIOCHEMISTRY, MOLECULAR BIOLOGY. Educ: Univ Calif, Davis, BS, 60; Univ Calif, Los Angeles, PhD(microbiol), 70. Prof Exp: Res asst, Gerbers Baby Foods, 60-62 & Univ Calif, Berkeley, 62-64; USPHS res fel molecular biol, Univ Calif, Davis, 71-73; ASST PROF BIOCHEM, UNIV CALIF, RIVERSIDE, 73- Res: Regulation of protein synthesis; protein modification enzymes. Mailing Add: Dept of Biochem Univ of Calif Riverside CA 92502

TRAUL, KARL ARTHUR, b Akron, Ohio, July 25, 41; c 1. IMMUNOLOGY, VIROLOGY. Educ: Univ Akron, BSc, 63; Ohio State Univ, MSc, 65, PhD(immunol), 69. Prof Exp: Staff immunologist, 69-71, lab head, 71-74, PROJ LEADER, DEPT CANCER RES, PFIZER INC, 74- Mem: Am Soc Microbiol; NY Acad Sci; AAAS. Res: Isolation and purification of antigens associated with Epstein-Barr Virus and characterization of same; development of new systems for propagation and purification of oncogenic viruses on a large scale. Mailing Add: Dept of Cancer Res Pfizer Inc 199 Maywood Ave Maywood NJ 07607

TRAUMANN, KLAUS FRIEDRICH, b Schweinfurt, Ger, Mar 23, 24; nat US; m 58. ORGANIC CHEMISTRY. Educ: Univ Heidelberg, PhD(org chem), 54. Prof Exp: Res chemist, Carothers Lab, 54-66, sr res chemist, Textile Res Lab, 67-73, DEVELOP ASSOC, TEXTILE RES LAB, E I DU PONT DE NEMOURS & CO, INC, 73- Mem: Am Chem Soc. Res: Synthetic organic fibers. Mailing Add: Textile Res Lab E I du Pont de Nemours & Co Inc Wilmington DE 19898

TRAURIG, HAROLD HENRY, b Chicago, Ill, July 28, 36; m 59; c 3. ANATOMY, ENDOCRINOLOGY. Educ: Mankato State Col, BS, 58; Univ Minn, PhD(anat), 63. Prof Exp: From instr to asst prof, 63-69, ASSOC PROF ANAT, MED CTR, UNIV KY, 69- Concurrent Pos: NIH res grants, 64-67 & 68-72; res fel neurochem, Ohio State Univ, 70-71. Mem: Am Asn Anatomists; Soc Exp Biol & Med; Soc Study Reproduction. Res: Cytology; neurobiology; radioautography; cell proliferation. Mailing Add: Dept of Anat Univ of Ky Med Ctr Lexington KY 40506

TRAURING, MITCHELL, b Brooklyn, NY, Mar 8, 22; m 43; c 2. OPERATIONS RESEARCH. Educ: Brooklyn Col, BA, 41; Johns Hopkins Univ, MA, 47. Prof Exp: Physicist, Nat Adv Comt Aeronaut, 41-46; optical engr, Bur Ships, US Dept Navy, 49; ballistician & sect head, Ballistics Res Labs, Ord Corps, US Dept Army, 49-53; sect head, Guided Missiles Div, Repub Aviation Corp, 53-57; asst sect head, Ground Systs Group, Hughes Aircraft Co, 57-59, sr staff physicist, Res Labs, 59-63, sr staff engr, Aerospace Corp, Calif, 63-68; sr mem tech staff, Data Systs Div, Litton Industs, Inc, 68-72; CONSULT OPERS ANAL, 72- Mem: Sigma Xi; Opers Res Soc Am. Res: Weapon, electronic and space systems; automatic recognition; business and international economics. Mailing Add: 1645 Comstock Ave Los Angeles CA 90024

TRAUT, ROBERT RUSH, b Utica, NY, Oct 21, 34; m 62; c 1. BIOCHEMISTRY, MOLECULAR BIOLOGY. Educ: Haverford Col, AB, 56; Rockefeller Univ, PhD(biochem), 62. Prof Exp: Res asst, Inst Molecular Biol, Univ Geneva, 64-68, Am Heart Asn estab investr, 68-70; ASSOC PROF BIOL CHEM, SCH MED, UNIV CALIF, DAVIS, 70- Concurrent Pos: Jane Coffin Childs Mem Fund fel molecular biol, Med Res Coun Lab Molecular Biol, Cambridge Univ, 62-64; Am Heart Asn estab investr, Univ Calif, Davis, 70-73. Mem: AAAS; Am Soc Microbiol; Am Soc Biol Chem. Res: Mechanism and regulation of protein synthesis; structure and function of ribosomes. Mailing Add: Dept of Biol Chem Sch of Med Univ of Calif Davis CA 95616

TRAUTH, CHARLES ARTHUR, JR, b Santa Fe, NMex, Mar 8, 37; m 60; c 1. MATHEMATICS. Educ: Kans State Teachers Col, BA, 58; Univ Mich, MS, 60, PhD(math), 63. Prof Exp: Staff mem, 62-66, tech supvr systs studies, 66-72, SUPVR BIOSYSTS STUDIES, SANDIA LABS, 72- Concurrent Pos: Mem, Subcomt Math Modeling, Spacecraft Sterilization Adv Comt, Am Inst Biol Sci, 67- Mem: AAAS; NY Acad Sci; fel Royal Soc Health; Am Soc Microbiol; Am Inst Biol Sci. Res: Biomathematics; microbiology; planetary quarantine; radiobiology; environmental biology. Mailing Add: 4505 Royene Ave NE Albuquerque NM 87110

TRAUTMAN, JACK CARL, b Cushing, Okla, Dec 7, 29; m 57; c 1. DAIRYING, BIOCHEMISTRY. Educ: Univ Idaho, BS, 51; Univ Calif, MS, 53; Univ Wis, PhD(dairy tech), 58. Prof Exp: Asst dairy indust, Univ Calif, 51-53; asst dairy & food indust, Univ Wis, 56-58, from instr to asst prof dairy indust, 58-59; asst prof dairy technol, Ohio State Univ, 59-60; supvr indust prod & processes, 60-72, MGR BIOL RES & PRES SCI PROTEIN LABS, OSCAR MAYER & CO, 72- Concurrent Pos: Mem sci adv comt, Fats & Protein Res Found. Mem: Am Chem Soc; Am Meat Sci Asn; Inst Food Technol. Res: Process development of animal biologicals. Mailing Add: Oscar Mayer & Co Res Div 910 Mayer Ave Madison WI 53701

TRAUTMAN, MILTON BERNHARD, b Columbus, Ohio, Sept 7, 99; m 40; c 1. ZOOLOGY. Hon Degrees: DSc, Col Wooster, 51. Prof Exp: Asst, Bur Sci Res, Ohio Div COnserv, 30-34; asst cur, Mus Zool, Univ Mich, 34-39; res biologist, Stone Lab, 39-40, res assoc, Stone Inst Hydrobiol, 40-55, lectr zool & cur vert collections, Dept Zool & Entom, 55-69, prof fac biol, Col Biol Sci, 69-72, EMER PROF ZOOL & EMER CUR BIRDS, OHIO STATE UNIV, 72-; RES ASSOC BIOL, JOHN CARROLL UNIV, 72- Concurrent Pos: Asst dir inst fisheries res, State Conserv Dept, Mich, 34-35, res assoc, 35-36; mem, Univ Mich Zool Exped, Yucatan, 36; mem, Ohio State Univ Exped, Inst Polar Studies, Alaska, 65. Honors & Awards: Wildlife Soc Award, 58. Mem: AAAS; Wildlife Soc; Wilson Ornith Soc (treas, 43-45); Am Soc Ichthyol & Herpet (vpres, 46-49); Am Ornith Union. Res: Life history and taxonomy of lampreys, freshwater fishes and birds of North America; factors affecting animal distribution and abundance; animal behavior. Mailing Add: Mus of Zool 1813 N High St Ohio State Univ Columbus OH 43210

TRAUTMAN, RODES, b Portsmouth, NH, Apr 7, 23; m 46; c 3. BIOPHYSICS. Educ: Yale Univ, BE, 44; Univ Calif, PhD(biophys), 50. Prof Exp: Res asst phys biochem, Donner Lab, Univ Calif, 50-54; asst phys chem, Med Lab, Rockefeller Inst, 54-57; CHIEF RES PHYSICIST, PLUM ISLAND ANIMAL DIS LAB, USDA, 57- Concurrent Pos: Res collabr, Brookhaven Nat Lab, 56- Mem: Biophys Soc; Sigma Xi. Res: Application of physics and mathematics to virus research; analytical and preparative ultracentrifugation; programming and utilization of digital computers; simulation of immunochemical reactions; information retrieval. Mailing Add: USDA Plum Island Animal Dis Ctr PO Box 848 Greenport NY 11944

TRAUTMAN, WILLIAM DEAN, b Cleveland, Ohio, Dec 22, 19; m 48; c 2. COMPUTER SCIENCES. Educ: Swarthmore Col, BA, 42; Case Inst Technol, MS, 50, PhD(chem), 51. Prof Exp: Chemist, Atlas Powder Co, 42-43, chem supvr, 43-44; res chemist, Gen Elec Co, 51 & 52-56, specialist liaison & recruiting, 56-59; proj adminr, Off Naval Res, 51-52; mgr lab admin, Brush Beryllium Co, Ohio, 59-71; consult, 71-72, mkt rep, 72-74, DIR EDUC SERV, CHI CORP, 74- Mem: AAAS; Am Chem Soc. Res: Physical inorganic chemistry; chemical metallurgy; protective coatings; plastics; high temperature reactions; computer applications. Mailing Add: 12 Pepperwood Lane Cleveland OH 44124

TRAUTMANN, WILLIAM LESTER, soil chemistry, water chemistry, see 12th edition

TRAVER, JANET HOPE, b Boston, Mass, Apr 18, 26. BIOCHEMISTRY. Educ: Cornell Univ, BS, 47; Mich State Univ, MS, 49. Prof Exp: Asst nutrit, Mich State Univ, 47-49; res assoc biochem, 50-65, RES BIOCHEMIST, STERLING-WINTHROP RES INST, 65- Res: Isolation of natural products. Mailing Add: 11 Brilan Ave East Greenbush NY 12061

TRAVERS, JOHN JOSEPH, b Philadelphia, Pa, May 20, 18; m 42; c 4. BIOCHEMISTRY. Educ: St Joseph's Col, Pa, BS, 39; Univ Detroit, MS, 41; Fordham Univ, PhD(biochem), 51. Prof Exp: Res chemist, Arlington Chem Co, 46-48; asst, Fordham Univ, 48-51, res assoc, 51, head org chemist sect, Eastern Res Div, Stauffer Chem Co, 53-55; sr res assoc, Res & Develop Div, Lever Bros Co, 55-62; mgr semibasic res dept, 62-64, dir res, 64-70, SR RES & DEVELOP ASSOC, AVON PROD, INC, 70- Mem: Am Cosmetic Chem. Soc; Soc Cosmetic Chem. Res: Biometrics; biocidal agents; toxicology; skin structure and function. Mailing Add: 38 Johnson's Lane New City NY 10956

TRAVERS, WILLIAM BRAILSFORD, b Long Beach, Calif, June 13, 34; m 58; c 3. GEOLOGY. Educ: Stanford Univ, BS, 56, MS, 59; Princeton Univ, PhD(geol), 72. Prof Exp: Geologist, Stand Oil Co, Calif, 59-61; geologist, Sante Fe Drilling Co, 61-63; chief geologist, Santa Fe Int, Inc, 63-67; asst instr geol, Princeton Univ, 67-71; ASST PROF GEOL SCI, CORNELL UNIV, 72- Concurrent Pos: Consult petrol geologist; vpres, Anacapa Oil Co, 67- Mem: AAAS; Geol Soc Am; Am Asn Petrol Geol; Am Geophys Union. Res: Problems of mountain building; deformation of continental margins; structural geology and sedimentology; continental rifting; tectonism in Italy, western United States and western Canada. Mailing Add: Dept of Geol Sci Cornell Univ Ithaca NY 14850

TRAVERSE, ALFRED, b Port Hill, Prince Edward Island, Sept 7, 25; nat US; m 51; c 4. PALYNOLOGY, PALEOBOTANY. Educ: Harvard Univ, SB, 46, AM, 48, PhD(paleobot), 51; Episcopal Theol Sem Southwest, MDiv, 65. Prof Exp: Coal technologist, Lignite Res Lab, US Bur Mines, NDak, 51-55, head fuels micros lab, Colo, 55; geologist, Shell Develop Co, 55-62; palynological consult, Tex, 62-65; asst prof geol, Univ Tex, 65-66; assoc prof, 66-70, PROF GEOL, PA STATE UNIV, UNIVERSITY PARK, 70- Concurrent Pos: Mem, Int Comn Palynology, 73- Mem: AAAS; Paleont Soc; Geol Soc Am; Bot Soc Am; Int Asn Plant Taxon. Res: Palynology of Cenozoic and older rocks; theory of palynology. Mailing Add: Dept of Geosci Deike 529 Pa State Univ University Park PA 16802

TRAVIS, BERNARD VALENTINE, b Uncompahgre, Colo, Mar 29, 07; m 30; c 3. MEDICAL ENTOMOLOGY, PARASITOLOGY. Educ: Colo State Col, BS, 30; Iowa State Col, MS, 32, PhD(entom), 37. Prof Exp: Asst entom, Iowa State Col, 30-34, wildlife exten specialist, 34-35; asst entomologist, Bur Entom & Plant Quarantine, USDA, 35-42, assoc entomologist, 42-47, res entomologist, 47-49; PROF MED ENTOM & PARASITOL, CORNELL UNIV, 49-, CHMN DEPT ENTOM, 70- Concurrent Pos: Teacher from Cornell Univ, Philippines Col Agr, 57-59; vis prof, Med Sch, La State Univ, 68-69 & Univ Costa Rica, 68-69. Mem: AAAS; Soc Protozool; Am Entom Soc; Am Soc Trop Med & Hyg; Am Soc Parasitol. Res: Mosquito and black fly ecology and control; insect repellents; flagellate protozoa; wildlife. Mailing Add: 162 Comstock Hall Cornell Univ Ithaca NY 14850

TRAVIS, DAVID M, b Nashville, Tenn, June 6, 26; m 53; c 3. INTERNAL MEDICINE, PHARMACOLOGY. Educ: Vanderbilt Univ, BA, 47, MD, 51; Am Bd Internal Med, dipl, 60. Prof Exp: Intern & resident med, Boston City Hosp, Harvard Univ, 51-54; from asst prof to assoc prof, 58-70, PROF PHARMACOL & MED, COL MED, UNIV FLA, 70- Concurrent Pos: Teaching fel, Harvard Med Sch, 52-54; Nat Heart Inst res fel, Peter Bent Brigham Hosp, Boston, 56-58; Nat Heart & Lung Inst sr res fel, Harvard Univ, 71-72; mem corp, Marine Biol Lab, Woods Hole, 62- Honors & Awards: Borden Res Award, 51. Mem: Am Fedn Clin Res; Am Col Physicians; Am Physiol Soc; Soc Gen Physiol; Am Soc Pharmacol & Exp Therapeut. Res: Respiratory physiology and pharmacology; biological role of respiratory gases in health and disease. Mailing Add: Dept of Pharmacol Univ of Fla Col of Med Gainesville FL 32610

TRAVIS, DENNIS MICHAEL, b Erie, Pa, Aug 14, 44; m 65; c 1. BOTANY, GENETICS. Educ: Edinboro State Col, BS, 66, MEd, 67; Miami Univ, PhD(bot), 74. Prof Exp: Asst biol, Edinboro State Col, 66-67; instr, Community Col Baltimore, 67-69; assoc, 6970, instr bot & coordr gen plant biol, 70-71, asst to dean, Col Arts & Sci, 71-73, instr bot, 7174, actg asst dean, 73-74, ASST DEAN, COL ARTS & SCI, MIAMI UNIV & ASST PROF BOT, 74- Mem: AAAS; Am Inst Biol Sci; Bot Soc Am; Genetics Soc Am. Res; Toxic and mutagenic effects of chemical mutagens in higher plants; genetic analysis of mutants, with particular regard to cytoplasmic inheritance; organelle DNA is biochemically analyzed. Mailing Add: Dept of Bot Col of Arts & Sci Miami Univ Oxford OH 45056

TRAVIS, DOROTHY FRANCES, b Atlanta, Ga, Dec 3, 20. CELL BIOLOGY. Educ: George Washington Univ, AB, 45; Radcliffe Col, AM, 50, PhD(biol), 51. Prof Exp: Lab asst zool, George Washington Univ, 42-44; preparator, Div Paleont, US Dept Interior, 44-45; jr zoologist, Div Zool, NIH, 45-46; AEC fel, Bermuda Biol Sta, 51-53; instr zool & physiol, Univ NH, 53-55, asst prof, 55-58; res fel biol, Harvard Univ, 58-60; res assoc & asst biologist, Orthop Res Labs, Mass Gen Hosp, 60-69; chief morphol sect, Geront Res Ctr, Nat Inst Child Health & Human Develop, 69-74, HEALTH SCI ADMINR & EXEC SECY, GEN RES SUPPORT ADV COMT, DIV RES RESOURCES, NIH, 74- Mem: AAAS; Am Soc Zoologists; Electron Micros Soc Am; NY Acad Sci; Am Soc Cell Biol. Res: Molecular and cell biology of mineralized tissues. Mailing Add: 733 Sligo Ave Apt 503 Silver Spring MD 20910

TRAVIS, HUGH FARRANT, b Grand Rapids, Mich, Sept 5, 22; m 48; c 2. ANIMAL NUTRITION, PHYSIOLOGY. Educ: Mich State Univ, BS, 47, MS, 52, PhD(poultry nutrit), 61. Prof Exp: Biologist, 48-60, physiologist, 61-62, res animal husbandman, 62-65, LEADER FUR ANIMAL INVESTS, AGR RES SERV, USDA, 65- Concurrent Pos: Assoc prof, Cornell Univ, 62- Mem: AAAS; Am Soc Animal Sci; Am Inst Nutrit. Res: Nutrition, physiology and husbandry of fur-bearing animals. Mailing Add: US Fur Animal Exp Sta Cornell Univ 321 Morrison Hall Ithaca NY 14850

TRAVIS, JAMES, b Winnipeg, Can, Nov 11, 35; m 60; c 4. BIOCHEMISTRY. Educ: Univ Man, BSc, 58, MSc, 60; Univ Minn, PhD(biochem), 64. Prof Exp: Fel biochem, Johns Hopkins Univ, 64-66; asst prof, Univ Md, 66-67; asst prof, 67-72, ASSOC PROF BIOCHEM, UNIV GA, 72- Concurrent Pos: USPHS career develop award, 72- Mem: Am Chem Soc; Am Soc Biol Chem. Res: Protein structure and function. Mailing Add: Dept of Biochem Univ of Ga Athens GA 30601

TRAVIS, JAMES ROLAND, b Iowa City, Iowa, Dec 20, 25; m 50. PHYSICS, EXPLOSIVES. Educ: Tufts Univ, BS, 49; Johns Hopkins Univ, PhD(physics), 56. Prof Exp: Res assoc spectros, Johns Hopkins Univ, 56-57; MEM STAFF, LOS ALAMOS SCI LAB, 57- Mem: Am Phys Soc; Sigma Xi; Combustion Inst. Res: Physics of detonation and shock waves; atomic and molecular spectroscopy. Mailing Add: 417 Estante Way Los Alamos NM 87544

TRAVIS, JOHN RICHARD, b Billings, Mont, Sept 3, 42; m 63; c 2. PHYSICS, FLUID DYNAMICS. Educ: Univ Wyo, BS, 65; Purdue Univ, MS, 69, PhD(nuclear eng), 71. Prof Exp: Asst scientist reactor anal & safety, Argonne Nat Lab, 71-73; STAFF MEM NUMERICAL FLUID DYNAMICS, LOS ALAMOS SCI LAB, 73- Mem: Am Soc Mech Engr; Am Nuclear Soc. Res: Numerical fluid dynamics. Mailing Add: Theoret Div T-3 MS 216 Los Alamos Sci Lab Los Alamos NM 87545

TRAVIS, LARRY DEAN, b Burlington, Iowa, July 29, 43. ASTROPHYSICS. Educ: Univ Iowa, BA, 65, MS, 67; Pa State Univ, University Park, PhD(astron), 71. Prof Exp: Asst prof physics, Pa State Univ, Worthington Scranton Campus, 71-74; MEM STAFF, INST SPACE STUDIES, 74- Mem: AAAS; Am Astron Soc; Am Geophys Union. Res: Stellar atmospheres; solar physics. Mailing Add: Inst for Space Studies 2880 Broadway New York NY 10025

TRAVIS, LUTHER BRISENDINE, b Atlanta, Ga, May 25, 31; m 54; c 4. PEDIATRICS, NEPHROLOGY. Educ: NGa Col, BS, 51; Med Col Ga, MD, 55. Prof Exp: Intern, Med Col Va, 55-56; resident pediat, Wyeth Labs, Col Med, Baylor Univ, 58-60; from asst prof to assoc prof pediat, 62-73, co-dir pediat nephrology, 64-71, PROF PEDIAT, UNIV TEX MED BR, GALVESTON, 73-, DIR PEDIAT NEPHROLOGY, 71- Concurrent Pos: Nat Inst Arthritis & Metab Dis fel pediat nephrology, Univ Tex Med Br, Galveston, 60-62; vis prof, William Beaumont Army Hosp, El Paso, Tex, 66-70. Mem: Am Soc Nephrology; Am Soc Pediat Nephrology; Am Acad Pediat; Am Fedn Clin Res; Soc Pediat Res. Res: Medical diseases of the kidney in children, particularly glomerulonephritis, nephrosis and pyelonephritis; juvenile Diabetes Mellitus. Mailing Add: Dept of Pediat Univ of Tex Med Br Galveston TX 77550

TRAVIS, RANDALL HOWARD, b Curdsville, Ky, July 11, 24; div; c 2. PHYSIOLOGY. Educ: Univ Chicago, BS, 47; Case Western Reserve Univ, MD, 52. Prof Exp: Intern, Univ Hosps, Cleveland, 52-53, jr asst resident, 53-54, asst resident, 54-55; sr instr physiol & med, 59-63, asst prof physiol, 63-68, ASSOC PROF PHYSIOL, CASE WESTERN RESERVE UNIV, 68-, ASST PROF MED, 63-; DIR ENDOCRINOL, CLEVELAND METROP GEN HOSP, 74- Concurrent Pos: Nat Heart Inst res fel, 55-57; Am Heart Asn res fel, 57-59; estab investr, Am Heart Asn, 59-64; asst physician, Univ Hosps, Cleveland, 59-; attend physician, Wade Park Vet Admin Hosp, 59-; assoc dir employees clin & consult endocrinol, Univ Hosps, Cleveland, 67-; asst phys, Cuyahoga County Hosp, 73-75, assoc, 75- Mem: Endocrine Soc. Res: Experimental endocrinology, adrenal and renal hormones relating to cardiovascular system and to nervous system. Mailing Add: 3395 Scranton Rd Cleveland OH 44109

TRAVIS, ROBERT VICTOR, b Ames, Iowa, Aug 6, 33; m 55; c 5. ENTOMOLOGY. Educ: Cornell Univ, BS, 55; Univ Md, MS, 57, PhD(entom), 61. Prof Exp: Horticulturist, Agr Res Serv, USDA, 55-60; teacher sci & chmn dept, Gwynn Park High Sch, Md, 60-63; assoc prof biol, Mansfield State Col, 63-66; ASSOC PROF BIOL, WESTMINSTER COL, PA, 66- Concurrent Pos: Owner, Garden Pest Control Co, 58-61. Mem: Nat Asn Biol Teachers; Entom Soc Am. Res: Mechanism of plant virus transmission; insect physiology and behavior; new methods of illustrating general biological principles using microorganisms and insects; biological clocks; insect diapause. Mailing Add: 620 S Market St New Wilmington PA 16142

TRAVIS, RUSSELL BURTON, b San Francisco, Calif, June 18, 18; m 40, 60; c 5. GEOLOGY. Educ: Colo Sch Mines, GeolE, 43; Univ Calif, PhD(geol), 51. Prof Exp: Geologist, Stand Oil Co Calif, 46; teaching asst, Univ Calif, 47-50; asst prof geol, Univ Idaho, 51; sr geologist, Int Petrol Co, Ltd, 51-53; asst prof geol, Colo Sch Mines, 53-56; sr geologist, Peru, 56-62, Fla, 62-63, Colombia, 63-67, Peru, 67-70, Colombia, 70-74, SR GEOLOGIST, INT PETROL CO, LTD, PERU, 74- Mem: AAAS; Geol Soc Am; Soc Econ Paleontologists & Mineralogists; Am Asn Petrol Geologists; Colombian Soc Petrol Geologists & Geophysicists (pres, 65-66). Res: Computer science in petrology, petrography and petroleum geology. Mailing Add: Petroperu Apto 3126 Lima Peru

TRAWICK, WILLIAM GEORGE, b Sandersville, Ga, Aug 16, 24; m 48; c 2. PHYSICAL CHEMISTRY, CLINICAL CHEMISTRY. Educ: Ga Inst Technol, BS, 48; PhD(phys chem), 55. Prof Exp: Chemist, Union Carbide Nuclear Co, Tenn, 54-58; assoc prof phys chem, La Polytech Inst, 58-61; chmn dept chem, 62-74, PROF CHEM, GA STATE UNIV, 61- Mem: Am Chem Soc; Sigma Xi; Am Asn Clin Chemists. Res: Innovations in teaching of college chemistry. Mailing Add: Dept of Chem Ga State Univ Atlanta GA 30303

TRAWINSKI, BENON JOHN, b Poland, Oct 20, 24; m; c 1. MATHEMATICAL STATISTICS. Educ: McMaster Univ, BSc, 58; Va Polytech Inst & State Univ, PhD(math statist), 61. Prof Exp: Asst prof statist, Va Polytech Inst & State Univ, 60-61; from asst prof to assoc prof biostatist, Tulane Univ, 61-65; assoc prof statist, Univ Ky, 65-66; ASSOC PROF BIOSTATIST, MED CTR, UNIV ALA, BIRMINGHAM, 66-, DIR GRAD PROG, 69- Concurrent Pos: NIH fel, 62; Nat Res Coun Can grant, 63; vis lectr, NSF, 71-; univ statist & math sci curric consult. Mem: AAAS; Inst Math Statist; Am Statist Asn; Am Math Soc; NY Acad Sci. Res: Theoretical and applied research in statistics, especially order and nonparametric statistics and decision theory. Mailing Add: Dept of Biostatist Univ of Ala Med Ctr Birmingham AL 35233

TRAWINSKI, IRENE PATRICIA MONAHAN, b Bayonne, NJ, Mar 17, 29; m 63; c 1. MATHEMATICAL STATISTICS. Educ: Rutgers Univ, BSc, 50; Univ Ill, MS, 51; Va Polytech Inst, PhD(math statist), 61. Prof Exp: From instr to prof math, Keuka Col, 51-63, head dept, 57-63; assoc prof, La State Univ, 64-66; BIOSTATIST, UNIV ALA, BIRMINGHAM, 66- Mem: Math Asn Am; Am Statist Asn; Biomet Soc; Inst Math Statist. Res: Multivariate analysis. Mailing Add: Dept of Biostat Univ of Ala Birmingham AL 35294

TRAXLER, JAMES THEODORE, b Minneapolis, Minn, Oct 17, 29; m 56; c 5. SYNTHETIC ORGANIC CHEMISTRY, PESTICIDE CHEMISTRY. Educ: St John's Univ, Minn, BA, 51; Univ Notre Dame, PhD(org chem, biochem), 56. Prof Exp: Res chemist, Cent Res Labs, Armour & Co, Ill, 55-60 & Am Cyanamid Co, Conn, 60-62; head org lab, Durkee Foods Div, Glidden Co, Ill, 62-66; sr res chemist, Peter Hand Found, Ill, 66-69; org res specialist, Growth Sci Ctr, Int Minerals & Chem Corp, Libertyville, 69-74; SR RES CHEMIST, VELSICOL CHEM CORP, CHICAGO, 74- Mem: AAAS; Am Chem Soc. Res: Amino acids and alcohols; pesticides; heterocycles; natural products; polycyclic aromatics; antimalarials. Mailing Add: 1630 Ashland Ave Evanston IL 60201

TRAXLER, RALPH NEWTON, chemistry, see 12th edition

TRAXLER, RICHARD WARWICK, b New Orleans, La, July 25, 28; m 52; c 3. BACTERIAL PHYSIOLOGY. Educ: Univ Tex, BA, 51, MA, 55, PhD(bact), 58. Prof Exp: Asst serologist, Port Arthur Health Dept, 49-52; asst, Univ Tex, 54-58; asst prof bact, Univ Southwestern La, 58-62, from assoc prof to prof microbiol, 62-71; PROF PLANT PATH & ENTOM & MICROBIOL & CHMN DEPT, UNIV RI, 71- Mem: Am Soc Microbiol; Soc Indust Microbiol; Biodeterioration Soc. Res: Microbial physiology, especially degradation; aliphatic hydrocarbons and related molecules. Mailing Add: Dept of Plant Path & Entom Univ of RI Kingston RI 02881

TRAYLOR, DONALD REGINALD, mathematics, see 12th edition

TRAYLOR, MELVIN ALVAH, b Chicago, Ill, Dec 16, 15; m 41; c 2. ORNITHOLOGY. Educ: Harvard Univ, AB, 37. Prof Exp: Assoc, Div Birds, 40-48, res assoc, 48-55, assoc cur, 55-62, CUR BIRDS, FIELD MUS NATURAL HIST, 72- Concurrent Pos: Mem expeds, Yucatan, Mex, 39-40, Galapagos Island, 41, US, 41, Mex, 48 & Africa, 61-62; pelagic fishing surv, Oper Crossroads, Bikini Atoll, 46. Mem: Wilson Ornith Soc; fel Am Ornith Union; Brit Ornith Union. Res: Taxonomy of Neotropical and African birds; biogeography of South America. Mailing Add: Field Mus of Natural Hist Chicago IL 60605

TRAYLOR, PATRICIA SHIZUKO, b San Francisco, Calif, Jan 21, 30; m 59; c 2. CHEMISTRY, BIOCHEMISTRY. Educ: Univ Calif, Berkeley, AB, 51; Univ Wis, MS, 53; Harvard Univ, PhD(chem), 63. Prof Exp: Res biochemist, Univ Calif, Berkeley, 53-55; chemist, Dow Chem Co, 55-59; NIH res fel, 63-66; asst prof, 66-70, ASSOC PROF CHEM, UNIV SAN DIEGO, 70- Mem: AAAS; Am Chem Soc; NY Acad Sci. Res: Mechanisms of reactions. Mailing Add: Dept of Chem Univ of San Diego San Diego CA 92110

TRAYLOR, TEDDY G, b Sulphur, Okla, May 21, 25; m 59; c 6. ORGANIC CHEMISTRY. Educ: Univ Calif, Los Angeles, AB, 49, PhD(chem), 52. Prof Exp: Sr res chemist, Dow Chem Co, Calif, 52-59; fel, Harvard Univ, 59-61, instr chem, 61; from asst prof to assoc prof, 61-68, PROF CHEM, UNIV CALIF, SAN DIEGO, 68- Concurrent Pos: Consult, Rohm and Haas Res Labs, Pa, 63-74; Guggenheim fel, 76. Mem: Am Chem Soc. Res: Organometallic chemistry; autoxidation; oxygen transport; electrophilic substitution. Mailing Add: Dept of Chem B-017 Univ of Calif San Diego La Jolla CA 92093

TRAYNELIS, VINCENT JOHN, b Cliffside Park, NJ, Oct 3, 28; m 53; c 5. ORGANIC CHEMISTRY. Educ: Rutgers Univ, BS, 50; Wayne State Univ, PhD(chem), 53. Prof Exp: Res assoc, Univ Minn, 53-54; from instr to assoc prof chem, Univ Notre Dame, 54-65, asst head dept, 63-64; chmn dept, 65-72, PROF CHEM, WVA UNIV, 65- Mem: Am Inst Chemists; Am Chem Soc. Res: Aromatic N-oxides; ylid chemistry; heterocyclic sulfur chemistry; oxidations and dehydrations in dimethylsulfoxide. Mailing Add: Dept of Chem WVa Univ Morgantown WV 26506

TRAYNHAM, JAMES GIBSON, b Broxton, Ga, Aug 5, 25; m 48; c 2. ORGANIC CHEMISTRY. Educ: Univ NC, BS, 46; Northwestern Univ, PhD(chem), 50. Prof Exp: Instr chem, Northwestern, 49-50; asst prof, Denison Univ, 50-53; from asst prof to assoc prof, 53-63, chmn dept, 68-73, PROF CHEM, LA STATE UNIV, BATON ROUGE, 63-, VCHANCELLOR ADVAN STUDIES & RES & DEAN GRAD SCH, 73- Concurrent Pos: Res assoc, Ohio State Univ, 51-53; Am Chem Soc Petrol Res Fund int award, Swiss Fed Inst Technol, 59-60; NATO sr fel sci, Univ Saarland, 72. Mem: Am Chem Soc; Sigma Xi. Res: Stereochemistry; mechanisms of reactions; alicyclic systems; olefin reactions. Mailing Add: Off of Advan Studies & Res La State Univ Baton Rouge LA 70803

TRAYNOR, LEE, b Flint, Mich, July 9, 38; m 60; c 2. ORGANIC CHEMISTRY, POLYMER CHEMISTRY. Educ: Mich State Univ, BS, 60; Univ Mich, PhD(org chem), 64. Prof Exp: Res chemist, 64-67, sr res chemist, 67-72, RES ASSOC, B F GOODRICH CO, 72- Mem: Am Chem Soc. Res: Organic reaction mechanisms; heterogeneous catalysis; new methods in vinyl polymerization. Mailing Add: 2824 Yellow Creek Rd Akron OH 44313

TRAYSER, KENNETH ALLISON, biochemistry, see 12th edition

TREADO, PAUL A, b Ironwood, Mich, Mar 6, 36; m 59; c 6. NUCLEAR PHYSICS. Educ: Univ Mich, BSE(math) & BSE(physics), 58, MS, 59, PhD(physics), 61. Prof Exp: Assoc prof, 62-73, PROF PHYSICS, GEORGETOWN UNIV, 73- Concurrent Pos: Consult, US Dept Navy, 62- Mem: Am Phys Soc; Sigma Xi. Res: Few nucleon physics; multiparticle breakup reactions; complex reactions and heavy-ion products from intermediate energy reactions. Mailing Add: Dept of Physics Georgetown Univ Washington DC 20057

TREADWAY, ROBERT HOLLAND, b Bloomington, Ind, Nov 10, 10; m 38; c 1. PHYSICAL CHEMISTRY. Educ: Ind Univ, AB, 32, AM, 34, PhD(gen & phys chem), 36. Prof Exp: Asst chem, Ind Univ, 33-36; res chemist, E I du Pont de Nemours & Co, Ohio, 36-40 & A E Staley Mfg Co, Ill, 40-41; chemist, Eastern Regional Res Lab, Bur Agr & Indust Chem, 41-53, chemist, Eastern Utilization Res & Develop Div, 53-61, PHYS SCI ADMINR, EASTERN UTILIZATION RES & DEVELOP DIV, AGR RES SERV, USDA, 61- Mem: Am Chem Soc. Res: Reaction mechanisms in nonaqueous solutions; adhesives and coatings from starch, proteins and resins; industrial utilization of potatoes; industrial developments using agricultural raw materials. Mailing Add: 8400 Hull Dr Philadelphia PA 19118

TREADWAY, WILLIAM JACK, JR, b Johnson City, Tenn, Feb 22, 49; m 71. IMMUNOCHEMISTRY. Educ: Univ Ill, Urbana-Champaign, BS, 72; Loyola Univ Chicago, PhD(biochem), 76. Prof Exp: Teaching asst biochem, Loyola Univ Chicago Med Ctr, 71-73, res asst, 73-75; RES ASSOC IMMUNOCHEM, JEFFERSON MED COL, 75- Mem: Am Chem Soc; AAAS. Res: Cellular immunology. Mailing Add: Dept Biochem Jefferson Med Col 1020 Locust St Philadelphia PA 19107

TREADWELL, CARLETON RAYMOND, b Burlington, Mich, Dec 28, 11; m 41; c 2. BIOCHEMISTRY. Educ: Battle Creek Col, AB, 34; Univ Mich, MS, 35, PhD(biochem), 39. Prof Exp: Asst biochem, Univ Mich, 35-39; from instr to assoc prof, Col Med, Baylor Univ, 39-43, assoc prof, Southwestern Med Found, 43-45; from asst prof to assoc prof, 45-52, PROF BIOCHEM, SCH MED, GEORGE WASHINGTON UNIV, 52-, CHMN DEPT, 59- Mem: AAAS; Am Soc Biol Chemists; Soc Exp Biol & Med; Am Inst Nutrit. Res: Cholesterol metabolism; composition of tissue lipids; fat metabolism; interrelationships of carbohydrate and fat metabolism; amino acids; lipotropism. Mailing Add: Dept of Biochem George Washington Univ Sch of Med Washington DC 20037

TREADWELL, GEORGE EDWARD, JR, b Selma, Ala, Dec 22, 41. BOTANY, BIOCHEMISTRY. Educ: King Col, BA, 64; Iowa State Univ, MS, 67, PhD(biochem, plant physiol), 70. Prof Exp: ASST PROF BIOL, EMORY & HENRY COL, 70- Mem: Am Chem Soc; Bot Soc Am. Res: Plant physiology; vitamin B-2; chromatography. Mailing Add: PO Drawer DDD Emory VA 24327

TREADWELL, PERRY EDWARD, b Chicago, Ill, June 18, 32; m 52; c 4. MICROBIOLOGY. Educ: Univ Calif, Los Angeles, AB, 55, PhD(microbiol), 58. Prof Exp: Fel ment health, Univ Calif, Los Angeles, 58-60; instr bact, Univ Minn, 60-62; from asst prof to assoc prof, Emory Univ, 62-74; DIR, EXP ENERGY ENVIRON, 74- Mem: Am Asn Immunol; Am Soc Microbiol. Res: Effect of physical and psychological stress on mice; cytology of mammalian cells in culture; immunology, pathology and enzymology of shock in mice, including role of lysosomes of the reticuloendothelial system; antigen-antibody complexes in human disease; alternative energy systems. Mailing Add: 428 Princeton Way Atlanta GA 30307

TREAGAN, LUCY, b Novosibirsk, Russia, July 20, 24; US citizen; m 42; c 2. MICROBIOLOGY. Educ: Univ Calif, Berkeley, AB, 45, PhD(bact), 60. Prof Exp: Lectr, Col Holy Names, Calif, 61-66; asst prof biol, 66-73, ASSOC PROF BIOL, UNIV SAN FRANCISCO, 73- Concurrent Pos: Lectr, Univ San Francisco, 62-66. Res: Viral inhibitors; immunological study of interferons; effect of metals on the immune response. Mailing Add: Dept of Biol Univ of San Francisco San Francisco CA 94117

TREANOR, CHARLES EDWARD, b Buffalo, NY, Oct 22, 24; m 50; c 5. PHYSICS, AERODYNAMICS. Educ: Univ Minn, BA, 47; Univ Buffalo, PhD(physics), 55. Prof Exp: Instr physics, Univ Buffalo, 53; physicist, Cornell Aeronaut Lab, Inc, 54-68, HEAD AERODYN RES DEPT, CALSPAN CORP, 68- Mem: Am Phys Soc; Am Inst Aeronaut & Astronaut; Combustion Inst. Res: High temperature gases; spectroscopy; hypersonic flows; molecular interactions. Mailing Add: Aerodyn Res Dept Calspan Corp Buffalo NY 14221

TREANOR, KATHERINE P, b Buffalo, NY, Apr 23, 27. GENETICS. Educ: D'Youville Col, AB, 48; Univ Buffalo, PhD(genetics), 62. Prof Exp: Lab asst biol, Canisius Col, 48-52; instr, D'Youville Col, 52-53; from instr to asst prof, 53-66, ASSOC PROF BIOL, CANISIUS COL, 66- Mem: Genetics Soc Am; Genetics Soc Can. Res: Enzyme deficiencies associated with genetic mutations. Mailing Add: Dept of Biol Canisius Col Buffalo NY 14208

TREAT, ASHER EUGENE, b Antigo, Wis, July 6, 07; m 39; c 1. ENTOMOLOGY. Educ: Univ Wis, BA, 29; Columbia Univ, PhD(physiol), 41. Prof Exp: From instr to prof, 30-66, EMER PROF BIOL, CITY COL NEW YORK, 66- Concurrent Pos: Res assoc entom, Am Mus Natural Hist, 66- Mem: AAAS; Am Soc Zoologists; Entom Soc Am; Acarological Soc Am. Res: Insect physiology; acarology; biology and taxonomy of mites associated with Lepidoptera. Mailing Add: Dept of Entom Am Mus of Natural Hist New York NY 10024

TREAT, CHARLES HERBERT, b Cambridge, Mass, Dec 9, 31; m 56; c 3. NUMERICAL ANALYSIS, HEAT TRANSFER. Educ: Purdue Univ, BSCE, 53, MSE, 59; Univ NMex, PhD(mech eng), 68. Prof Exp: Instr eng graphics, Purdue Univ, 56-59; staff mem heat transfer, Sandia Corp, 59-61; instr mech eng, Univ NMex, 61-68; asst prof eng sci, 68-72, asst prof comput & info sci, 72-75, ASSOC PROF COMPUT & INFO SCI, TRINITY UNIV, TEX, 75- Concurrent Pos: Res assoc, United Nuclear Corp, 63 & Los Alamos Sci Lab, 64; consult, Sch Aerospace Med, Brooks AFB, Tex, 68- & Q-Dot Corp, NMex, 69- Mem: Am Soc Mech Eng. Res: Radiation heat transfer. Mailing Add: Dept of Info & Comput Sci Trinity Univ San Antonio TX 78284

TREAT, DONALD FACKLER, b Hartford, Conn, Feb 14, 25; m 49; c 4. MEDICAL EDUCATION. Educ: Univ Mich, Ann Arbor, BA, 46, MD, 49. Prof Exp: Intern gen pract, Univ Hosp, Ann Arbor, Mich, 49-50, resident, 50-52; pvt pract, Springfield, Vt, 54-69; ASSOC PROF FAMILY MED, SCH MED & DENT, UNIV ROCHESTER, 69-, ASSOC DIR, FAMILY MED PROG & DIR GRAD EDUC IN FAMILY MED, 70- Mem: Soc Teachers Family Med. Res: Family medicine; medical audit and peer review; measurement of attitudinal change in residents; defining patterns of medical care. Mailing Add: 885 South Ave Rochester NY 14620

TREAT, JAY EMERY, JR, b Trinidad, Colo, Nov 16, 20; m 45; c 4. PHYSICS. Educ: Univ Ariz, BS, 42; Cornell Univ, PhD, 54. Prof Exp: Mem staff magnetrons, Radiation Lab, Mass Inst Technol, 42-45; asst gen physics, Cornell Univ, 45-48, cosmic rays and nuclear physics, 48-51; asst prof nuclear physics, 51-58, ASSOC PROF PHYSICS, UNIV ARIZ, 58- Mem: Am Phys Soc; Am Asn Physics Teachers. Res: Cosmic rays; electromagnetic theory; elementary particles. Mailing Add: Dept of Physics Univ of Ariz Tucson AZ 85721

TREAT, THEODORE ARTHUR, organic chemistry, see 12th edition

TREBLE, DONALD HAROLD, b Liverpool, Eng, Apr 14, 34; m 59; c 2. BIOCHEMISTRY. Educ: Bristol Univ, BSc, 55; Univ Liverpool, PhD(biochem), 59. Prof Exp: From asst prof to assoc prof, 63-75, PROF BIOCHEM, ALBANY MED COL, 75- Concurrent Pos: Res fel biochem, Inst Animal Physiol, Babraham, Eng, 58-59, Cambridge Univ, 59-61 & Harvard Med Sch, 61-63. Mem: Am Soc Biol Chemists. Res: Lipid metabolism. Mailing Add: Dept of Biochem Albany Med Col Albany NY 12208

TREECE, JACK MILAN, b Findlay, Ohio, Dec 19, 32; m 54; c 3. GENETICS, BIOCHEMISTRY. Educ: Ohio State Univ, BS, 54, MSc, 55, PhD(biochem genetics), 60. Prof Exp: Res asst biol res, Ohio Agr Exp Sta, 54-60; tech mgr genetics, Cent Ohio Breeding Asn, 60-62; Nat Acad Sci-Nat Res Coun res fel biochem, Animal Protein Pioneering Lab, Eastern Regional Res Utilization Lab, Agr Res Serv, USDA, 62-63; asst prof, Sch Vet Med, Univ Pa, 63-66; asst prof biochem, Univ Del, 66-70; DIR, BLOOD PLASMA & COMPONENTS, INC, 71- Mem: AAAS; Am Chem Soc; Am Asn Clin Chemists; NY Acad Sci. Res: Clinical biochemistry and related product development. Mailing Add: Blood Plasma & Components Inc Box 414 West Chester PA 19380

TREECE, ROBERT EUGENE, b Bluffton, Ohio, Oct 1, 27; m 55; c 3. ECONOMIC ENTOMOLOGY. Educ: Ohio State Univ, BS, 51, MS, 53; Cornell Univ, PhD(econ entom), 57. Prof Exp: Asst exten specialist entom, Rutgers Univ, 56-58; asst prof entom & asst entomologist, Agr Exp Sta, Ohio State Univ, 58-64, from assoc prof to prof, Ohio Agr Res & Develop Ctr, 64-73, ASSOC CHMN OHIO AGR RES & DEVELOP CTR, OHIO STATE UNIV, 73- Mem: Entom Soc Am; AAAS. Res: Bionomics and control of insect pests of livestock. Mailing Add: Dept of Entom Ohio Agr Res & Develop Ctr Wooster OH 44691

TREFETHEN, JOSEPH MUZZY, b Kent's Hill, Maine, May 27, 06; m 31; c 3. GEOLOGY. Educ: Colby Col, AB, 31; Univ Ill, MS, 32; Univ Wis, PhD(geol), 35. Prof Exp: Instr geol, Univ Mo, 35-38; from asst prof to prof, Univ Maine, 38-71; CONSULT GEOLOGIST, 71- Concurrent Pos: Field geologist, State Geol Surv, Maine, 29-32, dir, 42-55; dir, State Geol Surv, Wis, 35; vis lectr, Stephens Col, 35-36. Mem: Fel Geol Soc Am; Soc Econ Geol; Soc Econ Paleontologists & Mineralogists; Mineral Soc Am; Am Inst Mining, Metall & Petrol Engrs. Res: Structural and applied geology; economic geology of the non-metallics; engineering geology. Mailing Add: Friendship ME 04547

TREFETHEN, PARKER S, b Wilton, Maine, Aug 29, 19; m 54; c 4. FISHERIES, FORESTRY. Educ: Univ Maine, BS, 47. Prof Exp: Cruiser-scaler, St Regis Paper Co, 47-49; fishery aide herring invest, US Fish & Wildlife Serv, 49-51, fishery res biologist, 51-52, proj leader elec guiding, 52-54, sonic fish tracking, 54-59, chief fish tracking prog, Bur Com Fisheries, 59-61, res supvr fish passage res prog, 61-67, biol groundfish res prog, 68, asst chief br anadromous fish, 69, actg chief, 70, PROG MGR ECOL EFFECTS OF DAMS, NAT MARINE FISHERIES SERV, NAT OCEANIC & ATMOSPHERIC ADMIN, 71- Concurrent Pos: Consult, US Agency Int Develop, Korea, 68. Mem: Am Fisheries Soc; Am Inst Fisheries Res Biol. Res: Assess effects of dams and water resource developments of fish migration, survival and ecology of rivers; development of systems to protect fish from losses in hazardous areas; evaluate effectiveness of protective actions. Mailing Add: Nat Marine Fish Serv 2725 Montlake Blvd Nat Oceanic & Atmospheric Admin Seattle WA 98112

TREFFERS, HENRY PETER, b New York, NY, Aug 21, 12; m 37; c 2. MICROBIOLOGY. Educ: Columbia Univ, AB, 33, PhD(chem), 37. Hon Degrees: MA, Yale Univ, 50. Prof Exp: Asst chem, Univ Exten, Columbia Univ, 33-36, instr biochem, 36-42; asst prof comp path & biochem, Harvard Med Sch, 42-44; asst prof immunol, 44-46, assoc prof immunochem, 46-49, prof microbiol & Davenport Col fel, 49-69, chmn dept microbiol, 50-61, PROF PATH, SCH MED, YALE UNIV, 69- Concurrent Pos: Fulbright res scholar, Univ Otago, NZ, 54; USPHS spec fel, Wright-Fleming Inst, St Mary's Hosp, London, Eng, 61-62; consult, USPHS, 48-50, 58-61 & 66-69; asst, Presby Hosp, 36-42; assoc mem, Comn Immunization, US Army Epidemiol Bd, 47-55; vchmn, Sect Microbiol, Chem & Biol Coord Ctr, Nat Res Coun, 48-53; mem, Microbiol Panel, US Off Naval Res, 51-54; ed, J Am Asn Immunol, 52-57; vis instr, Univ Calif, San Diego, 69-70. Mem: Am Asn Immunologists. Res: Chemistry of non-aqueous solutions; quantitative chemistry of immune reactions; antibiotic resistance; bacterial genetics. Mailing Add: Dept of Microbiol Yale Univ Sch of Med New Haven CT 06510

TREFFNER, WALTER SEBASTIAN, b Freistadt, Austria, Oct 27, 21; US citizen; m 45; c 2. INORGANIC CHEMISTRY, PHYSICAL CHEMISTRY. Educ: Graz Tech Univ, Dipl Ing, 49. Prof Exp: Tech secy, Oesterr-Amerik Magnesit AG, 49-50, res engr, 50-54; res engr, 54-62, mgr res labs, 62-66, mgr prod develop, 66-73, MGR RES CTR, GEN REFRACTORIES CO, 73- Mem: Am Chem Soc; fel Am Ceramic Soc; fel Am Inst Chemists; Nat Inst Ceramic Eng; Am Soc Testing & Mat. Res: Petrography and applied research and development of refractory oxide materials; oxide microstructures; ceramics; research management. Mailing Add: Res Ctr Gen Refractories Co PO Box 1673 Baltimore MD 21203

TREFIL, JAMES S, b Chicago, Ill, Sept 10, 38; m 60; c 2. PARTICLE PHYSICS. Educ: Univ Ill, BS, 60; Oxford Univ, BA & MA, 62; Stanford Univ, MS & PhD(physics), 66. Prof Exp: Res assoc physics, Stanford Linear Accelerator Ctr, 66; Air Force Off Sci Res fel, Europ Ctr Nuclear Res, 66-67; res assoc, Mass Inst Technol, 67-68; asst prof, Univ Ill, Urbana, 68-70; assoc prof & fel, Ctr Advan Studies, 70-75, PROF PHYSICS, UNIV VA, 75- Res: Interactions of high energy particles with nuclei; magnetic monopoles; cosmic ray interaction; medical physics. Mailing Add: Dept of Physics Univ of Va Charlottesville VA 22901

TREFNY, JOHN ULRIC, b Greenwich, Conn, Jan 28, 42; m 67. PHYSICS. Educ: Fordham Univ, BS, 63, Rutgers Univ, New Brunswick, PhD(physics), 68. Prof Exp: Res assoc physics, Cornell Univ, 67-69; ASST PROF PHYSICS, WESLEYAN UNIV, 69- Concurrent Pos: Consult, Inst for Future, 70. Mem: Am Phys Soc; Sigma Xi. Res: Low temperature experimental studies of liquid and solid helium; certain areas in superconductivity including the surface sheath and the Josephson effect and general topics in low temperature physics. Mailing Add: Dept of Physics Wesleyan Univ Middletown CT 06457

TREFONAS, LOUIS MARCO, b Chicago, Ill, June 21, 31; m 57; c 6. STRUCTURAL CHEMISTRY, PHYSICAL BIOCHEMISTRY. Educ: Univ Chicago, BA, 51, MS, 54; Univ Minn, PhD, 59. Prof Exp: From asst prof to assoc prof, 59-66, PROF CHEM, UNIV NEW ORLEANS, 66-, CHMN DEPT, 64- Concurrent Pos: NIH spec fel, 72-73; hon res assoc, Harvard Univ, 72-73. Res: Molecular structure studies by x-ray diffraction; small ring nitrogen compounds; structures of biologically interesting compounds by x-ray diffraction techniques. Mailing Add: Dept of Chem Univ of New Orleans New Orleans LA 70122

TREGILLUS, LEONARD WARREN, b Toronto, Ont, Sept 28, 21; nat US; m 45; c 3. PHYSICAL CHEMISTRY. Educ: Antioch Col, BS, 44; Univ Calif, PhD(chem), 50. Prof Exp: RES ASSOC PHOTOCHEM, EASTMAN KODAK CO, 50- Mem: Am Chem Soc; Soc Photog Sci & Eng; Soc Motion Picture & TV Engrs. Res: Photographic film preparation and processing. Mailing Add: 297 Pinecrest Dr Rochester NY 14617

TREGUNNA, E BRUCE, b Neepawa, Man, Sept 20, 37; m 58; c 3. PLANT PHYSIOLOGY. Educ: Queen's Univ, Ont, BSc, 59, MSc, 61, PhD(bot), 64. Prof Exp: Lectr bot, McGill Univ, 63-64; from asst prof to assoc prof, 64-74, PROF BOT, UNIV BC, 74- Mem: AAAS; Can Soc Plant Physiol. Res: Influence of environment on metabolism, growth and differentiation of plants. Mailing Add: Dept of Bot Univ of BC Vancouver BC Can

TREHUB, ARNOLD, b Malden, Mass, Oct 19, 23; m 50; c 3. PSYCHOPHYSIOLOGY. Educ: Northeastern Univ, AB, 49; Boston Univ, MA, 51, PhD(clin psychol), 54. Prof Exp: Clin psychologist, 54-59, DIR PSYCHOL RES LAB, NORTHAMPTON VET ADMIN HOSP, 59- Concurrent Pos: Vis lectr, Univ Mass, 59-66; lectr, Clark Univ, 60-62; mem grad fac, Univ Mass, Amherst, 70-, adj prof, 71- Mem: AAAS; Am Psychol Asn; NY Acad Sci; Soc Neurosci. Res: Electrophysiology of brain; biomathematics; artificial intelligence; neurophysiology. Mailing Add: 145 Farview Way Amherst MA 01002

TREICHEL, PAUL MORGAN, JR, b Madison, Wis, Dec 4, 36; m 61; c 2. INORGANIC CHEMISTRY. Educ: Univ Wis, BS, 58; Harvard Univ, AM, 60, PhD(chem), 62. Prof Exp: Teaching asst chem, Harvard Univ, 61-62; NSF fel, Queen Mary Col, London, 62-63; from asst prof to assoc prof, 63-72, PROF CHEM, UNIV WIS-MADISON, 72- Mem: Am Chem Soc; The Chem Soc. Res: Organometallic chemistry, including metal carbonyls, cyclopentadienyls, alkyls and related compounds; organophosphorus and organoboron chemistry. Mailing Add: Dept of Chem Univ of Wis Madison WI 53706

TREICHLER, RAY, b Rock Island, Ill, Sept 10, 07; m 42. AGRICULTURAL CHEMISTRY, BIOLOGICAL CHEMISTRY. Educ: Pa State Col, BS & MS, 29; Univ Ill, PhD, 39. Prof Exp: Chemist, Exp Sta, Agr & Mech Col, Tex, 29-37; spec asst animal nutrit, Univ Ill, 37-38; chemist, Ext Sta, Agr & Mech Col, Tex, 39-40; vis scientist, Univ Ill, 40-41; chemist, US Fish & Wildlife Serv, 41-42, technologist, 42-43, chemist, 43-44; res & develop div & head biol activities sect, Off Qm Gen, US Dept Army, 45-52, head chem & biol br, 53; chief agts br, Res & Eng Command, US Army Chem Corps, 53-55, asst to dir med res, 55-58; scientist, US Air Force, 58-68; TECH SERV MGR, H D HUDSON MFG CO, 68- Mem: NY Acad Sci; Am Soc Trop Med & Hyg; Entom Soc Am; Am Chem Soc; Am Soc Microbiologists. Res: Chemistry and biology of natural products; environmental pollution; research and development of pesticides; application equipment for pesticides. Mailing Add: H D Hudson Mfg Co Suite 819 1625 I St NW Washington DC 20006

TREICK, RONALD WALTER, b Scotland, SDak, June 8, 34; m 55; c 2. MICROBIAL PHYSIOLOGY. Educ: Univ SDak, BA, 56, MA, 57; Ind Univ, PhD(bact), 65. Prof Exp: Med technician, Univ SDak, 57-58; res asst infectious dis, Upjohn Co, 58-62; asst prof, 65-69, ASSOC PROF MICROBIOL, MIAMI UNIV, 69- Concurrent Pos: Res grants, Miami Univ, 65-66, 68-69, 70-71 & 72- Mem: AAAS; Am Soc Microbiol. Res: Antimicrobial agents; effect of metabolic inhibitors on bacterial macromolecular synthesis; inhibition of bacterial luminescence; bacteria lipid metabolism. Mailing Add: Dept of Microbiol Miami Univ Oxford OH 45056

TREIMAN, LEONARD HAMILTON, physical chemistry, see 12th edition

TREIMAN, SAM BARD, b Chicago, Ill, May 27, 25; m 52; c 3. PHYSICS. Educ: Univ Chicago, PhD(physics), 52. Prof Exp: Res assoc physics, Univ Chicago, 52; from instr to assoc prof, 52-63, PROF PHYSICS, PRINCETON UNIV, 63- Mem: Nat Acad Sci; Am Phys Soc; Am Acad Arts & Sci. Res: Cosmic ray physics; fundamental particles; field theory. Mailing Add: Joseph Henry Labs Princeton Univ Princeton NJ 08540

TREITEL, SVEN, b Freiburg, Ger, Mar 5, 29; US citizen; m 58; c 4. GEOPHYSICS. Educ: Mass Inst Technol, BS, 53, MS, 55, PhD(geophys), 58. Prof Exp: Geophysicist, Stand Oil Co Calif, 58-60; res assoc commun theory, Pan Am Petrol Corp, 60-65, group supvr, 66-71, mgr res sect, 71-74, SR RES ASSOC, AMOCO PROD CO, 74- Mem: Soc Explor Geophys; Am Geophys Union; Europ Asn Explor Geophys; Seismol Soc Am. Res: Application of statistical communication theory to seismic analysis. Mailing Add: Amoco Prod Co Res Ctr PO Box 591 Tulsa OK 74102

TREITLER, THEODORE LEO, b New York, NY, Apr 18, 20; m 47; c 2. PHYSICAL CHEMISTRY. Educ: City Col New York, BS, 47; Stevens Inst Technol, MS, 50; Rutgers Univ, PhD(phys chem), 56. Prof Exp: Res chemist, Corning Glass Works, NY, 55-57, Colgate-Palmolive Co, 57-61 & Thomas A Edison Labs, 61-63; chemist, Inorg Res & Develop Dept, FMC Corp, 64-70; ASSOC DIR RES, RES & DEVELOP DEPT, BLOCK DRUG CO, INC, JERSEY CITY, 70- Concurrent Pos: Instr, Rutgers Univ, 54 & 62 & Fairleigh Dickinson Univ, 59-67. Mem: Am Chem Soc. Res: Electrochemistry; ion exchange; surface chemistry; fuel cells; inorganic chemical purification; inorganic synthesis; glass and ceramics; detergency; gas measurement; phosphate chemistry; barium chemistry; phase compatibility; nonprescription drugs, analgesics, dental products and medicated shampoos; denture adhesive systems, synthetic, natural and mixtures. Mailing Add: 744 Ridgewood Rd Millburn NJ 07041

TRELA, JOHN MICHAEL, microbial physiology, see 12th edition

TRELAWNY, GILBERT STERLING, b Cincinnati, Ohio, Nov 12, 29; m 50; c 3. MICROBIAL PHYSIOLOGY. Educ: Delaware Valley Col, BS, 57; Lehigh Univ, MS, 60, PhD(biol), 66. Prof Exp: From instr to asst prof microbiol, Delaware Valley Col, 57-66; PROF BIOL, MADISON COL, VA, 66-, HEAD DEPT, 71- Concurrent Pos: NSF res grant, 67-69. Mem: Mycol Soc Am. Res: Microbial metabolism and nutrition. Mailing Add: Dept of Biol Madison Col Harrisonburg VA 22801

TRELEASE, RICHARD DAVIS, b Chicago, Ill, Sept 23, 17; m 42; c 6. FOOD CHEMISTRY. Educ: Univ Ill, BS, 40. Prof Exp: Chemist, Swift & Co, 40-42, res chemist poultry, 43-44; food technician, Qm Food & Container Inst, 44-45; asst to vpres res, 48-50, head frozen food res, 50-63, mgr processed meats res, 63-69, meat res, 69-71, GEN MGR, PROCESSED MEATS RES, SWIFT & CO, 71- Mem: Am Chem Soc; Inst Food Technol. Res: Food processing and preservation; poultry products; frozen foods; cured meats. Mailing Add: Swift & Co Res & Develop Ctr 1919 Swift Dr Oak Brook IL 60521

TRELEASE, RICHARD NORMAN, b Las Vegas, Nev, Nov 6, 41; m 65; c 2. CELL BIOLOGY. Educ: Univ Nev, Reno, BS, 63, MS, 65; Univ Tex, Austin, PhD(cell biol), 69. Prof Exp: NIH fel, Univ Wis-Madison, 69-71; ASST PROF BIOL, ARIZ STATE UNIV, 71- Mem: AAAS; Am Soc Cell Biol; Am Soc Plant Physiol; Am Inst Biol Sci. Res: Diversity of the Glyoxylate cycle within glyoxysomes of plants and animals; application of electron cytochemistry: enzymology; cell fractionation. Mailing Add: Dept of Bot & Micros Ariz State Univ Tempe AZ 85281

TRELFORD, JOHN D, b Toronto, Ont, Feb 7, 31; US citizen; c 3. OBSTETRICS & GYNECOLOGY, ONCOLOGY. Educ: Univ Toronto, MD, 56; FRCS(C), 64. Prof Exp: Asst prof obstet & gynec, Med Sch, Ohio State Univ, 65-70; assoc prof, 70-75, PROF OBSTET & GYNEC, SCH MED, UNIV CALIF, DAVIS, 75- Concurrent Pos: Grants, Univ Calif, Davis, 71-72; dir, Oncol Serv, Sacramento Med Ctr, 72-; consult, Vet Admin Hosp, Martinez, Calif, 72-; dir, Am Cancer Soc, Yolo County. Mem: Fel Am Col Surgeons; Soc Obstet & Gynaec Can; fel Am Col Obstet & Gynec. Res: Antigenicity of the trophoblastic cell and its relationship to cancer. Mailing Add: Dept of Obstet & Gynec Sch of Med Univ Calif at Davis 4301 X St Rm 207 Sacramento CA 95817

TRELKA, DENNIS GEORGE, b Lorain, Ohio, June 11, 40; m 66; c 1. ANIMAL PHYSIOLOGY. Educ: Kent State Univ, BA, 67, MA, 68; Cornell Univ, PhD(animal physiol), 72. Prof Exp: Res asst invert zool, Kent State Univ, 67-68; res asst animal physiol, Cornell Univ, 68-72; ASST PROF ANIMAL PHYSIOL, WASHINGTON & JEFFERSON COL, 72- Mem: Zool Soc Am; AAAS; Sigma Xi. Res: The effects of selected pesticides on mites associated with Junipers; hemorrhagic shock studies on pigs; the effect of carotid ligation on changes in cerebral spinal fluid in rats. Mailing Add: Dept of Biol Washington & Jefferson Col Washington PA 15301

TRELOAR, ALAN EDWARD, b Melbourne, Australia, Sept 27, 02; nat US; m 29, 49; c 3. MEDICAL ANTHROPOLOGY, BIOMETRICS. Educ: Univ Sydney, BSc, 26; Univ Minn, MS, 29, PhD(agr biochem), 30. Prof Exp: Demonstr geol, Univ Sydney, 24-25; from instr to assoc prof biomet, Univ Minn, 29-47, prof biostatist, Sch Pub Health, 47-56; asst dir res, Am Hosp Asn, 56-59; chief statist & anal br, Div Res Grants, NIH, 59-61, spec asst to dir for biomet, Nat Inst Neurol Dis & Blindness, 61-66, chief reproduction anthropometry sect, Nat Inst Child Health & Human Develop, 66-74; DIR MENSTRUATION & REPRODUCTION HIST RES PROG, RUTH E GOYNTON HEALTH SERV, UNIV MINN, MINNEAPOLIS, 74- Concurrent Pos: Consult, USPHS, 53-58; dir, Hosp Res & Educ Trust, 57-59; mem, Human Biol Coun. Mem: AAAS; fel Am Statist Asn; Pop Asn Am; Am Fertil Soc. Res: Biometry of the menstrual cycle and gestation period; relationship of menstrual history to illness; effects of oral contraceptives. Mailing Add: Ruth E Boynton Health Serv Univ of Minn Minneapolis MN 55455

TRELSTAD, ROBERT LAURENCE, b Redding, Calif, June 16, 40; m 61; c 4. EMBRYOLOGY, PATHOLOGY. Educ: Columbia Univ, BA, 61; Harvard Univ, MD, 66. Prof Exp: Intern path, Mass Gen Hosp, 66-67; NIH res assoc embryol, 67-69; ASST PATHOLOGIST, MASS GEN HOSP, 72-, CHIEF PATH, SHRINERS BURNS INST, 75- Concurrent Pos: Helen Hay Whitney Found res fel path, Mass Gen Hosp, 69-72; Am Cancer Soc fac res award, 72-; asst prof path, Harvard Med Sch, 72- Mem: Soc Develop Biol; Am Soc Cell Biol; Am Soc Zoologists; Int Soc Develop Biol. Res: Biological function of connective tissues in normal growth and development and in disease. Mailing Add: Dept of Path Shriners Burns Inst MGH Boston MA 02114

TREMAIN, HENRY EARL, b Salina, Kans, Dec 8, 11; m 31; c 4. CHEMISTRY. Prof Exp: Chemist, Eller Co, 32-33; chief chemist, Western Star Milling Co, 33-43; asst to vpres res & develop, Wyandotte Chem Corp, 43-52, dir contract res, 52-54; dir contract res, Colgate-Palmolive Co, 54-57; vpres, W Alec Jordan Assocs, 57-67; PRES, REDACT, 67- Mem: Am Chem Soc; Soc Plastics Eng. Res: Plastics; research administration; laboratory design and construction. Mailing Add: 8 Kennedy Dr Flanders NJ 07836

TREMAINE, JACK H, b Galt, Ont, June 15, 28; m 56; c 3. PLANT VIROLOGY. Educ: McMaster Univ, BSc, 51, MSc, 53; Univ Pittsburgh, PhD(virol), 57. Prof Exp: Plant pathologist, 52-64, PLANT VIROLOGIST, CAN DEPT AGR, 64- Mem: Am Phytopath Soc. Mailing Add: Res Sta Res Br Can Dept Agr 6660 NW Marine Dr Vancouver BC Can

TREMAINE, MARY M, b Eagle Grove, Iowa, July 31, 12. MICROBIOLOGY. Educ: Univ Wyo, BA, 35; Univ Iowa, MS, 52, PhD(bact), 54. Prof Exp: Instr microbiol, State Univ NY Downstate Med Ctr, 54-60; asst prof, 60-65, ASSOC PROF MED MICROBIOL, UNIV NEBR MED CTR, OMAHA, 65-, DIR ANTIBIOTIC LAB, 65- Mem: Am Soc Microbiologists; AAAS. Res: Immunology; allergy; antibiotics. Mailing Add: Dept of Med Microbiol Univ of Nebr Med Ctr Omaha NE 68105

TREMAINE, PETER RICHARD, b Toronto, Ont, Dec 20, 47; m 72. PHYSICAL CHEMISTRY. Educ: Univ Waterloo, Ont, BSc, 69; Univ Alta, PhD(phys chem), 74. Prof Exp: Nat Res Coun Can fel surface chem, Pulp & Paper Res Inst Can, Montreal, 74-75; ASST RES OFFICER PHYS CHEM, WHITESHELL NUCLEAR RES ESTAB, ATOMIC ENERGY CAN, LTD, 75- Mem: Chem Inst Can. Res: Thermodynamics of high temperature, high pressure aqueous systems. Mailing Add: Atomic Energy of Can Ltd Whiteshell Nuclear Res Estab Pinawa MB Can

TREMBLAY, GEORGE CHARLES, b Pittsfield, Mass, Oct 13, 38; div; c 4. BIOCHEMISTRY. Educ: Mass Col Pharm, BS, 60; St Louis Univ, PhD(biochem), 65. Prof Exp: Am Cancer Soc fel biol chem, Harvard Univ, 65-66; from asst prof to assoc prof, 66-75, PROF BIOCHEM, UNIV RI, 75- Concurrent Pos: Res grants, Nat Inst Child Health & Human Develop, 67- & Nat Inst Arthritis & Metab Dis, 71- Mem: AAAS. Res: Regulatory mechanisms involved in the control of cellular metabolism. Mailing Add: Dept of Biochem Univ of RI Kingston RI 02881

TREMBLAY, GILLES, b Montreal, Que, Apr 18, 27; m 54; c 2. PATHOLOGY. Educ: Univ Montreal, BA, 48, MD, 53. Prof Exp: Resident path, Hotel-Dieu Hosp, 54-55; resident path, New Eng Deaconess Hosp, Boston, Mass, 55-57; pathologist, Hotel-Dieu Hosp, Montreal, 59-61; pathologist, Notre-Dame Hosp, 61-64; prof path & chmn dept, Fac Med, Univ Montreal, 64-70; PROF PATH, McGILL UNIV, 70- Concurrent Pos: Nat Res Coun Can med res fel, Hotel-Dieu Hosp, Montreal, 53-54; Can Cancer Soc Allan Blair Mem res fel histochem, Postgrad Med Sch, Univ London, 57-59. Mem: Histochem Soc; Am Asn Pathologists & Bacteriologists; Can Asn Path; Int Acad Path. Res: Experimental studies on tumor cell-host cell interaction; ultrastructural and cytochemical studies on breast carcinoma. Mailing Add: Dept of Path McGill Univ Montreal PQ Can

TREMBLAY, JEAN-LOUIS, biology, see 12th edition

TREMBLAY, LEO-PAUL, geology, see 12th edition

TREMBLAY, LOUIS-MARIE, physics, see 12th edition

TREMBLY, LYNN DALE, b Parma, Idaho, July 26, 39; m 68; c 2. EXPLORATION GEOPHYSICS. Educ: Eastern Ore Col, BS, 62; Ore State Univ, MS, 65, PhD(geophysics), 68. Prof Exp: From res asst to res assoc geophysics, Ore State Univ, 62-67; res scientist, 67-70, advan scientist geophys, 70-76, SR RES GEOPHYSICIST, MARATHON OIL CO, 76- Mem: Soc Explor Geophysicists; Sigma Xi; Am Geophys Union. Res: Origination and conducting of original and reative research through the application of both theoretical and experimental techniques in seismology. Mailing Add: Marathon Oil Co Denver Res Ctr Box 269 Littleton CO 80121

TREMELLING, MICHAEL, JR, b Rigby, Idaho, Oct 14, 45; m 66; c 1. PHYSICAL ORGANIC CHEMISTRY. Educ: Idaho State Univ, BA, 68; Yale Univ, MPhil, 70, PhD(chem), 72. Prof Exp: Res assoc, Calif Inst Technol, 72-73; ASST PROF CHEM, VASSAR COL, 74- Mem: Am Chem Soc. Res: Solution dynamics of free radicals; high temperature heterogenous reactions. Mailing Add: Dept of Chem Vassar Col Poughkeepsie NY 12601

TREMERE, ARNOLD WESLEY, biological sciences, see 12th edition

TREMMEL, CARL GEORGE, b Lakewood, Ohio, June 25, 33; m 54; c 3. ANALYTICAL CHEMISTRY. Educ: Kent State Univ, BS, 55; Iowa State Univ, MS, 58. Prof Exp: RES CHEMIST, EASTMAN KODAK CO, 58- Mem: Am Chem Soc. Res: Photographic chemistry; physical chemistry of color photography; electrochemistry. Mailing Add: Explor Color Photog Eastman Kodak Co Res Labs Rochester NY 14650

TREMOR, JOHN W, b East Aurora, NY, Jan 24, 32; m 59; c 2. COMPARATIVE PHYSIOLOGY, ENVIRONMENTAL BIOLOGY. Educ: Univ Buffalo, BA, 53, MA, 57; Univ Ariz, PhD(zool), 62. Prof Exp: Res asst biochem, Vet Admin Hosp, Buffalo, NY, 58; teaching asst radiation biol, NSF Inst Biol Sta, Mont State Univ, 59; instr biol & bot, Phoenix Col, 62; asst prof physiol & embryol, Humboldt State Col, 62-63; group leader gen biol, Biosatellite Proj, 63-72, DEP CHIEF EARTH SCI APPLNS OFF, AMES RES CTR, NASA, 72-, PROJ SCIENTIST, BIOMED EXP SCI SATELLITE & JOINT USSR/US BIOL SATELLITE PROG, 74- Mem: AAAS. Res: Development and implementation of biological experiments for space flight; developmental biology of amphibians. Mailing Add: Earth Sci Applns Off NASA Ames Res Ctr Moffett Field CA 94035

TRENCH, ROBERT KENT, b Belize City, Brit Honduras, Aug 3, 40; m 68. MARINE BIOLOGY, BIOCHEMISTRY. Educ: Univ West Indies, BSc, 65; Univ Calif, Los Angeles, MA, 67, PhD(invert zool), 69. Prof Exp: UK Sci Res Coun fel, Oxford Univ, 69-71; instr, 71-72, ASST PROF BIOL, YALE UNIV, 72- Mem: AAAS; Am Soc Limnol & Oceanog. Res: Coral reef biology and ecology; biochemical integration of plasmids in autotroph-heterotroph endosymbioses. Mailing Add: Dept of Biol Osborn Mem Lab Yale Univ New Haven CT 06520

TRENCH, WILLIAM FREDERICK, b Trenton, NJ, July 31, 31; m 54; c 4. MATHEMATICS. Educ: Lehigh Univ, BA, 53; Univ Pa, MA, 55, PhD(math), 58. Prof Exp: Instr, Moore Sch Elec Eng, Univ Pa, 53-56; mathematician, Gen Elec Co, Pa, 56-57; eng specialist, Philco Corp, Pa, 57-59; engr, Radio Corp Am, NJ, 59-64; assoc prof math, 64-67, PROF MATH, DREXEL UNIV, 67- Mem: Math Asn Am; Am Math Soc; Soc Indust & Appl Math. Res: Applied mathematics; differential equations; numerical analysis. Mailing Add: Dept of Math Drexel Univ Philadelphia PA 19104

TRENHAILE, ALAN STUART, b Ebbw Vale, SWales, Apr 14, 45; m 66. GEOMORPHOLOGY. Educ: Univ Wales, BS, 66, PhD(geog), 69. Prof Exp: Asst prof, 69-73, ASSOC PROF GEOG, UNIV WINDSOR, 73- Concurrent Pos: Mem, Brit Geomorphol Res Group. Mem: Inst Brit Geog; Can Asn Geogr; Asn Am Geogr. Res: Shore platforms and Pleistocene raised platforms; morphometry and spatial variation of glacial features, with particular reference to drumlins, eskers and cirques; beach processes and cliff erosion; air photography; shore platforms in Eastern Canada; glacial morphometry. Mailing Add: Dept of Geog Univ of Windsor Windsor ON Can

TRENHOLM, HAROLD LOCKSLEY, b Amherst, NS, July 24, 41; m 68. BIOCHEMISTRY, ENZYMOLOGY. Educ: McGill Univ, BSc, 63; Cornell Univ, PhD(physiol), 68. Prof Exp: Asst physiol, State Univ NY Vet Col, Cornell, 63-67; res scientist, Res Labs, Health Protection Br, 67-73, DIR RES BUR, NO NMED USE DRUGS DIRECTORATE, CAN DEPT HEALTH & WELFARE, 73- Mem: Am Chem Soc; NY Acad Sci; Am Acad Clin Toxicol. Res: Drug abuse; toxicology. Mailing Add: Nonmed Use of Drugs Directorate Can Dept of Health & Welfare Ottawa ON Can

TRENKLE, ALLEN H, b Alliance, Nebr, July 23, 34; m 56; c 3. NUTRITION, BIOCHEMISTRY. Educ: Univ Nebr, BSc, 56; Iowa State Univ, MSc, 58, PhD(nutrit), 60. Prof Exp: NIH res fel, Univ Calif, Berkeley, 61-62; from asst prof to assoc prof animal sci, 62-71, PROF ANIMAL SCI, IOWA STATE UNIV, 71- Mem: Am Soc Animal Sci; Am Chem Soc; Soc Exp Biol & Med; NY Acad Sci; Am Inst Nutrit. Res: Physiology of growth hormone and insulin secretion; endocrinology studies with ruminants; influence of hormones on growth and development of mammals. Mailing Add: 301 Kildee Hall Iowa State Univ Ames IA 50011

TRENT, DENNIS W, b Bend, Ore, Oct 17, 35; m 55; c 6. MICROBIOLOGY, BIOCHEMISTRY. Educ: Brigham Young Univ, BS, 59, MS, 61; Univ Okla, PhD(med sci), 64. Prof Exp: From asst prof to assoc prof bact, Brigham Young Univ, 67-69; from asst prof to assoc prof microbiol, Univ Tex Med Sch, San Antonio, 69-74; CHIEF IMMUNOCHEM BR, VECTOR-BORNE DIS DIV, CTR DIS CONTROL, USPHS, 74- Concurrent Pos: NIH res grants, 66-72. Mem: Am Soc Microbiol. Res: Arbovirus replication; immunology; genetics. Mailing Add: Vector-Borne Dis Div Ctr for Dis Control PO Box 2037 Ft Collins CO 80522

TRENT, JOHN ELLSWORTH, b Wabash, Ind, Sept 22, 42; m 68; c 2. ANALYTICAL CHEMISTRY. Educ: Manchester Col, BA, 64; Ohio State Univ, PhD(org chem), 70. Prof Exp: RES CHEMIST, STAND OIL CO, IND, 71- Mem: Am Chem Soc. Res: Analytical instrumentation. Mailing Add: Stand Oil Co of Ind PO Box 400 Naperville IL 60540

TRENT, WALTER RUSSELL, b Charleston, Mo, Sept 27, 08; m 31; c 6. CHEMISTRY. Educ: Drury Col, BS, 29. Prof Exp: Jr res chemist, Monsanto Chem Co, WVa, 29-30; asst org chem, Pa State Col, 30-34; sr org res chemist, Colgate-Palmolive-Peet Co, NJ, 34-46; dir res, Nat Home Prod Co, NY, 46-47; sr proj chemist, Colgate-Palmolive Co, 47-50, chief chemist, 50-62; chem dir, John T Stanley Co, 62-65; sr res chemist, Stauffer Chem Co, 65-67; dir appl res, J H Baxter Co, 67-70; INDUST CONSULT, 70- Mem: Am Chem Soc. Res: Organic chemical synthesis; synthetic detergents; soap processes; applications research. Mailing Add: 835 Bobolink Ave Eugene OR 97404

TRENTELMAN, GEORGE FREDERICK, b Amsterdam, NY, Apr 27, 44. PHYSICS. Educ: Clarkson Col Technol, BS, 66; Mich State Univ, MS, 68, PhD(physics), 70. Prof Exp: Res assoc nuclear physics, Mich State Univ, 70-71; ASST PROF PHYSICS, NORTHERN MICH UNIV, 71- Concurrent Pos: Vis physicist, Inst Nuclear Sci, Univ Tokyo, 72-73. Mem: Am Phys Soc. Res: Nuclear physics. Mailing Add: Dept of Physics Northern Mich Univ Marquette MI 49855

TRENTHAM, JIMMY N, b Dresden, Tenn, Jan 7, 36; m 65; c 2. MICROBIOLOGY. Educ: Univ Tenn, Martin, BS, 58; Vanderbilt Univ, PhD(microbiol), 65. Prof Exp: From asst prof to assoc prof microbiol, 65-73, chmn dept biol, 69-73, PROVOST, UNIV TENN, MARTIN, 73- Concurrent Pos: NSF res fel, 67-69. Mem: AAAS; Am Soc Microbiologists; Am Inst Biol Sci. Res: Microbial ecology, including qualitative and quantitative fluctuations of bacterial populations in freshwater and effects of temperature and nutritional factors on the structure of bacterial communities in freshwater. Mailing Add: Univ of Tenn Martin TN 38237

TRENTIN, JOHN JOSEPH, b Newark, NJ, Dec 15, 18; m 46; c 2. EXPERIMENTAL BIOLOGY. Educ: Pa State Univ, BS, 40; Univ Mo, AM, 41, PhD(endocrinol), 47. Prof Exp: Res asst, Univ Mo, 47-48; Childs fel anat, Sch Med, Yale Univ, 48-51, from instr to asst prof, 51-54; from assoc prof to prof, 54-60, actg chmn dept, 58-60, PROF EXP BIOL & HEAD DIV, BAYLOR COL MED, 60- Concurrent Pos: Assoc prof anat, Univ Tex Dent Br, 54-60; mem, Adv Comt Pathogenesis Cancer, Am Cancer Soc, 58-60, Comt Tissue Transplantation, Nat Res Coun, 60-70 & Comt Med Res & Educ, Vet Admin Hosp, Houston, 60-65; consult, Univ Tex M D Anderson Hosp & Tumor Inst, 59-62; mem, Bd Sci Counrs, Nat Cancer Inst, 63-65, chmn, 65-67; mem, Spec Animal Leukemia Ecol Studies Comt, NIH, 64-67; vis prof microbiol, Grad Sch, Tex A&M Univ, 65- Honors & Awards: Esther Langer-Bertha Teplitz Mem Award, Ann Langer Res Found, Chicago, 63; Golden Plate Award, Am Acad Achievement, 65. Mem: AAAS; Soc Exp Biol & Med; Radiation Res Soc; Am Soc Exp Path; Am Asn Cancer Res. Res: Cancer research; human cancer viruses; genetics; immunology; tissue and organ transplantation; hematology; cancer immunity; radiobiology; graft-versus-host diseases; immunological tolerance. Mailing Add: Div of Exp Biol Baylor Col of Med Tex Med Ctr Houston TX 77025

TREPAGNIER, JOSEPH HARDONCOURT, chemistry, see 12th edition

TREPANIER, DONALD L, pharmacy, pharmaceutical chemistry, see 12th edition

TREPKA, ROBERT DALE, b Crete, Nebr, Dec 16, 38; m 61; c 3. ORGANIC CHEMISTRY. Educ: Grinnell Col, BA, 61; Univ Calif, Los Angeles, PhD(org chem), 65. Prof Exp: NATO fel org chem, Munich, 65-66; RES SPECIALIST, CENT RES DEPT, 3M CO, 66- Mem: Am Chem Soc; Soc Photog Sci & Eng. Res: Physical organic chemistry; organic stereochemistry; photographic chemistry; synthetic organic chemistry. Mailing Add: 3081 Birchwood Rd St Paul MN 55119

TREPKA, WILLIAM JAMES, b Crete, Nebr, Apr 17, 33; m 55; c 3. POLYMER CHEMISTRY, RUBBER CHEMISTRY. Educ: Doane Col, BA, 55; Iowa State Univ, PhD(chem), 60. Prof Exp: RES CHEMIST, PHILLIPS PETROL CO, 60- Mem: Am Chem Soc. Res: Polymer and rubber synthesis. Mailing Add: 115 Ramblewood Rd Bartlesville OK 74003

TREPTOW, RICHARD S, b Chicago, Ill, Feb 8, 41; m 68; c 2. INORGANIC CHEMISTRY. Educ: Blackburn Col, BA, 62; Univ Ill, Urbana, MS, 64, PhD(chem), 66. Prof Exp: Staff chemist, Procter & Gamble Co, 66-72; asst prof chem, 72-75, ASSOC PROF CHEM, CHICAGO STATE UNIV, 75- Mem: Am Chem Soc. Res: Transition metal compounds; forensic chemistry; physical biochemistry. Mailing Add: Dept of Phys Sci Chicago State Univ Chicago IL 60628

TRESCOTT, PETER CHAPIN, b Evanston, Ill, Dec 12, 39; m 63; c 3. HYDROGEOLOGY. Educ: Williams Col, BA, 62; Univ Ill, Urbana, MS, 64, PhD(geol), 67. Prof Exp: Geologist, NS Dept Mines, Halifax, 67-70; RES

HYDROLOGIST, US GEOL SURV, 70- Mem: Am Asn Petrol Geologists; Am Geophys Union. Res: Areal variation in properties of porous media; digital simulation of flow through porous media; coupled flow in porous media. Mailing Add: US Geol Surv MS 431 Reston VA 22092

TRESHOW, MICHAEL, b Copenhagen, Denmark, July 14, 26; nat US; m 51; c 1. PLANT PATHOLOGY. Educ: Univ Calif, Los Angeles, BS, 50; Univ Calif, Davis, PhD(plant path), 54. Prof Exp: Sr lab technician, Univ Calif, 52-53; plant pathologist, Columbia-Geneva Steel Div, US Steel Corp, 53-61; from asst prof to assoc prof bot, 61-67, assoc prof biol, 67-70, PROF BIOL, UNIV UTAH, 70- Mem: Air Pollution Control Asn; Am Phytopath Soc; Mycol Soc Am; Bot Soc Am. Res: Environmental pathology, particularly diseases caused by air pollutants; diseases of fruit crops; environmental stress. Mailing Add: Dept of Biol Univ of Utah Salt Lake City UT 84112

TRESNER, HOMER DAVID, b Wis, Aug 18, 18; m 43; c 2. MYCOLOGY. Educ: Univ Wis, BS, 47, MS, 49, PhD(mycol), 52. Prof Exp: Res mycologist, Commercial Solvents Corp, Ind, 52-53; mycologist, Fed Govt, Washington, DC, 53-54; MICROBIOLOGIST & CUR CULT COLLECTION, LEDERLE LABS, AM CYANAMID CO, 54- Mem: Am Acad Microbiologists; Mycol Soc Am; Soc Indust Microbiol; Am Soc Microbiologists; Am Inst Biol Sci. Res: Taxonomy of Actinomycetes; ecology of soil microorganisms; antibiotic research. Mailing Add: Dept 922 Bldg 96A Am Cyanamid Co Lederle Labs Pearl River NY 10965

TRESSELT, HUGH BENJAMIN, biochemistry, see 12th edition

TRESSLER, DONALD KITELEY, b Cincinnati, Ohio, Nov 7, 94; m 19; c 2. FOOD CHEMISTRY. Educ: Univ Mich, AB, 13; Cornell Univ, PhD(agr & food chem), 18. Prof Exp: Scientist, Bur Soils, USDA, 17; instr chem, Ore State Col, 17-18; asst, US Bur Fisheries, 18-19; indust fel, Mellon Inst, 20-29; chief chemist, Birdseye Labs, Mass, 29-33; prof agr chem, Cornell Univ & chief res & head div chem, Exp Sta, 33-43, prof, Nutrit Sch, 42; mem food refrig staff, Gen Elec Co, 42-46; prof food technol, NY Univ, 47-50; sci dir, Qm Food Container Inst, 50-58; PRES & PUBLISHER, AVI PUBL CO, 56- Concurrent Pos: Vis prof, Sch Agr Fisheries, Laval Univ, 41, 43 & 45; consult & mgr, Food Res Labs, 43-50 & 58-73; consult & mgr, Sch Hotel Admin, Cornell Univ; consult, Off Civil Defense Mobilization, 58-64. Honors & Awards: Gold Medal, US Dept Com, Brazilian Centennial Expos, 23; Appert Award, 68; Paul Anderson Award, 71; Babcock-Hart Award, Inst Food Technologists, 75. Mem: Am Chem Soc; fel Am Soc Heating, Refrig & Air Conditioning Engrs (treas, 50-52); fel Am Pub Health Asn; Inst Food Technologists; Am Soc Bakery Engrs. Res: Food technology; soil fertility; fertilizers; preservation of fish; fish products; composition of fish; liquid glue; beet molasses; sodium glutamate; synthetic waxes; freezing of foods; fruit juices; wines; nutritive values of fruits and vegetables; military subsistence; rations; food dehydration. Mailing Add: Avi Publishing Co PO Box 831 Westport CT 06880

TRETSVEN, WAYNE I, dairy industry, see 12th edition

TRETTER, JAMES RAY, b Boone, Iowa, June 7, 33; m 56; c 6. ORGANIC CHEMISTRY. Educ: Loras Col, BA, 56; Univ Calif, Berkeley, PhD(org chem), 60. Prof Exp: DIR CHEM DEPT, PFIZER INC, 60- Mem: Am Chem Soc. Res: Medicinal chemistry; synthesis of organic chemicals as medicinal agents. Mailing Add: Chem Dept Pfizer Inc Groton CT 06340

TREU, JESSE ISAIAH, b New York, NY, Apr 10, 47; m 70; c 1. BIOPHYSICS, OPTICS. Educ: Rensselaer Polytech Inst, BS, 68; Princeton Univ, MA, 71, PhD(physics), 73. Prof Exp: Physicist optics & immunol, 73-75, LIAISON SCIENTIST COMPONENTS & MAT GROUP, GEN ELEC RES & DEVELOP CTR, 75- Mem: Am Phys Soc; Am Asn Physicists Med; Biphys Soc. Res: Development of clinical diagnostic tests based on immunology and the optical properties of proteins immobilized onto surfaces. Mailing Add: Gen Elec Res & Develop Ctr Schenectady NY 12301

TREUENFELS, PETER MARTIN, computer science, see 12th edition

TREUMANN, WILLIAM BORGEN, b Grafton, NDak, Feb 26, 16; m 45, 48; c 3. PHYSICAL CHEMISTRY. Educ: Univ NDak, BS, 42; Univ Ill, MS, 44, PhD(phys chem), 47. Prof Exp: Asst chem, Univ Ill, 42-46, asst math, 46; from asst prof to prof phys chem, NDak State Univ, 46-55; assoc prof, 60-62, assoc dean acad affairs, 68-70, PROF PHYS CHEM, MOORHEAD STATE UNIV, 62-, DEAN FAC MATH & SCI, 70- Mem: AAAS; Am Chem Soc; Am Inst Chemists. Res: Complex ion formation. Mailing Add: Off of Acad Affairs Moorhead State Univ Moorhead MN 56560

TREUPEL, HANS WILHELM, experimental physics, see 12th edition

TREUTING, THEODORE FRANCIS, internal medicine, psychiatry, deceased

TREUTING, WALDO LOUIS, b New Orleans, La, Apr 28, 11; m 36; c 3. PUBLIC HEALTH. Educ: Tulane Univ, BS, 30, MD, 34; Johns Hopkins Univ, MPH, 39. Prof Exp: Dir, Parish Health Units, La, 36-38; chief epidemiol, State Health Dept, La, 39-43, dir div prev med, 43-46, state health officer, 46-48; instr prev med, Sch Med, Tulane Univ, 45-47, lectr pub health, 47-48, prof pub health admin, 48-58, actg chmn dept trop med & pub health, 53-58; head dept, 58-74, PROF PUB HEALTH PRACT, GRAD SCH PUB HEALTH, UNIV PITTSBURGH, 58-, ASSOC DEAN, 74- Concurrent Pos: Consult, WHO, 53. Mem: Am Pub Health Asn; Am Col Prev Med. Res: Virus pneumonitis; pertussis immunization; public health administration. Mailing Add: Dept of Pub Health Pract Univ Pittsburgh Grad Sch Pub Hlth Pittsburgh PA 15213

TREVELYAN, BENJAMIN JOHN, b Beamsville, Ont, Nov 8, 22; m 53; c 4. PHYSICAL CHEMISTRY. Educ: McMaster Univ, BA, 44; McGill Univ, PhD(chem), 51. Prof Exp: Chemist, Aluminum Co Can, Ltd, Que, 44-47; asst res dir, Fraser Co, Ltd, NB, 51-53, res dir, 53-60; res scientist, WVa Pulp & Paper Co, 60-62; dir res, Celfibe Div, Johnson & Johnson, 62-65; mgr pulp res & eng, Kimberly-Clark Corp, 65-69, dir pulp & wood prep, Res & Eng, 69-73; INDUST LIAISON OFFICER, PULP & PAPER RES INST CAN, 73- Mem: Tech Asn Pulp & Paper Indust; Am Chem Soc; Can Pulp & Paper Asn. Res: Pulp and paper technology. Mailing Add: Pulp & Paper Inst of Can 570 St Johns Blvd Pointe Claire PQ Can

TREVES, GINO ROBERT, b Turin, Italy, Apr 6, 17; nat US; m 52; c 3. AGRICULTURAL CHEMISTRY. Educ: Univ Turin, DSc, 39; Cornell Univ, MS, 42. Prof Exp: Asst res chemist, Littauer Pneumonia Res Fund, Sch Med, NY Univ, 42-43; chief chemist, Res Labs, Schieffelin & Co, NY, 43-55; RES ASSOC, AGR CHEM DIV, FMC CORP, 56- Mem: AAAS; Am Chem Soc; fel Am Inst Chemists. Res: Herbicides; fungicides; insecticides; estrogens; antispasmodics; local anesthetics. Mailing Add: 9 Adams Dr Princeton NJ 08540

TREVES, JEAN FRANCOIS, b Brussels, Belg, Apr 23, 30; m 62; c 1. PURE MATHEMATICS. Educ: Univ Sorbonne, Lic, 53, Dr(math), 58. Prof Exp: Asst prof math, Univ Calif, Berkeley, 58-60; assoc prof, Yeshiva Univ, 61-64; prof, Purdue Univ, 64-71; PROF MATH, RUTGERS UNIV, NEW BRUNSWICK, 71- Concurrent Pos: Sloan fel, 60-64; vis prof, Univ Sorbonne, 65-67; mem, Mission Orgn Am States In Brazil, 61. Honors & Awards: Chauvenet Prize, Am Math Soc, 71. Mem: Am Math Soc; Math Soc France. Res: Partial differential equations; functional analysis. Mailing Add: Dept of Math Rutgers Univ New Brunswick NJ 08903

TREVES, SALVADOR, b Ramallo, Arg, Aug 3, 40; US citizen; m 66; c 2. NUCLEAR MEDICINE, PEDIATRICS. Educ: Nat Col III, Buenos Aires, BA, 59; Univ Buenos Aires, MD, 66. Prof Exp: Asst physician, Ctr Nuclear Med, Hosp Clin, Univ Buenos Aires, 66; res fel, Inst Med & Exp Surg, Univ Montreal, 67; resident nuclear med, Royal Victoria Hosp, McGill Univ, 67-68; clin fel & res assoc nuclear med & clin fel med, Yale-New Haven Hosp, Sch Med, Yale Univ, 68-70; instr, 70-73, ASST PROF RADIOL, HARVARD MED SCH, 73- Concurrent Pos: Head nuclear med, Childrens Hosp Med Ctr, 70-73, chief pediat nuclear med, 75- Mem: Soc Nuclear Med; Sigma Xi; Am Heart Asn; Soc Pediat Radiol. Res: Development of ultrashort lived radionuclide generators for angiocardiography in children; development of magnifying collimators for children; development of computer software for nuclear medicine functional studies. Mailing Add: Div of Nuclear Med Childrens Hosp Med Ctr Boston MA 02115

TREVES, SAMUEL BLAIN, b Detroit, Mich, Sept 11, 25; m 60; c 2. VOLCANOLOGY, PETROLOGY. Educ: Mich Col Mining & Technol, BS, 51; Univ Idaho, MS, 53; Ohio State Univ, PhD(geol), 58. Prof Exp: Geologist, Ford Motor Co, 51, State Bur Mines & Geol, Idaho, 52 & Otago Catchment Bd, NZ, 53-54; from instr to assoc prof geol, 59-55; chmn dept, 64-70, PROF GEOL, UNIV NEBR, LINCOLN, 66-, CHMN DEPT, 75- Concurrent Pos: Chief scientist, Antarctica Expeds, 60-75 & Greenland Exped, 62-64. Honors & Awards: Antarctic Serv Medal, 68. Mem: AAAS; fel Geol Soc Am; Nat Asn Geol Teachers; Royal Soc NZ. Res: Petrography; economic geology; antarctic geology. Mailing Add: 1710 B St Lincoln NE 68502

TREVILLYAN, ALVIN EARL, b Moline, Ill, Apr 12, 36; m 58; c 4. ORGANIC CHEMISTRY. Educ: Augustana Col, BA, 57; Purdue Univ, MS, 59, PhD(organometallics), 62. Prof Exp: Res chemist, Sinclair Res Inc, 62-69; RES CHEMIST, AMOCO CHEM, 69- Mem: Am Chem Soc. Res: Organometallics; petrochemicals. Mailing Add: 107 Hopi Ct Naperville IL 60540

TREVINO, DANIEL LOUIS, b Edinburg, Tex, May 15, 43; m 65; c 2. NEUROPHYSIOLOGY. Educ: Univ Tex, Austin, BA, 65, PhD(physiol), 70. Prof Exp: Instr, 72-73, ASST PROF PHYSIOL, SCH MED, UNIV NC, CHAPEL HILL, 73- Concurrent Pos: USPHS fel, Univ Tex Southwest Med Sch Dallas, 70 & Marine Biomed Inst, Galveston, 70-72; fel neurobiol, Sch Med, Univ NC, Chapel Hill, 72- Mem: AAAS; Int Asn Study Pain; Soc Neurosci. Res: Sensory neurophysiology; pain. Mailing Add: Dept of Physiol Univ NC Sch of Med Chapel Hill NC 27514

TREVINO, SAMUEL FRANCISCO, b San Antonio, Tex, Apr 2, 36; m 58; c 5. SOLID STATE PHYSICS. Educ: St Mary's Univ, Tex, BS, 58; Univ Notre Dame, PhD(physics), 63. Prof Exp: RES PHYSICIST, PICATINNY ARSENAL, DOVER, NJ, 64- Concurrent Pos: Guest physicist, Reactor Radiation Div, Nat Bur Stand, 71- Honors & Awards: Paul A Siple Award, Dept Army, 70. Res: Molecular spectroscopy using inelastic scattering of low energy neutrons, particularly vibrational properties of polymers and molecular crystals; low energy nuclear physics involving reactions producing polarized neutrons. Mailing Add: Reactor Radiation Div Nat Bur of Standards Washington DC 20234

TREVITHICK, JOHN RICHARD, b St Thomas, Ont, Nov 30, 38; c 1. BIOCHEMISTRY. Educ: Queen's Univ, Ont, BSc, 61; Univ Wis-Madison, PhD(physiol chem), 65. Prof Exp: Nat Res Coun fel, Univ BC, 65-67; ASST PROF BIOCHEM, UNIV WESTERN ONT, 67- Concurrent Pos: Ed, J Can Fedn Biol Socs, 71-72. Mem: Can Biochem Soc; Can Fedn Biol Socs; Am Chem Soc. Res: Biochemistry of development and differentiation, especially extracellular enzymes of microorganisms; histones and nuclear proteins; proteins of the lens. Mailing Add: Dept of Biochem Univ of Western Ont London ON Can

TREVOR, ANTHONY JOHN, b London, Eng, Dec 27, 34; m 63; c 4. BIOCHEMICAL PHARMACOLOGY, NEUROCHEMISTRY. Educ: Univ Southampton, BSc, 60; Univ London, PhD(biochem), 63. Prof Exp: Lectr, 64-65, asst prof, 65-70, ASSOC PROF PHARMACOL, SCH MED, UNIV CALIF, SAN FRANCISCO, 70- Concurrent Pos: NSF res grants, 72-74 & 75-; USPHS fel neuropharmacol, Sch Med, Univ Calif, San Francisco, 63-64, res grants, 66-75. Mem: AAAS; Am Soc Pharmacol & Exp Therapeut; Brit Biochem Soc. Res: Brain enzymes purification, acetylcholinesterase, mechanisms of action and biodisposition of anesthetics. Mailing Add: Dept of Pharmacol Univ of Calif Sch of Med San Francisco CA 94143

TREVORROW, LAVERNE EVERETT, b Moline, Ill, Nov 1, 28; m 50; c 2. INORGANIC CHEMISTRY, PHYSICAL CHEMISTRY. Educ: Augustana Col, Ill, AB, 50; Okla State Univ, MS, 52; Univ Wis, PhD(chem), 55. Prof Exp: Asst, Okla State Univ, 50-52; asst, Univ Wis, 52-55; asst chemist, 55-59, ASSOC CHEMIST, ARGONNE NAT LAB, 59- Mem: Am Chem Soc; Sigma Xi; The Chem Soc. Res: Chemistry of fluorine and metal fluorides; uranium, neptunium, plutonium and fission elements; molten salt batteries; disposal of radioactive wastes; technological economics. Mailing Add: Argonne Nat Lab 9700 S Cass Ave Argonne IL 60439

TREVOY, DONALD JAMES, b Saskatoon, Sask, Jan 27, 22; m 46; c 4. SOLID STATE CHEMISTRY, ENERGY CONVERSION. Educ: Univ Sask, BE, 44, MSc, 46; Univ Ill, PhD(chem eng), 49. Prof Exp: Chem engr, Nat Res Coun Can, 44-46; SR LAB HEAD, EASTMAN KODAK CO, 49- Mem: Am Chem Soc; AAAS; Int Solar Energy Soc. Res: Liquid-liquid extraction; thermal diffusion; high vacuum evaporation of liquids; physical chemistry of lithography; antistatic agents; photochemistry; organic semiconductors; conducting coatings; solar energy conversion. Mailing Add: 13 Countryside Rd Fairport NY 14450

TREVOY, LLOYD WOODBURY, organic chemistry, agricultural chemistry, see 12th edition

TREW, JOHN ALLAN, b Lemsford, Sask, Sept 6, 23; m 49; c 3. BIOCHEMISTRY. Educ: Univ Sask, BSA, 50, MSc, 51; Univ Western Ont, PhD(biochem), 57. Prof Exp: Biochemist, Regina Gen Hosp, 51-54 & 57-63; assoc prof, 65-73, PROF PATH, UNIV HOSP, UNIV SASK, 73- Mem: Can Soc Clin Chemists; Chem Inst Can; Asn Clin Scientists; Can Biochem Soc. Res: Identification of fungal enzymes; protein electrophoresis; cancer; lactic dehydrogenase; computer applications in clinical pathology. Mailing Add: Path Dept Univ Hosp of Sask Saskatoon SK Can

TREWILER, CARL EDWARD, b Sheridan, NY, Sept 17, 34; m 54; c 4. POLYMER

CHEMISTRY. Educ: Alfred Univ, BA, 56; Univ Akron, PhD(polymer sci), 66. Prof Exp: Resin chemist, Durez Plastics Div, Hooker Chem Co, 56-58; develop engr, Laminated Prod Bus Dept, Gen Elec Co, Ohio, 58-64; staff chemist addn polymerization, Akron Univ, 64-66; POLYMER CHEMIST, LAMINATED PROD BUS DEPT, GEN ELEC CO, 66- Mem: Am Chem Soc. Res: Thermosetting resins and polymers in laminate applications. Mailing Add: Laminated Prod Bus Dept Gen Elec Co S Second St Coshocton OH 43812

TREXLER, BRYSON DOUGLAS, JR, b Wadesboro, NC, Oct 29, 47; m 74. HYDROGEOLOGY. Educ: NC State Univ,BS, 70, MS, 74; Univ Idaho, PhD(geol), 75. Prof Exp: Teaching asst geol, NC State Univ, 70-72; res asst hydrogeol, Univ Idaho, 72-73, res assoc, 73-74, hydrogeologist mine hydrol, 74-75; ASST PROF HYDROGEOL, UNIV TOLEDO, 75- Mem: Soc Mining Engrs; Am Inst Mining Engrs; Nat Water Well Asn; Am Water Resources Asn. Res: Mining hydrogeology and hydrology; ground-water exploration, evaluation and development; dewatering and drainage design. Mailing Add: Dept of Geol Univ of Toledo Toledo OH 43606

TREXLER, DAVID WILLIAM, b Walla Walla, Wash, Apr 22, 20; m 44; c 3. GEOLOGY, PALEONTOLOGY. Educ: Southern Methodist Univ, BS, 41; Johns Hopkins Univ, PhD(geol), 55. Prof Exp: Instr geol, Southern Methodist Univ, 49; asst, Southwestern La Inst, 50-53; photogeologist, Geophoto Servs, Colo, 53-56; asst prof, 56-71, ASSOC PROF GEOL, COLO SCH MINES, 71- Mem: Geol Soc Am; Paleont Soc; Am Asn Petrol Geologists. Res: Mesozoic mollusks; coccoliths; photogeology. Mailing Add: Dept of Geol Colo Sch of Mines Golden CO 80401

TREXLER, FREDERICK DAVID, b Rahway, NJ, Feb 24, 42; m 64; c 2. PHYSICS. Educ: Houghton Col, BS, 64; Pa State Univ, University Park, PhD(solid state sci), 71. Prof Exp: Asst prof, 69-71, ASSOC PROF PHYSICS, HOUGHTON COL, 71- Mem: Am Asn Physics Teachers. Res: Teaching physics, computer applications thereof; solid state electronics. Mailing Add: Dept of Physics Houghton Col Houghton NY 14744

TREXLER, JOHN PETER, b Allentown, Pa, Nov 8, 26; m 50; c 2. GEOLOGY, STRATIGRAPHY. Educ: Lehigh Univ, BA, 50, MS, 53; Univ Mich, PhD(geol), 64. Prof Exp: Asst geologist, Lehigh Portland Cement Co, 50-52; geologist, Fuels Br, US Geol Surv, 53-59; assoc prof geol, 62-69, chmn dept, 62-69, chmn sci div, 67-70, PROF GEOL, JUNIATA COL, 69-, CHMN DEPT, 74- Mem: Fel Geol Soc Am; Am Asn Petrol Geologists. Res: Field geology; stratigraphy and structural geology of sedimentary rocks of central and eastern Pennsylvania, particularly in the anthracite region. Mailing Add: Dept of Geol Juniata Col Huntingdon PA 16652

TREYBIG, LEON BRUCE, b Yoakum, Tex, Aug 29, 31; m 57; c 3. MATHEMATICS. Educ: Univ Tex, BA, 53, PhD(math), 58. Prof Exp: Spec instr math, Univ Tex, 54-58; instr, Tulane Univ, 58-59, res assoc, 59-60, from asst prof to prof, 60-70; PROF MATH, TEX A&M UNIV, 70- Mem: Am Math Soc. Res: Point set topology; separability in metric spaces; continuous images of ordered compacta; knot theory. Mailing Add: Dept of Math Tex A&M Univ College Station TX 77843

TREYTL, WILLIAM JOSEPH, nuclear chemistry, analytical chemistry, see 12th edition

TREZISE, WILLARD JOSEPH, b Minersville, Pa, Mar 30, 10; m 36; c 1. BIOCHEMISTRY, BACTERIOLOGY. Educ: Lebanon Valley Col, BS, 31; Johns Hopkins Univ, MS, 32, PhD, 36. Prof Exp: Instr math, Frostburg State Col, Md, 36-38; from instr to assoc prof, 38-58, asst chmn dept sci, 38-60, vchmn, 60-62, chmn new sci bldg, 60, PROF BIOL, WEST CHESTER STATE COL, 58-, DEAN GRAD STUDIES, 62- Concurrent Pos: Head biochem res lab, Nat Foam Systs, Inc, 42-44; consult, R M Hollingshead, Inc, NJ, 44-45; prof, Pa Area Cols, Chester, 46-47 & Philadelphia, 47-49; consult, Govt Proj, Aeroprojs, Inc, 52-54. Honors & Awards: E Award, US Navy, 44. Mem: AAAS; Am Soc Microbiol; fel Am Inst Chem. Res: Light sensitivity and pigment migration in eye of beetle Dineutes nigroir; effect of bacteriostatic agents on industrial solutions; hydrogen ion concentration control; fermentation and protein decomposition; fire fighting foam from soybean, maize and grasses; non-inflammable paint removers; zoology. Mailing Add: 443 Caswallen Dr West Chester PA 19380

TRIANTAPHYLLOPOULOS, DEMETRIOS, b Athens, Greece, July 8, 20; m 54. PHYSIOLOGY, INTERNAL MEDICINE. Educ: Athens Sch Med, MD, 46. Prof Exp: From asst prof to assoc prof physiol, Univ Alta, 60-65; assoc prof, Wayne State Univ, 65-69; SR RES SCIENTIST & HEAD COAGULATION DEPT, AM NAT RED CROSS BLOOD RES LAB, 69- Concurrent Pos: Res assoc, Med Res Coun Can, 60-66. Mem: Am Physiol Soc; Can Physiol Soc; Am Soc Hemat; Int Soc Hemat; Am Heart Asn. Res: Blood coagulation; immunohematology. Mailing Add: Coagulation Dept Am Nat Red Cross Blood Res Lab Bethesda MD 20014

TRIANTAPHYLLOPOULOS, EUGENIE, b Astros, Greece, Nov 27, 21; Can citizen; m 54. MEDICINE, BIOCHEMISTRY. Educ: Nat Univ Athens, MD, 47; Univ Alta, PhD(biochem), 57. Prof Exp: Nat Cancer Inst Can res fel, Univ Alta, 57-60, res assoc blood coagulation, 60-62; Can Heart Found res fel, Univ Alta & Wayne State Univ, 62-67; res scientist, Mt Carmel Mercy Hosp, Detroit, Mich, 68-69; NIH grant & chief coagulation res, Washington Hosp Ctr, 69-74; MED OFFICER, BUR DRUGS, FOOD & DRUG ADMIN, 74- Concurrent Pos: Asst prof, Univ Alta, 65-66. Mem: AAAS; Am Physiol Soc; Can Biochem Soc; NY Acad Sci. Res: Blood coagulation; fibrinolysis; hemolysis. Mailing Add: Bur of Drugs Food & Drug Admin Rockville MD 20852

TRIANTAPHYLLOU, ANASTASIOS CHRISTOS, b Amaliapolis-Volou, Greece, Nov 30, 26; m 60; c 1. CYTOGENETICS, NEMATODES. Educ: Athens Superior Sch Agr, Greece, BS, 49, MS, 50; NC State Col, PhD(plant path, bot), 59. Prof Exp: Nematologist, Benaki Phytopath Inst, Greece, 50-60; asst geneticist, 60-62, from asst prof to assoc prof genetics, 62-68, PROF GENETICS, NC STATE UNIV, 68- Mem: AAAS; Genetics Soc Am; Soc Nematol; Am Inst Biol Sci. Res: Cytogenetics, evolution mode of reproduction and sexuality of nematodes; genetics and cytology of plant parasitic nematodes. Mailing Add: Dept of Genetics NC State Univ Raleigh NC 27607

TRIBBEY, BERT ALLEN, b Moorpark, Calif, Aug 8, 38; m 61; c 2. ZOOLOGY, ECOLOGY. Educ: Univ Calif, Santa Barbara, AB, 61; Univ Tex, PhD(zool), 65. Prof Exp: Asst prof, 65-69, ASSOC PROF BIOL, CALIF STATE UNIV, FRESNO, 69-, CHMN DEPT, 72- Mem: AAAS; Ecol Soc Am; Am Soc Limnol & Oceanog. Res: Structure and succession of aquatic communities; ecology of temporary ponds; physiological ecology of freshwater invertebrates. Mailing Add: Dept of Biol Calif State Univ Fresno CA 93710

TRIBBLE, LELAND FLOYD, b Oxnard, Calif, July 12, 23; m 52; c 4. ANIMAL SCIENCE. Educ: Univ Mo, BS, 49, MS, 50, PhD(agr), 56. Prof Exp: Instr animal husb, Univ Mo, 49-56, assoc prof, 56-57; PROF ANIMAL HUSB, TEX TECH UNIV, 67- Concurrent Pos: Vis prof, Kans State Univ, 65-66; nonruminant nutritionist, USDA, Washington, DC, 74-75. Mem: AAAS; Am Soc Animal Soc; Nutrition Today Soc. Res: Animal nutrition; swine production and management. Mailing Add: Dept of Animal Sci Tex Tech Univ Lubbock TX 79409

TRIBBLE, MERRELL THOMAS, physical organic chemistry, computer science, see 12th edition

TRIBBLE, ROBERT EDMOND, b Mexico, Mo, Jan 7, 47; m 69. NUCLEAR PHYSICS. Educ: Univ Mo, Columbia, BS, 69; Princeton Univ, PhD(nuclear physics), 73. Prof Exp: Instr physics, Princeton Univ, 73-75; ASST PROF PHYSICS, TEX A&M UNIV, 75- Mem: Am Phys Soc. Res: Experimental research to determine isospin mixing in light nuclei; nuclear weak interaction experiments that will determine the role of nuclear induced weak currents. Mailing Add: Cyclotron Inst Tex A&M Univ College Station TX 77843

TRICE, JAMES BUCKNER, b Hopkinsville, Ky, Oct 30, 20; m 47; c 7. RADIATION PHYSICS, ATOMIC SPECTROSCOPY. Educ: Univ Ky, BS, 43, MS, 47. Prof Exp: Exp physicist, Fairchild Engine & Airplane Corp, Tenn, 47-49, theoret physicist, 49-51; solid state physicist, Oak Ridge Nat Lab, 51-55; opers analyst, Hq, Strategic Air Command, Nebr, 55-57; consult physicist, Develop Labs, 57-61, consult physicist, Space Sci Lab, 61-65, group leader sensors & detectors res, 63-71, CONSULT PHYSICIST, Space Technol Prod, 71-75, consult, Nuclear Fuels Dept, 74-75; SYSTS ANALYST, ADVAN ENERGY PROGS, GEN ELEC CO, 75- Concurrent Pos: Consult, Oak Ridge Nat Lab, 56. Mem: Am Nuclear Soc; Am Geophys Union; Inst Elec & Electronics Engrs; Am Soc Testing & Mat. Res: Experimental nuclear physics of reactors, shields and radiation effect; mathematical methods; development of semiconductor radiation detectors; development of x-ray focusing diffraction systems; x-ray fluorescence elemental analyzers for industry. Mailing Add: Gen Elec Co Advan Energy Systs PO Box 8555 Philadelphia PA 19101

TRICE, WILLIAM HENRY, b Geneva, NY, Apr 4, 33; m 55; c 2. RESEARCH ADMINISTRATION. Educ: State Univ NY Col Forestry, BS, 55; Inst Paper Chem, MS, 60, PhD(phys chem), 63. Prof Exp: Res scientist, 63-66, group leader res & develop, 66-68, sect leader, 68-72, tech dir bleached div, 72-74, VPRES RES & DEVELOP & CORP TECH DIR, UNION CAMP CORP, 74- Concurrent Pos: Chmn res comt, Univ Maine Pulp & Paper Found, 76-77. Mem: Tech Asn Pulp & Paper Indust. Res: Fatty acid and terpene chemistry; pulp and paper chemistry; engineering in fields related to pulp and paper. Mailing Add: Union Camp Corp 1600 Valley Rd Wayne NJ 07470

TRICHE, TIMOTHY JUNIUS, b San Angelo, Tex, June 4, 44; m 71. PATHOLOGY. Educ: Cornell Univ, AB, 66; Tulane Univ, MD & PhD(path), 71. Prof Exp: Resident path, Barnes Hosp, St Louis, Mo, 71-74; SR STAFF FEL, LAB PATH, NAT CANCER INST, 74- Res: Membrane receptor sites; electron microscopic visualization of membrane bound lectins by freeze etch electron microscopy. Mailing Add: 8615 Lancaster Dr Bethesda MD 20014

TRICK, GORDON STAPLES, b Winnipeg,Winnipeg, Man, May 6, 27; m 52; c 4. PHYSICAL CHEMISTRY. Educ: McGill Univ, BS, 49, PhD(chem), 52; Univ Western Ont, MS, 50. Prof Exp: Ramsay Mem fel, London, 53; sci serv officer, Defence Res Bd Can, 53-56; sr res chemist & res scientist, Goodyear Tire & Rubber Co, 56-71; EXEC DIR MANITOBA RES COUN & DIR RES & TECHNOL BR, DEPT OF INDUST & COM, GOVT MANITOBA, 71- Concurrent Pos: Adj prof, Univ Manitoba, 72-; mem adv bd sci & tech info, Nat Res Coun Can, 72- Mem: Am Chem Soc; Chem Inst Can. Res: Gas and liquid phase kinetics; molecular weights of polymers; phase transitions in polymers; stress-strain properties; technology transfer; innovation development; science policy. Mailing Add: Manitoba Res Coun Govt of Manitoba Winnipeg MB Can

TRICKEL, RALPH JOSEPH, bacteriology, biochemistry, see 12th edition

TRICKEY, SAMUEL BALDWIN, b Detroit, Mich, Nov 28, 40; m 62; c 2. THEORETICAL PHYSICS, SOLID STATE PHYSICS. Educ: Rice Univ, BA, 62; Tex A&M Univ, MS, 66, PhD(physics), 68. Prof Exp: Physicist, Mason & Hanger, Silas Mason Corp, Tex, 62-64; vis asst prof physics, 68-70, asst prof, 70-73, ASSOC PROF PHYSICS, UNIV FLA, 73- Concurrent Pos: Vis staff mem, Los Alamos Sci Lab, 71; consult, Quantum Physics Group, Redstone Arsenal, Ala, 72-; consult phys sci, IBM Res Labs, San Jose, Calif, 75- Mem: Am Phys Soc; Am Asn Physics Teachers. Res: Theory of quantum crystals, rare gas solids, computational methods in solid state physics, energy band theory; lattice dynamics; quantum chemistry. Mailing Add: Quantum Theory Proj Univ of Fla Dept of Physics Gainesville FL 32611

TRICOLES, GUS P, b San Francisco, Calif, Oct 18, 31; m 53; 53; c 2. OPTICS, ELECTRICAL ENGINEERING. Educ: Univ Calif, BA, 55, MS, 62, PhD, 71; San Diego State Col, MS, 58. Prof Exp: Asst res engr, Gen Dynamics/Convair, 55, res engr, 55-59, sr res engr, 59; physicist, Smyth Res Assocs, 59-61 & Univ Calif, 62; PHYSICIST, ELECTRONICS DIV, GEN DYNAMICS CORP, 62- Mem: Inst Elec & Electronics Engrs; Optical Soc Am. Res: Microwave optics; diffraction; physical optics; holograms; antennas; waves. Mailing Add: 4868 Monroe Ave San Diego CA 92115

TRICOMI, VINCENT, b New York, NY, Sept 16, 21; m 49; c 6. OBSTETRICS & GYNECOLOGY. Educ: Syracuse Univ, AB, 42; State Univ NY Downstate Med Ctr, MD, 50. Prof Exp: From instr to assoc prof, 55-69, clin prof, 69-74, asst dean, 70-74, PROF OBSTET & GYNEC, STATE UNIV NY DOWNSTATE MED CTR, 74-, ASSOC DEAN, 70-; DIR MED AFFAIRS, BROOKLYN-CUMBERLAND MED CTR, 74- Concurrent Pos: Brooks scholar, 54-55; consult, Lutheran Med Ctr, 69- & Kings County Med Ctr, 71- Mem: Soc Gynec Invest; Sigma Xi. Res: Human cytogenetics. Mailing Add: Brooklyn Hosp 121 DeKalb Ave Brooklyn NY 11201

TRIEBWASSER, SOL, b New York, NY, Aug 16, 21; m 41; c 2. PHYSICS. Educ: Brooklyn Col, AB, 41; Columbia Univ, MA, 48, PhD, 52. Prof Exp: Instr physics, Brooklyn Col, 47-50; asst, Radiation Lab, Columbia, 51; mem tech staff, 52-69, ASST DIR APPL RES, RES CTR, IBM CORP, 69- Mem: Fel Am Phys Soc; Sigma Xi; fel Inst Elec & Electronics Engrs. Res: Atomic structure; microwave spectroscopy; solid state physics; ferroelectricity; photoconductivity; semiconductors. Mailing Add: IBM Corp Res Ctr PO Box 218 Yorktown NY 10598

TRIEFF, NORMAN MARTIN, b Brooklyn, NY, May 11, 29; m 60, 74; c 4. ENVIRONMENTAL CHEMISTRY, TOXICOLOGY. Educ: Polytech Inst Brooklyn, BS, 50; Univ Iowa, MS, 55; NY Univ, PhD(chem), 63. Prof Exp: Res asst biochem, Atran Labs, Mt Sinai Hosp, New York, 54-55; instr chem, Cooper Union, 55-58; res assoc biochem, Isaac Albert Res Ctr, Jewish Chronic Dis Hosp, Brooklyn, NY, 61-62; supvr blood res, Blood derivatives Sect, Mich Dept Pub Health, 63-65; asst prof chem, Drexel Inst, 65-68; asst prof toxicol & phys chem, 68-70, ASST PROF PREV MED & COMMUNITY HEALTH, UNIV TEX MED BR GALVESTON, 70- Concurrent Pos: Nat Ctr for Air Pollution Control grant, 66-68; consult, US Army Corps Engr, 67-68 & US Coast Guard, 70-71; Robert A Welch Found res grant, 70-; consult, Farner & Winslow, Inc, Houston, TX & Heat Systs Ultrasonics, Inc, Plainview, NY,

75- Mem: Fel Am Inst Chemists; Am Chem Soc; Air Pollution Control Asn; Am Indust Hyg Asn; NY Acad Sci. Res: Environmental chemistry; development of analytic methods of drugs, toxins and environmental pollutants; odor analysis and olfaction; chemical and biochemical kinetics. Mailing Add: Dept Prev Med & Commun Hlth Univ of Tex Med Br Galveston TX 77550

TRIER, JERRY STEVEN, b Frankfurt, Ger, Apr 12, 33; US citizen; m 57; c 3. INTERNAL MEDICINE, GASTROENTEROLOGY. Educ: Univ Wash, MD, 57. Prof Exp: Intern med, Univ Rochester, 57-58, asst resident, 58-59; clin assoc, Nat Cancer Inst, 59-61; trainee gastroenterol, Univ Wash, 61-63; asst prof med, Univ Wis, 63-67; assoc prof, Univ NMex, 67-69; assoc prof med & anat, Sch Med, Boston Univ, 69-73; ASSOC PROF MED, HARVARD MED SCH, 73- Concurrent Pos: USPHS grant, 64-67; USPHS grant, Univ NMex, 67-69; USPHS grants, Sch Med, Boston Univ, 69-; assoc physician, Univ Hosp, Boston & Boston City Hosp, 69-73; consult, Vet Admin Hosp, Boston, 69-, Chelsea Naval Hosp, 71-74 & US Vet Admin Cent Off, Washington, DC, 71-; physician & dir div gastroenterol, Peter Bent Brigham Hosp, Boston, 73- Mem: Am Soc Clin Invest; Am Gastroenterol Asn; Am Fedn Clin Res; Am Soc Cell Biol. Res: Functional morphology of the gastrointestinal tract of humans in health and disease; developmental morphology of the intestine; cell renewal in the alimentary tract. Mailing Add: Dept of Med Peter Bent Brigham Hosp Boston MA 02115

TRIFAN, DANIEL SIEGFRIED, b Cleveland, Ohio, Dec 23, 18; m 48; c 3. PHYSICAL CHEMISTRY, ORGANIC CHEMISTRY. Educ: Baldwin-Wallace Col, BS, 40; Western Reserve Univ, MS, 41; Harvard Univ, MA, 46, PhD(chem), 48. Prof Exp: Res fel phys & org chem, Univ Calif, Los Angeles, 48-49, instr org chem, 49-50; asst prof, Bowling Green State Univ, 50-51; asst prof phys, org & polymer chem & head chem sect, Plastics Lab, Princeton Univ, 51-64; PROF CHEM, FAIRLEIGH DICKINSON UNIV, 64- Mem: Am Chem Soc. Res: Organic reaction mechanisms; polymer chemistry. Mailing Add: Dept of Chem Fairleigh Dickinson Univ Rutherford NJ 07070

TRIFAN, DEONISIE, b Cleveland, Ohio, July 27, 15; m 48; c 1. APPLIED MATHEMATICS. Educ: Baldwin-Wallace Col, AB, 37; Univ Toledo, MA, 40; Brown Univ, PhD(appl math), 48. Prof Exp: Instr math, Bluffton Col, 40-41; res assoc appl math, Brown Univ, 47-48; instr math, Case Inst Technol, 48-49; from asst prof to assoc prof, 49-67, PROF MATH, UNIV ARIZ, 67- Concurrent Pos: Consult, Radio Corp Am, 57-58. Mem: Am Math Soc; Math Asn Am. Res: Elasticity and plasticity. Mailing Add: Dept of Math Univ of Ariz Tucson AZ 85721

TRIFARO, JOSE MARIA, b Mercedes, Arg, Nov 29, 36; m 64; c 2. PHARMACOLOGY. Educ: Liceo Militar Gen San Martin, BA, 54; Univ Buenos Aires, MD, 61. Prof Exp: Demonstr anat, Univ Buenos Aires, 57-58, instr pharmacol, 61-62, lectr physiol, 62-64; lectr, 67-68, asst prof, 68-72, ASSOC PROF PHARMACOL, McGILL UNIV, 72- Concurrent Pos: A Thyssen Found res fel, Arg, 62-63; res fel, Nat Res Coun, Arg, 63-64; Rockefeller Found res fel, US, 64-66; NIH res fel, 66-67; Med Res Coun Can scholar, 68. Mem: Am Soc Pharmacol & Exp Therapeut; Pharmacol Soc Can. Res: Cellular and molecular mechanism of hormone and neurotransmitter release. Mailing Add: Dept of Pharmacol & Therapeut McGill Univ Montreal PQ Can

TRIFUNAC, ALEXANDER DIMITRIJE, b Yugoslavia, July 29, 44; US citizen; m 67. PHYSICAL CHEMISTRY, MAGNETIC RESONANCE. Educ: Columbia Univ, BA, 66; Univ Chicago, PhD(chem), 71. Prof Exp: Res asst chem, Univ Chicago, 66-71; res assoc chem, Univ Notre Dame, 71-72 & Univ Chicago, 72; ASST SCIENTIST, CHEM DIV, ARGONNE NAT LAB, 74- Concurrent Pos: Presidential intern, Argonne Nat Lab, 72-73. Mem: Am Chem Soc; The Chem Soc; Int Soc Magnetic Resonance. Res: Chemically induced magnetic polarization in radiation chemistry and photochemistry; timeresolved electron paramagnetic resonance spectroscopy. Mailing Add: Chem Div Argonne Nat Lab 9700 S Cass Argonne IL 60439

TRIGG, GEORGE LOCKWOOD, b Washington, DC, Sept 30, 25; m 48; c 2. THEORETICAL PHYSICS. Educ: Washington Univ, AB, 47, AM, 50, PhD(physics), 51. Prof Exp: Asst physics, Washington Univ, 47-50; asst prof, Knox Col, 51-54, actg chmn dept, 51-52; from asst prof to assoc prof, Ore State Univ, 54-62; asst ed, 62-65, ED, PHYS REV LETTERS, AM PHYS SOC, 65- Concurrent Pos: NSF fel, 57-58; asst ed, Phys Rev & Phys Rev Letters, 58; consult, Funk & Wagnalls Dictionary, 66-71 & Am Heritage Dictionary, 67-70; mem comt symbols, units & terminology, Nat Res Coun, 71- Mem: AAAS; fel Am Phys Soc; Fedn Am Sci; Am Asn Physics Teachers. Res: Elementary particle theory; fundamentals of quantum theory; history of physics. Mailing Add: Phys Rev Letters Brookhaven Nat Lab Upton NY 11973

TRIGG, WILLIAM WALKER, b Little Rock, Ark, Dec 4, 31; m 57; c 2. INORGANIC CHEMISTRY. Educ: Univ Ark, BSChE, 56, MS, 60; La State Univ, PhD(inorg chem), 66. Prof Exp: Control chemist, Niagara Chem Div, Food Mach & Chem Corp, 56-57; from instr to assoc prof, 59-75, PROF CHEM, ARK POLYTECH COL, 75-, HEAD DEPT, 66- Mem: Am Chem Soc; Am Inst Chem Engrs. Res: Precipitation from homogeneous solution and studies of ion solvent effects in solvents of low dielectric constant. Mailing Add: Dept of Chem Ark Polytech Col Russellville AR 72801

TRIGGLE, DAVID J, b London, Eng, May 4, 35; m 59; c 1. PHARMACOLOGY, MEDICINAL CHEMISTRY. Educ: Univ Southampton, BSc, 56; Univ Hull, PhD(org chem), 59. Prof Exp: Fel org chem, Univ Ottawa, 59-61; res fel, Bedford Col, London, 61-62; asst prof biochem pharmacol, 62-65, assoc prof biochem pharmacol & theoret biol, 65-69, PROF BIOCHEM PHARMACOL & THEORET BIOL, SCH PHARM, STATE UNIV NY BUFFALO, 69-, CHMN DEPT BIOCHEM PHARMACOL, 71- Concurrent Pos: NIH res grants, 64- Mem: Am Chem Soc; The Chem Soc; Am Soc Pharmacol & Exp Therapeut. Res: Molecular pharmacology of adrenergic and cholinergic systems; organic reaction mechanisms; synthesis of organic heterocyclic systems; ion translocation and cell membranes; molecular basis of neurotransmitter action. Mailing Add: Dept of Biochem Pharmacol State Univ NY Sch of Pharm Buffalo NY 14214

TRIGLIA, EMIL J, b Lucca, Italy, Aug 26, 21; US citizen; m 48; c 5. ANALYTICAL CHEMISTRY. Educ: City Col New York, BS, 43. Prof Exp: Chemist, Ledoux & Co, Inc, 46-51; group leader anal chem, Chem Construct Corp & Am Cyanamid Co, 51-56; sr res chemist, Minerals & Chem, Philipp Corp, 56-61, res group supvr, 61-68; group leader anal & phys testing, 68-74, MGR ANAL & PHYS MEASUREMENTS, ENGELHARD MINERALS & CHEM CORP, MENLO PARK, 74- Mem: Am Chem Soc; Tech Asn Pulp & Paper Indust; Am Soc Testing & Mat. Res: Analyses of minerals, ores and rocks by chemical and instrumental methods. Mailing Add: 12 Sharon Ct Metuchen NJ 08840

TRILLING, GEORGE HENRY, b Bialystok, Poland, Sept 18, 30; nat US; m 55; c 2. ELEMENTARY PARTICLE PHYSICS. Educ: Calif Inst Technol, BS, 51, PhD(physics), 55. Prof Exp: Res fel physics, Calif Inst Technol, 55-56; Fulbright res fel, Polytech Sch, Paris, 56-57; from asst prof to assoc prof, Univ Mich, 57-60; assoc prof, 60-64, chmn dept, 68-72, PROF PHYSICS, UNIV CALIF, BERKELEY, 64- Concurrent Pos: NSF sr fel, Europ Orgn Nuclear Res, 66-67, Guggenheim fel, 73-74. Mem: Fel Am Phys Soc. Res: Properties of elementary particle produced by high energy accelerators. Mailing Add: Dept of Physics Univ of Calif Berkeley CA 94720

TRIMBERGER, GEORGE WILLIAM, b Neilsville, Wis, Dec 8, 09; m 38; c 3. DAIRY SCIENCE. Educ: Univ Wis, BS, 33; Univ Nebr, MS, 42, PhD(zool), 48. Prof Exp: Herd supt, Univ Nebr, 34-40, instr dairy prod, 40-44; from asst prof to assoc prof, 44-50, PROF DAIRY HUSB, CORNELL UNIV, 50- Concurrent Pos: Vis prof & proj leader, Univ Philippines, 55-57 & 66-67. Mem: AAAS; Am Soc Animal Sci; Am Diary Sci Asn; Am Genetic Asn. Res: Artificial insemination; reproduction and nutrition in dairy cattle. Mailing Add: Morrison Hall Cornell Univ Ithaca NY 14850

TRIMBLE, HAROLD CALLANDER, b Can, Mar 6, 14; nat US; m 38; c 3. MATHEMATICS. Educ: Univ Western Ont, BA, 35; Univ Wis, MA & PhD(math), 39. Prof Exp: Gen Educ Bd fel, Univ Chicago, 39, instr educ, 40; from instr to asst prof math, Iowa State Univ, 40-47; assoc prof, Fla State Univ, 47-51; from assoc prof to prof, Iowa State Univ, 51-63, head dept, 58-63; PROF MATH EDUC, OHIO STATE UNIV, 63- Concurrent Pos: Ford fel, 54-55. Mem: Math Asn Am. Res: Secondary mathematics curriculum. Mailing Add: 613 Farrington Dr Worthington OH 43085

TRIMBLE, MARY ELLEN, b Englewood, NJ, Nov 1, 36. PHYSIOLOGY, BIOLOGY. Educ: Wellesley Col, AB, 58; Syracuse Univ, MA, 59; Case Western Reserve Univ, PhD(biol), 69. Prof Exp: Res assoc develop biol, Brown Univ, 68-70; res assoc physiol, 70-71, ASST PROF PHYSIOL, STATE UNIV NY UPSTATE MED CTR, 72-; RES BIOLOGIST, VET ADMIN HOSP, SYRACUSE, 71- Mem: AAAS; Am Soc Zoologists; Am Physiol Soc; Int Soc Nephrology. Res: Renal physiology, particularly countercurrent mechanism, metabolic aspects of sodium transport and compensatory renal hypertrophy. Mailing Add: Dept of Physiol State Univ NY Upstate Med Ctr Syracuse NY 13210

TRIMBLE, ROBERT BOGUE, b Baltimore, Md, July 2, 43; m 69. BIOCHEMISTRY, GENETICS. Educ: Rensselaer Polytech Inst, BS, 65, MS, 67, PhD(microbiol), 69. Prof Exp: Health Res Inc fel, 69-70, res scientist, 70-72, SR RES SCIENTIST, DEVELOP BIOCHEM LABS, DIV LABS & RES, NY STATE DEPT HEALTH, 72- Mem: AAAS; Am Soc Microbiol. Res: Regulation of enzyme synthesis and activity; bacteriophage biochemistry; viral genetic expression; pyrimidine nucleotide interconversions; geomicrobiology. Mailing Add: Develop Biochem Labs Div of Labs NY Health Dept Albany NY 12101

TRIMBLE, RUSSELL FAY, b Montclair, NJ, Feb 23, 27; m 50; c 3. INORGANIC CHEMISTRY. Educ: Mass Inst Technol, BS, 48, PhD(chem), 51. Prof Exp: Instr chem, Univ Rochester, 51-54; from asst prof to assoc prof, 54-70, PROF CHEM, SOUTHERN ILL UNIV, CARBONDALE, 70- Concurrent Pos: Abstractor, Chem Abstr, 54-, tech translator, 59-; vis lectr, Univ Ill, 63-64. Mem: AAAS; Am Chem Soc; Am Inst Chemists; Am Translators Asn. Res: Coordination compounds; inorganic synthesis. Mailing Add: Dept of Chem & Biochem Southern Ill Univ Carbondale IL 62901

TRIMBLE, RUSSELL HAROLD, b Kingsport, Tenn, Oct 30, 43; m 63; c 1. CHEMICAL PHYSICS. Educ: Davidson Col, BS, 65; Mass Inst Technol, PhD(chem physics), 70. Prof Exp: SR RES CHEMIST, MOBIL FIELD RES LAB, MOBIL OIL CORP, 70- Mem: Am Inst Mining, Metall & Petrol Engrs. Res: Fluid flow, computer simulation of physical systems; statistical mechanics; non-equilibrium thermodynamics; numerical analysis. Mailing Add: Mobil Field Res Lab Mobil Oil Corp PO Box 900 Dallas TX 75221

TRIMBLE, VIRGINIA LOUISE, b Los Angeles, Calif, Nov 15, 43; m 72. ASTRONOMY, ASTROPHYSICS. Educ: Univ Calif, Los Angeles, BA, 64; Calif Inst Technol, MS, 65, PhD(astron), 68; Cambridge Univ, MA, 69. Prof Exp: Res fel astrophys, Inst Theoret Astron, Cambridge Univ, 68; asst prof astron, Smith Col & Four Cols Observ, 68-69; NATO sr fel, Inst Theoret Astron, Cambridge Univ, 69-70, vis fel, 70-71; asst prof, 71-74, ASSOC PROF PHYSICS, UNIV CALIF, IRVINE, 74- Concurrent Pos: Vis asst prof astron, Univ Md, College Park, 72-74, vis assoc prof, 74-; Sloan fel, 72-74; nat lectr, Sigma Xi, 74-76. Mem: Am Astron Soc; Royal Astron Soc; Europ Phys Soc; Int Astron Union; Int Soc Gen Relativity & Gravitation. Res: Late phases of stellar evolution; supernovae; white dwarfs; neutron stars; collapsed configurations; binary stars. Mailing Add: Dept of Physics Univ of Calif Irvine CA 92664

TRIMITSIS, GEORGE B, b Assiut, Egypt, Nov 28, 39; m 64; c 1. ORGANIC CHEMISTRY. Educ: Am Univ Cairo, BSc, 64; Va Polytech Inst & State Univ, PhD(org chem), 68. Prof Exp: Res assoc org res, Ohio State Univ, 68-69; ASST PROF ORG CHEM, WESTERN MICH UNIV, 69- Mem: Am Chem Soc. Res: Formation, study and synthetic applications of carbanions. Mailing Add: Dept of Chem Western Mich Univ Kalamazoo MI 49001

TRIMMER, JOHN DEZENDORF, b Washington, DC, Sept 19, 07; m 30; c 2. PHYSICS. Educ: Elizabethtown Col, AB, 26; Pa State Univ, MS, 33; Univ Mich, PhD(physics), 36. Hon Degrees: DSc, Elizabethtown Col, 53. Prof Exp: Asst prof aeronaut eng, Mass Inst Technol, 37-41, res assoc underwater sound, 41-43; physicist, Tenn Eastman Co, 43-46; prof physics, Univ Tenn, 46-57; prof, Univ Mass, 57-66, head dept, 57-63; prof & head dept, 66-73, EMER PROF PHYSICS, WASHINGTON COL, 73- Mem: AAAS; Am Phys Soc; Acoust Soc Am; Soc Gen Systs Res. Res: Instrumentation; principles of measurement; communication; physics of ionized fluids; cybernetics; general systems theory; optical physics. Mailing Add: Box 75A RD 1 Millington MD 21651

TRIMMER, ROBERT WHITFIELD, b Binghamton, NY, Dec 13, 37; m 65; c 2. INDUSTRIAL ORGANIC CHEMISTRY. Educ: Hope Col, Mich, AB, 60; Rensselaer Polytech Inst, PhD(org chem), 73. Prof Exp: Asst res polymer chem, Schenectady Chem Co, 60; asst res med chemist, Sterling Winthrop Res Inst, Sterling Drug, Inc, 64-69; SUPVR LAB PROD DEVELOP & ORG CHEMIST, SUMNER DIV, MILES LABS, INC, 73- Mem: Am Chem Soc; The Chem Soc; AAAS; Org Reactions Catalysis Soc; Am Sci Affil. Res: Hydrogenation technology; specialty products; pharmaceuticals; aryl and alkyl amines as intermediates and polymer catalysts; quaternary ammonium compounds, citric acid derivatives; heterocycles; organic photochemistry; development of new synthetic methods. Mailing Add: 23357 Delany Lane Elkhart IN 46514

TRIMNELL, DONALD, chemistry, see 12th edition

TRINDELL, ROGER THOMAS, b Newark, NJ, Jan 2, 32; m 57; c 1. CULTURAL GEOGRAPHY, HISTORICAL GEOGRAPHY. Educ: Montclair State Col, BA, 58; Univ Colo, Boulder, MA, 60; La State Univ, Baton Rouge, PhD(geog), 66. Prof Exp: Assoc prof geog & anthrop, Millersville State Col, 62-64; asst prof geog, Calif State Col, Hayward, 64-65; asst prof, State Univ NY Col, Plattsburg, 65-66; asst prof, Mich

State Univ, 66-70; PROF GEOG & REGIONAL PLANNING & CHMN DEPT, MANSFIELD STATE COL, 70- Mem: AAAS; Asn Am Geogrs; Am Geog Soc; Nat Coun Geog Educ. Res: Human geography: heuristic, applied and theoretical; environmental systems; spatial analysis; regional planning; geographic education. Mailing Add: Dept of Geog Mansfield State Col Mansfield PA 16933

TRINDLE, CARL OTIS, b Des Moines, Iowa, Aug 26, 41; m 62; c 1. THEORETICAL CHEMISTRY. Educ: Grinnell Col, BA, 63; Tufts Univ, PhD(phys chem), 67. Prof Exp: NSF fel theoret chem, Yale Univ, 67-68; res assoc, Argonne Nat Lab, 68-69; asst prof, 69-73, ASSOC PROF THEORET CHEM, UNIV VA, 73- Concurrent Pos: Consult, Argonne Nat Lab, 69-; vis asst prof, Mideast Tech Univ, 71; Sloan Found fel, 71-73. Mem: AAAS; Am Chem Soc. Res: Impact of orbital topology on organic stereo-chemistry; localized description of charge distributions; group theory of easily rearranged systems. Mailing Add: Dept of Chem Univ of Va Charlottesville VA 22901

TRINE, FRANKLIN DAWSON, b Cincinnati, Ohio, July 12, 30; m 48; c 4. MATHEMATICS. Educ: Wis State Univ, Platteville, BS, 53; Univ Wis, MS, 57, PhD(math educ), 65. Prof Exp: Instr high sch, Ill, 53-56 & 57-58; from instr to assoc prof, 58-66, PROF MATH, UNIV WIS-PLATTEVILLE, 66- Mem: Math Asn Am. Res: Foundations of mathematics; mathematics education, particularly at junior college and college level. Mailing Add: Dept of Math Univ of Wis Platteville WI 53818

TRINKAUS, ERIK, b New Haven, Conn, Dec 24, 48. PHYSICAL ANTHROPOLOGY. Educ: Univ Wis, BA, 70; Univ Pa, MA, 73, PhD(anthrop), 75. Prof Exp: ASST PROF ANTHROP, HARVARD UNIV, 75- Mem: Am Asn Phys Anthropologists. Res: Paleontological study of Middle and Upper Pleistocene hominids emphasizing the behavioral interpretation of fossil remains; reconstructing human evolutionary history using complex evolutionary models as well as comparative anatomy. Mailing Add: Dept Anthrop Peabody Mus Harvard Univ Cambridge MA 02138

TRINKAUS, JOHN PHILIP, b Rockville Center, NY, May 23, 18; m 63; c 3. DEVELOPMENTAL BIOLOGY. Educ: Wesleyan Univ, BA, 40; Columbia Univ, MA, 41; Johns Hopkins Univ, PhD(embryol), 48. Prof Exp: From instr to assoc prof, 48-72, PROF BIOL, YALE UNIV, 48- Concurrent Pos: Mem staff embryol, Marine Biol Lab, Yale Univ, 53-57, from fel to master, Branford Col, 51-73, dir grad studies biol, 65-66; Guggenheim fel, Lab Exp Embryol, Col France, 59-60. Mem: Soc Develop Biol; Am Soc Zoologists; Tissue Cult Asn; Am Soc Cell Biol; Int Inst Embryol. Res: Cytodifferentiation; mechanism of morphogenetic movements; teleost development; contact behavior and locomotion of normal and transformed cells; fine structure. Mailing Add: Dept of Biol Yale Univ New Haven CT 06520

TRINLER, WILLIAM A, b Louisville, Ky, Dec 24, 29; m 62; c 1. ORGANIC CHEMISTRY. Educ: Univ Louisville, BS, 55, PhD(org chem), 59. Prof Exp: Chemist, E I du Pont de Nemours & Co, 59-60; from asst prof to assoc prof, 60-74, PROF ORG CHEM, IND STATE UNIV, TERRE HAUTE, 74- Mem: Am Chem Soc; Am Inst Chemists. Res: Synthesis and polymerization of vinyl monomers; liquid chromatography. Mailing Add: Dept of Chem Ind State Univ Terre Haute IN 47809

TRIOLO, ANTHONY J, b Philadelphia, Pa, Aug 8, 32; m 59; c 3. PHARMACOLOGY. Educ: Philadelphia Col Pharm, BS, 59; Jefferson Med Col, MS, 62, PhD(pharmacol), 64. Prof Exp: From instr to asst prof pharmacol, 67-72, ASSOC PROF PHARMACOL, JEFFERSON MED COL, 72- Res: Neuropharmacological effects of tremorine on motor reflex activity; toxicological interactions between organochlorine or organophosphate insecticides and Benzo-(a)-pyrene carcinogenesis. Mailing Add: Dept of Pharmacol Jefferson Med Col Philadelphia PA 19107

TRIOLO, VICTOR ANTHONY, b New York, NY, May 31, 32; m 68. INFORMATION SCIENCE. Educ: Brooklyn Col, BS, 53; Univ Mass, Amherst, MA, 56; Univ Wis-Madison, PhD(biomed hist), 62; Columbia Univ, MS, 74. Prof Exp: Asst physiol, Univ Ill, Urbana, 55-56; asst chemother, Mem Sloan Kettering Cancer Ctr, NY, 56-57; proj asst oncol, Sch Med, Univ Wis, 57-59, res asst, 59-62; from instr to assoc prof behav sci, Sch Med, Temple Univ, 66-74, investr, Fels Res Inst, 66-74; ASSOC PROF, DIV LIBR SCI & INSTRNL TECHNOL, SOUTHERN CONN STATE COL, 75- Concurrent Pos: USPHS fel, McArdle Mem Lab, 62-66; Nat Cancer Inst res career develop award, Health Sci Ctr, Temple Univ, 66-71. Honors & Awards: Wellcome Trust Award, London, Eng, 70. Mem: Am Soc Info Sci; Am Libr Asn; Med Libr Asn. Res: Machine based information systems and audiovisual instructional applications in health science documentation. Mailing Add: Southern Conn State Col 501 Crescent St New Haven CT 06515

TRIONE, EDWARD JOHN, b Ill, Mar 10, 26; m 49; c 3. BIOCHEMISTRY. Educ: Chico State Col, BA, 50; Ore State Col, PhD(bot), 57. Prof Exp: Chem technician, Ore State Col, 53-54, asst, 54-57; res plant pathologist, Univ Calif, 57-59; PLANT BIOCHEMIST, ORE STATE UNIV, 59- Mem: Am Soc Plant Physiologists; Bot Soc Am; Am Phytopath Soc; Can Soc Plant Physiologists; Japanese Soc Plant Physiologists. Res: Biochemistry and physiology of reproduction in higher plants and fungi; biochemistry of host-pathogen interactions. Mailing Add: Dept of Bot & Plant Path Ore State Univ Corvallis OR 97331

TRIPARD, GERALD EDWARD, b Saskatoon, Sask, Apr 18, 40; m 63; c 2. NUCLEAR PHYSICS. Educ: Univ BC, BSc, 62, MSc, 64, PhD(physics), 67. Prof Exp: Nat Res Coun Can fel, Swiss Fed Inst Technol, 67-69; ASST PROF PHYSICS, WASH STATE UNIV, 69- Res: Neutron scattering; final state interactions; stopped pions. Mailing Add: Dept of Physics Wash State Univ Pullman WA 99163

TRIPATHI, KAMALA KANT, b Varanasi, India, July 10, 34; m 51; c 4. BIOCHEMISTRY, ANALYTICAL CHEMISTRY. Educ: Banaras Hindu Univ, BS, 54; Univ Bihar, DVM, 59; Univ Mo-Columbia, 64, PhD(biochem), 68. Prof Exp: Vet surgeon & exten officer, Govt Bihar, 59-63; res asst, Univ Nebr, Lincoln, 64-65; res asst bioanal chem, Univ Mo-Columbia, 65-68, NIH fel, 68-70; RES SCIENTIST CANCER CHEMOTHER, AM MED CTR DENVER, 70- Mem: Am Chem Soc. Res: Cancer biochemistry, proteins, fatty acids, amino acids and steroids; analytical, immunochemical, ion-exchange, electrophoretic, gas-liquid and thin-layer chromatographic methods and their applications. Mailing Add: Am Med Ctr at Denver 6401 W Colfax Ave Spivak CO 80214

TRIPATHI, UMA PRASAD, b Lumbini, Nepal, Apr 29, 45; m 70; c 1. INORGANIC CHEMISTRY, COSMETIC CHEMISTRY. Educ: Univ Gorakhpur, India, BSc, 63; Univ Allahabad, India, MSc, 65; Mont State Univ, PhD(inorg chem), 72. Prof Exp: Anal chemist, Nepal Bur Mines, Kathmandu, 65; asst prof chem, Trichandra Col, Kathmandu, 65-67; SR RES CHEMIST COSMETICS, CHESEBROUGH-POND'S INC, 72- Mem: Soc Cosmetic Chemists; Am Chem Soc. Res: Elucidation of molecular and crystal structures of organometallics; inorganic synthesis; inorganic reaction mechanisms; emulsion technology; surface active agents. Mailing Add: Chesebrough-Pond's Res Labs Trumbull Indust Park Trumbull CT 06611

TRIPATHY, DEOKI NANDAN, b Dwarahat, India, July 1, 33; m 69; c 2. VETERINARY MICROBIOLOGY. Educ: Utter Pradesh Agr Univ, India, BVS & AH, 64; Univ Ill, Urbana, MS, 67, PhD(vet microbiol), 70; Am Col Vet Microbiologists, dipl, 72. Prof Exp: Fel, 64-65, res asst, 65-70, assoc, 70-73, ASST PROF VET MICROBIOL, UNIV ILL, URBANA, 73- Mem: Am Vet Med Asn; Am Asn Avian Pathologists; Am Soc Microbiol; Am Col Vet Microbiologists; Am Leptospirosis Res Conf. Res: Pathogenesis, epidemiology and characterization of avian pox viruses; immune response of ducks to Pasteurella anatipestifer and duck hepatitis virus; immunization and evaluation of immune response to leptospirosis. Mailing Add: Dept of Vet Path & Hyg Univ of Ill Urbana IL 61801

TRIPLEHORN, CHARLES A, b Bluffton, Ohio, Oct 27, 27; m 49; c 2. ENTOMOLOGY. Educ: Ohio State Univ, BS, 49, MS, 52; Cornell Univ, PhD(entom), 57. Prof Exp: Asst prof entom, Univ Del, 52-54, Ohio Agr Exp Sta, 57-62 & Ohio State Univ, 62-64; entomologist, US AID, Brazil, 64-66; assoc prof, 66-67, PROF ENTOM, OHIO STATE UNIV, 67- Mem: Entom Soc Am; Soc Syst Zool; Coleopterists Soc (pres, 76). Res: Taxonomy of Coleoptera; animal ecology; herpetology. Mailing Add: Dept of Entom Ohio State Univ Columbus OH 43210

TRIPLEHORN, DON MURRAY, b Bluffton, Ohio, July 24, 34; m 57; c 3. GEOLOGY. Educ: Ohio Wesleyan Univ, BA, 56; Ind Univ, MA, 57; Univ Ill, PhD(geol), 61. Prof Exp: Instr geol, Col Wooster, 60-61; res geologist, Tulsa Res Ctr, Sinclair Oil & Gas Co, 61-69; ASSOC PROF GEOL, UNIV ALASKA, FAIRBANKS, 69- Mem: Geol Soc Am; Mineral Soc Am; Am Asn Petrol Geologists; Soc Econ Paleontologists & Mineralogists; Clay Minerals Soc. Res: Glauconite; clay mineralogy; shale petrology; diagenesis. Mailing Add: Dept of Geol Univ of Alaska Fairbanks AK 99701

TRIPLETT, EDWARD LEE, b Denver, Colo, July 14, 30; m 51; c 3. ZOOLOGY. Educ: Stanford Univ, BS, 51, PhD, 56. Prof Exp: Actg instr biol, Stanford Univ, 54-55; from instr to asst prof, 55-63, ASSOC PROF BIOL, UNIV CALIF, SANTA BARBARA, 63- Mem: AAAS; Soc Develop Biol. Res: Development of the nervous system; embryology and immune mechanisms of fishes; control of protein synthesis in developing systems. Mailing Add: Dept of Biol Sci Univ of Calif Santa Barbara CA 93106

TRIPLETT, GLOVER BROWN, JR, b Miss, June 2, 30; m 51; c 1. AGRONOMY. Educ: Miss State Univ, BS, 51, MS, 55; Mich State Univ, PhD(farm crops), 59. Prof Exp: From asst prof to assoc prof, 59-67, PROF AGRON, OHIO AGR RES & DEVELOP CTR, 67-, AGRONOMIST, 59- Mem: Am Soc Agron; Weed Sci Soc Am. Res: Crop production and management. Mailing Add: Ohio Agr Res & Dev Ctr Wooster OH 44691

TRIPODI, DANIEL, b Cliffside Park, NJ, May 13, 39; m 63. IMMUNOCHEMISTRY, MICROBIOLOGY. Educ: Univ Del, BS, 61, MS, 63; Temple Univ, PhD(microbiol, immunol), 66. Prof Exp: Asst microbiol, Univ Del, 63; sr scientist, Ortho Res Found, 66-69, dir div immunol & diag res, 69-74; res asst prof, 69-74, RES ASSOC PROF MICROBIOL, SCH MED, TEMPLE UNIV, 73-; GEN MGR, BIOMED DIV, NEW ENG NUCLEAR CORP, 74- Mem: Am Soc Microbiol; Am Asn Immunol. Res: Physical and chemical aspects of structures which display serological reactivity. Mailing Add: Biomed Div New Eng Nuclear Corp North Billerica MA 01862

TRIPP, MARENES ROBERT, b Poughkeepsie, NY, Aug 20, 31; m 55; c 4. INVERTEBRATE PATHOLOGY. Educ: Colgate Univ, AB, 53; Univ Rochester, MS, 56; Rutgers Univ, PhD(zool), 58. Prof Exp: Res fel trop pub health, Harvard Univ, 58-60; from asst prof to assoc prof, 60-71, PROF BIOL SCI, UNIV DEL, 71- Mem: AAAS; Am Soc Parasitol; Soc Invert Path; Am Soc Zoologists; Reticuloendothelial Soc. Res: Invertebrate defense mechanisms. Mailing Add: Dept of Biol Sci Univ of Del Newark DE 19711

TRIPP, RALPH HARRY, b Denton, Mont, Mar 11, 15; m; c 3. MATHEMATICS. Educ: Drake Univ, AB, 37; Iowa State Col, MS, 39, PhD(appl math), 42. Prof Exp: Instr math, Iowa State Col, 38-42; from vibrations engr to chief res, Grumman Aircraft Eng Corp, 42-48, chief instrumentation, 48-57, asst dir, Flight Test Div, 57-62, dir orbiting astron observ prog, 62-66, dir unmannned space progs, 66-68, lunar module prog mgr, 68-70, vpres lunar module & unmanned space progs, 70-72, dir corp systs, 72-74, dir, Training Systs & mgr, Great River Responsibility Ctr, 74-75, EXEC VPRES & DIR, GRUMMAN AEROSPACE CORP, 75- Mem: Fel AAAS; fel Instrument Soc Am (pres, 61); assoc fel Am Inst Aeronaut & Astronaut; Am Astron Soc; Am Mgt Asn. Res: Vibration and flutter; supersonic aerodynamics; transient loads; dynamic stability and control; data processing. Mailing Add: 9 Apex Rd Melville NY 11746

TRIPP, ROBERT D, b Oakland, Calif, Jan 9, 27; m 64; c 3. PHYSICS. Educ: Mass Inst Technol, BS, 49; Univ Calif, PhD, 55. Prof Exp: From asst prof to assoc prof, 60-66, PHYSICIST, LAWRENCE BERKELEY LAB, UNIV CALIF, 55-, PROF PHYSICS, UNIV CALIF, BERKELEY, 66- Concurrent Pos: Physicist, AEC, France, 59-60; NSF sr fel, Europ Orgn Nuclear Res, Switz, 64-65, vis scientist, 71-72. Res: Elementary particle physics. Mailing Add: Lawrence Berkeley Lab Univ of Calif Berkeley CA 94720

TRIPP, RUSSELL MAURICE, b Holton, Kans, July 12, 16; m 37; c 7. RADIOLOGY, GEOPHYSICS. Educ: Colo Sch Mines, GeolE, 39, MGeophysEng, 43; Mass Inst Technol, ScD(geol), 48. Prof Exp: Geophysicist, Geotech Corp, Tex, 36-41, asst to pres in-chg res, 43-46; instr geol & geophys, Colo Sch Mines, 41-43; sr scientist, Bur Ships, US Dept Navy, 46; consult geologist & geophysicist, 47-49; vpres & dir res, Res Inc, 49-53; pres, Explor, Inc, 52-58; PRES, TRIPP RES CORP, 55- Concurrent Pos: Managing partner, Tripp Lead & Zinc Co, 52-, pres, Tripp Prod, Inc, 62-; dir, Sonic Res Corp, 56-; consult, Bostwick Propecting Co, 56- & Archilithic Corp, 58-; pres, Skia Corp, 72- Mem: AAAS; Soc Econ Paleont & Mineral; Soc Explor Geophys; Am Asn Petrol Geologists; Am Inst Mining, Metall & Petrol Engrs. Res: Geophysical and geochemical exploration for minerals; relation between clay minerals, organic matter and radioelements in sediments; genesis of uranium ore bodies; mineral benefication; free viewing depth-perception radiography; nuclear medicine instrumentation; animal genetics; ophthalmic surgical instrumentation. Mailing Add: 15231 Quito Rd Saratoga CA 95070

TRIPPE, THOMAS GORDON, b Los Angeles, Calif, Nov 17, 39; div; c 3. EXPERIMENTAL HIGH ENERGY PHYSICS. Educ: Univ Calif, Los Angeles, PhD(physics), 68. Prof Exp: Physicist, Univ Calif, Los Angeles, 68-69; physicist, Europ Orgn Nuclear Res, 69-70; PHYSICIST, LAWRENCE BERKELEY LAB, 71- Concurrent Pos: NSF fel, 69-70. Mem: Am Phys Soc. Res: Experimental weak interactions; particle properties. Mailing Add: Lawrence Berkeley Lab Berkeley CA 94720

TRIPPEER, WILLIAM MOWBRAY, b Chagrin Falls, Ohio, Aug 23, 27; m 50; c 2. SPECTROSCOPY. Educ: Va Polytech Inst, BS, 51. Prof Exp: Draftsman, Appalachian Elec Power Co, Va, 48-50; engr, E I du Pont de Nemours & Co, Tex, 51-55, res engr, Eng Physics Lab, Del, 55-60, sr res engr, 60-61, res supvr, 61-65;

mgr, Com Div, Block Eng, Inc, Mass, 65; tech supvr, Textile Fibers Dept, Del, 66-67, sr tech supvr, Tech Div, Tenn, 67-69, supvr, Export Sales Instrument & Equip Div, Photo Prod Dept, 69-70, mgr int opers, Instrument Prod Div, 70-71, prod mgr, 71-73, nat serv mgr, 73-74, PLANNING MGR, INSTRUMENT PROD DIV, PHOTO PROD DEPT, E I DU PONT DE NEMOURS & CO, DEL, 74- Mem: Optical Soc Am; NY Acad Sci; Electron Micros Soc Am. Res: Process analysis; infrared instrumentation; automated laboratory analyses; specialized data analysis systems; general instrumentation development and application. Mailing Add: Instr Prod Div Photo Prod Dept E I du Pont de Nemours & Co Wilmington DE 19898

TRISCHKA, JOHN WILSON, b Bisbee, Ariz, Dec 30, 16; m 46; c 2. PHYSICS. Educ: Univ Ariz, BS, 37; Cornell Univ, PhD(physics), 43. Prof Exp: Test engr, Gen Elec Co, 37-38; asst physics, Cornell Univ, 39-42, instr, 42-45; res physicist, Los Alamos Sci Lab, Univ Calif, 45; assoc physics, Columbia Univ, 46-48; from asst prof to assoc prof, 48-56, PROF PHYSICS, SYRACUSE UNIV, 56- Mem: AAAS; Am Phys Soc; Am Asn Physics Teachers; Am Meteorol Soc. Res: Radio frequency spectroscopy of molecules; molecular beams; surface physics; atmospheric physics. Mailing Add: Dept of Physics Syracuse Univ Syracuse NY 13210

TRISCHLER, FLOYD D, b Pittsburgh, Pa, Aug 31, 29; m 51; c 6. ORGANIC CHEMISTRY, POLYMER CHEMISTRY. Educ: Univ Pittsburgh, BS, 51. Res chemist, Pa Indust Chem Corp, 53-55, asst lab mgr, 55-56; develop & tech chemist, Tar Prod Div, Koppers Co, 56-58, asst group leader, Res Dept, 58-63; sr res chemist, Narmco Res & Develop, Whittaker Corp, Calif, 63-65, prog mgr, 65-69; mgr mkt & admin asst to pres, Mat Systs Corp, Calif, 69-72; exec vpres, Taylor Bldg Corp, Ind, 72-73; PRES GUADALUPE BUILDERS, IND, 73- Mem: Am Chem Soc. Res: Polymer applications and research; pulp and paper; organic coatings; fluorine compounds; elastomers; adhesives; high performance polymers. Mailing Add: RR 1 Box 225L Mooresville IN 46158

TRISKA, FRANK JOHN, b Pittsburgh, Pa, Nov 17, 43. AQUATIC ECOLOGY, MICROBIAL ECOLOGY. Educ: Univ Pittsburgh, PhD(biol), 70. Prof Exp: RES ASSOC STREAM ECOL, ORE STATE UNIV, 73- Mem: Int Soc Limnol; Am Soc Limnol & Oceanog; Am Ecol Soc. Res: Decomposition processes; animal-microbial interactions in flowing water systems; general stream ecology. Mailing Add: Dept Fisheries & Wildlife Ore State Univ Corvallis OR 97330

TRISLER, JOHN CHARLES, b Eva, La, Dec 24, 33; m 53; c 2. ORGANIC CHEMISTRY. Educ: La Polytech Inst, BS, 56; Tex Tech Univ, PhD(org chem), 59. Prof Exp: From asst prof to assoc prof, 59-66, PROF CHEM, LA TECH UNIV, 66- Mem: Am Chem Soc. Res: Organic reaction mechanisms. Mailing Add: Dept of Chem La Tech Univ Ruston LA 71270

TRISTAN, THEODORE A, b Mexico City, Mex, Oct 5, 24; US citizen; m 47; c 4. RADIOLOGY. Educ: Univ Nebr, BS, 47, MD & MSc, 50; Univ Pa, MS, 58; Am Bd Radiol, dipl, 57. Prof Exp: Intern, Hosp, Univ Pa, 50-51; instr radiol, Med Ctr, Univ Rochester, 56-59, sr instr, 59; from asst prof to assoc prof, Univ Pa, 59-65, lectr, 65-70; CLIN PROF RADIOL & ANAT, HERSHEY MED CTR, PA STATE UNIV, 70- Concurrent Pos: Fel radiol, Univ Pa Hosp, 53-56; Am Cancer Soc fel, 53-54; NIH grants, Univ Rochester, 58-59 & Univ Pa, 59-63; Nat Res Coun Picker Found grant, Univ Pa, 58-65; mem & task group chmn, Int Comn Radiation Units & Measurements, 64-67; assoc prof radiol, Hahnemann Med Col, 70-73. Mem: Radiol Soc NAm; Am Roentgen Ray Soc; AMA; fel Am Col Radiol. Res: Image intensification, quality and information development, storage and retrieval; analysis of the function of motion in cinefluorography with regard to its value in development of diagnostic criteria in clinical radiology. Mailing Add: Dept of Radiol Polyclin Hosp Harrisburg PA 17105

TRITES, RONALD WILMOT, b Moncton, NB, July 17, 29; m 56; c 5. PHYSICAL OCEANOGRAPHY. Educ: Univ NB, BSc, 50; Univ BC, MA, 52, PhD(physics), 55. Prof Exp: Phys oceanogr, Fisheries Res Bd Can, 50-55 & 56-66 & Defence Res Bd Can, 55-56; head appl oceanog, Bedford Inst Oceanog, 66-70; adv Atlantic, Fisheries Res Bd Can, 70-71; head coastal oceanog div, Bedford Inst Oceanog, 71-75, SR OCEANOGR, MARINE ECOL LAB, BEDFORD INST OCEANOG, 75- Concurrent Pos: Hon lectr, Dalhousie Univ, 60- Mem: Can Asn Physicists. Res: Coastal and estuarine circulation, mixing and dispersion processes; role of physical processes in production of fish stocks. Mailing Add: Marine Ecol Lab Bedford Inst of Oceanog PO Box 1006 Dartmouth NS Can

TRITSCH, GEORGE LEOPOLD, b Vienna, Austria, Apr 8, 29; nat US; m 51; c 3. BIOCHEMISTRY. Educ: NY Univ, BS, 48; Univ Md, MS, 51; Purdue Univ, PhD(biochem), 54. Prof Exp: Res assoc, Med Col, Cornell Univ, 54-56, res assoc, Rockefeller Inst, 56-59; ASSOC CANCER RES SCIENTIST, ROSWELL PARK MEM INST, 59-; ASSOC PROF BIOCHEM, STATE UNIV NY BUFFALO, 68- Concurrent Pos: Asst res prof, State Univ NY Buffalo, 63-68; res prof, Niagara Univ, 71- Mem: Am Asn Cancer Res; Am Inst Chemists; Am Soc Biol Chemists; Am Inst Nutrit; Am Chem Soc. Res: Mechanisms of control of growth of mammalian cells in axenic culture; cancer chemotherapy; synthesis and evaluation of analogs of naturally occurring compounds in chemotherapy; endocrinology; enzyme structure and kinetics. Mailing Add: Roswell Park Mem Inst 666 Elm St Buffalo NY 14203

TRITSCHLER, LOUIS GEORGE, b St Louis, Mo, Jan 24, 27; m 47; c 2. VETERINARY MEDICINE, SURGERY. Educ: Univ Mo, BSAgr, 49, DVM, 60, MS, 62. Prof Exp: From instr to asst prof, 60-72, ASSOC PROF VET MED & SURG, UNIV MO-COLUMBIA, 72- Mem: Am Vet Med Asn; Am Asn Equine Practrs; Am Asn Vet Clinicians. Res: Use of estrone in the treatment of anestrus in cattle; evaluation of bulls for breeding soundness; wound treatment and fracture repair in equine. Mailing Add: Vet Hosp & Clin Univ of Mo-Columbia Columbia MO 65201

TRITZ, GERALD JOSEPH, b Sioux City, Iowa, Apr 12, 37; m 66; c 1. MICROBIOLOGY, GENETICS Educ: Utah State Univ, BS, 62; Colo State Univ, MS, 65; Univ Tex Med Sch Houston, PhD(biomed res), 70. Prof Exp: Res microbiologist, USPHS, 65-67; NIH fel microbiol, Univ Tex MD Anderson Hosp & Tumor Inst, Houston, 70; ASST PROF MICROBIOL, UNIV GA, 70- Concurrent Pos: NSF grant, 71- Mem: Am Soc Microbiol. Res: Pyridine nucleotide metabolism and its control. Mailing Add: Dept of Microbiol Univ of Ga Athens GA 30602

TRIVEDI, KISHOR SHRIDHARBHAI, b Bhavnagar, India, Aug 20, 46; m 73; c 1. COMPUTER SCIENCES. Educ: Indian Inst Technol, Bombay, BTech, 68; Univ Ill, Urbana-Champaign, MS, 72, PhD(comput sci), 74. Prof Exp: Assoc customer engr, Int Bus Mach World Trade Corp, Bombay, 68-70; res asst comput sci, Univ Ill, Urbana-Champaign, 70-74, res assoc, 74-75; ASST PROF COMPUT SCI, DUKE UNIV, 75- Mem: Am Asn Comput Mach. Res: Algorithms for construction and performance evaluation of computer operating systems; different methods of computer system organization and techniques to exploit the systems; new organizations of arithmetic units. Mailing Add: Dept of Comput Sci Duke Univ Durham NC 27706

TRIVELPIECE, ALVIN WILLIAM, b Stockton, Calif, Mar 15, 31; m 53; c 3. PLASMA PHYSICS. Educ: Calif State Polytech Col, BS, 53; Calif Inst Technol, MS, 55, PhD(elec eng), 58. Prof Exp: From asst prof to assoc prof elec eng, Univ Calif, Berkeley, 59-66; PROF PHYSICS, UNIV MD, COLLEGE PARK, 66- Concurrent Pos: Fulbright scholar, Delft Technol Univ, 58-59; consult to govt & indust, 61-; Guggenheim fel, 67-68; asst dir res, Div Controlled Thermonuclear Res, AEC, 73-75. Mem: Fel AAAS; Am Nuclear Soc; Sigma Xi; Am Phys Soc; Inst Elec & Electronics Engrs. Res: Plasma physics and controlled thermonuclear fusion research; particle accelerators; microwave devices; electromagnetic waves. Mailing Add: Dept of Physics & Astron Univ of Md College Park MD 20742

TRIVETT, TERRENCE LYNN, b Madison, Tenn, Oct 3, 40; m 65. BACTERIOLOGY. Educ: Southern Missionary Col, BS, 64; Univ Ore, PhD(microbiol), 69. Prof Exp: Asst prof, 69-74, ASSOC PROF BIOL, PAC UNION COL, 74- Mem: Am Soc Microbiol. Res: Carbohydrate metabolism of Listeria monocytogenes. Mailing Add: Dept of Biol Pac Union Col Angwin CA 94508

TRIVETTE, CHESTER DRAPER, JR, organic chemistry, rubber chemistry, see 12th edition

TRIVICH, DAN, b Jenkins, Ky, May 20, 16; m 43; c 3. PHYSICAL CHEMISTRY. Educ: Ohio State Univ, BA, 38, PhD(chem), 42. Prof Exp: Asst, Ohio State Univ, 38-42; res chemist, United Chromium, Inc, 42-48; from asst prof to assoc prof, 48-57, PROF CHEM, WAYNE STATE UNIV, 57- Concurrent Pos: Vis prof, Karlsruhe Tech Univ, 58-59; vis prof & Fulbright-Hays advan res scholar, Lab Physics, Ecole Normale Superieure, Paris, 65-66; vis prof & Fulbright lectr, Lab Physics Solids, Univ Paris, 72-73. Mem: Am Chem Soc; Electrochem Soc. Res: Chemistry of the solid state; chemical aspects of semiconductors; electrochemistry. Mailing Add: Dept of Chem Wayne State Univ Detroit MI 48202

TRIVISONNO, JOSEPH, JR, b Cleveland, Ohio, Feb 28, 33; m 57; c 4. SOLID STATE PHYSICS. Educ: John Carroll Univ, BS, 55, MS, 56; Case Western Reserve Univ, PhD(physics), 61. Prof Exp: Instr physics & math, John Carroll Univ, 55-58 & physics, Case Western Reserve Univ, 58-61; asst prof, 61-62, from asst prof to assoc prof, 63-69, PROF PHYSICS, JOHN CARROLL UNIV, 69- Mem: Am Asn Physics Teachers; Am Phys Soc. Res: Elastic constants of metals; magnetoacoustic studies; ultrasonics; low temperature physics; superconductivity. Mailing Add: Dept of Physics John Carroll Univ Cleveland OH 44118

TRIX, PHELPS, b Detroit, Mich, Apr 29, 21; m 46; c 3. ORGANIC CHEMISTRY. Educ: Olivet Col, BS, 43; Ind Univ, MA, 44; Pa State Col, PhD(chem), 49. Prof Exp: Jr res chemist, Parke, Davis & Co, 44-46; sect head org lab, Wyandotte Chem Corp, 49-53, mgr mkt develop, 53-59, dir prod develop, 59-69; assoc prof chem, 69-74, vpres acad affairs, 71-74, PROF CHEM, DETROIT INST TECHNOL, 74- Mem: Am Chem Soc. Res: Polyurethanes; nutritional biochemistry. Mailing Add: Dept of Chem Detroit Inst of Technol Detroit MI 48201

TRKULA, DAVID, b Monroeville, Pa, Aug 19, 27; m 54; c 3. BIOPHYSICS, VIROLOGY. Educ: Univ Pittsburgh, BS, 49, MS, 55, PhD(biophys), 59. Prof Exp: Instr physics, Johnstown Col, Univ Pittsburgh, 57-59; asst biophysicist, M D Anderson Hosp & Tumor Inst, 59-61; physicist, US Army Biol Labs, Ft Detrick, 61-68; ASST PROF BIOPHYS, BAYLOR COL MED, 68- Mem: AAAS; Biophys Soc. Res: Biophysics of viruses. Mailing Add: Div of Biochem Virol Baylor Col of Med Tex Med Ctr Houston TX 77025

TROBAUGH, FRANK EDWIN, JR, b West Frankfort, Ill, Oct 20, 20; m 51; c 4. MEDICINE. Educ: Univ Ill, AB, 40; Harvard Med Sch, MD, 43; Am Bd Internal Med, dipl, 55. Prof Exp: Asst physician, Barnes Hosp, St Louis, Mo, 52-54; DIR, BLOOD BANK & SECT HEMAT, DEPT MED, RUSH-PRESBY-ST LUKE'S MED CTR, 54-, ATTEND PHYSICIAN, 60-; PROF MED, RUSH COL MED, 72- Concurrent Pos: Assoc attend physician med, Presby-St Luke's Hosp, 59-60; assoc prof, Univ Ill, 59-65, prof, 65-72. Mem: Am Soc Hemat; fel Am Col Physicians; Am Fedn Clin Res; Int Soc Hemat; AMA. Res: Hematology. Mailing Add: Rush-Presby-St Luke's Med Ctr 1753 W Congress Pkwy Chicago IL 60612

TROCHU, LOUIS, b Montreal, Que, Feb 12, 21; m 50; c 2. BIOCHEMISTRY. Educ: Univ Montreal, MSc, 46, DSc(biochem), 53. Prof Exp: Dir biochem lab, Verdun Gen Hosp, Montreal, 53-63; consult, Orgn Coop Develop Econ, Paris, 63-66; dir biochem lab, Hosp de la Misericorde, 66-70; DIR BIOCHEM SERV, HOSP DU SACRE-COEUR, MONTREAL, 70- Mem: Chem Soc France. Res: Clinical and analytical chemistry. Mailing Add: Serv de Biochimie Hosp du Sacre-Coeur Montreal PQ Can

TROEGER, GARY LESLIE, b Santa Monica, Calif, July 12, 46; m 70. SOLID STATE PHYSICS. Educ: Univ Calif, Los Angeles, BS, 68; Univ Colo, MS, 70, PhD(physics), 75. Prof Exp: RES ASST PHYSICS, UNIV COLO, 70- Res: Investigation of magnetic properties of impurities in ternary semiconductors. Mailing Add: Dept of Physics Univ Colo Boulder CO 80302

TROEH, FREDERICK ROY, b Grangeville, Idaho, Jan 23, 30; m 51; c 3. SOIL SCIENCE. Educ: Univ Idaho, BSAgr, 51, MSAgr, 52; Cornell Univ, PhD(soil sci), 63. Prof Exp: Soil scientist, Soil Conserv Serv, 52-59; asst prof, 63-70, ASSOC PROF AGRON, IOWA STATE UNIV, 70- Mem: Am Soc Agron; Soil Sci Soc Am; Soil Conserv Soc Am. Res: Soil formation and classification; measuring the rate of soil creep; soil permeability relationships with microbial activity. Mailing Add: Dept of Agron Iowa State Univ Ames IA 50010

TROEN, PHILIP, b Portland, Maine, Nov 24, 25; m 53; c 3. MEDICINE. Educ: Harvard Univ, AB, 44, MD, 48; Am Bd Internal Med, dipl. Prof Exp: Intern, Boston City Hosp, 48-49, asst resident med, 49-50; chief med serv, US Army Hosp, Kobe, Japan, 50-52; asst, Harvard Univ, 53-54, instr, 56-59, assoc, 59-60, asst prof, 60-64; PROF MED, SCH MED, UNIV PITTSBURGH, 64-, ASSOC CHMN DEPT, 69-; PHYSICIAN-IN-CHIEF, MONTEFIORE HOSP, 64- Concurrent Pos: Teaching fel, Harvard Univ, 52-53, res fel, 55-56; Ziskind teaching fel, Beth Israel Hosp, 55-60; fel endocrinol & metab, Mayo Clin, 54-55, Kendall-Hench res fel, 55; teaching fel med, Tufts Univ, 53-54; Guggenheim fel, Stockholm, Sweden, 60-61; asst resident, Beth Israel Hosp, 50 & 52-53, resident, 53-54, asst, 55-56, assoc, 56-64, asst vis physician, 59-64; sr instr, Tufts Univ, 57-60; mem, Contract Rev Comt, Nat Inst Child Health & Human Develop. Mem: Am Soc Clin Invest; Am Fedn Clin Res; Endocrine Soc; Am Soc Biol Chemists; Soc Study Reproduction. Res: Endocrinology; internal medicine. Mailing Add: Montefiore Hosp 3459 Fifth Ave Pittsburgh PA 15213

TROESCH, BEAT ANDREAS, b Bern, Switz, Mar 2, 20; nat US; m 48; c 4. APPLIED MATHEMATICS. Educ: Swiss Fed Inst Technol, Zurich, dipl, 47, PhD(math), 52. Prof Exp: Asst mech & physics, Swiss Fed Inst Technol, Zurich, 47-52; res assoc appl math, Inst Math, NY Univ, 52-56; head appl math sect, Comput Ctr, Ramo-Wooldridge Corp, 56-58 & Space Tech Labs, 58-61; mgr comput sci dept, Aerospace Corp, 61-66; PROF AEROSPACE ENG & MATH, UNIV SOUTHERN CALIF, 66- Concurrent Pos: Consult, Aerospace Corp, 66- Mem: Math Asn Am; Soc Indust & Appl Math; Am Math Soc; Sigma Xi. Res: Applied mathematics and numerical analysis in hydrodynamics and gas dynamics; elliptic and hyperbolic partial differential equations. Mailing Add: 523 N Elm Dr Beverly Hills CA 90210

TROFIMENKO, SWIATOSLAW, b Lviv, Ukraine, Dec 15, 31; nat US; m 62; c 1. ORGANIC CHEMISTRY. Educ: Wesleyan Univ, BA, 55; Northwestern Univ, PhD(org chem), 58. Prof Exp: Res fel, Columbia Univ, 58-59; sr res chemist, 59-73, RES ASSOC, PLASTICS DEPT, E I DU PONT DE NEMOURS & CO, 73- Mem: Am Chem Soc; NY Acad Sci. Res: Heterocyclic chemistry; cyanocarbons; reaction mechanisms; polyhedral boranes; boron-pyrazole and organometallic chemistry; coordination chemistry and catalysis; polymer chemistry. Mailing Add: 101 Cheltenham Rd Oaklands Newark DE 19711

TROGDON, WILLIAM OREN, b Anadarko, Okla, Nov 1, 20; m 42; c 2 5SOIL5S Educ: Okla State Univ, BS, 42; Ohio State Univ, PhD(soil fertility), 49. Prof Exp: Asst agronomist, Agr Exp Sta, Univ Tex, 48; soil scientist, Res Div, Soil Conserv Serv, USDA, 49; chmn dept agr & dir soils lab, Midwestern Univ, 49-53; agronomist, Olin Mathieson Chem Corp, 53-58; prof agron & head dept, Tex A&M Univ, 58-63; exec vpres, Best Fertilizers Co, Tex, 63-65; dir agron & mkt develop, Occidental Agr Chem Corp, 65-66; PRES, TARLETON STATE UNIV, 66- Mem: Am Soc Agron; Crop Sci Soc Am; Soil Sci Soc Am; Am Chem Soc; Soil Conserv Soc Am. Res: Soil fertility and management, especially fertilizer usage; fertilizer technology; salinity control and water quality; polyphosphate fertilizers; academic administration. Mailing Add: Tarleton State Univ Stephenville TX 76402

TROIANO, MARLIN FRANK, b Monessen, Pa, Jan 29, 38; c 3. ORAL SURGERY. Educ: Kent State Univ, BA, 60; Med Col Va, DDS, 64; Ohio State Univ, MSc, 67, PhD(anat), 69; Am Bd Oral Surg, dipl. Prof Exp: Instr oral surg, Ohio State Univ, 66-67, from asst prof to assoc prof, 69-71, dir respiratory anat & oral surg, 69-71; PROF ORAL SURG, CHMN DEPT, DIR HOSP DENT SERV & ASSOC DEAN HOSP AFFAIRS, COL MED & DENT, NJ, 72- Honors & Awards: Outstanding Res Award, Am Soc Oral Surg, 66; Recognition Serv Award, Ohio State Univ, 71. Mem: Am Soc Oral Surg; fel Int Asn Oral Surgeons; Int Asn Dent Res; Am Acad Oral Path; Am Asn Anatomists. Mailing Add: Dept of Oral Surg Col of Med & Dent of NJ Jersey City NJ 07304

TROIANO, PAUL FRANCIS, b Boston, Mass, July 31, 37; m 60; c 4. PHYSICAL CHEMISTRY. Educ: Northeastern Univ, BS, 60; Mass Inst Technol, PhD(phys chem), 64. Prof Exp: Sr res chemist, 64-67, head anal res & serv sect, 67-68, tech mgr mat res, 68-70, mgr fine particle res sect, 70-72, DIR CAB-O-SIL RES & DEVELOP, CABOT CORP, BILLERICA, MASS, 72- Mem: Am Chem Soc; AAAS. Res: Infrared studies of absorbed molecules; surface chemistry of amorphous carbons; chemical and physical properties of fumed silica. Mailing Add: 19 Bonney Lane Norwood MA 02062

TROISI, RAPHAEL ANGELO, b Avellino, Italy, Feb 14, 04; nat US; m 31; c 3. PHYSICAL CHEMISTRY. Educ: Temple Univ, BA, 27; Univ Pa, AM, 30, MSc, 31, PhD(biochem), 33; Mass Inst Technol, MA, 37. Prof Exp: From instr biol to assoc prof physiol & histol, Temple Univ, 27-42; res & develop chemist, Rohm & Haas, Inc, 44-46; chemist, Coopers Creek Chem Corp, Pa, 46-54, vpres chg res & develop, 54-68; mgr, Laycee Consults, Pa, 68-72; RETIRED. Concurrent Pos: Consult & lectr, 72- Mem: Am Chem Soc; Am Soc Biol Chemists; Fedn Socs Paint Technol. Res: Biophysics; biochemistry; bituminology; physiology; cytology; histology. Mailing Add: Allison Apts 233 N Maple Ave Marlton NJ 08053

TROLAN, J KENNETH, b Madras, Ore, Jan 17, 17; m 41, 66; c 5. SURFACE PHYSICS. Educ: Linfield Col, BA, 39; Ore State Univ, MA, 48, PhD(physics), 50. Prof Exp: Instr physics, Univ Alaska, 41-42; mem staff, Radiation Lab, Mass Inst Technol, 42-45; instr & res asst, Ore State Univ, 45-48; from asst to assoc prof, Linfield Col, 48-57, asst dir res, Res Inst, 51-62; asst dir res & develop physics & eng, Field Emission Corp, 62-64; PROF PHYSICS, UNIV REDLANDS, 64- Concurrent Pos: Consult, NSF-AID Sci Educ Improv Prog, India, 66. Mem: Am Phys Soc; Am Asn Physics Teachers. Res: Basic field emission research and use of the pulsed field emission microscope to observe surface migration and dislocation in metals. Mailing Add: Dept of Physics Univ of Redlands Redlands CA 92373

TROLINGER, JAMES DAVIS, physics, optics, see 12th edition

TROLL, JOSEPH, b Paterson, NJ, May 5, 20; m 43; c 2. SOIL SCIENCE, AGROSTOLOGY. Educ: Univ RI, BS, 54, MS, 57; Univ Mass, PhD(nematol), 65. Prof Exp: Asst, Univ RI, 54-57; asst prof agron, 57-65, assoc prof plant & soil sci, 65-71, PROF PLANT & SOIL SCI, UNIV MASS, AMHERST, 71- Res: Turf management; plant pathology and nematology. Mailing Add: Dept of Plant & Soil Sci Univ of Mass Amherst MA 01002

TROLL, RALPH, b Reinheim, Ger, Oct 8, 32; US citizen; m 58; c 3. ZOOLOGY, BOTANY. Educ: Univ Ill, BS, 57, MS, 58; Univ Minn, PhD(parasitol), 65. Prof Exp: From instr to assoc prof, 59-72, PROF BIOL, AUGUSTANA COL, ILL, 72-, CHMN DEPT, 68- Mem: Am Inst Biol Sci; Hist Sci Soc. Res: Ecology of myxomycetes; abnormal development in vertebrates; the development of German naturphilosophie and its role in nineteenth century biology. Mailing Add: Dept of Biol Augustana Col Rock Island IL 61201

TROLL, WALTER, b Vienna, Austria, Oct 25, 22; nat US; m 44; c 2. BIOCHEMISTRY, ORGANIC CHEMISTRY. Educ: Univ Ill, BS, 44; Pa State Univ, MS, 46; NY Univ, PhD(biochem), 51. Prof Exp: Instr biochem, Univ Cincinnati, 51-52, asst prof, 52-54; assoc, Cancer Res Inst, New Eng Deaconess Hosp, 54-56; from asst prof to assoc prof, 56-70, PROF INDUST MED, SCH MED, NY UNIV, 70- Concurrent Pos: Asst dir, May Inst Med Res, Cincinnati, Ohio, 51-54. Mem: Am Soc Biol Chem; Am Chem Soc; NY Acad Sci. Res: Assay of amino acids; synthetic substrates for enzymes involved in blood clotting; metabolism of aromatic amines and its relation to carcinogenesis. Mailing Add: Inst of Environ Med NY Univ Col of Med 550 First Ave New York NY 10016

TROLLER, JOHN ARTHUR, b Hartford, Wis, Apr 17, 33; m 56; c 4. MICROBIOLOGY. Educ: Univ Wis, BS, 55, MS, 56, PhD(bact), 62. Prof Exp: GROUP LEADER MICROBIOL, WINTON HILL TECH CTR, PROCTER & GAMBLE CO, 62- Mem: AAAS; Am Soc Microbiol; Inst Food Technologists; Soc Indust Microbiol; Brit Soc Appl Bact. Res: Food technology and microbiology; water relations of microorganisms; mechanism of action of food preservatives; staphylococcal food poisoning and other food-borne diseases. Mailing Add: 314 Ritchie Ave Cincinnati OH 45215

TROMBA, ANTHONY JOSEPH, b Brooklyn, NY, Aug 10, 43. PURE MATHEMATICS. Educ: Cornell Univ, BA, 65; Princeton Univ, MA, 67, PhD(math), 68. Prof Exp: Asst prof math, Stanford Univ, 68-69; vis prof, Univ Pisa, Italy, 70; ASSOC PROF MATH, UNIV CALIF, SANTA CRUZ, 70- Concurrent Pos: Woodrow Wilson & NSF fels; vis prof, Univ Calif, Stony Brook, 74; mem, Inst Advan Study, 75. Res: Topological methods in non-linear analysis. Mailing Add: Dept of Math Univ of Calif Santa Cruz CA 95064

TROMBA, FRANCIS GABRIEL, b New York, NY, Oct 19, 20; m 43; c 4. PARASITOLOGY. Educ: Univ Md, PhD(zool), 53. Prof Exp: Parasitologist, USDA, Md, 53-55, parasitologist in-chg field sta, Ga, 55-56, actg proj leader helminths & dis swine, 56-58, proj leader, 58-61, prin res parasitologist & proj leader immunol res, Animal Dis & Parasite Res Div, 61-71, leader anti-parasitic invests, Nat Animal Parasite Lab, 71-72, CHIEF, NONRUMINANT HELMINTHIC DIS LAB, ANIMAL PARASITOL INST, 72- Concurrent Pos: Ed, Proceedings, Helminth Soc Wash, 66-71. Mem: Am Soc Parasitol. Res: Life histories, morphology, and host-parasite relationships of parasitic helminths; diagnosis, pathology, immunology, and biological control of helminthic diseases. Mailing Add: Animal Parasitol Inst Agr Res Ctr Beltsville MD 20705

TROMBETTA, LOUIS DAVID, b New York, NY, Sept 8, 46; m 73. HISTOLOGY. Educ: Fordham Univ, BS, 68, MS, 69, PhD(biol), 74. Prof Exp: RES ASSOC PATH, ISSAC ALBERT RES INST, 73- Mem: NY Acad Sci; Electron Micros Soc Am; Sigma Xi. Res: The study of insect development and endocrinology by electron microscopy and histochemistry. Mailing Add: 469 Wolf's Lane Pelham Manor NY 10803

TROMBKA, JACOB ISRAEL, b Detroit, Mich, Jan 7, 30; m 52; c 3. RADIATION PHYSICS, SPACE PHYSICS. Educ: Wayne State Univ, BS, 52, MS, 54; Univ Mich, PhD(nuclear sci), 62. Prof Exp: Res physicist, Oak Ridge Inst Nuclear Studies, 54-56; res assoc nuclear eng & fel gamma ray spectros, Univ Mich, 56-62; sr scientist, Jet Propulsion Lab, Calif Inst Technol, 62-64; prog scientist, Hq, NASA, 64-65, STAFF SCIENTIST, GODDARD SPACE FLIGHT CTR, NASA, 65- Concurrent Pos: Mem panel in-flight exp, NASA, 63-66, mem working group, Manned Space Flight Exp Bd, 64-66, secy & mem, Geochem Working Group Planetology Subcomt, 65, mem, Terrestrial Bodies Sci Working Group, 76-; adj prof, Law Sch, Georgetown Univ, 67-; co-investr, Apollo 15 & 16 x-ray, gamma ray & alpha particle spectrometer exp, 68-; mem Apollo sci working panel, 71-; prin investr, Apollo-Soyuz Crystal Activation Exp, 75-; vis prof, Dept Chem, Univ Md, 76- Honors & Awards: John Lindsay Mem Award, Goddard Space Flight Ctr, 72. Mem: Am Phys Soc; Am Nuclear Soc. Res: Gamma ray spectroscopy; techniques in activation analysis, dosimetry and tracer techniques; planetary physics; gamma ray astrophysics; gamma-ray, x-ray and neutron-gamma ray in situ and remote sensing methods. Mailing Add: Goddard Space Flight Ctr NASA Greenbelt MD 20770

TROMBLE, JOHN M, b Lincoln, Kans, Jan 26, 32; m 52; c 2. HYDROLOGY, SOIL SCIENCE. Educ: Utah State Univ, BS, 61; Univ Ariz, MS, 64, PhD(watershed hydrol), 73. Prof Exp: Res assoc watershed hydrol, Univ Ariz, 64-67; asst dir, Southwest Watershed Res Ctr, 67-74, RES HYDROLOGIST, JORNADA EXP RANGE, AGR RES SERV, USDA, 74- Mem: Crop Sci Soc Am; Soc Range Mgt; Am Soc Agron; Soil Sci Soc Am; Int Soil Sci Soc. Res: Watershed hydrological studies; runoff, erosion, infiltration, simultaneous transfer of heat and water in soils; evaluation of consumptive use of water by native vegetation. Mailing Add: Jornada Exp Range PO Box 698 Las Cruces NM 88001

TROMBLY, THELMA WOODHOUSE, b Marysville, Wash, Oct 23, 09; m 38; c 2. SPEECH PATHOLOGY, AUDIOLOGY. Educ: Kans Wesleyan Univ, AB, 32; Univ Mo, MA, 33, PhD, 58. Prof Exp: Instr French, Stephens Col, 34-37; instr, 37-38, asst, 51-52, instr speech, 52-58, asst prof speech path & speech pathologist, Med Ctr, 58-62, assoc prof speech, 62-69, PROF SPEECH, MED CTR, UNIV MO-COLUMBIA, 69-, ASSOC DIR SPEECH & HEARING CLIN, 58-, CO-DIR AREA SPEECH PATH-AUDIOL, 71- Concurrent Pos: Vis assoc prof, Univ Hawaii, 67. Mem: AAAS; Am Speech & Hearing Asn; Am Cleft Palate Asn. Res: Etiological factors in delayed speech development. Mailing Add: 107 Parker Hall Univ of Mo Med Ctr Columbia MO 65201

TROMMEL, JAN, b Rotterdam, Netherlands, July 16, 26; m 54; c 2. PHYSICAL CHEMISTRY. Educ: Univ Utrecht, BSc, 47, MA, 51, PhD(phys chem), 54. Prof Exp: Mgr res, Royal Dutch Explosive Co, Amsterdam, 54-64, vpres res, 64-67; MGR MAT CHARACTERIZATION, XEROX CORP, 67- Concurrent Pos: Mem, Adv Group Rockets, Dutch Govt, 57-67; Netherlands govt rep, AC 60 Group Experts Explosives & Propellants, NATO, 63-67. Mem: Am Chem Soc; NY Acad Sci; Royal Netherlands Chem Soc. Res: X-ray diffraction; scanning and transmission electron microscopy; electron microprobe; propellants; explosives; rocket propellants. Mailing Add: 515 Bay Rd Webster NY 14580

TROMMERSHAUSEN-SMITH, ANN L, b Portland, Ore, June 1, 43. GENETICS. Educ: Carleton Col, AB, 65; Univ Calif, Davis, PhD(genetics), 69. Prof Exp: Asst prof biol, Occidental Col, 69-73; ASST RES GENETICIST, UNIV CALIF, DAVIS, 73- Mem: AAAS; Sigma Xi; Am Genetics Soc; Int Soc Animal Blood Group Res. Res: Immunogenetics, cytogenetics and genetics of problems of genetic etiology encountered in clinical veterinary medicine. Mailing Add: Serol Lab Univ of Calif Davis CA 95616

TROPF, CHERYL GRIFFITHS, b Newark, NJ, Oct 15, 46; m 68. APPLIED MATHEMATICS. Educ: Col William & Mary, BS, 68; Univ Va, MAM, 72, PhD(appl math), 73. Prof Exp: SR MATHEMATICIAN, APPL PHYSICS LAB, JOHNS HOPKINS UNIV, 73- Concurrent Pos: Instr, Johns Hopkins Evening Col, 74- Mem: Soc Indust & Appl Math. Res: Application of analytical mathematical techniques to the modelling of physical systems. Mailing Add: 18213 Queen Elizabeth Dr Olney MD 20832

TROPF, WILLIAM JACOB, b Chicago, Ill, Jan 14, 47; m 68. OPTICS, MILITARY SYSTEMS. Educ: Col William & Mary, BS, 68; Univ Va, PhD(physics), 73. Prof Exp: PROJ SCIENTIST, B-K DYNAMICS, INC, ROCKVILLE, 73- Mem: Am Phys Soc; Optical Soc Am; Sigma Xi. Res: Analysis and modelling of new or improved technology as applied to military systems and determination of the impact of new developments on system capabilities. Mailing Add: 18213 Queen Elizabeth Dr Olney MD 20832

TROPP, BURTON E, b New York, NY, Aug 8, 40; m 65; c 3. BIOCHEMISTRY. Educ: Brooklyn Col, BS, 61; Harvard Univ, PhD(biochem), 66. Prof Exp: NIH fel bacteriol, Harvard Med Sch, 65-67; asst prof biochem, Richmond Col, NY, 67-70; asst prof, 70-72, ASSOC PROF BIOCHEM, QUEEN'S COL, NY, 72- Mem: AAAS; Am Chem Soc; NY Acad Sci; Am Soc Microbiol; Am Soc Biol Chemists. Res: Control and regulation of phospholipid biosynthesis. Mailing Add: Dept of Chem Queen's Col Flushing NY 11367

TROPP, HENRY S, b Chicago, Ill, July 15, 27; m 54; c 3. MATHEMATICS. Educ: Purdue Univ, BS, 49; Ind Univ, MS, 53. Prof Exp: Instr math, Mont Sch Mines, 55-57; from asst prof to assoc prof, Humboldt State Col, 57-72; prin investr, Comput Hist Proj, Smithsonian Inst, 71-74; PROF MATH, HUMBOLDT STATE UNIV, 74- Concurrent Pos: Vis lectr, Asn Comput Mach, 73-; mem, Prog Comt, Int Res Conf

Hist Comput. Mem: Math Asn Am; Hist Sci Soc; Can Soc Hist & Philos Math; Asn Comput Mach. Res: History of mathematics; history of computers. Mailing Add: Dept of Math Humboldt State Univ Arcata CA 95521

TROSETH, FRANK PATON, geophysics, see 12th edition

TROSKO, JAMES EDWARD, b Muskegon, Mich, Apr 2, 38; m 60; c 1. GENETICS, ONCOLOGY. Educ: Cent Mich Univ, BA, 60; Mich State Univ, MS, 62, PhD(radiation genetics), 63. Prof Exp: Fel, Oak Ridge Nat Lab, 63-64, Am Cancer Soc fel, 64-65; res scientist radiation biophys, Biol Div, Oak Ridge Nat Lab, 65-66; asst prof sci & philos, Dept Natural Sci, 66-70, ASSOC PROF CARCINOGENESIS & MED ETHICS, DEPT HUMAN DEVELOP, MICH STATE UNIV, 70- Concurrent Pos: Nat Cancer Inst career develop award, 72; consult, Oak Ridge Nat Lab, 70-72; vis prof oncol, McArdle Lab Cancer Res, Univ Wis-Madison, 72-73; consult, Wis Res & Develop Ctr Cognitive Learning, 73-74. Mem: AAAS; Genetics Soc Am; Tissue Cult Asn. Res: Molecular basis for genetic and environmental influences on carcinogenesis and aging; integration of science and human values. Mailing Add: Dept of Human Develop Col Human Med Mich State Univ East Lansing MI 48824

TROSMAN, HARRY, b Toronto, Ont, Dec 9, 24; nat US; m 52; c 3. PSYCHIATRY, PSYCHOANALYSIS. Educ: Univ Toronto, MD, 48. Prof Exp: Intern, Grace Hosp, Detroit, Mich, 48-49; resident, Psychopath Hosp, Iowa City, Iowa, 49-51; resident, Cincinnati Gen Hosp, Ohio, 51-52; from asst prof to assoc prof, 54-74, PROF PSYCHIAT, PRITZKER SCH MED, UNIV CHICAGO, 75- Concurrent Pos: Psychoanal training, Chicago Inst Psychoanal, 54-62, fac mem, 65-; consult, Chicago Police Dept, 61-, Ill State Psychiat Inst Psychoanal, 65. Honors & Awards: Franz Alexander Prize, Chicago Inst Psychoanal, 65. Mem: Am Psychiat Asn; Am Psychoanal Asn. Res: Dreams; applied psychoanalysis; creativity. Mailing Add: Dept of Psychiat Univ of Chicago Sch of Med Chicago IL 60637

TROSPER, TERRY LOUISE, b San Francisco, Calif, Oct 5, 39; m 73. BIOPHYSICS. Educ: Univ Calif, Berkeley, BA, 61, PhD(biophysics), 66. Prof Exp: Fel colloid sci, Univ Cambridge, 67-69; fel photosynthesis, Charles F Kettering Res Lab, 69-70; res assoc chem, Pomona Col, 70-72; res assoc biochem, Univ Southern Calif Sch Med, 72-74; fel biol, 74-75, RES ASSOC BIOL, UNIV CALIF, LOS ANGELES, 75- Concurrent Pos: Fel, Am Asn Univ Women, 67-69 & Molecular Biol Inst, Univ Calif, Los Angeles, 74-75. Res: Molecular architecture of biological membranes, especially those containing apparatus for photosynthesis and other processes involving energy conversion; physical properties of model and biological membrane systems. Mailing Add: Dept of Biol Univ of Calif Los Angeles CA 90024

TROSS, CARL HENRY, b Bad Kreuznach, Ger, June 24, 25; nat US; m 56; c 3. PHYSICS, MATHEMATICS. Educ: Univ Ill, BS, 50, MS, 52. Prof Exp: Asst, Univ Ill, 50-52; tech dir anal group, Lockheed Aircraft Corp, 52-56; dir comput ctr, Am Bosch Arma Corp, 56-58; proj mgr & mem sr staff, Aeronutronic Div, Philco-Ford Corp, 58-61; eng mgr, Univac Div, Sperry Rand Corp, 61-64; prog mgr, IBM Corp, 65-69; pres, Coun Technol & Indust Develop, 64-73; head plans anal & design off, 69-73, PHYS SCI ADMINR & DEP PROG MGR, NAVAL AIR SYSTS, DEPT NAVY, 73- Concurrent Pos: Ed, For Tech Notes, J Astronaut Sci. Mem: Fel Am Astronaut Soc (vpres, 69-71, treas, 68-71); Am Inst Aeronaut & Astronaut; Am Mgt Asn. Res: Ecology, aerodynamics; mission analysis; program management; hydrological resources and management; high speed surface transportation; communications; command and control; astrodynamics; computer technology; space applications; data management; management science; corporate planning; operations research; shock wave propagation. Mailing Add: PO Box 5878 Bethesda MD 20014

TROSS, RALPH G, b Bad Kreuznach, Ger, Jan 17, 23; US citizen; m 47; c 2. MATHEMATICAL PHYSICS. Educ: Sophia Univ, Japan, BS, 52; Mo Sch Mines, BS, 59; Univ Mo-Rolla, MS, 66, PhD(physics), 68. Prof Exp: NASA fel physics, Univ Mo-Rolla, 64-67, instr math, 67-68, res assoc physics, 68; asst prof, 68-69, actg chmn dept math, 70-71, ASSOC PROF MATH, UNIV OTTAWA, 69-, CHMN COMPUT COMT, 71- Concurrent Pos: Nat Res Coun grant, 68-; mem, Adv Comt, Algonquin Col, Ottawa, 71-; educ develop grant, Comt Ont Univs, 74. Mem: Am Phys Soc; Can Asn Physicists; Am Math Soc; Math Asn Am; Can Math Cong. Res: Statistical mechanics; Ising model; cooperative phenomena. Mailing Add: Dept of Math Univ of Ottawa Ottawa ON Can

TROST, BARRY M, b Philadelphia, Pa, June 13, 41. ORGANIC CHEMISTRY. Educ: Univ Pa, BA, 62; Mass Inst Technol, PhD(org chem), 65. Prof Exp: From asst prof to assoc prof, 65-69, PROF CHEM, UNIV WIS-MADISON, 69- Concurrent Pos: Assoc ed, J Am Chem Soc & adv, NSF Chem Sect, 73-; Sloan fel; Dreyfuss Found teacher-scholar grant; consult, E I du Pont de Nemours & Co. Mem: Am Chem Soc. Res: Development of new synthetic methods; synthesis of natural products and theoretically important systems; investigations of model biogenetic systems. Mailing Add: Dept of Chem Univ of Wis Madison WI 53706

TROST, CHARLES HENRY, b Erie, Pa, Apr 4, 34; m 60; c 1. VERTEBRATE ZOOLOGY, PHYSIOLOGICAL ECOLOGY. Educ: Pa State Univ, BS, 60; Univ Fla, MS, 64; Univ Calif, Los Angeles, PhD(zool), 68. Prof Exp: Grad fac grant, 69-70, ASST PROF BIOL, IDAHO STATE UNIV, 68- Mem: Am Ornithologists Union; Cooper Ornith Soc; Wilson Ornith Soc; Am Soc Zoologists; Am Soc Mammalogists. Res: Water balance and energetics of birds and mammals; relation of behavior to the adaptations of animals to their environment. Mailing Add: Dept of Biol Idaho State Univ Pocatello ID 83201

TROST, HENRY BIGGS, b Lancaster, Pa, Aug 18, 20; m 43; c 3. ORGANIC CHEMISTRY. Educ: Franklin & Marshall Col, BS, 42. Prof Exp: Anal chemist, Org Anal Group, 42-44, shift supvr, Acid Lab, Badger Ord Works, 44-45, Org Anal Lab, 45-46, Size & Solvents Anal Lab, Naval Stores Res Div, 55-61, MEM STAFF MAT RES DIV, HERCULES INC, 61- Mem: Am Chem Soc; Sigma Xi. Res: Product application, development, formulation and sales service type work on water soluble polymers and surface active agents. Mailing Add: Res Ctr Hercules Inc Wilmington DE 19899

TROST, WALTER RAYMOND, b Castor, Alta, Mar 2, 19; m 47; c 3. PHYSICAL CHEMISTRY. Educ: Univ Alta, BSc, 44; McGill Univ, PhD(chem), 47. Prof Exp: Fel atomic reactions, Nat Res Coun Can, 47; Royal Soc Can fel infrared spectros, Oxford Univ, 48; from asst prof to prof chem, Dalhousie Univ, 48-66, dean fac grad studies, 61-66; acad vpres, Univ Calgary, 66-69, actg pres, 69-70; CHMN, ENVIRON CONSERV AUTHORITY, EDMONTON, 70- Concurrent Pos: Pres, Laborde, Simat & Trost, 69-70. Mem: Fel Chem Inst Can; NY Acad Sci; Can Asn Chem Engrs. Mailing Add: 2100 College Plaza 8215 112 St Edmonton AB Can

TROTT, CHARLES EUGENE, b Columbus, Ohio, Oct 21, 40; m 60; c 5. URBAN GEOGRAPHY. Educ: Ohio State Univ, BSEd, 63, MA, 64, PhD(geog), 69. Prof Exp: Economist regional anal, Fed Reserve Bank Cleveland, 67-69; regional economist migration, Bur Econ Anal, US Dept Com, 69-72; ASSOC PROF GEOG, NORTHERN ILL UNIV, 72- Mem: Regional Sci Asn; Asn Am Geogrs. Res: Determinants of mobility of the labor force, population redistribution and inter-urban relationships in growth and migration; migration analysis. Mailing Add: Dept of Geog Northern Ill Univ De Kalb IL 60115

TROTT, GENE F, b Louisville, Ky, May 27, 29; m 55; c 4. RUBBER CHEMISTRY. Educ: Univ Louisville, BA, 54, MS, 68, PhD, 71. Prof Exp: Chemist, Pillsbury Co, 54-56; res chemist & group leader, Am Synthetic Rubber Corp, 56-66; chemist, Gen Elec Co, Louisville, Ky, 66-73; VPRES RES & DEVELOP, BURTON RUBBER PROCESSING CO, 73- Concurrent Pos: Adj asst prof, Sch Med, Univ Louisville. Mem: Am Chem Soc. Res: Polymer chemistry; polymer characterization; biopolymeric interactions; bone structure; flame retardancy; metal to rubber adhesion; radiation curing. Mailing Add: 1523 Briarwood Dr Clarksville IN 47130

TROTT, JOHN RICHARD, b Kasauli, India, Sept 2, 25; nat US; m 53; c 2. DENTISTRY, PERIODONTOLOGY. Educ: Univ Adelaide, BDS, 50, MDS, 66; Australian Col Dent Surg, FACDS, 67. Prof Exp: Tutor dent surg, St Mark's Col, Univ Adelaide, 51; house surgeon, London Hosp Med Col, London, 52, res asst periodont & path, 53, Royal Dent Hosp, Sch Dent Surg, 53-55; gen dent surgeon, Post-Grad Inst Dent Surg, Eastman Dent Hosp, 56-59; assoc prof, 59-67, PROF PERIODONT, FAC DENT, UNIV MAN, 67- Mem: Int Asn Dent Res. Res: Pathology, general and oral; histopathology of the oral and gingival epithelium at various stages of development. Mailing Add: Dept of Oral Path Univ of Man Fac of Dent Winnipeg MB Can

TROTT, LAMARR BRICE, b North Wilkesboro, NC, Jan 2, 35; m 57. MARINE ECOLOGY, ICHTHYOLOGY. Educ: Fla State Univ, BA, 57, MS, 60; Univ Calif, Los Angeles, PhD(zool), 67. Prof Exp: Biologist, Bur Sanit Eng, Fla State Bd Health, 57-58; res scientist, Inst Marine Sci, Univ Tex, 60; asst prof biol, Col Guam, 62-63; assoc zool, Univ Calif, Los Angeles, 64-65; dir marine sci lab, Chinese Univ Hong Kong, 67-73; RES ADMINR, NAT MARINE FISHERIES SERV, NAT OCEANIC & ATMOSPHERIC ADMIN, 73- Concurrent Pos: Fel, Chinese Univ Hong Kong Inst Sci & Technol, Chinese Univ Hong Kong, 67-70 & Asia Found, 70-72. Mem: AAAS; Ecol Soc Am; Am Soc Limnol & Oceanog; Am Soc Zoologists. Res: Biology of pearlfishes; marine pollution; marine benthic biology. Mailing Add: Nat Marine Fish Serv 418C 3300 Whitehaven St NW Washington DC 20035

TROTT, S M, b Barros, PR, Jan 22, 15; m 41; c 3. MATHEMATICS. Educ: US Naval Acad, SB, 36; Univ Tasmania, BSc, 52, PhD(math), 67; Univ Toronto, MA, 62. Prof Exp: Head dept math & physics, Hobart Tech Col, 52-55; sr lectr mech eng, Univ Tasmania, 55-63; assoc prof math, Univ Sask, Regina Campus, 63-65; ASSOC PROF MATH, UNIV TORONTO, 65- Res: Geometric algebra. Mailing Add: Dept of Math Fac of Arts & Sci Univ of Toronto Toronto ON Can

TROTT, WINFIELD JAMES, b Lockport, NY, Mar 11, 15; m 41; c 3. UNDERWATER ACOUSTICS. Educ: Hillsdale Col, BS, 38. Prof Exp: Lab instr physics, Columbia Univ, 39-40; res physicist, Res Div, Gen Motors Corp, 40-46; res physicist, Res Lab, Ford Motor Co, 46-47; head res dept, Underwater Sound Ref Div, Naval Res Lab, Fla, 48-67; prin scientist, Sci-Atlanta, Inc, Ga, 67-70; SUPVRY RES PHYSICIST, ACOUST DIV, NAVAL RES LAB, 70- Concurrent Pos: Liaison scientist, London Br Off, Off Naval Res, 66-67; chmn, Writing Subcomt Underwater Transducers, Am Nat Standards Inst; control agent, Working Group 7 Ultrasonics, Int Round-Robin Hydrophone Calibration Tech Comt, 29, Int Electrotech Comn. Mem: AAAS; fel Acoust Soc Am; Sigma Xi. Res: Instrument development; acoustics; underwater sound. Mailing Add: Acoust Div Naval Res Lab Code 8150 Washington DC 20375

TROTTER, ALLEN RICHARD, b New Iberia, La, Mar 12, 11; m 40; c 2. PLANT BREEDING. Educ: La State Univ, BS, 35; Cornell Univ, PhD(veg crops, genetics), 40. Prof Exp: Asst veg crops, Cornell Univ, 35-40; plant breeder, Assoc Seed Growers, Inc, 40-60; dir veg res, Conn, 60-71, DEP DIR RES, ASGROW SEED CO, MICH, 71- Mem: Am Soc Hort Sci; Genetics Soc Am. Res: Tomato breeding for wilt resistance; vegetable and corn breeding. Mailing Add: Asgrow Seed Co Kalamazoo MI 49001

TROTTER, GORDON TRUMBULL, b Washington, DC, Aug 27, 34; m 65; c 2. COMPUTER SCIENCES. Educ: Univ Md, BS, 56; Johns Hopkins Univ, MS, 72. Prof Exp: Mathematician, Nat Bur Stand, 56-58; from assoc mathematician to mathematician appl physics lab, Johns Hopkins Univ, 58-66; res mathematician & supvr comput opers, IIT Res Inst, 66-67; MATHEMATICIAN & SUPVR INFO PROCESSING PROG PROJ, APPL PHYSICS LAB, JOHNS HOPKINS UNIV, 67- Concurrent Pos: Asst, Grad Sch, Univ Md, 56-58. Mem: Asn Comput Mach. Res: Information storage and retrieval systems; text processing, software management; programming theory. Mailing Add: Appl Physics Lab Johns Hopkins Univ Laurel MD 20810

TROTTER, HALE FREEMAN, b Kingston, Ont, May 30, 31. MATHEMATICS. Educ: Queen's Univ, Ont, BA, 52, MA, 53; Princeton Univ, PhD(math), 56. Prof Exp: Fine instr math, Princeton Univ, 56-58; asst prof, Queen's Univ, Ont, 58-60; vis assoc prof, 60-62, assoc prof, 63-69, PROF MATH, PRINCETON UNIV, 69-, ASSOC DIR COMPUT CTR, 62- Mem: Am Math Soc; Can Math Cong; Math Asn Am; Asn Comput Mach. Res: Knot theory; computing. Mailing Add: Dept of Math Princeton Univ Princeton NJ 08540

TROTTER, JAMES, b Dumfries, Scotland, July 15, 33; m 57; c 2. PHYSICAL CHEMISTRY. Educ: Univ Glasgow, BSc, 54, PhD(chem), 57, DSc(chem), 63. Prof Exp: Asst lectr chem, Univ Glasgow, 54-57, Nat Res Coun Can fel physics, 57-59, Imp Chem Indust fel chem, 59-60; from asst prof to assoc prof, 60-65, PROF CHEM, UNIV BC, 65- Mem: AAAS; Am Crystallog Asn; Chem Inst Can; The Chem Soc; Royal Inst Chemists. Res: Chemistry; crystallography. Mailing Add: Dept of Chem Univ of BC Vancouver BC Can

TROTTER, JOHN ELLIS, b Chicago, Ill, May 18, 21. GEOGRAPHY. Educ: Univ Chicago, SM, 53, PhD(geog), 62. Prof Exp: From asst prof to assoc prof, 56-60, PROF GEOG, ILL STATE UNIV, 60-, HEAD DEPT GEOG & GEOL, 67- Mem: Asn Am Geogrs; Am Geog Soc. Res: Climatology; recreational resources. Mailing Add: Dept of Geog & Geol Ill State Univ Normal IL 61761

TROTTER, NANCY LOUISA, b Monaca, Pa, July 26, 34. CYTOLOGY, ELECTRON MICROSCOPY. Educ: Oberlin Col, AB, 56; Brown Univ, ScM, 58, PhD(cytol), 60. Prof Exp: From instr to asst prof histol, Col Physicians & Surgeons, Columbia Univ, 61-68; ASSOC PROF ANAT, JEFFERSON COL MED, THOMAS JEFFERSON UNIV, 68- Concurrent Pos: USPHS trainee, 60, fel, 61, res grant, 62-68. Mem: Am Asn Anatomists. Res: Hepatomas; liver cytology; partial hepatectomy. Mailing Add: Box 95 RD 1 Enon Valley PA 16120

TROTTER, PATRICK CASEY, b Longview, Wash, Jan 26, 35; div; c 2. PULP CHEMISTRY, PAPER CHEMISTRY. Educ: Ore State Univ, BS, 57; Inst Paper

Chem, MS, 59, PhD(chem), 61. Prof Exp: Sect leader, Paperboard & Coatings Group, Pulp & Paperboard Res Dept, 61-68, DEPT MGR, FIBER PROD RES & DEVELOP DIV, WEYERHAEUSER CO, 68- Mem: Tech Asn Pulp & Paper Indust. Res: Long range and basic research on pulping, bleaching, papermaking and properties of paper and paperboard. Mailing Add: Weyerhaeuser Co Tacoma WA 98401

TROTTER, PHILIP JAMES, b Jackson, Mich, Jan 31, 41; m 67. PHYSICAL CHEMISTRY. Educ: Ill Inst Technol, BS, 64; Univ Colo, PhD(phys chem), 67. Prof Exp: Fel molecular complexes, New Eng Inst & Univ Conn, 67-69; res chemist raman spectra, Shell Res, Holland, 69-72; RES CHEMIST RAMAN & INFRARED SPECTRA, EASTMAN KODAK CO, 73- Mem: Am Chem Soc; Soc Appl Spectros. Res: Laser-Raman and infrared spectroscopic applications. Mailing Add: Eastman Kodak Co 343 State St Rochester NY 14650

TROTTER, ROBERT RUSSELL, b Morgantown, WVa, Apr 23, 15; m 56. OPHTHALMOLOGY. Educ: WVa Univ, AB & BS, 40; Temple Univ, MD, 42. Prof Exp: Asst ophthal res, Howe Lab, Harvard Med Sch, 47, instr, 52-55, instr ophthal, 56-60; clin assoc prof surg, 61-63, assoc prof, 63-65, PROF SURG, MED CTR, W VA UNIV, 65-, CHMN DIV OPHTHAL, 61- Concurrent Pos: Fel, Harvard Med Sch, 48-49; resident ophthal, Mass Eye & Ear Infirmary, 49-51, dir glaucoma consult serv & asst to chief ophthal, 55-60. Mem: AMA; Am Acad Ophthal & Otolaryngol; Am Col Surgeons; Asn Res Vision & Ophthal. Res: Glaucoma; testing vision of preschool children. Mailing Add: Div of Ophthal WVa Univ Med Ctr Morgantown WV 26506

TROTTIER, CLAUDE HENRY, b Woonsocket, RI, Jan 13, 39; m 63; c 2. ORGANIC CHEMISTRY. Educ: Univ RI, BS, 60; Providence Col, MS, 62, PhD(org chem), 65. Prof Exp: Res chemist, Am Hoechst Corp, 66-67; NIH fel, Worcester Found Exp Biol, 67-68; sr develop chemist, Ciba-Geigy Corp, RI, 68-70, Ciba-Geigy Ltd, Switz, 70-71, group leader develop, Ciba-Geigy Corp, RI, 71-72, mfg mgr dyestuff & chem, 72-74, DEVELOP MGR, CIBA-GEIGY CORP, ALA, 75- Concurrent Pos: Consult, US Army Natick Labs, 67. Mem: Am Chem Soc. Res: Mixed hydride reduction of oximes. Mailing Add: 612 Tuthill Lane Mobile AL 36608

TROTZ, SAMUEL ISAAC, b Chattanooga, Tenn, Nov 6, 27; m 55; c 2. CHEMISTRY. Educ: Univ Chattanooga, BS, 48; Univ Tenn, MS, 51; Univ St Louis, PhD, 56. Prof Exp: Asst chem, Univ Tenn, 48-50 & Univ St Louis, 52-55; res chemist, Olin Mathieson Chem Corp, 55, sr res chemist & group leader, 55-59, proj supvr org div, 59-66, sect mgr, 66-70, TECH MGR CHEM GROUP, OLIN CORP, 70- Mem: AAAS; Am Chem Soc; Sigma Xi. Res: Synthesis; product development; oxyhalogens; water chemistry; antimicrobial agents; organic chemistry; organometallics; boranes; light metal hydrides; heterocyclics; high energy fuels; polymers. Mailing Add: Chem Group Res Olin Corp 275 Winchester Ave New Haven CT 06504

TROUBETZKOY, EUGENE SERGE, b Clamart, Seine, France, Apr 7, 31; US citizen; m 58; c 3. THEORETICAL PHYSICS. Educ: Univ Paris, B es Sc, 49, lic es SC, 53; Columbia Univ, PhD(physics), 58. Prof Exp: Res asst physics, Columbia Univ, 53-58; sr scientist, United Nuclear Corp, NY, 58-64, adv scientist, 64-69; sr res assoc, Div Nuclear Sci & Eng, Columbia Univ, 69-70; MEM STAFF, MAGI CORP, 70- Mem: Am Phys Soc. Res: Neutrons; radiation transport theory and calculations applied to shielding and reactor calculations. Mailing Add: Magi Corp 3 Westchester Plaza Elmsford NY 10523

TROUP, STANLEY BURTON, b Minneapolis, Minn, Feb 9, 25; m 49; c 2. INTERNAL MEDICINE, HEMATOLOGY. Educ: Univ Minn, Minneapolis, BS, 48, BM, 49, MD, 50; Mass Inst Technol, MS, 72; Am Bd Internal Med, dipl, 57. Prof Exp: Intern med, Strong Mem Hosp, Univ Rochester, 49-50, intern path, 50-51, asst resident med, 51-52; resident, Beth Israel Hosp, Harvard Univ, 52-53; from instr to prof med, Strong Mem Hosp, Univ Rochester, 58-74; PROF MED, DIR MED CTR & V PRES, UNIV CINCINNATI, 74- Concurrent Pos: Fel path, Strong Mem Hosp, Univ Rochester, 50-51, fel hemat, 55-58; NIH spec fel, Kocher Inst, Univ Bern, 61-62; Alfred P Sloan fel mgt, Mass Inst Technol, 71-72; chief med, Rochester Gen Hosp, 65-74; consult, Genesee, St Mary's & Highland Hosps, Rochester & Vet Admin Hosp, Bath, NY, 65-; consult spec ctrs res, NIH, 71-72. Mem: Fel Am Col Physicians; Am Fedn Clin Res; Am Soc Hemat. Res: Bleeding disorders; hemolytic anemia; medical education and management. Mailing Add: 234 Goodman St Cincinnati OH 45267

TROUSDALE, WILLIAM LATIMER, b Littleton, NH, Nov 10, 28; m 55; c 4. PHYSICS. Educ: Trinity Col, Conn, BS, 50; Rutgers Univ, PhD, 56. Prof Exp: Asst prof physics, Trinity Col, Conn, 55-61; res assoc, Univ Pa, 61-62; asst prof, 62-66, ASSOC PROF PHYSICS, WESLEYAN UNIV, 66- Concurrent Pos: Consult, United Aircraft Corp, 56-61; vis scientist, Brookhaven Nat Lab, 66-67. Mem: Am Phys Soc. Res: Mössbauer effect; magnetism; low temperature physics; physical electronics; holography. Mailing Add: Dept of Physics Wesleyan Univ Middletown CT 06457

TROUSE, ALBERT CHARLES, b Hanford, Calif, May 19, 21; m 47; c 4. AGRONOMY. Educ: Univ Calif, BS, 43, MS, 48; Univ Hawaii, PhD(soil physics), 64. Prof Exp: Soil scientist, USDA, Nev, Calif & Hawaii, 46-51; assoc agronomist, Exp Sta, Hawaiian Sugar Planters Asn, 51-57, sr agronomist, 58-63; SOIL SCIENTIST, NAT TILLAGE MACH LAB, AGR ENG RES DIV, AGR RES SERV, USDA, 64- Mem: Am Soc Agron; Am Soc Agr Eng; AAAS. Res: Soil physical properties; requirements for plant root bed and seed bed with respect to various crops and climatic situations, soil strength and aeration; interactions of roots of various species to each other. Mailing Add: PO Box 792 Auburn AL 36830

TROUT, DAVID LYNN, b Ann Arbor, Mich, Aug 2, 27; m 59; c 3. PHYSIOLOGY. Educ: Swarthmore Col, AB, 51; Duke Univ, MA, 54, PhD(physiol, pharmacol), 58. Prof Exp: Lab technician, Baxter Labs, 51-52; asst physiol & pharmacol, Duke Univ, 54-58; sr biochemist, Cent Ref Lab, Vet Admin Hosp, Durham, NC, 58-61; res physiologist, US Air Force Sch Aerospace Med, Brooks AFB, Tex, 61-66; res physiologist, Human Nutrit Res Div, Agr Res Serv, 66-71, RES PHYSIOLOGIST, NUTRIT INST, AGR RES SERV, USDA, 71- Mem: Fel AAAS; Am Physiol Soc; Am Inst Nutrit; Soc Exp Biol & Med. Res: Carbohydrate nutrition; physiological adaptation to diet; lipid transport and metabolism; isotope kinetics and tracer techniques. Mailing Add: 8905 Royal Ridge Lane Laurel MD 20811

TROUT, DENNIS ALAN, b Washington, DC, July 26, 47; m 75. AIR POLLUTION, METEOROLOGY. Educ: Pa State Univ, BS, 68, MS, 69, PhD(meteorol, air pollution), 73. Prof Exp: Res asst, Dept Meteorol, Pa State Univ, 68-70, res asst, Ctr Air Environ Studies, 70-71; environ/syst analyst, Environ Tech Appl Ctr, US Air Force, 71-73; STAFF METEOROLOGIST, BATTELLE-COLUMBUS LABS, BATTELLE MEM INST, 73- Concurrent Pos: Lectr air pollution & meteorol, Ohio State Univ, 75. Mem: Am Meteorol Soc; Air Pollution Control Asn; Am Geophys Union; AAAS; Sigma Xi. Res: Ambient air quality measurements and analysis; computer modeling of atmospheric dispersion of air pollutants; development of dynamic emission control strategies; assessment of trace contaminant emissions and resulting ambient concentrations. Mailing Add: Battelle-Columbus Labs Battelle Mem Inst 505 King Ave Columbus OH 43201

TROUT, EDRIE DALE, b Franklin, Ind, Nov 3, 01; m 46. PHYSICS. Educ: Franklin Col, BS, 22; Am Bd Health Physics, dipl. Hon Degrees: DSc, Franklin Col, 50. Prof Exp: Instr high sch, Ill, 22-28; mem, Educ Dept, Gen Elec X-ray Corp, 28-34, mem staff, 34-45, dir tech serv, 45-47, mgr therapeut prod dept, 47-52, consult radiation physicist, X-ray Dept, Gen Elec Co, 52-62; PROF RADIOL PHYSICS & DIR X-RAY SCI & ENG LAB, ORE STATE UNIV, 62- Concurrent Pos: Mem, Nat Coun Radiation Protection. Mem: Fel Am Phys Soc; Radiation Res Soc; assoc Am Roentgen Ray Soc; assoc Radiol Soc NAm; Am Radium Soc. Res: X-ray physics as applied to diagnosis, therapy and industrial applications; radiation physics. Mailing Add: X-ray Sci & Eng Lab Ore State Univ Corvallis OR 97331

TROUT, PAUL EUGENE, b Baker, Mont, Sept 17, 21; m 48; c 3. CHEMISTRY. Educ: Mont State Col, BS, 43; Lawrence Col, MS, 48, PhD(chem), 51. Prof Exp: Res asst, Am Box Bd Co, 50-54; vpres & tech dir, Waldorf Paper Prod Co, 54-61; mgr container bd res, 62-69, DIR ENVIRON CONTROL, CONTAINER CORP AM, 69- Mem: AAAS; Am Chem Soc; Tech Asn Pulp & Paper Indust. Res: Wet strength of paper; neutral sulfite pulping of hardwood; lignin investigations; air and water pollution abatement; paper and cellulose chemistry. Mailing Add: Container Corp Am 500 E North Ave Carol Stream IL 60187

TROUT, ROBERT OREN, geography, sociology, see 12th edition

TROUT, VERDINE ELZA, b Waurika, Okla, Sept 19, 24; m 52; c 1. SCIENCE EDUCATION, ECOLOGY. Educ: Southeastern State Col, BS, 49; Univ Okla, MEd, 53; Okla State Univ, EdD(biol & sci educ), 66. Prof Exp: Teacher high schs, Okla, 49-64; ASSOC PROF BIOL PHYSICS & SCI EDUC, CENT STATE UNIV, OKLA, 64- Mem: Am Asn Physics Teachers; Nat Sci Teachers Asn. Res: Human population biology. Mailing Add: 2902 Wanetta Edmond OK 73034

TROUT, WILLIAM EDGAR, JR, b Clifton Forge, Va, July 30, 03; m 34; c 2. CHEMISTRY. Educ: Johns Hopkins Univ, AB, 25, PhD(chem), 35. Prof Exp: Instr biol, Va Mil Inst, 26; prof natural sci & math, Md Col Women, 27-29; asst chem, Johns Hopkins Univ, 29-32, instr & jr instr, 32-35; prof, Mary Baldwin Col, 35-46; chmn dept, 53-59, prof Chem, 46-73; EMER PROF CHEM, UNIV RICHMOND, 73- Mem: AAAS; Am Chem Soc; fel Am Inst Chemists; The Chem Soc. Res: Heterogeneous catalysis; complex compounds; inorganic chemistry. Mailing Add: Dept of Chem Univ of Richmond Richmond VA 23173

TROUT, WILLIAM EDGAR, III, b Staunton, Va, Apr 21, 37. GENETICS. Educ: Univ Richmond, BS, 59; Ind Univ, AM, 64, PhD(genetics), 65. Prof Exp: USPHS res fel radiation genetics, Biol Div, Oak Ridge Nat Lab, 65-66; RES SCIENTIST, BIOL DEPT, CITY OF HOPE MED CTR, 66- Mem: AAAS; Genetics Soc Am; Nat Speleol Soc. Res: Behavior genetics of Drosophila melanogaster. Mailing Add: Dept of Biol City of Hope Med Ctr Duarte CA 91010

TROUTMAN, JOSEPH LAWRENCE, b Concordia, Ky, June 21, 21; m 45; c 4. PLANT PATHOLOGY. Educ: Univ Ky, BS, 50; Univ Wis, PhD(plant path), 57. Prof Exp: Asst agron, Univ Ky, 50-53; from asst prof to assoc prof plant path & physiol, Res Div, Va Polytech Inst, 57-68; assoc prof, 68-74, ASSOC PLANT PATHOLOGIST, AGR EXP STA, UNIV ARIZ, 74- Mem: AAAS. Am Phytopath Soc. Res: Diseases of vegetables and citrus. Mailing Add: Univ of Ariz Agr Exp Sta Rte 1 Box 587 Yuma AZ 85364

TROUTMAN, RICHARD CHARLES, b Columbus, Ohio, May 16, 22; m; c 3. OPHTHALMOLOGY. Educ: Ohio State Univ, BA, 42, MD, 45; Am Bd Ophthal, dipl, 51. Prof Exp: Intern ophthal, New York Hosp, 45-46; resident, Cornell Med Ctr, 48-50, from instr to asst prof, Med Col, Cornell Univ, 52-55; PROF OPHTHAL & CHMN DIV, STATE UNIV NY DOWNSTATE MED CTR, 55- Concurrent Pos: Instr, Manhattan Eye, Ear & Throat Hosp, 51-55, mem courtesy staff, 55-, surgeon dir, 61-; consult hosps, 54-; vis surgeon, Kings County Hosp, 55-; mem courtesy staff, Cornell Med Ctr, New York Hosp, 55-, attend surgeon, 71-; mem courtesy staff, New York Eye & Ear Infirmary, 55-; mem ophthal postgrad training comt, Nat Inst Neurol Dis & Blindness, 59-63; consult neurol & blindness div, Bur State Serv, Dept Health, Educ & Welfare, 63-67; mem, Bd Dirs, Baraquer Inst, 63-73. Mem: Am Asn Res Vision & Ophthal; fel Am Col Surgeons; fel Am Acad Ophthal & Otolaryngol; fel NY Acad Med; Am Ophthal Soc. Res: Orbital surgery and surgery of the anterior segment of the eye; microsurgery and stereotaxic ophthalmic surgery. Mailing Add: 755 Park Ave New York NY 10021

TROUTNER, DAVID ELLIOTT, b Eolia, Mo, Oct 11, 29; m 55; c 3. NUCLEAR CHEMISTRY. Educ: Washington Univ, AB, 52, PhD(chem), 59; Univ Mo-Rolla, MS, 56. Prof Exp: From asst prof to assoc prof, Univ Mo-Rolla, 59-61; assoc prof, 61-69, PROF CHEM, UNIV MO-COLUMBIA, 69-, CHMN DEPT, 71- Concurrent Pos: Vis scientist, Oak Ridge Nat Lab, 67-68. Mem: Am Chem Soc; Am Phys Soc. Res: Radiochemical studies of nuclear fission. Mailing Add: Dept of Chem Univ of Mo-Columbia Columbia MO 65201

TROW, JAMES, b Chicago, Ill, Apr 21, 22; m 47. GEOLOGY. Educ: Univ Chicago, SB, 43, SM, 45, PhD(geol), 48. Prof Exp: Asst surveying, Univ Chicago, 43-44; from asst prof to assoc prof geol, 47-59, PROF GEOL, MICH STATE UNIV, 59- Mem: Geol Soc Am; Am Geophys Union. Res: Structure, stratigraphy, sedimentology, igneous petrology, and metamorphism of Pre-Cambrian rocks and ores; regional tectonics. Mailing Add: Dept of Geol Mich State Univ East Lansing MI 48823

TROWBRIDGE, DALE BRIAN, b Glendale, Calif, May 17, 40; m 66; c 1. ORGANIC CHEMISTRY. Educ: Whittier Col, AB, 61; Univ Calif, Berkeley, MS, 64, PhD(org chem), 70. Prof Exp: Chemist, Aerojet Gen Corp, 61-62; teacher chem high sch, Calif, 64-66; asst prof, 69-72, ASSOC PROF CHEM, CALIF STATE COL, SONOMA, 72- Mem: Am Chem Soc. Res: Preparation and study of organo-phosphorus compounds of biological interest. Mailing Add: 10641 Barnett Valley Rd Sebastopol CA 95472

TROWBRIDGE, HENRY O, pathology, oral biology, see 12th edition

TROWBRIDGE, JAMES RUTHERFORD, b Glen Ridge, NJ, Sept 19, 20; m 49; c 3. SURFACE CHEMISTRY. Educ: Yale Univ, BS, 42, MS, 48, PhD(org chem), 50. Prof Exp: Group leader, Explor Org Div, 51-57, sr proj chemist, 57-62, RES ASSOC, CHEM RES SECT, COLGATE-PALMOLIVE RES CTR, 62- Mem: Am Chem Soc; Am Oil Chem Soc. Res: Surface active agents. Mailing Add: Colgate-Palmolive Res Ctr 909 River Rd Piscataway NJ 08854

TROWBRIDGE, LESLIE WALTER, b Curtiss, Wis, May 21, 20; m 46; c 4. SCIENCE EDUCATION. Educ: Wis State Univ, Stevens Point, BS, 40; Univ Chicago, MS, 48; Univ Wis, MS, 53; Univ Mich, PhD(sci educ), 61. Prof Exp: Teacher jr high sch, Wis, 41 & 46, instr high sch, 46-54; univ scholar, Univ Mich, 54-62; from asst prof to assoc prof, 62-70, chmn dept sci educ, 66-72, PROF SCI EDUC, UNIV

NORTHERN COLO, 70- Concurrent Pos: Fel, NY Univ, 69-70. Mem: Nat Sci Teachers Asn (pres, 73-74); Nat Asn Res Sci Teaching. Mailing Add: Dept of Sci Educ Univ of Northern Colo Greeley CO 80631

TROWER, WILLIAM PETER, b Rapid City, SDak, May 25, 35; m 57, 63; c 3. ELEMENTARY PARTICLE PHYSICS. Educ: Univ Calif, Berkeley, AB, 57; Univ Ill, Urbana, MS, 63, PhD(physics), 66. Prof Exp: Physicist, Lawrence Radiation Lab, Univ Calif, Berkeley, 60-62; res asst, Digital Comput Lab & Dept Physics, Univ Ill, Urbana, 62-66; asst prof, 66-70, ASSOC PROF PHYSICS, VA POLYTECH INST & STATE UNIV, 70-, COL ARCHIT, 73- Concurrent Pos: Chmn, Gordon Res Conf Multiparticle Prod Processes, 73. Honors & Awards: Bronze Medal, Int Film & TV Festival NY, 73. Mem: Fel Am Phys Soc; Sigma Xi; AAAS. Res: Experimental nuclear and particle physics; scientific computer applications. Mailing Add: 1105 Highland Circle SE Blacksburg VA 24060

TROXEL, ALLEN WENDELL, b Okla, Aug 19, 18; m 45; c 3. PLANT PATHOLOGY. Educ: Univ Calif, BS, 48, PhD(plant path), 54. Prof Exp: Lab asst plant path, Univ Calif, 48-50, asst, 51-54; from instr to asst prof bot & plant path, assoc prof bot, 71-73, ASSOC PROF PLANT PATH, OHIO STATE UNIV, 71- Mem: Am Phytopath Soc. Res: Botany, plant virology; bacterial diseases; plant virus multiplication, detection and chemotherapy. Mailing Add: Col Agr & Home Econ Ohio State Univ Dept Plant Path Columbus OH 43210

TROXEL, BENNIE WYATT, b Osawatomie, Kans, Aug 9, 20; m 46; c 2. GEOLOGY. Educ: Univ Calif, Los Angeles, BA, 51, MA, 58. Prof Exp: Geologist, Calif Div Mines & Geol, 52-71; sci ed, Geol Soc Am, 71-75; MEM STAFF, DIV MINES & GEOL, STATE CALIF, 75- Mem: Soc Econ Geologists; AAAS; Geol Soc Am; Asn Earth Sci Eds. Res: Geology of Death Valley regions; geologic factors that influence slope stability in urban areas of California; mineral resources; Precambrian stratigraphy and faults in California. Mailing Add: Calif Div Mines & Geol Rm 1341 1416 Ninth St Sacramento CA 95814

TROXELL, HARRY EMERSON, JR, b Northumberland, Pa, Sept 19, 21; wid; c 4. WOOD SCIENCE & TECHNOLOGY. Educ: Duke Univ, BS, 43, MF, 47, DF, 61. Prof Exp: Instr forest mgt, 47-49, from asst prof to assoc prof wood technol, 49-62, PROF WOOD SCI, COLO STATE UNIV, 62-, HEAD DEPT WOOD SCI & TECHNOL, COLO STATE UNIV, 74- Mem: Soc Am Foresters; Forest Prods Res Soc; Soc Wood Sci & Technol. Res: Wood products. Mailing Add: 624 Armstrong Ave Ft Collins CO 80521

TROXELL, TERRY CHARLES, b Allentown, Pa, Jan 1, 44; m 64; c 2. BIOPHYSICAL CHEMISTRY. Educ: Muhlenberg Col, BS, 65; Cornell Univ, PhD(biophys chem), 71. Prof Exp: Res assoc phys chem, Univ Ore, 71-74; SR PHYS CHEMIST, ELI LILLY & CO, 74- Mem: Am Chem Soc; AAAS. Res: Optical spectroscopies especially circular dichroism and linear dichroism; conformational analysis; binding studies; drug structure-activity relationships; drug-receptor interactions. Mailing Add: Eli Lilly & Co Indianapolis IN 46206

TROXLER, RAYMOND GEORGE, b New Orleans, La, Sept 21, 39; m 63; c 2. PATHOLOGY. Educ: Univ Southwestern La, BS, 64; La State Univ, MD, 64. Prof Exp: CHIEF CLIN PATH, CLIN SCI DIV, US AIR FORCE SCH AEROSPACE MED, 71- Concurrent Pos: Clin asst prof, Univ Tex Health Sci Ctr, San Antonio, 72-76. Res: Early detection of latent coronary artery disease by laboratory screening of blood and serum. Mailing Add: USAF Sch of Aerospace Med NGP Brooks AFB TX 78235

TROXLER, ROBERT FULTON, b Santa Monica, Calif, July 11, 38; m 64. PLANT PHYSIOLOGY. Educ: Grinnell Col, BS, 60; Pa State Univ, MS, 62; Univ Chicago, PhD(bot), 65. Prof Exp: Teaching asst plant physiol, Pa State Univ, 62; res assoc bot, Univ Chicago, 65-66; res assoc med, 66-68, asst prof biochem, 68-72, ASSOC PROF BIOCHEM, SCH MED, BOSTON UNIV, 72- Mem: AAAS; Am Soc Plant Physiologists. Res: Porphyrin and bile pigment biosynthesis and chemistry. Mailing Add: Dept of Biochem Boston Univ Sch of Med Boston MA 02118

TROY, DANIEL JOSEPH, b St Louis Co, Mo, Feb 2, 32; m 55; c 6. MATHEMATICS. Educ: St Louis Univ, BS, 53, MS, 58, PhD(math), 61. Prof Exp: Instr math, St Louis Univ, 58-61; asst prof, Ohio State Univ, 61-67; ASSOC PROF MATH, PURDUE UNIV, 67- Mem: Am Math Soc; Math Asn Am; Inst Math Statist. Res: Complex variable. Mailing Add: Dept Math Calumet Campus Purdue Univ Hammond IN 46323

TROY, FREDERIC ARTHUR, b Evanston, Ill, Feb 16, 37; m 59; c 2. BIOCHEMISTRY, ONCOLOGY. Educ: Washington Univ, BS, 61; Purdue Univ, West Lafayette, PhD(biochem), 66. Prof Exp: Am Cancer Soc res fel physiol chem, Sch Med, Johns Hopkins Univ, 66-68; asst prof, 68-74, ASSOC PROF BIOL CHEM, SCH MED, UNIV CALIF, DAVIS, 74- Concurrent Pos: USPHS res grant, Sch Med, Univ Calif, Davis, 69-81, cancer res grant, 71-78; Nat Cancer Inst career res develop award, 75-80; co-dir tumor biol training grant, Nat Cancer Inst, 72- Mem: Am Soc Microbiol; Am Inst Chemists; Am Soc Biol Chemists; Am Chem Soc; Brit Biochem Soc. Res: Relationship of virally determined membrane components to the proliferative transforming interaction of virus and lymphocytes; role of surface-mediated phosphorylation and sialylation reactions in lymphoid cell proliferation; chemistry and biosynthesis of complex microbial cell envelope polymers; the role of bacterial membranes in the synthesis of macromolecules. Mailing Add: Dept of Biol Chem Univ of Calif Sch of Med Davis CA 95616

TROY, WILLIAM CHRISTOPHER, b Rochester, NY, July 7, 47. APPLIED MATHEMATICS. Educ: St John Fisher Col, BS, 69; State Univ NY Buffalo, MA, 70, PhD(math), 74. Prof Exp: ASST PROF MATH, UNIV PITTSBURGH, 74- Concurrent Pos: NIH res grant, 75-77. Mem: Am Math Soc. Res: Application of the theory of differential equations to mathematical problems arising in biology, neurophysiology and chemistry which includes nerve conduction and the Belousov-Zhabotinskii chemical reaction. Mailing Add: Dept of Math Univ of Pittsburgh Pittsburgh PA 15260

TROYER, ALVAH FORREST, JR, plant breeding, genetics, see 12th edition

TROYER, JAMES RICHARD, b Goshen, Ind, Feb 26, 29; m 51; c 3. PLANT PHYSIOLOGY. Educ: DePauw Univ, BA, 50; Ohio State Univ, MS, 51; Columbia Univ, PhD(bot), 54. Prof Exp: Vis asst prof biol, Univ Ala, 54-55; instr plant physiol, Sch Forestry, Yale Univ, 55-57; asst prof bot, 57, assoc prof, 61-69, PROF BOT, NC STATE UNIV, 69- Concurrent Pos: Fel, Biomath Training Prog, NC State Univ, 64-66. Mem: AAAS; Am Soc Plant Physiol. Res: Mathematical plant physiology; flavonoid substances of plants. Mailing Add: Dept of Bot NC State Univ Raleigh NC 27607

TROYER, JOHN ROBERT, b Princeton, Ill, Feb 5, 28; m 56; c 4. ANATOMY. Educ: Syracuse Univ, AB, 49; Cornell Univ, PhD(histol, embryol), 55. Prof Exp: Asst histol & embryol, Cornell Univ, 49-54; from instr to assoc prof anat, 54-69, actg chmn dept, 71-72, PROF ANAT, SCH MED, TEMPLE UNIV, 69-, VCHMN DEPT IN CHG TEACHING, 73- Mem: AAAS; Am Asn Anat; Am Soc Mammal; NY Acad Sci. Res: Liver glycogen, porphyrin synthesis and neurosecretion in the hibernating bat; gross anatomy, normal development and abnormal development of the human heart. Mailing Add: Dept of Anat Temple Univ Sch of Med Philadelphia PA 19140

TROYER, ROBERT JAMES, b Sturgis, Mich, Sept 21, 28; m 54; c 3. MATHEMATICS. Educ: Ball State Univ, BS, 50; Ind Univ, MAT, 56, PhD(math), 60. Prof Exp: Instr math, Ind Univ, 60-62, asst prof, 62-65; vis fel, Dartmouth Col, 65-66; from asst prof to assoc prof, Univ NC, Chapel Hill, 66-68; vis assoc prof, 68-69, assoc prof, 69-70, PROF MATH, LAKE FOREST COL, 70- Concurrent Pos: Vis scholar, Northwestern Univ, 74-75. Mem: Math Asn Am. Res: Topology; geometry. Mailing Add: Dept of Math Lake Forest Col Lake Forest IL 60045

TROZZOLO, ANTHONY MARION, b Chicago, Ill, Jan 11, 30; m 55; c 6. ORGANIC CHEMISTRY, PHOTOCHEMISTRY. Educ: Ill Inst Technol, SB, 50; Univ Chicago, SM, 57, PhD(chem), 60. Prof Exp: Asst chemist, Chicago Midway Labs, 52-53; assoc chemist, Armour Res Found, 53-56; mem tech staff, Bell Tel Labs, 59-75; HUISKING PROF CHEM & MEM SR RES STAFF, RADIATION LAB, UNIV NOTRE DAME, IND, 75- Concurrent Pos: Adj prof, Columbia Univ, 71; Phillips Lectr, Univ Okla, 71; Reilly Lectr, Univ Notre Dame, 72; Brown Lectr, Rutgers Univ, 75; assoc ed, J Am Chem Soc, 75- Honors & Awards: Am Inst Chemists Award, 50. Mem: Fel AAAS; Am Chem Soc; Am Inst Chemists; fel NY Acad Sci. Res: Free radicals; carbenes; charge transfer complexes; electron spin resonance; organic solid state; singlet molecular oxygen; polymer stabilization; chemically-induced dynamic nuclear polarization. Mailing Add: Dept of Chem Univ of Notre Dame Notre Dame IN 46556

TRPIS, MILAN, b Mojsova Lucka, Czech, Dec 20, 30; US citizen; m 56; c 3. MEDICAL ENTOMOLOGY. Educ: Comenius Univ, Bratislava, Prom Biol, 56; Charles Univ, Prague, Dr rer nat(zool, med entom), 60. Prof Exp: Res asst entom, Faunistic Lab, Slovak Acad Sci, 53-56, sci asst, Dept Biol, 56-60, scientist, 60-62, independent scientist & head, Dept Ecol Physiol Insects, Inst Landscape Biol, 62-65; res assoc med entom, Univ Ill, Urbana, 66-67; res assoc, Can Dept Agr, Alta, 67-68; independent scientist & head, Dept Ecol Physiol Insects, Slovak Acad Sci, 68-69; entomologist-ecologist, EAfrica Aedes Res Unit, WHO, UN, Tanzania, 69-71; from asst fac fel to assoc fac fel, Vector Biol Labs, Dept Biol, Univ Notre Dame, Ind, 71-74; ASSOC PROF MED ENTOM, LABS MED ENTOM, DEPT PATHOBIOL, JOHNS HOPKINS UNIV, 74- Honors & Awards: First Prize Award, Slovak Acad Sci, Bratislava, 61. Mem: AAAS; Am Soc Trop Med & Hyg; Am Soc Parasitologists; Entom Soc Am; Am Mosquito Control Asn. Res: Parasitic insects, particularly their population dynamics, ecological genetics of populations, behavior and behavioral genetics; embryonic development of insects; biological and genetic control of vectors; ecology of vector-borne diseases. Mailing Add: Dept of Pathobiol Labs Med Entom Johns Hopkins Sch Hyg & Pub Hlth Baltimore MD 21205

TRUANT, ALDO PETER, pharmacology, deceased

TRUANT, JOSEPH PAUL, b San Martino Al Tagliamento, Italy, Aug 10, 23; US citizen; m 48; c 3. MICROBIOLOGY, IMMUNOLOGY. Educ: Univ Toronto, BSA, 45; Univ Western Ont, MSc, 52, PhD(microbiol), 54. Prof Exp: Lectr microbiol, Assumption & Holy Name Cols, 45-48, assoc prof, 48-51; res assoc, Med Sch, Univ Western Ont, 51-54; dir div bact mycol, Henry Ford Hosp, Detroit, Mich, 54-66; dir div microbiol, Providence Hosp, 66-71; DIR DIV MICROBIOL, ADVANCE MED & RES CTR, 72- Concurrent Pos: Guest lectr, Wayne State Univ, 54-66; consult, Oakland Med Ctr, 66- & Garden City Hosp, 66-; prof microbiol, Mich State Univ, 70-73. Mem: Am Soc Microbiol; Am Soc Pub Health; Can Soc Microbiol. Res: Chemotherapeutic agents for treatment of endocarditis, ocular and urinary tract infections; Truant's fluorescence procedure for staining of acid-fast bacilli; fluorescent procedures for beta-hemolytic streptococci. Mailing Add: 28060 New Bedford Farmington MI 48024

TRUAX, DONALD R, b Minneapolis, Minn, Aug 29, 27; m 50; c 3. MATHEMATICAL STATISTICS. Educ: Univ Wash, BS, 51, MS, 53; Stanford Univ, PhD(statist), 55. Prof Exp: Res fel math, Calif Inst Technol, 55-56; asst prof, Univ Kans, 56-59; from asst prof to assoc prof, 59-69, PROF MATH, UNIV ORE, 69- Concurrent Pos: Managing ed, Inst Math Statist, 75- Mem: Am Math Soc; Math Asn Am; Am Statist Asn; Inst Math Statist. Res: Testing statistical hypotheses; multiple decision problems. Mailing Add: Dept of Math Univ of Ore Eugene OR 97403

TRUAX, ROBERT LLOYD, b Gillett, Ark, May 8, 28; m 52; c 2. MATHEMATICS. Educ: Ark State Teachers Col, BSE, 50; Univ Miss, MA, 58; Okla State Univ, EdD(math educ), 64. Prof Exp: Coordr math, Pub Schs, Ark, 50-62; assoc prof math, Southern State Col, 63-65; assoc prof, Northeast La State Col, 65-67; ASSOC PROF MATH, UNIV MISS, 67- Mem: Math Asn Am. Res: Multivariable function approximations; statistical analysis of research related to paper industry. Mailing Add: Dept of Math Univ of Miss University MS 38677

TRUBATCH, JANETT, b New York, NY, Oct 13, 42; m 62; c 2. NEUROPHYSIOLOGY, BIOPHYSICS. Educ: Polytech Inst Brooklyn, BSc, 62; Brandeis Univ, MA, 64, PhD(physics), 68. Prof Exp: Asst prof physics, Calif State Univ, Los Angeles, 67-68; res fel biol, Calif Inst Technol, 68-74; ASST PROF PHYSIOL, NY MED COL, 74- Res: Mechanisms of synaptic transmission; synapse formation; mathematical modeling of biological systems. Mailing Add: Dept of Physiol New York Med Col Valhalla NY 10595

TRUBATCH, SHELDON L, b Brooklyn, NY, Mar 12, 42; m 62; c 2. THEORETICAL PHYSICS, BIOPHYSICS. Educ: Polytech Inst Brooklyn, BS, 62; Brandeis Univ, MA, 64, PhD(physics), 68. Prof Exp: ASSOC PROF PHYSICS, CALIF STATE UNIV, LONG BEACH, 67- Res: Non-relativistic field theory; sensory physiology. Mailing Add: Dept of Physics Calif State Univ Long Beach CA 90801

TRUBEK, MAX, b New York, NY, Nov 28, 98; m 37; c 1. MEDICINE. Educ: Johns Hopkins Univ, AB, 22; Univ Md, MD, 26. Prof Exp: House physician, Bellevue Hosp, 27-29; asst pathologist, Newark City Hosp, 29-31; assoc prof med, NY Univ-Bellevue Med Ctr, 44-56, PROF CLIN MED, MED SCH, NY UNIV, 56- Concurrent Pos: Vis physician, Bellevue Hosp, 46-; attend physician, Univ Hosp, 52- Res: Clinical medicine. Mailing Add: 121 E 60th St New York NY 10022

TRUBEY, DAVID KEITH, b Coldwater, Mich, Apr 23, 28; m 50; c 2. RADIATION PHYSICS. Educ: Mich State Univ, BS, 53. Prof Exp: Physicist, 53-54 & 55-56, mgr radiation shielding info ctr, 66-70, MEM STAFF, OAK RIDGE NAT LAB, 72- Concurrent Pos: Lectr, Oak Ridge Sch Reactor Technol, 60-62. Mem: AAAS; Am Nuclear Soc; Am Soc Info Sci. Res: Radiation shielding, transport and dosimetry. Mailing Add: Oak Ridge Nat Lab PO Box X Oak Ridge TN 37830

TRUBISKY, MICHAEL P, physical chemistry, see 12th edition

TRUBOWITZ, SIDNEY, b Brooklyn, NY, Aug 25, 11; m 49; c 2. HEMATOLOGY, CELL PHYSIOLOGY. Educ: Columbia Univ, AB, 31; Univ Chicago, MD, 36. Prof Exp: Chief hemat, Vet Admin Hosp, Staten Island, NY, 47-51, CHIEF HEMAT & HEMAT RES, VET ADMIN HOSP, EAST ORANGE, NJ, 52- Concurrent Pos: Fel, Nat Transfusion Ctr, France, 51-52; prof, Col Med, NJ, 65- Mem: Am Soc Hemat. Res: Structure and function of the marrow matrix and its role in marrow regeneration. Mailing Add: Hemat Res Lab Vet Admin Hosp East Orange NJ 07019

TRUBY, CHARLES PAUL, b Chicago, Ill, June 28, 37; m 60; c 2. MEDICAL SCIENCE. Educ: Hope Col, BS, 61; Ariz State Univ, MS, 63; Univ Houston, PhD(microbiol), 67. Prof Exp: Supvr microbiol, Brown & Root-Northrup Lunar Receiving Lab, Manned Spacecraft Ctr, NASA, 67-70, biol sect supvr, 70-72, lab mgr, 72-74; ASSOC DIR, SERAFY LABS, 74- Concurrent Pos: Lab dir, Am Bd Bioanal, 75- Mem: Am Soc Microbiol; Am Asn Bioanalysts; Am Acad Microbiol. Res: Management in all fields of biology, especially bacteriology; clinical, water, food and aerospace microbiology; quality control, proficiency testing. Mailing Add: Serafy Labs 205 W Levee Brownsville TX 78520

TRUBY, FRANK KEELER, b Painesville, Ohio, Nov 6, 24; m 48; c 3. CHEMICAL PHYSICS. Educ: Hiram Col, BA, 48; Case Inst Technol, MS, 50. Prof Exp: Physicist, Bausch & Lomb Optical Co, 50-51; assoc physicist, NMex Inst Mining & Technol, 51-55; proj leader, Southwest Res Inst, 55-59; MEM STAFF, SANDIA CORP, 59- Mem: Am Phys Soc. Res: Magnetic resonance spectroscopy; gaseous electronics; gas phase kinetics. Mailing Add: 6609 Loftus Ave NE Albuquerque NM 87109

TRUCCO, RAUL E, biochemistry, microbiology, see 12th edition

TRUCE, WILLIAM EVERETT, b Chicago, Ill, Sept 30, 17; m 40; c 2. CHEMISTRY. Educ: Univ Ill, BS, 39; Northwestern Univ, PhD(chem), 43. Prof Exp: Instr chem, Wabash Col, 43-44; res chemist, Swift & Co, Ill, 44-46; from asst prof to assoc prof, 46-55, PROF CHEM, PURDUE UNIV, WEST LAFAYETTE, 55- Concurrent Pos: Guggenheim fel, Oxford Univ, 57. Mem: Am Chem Soc; The Chem Soc. Res: Organic sulfur chemistry; acetylenes; vinylic halides; organic theory and its relationship to synthetic organic chemistry. Mailing Add: Dept of Chem Purdue Univ West Lafayette IN 47907

TRUCHARD, JAMES JOSEPH, b Sealy, Tex, June 25, 43; m 66; c 4. ACOUSTICS, ELECTRONIC ENGINEERING. Educ: Univ Tex, BS, 64, MA, 67, PhD(elec eng), 74. Prof Exp: Lab res asst, 63-65, RES SCIENTIST ACOUST ELECTRONICS, APPL RES LABS, UNIV TEX, AUSTIN, 65- Mem: Acoust Soc Am. Res: Transducer measurement systems; digital signal processing of acoustic signals; nonlinear acoustics; parametric receiving arrays for acoustic signals. Mailing Add: 10000 FM Rd 1325 Appl Res Labs PO Box 8029 Austin TX 78712

TRUCHELUT, GEORGE BURNETT, b Savannah, Ga, Sept 3, 16; div; c 2. PLANT PHYSIOLOGY. Educ: Emory Univ, AB, 41, MS, 42; Tex A&M Univ, PhD(plant physiol), 54. Prof Exp: Res chemist, Rayon Dept, E I du Pont de Nemours & Co, Va, 42-43 & Magnolia Petrol Co, Tex, 43-46; asst prof chem, Mercer Univ, 48; vis prof, Col Agr, Univ PR, 48-50; asst, Tex A&M Univ, 50-53, sr plant physiologist agr res, Dow Chem Co, 53-59, group leader, 59-60, adminr & agr res scientist, Field Res Sta, 60-63; res plant physiologist, US Army Biol Labs, Ft Detrick, 63-66, chief chem br, Crops Div, 67-68; CHMN DEPT CHEM, PALM BEACH JR COL, 68- Mem: Am Chem Soc. Res: Biochemistry and physiology of weed control; herbicides; defoliants; desiccants; general, organic and medically related chemistry; instrumentation. Mailing Add: Dept of Chem Palm Beach Jr Col Lake Worth FL 33460

TRUCKER, DONALD EDWARD, b Jamaica, NY, Jan 20, 26; m 49; c 3. ORGANIC CHEMISTRY. Educ: Polytech Inst Brooklyn, BS, 48, PhD(chem), 51; Purdue Univ, MS, 48. Prof Exp: Asst chem, Purdue Univ, 47-48; res chemist, Wyandotte Chem Corp, Mich, 51-57; TECH ASSOC, PHOTO & REPROD DIV, GAF CORP, 57- Honors & Awards: Serv Award, Soc Photog Scientists & Engrs, 73. Mem: Am Chem Soc; Soc Photog Scientists & Engrs; The Chem Soc. Res: Photographic chemistry; electrophotography; chlorination of hydrocarbons; alkylene oxides. Mailing Add: 115 Morgan Rd Binghamton NY 13903

TRUDEL, GERALD JOSEPH, b Ottawa, Ont, Dec 8, 31; m 56; c 3. RADIATION CHEMISTRY, POLYMER CHEMISTRY. Educ: McGill Univ, BSc, 55; Leeds Univ, PhD(chem), 59. Prof Exp: Res chemist, Nat Res Coun Can, 55-56; fel, Western Reserve Univ, 59-61; sr res chemist, Cent Res Lab, Can Res Lab, Can Industs, Ltd, 61-65; Asst prof, 65-74, ASSOC PROF CHEM, CONCORDIA UNIV, 74- Res: Effects of metal ions as free radical scavengers in radiation initiated polymerizations; study of electron transfer in aqueous solutions; gas phase kinetics of reaction deuterium with methane beta ray initiated; effects of radiation on the physical properties of polymers. Mailing Add: Dept of Chem Concordia Univ Montreal PQ Can

TRUDEN, JUDITH LUCILLE, b Duluth, Minn, Sept 29, 31. VIROLOGY. Educ: Wayne State Univ, BA, 53, MS, 55; Univ Miami, PhD(microbiol), 67. Prof Exp: Technician virol, Henry Ford Hosp, Detroit, Mich, 55-59; USPHS res fel, Pub Health Res Inst, New York, NY, 68-71; res instr, Med Col Wis, 71-74, res assoc, 74; SCHOLAR MOLECULAR BIOL, UNIV MICH, ANN ARBOR, 75- Mem: AAAS; Am Soc Microbiol. Res: Integration of polyoma virus genomes into the DNA of transformed cells. Mailing Add: Dept of Biol Chem Univ of Mich Ann Arbor MI 48109

TRUE, MERRILL ALLAN, b Waco, Tex, Jan 1, 40; m 63. MARINE BIOLOGY. Educ: Univ Aix-Marseille, cert oceanog, 62, PhD(biol oceanog), 66. Prof Exp: Res asst biol oceanog, Oceanog Inst, Univ Aix-Marseille, 62-66; res assoc marine ecol, Tulane Univ, La, 67-69; PRES, BIO-OCEANIC RES, INC, 69- Mem: AAAS; Am Fisheries Soc; Am Littoral Soc; Am Soc Limnol & Oceanog; Int Asn Prof Diving Scientists. Res: Benthic ecology; bio-fouling; oceanographic sampling techniques. Mailing Add: 7733 Freret St New Orleans LA 70118

TRUE, RENATE (SCHLENZ), b Porto Alegre, Brazil, Sept 19, 36; m 63. BIOLOGICAL OCEANOGRAPHY. Educ: Univ Sao Paulo, BS, 59, MS, 60; Univ Aix-Marseille, PhD(biol oceanog), 65. Prof Exp: Asst, investr & mem fac, Lab Marine Biol Sao Sebastio, Univ Sao Paulo, 60-61; teacher, Sao Paulo State, Brazil, 61; res assoc biol oceanog, Marine Sta Endoume, Univ Aix-Marseille, 61-64, teaching asst biol, Univ Aix-Marseille, 64-67; asst, Tulane Univ, La, 68-69, res asst pharmacol, Sch Med, 69-70; CHIEF MARINE BIOLOGIST, BIO-OCEANIC RES, INC, 70- Concurrent Pos: Chief scientist, Bio-Nica, Nicaragua, 73-75. Honors & Awards: Conserv Award, US Dept Interior, 70. Mem: Int Asn Prof Diving Scientists. Res: Marine benthic ecology; benthic communities distribution; pollution evaluation; pre and post spill evaluation. Mailing Add: 7733 Freret St New Orleans LA 70118

TRUE, WILLIAM WADSWORTH, b Rockland, Maine, Dec 27, 25; m 54; c 4. PHYSICS. Educ: Univ Maine, BS, 50; Univ RI, MS, 52; Ind Univ, PhD(physics), 57. Prof Exp: Instr physics, Princeton Univ, 57-60; from asst prof to assoc prof, 60-69, PROF PHYSICS, UNIV CALIF, DAVIS, 69- Mem: Am Phys Soc. Res: Theoretical nuclear physics. Mailing Add: Dept of Physics Univ of Calif Davis CA 95616

TRUEBLOOD, EMILY WALCOTT EMMART, b Baltimore, Md; m 49. CYTOLOGY. Educ: Goucher Col, AB, 22; Johns Hopkins Univ, MA, 24, PhD(cytol, genetics), 30. Prof Exp: Assoc prof biol, Western Md Col, 24-28; assoc entomologist, Bur Entom, USDA, Mexico City, Mex, 30-31; instr histol, Johns Hopkins Univ, 32-36; assoc cytologist, Div Pharmacol, NIH, 36-40, cytologist, 44-60, cytologist, Lab Exp Path, Nat Inst Arthritis & Metab Dis, 60-69; RES FEL, HUNT BOT LIBR, CARNEGIE MELLON UNIV, 70- Concurrent Pos: Res assoc, Smithsonian Inst, 70-71; hon fel, Bot Dept, Harvard Univ, 70-73. Mem: Fel AAAS; Soc Exp Biol & Med; Am Soc Pharmacol & Exp Therapeut; Am Soc Microbiol; Soc Indust Microbiol. Res: Cytology of chromosome pattern and induced liver cancer; chemotherapy of tuberculosis; antibiotics in therapy; fluorescence microscopy; cytological localization of antigens; cellular localization of streptococcal hyaluronidase, glyceraldehyde 3-phosphate dehydrogenase in muscle and kidney and myosin in skeletal muscle and the conduction bundle of the heart. Mailing Add: 7100 Armat Dr Bethesda MD 20034

TRUEBLOOD, KENNETH NYITRAY, b Dobbs Ferry, NY, Apr 24, 20; m 70. CHEMISTRY. Educ: Harvard Univ, AB, 41; Calif Inst Technol, PhD(chem), 47. Prof Exp: Asst chem, Calif Inst Technol, 43-46, res fel, 47-49; from instr to assoc prof, 49-60, dean, Col Lett & Sci, 71-74, chmn dept chem, 65-70, PROF CHEM, UNIV CALIF, LOS ANGELES, 60- Concurrent Pos: Fulbright award, 56-57; mem, US Nat Comt Crystallog, 60-65; vis prof, Ibadan, 64-65; vis scientist, Inst Elemento-Org Compounds, Moscow, 65. Mem: Am Chem Soc; Am Crystallog Asn (pres, 61). Res: X-ray studies of molecular and crystal structure. Mailing Add: Dept of Chem Univ of Calif Los Angeles CA 90024

TRUELOVE, BRYAN, b Bradford, Eng. WEED SCIENCE. Educ: Sheffield Univ, BSc, 55, PhD(plant physiol), 61. Prof Exp: Asst lectr bot, Manchester Univ, 60-62, lectr, 62-67; assoc prof bot, 68-75, PROF BOT & MICROBIOL, AUBURN UNIV, 75- Concurrent Pos: Vis asst prof, Univ Ill, 65. Mem: Soc Exp Biol & Med; Am Soc Plant Physiologists; Weed Sci Soc Am. Res: Mitochondrial metabolism, particularly in relation to their energy linked functions and the effects of aging in mitochondria; mode of action of herbicides and effects on plant physiological processes and metabolism. Mailing Add: Dept of Bot & Microbiol Auburn Univ Auburn AL 36830

TRUEMAN, THOMAS LAURENCE, b Media, Pa, Sept 24, 35; m 61; c 2. PHYSICS. Educ: Dartmouth Col, AB, 57; Univ Chicago, MS, 58, PhD(physics), 62. Prof Exp: Res assoc physics, 62-64, from asst physicist to physicist, 65-74, SR PHYSICIST & GROUP LEADER, BROOKHAVEN NAT LAB, 74- Concurrent Pos: Guggenheim fel, Oxford Univ, 72-73. Res: High energy theory; scattering theory; strong interactions. Mailing Add: Brookhaven Nat Lab Upton NY 11973

TRUEMPER, JOSEPH TUCKER, b Memphis, Tenn, Dec 25, 28; m 53; c 7. CHEMISTRY. Educ: Loyola Univ, La, BS, 50; La State Univ, MS, 57, PhD(chem), 59. Prof Exp: Sr chemist, Chem Eng Dept, Atlas Chem Industs, Inc, 58-63, proj leader, 63-72; MGR TECH SERV SECT, QUAL ASSURANCE DEPT, STUART PHARMACEUT DIV, ICI UNITED STATES, 72- Mem: Am Chem Soc; Am Soc Qual Control. Res: Analytical development; pharmaceutical analysis; quality control; adsorption at solid liquid, solid gas interfaces. Mailing Add: Stuart Pharmaceut Div ICI United States Wilmington DE 19899

TRUESDELL, ALFRED HEMINGWAY, b Washington, DC, Sept 10, 33; m 64. GEOLOGY, CHEMISTRY. Educ: Oberlin Col, AB, 57; Harvard Univ, AM, 61, PhD(geol), 62. Prof Exp: GEOCHEMIST, US GEOL SURV, 55- Concurrent Pos: Res assoc, Stanford Univ, 64- Mem: AAAS; Am Mineral Soc; Geochem Soc. Res: Application of physical chemistry to the study of geologic processes; electrochemistry of membranes; ion exchange equilibria and energetics; hot springs solution geochemistry. Mailing Add: US Geol Surv 345 Middlefield Rd Menlo Park CA 94025

TRUESDELL, CLIFFORD AMBROSE, III, b Los Angeles, Calif, Feb 18, 19; m 39, 51; c 1. MATHEMATICS. Educ: Calif Inst Technol, BS, 41, MS, 42; Brown Univ, cert, 42; Princeton Univ, PhD(math), 43. Hon Degrees: Dr Eng, Milan Polytech Univ, 64. Prof Exp: Asst math & hist, Calif Inst Technol, 41-42; asst mech, Brown Univ, 42; instr math, Princeton Univ, 42-43 & Univ Mich, 43-44; mem staff, Radiation Lab, Mass Inst Technol, 44-46; chief theoret mech subdiv, Naval Ord Lab, 46-48, head theoret mech sect, Naval Res Lab, 48-51; prof math, Ind Univ, 50-61; PROF RATIONAL MECH, JOHNS HOPKINS UNIV, 61- Concurrent Pos: From lectr to assoc prof, Univ Md, 46-50; consult, Naval Res Lab, 51-55, Nat Bur Standards, 59-62, Sandia Corp, 66, Ga Inst Technol, 73- & US Nuclear Regulatory Comn, 75-; ed, J Rational Mech & Anal, 52-56 & Arch Hist Exact Sci, 60-; Guggenheim fel, 57; ed, Arch Rational Mech & Anal, 57-66, co-ed, 67-; co-ed, Ergebnisse der Angewandten Math, 57-62; NSF sr res fel, Univ Bologna & Univ Basel, 60-61; ed, Springer Tracts Natural Philos, 62-66, co-ed, 67-; Walker-Ames prof, Univ Wash, 64; distinguished vis prof, Syracuse Univ, 65; 75th Anniversary lectr, Drexel Inst Technol, 66-67; Lincean prof, Ital Acad Lincei, Rome, 70, 73 & 74; lectr, Fed Univ Rio de Janeiro, 72. Honors & Awards: Bingham Medal, Soc Rheol, 63; Panetti Int Medal & Prize, Acad Sci Turin, 67. Mem: Soc Natural Philos (secy, 63-65 & 70-71, chmn, 67-68); Int Acad Hist Sci; hon mem, Ital Acad Sci; hon mem, Int Acad Philos Sci; Ital Math Soc. Res: Rational mechanics. Mailing Add: 119 La Trobe Johns Hopkins Univ Baltimore MD 21218

TRUESDELL, SUSAN JANE, b Oak Park, Ill, Mar 22, 45. MOLECULAR BIOLOGY. Educ: Mich State Univ, BS, 67; Univ Calif, Los Angeles, PhD(molecular biol), 71. Prof Exp: Am Cancer Soc fel, Univ Mich, 71-73; MICROBIOLOGIST MOLECULAR BIOL, PFIZER, INC, 73- Mem: Am Soc Microbiol; AAAS. Res: Genetics and physiology of penicillin production; viruses that infect Penicillium chrysogenum. Mailing Add: Fermentation Res & Develop Pfizer Inc Gordon CT 06340

TRUETT, WILLIAM LAWRENCE, b Nashville, Tenn, July 10, 22; m 61; c 4. CHEMISTRY. Educ: Emory Univ, BS, 43, MS, 47; Univ Va, PhD, 50. Prof Exp: Res chemist, Union Carbide Corp, 52-53; sr res chemist, E I du Pont de Nemours & Co, 53-66; res mgr, Wilks Sci Corp, 66-67; prod mgr, Picker Nuclear, NY, 67-69; staff chemist, Fabrics & Finishes Dept, Marshall Lab, E I du Pont de Nemours & Co, Inc, 69-76; PROD MGR, WILKS CORP, 76- Mem: Am Chem Soc. Res: Coordination polymerization of olefins; infrared spectroscopy; nuclear quadrupole resonance. Mailing Add: Wilks Corp 140 Water St Box 449 South Norwalk CT 06856

TRUEX, RAYMOND CARL, b Norfolk, Nebr, Dec 11, 11; m 38; c 2. ANATOMY. Educ: Nebr Wesleyan Univ, AB, 34; St Louis Univ, MS, 36; Univ Minn, PhD(anat), 39. Prof Exp: From instr to assoc prof anat, Col Physicians & Surgeons, Columbia Univ, 38-48; prof & head div, Hahnemann Med Col, 48-61; PROF ANAT, SCH MED, TEMPLE UNIV, 61- Concurrent Pos: USPHS award prof, 61-; consult, NIH, 62-66 & Nat Bd Med Examrs, 67-71. Honors & Awards: AMA Awards, 52 & 58. Mem: AAAS; Geront Soc; Am Asn Anatomists (pres, 71); Harvey Soc; Am Vet Med Asn. Res: Histological changes with age and pathology of the human nervous system;

histology and physiology of the conduction system and circulation of the heart. Mailing Add: Dept of Anat Temple Univ Sch of Med Philadelphia PA 19140

TRUEX, TIMOTHY JAY, b Goshen, Ind, June 11, 45. INORGANIC CHEMISTRY. Educ: Hanover Col, BS, 67; Mass Inst Technol, PhD(inorg chem), 72. Prof Exp: RES SCIENTIST INORG CHEM, FORD MOTOR CO, 72- Mem: Am Chem Soc. Res: Transition metal organometallic chemistry; catalysis chemistry; chemistry of surface coatings. Mailing Add: Sci Res Staff Ford Motor Co PO Box 2053 Dearborn MI 48121

TRUFANT, SAMUEL ADAMS, b New Orleans, La, May 24, 19; m 45; c 4. MEDICINE, NEUROLOGY. Educ: Tulane Univ, BS, 40, MD, 43; Am Bd Psychiat & Neurol, dipl, 51. Prof Exp: Asst gross anat, Tulane Univ, 40-41; Rockefeller fel neurol, Washington Univ, 47-49, USPHS res fel, 49-50; from asst prof to assoc prof, 50-62, from asst dean to assoc dean, Col Med, 51-62, PROF NEUROL, COL MED, UNIV CINCINNATI, 62- Concurrent Pos: Consult, Wright-Patterson AFB, Ohio, 52-; dir pediat neurol, Children's Hosp, 59-71; dir, Am Bd Psychiat & Neurol, 66-73, vpres, 71, pres, 72-; proj officer, India Neurol & Sensory Dis Serv Prog, USPHS, 66-; mem, Residency Rev Comt Psychiat & Neurol, 67-72, chmn, 70-72; ed, Trans, Am Neurol Asn, 68-73. Mem: Am Neurol Asn (vpres, 65, secy-treas, 68-73, pres, 75); Asn Res Nerv & Ment Dis; Asn Am Med Cols (asst secy, 59-64); Am Acad Neurol. Res: Clinical neurology; electroencephalography. Mailing Add: Dept of Neurol Univ of Cincinnati Col of Med Cincinnati OH 45267

TRUHLAR, DONALD GENE, b Chicago, Ill, Feb 27, 44; m 65. PHYSICAL CHEMISTRY, MOLECULAR PHYSICS. Educ: St Mary's Col, Minn, BA, 65; Calif Inst Technol, PhD(chem), 70. Prof Exp: Student aide chem, Argonne Nat Lab, 65; asst prof, 69-72, ASSOC PROF CHEM, UNIV MINN, MINNEAPOLIS, 72- Concurrent Pos: Sr vis fel, Battelle Mem Inst, Columbus & Sloan res fel, 73; vis fel, Joint Inst Lab Astrophysics, Boulder, Colo, 75-76. Mem: Am Phys Soc; Am Chem Soc. Res: Theory and computations for collision processes involving atoms, molecules and electrons; potential energy surfaces for triatomic systems; theory of molecular spectroscopy. Mailing Add: Dept of Chem Univ of Minn Minneapolis MN 55455

TRUITT, B PRICE, b Gainesville, Tex, Oct 30, 19; m 39; c 2. ORGANIC CHEMISTRY. Educ: NTex State Col, BS, 41, MS, 42; Univ Tex, PhD(chem), 44. Prof Exp: Anal chemist, NTex State Col, 41-42; instr chem, Univ Tex, 43-44; res & develop chemist, Gen Aniline Corp, NJ, 44-45; from asst prof to assoc prof, 45-51, PROF CHEM, N TEX STATE UNIV, 51- Mem: Am Chem Soc. Res: Organic reactions; pharmaceutical and dye chemistry. Mailing Add: Dept Chem NTex State Univ NT Sta Box 13707 Denton TX 76203

TRUITT, EDWARD BYRD, JR, b Norfolk, Va, Aug 23, 22; m 49; c 2. PHARMACOLOGY. Educ: Med Col Va, BS, 43; Univ Md, PhD(pharmacol), 50. Prof Exp: Asst prof pharmacol, Bowman Gray Sch Med, Wake Forest Col, 50-55; from assoc prof to prof, Sch Med, Univ Md, Baltimore City, 55-67; sr res fel, Columbus Labs, Battelle Mem Inst & prof pharmacol, Col Med, Ohio State Univ, 66-72; RES PROF PHARMACOL, SCH MED, GEORGE WASHINGTON UNIV, 72- Concurrent Pos: Robins Co fel, Bowman Gray Sch Med, Wake Forest Col, 50-55. Mem: Am Chem Soc; Am Soc Pharmacol & Exp Therapeut; Soc Exp Biol & Med; NY Acad Sci. Res: Neuropharmacology; psychopharmacology; drug metabolism; alcoholism; marijuana and drug abuse research. Mailing Add: Sch of Med Dept of Pharmacol George Washington Univ Med Ctr Washington DC 20037

TRUITT, ROBERT LINDELL, b Carbondale, Ill, July 26, 46; m 67; c 4. MICROBIOLOGY, TRANSPLANTATION IMMUNOLOGY. Educ: Southern Ill Univ, Carbondale, BA, 68, PhD(microbiol), 73. Prof Exp: Fel germfree syst, Lobund Lab, Univ Notre Dame, 72-74; RES ASSOC TUMOR IMMUNOL, WINTER RES LAB, MT SINAI MED CTR, 74- Concurrent Pos: Fel, United Cancer Coun, 72-73 & Damon Runyon Mem Fund Cancer Res, 73-75; NIH/Nat Cancer Inst res grant, 75. Mem: Sigma Xi; Am Soc Microbiol; Int Soc Exp Hemat. Res: Germfree animal systems; tumor immunology; bone marrow transplantation; virology. Mailing Add: Winter Res Lab Mt Sinai Med Ctr 950 N 12th St Milwaukee WI 53233

TRUJILLO, EDUARDO E, b Horconcitos, Panama, Apr 22, 30. PLANT PATHOLOGY. Educ: Univ Ark, BSA, 56, MS, 57; Univ Calif, PhD(plant path), 62. Prof Exp: Res asst plant path, Univ Calif, 57-62; asst plant pathologist, 62-65, asst prof plant path & asst specialist, 65-67, assoc prof, 67-74, ASSOC PROF PLANT PATH, 74-, ASSOC PLANT PATHOLOGIST, UNIV HAWAII, MANOA, 67- Concurrent Pos: Consult, Pac Southwest Forest & Range Exp Sta, US Forest Serv, 63- Mem: AAAS; Am Phys Soc; Am Soc Hort Sci. Res: Aspects of research dealing with soil borne pathogens, mainly ecology and epidemiology of Pythium and Phytophthoras in tropical environments; biology of Fusarium species. Mailing Add: Dept of Plant Path Univ of Hawaii at Manoa Honolulu HI 96822

TRUJILLO, PHILLIP M, b Chimayo, NMex, Feb 5, 21; m 45; c 3. AGRONOMY. Educ: NMex State Univ, BS, 47; Univ Md, MS, 51. Prof Exp: County exten agent, NMex State Univ, 47-48; instr farm training prog, State Dept Voc Agr, NMex, 48-49; asst supvr, Farmers Home Admin, USDA, 51-52; ASSOC PROF AGRON & SUPT, ESPANOLA VALLEY BR STA, N MEX STATE UNIV, 52- Mem: Am Soc Agron. Res: Improvement of agronomic crops. Mailing Add: Espanola Valley Br Sta NMex State Univ Dept Agron Alcalde NM 87511

TRUJILLO, RALPH EUSEBIO, b Embudo, NMex, Sept 22, 40; m 70; c 4. BIOCHEMISTRY. Educ: Univ NMex, BS, 62; Ind Univ, PhD(biochem), 67. Prof Exp: Mem Peace Corps, Ecuador, 62-64; USPHS fel biochem, Univ Tex M D Anderson Hosp & Tumor Inst, 67-69; MEM TECH STAFF, SANDIA LABS, 69- Res: Response of biological systems to thermal, chemical and radiation environments. Mailing Add: Div 5811 Sandia Labs Albuquerque NM 87115

TRULIO, JOHN GEORGE, b Brooklyn, NY, Feb 28, 29; c 3. CONTINUUM MECHANICS, PHYSICS. Educ: Harvard Univ, BA, 49; Columbia Univ, MS, 50, PhD(chem), 54. Prof Exp: Physicist, Lawrence Radiation Lab, 55-62; dir res, Northrop Ventura, 62-65; PRES, APPL THEORY, INC, 65- Concurrent Pos: Consult, Boeing Co, 55-62. Mem: AAAS; Am Phys Soc; Am Ord Asn; Am Geophys Union. Res: Classical physics; development and application of numerical and analytical techniques in the fields of continuum mechanics, electromagnetic wave propagation and neutronics. Mailing Add: Appl Theory Inc 1010 Westwood Blvd Los Angeles CA 90024

TRUM, BERNARD FRANCIS, b Natick, Mass, Dec 10, 09; m 36; c 4. VETERINARY MEDICINE. Educ: Boston Col, AB, 31; Cornell Univ, DVM, 35. Prof Exp: Prof zootechnol, Univ Mayor de San Simon, Bolivia, 49-50 & Univ Tenn, 51-56; vet, US AEC, 56-58; LECTR & DIR ANIMAL RES CTR, HARVARD MED SCH, 58-; DIR, NEW ENG REGIONAL PRIMATE RES CTR, 62- Concurrent Pos: Mem, Nat Coun Radiation Protection & Measurement. Mem: Am Asn Lab Animal Sci (pres, 65); Radiation Res Soc; Soc Exp Biol & Med; NY Acad Sci. Res: Effects of total body radiation; zootechnics of domestic and laboratory animals. Mailing Add: Washington St Sherborn MA 01770

TRUMAN, JAMES WILLIAM, b Akron, Ohio, Feb 5, 45; m 70. INSECT PHYSIOLOGY. Educ: Univ Notre Dame, BS, 67; Harvard Univ, MA, 69, PhD(biol), 70. Prof Exp: Harvard Soc Fels jr fel, Harvard Univ, 70-73; ASST PROF ZOOL, UNIV WASH, 73- Honors & Awards: Newcomb Cleveland Prize, AAAS, 70. Mem: Entom Soc Am; Am Soc Zoologists; Soc Gen Physiol. Res: Physiological aspects of circadian rhythms; interaction of hormones with the insect nervous system. Mailing Add: Dept of Zool Univ of Wash Seattle WA 98195

TRUMBO, BRUCE EDWARD, b Springfield, Ill, Dec 12, 37. STATISTICS. Educ: Knox Col, Ill, AB, 59; Univ Chicago, SM, 61, PhD(statist), 65. Prof Exp: Asst prof math, San Jose State Col, 63-64; from asst prof to assoc prof, 65-72, chmn dept, 70-75, PROF STATIST, CALIF STATE UNIV, HAYWARD, 72- Concurrent Pos: Consult, 64-70; vis assoc prof, Stanford Univ, 67-69 & 71; coun fel, Acad Admin Internship Prog, Am Coun Educ, 68-69; prog dir statist res, NSF, 74-75; mem, Post-Toussas-Trumbo Statist. Mem: Am Statist Asn; Inst Math Statist. Res: Application of statistical methods to social, behavioral and biological sciences; probability. Mailing Add: Dept of Statist Calif State Univ Hayward CA 94542

TRUMBORE, CONRAD NOBLE, b Denver, Colo, Feb 17, 31; m 55; c 2. PHYSICAL CHEMISTRY. Educ: Dickinson Col, BS, 52; Pa State Univ, PhD(chem), 55. Prof Exp: Fulbright grant, Inst Nuclear Res, Netherlands, 55-56; asst scientist, Argonne Nat Lab, 56-57; instr chem, Univ Rochester, 57-60; asst prof, 60-66, ASSOC PROF CHEM, UNIV DEL, 66- Concurrent Pos: USPHS spec fel, Inst Cancer Res, Sutton, Eng, 67-68. Mem: AAAS; Radiation Res Soc; Am Chem Soc. Res: Primary chemical processes in radiation chemistry of liquids; correlations between photochemistry and radiation chemistry; biological radiation chemistry; pulse radiolysis and flash photolysis. Mailing Add: 113 Dallas Ave Newark DE 19711

TRUMBORE, FORREST ALLEN, b Denver, Colo, Dec 28, 27; m 51; c 2. PHYSICAL CHEMISTRY. Educ: Dickinson Col, BS, 46; Univ Pittsburgh, PhD(chem), 50. Prof Exp: Aeronaut res scientist thermodyn alloys, Lewis Flight Propulsion Lab, Nat Adv Comt Aeronaut, 50-52; MEM TECH STAFF, BELL LABS, INC, 52- Mem: AAAS; Electrochem Soc. Res: Solubilities and electrical properties of impurities in semiconductors; crystal growth; photoluminescence and electroluminescence in semiconductors; battery materials. Mailing Add: 30 Glen Oaks Ave Summit NJ 07901

TRUMBORE, ROGER H, b Wilmington, Del, May 24, 34; m 55; c 3. PHYSIOLOGY. Educ: Univ Wis, BS, 55; Univ Md, PhD(physiol), 59. Prof Exp: Instr zool, Univ Md, 57-58; asst prof biol, Franklin & Marshall Col, 59-60; asst prof, Lawrence Univ, 60-63; ASSOC PROF BIOL & HEAD DEPT BIOL SCI, STATE UNIV NY BINGHAMTON, 63- Concurrent Pos: NIH fel physiol & biophys, Univ Ill, 63-64. Mem: AAAS; Am Soc Zool; Am Inst Biol Sci. Res: Cell physiology; metabolic effects of physiological and environmental shifts. Mailing Add: Dept of Biol Sci State Univ of NY Binghamton NY 13901

TRUMBULL, ELMER ROY, JR, b Lawrence, Mass, Apr 5, 24; m 54; c 3. ORGANIC CHEMISTRY. Educ: Dartmouth Col, AB, 44; Univ Ill, PhD(org chem), 47. Prof Exp: Asst chem, Univ Ill, 44-46; res assoc, Mass Inst Technol, 47-48, Du Pont fel, 51-52; instr, Tufts Col, 48-51; asst prof chem, Brown Univ, 52-58; from asst prof to assoc prof, 58-63, dir div natural sci & math, 64-70, PROF CHEM, COLGATE UNIV, 63-, CHMN DEPT, 70- Concurrent Pos: NSF fac fel, Univ Ariz, 66-67. Mem: Am Chem Soc; The Chem Soc. Res: Elimination reactions. Mailing Add: Dept of Chem Colgate Univ Hamilton NY 13346

TRUMBULL, RICHARD, b Johnstown, NY, Apr 6, 16; m 39; c 4. PSYCHOPHYSIOLOGY. Educ: Union Col, NY, AB, 37; Union Univ, NY, MS, 39; Syracuse Univ, PhD(psychol), 51. Prof Exp: Asst prof psychol, Green Mountain Jr Col, 39-41; lectr, Syracuse Univ, 41-43; aviation psychologist, US Navy, 43-46; chmn psychol, Green Mountain Jr Col, 46-49; chmn undergrad prog psychol, Syracuse Univ, 49-51; mem staff res, Sch Aviation Med, US Navy, 51-53; from asst head to head physiol psychol br, Off Naval Res, 53-61; dir psychol sci, 61-67; res dir, 67-68; dir res, 68-70; dep exec dir, AAAS, 70-74; EXEC DIR, AM INST BIOL SCI, ARLINGTON, VA, 74- Honors & Awards: Distinguished Civilian Serv Award, US Navy, 61, Sustained Super Accomplishment Award, US Navy, 66; Longacre Award, Aerospace Med Asn, 66. Mem: Am Inst Biol Sci; AAAS; Aerospace Med Asn (vpres, 60-64). Res: Research administration; translation and utilization; structure for management of research and development and development of policy and procedure. Mailing Add: 4708 N Chelsea Lane Bethesda MD 20014

TRUMMER, MAX JOSEPH, b Bogota, Colombia, Aug 12, 24; US citizen; m 45; c 1. THORACIC SURGERY. Educ: Univ Ill, MD, 48; Univ Pa, MS, 65; Am Bd Surg, dipl, 58; Bd Thoracic Surg, dipl, 61. Prof Exp: Resident thoracic surg, US Naval Hosp, Med Corps, US Navy, 58-60, chief thoracic surgeon, San Diego, 67-70; chief thoracic and cardiac surg, Los Angeles County-Olive View Med Ctr, 70-71; DIR SURG TEACHING PROG, MERCY HOSP & MED CTR, SAN DIEGO, 71- Mem: Fel Am Col Surgeons; fel Am Col Chest Physicians; Soc Thoracic Surg; Am Asn Thoracic Surg; Am Thoracic Soc. Res: Lung transplantation; open-heart surgery; cardiopulmonary physiology. Mailing Add: Mercy Hosp & Med Ctr 4077 Fifth Ave San Diego CA 92103

TRUMP, BENJAMIN FRANKLIN, b Kansas City, Mo, July 23, 32; m 61; c 2. PATHOLOGY, CELL BIOLOGY. Educ: Univ Mo-Kansas City, BA, 53; Univ Kans, MD, 57. Prof Exp: Intern path, Med Ctr, Univ Kans, 57-58, resident, 58-59; resident anat, Sch Med, Univ Wash, 59-60, resident-trainee, 60-61, investr exp path, Armed Forces Inst Path, 61-63; asst prof path, Sch Med, Univ Wash, 63-65; from assoc prof to prof, Med Ctr, Duke Univ, 65-70; PROF PATH & CHMN DEPT, SCH MED, UNIV MD, BALTIMORE CITY, 70- Concurrent Pos: Fel, Med Ctr, Univ Kans, 58-59; US Food & Drug Admin fel, Univ Assoc for Res & Educ in Path; NIH fel & training grant, Univ Md; mem bd dirs, Univ Assoc for Res & Educ in Path, 70-; mem, Md Post Mortem Exam Comn, 70-; consult, US Food & Drug Admin, 71- & Vet Admin Hosp, Baltimore, Md, 72-; docent, Dept Cell Biol, Univ Jyvaskyla, Finland, 73- Mem: AAAS; Am Asn Path & Bact; Am Soc Exp Path; Am Soc Cell Biol; Am Soc Microbiol. Res: Cellular and subcellular pathology; membrane structure and functions; lysosome structure and function; chemical carcinogenesis; kidney pathophysiology; fish physiology and pathology; environmental pathology. Mailing Add: Dept of Path Univ of Md Sch of Med Baltimore MD 21201

TRUMPLER, DONALD ALASTAIR, mathematics, see 12th edition

TRUMPOWER, BERNARD LEE, b Chambersburg, Pa, July 20, 43; m 64. BIOCHEMISTRY. Educ: Univ Pittsburgh, BS, 65; St Louis Univ, PhD(biochem), 69. Prof Exp: ASST PROF BIOCHEM, DARTMOUTH MED SCH, 72- Concurrent Pos: NIH fel, Cornell Univ, 69-71. Mem: AAAS. Res: Bioenergetics; membrane structure. Mailing Add: Dept of Biochem Dartmouth Med Sch Hanover NH 03755

TRUNNELL, JACK B, b Milledgeville, Ill, Oct 21, 18; m 42; c 6. MEDICINE. Educ: Brigham Young Univ, BA, 42; Univ Utah, MD, 45. Prof Exp: Asst, Sloan-Kettering Inst, 48-50; from asst prof to assoc prof med, Post-Grad Sch Med, Univ Tex, 50-58, head exp med, M D Anderson Hosp & Tumor Clin, 50-58; prof develop biol, Brigham Young Univ, 58-68, dean col family living, 58-61, dir ctr cell res, 61-68; CHIEF MED, GULF COAST HOSP & CLIN, 72- Concurrent Pos: Fel med, Sloan-Kettering Inst, 46-48; fel, Mem Hosp, New York, 47-50; asst res, Mem Hosp, New York, 46-47; lectr, US Naval Med Ctr, 46-50; consult, Brookhaven Nat Labs, 48-50; instr, Med Sch, Cornell Univ, 49-50; mem med staff, New York Hosp, 49-50; pvt pract, 69-; chief med, Baytown Med Ctr Hosp, 72. Mem: Soc Nuclear Med; Endocrine Soc; AMA; Am Thyroid Asn; Acad Psychosom Med. Res: Cancer; endocrinology; radioisotopes; psychosomatics; nutrition. Mailing Add: Gulf Coast Clin 2800 Garth Rd Baytown TX 77520

TRUONG, XUAN THOAI, b Saigon, French Cochin-Chine, Nov 17, 30; US citizen. PHYSICAL MEDICINE & REHABILITATION, PHYSIOLOGY. Educ: West Liberty State Col, BS, 52; Columbia Univ, MD, 56; Univ Louisville, PhD(physiol), 64; Am Bd Phys Med & Rehab, dipl, 65. Prof Exp: Intern surg, Ind Univ Med Ctr, 56-57; resident phys med & rehab, Univ Louisville Hosps, 58-61, instr, Sch Med, Univ, 62-64; clin & res consult, Inst Phys Med & Rehab, 64-68; asst prof, Baylor Col Med, 68-71; DIR RES & EDUC, INST PHYS MED & REHAB, 72- Concurrent Pos: Clin investr, Vet Admin Hosp, Houston, 69-71; asst prof, Peoria Sch Med, Univ Ill, 72- Mem: Am Physiol Soc; Am Acad Phys Med & Rehab; Inst Elec & Electronics Engrs; Am Asn Electromyog & Electrodiag. Res: Mechanical properties of muscle tissue; electrophysiology of neuro-muscular system; orthotic and prosthetic devices for neuro-muscular disabilities. Mailing Add: Inst of Phys Med & Rehab 619 NE Glen Oak Ave Peoria IL 61603

TRUPIN, JOEL SUNRISE, b Brooklyn, NY, Mar 15, 34; m 57; c 1. BIOCHEMISTRY, NUTRITION. Educ: Cornell Univ, BS, 54, MNS, 56; Univ Ill, PhD(biochem), 63. Prof Exp: Asst prof microbiol, Sch Med, St Louis Univ, 66-71; ASSOC PROF GENETICS & MOLECULAR MED, MEHARRY MED COL, 71- Concurrent Pos: Am Cancer Soc res fel biochem genetics, Nat Heart Inst, 63-66. Mem: AAAS; Am Chem Soc; Am Soc Microbiol. Res: Biochemistry and regulation of amino acid biosynthesis; transfer RNA; protein and amino acid nutrition. Mailing Add: Grad Studies Meharry Med Col Nashville TN 37208

TRUPP, CLYDE RULON, b St Anthony, Idaho, May 14, 41; m 64; c 2. PLANT BREEDING. Educ: Univ Idaho, BS, 63, MS, 65; Iowa State Univ, PhD(plant breeding), 69. Prof Exp: Asst prof hybrid wheat breeding, Mich State Univ, 69-75; PLANT BREEDER SUGAR BEETS, AMALGAMATED SUGAR CO, 75- Mem: Am Soc Agron; Crop Sci Soc Am; Sigma Xi. Res: Breeding improved hybrid varieties of sugar beets for commercial culture and processing. Mailing Add: Beet Seed Develop Amalgamated Sugar Co Nyssa OR 97913

TRURAN, JAMES WELLINGTON, JR, b Mt Kisco, NY, July 12, 40; m 65; c 3. ASTROPHYSICS. Educ: Cornell Univ, BA, 61; Yale Univ, MS, 63, PhD(physics), 65. Prof Exp: Resident res assoc, Goddard Inst Space Studies, NASA, NY, 65-67; res fel physics, Calif Inst Technol, 68-69; from assoc prof to prof physics, Belfer Grad Sch Sci, 70-73; PROF ASTRON, UNIV ILL, URBANA-CHAMPAIGN, 73- Concurrent Pos: Mem bd contribr, Comments Astrophys & Space Physics, 73-74; ed, Physics Letters B, 74- Mem: Am Phys Soc; Am Astron Soc. Res: Nucleosynthesis; nuclear reactions in stars; mechanisms of nova and supernova explosions; stellar evolution; galactic evolution; origin of cosmic rays; white dwarfs; binary evolution. Mailing Add: Dept of Astron Univ of Ill Urbana IL 61801

TRUSCOTT, BASIL LIONEL, b Chambers, Nebr, Aug 4, 16; m 48. BIOLOGY, ANATOMY. Educ: Drew Univ, BA, 39; Syracuse Univ, MA, 40; Yale Univ, MS, 42, PhD(exp embryol), 43, MD, 50; Am Bd Neurol, dipl, 59. Prof Exp: Instr anat, Georgetown Univ, 43-45; instr biol, Yale Univ, 46, instr anat, 47-51; asst prof, Sch Med, Univ NC, 51-54; from assoc prof to prof neurol, Albany Med Col, 60-68; PROF NEUROL, BOWMAN GRAY SCH MED, 68-, ASST DEAN STUDENT ADMIS, 73- Concurrent Pos: Chief neurol sect, Vet Admin Hosp, Albany, NY, 60-68; dir comprehensive stroke prog, NC Regional Med Prog, 68-73. Mem: Am Asn Anat; Am Acad Neurol; fel Am Col Physicians. Res: Hypophysial-gonadal interaction; vitamin A physiology and steroid interrelationships; epidemiology of stroke; physiology; pathology; zoology; chemistry; neurology. Mailing Add: 460 Briarlea Rd Winston-Salem NC 27104

TRUSCOTT, FREDERICK HERBERT, b Meredith, NY, Mar 16, 26; m 54; c 2. PLANT PHYSIOLOGY. Educ: State Univ NY Albany, AB, 50; Rutgers Univ, PhD(bot), 55. Prof Exp: Asst bot, Rutgers Univ, 52-55; res fel, Jackson Mem Lab, Bar Harbor, Maine, 55-56; instr bot, Univ RI, 56-58; from asst prof to assoc prof, 58-65, PROF BIOL, STATE UNIV NY ALBANY, 65-, CHMN DEPT, 72- Mem: Bot Soc Am; Am Soc Plant Physiologists; Am Inst Biol Sci. Res: Morphogenesis. Mailing Add: Dept of Biol State Univ of NY Albany NY 12222

TRUSCOTT, ROBERT BRUCE, b Winnipeg, Man, July 9, 28; m 53; c 5. VETERINARY MICROBIOLOGY. Educ: Univ Toronto, BSA, 50, MSA, 53, DVM, 62; Univ Waterloo, PhD(microbiol physiol), 66. Prof Exp: Supvr, Animal House, Univ Western Ont, 50; fermentation supvr, Merck & Co Ltd, Que, 50-51; res asst microbiol, Ont Agr Col, Guelph, 51-53; LECTR POULTRY PATH, ONT VET COL, UNIV GUELPH, 53-, ASSOC PROF VET MICROBIOL, 69- Concurrent Pos: Consult, Shaver Poultry Breeding Farms Ltd, Ont, 68- Mem: Can Soc Microbiol; Am Soc Microbiol; Poultry Sci Asn; Am Asn Avian Path; Can Vet Med Asn. Res: Avian pathology; avian and bovine mycoplasmas, detection, pathology and control. Mailing Add: Dept of Vet Microbiol & Immunol Univ of Guelph Guelph ON Can

TRUSELL, FRED CHARLES, b Kansas City, Mo, May 12, 31; m 57; c 2. ANALYTICAL CHEMISTRY. Educ: Univ Mo, Kansas City, BA, 52, BS, 56; Iowa State Univ, MS, 59, PhD(chem), 61. Prof Exp: Asst prof chem, Tex Tech Col, 61-64; RES CHEMIST, DENVER RES CTR, MARATHON OIL CO, 64- Mem: Am Chem Soc; Am Inst Chemists. Res: Composition and analysis of crude oils. Mailing Add: Denver Res Ctr Marathon Oil Co PO Box 269 Littleton CO 80120

TRUSK, AMBROSE, b DePue, Ill, Sept 16, 21. ANALYTICAL CHEMISTRY. Educ: St Mary's Col, Minn, BS, 43; Univ Minn, MA, 51; Univ Notre Dame, MS, 62, PhD(chem), 66. Prof Exp: Instr physics, St Mary's Col, Minn, 43-45; teacher high sch, Minn, 45-47 & Mo, 47-49; from instr to assoc prof, 49-67, PROF CHEM & CHMN DEPT, ST MARY'S COL, MINN, 67- Mem: Am Chem Soc. Res: Fluorometric methods of analysis; structure of inorganic compounds. Mailing Add: Dept of Chem St Mary's Col Winona MN 55987

TRUSSELL, MARGARET EDITH, b Alameda, Calif, Jan 26, 28. GEOGRAPHY. Educ: Univ Calif, Berkeley, AB, 49; Long Beach State Col, MA, 57; Univ Calif, Berkeley, MA, 60; Univ Ore, PhD(geog), 69. Prof Exp: Teacher social studies, Calif High Schs, 53-57; instr econ & US hist, San Jose City Col, 58; teaching asst geog, Univ Calif, Berkeley, 60-61; asst prof soc sci, Southwestern Ore Col, 61-66; from asst prof to assoc prof, 66-74, PROF GEOG, CALIF STATE UNIV, CHICO, 74- Mem: AAAS; Soc Woman Geogrs; Asn Am Geogrs; Am Geog Soc; Soc Econ Bot. Res: Pioneer agriculture and agricultural development, including origins and diffusion of land choice decisions and agricultural practices; distribution and human uses of native plants; man's modification of the environment. Mailing Add: Dept of Geog Calif State Univ Chico CA 95929

TRUSSELL, PAUL CHANDOS, b Vancouver, BC, July 4, 16; m 43; c 2. BACTERIOLOGY. Educ: Univ BC, BSA, 38; Univ Wis, MS, 42, PhD(agr bact), 43. Prof Exp: Chief res microbiologist, Ayerst, McKenna & Harrison, Ltd, Que, 44-47; head div appl biol, 47-61, DIR, BC RES COUN, 61- Concurrent Pos: Secy-gen, World Asn Indust & Technol Res Insts; dir, Fisheries Res Bd Can. Mem: Am Soc Microbiol; Inst Food Technologists. Res: Agricultural bacteriology; industrial fermentations; antibiotics; marine borer control; food spoilage; water pollution; bacteriological leaching of ores. Mailing Add: BC Research Coun 3650 Wesbrook Crescent Vancouver BC Can

TRUSSELL, RAY, b Toledo, Iowa, Feb 7, 14. MEDICINE, EPIDEMIOLOGY. Educ: Univ Iowa, BA, 35, MD, 41; Johns Hopkins Univ, MPH, 47. Prof Exp: Lab instr bact, Univ Iowa, 35-37, asst obstet & gynec, 36-41, assoc prev med, 42-43, instr, 46; asst prof, Albany Med Col, Union, NY, 47-48, prof, 48-50; clin prof, NY Univ, 51-55; De Lamar prof admin med, assoc dean pub health & exec officer, Sch Pub Health & Admin Med, Columbia Univ, 55-68; PROF ADMIN MED, MT SINAI SCH MED & GEN DIR, BETH ISRAEL MED CTR, NEW YORK, 68- Concurrent Pos: Epidemiologist, State Dept Health, NY, 47-48; dir, Hunterdon Med Ctr, Flemington, NJ, 50-55; comnr hosps, New York, 61-65. Mem: AMA; fel Am Pub Health Asn; fel NY Acad Med. Res: Trichomoniasis; epidemiology of various communicable diseases; rural chronic disease problems; medical administration. Mailing Add: Beth Israel Med Ctr 10 Nathan D Perlman Pl New York NY 10003

TRUST, TREVOR JOHN, b Melbourne, Australia, June 24, 42; m 69. MICROBIOLOGY. Educ: Univ Melbourne, BSc, 64, MSc, 66, PhD(microbiol), 69. Prof Exp: Lectr microbiol, Royal Melbourne Inst Technol, 69; asst prof, 69-73, ASSOC PROF BACT, UNIV VICTORIA, BC, 73-, CHMN DEPT BACT & BIOCHEM, 75- Mem: Can Soc Microbiol; Am Soc Microbiol; Brit Soc Gen Microbiol; Australian Soc Microbiol; Soc Appl Microbiol. Res: Bacteriology of fish and wood; antimicrobial agents; microbiology of fish and birds; environmental sources of pathogens. Mailing Add: Dept of Bact & Biochem Univ of Victoria Victoria BC Can

TRUSTY, JOSANN WATKINS, b Pittsburgh, Pa, May 22, 44; m 66. SURFACE PHYSICS. Educ: Ohio State Univ, BS, 66, MS, 67, PhD(physics), 70. Prof Exp: Lectr physics, Ohio State Univ, 70-72, vis asst prof, 72-73; res physicist, E I du Pont de Nemours & Co, 73-75; RES SCIENTIST, OWENS CORNING FIBERGLAS CORP, 76- Mem: Am Asn Physics Teachers; Am Phys Soc. Mailing Add: Owens Corning Tech Ctr Granville OH 43023

TRUTT, DAVID, b New York, NY, Mar 21, 38; m 65; c 2. MATHEMATICS. Educ: Lafayette Col, BS, 59; Brown Univ, MS, 62; Purdue Univ, PhD(math), 64. Prof Exp: Instr math, Purdue Univ, 64-65; asst prof, Lehigh Univ, 65-68; asst prof, Univ Va, 68-70; asst prof, 70-71, ASSOC PROF MATH, LEHIGH UNIV, 71- Mem: Am Math Soc. Res: Hilbert space operator theory; spaces of analytic functions. Mailing Add: Dept of Math Lehigh Univ Bethlehem PA 18015

TRUXAL, FRED STONE, b Great Bend, Kans, Feb 20, 22; m 43; c 2. ENTOMOLOGY. Educ: Univ Kans, AB, 47, MA, 49, PhD(entom), 52. Prof Exp: Agent, Bur Entom & Plant Quarantine, USDA, 42; asst instr biol & entom, Univ Kans, 47-52; cur entom, 52-61, CHIEF CUR LIFE SCI DIV, LOS ANGELES COUNTY MUS NATURAL HIST, 61- Concurrent Pos: Asst to state entomologist, Kans, 48-51; asst prof, Ottawa Univ, Kans, 51-52; adj prof, Univ Southern Calif, 57-; biol consult, Pac Horizons; mem expeds, Mex, Cent Am, Brazil & Peru. Mem: Fel AAAS; Entom Soc Am; Soc Study Evolution; Soc Syst Zool. Res: Biology, ecology and taxonomy of aquatic and semiaquatic Hemiptera. Mailing Add: Life Sci Div Los Angeles Co Mus of Nat Hist Los Angeles CA 90007

TRUXILLO, STANTON GEORGE, b New Orleans, La, June 23, 41; m 65; c 1. PHYSICS. Educ: Loyola Univ, La, BS, 63; La State Univ, Baton Rouge, PhD(physics), 69. Prof Exp: Fel, Coastal Studies Inst, La State Univ, Baton Rouge, 68-70; asst prof, 70-73, ASSOC PROF PHYSICS, UNIV TAMPA, 73- Mem: Am Phys Soc; Am Asn Physics Teachers. Res: Time-dependent rotating fluid dynamics; fluid dynamics of circulatory system; acoustics. Mailing Add: Dept of Physics Univ of Tampa Tampa FL 33606

TRYFIATES, GEORGE P, b Mesolongi, Greece, Feb 26, 35; US citizen; m 59; c 4. BIOCHEMISTRY, CANCER. Educ: Univ Toledo, BS, 58; Bowling Green State Univ, MA, 59; Rutgers Univ, PhD(biochem), 63. Prof Exp: Teaching & res asst biol, Bowling Green State Univ, 58-59; res asst biochem, Rutgers Univ, 59-62; res assoc, Grad Sch Med, Univ Pa & Sch Med, Temple Univ, 62-64; res biochemist, P Lorillard Co, NC, 64-66; instr pharmacol, Sch Med, Duke Univ, 66-67, assoc, 67; asst prof biochem, 67-72, ASSOC PROF BIOCHEM, SCH MED, W VA UNIV, 72- Concurrent Pos: USPHS trainee, Grad Sch Med, Univ Pa & Sch Med, Temple Univ, 62-64, USPHS grants, Nat Cancer Inst. Mem: AAAS; Am Chem Soc; Soc Exp Biol & Med; Sigma Xi; NY Acad Sci. Res: Enzyme regulation in vivo and in vitro; hormone action; molecular aspects of control of neoplasia; nutritional control of tumor growth and of expression of enzyme activity. Mailing Add: Dept of Biochem WVa Univ Med Ctr Morgantown WV 26506

TRYGGVASON, EYSTEINN, b Iceland, July 19, 24; m 54; c 3. GEOPHYSICS. Educ: Univ Oslo, Cand real, 51; Univ Uppsala, Fil lic, 61. Prof Exp: Head div geophys, Icelandic Meteorol Serv, 52-62; from asst prof to assoc prof, 62-68, PROF GEOPHYS, UNIV TULSA, 68- Mem: Am Geophys Union; Seismol Soc Am; Am Meteorol Soc; Soc Explor Geophys. Res: Seismology; structure of the earth; deformation of the earth's crust. Mailing Add: Dept of Earth Sci Univ of Tulsa Tulsa OK 74104

TRYON, CLARENCE ARCHER, b Niagara Falls, NY, Nov 15, 11; m 35; c 2. ZOOLOGY. Educ: Cornell Univ, BS, 35, PhD(mammal), 42. Prof Exp: Asst biol, State Conserv Dept, NY, 35-37; from instr to asst prof zool, Mont State Col, 37-42, assoc prof, 43-47, assoc zoologist, Exp Sta, 42-47; from asst prof to assoc prof, 47-53, PROF ZOOL & DIR PYMATUNING LAB ECOL, UNIV PITTSBURGH, 54- Mem: AAAS; Ecol Soc Am; Wildlife Soc; Am Ornith Union. Res: Ecology. Mailing Add: Dept of Biol Sci Univ of Pittsburgh Lab of Ecol Linesville PA 16424

TRYON, EARL HAVEN, b Yarmouth, Maine, Apr, 7, 13; m 40; c 2. SILVICULTURE. Educ: Univ NH, BS, 36; Ore State Col, MS, 40; Yale Univ, PhD(forestry), 45. Prof Exp: Asst technologist, Northeastern Forest Exp Sta, US Forest Serv, Conn, 36-37; jr pathologist, Div Forest Path, Univ Ore, 37-41; from asst prof to assoc prof silvicult, 45-52, PROF SILVICULT & SILVICULTURIST, AGR EXP STA, W VA UNIV, 52-

Mem: Soc Am Foresters. Res: Forest nursery disease; planting projects; stripmine planting; oak regeneration; frost damage to hardwoods; geobotany and wood properties. Mailing Add: Dept of Forestry WVa Univ Morgantown WV 26506

TRYON, EDWARD POLK, b Terre Haute, Ind, Sept 4, 40. THEORETICAL HIGH ENERGY PHYSICS, COSMOLOGY. Educ: Cornell Univ, AB, 62; Univ Calif, Berkeley, PhD(physics), 67. Prof Exp: Res assoc physics, Columbia Univ, 67-68, asst prof, 68-71; asst prof, 71-74, ASSOC PROF PHYSICS, HUNTER COL, CITY UNIV NEW YORK, 71- Mem: Am Phys Soc; NY Acad Sci; Sigma Xi. Res: Gravitational interactions; high energy theory; pion-pion interaction. Mailing Add: Dept of Physics Hunter Col 695 Park Ave New York NY 10021

TRYON, MAX, b Brentwood, Md, Mar 27, 21; m 48; c 4. INSTRUMENTATION, MATERIALS SCIENCE. Educ: Univ Md, BS, 43. Prof Exp: Chemist, 43-44 & 46-65, Polymerization Sect, 65-68, PHYS CHEMIST, BLDG RES DIV, NAT BUR STANDARDS, 68- Mem: Am Chem Soc; Nat Geog Soc. Res: Measurement of molecular weight and molecular weight distributions of polymers; structure of natural and synthetic elastomers; mechanism of degradation of polymeric substances, primarily polyhydrocarbons; analysis and identification of polymeric substances; plastics in building technology, plastic pipe, roofing, siding and coatings. Mailing Add: Building Research National Bur of Standards Washington DC 20234

TRYON, PETER VINCENT, b Ossining, NY, July 1, 41; m 64. STATISTICS. Educ: Pa State Univ, BS, 63, PhD(statist), 70; NY Univ, MS, 65. Prof Exp: Mem tech staff, Bell Tel Labs, 63-65; res asst, Pa State Univ, 65-70; STATIST CONSULT, NAT BUR STANDARDS, 70- Mem: Inst Math Statist; Am Statist Asn. Res: Mathematical statistics; applications of statistics in physical sciences. Mailing Add: 4005 Radio Bldg Nat Bur of Standards Boulder CO 80302

TRYON, ROLLA MILTON, JR, b Chicago, Ill, Aug 26, 16; m 45. BOTANY. Educ: Univ Chicago, BS, 37; Univ Wis, PhM, 38; Harvard Univ, MS, 40, PhD(bot), 41. Prof Exp: Lab technician, Chem Warfare Serv, Mass Inst Technol, 42; instr bot, Dartmouth Col, 42, lab technician, 43-44; instr bot, Univ Wis, 44-45; asst prof plant taxon, Univ Minn, 45-48; assoc prof, Wash Univ, St Louis, 48-57; CUR, HERBARIUM & CUR FERNS, GRAY HERBARIUM, HARVARD UNIV, 58-, PROF BIOL, 72- Concurrent Pos: Cur, Herbarium, Minneapolis, Minn, 46-48; asst cur herbarium, Mo Bot Garden, 48-57. Mem: Am Soc Plant Taxon; Am Fern Soc; Bot Soc Am. Res: Taxonomy of pteridophytes; Doryopteris; Pteridium; ferns and fern allies of Wisconsin, Minnesota and Peru. Mailing Add: Gray Herbarium Harvard Univ 22 Divinity Ave Cambridge MA 02138

TRYON, SAGER, b Strasburg, Ohio, Aug 7, 11; m 37; c 3. ORGANIC CHEMISTRY. Educ: Otterbein Col, BS, 34; Ohio State Univ, MS, 35, PhD(chem), 36. Prof Exp: Res chemist, Gen Chem Div, Allied Chem & Dye Corp, 36-44, tech supvr, 44-55; sr res chemist, Am Viscose Corp, 55-61 & Avisun Corp, 61-63; sr res chemist, Viscose Div, FMC Corp, 63-74; RETIRED. Mem: AAAS; Am Chem Soc. Res: Industrial and polymer chemistry. Mailing Add: 2605 Lincoln Ave Claymont DE 19703

TRYPHONAS, LEANDER, b Corfu, Greece, Mar 26, 31; Can citizen; m 63; c 1. VETERINARY PATHOLOGY, TOXICOLOGY. Educ: Univ Perugia, DVM, 56; Univ Sask, PhD(vet path), 68. Prof Exp: Pvt pract, vet med, 58-60; vet, Health Animals Br, Can Dept Agr, 60-62; pvt pract, 62-65; trainee vet path & comp neuropath, 65-69; from asst prof to assoc prof, 69-75, PROF VET PATH, WESTERN COL VET MED, UNIV SASK, 75-; RES SCIENTIST, CAN DEPT HEALTH & WELFARE, 75- Concurrent Pos: Can Med Res Coun fel, Comn Comp Neuropath, Bern, Switz, 68-69, grant, Western Col Vet Med, Univ Sask, 69-, Alta Agr Res Trust, 72-73. Mem: Am Soc Neuropath; Am Col Vet Pathologists; Europ Asn Pathologists; Can Vet Med Asn; Greek Vet Med Asn. Res: Brain poisons; pathology of alkylmercurial poisoning in pigs; lead poisoning in cattle; kernicterus in kittens; vitamin B6 poisoning in dogs; pathogenesis of lead poisoning in primates. Mailing Add: Health & Welfare Can HPB Tunney's Pasture Ottawa ON Can

TRYTTEN, GEORGE NORMAN, b Pittsburgh, Pa, Apr 21, 28; m 52; c 4. MATHEMATICS. Educ: Luther Col, Iowa, AB, 51; Univ Wis, MS, 53; Univ Md, PhD(math), 62. Prof Exp: Asst math, Univ Wis, 51-53; asst, Inst Fluid Dynamics & Appl Math, Univ Md, 53-57; mathematician, US Naval Ord Lab, 57-62; res assoc math, Inst Fluid Dynamics & Appl Math, Univ Md, College Park, 62-63, from res asst prof to res assoc prof, 63-69, assoc dean sponsored res & fels, Grad Sch, 67-69; vpres, Math Sci Group, Inc, 69-70; free-lance filmstrip producer, 70-71; prof math & chmn dept, Hood Col, 71-72; ASSOC PROF MATH, LUTHER COL, IOWA, 72- Mem: AAAS; Am Math Soc; Math Asn Am. Res: Partial differential equations; fluid dynamics; numerical solution of partial differential equations; celestial mechanics. Mailing Add: Dept of Math Luther Col Decorah IA 52101

TRYTTEN, MERRIAM HARTWICK, b Albert Lea, Minn, Jan 17, 94; m 20; c 2. PHYSICS. Educ: Luther Col, BA, 16; Univ Iowa, MS, 24; Univ Pittsburgh, PhD(physics), 28; Lutheran Col, LLD, 50; St Olaf Col, LLD, 51. Hon Degrees: DSc, Wesleyan Univ, 51, Carthage Col, 52, Drexel Inst, 56, Univ Pittsburgh, 62. Prof Exp: From instr to prof physics, Luther Col, 17-24; prof, Univ Pittsburgh, 24-41; tech aide, Off Sci Res & Develop, 41-43, specialist physics, War Manpower Comn, 43-44; dir officer sci personnel, Nat Acad Sci-Nat Res Coun, 44-67, consult to pres, Nat Acad Sci, 67-68, spec asst to pres, 68-70, CONSULT, OFF SCI PERSONNEL, NAT ACAD SCI, 70- Concurrent Pos: Consult, Eng Manpower Comn, 53-; chmn, Comt Int Exchange of Persons, 49-67; chmn, Civil Serv Adv Comn, 46-52; chmn, Selective Serv Adv Comn, 50-52; men & vchmn, Comn Human Resources; pres, Sci Manpower Comn. Mem: AAAS; fel Am Phys Soc; Am Asn Physics Teachers. Res: Magnetism; science education; paramagnetism of certain halides. Mailing Add: 9508 St Andrews Way Silver Spring MD 20901

TRYTTEN, ROLAND AAKER, b Tower City, NDak, Oct 15, 13; m 42; c 6. CHEMISTRY. Educ: St Olaf Col, AB, 35; Univ Wis, PhD(chem), 41. Prof Exp: Control chemist, Kimberly-Clark Corp, 41-42; instr chem, Ripon Col, 42-45; instr, Cent State Teachers Col, 45-51, chmn dept, 49-51; chmn dept, 51-72, PROF CHEM, UNIV WIS-STEVENS POINT, 51- Mem: Am Chem Soc. Res: Foams; foaming tendency of aqueous aliphatic alcohol solutions; sulfur dioxide determination in polluted air. Mailing Add: Dept of Chem Univ of Wis Stevens Point WI 54481

TRZCIENSKI, WALTER EDWARD, JR, b Montague City, Mass, Sept 19, 42; m 65. GEOLOGY. Educ: Bowdoin Col, AB, 65; McGill Univ, PhD(geol), 71. Prof Exp: NASA fel, Princeton Univ, 71; asst prof geol, Brooklyn Col, 71-72; ASST PROF GEOL, ECOLE POLYTECH, MONTREAL, 72- Mem: AAAS; Geol Soc Am; Mineral Soc Am; Mineral Asn Can; Geol Asn Can; Microbeam Soc. Res: Metamorphic and igneous petrology and mineralogy-geochemistry, especially in the northern Appalachians. Mailing Add: Dept of Geol Eng Ecole Polytech Montreal PQ Can

TRZECIAK, MAX JOSEPH, physical chemistry, see 12th edition

TSAGARIS, THEOFILOS JOHN, b Fernandina, Fla, June 27, 29; m 54; c 3. INTERNAL MEDICINE, CARDIOLOGY. Educ: Univ Fla, BS, 50; Emory Univ, MD, 54. Prof Exp: Chief cardiol, Vet Admin Hosp, Wood, Wis, 62-65; from asst prof to assoc prof, 65-75, PROF MED, UNIV UTAH, 75-; CHIEF CARDIOL, VET ADMIN HOSP, SALT LAKE CITY, 65- Concurrent Pos: Fel cardiol, Emory Univ, 59-60 & Univ Utah, 60-62; asst prof, Marquette Univ, 62-65. Mem: Am Fedn Clin Res. Res: Hemodynamics; coronary blood flow. Mailing Add: Vet Admin Hosp Salt Lake City UT 84112

TSAI, ALAN CHUNG-HONG, b Chang-hua Hsien, Taiwan, June 18, 43; m 69; c 1. NUTRITION. Educ: Taiwn Prov Chung Hsing Univ, BS, 66; Wash State Univ, MS, 69, PhD(nutrit), 72. Prof Exp: Res assoc nutrit, Mich State Univ, 72-73; ASST PROF NUTRIT, UNIV MICH, ANN ARBOR, 73- Res: Cholestdrol feeding associated metabolic alterations including enzyme activities; microsomal activities; tissue lipid peroxidation and insulin metabolism; nutrition of dietary fiber, its effect on cholesterol metabolism and mineral metabolism. Mailing Add: Human Nutrit Prog Sch Pub Health Univ of Mich Ann Arbor MI 48109

TSAI, BILIN PAULA, b Seattle, Wash, May 23, 49. CHEMICAL PHYSICS. Educ: Univ Chicago, BS, 71; Univ NC, Chapel Hill, PhD(chem physics), 75. Prof Exp: RES ASSOC CHEM PHYSICS, UNIV NEBR, LINCOLN, 75- Mem: Sigma Xi. Mailing Add: Dept of Chem Univ of Nebr Lincoln NE 68588

TSAI, CHESTER E, b Amoy, China, Mar 7, 35; m 62; c 2. ALGEBRA. Educ: Nat Taiwan Univ, BA, 57; Marquette Univ, MS, 61; Ill Inst Technol, PhD(math), 64. Prof Exp: Instr math, Ill Inst Technol, 62-64; asst prof, 64-68, ASSOC PROF MATH, MICH STATE UNIV, 68- Mem: Am Math Soc; Math Asn Am. Res: Non-associative algebra. Mailing Add: Dept of Math Mich State Univ East Lansing MI 48823

TSAI, CHIA-YIN, b Taichung, Taiwan, Dec 15, 37; m 67; c 2. GENETICS, BIOCHEMISTRY. Educ: Nat Taiwan Univ, BS, 60; Purdue Univ, Lafayette, PhD(genetics), 67. Prof Exp: Res asst genetics, 63-67, res assoc, 67-69, asst prof, 69-74, ASSOC PROF GENETICS, PURDUE UNIV, WEST LAFAYETTE, 74- Mem: AAAS; Genetics Soc Am; Am Soc Plant Physiologists; Crop Sci Soc Am. Res: Carbohydrate metabolism and storage protein synthesis in maize. Mailing Add: Dept of Bot & Plant Path Purdue Univ West Lafayette IN 47906

TSAI, CHISHIUN S, b Chia-yi, Taiwan, Dec 19, 33; m 65. BIOCHEMISTRY, ENZYMOLOGY. Educ: Nat Taiwan Univ, BS, 56; Purdue Univ, MS, 61, PhD(biochem), 63. Prof Exp: Fel chem, Cornell Univ, 63; fel biosci, Nat Res Coun Can, 63-64, asst res officer, 64-65; asst prof chem, 65-71, ASSOC PROF CHEM, CARLETON UNIV, 71- Mem: Am Chem Soc; Can Biochem Soc. Res: Function and reactivity of enzymes in relation with the structures of enzymes; substrates and inhibitors. Mailing Add: Dept of Chem Carleton Univ Ottawa ON Can

TSAI, DONALD HSI-NIEN, solid state physics, see 12th edition

TSAI, LIN, b Hong Kong, May 30, 22; US citizen. ORGANIC CHEMISTRY. Educ: Chinese Nat Southwest Assoc Univ, BSc, 46; Univ Ore, MA, 49; Fla State Univ, PhD(org chem), 54. Prof Exp: Res assoc chem, Ohio State Univ, 54-57; res scientist, Worcester Found Exp Biol, 57-59; vis scientist, 59-62, ORG CHEMIST, NAT HEART INST, 62- Mem: Am Chem Soc; The Chem Soc. Res: Syntheses, reactions and microbial degradations of heterocyclic compounds; stereochemistry of enzymic reactions. Mailing Add: NIH Bldg 3 Room 110 Bethesda MD 20014

TSAI, MING-JER, b Taipei, Taiwan, Nov 3, 43; m 71; c 1. BIOCHEMISTRY, MOLECULAR BIOLOGY. Educ: Nat Taiwan Univ, BS, 66; Univ Calif, Davis, PhD(biochem), 71. Prof Exp: Damon Runyon fel, Univ Tex M D Anderson Hosp & Tumor Inst Houston, 71-73; instr, 73-74, ASST PROF CELL BIOL, BAYLOR COL MED, 74- Res: Hormonal regulation of gene expression; chromatin structure; initiation of RNA synthesis by DNA-dependent RNA polymerases. Mailing Add: Dept of Cell Biol Baylor Col of Med Houston TX 77025

TSAI, MIN-SHEN CHEN, b Fukien, China. CARBOHYDRATE BIOCHEMISTRY. Educ: Nat Taiwan Univ, BS, 67; State Univ NY Buffalo, MA, 70, PhD(biochem), 72. Prof Exp: NIH fel biochem, Diabetes Ctr, New York Med Col, 72-73; res assoc, 73-75, NIH FEL, DEPT BIOCHEM, PURDUE UNIV, 75- Mem: Sigma Xi. Res: Glycolipid and glycoprotein metabolism in disease stage; biological active compound, such as antimetabolite or antibiotics. Mailing Add: Dept of Biochem Purdue Univ West Lafayette IN 47906

TSAI, TOM CHUNG HSIUNG, b Chang Hua, Taiwan, Apr 7, 35; m 59; c 4. POLYMER CHEMISTRY, COLLOID CHEMISTRY. Educ: Nat Taiwan Univ, BS, 57; NDak State Univ, PhD(polymer chem), 69. Prof Exp: Res chemist, Taiwan Fertilizer Corp, 59-61; res asst nuclear chem, Nat Tsing Hua Univ, Taiwan, 62-64; res chemist, Dow Corning Corp, Mich, 69-72; sr res chemist, 72-73, PROJ LEADER, COPOLYMER RUBBER & CHEM CORP, 73- Mem: AAAS; Am Chem Soc. Res: Emulsion polymerization; free radical, radiation, anionic and coordination polymerizations; polymer characterizations; room temperature vulcanizations; rubber chemistry plastics modifications; silicone chemistry; coatings; adhesives; utilization of silicones. Mailing Add: Res Dept Shade Ave Copolymer Rubber & Chem Corp Baton Rouge LA 70821

TSAI, TSUI HSIEN, b Taichung, Formosa, Nov 19, 35; m 63; c 1. PHARMACOLOGY. Educ: Nat Taiwan Univ, BS, 58; WVa Univ, MS, 63, PhD(pharmacol), 65. Prof Exp: Sect head autonomic pharmacol, Merrell-Nat Labs, Ohio, 65-71; SR RES PHARMACOLOGIST, WELLCOME RES LABS, 71- Concurrent Pos: Fel, Harvard Med Sch, 64-66. Mem: Am Soc Pharmacol & Exp Therapeut. Res: Autonomic and cardiovascular pharmacology. Mailing Add: Dept of Pharmacol Wellcome Res Labs Research Triangle Park NC 27709

TSAI, YUAN-HWANG, b Ping-Tung, Taiwan, Jan 28, 36; US citizen; m 64; c 3. MICROBIOLOGY. Educ: Nat Taiwan Univ, BS, 59; Univ Utah, MS, 64, PhD(parasitol), 67. Prof Exp: Lab instr clin parasitol, Med Sch, Nat Taiwan Univ, 61-62; asst biol, parasitol & med entom, Univ Utah, 62-66; parasitologist, Trop Dis Ctr, St Clare's Hosp, New York, 66-67; SR MICROBIOLOGIST, BRISTOL LABS, 68- Mem: Am Soc Microbiol. Res: Antimicrobial agents and chemotherapy; experimental bacterial meningitis and therapy; pharmacokinetics of antibiotics. Mailing Add: Dept of Microbiol Bristol Labs Thompson Rd Syracuse NY 13201

TSAI, YUNG SU, b Yuli, Taiwan, Feb 1, 30, US citizen; m 61; c 2. THEORETICAL PHYSICS, ELEMENTARY PARTICLE PHYSICS. Educ: Nat Taiwan Univ, BS, 54; Univ Minn, MS, 56, PhD(physics), 58. Prof Exp: Res assoc theoret physics, Stanford Univ, 59-61, asst prof, 61-63, SR STAFF MEM THEORET PHYSICS, STANFORD LINEAR ACCELERATOR CTR, STANFORD UNIV, 63- Mem: Am Phys Soc; Sigma Xi. Res: Passage of particles through matter; production of particles; energy loss and straggling due to ionization; bremsstrahlung and cerenkov radiation; radiative corrections to scatterings; new particles and properties of elementary particles. Mailing Add: Stanford Linear Accelerator Ctr Stanford Univ Stanford CA 94305

TSAN, ALICE TUNG-HUA, b Taiwan, Rep China, Dec 3, 49; m 73. NUMERICAL ANALYSIS, APPLIED MATHEMATICS. Educ: Nat Taiwan Univ, BS, 71; State Univ NY Stony Brook, MS, 73, PhD(numerical anal), 75. Prof Exp: VIS FEL RES MATH, NIH, 75- Res: Development of a multinephron model of the kidney that is adequate for the interpretation of experimental data. Mailing Add: Bldg 31 Rm 9A21 NIH Bethesda MD 20014

TSAN, MIN-FU, b Taiwan, Jan 27, 42; m 75. HEMATOLOGY, NUCLEAR MEDICINE. Educ: Nat Taiwan Univ, MB, 67; Harvard Univ, PhD(physiol), 71; Am Bd Internal Med, dipl, 75. Prof Exp: Intern med, Nat Taiwan Univ Hosp, 66-67; med officer, Chinese Navy, 67-68; med intern, Boston Vet Admin Hosp, 71-72; med resident, 72-73, fel hematol, Johns Hopkins Hosp, 73-75; ASST PROF MED, RADIOL & RADIOL SCI, MED SCH & ASST PROF ENVIRON HEALTH, SCH PUB HEALTH & HYG, JOHNS HOPKINS UNIV, 75- Mem: AAAS; Am Fedn Clin Res; Am Soc Hematol. Res: Metabolism and function of neutrophils, with particular emphasis on the control mechanism of the oxidative metabolism associated with phagocytosis. Mailing Add: Johns Hopkins Med Inst 615 N Wolfe St Baltimore MD 21205

TSANG, CHARLES PAK WAI, Can citizen. ENDOCRINOLOGY. Educ: McGill Univ, BSc, 61, MSc, 65, PhD(steroid biochem), 68. Prof Exp: Staff scientist, Worcester Found Exp Biol, 68-70; fel steroid bioinorganic assay, Queen Mary Hosp, Montreal, 70-71; RES SCIENTIST REPRODUCTIVE PHYSIOL, ANIMAL RES INST, 71- Res: Metabolism of steroid hormones in relation to pregnancy and control of parturition. Mailing Add: Animal Res Inst Ottawa ON Can

TSANG, KANG TOO, b Hong Kong. PLASMA PHYSICS. Educ: Chinese Univ Hong Kong, BSc, 70; State Univ NY Stony Brook, MA, 71; Princeton Univ, PhD(plasma physics), 74. Prof Exp: RES STAFF PLASMA PHYSICS, OAK RIDGE NAT LAB, 74- Mem: Am Phys Soc. Res: Theoretical investigation of microinstability and transports in thermonuclear plasma. Mailing Add: Oak Ridge Nat Lab Oak Ridge TN 37830

TSANG, REGINALD C, b Hong Kong, Sept 20, 40; UK citizen; m 66; c 2. NEONATOLOGY, NUTRITION. Educ: Univ Hong Kong, MBBS, 66. Prof Exp: Intern med & surg, Queen Mary Hosp, Hong Kong Univ, 64-65, resident pediat, 65-66; resident psychiat, Hong Kong Psychiat Hosp, 65; intern pediat & med, Michael Reese Hosp, Chicago, 66-67, resident pediat, 67-68, fel neonatology, 68-69; fel, Cincinnati Gen Hosp & Childrens Hosp, 69-71; asst prof pediat & obstet & gynec, 71-75, DIR, FELS DIV PEDIAT RES, UNIV CINCINNATI, 74-, ASSOC PROF PEDIAT & OBSTET & GYNEC, 75- Concurrent Pos: Attend pediatrician, Childrens Hosp, Cincinnati & NIH grant neonatal mineral metab, Nat Inst Child Health & Human Develop, 71-; attend pediat, Cincinnati Gen Hosp, 71-, pediat dir, Lipid Res Clin, 73- Mem: Am Fedn Clin Res; Soc Pediat Res; Endocrine Soc; Am Col Clin Nutrit. Res: Pathophysiology of disturbances in calcium-phosphate-magnesium homeostasis in the neonate; examination of parathyroid hormone, vitamin D, glucagon and calcitonin; pediatric hyperlipoproteinemia; identification and prevention of premature atherosclerosis. Mailing Add: Dept of Pediat Col of Med Univ of Cincinnati Cincinnati OH 45267

TSANG, SIEN MOO, b Shanghai, China, Apr 12, 12; nat US; m 46; c 2. PHOTOCHEMISTRY. Educ: St John's Univ, China, BS, 36; Cornell Univ, MS, 40, PhD(org chem), 44. Prof Exp: Chemist, H Z Synthetic Chem Industs, Ltd, China, 36-37; instr, St John's Univ, China, 38-39; asst, Nat Defense Res Comt Proj, Cornell Univ, 42-44, res assoc, B F Goodrich Co Proj, 44-45; res chemist, Am Cyanamid Co, 45-46; chemist, Wahca Chem Corp, 47-48; from res chemist to sr res chemist, Am Cyanamid Co, 49-63; Am Cyanamid Co sr res award, Univ Sheffield, 63-64; res assoc, 65-74, PRIN RES SCIENTIST, AM CYANAMID CO, 74- Honors & Awards: Naval Ord Develop Award, Bur Ord, Navy Dept, 46. Mem: Am Chem Soc. Res: Aromatic substitution reactions; organic photochemistry; photostabilizations of polymers. Mailing Add: Am Cyanamid Co Bound Brook NJ 08805

TSANG, TUNG, b Shanghai, China, Aug 17, 32; US citizen; m 57; c 2. PHYSICAL CHEMISTRY. Educ: Ta-Tung Univ, China, BS, 49; Univ Minn, MS, 52; Univ Chicago, PhD(chem), 60. Prof Exp: Chemist, Minneapolis-Honeywell Regulator Co, 52-55 & 56; asst chemist, Argonne Nat Lab, 60-64, assoc chemist, 64-67; phys chemist, Nat Bur Standards, 67-69; ASSOC PROF PHYSICS, HOWARD UNIV, 70- Mem: Am Phys Soc. Res: Magnetic resonance and susceptibility. Mailing Add: Dept of Physics Howard Univ Washington DC 20001

TSANG, WING, physical chemistry, see 12th edition

TSAO, CHEN-HSIANG, b Shanghai, China, Jan 21, 29; US citizen; m 57; c 2. PHYSICS. Educ: Univ Wash, Seattle, BS, 53, MS, 56, PhD(physics), 61. Prof Exp: Res assoc particle physics, Univ Wash, Seattle, 56-60, res instr, 60-61; res assoc high energy physics, Enrico Fermi Inst Nuclear Studies, Univ Chicago, 61-65; Nat Acad Sci-Nat Res Coun resident res assoc fel, 65-66, RES PHYSICIST LAB COSMIC RAY PHYSICS, US NAVAL RES LAB, 66- Mem: Am Phys Soc; Sigma Xi. Res: Elementary particle physics; high energy interactions; astrophysics. Mailing Add: Lab for Cosmic Ray Physics Code 7023 US Naval Res Lab Washington DC 20375

TSAO, CHIA KUEI, b China, Jan 14, 22; m 52; c 4. MATHEMATICAL STATISTICS. Educ: Univ Ore, MA, 50, PhD(math statist), 52. Prof Exp: Asst math, Univ Ore, 48-52; from instr to assoc prof, 52-63, PROF MATH, WAYNE STATE UNIV, 63- Mem: Am Math Soc; Math Asn Am; Inst Math Statist. Res: Nonparametric statistics. Mailing Add: Dept of Math Wayne State Univ Detroit MI 48202

TSAO, CHING HSI, b Hopei, China, Jan 24, 18; nat US; m 53; c 3. ENTOMOLOGY. Educ: Chekiang Univ, BS, 40; Univ Minn, MS, 48, PhD(entom), 51. Prof Exp: Asst entomologist, Ministry Agr & Forestry, China, 40-41; asst entom, Tsing Hwa, China, 41-43; sr asst entomologist, Inst Zool, China Acad Sci, 44-46; asst entom, Univ Minn, 49-52; US State Dept scholar, Entom Res Sect, USDA, Md, 52-54; entomologist, Univ Tex, 54-61; asst prof entom, 61-67, ASSOC PROF ENTOM, UNIV GA, 67- Mem: AAAS; Entom Soc Am. Res: Behavior of insects; laboratory and field ecology of insects; forest entomology; insect physiology. Mailing Add: Dept of Entom Univ of Ga Athens GA 30602

TS'AO, CHUNG-HSIN, b Nanking, China, 33; m 62; c 3. PHYSIOLOGY, EXPERIMENTAL PATHOLOGY. Educ: Tunghai Univ, Taiwan, BS, 60; Ind Univ, Bloomington, MA, 61; Yale Univ, PhD(physiol), 66. Prof Exp: Res assoc hemat, Montefiore Hosp & Med Ctr, 66-67; res assoc physiol, Sch Med, Univ Chicago, 67-72, asst prof path, 68-72; asst prof, 73-75, ASSOC PROF PATH, MED SCH, NORTHWESTERN UNIV, CHICAGO, 75- Concurrent Pos: Dir path, Coagulation Lab, Chicago Northwestern Mem Hosp, 73- Mem: AAAS; Am Soc Hemat; Int Soc Thrombosis & Haemostasis; Am Soc Exp Path. Res: Experimental thrombosis; vascular morphology and function; intravascular leukocytolysis. Mailing Add: 201 Riverside Dr Northfield IL 60093

TSAO, LIANG-CHI, b Taipei, Taiwan, Oct 14, 43; m 73. NUMBER THEORY. Educ: Nat Taiwan Univ, BSc, 66; Univ Chicago, MSc, 68, PhD(math), 72. Prof Exp: Vis asst prof math, Univ Mich, 72-73; mem, Inst Advan Study, 73-75; VIS LECTR MATH, UNIV ILL, URBANA, 75- Mem: Am Math Soc. Res: Fields of definition for fields of automorphic functions of several complex variables. Mailing Add: Dept of Math Univ of Ill Urbana IL 61801

TSAO, MAKEPEACE UHO, b Shanghai, China, Aug 28, 18; nat US; m 47; c 4. CHEMISTRY. Educ: Univ Tatung, BS, 37; Univ Mich, MS, 41, PhD(pharmaceut chem), 44. Prof Exp: Wm S Merrill Co fel, Univ Mich, 44-45, sr biochemist, 46-48, head biochemist, 48-52, from asst prof to assoc prof biochem, 52-67; PROF SURG, UNIV CALIF, DAVIS, 67- Mem: AAAS; Am Chem Soc; Am Soc Biol Chem; Biomet Soc; NY Acad Sci. Res: Synthetic medicinals; physiological chemistry of premature and newborn infants; biochemical analytical methods; multiple molecular forms of enzymes; carbohydrate metabolism; experimental diabetes; Neurospora crassa. Mailing Add: Dept of Surg Univ of Calif Davis CA 95616

TSAO, PAMELA WEN-CHAU WANG, b Shen-Yang, China, May 16, 30; US citizen; m 56; c 1. PLANT PATHOLOGY. Educ: Nat Taiwan Univ, BS, 53; Univ Wis, MS, 55, PhD(plant path), 62. Prof Exp: Asst specialist plant path, 61-66, RES ASSOC PLANT PATH, UNIV CALIF, RIVERSIDE, 66- Mem: Am Phytopath Soc; Bot Soc Am. Res: Virology; plant anatomy and histology; cytology. Mailing Add: Dept of Plant Path Univ of Calif Riverside CA 92502

TSAO, PETER, b Kiangsu, China, June 9, 26. PHYSICS, GEOPHYSICS. Educ: John Carroll Univ, BS, 51, MS, 52; Tex A&M Univ, PhD, 70. Prof Exp: Instr physics, Undergrad Div, Univ Ill, 56-58; asst prof, 58-72, ASSOC PROF PHYSICS, NORTHERN ILL UNIV, 72- Mem: Am Phys Soc; Am Asn Physics Teachers; Inst Elec & Electronics Engrs; Seismol Soc Am. Res: Ultraviolet-visible and Raman spectroscopy of small molecules. Mailing Add: Dept of Physics Northern Ill Univ De Kalb IL 60115

TSAO, PETER HSING-TSUEN, b Shanghai, China, Mar 22, 29; nat US; m 56; c 1. PLANT PATHOLOGY. Educ: Univ Wis, BA, 52, PhD(plant path), 56. Prof Exp: Jr plant pathologist, 56-58, asst plant pathologist, 58-64, assoc prof plant path, 64-70, PROF PLANT PATH, UNIV CALIF, RIVERSIDE, 70- Concurrent Pos: Guggenheim fel, 66-67; consult Thailand dept agr, UN Food & Agr Orgn, 73-74; res consult, Nat Sci Coun, Taiwan, 74-76. Mem: Am Phytopath Soc; Bot Soc Am; Mycol Soc Am; Brit Mycol Soc; Am Inst Biol Sci. Res: Ecology of soil fungi; citrus root diseases; antagonism and antibiotics. Mailing Add: Dept of Plant Path Univ of Calif Riverside CA 92502

TSAO, RHETT FENSHENG, applied statistics, economics, see 12th edition

TSAPOGAS, MAKIS JOAKIM, b Piraeus, Greece, May 7, 26; m 61. SURGERY, BIOMEDICAL SCIENCES. Educ: Univ Athens, MD, 54, MChir, 60. Prof Exp: Res fel, King's Col Hosp Med Sch, Univ London, 60-61; PROF SURG, RUTGERS MED SCH, 76- Concurrent Pos: Hunterian prof, Royal Col Surgeons, Eng, 64-; prof surg, Univ Athens, 68-; adj prof, Rensselaer Polytech Inst, 69-; adj prof surg, Albany Med Col, 70-; chief vascular surg, Raritan Hosp, Greenbrook, NJ; surgeon, Lyons Vet Admin Hosp, NJ; consult gen & vascular surg, Albany Med Ctr, Ellis, St Clare's, St Mary's & Vet Admin Hosps, Sunnyview & Glenridge Hosp, Albany, NY. Honors & Awards: Sci Prize, Brit Med Asn, 64. Mem: Soc Vascular Surg; Int Cardiovasc Soc; Brit Surg Res Soc; AMA; Brit Med Asn. Res: Various aspects of vascular research, including organization of thrombi, fibrinolytic therapy, new reconstructive arterial procedures, prevention and early detection and therapy of venous thrombosis and pulmonary embolism. Mailing Add: Dept of Surg Rutgers Med Sch Piscataway NJ 08854

TSAY, JIA-YEONG, b Taiwan. BIOSTATISTICS, MATHEMATICAL STATISTICS. Educ: Fu Jen Cath Univ, Taiwan, BS, 67; Bowling Green State Univ, MA, 69; Purdue Univ, MS, 71, PhD(statist), 74. Prof Exp: NIH fel biostatist, 74-76, ASST PROF BIOSTATIST, MED COL, UNIV CINCINNATI, 76- Mem: Am Statist Asn; Inst Math Statist. Res: Optimal experimental designs. Mailing Add: Dept of Environ Health Univ of Cincinnati Med Col Cincinnati OH 45267

TSCHABOLD, EDWARD EVERTT, b Wichita Falls, Tex, Dec 7, 34; m 70; c 3. PLANT PHYSIOLOGY. Educ: Midwestern Univ, BS, 57; Miami Univ, MA, 62; Colo State Univ, PhD(plant physiol), 67. Prof Exp: Sr plant physiologist, 6773, RES PLANT PHYSIOLOGIST, LILLY RES LABS, 73- Mem: Am Soc Hort Sci; Am Soc Plant Physiologists. Res: Basic and applied aspects of agricultural plant growth regulators. Mailing Add: Lilly Res Labs Greenfield Res Labs G785 Greenfield IN 46140

TSCHANZ, CHARLES MCFARLAND, b Mackay, Idaho, July 9, 26; m 58; c 3. GEOLOGY. Educ: Univ Idaho, BS, 49; Stanford Univ, MS, 51. Prof Exp: Geologist, Colo, 49-50, Pioche, Nev, 51-53, chief uranium-copper proj, NMex, 53-55, geochem researcher, 55-56, chief mapping proj, Lincoln County, Nev, 56-60, advisor, Bolivian Mineral Resources, US Opers Mission, USAID, 60-65, geol consult, Nat Mineral Inventory, Colombia, 65-69, proj chief, Boulder Mountains Mapping Proj, Idaho, 69-70, proj chief mineral eval, Sawtooth Nat Recreation Area, Idaho, 71-74, PROJ CHIEF, BOULDER MOUNTAINS, IDAHO, US GEOL SURV, 74- Mem: Geol Soc Am; Geochem Soc; Soc Econ Geol. Res: Regional mapping and economic evaluation as an integrated project; geochemistry, especially distribution of minor elements in igneous rocks; geology of eastern Nevada, Colorado Plateau, Bolivian Altiplano and Sierra Nevada of Santa Marta, Colombia. Mailing Add: US Geol Surv Bldg 25 Denver Fed Ctr Denver CO 80225

TSCHARNER, CHRISTOPHER J, b Oberhallau, Switz, July 15, 29; m 60; c 2. ORGANIC CHEMISTRY, PHYSICAL CHEMISTRY. Educ: Univ Zurich, PhD(carotenes), 60. Prof Exp: Instr gen org chem, Univ Zurich, 56-59; develop chemist, Geigy Chem Corp, 60-62, head of plant lab, 62-64, dir develop, 64-70, mgr process develop plastics & additives div, Ciba-Geigy Basel, 70-75, PROD MGR PLASTICS ADDITIVES, CIBA-GEIGY LTD, 75- Mem: Am Chem Soc; Swiss Chem Soc. Res: Research and development of industrial chemicals, optical brighteners and antioxidants. Mailing Add: Neuackerweg 3 CH-4105 Biel-Benken Basel Switzerland

TSCHIEGG, CARL EMERSON, b Bluffton, Ohio, Apr 14, 24; m 50; c 2. PHYSICS. Educ: Bowling Green State Univ, BS, 49. Prof Exp: PHYSICIST SOUND SECT, NAT BUR STANDARDS, 50- Honors & Awards: Meritorious Serv Award, US Dept Com, 61. Mem: Am Phys Soc; sr mem Inst Elec & Electronics Eng. Res: Ultrasonics; underwater sound. Mailing Add: Sound Sect B106 Bldg 233 Nat Bur of Standards Washington DC 20234

TSCHIRGI, ROBERT DONALD, b Sheridan, Wyo, Oct 9, 24. PHYSIOLOGY. Educ:

Univ Chicago, BS, 45, MS, 47, PhD, 49, MD, 50. Prof Exp: Asst physiol, Univ Chicago, 45-48, from instr to asst prof, 48-53; from assoc prof to prof, Sch Med, Univ Calif, Los Angeles, 53-66; vchancellor acad planning, 66-67, vchancellor acad affairs, 67-68, PROF NEUROSCI, UNIV CALIF, SAN DIEGO, 66- Concurrent Pos: Dir med educ study, Univ Hawaii, 63-64; univ dean planning, Univ Calif, 64-66; consult, NSF. Mem: Int Brain Res Orgn; fel AAAS; Am Physiol Soc; Biophys Soc. Res: Intracranial fluids and barriers; direct current potentials in central nervous system; neurophysiology of perception. Mailing Add: Dept of Neurosci Sch of Med Univ of Calif at San Diego La Jolla CA 92093

TSCHIRLEY, FRED HAROLD, b Ethan, SDak, Dec 19, 25; m 48; c 5. ECOLOGY. Educ: Univ Colo, BA, 51, MA, 54; Univ Ariz, PhD, 63. Prof Exp: Res asst, Univ Ariz, 52-53, instr, 53-54; range scientist, Crops Res Div, Agr Res Serv, USDA, 54-68, asst br chief, Crops Protection Res Br, 68-71, asst coordr environ qual activ sci & educ, 71-73, coordr environ qual activ, Off Secy, 73-74; PROF & CHMN DEPT BOT & PLANT PATH, MICH STATE UNIV, 74- Mem: AAAS; Weed Sci Soc Am; Soc Range Mgt; Ecol Soc Am. Res: Woody plant control; physiological ecology. Mailing Add: Dept of Bot & Plant Path Mich State Univ East Lansing MI 48224

TSCHOEGL, MICHOLAS WILLIAM, b Zidlochovice, Czech, June 4, 18; m 46; c 2. PHYSICAL CHEMISTRY. Educ: New South Wales, BSc, 54, PhD(chem), 58. Prof Exp: Sr res officer, Bread Res Inst, Australia, 58-61; proj assoc dept chem, Univ Wis, 61-63; sr phys chemist, Stanford Res Inst, 63-65; assoc prof mat sci, 65-67, PROF CHEM ENG, CALIF INST TECHNOL, 67- Concurrent Pos: Consult, Phillips Petrol Co, 67- Mem: Am Phys Soc; Am Chem Soc; Soc Rheol; Brit Soc Rheol; Royal Australian Chem Inst. Res: Polymer rheology; physical chemistry of macromolecules; mechanical properties of polymeric materials. Mailing Add: Dept of Chem Eng Calif Inst of Technol Pasadena CA 91125

TSCHUDI, WILBUR JAMES, b Baltimore, Md, July 27, 23; m 45; c 3. INORGANIC CHEMISTRY. Educ: Ohio Univ, BS, 45. Prof Exp: From jr chemist to sr chemist, 45-51, chief chemist, 52-56, tech dir, 57-62, tech dir chem sales & develop, 63-71, CHEM MKT MGR, CLIMAX MOLYBDENUM CO, GREENWICH, CONN, 72- Mem: Am Chem Soc; Am Soc Testing & Mat. Res: Sales and market development; molybdenum chemicals. Mailing Add: 309 Meadowbrook Ave Ridgewood NJ 07450

TSCHUDY, DONALD P, b Palmerton, Pa, Nov 8, 26; m 51; c 2. INTERNAL MEDICINE, BIOCHEMISTRY. Educ: Princeton Univ, AB, 46; Columbia Univ, MD, 50; Am Bd Internal Med, dipl, 61. Prof Exp: Intern med, Presby Hosp, New York, 51-52, asst resident, 52-53; asst resident, Francis Delafield Hosp, 53-54; clin assoc, Clin Ctr, NIH, 54-55, SR INVESTR, METAB SERV, CLIN CTR, NAT CANCER INST, 55- Mem: Am Soc Clin Invest; Am Fedn Clin Res; Am Soc Biol Chemists; AMA. Res: Clinical and biochemical research on porphyrin metabolism and the porphyrias; research on tumor-host relationships. Mailing Add: NIH Clin Ctr Nat Cancer Inst Bethesda MD 20014

TSCHUDY, ROBERT HAYDN, b Pocatello, Idaho, May 7, 08; m 34. PALEOBOTANY. Educ: Univ Wash, Seattle, BS, 32, MS, 34, PhD, 37. Prof Exp: Instr bot, Univ Wyo, 37-38; res biologist, Scripps Inst, Univ Calif, 39-41; assoc prof bot, Willamette Univ, 41-45; res biologist, Creole Petrol Corp, 45-49, asst res coordr, 49-50; dir, Palynological Res Lab, 50-61; MEM RES STAFF, PALEONT & STRATIG BR, US GEOL SURV, 61- Concurrent Pos: Lectr, Univ Colo, 55-56 & 64-66 & Univ Minn, 58. Mem: Bot Soc Am; Geol Soc Am. Res: Palynology. Mailing Add: US Geol Surv Denver Fed Ctr Bldg 25 Denver CO 80225

TSCHUIKOW-ROUX, EUGENE, b Kharkov, USSR, Jan 16, 36; US citizen; m 59. PHYSICAL CHEMISTRY. Educ: Univ Calif, Berkeley, BS, 57, PhD(chem), 61; Univ Wash, Seattle, MS, 58. Prof Exp: Sr scientist, Jet Propulsion Lab, Calif Inst Technol, 60-65; Nat Acad Sci-Nat Res Coun res assoc chem, Nat Bur Standards, 65-66; assoc prof, 66-71, PROF CHEM, UNIV CALGARY, 71-, CHMN DEPT, 73- Concurrent Pos: Vis scholar, Univ Calif, Santa Barbara, 72; consult, Jet Propulsion Lab, Calif Inst Technol, 73. Mem: Am Chem Soc; Am Phys Soc; Can Inst Chem. Res: Gas phase reaction kinetics; high temperature shock tube studies; kinetic isotope effects; photochemistry; reaction dynamics; unimolecular reactions; gas phase ion-molecule reactions. Mailing Add: Dept of Chem Univ of Calgary Calgary AB Can

TSCHUNKO, HUBERT F A, b Weidenau, Austria, Sept 9, 12; US citizen; m 46; c 2. PHYSICS, OPTICS. Educ: Darmstadt Tech Univ, Diplom-Ing, 35. Prof Exp: Develop engr aeronaut indust, Europe, 36-45; res assoc astron, Astron Observ, Heidelberg, 45-50; engr pvt indust, WGer, 51-57; physicist, US Air Force, Wright-Patterson AFB, 57-65; opticist electronics res ctr, 65-70, OPTICIST, NASA GODDARD SPACE FLIGHT CTR, 70- Honors & Awards: Apollo Achievement Award, NASA, 69. Mem: AAAS; Optical Soc Am; Ger Soc Aeronaut & Astronaut. Res: Wave, space and astronomical optics; large optical systems. Mailing Add: NASA Goddard Space Flight Ctr Code 722 Greenbelt MD 20771

TSE, ROSE (LOU), b Shanghai, China, July 27, 27; US citizen; m 53. ORGANIC CHEMISTRY, MEDICINE. Educ: St John's Univ, China, BS, 49; Mt Holyoke Col, MA, 50; Yale Univ, PhD(org chem), 53; Med Col Pa, MD, 60; Am Bd Internal Med, dipl & cert rheumatology. Prof Exp: Instr, Ohio State Univ, 53-55; res assoc, Univ Pa, 55-56; intern, Philadelphia Gen Hosp, 60-61, resident internal med, 61-64; assoc in med, 68-71, asst prof clin med, 71-75, ASSOC PROF MED, SCH MED, UNIV PA, 75- Concurrent Pos: Attend physician, Philadelphia Gen Hosp, 64-68, sr attend physician, 68-, assoc chief spec ward cardiol, 68-71, chief rheumatology sect, 71-; clin instr internal med, Med Col Pa, 64-68. Mem: Fel Am Col Physicians; fel Am Inst Chemists; fel Am Col Angiol; Am Heart Asn; Am Rheumatism Asn. Res: Reaction mechanisms; organic synthesis; electrocardiology; inflammatory mediators; non-steroidal inflammatory agents; cardiology; rheumatology; catecholamines; cyclic adenosine monophosphate; prostaglandin; crystal-induced synovitis. Mailing Add: 700 Civic Ctr Blvd Philadelphia PA 19104

TSE, WARREN W, b Hong Kong, Mar 13, 39; m 68; c 2. HUMAN PHYSIOLOGY. Educ: Univ Cincinnati, BS, 65; Univ Wis, MS, 67, PhD(physiol), 70. Prof Exp: Lectr physiol, Univ Wis, 69-70, fel physiol, 70-72; res assoc cardiol, Albany Med Col, 72-75; ASST PROF PHYSIOL, DEPT PHYSIOL & BIOPHYS, CHICAGO MED SCH, UNIV HEALTH SCI, 75- Mem: Am Physiol Soc. Res: Electrophysiological properties of single fibers of atrioventricular node and purkinje fibers of dog hearts. Mailing Add: Dept of Physiol & Biophys Chicago Med Sch Univ of Health Sci 2020 W Ogden Chicago IL 60612

TSEN, CHO CHING, b Chekiang, China, Oct 12, 22; nat US; m 52; c 4. BIOCHEMISTRY, NUTRITION. Educ: Nat Chekiang Univ, China, BS, 44, MS, 46; Univ Calif, PhD(biochem), 58. Prof Exp: Res chemist, Taiwan Sugar Exp Sta, 46-50; from instr to assoc prof biochem, Taiwan Prov Col Agr, 51-54; fel, Nat Res Coun Can, 58-59; chemist, Grain Res Lab, Bd Grain Comnrs Can, 59-65, scientist, 65-67; res group leader, Am Inst Baking, Ill, 67-69; PROF CEREAL CHEM & BAKING SCI, KANS STATE UNIV, 69- Mem: AAAS; Am Chem Soc; Am Asn Cereal Chemists; Inst Food Technologists; Int Union Food Sci & Technol. Res: Cereal and food chemistry; baking technology; enrichment and fortification of cereal products and nutrition of cereal products. Mailing Add: Dept Grain Sci & Indust Kans State Univ, Manhattan KS 66506

TSENG, CHARLES C, b Fuchow, Fukien, China, Dec 20, 32; m 65; c 2. PLANT ANATOMY, MORPHOLOGY. Educ: Taiwan Norm Univ, BS, 55; Taiwan Univ, MS, 57; Univ Calif, Los Angeles, PhD(plant sci), 65. Prof Exp: From asst prof to assoc prof bot, Windham Col, 65-75; ASSOC PROF BIOL, PURDUE UNIV CALUMET CAMPUS, 75- Mem: AAAS; Bot Soc Am; Am Inst Biol Sci; Am Soc Plant Taxon. Res: Floral anatomy of Umbelliferae and Araliaceae; palynology of angiosperms and gymnosperms; taxonomy of Umbelliferae. Mailing Add: Dept of Biol Purdue Univ Hammond IN 46323

TSENG, CHIEN KUEI, b Tao Yuan, Taiwan, Feb 21, 34; m 66; c 2. ORGANIC CHEMISTRY. Educ: Cheng Kung Univ, Taiwan, BS, 57; WVa Univ, MS, 64; Ill Inst Technol, PhD(chem), 68. Prof Exp: USPHS fels, Ill Inst Technol, 67-68; from res chemist to sr res chemist, 68-71, SUPVR ANAL CHEM, STAUFFER CHEM CO, 71- Mem: Am Chem Soc. Res: Nuclear magnetic resonance; structure determination; stereochemistry; phosphorus chemistry; infrared and mass spectroscopy. Mailing Add: Stauffer Chem Co 1200 S 47th St Richmond CA 94804

TSENG, HSIANG LEN, b Wukingfu via Swatow, China, Apr 24, 13; US citizen; m 52; c 4. PATHOLOGY. Educ: Hsiang Ya Sch Med, MD, 37; Univ Basel, MD, 48; Am Bd Path, cert anat path, 55, cert clin path, 66. Prof Exp: Vol asst, Inst Anat Path, Univ Basel, 47-48; vol asst internal med, Bürgerspital, Basel, 48; vol asst, Postgrad Med Sch, Univ London, 48; resident internal med, St Catherine Hosp, Brooklyn, NY, 49-51; resident path, Univ Hosp, New York, 51-52; asst pathologist, Ill Masonic Hosp, 52-55; chief pathologist, St Elizabeth Hosp, Washington, DC, 55-57; chief pathologist, Provident Hosp, Baltimore, 57-69; pathologist, 69-70, CHIEF ANAT PATH, ST ELIZABETH HOSP, WASHINGTON, DC, 70- Concurrent Pos: Fel, Malaria Inst India, Delhi, 42; clin instr, Col Med, Univ Ill, 54-55. Mem: Fel Am Col Path; Am Soc Cytol; Int Acad Path. Res: Anatomical and clinical pathology; myocarditis, both idiopathic and tuberculous type; acute sicklemia. Mailing Add: 1211 LaGrande Rd Silver Spring MD 20903

TSENG, SHU-TEN, b Kaohsiung, Taiwan, May 22, 33; m 70; c 1. AGRONOMY. Educ: Nat Taiwan Univ, BS, 57; Utah State Univ, MS, 67; Univ Mo-Columbia, PhD(plant breeding), 70. Prof Exp: Res assoc seed technol & crop mgt, Nat Taiwan Univ, 57-64; PLANT BREEDER, CALIF COOP RICE RES FOUND, INC, 70- Mem: Am Soc Agron; Crop Sci Soc Am; Am Genetic Asn. Res: Rice breeding and genetics. Mailing Add: Rice Exp Sta PO Box 306 Calif Coop Rice Res Found Inc Biggs CA 95917

TSERPES, NICOLAS A, b Messene, Greece, 1936; US citizen; m 68. MATHEMATICS, STATISTICS. Educ: Wayne State Univ, BS, 61, MA, 64, PhD(math), 68. Prof Exp: Res phys scientist statist, Corps Eng, US Lake Surv, Detroit Dist, 63-64 & 66-67; ASSOC PROF MATH, UNIV S FLA, 68- Concurrent Pos: Univ res coun grant, Univ SFla, 71-73. Mem: Am Math Soc; Math Asn Am. Res: Theory and measure on topological semigroups; probability structures and random walks on semigroups; mathematical statistics; stochastic equations. Mailing Add: Dept of Math Univ of SFla Tampa FL 33620

TSIAPALIS, CHRIS MILTON, b Sykaminea, Greece, Nov 18, 39; Can citizen; m 63; c 2. BIOCHEMISTRY. Educ: Univ Western Ont, BA, 64, MSc, 65, PhD(biochem), 69. Prof Exp: ASSOC BIOCHEMIST, LIFE SCI RES LABS, 74- Concurrent Pos: Nat Res Coun Can fel biochem, 69-70; res scholar biochem, Med Sch, Univ Ky, 71-73. Mem: AAAS; Am Soc Biol Chem; Can Biochem Soc. Res: Nucleic acid enzymology in differentiation and proliferation of eucaryotes; mechanisms of RNA and DNA synthesis. Mailing Add: Life Sci Res Labs 1509 112 49th St S St Petersburg FL 33707

TSIBRIS, JOHN CONSTANTINE MICHAEL, b Jannina, Greece, Dec 22, 36; m 69; c 2. BIOCHEMISTRY, BIOPHYSICS. Educ: Nat Univ Athens, BSc, 59; Cornell Univ, PhD(biochem), 65-68, vis scientist, 69-71; ASST PROF BIOCHEM, UNIV FLA, 71- Concurrent Pos: NIH trainee biophys chem, 67-68, grant, 69-71, career develop award, 69- Mem: Am Soc Biol Chem; Am Chem Soc. Res: Function and structure of flavin coenzymes; structure, function and regulation of hydroxylase enzyme systems. Mailing Add: Dept of Biochem Miller Health Ctr Univ of Fla Gainesville FL 32610

TSIEN, HSIENCHYANG, b Nanking, China, July 26, 39; m 66; c 2. MICROBIOLOGY. Educ: Nat Taiwan Univ, BS, 61; Cath Univ Louvain, Belg, DrS(microbiol & biochem), 67. Prof Exp: Asst, Cath Univ Louvain, 67-68 & 70-72; res instr microbiol, Temple Univ, 72-75; res fel, 68-70, ASST PROF MICROBIOL, UNIV MINN, 75- Mem: Am Soc Microbiol; AAAS. Res: Soil microbiology; microorganism-plant symbiotic nitrogen fixation; microbial ecology; microbial physiology; structure and function of bacterial cell membranes and cell walls. Mailing Add: Dept of Microbiol Univ of Minn Minneapolis MN 55455

TSIGDINOS, GEORGE ANDREW, b Trikkala, Greece, May 30, 29; nat US; m 52; c 3. INORGANIC CHEMISTRY. Educ: Univ Boston, AB, 53, AM, 55, PhD(inorg chem), 61. Prof Exp: Res chemist res & eng div spec projs dept, Monsanto Chem Co, 59-61, res chemist, Boston Labs, Monsanto Res Corp, 61-65; sr res chemist, 65-74, RES SUPVR, CLIMAX MOLYBDENUM CO, 74- Mem: Am Chem Soc. Res: Heteropoly compounds; synthesis and characterization of molybdenum, tungsten and vanadium compounds and their application to catalysis; inorganic fluorine chemistry; inorganic polymers; molybdenum flame retardants. Mailing Add: 1810 Traver Rd Ann Arbor MI 48105

TSINA, RICHARD VASIL, physical chemistry, see 12th edition

TSO, MARK ON-MAN, b Hong Kong, China, Oct 19, 36; US citizen; m 64; c 2. OPHTHALMOLOGY, PATHOLOGY. Educ: Univ Hong Kong, MB, BS, 61. Prof Exp: ASSOC RES PROF OPHTHAL, MED CTR, GEORGE WASHINGTON UNIV, 73-; RES ASSOC, ARMED FORCES INST PATH, 71- Concurrent Pos: Fel ophthal path, Armed Forces Inst Path, 67-68, res fel, 68-69. Honors & Awards: Distinguished Serv Award, Armed Forces Inst Path, 71. Mem: Fel Am Acad Ophthal & Otolayrngol; Asn Res Vision & Ophthal. Res: Experimental pathology; electron microscopy; tissue culture; ocular oncology. Mailing Add: Ophthalmic Path Br Armed Forces Inst of Path Washington DC 20306

TS'O, PAUL ON PONG, b Hong Kong, July 17, 29; m 55; c 3. BIOPHYSICAL CHEMISTRY. Educ: Lingnan Univ, China, BS, 49; Mich State Univ, MS, 51; Calif Inst Technol, PhD, 55. Prof Exp: Res fel biol, Calif Inst Technol, 55-61, sr res fel, 61-62; assoc prof biophys chem, 62-67, PROF BIOPHYS CHEM, JOHNS HOPKINS UNIV, 67, DIR DIV BIOPHYSICS SCH HYGIENE & PUB HEALTH, 72- Concurrent Pos: Consult, Nat Cancer Inst; assoc ed, Molecular Pharmacol, 66-, Biochem, 68-74, Biophys J, 69-72, Biochem Biophys Acta, 73- & Cancer Review, 74-

Mem: Am Chem Soc; Am Soc Biol Chem; Biophys Soc; Am Asn Cancer Res; Am Soc Microbiol. Res: Physical, organic and biochemistry of nucleic acids, nuclear magnetic resonance, electron spin resonance, on biological systems; molecular carcinogenesis of polycyclic hydrocarbons, in vitro neoplastic transformation, tissue culture and anti-viral substances. Mailing Add: Sch of Hygiene & Pub Health Johns Hopkins Univ 615 N Wolfe Baltimore MD 21205

TSO, TIEN CHIOH, b Hupeh, China, July 25, 17; nat US; m 49; c 2. PHYTOCHEMISTRY. Educ: Nanking Univ, China, BS, 41, MS, 44; Pa State Univ, PhD(agr biochem), 50. Prof Exp: Supt exp farm, Ministry Social Affairs, China, 44-46; secy, Tobacco Improv Bur, 46-47; chemist res lab, Gen Cigar Co, 50-51; chemist div tobacco & spec crops, USDA, 52; asst prof & res assoc agron, Univ Md, 53-59; res plant physiologist, Tobacco & Sugar Crops Res Br, 59-62, sr plant physiologist, 62-64, prin plant physiologist, 64-66, leader tobacco qual invests, Tobacco & Sugar Crops Res Br, 66-72, CHIEF TOBACCO LAB, BELTSVILLE AGR RES CTR, AGR RES SERV, USDA, 72- Concurrent Pos: Mem, Tobacco Chem Res Conf, Tobacco Workers Conf, World Conf Tobacco & Health, Tobacco Working Group, Lung Cancer Task Force & Int Tobacco Working Group. Honors & Awards: Super Serv Award, USDA. Mem: Fel AAAS; Am Agron Soc; Am Chem Soc; fel Am Inst Chem; Am Soc Plant Physiol. Res: Plant physiology; tobacco alkaloids; biochemistry; culture; radio elements; health related components; tobacco production research relating to smoking and health. Mailing Add: Agr Res Ctr W USDA Beltsville MD 20705

TS'O, TIMOTHY ON TO, b Hong Kong, Nov 9, 34; m 63; c 2. PHARMACOLOGY, PATHOLOGY. Educ: Univ Hong Kong, MB, BS, 59; Stanford Univ, PhD, 68. Prof Exp: House surgeon, Govt Surg Unit, Queen Mary Hosp, Hong Kong, 59, house physician, Univ Med Unit, 60; demonstr path, Univ Hong Kong, 60-62; res physician neuropharmacol, Biochem Res Lab, Dow Chem Co, 68-72, sr res specialist, Chem Biol Res, 72-74; RES PROF HUMAN BIOL, SAGINAW VALLEY COL, 74- Mem: AAAS; Biophys Soc; NY Acad Sci. Res: Bronchogenic carcinoma; congenital tumors; psychopharmacology and neuropharmacology, particularly central nervous stimulants; computer application in biomedical research; biomathematics. Mailing Add: Saginaw Valley Col 2250 Pierce Rd University Center MI 48710

TSOKOS, CHRIS PETER, b Greece, Mar 25, 37; US citizen; m 57; c 3. APPLIED MATHEMATICS. Educ: Univ RI, BS & MS, 61; Univ Conn, PhD(math statist & probability), 67. Prof Exp: Consult opers res anal bur naval weapons, US Naval Air Sta, 64; proj engr elec boat div, Gen Dynamics Corp, 61-63; asst prof math, Univ RI, 63-69; assoc prof statist, Va Polytech Inst, 69-71; PROF MATH & STATIST, UNIV S FLA, 71-, DIR GRAD PROG STATIST & STOCHASTIC SYSTS, 71- Concurrent Pos: RI Res Coun res grants 65-66 & 67-68; NSF lectr, Univ RI, 65-68; consult, US Army Electronics Command Ctr, Ft Monmouth, NJ. Mem: AAAS; Am Math Soc; fel Am Statist Asn; Opers Res Soc Am. Res: Statistical theory and applications; stochastic integral equations; stochastic systems theory; biomathematics; stochastic modeling; time series; stochastic differential games; Bayesian reliability theory and simulation. Mailing Add: Dept of Math Univ of SFla Tampa FL 33620

TSOKOS, JANICE OSETH, b Canacao, Philippines, June 29, 40; US citizen; m; c 3. BIOCHEMISTRY. Educ: Univ RI, BS, 62, PhD(biochem), 59. Prof Exp: Asst prof biol sci, Va Polytech Inst & State Univ, 69-72; res assoc biochem, 72-73, ASST PROF CHEM, UNIV S FLA, 73- Mem: Am Chem Soc; AAAS. Res: Investigations in cardiac and smooth muscle mitochondrial physiology; role of calcium in cell injury. Mailing Add: Dept of Chem Univ of S Fla Tampa FL 33620

TSOLAS, ORESTES, b Istanbul, Turkey; US citizen. MOLECULAR BIOLOGY, BIOCHEMISTRY. Educ: Robert Col, Istanbul, BSc, 54; Cambridge Univ, MA, 57; Albert Einstein Col Med, PhD(molecular biol), 67. Prof Exp: ASST PROF MOLECULAR BIOL, ALBERT EINSTEIN COL MED, 70-; ASST MEM, ROCHE INST MOLECULAR BIOL, 72- Concurrent Pos: Prof, Univ Sao Paulo, 69-; adj prof, Rutgers Univ, 74-; vis lectr, Rotterdam Med Fac, Neth, 72. Res: Structure, function, biology and technology of enzymes. Mailing Add: Roche Inst of Molecular Biol Hoffman-La Roche Inc Nutley NJ 07110

TSONG, TIAN YOW, b Taiwan, Sept 6, 34; m 71; c 1. BIOPHYSICAL CHEMISTRY, BIOCHEMISTRY. Educ: Chung Hsing Univ, Taiwan, BS, 64; Yale Univ, MS, 67, MPh, 68, PhD(phys biochem), 69. Prof Exp: Asst prof, 72-75, ASSOC PROF PHYSIOL CHEM, SCH MED, JOHNS HOPKINS UNIV, 75- Concurrent Pos: Fel, Stanford Univ, 70-72; NSF res grant, Sch Med, Johns Hopkins Univ, 73-; NIH res grant, 75- Mem: Am Chem Soc; Biophys Soc; Am Soc Biol Chemists; AAAS. Res: Physical chemistry of proteins, nucleic acids and membrane lipids and its correlation to biological functions. Mailing Add: Dept of Physiol Chem Johns Hopkins Univ Sch of Med Baltimore MD 21205

TSONG, TIEN TZOU, b Taiwan, China, Sept 6, 34; m 64; c 3. SOLID STATE PHYSICS. Educ: Taiwan Norm Univ, BSc, 59; Pa State Univ, MS, 64, PhD(physics), 66. Prof Exp: Res assoc physics, 67-69, from asst prof to assoc prof, 69-74, PROF PHYSICS, PA STATE UNIV, 75- Mem: Am Phys Soc. Res: Surface physics; field effect on metal surface; field ionization; field desorption and field ion microscopy. Mailing Add: 104 Davey Lab Pa State Univ University Park PA 16802

TSONG, YUN YEN, b Taiwan, China, Jan 15, 37; m 67; c 1. BIOCHEMISTRY, ORGANIC CHEMISTRY. Educ: Nat Taiwan Univ, BS, 60; Univ Wis-Madison, PhD(biochem), 68. Prof Exp: Sr med chemist, Smith Kline & French Labs, 68-70; res assoc biochem, 70-72, STAFF SCIENTIST BIOCHEM, POP COUN, ROCKEFELLER UNIV, 72- Mem: AAAS; Am Chem Soc; Endocrine Soc. Res: Mechanism of action of steroid and peptide hormones; metabolism and microbial transformation of steroids. Mailing Add: Pop Coun Rockefeller Univ 66th & York Ave New York NY 10021

TSOU, KWAN CHUNG, b Shanghai, China, Apr 5, 22; m 49; c 3. ORGANIC CHEMISTRY, BIOCHEMISTRY. Educ: Nat Cent Univ, China, BS, 44; Univ Nebr, PhD(chem), 50. Prof Exp: Res assoc chem, Harvard Univ, 50-55; dir res, Monomer-Polymer & Dajac Labs, Borden Chem Co, 55-56, develop mgr, 56-69, head cent res lab, 60-63; ASSOC PROF CHEM, HARRISON DEPT SURG RES, SCH MED, UNIV PA, 63- Concurrent Pos: Assoc, Beth Israel Hosp, Boston, Mass, 53-55. Mem: AAAS; Am Chem Soc. Res: Enzyme, polymer, synthetic and structural organic chemistry; histochemistry and chemotherapy. Mailing Add: Harrison Dept of Surg Res Univ of Pa Sch of Med Philadelphia PA 19104

TSUANG, MING TSO, b Tainan, Taiwan, Nov 16, 31; m 58; c 3. PSYCHIATRY. Educ: Nat Taiwan Univ, MD, 57; Univ London, PhD(psychiat), 65. Prof Exp: Lectr psychiat & sr psychiatrist, Dept Neurol & Psychiat, Nat Taiwan Univ Hosp, 61-63; vis res worker psychiat, Med Res Coun Psychiat Genetics Res Unit, Maudsley Hosp & Inst Psychiat, Univ London, 63-65; lectr psychiat & sr psychiatrist, Dept Neurol & Psychiat, Nat Taiwan Univ Hosp, 65-68, assoc prof psychiat & sr psychiatrist, 68-71; vis assoc prof psychiat & staff psychiatrist, Barnes & Renard Hosp, Sch Med, Wash Univ, 71-72; assoc prof psychiat & staff psychiatrist, 72-75, PROF PSYCHIAT & PSYCHIATRIST, IOWA PSYCHOPATHIC HOSP, COL MED, UNIV IOWA, 75- Concurrent Pos: Res fel, Nat Coun Sci Develop, Repub China, 60-70; fel, Sino-Brit Fel Trust, UK, 63-65; collab investr, Int Pilot Study Schizophrenia, WHO, Geneva, Switz, 66-71; consult psychiatrist, Vet Admin Hosp, Iowa City, Iowa, 72-; chief staff psychiatrist, EWard, Psychopathic Hosp, Univ Iowa, 73- Mem: Psychiat Res Soc; AAAS; Am Psychopath Asn; Behav Genetics Asn; Am Psychiat Asn. Res: Long-term follow-up and family studies of schizophrenia, mania, depression and atypical psychoses; diagnostic classification of mental disorder; psychiatric genetics; clinical psychopharmological research. Mailing Add: Dept of Psychiat Univ of Iowa Col of Med Iowa City IA 52242

TSUBOI, KENNETH KAZ, b Seno, Japan, Feb 7, 22; nat US; m 47; c 2. BIOCHEMISTRY. Educ: St Thomas Col, BS, 44; Univ Minn, MS, 46, PhD(biochem), 48. Prof Exp: Asst physiol chem, Univ Minn, 44-47; res assoc path, Washington Univ, 48; res assoc oncol, Univ Kans Med Ctr, Kansas City, 48-51; res assoc biochem, Columbia Univ, 51-55; asst prof, Med Col, Cornell Univ, 55-60; assoc prof pediat, 60-66, sr res assoc, 66-73, ADJ PROF PEDIAT, SCH MED, STANFORD UNIV, 73- Concurrent Pos: Estab investr, Am Heart Asn, 59-64; vis prof, Univ Tokyo, 67. Mem: Am Soc Biol Chemists; Biophys Soc; Am Asn Cancer Res. Res: Cellular and muscle biochemistry; enzymology. Mailing Add: Dept of Pediat Stanford Univ Stanford CA 94305

TSUCHIYA, HENRY MITSUMASA, b Seattle, Wash, Dec 9, 14; m 41; c 2. BACTERIOLOGY. Educ: Univ Wash, Seattle, BS, 36, MS, 38; Univ Minn, PlhD(bact), 42. Prof Exp: Asst bact, Univ Minn, 40-42, res fel, 42-44, res assoc, 44-47; microbiologist, USDA, 47-56; assoc prof chem eng, 56-63, PROF CHEM ENG, UNIV MINN, MINNEAPOLIS, 63-, PROF MICROBIOL, 66- Mem: AAAS; Am Chem Soc; Ecol Soc Am; Am Soc Limnol & Oceanog; Am Inst Biol. Res: Industrial fermentations; sulfonamide chemotherapy; treatment of industrial wastes; physiology of microorganisms; biological polymerization and depolymerization; bioengineering; dynamics of microbial populations; ecology. Mailing Add: Dept of Chem Eng Univ of Minn Minneapolis MN 55455

TSUCHIYA, MIZUKI, b Matsuyama, Japan, May 2, 29; m 56; c 2. PHYSICAL OCEANOGRAPHY. Educ: Univ Tokyo, BS, 53, ScD(oceanog), 62. Prof Exp: Res asst geophys, Univ Tokyo, 55-60; instr oceanog, Meteorol Col, Japan Meteorol Agency, 60-64; res assoc, Johns Hopkins Univ, 64-67; lectr, Univ Tokyo, 67-69; asst res oceanographer, 69-73, ASSOC RES OCEANOGRAPHER, INST MARINE RESOURCES, SCRIPPS INST OCEANOG, 73- Honors & Awards: Okada Takematsu Prize, Oceanog Soc Japan, 67. Mem: Am Geophys Union; Oceanog Soc Japan. Res: Circulation and distributions of water characteristics in the ocean. Mailing Add: Scripps Inst of Oceanog La Jolla CA 92093

TSUCHIYA, TAKUMI, b Oita-Ken, Japan, Mar 10, 23; m 53; c 2. PLANT CYTOLOGY, PLANT GENETICS. Educ: Gifu Univ, BAgr, 43; Kyoto Univ, BAgr, 47, DAgr(genetics), 60. Prof Exp: Asst prof biol, Beppu Univ, 50-57; cytogeneticist, Kihara Inst Biol Res, 57-63; Nat Res Coun Can fel plant cytogenetics, Univ Man, 63-64; cytogeneticist, Children's Hosp, Winnipeg, 64-65; res assoc plant cytogenetics, Univ Man, 65-68; assoc prof plant cytogenetics, 68-73, PROF PLANT CYTOGENETICS, COLO STATE UNIV, 73- Concurrent Pos: Rockefeller Found travel grant insts & univs, US & Can, 61; coordr genetic & linkage studies barley, Int Barley Genetics Symp, 70-; chmn, Int Comt Nomenclature & Symbolization Barley Genes, 70-; chmn barley genetics comt, Am Barley Res Workers' Conf, 71- Mem: Genetics Soc Am; Am Soc Agron; Crop Sci Soc Am; Am Genetics Asn; Genetics Soc Can. Res: Cytogenetics of barley; cytogenetic and evolutionary studies of species of Gramineae; breeding of sugarbeet by chromosome engineering. Mailing Add: Dept of Agron Colo State Univ Ft Collins CO 80523

TSUDA, ROY TOSHIO, b Honolulu, Hawaii, Dec 25, 39; m 59; c 3. PHYCOLOGY. Educ: Univ Hawaii, BA, 63, MS, 66; Univ Wis-Milwaukee, PhD(bot), 70. Prof Exp: Instr biol, 67-68, asst prof dept biol & marine lab, 68-70, assoc prof marine sci, Marine Lab, 70-74, assoc dir marine sci, 69-71, PROF MARINE SCI & DIR MARINE LAB, UNIV GUAM, 74- Concurrent Pos: Consult on Acanthaster, Trust Territory Pac Islands, 70-72; algal consult, Environ Protection Agency Water Pollution Grants to Univ Guam, 70-; gen ed, Micronesica, Univ Guam, 72- Mem: Am Soc Limnol & Oceanog; Asn Trop Biol; Phycol Soc Am. Res: Taxonomy and ecology of tropical marine algae; primary productivity. Mailing Add: Marine Lab Univ of Guam PO EK Agana GU 96910

TSUEI, YEONG GING, b China, Feb 25, 32; m 62; c 4. APPLIED MECHANICS. Educ: Cheng Kung Univ, Taiwan, BSCE, 56; Colo State Univ, MCE, 60, PhD(fluid mech), 63. Prof Exp: Asst, Cheng Kung Univ, Taiwan, 56-58; from instr to asst prof, 61-68, ASSOC PROF MECH, UNIV CINCINNATI, 68- Mem: Am Inst Aeronaut & Astronaut; Am Soc Civil Engr; Am Soc Eng Educ. Res: Fluid and engineering mechanics. Mailing Add: Dept of Mech Eng Univ of Cincinnati Cincinnati OH 45221

TSUI, DANIEL CHEE, b Honan, China, Feb 28, 39; US citizen; m 64; c 2. SOLID STATE PHYSICS. Educ: Augustana Col, BA, 61; Univ Chicago, MS & PhD(physics), 67. Prof Exp: MEM TECH STAFF SOLID STATE PHYSICS, BELL LABS, 68- Mem: Am Phys Soc. Res: Electronic properties of metals, surface properties of semiconductors; low temperature physics. Mailing Add: Bell Labs Murray Hill NJ 07974

TSUJI, FREDERICK ICHIRO, b Honolulu, Hawaii, Aug 23, 23. BIOCHEMISTRY. Educ: Cornell Univ, AB, 46, MS, 48, PhD(biochem), 50. Prof Exp: Asst biochem & nutrit, Cornell Univ, 48-49; res biochemist, Children's Fund Mich, 49-50; asst prof biochem & pharmacol, Duquesne Univ, 50-52; res asst biol, Princeton Univ, 52-55; tech dir res lab, Vet Admin Hosp, Pittsburgh, Pa, 55-72; HANCOCK FEL, UNIV SOUTHERN CALIF, 72-; BIOCHEMIST, BRENTWOOD VET ADMIN HOSP, LOS ANGELES, 72- Concurrent Pos: Res assoc, Mercy Hosp, 51-52; Anathan fel inst res, Montefiore Hosp, 52; investr, Marine Biol Lab, Woods Hole, 53-54; lectr, Univ Pittsburgh, 56-65, from adj assoc prof to adj prof, 65-72; sr scientist, Te Vega Exped Pac Ocean, Hopkins Marine Sta, Stanford Univ, 66; mem, Alpha Helix Biol Exped to New Guinea, Scripps Inst Oceanog, Univ Calif, San Diego, 69; vis prof dept med chem, Fac Med, Kyoto Univ, Japan, 74. Mem: Am Soc Biol Chem; Am Chem Soc; Biophys Soc; Am Asn Immunol; Soc Gen Physiol. Res: Bioluminescence; chemistry of antigen-antibody reactions. Mailing Add: Dept of Biol Sci Univ of Southern Calif Los Angeles CA 90007

TSUJI, GORDON YUKIO, b Honolulu, Hawaii, July 31, 42; m 67; c 1. SOIL PHYSICS. Educ: Univ Hawaii, BS, 65, MS, 67; Purdue Univ, Lafayette, PhD(soil physics), 71. Prof Exp: Asst soil scientist dept agron & soil sci, 71-74, PROJ ASSOC, UNIV HAWAII/US AID BENCHMARK SOILS PROJ, UNIV HAWAII, 74- Concurrent Pos: AID fel, Univ Hawaii, 71- Mem: Am Soc Agron; Sigma Xi. Res: Water movement in soils; infiltration of water into soils; water distribution under drip irrigation; United States soil taxonomy and agricultural development. Mailing Add: Dept of Agron & Soil Sci Univ of Hawaii 3190 Maile Way Honolulu HI 96822

TSUJI, KIYOSHI, b Kyoto, Japan, May 31, 31; m 58; c 3. MICROBIOLOGY,

ANALYTICAL CHEMISTRY. Educ: Kyoto Univ, BS, 54; Univ Mass, MS, 56, PhD(food technol), 60. Prof Exp: Fel food sci, Rutgers Univ, 59-60; staff microbiol, Nat Canners Asn, Calif, 60-63; res assoc food sci, Mass Inst Technol, 63-64; res assoc anal res & develop, 64-71, sr res scientist, 71-74, SR SCIENTIST, CONTROL ANAL RES & DEVELOP, UPJOHN CO, 74- Honors & Awards: William E Upjohn Prize, Upjohn Co, 71. Mem: Am Chem Soc; Am Soc Microbiol; Inst Food Technologists. Res: Analytical microbiology; microbioassay automation; analysis of antibiotics by gas-liquid chromatography; high-performance liquid chromatography; sterility test; chromatography of antibiotics and vitamins; endotoxin detection by Limulus amebocyte lymphatics; environmental control. Mailing Add: Upjohn Co 7831-41-1 Kalamazoo MI 49001

TSUK, ANDREW GEORGE, b Budapest, Hungary, July 11, 32; US citizen; m 63; c 2. PHYSICAL CHEMISTRY, POLYMER CHEMISTRY. Educ: Budapest Polytech Inst, dipl chem eng, 54; Polytech Inst Brooklyn, PhD(chem), 64. Prof Exp: Engr, Indust Fermentations, Hungary, 54-56; res chemist, Schwarz Bioresearch, Inc, 58-62, tech asst to pres, 64-65, dir radiochem div, 65-66; sr res chemist, W R Grace & Co, Md, 66-72; GROUP LEADER PHARMACEUT DEVELOP, AYERST LABS, INC, 72- Mem: Am Chem Soc. Res: Polymers in pharmaceutical dosage forms; physical chemistry of polymers; polyelectrolytes; biomedical materials; biological macromolecules; ion-exchange and radioactive tracers. Mailing Add: Pharmaceut Develop Ayerst Labs Inc Rouses Point NY 12979

TSUKADA, MATSUO, b Nagano, Japan, Jan 4, 30; m 56; c 2. ECOLOGY, PALEOECOLOGY. Educ: Shinshu Univ, Japan, BS, 53; Osaka City Univ, MA, 58, PhD(biol), 61. Prof Exp: Japan Acad Sci fel, Osaka City Univ, 61; Seessel fel, Yale Univ, 61-62, res assoc palynology, 62-66, lectr & res assoc biol, 66-68; assoc prof, 69-71, PROF BOT, UNIV WASH, 71-, DIR LAB PALEOECOL, 69- Concurrent Pos: Sigma Xi res grant, Yale Univ, 63-64, Am Philos Soc res grant, 64-65; prin investr NSF res grants, Univ Wash, 70- Mem: Ecol Soc Am; Am Soc Limnol & Oceanog; Am Quaternary Asn; Am Asn Stratig Palynologists; Bot Soc Japan. Res: Present and past environmental changes on a global scale, mainly by means of modern and fossil plants, including pollen and also animals, chemicals and heavy metals, such as lead and cadmium. Mailing Add: Dept Bot Quaternary Res Ctr Univ of Wash Seattle WA 98195

TSUTAKAWA, ROBERT K, b Seattle, Wash, Mar 28, 30; m 61; c 3. STATISTICS. Educ: Univ Chicago, BS, 56, MS, 57, PhD(statist), 63. Prof Exp: Res specialist, Boeing Co, 58-60 & 63-65; res assoc statist, Univ Chicago, 65-68; ASSOC PROF STATIST, UNIV MO-COLUMBIA, 68- Mem: Am Statist Asn; Inst Math Statist. Res: Statistical inference. Mailing Add: Dept of Statist Univ of Mo Columbia MO 65201

TSUTSUI, ETHEL ASHWORTH, b Geneva, NY, May 31, 27; m 56; c 1. BIOCHEMISTRY. Educ: Keuka Col, BA, 48; Univ Rochester, PhD(biochem), 54. Prof Exp: Res assoc med sch med & dent, Univ Rochester, 53-55; Nat Cancer Inst fel, Sloan-Kettering Inst Cancer Res, 55-56; lectr, Tokyo Med & Dent Univ, Japan, 56-57; asst prof biol & res biochemist, C F Kettering Found, Antioch Col, 57-60; res assoc Inst Cancer Res, Col Physicians & Surgeons, Columbia Univ, 60-63, res assoc dept biochem, 63-65; asst prof biol sci, Hunter Col, 65-69; assoc prof biol, 69-71, ASSOC PROF BIOCHEM & BIOPHYS, TEX A&M UNIV, 71- Mem: AAAS; Am Asn Cancer Res; Am Chem Soc; fel NY Acad Sci; Harvey Soc. Res: Enzymatic methylation of nucleic acids; biochemistry of cancer cells; biochemical control mechanisms. Mailing Add: Dept of Biochem & Biophysics Tex A&M Univ College Station TX 77843

TSUTSUI, MINORU, b Wakayama, Japan, Mar 31, 18; nat US; m 56; c 1. ORGANOMETALLIC CHEMISTRY. Educ: Gifu Univ, Japan, BA, 38; Univ Tokyo, MS, 41; Yale Univ, MS, 53, PhD(chem), 54; Nagoya Univ, DSc(chem), 60. Prof Exp: Asst prof chem, Univ Tokyo, 50-53, assoc pharmaceut inst, 56-57; vis res fel, Sloan-Kettering Inst Cancer Res, NY, 54-56; res chemist cent res lab, Monsanto Chem Co, 57-60; res scientist res div, NY Univ, 60-68, assoc prof chem, 66-68, lectr, 65; PROF CHEM, TEX A&M UNIV, 69- Concurrent Pos: Mem spec proj, Mass Inst Technol, 51; vis res fel, Brookhaven Nat Lab, 55; consult, Union Carbide Co, 63-73, Maruzen Oil Co, 63, Glidden Co, 64 & 65, Toyo Rayon Corp, 65, Kurashiki Rayon Co, 66, Tokuyama Soda Co, 67 & Celanese Co, 73-; mem bd, Southwest Catalysis Soc, 72-74; partic, US-Russia Cultural Exchange Prog; mem bd, Nard Inst, Japan. Honors & Awards: Morrison Award, NY Acad Sci, 60. Mem: AAAS; Am Chem Soc; NY Acad Sci (vpres, 65-66, pres-elect, 67, pres, 68); The Chem Soc; Chem Soc Japan. Res: Organotransition-metal chemistry; catalysis, conductors and biological aspects; diagnostic and chemotherapeutic tumor localizers. Mailing Add: Dept of Chem Tex A&M Univ College Station TX 77843

TSUYUKI, HIROSHI, biochemistry, see 12th edition

TSUZUKI, TOSHIO, b Japan, Dec 22, 24; nat US; m 51; c 3. PHYSICAL CHEMISTRY. Educ: Univ Tokyo, BS, 46; Univ Ill, PhD(chem), 57. Prof Exp: Asst phys chem, Univ Ill, 53-55; res fel, Purdue Univ, 55-57; res chemist, Corn Prods Co, 57-61; proj leader, Am Maize Prods Co, 61-65; sr scientist, 65-70, ASSOC DIR RES, DEVRO, INC, 70- Res: Surface and colloid chemistry; thermochemistry; carbohydrate; food; collagen. Mailing Add: Devro Inc Res Div Southside Ave Somerville NJ 08876

TTERLIKKIS, LAMBROS, b Beirut, Lebanon, Oct 17, 34; US citizen; m 60; c 2. PHYSICS. Educ: Walla Walla Col, BSc, 59; Univ Denver, MSc, 62; Univ Calif, Riverside, PhD(physics), 68. Prof Exp: Res asst solid state physics, Denver Res Inst, Univ Denver, 59-62; assoc physicist, IBM Corp, 62-63; NIH res assoc biophys, Inst Molecular Biophys, Fla State Univ, 68-70; asst prof, 70-74, ASSOC PROF PHYS PHARMACEUT, FLA A&M UNIV, 74- Mem: AAAS; Soc Nuclear Med; Am Pharmaceut Asn. Res: Pharmacokinetics of drug metabolism; physicochemical properties of drugs; solid state physics; optical properties of biopolymers. Mailing Add: Sch of Pharm Fla A&M Univ Tallahassee FL 32307

TU, ANTHONY T, b Taipei, Formosa, Apr 12, 30; US citizen; m 57; c 5. BIOCHEMISTRY. Educ: Nat Taiwan Univ, BS, 53; Univ Notre Dame, MS, 56; Stanford Univ, PhD(biochem), 60. Prof Exp: Res assoc biochem, Yale Univ, 61-62; asst prof, Utah State Univ, 62-67; assoc prof, 67-70, PROF BIOCHEM, COLO STATE UNIV, 70- Concurrent Pos: NIH career develop award, 69-73. Mem: Am Chem Soc; Am Soc Biol Chem; Int Soc Toxinology. Res: Snake venom toxins and enzymes; hemepeptides; metal-nucleotide interaction. Mailing Add: Dept of Biochem Colo State Univ Ft Collins CO 80521

TU, CHEN CHUAN, b Husin, Oct 5, 18; nat US; m 47; c 2. CHEMISTRY. Educ: Chinese Nat Col Pharm, dipl, 42; Purdue Univ, MS, 49, PhD, 51. Prof Exp: Asst, Chinese Nat Chekiang Univ, 42-47; asst, Purdue Univ, 47-51, res fel, 51-52; res asst, Inst Paper Chem, 52-56; res assoc, 56-57, SR SCIENTIST, EXP STA, HAWAIIAN SUGAR PLANTERS' ASN, 57- Mem: Am Chem Soc; Int Soc Sugarcane Technol. Res: Biochemistry; natural products; sugar technology; food science. Mailing Add: Exp Sta Hawaiian Sugar Planters' Asn Honolulu HI 96822

TU, CHIN MING, b Hsiuchu, Taiwan, Dec 14, 32; Can citizen; m 62; c 3. MICROBIOLOGY, BIOCHEMISTRY. Educ: Chung Hsing Univ, Taiwan, BSc, 56; Univ Sask, MSc, 63; Ore State Univ, PhD(microbiol), 66. Prof Exp: Asst org chem & soil fertil dept agr chem, Chung Hsing Univ, Taiwan, 58-60; asst soil sci, Univ Sask, 60-62; asst microbiol, Ore State Univ, 63-66; RES SCIENTIST RES INST, CAN DEPT AGR, 66- Mem: Am Soc Microbiol; Can Soc Microbiol. Res: Interaction between pesticides and soil microorganisms. Mailing Add: Res Inst Can Dept of Agr London ON Can

TU, CHINGKUANG, b Keichou, China, Feb 1, 44; m 72; c 2. BIOCHEMISTRY, PHARMACOLOGY. Educ: Chunghsin Univ, Taiwan, BS, 68; Univ Miami, PhD(chem), 72. Prof Exp: Fel chem, 72-73, fel biochem & pharmacol, 73-75, ASSOC INSTR BIOCHEM & PHARMACOL, UNIV FLA, 75- Mem: Am Chem Soc. Res: Study enzyme carbonic anhydrase, mechanism, kinetics and inhibition, by stable isotope technique; GC-MS-Data system application to medical science. Mailing Add: Dept of Pharmacol & Therapeut Univ Fla J267 JHM Hlth Ctr Gainesville FL 32610

TU, HSIN-YUAN, mineralogy, soil science, see 12th edition

TU, SHU-I, b Chungking, China, Jan 3, 43; m 69; c 1. BIOPHYSICAL CHEMISTRY. Educ: Nat Taiwan Univ, BS, 65; Yale Univ, MPhil, 68, PhD(chem), 69. Prof Exp: Res assoc biochem, Yale Univ, 69-72; res asst prof, State Univ NY Buffalo, 72-74; ASST PROF CHEM, STATE UNIV NY STONY BROOK, 74- Mem: Am Chem Soc; Biophys Soc; Sigma Xi. Res: Structure of inner mitochondrial membrane, mechanism of oxidative phosphorylation and photophosphorylation and conversion of light into electric energy in synthetic membrane system. Mailing Add: Dept of Chem State Univ NY Stony Brook NY 11794

TU, SHU-TUNG (ROBERT), organic chemistry, polymer chemistry, see 12th edition

TU, YIH-O, b Kiangsi, China, Jan 8, 20; m 60; c 1. APPLIED MATHEMATICS. Educ: Col Ord Eng, Chungking, China, BS, 46; Carnegie Inst Technol, MS, 54; Rensselaer Polytech Inst, PhD(math), 59. Prof Exp: Designer mech eng, Rockwell Mfg Co, Pa, 53-55; STAFF MATHEMATICIAN, IBM CORP, 59- Mem: Am Math Soc; Am Phys Soc; Am Soc Mech Eng; Soc Eng Sci; Soc Indust & Appl Math. Res: Fluid mechanics; elasticity; vibration and elastic stability; continuum mechanics. Mailing Add: IBM Res Lab 5600 Cottle Rd San Jose CA 95193

TUAN, DEBBIE FU-TAI, b Kiangsu, China, Feb 2, 30. PHYSICAL CHEMISTRY, CHEMICAL PHYSICS. Educ: Taiwan Univ, BS, 54, MS, 58; Yale Univ, MS, 60, PhD(chem), 61. Prof Exp: Teaching asst chem, Taiwan Univ, 54-55; NSF res fel, Yale Univ, 61-64; NASA res grant & proj assoc theoret chem inst, Univ Wis, 64-65; from asst prof to assoc prof, 65-73, summer res fels, 66, 68 & 71, PROF CHEM, KENT STATE UNIV, 73- Concurrent Pos: Res fel, Harvard Univ, 70. Mem: Am Phys Soc; Am Chem Soc; Sigma Xi. Res: Many electron theory of atoms and molecules; perturbation theory; other applications of quantum mechanics to chemical problems. Mailing Add: Dept of Chem Kent State Univ Kent OH 44242

TUAN, RENDEH, physical science, see 12th edition

TUAN, SAN FU, b Tientsin, China, May 14, 32; m 63; c 3. THEORETICAL PHYSICS, APPLIED MATHEMATICS. Educ: Oxford Univ, BA, 54, MA, 58; Univ Calif, PhD(appl math), 58. Prof Exp: Res assoc, Univ Chicago, 58-60; asst prof, Brown Univ, 60-62; assoc prof, Purdue Univ, 62-65; vis prof, Univ Hawaii, 65-66; mem inst adv study, Princeton Univ, 66; PROF THEORET PHYSICS, UNIV HAWAII, 66- Concurrent Pos: Mackinnon scholar, Magdalen Col, Oxford Univ, 51-54; consult, Argonne Nat Lab, 63-70; John S Guggenheim fel, 65-66; dir & co-ed proc, Second, Third, Fifth & Sixth Hawaii Topical Conf Particle Physics, 67, 69, 73 & 75; vis lectr, Bariloche Atomic Ctr, Argentina & Univ Buenos Aires, 69-70; vis lectr, US-China Sci Coop Prog, 70-71. Mem: Am Math Soc; fel Am Phys Soc. Res: Mathematical physics; theory of elementary particles; superconductivity; political science. Mailing Add: Dept of Physics & Astron Univ of Hawaii Honolulu HI 96822

TUAN, TAI-FU, b Tientsin, China, Sept 7, 29; US citizen; m 68. THEORETICAL PHYSICS. Educ: Cambridge Univ, BA, 51; La State Univ, MS, 53; Univ Pittsburgh, PhD(physics), 59. Prof Exp: Instr physics, Northwestern Univ, 59-60; univ res fel, Univ Birmingham, 61-64, Dept Sci & Indust Res res fel, 64-65; from asst prof to assoc prof, 65-71, PROF PHYSICS, UNIV CINCINNATI, 71- Concurrent Pos: US Air Force res grant, 70- Mem: Am Phys Soc; Am Geophys Union. Res: Scattering theory in atomic physics; atmospheric physics; research in airglow, gravity waves and magnetohydrodynamics models for magnetosphere. Mailing Add: Dept of Physics Univ of Cincinnati Cincinnati OH 45221

TUAN, YI-FU, b Tientsin, China, Dec 5, 30. CULTURAL GEOGRAPHY. Educ: Oxford Univ, BA, 51, MA, 55; Univ Calif, Berkeley, PhD(geog), 57. Prof Exp: Instr geog, Ind Univ, 56-58; fel statist, Univ Chicago, 58-59; from asst to assoc prof, Univ NMex, 59-65; assoc prof, Univ Toronto, 66-68; PROF GEOG, UNIV MINN, MINNEAPOLIS, 68- Concurrent Pos: John Simon Guggenheim fel, 67-68. Mem: AAAS; Asn Am Geogr; Asn Asian Studies; Am Geog Soc; Asn Study Man-Environ Rels. Res: Attitudes to environment; humanistic geography. Mailing Add: Dept of Geog Univ of Minn Minneapolis MN 55455

TUBB, RICHARD ARNOLD, b Weatherford, Okla, Dec 18, 31; m 57; c 2. LIMNOLOGY, FISHERIES. Educ: Okla State Univ, BS, 58, MS, 60, PhD(zool), 63. Prof Exp: Res asst aquatic biol lab, Okla State Univ, 60-62; asst prof biol, Univ NDak, 63-66; asst leader fisheries, SDak Coop Fishery Unit, 66-67; leader, Ohio Coop Fishery Unit, 67-75; PROF & HEAD DEPT FISHERIES & WILDLIFE MGT, ORE STATE UNIV, , 75- Mem: Am Soc Limnol & Oceanog; Am Fisheries Soc; Int Asn Theoret & Appl Limnol; Am Inst Fishery Res Biol. Res: Herbivorous insect population in oil refinery effluent holding pond series; investigations of whirling disease of trout; freshwater bivalves as stream monitors for pesticides; environmental impact of nuclear power plants and stream channelization. Mailing Add: Dept of Fisheries & Wildlife Mgt Ore State Univ Corvallis OR 97331

TUBBS, CARL H, forestry, forest ecology, see 12th edition

TUBBS, ELDRED FRANK, b Buffalo, NY, Mar 31, 24; m 49; c 3. OPTICS, ATOMIC SPECTROSCOPY. Educ: Carnegie Inst Technol, BS, 49; Johns Hopkins Univ, PhD(physics), 56. Prof Exp: Sr physicist res ctr, Am Optical Co, 55-58; res physicist microwave physics lab, Sylvania Elec Prod Inc, 58-60; res physicist, WCoast Br, Gen Tel & Electronics Labs, Inc, 60-63; from asst prof to assoc prof physics, 63-72, PROF PHYSICS, HARVEY MUDD COL, 72- Honors & Awards: Prize, Am Asn Physics Teachers, 67. Mem: Optical Soc Am; Am Asn Physics Teachers. Res: Optical instruments; interferometry; absolute f-values; teaching demonstrations. Mailing Add: Dept of Physics Harvey Mudd Col Claremont CA 91711

TUBBS, ROBERT KENNETH, b Gary, Ind, Nov 25, 36; m 56; c 3. COLLOID CHEMISTRY, SURFACE CHEMISTRY. Educ: Ohio State Univ, BS, 58, PhD(colloid chem), 62. Prof Exp: Res chemist electrochem dept, Del, 62-67, staff scientist, 67-68, res supvr, 68-70, gen tech supt indust chem dept, NY, 70-73, sr res supvr, Plastics Dept, 73-74, prod mgr, 74-75, DEVELOP MGR PLASTICS DEPT, E I DU PONT DE NEMOURS & CO, INC, 75- Mem: Am Chem Soc. Res: Structure and interactions of macromolecules; kinetics of polymerization; molecular biology; emulsion polymerization; coatings; adhesives. Mailing Add: Plastics Dept E I du Pont de Nemours & Co Inc Wilmington DE 19809

TUBIASH, HASKELL SOLOMON, b Boston, Mass, Feb 7, 13; m 54; c 1. MICROBIOLOGY. Educ: Univ Mass, BS, 37; Univ NC, MS, 47. Prof Exp: Jr bacteriologist, Mass State Dept Pub Health, 40-43; sanit bacteriologist, USPHS, Ohio, 47-48; bacteriologist, USDA, Md & Colo, 48-52, microbiologist, Plum Island Animal Dis Lab, 52-61, res microbiologist, US Bur Com Fisheries, 61-70, res microbiologist, 70, TECH ED, MID ATLANTIC COASTAL FISHERIES CTR, NAT OCEANIC & ATMOSPHERIC ADMIN, 70- Mem: AAAS; Am Soc Microbiol; NY Acad Sci. Res: Microbiology of foods; sanitary bacteriology; clostridial toxins; animal viruses and tissue culture; diseases of bivalve mollusks; technical editing. Mailing Add: Oxford Lab Nat Oceanic & Atmospheric Admin Oxford MD 21654

TUBIS, ARNOLD, b Pottstown, Pa, Mar 28, 32; m 59; c 2. THEORETICAL PHYSICS. Educ: Mass Inst Technol, BS, 54, PhD(theoret physics), 59. Prof Exp: Res asst, Mass Inst Technol, 54-57; asst prof physics, Worcester Polytech Inst, 58-60; res assoc, 60-62, from asst prof to assoc prof, 62-69, asst head dept, 66-73, PROF PHYSICS, PURDUE UNIV, 69- Concurrent Pos: Asst physicst, Brookhaven Nat Lab, 59; res assoc, Argonne Nat Lab, 61; vis physicist, Lawrence Radiation Lab, 63, Stanford Linear Accelerator Ctr, 71 & Los Alamos Sci Lab, 72. Mem: AAAS; Am Phys Soc; Am Asn Physics Teachers. Res: Theory of atomic structure; theory of interactions of nuclei and elementary particles. Mailing Add: Dept of Physics Purdue Univ West Lafayette IN 47906

TUBIS, MANUEL, b Philadelphia, Pa, July 14, 09; m 36; c 1. NUCLEAR MEDICINE. Educ: Philadelphia Col Pharm, BSc, 31; Univ Pa, MSc, 32; Univ Tokyo, PhD(pharmaceut sci), 66. Prof Exp: Chemist, US Food & Drug Admin, 35-44; pharmaceut res chemist, Wyeth, Inc, 44-46; res chemist, Dartell Labs, 46-47; tech dir, Am Biochem Corp, 47-48; biochemist, 48-70, CHIEF BIOCHEMIST NUCLEAR MED SERV, VET ADMIN HOSP, WADSWORTH CTR, 70- Concurrent Pos: Asst prof sch med, Univ Calif, Los Angeles, 53-68; consult nuclear med & radiopharm, Int Atomic Energy Agency, Vienna; adj prof biomed chem & co-dir radiopharm prog, Sch Pharm, Univ Southern Calif, 74-, consult radiopharmaceut & nuclear med, 74- Honors & Awards: Super Performance Award, US Vet Admin, 75. Mem: Fel & hon mem, Am Inst Chem; Am Chem Soc; Soc Nuclear Med; Japanese Soc Nuclear Med; Asn Latin Am Soc Biol & Nuclear Med. Res: Application of radioisotopes to biochemistry, medicine and pharmacy; research and development of new labeled radiopharmaceuticals; supervision of preparation, production and quality control. Mailing Add: Nuclear Med Serv Vet Admin Hosp Wadsworth Ctr Room 209 Bldg 114 Los Angeles CA 90073

TUCCI, ANTHONY FREDERICK, b Brooklyn, NY, Oct 23, 35; m 57; c 2. BIOCHEMISTRY, MICROBIOLOGY. Educ: City Col New York, BS, 59; Albert Einstein Col Med, PhD(biochem), 66. Prof Exp: Res assoc muscle biochem, Inst Muscle Dis, 65-66; from asst mem to assoc mem amino acid biosynthesis, 66-71, DIR METAB LAB, DIV PEDIAT, ALBERT EINSTEIN MED CTR, PHILADELPHIA, 71- Concurrent Pos: NSF grants, Albert Einstein Med Ctr, Philadelphia, 66-67, NIH grants, 67-73. Mem: AAAS; Am Soc Biol Chemists; Am Soc Microbiol; Am Chem Soc. Res: Amino acid biosynthesis and metabolism; inborn errors of amino acid metabolism; biosynthetic control mechanisms. Mailing Add: 237 Willets Ave West Hempstead NY 11552

TUCCI, EDMOND RAYMOND, b Pawtucket, RI, May 31, 33; m 63. PHYSICAL INORGANIC CHEMISTRY. Educ: RI Sch Design, BS, 55; St Louis Univ, MS, 57; Duquesne Univ, PhD(phys chem), 61. Prof Exp: Res chemist, Olin Mathieson Chem Corp, 57-58; chemist, Process Res Div, Gulf Res & Develop Co, 61-62, res chemist, 62-66, sr res chemist, 66-73; supvr, Engelhard Minerals & Chem Co, 73-75; MGR COMMERCIAL DEVELOP, MATTHEY BISHOP INC, 76- Mem: AAAS; Am Chem Soc; Am Asn Textile Chemists & Colorists; Am Inst Chemists. Res: Homogeneous and heterogeneous catalysis; organometallic compounds or metal carbonyls; petrochemical process development; energy programs in coal gasification, coal liquefaction, methanation and shift catalysis; synthetic natural gas production. Mailing Add: Admin Bldg Matthey Bishop Inc Malvern PA 19355

TUCCI, JAMES VINCENT, b Hollis, NY, Feb 13, 39; m 62; c 1. CHEMICAL PHYSICS. Educ: Hofstra Univ, BA, 62; Univ Mass, MS, 66, PhD(chem), 67. Prof Exp: Instr physics, 66-67, from asst prof to assoc prof, 67-73, actg chmn dept, 67-71, PROF PHYSICS, UNIV BRIDGEPORT, 73-, CHMN DEPT, 72- Concurrent Pos: NSF grant, 67-70; Conn Res Comn res grant, 68-70. Mem: NY Acad Sci; Optical Soc Am. Res: Laser Raman spectroscopy; vibrational spectroscopy. Mailing Add: Dept of Physics Univ of Bridgeport Bridgeport CT 06602

TUCCIARONE, JOHN PETER, b New York, NY, Apr 9, 40; m 63; c 3. MATHEMATICS, LAW. Educ: Fordham Univ, BS, 61; St John's Univ, NY, MA, 63, JD, 66; NY Univ, PhD(math), 69. Prof Exp: Asst prof math, St John's Univ, NY, 62-74; ASSOC PROF MATH, MERCY COL, 74-; DIR PROB ANAL CORP, 71- Concurrent Pos: Attorney pvt practice, NY, 70- Mem: Math Asn Am. Res: Numerical analysis; computer science; application of computers to instruction. Mailing Add: 92 Highland Rd Thornwood NY 10594

TUCCIO, SAM ANTHONY, b Rochester, NY, Jan 15, 39; m 72; c 3. LASERS. Educ: Univ Rochester, BS, 65. Prof Exp: Syst engr, Eastman Kodak Co, 65-68, res physicist, 68-73; staff physicist, Lawrence Livermore Labs, 73-75; STAFF PHYSICIST, ALLIED CHEM CORP, 75- Honors & Awards: IR-100 Award, Indust Res Mag, 71. Res: Selectively induced chemical reactions through the application of tunable lasers. Mailing Add: Mats Res Ctr Allied Chem Corp Morristown NJ 07960

TUCHINSKY, PHILIP MARTIN, b Philadelphia, Pa, June 17, 45. MATHEMATICS. Educ: Queens Col, BA, 66; Courant Inst, NY Univ, MS, 68, PhD(math), 71. Prof Exp: Instr math, NY Univ, 69-70; adj instr, Cooper Union Advan Sci & Art, 70-71; asst prof, Kalamazoo Col, 71-72; ASST PROF MATH, OHIO WESLEYAN UNIV, 72- Concurrent Pos: Great Lakes Col Asn teaching fel, Lilly Found Grant, 75. Mem: Am Math Soc; Math Asn Am; Am Asn Higher Educ. Res: Development of applications of mathematics to other disciplines for use in undergraduate teaching. Mailing Add: Dept of Math Ohio Wesleyan Univ Delaware OH 43015

TUCHMAN, ALBERT, b Brooklyn, NY, July 1, 35; m 57; c 4. PHYSICS. Educ: Yeshiva Univ, BA, 56; Mass Inst Technol, PhD(physics), 63. Prof Exp: Staff scientist res & develop div, 63-64, group leader propulsion, 64-66, sect chief plasma physics space systs div, 66-70, sr consult scientist systs div, 70-73, PRIN STAFF SCIENTIST PHYSICS, AVCO CORP, 73- Mem: Asn Orthodox Jewish Scientists. Res: Plasma, quantum and elementary particle physics; optical and infrared sources; infrared optical design; missile countermeasures; optical, infrared and laser countermeasures; electrode phenomena; re-entry simulation and ablation measurement; satellite detection. Mailing Add: Avco Corp Systs Div 201 Lowell St Wilmington MA 01887

TUCK, DENNIS GEORGE, b UK, Apr 8, 29; m 56; c 3. INORGANIC CHEMISTRY, RADIOCHEMISTRY. Educ: Univ Durham, BSc, 49, PhD(chem), 56, DSc, 71. Prof Exp: Brit Coun fel, Inst du Radium, Paris, France, 52-53; sci officer chem, Windscale Works, UK Atomic Energy Auth, Eng, 53-56; lectr inorg chem, Univ Nottingham, 59-65; from assoc prof to prof chem, Simon Fraser Univ, 66-72; PROF CHEM & HEAD DEPT, UNIV WINDSOR, 72- Concurrent Pos: Res fel chem, Univ Manchester, 56-59; res fel lab nuclear sci, Cornell Univ, 57-58; vis expert, Concepcion Univ, 64; mem chem grant selection comt, Nat Res Coun Can, 72-74. Mem: The Chem Soc; fel Royal Inst Chem; fel Chem Inst Can. Res: Coordination chemistry of non-transition metals, especially indium; complexes in aqueous solution; use of radiochemical methods in inorganic chemistry. Mailing Add: Dept of Chem Univ of Windsor Windsor ON Can

TUCK, GARY ALLEN, biostatistics, behavioral science, see 12th edition

TUCK, JAMES LESLIE, b Eng, Jan 9, 10; nat US; m 38; c 2. PHYSICS. Educ: Univ Manchester, BSc, 31, MSc, 32; Oxford Univ, MA, 48. Prof Exp: Demonstr phys chem, Univ Manchester, 34; res assoc inst nuclear studies, Univ Chicago, 49; mem staff theoret div, Los Alamos Sci Lab, Univ Calif, 50-54, assoc leader physics div, 54-73; RETIRED. Educ: Guggenheim fel, 62; consult, Los Alamos Sci Lab, 73-; Walker-Ames prof, Univ Wash, Seattle, 74; vis fel, Univ Waikato, NZ, 75; Regents' fel, Univ Calif, San Diego, 75. Honors & Awards: Order Brit Empire Medal, 44. Mem: Fel Am Phys Soc. Res: Thermonuclear physics; ball lightning; fluid dynamics of explosives; accelerators; nuclear cross sections. Mailing Add: 2502 35th St Los Alamos NM 87544

TUCK, LEO DALLAS, b San Francisco, Calif, Oct 12, 16; m 53; c 2. PHYSICAL CHEMISTRY. Educ: Univ Calif, AB, 39, PhD(chem), 48. Prof Exp: Res assoc radiation lab, Univ Calif, 42; res assoc chem dept, Univ Chicago, 442-43; res assoc radio res lab, Harvard Univ, 43-45; lectr & res asst chem, 48-50, instr, 50-51, from asst prof to assoc prof, 51-63, vchancellor acad affairs, 71-73, PROF CHEM & PHARMACEUT CHEM, SCH PHARM, UNIV CALIF, SAN FRANCISCO, 63- Mem: Fel AAAS; Am Chem Soc; Am Phys Soc; Am Pharmaceut Asn. Res: Thermodynamics and electrochemistry; thermodynamics of nonisothermal systems; chemistry of boron and uranium compounds; electrolytes in aqueous and nonaqueous solutions; microwave electronics; chemistry of free radicals; magnetic resonance spectroscopy. Mailing Add: Univ of Calif Sch of Pharm San Francisco Med Ctr San Francisco CA 94143

TUCK, NORMAN GORDON MAXWELL, b Lavoy, Alta, Can, Apr 13, 22; m 45; c 4. PULP CHEMISTRY. Educ: Univ Alta, BSc, 44, MSc, 45; McGill Univ, PhD, 47. Prof Exp: Res assoc, Pulp & Paper Res Inst Can, 47-48; chemist, Howard Smith Paper Mills, Ltd, 48-59, group leader fiber processing, 59-64; tech dir, Domtar Pulp & Paper Lts, 64-66, mkt mgr, Domtar Paper Mills Ltd, Eng, 66-67, develop mgr, Fine Papers Div, Domestic & Int Overseas, 67-69, TECH DEVELOP DIR, DOMTAR FINE PAPERS LTD, 69- Mem: Tech Asn Pulp & Paper Indust; Can Pulp & Paper Asn. Res: Fibre processing. Mailing Add: Domtar Fine Papers Ltd PO Box 7211 Montreal 111 PQ Can

TUCKER, ALAN CURTISS, b Princeton, NJ, July 6, 43; m 68. MATHEMATICS. Educ: Harvard Univ, BA, 65; Stanford Univ, MS, 67, PhD(math), 69. Prof Exp: Vis asst prof math, Math Res Ctr, Univ Wis-Madison, 69-70; ASST PROF APPL MATH & STATIST, STATE UNIV NY STONY BROOK, 70- Concurrent Pos: Res consult, Rand Corp, 64- Mem: Am Math Soc; Math Asn Am. Res: Extremal characterization problems in graph theory; zero-one matrices; network flows. Mailing Add: Dept Appl Math & Statist State Univ of NY Stony Brook NY 11794

TUCKER, ALBERT WILLIAM, b Oshawa, Ont, Nov 28, 05; nat US; m 64; c 3. MATHEMATICS. Educ: Univ Toronto, BA, 28, MA, 29; Princeton Univ, PhD(topol), 32. Hon Degrees: ScD, Dartmouth Col, 61. Prof Exp: Instr math, Princeton Univ, 29-31; Nat Res Coun fel, Cambridge Univ, Harvard Univ & Univ Chicago, 32-33; from instr to prof, 33-74, chmn dept, 53-63, EMER PROF MATH, PRINCETON UNIV, 74- Concurrent Pos: Ed, Princeton Math Series, 38- & Annals Math Studies, 40-49; res mathematician, Nat Defense Res Comt, 41-43; assoc dir appl math group, Off Sci Res & Develop, 44-45; dir logistics proj, US Off Naval Res, 48-72; vis prof, Stanford Univ, 49-50; Philips vis, Haverford Col, 53-54 & 58-59; mem basic sci prog comt, Alfred P Sloan Found, 55-59; chmn comn math, Col Entrance Exam Bd, 55-59; Fulbright lectr, Australia, 56; mem, Presidential Comt on Nat Medal Sci, 62-66; vis prof, Dartmouth Col, 63. Honors & Awards: Distinguished Serv Award, Math Asn Am, 68. Mem: AAAS (vpres, 57 & 66); Am Math Soc; Soc Indust & Appl Math; Math Asn Am (pres, 61-62); Can Math Cong. Res: Topology; differential geometry; theory of games and programming; linear algebra; combinatorics. Mailing Add: 37 Lake Lane Princeton NJ 08540

TUCKER, ALLEN BRINK, b Highland, Ind, Oct 12, 36; m 63; c 2. NUCLEAR PHYSICS. Educ: Mass Inst Technol, BS, 58; Stanford Univ, PhD(physics), 65. Prof Exp: Asst prof physics, San Jose State Col, 63-65 & Iowa State Univ, 65-70; ASSOC PROF PHYSICS, SAN JOSE STATE UNIV, 70- Mem: Am Phys Soc; Am Asn Physics Teachers. Res: Neutron scattering; nuclear spectroscopy; delayed neutrons. Mailing Add: Dept of Physics San Jose State Univ San Jose CA 95114

TUCKER, ALLEN BROWN, b Worcester, Mass, Feb 19, 42; m 65; c 1. MATHEMATICS. Educ: Wesleyan Univ, BA, 63; Northwestern Univ, MS, 69, PhD(appl math), 70. Prof Exp: Systs analyst, Norton Co, 63-67; asst prof comput sci, Univ Mo-Rolla, 70-71; ASST PROF COMPUT SCI, GEORGETOWN UNIV, 71-; VPRES, TABOR, INC-SYSTS CONSULT, 74- Concurrent Pos: Assoc ed, J Comput Lang, 75- Mem: Asn Comput Mach; Sigma Xi. Res: Formal languages; automata theory; programming languages; computer applications. Mailing Add: Dept of Math Georgetown Univ 37th & O St NW Washington DC 20007

TUCKER, ARTHUR SMITH, b Hopei, China, May 2, 13; m 45; c 2. MEDICINE, RADIOLOGY. Educ: Oberlin Col, AB, 35; Yale Univ, MD, 39. Prof Exp: Radiologist, Atomic Bomb Casualty Comn, Japan, 48-50; instr radiol, Univ Calif, 50-51; radiologist, Cleveland Clin, Ohio, 51-56; from asst prof to assoc prof, 56-72, PROF RADIOL, SCH MED, CASE WESTERN RESERVE UNIV, 72- Mem: Roentgen Ray Soc; Radiol Soc NAm; AMA. Res: Pediatric radiology. Mailing Add: Sch of Med Dept of Radiol Case Western Reserve Univ Cleveland OH 44106

TUCKER, BERT E, JR, physics, see 12th edition

TUCKER, BILLY BOB, b Cheyenne, Okla, Jan 13, 28; m 49; c 3. AGRONOMY. Educ: Okla State Univ, BS, 52, MS, 53; Univ Ill, PhD, 55. Prof Exp: Soil scientist,

Agr Res Serv, USDA, 55-56; from asst prof to assoc prof agron, 56-67, PROF AGRON, OKLA STATE UNIV, 67- Mem: Am Soc Agron; Soil Sci Soc Am; Soil Conserv Soc Am. Res: Soil management, especially improvment and maintenance of soil productivity; soil chemistry, plant nutrition and fertilizer technology. Mailing Add: Dept of Agron Okla State Univ Stillwater OK 74074

TUCKER, CHARLES EUGENE, b Montgomery, Ala, July 2, 33; m 58; c 2. BIOLOGY. Educ: Huntingdon Col, BA, 59; Univ Ala, MS, 65, PhD(biol), 67. Prof Exp: ASSOC PROF BIOL, LIVINGSTON UNIV, 67-, CHMN DIV NATURAL SCI & MATH, 70-, ASSOC DEAN HEALTH RELATED PROG, 75- Mem: Am Soc Ichthyol & Herpet; Sigma Xi. Res: Vertebrate field zoology; ichthyology; survey of fishes. Mailing Add: Sta 7 Livingston Univ Livingston AL 35470

TUCKER, CHARLES LEROY, JR, b Winston-Salem, NC, May 19, 21; m 49; c 2. PLANT CHEMISTRY. Educ: The Citadel, BS, 43; NC State Col, MS, 52. Prof Exp: Jr anal chemist div tests & mat, NC State Hwy Comn, 47-48, anal chemist dept conserv & develop water resources div, 48-50; chemist anal res, Liggett & Myers Tobacco Co, 52-58; res chemist, 58-67, mgr leaf & flavor res, 67-68, MGR PROD DEVELOP, LORILLARD RES CTR, P LORILLARD CO, 68- Mem: Am Chem Soc. Res: Chemical and physical constitution of tobacco, especially as related to final tobacco product characteristics; analytical methods peculiar to tobacco and tobacco smoke. Mailing Add: 903 Caswell Dr Greensboro NC 27408

TUCKER, CHARLES R, b Clinton, SC, Sept 2, 41; m 74; c 2. TEXTILE CHEMISTRY. Educ: Erskine Col, BS, 63. Prof Exp: Res chemist, Deering Milliken Res Corp, 63-66; prod mgr plastics, Warco Inc, Div Albany Felt Co, 66-67; sr res chemist, Fibers Div, Hercules Inc, 68-75; RES MGR FIBER FINISHES, QUAKER CHEM CO, 75- Concurrent Pos: Nat Health Serv training grant water qual mgt, 66. Mem: Am Asn Textile Chemists; Am Chem Soc. Res: Effects of chemical structure on fiber finish properties; nuclear magnetic resonance characterization of textile finishes. Mailing Add: Quaker Chem Co Conshohocken PA 19428

TUCKER, CHARLES THOMAS, b Laredo, Tex, Aug 6, 36; m 64. MATHEMATICS. Educ: Tex A&M Univ, BS & BA, 58; Univ Tex, MA, 62, PhD(math), 66. Prof Exp: Chem engr, Tracor, Inc, Tex, 60-62, engr & scientist, 63-66; asst prof math, 66-73, ASSOC PROF MATH, UNIV HOUSTON, 73- Mem: Am Math Soc; Math Asn Am. Res: Sonar signal processing; pure mathematics. Mailing Add: Dept of Math Univ of Houston Houston TX 77004

TUCKER, CHARLES WINFRED, JR, b East Orange, NJ, Nov 9, 17; m 51; c 2. PHYSICAL CHEMISTRY. Educ: Cooper Union, BChE, 41; Lehigh Univ, MS, 46, PhD(chem), 47. Prof Exp: Anal chemist, Bell Tel Labs, Inc, NY, 38-41; asst chem, Lehigh Univ, 41-47; res assoc, Knolls Atomic Power Lab, Gen Elec Co, 47-57, PHYS CHEMIST, GEN ELEC RES & DEVELOP CTR, 57- Mem: Fel AAAS; Am Phys Soc. Res: Stress measurement by x-ray diffraction; crystal structures; irradiation effects; x-ray scattering by point defects; surface chemistry by low energy electron diffraction. Mailing Add: Phys Chem Lab Gen Elec Res & Develop Ctr Schenectady NY 12345

TUCKER, DAVID PATRICK HISLOP, b Trinidad, West Indies, Oct 26, 34; m 66; c 1. HORTICULTURE. Educ: Univ Birmingham, BSc, 58; Univ Calif, PhD(plant sci), 66. Prof Exp: Agronomist, Dept Agr, Brit Honduras, 60-63; agronomist, 66-74, ASSOC PROF FRUIT CROPS & ASSOC HORTICULTURIST, AGR RES & EDUC CTR, UNIV FLA, 74- Mem: Am Soc Hort Sci. Res: All aspects of citrus production. Mailing Add: Agr Res & Educ Ctr Univ of Fla PO Box 1088 Lake Alfred FL 33850

TUCKER, DON, b Seligman, Mo, May 19, 24; m 54; c 3. BIOPHYSICS. Educ: Univ Ill, BS, 51; Fla State Univ, PhD(physiol), 61. Prof Exp: Res asst, Univ Ill, 51-54; jr elec engr, Raytheon Tel & Radio, 54; res assoc, 54-57, USPHS trainee, 61-64, res investr biol sci, 64-72, ASSOC RESEARCHER BIOL SCI, FLA STATE UNIV, 72- Mem: AAAS; Am Phys Soc. Res: Biophysics of chemical senses; biological effects of high intensity ultrasound; response characteristics of olfactory, vomeronasal and trigeminal receptor responses to odorants; chemoreceptor mechanisms, including taste; neural coding in chemical senses. Mailing Add: Dept of Biol Sci Fla State Univ Tallahassee FL 32306

TUCKER, DON HARRELL, b Brown County, Tex, Jan 21, 30; m 51; c 3. MATHEMATICS. Educ: Western Tex State Univ, BA, 51; Univ Tex, MA, 55, PhD(math), 58. Prof Exp: Res scientist mil physics res lab, Balcones Res Ctr, Univ Tex, 52-53, instr math, Univ, 53-58; from asst prof to assoc prof, 58-67, PROF MATH, UNIV UTAH, 67- Concurrent Pos: Vis prof, Cath Univ Am, 68-69; guest prof, Univ Marburg, 69-71; vis lectr, Math Asn Am, 73- Mem: Am Math Soc; Math Asn Am. Res: Functional analysis; abstract summability theory; differential equations. Mailing Add: Dept of Math Univ of Utah Salt Lake City UT 84112

TUCKER, E SCOTT, III, analytical chemistry, see 12th edition

TUCKER, EDMUND BELFORD, b NS, Can, May 6, 22; m 46; c 3. PHYSICS. Educ: Mt Allison Univ, BSc, 43; Oxford Univ, BA, 48; Yale Univ, MS, 49, PhD(physics), 51. Prof Exp: Res assoc physics, Univ Minn, 50-52, prof, 53, res assoc, 53-55; physicist res & develop ctr, 55-66, mgr personnel & admin info sci lab, 66-69, consult educ rels, NY, 69-71, consult, Conn, 71-75, MGR SCI & TECHNOL SUPPORT PROG, GEN ELEC CO, CONN, 75- Mem: AAAS; Am Phys Soc. Res: Linear accelerators; magnetic resonance; microwave ultrasonics; crystal fields; energy related problems. Mailing Add: 1285 Boston Ave Bridgeport CT 06602

TUCKER, EDWIN WALTER, medicine, psychiatry, see 12th edition

TUCKER, FREDERICK ROBERT, b Toronto, Ont, July 29, 12; m 40; c 2. ORTHOPEDIC SURGERY. Educ: Univ Man, MD, 36; Univ Liverpool, MCh, 46; FRCS(E), 39; FRCPS(C), 51. Prof Exp: Assoc prof, 60-66, PROF ORTHOP SURG, FAC MED, UNIV MAN, 66- Concurrent Pos: Chief orthop surg & dir teaching res, Winnipeg Gen Hosp, 60- Mem: Can Orthop Asn; Brit Orthop Asn. Res: Vascularity of head of femur and method of demonstration avascular necrosis by use of radioactive phosphorus. Mailing Add: Dept of Surg Univ of Man Winnipeg MB Can

TUCKER, GABRIEL FREDERICK, JR, b Bryn Mawr, Pa, June 18, 24; m 47; c 6. LARYNGOLOGY. Educ: Princeton Univ, AB, 47; Johns Hopkins Univ, MD, 51; Am Bd Otolaryngol, dipl, 58. Prof Exp: Asst prof pharmacol, Sch Med, Univ NC, 52-53; instr otolaryngol, Sch Med, Johns Hopkins Univ, 56; instr surg, Sch Med, Univ NC, 57-58; asst prof laryngol & otol, Sch Med, Johns Hopkins Univ & otolaryngologist, Johns Hopkins Hosp, 58-62; CLIN PROF LARYNGOL & BRONCHOESOPHAGOL, CHEVALIER JACKSON CLIN, SCH MED, TEMPLE UNIV, 62- Concurrent Pos: Fel bronchoesophagol & laryngol surg, Grad Sch Med, Univ Pa, 53-54; fel laryngol & otol, Sch Med, Johns Hopkins Univ, 54-56; vis otolaryngologist, Baltimore City Hosp, 59-62; consult, Clin Ctr, NIH, 60-62; chief bronchoesophagology, St Christopher's Hosp, 66-; consult-lectr laryngol, US Naval Hosp, Philadelphia, 67-; consult, Episcopal Hosp & Shriner's Hosp Crippled Children, Philadelphia; mem, Am Joint Comt Cancer Staging & End Results Reporting, Task Force on Larynx; lectr div grad med, Univ Pa, 68-, lectr grad bronchol, esophagol & laryngol surg, 71-; vis res prof laryngol bronchoesophagol, Hahnemann Med Col, 72-; consult otolaryngol, Lankenau Hosp, 74- Honors & Awards: Honor Award, Am Acad Ophthal & Otolaryngol, 69. Mem: AAAS; AMA; fel Am Col Chest Physicians; fel Am Acad Ophthal & Otolaryngol; fel Am Laryngol Asn. Res: Bronchoesophagology; laryngeal pathology; prevention of foreign bodies accidents. Mailing Add: Chevalier Jackson Clin Temple Univ Health Sci Ctr Philadelphia PA 19140

TUCKER, GAIL SUSAN, b New York, NY, Aug 30, 45. EXPERIMENTAL EMBRYOLOGY. Educ: Mercy Col, BA, 67; Univ Kans, Lawrence, PhD(develop biol), 73. Prof Exp: Res asst develop biol, 68-72, res asst mycology, Univ Kans, Lawrence, 72-73; instr, Mercy Col, 73-75; RES ASSOC, BASCOM PALMER EYE INST, UNIV MIAMI, 76- Concurrent Pos: Fel, Eye Inst, Col Physicians & Surgeons, Columbia Univ, 73-75. Mem: Am Res Vision & Ophthal; Am Asn Univ Prof; AAAS; Women in Cell Biol; Am Soc Zoologists. Res: Light and electron microscopy of retinal development and cytoarchitecture of the normal and visually deprived amphibian Xenopus laevis and mammalian cat retina; influence of neurotransmitters on retinal synaptogenesis. Mailing Add: Bascom Palmer Eye Inst Univ of Miami Miami FL 33152

TUCKER, GARDINER LUTTRELL, physics, see 12th edition

TUCKER, GARY EDWARD, b Michigan Valley, Kans, Aug 17, 41; m 60; c 2. BOTANY. Educ: Kans State Teachers Col, BA, 64; Univ NC, Chapel Hill, MA, 67; Univ Ark, PhD(bot), 76. Prof Exp: ASST PROF BIOL & CUR HERBARIUM, ARK POLYTECH COL, 66- Concurrent Pos: Consult, Ozark Nat Forest, USDA, 71-72 & Ark Dept Planning, 73-75. Mem: Int Asn Plant Taxonomists. Res: Rare and endangered plant species of Arkansas; woody flora of Arkansas; vascular plant family Lauraceae of southeastern US; atlas of Arkansas plant species. Mailing Add: Dept of Biol Sci Ark Polytech Col Russellville AR 72801

TUCKER, GARY JAY, b Cleveland, Ohio, May 6, 34; m 56; c 2. PSYCHIATRY. Educ: Oberlin Col, AB, 56; Western Reserve Univ, MD, 60. Prof Exp: Asst med dir, Acute Psychiat Inpatient Div, Yale-New Haven Hosp, Yale Univ, 67-68, from asst prof to assoc prof psychiat, Sch Med, 67-71, med dir, Psychiat Inpatient Div, Med Ctr, 68-71, attend psychiatrist, 69-71, asst chief psychiat, 70-71; assoc prof psychiat, 71-74, PROF PSYCHIAT, DARTMOUTH MED SCH, 74-, DIR RESIDENCY TRAINING, 71- Concurrent Pos: Fel psychiat, Sch Med, Yale Univ, 61-64; consult, Norwich State Hosp, Conn, 67-68, Univ Conn, 68-70, Off Aviation Med, Fed Aviation Admin, Dept Transp, 68-70, Vet Admin Hosp, West Haven, Conn, 70-71 & White River Junction, Vt, 71- Mem: Fel Am Psychiat Asn. Res: Behavioral implications of psychomotor functions; psychopathology and hospital psychiatry. Mailing Add: Dept of Psychiat Dartmouth Med Sch 9 Maynard St Hanover NH 03755

TUCKER, GEOFFREY THOMAS, b London, Eng, June 1, 43; m 67; c 2. PHARMACOLOGY, PHARMACY. Educ: Univ London, BPharm, 64, PhD(med chem), 67. Prof Exp: Res investr & dir lab anesthesiol res, 68-71, SR INVESTR, VIRGINIA MASON RES CTR, 71-; RES ASST PROF, DEPT ANESTHESIOL & ANESTHESIA RES CTR, SCH MED, UNIV WASH, 71- Mem: AAAS; Am Acad Pharmaceut Sci; NY Acad Sci. Res: Bioanalysis of drugs; pharmacokinetics; biopharmaceutics; computer applications; clinical pharmacology; local anesthesia. Mailing Add: Virginia Mason Res Ctr 1000 Seneca St Seattle WA 98101

TUCKER, HAROLD, b Nicholsville, Ala, Sept 10, 12; m 35; c 2. POLYMER CHEMISTRY. Educ: Ala Polytech Inst, BS, 35. Prof Exp: Chemist res ctr, B F Goodrich Co, 35-36, res chemist, 36-41 & 45-55, sect leader, 55-64, mgr res, Goodrich-Gulf Chem, Inc, 64-69, sect leader, B F Goodrich Res Ctr, 69-73, PROJ TECH MGR, B F GOODRICH RES CTR, 73- Mem: AAAS; Am Chem Soc. Res: Polymer chemistry, especially the emulsion and solution polymerization of dienes; directive polymerization using coordination catalysts; rubber technology. Mailing Add: B F Goodrich Res Ctr Brecksville OH 44141

TUCKER, HARVEY MICHAEL, b New Brunswick, NJ, Nov 27, 38; m 60; c 3. OTOLARYNGOLOGY, SURGERY. Educ: Bucknell Univ, BS, 60; Jefferson Med Col, MD, 64. Prof Exp: Resident, Jefferson Med Col, 69; asst otolaryngol, Barnes Hosp, Washington Univ, 69-70; assoc prof otolaryngol, State Univ NY Upstate Med Ctr, 70-75; CHMN, DEPT OTOLARYNGOL & COMMUN DIS, CLEVELAND CLIN FOUND, 75- Concurrent Pos: Fel head & neck surg, Barnes Hosp, Washington Univ, 69-70. Honors & Awards: S Macuen Smith Award, Jefferson Med Col, 64; Benjamin Shuster Award, Am Acad Plastic & Reconstruct Surg, 70. Mem: Fel Am Col Surgeons; fel Am Acad Facial Plastic & Reconstruct Surg; Am Soc Surg; fel Am Acad Head & Neck Surg; fel Am Acad Ophthal & Otolaryngol. Res: Laryngeal reinnervation and transplantation; cancer surgery of the head and neck. Mailing Add: 9500 Euclid Ave Cleveland OH 44106

TUCKER, HERBERT ALLEN, b Milford, Mass, Oct 25, 36; m 59; c 3. ANIMAL PHYSIOLOGY. Educ: Univ Mass, BS, 58; Rutgers Univ, MS, 60, PhD(animal physiol), 63. Prof Exp: From asst prof to assoc prof mammary physiol, 62-75, PROF MAMMARY PHYSIOL, MICH STATE UNIV, 75- Concurrent Pos: NIH spec fel, Univ Ill, 69. Mem: AAAS; Am Soc Animal Sci; Am Dairy Sci Asn; Soc Exp Biol & Med; Am Physiol Soc. Res: Endocrinology of mammary development and lactation; nucleic acid metabolism of mammary tissue; radioimmunoassay of hormones; endocrinology of reproduction; hormone binding to mammary cells. Mailing Add: Dept of Dairy Sci Mich State Univ East Lansing MI 48824

TUCKER, HOWARD GREGORY, b Lawrence, Kans, Oct 3, 22; m 46; c 4. MATHEMATICS. Educ: Univ Calif, AB, MA, 49, PhD(math), 55. Prof Exp: Instr math, Rutgers Univ, 52-53; asst prof, Univ Ore, 55-56; asst prof, Univ Calif, Riverside, 56-58; PROF MATH, UNIV CALIF, IRVINE, 68- Mem: Am Math Soc; Math Asn Am; Inst Math Statist. Res: Probability theory; mathematical statistics. Mailing Add: Dept of Math Univ of Calif Irvine CA 92664

TUCKER, IRWIN WILLIAM, b New York, NY, Oct 30, 14; m 73; c 1. ORGANIC CHEMISTRY, ENVIRONMENTAL ENGINEERING. Educ: George Washington Univ, BS, 39; Univ Md, PhD(org chem), 48. Prof Exp: Chemist, USDA, Washington, DC, 36-45; asst, Univ Md, 45-48; res chemist, Ligget & Meyers Tobacco Co, NC, 48-51; indust specialist, Indust Eval Bd, US Dept Com, 51-53; dir res, Brown & Williamson Tobacco Corp, 53-59, mem, Bd Dirs, 56-59; PROF ENG RES, UNIV LOUISVILLE, 66- Concurrent Pos: Dir, Inst Indust Res, 66-72; pres, Coun Environ Balance, 73- Mem: Am Chem Soc; Air Pollution Control Asn; Inst Food Technologists. Res: Synthetic and determination of structure; fermentation chemistry; enzyme chemistry; chemistry of natural products; air pollution and solid wastes. Mailing Add: 1810 Crossgate Lane Louisville KY 40222

TUCKER, JAMES, b Shamrock, Okla, July 17, 22; m 46; c 4. VETERINARY SCIENCE. Educ: Okla State Univ, BS, 47, MS, 48, DVM, 51. Prof Exp: Asst prof vet path, Univ Ga, 51-53; PROF VET SCI & HEAD DIV VET MED & MICROBIOL, UNIV WYO, 53- Mem: AAAS; Am Vet Radiol Soc; Am Vet Med Asn; Am Asn Vet Nutritionists; Am Col Vet Toxicol. Res: Diseases and toxicology of cattle and sheep. Mailing Add: Div of Vet Med & Microbiol Univ Wyo PO Box 3354 Univ Sta Laramie WY 82070

TUCKER, JOHN MAURICE, bYamhill Co, Ore, Jan 7, 16; m 42; c 3. BOTANY. Educ: Univ Calif, AB, 40, PhD(bot), 50. Prof Exp: Botanist, Univ Calif Exped, El Salvador, Cent Am, 41-42; asst bot, Univ Calif, Berkeley, 46-47; assoc, 47-49, instr & jr botanist, Exp Sta, 49-51, from asst prof & asst botanist to assoc prof & assoc botanist, 51-63, assoc, Exp Sta, 47-49, PROF BOT & BOTANIST, UNIV CALIF, DAVIS, 63-, DIR ARBORETUM, 72- Concurrent Pos: Guggenheim fel, 55-56. Mem: AAAS; Bot Soc Am; Am Inst Biol Sci; Soc Study Evolution; Am Soc Plant Taxon. Res: Systematics and evolution of oaks of North America; classification of Fagaceae of the New World. Mailing Add: Dept of Bot Univ of Calif Davis CA 95616

TUCKER, JOHN SHEPARD, b San Francisco, Calif, Aug 19, 29; m 53; c 2. COMPARATIVE PHYSIOLOGY, AQUATIC BIOLOGY. Educ: Pomona Col, BA, 51; Stanford Univ, PhD(biol), 61. Prof Exp: Asst prof zool, 59-63, assoc res prof, Pac Marine Sta, 61-69, asst dir, 61-63 & 67-69, ASST PROF BIOL, UNIV OF PAC, 63- Concurrent Pos: Partic guest, Lawrence Livermore Lab, 71- Mem: AAAS; Am Soc Zool; Am Soc Limnol & Oceanog; Marine Biol Asn UK. Res: Adaptive physiology, especially nutritive and reproductive, of aquatic invertebrates. Mailing Add: Dept of Biol Univ of the Pac Stockton CA 95211

TUCKER, KENNETH WILBURN, b Santa Barbara, Calif, Aug 8, 24; m 53. APICULTURE. Educ: Univ Calif, BS, 50, PhD(entom), 57. Prof Exp: Instr biol, Lake Forest Col, 60-63; asst res apiculturist, Univ Calif, Davis, 63-66; RES ENTOMOLOGIST, USDA, 66- Mem: AAAS; Entom Soc Am; Bee Res Asn; Genetics Soc Am; Am Genetic Asn. Res: Genetics of honey bees. Mailing Add: Bee Stock Res USDA Agr Res Serv Route 3 Box 82-B Baton Rouge LA 70808

TUCKER, LAWRENCE MANFRED, chemistry, see 12th edition

TUCKER, MARIE, b Cowden, Ill, Mar 8, 11. EMBRYOLOGY, BIOLOGY. Educ: Greenville Col, BA, 36; Univ Ill, Urbana, MS, 43, PhD(zool), 59. Prof Exp: Assoc prof biol & chmn dept, Dana Col, 46-51; instr, Univ Ill, Urbana, 52-53; assoc prof, Youngstown Univ, 53-61; prof & chmn dept, Dana Col, 61-65; ASSOC PROF BIOL, N CENT COL, 65- Mem: AAAS; Am Soc Zool; Am Genetic Asn. Res: Inhibitory control of regeneration in nemertean worms. Mailing Add: 125 S Brainard St Naperville IL 60540

TUCKER, NATHANIEL BEVERLEY, organic chemistry, see 12th edition

TUCKER, PAUL ARTHUR, b Albemarle, NC, May 14, 41; m 65. TEXTILES, MICROSCOPY. Educ: NC State Univ, BS, 63, MS, 66, PhD(fiber & polymer sci), 73. Prof Exp: Instr textiles, NC State Univ, 64-71, asst prof, 73; NATO vis fel, Dept Textile Indust, Univ Leeds, 74; ASST PROF TEXTILES, SCH TEXTILES, NC STATE UNIV, 75- Mem: Royal Micros Soc. Res: Seeking the basic materials science underlying fibrous materials and relating applied technology to this science; polymer fine structure; microscopy; yarn processing; particulate analyses. Mailing Add: Sch of Textiles NC State Univ Raleigh NC 27607

TUCKER, PAUL WILLIAM, b Liberty, Mo, Dec 21, 21; m 43; c 2. PETROLEUM, NATURAL RESOURCES. Educ: William Jewell Col, AB, 42; La State Univ, MS, 44; Univ Mo, PhD(chem), 48. Hon Degrees: LLD, William Jewell Col, 68. Prof Exp: Asst army specialized training prog, La State Univ, 42-44; chemist-spectroscopist, Manhattan Proj, Tenn Eastman Co, 44-46; chemist, Phillips Petrol Co, 48-49, tech rep, 49-60, asst dir pub affairs, Okla, 60-62, managing dir, Phillips Petrol-UK, Ltd, 62-68, vpres natural gas & natural gas liquids, Phillips Petrol Co Europe-Africa, 68-72, vpres natural gas, natural gas liquids & govt rels, 72-74, MGR INT GAS & GAS LIQUIDS, PHILLIPS PETROL CO EUROPE- AFRICA, 74- Mem: Am Chem Soc; Nat Soc Prof Eng; Brit Inst Petrol; Brit Inst Gas Eng. Res: Chemical spectroscopy; physical and chemical properties of rice hulls; organic synthesis; adsorption by silica gel; ultraviolet absorption; stibestrol compounds; light hydrocarbons; anhydrous ammonia; flammable liquids; ammonium nitrate; citizenship education and development; international natural gas and gas liquids. Mailing Add: Phillips Petrol Co 5A3 Phillips Bldg Bartlesville OK 74004

TUCKER, RAY EDWIN, b Somerset, Ky, Dec 31, 29; m 53; c 3. ANIMAL SCIENCE. Educ: Univ Ky, BS, 51, MS, 66, PhD(animal nutrit), 68. Prof Exp: Asst prof animal sci, Va Polytech Inst & State Univ, 68-69; ASST PROF ANIMAL SCI, UNIV KY, 69- Mem: Am Soc Animal Sci. Res: Ruminant nutrition; starch utilization; urea utilization; magnesium deficiency; vitamin A antagonists; poultry litter as a feedstuff for ruminants. Mailing Add: Dept of Animal Sci Univ of Ky Lexington KY 40506

TUCKER, ROBERT GENE, b Springfield, Mo, Sept 14, 18; m 46; c 3. NUTRITION. Educ: Southwest Mo State Col, AB, 40; Univ Minn, MS, 48, PhD(physiol hyg), 50. Prof Exp: Chemist, MFA Milling Co, 40-44; chemist, Tenn Eastman Co, 44-45; asst physiol hyg, Univ Minn, 46-50; instr biochem sch med, Vanderbilt Univ, 51-55; sr scientist, Smith Kline & French Labs, 55-60; MGR SCI SERVS, BAXTER LABS, INC, 60- Concurrent Pos: Biochemist, Thayer Vet Admin Hosp, Nashville, Tenn, 51-55. Mem: AAAS; Am Chem Soc; Am Inst Nutrit; Drug Info Sci. Res: Pharmaceuts; information science. Mailing Add: Baxter Labs Inc 6301 Lincoln Ave Morton Grove IL 60053

TUCKER, ROBERT H, b Clovis, NMex, Aug 3, 44; m; c 1. THEORETICAL PHYSICS. Educ: Univ Ariz, BS, 65; Iowa State Univ, PhD(physics), 70. Prof Exp: Asst ed physics, Phys Rev Lett, 71-74; ED PHYSICS, PHYS REV A, 74- Mem: Am Phys Soc. Res: Electrodynamcis; quantum optics; quantum electrodynaminсs. Mailing Add: 14 Rockhill Rd Rocky Point NY 11778

TUCKER, ROSS NORMAN, solid state chemistry, see 12th edition

TUCKER, ROY WILBUR, b Exeter, Calif, Jan 25, 27; m 54; c 2. MATHEMATICS. Educ: Stanford Univ, BS, 51, MA, 53, MS, 54. Prof Exp: Instr math, Colo Col, 54-55; instr, Modesto Jr Col, 55-59; assoc prof, 59-71, coordr comput ctr, 64-65 & consult, 65-66, PROF MATH, HUMBOLDT STATE UNIV, 71- Mem: Math Asn Am; Soc Indust & Appl Math; Asn Comput Mach. Res: Numerical analysis; linear algebra. Mailing Add: Dept of Math Humboldt State Univ Arcata CA 95221

TUCKER, RUTH EMMA, b Warrensburg, Ill, Feb 17, 01. NUTRITION. Educ: Univ Ill, AB, 23, MS, 25; Univ Chicago, PhD(nutrit, food chem), 48. Prof Exp: Asst, Univ Ill, 23-25; instr food & nutrit, Kans State Col, 25-37; prof home econ, Univ Alaska, 37-42; prof & res prof food & nutrit, 44-72, EMER PROF FOOD & NUTRIT, UNIV RI, 72- Mem: AAAS; fel Am Pub Health Asn; Am Dietetic Asn; Am Home Econ Asn. Res: Food chemistry. Mailing Add: 64 Linden Dr Kingston RI 02881

TUCKER, SHIRLEY COTTER, b St Paul, Minn, Apr 4, 27; m 53. BOTANY. Educ: Univ Minn, Minneapolis, BA, 49, MS, 51; Univ Calif, Davis, PhD, 56. Prof Exp: Scientist plant path, Univ Minn, 55-56, res fel bot, 57-60, instr, 60; res fel biol, Northwestern Univ, 61-63; res fel bot, Univ Calif, 63-66; asst prof, 68-71, ASSOC PROF BOT, LA STATE UNIV, BATON ROUGE, 71- Mem: Bot Soc Am; Am Bryol & Lichenological Soc; Int Soc Plant Morphol. Res: Developmental anatomy of flower and vegetative shoots; determinate growth; lichens; plant anatomy; morphology; lichenology. Mailing Add: Dept of Bot La State Univ Baton Rouge LA 70803

TUCKER, THOMAS CURTIS, b Hanson, Ky, Nov 1, 26; m 47; c 3. SOILS, PLANT NUTRITION. Educ: Univ Ky, BS, 49; Kans State Univ, MS, 51; Univ. Ill, PhD, 55. Prof Exp: Asst soils, Kans State Univ, 49-51, instr, 51; asst soil fertility, Univ Ill, 51-55; asst prof soils, Miss State Univ, 55-56; from assoc prof to prof agr chem & soils, 56-74, PROF SOILS, WATER & ENG & SOIL SCIENTIST, AGR EXP STA, UNIV ARIZ, 74- Concurrent Pos: Vis prof, NC State Univ, 66-67. Mem: AAAS; Soil Sci Soc Am; Am Soc Agron; Am Chem Soc. Res: Agronomy; soil fertility and chemistry; analytical chemistry; soil-plant relationships. Mailing Add: Dept of Soils Univ of Ariz Tucson AZ 85721

TUCKER, VANCE ALAN, b Niagara Falls, NY, Apr 4, 36. COMPARATIVE PHYSIOLOGY. Educ: Univ Calif, Los Angeles, BA, 58, PhD(zool), 63; Univ Wis, MS, 60. Prof Exp: NSF fel zool, Univ Mich, 63-64; from asst prof to assoc prof zool, 64-73, PROF ZOOL, DUKE UNIV, 73- Concurrent Pos: NSF res grants & Duke Univ Coun Res grants, 65-75. Mem: AAAS. Res: Vertebrate locomotion, respiration, circulation, energy metabolism; avian aerodynamics. Mailing Add: Dept of Zool Duke Univ Durham NC 27706

TUCKER, WALLACE HAMPTON, astrophysics, see 12th edition

TUCKER, WALTER EUGENE, JR, b Atlanta, Ga, Aug 7, 31; m 56; c 3. VETERINARY PATHOLOGY. Educ: Univ Ga, DVM, 56; Am Col Vet Pathologists, dipl, 62. Prof Exp: From resident to staff mem, Vet Path Div, Armed Forces Inst Path, 58-62; sr pathologist, Dow Chem Co, 62-68; mgr path sect, Wyeth Labs Inc, 68-74; HEAD DEPT TOXICOL & EXP PATH, BURROUGHS WELLCOME CO, 74- Mem: Int Acad Pathologists; Soc Pharmacol Environ Pathologists. Res: Research and development of pharmaceutical and agricultural chemicals with emphasis on characterization of toxicopathologic responses in laboratory and domestic animals of administration of candidate human and veterinary chemotherapeutic agents. Mailing Add: Burroughs Wellcome Co 3030 Cornwallis Rd Research Triangle Park NC 27709

TUCKER, WILLIAM BOOSE, b Peitaiho, China, Aug 17, 05; m 32; c 2. INTERNAL MEDICINE. Educ: Oberlin Col, AB, 29; Univ Chicago, MD, 34. Prof Exp: Instr biol, Bennington Col, 34-35; asst med, Col Med, Univ Chicago, 36-39; from instr to asst prof, 39-47; from assoc prof to prof, Med Univ Minn, 47-54; prof, Sch Med, Duke Univ, 54-56; dir tuberc serv, Cent Off, US Vet Admin, 56-59, dir pulmonary dis serv, 59-61, dir med serv, 61-69; PROF MED, COL MED, UNIV FLA, 70- Concurrent Pos: Mem teaching staff, Billings Hosp, Univ Chicago Clins, 36-47 & Univ Minn Hosps, Minneapolis, 47-54; chief tuberc serv, Vet Admin Hosp, Minneapolis, 47-54, chief pulmonary dis serv, Durham, 54-55, chief med serv, 55-57; consult, Vet Admin Hosp, Gainesville, 70- Honors & Awards: Trudeau Medal, 66. Mem: AAAS; Am Thoracic Soc (pres, 60-61); fel AMA; Am Lung Asn (pres, 70-71); Am Col Physicians. Res: Pulmonary diseases; chemotherapy of tuberculosis; non-tuberculosis pulmonary diseases; respiratory physiology; medical administration. Mailing Add: 2544 SW 14th Dr Gainesville FL 32608

TUCKER, WILLIAM ERIC, b Port Arthur, Tex, Mar 7, 30; m 54; c 2. NUCLEAR PHYSICS. Educ: Univ Tex, Austin, BS, 53, MA, 59, PhD(physics), 65. Prof Exp: Res scientist, Tex Nuclear Div, Nuclear-Chicago Corp, 57-65, from sr res scientist to prin res scientist, 65-74, sr group leader, Nuclear Physics & Appl Radiation Group, 67-74; TECHNOL UTILIZATION COORDR, ENERGY RES & DEVELOP ADMIN, 74- Concurrent Pos: Ex-officio mem US nuclear data comt & mem elastic & inelastic subcomt, AEC, 69- Mem: Am Phys Soc; Inst Elec & Electronics Engrs; Am Nuclear Soc. Res: Neutron interactions with complex nuclei; neutron and gamma-ray spectroscopy; nuclear structure, nuclear physics with electrostatic accelerators; radiation dosimetry; activation analysis; applications of nuclear methods to industrial and military problems. Mailing Add: Energy Res & Develop Admin Washington DC 20460

TUCKER, WILLIAM GOUGH, b Granby, Conn, Dec 14, 27; m 53, 72; c 4. ONCOLOGY. Educ: Mt Allison Univ, BS, 49; Queen's Univ, Ont, MD, CM, 58. Prof Exp: Intern, Akron City Hosp, 58-59; med resident, Henry Ford Hosp, 59-61; pvt practr, NS, 60-61; Navy residency med, Royal Can Navy, Can Forces Hosp, HMCS Stadacoma, 61-62; from med resident to chief med resident, Henry Ford Hosp, 62-65, assoc physician, Div Oncol, 65-69; DIR MIDWEST ONCOL CTR, BORGESS HOSP, KALAMAZOO, MICH, 69- Concurrent Pos: Prin investr, Southwest Oncol Group, 70- Mem: Am Soc Clin Oncol; AMA. Res: Clinical investigation of chemotherapeutic agents; microbiology tissue culture; biochemical investigation of metabolites of chemotherapeutic agents. Mailing Add: Midwest Oncol Ctr 1521 Gull Rd Kalamazoo MI 49001

TUCKER, WILLIAM PRESTON, b Louisville, Ky, Sept 23, 32; m 59; c 3. ORGANIC CHEMISTRY. Educ: Wake Forest Col, BS, 57; Univ NC, MA, 60, PhD(chem), 62. Prof Exp: NIH fel, Univ Ill, 62-63; from asst prof to assoc prof chem, 63-72, PROF CHEM, NC STATE UNIV, 72- Mem: Am Chem Soc. Res: Chemistry of organic compounds of divalent sulfur; natural products. Mailing Add: Dept of Chem NC State Univ Raleigh NC 27607

TUCKER, WILLIE GEORGE, b Tampa, Fla, Nov 26, 34. ORGANIC CHEMISTRY. Educ: Tuskegee Inst, BS, 56, MS, 58; Univ Okla, PhD(org chem), 62. Prof Exp: PROF CHEM, SAVANNAH STATE COL, 62-, HEAD DEPT, 69- Mem: AAAS; Am Chem Soc; Sigma Xi. Res: Chlorination with cupric chloride; halogenation of pyridine. Mailing Add: Dept of Chem Box 395 Savannah State Col Savannah GA 31404

TUCKER, WOODSON COLEMAN, JR, b Halsey, Ky, Sept 17, 08; m 32, 73; c 2. ACADEMIC ADMINISTRATION, PHYSICAL CHEMISTRY. Educ: Univ Fla, BS, 29, MS, 30, PhD(chem), 53. Prof Exp: Chemist, Superior Earth Co, 31-36; asst gen mgr in chg prod, United Prod Co, 36-38; chemist & tech adv to supt, Edgar Plastic Kaolin Co, 40-41; interim instr chem, Univ Fla, 46-51; instr, 51-52, from asst prof to assoc prof, 52-66, PROF CHEM, UNIV NEW ORLEANS, 66-, ASST VCHANCELLOR ACAD AFFAIRS, 69- Mem: Am Chem Soc. Res: Physical and thermodynamic properties of terpenes; chemical education. Mailing Add: Off of Acad Affairs Univ of New Orleans New Orleans LA 70122

TUCKERMAN, BRYANT, b Lincoln, Nebr, Nov 28, 15; m 53; c 3. MATHEMATICS. Educ: Antioch Col, BS, 39; Princeton Univ, MA, 46, PhD(math), 47. Prof Exp: Physicist, Off Sci Res & Develop Proj, Dept Terrestrial Magnetism, Carnegie Inst, 41-45; instr math, Cornell Univ, 47-49; asst prof, Oberlin Col, 49-52; mathematician electronic comput proj, Inst Adv Study, NJ, 52-57; MATHEMATICIAN RES CTR, IBM CORP, 57- Mem: Am Math Soc; Math Asn Am; Asn Comput Mach. Res: Computational number theory; cryptology; celestial mechanics; computational efficiency. Mailing Add: 121 Schrade Rd Briarcliff Manor NY 10510

TUCKERMAN, MURRAY MOSES, b Boston, Mass, July 19, 28; m 48; c 4. PHARMACEUTICAL CHEMISTRY. Educ: Yale Univ, BS, 48; Temple Univ, BS, 53; Rensselaer Polytech Inst, PhD(chem), 58. Prof Exp: Asst anal chem, Sterling-Winthrop Res Inst, 53-55, res assoc, 55-58; assoc prof chem, 58-62, proj dir & radiol health specialist training prog, 63-69, head dept, 61-72, PROF CHEM, SCH PHARM, TEMPLE UNIV, 62- Concurrent Pos: Resident res assoc, Argonne Nat Lab, 59; mem bd revision, US Pharmacopeia, 60-; consult clin ctr, NIH, 59-65; sci adv, Food & Drug Admin, 67-71, comn labs & servs for control of drugs, 64-; temp hon mem secretariat Europ pharmacopeia, Coun Europe, 74. Mem: Fel Am Inst Chem; Am Chem Soc; Am Acad Pharmaceut Sci; Int Pharmaceut Fedn; Am Mgt Asn. Res: Pharmaceutical analysis; drug standards; reference standards; pharmaceutical quality assurance. Mailing Add: Temple Univ Sch of Pharm 3307 N Broad St Philadelphia PA 19140

TUCKEY, STEWART LAWRENCE, b Browns Valley, Minn, Aug 24, 05; m 36; c 1. DAIRYING. Educ: Univ Ill, Urbana, BS, 28, MS, 30, PhD(dairy tech), 37. Prof Exp: Asst dairy mfg, 28-30, instr, 30-32, assoc, 32-37, from asst prof to assoc prof, 37-57, prof dairy technol, 57-72, EMER PROF DAIRY TECHNOL, UNIV ILL, URBANA, 72- Concurrent Pos: Sabbatical, Neth Inst Dairy Res, Ede, 67. Honors & Awards: Borden Award, 39; Charles E Pfizer Award, Am Dairy Sci Asn, 69. Mem: AAAS; Am Chem Soc; Am Dairy Sci Asn. Res: Biochemical and microbiological changes in cheese; microbiol clotting enzyme for cheese. Mailing Add: 919 W Charles St Champaign IL 61820

TUCKFIELD, RALPH GEORGE, JR, b Independence, Mo, Mar 3, 26; m 50; c 4. PHYSICS. Educ: William Jewell Col, BS, 49; Univ Iowa, MS, 54. Prof Exp: Mem res staff, Princeton Univ, 54-56, mem thermonuclear res staff, 56-57; mem res staff, Gulf Energy & Environ Systs, Inc, 57-71, head laser physics lab, 69-71; LASER PHYSICS & ELECTROOPTICS CONSULT, 71- Mem: Am Phys Soc; Am Vacuum Soc. Res: Thermonuclear research; plasma physics; laser physics and application. Mailing Add: 5034 February St San Diego CA 92110

TUCKSON, COLEMAN REED, JR, b Washington, DC, Sept 2, 23; m 49; c 2. DENTISTRY. Educ: Howard Univ, DDS, 47; Univ Pa, MScD, 53. Prof Exp: From instr to assoc prof, 48-64, PROF DENT RADIOL, COL DENT, HOWARD UNIV, 64-, CHMN DEPT ORAL DIAG & RADIOL, 68-, ASSOC DEAN COL, 74- Mem: AAAS; Am Acad Dent Radiol (pres, 71-72); Int Asn Dent Res. Res: Oral surgery, dental radiology. Mailing Add: Howard Univ Col of Dent 600 W St NW Washington DC 20059

TUDDENHAM, WILLIAM J, b Salt Lake City, Utah, Nov 15, 22; m 47; c 3. RADIOLOGY, ROENTGENOLOGY. Educ: Univ Utah, BA, 43; Univ Pa, MD, 50, MSc, 56; Am Bd Radiol, dipl, 55. Prof Exp: Asst biochem, Calif Inst Technol, 43-44; intern, Univ Hosp, 50-51, asst instr, Sch Med, 51-54, instr, 54-56, assoc, 56-57, from asst prof to assoc prof, 57-61, PROF RADIOL, SCH MED, UNIV PA, 61-, DIR DEPT RADIOL, PA HOSP, 62- Concurrent Pos: Resident, Univ Hosp, Univ Pa, 51-54, mem staff, 54-67, chief diag sect, 59-61. Honors & Awards: Bronze Medal, Roentgen Ray Soc, 56, Cert Merit, 58; Lindback Award, Univ Pa, 64. Mem: Roentgen Ray Soc; Radiol Soc NAm; AMA; Am Col Radiol. Res: Visual physiology of roentgen interpretation. Mailing Add: Pa Hosp Dept of Radiol Eighth & Spruce Sts Philadelphia PA 19107

TUDEN, ARTHUR, b Wilmington, Mass, Sept 2, 27; m 58; c 3. CULTURAL ANTHROPOLOGY. Educ: Yale Univ, BA, 51; Northwestern Univ, PhD, 62. Prof Exp: Instr anthrop, Princeton Univ, 55-56; Ford Found fel, Cent Africa, 56-58; from asst prof to assoc prof, 59-69, PROF ANTHROP, UNIV PITTSBURGH, 69- Concurrent Pos: Wenner-Gren Found fel, Conf Polit Anthrop, Burgh-Wartenstein, Austria, 63 & Conf Social Stratification, 66; ed, Ethnology, 69- Res: Political economy; social stratification. Mailing Add: Dept of Anthrop Univ of Pittsburgh Pittsburgh PA 15213

TUDOR, DAVID CYRUS, b Wildwood, NJ, May 10, 18; m 41; c 1. POULTRY PATHOLOGY. Educ: Rutgers Univ, BS, 40; Univ Pa, VMD, 51. Prof Exp: Instr high sch, NJ, 40-44; res asst poultry path, 51-59, assoc res specialist, 59-66, RES PROF POULTRY PATH, RUTGERS UNIV, 66- Concurrent Pos: Assoc ed, Poultry Sci, Poultry Sci Asn, 73-75. Mem: Am Vet Med Asn; Am Asn Avian Path; US Animal Health Asn; NY Acad Sci; World Poultry Sci. Res: Poultry science; infectious nephrosis; Salmonella and chronic respiratory disease control pox; mycoplasma; pox; pet bird diseases. Mailing Add: 29 Station Rd Cranbury NJ 08512

TUELLER, PAUL T, b Paris, Idaho, July 30, 34; m 63; c 3. PLANT ECOLOGY, RANGE MANAGEMENT. Educ: Idaho State Univ, BS, 57; Univ Nev, MS, 59. Prof Exp: From asst prof to assoc prof range sci, 62-73, PROF RANGE SCI, UNIV NEV, RENO, 73- Mem: Soc Range Mgt; Ecol Soc Am; Wildlife Soc; Am Soc Photogram; Am Inst Biol Sci. Res: Range ecology, especially vegetation-soil relationships; management of big game populations; remote sensing pf renewable natural resources. Mailing Add: Col of Agr Univ of Nev Reno NV 89507

TUEMMLER, WILLIAM BRUCE, organic chemistry, see 12th edition

TUESDAY, CHARLES SHEFFIELD, b Trenton, NJ, Sept 7, 27; m 52; c 3. RESEARCH ADMINISTRATION. Educ: Hamilton Col, NY, AB, 51; Princeton Univ, MA & PhD(phys chem), 55. Prof Exp: Res chemist panelyte div, St Regis Paper Corp, 50-51; res asst, Princeton Univ, 51-55; sr res chemist fuels & lubricants dept, 55-64, supvry res chemist, 64-67, from asst head to head, 67-72, head environ sci dept & actg head phys chem dept, 72-74, TECH DIR, RES LABS, GEN MOTORS CORP, 74- Concurrent Pos: Consult vapor-phase org air pollutants panel, Nat Res Coun, 72- Mem: AAAS; Am Chem Soc; Soc Automotive Eng. Res: Catalysis; corrosion; molecular energy exchange; air pollution control; atmospheric chemistry; polymers; metallurgy; analytical chemistry. Mailing Add: Res Labs Gen Motors Corp Gen Motors Tech Ctr Warren MI 48090

TUFARIELLO, JOSEPH JAMES, b Brooklyn, NY, Oct 3, 35; m 60; c 4. ORGANIC CHEMISTRY. Educ: Queens Col, NY, BS, 57; Univ Wis-Madison, PhD(chem), 62. Prof Exp: Assoc org chem, Purdue Univ, Lafayette, 62; NIH fel, Cornell Univ, 62-63; ASSOC PROF ORG CHEM, STATE UNIV NY BUFFALO, 63- Mem: AAAS; Am Chem Soc; The Chem Soc. Res: Organic synthesis; synthesis and reactivity of strained or otherwise unique carbocyclic systems; synthesis of natural products; organometallic chemistry. Mailing Add: Dept of Chem State Univ of NY Buffalo NY 14214

TUFF, DONALD WRAY, b San Francisco, Calif, May 4, 35; m 55; c 3. PARASITOLOGY, TAXONOMY. Educ: San Jose State Col, BA, 57; Wash State Univ, MS, 59; Tex A&M Univ, PhD(entomol), 63. Prof Exp: From asst prof to assoc prof biol, 63-73, PROF BIOL, SOUTHWEST TEX STATE UNIV, 73- Mem: Entom Soc Am; Soc Syst Zool. Res: Taxonomy of avian Mallophaga. Mailing Add: Dept of Biol Southwest Tex State Univ San Marcos TX 78666

TUFFLY, BARTHOLOMEW LOUIS, b Houston, Tex, Apr 9, 28; m 58; c 3. PHYSICAL CHEMISTRY, ANALYTICAL CHEMISTRY. Educ: Univ Tex, BA, 48, MA, 50, PhD(chem), 52. Prof Exp: Chemist, Carbide & Carbon Chem Co, 52-60; chemist, Rocketdyne Div, NAm Rockwell Corp, 60-73, CHEMIST, ROCKETDYNE DIV, ROCKWELL INT CORP, 73- Mem: Am Chem Soc; Water Pollution Control Fedn. Res: Mass spectrometry; pollution; water management; environmental control systems; rocket propellants. Mailing Add: 4709 Dunman Ave Woodland Hills CA 91364

TUFT, RICHARD ALLAN, b Newark, NJ, Oct 9, 40; m 63; c 3. QUANTUM OPTICS. Educ: Pa State Univ, BS, 63; Mass Inst Technol, MS, 66; Worcester Polytech Inst, PhD(physics), 71. Prof Exp: Engr optical physics, Aerospace Syst Div, RCA Corp, 63-68 & 70-71; ASST PROF PHYSICS, WORCESTER POLYTECH INST, 71- Concurrent Pos: Consult, Govt & Com Syst Div, RCA Corp, 71- Mem: Optical Soc Am; Sigma Xi. Res: Experimental investigations in light scattering spectroscopy using photon correlation spectrometer; development of holographic displays and information storage devices. Mailing Add: Dept of Physics Worcester Polytech Inst Worcester MA 01609

TUFTE, MARILYN JEAN, b Iron Mountain, Mich, Nov 20, 39; m 72. BACTERIOLOGY. Educ: Northern Mich Univ, AB, 61; Univ Wis, Madison, MS, 65, PhD(bact), 68. Prof Exp: Trainee & fel, Univ Wis, Madison, 68; asst prof biol, 68-71, ASSOC PROF BIOL, UNIV WIS-PLATTEVILLE, 71- Mem: Am Soc Microbiol. Res: Electron microscopic analysis of guinea pig peritoneal phagocytes infected with strains of Brucella abortus of different degrees of virulence. Mailing Add: Dept of Biol Univ of Wis Platteville WI 53818

TUFTE, OBERT NORMAN, b Northfield, Minn, May 30, 32; m 56; c 3. SOLID STATE PHYSICS. Educ: St Olaf Col, BA, 54; Northwestern Univ, PhD(physics), 60. Prof Exp: Asst physics, Northwestern Univ, 54-59; DEPT MGR RES CTR, HONEYWELL CORP, 60- Mem: Am Phys Soc; Inst Elec & Electronics Engrs. Res: Semiconductors; electrical and optical properties of solids; lasers and optical data recording. Mailing Add: Honeywell Corp Res Ctr 10701 Lyndale Ave Bloomington MN 55420

TUFTS, NORMAN ROYAL, veterinary medicine, epidemiology, see 12th edition

TUHOLSKI, JAMES MARTIN, b Henning, Tenn, Oct 21, 24; m 45; c 5. MEDICINE. Educ: Univ Tenn, MD, 48; Am Bd Pediat, dipl. Prof Exp: Pvt pract pediat, Tenn, 54-56; asst dir clin res, 56-58, assoc dir, 58, dir prod develop, 58-59, vpres, 59-60, pres labs div, 61-66, exec vpres, 66-69, PRES, MEAD JOHNSON & CO, 69-, VPRES, BRISTOL-MYERS CO, NY, 69- Concurrent Pos: Asst Col Med, Univ Tenn, 54. Mem: AMA; Am Acad Pediat. Res: Therapeutics. Mailing Add: Exec Admin Mead Johnson & Co 2404 Pennsylvania St Evansville IN 47721

TUHY, PETER MIRKO, b Plainfield, NJ, May 20, 49. ORGANIC BIOCHEMISTRY. Educ: Ga Inst Technol, BS, 70, PhD(chem), 75. Prof Exp: RES COLLABR BIOCHEM, BROOKHAVEN NAT LAB, 75- Concurrent Pos: NIH fel, 75-77. Mem: Am Chem Soc. Res: Mechanism of action and specificity of proteolytic enzymes; synthesis and interaction of substrate-like inhibitors; alteration of enzymatic activity by chemical modification. Mailing Add: Biol Dept Brookhaven Nat Lab Upton NY 11973

TUINSTRA, KENNETH EUGENE, b Des Moines, Iowa, Dec 22, 40; m 63; c 2. ECOLOGY, LIMNOLOGY. Educ: Univ Wyo, BS, 62; Mont State Univ, PhD(bot), 67. Prof Exp: Res assoc plant ecol, Mont State Univ, 67, Fed Water Pollution Control Admin fel & res assoc limnol, 67-68; ASST PROF BIOL, WESTMONT COL, 68- Mem: AAAS; Am Soc Limnol & Oceanog; Ecol Soc Am; Am Inst Biol Sci. Res: Ecology of freshwater phytoplankton and zooplankton; population ecology; systems ecology. Mailing Add: Dept of Biol Westmont Col Santa Barbara CA 93108

TUITE, JOHN F, b New York, NY, Nov 23, 27; m 54; c 6. PHYTOPATHOLOGY. Educ: Hunter Col, BA, 51; Univ Minn, MS, 53, PhD(plant path), 56. Prof Exp: From asst prof to assoc prof plant path, 56-69, PROF PLANT PATH, PURDUE UNIV, 69- Mem: Am Phytopath Soc; Bot Soc Am; Soc Indust Microbiol; Am Asn Cereal Chem; Sigma Xi. Res: Identification; ecology of fungi growing on stored grain; production and detection of mycotoxins. Mailing Add: Dept of Bot & Plant Path Lilly Hall Purdue Univ West Lafayette IN 47906

TUITE, ROBERT JOSEPH, b Rochester, NY, Aug 28, 34; m 58; c 3. RESEARCH ADMINISTRATION, PHOTOGRAPHIC CHEMISTRY. Educ: St John Fisher Col, BS, 56; Univ Ill, PhD(org chem), 60. Prof Exp: Asst chem, Univ Ill, 56-57, asst pub health serv, 59; from res chemist to sr research chemist, 59-67, res assoc res labs, 67-70, head color photo chem lab, 70-73, from head to sr head, Color Reversal Systs Lab, 73-74, ASST DIR COLOR PHOTOG DIV, EASTMAN KODAK CO, 74- Mem: Am Chem Soc; Soc Photog Scientists & Engrs; Soc Motion Picture & TV Engrs. Res: Applied photochemistry; color photographic imaging chemistry; color photographic systems design; parametrization of color photographic system response and correlation with molecular structure and other systems variables. Mailing Add: Res Labs Eastman Kodak Co Rochester NY 14650

TUITES, DONALD EDGAR, b Saginaw, Mich, Dec 27, 25; m 50; c 4. POLYMER CHEMISTRY. Educ: Univ Rochester, BS, 49; Clarkson Tech Univ, MS, 52; Cornell Univ, PhD(org chem), 56. Prof Exp: Chemist, NY, 55-63, chemist electrochem dept, Del, 63-71, CHEMIST PLASTICS DEPT, E I DU PONT DE NEMOURS & CO, INC, DEL, 72- Mailing Add: E I du Pont de Nemours & Co Inc Plastics Dept Chestnut Run Wilmington DE 19898

TUITES, RICHARD CLARENCE, b Rochester, NY, Oct 31, 33; m 54; c 4. ORGANIC CHEMISTRY, POLYMER CHEMISTRY. Educ: Univ Rochester, BS, 55; Univ Ill, PhD(org chem), 59. Prof Exp: Res chemist, E I du Pont de Nemours & Co, Inc, 58-62; from chemist to sr chemist, 62-72, RES ASSOC, EASTMAN KODAK CO, 72- Mem: Am Chem Soc; Soc Photog Scientists & Engrs. Res: Photographic chemistry. Mailing Add: Eastman Kodak Co Res Labs 343 State St Rochester NY 14650

TUKEY, HAROLD BRADFORD, JR, b Geneva, NY, May 29, 34; m 55; c 3. HORTICULTURE. Educ: Mich State Univ, BS, 55, MS, 56, PhD(plant physiol), 58. Prof Exp: AEC fel, Mich State Univ, 58; NSF fel, Calif Inst Technol, 58-59; from asst prof to assoc prof ornamental hort, 59-70, PROF FLORICULT & ORNAMENTAL HORT, CORNELL UNIV, 70- Concurrent Pos: Consult, PR Nuclear Ctr, 65-66; lectr, Int Atomic Energy Agency Radioisotope Sch, Hanover, Ger, 68; vis horticulturist, Univ Calif, Davis, 72; mem coun, Int Soc Hort Sci, 72- Mem: Am Soc Hort Sci; Am Soc Plant Physiol; Bot Soc Am; Asn Trop Biol; Int Plant Propagators

Soc. Res: Growth and development of plants; role of above-ground plant parts in nutrition; environmental effects on plant growth and behavior. Mailing Add: Dept of Floricult & Ornamental Hort Cornell Univ Ithaca NY 14850

TUKEY, JOHN WILDER, b New Bedford, Mass, June 16, 15; m 50. STATISTICS, STATISTICAL ANALYSIS. Educ: Brown Univ, ScB, 36, ScM, 37; Prineton Univ, MA, 38, PhD(math), 39. Hon Degrees: ScD, Case Inst Technol, 62, Brown Univ, 65, Yale Univ, 68 & Univ Chicago, 69. Prof Exp: Instr math, Princeton Univ, 39-41, res assoc, Fire Control Res Off, 41-45; mem tech staff, 45-58, asst dir res commun prin, 58-61, ASSOC EXEC DIR RES COMMUN PRIN DIV, BELL LABS, 61-; PROF STATIST, PRINCETON UNIV, 65-, DONNER PROF SCI, 76- Concurrent Pos: From asst prof to prof math, Princeton Univ, 41-65, chmn dept statist, 65-70; Guggenheim fel, 49-50; fel, Ctr Advan Study Behav Sci, 57-58; visitor, Commonwealth Sci & Indust Res Orgn, Canberra, Australia, 71 & Stanford Linear Accelerator Ctr, Calif, 72. Mem, US deleg, Tech Working Group 2 of Conf on Discontinuance of Nuclear Weapon Tests, Geneva, Switz, 59 & UN Conf on Human Environ, Stockholm, Sweden, 72; mem, President's Sci Adv Comt, Off Sci & Technol, 60-63, chmn, Panel on Environ Pollution, 64-65 & Panel on Chem & Health, 71-72; mem, President's Air Qual Adv Bd, 68-71 & President's Comn on Fed Statist, 70-71; mem, Sci Info Coun, NSF, 62-64; chmn anal adv comt, Nat Assessment Educ Progress, 63-73 & chmn sci panel of anal adv comt, 73-; mem coun, Nat Acad Sci, 69-71, chmn class III, 69-72 & chmn climatic impact comt, 75-; mem, Nat Adv Comt for Oceans & Atmosphere, 75- Honors & Awards: S S Wilks Medal, Am Statist Asn, 65; Nat Medal of Sci, 73; Hitchcock Prof, Univ Calif, Berkeley, 75. Mem: Nat Acad Sci; Am Philos Soc; Am Acad Arts & Sci; Int Statist Inst; hon mem Royal Statist Soc. Res: Theoretical, applied and mathematical statistics; point set topology; fire control equipment; military analysis. Mailing Add: Bell Labs 600 Mountain Ave Murray Hill NJ 07974

TUKEY, LOREN DAVENPORT, b Geneva, NY, Dec 4, 21; m 52; c 2. HORTICULTURE. Educ: Mich State Univ, BS, 43, MS, 47; Ohio State Univ, PhD(hort), 52. Prof Exp: Asst hort, Ohio State Univ, 47-50; from asst prof to assoc prof pomol, 50-66, PROF POMOL, PA STATE UNIV, 66- Concurrent Pos: Mem coop fruit res prog, Inst Nat Tech Agr, Argentina, 65-70. Mem: Fel AAAS; fel Am Soc Hort Sci; Bot Soc Am; Am Soc Plant Physiol; Am Pomol Soc. Res: Pomology; growth and development-plant growth regulators; plant-environmental relationships; tree form and mechanical harvesting. Mailing Add: Dept of Hort Pa State Univ Tyson Bldg University Park PA 16802

TUKEY, RONALD BRADFORD, b Hudson, NY, July 19, 24; m 51; c 6. POMOLOGY. Educ: Mich State Col, BS, 47, MS, 48; Cornell Univ, PhD(pomol), 52. Prof Exp: Asst pomol, NY State Agr Exp Sta, Geneva, 48-50; from asst prof to assoc prof hort, Purdue Univ, 50-65; EXTEN HORTICULTURIST, WASH STATE UNIV, 65- Mem: Am Soc Hort Sci; Int Soc Hort Sci. Res: Tree fruit physiology, especially dwarfing, propagation, plant nutrition varieties and production efficiency; soil physics and nutrient availability; production management and economics. Mailing Add: Dept of Hort Wash State Univ Pullman WA 99163

TULAGIN, VSEVOLOD, b Leningrad, Russia, June 16, 14; US citizen; m 38. CHEMISTRY. Educ: Calif Inst Technol, BS, 37; Univ Calif, Los Angeles, MA, 41, PhD(chem), 43. Prof Exp: Chemist, Wesco Water Paints, Inc, Calif, 37-39; Nat Defense Res Comt asst, Univ Calif, Los Angeles, 42-43; res chemist, Gen Aniline & Film Corp, Pa, 43-47, group leader, 47-52, res specialist & group leader, Ansco Div, 52-57; res supvr, Minn Mining & Mfg Co, 57-60, mgr res photo prod, 60-66; res mgr, 66-69, CHIEF SCIENTIST, XEROX CORP, 69- Honors & Awards: Kosar Mem Award, Soc Photog Sci & Eng. Mem: Fel Am Inst Chem; Royal Photog Soc; Am Chem Soc; Soc Photog Sci & Eng. Res: Paint technology; chemical synthesis; color photography; polychrome photoelectrophoresis. Mailing Add: 89 Shirewood Dr Rochester NY 14625

TULCZYJEW, WLODZIMIERZ MAREK, b Wlodawa, Poland, June 18, 31; m 75. MATHEMATICAL PHYSICS. Educ: Univ Warsaw, MS, 56, PhD(physics), 59, docent, 65. Prof Exp: From instr to assoc prof physics, Univ Warsaw, 56-68; ASSOC PROF MATH, UNIV CALGARY, 69- Concurrent Pos: Brit Coun fel, Imp Col, Univ London, 59-60; vis asst prof, Lehigh Univ, 60-61; US NSF sr foreign scientist fel, Boston Univ, 66. Mem: Am Math Soc; NY Acad Sci. Res: Differential geometry. Mailing Add: Dept of Math & Statist Univ of Calgary Calgary AB Can

TULECKE, WALTER, b Detroit, Mich, Feb 10, 24; m 46; c 4. BOTANY. Educ: Univ Mich, BA, 46, MS, 50, PhD(bot, 53. Prof Exp: Asst prof bot, Ariz State Col, 53-55; res assoc, Brooklyn Bot Garden, 55-57; res botanist, Chas Pfizer & Co, 57-59; assoc plant physiologist, Boyce Thompson Inst, 59-67; ASSOC PROF BIOL, ANTIOCH COL, 67- Mem: AAAS; Bot Soc Am; Am Soc Plant Physiol. Res: Plant science and nutrition. Mailing Add: Dept of Biol Antioch Col Yellow Springs OH 45387

TULEEN, DAVID L, b Oak Park, Ill, Sept 19, 36; m 60; c 4. ORGANIC CHEMISTRY. Educ: Wittenberg Univ, BS, 58; Univ Ill, PhD(org chem), 62. Prof Exp: Fel, Pa State Univ, 62-63; asst prof, 63-68, ASSOC PROF ORG CHEM, VANDERBILT UNIV, 68- Mem: Am Chem Soc; The Chem Soc. Res: Sulfur chemistry. Mailing Add: Dept of Chem Vanderbilt Univ Box 1613 Sta B Nashville TN 37203

TULENKO, JAMES STANLEY, b Holyoke, Mass, June 1, 36; m 65; c 3. NUCLEAR PHYSICS, APPLIED MATHEMATICS. Educ: Harvard Univ, BA, 58, MA, 60; Mass Inst Technol, MS, 63. Prof Exp: Mgr nuclear develop, United Nuclear Corp, NY, 63-70; mgr physics, Nuclear Mat & Equip Corp, 70-71; mgr physics, Nuclear Power Generation Div, 71-74, MGR NUCLEAR FUEL ENG, BABCOCK & WILCOX, 74- Mem: Am Nuclear Soc. Res: Mechanical and material design nuclear fuel; thermal hydraulic behavior of nuclear cores; reactor physics; fuel management of nuclear reactors; fuel cycle economics of nuclear power plants; plutonium recycle in thermal water reactors. Mailing Add: Babcock & Wilcox PO Box 1260 Lynchburg VA 24505

TULER, FLOYD ROBERT, b Chicago, Ill, May 24, 39; m 61; c 2. MATERIALS SCIENCE. Educ: Univ Ill, Urbana, BS, 60, MS, 62; Cornell Univ, PhD(mat sci & eng), 67. Prof Exp: Tech staff mem stress wave response-solids, Sandia Labs, 66-69; vpres tech dynamic response mat, Effects Technol Inc, 69-74; ASSOC PROF MAT SCI, HEBREW UNIV, ISRAEL, 74- Concurrent Pos: Panel mem nat mat adv bd ad hoc comt, Nat Acad Sci, 68-69. Mem: Am Soc Testing & Mat; Sigma Xi. Res: Mechanical properties of metals and reinforced composite materials; fracture; fatigue initiation and propagation; impact and impulsive loading; mechanical testing and nondestructive testing techniques; failure prediction and analysis. Mailing Add: Sch Appl Sci & Technol Mat Sci Div Hebrew Univ Jerusalem Israel

TULI, VIRENDRA, plant biochemistry, microbiology, see 12th edition

TULINSKY, ALEXANDER, b Philadelphia, Pa, Sept 25, 28; m 55; c 4. STRUCTURAL CHEMISTRY. Educ: Temple Univ, AB, 52; Princeton Univ, PhD(chem), 56. Prof Exp: Res assoc, Protein Struct Proj, Polytech Inst Brooklyn, 55-59; asst prof chem, Yale Univ, 59-65; assoc prof chem, 65-67 & chem & biochem, 67, PROF CHEM & BIOCHEM, MICH STATE UNIV, 68- Mem: Am Crystallog Asn; Am Chem Soc. Res: X-ray crystallographic structure determination of biological molecules; enzymes; electron density distributions. Mailing Add: Dept of Chem Mich State Univ East Lansing MI 48823

TULIS, JERRY JOHN, b Cicero, Ill, Apr 13, 30; m 54; c 3. MICROBIOLOGY, IMMUNOLOGY. Educ: Univ Ill, BS, 53; Loyola Univ Chicago, MS, 55; Cath Univ, PhD(radiation microbiol), 65; Nat Registry Microbiol, registered. Prof Exp: Res assoc immunochem, Merck Sharpe & Dohme Res Labs & Merck Inst Therapeut Res, 55-57; proj chief med microbiol, US Army Chem Corps Biol Labs, Ft Detrick, 57-68; DIR MICROBIOL SCI, BECTON DICKINSON RES CTR, 68- Concurrent Pos: Adj assoc prof, NC State Univ. Mem: AAAS; Soc Indust Microbiol; Am Asn Contamination Control; Am Soc Microbiol. Res: Development and evaluation of bacterial vaccines; pathogenesis of bacterial infections; radiobiological and radiotracer studies; sterilization research; cancer immunology; automated biological systems; immunochemical and immunodiagnostic research: antibacterial and biocidal materials research. Mailing Add: Becton Dickinson Res Ctr Box 12016 Research Triangle Park NC 27709

TULK, ALEXANDER STUART, b Hamilton, Ont, Feb 25, 18; m 46; c 4. APPLIED CHEMISTRY. Educ: McMaster Univ, BSc, 44, MSc, 45; Pa State Univ, PhD(inorg chem), 51. Prof Exp: Lectr, McMaster Univ, 45-47; asst, Pa State Univ, 47-50; sr engr, 50-52, engr in-chg, 52-55, adv develop engr, 55-56, eng specialist, 56-57, eng mgr, 57-67, eng mgr inorg mat, 67-71, sr eng specialist, Electronic Mat, 71-73, SR ENG SPECIALIST, SPEC PROJ, GTE SYLVANIA INC, 73- Honors & Awards: Sullivan Award, Am Chem Soc, 74. Mem: Am Chem Soc; Electrochem Soc; Chem Inst Can. Res: Chemical warfare; photosensitive materials; liquid bright gold; fluorocarbon chemistry; electroplating of precious metals; germanium; silicon and gallium arsenide preparation; semiconductor measurements; chemistry of tungsten, molybdenum, rare earths, tantalum and niobium. Mailing Add: 510 Poplar St Towanda PA 18848

TULL, JACK PHILLIP, b Jackson, Mich, Dec 2, 30; m 52; c 3. MATHEMATICS. Educ: Univ Ill, PhD(math), 57. Prof Exp: From instr to asst prof, 56-61, ASSOC PROF MATH, OHIO STATE UNIV, 61- Concurrent Pos: Vis sr lectr, Univ Adelaide, 63-64; prof & head dept, Univ Zambia, 70-71, dean humanities & social sci, 71. Mem: Am Math Soc; Math Asn Am; London Math Soc. Res: Analytic theory of numbers. Mailing Add: Dept of Math Ohio State Univ Columbus OH 43210

TULL, ROBERT GORDON, b Jackson, Mich, May 1, 29; m 52; c 4. ASTRONOMY. Educ: Univ Ill, BS, 52, MS, 57; Univ Mich, PhD(astron), 63. Prof Exp: Res scientist, 61, from instr to asst prof, 62-70, RES SCIENTIST ASTRON, UNIV TEX, AUSTIN, 70- Concurrent Pos: NSF grants, 63-66 & 74-; mem high resolution spectrograph instrument definition team, NASA Large Space Telescope, 73-; consult, Electronic Vision Co Div of Sci Appln Inc, 74- Mem: AAAS; Am Astron Soc; Int Astron Union; Optical Soc Am. Res: Stellar populations of galaxies; photoelectric spectrophotometry of astronomical sources; astronomical instrumentation; planetary atmospheres; development and application of multi-channel image detectors for astronomical spectrophotometry. Mailing Add: Dept of Astron Univ of Tex Austin TX 78712

TULL, ROGER JAMES, b Duluth, Minn, Jan 9, 22; m 45; c 5. ORGANIC CHEMISTRY. Educ: Mich State Univ, BS, 44; Univ Ill, PhD(chem), 49. Prof Exp: Chemist, Eastman Kodak Co, 44-46; chemist, 49-57, mgr process res, 57-69, DIR PROCESS RES, MERCK & CO, INC, RAHWAY, 69- Mem: AAAS; Am Chem Soc. Res: Photographic chemicals; steroids; veterinary and organic medicinals. Mailing Add: 48 Spring St Metuchen NJ 08840

TULLAR, PAUL EDGAR, pharmacology, physiology, deceased

TULLAR, RICHARD MONTGOMERY, b Tacoma, Wash, Aug 30, 10; m 36, 63; c 2. VERTEBRATE PALEONTOLOGY, POPULATION GENETICS. Educ: Univ Calif, Los Angeles, AB, 33; Univ Southern Calif, MS, 47. Prof Exp: Wildlife & range mgt specialist, US Forest Serv & US Soil Conserv Serv, 35-38; asst refuge supt, US Fish & Wildlife Serv, 38-43; chem & stress engr, Timm Aircraft Corp, Calif, 43-45; instr sci & chmn dept, High Sch, 45-47; from instr to assoc prof zool, 47-70, PROF ZOOL & CHMN DEPT LIFE SCI, LOS ANGELES PIERCE COL, 70- Mem: AAAS. Res: Thecodonts; theropods; mosasaurs; ichthyosaurs. Mailing Add: Dept of Zool Los Angeles Pierce Col Woodland Hills CA 91364

TULLER, ANNITA, b New York, NY, Dec 30, 10; m 38; c 2. MATHEMATICS. Educ: Hunter Col, BA, 29; Bryn Mawr Col, MA, 30, PhD(math), 37. Prof Exp: Substitute math, Hunter Col, 30-31; teacher high sch, NY, 31-35; from tutor to assoc prof math, Hunter Col, 37-68; prof, 68-71, EMER PROF MATH, LEHMAN COL, 71- Mem: AAAS; Am Math Soc; Hist Sci Soc; Math Asn Am; NY Acad Sci. Res: Differential geometry; ergodic theory. Mailing Add: 139-62 Pershing Crescent Jamaica NY 11435

TULLER, STANTON ERNEST, b Portland, Ore, July 5, 44. GEOGRAPHY, CLIMATOLOGY. Educ: Univ Ore, BA, 66; Univ Calif, Los Angeles, MA, 67, PhD(geog), 71. Prof Exp: ASSOC PROF GEOG, UNIV VICTORIA, BC, 69- Concurrent Pos: Vis lectr geog, Univ Canterbury, Christchurch, NZ, 75. Mem: AAAS; Asn Am Geog; Can Asn Geog; Am Meteorol Soc. Res: Applied microclimatology, including urban, agricultural and forest climatology. Mailing Add: Dept of Geog Univ of Victoria Victoria BC Can

TULLIER, PETER MARSHALL, JR, b New Orleans, La, Aug 15, 16; m 43; c 9. OPERATIONS RESEARCH, SYSTEMS ANALYSIS. Educ: Loyola Univ, New Orleans, BS, 38; La State Univ, Baton Rouge, MS, 40. Prof Exp: Instr high sch, La, 40-41; from instr to asst prof math, Loyola Univ, 45-52; mathematician, Dept Geophys, Calif Co, 52-53; assoc prof math, Southwestern La Inst, 53-60; opers anal, Naval Sci Dept, US Naval Acad, 60-67; opers analyst, Pac Tech Analysts, Inc, 67-71; dir opers res, Southeast Asia Comput Assocs, 71-74; PRIN ANALYST, OPERS RES, INC, 74- Concurrent Pos: Consult, Electronic Compatibility Anal Ctr; lectr math, Univ Md Far East Div, Saigon; lectr systs anal for Vietnamese analysts, Comput Ctr, Govt Vietnam. Mem: Opers Res Soc Am. Res: Computer software marketing; management information systems and design; data processing in developing countries. Mailing Add: 39 Murray Ave Annapolis MD 21401

TULLIO, VICTOR, b Philadelphia, Pa, May 29, 27; m 51; c 2. CHEMISTRY. Educ: Univ Pa, BS, 48; Univ Ill, PhD(org chem), 51. Prof Exp: Asst, Univ Ill, 48-49; res chemist, 51-61, supvr new dye eval, 64-69, tech asst textile dyes, 69-71, supvr new dye eval, 71-72, SR RES CHEMIST, E I DU PONT DE NEMOURS & CO, INC, 72- Mem: Am Asn Textile Chemists & Colorists; Am Chem Soc. Res: Organic chemistry; dyes; dyeing of synthetic fibers. Mailing Add: Jackson Lab E I du Pont de Nemours & Co Inc Wilmington DE 19898

TULLIS, EDGAR CECIL, plant pathology, see 12th edition

TULLIS, JAMES EARL, b Cincinnati, Ohio. GENETICS. Educ: Miami Univ, BS, 51; Ohio State Univ, MS, 54, PhD(genetics), 61. Prof Exp: Instr zool, Ohio Univ, 56-59; asst prof, Wash State Univ, 61-65; asst prof, 65-71, ASSOC PROF, IDAHO STATE UNIV, 71- Mem: AAAS; Genetics Soc Am. Mailing Add: Dept of Biol Idaho State Univ Pocatello ID 83209

TULLIS, JAMES LYMAN, b Newark, Ohio, June 22, 14; m 37; c 4. BIOCHEMISTRY. Educ: Duke Univ, MD, 40; Am Bd Internal Med, dipl, 48. Prof Exp: Intern med, Roosevelt Hosp, New York, 40-41, sr intern & resident physician, 41-42; from assoc dir to dir blood characterization & preservation lab, 51-55, RES ASSOC BIOCHEM, HARVARD MED SCH, 54-, SR INVESTR, PROTEIN FOUND, 56-, DIR CYTOL LABS, 60-, PROF MED, 75- Concurrent Pos: Donner Found res fel, Harvard Med Sch, 45-48; asst, Peter Bent Brigham Hosp, Boston, 46-50, assoc, 55-58, sr assoc, 58-; attend physician, West Roxbury Vet Admin Hosp, 48-; attend physician & hematologist, New Eng Deaconess Hosp, 49-, chief hemat & chemother clin, 57-, chmn gen med div, 60-, chmn dept med, 64-; consult, Panel Mil & Field Med, Div Med Sci, Res & Develop Bd, US Secy Defense, 52-54; consult, Cambridge City Hosp, 54-60; vpres & treas, Int Cong Hemat, 55-56; assoc clin prof med, Harvard Med Sch, 70-75. Honors & Awards: Glycerol Producers Award, 57; Hektoen Medal, AMA, 58; Katsunuma Award, Int Soc Hemat, 59. Mem: Fel Am Soc Hemat (pres, 58-59); fel AMA; fel Am Col Physicians; fel NY Acad Sci; Int Soc Hemat (vpres, 56-58, secy gen, western hemisphere, 58-). Res: Chemical interactions between blood cells and plasma proteins. Mailing Add: 110 Francis St Boston MA 02215

TULLIS, JULIA ANN, b Swedesboro, NJ, Feb 21, 43; m 65. STRUCTURAL GEOLOGY. Educ: Carleton Col, BA, 65; Univ Calif, Los Angels, PhD(geol), 71. Prof Exp: ASST PROF (RES) GEOL, BROWN UNIV, 71- Concurrent Pos: Mem, Am Geol Inst Women Geoscientist Comt, 73- Mem: Am Geophys Union. Res: High temperature and pressure experimental rock deformation; flow laws of crustal rocks; mechanisms of development of microstructures and preferred orientations; influence of non-hydrostatic stress and deformation on solid state processes in minerals. Mailing Add: Dept of Geol Sci Brown Univ Providence RI 02912

TULLIS, RICHARD EUGENE, b Long Beach, Calif, Apr 26, 36; m 62; c 3. COMPARATIVE PHYSIOLOGY, COMPARATIVE ENDOCRINOLOGY. Educ: Univ Wash, BS, 63; Univ Hawaii, MS, 68, PhD(zool), 72. Prof Exp: Res asst, Univ Hawaii, 70-71; ASST PROF PHYSIOL, CALIF STATE UNIV, HAYWARD, 72- Concurrent Pos: Partic guest, Biomed Div, Lawrence Livermore Lab, 73-; NSF sci equip grant, 74. Mem: Am Soc Zoologists; AAAS. Res: Neuroendocrine control of hydromineral regulation in crustaceans including isolation of neuroendocrine substances, enzyme regulation mechanisms and target organ identification; basic physiological invertebrate functions affected by environmental and man-made substances. Mailing Add: Dept of Biol Sci Calif State Univ Hayward CA 94542

TULLIS, TERRY EDSON, b Rapid City, SDak, July 21, 42; m 65. STRUCTURAL GEOLOGY, GEOPHYSICS. Educ: Carleton Col, AB, 64; Univ Calif, Los Angeles, MS, 67, PhD(struct geol), 71. Prof Exp: Actg instr geol, Univ Calif, Los Angeles, 69-70; ASST PROF GEOL, BROWN UNIV, 70- Concurrent Pos: Sloan res fel, Brown Univ, 73-75. Mem: Am Geophys Union; Geol Soc Am; AAAS. Res: Experimental rock deformation; tectonophysics; plate tectonics; origin of slaty cleavage and schistosity; thermodynamic systems under nonhydrostatic stress; rheology fo rocks at high temperature and pressure; study of in situ stress. Mailing Add: Dept of Geol Sci Brown Univ Providence RI 02912

TULLMAN, GERALD M, organic chemistry, physical chemistry, see 12th edition

TULLNER, WILLIAM W, b Sea Isle City, NJ, Oct 11, 14; m 47; c 4. ENDOCRINOLOGY. Educ: Temple Univ, BA, 37; George Washington Univ, MS, 53, PhD(physiol), 57. Prof Exp: Sr investr endocrinol, physiol & pharmacol, Nat Cancer Inst, 46-65; chief, 65-73, SR INVESTR, SECT ENDOCRINOL, REPRODUCTION RES BR, NAT INST CHILD HEALTH & HUMAN DEVELOP, 73- Concurrent Pos: Lectr endocrine chem, Georgetown Univ, 60-64; res consult endocrinol, Grad Coun, George Washington Univ, 68-; res assoc, Howard Univ, 70- Mem: Fel AAAS; Endocrine Soc. Res: Biology and biochemistry of reproduction of subhuman primates and laboratory animals; effects of hormones on fetal development. Mailing Add: Sect Endocrinol Reproduct Res Br Nat Inst Child Health & Human Dev NIH Auburn Bldg Rm 205 Bethesda MD 20014

TULLOCH, ALEXANDER PATRICK, b Garelochhead, Scotland, Nov 27, 27; m 57; c 4. ORGANIC CHEMISTRY. Educ: Univ Glasgow, BSc, 49, PhD(chem), 55, DSc, 72. Prof Exp: Tech off org chem, Imp Chem Industs, Ltd, 54-58; fel, 56-57, asst res off, 58-62, assoc res off, 62-69, SR RES OFF, FATS & OILS LAB, PRAIRIE REGIONAL LAB, NAT RES COUN CAN, 69- Mem: The Chem Soc; Am Oil Chemists Soc. Res: Organic chemistry of lipids, particularly glycolipids of microorganisms; gas chromatography of fats and waxes; nuclear magnetic resonance spectroscopy. Mailing Add: Prairie Regional Lab Nat Res Coun Can Saskatoon SK Can

TULLOCH, GEORGE SHERLOCK, b Bridgewater, Mass, Aug 3, 06; m 31; c 2. PARASITOLOGY. Educ: Mass Col, BS, 28; Harvard Univ, MS, 29, PhD(entom), 31. Prof Exp: Asst biol, Harvard Univ, 28-31; from instr to prof, 32-65, EMER PROF BIOL, BROOKLYN COL, 65-; RES DIR, GEORGE S TULLOCH & ASSOCS, 70- Concurrent Pos: Asst, Radcliffe Col, 29-30; assoc entomologist, PR Insect Pest Surv, 35-36; chief entomologist, State Mosquito Surv, Mass, 39; Rockefeller Found fel, Brazil, 40-41; consult, Arctic Aeromed Lab, US Air Force, 54-55, res scientist, 65-67; vis prof, Univ Queensland, 61; res scientist, US Air Force Sch Aerospace Med, 67-69; vis investr, Southwest Found Res & Educ, Tex, 70-; consult, Dept Path & Labs, Nassau County Med Ctr, NY, 71- Mem: AAAS; Am Soc Parasitol; Entom Soc Am. Res: Arthropod carried and parasitic diseases of man and animals; life cycle and therapy of canine dirofilariasis; collection of parasitic antigens for skin tests. Mailing Add: 4919 Pecan Grove 213 San Antonio TX 78222

TULLOCH, LYNN HARDYN, b Belton, Tex, Apr 23, 07. MATHEMATICS. Educ: Baylor Univ, AB, 28; Brown Univ, AM, 32; Univ Tex, PhD, 65. Prof Exp: Instr math, Baylor Univ, 28-30; head dept, Victoria Jr Col, 34-37 & San Antonio Jr Col, 38-46; assoc prof, 46-57, PROF MATH, SOUTHWEST TEX STATE UNIV, 57- Concurrent Pos: NSF fel, Univ Tex, 57-58. Mem: AAAS; Am Math Soc; Soc Indust & Appl Math; Math Asn Am; Can Math Cong. Res: Riesz-Lebesque and Stieltjes-Lebesque integrals; theory of integration; analytical mechanics. Mailing Add: Dept of Math Southwest Tex State Univ San Marcos TX 78666

TULLOCK, CHARLES WILLIAM, b Antigo, Wis, Nov 19, 12; m 54. ORGANIC CHEMISTRY. Educ: Univ Ill, BS, 34; Univ Wis, PhD(org chem), 38. Prof Exp: Instr chem, Univ Wis, 34-38; RES CHEMIST, EXP STA, E I DU PONT DE NEMOURS & CO, INC, 38-42 & 46- Mem: Am Chem Soc; The Chem Soc. Res: Fluorochemicals; heat stable fluids; olefin polymers. Mailing Add: Cent Res Dept Exp Sta EI de Pont de Nemours & Co Inc Wilmington DE 19898

TULLOCK, ROBERT JOHNS, b Atascadero, Calif, Oct 3, 40; m 62; c 3. SOIL CHEMISTRY. Educ: Calif State Polytech Col, San Luis Obispo, BS, 67; Purdue Univ, West Lafayette, MS, 70, PhD(soil chem), 72. Prof Exp: Asst prof soil sci, Univ Calif, Riverside, 72-74; ASST PROF SOIL SCI, ORE STATE UNIV, Mem: Am Soc Agron; Soil Sci Soc Am; Clay Minerals Soc. Res: Physicochemical properties of colloidal surfaces. Mailing Add: Dept of Soil Sci Ore State Univ Corvallis OR 97331

TULLY, EDWARD JOSEPH, JR, b Brooklyn, NY, Jan 22, 30. MATHEMATICS. Educ: Fordham Univ, AB, 51, MS, 52; Tulane Univ, PhD(math), 60. Prof Exp: Instr math, St John's Univ, Minn, 56-57; asst, Tulane Univ, 57-60; NSF fel, Calif Inst Technol, 60-61; lectr, Univ Calif, Los Angeles, 61-63; lectr, 63-64, asst prof, 64-68, ASSOC PROF MATH, UNIV CALIF, DAVIS, 68- Mem: Am Math Soc; Math Asn Am. Res: Algebraic theory of semigroups; ordered algebraic systems; application of algebra to linguistics. Mailing Add: Dept of Math Univ of Calif Davis CA 95616

TULLY, FRANK PAUL, b Hartford, Conn, Apr 27, 46; m 74. PHYSICAL CHEMISTRY. Educ: Clark Univ, BA, 68; Univ Chicago, MS, 69, PhD(chem), 73. Prof Exp: Res asst chem, Clark Univ, 65-68; res asst, Univ Chicago, 68-73; fel, Univ Toronto, 73-74; NSF FEL CHEM, MICH STATE UNIV, EAST LANSING, 74- Concurrent Pos: NSF energy related fel, 75. Mem: Am Phys Soc. Res: Use of the crossed molecular beam method in studies of elastic, inelastic and reactive scattering; photoconization; surface studies. Mailing Add: Dept of Chem Mich State Univ East Lansing MI 48824

TULLY, JOHN CHARLES, b New York, NY, May 17, 42; m 71. CHEMICAL PHYSICS. Educ: Yale Univ, BS, 64; Univ Chicago, PhD(chem), 68. Prof Exp: NSF fel chem, Univ Colo, 68-69 & Yale Univ, 69-70; MEM TECH STAFF, BELL LABS, 70- Mem: Am Phys Soc. Res: Theoretical studies of electron-molecule scattering and molecular photoionization; experimental and theoretical studies of molecular collisions. Mailing Add: Bell Labs Murray Hill NJ 07974

TULLY, JOHN PATRICK, b Brandon, Man, Nov 29, 06; m 38; c 3. OCEANOGRAPHY. Educ: Univ Man, BSc, 31; Univ Wash, Seattle, PhD(oceanog chem), 48. Prof Exp: Sci asst chem, Fisheries Res Bd Can, 31-36, hydrog, 36-46, oceanogr-in-chg, Pac Oceanog Group, 46-65, oceanog consult, 66-69; consult, 69-75; RETIRED. Concurrent Pos: Hon lectr, Univ BC, 50-53. Honors & Awards: Mem, Order of Brit Empire, 46; Commemorative Medal, Albert I of Monaco, 67, 70. Mem: AAAS (pres western div, 63); Am Soc Limnol & Oceanog; Am Geophys Union; Royal Soc Can. Res: Oceanographic chemistry; seawater structure; estuarine mechanisms; physical and chemical processes and submarine acoustics. Mailing Add: 2740 Fandell Ave Nanaimo BC Can

TULLY, JOSEPH GEORGE, JR, b Sterling, Colo, July 14, 25; m 57. MEDICAL MICROBIOLOGY. Educ: Portland Univ, BS, 49; Brigham Young Univ, MS, 51; Cincinnati Univ, PhD(microbiol), 55. Prof Exp: Asst prof microbiol, Col Med, Cincinnati Univ, 55-57; microbiologist, Walter Reed Army Inst Res, 57-61, chief, Dept Microbiol, 61-62; res microbiologist, 62-68, HEAD MYCOPLASMA SECT, NAT INST ALLERGY & INFECTIOUS DIS, 68- Concurrent Pos: China med bd fel, Cent Am, 56; attend microbiologist, Cincinnati Gen Hosp, Ohio, 56-57; mem bd for Food & Agr Orgn/WHO Prog on Comp Mycoplasmology, 69-, chmn bd, 72-; mem, Int Subcomt Taxon of Mycoplasmatales, 70- Mem: Fel AAAS; Am Asn Immunol; fel Am Acad Microbiol; Am Soc Microbiol; Soc Exp Biol & Med. Res: Bacillary dysentery; immunology and pathogenesis of enteric diseases; typhoid infection in primates; basic biology of the mycoplasmas; murine mycoplasmas. Mailing Add: Mycoplasma Sect Lab Microbiol Nat Inst Allergy & Infectious Dis Bethesda MD 20014

TULSKY, EMANUEL GOODEL, b Philadelphia, Pa, Dec 6, 23; m 50; c 2. RADIOLOGY. Educ: Jefferson Med Col, MD, 48. Prof Exp: Attend radiologist, Delafield Hosp, New York, 52-53; assoc radiologist, Sch Med & Univ Hosp, Temple Univ, 53-55; assoc radiol, Div Grad Med & asst prof radiol, Univ Pa, 55-65; RADIOLOGIST & DIR DIV RADIATION THER & NUCLEAR MED, ABINGTON MEM HOSP, PA, 65- Concurrent Pos: Assoc radiologist & dir, Tumor Clin, Hosp Univ Pa, 55-67; asst prof radiol, Med Col Pa, 62- Mem: Radiol Soc NAm; AMA; Am Col Radiol. Res: Intracavitary radiation dosimetry; isotopic studies of gastrointestinal absorption; cancer therapy; synergistic action of radiation and cytotoxics. Mailing Add: Dept of Radiol Abington Mem Hosp Abington PA 19001

TULUMELLO, ANGELO C, inorganic chemistry, see 12th edition

TUMA, DEAN J, b Howells, Nebr, Oct 20, 41; m 64; c 3. BIOLOGICAL CHEMISTRY. Educ: Creighton Univ, BS, 64, MS, 68; Univ Nebr, PhD(biochem), 73. Prof Exp: Instr, 73-75, ASST PROF INTERNAL MED & BIOCHEM, COL MED, UNIV NEBR, 75-; RES CHEMIST BIOCHEM, VET ADMIN HOSP, OMAHA, 64- Mem: Am Asn Study Liver Dis; Am Fedn Clin Res. Res: Investigation of the role of ethanol, drugs and nutrition in liver metabolism and liver disease. Mailing Add: 2223 S 161 Circle Omaha NE 68130

TUMA, HAROLD J, b Belleville, Kans, Feb 28, 33; m 56; c 3. MEAT SCIENCE, FOOD SCIENCE. Educ: Kans State Univ, BS, 55, MS, 58; Okla State Univ, PhD(food sci), 61. Prof Exp: Asst food sci, Okla State Univ, 58-61; asst prof animal sci, SDak State Univ, 61-65; assoc prof, 65-73, PROF ANIMAL SCI, KANS STATE UNIV, 73- Mem: Am Soc Animal Sci; Am Meat Sci Asn; Inst Food Technol. Res: Histological and physiological characteristics of muscle associated with meat quality. Mailing Add: Dept of Animal Sci Kans State Univ Manhattan KS 66502

TUMAN, VLADIMIR SHLIMON, b Kermanshah, Iran, May 21, 23; US citizen; m 51; c 3. PHYSICS, GEOPHYSICS. Educ: Univ Birmingham, BSc, 48; Univ London, DIC, 49; Stanford Univ, PhD(geophys), 64. Prof Exp: Geophysicist, Anglo Iranian Oil Co, SW Iran, 50-52; actg chief petrol, Nat Iranian Oil Co, 52-55, engr trainee, Europe & USA, 55-56; sr petrol physicist, Nat Iranian Consortium, 56-57; res physicist, Atlantic Refining Oil Co, Dallas, Tex, 57-59; assoc prof petrol eng, Univ Ill, Urbana, 59-62; res assoc geophys, Stanford Univ, 62-65, res physicist, Res Inst, 65-66; assoc prof physics, 66-67, chmn dept phys sci, 66-71, PROF PHYSICS, CALIF STATE COL, STANISLAUS, 67- Concurrent Pos: Calif Res Corp grant, 60-61; consult, Esso Res Lab, 61, Schlumberger Well Logging Co, 61 & Sinclair Oil Co, 61; Am Petrol Inst grants, 61-63; consult, Comput Symp, Stanford Univ, 63-64, res assoc, Physics Dept, 65-; lectr, Varian Assoc, Palo Alto, Calif, 66. Mem: fel Royal Astron Soc; Am Phys Soc; Am Asn Physics Teachers; Am Geophys Union; Soc Explor Geophys. Res: Development of cryogenic gravity meter to study earth eigen vibrations and look for gravitational radiation. Mailing Add: Dept of Physics Calif State Col at Stanislaus Turlock CA 95380

TUMASONIS, CASIMIR F, experimental embryology, entomology, see 12th edition

TUMBLESON, MYRON EUGENE, b Mountain Lake, Minn, Mar 13, 37; m 58; c 3. BIOCHEMISTRY, NUTRITION. Educ: Univ Minn, BS, 58, MS, 61, PhD(nutrit), 64. Prof Exp: From res assoc to asst prof animal sci, Univ Minn, 64-66; asst prof vet physiol & pharmacol, Univ Mo-Columbia & res assoc med biochem, Sinclair Comp Med Res, 66-69; ASSOC PROF VET ANAT & PHYSIOL, UNIV 66-69, MO-COLUMBIA & RES ASSOC SINCLAIR COMP MED RES, 69- Mem: Am Inst Nutrit; Soc Exp Biol & Med; Am Soc Neurochem; NY Acad Sci; Sigma Xi. Res: Protein-calórie malnutrition; alcoholism and aging, using miniature swine as biomedical research subjects. Mailing Add: Dept of Vet Anat-Physiol Univ of Mo Columbia MO 65201

TUMEN, HENRY JOSEPH, b Philadelphia, Pa, Apr 7, 02; m 26; c 1. MEDICINE. Educ: Univ Pa, AB, 22, MD, 25; Am Bd Internal Med, dipl, 52. Prof Exp: Prof clin gastroenterol, 54-60, prof med & chmn dept, 60-71, EMER PROF MED, GRAD SCH MED, UNIV PA, 71- Concurrent Pos: Consult, Walter Reed Army Med Ctr, DC, 58- & Albert Einstein Med Ctr, 60-; chmn, Am Bd Internal Med, 58-60. Mem: Am Soc Gastrointestinal Endoscopy; Am Gastroenterol Asn; Am Col Physicians. Res: Gastroenterology. Mailing Add: 135 S 18th St Philadelphia PA 19103

TUMILOWICZ, JOSEPH J, b South River, NJ, June 12, 30; m 55; c 1. VIROLOGY, IMMUNOLOGY. Educ: Rutgers Univ, BS, 55; Univ Pa, PhD(virol), 63. Prof Exp: Res asst virol, Ciba Pharmaceut Co, 55-58; asst mem virol, Inst Med Res, NJ, 64-71; asst mem, NY Blood Ctr, 71-72; HEAD DEPT VIROL & TISSUE CULT, VIRAL SCI LAB, ELECTRO-NUCLEONICS LABS, INC, 72- Concurrent Pos: NIH fel, 63-64. Mem: AAAS; Am Soc Microbiol; Tissue Cult Asn; NY Acad Sci. Res: Fractionation and analysis of poliovirus populations; in vitro properties of cells derived from mammalian tumors; cell-virus interactions. Mailing Add: Electro-Nucleonic Labs Inc Viral Sci Lab 12050 Tech Rd Silver Spring MD 20014

TUMULTY, PHILIP A, b Jersey City, NJ, Nov 4, 12; m 42; c 5. MEDICINE. Educ: Georgetown Univ, AB, 35; Johns Hopkins Univ, MD, 40; Am Bd Internal Med, dipl, 48. Prof Exp: Intern med, Johns Hopkins Hosp, 40-41, asst resident, 41-42 & 45-46, resident, 46-47, from instr to assoc prof, Johns Hopkins Univ, 46-53, from asst dir to dir, Med Clin, Johns Hopkins Hosp, 48-53, physician-in-chg, Med Surg Group Clin, 48-53; prof med & dir dept, Sch Med, St Louis Univ, 53-54; assoc prof, 55-63, PROF MED, SCH MED, JOHNS HOPKINS UNIV, 63-, CHMN PVT MED SERV, JOHNS HOPKINS HOSP, 55-, PHYSICIAN-IN-CHG, PVT PATIENT CLIN, 56- Concurrent Pos: Am Col Physicians clin fel, Johns Hopkins Hosp, 46-47; regional consult, Vet Admin, 54-55; consult, Baltimore City Hosp, 55-; spec consult, NIH, 59-; consult, Vet Admin & USPHS Hosps, 60- & Walter Reed Hosp, 61- Mem: Asn Am Physicians; Am Soc Internal Med; Am Clin & Climat Asn. Res: Bacterial endocarditis; collagen diseases. Mailing Add: Johns Hopkins Hosp 601 N Broadway Baltimore MD 21205

TUNA, NAIP, b Constanta, Romania, Aug 18, 21; m 49; c 2. INTERNAL MEDICINE, CARDIOVASCULAR DISEASES. Educ: Istanbul Univ, MD, 47; Univ Minn, PhD(med), 58. Prof Exp: Asst med, Therapeut Clin, Istanbul Univ, 49-52; resident med, St Joseph's Hosp, Lexington, Ky, 52-53; from instr to asst prof, 57-64, ASSOC PROF MED, MED SCH, UNIV MINN, MINNEAPOLIS, 64- Concurrent Pos: Am Heart Asn res fel, Univ Minn, Minneapolis, 58-59, advan res fel, 59-61. Mem: Fel Am Col Physicians; fel Am Col Cardiol; fel Am Heart Asn; Am Fedn Clin Res. Res: Electro and vector cardiography; cardiology. Mailing Add: Dept of Med Univ of Minn Minneapolis MN 55455

TUNG, FRED FU, b Manchouli, Heilungkiang. BIOCHEMISTRY. Educ: Taiwan Prov Col Agr, BS, 56; Univ Vt, MS, 63; Univ Mich, MS, 66; Univ Mo, PhD(biochem), 70. Prof Exp: Res asst agr chem, Taiwan Prov Col Agr, 58-59; fel biochem, State Univ NY Albany, 70-73; BIOCHEMIST, MICH DEPT PUB HEALTH, 74- Mem: Am Chem Soc; AAAS. Res: Use of plasmin to modify immune serum globulin for intravenous administration; isolation and preparation of blood derivatives. Mailing Add: Mich Dept Pub Health Bur of Labs 3500 N Logan St Lansing MI 48914

TUNG, JOHN SHIH-HSIUNG, b Keelung, Taiwan, July 19, 28; m 54; c 4. MATHEMATICS. Educ: Taiwan Norm Univ, BA, 50; Pa State Univ, MA, 60, PhD(math), 62. Prof Exp: Asst civil eng, Taihoku Imp Univ, Taiwan, 44-45; asst instr math, Taipei Inst Technol, 50-53; asst math & indust educ, Taiwan Norm Univ, 53-56, instr, 56-58; asst math, Pa State Univ, 58-60 & 61-62; asst prof, 62-66, ASSOC PROF MATH, MIAMI UNIV, OHIO, 66-, ASSOC PROF STATIST, 74- Mem: Math Asn Am; Am Math Soc. Res: Theory of functions of a complex variable; infinite and orthogonal series; matrix algebra. Mailing Add: Dept of Math Miami Univ Oxford OH 45056

TUNG, MARVIN ARTHUR, b Sask, Can, Nov 9, 37; m 61; c 3. FOOD SCIENCE. Educ: Univ BC, BSA, 60, teaching cert, 61, MSA, 67, PhD(food sci), 70. Prof Exp: Teacher sch bd, 61-70; ASSOC PROF FOOD SCI, UNIV BC, 70- Mem: Can Inst Food Sci & Technol; Inst Food Technologists; Brit Inst Food Sci & Technol; Can Soc Agr Eng; Micros Soc Can. Res: Food rheology; microstructure of food systems; food processing and packaging. Mailing Add: Dept of Food Sci Univ of BC Vancouver BC Can

TUNG, WU-KI, b Kunming, China, Oct 16, 39; m 63; c 2. THEORETICAL PHYSICS, ELEMENTARY PARTICLE PHYSICS. Educ: Univ Taiwan, BS, 60; Yale Univ, PhD(physics), 66. Prof Exp: Res assoc theoret physics, Inst Theoret Physics, State Univ NY Stony Brook, 66-68; mem staff, Inst Adv Study, 68-70; asst prof physics & mem staff, Enrico Fermi Inst, Univ Chicago, 70-75; ASSOC PROF PHYSICS, ILL INST TECHNOL, 75- Mem: Am Phys Soc. Res: High energy theoretical physics. Mailing Add: Dept of Physics Ill Inst Technol Chicago IL 60616

TUNHEIM, JERALD ARDEN, b Claremont, SDak, Sept 3, 40; m 63; c 1. SOLID STATE PHYSICS. Educ: SDak State Univ, BS, 62, MS, 64; Okla State Univ, PhD(physics), 68. Prof Exp: Asst prof, 68-72, ASSOC PROF PHYSICS, S DAK STATE UNIV, 72- Mem: Am Phys Soc; Am Asn Physics Teachers. Res: Alpha particle model of sulphur nucleus; electron spin resonance measurements of transition metal ions in stannic oxide; surface effects on conductivity of stannic oxide; application and development of models for remote sensing application. Mailing Add: Dept of Physics SDak State Univ Brookings SD 57006

TUNIK, BERNARD D, b New York, NY, Dec 22, 21; m 49; c 3. PHYSIOLOGY. Educ: Univ Wis, BA, 42; Columbia Univ, MA, 51, PhD(zool), 59. Prof Exp: Asst cytol, Sloan-Kettering Inst, 50-52; asst zool, Columbia Univ, 52-55; instr anat, Sch Med, Univ Pa, 58-60; dep chmn dept, 62-63 & 64-65, ASSOC PROF BIOL SCI, STATE UNIV NY STONY BROOK, 60- Concurrent Pos: Nat Inst Arthritis & Metab Dis spec fel, Dept Polymer Sci, Weizmann Inst, 66-67; vis scholar, Dept Zool, Univ Calif, Berkeley, 75. Mem: AAAS; Am Soc Cell Biol. Res: Cellular physiology; mechanochemical aspects of muscle contraction; triggers of muscle hypertrophy. Mailing Add: Dept of Cellular & Comp Biol State Univ NY Stony Brook NY 11794

TUNIS, MARVIN, b New York, NY, Apr 18, 25; m 52; c 4. BIOCHEMISTRY. Educ: Hunter Col, AB, 50; Univ Ill, MS, 51, PhD(biochem), 54. Prof Exp: Res assoc, Univ Ill, 54-55; USPHS fel, Col Physicians & Surgeons, Columbia Univ, 55-56; sr cancer res scientist, Roswell Park Mem Inst, 57-68; assoc prof, 68-74, ASSOC PROF CHEM, STATE UNIV NY COL BUFFALO, 74- Mem: Am Chem Soc; Am Soc Biol Chem; Am Asn Cancer Res. Res: Biochemistry and metabolism of nucleic acids, proteins, glycoproteins and mucopolysaccharides; enzymology. Mailing Add: Dept of Chem State Univ of NY Col Buffalo NY 14222

TUNIS, WILLIAM DAVID, b Northampton, Mass, Nov 11, 24; m 49; c 3. ENTOMOLOGY. Educ: Univ Mass, BS, 49, PhD(entom), 59; Univ Minn, MS, 51. Prof Exp: Asst prof entom, Univ Conn, 52-59; assoc prof, 59-72, PROF ENTOM, UNIV MASS, 72-, DEAN ADMIS & REC, 63- Res: Economic entomology; insect ecology. Mailing Add: Dept Entom & Plant Path Univ of Mass Amherst MA 01002

TUNKEL, NORMAN, organic chemistry, see 12th edition

TUNKELANG, BEN, physics, environmental sciences, see 12th edition

TUNNICLIFFE, PHILIP ROBERT, b Derby, Eng, May 3, 22; Can citizen; m 46; c 2. PHYSICS. Educ: Univ London, BSc, 42. Prof Exp: Res staff, Telecommun Res Estab, Malvern, Eng, 42-46, UK Atomic Energy Auth, Ont, Can, 46-49 & Atomic Energy Res Estab, Harwell, Eng, 49-51; res staff, 51-61, br head electronics, 61-63, br head appl physics, 63-67, BR HEAD ACCELERATOR PHYSICS, ATOMIC ENERGY CAN LTD, 67- Mem: Can Asn Physicists; Am Phys Soc. Res: Microwave tubes; neutron, low energy nuclear, reactor and accelerator physics; reactor control and instrumentation. Mailing Add: Atomic Energy of Can Ltd Chalk River ON Can

TUNSTALL, LUCILLE HAWKINS, b Thurber, Tex, Jan 17, 22; m 44; c 2. MICROBIOLOGY, IMMUNOLOGY. Educ: Univ Colo, BS, 43; Wayne State Univ, MS, 59, PhD(biol, microbiol), 63. Prof Exp: Med technologist, Med Sch, Univ Colo, 43-45, Presby Hosp Colo, 45-47, Evangel Deaconess Hosp, 50-52, Sinai Hosp Detroit, 52-55 & Brent Gen Hosp, 55-58; res & tech asst biol, Wayne State Univ, 58-62; asst prof, Delta Col, 62-65; assoc prof, Saginaw Valley Col, 65-67; prof & chmn dept, Bishop Col, 67-71; assoc dir, United Bd Col Develop, 71-72; PROF BIOL & DIR ALLIED HEALTH PROG, CLARK COL & PROF BIOL, ATLANTA UNIV, 72- Concurrent Pos: Consult, United Bd Col Develop, 72-; consult, Nat Urban League, Moton Consortium Admis & Financial Aid & Univ Assocs, 72-74; spec consult, Nat Inst Gen Med Sci, 75- Mem: AAAS; Am Soc Clin Path; Am Soc Microbiol; Am Soc Cell Biol; NY Acad Sci. Res: L-variation; frequency of occurrence and pathogenicity of organisms; immunological studies of blood and pleural fluid in patients with coronary thrombosis; cell wall-deficient bacteria in microbial ecology. Mailing Add: Allied Health Dept Clark Col Atlanta GA 30314

TUNTURI, ARCHIE ROBERT, b Portland, Ore, July 28, 17; m 48. ANATOMY. Educ: Reed Col, AB, 39; Univ Ore, MS, 43, MD & PhD(anat), 44: Prof Exp: Asst, 40-44, from instr to asst prof, 43-59, ASSOC PROF ANAT, SCH MED, UNIV ORE HEALTH SCI CTR, 59- Concurrent Pos: Dir contract, US Off Naval Res, 47-69. Mem: Int Brain Res Orgn; Am Asn Anatomists; Am Physiol Soc; sr mem Inst Elec & Electronics Engrs; Acoust Soc Am. Res: Physiology, biophysics and communication aspects of the auditory cortex in the brain; neuroanatomy; hearing and acoustics; use of computer techniques for the study of the nervous system. Mailing Add: Dept of Anat Sch of Med Univ Ore Hlth Sci Ctr Portland OR 97201

TUNZI, MILTON GEORGE, limnology, biometry, see 12th edition

TUOMI, DONALD, b Willoughby, Ohio, Sept 12, 20; m 45; c 2. PHYSICAL CHEMISTRY. Educ: Ohio State Univ, BS, 43, PhD(phys chem), 52. Prof Exp: Res scientist, SAM Lab, Columbia Univ, 43-45 & Carbide & Carbon Chem Corp, 45-46; res assoc photoemissive surfaces, Res Found, Ohio State Univ, 50-53; mem staff semiconductor devices, Lincoln Lab, Mass Inst Technol, 53-54; res chemist, Baird Assocs, Inc, 54-55 & Res Lab, McGraw-Edison Co, 55-61; MGR SOLID STATE PHYSICS, RES LAB, BORG-WARNER CORP, 61- Mem: AAAS; Electrochem Soc; Am Phys Soc; Am Chem Soc; Am Crystallog Asn. Res: Structural chemistry; semiconductors; photoemissive surfaces; electrochemistry; charge transfer phenomena; thermoelectric cooling devices and semiconductor alloys; polymer solid state structural chemistry. Mailing Add: R C Ingersoll Res Ctr Borg-Warner Corp Des Plaines IL 60018

TUONO, JOSEPH G, physical chemistry, organic chemistry, see 12th edition

TUPA, DIANNA LOU DOWDEN, b San Marcos, Tex, Jan 12, 45; m 70. PHYCOLOGY, MORPHOLOGY. Educ: Southwest Tex State Univ, BS, 67, MA, 68; Univ Tex, Austin, PhD(bot), 72. Prof Exp: ASST PROF BIOL, UNIV N DAK, 72- Concurrent Pos: Res scientist, Southwest Found Res & Educ, San Antonio, 75; sr res scientist, Southwest Res Inst, San Antonio, 75- Mem: AAAS; Phycol Soc Am; Am Inst Biol Sci; Bot Soc Am. Res: Use of cultural techniques to elucidate the morphology, life histories, taxonomy and ecology of freshwater algae. Mailing Add: Dept of Biol Univ of NDak Grand Forks ND 58201

TUPAC, JAMES DANIEL, b Chisholm, Minn, May 4, 27; m 47; c 2. MATHEMATICS, PHYSICS. Educ: Univ Minn, BS, 50, MS, 51. Prof Exp: Mathematician, US Naval Missile Test Ctr, Calif, 51-53; mathematician, Rand Corp, Calif, 53-59, head comput serv, 59-68; PRES, PRC COMPUT CTR INC, 68- Concurrent Pos: Lectr, Exten Div, Univ Calif, Los Angeles, 60-62; mem comput adv coun, Nat Ctr Atmospheric Res, 66-69. Mem: Asn Comput Mach. Res: Applied mathematical, numerical and computer systems analyses; computer programming; computer center management. Mailing Add: PRC Comput Ctr Inc 7670 Old Springhouse Rd McLean VA 22101

TUPIN, JOE PAUL, b Comanche, Tex, Feb 17, 34; m 55; c 3. PSYCHIATRY. Educ: Univ Tex, Austin, BS, 55; Univ Tex Med Br, Galveston, MD, 59. Prof Exp: Resident psychiatrist, Univ Tex Med Br, Galveston, 60-62; resident, NIMH, 63-64; NIMH career teaching award, Group Advan Psychiat, 64-66; assoc prof psychiat & assoc dean, Univ Tex Med Br, Galveston, 68-69; assoc prof, 69-71, PROF PSYCHIAT, SCH MED, UNIV CALIF, DAVIS, 71-, VCHMN DEPT, 70- Concurrent Pos: Fel, Group Advan Psychiat, 60-62; dir, Psychiat Consult Serv, Sacramento Med Ctr, 69-; consult, Calif Med Facil, Vacaville, 69- & Twin & Sibling Study, NIMH, 69-; consult, Dept Corrections, State of Calif, 71; chmn, NIMH Clin Psychopharmacol Rev Comt, 75-77. Mem: Fel Am Psychiat Asn; Soc Biol Psychiat; Am Col Psychiat; Soc Health & Human Values; Am Psychosomatic Soc. Res: Teaching of medical education; psychopharmacology; identification and treatment of violent behavior. Mailing Add: Dept Psychiat Univ Calif Davis Bldg Sacramento Med Ctr 2233 Stockton Blvd Sacramento CA 95817

TUPPER, CHARLES JOHN, b Miami, Ariz, Mar 7, 20; m 42; c 2. INTERNAL MEDICINE. Educ: San Diego State Col, BS, 43; Univ Nebr, MD, 48. Prof Exp: Asst prof internal med, Med Sch, Univ Mich, 56-59, secy med sch, 57, assoc prof & asst dean, 59-66; PROF MED & DEAN SCH MED, UNIV CALIF, DAVIS, 66- Concurrent Pos: Consult, St Joseph Mercy Hosp, 56- Honors & Awards: Billings Bronze Medal, AMA, 55. Mem: Am Soc Internal Med; AMA; Am Col Health Asn;

Asn Am Med Cols; fel Am Col Physicians. Res: Medical education; application of principles of preventive medicine to care of the individual patient through periodic health examination; evaluation of diagnostic procedures for effectiveness and reliability. Mailing Add: Sch of Med Univ of Calif Davis CA 95616

TUPPER, W R CARL, b New Glasgow, NS, Feb 15, 15; m 43; c 3. OBSTETRICS & GYNECOLOGY. Educ: Dalhousie Univ, BSc, 39, MD, CM, 43; FRCOG; FRCS(C). Prof Exp: PROF OBSTET & GYNEC & HEAD DEPT, DALHOUSIE UNIV, 59- Mem: Am Col Surgeons; Am Col Obstet & Gynec; Can Soc Obstet & Gynec; Int Col Surgeons. Mailing Add: Dept of Obstet & Gynec Dalhousie Univ Fac of Med Halifax NS Can

TUPPER, WILLIAM MACGREGOR, b New Glasgow, NS, Oct 7, 33; m 53; c 2. GEOLOGY, GEOCHEMISTRY. Educ: Mt Allison Univ, BSc, 53; Univ NB, MSc, 55; Mass Inst Technol, PhD(geochem), 59. Prof Exp: Geologist, NB Dept Lands & Mines, 55-58; from asst prof to assoc prof, 59-72, PROF GEOL, CARLETON UNIV, 72-, CHMN DEPT, 70- Concurrent Pos: Mem educ comt, Nat Adv Coun Geol Sci, Nat Res Coun Can, 61-63, assoc comt isotope geol & geochronology, 66-68; geologist, Geol Surv Can, 61-65; Nat Res Coun Can fel, Inst Geol Sci, 66-67. Mem: Can Inst Mining & Metall. Res: Geology and geochemistry of mineral deposits; application of chemical methods in search for mineral deposits; application of staple isotope studies to geological problems. Mailing Add: Dept of Geol Carleton Univ Ottawa ON Can

TURAN, THOMAS STEPHEN, analytical chemistry, see 12th edition

TURBAK, ALBIN FRANK, b New Bedford, Mass, Sept 23, 29; m 52; c 2. ORGANIC CHEMISTRY. Educ: Southeastern Mass Technol Inst, BS, 51; Inst Textile Tech, MS, 53; Ga Inst Technol, PhD, 57. Prof Exp: Res chemist, Esso Res Co, 57-63; res mgr, Teepak Inc, 63-72; RES SUPVR, ITT RAYONIER CO, 72- Mem: AAAS; Am Chem Soc; Am Asn Textile Chem & Colorists; Inst Food Technol; fel Royal Soc Dyers & Colorists. Res: Cellulose, protein and synthetic polymer research; polymer modification; new process and methods research; phosphorus chemistry; dyeing and finishing of textiles; food products research; paper products and wood research. Mailing Add: 7 Fairfield Dr Convent Station NJ 07961

TURBYFILL, CHARLES LEWIS, b Newland, NC, Feb 27, 33; m 55; c 2. RESEARCH ADMINISTRATION. Educ: Univ Ore, BA, 55, MS, 57; Univ Ga, PhD(zool), 64. Prof Exp: Prin investr, Worcester Found Exp Biol, 64-66 & Armed Forces Radiobiol Res Inst, 66-72; HEALTH SCI ADMINR, NAT HEART & LUNG INST, NIH, 72-, HEAD INSTNL TRAINING, NAT CANCER INST, NIH, 75- Mem: Am Soc Zool; Brit Soc Endocrinol; NY Acad Sci. Res: Primate cardiovascular physiology, atherosclerosis. Mailing Add: 20512 Clarksburg Rd Boyds MD 20720

TURCHECK, JOSEPH EDWARD, b Brownville, Pa, Nov 26, 42; m 71. APPLIED MATHEMATICS. Educ: Pa State Univ, BS, 64; Columbia Univ, MA, 65, PhD(math), 72. Prof Exp: Instr math, Clarkson Col Technol, 69-71; asst prof, Va Polytech Inst, 72-73; ASST PROF MATH, ST LAWRENCE UNIV, 73- Mem: Am Math Soc; Math Asn Am. Res: Abstract theory of asymptotic expansions for solutions of algebraic differential equations; application of the theory of abstract differential fields with order relations to problems in asymptotic expansions. Mailing Add: Dept of Math St Lawrence Univ Canton NY 13617

TURCHI, JOSEPH J, b Philadelphia, Pa, Feb 16, 33; m 59; c 3. INTERNAL MEDICINE, HEMATOLOGY. Educ: Univ Pa, BA, 54; Jefferson Med Col, MD, 58. Prof Exp: Head clin hemat & cancer chemother, US Naval Hosp, Bethesda, Md, 62-64; sr investr med, Hahnemann Med Col, 64-69; CLIN ASST PROF MED, THOMAS JEFFERSON UNIV, 69- Concurrent Pos: Nat Cancer Inst grant, Misericordia Hosp, 66-; assoc dept path, Hemat Sect & attend physician dept med, Misericordia Hosp, 64-, Nat Cancer Inst prin investr hemat res, 66- Mem: Am Col Physicians. Mailing Add: Township & Belfield Ave Havertown PA 19083

TURCHINETZ, WILLIAM ERNEST, b Winnipeg, Man, Nov 18, 28; m 54; c 2. PHYSICS. Educ: Univ Man, BSc, 52, MSc, 53, PhD(physics), 55. Prof Exp: Asst prof physics, Univ Man, 55-56; res fel, Australian Nat Univ, 56-59; Sloan Found fel, 59-69, mem res staff, 60-65, lectr, 65-68, SR RES SCIENTIST, MASS INST TECHNOL, 68-, HEAD OPERS, BATES LINEAR ACCELERATOR, 73- Concurrent Pos: Chmn, Gordon Conf Photonuclear Reactions, 69-71. Res: Nuclear physics; particle accelerators; radiation therapy; technical education. Mailing Add: Rm 26-409 Mass Inst Technol Cambridge MA 02139

TURCK, JOSEPH ABRAHAM VALENTINE, JR, b Providence, RI, Jan 10, 11; m 41; c 1. ORGANIC CHEMISTRY. Educ: Univ Ill, BS, 33; Iowa State Col, PhD(org chem), 38. Prof Exp: Org res chemist, Colgate-Palmolive-Peet Co, 38-41, chem engr, 41-49, div head, 49-56; develop mgr, Curon Div, Curtiss-Wright Corp, 56-60; dir res & develop, Fels & Co, 60-64; scientist assoc, Scott Paper Co, 64-71; RES & DEVELOP CONSULT, 71- Mem: Am Chem Soc; Am Inst Chem Eng; Soc Cosmetic Chem; Soc Invest Dermat. Res: Chemical engineering; synthesis and manufacture of detergents; development of first successful synthetic detergent toilet bar; urethane foams. Mailing Add: 5005 Dermond Rd Drexel Hill PA 19026

TURCK, MARVIN, b Chicago, Ill, June 13, 34; m 56; c 4. INTERNAL MEDICINE, INFECTIOUS DISEASE. Educ: Univ Ill, BS, 57, MD, 59; Am Bd Internal Med, dipl. Prof Exp: Intern med, Res & Educ Hosp, Ill, 59-60; fel, Univ Wash, 60-62; resident & asst, Res & Educ Hosp, Ill, 62-63, chief resident, Cook County Hosp Serv, 63-64; head, Div Infectious Dis & prog dir, Res Infecious Dis Lab, King County Hosp, 64-68; chief med, USPHS Hosp, 68-72; PROF MED, UNIV WASH, 72-, PHYSICIAN-IN-CHIEF DEPT MED, HARBORVIEW MED CTR, 72- Concurrent Pos: Instr, Univ Ill, 63-64; from asst prof to assoc prof, Univ Wash, 64-72; attend physician, King County Hosp, Wash, 64-; attend physician & consult, Univ Wash Hosp, 66; attend physician, USPHS Hosp, 66. Mem: Am Fedn Clin Res; Infectious Dis Soc Am; fel Am Col Physicians. Res: Laboratory and clinical aspects of pyelonephritis; investigation of new antibiotics. Mailing Add: Harborview Med Ctr 325 Ninth Ave Seattle WA 98104

TURCO, CHARLES PAUL, b Brooklyn, NY, Sept 23, 34; m 55; c 4. PARASITOLOGY, NEMATOLOGY. Educ: St John's Univ, NY, BS, 56, MS, 58, MS, 60; Tex A&M Univ, PhD(biol), 69. Prof Exp: Teacher, High Sch, NY, 56-64; dir univ develop, 71-74, ASSOC PROF BIOL, LAMAR UNIV, 65-, DIR RES & DEVELOP, 74- Concurrent Pos: Sigma Xi res award, 68. Mem: Am Soc Parasitol; Am Inst Biol Sci; Am Soc Nematol. Res: Nematodes of rice and associated insect pests; nematode parasites associated with man's domestic animals. Mailing Add: Off of Develop Lamar Univ Beaumont TX 77710

TURCO, RICHARD PETER, b New York, NY, Mar 9, 43; m 67; c 1. AERONOMY. Educ: Rutgers Univ, New Brunswick, BS, 65; Univ Ill, Urbana, MS, 67, PhD(elec eng), 71. Prof Exp: NSF res grant, Space Sci Div, Ames Res Ctr, NASA, Moffett Field, Calif, 71; SCIENTIST AERONOMY, R&D ASSOCS, 71- Mem: Am Geophys Union. Res: Dynamical photochemistry of the ambient and disturbed atmosphere; pollutant effects on the stratosphere; charged species in the lower ionosphere. Mailing Add: R&D Assocs PO Box 9695 Marina Del Rey CA 90291

TURCO, SALVATORE J, b Philadelphia, Pa, Mar 4, 32; m 57; c 2. PHARMACY. Educ: Philadelphia Col Pharm & Sci, BSc, 59, MSc, 66, PharmD, 67. Prof Exp: Instr sterile prod, Philadelphia Col Pharm & Sci, 66-67; from instr to asst prof, 67-73, ASSOC PROF PHARM, SCH PHARM, TEMPLE UNIV, 73- Concurrent Pos: Indust res grants, Temple Univ, 69, Roche award, 72, univ grant, 72-73. Mem: Am Pharmaceut Asn; Am Soc Hosp Pharmacists. Res: Parenteral products; particulate matter in parenterals; hospital pharmacy. Mailing Add: Dept of Pharm Temple Univ Sch of Pharm Philadelphia PA 19140

TURCOT, JACQUES, b Quebec, Ont, Oct 26, 14; m 43; c 6. SURGERY. Educ: Laval Univ, BA, 35, MD, 40; FRCS(C). Prof Exp: Assoc prof, 51-58, PROF SURG, FAC MED, LAVAL UNIV, 68- Concurrent Pos: Surgeon, L'Hotel-Dieu Hosp, Quebec, 47- Mem: Fel Am Col Surgeons. Res: General surgery; cancerology. Mailing Add: Dept of Surg Laval Univ Fac of Med Quebec PQ Can

TURCOTTE, DONALD LAWSON, b Bellingham, Wash, Apr 22, 32; m 57; c 2. FLUID MECHANICS, GEOPHYSICS. Educ: Calif Inst Technol, BS, 54, DPh(aeronaut eng), 58; Cornell Univ, MAeroE, 55. Prof Exp: Asst prof aeronaut eng, US Naval Postgrad Sch, 58-59; from asst prof to prof aeronaut eng, 59-72, PROF GEOL SCI, CORNELL UNIV, 72- Concurrent Pos: NSF fel, Oxford Univ, 65-66, Guggenheim fel, 72-73. Mem: Am Phys Soc; Am Inst Aeronaut & Astronaut; Am Geophys Union. Res: Mantle convection; geophysical heat transfer. Mailing Add: Kimble Hall Cornell Univ Ithaca NY 14850

TURCOTTE, EDGAR LEWIS, b Duluth, Minn, June 7, 29. PLANT GENETICS. Educ: Univ Minn, BA, 51, MS, 57, PhD(genetics), 58. Prof Exp: PLANT GENETICIST, USDA, 58- Mem: AAAS; Am Genetic Asn; Am Soc Agron; Genetics Soc Can. Res: Genetics and speciation of Gossypium barbadense. Mailing Add: Univ of Ariz Cotton Res Ctr 4207 E Broadway Phoenix AZ 85040

TURCOTTE, JEREMIAH G, b Detroit, Mich, Jan 20, 33; m 58; c 4. SURGERY. Educ: Univ Mich, BS, 55, MD, 57; Am Bd Surg, dipl, 65. Prof Exp: Resident, 58-63, from instr to assoc prof, 63-71, PROF SURG, MED SCH, UNIV MICH, ANN ARBOR, 71- Concurrent Pos: Co-investr, USPHS Res Grant, 64-; consult, Ann Arbor Vet Admin Hosp & Wayne County Gen Hosp. Mem: Fel Am Col Surg; Transplantation Soc; Soc Univ Surgeons; Soc Surg Alimentary Tract; Asn Acad Surg. Res: Portal hypertension; organ transplantation. Mailing Add: Dept of Surg Univ of Mich Med Ctr Ann Arbor MI 48104

TURCOTTE, JOSEPH GEORGE, b Boston, Mass, Dec 25, 36; m 62; c 5. ORGANIC CHEMISTRY, MEDICINAL CHEMISTRY. Educ: Mass Col Pharm, BS, 58, MS, 60; Univ Minn, PhD(med chem), 67. Prof Exp: Sr biochemist, Spec Lab Cancer Res & Radioisotope Serv, Vet Admin Hosp, Minneapolis, 65-67; asst prof med chem, 67-72, ASSOC PROF MED CHEM, COL PHARM, UNIV RI, 72- Concurrent Pos: Res comt grants, Univ RI, 67-69; res grants, Nat Cancer Inst, 67-70, RI Water Resources Ctr, 68-69, RI Heart Asn, 70-72 & Nat Heart & Lung Inst, 71- Mem: AAAS; NY Acad Sci; Am Chem Soc; Am Pharmaceut Asn. Res: Synthesis of potential medicinal agents, including phospholipids, anticancer agents, antihypertensives, molluscicides, parasympathomimetic and parasympatholytic agents. Mailing Add: Dept of Med Chem Univ of RI Col of Pharm Kingston RI 02881

TUREK, ANDREW, b Lemberg, Poland, Nov 11, 35; Can citizen; m 58; c 2. GEOCHEMISTRY. Educ: Univ Edinburgh, BSc, 57; Univ Alta, MSc, 62; Australian Nat Univ, PhD(geophys), 66. Prof Exp: Chemist, Scottish Agr Industs, 57-58; mine geologist, Lake Cinch Mines Ltd, Can, 58-60 & Sherritt-Gordon Mines, 62-63; res coordr geol, Can Dept Mines & Natural Resources, Man, 66-69; assoc prof, Northern Ill Univ, 69-71; ASSOC PROF GEOL, UNIV WINDSOR, 71-, ASSOC PROF CHEM, 73- Mem: Geol Asn Can; Australasian Inst Mining & Metall. Res: Economic geology; geochronology; geostatistics. Mailing Add: Dept of Geol Univ of Windsor Windsor ON Can

TUREK, FRED WILLIAM, b Detroit, Mich, July 31, 47; m 70. REPRODUCTIVE ENDOCRINOLOGY, PHOTOBIOLOGY. Educ: Mich State Univ, BS, 69; Stanford Univ, PhD(biol sci), 73. Prof Exp: Fel reproductive biol, Univ Tex, Austin, 73-75; ASST PROF REPRODUCTIVE ENDOCRINOL, NORTHWESTERN UNIV, 75- Concurrent Pos: NIH fel, 73. Mem: Am Soc Zoologists; AAAS; Soc Study of Reproduction; Am Physiol Soc; Endocrine Soc. Res: Study of how the photoperiod is involved in the regulation of the hypothalamo-pituitary-gonadal axis in birds and mammals. Mailing Add: Dept of Biol Sci Northwestern Univ Evanston IL 60201

TUREK, WILLIAM NORBERT, b St Paul, Minn, June 30, 31; m 66. SYNTHETIC ORGANIC CHEMISTRY. Educ: Col St Thomas, BS, 53; Univ Md, PhD(org chem), 58. Prof Exp: From asst prof to assoc prof, 63-75, PROF CHEM, ST BONAVENTURE UNIV, 75- Mem: Am Chem Soc. Res: Pyrrolines; substituted furans; vinyl heterocycles. Mailing Add: Dept of Chem St Bonaventure Univ St Bonaventure NY 14778

TUREKIAN, KARL KAREKIN, b New York, NY, Oct 25, 27; m 62; c 2. GEOCHEMISTRY. Educ: Wheaton Col, Ill, AB, 49; Columbia Univ, MA, 51, PhD, 55. Prof Exp: Lectr geol, Columbia Univ, 53-54, res assoc geochem, Lamont Geol Observ, 54-56; from asst prof to assoc prof, 56-65, PROF GEOL, YALE UNIV, 65- Concurrent Pos: Guggenheim fel, 62-63; consult, President's Comn Marine Sci Eng & Resources, 67-68 & Oceanog Panel, NSF, 68-71; ed, J Geophys Res, 69-75; mem, US Nat Comn Geochem, Nat Acad Sci, 70-73; group experts Sci Aspects Marine Pollution, UN; co-ed, Earth & Planetary Sci Lett, 75- Mem: AAAS; Geol Soc Am; Geochem Soc (pres, 75-76); Am Chem Soc; Meteoritical Soc. Res: Marine geochemistry; geochemistry of trace elements. Mailing Add: Dept of Geol & Geophys Yale Univ New Haven CT 06520

TUREL, FRANZISKA LILI MARGARETE, b Berlin, Ger, Jan 10, 24; nat Can. PLANT PHYSIOLOGY. Educ: Swiss Fed Inst Tech, dipl, 47, Dr sc nat, 52. Prof Exp: Asst to prof, Inst Appl Bot, Swiss Fed Inst Tech, 47-50; asst to dir, Swiss Fed Exp Sta, Waedenswil, 50-53; fel, Prairie Regional Lab, Nat Res Coun Can, 53-55, asst res officer, 55-58; res assoc bact, 58-59, lectr, 65-69, asst prof, 69-75, ASSOC PROF PLANT PHYSIOL, UNIV SASK, 75-, RES ASSOC PLANT PHYSIOL, 60- Mem: AAAS; Can Bot Asn. Res: Plant pathology; physiology of host-parasite relationships. Mailing Add: 844 Univ Dr Saskatoon SK Can

TURELL, ROBERT, b Lublin, Poland, Mar 25, 02; nat US. SURGERY. Educ: Univ Wis, BS, 26, MD, 28. Prof Exp: PROF CLIN SURG, ALBERT EINSTEIN COL MED, 56- Mem: AAAS; Asn Med Cols; Am Col Surgeons; NY Acad Sci; NY Acad Med. Res: Clinical proctology. Mailing Add: Dept of Clin Surg Albert Einstein Col of Med New York NY 10028

TURER, JACK, b New York, NY, Mar 18, 12; m 38; c 2. ORGANIC CHEMISTRY,

PHYSICAL CHEMISTRY. Educ: City Col New York, BS, 34; Fairleigh Dickinson Univ, MAS, 69. Prof Exp: Res chemist, US Pub Rds Admin, 36-39; res chemist, USDA, 39-41, Eastern Regional Res Labs, 41-45; chief chemist, Va-Carolina Chem Corp, 45-52; tech dir, Textile Chem, 52-66, tech mgr automotive lubricants & petrol prod, 66-72, TECH ADV LABELING GOVT REGULATIONS & CHEM PROBS, WITCO CHEM CORP, 72- Concurrent Pos: Abstr, Chem Abstracts, 53- Mem: Am Chem Soc; Am Soc Test & Mat; Am Asn Textile Chem & Colorists; Am Inst Chem; Am Soc Lubrication Eng. Res: Soils; chemurgy; electrochemistry; oils and fats; proteins for synthetic textile fibers; textile chemicals and finishes; automotive lubricants and petrochemicals; pollution control; labeling government regulations. Mailing Add: 15 Brentwood Dr Verona NJ 07044

TURESKY, SAMUEL SAUL, b Portland, Maine, Feb 22, 16; m 52; c 5. DENTISTRY. Educ: Harvard Univ, AB, 37; Tufts Col, DMD, 41. Prof Exp: Res assoc oral path & periodont, 47-55, from asst prof to assoc prof, 55-71, PROF PERIDONT, SCH DENT MED, TUFTS UNIV, 71- Mem: Int Asn Dent Res. Res: Histochemistry of gingiva; calculus and plaque formation and prevention. Mailing Add: Tufts Univ Sch of Dent Med 1 Kneeland St Boston MA 02111

TURETZKY, MELVIN N, organic chemistry, see 12th edition

TURGEON, ALFRED J, b White Plains, NY, Sept 13, 43; m 66; c 1. WEED SCIENCE, ECOLOGY. Educ: Rutgers Univ, New Brunswick, BS, 65; Mich State Univ, MS, 70, PhD(weed sci), 71. Prof Exp: ASST PROF HORT, UNIV ILL, URBANA, 71- Concurrent Pos: Mem, NCent Weed Control Conf. Mem: Weed Sci Soc Am; Am Soc Agron. Res: Turfgrass ecology; life history and control of annual bluegrass in turf; ecological factors affecting weed incidence in turf; physiological effects of herbicides on the turfgrass community. Mailing Add: 202 Ornamental Hort Bldg Univ of Ill Urbana IL 61801

TURGEON, JEAN, b Montreal, Que, May 8, 36. MATHEMATICS. Educ: Univ Toronto, MA, 65, PhD, 68. Prof Exp: Asst prof math, Univ Montreal, 69-73; ASSOC PROF MATH, CONCORDIA UNIV, 73- Mem: AAAS; Am Math Soc; Math Asn Am; Can Math Cong. Res: Geometry. Mailing Add: Dept of Math Concordia Univ Montreal PQ Can

TURGEON, JUDITH LEE, b Topeka, Kans, Mar 19, 42. NEUROENDOCRINOLOGY, REPRODUCTIVE PHYSIOLOGY. Educ: Washburn Univ, BA, 65; Univ Kans, PhD(anat), 69. Prof Exp: Res assoc physiol, Sch Med, Univ Md, 69-71, asst prof, 71-75; ASST PROF HUMAN PHYSIOL, SCH MED, UNIV CALIF, DAVIS, 75- Mem: Soc Study Reproduction; Endocrine Soc; Am Physiol Soc. Res: Hypothalamic control of gonadotrophin secretion by the anterior pituitary. Mailing Add: Dept of Human Physiol Univ of Calif Sch of Med Davis CA 95616

TURI, PAUL GEORGE, b Battonya, Hungary, Apr 16, 17; US citizen; m 41; c 1. INDUSTRIAL PHARMACY, ANALYTICAL CHEMISTRY. Educ: Pazmany Peter Univ, Budapest, MS, 40, PhD(pharm), 46. Prof Exp: Mgr, Szanto Pharm Labs, Budapest, 46-48; res coord, Pharmaceut Indust Ctr, 48-49; dep dir, Pharm Res Inst, 50-53, head pharm res & develop, 55-56; dep mgr qual control dept, Chinoin Chem Works, 53-55; anal chemist, Chase Chem Co, NJ, 57-59; sr scientist, 59-60, group leader anal res, 60-63, head anal labs, 63-70, mgr pharm res, 70-74, ASSOC SECT HEAD, SANDOZ PHARMACEUT, SANDOZ INC, 75- Concurrent Pos: Hon asst prof, Pazmany Peter Univ, Budapest, 46-48, hon adj prof, 48-56; lectr, Budapest Tech Univ, 52-55. Mem: Am Pharmaceut Asn; Acad Pharmaceut Sci; Int Pharmaceut Fedn; Am Chem Soc. Res: Pharmaceutical analysis; pharmacy research and development. Mailing Add: Pharm Res Dept Sandoz Inc Rte 10 East Hanover NJ 07939

TURIN, JOHN JOSEPH, physics, deceased

TURINO, GERARD MICHAEL, b New York, NY, May 16, 24; m 51; c 3. MEDICINE. Educ: Princeton Univ, AB, 45; Columbia Univ, MD, 48. Prof Exp: Mem staff, Div Med Sci, Nat Res Coun, 51-53; resident med, Bellevue Hosp, New York, 53-54; assoc, 56-60, from asst prof to assoc prof, 60-72, PROF MED, COL PHYSICIANS & SURGEONS, COLUMBIA UNIV, 72- Concurrent Pos: Nat Found Infantile Paralysis fel, Col Physicians & Surgeons, Columbia Univ, 54-56, NY Heart Asn sr fel, 56-60; asst physician, Presby Hosp, New York, 56-61, from asst attend physician to assoc attend physician, 61-72, dir, Cardiovasc Lab, 66-, attend physician, 72-; consult, Vet Admin Hosp, East Orange, NJ; mem, Career Invest Health Res Coun, New York, 61. Mem: AAAS; Harvey Soc; Am Fedn Clin Res; Am Soc Clin Invest; Am Physiol Soc. Res: Internal medicine; cardio-pulmonary physiology. Mailing Add: Col of Physicians & Surgeons Columbia Univ New York NY 10032

TURINSKY, JIRI, b Prague, Czech, Apr 9, 35; m 64; c 2. PHYSIOLOGY, BIOCHEMISTRY. Educ: Charles Univ, Prague, MD, 59, PhD(physiol), 62. Prof Exp: Instr & res assoc physiol, Med Sch, Charles Univ, Prague, 59-66, asst prof, 68-69; asst prof, 70-71, ASSOC PROF PHYSIOL, ALBANY MED COL, 71- Concurrent Pos: Res fel surg res, Med Sch, Univ Pa, 66-68. Res: Regulation of hormone secretion; endocrine control of metabolism. Mailing Add: Dept of Physiol Albany Med Col Albany NY 12208

TURK, AMOS, b New York, NY, Feb 28, 18; m 41; c 3. ORGANIC CHEMISTRY. Educ: City Col New York, BS, 37; Ohio State Univ, MA, 38, PhD(chem), 40. Prof Exp: Res assoc, Explosives Res Lab, Pa, 42-44 & Allegany Ballistics Lab, Md, 44-46; instr org chem, City Col New York, 46-49; dir res & develop, Connor Eng Corp, 49-54; from asst prof to assoc prof, 56-66, PROF CHEM, CITY COL NEW YORK, 67-; CONSULT CHEMIST, 54- Concurrent Pos: Consult, USPHS, 56-; lectr, Inst Indust Med, NY Univ, 56-62. Mem: Am Chem Soc; Am Soc Test & Mat; Am Indust Hyg Asn; Air Pollution Control Asn; NY Acad Sci. Res: Organic synthesis; activated carbon; air analysis and purification; odors. Mailing Add: 7 Tarrywile Lake Dr Danbury CT 06810

TURK, DONALD EARLE, b Dryden, NY, Sept 4, 31. NUTRITION, BIOCHEMISTRY. Educ: Cornell Univ, BS, 53, MNS, 57; Univ Wis, PhD(biochem, nutrit), 60. Prof Exp: Asst prof, 60-67, ASSOC PROF POULTRY SCI, CLEMSON UNIV, 67- Concurrent Pos: From vchmn to chmn, SC Nutrit Comt, 75-77. Mem: AAAS; Poultry Sci Asn; Am Chem Soc; Am Inst Nutrit; Sigma Xi. Res: Protein and energy relationships in the fowl; mineral metabolism in obese humans; trace mineral metabolism; digestive tract disease and nutrient absorption; trace mineral deficiency and nucleic acid metabolism. Mailing Add: Food Sci Dept Clemson Univ Clemson SC 29631

TURK, JESSIE ROSE, b Newark, NJ, June 27, 20. GEOGRAPHY OF LATIN AMERICA, POLITICAL GEOGRAPHY. Educ: Montclair State Col, BA, 42; Oberlin Col, AM, 47; Columbia Univ, EdD(geog), 64. Prof Exp: Asst geog, Oberlin Col, 42-44; instr, Montclair State Col, 47-49; from instr to assoc prof geog, 49-64, PROF GEOG, TRENTON STATE COL, 64- Mem: Asn Am Geog; Am Geog Soc; Nat Coun Geog Educ. Res: Historical geography of East Coast cities. Mailing Add: Dept of Geog Trenton State Col Trenton NJ 08625

TURK, KENNETH LEROY, b Mt Vernon, Mo, July 14, 08; m 34. DAIRY HUSBANDRY. Educ: Univ Mo, BS, 30; Cornell Univ, MS, 31, PhD(animal husb), 34. Prof Exp: Asst animal husb, Cornell Univ, 31-34, from exten instr to exten asst prof, 34-38; prof dairy husb, Univ Md, 38-44, head dept, 40-44; prof animal husb, 44-74, dir int agr develop, 63-74, head dept, 45-63, EMER PROF ANIMAL HUSB, CORNELL UNIV, 74- Concurrent Pos: Vis prof, Univ Philippines, 54-55; consult, Rockefeller Found, 58-62; mem, Expert Panel Dairy Educ, Food & Agr Orgn, 65-71, Expert Panel Animal Husb Educ, 69-71; tech adv comt, Inst Nutrit Cent Am & Panama, 67-69; mem, Working Group Agr Res, US-USSR Joint Comn Sci & Tech Coop, 72. Mem: AAAS; Am Soc Animal Sci; Am Dairy Sci Asn (pres, 59-60); Soc Int Develop. Res: Nutrition of dairy calves and cows; nutritive value of hay and pasture crops; dairy cattle breeding; international agricultural development; animal science. Mailing Add: 803 Hanshaw Rd Ithaca NY 14850

TURK, LELAND JAN, b Tulare, Calif, July 18, 38; m 56; c 3. GEOLOGY, HYDROLOGY. Educ: Fresno State Col, BA, 61; Stanford Univ, MS, 63 & 67, PhD(geol), 69. Prof Exp: Jr geologist, Mobil Oil Libya Ltd, Tripoli, Libya, 63-65, geologist, 65-66; asst prof, 68-72, ASSOC PROF GEOL, UNIV TEX, AUSTIN, 72- Concurrent Pos: Prof engr, Tex; ed-in-chief, Environ Geol, 74- Mem: Geol Soc Am; Asn Eng Geol; Am Geophys Union; Am Inst Prof Geol; Am Water Res Asn. Res: Hydrogeology; environmental geology; engineering geology. Mailing Add: Dept of Geol Sci Univ of Tex Austin TX 78712

TURKANIS, STUART ALLEN, b Everett, Mass, Dec 15, 36; m 64; c 2. PHARMACOLOGY. Educ: Mass Col Pharm, BS, 58, MS, 60; Univ Utah, PhD(pharmacol), 67. Prof Exp: ASST PROF PHARMACOL, COL MED, UNIV UTAH, 67- Concurrent Pos: USPHS fel, Univ Col, Univ London, 67-69. Mem: AAAS; Am Soc Pharmacol & Exp Therapeut; Soc Neurosci. Res: Pharmacology and physiology of synaptic transmission; mechanisms of action of antiepileptic drugs; pharmacology of marijuana and its surrogates. Mailing Add: Dept of Pharmacol Univ of Utah Med Ctr Salt Lake City UT 84112

TURKEL, RICKEY MARTIN, b New York, NY, Apr 12, 43; m 65; c 2. ORGANIC CHEMISTRY. Educ: Hofstra Univ, BA, 63; Mass Inst Technol, PhD(org chem), 68. Prof Exp: Fel org chem, Hebrew Univ, Jerusalem, 68-69 & Tulane Univ, 69-70; assoc abstractor, 70-71, ASSOC ED, CHEM ABSTR SERV, 71- Honors & Awards: Award, Am Chem Soc, 63. Mem: Am Chem Soc. Res: Synthetic organic chemistry; organometallic chemistry; literature chemistry. Mailing Add: 2980 Fair Ave Columbus OH 43209

TURKEL, SOLOMON HENRY, physics, see 12th edition

TURKEVICH, ANTHONY, b New York, NY, July 23, 16; m 48; c 2. NUCLEAR CHEMISTRY, SPACE CHEMISTRY. Educ: Dartmouth Col, BA, 37; Princeton Univ, PhD(phys chem), 40. Hon Degrees: DSc, Dartmouth Col, 71. Prof Exp: Res assoc molecular spectra, Dept Physics, Univ Chicago, 40-41, res chemist, Metall Lab, 43-45; res chemist, SAM Labs, Columbia Univ, 42-43; res physicist, Los Alamos Sci Lab, 45-46; from asst prof to assoc prof, 46-50, James Franck prof, 65-71, PROF CHEM, ENRICO FERMI INST & CHEM DEPT, UNIV CHICAGO, 50-, JAMES FRANCK DISTINGUISHED SERV PROF, 71- Concurrent Pos: NSF fel, Europ Orgn Nuclear Res, Switz, 61-62; J W Kennedy Mem lectr, Univ Wash, St Louis, 64; NSF fel, Orsay, France, 70; consult, Labs, US AEC. Honors & Awards: E O Lawrence Award, US AEC, 62; Atoms for Peace Award, 69; Nuclear Applns Award, Am Chem Soc, 72. Mem: Nat Acad Sci; AAAS; Am Chem Soc; Am Acad Arts & Sci; Royal Soc Arts. Res: Reactions of energetic particles with complex nuclei; radioactivity in meteorites; chemical composition of the moon and meteorites. Mailing Add: Enrico Fermi Inst Univ of Chicago 5640 S Ellis Ave Chicago IL 60637

TURKEVICH, JOHN, b Minneapolis, Minn, Jan 20, 07; m 36; c 2. PHYSICAL CHEMISTRY. Educ: Dartmouth Col, BS, 28, MA, 30; Princeton Univ, AM, 32, PhD(chem), 34. Prof Exp: Instr, Dartmouth Col, 28-31; from instr to prof, 36-55, EUGENE HIGGINS PROF CHEM, PRINCETON UNIV, 55- Concurrent Pos: Consult, M W Kellogg Co Div, Pullman, Inc, NY, 36-, Radio Corp Am Labs, 43-, Brookhaven Nat Lab, 47, US Energy Res & Develop Admin, 50- & US Dept State; chmn, US Deleg Educators, USSR, 58; rep, US Sci Am Nat Exhib, Moscow, 59; actg sci attache, US Embassy, Moscow, 60, sci attache, 61; lectr, US Army War Col & US Air War Col; Phi Beta Kappa vis scholar, 61. Honors & Awards: Award, Mfg Chemists Asn, 56. Mem: Am Chem Soc; Am Phys Soc; Electron Micros Soc Am (secy-treas). Res: Catalysis; molecular structure. Mailing Add: 109 Rollingmead Princeton NJ 08540

TURKINGTON, ROGER W, b Manchester, Conn, Jan 13, 36; m 60; c 1. MEDICINE, BIOCHEMISTRY. Educ: Wesleyan Univ, AB, 58; Harvard Med Sch, MD, 63; Am Bd Internal Med, dipl. Prof Exp: Intern med, Duke Univ Hosp, 63-64, resident, 64-65; res assoc, NIH, 65-67; res assoc, Sch Med, Duke Univ, 67-68, asst prof med, 68-69, asst prof med & biochem, 69-71; assoc prof med, Univ Wis-Madison, 71-73; CHIEF ENDOCRINOL, ST LUKES HOSP, 73- Concurrent Pos: Chief endocrinol, Vet Admin Hosp, Durham, NC, 67-71. Mem: Am Soc Clin Invest; Am Soc Cancer Res; Endocrine Soc; Am Soc Biol Chem; Am Fedn Clin Res. Res: Biochemistry of development and cancer; mechanisms of hormone action. Mailing Add: St Luke's Hosp 2900 W Oklahoma Ave Milwaukee WI 53215

TURKKI, PIRKKO REETTA, b Laitila, Finland, Aug 27, 34; m 57; c 2. NUTRITION, FOOD SCIENCE. Educ: Helsinki Home Econ Teacher's Col, dipl, 57; Univ Mass, Amherst, MS, 62; Univ Tenn, Knoxville, PhD(nutrit, food sci), 65. Prof Exp: Res assoc biochem, Univ Tenn, 65-66; NIH res fel, Albany Med Col, 66-68; asst prof, 68-72, ASSOC PROF NUTRIT, SYRACUSE UNIV, 72-, ASST DEAN, COL HUMAN DEVELOP, 75- Mem: Am Dietetic Asn; Soc Nutrit Educ; Nutrit Today Soc. Res: Phospholipid metabolism in choline deficiency; role of diet in hyperlipemia. Mailing Add: Slocum Hall Col Human Develop Syracuse Univ Syracuse NY 13210

TURKOS, ROBERT EDWARD, physical organic chemistry, see 12th edition

TURKOT, FRANK, b Woodlynne, NJ, Sept 29, 29; m 59; c 4. PHYSICS. Educ: Univ Pa, BA, 51; Cornell Univ, PhD(physics), 59. Prof Exp: Res assoc particle physics, Lab Nuclear Studies, Cornell Univ, 59-60; from asst physicist to sr physicist, Brookhaven Nat Lab, NY, 60-74; PHYSICIST, FERMILAB, 74- Mem: Fel Am Phys Soc. Res: Experiments in elementary particle physics to study photoproduction processes and high energy collisions of strongly-interacting particles; particle accelerator research. Mailing Add: Fermilab PO Box 500 Batavia IL 60510

TURLEY, HUGH PATRICK, b Queens, NY, Oct 1, 36. CELL PHYSIOLOGY. Educ: Marist Col, BA, 58; Fordham Univ, MS, 63; Catholic Univ, PhD(cellular physiol), 68. Prof Exp: Teacher, Bishop DuBois High Sch, NY, 58-61 & Mt St Michael Acad, NY, 61-63; ASSOC PROF BIOL, MARIST COL, 67- Mem: AAAS; Am Soc Microbiol; NY Acad Sci; Am Inst Biol Sci; Inst Soc Ethics & Life Sci. Res: Role of induction in differentiation. Mailing Add: Dept of Biol Marist Col Poughkeepsie NY 12601

TURLEY, JUNE WILLIAMS, b Boston, Mass, Apr 12, 29; m 50; c 3. X-RAY CRYSTALLOGRAPHY, ECONOMIC STATISTICS. Educ: Wilkes Col, BS, 50; Pa State Univ, MS, 51, PhD(biochem), 57. Prof Exp: Asst x-ray crystal struct anal, Pa State Univ, 53-56, res assoc, 56-57; from res chemist to sr res chemist, 57-71, res mgr, Anal Labs, 71-74, ECON PLANNER, ECON PLANNING & DEVELOP, DOW CHEM CO, 74- Mem: AAAS; Am Crystallog Asn; Sigma Xi; Women's Nat Sci Fraternity. Res: Crystal structure analysis; identification and analysis of compounds using x-ray methods; industrial projects correlating chemical process and business data for long range economic planning. Mailing Add: Econ Planning & Develop Bldg 2030 The Dow Chem Co Midland MI 48640

TURLEY, SHELDON GAMAGE, b Pa, June 13, 22; m 50; c 3. PHYSICS. Educ: Pa State Univ, BS, 50, MS, 51, PhD(physics), 57. Prof Exp: Res assoc physics of aerosols, Pa State Univ, 52-53; res physicist, 57-62, SR RES PHYSICIST, DOW CHEM CO, 62- Mem: AAAS; Sigma Xi; Am Phys Soc. Res: High polymer physics, especially with dynamic mechanical and electrical properties of polymers and their relationship to molecular structure. Mailing Add: Bldg 672 Dow Chem Co Midland MI 48640

TURLINGTON, BOYCE LYNN, applied mathematics, computer science. see 12th edition

TURMAN, ELBERT JEROME, b Granite, Okla, Jan 26, 24; m 49; c 2. ANIMAL SCIENCE. Educ: Okla State Univ, BS, 49; Purdue Univ, MS, 50, PhD(physiol), 53. Prof Exp: From instr to asst prof animal husb, Purdue Univ, 50-55; from asst prof to assoc prof, 55-63, PROF ANIMAL HUSB, OKLA STATE UNIV, 63- Mem: Soc Study Reproduction; Am Soc Animal Sci. Res: Physiology of reproduction of beef cattle, sheep and swine. Mailing Add: Dept of Animal Sci Okla State Univ Stillwater OK 74074

TURNBLOM, ERNEST WAYNE, b Boston, Mass, Nov 1, 46; m 73. ORGANIC CHEMISTRY. Educ: Worcester Polytech Inst, BS, 68; Columbia Univ, PhD(chem), 72. Prof Exp: Instr org chem, Princeton Univ, 72-74; RES CHEMIST, EASTMAN KODAK CO, 74- Mem: Am Chem Soc. Res: Organophosphorus chemistry; pentaalkylphosphoranes and related compounds; chemistry of other main group elements; electrophotographic processes and materials; novel imaging systems. Mailing Add: Eastman Kodak Co Res Labs 1669 Lake Ave Rochester NY 14650

TURNBULL, BRUCE FELTON, b Cleveland, Ohio, Mar 2, 28; m 51; c 3. PHYSICAL CHEMISTRY. Educ: Case Inst Technol, BS, 50; Faith Theol Sem, BD, 54; Western Reserve Univ, MS, 55, PhD(phys chem), 63. Prof Exp: Chemist, Gen Motors Corp, Ohio, 50-51; asst prof chem, Cedarville Col, 55-63; from asst prof to assoc prof, 63-70, PROF CHEM, CLEVELAND STATE UNIV, 70-, OMBUDSMAN, 71- Mem: AAAS; Am Chem Soc; Am Asn Physics Teachers. Res: Thermodynamics; molecular structure studies with infrared spectroscopy. Mailing Add: Dept of Chem Cleveland State Univ Cleveland OH 44115

TURNBULL, CHRISTOPHER JOHN, b Victoria, BC, Sept 25, 43; m 69; c 2. ANTHROPOLOGY. Educ: Univ BC, BA, 65; Univ Calgary, PhD(archaeol), 73. Prof Exp: PROV ARCHAEOLOGIST, HIST RESOURCES ADMIN, PROV NB, 70-, ASST DIR RES & DEVELOP BR, 74- Concurrent Pos: Lectr anthrop, Univ NB, 71-; chmn, Fed-Prov Comt Archaeol, 74- Mem: Can Archaeol Asn (vpres, 74-75); Soc Am Archaeol. Res: Cultural history and ecology of the Northeast, specializing on the maritime region of Canada; burial complexes in late archaic-woodland, the management of archaeological sites as a heritage resource. Mailing Add: PO Box 6000 Fredericton NB Can

TURNBULL, COLIN MACMILLAN, b Eng, Nov 23, 24; nat US. SOCIAL ANTHROPOLOGY. Educ: Oxford Univ, BA, 47, MA, 49, dipl social anthrop, 55, BLitt, 56, DPhil(anthrop), 64; London Univ, dipl educ, 49. Prof Exp: Assoc cur anthrop, Am Mus Natural Hist, 59-69; prof, Hofstra Univ, 69-72; PROF SOCIOL & ANTHROP, VA COMMONWEALTH UNIV, 72- Concurrent Pos: Corresp mem, Royal Mus Cent Africa, Tervuren, 57; res assoc, Am Mus Natural Hist, NY, 69. Honors & Awards: Lit Award, Nat Acad Arts & Lett, 75. Res: African social organization; hunter-gatherer societies; community research; interdisciplinary approaches to social problems. Mailing Add: Dept of Sociol & Antrhop Va Commonwealth Univ Richmond VA 23284

TURNBULL, DAVID, b Stark Co, Ill, Feb 18, 15; m 46; c 3. PHYSICAL CHEMISTRY. Educ: Monmouth Col, BS, 36; Univ Ill, PhD(phys chem), 39. Hon Degrees: ScD, Monmouth Col, 58. Prof Exp: Instr phys chem, Case Inst Technol, 39-43, asst prof, 43-46, res proj leader, 45-46; res assoc, Res Lab, Gen Elec Co, 46-51, mgr chem metall sect, 51-58, phys chemist, 58-62; GORDON McKAY PROF APPL PHYSICS, HARVARD UNIV, 62- Mem: Nat Acad Sci; AAAS; fel Am Acad Arts & Sci; Am Chem Soc; fel Am Phys Soc. Res: Thermionic emission; thermodynamic properties of gases at high pressures; corrosion in non-aqueous media; diffusion in metals; kinetics of nucleation in solid state transformation; solidification; theory of liquids; glass. Mailing Add: Div Eng & Appl Physics Pierce Hall Harvard Univ Cambridge MA 02138

TURNBULL, LENNOX BIRCKHEAD, b Richmond, Va, Mar 18, 25; m 63; c 2. ORGANIC CHEMISTRY. Educ: Davidson Col, BS, 47; Univ Va, PhD(org chem), 51. Prof Exp: Sr res chemist, Merck & Co, Inc, 51-55; res assoc pharmacol, Med Col Va, 56-63; sr res chemist, 63-72, ASSOC DIR DRUG METAB, A H ROBINS CO, 72- Mem: AAAS; Am Chem Soc; Toxicol Soc. Res: Synthetic medicinal chemistry; steroid hormones; tobacco chemistry; isolation of natural products; drug metabolism. Mailing Add: A H Robins Co 1211 Sherwood Ave Richmond VA 23220

TURNDORF, HERMAN, b Paterson, NJ, Dec 22, 30; m 59; c 2. ANESTHESIOLOGY. Educ: Oberlin Col, AB, 52; Univ Pa, MD, 56; Am Bd Anesthesiol, dipl, 62. Prof Exp: Asst dir, US Naval Hosp, Portsmouth, Va, 59-61; asst anesthetist, Mass Gen Hosp, 61-63; coordr clin affairs, Mt Sinai Hosp, New York, 63-70; PROF ANESTHESIOL & CHMN DEPT, SCH MED, W VA UNIV, 70- Concurrent Pos: From asst attend anesthetist to assoc attend anesthetist, Mt Sinai Hosp, 63-66. Mem: Fel Am Col Anesthesiol; fel Am Soc Anesthesiol; Am Thoracic Soc; AMA; fel Am Col Chest Physicians. Mailing Add: Dept of Anesthesiol WVa Univ Morgantown VA 26505

TURNEAURE, JOHN PAUL, b Yakima, Wash, Jan 16, 39; m 68; c 2. LOW TEMPERATURE PHYSICS. Educ: Univ Wash, BS, 61; Stanford Univ, PhD(physics), 67. Prof Exp: Res assoc, 66-69, res physicist, 69-75, actg asst prof, 70-73, SR RES ASSOC PHYSICS, STANFORD UNIV, 75- Mem: Am Phys Soc. Res: Study of radio frequency properties of superconductors, time variations of the fundamental physical constants and general relativistic effects; development of ultra-stable superconducting cavity oscillators and superconducting microwave structures for particle accelerators. Mailing Add: Dept of Physics Stanford Univ Stanford CA 94305

TURNER, ALICE WILLARD, b Norval, Ont. MATHEMATICS. Educ: McGill Univ, BA, 27, MA, 28; Univ Toronto, PhD, 32. Prof Exp: With Dom Bur Statist, 32-37 & Wood, Gundy & Co, 37-60; spec lectr, 60-61, from asst prof to assoc prof, 61-70, PROF MATH, YORK UNIV, 70- Concurrent Pos: Dir, Toronto Mutual Life Ins Co. Mem: Fel Royal Statist Soc. Res: Statistics; analysis; functions of a real variable; Fourier and Poisson integrals. Mailing Add: Dept of Math York Univ Downsview ON Can

TURNER, ALMON GEORGE, JR, b Detroit, Mich, June 9, 32; m 64; c 3. THEORETICAL CHEMISTRY, INORGANIC CHEMISTRY. Educ: Univ Mich, BS, 55; Purdue Univ, MS, 56, PhD(inorg chem), 58. Prof Exp: Assoc prof inorg chem, NDak Agr Col, 58-59; instr & fel chem, Carnegie Inst Technol, 59-61; asst prof inorg chem, Polytech Inst Brooklyn, 61-66; ASSOC PROF CHEM, UNIV DETROIT, 66- Mem: Am Chem Soc; Am Phys Soc. Res: Chemical bonding; electronic structure of molecules; polyhedral molecules; nitrogen-sulfur chemistry; prebiotic chemistry of sulfur. Mailing Add: Dept of Chem Univ of Detroit 4001 W McNichols Rd Detroit MI 48221

TURNER, ANDREW B, b Lock Haven, Pa, Dec 23, 40. ORGANIC CHEMISTRY. Educ: Franklin & Marshall Col, AB, 62; Bucknell Univ, MS, 65; Univ Va, PhD(chem), 68. Prof Exp: Interim asst prof chem, Univ Fla, 68-69; asst prof chem, Lycoming Col, 69-74; ASST PROF CHEM, ST JOHN FISHER COL, 74- Mem: Am Chem Soc. Res: Aziridines; synthetic tropane alkaloid analogs; nuclear magnetic resonance spectroscopy. Mailing Add: Dept of Chem St John Fisher Col Rochester NY 14618

TURNER, ANNE HALLIGAN, b Columbus, Ohio, Feb 3, 41; m 66; c 2. ANALYTICAL CHEMISTRY. Educ: Middlebury Col, AB, 63; Univ Rochester, PhD(org chem), 69. Prof Exp: Instr, USDA Grad Sch, 70-75; LECTR CHEM, PRINCE GEORGE'S COMMUNITY COL, 69- Mem: AAAS; Am Chem Soc. Res: Nuclear magnetic resonance spectra of sterically hindered systems. Mailing Add: Prince Georges Community Col 301 Largo Rd Div of Sci Largo MD 20870

TURNER, ARTHUR FRANCIS, b Detroit, Mich, Aug 8, 06; m 34; c 2. OPTICS, PHYSICS. Educ: Mass Inst Technol, BS, 29; Univ Berlin, PhD(physics), 35. Prof Exp: From asst to instr physics, Mass Inst Technol, 35-39; from physicist to head optical physics dept, Bausch & Lomb, Inc, 39-71; PROF OPTICAL SCI, UNIV ARIZ, 71- Concurrent Pos: Mem US Nat Comt, Int Comn Optics, 60-65. Honors & Awards: Frederic Ives Medalist, Optical Soc Am, 71. Mem: Fel Am Phys Soc; fel Optical Soc Am (pres, 68); Brit Inst Physics & Phys Soc. Res: Optical physics; evaporated thin films; infrared. Mailing Add: Optical Sci Ctr Univ of Ariz Tucson AZ 85721

TURNER, BARBARA HOLMAN, b Evergreen, Ala, Aug 31, 26; m 50; c 2. ECOLOGY. Educ: Miss State Col for Women, BS, 47; Univ Kans, MT, 48; Vanderbilt Univ, MA, 66, PhD(ecol), 72. Prof Exp: Asst prof biol, George Peabody Col, 66-72; ASST PROF MICRO-MED TECHNOL, MISS STATE UNIV, 73- Mem: Am Inst Biol Sci; Ecol Soc; Bot Soc; Sigma Xi; Am Soc Med Technologists. Res: Interactions between microorganisms and higher plants. Mailing Add: Dept of Microbiol Drawer MB Mississippi State MS 39762

TURNER, BARRY EARL, b Victoria, BC, Sept 8, 36; m 62. RADIO ASTRONOMY. Educ: Univ BC, BSc, 59, MSc, 62; Univ Calif, Berkeley, PhD(astron), 67. Prof Exp: Res off elec eng, Nat Res Coun Can, 62-64; from res assoc radio astron to assoc scientist, 67-74, SCIENTIST, NAT RADIO ASTRON OBSERV, 74- Mem: Int Astron Union; Am Astron Soc. Res: Theoretical and observational studies of interstellar molecules, interstellar chemistry, physics of the interstellar medium. Mailing Add: Nat Radio Astron Observ Charlottesville VA 22901

TURNER, BILLIE LEE, b Yoakum, Tex, Feb 22, 25; div; c 2. SYSTEMATIC BOTANY. Educ: Sul Ross State Col, BS, 49; Southern Methodist Univ, MS, 50; Wash State Univ, PhD(bot), 53. Prof Exp: From instr to assoc prof, 53-58, chmn dept, 67-74, PROF BOT & DIR HERBARIUM, UNIV TEX, AUSTIN, 59- Concurrent Pos: Vis prof & NSF sr fel, Univ Liverpool, 65-66; assoc investr ecol study African veg, 56-57. Honors & Awards: NY Bot Garden Award, 65. Mem: AAAS; Bot Soc Am (secy, 58-59 & 60-64, vpres, 65); Am Soc Plant Taxon; Soc Study Evolution; Int Asn Plant Taxon. Res: Plant geography; chromosomal studies of higher plants; flora of Texas and Mexico; biochemical systematics. Mailing Add: Dept of Bot Univ of Tex Austin TX 78712

TURNER, BILLIE LEE, II, b Texas City, Tex, Dec 22, 45; div; c 1. GEOGRAPHY, ANTHROPLOGY. Educ: Univ Tex, Austin, BA, 68, MA, 69; Univ Wis, Madison, PhD(geog), 74. Prof Exp: ASST PROF GEOG, UNIV MD, BALTIMORE COUNTY, 74- Concurrent Pos: Consult, Nat Geog Soc, 74-75; res assoc, Univ Okla, 75-76. Mem: Asn Am Geogr; Am Geog Soc; Conf Latin Am Geogr. Res: Cultural ecology in the tropical world with special interests in subsistence systems, agricultural development and the ecological and subsistence base of the Classic Maya civilization. Mailing Add: Dept of Geog Univ of Okla Norman OK 73069

TURNER, BYRON, b Weston, WVa, July 8, 12; m 49; c 1. CHEMISTRY. Educ: Glenville State Col, AB, 34; WVa Univ, MA, 40; Columbia Univ, EdD(sci educ), 52. Prof Exp: Teacher, Lewis County Schs, WVa, 34-42; PROF CHEM, GLENVILLE STATE COL, 46- Mailing Add: Dept of Chem Glenville State Col Glenville WV 26351

TURNER, CARLTON EDGAR, b Choctaw Co, Ala, Sept 13, 40. CHEMISTRY, PHARMACOGNOSY. Educ: Univ Southern Miss, BS, 66, MS, 69, PhD(org chem), 70. Prof Exp: Fel pharmacog res, 70-71, proj supvr & coord marihuana proj, Dept Pharmacog, 71-72, ASSOC DIR RES, INST PHARMACEUT SCI, SCH PHARM, UNIV MISS, 72- Mem: AAAS; Am Soc Pharmacog; Soc Econ Bot; Am Chem Soc. Res: Organo-silicone compounds hypotensive agents; chemistry of cannabinoids; gas chromatographic and thin layer chromatographic techniques for separation of natural products; vito chemistry. Mailing Add: Res Inst Pharmaceut Sci Univ of Miss Sch of Pharm University MS 38677

TURNER, CHRISTY GENTRY, II, b Columbia, Mo, Nov 28, 33; m 57; c 3. PHYSICAL ANTHROPOLOGY, DENTAL ANTHROPOLOGY. Educ: Univ Ariz, BA, 57, MA, 58; Univ Wis, Madison, PhD(phys anthrop), 67. Prof Exp: Actg asst prof phys anthrop, Univ Calif, Berkeley, 63-66; from asst prof to assoc prof, 66-75, PROF PHYS ANTHROP, ARIZ STATE UNIV, 75-, ASST DEAN GRAD COL, 72- Concurrent Pos: Am Dent Asn & NIH dent epidemiol & biomet trainee, Univ Cincinnati & Univ Chicago, 69; collabr phys anthrop, US Nat Park Serv, 69-; fel, Ctr Advan Study Behav Sci, Stanford Univ, 70-71; Nat Geog Soc grants, 72 & 73. Mem: Am Asn Phys Anthropologists; Sigma Xi; Soc Am Archaeol; Am Anthrop Asn. Res: Co-evolution of human biology and culture; dental morphology, genetics and related behavior; biology and culture of New World peoples, especially southwestern United States Indians and Alaskan Aleuts. Mailing Add: Dept of Anthrop Ariz State Univ Tempe AZ 85281

TURNER, CLAIR ELSMERE, public health education, deceased

TURNER, CLARENCE MARSHALL, b Economy, Ind, Aug 14, 11; m 42; c 2. PHYSICS. Educ: DePauw Univ, AB, 37; Univ Wis, PhD(physics), 43. Prof Exp: Res assoc, Nat Defense Res Lab, Isotron Proj, Princeton Univ, 42-43, Manhattan Proj, Los Alamos Sci Lab, 42-45 & Radiation Lab, Univ Calif, 45-49; from assoc physicist to sr physicist, Brookhaven Nat Lab, 49-75. Mem: Am Phys Soc; Am Vacuum Soc. Res: Electrostatic accelerator development; charging belt operation; high brightness pulsed ion sources; proton linear accelerator and cosmotron injection; cosmotron injector conversion to high intensity radiological research accelerator. Mailing Add: 8 Hawkins Rd PO Box 393 Stony Brook NY 11790

TURNER, DANIEL STOUGHTON, b Madison, Wis, Feb 8, 17; m 44; c 4. PETROLEUM, GEOLOGY. Educ: Univ Wis, PhB, 40, PhM, 42, PhD(geol), 48. Prof Exp: Field geologist, Buchans Mining Co, Nfld, 47; geologist, US AEC, Colo, 48; asst prof geol, Univ Wyo, 49-50; geologist, Carter Oil Co, Okla, 51 & Petrol Res Co, Colo, 52-53; div geologist, Wm R Whittaker Co, Ltd, 53; geol consult, 54-63; consult, Earth Sci Curric Proj, Boulder, 63-65; PROF GEOL, EASTERN MICH UNIV, 65- Mem: Geol Soc Am; Soc Econ Geol; Am Asn Petrol Geol. Res: Arctic geology and glaciation; permafrost; thermal activity of Yellowstone National Park; hydrodynamics of oil; Rocky Mountain petroleum exploration; earth science education. Mailing Add: Dept of Geog & Geol Eastern Mich Univ Ypsilanti MI 48197

TURNER, DAVID LEE, b Afton, Wyo, Nov 20, 49; m 69; c 1. STATISTICAL ANALYSIS. Educ: Colo State Univ, BS, 71, MS, 73, PhD(statist), 75. Prof Exp: From res asst to instr statist, Colo State Univ, 71-75; ASST PROF STATIST, UTAH STATE UNIV, 75- Mem: Sigma Xi; Am Statist Soc; Biomet Soc. Res: Tolerance intervals and bands for regression; analysis of unbalanced data; analysis of categorical data. Mailing Add: Dept of Appl Statist/Comput Sci Utah State Univ Logan UT 84322

TURNER, DEREK T, b London, Eng, Dec 19, 26; m 55; c 2. PHYSICAL CHEMISTRY, POLYMER CHEMISTRY. Educ: Northern Polytech Eng, BSc, 51; Nat Col Rubber Technol, Eng, AIRI, 52; Univ London, PhD(chem), 57. Prof Exp: Res chemist, Brit Insulated Callenders Cables, Ltd, Eng, 52-55; sr chemist, Brit Rubber Producers Res Asn, 55-63; res fel, Camille Dreyfus Lab, Res Triangle Inst, 62-63, sr chemist, 62-68; from assoc prof to prof metall eng, Drexel Univ, 68-70; prof oral biol, 70-74, PROF OPER DENT, UNIV NC, CHAPEL HILL, 74- Mem: Am Chem Soc; Am Phys Soc; Soc Plastics Engr; Int Asn Dent Res. Res: Radiation chemistry and photochemistry of polymers; solution properties of polymers; biomaterials. Mailing Add: Dent Res Ctr Univ of NC Chapel Hill NC 27514

TURNER, DONALD LLOYD, b Richmond, Calif, Dec 21, 37; c 3. GEOLOGY, GEOCHRONOLOGY. Educ: Univ Calif, Berkeley, AB, 60, PhD(geol), 68. Prof Exp: Nat Res Coun res assoc, Isotope Geol Br, US Geol Surv, Colo, 68-69, Calif, 69-70; ASSOC PROF GEOL, GEOPHYS INST & DEPT GEOL, UNIV ALASKA, FAIRBANKS, 70- Mem: Am Geophys Union. Res: Geochronology, K-Ar and fission track dating applied to problems of regional tectonics and radiometric calibration of paleontological time scales. Mailing Add: Geophys Inst Univ of Alaska Fairbanks AK 99701

TURNER, DOUGLAS HUGH, b Staten Island, NY, July 24, 46. BIOPHYSICAL CHEMISTRY. Educ: Harvard Col, AB, 67; Columbia Univ, PhD(phys chem), 72. Prof Exp: Fel biophys chem, Univ Calif, Berkeley, 73-74; ASST PROF CHEM, UNIV ROCHESTER, 74- Concurrent Pos: Tech collabr, Brookhaven Nat Lab, 70- Mem: Am Chem Soc; Am Phys Soc. Res: Structure and function of nucleic acids; laser temperature jump kinetics; florescence detected circular dichroism. Mailing Add: Dept of Chem Univ of Rochester Rochester NY 14627

TURNER, EARL WILBERT, b Bozeman, Mont, Feb 6, 18; m 39; c 5. BIOCHEMISTRY, PROTEIN CHEMISTRY. Educ: Mont State Univ, BS, 46, MS, 47; Univ Minn, PhD(biochem), 50. Prof Exp: Instr biochem & food chem, Univ Minn, 47-50; biochemist, Protein & Food Res, Oscar Mayer & Co, Wis, 50-51, chief chemist & head res lab, 51-55; chief animal prod br, Qm Food & Container Inst Armed Forces, Ill, 55-56; dir res, Int Packers, Ltd, Ill, 56-66; RES MGR, ITT CONTINENTAL BAKING CO, INC, 67- Mem: AAAS; Am Chem Soc; Inst Food Technol; Am Asn Cereal Chemists. Res: Meat science, cereal chemistry, lipid chemistry, microbiology, food chemistry; product and process development, meat products, frozen prepared foods, bakery products, food safety and microbiology; immunochemistry; food systems; edible fats and oils; special rations for armed forces. Mailing Add: Res Labs ITT Continental Baking Co Inc Rye NY 10580

TURNER, EDWARD FELIX, JR, b Newport News, Va, Apr 21, 20; m 45; c 3. PHYSICS. Educ: Washington & Lee Univ, BS & BA, 50; Mass Inst Technol, MS, 52; Univ Va, PhD(physics), 54. Prof Exp: Asst prof physics, George Washington Univ, 54-57; assoc prof, 57-59, PROF PHYSICS, WASHINGTON & LEE UNIV, 59-, CHMN DEPT, 61- Concurrent Pos: Consult & physicist, Diamond Ord Fuze Lab, 57-58; consult, US Off Educ, 65- Mem: AAAS; Am Phys Soc; Am Asn Physics Teachers. Res: Solid state physics; electrical properties of solids; electronics; astronomy. Mailing Add: Dept of Physics Washington & Lee Univ Lexington VA 24450

TURNER, EDWARD HARRISON, b Cleveland, Ohio, Dec 21, 20; m 45; c 4. MAGNETISM. Educ: Harvard Univ, SB, 42, AM, 47, PhD, 50. Prof Exp: Mem staff, Radiation Lab, Mass Inst Technol, 42-45; MEM TECH STAFF MICROWAVE PHYSICS, BELL TEL LABS, INC, 49- Mem: Am Phys Soc; sr mem Inst Elec & Electronics Eng. Res: Electrp-magnetic theory in gyrotropic media; spin waves in ferrimagnetic materials; interaction with lattice vibrations. Mailing Add: Rm 4B-415 Bell Tel Labs PO Box 400 Holmdel NJ 07733

TURNER, EDWARD V, b Belmont, Ohio, May 19, 13; m 39; c 1. PEDIATRICS. Educ: Ohio Univ, AB, 34; Harvard Med Sch, MD, 38. Prof Exp: From asst prof to assoc prof, 48-68, PROF PEDIAT, COL MED, OHIO STATE UNIV, 68- Mem: Fel Am Acad Pediat. Res: Clinical pediatrics; medical education. Mailing Add: Dept of Pediat Children's Hosp Columbus OH 43205

TURNER, EDWIN MORRIS, b Steele, Mo, Nov 28, 44; m 71. PHYSICAL CHEMISTRY. Educ: Mich State Univ, BS, 66; Univ Wis, Madison, PhD(phys chem), 70. Prof Exp: Lectr & grant chem, Univ Wis, Madison, 70-71; grant, Univ Ky, 71-72; DIR PHYS CHEM LAB, UNIV WIS, MADISON, 72- Mem: Am Chem Soc. Res: Dielectric properties of highly polar liquids. Mailing Add: Dept of Chem Univ of Wis Madison WI 53706

TURNER, ELLA VICTORIA, b Columbia, Mo, Jan 23, 46; m 66; c 1. IMMUNOLOGY. Educ: Univ Ark, BA, 67; Univ Louisville, PhD(microbiol), 73. Prof Exp: Res asst, 73-74, RES ASSOC, DEPT MICROBIOL & IMMUNOL, UNIV LOUISVILLE, 74- Mem: Am Soc Microbiol. Res: Immunologic parameters of tumor growth and rejection; effects of mediators of inflammation and of specific immunological responses on tumor growth; regulation of immune responses. Mailing Add: Dept of Microbiol & Immunol Health Sci Ctr Univ Louisville Louisville KY 40201

TURNER, ERNEST CRAIG, JR, b West Jefferson, NC, June 15, 27; m 53; c 3. ENTOMOLOGY. Educ: Clemson Col, BS, 48; Cornell Univ, PhD(entom), 53. Prof Exp: Asst econ entom, Cornell Univ, 48-53; assoc prof entom, 53-65, PROF ENTOM, VA POLYTECH INST & STATE UNIV, 65-, ASSOC ENTOMOLOGIST, AGR EXP STA, 53- Mem: Am Mosquito Control Asn; Entom Soc Am. Res: Medical and veterinary entomology. Mailing Add: 1413 Locust Ave Blacksburg VA 24060

TURNER, EUGENE BONNER, b Wolf Point, Mont, Oct 6, 22; m 46; c 3. PHYSICS. Educ: Mont State Col, BS, 44; Univ Mich, MS, 50, PhD(physics), 56. Prof Exp: Mem tech staff, Ramo-Wooldridge Corp, Calif, 56-58 & Space Tech Labs, Inc, 58-60; mem tech staff, 60-68, head lasers & optics dept, Electronics Res Lab, 68-69, sr staff engr, Develop Planning Div, 69-73, staff engr, 73-75, STAFF SCIENTIST, LABS DIV, AEROSPACE CORP, 75- Mem: Am Phys Soc; Optical Soc Am; Soc Photo-Optical Instrument Eng (pres, 70); Sigma Xi; Am Inst Aeronaut & Astronaut. Res: Plasma physics; spectroscopy; optical and photographic instrumentation; lasers; infrared systems; strategic space systems. Mailing Add: Labs Div Aerospace Corp PO Box 92957 Los Angeles CA 90009

TURNER, EVERETT EUGENE, animal science, see 12th edition

TURNER, FRANCIS JOHN, b Auckland, NZ, Apr 10, 04; nat US; m 30; c 1. PETROLOGY. Educ: Univ NZ, MSc, 26, DSc(geol), 34. Hon Degrees: DSc, Univ Auckland, 65. Prof Exp: From lectr to sr lectr geol, Otago Univ, 26-46; from assoc prof to prof, 46-72, EMER PROF GEOL, UNIV CALIF, BERKELEY, 72- Concurrent Pos: Sterling fel, Yale Univ, 38-39; Guggenheim fel, 51 & 59-60; Fulbright fel, 56. Honors & Awards: Hector Medal, Royal Soc NZ, 56; Lyell Medal, Geol Soc London, 70. Mem: Nat Acad Sci; fel Geol Soc Am; fel Mineral Soc Am; fel Geol Soc London; fel Royal Soc NZ. Res: Petrology of ultrabasic rocks in New Zealand; metamorphic and igneous petrology; fabric of deformed rocks and artificially deformed rocks and minerals; deformation of mineral crystals; metamorphic paragenesis and facies. Mailing Add: 2525 Hill Ct Berkeley CA 94708

TURNER, FRANK JOSEPH, b New York, NY, Nov 23, 25; m 51; c 3. MICROBIOLOGY. Educ: St John's Univ, NY, BS, 50; Rutgers Univ, MS, 58, PhD, 71. Prof Exp: Bacteriologist, Nepera Chem Co, Inc, 51-52; asst scientist, Warner Inst Therapeut Res & Warner-Chilcott Labs Div, 52-53, assoc scientist, Warner-Chilcott Labs Div, 53-56, scientist, 56-60, sr scientist, Warner-Lambert Res Inst, 61-64, sr res assoc, 64-72, DIR, DEPT MICROBIOL & IMMUNOL, WARNER-LAMBERT RES INST, WARNER-HUDNUT, INC, 72- Mem: AAAS; Am Soc Microbiol; NY Acad Sci. Res: Chemotherapy of bacterial, fungal and viral diseases; investigation of host defense mechanisms; immunology of delayed hypersensitivity; disinfecting properties of hydrogen peroxide. Mailing Add: Dept Microbiol & Immunol Warner-Lambert Res Inst Morris Plains NJ 07950

TURNER, FRED, JR, b Paris Crossing, Ind, Jan 13, 20; m 42; c 1. AGRICULTURAL CHEMISTRY, SOILS. Educ: Univ Ariz, BS, 48; State Col Wash, MS, 51; Mich State Univ, PhD(soil sci), 57. Prof Exp: Asst soils, State Col Wash, 48-50; soil scientist, Irrig Exp Sta, Agr Res Serv, USDA, Prosser, Wash, 50-51 & Northern Great Plains Exp Sta, NDak, 54; asst soils, Mich State Univ, 54-57; asst prof agr chem & soils & agr chemist, Univ Ariz, 57-61, supt & agr chemist, Safford Exp Sta, 61-66, adv soils, US AID-Univ Ariz Contract, Sch Agron, Univ Ceara, Brazil, 66-68, SUPT, SAFFORD EXP STA, UNIV ARIZ, 68-, AREA EXTEN SOIL SPECIALIST, 75- Mem: AAAS; Soil Sci Soc Am; Am Soc Agron; Coun Agr Sci & Technol. Res: Various salinity aspects of irrigation waters and soils; fertility and management of salt load in irrigation agriculture to minimize water pollution; fertility and management of tropical soils. Mailing Add: Univ of Ariz Safford Exp Sta PO Box 1015 Safford AZ 85546

TURNER, FRED ALLEN, b Chicago, Ill, Mar 16, 33; m 64; c 2. ORGANIC CHEMISTRY. Educ: Univ Ill, BS, 55, MS, 58, PhD(pharmaceut chem), 63. Prof Exp: Instr, Univ Ill, Chicago, 58-59 & 61-62; from instr to assoc prof, 62-74, PROF CHEM, ROOSEVELT UNIV, 74- Concurrent Pos: NSF fel, 59-61. Mem: Am Chem Soc; The Chem Soc. Res: Synthesis of medicinal compounds; study of organic halogenating agents; synthesis of heterocyclic systems. Mailing Add: Dept of Chem Roosevelt Univ 430 S Michigan Ave Chicago IL 60605

TURNER, FREDERICK BROWN, b Carlinville, Ill, Feb 4, 27; m 54. VERTEBRATE ZOOLOGY. Educ: Univ Calif, AB, 49, MA, 50, PhD(zool), 57. Prof Exp: Asst zool, Univ Calif, 50-52 & 55-56; instr, Ill Col, 52-53; seasonal park naturalist, Death Valley Nat Monument, Calif, 53-55; instr biol, Wayne State Univ, 56-60, univ res coun res fel, 60; vis asst prof, 60-61, MEM STAFF, LAB NUCLEAR MED & RADIATION BIOL, UNIV CALIF, LOS ANGELES, 61- Concurrent Pos: NSF partic, Inst Desert Biol, Ariz State Univ, 59. Mem: AAAS; Ecol Soc Am; Brit Ecol Soc. Res: Radioecology; population ecology of amphibians and reptiles; ecosystem analysis; cycling and effects of trace metals in desert environments. Mailing Add: Lab Nuclear Med & Radiation Biol Univ of Calif Los Angeles CA 90024

TURNER, GEORGE CLEVELAND, b Spokane, Wash, Jan 26, 25; m 52; c 1. ECOLOGY, SCIENCE EDUCATION. Educ: Stanford Univ, BA, 47; Utah State Univ, MS, 50; East Wash State Col, MEd, 52; Ariz State Univ, EdD(sci ed), 64. Prof Exp: Teacher high schs, Wash, 52-53 & Calif, 53-60; assoc prof, 60-67, PROF BIOL & SCI EDUC, CALIF STATE UNIV, FULLERTON, 67-, CHMN DEPT SCI EDUC, 64- Concurrent Pos: Lectr, Claremont Grad Sch, 57-60; dir, NSF grants for adv topics inst for high sch biol teachers, 65-67, intern-master's degree prog for sci teachers, 67-, Human Ecol Inst, 69- & Urban Sci Intern Teaching Proj, 72-; dir, Off Educ grant for biol sci curriculum study test eval team, 66-67; pres, Calif Intersci Coun, 73-75; dir, Energy and the Environ Inst, 75- Mem: Nat Sci Teachers Asn; Nat Asn Biol Teachers. Res: Ecology of rodents, Wasatch Mountains, Utah; scientific enquiry. Mailing Add: Dept of Sci Educ Calif State Univ Fullerton CA 92634

TURNER, GEORGE ROBERT, b Newark, NJ, Oct 11, 17; m 42; c 3. TEXTILES, CHEMISTRY. Educ: Lowell Textile Inst, BTC, 41. Prof Exp: Lab technician, E I du Pont de Nemours & Co, 41-43, Blue Prods Div, Manhattan Proj, 43-44; tech field rep, Nat Res Coun, 44-45; mem tech serv, 45-47, sales serv, NC, 47-57, sales, 58-65, mgr tech serv dye appln, Del, 65-73, SR TECH CONSULT, ORG CHEM DEPT, DYES & CHEM DIV, E I DU PONT DE NEMOURS & CO, INC, 73- Mem: Am Asn Textile Chem & Colorists. Res: Dyestuffs application and manufacturing. Mailing Add: Org Chem Dept E I du Pont de Nemours & Co Inc Wilmington DE 19803

TURNER, HAL RUSSELL, b Des Moines, Iowa, Sept 4, 46; m 66; c 2. MICROBIAL PHYSIOLOGY. Educ: Iowa State Univ, BS, 69, MS, 73, PhD(bact), 75. Prof Exp: SR RESEARCHER FUNDAMENTAL BIOCHEM, CPC INT INC, 75- Mem: Am Soc Microbiol; Sigma Xi. Res: All areas of microbial physiology with special interests in the areas of enzymology; natural product isolation and metabolic regulation. Mailing Add: CPC Int Inc Moffett Tech Ctr Argo IL 60501

TURNER, HARRY JACKSON, JR, b Ansonia, Conn, Jan 20, 15; m 40; c 2. WATER POLLUTION. Educ: Yale Univ, AB, 35, MS, 42. Prof Exp: Asst, Yale Univ, 35-37, lab asst zool, 39-42; instr biol, Springfield Col, 37-39; jr biologist, US Fish & Wildlife Serv, Mass, 42-44; res assoc, Woods Hole Oceanog Inst, 44-50, marine biologist, 50-64, assoc scientist, 64-69; asst prof sci, Nathaniel Hawthorne Col, NH, 69-71; WATER POLLUTION BIOLOGIST, NH WATER SUPPLY & POLLUTION CONTROL COMN, 71- Mem: Am Micros Soc; Am Soc Limnol & Oceanog. Res: Marine, salt water and fisheries biology; shellfish propagation; protozoology; ciliates; antifouling. Mailing Add: Sun Valley RFD 1 Antrim NH 03440

TURNER, HENRY FORD, b Dozier, Ala, Sept 8, 22; m 56; c 2. ZOOLOGY, PARASITOLOGY. Educ: Auburn Univ, BS, 48, MS, 50; Iowa State Col, PhD(parasitol), 58. Prof Exp: Instr zool, Auburn Univ, 50-54; asst, Iowa State Col, 54-56; asst prof zool, Auburn Univ, 56-61; PROF ZOOL, UNIV MONTEVALLO, 63- Mem: Am Soc Parasitol. Res: Life cycles of digenetic trematodes; physiology of helminths; biology curricula. Mailing Add: Dept of Biol Univ of Montevallo Montevallo AL 35115

TURNER, J HOWARD, b San Diego, Calif, Oct 16, 27; m 49; c 4. HUMAN GENETICS. Educ: Utah State Univ, BS, 57, MS, 59; Univ Pittsburgh, ScD(human genetics), 62. Prof Exp: Instr zool & res asst genetics, Utah State Univ, 57-59; res assoc cytogenetics, 61-62, asst prof human genetics, 62-66, assoc prof, 66-71, PROF BIOSTATIST, UNIV PITTSBURGH, 71- Concurrent Pos: Res assoc, Magee Womens Hosp, 64- Mem: AAAS; Soc Human Genetics; Am Genetic Asn; Genetics Soc Am; Am Pub Health Asn. Res: Developmental cytogenetics; somatic cell genetics; statistical genetics. Mailing Add: Rm 316 Grad Sch Pub Health Univ of Pittsburgh Pittsburgh PA 15213

TURNER, JACK ALLEN, b Milner, Colo, Feb 2, 42; m 66; c 2. ECOLOGY. Educ: Colo State Univ, BS, 68; SDak State Univ, MS, 71; Univ Okla, PhD(ecol), 74. Prof Exp: Asst bact, SDak State Univ, 69-72; ASST PROF BIOL, UNIV SC, SPARTANBURG, 74- Mem: Am Soc Microbiol; Ecol Soc Am. Res: The antimicrobial activity of various types of textile finishes. Mailing Add: 121 Greengate Lane Spartanburg SC 29302

TURNER, JAMES ALFRED, physical chemistry, see 12th edition

TURNER, JAMES E, b Roann, Ind, Oct 11, 26. BIOCHEMISTRY. Educ: Manchester Col, BS, 50; Univ Ind, MA, 53, PhD(chem), 58. Prof Exp: Nat Res Coun fel & resident res assoc, Argonne Nat Lab, 58-60; chemist, Northern Regional Lab, USDA, 60-61; SR RES ASSOC, RES INST, WARNER-LAMBERT PHARMACEUT CO, 61- Mem: AAAS. Res: Protein structure studies; isolation and characterization of cereal proteins; optical rotation of proteins; transmethylation in plants; peptide hormone structure and synthesis; development of clinical chemistry assays and controls. Mailing Add: Warner-Lambert Res Inst 170 Tabor Rd Morris Plains NJ 07950

TURNER, JAMES EDWARD, theoretical physics, see 12th edition

TURNER, JAMES ELDRIDGE, b Richmond, Va, Oct 1, 42; m 67; c 2. NEUROBIOLOGY, ELECTRON MICROSCOPY. Educ: Va Mil Inst, BA, 65; Univ Richmond, MS, 67; Univ Tenn, PhD(zool), 70. Prof Exp: Teaching asst zool, Univ Tenn, 67-69; NIH res trainee neurobiol & electron micros, Dept Anat, Sch Med, Case Western Reserve Univ, 71; asst prof biol, Va Mil Inst, 71-72; NIH res fel neurobiol & electron micros, Dept Anat, Sch Med, Case Western Reserve Univ, 72-74; ASST PROF ANAT, BOWMAN GRAY SCH MED, 74- Mem: AAAS; Am Inst Biol Sci; Am Soc Zool; Am Asn Anatomists; Sigma Xi. Res: Nerve injury, repair and regeneration; biology of neuroglia. Mailing Add: Dept of Anat Bowman Gray Sch of Med Winston-Salem NC 27103

TURNER, JAMES HENRY, b Stuart, Va, June 13, 22. PARASITOLOGY. Educ: Univ Md, BS, 47, MS, 52, PhD(zool), 57. Prof Exp: Entomologist, Div Insects, Dept Zool, US Nat Mus, 48; from jr parasitologist to sr res parasitologist, Animal Dis Parasite Res Div, Agr Res Serv, USDA, 48-62; prin res parasitologist, Beltsville Parasitol Lab & McMaster Health Lab, Commonwealth Sci & Indust Res Orgn, Australia, 62-64; HEALTH SCIENTIST ADMINSTR, IMMUNOBIOL STUDY SECT, DIV RES GRANTS, NIH, 64- Concurrent Pos: Tutorial lectr, Univ Md, 58-61; Fulbright res fel, Australia, 62-63. Honors & Awards: Ransom Mem Award, 60. Mem: Am Soc Parasitol; Am Soc Trop Med & Hyg; Transplantation Soc. Res: Pathogenesis and immunological aspects of parasitic infections. Mailing Add: Div of Res Grants NIH Bethesda MD 20014

TURNER, JAMES HENRY, b Skowhegan, Maine, Jan 22, 37; m 61; c 4. SOLID STATE PHYSICS. Educ: Bowdoin Col, AB, 61; Mass Inst Technol, BS & MS, 61, PhD(physics), 64. Prof Exp: Asst prof, 64-70, ASSOC PROF PHYSICS, BOWDOIN COL, 70- Mem: Am Phys Soc. Res: Physical properties of thin metal foils and wires. Mailing Add: 19 Potter St Brunswick ME 04011

TURNER, JAMES HOWARD, b Colorado Springs, Colo, July 18, 12. ANALYTICAL CHEMISTRY. Educ: Colo Col, AB, 33; Univ Iowa, MS, 36. Prof Exp: Asst chem, Colo Col, 31-33 & Calif Inst Technol, 33-34; anal chemist, SW Shattuck Chem Corp, Colo, 41-44; asst chem, Univ Iowa, 45; chemist, Colo Fuel & Iron Corp, 47-48; res chemist, Holly Sugar Corp, 48-54; anal chemist, Holloman Air Force Base, NMex, 55-59 & US Bur Mines, 59-61; ANAL CHEMIST, US GEOL SURV, 61- Mem: Soc Appl Spectros; Am Chem Soc; assoc Cooper Ornith Soc; Am Ornith Union. Res: Spectrophotometric methods of analysis; sugar analysis; sugar beet by-products; rarer metal analysis; infrared analysis. Mailing Add: 132 5 W Second Ave Denver CO 80223

TURNER, JAMES MARSHALL, b Washington, DC, Aug 20, 44; m 67; c 3. PLASMA PHYSICS. Educ: Johns Hopkins Univ, BA, 66; Mass Inst Technol, PhD(physics), 71. Prof Exp: Lectr physics, Lesley Col, 70-71; res staff mem, Mass Inst Technol, 66-71; asst prof, Southern Univ, 71-73; assoc prof, 73-76, PROF PHYSICS, MOREHOUSE COL, 76- Mem: Am Phys Soc; Am Geophys Union; Inst Elec & Electronics Engrs; Am Chem Soc; AAAS. Res: Instabilities in weakly ionized plasmas; MHD structures and plasma properties in the solar wind; structure of hemoglobin S. Mailing Add: Dept of Physics Morehouse Col Box 116 Atlanta GA 30314

TURNER, JAMES WILLIAM, b Sweetwater, Tex, July 9, 40; m 62; c 2. ANIMAL BREEDING, STATISTICS. Educ: Tex Tech Col, BS, 62; Okla State Univ, MS, 64, PhD(animal breeding), 66. Prof Exp: From asst prof to assoc prof animal sci, La State Univ, Baton Rouge, 65-72; PROF ANIMAL SCI & HEAD DEPT, MISS STATE UNIV, 72- Mem: Am Soc Animal Sci; Am Genetic Asn. Res: Beef cattle breeding, especially crossbreeding. Mailing Add: Dept of Animal Sci Miss State Univ Mississippi State MS 39762

TURNER, JAN ROSS, b Okla, Sept 25, 37; m 62; c 2. MICROBIOLOGY. Educ: Ore State Univ, BS, 61, MS, 63, PhD(microbial physiol), 65. Prof Exp: Res asst microbial physiol, Ore State Univ, 61-65; AEC fel microbiol, Biol Div, Pac Northwest Labs, Battelle Mem Inst, 65-67, sr res scientist, 67-73; SR MICROBIOLOGIST, LILLY RES LABS, 73- Mem: AAAS; Am Chem Soc; Am Soc Microbiol. Res: Microbial physiology; sterol biosynthesis; biochemical genetics; aromatic amino acid metabolism; medical mycology; clinical microbiology; antibiotic mechanisms. Mailing Add: Lilly Res Labs Indianapolis IN 46206

TURNER, JANICE BUTLER, b Lincolnton, Ga, Dec 1, 36; m 58; c 1. MOLECULAR SPECTROSCOPY. Educ: Ga State Col for Women, AB, 58; Emory Univ, MS, 59; Univ SC, PhD(chem), 70. Prof Exp: From instr to asst prof chem, 59-70, ASSOC PROF CHEM, AUGUSTA COL, 70- Mem: Sigma Xi; Coblentz Soc; Am Chem Soc. Res: Preparation and strucutre determination of organogermanes-microwave studies. Mailing Add: Dept of Chem & Physics Augusta Col 2500 Walton Way Augusta GA 30904

TURNER, JOHN D, organic chemistry, see 12th edition

TURNER, JOHN DEAN, b Pasadena, Calif, Oct 2, 30. MEDICINE. Educ: Univ Calif, Berkeley, BA, 52; McGill Univ, MD, CM, 56. Prof Exp: Intern med, Mass Gen Hosp, 56-57, asst resident, 57-58; attend physician, Nat Heart Inst, 61-65; asst prof med, Baylor Col Med, 65-71; ASSOC PROF MED, SCH MED, UNIV CALIF, SAN DIEGO, 71- Concurrent Pos: Paul Dudley White fel cardiol, Mass Gen Hosp, 59-60, teaching fels, 59-61; res fel med, Peter Bent Brigham Hosp, Harvard Univ, 58-59; staff assoc, President's Comn Heart Dis, Cancer & Stroke, 64; fel coun epidemiol, Am Heart Asn, 65- Mem: Am Heart Asn; fel Am Col Physicians; Am Oil Chem Soc; Am Pub Health Asn. Res: Cardiovascular hemodynamics and epidemiology; angiocardiography in the diagnosis of valvular and congenital heart disease; external scintillation counting in detection of intracardiac shunts; lipid composition in human erythrocytes and plasma; lipid and lipoprotein metabolism in human plasma. Mailing Add: Dept of Med Sch of Med Univ of Calif at San Diego La Jolla CA 92037

TURNER, JOHN K, b Fairfield, Ill, May 7, 23; m 46; c 5. ANIMAL PHYSIOLOGY. Educ: Millikin Univ, AB, 47; Univ Ill, MA, 49; Univ Wis, PhD(physiol), 60. Prof Exp: Physiologist, Army Med Res Lab, Ft Knox, Ky, 59-61 & Vet Admin Hosp, Long Beach, Calif, 61-64; ASSOC PROF BIOL SCI, WESTERN ILL UNIV, 65- Concurrent Pos: Lectr, Sch Med, Univ Calif, Los Angeles, 62-65. Res: Reflex regulation of respiration and circulation. Mailing Add: Dept of Biol Sci Western Ill Univ Macomb IL 61455

TURNER, KENNETH CLYDE, b Mt Vernon, Wash, Apr 6, 34; m 55; c 3. RADIO ASTRONOMY. Educ: Portland Univ, BS, 57; Princeton Univ, PhD(physics), 62. Prof Exp: Instr physics, Princeton Univ, 62; Carnegie fel, 62-64; res staff assoc, 64-68, RES STAFF MEM, CARNEGIE INST WASH DEPT TERRESTRIAL MAGNETISM, 68- Concurrent Pos: Dir, Arg Inst Radio Astron, 70-73; vis prof, Nat Univ La Plata, 72-73. Mem: AAAS; Am Phys Soc; Am Astron Soc; Int Union Radio Sci; Int Astron Union. Res: Experimental foundations of relativity; hydrogen line and galactic continuum radio astronomy; history of the galaxy-magellanic cloud system; radio astronomical instrumentation. Mailing Add: Carnegie Inst of Wash Dept Terrestrial Magnetism 5241 Broad Branch Rd NW Washington DC 20015

TURNER, LARRY WEBSTER, b Alhambra, Calif, Jan 5, 41; m 70. VERTEBRATE BIOLOGY. Educ: Portland State Univ, BS, 68; Univ Ariz, PhD(zool), 72. Prof Exp: Asst prof zool, Univ Minn, 71-73; asst dir, Malheur Environ Field Sta, Burns, Ore, 73-75; LECTR BIOL, LEWIS & CLARK COL, 76- Mem: AAAS; Am Soc Mammalogists; Animal Behav Soc. Res: Ecology and behavior of vertebrates, especially small mammals. Mailing Add: Dept of Biol Lewis & Clark Col Portland OR 97219

TURNER, LEAF, b Brooklyn, NY, Mar 23, 43; m 66; c 2. THEORETICAL PHYSICS. Educ: Cornell Univ, AB, 63; Univ Wis, Madison, MS, 64, PhD(theoret physics), 69. Prof Exp: Fel, Weizmann Inst Sci, Rehovot, Israel, 69; fel physics, Univ Toronto, 69-71; proj assoc, Univ Wis, Madison, 71-74; STAFF MEM, LOS ALAMOS SCI LAB, UNIV CALIF, 74- Mem: Am Phys Soc. Res: Current algebra; phenomenological lagrangians; interactions of mesons and baryons; quantum field theory; symmetries in high energy physics; high energy phenomenology; scattering theory; plasma kinetic theory; magnetohydrodynamics. Mailing Add: Los Alamos Sci Lab PO Box 1663 Los Alamos NM 87545

TURNER, LINCOLN HULLEY, b Chicago, Ill, June 30, 28. MATHEMATICS. Educ: Univ Chicago, MS, 48; Purdue Univ, PhD(math), 57. Prof Exp: Mathematician, Space Tech Labs, Inc, 58-60; prof math, Univ Minn, 60-63; PROF MATH, UNIV TENN, KNOXVILLE, 63- Mem: Am Math Soc; Math Asn Am; Soc Indust & Appl Math. Res: Real variables; calculus of variations; applied mathematics. Mailing Add: Dept of Math Univ of Tenn Knoxville TN 37916

TURNER, LUTHER P, analytical chemistry, see 12th edition

TURNER, MALCOLM ELIJAH, JR, b Atlanta, Ga, May 27, 29; m 48, 68; c 4. MATHEMATICAL BIOLOGY, STATISTICS. Educ: Duke Univ, BA, 52; NC State Univ, MES, 55, PhD(statist), 59. Prof Exp: Asst biostatist, Univ NC, 53-54; sr res assoc biomet, Med Sch, Univ Cincinnati, 55, asst prof, 56-58; assoc prof & Williams res fel, Med Col Va, 58-63, chmn div biomet, Dept Biophys & Biomet, 59-63; prof statist & biomet & chmn dept, Emory Univ, 63-69, prof math, 66-69; prof biostatist & biomath & assoc prof physiol & biophys, 70, PROF MATH, UNIV ALA, BIRMINGHAM, 72-, CHMN DEPT BIOMATH, 75- Concurrent Pos: Asst statistician, NC State Univ, 57-58; managing ed, Biometrics, 62-69; Merck lectr, Med Ctr, Univ Kans, 65, vis prof & chmn dept biomet, 68-69; consult, Southern Res Inst, 74- Mem: Fel AAAS; Soc Indust & Appl Math; hon fel Am Statist Asn; Soc Math Biol; Soc Comput Simulation. Res: Application of mathematics and statistics to biological research. Mailing Add: Dept of Biomath Univ of Ala Birmingham AL 35294

TURNER, MANSON DON, b Pleasanton, Tex, Nov 15, 28; m 53; c 3. PHYSIOLOGY. Educ: Baylor Univ, BS, 50, MS, 51; Univ Tenn, PhD(physiol), 55. Prof Exp: Lab instr, Baylor Univ, 50-51; res asst, Univ Tenn, 51-54, instr clin physiol, 54-55; res asst prof surg, Sch Med, Univ Miss, 55-57, asst prof biochem, 58-61, assoc prof res surg & asst prof physiol & biophys, 61-65; supvry res physiologist, US Air Force Sch Aerospace Med, 65; assoc prof surg & physiol, 65-69, ASSOC PROF PHYSIOL & BIOPHYS & RES PROF SURG, SCH MED, UNIV MISS, 69- Concurrent Pos: Attend in physiol, Vet Admin Hosp, Jackson, 58-; consult, Oak Ridge Inst Nuclear Studies, 62-65. Mem: Transplantation Soc; Cryobiol Soc; Am Heart Asn; Am Physiol Soc. Res: Organ preservation; cardiovascular physiology. Mailing Add: Dept of Surg Univ of Miss Sch of Med Jackson MS 39216

TURNER MATTHEW X, b Rockaway, NY, July 27, 28; m 58; c 5. CYTOLOGY, IMMUNOLOGY. Educ: Fordham Univ, BS, 56, MS, 62, PhD(cytol), 68. Prof Exp: Jr bacteriologist, NY State Conserv Dept, 56-58; jr isotope officer, State Univ NY Downstate Med Ctr, 58-60, radiation officer, 60-62, res asst anat & cytol, 62-66, NIH fel cytol & immunol, 67-68; microbiologist, US Naval Appl Sci Lab, NY, 66-67; asst prof, 68-72, ASSOC PROF BIOL, JERSEY CITY STATE COL, 72- Mem: Reticuloendothelial Soc; Am Soc Microbiol. Res: Effect of antigen upon the cellular and humoral responses in normal and immunized animals; determination of x-ray and drugs upon antigenic stimulation in normal and immunized mice. Mailing Add: Dept of Biol Jersey City State Col Jersey City NJ 07305

TURNER, MICHAEL D, b Weston, Eng, Oct 26, 27; m 57; c 4. MEDICINE, BIOCHEMISTRY. Educ: Bristol Univ, MB, ChB, 50, MD, 59; Univ Rochester, PhD(biochem), 64. Prof Exp: Tutor med, Postgrad Med Sch, Univ London, 54-57; lectr, Royal Free Hosp Med Sch, 60-63; consult med, 63-64, from assoc prof to prof, 64-72, SEGAL-WATSON PROF GASTROENTEROL, SCH MED & DENT, UNIV ROCHESTER, 72- Concurrent Pos: Lederle med fac fel, 63-66. Mem: Am Fedn Clin Res; Am Gasteroenterol Asn; Brit Soc Gastroenterol. Res: Liver diseases and portal hypertension; chemistry and immunology of gastric macromolecules. Mailing Add: Sch of Med & Dent Univ of Rochester Rochester NY 14642

TURNER, MORTIMER DARLING, b Greeley, Colo, Oct 24, 20; m 45, 65; c 4. ECONOMIC GEOLOGY, ENGINEERING GEOLOGY. Educ: Univ Calif, Berkeley, BS, 43, MS, 54; Univ Kans, PhD, 72. Prof Exp: Mech & elec engr, Aberdeen Proving Grounds, US Dept Army, 46; asst geol sci, Univ Calif, Berkeley, 48; from jr mining geologist to asst mining geologist, Calif State Div Mines, 48-54; state geologist, Econ Develop Admin, PR, 54-58; phys sci administr & asst to dir, Antarctic Res Prog, NSF, 59-61; res assoc geol, Univ Kans, 62-65; prog dir, Antarctic Earth Sci, Off Antarctic Progs, 65-70, PROG MGR, POLAR EARTH SCI, POLAR OCEANOG, OFF POLAR PROGS, NSF, 70-; ASSOC PROF LECTR GEOL, GEORGE WASHINGTON UNIV, 72- Concurrent Pos: Consult, 55-59; mem, Orgn Comt, 1st Conf Clays & Clay Technol, 52 & Caribbean Geol Conf, 59; mem, US Planning Comt, 2nd Int Conf Permafrost, Nat Acad Sci; mem, 19th, 20th (vpres), 21st, 22nd & 24th (US Govt deleg), Int Geol Congs. Mem: Geol Soc Am; Am Geog Soc; Brit Geol Asn; Geol Asn PR; Am Polar Soc. Res: Tectonics, economic geology and engineering geology of California, Puerto Rico, Caribbean area and Antarctica; geology of early man in North America; engineering geology of costal zones. Mailing Add: 3920 Rickover Rd Silver Spring MD 20902

TURNER, NATHAN JOE, plant pathology, biochemistry, see 12th edition

TURNER, NOEL HINTON, b Redlands, Calif, Dec 24, 40; m 66; c 2. PHYSICAL CHEMISTRY. Educ: Univ Calif, Berkeley, BS, 62; Univ Rochester, PhD(phys chem), 68. Prof Exp: RES CHEMIST, US NAVAL RES LAB, 68- Mem: AAAS; Am Chem Soc; Sci Res Soc Am. Res: Surface chemistry; gas-solid interactions with reinforcement fibers; electron spectroscopy for chemical analysis and auger electron spectroscopy; kinetics of gas phase reactions and gas-solid adsorption. Mailing Add: Chem Div Code 6173 US Naval Res Lab Washington DC 20390

TURNER, NURA DOROTHEA, b Gregor, Iowa, Mar 5, 05. MATHEMATICS. Educ: Univ Iowa, BA, 28, MS, 36. Prof Exp: High sch teacher, Iowa, 28-30 & Tex, 30-43; asst prof math, Nebr State Teachers Col, Kearney, 43-44; computer, Nat Adv Comt Aeronaut, Va, 44-45; from asst prof to prof, 46-70, EMER PROF MATH, STATE UNIV NY ALBANY, 70- Concurrent Pos: Asst, Ohio State Univ, 45; mem, Int Cong Math, Stockholm, 62, Moscow, 66 & Nice, France, 70; mem, Int Cong Math Educ, Exeter, Eng, 72. Mem: Am Math Soc; Math Asn Am; Am Statist Asn. Res: Statistics. Mailing Add: Dept of Math State Univ of NY Albany NY 12222

TURNER, RALPH B, b Lynchburg, Va, Mar 1, 31; m 52; c 2. BIOCHEMISTRY. Educ: Va Polytech Inst, BS, 52; Univ Tex, PhD(chem), 63. Prof Exp: Res biochemist, Entom Res Div, USDA, 63-68; RES BIOCHEMIST, DEPT BOT & ENTOM, N MEX STATE UNIV, 68- Mem: Fel AAAS; fel Am Inst Chem; Am Chem Soc; Sigma Xi; NY Acad Sci. Res: Insect nutrition and development; mode of action of chemosterilants; insect embryogenesis; proteolytic enzymes. Mailing Add: Box 3BE NMex State Univ Las Cruces NM 88001

TURNER, RALPH WALDO, b Blakely, Ga, Nov 9, 38. PHYSICAL CHEMISTRY, MATHEMATICS. Educ: J C Smith Univ, BS, 59; Univ Pittsburgh, PhD(phys chem), 65. Prof Exp: Res asst chem, Univ Pittsburgh, 60-64; res engr, Gen Tel & Electronics, 65-66, advan res engr, 66-67; prof sci educ, 67-68, prof chem & physics, 68-69, PROF CHEM & DIR, DIV BASIC STUDIES, FLA A&M UNIV, 69- Concurrent Pos: US Off Educ grant, Fla A&M Univ, 62-; dir 13 col prog, Fla A&M Univ, 68-69; consult, Inst Servs Educ, 68-72. Mem: AAAS; Am Chem Soc; Am Crystallog Asn; Sigma Xi. Res: Structure of metal ion aromatic complexes using x-ray crystallography. Mailing Add: Dept of Chem Fla A&M Univ Tallahassee FL 32307

TURNER, RAYMOND MARRINER, b Salt Lake City, Utah, Feb 25, 27; m 49; c 3. ECOLOGY. Educ: Univ Utah, BS, 48; State Col Wash, PhD(plant ecol), 54. Prof Exp: Instr range mgt, Univ Ariz, 54-56, asst prof, 56, instr bot, 56-57, asst prof, 57-62; RES BOTANIST, US GEOL SURV, 62- Concurrent Pos: NSF res grant, 57-60. Mem: AAAS; Ecol Soc Am; Bot Soc Mex; Am Soc Photogram. Res: Ecology of arid and semi-arid regions; remote sensing. Mailing Add: US Geol Surv 301 W Congress Tucson AZ 85701

TURNER, REX HOWELL, b Birmingham, Ala, Aug 22, 38; m 60; c 2. CELLULOSE CHEMISTRY. Educ: Univ S Ala, BS, 68; Univ Ga, PhD(chem), 73. Prof Exp: Res asst org chem, Univ Ga, 68-73; develop chemist, Millmaster Onyx Corp, 73-74; SR RES CHEMIST, INT PAPER CO, 74- Mem: Am Chem Soc; Am Asn Textile Chemists & Colorists. Res: Isolation and enrichinhomogeneities; macrocyclic synthesis via thermochemical and photochemical decomposition of ketone peroxides. Mailing Add: Int Paper Co PO Box 2328 Mobile AL 36601

TURNER, RICHARD JOSEPH, chemistry, see 12th edition

TURNER, ROBERT CHAPMAN, b Dorchester, NB, Apr 7, 10; m 40; c 2. PHYSICAL CHEMISTRY. Educ: McGill Univ, BSA, 32, MSc, 49, PhD(chem), 51. Prof Exp: RES SCIENTIST, SOIL RES INST, RES BR, CAN DEPT AGR, 51- Mem: Am Soc Soil Sci; Can Soc Soil Sci; Agr Inst Can; Int Soc Soil Sci. Res: Soil chemistry. Mailing Add: Soil Res Inst Res Br Can Dept Agr Ottawa ON Can

TURNER, ROBERT E L, b Montclair, NJ, Nov 15, 36; m 60; c 3. MATHEMATICS. Educ: Cornell Univ, BEngPhys, 59; NY Univ, PhD(math), 63. Prof Exp: From asst prof to assoc prof, 63-71, PROF MATH, UNIV WIS-MADISON, 71- Mem: Am Math Soc. Res: Functional analysis. Mailing Add: Dept of Math Univ of Wis Madison WI 53706

TURNER, ROBERT ELWOOD, b Covington, Ky, Dec 8, 37. ATMOSPHERIC PHYSICS, THEORETICAL ASTROPHYSICS. Educ: Univ Cincinnati, BS, 59, MS, 60; Columbia Univ, MA, 63; Wash Univ, PhD(physics), 70. Prof Exp: Res asst physics, Wash Univ, 64-69; res physicist, Univ Mich, Ann Arbor, 70-73; RES PHYSICIST, ENVIRON RES INST MICH, 73- Concurrent Pos: Astronomer, McDonnell Planetarium, 65-68; lectr, NASA, 75- Mem: AAAS; Am Inst Physics; Am Asn Physics Teachers; Am Astron Soc. Res: Radiative transfer in planetary atmospheres; remote sensing of earth's resources; atmospheric optics; aerosol physics; cosmic ray astrophysics. Mailing Add: Environ Res Inst of Mich PO Box 618 Ann Arbor MI 48107

TURNER, ROBERT JAMES, b Loda, Ill, Nov 15, 21; m 47; c 2. ORGANIC CHEMISTRY. Educ: Univ Ill, BS, 47; Univ Wis, MS, 49, PhD(org chem), 50. Prof Exp: Asst, Alumni Res Found, Univ Wis, 47-49; res chemist, Mallinckrodt Chem Works, 50-55, group leader, 55; group leader, 56-58, res supvr, 58-64, asst dir res, 64-68, DIR RES, MORTON CHEM CO, MORTON-NORWICH PROD, INC, 68- Mem: Am Chem Soc. Res: Synthetic organic chemistry; hydrogenation; hydroformylation; polymers; surface coatings; dyes. Mailing Add: Morton Chem Co Res Dept 1275 Lake Ave Woodstock IL 60098

TURNER, ROBERT LAWRENCE, b Chicago, Ill, Nov 4, 45; m 66; c 2. ORGANIC POLYMER CHEMISTRY. Educ: Albion Col, BA, 67; Mich State Univ, PhD(org chem), 72. Prof Exp: Teaching asst org chem, Mich State Univ, 67-69, res asst, 71-72; RES CHEMIST, E I DU PONT DE NEMOURS & CO, INC, 72- Mem: Am Chem Soc. Res: Development of new polymer-catalyst systems for use in low energy and nonpolluting protective organic coatings for industrial use. Mailing Add: E I du Pont de Nemours & Co Inc Fabrics & Finishes Dept Exp Sta Wilmington DE 19898

TURNER, ROBERT SCOTT, JR, b Seattle, Wash, May 18, 42; m 72; c 3. DEVELOPMENTAL BIOLOGY. Educ: Seattle Univ, BS, 64; Univ Ore, PhD(develop biol), 71. Prof Exp: Fel, Dept Biol Chem, Princeton Univ, 71-72, Dept Biochem, Univ Basel, Switz, 72-74; res fel, Dept Biol Chem, Harvard Med Sch, 74-75; ASST PROF BIOL, WESLEYAN UNIV, 75- Mem: Am Soc Cell Biol; Soc Develop Biol. Res: Genetic control of the synthesis and structure of cell membrane components involved in cell interactions during development and neoplasia; the nature and control of the interactions between these surface components. Mailing Add: Dept of Biol Wesleyan Univ Middletown CT 06457

TURNER, ROBERT STUART, b Red Oak, Iowa, June 25, 12; m 42; c 2. NEUROANATOMY. Educ: Dartmouth Col, AB, 33; Yale Univ, PhD(anat), 38. Prof Exp: Instr biol, Dartmouth Col, 33-35; asst resident in neuroanat, Med Sch, Yale Univ, 37-38; from instr to assoc prof anat, 38-55, PROF ANAT, STANFORD UNIV, 55- Concurrent Pos: NIH spec fel, 48-49; consult, Agnews State Hosp & Vet Admin Hosp, 59- Mem: AAAS; Am Asn Anat; corresp mem Mex Soc Anat; Pan Am Soc Anat. Res: Functional anatomy of nervous system; comparative neurology and neurophysiology. Mailing Add: Dept of Anat Stanford Univ Stanford CA 94305

TURNER, RUTH DIXON, b Melrose, Mass, Dec 7, 14. MALACOLOGY. Educ: Bridgewater Teachers Col, BS, 36; Cornell Univ, MA, 43; Radcliffe Col, PhD, 54. Prof Exp: Teacher high sch, Vt, 36-37 & jr high sch, Mass, 37-40; asst ed, Boston Soc Natural Hist, 40-42, asst cur birds, 41-42; instr ornith, Vassar Col, 43-44; biologist, William F Clapp Labs, Mass, 44-45; res mollusks, Mus Comp Zool, 45-55, res assoc malacol & Agassiz fel oceanog & zool, 55-75, PROF BIOL, HARVARD UNIV & CUR MALACOL, MUS COMP ZOOL, 75- Concurrent Pos: Res assoc, Inst Marine Biol, PR, 56-; consult, William F Clapp Labs, 57- Mem: Am Soc Limnol & Oceanog; Soc Syst Zool; Am Malacol Union. Res: Marine boring and fouling mollusks; taxonomy and biology of mollusks, particularly Western Atlantic marine and North American freshwater. Mailing Add: Mollusk Dept Mus Comp Zool Harvard Univ Cambridge MA 02138

TURNER, STANLEY EUGENE, b Greenville, Tex, Sept 3, 26; m; c 4. NUCLEAR CHEMISTRY. Educ: Univ SC, BS, 47; Univ Tex, PhD(chem), 51. Prof Exp: Chemist, US Naval Radiol Defense Lab, 51-52; sr res chemist, Field Res Labs, Magnolia Petrol Co, 52-57; staff physicist, Gen Nuclear Eng Corp, 57-64; vpres physics, Southern Nuclear Eng, Inc, 64-73; EXEC SCIENTIST, NUS CORP, 73- Mem: Am Chem Soc; Am Nuclear Soc. Res: Nuclear transformations; reactor design; physics calculations; reactor physics. Mailing Add: 2536 Countryside Blvd Clearwater FL 33515

TURNER, TERRY EARLE, b NS, Can, Jan 18, 21; nat US; m 42; c 2. PHYSICS. Educ: Acadia Univ, BSc, 43, BA, 44; McGill Univ, PhD(physics), 48. Prof Exp: Jr res physicist optics, Nat Res Coun Can, 44-45; head physics lab, Sprague Elec Co, 48-51; chief combustion br, Ballistic Res Labs, Aberdeen Proving Ground, Md, 51-56; group leader, Spark Heated Wind Tunnel, Missiles & Space Div, Lockheed Aircraft Corp, 56-59; develop prog mgr, Boeing Co, 59-72; CONSULT PHYSICIST, 72- Mem: Am Inst Aeronaut & Astronaut; Am Phys Soc; Combustion Inst. Res: High temperature physics; spectroscopy; combustion. Mailing Add: 18161 Brittany Dr SW Seattle WA 98166

TURNER, THOMAS BOURNE, b Prince Frederick, Md, Jan 28, 02; m 27; c 2. MICROBIOLOGY. Educ: St John's Col, Md, BS, 21; Univ Md, MD, 25. Hon Degrees: ScD, Univ Md, 66. Prof Exp: Intern, Hosp for Women of Md, 25-26; resident, Mercy Hosp, 26-27; Loeb fel, Sch Med, Johns Hopkins Univ, 27-28, instr med, 28-31, assoc, 31-32; mem staff, Int Health Div, Rockefeller Found, 32-39, clin dir, Jamaica Yaws Comn, 32-34, labs, 34-36; lectr med & pub health admin, Sch Hyg & Pub Health, 36-39, prof microbiol, Sch Med, 39-68, dean, 57-68, EMER DEAN, SCH MED, JOHNS HOPKINS UNIV, 68- Concurrent Pos: Consult, Surgeon Gen, US Army; vchmn comt virus res & epidemiol, Nat Found, 49-67; coord, Regional Med Prog for Md, 67-68; physician, out-patient dept, Johns Hopkins Hosp; mem bd visitors, St John's Col, Md. Mem: Am Soc Clin Invest; Asn Am Physicians; Asn Am Med Cols (past pres); Am Venereal Dis Asn (past pres). Res: Spirochetal diseases; poliomyelitis; tetanus; internal medicine. Mailing Add: Johns Hopkins Univ Sch Med 720 Rutland Ave Baltimore MD 21205

TURNER, THOMAS JENKINS, b Albany, Ga, Sept 11, 26; m 48; c 4. SOLID STATE PHYSICS. Educ: Univ NC, BS, 47; Clemson Col, MS, 49; Univ Va, PhD(physics), 51. Prof Exp: Instr physics, Clemson Col, 47-49; asst prof, Univ NH, 52; asst prof, 53-56, chmn dept, 56-74, PROF PHYSICS, WAKE FOREST UNIV, 56- Mem: Fel Am Phys Soc; Am Asn Physics Teachers. Res: Defects in crystalline materials; color centers; internal friction. Mailing Add: Dept of Physics Wake Forest Univ Winston-Salem NC 27109

TURNER, VERAS D, b Tompkinsville, Ky, Oct 19, 25; m 50; c 2. MATHEMATICS. Educ: Northwestern Univ, BS, 46; Univ Ill, MA, 49; Univ Okla, PhD, 68. Prof Exp: Teacher math, Moark Col, 49-51 & high sch, Ill, 53-56; PROF MATH, MANKATO STATE UNIV, 56-, CHMN DEPT MATH, ASTRON & STATIST, 73- Mem: Math Asn Am. Mailing Add: Dept of Math Mankato State Univ Mankato MN 56001

TURNER, VERNON LEE, JR, b New Orleans, La, Apr 17, 24; m 47; c 4. CHEMISTRY. Educ: Clemson Col, BS, 47; Canisius Col, MS, 52. Prof Exp: Res chemist, Spencer Kellogg & Sons, 47-48; res chemist, Textile Fibers Dept, Pioneering Res Div, 48-52, Indust & Biochem Dept, Res Div, 52-62, RES CHEMIST, FILM DEPT, RES DIV, E I DU PONT DE NEMOURS & CO, INC, 62- Mem: Am Chem Soc; Sigma Xi; Am Soc Testing & Mat. Res: Catalytic hydrogenation and deodorization of vegetable oils; instrumentation and application of radioisotopes in analytical chemistry; general physical and chemical development in the analytical field; thin film deposition; high vacuum technology. Mailing Add: E I du Pont de Nemours & Co Wilmington DE 19898

TURNER, WILLIAM JR, b Bell, Calif, June 27, 40; m 62; c 2. ENTOMOLOGY. Educ: Univ Calif, Berkeley, AB, 63, MS, 66, PhD(entom), 71. Prof Exp: Lab technician entom, Univ Calif, Berkeley, 63-64, from asst to assoc entomologist, 64-67, NIH trainee, 67-70; ASST PROF ENTOM & ZOOL & ASST ENTOMOLOGIST, WASH STATE UNIV, 70- Mem: Entom Soc Am; Entom Soc Can; Soc Syst Zool. Res: Insect biosystematics; zoogeography, phylogeny and biology of Diptera; swarming behavior in insects; medical entomology, especially biting flies as vectors of pathogens. Mailing Add: Dept of Entom Wash State Univ Pullman WA 99163

TURNER, WILLIAM JOSEPH, b Canandaigua, NY, Mar 7, 27; m 51; c 4. SOLID STATE PHYSICS. Educ: Villanova Univ, BS, 49; Cath Univ, PhD(physics), 55. Prof Exp: Physicist, Naval Res Lab, DC, 51-52 & Nat Bur Stands, 52-56; assoc physicist, Phys Res Dept, 56-57, staff physicist, Res Lab, 57-58, proj physicist, 58-59, develop physicist & res staff mem semiconductor physics, 59-61, res staff mem optical properties of semiconductors, Thomas J Watson Res Ctr, 61-64, MGR RES STAFF OPERS, THOMAS J WATSON RES CTR, IBM CORP, 64- Mem: Fel Am Phys Soc. Res: Use of optical absorption, reflection and luminescence measurements to study the intrinsic, lattice and extrinsic properties of semiconductors. Mailing Add: Thomas J Watson Res Ctr PO Box 218 IBM Yorktown Heights NY 10598

TURNER, WILLIAM RICHARD, b Drexel Hill, Pa, June 26, 36; m 60; c 3. ANALYTICAL CHEMISTRY. Educ: Philadelphia Col Pharm, BS, 58; Univ Conn, MS, 61, PhD(anal chem), 64. Prof Exp: Chemist, Borden Chem Co, 58-59; fel, Univ Mich, 64; RES CHEMIST, ICI AM, 65- Mem: Am Chem Soc. Res: Liquid chromatography; electroanalytical chemistry, especially organic polarography, votammetry at solid electrodes, coulometric titrimetry and amperometric titrimetry. Mailing Add: Specialty Chem Res Dept ICI Am Wilmington DE 19899

TURNER, WILLIAM RUSSEL, b Gordon, Pa, July 27, 11; m 40, 69; c 6. PETROLEUM CHEMISTRY, POLYMER CHEMISTRY. Educ: Univ Pa, AB, 42. Prof Exp: Jr chemist, Atlantic Ref Co, 42-45, from asst to assoc chem engr, 45-49, sr chem engr, 49-54, supv engr, 54-60; sr res chemist, Arco Chem Co Div, Atlantic Richfield Co, 60-71; CONSULT, W R TURNER CO, 71- Mem: AAAS; Am Chem Soc; Am Inst Chem. Res: Lubricants; process oils; petroleum waxes; rust preventives; synthetic additives for lubricants and waxes; paper coatings; polymers and plastics; petrochemicals; synthetic foods; synthetic food and feed supplements. Mailing Add: W R Turner Co 629 Paddock Rd Havertown PA 19083

TURNER, WILLIE, b Suffolk, Va, Feb 1, 35; m 64; c 2. VIROLOGY, IMMUNOLOGY. Educ: Md State Col, BS, 57; Ohio State Univ, MS, 59, PhD(microbiol), 61. Prof Exp: NIH fel, Naval Med Res Inst, 61-62, Nat Inst Allergy & Infectious Dis grant, 62-64; asst prof microbiol, Meharry Med Col, 62-66; head microbiol sect, Viral Biol Br, Nat Cancer Inst, 70-71; PROF MICROBIOL & CHMN DEPT, COL MED, HOWARD UNIV, 71- Concurrent Pos: NIH staff fel oncol virol, Nat Cancer Inst, 66-69, sr fel, 69-70. Mem: AAAS; Am Soc Microbiol; Am Asn Immunologists; Am Asn Cancer Res; Soc Exp Biol & Med. Res: Oncogenic virology, especially interaction of oncogenic and nononcogenic viruses in vitro and in vivo; immunology of murine oncogenic virus as well as the immunology involved with tumors induced by these agents in vivo. Mailing Add: Dept of Microbiol Howard Univ Col of Med Washington DC 20001

TURNEY, LAWRENCE JOSEPH, b Charleroi, Pa, Sept 29, 13; m 46; c 2. FOOD SCIENCE. Educ: Univ Pittsburgh, BS, 37. Prof Exp: Chemist, H J Heinz Co, Pa, 40-44; res dir, H W Madison Co, Div, J C Smucker Co, 44-72; DIR TECH SERV, PARAMOUNT FOODS DIV, HIRSCH BROS CO, 72- Concurrent Pos: Consult, Kellogg Co, 70-72. Mem: Am Chem Soc; Am Oil Chem Soc; Fel Am Inst Chem; Inst Food Technol. Res: Food chemistry and fermentation; emulsions; vinegar organisms and nutrients; pasteurization; waste treatment; canned meat and related products; brine recycling; pure culture fermentations; controlled fermentations. Mailing Add: Paramount Foods Piccadilly Sta Louisville KY 40213

TURNEY, TULLY HUBERT, b Lakewood, Ohio, Sept 7, 36; m 62; c 1. ZOOLOGY. Educ: Oberlin Col, AB, 58; Univ NC, PhD(zool), 63. Prof Exp: Fel, Oak Ridge Nat Labs, 63-64; instr zool, Univ NC, 64-65; assoc prof biol, 65-74, PROF BIOL, HAMPDEN-SYDNEY COL, 74-, CHMN DEPT, 67- Mem: AAAS; Am Inst Biol Sci. Res: Cell control mechanisms; molecular biochemistry; physiology. Mailing Add: Dept of Biol Hampden-Sydney Col Hampden-Sydney VA 23943

TURNIPSEED, GLYN D, b Hazlehurst, Miss, Dec 19, 42; m 67; c 1. PLANT PHYSIOLOGY. Educ: Delta State Univ, BS, 66; Miss State Univ, PhD(bot), 73. Prof Exp: Sci teacher biol chem, Jackson Pub Schs & Wingfield High Sch, 66-70; asst bot, Miss State Univ, 70-73; ASST PROF BIOL, ARK POLYTECH COL, 73- Mem: Soc Study of Amphibians & Reptiles; Bot Soc Am; Am Hibiscus Soc. Res: Nitrogen source preference of Oophila ambystomatis; feeding efficiency of larval anurans; population density of larval anurans. Mailing Add: Dept of Biol Ark Polytech Col Russellville AR 72801

TURNIPSEED, MARVIN ROY, b Carrollton, Miss, Nov 11, 34; m 57; c 2. REPRODUCTIVE PHYSIOLOGY, BIOCHEMISTRY. Educ: Miss State Univ, BS, 56, MEd, 63; Univ Ga, PhD(zool), 69. Prof Exp: NIH fel steroid biochem, 69-71, RES ASSOC STEROID BIOCHEM, MED SCH, UNIV MINN, MINNEAPOLIS, 71- Mem: AAAS. Res: Effects of gamma irradiation on reproduction in female cotton rats; ovarian production of steroid hormones in response to trophic hormones; plasma and urinary estrogens and adrenal metabolism in newborn babies. Mailing Add: Univ of Minn Dept of Pediat Box 391 Mayo Mem Bldg Minneapolis MN 55455

TURNIPSEED, SAMUEL GUY, b Mobile, Ala, Jan 19, 34; m 55; c 3. ENTOMOLOGY. Educ: Univ NC, AB, 56; Clemson Univ, MS, 58; NC State Univ, PhD(entom), 61. Prof Exp: From asst prof to assoc prof, 61-71, PROF ENTOM, EDISTO EXP STA, CLEMSON UNIV, 71-, PROF ECON ZOOL, 74- Mem: Entom Soc Am; Am Soybean Asn. Res: Economic injury thresholds for insects of soybeans; development of integrated control methods, with emphasis on host plant resistance, predator manipulation, microbial pathogens and selected chemicals. Mailing Add: Dept Entom Edisto Exp Sta Clemson Univ Clemson SC 29631

TURNOCK, WILLIAM JAMES, b Winnipeg, Man, May 17, 29; m 58; c 3. POPULATION ECOLOGY, BIOLOGICAL CONTROL. Educ: Univ Man, BSA, 49; Univ Minn, MS, 51, PhD(entom), 59. Prof Exp: Res scientist, Div Forest Biol, Can Dept Agr, 49-61 & Forest Entom & Path Br, Can Dept Forestry, Man, 61-70; sci adv, Can Ministry State for Sci & Technol, 70-72; SECT HEAD INTEGRATED PEST CONTROL, RES BR, CAN DEPT AGR, 72- Concurrent Pos: Hon prof, Grad Sch, Univ Man, 65-70; Can del, Int Coord Coun Man & Biosphere, UNESCO, 71. Mem: Ecol Soc Am; Entom Soc Can. Res: Integrated and biological control of agriculture pests. Mailing Add: Agr Res Sta 25 Dafoe Rd Winnipeg MB Can

TURNQUEST, BYRON W, b Chicago, Ill, Dec 16, 24; m 51; c 4. ORGANIC CHEMISTRY. Educ: DePaul Univ, BS, 49, MS, 51; Ill Inst Technol, PhD(phys org chem), 57. Prof Exp: Res chemist, Sinclair Res Inc, Harvey, 56-60, sr res chemist, 60-66, SR RES CHEMIST & SECT LEADER, ARCO, 66-, MGR PROD RES & DEVELOP, 73- Res: Physical and synthetic organic chemistry; engine oils; engine oil additives; greases; catalysis. Mailing Add: Harvey Res Ctr ARCO 400 E Sibley Blvd Harvey IL 60426

TURNQUIST, ORRIN CLINTON, b Minneapolis, Minn, Apr 29, 13; m 42; c 2. HORTICULTURE. Educ: Univ Minn, BS, 37, MS, 40, PhD(hort, plant genetics), 51. Prof Exp: Asst hort, Univ Minn, 43-45, instr & horticulturist, Northwest Agr Exp Sta, 45-47; horticulturist, USDA, 48-49; instr, 49-52, assoc prof, 52-58, PROF HORT, UNIV MINN, ST PAUL, 58-, EXTEN HORTICULTURIST, 49- Mem: Am Soc Hort Sci; Potato Asn Am (vpres, 58, pres elect, 59). Res: Potato culture and vegetable corps, particularly potato variety testing. Mailing Add: 342 Hort Sci Univ of Minn St Paul MN 55101

TURNQUIST, RICHARD LEE, b Rugby, NDak, Aug 12, 44; m 66. INSECT TOXICOLOGY, BIOCHEMISTRY. Educ: Concordia Col, BA, 66; Utah State Univ, PhD(physiol), 71. Prof Exp: Fel biochem, Utah State Univ, 71-74; ASST PROF BIOL, AUGUSTANA COL, 74- Mem: AAAS; Sigma Xi. Res: Ultrastructural changes and responses during detoxication of xenobiotics in mammals and insects. Mailing Add: Dept of Biol Augustana Col Rock Island IL 61264

TURNQUIST, TRUMAN DALE, b Kipling, Sask, Apr 8, 40; m 64; c 2. ANALYTICAL CHEMISTRY. Educ: Bethel Col, Minn, BA, 61; Univ Minn, PhD(anal chem), 65. Prof Exp: ASSOC PROF CHEM, MT UNION COL, 65- Mem: Am Chem Soc. Res: Metal complex formation; solvent extraction of metal complexes; spectrophotometry. Mailing Add: Dept of Chem Mt Union Col Alliance OH 44601

TURNROSE, BARRY EDMUND, b New Britain, Conn, June 3, 47; m 70; c 2. ASTRONOMY. Educ: Wesleyan Univ, BA, 69; Calif Inst Technol, PhD(astron), 76. Prof Exp: RESIDENT RES ASSOC ASTRON, NASA-JOHNSON SPACE CTR, HOUSTON, 75- Mem: Am Astron Soc. Res: Extragalactic astronomy; spectrophotometry of galaxies; stellar content of galaxies; surface photometry of extragalactic objects. Mailing Add: Code TC NASA-Johnson Space Ctr Houston TX 77058

TUROFF, MURRAY, b San Francisco, Calif, Feb 13, 36; m 61; c 1. COMPUTER SCIENCE, OPERATIONS RESEARCH. Educ: Univ Calif, Berkeley, BA, 58; Brandeis Univ, PhD(physics), 65. Prof Exp: Syst engr, IBM Corp, 61-64; mem prof staff syst anal, Inst Defense Anal, 64-68; opers res & info systs, Off Emergency Preparedness, 68-73; PROF COMPUT & INFO SCI, NJ INST TECHNOL, 73- Concurrent Pos: Lectr, Am Univ, 70-73. Mem: AAAS; Inst Mgt Sci; Opers Res Soc Am; Asn Comput Mach; Am Soc Cybernet. Res: Delphi design; computerized conferencing systems; information systems design; technology assessment and forecasting; gaming, simulation and modeling; policy analyses. Mailing Add: NJ Inst of Technol 323 High St Newark NJ 07102

TUROFF, ROBERT DAVID, b New York, NY, May 29, 35; m 58; c 2. SOLID STATE PHYSICS. Educ: Queens Col, NY, BS, 57; Rutgers Univ, MS, 60, PhD(physics), 64. Prof Exp: Asst prof, San Francisco State Col, 64-65; ASST PROF PHYSICS, ANTIOCH COL, 65- Mem: Am Phys Soc; Am Asn Physics Teachers. Res: Magnetic resonance; low temperature magnetism. Mailing Add: Dept of Physics Antioch Col Yellow Springs OH 45387

TURPENING, ROGER MUNSON, b Detroit, Mich, Oct 8, 39; m 66. SEISMOLOGY. Educ: Univ Mich, Ann Arbor, BS, 61, MS, 63, PhD(geophys), 66; Mass Inst Technol, 67. Prof Exp: Asst analyst seismol, Geotech Corp, 62-63; res asst, Geophys Lab, Univ Mich, Ann Arbor, 63-66, res assoc, 66; res assoc, Mass Inst Technol, 66-67; assoc res geophysicist, 67-70, RES GEOPHYSICIST, GEOPHYS LAB, UNIV MICH, ANN ARBOR, 70- Mem: Seismol Soc Am; Am Geophys Union; Soc Explor Geophys; Europ Asn Explor Geophys; Soc Automotive Eng. Res: Crustal studies by means of long refraction profiles; seismic signal processing by means of three-component analysis; seismic source function studies by means of seismic wave analysis. Mailing Add: Geophys Lab PO Box 618 Environ Res Inst of Mich Ann Arbor MI 48107

TURPIN, FRANK THOMAS, b Troy, Kans, June 4, 43; m 70; c 1. ECONOMIC ENTOMOLOGY. Educ: Washburn Univ, BS, 65; Iowa State Univ, PhD(entom), 71. Prof Exp: ASST PROF ENTOM, PURDUE UNIV, 71- Concurrent Pos: Consult, US Environ Protection Agency, 73- Mem: Entom Soc Am. Res: Biology; ecology; population dynamcis and control of insects attacking corn. Mailing Add: Entom Hall Purdue Univ West Lafayette IN 47907

TURRELL, BRIAN GEORGE, b Shoreham-by-Sea, Eng, May 6, 38; m 62; c 3. PHYSICS. Educ: Oxford Univ, BA, 59, MA & DPhil(nuclear orientation), 63. Prof Exp: Asst lectr physics, Univ Sussex, 63-64; asst prof, 64-70, ASSOC PROF PHYSICS, UNIV BC, 70- Res: Nuclear orientation; nuclear magnetic resonance; use of these techniques to study hyperfine interactions in magnetic materials. Mailing Add: Dept of Physics Univ of BC Vancouver BC Can

TURRELL, EUGENE SNOW, b Hyattsville, Md, Feb 27, 19; m 42; c 2. PSYCHIATRY. Educ: Ind Univ, BS, 39, MD, 47; Am Bd Psychiat & Neurol, dipl, 53. Prof Exp: Asst physiol, Ind Univ, 39-42; res assoc, Fatigue Lab, Harvard Univ, 42-43; res assoc physiol, Ind Univ, 43-44, asst biochem, Sch Med, 45-47; med house officer, Peter Bent Brigham Hosp, Boston, 47-48; resident psychiat, Kankakee State Hosp, Ill, 48-49; clin asst, Sch Med, Univ Calif, 49-52; asst prof, Sch Med, Ind Univ, 52-53; assoc prof, Sch Med, Univ Colo, 53-58, asst dean sch, 57; prof & chmn dept, Sch Med, Marquette Univ, 58-63, clin prof, 63-69; sr psychiatrist, 69-72, dir, Ctr Spec Probs, Community Ment Health Serv, City & County of San Francisco, 72-75; STAFF PSYCHIATRIST, MIDTOWN COMMUNITY MENT HEALTH CTR, WISHARD MEM HOSP, 75-; ASSOC PROF, DEPT PSYCHIAT, IND UNIV SCH MED, 75- Concurrent Pos: Resident, Langley Porter Clin, 49-50; chief psychiat consult serv, Robert W Long Hosp, Indianapolis, Ind, 52-53; med dir, Colo Psychopath Hosp, 53-54; assoc attend, Denver Gen Hosp, 53-57, dir psychiat serv, 57-58; attend, Vet Admin Hosp, Denver, 53-58; mem staff psychosom div, Colo Gen Hosp, 54-57; dir psychiat serv, Milwaukee Sanitarium Found, 58-65; consult, Hosp Ment Dis, Milwaukee, 58-69, Vet Admin Hosp, Wood, Wis, 58-69, Columbia Hosp, 59-69 & Milwaukee Children's Hosp, 60-69. Mem: AAAS; Am Psychiat Asn; AMA. Res: Psychosomatic medicine; psychotherapy. Mailing Add: 1001 W Tenth St Indianapolis IN 46202

TURRELL, FRANKLIN MARION, b Pittsburgh, Pa, Mar 21, 05; m 29; c 1. PLANT PHYSIOLOGY. Educ: Eastern Ill Univ, BE, 29; Univ Iowa, MS, 32, PhD(exp morphol, physiol, anal chem), 35. Prof Exp: Instr plant physiol, Univ Cincinnati, 35-36; from jr plant physiologist to assoc plant physiologist, 36-55, radiol adv, 49-54, dir AEC Proj, 50-55, plant physiologist, 55-72, prof biochem, 61-72, EMER PLANT PHYSIOLOGIST, UNIV CALIF, RIVERSIDE, 72- Concurrent Pos: Grants-in-aid,

US Nat Youth Admin & Fed Security Agency, Univ Calif, Riverside, 38-42, Tex Gulf Sulphur Co, 39-46, US Govt Atomic Comn, 50-55, Pest Control Ltd, 52-53, Leffingwell Chem Co, 57-59, Nat Frost Protection, Inc, 57-67, Baird Atomic, Inc, 63-66 & Mobil Oil Co, 67-70; chmn radiol safety comt, AEC, 65-72; consult remote sensing of environ proj, NASA-US Geol Surv, 66- Mem: Fel AAAS; Am Soc Plant Physiol; Bot Soc Am; Am Chem Soc; Am Soc Hort Sci. Res: Measurement and physiology of plant surfaces; extent and growth of leaf internal surface and intercellular space; heat exchange; thermal, sulfur and freeze injury; physiological diseases of subtropical trees; radioactive tracers; logarithmic parameters. Mailing Add: Dept of Biochem Univ of Calif Riverside CA 92502

TURRELL, GEORGE CHARLES, b Portland, Ore, June 19, 31; m 54; c 4. PHYSICAL CHEMISTRY. Educ: Lewis & Clark Col, BA, 50; Ore State Univ, MS, 52, PhD(phys chem), 54. Prof Exp: Asst, Ore State Univ, 50-52; mem tech staff, Electron Device Dept, Bell Tel Labs, Inc, 54-56; res assoc, Metcalf Res Lab, Brown Univ, 56-57, instr chem, 57-58; Guggenheim fel, Bellevue Labs, Nat Ctr Sci Res, France, 58-59; asst prof chem, Howard Univ, 59-62, assoc prof, 62-67; exchange prof & Fulbright fel, Infrared Spectros Lab, Univ Bordeaux, 66-67, vis prof, 67-70; prof, Nat Univ Zaire, Kisangani, 70-71 & Kinshasa, 71-72; actg ed, Can Jour Spectroscopy, 72; vis prof chem, Univ Montreal, 72-74 & McGill Univ, 73-75; RES PROF CHEM, UNIV LAVAL, QUEBEC, 75- Mem: Am Phys Soc; Coblentz Soc. Res: Infrared and Raman spectra; inter-molecular forces. Mailing Add: Dept Chem Univ of Laval Quebec PQ Can

TURRELL, SYLVIA JONES, b Washington, DC, Oct 10, 43; m 70; c 2. PHYSICAL CHEMISTRY. Educ: Howard Univ, BS, 66; Univ Paris, DES, 67; Univ Bordeaux, Doctorat(struct chem), 70. Prof Exp: Asst prof chem, Univ Zaire, 70-72; res assoc phys chem, Univ Que, Montreal, 72-73 & 74-75; prof chem, Col Snowden, 73-74; RES ASSOC PHYS CHEM, UNIV LAVAL, 75- Mem: Chem Inst Can; Spectros Soc Can. Res: Vibrational studies of biologically-active molecules and their metal complexes; structure of aqueous and nonaqueous solutions. Mailing Add: Dept of Chem Fac of Sci Univ Laval Quebec PQ Can

TURRENT, ANTONIO, soil fertility, see 12th edition

TURRITTIN, HUGH LONSDALE, b Rice, Minn, Apr 24, 06; m 32; c 3. MATHEMATICS. Educ: Univ Minn, BS, 27; Univ Wis, MS, 32, PhD(math), 33. Prof Exp: Instr mech, Univ Wis, 28-29 & math, 31-33; asst prof math & eng, Col Mines & Metall, Univ Tex, 33-39 & math & mech, Univ Minn, 39-44; opers analyst, US War Dept, 44-45; from asst prof to assoc prof math & mech, 45-50, prof, 50-72, EMER PROF MATH, UNIV MINN, MINNEAPOLIS, 72- Concurrent Pos: Distinguished Vis Prof, Rockford Col, 74. Mem: Am Math Soc; Math Asn Am. Res: Solution of ordinary differential equations. Mailing Add: Dept of Math Univ of Minn Minneapolis MN 55455

TURRO, NICHOLAS JOHN, b Middletown, Conn, May 18, 38; m 60; c 2. ORGANIC CHEMISTRY. Educ: WVa Wesleyan Col, BA, 60; Calif Inst Technol, PhD(chem), 63. Prof Exp: Instr, 64-65, from asst prof to assoc prof, 65-69, PROF ORG CHEM, COLUMBIA UNIV, 69- Concurrent Pos: NSF fel, Harvard Univ, 63-64; consult, E I du Pont de Nemours & Co, 64-; vis prof, Pa State Univ, 66; Sloan Found fel, 66-68; mem ed bd, J Org Chem, J Molecular Photochem, 75- Honors & Awards: Pure Chem Award, Am Chem Soc, 74. Mem: AAAS; Am Chem Soc; The Chem Soc; NY Acad Sci; Am Soc Photobiol. Res: Photochemistry; electronic energy transfer in fluid solution; mass spectrometry; cycloaddition reactions; thermal rearrangements; cyclopropane chemistry; mechanism of the Favorskii rearrangement; application of laser techniques to organic photochemistry. Mailing Add: 125 Downey Dr Tenafly NJ 07670

TURSE, RICHARD S, b Jersey City, NJ, Mar 24, 35; m 64; c 3. ANALYTICAL CHEMISTRY, SPECTROCHEMISTRY. Educ: Rutgers Univ, BS, 56, MS, 58, PhD(anal chem), 60. Prof Exp: Instr anal chem, Rutgers Univ, 56-58; SR RES CHEMIST, COLGATE-PALMOLIVE CO, 60- Mem: Acad Pharmaceut Sci; Am Pharmaceut Asn. Res: Development of atomic absorption methods for determination of trace metals in consumer products for safety and environmental reasons. Mailing Add: Colgate-Palmolive Co 909 River Rd Piscataway NJ 08854

TURYN, RICHARD JOSEPH, mathematics, see 12th edition

TUSHAUS, LEONARD A, organic chemistry, polymer chemistry, see 12th edition

TUSING, THOMAS WILLIAM, b New Market, Va, Feb 2, 20; m 49; c 5. PHARMACOLOGY. Educ: George Washington Univ, BS, 42; Med Col Va, MD, 50. Prof Exp: Med dir, Hazleton Labs, Inc, 51-58, dir res & vpres, 58-69; MED DIR PHARMACEUT RES & DEVELOP DIV, MALLINCKRODT INC, 69- Mem: AMA; Indust Med Asn; Soc Toxicol. Mailing Add: 13561 Featherstone Dr St Louis MO 63131

TUSTANOFF, EUGENE RENO, b Windsor, Ont, Jan 30, 29; m 54; c 4. BIOCHEMISTRY. Educ: Assumption Col, BA, 52; Detroit Univ, MS, 54; Western Ont Univ, PhD(biochem), 59. Prof Exp: Res assoc biochem, Western Ont Univ, 55-59; fel, Western Reserve Univ, 59-61; Life Ins Med Res fel, Oxford Univ, 61-62; asst scientist, Hosp for Sick Children, Univ Toronto, 62-64; assoc biochem, Univ Toronto, 63-65, asst prof pharmacol, 64-65; assoc prof biochem, McMaster Univ, 65-67; assoc prof path chem, 67-72, assoc prof clin biochem, 72-75, PROF CLIN BIOCHEM, UNIV WESTERN ONT, 75- Mem: Can Biochem Soc; Brit Biochem Soc; Am Soc Biol Chem; Am Asn Clin Chem. Res: Biogenesis and control of mitochondria; biogenesis of membranes; biochemistry of tumour model systems. Mailing Add: Dept of Biochem Victoria Hosp Univ of Western Ont London ON Can

TUTAS, DANIEL JOSEPH, b Niagara, Wis, Nov 1, 35. CHEMISTRY, BIOCHEMISTRY. Educ: Univ Wis-Madison, BS, 60; Georgetown Univ, MS, 64; State Univ NY Buffalo, PhD(biochem), 72. Prof Exp: Chemist, Naval Propellant Plant, 60-61 & Nat Bur Standards, 61-64; instr chem, Holyoke Community Col, 65-67; res assoc biochem, Univ Minn, Minneapolis, 72-74; BIOCHEMIST, DIV OF LABS AND RES, NY STATE DEPT HEALTH, 74- Mem: AAAS; Am Chem Soc. Res: Enzymology. Mailing Add: 120 New Scotland Ave Albany NY 12201

TUTHILL, HARLAN LLOYD, b Fillmore, NY, Nov 24, 17; m 41; c 2. PHARMACEUTICS, Rohm & Haas Co, 43-46; head phys chem sect, Smith Kline & French Labs, 46-48, tech dir, 48-54, sci dir int div, 54-57, asst dir res & develop labs, 57-62, vpres, Smith Kline Instrument Co, 62-65; dir prod develop & assoc dir, Squibb Inst Med Res, 65-70; DIR HEALTH & MED RES, INT PAPER CO, 70- Mem: AAAS; Am Chem Soc; Acad Pharmaceut Sci. Res: Properties and applications of polymers; analysis of drugs and dosage forms; design, development and biological and clinical evaluation of drugs and devices; pharmaceutical research and engineering; medical instrumentation. Mailing Add: Int Paper Co Box 797 Tuxedo Park NY 10987

TUTHILL, RICHARD LOVEJOY, b Suffern, NY, Apr 29, 11; m 34; c 2. GEOGRAPHY OF AFRICA, ECONOMIC GEOGRAPHY. Educ: Bates Col, AB, 33; Columbia Univ, AM, 38, EdD, 46. Prof Exp: Teacher social studies, Park Ridge High Sch, NJ, 34-40; instr geog, Teachers Col, Columbia Univ, 40-41; chief res analyst, Gen Staff Corps, US Army, 42-44; asst dir res, Pan Am World Airways, 45-46; prof geog, Univ Ky, 46-53; PROF GEOG, DUKE UNIV, 54- Concurrent Pos: Specialist, In-serv Teacher Training, NC Dept Pub Instr, 60-; vis prof, Ahmadu Bello Univ, Nigeria, 69-70. Mem: Fel Am Geog Soc; Asn Am Geog; Nigerian Geog Asn; African Studies Asn. Res: Commodity processing and production, transportation and communication and cultural pluralism in West Africa, principally in Nigeria; Canadian area studies; resource inventories and initial processing. Mailing Add: Dept of Geog Duke Univ PO Box 6537 Col Sta Durham NC 27708

TUTHILL, SAMUEL JAMES, b San Diego, Calif, Sept 6, 25; m 52; c 3. GEOLOGY, PALEOECOLOGY. Educ: Drew Univ, AB, 51; Syracuse Univ, MS, 60; Univ NDak, MA, 63, PhD(geol), 69. Prof Exp: Geologist, NDak Geol Surv, 63-64; asst prof geol, Muskingum Col, 64-68; asst state geologist, 69-, DIR & STATE GEOLOGIST, IOWA GEOL SURV, 69- Concurrent Pos: NSF grants, Muskingum Col expeds Alaska, 65-66, 67-69; adj prof, Univ Iowa, 69-; adminr, Iowa Oil & Gas Admin, 69-; mem, Iowa Natural Resources Coun, 70-; mem, Iowa Land Rehab Coun, 70-; secy, Iowa State Map Adv Coun, 72- Mem: AAAS; fel Geol Soc Am; Asn Am State Geol; Am Water Works Asn. Res: Research management; water; minerals; environmental protection; landuse planning; remote sensing; resources managemental development. Mailing Add: Iowa Geol Surv 16 W Jefferson Iowa City IA 52240

TUTHILL, SAMUEL MILLER, b Rocky Point, NY, Jan 7, 19; m 41; c 3. ANALYTICAL CHEMISTRY. Educ: Wesleyan Univ, BA, 39, MA, 41; Ohio State Univ, PhD(chem), 48. Prof Exp: Lab asst chem, Wesleyan Univ, 39-41; anal chemist, Mallinckrodt Chem Works, 41-45; lab asst chem, Ohio State Univ, 45-46, from asst instr to instr, 47-48; anal chemist, 48-56, dir qual control, 56-70, corp dir qual control, 70-72, CORP DIR QUAL ASSURANCE, MALLINCKRODT, INC, 72- Concurrent Pos: Mem comt revision, US Pharmacopeia. Mem: Am Chem Soc; Am Soc Qual Control; Pharmaceut Mfrs Asn; Acad Pharmaceut Sci; Food Chem Codex. Res: Methods of analysis of pharmaceutical and reagents chemicals; separation by electrodeposition; instrumental methods of analysis; separation of rhodium from iridium by electrolysis with control of the cathode potential; determination of rare earths in steels; analysis of opium and narcotics; good manufacturing practice in manufacture of drugs. Mailing Add: Mallinckrodt Inc Mallinckrodt & Second St St Louis MO 63147

TUTIHASI, SIMPEI, b Tokyo, Japan, Mar 2, 22; US citizen; m 47; c 2. SOLID STATE PHYSICS. Educ: Kyoto Univ, BSc, 46, DSc(physics), 56. Prof Exp: Res assoc solid state physics, Inst Optics, Univ Rochester, 56-59; sr engr, Sylvania Elec Prods, Inc, 59-64; SCIENTIST, RES LABS, XEROX CORP, 64- Mem: Am Phys Soc; Optical Soc Am. Res: Solid state spectroscopy; spectroscopy of ordered crystals, photoconductivity, luminescence Mailing Add: Res Lab Xerox Corp Rochester NY 14644

TUTTE, WILLIAM THOMAS, b Newmarket, Eng, May 14, 17; m 49. MATHEMATICS. Educ: Cambridge Univ, PhD(math), 48. Prof Exp: Lectr, Univ Toronto, 48-52, from asst prof to assoc prof, 52-62; PROF MATH, UNIV WATERLOO, 62- Honors & Awards: Henry Marshall Tory Medal, Royal Soc Can, 75. Mem: Am Math Soc; Math Asn Am; fel Royal Soc Can; Can Math Cong; London Math Soc. Res: Graph theory; matroid theory. Mailing Add: Dept of Math Univ of Waterloo Waterloo ON Can

TUTTLE, DONALD MONROE, b Bay City, Mich, Feb 1, 17; m 47; c 3. ENTOMOLOGY. Educ: Mich State Col, BS, 40, MS, 47; Univ Ill, PhD(entom), 52. Prof Exp: Lab asst entom, Mich State Col, 40; instr, Univ Maine, 47-49; asst, Univ Ill, 49-52; RES ENTOMOLOGIST, UNIV ARIZ, 52- Mem: Entom Soc Am; Acarological Soc Am. Res: Citrus, alfalfa, melon and turf insects; systematics and biology of Tetranychoidea. Mailing Add: Univ of Ariz Farm Route 1 Box 587 Yuma AZ 85364

TUTTLE, ELBERT P, JR, b Ithaca, NY, Sept 1, 21; m 52; c 5. PHYSIOLOGY, INTERNAL MEDICINE. Educ: Princeton Univ, AB, 42; Harvard Univ, MD, 51. Prof Exp: Asst med, Harvard Med Sch & Mass Gen Hosp, 54-56; from asst prof to assoc prof, 57-66, PROF MED, SCH MED, EMORY UNIV, 66- Concurrent Pos: Nat Heart Inst res fel, 53-56; Am Heart Asn res fel, 57; chair cardiovasc res, Ga Heart Asn, 58-72. Mem: Am Fedn Clin Res. Res: Inorganic metabolism; renal and circulatory physiology; hypertension; nephrology. Mailing Add: Dept of Med Emory Univ Sch of Med Atlanta GA 30303

TUTTLE, ELIZABETH R, b Boston, Mass, Dec 5, 38. THEORETICAL PHYSICS. Educ: Univ NH, BS, 60; Univ Colo, MS, 61, PhD(physics), 64. Prof Exp: Asst prof, 64-68, ASSOC PROF PHYSICS, UNIV DENVER, 68- Mem: Am Phys Soc; Am Asn Physics Teachers; Nat Soc Prof Engrs. Res: Theory of quantum fluids at very low temperatures with applications to liquid He4, liquid He3 and He3-He4 mixtures. History of science and technology. Mailing Add: Dept of Physics Univ of Denver Denver CO 80210

TUTTLE, LAWRENCE WESSELL, biochemistry, radiation biology, see 12th edition

TUTTLE, MERLIN DEVERE, b Honolulu, Hawaii, Aug 26, 41. POPULATION ECOLOGY, MAMMALOGY. Educ: Andrews Univ, BA, 65; Univ Kans, MA, 69, PhD(pop ecol), 74. Prof Exp: Co-dir, Smithsonian Venezuelan Res Proj, Smithsonian Inst, 65-67; res assoc pop ecol, Univ Minn, 72; CUR MAMMALS, MILWAUKEE PUB MUS, 75- Concurrent Pos: Consult endangered bats, Tenn Valley Authority, 76- Mem: Am Soc Mammalogists; Am Soc Naturalists; Ecol Soc Am; Soc Study Evolution; Nat Speleol Soc. Res: Niche breadth, intra- and inter-specific competition, reproductive and survival strategies; foraging behavior in refuging species and the energetics of thermoregulation, hibernation and migration. Mailing Add: Milwaukee Pub Mus Vert Div 800 W Wells St Milwaukee WI 53233

TUTTLE, O FRANK, b Olean, NY, June 25, 16; m 41; c 2. GEOLOGY. Educ: Pa State Col, BS, 39, MS, 40; Mass Inst Technol, PhD(petrol), 48. Prof Exp: Asst, Off Sci Res & Develop, Mass Inst Technol, 42, phys chemist, Geophys Lab, 42-45; phys chemist, US Naval Res Lab, Washington, DC, 45-47, petrologist, Geophys Lab, 47-53; prof geochem & chmn div earth sci, Pa State Univ, 53-65, dean col mineral industs, 59-60; prof geochem, Stanford Univ, 65-70. Honors & Awards: Mineral Soc Award, Mineral Soc, 49; Day Medal, Geol Soc Am, 67; Roebling Medal, Mineral Soc, 75. Mem: Nat Acad Sci; fel Geol Soc Am; fel Mineral Soc Am; Brit Mineral Soc. Res: Application of phase equilibria in silicate systems to petrology; mineralogy of rock forming minerals; structural petrology; high temperature and high pressure syntheses of minerals and rocks; growth of large single crystals. Mailing Add: 4850 W Lazy C Dr Tucson AZ 85705

TUTTLE, RICHARD SUNESON, b Pottsville, Pa, Aug 18, 30; m 60; c 1. PHYSIOLOGY, PHARMACOLOGY. Educ: State Univ NY, PhD(pharm), 60. Prof

Exp: Res fel, 60-64, RES ASSOC NEUROPHYSIOL, MASONIC MED RES FOUND, 64- Concurrent Pos: USPHS fel, 60-61; NIH grants, 63- Mem: Am Physiol Soc; Am Soc Pharmacol Exp Therapeut. Res: Pharmacology of cardiac glycosides and electrolytes; neurophysiology of vasomotor regulation; centrally evoked histamine release; role of histamine in control of cardiovascular tone; cardiovascular effects of imidazoles and mesenteric Pacinian baroreceptors. Mailing Add: Masonic Med Res Found Bleeker St Utica NY 15301

TUTTLE, ROBERT LEWIS, b Boston, Mass, July 26, 22; m 42; c 2. MICROBIOLOGY. Educ: Univ NH, BS, 43; Univ Rochester, MD, 47. Prof Exp: Asst trop med, Bowman Gray Sch Med, 48-50, from instr to assoc prof microbiol, 50-70, chmn dept, 55-62, assoc dean, 62-69, acad dean, 69-70; PROF MICROBIOL & ASSOC DEAN ACAD AFFAIRS, UNIV TEX MED SCH HOUSTON, 70- Mailing Add: Univ of Tex Med Sch 6400 W Cullen Houston TX 77025

TUTTLE, RONALD RALPH, b Colorado Springs, Colo, July 10, 36; m 63; c 1. PHARMACOLOGY, PHYSIOLOGY. Educ: Colo Col, BA, 60; Univ Man, MS, 64, PhD(pharmacol), 66. Prof Exp: Fel pharmacol, Emory Univ, 66-67; sr pharmacologist, 67-71, res scientist, 71-74, RES ASSOC PHARMACOL, LILLY RES LABS, 74- Mem: Am Soc Pharmacol & Exp Therapeut. Res: Cardiovascular pharmacology. Mailing Add: Lilly Res Labs Indianapolis IN 46206

TUTTLE, RUSSELL H, b Marion, Ohio, Aug 18, 39; m 68; c 1. ANTHROPOLOGY, PRIMATOLOGY. Educ: Ohio State Univ, BSc, MA, 62; Univ Calif, Berkeley, PhD(anthrop), 65. Prof Exp: Instr anat, 64-65, from inst to asst prof anat & anthrop, 65-69, asst prof anthrop, 69-70, ASSOC PROF ANTHROP & EVOLUTIONARY BIOL, UNIV CHICAGO, 70- Concurrent Pos: Wenner-Gren Found grants, 65, 66, 69 & 70; NSF res grants, 65-74. Honors & Awards: NIH career develop award, 68-73. Mem: AAAS; Am Anthrop Asn; Am Asn Phys Anthrop; Am Soc Nat; Am Soc Mammal. Res: Functional comparative morphology, behavior, ecology and evolution of primates, especially those features that may provide insights into the evolutionary history of man and the human condition. Mailing Add: Dept of Anthrop Univ of Chicago 1126 E 59th St Chicago IL 60637

TUTTLE, SHERWOOD DODGE, b Medford, Mass, June 8, 18; m 41; c 3. GEOMORPHOLOGY. Educ: Univ NH, BS, 39; Wash State Univ, MS, 41; Harvard Univ, MA & PhD(geol), 53. Prof Exp: Instr geol, Wash State Univ, 41, 46-48; from asst prof to assoc prof, 52-62, chmn dept geol, 63-68, PROF GEOL, UNIV IOWA, 62-, ASSOC DEAN COL LIB ARTS, 70- Concurrent Pos: Res assoc, Woods Hole Oceanog Inst, 59-66; Fulbright lectr, Chinese Univ Hong Kong, 68-69. Mem: Col Fel Geol Soc Am; Nat Asn Geol Teachers. Res: Glacial geology; shorelines. Mailing Add: Dept of Geol Univ of Iowa Iowa City IA 52240

TUTTLE, THOMAS R, JR, b Somerville, Mass, Mar 28, 28; m 54; c 3. PHYSICAL CHEMISTRY. Educ: Northeastern Univ, BS, 53, MS, 55; Wash Univ, St Louis, PhD, 57. Prof Exp: Actg asst prof chem, Stanford Univ, 57-60; asst prof, 60-64, ASSOC PROF CHEM, BRANDEIS UNIV, 64- Mem: Am Chem Soc; Am Phys Soc. Res: Electron distribution in ion radicals by electron spin-resonance; molecular motions in solutions; properties of metal solutions in ammonia and other solvents. Mailing Add: Dept of Chem Brandeis Univ Waltham MA 02154

TUTTLE, WARREN WILSON, b Fulton, Mo, Aug 2, 30; m 52; c 4. NEUROPHARMACOLOGY. Educ: Univ Mo, BA, 52; Univ Kans City, BS, 58, MS, 60; Univ Calif, San Francisco, PhD(pharmacol), 66. Prof Exp: Asst prof pharmacol, Univ Mo-Kans City, 65-68; asst prof med pharmacol & therapeut, Sch Med, Univ Calif, Irvine, 68-69; asst prof pharmacol, Univ Mo-Kans City, 69-72; actg chmn dept, 72-73, ASSOC PROF PHARMACOL, KANS CITY COL OSTEOP MED, 72-, CHMN DEPT, 73- Res: Drug metabolism; electroencephalographic investigation into the sites of action of various drugs in the central nervous system. Mailing Add: Dept of Pharmacol Kans City Col of Osteop Med Kansas City MO 64124

TUTUPALLI, LOHIT VENKATESWARA, b Guntur, Andhra Pradesh, India, Aug 10, 45; m 74. PHARMACOGNOSY, PHYTOCHEMISTRY. Educ: Andhra Univ, BS, 63; Bombay Univ, BS, 66, MS, 68; Univ of the Pac, PhD(pharmacog), 74. Prof Exp: Res pharmacist product develop, M/S Pfizer (India), Ltd, Bombay, 68-69; from teaching asst to instr pharmacog, Univ of the Pac, 69-74; RES CHEMIST, CALIF CEDAR PROD RES LAB, 74- Mem: Am Pharmaceut Asn; Am Chem Soc; Am Asn Col Pharm; AAAS; Forest Prod Res Soc. Res: Investigating the medicinal and economic uses of natural products. Mailing Add: Calif Cedar Prod Res Lab PO Box 8449 Stockton CA 95205

TUTWILER, FRANK BRYAN, b Richmond, Va, Oct 6, 24; m 48; c 4. PHYSICAL CHEMISTRY, ORGANIC CHEMISTRY. Educ: George Washington Univ, BS, 44, MS, 46; Univ NC, Chapel Hill, PhD(chem), 51. Prof Exp: Asst chem, Univ NC, 46-49; prof, King Col, 50-53, head dept, 52-53; instr, Williams Col, Mass, 53-55; chmn dept chem & physics, 59-72, PROF CHEM, WINTHROP COL, 55- Mem: AAAS; Am Chem Soc. Res: Synthesis of organic medicinals; preparation and properties of halogenated ethers; organic reaction kinetics; theoretical organic chemistry. Mailing Add: Dept of Chem & Physics Winthrop Col Rock Hill SC 29730

TUTWILER, GENE FLOYD, b Peoria, Ill, Sept 19, 45; m 68; c 1. BIOCHEMISTRY, ENDOCRINOLOGY. Educ: Western Ill Univ, BS, 67; Univ Mich, Ann Arbor, PhD(biochem), 70. Prof Exp: Lab asst quant anal, Western Ill Univ, 66-67; sr scientist, 70-74, GROUP LEADER, McNEIL LABS INC, JOHNSON & JOHNSON, 74- Concurrent Pos: Teaching fel biochem, Univ Mich, Ann Arbor, 67-70; vis asst prof, Temple Univ, 73-74 & Bucks County Community Col, 74- Mem: AAAS; Am Diabetes Asn; Soc Exp Biol & Med. Res: Diabetes; obesity; protein purification; isolation pituitary proteins; free fatty acid metabolism; carbohydrate metabolism; atherosclerosis. Mailing Add: Biochem Dept McNeil Labs Inc Ft Washington PA 19034

TUUL, JOHANNES, b Tarvastu, Estonia, May 23, 22; US citizen; m 57; c 2. PHYSICS. Educ: Stockholm Univ, BS, 55, MA, 56; Brown Univ, ScM, 57, PhD(physics), 60. Prof Exp: Instr elec eng, Stockholm Tech Inst, 47-49; res engr, Elec Prospecting Co, Sweden, 49-53; elec engr, L M Ericsson Tel Co, 54-55; res asst physics, Brown Univ, 55-57, 58-60, res assoc, 60; res physicist, Stamford Res Labs, Am Cyanamid Co, Conn, 60-62; sr res physicist, Bell & Howell Res Ctr, Calif, 62-65; asst prof, 65-68, assoc prof, 68-74, chmn dept physics & earth sci, 71-75, PROF PHYSICS, CALIF STATE POLYTECH UNIV, 74- Concurrent Pos: Consult, Bell & Howell Res Ctr, 65 & Teledyne Inc, 68; vis assoc prof, Pahlavi Univ, Iran, 68-70. Mem: AAAS; Am Phys Soc; Am Chem Soc; Am Vacuum Soc; Am Asn Physics Teachers. Res: Adsorption of gases on solids; low-energy electron diffraction studies of effects of adsorption and ion bombardment on initially clean surfaces; ultra-high vacuum technology; mass spectrometry. Mailing Add: Dept of Physics Calif State Polytech Univ Pomona CA 91768

TUVE, MERLE ANTONY, b Canton, SDak, June 27, 01; m 27; c 2. PHYSICS. Educ: Univ Minn, BS, 22, AM, 23; Johns Hopkins Univ, PhD(physics), 26. Hon Degrees: DSc, Case Western Reserve Univ, 48, Kenyon Col, 49, Williams Col, 49, Johns Hopkins Univ, 50, Augustana Col, 52, Alaska, 53; LLD, Carleton Col, 61. Prof Exp: Instr physics, Princeton Univ, 23-24 & Johns Hopkins Univ, 24-26; assoc physicist, Carnegie Inst Dept Terrestrial Magnetism, 26-28, physicist, 28-38, chief physicist, 38-46, dir, 46-66, DISTINGUISHED SERV MEM, CARNEGIE INST WASHINGTON, 66- Concurrent Pos: Dir appl physics lab, Johns Hopkins Univ, 42-46; ed, J Geophys Res, 49-53; mem nat comt, Int Geophys Year, 54-59; chmn geophys res bd, Nat Acad Sci-Nat Res Coun, 60-70. Honors & Awards: Presidential Medal for Merit, 46; Award, Res Corp, 47; Scott Award, 48; Comstock Prize, Nat Acad Sci, 49; Potts Medal, Franklin Inst, 50; Barnard Medal, Columbia, 55; Bowie Medal, Am Geophys Union, 63; Award, Cosmos Club, 66. Mem: Nat Acad Sci (home secy, 66-71); AAAS (vpres, 47); fel Am Phys Soc; Am Philos Soc; Am Acad Arts & Sci. Res: Nuclear physics; geophysics; echo investigations of Kennelly-Heaviside layer; high voltage vacuum tubes; high-speed protons; transmutations of atomic nuclei; nuclear forces; deep seismic shots; earth's crust; radio astronomy. Mailing Add: Carnegie Inst 5241 Broad Branch Rd NW Washington DC 20015

TUVE, RICHARD LARSEN, b Canton, SDak, Feb 1, 12; m 36; c 2. PHYSICAL CHEMISTRY, INORGANIC CHEMISTRY. Educ: Am Univ, BA, 35. Hon Degrees: DSc, Carleton Col, 61. Prof Exp: Asst chemist, Res Assocs, Inc, 35-38; asst chemist, US Naval Res Lab, 38-40, assoc phys chemist, 40-41, head, Eng Res Br, 41-70; CONSULT, APPL PHYSICS LAB, JOHNS HOPKINS UNIV, SILVER SPRING, 70- Concurrent Pos: Mem comt fire res conf, Nat Acad Sci-Nat Res Coun, 55-68. Honors & Awards: Meritorious civilian serv award, US Dept Navy, 46; superior accomplishment award, US Dept Defense, 59. Mem: Fel AAAS; Am Chem Soc; Am Inst Chem Eng; fel Am Inst Chem; Inst Chem Eng. Res: Chemistry of fire extinguishment; foam extinguishment methods and materials; flame propagation; surface chemistry; special explosives; fire fighting equipment design; combustion inhibition. Mailing Add: 9211 Crosby Rd Silver Spring MD 20910

TUVESON, ROBERT WILLIAMS, b Chicago, Ill, Aug 30, 31. GENETICS, BOTANY. Educ: Univ Ill, BS, 54, MS, 56; Univ Chicago, PhD(bot), 59. Prof Exp: Asst prof biol, Wayne State Univ, 59-61; asst prof bot, Univ Chicago, 61-68; ASSOC PROF BOT, UNIV ILL, URBANA-CHAMPAIGN, 68- Concurrent Pos: NIH spec fel, Dartmouth Col, 65-66. Mem: AAAS; Mycol Soc Am; Bot Soc Am; Genetics Soc Am. Res: Fungal genetics; viruses of fungi; radiation sensitivity. Mailing Add: Dept of Bot Univ of Ill Urbana IL 61801

TUZAR, JAROSLAV, b Czech, Mar 25, 15; nat US; m 48; c 1. MATHEMATICS. Educ: Charles Univ, Prague, MA, 39 & 45, ScD(math), 48. Prof Exp: Asst prof, State Tech Col, Prague, 45-48; Rockefeller Found fel, Univ Chicago, 48-50; dir control & res lab, Salerno-Megowen Biscuit Co, 50-70; ASSOC PROF MATH, NORTHEASTERN ILL UNIV, 70- Concurrent Pos: Lectr, Northwestern Univ, 60-70. Mem: AAAS; Math Asn Am; Am Statist Asn. Res: Mathematical probability and statistics; pedagogy of mathematics. Mailing Add: Dept of Math Northeastern Ill Univ Chicago IL 60625

TUZZOLINO, ANTHONY J, b Chicago, Ill, July 1, 31; m 54; c 2. PHYSICS, SOLID STATE PHYSICS. Educ: Univ Chicago, MS, 55, PhD(physics), 58. Prof Exp: PHYSICIST, UNIV CHICAGO, 58- Res: Semiconductor nuclear particle and photon detectors. Mailing Add: 6615 N Knox Lincolnwood IL 60646

TVARUSKO, ALADAR, electrochemistry, physical chemistry, see 12th edition

TVEEKREM, JAMES OLAF, analytical chemistry, physical chemistry, see 12th edition

TVETEN, JOHN LOWELL, organic chemistry, see 12th edition

TWARDOCK, ARTHUR ROBERT, b Normal, Ill, July 20, 31; m 54; c 4. VETERINARY PHYSIOLOGY. Educ: Univ Ill, BS, 54, DVM, 56; Cornell Univ, PhD(animal physiol), 61. Prof Exp: Vet practioner, Hillcrest Animal Hosp, 56-57; res asst radiobiol, Cornell Univ, 57-60, res assoc phys biol, 60-62; from asst prof to assoc prof, 62-70, PROF, VET PHYS, UNIV ILL, URBANA, 70-, ASSOC DEAN ACAD AFFAIRS, 73-, ACTG HEAD, DEPT OF VET PHYSIOL & PHARM, 74- Mem: AAAS; Am Physiol Soc; Am Soc Vet Physiol & Pharmacol; Am Vet Med Asn; Conf Res Workers Animal Diseases. Res: Mineral metabolism; placental transfer of mineral elements; mechanisms of ion transport; applications of radioisotope techniques in veterinary medicine. Mailing Add: Col of Vet Med Univ of Ill Urbana IL 61801

TWAROG, BETTY MACK, b New York, NY, Aug 28, 27; m 47; c 1. PHYSIOLOGY. Educ: Swarthmore Col, AB, 48; Tufts Col, MS, 49; Radcliffe Col, PhD(biol), 52. Prof Exp: Asst, Harvard Univ, 52 & Res Div, Cleveland Clin, 52-53; res assoc & instr, Tufts Col, 53-55; res fel, Harvard Univ, 55-58; instr, 58-60; USPHS trainee, Oxford Univ, 60-61; asst prof physiol & biophys, Sch Med, NY Univ, 61-65; res fel, Harvard Univ, 65-66; from asst prof to assoc prof, 66-70, prof biol, Tufts Univ, 70-75; PROF BIOL, STATE UNIV NY, STONY BROOK, 75- Concurrent Pos: John Simon Guggenheim Mem fel, 72-73. Mem: Soc Gen Physiologists; Am Physiol Soc. Res: Physiology and pharmacology of smooth muscle; neurophysiology; neuropharmacology. Mailing Add: Dept of Anat Sci Health Sci Ctr State Univ NY Stony Brook NY 11794

TWAROG, ROBERT, b Lowell, Mass, Mar 17, 35; m 58; c 2. MICROBIAL BIOCHEMISTRY. Educ: Univ Conn, BS, 56, MS, 58; Univ Ill, PhD(microbiol), 62. Prof Exp: Lab officer, US Air Force Epidemiol Lab, San Antonio, Tex, 62-65; asst prof, 65-74, ASSOC PROF BACT, SCH MED, UNIV NC, CHAPEL HILL, 74- Mem: AAAS; Am Soc Microbiol. Res: Control mechanisms of microbial processes; bacteriophage biochemistry. Mailing Add: Dept of Bact & Immunol Univ of NC Sch of Med Chapel Hill NC 27514

TWEDDELL, COLIN ELLIDGE, b Melbourne, Australia, Mar 19, 99; US citizen; m 33; c 3. CULTURAL ANTHROPOLOGY, ANTHROPOLOGICAL LINGUISTICS. Educ: Univ Wash, BA, 45; Univ Wash, MA, 48; Melbourne Bible Inst, Australia, dipl, 48; Univ Wash, PhD(ling), 58. Prof Exp: Instr Chinese, Univ Wash, 42-45; sr missionary, Overseas Missionary Fel, 24-63; assoc prof anthrop, Western Wash State Col, 65-66, LECTR ANTHROP & LING, WESTERN WASH STATE COL, 66- Concurrent Pos: Instr phonetics, Simpson Bible Col, 51-53; coord & linguistic in charge, Cuyonon Transl Proj, Philippine Bible House, Manila, Philippines, 58-63; consult in Cuyonon orthography, Div Supt, Palawan Prov Div Pub Schs, Philippines, 60-63; mem bd, Southwest Regional Sch Arts & Trades, Cuyo, Philippines, 62-63; acad adv & mem fac anthrop, Mountlake Col, 64-65, mem bd dirs, 64-67; consult-instr, Sem Honor Sr, Ferndale Sch Dist, Wash, 68-73; vis lectr, Vancouver Bible Col, 75-; vis lectr anthrop, Trinity Western Col, 75- Mem: AAAS; Ling Soc Am; fel Am Anthrop Asn; Am Ethnol Soc; Asn Social Anthrop Oceania. Res: Areal studies; ethno-history; peoples and institutions; religion of China; Southeast Asia; Oceania; Pacific Northwest Coast of America; Sinitic, Philippine and Salishan linguistics; descriptive linguistics; ethnic identity and intergroup relations. Mailing Add: Dept of Sociol-Anthrop Western Wash State Col Bellingham WA 98225

TWEDT, ROBERT MADSEN, b Rochester, Minn, July 4, 24; m 72. MICROBIOLOGY. Educ: Univ Minn, BS, 45; Univ Colo, MS, 49, PhD(microbiol), 52. Prof Exp: Res assoc microbiol, Western Reserve Univ, 53-54; asst scientist, Univ Minn, 54-56, res fel physiol chem, 56-59; from asst prof to assoc prof biol, Univ Detroit, 59-67; res microbiologist, Nat Ctr Urban & Indust Health, 67-69, ASST CHIEF, FOOD MICROBIOL BR, FOOD & DRUG ADMIN, USPHS, 69- Concurrent Pos: Am Cancer Soc fel, 52-53; adj prof, Univ Detroit, 68-69. Mem: Am Soc Microbiol; Am Acad Microbiol; AAAS; Int Asn Milk, Food & Environ Sanitarians; Sigma Xi. Res: Research and field investigations to identify, evaluate and resolve microbiological problems of public health significance associated with foods. Mailing Add: Food Microbiol Br Food & Drug Adm 1090 Tusculum Ave Cincinnati OH 45226

TWEED, JOHN, b Greenock, Scotland, Mar 29, 42; m 66; c 1. APPLIED MATHEMATICS. Educ: From asst lectr to lectr math, Univ Glasgow, 65-69; vis asst prof, NC State Univ, 69-70; lectr, Univ Glasgow, 70-73; vis prof, NC State Univ, 73-74; ASSOC PROF MATH, OLD DOMINION UNIV, 74- Mem: Soc Indust & Appl Math; assoc fel Brit Inst Math & Appln. Res: Applications of transform techniques and integral equations to the solution of mixed boundary value problems in fracture mechanics Mailing Add: Dept of Math & Comput Sci Old Dominion Univ Norfolk VA 23508

TWEED, PAUL BASSET, b Zvinigorodka, Russia, Sept 22, 13; US citizen; m 42; c 2. CHEMISTRY. Educ: Rensselaer Polytech Inst, BS, 34, MS, 35; NY State Col Teachers, MA, 37; Am Inst Chemists, cert, 69. Prof Exp: Chemist, Am Hard Rubber Co, 36-40 & Picatinny Arsenal, 40-62; engr, Avco Corp, 62-66; ENGR, MARTIN MARIETTA CORP, 66- Concurrent Pos: Chmn explosive subcomt, Ord Handbooks, 55-62; mem, Mutual Weapons Develop Team, 58 & 60. Mem: Am Defense Preparedness Asn; Am Inst Chemists. Res: Explosive loading; colored smokes; electroexplosive devices; exploding bridgewire systems; shaped charges. Mailing Add: 4624 Tinsley Dr Orlando FL 32809

TWEEDDALE, MARTIN GEORGE, b Bristol, Eng, Aug 22, 40. CLINICAL PHARMACOLOGY. Educ: King's Col, Univ London, BSc, 62, PhD(pharmacol), 65; Westminister Med Sch, Univ London, MB, BS, 67; FRCPS(C), 72. Prof Exp: ASST PROF MED CLIN PHARMACOL, FAC MED, MEM UNIV NFLD, 73-; ASSOC PHYSICIAN INTERNAL MED, GEN HOSP, ST JOHN'S, NFLD, 73- Concurrent Pos: Develop grant, Can Found Advan Clin Pharmacol, 73-77. Mem: Can Soc Clin Invest; Can Pharmacol Soc; Am Soc Clin Pharmacol & Therapeut. Res: In vivo and in vitro study of drug interactions and pharmacokinetics, in particular the mechanism of antagonism between acetylsalicylic acid and spironolactone in man and the development of hydraulic pharmacokinetic models. Mailing Add: Fac of Med Mem Univ of Nfld St John's NF Can

TWEEDELL, KENYON STANLEY, b Sterling, Ill, Mar 28, 24; m 56; c 5. ZOOLOGY. Educ: Univ Ill, BS, 47, MS, 49, PhD(zool,physiol), 53. Prof Exp: Res assoc biol, Control Systs Lab, Univ Ill, 51-54; instr zool, Univ Maine, 54-56, asst prof, 56-58; from asst prof to assoc prof, 58-68, PROF BIOL, UNIV NOTRE DAME, 68- Mem: AAAS; Soc Exp Biol & Med; Am Soc Zool; Soc Develop Biol. Res: Experimental pathology; transmissable tumors of Amphibia; developmental biology, especially oogenesis, ovulation, regeneration in invertebrates, cytodifferentiation in normal and malignant cells. Mailing Add: Dept of Biol Box 369 Univ of Notre Dame Notre Dame IN 46556

TWEEDIE, ADELBERT THOMAS, b Saginaw, Mich, Jan 5, 31; m 53; c 3. POLYMER CHEMISTRY. Educ: Univ Mich, BSCh, 53; Univ Ill, PhD(org chem), 56. Prof Exp: Res chemist, Dow Chem Co, Mich, 56-58 & Aerojet-Gen Corp, Calif, 58-62; tech supvr rocket propellants, Union Carbide Corp, WVa, 62-64; MGR MAT ENG, SPACE DIV, GEN ELEC CO, KING OF PRUSSIA, 64- Mem: Am Chem Soc. Res: Behavior of materials in the space environment; vibration damping; application of materials to spacecraft; polymers; solar energy. Mailing Add: 24 Harvey Lane Malvern PA 19355

TWEEDIE, STEPHEN WILLIAM, b Walton, NY, Feb 18, 38; m 66; c 2. CULTURAL GEOGRAPHY. Educ: Cornell Univ, BA, 59, MEd, 60; Syracuse Univ, PhD(geog), 69. Prof Exp: Asst prof geog, Eastern Wash State Col, 69-71; ASST PROF GEOG, OKLA STATE UNIV, 71- Mem: Asn Am Geogr; Am Geog Soc. Res: Geography of religion; geography of popular culture; use of computers in geographical teaching and research. Mailing Add: Dept of Geog Okla State Univ Stillwater OK 74074

TWEEDIE, VIRGIL LEE, b Norborne, Mo, Feb 18, 18; m 43; c 3. ORGANIC CHEMISTRY. Educ: Univ Mo, AB, 41, MA, 43; Univ Tex, PhD(chem), 51. Prof Exp: Res chemist, Commercial Solvents Corp, 42-46; asst prof, 46-48, assoc prof, 50-53, PROF CHEM, BAYLOR UNIV, 53- Mem: AAAS; Am Chem Soc. Res: Allylic compounds; organometallics; complex metal hydrides and alkides; hydrogenolysis. Mailing Add: 7720 Tallahassee Rd Waco TX 76710

TWEEDLE, CHARLES DAVID, b Astoria, Ore, Jan 22, 44. NEUROBIOLOGY. Educ: Univ Ore, BA, 66, MA, 67, PhD(zool), 70. Prof Exp: NIH fel, Yale Univ, 70-72; NIH Res Grant & Res Assoc, 72-73; ASST PROF, DEPT BIOMECH & ZOOL, MICH STATE UNIV, 73- Mem: AAAS; Soc Neurosci; Am Soc Zool; Am Asn Anatomists. Res: Developmental neurobiology; trophic effects of nerves; neuromuscular development; nerve regeneration; dendritic stability and development; neurophysiology of simple nervous systems. Mailing Add: Dept of Biomech Mich State Univ East Lansing MI 48823

TWEEDY, BILLY GENE, b Cobden, Ill, Dec 31, 34; m 57; c 3. PLANT PATHOLOGY. Educ: Univ Southern Ill, BS, 56; Univ Ill, MS, 59, PhD(plant path), 61. Prof Exp: Asst plant pathologist, Royce Thompson Inst Plant Res, 61-65; asst prof plant path, Univ Mo-Columbia, 65-69, from assoc prof to prof, 69-73; MGR RESIDUE INVEST, CIBA-GEIGY CORP, 73- Mem: AAAS; Am Phytopath Soc; Am Chem Soc; Am Soc Plant Physiol; Weed Sci Soc Am. Res: Degradation of pesticides; fungus physiology; fruit pathology. Mailing Add: 111 Cresthill Dr Jamestown NC 27282

TWEEDY, JAMES ARTHUR, b Cobden, Ill, Nov 29, 39; m 64; c 1. HORTICULTURE. Educ: Southern Ill Univ, BS, 62; Mich State Univ, MS, 64, PhD(hort), 66. Prof Exp: From asst prof to assoc prof, 66-74, asst dean, Sch of Agr, 74-75, PROF PLANT INDUST, PLANT & SOIL SCI DEPT, SOUTHERN ILL UNIV, 74- Mem: Weed Sci Soc Am. Res: Influence of herbicides on plant physiological processes; evaluation of herbicides for weed control in horticultural crops. Mailing Add: Dept of Plant & Soil Sci Southern Ill Univ Carbondale IL 62901

TWEET, ARTHUR GLENN, b Aberdeen, SDak, Sept 20, 27; m 50; c 3. SOLID STATE PHYSICS. Educ: Harvard Univ, AB, 48; Univ Wis, MS, 49, PhD(physics), 53. Prof Exp: Physicist, Semiconductor Sect, Res Labs, Gen Elec Corp, 53-59, liaison scientist, 59-61, physicist, Biol Studies Sect, 61-64; mgr phys imaging br, Advan Imaging Technol Lab, 64-68, mgr, Imaging Res Lab, 68-72, mgr technol planning, Info Technol Group, 72-75, MGR, TECH STRATEGIC PLANNING, XEROX CORP, 75- Concurrent Pos: Adj prof, Rensselaer Polytech Inst, 62. Mem: Fel Am Phys Soc; Am Chem Soc; Soc Photog Sci & Eng. Res: Energy transfer in excited molecules; unconventional photographic systems; dye sensitized reactions; electrical and surface properties of polymers; surface chemistry of chromophores; optical properties of large molecules; imperfections in semiconductors and insulators; nonaqueous electrochemistry; decision analysis; technological forecasting; operations analysis and modelling. Mailing Add: Xerox Corp Xerox Sq Rochester NY 14603

TWEIT, ROBERT CHRISTOPHER, b Orange, NJ, July 26, 28; m 52; c 2. MEDICINAL CHEMISTRY. Educ: Mass Inst Technol, SB, 50; Univ Calif, PhD(chem), 53. Prof Exp: Asst chem, Univ Calif, 50-52; res chemist, 55-70, GROUP LEADER, G D SEARLE & CO, 70- Concurrent Pos: Vis scholar, Cambridge Univ, 74. Mem: Am Chem Soc; Am Ornithologist Union; Brit Trust for Ornith; Inland Bird Banding Asn. Res: Organic sulfur compounds; macrocyclic antifungal antibiotics; antiviral agents; substituted carbohydrates; bird and plant population studies. Mailing Add: 2104 Birchwood Wilmette IL 60091

TWELVES, ROBERT RALPH, b Chicago, Ill, Nov 4, 27; m 53; c 3. ORGANIC CHEMISTRY. Educ: Univ Utah, BS, 50, MA, 52; Univ Minn, PhD(org chem), 57. Prof Exp: Asst, Univ Utah, 50-51; process develop chemist, Merck & Co, 52-53; asst, Univ Minn, 53-55; RES CHEMIST, E I DU PONT DE NEMOURS & CO, 57- Mem: Am Chem Soc. Res: Fluorochemicals; dyes and intermediates; synthetic organic chemistry. Mailing Add: Chamber Works Bldg 1155 E I du Pont de Nemours & Co Deepwater NJ 08023

TWENHOFEL, WILLIAM STEPHENS, b Madison, Wis, Aug 30, 18; m 51; c 5. GEOLOGY. Educ: Univ Wis, BA, 40, PhD(geol), 52. Prof Exp: Geologist, US Geol Surv, Washington, DC, 42-45; physicist, US Naval Res Lab, 45-46; geologist, 47-64, CHIEF, BR SPEC PROJS, US GEOL SURV, 64- Honors & Awards: Meritorious Serv Award, US Dept Interior, 73- Mem: AAAS; Geol Soc Am; Asn Eng Geol. Res: Geology of Alaska; growth of artificial crystals; structural and uranium geology; engineering geology of underground nuclear explosions. Mailing Add: US Geol Surv Mailstop 954 Box 25046 Denver CO 80215

TWENTE, JANET, b Geneva, Switz, Feb 5, 29; US citizen; m 53; c 1. PHYSIOLOGY, ZOOLOGY. Educ: Swarthmore Col, BA, 49; Univ Mich, MS, 52, PhD(zool), 55. Prof Exp: Res assoc cancer, Northwestern Univ & Hines Vet Admin, 56-57; trainee steroid biochem, Univ Utah, 57-58, res assoc radiobiol, 58-61, asst res prof molecular & gen biol, 62-66; RES ASSOC SPACE SCI, UNIV MO-COLUMBIA, 66- Mem: AAAS; Am Soc Zool; Radiation Res Soc. Res: Hibernation physiology; physiology of arousal from hibernating periods; neuroendocrinological and other hormonal factors and their effects upon hibernation. Mailing Add: Dalton Res Ctr Univ of Mo Columbia MO 65201

TWENTE, JOHN W, b Lawrence, Kans, Dec 18, 26; m 53; c 1. ZOOLOGY. Educ: Univ Kans, AB, 50; Univ Mich, MS, 52, PhD(zool), 54. Prof Exp: Interim instr biol, Univ Fla, 54-55; instr zool, Col Pharm, Univ Ill, 55-56; instr, Univ Utah, 57-58, asst prof, 58-62, res biologist, 62-66; ASSOC PROF BIOL & INVESTR, DALTON RES CTR, UNIV MO-COLUMBIA, 66- Mem: Am Soc Zool; Ecol Soc Am; Am Soc Mammal. Res: Physiological ecology and behavior; hibernation physiology. Mailing Add: Dalton Res Ctr Univ of Mo Columbia MO 65201

TWERSKY, VICTOR, b Poland, Aug 10, 23; nat US; m 50; c 3. MATHEMATICAL PHYSICS. Educ: City Col New York, BS, 47; Columbia Univ, AM, 48; NY Univ, PhD(physics), 50. Prof Exp: Assoc, Guid Device Proj, Biol Dept, City Col New York, 46-49; asst physics, NY Univ, 49, res assoc electromagnetic theory, Inst Math Sci, 50-53; assoc, Nuclear Develop Assocs, 51-53; specialist theoret physics, Electronic Defense Labs, Sylvania Electronic Systs-West, Sylvania Elec Prod, Inc Div, Gen Tel & Electronics Corp, 53-58, sr specialist & lab consult, 58-60, sr scientist, 60-66, head res, Electronics Defense Labs, 58-66 & Sylvania Electronic Systs-West, 64-66; PROF MATH, UNIV ILL, CHICAGO CIRCLE, 66- Concurrent Pos: Mem tech res comt, Am Found Blind, 47-49; lectr math, Stanford Univ, 56-58; consult, Sylvania Elec Prod Inc Div, Gen Tel & Electronics Corp, 66; NSF grants, 67-75; mem ctr advan study, Univ Ill, 69-70; Guggenehim fel, 72-73; mem at large, Conf Bd Math Sci, 75-; mem US Comn VI, Int Union Radio Sci. Mem: Fel AAAS; fel Am Phys Soc; fel Acoust Soc Am; Am Math Soc; NY Acad Sci. Res: Multiple scattering of electromagnetic and acoustic waves; rough surfaces; gratings; scattering and propagation in random distributions; radiative diagnostics of biological cells; relativistic scattering; applied mathematics; obstacle perception by the blind. Mailing Add: Dept of Math Univ of Ill Box 4348 Chicago IL 60680

TWETO, JOHN HALVOR, b Boulder, Colo, Apr 16, 45; m 69; c 1. BIOLOGICAL CHEMISTRY. Educ: Univ Colo, BA, 67; Univ Ore, PhD(biochem), 72. Prof Exp: Asst cancer res scientist, 72-74, NIH res fel, 74-75, CANCER RES SCIENTIST, ROSWELL PARK MEM INST, 75- Res: Control of levels of proteins in eukaryotic cells with particular emphasis on membrane organization and biogenesis. Mailing Add: Roswell Park Mem Inst 666 Elm St Buffalo NY 14263

TWETO, OGDEN, b Abercrombie, NDak, June 10, 12; m 40; c 2. ECONOMIC GEOLOGY. Educ: Univ Mont, AB, 34, MA, 37; Univ Mich, PhD(geol), 47. Prof Exp: Instr, Univ NC, 39-40; geologist, 40-61, chief, South Rockies Br, 61-65, asst chief geologist & chief off econ geol, 65-68, RES GEOLOGIST, US GEOL SURV, 68- Mem: AAAS; Geol Soc Am; Soc Econ Geol; Mineral Soc Am; Am Inst Mining, Metall & Petrol Eng. Res: Geology and mineral deposits of Southern Rocky Mountains. Mailing Add: US Geol Surv Bldg 25 Federal Ctr Denver CO 80225

TWIEST, GILBERT LEE, b Grand Rapids, Mich, Apr 23, 37; m 58; c 3. ORNITHOLOGY, SCIENCE EDUCATION. Educ: Mich State Univ, BS, 61, MS, 63; Univ Toledo, PhD(sci educ), 68. Prof Exp: Instr biol, Kellogg Community Col, 63-65 & Univ Toledo, 66-68; ASSOC PROF SCI EDUC, CLARION STATE COL, 68- Mem: Nat Asn Res Sci Teaching; Nat Sci Teachers Asn; Am Inst Biol Sci; Am Ornith Union; Wilson Ornith Soc. Res: Methods of teaching students the processes needed to solve problems. Mailing Add: Dept of Biol Clarion State Col Clarion PA 16214

TWIGG, BERNARD ALVIN, b Cumberland, Md, Oct 15, 28; m 51; c 4. HORTICULTURE, FOOD SCIENCE. Educ: Univ Md, BS, 52, MS, 54, PhD, 59. Prof Exp: Asst hort, 52-54, from instr to assoc prof, 54-69, PROF HORT, UNIV MD, COLLEGE PARK, 69-, CHMN DEPT, 75- Mem: Am Soc Hort Sci; Inst Food Technologists. Res: Objective evaluation of food products; statistical quality control; food chemistry and physics; horticultural food processing. Mailing Add: Dept of Hort Univ of Md College Park MD 20740

TWIGG, HOMER LEE, b Westminster, Md, Apr 10, 26; m 55; c 6. MEDICINE, RADIOLOGY. Educ: Univ Md, MD, 51; Am Bd Radiol, dipl, 56. Prof Exp: Intern med, USPHS Hosp, Boston, 51-52, intern surg serv outpatient clin, 52, resident radiol,

Baltimore, 55-56, chief radiol, Detroit, 56-57; actg chmn & dir dept, 67-69, from asst prof to assoc prof, 57-70, PROF RADIOL, GEORGETOWN UNIV HOSP, 70-, CHMN DEPT, 69- Concurrent Pos: Spec assignment, US Dept Interior, 52; consult, Vet Admin Hosp, DC, 60-, St Elizabeth's Hosp, 65- & NIH, 65- Mem: AMA; fel Am Col Radiol; Am Roentgen Ray Soc; Radiol Soc NAm. Mailing Add: Dept of Radiol Georgetown Univ Hosp Washington DC 20007

TWILLEY, IAN CHARLES, b London, Eng, Apr 4, 27; m 54; c 4. CHEMISTRY. Educ: Univ London, BSc, 53, FRIC, Royal Inst Chem, 62. Prof Exp: Sr chemist, Nelsons Silk Ltd, Eng, 53-56 & Micanite & Insulators, Ltd, 56-57; sr develop chemist, Textile Fibers Div, Du Pont of Can, 57-59; sect leader moulding polymers, Nat Aniline Div, 59-60, group leader polymer res, 60-61, res supvr, 61-65, process develop supvr, 65-66, mgr systs eng, 66-68, mgr eng res, Fibers Div, 68-69, asst chief engr, 69, tech dir polyester, 69-72, TECH DIR HOME FURNISHINGS, FIBERS DIV, ALLIED CHEM CORP, HOPEWELL, 72- Mem: Am Chem Soc; Am Inst Chem Eng; Can Soc Chem Eng; Chem Inst Can; Royal Inst Chem. Res: Polymeric and textile processes and products; production and design problems; polymer and textile chemistry; polyamide and polyester technology. Mailing Add: 12625 Merry Dr Chester VA 23831

TWISS, PAGE CHARLES, b Columbus, Ohio, Jan 2, 29; m 54; c 3. GEOLOGY. Educ: Kans State Univ, BS, 50, MS, 55; Univ Tex, PhD(geol), 59. Prof Exp: Instr, 53-55, from asst prof to assoc prof, 59-69, PROF GEOL, KANS STATE UNIV, 69-, HEAD DEPT, 68- Concurrent Pos: Co-investr, NSF grants, 60-62, 66, 67-68; res scientist, Univ Tex, 66-67. Mem: AAAS; Geol Soc Am; Am Asn Petrol Geol; Soc Econ Paleont & Mineral; Clay Minerals Soc. Res: Sedimentary and igneous petrology; clay mineralogy; stratigraphy, tectonics, petrology and geochemistry of Mesozoic and Cenozoic rocks of Trans-Pecos Texas and Chihuahua, Mexico; petrology of recent dust deposits. Mailing Add: 2327 Bailey Dr Manhattan KS 66502

TWISS, ROBERT JOHN, b Baltimore, Md, May 12, 42; m 68; c 1. GEOLOGY. Educ: Yale Univ, BS, 64; Princeton Univ, MA, 68, PhD(geol), 71. Prof Exp: NATO fel geol, Australian Nat Univ, 71-72; ASST PROF GEOL, UNIV CALIF, DAVIS, 72- Mem: AAAS; Am Geophys Union; Geol Soc Am. Res: Continuum mechanics theory applied to understanding the behavior of geologic materials; deformation mechanisms in silicates; structural analysis of tectonites. Mailing Add: Dept of Geol Univ of Calif Davis CA 95616

TWISS, SUMNER BARNES, physical chemistry, organic chemistry, see 12th edition

TWITCHELL, PAUL F, b Somerville, Mass, Mar 7, 32; m 56; c 4. METEOROLOGY. Educ: Boston Col, BS, 53, MS, 62; Pa State Univ, BS, 54. Prof Exp: Weather officer, US Air Force, 53-57; res engr, Res Dept, Melpar, Inc, Mass, 57-60; sr res engr, Appl Sci Div, 60-62; phys sci coordr, 62-72, PHYS SCI ADMINR, BOSTON BR, OFF NAVAL RES, 72- Concurrent Pos: Student, Univ Wis, 67- Mem: Am Meteorol Soc; Am Geophys Union. Res: Physical processes in the terrestrial atmosphere related to extraterrestrial events. Mailing Add: Off of Naval Res 495 Summer St Boston MA 02210

TWITCHELL, THOMAS EVANS, b Springfield, Ohio, Sept 4, 23; m 56; c 4. NEUROLOGY. Educ: Univ Mich, MD, 46. Prof Exp: Res fel physiol, Med Sch, Yale Univ, 47; intern neurol, Boston City Hosp, 47-48, res fel, Harvard Med Sch & Boston City Hosp, 48-49; asst resident med, New Eng Ctr Hosp, 54, chief resident neurol, 54-55; from instr to asst prof, 55-63, ASSOC PROF NEUROL, SCH MED, TUFTS UNIV, 63- Concurrent Pos: USPHS res fel, Yale Univ, 49-51; res assoc, Mass Inst Technol, 63-; neurologist, New Eng Med Ctr Hosps, 63- Mem: AAAS; Asn Res Nerv & Ment Dis; Am Neurol Asn; AMA; Am Fedn Clin Res. Res: Neurophysiology of primate motor function; physiologic nature of development of behavior in infants; sensory mechanisms in movement; clinical neurology; applied neurophysiology; neuropsychology. Mailing Add: Dept of Neurol Tufts Univ Sch of Med Boston MA 02111

TWOHY, DONALD WILFRED, b Clackamas, Ore, Sept 9, 24; m 55; c 1. PARASITOLOGY. Educ: Ore State Col, BS, 48, MS, 51; Johns Hopkins Univ, ScD, 55. Prof Exp: Aquatic biologist, Ore State Game Comn, 49-51; res asst, Sch Hyg & Pub Health, Johns Hopkins Univ, 55-56; asst prof zool, Okla State Univ, 56-60; instr, 60-62, ASST PROF MICROBIOL, MICH STATE UNIV, 62- Mem: Am Soc Parasitologists; Am Soc Trop Med & Hyg; Soc Protozool. Res: Parasitic protozoa; cellular immunity. Mailing Add: Dept of Microbiol & Pub Health Mich State Univ East Lansing MI 48823

TWOMBLY, GRAY HUNTINGTON, b Newton, Mass, Apr 27, 05; m 34; c 3. OBSTETRICS & GYNECOLOGY. Educ: Franklin & Marshall Col, AB, 25; Harvard Univ, MD, 29; Am Bd Surg, dipl, 40; Am Bd Obstet & Gynec, dipl, 54. Hon Degrees: DSC, Franklin & Marshall Col, 49. Prof Exp: Intern surg, Mass Gen Hosp, 30-32; asst resident, Free Hosp Women, 32; intern, Mem Hosp, 33-37; asst prof cancer res, Columbia Univ, 42-48, asst prof clin obstet & gynec, Col Physicians & Surgeons, 48-52; prof gynec, Sch Med, NY Univ & assoc dir gynec, Univ Hosp, 52-73; RETIRED. Concurrent Pos: Fel, Mem Hosp, New York, 33-37, from clin asst to asst surgeon, 37-48; asst attend gynecologist, Sloane Hosp Women, 48-52 & Delafield Hosp, '50-52; consult, Yonkers Gen Hosp, 59-72, Holy Name Hosp, Teaneck, NJ, 59-72, Queen's Gen Hosp, 62-69 & St Luke's Hosp, 66- Mem: Am Gynec Soc; Am Radium Soc (vpres, 59); Am Asn Cancer Res; Am Cancer Soc; Soc Pelvic Surg (pres, 71-72). Res: Cancer and steroid research. Mailing Add: 450 Riverside Dr New York NY 10027

TWOMEY, ARTHUR CORNELIUS, b Midland, Ont, Oct 15, 08; nat US; m 34; c 1. ZOOLOGY. Educ: Univ Alta, BS, 33; Univ Ill, MSc, 35, PhD(zool), 37. Prof Exp: Asst zool, Univ Ill, 33-37; asst cur & field collector, Dept Ornith, Carnegie Mus, 37-44, lectr, 43-44, DIR FIELD EXPEDS & CUR ORNITH, CARNEGIE MUS, CARNEGIE INST, 44-, DIR DEPT EDUC, INST, 48- Concurrent Pos: Asst paleot exped, 27; conductor expeds, Can, 28-34, 46-47, 54, Arctic Regions & Hudson Bay, 31-34, 36, 38, Pac Coast, 33, Utah, 37, NMex, 39, Ariz, 40, Ore, 41, Peru, Chile, Tierra del Fuego & Galapagos Islands, 39, Bahama Islands, 42, Delta of Mackenzie River, 42-43, 55, 66, Isles of Bahia, Honduras, 47, Repub of Honduras, 48-51, EAfrica, Kenya, Uganda, Tansania, 60, 63, 64, 65, Mozambique, Batswana, 65, Pamir Mt, Afghanistan, 69, Gobi Desert & Alti Mt, Outer Mongolia, 70, Caucasus Mt, USSR, 71. Mem: AAAS; Soc Syst Zool; Ecol Soc Am; Wilson Ornith Soc; Am Ornith Union. Res: Animal ecology; general, economic, life history systematics and physiological ornithology; climatology relating to ornithology. Mailing Add: Carnegie Mus 4400 Forbes Ave Pittsburgh PA 15213

TYAGI, AVDHESH KUMAR, b Bijnor, India, May 10, 44; US citizen; m 67; c 1. HYDROLOGY. Educ: Univ Allahabad, BS, 65; Univ Roorkee, MS, 67; Univ Calif, Berkeley, PhD(hydrol, water resources eng), 70. Prof Exp: Mem faculty hydrol & water resources, Univ Ariz, 70-74, off water resources res grant, 71-75, CHIEF HYDROLOGIST, CTR ADVAN RES HYDROLOGY & WATER RESOURCES, 74- Concurrent Pos: Consult, Ford Found, Mex, 71-73; adv Nat Sch Agr, Mex, 71-73; consult, Off Water Res & Technol, 71-75 & Sperry Rand Corp, 75-; Sr Scientist, NSF, 75-; consult, Environ Protection Agency, 75- Res: Numerical methods in hydrologic systems; water quality modeling of surface and ground waters; stochastic and decision-making processes in hydrology; water resources system optimization. Mailing Add: 5129 E Monte Vista Phoenix AZ 85008

TYAN, MARVIN L, b Los Angeles, Calif, Nov 29, 26; m 50; c 2. INTERNAL MEDICINE, EXPERIMENTAL BIOLOGY. Educ: Univ Calif, Berkeley, BA, 49; Univ San Francisco, MD, 52. Prof Exp: Intern med, Boston City Hosp, Mass, 52-53; resident, Boston Vet Admin Hosp, 53-54; sr asst resident, San Francisco County Hosp, Calif, 54-55; pvt pract, 56-61; sr investr exp path, US Naval Radiol Defense Lab, 61-68 & Stanford Res Inst, Calif, 68-71; PROF BACT, IMMUNOL & ORAL BIOL, DENT RES CTR, UNIV NC, CHAPEL HILL, 71- Concurrent Pos: Fel hemat, Stanford Lane Hosp, San Francisco, 55-56; fel, Tumor Biol Inst, Karolinska Inst, Sweden, 63-64. Honors & Awards: Gold Medal Sci Achievement, US Naval Radiol Defense Lab, 64; Superior Civil Serv Award, Bur Ships, US Navy. Mem: AAAS; NY Acad Sci. Res: Ontogeny of immune system of the mouse; processes involved in transplantation immunity. Mailing Add: Dent Res Ctr Univ of NC Chapel Hill NC 27514

TYBERG, JOHN VICTOR, b Grantsburg, Wis, May 4, 38; m 60; c 1. CARDIOVASCULAR PHYSIOLOGY. Educ: Bethel Col, Minn, BA, 60; Univ Minn, PhD(physiol), 67, MD, 72. Prof Exp: Res assoc med, Harvard Med Sch, 69; res physiologist, Riverside Res Inst, 69-71; res scientist cardiol, Cedars-Sinai Med Ctr, 71-73; ASST PROF MED & PHYSIOL & STAFF MEM, CARDIOVASC RES INST, MED CTR, UNIV CALIF, SAN FRANCISCO, 74- Concurrent Pos: Lectr, Med Ctr, Univ Calif, San Francisco, 69-71; mem, Basic Sci Coun, Am Heart Asn, 74- Mem: Am Heart Asn; Am Physiol Soc. Res: Mechanics and metabolism of ischemic myocardium; coronary flow. Mailing Add: Dept of Med Div of Cardiol Univ of Calif Med Ctr San Francisco CA 94143

TYBOR, PHILIP THOMAS, b Fredericksburg, Tex, Oct 3, 48; m 69; c 3. FOOD SCIENCE. Educ: Tex A&M Univ, BS, 70, PhD(food sci), 73. Prof Exp: SCIENTIST PROTEIN RES, CENT SOYA CO, INC, 73- Mem: Inst Food Technologists; Am Asn Cereal Chemists; Am Chem Soc. Res: Development of protein products which have nutritional and functional properties necessary for utilization in food applications. Mailing Add: Cent Soya Co Inc 1825 N Laramie Ave Chicago IL 60639

TYCE, FRANCIS ANTHONY, b South Wales, Eng, Oct 31, 17; US citizen; m 52; c 1. PSYCHIATRY. Educ: Univ Durham, BS & MD, 52; Univ Minn, MS, 64; Am Bd Psychiat & Neurol, dipl, 64. Prof Exp: House surgeon, Teaching Hosp, Durham, Eng, 52-53; rotating intern, St Vincents Hosp, Erie, Pa, 53-54; gen pract, Seaham Harbor, 54-56; actg supt, Rochester State Hosp, 60-61; asst prof psychiat, Mayo Grad Sch Med, Univ Minn, 69-73, ASSOC PROF PSYCHIAT, MAYO MED SCH, ROCHESTER, MINN, 73-; SUPT, ROCHESTER STATE HOSP, 71- Concurrent Pos: Fel psychiat, Mayo Clin, 56-60; lectr, Mayo Found, 65-; consult, WHO, 67-; vpres, Zumbro Valley Med Soc, Rochester, Minn, 71-72, pres, 72-73; task force comt, Psychiat Rehab in Correctional Systs, 73-; Field Rep Accreditation Coun for Psychiat Facil-Jt Comn on Accreditation of Hosps, 73-; mem, Juvenile Delinquency-Nat Adv Comt on Criminal Justice Stand & Goals, 75- Mem: Fel Psychiat Soc; Asn Med Supt Ment Hosp (pres elect, 67-68 & pres, 68-); chmn, Am Psychiat Asn. Res: Neurophysiology, especially electrical stimulation of the brain in rats; psychiatric program design in mental hospitals. Mailing Add: 1520 E Center Rochester MN 55901

TYCE, GERTRUDE MARY, b Wark, Eng, Mar 26, 27; m 52; c 1. BIOCHEMISTRY. Educ: Univ Durham, BSc, 48, PhD(plant physiol & biochem), 52. Prof Exp: Instr chem, Villa Maria Col, Pa, 52-53; instr biol & chem, Nottingham & Dist Tech Col Eng, 54-56; res asst biochem, Mayo Clin, 58-63, res assoc, 63-71, ASSOC CONSULT BIOCHEM, MAYO CLIN & MAYO FOUND, 71- Mem: AAAS; Am Chem Soc; Am Soc Exp Path. Res: Metabolism of glucose, amino acids and biogenic amines in brain and liver. Mailing Add: Mayo Clin & Mayo Found Rochester MN 55901

TYCHSEN, PAUL C, b Chicago, Ill, Nov 1, 16; m 43; c 2. GEOLOGY. Educ: Carleton Col, BA, 41; Univ Nebr, MSc, 49, PhD(geol), 54. Prof Exp: Geolgist, US Geol Surv, 46-49; Proj rep Australia, Earth Sci Curric Proj, Am Geol Inst, 72-73; CHMN DEPT GEOL, UNIV WIS-SUPERIOR, 52- Honors & Awards: Johnson Wax Found teaching award, 65. Mem: AAAS; fel Geol Soc Am; Nat Asn Geol Teachers. Res: Stratigraphy; field-mapping; geomorphology. Mailing Add: Dept of Geosci Univ of Wis Superior WI 54880

TYCKO, DANIEL H, b Los Angeles, Calif, Nov 14, 27; m 52; c 3. COMPUTER SCIENCES, EXPERIMENTAL HIGH ENERGY PHYSICS. Educ: Univ Calif, Los Angeles, BA, 50; Columbia Univ, PhD(physics), 57. Prof Exp: Sr res assoc, Nevis Labs, Columbia Univ, 57-66; assoc prof physics, Rutgers Univ, 66-67; eng, 67-70, PROF COMPUT SCI, STATE UNIV NY, STONY BROOK, 70- Concurrent Pos: Res collab, Saclay Nuclear Res Ctr, France, 61-62; consult, Technicon Instruments Corp, 70- Mem: Am Phys Soc; Asn Comput Mach; Inst Elec Electronic Engrs; AAAS. Res: Image processing by computers; pattern recognition; applications to cytology and hematology. Mailing Add: Dept of Comput Sci State Univ of NY Stony Brook NY 11794

TYCZKOWSKI, EDWARD ALBERT, b Providence, RI, May 15, 24; m 50. ORGANIC CHEMISTRY. Educ: Brown Univ, ScB, 49; Duke Univ, PhD(chem), 53. Prof Exp: Res assoc, US Army Off Ord Res, Duke Univ, 52-53; res chemist, Gen Chem Div, Allied Chem & Dye Corp, 53-56 & Pennsalt Chem Corp, 56-62; process chemist, Air Prod & Chem Corp, 62-63; vpres, Hynes Chem Res Corp, 63-66; sr res chemist, Fibers Div, Beaunit Corp, 66-67, res assoc, 67-72; PRES, ARMAGEDDON CHEM CO, 72- Mem: AAAS; Am Chem Soc. Res: Organic fluorine chemistry; reactions of elementary fluorine with organic compounds; flame reactions; explosions; reactor design; organic synthesis; polymer chemistry; fiber structure. Mailing Add: 824 E Forest Hills Blvd Durham NC 27707

TYE, ARTHUR, b Rutherglen, Australia, Sept 13, 09; nat US; m 41, 65. PHARMACOLOGY. Educ: Ohio State Univ, PhD(pharm), 50. Prof Exp: Asst & lectr pharm, Peiping Union Med Col, China, 32-38, head dept, 39-41; head pharmaceut prods div, Oriental Corp, China, 42-47; from asst prof to assoc prof, 50-60, prof, 60-70, EMER PROF PHARM, COL PHARM, OHIO STATE UNIV, 70- Concurrent Pos: Adminr, Redwood Health Consortium, 72- Mem: AAAS; Am Pharmaceut Asn; Am Soc Pharmacol & Exp Therapeut. Res: Autonomic nervous system; steric aspects of adrenergic drugs; structure-action relationships; chemical pharmacology; modular systems approach to allied health education. Mailing Add: 219 Mocking Bird Circle Santa Rosa CA 95405

TYERYAR, FRANKLIN JOSEPH, JR, b Frederick, Md, Apr 29, 35; m 59; c 3. MEDICAL MICROBIOLOGY. Educ: Univ Md, BS, 60, MS, 62, PhD(microbiol), 68. Prof Exp: Microbiologist, US Bur Mines, US Dept Interior, 62-63, US Dept Army, Ft Detrick, 63-71 & Dept Microbiol, Naval Med Res Inst, 71-73; MICROBIOLOGIST, NAT INST ALLERGY & INFECTIOUS DIS, 73- Mem: Am Soc Microbiol. Res:

Microbial genetics and physiology, specifically gene exchange mechanisms by transformation and transduction; development and testing of viral vaccines for clinical use; clinical evaluation of viral vaccines for efficacy; persistent viral infections; prevention and control of virus caused respiratory disease. Mailing Add: Nat Inst of Allergy & Infectious Dis Nat Insts of Health Bethesda MD 20014

TYHURST, JAMES STEWART, b Victoria, BC, Feb 24, 22; m 49; c 2. PSYCHIATRY. Educ: McGill Univ, BSc, 41, MD, CM, 44. Prof Exp: Res asst & demonstr psychiat, McGill Univ, 48-49, sr res psychiatrist & demonstr, 49-50; asst prof, Col Med & asst, Cornell Univ, 50-53; asst prof, McGill Univ, 53-60; head dept, 60-70, PROF PSYCHIAT, UNIV BC, 60- Concurrent Pos: Lectr psychiat, Dalhousie Univ, 50-51; dir, Psychiat Clin, Digby, NS, 51-53; assoc psychiatrist, Royal Victoria Hosp, Montreal, Que, 53-60; consult, Can Dept Vet Affiars, 53-; mem, Adv Bd, Nat Lab Group Develop, 54- Mem: AAAS; fel Am Psychiat Asn; Can Psychiat Asn; Can Med Asn. Res: Psychopharmacology; social, clinical, community and industrial psychiatry; behavior of individuals and groups in emergencies; ageing and retirement; displacement and migration. Mailing Add: Dept of Psychiat Univ of BC Vancouver BC Can

TYKODI, RALPH JOHN, b Cleveland, Ohio, Apr 18, 25; m 55; c 3. PHYSICAL CHEMISTRY. Educ: Northwestern Univ, BS, 49; Pa State Univ, PhD, 54. Prof Exp: Instr, Ill Inst Technol, 55-57, from asst prof to assoc prof, 57-65; assoc prof chem, 65-68, assoc dean, Col Arts & Sci, 69-72, PROF, SOUTHEASTERN MASS UNIV, 68- Mem: Am Chem Soc; Am Phys Soc. Res: Equilibrium and non-equilibrium thermodynamics. Mailing Add: Dept of Chem Southeastern Mass Univ North Dartmouth MA 02747

TYLER, ALBERT VINCENT, b Philadelphia, Pa, June 25, 38; m 60; c 2. FISHERIES. Educ: Univ Pa, BA, 60; Univ Toronto, MA, 64, PhD(synecol), 68. Prof Exp: Scientist, Fisheries Res Bd Can, 64-74, ASSOC PROF FISHERIES, ORE STATE UNIV, 74- Mem: Am Fisheries Soc; Can Soc Zool. Res: Physiological and ecological energetics; competitive and predatory relationships among fishes; population dynamics. Mailing Add: Marine Sci Ctr Ore State Univ Newport OR 97365

TYLER, BONNIE MORELAND, b New York, NY, Jan 5, 41; m 63; c 3. MICROBIAL PHYSIOLOGY, MOLECULAR BIOLOGY. Educ: Wheaton Col, BA, 62; Mass Inst Technol, PhD(biol), 68. Prof Exp: Res assoc bact, Univ Calif, Davis, 68-70; instr, 70-73, ASST PROF BIOL, MASS INST TECHNOL, 73- Mem: Sigma Xi. Res: Regulation of transcription and translation in bacteria; studies with whole cells and with purified transcription systems on factors regulating transcription. Mailing Add: Dept of Biol Mass Inst Technol Cambridge MA 02139

TYLER, CHAPLIN, b Washington, DC, Mar 28, 98; m 25; c 3. CHEMISTRY. Educ: Northeastern Univ, BChE, 20, hon ScD, 61; Boston Univ, BBA, 22; Mass Inst Technol, SM, 23. Prof Exp: Asst appl chem, Mass Inst Technol, 20-22, res assoc, 23-24; asst ed, Chem & Metall Eng, 24-27; chemist, E I du Pont de Nemours & Co, 27-35, asst dir, 35-41, mem develop dept, 42-62; dir, Sterling Inst, Inc, 71-75; DIR, ROGER WILLIAMS TECH & ECON SERV, INC, 63- Concurrent Pos: Assoc, Columbia Univ, 26-27; spec lectr, Univ Del, 46-48; consult, President's Mat Policy Comn, 51-52; consult, 63-; Trustee, Northeastern Univ, 66- Mem: Am Chem Soc; Am Inst Chem Eng. Res: Economics of chemical industry; management. Mailing Add: Apt 1014 Devon Apts 2401 Pennsylvania Ave Wilmington DE 19806

TYLER, DAVID BERNARD, b New York, NY, Jan 13, 05; m 29. PHYSIOLOGY, PHARMACOLOGY. Educ: Univ Southern Calif, AB, 33, PhD(physiol, biochem), 37. Prof Exp: Teaching asst physiol, Sch Med, Univ Southern Calif, 33-37, lectr, 37-47; asst chief physiolo- gist, Med Div, Army Chem Ctr, 47; mem staff, Carnegie Inst Dept Embryol & Johns Hopkins Univ, 47-51; prof pharmacol & dir dept, Sch Med & Dent, Univ PR, 51-64; sect head physiol processes, Biomed Sci Div, NSF, 61-72; PROF PHARMACOL, COL MED, UNIV S FLA, 72- Concurrent Pos: Hixon Fund fel, Calif Inst Technol, 40-46; res assoc, Calif Inst Technol, 42-46; mem comt motion sickness, Off Sci Res & Develop, 42-46; vis prof, Med Col, Trivandrum, India, 59-60; foreign serv officer, Tech Coop Mission, India, 59-60; panel regulatory biol, NSF, 61. Mem: Am Physiol Soc; Soc Exp Biol & Med; Am Soc Pharmacol & Exp Therapeut. Res: Brain and tissue metabolism; brain damage; experimental psychoses; prolonged wakefulness; fatigue; carbohydrate and insulin metabolism; motion sickness. Mailing Add: Dept of Pharmacol Univ of S Fla Col of Med Tampa FL 33620

TYLER, DAVID E, b Carlisle, Iowa, July 12, 28; m 52; c 2. VETERINARY PATHOLOGY. Educ: Iowa State Univ, BS, 53, DVM, 57, PhD(vet path), 63; Purdue Univ, MS, 60. Prof Exp: Instr, Purdue Univ, 57-60; asst prof, Iowa State Univ, 60-64, assoc prof, 64-66; PROF VET PATH & HEAD DEPT, COL VET MED, UNIV GA, 66- Mem: Am Vet Med Asn; Am Col Vet Path; Am Asn Vet Med Educr; Conf Res Workers Animal Diseases. Res: Epidemiology, pathology and immunology of the bovine mucosal disease-virus diarrhea complex; respiratory complex of calves—epidemiology and pathology. Mailing Add: Dept of Path Col of Vet Med Univ of Ga Athens GA 30602

TYLER, EDWARD ALTON, b Portsmouth, Va, Jan 17, 21; c 3. MEDICINE, PSYCHIATRY. Educ: Univ Va, MD, 44; Am Bd Psychiat & Neurol, dipl & cert psychiat, 53, cert child psychiat, 60. Prof Exp: Resident psychiat, Med Sch, Duke Univ, 45-47; fel child psychiat, Med Sch, Univ Md, 47-48; asst prof, Med Sch, Univ Pittsburgh, 50-53; asst prof, Dartmouth Med Col, 55-58; from assoc prof to prof, Med Sch, Ind Univ, Indianapolis, 58-72, dir undergrad psychiat educ & co-chmn curric comt, 64-68, psychotherapist for med students, 68-70, chmn sex educ comt, 68-72, asst dean student affairs, 70-72; vis prof family pract, Sch Med, Univ Wis-Madison, 72-73; ASSOC DEAN, MED SCH, NORTHWESTERN UNIV, CHICAGO, 73- Concurrent Pos: Mem staff, Pittsburgh Child Guid Ctr, 49-53 & Hitchcock Clin, NH, 55-58; dir, Child Guid Clin, 58-64. Res: Motivation of human social and psychological behavior; curriculum innovation; human sexual behavior. Mailing Add: Northwestern Univ Med Sch Chicago IL 60611

TYLER, EDWARD TITLEBAUM, medicine, deceased

TYLER, FRANK HILL, b Villisca, Iowa, Jan 5, 16; m 41; c 3. MEDICINE. Educ: Willamette Univ, BA, 38; Johns Hopkins Univ, MD, 42. Prof Exp: Intern med, Johns Hopkins Hosp, 42-43; from asst resident to resident, Peter Bent Brigham Hosp, Boston, 43-47; research instr, 47-54, from asst prof to assoc prof, 50-59, PROF MED, MED SCH, UNIV UTAH, 59- Mem: Am Soc Clin Invest; Endocrine Soc; Am Fedn Clin Res; Asn Am Physicians. Res: Disease of the muscle; human inheritance; metabolism of steroids and metabolic disorders. Mailing Add: 50 N Medical Dr Salt Lake City UT 84132

TYLER, GEORGE WILLIAM, b Smyth Co, Va, Oct 16, 08; m 33; c 3. MATHEMATICS. Educ: Emory & Henry Col, BS, 30; Duke Univ, MA, 35; Va Polytech Inst, PhD(statist), 49. Prof Exp: Teacher schs, Va, 30-40; from instr to asst prof math, Va Polytech Inst, 40-44, assoc prof, 46-48; assoc physicist, Div War Res, Univ Calif, 44-46; consult, Navy Electronics Lab, 48-51; opers analyst & team chief, Hqs, US Air Force, 51-61; scientist, Inst Defense Anal, 61-63 & Ctr Naval Anal, 63-66; sr scientist, Tech Ctr, Supreme Hqs, Allied Powers, Europe, 66-69; SCI CONSULT, 69- Honors & Awards: Meritorious civilian serv award, US Air Force, 58. Mem: Opers Res Soc Am; Math Asn Am; Am Statist Asn; Inst Math Statist. Res: Experimental design; sampling; variance analysis; error control; systems simulation and analysis; weapons effects and arms control. Mailing Add: 3331 Bayshore Blvd NE St Petersburg FL 33703

TYLER, JACK D, b Snyder, Okla, July 18, 40; m 69. ORNITHOLOGY, ECOLOGY. Educ: Southwestern State Col, BS, 62; Okla State Univ, 65; Univ Okla, PhD(zool), 68. Prof Exp: Teaching asst gen zool, Univ Okla, 64-66; from instr to asst prof biol, 67-70, ASSOC PROF BIOL, CAMERON UNIV, 70- Mem: Am Soc Mammal; Wilson Ornith Soc. Res: Ecological relationships between certain birds in southwest Oklahoma and among vertebrates in prairie dog towns. Mailing Add: Dept of Biol Cameron Univ Lawton OK 73501

TYLER, JAMES CHASE, b Shanghai, China, Mar 31, 35; US citizen; m 58; c 2. ICHTHYOLOGY. Educ: George Washington Univ, BS, 57; Stanford Univ, PhD(biol), 62. Prof Exp: Actg instr gen biol, Stanford Univ, 61; asst curator ichthyol, Acad Natural Sci Philadelphia, 62-66, assoc curator, 67-72; asst dir, Lerner Marine Lab, Am Mus Natural Hist, 72-73, dir, 73-75; FISHERY BIOL, NAT MARINE FISHERIES SERV, WASHINGTON, DC, 75- Concurrent Pos: NSF grants, 63-72. Mem: Am Soc Ichthyol & Herpet. Res: Ichthyology, especially the anatomy and phylogeny of plectognath fishes and their classification; behavior and ecology of coral reef fishes. Mailing Add: Office of Resource Res Nat Marine Fisheries Serv Washington DC 20235

TYLER, JEAN MARY, b Sheffield, Eng, Apr 7, 28. ENDOCRINOLOGY, IMMUNOCHEMISTRY. Educ: Univ London, BSc, 49, PhD(org chem), 55. Prof Exp: Asst lectr org chem, Univ London, 51-55; Imp Chem Industs fel, Univ Edinburgh, 55-59; res fel, Univ West Indies, 59-60; res assoc immunochem, Inst Microbiol, Rutgers Univ, 60-64; NATO res fel org chem, Univ Newcastle, 64-65; sr hosp biochemist, Dept Med, Royal Free Hosp, London, Eng, 65-66; res fel chem, Imp Col Sci & Technol, Univ London, 66-68; asst res prof med, 68-74, ASSOC RES PROF MED, MED COL GA, 74- Honors & Awards: Commonwealth Award, Royal Soc & Nuffield Found, 59; NATO Award, 64. Mem: The Chem Soc; Endocrine Soc; Am Diabetes Asn. Res: Carbohydrate chemistry; immunochemistry of microbial polysaccharide antigens; radioimmunoassay of peptide hormones; diabetes. Mailing Add: Div of Endocrinol & Metabolic Dis Dept of Med Med Col of Ga Augusta GA 30902

TYLER, JOHN EDWARDS, b Boston, Mass, Nov 11, 11; m 40; c 4. PHYSICS. Educ: Mass Inst Technol, BS, 40; Stevens Inst Technol, 47-49. Prof Exp: Asst physics, Mass Inst Technol, 40-41, res assoc, 41-42; res physicist, Nat Res Corp, Mass, 42-44 & Interchem Corp, 44-52; res physicist, 52-75, EMER RES PHYSICIST, VISIBILITY LAB, SCRIPPS INST OCEANOG, UNIV CALIF, SAN DIEGO, 75- Concurrent Pos: Chief Scientist, SCOR Discoverer Exped, 70. Mem: Fel AAAS; fel Optical Soc Am; Sigma Xi(treas, 75); Am Soc Limnol Oceanog. Res: Development of instruments for measurement of light in the ocean; collection of fundamental data on ocean optical properties; application of ocean optical data and measuring techniques to the determination of the growth and concentration of oceanic phytoplankton. Mailing Add: Scripps Inst of Oceanog Box 1529 La Jolla CA 92037

TYLER, JOHN HOWARD, b Madison, Wis, Aug 29, 35; m 66. GEOLOGY. Educ: Univ Wis, BS, 58; Va Polytech Inst, MS, 60; Univ Mich, PhD(geol), 63. Prof Exp: Res asst geol, Va Polytech Inst, 58-60; res asst geol & paleont, Univ Mich, 60-63; tech asst, US Geol Surv, Calif, 63-64; post-doctoral fel, Univ of Wales, Swansea, 64-65; engr, Itek Corp, Calif, 65-66 & Mark Systs Inc, 66; asst prof, 66-72, ASSOC PROF GEOL, CALIF STATE UNIV, SAN FRANCISCO, 72- Mem: Geol Soc Am; Soc Econ Paleont & Mineral. Res: Structural geology and tectonics; stratigraphy; sedimentology; photogeology. Mailing Add: Dept of Geol Calif State Univ San Francisco CA 94132

TYLER, JOHN MASON, medicine, deceased

TYLER, LESLIE JUNIOR, organic chemistry, see 12th edition

TYLER, LYNN DOLAN, fluid mechanics, solid mechanics, see 12th edition

TYLER, MAX EZRA, b Groveland, NY, June 1, 16; m 45; c 2. MICROBIAL ECOLOGY. Educ: Cornell Univ, BS, 38; Ohio State Univ, MS, 40, PhD(bact), 48. Prof Exp: Asst bacteriologist, Ohio State Univ, 38-41; instr, Colo State Col, asst, Exp Sta, 41-42; bacteriologist, US War Dept, 46-53; head dept, 53-71, PROF BACT, UNIV FLA, 53- Concurrent Pos: Mem, Conf State & Prov Lab Dirs. Mem: Fel AAAS; Am Soc Microbiol; fel Am Pub Health Asn; Brit Soc Gen Microbiol. Res: Ecology; physiology; taxonomy marine, estuarine bacteria; water pollution; bacterial sampling from air. Mailing Add: Dept of Microbiol 1053 McCarty Univ Fla Gainesville FL 32601

TYLER, STANLEY WARREN, b Lynn, Mass, Oct 24, 11; m 34; c 2. PHYSIOLOGICAL CHEMISTRY. Educ: Univ Mass, BS, 33; Staley Col Spoken Word, BS, 50, MA, 51, hon DA, 52. Prof Exp: Teacher high sch, Mass, 33-35; chemist, Forbes Lithograph Co, 35-36; res chemist, Wirthmore Res Lab, Chas M Cox Co, 36-48; prof math, Staley Col Spoken Word, 48-52; inspector gen, US Army, 52-54 & Ord Corps, 54-55, mil adv to Korean Army, 55-57, res & develop chem engr, Frankford Arsenal, 57-61, Mil Assistance Adv Group, Ethiopia, 61-64, Missile Command, Redstone Arsenal, Ala, 64-66, Off Joint Chiefs of Staff, Dept Defense, Washington, DC, 66 & Picatinny Arsenal, 66-68; res consult, 68-69; CHEM ENGR, US ARMY PICATINNY ARSENAL, DOVER, 69- Mem: Fel AAAS; fel Am Inst Chem. Res: Vitamins; amino acids; nutritional requirements of bacteria; trace minerals in nutrition; biotics and antibiotics; biochemistry; explosives. Mailing Add: 75 Summit Trail Sparta NJ 07871

TYLER, STEPHEN ALBERT, b Hartford, Iowa, May 8, 32; m 62. ANTHROPOLOGY, ANTHROPOLOGICAL LINGUISTICS. Educ: Simpson Col, BA, 57; Stanford Univ, MA, 62, PhD(anthrop), 64. Prof Exp: Asst prof anthrop, Univ Calif, Davis, 64-67; Ford Found res grant, Univ Calif, Berkeley, 67-69; assoc prof, Tulane Univ, 67-70, Am Philos Soc res grant, 70-71; PROF ANTHROP & LING, RICE UNIV, 70-, CHMN DEPT ANTHROP, 71- Concurrent Pos: Spec lectr anthrop, Univ La, Baton Rouge, fall 68; assoc ed, Ann Revs, Inc, 70-; mem adv panel, NSF, 72- Mem: Fel Am Anthrop Asn; Ling Soc Am; Asn Asian Studies. Res: Dravidian linguistics, language and culture; kinship and social organization; cognitive anthropology; India. Mailing Add: Dept of Anthrop Rice Univ Houston TX 77001

TYLER, TIPTON RANSOM, b Milwaukee, Wis, Jan 3, 41; m 62; c 3. BIOCHEMISTRY, PHARMACOLOGY. Educ: Colo State Univ, BS, 63, PhD(nutrit), 68; NC State Univ, MS, 65. Prof Exp: Sr res chemist, Merck Sharp & Dohme Res Labs, 68-74; ASST PROF ANIMAL SCI, UNIV ILL, URBANA-CAMPAIGN, 74-

Mem: AAAS: Am Chem Soc. Res: Drug metabolism; residues in tissues; disposition and clearance from animals. Mailing Add: 124 Animal Sci Lab Univ of Ill Urbana IL 61801

TYLER, VARRO EUGENE, b Auburn, Nebr, Dec 19, 26; m 47; c 2. PHARMACOGNOSY. Educ: Univ Nebr, BS, 49; Univ Conn, MS, 51, PhD(pharmacog), 53. Prof Exp: Assoc prof pharmacog, Univ Nebr, 53-57; assoc prof pharmacog, Univ Wash, 57-61, prof, 61-66, chmn dept pharmacog & dir drug plant gardens, 57-66; DEAN, SCH PHARM & PHARMACAL SCI, PURDUE UNIV, 66- Honors & Awards: Found res achievement award, Am Pharmaceut Asn, 66. Mem: Am Asn Cols Pharm (pres, 70-71), Am Soc Pharmacog (pres, 59-61); Am Pharmaceut Asn; Am Coun Pharmaceut Educ (pres, 74-). Res: Alkaloid biosynthesis; drug plant cultivation; phytochemical analysis; medicinal and toxic constituents of higher fungi. Mailing Add: Sch of Pharm and Pharmacal Sci Purdue Univ Lafayette IN 47907

TYLER, WALTER STEELE, b Caspian, Mich, Nov 2, 25; m 49; c 2. ANATOMY. Educ: Mich State Col, DVM, 51; Univ Calif, Davis, PhD(comp path), 56. Prof Exp: Lectr vet sci, 52-57, from asst prof to assoc prof vet med, 57-67, actg chmn dept, 65-67, chmn dept, 67-70, jr vet, Exp Sta, 52-56, asst vet, 56-62, PROF VET MED, UNIV CALIF, DAVIS, 67-, ANATOMIST, 62-, DIR, CALIF PRIMATE RES CTR, 72- Concurrent Pos: Fel, Postgrad Med Sch, Univ London, 63-64. Mem: Am Asn Anatomists; Am Asn Vet Anat (secy-treas, 66-67); Am Physiol Soc; Am Soc Zool; Am Vet Med Asn. Res: Relationship of structure to function in health and disease; pulmonary anatomy; emphysema; air pollution; histochemistry; scanning and transmission; electron microscopy; comparative anatomy; primate morphology. Mailing Add: Calif Primate Res Ctr Univ of Calif Davis CA 95616

TYLER, WILBER JONES, dairy science, genetics, see 12th edition

TYLER, WILLARD PHILIP, b Newton, Mass, Nov 8, 09; m 35, 55; c 3. ANALYTICAL CHEMISTRY. Educ: Ore State Col, BS, 31, MS, 33; Univ Ill, PhD(anal chem), 38. Prof Exp: Instr, Clark Jr Col, 34-36; from res analyst, 36-38, SECT LEADER, RES CTR, B F GOODRICH CO, 38- Mem: Am Chem Soc. Res: Classical and instrumental chemical analysis; absorption spectroscopy; x-ray diffraction; gas chromatography; electroanalytical methods; high polymers. Mailing Add: 8471 Whitewood Rd Brecksville OH 44141

TYLER, WINFIELD WARREN, b Batavia, NY, Sept 8, 20; m 49; c 5. SOLID STATE PHYSICS. Educ: Cornell Univ, AB, 43, PhD(physics), 50. Prof Exp: Engr, US Naval Res Lab, 43-45; res assoc, Knolls Atomic Power Lab, Gen Elec Co, 50-52, res lab, 52-59, mgr light prod studies sect, 59-62; mgr, light prod studies sect, Xerox Corp, 62-64, dir fundamental res lab, 64-65, asst vpres, 65-66, res div, 66-67, vpres res labs, 67-74; MEM STAFF CORNELL RES FOUND, CORNELL UNIV, 74- Mem: Fel Am Phys Soc. Res: Crystal growth; low temperature physics; photoconductivity; semiconductor studies; luminescence. Mailing Add: Cornell Res Found 476 Uris Hall Cornell Univ Ithaca NY 14853

TYLUTKI, EDMUND EUGENE, b Chicago, Ill, Nov 6, 26; m 56; c 6. MYCOLOGY. Educ: Univ Ill, BS, 51, MS, 52; Mich State Univ, PhD(mycol), 55. Prof Exp: Asst bot, Univ Ill, 51-52; bot & plant path, Mich State Univ, 52-55; actg asst prof plant path & actg asst plant pathologist, Wash State Univ, 56; asst prof, 56-65, ASSOC PROF BOT, UNIV IDAHO, 65- Concurrent Pos: Ed-in-chief, Jour Idaho Acad Sci, 65-, pres, 70-71; actg chmn, Dept Biol Sci, Univ Idaho, 75-76. Mem: Agaricales of Idaho; Classifaction Soc; Int Asn Plant Taxonomists. Res: Computer applications to fungal taxonomy; taxonomy of fleshy fungi. Mailing Add: Dept of Biol Sci Univ of Idaho Moscow ID 83843

TYMAN, JOHN LANGTON, b Gravesend, Eng, Aug 24, 35; Can citizen; m 62; c 2. GEOGRAPHY. Educ: Oxford Univ, BA, 59, MA, 63; McGill Univ, MA, 61. Prof Exp: Lectr geog, 62, from asst prof to assoc prof, 62-71, head dept, 62-74, PROF GEOG, BRANDON UNIV, 71- Concurrent Pos: Vis prof, Univ Western Australia, 73. Mem: Asn Am Geogr; Can Asn Geogr. Res: Historical and cultural geography, especially pioneer settlement. Mailing Add: Dept of Geog Brandon Univ Brandon MB Can

TYMCHATYN, EDWARD DMYTRO, b Leoville, Sask, Nov 11, 42. MATHEMATICS. Educ: Univ Sask, BA, 63, Hons, 64; Univ Ore, MA, 65, PhD(math), 68. Prof Exp: Asst prof, 68-71, ASSOC PROF MATH, UNIV SASK, 71- Concurrent Pos: Nat Res Coun res grant, Univ Sask, 69- Mem: Am Math Soc. Res: General topology and point set topology; continua; low dimensional spaces; partially ordered spaces and topological semigroups. Mailing Add: Dept of Math Univ of Sask Saskatoon SK Can

TYNAN, EUGENE JOSEPH, b Middletown, Conn, Sept 15, 24. PALYNOLOGY. Educ: Univ Conn, BA, 53; Univ Mass, MS, 56; Univ Okla, PhD, 62. Prof Exp: Instr, 59-62, asst prof, 62-70, actg chmn dept, 66-70, ASSOC PROF GEOL, UNIV RI, 70- Mem: Sigma Xi; Am Asn Stratig Palynologists. Res: Taxonomy, ecology and stratigraphic application of spores, pollen and other lesser known groups of microfossils. Mailing Add: Dept of Geol Univ of RI Kingston RI 02881

TYNDALL, JESSE PARKER, b Jones Co, NC, Jan 9, 25; m 65; c 1. BIOLOGY, SCIENCE EDUCATION. Educ: Atlantic Christian Col, AB, 45; Univ NC, Chapel Hill, MA, 49; Univ Fla, EdD, 56. Prof Exp: Teacher, Jones County Bd Educ, NC, 45-47; from instr to assoc prof biol & sci educ, 49-52, prof sci educ, 52-73, PROF BIOL, ATLANTIC CHRISTIAN COL, 52-, CHMN DEPT SCI, 54- Concurrent Pos: Mem bd dirs, Joint Comn Nursing Educ, NC Bd Educ & Bd Gov, 69-72, chmn, 72-73. Mem: Fel AAAS. Res: Genetics; nursing education. Mailing Add: Dept of Sci Atlantic Christian Col Wilson NC 27893

TYNDALL, RICHARD LAWRENCE, microbiology, virology, see 12th edition

TYNER, DAVID ANSON, b Berrien Co, Mich, Feb 19, 22; m 49; c 4. ORGANIC CHEMISTRY. Educ: Univ Mich, BS, 44, MS, 49, PhD(org chem), 52. Prof Exp: RES CHEMIST, G D SEARLE & CO, 52- Concurrent Pos: Civilian res chemist, Manhattan Proj, 44-45. Mem: Am Chem Soc. Res: Total and partial synthesis of steroids; peptides. Mailing Add: 909 Glendale Rd Glenview IL 60025

TYNER, EDWARD HENRY, b Chicago, Ill, June 15, 07; m 32; c 3. SOILS. Educ: Univ Nebr, BS, 30; Univ Wis, MS, 32, PhD(soils), 34. Prof Exp: Instr soils, Univ Nebr, 34-35; asst soil scientist, Bur Chem & Soils, USDA, 35-36; prof soils, NDak Col, 36-38; asst prof agron & asst agronomist, Exp Sta, Univ WVa, 38-43, assoc prof & assoc agronomist, 43-48, prof, 48-50; PROF SOIL FERTILITY, UNIV ILL, URBANA, 50- Concurrent Pos: Vis prof, Univ Philippines, 58-59; soil scientist, Rockefeller Found, 62-63; Fulbright lectr, Imp Col Trop Agr, Trinidad, 65; consult, IRI Res Inst, Brazil, 66. Mem: Soil Sci Soc Am; Am Soc Agron; Am Chem Soc; Int Soil Sci Soc. Res: Plant nutrition; tree-soil relationships in the Great Plains; soil dispersion method; strip mine reclamation; soil phosphate chemistry; tropical agriculture; soil-economic studies; electron microprobe analysis techniques; sulfur chemistry of tidal marshlands and upland tropical soils. Mailing Add: Dept of Agron Univ of Ill Urbana IL 61803

TYNER, GEORGE S, b Omaha, Nebr, Oct 9, 16; m; c 2. SURGERY, OPHTHALMOLOGY. Educ: Univ Nebr, BS, 40, MD, 42; Univ Pa, MS, 52; Am Bd Ophthal, dipl, 50. Prof Exp: Intern, Philadelphia Gen Hosp, 42-43, resident ophthal, 47-48; resident, Hosp Univ Pa, 48-51, asst instr ophthal, Sch Med, Univ Pa, 48-52; from asst instr to asst clin prof, Sch Med, Univ Colo, 52-61, assoc dean & asst to vpres med affairs, 63-71, chief glaucoma clin & assoc prof ophthal, 64-71; PROF OPHTHAL, SCH MED, TEX TECH UNIV, 71-, DEAN, 74- Concurrent Pos: Res fel, Univ Pa, 48-51; asst abstr ed, Am J Ophthal, 52-57; pvt pract, Colo, 52-61; mem, Colo State Bd Basic Sci Exam, 62-67; mem consult staff, Children's Hosp & Denver Gen Hosp; mem courtesy staff, St Luke's Hosp; mem assoc staff, St Mary's Hosp, 73. Mem: Fel Am Col Surg; Am Acad Ophthal & Otolaryngol; AMA. Mailing Add: Off of the Dean Tex Tech Univ Sch of Med Lubbock TX 79409

TYNER, HOWARD DALE, physics, see 12th edition

TYNES, ARTHUR RICHARD, b Great Falls, Mont, Oct 1, 26; m 46; c 5. OPTICS. Educ: Mont State Univ, BS, 50; Ore State Univ, MS, 53, PhD(physics), 63. Prof Exp: Physicist, US Bur Mines, 51-54; instr physics, Ore State Univ, 54-61; MEM TECH STAFF, BELL TEL LABS, 61- Mem: Optical Soc Am. Res: Spectroscopy; plasma diagnositics; physical optics; applications of lasers to optical measurements; light scattering and light transmission; fiber optics. Mailing Add: 120 Bruce Rd Red Bank NJ 07701

TYOR, MALCOLM PAUL, b New York, NY, Apr 20, 23; m 47; c 4. MEDICINE. Educ: Univ Wis, AB, 44; Duke Univ, MD, 46; Am Bd Internal Med, dipl, 57. Prof Exp: Intern, Madison Gen Hosp, Univ Wis-Madison, 46-47; resident med, Bowman Gray Sch Med, Wake Forest Col, 49-51, fel gastroenterol, 51-52; clinician, Med Div, Oak Ridge Inst Nuclear Studies, 52-54; physician, pvt pract, 54-55; assoc, 55-57, from asst prof to assoc prof, 57-62, PROF MED, MED SCH, DUKE UNIV, 62-, CHIEF DIV GASTROENTEROL, MED CTR, 65- Concurrent Pos: Asst chief med serv & chief radioisotope serv & gastroenterol, Vet Admin Hosp, Durham, 55- Mem: AAAS; AMA; Am Gastroenterol Asn; Am Fedn Clin Res; Am Soc Clin Invest. Res: Gastroenterology. Mailing Add: Div of Gastroenterol Duke Univ Med Ctr Durham NC 27706

TYPPO, JOHN TAUNO, human nutrition, biochemistry, see 12th edition

TYREE, SHEPPARD YOUNG, JR, b Richmond, Va, July 4, 20; m 43; c 5. INORGANIC CHEMISTRY. Educ: Mass Inst Technol, BS, 42, PhD(inorg chem), 46. Prof Exp: From asst to instr chem, Off Sci Res & Develop, Mass Inst Technol, 42-46; from asst prof to assoc prof, 46-57, prof, Univ NC, 57-66; chmn dept, 68-73, PROF CHEM, COL WILLIAM & MARY, 66- Concurrent Pos: Lectr, Morehead Planetarium, 51-52; vis prof, NC Col Durham, 52 & Puerto Rico, 63; sci off, US Off Naval Res, 54-55, liaison scientist, London, 65-66; chmn, Gordon Res Conf Inorg Chem, 57. Honors & Awards: Herty Medal, Am Chem Soc, 64- Mem: Fel AAAS; Am Chem Soc; The Chem Soc. Res: Metal halides; solution chemistry of inorganic ions. Mailing Add: Dept of Chem Col of William & Mary Williamsburg VA 23185

TYREY, LEE, b Chicago, Ill, Oct 26, 37; m 61; c 3. NEUROENDOCRINOLOGY, PHYSIOLOGY. Educ: Univ Ill, Urbana, BSc, 63, MSc, 64, PhD(physiol), 69. Prof Exp: Assoc, 70-72, ASST PROF ANAT & ASSOC PROF OBSTET & GYNEC, MED CTR, DUKE UNIV, 76-, DIR OBSTET-GYNEC ENDOCRINE LAB, 70- Concurrent Pos: NIH res fel, Med Ctr, Duke Univ, 69-70, Duke Endowment res grant, 71-73, Pop Coun res grant, 74-76; NC United Community Serv grant, 75-76. Res: Neural control of gonadotropin secretion; radioimmunoassay of protein hormones; gonadotropin receptor sites. Mailing Add: Dept of Obstet-Gynec Duke Univ Med Ctr Box 3244 Durham NC 27710

TYRL, RONALD JAY, b Lawton, Okla, June 16, 43; m 65; c 3. PLANT TAXONOMY. Educ: Park Col, BA, 64; Ore State Univ, MS, 67, PhD(syst bot), 69. Prof Exp: Herbarium asst taxon, Ore State Univ, 65-69; asst prof biol, Park Col, 70-72; ASST PROF BOT & CURATOR HERBARIUM, OKLA STATE UNIV, 72- Mem: Am Soc Plant Taxon. Res: Plant biosystematics; evolutionary mechanisms; cytogenetic patterns. Mailing Add: Sch Biol Sci Okla State Univ Stillwater OK 74074

TYROLER, JESSE FRANKLIN, physics, see 12th edition

TYRRELL, DAVID, b Doncaster, Gt Brit, Aug 4, 40; m 65. BIOCHEMISTRY. Educ: Univ Birmingham, BSc, 61; Imp Col, dipl & Univ London, PhD(biochem), 64. Prof Exp: Demonstr biochem, Imp Col, Univ London, 61-64; RES SCIENTIST, INSECT PATH RES INST, 64- Mem: Can Soc Microbiol; Am Soc Microbiol. Res: Fungal biochemistry and physiology. Mailing Add: Insect Path Res Inst PO Box 490 Sault Ste Marie ON Can

TYRRELL, ELIZABETH ANN, b Pittsfield, Mass, Oct 16, 31. MICROBIOLOGY. Educ: Simmons Col, BS, 53; Univ Mich, MS, 56, PhD(bact), 62. Prof Exp: Res asst virol, Parke, Davis & Co, Mich, 53-55; from inst to asst prof, 60-71, ASSOC PROF MICROBIOL, SMITH COL, 71- Mem: AAAS; Am Soc Microbiol. Res: Concentrated culture of microorganisms; autolysis in bacteria. Mailing Add: Clark Sci Ctr Smith Col Northampton MA 01060

TYRRELL, HENRY FLANSBURG, b Gloversville, NY, Aug 4, 37; m 69. NUTRITION, BIOMETRY. Educ: Iowa State Univ, BS, 59; Cornell Univ, MS, 64, PhD(nutrit), 66. Prof Exp: Asst prof animal sci, Cornell Univ, 66-69; res dairy husbandman, Energy Metab Lab, Animal Husb Res Div, Agr Res Serv, USDA, 69-72, RES ANIMAL SCIENTIST, RUMINANT NUTRIT LAB, NUTRIT INST, 72- Mem: Am Dairy Sci Asn; Am Soc Animal Sci. Res: Utilization of energy by domestic animals; nitrogen utilization by ruminant animals; energy requirements for growthl lactation in cattle. Mailing Add: Ruminant Nutrit Lab Nutrit Inst Agr Res Ctr Beltsville MD 20705

TYRRELL, JAMES, b Kilsyth, Scotland, Apr 19, 38; m 64; c 1. THEORETICAL CHEMISTRY. Educ: Univ Glasgow, BS, 60, PhD(chem), 63. Prof Exp: Teaching fel chem, McMaster Univ, 63-65; fel, Div Pure Physics, Nat Res Coun, 65-67; asst prof, 67-72, ASSOC PROF CHEM, SOUTHERN ILL UNIV, 72- Mem: Am Chem Soc. Res: High resolution spectroscopy in the ultraviolet, visible and infrared of vapors; theoretical calculations on atoms and molecules. Mailing Add: Dept of Chem Southern Ill Univ Carbondale IL 62901.

TYRRELL, WILLIS W, JR, b Mobile, Ala, Feb 12, 30; m 52; c 4. GEOLOGY. Educ: Fla State Univ, BS, 52; Yale Univ, MS, 54, PhD(geol), 57. Prof Exp: Asst, Fla State Univ, 52-53; field asst, Texaco, 54; geologist, Pan Am Petrol Corp, 55-64, res group supvr, 64-68, SR STAFF GEOLOGIST, AMOCO PROD CO, 68- Mem: Geol Soc Am; Am Asn Petrol Geol; Soc Econ Paleont & Mineral. Res: Stratigraphy;

sedimentary petrology; carbonate petrology and sedimentation. Mailing Add: Amoco Prod Co Box 50879 New Orleans LA 70150

TYSON, BRUCE CARROLL, JR, b Greenville, NC, Aug 17, 36. ANALYTICAL CHEMISTRY. Educ: Duke Univ, BS, 58; Princeton Univ, MA, 60; Univ Del, PhD(anal chem), 69. Prof Exp: Res chemist, US Army Edgewood Arsenal, Md, 63-67; res assoc anal chem, 68-75, DIR ANAL RES, A H ROBINS CO, 75- Mem: Am Chem Soc. Res: Drug purity evaluation; identification of trace impurities; polarography; thermometric titrations. Mailing Add: A H Robins Co 1211 Sherwood Ave Richmond VA 23220

TYSON, GEORGE NOBLIT, JR, b Philadelphia, Pa, Jan 7, 06; m 39; c 2. INORGANIC CHEMISTRY. Educ: Pomona Col, AB, 35; Univ Southern Calif, MS, 39, PhD(chem), 41. Prof Exp: Asst chem, Pomona Col, 35-36, instr, 38-42, from asst prof to assoc prof, 42-44; res chemist, Union Oil Co, Calif, 36-38; dir res, Pac Coast Borax Co, 44-51; asst to pres, Electro Circuits, Inc, 51-52; res coordinator, Olin Mathieson Chem Corp, 52-55, res mgr, 55-58, assoc res dir, 58-59; adminstr dir, Nat Eng Sci Co, 59-61; mgr customer rels, Aerojet-Gen Corp, 61-65; exec vpres, Nesbitt Food Prods, 65-68; PRES, TRANS NUCLEAR CHEM CO, 68- Mem: Am Chem Soc; Sigma Xi. Res: Magnetic susceptibility and configuration studies of complex compounds; nature of catalysis. Mailing Add: 4950 Live Oak Canyon Rd La Verne CA 91750

TYSON, GRETA E, b Medford, Mass, Nov 2, 33. ZOOLOGY. Educ: State Teachers Col Bridgewater, BS, 55; Univ NH, MS, 57; Univ Calif, Berkeley, PhD(zool), 67. Prof Exp: NIH fel biol struct, Univ Wash, 67-69, NIH fel path, 69-70, instr, 70-72; ASST PROF, UNIV MD, BALTIMORE, 72- Mem: Am Asn Anat; Am Soc Cell Biol; Am Soc Zool; NY Acad Sci; Soc Invert Path. Res: Comparative renal morphology and physiology; structure and function of microtubules and microfilaments. Mailing Add: Dept of Path 369 Howard Hall Univ of Md Baltimore MD 21201

TYSON, J ANTHONY, b Pasadena, Calif, Apr 5, 40; Stanford Univ, BS, 62; Univ Wis, MS, 64, PhD(physics), 67. Prof Exp: Nat Res Coun-Air Force Off Sci Res fel, Univ Chicago, 67-69; MEM TECH STAFF, BELL TEL LABS, 69- Mem: Am Astron Soc; Am Phys Soc. Res: Experimental gravitation and relativity; gravitational radiation; astrophysics; infrared astronomy; optical astronomy; x-ray astronomy; radio astronomy. Mailing Add: Bell Tel Labs Murray Hill NJ 07974

TYSON, RALPH ROBERT, b Philadelphia, Pa, Dec 14, 20; m 45; c 3. SURGERY. Educ: Dartmouth Col, AB, 41; Univ Pa, MD, 44; Am Bd Surg, dipl, 52; Pan-Am Med Asn, dipl. Prof Exp: From instr to assoc prof surg, 52-62, chief sect vascular surg, 62-73, PROF SURG, SCH MED, TEMPLE UNIV, 62-, CHMN DEPT & DIV, 73- Concurrent Pos: Asst attend surgeon, St Christopher's Hosp, Philadelphia; consult surg, Chestnut Hill Hosp, Philadelphia & Wilkes Barre Vet Admin Hosp; mem Nat Bd Med Exam. Mem: AMA; fel Am Col Surg; NY Acad Sci; Am Fedn Clin Res; Int Cardiovasc Soc. Res: Variations in blood supply to the liver; 9-amino-arcidine in wound infections; venography; venous pressures; large bowel preparation; cross circulation in dogs; staphyloccus infections. Mailing Add: Dept of Surg Temple Univ Sch of Med Philadelphia PA 19140

TYSVER, JOSEPH BRYCE, b Hazen, NDak, Mar 2, 18; m 45; c 1. APPLIED STATISTICS, OPERATIONS RESEARCH. Educ: Wash State Univ, BA, 42, MA, 48; Univ Mich, PhD(statist), 57. Prof Exp: Assoc res engr, Eng Res Inst, Univ Mich, 51-57; res specialist, Boeing Airplane Co, 57-63; statistician, Stanford Res Inst, 63-67; ASSOC PROF STATIST, NAVAL POSTGRAD SCH, 67-, ASSOC PROF OPER RES, 73- Concurrent Pos: Statistician, Litton Sci Support Lab, 68. Mem: Inst Math Statist; Am Math Soc; Math Asn Am; Inst Indust & Appl Math. Res: Statistical applications for engineering and military systems; sensitivity testing. Mailing Add: Dept Opers Res & Admin Sci Naval Postgrad Sch Monterey CA 93940

TYTELL, ALFRED A, b New York, NY, Aug 1, 15; m 45; c 2. BIOCHEMISTRY. Educ: Tufts Col, BS, 36; Mass Inst Technol, PhD(biochem), 40. Prof Exp: Instr biochem, Col Med, Univ Cincinnati, 40-44, from asst prof to assoc prof, 44-52; res assoc, 52-58, sr investr & dir cell biol, Virus & Cell Biol Res Div, 58-75, SR SCI & DIR CELL BIOL, MERCK, SHARP & DOHME RES LABS, 75- Concurrent Pos: Civilian with Off Sci Res & Develop & USPHS, 44. Mem: Am Asn Immunol; NY Acad Sci. Res: Bacterial toxins and toxoids; biochemistry and nutrition of microorganisms and mammalian cells; enzymes and microorganisms; antibiotics; virus chemistry; virus and cell biology. Mailing Add: Merck Sharp & Dohme Res Labs Virus & Cell Biol Res West Point PA 19486

TYZNIK, WILLIAM JOHN, b Milwaukee, Wis, Apr 26, 27; m 50; c 5. ANIMAL NUTRITION. Educ: Univ Wis, BS, 48, MS, 49, PhD(nutrit), 51. Prof Exp: Asst nutrit, Univ Wis, 48-51; from asst prof to assoc prof animal nutrit, 51-59, PROF ANIMAL SCI & VET PREV MED, OHIO STATE UNIV, 59- Mem: Am Soc Animal Sci; Am Dairy Sci Asn. Res: Ruminant and monogastric nutrition. Mailing Add: Dept of Animal Sci Ohio State Univ Columbus OH 43210

TZAGOURNIS, MANUEL, b Youngstown, Ohio, Oct 20, 34; m 58; c 5. MEDICINE, ENDOCRINOLOGY. Educ: Ohio State Univ, BS, 56, MD, 60, MMS, 65. Prof Exp: Intern med, Philadelphia Gen Hosp, Pa, 60-61; resident internal med, Univ Hosp, 61-62 & 64-65, chief resident internal med, 66-67, asst prof med, Col Med, 67-70, assoc prof med, 70-74, PROF MED, COL MED, OHIO STATE UNIV, 74-, ASSOC DEAN, 75- Concurrent Pos: USPHS fel endocrinol & metab, Ohio State Univ Hosp, 65-66. Mem: Am Fedn Clin Res; Am Diabetes Asn; AMA. Res: Diabetes, glucose metabolism and insulin secretion, especially as they relate to lipid disorders and coronary atherosclerosis. Mailing Add: Ohio State Univ Hosp Columbus OH 43210

TZIANABOS, THEODORE, b Manchester, NH, Feb 12, 33. VIROLOGY. Educ: Univ NH, BA, 55, MS, 59; Univ Mass, PhD(microbiol), 65; Am Bd Med Microbiol, dipl. Prof Exp: Res instr microbiol, Dept Poultry Dis, Univ NH, 55-57; microbiologist, Diagnostic Virol, State Mass, 59-60; res instr microbiol, Dept Vet Sci, Univ Mass, 60-65; resident, Ctr Dis Control, 65-67; res microbiologist, Med Sci Div, Ft Detrick, 67-70; microbiologist, Beckman Instruments, 70-71; RES MICROBIOLOGIST VIROL, CTR DIS CONTROL, 71- Mem: Am Soc Microbiologists; Am Soc Trop Med & Hyg. Res: Development and research on rickettsial products involving serologic tests, including fluorescent microscopy; purification and protein composition of rickettsiae. Mailing Add: Ctr for Dis Control 1600 Clifton Rd Atlanta GA 30333

U

U, RAYMOND, b Pyoung-Yang, Korea, Dec 27, 36; US citizen; m 69; c 1. GENETICS, RADIOBIOLOGY. Educ: Northern Ill Univ, BS, 63, MS, 65; Kyoto Univ, PhD(radiobiol, genetics), 70. Prof Exp: Res assoc radiation genetics, Northern Ill Univ, 65-66; instr radiol, Albert Einstein Col Med, 66-67; assoc radiol, 67-70, ASSOC PROF RADIOL, SCH MED, DUKE UNIV, 71- Concurrent Pos: Consult res biologist, US Vet Admin Hosp, 67-; fel, US-Japan Coop Cancer Res Prog on high linear energy transfer radiother, 76. Mem: AAAS; Genetics Soc Am; Am Inst Biol Sci; Radiation Res Soc; Tissue Cult Asn. Res: Cancer; low and high linear energy transfer radiations and combined therapy with chemotherapeutic chemicals; chemical modification of repair and recovery of radiation-induced chromosome aberration; development of radiation damage and post-radiation repair processes at the chromosomal level in mammalian cells. Mailing Add: Dept of Radiol Duke Univ Med Ctr Durham NC 27710

UBELAKER, DOUGLAS HENRY, b Horton, Kans, Aug 23, 46; m 75. PHYSICAL ANTHROPOLOGY. Educ: Univ Kans, Lawrence, BA, 68, PhD(anthrop), 73. Prof Exp: ASSOC CUR PHYS ANTHROP, SMITHSONIAN INST, 73-; ASSOC PROF ANTHROP, AM UNIV, 75- Mem: AAAS; Am Asn Phys Anthrop; Am Anthrop Asn; Am Acad Forensic Sci; Soc Am Archaeol. Res: Physical anthropology of North America, Latin America; skeletal biology; prehistoric demography; forensic problems in osteology. Mailing Add: Nat Mus of Natural Hist Smithsonian Inst Washington DC 20560

UBELAKER, JOHN E, b Everest, Kans, Mar 21, 40. PARASITOLOGY. Educ: Univ Kans, BA, 62, MA, 65, PhD(zool), 67. Prof Exp: Fel parasitol, Emory Univ, 67-68; asst prof biol, 68-71, ASSOC PROF BIOL, SOUTHERN METHODIST UNIV, 71- Mem: Am Soc Parasitol; Am Soc Zool; Am Micros Soc; Wildlife Dis Asn. Res: Helminthology. Mailing Add: Dept of Biol Southern Methodist Univ Dallas TX 75222

ÜBERALL, HERBERT MICHAEL, b Neunkirchen, Austria, Oct 14, 31; US citizen; m 57; c 2. THEORETICAL NUCLEAR PHYSICS, ACOUSTICS. Educ: Univ Vienna, PhD(theoret physics), 53; PhD(theoret physics), 56. Prof Exp: Res fel physics, Univ Liverpool, 56-57; Ford Found fel, European Org Nuclear Res, Geneva, Switz, 57-58; res physicist, Carnegie Inst Technol, 58-60; asst prof, Univ Mich, 60-64; assoc prof, 64-65, PROF PHYSICS, CATH UNIV AM, 65- Concurrent Pos: Sr res physicist, Conductron Corp, Mich, 61-64; consult, Naval Res Lab, Washington, DC, 66- Mem: Fel Am Phys Soc; Acoust Soc Am. Res: Theoretical nuclear and elementary particle physics; scattering and radiation theory; theory of muons, neutrinos, and nuclear photo excitation and electroexcitation; electromagnetic and acoustic waves; underwater acoustics. Mailing Add: Dept of Physics Cath Univ of Am Washington DC 20064

UCCI, POMPELIO ANGELO, b Warwick, RI, Jan 15, 22; m 49; c 4. PHYSICAL CHEMISTRY, MATHEMATICS. Educ: Univ RI, BS, 43. Prof Exp: Res chemist, Celanese Corp Am, 43-44; 46-52; sr res chemist, 52-54, group leader synthetic fibers, 54-58, sect head, 58-69, site mgr, New Enterprise Div, 69-71, sr res specialist, 71-75, ENG FEL, TEXTILES DIV, MONSANTO CO, 75- Mem: AAAS; Am Chem Soc. Res: Solution and melt properties of natural and synthetic fiber forming polymers; fundamental mechanical and engineering properties; statistics and quality control; paper making from synthetic fibers; testing equipment and procedures. Mailing Add: Textiles Div Monsanto Co Box 12830 Pensacola FL 32575

UCHIDA, HENRY SHIGETOMI, physical chemistry, see 12th edition

UCHIDA, IRENE AYAKO, b Vancouver, BC, Apr 8, 17. GENETICS. Educ: Univ Toronto, PhD(human genetics), 51. Prof Exp: Res assoc, Hosp Sick Children, Toronto, Ont, 51-59; proj assoc, Univ Wis, 59-60; lectr pediat, Univ Man, 60-62, from asst prof to assoc prof, 63-69, asst prof anat, 67-69; PROF PEDIAT & DIR REGIONAL CYTOGENETICS LAB, MED CTR, McMASTER UNIV, 69- Concurrent Pos: Dir med genetics, Children's Hosp, Winnipeg, 60-69; mem, Sci Coun Can, 70-73. Mem: AAAS; Am Inst Biol Sci; Am Soc Human Genetics (pres, 68); Genetics Soc Am; Genetics Soc Can. Res: Human genetics; cytogenetics. Mailing Add: Dept of Pediat McMaster Univ Hamilton ON Can

UCHIDA, RICHARD NOBORU, b Honolulu, Hawaii, Sept 4, 29; m 55; c 3. MARINE SCIENCES. Educ: Univ Wash, Seattle, BS, 51; FISHERY RES BIOLOGIST, US DEPT OF COM, NOAA, NAT MARINE FISHERIES SERV, SOUTHWEST FISHERIES CTR, HONOLULU LAB, 54- Concurrent Pos: Counr, Hawaiian Acad Sci, 75-76. Honors & Awards: Superior performance award, Nat Marine Fisheries Serv, 62, 69, Letter of Commendation, 67, Spec Achievement Award, 74. Mem: AAAS; Am Fisheries Soc; Am Inst Fishery Res Biol; Smithsonian Inst; Int Oceanog Found. Res: Tuna baitfish research; studies on catch and effort in the Hawaiian skipjack fishery; behavior of skipjack tuna schools; studies dealing with skipjack stocks in Pacific and Indian Oceans. Mailing Add: Nat Marine Fisheries Serv Honolulu Lab PO Box 3830 Honolulu HI 96812

UCHTMAN, VERNON ALBERT, b Cincinnati, Ohio, Oct 18, 41; m 68; c 3. PHYSICAL INORGANIC CHEMISTRY. Educ: Univ Cincinnati, BS, 63; Univ Wis-Madison, PhD(inorg chem), 68. Prof Exp: RES CHEMIST, WINTON HILL TECH CTR, PROCTER & GAMBLE CO, 68- Mem: AAAS; Am Chem Soc; Sigma Xi. Res: Molecular structure determination; x-ray crystallography; detergency; metal ion binding and control in aqueous solutions. Mailing Add: Winton Hill Tech Ctr Procter & Gamble Co Cincinnati OH 05224

UCHUPI, ELAZAR, b New York, NY, Oct 31, 28. GEOLOGY. Educ: City Col New York, BS, 52; Univ Southern Calif, MS, 54, PhD(geol), 62. Prof Exp: Res asst geol, Univ Southern Calif, 55-60; ASSOC SCIENTIST, WOODS HOLE OCEANOG INST, 62- Mem: Geol Soc Am; Am Asn Petrol Geol; Soc Econ Paleont & Mineral; Am Geophys Union; Am Soc Limnol & Oceanog. Res: Sedimentation; submarine topography; structure; recent marine sediments; topography and structure of continental margins. Mailing Add: Dept of Geol & Geophys Woods Hole Oceanog Inst Woods Hole MA 02543

UCKO, DAVID A, b New York, NY, July 9, 48; m 72. INORGANIC CHEMISTRY, BIOINORGANIC CHEMISTRY. Educ: Columbia Col, BA, 69; Mass Inst Technol, PhD(chem), 72. Prof Exp: Asst prof chem, Hostos Community Col, 72-76; ASST PROF CHEM, ANTIOCH COL, 76- Concurrent Pos: NIH fel, Columbia Univ, 72; fac res fel, Res Found, State Univ NY, 75. Mem: Am Chem Soc; Sigma Xi; The Chem Soc; NY Acad Sci. Res: Interaction of transition metal complexes with nucleic acids and other molecules of biological interest. Mailing Add: 646 Omar Circle Yellow Springs OH 45387

UDALL, JOHN ALFRED, b Holbrook, Ariz, Sept 11, 29; m 50; c 3. INTERNAL MEDICINE, CARDIOLOGY. Educ: Brigham Young Univ, BS, 51; Temple Univ, MD, 58. Prof Exp: Resident internal med & cardiol, Med Ctr, Univ Calif, San Francisco, 59-62; dir med educ, Maricopa County Gen Hosp, Phoenix, Ariz, 62-66; asst prof, 66-69, ASSOC PROF MED, COL MED, UNIV CALIF, IRVINE, 69- Concurrent Pos: Consult cardiovasc dis, Long Beach Vet Admin Hosp, Calif, 66- & Fairview State Hosp Retarded Children, Costa Mesa, 69-; dir med serv, Orange County Med Cyr, Calif, 71- Mem: AAAS; fel Am Col Cardiol; fel Am Col Physicians; Am Fedn Clin Res; Asn Hosp Med Educ. Res: Pervenous pacemaker electrode endocardial implantation; clinical research in all aspects of oral anticoagulant therapy, especially the problem of the lack of stability of long-term therapy. Mailing Add: Dept of Med Univ of Calif Col of Med Irvine CA 92664

UDALL, ROBERT HOVEY, b Ithaca, NY, Aug 4, 16; m 43. ANIMAL NUTRITION. Educ: Cornell Univ, AB, 38, DVM, 41, PhD(animal nutrit), 51. Prof Exp: Scripps fel, San Diego Zool Soc, 41-42; asst vet, Agr Exp Sta, Univ Ky, 46-47; assoc prof physiol & vet physiol, 51-54, assoc prof vet path, 54-59, PROF VET PATH, COLO STATE UNIV, 59- Concurrent Pos: Chief of party, Colo State Univ-US AID, Univ Col Nairobi, Kenya, 65-70. Mem: Am Soc Animal Sci; Am Vet Med Asn. Res: Metabolic diseases; biochemical pathology. Mailing Add: Dept of Path Colo State Univ Ft Collins CO 80521

UDELHOFEN, JOHN HENRY, b Chicago, Ill, Nov 28, 31; m 55; c 5. PETROLEUM CHEMISTRY. Educ: St Joseph Col, BS, 53; Iowa State Univ, PhD(chem), 58. Prof Exp: Proj chemist, Am Oil Co, 58-63, sr proj chemist, 64-69, sr res chemist, 69-73, RES SUPVR, AMOCO CHEM CORP, 73- Mem: Am Chem Soc. Res: Synthesis and development of petrochemicals. Mailing Add: Amoco Chem Corp Naperville IL 60540

UDEM, STEPHEN ALEXANDER, b New York, NY, Apr 4, 44. VIROLOGY, INFECTIOUS DISEASES. Educ: City Col New York, BS, 64; Albert Einstein Col Med, PhD(genetics), 71, MD, 72. Prof Exp: Intern internal med, Bronx Munic Hosp Complex, 72-73, resident, 73-74; NIH fel infectious dis, Montefiore/Albert Einstein/Jacobi Hosps & Albert Einstein Col Med, 74-76; ASST PROF MED, DEPT CELL BIOL, ALBERT EINSTEIN COL MED, 76- Mem: Am Soc Microbiol. Res: Investigation of persistent viral infections and their relationship to the production of chronic disease, particularly chronic neurological and rheumatic diseases. Mailing Add: Dept of Cell Biol Albert Einstein Col of Med New York NY 10461

UDEN, PETER CHRISTOPHER, b Southampton, Eng, May 19, 39; m 67; c 2. ANALYTICAL CHEMISTRY. Educ: Bristol Univ, BSc, 61, PhD(chem), 64. Prof Exp: Instr chem, Univ Ill, Urbana, 65-66; ICI fel, Univ Birmingham, 66-67, from lectr to asst prof, 67-72, ASSOC PROF ANAL CHEM, UNIV MASS, AMHERST, 72- Concurrent Pos: Mallinckrodt Chem Corp res assoc, Univ Ill, Urbana, 64-66. Mem: Am Chem Soc; The Chem Soc; Brit Inst Petrol. Res: Analytical and inorganic chemistry; separation and thermal methods; gas and liquid chromatography; mass spectrometry; metal complexes. Mailing Add: Dept of Chem Univ of Mass Amherst MA 01002

UDENFRIEND, SIDNEY, b New York, NY, Apr 5, 18; m 43; c 2. BIOCHEMISTRY. Educ: City Col New York, BS, 39; NY Univ, MS, 42, PhD(biochem), 48. Hon Degrees: DSc, New York Med Col, 74. Prof Exp: Lab asst bact, City Dept Health, New York, 40-42; asst chemist, Res Div, NY Univ, 42-45, asst biochem, Col Med, 45-48; instr, Sch Med, Wash Univ, 48-50; chief sect cellular pharmacol, Chem Pharmacol Lab, Nat Heart Inst, 50-56, chief lab clin-biochem, 56-58; DIR, ROCHE INST MOLECULAR BIOL, 68- Concurrent Pos: NIH spec fel, St Mary's Hosp Med Sch, Univ London, 57. Honors & Awards: Cert merit, Am Med Asn, 56; Flemming Award, 58; superior serv award, Dept Health, Educ & Welfare, 65, distinguished serv award, 66; Van Slyke Award, 67; Gairdner Award, 67; Hillebrand Award, Am Chem Soc, 62; Ames Award, Am Asn Clin Chem, 69. Mem: Nat Acad Sci; AAAS; Am Chem Soc; Am Soc Biol Chem; Am Soc Pharmacol & Exp Therapeut(secy, 62-64); Soc Exp Biol & Med. Res: Metabolism of aromatic amino acids; biosynthesis and metabolism of epinephrine and serotonin. Mailing Add: Roche Inst of Molecular Biol Nutley NJ 07110

UDIPI, KISHORE, b Udipi, SIndia, May 19, 40; m 73; c 1. POLYMER CHEMISTRY. Educ: Univ Bombay, BSc Hons, 59, MSc, 63; Univ Akron, PhD(polymer chem), 72. Prof Exp: Works mgr paints & polymers, Bombay Paints, India, 63-68; fel, Princeton Univ, 72-73; RES CHEMIST POLYMER CHEM, PHILLIPS PETROL CO, 73- Mem: Am Chem Soc. Res: Polymer synthesis; study of polymer microstructure and chemical modifications of polymers. Mailing Add: 365 RBI Phillips Res Ctr Bartlesville OK 74004

UDOVIC, JOSEPH DANIEL, b Cleveland, Ohio, July 9, 47. POPULATION BIOLOGY. Educ: Univ Tex, Austin, BA, 70; Cornell Univ, PhD(entom), 74. Prof Exp: ASST PROF BIOL, UNIV ORE, 73- Mem: AAAS; Ecol Soc Am; Soc Study Evolution. Res: Population biology, with particular emphasis on mathematical models of coevolution and speciation, and on plant-insect interactions. Mailing Add: Dept of Biol Univ of Ore Eugene OR 97403

UDVARDY, MIKLOS DEZSO FERENC, b Debrecen, Hungary, Mar 23, 19; nat US; m 51; c 3. ZOOLOGY. Educ: Debrecen Univ, PhD, 42. Prof Exp: Asst biologist, Hungarian Inst Ornith, 42-45; res assoc, Biol Res Inst, Hungarian Acad Sci, 45-48; res fel zool, Univ Helsinki, 48-49 & Univ Uppsala, 49-50; asst cur, Swedish Mus Natural Hist, 51; vis lectr ecol, Univ Toronto, 51-52; lectr zool, Univ BC, 52-53, from asst prof to assoc prof, 53-66; PROF BIOL SCI, CALIF STATE UNIV, SACRAMENTO, 66- Concurrent Pos: Asst scientist, Fisheries Res Bd Can, 52-55; vis prof, Univ Hawaii, 58-59; vis spec lectr, Univ Calif, Los Angeles, 63-64; vis prof, Univ Bonn, 70-71; Fulbright lectr, Honduras, 71-72; mem, Int Protecting Bd, Biol Sta Wilhelminberg, Austria & Point Reyes Bird Observ. Mem: Am Ornith Union; Ecol Soc Am; Cooper Ornith Soc; Nat Audubon Soc; Wilson Ornith Soc. Res: Zoogeography, especially distributional and ecological zoogeography; animal ecology and behavior; ornithology; biology and distribution of Appendicularia. Mailing Add: Dept of Biol Sci Calif State Univ Sacramento CA 95819

UDVARHELYI, GEORGE BELA, b Budapest, Hungary, May 14, 20; US citizen; m 56; c 3. NEUROSURGERY. Educ: St Stephan Col, Hungary, BA, 38; Pazmany Peter Univ, MD, 44; Univ Buenos Aires, MD, 52. Prof Exp: Resident neurol, Univ Budapest Hosp, 44-46; fel, Univ Vienna, 46-47; fel neuropath & psychiat, Univ Berne, 47-48; resident neurosurg, Hosp Espanol, Cordoba, 48-50; resident & surgeon, Univ Buenos Aires, 50-52; registr, Univ Edinburgh, 53-55; from instr neurosurg to asst prof neurosurg & radiol, 56-63, ASSOC PROF RADIOL, SCH MED, JOHNS HOPKINS UNIV, 63-, PROF NEUROSURG, 69- Concurrent Pos: WGer fel neurosurg, Univ Cologne, 52-53; Brit Coun scholar, Univ Edinburgh, 53-55; NIH fel neurosurg, Sch Med, Johns Hopkins Univ, 55-56; NIH res fel, 57-58; consult, Baltimore City Hosp, 58-, Danville Hosp, 58-71 & Harrisburg State Hosp, 59-71. Mem: AMA; Am Asn Neuropath; Am Asn Neurol Surg; Cong Neurol Surg; Soc Fr Speaking Neurosurgeons. Res: Clinical neurosurgery; neuroradiology; cerebral circulation; pediatric neurosurgery; pituitary surgery. Mailing Add: Dept of Neurosurg & Radiol Johns Hopkins Hosp Baltimore MD 21205

UEBEL, JACOB JOHN, b Chicago, Ill, Dec 25, 37; m 58; c 3. ORGANIC CHEMISTRY. Educ: Carthage Col, BA, 59; Univ Ill, MA, 62, PhD(chem), 64. Prof Exp: Res assoc, Univ Mich, 64; from asst prof to assoc prof, 64-73, PROF ORG CHEM, UNIV NH, 73- Concurrent Pos: Vis prof, Univ Calif-Riverside, 71-72. Mem: Am Chem Soc. Res: Organic reaction mechanisms and conformational analysis. Mailing Add: Dept of Chem Univ of NH Durham NH 03824

UEBELE, CURTIS EUGENE, b Kenosha Co, Wis, Dec 3, 35; m 58; c 6. POLYMER CHEMISTRY. Educ: Carroll Col, Wis, BS, 58; Univ Kans, PhD(chem), 65. Prof Exp: PROJ LEADER POLYMER RES, STANDARD OIL CO OHIO, CLEVELAND, 65- Mem: Am Chem Soc. Res: Formulation and evaluation of gas barrier resins. Mailing Add: 128 Eldred Ave Bedford OH 44146

UECKER, FRANCIS AUGUST, b Ft Wayne, Ind, Dec 18, 30; m 61; c 3. MYCOLOGY. Educ: Quincy Col, BS, 56; Univ Ill, Urbana, MS, 59, PhD(bot), 62. Prof Exp: Asst prof bot, Univ Ill, Urbana, 62-63, biologist, 63; asst prof, Winona State Col, 63-65; RES MYCOLOGIST, AGR RES CTR, PLANT PROTECTION INST, USDA, 65- Mem: AAAS; Mycol Soc Am. Res: Cytology and cytotaxonomy of fungi, especially Tuberales and pyrenomycetes. Mailing Add: 4611 Barbara Dr Beltsville MD 20705

UECKERT, DARRELL NEAL, b Merkel, Tex, May 25, 44; m 64; c 2. RANGE ECOLOGY, INSECT ECOLOGY. Educ: Tex Technol Col, BS, 66; Colo State Univ, MS, 68, PhD(range sci, entom), 69. Prof Exp: Asst prof entom, 69-71, biol control & range mgt, 69-71; asst prof, 71-73, ASSOC PROF, RANGE & WILDLIFE MGT DEPT, TEX TECH UNIV, 73- Concurrent Pos: Ed bd, J Range Mgt, Soc Range Mgt, 75-77. Mem: Soc Range Mgt; Entom Soc Am; Ecol Soc Am; Int Orgn Biol Control. Res: Biological control of noxious range plants; insect and range plant ecology; rangeland entomology. Mailing Add: Dept of Range & Wildlife Mgt Tex Tech Univ Lubbock TX 79409

UEHARA, HIROSHI, b Kobe City, Japan, Mar 7, 23; m 47; c 2. MATHEMATICS. Educ: Univ Tokyo, MS, 49; Osaka Univ, DSc, 54. Prof Exp: Instr math, Nagoya Univ, 49-51; asst prof, Math Inst, Kyushu Univ, 53-56; prof, Univ of the Andes & Nat Univ Colombia, 56-58; asst prof, Univ Southern Calif, 58-60; assoc prof, Univ Iowa, 60-64; PROF MATH, OKLA STATE UNIV, 64- Concurrent Pos: Lectr, Nat Univ Mex, 58. Mem: Am Math Soc; Math Soc France; Math Soc Japan. Res: Algebraic topology. Mailing Add: Dept of Math Okla State Univ Stillwater OK 74074

UELAND, KENT, b Chicago, Ill, May 27, 31; m 54; c 4. OBSTETRICS & GYNECOLOGY. Educ: Carleton Col, BA, 53; Univ Ill, BS & MD, 57; Am Bd Obstet & Gynec, dipl, 67. Prof Exp: Intern, Med Sch, Univ Ore, 57-58; asst resident obstet & gynec, King County & Univ Hosps, 60-61; from asst instr to assoc instr, 61-63, from instr to assoc prof, 63-74, PROF OBSTET & GYNEC, UNIV WASH, 74-, DIR OBSTET, UNIV HOSP, 68- Concurrent Pos: Res fel cardiovasc res, Univ Wash, 63; res fel med, Univ Ore, 63-64; resident, King County & Univ Hosps, 61-62, chief resident, 63; consult, USPHS Hosp, Seattle, Wash, Harborview Med Ctr, Madigan Gen Hosp, US Army, Tacoma & Nat Found March of Dimes. Mem: AMA; Am Col Obstet & Gynec; Soc Gynec Invest. Res: Pregnancy and cardiovascular dynamics; toxemia of pregnancy; pregnancy and heart disease. Mailing Add: Dept of Obstet & Gynec Univ of Wash Seattle WA 98195

UELTZ, HERBERT FRANK, b Morristown, NJ, Dec 14, 20. HIGH TEMPERATURE CHEMISTRY. Educ: Rutgers Univ, BS, 42, MS, 43, PhD(ceramics), 49. Prof Exp: Res assoc ceramics, Rutgers Univ, 46-49; res assoc, Norton Co, 49-62; dir res, 62-71, VPRES RES & DEVELOP, GEN ABRASIVE CO, 71- Mem: Fel Am Inst Chem; Electrochem Soc. Res: Abrasive and refractory substances. Mailing Add: 140 William St Youngstown NY 14174

UELZMANN, HEINZ, organic chemistry, see 12th edition

UFFEN, ROBERT JAMES, b Toronto, Ont, Sept 21, 23; m 49; c 2. GEOPHYSICS. Educ: Univ Toronto, BASc, 49, MA, 50; Univ Western Ont, PhD(physics), 52. Hon Degrees: Hon DSc, Univ Western Ont, 70; hon DSc, Queen's Univ, Ont, 67. Prof Exp: Lectr physics & geol, Univ Western Ont, 51-53, from asst prof to assoc prof geophys, 53-58, prof & head dept, 58-61, actg head dept physics, 60-61, asst prin, Univ Col Arts & Sci, 60-61, prin, 61-65, dean col sci, 65-66; mem, Defence Res Bd Can, 63-66, vchmn, 66-67, chmn, 67-69; chief sci adv to cabinet, Can Govt, 69-71; DEAN FACULTY APPL SCI, QUEEN'S UNIV, ONT, 71- Concurrent Pos: Consult, Kennco Explor Ltd, 52-59; res fel, Inst Geophys Univ Calif, Los Angeles, 53; Can deleg, Int Union Geod & Geophys, Rome, 54, Toronto, 57, Helsinki, 60, Tokyo, 62 & Int Union Geol Sci, New Delhi, 64; consult, Utah Construct Co, 55-58; mem Nat Adv Comt Res Geol Sci, 58-61; adv comt geod & geophys, Nat Res Coun Can, mem, 63-66; mem, Nat Feasibility Comt Proj Oilsand, 59; chmn, Can Sci Comt, Int Upper Mantle Proj, 60-65; consult, Stanford Res Inst, 61-62; ed, Earth & Planetary Sci Letters, 65-69; mem, Coun Regents for Cols Appl Arts & Technol, Prov Ont, 66-69; chmn, Can Eng Manpower Coun, 72-; mem, Ctr Resource Studies, 73-; counr, Assoc Prof Engrs Ont, 75-; mem, Fisheries Res Bd Can, 75-; vchmn bd, Ont Hydro Electric Power Corp, 74- Mem: AAAS; Soc Explor Geophys; Am Geophys Union; fel Geol Soc Am; Am Inst Mining, Metall & Petrol Eng. Res: Geothermometry; internal constitution of the earth; paleomagnetism; science policy. Mailing Add: Fac of Appl Sci Queen's Univ Kingston ON Can

UFFEN, ROBERT L, b Oxnard, Calif, Dec 2, 37. MICROBIOLOGY. Educ: Stanford Univ, BA, 62; Univ Mass, Amherst, MA, 64, PhD(microbiol), 68. Prof Exp: NIH fel, Univ Ill, Urbana, 68-70; ASST PROF MICROBIOL, MICH STATE UNIV, 70- Mem: AAAS; Am Soc Microbiol. Res: General microbiology and microbial physiology; regulation during cell differentiation; bacterial photosynthesis; ecology. Mailing Add: Dept of Microbiol Mich State Univ East Lansing MI 48823

UGARTE, EDUARDO, b Santa Ana, El Salvador, Oct 22, 35; m 63; c 4. BIOCHEMISTRY, MICROBIOLOGY. Educ: Nat Univ Mex, lic biochem, 62; Univ El Salvador, Dr(biochem), 65; Univ Rio de Janeiro, dipl oral microbiol, 66; Inter-Am Inst Agr Sci, El Salvador, dipl agr sci educ, 69. Prof Exp: Assoc prof biochem, Sch Dent, Univ El Salvador, 65-67, secy dept basic sci, 66, secy curriculum comn, 66-67, prof chem, Sch Agron Sci, 67-68, chief prof biochem, 68-70; researcher biochem, United Med Labs, 70-74; dir, Analytico Lab, 75; VPRES, PAGE BIOCHEM LABS, INC, 75- Concurrent Pos: Fel bact & virol, Life Labs, Ecuador, 63-64; fel oral microbiol, Pan-Am Health Orgn, Brazil, 66. Mem: Am Chem Soc; Microbiol Soc El Salvador. Res: Relation of the mechanism of different hormones between pituitary and thyroid glands; development of the radioimmunoassays for these hormones; radiolabeling materials for different fractions. Mailing Add: 5235 Gulf Blvd St Petersburg Beach FL 33706

UGINCIUS, PETER, b Geniai, Lithuania, Feb 23, 36; US citizen; m 62; c 4. GEODESY. Educ: Kalamazoo Col, BA, 58; Ind Univ, MS, 61; Cath Univ Am, PhD(physics), 68. SUPVRY GEODESIST, NAVAL SURFACE WEAPONS CTR, 72- Concurrent Pos: Adj prof, Va Polytech Inst & State Univ, 72- Mem: Am Phys Soc. Res: Physical geodesy; earth gravitational field. Mailing Add: Naval Surf Weapons Ctr Ocean Geodesy Br DK-12 Dahlgren VA 22448

UGLEM, GARY LEE, b Grand Forks, NDak, Sept 19, 41; m 64; c 2. PARASITOLOGY, ZOOLOGY. Educ: Univ NDak, BS, 66, MS, 68; Univ Idaho, PhD(zool), 72. Prof Exp: Instr parasitol, Univ Idaho, 71; NIH fel, Rice Univ, 72-74; ASST PROF BIOL SCI, UNIV KY, 74- Mem: Am Micros Soc; Am Soc

Parasitologists; Am Inst Biol Sci. Res: Physiology of host-parasite relations; Acanthocephalan life histories. Mailing Add: Sch of Biol Sci Univ of Ky Lexington KY 40506

UGLUM, JOHN RAYMOND, elementary particle physics, see 12th edition

UGOLINI, FIORENZO CESARE, b Florence, Italy, Jan 16, 29; m 63; c 2. SOILS. Educ: Rutgers Univ, BS, 57, PhD(soils), 60. Prof Exp: Arctic Inst NAm fel, Rutgers Univ, 60-61, asst prof soils, 61-64; asst prof & res assoc, Ohio State Univ, 64-66; assoc prof, 66-73, PROF SOILS, UNIV WASH, 73- Concurrent Pos: NATO prof, Univ Milan. Mem: AAAS; Am Polar Soc; Am Soc Agron; Int Soc Soil Sci; Arctic Inst NAm; Int Union Quaternary Res. Res: Soil formation and weathering in the cold regions, including Arctic, Antarctica and Alpine environments; soil development and the impact of time on glacial deposits, chronologically different; forest soils; paleosoils. Mailing Add: Col of Forest Resources AR-10 Univ of Wash Seattle WA 98195

UGRO, JOSEF VENDAL, JR, physical chemistry, see 12th edition

UHERKA, DAVID JEROME, b Wagner, SDak, June 2, 38; m 64. NUMERICAL ANALYSIS. Educ: SDak Sch Mines & Technol, BS, 60; Univ Utah, MA, 63, PhD(math), 64. Prof Exp: Mathematician, US Naval Radiol Defense Lab, Calif, 62; mathematician & programmer, US Army Natick Labs, Mass, 64-66; asst prof, Ariz State Univ, 66-68; ASSOC PROF MATH, UNIV N DAK, 68- Concurrent Pos: Consult, US Army Natick Labs, 66-68. Mem: Math Asn Am; Am Math Soc. Res: Functional and numerical analysis; computer applications. Mailing Add: Dept of Math Univ of ND Grand Forks ND 58202

UHL, CHARLES HARRISON, b Schenectady, NY, May 28, 18; m 45; c 4. BOTANY. Educ: Emory Univ, BA, 39, MS, 41; Cornell Univ, PhD(bot), 47. Prof Exp: Asst, 41-42, 45-46, from instr to asst prof, 46-52, ASSOC PROF BOT, CORNELL UNIV, 52- Mem: AAAS; Bot Soc Am; Genetics Soc Am; Soc Study Evolution; Int Asn Plant Taxon. Res: Chromosomes and evolution, especially of Crassulaceae. Mailing Add: Dept of Bot Cornell Univ Ithaca NY 14850

UHL, DALE LYNDEN, b Monroe, Mich, May 21, 43; m 68; c 1. NUCLEAR CHEMISTRY. Educ: Eastern Mich Univ, BA, 66; Purdue Univ, PhD(nuclear chem), 70. Prof Exp: Engr, Westinghouse Elec Corp, 70-72; Nuclear chemist, Lynchburg Res Ctr, 72-73, supvr radiochem, 73-75, SUPVR CHEMONUCLEAR TECHNOL, BABCOCK & WILCOX, 75- Mem: Am Nuclear Soc; Am Chem Soc; Am Nat Standards Inst. Res: Nuclear and radiochemistry, fission, iodine chemistry; pressurized water reactor chemistry and chemistry training; PWR Radwaste system; PWR primary system corrosion products; decontamination. Mailing Add: Lynchburg Res Ctr Babcock & Wilcox PO Box 1260 Lynchburg VA 24505

UHL, EDWARD GEORGE, b Elizabeth, NJ, Mar 24, 18; m 42, 66; c 4. APPLIED PHYSICS. Educ: Lehigh Univ, BS, 40. Prof Exp: With Nat Carbon Co, Md & Fla, 46-59; vpres, Teledyne Ryan Aeronaut Co, Calif, 59-61; PRES & CHIEF EXEC OFFICER, FAIRCHILD INDUSTS, INC, 61- Concurrent Pos: Directorships, Am Satellite Corp, Fairchild Industs, Inc, Md Nat Corp & Md Nat Bank; mem bd trustees, Hood Col; mem adv comt, Sch Eng & Archit, Cath Univ Am; mem adv coun, Sch Eng & Appl Sci, George Washington Univ; mem vis comt, Dept Mech Eng & Mech, Lehigh Univ; mem indust adv coun, Off Secy Defense. Mem: Fel Am Inst Aeronaut & Astronaut; Am Ord Asn; Soc Automotive Eng. Res: Aircraft and spacecraft design. Mailing Add: Fairchild Industs Inc Sherman Fairchild Technol Ctr Germantown MD 20767

UHL, HENRY STEPHEN MAGRAW, b Wilkes-Barre, Pa, July 23, 21; m 46; c 1. INTERNAL MEDICINE. Educ: Princeton Univ, AB, 43; Harvard Univ, MD, 47. Prof Exp: Asst path, Sch Med, Johns Hopkins Univ, 47-48, instr anat & path, 48-49; instr med, Med Sch, Wayne State Univ, 52-53; dir med educ, Worcester City Hosp, Mass, 53-58; dir med educ, Springfield Hosp, 58-60; from asst prof to assoc prof postgrad med, Albany Med Col, 60-66; PROF MED SCI, BROWN UNIV, 66-, DIR CONTINUING MED EDUC, 68- Concurrent Pos: Consult, Postgrad Med Inst, Boston, 55-60. Mem: Asn Hosp Med Educ (pres, 58-60); Am Col Physicians; Asn Am Med Cols; Am Fedn Clin Res. Res: Medical education; experimental atherosclerosis; research and development in continuing medical education and in medical care research. Mailing Add: Div of Biol & Med Sci Brown Univ Providence RI 02912

UHL, JOHN JERRY, JR, b Pittsburgh, Pa, June 27, 40. MATHEMATICS. Educ: Col William & Mary, BS, 62; Carnegie Inst Technol, MS, 64, PhD(math), 66. Prof Exp: Asst prof, 68-72, ASSOC PROF MATH, UNIV ILL, URBANA-CHAMPAIGN, 72- Mem: Am Math Soc; Math Asn Am. Res: Functional analysis and integration theory; interrelations among martingales, vector measures and operator theory. Mailing Add: 273 Altgeld Hall Univ of Ill Dept of Math Urbana IL 61801

UHL, NATALIE W, plant anatomy, plant morphology, see 12th edition

UHLE, CHARLES AUGUSTUS WOERWAG, b Philadelphia, Pa, Nov 20, 04; m 33; c 2. UROLOGY. Educ: Univ Pa, AB, 26, MD, 30. Prof Exp: Intern, Lankenau Hosp, Philadelphia, 30-32; assoc, Sch Med, 35-46, assoc prof, Div Grad Med, 48-52, PROF UROL, DIV GRAD MED, UNIV PA, 52- Concurrent Pos: Bishop Mathews fel urol, Sch Med, Univ Pa, 33-35; asst urologist, Lankenau Hosp, 35-43, urologist, 43-70, sr consult, 70-; from asst urologist to urologist, Philadelphia Gen Hosp, 35-52; consult, US Naval Hosp, Philadelphia, 37-; urologist, Chestnut Hill Hosp, 49-54, consult, 54-; from asst urologist to urologist, Grad Hosp, Univ Pa, 52-57, consult, 57- Mem: Fel Am Urol Asn; fel AMA; fel Am Asn Genito-Urinary Surg; fel Am Col Surg; Int Soc Urol. Res: Surgery of cancer of the urologic organs; physiology of the urinary bladder; surgical techniques for correction of urethral traumatic strictures, incontinence and fistulae; post-prostatactomy vesical neck sclerosis; retention urine following ingestion of antihistamines; urinary tract allergy. Mailing Add: Lankenau Hosp Philadelphia PA 19151

UHLENBECK, GEORGE EUGENE, b Batavia, Java, Dec 6, 00; nat US; m 27; c 1. PHYSICS. Educ: State Univ Leiden, PhD, 27. Hon Degrees: ScD, Univ Notre Dame, 55, Case Inst Technol, 59, Univ Colo, 68, Yeshiva Univ, 69. Prof Exp: Asst theoret physics, State Univ Leiden, 25-27; from instr to assoc prof, Univ Mich, 27-35; prof, State Univ Utrecht, 35-39; Henry S Carhart prof, Univ Mich, 39-60; prof physics & physicist, 60-74, EMER PROF PHYSICS, ROCKEFELLER UNIV, 74- Concurrent Pos: Mem staff, Radiation Lab, Mass Inst Technol, 43-45; Lorentz prof, State Univ Leiden, 54-55; Van der Waals prof, Univ Amsterdam, 63-64. Honors & Awards: Max Planck Medal, 64; Lorentz Medal, 70. Mem: Nat Acad Sci; fel Am Phys Soc (pres, 59); Am Philos Soc; Netherlands Phys Soc; Nat Acad Lincei. Res: Theory of atomic structure and quantum mechanics; statistical mechanics and the kinetic theory of matter; nuclear physics. Mailing Add: Rockefeller Univ York Ave & E 66th St New York NY 10021

UHLENBECK, KAREN K, b Cleveland, Ohio, Aug 24, 42; m 65. MATHEMATICS. Educ: Univ Mich, BA, 64; Brandeis Univ, PhD(math), 68. Prof Exp: Instr math, Mass Inst Technol, 68-69; lectr, Univ Calif, Berkeley, 69-71; ASST PROF MATH, UNIV ILL, URBANA, 71- Mem: Am Math Soc. Res: Infinite dimensional differential topology as applied to nonlinear or global analysis. Mailing Add: 273 Altgeld Hall Univ of Ill Dept of Math Urbana IL 61801

UHLENBECK, OLKE CORNELIS, b Ann Arbor, Mich, Apr 20, 42. BIOPHYSICAL CHEMISTRY. Educ: Univ Mich, Ann Arbor, BS, 64; Harvard Univ, PhD(biophys), 69. Prof Exp: Miller fel, Univ Calif, Berkeley, 69-71; ASST PROF BIOCHEM & CHEM, UNIV ILL, URBANA, 71- Concurrent Pos: NIH res grant, Univ Ill, Urbana, 71- Res: Nucleic acid interactions; structure and function of RNA. Mailing Add: Dept of Biochem Univ Ill at Urbana-Champaign Urbana IL 61801

UHLENBROCK, DIETRICH A, b Schweinfurt, Ger, Oct 13, 37. APPLIED MATHEMATICS, MATHEMATICAL PHYSICS. Educ: Univ Cologne, Vordiplom, 59; NY Univ, MS, 62, PhD(physics), 63. Prof Exp: Adj asst prof theoret physics, Univ & res assoc, Courant Inst, NY Univ, 63-64; vis mem, Inst Adv Study, NJ, 64-66; asst prof, 66-68, ASSOC PROF MATH & PHYSICS, UNIV WIS-MADISON, 68- Concurrent Pos: Prof math & physics, Free Univ Berlin, 73-75. Res: Classical and quantum statistical physics; quantum field theory. Mailing Add: Dept of Math Univ of Wis Madison WI 53706

UHLENHOPP, ELLIOTT LEE, b Hampton, Iowa, Dec 8, 42; m 67; c 2. BIOCHEMISTRY. Educ: Carleton Col, BA, 65; Columbia Univ, PhD(biochem), 71. Prof Exp: Res chemist, Univ Calif, San Diego, 73-74; ASST PROF CHEM, WHITMAN COL, 75- Concurrent Pos: Fel, Damon Runyon Mem Fund, Cancer Res Inc, 71-73. Mem: Sigma Xi; Biophys Soc; Am Chem Soc; AAAS. Res: Viscoelastic characterization of high molecular weight native and denatured DNA from bacterial and eukaryotic cells. Mailing Add: Dept of Chem Whitman Col Walla Walla WA 99362

UHLENHUTH, EBERHARD HENRY, b Baltimore, Md, Sept 15, 27; m 52; c 3. PSYCHIATRY, PSYCHOPHARMACOLOGY. Educ: Yale Univ, BS, 47; Johns Hopkins Univ, MD, 51. Prof Exp: USPHS fel psychiat, Johns Hopkins Univ, 52-56, from instr to assoc prof, 56-68; assoc prof, 68-74, PROF PSYCHIAT & CHIEF ADULT PSYCHIAT CLIN, UNIV CHICAGO, 74- Concurrent Pos: Consult, Patuxent Inst, 56-57; asst psychiatrist chg, Outpatient Dept, Johns Hopkins Hosp, 56-61, consult, Div Plastic Surg, 59-60, psychiatrist chg, Outpatient Dept, 61-62; USPHS career teacher trainee, Johns Hopkins Univ, 57-59, USPHS career res develop awards, 62-68; consult, Ill State Psychiat Inst, 68, Clin Psychopharmacol Res Rev Comt, Psychopharmacol Br, NIMH, 68-72, Ment Health Task Force, Mid-Southside Planning Orgn, 70, Woodlawn Ment Health Ctr, 71 & US Food & Drug Admin, 71. Honors & Awards: Assoc Clin Psychiatrists' Award, 58. Mem: Am Col Neuropsychopharmacol; Am Psychiat Asn; Soc Gen Systs Res; Psychiat Res Soc; Group Advan Psychiat. Res: Clinical psychopharmacology; anxiety and depression; life stress, dysfunction and recovery processes; evaluation of treatments. Mailing Add: Pritzker Sch of Med Univ of Chicago Chicago IL 60637

UHLER, LOWELL DOHNER, b Conemaugh, Pa, May 10, 14; m 44; c 2. INSECT ECOLOGY. Educ: Pa State Teachers Col, Ind, BS, 35; Cornell Univ, MS, 41, PhD(entom), 48. Prof Exp: Teacher pub sch, Pa, 35-39; asst entom, Cornell Univ, 40-42; med entomologist, Douglas Aircraft Co, 42-44; asst entom, 46-48, from asst prof to assoc prof, 48-58, PROF BIOL, CORNELL UNIV, 58- Concurrent Pos: Vis prof, Univ Philippines, 66-68. Mem: Fel AAAS; Entom Soc Am; Ecol Soc Am; Philippine Asn Entomologists. Res: Insect ecology; rodent control; microtechniques. Mailing Add: 4 Redwood Lane Ithaca NY 14850

UHLER, ROBERT LOWELL, plant nutrition, see 12th edition

UHLER, ROGER OSCAR, organic chemistry, polymer chemistry, see 12th edition

UHLHORN, KENNETH W, b Mankato, Minn, Sept 24, 33; m 54; c 3. SCIENCE EDUCATION. Educ: Mankato State Col, BS, 54; Univ Minn, MA, 60; Univ Iowa, PhD(biol, sci ed), 63. Prof Exp: Teacher pub sch, Minn, 56-60; instr biol & physics, Univ Iowa, 60-63, asst prof biol, 63; assoc prof, 63-68, PROF SCI EDUC, IND STATE UNIV, TERRE HAUTE, 68-, DIR SCI TEACHING CTR, 66- Concurrent Pos: Mem, Ind State Sci Adv Bd. Mem: Nat Asn Res Sci Teaching; Nat Sci Teachers Asn. Res: Preparation, use and application of science experience inventories and their role in teaching science; science for elementary education majors; children's concept of science and scientists based on their drawings of scientists. Mailing Add: Sci Teaching Ctr Ind State Univ Terre Haute IN 47809

UHLIG, HANS GERD, b Oberhohndorf, Ger, Nov 3, 22; nat US; m; c 3. BIOLOGY. Educ: Iowa State Univ, BS, 45; Ore State Col, MS, 47. Prof Exp: Biol aide fish res, State Div Fisheries & Game, Mass, 47; aide waterfowl res, State Natural Hist Serv, Ill, 47; biologist & proj leader small game, WVa Conserv Comn, 47-56; biologist, 56-67, RECREATION SPECIALIST, SOIL CONSERV SERV, USDA, 67- Honors & Awards: Nash award, 56; Pearce Mem award, Northeast Wildlife Soc. Mem: Nat Recreation & Park Asn; Northeast Wildlife Soc. Res: Small game, particularly gray squirrel; resource development; recreation development on private lands. Mailing Add: Soil Conserv Serv USDA 7600 West Chester Pike Upper Darby PA 19082

UHLRICH, HELEN MARIE, b Bozeman, Mont, Mar 29, 13. BIOLOGY. Educ: St Louis Univ, BA, 40; Fordham Univ, MS, 45, PhD(biol), 47. Prof Exp: Instr biol, Viterbo Col, 46-65; assoc prof, 65-72, PROF BIOL, CARLOW COL, 72- Mem: AAAS; Am Inst Biol. Res: Effect of hydrogen ion concentration on mitosis in Allium cepa; botany. Mailing Add: Dept of Biol Carlow Col 3333 Fifth Ave Pittsburgh PA 15213

UHR, JONATHAN WILLIAM, b New York, NY, Sept 8, 27; m 54; c 2. MEDICINE. Educ: Am Bd Internal Med, dipl, 60. Prof Exp: PROF MED & MICROBIOL & CHMN DEPT MICROBIOL, UNIV TEX HEALTH SCI CTR DALLAS, 72- Concurrent Pos: Dept dir comn immunization, Armed Forces Epidemiol Bd, 69- Mem: Am Asn Immunol; Am Soc Clin Invest; Am Rheumatism Asn; Am Soc Exp Path; Asn Am Physicians. Res: Immunology. Mailing Add: Dept of Microbiol Univ of Tex Southwestern Med Sch Dallas TX 75235

UHR, LEONARD MERRICK, b Philadelphia, Pa, June 26, 27; m 49; c 2. COMPUTER SCIENCE, PSYCHOLOGY. Educ: Princeton Univ, BA, 49; Johns Hopkins Univ, MA, 51; Univ Mich, PhD(psychol), 57. Prof Exp: Assoc prof psychol, Univ Mich & res psychologist & coordr psychol sci, Ment Health Res Inst, 57-65; PROF COMPUT SCI, UNIV WIS-MADISON, 65- Concurrent Pos: Vis distinguished scholar, Ed Testing Serv, 67. Mem: Am Psychol Asn; Asn Comput Mach; Inst Elec & Electronics Eng. Res: Dynamic computer models of perceptual and cognitive processes; perception; learning; computers and education; behavioral psychopharmacology. Mailing Add: Dept of Comput Sci Univ of Wis Madison WI 53706

UHRICH, DAVID LEE, b Buffalo, NY, Jan 5, 39; m 61; c 3. PHYSICS. Educ: Canisius Col, BS, 60; Univ Pittsburgh, PhD(physics), 65. Prof Exp: Fels, Univ Pittsburgh, 66 & Iowa State Univ, 66-67; asst prof, 67-71, ASSOC PROF PHYSICS, KENT STATE UNIV, 71- Mem: Am Phys Soc. Res: Mössbauer effect studies of liquid crystals and solids. Mailing Add: 707 Berkeley Dr Kent OH 44240

UHRICH, JACOB, b McCook, Nebr, Aug 15, 09; m 43; c 4. ZOOLOGY. Educ: Doane Col, BA, 32; Univ Nebr, MA, 34; Univ Chicago, PhD(zool), 37. Prof Exp: Asst prof biol, Kans State Teachers Col, Pittsburg, 37-46; chmn dept, 46-61, PROF BIOL, TRINITY UNIV, TEX, 46- Mem: AAAS; Am Soc Limnol & Oceanog. Res: Social behavior of mice; ecological succession on strip mines; parasites of birds and mammals; animal ecology of the southwest; biology of man-made lakes. Mailing Add: Dept of Biol Trinity Univ 715 Stadium Dr San Antonio TX 78284

UITTO, JOUNI JORMA, b Helsinki, Finland, Sept 15, 43; m 65; c 1. DERMATOLOGY, BIOCHEMISTRY. Educ: Univ Helsinki, BM, 65, MD & PhD(med biochem), 70. Prof Exp: Intern med, surg, med biochem, Univ Helsinki Cent Hosp, 69; instr med biochem, Univ Helsinki, 70-71; clin asst dermat, Univ Cent Hosp, Univ Copenhagen, 71; instr, 72-73, ASST PROF BIOCHEM, RUTGERS MED SCH, COL MED & DENT NJ, 73- Concurrent Pos: Fel, Gen Clin Res Ctr, Philadelphia Gen Hosp & Dept Dermat, Univ Pa, 71-72. Mem: Soc Invest Dermat; Europ Soc Dermat Res; Am Chem Soc. Res: Biochemistry of connective tissues; collagen metabolism; investigative dermatology. Mailing Add: Dept of Biochem Rutgers Med Sch Piscataway NJ 08854

UKELES, RAVENNA, b New York, NY, Aug 1, 29. MICROBIOLOGY. Educ: Hunter Col, BS, 49; NY Univ, MSc, 56, PhD(biol), 60. Prof Exp: RES MICROBIOLOGIST, LAB EXP BIOL, NAT MARINE FISHERIES SERV, 59- Mem: Soc Protozool; Phycol Soc Am; Am Soc Microbiol; NY Acad Sci. Res: Nutrition of protozoa and algae; growth of organisms in mass culture using algae as the potential food sources. Mailing Add: Lab for Exp Biol Nat Marine Fisheries Serv Milford CT 06460

ULAGARAJ, MUNIYANDY SEYDUNGANALLUR, b Tuticorin, India, Mar 6, 44; m 70. ENTOMOLOGY. Educ: Madras Univ, India, BScAg, 64; Indian Agr Res Inst, MSc, 70; Univ Fla, Gainesville, PhD(entom), 74. Prof Exp: Res asst entom, Dept Agr, Coimbatore, India, 64-68; res & teaching asst entom, Univ Fla, Gainesville, 71-74, assoc entom, 74-75; RES ASSOC, McGILL UNIV, MACDONALD CAMPUS, QUE, 75- Concurrent Pos: Assoc entom, Univ Fla, Gainesville, 74; res assoc, Purdue Univ, WLafayette, Ind, 75. Mem: AAAS; Entom Soc Am; Entom Soc Can; Sigma Xi. Res: Ecology, behavior and systematics of Orthoptera; acoustical communication of insects; behavioral biology of soil arthropods; radiation biology of agricultural insect pests. Mailing Add: Lyman Entom Mus & Res Lab McGill Univ MacDonald Campus Ste Anne de Bellevue PQ Can

ULAM, STANISLAW MARCIN, b Lwow, Poland, Apr 3, 09; m 41; c 1. MATHEMATICS, MATHEMATICAL PHYSICS. Educ: Polytech Inst, Poland, MA, 32, Dr Math Sc, 33; Univ NMex, PhD, 65. Prof Exp: Mem, Inst Adv Study, 36; lectr math, Harvard Univ, 39-40; asst prof, Univ Wis, 41-43; from mem staff to res adv, Los Alamos Sci Lab, 43-65; prof math, Univ Colo, Boulder, 65-74; GRAD RES PROF, UNIV FLA, 74- Concurrent Pos: Assoc prof, Univ Southern Calif, 45-46; vis lectr, Harvard Univ, 51; vis prof, Mass Inst Technol, 56-57; chmn math sect, AAAS & mem bd govs, Weizmann Inst Sci Behav, Israel, 75- Honors & Awards: Polish Millenium Prize, Zurzykowski Foundation. Mem: Nat Acad Sci; Am Math Soc; Am Phys Soc; Math Asn Am. Res: Set theory; functions of real variable; mathematical logic; thermonuclear reactions; topology; Monte Carlo method. Mailing Add: 775 Pleasant St Boulder CO 80302

ULANOWICZ, ROBERT EDWARD, b Baltimore, Md, Sept 17, 43; m 67. THEORETICAL BIOLOGY. Educ: Johns Hopkins Univ, BES, 64, PhD(chem eng), 68. Prof Exp: Res asst phys chem, Univ Göttingen, 64; res asst chem eng, Johns Hopkins Univ, 64-68; asst prof, Cath Univ Am, 68-70; res asst prof, Natural Resources Inst, 70-75, RES ASSOC PROF, CTR ENVIRON & ESTUARINE STUDIES, UNIV MD, 75- Mem: Atlantic Estuarine Res Soc; Estuarine Res Fedn. Res: Mass and energy transfer in ecosystems; ecological modelling; thermodynamics of ecosystems; hydrographical modeling. Mailing Add: Ctr Environ & Estuarine Studies Univ of Md PO Box 38 Solomons MD 20688

ULBERG, LESTER CURTISS, b Wis, Dec 2, 17; m 45; c 1. REPRODUCTIVE PHYSIOLOGY. Educ: Univ Wis, BS, 48, MS, 49, PhD(reprod physiol), 52. Prof Exp: Asst reprod physiol, Univ Wis, 47-50, instr, 50-52; instr animal husb, Miss State Col, 52-55, assoc prof, 55-57; assoc prof, 57-60, PROF ANIMAL HUSB, NC STATE UNIV, 60- Concurrent Pos: Agent, USDA, 50-52. Mem: AAAS; Am Soc Animal Sci; Am Dairy Sci Asn; Brit Soc Study Fertil. Res: Early embryonic development; hormone control of ovarian activity. Mailing Add: Dept of Animal Sci NC State Univ Raleigh NC 27607

ULBRICH, CARLTON WILBUR, b Meriden, Conn, Oct 1, 32; m 62; c 2. LOW TEMPERATURE PHYSICS. Educ: Univ Conn, BSME, 60, MS, 62, PhD(physics), 65. Prof Exp: Res asst physics, Univ Conn, 61-65; asst prof, Wittenberg Univ, 65-66; asst prof, 66-74, ASSOC PROF PHYSICS, CLEMSON UNIV, 74- Mem: Am Phys Soc. Res: Thermal, electric and magnetic properties of superconductors. Mailing Add: Dept of Physics Clemson Univ Clemson SC 29631

ULERY, HARRIS ELLSWORTH, organic chemistry, see 12th edition

ULETT, GEORGE ANDREW, b Needham, Mass, Jan 10, 18; m 43; c 3. PSYCHIATRY. Educ: Stanford Univ, BA, 40; Univ Ore, MS, 43, PhD(anat) & MD, 44; Am Bd Psychiat & Neurol, dipl & cert psychiat. Prof Exp: Instr anat, Med Sch, Univ Ore, 43-44; asst neurol, Harvard Univ, 45-46; asst neuropsychiat, Sch Med, Washington Univ, 48-49, from asst prof to prof, 50-64; prof psychiat & chmn dept, Mo Inst Psychiat, Univ Mo-St Louis, 64-73; DIR NEUROPSYCHIAT SERV & DIR PSYCHOSOM RES LABS, DEACONESS HOSP, ST LOUIS, 74- Concurrent Pos: USPHS fel psychiat, Washington Univ, 49-50; asst psychiatrist, Barnes & St Louis Children's Hosps, 50-64; vis physician, Malcolm Bliss Ment Health Ctr, 51-56, med dir, 56-61; dir psychiat serv, City Hosp Div, 59-61; dir, Mo State Div Ment Dis, 61-72; physician, Pvt Pract, 74- Mem: Am Electroencephalog Soc (secy); Am Soc Med Psychiat (past pres); fel Am Psychiat Asn; AAAS; Soc Exp Biol & Med. Res: Electroencephalography; psychopharmacology; neurophysiology. Mailing Add: 1144 Hampton Ave St Louis MO 63110

ULFELDER, HOWARD, b Mexico City, Mex, Aug, 15, 11; US citizen; m 32; c 4. OBSTETRICS & GYNECOLOGY, SURGERY. Educ: Harvard Univ, BA, 32, MD, 36. Prof Exp: Intern surg, Mass Gen Hosp, 37-39; asst, 40-46, instr, 46-48, clin assoc, 48-49, clin assoc gynec, 49-53, from asst clin prof to clin prof, 53-62, JOE V MEIGS PROF GYNEC, HARVARD MED SCH, 62- Concurrent Pos: Resident surg, Mass Gen Hosp, 39-41, asst, 41-46, asst surgeon, 46-55, vis surgeon & chief gynec serv, 55-; physician, Pondville Hosp, 48-; consult surgeon, Mass Eye & Ear Infirmary, 48-; chief of staff, Vincent Mem Hosp, 55- Mem: AAAS; Am Gynec Soc; Soc Univ Surg; Soc Pelvic Surg (secy-treas, 54, pres, 59); fel AMA. Res: Carcinoma of the cervix; biological mechanics of pelvic support; stilbestrol, adenosis carcinoma syndrome. Mailing Add: Mass Gen Hosp 275 Charles St Boston MA 02114

ULICH, BOBBY LEE, b Bryan, Tex, Aug 13, 47; m 65; c 1. RADIO ASTRONOMY. Educ: Tex A&M Univ, BS, 69; Calif Inst Technol, MS, 70; Univ Tex, Austin, PhD(elec eng), 73. Prof Exp: HEAD TELESCOPE OPER, TUCSON DIV, NAT RADIO ASTRON OBSERV, ASSOC UNIVS INC, 73- Concurrent Pos: Mem, US Nat Comt Int Union Radio Sci, Comn J. Mem: Inst Elec & Electronic Engrs; Am Astron Soc. Res: Millimeter wavelength instrumentation and calibration techniques; solar system astronomy; interstellar molecules. Mailing Add: Nat Radio Astron Observ Ste 100 2010 N Forbes Blvd Tucson AZ 85705

ULINSKI, PHILIP STEVEN, b Detroit, Mich, Feb 17, 43. NEUROANATOMY. Educ: Mich State Univ, BS, 64, MS, 67, PhD(zool), 69. Prof Exp: Asst prof biol, Oberlin Col, 69-70; asst prof anat, Sch Dent, Loyola Univ Chicago, 70-74; ASST PROF ANAT, UNIV CHICAGO, 75- Concurrent Pos: NIH fel, Univ Chicago, 75-78. Mem: AAAS; Am Soc Zool; Soc Neurosci; Am Asn Anat. Res: Comparative anatomy of reptilian nervous systems. Mailing Add: Dept of Anat Univ of Chicago Chicago IL 60637

ULISS, DAVID BARRY, organic chemistry, mass spectrometry, see 12th edition

ULLERY, CHARLES HOWARD, b Dayton, Ohio, Feb 7, 42; m 64; c 2. SOIL SCIENCE, AGRONOMY. Educ: Ohio State Univ, BS, 64, MS, 67; Colo State Univ, PhD(agron), 71. Prof Exp: Res asst agron, Ohio State Univ, 65-67 & Colo State Univ, 67-70; ASST PROF SOIL SCI, ORE STATE UNIV, 70- Mem: Soil Sci Soc Am; Am Soc Agron. Res: Soil-plant-water relationships; influence of soil management practices on plant growth. Mailing Add: Dept of Soil Sci Ore State Univ Corvallis OR 97331

ULLMAN, ARTHUR WILLIAM JAMES, b New York, NY, Dec 30, 36. MATHEMATICS. Educ: Univ Miami, BS, 57; Rice Univ, MS, 64, PhD(math), 66. Prof Exp: Instr math, Rice Univ, 61-62; asst prof, Tex Christian Univ, 64-66; asst prof philos, Tex A&M Univ, 66-67; prof & chmn dept, State Univ NY Col New Paltz, 68-71, resident dir, State Univ NY prog, Int Documentation Ctr, Cuernavaca, Mex, 71-72, PROF MATH, STATE UNIV NY COL NEW PALTZ, 72- Concurrent Pos: Founder & pres, MH Hedging Analysts, Inc, 74; contract to produce educ modules on health, Empire State Col, State Univ NY, 75. Mem: Am Math Soc; Asn Symbolic Logic. Res: Logic applying algebraic and analytic techniques to the study of logical theories; mathematical analysis of various commodity markets; potential conflict of interest between the medical profession and the public interest. Mailing Add: Dept of Math State Univ NY New Paltz NY 12561

ULLMAN, BETTY M, b Jersey City, NJ. BIOSTATISTICS. Educ: Brown Univ, AB, 45, ScM, 46; Univ Mich, MPH, 66, PhD(biostatist), 69. Prof Exp: Asst prof biostatist, 69-72, ASSOC PROF BIOSTATIST, SCH PUB HEALTH, UNIV MICH & ASSOC DIR, CTR RES DIS OF HEART, 72- Concurrent Pos: Mem, Clin Applns & Prev Adv Comt, Div Heart & Vascular Dis, Nat Heart & Lung Inst, 75-76. Mem: Am Statist Asn; Biomet Soc; Soc Epidemiol Res; Am Heart Asn; Am Pub Health Asn. Res: Epidemiological studies related to heart disease; longitudinal studies. Mailing Add: Apt 3D 1050 Wall St Ann Arbor MI 48105

ULLMAN, EDWARD LOUIS, b Chicago, Ill, July 24, 12; m 42, 68; c 2. GEOGRAPHY. Educ: Univ Chicago, SB, 34, PhD(geog), 42; Harvard Univ, AM, 35. Prof Exp: Instr, Wash State Col, 35-37 & Ind Univ, 41; chief transport sect, Off Strategic Serv, 42-43 & US Maritime Comn, 46; from asst prof to assoc prof regional planning, Harvard Univ, 46-51; assoc dean, Grad Sch, 61-65, PROF GEOG, UNIV WASH, 51- Concurrent Pos: Consult, Boston Globe, 48-50 & res & develop bd, US Dept Defense, 49-52; mem, Am Coun Learned Socs, 50-54; consult, Stanford Res Inst, Pl, 56 & European Prod Agency, OEEC, Italy, 57; Fulbright res prof, Univ Rome, 56-57; lectr, Univ Col, London & Vienna Univs, 57; consult, US Dept Com, Calif, 58-59; vis prof, Wash Univ, 59-61; mem,, US del Sweden, 60 & Int Geog Union; dir, Meramec Basin Res Proj, 59-62; vis prof, Moscow State Univ, 65 & Salzburg Seminar Am Studies, Austria, 65; pres, Wash Ctr Metrop Studies, DC, 65-66; consult, Asian Develop Bank, 68; mem behav sci div, Nat Acad Sci, 68-71; vis prof, Hebrew & Haifa Univs, Israel, 73; mem bd dirs, AMTRAK, 74- Honors & Awards: Citation, Asn Am Geog, 58; Citation, Italian Soc Geog, 59. Mem: Asn Am Geog; Regional Sci Asn (pres, 60-61); corresp mem, Italian Soc Geog. Res: Transportation and urban geography; area development; geographic theory. Mailing Add: Dept of Geog Univ of Wash Seattle WA 98105

ULLMAN, EDWIN FISHER, b Chicago, Ill, July 19, 30; m 54; c 2. ORGANIC CHEMISTRY. Educ: Reed Col, AB, 52; Harvard Univ, AM, 53, PhD(org chem), 56. Prof Exp: Res chemist, Lederle Labs, Am Cyanamid Co, 55-60, group leader, Cent Res Div, 60-66; sci dir, Synvar Assocs, 66-70, VPRES & DIR RES, SYVA CORP, 70- Concurrent Pos: Adv bd, J Org Chem, 70-75. Mem: AAAS; Am Chem Soc; The Chem Soc; Am Asn Clin Chemists. Res: Reaction mechanisms; chemistry of stable radicals; photochemistry; enzyme chemistry; immunochemical assay methods. Mailing Add: Syva Res Inst 3221 Porter Dr Palo Alto CA 94304

ULLMAN, FRANK GORDON, b New York, NY, Dec 14, 26; m 51; c 3. SOLID STATE PHYSICS. Educ: NY Univ, BA, 49; Polytech Inst Brooklyn, PhD(physics), 58. Prof Exp: Jr engr, Sylvania Elec Prod, Inc, 51-54; asst physics, Polytech Inst Brooklyn, 54-57, res assoc, 57-58; sr physicist, Nat Cash Register Co, 58-66; PROF ELEC ENG, UNIV NEBR, LINCOLN, 66- Mem: Am Phys Soc. Res: Thin films and surfaces; photoconduction; luminescence; ferroelectricity. Mailing Add: Dept of Elec Eng Univ of Nebr Lincoln NE 68588

ULLMAN, JACK DONALD, b Chicago, Ill, Sept 5, 29; m 54. NUCLEAR PHYSICS. Educ: Univ Ill, BS, 51, MS, 56, PhD(physics), 60. Prof Exp: Physicist, Bur Ships, Dept Navy, 51-53 & US Bur Stand, 60-61; res assoc, Dept Nuclear Physics, Univ Strasbourg, 61-62; res assoc physics, Columbia Univ, 62-70; ASSOC PROF PHYSICS, LEHMAN COL, 70- Mem: Am Phys Soc. Res: Nuclear structure physics; weak interactions. Mailing Add: Dept of Physics & Astron Lehman Col Bedford Park Blvd Bronx NY 10468

ULLMAN, JOHN SUMNER, biochemical genetics, see 12th edition

ULLMAN, JOSEPH LEONARD, b Buffalo, NY, Jan 30, 23. MATHEMATICS. Educ: Univ Buffalo, BA, 42; Stanford Univ, PhD(math), 49. Prof Exp: From instr to assoc prof, 49-66, PROF MATH, UNIV MICH, ANN ARBOR, 66- Mem: Am Math Soc; Math Asn Am. Res: Complex variable theory; potential theory; approximation theory in the complex plane. Mailing Add: Dept of Math Univ of Mich Ann Arbor MI 48104

ULLMAN, MONTAGUE, b New York, NY, Sept 9, 16; m 41; c 3. PSYCHIATRY, PARAPSYCHOLOGY. Educ: City Col New York, BS, 34; NY Univ, MD, 38. Prof Exp: Intern med, Morrisania City Hosp, New York, 36-41; resident neurol,

Montefiore Hosp, 41-42; resident psychiat, NY State Psychiat Inst, 42; PROF PSYCHIAT, STATE UNIV NY DOWNSTATE MED CTR, 62-; DIR, DIV PARAPSYCHOL & PSYCHOPHYS, MAIMONIDES MED CTR, 62- Concurrent Pos: Ittleson Found grant, Maimonides Med Ctr, 62-65, Shanti Found grant, 66-70 & NIMH grant, 72-74; training analyst, Comprehensive Course in Psychoanal, NY Med Col, 50-62; lectr, NY Skin & Cancer Unit, 50-62; dir, Putnam County Clin Retarded Children, 51-61; consult ment hyg, Skidmore Col, 57-71; dir, Maimonides Community Ment Health Ctr, 61-73; dir psychiat, Maimonides Med Ctr, 61-74; lectr psychother, Psykoteripikiniken, Göteburg, Sweden, 74-76. Honors & Awards: Parapsychol Found Essay Prize, 67. Mem: Soc Med Psychoanal (pres, 57-58); fel Am Acad Psychoanal; life fel Am Psychiat Asn; Am Soc Psychical Res (pres, 71-); Parapsychol Asn (pres, 66-67). Res: Studies on suggestion and warts; behavioral changes in patients with strokes and studies on dreams and extra-senosry communication. Mailing Add: 55 Orlando Ave Ardsley NY 10502

ULLMAN, NELLY SZABO, b Vienna, Austria, Aug 11, 25; US citizen; m 47; c 4. APPLIED MATHEMATICS, BIOSTATISTICS. Educ: Hunter Col, BA, 45; Columbia Univ, MA, 48; Univ Mich, Ann Arbor, PhD(biostatist), 69. Prof Exp: Res assoc, Radiation Lab, Mass Inst Technol, 45; res assoc, Microwave Res Inst, Polytech Inst Brooklyn, 45-46, instr math, 45-63; instr, 63-64, asst prof, 64-66, 68-71, ASSOC PROF MATH, EASTERN MICH UNIV, 71- Mem: Am Asn Univ Prof (treas, 71-); Am Math Asn; Am Statist Soc; Biomet Soc. Res: Integral equations; mathematical models in biological data; nonlineal parameter estimation; mathematical statistics. Mailing Add: 1430 Granger Ave Ann Arbor MI 48104

ULLMAN, ROBERT, b New York, NY, Nov 21, 20; m 47; c 4. POLYMER CHEMISTRY, PHYSICAL CHEMISTRY. Educ: City Col New York, BS, 41; Polytech Inst Brooklyn, MS, 46, PhD(chem), 50. Prof Exp: Res chemist, Ridbo Lab, 41-42, Columbia Univ, 42-45 & Carbide & Chem Co, 45-46; asst & instr chem, Polytech Inst Brooklyn, 46-47, from instr to assoc prof math, 47-59, assoc prof chem, 59-63; STAFF SCIENTIST, FORD MOTOR CO, 63- Concurrent Pos: Fulbright fel, Univ Groningen, 52; assoc prof, Univ Strasbourg, 61-62, Guggenheim fel, Ctr Res on Macromolecules, Strasbourg, France, 61-62; adj lectr nuclear eng, Univ Mich, 74- Mem: AAAS; Am Chem Soc; fel Am Phys Soc; Sigma Xi. Res: Macromolecules viscosity; chain statistics; surface chemistry; light scattering; magnetic resonance and relaxation; rheology; molecular motion in liquids and solutions; neutron scattering from macromolecules. Mailing Add: Sci Res Staff Ford Motor Co Dearborn MI 48121

ULLMANN, WILLIAM W, microbiology, analytical chemistry, see 12th edition

ULLOM, STEPHEN VIRGIL, b Washington, DC, Nov 9, 38; m 66; c 2. MATHEMATICS. Educ: Am Univ, BA, 62; Harvard Univ, MA, 64; Univ Md, PhD(math), 68. Prof Exp: Asst prof math, 70-74, ASSOC PROF MATH, UNIV ILL, URBANA, 74- Concurrent Pos: NSF fel, Math Inst, Karlsrube, WGer & King's Col, Univ London, 68-69; mem, Inst Advan Study, 69-70. Mem: Am Math Soc. Res: Algebraic number theory; class groups of integral group rings; K-theory of groups and orders. Mailing Add: Dept of Math Univ Ill Urbana IL 61801

ULLREY, DUANE EARL, b Niles, Mich, May 27, 28; m 61. ANIMAL NUTRITION. Educ: Mich State Univ, BS, 50, MS, 51; Univ Ill, PhD(animal nutrit), 54. Prof Exp: Asst animal sci, Univ Ill, 51-54; instr physiol & pharmacol, Okla State Univ, 54-55; from asst prof to assoc prof, 56-68, PROF ANIMAL HUSB, MICH STATE UNIV, 68-, PROF FISH & WILDLIFE, 73- Concurrent Pos: Moorman fel nutrit res, 70; mem comt animal nutrit, Nat Res Coun. Honors & Awards: Am Feed Mfrs Asn Nutrit Res Award, 67; G Bohstedt Mineral Res Award, 69. Mem: AAAS; Am Soc Animal Sci; Am Inst Nutrit. Res: Nutrient requirements of swine; normal development of the swine fetus; hematology of domestic animals; nutrition of exotic animals. Mailing Add: Dept of Animal Husb Mich State Univ East Lansing MI 48823

ULLRICH, DAVID FREDERICK, b Waterbury, Conn, Sept 10, 37. MATHEMATICS. Educ: Rensselaer Polytech Inst, BS, 59; Case Western Reserve Univ, MS, 62; Carnegie-Mellon Univ, PhD(differential equations), 67. Prof Exp: ASST PROF MATH, NC STATE UNIV, 66- Mem: Am Math Soc; Math Asn Am. Res: Nonlinear ordinary differential equations. Mailing Add: Dept of Math NC State Univ Raleigh NC 27607

ULLRICH, GEORGE WERNER, b Augsburg, WGer, May 1, 47; US citizen; m 71; c 1. THEORETICAL PHYSICS. Educ: Drexel Univ, BS, 69, MS, 71, PhD(theoret biophys), 74. Prof Exp: Engr, Res & Environ Systs Div, Gen Elec Corp, Pa, 69-70; teaching asst physics, Drexel Univ, 70-71; PHYSICIST, US ARMY MOBILITY EQUIP RES & DEVELOP CTR, 72- Mem: Am Phys Soc. Res: Biomathematics; nonlinear partial differential equations; aerosol diffusion and dispersion; hypervelocity impact phenomena; spallation in metals. Mailing Add: Counter Mine/Counter Intrusion US Army Mobil Equip R&D Ctr Ft Belvoir VA 22060

ULLRICK, WILLIAM CHARLES, b Evanstcn, Ill, June 6, 24; m 48; c 3. PHYSIOLOGY. Educ: Northwestern Univ, BS, 49; Univ Ill, MS, 51, PhD(physiol), 55. Prof Exp: Lab asst zool, comp anat & embryol, Northwestern Univ, 48-49; asst physiol, Col Med, Univ Ill, 50-54; from instr to assoc prof, 54-63, PROF PHYSIOL, SCH MED, BOSTON UNIV, 63- Concurrent Pos: USPHS career res develop awards, 59-; mem bd dirs, Harvard Apparatus Co, Mass, 58-61. Mem: AAAS; Biophys Soc; Am Physiol Soc. Res: Muscle physiology and biochemistry. Mailing Add: Dept of Physiol Boston Univ Sch of Med Boston MA 02118

ULLYOT, GLENN EDGAR, b Clark Co, SDak, Mar 11, 10; m 35. ORGANIC CHEMISTRY. Educ: Univ SDak, BS, 29; Univ Ill, MS, 35, PhD(org chem), 38. Prof Exp: Res chemist, Smith Kline & French Labs, 37-45, head org chem sect, 45-50, dir chem labs, 50-57, assoc dir res, 57-60, res & develop, 60-67, dir sci liaison res & develop div, 67-75; RETIRED. Concurrent Pos: Mem ad hoc comt anticonvulsants, Nat Inst Neurol Dis & Stroke, 69-, epilepsy adv comt, 72- Mem: AAAS; Am Chem Soc; fel Am Inst Chem; Am Soc Pharmacol & Exp Therapeut; Acad Pharmaceut Sci. Res: Research and development administration; synthetic medicinal agents; central nervous system active agents; diuretics; structure-biological activity relation; alkaloids. Mailing Add: Box 77 Kimberton PA 19442

ULM, EDGAR H, b McKeesport, Pa, July 23, 42; m 65; c 2. BIOCHEMISTRY. Educ: Ind Univ Pa, BA, 65; Ohio Univ, MS, 67; Purdue Univ, PhD(biochem), 72. Prof Exp: Res assoc biochem, St Louis Univ Med Sch, 71-73; SR RES BIOCHEMIST, MERCK INST, 73- Concurrent Pos: NIH res fel, 72-73. Mem: Am Chem Soc; AAAS. Res: Prostaglandin metabolism; biochemistry of hypertension; rational drug design based on specific alterations of enzymatic activities. Mailing Add: Merck Inst Therapeut Res West Point PA 19486

ULMER, DAVID D, b McCall, Idaho, Nov 10, 29; m 54; c 2. INTERNAL MEDICINE, PHYSICAL CHEMISTRY. Educ: Univ Idaho, BS, 50; Wash Univ, MD, 54. Prof Exp: Intern, Peter Bent Brigham Hosp, 54-55, asst resident physician, 55-56; res fel med, Harvard Med Sch, 58-61, from instr to assoc prof, 61-72; PROF MED & CHMN DEPT, CHARLES R DREW POSTGRAD MED SCH, 72-; PROF, UNIV SOUTHERN CALIF, 72- Concurrent Pos: Nat Found fel, 58-60; Nat Inst Arthritis & Metab Dis fel, 60-61; NIH res career develop award, 61-71; asst med, Peter Bent Brigham Hosp, 58-61, jr assoc, 61-63, chief resident physician, 63-64, assoc med, 64-67, sr assoc, 67-71, physician, 71-72; chief internal med, Martin Luther King Hosp, Los Angeles, 72- Mem: Am Chem Soc. Res: Chemical basis of enzyme specificity; physical chemistry of proteins; spectropolarimetric properties of proteins. Mailing Add: Los Angeles County-King Hosp 12021 S Wilmington Ave Los Angeles CA 90059

ULMER, DAVID HEADING BARTINE, JR, b Morrestown, NJ, Mar 31, 16; div; c 1. PHYSIOLOGY, MICROBIOLOGY. Educ: Duke Univ, AB, 46, PhD(bot), 55. Prof Exp: Asst med mycol, Sch Med, Duke Univ, 52-53; asst serol, State Dept Health, Mich, 54; res dir vet pharmaceut, LeGear Med Co, Mo, 54-56; prod mgr, St Louis Brewers Yeast Corp, 57; asst dir, Seafood Processing Lab, Univ Md, 57-65; assoc prof biol, Keuka Col, 65-67; ASSOC PROF BIOL, STATE UNIV NY COL GENESEO, 67- Mem: AAAS; Am Inst Biol Sci; Am Soc Microbiol. Res: Antagonism, synergism, symbiosis, commensalism and other simple interactive competition, both among microorganisms and higher organisms and between the two groups; environmental microbiology; ecology. Mailing Add: Dept of Biol State Univ NY Col Geneseo NY 14454

ULMER, GENE CARLETON, b Cincinnati, Ohio, Jan 28, 37; m 60; c 4. GEOCHEMISTRY. Educ: Univ Cincinnati, BS, 58; Pa State Univ, PhD(geochem), 64. Prof Exp: Asst geochem, Pa State Univ, 59-62, staff res asst, Col Mineral Industs, 62-64; engr, Homer Res Labs, Bethlehem Steel Corp, Pa, 64-69; ASSOC PROF GEOL, TEMPLE UNIV, 69- Mem: AAAS; Mineral Soc Am; Am Ceramic Soc; Am Geophys Union. Res: High temperature phase equilibria; oxide systems; oxidation reduction reactions, equilibria and kinetics; experimental petrology; materials research, especially spinels and silicates, ultramafics in Africa and basalts in Idaho-Oregon; high pressure-high temperature research. Mailing Add: Rm 307 Buery Hall Temple Univ Dept of Geol Philadelphia PA 19122

ULMER, GILBERT, b Alexandria, Ind, Oct 27, 03; m 37; c 6. MATHEMATICS. Educ: Butler Univ, AB, 31; Univ Kans, AM, 33, PhD(educ), 39. Prof Exp: Asst instr math, 31-34, instr educ, 34-39, asst prof math & educ, 39-41, asst dean, Col Lib Arts, 41-73, assoc prof math, 44-56, PROF MATH, UNIV KANS, 56- Concurrent Pos: Mem, Nat Coun Teachers Math. Mem: Am Math Soc; Math Asn Am. Res: Projective differential geometry. Mailing Add: Dept of Math Univ of Kans Lawrence KS 66044

ULMER, MARTIN JOHN, b US, June 4, 20; m 46; c 4. HELMINTHOLOGY. Educ: Univ Mich, BS, 42, MS, 43, PhD(zool), 50. Prof Exp: From asst prof to assoc prof, 50-59, PROF ZOOL, IOWA STATE UNIV, 59-, GRAD FAC, 53-, ASSOC DEAN GRAD COL, 71- Concurrent Pos: Univ Va-Mt Lake Biol Sta, 60, 63. Honors & Awards: Ward Medal, Am Soc Parasitol, 66. Mem: Am Soc Parasitol (treas, 59-62); Am Micros Soc; Am Soc Zool; Wildlife Dis Asn. Res: Life cycles and host-parasite relationships of helminths. Mailing Add: Dept of Zool Iowa State Univ Ames IA 50010

ULMER, MELVILLE PAUL, b Washington, DC, Mar 12, 43; m 68; c 1. X-RAY ASTRONOMY. Educ: Johns Hopkins Univ, BA, 65; Univ Wis, PhD(physics), 70. Prof Exp: Res assoc physics, Univ Calif, San Diego, 70-74; ASTROPHYSICIST, SMITHSONIAN ASTROPHYS OBSERV, 74- Mem: Am Astron Soc; Am Phys Soc. Res: Measurement of spectra and positions of galactic and extragalactic x-ray sources; application of x-ray astronomy to studies in cosmology, galactic structure and the interstellar medium. Mailing Add: Ctr for Astrophys SAO-HCO 60 Garden St Cambridge MA 02138

ULMER, MILTON DON, mathematics, see 12th edition

ULMER, RICHARD CLYDE, b Lancaster, Ohio, July 4, 09; m 36; c 1. PHYSICAL CHEMISTRY. Educ: Ohio State Univ, AB, 30, PhD(chem), 36. Prof Exp: Chief chemist, Columbus and Southern Ohio Elec Co, 30-33; asst head chem div, Res Dept, Detroit Edison Co, 36-45; tech dir, E F Drew & Co, Inc, 45-53; mgr res, Combustion Eng, Inc, Windsor, 53-66, exec engr, 66-75; RETIRED. Mem: Am Chem Soc; Am Soc Test & Mat; Am Soc Mech Eng; Am Asn Corrosion Eng. Res: Water treatment and corrosion in the power boiler and electric utility fields. Mailing Add: 17 Latimer Lane Simsbury CT 06070

ULOTH, ROBERT HENRY, b Valley City, NDak, Mar 17, 27; m 50; c 2. ORGANIC CHEMISTRY. Educ: Valley City State Col, BS, 49; Univ NDak, MSc, 54. Prof Exp: Assoc chemist, 54-58, chemist, 59-60, sr scientist, 60-68, res assoc, 68-69, patent coordr, 69-70, PATENT AGENT, MEAD JOHNSON & CO, 70- Mem: Am Chem Soc. Res: Synthetic pharmaceutical drugs. Mailing Add: Mead Johnson Res Ctr Evansville IN 47721

ULREY, STEPHEN SCOTT, b Wilmington, Del, July 29, 46; m 69; c 2. INDUSTRIAL ORGANIC CHEMISTRY. Educ: WVa Univ, BS, 68; Ohio State Univ, PhD(org chem), 73. Prof Exp: DEVELOP CHEMIST, AM CYCNAMID CO, 74- Mem: Am Chem Soc. Res: Catalysis in organic chemistry, especially enzyme model compounds; industrial process development. Mailing Add: Am Cyanamid Co Willow Island WV 26190

ULRICH, AARON JACK, b Benton, Ill, Feb 27, 21; m; c 2. REACTOR PHYSICS. Educ: Univ Chicago, BS, 43, MS, 50. Prof Exp: Instr physics, Univ Chicago, 43-44; jr physicist, Oak Ridge Nat Lab, 44-46; asst physicist, 50-51, assoc physicist, 51-72, PHYSICIST, ARGONNE NAT LAB, 72- Mem: Am Phys Soc; Am Nuclear Soc; Sigma Xi. Res: Theory of nuclear power reactors, planning and analysis of critical experiments and reactor safety tests; energy conversion; plasma physics. Mailing Add: Appl Physics Div Argonne Nat Lab 9700 Cass Ave Argonne IL 60439

ULRICH, ALBERT, plant physiology, see 12th edition

ULRICH, ARLENE LOUISE, b Manhattan, Kans, Sept 1, 29. MICROBIOLOGY. Educ: Univ Kans, AB, 51; Univ Tex, Austin, MA, 58; Kans State Univ, PhD(bact), 67. Prof Exp: Med technologist, Lake Charles Mem Hosp, La, 52-53; head, Old Hermann Lab, Hermann Hosp, Tex, 53-55; head bact dept, Austin State Hosp, Tex, 57-63; res assoc, Brookhaven Nat Lab, 67-68; ASST PROF BIOL, EMPORIA KANS STATE COL, 68- Mem: AAAS; Am Soc Microbiol. Res: Phage-host relationships in bacterial hybrids. Mailing Add: Div of Biol Sci Emporia Kans State Col Emporia KS 66801

ULRICH, BRUCE T, b May 7, 40. LOW TEMPERATURE PHYSICS. Educ: Calif Inst Technol, BS, 61; Cornell Univ, PhD(physics), 64. Prof Exp: Res fel, Calif Inst Technol, 64-66; res scientist, Ford Sci Lab, Newport Beach, Calif, 67-68; ASST PROF ASTRON & PHYSICS, UNIV TEX, AUSTIN, 68- Concurrent Pos: Chief res, Normal Sec Sch, Paris, France, 70-71; sr Fulbright lectr comt int exchange scholars, Univ Moscow, 75; vis McKay prof, Univ Calif, Berkeley, 76; mem comn instruments & tech, Int Astron Union. Mem: Am Phys Soc; Int Astron Union; Am Astron Soc; fel

AAAS. Res: Superconductivity; applications of the Josephson effect to microwave detection and studies of nonlinear oscillatory phenomena. Mailing Add: Astron Dept Univ of Tex Austin TX 78712

ULRICH, DALE V, b Wenatchee, Wash, Mar 1, 32; m 53; c 3. PHYSICS. Educ: La Verne Col, BA, 54; Univ Ore, MS, 56; Univ Va, PhD(physics), 64. Prof Exp: Instr, 58-61, from asst prof to assoc prof, 64-67, PROF PHYSICS & DEAN COL, BRIDGEWATER COL, 67- Concurrent Pos: NSF res grant, 64-68. Mem: Am Asn Physics Teachers; Am Phys Soc. Res: Partial specific volume studies on biological macromolecules; density studies in the critical region of single component systems. Mailing Add: Bridgewater Col Bridgewater VA 22812

ULRICH, FLOYD SEYMOUR, b Huron, SDak, Feb 22, 15; m 41; c 4. MEDICINAL CHEMISTRY, ANALYTICAL CHEMISTRY. Educ: Huron Col, SDak, BA, 40; Kans State Univ, MS, 51; Okla State Univ, PhD(chem), 56; Southwestern State Col, BS, 68. Prof Exp: Asst prof chem, Huron Col, SDak, 46-52; tech asst to chief chemist, Celanese Chem Co, 55-65; instr pharm, 65-67, ASSOC PROF MED CHEM, SCH PHARM, SOUTHWESTERN STATE COL, 67- Mem: Am Chem Soc; Acad Pharmaceut Sci. Res: Nonaqueous high resistance, organic, polarographic analysis. Mailing Add: 1021 E Kee Weatherford OK 73096

ULRICH, FRANK, b Frankfurt-am-Main, Ger, Aug 30, 26; nat US; m 57; c 3. CELL PHYSIOLOGY. Educ: Univ Calif, BA, 48, PhD, 52. Prof Exp: Jr res endocrinologist, Univ Calif, 52-53; estab investr, Am Heart Asn, 57-62; asst prof, Yale Univ, 62-67; sr res assoc, Grad Sch Nutrit, Cornell Univ, 67-69; ASSOC PROF PHYSIOL, DEPT SURG, SCH MED, TUFTS UNIV, 69- Concurrent Pos: Brown Mem fel physiol, Sch Med, Yale Univ, 53-54; Nat Cancer Inst fel, 54-56; Arthritis & Rheumatism Found fel, 56-57. Mem: Am Soc Cell Biol; Am Physiol Soc; Brit Biochem Soc. Res: Ion transport in mitochondria, ions and cell respiration; enzyme kinetics; macrophage physiology. Mailing Add: Surg Unit Vet Admin Hosp 150 S Huntington Ave Boston MA 02130

ULRICH, GEORGE ERWIN, b Cortland, NY, Jan 8, 34; m 55; c 2. GEOLOGY. Educ: Brown Univ, BA, 55; Univ Colo, PhD(geol), 63. Prof Exp: GEOLOGIST, CTR ASTROGEOL, US GEOL SURV, 63- Concurrent Pos: Coinvestr, Apollo Lunar Geol Invest Team, Apollo, 16 & 17. Mem: AAAS; fel Geol Soc Am; Am Geophys Union. Res: Geologic training of astronaut crews; geologist traverse planning; analysis of geological data returned from the lunar surface. Mailing Add: Ctr of Astrogeol US Geol Surv 601 E Cedar Ave Flagstaff AZ 86001

ULRICH, HENRI, b Rheinsberg, Ger, May 4, 25; nat US; m 54; c 4. ORGANIC POLYMER CHEMISTRY. Educ: Univ Berlin, dipl, 52, Dr rer nat, 54. Prof Exp: Instr org chem, Univ Berlin, 53-54; res assoc, Res Found, Ohio State Univ, 55-56; res chemist, Olin Mathieson Chem Corp, Ohio, 56-59; group leader org res, Carwin Co, 59-62, head org res, Donald S Gilmore Res Lab, 62-65, mgr chem res & develop, 65-76, DIR CHEM RES & DEVELOP, UPJOHN CO, 76- Mem: AAAS; Am Chem Soc; Soc Ger Chem. Res: Carbodiimides; polyurethanes; agricultural chemicals; light sensitive chemicals. Mailing Add: Upjohn Co 410 Sackett Pt Rd North Haven CT 06473

ULRICH, JOHN AUGUST, b St Paul, Minn, May 15, 15; m 40; c 6. MICROBIOLOGY, BACTERIOLOGY. Educ: St Thomas Col, BS, 38; Univ Minn, PhD(bact), 47; Am Bd Microbiol, dipl, 63. Prof Exp: Teacher, High Sch, Minn, 38-41; asst bact, 41-45 & Hormel Inst, 45-46; first asst, Mayo Clin, 49-50, consult, 50-65, assoc prof bact, Mayo Grad Sch Med, Univ Minn, 65-69, assoc prof microbiol, Univ, 66-69; PROF MICROBIOL, MED SCH, UNIV N MEX, 69- Concurrent Pos: Res fel, Hormel Inst, Univ Minn, 46-49; consult, Econ Labs, Mo, 45, Hormel Packing Plant, Minn, 47-, NIH, 56-, Nat Commun Dis Ctr, 63-, NASA, 65-, Vet Admin, 69-, Sandia Labs, 70- & Midwest Res Inst, 71- Mem: AAAS; Am Soc Microbiol; Mycol Soc Am; Am Chem Soc; Am Acad Microbiol. Res: Skin bacteriology; hospital epidemiology; surgery air recirculation; infected wounds; chemotherapy; food preservation; low temperature; bacterial metabolism; medical mycology and bacteriology. Mailing Add: Dept of Microbiol Univ of NMex Med Sch Albuquerque NM 87131

ULRICH, JOHN PAUL, physics, mathematics, see 12th edition

ULRICH, MARIE-HELENE DEMOULIN, b Marseille, France, Mar 10, 39; m 70. ASTRONOMY. Educ: Univ Marseille, MA, 63; Univ Paris, PhD(radio galaxy), 69. Prof Exp: Asst astron, Univ Marseille, 63-66; res physicist, Univ Calif, San Diego, 67-68; in charge res, Nat Sci Res Ctr Observ Paris, Meudon, 69-70; res assoc, 70-73, ASST PROF ASTRON, UNIV TEX, AUSTIN, 74- Concurrent Pos: Mem user's comt, Kitt Peak Nat Observ, Tucson, 74-75; NSF grant, 71, 73 & 75. Mem: Int Astron Union; Am Astron Soc; AAAS. Res: Optical observation of radio galaxies and quasars aimed to make progress toward understanding the phenomenon producing radio sources in nuclei of galaxies and in quasars. Mailing Add: Dept of Astron Univ of Tex Austin TX 78712

ULRICH, MERWYN GENE, b Norfolk, Nebr, July 14, 36; m 58. ZOOLOGY. Educ: Westmar Col, BA, 58; Univ SDak, MA, 62; Univ Southern Ill, PhD(zool), 66. Prof Exp: Asst prof, 66-68, ASSOC PROF BIOL, WESTMAR COL, 68- Mem: Am Fisheries Soc. Res: Fisheries management and culture. Mailing Add: Dept of Biol Westmar Col Le Mars IA 51031

ULRICH, PETER B, b New Haven, Conn, July 4, 37; m 63; c 2. OPTICAL PHYSICS. Educ: Yale Univ, BS, 59; Mass Inst Technol, PhD(physics), 66. Prof Exp: Comput scientist, Am Sci & Eng, Inc, Mass, 66-67; RES PHYSICIST, US NAVAL RES LAB, 67- Concurrent Pos: Consult, Phelps-Dodge Electronic Prod, Inc, 66-67. Mem: Optical Soc Am. Res: Laser propogation in fluids; computer studies of non-linear partial differential equations. Mailing Add: Optical Sci Div US Naval Res Lab Washington DC 20375

ULRICH, RENEE SANDRA, b New York, NY, Aug 26, 40; m 71. NEUROENDOCRINOLOGY. Educ: Univ Calif, Los Angeles, BA, 60, PhD(anat), 66. Prof Exp: USPHS fels, Sch Med, Univ Southern Calif & Med Res Div, Long Beach Vet Admin Ctr, Long Beach, 66-67; fel neurobiochem, Univ Calif, Los Angeles, 68; RES ASSOC NEUROBIOCHEM LAB, VET ADMIN CTR, WEST LOS ANGELES, 68-; ASST RES NEUROENDOCRINOLOGIST, CTR HEALTH SCI, UNIV CALIF, LOS ANGELES, 71-, ASSOC RES NEUROENDOCRINOLOGIST, DEPT PSYCHIAT, 75- Concurrent Pos: Consult, Vet Admin Ctr, Los Angeles, 68-; asst prof, San Fernando Valley State Col, 69-70. Mem: AAAS; Endocrine Soc; Soc Neurochem; Soc Neurosci; NY Acad Sci. Res: Neuroendocrine regulation of adrenocorticotropic hormone; developmental consequences of neonatal hyrocortisone treatment on the pituitary-adrenal axis. Mailing Add: Neurobiochem Lab 601 T-85 Vet Admin Ctr West Los Angeles CA 90073

ULRICH, STEPHEN EDGAR, b Hartford, Conn, Jan 8, 44; m 69. CHEMISTRY. Educ: Pa State Univ, University Park, BS, 65; Cornell Univ, PhD(inorg chem), 71. Prof Exp: Fel chem, Univ BC, 71-74; SPECTROSCOPIST, SCH CHEM SCI, UNIV ILL, URBANA, 74- Mem: Am Chem Soc. Res: High resolution and broadline nuclear magnetic resonance. Mailing Add: 148 Roger Adams Lab Univ of Ill Urbana IL 61801

ULRICH, STEPHEN ELMER, organic chemistry, see 12th edition

ULRICH, VALENTIN, b Palmerton, Pa, Aug 5, 26. GENETICS. Educ: Rutgers Univ, AB, 53, PhD, 61. Prof Exp: From asst prof to assoc prof, 57-68, PROF GENETICS, W VA UNIV, 68-, CHMN DEVELOP BIOL FAC, 74- Concurrent Pos: Consult, Nat Tech Adv Comt Pesticides in Water Environ, 70 & Environ Protection Agency, 70-75. Mem: AAAS; Genetic Soc Am. Res: Biochemistry and genetics of heterosis. Mailing Add: Plant Sci Div WVa Univ Morgantown WV 26506

ULRICH, WILLIAM FREDERICK, b Pinckneyville, Ill, Nov 4, 26; m 51; c 3. INORGANIC CHEMISTRY, ANALYTICAL CHEMISTRY. Educ: Southern Ill Univ, BS, 49; Univ Ill, PhD, 52. Prof Exp: Supvr spectrochem group, Shell Develop Co, 52-56; supvr appln eng, 55-66, mgr appl res, Sci Instruments Dir, 66-74, MGR CLIN MKT DEVELOP, BECKMAN INSTRUMENTS, INC, 74- Mem: Am Chem Soc; Soc Appl Spectros (treas, 67-69); Spectros Soc Can; Am Asn Clin Chemists. Res: Analytical instrumentation, ultraviolet and infrared spectrophotometry; atomic absorption; gas chromatography; electrochemistry; radioimmunoassay. Mailing Add: Sci Instruments Div Beckman Instruments Inc Box D-W Irvine CA 92664

ULRYCH, TADEUSZ JAN, b Warsaw, Poland, Aug 9, 35; Can citizen; m 58; c 2. GEOPHYSICS. Educ: Univ London, BSc, 57; Univ BC, MSc, 61, PhD(geophys), 63. Prof Exp: Asst prof geophys, Univ Western Ont, 61-64; Nat Res Coun fel, Oxford Univ, 64-65 & Bernard Price Inst Geophys, 65; asst prof, Univ BC, 65-67; vis prof, Petrobras, Salvador, Brazil, 67-68; assoc prof, 68-74, PROF GEOPHYS, UNIV BC, 74- Concurrent Pos: Vis lectr, B A Oil, Alta, Can, 66. Res: Applications of communication theory to geophysics and astronomy; isotope geophysics. Mailing Add: Dept of Geophys & Astron Univ of BC Vancouver BC Can

ULSAMER, ANDREW GEORGE, JR, b Yonkers, NY, Nov 13, 41; m 65; c 2. BIOCHEMISTRY. Educ: Siena Col, BS, 63; Albany Med Col, PhD(biochem), 67. Prof Exp: USPHS fel, Nat Heart Inst, Md, 67-68, staff fel, 68-70; res chemist, Div Toxicol, US Food & Drug Admin, 70-73, RESEARCH BIOCHEMIST, BUR BIOMED SCI, CONSUMER PROD SAFETY COMN, 73- Mem: AAAS; Am Chem Soc. Res: Effects of toxic substances on cellular biochemistry with particular emphasis on those effects induced by inhalation of respirable substances. Mailing Add: Consumer Prod Safety Comn 200 C St SW Washington DC 20204

ULSHAFER, PAUL R, b Kellersville, Pa, Aug 9, 14; m 57; c 2. ORGANIC CHEMISTRY. Educ: Pa State Univ, BS, 40, MS, 41, PhD(chem), 44. Prof Exp: Asst, Pa State Univ, 40-41 & 43; from res chemist to sr res chemist, Ciba Pharmaceut Co, 44-70, sr biochemist, 70-71, sr molecular biologist, Pharmaceut Div, Ciba-Geigy Corp, 71-72, SR STAFF SCIENTIST, PHARMACEUT DIV, CIBA-GEIGY CORP, 72- Mem: Am Chem Soc; fel Am Inst Chemists; Am Soc Microbiol; Am Inst Biol Sci; NY Acad Sci. Res: Steroid sapogenins; sex hormones; sterols; adrenal cortical hormones; alkaloids; methods of separation and purification; molecular biology. Mailing Add: Pharmaceut Div Ciba-Geigy Corp Summit NJ 07901

ULSTROM, ROBERT, b Minneapolis, Minn, Feb 23, 23; m 46; c 3. MEDICINE. Educ: Univ Minn, BS, 44, MS, 46. Prof Exp: Intern & resident, Strong Mem Hosp, Rochester, NY, 46-48; from instr to asst prof pediat, Univ Minn, 50-53; asst prof, Univ Calif, Los Angeles, 53-56; from assoc prof to prof, Sch Med, Univ Minn, 56-64; prof & chmn dept, Sch Med, Univ Calif, Los Angeles, 64-67; assoc dean med sch, 67-70, PROF PEDIAT, MED SCH, UNIV MINN, MINNEAPOLIS, 67- Concurrent Pos: Markle scholar, 54-59; consult, Hennepin County Gen Hosp, 56-; mem study sect, NIH, 64-68; examr, Am Bd Pediat, 71- Mem: AAAS; Soc Pediat Res; Endocrine Soc; Lawson Wilkins Pediat Endocrine Soc; Am Pediat Soc. Res: Metabolism of children, particularly the endocrine aspects of the neo-natal period; developmental endocrinology. Mailing Add: Dept of Pediat Univ of Minn Med Sch Minneapolis MN 55455

ULTEE, ARNOLDUS JOHANNES, JR, organic chemistry, see 12th edition

ULTEE, CASPER JAN, b Noordwyk, Netherlands, Apr 5, 28; nat US; m 50; c 4. MOLECULAR SPECTROSCOPY, PHYSICAL CHEMISTRY. Educ: Hope Col, BA, 50; Purdue Univ, PhD(chem), 54. Prof Exp: Res chemist, Linde Co Div, Union Carbide Corp, 54-60; chemist, Res & Adv Develop Div, Avco Corp, 60-61; sr res scientist, 61-67, PRIN SCIENTIST, RES LAB, UNITED AIRCRAFT CORP, 67- Mem: Am phys Soc; Am Chem Soc. Res: Infrared, Raman and electron spin resonance spectroscopy; molecular structure; chemical kinetics; spectroscopy of high temperature arcs; gas phase and chemical lasers. Mailing Add: 55 Harvest Lane Glastonbury CT 06033

ULTMANN, JOHN ERNEST, b Vienna, Austria, Jan 6, 25; US citizen; m 52; c 3. HEMATOLOGY, ONCOLOGY. Educ: Columbia Univ, MD, 52; Am Bd Internal Med, dipl. Prof Exp: Intern, NY Hosp-Cornell Med Ctr, 52-53, asst resident med, 53-54, asst med & resident hemat, 54-55; instr med, Col Physicians & Surgeons, Columbia Univ, 56-61, assoc, 61-62, asst prof, 62-68; assoc prof, 68-70, PROF MED, SCH MED, UNIV CHICAGO, 70-, DIR CANCER RES CTR, 73-; DIR CLIN ONCOL, FRANKLIN McLEAN MEM RES INST, 68- Concurrent Pos: Nat Cancer Inst trainee, NY Hosp-Cornell Med Ctr, 53-55; Am Cancer Inst Soc fel hemat, Col Physicians & Surgeons, Columbia Univ, 55-56; from asst vis physician to vis physician, Francis Delafield Hosp, 56-68; career scientist, Health Res Coun City New York, 59-; asst physician, Presby Hosp, 59-65, asst attend physician, 65-68; clin asst vis physician, 1st Med Div, Bellevue Hosp, 61-62, asst vis physician, 63-68; consult, Harlem Hosp, 66-68. Mem: Fel Am Col Physicians; Am Asn Cancer Res; Am Soc Hemat; NY•Acad Sci; Int Soc Hemat. Res: Chemotherapy cancer, lymphoma and leukemia; pathophysiology of anemia of cancer; immune defects of patients with lymphoma. Mailing Add: 950 E 59th St Chicago IL 60637

ULVEDAL, FRODE, b Oslo, Norway, Nov 20, 32; US citizen; m 57; c 1. PHYSIOLOGY, ENDOCRINOLOGY. Educ: St Svithun's Col, Norway, BS, 51; Drew Univ, BA, 55; Emory Univ, PhD(physiol), 59. Prof Exp: Chief physiol support div, Laughlin AFB, Tex, 59-60, aviation physiologist, US Air Force Sch Aerospace Med, 60-62, chief adv res unit, SMBE, 62-65, chief chem sect, 65-66, chief sealed environ sect & task scientist, 66-68, chief sealed environ br, 68-72; chief pulmonary scor br, Nat Heart & Lung Inst, 72-73; ACTG DIR, HEALTH EFFECTS DIV, ENVIRON PROTECTION AGENCY, 72- Concurrent Pos: Fel, Emory Univ, 66. Mem: Am Physiol Soc; Endocrine Soc. Res: Effects of altered atmospheric conditions like altitude and gaseous composition on man and other animals during prolonged exposures in space cabin environments; oxygen toxicity at decreased pressure, especially in regard to endocrinology, hematology and biochemistry. Mailing Add: Health Effects Div Environ Protection Agency 401 M St SW Washington DC 20460

UMANS, ROBERT SCOTT, b New York, NY, Dec 17, 41; m 69. BIOPHYSICAL CHEMISTRY. Educ: Columbia Univ, AB, 62; Yale Univ, MS, 63, PhD(chem), 66. Prof Exp: Res assoc biophys chem, Johns Hopkins Univ, 66-68, NIH fel, 67-68; res assoc, Inst Biophys & Biochem, Paris, 68-69; Mass Div, Am Cancer Soc res grant, 71-72, ASST PROF CHEM, BOSTON UNIV, 69- Mem: Am Chem Soc. Res: Physicochemical studies of influence of carcinogenic hydrocarbons and steroid hormones on deoxyribonucleoprotein control mechanisms. Mailing Add: Dept of Chem Boston Univ 675 Commonwealth Ave Boston MA 02215

UMANZIO, CARL BEEMAN, b Thompsonville, Conn, Apr 9, 07; m 29; c 1. MICROBIOLOGY. Educ: Univ Boston, AM, 35; Wash Univ, St Louis, PhD, 50. Prof Exp: Asst med mycol, Sch Med, Univ Boston, 34-35; educ adv, Civilian Conserv Corps, 35-37; teacher, High Sch, Mass, 37-39; mem staff pharm, Franklin Tech Inst, 39-47; prof bact & parasitol & chmn dept microbiol, 47-74, EMER PROF MICROBIOL, KIRKSVILLE COL OSTEOP MED, 74- Concurrent Pos: Mem staff pharm, Cambridge Jr Col, 42-47; consult, 47- Mem: Am Soc Microbiol; Mycol Soc Am; Sigma Xi; Am Soc Trop Med & Hyg; Nat Asn Biol Teachers. Res: Medical mycology; parasitology; hypersensitivity; antibiotics. Mailing Add: Dept of Microbiol Kirksville Col of Osteop Med Kirksville MO 63501

UMBERGER, ERNEST JOY, b Burke, SDak, Aug 5, 09; m 32; c 3. BIOCHEMISTRY. Educ: George Washington Univ, BS, 37, MA, 41; Georgetown Univ, PhD(biochem), 48. Prof Exp: Asst sci aide fermentation sect, Bur Agr Chem & Eng, USDA, 37-39, jr chemist, Div Allergen Invests, 39-42, asst chemist, Naval Stores Res Div, 42-43; asst chemist, Div Pharmacol, US Food & Drug Admin, 43-44, pharmacologist, 44-47, chief endocrine sect, Drug Pharmacol Br, 57-67, chief drug anal br, Div Pharmaceut Sci, Bus Sci Consumer Protection & Environ Health Serv, 67-70, dir div drug biol, Off Pharmaceut Res & Testing, Bur Drugs, 70-71; RES CONSULT ENDOCRINE PHYSIOL, GRAD SCH ARTS & SCI, GEORGE WASHINGTON UNIV, 71- Mem: Am Chem Soc; Soc Exp Biol & Med; Endocrine Soc. Res: Fermentation; chemistry of allergenic proteins; rosin esters; absorption of calomel from ointments; bioassay of estrogens; androgens, gonadotropins and adrenocorticotropic hormone; metabolism of steroid hormones; central nervous systems endocrine relationships. Mailing Add: Grad Sch Arts & Sci George Washington Univ Washington DC 20006

UMBERGER, JACOB QUENTIN, physical chemistry, see 12th edition

UMBREIT, GERALD ROSS, b Minneapolis, Minn, June 17, 30; m 53; c 3. ANALYTICAL CHEMISTRY. Educ: Augustana Col, SDak, BA, 54; Iowa State Univ, PhD(anal chem), 57. Prof Exp: Res asst, Ames Lab, AEC, Iowa, 54-57; res assoc, Upjohn Co, Mich, 58-63; res scientist, Lockheed Missiles & Space Co, Calif, 63-64; appln lab mgr, F & M Sci Div, Hewlett-Packard Co, Pa, 64-66; PRES, GREENWOOD LABS, INC, 66- Mem: Am Chem Soc; Am Inst Chem. Res: Chromatography; ion exchange. Mailing Add: Greenwood Labs Inc PO Box 187 Kennett Square PA 19348

UMBREIT, WAYNE WILLIAM, b Marksan, Wis, May 1, 13; m 37; c 3. BACTERIOLOGY, BIOCHEMISTRY. Educ: Univ Wis, BA, 34, MS, 36, PhD(bact, biochem), 39. Prof Exp: Asst bact & biochem, Univ Wis, 34-37, instr bact & chem, 38-41, asst prof bact, 41-44; instr soil microbiol, Rutgers Univ, 37-38; assoc prof bact, Cornell Univ, 44-46, prof, 46-47; head enzyme chem dept, Merck Inst Therapeut Res, 47-56, assoc dir, 56-58; prof bact & head dept, Rutgers Univ, 58-75. Honors & Awards: Lilly Award, 47; Waksman Award, 57; Carski Award, 68. Concurrent Pos: Mem, Am Bd Microbiol, 60- Mem: AAAS; Am Soc Microbiol; Am Chem Soc; Am Soc Biol Chem; fel Am Acad Microbiol (vpres, 60). Res: Mode of action of antibiotics; nature of autotrophic bacteria; transformations of morphine by microorganisms. Mailing Add: Dept of Microbiol Nelson Biol Labs Rutgers Univ New Brunswick NJ 08903

UMEN, MICHAEL JAY, b Jamaica, NY, Feb 10, 48; m 69; c 1. ORGANIC CHEMISTRY, MEDICINAL CHEMISTRY. Educ: Queens Col, BA, 69; Mass Inst Technol, PhD(org chem), 73. Prof Exp: Res scientist, 73-74, SR SCIENTIST ORG MED CHEM, McNEIL LABS INC, JOHNSON & JOHNSON, 74- Mem: Am Chem Soc. Res: Design and synthesis of organic moleucles to solve fundamental questions of biological interest and as sources of new drugs; calcium in biological systems; ionophores as chemical and biological tools. Mailing Add: Dept of Chem Res McNeil Labs Inc Camp Hill Rd Ft Washington PA 19034

UMEZAWA, HIROOMI, b Saitama-Ken, Japan, Sept 20, 24; m 58; c 2. THEORETICAL PHYSICS. Educ: Nagoya Univ, BS, 46, PhD(physics), 51. Prof Exp: Assoc prof physics, Nagoya Univ, 52-56; from assoc prof to prof, Univ Tokyo, 56-65; prof, Univ Naples, 65-66; prof, Univ Wis-Milwaukee, 66-67, distinguished prof, 67-75; KILLAM MEM PROF PHYSICS, UNIV ALTA, 75- Concurrent Pos: Imp Chem Indust fel, Univ Manchester, 53-55; vis prof, Univ Wash, 56, Univ Md, 57, Univ Iowa, 57 & Univ Aix Marseille, 59-60; leading mem Naples group of struct of matter, Ctr Nat Res, Italy, 65-66. Mem: Am Phys Soc; Phys Soc Japan; Italian Phys Soc. Res: Theoretical research on quantum field theory, high energy particle physics and many-body-problems. Mailing Add: Dept of Physics Univ of Alta Edmonton AB Can

UMLAND, JEAN BLANCHARD, organic chemistry, see 12th edition

UMMINGER, BRUCE LYNN, b Dayton, Ohio, Apr 10, 41; m 66; c 2. COMPARATIVE PHYSIOLOGY, ENVIRONMENTAL PHYSIOLOGY. Educ: Yale Univ, BS, 63, MS, 66, MPhil, 68, PhD(biol), 69; Univ Cincinnati, Cert-Univ Admin Mgt Training Prog, 75. Prof Exp: From asst prof to assoc prof, 69-75, actg head dept, 73-75, admin intern, Off of Develop, 74-75, PROF BIOL SCI, UNIV CINCINNATI, 75- Concurrent Pos: NSF grant, Univ Cincinnati, 71-76; trainee, Nat Aeronaut & Space Admin, 64-67; secy div comp physiol & biochem, Am Soc Zoologists, 76-77. Mem: Fel AAAS; Am Soc Zoologists; Am Inst Biol Sci; Am Soc Ichthyol & Herpet; Ecol Soc Am. Res: Comparative physiology, biochemistry and endocrinology of fish acclimated to environmental extremes of temperature and salinity; carbohydrate metabolism and osmoregulation in fish; stress physiology of fish; low temperature biology. Mailing Add: Dept of Biol Sci Univ of Cincinnati Cincinnati OH 45221

UMPHLETT, CLYDE JEFFERSON, b Suffolk, Va, July 11, 28; m 50; c 2. BOTANY. Educ: Va Polytech Inst & State Univ, BS, 51, MS, 57; Univ NC, PhD(mycol), 61. Prof Exp: Asst bot, Va Polytech Inst & State Univ, 55-57; asst, Univ NC, Chapel Hill, 57-61, res assoc aquatic fungi, 61, from instr to assoc prof mycol, 61-70; PROF BOT & HEAD DEPT, CLEMSON UNIV, 70- Mem: Bot Soc Am; Mycol Soc Am. Res: Ecology, systematics, morphology and cytology of aquatic fungi, especially Chytridiomycetes and parasites of mosquito larvae. Mailing Add: Dept of Bot Clemson Univ Clemson SC 29631

UN, HOWARD HO-WEI, b Hong Kong, June 8, 38; m 67; c 2. ORGANIC POLYMER CHEMISTRY. Educ: Beloit Col, BS, 60; Univ Mich, MSCh, 63, PhD(org chem), 65. Prof Exp: Res chemist, Exp Sta Lab, 65-69, tech rep, Chestnut Run Lab, 69-71, tech rep, Fluorocarbons Mkt, 71-74, MKT REP, PLASTICS DEPT, FLUORCARBONS DIV, E I DU PONT DE NEMOURS & CO, INC, 74- Mem: Am Chem Soc. Res: Fluorocarbon chemistry and polymers; thermally stable polymers. Mailing Add: Plastics Dept Farmers Bank Bldg E I Du Pont de Nemours & Co Inc Wilmington DE 19898

UNAKAR, NALIN J, b Karachi, Pakistan, Mar 26, 35; m 62; c 2. CELL BIOLOGY. Educ: Gujarat Univ, India, BSc, 55; Univ Bombay, MSc, 61; Brown Univ, PhD(biol), 65. Prof Exp: Res asst biol, Indian Cancer Res Ctr, 55-61; res assoc path, Univ Toronto, 65-66; asst prof biol, 66-69, assoc prof biol sci, 69-74, PROF & CHMN DEPT BIOL SCI, OAKLAND UNIV, 74- Concurrent Pos: Nat Cancer Inst Can fel, 65-66; NIH res grant, 71- Mem: AAAS; Am Soc Cell Biol; Asn Res Vision & Opthal. Res: Cell ultrastructure and function; inhibition of hepatomas induced by chemical carcinogens; control of cell division; wound healing. Mailing Add: Dept of Biol Sci Oakland Univ Rochester MI 48063

UNANGST, PAUL CHARLES, b Fountain Hill, Pa, Apr 19, 44; m 69. MEDICINAL CHEMISTRY. Educ: Lehigh Univ, BS, 65; Carnegie-Mellon Univ, MS, 68, PhD(org chem), 70. Prof Exp: Res chemist, Ozone Systs Div, Welsbach Corp, 70-72; assoc chem, Lehigh Univ, 72-73; asst prof, East Stroudsburg State Col, 72-73; SCIENTIST, WARNER-LAMBERT RES INST, 73- Mem: Am Chem Soc. Res: Imidates; conjugated nucleophilic addition; 1, 5-benzo-diazepenes; quinaldines; antimalarials; anti-allergy agents; CNS agents. Mailing Add: Warner-Lambert Res Inst 170 Tabor Rd Morris Plains NJ 07950

UNANUE, EMIL R, b Havana, Cuba, Sept 13, 34; m 65; c 3. IMMUNOLOGY. Educ: Inst Sec Educ, BSc, 52; Univ Havana Sch Med, MD, 60; Harvard Univ, MA, 74. Prof Exp: Assoc exp path, Scripps Clin & Res Found, 60-70; intern path, Presby Univ Hosp, Pittsburgh, Pa, 61-62; res fel exp path, Scripps Clin & Res Found, 62-65; res fel immunol, Nat Inst Med Res London, 66-68; from asst prof to assoc prof path, 71-74, PROF IMMUNOPATH, HARVARD MED SCH, 74- Concurrent Pos: Prof path, Harvard Med Sch. Honors & Awards: T Duckett Jones Award, Helen Hay Whitney Found, 68; Parke-Davis Award, Am Soc Exp Path, 73. Mem: Am Soc Exp Path; Am Soc Immunologists; Brit Soc Immunol; Reticuloendothelial Soc. Res: The cellular basis of the immune response; regulatory mechanisms in immunity. Mailing Add: Dept of Path Harvard Med Sch 25 Shattuck St Boston MA 02115

UNBEHAUN, LARAINE MARIE, b Kearney, Nebr, May 4, 40; m 65. PLANT PATHOLOGY. Educ: Kearney State Col, BAEd, 61; Univ Northern Colo, MA, 64; Va Polytech Inst & State Univ, PhD(plant path), 69. Prof Exp: Teaching assoc biol, Univ Colo, Boulder, 64-65; asst prof, 69-71, ASSOC PROF BIOL, UNIV WIS-LA CROSSE, 71- Mem: Am Phytopath Soc; Sigma Xi. Res: Pectic enzyme production by Thielaviopsis basicola grown on synthetic and natural media; enzyme purification; characterization of pectic enzymes produced in black root rot diseased tobacco. Mailing Add: Dept of Biol Univ of Wis La Crosse WI 54601

UNDEEN, ALBERT HAROLD, b San Francisco, Calif, Oct 8, 37; m 64; c 4. PARASITOLOGY. Educ: Western Wash State Col, BA, 66; Univ Ill, Urbana, MS, 69, PhD(zool), 73. Prof Exp: Res assoc parasitol, Univ Ill, Urbana, 73-75; DEP DIR RES UNIT VECTOR PATH, MEM UNIV NFLD, 76- Mem: Soc Invert Path; Am Soc Parasitologists; Soc Protozoologists. Res: Biological control of mosquitoes and blackflies with Nosema and the relationships between microsporidian parasites and their hosts. Mailing Add: PO Box 147 Mansfield IL 61854

UNDERBRINK, ALAN GEORGE, b Quincy, Ill, Jan 18, 37. PLANT CYTOLOGY, RADIOBIOLOGY. Educ: Quincy Col, BS, 59; Southern Ill Univ, MS, 61, PhD(bot), 65. Prof Exp: Res assoc, 65-67, asst cytologist, 67-68, RES COLLABR, BROOKHAVEN NAT LAB, 68-; ASSOC RADIOL, COLUMBIA UNIV, 68- Mem: Radiation Res Soc; Am Soc Cell Biol. Res: Plant cytology and fine structure, especially mitotic disturbances; chromosome structure; chemical mutagenesis. Mailing Add: Dept of Biol Brookhaven Nat Lab Upton NY 11973

UNDERDAHL, NORMAN RUSSELL, b Minn, June 5, 18; m 48; c 1. BACTERIOLOGY, VIROLOGY. Educ: St Olaf Col, BA, 41; Univ Minn, MS, 48. Prof Exp: Asst scientist, Hormel Inst, Univ Minn, 46-55; from asst prof to assoc prof, 55-68, PROF VET SCI, UNIV NEBR, LINCOLN, 68- Honors & Awards: Serv Award, Nat SPF Swine Accrediting Agency, 74. Mem: Assoc Am Vet Med Asn; Am Soc Microbiol. Res: Elimination of swine diseases by repopulation with disease-free pigs; isolation of causative agents of swine diseases using antibody-devoid, disease-free pig, obtained by hysterectomy, as the host animal. Mailing Add: Dept of Vet Sci Univ of Nebr Lincoln NE 68583

UNDERDOWN, BRIAN JAMES, b Montreal, Que, Mar 23, 41; m 65; c 1. IMMUNOLOGY. Educ: McGill Univ, BSc, 64, PhD(immunol), 68. Prof Exp: Med Res Coun Can fel, Sch Med, Washington Univ, 68-70; ASST PROF MED & IMMUNOL, UNIV TORONTO, 70- Mem: Can Soc Immunol. Res: Studies of the IgA immune response; studies of the structure of antibody molecules. Mailing Add: Dept of Cell Biol Univ of Toronto Toronto ON Can

UNDERHILL, ADNA HEATON, b Jersey City, NJ, June 8, 14; m 43; c 5. ICHTHYOLOGY. Educ: Dartmouth Col, AB, 36; Cornell Univ, PhD(fisheries), 48. Prof Exp: Dist game mgr, NY Conserv Dept, 40, 46-48; exec dir, Mass Fish & Game Asn, 49-50; dir, NJ Div Fish & Game, 50-62; ASST DIR, BUR OUTDOOR RECREATION, 62- Mem: Am Fisheries Soc; Int Asn Game, Fish & Conservation Comnrs. Res: Administration of natural resources in relation to human needs; development and management of fisheries and wildlife resources; measuring outdoor recreation demand and projecting needs. Mailing Add: 3428 N Glebe Rd Arlington VA 22207

UNDERHILL, ANNE BARBARA, b Vancouver, BC, June 12, 20. ASTROPHYSICS. Educ: Univ BC, BA, 42, MA, 44; Univ Chicago, PhD(astrophys), 48. Hon Degrees: DSc, York Univ, 69. Prof Exp: Nat Res Coun Can fel, Copenhagen Observ, 48-49; astrophysicist, Dom Astrophys Observ, 49-62; prof astrophys, State Univ Utrecht, 62-70; CHIEF LAB OPTICAL ASTRON, GODDARD SPACE FLIGHT CTR, 70- Concurrent Pos: Vis lectr, Harvard Univ, 55-56; vis prof, Univ Colo, 67. Mem: Am Astron Soc; Royal Astron Soc Can; Royal Astron Soc. Res: Atmospheres of hot stars; shell stars; model atmospheres; ultraviolet spectra of stars. Mailing Add: Code 670 Goddard Space Flight Ctr Greenbelt MD 20771

UNDERHILL, DONALD KRAFT, b Duluth, Minn, Dec 13, 33; m 52; c 4. ZOOLOGY. Educ: Univ Minn, BA, 59; Univ Ill, MA, 63, PhD(zool), 67. Prof Exp: Instr life sci, Univ Ill, 66-67; asst prof zool, Rutgers Univ, 67-74; PVT PRACT, 74- Mem: AAAS; Am Inst Biol Sci; Genetics Soc Am; Soc Study Evolution; Am Soc Ichthyol & Herpet. Res: Population genetics and evolution of vertebrates. Mailing Add: 23 Tyndall Rd Kendall Park NJ 08824

UNDERHILL, EDWARD WESLEY, b Regina, Sask, Jan 28, 31; m 54; c 3. PLANT BIOCHEMISTRY. Educ: Univ Sask, BScP, 54, MSc, 56; Univ RI, PhD(pharmacog), 60. Prof Exp: Lectr pharm, Univ Sask, 54-55, asst prof, 60-61; from asst res off to assoc res off, 61-74, SR RES OFF, NAT RES COUN CAN, 74- Mem: Am Soc Plant

Physiol; Phytochem Soc NAm; Can Soc Plant Physiol. Res: Plant biosynthesis studies using labeled compounds and enzyme techniques with special emphasis on the formation of glucosinolat- es; insect attractants in plants; insect sex pheromones. Mailing Add: Prairie Regional Lab Nat Res Coun Can Saskatoon SK Can

UNDERHILL, GLENN, b Trenton, Nebr, Oct 30, 25; m 58; c 5. THEORETICAL PHYSICS, ASTRONOMY. Educ: Nebr State Col, BS, 55; Univ Nebr, MA, 57, PhD(physics), 63. Prof Exp: Instr physics, Univ Nebr, 59-61; assoc prof, 63-68, PROF PHYSICS, KEARNEY STATE COL, 68-, HEAD DEPT PHYSICS & PHYS SCI, 71- Mem: Am Phys Soc; Am Asn Physics Teachers; Am Sci Affiliation. Res: Structure of beryllium-9 nucleus; interaction of radiation with matter. Mailing Add: Dept of Physics & Phys Sci Kearney State Col Kearney NE 68847

UNDERHILL, JAMES CAMPBELL, b Duluth, Minn, June 8, 23; m 43; c 3. ZOOLOGY. Educ: Univ Minn, BA, 49, MA, 52, PhD(zool), 55. Prof Exp: From asst prof to assoc prof zool, Univ SDak, 55-59; from asst prof to assoc prof, 59-69, PROF ZOOL, UNIV MINN, MINNEAPOLIS, 69-, COORDR GEN ZOOL PROG, 70- Mem: Am Soc Ichthyol & Herpet; Ecol Soc Am; Soc Study Evolution; Am Fisheries Soc; Am Soc Limnol & Oceanog. Res: Ecology of minnows and darters; variation in fishes; aquatic ecology. Mailing Add: 1262 Raymond Ave St Paul MN 55108

UNDERHILL, RAYMOND ALDEN, b Seattle, Wash, Jan 8, 19; m 42; c 2. BIOLOGY. Educ: Walla Walla Col, BA, 42; Agr & Mech Col, Tex, MS, 46; Ore State Col, PhD, 51. Prof Exp: Dean men & instr biol, Champion Acad, Colo, 42-44; instr biol, Southwestern Jr Col, 44-46; from asst prof to assoc prof biol, Walla Walla Col, 46-56; acad dean, Southern Missionary Col, Tenn, 56-58; instr life sci, 58-64, CHMN DEPT LIFE SCI, SIERRA COL, 65- Res: Economic entomology; insects affecting man and animals; forest insects; marine biology and entomology. Mailing Add: Dept of Life Sci Sierra Col Rocklin CA 95677

UNDERKOFLER, LELAND ALFRED, b Fairfield, Nebr, Apr 24, 06; m 32; c 2. BIOPHYSICAL CHEMISTRY. Educ: Nebr Wesleyan Univ, AB, 28, DSc(biochem), 53; Iowa State Col, PhD(biophys chem), 34. Prof Exp: Asst chem, Iowa State Col, 28-33, from instr to prof, 35-55; prof, Westminster Col, Utah, 33-35; res dir, Miles Labs, Inc, Ind, 55-71; CONSULT, 71- Concurrent Pos: Chemist, Atchison Agrol Co, Kans, 36-37; consult, Coord Interam Affairs, 42-43; chief chemist, Farm Crops Co, Nebr, 44-45. Honors & Awards: Charles Thom Award, Soc Indust Microbiol, 71. Mem: Am Chem Soc; Am Soc Biol Chem; Soc Indust Microbiol (pres, 67); Am Asn Cereal Chem; Soc Indust Microbiol. Res: Fermentation; microbial enzyme production; microbiology; microbial nutrition; enzyme composition and properties. Mailing Add: Crestmoor Village Apt 3A 2200 W Pierce St Carlsbad NM 88220

UNDERKOFLER, WILLIAM LELAND, b Ames, Iowa, Nov 10, 36; m 61; c 3. ANALYTICAL CHEMISTRY. Educ: Iowa State Univ, BS, 58; Univ Wis, PhD(anal chem), 64. Prof Exp: Staff chemist, 63-71, ADV CHEMIST, IBM CORP, 71- Mem: Am Chem Soc. Res: Electrochemistry; electrochemical analysis; electroplating; general chemical analysis. Mailing Add: Dept T42 IBM Corp Endicott NY 13760

UNDERWOOD, ARTHUR LOUIS, b Rochester, NY, May 18, 24; m 48; c 3. BIOCHEMISTRY. Educ: Univ Rochester, BS, 44, PhD(biochem), 51. Prof Exp: Res assoc, Atomic Energy Proj, Univ Rochester, 46-51; res assoc anal chem, Mass Inst Technol, 51-52; from asst prof to assoc prof, 52-62, PROF CHEM, EMORY UNIV, 62- Concurrent Pos: Res fel biochem, Univ Rochester, 48-51; res assoc, Cornell Univ, 59-60. Mem: Fel AAAS; Am Chem Soc. Res: Electrochemistry of biological compounds solubilized in micelles; electrochemical studies of pyridine coenzymes, nicotinamide model compounds and other substituted pyridinium salts; plant phenol oxidase systems. Mailing Add: Dept of Chem Emory Univ Atlanta GA 30322

UNDERWOOD, BARBARA ANN, b Santa Ana, Calif, Aug 24, 34. NUTRITION, BIOCHEMISTRY. Educ: Univ Calif, Santa Barbara, BA, 56; Cornell Univ, MS, 58; Columbia Univ, PhD(nutrit biochem), 62. Prof Exp: Res asst nutrit, Cornell Univ, 58-59; res asst nutrit biochem, Columbia Univ & St Luke's Hosp, 59-61; res assoc, Inst Int Med, Univ Md, 62-64, asst prof, 64-66; res assoc, Columbia Univ, 66-68, asst prof nutrit sci, 68-72; ASSOC PROF NUTRIT, PA STATE UNIV, 72-, DIR DIV BIOL HEALTH, 74- Concurrent Pos: Nutrit Found future leaders grant, 67-69; adv comt mem, Am Found Overseas Blind, 73-; mem nutrit panel, US-Japan Med Res Comt, 74- Mem: Am Inst Nutrit; Am Pub Health Asn; NY Acad Sci. Res: Malnutrition children; lipid metabolism; absorption and metabolism fat soluble vitamins in cystic fibrosis; vitamin A. Mailing Add: Div Biol Health Col Human Develop 118 Human Develop Pa State Univ University Park PA 16802

UNDERWOOD, DONALD LEE, b Grand Rapids, Mich, Apr 20, 28; m 56; c 2. PHYSICAL CHEMISTRY. Educ: Wheaton Col, BS, 50; Princeton Univ, MA, 53. Prof Exp: Res chemist, Personal Care Div, Gillette Co, 54-55, from res supvr to sr res supvr, 55-73, PRIN ENGR, GILLETTE ADVAN TECHNOL LAB, 73- Mem: Am Chem Soc; Soc Cosmetic Chemists; Am Inst Chemists. Res: Sorption and diffusion of salt, acid and water in human hair; hair cosmetics; physics; appliance engineering and development. Mailing Add: Gillette Advan Technol Lab 83 Rogers St Cambridge MA 02142

UNDERWOOD, DOUGLAS HAINES, b Ravenna, Ohio, Nov 29, 34; m 58; c 2. MATHEMATICS. Educ: Case Inst Technol, BS, 56; Univ Calif, Berkeley, MA, 58; Univ Wis-Madison, PhD, 68. Prof Exp: Assoc prof, 58-74, PROF MATH, WHITMAN COL, 74- Mem: Am Math Soc; Math Asn Am. Res: Commutative rings. Mailing Add: Dept of Math Whitman Col Walla Walla WA 99362

UNDERWOOD, FRANCES WENRICH, b Philadelphia, Pa, Apr 10, 17; m 42; c 1. ANTHROPOLOGY. Educ: Univ Pa, BA, 38; Yale Univ, PhD(anthrop), 48. Prof Exp: Instr sociol & anthrop, Univ Conn, 48-51; lectr anthrop, Stanford Univ, 53-64; asst prof, 64-70, ASSOC PROF ANTHROP, SAN JOSE STATE UNIV, 70- Mem: Fel Am Anthrop Asn. Res: Peasant culture, especially Bangladesh; culture and personality, particularly cross-cultural socialization. Mailing Add: Dept of Anthrop San Jose State Univ San Jose CA 95192

UNDERWOOD, GERALD EMERSON, b Wellsburg, WVa, Nov 30, 21; m 44; c 3. VIROLOGY. Educ: Mt Union Col, BS, 43; Ohio State Univ, MSc, 46, PhD(biochem), 51. Prof Exp: Res assoc chem, Ohio State Univ, 45-47; res chemist, Babcock & Wilcox Co, 47-49; res assoc, 52-67, RES HEAD, UPJOHN CO, 67- Mem: AAAS; Am Soc Microbiol. Res: Animal viruses and antiviral chemotherapy. Mailing Add: Exp Biol Res Upjohn Co Kalamazoo MI 49001

UNDERWOOD, HERBERT ARTHUR, JR, b Austin, Tex, Sept 4, 45. BIOLOGICAL RHYTHMS. Educ: Univ Tex, Austin, BA, 67, MA, 68, PhD(zool), 72. Prof Exp: Fel zool, Max-Planck Inst, 72-73; fel, Univ Tex, Austin, 73-75; ASST PROF ZOOL, NC STATE UNIV, 75- Mem: Am Soc Zoologists; AAAS. Res: Role of the eyes and extraretinal photoreceptors in the control of the biological clock of lizards; involvement of the pineal system in lizard circadian rhythms. Mailing Add: Dept of Zool NC State Univ Raleigh NC 27607

UNDERWOOD, JAMES HENRY, b Minster, Eng, Apr 18, 38; m 65. SOLAR PHYSICS, X-RAY ASTRONOMY. Educ: Univ Leicester, BSc, 59, PhD(physics), 63. Prof Exp: Res assoc space res, Nat Acad Sci, NSF, 63-66; aerospace scientist, NASA/Goddard Space Flight Ctr, 66-72; STAFF SCIENTIST SPACE RES, AEROSPACE CORP, 72- Mem: Int Astron Union; Am Astron Soc; Optical Soc Am. Res: Application of x-ray techniques, in particular x-ray optics and crystal spectroscopy, to the study of the solar coronal plasma and other celestial x-ray sources. Mailing Add: Aerospace Corp Mail Sta 120/1009 PO Box 92957 Los Angeles CA 90009

UNDERWOOD, JAMES ROSS, JR, b Austin, Tex, May 15, 27; m 61; c 3. STRUCTURAL GEOLOGY, PETROLEUM ENGINEERING. Educ: Univ Tex, Austin, BS, 48, BS, 49, MA, 56, PhD(geol), 62. Prof Exp: Petrol eng trainee, Sohio Petrol Co, 49-50, jr petrol engr, 50-51, petrol engr, 53-54; instr geol, Univ Tex, 56-57, Univ-Agency Int Develop asst prof, Univ Baghdad, 62-65; asst prof, Univ Fla, 65-67; assoc prof, 67-74, PROF GEOL, WTEX STATE UNIV, 74- Concurrent Pos: On leave as Exxon-sponsored prof, Univ Libya, 69-71. Mem: Geol Soc Am; Am Asn Petrol Geol; Soc Econ Paleontologists & Mineralogists; Nat Asn Geol Teachers; Am Inst Mining, Metall & Petrol Eng. Res: Structural geology; geomorphology; planetology, especially Mars; geology of Trans-Pecos Texas, northern Chihuahua and the Middle East; petroleum engineering, especially drilling and production. Mailing Add: Dept of Geol WTex State Univ Canyon TX 79016

UNDERWOOD, JANE HAINLINE, b Ft Bliss, Tex, Oct 30, 31; m 47, 68; c 3. ANTHROPOLOGY. Educ: Univ Calif, Riverside, BA, 60; Univ Calif, Los Angeles, MA, 62, PhD(anthrop), 64. Prof Exp: Res asst anthrop, Univ Calif, Los Angeles, 62; asst prof, Univ Calif, Riverside, 62-68; assoc prof, 68-73, PROF ANTHROP, UNIV ARIZ, 73- Concurrent Pos: NSF grant, 65-66; Wenner-Gren Found res grant, Australia, Fiji Islands & Hawaii, 74-75. Mem: Fel AAAS; fel Am Anthrop Asn; Am Asn Physical Anthropologists; Brit Soc Study Human Biol; Asn Social Anthrop Oceania. Res: Population genetics, ecology and historical demography of Micronesian societies. Mailing Add: Dept of Anthrop Univ of Ariz Tucson AZ 85721

UNDERWOOD, LAWRENCE STATTON, b Kansas City, Kans, July 29, 36; m 64; c 2. PHYSIOLOGICAL ECOLOGY. Educ: Univ Kans, BA, 59; Syracuse Univ, MS, 68; Pa State Univ, University Park, PhD(zool), 71. Prof Exp: Teacher pub schs, Kans, Mo & Alaska, 57-67; asst biol, Pa State Univ, University Park, 68-71; asst prof zool, Univ Conn, West Hartford Br, 71-73; ASST DIR SCI, NAVAL ARCTIC RES LAB, 73- Concurrent Pos: Int Biol Prog, NSF & Arctic Inst NAm grants, Naval Arctic Res Lab, 71-72. Mem: AAAS; Am Inst Biol Sci; Ecol Soc Am; fel Arctic Inst NAm; Int Soc Biometeorol. Res: Physiological and behavioral adjustments in cold acclimatized mammals to varying environmental conditions; population energetics of the arctic fox in northern Alaska; study of nutrition in the field. Mailing Add: Naval Arctic Res Lab Barrow AK 99723

UNDERWOOD, LOUIS EDWIN, b Danville, Ky, Feb 20, 37; m 60; c 3. PEDIATRICS, ENDOCRINOLOGY. Educ: Univ Ky, AB, 58; Vanderbilt Univ, MD, 61. Prof Exp: From intern to asst resident pediat, Vanderbilt Univ, 61-63; asst resident, Univ NC, 63-64; instr, Vanderbilt Univ, 64-65; attend pediat, US Naval Hosp, Chelsea, Mass, 65-67; instr, 69-70, asst prof, 70-75, ASSOC PROF PEDIAT, SCH MED, UNIV NC, CHAPEL HILL, 75- Concurrent Pos: USPHS fel endocrinol, 67-70. Mem: AAAS; Am Fedn Clin Res; Endocrine Soc; Soc Pediat Res; Lawson Wilkins Pediat Endocrine Soc. Res: Pediatric endocrine diseases; growth problems and hormonal control of growth. Mailing Add: Dept of Pediat Univ of NC Sch of Med Chapel Hill NC 27514

UNDERWOOD, NEWTON, b Atlanta, Ga, Nov 19, 06; m 34; c 3. PHYSICS. Educ: Emory Univ, BS, 28; Brown Univ, MS, 30, PhD(physics), 34. Prof Exp: Instr physics, Hood Col, 32-36; instr, Vanderbilt Univ, 36-50; prof, NC State Col, 50-63; PROF PHYSICS, SCH PUB HEALTH, UNIV NC, CHAPEL HILL, 63- Concurrent Pos: Mem sci staff, Nat Defense Res Comt, Columbia Univ, 41-42; mem res staff, Labs, Carbide & Carbon Chems Corp, Tenn, 48-49; consult, Lockheed Aircraft Corp & US Dept Health, Educ & Welfare. Mem: Am Phys Soc; Health Phys Soc. Res: Electronics; ultra-sonics; behavior of gases; photoelectricity; electron diffraction; polarized light; bacteria and virus measurements; environmental radioactivity; thermoluminescence; stopping power. Mailing Add: Dept of Environ Sci Univ NC Sch Pub Health Chapel Hill NC 27514

UNDERWOOD, REX J, b Eugene, Ore, Nov 13, 26; m 50; c 2. MEDICINE. Educ: Stanford Univ, AB, 50; Univ Ore, MS & MD, 55. Prof Exp: From instr to assoc prof anesthesiol, Med Sch, Univ Ore, 58-67; asst dir anesthesiol, 67-68, DIR ANESTHESIOL, BESS KAISER HOSP, 68- Mem: AMA; Am Soc Anesthesiol. Res: Anesthesiology, particularly related to cardiovascular physiology. Mailing Add: Bess Kaiser Hosp 5055 Greeley Ave Portland OR 97217

UNDERWOOD, ROBERT MARSHALL, b Ravenna, Ohio, Mar 27, 24; m 46; c 2. ECONOMIC GEOGRAPHY, GEOGRAPHY OF THE SOVIET UNION. Educ: Houghton Col, AB, 49; Univ Ky, MA, 50; Univ Northern Colo, EdD(geog), 64. Prof Exp: Prin, Mt Tabor Christian High Sch, WVa, 52-53; asst prof soc sci, Cedarville Col, 53-61, dir admis & registr, 57-61; teacher & supvr geog & soc sci, Bryan High Sch, Ohio, 61-62; prof soc sci, Bethel Col, 64-68; PROF GEOG, UNIV WIS-WHITEWATER, 68-, CHMN DEPT GEOG/GEOL, 70- Concurrent Pos: Partic, Advan Study Inst, Univ Ill, 68; Danforth assoc, Danforth Found, 70. Mem: Asn Am Geog; Nat Coun Geog Educ. Res: Geographic education. Mailing Add: Dept of Geog/Geol Univ of Wis Whitewater WI 53190

UNDESSER, KARL, physics, mathematics, see 12th edition

UNDEUTSCH, WILLIAM CHARLES, b Hamilton, Ohio, Oct 6, 25; m 54; c 4. ORGANIC CHEMISTRY. Educ: Univ Cincinnati, BS, 48, MS, 50; Univ Del, PhD(chem), 53. Prof Exp: Lab technician, Children's Hosp Res Found, 48, 50; res chemist, Photo Prods Dept, E I du Pont de Nemours & Co, Inc, 53-58; SR ED CHEM ABSTR SERV, OHIO STATE UNIV, 58- Mem: Am Chem Soc. Res: Chemical literature. Mailing Add: Chem Abstr Serv Ohio State Univ Columbus OH 43210

UNGAR, ANDREW, b Romania, Oct 17, 22; nat US; m 50; c 2. APPLIED STATISTICS. Educ: City Col, BChE, 47; Univ Chicago, MS, 55. Prof Exp: Statistician, IIT Res Inst, 48-59, asst supvr comput appln & opers res, 59-62, mgr opers res, 62-64, sr scientist, 64-66; sr analyst, Matson Res Corp, Calif, 66-68, assoc tech dir, 68-70; prog dir, Manalytics, Inc, 70-73; SR QUANTITATIVE ANALYST, METROP TRANSP COMN, 73- Mem: Am Statist Asn; Opers Res Soc Am. Res: Combinatorial analysis; digital computer applications; probability; operations research. Mailing Add: Metrop Transp Comn Claremont Hotel Berkeley CA 94705

UNGAR, FRANK, b Cleveland, Ohio, Apr 30, 22; m 48; c 4. BIOCHEMISTRY, ENDOCRINOLOGY. Educ: Ohio State Univ, BA, 43; Western Reserve Univ, MSc, 48; Tufts Univ, PhD(biochem, physiol), 52. Prof Exp: Res staff mem, Cleveland Clin,

47-48; res staff mem, Worcester Found Exp Biol, 51-58; assoc prof, 58-67, PROF BIOCHEM, MED SCH, UNIV MINN, MINNEAPOLIS, 68- Concurrent Pos: Fulbright sr scholar, Univ Col, Cork, 74-75; asst vis prof chem, Clark Univ, 56-58; consult cancer chemother group, NIH, 65-72. Mem: AAAS; Am Chem Soc; Endocrine Soc; Am Soc Biol Chem. Res: Regulation of hormone action; metabolism of steroid hormones. Mailing Add: Dept of Biochem Univ of Minn Med Sch Minneapolis MN 55455

UNGAR, GEORGES, b France, Mar 30, 06; nat US; m 37; c 1. PHARMACOLOGY, NEUROCHEMISTRY. Educ: Univ Paris, DSc, 34, MD, 39. Prof Exp: Asst prof physiol, Univ Paris, 34-37, head lab exp med, 37-39; lectr physiol, Oxford Univ, 41-44; lectr chem path, Univ London, 45-47; Claude Bernard prof, Univ Montreal, 48; res assoc, Northwestern Univ, 48-53 & Univ Ill, 53-54; dir dept pharmacol, US Vitamin & Pharmaceut Corp, 54-62; dir, Inst Comp Biol, San Diego, 62-63; PROF PHARMACOL, BAYLOR COL MED, 63- Concurrent Pos: Sci adv, French Embassy, Eng, 45-47. Mem: AAAS; Am Physiol Soc; Soc Exp Biol & Med; Soc Gen Syst Res; Am Fedn Clin Res. Res: Reaction to injury; mechanism of cellular excitation and drug action; protein structure and biological function; information theory in biology; molecular processes in neural function, learning and memory. Mailing Add: 1800 Holcombe Blvd Houston TX 77025

UNGAR, GERALD S, b Wilkes-Barre, Pa, Jan 27, 41; m 59; c 3. MATHEMATICS. Educ: Franklin & Marshall Col, BA, 61; Rutgers Univ, MS, 63, PhD(topol), 66. Prof Exp: Asst prof math, La State Univ, Baton Rouge, 66-68 & Case Western Reserve Univ, 68-70; ASSOC PROF MATH, UNIV CINCINNATI, 70- Mem: Am Math Soc. Res: Fiber maps; local homogeneity. Mailing Add: 774 Avon Fields Lane Cincinnati OH 45229

UNGAR, IRWIN A, b New York, NY, Jan 21, 34; m 59; c 3. PLANT ECOLOGY. Educ: City Col New York, BS, 55; Univ Kans, MA, 57, PhD(bot), 61. Prof Exp: Instr bot, Univ RI, 61-62; asst prof, Quincy Col, 62-66; from asst prof to assoc prof, 66-73, PROF BOT, OHIO UNIV, 74- Concurrent Pos: Sigma Xi res awards, 59 & 66; NSF grants, 64-65, 67-69 & 74-75, panelist, 66; res grant, Ohio Univ, 66-68, John C Baker res grant, 72-73; res grant, Ohio Biol Surv, 70-71; res assoc, Nat Ctr Sci Res grant, 72-73; Res Inst fel, Ohio Univ, 74. Mem: AAAS; Ecol Soc Am; Bot Soc Am; Am Inst Biol Sci. Res: Vegetation-soil relations on acid forest and river bottom soils; ecology of halophytes; studies in salt tolerance and behavior of species under field conditions. Mailing Add: Dept of Bot Ohio Univ Athens OH 45701

UNGEFUG, GARY ALLAN, organic chemistry, biochemistry, see 12th edition

UNGER, HILBERT JOHN, physics, see 12th edition

UNGER, ISRAEL, b Tarnow, Poland, Mar 30, 38; Can citizen; m 64. CHEMISTRY. Educ: Sir George Williams Univ, BSc, 58; Univ NB, MSc, 60, PhD(chem), 63. Prof Exp: Fel, Univ Tex, 63-65; from asst prof to assoc prof, 65-74, PROF CHEM, UNIV NB, FREDERICTON, 74- Mem: Chem Inst Can. Res: Kinetics and photochemistry of small organic molecules; photochemistry of pesticides. Mailing Add: Dept of Chem Univ of NB Fredericton NB Can

UNGER, JAMES WILLIAM, b Marshfield, Wis, Apr 1, 21; m 47; c 1. PLANT MORPHOLOGY. Educ: Wis State Col, Stevens Point, BS, 42; Univ Wis, MS, 47, PhD(bot), 53. Prof Exp: Asst prof bot, Hope Col, 47-51; prof bot & chmn dept biol, 53-67, PROF BIOL, UNIV WIS-OSHKOSH, 67- Mem: Am Inst Biol Sci; Bot Soc Am. Res: Anatomical considerations of gymnosperm tissue cultures; anatomical studies of stem apices and stem to root vascular transitions; membrane permeability studies; tissue culture of orange. Mailing Add: Dept of Biol Univ of Wis Oshkosh WI 54901

UNGER, JOHN DUEY, b Harrisburg, Pa, Mar 2, 43; m 66; c 2. GEOPHYSICS, SEISMOLOGY. Educ: Mass Inst Technol, BS & MS, 67; Dartmouth Col, PhD(geol), 69. Prof Exp: Geophysicist, Hawaiian Volcano Observ, US Geol Surv, Hawaii Nat Park, 69-74, GEOPHYSICIST, NAT CTR FOR EARTHQUAKE RES, US GEOL SURV, 74- Mem: AAAS; Geol Soc Am; Am Geophys Union; Seismol Soc Am. Res: Volcanic seismology; deformation studies of active volcanoes; general microearthquake seismology. Mailing Add: US Geol Surv 345 Middlefield Rd Menlo Park CA 94025

UNGER, LLOYD GEORGE, b Stickney, SDak, Feb 24, 18; m 47; c 4. PHYSICAL CHEMISTRY. Educ: Yankton Col, BA, 39; Pa State Univ, MS, 41, PhD(phys chem), 45. Prof Exp: Lab asst, Pa State Univ, 39-44; res chemist, CPC Int Inc, Ill, 44-68; ASSOC PROF PHYS SCI & CHEM, WRIGHT COL, 68- Mem: Fel AAAS; Am Chem Soc. Res: Cereal proteins; textile chemicals. Mailing Add: Dept of Chem & Physics Sci Wright Col Chicago IL 60634

UNGER, PAUL WALTER, b Winchester, Tex, Sept 10, 31; m 60; c 6. SOIL SCIENCE. Educ: Tex A&M Univ, BS, 61; Colo State Univ, MS, 63, PhD(soil sci), 66. Prof Exp: SOIL SCIENTIST, AGR RES SERV, USDA, 65- Mem: Am Soc Agron; Soil Sci Soc Am; Soil Conserv Soc Am. Res: Soil management and moisture conservation, especially tillage and crop residue management as they relate to soil structure, water movement and water storage in the soil. Mailing Add: Southwestern Gt Plains Res Ctr USDA Bushland TX 79012

UNGER, ROGER HAROLD, b New York, NY, Mar 7, 24; m 46; c 3. INTERNAL MEDICINE. Educ: Yale Univ, BS, 44; Columbia Univ, MD, 47; Am Bd Internal Med, dipl, 56. Hon Degrees: Dr, Univ Geneva, 76. Prof Exp: From intern to resident med, Bellevue Hosp, New York, 47-51; dir, Dallas Diabetes Unit, Tex, 51-52; clin instr, 52-59, from asst prof to assoc prof, 59-70, PROF MED, UNIV TEX HEALTH SCI CTR DALLAS, 71- Concurrent Pos: Clin instr, Postgrad Med Sch, NY Univ, 51-56; chief gastroenterol sect, Vet Admin Hosp, Dallas, 58-64, chief metab sect, 64-74, dir res, 65-75; vis prof, Univ Geneva, 72-73. Honors & Awards: Lilly Award, Am Diabetes Asn, 64; Tinsley Harrison Award, 67; Middleton Award, Vet Admin, 69; David Rumbaugh Award, Juvenile Diabetes Found, 75; Banting Medal, Am Diabetes Asn, 75. Mem: Am Diabetes Asn; Am Fedn Clin Res; Am Soc Clin Invest; Endocrine Soc; Asn Am Physicians. Res: Diabetes. Mailing Add: 4500 S Lancaster Rd Dallas TX 75216

UNGER, VICTOR HERMAN, b New Haven, Conn, Jan 7, 28; m 53; c 3. AGRONOMY. Educ: Univ Pa, AB, 49; Univ Conn, MS, 52. Prof Exp: Res scientist, 51-66, lab head, 66-72, res supvr, 72-75, DIR AGR PROD RES, ROHM AND HAAS CO, 75- Mem: Am Agron Soc; Weed Sci Soc Am. Res: Discovery and development of new chemicals to improve crop production, such chemicals to be herbicides, growth regulators, insecticides, fungicides, or other. Mailing Add: Mechanicsville Rd Mechanicsville PA 18934

UNGLAUBE, JAMES M, b Milwaukee, Wis, Apr 13, 42; m 64; c 1. ORGANIC CHEMISTRY. Educ: Carthage Col, BA, 63; Univ Iowa, MS, 66, PhD(org chem), 68. Prof Exp: Teaching asst org chem, Univ Iowa, 63-64; asst prof chem, 67-70, ASSOC PROF CHEM & ACAD DEAN, LENOIR RHYNE COL, 70- Mem: AAAS; Am Asn Higher Educ; Am Chem Soc. Res: Chemical education; organic chemistry syntheses, including heterocyclic nitrogen compounds. Mailing Add: Lenoir Rhyne Col Hickory NC 28601

UNGRIN, JAMES, b Man, Can, Feb 25, 42; m 72. NUCLEAR PHYSICS. Educ: Univ Man, BSc, 64, MSc, 65; McMaster Univ, PhD(nuclear physics), 68. Prof Exp: Nat Res Coun Can fel, Niels Bohr Inst, Copenhagen, Denmark, 68-70; RES OFFICER, ACCELERATOR PHYSICS, ATOMIC ENERGY CAN LTD, 70- Res: Accelerator physics. Mailing Add: Atomic Energy of Can Ltd Chalk River ON Can

UNIACKE, CHARLES ALLYN, b Washington, DC, Sept 21, 45; m 67; c 2. VISUAL PHYSIOLOGY, OPTOMETRY. Educ: Ohio State Univ, BS, 67, MS & OD, 71, PhD(physiol optics), 73. Prof Exp: Instr optom, 72-73, ASST PROF PHYSIOL OPTICS & OPTOM, COL OPTOM, OHIO STATE UNIV, 73- Mem: Am Acad Optom; Am Optom Asn. Res: Color vision; binocular vision; low vision. Mailing Add: Col of Optom Ohio State Univ Columbus OH 43210

UNIK, JOHN PETER, b Chicago, Ill, May 18, 34; m 57; c 2. NUCLEAR CHEMISTRY. Educ: Ill Inst Technol, BS, 56; Univ Calif, Berkeley, PhD(chem), 60. Prof Exp: From asst chemist to assoc chemist, 60-74, sect head, 73-74, SR CHEMIST & ASSOC DIR CHEM DIV, ARGONNE NAT LAB, 74- Concurrent Pos: Consult, Oak Ridge Nat Lab, 74- Mem: Am Phys Soc; Am Chem Soc. Res: Nuclear fission and heavy ion reactions. Mailing Add: Argonne Nat Lab Bldg 200 9700 S Cass Ave Argonne IL 60439

UNKLESBAY, ATHEL GLYDE, b Byesville, Ohio, Feb 11, 14; m 40; c 4. GEOLOGY. Educ: Marietta Col, AB, 38; Univ Iowa, MS, 40, PhD(paleont), 42. Prof Exp: Lab asst, State Geol Surv, Iowa, 38-39, geologist, 45-46; asst geol, Univ Iowa, 39-42; jr geologist, US Geol Surv, Fla, 42-44, asst geologist, 44-45; instr geol, Colgate Univ, 46-47; from asst prof to prof, 47-67, VPRES ADMIN, UNIV MO-COLUMBIA, 67- Mem: Fel Geol Soc Am; Soc Econ Paleont & Mineral; Paleont Soc; Am Asn Petrol Geol. Res: Stratigraphy; Paleozoic nautiloids; ground water geology. Mailing Add: Univ Hall Univ of Mo Columbia MO 65201

UNKLESBAY, NAN F, b North Vancouver, BC, May 28, 44; m 74. FOOD SCIENCE, NUTRITION. Educ: Univ BC, BHE, 66; Univ Wis, Madison, MS, 71, PhD(food sci), 73. Prof Exp: Dietary consult, Dept Health, Govt Nfld & Labrador, 67-70; ASST PROF FOOD SYSTS MGT, UNIV MO, COLUMBIA, 73- Concurrent Pos: Consult, NSF, 75-76. Mem: Inst Food Technologists; Am Dietetic Asn; Can Dietetic Asn. Res: Major amounts of energy utilization within the food industry; optimization of resource utilization in food services while maintaining microbial safety, quality and nutritional value. Mailing Add: 1504 W Lexington Circle Columbia MO 65201

UNLAND, MARK LEROY, b Jacksonville, Ill, Mar 17, 40; m 62; c 2. PHYSICAL CHEMISTRY. Educ: MacMurray Col, AB, 62; Univ Ill, Urbana, MS, 64, PhD(chem), 66. Prof Exp: RES CHEMIST, MONSANTO CO, 66- Res: Molecular structure; microwave and Mossbauer effect of spectroscopy; fundamental studies of heterogeneous catalysts and catalysis mechanisms. Mailing Add: Corp Res Dept Monsanto Chem Co 800 N Lindbergh Blvd St Louis MO 63166

UNNA, KLAUS ROBERT, b Hamburg, Ger, July 30, 08; nat US; m 39; c 1. PHARMACOLOGY. Educ: Univ Freiburg, MD, 30. Prof Exp: Asst surg, Univ Hamburg, 32, asst pharmacol, 32-33; asst, Univ Vienna, 33-37; sr pharmacologist, Merck Inst Therapeut Res, 37-44; instr pharmacol, Univ Pa, 44-45; from asst prof to assoc prof, 45-50, head dept, 55-74, PROF PHARMACOL, UNIV ILL COL MED, 50- Concurrent Pos: Mem, Int Cong Physiol. Mem: AAAS; Am Physiol Soc; Soc Exp Biol & Med; Am Soc Pharmacol & Exp Therapeut; Am Inst Nutrit. Res: Vitamins; autonomic agents; skeletal muscle depressants; site of action of drugs in the central nervous system. Mailing Add: Dept of Pharmacol Univ of Ill Col of Med Chicago IL 60680

UNOWSKY, JOEL, b St Paul, Minn, Dec 11, 38; m 75; c 2. INDUSTRIAL MICROBIOLOGY, MICROBIAL GENETICS. Educ: Univ Minn, Minneapolis, BA, 61; Northwestern Univ, Evanston, PhD(bact), 66. Prof Exp: Res fel, E I du Pont de Nemours & Co, Inc, 65-66; SR MICROBIOLOGIST, HOFFMANN-LA ROCHE INC, 66- Mem: AAAS; Am Soc Microbiol; Sigma Xi. Res: Resistance transfer factor; antibiotic strain development; antibiotic screening; mutation and genetics. Mailing Add: Hoffmann-La Roche Inc Nutley NJ 07110

UNRATH, CLAUDE RICHARD, b Benton Harbor, Mich, Nov 29, 41; m 62; c 2. HORTICULTURE. Educ: Mich State Univ, BS, 63, MS, 66, PhD(hort), 68. Prof Exp: Asst prof, 68-73, ASSOC PROF TREE FRUIT PHYSIOL, N C STATE UNIV, 73- Mem: Am Soc Hort Sci; Am Pomol Soc. Res: Applied tree fruit physiology, apple research, growth regulator physiology, environmental control and modification and cultural improvement and efficiency. Mailing Add: Dept of Hort Sci 250 Kilgore Hall NC State Univ Raleigh NC 27607

UNRAU, ABRAHAM MARTIN, b Orenbourg, Can, Feb 16, 26; m 52; c 3. BIOCHEMISTRY, PLANT PHYSIOLOGY. Educ: Univ BC, BSA, 52, MSA, 53; Univ Minn, PhD(hort sci), 56, PhD(biochem), 60. Prof Exp: Asst hort, Exp Farms, Can Dept Agr, 50; asst & instr biol, genetics & hort, Univ BC, 51-53, fel chem, 61; asst hort, Univ Minn, 53-56, res asst & res fel, 56-59; asst chemist, Univ Hawaii, 59-61; assoc prof plant chem, Univ Man, 61-66; PROF ORG CHEM, SIMON FRASER UNIV, 66- Mem: Am Chem Soc; Agr Inst Can. Res: Carbohydrate chemistry; plant biochemistry; synthesis of biologically important organic compounds. Mailing Add: Dept of Chem Simon Fraser Univ Burnaby BC Can

UNRAU, DAVID GEORGE, b Leamington, Ont, July 21, 38; m 62; c 4. BIOCHEMISTRY. Educ: Univ Toronto, BSA, 62; Purdue Univ, MS, 65, PhD(biochem), 67. Prof Exp: Res scientist, Union Camp Corp, NJ, 67-71; RES CHEMIST, ITT RAYONIER INC, WHIPPANY, 71-, RES GROUP LEADER, 74- Mem: Am Chem Soc. Res: Carbohydrate modification for industrial uses; flame retardants for cellulosics; viscose; new rayon fiber development. Mailing Add: 20 Glenside Dr Budd Lake NJ 07828

UNRUH, HENRY, JR, b Greensburg, Kans, Dec 31, 26; m 50; c 2. PHYSICS. Educ: Wichita State Univ, AB, 50; Kans State Univ, MS, 52; Case Inst Technol, PhD(physics), 60. Prof Exp: Instr physics, Fenn Col, 54-57; asst prof, Colo State Univ, 59-61; assoc prof, 61-72, PROF PHYSICS, WICHITA STATE UNIV, 72- Mem: Am Phys Soc. Res: Magnetic properties of solids. Mailing Add: Dept of Physics Wichita State Univ Wichita KS 67208

UNRUH, JERRY DEAN, b Colorado Springs, Colo, Nov 4, 44; m 66; c 1. INDUSTRIAL ORGANIC CHEMISTRY. Educ: Colo State Univ, BS, 66; Ore State Univ, PhD(org chem), 70. Prof Exp: Res chemist, 70-73, SR RES CHEMIST, CORPUS CHRISTI TECH CTR, CELANESE CHEM CO, 73- Mem: Am Chem Soc. Res: Organic free radicals; linear free energy relationships; molecular orbital theory; homogeneous catalysis. Mailing Add: Box 9077 Corpus Christi TX 78403

UNRUH, WILLIAM GEORGE, b Winnipeg, Man, Aug 28, 45; m 74. THEORETICAL PHYSICS. Educ: Univ Man, BSc, 67; Princeton Univ, MA, 69, PhD(physics), 71. Prof Exp: Nat Res Coun Can fel physics, Birkbeck Col, Univ London, 71-72; Miller fel physics, Miller Inst Basic Res & Univ Calif, Berkeley, 73-74; ASST PROF APPL MATH, McMASTER UNIV, 74- Res: Relation between quantum mechanics and gravitation, and other effects of strong gravitational fields. Mailing Add: Dept of Appl Math McMaster Univ Hamilton ON Can

UNSWORTH, BRIAN RUSSELL, b London, Eng, July 30, 37; m 66; c 2. BIOCHEMISTRY. Educ: Univ London, BSc, 61, PhD(biochem), 65. Prof Exp: Proj assoc biochem, Univ Wis-Madison, 65-67; USPHS fel biol, Univ Calif, San Diego, 67-69; ASST PROF BIOL, MARQUETTE UNIV, 69- Mem: Soc Develop Biol. Res: Control of organogenesis during mammalian embryonic development; tissue culture and biochemical analysis of embryonic organs differentiating in vitro; relationship between differentiation and cancer. Mailing Add: Dept of Biol Marquette Univ Milwaukee WI 53233

UNTCH, KARL GEORGE, b Cleveland, Ohio, Apr 24, 31; m 53; c 2. ORGANIC CHEMISTRY. Educ: Oberlin Col, BA, 53; Univ NDak, MS, 55; Columbia Univ, MA, 57, PhD(org chem), 59. Prof Exp: Du Pont teaching fel, Columbia Univ, 57-58; fel org chem, Univ Wis, 58-60; Fundamental Res staff fel, Mellon Inst, 60-66; assoc prof chem, Belfer Grad Sch Sci, Yeshiva Univ, 66-68; DEPT HEAD, SYNTEX RES, 68- Concurrent Pos: Alfred P Sloan fel, 67-69. Mem: Am Chem Soc; The Chem Soc. Res: Aromaticity; chemistry of unsaturated medium sized ring compounds; synthesis and structure of natural products; conformational analysis; synthesis of heterocyclic systems; total synthesis of prostaglandins; synthesis of anti-infammatory and cardiovascular agents. Mailing Add: Syntex Res 3401 Hillview Ave Palo Alto CA 94304

UNTERBERGER, ROBERT RUPPE, b New York, NY, Apr 27, 21; m 44; c 3. PHYSICS. Educ: State Univ NY Col Forestry, Syracuse, BS, 43; Syracuse Univ, BS, 43; Duke Univ, PhD(physics), 50. Prof Exp: Lab instr physics, Syracuse Univ, 43; electronics engr, Watson Labs, NJ, 46; res physicist, Chevron Res Co, Standard Oil Co, Calif, 50-52, sr res physicist, 52-54, tech asst mgr, 54-56, res assoc, 56-59, supvr res physicist, 59-65, sr res assoc, 65-68; PROF GEOPHYS, TEX A&M UNIV, 68- Concurrent Pos: Electronic engr, White Sands Proving Ground, 46; consult to many salt and potash companies in US, Can & Europe. Mem: Am Phys Soc; Inst Elec & Electronics Eng. Res: Geophysics; acoustics; electronic instrumentation; microwave spectroscopy; structure of molecules; secondary frequency standards for K-band and higher frequencies; electron spin resonance; optical pumping; high sensitivity magnetometry; electromagnetic wave propagation in rocks Mailing Add: Dept of Geophys Tex A&M Univ College Station TX 77843

UNTERHARNSCHEIDT, FRIEDRICH J, b Essen, Ger, July 17, 26; m 57. NEUROPATHOLOGY, NEUROLOGY. Educ: Univ Münster, MD, 53, Venia Legendi, 61; Ger Bd Neurol & Psychiat, cert, 57. Prof Exp: Asst neurol & psychiat, Univ Hosp Neurol & Psychiat, Bonn, 52-57; assoc, Neuropath Inst, Univ Bonn, 57-61; res assoc, Ger Res Inst Psychiat, Max Planck Inst, Munich, 61-66; res prof neuropath & chief div neuropath & exp neurol, Univ Tex Med Br Galveston, 66-72; WITH NAVAL AEROSPACE MED RES LAB, DETACHMENT, 72- Mem: Ger Neurol Soc; Ger Asn Neuropath; Japanese Soc Neurol & Psychiat; Ger Soc Scientists & Physicians; fel Royal Soc Trop Med & Hyg. Res: Psychiatry; mechanics and pathomorphology of the central nervous system traumas; virus-induced tumors; neurovirology, especially safety tests of polio and measle vaccines; general and special neuropathology; malformations of the nervous system; diseases of the spinal cord. Mailing Add: NAMRL Detachment Box 29407 Michoud Sta New Orleans LA 70189

UNTERLEITNER, FRED CHARLES, solid state physics, see 12th edition

UNTERSTEINER, NORBERT, b Merano, Italy, Feb 24, 26. GLACIOLOGY. Educ: Innsbruck Univ, PhD(geophys), 50. Prof Exp: Asst prof meteorol, Univ Vienna, 51-56, res meteorologist, Cent Estab Meteorol & Geodyn, Vienna, Austria, 57-62; res assoc prof glaciol, 63-67, PROF ATMOSPHERIC SCI & GEOPHYS, UNIV WASH, 67-, AIDJEX COORDR, DIV MATH RESOURCES, 73- Concurrent Pos: Docent, Univ Vienna, 61; consult, Rand Corp, Calif, 65-72; mem, Int Comn Polar Meteorol, World Meteorol Orgn, 66-; mem comt Polar res, Nat Acad Sci, 70-; vpres, Int Comn Snow & Ice, Asn Sci Hydrol, Int Union Geod & Geophys, 71- Honors & Awards: Austrian Hon Cross Arts & Sci, 60. Mem: AAAS; Am Geophys Union. Res: Heat and mass budget of glaciers; physical properties of sea ice; sea-air interactions in polar regions. Mailing Add: Dept of Atmospheric Sci Univ of Wash Seattle WA 98105

UNTI, THEODORE WAYNE JOSEPH, b Kenosha, Wis, Mar 11, 31. PLASMA PHYSICS, OPTICS. Educ: Marquette Univ, BS, 54; Univ Pittsburgh, MS, 60, PhD(gen relativity physics), 64. Prof Exp: Sr physicist, Am Optical Co, Pa, 64-66; SR PHYSICIST, JET PROPULSION LABS, 66- Concurrent Pos: Physicist & consult optics, Fairchild Space-Sci Div, Calif, 66. Res: Solar wind; shape of magnetósphere; diffraction theory; lens design; statistical and spectral analysis; mathematics peripheral to lens design. Mailing Add: PO Box 178 Altadena CA 91001

UNZICKER, JOHN DUANE, b Harvey, Ill, May 8, 38; m 71; c 1. ARACHNOLOGY, ENTOMOLOGY. Educ: Univ Ill, BS, 62, MS, 63, PhD(entom), 66. Prof Exp: Res assoc Trichoptera, 66-67, ASST TAXONOMIST, ILL NATURAL HIST SURV, 67-; ASST PROF ZOOL, UNIV ILL, URBANA, 72- Mem: Soc Study Evolution; Soc Syst Zool; Entom Soc Am; Am Inst Biol Sci. Res: Biosystematics of North American Trichoptera and Araneae, especially Mimetidae. Mailing Add: Faunistic Surv Sect Ill Natural Hist Surv Urbana IL 61801

UOTA, MASAMI, plant physiology, see 12th edition

UOTILA, URHO ANTTI, b Pöytyä, Finland, Feb 22, 23; nat US; m 49; c 6. GEODESY. Educ: Finland Inst Tech, BS, 46, MS, 49; Ohio State Univ, PhD(geod), 59. Prof Exp: Surveyor & geodesist, Finnish Govt, 44-46 & 46-51; from res asst to res assoc, 52-58, from lectr to assoc prof geol, 55-65, PROF GEOL, OHIO STATE UNIV, 65-, CHMN DEPT, 64-, RES SUPVR, 59- Concurrent Pos: Geodesist, Swedish Govt, 46; mem, Solar Eclipse Exped Greenland, 54; gravity comt mem, Int Asn Geod, 60-, pres spec study group 5.30, 67-71, sect V, 71-; mem geod adv panel, Nat Acad Sci to US Coast & Geod Surv, 64-66; geod & cartog working group, Space Sci Steering Comt, NASA, 65-67, geod & cartog adv subcomt, 67-; Nat Acad Sci-Nat Acad Eng ad hoc comt NAm datum, Div Earth Sci, Nat Res Coun, 68-70. Honors & Awards: Kaarina & W A Heiskanen Award, 62; Apollo Achievement Award, NASA, 69. Mem: Fel Am Geophys Union (vpres geod sect, 64-68, pres, 68-70); Am Cong Surv & Mapping; Am Soc Photogram; Can Inst Surv; foreign mem Finnish Nat Acad Sci. Res: Geometric and physical geodesy and statistical analysis of data. Mailing Add: Dept of Geodetic Sci Ohio State Univ 1958 Neil Ave Columbus OH 43210

UPADHYAY, JAGDISH M, b Jambusar, Gujerat, India, July 2, 31; m 63; c 1. MICROBIOLOGY, BIOCHEMISTRY. Educ: Gujerat Univ, India, BPharm, 51; Univ Mich, MS, 57; Wash State Univ, PhD(bact), 63. Prof Exp: Chemist, Sarabhai Chem, India, 51-55; grant, Univ Tex, 63-65; asst prof microbiol, 65-67, ASSOC PROF MICROBIOL, LOYOLA UNIV, LA, 68- Concurrent Pos: NIH grants, Schlieder Found, 70-72. Mem: Am Soc Microbiol; Brit Soc Gen Microbiol. Res: Growth and metabolism of psychrophilic microorganisms and soil amebas; lytic enzymes; cell-wall composition; thermophilic microorganisms; carotenoid pigments. Mailing Add: Dept of Biol Sci Loyola Univ New Orleans LA 70118

UPADHYAYA, RAJARAMA BELLE, b Uchila, India, Aug 20, 35; m 69; c 1. GENETICS. Educ: Univ Madras, BS, 57; Univ Poona, MSc, 60; Univ Minn, St Paul, PhD(genetics), 67. Prof Exp: Plant breeder & geneticist, Manomin Develop Co, 67-73, gen mgr admin & plant breeding, 68-73; PRES BELMONT PROD, INC, 74- Concurrent Pos: Pres, Greenland Co, Inc, Calif, 74- Res: Plant breeding; barley; wild rice. Mailing Add: Belmont Prods Inc Aitkin MN 56431

UPCHURCH, DONALD GENE, b Benton, Ill, Aug 4, 37; m 60; c 2. INORGANIC CHEMISTRY. Educ: Southern Ill Univ, BS, 58, MA, 62; Univ Tex, PhD(chem), 66. Prof Exp: Teacher high sch, Ill, 58-63; res chemist, 66-69, field tech serv rep, Plastics Dept, 69-73, STAFF SCIENTIST, FILM DEPT, E I DU PONT DE NEMOURS & CO, INC, 73- Mem: Am Chem Soc; Soc Plastic Engrs; Sigma Xi. Res: Coordination chemistry and catalysis. Mailing Add: Film Dept E I du Pont de Nemours & Co Inc Wilmington DE 19898

UPCHURCH, ROBERT PHILLIP, b Raleigh, NC, Feb 9, 28; m 48; c 3. PLANT PHYSIOLOGY, WEED SCIENCE. Educ: BS & MS, NC State Univ, 49; Univ Calif, PhD(plant physiol), 53. Prof Exp: Instr crop sci, NC State Univ, 49-51, from asst prof to prof, 53-65; sr res group leader, Monsanto Co, 65-70, res mgr, 70-73, mgr res, 73-75; HEAD DEPT PLANT SCI, UNIV ARIZ, 75- Concurrent Pos: Consult, Shell Develop Co, 62-65; Sigma Xi res award, NC State Univ, 63; mem, Weeds Subcomt, Nat Acad Sci, 64-68; ed, Southern Weed Conf, 66-69. Mem: Am Soc Plant Physiol; Am Soc Agron; Crop Sci Soc Am; Weed Sci Soc Am (pres, 72-73). Res: Response of plants to phytoactive chemicals and the influence of soil and climate factors on the expression of such responses. Mailing Add: Dept of Plant Sci Col of Agr Univ of Ariz Tucson AZ 85721

UPCHURCH, SAM BAYLISS, b Murfreesboro, Tenn, June 30, 41; m 64. ENVIRONMENTAL GEOLOGY, SEDIMENTOLOGY. Educ: Vanderbilt Univ, AB, 63; Northwestern Univ, Evanston, MS, 66, PhD(geol), 70. Prof Exp: Resident in res marine geol, Northwestern Univ, Evanston, 67-68; res phys scientist chem limnol, Lake Surv Ctr, Nat Oceanic & Atmospheric Admin, 68-71; asst prof geol, Mich State Univ, 71-74; ASSOC PROF GEOL, UNIV SOUTH FLA, 74- Concurrent Pos: Mem limnol work group, Great Lakes Basin Comn, 68-75, mem bd tech adv, 69-75; chmn mineral & geol sect, Mich Acad Sci, Arts & Letters, 74. Mem: Soc Econ Paleont & Mineral; Geol Soc Am; Am Water Resources Asn; Int Asn Math Geol. Res: Trace element-sediment interaction; coastal sedimentation; carbonate sediment genesis and diagenesis; environmental geology; mathematical geology; geohydrology; land-use planning. Mailing Add: Dept of Geol Univ of South Fla Tampa FL 33620

UPCHURCH, WILLIAM JOSEPH, soils, see 12th edition

UPDEGRAFF, DAVID MAULE, b Woodstock, NY, Dec 19, 17; m 43; c 3. MICROBIOLOGY. Educ: Univ Calif, Los Angeles, AB, 41; Univ Calif, PhD(microbiol), 47. Prof Exp: Res assoc & actg dir, Am Petrol Inst Res Proj, Scripps Inst, Univ Calif, 46; sr res chemist, Field Res Labs, Magnolia Petrol Co, 47-50, sr res technologist, 50-55; res microbiologist, Cent Res Dept, Minn Mining & Mfg Co, Minn, 55-68; head microbiol sect, Chem Div, Denver Res Inst, Univ Denver, 68-72; HEAD BASIC RES DEPT, CAWTHRON INST, 72- Mem: AAAS; Am Soc Microbiol; Soc Indust Microbiol. Res: Biochemistry of carotenoid pigments; bacterial physiology; marine and petroleum microbiology; applied microbiology; fermentations; ecology of water pollution and waste treatment. Mailing Add: Basic Res Dept Cawthron Inst Box 175 Nelson New Zealand

UPDEGRAFF, IVOR HEBERLING, b Provincetown, Mass, July 31, 13; m 41; c 2. ORGANIC CHEMISTRY. Educ: Univ Calif, BS, 38; Yale Univ, PhD, 50. Prof Exp: Res chemist, Calif Res Corp, 38-46; instr gen chem, Yale Univ, 46-49; res chemist, E I du Pont de Nemours & Co, Pa, 49-50; res chemist, 50-55, GROUP LEADER, AM CYANAMID CO, 55- Mem: AAAS; Am Chem Soc. Res: Polymers; resins; plastics. Mailing Add: Am Cyanamid Co 1937 W Main St Stamford CT 06904

UPDEGRAFF, WILLIAM EDWARD, b Williamsport, Pa, Sept 8, 37; m 63; c 3. NUCLEAR PHYSICS, COMPUTER SCIENCE. Educ: Dickinson Col, BA, 59; Pa State Univ, University Park, MS, 62; Ohio Univ, PhD(physics), 69. Prof Exp: From instr to asst prof physics, Lycoming Col, 62-65; assoc prof & head dept, Simpson Col, 69-74; ASSOC DEAN, FAC NATURAL SCI, STATE UNIV COL AT BUFFALO, 74- Concurrent Pos: Instr, Iowa Methodist Hosp, Des Moines, Iowa, 70-72; dir comput ctr, Simpson Col, 72-74. Mem: Am Asn Physics Teachers; Am Asn Higher Educ. Res: Computer oriented physics development at the undergraduate level; dynamic collective model of nuclear photofission. Mailing Add: State Univ NY 1300 Elmwood Ave Buffalo NY 14222

UPESLACIS, JANIS, b Bad-Rothenfelde, Ger, Jan 12, 46; US citizen; m 68; c 2. PHARMACEUTICAL CHEMISTRY. Educ: Univ Nebr-Lincoln, BS, 67; Harvard Univ, MA, 69, PhD(org chem), 75. Prof Exp: Asst nuclear physics, Walter Reed Army Med Ctr, US Army, 69-71; RES CHEMIST, LEDERLE LABS, AM CYANAMID, 75- Mem: Am Chem Soc. Res: Synthetic applications of carbohydrates. Mailing Add: Lederle Labs Pearl River NY 10965

UPGREN, ARTHUR REINHOLD, JR, b Minneapolis, Minn, Feb 21, 33; m 67. ASTRONOMY. Educ: Univ Minn, BA, 55; Univ Mich, MS, 58; Case Inst Technol, PhD(astron), 61. Prof Exp: Res assoc astron, Swarthmore Col, 61-63; astronr, US Naval Observ, 63-66; asst prof astron, 66-74, actg chmn dept, 68-74, ADJ ASSOC PROF ASTRON, WESLEYAN UNIV, 74-, ACTG DIR, VAN VLECK OBSERV, 68- Concurrent Pos: Vis lectr, Univ Md, 64-66; exec officer, Fund Astrophys Res, Inc, 72- Mem: Am Astron Soc; Royal Astron Soc; Int Astron Union. Res: Galactic structure; photographic astrometry. Mailing Add: Dept of Astron Wesleyan Univ Middletown CT 06457

UPHAM, ROY HERBERT, b Boston, Mass, Feb 9, 20; m 45; c 8. ORGANIC CHEMISTRY. Educ: Boston Col, BS, 41, MS, 48. Prof Exp: Chemist, Howe & French, Inc, 41-42, Cities Serv Oil Co, 42, Rock Island Arsenal, 44-45 & E I du Pont de Nemours & Co, Inc, 47-49; PROF CHEM, ST ANSELM'S COL, 49- Mem: Am Chem Soc. Res: Reaction mechanisms; teaching methods; computer applications. Mailing Add: Dept of Chem St Anselm's Col Manchester NH 03102

UPHAM, ROY WALTER, b Ogden, Kans, Apr 11, 20; m 69. VETERINARY MEDICINE, FOOD TECHNOLOGY. Educ: Kans State Univ, DVM, 43; Mass Inst Technol, MS, 60; Am Bd Vet Pub Health, dipl. Prof Exp: Instr food technol, US Army Med Serv Sch, 54-56, proj off radiation of foods prog, US Army, 56-58, mil adv, Food Prog for Vietnam, 62-63, chief lab br food testing, Defense Personnel

Supply Command, 63-65, chief standardization br, Mil Specifications, Natick Army Lab, 65-66; CHIEF DIV FOOD & DRUGS, REGULATORY AGENCY, ILL DEPT PUB HEALTH, SPRINGFIELD, 66- Mem: Inst Food Technologists. Res: Application of controlled food processing to safeguard and protect public health. Mailing Add: Div of Food & Drugs Dept Pub Health 535 W Jefferson Springfield IL 62706

UPHAUS, ROBERT ALAN, physical chemistry, see 12th edition

UPHOFF, DELTA EMMA, b Brooklyn, NY, Jan 23, 22. GENETICS. Educ: Russell Sage Col, AB, 44; Univ Rochester, MS, 47. Prof Exp: Asst radiation genetics, Manhattan Proj, Rochester, 46-47; zoologist, Radiation Lab, Univ Calif, 47-48; geneticist, Mound Lab, Monsanto Chem Co, Ohio, 48-49; BIOLOGIST, LAB PHYSIOL, NAT CANCER INST, 49- Mem: Fel AAAS; Genetics Soc Am; Radiation Res Soc; Am Genetic Asn; Am Asn Cancer Res. Res: Immunogenetics. Mailing Add: Lab of Physiol Nat Cancer Inst Bethesda MD 20014

UPHOLT, WILLIAM BOYCE, b Orlando, Fla, Sept 14, 43. MOLECULAR BIOLOGY. Educ: Pomona Col, BA, 65; Calif Inst Technol, PhD(chem), 71. Prof Exp: Res fel molecular biol, Damon Runyon Mem Fund Cancer Res, Biochem Lab, Univ Amsterdam, 71-73; res fel, Dept Embryol, Carnegie Inst Wash, Baltimore, 73-75; RES ASSOC MOLECULAR BIOL, DEPT PEDIAT & BIOCHEM, UNIV CHICAGO, 75- Res: Physical chemistry of nucleic acids, mitochondrial biogenesis, DNA replication, organization of genetic material, developmental biology, control of gene expression. Mailing Add: Univ of Chicago Hosp Box 413 5825 S Maryland Ave Chicago IL 60637

UPHOLT, WILLIAM MARTIN, b Grand Rapids, Mich, July 16, 14; m 40; c 2. SCIENCE ADMINISTRATION. Educ: Univ Calif, BS, 35, MS, 36, PhD(insect toxicol), 39. Prof Exp: Asst entomologist, Exp Sta, Clemson Agr Col, 39-42; entomologist, Calif Spray-Chem Corp, Fla, 42-44; asst sanitarian, USPHS, 44-46, asst chief tech develop div, Commun Dis Ctr, 46-53, chief, Wenatchee Field Sta, 53-57, regional chief, Commun Dis Ctr Servs, 57-61, chief res & training grants br off resource develop, Bur State Serv, 61-65, exec secy fed comt pest control, 65-70, staff dir secy's pesticide adv comt, 70; dep asst admin pesticides progs, 71-72, SR SCI ADV, WATER & HAZARDOUS MATS, ENVIRON PROTECTION AGENCY, 72- Mem: AAAS; Entom Soc Am; Soc Toxicol; Am Inst Biol Sci. Res: Chemical control of agricultural pests, rodents and arthropods of public health importance; toxicology and regulation of pesticides; estimation of risk and evaluation of risk-benefit. Mailing Add: 525 E Indian Spring Dr Silver Spring MD 20901

UPJOHN, EVERETT GIFFORD, b Kalamazoo, Mich, Oct 7, 04; m 27; c 2. MEDICINE. Educ: Univ Mich, BS, 25, MD, 28. Hon Degrees: DSc, Philadelphia Col Pharm & St Louis Col Pharm, 61; LLD, Western Mich Univ, 64. Prof Exp: Mem staff, Upjohn Co, 30-43, dir, 37-73, vpres & med dir, 43-51, exec vpres, 51-53, pres, 53-61, chmn, Bd Dirs, 61-69; RETIRED. Concurrent Pos: Pres, Nat Vitamin Found, 48-52 & Am Found Pharmaceut Educ, 66-67; trustee, W K Kellogg Found. Mem: Pharmaceut Mfrs Asn. Mailing Add: 2230 Glenwood Dr Kalamazoo MI 49008

UPPER, CHRISTEN D, biochemistry, see 12th edition

UPPULURI, V R RAO, b Machilipattanam, India, Feb 22, 31; m 60; c 1. MATHEMATICAL STATISTICS. Educ: Andhra Univ, India, MA, 54; Ind Univ, Bloomington, PhD(math), 63. Prof Exp: Res asst math & statist, Tata Inst Fundamental Res, 54-57; res asst math, Ind Univ, Bloomington, 57-61; res assoc statist, Mich State Univ, 61-62, biophys & statist, 62-63; sr math statistician, Oak Ridge Nat Lab, 63-74; MEM STAFF, NUCLEAR DIV, UNION CARBIDE CORP, 74- Concurrent Pos: Lectr, Oak Ridge Traveling Lect Prog, Oak Ridge Assoc Univs & Oak Ridge Nat Lab, 64-; sr engr & scientist, Douglas Aircraft Co, 66; vis lectr prog, Comt Statist Southern Regional Educ Bd, 67-; adj prof, Univ Tenn, 67-; consult, Syst Develop Corp, 67- & med div, Oak Ridge Assoc Univ, 67-; vis prof, Univ Sao Paulo, 70 & Univ Minn, 71. Mem: Fel AAAS; fel Am Statist Asn; Am Math Soc. Res: Probability; statistics; stochastic approach for a better understanding of the structure of physical and natural phenomena; limit theorems in random difference equations and applications of probability. Mailing Add: Nuclear Div PO Box Y Union Carbide Corp Oak Ridge TN 37830

UPSON, DAN W, b Hutchinson, Kans, July 30, 29; m 59; c 3. PHARMACOLOGY, PHYSIOLOGY. Educ: Kans State Univ, DVM, 52, MS, 62, PhD(physiol), 69. Prof Exp: Vet, Pvt Pract, 52-59; instr pharmacol & physiol, 59-69, assoc prof pharmacol, 69-73, asst dean, 72-73, PROF PHARMACOL, COL VET MED, KANS STATE UNIV, 73-, DIR TEACHING RESOURCES, 75- Concurrent Pos: Consult, Tevcon Ind Inc, 70- Res: Veterinary pharmacology and clinical pharmacology. Mailing Add: Vet Teaching Bldg Col Vet Med Kans State Univ Manhattan KS 66506

UPSON, JOSEPH EDWIN, geology, see 12th edition

UPSON, ROBERT WILLIAM, b Odell, Nebr, Aug 7, 13; m 39; c 4. ORGANIC CHEMISTRY, PHOTOGRAPHIC CHEMISTRY. Educ: Univ Nebr, BS, 37, MS, 38; Univ Ill, PhD(org chem), 41. Prof Exp: Rohm & Haas asst, Univ Ill, 38-39; res chemist, Cent Res Dept, 41-52, res supvr, Photo Prod Dept, 52-56, asst lab dir, 56-57, tech mgr, 57-65, asst dir, Res & Develop, 65-68, dir, 68-74, GEN DIR RES & DEVELOP, PHOTO PROD DEPT, E I DU PONT DE NEMOURS & CO, INC, 74- Mem: Am Chem Soc; Soc Photog Scientists & Engrs; Soc Chem Indust; Am Inst Chemists. Res: Bisbenzimidazoles; synthetic antimalarials; organic and organo-inorganic polymers; photographic products and processes, including both silver and non-silver systems; research on photographic film base, gelatin, sensitizing dyes, etc. Mailing Add: Photo Prods Dept E I du Pont de Nemours & Co Inc Wilmington DE 19898

UPSON, U LAYTON, b Grants Pass, Ore, Mar 26, 11; m 44; c 3. ANALYTICAL CHEMISTRY. Educ: Ore State Col, BS, 38, 46; Univ Idaho, MS, 40. Prof Exp: Res chemist, Consol Water Power & Paper Co, Wis, 41-42; chemist, Williamette Valley Wood Chem Co, Ore, 47; res chemist, Hanford Atomic Prod Oper, Gen Elec Co, 48-64; sr res scientist, Pac Northwest Labs, Battelle Mem Inst, 65-71; PRES UPSON ENG CO, 71- Mem: AAAS; Am Chem Soc; Am Nuclear Soc; fel Am Inst Chem. Res: Instrumental analysis, especially radiochemical and physical methods; instrumentation for process control; analytical instrumentation and system design and evaluation; environmental analysis; instrumentation design and evaluation. Mailing Add: 632 Basswood Ave Richland WA 99352

UPTON, ARTHUR CANFIELD, b Ann Arbor, Mich, Feb 27, 23; m 46; c 3. EXPERIMENTAL PATHOLOGY. Educ: Univ Mich, BA, 44, MD, 46. Prof Exp: Intern, Univ Hosp, Univ Mich, 47, resident path, Med Sch, 48-50, instr, 50-51; pathologist, Biol Div, Oak Ridge Nat Lab, 51-54, chief, Path-Physiol Sect, 54-69; chmn dept path, Health Sci Ctr, 69-70, dean sch basic health sci, 70-75, PROF PATH, STATE UNIV NY STONY BROOK, 69-; ATTEND PATHOLOGIST, MED DEPT, BROOKHAVEN NAT LAB, 69- Concurrent Pos: Mem subcomts biol effectiveness of radiation & long term effects of radiation, Nat Acad Sci-Nat Res Coun, 57-; Ciba Found lectr, Univ London, 59; rep, USA Nat Comt on the Int Union against Cancer, 72-; mem, Int Comn Radiol Protection, 74-; chmn, Int Comn Radiol Protection, 73-; mem, Nat Coun Radiation Protection & Measurements, 65-, adv comt, Ctr on Human Radiobiol, Argonne Nat Lab, 72-, sci adv group, US-Japan Coop Cancer Res Prog, 74-, sci adv bd, Nat Ctr for Toxological Res, 74-, sci coun, Int Agency for Res on Cancer, WHO, 75- Honors & Awards: Lawrence Award, 65. Mem: AAAS; Sigma Xi; Radiation Res Soc (pres, 65-66); Am Soc Exp Path (pres, 67-68); Am Asn Cancer Res (pres, 63-64). Res: Pathology of radiation injury and endocrine glands; cancer; carcinogenesis; experimental leukemia; aging. Mailing Add: Sch of Basic Health Sci State Univ of NY Health Sci Ctr Stony Brook NY 11790

UPTON, G VIRGINIA, b New Haven, Conn, Oct 17, 29; wid; c 3. PHYSIOLOGY, BIOCHEMISTRY. Educ: Albertus Magnus Col, BA, 51; Yale Univ, MS, 61, PhD(physiol), 64. Prof Exp: NIH fel peptide chem, Yale Univ, 63-66, chief endocrine & polypeptide lab, Vet Admin Hosp, Yale Univ, 66-74, sr res assoc med, Sch Med, 71-74; ASSOC DIR CLIN RES, WYETH INT LTD, 74- Concurrent Pos: Asst prof comp endocrinol, Eve Div, South Conn State Col, 66-67. Mem: AAAS; Endocrine Soc; NY Acad Sci; Int Soc Neuroendocrinol; Am Physiol Soc. Res: Neuroendocrinology; hypothalamic-pituitary-adrenal relationships; isolation of pituitary peptides and tumor peptides with hormonal activity; relationship between endocrine disorders and hypothalamic dysfunction. Mailing Add: Wyeth Int Ltd PO Box 8616 Philadelphia PA 19101

UPTON, RONALD P, b Boston, Mass, May 6, 41; m 65; c 2. ANALYTICAL CHEMISTRY. Educ: New Bedford Inst Tech, BS, 63; Univ Del, PhD(anal chem), 68. Prof Exp: Staff chemist, Res Lab, Chas Pfizer & Co, 67-70, Univ Del, 68, MGR QUAL CONTROL, PFIZER INC, 70- Mem: Am Chem Soc; NY Acad Sci. Res: Quality control; instrumental analysis; infrared, ultraviolet, atomic absorption and nuclear magnetic spectroscopies; gas and high pressure liquid chromatography and pharmaceutical analysis. Mailing Add: Qual Control Pfizer Inc Groton CT 06340

UPTON, WILSON VINCENT, organic chemistry, see 12th edition

URBACH, FREDERICK, b Vienna, Austria, Sept 6, 22; nat US; m 52; c 3. DERMATOLOGY. Educ: Univ Pa, BS, 43; Jefferson Med Col, MD, 46; Am Bd Dermat, dipl, 53. Prof Exp: Asst instr dermat, Med Sch, Univ Pa, 49-50, instr, 50-52, assoc, 52-54; chief dermat serv, Roswell Park Mem Inst, 54-58; assoc prof dermat, 58-60, prof res dermat, 60-67, PROF DERMAT & CHMN DEPT, SCH MED, TEMPLE UNIV, 67-, MED DIR, SKIN & CANCER HOSP, 67- Concurrent Pos: Fel, Hosp Univ Pa, 49-52; Damon Runyon res fel clin cancer, Univ Pa, 51-53; asst vis physician, Philadelphia Gen Hosp, 51-54; mem, Int Cong Dermat, London, 52; from asst med dir to assoc med dir, Skin & Cancer Hosp, Temple Univ, 58-67; mem, US Nat Comt on photobiol of Nat Res Coun, 73- Mem: Fel AAAS; Soc Invest Dermat; Soc Exp Biol & Med; Am Asn Cancer Res; Am Acad Dermat. Res: Blood supply of cancer; biologic effects of ultraviolet radiation; photobiology; epidemiology of cancer. Mailing Add: Skin & Cancer Hosp 3322 N Broad St Philadelphia PA 19140

URBACH, FREDERICK LEWIS, b New Castle, Pa, Nov 21, 38; m 60; c 2. INORGANIC CHEMISTRY. Educ: Pa State Univ, University Park, BS, 60; Mich State Univ, PhD(chem), 64. Prof Exp: Res assoc & fel, Ohio State Univ, 64-66; asst prof, 66-74, ASSOC PROF CHEM, CASE WESTERN RESERVE UNIV, 74- Mem: Am Chem Soc. Res: Chemistry of metal chelates containing multidentate ligands; stereochemistry and optical activity; role of transition metal ions in biological processes. Mailing Add: Dept of Chem Case Western Reserve Univ Cleveland OH 44106

URBACH, JOHN C, b Vienna, Austria, Feb 18, 34; US citizen; m 56; c 3. OPTICS. Educ: Univ Rochester, BS, 55, PhD(optics), 62; Mass Inst Technol, MS, 57. Prof Exp: Assoc physicist, Int Bus Mach Corp, 57-58; NATO fel sci, Royal Inst Tech, Sweden, 61-62; scientist, NY, 63-66, sr scientist, 66-67, mgr optical & imaging anal br, 67-68, mgr optical sci br, 68-70, mgr optical sci area, 70-75, MGR OPTICAL SCI LAB, PALO ALTO RES CTR, XEROX CORP, 75- Concurrent Pos: Vchmn tech group info processing & holography, Optical Soc Am, 74-; mem, US Nat Comt, Int Comn Optics, 75- Mem: Fel Optical Soc Am; Soc Photog Sci & Eng; Soc Photo-Optical Instrument Eng. Res: Optical techniques for information storage and retrieval; effects of recording materials upon holographic imaging; unconventional photography, especially electrophotography; evaluation of optical and photographic image quality; displays; optical communication. Mailing Add: 142 Crescent Ave Portola Valley CA 94025

URBACH, JOHN ROBERT, b Vienna, Austria, Aug 2, 24; US citizen; m 55; c 2. MEDICINE. Educ: Univ Pa, MD, 47; Am Bd Internal Med, dipl, 54. Prof Exp: Intern, Philadelphia Gen Hosp, Pa, 47-48, resident med & cardiol, 48-50; asst prof physiol & biophys & assoc prof med, 65-72, CLIN PROF MED & ASSOC PROF PHYSIOL & BIOPHYS, MED COL PA, 72- Mem: Am Heart Asn. Res: Cardiology; physiology; medical instrumentation. Mailing Add: Med Col of Pa 3300 Henry Ave Philadelphia PA 19129

URBACH, KARL FREDERIC, b Vienna, Austria, Nov 9, 17; nat US; m 52; c 2. HOSPITAL ADMINISTRATION. Educ: Reed Col, BA, 42; Northwestern Univ, PhD(chem), 46, MD, 51. Prof Exp: Asst, Eve Sch, Northwestern Univ, 42-46, asst chem, Dent Sch, 43-45, asst pharmacol, Med Sch, 45-47, lectr chem, Univ, 46-47, instr chem & pharmacol, Univ & Med Sch, 47-50; resident anesthesiol, USPHS Hosp, Staten Island, NY, 52-54; actg chief, USPHS Hosp, San Francisco, 54-55; chief anesthesiol, USPHS Hosp, Staten Island, 55-69; chief med educ & res, 69-70, DIR, USPHS HOSP, SAN FRANCISCO, 70- Concurrent Pos: NIH res fel, USPHS, 51-52. Mem: AAAS; Soc Exp Biol & Med; AMA; Am Soc Anesthesiol; Asn Mil Surg US. Res: Synthesis of vasopressors and testing, local anesthetics and testing; histamine methods for identification; actions; metabolism; pharmacology; evaluation of coronary dilators; anesthesia; hypnotics. Mailing Add: USPHS Hosp San Francisco CA 94118

URBAIN, WALTER MATHIAS, b Chicago, Ill, Apr 8, 10; m 39; c 2. PHYSICAL CHEMISTRY, FOOD SCIENCE. Educ: Univ Chicago, SB, 31, PhD(chem), 34. Prof Exp: Phys chemist, Swift & Co, 33-50, assoc dir res, 50-59, dir eng res & develop dept, 59-65; prof, 65-75, EMER PROF FOOD SCI, MICH STATE UNIV, 75- Concurrent Pos: Mem comn radiation preservation of foods, Nat Res Coun, 56-62, chmn, 59-62, mem adv bd mil personnel supplies, 62-; mem adv comt isotopes & radiation develop, AEC, 64-66, chmn adv comt radiation pasteurization of foods, Am Inst Biol Sci, 71-; sci ed, Food Technol & J Food Sci, Inst Food Technologists, 66-70; consult, US Food & Drug Admin, 74. Honors & Awards: Outstanding Civilian Serv Medal, US Army, 62; Indust Achievement Award, Inst Food Technologists, 63. Mem: Am Chem Soc; Am Inst Chem Eng; Inst Food Technologists; Optical Soc Am; AAAS. Res: Activity coefficients; detergent action of soaps; meat pigments; color standards; egg processing; meat packaging and processing; spectrochemical analysis of foods; x-ray diffraction; instrumentation; microwave heating; treatment of foods with ionizing radiation; spun protein foods. Mailing Add: 10645 Welk Dr Sun City AZ 85351

URBAN, JAMES BARTEL, b Rush Springs, Okla, June 20, 33; m 55; c 2. GEOLOGY, PALYNOLOGY. Educ: Univ Okla, BS, 57, MS, 60, PhD(geol), 62. Prof Exp: Res geologist, Continental Oil Co, 62-63; chmn dept geol, Cent Mo State Col, 63-67; res scientist, Southwest Ctr Advan Studies, 67-68, asst prof geol, 68-72, ASSOC PROF GEOL, UNIV TEX, DALLAS, 72-, HEAD GRAD PROG SCI EDUC, 73- Mem: Am Asn Petrol Geologists. Res: Paleoecology and stratigraphy using palynology as the method of research. Mailing Add: Dept of Geol Univ of Tex PO Box 30365 Dallas TX 75230

URBAN, JAMES EDWARD, b Dime Box, Tex, Jan 5, 42; m 63; c 2. MICROBIAL PHYSIOLOGY. Educ: Univ Tex, Austin, BA, 65, PhD(microbiol), 68. Prof Exp: NSF fel, 68-70, ASST PROF BIOL, KANS STATE UNIV, 70- Mem: Am Soc Microbiol. Res: Regulation of cell division; medium and growth rate influences on the bacterial cell cycle. Mailing Add: Div of Biol Ackert Hall Kans State Univ Manhattan KS 66506

URBAN, JOHN, b Tisovec, Slovakia, May 31, 09; nat US; m 39; c 3. SCIENCE EDUCATION, BIOLOGY. Educ: Kent State Univ, BS, 30; Columbia Univ, MA, 35, PhD(sci educ), 43. Prof Exp: Teacher high sch, Ohio, 30-35; teacher & head dept sci, NJ, 36-46; fac senator, 66-72, PROF SCI & CHMN DEPT, STATE UNIV NY COL BUFFALO, 46- Concurrent Pos: Consult, State Dept Educ NC, 55. Mem: Fel AAAS; Am Inst Biol Sci. Mailing Add: Dept of Biol State Univ NY Col 1300 Elmwood Ave Buffalo NY 14222

URBAN, PAUL, b Philadelphia, Pa, June 11, 38. CELL BIOLOGY. Educ: Washington Univ, AB, 62; Tufts Univ, PhD(biol), 67. Prof Exp: Instr biol, Yale Univ, 67-68; asst prof biol sci, Union Col, NY, 68-74; MEM, ELECTRON MICROS UNIT, VET ADMIN HOSP, 74- Mem: Bot Soc Am; Phycol Soc Am; Am Soc Cell Biol. Res: Fine structure of reproduction in plants; gametogenesis; sporogenesis; gamete fusion; pronuclear fusion and early development. Mailing Add: Electron Micros Unit 113A Vet Admin Hosp Albany NY 11208

URBAN, THEODORE JOSEPH, b Chicago, Ill, May 29, 26; m 56; c 1. PHYSIOLOGY. Educ: Northwestern Univ, BS, 49, MS, 50; Purdue Univ, PhD(biol), 54. Prof Exp: Instr biol, Berea Col, 53-54; assoc prof, 54-68, PROF BIOL, CREIGHTON UNIV, 68-, ASST DEAN DENT SCH, 69- Mem: AAAS; Am Soc Zool; Int Asn Dent Res. Res: Minor element metabolism; salivary gland hormones and factors; cleft palate research. Mailing Add: Dept of Oral Biol Creighton Univ Dent Sch Omaha NE 68178

URBAN, WILLARD EDWARD, JR, b Chicago, Ill, Sept 16, 36; m 57; c 3. BIOMETRICS, ANIMAL BREEDING. Educ: Va Polytech Inst & State Univ, BS, 58; Iowa State Univ, MS, 60, PhD(animal breeding), 63. Prof Exp: Animal husbandman, Animal Husb Res Div, Agr Res Serv, USDA, 58-63; asst prof, 63-68, ASSOC PROF BIOMET, UNIV N H, 68-, STATISTICIAN, AGR EXP STA, 63-, ASST DIR, 72- Mem: AAAS; Biomet Soc; Am Soc Animal Sci. Res: Role of heredity and environment in economic traits of livestock; statistical methods for analyzing non-orthogonal data. Mailing Add: 203 Taylor Hall Univ of NH Durham NH 03824

URBANEK, VINCENT EDWARD, b Chicago, Ill, Jan 2, 27; m 49; c 2. PROSTHODONTICS, DENTISTRY. Educ: Northwestern Univ, Evanston, BS, 51, MA, 52; Univ Ill, Chicago, DDS, 57. Prof Exp: Instr removable prosthodontics, Col Dent, Univ Ill Med Ctr, 57-63, asst prof, 63-67, assoc prof oral diag & oral med, 67-70; PROF REMOVABLE PROSTHODONTICS, SCH DENT, MED COL GA, 70- Concurrent Pos: Mem attend staff, Eugene Talmadge Mem Hosp, 70- Mem: Fel Am Col Dent; Am Dent Asn; Am Prosthodont Soc; Am Acad Oral Med. Res: Removable prosthodontics; oral diagnosis; mandibular dysfunction as related to the temporomandibular joints, neuromuscular components and dental occlusion; effects of corticosteroids on vesiculobullous lesions of the oral mucosa. Mailing Add: Dept of Prosthodontics Med Col of Ga Sch of Dent Augusta GA 30902

URBANIC, ANTHONY JOSEPH, b Cleveland, Ohio, July 10, 14; m 42; c 3. COLLOID CHEMISTRY, POLYMER CHEMISTRY. Educ: Oberlin Col, AB, 38; Case Western Reserve Univ, MA, 40, PhD, 42. Prof Exp: Asst, Case Western Reserve Univ, 39-42; res chemist, B F Goodrich Co, 42-51; asst dir res, Gen Tire & Rubber Co, 51-62, coordr res & develop, 62-68, mgr plastics res & develop, 68-75; RETIRED. Mem: Am Chem Soc; Soc Plastics Indust; Soc Plastics Eng. Res: Synthetic polymers, especially theory; preparation, properties, evaluation and utilization. Mailing Add: 2255 Woodpark Rd Akron OH 44313

URBANIK, ARTHUR RONALD, b Union City, NJ, Apr 17, 39; m 66; c 2. ORGANIC CHEMISTRY, TEXTILE CHEMISTRY. Educ: St Vincent Col, BS, 61; WVa Univ, MS, 63, PhD(org chem), 67. Prof Exp: Sr res chemist, Bjorksten Res Labs, 67; SR RES CHEMIST, DAN RIVER, INC, 67- Concurrent Pos: Lectr chem, Stratford Col, 72-73. Mem: Am Chem Soc; Am Asn Textile Chem & Colorists. Res: Textile chemicals and finishes; dyeing auxiliaries; applied dyeing theory. Mailing Add: 709 Brightwell Dr Danville VA 24541

URBANYI, TIBOR, b Bikas, Hungary, Jan 8, 15; US citizen; m 48. CHEMISTRY. Educ: Pazmany Peter Univ, Budapest, MS, 37, PhD(anal chem), 40. Prof Exp: Mgr, Mercur Müszaki es Vegyipari Rt, Hungary, 40-45; chemist, Merck & Co, NJ, 52-56; head automated methods develop, Ciba Pharmaceut Co, 56-72; HEAD ANAL RES & DEVELOP, ALZA CORP, 72- Mem: Acad Pharmaceut Sci; Am Chem Soc. Res: New method development by ultraviolet and infrared spectroscopy. Mailing Add: Alza Corp 950 Page Mill Rd Palo Alto CA 94304

URBAS, BRANKO, b Zagreb, Yugoslavia, July 24, 29; m 56; c 1. ORGANIC CHEMISTRY, BIOCHEMISTRY. Educ: Univ Zagreb, Diplom Chem, 53, DSc(org chem), 60. Prof Exp: Group leader, Synthetic Org Chem, Pliva Chem & Pharmaceut Works, Zagreb, Yugoslavia, 52-60; sect leader polymer chem, Org Chem Indust, 60-61; Nat Res Coun Can fel carbohydrates, Ottawa Univ, Ont, 61-63; asst prof & NIH grant dept biochem, Purdue Univ, Lafayette, 63-65; res scientist, Res Br, Can Dept Agr, 65-68; SR RES CHEMIST, MOFFETT TECH CTR, CPC INT, INC, ARGO, 68- Mem: Am Chem Soc; Croatian Chem Soc. Res: Synthetic organic and carbohydrate chemistry. Mailing Add: 813 Belair Dr Darien IL 60559

URBATSCH, LOWELL EDWARD, b Osage, Iowa, July 5, 42; m 65; c 2. SYSTEMATIC BOTANY. Educ: Univ Northern Iowa, BA, 64; Univ Ga, PhD(bot), 70. Prof Exp: Asst prof bot, Chadron State Col, 70-71; ASST PROF PLANT SYST & BIOL, UNIV TEX, AUSTIN, 71- Mem: Int Asn Plant Taxon; Am Soc Plant Taxon; Soc Study Evolution. Res: Cytological and biochemical systematics of genera in the Compositae. Mailing Add: Dept of Bot Univ of Tex Austin TX 78712

URBIN, MATTHEW CHARLES, biochemistry, see 12th edition

URBSCHEIT, NANCY LEE, b Viroqua, Wis, Sept 7, 46. RESPIRATORY PHYSIOLOGY, PHYSICAL MEDICINE & REHABILITATION. Educ: State Univ NY Buffalo, BS, 68, MA, 70, PhD(physiol), 73. Prof Exp: Instr, 73-75, ASST PROF PHYSIOL, STATE UNIV NY BUFFALO, 75-, ASST TO THE V PRES HEALTH SCI, 73- Res: Mapping the activity of intercostal muscles in anesthetized cats during mechanical loading, in particular, positive and negative pressure breathing, threshold loading and elastic loading and chemical loading of respiration. Mailing Add: Dept of Physiol Sherman Hall State Univ NY Buffalo NY 14214

URBSCHEIT, PETER, b Berlin, Ger, Sept 29, 42; m 68; c 1. GEOGRAPHY. Educ: Univ Heidelberg, Dipl geog, Eng & polit sci & MA, 68; Univ Waterloo, PhD(geog), 73. Prof Exp: ASST PROF GEOG, SLIPPERY ROCK STATE COL, 72- Mem: Asn Am Geog. Res: Geography of boundaries and borderlands, in the context of interregional cross-boundary cooperation and supranational frameworks. Mailing Add: Dept of Geog Slippery Rock State Col Slippery Rock PA 16057

URDANG, ARNOLD, b Brooklyn, NY, Feb 10, 28; m 52; c 2. PHARMACEUTICAL CHEMISTRY, PHARMACY. Educ: Long Island Univ, BS, 49; Columbia Univ, MS, 52; Univ Conn, PhD(pharmaceut chem), 55. Prof Exp: Asst prof phys pharm, Long Island Univ, 55-58; res dir pharm, E Fougera & Co, NY, 58-64; asst dir clin res, Winthrop Prod Inc, Div, Sterling Drug Inc, 64-67, asst dir new prod, 67-69, DIR TECH COORD, WINTHROP LABS, 69- Concurrent Pos: Res Corp grant, 56-58. Mem: AAAS; Am Pharmaceut Asn; Am Chem Soc. Res: Development of new pharmaceuticals both from the aspect of new product development as well as the investigation of potential new chemical compounds. Mailing Add: Winthrop Labs 90 Park Ave New York NY 10016

URDY, CHARLES EUGENE, b Georgetown, Tex, Dec 27, 33; m 62; c 1. X-RAY CRYSTALLOGRAPHY, INORGANIC CHEMISTRY. Educ: Huston-Tillotson Col, BS, 54; Univ Tex, Austin, PhD(chem), 62. Prof Exp: Prof chem, Huston-Tillotson Col, 61-62; assoc prof, NC Col, Durham, 62-63; prof, Prairie View Agr & Mech Col, 63-72; PROF CHEM, HUSTON-TILLOTSON COL, 72- Concurrent Pos: Robert A Welch Found fel, Univ Tex, Austin, 62. Mem: Fel Am Inst Chem; Am Crystallog Asn; Am Chem Soc; Sigma Xi. Res: Determination of the crystal structures of coordination compounds of the transition metals by x-ray diffraction methods. Mailing Add: Dept of Chem Huston-Tillotson Col Austin TX 78702

URE, ROLAND WALTER, JR, b New York, NY, June 22, 25; m 49; c 4. SEMICONDUCTORS. Educ: Univ Mich, BSE, 47; Calif Inst Technol, MS, 48; Univ Chicago, PhD(physics), 57. Prof Exp: Eng aide underwater acoustics, US Naval Ord Lab, 43-45; jr engr, Res Lab, Westinghouse Elec Corp, 48-50, res physicist, 55-63, fel physicist, 63-69; PROF MAT SCI & ELEC ENG, COL ENG, UNIV UTAH, 69- Mem: Am Phys Soc; sr mem Inst Elec & Electronics Eng. Res: Solid state physics; ionic crystals; transport properties of semiconductors; thermoelectric and microwave semiconductor devices; quantum effects in semiconductors; electrical engineering; positron annihilation; photodiodes; materials for energy conversion. Mailing Add: Col of Eng Univ of Utah Salt Lake City UT 84112

URELES, ALVIN L, b Rochester, NY, Aug 8, 21; m 53; c 3. MEDICINE. Educ: Univ Rochester, MD, 45; Am Bd Internal Med, dipl; Am Bd Nuclear Med, dipl. Prof Exp: Intern, Beth Israel Hosp, Boston, 45-46, from resident to chief resident med, 48-51; from clin instr radiol & med to clin asst prof med, 51-64, assoc prof med, 64-67, clin assoc radiol, 61-67, PROF MED, SCH MED & DENT, UNIV ROCHESTER, 69-; ASSOC HEAD ENDOCRINOL & METAB DIV & CHIEF MED, GENESEE HOSP, 67- Concurrent Pos: Asst physician, Strong Mem Hosp, Rochester, NY, 51-58, sr assoc physician, 64- Mem: Fel Am Col Physicians; Int Soc Internal Med; Am Soc Internal Med; Endocrine Soc; AMA. Res: Thyroid disease; breast cancer. Mailing Add: Genesee Hosp 224 Alexander St Rochester NY 14607

URENOVITCH, JOSEPH VICTOR, inorganic chemistry, see 12th edition

URETSKY, MYRON, b New York, NY, May 28, 40; m 69; c 4. COMPUTER SCIENCE, DATA PROCESSING, Educ: City Col NY, BBA, 61; Ohio State Univ, MBA, 62, PhD(acct), 62. Prof Exp: Asst prof acct, Univ Ill, 64-67; assoc prof bus, Columbia Univ, 67-70; PROF INFO SYST, NY UNIV, 70- Mem: Inst Mgt Sci; Am Inst Cert Pub Acct; Asn Comput Mach. Res: Impact of computers on society; management fraud; simulation and gaming; East-West trade. Mailing Add: Grad Sch of Bus Admin NY Univ 100 Trinity Pl New York NY 10006

URETZ, ROBERT BENJAMIN, b Chicago, Ill, June 27, 24; m 55; c 2. BIOPHYSICS. Educ: Univ Chicago, BS, 47, PhD(biophys), 54. Prof Exp: Asst cosmic rays, 48-50, from instr to assoc prof biophys, 54-64, chmn dept biophys, 66-69; assoc dean, 69-70, PROF BIOPHYS, UNIV CHICAGO, 64-, DEP DEAN DIV BIOL SCI & PRITZKER SCH MED, 70- Mem: Radiation Res Soc; Biophys Soc; Am Soc Cell Biol. Res: Mechanism of biological effects of various radiations; optical analysis of biological structure. Mailing Add: Dept of Biophys Univ of Chicago Chicago IL 60637

UREY, HAROLD CLAYTON, b Walkerton, Ind, Apr 29, 93; m 26; c 4. CHEMISTRY. Educ: Univ Mont, BS, 17; Univ Calif, PhD(chem), 23. Hon Degrees: Twenty-three from US, Can & foreign univs, 35-59. Prof Exp: Res chemist, Barrett Chem Co, Md, 18-19; instr chem, Univ Mont, 19-21; Am-Scandinavian Found fel, Copenhagen Univ, 23-24; assoc, Johns Hopkins Univ, 24-29; from assoc prof to prof, Columbia Univ, 29-45, exec officer dept, 39-42, dir war res, S A M Lab, 40-45; prof chem, Inst Nuclear Studies, Univ Chicago, 45-52, Ryerson prof, 52-58; prof-at-large, 58-70, EMER PROF CHEM, UNIV CALIF, SAN DIEGO, 70- Concurrent Pos: Ed, J Chem Physics, 33-40; Adams fel, Columbia Univ, 33-36; Bownocker lectr, Ohio State Univ, 50; Silliman lectr, Yale Univ, 51; Montgomery lectr, Univ Nebr, 52; Hitchcock lectr, Univ Calif, 53; Eastman vis prof, Oxford Univ, 56-57; hon fel, Weizmann Inst, 57; Scott lectr, Cambridge Univ, 57; Priestley lectr, Pa State Univ, 63; mem bd dirs, Am-Scandinavian Found. Honors & Awards: Nobel Prize in Chem, 34; Gibbs Medal, Am Chem Soc, 34, Remsen Award, 63, Priestley Award, 73; Davy Medal, Royal Soc, 40; Franklin Medal, Franklin Inst, 43; Medal for Merit, 46; Cordoza Award, 54; Priestley Award, Dickinson Col, 55; Hamilton Award, Columbia Univ, 61; Univ Paris Medal, 64; Nat Medal Sci, 64; Am Acad Achievement Award, 66; Gold Medal, Royal Astron Soc, 66; Leonard Medal, Meteoritical Soc, 69; Linus Pauling Award, 70; Johann Kepler Medal, 71; Gold Medal, Am Inst Chemists, 72. Mem: Nat Acad Sci; Am Chem Soc; fel Am Phys Soc; Am Acad Arts & Sci; Royal Astron Soc. Res: Entropy of gases; atomic structure; absorption spectra and structure of molecules; discovery of deuterium; properties and separation of isotopes; exchange reactions; measurement of paleotemperature; chemical problems of the origin of the earth, meteorites, moon and solar system. Mailing Add: 7890 Torrey Lane La Jolla CA 92037

UREY, JOHN CLAYTON, biochemical genetics, see 12th edition

URIBE, ERNEST GILBERT, b Sanger, Calif, Nov 25, 35; m 57; c 3. PLANT BIOCHEMISTRY. Educ: Fresno State Col, AB, 57; Univ Calif, Davis, MS, 62, PhD(plant physiol), 65. Prof Exp: Res asst bot & biochem, Univ Calif, Davis, 58-65; res assoc biochem, Johns Hopkins Univ, 65-66 & Cornell Univ, 66-67; asst prof biol, Yale Univ, 67-74; ASSOC PROF BOT, WASH STATE UNIV, 74- Concurrent Pos: Res fels, NSF, 65-66 & NIH, 66-67. Mem: AAAS; Am Soc Plant Physiologists; Am Soc Biol Chemists. Res: Intermediary metabolism of phenolic compounds in higher plants; energy conversion in photosynthesis mechanism of photosynthetic phosphorylation. Mailing Add: Dept of Biol Wash State Univ Pullman WA 99163

URICCHIO, WILLIAM ANDREW, b Hartford, Conn, Apr 21, 24; m 50; c 5. BIOLOGY, HELMINTHOLOGY. Educ: Catholic Univ, BA, 49, MS, 51, PhD(zool), 53. Prof Exp: Asst prof, 53, PROF BIOL & CHMN DEPT, CARLOW COL, 53- Concurrent Pos: Guest prof, Carnegie-Mellon Univ, 63-68, NSF vis prof, 67-68. Honors & Awards: Bishop-Wright Award, 62; Knight of St Gregory the Great. Mem: Fel AAAS; Am Soc Zool; Nat Asn Biol Teachers; Am Inst Biol Sci; Nat Asn Sci Teachers. Res: Immunity to coccidiosis in chickens by physical and chemical means; science education; family planning. Mailing Add: 1402 Murray Ave Pittsburgh PA 15217

URICK, ROBERT JOSEPH, b Brooklyn, NY, Apr 1, 15; m 55; c 3. UNDERWATER ACOUSTICS. Educ: Brooklyn Col, BS, 35; Calif Inst Technol, MS, 39. Prof Exp: Asst seismologist, Shell Oil Co, Tex, 36-38; chief computer, Tex Co, 39-42; physicist, Radio & Sound Lab, US Dept Navy, Calif, 42-45, Naval Res Lab, Wash, DC, 45-55; physicist, Ord Res Lab, Pa State Univ, 55-57, Mine Defense Lab, Fla, 57-60, US Naval Ord Res Lab, 60-74 & Naval Surface Weapons Ctr, 74-75; ACOUSTIC CONSULT, TRACOR, INC, 75- Honors & Awards: Distinguished Civilian Serv Award, Navy Dept, 75. Mem: Fel Acoust Soc Am. Res: Underwater sound; sonar; sound propagation. Mailing Add: 11701 Berwick Rd Silver Spring MD 20904

URITAM, REIN AARNE, b Tartu, Estonia, Apr 11, 39; US citizen; m 70. ELEMENTARY PARTICLE PHYSICS. Educ: Concordia Col, Moorehead, Minn, BA, 61; Oxford Univ, BS, 63; Princeton Univ, MA, 65, PhD(physics), 68. Prof Exp: Res assoc physics, Princeton Univ, 67-68; asst prof, 68-74, ASSOC PROF PHYSICS, BOSTON COL, 74- Mem: Am Phys Soc; Philos Sci Asn; Hist Sci Soc. Res: Theory of elementary particles; weak interactions; current algebra; high-energy hadron collisions; history and philosophy of science. Mailing Add: Dept of Physics Boston Col Chestnut Hill MA 02167

URIU, KIYOTO, b Berryessa, Calif, May 25, 17; m 49; c 4. POMOLOGY, PLANT PHYSIOLOGY. Educ: Univ Calif, BS, 48, MS, 50, PhD(plant physiol), 53. Prof Exp: Prin lab technol, 53-55, jr pomologist, 55-56, asst pomologist, 56-63, assoc specialist, 63-64, assoc pomologist, 64-70, POMOLOGIST, UNIV CALIF, DAVIS, 70-, LECTR, 62- Mem: Am Soc Hort Sci; Am Soc Plant Physiol. Res: Mineral nutrition, especially microelements of deciduous fruit trees; water relations of deciduous fruit trees. Mailing Add: Dept of Pomology Univ of Calif Davis CA 95616

URIVETZKY, MORTON M, chemistry, biology, see 12th edition

URLING, GERARD PHELPS, b Pittsburgh, Pa, Oct 23, 24; m 49; c 3. WOOD SCIENCE & TECHNOLOGY. Educ: Univ Fla, BSF, 50, MSF, 51. Prof Exp: Asst wood tech, Univ Fla, 50-51; res wood technologist, Timber Eng Co, 51-55, asst to dir res, 55-57, asst lab mgr, 57-59; res scientist, 59-64, group leader, 64-68, SECT MGR, WEYERHAEUSER CO, 68- Mem: Forest Prod Res Soc; Soc Wood Sci & Technol. Res: Wood utilization; particleboard; hardboard; fiberboard; molded products; plywood; wood anatomy; adhesives and gluing; fasteners; finishes; overlays; wall and roof construction; doors; furniture; acoustics; evaluation procedures. Mailing Add: Weyerhauser Co Res & Develop Div Tech Ctr Longview WA 98632

URNESS, PHILIP JOEL, zoology, plant ecology, see 12th edition

URONE, PAUL, b Pueblo, Colo, Nov 29, 15; m 43; c 2. CHEMISTRY. Educ: Western State Col Colo, AB, 38; Ohio State Univ, MS, 47, PhD, 54. Prof Exp: Teacher, High Sch, Colo, 38-42; chemist, Colo Fuel & Iron Corp, 42-45; chief chemist, Div Indust Hyg, State Dept Health, Ohio, 47-55; from asst prof to prof chem, Univ Colo, Boulder, 55-70; prof atmospheric chem, 71-74, PROF CHEM & ENVIRON ENG SCI, UNIV FLA, 74- Concurrent Pos: Consult, Martin Co, 60-61; Univ Colo fac fel, Univ Calif, Los Angeles, 61-62; sci adv, Food & Drug Admin, 66-; Dept Health, Educ & Welfare air pollution fel, Univ Fla, 67; mem, Air Pollution Nat Manpower Adv Comt; sulfur oxides subcomt, Intersoc Comt of Methods of Sampling & Anal. Mem: Am Chem Soc; Am Indust Hyg Asn; Am Conf Govt Indust Hygienists; Air Pollution Control Asn. Res: Polarography of hydrocarbon combustion products; chemical analysis of air contaminants in industrial hygiene and air pollution; theoretical and applied gas chromatography; reaction of sulfur dioxide in air. Mailing Add: Dept of Environ Eng Univ of Fla Gainesville FL 32601

URONE, PAUL PETER, b Pueblo, Colo, Feb 11, 44; m 65. NUCLEAR PHYSICS. Educ: Univ Colo, BA, 65, PhD(physics), 70. Prof Exp: Teaching asst, Dept Physics, Univ Wash, 65-66; res asst nuclear physics, Univ Colo, 66-70; staff physicist, Kernfysisch Versneller Inst, Univ Groningen, Neth, 70-71; nuclear info res assoc, State Univ NY Stony Brook, 71-73; ASST PROF PHYSICS, CALIF STATE UNIV, SACRAMENTO, 73- Concurrent Pos: Consult, Calif State Univ & Cols, 74-75. Mem: Am Phys Soc. Res: Optical model of nucleus, nuclear data compilations, basic and applied neutron physics research. Mailing Add: Dept of Physics Calif State Univ Sacramento CA 95819

URQUHART, FREDERICK ALBERT, b Toronto, Ont, Dec 13, 11; m 45; c 1. ZOOLOGY. Educ: Univ Toronto, BA, 35, MA, 38, PhD(zool), 40. Prof Exp: Insect pest investr, Ont Dept Agr, 35-37; lectr & actg cur insects, Div Life Sci, Royal Ont Mus, 37-41, cur, 41-48, asst prof syst entom, Univ, 48-61, assoc prof zool, 61-72, PROF ZOOL, UNIV TORONTO, 71- Concurrent Pos: Asst dir div life sci, Royal Ont Mus, 48-49,, dir, 49-55, head, 55-61. Res: Insect migration. Mailing Add: Dept of Zool & Scarborough Col Univ of Toronto Toronto ON Can

URQUHART, JOHN, III, b Pittsbrugh, Pa, Apr 24, 34; m 57; c 3. PHYSIOLOGY, ENDOCRINOLOGY. Educ: Rice Univ, BA, 55; Harvard Univ, MD, 59. Prof Exp: Intern surg, Mass Gen Hosp, 59-60, asst resident, 60-61; investr exp cardiovasc dis, NIH, 61-63; from asst prof to prof physiol, Sch Med, Univ Pittsburgh, 63-70; prof biomed eng, Univ Southern Calif, 70-71; PRIN SCIENTIST & DIR BIOL RES, ALZA CORP, 71- Concurrent Pos: Josiah Macy, Jr fel obstet, Harvard Med Sch, 59-61; USPHS res career develop award, Nat Heart Inst, 63-70; NIH grants, 63-71 & 66-71; consult, Physiol Training Comt, NIH, 70- Honors & Awards: Borden Prize, Harvard Univ, 59; Upjohn Award, Endocrine Soc, 62. Mem: Am Physiol Soc; Endocrine Soc; Biomed Eng Soc. Res: Dynamics of drug and hormone action. Mailing Add: Alza Corp 950 Page Mill Rd Palo Alto CA 94304

URQUHART, N SCOTT, b Columbia, SC, Mar 15, 40; m 59; c 7. STATISTICS. Educ: Colo State Univ, BS, 61, MS, 63, PhD(statist), 65. Prof Exp: Asst prof statist, Cornell Univ, 65-69, assoc prof biol statist, 69-70; assoc prof, 70-75, PROF EXP STATIST, N MEX STATE UNIV, 75- Mem: Am Statist Asn; Inst Math Statist; Biomet Soc. Res: Development and dissemination of statistical methods used in biological research; teaching and development of teaching techniques for statistical methods. Mailing Add: Dept of Exp Statist Box 3130 NMex State Univ Las Cruces NM 88003

URQUILLA, PEDRO RAMON, b San Miguel, El Salvador, July 28, 39; m 64; c 2. PHARMACOLOGY. Educ: Cath Inst of the East, BA, 57; Univ El Salvador, MD, 65. Prof Exp: Instr pharmacol, Univ El Salvador, 64-65, Pan Am Health Orgn fel, 66-68, NIH fel, 68-69, assoc prof, 69-72, asst dean, Sch Med, 71-72; assoc prof pharmacol, Univ Madrid, 72-73; ASST PROF PHARMACOL, W VA UNIV, 73- Res: Analysis of the pharmacological receptors of cerebral arteries; pharmacological studies on cerebral vasospasm. Mailing Add: Dept of Pharmacol WVa Univ Med Ctr Morgantown WV 26505

URRY, DAN WESLEY, b Salt Lake City, Utah, Sept 14, 35; m; c 3. MOLECULAR BIOPHYSICS, BIOCHEMISTRY. Educ: Univ Utah, BA, 60, PhD(phys chem), 64. Prof Exp: Fel, Univ Utah, 64; vis investr, Chem Biodynamics Lab, Univ Calif, 65-66; prof lectr, Dept Biochem, Univ Chicago, 67-70; dir div molecular biophys, Lab Molecular Biol, 70-72, PROF BIOCHEM, UNIV ALA, BIRMINGHAM, 70-, DIR LAB MOLECULAR BIOPHYS, 72- Concurrent Pos: Assoc mem, Inst Biomed Res, AMA, 65-69, mem, 69-70. Mem: AAAS; Am Soc Biol Chem; Am Chem Soc; Biophys Soc; Am Inst Biol Chem. Res: Methods of absorption, optical rotation and nuclear magnetic resonance spectroscopies to study polypeptide conformation and its relation to biological function; emphasis on membrane structure, mechanism of ion transport, membrane active polypeptides, elastin, and atherosclerosis. Mailing Add: Lab of Molecular Biophys Univ of Ala Birmingham AL 35294

URRY, GRANT WAYNE, b Salt Lake City, Utah, Mar 12, 26; m 46; c 4. INORGANIC CHEMISTRY. Educ: Univ Chicago, SB, 47, PhD(chem), 53. Prof Exp: Asst bot, Univ Chicago, 46-47, res assoc, 47-48, asst chem, 49-52, res assoc, 53-55; asst prof, Wash Univ, 55-58; from assoc prof to prof, Purdue Univ, 58-68; prof, 68-70, chmn dept, 68-73, ROBINSON PROF CHEM, TUFTS UNIV, 70- Concurrent Pos: Sloan fel, 56-58; consult, E I du Pont de Nemours & Co, Inc. Mem: AAAS; Am Chem Soc; Fedn Am Sci. Res: Chemistry of convalently bonded inorganic compounds and electron spin resonance. Mailing Add: 2 Black Horse Terr Winchester MA 01890

URRY, RONALD LEE, b Ogden, Utah, June 5, 45; m 71; c 1. REPRODUCTIVE PHYSIOLOGY, UROLOGY. Educ: Weber State Col, BS, 70; Utah State Univ, MS, 72, PhD(physiol), 73. Prof Exp: Teaching & res asst physiol, Dept Biol, Utah State Univ, 70-72, NDEA fel, 72-73; DIR UROL RES LAB & ASST PROF UROL SURG, SCH MED & DENT, UNIV ROCHESTER, 73- Mem: Soc Study Reprod; AAAS; Am Fertil Soc; Endocrine Soc; Am Andrology Soc. Res: Relationship of stress and biogenic amines to male reproduction; testicular tissue culture, testicular perfusion, male infertility studies, vasectomy and vasovasostomy, and testicular physiology and endocrinology. Mailing Add: Div of Urol Sch Med & Dent Univ of Rochester Rochester NY 14642

URRY, WILBERT HERBERT, b Salt Lake City, Utah, Nov 5, 14; m 36; c 4. CHEMISTRY. Educ: Univ Chicago, BS, 38, PhD(chem), 46. Prof Exp: Lectr sci, Mus Sci & Indust, Chicago, 37-40, cur & consult chem, 40-44; instr, 44-48, from asst prof to assoc prof, 48-58, PROF CHEM, UNIV CHICAGO, 58- Concurrent Pos: Daines Mem lectr, Univ Kans, 54; Carbide & Carbon Prof fel, 56; vis prof, Univ Calif, 58-59; consult, Monsanto Co, Commercial Solvents Corp, Wyandotte Chem Corp & US Naval Weapons Ctr, Calif, 51- Mem: AAAS; Am Chem Soc. Res: Reaction of free radicals in solution; rearrangements of free radicals and anions; organic photochemistry; homogeneous catalysis via complex ions; chemistry of hydrazines; structure and synthesis of natural products. Mailing Add: Dept of Chem G D Searle Lab Univ of Chicago Chicago IL 60637

URSE, VLADIMIR GEORGE, b Chicago, Ill, Jan 18, 06; m 33; c 2. NEUROPSYCHIATRY. Educ: Northwestern Univ, BSm, 29, MD, 30; Am Bd Psychiat & Neurol, dipl, 39. Prof Exp: Clin prof neurol & psychiat, Stritch Sch Med, Loyola Univ, Ill, 53-60; clin prof, 61-72, PROF PSYCHIAT, UNIV ILL COL MED, 72- Concurrent Pos: Asst to health comnr ment health, Chicago Bd Health, 70-, actg dir, Div Ment Health, 72-; lectr, Sch Social Work, Loyola Univ, Ill; supt & med dir, Cook County Psychopath Hosp; mem attend staff, Presby-St Luke's Hosp; consult, Ill Cent Hosp; sr consult, Vet Admin Hosp, Hines. Mem: AMA; Am Psychiat Asn; Am Psychopath Asn; Am Acad Neurol. Res: Biochemical distrubances in schizophrenia; medico-legal problems in psychiatry and forensic psychiatry. Mailing Add: 1447 Keystone Ave River Forest IL 60305

URSELL, JOHN HENRY, b Leeds, Eng, June 9, 38. MATHEMATICS. Educ: Oxford Univ, BA, 59, MA & DPhil(math), 63. Prof Exp: Asst prof math, Pa State Univ, 62-63; assoc prof, State Univ NY Col Fredonia, 63-64; ASST PROF MATH, QUEEN'S UNIV, ONT, 64- Mem: Am Math Soc; Can Math Cong; Math Asn Am; fel Royal Asiatic Soc Gt Brit & Ireland. Res: Topological semigroups; algebra; graph theory; comparative religions; mathematical sociology; statistics. Mailing Add: Dept of Math Queen's Univ Kingston ON Can

URSENBACH, WAYNE OCTAVE, b Lethbridge, Alta, Dec 4, 23; m 44; c 7. CHEMISTRY. Educ: Brigham Young Univ, BSc, 47, MSc, 48. Prof Exp: Asst chemist, Dept Agr Res, Am Smelting & Refining Co, 42-43 & 46-51; lab supvr health physics, Dow Chem Co, Colo, 51-52; asst, Explosives Res Group, Univ Utah, 52-55, res assoc, 55-59, asst res prof, Inst Metals & Explosives, 59-61; mgr prod & res develop, Inter-Mountain Res & Eng Co, 61-65, asst res dir, 65-66; prod mgr, Ireco Chem, 66-69, mgr planning, 69-70, dir res, 70-71; RES ASSOC, UTAH ENG EXP STA, UNIV UTAH, 71-, ASST GEN MGR, RES INST, 74- Concurrent Pos: Consult, 71- Res: Air pollution; agricultural chemistry; health physics; theory of detonation; explosion and long range blast effects; terminal ballistics; seismology. Mailing Add: 4635 S 1175 East Salt Lake City UT 84117

URSIC, STANLEY JOHN, b Milwaukee, Wis, Apr 2, 24; m 50; c 3. WATERSHED MANAGEMENT, FOREST HYDROLOGY. Educ: Univ Minn, BS, 49; Yale Univ, MF, 50. Prof Exp: Res forester, Univ Ill, 50-51; PROJ LEADER, SOUTHERN FOREST EXP STA, US FOREST SERV, 51- Concurrent Pos: Mem, Int Union Forestry Res Orgn. Mem: Soc Am Foresters. Res: Effects of forestry practices, including rehabilitation of eroding lands, on water quality, yields and distribution; flow processes on forested lands. Mailing Add: US Forest Serv Forest Hydrol Lab PO Box 947 Oxford MS 38655

URSILLO, RICHARD CARMEN, b Lawrence, Mass, Oct 26, 26; m 66. PHARMACOLOGY. Educ: Tufts Univ, BS, 49, PhD(pharmacol), 54; Univ Calif, MS, 52. Prof Exp: From instr to asst prof pharmacol, Univ Calif, Los Angeles, 54-62; sect head pharmacol, Lakeside Labs, 62-66, dir pharmacol dept, 66-75; HEAD DEPT PHARMACOL II, MERRELL-NAT LABS, 75- Concurrent Pos: USPHS spec fel, Inst Sanita, Italy, 60-61. Mem: Am Soc Pharmacol & Exp Therapeut. Res: Pharmacology of autonomic and central nervous systems. Mailing Add: Dept of Pharmacol II Merrell-Nat Labs 110 E Amity Rd Cincinnati OH 45215

URSINO, DONALD JOSEPH, b Toronto, Ont, Nov 11, 35; m 60; c 3. PLANT PHYSIOLOGY, RADIATION BOTANY. Educ: Pomona Col, BA, 56; Queen's Univ, Ont, MSc, 64, PhD(biol), 67. Prof Exp: Teacher sci, High Sch, 57-63; Nat Res Coun Can fel, Milan, 67-69; asst prof biol, 69-72, ASSOC PROF BIOL, BROCK UNIV, 72-

, CHMN DEPT BIOL SCI, 74- Mem: AAAS; Am Soc Plant Physiol; Can Soc Plant Physiol. Res: Effects of ionizing radiation on plant metabolism and development; translocation of photosynthate; radiation research. Mailing Add: Dept of Biol Sci Brock Univ St Catherines ON Can

URSINO, JOSEPH ANTHONY, b Brooklyn, NY, Feb 28, 39. ORGANIC CHEMISTRY. Educ: St John's Univ, NY, BS, 60, MS, 62, PhD(chem), 67. Prof Exp: From asst prof to assoc prof, 66-71, PROF CHEM, STATE UNIV NY AGR & TECH COL FARMINGDALE, 71- Mem: Am Chem Soc; NY Acad Sci; Am Soc Eng Educ. Res: Synthesis and properties of heterocyclic organotin compounds. Mailing Add: Dept of Chem State Univ NY Agr & Tech Col Farmingdale NY 11735

URSO, PAUL, b Sicily, Italy, Aug 3, 25; US citizen; m 52; c 2. IMMUNOLOGY, ZOOLOGY. Educ: St Francis Col, NY, BS, 50; Marquette Univ, MS, 52; Univ Tenn, PhD(zool), 61. Prof Exp: Asst zool, Marquette Univ, 50-52; instr biol, Cardinal Stritch Col, 52-53; biologist, Nat Cancer Inst, 53-55; jr biologist, Oak Ridge Nat Lab, 55-57, assoc biologist, 58-59; from asst prof to assoc prof biol, Seton Hall Univ, 61-71; SR SCIENTIST, MED DIV, OAK RIDGE ASSOC UNIVS, 71- Concurrent Pos: Consult, Biol Div, Oak Ridge Nat Lab, 61-63; res partic, Oak Ridge Inst Nuclear Studies, 63- Mem: Transplantation Soc; Radiation Res Soc; Am Soc Zool; Reticuloendothelial Soc; Soc Exp Hemat. Res: Antibody formation; immune cell interactions; immunology recovery of chimeras; transplantation in chimeras. Mailing Add: Med Div Oak Ridge Assoc Univs PO Box 117 Oak Ridge TN 37830

URSPRUNG, JOSEPH JOHN, b Godre, Hungary, Mar 19, 24; Can citizen; m 55; c 2. ORGANIC CHEMISTRY, MEDICINAL CHEMISTRY. Educ: Queen's Univ, Ont, BA, 51, MA, 52; Univ Ill, Urbana, PhD(org chem), 55. Prof Exp: Res chemist, Chas Pfizer & Co, 55-59; res assoc med chem, 59-63, HEAD DEPT CARDIOVASC DIS RES, UPJOHN CO, 63- Mem: Am Chem Soc. Res: Organic synthesis; heterocyclic chemistry; cardiovascular diseases. Mailing Add: Dept of Cardiovasc Dis Res Upjohn Co Kalamazoo MI 49001

URTASUN, RAUL C, Can citizen. ONCOLOGY, EXPERIMENTAL RADIOBIOLOGY. Educ: Univ Buenos Aires, MD, 60; FRCP(C), 67; Am Bd Radiol, dipl radiother, 67. Prof Exp: Res fel oncol, Harvard Med Sch, 63-64; instr radiation oncol, Johns Hopkins Univ, 66-68; asst prof, McGill Univ, 68-70; ASSOC PROF RADIATION ONCOL, UNIV ALTA, 70- Mem: Am Soc Therapeut Radiologists; Am Soc Clin Oncologists; Am Soc Cancer Res; Am Soc Nuclear Res; Royal Col Physicians & Surgeons Can. Res: Clinical radiobiology; radiosensitizers; combined modalities in the treatment of cancer; high linear energy transfer particle radiation. Mailing Add: 26 Wellington Crescent Edmonton AB Can

URTIEW, PAUL ANDREW, b Nish, Yugoslavia, Feb 23, 31; US citizen; m 61; c 2. THERMODYNAMICS, AERONAUTICAL ENGINEERING. Educ: Univ Calif, Berkeley, BS, 55, MS, 59, PhD(mech eng), 64. Prof Exp: Res engr, Detonation Lab, 59-64, asst res engr, Propulsion Dynamics Lab, 64-67, engr, Physics Dept, 67-73, ENGR, CHEM & MAT SCI DEPT, LAWRENCE LIVERMORE LAB, UNIV CALIF, 73- Concurrent Pos: Consult, Hiller Aircraft Corp, Calif, 63-64 & MB Assocs, 66. Mem: Combustion Inst; Am Inst Aeronaut & Astronaut. Res: High pressure physics of shocked solid materials; nonsteady wave dynamics; wave interaction processes in reactive and nonreactive media; graphical and experimental techniques applicable in research. Mailing Add: Chem Dept L-402 Lawrence Livermore Lab PO Box 808 Livermore CA 94550

URY, HANS KONRAD, b Berlin, Ger, Nov 4, 24; US citizen; m 55. BIOSTATISTICS, MATHEMATICAL STATISTICS. Educ: Univ Calif, Berkeley, AB, 45 & 55, MA, 64, PhD(statist), 71. Prof Exp: Res asst statist, Inst Eng Res, Univ Calif, Richmond, 61-62; res assoc, Stanford Univ, 62-63; biostatistician, Calif State Dept Pub Health, 63-66; statist consult, Comput Ctr, San Francisco Med Ctr, Univ Calif, 67; spec consult, Calif State Dept Pub Health, 67-68; res specialist biostatist, 68-71; BIOSTATISTICIAN & CONSULT STATIST, MED METHODS RES DEPT, PERMANENTE MED GROUP, 71- Concurrent Pos: Instr, Exten Div, Univ Calif, Berkeley, 67-; statist consult, Environ Resources, Inc, Calif, 68-70; chmn, San Francisco Bay Area Biostatist Colloquium, 74-75; special consult biostatist, Inst Res Social Behav, 75- Mem: AAAS; Am Statist Asn; Biomet Soc; Inst Math Statist; fel Royal Statist Soc. Res: Nonparametric statistics; statistical techniques for evaluating adverse drug reactions; chronic disease epidemiology; application of computers to biostatistics; multiple comparison methods; statistical efficiency comparisons. Mailing Add: 2050 Drake Dr Oakland CA 94611

USAMI, SHUNICHI, b Tokyo, Japan, Jan 5, 24; m 52; c 2. PHYSIOLOGY. Educ: Kyoto Prefectural Med Col, MD, 49, PhD(physiol), 57. Prof Exp: Instr physiol, Kyoto Prefectural Med Col, 50-56, lectr, 56-59; asst prof, Sch Med, Mie Prefectural Univ, 59-64; res assoc, 64-66, asst prof physiol, 66-72, SR RES ASSOC, COL PHYSICIANS & SURGEONS, COLUMBIA UNIV, 72- Mem: AAAS; NY Acad Sci; Japan Physiol Soc; Int Soc Biorheol; Microcirc Soc. Res: Circulation; shock; blood rheology. Mailing Add: Lab of Hemorheology Columbia Univ New York NY 10032

USBORNE, WILLIAM RONALD, b Rochester, NY, Nov 22, 37; m 62; c 2. MEAT SCIENCES. Educ: Cornell Univ, BS, 59; Univ Ill, Urbana, MS, 61; Univ Ky, PhD(meat & animal sci), 67. Prof Exp: Res asst meat sci, Univ Ill, 59-61; res assoc, Cornell Univ, 61-62; res asst, Univ Ky, 63-64, instr, 64-65, res asst, 65-66; fel meat chem, Tex A&M Univ, 66-67; asst prof meat sci, Univ Minn, St Paul, 68-69; ASSOC PROF MEAT SCI, UNIV GUELPH, 69- Concurrent Pos: Welch Found fel, 67-68; fel, Tex A&M Univ, 68. Mem: Am Soc Animal Sci; Am Meat Sci Asn; Am Inst Food Technologists; Can Inst Food Sci & Technol; Agr Inst Can. Res: Meat chemistry and technology; meat animal evaluation techniques; meat processing and quality. Mailing Add: Meat Sci Lab Dept Anml & Pltry Sci Univ of Guelph Guelph ON Can

USCAVAGE, JOSEPH PETER, b Middleport, Pa, Nov 12, 25. BACTERIOLOGY, MYCOLOGY. Educ: Pa State Univ, BS, 49; Univ Ill, MS, 50, PhD, 53. Prof Exp: Asst bact, Univ Ill, 50-53; technician parasitol & bact, Hektoen Inst Med Res, Cook County Hosp, Chicago, 51; instr bact, NY Med Col, 54-58; assoc microbiol, Sch Vet Med, Univ Pa, 58-61; MEM STAFF, RES DIV, WILLIAM H RORER, INC, 61- Mem: AAAS; Am Soc Microbiol; Am Chem Soc; NY Acad Sci. Res: Medical bacteriology and mycology; bacterial physiology. Mailing Add: William H Rorer Inc 500 Virginia Dr Ft Washington PA 19034

USCHOLD, RICHARD L, b Buffalo, NY, Sept 10, 28; m 51; c 9. MATHEMATICS. Educ: Canisius Col, BS, 53; Univ Notre Dame, MS, 55; State Univ NY Buffalo, PhD(math), 63. Prof Exp: Instr math, Nazareth Col, NY, 55-56; from instr to asst prof, 56-64, chmn dept, 66-72, ASSOC PROF MATH, CANISIUS COL, 64- Mem: Am Math Soc; Math Asn Am; Hist Sci Soc. Mailing Add: Dept of Math Canisius Col Buffalo NY 14208

USDIN, EARL, b Brooklyn, NY, Mar 6, 24; m 49; c 4. PSYCHOPHARMACOLOGY, PHARMACOLOGY. Educ: Johns Hopkins Univ, AB, 43; Ohio State Univ, PhD(org chem), 51. Prof Exp: Res assoc, Inst Cancer Res, Philadelphia, 51-59; from assoc res prof to res prof, NMex Highlands Univ, 59-62; head biochem res dept, Inst Psychosom & Psychiat Res, Michael Reese Hosp, 62; head biochem & immunochem br, Melpar, Inc, 62-66; sr res biochemist, Life Systs Div, Hazelton Lab, 66; sr scientist, Atlantic Res Corp, Va, 66-68; exec secy, Preclin Psychopharmacol Res Rev Comt, 68-74, CHIEF PHARMACOL SECT, PSYCHOPHARMACOL BR, NIMH, 72- Concurrent Pos: Res fel, Univ Pa, 51; Am Cancer Soc fel, Univ Uppsala & Royal Inst Technol, Sweden, 55-56; dir joint ment health res proj, NMex State Hosp-NMex Highlands Univ, 59-62; lectr, Nobel Found & Univ Göttingen, 61; ed, Psychopharmacol Commun. Mem: AAAS; Am Asn Cancer Res; Am Chem Soc; Am Soc Biol Chem; Biophys Soc. Res: Drug metabolism; drug assay methodology; natural products; enzymology and purification techniques; mechanism of drug action; cholinesterases and cholinesterase inhibitors; folic acid-active substances. Mailing Add: Psychopharmacol Res Br Nat Inst of Ment Health Rockville MD 20852

USDIN, VERA RUDIN, b Vienna, Austria, May 31, 25; nat US; m 49; c 4. BIOCHEMISTRY. Educ: Sterling Col, BS, 45; Duke Univ, MA, 47; Ohio State Univ, PhD(biochem), 51. Prof Exp: Res assoc physiol chem, Grad Sch Med, Univ Pa, 51-56; chemist, Res Labs, Rohm and Haas Co, 56-59; assoc res prof physiol chem, NMex Highlands Univ, 59-62; head physiol chem br, Melpar, Inc, Va, 62-67; proj supvr, 67-69, res supvr, 69-73, GROUP LEADER, GILLETTE RES INST, INC, 73- Mem: Am Chem Soc; Am Soc Cell Biol; Int Asn Dent Res. Res: Biochemistry of skin; enzyme inhibition; salivary proteins; dental plaque. Mailing Add: Gillette Res Inst Inc 1413 Research Blvd Rockville MD 20850

USEEM, JOHN, b Buffalo, NY, Oct 15, 10; m 40; c 3. ANTHROPOLOGY. Educ: Univ Calif, Los Angeles, BA, 34; Univ Wis, PhD(sociol, anthrop), 39. Prof Exp: Prof sociol, Univ SDak, 39-43; vis lectr, Barnard Col, Columbia Univ, 45-46; assoc prof sociol & anthrop, Univ Wis, 46-49; head dept sociol & anthrop, 58-66, PROF SOCIOL & ANTHROP, MICH STATE UNIV, 49- Concurrent Pos: Vis lectr & consult, Foreign Serv Inst, US Dept State; soc sci analyst, Off Indian Affairs, 42; dir Palau Islands field study, Pac Sci Bd, Nat Res Coun, 48; co-dir, Hazen Found Study, India, 52-53 & 58-59; consult, Conf Bd Assoc Res Coun, 64- & Soc Sci Res Coun, 64- Mem: Am Sociol Asn; fel Am Anthrop Asn; Soc Appl Anthrop. Res: Comparative sociology; crosscultural relations; American culture. Mailing Add: Dept of Sociol Mich State Univ East Lansing MI 48823

USENIK, EDWARD A, b Eveleth, Minn, Jan 16, 27; m 55; c 2. VETERINARY MEDICINE. Educ: Univ Minn, BS, 50, DVM, 52, PhD(vet med, path), 57. Prof Exp: Instr vet surg & radiol, Col Vet Med, Univ Minn, 52-57, asst prof, 57-59; med assoc exp path, Med Dept, Brookhaven Nat Lab, 59-60; assoc prof vet surg & radiol, 60-64, PROF VET SURG & RADIOL, COL VET MED, UNIV MINN, ST PAUL, 64- Concurrent Pos: Collabr med dept, Brookhaven Nat Lab, 60-; Rockefeller consult, Col Vet Med, Lima, 65-66; mem adv coun, Inst Lab Animal Resources, Nat Res Coun-Nat Acad Sci; on leave to fac vet sci, Nat Univ, Neirobi, Kenya, 72-74. Res: Gastrointestinal diseases; anesthesia in veterinary medicine. Mailing Add: Col of Vet Med Univ of Minn St Paul MN 55101

USHER, DAVID ANTHONY, b Harrow, Eng, Nov 1, 36; m 60. BIO-ORGANIC CHEMISTRY. Educ: Victoria Univ, NZ, BSc, 58, MSc, 60; Univ Cambridge, PhD(chem), 63. Prof Exp: Res fel chem, Harvard Univ, 63-65; asst prof, 65-70, ASSOC PROF CHEM, CORNELL UNIV, 70- Concurrent Pos: NIH career develop award, 68-73; vis prof, Oxford Univ, 71-72. Honors & Awards: NZ Inst Chem Prize, 58. Mem: Am Chem Soc. Res: Models of enzyme action; biologically important organophosphates. Mailing Add: Dept of Chem Cornell Univ Ithaca NY 14850

USHER, JOHN LESLIE, stratigraphy, paleontology, see 12th edition

USHER, PETER DENIS, b Bloemfontein, SAfrica, Oct 27, 35; US citizen; m 61; c 1. ASTRONOMY. Educ: Univ of the Orange Free State, BS, 56, MS, 59; Harvard Univ, PhD(astron), 66. Prof Exp: Fel, Harvard Col Observ, 66-67; sr scientist, Am Sci & Eng, Inc, Mass, 67-68; asst prof, 68-73, ASSOC PROF ASTRON, PA STATE UNIV, UNIVERSITY PARK, 73- Mem: Int Astron Union; Am Astron Soc; Royal Astron Soc. Res: Perturbation theory; variable galaxies; stellar structure. Mailing Add: Dept of Astron 507 Davey Lab Pa State Univ University Park PA 16802

USHER, WILLIAM MACK, b Devol, Okla, Nov 10, 27; m 52. MATHEMATICAL STATISTICS. Educ: Okla State Univ, BS, 52, MS, 58. Prof Exp: Asst registr, Okla State Univ, 54-58; statistician, Tex Instruments, Inc, 58-59, opers res analyst, 59-61, mgr, 61-63, corp systs develop mgr, 63-67; dir instnl res, 67-69, DIR COMPUT & INFO SYST, OKLA STATE UNIV, 69- Mem: Am Statist Asn; Asn Comput Mach; Asn Instnl Res. Res: Development of management information systems. Mailing Add: Dept Comput & Info Syst Okla State Univ Stillwater OK 74074

USHERWOOD, NOBLE RANSOM, b Atlanta, Ill, Jan 13, 38; m 63; c 1. SOIL FERTILITY, PLANT NUTRITION. Educ: Southern Ill Univ, BS, 59, MS, 60; Univ Md, PhD(soils, plant physiol), 66. Prof Exp: Res asst soil fertil & test correlation, Univ Md, 60-66; asst prof soil fertil & plant nutrit, Univ Del, 66-67; midwest agronomist, Ill, 67-69, dir, Potash Res Asn Northern Latin Am, Guatemala, 69-71, DIR FLA & LATIN AM, POTASH INST, GA, 71- Concurrent Pos: Assoc ed, J Agron Educ. Honors & Awards: Award of Merit, Ministry Agr, Guatemala. Mem: Am Soc Agron; Soil Sci Soc Am; Latin Am Asn Earth Sci; Brazilian Soc Sci; Latin Am Asn Phytotechnol. Res: Nitrogen, phosphorus, potassium, magnesium and manganese soil fertility and plant nutrition; sub-surface irrigation feasability studies. Mailing Add: Potash Inst 1649 Tullie Circle NE Atlanta GA 30329

USHIJIMA, RICHARD N, b Molokai, Hawaii, Sept 29, 29; m 55; c 3. MICROBIOLOGY, VIROLOGY. Educ: Mont State Univ, BS, 53, MS, 57; Univ Utah, PhD(microbiol), 61. Prof Exp: Serologist, Univ Utah, 57-58, virologist, 62; asst scientist, Ore Regional Primate Res Ctr, 62-66; from asst prof to assoc prof virol & microbiol, 66-73, PROF VIROL & MICROBIOL, UNIV MONT, 73- Concurrent Pos: Consult, Battelle Mem Inst & Children's Diag Lab, Ore. Mem: AAAS; Am Soc Microbiol. Res: Microbial physiology; oncogenic and latent viruses; tumor immunology; cytogenetics. Mailing Add: Dept of Microbiol Univ of Mont Missoula MT 59801

USHIODA, SUKEKATSU, b Tokyo, Japan, Sept 18, 41; m 68. PHYSICS. Educ: Dartmouth Col, AB, 64; Univ Pa, MS, 65, PhD(physics), 69. Prof Exp: Asst prof, 69-74, ASSOC PROF PHYSICS, UNIV CALIF, IRVINE, 74- Mem: AAAS; Am Phys Soc; Phys Soc Japan. Res: Solid state physics; optical properties of solids. Mailing Add: Dept of Physics Univ of Calif Irvine CA 92664

USMANI, RIAZ AHMAD, b Farrukhabad, India, Nov 1, 34; m 54; c 3. NUMERICAL ANALYSIS. Educ: Aligarh Muslim Univ, India, BSc, 56, MSc, 57; Univ BC, PhD(numerical anal), 67. Prof Exp: Lectr math, Col Eng & Technol, Aligarh Muslim Univ, Indian, 57-61; teaching asst, Univ BC, 61-65; sessional lectr, Univ Calgary, 65-66; asst prof comput sci, 66-75, ASSOC PROF COMPUT SCI, UNIV MAN, 75- Res: Numerical integration of differential equations; initial and boundary value problems in ordinary differential equations. Mailing Add: Dept of Comput Sci Univ of Man Winnipeg MB Can

UTERMOHLEN, VIRGINIA, b New York, NY, June 17, 43; m 72; c 1. IMMUNOLOGY. Educ: Wash Univ, BS, 64; Columbia Univ, MD, 68. Prof Exp: Intern, resident & chief resident pediat, St Luke's Hosp, New York, 68-71; fel immunol, Rockefeller Univ, 71-74; ASST PROF BIOCHEM, CORNELL UNIV, 74- Concurrent Pos: NIH fel, 71-72; NY Heart Asn fel, 72-74; guest investr immunol, Rockefeller Univ, 74- Mem: Harvey Soc; Am Med Women's Asn. Res: Cell-mediated immunity in multiple sclerosis. Mailing Add: Dept of Biochem Cornell Univ Ithaca NY 14853

UTERMOHLEN, WILLIAM PERRY, JR, chemistry, see 12th edition

UTGAARD, JOHN EDWARD, b Anamoose, NDak, Jan 22, 36; m 61; c 4. GEOLOGY, PALEOZOOLOGY. Educ: Univ NDak, BS, 58; Ind Univ, AM, 61, PhD(geol), 63. Prof Exp: Res assoc paleont, US Nat Mus, Smithsonian Inst, 63-65; from asst prof to assoc prof geol, 65-73, PROF GEOL, SOUTHERN ILL UNIV, CARBONDALE, 73- Concurrent Pos: Smithsonian fel evolutionary & syst biol, Smithsonian Inst, 72. Mem: Geol Soc Am; Am Asn Petrol Geol; Paleont Soc; Brit Palaeont Asn; Soc Econ Paleont & Mineral. Res: Fossil bryozoans; carboniferous paleoecology and depositional environments; paleobiology of Paleozoic bryozoans; paleoecology of Late Paleozoic fossil communities. Mailing Add: Dept of Geol Southern Ill Univ Carbondale IL 62901

UTGARD, RUSSELL OLIVER, b Star Prairie, Wis, July 30, 33; m 56; c 3. GEOLOGY, SCIENCE EDUCATION. Educ: Wis State Col, River Falls, BS, 57; Univ Wis, MS, 58; Ind Univ, Bloomington, MAT, 66, EdD(sci), 69. Prof Exp: Instr geol, Joliet Jr Col, Ill, 58-67; asst prof, 69-72, ASSOC PROF GEOL, OHIO STATE UNIV, 72- Concurrent Pos: Teaching asst, Ind Univ, Bloomington, 65-66. Mem: AAAS; Nat Asn Geol Teachers; Nat Sci Teachers Asn; Geol Soc Am. Res: Environmental geology. Mailing Add: Dept of Geol Ohio State Univ Columbus OH 43210

UTHE, JOHN FREDERICK, b Saskatoon, Sadsk, Feb 27, 38; m 63; c 3. FISHERIES. Educ: Univ Sask, BA, 59, Hons, 60, MA, 61; Univ Western Ont, PhD(biochem), 68. Prof Exp: RES SCIENTIST, FRESH WATER INST, FISHERIES RES BD CAN, 63- Res: Analytical chemistry applied to toxic residues and the biochemical effects of such residues on fish. Mailing Add: 32 Simcoe Pl Halifax NS Can

UTKE, ALLEN R, b Moline, Ill, Feb 5, 36; m 57; c 3. INORGANIC CHEMISTRY. Educ: Augustana Col, Ill, BS, 58; Univ Iowa, MS, 61, PhD(inorg chem), 63. Prof Exp: Sr res chemist, Chem Div, Pittsburgh Plate Glass Co, Tex, 62-64; ASSOC PROF CHEM, UNIV WIS-OSHKOSH, 64- Mem: Am Chem Soc. Res: Chemistry of the alkali and alkaline earth metals and their reactions in liquid ammonia. Mailing Add: Dept of Chem Univ of Wis Oshkosh WI 54902

UTLEY, JAMES WILLIAM, physical chemistry, organic chemistry, see 12th edition

UTLEY, PHILIP RAY, b Ill, Dec 18, 41; m 63; c 2. ANIMAL SCIENCE. Educ: Southern Ill Univ, Carbondale, BS, 64; Univ Mo-Columbia, MS, 67; Univ Ky, PhD(animal sci), 69. Prof Exp: Asst instr animal sci, Southern Ill Univ, Carbondale, 63-65; ASSOC PROF ANIMAL SCI, UNIV GA COASTAL PLAINS EXP STA, TIFTON, 70- Mem: Am Soc Animal Sci. Res: Beef cattle nutrition and management. Mailing Add: Univ of Ga Coastal Plain Exp Sta Tifton GA 31794

UTTER, FRED MADISON, b Seattle, Wash, Nov 25, 31; m 58; c 3. BIOCHEMICAL GENETICS. Educ: Univ Puget Sound, BSc, 54; Univ Wash, MSc, 64; Univ Calif, Davis, PhD(genetics), 69. Prof Exp: Serologist, 59-60, CHEMIST, BIOL LAB, US BUR COM FISHERIES, 60- Concurrent Pos: Affiliate asst prof, Univ Wash, 71. Mem: AAAS; Genetics Soc Am; Am Genetic Asn; Am Inst Fishery Res Biol; Europ Soc Animal Blood Group Res. Res: Use of biochemical methods for the detection of genetic variations in fish for use in studies of fish populations. Mailing Add: 2725 Montlake Blvd Seattle WA 98112

UTTER, MERTON FRANKLIN, b Westboro, Mo, Mar 23, 17; m 39; c 1. BIOCHEMISTRY. Educ: Simpson Col, BA, 38; Iowa State Univ, PhD(bact, physiol), 42. Prof Exp: Instr & res assoc bact exp sta, Iowa State Univ, 42-44; asst prof physiol chem, Univ Minn, 44-46; assoc prof biochem, 46-57, PROF BIOCHEM, CASE WESTERN RESERVE UNIV, 56-, CHMN DEPT, 65- Concurrent Pos: Fulbright sr res scholar, Australia, 53-54; NSF sr res fels, Oxford Univ, 60-61 & Univ Leicester, 68-69. Honors & Awards: Paul-Lewis Award, Am Chem Soc, 56. Mem: Am Chem Soc; fel Am Acad Arts & Sci; Am Soc Biol Chem; Am Soc Microbiol; Brit Biochem Soc. Res: Structure, mechanism of action, control and metabolic role of enzymes of carbohydrate and oxidative metabolism. Mailing Add: Dept of Biochem Case Western Reserve Univ Cleveland OH 44106

UTTERBACK, NYLE GENE, b Oskaloosa, Iowa, Jan 19, 31; m 58; c 3. EXPERIMENTAL PHYSICS. Educ: Iowa State Univ, BS, 53, PhD(physics), 57. Prof Exp: Res asst, Ames Lab, AEC, 51-57; Fulbright scholar, Ger, 57-58; sr physicist, Ord Res Lab, Univ Va, 58-59; asst prof & res physicist, Denver Res Inst, 59-63; staff scientist, Gen Motors Corp, Calif, 63-72; staff scientist, Mission Res Corp, Santa Barbara, 72-75; CONSULT, 75- Concurrent Pos: Consult, Sansum Med Res Found, Santa Barbara, 74- & TRW, Inc, Redondo Beach, 75- Mem: Am Phys Soc. Res: Laser technology; atomic and molecular reaction kinetics; diabetic microangiopathy. Mailing Add: 718 Willowglen Rd Santa Barbara CA 93105

UTTERBACK, ROBERT ALLEN, neurology, neuropathology, deceased

UTTON, DONALD BRIAN, b London, Eng, June 10, 38; m 62; c 5. SOLID STATE PHYSICS. Educ: Univ Durham, BSc, 61; Univ Southampton, PhD(physics), 64. Prof Exp: Physicist, Nat Bur Standards, 64-68; res fel, Atomic Energy Res Estab, Harwell, Eng, 68-70; PHYSICIST, NAT BUR STANDARDS, 70- Concurrent Pos: Assoc mem, Brit Inst Physics. Res: Nuclear magnetic resonance; nuclear quadrupole resonance; applications to the solid state; thermometry. Mailing Add: Cryogenic Physics Sect Heat Div Nat Bur Stand Inst Basic Stand Washington DC 20234

UTZ, JOHN PHILIP, b Rochester, Minn, June 9, 22; m 47; c 5. MEDICINE. Educ: Northwestern Univ, BS, 43, MD, 47; Georgetown Univ, MS, 49. Prof Exp: Physician, Lab Clin Invest, Nat Inst Allergy & Infectious Dis, 47-49, chief infectious dis serv, 52-65; prof med & chmn div immunol & infectious dis, Med Col Va, 65-73; PROF MED & DEAN FAC, SCH MED, GEORGETOWN UNIV, 73- Concurrent Pos: Fel, Mayo Found, 49-52; intern, Evans Mem Hosp, Boston, 46-47; consult, Clin Ctr, NIH, US Vet Admin & Hoffmann-La Roche Pharmaceut Co; pres, Nat Found Infectious Dis, 72- Mem: Am Fedn Clin Res; Am Col Physicians; Am Col Chest Physicians; Soc Exp Biol & Med; Am Thoracic Soc. Res: Clinical investigations in infectious diseases. Mailing Add: 3257 N St NW Washington DC 20007

UTZ, WINFIELD ROY, JR, b Boonville, Mo, Nov 17, 19; m 41; c 3. MATHEMATICAL ANALYSIS. Educ: Cent Col, Mo, AB, 41; Univ Mo, MA, 42; Univ Va, PhD(math), 48. Prof Exp: Asst instr math, Univ Mo, 42; instr, Univ Notre Dame, 42-43; instr, Univ Va, 44-48; instr, Univ Mich, 48-49; from asst prof to assoc prof, 49-69, PROF MATH, UNIV MO-COLUMBIA, 69-, CHMN DEPT, 70- Concurrent Pos: Mem, Inst Advan Study, Princeton Univ, 55-56; vis scholar, Univ Calif, Berkeley, 62-63; vis prof, Brown Univ, 69. Mem: Am Math Soc; Math Asn Am; London Math Soc. Res: Surface dynamics; topological dynamics; differential equations. Mailing Add: Dept of Math Univ of Mo Columbia MO 65201

UY, OSCAR MANUEL, b Cadiz City, Philippines, Mar 8, 39; m 68; c 3. PHYSICAL CHEMISTRY, HIGH TEMPERATURE CHEMISTRY. Educ: La Salle Col, Philippines, BS, 61 & 62; Case Inst Technol, PhD(chem), 67. Prof Exp: Fel high temperature chem, Rice Univ, 67-68; fel, Lab Molecular Phys Chem, Free Univ Brussels, 68-70; sr scientist, Space Sci, Inc, Calif, 70-71; chief lab air pollution, Div Air Pollution Control, Ohio, 71-72; SR ENGR, LAMP DIV, GEN ELEC CO, 72- Mem: Am Chem Soc; The Chem Soc; Sigma Xi. Res: Mass spectrometry; air pollution; molecular discharges; emission spectroscopy. Mailing Add: Advan Eng Dept HIQ Dept Gen Elec Co Twinsburg OH 44087

UYEDA, CARL KAORU, b San Bernardino, Calif, July 11, 22; m 53; c 2. ANATOMY, CYTOPATHOLOGY. Educ: Syracuse Univ, BA, 47, MS, 49; Univ Md, PhD(anat), 66. Prof Exp: Div head cancer cytol dept, Md State Dept Health, 50-58; instr cytopath, Sch Med, Univ Md, 58-64, instr anat, 62-67; asst prof path & anat, 67-73, ASSOC DIR SCH CYTOTECHNOL & DIR CYTOPATH LAB, MED SCI, UNIV ARK, LITTLE ROCK, 67-, ASSOC PROF PATH, 73- Concurrent Pos: Res assoc, Sch Med, Johns Hopkins Univ, 56-65, sr cytologist, 65-67; sr cytologist, Ark State Dept Health, 65-67; contractor, Nat Ctr Toxicol Res, Food & Drug Admin, Ark, 71-; consult, Vet Admin Hosp, Little Rock, 72- Mem: Am Soc Cytol; Int Acad Cytol; Pan-Am Cancer Cytol Soc; Am Asn Anat; NY Acad Sci. Res: Cytogenetics; abnormal cytogenetic changes; clinical carcinoma and congenital anomalies; spontaneous leukemic C3H and C57 mice; cytopathology, refinement of interpretation in structural change of cancer; neoplasm of mice bladder; circadian rhythmicity in bronchogenic carcinoma and mice tissue. Mailing Add: Dept of Path Univ of Ark Med Sci Little Rock AR 72201

UYEDA, KOSAKU, b Kokawa Naga-gun, Japan, Mar 15, 32; US citizen; m 57; c 2. BIOCHEMISTRY. Educ: Ore State Univ, BS, 55, MS, 57; Univ Calif, Berkeley, PhD(biochem), 62. Prof Exp: Asst prof, 67-72, ASSOC PROF BIOCHEM, UNIV TEX HEALTH SCI CTR DALLAS, 72-; RES CHEMIST, VET ADMIN HOSP, 67-, CHIEF CELLULAR REGULATION, 71- Concurrent Pos: Fel, Univ Calif, Berkeley, 62; NIH fel, Pub Health Res Inst NY, 62-64; scholar, Univ Calif, Berkeley, 64-67. Mem: Am Chem Soc; Am Soc Biol Chem. Res: Elucidation of the mechanism of action of enzymes and allosteric enzymes and their roles in regulation of carbohydrate metabolism. Mailing Add: Gen Med Res Vet Admin Hosp 4500 S Lancaster Dallas TX 75216

UYEKI, EDWIN M, b Seattle, Wash, Mar 12, 28; m 51; c 3. PHARMACOLOGY, RADIOBIOLOGY. Educ: Kenyon Col, AB, 49; Univ Chicago, PhD(pharmacol), 53. Prof Exp: Instr pharmacol, Univ Chicago, 53-54; instr, Sch Med, Western Reserve Univ, 54-60, sect assoc radiation biol, 54-60; sr scientist, Hanford Labs Gen Elec Co, 60-65; assoc prof, 65-70, PROF PHARMACOL, MED CTR, UNIV KANS, 70- Mem: AAAS; Am Soc Pharmacol & Exp Therapuet; Am Soc Cell Biol; Radiation Res Soc. Res: Immunopharmacology; immunosuppressants on antibody formation; bone marrow transplantation in radiation chimeras; radiation effects; short term tissue culture. Mailing Add: Dept of Pharmacol Univ of Kans Med Ctr Kansas City KS 66103

UYEMOTO, JERRY KAZUMITSU, b Fresno, Calif, May 27, 39; m 65; c 1. PLANT PATHOLOGY, PLANT VIROLOGY. Educ: Univ Calif, Davis, BS, 62, MS, 64, PhD(plant path), 68. Prof Exp: Lab technician, Univ Calif, Davis, 63-67; ASST PROF VIROL, CORNELL UNIV, 68- Mem: Am Phytopath Soc. Res: Interaction between unlike virus particles, especially tobacco necrosis and satellite viruses; serological properties of plant viruses in general. Mailing Add: Dept of Plant Path NY State Exp Sta Cornell Univ Geneva NY 14456

UYENO, EDWARD TEISO, b Vancouver, BC, Apr 5, 21; m 69. PHARMACOLOGY. Educ: Univ Toronto, BA, 47, MA, 52, PhD(psychol), 58. Prof Exp: Res assoc psychol, Stanford Univ, 58-61, RES PSYCHOLOGIST & PHARMACOLOGIST, STANFORD RES INST, 61- Concurrent Pos: NIH grants, Stanford Res Inst, 63-66, 68-70 & 71-73; mem, Int Cong Primatol; Nat Inst Ment Health grant, 75-78. Mem: AAAS; Am Psychol Asn; Psychonomic Soc; Behav Genetics Asn; Am Soc Pharmacol & Exp Therapeut. Res: Comparative, developmental and experimental social psychology; behavioral psychopharmacology; effects of drugs on learning, retention, social behavior and reproduction of animals; interaction effects of non-medical drugs and medical drugs; self-administration of alcohol and narcotics by rats. Mailing Add: Life Sci Div Stanford Res Inst Menlo Park CA 94025

UZELMEIER, CHRISTOPHER WILLIAM, organic chemistry, see 12th edition

UZES, CHARLES ALPHONSE, b Downey, Calif, Dec 14, 39; m 67; c 1. THEORETICAL PHYSICS. Educ: Calif State Univ, Long Beach, BS, 62; Univ Calif, Riverside, MA, 64, PhD(physics), 67. Prof Exp: Asst prof, 67-73, ASSOC PROF PHYSICS, UNIV GA, 73- Mem: Am Phys Soc. Res: The use of non-perturbative calculational methods in nonlinear quantum mechanical problems in field and many body theory, and in solid state physics. Mailing Add: Dept of Physics Univ of Ga Athens GA 30602

UZGIRIS, EGIDIJUS E, b Lithuania, Jan 11, 41; m 67. PHYSICS. Educ: Univ Ill, BS, 62; Harvard Univ, MS, 64, PhD(physics), 68. Prof Exp: Res assoc, Harvard Univ, 68-69; res assoc, Joint Inst Lab Astrophys, Univ Colo, 69-70; PHYSICIST, GEN ELEC RES & DEVELOP CTR, 70- Mem: Am Phys Soc; Biophys Soc; AAAS. Res: Biophysics; atomic masers; frequency and wavelength standards; nonlinear absorption spectroscopy; charged particle tracks in solids; light scattering. Mailing Add: Gen Elec Res & Develop Ctr PO Box 8 Schenectady NY 12301

UZIEL, MAYO, b Seattle, Wash, May 3, 30; m 67; c 2. BIOLOGICAL CHEMISTRY. Educ: Univ Wash, BSc, 52, PhD(biochem), 55. Prof Exp: Nat Found Infantile Paralysis fel, Rockefeller Inst, 55-57; asst prof biochem, Sch Med, Tufts Univ, 57-62; biochemist, Mass Eye & Ear Infirmary, 62-64; BIOCHEMIST, BIOL DIV, OAK RIDGE NAT LAB, 64- Concurrent Pos: Mem subcomt specification & criteria nucleotides & related compounds, Nat Sci Found-Nat Res Coun, 68-; prof, Univ Tenn, 71- Mem: AAAS; Am Soc Biol Chem; Am Chem Soc. Res: Structure and function of biological macromolecules. Mailing Add: Oak Ridge Nat Lab Box Y Biol Div Oak Ridge TN 37830

UZMAN, BETTY GEREN, b Ft Smith, Ark, Nov 17, 22; m 55; c 1. NEUROBIOLOGY, PATHOLOGY. Educ: Univ Ark, Fayetteville, BS, 42; Wash

Univ, MD, 45. Hon Degrees: MA, Harvard Univ, 67. Prof Exp: Am Cancer Soc res fel biophys, Mass Inst Technol, 48-50; mem fac path, Harvard Med Sch, 50-67, from assoc prof to prof, 67-72; head res dept, Sparks Regional Med Ctr & Cancer Prog Coord, St Edward Mercy Hosp, Ft Smith, 72-74; PROF PATH, LA STATE UNIV MED CTR, SHREVEPORT, 74- Concurrent Pos: Dept chief ultrastruct, Children's Cancer Res Found, 50-71; mem path A study sect, NIH, 72-76, chmn, Path A Study, 73-76; assoc chief of staff for res, Vet Admin Hosp, Shreveport, 74-; distinguished vis investr, Venezuelan Inst Sci Res. Honors & Awards: Weinstein Award, United Cerebral Palsy Asn, 64; Order of Andres Bello, 3rd Class, Republic of Venezuela, 73. Mem: AAAS; Soc Develop Biol; Am Asn Cancer Res; Am Soc Cell Biol; Int Acad Path. Res: Myelin formation, control mechanisms; epidemiology of breast cancer in western Arkansas. Mailing Add: Vet Admin Hosp E Stoner Ave Shreveport LA 71130

UZNANSKI-BOTTEI, RITA MARLENE, b Chicago, Ill, Aug 13, 33; m 60; c 4. ANALYTICAL CHEMISTRY. Educ: Rosary Col, BA, 54; Univ Ill, Urbana, PhD(anal chem), 58. Prof Exp: Teaching asst chem & chem eng, Univ Ill, Urbana, 54-58; from instr to asst prof chem, St Ambrose Col, 58-60; vis lectr, St Mary's Col, Ind, 60-61; ASSOC LECTR CHEM, IND UNIV, SOUTH BEND, 71- Concurrent Pos: Consult, Ultrasonic Lab, Pioneer Cent Div, Bendix Aviation Corp, Iowa, 58-60. Mem: Am Chem Soc. Res: X-ray analysis; radiochemical analysis. Mailing Add: 19341 Wedgewood Dr South Bend IN 46637

UZODINMA, JOHN E, b Onitsha, Nigeria, July 26, 29; m 57; c 4. PREVENTIVE MEDICINE, MICROBIOLOGY. Educ: Grinnell Col, BA, 54; Univ Iowa, MS, 56, PhD(prev med, microbiol), 65. Prof Exp: Bacteriologist, State Hyg Labs, Iowa, 57-58; microbiologist, Broadlawns County Hosp, Des Moines, 58-61; PROF BIOL, JACKSON STATE COL, 64-, CHMN DEPT, 67- Mem: AAAS; Am Soc Clin Path; Am Inst Biol Sci; Am Soc Microbiol; Royal Soc Trop Med & Hyg. Res: Host-parasite relationships; effect of insulin, serotonin and thyroxine on the penetration of tissue culture cells by Toxoplasma gondii; effect of certain drugs on Trypanosoma equiperdum infections in mice. Mailing Add: Dept of Biol Jackson State Col Box 17111 Jackson MS 39209

UZZELL, THOMAS MARSHALL, JR, b Charleston, SC, Apr 6, 32. SYSTEMATIC BIOLOGY, VERTEBRATE BIOLOGY. Educ: Univ Mich, BA, 53, MS, 58, PhD(zool), 62. Prof Exp: From instr to asst prof biol, Univ Chicago, 62-67; asst prof, Yale Univ, 67-72, asst curator herpet, Peabody Mus & fel, Berkeley Col, 70-72; ASSOC CURATOR HERPET, ACAD NATURAL SCI, 72- Concurrent Pos: Adj assoc prof, Univ Pa, 74- Mem: Am Soc Naturalists; AAAS; Am Soc Ichthyol & Herpet; Soc Study Evolution; Soc Syst Zool. Res: Evolution of salamanders of the family Ambystomatidae; determination of the generic and specific limits of South American lizards of the family Teiidae; evolution of hybrid species of vertebrates. Mailing Add: 2424 Gulf Rd Philadelphia PA 19131

V

VAALA, ALLEN RICHARD, b Wilmington, Del, Jan 2, 44; m 67. EXPERIMENTAL PHYSICS. Educ: Col Wooster, BA, 65; Pa State Univ, PhD(physics), 71. Prof Exp: SR RES PHYSICIST, RES LABS, EASTMAN KODAK EASTMAN KODAK CO, 71- Res: Thermally stimulated current and conductivity techniques; thermoluminescent studies; ultra-low current measurement; crystal growth. Mailing Add: Res Labs Eastman Kodak Co 1669 Lake Ave Rochester NY 14650

VAALA, GORDON THEODORE, b Madison, Minn, Sept 11, 08; m 40; c 4. CHEMISTRY. Educ: St Olaf Col, AB, 30; Mass Inst Technol, PhD(org chem), 36. Prof Exp: Asst, Mass Inst Technol, 30-31; res chemist, Exp Sta, 36-39, group leader, 39-45, asst dir, Fairfield Lab, 45-49 & Newburgh Lab, 49-52, mgr trade prod sales, NY, 52-53, dir sales, Fabrics Div, 53-59, MGR, PLANNING DIV, E I DU PONT DE NEMOURS & CO, INC, 59- Res: High polymers; plasticizers; amino acids; coated fabrics; synthetic rubbers. Mailing Add: 1609 Shipley Rd Wilmington DE 19803

VAALER, LUTHER EVEN, electrochemistry, see 12th edition

VAARTAJA, OLLI, b Viipuri, Finland, Mar 19, 17; m 44; c 2. FOREST BIOLOGY. Educ: Univ Helsinki, BFA, 40, MFA, 48, lic & PhD, 51. Prof Exp: Asst plant path, Univ Helsinki, 47-52; forest biologist, Can Dept Agr, 52-60; sr lectr forest path, Waite Agr Res Inst, Univ Adelaide, 60-63; RES SCIENTIST, FOREST ECOL INST, CAN DEPT ENVIRON, OTTAWA, 63- Mem: Fel AAAS; Am Phytopath Soc; Mycol Soc Am; Ecol Soc Am; NY Acad Sci. Res: Ecology of tree seedlings; germination of tree seeds; diseases in forest nurseries; ecology and physiology of soil fungi; the genus Pythium; soil fungistasis. Mailing Add: Ottawa ON Can

VAARTNOU, HERMAN, b Estonia, Nov 23, 17; Can citizen; m 44; c 3. PLANT PATHOLOGY. Educ: Agronom, Royal Agr Col, Sweden, Agronom, 49; Univ BC, MSA, 53; Ore State Univ, PhD(farm crops), 67. Prof Exp: Res asst plant sci, Royal Agr Col, Sweden, 44-49; asst, Univ BC, 50-59, supvr landscaping, 59-66; res scientist, Res Br, Can Dept Agr, Beaverlodge, Alta, 67-68; supvr plant path, 68-71, HEAD BOT SECT, PLANT INDUST LAB, ALTA DEPT AGR, EDMONTON, 71- Mem: Agr Inst Can. Res: Turf management; forage ecology and taxonomy; effect of growth regulators on forage corps; occurrence and severity of plant diseases in Alberta; revegetation for reclamation. Mailing Add: Alta Dept Agr Longman Lab Bldg 6909 116 St Box 4370 Edmonton AB Can

VACHON, RAYMOND NORMAND, b Lawrence, Mass, Jan 14, 40; m 70. ORGANIC CHEMISTRY, POLYMER CHEMISTRY. Educ: Lowell Technol Inst, BS, 63; Princeton Univ, PhD(chem), 67. Prof Exp: Res grant, Inst Sci & Technol, Univ Manchester, 67-69; sr res chemist, Burlington Industs, Inc, 69-72; res chemist, 72-74, SR RES CHEMIST, TENN EASTMAN CO, 74- Mem: Am Asn Textile Chemists & Colorists; Am Chem Soc. Res: Chemical modification of natural and synthetic fibers. Mailing Add: Tenn Eastman Co Bldg 150-A PO Box 511 Kingsport TN 37662

VACIK, DOROTHY NOBLES, b Memphis, Tenn, Dec 16, 38; m 67; c 3. ORGANIC CHEMISTRY, PHARMACEUTICAL CHEMISTRY. Educ: Memphis State Univ, BS, 60; Univ Miss, PhD(pharmaceut chem), 65. Prof Exp: Asst prof pharmaceut chem & bionucleonics, Univ, 65-68, lab supvr & gen analyst, Dept Animal Sci, 71-72, LAB DIR, DRUG CONTROL LAB, DEPT PHARM & VET ADMIN HOSP, N DAK STATE UNIV, 72- Mem: Am Chem Soc; Am Pharmaceut Asn. Res: Development of analytical techniques for drug determination in human therapy. Mailing Add: Dept of Pharm Chem NDak State Univ Fargo ND 58102

VACIK, JAMES P, b North Judson, Ind, Nov 30, 31; m 67; c 4. PHARMACEUTICAL CHEMISTRY, BIONUCLEONICS. Educ: Purdue Univ, BS, 55, MS, 57, PhD(bionucleonics), 59. Prof Exp: Asst prof & res fel bionucleonics, Purdue Univ, 59-60; assoc prof pharmaceut chem & chmn dept, 60-63, PROF PHARMACEUT CHEM & BIONUCLEONICS & CHMN DEPT, N DAK STATE UNIV, 63- Mem: AAAS; Health Physics Soc; Am Pharmaceut Asn; Am Pub Health Asn. Res: Bionucleonics including metabolism, uptake and distribution of radioisotope tracers and large animal biosynthesis; synthesis of benzodioxans. Mailing Add: Dept of Pharmaceut Chem & Bionucleo NDak State Univ Col Pharm Fargo ND 58102

VACIRCA, SALVATORE JOHN, b Bronx, NY, July 20, 22; c 2. RADIATION PHYSICS, HEALTH PHYSICS. Educ: City Col New York, BS, 48; NY Univ, MS, 54, PhD(radiation physics), 70; Am Bd Health Physics, dipl, 60. Prof Exp: Res asst nuclear electronics instrumentation, Sloan Kettering Inst Cancer Res, 48-54, asst physicist, Mem-Sloan Kettering Ctr Cancer & Allied Dis, 54-56, asst attend physicist, 56-61; co-dir nuclear med lab, 65-70, chmn radiation technol prog, 69-71, asst prof, 61-73, ASSOC PROF RADIOL, STATE UNIV NY DOWNSTATE MED CTR, 73-, DIR RADIATION PHYSICS LAB & RADIATION SAFETY OFF, 61- Concurrent Pos: Instr & res assoc, Sloan Kettering Div, Med Col, Cornell Univ, 55-61; dir radiol physics & radiation safety off, Kings County Hosp Ctr, 61-, sr med physicist, Prof Staff, 72; consult physicist, Col Health Related Professions, State Univ NY, 71-; consult physicist, Coney Island Hosp, 72; adj assoc prof, York Col, NY, 72 & City Col New York, 74. Mem: Am Asn Physicists in Med; Health Physics Soc. Res: Radiation dosimetry as applied to therapy; film and thermoluminescent synergistic dosimetry system used as a method of mapping dose distribution; health physics problems as applied to hospital environment. Mailing Add: Dept of Radiol State Univ NY Downstate Med Ctr Brooklyn NY 11203

VACQUIER, VICTOR, b Leningrad, Russia, Oct 13, 07; nat US; m 31; c 2. GEOPHYSICS. Educ: Univ Wis, BS, 27, MA, 28. Prof Exp: Asst instr physics, Univ Wis, 27-30; geophysicist, Gulf Res & Develop Co, Pa, 30-42; mem staff airborne instruments lab, Columbia Univ, 42-44; marine instruments engr, Sperry Gyroscope Co, 44-53; prof geophys & prin geophysicist, NMex Inst Mining & Technol, 53-57; prof geophys, Scripps Inst Oceanog, 57-74, EMER PROF GEOPHYS, UNIV CALIF, SAN DIEGO, 74- Mem: Soc Explor Geophys; Geol Soc Am; Am Geophys Union. Res: Geomagnetism; airborne magnetometry; construction of gyroscopic instruments; terrestrial heat flow. Mailing Add: Scripps Inst of Oceanog Univ of Calif San Diego La Jolla CA 92037

VADAS, ROBERT LOUIS, b New Brunswick, NJ, Aug 5, 36; m 61; c 3. MARINE ECOLOGY, PHYCOLOGY. Educ: Utah State Univ, BS, 62; Univ Wash, PhD(bot), 68. Prof Exp: Asst prof bot, 67-72, asst prof zool, 68-72, ASSOC PROF BOT, OCEANOG & ZOOL, UNIV MAINE, ORONO, 72- Concurrent Pos: Maine Yankee Nuclear Atomic Power Co study grant, 69-74; Off Water Resources grants, 70-72 & 72-75. Mem: Ecol Soc Am; Am Soc Naturalists; Phycol Soc Am; Brit Phycol Soc; Int Phycol Soc. Res: Ecology of kelp communities; marine plant-herbivore interactions; algal distributions; culture and physiology of algae; thermal enrichment in marine communities; influence of pesticides on marine microalgae. Mailing Add: Dept of Bot Univ of Maine Orono ME 04473

VADHWA, OM PARKASH, b Mandi Maklot Ganj, India, May 10, 41; m 67; c 1. AGRONOMY. Educ: Rajasthan Univ, India, BS, 61; Punjab Agr Univ, MS, 63; Utah State Univ, PhD(agron), 71. Prof Exp: Lectr agron, Punjab Agr Univ, Hissar Campus, 63-65; fel agron & plant sci, Utah State Univ, 70-71; assoc prof natural resources, Ala A&M Univ, 71-72; ASST PROF AGRON, ALCORN STATE UNIV, 72- Mem: Am Soc Agron; Am Asn Univ Prof. Res: Forage crops; crop production; soil fertility and plant nutrition; vegetable crops. Mailing Add: Dept of Agron Alcorn State Univ Lorman MS 39096

VAGELOS, P ROY, b Westfield, NJ, Oct 8, 29; m 55; c 4. BIOCHEMISTRY. Educ: Univ Pa, AB, 50; Columbia Univ, MD, 54. Prof Exp: Intern med, Mass Gen Hosp, Boston, 54-55, asst resident, 55-56; sr asst surgeon, Lab Cellular Physiol, Nat Heart Inst, 56-59, surgeon, 59-61, actg chief sect enzymes, 59-60, sr surgeon, Lab Biochem, 61-62; sr surgeon, Pasteur Inst, Paris, 62-63; sr surgeon & res chemist, Lab Biochem, Nat Heart Inst, 63-64, head sect comp biochem, 64-66; chmn dept biol chem, Sch Med, Wash Univ, 66-75, dir div biol & biomed sci, 74-75; SR VPRES, MERCK SHARP & DOHME RES LABS, MERCK & CO, INC, 75- Concurrent Pos: NIH & NSF grants. Mem: Nat Acad Sci; Nat Inst Med; AAAS; Am Chem Soc; Am Soc Biol Chem. Res: Mechanism of lipid biosynthesis; involvement of acyl carrier protein in fatty acid biosynthesis. Mailing Add: Merck Sharp & Dohme Res Labs Merck & Co Inc Rahway NJ 07065

VAGNINA, LIVIO L, b North Bergen, NJ, Apr 26, 17; m 49; c 3. CHEMISTRY. Educ: Fordham Col, BS, 38. Prof Exp: Chemist, H A Wilson Co Div, Englehard Industs, Inc, NJ, 40-42; chief forensic chemist, US Army Criminal Invest Lab, France, 44-46, chief chemist, US Army Graves Regist Lab, Belg, 46-48, chief forensic chemist, Ger, 48-60; microanalyst, US Food & Drug Admin, Washington, DC, 60-63; sr chemist, Cent Intel Agency, 63-73; tech staff mem, Mitre Corp, Va, 73-75; CONSULT CRIMINALIST, 75- Mem: Fel Am Inst Chem; Am Chem Soc; Asn Off Anal Chem; fel Am Acad Forensic Sci; Int Soc Forensic Toxicol. Res: Forensic chemistry; microchemistry; serology of dried blood factors; analysis of narcotics; optical crystallography of drugs; microanalysis of foods and drugs; x-ray spectrometry. Mailing Add: 1034 Dead Run Dr McLean VA 22101

VAGNUCCI, ANTHONY HILLARY, b Terni, Italy, July 9, 28; US citizen; m 62; c 3. MEDICINE, PHYSIOLOGY. Educ: Univ Genoa, MD, 54. Prof Exp: Intern med, Wesson Mem Hosp, Springfield, Mass, 57-58; resident, NY Univ-Bellevue Med Ctr, 58-60; Am Heart Asn res fel renal physiol, Med Sch, NY Univ, 60-62; advan res fel, Hypertension Unit, Dept Med, Peter Bent Brigham Hosp, Boston, 62-63 & advan res fel endocrinol, 63-64; jr assoc med & assoc dir endocrinol metab unit, Peter Bent Brigham Hosp, 64-65; asst prof, 65-70, ASSOC PROF MED, SCH MED, UNIV PITTSBURGH, 70-; HEAD ADRENAL UNIT, MONTEFIORE HOSP, 65- Concurrent Pos: Res assoc, Harvard Med Sch, 64-65. Mem: Am Fedn Clin Res; Endocrine Soc; NY Acad Sci. Res: Diurnal aspect of electrolyte homeostasis; adrenal physiopathology; hypertensive disease. Mailing Add: Dept of Med Univ of Pittsburgh Pittsburgh PA 15213

VAGVOLGYI, JOSEPH, systematics, evolution, see 12th edition

VAHALA, GEORGE MARTIN, b Tabor, Czech, Mar 26, 46; Australian citizen; m 70. MAGNETOHYDRODYNAMICS, PLASMA PHYSICS. Educ: Univ Western Australia, BSc Hons, 67; Univ Iowa, MS, 69, PhD(physics), 72. Prof Exp: Res assoc plasma physics, Univ Tenn, Knoxville, 72; res scientist magnetohydrodynamics, Courant Inst Math Sci, NY Univ, 72-74; ASST PROF PHYSICS, COL WILLIAM & MARY, 74- Res: Magnetohydrodynamics and guiding-center stability of containment devices; spectral theory and its interpretation in plasma physics as well as in magnetohydrodynamics; transport effects in plasmas. Mailing Add: Dept of Physics Col of William & Mary Williamsburg VA 23185

VAHEY, DAVID WILLIAM, b Youngstown, Ohio, Nov 21, 44. OPTICAL PHYSICS, ELECTROOPTICS. Educ: Mass Inst Technol, BS, 66; Calif Inst Technol, MS, 67,

PhD(elec eng), 73. Prof Exp: Fel physics, 73-74, res scientist, 74-75, RES SCIENTIST PHYSICS, COLUMBUS LABS, BATTELLE MEM INST, 75- Mem: Am Phys Soc. Res: Integrated optics; photorefractive phenomena; optical data processing; laser communications; optical properties of organic dyes. Mailing Add: Battelle Columbus Labs 505 King Ave Columbus OH 43201

VAHOUNY, GEORGE V, b New York, NY, Feb 22, 32; m 55; c 3. CARDIOVASCULAR DISEASES. Educ: George Washington Univ, BS, 53, MS, 55, PhD, 58. Prof Exp: From instr to assoc prof, 56-69, PROF BIOCHEM, GEORGE WASHINGTON UNIV, 69- Concurrent Pos: Lectr, Univ Tex Southwestern Med Sch Dallas, 57. Honors & Awards: William B Peck Sci Res Award, Interstate Postgrad Med Asn, 66. Mem: Soc Exp Biol & Med; Am Soc Biol Chem; Am Inst Nutrit. Res: Absorption and metabolism of lipids, especially cholesterol; cholesterol esterase systems; effects of hormones on fat absorption from the small intestine. Mailing Add: Dept of Biochem George Washington Univ Washington DC 20005

VAIDHYANATHAN, V S, b Madras, India, Dec 15, 33; m 65; c 1. BIOPHYSICS. Educ: Annamalai Univ, Madras, BSC, 53, MA, 54; Ill Inst Technol, PhD(chem), 61. Prof Exp: Res assoc chem, Univ Kans, 60-62; chief math & statist sect, Southern Res Support Ctr, 62-63, chief theoret sci sect, 63-66; assoc prof theoret biol, 66-70, assoc prof biophys, 67-72, ASSOC PROF PHARMACEUT & BIOPHYS, STATE UNIV NY BUFFALO, 72- Concurrent Pos: Consult, Vet Admin Hosp, New Orleans; vis prof theoret biol, State Univ NY Buffalo, 65; Europ Molecular Biol Orgn fel, 69. Mem: AAAS; Am Chem Soc; Biophys Soc; Int Soc Cell Biol. Res: Statistical mechanics; active transport; nerve potentials; biophysics of membranes. Mailing Add: Dept of Biophys Sci State Univ NY at Buffalo Amherst NY 14226

VAIL, CHARLES BROOKS, b Bessemer, Ala, Apr 29, 23; m 44; c 2. PHYSICAL CHEMISTRY. Educ: Birmingham-Southern Col, BS, 45; Emory Univ, MS, 47, PhD(chem), 51. Prof Exp: Instr chem, Armstrong Col, 48-49; chemist, Southern Res Inst, 51-53; prof phys sci, Coker Col, 53-56; assoc prof chem, Agnes Scott Col, 56-57; prof chem & acad dean, Hampden-Sydney Col, 57-65; assoc exec secy, Comn on Cols, 65-68; dean, Sch Arts & Sci, Ga State Univ, 68-73; PRES, WINTHROP COL, 73- Mem: Am Chem Soc. Res: Thermal diffusion in liquids; heats of vaporization; photochemistry. Mailing Add: Winthrop Col Rock Hill SC 29733

VAIL, DERRICK TILTON, ophthalmology, deceased

VAIL, EDWIN GEORGE, b Toledo, Ohio, July 25, 21; m 46; c 5. MEDICAL PHYSIOLOGY, BIOENGINEERING. Educ: Univ Toledo, BSc, 47; Ohio State Univ, MSc, 48, PhD(aviation physiol), 53. Prof Exp: Proj engr, Aerospace Med Lab, Wright Air Develop Ctr, Ohio, 51-53, chief respiration sect, 53-54, proj scientist, 54-60, chief personnel protection equip & crew escape group X-20 syst prog officer, 60-62, asst chief bioastronaut div, 63-64; chief human eng & space suit res & develop, Hamilton Standard, United Aircraft Corp, Conn, 64-69; MEM STAFF PHYSIOL, NAVAL COASTAL SYSTS LAB, PANAMA CITY, 70-; PRES, VAIL APPL RES CO, INC, 73- Honors & Awards: Award, Air Force Systs Command, 64. Mem: Aerospace Med Asn; Undersea Med Soc; Sigma Xi. Res: Aerospace and oceanographic physiology, including respiratory, cardiovascular, environmental stress tolerance; space-pressure suits, diving equipment and life support system research and development of medical devices. Mailing Add: 4502 Vista Ln Lynn Haven FL 32444

VAIL, JAMES MONTGOMERY, cell physiology, biochemistry, see 12th edition

VAIL, JOHN MONCRIEFF, b Winnipeg, Man, Oct 17, 31; m 54; c 1. SOLID STATE PHYSICS. Educ: Univ Man, BSc, 55, MSc, 56; Brandeis Univ, PhD(physics), 60. Prof Exp: IBM res asst, Brandeis Univ, 57-59; Nat Res Coun Can fel, McGill Univ, 60-61; Leverhulme fel, Univ Liverpool, 61-62; from asst prof math physics to assoc prof physics, 62-72, PROF PHYSICS, UNIV MAN, 72- Mem: Can Asn Physicists; Am Phys Soc; Brit Inst Physics. Res: Solid state theory, including point defects in general and color centers in particular; ionic crystals. Mailing Add: Dept of Physics Univ of Man Winnipeg MB Can

VAIL, OAKLEY R, organic chemistry, see 12th edition

VAIL, SIDNEY LEE, b New Orleans, La, Aug 10, 28; m 53; c 4. ORGANIC CHEMISTRY, TEXTILES. Educ: Tulane Univ, BS, 49, PhD(org chem), 65; La State Univ, MS, 51. Prof Exp: Org chemist, Dow Chem Co, 51-53; sr chemist, Am Cyanamid Co, 55-59; proj leader, 59-72, RES LEADER COTTON TEXTILES, SOUTHERN REGIONAL RES CTR, USDA, 72- Concurrent Pos: Exchange scientist, Shirley Inst, Eng, 65-66. Mem: Am Chem Soc; Am Asn Textile Chemists & Colorists; Sigma Xi. Res: Petrochemicals, synthesis and process chemistry; organic synthesis and mechanisms of textiles; nuclear magnetic resonance; chemical modification of cotton. Mailing Add: Southern Regional Res Ctr 1100 Robert E Lee Blvd New Orleans LA 70119

VAIL, WILLIAM JERALD, b July 9, 36; US citizen; m 58; c 4. BIOPHYSICS, CELL BIOLOGY. Educ: WVa Univ, BA, 60, MS, 63, PhD(microbiol), 65. Prof Exp: Instr biol, WVa Univ, 64-65; assoc prof biol, Ind Univ Pa, 65-67; trainee biochem, Univ Wis-Madison, 67-70; asst prof, 70-76, ASSOC PROF MICROBIOL, UNIV GUELPH, 76- Concurrent Pos: Consult, Hosp for Sick Children, Toronto, 74- Mem: AAAS; Sigma Xi; Biophys Soc; Am Soc Cell Biol. Res: Structure and function of biological membranes; bioenergetics; phospholipid vesicles. Mailing Add: Dept of Microbiol Univ Guelph Guelph ON Can

VAILE, JOSEPH EDWIN, b Kokomo, Ind, Jan 13, 02; m 30. HORTICULTURE, POMOLOGY. Educ: Univ Ill, BS, 28, MS, 29, PhD(hort), 33. Prof Exp: Instr hort & asst horticulturist, Univ Del, 29-30; asst plant breeding, Univ Ill, 31-32; from asst prof to prof, 36-69, from assoc horticulturist to horticulturist, 43-69, EMER PROF HORT, UNIV ARK, FAYETTEVILLE, 69- Concurrent Pos: Garden supvr, Ill Emergency Relief Comn, 34, state field rep, Subsistence Garden Prog, 35. Mem: Am Soc Hort Sci; Am Pomol Soc. Res: Influence of environment on the development of flower parts; hardiness and drought resistance; uneven ripening of American grapes; variety testing and selection; breeding; strawberry and blackberry production. Mailing Add: 306 Adams Fayetteville AR 72701

VAILLANCOURT, DE GUISE, b Montreal, Que, Dec 11, 20; m 52; c 5. INTERNAL MEDICINE, RHEUMATOLOGY. Educ: Univ Montreal, BA, 41, MD, 47; Columbia Univ, DSc(med), 53; FRCPS(C), 55. Prof Exp: PROF MED, FAC MED, UNIV MONTREAL, 55-, DIR CONTINUING MED EDUC, 56-, V DEAN FAC MED, 68- Concurrent Pos: Attend physician, Hotel-Dieu Hosp, 52- Mem: Fel Am Col Physicians; Am Rheumatism Asn; Can Med Asn; Can Rheumatism Asn (pres, 67); Can Asn Continuing Med Educ. Res: Continuing medical education; clinical medicine. Mailing Add: Univ of Monteal Fac of Med PO Box 6207 Sta A Montreal PQ Can

VAILLANCOURT, REMI ETIENNE, b Maniwaki, Que, June 16, 34. MATHEMATICS. Educ: Univ Ottawa, BA, 57, BSc, 61, BTh, 63, MSc, 64, MTh, 65; NY Univ, PhD(math), 69. Prof Exp: Instr, NY Univ, 68-69; Off Naval Res res assoc, Univ Chicago, 69-70; ASSOC PROF MATH, UNIV OTTAWA, 70-, CHMN DEPT, 72- Mem: Am Math Soc; Math Asn Am; Can Math Cong; Fr-Can Asn Advan Sci. Res: Partial differential equations; pseudo-differential operators; finite difference and finite element methods. Mailing Add: Dept of Math Univ of Ottawa Ottawa ON Can

VAILLANT, HENRY WINCHESTER, b New York, NY, Dec 17, 36; m 58; c 3. POPULATION BIOLOGY. Educ: Harvard Univ, AB, 58, MD, 62, SMHyg, 69. Prof Exp: Intern med, Boston City Hosp, 62-63, resident, 63-64; res assoc, Nat Inst Child Health & Human Develop, 64-66; resident, Boston City Hosp, 66-67; res fel obstet & gynec, Harvard Med Sch, 67-68, ASST PROF POP STUDIES, SCH PUB HEALTH, HARVARD UNIV, 68- Concurrent Pos: Consutl cancer control prog, USPHS, 68. Mem: Am Pub Health Asn. Res: Clinical human reproductive physiology. Mailing Add: Harvard Univ Sch Pub Health 665 Huntington Ave Boston MA 02115

VAISNYS, JUOZAS RIMVYDAS, b Kaunas, Lithuania, Mar 12, 37; US citizen. PHYSICAL CHEMISTRY. Educ: Yale Univ, BS, 56; Univ Calif, Berkeley, PhD(chem), 60. Prof Exp: ASST PROF APPL SCI, YALE UNIV, 67- Mem: Am Phys Soc; Am Chem Soc. Res: Mechanical processes in the earth's mantle; ecology; population dynamics. Mailing Add: Kline Geol Lab Yale Univ New Haven CT 06520

VAISRUB, SAMUEL, b Uman, Russia, Sept 15, 06; nat Can; m 49; c 2. INTERNAL MEDICINE. Educ: Univ Man, MD, 32; Can Bd Internal Med, dipl, 47; MRCP, 48; FRCP(C), 52. Prof Exp: Pvt pract, 33-65; SR ED, JAMA, 65-; ASSOC ED, ARCH INTERNAL MED, 68- Concurrent Pos: Attend physician, Deer Lodge Vet Hosp, 48-65, St Boniface Gen Hosp, 50-65 & Misericordia Gen Hosp, 57-65; ed, Man Med Rev, 55-65; asst prof med, Univ Man, 58-65; clin assoc prof, Chicago Med Sch, 67- Mem: Am Diabetic Asn; Am Heart Asn; Am Med Writers' Asn; fel Am Col Physicians; Am Soc Internal Med. Res: Medical writing and editing. Mailing Add: Am Med Asn 535 N Dearborn St Chicago IL 60610

VAITKEVICIUS, VAINUTIS K, b Kaunas, Lithuania, Jan 12, 27; US citizen; m 51; c 6. ONCOLOGY. Educ: Univ Frankfurt, MD, 51. Prof Exp: Intern med, Grace Hosp, Detroit, 51-52, resident, 55-56; resident internal med, Detroit Gen Hosp, 56-58; assoc physician, Henry Ford Hosp, Detroit, 59-62; clin dir oncol, Detroit Inst Cancer Res, 62-66; from assoc prof to prof med & dir oncol, 66-73, PROF ONCOL & CHMN DEPT, SCH MED, WAYNE STATE UNIV, 73- Concurrent Pos: Fel cancer res, Detroit Inst Cancer Res, Mich, 58-59; consult, Detroit-Macomb Hosps Asn, Harper Hosp, Mt Carmel Mercy Hosp, Sinai Hosp & Vet Admin Hosp. Mem: AAAS; Am Col Physicians; Am Asn Cancer Res; Am Asn Cancer Educ; Am Soc Hemat. Res: Mechanism of metastases; pharmacology of cytostatic drugs. Mailing Add: Dept of Oncol Wayne State Univ Sch of Med Detroit MI 48201

VAITUZIS, ZIGFRIDAS, b Lithuania, Jan 31, 37; US citizen; m 62; c 2. MEDICAL MICROBIOLOGY. Educ: Univ Conn, AB, 59; Univ Md, College Park, MS, 65, PhD(microbiol), 69. Prof Exp: Res asst microbiol, Leonard Wood Mem Found, Nat Inst Arthritis, Metabolic & Digestive Dis, Bethesda, Md, 62-63; from instr to asst prof, Univ Md, College Park, 65-75; CHIEF MICROBIOLOGY, OSCAR B HUNTER MEM LAB & SCH MED TECHNOL, 75- Mem: AAAS; Am Soc Microbiol; Electron Micros Soc Am. Res: Bacterial motility systems; ultrastructure of viruses and procaryotic cells; development of rapid methods for identification of bacterial and viral pathogens. Mailing Add: Hunter Labs 915 19th St NW Washington DC 20006

VAJDA, GEZA LASZLO, b Gyulavari, Hungary, June 27, 98; US citizen; m 30. PHYSICS, ENGINEERING. Educ: Budapest Tech Univ, ME, 20; Acad Indust & Tech, Kassa, Hungary, ME, 21; Sorbonne, BS, 26, PhD, 27; Kaiser Wilhelm Inst, PhD, 29. Prof Exp: From engr to chief engr, Edison Radio & Electro Res, Hungary, 27-37; tech counselor, Royal Hungarian Ministry, Budapest, 37-43; pres, Wood Mach Factory, 43-44; engr, Mach Factory Fahr, Ger, 44-45; radio engr, Radio Stuttgart, Ger, 46-51; mech design engr, Scaife Co, Pa, 51-52; engr, Pittsburgh Waterproof Co, 52-54 & Nider Tool Co, Inc, 54-57; design engr, Top Indust Inc, 57-58 & Vironic Res Corp, 58; sr physicist, Acoustica Assocs, Inc, 58-59; chief physicist, Halex Inc, 59-60; pres & chief physicist, Advan Sci Trends Assocs, 60-65; OWNER, VAJDA SCI RES LAB, 65- Concurrent Pos: Tech counsr, Govt Comn, Karpatland, Hungary, 37-43; tech counsr city eng, Hust, 43-44; instr, Ursuline Acad, Pittsburgh, 52-54; construct engr, Am Indust, 58; assoc prof space physics, Univ Calif, Los Angeles, 58-68; founder, Vajda Scholar Found, 69; west coast ed, Am Hungarian Life Weekly; NAm pres, Int Comn Revision Postulates Theoret Physics; Founder & pres, St Stephen Sci Acad, San Francisco. Honors & Awards: Hon life mem, Kaiser Wilhelm Max Planck Inst; Knight, Int Constantinian Order. Mem: AAAS; Acoust Soc Am; Am Crystallog Asn; Am Geophys Union; Am Inst Physics. Res: Mechanical science; solid state; theoretical physics; electronics; electrics; mathematical science; atomic physics. Mailing Add: 4561 W 160th St Lawndale CA 90260

VAJK, JOSEPH PETER, b Budapest, Hungary, Aug 3, 42; US citizen; m 70; c 3. PHYSICS. Educ: Cornell Univ, AB, 63; Princeton Univ, MA, 65, PhD(physics), 68. SR PHYSICIST, LAWRENCE LIVERMORE LAB, UNIV CALIF, 68- Mem: Am Phys Soc. Res: Relativistic astrophysics and cosmology; general relativity theory; evolution of relativistic cosmological models; theory of electromagnetic pulses from nuclear explosions; world dynamics; socioeconomic benefits of space industrilization and colonization. Mailing Add: 57 Oakdene Ct Walnut Creek CA 94596

VAJK, RAOUL, b Vajdajunyad, Hungary, Dec 19, 96; nat US; m 32; c 3. GEOPHYSICS. Educ: Univ Sci, Kolozsvar, DrPolitSci, 22; Univ Sci, Budapest, PhD(geophys), 32. Prof Exp: Geophys interpreter, Torsion Balance Explor Co, Tex, 28-31; head geophysicist, Hungarian Oil Indust Co, Hungary, 33-46; res geophysicist, Standard Oil Co, NJ, 47-53, geophys adv, 54-56, assoc geophys adv, 57-61; res scientist, Lamont Geol Observ, Columbia Univ, 62-64, res assoc, 65; res geophysicist, Alpine Geophys Assocs, NJ, 65-68; assoc prof geol, 68-70, PROF GEOL, CALIFORNIA STATE COL, PA, 70- Concurrent Pos: Assoc prof, Univ Sci, Budapest, 39-46; adj assoc prof, NY Univ, 65-66. Honors & Awards: Soc Explor Geophys Medal Award, 61. Mem: Seismol Soc Am; Soc Explor Geophys; Nat Asn Geol Teachers; Am Asn Petrol Geologists; Am Geophys Union. Res: Gravity; magnetic and seismic methods; physical oceanography. Mailing Add: Dept of Geol California State Col California PA 15419

VAKILI, NADER GHOLI, b Bushir, Iran, Jan 14, 27; nat US; m 53; c 6. PLANT PATHOLOGY, PLANT GENETICS. Educ: Northwestern Univ, BS, 52; Univ Chicago, MS, 53; Purdue Univ, PhD, 58. Prof Exp: Pathologist, United Fruit Co, 53-65; asst plant pathologist, Everglades Exp Sta, Univ Fla, 65-67; area agron adv, US Agency Int Develop, Vietnam, 67-69; mem staff, 69-73, PLANT PATHOLOGIST, FED EXP STA, AGR RES SERV, USDA, PR, 73- Mem: Am Phytopath Soc; Bot Soc Am. Res: Genetics of pathogenicity; taxonomy and genetics of disease resistance in musa; vegetable diseases. Mailing Add: Fed Exp Sta PO Box 70 USDA Mayaguez PR 00708

VAKILZADEH, JAVAD, b Esfahan, Iran, June 26, 27. PUBLIC HEALTH, EPIDEMIOLOGY. Educ: Sharaf Col, Iran, BS, 48; Univ Teheran, DVM, 52; Univ

Pittsburgh, CPH, 58; Univ NC, MPH, 59. Prof Exp: Epidemiologist, Int Coop Admin, Iran, 53-57; res scientist, NC Sanitorium Syst, 59-63, from asst to actg dir res respiratory dis, 63-68; mem fac, Med Ctr, Duke Univ, 68-72; EPIDEMIOLOGIST, INT FERTILITY RES PROG, POP CTR, UNIV NC, CHAPEL HILL, 72- Concurrent Pos: Consult epidemiologist, Haitian-Am Tuberc Inst, 63- Mem: Fel Am Pub Health Asn; Am Vet Med Asn; Am Vet Epidemiol Soc. Res: Research and teaching of epidemiology; environmental health; communicable disease control. Mailing Add: 1305 Wildwood Dr Chapel Hill NC 27514

VALA, MARTIN THORVALD, JR, b Brooklyn, NY, Mar 28, 38; m 66; c 2. PHYSICAL CHEMISTRY, SPECTROCHEMISTRY. Educ: St Olaf Col, BA, 60; Univ Chicago, SM, 62, PhD(chem), 64. Prof Exp: NSF fel chem, Copenhagen Univ, 65-66; US-Japan Coop Sci Prog fel, Univ Nagoya, 66-67; asst prof, 68-72, ASSOC PROF CHEM, UNIV FLA, 72- Concurrent Pos: Merck Found fel, 70-71; vis prof, Advan Sch Physics & Chem, 73-74; NATO fel, 73-74; Fulbright sr fel, Franco-Am Scholar Exchange Comn, 73-74. Mem: Am Phys Soc; Am Chem Soc; Am Inst Chem. Res: Optical properties of organic molecules and transition metal complexes. Mailing Add: Dept of Chem Univ of Fla Gainesville FL 32601

VALADARES, JOSEPH R E, b Ahmadabad, India, Oct 3, 31; Can citizen. PHARMACOLOGY, BIOCHEMISTRY. Educ: Karnatak Univ, India, BSc, 52; Univ Bombay, MS, 54; Univ Okla, PhD(med sci), 60. Prof Exp: Lectr pharmacol, Sch Nursing, Univ Okla, 59-62; res assoc, Univ BC, 63-65; lectr pharmacol, Fac Med, Univ Ottawa, 65-66, asst prof, 67-69; SCI ADV, BUR DRUGS, HEALTH PROTECTION BR, NAT HEALTH & WELFARE, 69- Concurrent Pos: Res fel biochem, Haffkine Inst, Bombay, India, 54-55; res fel pharmacol, Univ Okla, 60-62; clin consult, St Anthony Hosp, Okla, 57-60; guest lectr, Hillcrest Med Ctr, Okla, 60-62. Mem: NY Acad Sci; Can Soc Cell Biol; Brit Biochem Soc; Int Soc Cybernet Med. Res: Ion transport; biochemical mechanisms of hormone and drug action; effects of drugs on enzyme systems in the central nervous system; mechanisms of drug action in the cardiovascular system. Mailing Add: 170 Viewmount Dr Ottawa ON Can

VALANCE, WILLIAM GEORGE, physical chemistry, see 12th edition

VALBERG, LESLIE S, b Churchbridge, Sask, June 3, 30; m 54; c 3. MEDICINE. Educ: Queen's Univ, Ont, MD, 54, MSc, 58; FRCPS(C), 60. Prof Exp: Lectr med, Queen's Univ, Ont, 60-61; res assoc, Med Res Coun, Can, 61-65; from asst prof to prof med, Queen's Univ, Ont, 61-75; PROF MED & CHMN DEPT, UNIV WESTERN ONT, 75- Concurrent Pos: Dir spec invest unit, Kingston Gen Hosp, 65-75; consult, Univ Hosp, London, 75- Mem: Am Gastroenterol Asn; Am Fedn Clin Res; fel Am Col Physicians; fel Royal Col Physicians & Surgeons Can; Can Soc Clin Invest. Res: Iron metabolism; absorption of metals. Mailing Add: Dept of Med Univ Hosp Univ of Western Ont London ON Can

VALCARCE, ARLAND CASIANO, b Brigham City, Utah, Mar 11, 23; m 50; c 5. ENTOMOLOGY. Educ: Univ Utah, BS, 49; Utah State Univ, MS, 53. Prof Exp: Entomologist & supvr grasshopper control proj, USDA, 50-51, inspector, Plant Quarantine Div, 52-54, res entomologist, Entom Res Div, 54-62; ENTOMOLOGIST, US FOREST SERV, 62- Res: Biology, ecology and cultural, biological and chemical control of forest insect and disease pests. Mailing Add: 3473 Manchester St Boise ID 83704

VALCOURT, ALFRED J, biochemistry, see 12th edition

VALDES-DAPENA, MARIE A, b Pottsville, Pa, July 14, 21; m 45; c 11. PEDIATRICS, PATHOLOGY. Educ: Immaculate Col, Pa, BS, 41; Temple Univ, MD, 44. Prof Exp: Instr path, Sch Med, Univ Pa, 45-49 & Sch Dent, 47; from instr to assoc prof surg, Med Col Pa, 47-59; from asst prof to assoc prof, 59-67, PROF PATH & PEDIAT, SCH MED, TEMPLE UNIV, 67-; ASSOC PATHOLOGIST, ST CHRISTOPHER'S HOSP CHILDREN, 59- Concurrent Pos: St Christopher's Hosp Children grant, 58-; instr, Grad Sch Med, Univ Pa, 48-55, vis lectr, 60-; consult pediat path, Div Med Exam, Dept Pub Health, Philadelphia, 67-70; mem, Perinatal Biol & Infant Mortality Res & Training Comt Nat Inst Child Health & Human Develop, 71-75; consult, Lankenau Hosp, Philadelphia, 71-; consult & lectr, US Naval Hosp, Philadelphia, 72. Mem: Am Asn Pathologists & Bacteriologists; Int Acad Path; Path Soc Gt Brit & Ireland. Res: Causes of neonatal mortality; sudden unexpected death in infancy; gynecologic pathology in infancy and childhood. Mailing Add: St Christopher's Hosp for Children Philadelphia PA 19133

VALDIVIESO, DARIO, b Fusagasuga, Colombia, Dec 12, 36; US citizen. MICROBIOLOGY. Educ: Univ Andes, Colombia, BS, 60, MS, 62; Univ PR Sch Med, PhD(med zool), 67. Prof Exp: Coordr biochem, Seneca Col, Toronto, 68-70; resident med microbiol, US Nat Ctr Dis Control, Atlanta, 70-72; instr microbiol, Univ Tex Sch Med, San Antonio, 72-73; RES ASSOC MAMMAL, ROYAL ONT MUS, 73- Concurrent Pos: Specialist microbiologist pub health & med lab microbiol, Am Acad Microbiol, 74- Mem: Am Soc Microbiologists; Am Soc Trop Med & Hyg. Res: Comparative biochemistry of proteins; microbiology of human pathogens; immunochemistry of human parasitic and mycotic diseases. Mailing Add: Dept of Mammal Royal Ont Mus 100 Queen's Park Toronto ON Can

VALDSAAR, HERBERT, b Tallinn, Estonia, Dec 6, 25; nat US; m 59; c 2. HIGH TEMPERATURE CHEMISTRY. Educ: Aachen Tech Univ, Dipl, 50; Univ Maine, MS, 52; Univ Fla, PhD(chem), 56. Prof Exp: Res chemist, 56-71, SR RES CHEMIST, PIGMENTS DEPT, E I DU PONT DE NEMOURS & CO, INC, 71- Mem: Am Ceramic Soc. Res: High temperature inorganic reactions; preparation of high purity silicon; high temperature refractories. Mailing Add: Exp Sta E I du Pont de Nemours & Co Inc Wilmington DE 19898

VALEGA, THOMAS MICHAEL, b Linden, NJ, May 23, 27; m 58; c 4. ORGANIC CHEMISTRY. Educ: Rutgers Univ, BS, 59, PhD(org chem), 63. Prof Exp: Chemist, Pesticide Chem Res Br, Entom Res Div, Agr Res Serv, USDA, 63-67; grants assoc, NIH, 67-68, health scientist adminr, Nat Inst Environ Health Sci, 68-69, coordr contracts artificial kidney-chronic uremia prog, Nat Inst Arthritis & Metab Dis, 69-72; prog analyst, Off Categorical Progs, Prog Planning & Eval, Environ Protection Agency, DC, 72; health scientist adminr, 72-74, CHIEF RESTORATIVE MATERIALS PROG, EXTRAMURAL PROGS, NAT INST DENT RES, 74- Mem: AAAS; Am Chem Soc; Nat Audubon Soc; Soc Biomaterials. Res: Peroxide and carbamate chemistry; insecticide and insecticide synergist chemistry; insect attractant and insect pheromone chemistry; medicinal chemistry; bioengineering; biomaterials; health sciences administration; dental chemistry. Mailing Add: 19005 Willow Grove Rd Olney MD 20832

VALENCIA, RUBY M, radiation genetics, see 12th edition

VALENCICH, TRINA J, b Long Beach, Calif, Feb 3, 43; c 1. PHYSICAL CHEMISTRY. Educ: Univ Calif, Irvine, BA, 68, PhD(chem), 74. Prof Exp: ADJ ASST PROF CHEM, UNIV CALIF, LOS ANGELES, 73- Mem: Am Chem Soc; Am Physics Soc. Res: Classical trajectory simulation of microscopic physical and chemical processes. Mailing Add: Dept of Chem Univ of Calif Los Angeles CA 90024

VALENSTEIN, ELLIOT SPIRO, b New York, NY, Dec 9, 23; m 47; c 2. NEUROSCIENCES, PSYCHOLOGY. Educ: City Col New York, BS, 49; Univ Kans, MA, 53, PhD(psychol), 54. Prof Exp: Asst anat & psychol, Univ Kans, 51, asst, Endocrinol Lab, 53-54, USPHS res fel anat, 54-55; chief lab neuropsychol, Walter Reed Army Inst Res, Walter Reed Army Med Ctr, 59-61; from assoc to prof psychol & sr res assoc, Fels Res Inst, Antioch Col, 61-70; PROF PSYCHOL, NEUROSCI LAB, UNIV MICH, ANN ARBOR, 70- Concurrent Pos: Mem, Exp Psychol Study Sect, Nat Sci Adv Bd, 64-66; vis prof, Univ Calif, Berkeley, 69-70; mem, Neurobiol Rev Panel, NSF, 71-72; mem, Exp Psychol Study Sect, NIH, 75- Mem: Fel AAAS; fel Am Psychol Asn (pres, Div Comp & Physiol Psychol, 75-76); Int Brain Res Orgn; NY Acad Sci; Soc Exp Psychol. Res: Hormones and behavior; development of behavioral capacities; physiological and comparative psychology; nervous system and motivation. Mailing Add: Neurosci Lab Univ of Mich Ann Arbor MI 48104

VALENTA, LUBOMIR JAN-VACLAV, b Czech, Feb 29, 32; m 70; c 1. MEDICINE, ENDOCRINOLOGY. Educ: Col Podebrady, BS, 50; Charles Univ, Prague, MD, 56, PhD(biochem), 66. Prof Exp: Instr surg, Univ Kosica, Czech, 56-57; family pract, Health Ctr Revuca & Kladno, 57-62; investr endocrinol, Res Inst Endocrinol, Prague, 62-68; assoc prof med, Sch Med, Mich State Univ, 72-74; ASSOC PROF MED & HEAD DIV ENDOCRINOL, DEPT MED, UNIV CALIF, IRVINE, 75- Concurrent Pos: Fels, Hosp Nestle & Univ Lausanne, 68-69, Univ Marseille, 69-71 & Mass Gen Hosp & Harvard Med Sch, 71-72; adv, Great Soviet Encycl, 73-; reviewer, J Clin Endocrinol & Metab, 73-; mem adv bd, J Cancer, 73- Mem: AMA; Am Thyroid Asn; Endocrine Soc. Res: Mechanism of hormone action; structure-function relationship of protein hormones. Mailing Add: Dept of Med Div of Endocrinol Univ of Calif Irvine CA 92650

VALENTA, ZDENEK, b Havlickuv Brod, Czech, June 14, 27; Can citizen; m 57; c 3. ORGANIC CHEMISTRY. Educ: Swiss Fed Inst Technol, Dipl Ing Chem, 50; Univ NB, MSc, 52, PhD(chem), 53. Prof Exp: Spec lectr chem, Univ NB, 53-54, lectr, 54-56; Univ NB fel & res assoc, Harvard Univ, 56-57; from asst prof to assoc prof chem, 57-63, chmn dept, 63-72, PROF CHEM, UNIV NB, 63- Honors & Awards: Merck, Sharp & Dohme Lect Award, Chem Inst Can, 67. Mem: Chem Inst Can; fel Royal Soc Can. Res: Total synthesis of organic molecules of biological and pharmaceutical interest; study of organic reactions and stereochemistry. Mailing Add: Dept of Chem Univ of NB Fredericton NB Can

VALENTE, FRANK ANTHONY, b Padula, Italy, Jan 22, 99; nat US; m 26. REACTOR PHYSICS, NUCLEAR PHYSICS. Educ: NY Univ, BS, 22, MS, 24, PhD(nuclear physics), 39. Prof Exp: Instr physics, NY Univ, 24-25; instr, High Sch, NY, 25-26; asst physicist, Picatinny Arsenal, US Dept War, 26-28 & Nat Bur Standards, 28-30; physicist, Devoe & Raynolds Co, 30; res physicist, Westinghouse Elec & Mfg Co, 30-31; physicist, Socony-Vacuum Oil Co, Inc, 31-42; asst officer in charge crystal br, Signal Lab, US Army, NJ, 42-43, asst opers officer, Manhattan Dist, Chicago, 43 & Clinton Lab, Tenn, 43-44, chief prod units, Manhattan Dist, Hanford Eng Works, 44-47; chief nuclear energy div, Cent Intel Agency, 55-56; prof physics, Rensselaer Polytech Inst, 56-60, prof nuclear eng & sci, 60-65, dir sub-critical reactor lab, 57-66; prof physics, 66-72, EMER PROF PHYSICS, SEATTLE UNIV, 72-, DIR NUCLEAR FACILITY, 74-; EMER PROF NUCLEAR ENG & SCI, RENSSELAER POLYTECH INST, 65- Concurrent Pos: Mem staff tech dir, Oper Sandstone, 48, Oper Greenhouse, Eniwetok Atoll & Oper Ranger, Nev, 51; lectr, Georgetown Univ, 47-56; consult, US Govt, 56-66. Mem: Fel AAAS; fel Am Phys Soc; Am Nuclear Soc; NY Acad Sci. Res: Calorimetry of propellants; spectroscopy; electricity and magnetism; rheology; x-rays; heat transfer; nuclear reactors of both the fusion and fission types. Mailing Add: 10240 SE 13th St Bellevue WA 98004

VALENTEKOVICH, MARIJA NIKOLETIC, b Dubrovnik, Yugoslavia, Feb 5, 32; m 62; c 2. CHEMISTRY. Educ: Univ Zagreb, MSChE, 57, PhD(chem), 63. Prof Exp: Res assoc, Rudjer Boskovic Inst, Zagreb, Yugoslavia, 57-65; fel & res assoc, Radiocarbon Lab, Univ Ill, Urbana, 65-67 & Univ Southern Calif, 67-68; sr chemist, Cyclo Chem Co, 68-69, dir qual control, 69-73; head dept radioisotopes, Curtis Nuclear Co, 73-74; DIR QUAL CONTROL, NICHOLS INST, 74- Mem: Am Chem Soc; Am Soc Qual Control. Res: Clinical diagnostics, particularly radioimmunoassays; readiolabelling of peptides and hormones; development and evaluation of new radioimmunoassay techniques. Mailing Add: 33 Silver Spring Dr Rolling Hills Estates CA 90274

VALENTI, CARLO, b Verona, Italy, Jan 18, 25; US citizen; m 58; c 1. MEDICINE. Educ: Univ Turin, MD, 49. Prof Exp: Assoc prof obstet & gynec, Univ Turin, 57-60; from instr to assoc prof, 61-70, PROF OBSTET & GYNEC, STATE UNIV NY DOWNSTATE MED CTR, 70- Concurrent Pos: Fulbright fel, 53-55; res fel obstet & gynec, State Univ NY, 56-57; USPHS trainee, Hahnemann Col Med, 60-61; Health Res Coun New York, res grant, 65-68. Mem· Tissue Cult Asn; Soc Gynec Invest; Harvey Soc. Res: Human cytogenetics; transplantation of tissue cultures. Mailing Add: Dept of Obstet State Univ NY Downstate Med Ctr Brooklyn NY 11203

VALENTINE, BARRY DEAN, b New York, NY, June 6, 24; m 53; c 2. ENTOMOLOGY, HERPETOLOGY. Educ: Univ Ala, BS, 51, MS, 54; Cornell Univ, PhD(entom), 60. Prof Exp: Asst prof biol, Miss Southern Col, 55-57, actg head dept, 57; from asst prof to assoc prof zool & entom, 60-74, PROF ZOOL & ENTOM, OHIO STATE UNIV, 74- Concurrent Pos: Consult, Standard Fruit Co, 56, Lerner Marine Lab, Am Mus Natural Hist, 65, Dames & Moore Inc, 72 & US Army Corps Engr, 72; entomologist zool expeds, Haiti & Jamaica, 56, Cent Am, 56, Mexico, 59, Bahama Islands, 65 & 72, Kenya & Tanzania, 71 & 74; Entom Soc Am travel grant, London, Eng, 64; Ohio State Univ develop fund travel grant, London, Copenhagen, Stockholm & Paris, 70. Mem: Entom Soc Am; Am Soc Ichthyologists & Herpetologists; Soc Study Amphibians & Reptiles. Res: Theory and practice of systematics and zoogeography, especially the weevil family Anthribidae of the world and salamanders of Eastern United States; comparative behavior of Coleoptera. Mailing Add: Dept of Zool Ohio State Univ Columbus OH 43210

VALENTINE, BOB LEON, b Dry Prong, La, Feb 5, 29; m 55; c 2. MEDICAL MICROBIOLOGY, BIOCHEMISTRY. Educ: La Col, BA, 54; La State Univ, MS, 56; Purdue Univ, PhD(microbiol), 63. Prof Exp: Clin lab technologist, Baptist Hosp, 46-47 & Murrells Clin Hosp, 51-53; res asst microbiol, La State Univ, 53-56; microbiologist, US Govt, 56-57 & Miles Labs, Inc, 57-61; Instr microbiol, Purdue Univ, 61-63; assoc prof, Miss State Univ, 63-67; DIR BIOL SAFETY, SHERWOOD MED INDUSTS, INC, 67- Concurrent Pos: Consult poultry & cattle indust, Miss State Univ, 64-67. Mem: AAAS; Am Soc Microbiol. Res: Diseases of animals and humans; safety of medical devices. Mailing Add: Sherwood Med Industs Inc 11802 Westline Industrial Dr St Louis MO 63141

VALENTINE, CHARLES A, anthropology, see 12th edition

VALENTINE, DONALD H, JR, b Orange, NJ, Nov 7, 40; m 66.

ORGANOMETALLIC CHEMISTRY, PHOTOCHEMISTRY. Educ: Wesleyan Univ, BA, 62; Calif Inst Technol, PhD(photochem), 66. Prof Exp: NSF fel, Stanford Univ, 65-66; asst prof chem, Princeton Univ, 66-71; sr chemist, 71-74, res fel, 74, GROUP CHIEF, HOFFMANN-LA ROCHE, INC, 75- Concurrent Pos: Lectr, Bell Tel Labs, 70-71; lectr, Exten Div, Rutgers Univ, 72; ed, Molecular Photochem, 72- Res: Redox reactions; spectroscopy; homogeneous catalysis; asymmetric synthesis. Mailing Add: Res Div Hoffmann-La Roche Inc Nutley NJ 07110

VALENTINE, FRANK ROSSITER, b Woodbridge, NJ, Dec 6, 15; m 40; c 2. ORGANIC CHEMISTRY. Educ: Yale Univ, BS, 37, PhD(org chem), 41. Prof Exp: Asst chem, Rutgers Univ, 37-38 & Yale Univ, 38-41; res chemist, Naugatuck Chem Div, US Rubber Co, 41-48; chemist, 49-60, TECH DIR, HOMOSOTE CO, TRENTON, 60- Mem: Am Chem Soc; Tech Asn Pulp & Paper Indust. Res: Sterols of sponges; agricultural and rubber chemicals; process research in resins; diarylamine carbonates as rubber antioxidants. Mailing Add: Fiddlers Creek Rd Titusville NJ 08560

VALENTINE, FRED TOWNSEND, b Detroit, Mich, Sept 1, 34; m 64; c 2. IMMUNOLOGY, INFECTIOUS DISEASES. Educ: Harvard Univ, AB, 56, MD, 60. Prof Exp: Asst prof, 69-75, ASSOC PROF MED, SCH MED, NY UNIV, 75- Concurrent Pos: Attend med, Manhattan Vet Admin Hosp, 70-; assoc attend med, Univ Hosp & assoc attend physician, Bellevue Hosp, New York, 76- Mem: Am Asn Immunologists; Infectious Dis Soc Am; Transplant Soc; Harvey Soc. Res: Cellular immunology, immunological defenses against infectious agents and against neoplasia. Mailing Add: Dept of Med Sch of Med NY Univ 550 First Ave New York NY 10016

VALENTINE, FREDERICK ALBERT, b Portland, Ore, May 8, 11; m; c 3. MATHEMATICS. Educ: Reed Col, AB, 33; Univ Chicago, MS, 34, PhD(math), 37. Prof Exp: Instr math, Univ Tenn, 36-37; from instr to assoc prof, 37-55, PROF MATH, UNIV CALIF, LOS ANGELES, 55- Mem: Am Math Soc; Math Asn Am. Res: Differential equations; mechanics; convex sets; linear spaces. Mailing Add: Dept of Math Univ of Calif Los Angeles CA 90024

VALENTINE, FREDRICK ARTHUR, b Detroit Lakes, Minn, June 26, 26; m 66; c 3. FOREST GENETICS. Educ: St Cloud State Teachers Col, BS, 49; Univ Wis, MS, 53, PhD(genetics), 57. Prof Exp: Instr genetics, Univ Wis, 54-56; from asst prof to assoc prof forest bot, 56-69, PROF FOREST BOT, STATE UNIV NY COL ENVIRON SCI & FORESTRY, 69- Mem: Genetics Soc Am; Am Genetic Asn. Res: Genetic control of growth and wood properties in Populus tremuloides, the genetics of Hypoxylon mammatum susceptibility to canker in Populus spp; genetics of urban trees. Mailing Add: Dept Bot & Forest Path State Univ NY Col Env Sci & For Syracuse NY 13210

VALENTINE, JAMES WILLIAM, b Los Angeles, Calif, Nov 10, 26; m 57; c 2. GEOLOGY. Educ: Phillips Univ, BA, 51; Univ Calif, Los Angeles, MA, 54, PhD(geol), 58. Prof Exp: Asst geol, Univ Calif, Los Angeles, 52-55, asst geophys, 57-58; from asst prof to assoc prof geol, Univ Mo, 58-64; assoc prof, 64-68, PROF GEOL, UNIV CALIF, DAVIS, 68- Concurrent Pos: Fulbright res scholar, Australia, 62-63; Guggenheim fel. Mem: AAAS; Geol Soc Am; Ecol Soc Am; Paleont Soc (pres, 73-74); Soc Econ Paleont & Mineral; Soc Study Evolution. Res: Evolutionary paleoecology. Mailing Add: Dept of Geol Univ of Calif Davis CA 95616

VALENTINE, JOAN SELVERSTONE, b Auburn, Calif, Mar 15, 45; m 66. BIOINORGANIC CHEMISTRY. Educ: Smith Col, AB, 67; Princeton Univ, PhD(chem), 71. Prof Exp: Instr chem, Princeton Univ, 71-72; ASST PROF CHEM, DOUGLASS COL, RUTGERS UNIV, NEW BRUNSWICK, 72- Mem: Am Chem Soc; The Chem Soc. Res: Synthesis of models for metal-containing biological materials; bioinorganic chemistry of oxygen and superoxide. Mailing Add: Dept of Chem Douglass Col Rutgers Univ New Brunswick NJ 08903

VALENTINE, JOSEPH EARL, b Kansas City, Kans, Apr 6, 33; m 55; c 2. MATHEMATICS. Educ: Southwest Mo State Col, BSEd, 58; Univ Ill, Urbana, MS, 60; Univ Mo-Columbia, PhD(distance geom), 67. Prof Exp: Teacher high sch, Ill, 58-59; teacher & prin high sch, Mo, 60-61; from instr to asst prof math, Southwest Mo State Col, 61-68; asst prof, 68-70, ASSOC PROF MATH, UTAH STATE UNIV, 70- Concurrent Pos: Fulbright fel, Univ Jordan, 71-72. Mem: Am Math Soc; Math Asn Am. Res: Distance geometry; non-euclidean geometry. Mailing Add: Dept of Math Utah State Univ Logan UT 84321

VALENTINE, RAYMOND CARLYLE, b Piatt Co, Ill, Sept 20, 36; m 58; c 1. BIOCHEMISTRY. Educ: Univ Ill, Urbana, BS, 58, MS, 60, PhD(microbiol), 62. Prof Exp: Asst microbiol, Univ Ill, Urbana, 58-62; fel, Rockefeller Inst, 62-64; asst prof biochem, Univ Calif, Berkeley, 64-70; asst prof in residence microbial biochem, Univ Calif, San Diego, 72-74; MEM STAFF, PLANT GROWTH LAB, UNIV CALIF, DAVIS, 74- Mem: Am Soc Microbiol; fel Am Soc Biol Chemists. Res: Nitrogen fixation; ferredoxin; microbial biochemistry and genetics. Mailing Add: Plant Growth Lab Univ of Calif Davis CA 95616

VALENTINE, WILBUR GOODRICH, b Meredith, NY, Oct 10, 03; m 28, 44; c 4. PETROLOGY, MINERALOGY. Educ: Univ Rochester, AB, 24, MS, 26; Columbia Univ, PhD(geol), 36. Prof Exp: Asst geologist, Cananea Consol Copper Co, 28-31; from tutor to prof geol & chmn dept, 33-72, EMER PROF GEOL, BROOKLYN COL, 72- Concurrent Pos: Instr, Cooper Union, 33-40 & NY Univ, 35-38; ed, NY Acad Sci Jour, 42; from asst engr to engr, Union Mines Develop Corp, 43-46. Mem: AAAS; Geol Soc Am; NY Acad Sci. Res: Quantitative microscopic analysis of rocks; natural sediments; artificial products; planetology. Mailing Add: 395 Ocean Ave Brooklyn NY 11226

VALENTINE, WILLIAM NEWTON, b Kans City, Mo, Sept 29, 17; m 40; c 3. MEDICINE. Educ: Tulane Univ, MD, 42; Am Bd Internal Med, dipl, 49. Prof Exp: Intern med, Strong Mem Hosp, Rochester, NY, 42-43, asst resident, 43, chief resident, 43-44; instr, Sch Med, Univ Rochester, 47-48, head sect hemat, AEC Proj, 47-53; asst clin prof, 49-50, from asst prof to assoc prof, 50-57, chmn dept, 63-71, PROF MED, SCH MED, UNIV CALIF, LOS ANGELES, 57- Concurrent Pos: Assoc, St John's Hosp, Santa Monica, Calif, 47, hon consult, 52; sr attend, Harbor Hosp, Torrance, 50; consult, AEC Proj, 53; consult, Hemat Study Sect, NIH, 55-58, mem coun, Inst Arthritis & Metab Dis, 66-70; mem, Am Bd Internal Med, 64-67. Mem: AAAS; Am Soc Clin Invest; Asn Am Physicians; master Am Col Physicians. Res: Hematology. Mailing Add: Dept of Med Univ of Calif Ctr for Health Sci Los Angeles CA 90024

VALENTY, VIVIAN BRIONES, b Concepcion, Philippines, Dec 15, 44; US citizen; m 69; c 2. BIO-ORGANIC CHEMISTRY. Educ: Mapua Inst Technol, BS, 64; Pa State Univ, PhD(chem), 71. Prof Exp: Res asst cereal chem, Int Rice Res Inst, 64-66; ASST PROF CHEM, SKIDMORE COL, 75- Mem: Am Chem Soc; Sigma Xi; AAAS. Res: Study of the enzymes involved in nucleic acid biosynthesis; isolation, purification, characterization and elucidation of the mechanism of reaction. Mailing Add: Dept of Chem & Physics Skidmore Col Saratoga Springs NY 12866

VALENZUELA, RAFAEL, b Montefrio, Spain, May 21, 46; m 71; c 2. IMMUNOPATHOLOGY, PATHOLOGY. Educ: Univ Seville, BS, 63, MD, 69. Prof Exp: Intern & resident internal med, Cadiz & Malaga, Spain, 70-71; resident anat path, Inst Path, Case Western Reserve Univ, 71-72, teaching fels, 72-75; SPEC FEL IMMUNOPATH, CLEVELAND CLIN FOUND, 75- Mem: Col Am Pathologists; Am Soc Clin Pathologists; Int Acad Path. Res: Ultrastructural aspects of immunopathology. Mailing Add: Dept of Immunopath Cleveland Clin Found 9500 Euclid Ave Cleveland OH 44106

VALEO, ERNEST JOHN, b New London, Conn, Aug 6, 45; m 70. PLASMA PHYSICS. Educ: Rensselaer Polytech Inst, BS, 67; Princeton Univ, MA, 69, PhD(astrophys sci), 71. Prof Exp: Res assoc plasma physics lab, Princeton Univ, Mem: Am Mem: Am Phys Soc. Res: Theoretical plasma physics, especially as related to controlled thermonuclear fusion research. Mailing Add: Lawrence Livermore Lab L545 Livermore CA 94550

VALERINO, DONALD MATTHEW, b Syracuse, NY, June 23, 41; m 63; c 1. PHARMACOLOGY, TOXICOLOGY. Educ: Rensselaer Polytech Inst, BS, 63; Univ Vt, PhD(pharmacol), 70. Prof Exp: Res asst pharmacol, Hazelton Labs, 63-65; asst prof pharmacol, Hershey Med Sch, Pa State Univ, 72-75; CRITERIA MGR, NAT INST OCCUP SAFETY & HEALTH, 75- Concurrent Pos: Staff fel, Nat Cancer Inst, 70-72; consult, Paul deHaen Co, 71-72. Res: Biochemical pharmacology-metabolism of antitumor drugs, pargyline and nitrogen heterocycles, studies of microsomal enzyme induction, toxicology of drug metabolites and criteria document preparation. Mailing Add: Nat Inst Occup Safety & Health 5600 Fishers Lane Rockville MD 20852

VALERIO, DAVID ALLEN, b Norway, Mich, May 23, 37; m 60; c 2. LABORATORY ANIMAL MEDICINE, RESEARCH ADMINISTRATION. Educ: Mich State Univ, BS, 59, DVM, 61; Am Col Lab Animal Med, dipl. Prof Exp: Intern small animal surg & med, Univ Pa, 61-62; vet, Harvey Mem Animal Hosp, 62-63; staff vet, Bionetics Res Labs Div, Litton Industs, 65-68, dir dept lab animal med, 68-70, vpres, Biomed Res Div, Litton Bionetics, Inc, 70-73; VPRES & SCI DIR, LIFE SCI DIV, HAZLETON LABS AM, INC, 73- Concurrent Pos: Chmn subcomt revision of nonhuman primate standards, Nat Acad Sci-Nat Res Coun, mem comt on standards, Inst Lab Animal Resources. Mem: Am Vet Med Asn; Am Asn Lab Animal Sci; NY Acad Sci; Int Primatol Soc; Soc Study Reproduction. Res: Administrative management of research in laboratory animal medicine; reproductive physiology; experimental oncology; microbiology, pathology, immunology and transplantation; biochemistry; pharmacology-toxicology; mutagenesis; teratology; carcinogenesis; cell biology; primatology. Mailing Add: Hazleton Labs Inc Life Sci Div 9200 Leesburg Tpk Vienna VA 22180

VALERIO, JOHN I, nuclear physics, space physics, see 12th edition

VALERIOTE, FREDERICK AUGUSTUS, b Montreal, Que, May 19, 41; m 66; c 2. BIOPHYSICS. Educ: Univ Toronto, BSc, 62, MA, 64, PhD(med biophys), 66. Prof Exp: Can Cancer Soc fel, Ont Cancer Inst, 66-67; Med Res Coun Can fel, NIH, 67-68, USPHS vis fel, 68-69; ASSOC PROF RADIOL, EDWARD MALLINCKRODT INST RADIOL, MED SCH, WASHINGTON UNIV, 69- Mem: Am Asn Cancer Res; Am Asn Cancer Educ. Res: Cancer research; experimental cancer chemotherapy; cell population kinetics. Mailing Add: 520 Warren Ave University City MO 63130

VALERO, FRANCISCO PEDRO JORGE, b Cordoba, Arg, Mar 12, 36; US citizen; m 62; c 3. ATOMIC SPECTROSCOPY, MOLECULAR SPECTROSCOPY. Educ: Univ La Plata, Arg, Licenciado, 60, PhD(physics), 65. Prof Exp: Instr physics, Univ La Plata, Arg, 60-66, assoc prof, 66-67; resident res assoc, Nat Acad Sci, 68-69; SR SCIENTIST PHYSICS, AMES RES CTR, NASA, 69- Concurrent Pos: Res assoc, Nat Acad Sci-Nat Res Coun, 68-69; mem comt on line spectra elements, Nat Acad Sci, 71- Mem: Optical Soc Am. Res: Spectra of neutral and highly ionized elements; spectra of laser generated plasmas; molecular spectroscopy of astrophysical interest. Mailing Add: Ames Res Ctr NASA Moffett Field CA 94035

VALI, GABOR, b Budapest, Hungary, Oct 22, 36; Can citizen; m 56; c 3. ATMOSPHERIC PHYSICS. Educ: Sir George Williams Univ, BSc, 61; McGill Univ, MSc, 64, PhD(physics), 68. Prof Exp: Lectr agr physics, Macdonald Col, McGill Univ, 65-68, asst prof, 68-69; asst prof atmospheric sci, 69-72, ASSOC PROF ATMOSPHERIC SCI, UNIV WYO, 72- Mem: Am Meteorol Soc; Can Meteorol Soc; Royal Meteorol Soc; Am Asn Physics Teachers. Res: Mechanisms and the role of ice nucleation in the atmosphere; development of ice elements in clouds; weather modification; atmospheric aerosols; human impact. Mailing Add: Univ Wyo Dept Atmospheric Sci PO Box 3038 University Sta Laramie WY 82070

VALIAVEEDAN, GEORGE DEVASIA, b Erattupetta, India, Mar 29, 32; m 65; c 2. ORGANIC CHEMISTRY. Educ: Univ Madras, BSc, 52; Georgetown Univ, MS, 60, PhD(org chem), 62. Prof Exp: Lab instr chem, St Joseph's Col, India, 52-53; high sch teacher, Ceylon, 54-55; instr, St Sylvester's Jr Col, 56-57; res asst steroid & carbohydrate chem, Georgetown Univ, 60-62; res fel, Sch Chem, Univ Minn, Minneapolis, 62-65; res chemist, 65-74, SR RES CHEMIST, PHOTO PROD DEPT, E I DU PONT DE NEMOURS & CO, INC, PARLIN, 74- Mem: Am Chem Soc. Res: Steroids; carbohydrates; microbial metabolites; aromatic hydrocarbons; polymer chemistry; dyes and pigments; photochemistry; photographic processes. Mailing Add: 9 Coventry Dr Freehold NJ 07728

VALK, HENRY SNOWDEN, b Washington, DC, Jan 26, 29. THEORETICAL NUCLEAR PHYSICS. Educ: George Washington Univ, BS, 53, MS, 54; Wash Univ, PhD(physics), 57. Prof Exp: Asst, Wash Univ, 54-56; asst prof physics, Univ Ore, 57-59; asst prog dir physics, NSF, 59-60; from asst prof to prof physics, Univ Nebr, Lincoln, 60-70, chmn dept, 66-70; DEAN, COL SCI & LIBERAL STUDIES, GA INST TECHNOL, 70- Concurrent Pos: Prog dir theoret physics, NSF, 65-66; vis prof, Univ Frankfurt, 70. Mem: Fel Am Phys Soc; Am Math Soc; Am Asn Physics Teachers; Math Asn Am. Res: College administration. Mailing Add: Col Sci & Liberal Studies Ga Inst Technol Atlanta GA 30332

VALK, WILLIAM LOWELL, b Muskegon, Mich, Aug 23, 09; m 37; c 2. SURGERY. Educ: Univ Mich, AB, 34, MD, 37. Prof Exp: Instr surg, Med Sch, Univ Mich, 40-43; assoc prof, 46-47, PROF SURG, UNIV KANS MED CTR, 47- Mem: Soc Univ Surg; Clin Soc Genito-Urinary Surg; Am Surg Asn; Am Urol Asn; AMA. Res: Urological surgery; physiology of kidney. Mailing Add: 5401 W 81st Prairie Village KS 66208

VALKO, EMERY IMRE, physical chemistry, deceased

VALKOVIC, VLADO, b Draga Baska, Yugoslavia, July 19, 39; m 62; c 2. EXPERIMENTAL NUCLEAR PHYSICS, APPLIED PHYSICS. Educ: Univ Zagreb, Yugoslavia, BA, 61, MA, 63, PhD(nuclear physics), 64. Prof Exp: Res assoc nuclear physics, Rudjer Boskovic Inst, Zagreb, Yugoslavia, 64-65; res assoc, Rice Univ, 65-67;

res assoc, Rudjer Boskovic Inst, 67-68, head nuclear reaction lab, 68-70; asst prof physics, 70-71, ASSOC PROF PHYSICS, RICE UNIV, 71- Concurrent Pos: Sr sci assoc, Rudjer Boskovic Inst, Zagreb, 71-; adj prof, Univ Houston, 71-; prof physics, Univ Rijeka, Yugoslavia, 73- Honors & Awards: Rudjer Boskovic Prize, Achievements in Nuclear Physics, Croatia, Yugoslavia, 69. Mem: Europ Phys Soc; Am Phys Soc. Res: Experimental nuclear physics and its applications; nuclear reactions with neutrons and charged particles; trace element analysis and applications to biology, medicine and environmental studies. Mailing Add: T W Bonner Nuclear Lab Rice Univ Houston TX 77001

VALLARTA, MANUEL SANDOVAL, b Mexico City, Mex, Feb 11, 99; m 33. THEORETICAL PHYSICS. Educ: Mass Inst Technol, SB, 21, ScD(physics), 24. Hon Degrees: PhD, Nat Univ Mex, 33; ScD, Univ Michoacan, 42; LLD, Univ of the Americas, 65. Prof Exp: Res assoc physics, Mass Inst Technol, 23-26, from asst prof to prof, 26-46; PROF PHYSICS, NAT COL, MEX, 43-; RES PROF, NAT UNIV MEX, 47- Concurrent Pos: Guggenheim fel, Berlin & Leipzig, 27-28; hon prof, Nat Univ Mex, 31-46 & Univ San Andres, Bolivia, 55; Comn Relief Belg Educ Found vis prof, Cath Univ Louvain, 35-36; ed, J Math & Physics, 37-46, J Geophys Res, 48 & Metrologia, 68-; gen dir, Nat Polytech Inst, 43-47; in charge phys res, Comn Impulsory Coord Sci Invest, 43-50; Mex rep & chmn, UN AEC, NY, 46; Mex deleg, UNESCO, 47; res assoc, Carnegie Inst, 48; vis prof, Tata Inst Fundamental Res, India, 48; guest, Inst Advan Study, 49; Pupin lectr, Columbia Univ, 49; undersecy educ, Mex, 54-58; comnr, Nat Comn Nuclear Energy, 56; mem, Int Comn Weights & Measures, Sevres, 60-; chmn sci coun, Int Ctr Theoret Physics, Trieste, 64-; governor for Mex, Int Atomic Energy Agency, Vienna, 66- Honors & Awards: Camacho Prize, 46; Nat Prize, Mex, 61. Mem: Fel Am Phys Soc; Am Math Soc; Am Philos Soc; fel Am Acad Arts & Sci; fel Mex Acad Sci (pres, 44-46). Res: Cosmic radiation; relativity. Mailing Add: Insurgentes Sur 1079 3er Piso Mexico 18 DF Mexico

VALLBONA, CARLOS, b Barcelona, Spain, July 29, 27; m 56; c 4. PEDIATRICS, PHYSIOLOGY. Educ: Univ Barcelona, BA & BS, 44, MD, 50. Prof Exp: Physician, Sch Child Health, Spain, 51-52; intern & resident, Sch Med, Univ Louisville, 53-55; from instr to assoc prof pediat & physiol, 56-67, from instr to assoc prof rehab, 57-67, PROF REHAB, BAYLOR COL MED, 67-, CHMN DEPT COMMUNITY MED, 69- Concurrent Pos: Fel, Children's Int Ctr, Univ Paris, 52-53. Mem: AAAS; Soc Pediat Res; Am Col Chest Physicians; Am Cong Rehab Med; AMA. Res: Pediatric rehabilitation; cardiorespiratory physiology in disabled persons and the newborn; application of electronic data processing techniques in health care; community medicine. Mailing Add: Dept of Community Med Baylor Col of Med Houston TX 77025

VALLEAU, JOHN PHILIP, b Toronto, Ont, Jan 17, 32; m 57; c 2. STATISTICAL MECHANICS, CHEMICAL PHYSICS. Educ: Univ Toronto, BA, 54, MA, 55; Cambridge Univ, PhD(theoret chem), 58. Prof Exp: Nat Res Coun Can fel, 58-60; from asst prof to assoc prof, 61-74, PROF CHEM, UNIV TORONTO, 74- Concurrent Pos: Res vis, Faculty Sci, Orsay, France, 68-69. Res: Theory of liquids and phase changes and of solutions; Monte Carlo and molecular dynamic computations; ultrasonic studies in liquids. Mailing Add: Lash Miller Lab Univ of Toronto Toronto ON Can

VALLEE, BERT L, b Hemer, WGer, June 1, 19; nat US; m 47. BIOCHEMISTRY, BIOPHYSICS. Educ: Univ Bern, BS, 38; NY Univ, MD, 43. Hon Degrees: MA, Harvard Univ, 60. Prof Exp: Res fel med, 45-49, from res assoc to assoc, 49-55, from asst prof to prof, 55-65, PAUL C CABOT PROF BIOL CHEM, HARVARD MED SCH, 65- Concurrent Pos: Nat Res Coun sr fel, Mass Inst Technol, 48-51; Hughes fel, Harvard Med Sch, 51-64; mem staff, Div Indust Coop, Mass Inst Technol, 45-48, res assoc biol, 48-; Merck Sharpe & Dohme prof, Univ Wash, 62; Arthur Kelley lectr, Purdue Univ, 63; mem adv bd, La Trinidad Health Care Facil, Caracas, Venezuela, 70- & Metrop Univ, Caracas, 70-; DuPont lectr, Univ SC, 71; Venable lectr, Univ NC, 72; mem bd gov, Tel Aviv Univ, 72- Honors & Awards: Warner-Chilcott Award, Am Asn Clin Chem, 69. Mem: Nat Acad Sci; Biochem Soc; Am Soc Clin Invest; Am Chem Soc; Optical Soc Am. Res: Composition, conformation, structure, function and mechanism of action of metalloenzymes; local conformation of enzymes; enzyme kinetics; physical chemistry; emission; atomic absorption; absorption spectroscopy; circular dichroism; magnetic circular dichroism; physics of spectrographic sources. Mailing Add: Biophys Res Lab Peter Bent Brigham Hosp Boston MA 02115

VALLEE, LIONEL, anthropology, see 12th edition

VALLEE, RICHARD BERT, b New York, NY, Aug 27, 45. BIOCHEMISTRY. Educ: Swarthmore Col, BA, 67; Yale Univ MPh & PhD(biol), 74. Prof Exp: FEL MOLECULAR BIOL, LAB MOLECULAR BIOL, UNIV WIS-MADISON, 74- Mem: Biophys Soc; AAAS. Res: The composition and function of intermediates in the formation of cytoplasmic microtubles. Mailing Add: Lab of Molecular Biol Univ of Wis Madison WI 53706

VALLEE, RICHARD EARL, b Cincinnati, Ohio, June 21, 28; m 51. PHYSICAL CHEMISTRY, INORGANIC CHEMISTRY. Educ: Univ Cincinnati, BS, 51, MS, 52, PhD(chem), 62. Prof Exp: Chemist, Procter & Gamble Co, 47-48; asst, Univ Cincinnati, 51; chemist, Monsanto Chem Co, 52-59, group leader, 59-60, ASF fel, 61-62, group leader, Monsanto Res Corp, 62-63, sect mgr, 63-67, mgr nuclear technol, 67-69, mgr non-weapons progs, 69-72, MGR TECHNOL APPLN & DEVELOP, MONSANTO RES CORP, 72- Mem: AAAS; Am Chem Soc. Res: Preparation, evaluation and handling of radioactive compounds; high temperature compounds; vacuum technology; isotope separation. Mailing Add: 619 S Bourbon St Blanchester OH 45107

VALLENTINE, JOHN FRANKLIN, b Ashland, Kans, Aug 1, 31; m 50; c 3. RANGE SCIENCE. Educ: Kans State Univ, BS, 52; Utah State Univ, MS, 53; Tex A&M Univ, PhD(range mgt, animal nutrit), 59. Prof Exp: Res aide, Rocky Mountain Forest & Range Exp Sta, US Forest Serv, 52; range conservationist, US Bur Land Mgt, 55-56; res asst range mgt, Exp Sta, Tex A&M Univ, 56-58; exten range specialist, Utah State Univ, 58-62; assoc prof range exten & res, Univ Nebr, 62-68; PROF RANGE SCI, BRIGHAM YOUNG UNIV, 68- Mem: Soc Range Mgt; Am Soc Animal Sci. Res: Range development and improvement; range animal nutrition and management; ranch management; forage production and utilization. Mailing Add: 114 Range Lab (B-49) Brigham Young Univ Provo UT 84602

VALLENTYNE, JOHN REUBEN WAY, limnology, organic geochemistry, see 12th edition

VALLETTA, ROBERT M, b Waterbury, Conn, Nov 10, 31; m 55; c 4. PHYSICAL CHEMISTRY. Educ: Univ Conn, BA, 53, MS, 56; Iowa State Univ, PhD(phys chem), 59. Prof Exp: Res physicist, Cent Res Lab, Am Mach & Foundry Co, 59-63; staff chemist, Components Div, Vt, 63-68, develop engr, 68-69, sr engr, 69-75, SR ENGR, GEN PROD DIV, IBM CORP, CALIF, 75- Mem: Am Chem Soc. Res: Solid state chemistry and physics. Mailing Add: IBM Gen Prod Div Monterey & Cottle Rds San Jose CA 95138

VALLEY, GEORGE EDWARD, JR, b New York, NY, Sept 5, 13; div; c 3. PHYSICS. Educ: Mass Inst Technol, SB, 35; Univ Rochester, PhD(physics), 39. Prof Exp: Optical engr, Bausch & Lomb Optical Co, 35-36; asst, Univ Rochester, 36-39; res assoc mass spectros, Harvard Univ, 39-40, Nat Res Coun fel nuclear physics, 40-41; mem sr staff, Radiation Lab, Mass Inst Technol, 41-45, from asst prof to prof physics, 46-75, Inst, 46-75, founded Lincoln Lab, 49, asst dir lab, 51-53, assoc dir, 53-57; GEN MGR, SUPER PARADIGM CO, 75- Concurrent Pos: Mem sci adv bd, US Air Force, 45-64, chmn air defense syst, Eng Comt, Chief Staff, 50-51, chief scientist, 57-58. Honors & Awards: Presidential Cert of Merit, 48; Except Serv Medal, US Air Force, 56, 58 & 64. Mem: Fel Am Phys Soc; fel Inst Elec & Electronics Eng. Res: Nuclear physics; cosmic radiation; electronics; research and engineering management; systems engineering. Mailing Add: 607 Main Concord MA 01742

VALLEY, KARL ROY, b Canton, Ohio, Aug 20, 43; m 72. INSECT TAXONOMY. Educ: Kent State Univ, BS, 65, MS, 68; Cornell Univ, PhD(entom), 74. Prof Exp: ENTOMOLOGIST, PA DEPT AGR, 72- Mem: Entomol Soc Am; Soc Syst Zool. Res: Biology and taxonomy of adults and immature stages of various families of acalyptrate Diptera. Mailing Add: Bur of Plant Indust Pa Dept of Agr Harrisburg PA 17120

VALLEY, LEONARD MAURICE, b Little Falls, Minn, July 3, 33; m 58; c 3. MOLECULAR PHYSICS. Educ: St John's Univ, Minn, BA(physics) & BA(math), 55; Iowa State Univ, PhD(physics), 60. Prof Exp: From asst prof to assoc prof physics, 60-72, PROF PHYSICS, ST JOHN'S UNIV, MINN, 72-, CHMN DEPT, 70- Concurrent Pos: Vis assoc prof, Univ Denver, 67-68. Mem: Am Phys Soc. Res: Vibrational, rotational and translational relaxation in gas molecules undergoing collisions; relaxation time. Mailing Add: Dept of Physics St John's Univ Collegeville MN 56321

VALLI, VICTOR EDWIN OSWALD, pathology, veterinary medicine, see 12th edition

VALLIER, TRACY L, b Oakland, Iowa, Sept 19, 36; m 57; c 4. MARINE GEOLOGY. Educ: Iowa State Univ, BS, 62; Ore State Univ, PhD(geol), 67. Prof Exp: Assoc prof geol, Ind State Univ, 66-72; geologist, Deep Sea Drilling Proj, Scripps Inst Oceanog, 72-75; MARINE GEOLOGIST, US GEOL SURV, 75- Mem: Geol Soc Am; Soc Econ Paleont & Mineral; Am Geophys Union. Res: Geology of the Bering Sea; deep sea sedimentation; basalt petrology; geology of northeastern Oregon and western Idaho. Mailing Add: US Geol Surv 345 Middlefield Rd Menlo Park CA 94025

VALLOTTON, WILLIAM WISE, b Valdosta, Ga, Nov 26, 27; m 50; c 4. MEDICINE. Educ: Duke Univ, AB, 47; Med Col Ga, MD, 52. Prof Exp: Intern, Univ Wis Hosp, 52; resident ophthal, Duke Univ Hosp, 53-56; instr, Duke Univ, 54-55, assoc, 55-56; chief, Eye, Ear, Nose & Throat Dept, US Naval Hosp, Beaufort, SC, 56-58; assoc prof, 58-65, PROF OPHTHAL, MED UNIV SC, 65-, CHMN DEPT, 66- Concurrent Pos: Consult, Vet Admin Hosp, Durham, NC, 55-56; consult, Vet Admin Hosp, Charleston, US Naval Hosp, SC State Hosp, Charleston County Hosp, St Francis Xavier Hosp & Williamsburg County Mem Hosp; mem, Res Prev Blindness, Inc & SC Eye Bank, Inc; counr, Southern Med Asn, 75-80. Honors & Awards: Honors Award, Am Acad Ophthal & Otolaryngol, 68. Mem: Am Acad Ophthal & Otolaryngol; Am Asn Ophthal; Am Ophthal Soc; Asn Res Vision & Ophthal; AMA. Res: Ophthalmology. Mailing Add: Dept of Ophthal Med Univ of SC Charleston SC 29401

VALLOWE, HENRY HOWARD, b Pittsburgh, Pa, Nov 18, 24; m 48. ENDOCRINOLOGY. Educ: Pa State Teachers Col, BS, 49; Univ Chicago, MS, 50, PhD(zool), 54. Prof Exp: Instr biol, Wright Jr Col, Ill, 52-56; assoc prof zool, Ohio Univ, 56-67; PROF BIOL, INDIANA UNIV, PA, 67- Mem: AAAS; Am Soc Zoologists. Res: Endocrines of poikilotherms; sexual physiology; phylogeny of endocrines; contraceptives of biological origin. Mailing Add: PO Box 294 Indiana PA 15701

VALSAMAKIS, EMMANUEL, b Istanbul, Turkey, May 11, 33; US citizen; m 60; c 2. SOLID STATE ELCTRONICS. Educ: Robert Col, Istanbul, BSEE, 55; Rensselaer Polytech Inst, MEE, 58, PhD(plasma physics), 63. Prof Exp: Instr elec eng, Rensselaer Polytech Inst, 56-61, res asst, 61-62; res scientist, Grumman Aircraft Eng Corp, 62-67; adv physicist, 67-76, MEM RES STAFF, IBM CORP, 76- Mem: Inst Elec & Electronics Engrs; Am Phys Soc. Res: Cryogenic tunneling device design and circuit analysis; experimental investigation of plasmas from pulsed plasma sources; bipolar, field effect transistor and cryogenic tunneling device design and circuit analysis for memory and logic applications. Mailing Add: T J Watson Res Ctr IBM Corp PO Box 218 Yorktown Heights NY 10598

VALTIN, HEINZ, b Hamburg, Ger, Sept 23, 26; nat US; m 53; c 2. PHYSIOLOGY, INTERNAL MEDICINE. Educ: Swarthmore Col, AB, 49; Cornell Univ, MD, 53. Prof Exp: From instr to prof, 57-73, ANDREW C VAIL PROF PHYSIOL, DARTMOUTH MED SCH, 73- Concurrent Pos: Consult, Hitchcock Clin, 61- Mem: Am Physiol Soc; Am Fedn Clin Res; Am Soc Clin Invest; Am Soc Nephrology; Int Soc Nephrology. Res: Kidney, electrolyte and water metabolism; neuroendocrinology. Mailing Add: Dept of Physiol Dartmouth Med Sch Hanover NH 03755

VALVANI, SHRI CHAND, b Mar 20, 40; Indian citizen. PHARMACEUTICAL CHEMISTRY, PHARMACEUTICS. Educ: Univ Saugar, BPharm, 65; Univ Mich, Ann Arbor, MS, 69, PhD(pharmaceut chem), 71. Prof Exp: RES SCIENTIST, UPJOHN CO, PHARM RES, 70- Mem: Acad Pharmaceut Sci; Am Chem Soc; Am Pharmaceut Asn. Res: Thermodynamics of solution process and its effect on drug design, physico-chemical parameters influencing performance of various drug dosage forms, computer applications in a typical pharmacy research laboratory or environment. Mailing Add: Upjohn Co Pharm Res Kalamazoo MI 49001

VALVASSORI, GALDINO E, b Milan, Italy, July 16, 26; US citizen; m 55; c 4. MEDICINE, RADIOLOGY. Educ: Univ Milan, MD, 50; Am Bd Radiol, dipl, 59. Prof Exp: Resident radiol, Univ Milan, 51-53; resident, Mem Hosp, New York, 54-56; asst prof, Univ Chicago, 60-65; assoc prof, 65-67, PROF RADIOL, UNIV ILL MED CTR, 67- Concurrent Pos: Dir dept radiol, Ill Eye & Ear Infirmary, Chicago, 65-; consult, Hinsdale Sanitarium & Hosp, 65, Grant Hosp, Chicago, 66, Henrotin Hosp, Chicago, 68 & Mercy Hosp, Chicago, 71. Mem: AMA; Am Col Radiol; Am Roentgen Ray Soc; Radiol Soc NAm; Am Acad Ophthal & Otolaryngol. Res: Radiology of the head and neck; development and refinement of new radiographic techniques for study of temporal bone in pathological conditions of the ear. Mailing Add: 55 E Washington Chicago IL 60602

VALYOCSIK, ERNEST WILLIAM, physical chemistry, photochemistry, see 12th edition

VAM CLITTERS, ROBERT L, b Alton, Iowa, Jan 20, 26; m 49; c 4. CARDIOVASCULAR PHYSIOLOGY. Educ: Univ Kans, AB, 49, MD, 53. Prof Exp: Intern, Med Ctr, Univ Kans, 53-55, resident internal med, 57-58; res assoc, Scripps Clin Res Found, Univ Calif, 61-62; asst prof cardiovasc physiol, 63-64, assoc prof physiol & biophys 63-68, Robert L King chair cardiovasc res, 63-68, mem staff,

Regional Primate Res Ctr, 64-68, assoc dean, Sch Med, 68-70, PROF PHYSIOL, BIOPHYS & MED, & DEAN, SCH MED, UNIV WASH, 70- Concurrent Pos: Nat Heart Inst res fel, Univ Kans, 55-56 & trainee, 56-57, spec res fel, 58-59; res fel, Sch Med, Univ Wash, 59-62; NIH career res award, 62; res grants, Nat Heart Inst, 62-67, Am Heart Asn, 62-67, US Air Force res grant, 67; exchange scientist, Joint US-USSR Sci Exchange, 62; mem, Myocardial Infarction Comt, Nat Heart & Lung Inst, 71-73; mem, Bd Trustees, Wash State Heart Asn. Mem: AAAS; Am Physiol Soc; Am Heart Asn; Am Fedn Clin Res; NY Acad Sci. Res: Cardiovascular physiology, left ventricular function and control, regional flow distribution, exercise and diving; development of instrumentation and techniques for studying cardiovascular dynacmis in healthy subjects during spontaneous activity. Mailing Add: Off of the Dean Univ of Wash Sch of Med Seattle WA 98195

VAMPOLA, ALFRED LUDVIK, b Dwight, Nebr, July 10, 34; m 56; c 8. SPACE PHYSICS. Educ: Creighton Univ, BS, 56; St Louis Univ, MS, 58, PhD(physics), 61. Prof Exp: Sr physicist, Convair Div, Gen Dynamics Corp, 61-62; STAFF SCIENTIST SPACE PHYSICS, AEROSPACE CORP, 62- Mem: Am Geophys Union. Res: Magnetospheric physics; solar particles. Mailing Add: Aerospace Corp PO Box 92957 El Segundo CA 90245

VAN ABEELE, FREDERICK RICHARD, b Chicago, Ill, Oct 19, 16; m; c 5. BIOCHEMISTRY. Educ: Purdue Univ, BS, 39; Univ Ill, PhD(biochem), 48. Prof Exp: Supvr prod control, US Rubber Co, 39-40; lab asst, Eli Lilly & Co, 40-45, fr chemist, 45-46, res chemist, 48-50, mgr biochem res, 50-53, mgr antibiotics purification & develop, 53-56, dir antibiotics opers, 56-63, dir chem res, 63-64, dir res, 65; pres, Elanco Prod Co, Ind, 67-68; vpres pharmaceut res & develop, Eli Lilly & Co, 68-72; RETIRED. Mem: Am Chem Soc. Res: Antibiotic research, development, manufacturing, vitamins, proteins and peptides; research administration. Mailing Add: State Rd 37 N Martinsville IN 46151

VANABLE, JOSEPH WILLIAM, JR, b Providence, RI, May 29, 36; m 62; c 2. DEVELOPMENTAL BIOLOGY. Educ: Brown Univ, AB, 58; Rockefeller Inst, PhD(biol), 62. Prof Exp: Asst prof, 61-72, ASSOC PROF BIOL, PURDUE UNIV, WEST LAFAYETTE, 71- Concurrent Pos: NIH spec fels, Univ Ore, 69 & Yale Univ, 69-70. Mem: AAAS; Soc Develop Biol; Am Soc Zool. Res: Neurogenesis; the specification of neural pathways during development; visual mutants; regeneration. Mailing Add: Dept of Biol Sci Purdue Univ West Lafayette IN 47906

VAN ALLEN, JAMES ALFRED, b Mt Pleasant, Iowa, Sept 7, 14; m 45; c 5. PHYSICS. Educ: Iowa Wesleyan Col, BS, 35; Univ Iowa, MS, 36, PhD(physics), 39. Hon Degrees: ScD, Iowa Wesleyan Col, 51, Grinnell Col, 57, Coe Col, 58, Cornell Col, 59, Univ Dubuque, 60, Univ Mich, 61, Northwestern Univ, 61, Ill Col, 63, Butler Col, 66, Boston Col, 66, Southampton Col, 67 & Augustana Col, 69. Prof Exp: Carnegie res fel nuclear physics, Dept Terrestrial Magnetism, Carnegie Inst, 39-41, physicist, 41-42; physicist appl physics lab, Johns Hopkins Univ, 42 & 46-50; prof physics, 51-72, CARVER PROF PHYSICS, UNIV IOWA, 72-, HEAD DEPT PHYSICS & ASTRON, 51- Concurrent Pos: Mem, Int Sci Radio Union; mem, Rocket & Satellite Res Panel, 46-, chmn, 47-48, mem exec comt, 58-; mem subcomt upper atmosphere, Nat Adv Comt Aeronaut, 48-52; leader sci exped, Cent Pac, 49, Gulf Alaska, 50, Arctic, 52, Int Geophys Year, Arctic, Atlantic, Cent Pac, SPac & Antarctic, 57; Guggenheim Mem Found fel, Brookhaven Nat Lab, 51; res assoc, Princeton Univ, 53-54; mem tech panel earth satellite prog, Int Geophys Year, 55-58, chmn working group internal instrumentation, 56-58, tech panel rocketry, 55-58, tech panel cosmic rays, 56-58, tech panel aurora & airglow, 57-58; adv comt nuclear physics, Off Naval Res, 57-; adv comt physics, Nat Sci Found, 57-60; mem space sci bd, Nat Acad Sci, 58-70, chmn Ad Hoc Panel Small Planetary Probes, 66; mem panel sci & technol, Comt Sci & Astronaut, US House Rep, 61-; consult, President's Sci Adv Comt; consult particles & fields subcomt, Nat Aeronaut & Space Admin, 61-, mem ad hoc sci adv comt, 66; chmn, Iowa's Int Coop Year Comt Sci & Advan Technol, 65; lectr, NATO Conf, Bergen, Norway, 65; mem, Planetary Missions Bd, 67. Honors & Awards: Hickman Medal, Rocket Soc, 49, First Annual Res Award, 61, Hill Award, Inst Aerospace Sci, 60, Am Inst Aeronaut & Astronaut; Physics Award, Wash Acad Sci, 49; Space Flight Award, Am Astronaut Soc, 58; Distinguished Civilian Serv Medal, US Army, 59; Space Flight Award, Int Acad Astronaut, 61; First Iowa Award Sci, 61; Elliot Cresson Medal, Franklin Inst, 61; Golden Omega Award, Elec Insulation Conf, 63; John A Fleming Awards, 63 & 64, Am Geophys Union; Comdr Order du Merite pour la Recherche et L'Invention, 64; Iowa Broadcasters Asn Award, 64. Mem: Nat Acad Sci; fel Am Phys Soc; fel Am Geophys Union; fel Inst Elec & Electronics Engrs; fel Am Astronaut Soc. Res: Nuclear physics; cosmic rays; use of rockets in physical research; satellites and space probes in planetary and solar physics. Mailing Add: Dept of Physics & Astron Univ of Iowa Iowa City IA 52240

VAN ALLEN, MAURICE WRIGHT, b Mt Pleasant, Iowa, Apr 3, 18; m 49; c 4. NEUROLOGY. Educ: Iowa Wesleyan Col, BA, 39; Univ Iowa, MD, 42; Am Bd Neurol Surg, dipl, 54; Am Bd Psychiat & Neurol, dipl, 63. Prof Exp: Chief neurol serv, Vet Admin Hosp, Iowa City, 54-59, asst dir prof serv for res, 57-59; assoc prof, 59-65, PROF NEUROL, COL MED, UNIV IOWA, 65-, HEAD DEPT, 74- Concurrent Pos: Attend physician & consult physician, Vet Admin Hosp; prog dir, Neurosensory Ctr, Nat Inst Neurol Dis & Stroke, 61; mem, Nat Adv Dent Res Coun, 75. Mem: Fel Am Acad Neurol; Am Neurol Asn; Harvey Cushing Soc; AMA; Asn Res Nerv & Ment Dis. Res: Disorders of ocular movement and muscles; reaction time in cerebral disease; visual perceptive disorders. Mailing Add: Dept of Neurol Univ Hosp Iowa City IA 52240

VAN ALLER, ROBERT THOMAS, b Mobile, Ala, June 18, 33; m 59; c 2. ORGANIC CHEMISTRY, BIOCHEMISTRY. Educ: Univ Ala, BS, 60, MS, 62, PhD(sulfonyl halides), 65. Prof Exp: Res assoc chem, Univ Miss, 65-67; aerospace technologist, Marshall Space Flight Ctr, NASA, 67-68; chmn dept chem, 68-70, dean, Col Sci, 70-71, DEAN, GRAD SCH, UNIV SOUTHERN MISS, 71- Mem: Am Chem Soc. Res: Reactions of aliphatic sulfonyl halides; biosynthesis of phytosterols and monocyclic monoterpenes. Mailing Add: Grad Sch Univ of Southern Miss Hattiesburg MS 39401

VAN ALSTINE, RALPH ERSKINE, b Paterson, NJ, May 18, 14; m 45; c 4. ECONOMIC GEOLOGY. Educ: Hamilton Col, NY, AB, 36; Northwestern Univ, MS, 38; Princeton Univ, PhD(geol), 44. Prof Exp: Asst geol, Northwestern Univ, 36-38; instr, Univ Colo, 38-39; asst geol, Princeton Univ, 39-41; GEOLOGIST, US GEOL SURV, 42- Concurrent Pos: Geologist & dir field studies, Nfld Geol Surv, 39 & 40. Mem: Assoc Soc Econ Geol; Geol Soc Am. Res: Geology of fluorspar deposits of the United States. Mailing Add: US Geol Surv Reston VA 22092

VAN ALSTYNE, JOHN PRUYN, b Albany, NY, Sept 12, 21; m 44; c 3. MATHEMATICS. Educ: Hamilton Col, BS, 44; Columbia Univ, MA, 52. Prof Exp: Vis instr math, Hamilton Col, 43-44, from instr to assoc prof, 48-61; assoc prof, 61-66, actg head dept, 68-71, PROF MATH, WORCESTER POLYTECH INST, 66-, DEAN ACAD ADVISING, 71- Mem: Am Math Soc; Math Asn Am. Res: Functional equations; linear algebra. Mailing Add: Boynton Hall Worcester Polytech Inst Worcester MA 01609

VAN ALTEN, LLOYD, b East Grand Rapids, Mich, Jan 2, 24; m 51; c 2. INORGANIC CHEMISTRY. Educ: Calvin Col, AB, 45; Purdue Univ, West Lafayette, MS, 48; Univ Wash, PhD(chem), 54. Prof Exp: Teacher, Lynden Christian High Sch, 48-50; Olin-Mathieson fel, Univ Wash-Boron Chem, 54-55; from asst prof to assoc prof, 55-68, PROF CHEM, SAN JOSE STATE UNIV, 68- Mem: Am Chem Soc; Sigma Xi. Res: Boranes; carboranes; absolute intensities in infrared spectroscopy; environmental mercury. Mailing Add: 2991 Fireside Dr San Jose CA 95128

VAN ALTEN, PIERSON JAY, b Grand Rapids, Mich, Feb 21, 28; m 53; c 2. EMBRYOLOGY, IMMUNOLOGY. Educ: Calvin Col, AB, 50; Mich State Univ, MS, 55, PhD(zool, physiol), 58. Prof Exp: From asst prof to assoc prof, 60-73, PROF ANAT, UNIV ILL MED CTR, 73- Concurrent Pos: NIH fel exp embryol & immunobiol, Univ Calif, Los Angeles, 58-60; Am Cancer Soc scholar, Univ Bern, 73-74; vis assoc prof pediat, Univ Minn, 66-67; guest prof immunobiol, Univ Bern, 73-74. Mem: AAAS; Am Soc Zool; Am Asn Immunol; Am Asn Anat; Soc Develop Biol. Res: Anatomy; zoology; experimental embryology and immunobiology; development of the digestive tract; graft-versus-host disease and immunological competence of chicken embryo; development of brain antigens in hamsters; culture of lymphocytes and human myeloma cells. Mailing Add: Dept of Anat Univ of Ill at the Med Ctr Chicago IL 60680

VAN ALTENA, WILLIAM F, b Hayward, Calif, Aug 15, 39; m 61; c 2. ASTRONOMY. Educ: Univ Calif, Berkeley, BA, 62, PhD(astron), 66. Prof Exp: From asst prof to assoc prof astron, Yerkes Observ, Univ Chicago, 66-74, dir observ, 72-74; PROF ASTRON, YALE UNIV, 74-, CHMN DEPT, 75- Concurrent Pos: Mem comt photographic plates & films, Am Nat Standards Inst. Mem: Am Astron Soc; Int Astron Union. Res: Trigonometric parallaxes and proper motions. Mailing Add: Yale Univ Observ PO Box 2023 Yale Sta New Haven CT 06437

VANAMAN, SHERMAN BENTON, b Lexington, Ky, July 25, 28; m 55; c 2. MATHEMATICS. Educ: Univ Louisville, BA, 49; Univ Ky, MS, 51; Univ Md, PhD(math educ), 67. Prof Exp: Instr math, Univ Ky, 51-55; assoc prof & actg head dept, 56-66, PROF MATH & CHMN DEPT, CARSON-NEWMAN COL, 66- Mem: Math Asn Am. Res: Learning theory, especially mathematics-education. Mailing Add: Dept of Math Carson-Newman Col Jefferson City TN 37760

VANAMAN, THOMAS CLARK, b Louisville, Ky, Aug 12, 41; m 62; c 1. BIOCHEMISTRY, MICROBIOLOGY. Educ: Univ Ky, BS, 64; Duke Univ, PhD(biochem), 68. Prof Exp: Asst prof, 70-73, ASSOC PROF MICROBIOL & IMMUNOL, MED CTR, DUKE UNIV, 73- Concurrent Pos: Am Chem Soc fel, Med Ctr, Stanford Univ, 69-70, NSF res grant, 71-73 & NIH res grant, 71-77. Mem: Am Soc Biol Chemists; Am Soc Microbiologists. Res: Study of the structure, function and evolution of proteins; structure and function of nervous system specific proteins; viral proteins and their roles in viral infection, replication and transformation. Mailing Add: Dept of Microbiol & Immunol Duke Univ Med Ctr Durham NC 27710

VANAMBURG, GERALD LEROY, b Hunter, Kans, Dec 17, 41; m 63; c 2. PLANT ECOLOGY, BOTANY. Educ: Ft Hays Kans State Col, BS, 64, MS, 65; Tex A&M Univ, PhD(plant ecol), 69. Prof Exp: ASST PROF BIOL, CONCORDIA COL, MOORHEAD, MINN, 69- Concurrent Pos: Expert, Int Atomic Energy Agency, 70-71. Mem: Ecol Soc Am; Am Inst Biol Sci; Soc Range Mgt. Res: Soil-vegetation relationships; biogeochemical cycling. Mailing Add: Dept of Biol Concordia Col Moorhead MN 56560

VANAMEE, PARKER, b Portland, Maine, Aug 9, 19; m 53; c 6. PHYSIOLOGY. Educ: Yale Univ, BS, 42; Cornell Univ, MD, 45. Prof Exp: Intern, RI Hosp, Providence, 45-46; res fel, Sloan-Kettering Inst Cancer Res, 51-54, res assoc, 55-57, assoc, 58-60; asst prof, 56-61, ASSOC PROF MED, MED COL, CORNELL UNIV, 61-; ASSOC MEM, SLOAN-KETTERING INST CANCER RES, 60- Concurrent Pos: From asst resident to chief resident med, Mem Hosp Cancer & Allied Dis, 51-54, clin asst, 54-57, from asst attend physician to assoc attend physician, 57-69, attend physician, 69-, chief clin physiol & renal serv, Dept Med, 70-; asst vis physician, James Ewing Hosp, New York, 57-61, assoc vis physician, 61-; consult, Urol Dept, USPHS Hosp, Staten Island, NY. Mem: AAAS; fel Am Col Physicians; Am Fedn Clin Res; AMA; Am Soc Clin Nutrit. Res: Medicine; clinical physiology. Mailing Add: Mem Sloan-Kettering Cancer Ctr 1275 York Ave New York NY 10021

VAN ANDEL, TJEERD HENDRIK, b Rotterdam, Netherlands, Feb 15, 23; US citizen; m 62; c 6. MARINE GEOLOGY. Educ: State Univ Groningen, BSc, 46, MSc, 48, PhD(geol), 50. Prof Exp: Asst prof geol, State Agr Univ, Wageningen, 48-50; sedimentologist, Royal Dutch Shell Res Lab, 50-53; sr sedimentologist, Cia Shell de Venezuela, 53-56; assoc res geologist, Scirpps Inst Oceanog, Univ Calif, 57-64, res geologist, 64-68, lectr geol, 57-68; PROF GEOL, SCH OCEANOG, ORE STATE UNIV, 68- Concurrent Pos: Vis prof, Univ Calif, Berkeley, 63; sr fel, Woods Hole Oceanog Inst, 63; sci adv, Deep Sea Drilling Proj, 64-68; res assoc geol, Scripps Inst Oceanog, Univ Calif, 68-72; mem geodynamics comt, Nat Acad Sci, 70-75; managing consult, Int Ocean Explor, NSF, 71-72; vis prof geophys, Stanford Univ, 74-75; group chmn, Sci Comt Ocean Res, UNESCO, 75- Mem: AAAS; Am Asn Petrol Geol; Soc Econ Paleont & Mineral; Geol Soc Am; Am Geophys Union. Res: Recent sediments of continents and oceans; origin and nature of the continental shelf; geology and geophysics of mid-ocean ridges; paleoceanography. Mailing Add: Sch Oceanog Ore State Univ Corvallis OR 97331

VAN ANTWERP, WALTER ROBERT, b Franklin, Ind, Aug 16, 25; m 46; c 2. SOLID STATE PHYSICS. Educ: Ind Univ, AB, 49; Univ Md, MS, 58. Prof Exp: Res physicist, Chem Res & Develop Lab, US Dept Army, 50-58 & Nuclear Defense Lab, Edgewood Arsenal, 58-60, chief solid state physics br, 60-64, chief nuclear physics div, 64-69; CHIEF EXP PHYSICS BR, BALLISTICS RES LABS, ABERDEEN PROVING GROUND, 69- Mem: AAAS; Am Phys Soc; Electron Micros Soc Am; Soc Appl Spectros. Res: Solid state and nuclear physics; radiation damage and detection; gamma-ray spectroscopy; thin films. Mailing Add: 110 Woodland Dr Bel Air MD 21014

VAN ARMAN, CLARENCE GORDON, b Detroit, Mich, Dec 29, 17; m 43, 69; c 3. PHARMACOLOGY. Educ: Univ Chicago, SB, 39; Northwestern Univ, MS, 48, PhD(pharmacol), 49. Prof Exp: Chemist, Price Extract Co, 39-40; chemist, G D Searle & Co, Ill, 40-41, pharmacologist, 50-60; res assoc pharmacol, Med Sch Northwestern Univ, 49-50; chief pharmacologist, Chem Therapeut Res Labs, Miles Labs, Inc, 60-61; mgr pharmacol eval sect, Wyeth Labs, Inc, 61-63; dir pharmacol res dept, Chas Pfizer & Co, 63-65; sr res fel, 65-75; SR INVESTR, MERCK INST THERAPEUT RES, 75- Concurrent Pos: Lectr, Med Sch, Northwestern Univ, 54-61. Mem: AAAS; Am Soc Pharmacol & Exp Therapeut; Soc Exp Biol & Med; Am Soc Clin Pharmacol & Therapeut; Am Rheumatism Asn. Res: Inflammation; polypeptides; analgesics; diuretics; cardiac drugs; glomerulomephritis. Mailing Add: Merck Inst for Therapeut Res West Point PA 19486

VAN ARSDEL, PAUL PARR, JR, b Indianapolis, Ind, Nov 4, 26; m 50; c 2. MEDICINE. Educ: Yale Univ, BS, 48; Columbia Univ, MD, 51. Prof Exp: Intern med, Presby Hosp, New York, 51-52, asst resident, 52-53; asst, Sch Med, Univ Wash, 53-55; asst, Presby Hosp, New York, 55-56; from instr to assoc prof, 56-69, PROF MED, SCH MED, UNIV WASH, 69-, HEAD DIV ALLERGY, 56- Concurrent Pos: Res fel med, Sch Med, Univ Wash, 53-55, Mass Mem Hosp & Sch Med, Boston Univ, 55 & Presby Hosp, New York, 55-56; mem, Growth & Develop Training Comt, Nat Inst Child Health & Human Develop, 70-74; consult, Vet Admin Hosp, USPHS Hosp, King County Hosp, Children's Hosp & Univ Hosp. Mem: AAAS; fel Am Col Physicians; Am Asn Immunol; Asn Am Med Cols; Am Acad Allergy (secy, 64-68, pres, 71-72). Res: Hypersensitivity; human immunology; histamine release and metabolism; drug sensitivity; autoantibodies. Mailing Add: BB 1333 Rm 13 Dept of Med Univ Hosp Seattle WA 98195

VAN ARSDEL, WILLIAM CAMPBELL, III, b Indianapolis, Ind, June 27, 20; div. PHARMACOLOGY. Educ: Ore State Col, BS, 49, MS, 51, PhD(physiol, zool), 59; Univ Ore, MS, 54. Prof Exp: Mem staff, Prod Control Lab, US Rubber Co, Ind, 41-45; res asst animal husb, Ore State Univ, 54-59, jr animal physiologist, 59-60, asst in animal physiol, 60-63; PHARMACOLOGIST, BUR DRUGS, US FOOD & DRUG ADMIN, 63- Mem: AAAS; NY Acad Sci; Sigma Xi; Soc Exp Biol & Med. Res: Toxicology; teratology; electrocardiology; physiology; marine biology; anthropometry. Mailing Add: Cardio-Renal Drug Prod Bur Drugs US Food & Drug Admin Rockville MD 20852

VAN ARTSDALEN, ERVIN ROBERT, b Doylestown, Pa, Nov 13, 13; m 45. PHYSICAL CHEMISTRY, INORGANIC CHEMISTRY. Educ: Lafayette Col, BS, 35; Harvard Univ, AB, 39, PhD(phys chem), 41. Prof Exp: Asst chem, Harvard Univ, 36-40; from instr to asst prof, Lafayette Col, 41-45; res scientist, Los Alamos Sci Lab, 45-46; asst prof chem, Cornell Univ, 46-51; prin chemist, Oak Ridge Nat Lab, 51-56; asst dir res, Parma Res Ctr, Union Carbide Corp, 56-63; John W Mallet Prof chem & chmn dept, Univ Va, 63-68; chmn dept, 68-72, PROF CHEM, UNIV ALA, 68- Concurrent Pos: Res assoc, Nat Defense Res Comt, Sch Med, Johns Hopkins Univ, 43-44; res assoc, Carnegie Inst Technol, 45; lectr, Western Reserve Univ, 59; mem coun, Oak Ridge Assoc Univs, 63-69, mem bd dir, 69-75, staff mem, Inst Energy Anal, 75-76. Mem: AAAS; Am Chem Soc; Am Phys Soc; fel Am Inst Chemists. Res: Energy analysis; adsorption indicators; photochemistry; bond strengths; reaction kinetics; thermodynamics and structure of inorganic systems; high temperature chemistry; fused salts; atomic energy; Mössbauer spectroscopy; radiochemistry; nuclear chemistry. Mailing Add: Dept of Chem Univ of Ala University AL 35486

VANAS, DON WOODRUFF, b Rochester, NY, June 11, 20; m 45. PHYSICAL CHEMISTRY. Educ: State Univ NY, BA, 43; Univ Rochester, PhD(phys chem), 46. Prof Exp: Res chemist rayon dept, E I du Pont de Nemours & Co, 46-52; res chemist biochem dept, Bristol Labs, Inc, 52-53; sr chemist, 53-55, res assoc, 55-71, ASST HEAD CHEM DIV, EASTMAN KODAK CO, 71- Mem: Am Chem Soc. Res: Physics and physical chemistry of polymeric melts, solutions and solids. Mailing Add: Research Labs Eastman Kodak Co 1669Lake Ave Rochester NY 14650

VAN ASDALL, WILLARD, b Knox, Ind, Apr 29, 34. PLANT ECOLOGY. Educ: Valparaiso Univ, AB, 56; Purdue Univ, SM, 58; Univ Chicago, PhD(bot), 61. Prof Exp: Instr biol, Knox Col, Ill, 61; asst prof, Duquesne Univ, 62-63; asst prof bot, 62-65, ASSOC PROF BOT, UNIV ARIZ, 65- Mem: Ecol Soc Am; Bot Soc Am. Res: Physiological ecology of desert plant species, especially winter-spring desert ephemerals. Mailing Add: Dept of Biol Sci Univ of Ariz Tucson AZ 85721

VAN ATTA, FLOYD AUGUST, b Vancouver, Wash, Sept 30, 06; m; c 3. INDUSTRIAL HYGIENE. Educ: Univ Ore, BS, 28, MS, 29; Northwestern Univ, PhD(phys chem), 36. Prof Exp: Instr, Armour Inst Technol, 34-39; state indust hygienist, Ill, 40-43; hygienist, Nat Safety Coun, 44-58; hygienist, United Automobile Workers-Am Fedn Labor-Cong Indust Orgns, 59-63; mem staff, US Dept Labor, Occup Safety & Health Admin, 63-75; PROF INDUST HYGIENE, QUINNIPIAC COL, 76- Concurrent Pos: Vis lectr, Sch Pub Health, Univ Mich, 58- Mem: AAAS; Am Chem Soc; Am Indust Hyg Asn; Am Acad Indust Hyg; Am Conf Govt Indust Hygienists. Res: Temperature coefficient of the dielectric constant of water. Mailing Add: Quinnipiac Col Hamden CT 06518

VANATTA, JOHN CROTHERS, III, b Lafayette, Ind, Apr 22, 19; m 44; c 2. PHYSIOLOGY. Educ: Ind Univ, AB, 41, MD, 44; Am Bd Internal Med, dipl, 53. Prof Exp: Intern, Wayne County Gen Hosp, 44-45, asst resident med, 46-47; instr physiol & pharmacol, 49-50, from asst prof to assoc prof physiol, 50-57, PROF PHYSIOL, UNIV TEX HEALTH SCI CTR DALLAS, 57- Concurrent Pos: Fel physiol & pharmacol, Univ Tex Southwestern Med Sch Dallas, 47-48 & fel exp med, 48-49; consult, Div Nuclear Educ & Training, USAEC, 64-67; adj prof physiol, Inst Technol, Southern Methodist Univ, 69- Mem: Am Physiol Soc; Soc Exp Biol & Med; AMA. Res: Sodium metabolism; transport functions of urinary bladder of toad. Mailing Add: Dept of Physiol Univ of Tex Health Sci Ctr Dallas TX 75235

VAN ATTA, JOHN R, b Boston, Mass, May 16, 39; m 60; c 2. BIOPHYSICS. Educ: Univ Calif, Santa Barbara, BA, 62; Univ Mich, PhD(biophys), 68. Prof Exp: Res scientist, 68-72, dir basic res, 72-74, DIR PROD DVELOP, CARNATION RES LABS, 74- Mem: Inst Food Technologists; Am Chem Soc. Res: Protein chemistry, especially protein structure. Mailing Add: 601 Toyopa Dr Pacific Palisades CA 90272

VAN ATTA, ROBERT ERNEST, b Ada, Ohio, Feb 29, 24; m 46; c 2. ANALYTICAL CHEMISTRY. Educ: Ohio Northern Univ, BA, 48; Purdue Univ, MS, 50; Pa State Col, PhD(chem), 52. Prof Exp: Instr chem, Ohio Northern Univ, 47-48; instr, Pa State Col, 51-52; assoc prof & head dept, Ohio Northern Univ, 52-54, dir, Div Natural Sci, 53-54; from asst prof to prof chem, Southern Ill Univ, 54-69; PROF CHEM & HEAD DEPT, BALL STATE UNIV, 69- Mem: Fel Am Inst Chem; Am Chem Soc. Res: Polarography of organic compounds; electrochemical measurements; flame photometry; kinetics; gas chromatography; organic reaction mechanisms; infrared spectroscopy; natural waters analysis; inexpensive instrumentation. Mailing Add: Dept of Chem Ball State Univ Muncie IL 47306

VAN AUSDAL, RAY GARRISON, b Cincinnati, Ohio, Sept 16, 43. ATOMIC PHYSICS, MUSICAL ACOUSTICS. Educ: Miami Univ, AB, 64, MA, 66; Univ Mich, Ann Arbor, PhD(physics), 72. Prof Exp: Asst prof physics, Northern Mich Univ, 72-73 & Kalamazoo Col, 73-74; ASST PROF PHYSICS, UNIV PITTSBURGH, JOHNSTOWN, 74- Res: Development of effective teaching methods and materials in physics. Mailing Add: Dept of Physics Univ of Pittsburgh Johnstown PA 15905

VAN BAVEL, CORNELIUS H M, b Breda, Neth, Sept 15, 21; nat US; m 47; c 8. AGRONOMY, BIOLOGY. Educ: State Agr Univ, Wageningen, MS, 45; Iowa State Univ, MSc, 46, PhD(soil physics), 49. Prof Exp: Instr physics, State Agr Univ, Wageningen, 41-43; soil surveyor, Neth Govt, 44; consult, Bolivia, SAm, 47; res assoc, Agr Exp Sta, Iowa State Univ, 48-50; assoc prof agron, NC State Univ, 50-54; soil scientist, Agr Res Serv, USDA, 54-61, chief physicist, Water Conserv Lab, 61-67; PROF AGRON & BIOL, TEX A&M UNIV, 67- Concurrent Pos: Consult, Int Atomic Energy Agency, 62 & 64. Honors & Awards: Superior Serv Award, USDA, 63; Horton Award, Am Geophys Union, 67. Mem: Soil Sci Soc Am; fel Am Soc Agron; Am Geophys Union; Int Soil Sci Soc (Am secy, 59-60); Int Asn Sci Hydrol. Res: Hydraulics of soils; gaseous diffusion in soils; micrometeorology; nuclear radiation methods in soil-plant systems; water resources; arid land problems; evaporation; environmental biology; irrigation; plant water balance. Mailing Add: Dept of Soil & Crop Sci 427 Biol Sci Tex A&M Univ College Station TX 77843

VAN BEAUMONT, KAREL WILLIAM, b Amsterdam, Netherlands, Sept 26, 30; US citizen; m 59; c 2. PHYSIOLOGY. Educ: Acad Phys Educ, The Hague, BS, 55; Cath Univ Louvain, MS, 57; Univ Ill, MS, 62; Ind Univ, PhD(physiol), 65. Prof Exp: Instr physiol, Univ Ind, 64-66; res physiologist, Miami Valley Labs, Proctor & Gamble Co, Ohio, 66-68; asst prof, 68-73, ASSOC PROF PHYSIOL, SCH MED, ST LOUIS UNIV, 73- Res: Temperature regulation; neural control systems; high altitude physiology; physiology of exercise; acceleration stress; biometeorology; body fluids and electrolytes; hematology. Mailing Add: Dept of Physiol St Louis Univ Sch of Med St Louis MO 63104

VAN BECKUM, WILLIAM GEORGE, b Wrightstown, Wis, Feb 22, 14; m 41; c 6. CHEMISTRY. Educ: St Norbert Col, BA, 36; Univ Wis, MA, 37. Prof Exp: Chief chem div, Wood Conversion Co, 39-41; chief chem div, Weyerhaeuser Co, Wash, 41-46, asst mgr develop dept, 46-48, asst & sales mgr tech serv, Spec Prod Div, 48-52; dir res & develop, 52-61, VPRES RES & DEVELOP, PAC LUMBER CO, 61-, SECY, 69- Concurrent Pos: Pres, Ideal Brushes, Inc, 73- Mem: Am Chem Soc; Forest Prod Res Soc; Tech Asn Pulp & Paper Indust. Res: Forest products. Mailing Add: Pac Lumber Co 1111 Columbus Ave San Francisco CA 94133

VAN BELLE, GERALD, b Enschede, Netherlands, July 23, 36; Can citizen; m 63; c 2. STATISTICS. Educ: Univ Toronto, BA, 62, MA, 64, PhD(math), 67. Prof Exp: Statistician, Connaught Med Res Labs, Univ Toronto, 57-62; from asst prof to assoc prof statist, Fla State Univ, 67-74, dir statist consult ctr, 71-74; vis assoc prof biostatist, 74-75, ASSOC PROF BIOSTATIST, UNIV WASH, 75- Mem: Am Statist Asn; Biomet Soc. Res: Application of statistics to biological and health-related problems. Mailing Add: Dept of Biostatist Univ of Wash Seattle WA 98195

VAN BERGEN, FREDERICK HALL, b Minneapolis, Minn, Sept 21, 14; c 4. ANESTHESIOLOGY. Educ: Univ Minn, MB, 41, MD, 42, MS, 52. Prof Exp: From instr to assoc prof, 48-57, assoc dir, 53-54, actg dir, 54-55, PROF ANESTHESIOL, MED SCH, UNIV MINN, MINNEAPOLIS, 57-, HEAD DEPT, 55- Mem: Am Soc Anesthesiol; Int Anesthesia Res Soc; AMA; Acad Anesthesiol. Res: Development and testing of respirators and respiratory assistors; evaluation of effects of respiratory patterns upon cardiovascular function; evaluation of pulmonary compliance under conditions of anesthesia; gas mass spectrometer. Mailing Add: Dept of Anesthesiol Univ of Minn Med Sch Minneapolis MN 55455

VAN BIESBROECK, GEORGE, astronomy, deceased

VAN BREEMEN, VERNE LEROY, b Zion, Ill, Oct 19, 21; m 49; c 3. HISTOLOGY, CYTOLOGY. Educ: Univ Iowa, BS, 46, MS, 48, PhD(cytol), 51. Prof Exp: Asst cytol, Univ Iowa, 47-51; res assoc anat, Univ Calif, Los Angeles, 51-53; instr sch med, Univ Colo, 53-54, asst prof, 54-59; dir, Mercy Inst Bio-Med Res, 59-64; res biol, US Air Force Sch Aerospace Med, 64-66; head dept biol, 66-74, PROF BIOL, SALISBURY STATE COL, 66- Concurrent Pos: Lederle Med fac award, 54-56; pres-elect, Wicomico Environ Trust, 72- Mem: Am Asn Anatomists; Am Soc Zoologists; Electron Micros Soc Am; Soc Exp Biol & Med; Biol Stain Comn. Res: Electron microscopy. Mailing Add: Dept of Biol Salisbury State Col Salisbury MD 21801

VAN BRUGGEN, JOHN TIMOTHY, b Chicago, Ill, Aug 12, 13; m 36; c 3. BIOCHEMISTRY. Educ: Linfield Col, BA, 37; Univ Ore, MA, 39; St Louis Univ, PhD(biochem), 44. Prof Exp: Instr biochem, Sch Med & biochemist, St Mary's Hosp, St Louis Univ, 43-45; res assoc, 45-47, from asst prof to assoc prof, 47-62, PROF BIOCHEM, MED SCH, UNIV ORE HEALTH SCI CTR, 62-, DIR TEACHING LABS & COORD FIRST YR MED EDUC, 73- Concurrent Pos: USPHS sr fel, Univ Copenhagen, 59-60; biochemist-biophysicist, Med Br, Div Biol & Med, USAEC, 67-68. Mem: AAAS; Am Chem Soc; Am Soc Biol Chem. Res: Metabolism of estrogens; isolation of penicillin B; clinical procedures; biochemistry of Vitamin A; lipid metabolism; metabolism of the diabetic; membrane transport and metabolism. Mailing Add: Dept of Biochem Univ of Ore Med Sch Portland OR 97201

VAN BRUGGEN, THEODORE, b Hawarden, Iowa, Jan 26, 26; m 48; c 2. BOTANY. Educ: Buena Vista Col, BS, 48; Univ SDak, MA, 50; Univ Iowa, PhD(bot), 58. Prof Exp: Instr biol, Northwestern Col, 50-55; from asst prof to prof bot & chmn dept, 58-59, ASSOC DEAN COL ARTS & SCI, UNIV SDAK, VERMILLION, 69- Concurrent Pos: Asst prog dir, NSF, 65-66. Mem: AAAS; Bot Soc Am; Mycol Soc Am; Am Soc Plant Taxon. Res: Plant systematics; flora of South Dakota, especially identification of vascular plants. Mailing Add: Col Arts & Sci Univ of SDak Vermillion SD 57069

VAN BRUNT, RICHARD JOSEPH, b Jersey City, NJ, May 11, 39; m 72; c 1. ATOMIC PHYSICS, MOLECULAR PHYSICS. Educ: Univ Fla, BS, 61, MS, 64; Univ Colo, Boulder, PhD(physics), 69. Prof Exp: Res asst atomic & molecular physics, Univ Fla, 64 & Joint Inst Lab Astrophys, Univ Colo, Boulder, 64-69; res assoc, 69-71, ASST PROF ATOMIC & MOLECULAR PHYSICS, UNIV VA, 71- Concurrent Pos: Univ grant, Univ Va, 72-73; vis mem, Joint Inst Lab Astrophys, Univ Colo, Boulder, 75-77. Mem: AAAS; Am Phys Soc; Am Asn Physics Teachers. Res: Experimental and theoretical studies of ionization and excitation processes produced by electron impact and x-ray absorption in simple molecular and atomic gases. Mailing Add: Dept of Physics Univ of Va Charlottesville VA 22901

VAN BUIJTENEN, JOHANNES PETRUS, b Netherlands, May 8, 28; nat US; m 63; c 3. FOREST GENETICS. Educ: State Agr Univ, Wageningen, BS, 52; Univ Calif, Berkeley, MS, 55; Tex A&M Univ, PhD(genetics), 56. Prof Exp: Forest geneticist, Inst Paper Chem, 56-60, Tex Forest Serv, 60-66 & Northeastern Forest Exp Sta, NH, 66-68; assoc prof, 68-71, PROF FOREST GENETICS, TEX A&M UNIV, 71-, PRIN GENETICIST, TEX FOREST SERV, TEX A&M UNIV, 68- Concurrent Pos: NSF travel grant, 63; consult, Tex Forest Serv, 66-68. Mem: AAAS; Tech Asn Pulp & Paper Indust; Soc Am Foresters. Res: Genetic improvement of forest trees for drought resistance, wood quality, insect resistance; physiology of forest trees as related to forest tree improvement. Mailing Add: Forest Genetics Lab Tex Forest Serv College Station TX 77843

VAN BUREN, ARNIE LEE, b Reynoldsburg, Ohio, Nov 28, 39; m 69. ACOUSTICS. Educ: Birmingham-Southern Col, BS, 61; Univ Tenn, PhD(physics), 67. Prof Exp: Res assoc acoustics, Univ Tenn, 67-68; RES PHYSICIST, NAVAL RES LAB, 68- Mem: Acoustical Soc Am. Res: Acoustic radiation; transducers; nonlinear acoustics. Mailing Add: Code 8150 Naval Res Lab Washington DC 20375

VAN BUREN, JEROME PAUL, b Brooklyn, NY, Oct 17, 26; m 53; c 3. BIOCHEMISTRY. Educ: Cornell Univ, BS, 50, MNS, 51, PhD, 54. Prof Exp: Proj leader cereal chem, Gen Mills, Inc, 54-57; from asst prof to assoc prof biochem, 57-

69, PROF BIOCHEM, CORNELL UNIV, 69- Concurrent Pos: Consult food & agr, USPHS, 62-65; vis prof, Swiss Fed Inst Technol, 64-65; vis prof, Agr Univ, Holland, 71-72. Mem: Am Chem Soc; Asn Cereal Chemists; Inst Food Technologists. Res: Protein interactions in food; anthocyanins and polyphenols; wine chemistry; effects of calcium on vegetable texture; pectic substances; food color and pigments. Mailing Add: Dept of Food Sci & Technol Cornell Univ Geneva NY 14456

VAN BUREN, JOHN MILLER, b Ridgewood, NJ, Apr 9, 23; m 46; c 3. NEUROPHYSIOLOGY, NEUROANATOMY. Educ: Dartmouth Col, AB, 43; Columbia Univ, MD, 47; McGill Univ, MSc, 50; George Washington Univ, PhD(neuroanat), 61. Prof Exp: Intern, Mary Hitchcock Hosp, Hanover, NH, 47-48, resident orthop surg, 48-49; asst resident neurosurg, Montreal Neurol Inst, 51-52; clin clerk neurol, Nat Hosp, Queen Sq, London, 52-53; clin assoc surg neurol, 54-55, assoc neurosurgeon, 55-70, actg chief surg neurol, 70-72, CHIEF SURG NEUROL, NAT INST NEUROL DIS & STROKE, 72-, ACTG CLIN DIR, 70- Concurrent Pos: Fel neurophysiol, McGill Univ, 49-50; sr fel neuropath, Montreal Neurol Inst, 50-51 & fel electroencephalog, 53; sr fel neurosurg, Lahey Clin, Boston, Mass, 53-54. Mem: Harvey Cushing Soc; Cong Neurol Surg; Am Asn Anat; Am Acad Neurol. Res: Anatomy of visual system and pituitary relationships to hormonal status; anatomical studies in thalamus and hypothalamus; physiological observations concerning autonomic changes in epilepsy; studies of the arrest response and speech in man. Mailing Add: Nat Inst of Neurol Dis & Stroke Rm 4N-236 Bldg 10 Bethesda MD 20014

VAN BURKALOW, ANASTASIA, b Buchanan, NY, Mar 16, 11. GEOMORPHOLOGY. Educ: Hunter Col, AB, 31; Columbia Univ, MA, 33, PhD(geomorphol), 44. Prof Exp: Asst geomorphol, Columbia Univ, 34-37; tutor geol & geog, Hunter Col, 38-41, instr, 41-45; asst med geog, Am Geog Soc, 45-48; from instr to assoc prof geol & geog, 48-61, chmn dept, 61-73, PROF GEOL & GEOG, HUNTER COL, 61- Concurrent Pos: Ed, Nat Asn Geol Teachers Jour, 54-56. Mem: Fel AAAS; fel Am Geog Soc; fel Geol Soc Am; Nat Asn Geol Teachers; fel NY Acad Sci. Res: Medical geography; fluorine in United States water supplies; angle of repose and sliding friction; statistics of petroleum utilization; medical geography of swimmers' itch; geography of New York City and metropolitan area. Mailing Add: Dept of Geol & Geog Hunter Col New York NY 10021

VAN BUSKIRK, FREDERICK WILLIAM, b Pottstown, Pa, Sept 17, 07; m 37; c 4. RADIOLOGY. Educ: Univ Pa, AB, 30, MD, 33. Prof Exp: Radiologist, Pottstown Hosp, Pa, 37-40; from asst prof to assoc prof, 46-70, PROF CLIN RADIOL, COL MED, UNIV VT, 70- Concurrent Pos: Attend radiologist, De Goesbriand Mem Hosp, Burlington, 46-69 & Fanny Allen Hosp, Winooski, 46-; assoc radiologist, Mary Fletcher Hosp, Burlington, 46-69; attend radiologist, Med Ctr Hosp, Vt, 69- Mem: AAAS; Am Soc Nuclear Med; Radiol Soc NAm; fel Am Col Radiol. Res: Diagnostic radiology; radiologic aspects of urticaria pigmentosa; cross indexing roentgen diagnoses; inquiry into familial aspects of infantile cortical hypertosis; nuclear medicine. Mailing Add: Dept of Radiol Univ of Vt Col of Med Burlington VT 05401

VAN CAESEELE, LAWRENCE ALOYSIUS, microbiology, see 12th edition

VAN CAMP, HARLAN LARUE, b Detroit, Mich, Dec 28, 43. BIOPHYSICS. Educ: Wayne State Univ, BS, 66, PhD(physics), 74. Prof Exp: RES ASSOC BIOPHYSICS, RES FOUND STATE NY, 75- Mem: Am Phys Soc; Am Asn Physics Teachers. Res: Electronic fine structure of heme model compounds and related metalloporphyrin proteins via electron paramagnetic and electron nuclear double resonance. Mailing Add: Dept of Physics State Univ NY Albany NY 12222

VAN CAMPEN, DARRELL R, b Two Buttes, Colo, July 15, 35; m 58; c 2. NUTRITION, BIOCHEMISTRY. Educ: Colo State Univ, BS, 57; NC State Univ, MS, 60, PhD(nutrit), 62. Prof Exp: NIH fel biochem, Cornell Univ, 62-63; RES CHEMIST, US PLANT, SOIL & NUTRIT LAB, USDA, 63-; ASST PROF ANIMAL NUTRIT, CORNELL UNIV, 72- Mem: AAAS; Am Soc Animal Sci; Am Chem Soc. Res: Mineral metabolism; absorption and utilization of trace minerals. Mailing Add: Plant Soil & Nutrit Lab USDA Tower Rd Ithaca NY 14850

VANCE, BENJAMIN DWAIN, b Cave City, Ark, May 7, 32; m 52; c 4. PLANT PHYSIOLOGY. Educ: Tex Tech Col, BS, 58; Univ Mo, AM, 59, PhD(bot), 62. Prof Exp: Asst prof biol, Tex Tech Col, 62-63; asst prof, 63-70, ASSOC PROF BIOL, N TEX STATE UNIV, 70- Concurrent Pos: Res grants, NTex Fac Res-Tex Col & Coord Bd, 66-67 & Nat Commun Dis Ctr, 66-68. Mem: AAAS; Bot Soc Am; Am Soc Plant Physiol; Phycol Soc Am. Res: Physiology of blue-green algae; phytohormones; action of phytochrome. Mailing Add: Dept of Biol NTex State Univ Denton TX 76203

VANCE, DENNIS E, biochemistry, see 12th edition

VANCE, DENNIS WILLIAM, b Quincy, Ill, Nov 20, 38; m 61; c 1. ELECTROOPTICS. Educ: St Lawrence Univ, BS, 60; Univ Fla, MS, 62, PhD(physics), 65. Prof Exp: SCIENTIST, XEROX RES LABS, 65-, MGR DISPLAY TECHNOL AREA, 75- Mem: Am Phys Soc. Res: Surface physics; display systems engineering and technology. Mailing Add: Xerox Corp Palo Alto Res Ctr 3333 Coyote Hill Rd Palo Alto CA 94304

VANCE, ELBRIDGE PUTNAM, b Cincinnati, Ohio, Feb 7, 15; m 39; c 4. MATHEMATICS. Educ: Col Wooster, AB, 36; Univ Mich, MA, 37, PhD(math), 39. Prof Exp: Dir statist lab, Univ Mich, 38; from instr to asst prof math, Univ Nev, 39-43; lectr, 43-46, from asst prof to assoc prof, 46-54, actg dean faculty, 66-70, PROF MATH, OBERLIN COL, 54-, CHMN DEPT, 48- Concurrent Pos: NSF fel, Stanford Univ, 60-61; Columbia Univ & US AID consult, Ranchi Univ, India, 65; math assoc, Univ Auckland, 67; ed, Am Math Monthly. Mem: AAAS; assoc Am Math Soc; assoc Math Asn Am. Res: Continuous transformations; foundations of mathematics; topology. Mailing Add: Dept of Math Oberlin Col Oberlin OH 44074

VANCE, HUGH GORDON, b Forest, Ont, Sept 18, 24; m 53; c 1. ANALYTICAL CHEMISTRY, BIOCHEMISTRY. Educ: Univ Western Ont, BSc, 50, PhD(path chem), 56. Prof Exp: Nat Res Coun Can fel, Dept Anat, McGill Univ, 55-57; res assoc biochem, Sinai Hosp of Baltimore, Ind, Md, 57-61, Nat Cancer Inst fel dept med, 60-61; asst prof chem, 61-65, ASSOC PROF CHEM, MORGAN STATE COL, 65- Mem: Am Chem Soc. Res: Investigations of the nature and content of mucopolysaccharides in skin; structural determination of the carbohydrate moiety of glycoporteins. Mailing Add: Morgan State Col Baltimore MD 21212

VANCE, IRVIN ELMER, b Mexico, Mo, Apr 8, 28; m 58; c 3. MATHEMATICS EDUCATION. Educ: Wayne State Univ, BS, 57; Washington Univ, MA, 59; Univ Mich, Ann Arbor, DEduc(math), 67. Prof Exp: From asst prof to assoc prof math, Mich State Univ, 66-71; ASSOC PROF MATH, NMEX STATE UNIV, 71- Concurrent Pos: Consult, Morel Lab Math Proj, 67; asst dir, Grand Rapids Math Lab Proj, 68-69; dir inner city math proj, Mich State Univ, 69-72. Mem: AAAS; Am Math Soc; Math Asn Am. Res: Finite projective planes; inductive learning and teaching of mathematics; laboratory techniques at school level and individualized instruction at the college level. Mailing Add: 207 Walden Hall Dept Math Sci NMex State Univ Las Cruces NM 88001

VANCE, JAMES ELMON, JR, b Natick, Mass, Dec 2, 25; m 54. GEOGRAPHY. Prof Exp: Res geogr cent bus dist study, Off Naval Res, Clark Univ, 52-53; instr geog, Univ Ark, 53-55; asst prof, Univ Wyo, 55-57 & Univ Nebr, 57-58; from asst prof to assoc prof, 58-70, PROF GEOG, UNIV CALIF, BERKELEY, 70- Mem: Asn Am Geogr; Am Geog Soc; Royal Geog Soc; Inst Brit Geog. Res: Urban and transportation geography. Mailing Add: Dept of Geog Univ of Calif Berkeley CA 94720

VANCE, JOHN EDWARD, analytical chemistry, physical chemistry, deceased

VANCE, JOSEPH ALAN, b Aberdeen, Wash, Mar 15, 30; m 49; c 3. GEOLOGY. Educ: Univ Wash, BSc, 51, PhD, 57. Prof Exp: Asst prof geol, 57-68, ASSOC PROF GEOL, UNIV WASH, 68- Mem: Geol Soc Am; Mineral Soc Am. Res: Igneous and metamorphic petrology; structure and stratigraphy; geology of the Pacific Northwest. Mailing Add: Dept of Geol Univ of Wash Seattle WA 98195

VANCE, JOSEPH FRANCIS, b Kansas City, Mo, July 24, 37. MATHEMATICS, STATISTICS. Educ: Southwest Tex State Col, BS, 59; Univ Tex, Austin, MA, 62, PhD(math), 67. Prof Exp: Teacher high sch, Tex, 60-61; spec instr math, Univ Tex, Austin, 66-67, asst prof, 67-68; from asst prof to assoc prof, 68-74, PROF MATH, ST MARY'S UNIV, TEX, 74- Mem: Am Math Soc. Res: Analysis; specialty, integration theory. Mailing Add: St Mary's Univ San Antonio TX 78284

VANCE, MILES ELLIOTT, b Findlay, Ohio, Jan 2, 32; m 55; c 3. MEDICAL TECHNOLOGY. Educ: Bowling Green State Univ, BA, 53; Ohio State Univ, PhD(physics), 62. Prof Exp: Instr physics, Ohio State Univ, 61-62; res physicist, Res & Develop Lab, NY, 62-68, sr res physicist, Electronics Res Lab, NC, 68-73, SR RES PHYSICIST, BIOMED TECH CTR, CORNING GLASS WORKS, 73- Mem: Optical Soc Am; Inst Elec & Electronics Engrs. Res: Cytopathology; microscopy; clinical instruments; applied optics. Mailing Add: Biomed Tech Ctr Corning Glass Works 3800 Electronics Dr Raleigh NC 27604

VANCE, RICHARD RAMEY, zoology, see 12th edition

VANCE, ROBERT FLOYD, b Columbus, Ohio, May 12, 26; m 53; c 3. INORGANIC CHEMISTRY. Educ: Otterbein Col, BS, 49; Univ Ill, MS, 50, PhD(chem), 52. Prof Exp: Res chemist, Battelle Mem Inst, 52-58; develop supvr, Girdler Catalysts Div, Chemetron Corp, 58-60; DEVELOP SUPVR, GEN ELEC CO, 60- Mem: Am Chem Soc. Res: Detergency; surface studies; water pollution; air pollution. Mailing Add: 7502 Tudor Ct Louisville KY 40222

VANCE, VELMA JOYCE, b Wilder, Idaho, May 13, 29. VERTEBRATE ZOOLOGY. Educ: Col Idaho, BS, 51; Univ Ariz, MS, 53; Univ Calif, Los Angeles, PhD(zool), 59. Prof Exp: Asst, Crookham Co, 52-54 & Univ Calif, Los Angeles, 54-58; instr biol, Occidental Col, 59; from asst prof to assoc prof zool, 59-71, PROF ZOOL, CALIF STATE UNIV, LOS ANGELES, 71- Concurrent Pos: Animal behavior; vertebrate biology. Mailing Add: Dept of Biol Calif State Univ Los Angeles CA 90032

VANCKO, ROBERT MICHAEL, b Johnson City, NY, Sept 15, 42; m 66; c 2. MATHEMATICS. Educ: Pa State Univ, BA, 64, MA, 65, PhD(math), 69. Prof Exp: Lectr math, Univ Man, 67-69; ASST PROF MATH, OHIO UNIV, 69- Mem: Am Math Soc. Res: Universal algebra; general algebraic systems; topological algebraic systems. Mailing Add: Dept of Math Ohio Univ Athens OH 45701

VANCLEAVE, ALLAN BISHOP, b Medicine Hat, Alta, Aug 19, 10; m 34; c 4. PHYSICAL CHEMISTRY. Educ: Univ Sask, BSc, 31, MSc, 33; McGill Univ, PhD, 35; Cambridge Univ, PhD, 37. Prof Exp: 1851 Exhib scholar, Cambridge Univ, 35-37; from asst prof to prof chem, Univ Sask, 37-62; chmn div natural sci, 62-69, dir sch grad studies, 65-69, DEAN GRAD STUDIES & RES, UNIV REGINA, 69- Mem: Fel Chem Inst Can; The Chem Soc; Royal Soc Can. Res: Active hydrogen; viscosity of gases; catalysis; accomodation coefficient method of measuring gas adsorption; reactions of cyanogen halides; radiation chemistry; properties of Saskatchewan volcanic ashes; beneficiation of low grade uranium ores; flotation characteristics of minerals; x-ray fluorescence analysis. Mailing Add: Fac of Grad Studies & Res Univ of Regina Regina SK Can

VAN CLEAVE, CHARLES DURWARD, b Holbrook, Nebr, Apr 8, 04; m 33. ANATOMY. Educ: Univ Colo, AB, 25; Univ Chicago, PhD(zool), 28. Prof Exp: Instr anat, Sch Med, Univ Pa, 28-38; instr, Med Col, Cornell Univ, 38-40; from asst prof to assoc prof, 40-61, PROF ANAT, SCH MED, UNIV NC, CHAPEL HILL, 61- Concurrent Pos: Mem staff, Med Res Br, Div Biol & Med, US AEC, 57-59. Mem: Am Asn Anatomists. Res: Regeneration in invertebrates; physiology of development; distribution studies of radioisotopes; late somatic effects of radiation. Mailing Add: Dept of Anat Univ of NC Sch of Med Chapel Hill NC 27515

VAN CLEAVE, HORACE WILLIAM, b Cherryvale, Kans, July 9, 31; m 56; c 2. ENTOMOLOGY. Educ: Tex A&M Univ, BS, 52, MS, 58; Okla State Univ, PhD(entom), 69; Am Registry Cert Entom, cert. Prof Exp: Teacher high sch, Tex, 54-56; surv entomologist, Okla State Univ, 58-61, instr, 62-64; asst prof entom, 64-71, ASSOC PROF ENTOM, TEX A&M UNIV, 71- Mem: Entom Soc Am. Res: Insect pests of pecans; taxonomy of aphids; economic entomology. Mailing Add: Dept of Entom Tex A&M Univ College Station TX 77843

VAN CLEVE, JOHN WOODBRIDGE, b Kansas City, Mo, Nov 22, 14; m 47; c 3. CARBOHYDRATE CHEMISTRY. Educ: Antioch Col, BS, 37, 37; Univ Minn, PhD(biochem), 51. Prof Exp: Jr res chemist, Aluminum Co Am, 43-45, assoc res chemist, 45-48; RES CHEMIST, CEREAL CROPS LAB, NORTHERN REGIONAL RES LAB, USDA, 51- Mem: AAAS; Am Chem Soc. Res: Carbohydrate chemistry. Mailing Add: USDA Northern Regional Res Lab 1815 N University St Peoria IL 61604

VAN CLEVE, RICHARD, b Seattle, Wash, Mar 6, 06; m 30; c 2. FISHERIES. Educ: Univ Wash, BS, 27, PhD(fisheries, zool), 36. Prof Exp: Scientist, Int Fisheries Comn, 26-31; lectr, Univ Wash, 35-41; chief bur marine fisheries, Calif State Div Fish & Game, 41-46; chief biologist, Int Pac Salmon Fisheries Comn, 46-48; actg dir, Col Fisheries, 48-50, dir, 50-58, actg dean, 58-59, dean, 59-71, PROF FISHERIES, UNIV WASH, 48- Concurrent Pos: US Dept State tech rep, Tuna Treaty Negotiations, Mex, 48; consult, Resources Sect, Gen Hq, Supreme Comdr Allied Powers, Japan, 51; consult, Int Coop Admin, India, 57, Bur Com Fisheries, Alaska Div, US Fish & Wildlife Serv, 57- & Calif Dept Fish & Game, 58-; US Govt tech rep, Pink Salmon Treaty Negotiations, Can, 56. Mem: AAAS; Am Soc Ichthyologists & Herpetologists; Am Soc Limnol & Oceanog; fel Am Int Fisheries Res Biol (pres, 66-); Biomet Soc. Res: Early life history of halibut; hydrography; general fisheries biology; biometry. Mailing Add: 1010 NW Innis Arden Dr Seattle WA 98177

VAN COTT, HARRISON CORBIN, b Schenectady, NY, Mar 13, 20; m 43; c 2. MINERAOLOGY, MICROSCOPY. Educ: Univ Rochester, AB, 42; Columbia Univ, MA, 63. Prof Exp: Petrographer, US Bur Mines, 43-45; MINERALOGIST, CORNING GLASS WORKS, 46- Mem: Mineral Soc Am; Am Ceramic Soc; Brit Soc Glass Technol; Mineral Asn Can; Brit Ceramic Soc. Res: Composition and crystallization of ceramics and glasses; microscopy. Mailing Add: Exp Melting Dept Corning Glass Works Corning NY 14830

VAN COUVERING, JOHN ANTHONY, b Long Beach, Calif, Aug 2, 31; m 56; c 4. STRATIGRAPHY, GEOCHRONOLOGY. Educ: Univ Calif, Los Angeles, BA, 56, MA, 62; Univ Cambridge, PhD(geol), 73. Prof Exp: Geologist, Water Resources Div, US Geol Surv, 60-62 & Vanderbilt Gold Corp, 63-65; instr phys geol, Los Angeles City Col, 66; asst engr geol, Calif Div Water Resources, 66; res geologist, Centre Prehist, Nairobi, Kenya, 67; instr geol, Univ Md Overseas Educ, 68; fel, Univ Calif, Berkeley, 70-71; geologist, Fluospar Co Kenya, 72; RES ASSOC, UNIV COLO MUS, BOULDER, 72- Concurrent Pos: Assoc, Isotope Geol Br, US Geol Surv, Denver, 74-75. Mem: Am Asn Petrol Geologists; Geol Soc Am; Int Asn Volcanology & Chem Earth's Interior; Geol Soc London; Soc Vertebrate Paleont. Res: Synthesis and refinement of the Cenozoic Time Scale, emphasizing continental geochronology; mammalian biochronology and marine-nonmarine biostratigraphic correlation in the Old World. Mailing Add: Univ of Colo Mus Boulder CO 80309

VAN COUVERING, JUDITH ANNE HARRIS, b Tulsa, Okla, Feb 20, 38; m 56; c 4. PALEONTOLOGY. Educ: Univ Calif, Berkeley, BA, 60; Univ Cambridge, PhD(geol), 72. Prof Exp: CUR FOSSIL VERTEBRATES & ASST PROF NATURAL HIST, UNIV COLO MUS, 72-, ASST PROF GEOL, UNIV COLO, BOULDER, 72- Mem: Soc Vertebrate Paleont; fel Linnean Soc; Soc Ichthyologists & Herpetologists; Paleont Soc. Res: African tertiary faunas and paleoenvironments; origin of modern terrestrial communities; biogepgraphy; evolutionary problems. cichlid fish evolution. Mailing Add: Univ of Colo Mus Boulder CO 80309

VANDAM, LEROY DAVID, b New York, NY, Jan 19, 14. ANESTHESIOLOGY. Educ: Brown Univ, PhB, 34; NY Univ, MD, 38. Prof Exp: Fel surg, Sch Med, Johns Hopkins Univ, 45-47; asst prof anesthesia, Sch Med, Univ Pa, 52-54; from assoc clin prof to clin prof, 54-67, PROF ANESTHESIA, HARVARD MED SCH, 67- Concurrent Pos: Consult, Valley Forge Army Hosp & Philadelphia Naval Hosp, 52-54, Children's Boston Lying In, Chelsea Naval, West Roxbury Vet Admin, Rutland Vet Admin, Winchester, Burbank & Nantucket Cottage Hosps, 54-; chmn adv panel anesthesiol, US Pharmacopoeia, 54-60; ed, J Anesthesiol, 64; chmn comn anesthesia, Nat Acad Sci-Nat Res Coun, 65. Mem: Am Soc Anesthesiol; AMA. Res: Pharmacology, physiology and biochemistry of surgery and anesthesia. Mailing Add: Boston MA

VAN DAM, M, b Amersfoort, Netherlands, Nov 13, 15; US citizen; m 46; c 2. PHOTOGRAPHY, ORGANIC CHEMISTRY. Educ: State Univ Utrecht, BS, 39, ChemDrs, 46. Prof Exp: Res chemist, Erdal Fabriek, Netherlands, 46-50; chief org lab, Netherlands Photog Indust, 50-52, tech coordr photog emulsions, 52-53; lab dir photog prod, Anken Chem & Film Corp, NJ, 53-68; staff scientist, Olivetti Underwood, 68-69, mgr imaging dept, Olivetti Corp Am, 69-72; sr photog chemist, 72-73, DIR RES, POLYCHROME CORP, 74- Mem: Am Chem Soc; Soc Photog Scientist & Engr; Royal Netherlands Chem Soc. Res: Photographic emulsions; photocopying and duplicating; synthesis of special chemicals for photographic industry. Mailing Add: 179 Beechwood Rd Oradell NJ 07649

VANDE BERG, JERRY STANLEY, physiology, entomology, see 12th edition

VANDE BERG, WARREN JAMES, b Orange City, Iowa, Sept 28, 43; m 62; c 3. PHYCOLOGY. Educ: Iowa State Univ, BS, 66; Ind Univ Univ, Bloomington, PhD(phycol), 70. Prof Exp: Asst prof, 70-75, ASSOC PROF BIOL, NORTHERN MICH UNIV, 75- Mem: Phycol Soc Am. Res: Control of cellular development in Volvox. Mailing Add: Dept of Biol Northern Mich Univ Marquette MI 49855

VAN DE CASTLE, JOHN F, b New York, NY, Sept 30, 33; m 57; c 4. POLYMER CHEMISTRY, PETROLEUM CHEMISTRY. Educ: St John's Col, BS, 55; Univ Md, PhD(org chem), 60. Prof Exp: Chemist, Hoffmann-La Roche, Inc, 55, Nat Bur Standards, 56 & Am Cyanamid Co, 57; group leader elastomers, Esso Res & Eng Co, NJ, 59-65, investment planning & mkt adv, Esso Chem Co, Inc, NY, 65-71; asst to vpres, 71-72, MGR COMM DEVELOP, ENGLEHARD MINERALS & CHEM CO, 72- Mem: Am Chem Soc. Res: Synthesis and characterization of ethylene propylene copolymers and terpolymers, polybutadienes; chemical modification of polymers; Ziegler and organometallic catalytic studies; technology licensing for petroleum and petrochemical industries; petroleum and petrochemical processing. Mailing Add: Engelhard Minerals & Chem Co 430 Mountain Ave Berkeley Heights NJ 07922

VANDEGAER, JAN EDMOND, b Tienen, Belg, July 28, 27; nat US; m 51; c 4. PHYSICAL CHEMISTRY, ORGANIC CHEMISTRY. Educ: Cath Univ Louvain, BS, 48, MS, 50, PhD(phys chem), 52. Prof Exp: Asst res proteins, Cath Univ Louvain, 52-54; fel, Nat Res Coun Can, 54-55; res chemist, Dow Chem Co, 55-59; sr chemist & proj leader, J T Baker Chem Co, 59-60, sr chemist & group leader, 60-62; mgr polymer res, Wallace & Tiernan, Inc, 62-65, dir cent res, 65-70; dir lab res, Chem Group, Dart Industs Inc, 70-74; DIR RES, ALCOLAC INC, 74-, VPRES, 75- Concurrent Pos: Scholar, Biochem Inst, Finland, 53. Mem: AAAS; Asn Res Dirs; Soc Plastics Engrs; Am Chem Soc. Res: Physical chemistry of proteins; polymer chemistry; correlation of physical properties of high polymers and molecular structure; plastics engineering; microencapsulation; polyolefins, surfactants, cosmetics and emulsion polymerization; research administration; technical planning. Mailing Add: 1206 Sillery Bay Rd Pasadena MD 21122

VAN DE GRAAFF, KENT MARSHALL, b Ogden, Utah, May 21, 42; m 62; c 4. GROSS ANATOMY, MAMMALOGY. Educ: Weber State Col, BS, 65; Univ Utah, MS, 69; Northern Ariz Univ, PhD(zool), 73. Prof Exp: Asst prof vet sci, Univ Minn, St Paul, 73-75; ASST PROF HUMAN ANAT, BRIGHAM YOUNG UNIV, 75- Honors & Awards: A Brazier Howell Honoriarum, Am Soc Mammalogists, 72. Mem: Am Soc Mammalogists; Am Soc Zoologists; Am Soc Vet Anatomists. Res: Functional morphological aspects of mammalian posture and locomotion. Mailing Add: Dept of Zool 575 Widtsoe Bldg Brigham Young Univ Provo UT 84602

VANDEGRIFT, VAUGHN, b Jersey City, NJ, Dec 7, 46; m 69; c 2. BIOCHEMISTRY. Educ: Montclair State Col, BA, 68, MA, 70; Ohio Univ, PhD(biochem), 74. Prof Exp: Teacher chem, River Dell Regional High Sch, Oradell, NJ, 68-70; asst prof chem & biol, Ill State Univ, 74-76; ASST PROF CHEM, MURRAY STATE UNIV, 76- Mem: Sigma Xi; Am Chem Soc; Biophys Soc. Res: Role of DNA associated histone proteins in chromosomal structure and funciton; use of circular dichroism in the study of nucleoprotein complexes. Mailing Add: Dept of Chem & Geol Murray State Univ Murray KY 42071

VANDEHEY, ROBERT C, b Hollandtown, Wis, July 19, 24. ENTOMOLOGY. Educ: Univ Notre Dame, MS, 57, PhD(biol), 61. Prof Exp: Instr high sch, Pa, 49-55; from instr to asst prof, 55-69, ASSOC PROF BIOL, ST NORBERT COL, 69- Concurrent Pos: NIH res fel, Inst Genetics, Johannes Gutenberg Univ, Mainz, 63-64; res assoc, Mt St Mary's Col, Calif, 71-72. Mem: AAAS; Entom Soc Am; Am Mosquito Control Asn; Sigma Xi. Res: Genetic variability and control of the mosquitoes, especially Aedes aegypti and Culex pipiens. Mailing Add: Dept of Biol St Norbert Col De Pere WI 54115

VAN DE KAMP, PETER, b Kampen, Holland, Dec 26, 01; nat US; m 47; c 1. ASTRONOMY. Educ: Univ Utrecht, Drs, 22; Univ Calif, PhD(astron), 25; Univ Groningen, DrPhil, 26. Prof Exp: Asst, Kapteyn Astron Lab, Univ Groningen, 22-23; res assoc astron, McCormick Observ, Univ Va, 23-24, from instr to asst prof, 25-37; from assoc prof to prof, 37-72, dir observ, 37-72, RES ASTRONOMER, SPROUL OBSERV, SWARTHMORE COL, 72- Concurrent Pos: Fulbright prof, France, 49, Netherlands, 74; prog dir astron, NSF, 54-55; leader, Shetland Site, Georgetown eclipse exped, 54; dir at large, Kitt Peak Nat Observ, Assoc Univs Res in Astron, 57-61; pres, Comt Double Stars, Int Astron Union, 58-64. Honors & Awards: Univ Va Pres & Vis Prize, 27, 37, 38; Glover Award, Dickinson Col, 61; Nason Award, Swarthmore Col, 63; Rittenhouse Soc Medal, 65. Mem: Am Astron Soc; corresp, Royal Netherlands Acad Sci; Int Astron Union. Res: Statistical astronomy; photographic astrometry; double stars; unseen companions of nearby stars. Mailing Add: Sproul Observ Swarthmore Col Swarthmore PA 19081

VAN DE KAMP, PETER CORNELIS, b Plainfield, NJ, Aug 25, 40; m 64; c 2. GEOLOGY, GEOCHEMISTRY. Educ: Lehigh Univ, BA, 62; McMaster Univ, MSc, 64; Univ Bristol, PhD(geochem), 67. Prof Exp: Geologist, Shell Develop Co, Calif, 67-71 & Shell Oil Co, Colo, 71-73; res prof, Univ Man, 73-74; V PRES, GEO-LOGIC, INC, 74- Mem: Soc Econ Paleont & Mineral; Geol Soc Am; Geol Soc London. Res: Stratigraphy; sedimentary, metamorphic and igneous petrology and geochemistry. Mailing Add: Geo-Logic Inc 440 W Colorado St Glendale CA 91204

VANDE KIEFT, LAURENCE JOHN, b Grand Rapids, Mich, May 14, 32; m 62; c 2. SOLID STATE PHYSICS, OPTICAL PHYSICS. Educ: Calvin Col, BA, 53; Univ Conn, MS, 55, PhD(physics), 68. Prof Exp: Sr physicist, Bendix Res Labs, 58-62; teaching & res asst, Univ Conn, 62-63 & 64-68; res physicist, Signature & Propagation Lab, 68-72, CHIEF OPTICAL & MICROWAVE SYSTS BR, CONCEPTS ANAL LAB, BALLISTIC RES LABS, US ARMY ABERDEEN RES & DEVELOP CTR, 72- Mem: Am Phys Soc. Res: Electron paramagnetic resonance investigations of radiation effects in single crystals; laser interactions with the atmosphere; laser semiactive terminal homing; computer simulation. Mailing Add: US Army Ballistic Res Labs Attn AMXBR-CA Aberdeen Proving Ground MD 21005

VAN DELL, ROBERT DUANE, b Detroit, Mich, Mar 14, 34; m 57; c 2. SURFACE CHEMISTRY, COLLOID CHEMISTRY. Educ: Wayne State Univ, BS, 57, MS, 62, PhD(phys chem), 66. Prof Exp: Res chemist, 67-69, RES SPECIALIST & PROJ MGR, DOW CHEM CO, 69- Concurrent Pos: Heart grant, Wayne State Univ, 66-67. Mem: Am Chem Soc. Res: Surface and colloid chemistry relating to the adsorption of polyelectrolytes, proteins and solution and film properties, especially monolayers; characterization of water soluble polymers such as molecular weight as determined by gel permeation chromatography and adsorption. Mailing Add: Dow Chem Co Midland MI 48640

VAN DELLEN, THEODORE ROBERT, b Chicago, Ill, Aug 15, 11; m 35; c 3. INTERNAL MEDICINE. Educ: Northwestern Univ, BS, 33, MD, 36, MS, 39. Prof Exp: Macy fel med, NY Postgrad Med Sch & Hosp, 36-37; ASSOC PROF MED & ASST DEAN, SCH MED, NORTHWESTERN UNIV, CHICAGO, 49- Concurrent Pos: Sr consult, Hines Vet Admin Hosp, 45- Honors & Awards: Am Med Writers Asn Award, 58. Mem: AAAS; AMA; Am Med Writers Asn (vpres, 58, pres, 60); Am Heart Asn; fel Am Col Physicians; Am Fedn Clin Res. Res: Cardiovascular disease. Mailing Add: 435 N Michigan Ave Chicago IL 60611

VAN DEMARK, DUANE R, b Elida, Ohio, Feb 26, 36; m 62; c 2. SPEECH PATHOLOGY. Educ: Hiram Col, BA, 58; Univ Iowa, MA & PhD(speech path, audiol), 62. Prof Exp: Instr speech path, Ind Univ, 62-65; asst prof, 65-68, assoc prof speech path & otolaryngol, 68-75, PROF SPEECH PATH & OTOLARYNGOL & MAXILLOFACIAL SURG, UNIV IOWA, 75- Concurrent Pos: Ind Univ Found res grant, 65; Am-Scand Found George Marshall fel, 70; Nat Inst Dent Res spec fel, 70; partic, Int Cong Cleft Palate, 67-69 & 73, mem prog comt, 68, secy & asst to secy gen, 69; sect ed, Cleft Palate J. Mem: Am Cleft Palate Educ Found; Am Speech & Hearing Asn; Am Cleft Palate Asn. Res: Cleft palate research. Mailing Add: Dept of Otolaryngol Univ of Iowa Iowa City IA 52242

VANDEMARK, J S, b Fairgrove, Mich, Mar 24, 19; m 45; c 1. HORTICULTURE, VEGETABLE CROPS. Educ: Mich State Univ, BS, 41, MS, 46; Univ Ill, PhD(hort), 60. Prof Exp: From asst prof to assoc prof hort, Purdue Univ, 47-57; commodity dir, Am Farm Fedn, 57-58; dir res, Cent Aquirre Sugar Co, 58-60; assoc prof hort, 60-63, PROF HORT, UNIV ILL, URBANA, 63- Mem: Am Soc Hort Sci. Res: Commercial production and fertility of sugar cane and vegetable crops. Mailing Add: 208 Vegetable Crops Univ of Ill Urbana IL 61801

VAN DE MARK, MILDRED S, b Alhambra, Calif, Aug 18, 10; m 43; c 1. NUTRITION. Educ: Ala Polytech Inst, BS, 34, MS, 39. Prof Exp: Specialist foods & nutrit, Exten Serv, 38-43, hort specialist foods, Ala Agr Sta, 43-46, from asst prof to prof, Sch Home Econ, 48-66, head dept, 55-64, head sch, 64-66, actg dean, 66-68, head nutrit & foods, 69-73, EMER PROF FOODS & NUTRIT, AUBURN UNIV, 73- Concurrent Pos: Mem, President's White House Conf Foods & Nutrit & Conf on Aging. Mem: Am Dietetic Asn; Inst Food Technol; Am Inst Chem. Res: Food technology and marketing research; decision making factors in good marketing; hemoglobin and folate levels of pregnant teen-agers as related to dietary levels; chemical factors affecting meat qualities. Mailing Add: Dept of Nutrit & Foods Auburn Univ Auburn AL 36830

VANDEMARK, NOLAND LEROY, b Columbus Grove, Ohio, July 6, 19; m 40; c 3. PHYSIOLOGY. Educ: Ohio State Univ, BS, 41, MS, 42; Cornell Univ, PhD, 48. Prof Exp: Asst animal husb, Ohio State Univ, 41-42; vitamin chemist, State Dept Agr, Ohio, 42; asst animal husb, Cornell Univ, 42-44, 48; livestock specialist, US Dept Army, Austria, 46-47; from asst prof to prof physiol, Univ Ill, Urbana, 48-64; prof dairy sci & chmn dept, Col Agr & Home Econ, Ohio State Univ & Ohio Agr Res & Develop Ctr, 64-73, mem fac, Coop Exten Serv, Univ, 64-73; DIR RES, NY STATE COL OF AGR & LIFE SCI, 74-; DIR AGR EXP STA, CORNELL UNIV, 74- Honors & Awards: Borden Award, 59; Gold Medal, Ital Govt, 64; Citation, Ital Soc Animal Sci, 72. Mem: AAAS; Am Soc Animal Sci; Am Dairy Sci Asn; Am Physiol Soc; Brit Soc Study Fertil. Res: Physiology and biochemistry of reproductive processes in cattle, especially semen production; sperm metabolism; female reproductive processes; artificial insemination and fertility-sterility problems. Mailing Add: 292 Roberts Hall Cornell Univ Ithaca NY 14853

VAN DEMARK, PAUL JOHN, microbiology, bacteriology, see 12th edition

VAN DENACK, JULIA MARIE, b New Franken, Wis, Sept 26, 23. BIOLOGY. Educ: Holy Family Col, Wis, BA, 52; Cath Univ, MS, 59, PhD(ecol, bact), 61. Prof Exp: Teacher, Cath Mem High Sch, Wis, 50-57; prof biol, 60-72, CHMN DEPT NATURAL SCI & MATH, SILVER LAKE COL, 70- Concurrent Pos: Mem adv curric comt sci, Nat Cath Educ Asn, 62-64, mem rev comt biol sci curric study, 63. Mem: AAAS; Ecol Soc Am. Res: Ecology of the sand dune complex at Point Beach State Forest, Two Rivers, Wisconsin; teacher preparation for secondary biology courses; environmental education. Mailing Add: Dept of Natural Sci & Math Silver Lake Col Manitowoc WI 53220

VAN DEN AKKER, JOHANNES ARCHIBALD, b Los Angeles, Calif, Dec 5, 04; m 30, 58; c 1. THERMODYNAMICS, OPTICS. Educ: Calif Inst Technol, BS, 26, PhD(physics), 31. Prof Exp: Instr physics, Wash Univ, 30-35; res assoc & chmn dept, 35-56, sr res assoc physics & chmn dept physics & math, 56-70, res counsr, 65-70, EMER PROF PHYSICS, INST PAPER CHEM, LAWRENCE UNIV, 70-; CONSULT RES & DEVELOP, AM CAN CO, NEENAH, 71- Concurrent Pos: Sr Fulbright lectr, Univ Manchester Inst Sci & Technol, 61-62. Honors & Awards: Res & Develop Award, Tech Asn Pulp & Paper Indust, 67, Gold Medal, 68. Mem: Fel AAAS; fel Am Phys Soc; fel Optical Soc Am; Am Asn Physics Teachers; fel Tech Asn Pulp & Paper Indust. Res: Spatial distribution of x-ray photoelectrons; optical properties of paper; spectrophotometry and color measurement; instrumentation of all properties of paper; paper and fiber physics. Mailing Add: One Brokaw Pl Appleton WI 54911

VANDENBELT, JOHN MELVIN, b Holland, Mich, Aug 9, 12; m 40; c 2. PHYSICAL CHEMISTRY. Educ: Hope Col, AB, 34; Boston Univ, AM, 36; Mich State Univ, PhD, 40. Prof Exp: Res chemist, 40-42, head phys chem labs, 42-55, lab dir phys chem, 55-70, SR RES SCIENTIST, PARKE, DAVIS & CO, 70- Concurrent Pos: Chmn, Gordon Instrumentation Conf, 61. Mem: Fel AAAS; Am Chem Soc; Instrument Soc Am; Optical Soc Am. Res: Physical measurements of biologically active compounds, absorption spectroscopy, spectrum-structure correlations; spectrophotometric standards; holmium wavelength filter; assay by group absorption; protein binding; dissociation constants. Mailing Add: 2645 Overridge Dr Ann Arbor MI 48104

VANDENBERG, EDWIN JAMES, b Hawthorne, NJ, Sept 13, 18; m 50; c 2. POLYMER CHEMISTRY. Educ: Stevens Inst Technol, ME, 39. Hon Degrees: DrEng, Stevens Inst Technol, 65. Prof Exp: Res chemist, 39-44, asst shift supvr, Sunflower Ord Works, Kans, 44-45, from res chemist to sr res chemist, 45-65, RES ASSOC, RES CTR, HERCULES INC, 65- Concurrent Pos: Mem adv bd, J Polymer Sci, 66- Honors & Awards: Indust Res 100 Award, 65. Mem: Am Chem Soc. Res: Paper sizing; soil stabilization; emulsion polymerization; peroxide reactions; polymers; Ziegler, viny ether, epoxide and graft polymerizations; mechanism of polymerization. Mailing Add: Hercules Inc Res Ctr Wilmington DE 19899

VAN DEN BERG, L, b Hattem, Neth, Mar 14, 29; Can citizen; m 53; c 4. FOOD SCIENCE. Educ: State Agr Univ, Wageningen, MSc, 53; Univ Man, MSc, 55. Prof Exp: SR RES OFF FOOD TECHNOL, DIV BIOL SCI, NAT RES COUN CAN, 56- Mem: Inst Food Technologists; Can Inst Food Sci & Technol. Res: Application of refrigeration to food preservation, including freezing and frozen storage of meat and vegetables and storage of fresh vegetables; anaerobic digestion of food plant waste. Mailing Add: Div Biol Sci Nat Res Coun Can Ottawa ON Can

VANDENBERG, STEVEN GERRITJAN, b Den Helder, Netherlands, July 7, 15; nat US, m 48. BEHAVIORAL GENETICS, PSYCHOLOGY. Educ: Univ Groningen, DrsJur, 46; Univ Mich, PhD, 55. Prof Exp: Psychologist, Ionia State Hosp for Criminally Insane, 47-48; asst psychol, Univ Mich, 48-50, psychologist, Inst Human Biol, 51-57, assoc dir schizophrenia study, Ment Health Res Inst, 57-60; psychologist, Sch Med, Univ Louisville, 60-67; PROF PSYCHOL, UNIV COLO, BOULDER, 67- Concurrent Pos: Psychologist, State Child Guid Clin, Mich, 49-50; lab schs, Eastern Mich Univ & pub schs, Mich, 51; exec ed, Behav Genetics; Nat Inst Ment Health res career develop award, 62-67; vis prof genetics, Univ Hawaii, 74-75. Mem: Am Soc Human Genetics; Soc Res Child Develop; Soc Personality Assessment; Am Psychol Asn; Brit Psychol Soc. Res: Objective measures of personality; behavior genetics; factor analysis and test theory; computer applications in the behavior sciences. Mailing Add: Dept of Psychol Univ of Colo Boulder CO 80302

VANDENBERGH, JOHN GARRY, b Paterson, NJ, May 5, 35; m 58; c 2. ANIMAL BEHAVIOR, ENDOCRINOLOGY. Educ: Montclair State Col, AB, 57; Ohio Univ, MA, 59; Pa State Univ, PhD(zool), 62. Prof Exp: Res biologist, Nat Inst Neurol Dis & Blindness, 62-65; RES SCIENTIST, NC DEPT MENT HEALTH, 65- Concurrent Pos: NIMH grant, 67-; mem, Nat Primate Adv Comt, NIH; mem psychobiol panel, NSF; mem comt conserv non-human primates, Nat Res Coun-Nat Acad Sci; adj prof zool, Univ NC & NC State Univ. Mem: Am Soc Zoologists; Animal Behav Soc; Am Soc Mammal; Soc Study Reproduction; Int Primatol Soc. Res: Environmental control of reproduction; endocrine basis of behavior; rodent and primate social behavior. Mailing Add: NC Dept of Ment Health Anderson Hall Box 7599 Raleigh NC 27611

VAN DEN BERGH, SIDNEY, b Wassenaar, Holland, May 20, 29; c 3. ASTRONOMY. Educ: Princeton Univ, AB, 50; Ohio State Univ, MSc, 52; Univ Göttingen, Drrernat(astron), 56. Prof Exp: Asst prof astron, Ohio State Univ, 56-58; from lectr to assoc prof, 58-66, PROF ASTRON, UNIV TORONTO, 66- Mem: Am Astron Soc; Royal Soc Can; Royal Astron Soc; Int Astron Union. Res: Extragalactic nebulae; star clusters; variable stars; supernovas. Mailing Add: David Dunlap Observ Richmond Hill Toronto ON Can

VAN DEN BERGHE, JOHN, b Rotterdam, Netherlands, Jan 30, 23; m 50; c 9. ORGANIC CHEMISTRY, POLYMER CHEMISTRY. Educ: Univ Utah, BA, 47, MS, 48; Univ Wis, PhD(org chem), 52. Prof Exp: SR RES CHEMIST, EASTMAN KODAK CO, 52- Mem: Am Chem Soc. Res: Polyesters; polycarbonates; polyesteramides; substituted polystyrenes and other polyolefins; Diels-Alder reaction; pyrolysis reactions; emulsion polymerization; identification of copolymer and terpolymer compositions; non-aqueous titrimetry; molecular weight distribution of polymers. Mailing Add: 34 Tanglewood Dr Rochester NY 14616

VAN DEN BOLD, WILLEM AALDERT, b Amsterdam, Neth, Mar 30, 21; m 50; c 5. GEOLOGY, PALEONTOLOGY. Educ: State Univ Utrecht, PhD, 46. Prof Exp: Micropaleontologist, Royal Dutch Shell Group, 46-58; assoc prof geol, 58-59, PROF GEOL, LA STATE UNIV, BATON ROUGE, 59- Mem: Paleont Res Inst; Am Geophys Union; Geol Soc Am; Soc Econ Paleontologists & Mineralogists. Res: Post-Paleozoic ostracoda; planktonic Foraminifera. Mailing Add: Dept of Geol La State Univ Baton Rouge LA 70803

VANDEN BORN, WILLIAM HENRY, b Rhenen, Netherlands, Nov 17, 32; Can citizen; m 58; c 5. WEED SCIENCE, PLANT PHYSIOLOGY. Educ: Univ Alta, BSc, 56, MSc, 58; Univ Toronto, PhD(plant physiol), 61. Prof Exp: Fel, Dept Plant Sci, 60-61, from asst prof to assoc prof, 61-72, chmn dept, 70-75, PROF WEED SCI & CROP ECOL, UNIV ALTA, 72- Mem: AAAS; Weed Sci Soc Am; Can Soc Plant Physiol; Can Soc Agron; Am Sci Affil. Res: Physiological plant ecology; plant growth regulators; herbicide translocation. Mailing Add: Dept of Plant Sci Univ of Alta Edmonton AB Can

VAN DEN BOS, JAN, b The Hague, Neth, Nov 1, 39. COMPUTER SCIENCE, ATOMIC PHYSICS. Educ: State Univ Leiden, BSc, 61; Univ Amsterdam, MSc, 64, PhD(atomic physics), 67. Prof Exp: Res assoc atomic physics, FOM-Inst Atomic Physics, Amsterdam, Neth, 64-67; res assoc, Joint Inst Lab Astrophys, Nat Bur Stand & Univ Colo, 67-68; res assoc, 68, asst prof, 69-72, ASSOC PROF COMPUT SCI, UNIV NEBR, LINCOLN, 72- Mem: Asn Comput Mach. Res: Computer graphics; systems programming; minicomputers; application of computers to physics problems; theoretical atomic physics. Mailing Add: Dept of Comput Sci Univ of Nebr Lincoln NE 68508

VAN DEN BOSCH, FRANK JOSEPH GERARD, b Antwerp, Belg, July 9, 04; US citizen; m 49. BIOPHYSICS, BIOENGINEERING. Educ: Univ Paris, DSc(physics), 27; Univ London, PhD(physiol), 30. Prof Exp: Sr physicist, Vacuum Sci Prod, Ltd, London, Eng, 30-39; dir res, Romac Radio, Ltd, Hendon, London, 39-45; sci adv, Belg Govt, 46-48; dir elec & phys labs, Inst Bunge, Antwerp, Belg, 48-57; founder, Electro-Microscopic Ctr, Antwerp, Belg, 57-61; mem staff, Audiotronics, Dayton, Ohio, 61-62; biophysicist, Marlboro State Hosp, 63-66; PROF ELECTRONIC ENG IN SURG RES, STATE UNIV NY DOWNSTATE MED CTR, 67- Concurrent Pos: Prof exp physics, Free Univ Antwerp, 47-56; dir res, Bootz Mfg Co, Evansville, Ind & Proj See, Biomed Inst, San Diego, Calif, 63-66. Honors & Awards: Sci Merit Medal, Brussels Worlds Fair, 58; Achievement Award, United Inventors & Scientists of Am. Res: Electronic and television development; biophysical mechanisms of vascular thrombosis; biophysical approach to live hematological studies; electron and ultraviolet microscopy; biophysical measuring instrumentation; biological power supplies and artificial hearts. Mailing Add: State Univ NY Downstate Med Ctr 450 Clarkson Ave Box 40 Brooklyn NY 11203

VAN DEN BOSCH, ROBERT, b Martinez, Calif, Mar 31, 22; m 44. ENTOMOLOGY. Educ: Univ Calif, AB, 43, PhD(entom), 50. Prof Exp: Asst entomologist, Univ Hawaii, 49-51; from asst entomologist to assoc entomologist, Univ Calif, Riverside, 51-64; ENTOMOLOGIST, UNIV CALIF, BERKELEY, 64-, PROF ENTOM, 67- Concurrent Pos: Guggenheim fel, 58. Mem: AAAS; Entom Soc Am; Ecol Soc Am. Res: Utilization of predatory and parasitic insects in control of pest insects; biology, ecology, behavior of entomophagous insects; integrated control of insect pests. Mailing Add: 4 Kensington Ct Kensington CA 94707

VANDENBOSCH, ROBERT, b Lexington, Ky, Dec 12, 32; m 56; c 2. NUCLEAR CHEMISTRY. Educ: Calvin Col, AB, 54; Univ Calif, PhD(chem), 57. Prof Exp: From asst chemist to assoc chemist, Agronne Nat Lab, 57-63; PROF CHEM, UNIV WASH, 63- Mem: Am Phys Soc; Sigma Xi. Res: Heavy ion nuclear reactions and nuclear fission. Mailing Add: Dept of Chem Univ of Wash Seattle WA 98195

VANDEN BOUT, PAUL ADRIAN, b Grand Rapids, Mich, June 16, 39; m 61; c 2. PHYSICS. Educ: Calvin Col, AB, 61; Univ Calif, Berkeley, PhD(physics), 66. Prof Exp: Fel physics, Lawrence Radiation Lab, 66-67; res assoc, Columbia Radiation Lab, 67-68, instr, Columbia Univ, 68-69, asst prof, 69-70; ASST PROF ASTRON, UNIV TEX, AUSTIN, 70- Mem: Am Phys Soc; Am Astron Soc; Royal Astron Soc. Res: X-ray astronomy and the interstellar medium. Mailing Add: Dept of Astron Univ of Tex Austin TX 78712

VANDEN EYNDEN, CHARLES LAWRENCE, b Cincinnati, Ohio, June 25, 36. NUMBER THEORY. Educ: Univ Cincinnati, BS, 58; Univ Ore, MA, 60, PhD(math), 62. Prof Exp: NSF fel, Univ Mich, 62-63; asst prof math, Univ Ariz, 63-65 & Miami Univ, 65-67; vis asst prof, Pa State Univ, 67-68; asst prof, Ohio Univ, 68-69; assoc prof, 69-74, PROF MATH, ILL STATE UNIV, 74- Mem: Am Math Soc; Math Asn Am. Res: Diophantine approximation; elementary number theory; sequences of integers. Mailing Add: Dept of Math Ill State Univ Normal IL 61761

VAN DEN HENDE, JAN H, crystallography, information science, see 12th edition

VAN DENHEUVEL, FRANZ AIME, b Brussels, Belg, Apr 30, 13; nat US; m 36. ORGANIC CHEMISTRY, BIOCHEMISTRY. Educ: Free Univ Brussels, BSc, 32, MSc, 34; Univ London, PhD(chem), 38, Imp Col, dipl, 38. Prof Exp: Res chemist, Belg Soc Carbochem, 38-47; head org chem, Fisheries Res Bd Can, 47-60; RES OFF, RES BR, ANIMAL RES INST, CAN DEPT AGR, 60- Mem: Chem Inst Can. Res: Methods of separation, identification, structural determination, chromatography and molecular distillation of lipids; synthesis; reaction kinetics; blood lipid chemistry and metabolism; lipoprotein systems; biological membrane structure and functions; steroid hormones and metabolites. Mailing Add: Physiol Sect Animal Res Inst Can Dept of Agr Ottawa ON Can

VANDEN HEUVEL, WILLIAM JOHN ADRIAN, III, b Brooklyn, NY, Mar 7, 35; m 60; c 3. DRUG METABOLISM. Educ: Princeton Univ, AB, 56, AM, 58, PhD(org chem), 60. Prof Exp: Instr chem, Lipid Res Ctr, Col Med, Baylor Univ, 62, asst prof, Dept Biochem, 62-64; sr res biochemist, 64-67, res fel, Dept Biochem, 67-72, SR RES FEL, DEPT DRUG METAB, MERCK SHARP & DOHME RES LABS, 72- Concurrent Pos: Sr asst scientist, Nat Heart Inst, 60-62. Mem: AAAS; Am Chem Soc; Am Soc Mass Spectrometry. Res: Development of gas chromatographic and mass spectrometric methods for the identification and quantification of drugs, metabolites and natural products; derivatization techniques; organic mass spectrometry; use of radioactive and stable isotopes. Mailing Add: Merck Sharp & Dohme Res Labs Rahway NJ 07065

VAN DEN NOORT, STANLEY, b Lynn, Mass, Sept 8, 30; m 54; c 5. NEUROLOGY. Educ: Dartmouth Col, AB, 51; Harvard Univ, MD, 54. Prof Exp: Asst prof neurol, Sch Med, Case Western Reserve Univ, 65-71; PROF MED (NEUROL) & DEAN, COL MED, UNIV CALIF, IRVINE, 71- Mailing Add: 101 City Dr S Orange CA 92668

VANDE NOORD, EDWIN LEE, b Pella, Iowa, Sept 10, 38; m 62; c 2. SPACE PHYSICS. Educ: Grinnell Col, BA, 60; Univ NMex, MS, 63, PhD(physics), 68. Prof Exp: Res asst physics, Grinnel Col, 60-61; asst, Univ NMex, 61-68, res assoc, 68-69; res scientist, Douglas Advan Res Labs, Calif, 69-70; staff scientist, 70-75, MGR ADVAN PROGS, BALL BROS RES CORP, 75- Mem: Am Astron Soc; Am Geophys Union; Optical Soc Am; Am Inst Aeronaut & Astronaut; Sigma Xi. Res: Photometry of zodiacal light; interplanetary dust; infrared Fourier transform spectroscopy; space instrumentation; remote sensing; earth radiation budget instrumentation. Mailing Add: Ball Bros Res Corp PO Box 1062 Boulder CO 80302

VAN DE POEL, JOSEPHUS, b Groenlo, Holland, Sept 15, 25; m 62; c 1. ORGANIC CHEMISTRY, ANALYTICAL CHEMISTRY. Educ: Univ Leiden, BS, 47, MS, 52, PhD(org chem), 56. Prof Exp: Instr anal chem, Univ Leiden, 53-56; res chemist, E I du Pont de Nemours & Co, 56-58; from asst prof to assoc prof, 58-68, PROF ANAL CHEM, WELLS COL, 68- Concurrent Pos: NSF res grant, 63-65; fel, Ga Inst

Technol, 67. Mem: Am Chem Soc. Res: Radiochemistry; stability and radiocyanide exchange of eight-coordinate molybdenum and tungsten cyanide complexes; tetracyano complexes of molybdenum; synthesis of potassium octacyanomolybdate (IV). Mailing Add: Dept of Chem Wells Col Aurora NY 13026

VANDEPUTTE, JOHN, b Catasauqua, Pa, May 3, 18; m 41; c 3. ORGANIC CHEMISTRY. Educ: Albright Col, BS, 43; Rutgers Univ, MS, 45, PhD(chem), 51. Prof Exp: Instr org chem, Rutgers Univ, 48-51; res assoc, 51-68, MGR RES PROJ PLANNING & COORD, SQUIBB INST MED RES, 68- Concurrent Pos: Mem, Proj Mgt Inst. Mem: Am Chem Soc. Res: Isolation, purification and identification of natural products and development of processes for their production. Mailing Add: Prog Planning & Coord PO Box 4000 Princeton NJ 08540

VANDER, ARTHUR J, b Detroit, Mich, Dec 28, 33; m 55; c 3. PHYSIOLOGY. Educ: Univ Mich, BA, 55, MD, 59. Prof Exp: Intern med, New York Hosp-Cornell Med Ctr, 59-60; from instr to assoc prof, 60-69, PROF PHYSIOL, UNIV MICH, ANN ARBOR, 69- Concurrent Pos: Mem, Pub Educ Comn, Nat Kidney Dis Found, 66- Mem: AAAS; Am Physiol Soc; Am Soc Nephrol; Soc Exp Biol & Med. Res: Renal physiology. Mailing Add: Dept of Physiol Univ of Mich Ann Arbor MI 48104

VANDERBEEK, LEO CORNELIS, b The Hague, Netherlands, Aug 28, 18; nat US; m 44; c 4. PLANT PHYSIOLOGY. Educ: Western Mich Univ, AB, 52; Univ Mich, MS, 53, PhD(bot), 57. Prof Exp: Asst, Univ Mich, 52-53, 54-55; from asst prof to assoc prof, 56-62, PROF BIOL, WESTERN MICH UNIV, 62- Mem: AAAS; Am Soc Plant Physiol; Am Soc Photobiol; Am Inst Biol Sci. Res: Plant growth regulators; effect of audible sound on plant development. Mailing Add: Dept of Biol Western Mich Univ Kalamazoo MI 49008

VANDERBEKE, LOIS M, mathematics, see 12th edition

VANDERBERG, JEROME PHILIP, b New York, NY, Feb 5, 35. MEDICAL ENTOMOLOGY, CELL PHYSIOLOGY. Educ: City Col New York, BS, 55; Pa State Univ, MS, 57; Cornell Univ, PhD(med entom), 61. Prof Exp: Fel biol, Johns Hopkins Univ, 62-63; from asst prof to assoc prof, 63-74, PROF PARASITOL SCH MED, NY UNIV, 74- Mem: Am Soc Trop Med & Hyg; Am Mosquito Control Asn; Entom Soc Am. Res: Cellular physiology of insects and of host-parasite complex; biology of mosquitoes; malariology. Mailing Add: Dept of Prev Med NY Univ New York NY 10016

VANDERBIE, JAN H, ergonomics, psychology, see 12th edition

VAN DER BIJL, WILLEM, b Alphen aan den Rijn, Netherlands, Aug 15, 20; nat US; m 46; c 2. METEOROLOGY. Educ: Vrije Univ, Netherlands, BSc, 41, MSc, 43; Univ Utrecht, PhD(meteorol), 52. Prof Exp: Res assoc climat, Royal Netherlands Meteorol Inst, 46-56; assoc prof physics & meteorol, Kans State Univ, 56-61; ASSOC PROF METEOROL, NAVAL POSTGRAD SCH, 61- Concurrent Pos: Fel statist & meteorol, Univ Chicago, 54-55. Mem: Am Meteorol Soc; Am Geophys Union. Res: Physics of the atmosphere; statistical treatment of data; statistical analysis of geophysical data. Mailing Add: 791 Toyon Dr Monterey CA 93940

VANDERBORGH, NICHOLAS E, analytical chemistry, see 12th edition

VANDER BROOK, MILTON JOHN, b Kalamazoo, Mich, Aug 5, 10; m 41; c 4. PHARMACOLOGY. Educ: Western Mich Univ, AB, 31; Kalamazoo Col, AM, 32; Univ Chicago, PhD(pharmacol), 39. Prof Exp: Asst biochem res, Upjohn Co, 33-36, pharmacologist, 39-50, head dept pharmacol, 59-62, mgr med develop, 62-64, admin asst to dir sci rels, 64-76; RETIRED. Mem: Am Soc Pharmacol & Exp Therapeut. Res: Psychopharmacology; autonomic pharmacology; chemotherapy. Mailing Add: 14313 Cathead Bay Dr Northport MI 49670

VAN DER BURG, SJIRK, b Makkum, Netherlands, Mar 23, 26; nat US; m; c 3. ORGANIC CHEMISTRY. Educ: Univ Groningen, Drs, 55. Prof Exp: Asst, Univ Groningen, 52-55; res chemist, Rubber Found, Delft Univ Technol, 55-56 & Res Ctr, US Rubber Co, NJ, 56-61; mgr mat res, US Rubber Tire Co, 61-66, develop mgr, Uniroyal Europ Tire Develop Ctr, Ger, 66-67, dir, 67-74, DIR RES & DEVELOP, INT DIV, UNIROYAL INC, 74- Mem: Am Chem Soc; Royal Netherlands Chem Soc. Res: Rubber chemistry and technology; plastics. Mailing Add: Uniroyal Europ Tire Develop Ctr Huettenstrasse 7 D-5100 Aachen-Rothe Erde Germany

VANDERBURG, VANCE DILKS, b Grand Rapids, Mich, July 22, 37; m 66. HIGH ENERGY PHYSICS. Educ: Syracuse Univ, BS, 60; Purdue Univ, MS, 63, PhD(high energy physics), 65. Prof Exp: Physicist, US AEC, 65-67; asst physicist, Brookhaven Nat Lab, 67-69, assoc physicist, 69-75; MEM STAFF, D C COOK NUCLEAR PLANT, BRIDGEMAN, MICH, 75- Mem: Am Phys Soc. Res: Experimental particle physics. Mailing Add: D C Cook Nuclear Plant Bridgeman MI 49106

VANDER BURGH, LEONARD F, b Chandler, Minn, Oct 10, 31; m 70; c 1. PHYSICAL ORGANIC CHEMISTRY. Educ: Hamline Univ, BS, 53; Univ Wash, PhD(org chem), 58. Prof Exp: Res chemist rubber, Shell Oil Co, 57-61; res chemist, polymers, 62-63, sr chemist paper, 63-66, group leader, 66-71, SR SCIENTIST STARCH, A E STALEY MFG CO, 71- Mem: Am Chem Soc. Res: New starch derivatives, new processes for modifying starch and starch-synthetic polymer combinations. Mailing Add: Res Ctr A E Staley Mfg Co 2200 E Eldorado St Decatur IL 62525

VAN DER ELST, DIRK H, b Dordrecht, Nath, June 15, 33; US citizen; m 65. BEHAVIORAL ANTHROPOLOGY, CULTURAL ANTHROPOLOGY. Educ: Univ Utah, BA, 60, MA, 61; Northwestern Univ, MA, 64, PhD(anthrop), 70. Prof Exp: Instr anthrop, Univ Nev, Reno, 64-65; asst prof, Marietta Col, 65-69; ASSOC PROF ANTHROP, CALIF STATE UNIV, FRESNO, 69-, CHMN DEPT, 74- Mem: Am Anthrop Asn. Res: Culture and choice-making; ideology and social organization; Negro in the New World; black tribesmen of Surinam; anthropology of war. Mailing Add: Dept of Anthrop Calif State Univ Fresno CA 93710

VAN DEREN, JOHN MEDEARIS, JR, b Cynthiana, Ky, July 12, 29; m 53; c 4. ANALYTICAL CHEMISTRY. Educ: Univ Cincinnati, BS, 50; La State Univ, MS, 53, PhD(biochem), 59. Prof Exp: Asst, La State Univ, 58-59; asst food technologist & biochemist, Clemson Col, 59-61; res biochemist, Monsanto Co, Mo, 61-68; ANAL RES CHEMIST, RES DEPT, BUCKMAN LABS, INC, 68- Res: Pesticide residue analysis; methods development; food additives analysis; herbicide metabolism. Mailing Add: Res Dept Buckman Labs Inc 1256 N McLean Blvd Memphis TN 38108

VANDERFORD, HARVEY BIRCH, b Maben, Miss, Feb 18, 10; m 39; c 1. AGRONOMY. Educ: Miss State Col, BS, 34, MS, 39; Univ Mo, PhD(soils), 42. Prof Exp: Asst agriculturist, Perkinston Jr Col, 34-37; from asst agronomist to assoc agronomist, 37-72, PROF AGRON, MISS STATE UNIV, 72- Res: Plant breeding; soil classification and survey. Mailing Add: Dept of Agron Miss State Univ Mississippi State MS 39762

VANDERGRAAF, TJALLE T, b Gravemoer, Netherlands, Sept 3, 36; Can citizen; m 66; c 2. ANALYTICAL CHEMISTRY, RADIOCHEMISTRY. Educ: Calvin Col, BS, 63; Pa State Univ, PhD(chem), 69. Prof Exp: RES SCIENTIST NUCLEAR FUEL ANAL, ANAL SCI BR, WHITESHELL NUCLEAR RES ESTAB, ATOMIC ENERGY CAN LTD, 69- Mem: Chem Inst Can. Res: Chemical analysis of highly radioactive materials; chemistry of nuclear fuels; neutron activation analysis; radiochemical separations; burnup determinations of nuclear fuels by destructive chemical analyses. Mailing Add: Anal Sci Br Atomic Energy of Can Whiteshell Nuclear Res Estab Pinawa MB Can

VANDER HAAR, ROY WILLIAM, b Sioux City, Iowa, Apr 26, 24; m 44; c 4. ANALYTICAL CHEMISTRY. Educ: Wesleyan Univ, AB, 47; Iowa State Col, PhD(anal chem), 52. Prof Exp: Res chemist, Am Oil Co, Ind, 52-70; res chemist, 70-75, SR RES CHEMIST, AMOCO RES CTR, STANDARD OIL CO, IND, 75- Mem: Am Chem Soc; Am Soc Mass Spectrometry. Res: Mass spectrometry; instrumental analytical chemistry. Mailing Add: Standard Oil Co Ind Amoco Res Ctr PO Box 400 Naperville IL 60540

VANDER HART, DAVID LLOYD, b Rehoboth, NMex, May 20, 41; m 64; c 1. PHYSICAL CHEMISTRY, POLYMER CHEMISTRY. Educ: Calvin Col, AB, 63; Univ Ill, Urbana, PhD(phys chem), 68. Prof Exp: Fel microwave spectros, Univ Ill, 68-69; RES CHEMIST, NAT BUR STANDARDS, 69- Mem: Am Phys Soc; Am Chem Soc. Res: Nuclear magnetic resonance, particularly carbon-13, application to the characterization of polymer solids. Mailing Add: Div 311.03 Nat Bur Standards Washington DC 20234

VAN DER HEEM, PETER, b Kampen, Neth, June 25, 20; US citizen; m 53; c 2. PHYSICAL CHEMISTRY, CHEMICAL ENGINEERING. Educ: Delft Univ Technol, MS, 47; Georgetown Univ, PhD(phys chem), 62. Prof Exp: Asst prod mgr, Electro, Neth, 47-52; construct engr, Royal Dutch Blast Furnaces, 52-57; asst sci attache, Neth Embassy, DC, 57-60; res fel, Georgetown Univ, 60-62; res group leader, Koppers Co, Pa, 62-65; SECT LEADER RES MGT, J M HUBER CORP, 65- Mem: Am Chem Soc; Royal Neth Chem Soc. Res: Water purification; chemistry of anesthesia; silicate chemistry; pollution control; pigment research. Mailing Add: Rte 1 Box 40B Perryville MD 21903

VANDERHEIDEN, BERNARDO S, b Bogota, Colombia, June 9, 26; nat US; m 53; c 3. BIOCHEMISTRY. Educ: Mich State Univ, BS, 47, MS, 49; Univ Wash, PhD(biochem), 56. Prof Exp: Res assoc biochem, Sch Med, Univ Wash, 56-58; sr res fel, 59-62, MED RES SCIENTIST, EASTERN PA PSYCHIAT INST, 62-; RES ASST PROF PSYCHIAT & HUMAN BEHAV (BIOCHEM), JEFFERSON MED COL, THOMAS JEFFERSON UNIV, 75- Mem: AAAS; Am Chem Soc; Int Soc Hemat; Am Soc Biol Chemists. Res: Glycolytic enzymes; phosphate esters and nucleotides; erythrocyte metabolism. Mailing Add: Eastern Pa Psychiat Inst Philadelphia PA 19129

VAN DER HELM, DICK, b Velsen, Netherlands, Mar 16, 33; m 60; c 5. PHYSICAL CHEMISTRY. Educ: Univ Amsterdam, Drs, 56, DSc(x-ray diffraction), 60. Prof Exp: Res assoc x-ray diffraction, Ind Univ, 57-59 & Inst Cancer Res, Philadelphia, 59-62; from asst prof to assoc prof, 62-69, PROF PHYS CHEM, UNIV OKLA, 69- Concurrent Pos: NIH Develop Award, 69-74. Mem: Am Chem Soc; Am Crystallog Asn; Royal Netherlands Chem Soc. Res: Molecular structure determination by means of x-ray diffraction of natural products and peptide chelates. Mailing Add: Dept of Chem 620 Parrington Oval Univ of Okla Norman OK 73069

VANDERHILL, BURKE GORDON, b Laporte, Ind, Jan 15, 20; m 52; c 2. GEOGRAPHY. Educ: Mich State Univ, BS, 41; Univ Nebr, MA, 47; Univ Mich, PhD(geog), 56. Prof Exp: Instr geog, Mich State Univ, 47-48; instr, Easterm Mich Univ, 50; from asst prof to assoc prof, 50-65, PROF GEOG, FLA STATE UNIV, 65- Concurrent Pos: Actg chmn dept geog, Fla State Univ, 71-72. Honors & Awards: J Geog Award, Nat Coun Geog Educ, 59 & 72. Mem: Asn Am Geog; Am Soc Can Studies in US; Arctic Inst NAm. Res: Land settlement both current and historical, with specific reference to boreal America. Mailing Add: Dept of Geog Fla State Univ Tallahassee FL 32306

VANDERHOEF, LARRY NEIL, b Frazee, Minn, Mar 20, 41; m 63; c 2. PLANT PHYSIOLOGY, MORPHOGENESIS. Educ: Univ Wis-Milwaukee, BS, 64, MS, 65; Purdue Univ, Lafayette, PhD(plant physiol), 69. Prof Exp: Nat Res Coun fel, Univ Wis-Madison, 69-70; ASSOC PROF PLANT DEVELOP, UNIV ILL, URBANA, 70-, HEAD BIOL PROGS, 74- Mem: AAAS; Am Soc Plant Physiol; Scand Soc Plant Physiol; Am Soc Limnol & Oceanog. Res: Plant hormones and nucleic acid metabolism; nitrogen fixation. Mailing Add: Dept of Bot Morrill Hall Univ of Ill Urbana IL 61801

VANDERHOEK, JACK YEHUDI, b Hilversum, Neth, Jan 1, 41; US citizen; m 66. BIOCHEMISTRY, ORGANIC CHEMISTRY. Educ: City Col New York, BS, 60; Mass Inst Technol, PhD(org chem), 66. Prof Exp: Sr res chemist, Gen Mills, Inc, 66-68, group leader vitamin E and sterol synthesis, 68-69; NIH spec fel, Princeton Univ, 69-70 & Univ Fla, 71; res assoc biochem, Univ Mich, Ann Arbor, 71-72, instr, 72-74; LECTR, HADASSAH UNIV HOSP, HEBREW UNIV MED SCH, JERUSALEM, ISRAEL, 74- Mem: AAAS. Res: Prostaglandin metabolism; photobiology. Mailing Add: c/o C Myerowitz Middle Rd Ellington CT 06029

VAN DER HOEVEN, BERNARD JACOB CORNELIS, JR, physics, see 12th edition

VAN DER HOEVEN, THEO A, b Indonesia, Feb 13, 33; US citizen; m 60; c 3. BIOCHEMISTRY. Educ: Brooklyn Col, BS, 65; Columbia Univ, PhD(biochem), 71. Prof Exp: Instr biochem, Univ Mich, 71-74; ASST PROF MED CHEM, UNIV MD, BALTIMORE, 74- Mem: Sigma Xi. Res: Steroid biochemistry; membrane bound enzymes; detoxication of xenobiotics. Mailing Add: Dept of Med Chem Sch of Pharm Univ of Md Baltimore MD 21201

VAN DER HOFF, BERNARD MARIA EUPHEMIUS, b Utrecht, Netherlands, Aug 19, 19; Can citizen; m 49; c 4. POLYMER CHEMISTRY, POLYMER PHYSICS. Educ: Tech Inst Amsterdam, Ing, 41; Delft Univ Technol, Ir, 50. Prof Exp: Sci officer electrochem anal, Delft Univ Technol, 50-51; fel surface energy, Nat Res Coun Can, 51-53; sr chemist, Polymer Corp, Ltd, 53-57, res assoc, 57-66; PROF POLYMER CHEM & PHYSICS, UNIV WATERLOO, 67- Concurrent Pos: Consult, Dunlop Res Ctr, 69- Mem: Am Chem Soc; Chem Inst Can. Res: Emulsion polymerization; rheology; viscoelasticity; structural characterization of elastomers; degradation of polymers; mechanical properties of elastomer vulcanizates. Mailing Add: Dept of Chem Eng Univ of Waterloo Waterloo ON Can

VANDERHOFF, JOHN W, b Niagara Falls, NY, Aug 2, 25; m 50; c 2. POLYMER CHEMISTRY, COLLOID CHEMISTRY. Educ: Niagara Univ, BS, 47; Univ Buffalo, PhD(phys chem), 51. Prof Exp: Chemist, Phys Res Lab, Dow Chem Co, 50-56, proj leader, 56-58, assoc scientist, 58-70, Plastics Dept Res Lab, 63-70; assoc prof, 70-74, DIR, NAT PRINTING INK RES INST, ASSOC DIR COATINGS, CTR SURFACE

& COATINGS RES, LEHIGH UNIV, 70-, PROF CHEM, 74-, CO-DIR EMULSION POLYMERS INST, 75- Concurrent Pos: Participant, Career Scientist Assignment Prog, van't Hoff Lab, Utrecht, 65-66. Honors & Awards: Res award, Union Carbide Chem, 64-65. Mem: AAAS; fel Am Inst Chem; Am Chem Soc. Res: Polymerization kinetics; solution properties of polymers; mechanism of emulsion polymerization; latex properties; foamed plastics; mechanism of latex film formation; colloidal properties of latexes; monodisperse latexes; printing inks; deinking of wastepaper. Mailing Add: Ctr for Surface & Coatings Res Lehigh Univ Bethlehem PA 18015

VANDERHOOF, ELLEN RUTH (MRS WALTER L PETERSON), b Plymouth, Wis, July 26, 30; m 62; c 1. PHYSIOLOGY. Educ: Univ Wis-La Crosse, BS, 53; Univ Iowa, MA, 56, PhD(physiol), 60. Prof Exp: Asst physiol, Univ Iowa, 56-57, from instr to asst prof, 57-62, res assoc psychiat, 62-65; INSTR PHYSIOL & ANAT, SOUTHEASTERN COMMUNITY COL, 70- Mailing Add: PO Box 252 Burlington IA 52601

VANDERHORST, PHILIP JOHN, organic chemistry, see 12th edition

VAN DER HOVEN, ISAAC, b Rotterdam, Netherlands, July 13, 23; US citizen; m 56; c 4. METEOROLOGY. Educ: Pa State Univ, BS, 51, PhD(meteorol), 56; Mass Inst Technol, MS, 52. Prof Exp: Res assoc meteorol, Pa State Univ, 52-55; res meteorologist, Brookhaven Nat Lab, US Weather Bur, 55-57, Nev Test Site, 57-61, Off Meteorol Res, DC, 61-65; CHIEF AIR RESOURCES ENVIRON LAB, NAT OCEANIC & ATMOSPHERIC ADMIN, 65- Mem: Am Meteorol Soc. Res: Atmospheric turbulence and diffusion; micrometeorology; meteorology as related to use of atomic energy. Mailing Add: Air Resources Environ Lab Nat Oceanic & Atmospheric Admin Silver Spring MD 20910

VAN DERIPE, DONALD R, b Lafayette, Ind, Feb 13, 34; m 64; c 1. PHARMACOLOGY. Educ: Purdue Univ, BS, 56, MS, 58; Northwestern Univ, PhD(pharmacol), 63. Prof Exp: Fel cardiovasc pharmacol, Emory Univ, 63-65; RES PHARMACOLOGIST, MALLINCKRODT CHEM WORKS, 65- Res: Radiopaque diagnostic agents; cardiovascular and radionuclide pharmacology. Mailing Add: Mallinkrodt Chem Works 3600 N Second St St Louis MO 63160

VANDER JAGT, DAVID LEE, b Grand Rapids, Mich, Jan 13, 42; m 67. BIOCHEMISTRY. Educ: Calvin Col, AB, 63; Purdue Univ, Lafayette, PhD(chem), 67. Prof Exp: NIH fel biochem, Northwestern Univ, 67-69; asst prof, 69-74, ASSOC PROF BIOCHEM, UNIV N MEX, 74- Concurrent Pos: Res career develop award, Nat Cancer Inst, 74- Mem: AAAS; Am Chem Soc: Am Soc Biol Chem. Res: Enzyme, coenzyme reaction mechanisms of glutathione requiring enzymes, especially glyoxalase; metabolic role of methylglyoxal; biomedical applications of 13-C; metabolism of chemical carcinogens. Mailing Add: Dept of Biochem Univ of N Mex Sch of Med Albuquerque NM 87106

VANDERJAGT, DONALD W, b Muskegon, Mich, Feb 25, 38; m 58; c 4. COMBINATORICS. Educ: Hope Col, AB, 59; Fla State Univ, MS, 61; Western Mich Univ, PhD(math), 73. Prof Exp: Instr math, Cent Univ Iowa, 62-64; from asst prof to assoc prof, 64-75, PROF MATH, GRAND VALLEY STATE COLS, 75- Mem: Am Math Soc; Math Asn Am; Asn Comput Mach; Nat Coun Teachers Math. Res: Graph theory, local properties, degree sets, Hamiltonian properties, generalized Ramsey theory. Mailing Add: Dept of Math Grand Valley State Cols Allendale MI 49401

VAN DER KLOOT, ALBERT PETER, b Chicago, Ill, Jan 22, 21; m 48; c 2. FOOD CHEMISTRY. Educ: Mass Inst Technol, SB, 42. Prof Exp: Chemist flower preserv, Flower Foods, Inc, 46-47; chemist biscuits & crackers, Independent Biscuit Mfg Tech Inst, 47-52; chief chemist brewing & foods, 52-72, PRES, WAHL-HENIUS INST, INC, 72- Mem: Am Soc Brewing Chem; Am Chem Soc; Instrument Soc Am; Inst Food Technol; AAAS. Res: Commercial application of plant tissue culture; soda cracker fermentation; freeze drying of microorganisms; vapor pressure-moisture relationships; statistical analysis of brewing process; gas chromatography; alcoholic beverages; trace components of food. Mailing Add: Wahl-Henius Inst Inc 4206 N Broadway Chicago IL 60613

VAN DER KLOOT, WILLIAM GEORGE, b Chicago, Ill, Feb 18, 27; m 63; c 2. PHYSIOLOGY. Educ: Harvard Univ, SB, 48, PhD(biol), 52. Prof Exp: Nat Res Coun fel, Cambridge Univ, 52-53; instr biol, Harvard Univ, 53-56; from asst prof to assoc prof zool, Cornell Univ, 56-58; prof pharmacol & chmn dept, Sch Med, NY Univ, 58-61, prof physiol & chmn dept physiol & biophys, 61-71; PROF PHYSIOL & BIOPHYS & CHMN DEPT, STATE UNIV NY STONY BROOK, 71- Concurrent Pos: Consult, NSF, 59-65 & NIH, 68-72. Mem: AAAS; Am Soc Zool; Soc Exp Biol & Med; Am Physiol Soc. Res: Comparative physiology and pharmacology. Mailing Add: Dept of Physiol & Biophys Health Sci Ctr State Univ of NY Stony Brook NY 11796

VANDERKOOI, NICHOLAS, JR, b Paterson, NJ, Jan 7, 32; m 55; c 4. PHYSICAL CHEMISTRY. Educ: Calvin Col, AB, 53; Wayne State Univ, PhD(phys chem), 57. Prof Exp: SR RES CHEMIST, SPECIALTY CHEM DIV, ALLIED CHEM CORP, 57- Mem: Am Chem Soc. Res: Electron spin resonance; electrodeposition; instrumental analysis; fluorine chemistry; inorganic synthesis; polymer characterization; polymerization; monomers process development. Mailing Add: Allied Chem Corp PO Box 1087R Morristown NJ 07960

VANDERKOOI, WILLIAM NICHOLAS, b Paterson, NJ, Dec 19, 29; m 54; c 3. INDUSTRIAL CHEMISTRY. Educ: Calvin Col, AB, 51; Purdue Univ, MS, 53, PhD(phys chem), 55. Prof Exp: Lab asst chem & physics, Calvin Col, 48-51, asst, Off Naval Res, 50-51; chem, Purdue Univ, 51-52, US Air Force contract, 52, instruments, 52-53; res chemist, C C Kennedy Res Lab, 55-64, group leader, Polymer & Chem Res Lab, 64-69, ASSOC SCIENTIST, HYDROCARBON & MONOMERS RES LAB, DOW CHEM CO, 69- Mem: Sigma Xi; Am Chem Soc. Res: Polymerization kinetics; radiation grafting; polymer synthesis and properties; metal chelation and purification; hydrocarbon analyses; high temperature reactions; hydrocarbon pyrolysis; petrochemical processes; catalysis; alkylation processes. Mailing Add: Dow Chemical Co 677 Bldg Midland MI 48640

VANDERKOOY, JOHN, b Neth, Jan 1, 41; Can citizen; m 65; c 1. SOLID STATE PHYSICS, OPTICS. Educ: McMaster Univ, BEng, 63, PhD(physics), 67. Prof Exp: Nat Res Coun Can fel physics, Cambridge Univ, 67-69; res assoc, 69-70, ASST PROF PHYSICS, UNIV WATERLOO, 70-, NAT RES COUN CAN GRANTS, 69- Mem: Can Asn Physicists. Res: Low temperature solid state physics of metals; optics and theories of light. Mailing Add: Dept of Physics Univ of Waterloo Waterloo ON Can

VANDERLAAN, WILLARD PARKER, b Muskegon, Mich, June 5, 17; m 44; c 3. ENDOCRINOLOGY. Educ: Harvard Med Sch, MD, 42. Prof Exp: Intern med, Boston City Hosp, 42-43, asst resident path, 43; instr pharmacother, Harvard Med Sch, 44-45; fel endocrinol, J H Pratt Diag Hosp, Boston, 45-47; from asst prof to assoc prof med, Med Sch, Tufts Univ, 47-56; HEAD ENDOCRINOL, SCRIPPS CLIN & RES FOUND, 56-; PROF MED, UNIV CALIF, SAN DIEGO, 68- Concurrent Pos: Fel med, Thorndike Mem Lab, Boston City Hosp, 44; consult, Boston Vet Admin Hosp, 54-56; mem endocrinol study sect, NIH, 71-75, chmn, 74-75. Mem: Am Thyroid Asn; Endocrine Soc; Am Soc Clin Invest. Res: Growth hormone; prolactin; thyroid physiology. Mailing Add: Scripps Clin & Res Found 476 Prospect St La Jolla CA 92037

VANDERLIN, CARL JOSEPH, JR, b Williamsport, Pa, Apr 28, 26; m 50; c 6. MATHEMATICS. Educ: Univ Chicago, PhB, 48, SB, 50; Univ Wis, MS, 52. Prof Exp: Asst math, Univ Wis, 51-54, instr, 54-55; from instr to assoc prof, Wis State Col, Whitewater, 55-61, actg chmn dept, 59-60; LECTR MATH, UNIV WIS-EXTEN, 61- Concurrent Pos: Univ Wis-US Agency Int Develop consult, Inst Adult Studies, Univ Nairobi, 67, vis lectr, 68-71, chief of party, 70-71. Mem: Math Asn Am; Nat Coun Teachers Math; Nat Univ Exten Asn; US Metric Asn; Can Metric Asn. Mailing Add: 845 Exten Bldg Univ of Wis-Exten Madison WI 53706

VANDERLINDE, RAYMOND E, b Newark, NY, Feb 28, 24; m 48; c 3. BIOCHEMISTRY. Educ: Syracuse Univ, AB, 44, MS(educ), 45, MS(chem), 47, PhD(biochem), 50; Am Bd Clin Chem, dipl, 60. Prof Exp: Teacher high sch, NY, 45-46; from asst prof to assoc prof biochem, Sch Med, Univ Md, 50-57; asst prof, Col Med, State Univ NY Upstate Med Ctr, 57-62; assoc lab dir & clin chemist, Mem Hosp Cumberland, Md, 62-65; DIR LABS CLIN CHEM, DIV LABS & RES, NY STATE DEPT HEALTH, 65- Concurrent Pos: Lab admin dir & clin biochemist, Syracuse Mem Hosp, 57-62; consult, Madison County Lab, NY, 59-62, Rome City Lab, 61-62 & Meyersdale Community Hosp Lab, Pa, 64-65; clin asst prof, Med Ctr, Univ WVa, 64-65; adj assoc prof, Albany Med Col, 70-; mem diag prods adv comt, Food & Drug Admin, 72-75. Mem: Am Chem Soc; fel Am Asn Clin Chem; Asn Clin Sci; assoc Am Soc Clin Path; Am Pub health Asn. Res: Clinical chemistry; clinical enzymology and standardization. Mailing Add: Labs for Clin Chem Div of Labs & Res NY State Dept of Health Albany NY 12201

VAN DER LINDE, REINHOUD H, b Amsterdam, Holland, July 14, 29; US citizen; m 58; c 5. MATHEMATICS. Educ: NY Univ, BA, 53, MS, 56; Rensselaer Polytech Inst, PhD(math), 68. Prof Exp: Res asst, NY Univ, 54-56; PROF MATH, BENNINGTON COL, 56- Mem: Math Asn Am; NY Acad Sci. Res: Random Eigenvalue problems; differential equations; functional analysis. Mailing Add: Dept of Math Bennington Col Bennington VT 05201

VANDERLIP, RICHARD L, b Woodston, Kans, May 6, 38; m 60; c 2. AGRONOMY. Educ: Kans State Univ, BS, 60; Iowa State Univ, MS, 62, PhD(agron), 65. Prof Exp: Asst prof, 64-70, ASSOC PROF AGRON, KANS STATE UNIV, 70- Mem: AAAS; Am Soc Agron; Soil Sci Soc Am; Crop Sci Soc Am. Res: Ecology of crop plants, especially climatic interrelationships. Mailing Add: Dept of Agron Waters Hall Kans State Univ Manhattan KS 66502

VANDER LUGT, KAREL L, b Pella, Iowa, Apr 25, 40; m 64; c 1. SOLID STATE PHYSICS. Educ: Hope Col, BA, 62; Wayne State Univ, PhD(exp solid state physics), 67. Prof Exp: Nat Res Coun assoc physics, Naval Res Lab, Washington, DC, 67-68; ASST PROF PHYSICS, AUGUSTANA COL, SDAK, 69- Mem: AAAS; Am Asn Physics Teachers; Hist Sci Soc. Res: Radiation damage in crystals; experimental solid state physics. Mailing Add: Dept of Physics Augustana Col Sioux Falls SD 57102

VAN DER MAATEN, MARTIN JUNIOR, b Alton, Iowa, Aug 6, 32; m 56; c 2. VETERINARY VIROLOGY. Educ: Iowa State Univ, DVM, 56, PhD(vet bact), 64. Prof Exp: Vet practice, 58-60; res asst vet virol, Iowa State Univ, 60-61, Nat Inst Allergy & Infectious Dis fel, 61-64, asst prof, 64-67; VET LAB OFFICER, NAT ANIMAL DIS LAB, US DEPT AGR, 67- Mem: Am Vet Med Asn; Am Soc Microbiol; Vet Cancer Soc. Res: Virological and serological studies of bovine lymphosarcoma and bovine leukemia virus. Mailing Add: Nat Animal Dis Lab US Dept of Agr PO Box 70 Ames IA 50010

VANDERMEER, CANUTE, b Kulangsu, Fukien, China, Apr 2, 30; US citizen; m 55; c 2. GEOGRAPHY. Educ: Hope Col, BA, 50; Univ Mich, MA, 56, PhD(geog), 62. Prof Exp: Teacher geog & Eng, Montague High Sch, Mich, 52-54 & geog, Univ Mich, Ann Arbor, 60-61; from instr to assoc prof geog, Univ Wis-Milwaukee, 61-73; PROF & CHMN DEPT GEOG, UNIV VT, 73- Concurrent Pos: Consult, Peace Corps, 65; Soc Sci Res Coun grant, Univ Wis-Milwaukee, 63; NSF fel, Joint Comn Rural Reconstruct, Taiwan, 65-66 & Johnson Found grant, 69; vis assoc prof geog, Univ Hawaii, 72. Mem: Asn Am Geog; Asn Asian Studies. Res: Irrigation water control methods and land ownership conditions in east Asia. Mailing Add: Dept of Geog Univ of Vermont Burlington VT 05401

VANDERMEER, JOHN H, b Chicago, Ill, July 21, 40; m 69; c 1. POPULATION BIOLOGY. Educ: Univ Ill, Urbana, BS, 61; Univ Kans, MA, 64; Univ Mich, Ann Arbor, PhD(ecol), 69. Prof Exp: Sloan Found fel, Univ Chicago, 69-70; asst prof ecol, State Univ NY Stony Brook, 70-71; asst prof, 71-74, ASSOC PROF ZOOL, UNIV MICH, ANN ARBOR, 74- Mem: Am Soc Ichthyologists & Herpetologists; Am Soc Naturalists; Ecol Soc Am. Res: Role of population processes as determiners of the structure of biological communities. Mailing Add: Dept of Zool Univ of Mich Ann Arbor MI 48104

VANDER MEER, PAUL, b Kulangsu, Amoy, China, Nov 6, 32; US citizen; m 64; c 2. CULTURAL GEOGRAPHY. Educ: Hope Col, BA, 54; Univ Mich, Ann Arbor, MA, 61, PhD(geog), 73. Prof Exp: Instr geog, Temple Univ, 67-71; ASST PROF GEOG, CALIF STATE UNIV, FRESNO, 71- Concurrent Pos: NSF res award, 76. Mem: China Acad; Asn Am Geogrs. Res: The relationship of land to the socioeconomic structure of Chinese villages. Mailing Add: Dept of Geog Calif State Univ Fresno CA 93740

VAN DER MEULEN, JOSEPH PIERRE, b Boston, Mass, Aug 22, 29; m 60; c 3. NEUROLOGY, NEUROPHYSIOLOGY. Educ: Boston Col, AB, 50; Boston Univ, MD, 54. Prof Exp: Intern med, Cornell Med Div, Bellevue Hosp, New York, 54-55, asst resident, 55-56; asst resident neurol, Harvard Neurol Unit, Boston City Hosp, 58-59, resident, 59-60; Nat Inst Neurol Dis & Blindness fel, Nobel Inst Neurophysiol, Karolinska Inst, Sweden, 60-62; instr, Harvard Neurol Unit, Boston City Hosp, 62-66, assoc, 66-67; asst prof, Sch Med, Case Western Reserve Univ, 67-69, assoc prof neurol & biomed eng, 69-71; PROF NEUROL & CHMN DEPT, SCH MED, UNIV SOUTHERN CALIF, 71-, DIR DEPT NEUROL, LOS ANGELES COUNTY-UNIV SOUTHERN CALIF MED CTR, 71- Concurrent Pos: Fel, Harvard Neurol Unit, Boston City Hosp, Mass, 62-66. Mem: Am Acad Neurol. Res: Neurophysiology of abnormalities of posture and movement; motor control systems in humans. Mailing Add: Dept of Neurol LAC-USC Med Ctr Los Angeles CA 90033

VANDER NOOT, GEORGE WARD, b Westchester, NY, Nov 21, 09; m 41; c 2. ANIMAL NUTRITION. Educ: Rutgers Univ, BS, 37, MEd, 41. Prof Exp: Asst animal husb, 37-39, from instr to assoc prof, 39-59, PROF ANIMAL HUSB, RUTGERS UNIV, 59-, ADJ PROF NUTRIT, 63- Concurrent Pos: Guest lectr, Sch Pub Health, Columbia Univ, 63- Honors & Awards: Recognition Award, Am Soc

Animal Sci, 62, Hon Fel Award, 69, Distinguished Serv Award, 72; Lindback Award, Rutgers Univ, 69. Mem: Am Inst Nutrit; Am Soc Animal Sci; NY Acad Sci; Animal Nutrit Res Coun. Mailing Add: Dept of Animal Sci Rutgers Univ Col of Agr & Environ Sci New Brunswick NJ 08903

VANDERPLOEG, HENRY ALFRED, b Chicago, Ill, Aug 23, 44; m 68. AQUATIC ECOLOGY. Educ: Mich Technol Univ, BS, 66; Univ Wis-Madison, MS, 68; Ore State Univ, PhD(biol oceanog), 72. Prof Exp: Aquatic ecologist modelling radionuclide cycling in aquatic systs, Environ Sci Div, Oak Ridge Nat Lab, 72-74; BIOL OCEANOGR PLANKTON ECOL GREAT LAKES, GREAT LAKES ENVIRON RES LAB, NAT OCEANIC & ATMOSPHERIC ADMIN, 74- Mem: AAAS; Am Soc Limnol & Oceanog; Ecol Soc Am; Sigma Xi. Res: Ecology of slective feeding of zooplankton; dynamics of seasonal succession of Great Lakes plankton. Mailing Add: Great Lakes Environ Res Lab 2300 Washtenaw Ave Ann Arbor MI 48104

VANDERRYN, JACK, b Groningen, Netherlands, Apr 14, 30; nat US; m 56; c 4. PHYSICAL CHEMISTRY, RESEARCH ADMINISTRATION. Educ: Lehigh Univ, BA, 51, MS, 52, PhD(chem), 55. Prof Exp: Asst, Lehigh Univ, 51-55; asst prof chem, Va Polytech Inst, 55-58; chemist, Res & Develop Div, US Atomic Energy Comn, Tenn, 58-62, tech adv to asst gen mgr res & develop, Washington, DC, 62-67; US Dept State sr sci adv to US Mission, Int Atomic Energy Agency, 67-71; tech asst, Off Gen Mgr, US Atomic Energy Comn, Washington, DC, 71-72, tech asst to dir, Div Appl Technol, 72-74, chief, Energy Technol Br, Div Appl Technol, 74-75, actg dir div energy storage, 75, DIR, OFF INT RES & DEVELOP PROGS, US ENERGY RES & DEVELOP ADMIN, 75- Mem: AAAS. Res: Applications of nuclear technology; science policy development; energy technology development; research and development administration; international cooperation in energy research and development. Mailing Add: 8112 Whittier Blvd Bethesda MD 20034

VAN DERSAL, WILLIAM RICHARD, b Portland, Ore, Apr 6, 07; m 49; c 5. PLANT ECOLOGY. Educ: Reed Col, AB, 29; Univ Pittsburgh, MS, 31, PhD(plant ecol), 33. Prof Exp: Asst, Univ Pittsburgh, 29-35; assoc biologist, Soil Conserv Serv, US Dept Agr, Pa, 35-36, biologist, 36-38, chief biol div, 38-42, personnel mgt div, 42-48, opers, Pac Region, 48-53, asst administr for mgt, 54-62, dep adminstr for mgt, 62-72; MGT CONSULT, 72-; DEAN MGT COL, NAT GRAD UNIV, 74- Concurrent Pos: Teacher, US Dept Agr Grad Sch, 57-; prof, Nat Grad Univ. Honors & Awards: Rockefeller Award, 57. Mem: Ecol Soc Am; AAAS; Am Soc Plant Taxon; Soil Conserv Soc Am; Am Geog Soc. Res: Plant and animal ecology; plant taxonomy; horticulture; natural resources. Mailing Add: 6 S Kensington St Arlington VA 22204

VANDERSALL, JOHN HENRY, b Helena, Ohio, July 20, 28; m 63; c 2. ANIMAL NUTRITION. Educ: Ohio State Univ, BS, 50, MS, 54, PhD(dairy sci), 59. Prof Exp: Instr dairy sci, Agr Exp Sta, Ohio State Univ, 57-59; from asst prof to assoc prof, 59-71, PROF DAIRY SCI, UNIV MD, COLLEGE PARK, 71- Mem: Fel AAAS; Am Soc Animal Sci; Am Dairy Sci Asn. Res: Effects of forages on milk production and growth and physiological bases for differences; effects of feeds upon the composition of milk. Mailing Add: Dept of Dairy Sci Univ of Md College Park MD 20740

VAN DER SCHALIE, HENRY, b Amsterdam, Netherlands, Jan 8, 07; nat US, m 36; c 4. MALACOLOGY, MEDICAL PARASITOLOGY. Educ: Calvin Col, AB, 29; Univ Mich, MS, 31, PhD(zool), 34. Prof Exp: Asst, Mus Zool, 29-34, from instr to assoc prof, 34-56, asst cur, 34-44, PROF ZOOL, UNIV MICH, ANN ARBOR, 56-, CUR MOLLUSKS, MUS ZOOL, 44- Concurrent Pos: Exchange prof, Univ PR, 40-41; sr adv, Govt Egypt, WHO, 52-53, 54; consult, Sudan Minister Health, 54 & 406 Med Lab, Japan & WHO Team, Leyte, Philippines, 55; prof, Cornell Univ, 55; rep comn parasitic dis, WHO Conf Bilharziasis, Brazzaville, French Equatorial Africa, 56, mem expert panel, WHO; exchange prof, Univ Hawaii, 59, vis prof, 68; mem ecol sect, AEC, Oak Ridge Nat Lab, 61; NSF sponsored study mollusks, Taiwan, 62; site-visit Am Found Trop Med, Liberian Inst, Liberia, 64; Tenn Valley Auth-du Pont & Limnol Dept, Phila Acad Nat Sci, mem surv mussels, Kentucky Lake, 64-65; partic panel US-Japan Health Agreement, Tokyo, 66; conserv, Found Conf Ecol Aspects of Int Develop, Washington, DC, 68; mem comn parasitic dis, Armed Forces Epidemiol Bd; adj prof Eastern Mich Univ, 69; consult, Smithsonian Inst in site-visit surv snailborne dis in southeast Asia, 70; external examr, Univ Ibadan, Nigeria, 75. Mem: Fel AAAS; Am Malacol Union (vpres, 41-45, pres, 45-46); Am Micros Soc; Am Soc Trop Med & Hyg; Am Soc Zool. Res: Systematics, ecology and life histories of mollusks; parasitic diseases; parasitic diseases, particularly in relation to medical malacology; zoogeography of mollusks. Mailing Add: 1422 Iroquois Dr Ann Arbor MI 48104

VANDERSCHMIDT, GEORGE FREDERICK, physics, see 12th edition

VANDERSLICE, JOSEPH THOMAS, b Philadelphia, Pa, Dec 21, 27; m 54; c 7. PHYSICAL CHEMISTRY. Educ: Boston Col, BS, 49; Mass Inst Technol, PhD(phys chem), 53. Prof Exp: From instr to asst prof chem, Catholic Univ, 52-56; from asst prof to assoc prof, 56-62, dir inst molecular physics, 67-68, PROF MOLECULAR PHYSICS, UNIV MD, COLLEGE PARK, 62-, HEAD DEPT CHEM, 68- Mem: AAAS; Am Chem Soc; fel Am Phys Soc. Res: Intermolecular forces; thermodynamic temperature scale; interpretations of molecular beam experiments; transport properties of high temperature gases; franck-condon factors and interpretation of spectroscopic data on diatomic molecules; auroral spectroscopy. Mailing Add: Dept of Chem Univ of Md College Park MD 20740

VANDERSLICE, THOMAS AQUINAS, b Philadelphia, Pa, Jan 8, 32; m 56; c 4. PHYSICAL CHEMISTRY. Educ: Boston Col, BA, 53; Catholic Univ, PhD(phys chem), 56. Prof Exp: Asst, Catholic Univ, 53-54, Fulbright fel, 56; res assoc, Res & Develop Ctr, 56-62, mgr eng, Vacuum Prod Oper, 62-64, mgr, Vacuum Prod Bus Sect, 64-66, gen mgr, Memory Equip Dept, Okla, 66-68, Info Devices Dept, 68, dep div gen mgr, Info Systs Sales & Serv, Ariz, 68-70, vpres & div gen mgr, Electronic Components Bus Div, 70-72, VPRES & GROUP EXEC, SPECIAL SYSTS & PROD GROUP, GEN ELEC CO, 72- Concurrent Pos: Mem sci adv bd, Boston Col, 65- Mem: Am Chem Soc; Am Phys Soc; Am Vacuum Soc. Res: Surface chemistry; mass spectrometry; high vacuum technology; gaseous discharges. Mailing Add: Gen Elec Co Fairfield CT 06431

VANDER SLUIS, KENNETH LEROY, b Holland, Mich, Dec 19, 25; m 52; c 3. PHYSICS. Educ: Baldwin-Wallace Col, BS, 47; Pa State Univ, MS, 50, PhD(physics), 52. Prof Exp: PHYSICIST, OAK RIDGE NAT LAB, 52- Concurrent Pos: Res guest, Spectros Lab, Mass Inst Technol, 60-61; mem comt line spectra of the elements, Nat Res Coun, 61- Mem: Am Phys Soc; Optical Soc Am; Sigma Xi. Res: Experimental atomic spectroscopy, echelle gratings, interferometry, gas laser systems. Mailing Add: 954 W Outer Dr Oak Ridge TN 37830

VANDERSPURT, THOMAS HENRY, b Lawrence, Mass, Apr 1, 46; m 71; c 1. PHYSICAL INORGANIC CHEMISTRY. Educ: Lowell Technol Inst, BS, 67; Princeton Univ, MA & PhD(chem), 72. Prof Exp: Fel catalysis chem, Princeton Univ, 72-73; RES CHEMIST CATALYSIS CHEM, CELANESE RES CO, CELANESE INC, 73- Mem: Am Chem Soc; NAm Catalysis Soc. Res: Supported metal alloy catalysts; metal/metal oxide selective oxidation catalysts; supported hydroformylation catalysis. Mailing Add: 203 Gates Ave Gillette NJ 07933

VANDER STOUW, GERALD GORDON, b Rochester, NY, May 15, 37; m 60; c 2. ORGANIC CHEMISTRY, INFORMATION SCIENCE. Educ: Calvin Col, AB, 58; Ohio State Univ, PhD(org chem), 64. Prof Exp: Res asst chem, 64-70, group leader, Chem Info Sci Dept, 70-74, CHEM SUBSTANCE PROJ MGR, CHEM ABSTR SERV, 74- Mem: AAAS; Am Chem Soc; Am Soc Info Sci. Res: Mechanized indexing and retrieval of chemical information; computer-based processing and searching of chemical information, especially chemical structures and nomenclature. Mailing Add: Chem Abstr Serv Ohio State Univ Columbus OH 43210

VAN DER VAART, HUBERTUS ROBERT, b Makassar, Celebes, Indonesia, Mar 2, 22; m 53; c 3. MATHEMATICS, STATISTICS. Educ: Univ Leiden, Drs, 50, PhD(theoret biol), 53. Prof Exp: Sci officer, Univ Leiden, 50-57; vis assoc prof exp statist, NC State Col, 57-58 & statist, Univ Chicago, 58; extraordinary prof theoret biol, Univ Leiden, 58-60, prof, 60-62, dir inst theoret biol, 58-62, dir cent comput inst, 61-62; assoc prof, 62-63, PROF EXP STATIST & MATH, N C STATE UNIV, 63- Concurrent Pos: Co-ed, Acta Biotheoretica, 53-75; Netherlands Orgn Pure Res fel, 57-58; co-ed, Statistica Neerlandica, 60-74; mem, Panel Life Sci, Comt Undergrad Prog Math, 67-70; mem, Panel Instructional Mats Appl Math, Comt Undergrad Prog Math, Math Asn Am, 74- Mem: Biomet Soc; Am Math Soc; Math Asn Am; Inst Math Statist; Soc Indust & Appl Math. Res: Mathematical statistics; probability; stochastic processes; theoretical biology; principles of scientific method; mathematical models for biosystems. Mailing Add: Inst of Statist NC State Univ PO Box 5457 Raleigh NC 27607

VANDER VALK, PAUL DAVID, b Passaic, NJ, Jan 15, 47; m 70. ORGANIC CHEMISTRY. Educ: Iowa Wesleyan BS, 69; Univ Iowa, MS, 71, PhD(org chem), 74. Prof Exp: RES CHEMIST ORG SYNTHESIS, EASTMAN KODAK CO, 74- Mailing Add: Res Labs Eastman Kodak Co Rochester NY 14650

VAN DER VEEN, JAMES MORRIS, b Chicago, Ill, Sept 19, 31; m 59; c 3. ORGANIC CHEMISTRY, X-RAY CRYSTALLOGRAPHY. Educ: Swarthmore Col, BA, 53; Harvard Univ, AM, 56, PhD(org chem), 59. Prof Exp: Fels, Ga Inst Technol, 58-59 & Argonne Nat Lab, 59-60; res assoc nuclear magnetic resonance spectros, Retina Found, 60-61; asst prof, 61-65, ASSOC PROF ORG CHEM, STEVENS INST TECHNOL, 65- Concurrent Pos: Consult, Picatinny Arsenal, NJ, 63-68 & Exxon Corp, 74- Mem: AAAS; Am Crystallog Asn; Am Chem Soc; Sigma Xi; The Chem Soc. Res: Physical organic chemistry; nuclear magnetic resonance spectroscopy; mechanism of enzyme action. Mailing Add: Dept of Chem & Chem Eng Stevens Inst of Technol Hoboken NJ 07030

VANDERVEEN, JOHN EDWARD, b Prospect Park, NJ, May 13, 34; m 67; c 2. NUTRITION, CHEMISTRY. Educ: Rutgers Univ, BS, 56; Univ NH, PhD(chem, nutrit), 61. Prof Exp: Res chemist, US Air Force Sch Aerospace Med, 64-75; DIR DIV NUTRIT, BUR FOODS, FOOD & DRUG ADMIN, DEPT HEALTH, EDUC & WELFARE, 75- Mem: Am Inst Nutrit; Am Chem Soc; Am Inst Clin Nutrit; Inst Food Technologists; Am Dairy Sci Asn. Res: Nutritional requirements of the American population; assessment of the nutritional quality of the national food supply; energy and mineral requirements and effects of excess nutrient intakes. Mailing Add: HFF 260 FDA 200 C St Washington DC 20204

VANDER VELDE, EDWARD JAY, b Royal Oak, Mich, July 1, 37; m 62; c 2. CULTURAL GEOGRAPHY. Educ: Univ Mich, Ann Arbor, BA, 59, MA, 63, PhD(geog), 72. Prof Exp: Instr geog, Mich State Univ, 67-71; lectr, 71-72, ASST PROF GEOG, STATE UNIV NY BINGHAMTON, 72- Mem: Asn Am Geogrs; Asn Asian Studies. Res: Irrigation development and socio-cultural change in South Asia; socio-cultural impact of coal strip mining in Appalachia. Mailing Add: Dept of Geog State Univ of NY Binghamton NY 13901

VANDER VELDE, JOHN CHRISTIAN, b Mich, Sept 25, 30; m 53; c 3. PHYSICS. Educ: Hope Col, AB, 52; Univ Mich, MA, 53, PhD(physics), 58. Prof Exp: From instr to assoc prof physics, 58-67, PROF PHYSICS, UNIV MICH, ANN ARBOR, 67- Concurrent Pos: Assoc in res, LePrince-Ringuet Lab, Polytech Sch, Paris, 66-67; mem prog comt, Argonne Nat Lab, 71-74; mem users exec comt, Nat Accelerator Lab, 72-74. Mem: Am Phys Soc. Res: Elementary particle physics. Mailing Add: Dept of Physics Univ of Mich Ann Arbor MI 48104

VANDERVEN, NED STUART, b Ann Arbor, Mich, July 15, 32; m 61; c 2. PHYSICS. Educ: Harvard Col, AB, 55; Princeton Univ, PhD(physics), 62. Prof Exp: Instr physics, Princeton Univ, 59-61; from instr to asst prof, 61-68, ASSOC PROF PHYSICS, CARNEGIE-MELLON UNIV, 68- Mem: Am Phys Soc. Res: Magnetic resonance; solid state physics. Mailing Add: Dept of Physics Carnegie-Mellon Univ Pittsburgh PA 15213

VAN DER VOO, ROB, b Zeist, Netherlands, Aug 4, 40; m 66; c 2. GEOLOGY, GEOPHYSICS. Educ: State Univ Utrecht, BSc, 61, Drs, 65 & 69, PhD(geol, geophys), 69. Prof Exp: Res assoc paleomagnetism, State Univ Utrecht, 65-70; vis asst prof, 70-72, asst prof, 72-75, ASSOC PROF GEOPHYS, UNIV MICH, ANN ARBOR, 75- Mem: Am Geophys Union; Geol Soc Am; Seismol Soc Am; Ger Geol Asn; Royal Netherlands Geol & Mining Soc. Res: Paleomagnetism, plate tectonics of the Atlantic Ocean and Mediterranean Sea; stratigraphy and tectonics of Pyrenean-Alpine Mountain belt; surface waves of earthquakes and nuclear explosions. Mailing Add: Dept of Geol & Mineral Univ of Mich Ann Arbor MI 48104

VAN DER VOORN, PETER C, b Haarlem, Netherlands, Feb 25, 40; US citizen; m 65. ELECTROPHOTOGRAPHY. Educ: Wichita Univ, BS, 61; Univ Ill, PhD(inorg chem), 65. Prof Exp: RES CHEMIST, EASTMAN KODAK CO, 65- Mem: Am Chem Soc; The Chem Soc. Res: Inorganic photoconductors; solid state chemistry. Mailing Add: Eastman Kodak Co Kodak Park Rochester NY 14617

VANDERVOORT, PETER OLIVER, b Detroit, Mich, Apr 25, 35; m 56; c 2. ASTRONOMY. Educ: Univ Chicago, AB, 54, SB, 55, SM, 56, PhD(physics), 60. Prof Exp: Vis res assoc, Nat Radio Astron Observ, Assoc Univs, Inc, WVa, 60; NSF fel, Princeton Univ Observ, 60-61; asst prof, 61-65, ASSOC PROF ASTRON, UNIV CHICAGO, 65- Concurrent Pos: NSF sr fel, Leiden Observ, Netherlands, 67-68. Mem: Int Astron Union; Am Astron Soc; Am Phys Soc; Royal Astron Soc. Res: Hydrodynamic stability; gas dynamics; interstellar matter; stellar dynamics; galactic structure. Mailing Add: Dept of Astron & Astrophys Univ of Chicago 1100 E 58th St Chicago IL 60637

VANDER WALL, EUGENE, b Munster, Ind, Feb 8, 31; m 53; c 4. PHYSICAL INORGANIC CHEMISTRY. Educ: Calvin Col, BS, 52; Univ Colo, PhD(chem), 57. Prof Exp: Res chemist, Atomic Energy Div, Phillips Petrol Corp, Idaho, 56-60, supvr chemist, 60-63; mem sci staff, Aerojet-Gen Corp, Sacramento, 63-64, supvr liquid propellant res, 64-67, tech supvr liquid propellant res & develop, 67-71, supvr chem res lab, 71-74, MGR CHEM PROCESSES, AEROJET LIQUID ROCKET CO, 74-

Concurrent Pos: Nat Reactor Testing Sta prof, Univ Idaho, 57-58 & 60-61; mem bd trustees, Bethesda Hosp, Denver, Colo. Mem: Fel Am Inst Chem; Am Chem Soc. Res: Gelation of liquids; characterization of hazardous chemicals; material-fluid compatibility evaluation; liquid propellant research; development of chemical processes for disposal of hazardous wastes. Mailing Add: 5552 Wildwood Way Citrus Heights CA 95610

VANDER WENDE, CHRISTINA, b Paterson, NJ, June 12, 30. BIOCHEMISTRY, PHARMACOLOGY. Educ: Upsala Col, BS, 52; Rutgers Univ, MS, 56, PhD(biochem), 59. Prof Exp: Jr pharmacologist, Wallace & Tiernan, Inc, 52-53 & Schering Corp, 53-55; pharmacologist, Maltbie Labs, 55-56; asst physiol & biochem, Rutgers Univ, 56-59; sr res scientist, E R Squibb & Sons, 59-60; sr res scientist, Vet Admin Hosp, 60-64; assoc prof pharmacol & biochem, 64-69, PROF PHARMACOL, RUTGERS UNIV, NEW BRUNSWICK, 69- Concurrent Pos: Nat Cancer Inst res grant, 61-67; Eastern Leukemia Asn Inc res scholar, 67-69; Epilepsy Found grants, 69-70 & 72-73; Pharmaceut Mfrs Asn grant, 70-72; lectr, Exten Serv, Rutgers Univ, 61-63; dir undergrad res, Upsala Col, 61-64; lectr, All Souls Hosp, Morristown, NJ, 62. Honors & Awards: Sci Achievement Award, Upsala Col, 68. Mem: AAAS; Am Chem Soc; Am Soc Pharmacol & Exp Theraput; Soc Neurosci; Acad Pharmaceut Sci. Res: Biochemical pharmacology of central nervous system metabolism and enzymology. Mailing Add: Col of Pharm Rutgers Univ New Brunswick NJ 08903

VANDERWERF, CALVIN ANTHONY, b Friesland, Wis, Jan 2, 17; m 42; c 6. ORGANIC CHEMISTRY. Educ: Hope Col, AB, 37; Ohio State Univ, PhD, 41; Hope Col, ScD, 63; St Benedict's Col, LLD, 66. Prof Exp: From instr to prof chem, Univ Kans, 41-63; prof chem & pres, Hope Col, 63-71; PROF CHEM & DEAN COL ARTS & SCI, UNIV FLA, 71- Concurrent Pos: Consult, Smith, Kline & French Co, 47-63 & Pan Am Oil Co, 58-63; dir, Kativo Chem Co, Ltd, Costa Rica, 63-; trustee, Res Corp, 65-; consult ed, Burgess Publ Corp, 72- Mem: Am Chem Soc; fel NY Acad Sci; The Chem Soc. Res: Aromatic fluorine compounds; phase diagrams; unsaturated lactones; epoxides; synthesis of medicinals; mechanisms of organic reactions; nitrogen containing constituents of petroleum; organo-boron and organo-phosphorous compounds. Mailing Add: Col of Arts & Sci Univ of Fla Gainesville FL 32601

VANDERWERFF, WILLIAM D, b Philadelphia, Pa, Dec 31, 29; m 58; c 4. ORGANIC CHEMISTRY. Educ: Univ Pa, BS, 51, PhD(org chem), 60. Prof Exp: Prod supvr explosives, E I du Pont de Nemours & Co, 51-53; res chemist, 59-70, CHIEF NEW PRODS RES, CORP RES DIV, SUN OIL CO, 70- Mem: Am Chem Soc. Res: Auto-oxidation; hydrocarbon chemistry; petrochemicals; polymer technology. Mailing Add: Res & Develop Bldg Sun Oil Co PO Box 1135 Marcus Hook PA 19061

VANDER WEYDEN, ALLEN JOSEPH, b Cripple Creek, Colo, July 21, 18; m 44; c 4. CHEMISTRY. Educ: Colo Col, AB, 40; Mass Inst Technol, PhD(inorg chem), 44. Prof Exp: Tech asst, Indust Coop Div, US Dept Navy, 42-43; res chemist, Rohm & Haas Co, Pa, 43-46; prof chem, Univ Denver, 46-50; mem tech staff, US AEC, 50-65; DIR ADVAN SYSTS & TECHNOL, McDONNELL DOUGLAS ASTRONAUT CO, 65- Honors & Awards: AEC Distinguished Serv Award, 65. Res: Chemistry and metallurgy of rare elements; nuclear reactor technology; advanced missile & space technology. Mailing Add: McDonnell Douglas Astronaut Co 5301 Bolsa Ave Huntington Beach CA 92647

VANDERWOLF, CORNELIUS HENDRIK, b Edmonton, Alta, Dec 13, 35; m 62. NEUROSCIENCES. Educ: Univ Alta, BSc, 58; McGill Univ, MSc, 59, PhD(psychol), 62. Prof Exp: Res fel biol, Calif Inst Technol, 62-63; Nat Res Coun fel, Brain Res Inst, Switzerland, 63-64; from asst prof to assoc prof psychol, McMaster Univ, 64-68; assoc prof, 68-73, PROF PSYCHOL, UNIV WESTERN ONT, 73- Concurrent Pos: Nat Res Coun res grants, 64- Mem: AAAS; Animal Behav Soc; Can Psychol Asn; Can Physiol Soc; Soc Neurosci; Psychonomic Soc. Res: Role of forebrain structures in patterning and control of motor activity. Mailing Add: Dept of Psychol Univ of Western Ont London ON Can

VAN DER WOUDE, WILLIAM JAN, b Hammond, Ind, Dec 6, 42; m 69; c 2. PLANT CYTOLOGY, PLANT PHYSIOLOGY. Educ: Purdue Univ, West Lafayette, BS, 64, MS, 69, PhD(plant cytol), 72. Prof Exp: ASST PROF BIOL, UNIV CALIF, RIVERSIDE, 72- Mem: Sigma Xi; Am Soc Plant Physiologists; Electron Micros Soc Am; Bot Soc Am; Am Inst Biol Sci. Res: Biochemistry and cytology of the plant plasma membrane and its function in cell wall biogenesis; cellulose microfibril synthesis and deposition; cell morphogenesis and cell interaction. Mailing Add: Dept of Biol Univ of Calif Riverside CA 92502

VANDER WYK, JAMES COLBY, b Waltham, Mass, June 22, 44; m 75. RESEARCH ADMINISTRATION. Educ: Mass Col Pharm, BS, 67; Univ Mass, PhD(microbiol), 73. Prof Exp: NIH fel, Univ Ill, 73-75; head dept microbiol, 75, DIR RES, PHARMACO, INC, 75- Concurrent Pos: Consult, Zimmerman & Assocs, 75- Mem: Am Soc Microbiol. Res: Testing of film-immobilized antimicrobial and antiviral compounds for activity; treatment of animal wastes as substrates for the production of single cell protein. Mailing Add: Pharmaco Inc Interstate Res Park Champaign IL 61820

VANDER WYK, RAYMOND WINSTON, b Waltham, Mass, June 26, 16; m 43; c 3. MICROBIOLOGY. Educ: Mass Col Pharm, BS, 37, PhC, 39, MS, 42; Boston Univ, AM, 42; Harvard Univ, PhD, 50. Prof Exp: Asst pharmacog & biol, 39-45, from instr pharmacog, plant anat & microtech to assoc prof biol, 45-55, PROF MICROBIOL, MASS COL PHARM, 55-, CHMN DEPT BIOL SCI, 68- Concurrent Pos: Bacteriologist, 7th Gen Hosp, 42-46; pres, Pharmaceut Res Assocs, Inc, 68-; consult, Purdue-Frederick Res Ctr & John H Breck, Inc. Mem: Soc Cosmetic Chem; Am Pharmaceut Asn; Am Soc Microbiol. Res: Microbiological flora of the human scalp; dandruff causes; cosmetic chemistry; antibiotics; dermatology. Mailing Add: Dept of Biol Sci Mass Col of Pharm 179 Longwood Ave Boston MA 02115

VANDERZANT, CARL, b Nymegen, Neth, Sept 7, 25; nat US; m 52; c 2. FOOD MICROBIOLOGY. Educ: State Agr Univ, Wageningen, BS, 47, MS, 49; Iowa State Univ, MS, 50, PhD(dairy bact), 53. Prof Exp: Asst prof dairy husb, 53-56, assoc prof dairy sci, 56-62, PROF DAIRY SCI, TEX A&M UNIV, 62- Mem: Am Soc Microbiol; Am Dairy Sci Asn; Inst Food Technologists. Res: Bacteriological problems of foods, especially in dairy products. Mailing Add: Dept of Animal Sci Tex A&M Univ College Station TX 77843

VANDERZANT, ERMA SCHUMACHER, b Elwood, Ill, Jan 30, 20; m 52; c 2. BIOCHEMISTRY. Educ: Iowa State Univ, BS, 42, PhD(biochem), 53. Prof Exp: BIOCHEMIST, COTTON INSECTS BR, ENTOM RES DIV, AGR RES SERV, USDA, TEX A&M UNIV, 54- Honors & Awards: Super Serv Award, USDA, 59; J Everett Bussart Mem Award, Entom Soc Am, 71. Mem: Am Chem Soc; Entom Soc Am; Am Inst Nutrit. Res: Chemically defined diets for insects; nutrition and metabolism of growth factors, lipides and amino acids. Mailing Add: Dept Biochem & Biophys Tex A&M Univ College Station TX 77843

VANDERZEE, CECIL EDWARD, b Wetonka, SDak, Apr 26, 12; m 44. PHYSICAL CHEMISTRY. Educ: Jamestown Col, BS, 38; Univ Iowa, PhD(chem), 49. Prof Exp: Instr, Jamestown Col, 39 & high sch, SDak, 39-42; from instr to assoc prof, 49-58, vchmn dept, 65-70, PROF CHEM, UNIV NEBR, LINCOLN, 58- Concurrent Pos: Mem bd dirs, Calorimetry Conf, 64-66, chmn elec, 67-68, chmn, 68-69. Honors & Awards: Huffman Mem Award, Calorimetry Conf, 75. Mem: Am Chem Soc. Res: Thermodynamics; calorimetry. Mailing Add: Dept of Chem Univ of Nebr Lincoln NE 65888

VAN DER ZIEL, ALDERT, b Zandeweer, Netherlands, Dec 12, 10; nat US; m 35; c 3. PHYSICS. Educ: Univ Groningen, BA, 30, MA, 33, PhD(physics), 34. Prof Exp: Electronics physicist, N V Philips' Incandescent Lamp Works, Netherlands, 34-47; assoc prof physics, Univ BC, 47-50; PROF ELEC ENG, UNIV MINN, MINNEAPOLIS, 50- Concurrent Pos: Grad res prof, Univ Fla, 68. Honors & Awards: Vincent Bendix Award, Am Soc Eng Educ, 75. Mem: Am Phys Soc; fel Inst Elec & Electronics Eng; Netherlands Phys Soc. Res: Noise in a wide variety of solid state devices, such as transistors, field effect transistors, cryogenic devices and radiation detectors; theory of such devices. Mailing Add: Dept of Elec Eng Univ of Minn Minneapolis MN 55455

VAN DER ZIEL, JAN PETER, b Eindhoven, Netherlands, Aug 17, 37; US citizen; m 65; c 1. SOLID STATE PHYSICS. Educ: Univ Minn, BS, 59; Harvard Univ, MS, 61, PhD(appl physics), 64. Prof Exp: Res fel appl physics, Harvard Univ, 64-65; MEM TECH STAFF, BELL LABS, 65- Mem: Am Phys Soc. Res: Lasers and nonlinear optics; optical spectroscopy. Mailing Add: 1D-465 Bell Labs Murray Hill NJ 07974

VANDER ZWAAG, ROGER, b Holland, Mich, Dec 27, 38; m 64; c 3. BIOSTATISTICS. Educ: Hope Col, AB, 60; Purdue Univ, MS, 62; Johns Hopkins Univ, PhD(biostat), 68. Prof Exp: ASST PROF BIOSTAT, SCH MED, VANDERBILT UNIV, 68- Mem: Am Statist Asn. Res: Epidemiology. Mailing Add: Dept of Preventive Med Sch of Med Vanderbilt Univ Nashville TN 37232

VAN DER ZWET, TOM, b Borneo, Indonesia, Apr 7, 32; nat US; m 55; c 3. PLANT PATHOLOGY. Educ: Col Trop Agr, Deventer, Netherlands, BS, 52; La State Univ, BS, 55, MS, 57, PhD(plant path), 59. Prof Exp: Plant pathologist, Tung Res Lab, Plant Sci Res Div, Bogalusa, La, 59-65, PLANT PATHOLOGIST, FRUIT & NUT CROPS RES BR, PLANT GENETICS & GERMPLASM INST, US DEPT AGR, AGR RES CTR, 65- Mem: Am Phytopath Soc; Int Soc Hort Sci; Int Soc Plant Path. Res: Diseases of tropical and subtropical crops; soil and leaf spot fungi; fire blight of pome fruit. Mailing Add: Plant Genetics & Germplasm Inst US Dept of Agr Agr Res Ctr West Beltsville MD 20705

VAN DE STEEG, GARET EDWARD, b Minneapolis, Minn, Feb 8, 40; m 65; c 2. RADIOCHEMISTRY, PHYSICAL CHEMISTRY. Educ: Marquette Univ, BS, 62; Univ NMex, PhD(chem), 68. Prof Exp: Sr res chemist, 68-75, PROJ RES CHEMIST, KERR-McGEE CORP, 75- Mem: Am Chem Soc. Res: Dilute solution chemistry of actinides, lanthanides and fission products; solvent extraction of boron, actinides, lanthanides and transition metals; health physics; analytical methods development. Mailing Add: 2312 NW 113th Pl Oklahoma City OK 73120

VAN DEUSEN, HOBART MERRITT, mammalogy, ornithology, see 12th edition

VAN DEUSEN, JAMES LOWELL, b Champaign, Ill, Sept 12, 28; m 51; c 1. FORESTRY. Educ: Iowa State Univ, BS, 55, MS, 57. Prof Exp: RES FORESTER, ROCKY MT FOREST & RANGE EXP STA, US FOREST SERV, 57- Res: Improvement of tree species for the northern Great Plains. Mailing Add: USFS Shelterbett Lab First and Brander Bottineau ND 58318

VAN DEUSEN, RICHARD L, b Norwich, NY, Jan 31, 26; m 52; c 3. POLYMER CHEMISTRY. Educ: Muhlenberg Col, BS, 50; Univ Miami, MS, 55; Univ Buffalo, PhD(chem), 60. Prof Exp: Cytotechnologist, Cancer Inst Miami, 51-54; chemist, Olin-Mathieson Chem Corp, 56; instr chem, Univ Buffalo, 56-59; res chemist, Ansco Div, Gen Aniline & Film Co, 59-60; res chemist, 60-70, CHIEF POLYMER BR, US AIR FORCE MAT LAB, 70- Mem: Am Chem Soc; AAAS; Sigma Xi; Am Inst Chem; Int Soc Heterocyclic Chem. Res: Photopolymerization; cyclopolymerization kinetics; solid state polymerization; polycondensation; thermally stable plastics, elastomers and fibrous materials. Mailing Add: Polymer Br AF Materials Lab MBP Wright-Patterson AFB Dayton OH 45433

VAN DE VAART, HERMAN, b Arnhem, Netherlands, Apr 11, 34; m 60; c 2. SOLID STATE ELECTRONICS. Educ: Delft Univ Technol, Ing, 58, PhD(tech sci), 69. Prof Exp: Res asst elec eng, Delft Univ Technol, 56-58; res engr, Transitron Electronic Corp, Mass, 60-62; res asst, 62-65, res staff mem, 65-73, MGR SOLID STATE DEVICES DEPT, SPERRY RES CTR, SPERRY RAND CORP, 73- Mem: Am Phys Soc; Inst Elec & Electronics Eng. Res: Semiconductor technology; quadrupole and ferro magnetic resonance; microwave magnetics; solid state delay lines; ferrites. Mailing Add: Sperry Rand Res Ctr 100 North Rd Sudbury MA 01776

VAN DEVENTER, WILLIAM CARL, b Salisbury, Mo, Oct 22, 08; m 34; c 1. BIOLOGY. Educ: Cent Methodist Col, AB, 30; Univ Ill, AM, 32, PhD(field biol), 35. Prof Exp: Asst biol, Cent Methodist Col, 28-30; asst zool, Univ Ill, 30-34; biologist, Monroe County Parks, NY, 34-35; prof biol, St Viator Col, 35-38; instr zool, Stephens Col, 38-43, prof biol, 43-53; head dept, 53-63, PROF BIOL, WESTERN MICH UNIV, 53- Mem: AAAS; Ecol Soc Am; Nat Sci Teachers Asn; Nat Asn Res Sci Teaching (vpres, 54, pres, 55). Res: Biology of Crustacea; ecology of birds; field biology; human ecology; science education. Mailing Add: Dept of Biol Western Mich Univ Kalamazoo MI 49008

VANDEVOORDE, JACQUES PIERRE, b Tourcoing, France, Apr 19, 30; US citizen; m 60; c 2. BIOCHEMISTRY. Educ: Loyola Univ, La, BA, 51; Tulane Univ, MS, 61, PhD(biochem), 68. Prof Exp: Asst biochemist, Touro Infirmary, 61-68; sr biochemist, 68-74, GROUP CHIEF, CANCER DETECTION RES LAB, HOFFMANN-LA ROCHE INC, 74- Mem: Reticuloendothelial Soc; Tissue Cult Asn; Am Asn Clin Chem. Res: Biochemical immunology; chemotherapy; clinical research; cancer detection. Mailing Add: Cancer Detection Res Lab Hoffmann-La Roche Inc Nutley NJ 07110

VANDE VUSSE, FREDERICK JOHN, b Holland, Mich, Sept 26, 39; m 61; c 3. PARASITOLOGY. Educ: Hope Col, BA, 61; Iowa State Univ, MS, 64, PhD(zool), 67. Prof Exp: ASSOC PROF BIOL, GUSTAVUS ADOLPHUS COL, 67- Concurrent Pos: Vis prof, Silliman Univ, Philippines, 70-71. Mem: Am Micros Soc; Am Soc Parasitol; Wildlife Dis Asn. Res: Parasites of wildlife; systematics of Trematoda and Hirudinea. Mailing Add: Dept of Biol Gustavus Adolphus Col St Peter MN 56082

VANDEWATER, STUART LESLIE, b Campbellford, Ont, Nov 16, 24; m 50; c 2. ANESTHESIOLOGY. Educ: Univ Toronto, MD, 47; FRCPS(C), 53; Am Bd Anesthesiol, dipl, 64. Prof Exp: R S McLaughlin fel, Toronto Gen Hosp, Univ Toronto, 52-53, clin teacher anesthesia, Fac Med, 53-59; head dept, 60-71, PROF

ANESTHESIA, FAC MED, QUEEN'S UNIV, ONT, 60-, ASSOC DEAN & SECY, 71- Concurrent Pos: Sr anesthetist, Kingston Gen Hosp, 60-71; consult, Can Armed Med Servs, Ongwanada Sanatorium & Ont Hosp, Rockwood, 60- & Can Armed Forces Med Comt, 64-69; examr, Royal Col Physicians & Surgeons Can, 63-70. Mem: Am Soc Anesthesiol; Acad Anesthesiol; Can Anesthetists Soc (vpres, 65-67, pres, 67-68, hon secy, 72-); Royal Soc Med; NY Acad Sci. Res: Clinical and laboratory anesthesia. Mailing Add: Fac of Med Queen's Univ Kingston ON Can

VAN DE WETERING, RICHARD LEE, b Bellingham, Wash, Aug 2, 28; m 60; c 2. MATHEMATICS. Educ: Univ Wash, Seattle, BS, 50; Western Wash State Col, EdM, 55; Stanford Univ, PhD(math), 60. Prof Exp: From asst prof to assoc prof, 60-67, PROF MATH, CALIF STATE UNIV, SAN DIEGO, 67- Concurrent Pos: Res grant, Delft Univ Technol, 66-67; res assoc, Math Inst, Univ Groningen, 73-74. Mem: Am Math Soc; Math Asn Am. Res: Ordinary differential equations and integral transforms. Mailing Add: Dept of Math Calif State Univ San Diego CA 92115

VANDEWIELE, RAYMOND LAURENT, b Kortryk, Belg, Oct 2, 22; m 54; c 3. OBSTETRICS & GYNECOLOGY. Educ: Cath Univ Louvain, MD, 47. Prof Exp: Fel, Col Physicians & Surgeons, Columbia Univ, 52-54 & Yale Univ, 54-55; from instr to prof gynec, 55-68, WILLARD C RAPPLEYE PROF OBSTET & GYNEC & CHMN DEPT, COL PHYSICIANS & SURGEONS, COLUMBIA UNIV, 71-; DIR OBSTET & GYNEC SERV, COLUMBIA-PRESBY MED CTR, 71- Concurrent Pos: From asst to attend, Columbia-Presby Med Ctr, 58-68; dir, Int Inst Study Human Reproduction, 70-; consult, Clin Ctr, NIH, 71- Mem: Soc Gynec Invest; Endocrine Soc. Res: Reproductive physiology; biochemistry of endocrine organs, especially adrenal, testis and ovary. Mailing Add: Dept of Obstet & Gynec Columbia-Presby Med Ctr New York NY 10032

VANDE WOUDE, GEORGE, b Brooklyn, NY, Dec 25, 35; m 59; c 4. BIOCHEMISTRY, VIROLOGY. Educ: Hofstra Col, BA, 59; Rutgers Univ, MS, 62, PhD(biochem), 64. Prof Exp: Agr Res Serv-Nat Acad Sci res assoc, 64-65; res chemist, Plum Island Animal Dis Lab, USDA, 65-72; HEAD HUMAN TUMOR STUDIES SECT, VIRAL BIOL BR, NAT CANCER INST, 72- Mem: AAAS; Am Chem Soc. Res: Chemical and physical properties of viruses; virus-host cell interactions. Mailing Add: Human Tumor Studies Sect Bldg 41 Viral Biol Br Nat Cancer Inst Bethesda MD 20014

VAN DIEPEN, JAN RUTGER, plant pathology, mycology, see 12th edition

VAN DIJK, CHRISTIAAN PIETER, b Amsterdam, Neth, Nov 10, 15; US citizen; m 53; c 3. PHYSICAL ORGANIC CHEMISTRY, CHEMICAL ENGINEERING. Educ: Amsterdam Munic Univ, BS, 36, MS, 40; Delft Univ Technol, PhD(org chem), 46. Prof Exp: Lab mgr org chem, Delft Univ Technol, 40-48; asst dept head, Shell Lab Amsterdam, Royal Dutch Shell Co, Neth, 48-57; sr chemist, Dow Chem Co, 57-60; sr scientist, M W Kellogg Co, NJ, 60-75, SR SCIENTIST, PULLMAN KELLOGG RES & DEVELOP CTR, 75- Concurrent Pos: Chmn, Nat Exam Comt towards Qual Sci Coworkers, 43-56; consult, Chemische Fabriek Rotterdam, Neth, 46-48. Honors & Awards: Chem Pioneer Award, Am Inst Chemists, 76. Mem: Am Chem Soc; Royal Neth Chem Soc; Sigma Xi. Res: Inventing new process forms for conversions in the petroleum, petrochemical and polymer field; combined information from the fields of organic and physical chemistry with design-engineering calculations. Mailing Add: Pullman Kellogg R&D Ctr PO Box 79513 Houston TX 77079

VAN DILLA, MARVIN ALBERT, b New York, NY, June 18, 19; m 51; c 4. PHYSICS. Educ: Mass Inst Technol, PhD(physics), 51. Prof Exp: Asst, Mass Inst Technol, 46-51; asst res prof physics, Radiobiol Lab, Univ Utah, 51-57; mem staff, Biomed Res Group, Los Alamos Sci Lab, 57-72; GROUP LEADER, BIOMED & ENVIRON RES DIV, LAWRENCE LIVERMORE LAB, 72- Mem: Biophys Soc. Res: Radiation detection and measurement; biological effects of radiation; cell analysis and sorting by high speed flow methods; flow microfluorometry; cellular DNA and life cycle analysis; flow chromosome analysis and sorting. Mailing Add: Lawrence Livermore Lab Livermore CA 94550

VANDIVER, BRADFORD B, b Orlando, Fla, Mar 7, 27; m 58; c 2. GEOLOGY. Educ: Univ Colo, BA, 57, MS, 58; Univ Wash, Seattle, PhD(geol), 64. Prof Exp: Explor geologist, Tenn Gas Transmission Co, 58-60; asst prof geol, Univ Ore, 64-65; PROF GEOL, STATE UNIV NY COL POTSDAM, 65- Mem: Fel Geol Soc Am; Sigma Xi. Res: Metamorphic petrology and structural geology, Cascades, Rocky Mountains, Alps, Adirondacks, Odenwald, (Germany); environmental geology. Mailing Add: Dept of Geol Sci State Univ of NY Potsdam NY 13676

VANDIVIERE, H MAC, b Dawsonville, Ga, Mar 26, 21; m 41; c 2; m 68. PUBLIC HEALTH. Educ: Mercer Univ, AB, 43, MA, 44; Univ NC, MD, 60. Prof Exp: From instr to asst prof biol, Mercer Univ, 42-48; chief spec serv, Res Lab, State Dept Pub Health, Ga, 48-51; res bacteriologist, NC Sanatorium Syst, 51-53, dir dept res, 53-67; assoc prof, 67-72, PROF COMMUNITY MED, COL MED, UNIV KY, 72- Concurrent Pos: Am Pub Health Asn fel epidemiol; instr bact, Univ Mich, 44-46; dir labs & res, Gravely Sanatorium, 53-62; med dir, Haitian Am Tuberc Inst, Jeremie, Haiti, 62-; asst prof community health sci, Sch Med, Duke Univ, 65-67; consult, Dept Pub Health & adv, President's Comt Control Tuberc, Repub Haiti, 65-; clin assoc prof, Sch Pub Health, Univ NC, 69-; dir div tuberc & fungal dis, State Health Dept, Ky, 72-73, dir div remedial health servs, 73- Mem: Am Thoracic Soc; Am Pub Health Asn; AMA. Res: Antituberculosis vaccination; purification of tuberculo-proteins, diagnostic testing methods; epidemiology. Mailing Add: Dept of Community Med Col of Med Univ of Ky Lexington KY 40506

VAN DOLAH, ROBERT WAYNE, b Cheyenne, Wyo, Feb 1, 19; m 42; c 3. CHEMISTRY. Educ: Whitman Col, AB, 40; Ohio State Univ, PhD(org chem), 43. Prof Exp: Asst chem, Ohio State Univ, 40-42; asst to sci dir, William S Merrell Co, Ohio, 43-44, res chemist & group leader, 44-46; actg head org chem br, US Naval Ord Test Sta, Calif, 46-48, head chem div, 53-54; chief, Explosives Res Lab, 54-71, RES DIR, PITTSBURGH MINING & SAFETY RES CTR, US BUR MINES, 71- Honors & Awards: Nitro Nobel Medal, 67; Distinguished Serv Award, US Dept Interior, 65. Mem: Fel AAAS; fel Am Inst Chem; Am Chem Soc; Am Inst Aeronaut & Astronaut; Sigma Xi. Res: Propellants and explosives; combustion; mine safety; industrial safety. Mailing Add: Pittsburgh Mining & Safety Res US Bur of Mines 4800 Forbes Ave Pittsburgh PA 15213

VAN DOMELEN, BRUCE HAROLD, b Shelby, Mich, May 27, 33; m 57; c 4. PHYSICS. Educ: Kalamazoo Col, BA, 55; Univ Wis, MA, 57, PhD(physics), 60. Prof Exp: Staff mem, 60-62, sect supvr phys metall, 62-65, div supvr anal physics, 65, tech adv syst res, 65-69, DIV SUPVR EXPLOR POWER SOURCES, SANDIA LABS, 69- Concurrent Pos: Actg chmn, Govenor's Sci Adv Comt, NMex, 66-70, Governor's sci adv, 66-75; NMex mem, Western Interstate Nuclear Bd, 67-, chmn, 71-73; mem Nat Governors' Coun Sci & Technol, 70-75. Mem: Am Phys Soc. Res: Electrochemistry; physical chemistry. Mailing Add: Explor Power Sources Div Sandia Labs 2523 Albuquerque NM 87115

VAN DONGEN, CORNELIS GODEFRIDUS, b Geertruidenberg, Netherlands, Mar 20, 34; m 69. ANIMAL HUSBANDRY, REPRODUCTIVE PHYSIOLOGY. Educ: Wageningen State Agr Univ, BS, 57, MS, 59; Univ Ill, Urbana, MS, 62, PhD(dairy sci), 64. Prof Exp: Res fel pharmacol, Harvard Univ, 64-65; res assoc physiol, Brown Univ, 66; asst prof pharmacol, NY Med Col, 66-67; RES ASSOC, BIO-RES INST, INC, 67- Concurrent Pos: Grants, NIH; investr, US Dept Agr Contract; consult, Bio Res Consults. Mem: Am Dairy Sci Asn; Am Soc Animal Sci; Brit Soc Study Fertil; Soc Study Reproduction. Res: Husbandry and reproductive physiology of inbred and hybrid hamsters; sex ratio shifts; pharmacological and toxicological effects on mammalian systems in vivo; sperm cell physiology. Mailing Add: Bio-Res Inst 9 Commercial Ave Cambridge MA 02141

VAN DOORNE, WILLIAM, b Utrecht, Neth, Dec 12, 37; US citizen; m 61; c 3. INORGANIC CHEMISTRY. Educ: Calvin Col, BS, 60; Univ Mich, MS, 62, PhD, 65. Prof Exp: From asst prof to assoc prof chem, 66-74, PROF CHEM, CALVIN COL, 74- Concurrent Pos: Vis assoc prof, Univ Hawaii, 72-73. Mem: Am Chem Soc. Res: Phophorus-nitrogen compounds; synthetic inorganic chemistry; crystallography. Mailing Add: Calvin Col 1801 E Beltline Ave SE Grand Rapids MI 49506

VANDOR, SANDOR LASZLO, b Baja, Hungary, Feb 1, 37; US citizen; m 64; c 2. BIOCHEMICAL PHARMACOLOGY. Educ: Augustana Col, SDak, BA, 62; SDak State Univ, MA, 64; Ohio State Univ, PhD(physiol chem), 67. Prof Exp: Asst chem, SDak State Univ, 62-64; asst physiol chem, Ohio State Univ, 64-67; res assoc biochem, Mich State Univ, 67-69; sr scientist, Dept Biochem, Mead Johnson Res Ctr, 69-75; ASST PROF PHARMACOL, OHIO NORTHERN UNIV, 75- Mem: AAAS; Am Chem Soc; Sigma Xi; Am Asn Col Pharm. Res: Enzymes, isolation and characterization; phospholipid metabolism; intermediary metabolism. Mailing Add: Dept of Pharmacol Ohio Northern Univ Ada OH 45810

VAN DOREN, CORNELIUS AUSTIN, agronomy, soil science, see 12th edition

VAN DOREN, DAVID MILLER, JR, b Urbana, Ill, Aug 16, 32; m 55; c 4. AGRONOMY, SOIL PHYSICS. Educ: Univ Ill, BS, 54; Mich State Univ, MS, 55, PhD(soil sci), 58. Prof Exp: From asst prof to assoc prof, 58-65, PROF AGRON, OHIO STATE UNIV & RES & DEVELOP CTR, 65- Mem: Soil Sci Soc Am; fel Am Soc Agron. Res: Soil structure and plant growth; tillage and land management. Mailing Add: Dept of Agron Ohio Agr Res & Develop Ctr Wooster OH 44691

VAN DORN, WILLIAM GEORGE, oceanography, see 12th edition

VAN DREAL, PAUL ARTHUR, b Chicago, Ill, Feb 15, 32; m 57; c 2. BIOCHEMISTRY, CYTOLOGY. Educ: Calvin Col, BS, 57; Mich State Univ, PhD(bot, biochem, cytol), 61. Prof Exp: Clin biochemist, St Lawrence Hosp, 61-63; NIH fel, Biol Div, Oak Ridge Nat Lab, 63-64; asst prof biochem, Med Sch, Univ Ore, 64-66; from asst prof to assoc prof clin chem, Med Sch, Univ Wash, 66-71, dir lab computer div, 69-71; vpres & dir res, Hycel Inc, 71-72; tech mgr radioimmunoassay develop, Corning Glass Works, Inc, 72-73; assoc prof path & clin chem, Med Sch, Univ Ky, 73-74; lab dir & tech dir, Nat Health Labs, Inc, 74-75; VPRES & LAB DIR, HERNER ANALYTICS, 75- Concurrent Pos: Lectr biochem, Mich State Univ, 62-63; Nat Cancer Inst fel carcinogenesis, Med Sch, Univ Ore, 64-66; consult, Ortec Div, EG & G, 67-71, Clin Instruments Div, Beckman Instruments, 67-, Pesticide Res Lab, Dept of Health, State of Wash, 68-69, Bausch & Lomb Inc, 69- & Corning Glass Inc, 74-; Bausch & Lomb Grant electrophoresis, Med Sch, Univ Wash, 69-70. Mem: AAAS; Am Chem Soc; fel Am Asn Clin Chemists; Am Soc Clin Path; Asn Clin Scientists. Res: New clinical laboratory diagnostic techniques; aging; patient normals as a function of age and disease onset; biochemistry of the cell cycle. Mailing Add: Herner Analytics 1500 E Jefferson St Rockville MD 20852

VAN DRIEST, EDWARD REGINALD, b Cleveland, Ohio, Sept 16, 13; m 42; c 3. FLUID DYNAMICS. Educ: Case Inst Technol, BS, 36; Univ Iowa, MS, 37; Calif Inst Technol, PhD, 40; Swiss Fed Inst Technol, ScD, 48. Prof Exp: Instr mech, Cornell Univ, 40-41; asst engr hydraul, Engr Off, USDA, 41-42; assoc prof civil eng, Univ Conn, 42-44; asst prof mech eng, Mass Inst Technol, 44-46; chief scientist, Missile Div, NAm Aviation, Inc, 48-58, dir space sci lab, dir space & info systs div, 58-65, chief scientist ocean systs opers, Autonetics Div, Calif, 65-70; prof aerospace eng, Va Polytech Inst & State Univ, 70-71; scientist, Phys Sci Dept, Rand Corp, 71-74; DIR, FLUID MECH & ACOUST LAB & ADJ PROF AEROSPACE ENG, UNIV SOUTHERN CALIF, 74- Concurrent Pos: Adj prof, Va Polytech Inst & State Univ, 71-; resident consult, Rand Corp, 74-; consult, Aerospace Corp, 74- Mem: Am Astronaut Soc; fel Am Inst Aeronaut & Astronaut; Ger Soc Appl Math & Mech. Res: Astronomy; space mechanics; electrophysics; geology; life sciences; fluid dynamics; supersonic aerodynamics; magnetohydrodynamics; boundary layer theory; aerodynamic heating; marine hydrodynamics. Mailing Add: Dept of Aerospace Eng Univ of Southern Calif Los Angeles CA 90007

VANDRUFF, LARRY WAYNE, b Elmira, NY, Apr 28, 42; m 66; c 2. WILDLIFE BIOLOGY. Educ: Mansfield State Col, BS, 64; Cornell Univ, MS, 66, PhD(wildlife ecol), 71. Prof Exp: Res asst vert ecol & genetics, Cornell Univ, 64-66, res asst wildlife ecol, NY Wildlife Res Unit, 66-70; ASST PROF VERT BIOL, STATE UNIV NEW YORK COL ENVIRON SCI & FORESTRY, 70- Mem: Am Soc Mammal; Wildlife Soc; Ecol Soc Am; Am Nature Study Soc. Res: Field studies in the ecology of urban wildlife species; waterfowl biology and wetland ecology; dynamics of homeotherm populations and wildlife; habitat relationships. Mailing Add: Dept of Zool SUNY Col Environ Sci & Forestry Syracuse NY 13210

VAN DUSEN, CLARENCE RAYMOND, b Elkhart, Ind, Nov 9, 07; m 41; c 1. SPEECH PATHOLOGY. Educ: Ind Univ, AB, 31; Univ Mich, AM, 32, ScD(speech & gen linguistics), 37. Prof Exp: Dir speech clin, Mich State Col, 37-42; dir speech clin, Univ Miami, 46-59, chmn dept speech, 59-61; coordr basic studies, Brevard Jr Col, 61-62, dean instr, 62-64; pres, Golden Hills Acad, 64-65; chmn fac, Fla Keyes Jr Col, 65-66; dir speech & hearing clin, Miss State Col Women, 66-68, coordr grad studies speech, 69-75; PVT PRACT & WRITING, 75- Concurrent Pos: Consult, Regional Rehab Ctr, Tupelo, Miss, 66-75. Mem: Fel Am Speech & Hearing Asn; Speech Commun Asn. Res: Speech development in the mentally retarded. Mailing Add: PO Box 3 Columbus MS 39701

VAN DUSEN, WILLIAM, JR, physical chemistry, analytical chemistry, see 12th edition

VAN DUSER, ARTHUR L, b Appleton, Wis, Dec 6, 11; m 39; c 2. PREVENTIVE MEDICINE, PUBLIC HEALTH. Educ: Univ Wis, BS, 36, MD, 38; Univ Mich, MS, 47; Am Bd Prev Med, cert pub health & gen prev med, 50. Prof Exp: Dist health officer, 41-46, dir cancer control div, 47-72, dir venereal dis control div, 49-69, DIR LAB EVAL DIV, WIS STATE BD HEALTH, 60-, DEP DIR BUR PREV DIS, 69- Mem: Am Pub Health Asn; AMA; Pub Health Cancer Asn Am; Am Social Health Asn. Res: Application of social-medical knowledge to the prevention and detection of cancer; improvement in qualitative and quantitative medical laboratory services, including methods, evaluations, standards and approving systems. Mailing Add: Wis Dept Health & Soc Serv 1 W Wilson St Madison WI 53701

VAN DUUREN, BENJAMIN LOUIS, b SAfrica, May 5, 27; nat US; div; c 2. ORGANIC CHEMISTRY. Educ: Univ SAfrica, BS, 46 & 48, MS, 49; Univ Orange Free State, ScD, 51. Prof Exp: Res assoc, Univ Ill, 51-53 & Univ Calif, Los Angeles, 53-54; res chemist, E I du Pont de Nemours & Co, NY, 54-55; from instr chem to assoc prof, 55-69, PROF ENVIRON MED, NY UNIV MED CTR, 69- Concurrent Pos: Mem special study sect, NIH, 75-; mem, Food Protection Comt, Nat Acad Sci, 75- Mem: Am Chem Soc; Am Asn Cancer Res; The Chem Soc; Soc Toxicol. Res: Chemistry of pyrrolizidine alkaloids and benzofuran fish toxic compounds; infrared spectroscopy; fluorescence spectroscopy of aromatic compounds; environmental carcinogens; carcinogenesis and metabolism of carcinogens; tobacco and cancer. Mailing Add: Dept of Environ Med NY Univ Med Ctr 550 First Ave New York NY 10016

VAN DUYNE, FRANCES OLIVIA, b Newark, NJ, Sept 16, 12. FOOD CHEMISTRY. Educ: Vassar Col, BA, 34, MA, 36; Columbia Univ, PhD(chem), 40. Prof Exp: Asst physiol, Vassar Col, 34-36; asst chem, Columbia Univ, 36-40; assoc home econ, 40-45, from asst prof to assoc prof, 45-53, PROF FOODS, UNIV ILL, URBANA, 53- Mem: AAAS; Am Chem Soc; Am Home Econ Asn; Inst Food Technol; Am Dietetic Asn. Res: Effects of home preparation and preservation on vitamin content of food; effect of freezing and freezer storage on quality of fruits, vegetables and cooked and prepared foods. Mailing Add: Sch of Human Resources & Family Univ of Ill Urbana IL 61801

VAN DUYNE, RICHARD PALMER, b Orange, NJ, Oct 28, 45. ANALYTICAL CHEMISTRY, CHEMICAL PHYSICS. Educ: Rensselaer Polytech Inst, BS, 67; Univ NC, PhD(anal chem), 71. Prof Exp: ASST PROF ANAL CHEM, NORTHWESTERN UNIV, EVANSTON, 71- Concurrent Pos: Fel, Alfred P Sloan Found, 74- Mem: Am Chem Soc; Electrochem Soc; AAAS. Res: Radical ion chemistry; electrochemiluminescence; time-resolved fluorescence spectroscopy; tunable dye laser resonance Raman spectroscopy; laboratory computer systems. Mailing Add: Dept of Chem Northwestern Univ Evanston IL 60201

VAN DYK, JOHN WILLIAM, b Paterson, NJ, May 2, 28; m 51; c 3. PHYSICAL CHEMISTRY. Educ: Rutgers Univ, AB, 50; Columbia, AM, 51, PhD(chem), 54. Prof Exp: Asst chem, Columbia Univ, 50-52; res chemist, Polychems Dept, 54-64, STAFF CHEMIST, FABRICS & FINISHES DEPT, E I DU PONT DE NEMOURS & CO, INC, 64- Mem: Am Chem Soc. Res: Polymerization kinetics; surface chemistry; polymer chemistry; paint chemistry; color science; visual perception. Mailing Add: Fabrics & Finishes Dept E I du Pont de Nemours & Co Inc Wilmington DE 19898

VAN DYKE, CECIL GERALD, b Effingham, Ill, Feb 4, 41; m 69; c 2. PLANT PATHOLOGY. Educ: East Ill Univ, BSEd, 63; Univ Ill, Urbana, MS, 66, PhD(plant path), 68. Prof Exp: NIH res assoc, Univ Ill, Urbana, 68; res assoc, 68-69, instr bot, 69, ASST PROF BOT & PLANT PATH, NC STATE UNIV, 69- Mem: Am Phytopath Soc. Res: Ultrastructure of fungi and fungus-host pathological interactions. Mailing Add: Dept of Bot NC State Univ Raleigh NC 27607

VAN DYKE, CHARLES H, b Rochester, Pa, Sept 19, 37; m 66; c 1. INORGANIC CHEMISTRY, ORGANOMETALLIC CHEMISTRY. Educ: Geneva Col, BS, 59; Univ Pa, PhD(inorg chem), 64. Prof Exp: Asst prof chem, 63-70, ASSOC PROF CHEM, CARNEGIE-MELLON UNIV, 70- Mem: Am Chem Soc; The Chem Soc. Res: Synthesis and study of volatile hydride derivatives of the Group IV elements. Mailing Add: Dept Chem Carnegie-Mellon Univ 4400 Fifth Ave Pittsburgh PA 15213

VAN DYKE, HENRY, b Pittsburgh, Pa, Oct 1, 21; m 43; c 4. MICROBIOLOGY. Educ: Western Reserve Univ, BS, 47; Univ Mich, MA, 49, PhD(zool), 55. Prof Exp: Malariologist, US Pub Health Serv, 52; instr zool, Univ Mich, 52-53; asst prof biol, Carleton Col, 53-60 & Ore Col Educ, 60-63; ASSOC PROF BIOL, ORE STATE UNIV, 63- Mem: AAAS; Soc Protozool; Am Soc Microbiol; Marine Biol Asn UK, Am Soc Limnol & Oceanog. Res: Ecology, physiology, culture and photobiology of marine communities. Mailing Add: 3300 NW Van Buren Corvallis OR 97331

VAN DYKE, JOHN HOWARD, b Palisades Park, NJ, Sept 13, 11; m 35; c 7. ANATOMY. Educ: Colgate Univ, AB, 35; Univ NH, MS, 37; Cornell Univ, PhD(anat), 41. Prof Exp: Asst zool, Univ NH, 36-37; asst histol & embryol, Cornell Univ, 37-41; res histologist, Plant, Soil & Nutrit Lab, USDA, NY, 41-42; from instr to asst prof anat, Washington Univ, 42-47; assoc prof, Sch Med, Ind Univ, 47-51; actg chmn dept, 61-63, PROF ANAT, HAHNEMANN MED COL, 51- Mem: Fel AAAS; fel Geront Soc; Am Soc Zoologists; Am Asn Anatomists; Am Asn Cancer Res. Res: Gross anatomy; embryology; histology; neuroanatomy; experimental pathology; congenital anomalies; experimental metaplasia related to neoplasia; mechanisms of calcitonin gland; thyroid and parathyroid; thymic neoplasias; studies in biomagnetics equivalent to lunar environment. Mailing Add: Dept of Anat Hahnemann Med Col Philadelphia PA 19102

VAN DYKE, JOHN WILLIAM, JR, b Holland, Mich, Nov 15, 35; m 59; c 3. ORGANIC CHEMISTRY. Educ: Hope Col, AB, 58; Univ Ill, PhD(org chem), 62. Prof Exp: RES CHEMIST, THERAPEUT RES LAB, MILES LABS, INC, 62- Mem: Am Chem Soc. Res: Organic synthesis of pharmacologically active compounds. Mailing Add: Miles Labs Inc 1127 Myrtle St Elkhart IN 46514

VAN DYKE, KNOX, b Chicago, Ill, June 23, 39; m 71; c 4. PHARMACOLOGY, BIOCHEMISTRY. Educ: Knox Col, AB, 61; St Louis Univ, PhD(biochem), 66. Prof Exp: Res assoc pharmacol, 66-68, sr res pharmacologist, 68-69, asst prof, 69-73, ASSOC PROF PHARMACOL, MED CTR, WVA UNIV, 73- Concurrent Pos: WHO grant, 70-; WVa Heart Asn grant, 70-; NIH instnl cancer & gen res support grant, 70-; WVa rep, Oak Ridge Assoc Univs, 72; consult, Walter Reed Workshop on Malaria, 72. Mem: AAAS; Am Chem Soc; Am Soc Pharmacol & Exp Therapeut; Int Soc Biochem Pharmacol. Res: Reproductive physiology; protein and nucleic acid synthesis; malariology and mechanism of drug resistance; automated analysis of enzyme and nucleic acid systems; radioimmunoassay; mechanisms of antimalarial drugs; adrenergic transmitter-energy complexes; adenosine utilization and syntheses; measurement of bioluminescent and chemiluminescent reaction. Mailing Add: Dept of Pharmacol WVa Univ Med Ctr Morgantown WV 26506

VAN DYKE, ROSS EDWIN, chemistry, see 12th edition

VAN DYKE, RUSSELL AUSTIN, b Rochester, NY, Feb 8, 30; m 56; c 2. BIOCHEMISTRY, MICROBIOLOGY. Educ: Hope Col, BS, 51; Univ Mich, MS, 53; Univ Ill, PhD(animal biochem), 60. Prof Exp: Asst animal biochem, Univ Ill, 56-60; res assoc radio biochem, Univ Colo, 60-61; sr res biochemist, Dow Human Health Res, Dow Chem Co, Mich, 61-68; ASSOC PROF BIOCHEM, MAYO MED SCH, 68-, CONSULT ANESTHESIA RES, MAYO CLIN, 68- Mem: AAAS; Am Chem Soc; Am Soc Pharmacol & Exp Therapeut. Res: Biochemical pharmacology; drug metabolism; cell membrane and transport; enzymatic dechlorination in mammals. Mailing Add: Dept of Anesthesiol Mayo Clin Rochester MN 55901

VAN DYKEN, ALEXANDER ROBERT, chemistry, see 12th edition

VAN DYNE, GEORGE M, b Pueblo, Colo, Sept 6, 32; m; c 4. ECOLOGY, NUTRITION. Educ: Colo Agr & Mech Col, BS, 54; SDak State Univ, MS, 56; Univ Calif, Davis, PhD(nutrit), 63; Oak Ridge Nat Lab, cert, 64; Univ Mich, cert, 64. Prof Exp: Res asst range ecol & nutrit, Agr Exp Sta, SDak State Univ, 54-56; instr, Range Mgt, Colo State Univ, 56-57; asst prof animal & range sci, Mont State Col, 57-61; asst res nutritionist, Univ Calif, Davis, 61-64; health physicist, Oak Ridge Nat Lab, 64-66; assoc prof, 66-68, PROF BIOL, COLO STATE UNIV, 68- Concurrent Pos: Ford Found prof, Dept Bot, Univ Tenn, 64-66; mem, Nat Acad Sci Comt Terrestrial Prod, Int Biol Prog, 65-, dir grassland biome study, 67-74; consult, Ecolog Systs Anal & Natural Resource Mgt & Res. Mem: AAAS; Am Soc Animal Sci; Ecol Soc Am; Soc Range Mgt; Wildlife Soc. Res: Systems ecology; soil-plant-animal energy and nutrient transfer; nutrition of range plants and animals; forage evaluation; modeling, analysis, simulation and computer applications to biological problems. Mailing Add: Col Forestry & Nat Resources Colo State Univ Ft Collins CO 80521

VANE, FLOIE MARIE, b Dawson, Minn, Nov 25, 37. DRUG METABOLISM. Educ: Gustavus Adolphus Col, BS, 59; Mich State Univ, PhD(org chem), 63. Prof Exp: Sr chemist, 64-72, GROUP CHIEF, HOFFMANN-LA ROCHE, INC, 73- Concurrent Pos: NIH fel, Mass Inst Technol, 63-64; vis asst prof, Baylor Col Med, 68-69. Mem: Am Chem Soc; The Chem Soc; Am Soc Mass Spectrometry; Am Soc Pharmacol & Exp Therapeut. Res: Structure determination of organic compounds by spectroscopic methods such as nuclear magnetic resonance and mass spectroscopy; structure identification of drug metabolites. Mailing Add: Hoffmann-La Roche Inc Nutley NJ 07110

VAN ECK, EDWARD ARTHUR, b Grand Rapids, Mich, May 26, 16; m 46; c 2. MICROBIOLOGY. Educ: Hope Col, BA, 38; Univ Mich, MSc, 41, PhD(bact), 50. Prof Exp: Instr bact, Univ Mich, 48-50; asst prof, Univ Kans, 50-53; purchasing agent, Stand Grocer Co, Mich, 53-58; lectr & reader microbiol, Christian Med Col, Vellore, India, 58-63, assoc prof, 62-63; chmn dept, 63-76, PROF NATURAL SCI & BIOL, NORTHWESTERN COL, IOWA, 63- Concurrent Pos: Vis mem staff, Univ Kans, 72-73. Mem: AAAS; Am Soc Microbiol; Am Inst Biol Sci. Res: Antigenic constitution of the Salmonellae; serological survey for presence of leptospirosis in South India; tumor immunology. Mailing Add: Dept of Biol Northwestern Col Orange City IA 51041

VAN ECK, WILLEM ADOLPH, b Wageningen, Netherlands, July 27, 28; nat US; m 56; c 3. AGRONOMY, SOIL SCIENCE. Educ: Wageningen State Agr Univ, BSc, 51; Mich State Univ, MSc, 54, PhD(soil sci), 58. Prof Exp: Asst plant ecol, Wageningen State Agr Univ, 50, soil surv, 51; forester, Gold Coast Govt Surv Team, 52; asst soil fertil, Mich State Univ, 52-53, soil surv & forest soils, 52-56; from asst prof to assoc prof, 57-66, PROF SOIL SCI & STATE EXTEN SPECIALIST, W VA UNIV, 66- Concurrent Pos: Vis sr lectr, Makerere Univ, Uganda, 66-72; pres, Acad Assocs, Econ-Environ Consults. Mem: AAAS; Soil Sci Soc Am; Ecol Soc Am; Soc Am Foresters; Soil Conserv Soc Am. Res: Effect of environment on soil and vegetation development, especially as applied to forest and watershed management; relation of soil morphology and pedology to soil and water conservation and to physical land use planning; assessment of soil fertility in agronomy and forestry. Mailing Add: Div of Plant Sci W Va Univ Morgantown WV 26506

VAN EEDEN, CONSTANCE, b Delft, Netherlands, Apr 6, 27; m 60; c 1. MATHEMATICS. Educ: Univ Amsterdam, BSc, 49, MA, 54, PhD, 58. Prof Exp: Res assoc, Math Ctr, Univ Amsterdam, 54-60; vis assoc prof, Mich State Univ, 60-61; res assoc, Univ Minn, Minneapolis, 61-64; assoc prof, 65-68, PROF MATH, UNIV MONTREAL, 68- Concurrent Pos: Assoc prof & actg dir statist ctr, Univ Minn, Minneapolis, 64-65; res mem, Math Res Ctr, Univ Wis-Madison, 69. Mem: AAAS; NY Acad Sci; Inst Math Statist; Am Statist Asn; Can Math Cong. Res: Mathematical statistics. Mailing Add: Dept of Math Univ of Montreal PO Box 6128 Montreal PQ Can

VAN EENAM, DONALD NEIL, organic chemistry, see 12th edition

VANELLI, RONALD EDWARD, b Quincy, Mass, July 5, 19; m 53; c 2. ORGANIC CHEMISTRY. Educ: Harvard Univ, AB, 41, MA & PhD(chem), 50. Prof Exp: Sr res chemist, Photo Prods Dept, E I du Pont de Nemours & Co, 50-51; DIR, CHEM LABS & LECTR CHEM, HARVARD UNIV, 51-, DIR SCI CTR, 72- Mem: Am Chem Soc. Res: Isonorcamphor; color and constitution. Mailing Add: Harvard Univ Chem Lab 12 Oxford St Cambridge MA 02138

VAN ELSWYK, MARINUS, JR, b Madera, Calif, July 30, 29; m 50; c 3. PLANT BREEDING, GENETICS. Educ: Fresno State Col, BS, 56; Univ Calif, Davis, MEd, 57; Univ Ariz, PhD(plant sci, agron), 67. Prof Exp: From instr to assoc prof plant sci & agron, 57-70, PROF PLANT BREEDING, STATIST & AGRON, CALIF STATE UNIV, FRESNO, 70- Mem: Am Soc Agron; Crop Sci Soc Am. Res: Agronomy; field plot techniques. Mailing Add: Dept of Plant Sci Sch of Agr Sci Calif State Univ Fresno CA 93740

VAN EMDEN, MAARTEN HERMAN, b Rheden, Netherlands. COMPUTER SCIENCE. Educ: Delft Univ Technol, MEng, 66; Univ Amsterdam, DSc(math & natural sci), 71. Prof Exp: Res assoc comput, Math Ctr, Amsterdam, 66-71; fel, IBM Thomas J Watson Res Ctr, 71-72; res fel artificial intel, Univ Edinbrugh, 72-75; ASST PROF COMPUT SCI, UNIV WATERLOO, 75- Res: Methods of programming and of problem solving in general; development of program languages; theory of computation; applications of firts-order predicate logic. Mailing Add: Dept of Comput Sci Univ of Waterloo Waterloo ON Can

VAN ENGEL, WILLARD ABRAHAM, b Milwaukee, Wis, June 3, 15. MARINE BIOLOGY. Educ: Univ Wis, PhB, 37, PhM, 40. Prof Exp: Asst limnol, Univ Wis, 38-41, zool, 41-42, 46; from asst biologist to assoc biologist, 46-60, SR MARINE SCIENTIST, VA INST MARINE SCI, 61- Concurrent Pos: Prof marine sci, Col William & Mary, 61-; asst prof, Univ Va, 63- Mem: AAAS. Res: Biology and ecology of the blue crab; vital statistics of aquatic animal populations; Crustacea of Chesapeake Bay and adjacent continental-shelf waters; biology and management of American lobster off Virginia. Mailing Add: Va Inst of Marine Sci Gloucester Point VA 23062

VAN ENGEN, HENRY, mathematics, see 12th edition

VAN ENKEVORT, RONALD LEE, b Escanaba, Mich, Dec 20, 39; m 62; c 1. MATHEMATICS. Educ: Univ Wash, BS, 62; Ore State Univ, MS, 66, PhD(math), 72. Prof Exp: High sch teacher, 62-67; ASST PROF MATH, UNIV PUGET SOUND, 71- Mem: Am Math Soc. Res: Additive number theory. Mailing Add: Dept of Math Univ of Puget Sound Tacoma WA 98416

VAN EPPS, DENNIS EUGENE, b Rock Island, Ill, Nov 26, 46; m 73. IMMUNOLOGY. Educ: Western Ill Univ, BS, 68; Univ Ill, PhD(microbiol), 72. Prof Exp: NIH fel immunol, 72-74, ASST PROF MED & MICROBIOL, UNIV N MEX, 74- Concurrent Pos: Arthritis Found fel, 74- Honors & Awards: Young Investr Pulmonary Res Award, Nat Heart & Lung Inst, 74. Mem: Am Soc Microbiol; Am Asn Immunologists; Am Fedn Clin Res. Res: Normal and abnormal phagocytic cell function and humoral factors which may alter this function. Mailing Add: Dept of Med & Microbiol NMex Sch of Med Albuquerque NM 87131

VAN EPPS, GORDON ALMON, b Salt Lake City, Utah, Apr 1, 20; m 45; c 4. FIELD CROPS. Educ: Utah State Univ, BS, 42, MS, 48. Prof Exp: Asst prof field crops, Calif State Polytech Col, 48-51; asst prof agron, Snow Col, 52-54, asst prof, 54-64, ASSOC PROF AGRON, UTAH STATE UNIV, 64- Concurrent Pos: Adv under contract with Utah State, Iran, 61-64 & Bolivia, 66-68. Mem: Soc Range Mgt. Res: Rehabilitate with vegetation disturbed sites in the northern desert; seed production of indigenous browse shrubs and forbs on crop land. Mailing Add: Snow Field Sta Utah State Univ Ephraim UT 84627

VAN ERT, MARK DEWAYNE, b Los Angeles, Calif, June 14, 47; m 68; c 3. INDUSTRIAL HYGIENE. Educ: Calif State Univ, Chico, BA, 69; Univ NC, Chapel Hill, MS, 71, PhD(indust hyg & toxicol), 74. Prof Exp: INDUST HYGIENIST & LAB DIR, OCCUP HEALTH STUDIES GROUP, DEPT ENVIRON SCI & ENG, UNIV NC, CHAPEL HILL, 74- Res: Assessment of work environments for potentially hazardous exposures, relating such exposures to health status; development of methodology for environmental assessment; determining the toxicological properties of environmental insults. Mailing Add: Occup Health Studies Group Suite 32 NCNB Plaza Chapel Hill NC 27514

VAN ESELTINE, WILLIAM PARKER, b Syracuse, NY, Aug 21, 24; m 48; c 2. BACTERIOLOGY. Educ: Oberlin Col, AB, 44; Cornell Univ, MS, 47, PhD(bact), 49. Prof Exp: Asst bact, NY State Agr Exp Sta, 44-45 & Cornell Univ, 46-48; assoc prof, Clemson Col, 48-52; asst prof vet hyg, 52-59, assoc prof vet microbiol & prev med, 59-67, PROF MED MICROBIOL, COL VET MED, UNIV GA, 67- Mem: AAAS; Am Soc Microbiol; Am Inst Biol Sci. Res: Microbiology of foods; bactericidal and bacteriostatic agents; physiology and taxonomy of bacteria, especially animal pathogens. Mailing Add: Dept of Med Microbiol Col of Vet Med Univ of Ga Athens GA 30601

VAN ETTEN, CECIL HERMAN, b Miller, SDak, Apr 18, 10; m 35; c 5. NATURAL PRODUCTS CHEMISTRY. Educ: Univ Ill, BA, 35. Prof Exp: Technician biochem, George Washington Med Sch, Washington, DC, 35-37; jr chemist animal nutrit, Bur Animal Ind, USDA, Beltsville, Md, 38-41; asst chemist org micro chem anal, 41-53, PRIN CHEMIST ANAL & ISOLATION NATURAL PROD, NORTHERN REGIONAL RES LABS, AGR RES SERV, USDA, PEORIA, ILL, 53- Honors & Awards: Distinguished Serv Award, USDA, 47 & Super Serv Award, 58. Mem: AAAS; Am Chem Soc; Soc Econ Bot; Am Asn Cereal Chemists. Res: Nutrition of men and animals as related to plant protein and amino acids; natural toxicants in plants with emphasis on glucosinolates and their products, their analysis and isolation. Mailing Add: Northern Regional Res Ctr USDA 1815 N University St Peoria IL 61604

VAN ETTEN, HANS D, b Peoria, Ill, Sept 16, 41; m 63; c 2. PLANT PATHOLOGY. Educ: Wabash Col, BA, 63; Cornell Univ, MS, 66, PhD(plant path), 70. Prof Exp: ASST PROF PLANT PATH, CORNELL UNIV, 70- Mem: AAAS; Am Phytopath Soc; NAm Photochem Soc. Res: Physiology of disease. Mailing Add: Dept of Plant Path Cornell Univ Ithaca NY 14850

VAN ETTEN, JAMES L, b Cherrydale, Va, Jan 7, 38; m 60; c 1. MICROBIAL PHYSIOLOGY. Educ: Carleton Col, BA, 60; Univ Ill, MS, 63, PhD(path), 65. Prof Exp: NSF fel microbiol, Univ Pavia, 65-66; from asst prof to assoc prof, 66-74, PROF PLANT PATH, UNIV NEBR, LINCOLN, 74- Mem: AAAS; Am Phytopath Soc; Soc Gen Microbiol; Am Soc Microbiol. Res: Biochemistry of fungal spore germination and bacteriophage; biochemistry. Mailing Add: Dept of Plant Path Univ of Nebr Lincoln NE 68583

VAN ETTEN, ROBERT LEE, b Evergreen Park, Ill, June 11, 37; c 3. CHEMISTRY. Educ: Univ Chicago, BS, 59; Univ Calif, Davis, 63-64, PhD(chem), 65. Prof Exp: Technician, Ben May Labs, Cancer Res, Univ Chicago, 57-59; teaching asst, Univ Calif, Davis, 60-63; NIH fel Northwestern Univ, 65-66; asst prof, 66-70, ASSOC PROF CHEM, PURDUE UNIV, LAFAYETTE, 70- Concurrent Pos: Res career develop award, NIH, 69-73; Alexander von Humboldt fel, Marburg, Germany, 75-76. Mem: Am Chem Soc; Am Soc Biol Chem; Am Crystallog Asn; Am Asn Univ Prof; NY Acad Sci. Res: Mechanisms of enzymatic catalysis; kinetics and mechanisms of solution reactions; enzyme models; clinical chemistry of phosphatases and sulfatases; relationship between structures in solution and in crystalline state. Mailing Add: Dept of Chem Purdue Univ Lafayette IN 47907

VAN EYS, JAN, b Hilversum, Neth, Jan 25, 29; nat US; m 55; c 2. BIOCHEMISTRY. Educ: Vanderbilt Univ, PhD, 55; Univ Wash, MD, 66. Prof Exp: Fel biochem, McCollum-Pratt Inst, Johns Hopkins Univ, 55-57; from asst prof to prof biochem, Sch Med, Vanderbilt Univ, 57-73, from asst prof to prof pediat, 68-73; PROF PEDIAT & HEAD DEPT, UNIV TEX SYST CANCER CTR, M D ANDERSON HOSP & TUMOR INST, 73- Concurrent Pos: Investr, Howard Hughes Med Inst, 57-66. Mem: Am Soc Biol Chemists; Am Inst Nutrit; NY Acad Sci. Res: Metabolism and enzymology in glycolysis, tetrahymena and red blood cells. Mailing Add: Dept Pediat Univ Tex Syst Can Ctr M D Anderson Hosp & Tumor Inst Houston TX 77025

VAN FAASEN, PAUL, b Holland, Mich, June 6, 34; m 58; c 2. PLANT TAXONOMY. Educ: Hope Col, BA, 56; Mich State Univ, MS, 62, PhD(bot), 71. Prof Exp: Chemist, Parke-Davis Co, 56-57; instr biol warfare, US Army, 57-58, technician comp physiol sect, 58-59; instr biol, Lake rest Col, 62-63; from instr to asst prof, 63-72, ASSOC PROF BIOL, HOPE COL, 72- Mem: Bot Soc Am; Am Soc Plant Taxonomists; Int Asn Plant Taxonomists; Am Inst Biol Sci; Sigma Xi. Res: Biosystematics of Aster, especially those of northeast United States; biology of weeds. Mailing Add: Dept of Biol Hope Col Holland MI 49423

VAN FLANDERN, THOMAS CHARLES, b Cleveland, Ohio, June 26, 40; m 63; c 4. CLESTIAL MECHANICS. Educ: Xavier Univ, Ohio, BS, 62; Yale Univ, PhD(astron), 69. Prof Exp: ASTRONOMER, US NAVAL OBSERV, 63- Concurrent Pos: Consult, Jet Propulsion Lab, 71. Mem: Am Astron Soc; Am Geophys Union; Int Astron Union; AAAS; Am Phys Soc. Res: Lunar motion; occulations; cosmology; gravitation; solar system astronomy. Mailing Add: US Naval Observ Washington DC 20390

VAN FLEET, DICK SCOTT, b Trenton, NJ, Mar 23, 12; m 39; c 2. BOTANY. Educ: Ind Univ, AB, 35, MA, 36, PhD(bot), 40. Prof Exp: Instr bot & geol, Heidelberg Col, 40-42, asst prof biol, 42-43; Sterling res fel, Yale Univ, 43-44; res assoc bot, Univ Mo, 44-45, from asst prof to prof, 45-57; prof bot & chmn dept, Univ Toronto, 57-59; prof bot & head dept, Univ Mass, 59-63; head dept bot, 63-69, PROF BOT, UNIV GA, 63- Mem: Bot Soc Am; Brit Phytochem Soc. Res: Enzyme localization; redox relationships; unsaturated lipid systems; phloem heme; enzymes of endodermis; gene linkage groups in polyenes and polyacetylenes in plants. Mailing Add: Dept of Bot Univ of Ga Athens GA 30601

VANFLEET, HOWARD BAY, b Salt Lake City, Utah, June 5, 31; m 54; c 7. SOLID STATE PHYSICS. Educ: Brigham Young Univ, BS, 55; Univ Utah, PhD(physics), 61. Prof Exp: Asst physics, Univ Utah, 56-60; from asst prof to assoc prof, 60-69, PROF PHYSICS, BRIGHAM YOUNG UNIV, 69- Concurrent Pos: Res grants, US Air Force Off Sci Res, Brigham Young Univ, 62-66, NSF, 69-73; phys scientist, US Army Electronics Command, 66-67; vis prof physics, Am Univ, Cairo, 73-74; consult, Codevintec Pac, Inc, 75- Mem: Am Phys Soc. Res: Ultra high pressure solid state physics; particular phenomena, such as solid state diffusion, Mössbauer effects, melting and high pressure calibration. Mailing Add: Dept of Physics Brigham Young Univ Provo UT 84601

VAN FOSSAN, DONALD DUANE, b El Paso, Tex, Jan 5, 29; m 49; c 3. BIOCHEMISTRY. Educ: Sul Ross State Col, BS, 49; Univ Tex, MA, 52, PhD(biochem), 54, MD, 61. Prof Exp: Asst, Univ Tex, 52-54; res biochemist, Air Force Sch Aviation Med, 54-57, head lab sect, Dept Phyiol-Biophys, 56-57; instr clin path & consult clin labs, Hosp, Univ Tex Med Br, 57-61; intern, St Joseph Hosp, Ft Worth, Tex, 61-62, resident path, Univ Tex, 62-66; dir clin path, Med Ctr, Baylor Univ, 66-69; CLIN PROF PATH, SOUTHERN ILL UNIV, 69-, ASST CHMN DEPT, 73-; ASSOC DIR CHEM, ST JOHN'S HOSP, 74- Concurrent Pos: Instr anal chem, Trinity Univ, 56; dir clin path, St John's Hosp, 68-74. Mem: AAAS; Soc Exp Biol & Med; AMA. Res: Analytic biochemistry; pathology; endocrinology. Mailing Add: 2011 Briarcliff Springfield IL 62704

VAN FOSSEN, PAUL, b Mansfield, Ohio, Aug 21, 23; m 47; c 2. ORGANIC CHEMISTRY. Educ: Univ Cincinnati, AB, 45; Princeton Univ, MA, 47, PhD(chem), 49. Prof Exp: Res chemist cellulose derivatives & org nitrogen compounds, 49-52, tech rep cellulose derivatives & org chem, 52-56, supt chem prod, Sales Develop Lab, 56-60, asst dir, Carney's Point Develop Lab, 60-61, mgr mkt serv, Chem Prod Sales Div, 61-65, mgr admin serv, 65-69, mgr indust prod, 69-72, prod mgr, 72-75, NEW PROD MGR, E I DU PONT DE NEMOURS & CO, INC, 75- Res: Electric blasting caps; blasting supplies and accessories. Mailing Add: 2007 Marsh Rd Wilmington DE 19810

VAN FRANK, RICHARD MARK, b Lansing, Mich, Oct 11, 30; m 54; c 2. CELL BIOLOGY, MOLECULAR PHARMACOLOGY. Educ: Mich State Univ, BS, 52, MS, 56. Prof Exp: Officer in-chg lab, US Naval Damage Control Training Ctr, Philadelphia, 52-54; SR SCIENTIST, BIOL RES DIV, LILLY RES LABS, ELI LILLY & CO, 57- Mem: AAAS. Res: Development of methodology for fractionation of cells and isolation of subcellular particles; effect of drugs at subcellular level. Mailing Add: Biol Res Div Lilly Res Labs Eli Lilly & Co Indianapolis IN 46206

VAN GEET, ANTHONY LEENDERT, b Rotterdam, Neth, July 24, 29; US citizen; m 56; c 3. PHYSICAL CHEMISTRY, ANALYTICAL CHEMISTRY. Educ: Delft Univ Technol, ChemEng, 55; Univ Southern Calif, PhD(phys chem), 61. Prof Exp: Res assoc chem, Mass Inst Technol, 61-63; asst prof, State Univ NY Buffalo, 63-69; assoc prof, Oakland Univ, 69-70; ASSOC PROF CHEM, STATE UNIV NY COL OSWEGO, 70- Mem: AAAS; Am Chem Soc; Royal Neth Chem Soc. Res: Nuclear magnetic resonance of protons, lithium, sodium and fluorine in solution; hydration, complexation and ion-pairing of monovalent ions; determination of trace amounts of heavy metals in lake water; instrumentation. Mailing Add: Dept of Chem State Univ of NY Oswego NY 13126

VAN GELDER, GARY ARTHUR, toxicology, biomedical engineering, see 12th edition

VAN GELDER, NICO MICHEL, b Sumatra, Netherlands E Indies, Dec 24, 33; Can citizen; m 59; c 2. BIOCHEMISTRY, PHYSIOLOGY. Educ: McGill Univ, BSc, 55, PhD(biochem), 59. Prof Exp: Life Ins Med Res Fund fel, Cambridge Univ, Eng, 59-60; res fel neurophysiol & neuropharmacol, Harvard Med Sch, 60-62; asst prof pharmacol, Sch Med, Tufts Univ, 62-67; ASSOC PROF PHYSIOL, UNIV MONTREAL, 67-, BD ADV, NEUROCHEM RES, 75- Concurrent Pos: Res grants, Nat Inst Neurol Dis & Blindness, 62-66, Nat Multiple Sclerosis Soc, 66- & Med Res Coun Can Neurol Sci Group, 67- Mem: Am Soc Neurochem; Int Soc Neurochem; Can Biochem Soc; Int Brain Res Orgn. Res: Biochemistry of epilepsy; structure-activity relationships; biochemistry of brain damage; brain-barrier systems; genetic contribution to epilepsy; function of taurine. Mailing Add: Dept of Physiol Univ of Montreal Montreal PQ Can

VAN GELDER, RICHARD GEORGE, b New York, NY, Dec 17, 28; m 62; c 3. MAMMALOGY. Educ: Colo Agr & Mech Col, BS, 50; Univ Ill, MS, 52, PhD(zool), 58. Prof Exp: Asst zool, Colo Agr & Mech Col, 47-50 & Univ Ill, 50-53; from asst to instr mammal, Univ Kans, 54-56; from asst cur to assoc cur mammals, 56-69, chmn dept, 59-74, CUR MAMMALS, AM MUS NATURAL HIST, 69- Concurrent Pos: Lectr, Columbia Univ, 58-59, asst prof, 59-63; mem bd dirs, Archbold Exped, Inc, 64-74; prof lectr, State Univ NY Downstate Med Ctr, 70-73. Mem: AAAS; Am Soc Mammalogists (vpres, 67-68, pres, 68-70); Wildlife Soc; Soc Syst Zool. Res: Mammalian taxonomy, evolution and ecology. Mailing Add: Am Mus of Natural Hist Central Park W at 79th St New York NY 10024

VAN GELUWE, JOHN DAVID, b Rochester, NY, Sept 18, 16; m 53; c 2. ENTOMOLOGY. Educ: State Univ NY, BS, 39. Prof Exp: Asst, Exten Serv, State Univ NY Col Agr, Cornell Univ, 39-44; dir res & develop, Soil Bldg Div, Coop GLF Exchange, Inc, 44-64; mgr prod develop, Agr Chem Div, 64-67, tech dir, 67-70, ASST DIR FIELD & FARM RES, CIBA-GEIGY CORP, 70- Mem: Fel Am Soc Hort Sci; fel Entom Soc Am; fel Am Phytopath Soc; Weed Sci Soc Am. Res: Formulations, basic laboratory evaluation and field testing of insecticides, fungicides and herbicides. Mailing Add: Ciba-Geigy Corp Agr Div PO Box 11422 Greensboro NC 27409

VAN GILS, GERARD EDUARD, colloid chemistry, see 12th edition

VAN GINNEKEN, ANDREAS J M, b Wynegem, Belg, Jan 1, 35. PHYSICS. Educ: Univ Chicago, MSc, 59, PhD(chem), 66. Prof Exp: Res assoc physics, McGill Univ, 66-70; PHYSICIST, FERMI NAT LAB, 70- Mem: Am Phys Soc; Am Asn Physics Teachers. Res: Nuclear chemistry; nuclear and particle physics; radiation physics. Mailing Add: Fermi Nat Lab Batavia IL 60510

VAN GULICK, NORMAN MARTIN, b Los Angeles, Calif, July 1, 26; m 48; c 2. ORGANIC CHEMISTRY. Educ: Univ Colo, AB, 48; Univ Southern Calif, PhD, 54. Prof Exp: Asst prof chem, Univ Ore, 56-60; RES CHEMIST, ELASTOMER CHEM DEPT, E I DU PONT DE NEMOURS & CO, INC, 60- Mem: Am Chem Soc; Sigma Xi. Res: Organic polymer chemistry; organometallic and organic fluorine chemistry;

reaction mechanisms. Mailing Add: Exp Sta E I du Pont de Nemours & Co Inc Wilmington DE 19898

VAN GUNDY, SEYMOUR DEAN, b Whitehouse, Ohio, Feb 24, 31; m 54; c 2. PLANT PATHOLOGY, NEMATOLOGY. Educ: Bowling Green State Univ, BA, 53; Univ Wis, PhD, 57. Prof Exp: Res assoc, Univ Wis, 53-57; from asst nematologist to assoc nematologist, 57-68, assoc dean res, 68-71, asst vchancellor res, 71-72, PROF NEMATOL, UNIV CALIF, RIVERSIDE, 68-, CHMN DEPT, 72- Concurrent Pos: NSF sr fel, Australia, 65-66; ed-in-chief, J Nematol, 67-71. Mem: Fel AAAS; Am Phytopath Soc; Soc Nematol (vpres, 72-73, pres, 73-74); Soc Europ Nematol; Am Inst Biol Sci. Res: Biology and control of nematodes. Mailing Add: Dept of Nematol Univ of Calif Riverside CA 92502

VAN HALL, CLAYTON EDWARD, b Grand Rapids, Mich, Apr 24, 24; m 51; c 2. ANALYTICAL CHEMISTRY. Educ: Hope Col, AB, 49; Mich State Univ, MSc, 54, PhD(anal chem), 56. Prof Exp: Asst, Mich State Univ, 52-56; chemist, 56-58, anal chemist, 58-61, anal specialist, 61-65, anal res specialist, 65-72, ASSOC SCIENTIST, DOW CHEM CO, 72- Mem: Am Chem Soc. Res: Instrumental methods; trace gas methods; trace element methods; purity of inorganic compounds; primary standards; water analysis. Mailing Add: 3712 Wintergreen Dr Midland MI 48640

VAN HANDEL, EMILE, b Rotterdam, Holland, Mar 29, 18; nat US; m 46; c 2. ORGANIC CHEMISTRY. Educ: State Univ Leiden, BS, 38, MS, 41; State Inst Technol, Delft, MS, 41; Univ Amsterdam, PhD(biochem), 54. Prof Exp: Indust chemist, 45-54; res biochemist, St Anthon's Hosp, Voorburg, Holland, 54-55, asst prof physiol, Univ Tenn, 55-58; HEAD DEPT BIOCHEM, ENTOM RES CTR, FLA STATE DIV HEALTH, 58- Mem: Am Heart Asn; Am Soc Biol Chemists. Res: Lipid and carbohydrate chemistry and metabolism; insect biochemistry; atherosclerosis. Mailing Add: Entom Res Ctr Fla State Div of Health Vero Beach FL 32960

VAN HARN, GORDON L, b Grand Rapids, Mich, Dec 30, 35; m 58; c 3. PHYSIOLOGY. Educ: Calvin Col, AB, 57; Univ Ill, MS, 59, PhD(physiol), 61. Prof Exp: Assoc prof biol, Calvin Col, 61-68 & Oberlin Col, 68-70; PROF BIOL, CALVIN COL, 70-; RES ASSOC, BLODGETT MEM HOSP, 70- Mem: AAAS; Am Sci Affil. Res: Smooth muscle contractile and electrical activity; intestinal smooth muscle; cardiac muscle. Mailing Add: Dept of Biol Calvin Col Burton St Grand Rapids MI 49506

VAN HARREVELD, ANTHONIE, b Haarlem, Neth, Feb 16, 04; nat US; m 28; c 2. PHYSIOLOGY. Educ: Univ Amsterdam, BA, 25, MA, 28, PhD(animal physiol), 29, MD, 30. Prof Exp: Asst physiol, Univ Amsterdam, 27-31; chief asst, State Univ Utrecht, 31-34; asst, 34-35, from instr to prof, 35-74, EMER PROF PHYSIOL, CALIF INST TECHNOL, 74- Concurrent Pos: Consult, Los Angeles County Gen Hosp, 43-55; res assoc, Univ Calif, 44. Mem: AAAS; Am Physiol Soc. Res: Electrophysiology of nerve muscle and central nervous system; nerve regeneration; metabolism; segmental physiology; water and electrolyte distribution in central nervous system. Mailing Add: Div Biol Calif Inst of Technol 1201 E California Blvd Pasadena CA 91109

VAN HASSEL, HENRY JOHN, b Paterson, NJ, May 2, 33; m 60. DENTISTRY, PHYSIOLOGY. Educ: Maryville Col, BA, 54; Univ Md, DDS, 63; Univ Wash, MSD, 64, PhD(physiol), 69. Prof Exp: Asst prof endodont & physiol, 69-71, ASSOC PROF PHYSIOL, MED SCH, UNIV WASH, 71-, ASSOC PROF ENDODONT, DENT SCH, 71-, RES ASSOC PHYSIOL, REGIONAL PRIMATE RES CTR, 69-; DEP CHIEF DENT SERV & DIR ENDODONT RESIDENCY, USPHS HOSP, SEATTLE, 71- Concurrent Pos: Consult, US Army, Ft Lewis, Wash, 69-; chmn, Nat Workshop Pulp Biol, 71-; vchmn, Sect Physiol, Am Asn Dent Schs, 72- Honors & Awards: First recipient Carl A Schlack Award, Asn Mil Surgeons US, 71. Mem: Am Den Asn; Am Asn Endodont; Int Ans Asn Dent Res. Res: Oral physiology; psychophysiology; neurophysiology of pain. Mailing Add: 7822 NE 14th Bellevue WA 98004

VAN HATTUM, ROLLAND JAMES, b Grand Rapids, Mich, July 14, 24; m 49; c 5. SPEECH PATHOLOGY, AUDIOLOGY. Educ: Western Mich Univ, BS, 50; Pa State Univ, MS, 52, PhD(speech path), 54. Prof Exp: Instr speech path, Pa State Univ, 50-54; dir spec educ, Kent County, Mich, 58-63; PROF COMMUN DIS & CHMN DEPT, STATE UNIV NY COL BUFFALO, 63-; CONSULT SPEECH, HEARING & LANG, BUFFALO CHILDREN'S HOSP, 65- Concurrent Pos: Consult speech & hearing, Children's Hosp, Grand Rapids, Mich, 57-63; lectr spec educ, Univ Mich & Mich State Univ, 58-63; res assoc, Eastman Dent Ctr, 63-68; audiologist, Joel Bernstein, MD, 71- Honors & Awards: Citation, Mich Asn Retarded Children, 63; Honors, NY State Speech & Hearing Asn, 74. Mem: Fel Am Speech & Hearing Asn (vpres, 71-74, pres-elect, 75). Res: Communication programs for mildly, moderately and severely retarded; automated programs. Mailing Add: Dept of Commun Dis State Univ NY Col Buffalo NY 14222

VAN HAVERBEKE, DAVID F, b Eureka, Kans, July 15, 28. TAXONOMY, SILVICULTURE. Educ: Kans State Univ, BS, 50; Colo State Univ, MS, 59; Univ Nebr, PhD(bot), 67. Prof Exp: Res forester, Rocky Mountain Forest & Range Exp Sta, 58 & Southeastern Forest Exp Sta, 59-62, RES FORESTER, ROCKY MOUNTAIN FOREST & RANGE EXP STA, USDA, 62- Concurrent Pos: Assoc prof forestry, Univ Nebr-Lincoln. Mem: Soc Am Foresters. Res: Forest botany and genetics; forest tree improvement; noise abatement; shelterbelt management. Mailing Add: Rocky Mountain Forest & Range Exp Sta Univ of Nebr Lincoln NE 68503

VAN HECKE, GERALD RAYMOND, b Evanston, Ill, Nov 1, 39. PHYSICAL CHEMISTRY. Educ: Harvey Mudd Col, BS, 61; Princeton Univ, AM, 63, PhD(phys chem), 66. Prof Exp: Chemist, Shell Develop Co, 66-70; asst prof, 70-74, ASSOC PROF CHEM, HARVEY MUDD COL, 74- Concurrent Pos: Shell Found fac improv grant, 71-72. Mem: AAAS; Am Chem Soc; The Chem Soc; Sigma Xi. Res: Nuclear magnetic resonance studies of paramagnetic transition metal complexes; physical properties of liquid crystals; calorimetry, dilatometry, firefringence. Mailing Add: Dept of Chem Harvey Mudd Col Claremont CA 91711

VAN HEERDEN, PIETER JACOBUS, b Utrecht, Neth, Apr 14, 15; nat US; m 49; c 2. PHYSICS. Educ: Univ Utrecht, PhD(physics), 45. Prof Exp: Res physicist, Bataafse Petrol Co, Neth, 44-45; vis lectr, Harvard Univ, 48-49, res fel nuclear physics, 49-53; res assoc, Gen Elec Res Lab, NY, 53-62; PHYSICIST, POLAROID RES LABS, 62- Mem: Am Phys Soc; Neth Phys Soc. Res: Experimental nuclear and solid state physics; foundation of physics; foundation of scientific knowledge and intelligence. Mailing Add: Polaroid Res Labs 750 Main St Cambridge MA 02139

VAN HEININGEN, JAN JACOB, organic chemistry, see 12th edition

VAN HEUVELEN, ALAN, b Buffalo, Wyo, Dec 15, 38; m 62; c 1. PHYSICS. Educ: Rutgers Univ, BA, 60; Univ Colo, PhD(physics), 64. Prof Exp: Assoc prof physics, 64-74, PROF PHYSICS, NMEX STATE UNIV, 74- Mem: Am Phys Soc. Res: Biophysics using electron spin resonance to study enzymes. Mailing Add: Dept of Physics NMex State Univ Las Cruces NM 88001

VAN HEYNINGEN, EARLE MARVIN, b Chicago, Ill, Oct 15, 21; m 51; c 4. ORGANIC CHEMISTRY. Educ: Calvin Col, AB, 43; Univ Ill, PhD(org chem), 46. Prof Exp: Res chemist, 46-65, res scientist, 65-66, res assoc, 66-69, dir agr chem, Greenfield Labs, 69-72, DIR CHEM, GREENFIELD LABS, ELI LILLY & CO, 72- Mem: Am Chem Soc. Res: Synthesis of barbituric acids; antimalarials, anti-arthritics; cholesterol lowering agents; cephalosporin antibiotics. Mailing Add: Lilly Res Labs Chem Res Indianapolis IN 46206

VAN HEYNINGEN, ROGER, b Chicago, Ill, Oct 2, 27; m 51; c 3. SOLID STATE PHYSICS. Educ: Calvin Col, AB, 51; Univ Ill, MS, 55, PhD, 58. Prof Exp: Sr physicist, 58-62, res assoc, 62-66, asst div head, 66-67, PHYSICS DIV DIR, EASTMAN KODAK CO, 67- Mem: Am Phys Soc; Optical Soc Am. Res: Electronic and optical properties of insulating and semiconducting solids. Mailing Add: Res Labs Eastman Kodak Co Rochester NY 14650

VAN HISE, JAMES R, b Tracy, Calif, Aug 11, 37; m 64; c 2. NUCLEAR CHEMISTRY, PHYSICAL CHEMISTRY. Educ: Walla Walla Col, BS, 59; Univ Ill, PhD(phys chem), 63. Prof Exp: Res assoc nuclear chem, Oak Ridge Nat Lab, 63-65; from asst prof to assoc prof chem & physics, Andrews Univ, 65-69; prof physics & chmn dept, Tri-State Col, 69-72; PROF CHEM, PAC UNION COL, 72- Concurrent Pos: Consult, Lawrence Livermore Lab, 74-; prof chem sci & chmn div, World Open Univ, 75- Mem: AAAS; Am Chem Soc; Am Phys Soc. Res: Nuclear photodisintegration; alpha, beta and gamma ray spectroscopy; positron annihilation in organic media. Mailing Add: 566 Sunset Dr Angwin CA 94508

VAN HOEVEN, WILLIAM, organic chemistry, polymer chemistry, see 12th edition

VAN HOLDE, KENSAL EDWARD, b Eau Claire, Wis, May 14, 28; m 50; c 4. PHYSICAL CHEMISTRY. Educ: Univ Wis, BS, 49, PhD(chem), 52. Prof Exp: Res chemist textile fibers dept, E I du Pont de Nemours & Co, 52-55; res assoc, Univ Wis, 55-56, asst prof chem, Univ Wis-Milwaukee, 56-57; from asst prof to prof, Univ Ill, Urbana, 57-67; PROF BIOPHYS, ORE STATE UNIV, 67- Mem: Am Soc Biol Chem. Res: Physical chemistry of macromolecules; biophysical chemistry. Mailing Add: Dept of Biochem & Biophys Ore State Univ Corvallis OR 97331

VAN HOOK, ANDREW, b Paterson, NJ, June 3, 07; m 34; c 5. PHYSICAL CHEMISTRY. Educ: Polytech Inst Brooklyn, BS, 31; NY Univ, PhD(phys chem), 34. Prof Exp: Indust & consult chem, 26-28; metall engr, Westinghouse Lamp Co, 28-29; instr, NY Univ, 31-34; res chemist, Autoxygen Co, 34-36; instr phys chem, Lafayette Col, 36-38; asst prof, Univ Idaho, 38-42 & Lafayette Col, 42-44; assoc prof, Univ Wyo, 44-46; PROF PHYS CHEM, COL OF THE HOLY CROSS, 46- Mem: AAAS; Am Chem Soc; assoc Am Inst Chem Engrs. Res: Kinetics; crystallization; sugar technology; theory of liquids and concentrated solutions. Mailing Add: Dept of Chem Col of the Holy Cross Worcester MA 01610

VAN HOOK, HARRY JERROLD, geochemistry, see 12th edition

VAN HOOK, JAMES PAUL, b Paterson, NJ, Oct 16, 31; m 57; c 5. PHYSICAL CHEMISTRY, ENGINEERING MANAGEMENT. Educ: Col of the Holy Cross, BS, 53; Princeton Univ, PhD(chem), 58. Prof Exp: Res chemist, M W Kellogg Co Div, Pullman, Inc, 57-62, supvr, 62-65, sect head process res, 65-74; PROCESS DEVELOP MGR, CORP ENG DEPT, ALLIED CHEM CORP, 74- Mem: AAAS; Am Chem Soc; Sigma Xi; Am Inst Chem Engrs. Res: Catalytic oxidation for chlorine production; steam-hydrocarbon reactions for production of synthesis gas, hydrogen or synthetic natural gas; coal gasification; air pollution control. Mailing Add: 102 Harrison Brook Dr Basking Ridge NJ 07920

VAN HOOK, WILLIAM ALEXANDER, b Paterson, NJ, Jan 14, 36; m 62; c 3. PHYSICAL CHEMISTRY. Educ: Col of the Holy Cross, BS, 57; Johns Hopkins Univ, MA, 59, PhD(chem), 61. Prof Exp: Res assoc phys chem, Brookhaven Nat Lab, 61-62; from asst prof to assoc prof, 62-72, PROF PHYSICS, UNIV TENN, KNOXVILLE, 72- Concurrent Pos: Fulbright res fel, Belg, 67-68; Nat Acad Sci exchange fel, Yugoslavia, 71. Mem: AAAS; Am Chem Soc. Res: Isotope effects on chemical and physical properties of molecular systems; solutions. Mailing Add: Dept of Chem Univ of Tenn Knoxville TN 37916

VAN HOOSIER, GERALD L, JR, b Weatherford, Tex, June 4, 34; m 59; c 2. LABORATORY ANIMAL SCIENCE. Educ: Agr & Mech Col Tex, DVM, 57. Prof Exp: Head animal test sect, Div Biol Standards, NIH, 57-59, head in serv training, Viral & Rickettsial Dis Lab, Calif State Dept Health, 59-60, head appl virol sect, Div Biol Standards, 60-62; from instr to assoc prof exp biol, Baylor Col Med, 62-69; from asst prof to assoc prof vet path & dir lab animal resources, Wash State Univ, 69-75; DIR DIV ANIMAL MED, UNIV WASH, 75- Concurrent Pos: Resident path, Baylor Col Med, 69-70 & Wash State Univ, 70-71; mem animal resources adv comt, Animal Res Bd, NIH, 74-78. Mem: AAAS; Am Asn Lab Animal Sci; Am Soc Exp Path; Am Vet Med Asn. Res: Laboratory animal disease and medicine; comparative pathology; animal virology. Mailing Add: Div of Animal Med SE-20 Univ of Wash Seattle WA 98195

VAN HORN, DAVID DOWNING, b Rochester, NY, Apr 23, 21; m 45. METAL PHYSICS, MATHEMATICS. Educ: Univ Rochester, BA, 42; Case Inst Technol, PhD(physics), 49. Prof Exp: Instr physics, Univ Rochester, 43-44; jr physicist, Clinton Eng Works, 44-46; instr physics, Case Inst Technol, 46-49; res assoc metall, Knolls Atomic Power Lab, 49-57, GROUP LEADER CHEM & METALL ENG, INCANDESCENT LAMP DEPT, GEN ELEC CO, 57- Mem: Am Phys Soc; Am Soc Metals; Am Asn Physics Teachers; Math Asn Am; Am Inst Mining, Metall & Petrol Engrs. Res: Solid state diffusion; mechanical properties; heat transfer; tungsten; incandescent lamps; radiation measurements. Mailing Add: Incandescent Lamp Dept 3437 Gen Elec Co Cleveland OH 44112

VAN HORN, DIANE LILLIAN, b Waukesha, Wis, Aug 21, 39; m 72; c 3. PHYSIOLOGY, ELECTRON MICROSCOPY. Educ: Univ Wis-Madison, BS, 61; Marquette Univ, MS, 66, PhD(physiol), 68. Prof Exp: Res assoc, Wood Vet Admin Ctr, 66-67, supvry scientist, Electron Micros Lab, 69-75; from instr to asst prof, 68-72, ASSOC PROF PHYSIOL & OPHTHAL, MED COL WIS, 72-; CHIEF ELECTRON MICROS SECT, WOOD VET ADMIN CTR, 75- Concurrent Pos: Seeing Eye Inc res grant, Med Col Wis, 69-72, Nat Eye Inst res grant, 72-75. Mem: Am Physiol Soc; Asn Res Vision & Ophthal; Electron Micros Soc Am. Res: Corneal physiology and ultrastructure; electron microscopy of ocular and other tissues. Mailing Add: Electron Micros Lab Wood Vet Admin Ctr Wood WI 53193

VAN HORN, DONALD H, b Hinsdale, Ill, Oct 9, 28; m 59; c 2. ECOLOGY. Educ: Kalamazoo Col, BA, 50; Univ Ill, MS, 52; Univ Colo, PhD(zool), 61. Prof Exp: Asst prof biol, Lake Forest Col, 61-62 & Utica Col, 62-65; vis asst prof, 65-71, assoc prof, 71-74, PROF BIOL & CHMN DEPT, UNIV COLO, COLORADO SPRINGS CTR,

74- Mem: AAAS; Am Ornith Union; Ecol Soc Am; Am Soc Zoologists. Res: Terrestrial ecology, especially community and population analysis of mountain animals. Mailing Add: Dept of Biol Univ of Colo Cragmor Rd Colorado Springs CO 80907

VAN HORN, GENE STANLEY, b Oakland, Calif, June 26, 40; m 62; c 3. SYSTEMATIC BOTANY. Educ: Humboldt State Univ, AB, 63; Univ Calif, Berkeley, PhD(bot), 70. Prof Exp: Vis asst prof biol, Tex Tech Univ, 70-71; ASST PROF BIOL, UNIV TENN, CHATTANOOGA, 71- Concurrent Pos: Tex State Inst Funds grant, Tex Tech Univ, 71; Univ Chattanooga Found grant, Univ Tenn, Chattanooga, 72-73. Mem: Bot Soc Am; Inst Asn Plant Taxonomists; Am Soc Plant Taxonomists; Torrey Bot Club. Res: Biosystematics and evolution of angiosperms, especially Asteraceae; floristics. Mailing Add: Dept of Biol Univ of Tenn Chattanooga TN 37401

VAN HORN, HAROLD H, JR, b Pomona, Kans, Jan 13, 37; m 58; c 3. DAIRYING. Educ: Kans State Univ, BS, 58, MS, 59; Iowa State Univ, PhD(dairy nutrit), 62. Prof Exp: Assoc prof dairy nutrit & mgt & exten dairyman, Iowa State Univ, 61-70; PROF DAIRY SCI & CHMN DEPT, UNIV FLA, 70-, ANIMAL NUTRITIONIST, 74- Mem: Am Dairy Sci Asn; Am Soc Animal Sci. Res: Improved dairy feeding and management practices; nutrition research in the use of urea in dairy rations. Mailing Add: Dept of Dairy Sci Univ of Fla Gainesville FL 32601

VAN HORN, HUGH MOODY, b Williamsport, Pa, Mar 5, 38; m 60; c 3. ASTROPHYSICS. Educ: Case Inst Technol, BS, 60; Cornell Univ, PhD(astrophys), 66. Prof Exp: Res assoc, 65-67, asst prof, 67-72, ASSOC PROF ASTROPHYS, UNIV ROCHESTER, 72- Concurrent Pos: Vis fel, Joint Inst Lab Astrophys, Univ Colo, Boulder, 73-74. Mem: Int Astron Union; Am Astron Soc. Res: Structure, evolution, and oscillations of degenerate dwarfs; rotation and magnetic fields in degenerate dwarfs; nuclear reactions in stars; statistical physics of astronomical systems. Mailing Add: Dept of Physics & Astron Univ of Rochester Rochester NY 14627

VAN HORN, JOHN A, b Butte, Mont, Aug 13, 15; m 45. PHYSICS. Educ: Chadron State Col, BS, 36; Univ Nebr, MS, 38; Pa State Univ, PhD, 49. Prof Exp: Instr physics, Mt St Mary's Col, 44-46; physicist, Airadio, Inc, 46-48; sr physicist, Hamilton Watch Co, 49-51, chief physicist, 51-54, dir res, 54-66; lectr physics, Bucknell Univ, 66-67; assoc prof, 67-69, PROF PHYSICS, MILLERSVILLE STATE COL, 69- Honors & Awards: Distinguished Achievement Award, United Horology Asn Am, 58. Mem: Am Phys Soc. Res: Horology; magnetism. Mailing Add: Millersville State Col Millersville PA 17551

VAN HORN, LESTER MILTON, b Brookfield, NY, Feb 26, 11; m 33, 67; c 6. ZOOLOGY, PHYSIOLOGY. Educ: Salem Col, WVa, BA & BS, 35; WVa Univ, MS, 37. Prof Exp: Asst biol, Salem Col, WVa, 34-35; asst nutrit, WVa Univ, 35-36, asst physiol, 36-37; dean, 56-64, PROF BIOL & HEAD DEPT, MILTON COL, 37- Concurrent Pos: Asst, Univ Wis, 49-51 & 55-56. Mem: Am Inst Biol Sci. Res: Bird pituitary-adrenal physiology during environmental stress. Mailing Add: Dept of Biol Milton Col Milton WI 53563

VAN HORN, MAURICE H, physical chemistry, deceased

VAN HORN, RICHARD NORMAN, b Milton, Wis, Feb 1, 40; m 61; c 2. PRIMATOLOGY, REPRODUCTIVE BIOLOGY. Educ: Beloit Col, BA, 63; Univ Calif, Berkeley, MA, 68, PhD(physical anthrop), 69. Prof Exp: Asst specialist primate behavior, Univ Calif, Berkeley, 67-69; NSF fel & res assoc cutaneous biol, 70-71, ASST SCIENTIST PRIMATE BEHAVIOR, ORE REGIONAL PRIMATE RES CTR, 71- Res: Primate sexual behavior and reproductive physiology with emphasis on the environmental regulation of breeding seasons and the social and hormonal determinants of sexual behavior in prosimians. Mailing Add: Primate Behav Ore Regional Primate Res Ctr Beaverton OR 97005

VAN HORN, RUTH WARNER, b Waterloo, Iowa, Mar 24, 18; m 45. ORGANIC CHEMISTRY. Educ: Univ Calif, Los Angeles, BA, 39, MA, 40; Pa State Univ, PhD(org chem), 44. Prof Exp: Org chemist, Am Cyanamid Co, 44-48; instr chem, Hunter Col, 48-49; from asst prof to assoc prof, 49-64, PROF CHEM, FRANKLIN & MARSHALL COL, 64- Mem: AAAS; Am Chem Soc. Res: Synthesis. Mailing Add: Dept of Chem Franklin & Marshall Col Lancaster PA 17604

VAN HORNE, ROBERT LOREN, b Malvern, Iowa, Dec 26, 15; m 41, 63; c 5. PHARMACOGNOSY, PHARMACY. Educ: Univ Iowa, BS, 41, MS, 47, PhD, 49. Prof Exp: Instr pharm, Univ Iowa, 49-51, from asst prof to assoc prof pharmacog, 51-56; dean sch pharm, 56-75, PROF PHARM, 56-, DIR CONTINUING EDUC, SCH PHARM, 75- Concurrent Pos: Mem fac adv coun, Gov of Mont; dir, Western Area Alcohol Educ & Training Prog, Nev, 74- Mem: Am Asn Cols Pharm; Am Pharmaceut Asn. Res: Water soluble embedding materials for microtechnique; polyethylene glycols as substitutes for glycerin and ethanol in pharmaceutical preparations; surfactants in the preparation of coal tar lotions; anionic exchange resins for alkaloid separation; phytochemistry of mistletoe species. Mailing Add: Sch of Pharm Univ of Mont Missoula MT 59801

VAN HORNE, WILLIAM LESLIE, chemistry, see 12th edition

VAN HOUTEN, FRANKLYN BOSWORTH, b New York, NY, July 14, 14; m 43; c 3. GEOLOGY. Educ: Rutgers Univ, BS, 36; Princeton Univ, PhD(geol), 41. Prof Exp: Instr geol, Williams Col, 39-42; from asst prof to assoc prof, 47-55, PROF GEOL, PRINCETON UNIV, 55- Concurrent Pos: Consult, 41-; geologist, US Geol Surv, 48- & Geol Surv Can, 53; vis prof, Univ Calif, Los Angeles, 63, State Univ NY Binghamton, 71 & Univ Basel, 71. Mem: Fel Geol Soc Am; Soc Econ Paleontologists & Mineralogists; Am Asn Petrol Geologists; Int Asn Sedimentol. Res: Sedimentology; clay minerals; zeolites; iron oxides; red beds; Triassic rocks, eastern North America and northwestern Africa, continental drift reconstructions; Cenozoic nonmarine deposits, western United States and northern South America; modern marine sediments; Molasse facies in orogenic belts. Mailing Add: Dept of Geol & Geophys Sci Princeton Univ Princeton NJ 08540

VAN HOUWELING, CORNELIUS DONALD, b Mahaska County, Iowa, July 19, 18; m 54; c 5. VETERINARY MEDICINE. Educ: Iowa State Univ, DVM, 42, MS, 66. Prof Exp: Vet, Springfield, Ill, 42-43; dir vet med rels, Ill Agr Asn, 46-48; dir prof rels & asst exec secy, Am Vet Med Asn, 48-53; instr, Col Vet Med, Univ Ill, 53-54; dir livestock regulatory progs, Agr Res Serv, USDA, 54-56, asst adminr, 56-61, asst dir regulatory labs, Nat Animal Dis Lab, Iowa, 61-66; DIR BUR VET MED, US FOOD & DRUG ADMIN, ROCKVILLE, MD, 67- Concurrent Pos: Mem comn vet educ, Southern Regional Educ Bd; mem subcomt laws, rules & regulations animal health, Nat Res Coun; mem exec bd, Nat Health Coun; mem, Nat Brucellosis Comt; chmn adv comt humane slaughter, US Secy Agr; mem & chmn, Coun Pub Health & Regulatory Vet Med. Honors & Awards: Award, Am Mgt Asn, 57. Mem: AAAS; Am Vet Med Asn; US Animal Health Asn; Conf Pub Health Vets; World Asn Vet Food Hyg. Res: Regulatory veterinary medicine. Mailing Add: Bur Vet Med US FDA 5600 Fishers Lane Rockville MD 20852

VAN HOVEN, GERARD, b Los Angeles, Calif, Nov 23, 32; m 56; c 2. PLASMA PHYSICS, SOLAR PHYSICS. Educ: Calif Inst Technol, BS, 54; Stanford Univ, PhD(physics), 63. Prof Exp: Mem tech staff, Bell Tel Labs, 54-56; electron physicist, Gen Elec Co, 56-63; res assoc, W W Hansen Labs Physics, Stanford Univ, 63-65, res physicist, Inst Plasma Res, 65-68; asst prof, 68-74, ASSOC PROF PHYSICS, UNIV CALIF, IRVINE, 74- Concurrent Pos: Fulbright fel, Vienna Tech Univ, 63-64; consult, Gen Elec Co, 63-65, Varian Assocs, 65-68 & Aerospace Corp, 75-; Langley-Abbot vis scientist, Ctr Astrophys, Harvard Univ, 75-76. Mem: Am Phys Soc; Am Astron Soc. Res: Solar-terrestrial activity magnetohydrodynamics, especially magnetic field reconnection instabilities; nonlinear plasma wave interactions, including instabilities and radiation. Mailing Add: Dept of Physics Univ of Calif Irvine CA 92664

VAN HULLE, GLENN JOSEPH, b Oconto, Wis, Dec 20, 37; m 63; c 3. FOOD SCIENCE. Educ: Univ Wis-Madison, BS, 64, MS, 65, PhD(food sci), 69. Prof Exp: SR FOOD SCIENTIST PROD DEVELOP, GEN MILLS INC, 69- Mem: Inst Food Technologists. Res: New product and process development in meat, dairy and soy based products. Mailing Add: 9000 Plymouth Ave N Minneapolis MN 55427

VAN HUYSTEE, ROBERT BERNARD, b Amsterdam, Holland, Sept 29, 31; Can citizen; m 58; c 1. BIOCHEMISTRY, BOTANY. Educ: Univ Sask, BA, 59, MA, 61; Univ Minn, St Paul, PhD(hort), 64. Prof Exp: Fel physiol, Purdue Univ, 64-66; asst prof radiobiol, 66-69, ASSOC PROF PLANT SCI, UNIV WESTERN ONT, 69- Mem: Am Soc Plant Physiol; Radiation Res Soc; Can Soc Plant Physiol. Res: Process of cold acclimation and its metabolism in plants; nucleic acid metabolism in photo induction and as related to ionizing radiation effects. Mailing Add: Dept of Plant Sci Univ of Western Ont London ON Can

VAN HYNING, JACK M, fishery biology, zoology, see 12th edition

VANICEK, C DAVID, b Waterloo, Iowa, Oct 12, 39. FISHERIES MANAGEMENT. Educ: Iowa State Univ, BS, 61, MS, 63; Utah State Univ, PhD(fishery biol), 67. Prof Exp: Fishery biologist, US Fish & Wildlife Serv, 63-67; asst prof, 67-73, ASSOC PROF BIOL, CALIF STATE UNIV, SACRAMENTO, 73- Mem: Am Fisheries Soc; Am Inst Biol Sci; Am Inst Fishery Res Biol. Res: Freshwater fishery biology and management. Mailing Add: Dept of Biol Sci Calif State Univ Sacramento CA 95819

VANICEK, PETR, b Susice, Czech, July 18, 35; m 60; c 3. GEODESY, GEOPHYSICS. Educ: Prague Tech Univ, Dipl Ing, 59; Czech Acad Sci, PhD(geophys), 68. Prof Exp: Div head land surv, Prague Inst Surv & Cartog, 59-63; consult numerical anal & comput prog, Fac Tech & Nuclear Physics, Prague Tech Univ, 63-67; sr sci officer, Inst Coastal Oceanog & Tides, Nat Environ Res Coun Gt Brit, 68-69; Nat Res Coun Can fel, Dept Energy, Mines & Resources, 69-71; ASSOC PROF GEOD, UNIV NB, FREDERICTON, 71-, DIR GRAD STUDIES, DEPT SURV ENG, 75- Concurrent Pos: Mem, NAm Working Group, Comn Recent Crustal Movements, Int Union Geod & Geophys, 72-; secy subcomt geod, Nat Res Coun Can, 72-74; mem, Can Subcomt Geodynamics, 75- Mem: Am Geophys Union; Can Inst Surv; fel Geol Asn Can. Res: Physical and dynamical geodesy; earth tides, crustal movements and mean sea level; applied mathematics, especially spectral analysis and mechanics. Mailing Add: Dept of Surv Eng Univ of NB Fredericton NB Can

VANIER, JACQUES, b Dorion, Que, Jan 4, 34; m 61; c 2. QUANTUM ELECTRONICS. Educ: Univ Montreal, BA, 55, BSc, 58; McGill Univ, MSc, 60, PhD(physics), 63. Prof Exp: Lectr physics, McGill Univ, 61-63; physicist, Quantum Electronics Div, Varian Assocs, Mass, 63-67; assoc prof, 68-71, PROF ELEC ENG, LAVAL UNIV, 71-; PHYSICIST, HEWLETT-PACKARD CO, 67- Concurrent Pos: Lectr, Univ Montreal, 62; prof, Univ Laval, 67. Mem: Can Asn Physicists; sr mem Inst Elec & Electronics Engrs; Am Phys Soc. Res: Electron paramagnetic resonance; nuclear magnetic resonance; optical pumping; masers; frequency standards; atomic clocks. Mailing Add: Quantum Electronics Lab Dept Elec Eng Laval Univ Quebec PQ Can

VAN ITALLIE, THEODORE BERTUS, b Hackensack, NJ, Nov 8, 19; m 48; c 5. MEDICINE. Educ: Harvard Univ, SB, 41; Columbia Univ, MD, 45; Am Bd Internal Med, dipl, 54. Prof Exp: Intern med, St Luke's Hosp, New York, 45-46, from asst resident to resident, 48-50; res fellow nutrit, Sch Pub Health, Harvard Univ, 50-51, res assoc, 51-52, asst prof clin nutrit, Schs Med & Pub Health, 55-57; instr, 52-55, from assoc clin prof to clin prof, 57-71, PROF MED, COL PHYSICIANS & SURGEONS, COLUMBIA UNIV, 71-, ASSOC DIR INST HUMAN NUTRIT, 67- Concurrent Pos: From asst to assoc, Peter Bent Brigham Hosp, Boston, 50-57; dir lab nutrit res, St Luke's Hosp, 52-55, asst attend physician, 53-55, attend physician, 57-, dir med, 57-75; vis lectr, Sch Pub Health, Harvard Univ, 57-60; mem gastroenterol & nutrit training comt, NIH, 69-73; mem food & nutrit bd, Nat Acad Sci, 70- Mem: Soc Exp Biol & Med; Am Clin & Climat Asn; Am Fedn Clin Res; fel Am Col Physicians; Am Soc Clin Nutrit (pres-elect, 75-76). Res: Carbohydrate and lipid physiology and biochemistry; nutrition; metabolism; regulation of food intake. Mailing Add: St Luke's Hosp Ctr 421 W 113th St New York NY 10025

VAN KAMPEN, KENT RIGBY, b Brigham City, Utah, July 30, 36; m 59; c 3. VETERINARY PATHOLOGY. Educ: Utah State Univ, BS, 61; Colo State Univ, DVM, 67; Univ Calif, Davis, PhD(comp path), 67. Prof Exp: Vet pathologist, Poisonous Plant Res Lab, Agr Res Serv, USDA, Utah, 67-70; assoc prof vet sci, Utah State Univ, 68-70; DIR RES, INTERMOUNTAIN LABS, INC, 69- Concurrent Pos: Adj assoc prof, Col Med, Univ Utah, 69- Mem: AAAS; Am Col Vet Path; Am Vet Med Asn; fel Am Col Vet Toxicol. Res: Pathogenesis of animal diseases related to similar disorders in man; pathology of natural and man made toxicants in animals; mechanisms of carcinogenesis. Mailing Add: 6174 Rodeo Lane Salt Lake City UT 84121

VAN KANN, FRANK JOACHIM, b Köln, WGer, Mar 9, 47; Australian citizen; m 72; c 2. LOW TEMPERATURE PHYSICS. Educ: Univ Western Australia, BSc, 68, PhD(physics), 75. Prof Exp: FEL PHYSICS, STANFORD UNIV, 74- Res: Test general relativity theory by measuring precession of earth orbiting cryogenic gyroscope mounted on telescope referenced to fixed star using SQUID magnetometry for London Moment readout of spin axis orientation. Mailing Add: Dept of Physics Stanford Univ Stanford CA 94305

VANKERKHOVE, ALAN PAUL, b Rochester, NY, Mar 18, 33; m 56; c 8. OPTICAL PHYSICS. Educ: Univ Rochester, BS, 64, MS, 71. Prof Exp: SR RES PHYSICIST OPTICAL PHYSICS, EASTMAN KODAK CO, 64- Res: Optical studies related to other disciplines such as organic chemistry and materials science; application of geometric and physical optics to consumer products. Mailing Add: Eastman Kodak Co Kodak Park Bldg Rochester NY 14650

VAN KEUREN, ROBERT W, b Virginia, Minn, Jan 2, 22; m 49; c 3. AGRONOMY. Educ: Wis State Univ, BS, 43; Univ Wis, MS, 52, PhD, 54. Prof Exp: Instr, Wis Pub Schs, 46-50; res asst, Univ Wis, 51-54; asst & assoc agronomist, Wash State Univ, 54-62; assoc prof, 62-70, PROF AGRON, OHIO STATE UNIV & OHIO AGR RES &

DEVELOP CTR, 70- Mem: Am Soc Agron; Crop Sci Soc Am; Am Soc Animal Sci. Res: Ecological and physiological factors associated with adaptation and utilization of forage plants; production, management and utilization of forage crops and pastures. Mailing Add: Dept of Agron Ohio Agr Res & Develop Ctr Wooster OH 44691

VANKIN, GEORGE LAWRENCE, b Baltimore, Md, Apr 22, 31; m 56; c 2. CELL BIOLOGY. Educ: NY Univ, BS, 54, PhD(zool), 52; Wesleyan Univ, MA, 56. Prof Exp: Res asst genetics, Wesleyan Univ, 56; teaching fel biol, NY Univ, 56-59, res asst embryol, 59-62, lectr biol, 61-62; from asst prof to assoc prof, 62-75, PROF BIOL, WILLIAMS COL, 75- Concurrent Pos: Vis asst prof, Med Col, Cornell Univ, 68. Mem: AAAS; Am Soc Zoologists; Soc Develop Biol. Res: Ultrastructure of differentiating amphibian red blood cells; fine structure of vertebrate smooth muscle; electron microscopy of post-implantation mouse embryos. Mailing Add: Dept of Biol Williams Col Williamstown MA 01267

VAN KLEY, HAROLD, b Chicago, Ill, Mar 7, 32; m 59; c 2. BIOCHEMISTRY, PROTEIN CHEMISTRY. Educ: Calvin Col, AB, 53; Univ Wis, MS, 55, PhD(biochem), 58. Prof Exp: From instr to sr instr, 58-61, ASST PROF BIOCHEM, SCH MED, ST LOUIS UNIV, 61-; DIR BIOCHEM RES, ST MARY'S HEALTH CTR, 67- Concurrent Pos: Mem Am pancreatic study group. Mem: AAAS; Am Chem Soc; Am Soc Biol Chemists. Res: Protein structure, especially as related to biological function and regulatory mechanisms in metabolism; protein changes in neoplasia; pancreatic enzymes. Mailing Add: Lab for Biochem Res St Mary's Health Ctr St Louis MO 63117

VANKO, MICHAEL, b Binghamton, NY, June 17, 29; m 52; c 2. BIOCHEMISTRY. Educ: Rensselaer Polytech Inst, BS, 52; Union Univ, NY, MS, 55, PhD(biochem), 59. Prof Exp: Asst prof, 59-68, ASSOC PROF BIOCHEM, ALBANY MED COL, 68-, DIR CLIN CHEM, ALBANY MED CTR, 59- Mem: AAAS; Am Chem Soc; Am Asn Clin Chem. Res: Clinical chemistry. Mailing Add: Dept of Biochem Albany Med Col Albany NY 12308

VAN KRANENDONK, JAN, b Delft, Neth, Feb 8, 24; m 52; c 3. THEORETICAL PHYSICS. Educ: Univ Amsterdam, PhD(physics), 52. Prof Exp: Res asst, Univ Amsterdam, 50-54; lectr, State Univ Leiden, 55-58; assoc prof, 58-60, PROF PHYSICS, UNIV TORONTO, 60- Concurrent Pos: Neth Orgn for Pure Res fel, Harvard Univ, 53-54. Honors & Awards: Steacie Prize, 64. Mem: Am Phys Soc; fel Royal Soc Can; Can Asn Physicists. Res: Molecular and solid-state physics. Mailing Add: Dept of Physics Univ of Toronto Toronto ON Can

VAN KREY, HARRY PETER, b Combined Locks, Wis, Oct 2, 31; m 52; c 2. PHYSIOLOGY, AGRICULTURE. Educ: Univ Calif, Davis, BS, 60, PhD(animal physiol), 64. Prof Exp: Fel, Univ Wis, 64-65; ASSOC PROF AVIAN PHYSIOL, VA POLYTECH INST & STATE UNIV, 65- Mem: AAAS; Sigma Xi; Poultry Sci Asn; World Poultry Sci Asn; Soc Study Reproduction. Res: Avian reproductive physiology; poultry science. Mailing Add: Dept of Poultry Sci Va Polytech Inst & State Univ Blacksburg VA 24061

VAN LAAN, GORDON JAMES, b Bay City, Mich, Aug 6, 22. ORNAMENTAL HORTICULTURE. Educ: Mich State Univ, BS, 48, MS, 49; Wash State Univ, PhD(hort), 53. Prof Exp: Agr missionary, Am Bd Comnrs For Missions, Angola, Port W Africa, 53-59; park foreman, City Walnut Creek, Calif, 60; nurseryman, Navlet's Nursery, 61; from asst prof to assoc prof, 61-73, PROF ORNAMENTAL HORT, CALIF STATE UNIV, CHICO, 73- Mem: Am Soc Hort Sci; Am Hort Soc. Res: Management practices in California nurseries with special attention to employee relations, merchandising and management problems; propagation of difficult to propagate ornamental shrubs and trees. Mailing Add: Dept of Plant & Soil Sci Calif State Univ Chico CA 95929

VAN LANCKER, JULIEN L, b Auderghem, Belg, Aug 14, 24; m 49; c 3. PATHOLOGY. Educ: Cath Univ Louvain, MD, 50. Prof Exp: Asst path, Cath Univ Louvain, 50-53; vis instr, Univ Kans, 53-54; Runyon fel oncol, Univ Wis, 54-55; asst path, Cath Univ Louvain, 55-56; asst prof path, Univ Utah, 56-60; assoc prof path & chief path sect, Primate Res Ctr, Univ Wis, 60-66; prof med sci, Brown Univ, 66-70; PROF PATH & CHMN DEPT, UNIV CALIF, LOS ANGELES, 70- Mem: Radiation Res Soc; Am Soc Exp Path; Am Soc Biol Chemists; NY Acad Sci; Int Acad Path. Res: Cell biology; chemical pathology; molecular mechanisms in disease. Mailing Add: Dept Path Ctr for Health Sci Univ of Calif Los Angeles CA 90024

VAN LANDINGHAM, AUDREY HOWARD, b Grady Co, Ga, Oct 19, 05; m 30; c 3. AGRICULTURAL CHEMISTRY. Educ: SGa Agr & Mech Col, BS, 29; WVa Univ, MS, 31, PhD(agr chem), 34. Prof Exp: Asst agr chem, Exp Sta, WVa Univ, 31-34, asst chemist, 34-41, assoc chemist, 41-46, biochemist, 46-50, asst dean & dir, Col Agr, Forestry & Home Econ, 50-57, actg dean col & actg dir agr exp sta, 57-59, assoc dean col & dir agr exp sta, 59-71, EMER PROF AGR BIOCHEM, COL AGR, FORESTRY & HOME ECON, WVA UNIV, 71- Mem: Am Chem Soc; Am Dairy Sci Asn. Res: Nutritional requirements of farm animals; composition of feeds and rations; human nutrition; nutritional status; reproduction and lactation. Mailing Add: 201 Gordon St Morgantown WV 26505

VAN LANDINGHAM, JOHN W, b Portland, Ore, Jan 25, 19; m 54; c 5. ENVIRONMENTAL CHEMISTRY. Prof Exp: Anal chemist, Testing & Res Labs, Deere & Co, 40-43; res technician, Pioneer Mill Co, Ltd, Hawaii, 46-47; asst chemist, Kekaha Sugar Co, Ltd, 47-50; anal chemist, Hawaiian Sugar Planters Exp Sta, 51-54; phys sci technician, Biol Lab, Bur Commercial Fisheries, 54-59; lab technician, Scripps Inst, Univ Calif, 59-62; phys sci technician, Trop Atlantic Biol Lab, Bur Commercial Fisheries, 62-65, phys scientist, 65-71, phys scientist, Miami Lab, Southeast Fisheries Res Ctr, Nat Oceanic & Atmospheric Admin, 71-73; LAB MGR, CONNELL, METCALF & EDDY SUBSID, RES COTTRELL, 73- Mem: Am Ord Asn. Res: Application of radio isotope techniques to ocean research problems; analytical chemical methodology; chemical and physical interaction of sea nutrients; infrared analysis of hydrocarbons in sea water; trace metals in sea water. Mailing Add: 3751 SW 124 Ct Miami FL 33165

VANLANDINGHAM, SAMUEL LEIGHTON, b Iraan, Tex, Aug 25, 35; m 67. BOTANY, PHYCOLOGY. Educ: Tex Tech Col, BS, 58; Univ Kans, BA, 60, MA, 63; Univ Louisville, PhD(bot), 66. Prof Exp: Asst prof biol, Northeast La Univ, 67-70; tech & biol consult, 70-72; phycologist, Water Qual Appraisal Sect, Biol Water Resources Comn, Mich Dept Natural Resources, 72; res assoc, Dept Geol, Calif Acad Sci, 72-73, res diatomist, 73-75; CONSULT BIOLOGIST, 75- Concurrent Pos: NSF res partic, Trop Atlantic Biol Lab, Nat Marine Fisheries Serv & Rosenstiel Sch Marine & Atmospheric Sci, 69, acad year exten, 70; consult, Environ Protection Agency, indust firms & govt agencies, 72-75 & telecommun firm, 74; res investr, NSF, 72-75. Mem: Phycol Soc Am; Int Phycol Soc; Brit Phycol Soc; Plankton Soc Japan. Res: Taxonomy, morphology and ecology of bacillariophyta; paleoecology; general phycology; paleobotany of vascular plants; stratigraphy of diatoms; plant geography and anatomy; exobiology; ecology of algae; evaluation of water quality with algae. Mailing Add: 3741 Woodsong Dr Cincinnati OH 45239

VAN LANDUYT, DENNIS CLARKE, b Geneva, Ill, Aug 27, 42; m 66; c 1. ORGANIC CHEMISTRY. Educ: Univ Wyo, BS, 66; Auburn Univ, PhD(chem), 75. Prof Exp: CHEMIST, ROHM AND HAAS CO, 75- Res: The development of polymers and organic chemicals used in the paper industry. Mailing Add: Res Labs Rohm and Haas Co Spring House PA 19477

VAN LANEN, JAMES MARVIN, bacteriology, see 12th edition

VAN LANEN, ROBERT JEROME, b Green Bay, Wis, July 30, 43. BIO-ORGANIC CHEMISTRY, PHYSICAL ORGANIC CHEMISTRY. Educ: St Norbert Col, BS, 65; Univ Colo, Boulder, PhD(org chem), 71. Prof Exp: NIH fel chem, Univ Wis-Madison, 71-73; ASST PROF CHEM, ST XAVIER COL, 73- Mem: AAAS; Am Chem Soc; Soc Social Responsibility in Sci. Res: Mechanisms of enzyme-catalyzed reactions; model systems; chemistry of small ring compounds. Mailing Add: St Xavier Col 103rd & Central Park Ave Chicago IL 60655

VAN LEAR, DAVID HYDE, b Clifton Forge, Va, Dec 1, 40. FORESTRY, SOILS. Educ: Va Polytech Inst, BS, 63, MS, 65; Univ Idaho, PhD(forest sci), 69. Prof Exp: Fel sch forestry, Univ Fla, 68-69; soil scientist, US Forest Serv, 69-71; ASSOC PROF SILVICULT, CLEMSON UNIV, 71- Mem: Soc Am Foresters; Soil Sci Soc Am. Res: Hardwood silviculture; environmental forestry; forest fertilization; soil-site relationships; strip mining. Mailing Add: Dept of Forestry Clemson Univ Clemson SC 29631

VAN LEAR, GEORGE EDWARD, organic chemistry, see 12th edition

VAN LEER, JOHN CLOUD, b Washington, DC, Feb 14, 40; m 62; c 3. PHYSICAL OCEANOGRAPHY, OCEAN ENGINEERING. Educ: Case Inst Technol, BSME, 62; Mass Inst Technol, ScD(phys oceanog), 71. Prof Exp: Engr, Draper Lab, Mass Inst Technol, 62-65, from res asst to res assoc phys oceanog, Mass Inst Technol, 65-71; ASST PROF PHYS OCEANOG, ROSENSTIEL SCH MARINE & ATMOSPHERIC SCI, UNIV MIAMI, 71- Mem: Sigma Xi; Am Geophys Union. Res: Physical oceanographic research on continental shelves and deep ocean; response to wind forcing, surface and bottom boundary layers; development of ocean instruments, notably the cyclesonde, an automatic oceanographic radiosonde.

VAN LEEUWEN, GERARD, b Hull, Iowa, July 4, 29; m 52; c 2. PEDIATRICS. Educ: Calvin Col, BA, 50; Univ Iowa, MD, 54. Prof Exp: Intern, Butterworth Hosp, Mich, 54-55; resident pediat, Univ Mo-Columbia, 57-59, from instr to asst prof, 62-69; PROF PEDIAT & CHMN DEPT, UNIV NEBR MED CTR, OMAHA, 69- Concurrent Pos: NIH trainee, 62-64; Am Thoracic Soc fel, 64-66; Nat Found Birth Defects grant, 66; proj consult, Head Start, 65; dir, Nat Found Treatment Ctr, 66. Mem: Am Acad Pediat; Am Pediat Soc. Res: Neonatal physiology and hypogylcemia; hyaline membrane syndrome; teratology. Mailing Add: Dept of Pediat Univ of Nebr Med Ctr Omaha NE 68105

VAN LENTE, KENNETH ANTHONY, b Holland, Mich, Mar 29, 03; m 29; c 4. PHYSICAL CHEMISTRY. Educ: Hope Col, AB, 25, MS, 26; Univ Mich, PhD, 31. Prof Exp: Asst, Univ Mich, 27-31; from asst prof to prof, 31-71, EMER PROF PHYS CHEM, SOUTHERN ILL UNIV, CARBONDALE, 71- Mem: Am Chem Soc. Res: Liquid junction potentials; constant temperature baths; composition of plating baths; chemical education. Mailing Add: 1209 W Chautauqua Carbondale IL 62901

VAN LEUWEN, BRUCE GUNN, organic chemistry, see 12th edition

VAN LIER, JAN ANTONIUS, b Ginneken, Neth, Nov 17, 24; m 54; c 4. PHYSICAL CHEMISTRY. Educ: Univ Utrecht, BS, 51, MS, 54, PhD(phys chem), 59. Prof Exp: Res assoc mineral eng, Mass Inst Technol, 55-58; instr colloid chem, Univ Utrecht, 58-59; res chemist, Philips Electronics, Holland, 59-60; sr res chemist, 60-74, STAFF RES CHEMIST, PARMA TECH CTR, UNION CARBIDE CORP, 74- Mem: Am Chem Soc; Royal Neth Chem Soc. Res: Microbalance techniques; solubility of quartz; battery materials; differential and thermogravimetric analysis; solid electrolytes; basic electrochemistry; ternary phase diagrams; surface chemistry; wetting phenomena. Mailing Add: Parma Tech Ctr Union Carbide Corp Parma OH 44130

VAN LIER, JOHANNES ERNESTINUS, b Amsterdam, Neth, May 26, 42; m 66. BIOCHEMISTRY. Educ: Delft Univ Technol, Ir, 66; Univ Tex Med Br Galveston, PhD(biochem), 69. Prof Exp: Res assoc, Univ Tex Med Br Galveston, 66-69, instr, 69-70; asst prof, 70-75, ASSOC PROF NUCLEAR MED RADIOBIOL, MED CTR, UNIV SHERBROOKE, 75- Concurrent Pos: Med Res Coun Que res grant, Med Ctr, Univ Sherbrooke, 70-72; Banting Res Found res grant, 75-76; Med Res Coun Can res grant, 70- Mem: AAAS; Am Chem Soc; Can Fedn Biol Socs; Royal Neth Chem Soc. Res: Mechanism of oxidative metabolism of steroids; lipids of human atherosclerotic aorta and brain; synthesis and evaluation of potential anticancer agents. Mailing Add: Dept of Nuclear Med Univ of Sherbrooke Med Ctr Sherbrooke PQ Can

VAN LIEW, HUGH DAVENPORT, b Spokane, Wash, Jan 28, 30; m 59; c 3. MEDICAL PHYSIOLOGY. Educ: State Col Wash, BS, 51; Univ Rochester, MS, 53, PhD(physiol), 56. Prof Exp: Res fel, Sch Pub Health, Harvard Univ, 59-61; asst prof physiol, Stanford Univ, 61-63; from asst prof to assoc prof, 63-74, PROF PHYSIOL, STATE UNIV NY BUFFALO, 74- Mem: AAAS; Am Physiol Soc; Undersea Med Soc. Res: Diffusion of gases through tissues; gas tensions in the tissues; subcutaneous gas pockets as models of decompression sickness bubbles; diffusion and convection in the lung. Mailing Add: Dept of Physiol State Univ of NY Buffalo NY 14214

VAN LIEW, JUDITH BRADFORD, b Boston, Mass, Jan 22, 30; m 59; c 3. PHYSIOLOGY. Educ: Bates Col, BS, 51; Univ Wash, MS, 54; Univ Rochester, PhD(physiol), 58. Prof Exp: Res assoc biochem, Woman's Med Col Pa, 58-60; res assoc med, 64-70, res asst prof, 70-74, ASST PROF PHYSIOL, SCH MED, STATE UNIV NY BUFFALO, 74-; RES PHYSIOLOGIST, VET ADMIN HOSP, BUFFALO, 73- Mem: Am Physiol Soc; Am Soc Nephrology. Res: Physiology of normal and abnormal proteinuria. Mailing Add: Vet Admin Hosp 3495 Bailey Ave Buffalo NY 14215

VAN LIGTEN, RAOUL FREDRIK, b Bandoeng, Java, Dutch E Indies, Sept 27, 32; m 75; c 1. OPTICS. Educ: Delft Univ Technol, MSc, 57; Univ Paris, PhD, 72. Prof Exp: Res physicist, Cent Orgn Appl Sci Res in Neth, 56-60 & Res Dept, Am Optical Co, 60-61; res physicist, Res Dept, Itek Corp, 61-62, head res group optical metrol, 62-63; sr res physicist, 63-67, chief coherent optics, 67-72, chief phys optics, 72-73, DIR RES & DEVELOP, INT DIV, AM OPTICAL CO, 73- Honors & Awards: Karl Fairbanks Mem Award, Soc Photo-Optical Instrument Engrs, 69. Mem: Fel Optical Soc Am; fel Royal Micros Soc; NY Acad Sci. Res: Physical and geometrical optics; holography; optical metrology; lens design; holographic, phase and interference microscopy; optical testing; ophthalmic instrumentation and optics. Mailing Add: 30 Rue Du Melon F67400 Ostwald France

VAN LINT, VICTOR ANTON JACOBUS, b Samarinda, Indonesia, May 10, 28; US citizen; m 50; c 4. PHYSICS. Educ: Calif Inst Technol, BS & PhD(physics), 54. Prof

Exp: Instr physics, Princeton Univ, 54-55; physicist, Gen Atomic Div, Gen Dynamics Corp, 57-65, assoc dir spec nuclear effects lab, 65-69, mgr defense sci dept, Gulf Radiation Technol, 69-70, vpres, Gulf Energy & Environ Systs & mgr, Gulf Radiation Technol Div, 70-73; pres, 73-74, CONSULT, INTELCOM RADIATION TECHNOL, 74- Mem: Am Phys Soc; sr mem Inst Elec & Electronics Engrs. Res: Radiation effects, including solid state physics, atomic physics, and electronic systems analysis. Mailing Add: 1032 Skylark Dr La Jolla CA 92037

VAN LOON, EDWARD JOHN, b Danville, Ill, Dec 3, 11; m 54; c 1. BIOCHEMISTRY. Educ: Univ Ill, AB, 36; Rensselaer Polytech Inst, MS, 37, PhD(chem), 39. Prof Exp: Res assoc & instr physiol & pharmacol, Albany Med Col, Union NY, 39-43; instr & asst prof biochem, Mich State Univ, 46; asst & assoc prof, Sch Med, Univ Louisville, 46-49; chief, Med Res Lab, Vet Admin Hosp, 49-55; group leader & head biochem sect, Smith Kline & French Labs, 55-67; CHIEF, SPEC PHARMACOL ANIMAL LAB, FOOD & DRUG ADMIN, 67- Concurrent Pos: Mem, Am Bd Clin Chemists. Mem: Am Soc Clin Invest; Am Inst Nutrit; Am Soc Pharmacol & Exp Therapeut; Am Chem Soc; Am Soc Biol Chem. Res: Clinical biochemistry; intermediary metabolism; biochemical pharmacology and drug metabolism; clinical biochemistry. Mailing Add: Div Toxicol Bur Foods FDA 200 C St SW Washington DC 20204

VAN LOON, JON CLEMENT, b Hamilton, Ont, Jan 9, 37; m 61; c 1. GEOLOGY, CHEMISTRY. Educ: McMaster Univ, BSc, 59; Univ Toronto, PhD(anal chem), 64. Prof Exp: Asst prof anal geochem, 64-69, ASSOC PROF ANAL GEOCHEM, UNIV TORONTO, 69- Res: Application of modern analytical methods to the analysis of natural products, with particular emphasis on environmental samples. Mailing Add: 38 Paultiel Dr Willowdale ON Can

VAN LOPIK, JACK RICHARD, b Holland, Mich, Feb 25, 29; m 52; c 1. RESEARCH ADMINISTRATION, MARINE SCIENCES. Educ: Mich State Univ, BS, 50; La State Univ, MS, 53, PhD(geol), 55. Prof Exp: Field investr, Coastal Studies Inst, La State Univ, 51-54, instr geol, 54; geologist, Waterways Exp Sta, Corps Engrs, US Army, 54-57, asst chief & chief geol br, 57-61; res scientist, chief area eval sect & mgr space & environ sci prog, Geosci Opers, Tex Instruments Inc, 61-66, tech requirements dir, 66-68; chmn dept marine sci, 68-74, dir, Ctr Wetland Resources, 70, PROF MARINE SCI & DIR SEA GRANT DEVELOP, LA STATE UNIV, BATON ROUGE, 68- Concurrent Pos: Mem, Nat Res Coun Earth Sci Div, Nat Acad Sci-Nat Res Coun, 67-72, chmn panel geog & human & cult resources, Comt Remote Sensing Progs for Earth Resources Surv, 69-; mem, La Adv Comn Coastal & Marine Resources, 71-73; mem bd dirs, Gulf South Res Inst, 74- Mem: Fel Geol Soc Am; Am Mgt Asn; Soc Res Adminr; World Future Soc; sr mem Am Astronaut Soc. Res: Photogeology and remote sensing; deltaic and arid zone geomorphology and sedimentation; military and engineering geology; terrain analysis and quantification; lunar and earth-orbiting-satellite exploration; coastal zone management. Mailing Add: Ctr for Wetland Resources La State Univ Baton Rouge LA 70803

VAN MAANEN, EVERT FLORUS, b Harderwyk, Neth, Sept 17, 18. PHARMACOLOGY. Educ: State Univ Utrecht, BS, 38, Phil Drs, 45; Harvard Univ, PhD(pharmacol), 49. Prof Exp: From instr to assoc prof, 49-66, PROF PHARMACOL, COL MED, UNIV CINCINNATI, 66- Concurrent Pos: Dir biol sci, Wm S Merrell Co Div, Richardson-Merrell, Inc, 55-62. Mem: AAAS; Am Soc Pharmacol & Exp Therapeut; Soc Exp Biol & Med. Res: Neuromuscular transmission; cholinesterase inhibitors; cardiovascular agents; autonomic drugs; theories of drug action. Mailing Add: Dept of Pharmacol Univ Cincinnati Col of Med Cincinnati OH 45267

VAN MARTHENS, EDITH, b Vienna, Austria, Nov 17, 40; US citizen. VETERINARY MEDICINE, NUTRITION. Educ: Vet Univ, Vienna, DVM, 61. Prof Exp: Res assoc physiol & microcirculation, NY Univ, 62-65; SR VET, NUTRIT, UNIV CALIF, LOS ANGELES, 65-, ASST RES VET, NUTRIT & PRENATAL BRAIN DEVELOP, 73- Mem: Am Asn Zool Vets; Am Asn Zool Parks & Aquariums. Res: Effects on prenatal malnutrition and brain development. Mailing Add: Univ of Calif NPI/MR Rm 48-149 760 Westwood Plaza Los Angeles CA 90024

VAN METER, CLARENCE TAYLOR, b Pittsburgh, Pa, Sept 1, 05; m 30; c 2. MEDICINAL CHEMISTRY, PHARMACEUTICAL CHEMISTRY. Educ: Pittsburgh Col Pharm, BS, 32; Univ Pittsburgh, PhD(chem), 41. Prof Exp: Instr chem, Sch Pharm, Univ Pittsburgh, 27-35, asst prof chem & physics, 35-44; dir res & control, Reed & Carnrick, 44-53; sr res investr & dir res projs, Off Eng Res, Univ Pa, 53-70; RETIRED. Concurrent Pos: Consult, US Dept Defense; adv, WHO; ed, Remington's Pharmaceut Sci; collabr & consult rev, US Pharmacopoeia & Nat Formulary, 35-, AMA & McGraw-Hill, Inc. Honors & Awards: US Pharmacopeia Award, 75. Mem: Am Chem Soc; Am Pharmaceut Asn. Res: Structure and systematic nomenclature of chemical compounds. Mailing Add: 5216 Oleander Rd Drexel Hill PA 19026

VAN METER, DAVID, b Southampton, NY, Mar 27, 19; m 50. APPLIED PHYSICS, INFORMATION SCIENCES. Educ: Mass Inst Technol, BS & SM, 43; Harvard Univ, MA, 53, PhD, 55. Prof Exp: Mem tech staff, Bell Tel Labs, 43-46; asst prof elec eng, Pa State Univ, 46-52; lab mgr, Melpar, Inc, 55-60 & Litton Systs, Inc, 60-66; chief comput res lab, Electronics Res Ctr, NASA, 66-70; CHIEF INFO SCI DIV, TRANSP SYTS CTR, US DEPT TRANSP, 70- Concurrent Pos: Lectr, Harvard Univ, 55 & 59; ed, Transp Info Theory, 64-67. Mem: Inst Elec & Electronics Engrs; Am Inst Aeronaut & Astronaut; Asn Comput Mach. Res: Computer science; information theory. Mailing Add: 10 Byron St Boston MA 02108

VAN METER, DONALD EUGENE, b Ashtabula, Ohio, Aug 30, 42; m 64; c 1. SOIL CONSERVATION. Educ: Purdue Univ, Lafayette, BS, 64; Mich State Univ, MS, 65; Ind Univ, Bloomington, DEduc, 71. Prof Exp: County agt agr, Coop Exten Serv, Purdue Univ, 65-68; ASSOC PROF NATURAL RESOURCES, BALL STATE UNIV, 69- Mem: Soil Conserv Soc Am; Am Soc Agron. Res: Natural resource management; agriculture extension education in developing nations. Mailing Add: Dept of Natural Resources Ball State Univ Muncie IN 47306

VAN METER, JAMES P, b Tiffin, Ohio, Dec 8, 36; m 61; c 1. ORGANIC CHEMISTRY. Educ: Ohio State Univ, BSc, 58, PhD(chem), 64. Prof Exp: RES CHEMIST, EASTMAN KODAK CO, 64- Mem: Am Chem Soc. Res: Organic synthesis in small ring compounds; heterocyclic chemistry; photochemistry; photographic research; synthesis of liquid crystals. Mailing Add: Eastman Kodak Co Res Lab 343 State St Rochester NY 14650

VAN METER, JOHN CONNELL, b Columbus, Ohio, Mar 29, 16; m 42; c 5. BIOCHEMISTRY. Educ: Ohio State Univ, BA, 39, MSc, 40. Prof Exp: Res chemist, Lederle Labs, Am Cyanamid Co, NY, 45-68; DIR LABS, NEW CASTLE STATE HOSP, 68- Mem: Am Chem Soc; fel Am Inst Chemists; Am Asn Clin Chemists; Sigma Xi. Res: Determination of drugs and their metabolites in body fluids. Mailing Add: New Castle State Hosp Lab New Castle IN 47362

VAN METER, WAYNE PAUL, b Fresno, Calif, Feb 16, 26; m 48; c 4. INORGANIC CHEMISTRY. Educ: Ore State Col, BS, 50, MS, 52; Univ Wash, PhD(inorg chem), 59. Prof Exp: Chemist, Hanford Atomic Prod Oper, Gen Elec Co, Wash, 51-56; from asst prof to assoc prof, 59-71, PROF CHEM, UNIV MONT, 71- Mem: Am Chem Soc. Res: Measurement of trace concentrations of metals in biological systems; development of sample processing techniques and instrumentation for atomic absorption spectrometry. Mailing Add: Dept of Chem Univ of Mont Missoula MT 59801

VAN METRE, THOMAS EARLE, JR, b Newport, RI, Jan 11, 23; m 47; c 5. MEDICINE. Educ: Harvard Univ, BS, 43, MD, 46; Am Bd Internal Med, dipl, 55. Prof Exp: Intern med, Johns Hopkins Hosp, 46-47, asst resident, 47-48, 50 & 51-52, Am Cancer Soc fel, 52-53; asst prof internal med, Sch Med, St Louis Univ, 53-54; from instr to asst prof, 56-70, ASSOC PROF MED, MED SCH, JOHNS HOPKINS UNIV, 70-, PHYSICIAN, JOHNS HOPKINS HOSP, 56-, PHYSICIAN-IN-CHG ADULT ALLERGY CLIN, 66- Concurrent Pos: Pvt pract, 54-; mem attend staff, Baltimore City Hosp, 54- & Union Mem Hosp, 59- Mem: AMA; Am Fedn Clin Res; Am Col Physicians; Am Acad Allergy; Am Clin & Climat Asn. Res: Allergy; effect of corticosteroids on growth; uveitis; asthma. Mailing Add: 11 E Chase St Baltimore MD 21202

VAN MIDDELEM, CHARLES HENRY, b Bruges, Belg, Aug 6, 19; nat US; m 43; c 1. BIOCHEMISTRY. Educ: Cornell Univ, BS, 44, PhD(hort), 52. Prof Exp: Asst biochemist, Dept Hort, Agr Exp Sta, Univ Fla, 52-57, assoc biochemist, Dept Food Technol & Nutrit, 57-62, biochemist, Dept Food Sci, 62-74, prof biochem, Inst Food & Agr Sci, 69-74; DIR DIV CHEM, FDAC, 74- Concurrent Pos: Mem toxicol study sect, NIH, 61-65, chmn pesticide subdiv, Agr & Food Chem Div, 61, consult & in charge pesticide res lab, 65- Mem: Am Chem Soc. Res: Pesticide residue methodology; fate of pesticides in plant soils and animal systems; use of gas and thin layer chromatography; radiolabeled pesticides. Mailing Add: Div of Chem FDAC Mayo Bldg Tallahassee FL 32304

VAN MIDDLESWORTH, LESTER, b Washington, DC, Jan 13, 19; m 48; c 4. PHYSIOLOGY, MEDICINE. Educ: Univ Va, BS, 40, MS, 42 & 44; Univ Calif, Berkeley, PhD(physiol), 47; Univ Tenn, MD, 51. Prof Exp: Chief chemist, Piedmont Apple Prod Corp, Va, 39-44; res assoc, Radiation Lab & teaching asst physiol, Univ Calif, 44-46; from instr to assoc prof, 46-59, PROF PHYSIOL & BIOPHYS, CTR HEALTH SCI, UNIV TENN, MEMPHIS, 59-, PROF MED, 74- Concurrent Pos: Res asst physiol, Univ Va, 42-44; intern, John Gaston Hosp, 51-52; USPHS career res award, 61- Mem: Am Chem Soc; Am Physiol Soc; Endocrine Soc; Am Thyroid Asn. Res: Vapor phase catalysis; hormone synthesis; carbohydrate metabolism; aviation medicine; anoxia; metabolism of plutonium, radium, iodide, thiocyanate and thyroxine; thyroid physiology; goiter; radioactive fallout. Mailing Add: Dept of Physiol & Biophys Univ Tenn Ctr for Health Sci Memphis TN 38103

VAN MIEROP, LODEWYK H S, b Surabaya, Java, Mar 31, 27; US citizen; m 54; c 5. MEDICINE. Educ: State Univ Leiden, MD, 52; Am Bd Pediat, cert pediat cardiol. Prof Exp: Lectr anat, McGill Univ, 61-62; from asst prof to assoc prof pediat, Albany Med Col, 62-66; assoc prof, 66-68, PROF PEDIAT & PATH, COL MED, UNIV FLA, 68- Concurrent Pos: NIH res career develop award, 64-73; res assoc, Mt Sinai Hosp, New York, 63-66; mem comt nomenclature of heart, NIH, 67-68; mem southern regional res rev comt, Am Heart Asn, 68-72. Mem: Am Pediat Soc; Am Heart Asn; Am Asn Anat; fel Am Acad Pediat; fel Am Col Cardiol. Res: Pediatric cardiology; pathology and pathogenesis of congenital heart disease; cardiac embryology. Mailing Add: Dept of Pediat Univ of Fla Col of Med Gainesville FL 32601

VANN, DOUGLAS CARROLL, b Coronado, Calif, May 3, 39; m 61; c 1. IMMUNOBIOLOGY. Educ: Univ Calif, Berkeley, AB, 60; Univ Calif, Santa Barbara, PhD(biol), 66. Prof Exp: Jr scientist, Inter-Am Trop Tuna Comn, 60-62; res assoc immunol, Biol Div, Oak Ridge Nat Lab, 66-68; USPHS training grant, Scripps Clin & Res Found, 68-70; asst prof, 70-73, ASSOC PROF GENETICS, UNIV HAWAII, HONOLULU, 73- Concurrent Pos: Prin investr, USPHS res grant, 71-77. Mem: AAAS. Res: Cellular basis of immune responses. Mailing Add: Dept of Genetics Univ of Hawaii Honolulu HI 96822

VANN, JOHN HERMAN, b Shreveport, La, Aug 29, 21. GEOGRAPHY. Educ: La State Univ, BA, 43, MS, 48; Univ Calif, PhD, 60. Prof Exp: Instr geog, La State Univ, 47-48, San Francisco State Col, 50-51 & Univ Calif, 51-52; instr, La State Univ, 53-58, asst prof, 59-63; from assoc prof to prof, State Univ NY Col Buffalo, 63-70; PROF GEOG, CALIF STATE UNIV, HAYWARD, 70-, CHMN DEPT, 74- Concurrent Pos: Vis prof, Univ Ariz, 67-68. Mem: AAAS; Asn Am Geogrs; Am Geog Soc. Res: Geomorphology; climatology; physical and cultural geography of Latin America and the Tropics. Mailing Add: Dept of Geog Calif State Univ 25800 Hillary St Hayward CA 94542

VANN, ROBERT LEE, b Wake Forest, NC, Mar 17, 22; m 46; c 3. PEDIATRICS, PHARMACEUTICS. Educ: Wake Forest Univ, BS, 42; Bowman Gray Sch Med, MD, 45. Prof Exp: Intern, Barnes Hosp, St Louis, Mo, 45-46; pvt pract, 48-51; resident pediat, NC Baptist Hosp, 51-53; mem staff, Pvt Diag Clin, Bowman Gray Sch Med, 53-61; pvt pract, 61-68; assoc med develop dir, E R Squibb & Sons, 68-73; CLIN RES DIR, BEECHAM LABS, 73- Mem: AMA; Am Acad Pediat; Am Soc Microbiol; Am Soc Clin Pharmacol & Therapeut. Mailing Add: Beecham Labs 501 Fifth St Bristol TN 37620

VANNEMAN, CLINTON ROSS, organic chemistry, see 12th edition

VAN NESS, JOHN WINSLOW, b McLean Co, Ill, Aug 16, 36; m 64; c 2. STATISTICS. Educ: Northwestern Univ, BS, 59; Brown Univ, PhD(appl math), 64. Prof Exp: Vis asst prof statist, Stanford Univ, 64-65, actg asst prof, 65-66; asst prof math, Univ Wash, 66-71; assoc prof statist, Carnegie-Mellon Univ, 71-73; assoc prof, 73-75, HEAD PROG MATH SCI, UNIV TEX, DALLAS, 73-, PROF, 75- Mem: Inst Math Statist; Am Statist Asn; Int Continence Soc. Res: Theoretical and applied statistics; biostatistics; multivariate and time series analysis; analysis of spectra and polyspectra; stochastic processes and their applications. Mailing Add: Prog in Math Sci Univ of Tex at Dallas PO Box 688 Richardson TX 75080

VAN NIEL, CORNELIS BERNARDUS, b Haarlem, Netherlands, Nov 4, 97; nat US; m 25; c 3. BIOCHEMISTRY. Educ: Delft Univ Technol, Chem E, 23, DSc, 28. Hon Degrees: DSc, Princeton Univ, 46; LLD, Univ Calif, 68. Prof Exp: Conservator, Microbiol Lab, Delft Univ Technol, 24-28; assoc prof microbiol, 28-35, prof, 35-46, Herstein prof biol, 46-63, EMER HERSTEIN PROF BIOL, HOPKINS MARINE STA, STANFORD UNIV, 63- Concurrent Pos: Rockefeller fel, 35-36; Guggenheim fel, 46 & 55-56; chmn adv comt biochem, Off Naval Res, 50-53; vis prof, Univ Calif, Santa Cruz, 64-68. Honors & Awards: Nat Medal Sci, 64; Emil Christian Hansen Medal, Denmark, 64; Rumford Medal, Am Acad Arts & Sci, 67; Antonie Van Leeuwenhoek Medal, Royal Netherlands Acad Sci, 70. Mem: Nat Acad Sci; AAAS; hon mem Am Soc Microbiol (pres, 54); Am Acad Arts & Sci; hon mem Brit Soc Gen Microbiol. Res: General microbiology; biochemistry of microorganisms;

photosynthesis. Mailing Add: Hopkins Marine Sta Stanford Univ Pacific Grove CA 93950

VANNIER, WILTON EMILE, b Pasadena, Calif, June 6, 24; m 53; c 2. IMMUNOCHEMISTRY, BIOCHEMISTRY. Educ: Univ Calif, San Francisco, MD, 48; Calif Inst Technol, PhD(immunochem), 58. Prof Exp: Instr exp med, Sch Pub Health, Univ NC, 51-54; immunochemist, Lab Immunol, Nat Inst Allergy & Infectious Dis, Md, 58-64, head, Immunochem Sect, 64-68; assoc prof biochem, Sch Med, Univ Southern Calif, 68-70; RES MED OFFICER, NAVAL MED RES INST, NAT NAVAL MED CTR, 70- Concurrent Pos: Res fel, Calif Inst Technol, 57-60. Mem: Fel AAAS; Am Chem Soc; Am Asn Immunol; Am Acad Allergy. Res: Parasite immunology; chemistry of antibodies and antigen-antibody reactions. Mailing Add: Naval Med Res Inst Nat Naval Med Ctr Bethesda MD 20014

VAN NORMAN, GILDEN RAMON, b Jamestown, NY, Dec 11, 32; m 58. PHOTOGRAPHIC CHEMISTRY, POLYMER CHEMISTRY. Educ: Univ Rochester, BS, 54; Mass Inst Technol, PhD(org chem), 57. Prof Exp: TECH ASSOC, EASTMAN KODAK CO, 57- Mem: AAAS; Am Chem Soc. Mailing Add: 520 Churchill Dr Rochester NY 14616

VAN NORMAN, JOHN DONALD, b Jamestown, NY, Sept 11, 34; m 58; c 1. ANALYTICAL CHEMISTRY, CLINICAL CHEMISTRY. Educ: Univ Rochester, BS, 55; Rensselaer Polytech Inst, PhD(anal chem), 59. Prof Exp: Res assoc chem, Brookhaven Nat Lab, 59-61, from asst chemist to chemist, 61-69; ASSOC PROF CHEM, YOUNGSTOWN STATE UNIV, 69- Mem: AAAS; Am Chem Soc; Am Asn Clin Chemists; NY Acad Sci. Res: Spectrophotometry and electroanalytical chemistry of bile pigments; radioimmunoassay; atomic absorption. Mailing Add: Dept of Chem Youngstown State Univ Youngstown OH 44555

VAN NORMAN, RICHARD WAYNE, b Spencer, Iowa, Mar 12, 23; m 46. PLANT PHYSIOLOGY. Educ: Iowa State Teachers Col, BA, 46; Univ Minn, PhD(bot), 50. Prof Exp: Asst bot, Univ Minn, 46-49; asst prof, Pa State Univ, 50-55; asst prof exp biol, 55-58, actg head dept, 58-60, assoc prof, 58-65, PROF BIOL, UNIV UTAH, 65- Concurrent Pos: Writer, Biol Sci Curriculum Study, 61-63; asst prog dir, Course Content Improv Sect, NSF, 63-64; AID-NSF consult, Gujarat Univ, India, 68 & Ravishankar Univ, India, 69; mem comt examr, Grad Rec Exam Biol Test, 72- Mem: Fel AAAS; Am Soc Plant Physiologists; Am Inst Biol Sci. Res: Photosynthetic pigments; cellular metabolism; biological education. Mailing Add: Dept of Biol Univ of Utah Salt Lake City UT 84112

VAN NORTON, ROGER NORMAN, b Kenosha, Wis, Apr 23, 29; m 52; c 1. APPLIED MATHEMATICS, COMPUTER SCIENCE. Educ: Univ Wis, BS, 52, MS, 53; NY Univ, PhD(math), 60. Prof Exp: Assoc res scientist, Appl Math & Comput Ctr, AEC, NY Univ, 54-62; asst chmn appl math dept, Brookhaven Nat Lab, 62-64; res scientist, Appl Math & Comput Ctr, AEC, NY Univ, 64-67; prof systs eng & dir comput ctr, Univ Ariz, 67-72; assoc prog dir, NSF, 72-73; ASST DEP CHANCELLOR, NY UNIV, 73- Mem: AAAS; Am Math Soc; Asn Comput Mach; Soc Indust & Appl Math. Mailing Add: NY Univ 251 Mercer St New York NY 10012

VAN NOSTRAND, ROBERT GAIGE, b Oneida, NY, Nov 28, 18; m 46; c 6. GEOPHYSICS. Educ: Univ Mo, BS, 42, MS, 49; Univ NC, PhD(physics), 53. Prof Exp: Instr physics & geophys, Mo Sch Mines, 47-49; instr geol, Univ NC, 49-50; sr res physicist, Magnolia Petrol Co, 52-56; staff geophysicist, Mobil Overseas Oil Co, 56-57; chief geophysicist, Soc Petrol Alsace, 57-61; gen mgr, Explor Geophys Rogers, 61-62; mgr res dept, Teledyne Earth Sci, 62-69, VPRES, TELEDYNE GEOTECH, 70- Concurrent Pos: Geophysicist, US Geol Surv, 48-53; asst ed, J Soc Explor Geophys, 55-56, ed, 66-67. Mem: Am Geophys Union; Soc Explor Geophys; Europ Asn Explor Geophys. Res: Exploration geophysics; applied and theoretical seismology; computer science. Mailing Add: 1424 Kingston Ave Alexandria VA 22302

VANNOTE, ROBIN L, b Summit, NJ, Aug 12, 34; m 59; c 2. AQUATIC ECOLOGY. Educ: Univ Maine, BS, 57; Mich State Univ, MS, 62, PhD(limnol), 63. Prof Exp: Biologist, Water Qual Br, Tenn Valley Authority, 63-65, chief biol sect, 65-66; DIR, STROUD WATER RES CTR, ACAD NATURAL SCI PHILADELPHIA, 66- Concurrent Pos: Mem pesticide monitoring subcomt, Fed Comt Pest Control, 64-66; adj prof entom & appl ecol, Univ Del, 74- Mem: AAAS; Am Fisheries Soc; Am Soc Limnol & Oceanog; Ecol Soc Am. Res: Ecology of streams and rivers; interactions with terrestrial landscapes, production ecology; detrital systems, energy flow and nutrient budgets, geomorphology of streams; effects of channel modifications, drainage, and rural runoff on biotic productivity and stability. Mailing Add: Stroud Water Res Ctr Acad of Natural Sci RD 1 Box 512 Avondale PA 19311

VAN OERS, WILLEM THEODORUS HENDRICUS, b Amsterdam, Neth, Mar 17, 34; m 60; c 2. NUCLEAR PHYSICS. Educ: Univ Amsterdam, PhD(math, physics), 63. Prof Exp: From res asst to res assoc, Inst Nuclear Physics Res, Amsterdam, Neth, 57-64; asst res physicist, Cyclotron Lab, Univ Calif, Los Angeles, 64-66, asst prof physics, 66-67; assoc prof, 67-74, PROF PHYSICS, UNIV MAN, 67- Concurrent Pos: Vis assoc prof, Univ Calif, Los Angeles, 71-72; vis scientist, Nuclear Res Ctr, Saclay, 75-76. Mem: Am Phys Soc; Can Asn Physicists; Neth Phys Soc. Res: Nuclear reactions induced by various types of particle beams at low and intermediate energies; phenomenological and theoretical analyses of few-nucleon problems; nuclear optical model. Mailing Add: Dept of Physics Univ of Man Winnipeg MB Can

VAN OLPHEN, HENDRIK, b Arnhem, Neth, Nov 4, 12; m 38; c 5. COLLOID CHEMISTRY. Educ: Univ Utrecht, Neth, Drs, 36; Delft Technol Univ, PhD(chem), 51. Prof Exp: Res chemist, Koninklijke Shell Lab, Neth, 38-51 & Explor & Prod Res Div, Shell Develop Co, Tex, 51-66; asst dir, Off Critical Tables, 66-69, EXEC SECY, NUMERICAL DATA ADV BD, NAT ACAD SCI, 69- Concurrent Pos: Ed, J Clay Minerals Soc, 68-; vis prof, Pa State Univ, 69-70 & Bristol Univ, 72-73. Mem: Am Chem Soc; Clay Minerals Soc; Am Soc Testing & Mat. Res: Lubricants; clays; colloid chemistry of clay systems; scientific information and data evaluation in physics and chemistry. Mailing Add: Nat Acad of Sci 2101 Constitution Ave NW Washington DC 20418

VAN ORDEN, HARRIS O, b Smithfield, Utah, Oct 6, 17; m 48; c 1. ORGANIC CHEMISTRY. Educ: Utah State Agr Col, BS, 38; Wash State Univ, MS, 42; Mass Inst Technol, PhD(org chem), 51. Prof Exp: From asst prof to assoc prof chem, Utah State Agr Col, 46-52; NIH spec fel, Univ Utah, 53; assoc prof, 54-59, actg head dept, 58-59, PROF CHEM, UTAH STATE UNIV, 60- Mem: Fel AAAS; Am Chem Soc; Sigma Xi. Res: Synthetic organic and bio-organic chemistry; protein sequence studies; synthesis of peptides; enzyme specificity studies. Mailing Add: Dept of Chem Utah State Univ Logan UT 84321

VAN ORDEN, LUCAS SCHUYLER, III, b Chicago, Ill, Nov 3, 28; m 53; c 4. NEUROBIOLOGY, PSYCHOPHARMACOLOGY. Educ: Northwestern Univ, BS, 50, MS, 52, MD, 56; Yale Univ, PhD(pharmacol), 66. Prof Exp: Intern, Harper Hosp, Detroit, 56-57; asst resident surg, Med Ctr, Yale Univ, 61-62; Nat Inst Neurol Dis & Blindness spec fel neuroanat, Dept Anat, Harvard Med Sch, 66-67; from asst prof to assoc prof, 67-73, PROF PHARMACOL, COL MED, UNIV IOWA, 73- Concurrent Pos: Resident psychiat, Univ of Iowa, 75-76. Mem: Am Soc Neurochem; Am Med Soc Alcoholism; Am Soc Pharmacol & Exp Therapeut; Soc Neurosci. Res: Neuropharmacology; autonomic nervous system fine structure and histochemistry of adrenergic transmitter; quantitative cytochemistry, immunocytochemistry; alcohol and drug abuse. Mailing Add: Depts Pharmacol & Psychiat Univ of Iowa Col of Med Iowa City IA 52242

VAN ORDER, ROBERT BRUCE, b Glenvale, Ont, Mar 19, 15; nat US; m 42; c 3. ORGANIC CHEMISTRY. Educ: Queen's Univ, Can, BA, 38, MA, 39; NY Univ, PhD(org chem), 42. Prof Exp: Asst, NY Univ, 39-42; res org chemist, Stamford Res Labs, Am Cyanamid Co, 42-46, plant chemist, Calco Chem Div, 46-53, asst chief chemist, Org Chem Div, Bound Brook Lab, 53-55 & Mkt Develop Dept, 55-62; tech dir, Pearsall Chem Co, 62-63; mgr sales develop, Bound Brook, 63-69, MGR ORD CHEM RES & DEVELOP, AM CYANAMID CO, WAYNE, 69- Mem: Fel Am Inst Chemists; Am Chem Soc. Res: Structure and synthesis of antibiotics; inorganic pigments and chemicals; dyestuffs and pigment dispersions; market development and sales development of new chemical products. Mailing Add: 58 Sycamore Ave Berkeley Heights NJ 07922

VAN OSDALL, THOMAS CLARK, b Ashland, Ohio, July 19, 11; m 39; c 1. INORGANIC CHEMISTRY. Educ: Ashland Col, BS, 32; Ohio State Univ, MS, 39. Prof Exp: Prin, Hayesville High Sch, 32-36; instr chem & chmn dept sci, Santa Ana Col, 39-41; res chemist, Goodyear Tire & Rubber Co, 43-47; assoc prof chem, Ashland Col, 47-48; instr chem & chmn dept sci, Santa Ana Col, 48-57; PROF CHEM, ASHLAND COL, 57-, CHMN DEPT, 67- Concurrent Pos: Guest lectr, Orange Coast Col, 58; lectr art-sci prog, Sussex Univ, 66 & interdisciplinary arts-sci prog, Ashland Col, 66- Mem: Am Chem Soc. Res: Separation of rare elements; polymerization of synthetic rubber; organosols and plastisols; high vacuum techniques; industrial waste problems; water chemistry and water pollution problems. Mailing Add: Dept of Chem Ashland Col Ashland OH 44805

VAN OSS, CAREL J, b Amsterdam, Neth, Sept 7, 23; m 51; c 3. IMMUNOCHEMISTRY, PHYSICAL BIOCHEMISTRY. Educ: Univ Paris, PhD(phys biochem), 55. Prof Exp: Fel colloid chem, Van't Hoff Lab, Univ Utrecht, 55-56; fel phys chem, Ctr Electrophoresis, Sorbonne, 56-57; dir lab phys biochem, Nat Vet Col Alfort, 57-63; asst head dept microbiol, Montefiore Hosp, NY, 63-65; assoc prof biol, Marquette Univ, 66-68; assoc prof, 68-72, PROF MICROBIOL, SCH MED, STATE UNIV NY BUFFALO, 72-, HEAD IMMUNOCHEM LAB, 68- Concurrent Pos: French Ministry Agr res fel, 55-57; master of res, French Nat Agron Inst, 62-63; consult, Amicon Corp, 64-67; dir serum & plasma depts, Milwaukee Blood Ctr, 65-68; prin investr, USPHS res grant, 66-75; consult, Gen Elec Co, 67; mem consult comt electrophoresis & other chem separation processes in outer space, NASA, 71-; exec ed, Preparative Biochem, 71-; consult mem immunol subcomt, Diag Prod Adv Comt, Food & Drug Admin, 75- Mem: Am Chem Soc; fel Am Inst Chemists; Am Asn Immunologists; Am Soc Microbiol; Reticuloendothelial Soc. Res: Membrane separation methods; precipitation in immunochemical, organic and inorganic systems; diffusion; sedimentation; physical surface properties of cells; mechanism of phagocytic engulfment; cell separation methods; function of glycoproteins; opsonins. Mailing Add: Immunochem Lab Dept of Microbiol State Univ of NY Sch of Med Buffalo NY 14214

VAN OSTENBURG, DONALD ORA, b East Grand Rapids, Mich, July 19, 29; m 51; c 2. SOLID STATE PHYSICS. Educ: Calvin Col, BS, 51; Mich State Univ, MS, 53, PhD(physics), 56. Prof Exp: Assoc physicist, Armour Res Found, Ill Inst Technol, 56-59; from asst physicist to assoc physicist, Argonne Nat Lab, 59-70; PROF PHYSICS, DE PAUL UNIV, 70- Mem: Am Phys Soc; Am Sci Affiliation. Res: Static electrification; electron paramagnetic resonance; lattice dynamics; magnetism; nuclear magnetic resonance; electronic structure of metals and alloys; biophysics. Mailing Add: Dept of Physics De Paul Univ Chicago IL 60614

VAN OVERBEEK, JOHANNES, b Schiedam, Holland, Jan 2, 08; nat US; m 32, 48; c 6. PLANT PHYSIOLOGY, BIOLOGY. Educ: State Univ Leiden, BS, 28; Univ Utrecht, MS, 32, PhD(bot), 33. Hon Degrees: Dr, Univ Belg, 60. Prof Exp: Asst bot, Univ Utrecht, 33-34; asst plant hormones, Calif Inst Technol, 34-37, instr, 37-39, asst prof plant physiol, 39-43; plant physiologist, Inst Trop Agr, PR, 43-44, head dept, 44-46, asst dir, 46-47; chief plant physiologist & head dept plant physiol, Agr Lab, Shell Develop Co, 47-67; head dept, 67-73, PROF BIOL, TEX A&M UNIV, 67- Concurrent Pos: Hon prof, Col Agr, Univ PR, 44-47; ed, Plant Physiol, 67-; mem, Gov Adv Panel Use of Agr Chem, Tex, 70. Honors & Awards: Award, Am Soc Agr Sci, PR, 48. Mem: AAAS; Charles Reed Barnes hon mem Am Soc Plant Physiologists; Bot Soc Am; Soc Gen Physiol; Am Inst Biol Sci. Res: Plant hormones; physiology of growth; applied plant physiology. Mailing Add: Dept of Biol Tex A&M Univ Col of Sci College Station TX 77843

VAN PATTER, DOUGLAS MACPHERSON, b Montreal, Que, July 4, 23; m 50; c 4. NUCLEAR PHYSICS. Educ: Queen's Univ, Can, BSc, 45; Mass Inst Technol, PhD(physics), 49. Prof Exp: Jr physicist, Nat Res Coun Can, 45-46; asst physics, Mass Inst Technol, 46-49, res assoc, 49-52; res assoc, Univ Minn, 52, asst prof, 52-54; PHYSICIST, BARTOL RES FOUND, FRANKLIN INST, 54- Concurrent Pos: Chmn subcomt nuclear constants, Nat Acad Sci-Nat Res Coun, 59-64; vis prof, Inst Nuclear Physics, Univ Frankfurt, 72. Mem: Fel Am Phys Soc. Res: Nuclear reactions using electrostatic accelerators; inelastic proton scattering; radioactive decay schemes; lifetimes of nuclear states. Mailing Add: Bartol Res Found Franklin Inst Swarthmore PA 19081

VANPEE, MARCEL, b Hasselt, Belg, Dec 4, 16; US citizen; m 48; c 3. PHYSICAL CHEMISTRY. Educ: Cath Univ Louvain, BS, MS & PhD(phys chem), 40, Agrege de l'Enseignement Superieur, 56. Prof Exp: Nat Res fel radio & photochem, Nat Found Sci Res, Belg, 40-45; head lab sci res, Nat Inst Mines, Belg, 46-56; prof physics, Univ Leopoldville, Congo, 56-57; supvry chemist, US Bur Mines, Pa, 57-60; sr scientist, Reaction Motor Div, Thiokol Chem Corp, 60-68; PROF CHEM ENG, UNIV MASS, AMHERST, 68- Concurrent Pos: Lectr & researcher, Cath Univ Louvain, 40-45; mem, Nat Found Sci Res, Belg, 40-; res fel, Univ Minn, Minneapolis, 48; staff physicist, Edsel B Ford Inst, Mich, 50. Honors & Awards: Awards, Prix Jean Stass, 40, Prix Louis Empain, Belg Acad Sci, 43 & Prix Frederic Swartz, 54. Mem: Combustion Inst; Am Inst Aeronaut & Astronaut. Res: Radiochemistry; photochemistry; kinetics of combustion reactions; cool flames; ignition; flame spectroscopy and structure; high energy fuel and oxidizers; rocket exhaust radiation; hypergolic ignition; atomic and chemiluminescent reactions; reentry observables. Mailing Add: Goessmann Lab Univ of Mass Amherst MA 01002

VAN PEENEN, PETER FRANZ DIRK, b Pensacola, Fla, Sept 18, 32; m 58; c 4. PREVENTIVE MEDICINE. Educ: Princeton Univ, AB, 53; Univ Calif, MS, 56, MD, 57; Johns Hopkins Univ, MPH, 59, DrPH, 60; Am Bd Prev Med, dipl, 65. Prof Exp: Med Corps, US Navy, 56-, asst pharmacol, Univ Calif, 56-57, intern, San Diego Naval Hosp, Calif, 57-58, head dept epidemiol, Med Res Unit 3, Cairo, 60-63, officer in chg

Prev Med Unit 5, Calif, 63-65, officer in chg Prev Med Unit, Danang, Vietnam, 65-66, mem res staff, Dept Clin Invest, Naval Med Res Inst, Md, 66-70, officer in chg, Nat Med Res Unit 2, Detachment Djakarta, 70-74, CMNDG OFFICER, NAVAL MED RES UNIT 2, MED CORPS, US NAVY, 74- Mem: Soc Protozool; Am Soc Parasitol; Am Soc Mammal. Res: Hemoprotozoology; preventive medicine; zoonotic diseases. Mailing Add: Nav Med Res Unit 2 Box 14 APO San Francisco CA 96263

VAN PELT, ARNOLD FRANCIS, JR, b Orange, NJ, Sept 24, 24; m 47; c 2. ECOLOGY. Educ: Swarthmore Col, BA, 45; Univ Fla, MS, 47, PhD(biol), 50. Prof Exp: Assoc prof biol, Appalachian State Teachers Col, 50-54; prof, Tusculum Col, 54-57, prof biol & chem, 57-63; chmn dept sci & math, 64-71, PROF BIOL, GREENSBORO COL, 63- Concurrent Pos: Sewell grant, Highlands Biol Sta, 53-54; NSF grants, 55-56 & 62; consult, Univ Ga ecol team, Savannah River Plant, AEC, 60; mem conf plant biochem, Inst Paper Chem, 61; Piedmont Univ Ctr grants, 64-67, 71 & 72; Res Corp Brown-Hazen Fund grant, 65-67; mem radiation biol conf, Oak Ridge Inst Nuclear Studies, 65; mem conf molecular genetics, S W Ctr Advan Studies, 68. Mem: AAAS; Soc Syst Zoologists; Entom Soc Am. Res: Ecology of ants; mouse genetics; guinea pig leukemia. Mailing Add: Dept of Biol Greensboro Col Greensboro NC 27420

VAN PELT, ROLLO WINSLOW, JR, b Chicago, Ill, Dec 14, 29; m 54; c 2. PATHOLOGY. Educ: Wash State Univ, BA, 54, DVM, 56; Mich State Univ, MS, 61, PhD(path), 65. Prof Exp: Vet, Rose City Vet Hosp, Ore, 56-57, Tigard Vet Hosp, 57 & Willamette Vet Hosp, 57-59; res asst arthrology, Mich State Univ, 59-60, res assoc, 60-62, Nat Inst Arthritis & Metab Dis fel, 62-64, spec fel, 64-65, from asst prof to assoc prof path, 65-70; vis assoc prof zoophysiol & path, 70-71, ASSOC PROF ZOOPHYSIOL & PATH, INST ARCTIC BIOL, UNIV ALASKA, FAIRBANKS, 71- Concurrent Pos: Upjohn Co grant-in-aid, 62-64; All Univ res grant, Mich State Univ, 63-65. Honors & Awards: Sigma Xi Award, 62. Mem: Am Vet Med Asn; Am Col Vet Path; Am Vet Radiol Soc; fel Royal Soc Health; Int Acad Path. Res: Comparative arthrology in man and animals. Mailing Add: Inst of Arctic Biol Univ of Alaska Fairbanks AK 99701

VAN PELT, WESLEY RICHARD, b Passaic, NJ, Oct 17, 43; m 65; c 1. HEALTH PHYSICS, ENVIRONMENTAL HEALTH. Educ: Rutgers Univ, New Brunswick, BA, 65, MS, 66; NY Univ, PhD(nuclear eng), 71; Am Bd Health Physics, cert, 73. Prof Exp: Asst res scientist aerosol physics, Med Ctr, NY Univ, 67-71; environ scientist, Environ Analysts, Inc, NY, 71-72; HEALTH PHYSICIST, RADIATION SAFETY OFF, HOFFMANN-LA ROCHE INC, 72- Mem: Sigma Xi; Health Physics Soc; Am Nuclear Soc; Am Indust Hyg Asn; NY Acad Sci. Res: Applied health physics; radiation protection program development; interaction of airborne radioactivity with natural aerosols; measurement and study of the natural ionizing radiation background. Mailing Add: Radiation Safety Off Hoffmann-La Roche Inc Nutley NJ 07110

VAN PERNIS, PAUL ANTON, b Owyhee, Nev, Sept 7, 14; m 39; c 8. MEDICINE. Educ: Hope Col, AB, 35; Rush Med Col, MD, 39; Am Bd Path, dipl, 46. Prof Exp: Intern, St Mary's Hosp, Grand Rapids, Mich, 39-40; resident, St Luke's Hosp, Chicago, Ill, 40-42; dir labs, Butterworth Hosp, Grand Rapids, 46-51; instr path, Sch Med, Univ Ill, 51-53, clin asst prof, 53-64; dir labs, Swedish-Am Hosp, Ill, 59-72; ASST DIR, DEPT GRAD MED EDUC, AM MED ASN, 72- Concurrent Pos: Instr, Univ Chicago, 40-41; Borland fel, St Luke's Hosp, Chicago, Ill, 40-42; mem, Governor's Adv Comt Blood Banks & Clin Labs, Ill State Health Dept. Honors & Awards: Foster Welfare Found Prize, 53-64. Mem: AMA; fel Am Col Physicians. Res: Medical mycology; anatomic pathology; blood banking; diagnostic radioisotopes. Mailing Add: 240 Longfellow Dr Wheaton IL 60187

VAN PETTEN, GARRY R, b Camrose, Alta, Nov 9, 36; m 57; c 3. PHARMACOLOGY. Educ: Univ Alta, BSc, 57, MSc, 59; Glasgow Univ, PhD(pharmacol), 63. Prof Exp: Assoc prof pharmacol, Ont Vet Col, Guelph, 63-65; res scientist, Food & Drug Directorate, Dept Nat Health & Welfare, Can, 65-67, head pharmacol sect, Res Labs, Ont, 67-70; actg head, Div Pharmacol & Therapeut, 74-75, assoc prof, 70-75, PROF PHARMACOL & THERAPEUT, FAC MED, UNIV CALGARY, 75-, DIR MED ANIMAL CARE, 74- Concurrent Pos: Res grants, Ont Heart Found, 63-65, Nat Res Coun Can, 65, Med Res Coun Can, 70-75 & Alta Heart Found, 71-75. Mem: Pharmacol Soc Can; Can Biochem Soc. Res: Foetal effects of cardiovascular and central nervous system drugs; kinetics of placental transfer of drugs; drug metabolism. Mailing Add: Div of Pharmacol & Therapeut Univ Calgary Fac of Med Calgary AB Can

VAN PEURSEM, RALPH LAWRENCE, organic chemistry, see 12th edition

VAN PILSUM, JOHN FRANKLIN, b Prairie City, Iowa, Jan 28, 22; m 58; c 6. BIOCHEMISTRY. Educ: Univ Iowa, BS, 43, PhD, 49. Prof Exp: Instr biochem, Long Island Col Med, 49-51; asst prof, Univ Utah, 51-54; from asst prof to assoc prof, 54-63, PROF BIOCHEM, MED SCH, UNIV MINN, MINNEAPOLIS, 63- Mem: Am Soc Biol Chemists; Am Inst Nutrit. Res: Guanidinium compound metabolism. Mailing Add: Dept Biochem 227 Millard Hall Univ Minn Col of Med Sci Minneapolis MN 55455

VAN PUTTEN, JAMES D, JR, b Grand Rapids, Mich, Apr 14, 34; m 59; c 2. NUCLEAR PHYSICS. Educ: Hope Col, AB, 55; Univ Mich, AM, 57, PhD(physics), 60. Prof Exp: Instr physics, Univ Mich, 60-61; NATO fel, Europ Orgn Nuclear Res, Geneva, 61-62; asst prof, Calif Inst Technol, 62-67; assoc prof, 67-70, PROF PHYSICS, HOPE COL, 70-, CHMN DEPT, 75- Concurrent Pos: Consult, Electro-Optical Systs, 63-67, Teledyne Corp, 67-70, Donnelly Mirrors, 67- & Maes, Inc, 68- Mem: AAAS; Am Phys Soc. Res: Bubble chambers; counter and spark chamber techniques; satellite bourn space physics experiments; nuclear charge structure; use of microcomputers in process control. Mailing Add: Dept of Physics Hope Col Holland MI 49423

VAN RAALTE, JOHN A, b Copenhagen, Denmark, Apr 10, 38; US Citizen; m 63; c 2. RESEARCH ADMINISTRATION. Educ: Mass Inst Technol, SB & SM, 60, EE, 62, PhD(solid state physics, elec eng), 64. Prof Exp: Res asst lab insulation res, Mass Inst Technol, 60-64; mem tech staff, 64-70, HEAD DISPLAYS & DEVICE CONCEPTS RES, RCA RES LABS, 70- Mem: AAAS; sr mem Inst Elec & Electronics Engrs; Am Phys Soc. Res: Materials; dielectrics; electro-optic materials; lasers; displays. Mailing Add: RCA Res Labs Princeton NJ 08540

VAN REEN, ROBERT, b Paterson, NJ, June 12, 21. BIOCHEMISTRY. Educ: NJ State Teachers Col, Montclair, AB, 43; Rutgers Univ, PhD(biochem), 49. Prof Exp: Assoc biochemist, Brookhaven Nat Lab, 49-51; res assoc, McCollum-Pratt Inst, Johns Hopkins Univ, 51-53, asst prof biol, 53-56; supv chemist & assoc head dent div, Naval Med Res Inst, 56-61, head nutrit biochem div, 61-70; PROF FOOD & NUTRIT SCI & CHMN DEPT, UNIV HAWAII, HONOLULU, 70- Honors & Awards: McLester Award, Asn Mil Surgeons of US, 59. Mem: Am Chem Soc; Soc Nutrit Educ; Am Soc Biol Chemists; Int Asn Dent Res; Am Inst Nutrit. Res: Mammalian and avian nutrition; requirements and functions of vitamins and trace elements; interrelationships between trace elements and enzyme systems; experimental dental caries; metabolism in calcified tissues; nutrition and urolithiasis. Mailing Add: Dept of Food & Nutrit Sci Univ of Hawaii Honolulu HI 96822

VAN REMOORTERE, EMILE C, b Herstal, Belg, July 5, 21; m 48; c 3. CARDIOVASCULAR PHYSIOLOGY, PHARMACOLOGY. Educ: State Univ Liege, BNatural & MedSci, 41, DrMed, 45. Prof Exp: Asst physiopath, State Univ Liege, 48-54, assoc, 54-57; prof physiol, State Univ Belg Congo, 57-63, dean Med Sch, 59-60; head dept cardiovasc pharmacol, Belg Chem Union, Brussels, 63-69; PROF PHARMACOL, SCH MED SCI, UNIV NEV, RENO, 70- Concurrent Pos: Belg-Am Educ Found fel, Cardiovasc Dept, Michael Reese Hosp, Chicago, 46-48. Res: Cardiac electrophysiology; general electrocardiography. Mailing Add: Sch of Med Sci Unvi of Nev Reno NV 89507

VAN RHEENEN, VERLAN H, b Oskaloosa, Iowa, Feb 15, 39; m 62; c 2. ORGANIC CHEMISTRY. Educ: Cent Col, Iowa, BA, 61; Univ Wis, PhD(org chem), 66. Prof Exp: SR SCIENTIST, UPJOHN CO, 66- Mem: Am Chem Soc. Res: Total synthesis of natural products; new synthetic methods; steroid chemistry; prostaglandin synthesis. Mailing Add: 2112 Vanderbilt Rd Kalamazoo MI 49002

VAN RIJ, WILLEM IDANIEL, b Brielle, Neth, Apr 19, 42; US citizen; m 68; c 1. COMPUTER SCIENCES, PLASMA PHYSICS. Educ: Univ Auckland, BS, 64, MS, 66; Fla State Univ, PhD(physics), 70. Prof Exp: Res assoc theoret nuclear physics, Brookhaven Nat Lab, 70-72 & Univ Wash, 72-74; COMPUT PHYSICIST, OAK RIDGE NAT LAB, 74- Mem: Am Phys Soc. Res: Computer simulations of plasmas for controlled thermonuclear research. Mailing Add: Oak Ridge Nat Lab PO Box X Oak Ridge TN 37830

VAN RIPER, GORDON EVERETT, b Flat Rock, Mich, Dec 7, 17; m 43; c 1. AGRONOMY. Educ: Mich State Univ, BS, 55; Univ Wis, MS, 57, PhD(agron), 58. Prof Exp: From asst prof to assoc prof agron, Univ Nebr, 58-64; mgr, Dept Agron, Deere & Co, 64-69, mgr, Dept Res Coord, 69-73; PRES, JAY DEE EQUIP, INC, 73- Concurrent Pos: Vpres, Agr Res Inst, 71-72; pres, Am Forage & Grassland Coun, 72. Mem: Am Soc Agron; Crop Sci Soc Am. Res: Crop physiology. Mailing Add: Jay Dee Equip Inc Box 210 Kewanee IL 61443

VAN RIPER, JOSEPH EDWARDS, b Champion, Mich, Nov 9, 10; m 35; c 1. GEOGRAPHY. Educ: Univ Mich, AB, 32, PhD(geog), 39; Syracuse Univ, AM, 34. Prof Exp: Specialist, State Planning Bd, NY, 34-35; instr geol, Colgate Univ, 35-37; asst prof geog, Southern Ill State Norm Univ, 39-42; geogr, Mil Intel Serv, US War Dept, 42-46; assoc prof geog & chmn dept geol & geog, Triple Cities Col, Syracuse Univ, 46-50; chmn dept, 50-71, PROF GEOG, STATE UNIV NY BINGHAMTON, 50- Concurrent Pos: Smith-Mundt vis prof, Am Univ, Beirut, 60-62. Mem: Fel Am Geog Soc; fel Asn Am Geogr; Mid East Inst. Res: Military terrain intelligence; Southwest Asia. Mailing Add: Dept of Geog State Univ of NY Binghamton NY 13901

VAN ROGGEN, AREND, b Nijmegen, Neth, Jan 2, 28; m 52; c 1. ELECTRODYNAMICS, COMPUTER SCIENCE. Educ: State Univ Leiden, Drs(phys chem), 53; Duke Univ, PhD(physics), 56. Prof Exp: Asst, Lab Phys Chem, State Univ Leiden, 48-54, sci clerk, 56; asst physics, Duke Univ, 54-56; from res physicist to sr res physicist, 56-72, sr res specialist, 72-75, RES ASSOC, E I DU PONT DE NEMOURS & CO, INC, 75- Concurrent Pos: Mem conf elec insulation, Nat Acad Sci-Nat Res Coun. Mem: Am Phys Soc; Am Soc Testing & Mat; NY Acad Sci. Res: Electronic and magnetic structure of matter; paramagnetic resonance; dielectrics; physics instrumentation; interaction of electric fields and materials; computational aspects of above. Mailing Add: Exp Sta E I du Pont de Nemours & Co Inc Wilmington DE 19898

VAN ROOSBROECK, WILLY WERNER, b Antwerp, Belg, Aug 10, 13; nat US; m 45. SEMICONDUCTORS. Educ: Columbia Univ, AB, 34, MA, 37. Prof Exp: RES PHYSICIST, BELL LABS, 37- Mem: AAAS; fel Am Phys Soc. Res: Mathematical physics of semiconductors; theory of current-carrier injection and transport and of amorphous and relaxation semiconductors; rectification, trapping, recombination, radiation, high-field and space-charge effects. Mailing Add: Bell Labs Murray Hill NJ 07974

VAN ROSSUM, GEORGE DONALD VICTOR, b London, Eng, Dec 13, 31; m 59; c 4. CELL PHYSIOLOGY, BIOCHEMICAL PHARMACOLOGY. Educ: Oxford Univ, MA, 59, DPhil(biochem), 60. Prof Exp: NATO fel physiol chem, Univ Amsterdam, 60-62; NIH fel, Johnson Res Found, Univ Pa, 62-63, res assoc phys biochem, Sch Med, 63-64; reader biochem, Christian Med Col, Vellore, India, 65-67; vis asst prof phys biochem, Johnson Res Found, Sch Med, Univ Pa, 67-69; actg chmn dept, 73-75, PROF PHARMACOL, SCH MED, TEMPLE UNIV, 69- Concurrent Pos: NATO vis prof, Inst Gen Path, Cath Univ, Rome, 71. Mem: Brit Biochem Soc; Am Soc Biol Chemists; Am Soc Pharmacol & Exp Therapeut. Res: Ion and water transport; tissue electrolytes in cancer; control of energy metabolism. Mailing Add: Dept of Pharmacol Temple Univ Sch of Med Philadelphia PA 19160

VAN ROYEN, PIETER, b Lahat, Indonesia, Jan 8, 23; m 56; c 8. BIOGEOGRAPHY, PLANT TAXONOMY. Educ: State Univ Utrect, BSc, 47, MSc, 49, PhD(bot), 51. Prof Exp: Res asst systs, Bot Mus & Herbarium, Utrecht, Neth, 45-51; sr res asst, Nat Herbarium, Leiden, Neth, 51-61; sr botanist, Div Bot, Dept Forests Lae, Australian New Guinea, 61-65; botanist, Queensland Herbarium, Brisbane, Australia, 65-67; CHMN & SR BOTANIST, B P BISHOP MUS, HONOLULU, 67- Concurrent Pos: Assoc, Lyone Arboretum & Univ Hawaii, 67-, sci adv, Pac Trop Bot Garden, 70-; NSF grant alpine studies, New Guinea, 70-72 & 74-76. Honors & Awards: French Resistance Medal, Legion d'Honneur. Mem: Asn Trop Biol; Am Asn Plant Taxonomists. Res: Alpine flora of New Guinea; plant systematics and geography of the Pacific basin; alpine research of Southwest Pacific area, Japan and the Americas. Mailing Add: Dept of Bot B P Bishop Mus Honolulu HI 96817

VAN RYSSELBERGHE, PIERRE, b Brussels, Belg, May 18, 05; nat US; m 30; c 3. ELECTROCHEMISTRY. Educ: Univ Brussels, Engr, 27; Stanford Univ, AM, 28, PhD(phys chem), 29. Prof Exp: From instr to asst prof chem, Stanford Univ, 29-41; from asst prof to prof, Univ Ore, 41-56; lectr & res assoc, Stanford Univ, 56-72, lectr chem eng, 61-72; RETIRED. Concurrent Pos: Vis Comn Relief Belg Educ lectr, Belg, 35-36; supvr, Proj Polarog of Corrosion, Off Naval Res, 46-51, naval technician, Europe, 48; pres, Int Comt Electrochem Thermodyn & Kinetics, 49-54; Fulbright vis lectr, Italy, 50-51; supvr, AEC Proj Corrosion of Zirconium, 52-56; mem comn electrochem, Int Union Pure & Appl Chem, 52-61, chmn, 61-67. Mem: AAAS; Am Chem Soc; Electrochem Soc; fel Am Inst Chemists; fel NY Acad Sci. Res: Solutions of electrolytes; polarography; corrosion; chemical thermodynamics and thermodynamics of irreversible processes. Mailing Add: 551 Santa Rita Ave Palo Alto CA 94301

VAN RYSWYK, ALBERT LEONARD, b Castor, Alta, May 10, 28; m 58; c 2. SOIL FERTILITY, SOIL MORPHOLOGY. Educ: Univ BC, BS, 50, MSc, 55; Wash State

Univ, PhD(soils), 69. Prof Exp: Soil surveyor, Doukhobor Lands, Univ BC, 50-54 & Soil Surv Br, BC Dept Agr, 54-58; SOILS RES RANGELANDS, RES BR, CAN DEPT AGR, 58- Concurrent Pos: Mem, Can Comt Soil Fertility, 75-77. Mem: Can Soc Soil Sci; Int Soc Soil Sci; Soil Sci Soc Am; Agr Inst Can; Soc Range Mgt. Res: Soil-test calibration for yield response of forage crops grown on wetlands and under irrigation; soil moisture and vegetation relationships affecting yield of rangeland grasses. Mailing Add: Res Sta Agr Can Box 940 Kamloops BC Can

VAN RYZIN, JOHN R, b Milwaukee, Wis, May 5, 35; m 59; c 4. MATHEMATICAL STATISTICS, STATISTICS. Educ: Marquette Univ, BS, 57, MS, 59; Mich State Univ, PhD(statist), 64. Prof Exp: Instr statist, Mich State Univ, 62-63; asst mathematician, Argonne Nat Lab, 63-66; res assoc statist, Stanford Univ, 66-67; assoc prof math, Univ Wis-Milwaukee, 67-69, assoc prof statist, 69-73, PROF STATIST, UNIV WIS-MADISON, 73- Concurrent Pos: NSF grant, Stanford Univ, 66-67. Mem: Fel Inst Math Statist; Am Statist Asn; Royal Statist Soc; Int Asn Statist in Phys Sci; Am Math Soc. Res: Inference procedures in statistics; empirical Bayes theory; decision theory; classification and pattern recognition procedures. Mailing Add: Dept of Statist Univ of Wis 1210 W Dayton St Madison WI 53706

VAN RYZIN, MARTINA, b Appleton, Wis, June 10, 23. HISTORY OF SCIENCE, MATHEMATICS. Educ: Holy Family Col, Wis, BA, 46; Marquette Univ, MS, 56; Univ Wis, PhD(hist of sci), 60. Prof Exp: Head dept math, 57-70, from instr to assoc prof math, 60-69, PROF MATH, SILVER LAKE COL, 69-, ACAD DEAN, 70- Mem: Math Asn Am; Hist Sci Soc. Res: History of mathematics, especially Medieval period; Arabic-Latin tradition of Euclid's elements in the 12th century. Mailing Add: Off of the Acad Dean Silver Lake Col Manitowoc WI 54220

VAN SAMBEEK, JEROME WILLIAM, b Milbank, SDak, Aug 1, 47; m 72; c 1. PLANT PHYSIOLOGY. Educ: SDak State Univ, BS, 59; Washington Univ, PhD(plant physiol), 75. Prof Exp: Fel plant path, Univ Mo, 75; RES PLANT PHYSIOL, SOUTHERN FOREST EXP STA, 75- Mem: Am Soc Plant Physiologists; Am Phytopathological Soc. Res: Stress physiology of higher plants in relationship to wounding and host-pathogen interactions. Mailing Add: Southern Forest Exp Sta 2500 Shreveport Hwy Pineville LA 71360

VAN SCHAACK, EVA BLANCHE, b Coxsackie, NY, July 19, 04. BOTANY. Educ: Hope Col, AB, 29; Johns Hopkins Univ, PhD(bot), 37. Prof Exp: Instr bot, Teachers Col, Johns Hopkins Univ, 35-40; asst prof, Kalamazoo Col, 46-47; asst prof plant sci, Mt Holyoke Col, 48-53; assoc prof biol, Wheaton Col, 53-56; from assoc prof to prof, 56-69, EMER PROF BIOL, HOPE COL, 69- Mem: Bot Soc Am; Mycol Soc Am. Res: Mayapple rust; cytology and taxonomy of fungi. Mailing Add: 250 College Ave Holland MI 49423

VAN SCHAACK, GEORGE BOOTH, b Coxsackie, NY, Sept 13, 03. MATHEMATICS, PLANT TAXONOMY. Educ: Harvard Univ, SB, 29, AM, 32, PhD(math), 35. Prof Exp: Instr math, Harvard Univ, 29-32 & 33-35; math clerk, Equitable Life Assurance Soc, 35-36; instr math, Univ Rochester, 36-38; from instr to asst prof, Mich State Col, 38-43; asst prof, Union Univ, NY, 46-47; from asst prof to assoc prof, Washington Univ, 47-60, actg cur herbarium, Mo Bot Garden, 55-58, cur grasses, 58-66, librn, 58-67; bibliog consult, Morton Arboretum, Ill, 67-72; RES ASSOC BIOL & HON CUR HERBARIUM, MUS NATURAL HIST, UNIV ORE, 74- Mem: AAAS; Int Asn Plant Taxonomists. Res: Grass taxonomy; botanical bibliography and illustration. Mailing Add: 1964 Harris St Eugene OR 97405

VAN SCHAIK, PETER HENDRIK, b Arnhem, Neth, Apr 18, 27; US citizen; m 54; c 5. PLANT BREEDING. Educ: Ont Agr Col, Univ Guelph, BSA, 52; Univ Toronto, MSA, 54; Purdue Univ, PhD(plant breeding), 56. Prof Exp: Res agronomist cotton, Agr Res Serv, USDA, Brawley, Calif, 57-64, coord res agronomist food legumes, Tehran, Iran & New Delhi, India, 64-70, res agronomist peanut res, Holland, Va, 70-72, ASST AREA DIR, AGR RES SERV, USDA, FRESNO, CALIF, 72- Concurrent Pos: Consult, Ford Found, 70, Rockefeller Found, 70 & Experience Inc, 71. Mem: Am Soc Agron. Res: Crop breeding; agronomy; foreign development; tropical agriculture. Mailing Add: Agr Res Serv USDA PO Box 8143 Fresno CA 93727

VAN SCHMUS, WILLIAM RANDALL, b Aurora, Ill, Oct 4, 38; m 61; c 3. GEOLOGY, METEORITICS. Educ: Calif Inst Technol, BS, 60; Univ Calif, Los Angeles, PhD(geol), 64. Prof Exp: From asst prof to assoc prof, 67-75, PROF GEOL, UNIV KANS, 75- Mem: Am Geophys Union; Geochem Soc; Meteoritical Soc; Mineral Soc Am; Geol Soc Am. Res: Geochronology and geochemistry of Precambrian rocks in North America; mineralogy and petrology of meteorites. Mailing Add: Dept of Geol Univ of Kans Lawrence KS 66045

VAN SCIVER, WESLEY J, b Philadelphia, Pa, Sept 6, 17; m 59; c 3. PHYSICS, OCEANOGRAPHY. Educ: Mass Inst Technol, BS, 40; Stanford Univ, PhD(physics), 55. Prof Exp: Dir res, scintillation crystals, Radiation at Stanford, Inc, 54-60; vis prof physics, Univ PR, 61-62; assoc prof, 62-65, PROF PHYSICS, LEHIGH UNIV, 65- Mem: Am Phys Soc; Optical Soc Am. Res: Electronic processes in insulating solids; time resolved optical spectroscopy of solids; physical and optical spectroscopy of solids; scintillation and ultraviolet luminescence; physical and optical oceanography. Mailing Add: Dept of Physics Lehigh Univ Bethlehem PA 18015

VAN SCOTT, EUGENE JOSEPH, b Macedon, NY, May 27, 22; m 48; c 3. DERMATOLOGY. Educ: Univ Chicago, BS, 45, MD, 48. Prof Exp: Intern, Millard Fillmore Hosp, Buffalo, NY, 48-49; resident physician dermat, Univ Chicago, 49-52; assoc, Univ Pa, 52-53; chief dermat br, Nat Cancer Inst, 53-68, sci dir gen labs & clins, 66-68; PROF DERMAT, HEALTH SCI CTR, TEMPLE UNIV, 68-; ASSOC DIR SKIN & CANCER HOSP PHILADELPHIA, 68- Honors & Awards: Taub Int Mem Award, 64; Clarke White Award, 65; Albert Lasker Award, 72; Stephen Rothman Award, Soc Invest Dermat, 75. Mem: Hon mem Can Dermat Asn; Soc Invest Dermat; Am Dermat Asn; Am Asn Cancer Res; Am Acad Dermat. Res: Biology and physiology of epithelial growth; differentiation and neoplasia; pathogenesis of psoriasis; biology; immunologic aspects and clinical management of cutaneous lymphomas. Mailing Add: Skin & Cancer Hosp 3322 N Broad St Philadelphia PA 19140

VANSELOW, CLARENCE HUGO, b Syracuse, NY, Sept 30, 28; m 51; c 7. PHYSICAL CHEMISTRY. Educ: Syracuse Univ, BS, 50, MS, 51, PhD, 58. Prof Exp: Instr chem, Colgate Univ, 55-56; prof, Thiel Col, 56-64; ASSOC PROF CHEM, UNIV NC, GREENSBORO, 64- Mem: Am Chem Soc. Res: Gas phase radiation chemistry; kinetic theory of precipitation processes. Mailing Add: Dept of Chem Univ of NC Greensboro NC 27412

VANSELOW, NEAL A, b Milwaukee, Wis, Mar 18, 32; m 58; c 2. MEDICAL ADMINISTRATION, ALLERGY. Educ: Univ Mich, AB, 54, MD, 58, MS, 63. Prof Exp: From instr to assoc prof internal med, Univ Mich, Ann Arbor, 63-74, from assoc prof to prof postgrad med & chmn dept, 67-74; PROF INTERNAL MED & DEAN COL MED, UNIV ARIZ, 74- Concurrent Pos: Consult, Ann Arbor Vet Admin Hosp, 64-74. Mem: Am Acad Allergy; Am Col Physicians. Res: Mechanisms of aspirin sensitivity; immunosuppression and immunologic adjuvants. Mailing Add: Col of Med Univ of Ariz Tucson AZ 85721

VANSELOW, RALF W, b Berlin, Ger, July 12, 31; m 61. PHYSICAL CHEMISTRY, SURFACE CHEMISTRY. Educ: Tech Univ Berlin, BS, 57, Dipl Ing, 62, Dr Ing, 66. Prof Exp: Res asst field emission micros, Fritz Haber Inst, Max Planck Soc, 52-66, res assoc, 66-68; asst prof, 68-74, ASSOC PROF CHEM, UNIV WIS-MILWAUKEE, 74- Mem: Ger Chem Soc; Am Chem Soc; Ger Vacuum Soc; Am Vacuum Soc. Res: Studies of metal surfaces by means of field electron and field ion microscopy; adsorption; surface migration; epitaxial growth. Mailing Add: Dept of Chem Univ of Wis Milwaukee WI 53201

VANSELOW, ROBERT DALE, physical chemistry, see 12th edition

VAN SICKLE, DALE ELBERT, b Ft Collins, Colo, Oct 8, 32; m 62; c 2. PHYSICAL ORGANIC CHEMISTRY. Educ: Colo State Univ, BS, 54; Univ Utah, MS, 56; Univ Calif, PhD, 59. Prof Exp: Asst chem, Univ Utah, 54-55 & Univ Calif, 56-59; org chemist, Stanford Res Inst, 59-69; SR RES CHEMIST, TENN EASTMAN CO, 69- Mem: AAAS; Am Chem Soc; Sigma Xi; Am Inst Chemists. Res: Mechanisms and kinetics of reactions; oxidation of hydrocarbons; thermodynamics of fluid phase equilibria. Mailing Add: Res Labs Tenn Eastman Co Kingsport TN 37662

VAN SICKLE, DAVID C, b Des Moines, Iowa, Jan 9, 34; m 56. HISTOLOGY, PATHOBIOLOGY. Educ: Iowa State Univ, DVM, 57; Purdue Univ, PhD(develop anat), 66. Prof Exp: Gen pract, Ill, 57-58 & 60-61; from instr to assoc prof, 61-75, PROF HISTOL & EMBRYOL, PURDUE UNIV, WEST LAFAYETTE, 75- Concurrent Pos: Morris Animal Found fel, 64-66; adj assoc prof anat, Sch Med, Ind Univ, 73- Mem: Sigma Xi; NY Acad Sci. Res: Osteogenesis and abnormalities associated with errors in osteogenesis; orthopedic pathobiology in arthritis, hip dysplasia, and osteochondritis. Mailing Add: Dept of Vet Anat Purdue Univ West Lafayette IN 47906

VAN SICLEN, DEWITT CLINTON, b Carlisle, Pa, Oct 25, 18; m 49; c 4. GEOLOGY. Educ: Princeton Univ, AB, 40, MA, 47, PhD(geol), 51; Univ Ill, MS, 41. Prof Exp: Jr geologist, Peoples Natural Gas Co, Pa, 41-42 & 46; field geologist, Drilling & Explor Co, Inc, 47-50; consult, 50-51; exec officer, Off Sci Res, US Dept Air Force, 51-52; res geologist, Pan-Am Prod Co, 52-56; sr geologist, Pan-Am Petrol Corp, 56-59; assoc prof, 59-65, chmn dept, 60-67, PROF GEOL, UNIV HOUSTON, 65- Mem: Am Asn Petrol Geologists; Nat Asn Geol Teachers; Am Inst Prof Geol; Soc Petrol Engrs. Res: Deposition, deformation and alteration of sedimentary rocks; behavior of subsurface fluids; petroleum geology; surficial geology and active faults of Texas-Louisiana coastal plain. Mailing Add: Dept of Geol Univ of Houston Houston TX 77004

VAN SLUYTERS, RICHARD CHARLES, b Chicago, Ill, June 12, 45; m 68. PHYSIOLOGICAL OPTICS, NEUROPHYSIOLOGY. Educ: Ill Col Optom, BS, 67, OD, 68; Ind Univ, Bloomington, PhD(physiol optics), 72. Prof Exp: ASST PROF OPTOM-PHYSIOL OPTICS, SCH OPTOM, UNIV CALIF, BERKELEY, 75- Concurrent Pos: Am Optom Found Res fel, 68-69; Nat Eye Inst fel, 69-71 & spec res fel, 72-74; NSF fel, 71-72; Miller Inst Basic Res Sci fel, Univ Calif, Berkeley, 74-76. Mem: Sigma Xi; Optical Soc Am; Asn Res Vision & Ophthal; fel Am Acad Optom. Res: Neurophysiology of developing mammalian visual systems. Mailing Add: Sch of Optom Univ of Calif Berkeley CA 94720

VAN SLYKE, ARTHUR LAWTON, b Perth, Ont, June 13, 25; m 50; c 2. FORESTRY, FOREST BIOMETRY. Educ: Univ NB, BSc, 49; Univ Mich, MF, 52. Prof Exp: Jr forester, NB Int Paper Co Div, Can Int Paper Co, 49-50, res forester, Causapscal Forest Res Sta, 50-56, supt, 56 & 57; PROF FORESTRY, UNIV NB, FREDERICTON, 57- Res: Tree crown measures to evaluate growth rates and to assist in management planning; inventory for stand management; stand area estimates. Mailing Add: Fac of Forestry Univ of NB Fredericton NB Can

VAN SLYKE, RICHARD M, b Manila, Philippines, Aug 17, 37; U S citizen; m 69. OPERATIONS RESEARCH. Educ: Stanford Univ, BS, 59; Univ Calif, Berkeley, PhD(opers res), 65. Prof Exp: Asst prof elec & indust eng, Univ Calif, Berkeley, 65-69; VPRES, NETWORK ANAL CORP, 69- Concurrent Pos: Consult, Rand Corp, 62-66 & Crown Zellerbach Corp, 63-65. Mem: Soc Indust & Appl Math; Opers Res Soc Am; Math Asn Am. Res: Mathematical techniques for optimization. Mailing Add: Network Analysis Corp Old Tappan Rd Beechwood Glen Cove NY 11542

VANSOEST, PETER JOHN, b Seattle, Wash, June 30, 29; m 59; c 3. ANIMAL NUTRITION. Educ: Wash State Univ, BS, 51, MS, 52; Univ Wis, PhD(nutrit), 55. Prof Exp: Biochemist, Agr Res Serv, USDA, 57-68; assoc prof animal nutrit, 68-73, PROF ANIMAL NUTRIT, CORNELL UNIV, 73- Honors & Awards: Am Feed Mfrs Award, 67; Hoblitzelle Nat Award Agr, 68; Am Grassland Coun Merit Cert, 68. Mem: AAAS; Am Dairy Sci Asn; Am Soc Animal Sci; Am Chem Soc; Asn Off Anal Chem. Res: Ruminant digestion and metabolism; chemistry of fibrous feedstuffs and methods of analysis; forage chemistry. Mailing Add: 120 Homestead Circle Ithaca NY 14850

VANSPEYBROECK, LEON PAUL, b Wichita, Kans, Aug 27, 35; m 59; c 3. X-RAY ASTRONOMY. Educ: Mass Inst Technol, BS, 57, PhD, 65. Prof Exp: Res assoc high energy physics, Mass Inst Technol, 65-67; staff scientist x-ray astron, Am Sci & Eng, Inc, Mass, 67-74; STAFF SCIENTIST, CTR FOR ASTROPHYS, 74- Res: Solar and stellar astronomy; physics of the solar corona. Mailing Add: Ctr for Astrophys 60 Garden St Cambridge MA 02138

VAN STEE, ETHARD WENDEL, b Traverse City, Mich, July 17, 36; m 60; c 2. PHARMACOLOGY, TOXICOLOGY. Educ: Mich State Univ, BS, 58, DVM, 60; Ohio State Univ, MS, 66, PhD(vet physiol, pharmacol), 70. Prof Exp: Res pharmacologist, 6570 Aerospace Med Res Lab, Wright-Patterson AFB, Ohio, 67-75; PHYSIOLOGIST, NAT INST ENVIRON HEALTH SCI, 75- Concurrent Pos: Adj assoc prof pharmacol, Univ NC, 75- Mem: Soc Toxicol; Am Soc Pharmacol & Exp Therapeut; Am Vet Med Asn; AAAS; Am Soc Vet Physiologists & Pharmacologists. Res: Inhalation toxicology; halogenated alkanes, anesthetics; cardiovascular pharmacology. Mailing Add: Nat Inst Environ Hlth Sci PO Box 12233 Research Triangle Park NC 27709

VAN STEENBERGEN, ARIE, b Vlaardingen, Neth, Feb 26, 28; m 53; c 4. PHYSICS. Educ: Delft Univ Technol, MSc, 52; McGill Univ, PhD(physics, math), 57. Prof Exp: Res asst electron beam optics, Delft Univ Technol, 50-53; res physicist, Nat Defense Res Lab, The Hague, Neth, 53-54; res assoc nuclear magnetic resonance, McGill Univ, 54-57; from asst physicist to assoc physicist, 57-75, head alternating gradient synchrotron div, 65-74, head booster synchrotron sect, 67-68, SR PHYSICIST, BROOKHAVEN NAT LAB, 75- Concurrent Pos: Consult, Radiation Dynamics, Inc, 60- Mem: Am Phys Soc; Europ Phys Soc. Res: High energy particle accelerators and storage rings; particle beam dynamics; ion sources. Mailing Add: Brookhaven Nat Lab Upton NY 11973

VANSTONE, J R, b Owen Sound, Ont, Aug 12, 33; m 56; c 3. MATHEMATICS. Educ: Univ Toronto, BA, 55, MA, 56; Univ Natal, PhD(math), 59. Prof Exp: Lectr math, 59-61, asst prof, 61-65, ASSOC PROF MATH, UNIV TORONTO, 65- Mem: Can Math Cong; Math Asn Am; Am Math Soc; Soc Indust & Appl Math. Res: Differential geometry. Mailing Add: Dept of Math Univ of Toronto Toronto ON Can

VAN STONE, JAMES MORRIL, b Bridgeport, Conn, Jan 19, 24; m 49; c 2. ZOOLOGY. Educ: Princeton Univ, PhD, 54. Prof Exp: From asst prof to assoc prof biol, 54-65, PROF BIOL, TRINITY COL, CONN, 65- Mem: Am Soc Zoologists; Am Asn Anatomists. Res: Amphibian regeneration; experimental zoology. Mailing Add: Dept of Biol Trinity Col Hartford CT 06106

VANSTONE, SCOTT ALEXANDER, b Chatham, Ont, Sept 14, 47; m 70; c 1. MATHEMATICS. Educ: Univ Waterloo, BMath, 70, MMath, 71, PhD(math), 74. Prof Exp: ASST PROF MATH, UNIV ST JEROME'S COL, 74- Res: The existence and construction of balanced incomplete block designs and regular pairwise balanced designs which are closely related to finite linear spaces and balanced equidistant codes. Mailing Add: Univ of St Jerome's Col Waterloo ON Can

VAN STRATEN, MARY PETRONIA, b Mar 3, 13. MATHEMATICS. Educ: Mt Mary Col, Wis, BA, 44; Univ Notre Dame, PhD(math), 47. Prof Exp: Elem teacher, Wis, 32-41 & St Mary's Sch, 41-42; assoc prof math, 47-69, PROF MATH, MT MARY COL, 69-, CHMN DEPT, 65- Mem: Math Asn Am. Res: Modern algebra. Mailing Add: Dept of Math Mt Mary Col N 92nd St Milwaukee WI 53222

VAN STRIEN, RICHARD EDWARD, b Battle Creek, Mich, Sept 17, 20; m 42; c 3. ORGANIC CHEMISTRY. Educ: Hope Col, AB, 42; Univ Pa, MS, 44, PhD(org chem), 48. Prof Exp: Res chemist, Standard Oil Co, Ind, 47-60, sect leader, Res & Develop Dept, 60-61, sect leader in-chg prod appln new chem, Amoco Chem Corp, 61-67, asst dir polymers & plastics div, Ind, 67-69, DIV DIR, CONDENSATION POLYMERS DIV, AMOCO CHEM CORP, 69- Mem: Am Chem Soc; Am Soc Testing & Mat. Res: Synthetic detergents; gelling agents; surface coatings; reinforced plastics; adhesives. Mailing Add: Res & Develop Dept Amoco Chem Corp PO Box 400 Naperville IL 60540

VAN SWAAY, MAARTEN, b The Hague, Neth, Aug 1, 30; m 54; c 4. ANALYTICAL CHEMISTRY. Educ: State Univ Leiden, BS, 53, Drs(chem), 56; Princeton Univ, PhD(chem), 56. Prof Exp: Sr res asst phys chem, State Univ Leiden, 56-59; res assoc instr anal, Eindhoven Technol Univ, 59-63; asst prof, 63-69, ASSOC PROF ANAL CHEM, KANS STATE UNIV, 69- Concurrent Pos: Consult, Medi-Comput Corp. Mem: Am Chem Soc. Res: Chemical instrumentation; computer interfacing and control. Mailing Add: Dept of Chem Kans State Univ Manhattan KS 66506

VAN TAMELEN, EUGENE EARL, b Zeeland, Mich, July 20, 25; m 51; c 3. CHEMISTRY. Educ: Hope Col, AB, 47; Harvard Univ, MA, 49, PhD(chem), 50. Hon Degrees: DSc, Hope Col & Bucknell Univ, 71. Prof Exp: From instr to prof org chem, Univ Wis, 50-61, Adkins prof chem, 61-62; PROF CHEM, STANFORD UNIV, 62-, CHMN DEPT, 74- Concurrent Pos: Guggenheim fels, 65 & 73; prof extraordinarius, Neth, 67-74; mem adv bd, Chem & Eng News, 68-70, Synthesis, 69- & Accounts of Chem Res, 70-73; mem adv bd & ed, Bioorg Chem, 71- Honors & Awards: Award in Pure Chem, Am Chem Soc, 61 & Award for Creative Work in Synthetic Org Chem, 70; Baekeland Award, 65. Mem: Nat Acad Sci; Am Acad Arts & Sci; Am Chem Soc. Res: Chemistry of natural products including structure, synthesis and biosynthesis; new reactions; organic-inorganic chemistry. Mailing Add: Dept of Chem Stanford Univ Stanford CA 94305

VAN TASSEL, JAMES HENRY, inorganic chemistry, analytical chemistry, see 12th edition

VAN TASSEL, ROGER A, b Orange, NJ, Oct 19, 36. AERONOMY. Educ: Wesleyan Univ, AB, 58; Northeastern Univ, MS, 68, PhD, 72. Prof Exp: Chemist, Lunar-Planetary Lab, 61-67, RES CHEMIST, AERONOMY LAB, AIR FORCE CAMBRIDGE RES LABS, 67- Mem: AAAS; Sigma Xi; Am Geophys Union; Optical Soc Am. Res: Atomic spectra in the vacuum ultraviolet; oscillator strengths of atomic transitions; ultraviolet airglow originating in the upper atmosphere. Mailing Add: Air Force Cambridge Res Labs Aeronomy Lab L G Hanscom AFB MA 01731

VAN TASSELL, MORGAN HOWARD, b Johnson City, NY, Oct 31, 23; m 53; c 1. MICROBIOLOGY. Educ: Univ Wis, BS, 50. Prof Exp: Plant bacteriologist, Commercial Solvents Corp, Ill, 50-52; res microbiologist, Cent Res Dept, Anheuser-Busch, Inc, 53-57, supvr microbiologist, Fermentation Pilot Lab, 57-60; asst plant mgr, Sheffield Chem Div, Nat Dairy Prod Corp, 60-62, plant mgr, Kraftco Corp, 62-75; DIR WATER SERV, CITY OF ONEONTA, NY, 75- Mem: Am Soc Microbiol. Res: Industrial microbiology; applied and developmental research of fermentations; mutational development and selection of cultures; pilot scale equipment; automated continuous fermentations; enzymatic hydrolysis of proteins. Mailing Add: City of Oneonta Dept of Water 110 East St Oneonta NY 13820

VANTERPOOL, THOMAS CLIFFORD, b W Indies, Apr 22, 98; Can citizen; m 26; c 2. PHYTOPATHOLOGY. Educ: McGill Univ, BSA, 23, MSc, 25; Univ Sask, DSc, 68. Prof Exp: Asst bot, Macdonald Col, McGill Univ, 23-25, lectr, 26-28; asst prof, 28-31, prof, 31-65, EMER PROF PLANT PATH, UNIV SASK, 65- Mem: AAAS; Am Phytopath Soc; Mycol Soc Am; Can Phytopath Soc (pres, 44-45); fel Royal Soc Can. Res: Plant diseases; tomato streak; Tilletia caries; Pythium root rot of cereals and grasses; flax, oil-seed rape, longevity of some of its root-rotting organisms and cereal diseases. Mailing Add: 1850 Penshurst Rd Victoria BC Can

VAN THIEL, MATHIAS, chemistry, physics, see 12th edition

VAN'T HOF, JACK, b Grand Rapids, Mich, Apr 11, 32; m 52; c 2. CELL BIOLOGY, RADIOBIOLOGY. Educ: Calvin Col, AB, 57; Mich State Univ, PhD(bot), 61. Prof Exp: Biologist, Hanford Labs, Gen Elec Corp, 61-62; res assoc radiobiol & fel, Biol Dept, Brookhaven Nat Lab, 62-64, asst cytologist, 64-65; asst prof cytol, Dept Bot, Univ Minn, 65-66; CYTOLOGIST, BIOL DEPT, BROOKHAVEN NAT LAB, 66- Mem: Am Soc Cell Biol; Bot Soc Am; Am Soc Plant Physiol; Radiation Res Soc. Res: Study of cell population kinetics in complex plant tissue emphasizing physiological, cytochemical and radiological events that occur in and govern the duration of the mitotic cycle of proliferating cells. Mailing Add: Biol Dept Brookhaven Nat Lab Upton NY 11973

VANT-HULL, LORIN LEE, b Sioux Co, Iowa, June 26, 32; m 55; c 3. PHYSICS. Educ: Univ Minn, Minneapolis, BS, 54; Univ Calif, Los Angeles, MS, 55; Calif Inst Technol, PhD(physics), 67. Prof Exp: Res engr, Res Lab, Hughes Aircraft Co, 54-58; sr res scientist cryogenic devices, Sci Lab, Ford Motor Co, 66-69; ASSOC PROF PHYSICS, UNIV HOUSTON, 69- Concurrent Pos: Consult, Lawrence Berkeley Lab, Univ Calif, 70-71 & Manned Spacecraft Ctr, NASA, 71-72; prog mgr, Univ Houston Solar Energy Lab, 74- Mem: Am Inst Physics; Sigma Xi. Res: Superconductivity; quantum interference effects involving Josephson junctions and their use in instrumentation; solar energy; large scale efficient collection of solar energy; solar tower central receiver systems. Mailing Add: Dept of Physics Univ of Houston Houston TX 77004

VAN TIENHOVEN, ARI, b The Hague, Neth, Apr 22, 22; nat US; m 50; c 3. ANIMAL PHYSIOLOGY. Educ: Univ Ill, MS & PhD(animal sci), 53. Prof Exp: Asst prof poultry husb, Miss State Col, 53-55; from asst prof to assoc prof avian physiol, 55-69, PROF ANIMAL PHYSIOL, COL AGR & LIFE SCI, CORNELL UNIV, 69- Concurrent Pos: NATO fel, 61-62; assoc ed, Biol of Reproduction, 74- Honors & Awards: Am Soc Agr Engrs Paper Award, 71. Mem: AAAS; Am Soc Zoologists; Poultry Sci Asn; Am Asn Anat; Soc Study Reproduction. Res: Neuroendocrinology; reproductive physiology; temperature regulation of birds. Mailing Add: Dept of Poultry Sci Cornell Univ Ithaca NY 14853

VAN TILL, HOWARD JAY, b Ripon, Calif, Nov 28, 38; m 58; c 4. ASTRONOMY. Educ: Calvin Col, BS, 60; Mich State Univ, PhD(physics), 65. Prof Exp: Res scientist physics, Univ Calif, Riverside, 65-66; asst prof, Univ Redlands, 66-67; PROF PHYSICS, CALVIN COL, 67- Concurrent Pos: Res scientist, Dept Astron, Univ Tex, Austin, 74. Mem: Am Astron Soc; Am Phys Soc; Am Asn Physics Teachers. Res: Study of interstellar molecular clouds using millimeter-wave techniques. Mailing Add: Dept of Physics Calvin Col Grand Rapids MI 49506

VAN'T RIET, BARTHOLOMEUS, b Arnhem, Neth, June 25, 22; nat US; m 55; c 3. ANALYTICAL CHEMISTRY. Educ: Vrye Univ, Neth, BSc, 50; Univ Minn, PhD(anal chem), 57. Prof Exp: Asst anal chem, Vrye Univ, Neth, 48-51; from asst to instr, Univ Minn, 51-57; instr chem, Univ Va, 58-64; ASSOC PROF CHEM, MED COL VA, 64- Honors & Awards: Chem Pioneer Award, Am Inst Chemists, 73. Mem: Am Chem Soc. Res: Application of complexing agents in mammals; dispersion of calculi by surface reactions; drug analysis. Mailing Add: Dept of Pharmaceut Chem Med Col of Va Richmond VA 23298

VAN TUYL, ANDREW HEUER, b Fresno, Calif, July 6, 22; m 55; c 4. MATHEMATICS. Educ: Fresno State Col, AB, 43; Stanford Univ, MA, 46, PhD(math), 47. Prof Exp: Asst chem, Stanford Univ, 43-44, asst elec eng, 44-45, asst physics, 46, asst math, 46-47; MATHEMATICIAN, NAVAL SURFACE WEAPONS CTR, 47- Concurrent Pos: Res assoc, Ind Univ, 53. Mem: Fel AAAS; Am Inst Aeronaut & Astronaut; Am Math Soc; Soc Indust & Appl Math; NY Acad Sci. Res: Potential theory; special functions; hydrodynamics; gas dynamics. Mailing Add: Naval Surface Weapons Ctr White Oak Silver Spring MD 20910

VAN TUYL, HAROLD HUTCHISON, b Ft Worth, Tex, Oct 13, 27; m 52; c 4. RADIOCHEMISTRY. Educ: Agr & Mech Col, Tex, BS, 48. Prof Exp: Res chemist, Gen Elec Co, 48-65; res chemist, 65-70, MGR APPL CHEM SECT, PAC NORTHWEST LABS, BATTELLE MEM INST, 70- Mem: Am Nuclear Soc; Am Chem Soc. Res: Fission product recovery; dose rate and shielding calculations; fission product and transuranics generation calculations; transuranic element separations; nuclear chemistry. Mailing Add: Pac Northwest Labs Battelle Mem Inst Richland WA 99352

VAN TUYLE, ROBERT (WOODING), organic chemistry, see 12th edition

VAN TYLE, WILLIAM KENT, b Frankfort, Ind, Feb 10, 44. PHARMACOLOGY. Educ: Butler Univ, BS, 67; Ohio State Univ, MSc, 69, PhD(pharmacol), 72. Prof Exp: Asst prof, 72-75, ASSOC PROF PHARMACOL, BUTLER UNIV, 75- Concurrent Pos: Consult pharmacol, Vet Admin Hosp, Indianapolis, Ind, 73-; secy-treas, Dist 4, Am Asn Cols Pharm-Nat Asn Bds Pharm, 75- Mem: Am Pharmaceut Asn. Res: Bioavailability of drugs to central nervous system and drug effects on central neurotransmitters. Mailing Add: Butler Univ 4600 Sunset Ave Indianapolis IN 46208

VAN UMMERSEN, CLAIRE ANN, b Chelsea, Mass, July 28, 35; m 58; c 2. DEVELOPMENTAL BIOLOGY, ANIMAL PHYSIOLOGY. Educ: Tufts Univ, BS, 57, MS, 60, PhD(biol), 63. Prof Exp: Res asst radiobiol, Tufts Univ, 57-60, res assoc, 60-67, lectr biol, 67-68; asst prof, 68-74, ASSOC PROF BIOL, UNIV MASS, BOSTON, 74- Concurrent Pos: Fel, Tufts Univ, 63-67; mem teaching fac, Lancaster Courses in Ophthal, Colby Col, 62- Mem: AAAS; Am Soc Zoologists; Soc Develop Biol. Res: Biological effects of microwave radiation on the eye and the developing embryo. Mailing Add: Dept of Biol Univ of Mass 100 Arlington St Boston MA 02125

VAN VALEN, LEIGH, b Albany, NY, Aug 12, 35; m 59; c 2. EVOLUTIONARY BIOLOGY. Educ: Miami Univ, BA, 56; Columbia Univ, MA, 57, PhD(zool), 61. Prof Exp: Boese fel, Columbia Univ, 61-62; NATO fel, Univ Col, London, 62-63; res fel vert paleont, Am Mus Natural Hist, 63-66; asst prof anat, 67-68, from asst prof to assoc prof evolutionary biol, 68-73, ASSOC PROF BIOL, UNIV CHICAGO, 73- Mem: Soc Study Evolution (vpres, 73); Soc Vert Paleont; Ecol Soc Am; Philos Sci Asn; Am Soc Naturalists (treas, 69-72, vpres, 74-75). Res: Extinction; ecological control of large-scale evolutionary patterns; analytical paleoecology; energy and evolution; the phenotype; competition; natural selection of plants and animals; evolutionary theory; biological variation; evolution of development; basal radiation of placental mammals. Mailing Add: Dept of Biol Univ of Chicago 103 E57th St Chicago IL 60637

VAN VALEN, PHEBE, embryology, genetics, see 12th edition

VAN VALKENBURG, JEPTHA WADE, JR, b Ann Arbor, Mich, Mar 26, 25; m 49; c 3. COLLOID CHEMISTRY, SURFACE CHEMISTRY. Educ: Kalamazoo Col, BS, 49; Univ Wis, MS, 51; Univ Mich, PhD(phys chem), 55. Prof Exp: Chemist, Dow Chem Co, Mich, 54-58, proj leader pesticidal formulations res, 58-60, group leader, 60-66, mgr patent admin, 66-68; supvr phys chem res, 68-71, info scientist specialist, 71-73, MEM PATENT LIAISON STAFF, 3M CO, 73- Mem: Am Chem Soc. Res: Patents; pesticidal formulations; surfactants; emulsions; diffusion; kinetics; encapsulation and surface coatings; biological correlations. Mailing Add: 3M Co PO Box 33221 St Paul MN 55133

VAN VECHTEN, JAMES ALDEN, b Washington, DC, July 29, 42. THEORETICAL SOLID STATE PHYSICS, SEMICONDUCTORS. Educ: Univ Calif, Berkeley, AB, 65; Univ Chicago, PhD(physics), 69. Prof Exp: Infrared res officer semiconductors, US Naval Res Lab, Wash, 69-71; mem tech staff electro optic res, Bell Tel Lab, Murray Hill, 71-74; RES STAFF MEM SEMICONDUCTOR PHYSICS, THOMAS J WATSON RES CTR, IBM, 74- Mem: Fel Am Phys Soc; Electrochem Soc. Res: Theoretical study of covalently bonded solids, their electronic, optical, mechanical and thermochemical properties; co-developer of the dielectric scale of electronegativity. Mailing Add: Thomas J Watson Res Ctr IBM PO Box 218 Yorktown Heights NY 10598

VAN VELD, ROBERT DALE, b Killduff, Iowa, Jan 31, 24; m 46; c 3. TEXTILE PHYSICS. Educ: Purdue Univ, BS, 49, MS, 51, PhD(physics), 55. Prof Exp: Asst, Purdue Univ, 55; res engr, 56-60, res proj engr, Eng Res Lab, Exp Sta, 60-61, sr res physicist, Kinston Plant, 61-65, res assoc, 65-66, tech supvr, 66-68, SUPVR RES &

DEVELOP, E I DU PONT DE NEMOURS & CO, INC, 68- Mem: Fiber Soc. Res: Microscopy. Mailing Add: E I du Pont de Nemours & Co Inc Kinston NC 28501

VAN VELDHUIZEN, PHILIP ANDROCLES, b Hospers, Iowa, Nov 6, 30; m 52; c 3. MATHEMATICS, STATISTICS. Educ: Cent Col, BA, 52; Univ Iowa, MS, 60. Prof Exp: Teacher jr high sch, 54; instr, Exten Ctr, Univ Ga, 54-56; instr math, Cent Col, 56-59; asst prof, Sacramento State Col, 60-63; assoc prof, 63-74, PROF MATH, UNIV ALASKA, FAIRBANKS, 74- Concurrent Pos: Spec lectr & resource personnel, Mod Math Prog, Fairbanks, Anchorage & Kodiak, Alaska, 64-67; mem adv bd every pupil eval prog, Northwest Regional Lab, State Dept of Educ, 75- Mem: Math Asn Am. Res: Algebraic structures and their relation to the mathematics program in the elementary and secondary schools; hypothesis testing. Mailing Add: Dept of Math Univ of Alaska Fairbanks AK 99701

VAN VERTH, JAMES EDWARD, b Huntington, WVa, Jan 26, 28; m 65; c 2. ORGANIC CHEMISTRY. Educ: Xavier Univ, Ohio, BS, 50; Univ Detroit, MS, 52; Ind Univ, PhD(org chem), 57. Prof Exp: Sr res chemist, Monsanto Chem Co, 56-61; asst, Yale Univ, 61-63; asst prof, 63-69, ASSOC PROF CHEM, CANISIUS COL, 69- Mem: Am Chem Soc. Res: Organic synthesis and mechanisms; anionic rearrangements; reactions of singlet oxygen. Mailing Add: Dept of Chem Canisius Col Buffalo NY 14208

VAN VLECK, DAVID B, vertebrate ecology, anatomy, see 12th edition

VAN VLECK, FRED SCOTT, b Clearwater, Nebr, Dec 12, 34; m 60; c 5. MATHEMATICS. Educ: Univ Nebr, BSc, 56, MA, 57; Univ Minn, PhD(math), 60. Prof Exp: Instr math, Mass Inst Technol, 60-62; from asst prof to assoc prof, 62-68, PROF MATH, UNIV KANS, 68- Concurrent Pos: Vis prof, Univ Colo, 71-72. Mem: Math Asn Am; Am Math Soc. Res: Optimal control theory; measurable multiplevalued functions; ordinary differential equations. Mailing Add: Dept of Math Univ of Kans Lawrence KS 66044

VAN VLECK, JOHN HASBROUCK, b Middletown, Conn, Mar 13, 99; m 27. PHYSICS. Educ: Univ Wis, AB, 20; Harvard Univ, AM, 21, PhD, 22. Hon Degrees: ScD, Numerous from US & foreign univs, 36-71. Prof Exp: Instr physics, Harvard Univ, 22-23; from asst prof to prof, Univ Minn, 23-28; prof, Univ Wis, 28-34; prof physics, 34-51, Hollis prof math & natural philos, 51-69, head theory group, Radio Res Lab, 43-45, chmn dept physics, 45-49, dean eng & appl physics, 51-57, EMER PROF MATH & NATURAL PHILOS, HARVARD UNIV, 69- Concurrent Pos: Guggenheim fel, 30; vis prof, Stanford Univ, 27, 34 & 41, Univ Mich, 33, Columbia Univ, 34 & Princeton Univ, 37; Lorentz guest prof, State Univ Leiden, 60; Eastman prof, Oxford Univ, 61-62. Honors & Awards: Albert A Michelson Prize, Case Inst Technol, 63; Irving Langmuir Prize, Am Phys Soc, 65; Nat Medal of Sci, 66; Cresson Medal, Franklin Inst, 71; Lorentz Medal, Royal Neth Acad Sci, 74. Mem: Nat Acad Sci; AAAS (vpres, 60); Am Phys Soc (pres, 52); Am Math Soc; Am Acad Arts & Sci (vpres, 56). Res: Quantum theory of atomic structure and magnetism. Mailing Add: Lyman Lab of Physics Harvard Univ Cambridge MA 02138

VAN VLECK, LLOYD DALE, b Clearwater, Nebr, June 11, 33; m 58; c 2. GENETICS, ANIMAL SCIENCE. Educ: Univ Nebr, BS, 54, MS, 55; Cornell Univ, PhD(animal breeding), 60. Prof Exp: Res assoc animal breeding, 59-60, res geneticist, 60-62, from asst prof to assoc prof animal genetics, 62-73, PROF ANIMAL GENETICS, CORNELL UNIV, 73- Concurrent Pos: Vis prof, Univ Nebr, Lincoln, 73. Honors & Awards: Am Soc Animal Sci Award, 72; Nat Asn Animal Breeders Award, Am Dairy Sci Asn, 74. Mem: Biomet Soc; Am Dairy Sci Asn; Am Soc Animal Sci. Res: Methods of improving genetic value of large animals using genetic theory, statistical technique for unbalanced data, and computer processing of large numbers of records. Mailing Add: Dept of Animal Sci Cornell Univ Ithaca NY 14850

VAN VLEET, JOHN F, b Lodi, NY, Mar 23, 38; m 61; c 2. VETERINARY PATHOLOGY. Educ: Cornell Univ, DVM, 62; Univ Ill, MS, 65, PhD(vitamin E deficiency), 67. Prof Exp: Asst vet, 62-63; USPHS trainee vet path, Univ Ill, 63-66, instr, 66-67; asst prof, 67-70, ASSOC PROF VET PATH, PURDUE UNIV, WEST LAFAYETTE, 70- Mem: Vet Med Asn; Int Acad Path; Am Col Vet Path. Res: Ultrastructural and nutritional pathology. Mailing Add: Dept of Vet Microbiol & Path Sch of Vet Sci & Med Purdue Univ West Lafayette IN 47907

VAN VLIET, ANTONE CORNELIS, b San Francisco, Calif, Jan 11, 30; m 53; c 4. WOOD SCIENCE, COMMUNICATIONS. Educ: Ore State Univ, BS, 53, MS, 58; Mich State Univ, PhD, 70. Prof Exp: Instr forest prod, Ore State Univ, 55-59; asst to plant mgr plywood prod, Bohemia Lumber Co, 59-60; asst prof forest prod, 63-65, asst prof wood prod, Exten, 63-71, ASSOC PROF FOREST PROD, ORE STATE UNIV, 65-, ASSOC DIR, OFF CAREERS, PLANNING & PLACEMENT, 71- Mem: Forest Prod Res Soc. Res: Plywood production; wood anatomy and utilization; company educational programs; behavioral aspect of communications; management science. Mailing Add: Forest Prods Dept Ore State Univ Corvallis OR 97331

VAN VLIET, CAREL (KAREL) M, b Dordrecht, Neth, Dec 27, 29; US citizen; m 53; c 4. PHYSICS, ELECTRICAL ENGINEERING. Educ: Free Univ, Amsterdam, BS, 49, MA, 53, PhD(physics), 56. Prof Exp: Fel elec eng, Univ Minn, Minneapolis, 56-57, asst prof, 57-58; asst dir physics lab, Free Univ, Amsterdam, 58-60; from assoc prof to prof elec eng, Univ Minn, Minneapolis, 60-66, prof elec eng & physics, 66-69; PROF ELEC ENG & PHYSICS, MATH RES CTR, UNIV MONTREAL, 69- Concurrent Pos: Vis prof, Univ Fla, 74. Mem: Am Phys Soc; Am Sci Affil; Europ Phys Soc; Neth Phys Soc; sr mem Inst Elec & Electronics Engrs. Res: Solid state physics; statistical mechanics; solid state electronics. Mailing Add: 30 Normandy Dr Montreal PQ Can

VAN VOORHIS, JOHN JAY, physical chemistry, see 12th edition

VAN VOROUS, TED, b Billings, Mont, Jan 6, 29; m 51; c 5. ANALYTICAL CHEMISTRY. Educ: Mont State Col, BS, 53, MS, 54. Prof Exp: Anal chemist, Dow Chem Co, 54-56, scientist-chemist, 56-59, sr chemist, 59-60, group supvr, 60-62, res group mgr, 62-69; PRES, VTA Inc, 69- Mem: Am Vacuum Soc; Geochem Soc; Sigma Xi. Res: High vacuum research; evaporation processes; ionphenomena; electron microsopy-metallurgy; electron microprobe analysis; instrumentation development; x-ray and emission spectroscopy; high temperature materials; epitaxial structures. Mailing Add: VTA Inc 2125 Pearl St Boulder CO 80302

VAN VUNAKIS, HELEN, b New York, NY, June 15, 24; m 58; c 2. BIOCHEMISTRY. Educ: Hunter Col, BA, 46; Columbia Univ, PhD(biochem), 51. Prof Exp: USPHS fel & res assoc, Johns Hopkins Univ, 51-54; sr res scientist, State Dept Health, NY, 54-58; from asst prof to assoc prof, 58-74, PROF BIOCHEM, BRANDEIS UNIV, 74- Concurrent Pos: NIH career res award. Res: Structure of proteins and nucleic acids; interaction of pharmacologically active compounds with specific antibodies and cellular receptor sites. Mailing Add: Dept of Biochem Brandeis Univ Waltham MA 02154

VAN WAGNER, CHARLES EDWARD, b Montreal, Que, Dec 9, 24; m 55; c 3. FORESTRY, CHEMICAL ENGINEERING. Educ: McGill Univ, BEng, 46; Univ Toronto, BScF, 61. Prof Exp: Chief chemist, Can Pittsburg Industs, 46-58; RES SCIENTIST, CAN DEPT ENVIRON, 60- Mem: Can Inst Forestry. Res: Forest fire; measurement and theory of fire behavior; variation in moisture content of forest fuel with weather; use of prescribed fire in forest management; ecological effects of forest fire. Mailing Add: Petawawa Forest Exp Sta Chalk River ON Can

VAN WAGTENDONK, JAN WILLEM, b Palo Alto, Calif, Feb 21, 40; m 68; c 2. FOREST ECOLOGY. Educ: Ore State Univ, BS, 63; Univ Calif, Berkeley, MS, 68, PhD(wildland res sci), 72. Prof Exp: RES SCIENTIST FIRE ECOL, YOSEMITE NAT PARK, NAT PARK SERV, 72- Mem: Soc Am Foresters; Ecol Soc Am. Res: Ecological role of fire in the Sierra Nevada ecosystems; recreational carrying capacities for wilderness areas. Mailing Add: Nat Park Serv PO Box 577 Yosemite Nat Park CA 95389

VAN WAGTENDONK, WILLEM JOHAN, b Jakarta, Indonesia, Apr 10, 10; nat US; m 37; c 3. BIOCHEMISTRY. Educ: State Univ Utrecht, AB, 31, MA, 34, PhD(biochem), 37. Prof Exp: Instr org chem, State Univ Utrecht, 35-37; res chemist, N V Polaks, Frutal Works, Neth, 37-39; res assoc biol, Stanford Univ, 39-41; from asst prof to assoc prof biochem, Ore State Col, 41-46; assoc prof zool, Ind Univ, 46-60; prof biochem, Sch Med, Univ Miami, 60-71; PROF BIOCHEM, TALLADEGA COL, 71- Concurrent Pos: NSF & NIH grants, 48-; scientist, Vet Admin Hosp, Coral Gables, Fla, 60-62, chief basic res, 62-71; liaison off, Study Sect, NIH, 64-67, mem, 67-71; chmn basic sci prog comt, Vet Admin, 66-71; Nat Acad Sci-Polish Acad Sci exchange prof, M Nencki Inst Exp Biol, Warsaw, Poland, 72-73. Mem: Fel AAAS; Am Soc Biol Chem; Soc Protozool; fel NY Acad Sci. Res: Biochemistry and nutrition of Paramecium aurelia and its endosymbiotes. Mailing Add: PO Box 206 Cedar Mountain NC 28718

VAN WAZER, JOHN ROBERT, b Chicago, Ill, Apr 11, 18; m 40; c 1. CHEMISTRY. Educ: Northwestern Univ, BS, 40; Harvard Univ, AM, 41, PhD(phys chem), 42. Prof Exp: Phys chemist, Eastman Kodak Co, 42-44; res group leader, Clinton Eng Works, Tenn, 44-46; phys chemist, Rumford Chem Works, RI, 46-49; head physics res, Great Lakes Carbon Corp, 49-50; sr scientist, Monsanto Co, 50-68; PROF CHEM, VANDERBILT UNIV, 68- Concurrent Pos: Asst res dir, Monsanto Co, 51-60, dir chem dynamics res, 64-67. Mem: AAAS; Am Chem Soc; Soc Rheol; NY Acad Sci; Ger Chem Soc. Res: Chemistry of phosphorus compounds; substituent-exchange or redistribution reactions; inorganic chemistry; rheology; applied nuclear-magnetic resonance; photoelectron spectroscopy; applied quantum mechanics; nuclear-medical chemistry; nutrition. Mailing Add: Dept of Chem Vanderbilt Univ Box 1521 Sta B Nashville TN 37235

VAN WEEL, PIETER BOUDEWIJN, b Amboina, Indonesia, Mar 20, 10; nat US; m 41; c 1. PHYSIOLOGY. Educ: State Univ Utrecht, Drs, 35, PhD, 37. Prof Exp: Asst zool, State Univ Utrecht, 32-35; asst microbiol, Univ Amsterdam, 38; chief asst, Dept Biochem & Histol, Med High Sch, Batavia, 39-48; assoc prof zool & histol, Med Fac, Univ Indonesia, 48-50; prof, 50-75, EMER PROF ZOOL, UNIV HAWAII, 75- Concurrent Pos: Donders Found scholar, Cambridge & Dutch Govt scholar, Naples, 39. Mem: AAAS; Am Soc Zool; Soc Gen Physiol; Neth Royal Zool Soc. Res: Comparative physiology; digestion; vital staining; histophysiology of pancreas, duodenum, digestive glands and liver; electrophysiology sense organs. Mailing Add: 1350 Ala Moana Honolulu HI 96814

VAN WIJNGAARDEN, ARIE, b Holland, Apr 8, 33; Can citizen; m 57; c 3. PHYSICS. Educ: McMaster Univ, PhD(physics), 62. Prof Exp: Teacher physics, 62-70, assoc prof, 70-73, PROF PHYSICS, UNIV WINDSOR, 73- Concurrent Pos: Nat Res Coun-Ont Res Found res grants, 62- Res: Study of fast atomic particles through matter. Mailing Add: Dept of Physics Univ of Windsor Windsor ON Can

VAN WILLIGEN, JOHN GILBERT, b Milwaukee, Wis, Dec 4, 39; m 64; c 2. APPLIED ANTHROPOLOGY, ETHNOGRAPHY. Educ: Univ Wis, BS, 64; Univ Ariz, MA, 68, PhD(anthrop, oriental studies), 71. Prof Exp: Dir community develop, Papago Tribe, Ariz, 68-70; asst prof anthrop, Univ Wis-Parkside, 70-74; ASSOC PROF ANTHROP & DIR GRAD STUDIES, UNIV OF KY, 74- Concurrent Pos: NSF fel & univ fel undergrad curric develop, Univ Wis-Parkside, 72-73. Mem: Fel AAAS; fel Soc Appl Anthrop; fel Am Anthrop Asn. Res: Dynamics of developing communities; south Asian urban research; Japanese industrial systems in the United States. Mailing Add: Dept of Anthrop Univ of Ky Lexington KY 40506

VAN WINKLE, QUENTIN, b Grand Forks, NDak, Mar 10, 19; m 41; c 4. CHEMISTRY. Educ: SDak Sch Mines & Technol, BS, 40; Ohio State Univ, PhD(chem), 47. Prof Exp: Asst chem, Ohio State Univ, 40-43, res assoc eng, Exp Sta, 44; asst chemist metall lab, Univ Chicago, 44-46; res assoc chem, 46-48, from asst prof to assoc prof, 48-59, PROF CHEM, OHIO STATE UNIV, 59- Concurrent Pos: Consult, E I du Pont de Nemours & Co, 57- Mem: AAAS; Am Chem Soc. Res: Physico-chemical properties of proteins, high polymers; nucleic acids; surface chemistry. Mailing Add: 280 Dixon Ct Columbus OH 43214

VAN WINKLE, WALTON, JR, b Seattle, Wash, July 16, 10; m 38; c 1. BIOCHEMISTRY. Educ: Stanford Univ, AB, 33, MD, 38. Prof Exp: From asst to instr pharmacol, Sch Med, Stanford Univ, 36-41; from sr med officer to prin med officer, Food & Drug Admin, Fed Security Agency, 42-46; secy comt res, AMA, 46-51; vpres res, Ethicon, Inc, NJ, 51-65, vpres med affairs, 65-70; PROF SURG BIOL, SCH MED, UNIV ARIZ, 70- Concurrent Pos: Prof lectr, Univ Ill, 46-51; from assoc pharmacologist to pharmacologist, Food & Drug Admin, Fed Security Agency, 51-52. Mem: Soc Exp Biol & Med; Am Soc Pharmacol & Exp Therapeut; Endocrine Soc; Am Thoracic Soc; fel NY Acad Sci. Res: Experimental syphilis, bismuth compounds; pharmacology of glycols; steroids and cancer; biochemistry of collagen; irradiation sterilization; research administration; wound healing. Mailing Add: Div of Surg Biol Univ of Ariz Med Ctr Tucson AZ 85724

VAN WINKLE, WEBSTER, JR, b Plainfield, NJ, Nov 18, 38; m 61; c 3. ENVIRONMENTAL SCIENCES. Educ: Oberlin Col, BA, 61; Rutgers Univ, New Brunswick, PhD(zool), 67. Prof Exp: Res assoc, Shellfish Res Lab, Rutgers Univ, 66-67; asst prof biol, Col William Mary, 67-70; USPHS fels, NC State Univ, 70 & 72; res assoc, 72-75, RES STAFF MEM, ENVIRON SCI DIV, OAK RIDGE NAT LAB, 75- Concurrent Pos: NSF fel, Marine Lab, Duke Univ, 69; NSF sci fac fel, NC State Univ, 71-72. Mem: AAAS; Ecol Soc Am; Am Fisheries Soc; Atlantic Estuarine Res Soc. Res: Assessment of environmental impacts on aquatic ecosystems; fish population modeling; spectral analysis of environmental time series; data analysis. Mailing Add: Environ Sci Div Oak Ridge Nat Lab Oak Ridge TN 37830

VAN WINTER, CLASINE, b Amsterdam, Netherlands, Apr 8, 29. MATHEMATICAL PHYSICS. Educ: State Univ Groningen, BSc, 50, MSc, 54, PhD(physics), 57. Prof Exp: Res asst physics, State Univ Groningen, 51-58, sci officer, 58-68; PROF MATH & PHYSICS, UNIV KY, 68- Concurrent Pos: Fel physics, Univ Birmingham, 57 & Niels Bohr Inst Theoret Physics, Univ Copenhagen, 63; vis assoc prof physics, Ind

Univ, Bloomington, 67-68. Mem: Am Math Soc; Am Phys Soc. Res: Three-and more-body problem in quantum mechanics; quantum scattering theory; functional analysis; complex variables. Mailing Add: Dept of Math Univ of Ky Lexington KY 40506

VAN WOERT, MELVIN H, b Brooklyn, NY, Nov 3, 29; m 55. INTERNAL MEDICINE. Educ: Columbia Univ, BA, 51; NY Med Col, MD, 56. Prof Exp: From intern to resident internal med, Univ Chicago, 56-60, res asst gastroenterol, 62-63; from asst scientist to assoc scientist, Brookhaven Nat Lab, 63-67; from asst prof to assoc prof med & pharmacol, Sch Med, Yale Univ, 67-74; PROF PHARMACOL & NEUROL & HEAD SECT CLIN PHARMACOL, MT SINAI SCH MED, 74- Mem: AAAS; fel Am Col Physicians; Soc Exp Biol & Med; Am Fedn Clin Res; Am Soc Pharmacol & Exp Therapeut. Res: Catecholamine and melanin metabolism, particularly in relationship to Parkinson's disease; serotonin metabolism and myoclonus. Mailing Add: Dept of Neurol Mt Sinai Sch of Med New York NY 10029

VAN WORMER, MARVIN CLINTON, animal science, biochemistry, see 12th edition

VAN WYK, JUDSON JOHN, b Maurice, Iowa, June 1, 21; m 44; c 4. PEDIATRICS, ENDOCRINOLOGY. Educ: Hope Col, AB, 43; Johns Hopkins Univ, MD, 48; Am Bd Pediat, dipl. Prof Exp: Intern & asst resident pediat, Johns Hopkins Hosp, 48-50; resident, Cincinnati Children's Hosp, Ohio, 50-51; investr metab, Nat Heart Inst, 51-53; fel pediat endocrinol, Johns Hopkins Univ, 53-55; from asst prof to assoc prof pediat, 55-62, prof, 62-75, KENAN PROF PEDIAT, SCH MED, UNIV NC, CHAPEL HILL, 75- Concurrent Pos: Attend physician, NC Mem Hosp, 55-; Markle scholar med sci, 56-61; USPHS res career award, 62-; mem training grants comt in diabetes & metab, NIH, 67-71 & endocrine study sect, 71-75; vis scientist, Karolinska Inst, Sweden, 68-69; consult, Womack Army Hosp, Ft Bragg. Mem: Endocrine Soc; Soc Pediat Res; fel Am Acad Pediat; Am Pediat Soc; Lawson Wilkins Pediat Endocrine Soc (pres, 76-77). Res: Human sex differentiation; pituitary function and hormonal control of growth and sexual maturation; isolation and physiologic role of somatomedin. Mailing Add: Dept of Pediat Univ of NC Sch of Med Chapel Hill NC 27514

VAN WYK, RODNEY, b Platte, SDak, Jan 20, 38; m 59; c 3. MATHEMATICS. Educ: Univ SDak, BA, 60; Iowa State Univ, MS, 61. Prof Exp: Scientist, Lockheed Missiles Space & Co, 61-63; sr physicist, Rocketdyne Div, NAm Aviation, Inc, 63-67, mgr math statist, 67-70; MGR SYSTS ANAL, WINCHESTER GROUP RES, OLIN CORP, NEW HAVEN, 70- Concurrent Pos: Consult, 65-66. Mem: Soc Indust & Appl Math. Res: Numerical analysis; differential equations. Mailing Add: 250 Stonehedge Lane Guilford CT 06437

VAN ZANDT, GERTRUDE, b Ft Worth, Tex, Aug 23, 10. CHEMISTRY. Educ: Tex Christian Univ, BS, 31; Tulane Univ, MS, 35; Univ Tex, PhD(org chem), 50. Prof Exp: Teacher high sch, Tex, 32-34 & 35-37; res chemist, Eastman Kodak Co, 37-44; instr chem, Univ of the South, 44-45, assoc prof, 47-56; assoc prof, Middle Tenn State Col, 57-61; assoc prof, 61-74, PROF CHEM, TEX WESLEYAN COL, 74-, HEAD DEPT, 61- Mem: AAAS; Am Chem Soc. Res: Azo dyes; photographic sensitizers; oxidation of cholesterol; application of chemical theories in college teaching. Mailing Add: Dept of Chem Tex Wesleyan Col Ft Worth TX 76105

VAN ZANDT, LONNIE L, b Bound Brook, NJ, Sept 29, 37; m 61; c 3. SOLID STATE PHYSICS. Educ: Lafayette Col, BS, 58; Harvard Univ, AM, 59, PhD(physics), 64. Prof Exp: Mem staff, Sci Lab, Ford Motor Co, 62-64 & Lincoln Lab, Mass Inst Technol, 64-67; asst prof, 67-70, ASSOC PROF PHYSICS, PURDUE UNIV, WEST LAFAYETTE, 70- Res: Degenerate electron gases; transition metal oxides. Mailing Add: Dept of Physics Purdue Univ West Lafayette IN 47907

VAN ZANDT, PAUL DOYLE, b Vandalia, Ill, Dec 29, 27; m 54; c 1. PARASITOLOGY. Educ: Greenville Col, AB, 52; Univ Ill, MS, 53; Univ NC, MSPH, 55, PhD(parasitol), 60. Prof Exp: Asst pub health, Univ NC, 58-61; from asst prof to assoc prof biol, 61-69, PROF BIOL, YOUNGSTOWN STATE UNIV, 69- Mem: Fel AAAS; Am Soc Parasitol; Am Soc Trop Med & Hyg; Royal Soc Trop Med & Hyg. Res: Immunology of animal parasites; medical parasitology and microbiology. Mailing Add: Dept of Biol Youngstown State Univ Youngstown OH 44503

VAN ZANDT, THOMAS EDWARD, b Highland Park, Mich, July 10, 29; m 61; c 2. AERONOMY, ATMOSPHERIC PHYSICS. Educ: Duke Univ, BS, 50; Yale Univ, PhD, 55. Prof Exp: Physicist, Sandia Corp, 54-57; PHYSICIST, AERONOMY LAB, NAT OCEANIC & ATMOSPHERIC ADMIN, 57- Concurrent Pos: Vis lectr, Univ Colo, 61-70, adj prof, 70- Mem: Am Geophys Union; Int Union Radio Sci. Mailing Add: Aeronomy Lab Nat Oceanic & Atmospheric Admin Boulder CO 80302

VAN ZEGGEREN, FREDERIK, b Amsterdam, Neth, June 7, 33; m 59; c 2. PHYSICAL CHEMISTRY. Educ: Univ Amsterdam, BSc, 52, PhD, 55, DSc, 56. Prof Exp: Res chemist, Appl Sci Res, Inc, Neth, 55; res chemist, 57-64, res leader, 64-69, res mgr, 69-73, physics group mgr, 73-75, MGR, EXPLOSIVES RES LAB, CAN INDUSTS, LTD, 75- Mem: Can Inst Mining & Metall; fel Chem Inst Can. Res: Thermodynamics, explosives and propellants technology; numerical analysis; blasting physics; technology of explosives, explosives accessories and propellants. Mailing Add: Explosives Res Lab Can Industs Ltd McMasterville PQ Can

VAN ZWALENBERG, GEORGE, b Neth, Sept 7, 30; US citizen; m 53; c 3. MATHEMATICS. Educ: Calvin Col, BS, 53; Univ Fla, MA, 55; Univ Calif, Berkeley, PhD(math), 68. Prof Exp: Instr math, Bowling Green State Univ, 59-60; vis lectr, Calvin Col, 60-61, asst prof, 61-63; asst prof, Calif State Univ, Fresno, 63-67; PROF MATH, CALVIN COL, 68-, CHMN DEPT, 74- Concurrent Pos: Math Asn Am vis lectr, High Schs, 65-66. Mem: Am Math Soc; Math Asn Am. Res: Complex variables. Mailing Add: Dept of Math Calvin Col Grand Rapids MI 49506

VAN ZWIETEN, MATTHEW JACOBUS, b Zeist, Netherlands, Apr 6, 45; US citizen; m 66; c 2. VETERINARY PATHOLOGY. Educ: Univ Calif, Davis, BS, 67, DVM, 69; Am Col Vet Pathologists, dipl, 74. Prof Exp: Sci investr cell biol & path, Med Res Inst Infectious Dis, US Army, 69-71; res fel, 71-75, ASSOC PATH, NEW ENG REGIONAL PRIMATE RES CTR & ANIMAL RES CTR, HARVARD MED SCH, 75- Concurrent Pos: Res assoc path, Angell Mem Animal Hosp, 71-74, consult, 75-; res assoc path, Children's Hosp Med Ctr, 73-75; head diag procedures, New Eng Regional Primate Res Ctr & Animal Res Ctr, Harvard Med Sch, 75- Mem: Am Col Vet Pathologists; Int Acad Path; Am Asn Lab Animal Sci; Am Vet Med Asn. Res: Studies on the pathogenesis of several forms of immunologically mediated renal disease in laboratory animals; identification and development of spontaneous animal diseases as models for their human counterparts. Mailing Add: Animal Res Ctr 25 Shattuck St Harvard Med Sch Boston MA 02115

VAN ZYTVELD, JOHN BOS, b Hammond, Ind, Nov 12, 40; m 61; c 3. SOLID STATE PHYSICS. Educ: Calvin Col, AB, 62; Mich State Univ, MS, 64, PhD(physics), 67. Prof Exp: Fel physics, Univ Sheffield, 67-68; asst prof, 68-72, ASSOC PROF PHYSICS, CALVIN COL, 72- Concurrent Pos: Res physicist, Battelle Mem Inst, Ohio, 69; sr fel, Dept Physics, Univ Leicester, Eng, 74-75. Mem: AAAS; Am Phys Soc; Am Asn Physics Teachers; Am Sci Affiliation. Res: Electron transport properties of solid and liquid metals, alloys and semiconductors. Mailing Add: Dept of Physics Calvin Col Grand Rapids MI 49506

VARADARAJAN, KALATHOOR, b Bezwada, India, Apr 13, 35; m 61; c 2. MATHEMATICS, TOPOLOGY. Educ: Loyola Col, Madras, India, BA, 55; Columbia Univ, PhD(topology), 60. Prof Exp: Res fel math, Tata Inst Fundamental Res, India, 60-61, fel, 61-67; vis assoc prof, Univ Ill, Urbana, 67-69; reader, Tata Inst Fundamental Res, India, 69-71; vis prof, Ramanujan Inst, Madras, 71; assoc prof, 71-73, PROF MATH, UNIV CALGARY, 73- Mem: Am Math Soc; Can Math Cong. Res: Algebraic and differential topology; homological algebra. Mailing Add: Dept Math Statist & Comput Sci Univ of Calgary 2920 24th Ave NW Calgary AB Can

VARADI, PETER FERENCZ, b Szeged, Hungary, July 7, 26; US citizen; m 51; c 1. PHYSICAL CHEMISTRY. Educ: Univ SZeged. MS & PhD(Phys chem), 49. Prof Exp: Group leader, Res Inst Telecommun, Budapest, Hungary, 49-56; staff mem, Tube Develop Lab, Telefunken GmbH, Ulm-Donan, WGer, 56-58; engr, Machlett Labs Inc, 58-61, sr scientist, Machlett Labs Inc Div, Raytheon Co, Conn, 61-67, sect head advan develop, 67-68; mgr mat technol br, Comsat Labs, 68-72, mgr mat sci dept, 72-74; WITH SOLAREX CORP, 74- Mem: Fel Am Inst Chem; sr mem Am Vacuum Soc; Soc Appl Spectros; Am Soc Test & Mat; Fr Soc Vacuum Eng & Tech. Res: Materials technology; x-ray, electron and mass spectroscopy; analytical chemistry; space materials; chemical fabrication of electronic circuits. Mailing Add: Solarex Corp 1335 Piccard Dr Rockville MD 20850

VARADY, JOHN CARL, b Niagara Falls, NY, Feb 26, 35; m 66. BIOSTATISTICS. Educ: Calif Inst Technol, BS, 56; Univ Wash, MA, 58; Univ Calif, Los Angeles, PhD(biostatist), 65. Prof Exp: Opers res analyst, Radioplane Div, Northrop Corp, 57-58; sr mathematician, Systs Develop Corp, 58-65; chief biostatist, Calif Dept Ment Hyg, 65-66; dir comput servs, Univ Cincinnati, 66-70; SR BIOSTATISTICIAN, SYNTEX RES, 70- Mem: Inst Math Statist; Am Statist Asn; Asn Comput Mach. Mailing Add: Dept of Biostatist Syntex Res 3401 Hiliview Palo Alto CA 94304

VARANASI, PRASAD, b Vijayavada, India, Dec 20, 38; m 72. PLANETARY ATMOSPHERES, SPECTROSCOPY. Educ: Andhra Univ, India, BSc, 57; Indian Inst Sci, Bangalore, MSc, 61; Mass Inst Technol, SM, 62; Univ Calif, San Diego, PhD(eng physics), 67. Prof Exp: Asst prof, 67-71, ASSOC PROF ENG PHYSICS, STATE UNIV NY STONY BROOK, 71- Concurrent Pos: NASA grant; assoc ed, J Quant Spectros & Radiative Transfer, 73-78. Mem: AAAS; Am Phys Soc. Res: Infrared spectroscopy as applied to planetary atmospheres; experimental work on collision broadening of spectral lines and molecular structure. Mailing Add: Lab Planetary Atmospheric Res State Univ of NY Stony Brook NY 11794

VARANASI, USHA SURYAM, b Bassien, Burma. BIOCHEMISTRY, CHEMISTRY. Educ: Univ Bombay, BSc, 61; Calif Inst Technol, MS, 64; Univ Wash, PhD(chem), 68. Prof Exp: Res assoc lipid biochem, Oceanic Inst, Oahu, Hawaii, 69-71; assoc res prof, 71-75, RES PROF LIPID BIOCHEM, SEATTLE UNIV, 75-; RES CHEMIST, NORTHWEST FISHERIES CTR, NAT MARINE FISHERIES SERV, NAT OCEANIC & ATMOSPHERIC ADMIN, 75- Concurrent Pos: Vis scientist, Pioneer Res Unit, Northwest Fisheries Ctr, Nat Marine Fisheries Serv, Nat Oceanic & Atmospheric Admin, Wash, 69-72. Mem: AAAS. Res: Lipid structure and metabolism; chemistry of bioacoustics; reaction kinetics in binary solvent systems; adaptive mechanisms in marine organisms; environmental conservation research in biochemistry. Mailing Add: Dept of Chem Seattle Univ 12th & EColumbia Seattle WA 98122

VARANDANI, PARTAB T, b Karachi, India, Sept 5, 29; US citizen; m 62; c 3. BIOCHEMISTRY, ENDOCRINOLOGY. Educ: Agra Univ, BSc, 50, MSc, 52; Univ Ill, PhD(biochem, animal nutrit), 59. Prof Exp: Res asst biochem, Indian Vet Res Inst, 52-56; res assoc, Radiocarbon Lab, Univ Ill, 59-61 & Roswell Park Mem Inst, 61-63; sr investr, 63-72, SR SCIENTIST & CHIEF, SECT ENDOCRINOL, FELS RES INST, 72- Mem: Am Chem Soc; Am Soc Biol Chemists; Endocrine Soc; Am Diabetes Asn; Am Asn Immunol. Res: Carbohydrate and fat metabolism; metabolism and function of vitamin A; immunochemistry, structure, metabolism and regulation of enzymes and hormones; pepsin, glutathione-insulin-transhydrogenase, insulin, growth hormone and glucagon. Mailing Add: Sect of Endocrinol Fels Res Inst Yellow Springs OH 45387

VARBERG, DALE ELTHON, b Forest City, Iowa, Sept 9, 30; m 55; c 3. MATHEMATICS. Educ: Univ Minn, BA, 54, MA, 57, PhD(math), 59. Prof Exp: From asst prof to assoc prof, 58-65, PROF MATH, HAMLINE UNIV, 65- Concurrent Pos: NSF fel, Inst Advan Study, 64-65; sci fac fel, Univ Wash, 71-72. Mem: Am Math Soc; Math Asn Am. Res: Stochastic and Gaussian processes; measure theory; convexity theory. Mailing Add: Dept of Math Hamline Univ St Paul MN 55104

VARCO, RICHARD LYNN, b Fairview, Mont, Aug 14, 12; m 40; c 7. SURGERY. Educ: Univ Minn, MB, 36, MD, 37, PhD, 44; Am Bd Surg, dipl. Prof Exp: Instr physiol, 40-41, sr resident surg, 42-43, from instr to prof, 43-74, REGENTS' PROF SURG, UNIV MINN, MINNEAPOLIS, 74- Concurrent Pos: Am Bd Surg rep, Am Bd Med Specialities. Mem: Am Thoracic Soc; Soc Univ Surgeons; AMA; Am Asn Thoracic Surg; Am Col Surgeons. Res: Cardiovascular problems; immunity; gastrointestinal physiology. Mailing Add: Univ of Minn Hosps 412 Union St Box 495 Minneapolis MN 55455

VARDANIS, ALEXANDER, b Athens, Greece, Mar 13, 33; Can citizen; m 59; c 2. BIOCHEMISTRY. Educ: Univ Leeds, BSc, 55; McGill Univ, MSc, 58, PhD(biochem), 60. Prof Exp: RES OFF BIOCHEM, RES INST, CAN DEPT AGR, UNIV WESTERN ONT, 61- Concurrent Pos: Nat Res Coun Can fel, 59-61. Mem: Can Biochem Soc. Res: Intermediary metabolism of carbohydrates, particularly glycogen metabolism; metabolism of toxic chemicals. Mailing Add: Res Inst Can Dept Agr University Sub PO London ON Can

VARDEMAN, STEPHEN BRUCE, b Louisville, Ky, Aug 27, 49; m 70. MATHEMATICAL STATISTICS. Educ: Iowa State Univ, BS, 71, MS, 73; Mich State Univ, PhD(statist), 75. Prof Exp: ASST PROF STATIST, PURDUE UNIV, WEST LAFAYETTE, 75- Mem: Inst Math Statist. Res: Compound decision problems, pattern recognition problems. Mailing Add: Dept of Statist Math Sci Bldg Purdue Univ West Lafayette IN 47907

VARELA-DIAZ, VICTOR M, parasitology, immunoparasitology, see 12th edition

VARGA, CHARLES E, b Philadelphia, Pa, Sept 13, 45; m 68; c 2. PHYSICAL ORGANIC CHEMISTRY. Educ: St Joseph's Col, Pa, BS, 67, MS, 69. Prof Exp: Develop chemist packaging, Single Serv Div Lab, Int Paper Co, 69-73; ASST LAB MGR & CHEMIST COLD ROLLING LUBRICANTS, METALS DIV LAB, QUAKER CHEM CORP, 73- Mem: Am Chem Soc. Res: Lubrication theory; metal organic bonding and chemisorption; surface activity; bidentate molecules. Mailing Add: Quaker Chem Corp Elm & Lime Sts Conshohocken PA 19428

VARGA, GIDEON MICHAEL, JR, inorganic chemistry, see 12th edition

VARGA, LOUIS P, b Portland, Ore, Mar 25, 22; m 48; c 4. ANALYTICAL CHEMISTRY, RADIOCHEMISTRY. Educ: Reed Col, BA, 48; Univ Chicago, MS, 50; Ore State Univ, PhD(anal chem), 60. Prof Exp: Chemist, Hanford Labs, 50-53; instr & res assoc, Reed Col, 53-57; res assoc anal chem, Mass Inst Technol, 60-61; asst prof chem, 61-67, ASSOC PROF CHEM, OKLA STATE UNIV, 67- Concurrent Pos: Vis staff mem, Los Alamos Sci Lab, 68- Mem: Am Chem Soc. Res: Analytical instrumentation; water analysis; rare earth spectra. Mailing Add: Dept of Chem Okla State Univ Stillwater OK 74074

VARGA, RICHARD S, b US, Oct 9, 28; m 51; c 1. MATHEMATICS. Educ: Case Inst Technol, BS, 50; Harvard Univ, AM, 51, PhD(math), 54. Prof Exp: Adv mathematician, Bettis Atomic Power Lab, Westinghouse Elec Co, 54-60; prof math, Case Western Reserve Univ, 60-69; UNIV PROF MATH, KENT STATE UNIV, 69- Concurrent Pos: Consult, Gulf Res & Develop Co, 60, Argonne Nat Lab, 61 & Los Alamos Sci Lab, 68-; Guggenheim fel, Harvard Univ & Calif Inst Technol, 63. Mem: Am Math Soc; Soc Indust & Appl Math; Math Asn Am. Res: Numerical analysis. Mailing Add: Dept of Math Kent State Univ Kent OH 44242

VARGAS, JOSEPH MARTIN, JR, b Fall River, Mass, Mar 11, 42; m 63; c 2. PLANT PATHOLOGY. Educ: Univ RI, BS, 63; Okla State Univ, MS, 65; Univ Minn, Minneapolis, PhD(plant path), 68. Prof Exp: Asst prof, 68-74, ASSOC PROF BOT & PLANT PATH, MICH STATE UNIV, 74- Mem: Am Phytopath Soc. Res: Turfgrass pathology; resistance to fungicides; pesticide degradation. Mailing Add: Dept of Bot & Plant Path Mich State Univ East Lansing MI 48823

VARGAS, LESTER LAMBERT, b Providence, RI, May 20, 21; m 48. CARDIOVASCULAR SURGERY. Educ: Brown Univ, AB, 43; George Washington Univ, MD, 45. Prof Exp: Asst surg, Col Physicians & Surgeons, Columbia Univ, 52-55; asst clin prof, Sch Med, Tufts Univ, 59-62; prof surg, 63-70, PROF MED SCI, DIV BIOL & MED SCI, BROWN UNIV, 70- Concurrent Pos: Lectr, Sch Nursing, Brown Univ, 42; dir cardiac surg & Cardiovasc Surg Res Lab, RI Hosp, 56-70, surgeon-in-chief, 63-69, assoc surgeon-in-chief, 69-; lectr, Sch Med, Tufts Univ, 63- Mem: Am Col Cardiol; Am Col Chest Physicians; Am Col Surgeons; Am Thoracic Soc; Int Cardiovasc Soc. Mailing Add: 110 Lockwood St Providence RI 02903

VARGHESE, SANKOORIKAL LONAPPAN, b Narakal, Kerala, Mar 13, 43; Indian citizen; m 71; c 2. EXPERIMENTAL ATOMIC PHYSICS. Educ: Kerala Univ, BSc, 63, MSc, 65; Univ Louisville, MS, 67; Yale Univ, PhD(physics), 74. Prof Exp: Res asst physics, Yale Univ, 68-74, res staff physicist, 74; RES ASSOC PHYSICS, KANS STATE UNIV, 74- Mem: Am Phys Soc; Sigma Xi. Res: Positron and positronium research, first observation of the n—2 state of positronium; accelerator based atomic physics, first direct lifetime measurement of x-ray emitters in the pico-second range; Mössbauer studies. Mailing Add: Dept of Physics Kans State Univ Manhattan KS 66506

VARGO, ROBERT ALLEN, b Cleveland, Ohio, Jan 29, 40; m 65; c 2. PHYSIOLOGY. Educ: Augusta Col, BS, 67; Med Col Ga, PhD(physiol), 72. Prof Exp: Instr, 72-74, ASST PROF PHYSIOL, MED COL GA, 74- Mailing Add: Dept of Physiol Med Col of Ga Augusta GA 30902

VARGO, STEVEN WILLIAM, b Whiting, Ind, Sept 9, 31; m 56; c 3. AUDIOLOGY. Educ: Ind State Univ, BS, 54; Purdue Univ, MS, 57; Ind Univ, PhD(audiol), 65. Prof Exp: Assoc prof audiol, Ill State Univ, 65-71; Nat Inst Neurol Dis & Stroke spec res fel, Auditory Res Lab, Northwestern Univ, Evanston, 71-73; ASSOC PROF SURG, HERSHEY MED CTR, PA STATE UNIV, 73- Mem: Am Speech & Hearing Asn; Acoust Soc Am. Res: Scientific study of communication behavior with primary emphasis on the auditory mechanism of both normal and abnormal systems. Mailing Add: Hearing & Speech Clin Pa State Univ Hershey Med Ctr Hershey PA 17033

VARIMBI, JOSEPH, b Philadelphia, Pa, Nov 27, 27; m 61. ELECTROCHEMISTRY. Educ: Univ Pa, BA, 48, MS, 50, PhD, 53. Prof Exp: Res chemist, Edison Lab, McGraw-Edison Co, NJ, 52-55; Du Pont fel chem, Yale Univ, 55-56; asst prof, Lafayette Col, 56-57; asst prof, 57-73, ASSOC PROF CHEM, BRYN MAWR COL, 73- Mem: Am Chem Soc; Electrochem Soc. Res: Electrolyte solutions; nonaqueous solvents; electrochemistry of solutions of sulfur in amines. Mailing Add: Dept of Chem Bryn Mawr Col Bryn Mawr PA 19010

VARIN, ROGER ROBERT, b Bern, Switz, Feb 15, 25; nat US; m 51; c 3. PHYSICAL CHEMISTRY. Educ: Univ Bern, PhD(chem), 51. Prof Exp: Fel phys chem, Harvard Univ, 51-52; res chemist, E I du Pont de Nemours & Co, 52-57, res assoc, 57-62; dir res, Riegel Textile Corp, 62-71; PRES, VARINIT CORP, 71- Concurrent Pos: Pres, Technol Assocs, Greenville, SC, 71- & Varinit S A, Fribourg, Switz, 74- Mem: AAAS; Am Chem Soc; Am Asn Textile Chem; Fiber Soc; Am Asn Textile Chem & Colorists. Res: Fiber physics and chemistry; textile technology; polymer physics and chemistry; rheology. Mailing Add: 4 Barksdale Rd Greenville SC 29607

VARINEAU, VERNE JOHN, b Escanaba, Mich, Mar 11, 15; m 45; c 4. MATHEMATICS. Educ: Col St Thomas, BS, 36; Univ Wis, AM, 38, PhD(math), 40. Prof Exp: Asst, Univ Wis, 36-39; from instr to assoc prof, 40-72, PROF MATH, UNIV WYO, 72- Concurrent Pos: NSF sci fac fel, Stanford Univ, 63-64. Mem: Am Math Soc; Math Asn Am. Res: Matrices with elements in a principal ideal ring. Mailing Add: Dept of Math Univ of Wyo Laramie WY 82070

VARKEY, THANAKAMMA EAPEN, b Palai, India, Oct 31, 36; m 64; c 1. ORGANIC CHEMISTRY. Educ: Kerala Univ, BSc, 56, MSc, 57; Temple Univ, PhD(chem), 74. Prof Exp: Lectr chem, Kerala Univ, 57-64; prof, Alphonsa Col, India, 64-68; FEL CHEM, DREXEL UNIV, 74- Mem: Am Chem Soc; Sigma Xi. Res: Synthesis and conformational analysis of steroids; synthesis, structure and stereochemistry of nitrogen-sulfur ylides (iminosulfuranes). Mailing Add: Dept of Biol Sci Drexel Univ Philadelphia PA 19104

VARLASHKIN, PAUL, b San Antonio, Tex, Aug 28, 31; m 54; c 3. SOLID STATE PHYSICS. Educ: Univ Tex, BS, 52, MA, 54, PhD(physics), 63. Prof Exp: Asst, Defense Res Lab, Univ Tex, 51-52, physicist, 52-53; from res scientist to chief res & develop, Electro-Mech Co, 53-63; fel & res assoc, Univ NC, 66-66; asst prof physics, La State Univ, Baton Rouge, 66-72; ASST PROF PHYSICS, ECAROLINA UNIV, 72- Concurrent Pos: Res physicist, White Sands Proving Grounds, 52. Mem: Am Phys Soc; Sigma Xi; Am Asn Physics Teachers. Res: Positron annihilation; liquid metals; positronium formation; solid state physics; chemical physics; metal-ammonia solutions. Mailing Add: Dept of Physics ECarolina Univ Greenville NC 27834

VARMA, ARUN KUMAR, b Faizabad, India. MATHEMATICS. Educ: Banaras Hindu Univ, BSc, 55; Univ Lucknow, MSc, 58; Univ Alta, PhD(math), 64. Prof Exp: Lectr math, Univ Rajasthan, 64-66; fel, Univ Alta, 66-67; asst prof, 67-69, ASSOC PROF MATH, UNIV FLA, 69- Mem: Am Math Soc. Res: Interpolation theory; approximation theory; numerical analysis. Mailing Add: Dept of Math Univ of Fla Gainesville FL 32601

VARMA, DAYA RAM, pharmacology, see 12th edition

VARMA, RAJENDRA, b India, Dec 26, 42; US citizen; m 69. ORGANIC CHEMISTRY, BIOCHEMISTRY. Educ: Univ Delhi, BS, 58, MS, 60; Univ NSW, PhD(chem), 66. Prof Exp: Lectr & res worker chem, Univ Delhi, 60-62; lectr, Sydney Tech Col, Australia, 63-66; fel biochem, Iowa State Univ, 67 & Purdue Univ, Lafayette, 68; asst prof chem, Alliance Col, 68-69; assoc prof, Edinboro State Col, 69-71; MED RES SCIENTIST, BIOCHEM RES DEPT, WARREN STATE HOSP, 71- Mem: Am Chem Soc; Sigma Xi. Res: Carbohydrate chemistry; biochemistry of glycoproteins and mucopolysaccharides from eye, brain and biological fluids. Mailing Add: Biochem Res Dept Warren State Hosp Warren PA 16365

VARMA, RANBIR S, b Bhera, Pakistan, July 18, 38; US citizen; m 69; c 2. CLINICAL BIOCHEMISTRY, MEDICAL RESEARCH. Educ: Govt Med Col, BS, 65; Panjab Univ, BS Hons, 66, MS Hons, 68; State Univ NY Buffalo, MA, 71; Am Bd Bioanalysis, cert, 75. Prof Exp: Res asst pharmacol, All India Inst Med Sci, 61-62, asst res officer anesthesiol, 68; res scholar radiation biol, Univ Jaipur, India, 69; clin chem trainee, Sch Med, State Univ NY Buffalo, 69-71; CLIN BIOCHEMIST, WARREN STATE HOSP, PA, 71- Mem: Am Asn Clin Chem; Am Asn Bioanalysts; Biochem Soc UK; Asn Clin Biochem UK; Sigma Xi. Res: Glycoproteins and acid mucopolysaccharides of tissues and biological fluids. Mailing Add: Biochem Dept Warren State Hosp Warren PA 16365

VARMA, RAVI KANNADIKOVILAKOM, b Tripunithura, India, Dec 23, 37; m 65; c 2. MEDICINAL CHEMISTRY. Educ: Maharaja's Col, Ernakulam, India, BSc, 57, MSc, 59; Univ Poona, PhD(org chem), 65. Prof Exp: Sci asst chem, Nat Chem Labs, Poona, India, 59-65; res fel org chem, Purdue Univ, Lafayette, 65-66; staff scientist bio-org chem, Worcester Found, Mass, 66-69; res fel org chem, Harvard Univ, 70-72; SR RES INVESTR, E R SQUIBB & SONS, INC, 72- Mem: Am Chem Soc; The Chem Soc. Res: Organic synthesis; chemistry of biologically active molecules; mechanism of action of drugs. Mailing Add: Dept of Org Chem E R Squibb & Sons Inc Princeton NJ 08540

VARNELL, THOMAS RAYMOND, b Whiteriver, Ariz, Jan 27, 31; m 57; c 2. BIOCHEMISTRY, PHYSIOLOGY. Educ: Univ Ariz, BS, 57, MS, 58, PhD(agr biochem), 60. Prof Exp: Instr animal nutrit, Univ Wyo, 60-62; res biochemist, Dept Health, Educ & Welfare, Food & Drug Admin, 62-63; asst prof animal nutrit, 63-65, from asst prof to assoc prof animal physiol, 65-73, PROF ANIMAL PHYSIOL, UNIV WYO, 73- Concurrent Pos: Stanford Res Inst grant, 61-64. Mem: Am Chem Soc; Animal Nutrit Res Coun; AAAS; Am Soc Animal Sci. Res: Metabolism of vitamin A and carotene; lipid metabolism; intestinal transport and metabolism of amino acids. Mailing Add: Div of Animal Sci Univ of Wyo Laramie WY 82070

VARNER, JOSEPH ELMER, b Nashport, Ohio, Oct 7, 21; div; c 4. PLANT PHYSIOLOGY. Educ: Ohio State Univ, BSc, 42, MSc, 43, PhD(biochem), 49. Prof Exp: Chemist, Owens-Corning Fiberglas Corp, 43-44; res engr, Battelle Mem Inst, 46-47; res assoc, Res Found, Ohio State Univ, 49-50, asst prof agr biochem, 50-53; res fel, Calif Inst Technol, 53-54; from assoc prof to prof biochem, Ohio State Univ, 54-61; prof, Res Inst Advan Study, 61-65 & Mich State Univ, 65-73; PROF BIOCHEM, WASHINGTON UNIV, 73- Concurrent Pos: NSF fel, Cambridge Univ, 59-60 & Univ Wash, 71-72. Mem: AAAS; Am Chem Soc; Am Soc Biol Chem; Am Soc Plant Physiol; NY Acad Sci. Res: Plant biochemistry; biochemistry of aging cells; action mechanism of plant hormones. Mailing Add: Dept of Biol Washington Univ St Louis MO 63130

VARNER, LARRY WELDON, b San Antonio, Tex, June 25, 44; m 67; c 1. ANIMAL NUTRITION, WILDLIFE RESEARCH. Educ: Abilene Christian Col, BS, 66; Univ Nebr, Lincoln, MS, 68, PhD(nutrit), 70. Prof Exp: Asst animal sci, Univ Nebr, Lincoln, 66-69, res assoc animal nutrit, 69-70, asst prof, 70-71; res scientist, USDA, 71-74, ASST PROF ANIMAL NUTRIT, TEX A&M UNIV, 74- Mem: Am Soc Animal Sci; Soc Range Mgt. Res: Ruminant nutrition; nitrogen and energy metabolism; wildlife nutrition. Mailing Add: PO Drawer 1051 Uvalde TX 78801

VARNER, REED WILLIAM, b Columbus, Ohio, Oct 29, 16; m 46; c 3. PLANT PATHOLOGY, FORESTRY. Educ: Univ Mich, BS, 39, MS, 41, PhD(plant path), 51. Prof Exp: Asst forestry, Univ Mich, 41-43 & 46-48; biologist, 48-51, tech field specialist, 51-53, res investr pesticides, 53-55, res supvr, 55-67, RES MGR INDUST & BIOCHEMS, E I DU PONT DE NEMOURS & CO, INC, 68- Mem: Am Phytopath Soc. Res: Pesticide evaluation and development; silviculture; conservation. Mailing Add: E I du Pont de Nemours & Co Inc Exp Sta Bldg 268 Wilmington DE 19898

VARNERIN, ROBERT E, b Boston, Mass, May 11, 26; m 70; c 2. PHYSICAL CHEMISTRY. Educ: Boston Col, AB, 49, MA, 50; Cath Univ Am, PhD(chem), 54. Prof Exp: Res assoc, Cath Univ Am, 53-54; lectr chem, Weston Col, 54-58; from asst prof to assoc prof, Fairfield Univ, 59-70; MGR EDUC, MFG CHEMISTS ASN, 70- Concurrent Pos: Vis scientist, Cath Univ Am, 66-67; chmn dept chem, Fairfield Univ, 67-70. Mem: AAAS; Am Chem Soc. Res: Mechanism of organic reactions; mass spectrometry; aliphatic free radicals; photochemistry. Mailing Add: Mfg Chemists Asn 1825 Connecticut Ave NW Washington DC 20009

VARNES, ARTHUR WAYNE, analytical chemistry, see 12th edition

VARNES, DAVID JOSEPH, b Howe, Ind, Apr 5, 19; m 43, 66; c 2. ENGINEERING GEOLOGY. Educ: Calif Inst Technol, BS, 40. Prof Exp: Lab instr geol, Northwestern Univ, 40-41; recorder & jr geologist, 41, asst geologist, 43-45, assoc geologist, 45-48, chief br eng geol, 61-64, GEOLOGIST, US GEOL SURV, 48- Concurrent Pos: Mem comts, Hwy Res Bd, Nat Acad Sci-Nat Res Coun. Honors & Awards: E B Burwell Jr Award, Geol Soc Am, 70; Distinguished Serv Award, US Dept Interior, 75. Mem: Asn Eng Geologists; Geol Soc Am; Am Soc Test & Mat. Res: Geologic studies of Lake Bonneville; landslides; mechanics of soil and rock deformation; logic of mapping. Mailing Add: US Geol Surv Denver Fed Ctr Denver CO 80225

VARNEY, CHARLES BROADWELL, b Bennet, Nebr, Nov 19, 16; m 48; c 5. GEOGRAPHY. Educ: Univ Colo, BA, 50; Clark Univ, AM, 53, PhD(geog), 63. Prof Exp: Instr geog, State Univ NY Col Fredonia, 52-53; asst prof geog & geol, Univ Tampa, 54-57; from instr to asst prof geog & soc sci, Univ Fla, 57-63; assoc prof geog, 63-64, PROF GEOG, UNIV WIS-WHITEWATER, 64- Concurrent Pos: Dir NSF Acad Yr In-Serv Inst Phys Geog, 67-68; sr lectr, Chinese Univ Hong Kong, 68-70. Mem: Asn Am Geog; Nat Coun Geog Educ. Res: Environmental quality improvement and education. Mailing Add: Dept of Geog & Geol Univ of Wis Whitewater WI 53190

VARNEY, EUGENE HARVEY, b South Egremont, Mass, Dec 25, 23; m 56; c 3. PLANT PATHOLOGY. Educ: Univ Mass, BS, 49; Univ Wis, PhD, 53. Prof Exp: Plant pathologist, USDA, 53-56; asst res specialist, 56-59, assoc prof plant path, 59-64, PROF PLANT PATH, RUTGERS UNIV, NEW BRUNSWICK, 64- Mem: AAAS; Am Phytopath Soc; Mycol Soc Am. Res: Diseases of small fruits; plant virology; mycology. Mailing Add: Dept of Plant Biol Cook Col Rutgers Univ New Brunswick NJ 08903

VARNEY, ROBERT NATHAN, b San Francisco, Calif, Nov 7, 10; m 48; c 2. MOLECULAR PHYSICS. Educ: Univ Calif, AB, 31, MA, 32, PhD(physics), 35. Prof Exp: Instr physics, Univ Calif, 35-36 & NY Univ, 36-38; asst prof, Washington Univ, 38-41, from assoc prof to prof, 46-64; sr mem & sr consult scientist, Lockheed Palo Alto Res Lab, 64-75; NSF SR FEL, US ARMY BALLISTIC RES LABS, 75- Concurrent Pos: Vis mem tech staff, Bell Tel Labs, NJ, 51-52; mem exec comt, Gaseous Electronics Conf, 51-53, 60-62 & 65-, secy, 67; NSF sr fel, Royal Inst Technol, Sweden, 58-59; mem, Gov Sci Adv Comt, Mo, 61-64; Fulbright lectr, Inst Atomic Physics, Innsbruck Univ, 71-72. Mem: Fel AAAS; fel Am Phys Soc; Am Asn Physics Teachers. Res: Ion-molecule reactions; collisions of positive ions in gases; spark breakdown; secondary electron emission.

VARNEY, ROGER FRANKLIN, biochemistry, see 12th edition

VARNEY, WILLIAM YORK, b Forest Hills, Ky, Apr 1, 17; m 40; c 2. ANIMAL HUSBANDRY. Educ: Univ Ky, BS, 51, MS, 52; Mich State Univ, PhD, 60. Prof Exp: Prin pub schs, Ky, 39-43; dist mgr, Southern States Coop, Va, 52-53; sales promoter, Swift & Co, Ill, 53-54; instr animal husb, 54-56, from asst prof to assoc prof, 56-69, PROF ANIMAL HUSB, UNIV KY, 69- Mem: Am Soc Animal Sci; Inst Food Technologists. Res: Carcass studies of beef, pork and lamb. Mailing Add: Div of Animal Sci Univ of Ky Col of Agr Lexington KY 40506

VARNHORN, MARY CATHERINE, b Baltimore, Md, July 8, 14. MATHEMATICS. Educ: Col Notre Dame, Md, AB, 36; Cath Univ Am, AM, 37, PhD(math), 39. Prof Exp: Instr math, Col Notre Dame, Md, 39-40; from instr to assoc prof, 40-72, PROF & CHMN DEPT, TRINITY COL, DC, 72- Mem: Math Asn Am; Am Math Soc. Res: Modern algebra; properties of quartic functions of one variable. Mailing Add: 6111 Northdale Rd Baltimore MD 21228

VARON, ALBERT, b New York, NY, July 14, 36. ANALYTICAL CHEMISTRY. Educ: City Col New York, BS, 58; Rutgers Univ, PhD(anal chem), 64. Prof Exp: CHEMIST, E I DU PONT DE NEMOURS & CO, INC, 63- Mem: Am Chem Soc. Res: Liquid and gas chromatography; analytical methods development. Mailing Add: Gen Anal Lab Chambers Works E I du Pont de Nemours & Co Inc Deepwater NJ 08069

VARON, MYRON IZAK, b Chicago, Ill, Aug 20, 30; m 59; c 3. RADIOBIOLOGY, MEDICINE. Educ: Univ Chicago, PHB, 50; Northwestern Univ, BSM, 52, MD, 55; Univ Rochester, MS, 63, PhD(radiation biol), 65. Prof Exp: Intern, Cook County Hosp, Chicago, 55-56; Med Corps, US Navy, 56-, med officer, USS Lenawee, 56-58 & Armed Forces Spec Weapons Proj, 58-59, asst to mgr naval reactor br, Idaho Br Off, AEC, 59-60, sr med officer & radiation safety officer, USS Long Beach, 60-62, assoc surg, Univ Rochester, 62-65, med dir radiation biol, Naval Radiol Defense Lab, 65-67, from asst dep sci dir to dir, Armed Forces Radiobiol Res Inst, 67-75, DEP COMNDG OFFICER, NAV MED RES & DEVELOP COMMAND, MED CORPS, US NAVY, 75- Concurrent Pos: Lectr, US Naval Hosp, Oakland, Calif, 66-67. Mem: Health Physics Soc; Radiation Res Soc; Asn Mil Surg US; NY Acad Sci; Soc Nuclear Med. Res: Radiation safety of nuclear reactors; labeled antibody localization for cancer therapy; nuclear weapons effects; radiation behavioral effects; radiation recovery and residual injury; experimental pathology. Mailing Add: Naval Med Res & Develop Command Med Corps US Navy Bethesda MD 20014

VARON, SILVIO SALOMONE, b Milan, Italy, July 25, 24; m 60 & 67; c 2. NEUROCHEMISTRY, NEUROBIOLOGY. Educ: Univ Lausanne, EngD, 45; Univ Milan, MD, 59. Prof Exp: Resident asst prof neurochem, Inst Psychiat, Univ Milan, 60-63; res assoc, Dept Biochem, City of Hope Med Ctr, Duarte, Calif, 61-63; res assoc neurobiol, Dept Biol, Wash Univ, 63-64, assoc prof, 64-65; vis assoc prof, Dept Genetics, Sch Med, Stanford Univ, 65-67; assoc prof, 67-72, PROF NEUROBIOL, DEPT BIOL, MED SCH, UNIV CALIF, SAN DIEGO, 72- Mem: Int Soc Neurochem; Am Soc Neurochem; Soc Neurosci; Am Soc Cell Biol. Res: Binding and transport in neural cells and subcellular particles; structure and properties of the nerve growth factor protein; dissociation fractionation and culture of cells from nervous tissues. Mailing Add: Dept of Biol Univ Calif San Diego Med Sch La Jolla CA 92037

VARRICCHIO, FREDERICK, b Brooklyn, NY, May 18, 38; m 62; c 2. BIOCHEMISTRY, DEVELOPMENTAL BIOLOGY. Educ: Univ Maine, Orono, BS, 60; Univ NDak, MS, 64; Univ Md, Baltimore, PhD(biochem), 66. Prof Exp: Asst biochem, Univ Freiburg, 66-67; researcher molecular biol, Nat Ctr Sci Res, France, 67-69; fel internal med, Yale Univ, 69-72; ASSOC, MEM SLOAN-KETTERING INST CANCER RES, 72- Concurrent Pos: Am Cancer Soc fel, Univ Freiburg, 66-67. Mem: Am Soc Biol Chemists; Am Chem Soc; Am Soc Microbiol; Ger Soc Biol Chemists. Res: Transfer RNA; chromatin. Mailing Add: Mem Sloan-Kettering Inst Cancer Res 1275 York Ave New York NY 10021

VARRO, STEPHEN, JR, b Budapest, Hungary, Oct 3, 09; US citizen; m 34; c 1. ECOLOGY, PLANT NUTRITION. Educ: Univ Frankfurt, PhD(econ), 34. Prof Exp: Pres, Int Equip Consult, Inc, NY, 47-57; vpres, Nat Waste Conversion Corp, 57-69; PRES, ECOL, INC, 69- Mem: AAAS; Am Chem Soc. Res: Conversion of organic wastes through biological oxidation into organic plant nutrients to improve soil productivity and plant health. Mailing Add: 425 E51st St New York NY 10022

VARS, HARRY MORTON, b Edelstein, Ill, July 2, 03; m; c 3. PHYSIOLOGICAL CHEMISTRY. Educ: Univ Colo, AB, 24; Yale Univ, PhD(physiol chem), 29. Prof Exp: Asst chem, Cornell Univ, 24-25; asst physiol chem, Yale Univ, 26-28, instr, 28-31; res assoc biol, Princeton Univ, 31-34; Merck fel physiol, 34-36, assoc, Harrison Dept Surg Res, 36-40, asst prof biochem in surg res, 40-48, assoc prof, 48-54, PROF BIOCHEM IN SURG RES, HARRISON DEPT SURG RES, SCH MED, UNIV PA, 54- Concurrent Pos: Responsible investr, Comt Med Res, Off Emergency Mgt, 42-44; res chemist, Bryn Mawr Hosp, 44-46. Mem: Am Chem Soc; Am Soc Biol Chemists; Soc Exp Biol & Med; Am Inst Nutrit. Res: Protein, amino acid and liver metabolism; adrenal cortical hormones; hepatotoxic agents; liver regeneration; enzymes; dietary factors. Mailing Add: Harrison Dept of Surg Res Univ of Pa Sch of Med Philadelphia PA 19104

VARSA, EDWARD CHARLES, b Marissa, Ill, Oct 18, 38; m 65; c 2. SOIL FERTILITY. Educ: Southern Ill Univ, Carbondale, BS, 61; Univ Ill, Urbana, MS, 65; Mich State Univ, PhD(soil sci), 70. Prof Exp: Asst agron, Univ Ill, 64-65; asst soil sci, Mich State Univ, 65-66, instr, 66-68, teaching asst, 68-70; ASST PROF SOILS, SOUTHERN ILL UNIV, CARBONDALE, 70- Mem: Am Soc Agron; Soil Sci Soc Am. Res: Soil fertility research on, and the fate of, applied fertilizer nitrogen in soils. Mailing Add: Dept of Plant & Soil Sci Southern Ill Univ Carbondale IL 62901

VARSAMIS, IOANNIS, b Alexandria, UAR, July 24, 32; Can citizen; m 58. PSYCHIATRY. Educ: Univ Alexandria, MB, ChB, 57; Conjoint Bd, London, Eng, dipl psychol med, 61; Univ Man, dipl psychiat, 64; FRCP(C), 65. Prof Exp: Psychiatrist, Winnipeg Psychiat Inst, 64-68, med supt, 68-73; PSYCHIATRIST, GRACE GEN HOSP, 74-; ASSOC PROF PSYCHIAT, UNIV MAN, 67- Mem: Can Med Asn; Can Psychiat Asn. Res: Phenomenology of schizophrenia; geriatric psychiatry. Mailing Add: Dept of Psychiat Grace Gen Hosp Winnipeg MB Can

VARSEL, CHARLES JOHN, b Fayette City, Pa, Jan 11, 30; m 50; c 6. FOOD CHEMISTRY, FOOD BIOCHEMISTRY. Educ: St Vincent Col, BA, 54; Univ Richmond, MS, 58; Med Col Va, PhD, 70. Prof Exp: Chemist, Linde Air Prod Co, Union Carbide & Carbon Corp, 54-55 & Res & Develop Dept, Philip Morris, Inc, 55-59; res assoc fuel technol, Pa State Univ, 59-60; from res chemist to sr res chemist, Res & Develop Dept, Philip Morris, Inc, Va, 60-69; prin chemist, Food Div, Citrus Res & Develop, 69-70, mgr chem, 70-73, dir citrus res & develop, 73, DIR RES & DEVELOP, FOODS DIV, COCA-COLA CO, 73- Mem: AAAS; Am Chem Soc; Inst Food Technol. Res: Instrumental analysis; citrus chemistry; essential oils; mass spectroscopy; chemistry of natural products. Mailing Add: Coca Cola Co Foods Div Dept Res & Develop PO Box 2079 Houston TX 77001

VARSHNI, YATENDRA PAL, b Allahabad, India, May 21, 32. ASTROPHYSICS, SOLID STATE PHYSICS. Educ: Univ Allahabad, BSc, 50, MSc, 52, PhD(physics), 56. Prof Exp: Asst prof physics, Univ Allahabad, 55-60, fel, Nat Res Coun Can, 60-62; from asst prof to assoc prof, 62-69, PROF PHYSICS, UNIV OTTAWA, 69- Mem: Am Phys Soc; Can Asn Physicists; Brit Inst Physics; Brit Interplanetary Soc; Am Astron Soc. Res: Molecular and nuclear structure; quasi-stellar objects; lattice dynamics; energy levels of nuclei. Mailing Add: Dept of Physics Univ of Ottawa Ottawa ON Can

VARTY, ISAAC WILLIAM, b Consett, Eng, Feb 9, 24; m 52; c 3. FOREST ENTOMOLOGY. Educ: Aberdeen Univ, BSc, 50, PhD(entom), 54. Prof Exp: Asst forest zool, Aberdeen Univ, 50-54; dist forest officer, Forestry Comn, Scotland, 54-58; FOREST RES SCIENTIST, MARITIMES FOREST RES CTR, CAN FORESTRY SERV, 58- Mem: Entom Soc Can; Can Inst Forestry. Res: Environmental impact of forest spraying; introduction of exotic parasites for control of forest pests. Mailing Add: Maritimes Forest Res Ctr Can Forestry Serv Box 4000 Fredericton NB Can

VAS, STEPHEN ISTVAN, b Budapest, Hungary, June 4, 26; Can citizen; m 53. MEDICAL MICROBIOLOGY, IMMUNOLOGY. Educ: Pazmany Peter Univ, Budapest, MD, 50, PhD(microbiol), 56. Prof Exp: Lectr microbiol, Pazmany Peter Univ, 48-49, asst prof, 49-50; Rockefeller res fel microbiol, 57-59, asst virologist, 59-60, from asst prof to assoc prof immunol, 60-69, PROF IMMUNOL, McGILL UNIV, 69-, CHMN DEPT, 72- Concurrent Pos: Microbiologist-in-chief & sr physician, Royal Victoria Hosp. Mem: Am Soc Microbiol; Can Soc Microbiol; Can Soc Immunol; Can Med Asn; Can Asn Med Microbiol. Res: Antibody synthesis; synthesis of complement; effect of antibodies on bacteria and on tissue cells. Mailing Add: Dept of Microbiol & Immunol McGill Univ Montreal PQ Can

VASAVADA, KASHYAP V, b Ahmedabad, India, July 25, 38; m 69; c 1. THEORETICAL PHYSICS. Educ: Univ Baroda, BS, 58; Univ Delhi, MS, 60; Univ Md, PhD(physics), 64. Prof Exp: Nat Acad Sci fel physics, Goddard Space Flight Ctr, NASA, 64-66; asst prof, Univ Conn, 66-70; assoc prof, 70-74, PROF PHYSICS, IND UNIV-PURDUE UNIV, INDIANAPOLIS, 74- Mem: Am Phys Soc. Res: High energy physics; scattering theory; atomic physics. Mailing Add: Dept of Physics Ind Univ-Purdue Univ Indianapolis IN 46205

VASEK, FRANK CHARLES, b Maple Heights, Ohio, May 9, 27; m 54; c 2. BOTANY. Educ: Ohio Univ, BS, 50; Univ Calif, Los Angeles, PhD(bot), 54. Prof Exp: Asst bot, Univ Calif, Los Angeles, 50-54; from instr to assoc prof, 54-68, PROF BOT, UNIV CALIF, RIVERSIDE, 68- Mem: Am Soc Naturalists; Bot Soc Am; Ecol Soc Am; Soc Study Evolution; Am Soc Plant Taxon. Res: Plant taxonomy; evolution population dynamics. Mailing Add: Dept of Biol Univ of Calif Riverside CA 92502

VASEY, EDFRED H, b Mott, NDak, Aug 29, 33; m 55; c 4. SOIL FERTILITY. Educ: NDak State Univ, BS, 55, MS, 57; Purdue Univ, PhD(plant nutrit), 62. Prof Exp: From asst prof to assoc prof, 61-69, exten soils specialist, Exten Serv, 67-71, PROF SOILS, NDAK STATE UNIV, 69-, HEAD PLANT SCI SECT, NDAK COOP EXTEN SERV, 71- Mem: Am Soc Agron; Sigma Xi; Soil Conservation Soc Am. Res: Fertility needs of crops grown on North Dakota soils. Mailing Add: Rm 203B Waldron Hall NDak State Univ Sta Fargo ND 58102

VASICEK, OLDRICH ALFONSO, b Prague, Czech, June 3, 41; US citizen; m 65; c 3. MATHEMATICAL STATISTICS. Educ: Czech Tech Univ, Prague, MS, 64; Charles Univ, Prague, PhD(probability theory), 68. Prof Exp: Res asst probability theory, Inst Info Theory, Charles Univ, 64-67, res asst math, Tech-Econ Inst, 67-68; res assoc, Mgt Sci Dept, Wells Fargo Bank, 69-74; ASST PROF QUANT METHODS, GRAD SCH MGT, UNIV ROCHESTER, 74- Honors & Awards: Graham & Dodd Award, Financial Analysts Fedn, 73. Mem: Inst Math Statist. Res: Stochastic processes; statistical decision theory; non-parametric methods; operations research. Mailing Add: Grad Sch of Mgt Univ of Rochester Rochester NY 14627

VASICKA, ALOIS, b Czech, Sept 23, 17; US citizen. OBSTETRICS & GYNECOLOGY. Educ: Masaryk Univ, MD, 45; Am Bd Obstet & Gynec, dipl, 60. Prof Exp: Resident obstet & gynec, Univ Hosp, Brno, Czech, 45-50; intern, Beverly Hosp, Mass, 52; asst resident obstet & gynec, Boston Lying-in Hosp, 52-53; resident gynec path, Free Hosp Women, Brookline, 53-54; asst gynec, Peter Bent Brigham Hosp, Boston, 54-55; resident obstet & gynec, Cleveland Metrop Gen Hosp, Ohio, 55-56, asst, 56-57, asst vis obstetrician & gynecologist, 57-61; obstetrician & gynecologist, John Sealy Hosp, Galveston, Tex, 61-65; CHIEF OBSTET & GYNEC, CONEY ISLAND HOSP, MAIMONIDES MED CTR, 65-; PROF OBSTET & GYNEC, STATE UNIV NY DOWNSTATE MED CTR, 65- Concurrent Pos: Res fels, Peter Bent Brigham Hosp, Boston & Harvard Med Sch, 54-55; demonstr, Sch Med, Western Reserve Univ, 56-57, from instr to asst prof, 57-61; prof, Univ Tex, 61-65. Mem: Soc Gynec Invest; Am Col Obstet & Gynec; AMA; Am Fertil Soc. Res: Physiology of human obstetrics; uterine and fetal physiology; uterine contractility and myometrial function at its cellular level. Mailing Add: Coney Island Hosp Brooklyn NY 11235

VASIL, INDRA KUMAR, b Basti, India, Aug 31, 32; m 59; c 2. BOTANY. Educ: Banaras Hindu Univ, BSc, 52; Univ Delhi, MSc, 54, PhD(bot), 58. Prof Exp: Res asst bot, Univ Delhi, 54-58, asst prof, 59-63; res assoc, Univ Wis, 63-65; scientist, Indian Agr Res Inst, 65-67; assoc prof, 67-74, PROF BOT, UNIV FLA, 74- Concurrent Pos: Res assoc, Univ Ill, 62-63. Honors & Awards: Sr US Scientist Award for Res, Fed Repub Ger, 74. Mem: AAAS; Int Asn Plant Tissue Cult; Bot Soc Am; Tissue Cult Asn; Int Soc Plant Morphol. Res: Developmental morphology; physiology of reproduction in flowering plants; morphogenesis and differentiation in higher plants;

plant tissue and organ culture. Mailing Add: Dept of Bot Univ of Fla Gainesville FL 32611

VASILE, MICHAEL JOSEPH, b Newton, NJ, May 11, 40. PHYSICAL CHEMISTRY. Educ: Rutgers Univ, BS, 62; Princeton Univ, MA, 64, PhD(chem), 66. Prof Exp: Res fel chem, Nat Res Coun Can, 66-68; MEM TECH STAFF, BELL LABS, 68- Mem: Am Chem Soc; Am Soc Mass Spectrometry; Am Vacuum Soc. Res: Mass spectrometry of inorganic fluorides; plasma chemistry; secondary ion mass spectrometry. Mailing Add: Bell Labs 1A259 Mountain Ave Murray Hill NJ 07974

VASILEFF, VASIL, b Madison, Ill, Jan 1, 14; m 41; c 3. DENTISTRY. Educ: Washington Univ, AB, 35; St Louis Univ, DDS, 50. Prof Exp: Instr dent mat, Sch Dent, St Louis Univ, 49-50, from instr to asst prof prosthetic dent, 50-64, assoc clin prof orthodont, 64-72; ASSOC PROF DENT, SCH DENT MED, SOUTHERN ILL UNIV, EDWARDSVILLE, 72- Honors & Awards: Cert Merit, Am Acad Dent Med, 50. Mem: Int Asn Dent Res; Am Acad Oral Med; Am Col Dent. Res: Reaction of living tissue to dental restorative materials. Mailing Add: Dept of Dent Southern Ill Univ Sch of Dent Med Edwardsville IL 62025

VASILEVSKIS, STANISLAUS, b Latvia, July 20, 07; nat US; m 29; c 2. ASTRONOMY. Educ: Univ Latvia, PhD(math), 39. Prof Exp: Asst astron, Univ Latvia, 28-32, from instr to asst prof, 33-44; res assoc, Leipzig Observ, 45; assoc prof, UNRRA Univ, Munich, 46-47; asst, 49-54, lectr, 54, from asst astronr to assoc astron, 54-64, prof astron, Univ, 66-74, EMER PROF ASTRON, UNIV CALIF, SANTA CRUZ, 74-, ASTRONR, LICK OBSERV, 64- Concurrent Pos: Vis prof astron, Leiden Univ, Neth, 75-76. Mem: Am Astron Soc. Res: Astrometry, particularly stellar proper motions and trigonometry parallexes. Mailing Add: Lick Observ Univ of Calif Santa Cruz CA 95064

VASILIADES, JOHN, analytical chemistry, see 12th edition

VASILIAUSKAS, EDMUND, b Lithuania, June 18, 38; US citizen; m 70; c 3. ORGANIC CHEMISTRY. Educ: Rochester Inst Technol, BS, 63; Loyola Univ Chicago, PhD(org chem), 70. Prof Exp: Chemist, Olin Corp, 63-65 & Witco Chem Corp, 70-71; instr, 71-74, ASST PROF CHEM, MORAINE VALLEY COMMUNITY COL, 74- Mem: Am Chem Soc. Res: Study of the derivatives of benzonorbornene. Mailing Add: Dept of Chem Moraine Valley Community Col Palos Hills IL 60465

VASILOS, THOMAS, b New York, NY, Oct 18, 29; m 54; c 2. CERAMICS, CHEMISTRY. Educ: Brooklyn Col, BS, 50; Mass Inst Technol, DSc(ceramics), 54. Prof Exp: Asst ceramics, Mass Inst Technol, 50-53; res engr, Ford Motor Co, 53; mgr, Ceramics Res Dept, Corning Glass Works, 55-57; sect chief metals & ceramics, Res & Adv Develop Div, 57-66, MGR MAT SCI DEPT, AVCO SYSTS DIV, AVCO CORP, 66- Concurrent Pos: Consult, Mat Adv Bd, Nat Acad Sci, 61- Honors & Awards: Ross Coffin Purdy Award, Am Ceramic Soc, 67. Mem: Am Ceramic Soc. Res: Thermal and mechanical properties of ceramics; crystal growing; ferroelectric ceramics and crystals; nuclear fuel ceramics; ceramic coatings; metal reinforced ceramics; diffusion in crystals; plastic-ceramic composites; composite formulation and properties; refractory metals. Mailing Add: 92 Bartlett Rd Winthrop MA 02152

VASINGTON, FRANK D, b Norwich, Conn, Nov 3, 28; m 52; c 4. BIOCHEMISTRY. Educ: Univ Conn, AB, 50, MS, 52; Univ Md, PhD(biochem), 55. Prof Exp: Asst prof biochem, Sch Med, Univ Md, 55-57; Nat Found res fel, McCollum-Pratt Inst, Johns Hopkins Univ, 57-59, asst prof physiol chem, Sch Med, 59-64; assoc prof, 64-68, exec officer biol sci group, 67-71, PROF BIOCHEM, UNIV CONN, 68-, HEAD BIOCHEM & BIOPHYS SECT, 67- Mem: AAAS; Am Chem Soc; Am Soc Biol Chem. Res: Mitochondrial metabolism; active transport processes in mitochondria and bacteria. Mailing Add: Biochem & Biophys Sect Biol Sci Group Univ of Conn Storrs CT 06268

VASINGTON, PAUL JOHN, b Norwich, Conn, June 6, 27; m 54; c 3. MICROBIOLOGY, VIROLOGY. Educ: Univ Conn, BA, 56; Univ Md, MS, 59, PhD(microbiol), 61. Prof Exp: Asst prof virol & tissue cult, St John's Univ, NY, 61-62; dir virol, Flow Labs Inc, Md, 62-64; mgr biol control, Lederle Labs, 64-67, asst mgr biol prod, 67-68, mgr biol sect, 68-70, MGR LEDERLE DIAG, AM CYANAMID CO, 70- Mem: AAAS; Am Soc Microbiol; NY Acad Sci. Res: Respiratory virus effects in non-human primates; tissue culture toxicity studies of rubbers and plastics; clinical chemistry. Mailing Add: Lederle Diag Am Cyanamid Co Pearl River NY 10965

VASKA, LAURI, b Rakvere, Estonia, May 7, 25; nat US; m 54; c 5. INORGANIC CHEMISTRY. Educ: Univ Göttingen, BS, 49; Univ Tex, PhD(chem), 56. Prof Exp: Fel Magnetism & chemisorption, Northwestern Univ, 56-57; res fel inorg chem, Mellon Inst, 57-64; assoc prof, 64-67, PROF INORG CHEM, CLARKSON COL TECHNOL, 67- Concurrent Pos: Fulbright-Hays fel, Univ Helsinki, 72. Honors & Awards: Boris Pregel Award, NY Acad Sci, 71. Mem: AAAS; Am Chem Soc; Catalysis Soc; fel NY Acad Sci; The Chem Soc. Res: Coordination chemistry; homogeneous catalysis; oxygen-carrying complexes; noble metal chemistry. Mailing Add: Dept of Chem Clarkson Col of Technol Potsdam NY 13676

VASKO, JOHN STEPHEN, b Cleveland, Ohio, Mar 17, 29; m 52; c 6. CARDIOVASCULAR & THORACIC SURGERY. Educ: Ohio State Univ, DDS, 54, MD, 58. Prof Exp: Instr anat & physiol, Cols Med & Dent, Ohio State Univ, 54-55, instr oper dent & consult prosthetic dent, Col Dent, 54-58; intern surg, Johns Hopkins Univ Hosp, 58-59, asst instr cardiac surg, Sch Med, Johns Hopkins Univ, 59-60; asst resident surgeon, Vanderbilt Univ Hosp, chief resident surgeon & chief labs exp surg, Nat Heart Inst, 64-66; assoc prof, 66-73, PROF THORACIC & CARDIOVASC SURG, COL MED, OHIO STATE UNIV, 73- Concurrent Pos: Halsted res fel cardiac surg, Sch Med, Johns Hopkins Univ, 59-60; asst resident surgeon cardiac surg, Johns Hopkins Hosp, 59-61; mem, Tech Rev Comt, Artificial Heart Prog, NIH & mem, Adv Comt, Coun Cardiovasc Surg, Am Heart Asn, 65-; partic, NIH Grad Prog, 65. Mem: AAAS; Asn Acad Surg; Soc Thoracic Surg; AMA; Am Col Surgeons. Res: Cardiovascular physiology. Mailing Add: Dept of Surg Ohio State Univ Col of Med Columbus OH 43210

VASLOW, DALE FRANKLIN, b Chicago, Ill, Aug 27, 45; m 69. PLASMA PHYSICS. Educ: Univ Wis, BS, 67, MS, 69, PhD(elec engr), 73. Prof Exp: SR SCIENTIST PLASMA PHYSICS, GEN ATOMIC CO, 74- Mem: Inst Elec & Electronics Engrs; Sigma Xi. Res: Ablation of hydrogen fuel pellet in a thermo nuclear plasma; optimization of power output and power gain for a tokamak fusion power reactor; chemical production utilizing fusion neutrons. Mailing Add: Box 81608 San Diego CA 92138

VASLOW, FRED, physical chemistry, see 12th edition

VASQUEZ, ALPHONSE THOMAS, b Boston, Mass, Apr 19, 38; m 65. MATHEMATICS. Educ: Mass Inst Technol, BS, 59; Univ Calif, Berkeley, PhD(math), 62. Prof Exp: Mem, Inst Advan Study, 62-64; res assoc math, Brandeis Univ, 64-65, asst prof, 65-67; ASSOC PROF MATH, GRAD DIV, CITY UNIV NEW YORK, 67- Concurrent Pos: NSF fel, 62-63. Mem: Am Math Soc. Res: Algebraic and differential topology; homological algebra. Mailing Add: Grad Div City Univ of NY 33 W 42nd St New York NY 10036

VASSALLE, MARIO, b Viareggio, Italy, May 26, 28; m 59; c 4. PHYSIOLOGY. Educ: Univ Pisa, MD, 53. Prof Exp: Fulbright travel grant, 58-62; NIH trainee, Med Col Ga, 59-60; NY Heart Asn fel, State Univ NY Downstate Med Ctr, 60-62; NIH fel, Physiol Inst, Bern, Switz, 62-64; from asst prof to assoc prof, 65-71, PROF PHYSIOL, STATE UNIV NY DOWNSTATE MED CTR, 71- Concurrent Pos: Mem coun basic sci, Am Heart Asn, 69- & Nat Conf Cardiovasc Dis, 69. Mem: AAAS; Am Physiol Soc; Harvey Soc; NY Acad Sci; Am Heart Asn. Res: Cardiac electrophysiology, particularly cardiac automaticity and its control. Mailing Add: Dept of Physiol State Univ NY Downstate Med Ctr Brooklyn NY 11203

VASSALLO, DONALD ARTHUR, b Waterbury, Conn, June 7, 32; m 60; c 4. POLYMER CHEMISTRY. Educ: Univ Conn, BA, 54; Univ Ill, MS, 56, PhD(anal chem), 58. Prof Exp: RES ASSOC, PLASTICS DEPT, E I DU PONT DE NEMOURS & CO, INC, 58- Mem: AAAS; Am Chem Soc; Soc Plastics Eng. Res: Automatic nonaqueous titrations; thermogravimetry; differential thermal analysis, especially polymers; polymer melt rheology; polymer structure-property correlations, especially polyolefins. Mailing Add: Plastics Dept Exp Sta E I du Pont de Nemours & Co Inc Wilmington DE 19898

VASSALLO, GODFREY, physics, deceased

VASSAMILLET, LAWRENCE FRANCOIS, b Elizabethville, Congo, Sept 14, 24; nat US; m 53; c 2. SOLID STATE PHYSICS. Educ: Mass Inst Technol, BSc, 46, MS, 50; Univ Liege, DSc(phys sci), 52; Carnegie Inst Technol, PhD(physics), 57. Prof Exp: Jr physicist, Monsanto Chem Co, Ohio, 47-48; physicist, Nat Carbon Co Div, Union Carbide Corp, 53; fel, 57-63, sr fel, 63-67, ASSOC PROF METALL & MAT SCI, CARNEGIE-MELLON UNIV, 67-, SR FEL, MELLON INST SCI, 74- Mem: Am Phys Soc; Electron Micros Soc Am; Electron Probe Anal Soc Am; Metall Soc. Res: X-ray diffraction; inperfections in crystals; electron probe microanalysis; electron microscopy. Mailing Add: Carnegie-Mellon Univ 4400 Fifth Ave Pittsburgh PA 15213

VASSAR, PHILIP STANLEY, b London, Eng, July 30, 24; m 50. PATHOLOGY. Educ: Univ London, MB, BS, 48; FRCPath. Prof Exp: Instr path, Columbia Univ, 50-53; resident pathologist, Royal Cancer Hosp, London, Eng, 54-56; from instr to assoc prof path, 56-66, PROF PATH, FAC MED, UNIV BC, 66- Concurrent Pos: resident, Presby Hosp, NY, 50-53; assoc surg pathologist, Vancouver Gen Hosp, 56-; consult, BC Cancer Inst, 56-, Can Tumour Registry, 57 & Nat Cancer Inst Can. Mem: Fel Col Am Path; Can Asn Path; Brit Med Asn; Int Acad Path. Res: Cell surfaces. Mailing Add: Dept of Path Univ of BC Fac of Med Vancouver BC Can

VASSEL, BRUNO, b Allahabath, Brit India, Oct 17, 08; nat US; m 36; c 3. BIOCHEMISTRY. Educ: Yale Univ, BS, 36; Univ Mich, MS, 37, PhD(biochem), 39. Prof Exp: Lab asst biochem, Univ Mich, 37-39; res biochemist, Am Cyanamid Co, 39-43; assoc prof biochem & res agr chemist exp sta, NDak Col, 43-46; supvr org & biochem res, Int Minerals & Chem Corp, 46-55; DIR RES, JOHNSON & JOHNSON OF BRAZIL, 55-, MEM EXEC COMT, 59- Honors & Awards: Robert W Johnson Medal Res & Develop, 64. Mem: AAAS; Am Chem Soc; Am Oil Chem Soc; Am Soc Sugar Beet Technol. Res: Protein isolations; monosodium glutamate processes; amino acid analyses and syntheses; pharmaceuticals; polarograph; flotation reagents; detergents; starch derivatives; surgical and pharmaceutical products. Mailing Add: 1630 Montia Ct Punta Gorda FL 33950

VASSELL, MILTON O, b Jamaica, West Indies, May 8, 31; m 58; c 2. THEORETICAL PHYSICS. Educ: NY Univ, BA, 58, PhD(physics), 64. Prof Exp: RES SCIENTIST, GTE LABS, INC, 64- Mem: Am Phys Soc; Sigma Xi. Res: Many particles physics; transport theory in semiconductors and metals; nonlinear optics; physics of lasers; acoustic surface wave propagation; acousto-electric effects; electron optics; integrated optics; optical guided wave propagation; holography. Mailing Add: GTE Labs Inc 40 Sylvan Rd Waltham MA 02154

VASSILIADES, ANTHONY E, b Chios, Greece, Nov 26, 33; US citizen; m 57; c 2. PHYSICAL CHEMISTRY, POLYMER CHEMISTRY. Educ: Wagner Col, BS, 56; Syracuse Univ, MS, 58; Polytech Inst Brooklyn, PhD(phys chem), 62. Prof Exp: From asst prof to assoc prof chem, Wagner Col, 61-66; assoc dir res, Champion Papers, Inc, 66-68, dir res, Champion Papers Group, US Plywood-Champion Papers, Inc, 68-70, VPRES & DIR RES & DEVELOP, CHAMPION INT CORP, 70- Concurrent Pos: Consult, Champion Papers, Inc, 64-66. Mem: Am Chem Soc; NY Acad Sci. Res: Coacervation of charged colloidal systems; transport phenomena in liquids; physical chemistry of high polymers; surface phenomena. Mailing Add: Champion Int Inc 130 N Franklin St Chicago IL 60606

VASSILIOU, ANDREAS H, b Ora, Larnaca, Cyprus, Nov 30, 36; m 65; c 1. MINERALOGY. Educ: Columbia Univ, BS, 63, MA, 65, PhD(mineral), 69. Prof Exp: Asst prof, 69-75, ASSOC PROF GEOL, RUTGERS UNIV, 75- Mem: Geol Soc Am; Mineral Soc Am. Res: Mineralogy of uranium, particularly the nature of urano-organic deposits in the Colorado Plateau and elsewhere. Mailing Add: Dept of Geol Rutgers Univ Newark NJ 07102

VASSILIOU, EUSTATHIOS, b Athens, Greece, Aug 22, 34; m 60; c 3. PHYSICAL CHEMISTRY, PLASTICS. Educ: Nat Tech Univ Athens, BScChE, 58; Univ Manchester, PhD(chem), 64. Prof Exp: Res chemist, Nuclear Res Ctr, Democritus, Greece, 64-66; res fel solid state physics, Harvard Univ, 66-67; res chemist, 67-72, STAFF CHEMIST, MARSHALL LAB, E I DU PONT DE NEMOURS & CO, INC, 72- Mem: Am Chem Soc. Res: Inorganic physical chemistry; physics and chemistry of glasses; vanadium; thermistors; anodic films; luminescence; ozone physics; semiconductor surface phenomena; magnetic phenomena; lubrication; structural plastics; fluorocarbon and other high temperature resistant coatings. Mailing Add: 12 S Townview Lane Newark DE 19711

VASSOS, BASIL HARILAOS, b Urziceni, Romania, Sept 3, 30; m 66; c 2. CHEMICAL INSTRUMENTATION, ANALYTICAL CHEMISTRY. Educ: Univ Bucharest, BSc, 52; Univ Mich, MSc, 62, PhD(chem), 65. Prof Exp: Lectr chem, Univ Bucharest, 52-60; res nuclear chem, Democritos Nuclear Res Ctr, Athens, 61-62; Alex von Humboldt Found fel, Max Planck Inst Metall Res, 67-68; asst prof anal chem, Seton Hall Univ, 68-72; fel, Colo State Univ, 72-73; ASSOC PROF ANAL CHEM, UNIV PR, 73- Mem: Am Chem Soc. Res: Trace analysis of pollutants. Mailing Add: Dept of Chem Univ of PR Rio Piedras PR 00931

VASTA, BRUNO MORREALE, b Washington, DC, Apr 1, 33; m 56; c 4. INFORMATION SCIENCE, BIOCHEMISTRY. Educ: Georgetown Univ, BS, 54; George Washington Univ, MS, 57. Prof Exp: Biochemist, Nat Heart Inst, 54-57; plant biochemist, USDA, 57-60; sr chemist, Life Sci Dept, Melpar, Inc, Va, 60-64; pesticide res chemist, Bur Sci, Food & Drug Admin, 64-66, chief chemist & sci info off, Bur Vet Med, 66-68, chief chem info, Sci Info Facil, 68-70; CHIEF, TOXICOL INFO SERV, NAT LIBR MED, 70- Honors & Awards: Merit Award, Food & Drug Admin, 70. Mem: AAAS; Am Chem Soc; Drug Info Asn; Chem Notation Asn. Res: Chemical substructure searching techniques; structure display; toxicology data; files building; computerized information storage and retrieval. Mailing Add: Nat Libr Med 8600 Rockville Bethesda MD 20014

VASTANO, ANDREW CHARLES, b New York, NY, Feb 26, 36; m 58; c 4. PHYSICAL OCEANOGRAPHY, NUMERICAL ANALYSIS. Educ: NC State Univ, BS, 56; Univ NC, MS, 60; Tex A&M Univ, PhD(phys oceanog), 67. Prof Exp: Anal engr, Pratt & Whitney Aircraft Corp, 56-57; sonar engr, Western Elec Co, Bell Tel Co, 60-62; instr physics, Tex A&M Univ, 62-63, res scientist, 63-66; assoc prof phys oceanog, Univ Fla, 66-67; asst scientist, Woods Hole Oceanog Inst, 67-69; instr, 64, ASSOC PROF OCEANOG, TEX A&M UNIV, 69- Mem: Am Geophys Union. Res: Numerical studies of tsunami and storm surges; theory of gravity waves; mesoscale ocean dynamics; topographic interaction of current systems. Mailing Add: Dept of Oceanog Tex A&M Univ College Station TX 77843

VASTINE, FREDERICK DAVIDSON, b Danville, Pa, Dec 31, 40; m 68. ORGANIC CHEMISTRY, ORGANOMETALLIC CHEMISTRY. Educ: Ursinus Col, BS, 62; Univ NC, Chapel Hill, PhD(org chem), 67. Prof Exp: Sr chemist, Arco Chem Co, Div Atlantic Richfield Co, 67-71; tech consult, Oblon, Fisher, Spivak & McClelland, 71-73, PATENT AGENT, OBLON, FISHER, SPIVAK, MCCLELLAND & MAIER, 73- Mem: Am Chem Soc; AAAS. Res: Synthesis of organometallic compounds; coal hydrogenation; petroleum chemistry. Mailing Add: 9119 Rockefeller Lane Springfield VA 22153

VASTOKAS, JOAN M, b Toronto, Ont, Apr 20, 38; m 59. ANTHROPOLOGY. Educ: Univ Toronto, BA, 60, MA, 61; Columbia Univ, PhD(art hist, archaeol), 66. Prof Exp: Lectr anthrop, Univ Toronto, 65-66, lectr anthrop & fine art, 66-67, asst prof, 67-70; asst prof anthrop, 70-74, ASSOC PROF ANTHROP & CHMN DEPT, TRENT UNIV, 74- Concurrent Pos: Res assoc ethnol, Royal Ont Mus, Toronto, 67-; consult, Nat Gallery Can, 67-; Can Coun Soc Sci & Humanities res grant, 70-71. Mem: Am Soc Ethnohist; Univ Art Asn Can; fel Royal Anthrop Inst Gt Brit & Ireland. Res: Art and architecture of the Northwest Coast Indians; petroglyphs of the Canadian shield; arts of Africa, Oceania and Americas. Mailing Add: Dept of Anthrop Trent Univ Peterborough ON Can

VASTOLA, EDWARD FRANCIS, b Waterbury, Conn, June 1, 24; m 49; c 2. MEDICINE, NEUROLOGY. Educ: Yale Univ, BS, 45; Columbia Univ, MD, 47; Am Bd Psychiat & Neurol, dipl. Prof Exp: Instr neurol, Sch Med, Washington Univ, 53-55; from asst prof to assoc prof, 55-70, PROF NEUROL, COL MED, STATE UNIV NY DOWNSTATE MED CTR, 70-, HEAD DEPT, 59-; DIR DEPT NEUROL, KINGS COUNTY HOSP, 59- Concurrent Pos: Consult, Vet Admin Hosp, Brooklyn; vis neurologist, Ft Hamilton Vet Admin Hosp; mem neurol staff, Huntington Hosp, NY. Res; Neurophysiology; clinical neurology. Mailing Add: Dept of Neurol Col of Med State Univ NY Downstate Med Ctr Brooklyn NY 11203

VASTOLA, FRANCIS J, b Buffalo, NY, Feb 22, 28; m 69. PHYSICAL CHEMISTRY. Educ: Univ Buffalo, BA, 50; Pa State Univ, PhD(fuel technol), 59. Prof Exp: Chemist, Nat Bur Standards, Washington, DC, 50-52; asst & res assoc, 56-59, from asst prof to assoc prof fuel sci, 59-72, PROF FUEL SCI, PA STATE UNIV, 72- Mem: Am Chem Soc. Res: Mass spectrometry; solid and gaseous combustion; kinetics and instrumentation. Mailing Add: Dept of Mat Sci Pa State Univ University Park PA 16802

VASU, BANGALORE SESHACHALAM, b Bangalore, India, May 20, 29; Indian citizen; m 65; c 1. BIOLOGY. Educ: Univ Madras, BSc, 49, MSc, 62; Stanford Univ, PhD(biol), 65. Prof Exp: Asst prof zool, Pachaiyappa's Col, Madras Univ, 50-59, lectr, Zool Res Lab, 59-62; Fulbright res fel biol, Standord Univ, 62-65; AEC fel, Univ Notre Dame, 65-67; asst prof zool, Ohio Wesleyan Univ, 67-68; UNESCO specialist biol, Univ Zambia, 68-73; ASST PROF BIOL, CALIF STATE UNIV, CHICO, 74- Concurrent Pos: Fulbright grant, US Educ Found India, 62. Mem: Sigma Xi; Am Inst Biol Sci; AAAS; Radiation Res Soc; Marine Biol Asn UK. Res: Age-related changes at the cellular and moleuclar level. Mailing Add: Dept of Biol Sci Calif State Univ Chico CA 95926

VATNE, ROBERT DAHLMEIER, b Pipestone, Minn, Oct 21, 34; m 63; c 1. PARASITOLOGY. Educ: Augustana Col, SDak, BA, 56; Kans State Univ, MS, 58, PhD(parasitol), 63. Prof Exp: Asst prof biol, St Cloud State Col, 63-64; assoc scientist, Salsbury Labs, Iowa, 65-66, scientist, 67-69; res specialist, 70-73, REGIST SPECIALIST, DOW CHEM CO, USA, 74- Mem: Am Soc Parasitol. Res: Etiology and chemotherapy of histomoniasis, coccidiosis and helminthiasis. Mailing Add: Ag-Org Dept Dow Chem Co USA PO Box 1706 Midland MI 48640

VATSIA, MISHRI LAL, physics, optics, see 12th edition

VATTHAUER, RICHARD JAMES, animal nutrition, microbiology, see 12th edition

VATUK, SYLVIA DUTRA, b Providence, RI, Mar 1, 34; m 57; c 4. ANTHROPOLOGY. Educ: Cornell Univ, BS, 55; Univ London, MA, 58; Harvard Univ, PhD(anthrop), 70. Prof Exp: ASSOC PROF ANTHROP, UNIV ILL, CHICAGO CIRCLE, 70- Concurrent Pos: NIMH res fel, 74-76. Mem: Am Anthrop Asn; Royal Anthrop Inst Gt Brit & Ireland; Asn Asian Studies. Res: Social organization, family and kinship in India; social and cultural dimensions of aging in India. Mailing Add: Dept of Anthrop Univ of Ill Chicago Circle Chicago IL 60680

VAUDO, ANTHONY FRANK, b Brooklyn, NY, Jan 21, 46; m 67; c 1. PHYSICAL CHEMISTRY. Educ: City Col New York, BS, 66; Mass Inst Technol, PhD(phys chem), 70. Prof Exp: NSF fel, Boston Univ, 70-71; sr develop chemist, 71-74, TECH DIR LAMINATED & COATED PRODS DIV, ST REGIS PAPER CO, 74- Mem: Am Chem Soc; Soc Photog Sci & Eng; Tech Asn Pulp & Paper Indust. Res: Photochemistry and its applications to imaging materials. Mailing Add: St Regis Paper Co 55 Starkey Ave Attleboro MA 02703

VAUGHAN, ALFRED LELAND, b Veedersburg, Ind, Sept 3, 06; m 33; c 2. MOLECULAR PHYSICS, MASS SPECTROSCOPY. Educ: DePauw Univ, BA, 29; Univ Minn, PhD(physics), 34. Prof Exp: Asst physics, Univ Minn, Minneapolis, 29-35, from instr to prof phys sci, 35-75, admin officer, 40-46, from asst dean to assoc dean gen col, 46-66, dir, Univ Col, 65-71, from actg dean to dean gen col, 66-75; RETIRED. Mem: AAAS; Am Phys Soc; Am Asn Physics Teachers. Res: Ionization potentials; dissociation phenomena; isotopic abundance ratios; collision phenomena of electrons in gases; sound transmission. Mailing Add: 1269 N Cleveland Ave St Paul MN 55108

VAUGHAN, ARTHUR HARRIS, JR, b Salem, Ohio, July 19, 34; m 59; c 1. ASTRONOMY. Educ: Cornell Univ, BEngPhys, 58; Univ Rochester, PhD(physics, astron), 65. Prof Exp: Jr res scientist, Avco-Everett Res Lab, Mass, 58-59; fel, 64-65, asst astronr, 65-66, staff assoc, 66-67, STAFF MEM, CARNEGIE INST WASH HALE OBSERV, 67- Honors & Awards: Sci Award, Eastman Kodak, 63. Mem: Am Astron Soc; Royal Astron Soc; Int Astron Union. Res: Spectroscopy; interstellar medium; planetary nebulae; chromospheres; astronomical instrumentation; interferometry. Mailing Add: Carnegie Inst Wash Hale Observ 813 Santa Barbara St Pasadena CA 91106

VAUGHAN, BURTON EUGENE, b Santa Rosa, Calif, May 31, 26; m 49; c 2. PHYSIOLOGY, BIOPHYSICS. Educ: Univ Calif, Berkeley, AB, 49, PhD, 55. Prof Exp: Vis scientist, White Mountain High Altitude Res Sta, Calif, 53-54; proj leader, Oper Deepfreeze I, US Exped to Antarctic, 55; staff scientist biophys br, US Naval Radiol Defense Lab, 56-61, br head, 62-69; MGR ECOSYSTS DEPT, PAC NORTHWEST LABS, BATTELLE MEM INST, 69- Concurrent Pos: Actg instr, Sch Med, Stanford Univ, 57, Lectr, 60-63, res assoc, 63-70; consult, Clin Invest Ctr, Oakland Naval Hosp, 60-61 & US Naval Med Res Unit 2, Taipei, Taiwan, 61, 64 & 65; trustee, Independent Sch Dist, 63, presiding off, 64-65; rep, County Comt Sch Dist Orgn, 65; mem bd educ, Castro Valley Unified Sch Dist, 65, pres, 66-67; consult, Govt Health Facil, Manila, Philippines; sci coun mem, Pac Sci Ctr, Seattle. Mem: Am Soc Plant Physiol; Am Physiol Soc; Biophys Soc; NY Acad Sci; Radiation Res Soc. Res: Electrolyte and other absorption processes, including permeability, metabolically coupled transport; gastro-intestinal irradiation injury in the mammal; ionic uptake, discrimination and fixation by terrestrial plant and marine algal tissues; radiocontamination processes in organisms and ecological systems; plant and animal physiology. Mailing Add: Eco-Systs Dept Pac Northwest Labs Box 999 Richland WA 99352

VAUGHAN, CHARLES EDWIN, b Mathiston, Miss, Jan 18, 33; m 57; c 1. AGRONOMY. Educ: Miss State Univ, BS, 55, MS, 62; NC State Univ, PhD(agron), 69. Prof Exp: Asst agronomist & asst prof agron, 59-69, ASSOC AGRONOMIST & ASSOC PROF AGRON, MISS STATE UNIV, 69- Mem: Am Soc Agron; Asn Off Seed Anal. Res: Seed physiology; seed deterioration; quality control in seed programs. Mailing Add: Seed Technol Lab Miss State Univ Mississippi State MS 39762

VAUGHAN, DAVID ARTHUR, b Mattoon, Wis, Mar 5, 23; m 51; c 1. NUTRITION. Educ: Univ Calif, BA, 49; Univ Ill, MA, 54, PhD(animal nutrit), 55. Prof Exp: Res physiologist, Arctic Aeromed Lab, US Dept Air Force, 55-59, supvry chemist, 59-68, res biochemist, US Air Force Sch Aerospace Med, 68-69; RES BIOCHEMIST, NUTRIT INST, AGR RES SERV, USDA, 69- Mem: Am Physiol Soc; Am Inst Nutrit; Arctic Inst NAm. Res: Survival nutrition in the Arctic; vitamin B requirements and intermediary metabolism during stress; biochemistry of hibernation; protein nutrition. Mailing Add: Nutrit Inst Agr Res Serv USDA Beltsville MD 20705

VAUGHAN, DAVID EVAN WILLIAM, b Hull, Yorkshire, Eng, Feb 4, 39; m 63; c 3. CHEMISTRY, GEOCHEMISTRY. Educ: Univ Durham, BSc, 61; Pa State Univ, MS, 63; Univ London, DIC & PhD(phys chem), 67. Prof Exp: Sr res asst chem, Imp Col, Univ London, 63-67; from res chemist to sr res chemist, 67-71, RES SUPVR, W R GRACE & CO, 72- Mem: The Chem Soc; Am Chem Soc; Am Mineral Soc. Res: Sorption and catalytic properties of molecular sieves; geology and geochemistry of zeolites. Mailing Add: Davidson Chem Div W R Grace & Co Clarksville MD 21044

VAUGHAN, DEBORAH WHITTAKER, b Concord, NH, Nov 30, 43; m 66. NEUROANATOMY. Educ: Univ Vt, BA, 66; Boston Univ, PhD(biol), 71. Prof Exp: USPHS fel, 71-72, RES ASST PROF NEUROANAT, SCH MED, BOSTON UNIV, 72- Res: Electron microscopic analysis of neocortex, primarily of rat, in regards to the effects of aging on the brain. Mailing Add: Dept of Anat Boston Univ Sch of Med Boston MA 02118

VAUGHAN, EDWARD KEMP, b East Las Vegas, NMex, Nov 16, 08; m 32; c 2. PLANT PATHOLOGY. Educ: NMex State Univ, BS, 29; Ore State Univ, MS, 32; Univ Minn, PhD(plant path), 42. Prof Exp: Jr plant quarantine inspector, USDA, 29-30; instr high sch, NMex, 33-34; asst horticulturist, Soil Conserv Serv, USDA, 34-36; instr plant path, Univ Minn, 36-37; agent, Bur Plant Indust, USDA, 37-41; assoc prof, Va Polytech Inst, 41-43, exten plant pathologist, 43-44; plant pathologist, USDA, 44-47; plant pathologist, Exp Sta, 47-74, EMER PROF PLANT PATH, ORE STATE UNIV, 74- Concurrent Pos: Guggenheim Found fel, Inst Phytopath Res, Neth, 54-55; NZ Dept Sci & Indust Res sr res fel soil microbiol, Plant Dis Div, Univ Auckland, NZ, 64-65; res adv plant path, Develop & Res Corp, Andimeshk, Iran, 74-76. Mem: Fel AAAS; Am Phytopath Soc; Brit Asn Appl Biol; Neth Soc Plant Path; Iranian Phytopathol Soc. Res: Vegetable diseases; physiology of disease resistance. Mailing Add: 1606 NW Alta Vista Dr Corvallis OR 97330

VAUGHAN, FRANCIS EDWARD, b Ludden, NDak, Sept 12, 89; m 20; c 3. GEOLOGY. Educ: Univ Calif, BS, 12, MS, 16, PhD(geol), 18. Prof Exp: Mining engr, Ariz, 12-14; geologist, NAm Oil Co, Calif, 14-15 & Royal Dutch-Shell Combine, 18-30; consult geophysicist, 31-33; geologist, Shell Oil Co, 33-38; consult geophysicist, 39; pres & sr partner, V & E Mfg Co, 40-62, PRES, VEMCO CORP, 62- Mem: AAAS; Am Asn Petrol Geol; Am Inst Mining, Metall & Petrol Eng. Res: Petroleum geology; geophysics; gravimetric and seismometric methods; development and manufacture of geophysical apparatus, drafting equipment and special precision instruments. Mailing Add: 3476 E Lombardy Rd Pasadena CA 91107

VAUGHAN, HERBERT EDWARD, b Ogdensburg, NY, Feb 18, 11; m 47. MATHEMATICS. Educ: Univ Mich, BS, 32, AM, 33, PhD(math), 35. Prof Exp: Instr math, Brown Univ, 35-36; Lloyd fel, Univ Mich, 36-37; instr, 37-41, assoc, 41-45, from asst prof to assoc prof, 45-60, PROF MATH, UNIV ILL, URBANA, 60-, MEM UNIV COMT ON SCH MATH PROJ EDUC, 56- Mem: Am Math Soc; Math Asn Am; Asn Symbolic Logic. Res: Topology; abstract spaces. Mailing Add: Dept of Math Univ of Ill Urbana IL 61801

VAUGHAN, JAMES HERBERT, JR, b Norfolk, Va, July 28, 27; m 50; c 2. ANTHROPOLOGY. Educ: Univ of the South, BA, 50; Univ NC, MA, 52; Northwestern Univ, PhD, 60. Prof Exp: Asst prof anthrop, Rockford Col, 61 & Univ Cincinnati, 61-67; assoc prof, 67-71, chmn dept, 70-75, PROF ANTHROP, IND UNIV, BLOOMINGTON, 71- Concurrent Pos: Ford Found fel, 59-60; mem, President's Task Force, Africa, 60. Mem: AAAS; Am Anthrop Asn; African Studies Asn; Int African Inst. Res: Cultural theory; political anthropology; ethnography of Africa, especially the Mandara Mountains and Nigeria. Mailing Add: Dept of Anthrop Ind Univ Bloomington IN 47401

VAUGHAN, JAMES ROLAND, b Allentown, Pa, June 7, 28; m 50; c 3. MICROBIOLOGY, BIOCHEMISTRY. Educ: Muhlenberg Col, BS, 52; Lehigh Univ, MS, 54, PhD(biol), 61. Prof Exp: From instr to assoc prof microbiol, 56-67, PROF MICROBIOL, MUHLENBERG COL, 67-, HEAD DEPT, 65- Mem: AAAS; Am Soc Microbiol; Am Chem Soc; Am Soc Cell Biol. Res: Microbial physiology, aerobiology and biochemistry of mucopolysaccharides and bacterial pigments. Mailing Add: Dept of Biol Muhlenberg Col Allentown PA 18104

VAUGHAN, JERRY EUGENE, b Gastonia, NC, Oct 30, 39; m 69. MATHEMATICS, TOPOLOGY. Educ: Davidson Col, BS, 61; Duke Univ, PhD(math), 65. Prof Exp: Teaching asst math, Duke Univ, 62-63; assoc prof, Eve Div, Univ Md, 66-67; asst prof, Univ NC, Chapel Hill, 67-73; ASSOC PROF MATH, UNIV NC, GREENSBORO, 73- Mem: Am Math Soc; Math Asn Am. Res: General topology; generalized metric spaces; product spaces; cardinal invariant properties. Mailing Add: Dept of Math Univ of NC Greensboro NC 27412

VAUGHAN, JOHN DIXON, b Clarksville, Va, Mar 3, 25; m 62. PHYSICAL CHEMISTRY. Educ: Col William & Mary, BS, 50; Univ Ill, PhD(chem), 54. Prof Exp: Asst phys chem, Univ Ill, 50-51, AEC, 51-54; res chemist, Chem Dept, Exp Sta, E I du Pont de Nemours & Co, 54-58; asst prof phys chem, Va Polytech Inst, 59-62 & Univ Hawaii, 62-64; assoc prof, 64-71, PROF PHYS CHEM, COLO STATE UNIV, 71- Mem: Am Chem Soc. Res: Quantum chemistry; homogeneous and heterogeneous catalysis; reactivities of heterocycles. Mailing Add: Dept of Chem Colo State Univ Ft Collins CO 80521

VAUGHAN, JOHN HEATH, b Richmond, Va, Nov 11, 21; m 46; c 4. IMMUNOLOGY, MEDICINE. Educ: Harvard Univ, AB, 42, MD, 45. Prof Exp: Intern med, Peter Bent Brigham Hosp, 45-56, resident, 48-51; Nat Res Coun fel med sci, Col Physicians & Surgeons, Columbia Univ, 51-53; asst prof, Med Col Va, 53-58; assoc prof med & asst prof bact, Sch Med & Dent, Univ Rochester, 58-63, prof med & head div immunol & infectious dis, 63-70; chmn clin div, 70-74, CHMN DEPT CLIN RES, SCRIPPS CLIN & RES FOUND, 74- Concurrent Pos: Fel, Peter Bent Brigham Hosp, Boston, 48-51; consult, NIH, 56-63, mem bd sci counr, Nat Inst Allergy & Infectious Dis, 68-72. Mem: Am Soc Clin Invest; Asn Am Physicians; Am Asn Immunol; Am Rheumatism Asn (pres, 70-); Am Acad Allergy (pres, 66-67). Res: Immunological phenomena in internal medicine; allergy; arthritis; hemolytic disease. Mailing Add: Scripps Clin & Res Found 476 Prospect Lane La Jolla CA 92037

VAUGHAN, JOHN THOMAS, b Tuskegee, Ala, Feb 6, 32; m 56; c 3. VETERINARY MEDICINE. Educ: Auburn Univ, DVM, 55, MS, 63. Prof Exp: From instr to assoc prof large animal surg & med, Auburn Univ, 55-70; prof vet surg & dir large animal hosp, NY State Vet Col, Cornell Univ, 70-74; CHMN DEPT LARGE ANIMAL SURG & MED, AUBURN UNIV, 74- Mem: Am Vet Med Asn; Am Asn Equine Practitioners; Am Asn Vet Clinicians; Am Col Vet Surg. Res: Large animal and equine surgery and medicine; general surgery of the equine system with emphasis on urogenital and gastrointestinal surgery. Mailing Add: Dept of Large Animal Surg & Med Auburn Univ Auburn AL 36820

VAUGHAN, LOY OTTIS, JR, b Birmingham, Ala, June 30, 45; m 66; c 1. MATHEMATICS. Educ: Fla State Univ, BA, 66; Univ Ala, MA, 67, PhD(math), 70. ASST PROF MATH, UNIV ALA, BIRMINGHAM, 69- Mem: Am Math Soc; Math Asn Am. Res: General topology, fixed and almost fixed point theory of continua. Mailing Add: Dept of Math Univ of Ala University Sta Birmingham AL 35294

VAUGHAN, MARTHA, b Dodgeville, Wis, Aug 4, 26; m 51; c 3. BIOCHEMISTRY. Educ: Univ Chicago, PhB, 44; Yale Univ, MD, 49. Prof Exp: Asst instr res med, Univ Pa, 51-52; Nat Res Coun fel, 52-54, from sr asst surgeon to sr surgeon, 54-62, MED DIR, NAT HEART INST, 62- Concurrent Pos: Fel res med, Univ Pa, 51-52; mem metab study sect, Div Res Grants, USPHS, 65-68, head sect metab, 68-74, chief, Lab Cellular Metab, 74- Mem: Am Soc Biol Chemists; Am Soc Clin Invest; Asn Am Physicians. Res: Adipose tissue metabolism; mechanism of hormone action. Mailing Add: Rm 5N-307 Bldg 10 NIH Bethesda MD 20014

VAUGHAN, NICK HAMPTON, b Graham, Tex, Feb 11, 23; m 60; c 2. MATHEMATICS. Educ: NTex State Univ, BS, 47, MS, 48; La State Univ, PhD(math), 68. Prof Exp: Mathematician, US Naval Ord Plant, Ind, 51-53; res assoc math, Statist Lab, Purdue Univ, Lafayette, 53-55; sr aerophysics engr, Gen Dynamics, Tex, 55; instr math, La State Univ, 55-58; asst prof, NTex State Univ, 58-65; instr, La State Univ, 65-68; ASSOC PROF MATH, NTEX STATE UNIV, 68- Mem: Am Math Soc; Math Asn Am. Res: Commutative rings; ideal theory; algebraic number theory. Mailing Add: Dept of Math NTex State Univ Denton TX 76203

VAUGHAN, PHILIP ALFRED, b Palo Alto, Calif, Sept 21, 23; m 45; c 2. PHYSICAL CHEMISTRY. Educ: Pomona Col, BA, 43; Calif Inst Technol, PhD(chem), 49. Prof Exp: Chemist, Union Oil Co Calif, 43-46; Hale fel, Calif Inst Technol, 49-50; from asst prof to prof chem, Rutgers Univ, 50-68; vpres, V & E Mfg Co, Calif, 68, EXEC VPRES, VEMCO CORP, 68- Res: Crystal and molecular structure; x-ray crystallography; structure and properties of solutions. Mailing Add: Vemco Corp 766 SFair Oaks Pasadena CA 91105

VAUGHAN, TERRY ALFRED, b Los Angeles, Calif, May 5, 28; m 50; c 2. VERTEBRATE ZOOLOGY. Educ: Pomona Col, BA, 50; Claremont Cols, MA, 52; Univ Kans, PhD, 58. Prof Exp: Asst zool, Univ Kans, 52-54 & 56-58; asst biologist, Colo State Univ, 58-64, assoc biologist, 64-70; PROF ZOOL, NORTHERN ARIZ UNIV, 70- Mem: Am Soc Mammalogists; Cooper Ornith Soc. Res: Chiropteran and rodent ecology; functional morphology. Mailing Add: Dept of Zool Northern Ariz Univ Flagstaff AZ 86001

VAUGHAN, THERESA PHILLIPS, b Kearney, Nebr, Oct 13, 41; m 69. ALGEBRA. Educ: Antioch Col, BA, 64; Am Univ, MA, 68; Duke Univ, PhD(math), 72. Prof Exp: Math programmer, Naval Ship Res & Develop Ctr, 64-69; asst prof math, NC Wesleyan Col, Rocky Mt, 72-73; LECTR MATH, UNIV NC, GREENSBORO, 74- Mem: Am Math Soc. Res: Polynomials and linear structure of finite fields; enumeration of sequences of integers by patterns; 2x2 matrices with positive integer entries. Mailing Add: Dept of Math Univ of NC Greensboro NC 27412

VAUGHAN, VICTOR CLARENCE, III, b Toledo, Ohio, July 19, 19; m 41; c 3. PEDIATRICS. Educ: Harvard Univ, AB, 39, MD, 43; Am Bd Pediat, dipl, 51, cert pediat allergy, 60. Prof Exp: Instr pediat, Sch Med, Yale Univ, 45-47 & 49-50, asst prof, 50-52; assoc prof, Sch Med, Temple Univ, 52-57; prof & chmn dept, Med Col Ga, 57-64; PROF PEDIAT & CHMN DEPT, SCH MED, TEMPLE UNIV, 64-; MED DIR, ST CHRISTOPHER'S HOSP CHILDREN, 64- Concurrent Pos: Res fel, Harvard Med Sch, 47-49; mem bd, Am Bd Pediat, 60-65 & 68-73, secy, 63-65, pres, 73. Mem: Fel Am Acad Allergy; Soc Pediat Res (vpres, 64-65); Am Acad Pediat; Am Fedn Clin Res; Am Pediat Soc. Res: Hemolytic disease of newborn; human genetics; allergic disorders of children; human growth and development. Mailing Add: Christophers Hosp Children 2600 N Lawrence St Philadelphia PA 19133

VAUGHAN, WILLIAM WALTON, b Clearwater, Fla, Sept 7, 30; m 51; c 4. METEOROLOGY, AEROSPACE SCIENCES. Educ: Univ Fla, BS, 51; Fla State Univ, cert meteorol, 52. Prof Exp: Meteorologist, US Air Force, 52-55, res & develop meteorologist, Air Force Armament Ctr, 55-57, tech asst meteorol, Army Ballistic Missile Agency, 58-60; chief, Aerospace Environ Off, 60-65, CHIEF, AEROSPACE ENVIRON DIV, MARSHALL SPACE FLIGHT CTR, NASA, 65- Honors & Awards: Exceptional Serv Medal, NASA, 69. Mem: Am Meteorol Soc; Am Inst Aeronaut & Astronaut. Res: Applied research in aerospace sciences and especially meteorology relative to space system and spacecraft experiment development. Mailing Add: 5606 Alta Dena Dr Huntsville AL 35802

VAUGHAN, WORTH E, b New York, NY, Feb 1, 36; m 69; c 4. PHYSICAL CHEMISTRY. Educ: Oberlin Col, AB, 57; Princeton Univ, AM, 59, PhD(phys chem), 60. Prof Exp: Res assoc phys chem, Princeton Univ, 60-61; asst prof chem, 61-71, ASSOC PROF CHEM, UNIV WIS-MADISON, 71- Mem: AAAS; Am Phys Soc; Am Chem Soc. Res: Dielectric and nuclear magnetic relaxation in liquids; irreversible statistical mechanics. Mailing Add: Dept of Chem Univ of Wis Madison WI 53706

VAUGHAN, WYMAN RISTINE, b Minneapolis, Minn, Oct 28, 16; m 43; c 2. CHEMISTRY. Educ: Dartmouth Col, AB, 39, AM, 41; Harvard Univ, AM, 42, PhD(org chem), 44. Prof Exp: Instr chem, Dartmouth Col, 39-41; res assoc, Harvard Univ, 42-44; res assoc, Dartmouth Col, 44-46; res assoc, Univ Mich, 46-47, from instr to prof, 47-66; PROF CHEM & HEAD DEPT, UNIV CONN, 66- Mem: Am Chem Soc; NY Acad Sci. Res: Synthetic organic chemistry; Diels-Alder reaction; sterochemistry; reaction mechanisms; potential anticancer agents; molecular rearrangements. Mailing Add: Dept of Chem Univ of Conn Storrs CT 06268

VAUGHN, CHARLES MELVIN, b Deadwood, SDak, Nov 23, 15; m 41; c 2. PARASITOLOGY, PROTOZOOLOGY. Educ: Univ Ill, BA, 39, MA, 40; Univ Wis, PhD(invert zool), 43. Prof Exp: Asst zool, Univ Ill, 39-40 & Univ Wis, 40-43; from asst prof to assoc prof, Miami Univ, 46-51; assoc dir, Field Serv Unit, Am Found Trop Med, 52-53; prof zool & chmn dept, Univ SDak, 53-65; prof zool & physiol & chmn dept, 65-71, actg dean res, 69-71, PROF ZOOL & CHMN DEPT, MIAMI UNIV, 71- Concurrent Pos: Asst prof parasitol & sr parasitologist, Inst Trop Med, Bowman Gray Sch Med, Wake Forest Col, 50-52, assoc dir, 51-52; dir, NSF Acad Year Inst, Univ SDak, 58-64, prog dir, NSF Col & Elem Prog, Res Training & Acad Year Study Prog, 64-65; consult, NSF, 65-69; biologist, Ohio State Univ-US AID Indian Educ Proj, 66 & NSF-US AID India Educ Proj, 67. Mem: AAAS; Am Soc Parasitol; Am Micros Soc (pres, 73); Soc Protozool; Am Soc Trop Med & Hyg. Res: Human and animal parasitology; malaria parasite and vector surveys; intestinal parasite surveys; schistosomiasis survey and control; physiology and ecology of gastropoda; life history and culture methods in protozoa, other invertebrates. Mailing Add: Dept of Zool 282 Upham Hall Miami Univ Oxford OH 45056

VAUGHN, CLARENCE BENJAMIN, b Philadelphia, Pa, Dec 14, 28; m 53; c 4. ONCOLOGY, PHYSIOLOGICAL CHEMISTRY. Educ: Benedict Col, BS, 51; Howard Univ, MS, 55, MD, 57; Wayne State Univ, PhD(physiol chem), 65. Prof Exp: Res fel oncol, Detroit Inst Cancer Res, 65-67; ASSOC BIOCHEM & ASST PROF ONCOL, WAYNE STATE UNIV, 67-; DIR ONCOL, PROVIDENCE HOSP, 73- Concurrent Pos: Mem consult staff, Depts Med, Blvd Gen Hosp & Kirwood Gen Hosp, 67-, Oakwood Hosp, 68-, Detroit Mem Hosp, 70- & Crittenton Hosp, Rochester, Mich, 72-; clin dir, Milton A Darling Mem Ctr, Mich Cancer Found, 70-72; mem, Coop Breast Cancer Study Group; comdr, 927 Tactical Air Command Hosp & dir med serv, Selfridge AFB, Mich. Mem: AMA; Am Col Physicians; NY Acad Sci. Res: Organic acid metabolism; prophyria metabolism; estrogen metabolism. Mailing Add: Providence Hosp 16001 W Nine Mile Rd Southfield MI 48075

VAUGHN, GWENYTH RUTH, b Lander, Wyo, June 8, 17; c 1. SPEECH PATHOLOGY, AUDIOLOGY. Educ: Univ Denver, BA, 39, MA, 42, PhD(speech path, audiol), 59. Prof Exp: Dean of personnel & head dept speech, Colo Womens Col, 41-44; dean of women & head dept speech, Univ of the Americas, 46-49; dir & founder, Johnson Sch Handicapped Children & Sch Spec Educ, Mexico City, 44-57; assoc prof speech path & audiol, head dept & dir speech & hearing clin, Idaho State Univ, 60-65; dir med ctr speech lang clin, head sect speech pathophysiol & dir clin communicology & asst dir training, Res & Training Ctr, 66, PROF SPEECH PATH & AUDIOL & DIR VET ADMIN DIV, DEPT BIOCOMMUN, SCHS MED & DENT, UNIV ALA, BIRMINGHAM, 66-; VET ADMIN MED DIST COORDR AUDIOL & SPEECH PATH, BIRMINGHAM VET ADMIN HOSP, 67- Concurrent Pos: Dir, Voc Rehab Admin Proj, Idaho, 62-65, Title III Proj, Birmingham Pub Schs, 69-72, Univ Ala Sch Med & Vet Admin Proj, Birmingham, 72-73 & Vet Admin Exchange Med Info Proj, Birmingham, 73-76. Honors & Awards: Adminr Commendation, Vet Admin, 73, Outstanding Hamdicapped Fed Employee of the Yr, Vet Admin, President of US, 73. Mem: Fel Am Speech & Hearing Asn; Acad Rehab Audiol; fel Am Asn Ment Deficiency. Res: Health care delivery systems to persons with communicative disorders; programs for the hearing impaired. Mailing Add: Rte 11 Box 499-700 N Birmingham AL 35210

VAUGHN, HOWARD ALTON, JR, organic chemistry, see 12th edition

VAUGHN, JACK C, b Burbank, Calif, July 4, 37; m 63; c 3. BIOCHEMICAL CYTOLOGY. Educ: Univ Calif, Los Angeles, BA, 60; Univ Tex, PhD(bot), 64. Prof Exp: Asst zool, Univ Calif, Los Angeles, 60-61; USPHS fel, Univ Wis, 64-66; asst prof, 66-70, ASSOC PROF ZOOL, MIAMI UNIV, 70- Concurrent Pos: NSF res grant, 67-69 & 71. Mem: AAAS; Am Inst Biol Scientist; Am Soc Cell Biol; Am Soc Zoologists. Res: Cell biology; histone metabolism; satellite DNA's; cellular differentiation. Mailing Add: Dept of Zool Miami Univ Oxford OH 45056

VAUGHN, JAMES E, JR, b Kansas City, Mo, Sept 17, 39; m 61; c 2. ANATOMY. Educ: Westminster Col, BA, 61; Univ Calif, Los Angeles, PhD(anat), 65. Prof Exp: Fel brain res, Univ Edinburgh, 65-66; asst prof anat, Boston Univ, 66-70; HEAD SECT NEUROANAT & ULTRASTRUCT, DIV NEUROSCI, CITY OF HOPE MED CTR, 70- Mem: Am Asn Anat; Am Soc Cell Biol; Soc Neurosci. Res: Fine structure of central nervous tissue. Mailing Add: City of Hope Med Ctr 1500 E Duarte Rd Duarte CA 91010

VAUGHN, JAMES L, b Marshfield, Wis, Mar 2, 34; m 55; c 4. INSECT PATHOLOGY. Educ: Univ Wis, BS, 57, MS, 59, PhD(zool), 62. Prof Exp: Res officer tissue cult virol, Insect Path Res Inst, Can Dept Forestry, 61-65; RES MICROBIOLOGIST, INSECT PATH LAB, USDA, 65- Concurrent Pos: Mem adv comt collection animal cell cult, Am Type Cult Collection. Mem: Am Soc Microbiol; Soc Invert Path; Tissue Cult Asn. Res: Methods for growth of insect tissue in vitro; study of processes of virial infection and development in in vitro insect systems. Mailing Add: Insect Path Lab Rm 214 Bldg 011A Agr Res Ctr-W USDA Beltsville MD 20705

VAUGHN, JOE WARREN, b Otterbein, Ind, Oct 8, 33; m 55; c 4. INORGANIC CHEMISTRY, PHYSICAL CHEMISTRY. Educ: DePauw Univ, BA, 55; Univ Ky, MS, 57, PhD(phys chem), 59. Prof Exp: Welch Found fel, Univ Tex, 59-61; from asst prof to assoc prof chem, 61-70, PROF CHEM, NORTHERN ILL UNIV, 70- Mem: Am Chem Soc. Res: Nonaqueous solvent; fluoro complexes of trivalent chromium; stereochemistry of coordination compounds; synthesis of cis-trans isomers. Mailing Add: Dept of Chem Northern Ill Univ De Kalb IL 60115

VAUGHN, JOHN B, b Birmingham, Ala, Mar 3, 24; m 45; c 6. EPIDEMIOLOGY. Educ: Auburn Univ, DVM, 49; Tulane Univ, MPH, 56. Prof Exp: From instr to asst prof, 57-60, ASSOC PROF EPIDEMIOL, SCH MED, TULANE UNIV, 70- Mem:

Am Pub Health Asn; Am Vet Med Asn. Res: Zoonotic epidemiology. Mailing Add: Dept Epidemiol Sch Pub Health & Trop Med Tulane Univ New Orleans LA 70112

VAUGHN, MICHAEL THAYER, b Chicago, Ill, Aug 6, 36. PHYSICS. Educ: Columbia Univ, AB, 55; Purdue Univ, PhD(physics), 60. Prof Exp: Res assoc physics, Univ Pa, 59-62; asst prof, Ind Univ, 62-64; assoc prof, 64-73, PROF PHYSICS, NORTHEASTERN UNIV, 73- Concurrent Pos: Vis prof, Tex A&M Univ, 75. Mem: Am Phys Soc. Res: High energy physics; elementary particles; scattering theory; quantum field theory. Mailing Add: Dept of Physics Northeastern Univ Boston MA 02115

VAUGHN, MOSES WILLIAM, b Rock Hill, SC, Nov 2, 13; m 42; c 3. FOOD TECHNOLOGY. Educ: WVa State Col, BS, 38; Mich State Univ, MS, 42 & 49; Univ Mass, PhD(food technol), 51. Prof Exp: Prof food technol, 46-69, PROF AGR, UNIV MD, EASTERN SHORE, 69- Mem: AAAS; Inst Food Technologists. Res: Combination jellies and jams; chemical methods for detecting meat spoilage; nitrogen partitions in fish meal; bacteriological effects of ionizing radiations in fishery products; bacterial flora of bottom muds; effects of washing shellstock and shucked oysters to bacterial quality, microwave opening of oysters; nutritive value of selected pork products. Mailing Add: Dept of Agr Univ of Md Eastern Shore Princess Anne MD 21853

VAUGHN, PETER PAUL, b Altoona, Pa, Aug 22, 28; m 48; c 3. VERTEBRATE PALEONTOLOGY. Educ: Brooklyn Col, BA, 50; Harvard Univ, MA, 52, PhD(biol), 54. Prof Exp: Instr anat, Univ NC, 54-56; asst prof zool & asst cur fossil vert, Univ Kans, 56-57; assoc cur vert paleont, US Nat Mus, 57-58; from asst prof to assoc prof, 59-67, PROF ZOOL, UNIV CALIF, LOS ANGELES, 67- Concurrent Pos: Res assoc, Los Angeles County Mus, 61- Mem: Soc Vert Paleont; Paleont Soc. Res: Comparative vertebrate anatomy and paleontology; Paleozoic tetrapods; late Paleozoic vertebrate faunas and paleobiogeography. Mailing Add: Dept of Biol Univ of Calif Los Angeles CA 90024

VAUGHN, REESE HASKELL, b Farragut, Iowa, Oct 1, 08; m 35; c 5. FOOD MICROBIOLOGY. Educ: Simpson Col, AB, 30; Iowa State Col, MS, 32, PhD(bact), 35. Prof Exp: Lab asst biol, Simpson Col, 28-30; asst bact, Iowa State Col, 35-36; from instr to assoc prof food technol, Univ Calif, Berkeley, 36-52; from jr bacteriologist to assoc bacteriologist, 36-52, PROF FOOD TECHNOL, UNIV CALIF, DAVIS, 52-, FOOD TECHNOLOGIST, 56- Concurrent Pos: Rockefeller Found fel, 57; Fulbright lectr, 67. Mem: Am Soc Microbiol; Am Soc Enol; Am Pub Health Asn; Inst Food Technol; Royal Soc Health. Res: Food fermentations and bacteria causing them; spoilage of foods; bacterial indices of sanitation; food plant waste disposal; pectolytic activity of microbes. Mailing Add: Dept of Food Sci & Technol Univ of Calif Davis CA 95616

VAUGHN, THOMAS HUNT, b Clay, Ky, Nov 11, 09; m 30; c 3. ORGANIC CHEMISTRY. Educ: Univ Notre Dame, BS, 31, MS, 32, PhD(org chem), 34. Prof Exp: Vpres, Vitox Labs, Ind, 33-34; head dept org res, Union Carbide & Carbon Res Labs, Inc, NY, 34-39; dir org res, Mich Alkali Co, 39-41, asst dir res, 41-43; asst dir, Wyandotte Chem Corp, 43-45, dir, 45-48, vpres, 48-53; vpres res & develop, Colgate-Palmolive Co, 53-57; exec vpres, Pabst Brewing Co, 57-60; PRES, THOMAS H VAUGHN & CO, WARETOWN, 60- Concurrent Pos: Consult to Secy Navy & Qm Gen, US Army, 43-53; mem Scand Res & Indust Tour, 46; lectr, Prog Bus Admin for Top Level Ger Indust Leaders, Mutual Security Agency Sem, 52; pres, Tech Sect, World Conf Surface Active Agents, Paris, 54; dir, Indust Res Inst, 54-58, from vpres to pres, 57-58; chmn, Conf Admin Res, 55; mem adv bd, Off Critical Tables, Nat Acad Sci, 56; mem adv bd, Col Sci, Univ Notre Dame, 56-; ed, Res Mgt, 58-69; pres, Myzon Labs, Chicago, Ill, 60-62; tech consult, UN, 70- Mem: AAAS; fel Am Inst Chem; Am Inst Chem Eng; Am Chem Soc; Soc Indust Chem. Res: Unsaturated compounds and their derivatives; polyols; detergents; polymers; research and business administration. Mailing Add: Box 577 Waretown NJ 08758

VAUGHN, WILLIAM KING, b Denison, Tex, July 28, 38; m 60; c 2. BIOSTATISTICS. Educ: Tex Wesleyan Col, BS, 60; Southern Methodist Univ, MS, 65; Tex A&M Univ, PhD(statist), 70. Prof Exp: Med technologist, St Joseph Hosp, Ft Worth, Tex, 61-63; res analyst biostatist, Univ Tex MD Anderson Hosp & Tumor Inst, 65-67; ASST PROF BIOSTATIST, SCH MED, VANDERBILT UNIV, 70- Mem: Am Statist Asn; Sigma Xi. Res: Clinical trials; simulation and Monte Carlo studies; design of experiments; statistical methodology in toxicology; biometrics. Mailing Add: Dept of Prev Med Sch of Med Vanderbilt Univ Nashville TN 37232

VAUGHT, DAVID MITCHELL, chemical physics, computer science, see 12th edition

VAUGHT, ROBERT L, b Alhambra, Calif, Apr 4, 26; m 55; c 2. MATHEMATICS. Educ: Univ Calif, AB, 45, PhD(math), 54. Prof Exp: From instr to asst prof math, Univ Wash, 54-58; from asst prof to assoc prof, 58-63, PROF MATH, UNIV CALIF, BERKELEY, 63- Concurrent Pos: Fulbright scholar, Univ Amsterdam, 56-57; NSF fel, Univ Calif, Los Angeles, 63-64; Guggenheim fel, 67. Mem: Am Math Soc; Asn Symbolic Logic. Res: Foundations of mathematics. Mailing Add: Dept of Math Univ of Calif Berkeley CA 94720

VAULES, DAVID WILSON, b Great Neck, NY, Feb 14, 38; m 59; c 3. INTERNAL MEDICINE, CARDIOLOGY. Educ: Dartmouth Col, AB, 60; Dartmouth Med Sch, BMS, 61; Univ Rochester, MD, 63; Am Bd Internal Med, dipl, cert cardiovasc dis, 74. Prof Exp: From intern to asst resident med, Strong Mem Hosp, 63-65, from assoc resident to chief resident internal med, 67-69; clin investr training prog, 68-70, instr internal med, 69-71, ASST PROF MED, SCH MED & DENT, UNIV ROCHESTER, 71- Concurrent Pos: Trainee cardiol, Strong Mem Hosp, 69-71, assoc physician, 71-72; assoc physician, Mary Imogene Bassett Hosp, 72-; asst prof, Columbia Univ, 72-; mem coun clin cardiol, Am Heart Asn, 75. Mem: Fel Am Heart Asn; fel Am Col Cardiol. Res: Regulation of myocardial protein synthesis. Mailing Add: Dept of Med Mary Imogene Bassett Hosp Cooperstown NY 13326

VAULX, RANN L, organic chemistry, see 12th edition

VAUN, WILLIAM STRATIN, b Hartford, Conn, Oct 10, 29; m 61; c 1. MEDICINE. Educ: Trinity Col, Conn, BS, 51; Univ Pa, MD, 55. Prof Exp: Resident internal med, Hartford Hosp, Conn, 55-57 & 59-61; asst dir med, St Luke's Hosp, Cleveland, Ohio, 62-65; assoc prof med, 69-72, actg dean, Sch Continuing Educ, 72-73, PROF MED, HAHNEMANN MED COL, 72-; DIR MED EDUC, MONMOUTH MED CTR, 65- Concurrent Pos: Consult, Am Bd Internal Med, 73- & Nat Acad Sci, 75-76; chmn adv coun, Off Consumer Health Educ, Rutgers Med Sch, 73-75. Mem: Asn Hosp Med Educ (vpres, 72-73). Mailing Add: Monmouth Med Ctr Long Branch NJ 07740

VAUPEL, MARTIN ROBERT, b Evansville, Ind, July 24, 28. ANATOMY, EMBRYOLOGY. Educ: Ind Univ, AB, 49; Tulane Univ, PhD(anat), 54. Prof Exp: Res asst, 49-50, grad asst, 50-54, from instr to asst prof, 54-71, ASSOC PROF ANAT, TULANE UNIV, 71- Res: Congenital anomalies of nervous system; teratology. Mailing Add: Dept of Anat Tulane Univ Sch of Med New Orleans LA 70112

VAURIE, CHARLES, ornithology, deceased

VAUX, HENRY JAMES, b Bryn Mawr, Pa, Nov 6, 12; m 37; c 2. FORESTRY. Educ: Haverford Col, BS, 33; Univ Calif, MS, 35, PhD(agr econ), 48. Prof Exp: Instr forestry, Ore State Col, 37-42; asst economist, La Agr Exp Sta, 42-43; assoc economist, US Army, 43; assoc economist, US Forest Serv, 46-48; from lectr to assoc prof forestry, Sch Forestry, 48-53, dean, Sch Forestry & assoc dir, Agr Exp Sta, 55-65, PROF FORESTRY, SCH FORESTRY, UNIV CALIF, BERKELEY, 53- Concurrent Pos: Consult ed, McGraw-Hill Bk Co, 54- Mem: AAAS; fel Soc Am Foresters; Forest Hist Soc; hon mem Soc Foresters Finland. Res: Long term timber supply; price behavior and market structures for forest products. Mailing Add: 622 San Luis Rd Berkeley CA 94707

VAUX, JAMES EDWARD, JR, b Pittsburgh, Pa, June 13, 32; m 54; c 3. CHEMISTRY, STATISTICS. Educ: Carnegie Inst, BS, 52, MS, 64, PhD(chem), 67. Prof Exp: Chemist, E I du Pont de Nemours & Co, Inc, 53-57; teacher, Shady Side Acad, 57-64; asst dir, 64-67, EXEC DIR, FOUND STUDY CYCLES, 67-; LECTR CHEM, UNIV PITTSBURGH, 65- Mem: Am Chem Soc; Am Statist Asn. Res: Organic and analytical chemistry; statistics. Mailing Add: Dept of Chem Univ of Pittsburgh Pittsburgh PA 15260

VAVICH, MITCHELL GEORGE, b Miami, Ariz, Aug 24, 16; m 37; c 1. BIOCHEMISTRY. Educ: Univ Ariz, BS, 38, MS, 40; Pa State Univ, PhD(biochem), 43. Prof Exp: Asst physiol chem, Pa State Univ, 40-42, instr biochem, 43-46; from assoc prof to prof agr biochem, 46-75, head dept, 69-75, PROF NUTRIT & FOOD SCI, UNIV ARIZ, 75- Concurrent Pos: Spec field staff mem, Rockefeller Found, 65-67; chmn, Comt Agr Biochem & Nutrit, 69- Mem: AAAS; Am Chem Soc; Am Inst Nutrit. Res: Fluorides, ascorbic acid; vitamins in canned foods; interrelationships of vitamins and other food components; carotenes and vitamin A; nutritional status; biochemistry of cyclopropenoid fatty acids; nutrient value of dietary proteins. Mailing Add: Dept of Nutrit & Food Sci Univ of Ariz Tucson AZ 85721

VAVRA, JAMES JOSEPH, b Boulder, Colo, Aug 30, 29; m 51; c 5. INFECTIOUS DISEASES. Educ: Univ Colo, BA, 51; Univ Wis, MS, 53, PhD, 55. Prof Exp: Res assoc, Dept Biochem, 55-57, res assoc & proj leader, Dept Microbiol, 57-62, sr res scientist, Dept Clin Res, 62-64 & Dept Microbiol, 64-68, HEAD DEPT INFECTIOUS DIS RES, UPJOHN CO, 68- Mem: AAAS; Am Soc Microbiol; NY Acad Sci. Res: Fermentation biochemistry, especially metabolism; antibiotic-pathogen relationships, especially resistance development. Mailing Add: Dept of Infectious Dis Upjohn Co Kalamazoo MI 49001

VAVRA, JOSEPH PETER, soil chemistry, soil fertility, see 12th edition

VAWTER, SPENCER MAX, b Morgan Co, Ind, Feb 18, 37; m 72; c 3. PHYSICS. Educ: Franklin Col, BA, 59; DePaul Univ, MS, 67; Univ Mich, Ann Arbor, cert physiol, 68; Univ Southern Calif, cert comput systs, 70. Prof Exp: Student, Defense Projs Div, Western Elec Co at Lincoln Lab, Mass Inst Technol, 59-60; systs planning & develop engr, Western Elec Co, 60-63; dir ionizing radiation sect, AMA, Chicago, 63-66 & med physics sect, 66-70, assoc dir dept med instrumentation, 70-73; asst to pres, 73-74, VPRES, BIO-DYNAMICS, INC, 74- Mem: Am Phys Soc; Instrument Soc Am; Asn Advan Med Instrumentation. Res: Interaction of electromagnetic energy with human tissue, especially the effect of laser energy on the human eye; application of physical principles to medical practice; application of physical sciences to health care. Mailing Add: Bio-Dynamics Inc PO Box 50528 Indianapolis IN 46250

VAYDA, ANDREW PETER, b Budapest, Hungary, Dec 7, 31; US citizen; m 62; c 1. HUMAN ECOLOGY, ANTHROPOLOGY. Educ: Columbia Univ, BA, 52, PhD(anthrop), 56. Prof Exp: Soc Sci Res Coun res training fel, Coral Atolls Northern Cook Islands, Polynesia, 56-57; lectr anthrop, Univ BC, 58-60; from asst prof to prof, Columbia Univ, 60-72; prof anthrop & actg assoc dean, 72-74, PROF ANTHROP & ECOL, COOK COL, RUTGERS UNIV, NEW BRUNSWICK, 74- Concurrent Pos: Am Coun Learned Socs grant, Columbia Univ, 61-62; NSF grants, Bismarck Mountains, New Guinea, 62-; mem bd dirs, Soc Sci Res Coun, 70-72; consult zoonoses, WHO, Switz, 70; ed, Human Ecol: An Interdisciplinary J, Plenum Press, 71-; consult & US directorate mem, UNESCO'S Man & Biosphere Prog, 74- Mem: Fel AAAS; fel Am Anthrop Asn; Polynesian Soc. Res: War and peace as an ecological process; interrelations of demographic and cultural processes; individual and social responses to environmental problems. Mailing Add: Dept of Human Ecol Cook Col Rutgers Univ PO Box 231 New Brunswick NJ 08903

VAYO, HARRIS WESTCOTT, b Chicago, Ill, Nov 15, 35; m 62; c 1. APPLIED MATHEMATICS. Educ: Culver-Stockton Col, BA, 57; Univ Ill, MS, 59, PhD(math), 63. Prof Exp: Asst math, Univ Ill, 57-62; res fel biomath, Harvard Univ, 63-65; from asst prof to assoc prof, 65-74, PROF MATH, UNIV TOLEDO, 74- Concurrent Pos: Mem coun basic sci, Am Heart Asn, 67-; vis prof, McGill Univ, 76. Mem: Math Asn Am; Am Math Soc; Am Acad Mech. Res: Applications of mathematics to biological and medical problems. Mailing Add: Dept of Math Univ of Toledo Toledo OH 43606

VAZAKAS, ARISTOTLE JOHN, b Haverhill, Mass, July 9, 22; m 56; c 2. MEDICINAL CHEMISTRY. Educ: Mass Col Pharm, BS, 43, MS, 49; Purdue Univ, PhD(pharmaceut chem), 52. Prof Exp: From asst prof to assoc prof chem, Sch Pharm, Temple Univ, 52-58; sr res chemist, Nat Drug Co, 58-64; chemist, Div New Drugs, Bur Med, US Food & Drug Admin, 64-66; asst mgr regulatory affairs, 66-69, MGR REGULATORY COMPLIANCE, JOHNSON & JOHNSON, 69- Mem: Am Chem Soc; Am Pharmaceut Asn. Res: Synthetic organic medicinal chemistry. Mailing Add: Johnson & Johnson Res Ctr New Brunswick NJ 08903

VAZQUEZ, ALFREDO JORGE, b Buenos Aires, Arg, Jan 21, 37; m 62; c 1. NEUROPHARMACOLOGY, ELECTROPHYSIOLOGY. Educ: Bernadino Rivadavia Col, Arg, BS, 54; Univ Buenos Aires, MD, 62. Prof Exp: From intern to resident med, Tigre Hosp, Arg, 59-61; instr pharmacol, Chicago Med Sch, 62-65, assoc, 65-67; asst prof, Fac Med, Univ Man, 67-71; ASSOC PROF PHARMACOL, CHICAGO MED SCH, 71- Concurrent Pos: Res assoc, Lab Psychopharmacol, Nat Neuropsychiat Inst, Buenos Aires, 60-62. Mem: Pharmacol Soc Can; Soc Biol Psychiat; Soc Neurosci; Am Soc Pharmacol & Exp Therapeut. Res: Physiology and pharmacology of the cerebral cortex; epilepsy; drug abuse and hallucinogenic drugs. Mailing Add: Dept of Pharmacol Chicago Med Sch Chicago IL 61612

VAZQUEZ, JACINTO JOSEPH, b Havana, Cuba, Aug 17, 23; nat US; m 48; c 2. PATHOLOGY. Educ: Inst Havana, BAS, 40; Univ Havana, MD, 48; Ohio State Univ, MS, 54. Prof Exp: Instr & chief resident path, Ohio State Univ, 52-55; from instr to assoc prof, Sch Med, Univ Pittsburgh, 55-61; assoc mem, Scripps Clin & Res Found, 61-63; from assoc prof to prof path, Med Ctr, Duke Univ, 63-70; path & chmn dept, Med Ctr, Univ Ky, 70-74; MEM & ASST DIR, SCRIPPS CLIN & RES FOUND, 74- Mem: Am Soc Exp Path; Am Asn Path & Bact; Am Asn Immunol; NY

Acad Sci; Int Acad Path. Res: Experimental pathology; immunology; immunopathology; antibody formation; immunopathology antigen-antibody reactions; immunohematology. Mailing Add: Scripps Clin & Res Found 476 Prospect Lane La Jolla CA 92037

VAZQUEZ, PEDRO C, general chemistry, analytical chemistry, see 12th edition

VAZQUEZ, ROBERTO, soil physics, irrigation, see 12th edition

VEAL, BOYD WILLIAM, JR, b Chance, SDak, May 21, 37; m 62; c 2. SOLID STATE PHYSICS. Educ: SDak State Univ, BS, 59; Univ Pittsburgh, MS, 62; Univ Wis, PhD(physics), 69. Prof Exp: Engr, Westinghouse Res Labs, 59-63; asst physicist, 69-73, PHYSICIST, ARGONNE NAT LAB, 73- Mem: Am Phys Soc. Res: Electronic properties of solids, particularly optical properties and band structure studies. Mailing Add: Argonne Nat Lab Argonne IL 60439

VEAL, DEAN JOHNSON, b Birmingham, Ala, Apr 4, 16; m 41; c 3. ANALYTICAL CHEMISTRY. Educ: Tex Tech Col, BS, 38; Ohio State Univ, MS, 50. Prof Exp: Qual supvr, US Gypsum Co, Tex, 38-41; chemist, Ansul Chem Co, Wis, 45-50; res chemist, Phillips Petrol Co, 51-67; res physicist, US Bur Mines, Okla, 67-70; SUPVRY PHYS SCIENTIST, FELTMAN RES LAB, US ARMY MUNITIONS COMMAND, 70- Mem: AAAS; Am Chem Soc; fel Am Inst Chem. Res: Development of instrumental microanalytical techniques; activation analysis with accelerators; radiochemical analyses; organic geochemistry; explosives; propellants; pyrotechnics. Mailing Add: Feltman Res Lab US Army Mun Com Picatinny Arsenal Bldg 3124 Dover NJ 07801

VEAL, DONALD L, b Chance, SDak, Apr 17, 31; m 53; c 2. METEOROLOGY. Educ: SDak State Univ, BS, 53; Univ Wyo, MS, 60, PhD, 64. Prof Exp: Instr civil eng, SDak State Univ, 57-58; from instr to assoc prof, 58-67, asst dir natural resources res inst, 67-70, PROF ATMOSPHERIC RESOURCES, UNIV WYO, 67-, HEAD DEPT, 70- Mem: Am Meteorol Soc; Royal Meteorol Soc; Am Geophys Union; Am Soc Eng Educ; Am Soc Civil Engrs. Mailing Add: Univ of Wyo Laramie WY 82071

VEALE, WARREN LORNE, b Antler, Sask, Mar 13, 43; m 66; c 1. PHYSIOLOGY, NEUROPSYCHOLOGY. Educ: Univ Man, BSc, 64; Purdue Univ, West Lafayette, MSc, 68, PhD(neuropsychol), 71. Prof Exp: Instr psychol, Brandon Univ, 64-66, lectr, 66-67; from asst prof to assoc prof, 70-76, PROF MED PHYSIOL, FAC MED, UNIV CALGARY, 76-, ASSOC DEAN RES & ADMINR GRAD STUDIES PROG, 74- Concurrent Pos: Vis scientist, Nat Inst Med Res, London, Eng, 69; mem sci adv comt non-med use of drugs, Nat Health & Welfare-Med Res Coun, 73-74, chmn comt, 74-; regional ed, Pharmacol, Biochem & Behav. Mem: Can Psychol Soc; Soc Neurosci; Can Biochem Soc; NY Acad Sci; Am Physiol Soc. Res: Central nervous systems' involvement in temperature regulation, fever and action of antipyretics; neurohumoral changes in brain related to alcoholism. Mailing Add: Div of Med Physiol Univ of Calgary Calgary AB Can

VEATCH, COLLINS, b Liscomb, Iowa, Jan 23, 04; m 34; c 2. AGRONOMY. Educ: Va Polytech Inst, BS, 26; Iowa State Col, MS, 26; Univ Ill, PhD(agron, plant breeding), 29. Prof Exp: Asst plant breeding, Univ Ill, 26-28; agronomist, Compania Agricola Dominicana C por A, Santo Domingo, 30-33; res chemist, Corn Prod Refining Co, Ill, 33-43; agronomist, Inst Inter-Am Affairs, USDA, Peru, 43-44; from assoc prof to prof agron, 45-71, from assoc agronomist to agronomist, Exp Sta, 45-71, actg chmn dept agron & genetics, Univ, 68-70, EMER PROF AGRON, WVA UNIV, 71- Concurrent Pos: Res assoc, Univ Calif, Davis, 65. Mem: Am Soc Agron; Weed Sci Soc Am; Am Genetic Asn. Res: Weed control in field crops and pasture; field crop production, temperate and tropical; weed life history and herbicidal action. Mailing Add: 1180 Parkview Dr Morgantown WV 26505

VEATCH, FRANKLIN, chemistry, see 12th edition

VEATCH, RALPH WILSON, b Ringwood, Okla, May 1, 00; m 32; c 2. MATHEMATICS. Educ: Univ Tulsa, AB, 25; Northwestern Univ, MA, 27. Prof Exp: From instr to asst prof math, Ursinus Col, 27-30; from asst prof to prof, 30-70, head dept, 47-65, EMER PROF MATH, UNIV TULSA, 70- Mem: Math Asn Am; Sigma Xi. Res: Physics; applied mathematics; tensor analysis; astronomy; rotation of a rigid body passing through a swarm of meteors. Mailing Add: 332 NSanta Fe Tulsa OK 74127

VEAZEY, SIDNEY EDWIN, b Wilmington, NC, Sept 18, 37; m 62; c 4. PHYSICS. Educ: US Naval Acad, BS, 59; Duke Univ, PhD(physics), 65. Prof Exp: US Navy, 55-, electronics mat officer & main propulsion asst, USS Pollack, 66-68, navigator-opers off, USS Ulysses S Grant, 68-71 & USS James Madison, 71-72, dep dir trident submarine design develop proj, Naval Ship Eng Ctr, 72, mem Admiral Kidd's Combat Systs Adv Group, Naval Mat Command, 72-73, exec asst to Admiral Davies, 73-74, DESIGN MGR NUCLEAR POWERED SUBMARINES, NAVAL SHIP ENG CTR, US NAVY, 74- Mem: Am Phys Soc; Am Soc Naval Engrs. Res: Microwave spectroscopy of the alkali fluorides; application of lasers to communication from ships; nuclear power; submarines. Mailing Add: 1611 Crofton Pkwy Crofton MD 21114

VEAZEY, THOMAS MABRY, b Paris, Tenn, Jan 13, 20; m 38; c 3. ORGANIC CHEMISTRY. Educ: Murray State Univ, BS, 40; Univ Ill, PhD(org chem), 53. Prof Exp: Chemist & job instr supvr, E I du Pont de Nemours & Co, Inc, 41-45; res chemist, Devoe & Raynolds Co, 45-50; res chemist, Chemstrand Corp, 53-55, develop group leader synthetic fibers, 55-58, supvr develop, 58-63; mgr patent liaison, Monsanto Textiles Div, 63-70, SR DEVELOP ASSOC, MONSANTO TEXTILES CO, 71- Mem: Am Chem Soc. Res: Chemical and spinning process development of synthetic textile fibers. Mailing Add: 2026 Woodland St SE Decatur AL 35601

VEBER, DANIEL FRANK, b New Brunswick, NJ, Sept 9, 39; m 59; c 2. CHEMISTRY. Educ: Yale Univ, BA, 61, MS, 62, PhD(org chem), 64. Prof Exp: Sr chemist, 64-66, res fel, 66-72, SR RES FEL, MERCK SHARP & DOHME RES LABS, 72- Mem: Am Chem Soc. Res: Protein and peptide synthesis; new methods and protecting groups in peptide synthesis; chemical and biological properties of enzymes and peptide hormones; chemistry of heterocyclic compounds. Mailing Add: Merck Sharp & Dohme Res Labs West Point PA 19486

VEDAM, KUPPUSWAMY, b Vedharanyam, India, Jan 15, 26; m 56; c 2. PHYSICS, MATERIALS SCIENCE. Educ: Univ Nagpur, BSc, 46, MSc, 47; Univ Saugor, PhD(physics), 51. Prof Exp: Lectr physics, Indian Govt Educ Serv, 46-47; lectr, Univ Saugor, 47-48 & 51-53; sr res asst, Indian Inst Sci, 53-56; res assoc, Pa State Univ, 56-57, asst prof, 57-59; sr res officer, Atomic Energy Estab, Bombay, India, 60-62; sr res assoc, 62-64, assoc prof, 63-70, PROF PHYSICS, PA STATE UNIV, 70- Concurrent Pos: Mem panel piezoelec transducers, Indian Stand Inst, 61-62. Mem: Am Phys Soc; Optical Soc Am; Phys Soc Japan. Res: Crystal physics; optics; ferroelectricity; x-ray and neutron diffraction; high pressure physics, physics of surfaces and materials characterization. Mailing Add: Mat Res Lab Pa State Univ University Park PA 16802

VEDAMUTHU, EBENEZER RAJKUMAR, b Tamilnadu, India, June 23, 32; m 63; c 1. FOOD SCIENCE, FOOD TECHNOLOGY. Educ: Univ Madras, BSc, 53; Nat Dairy Res Inst, India, dipl, 56; Univ Ky, MS, 61; Ore State Univ, PhD(microbiol), 65. Prof Exp: Dairy asst, Govt Madras, Animal Husb Serv India, 57-59; assoc microbiol, Ore State Univ, 65-66; Dept Health, Educ & Welfare trainee dairy microbiol, Iowa State Univ, 66-67, asst prof food technol, 67-71; asst prof microbiol, Ore State Univ, 71-72; sr microbiologist, 72-75, CHIEF RES MICROBIOLOGIST, MICROLIFE TECHNICS, SARASOTA, 75- Concurrent Pos: Cheese technol consult, 67-; mem chapter revision comt stand methods examination dairy prod, Am Pub Health Asn, 75- Mem: AAAS; Int Asn Milk, Food & Environ Sanit; Inst Food Technologists; Am Dairy Sci Asn; Am Soc Microbiol. Res: Microbiology of dairy and food products, especially starter and spoilage flora. Mailing Add: 3505-27th Ave W Bradenton FL 33505

VEDDER, JAMES FORREST, b Pomona, Calif, June 3, 28; m 70; c 2. PLANETARY SCIENCES. Educ: Pomona Col, BA, 49; Univ Calif, PhD(nuclear physics), 58. Prof Exp: Asst nuclear physics, Radiation Lab, Univ Calif, 51-58; res scientist, Missiles & Space Co, Lockheed Aircraft Corp, 58-63; RES SCIENTIST, AMES RES CTR, NASA, 63- Mem: Am Phys Soc; Am Geophys Union; Sigma Xi. Res: Nuclear physics; beta decay; space physics; meteoroids; microparticle accelerators; remote sensing of soil moisture; measurement of stratospheric halocarbons. Mailing Add: NASA Ames Res Ctr Moffett Field CA 94035

VEDDER, WILLEM, b Ouder-Amstel, Neth, July 17, 24; nat US; m 55. CHEMISTRY. Educ: Univ Amsterdam, PhD, 58. Prof Exp: Mem sci staff, Dept Chem, Univ Amsterdam, 55-58; res assoc, Princeton Univ, 58-60; phys chemist, Metall & Ceramics Lab, 60-68, mgr phys chem lab, 68-72, MGR TECHNOL EVAL OPER, CORP RES & DEVELOP, GEN ELEC CO, 74- Mem: Am Phys Soc; Am Chem Soc. Res: Solid state chemistry; energy science and technology; technology assessment. Mailing Add: Gen Elec Corp Res & Develop PO Box 8 Schenectady NY 12301

VEDEJS, EDWIN, b Riga, Latvia, Jan 31, 41; US citizen. ORGANIC CHEMISTRY. Educ: Univ Mich, Ann Arbor, BS, 62; Univ Wis-Madison, PhD(chem), 66. Prof Exp: Nat Acad Sci-Air Force Off Sci Res fel chem, Harvard Univ, 66-67; ASSOC PROF CHEM, UNIV WIS-MADISON, 67- Concurrent Pos: A P Sloan fel, 71-73. Mem: Am Chem Soc. Res: Synthetic organic and organophosphorus chemistry; thermal rearrangements. Mailing Add: Dept of Chem Univ of Wis 1101 University Ave Madison WI 53706

VEDROS, NEYLAN ANTHONY, b New Orleans, La, Oct 6, 29; m 55; c 2. MICROBIOLOGY, IMMUNOLOGY. Educ: La State Univ, BSc, 51, MSc, 57; Univ Colo, PhD(microbiol), 60. Prof Exp: Nat Inst Allergy & Infectious Dis fel, Med Sch, Univ Ore, 60-62; chief bact div, Naval Med Res Inst, Bethesda, Md, 62-66; res microbiologist, Biol Lab, 66-68, DIR NAVAL BIOMED RES LAB & PROF MED MICROBIOL & IMMUNOL, UNIV CALIF, BERKELEY, 68- Honors & Awards: Lab Sect Award, Am Pub Health Asn, 66. Mem: AAAS; Am Soc Microbiol; Am Asn Immunol; Soc Exp Biol & Med; Asn Mil Surg US. Res: Immunochemistry of Neisseria Meningitis; host-parasite studies in marine pinnipeds; ecology of terrestrial and marine fungi. Mailing Add: Naval Biosci Lab Naval Supply Ctr Oakland CA 94625

VEECH, JOSEPH A, b Passaic, NJ, June 2, 39; m 62; c 3. PLANT PATHOLOGY, PLANT PHYSIOLOGY. Educ: La Polytech Inst, BS, 62; Univ Ga, MS, 64, PhD(bot, plant physiol), 67. Prof Exp: Instr bot & plant physiol, Univ Ga, 64-67; Nat Acad Sci-Nat Res Coun fel nematol, Agr Res Serv, USDA, 67-68, res physiologist, Crops Protection Res Br, Crops Res Div, Plant Indust Sta, Beltsville, 68-73, RES PLANT PHYSIOLOGIST, NAT COTTON PATH LAB, AGR RES SERV, USDA, 73- Mem: Phytochem Soc NAm; Am Phytopath Soc; Soc Nematol; Can Soc Plant Physiol. Res: Physiology of parasitism and mechanisms of host resistance to plant diseases and histochemical, biochemical and physiological studies of these areas. Mailing Add: Nat Cotton Path Lab PO Drawer JF Agr Res Serv USDA College Station TX 77801

VEECH, RICHARD L, b Decatur, Ill, Sept 19, 35; m 65; c 3. BIOCHEMISTRY, MEDICINE. Educ: Harvard Univ, BA, 57, MD, 62; Oxford Univ, PhD(biochem), 69. Prof Exp: From intern to resident med, NY Hosp-Cornell Med Ctr, 62-64; from clin assoc to staff assoc, NIMH, 64-66, RES BIOCHEMIST, DIV SPEC MENT HEALTH RES, NIMH, ST ELIZABETHS HOSP, 69- Concurrent Pos: Consult, NIMH, 69. Mem: Brit Biochem Soc. Res: Control of the redox state and phosphorylation potential in mammalian cells and alteration of these states in certain metabolic diseases. Mailing Add: Div of Spec Ment Health Res St Elizabeths Hosp Washington DC 20032

VEECH, WILLIAM AUSTIN, b Detroit, Mich, Dec 24, 38; m 65; c 1. MATHEMATICS. Educ: Dartmouth Col, AB, 60; Princeton Univ, PhD(math), 63. Prof Exp: H B Fine instr math, Princeton Univ, 63-64, Higgins lectr, 64-66; asst prof, Univ Calif, Berkeley, 66-69; assoc prof, 69-72, PROF MATH, RICE UNIV, 72- Concurrent Pos: Mem math, Inst Advan Study, Princeton Univ, 68-69 & 72; NSF grant, Rice Univ, 69-, Alfred P Sloan fel, 71-73. Mem: Am Math Soc. Res: Topological dynamics; ergodic theory; probability theory; functional analysis; almost periodic functions; number theory. Mailing Add: Dept of Math Rice Univ Houston TX 77001

VEEN-BAIGENT, MARGARET JOAN, b Toronto, Ont, Dec 23, 33; m 69; c 2. NUTRITION. Educ: Univ Toronto, BA, 55, MA, 56, PhD(nutrit), 64. Prof Exp: From lectr to asst prof, 56-71, ASSOC PROF NUTRIT, SCH HYG, UNIV TORONTO, 71- Mem: Nutrit Soc Can; Brit Nutrit Soc; Am Inst Nutrit. Res: Metabolism of vitamin A; lipid metabolism. Mailing Add: Dept of Nutrit & Food Sci Fac of Med Univ of Toronto Toronto ON Can

VEENEMA, RALPH J, b Prospect Park, NJ, Dec 13, 21; m 44; c 4. UROLOGY. Educ: Calvin Col, AB, 42; Jefferson Med Col, MD, 45; Am Bd Urol, dipl, 57. Prof Exp: Asst resident, Vet Admin Hosps, Alexandria, La, 46 & Jackson, Miss, 47; asst resident path, Paterson Gen Hosp, 48 & surg path, Col Physicians & Surgeons, Columbia Univ, 49; asst resident urol, Vet Admin Hosp, Bronx, 49; asst resident & resident, Columbia-Presby Med Ctr, 50-52, from asst to assoc, 53-58, asst clin prof, 58-60, from asst prof to assoc prof clin urol, 60-68, PROF CLIN UROL, COL PHYSICIANS & SURGEONS, COLUMBIA UNIV, 68- Concurrent Pos: Assoc urologist, St Joseph Hosp, Paterson, NJ, 53-56; chief, Urol Outpatient Clin, Columbia-Presby Med Ctr, 55-60, from asst attend urologist to assoc attend urologist, 55-68, attend urologist, 68-, chief urol, Francis Delafield Hosp, Cancer Res Inst, 60-75; attend urologist & chief urol serv, Valley Hosp, Ridgewood, NJ, 56-60, consult, 60-; consult, USPHS Hosp, Staten Island, NY, 61- & Harlem Hosp, New York, 62- Honors & Awards: Am Urol Asn 2nd Prize, 62, 1st Prize, 64. Mem: Am Asn Genito-Urinary Surg; Am Urol Asn; fel Am Col Surgeons; fel AMA; NY Acad Med (secy, 60-61). Res: Pathophysiology of genitourinary neoplasms. Mailing Add: Dept of Clin Urol Columbia Univ Col Phys & Surg New York NY 10032

VEENING, HANS, b Neth, May 7, 31; nat US; m 57. ANALYTICAL CHEMISTRY. Educ: Hope Col, AB, 53; Purdue Univ, MS, 55, PhD, 59. Prof Exp: From instr to assoc prof, 58-72, PROF CHEM, BUCKNELL UNIV, 72- Concurrent Pos: NSF fac fel with Dr J F K Huber, Univ Amsterdam, 66-67; NIH spec res fel, Biochem Separations Sect, Oak Ridge Nat Lab, 72-73; NSF grants, 68-72; Petrol Res Fund grants, 68-75 & 75-77; prof in charge short course on automated anal, Am Chem Soc, 76- Mem: AAAS; Am Chem Soc. Res: Liquid and gas chromatography of physiological fluids; liquid chromatography of organometallic compounds; automated methods of analysis. Mailing Add: Dept of Chem Bucknell Univ Lewisburg PA 17837

VEENSTRA, MAURICE ARNOLD, plant pathology, see 12th edition

VEENSTRA, ROBERT J, veterinary medicine, see 12th edition

VEESER, LYNN RAYMOND, 8b Sturgeon Bay, Wis, Sept 18, 42; m 65. NUCLEAR PHYSICS. Educ: Univ Wis-Madison, BS, 64, MS, 65, PhD(physics), 68. Prof Exp: STAFF MEM PHYSICS, LOS ALAMOS SCI LAB, 67- Mem: Am Phys Soc. Res: Low energy nuclear physics. Mailing Add: Los Alamos Sci Lab Los Alamos NM 87544

VEGH, EMANUEL, b New York, NY, Nov 20, 36; m 60; c 3. MATHEMATICS. Educ: Univ Del, BA, 58, MA, 60; Univ NC, PhD(math), 65. Prof Exp: Lectr math, Univ Del, 58-60 & Univ NC, 60-63; RES MATHEMATICIAN, US NAVAL RES LAB, 63- Concurrent Pos: Assoc prof lectr, George Washington Univ, 65-67 & Univ Md, 67- Honors & Awards: Res Publ Award, US Naval Res Lab, 69. Mem: Math Asn Am; Am Math Soc; London Math Soc; Sigma Xi. Res: Number theory; algebra; combinatories. Mailing Add: US Naval Res Lab 4555 Overlook Ave SW Washington DC 20390

VEGORS, HALSEY HUGH, b Lehigh, Iowa, July 2, 16; m 36; c 3. ANIMAL PARASITOLOGY. Educ: Univ Calif, Los Angeles, BA, 39. Prof Exp: Jr engr chem, Dept Com, Nat Inventors Coun, 41-42, asst engr, 42-45; jr zoologist, Bur Animal Indust, US Dept Agr, 45-47, asst parasitologist, 47-49, assoc parasitologist, 49-53, parasitologist animal dis & parasite res div, Agr Res Serv, 53-67, prin res zoologist, Parasitol Lab, Agr Res Ctr, Md, 67-71, asst animal scientist, Allegheny Highlands Proj, WVa Univ, 71-74; CONSULT, 74- Mem: Am Soc Parasitol; World Asn Advan Vet Parasitol; Conf Res Workers Animal Dis; Am Inst Biol Scientists. Res: Liver fluke in ruminants; strongyloidiasis in ruminants, especially dairy calves; control of internal parasites in beef cattle; immunology of cattle lungworm; ecology, epidemiology and chemotherapeutic treatment of ruminant parasites; life cycles; epizootiology; pathogenesis control and treatment of parasites and parasitic diseases of sheep. Mailing Add: 4100 Stoconga Dr Beltsville MD 20705

VEGORS, STANLEY H, JR, b Detroit, Mich, Jan 5, 29; m 51; c 3. NUCLEAR PHYSICS. Educ: Middlebury Col, BA, 51; Mass Inst Technol, BS, 51; Univ Ill, MS, 52, PhD(physics), 55. Prof Exp: Res assoc physics, Univ Ill, 55-56; physicist, Phillips Petrol Co, 56-58; assoc prof physics, 58-61, head dept, 58-65, PROF PHYSICS, IDAHO STATE UNIV, 61- Mem: Am Phys Soc; Am Asn Physics Teachers. Res: Radioactivity, especially high Z nuclei and isomeric transitions with half-lives in the microsecond and millisecond region; radioactivity, solar energy, nuclear safeguards. Mailing Add: Dept of Physics Idaho State Univ Pocatello ID 83201

VEGOTSKY, ALLEN, b New York, NY, Mar 2, 31; m 67; c 1. BIOLOGICAL CHEMISTRY. Educ: City Col New York, BS, 52; Fla State Univ, MS, 57, PhD(chem), 61. Prof Exp: Asst biochem, NY Univ, 52 & Fla State Univ, 55-60; NIH fel pub health, Purdue Univ, 60-63; asst prof biol & chem, Wheaton Col, Mass, 63-69; assoc prof, Wells Col, 69-74; BIOSCI COORDR, BIOMED INTERDISCIPLINARY CURRIC PROJ, 74- Mem: AAAS; Am Inst Biol Scientists; Nat Sci Teachers Asn; Nat Asn Biol Teachers; Sigma Xi. Res: Microbial genetics; ultrasonic reactions; curriculum development. Mailing Add: 2150 Santa Clara Ave Alameda CA 94501

VEHSE, ROBERT CHASE, b Morgantown, WVa, Sept 9, 36; m 61; c 2. SOLID STATE PHYSICS. Educ: WVa Univ, BA, 58; Univ Tenn, Knoxville, PhD(physics), 64. Prof Exp: Mem tech staff compound semiconductor mat, Bell Tel Labs, 68-72, SUPVR COMPOUND SEMICONDUCTOR MAT GROUP, BELL LABS, 73- Mem: Am Phys Soc; Electrochem Soc. Res: Development of processes useful for production of epitaxial layers of semiconductor materials. Mailing Add: Dept 2354 Bell Labs 2525 N 11th St Reading PA 19604

VEHSE, WILLIAM E, b Morgantown, WVa, Apr 28, 32; m 56; c 4. PHYSICS. Educ: WVa Univ, BA, 55; Carnegie Inst, MS, 59, PhD(physics), 62. Prof Exp: From asst prof to assoc prof physics, 61-72, PROF PHYSICS, W VA UNIV, 72-, CHMN DEPT, 75- Mem: Am Phys Soc; Am Asn Physics Teachers. Res: Nuclear and electron resonance in metals; optics. Mailing Add: Dept of Physics WVa Univ Morgantown WV 26506

VEIDIS, MIKELIS VALDIS, b Riga, Latvia, Jan 25, 39; m 63; c 2. CHEMISTRY. Educ: Univ Queensland, BSc, 63, MSc, 67; Univ Waterloo, PhD(chem), 69. Prof Exp: Chemist, Queensland Govt Chem Lab, 63-67; res fel chem, Harvard Univ, 69-71; metallurgist, 71-75, VPRES RES & DEVELOP, WAKEFIELD BEARING CORP, 75- Mem: Am Chem Soc; Royal Australian Chem Inst. Res: Chemistry of metal surfaces as related to sintering phenomena; crystallographic studies of structures of inorganic complexes and compounds of biological interest. Mailing Add: Wakefield Bearing Corp Wakefield MA 01880

VEIGEL, JON MICHAEL, b Mankato, Minn, Nov 10, 38; m 62. RESEARCH ADMINISTRATION, ENVIRONMENTAL SCIENCES. Educ: Univ Wash, BS, 60; Univ Calif, Los Angeles, PhD(phys inorg chem), 65. Prof Exp: Res chemist, Jackson Lab, E I du Pont de Nemours & Co, Inc, Del, 65; asst prof phys inorg chem & res chemist, F J Seiler Res Lab, US Air Force Acad, 65-68; asst prof, Joint Sci Dept, Claremont Cols, 68-73; assoc prof energy & environ, Calif State Col, Dominguez Hills, Calif, 73-74; dir energy prog, Off Technol Assessment, US Cong, 74-75; ADMINR RES & DEVELOP, ENERGY COMN, SACRAMENTO, CALIF, 75- Concurrent Pos: Consult, Statewide Air Pollution Res Ctr & Churchill Films, Calif; cong scientist fel, AAAS, 74-75 & mem sci & pub policies comt, AAAS, 75-78; mem synthesis panel, Nat Acad Study Nuclear Power & Alternative Systs, 75-76. Mem: AAAS. Res: National energy policy; technology assessment; environmental energy policies. Mailing Add: Energy Comn 1111 Howe Ave Sacramento CA 95825

VEIGELE, WILLIAM JOHN, b New York, NY, June 18, 25; m 56; c 4. PHYSICS. Educ: Hofstra Col, BA, 49, MA, 51; Univ Colo, PhD(physics), 60. Prof Exp: Engr, New York Testing Labs, Inc, 49-50; instr physics, Williams Col, 51-52; instr physics & eng, Hofstra Col, 52-57; instr physics, Univ Colo, 57-58; physicist, Nat Bur Standards, 58-59; prof physics & head dept, Parsons Col, 60-61; assoc res scientist, Martin Co, 61-64; sr res scientist & mgr environ progs, Kaman Sci Corp, 64-74; PRES, RESOURCE SCI INC, 74- Concurrent Pos: Consult, Fairchild Camera & Instrument Corp & Hazeltine Electronics Div, Hazeltine Corp, 53-54; lectr physics, Univ Colo, 65- Mem: Am Phys Soc; Am Asn Physics Teachers; Air Pollution Control Asn. Res: Atomic, nuclear and solid state physics; radiation effects; magnetic resonance; thermodynamics; environmental and resource sciences. Mailing Add: 3003 Chelton Dr Colorado Springs CO 80909

VEILLON, CLAUDE, b Church Point, La, Jan 11, 40; m 66. ANALYTICAL CHEMISTRY, SPECTROSCOPY. Educ: Univ Southwestern La, BS, 62; Univ Fla, MS, 63, PhD(anal chem), 65. Prof Exp: Lab technician, Cabot Carbon Co, 60; res asst, Univ Southwestern La, 60-62; asst in res, Univ Fla, 62-65; res chemist, Nat Bur Standards, 65-67; from asst prof to assoc prof anal chem, Univ Houston, 67-74; VIS SCIENTIST, HARVARD MED SCH, 74- Concurrent Pos: Nat Acad Sci-Nat Res Coun res assoc, 65-67; res fel, NIH/Nat Cancer Inst, 74-76. Mem: Am Chem Soc; Soc Appl Spectros; Optical Soc Am; Am Inst Physics. Res: Atomic fluorescence, absorption and emission spectrometry; interferometry; photon counting; isotopic analysis; analytical instrumentation; optics; non-flame atomization systems; analytical biochemistry; enzymology; metabolism; cancer; forensic science. Mailing Add: Biophys Res Lab Harvard Med Sch Peter Bent Brigham Hosp Boston MA 02115

VEINOTT, ARTHUR FALES, JR, b Boston, Mass, Oct 12, 34; m 60; c 2. OPERATIONS RESEARCH. Educ: Lehigh Univ, BS & BA, 56; Columbia Univ, EngScD(indust eng), 60. Prof Exp: Asst prof indust eng, 62-64, assoc prof, 64-67, PROF OPERS RES, STANFORD UNIV, 67-, CHMN DEPT, 75- Concurrent Pos: Western Mgt Sci Indst grant, 64-65; Off Naval Res contract, 64-; consult, Rand Corp, 65- & IBM Res Ctr, 68-69; NSF grant, 67-, mem res initiation grant panel, 71; vis prof, Yale Univ, 72-73; ed, J Math Opers Res, 74- Mem: Inst Mgt Sci; Opers Res Soc Am; Math Prog Soc; Economet Soc; fel Inst Math Statist. Res: Development of lattice programming, a qualitative theory of optimization for predicting the direction of change of optimal decisions resulting from alteration of problem parameters; structure and computation of optimal policies for inventory systems and dynamic programs. Mailing Add: Dept of Opers Res Stanford Univ Stanford CA 94305

VEIRS, CARROLL EUGENE, b Olathe, Colo, May 3, 18; m 48; c 1. WATER POLLUTION. Educ: Western State Col, BA, 40. Prof Exp: Teacher high sch, Colo, 40-42; instr electronics, US Army Air Force Tech Training Command, Sioux Falls, SDak, 42-43; sci aide, Bur Reclamation, US Dept Interior, Salt Lake City, Utah, 46, soils analyst, 46-49, head regional lab, 49-64, soil scientist consult, 64-66; phys scientist adminr, Fed Water Pollution Control Admin, 67-71, chief water resources sect, Air & Water Div, 71-73, AGR SPECIALIST, WATER DIV, ENVIRON PROTECTION AGENCY, 73- Concurrent Pos: Mem subcomt agr, Nat Tech Adv Comt to Secy Interior, 67. Res: Tracing of ground water through evaluation of water quality; prediction model of surface water quality; water management planning for water quality. Mailing Add: 16931 NE 32nd Bellevue WA 98008

VEIRS, VAL RHODES, b Allegan, Mich, Sept 20, 42; m 64; c 1. ATMOSPHERIC PHYSICS. Educ: Case Inst Technol, BS, 64; Ill Inst Technol, PhD(physics), 69. Prof Exp: Res physicist, Zenith Radio Corp, 64-65; asst prof physics, Ill Inst Technol, 69-71; ASST PROF PHYSICS, COLO COL, 71- Mem: AAAS; Am Asn Physics Teachers. Res: Numerical and urban diffusion modeling. Mailing Add: Dept of Physics Colo Col Colorado Springs CO 80903

VEIS, ARTHUR, b Pittsburgh, Pa, Dec 23, 25; m 51; c 3. BIOCHEMISTRY, PHYSICAL CHEMISTRY. Educ: Univ Okla, BS, 47; Northwestern Univ, PhD(phys chem), 51. Prof Exp: Instr phys chem, Univ Okla, 51-52; res chemist, Dept Phys Chem, Armour & Co, 52-60, head dept, 59-60; assoc prof biochem, 60-65, asst dean grad affairs, 68-70, assoc dean med & grad schs, 70-76, PROF BIOCHEM, SCH MED, NORTHWESTERN UNIV, CHICAGO, 65- Concurrent Pos: Spec instr, Crane Jr Col, 55-56 & Loyola Univ, 57-58; Guggenheim fel, 67. Mem: Am Chem Soc; Am Soc Biol Chemists; Biophys Soc; NY Acad Sci; Int Asn Dent Res. Res: Physical chemistry and biology of the connective tissue systems; colloid chemistry. Mailing Add: Dept of Biochem Northwestern Univ Med Sch Chicago IL 60611

VEIT, JIRI JOSEPH, b Prague, Czech, Apr 15, 34; m 61. ATOMIC PHYSICS, NUCLEAR PHYSICS. Educ: Univ London, BSc, 55, PhD(nuclear physics), 59; Univ Birmingham, MSc, 56. Prof Exp: Instr physics, Univ BC, 59-62; lectr, Univ London, 62-63; from asst prof to assoc prof, 63-71, PROF PHYSICS, WESTERN WASH STATE COL, 71- Mem: Am Phys Soc; Am Asn Physics Teachers. Res: Positronium; stripping and pickup reactions; nuclear reaction mechanisms; nuclear structure. Mailing Add: Dept of Physics Western Wash State Col Bellingham WA 98225

VEITCH, FLETCHER PEARRE, b College Park, Md, Dec 21, 09; m 39; c 2. BIOCHEMISTRY. Educ: Univ Md, BS, 31, MS, 33, PhD(org chem), 35. Prof Exp: Instr org & physiol chem, Univ Md, 31-35; res chemist, Nat Canners Asn, 35-37; from instr to asst prof biochem, Sch Med, Georgetown Univ, 37-47; assoc prof, 47-55, PROF BIOCHEM, UNIV MD, COLLEGE PARK, 55- Mem: AAAS; Am Chem Soc; NY Acad Sci. Res: Sex hormones; enzymes; polymerization of olefins. Mailing Add: Dept of Chem Univ of Md College Park MD 20742

VEITH, DANIEL A, b Metairie, La, Apr 18, 36; m 56; c 2. SOLID STATE PHYSICS. Educ: Tulane Univ, BS, 56, PhD(physics), 63; Univ Calif, Los Angeles, MS, 58. Prof Exp: Mem tech staff, Hughes Aircraft Co, 56-59; sci specialist space div, Chrysler Corp, 63-67; PROF PHYSICS & DEPT HEAD, NICHOLLS STATE UNIV, 67- Res: Nucleation and growth of thin crystalline films. Mailing Add: Dept of Physics Nicholls State Univ Thibodaux LA 70301

VEITH, FRANK JAMES, b New York, NY, Aug 29, 31; m 54; c 4. SURGERY, TRANSPLANTATION BIOLOGY. Educ: Cornell Univ, AB, 52, MD, 55. Prof Exp: NIH fel, Harvard Med Sch, 63-64; asst prof surg, Cornell Univ, 64-67; assoc prof, 67-71, PROF SURG, ALBERT EINSTEIN COL MED, 71-; CO-DIR KIDNEY TRANSPLANT UNIT, MONTEFIORE HOSP, 67-, ATTEND SURG & CHIEF VASCULAR SURG, 72- Concurrent Pos: Markle scholar acad med, Cornell Univ, Albert Einstein Col Med & Montefiore Hosp, 64-69; career scientist award, Health Res Coun, City New York & Montefiore Hosp, 65-72; assoc attend surgeon, Montefiore Hosp, 67-71; consult, Heart-Lung Proj Comt, 71- Mem: Soc Univ Surgeons; Soc Vascular Surg; Am Asn Thoracic Surg; Am Surg Asn; Transplantation Soc. Res: Lung transplantation, pulmonary physiology, kidney transplantation and vascular surgery. Mailing Add: Dept of Surg Montefiore Hosp & Med Ctr New York NY 10467

VEITH, ILZA, b Ludwigshafen, Ger, May 13, 15; nat US; m 35. HISTORY OF MEDICINE. Educ: Johns Hopkins Univ, MA, 44, PhD(hist of med), 47; Juntendo Univ, Tokyo, Igaku hakase, 75. Prof Exp: Am Asn Learned Soc grant, 48; lectr hist of med, Univ Chicago, 49-51, from asst prof to assoc prof, 51-64; prof hist of health sci & vchmn dept, 64-67, PROF HIST OF PSYCHIAT, SCH MED, UNIV CALIF, SAN FRANCISCO, 67- Concurrent Pos: NSF res grant, 53-56; consult, Armed Forced Med Libr, 47-56 & NIH, 59-62; D J Davis mem lectr, Col Med, Univ Ill, 58; vis prof, Univ Calif, Los Angeles, 58; Kuntz mem lectr, Sch Med, St Louis Univ, 61; Alfred P Sloan vis prof, Menninger Sch Psychiat, Kans, 63, 64 & 66; vis prof, Med Sch, Univ Chicago, 68-; chmn hist res comt, World Fedn Neurol, 68-; Logan Clendening lectr, Sch Med, Univ Kans, 71; All-Univ Lectr, Univ Calif, Santa Barbara, 72; hon mem

Inst Hist of Med & Med Res, India. Mem: Hist Sci Soc; Am Asn Hist Med (secy-treas, 55); hon fel Am Psychiat Asn; Int Soc Hist Med; fel Royal Soc Med. Res: History of Chinese and Japanese medicine; history of psychiatry, particularly hysteria. Mailing Add: Dept of Hist of Hlth Sci Univ of Calif Med Ctr San Francisco CA 94122

VEJVODA, EDWARD, b New York, NY, Apr 18, 24; m 49; c 3. INDUSTRIAL CHEMISTRY. Educ: Univ Northern Colo, BA, 49, MA, 51. Prof Exp: Anal chemist, Anal Labs, Dow Chem Co, 52-56, res chemist, 56-60, sr res chemist res & develop labs, 60-62, anal supvr, 62-64, anal proj supvr, Dow Chem Int, Ger, 64-65, res staff asst, Chem-Physics Res & Develop Labs, 65-68, res mgr, Chem Res & Develop, 68-72, sr res mgr, 72-75; CHEM OPERS DIR, ROCKWELL INT, 75- Mem: Am Chem Soc; Soc Appl Spectros; Sigma Xi; Am Electroplaters Soc. Res: Actinide chemistry; development of analytical methods for the assay and impurity analysis of the actinide elements, especially optical emission spectroscopy for the impurity analysis of plutonium and americium compounds; process development for the separation and purification of plutonium compounds. Mailing Add: Rockwell Int Rocky Flats Plant Golden CO 80401

VELA, ADAN RICHARD, b Laredo, Tex, Oct 28, 30; m 55; c 4. PHYSIOLOGY. Educ: Baylor Univ, BS, 52; Univ Tenn, PhD(physiol) 62. Prof Exp: Res assoc data anal, Comput Ctr, Univ Tenn, 63; asst prof surg, 63-71, ASSOC PROF SURG, SCH MED, LA STATE UNIV MED CTR, 71- Res: Gastrointestinal physiology; esophageal motility; spinal fluid pressure and production; Escherichia coli endotoxin. Mailing Add: Dept of Surg La State Univ Sch of Med New Orleans LA 70112

VELA, GERARD ROLAND, b Eagle Pass, Tex, Sept 18, 27; m 53; c 4. MICROBIOLOGY. Educ: Univ Tex, BA, 50, MA, 51, PhD(microbiol), 64. Prof Exp: Res asst, Univ Tex, 50-51; res asst biochem, Southwest Found Res, 52-54; res asst immunol, Sch Med, Harvard Univ, 54-57; head clin chemist, Santa Rosa Hosp, San Antonio, Tex, 57-59; res microbiologist, US Air Force Sch Aerospace Med, 59-65; from asst prof to assoc prof microbiol, 65-72, PROF MICROBIOL, N TEX STATE UNIV, 72- Concurrent Pos: Fulbright lectr, Bogota, Colombia, 72; ed, Tex J Sci, 75- Res: Nature of microorganisms in their natural habitat; biochemical interrelationships in mixed cultures of microorganisms; physiology of azotobacter; microbiology of industrial waste-waters. Mailing Add: Dept of Biol NTex State Univ Denton TX 76203

VELARDO, JOSEPH THOMAS, b Newark, NJ, Jan 27, 23; m 48. ANATOMY. Educ: Northern Colo Univ, AB, 48; Miami Univ, SM, 49; Harvard Univ, PhD(biol, physiol, endocrinol), 52. Prof Exp: Asst org & inorg chem, Northern Colo Univ, 48; asst & instr zool & human heredity, Miami Univ, 48-49; res fel biol & endocrinol, Harvard Univ, 52-53, res assoc path, Sch Med, 53-54, res assoc surg, 54-55; asst prof anat, Sch Med, Yale Univ, 55-61; prof & chmn dept, NY Med Col, 61-62; dir, Inst Study Human Reproduction & dir educ prog, 62-67; chmn dept, 67-73, PROF ANAT, STRITCH SCH MED, LOYOLA UNIV, CHICAGO, 67- Concurrent Pos: Asst surg, Peter Bent Brigham Hosp, Boston, 54-55; Lederle med fac award, 55-58; prof biol, John Carroll Univ, 62-67; US del, Int Cong Reproduction, Vatican, 64; head dept res, St Ann Hosp, Cleveland, 64-67. Honors & Awards: Rubin Award, Am Soc Study Sterility, 55. Mem: Am Soc Zoologists; Endocrine Soc; fel Geront Soc; Am Asn Anat; fel NY Acad Sci. Res: Endocrinology of reproduction; anatomy, physiology, biochemistry, histochemistry and cytochemistry of reproductive organs. Mailing Add: Stritch Sch of Med Loyola Univ of Chicago Maywood IL 60153

VELASQUEZ, CARMEN C, b Bayambang, Philippines, Aug 7, 13; m 35; c 3. ZOOLOGY, PARASITOLOGY. Educ: Univ Philippines, BS, 34, PhD(zool, parasitol), 54; Univ Mich, MS, 37. Prof Exp: Instr zool, 39-41 & 48-53, from asst prof to assoc prof, 54-65, prof chair zool, 73-77, PROF ZOOL, UNIV PHILIPPINES, 66-, CHMN DEPT, COL ARTS & SCI, 73- Concurrent Pos: Guggenheim fel, 57-58 & 63; NIH res grant, 59-63, with Dr I E Wallen, Smithsonian Inst, 65-70; mem comn sci, Int Cong on Rizal, 61; Nat Sci Develop Bd Surv Team for Upgrading Sci in Philippines, 61-63; secy sects zool & parasitol, Nat Res Coun Philippines, 63- Honors & Awards: Presidential Distinguished Serv Medal & Dipl of Honor, Philippines, 65; Nat Sci Develop Bd Award, 73; Philippines Outstanding Woman in Sci, UNESCO Ninth Biennial Conf, 75. Mem: AAAS; fel Indian Acad Sci; Philippines Soc Advan Res; Philippine Soc Parasitol (secy, 55-56 & 63-); Am Soc Parasitologists. Res: Parasitic helminths of Philippine fishes—taxonomy, life cycles, distribution, ecology and zoonoses. Mailing Add: Dept of Zool Col Arts & Sci Univ of Philippines Syst Diliman Quezon City 3004 Philippines

VELAZQUEZ, THOMAS, pathology, see 12th edition

VELDHUIS, BENJAMIN, b Belgrade, Mont, Dec 28, 15; m 43; c 2. ORGANIC CHEMISTRY. Educ: Mont State Col, BS, 38; Univ Wis, PhD(chem), 42. Prof Exp: Mem staff, Allied Chem Corp, 42-70; MEM STAFF, OFF OF STATE MED EXAMR, NEWARK, 71- Mem: Am Chem Soc. Res: Surfactants; sulfonation; fluorochemicals; pesticides chlorination; forensic toxicology; explosives. Mailing Add: 34 Center Ave Morristown NJ 07960

VELECKIS, EWALD, b Kybartai, Lithuania, Aug 1, 26; US citizen; m 55; c 1. CHEMISTRY. Educ: Univ Ill, BS, 53; Ill Inst Technol, MS, 57, PhD(chem), 60. Prof Exp: ASSOC CHEMIST, ARGONNE NAT LAB, 59- Mem: Am Chem Soc. Res: Phase equilibria in inorganic systems; transition metal hydrides; molecular beams; metallic solutions and liquid state; thermodynamics. Mailing Add: Argonne Nat Lab 9700 S Cass Ave Argonne IL 60439

VELESZ, DUNSTAN GEORGE, b Mancelona, Mich, Apr 26, 09. MATHEMATICS. Educ: St Joseph's Sem, AB, 31; DePaul Univ, MS, 41. Prof Exp: Teacher, Quincy Col Acad, 36-40, instr math, 42-50, from asst prof to assoc prof, 50-60, registr, 50-60, PROF MATH, QUINCY COL, 60- Mem: Am Math Soc; Math Asn Am. Mailing Add: Dept of Math Quincy Col Quincy IL 62301

VELEZ, ANTONIO, b San Lorenzo, PR, Feb 20, 37; m 63; c 2. HORTICULTURE. Educ: Univ PR, BS, 60; La State Univ, MS, 70, PhD(hort), 75. Prof Exp: Sales supvr feeds, Quaker Oats Co, 61-62; agronomist, C Brewer PR, 62-65, dir res, 66-67; asst agronomist, 67-75, ACTG HEAD DEPT AGRON & SOILS, AGR EXP STA, UNIV PR, 76- Mem: Am Soc Agr Sci; Caribbean Food Crops Soc; Latin Am Soc Control Weeds. Res: Evaluation of herbicidal materials, both registered and unregistered, for weed control in sugarcane, tobacco, sweet potatoes, Tanier, yams, cassava and pasture lands. Mailing Add: Dept of Agron & Soils Box H Agr Exp Sta Univ PR Rio Piedras PR 00928

VELEZ, SAMUEL JOSE, b San Juan, PR, July 19, 45; m 67; c 2. NEUROPHYSIOLOGY. Educ: Univ PR, BS, 66, MS, 69; Yale Univ, PhD(neurophysiol), 74. Prof Exp: Instr biol, Univ PR, 66; biologist, US Naval Sta, San Juan, PR, 66 & 67-68; res asst pharmacol, Sch Med, Univ PR, 67-68; teaching asst neurophysiol, Yale Univ, 70 & 71; NIH fel zool, Univ Tex, Austin, 74-76; ASST PROF BIOL SCI, DARTMOUTH COL, NH, 76- Mem: AAAS. Res: Patterns of neuronal connections; rules of connectivity in neuromuscular connections; nerve-muscle trophic interactions; facilitation at the neuromuscular junction; regeneration of neuromuscular connections; developmental neurobiology. Mailing Add: Dept of Biol Sci Dartmouth Col Hanover NH 03755

VELEZ, WILLIAM YSLAS, b Tucson, Ariz, Jan 15, 47; m 68; c 2. MATHEMATICS. Educ: Univ Ariz, BS, 68, MS, 72, PhD(math), 75. Prof Exp: MEM TECH STAFF MATH, SANDIA LABS, 75- Mem: Am Math Soc; Math Asn Am. Res: Coding theory, number theory and combinatorics and their application to industrial problems. Mailing Add: Sandia Labs Albuquerque NM 87115

VELICK, SIDNEY FREDERICK, b Detroit, Mich, May 3, 13; m 41; c 2. BIOCHEMISTRY. Educ: Wayne State Univ, BS, 35; Univ Mich, MS, 36, PhD(biol chem), 38. Prof Exp: Rockefeller Found fel, Johns Hopkins Univ, 39-40; Int Cancer Res Found fel, Yale Univ, 41-45; from asst prof to prof biol chem, Sch Med, Wash Univ, 45-64; PROF BIOL CHEM & HEAD DEPT, COL MED, UNIV UTAH, 64- Concurrent Pos: Mem biochem study sect, NIH, 65-69. Mem: Am Soc Biol Chemists; Am Chem Soc. Res: Bacterial lipids; protein chemistry and metabolism; mechanism of enzyme action. Mailing Add: Dept of Biochem Univ of Utah Col of Med Salt Lake City UT 84112

VELIKY, IVAN ALOIS, b Zilina, Czech, Mar 23, 29; m 52; c 2. BIOLOGICAL CHEMISTRY. Educ: Slovak Tech Univ, Bratislava, EngC, 50, DiplEng, 52; Slovak Acad Sci, PhD, 60. Prof Exp: From asst prof to assoc prof biochem, Slovak Tech Univ, Bratislava, 52-65; fel, Prairie Regional Lab, 65-67, assoc res officer, 67-75, SR RES OFFICER, DIV BIOL SCI, NAT RES COUN CAN, 75- Mem: Chem Inst Can; Can Biochem Soc; Int Asn Plant Tissue Cult. Res: Cell physiology and biochemistry; physiology of cell growth in suspension cultures (fermentors); biosynthesis of secondary metabolites and biotransformation of precursors by cell cultures. Mailing Add: Div of Biol Sci Nat Res Coun Ottawa ON Can

VELLA, FRANCIS, b Malta, July 24, 29; m 56; c 4. BIOCHEMISTRY, GENETICS. Educ: Royal Univ Malta, BSc, 49, MD, 52; Oxford Univ, BA, 54, MA, 58; Univ Singapore, PhD(biochem), 62. Prof Exp: Asst lectr biochem, Univ Singapore, 56-57, lectr, 57-60; sr lectr, Univ Khartoum, 60-64, reader biochem genetics, 64-65; vis assoc prof biochem, 65-66, assoc prof, 66-71, PROF BIOCHEM, UNIV SASK, 71- Concurrent Pos: Tutor, WHO Lab Course in Abnormal Hemoglobins, Ibadan, Nigeria, 63; lectr, NATO Advan Course in Pop Genetics, Rome, Italy, 64; vis prof biochem, Univ Cambridge, 73-74. Honors & Awards: Chevalier, Order of St Sylvester, Vatican City, Italy, 65. Mem: Chem Inst Can; fel Royal Inst Chem; Royal Col Path. Res: Molecular genetics; abnormal human hemoglobins; hereditary enzyme deficiencies in man; teaching methods in biochemistry. Mailing Add: Dept of Biochem Univ of Sask Saskatoon SK Can

VELLA, PHILIP PETER, b Syracuse, NY, Aug 18, 37; m 59; c 6. MEDICAL BACTERIOLOGY, IMMUNOLOGY. Educ: Univ Notre Dame, BS, 59, PhD(biol), 65. Prof Exp: Res asst microbiol, Bristol Labs, 59-61; sr res virologist, 64-68, res fel, 68-71, sr res fel, 71-75, SR INVESTR, MERCK & CO, INC, 75- Mem: Am Soc Microbiol. Res: Research and development of vaccines against human, bacterial diseases. Mailing Add: Dept of Virus & Cell Biol Merck Sharp & Dohme Res Labs Merck & Co Inc West Point PA 19486

VELLACCIO, FRANK, b New Haven, Conn, Sept 24, 48; m 70; c 1. BIO-ORGANIC CHEMISTRY. Educ: Fordham Univ, BS, 70; Mass Inst Technol, PhD(org chem), 74. Prof Exp: ASST PROF CHEM, COL HOLY CROSS, 74- Mem: Sigma Xi; Am Chem Soc. Res: Synthetic methods for peptide synthesis; intromolecular acyl transfers. Mailing Add: Dept of Chem Col of the Holy Cross Worcester MA 01610

VELLA-COLEIRO, GEORGE, b Malta, Mar 15, 41. PHYSICS. Educ: Royal Univ Malta, BSc, 61; Oxford Univ, MA, 63, DPhil(physics), 67. Prof Exp: MEM TECH STAFF PHYSICS, BELL TEL LABS, 67- Mem: Am Phys Soc. Res: Magnetism; semiconductors. Mailing Add: Bell Tel Labs Rm 2D313 Murray Hill NJ 07974

VELLIOS, FRANK, b St Louis, Mo, Sept 8, 22; m 50. PATHOLOGY. Educ: Wash Univ, MD, 46; Am Bd Path, dipl. Prof Exp: Instr path, Wash Univ, 51-52; from asst prof to prof, Sch Med, Ind Univ, Indianapolis, 52-68, chmn dept, 65-68; prof, Case Western Reserve Univ, 68-69; PROF PATH, SOUTHWESTERN MED SCH, UNIV TEX HEALTH SCI CTR DALLAS, 69- Concurrent Pos: Ed, Jour, Am Soc Clin Path, 65- Mem: Am Soc Clin Path; Am Asn Path & Bact; Am Med Asn; Am Soc Cytol; Int Acad Path. Res: Surgical pathology, especially human neoplasms. Mailing Add: Southwestern Med Sch Univ Tex Hlth Sci Ctr Dallas TX 75235

VELLTURO, ANTHONY FRANCIS, b Ansonia, Conn, Dec 3, 36. APPLIED CHEMISTRY, ORGANIC CHEMISTRY. Educ: Yale Univ, BS, 58, MS, 59, PhD(org chem), 62. Prof Exp: NIH fel, Tulane Univ, La, 64-65; sr res chemist, Techni-Chem Co, Conn, 65-70; sr develop chemist, 70-73, GROUP LEADER, CIBA-GEIGY CHEM CO, 73- Mem: Am Chem Soc. Res: Reaction mechanisms; synthetic organic chemistry. Mailing Add: Ciba-Geigy Corp 180 Mill St Cranston RI 02905

VELTMAN, PRESTON LEONARD, b Grand Rapids, Mich, July 18, 12; m 38; c 2. CHEMISTRY. Educ: Mich Col Mining & Technol, BS, 34, MS, 35; Univ Wis, PhD(phys chem), 38. Prof Exp: Asst dir fuels res, Tex Co, NY, 38-43; plant engr, Manhattan Dist, Los Alamos Sci Lab, NMex, 43-45; mgr res, Davison Chem Corp, 45-50, mgr res & develop, Curtis Bay Labs, 50-54; res dir, 55-59, VPRES RES DIV & DIR RES LIAISON, W R GRACE & CO, 60- Concurrent Pos: Mem, Governor's Sci Adv Coun, Md. Mem: Am Chem Soc; Am Inst Chem Engrs. Res: Fuels synthesis; metallurgy; catalysis; adsorption phenomena; fluorides; fertilizers; plant foods; nitrogen chemistry; marine protein; construction materials; food processing. Mailing Add: Res Div W R Grace & Co Columbia MD 21029

VELTRI, ROBERT WILLIAM, b McKeesport, Pa, Dec 1, 41; m 62; c 2. MICROBIOLOGY. Educ: Youngstown Univ, BA, 63; WVa Univ, MS, 65, PhD(microbiol), 68. Prof Exp: Asst prof microbiol, 68-72, ASSOC PROF MICROBIOL & OTOLARYNGOL, MED CTR, WVA UNIV, 72-, DIR OTOLARYNGIC RES, DIV OTOLARYNGOL, 68- Concurrent Pos: Immunol consult, Dent Sci Inst, Univ Tex, Houston, 72- Mem: AAAS; assoc fel Am Acad Ophthal & Otolaryngol; Soc Gen Microbiol; Am Soc Microbiol. Res: Role of tonsils in immunobiology; virology and immunology of herpesvirus infections; microbiology and immunology of otolaryngic infections; isolation and identification of human tumor-associated antigens; Epstein-Barr virus-host relationships. Mailing Add: Div of Otolaryngol Rm 2156 BSB WVa Univ Med Ctr Morgantown WV 26506

VENA, JOSEPH AUGUSTUS, b Jersey City, NJ, Apr 18, 31; m 56; c 2. CYTOLOGY. Educ: St Peter's Col, BS, 52; Fordham Univ, MS, 55, PhD(cytol), 63. Prof Exp: Teacher high sch, 53-57; from instr to asst prof biol, Fordham Univ, 57-63; assoc prof, 63-66, PROF BIOL, TRENTON STATE COL, 66- Concurrent Pos: Sigma Xi res grant, 66-67; res consult, Univ Calif, Berkeley; consult, USPHS, 73-; NSF grant, 74. Mem: AAAS; Sigma Xi; Am Soc Cell Biol. Res: Coordinated studies of ultrastructural changes and electrophysiological properties in conduction system of canine heart during pharmacohogically induced alterations. Mailing Add: Dept of Biol Trenton State Col Trenton NJ 08625

VENABLE, DOUGLAS, b Charleston, WVa, Aug 17, 20; m 43; c 1. PHYSICS. Educ: Hampden-Sydney Col, BS, 42; Univ Va, MS, 47, PhD(physics), 50. Prof Exp: Design engr indust electronics div, Westinghouse Elec Corp, 42-46; mem staff, 50-57, alt group leader, 57-65, group leader, 65-72, ALT DIV LEADER, LOS ALAMOS SCI LAB, UNIV CALIF, 72- Concurrent Pos: Prof, Los Alamos Grad Ctr, Univ NMex, 57, 58 & 61. Mem: Fel AAAS; Am Phys Soc. Res: Crystal physics; electron beam dynamics; electron linear accelerators, gaseous discharges; flash radiography; hydrodynamics and shock wave phenomena. Mailing Add: 118 Aztec Ave Los Alamos NM 87544

VENABLE, JOHN HEINZ, b Atlanta, Ga, Dec 5, 08; m 34; c 2. PUBLIC HEALTH. Educ: Emory Univ, BS, 29, MD, 33; Tulane Univ, MPH, 51. Prof Exp: From instr to actg chmn dept, Emory Univ, 30-46; comnr health, Whitfield & Murray Counties, Ga, 47-50 & Spalding, Pike & Lamar Counties, 50-52; comnr health, State Dept Pub Health, Ga, 52-59, dir, 60-72, dir div phys health, Ga Dept Human Resources, 72-74; RETIRED. Mem: Am Pub Health Asn; AMA; Asn State & Territorial Health Officers (pres, 66-67). Mailing Add: 2418 Howell Mill Rd NW Atlanta GA 30318

VENABLE, JOHN HEINZ, JR, b Atlanta, Ga, June 9, 38; m 62. MOLECULAR BIOPHYSICS. Educ: Duke Univ, BS, 60; Yale Univ, MS, 63, PhD(biophys), 65. Prof Exp: Vis scientist, King's Col, Univ London, 65-67; asst prof molecular biol, 67-72, ASSOC PROF MOLECULAR BIOL, VANDERBILT UNIV, 72- Concurrent Pos: NSF fel, 65-66; USPHS fel, 66-67. Mem: AAAS; Biophys Soc; Sigma Xi. Res: Macromolecular structure; biophysical chemistry; transition-metal complexes; x-ray diffraction; electron paramagnetic resonance. Mailing Add: Dept of Molecular Biol Vanderbilt Univ Nashville TN 37235

VENABLE, JOHN HOWARD, b Oklahoma City, Okla, Oct 16, 29; m 50; c 2. VETERINARY ANATOMY, HISTOLOGY. Educ: Okla State Univ, DVM, 53, MS, 56; Harvard Univ, PhD(anat), 65. Prof Exp: Instr vet anat, Okla State Univ, 54-56; from asst prof to prof, 56-71, prof physiol sci & head dept, 71-75; PROF ANAT & CHMN DEPT, COL VET MED & BIOMED SCI, COLO STATE UNIV, 76- Concurrent Pos: NSF sci fac fel, Harvard Med Sch, 61-62, univ res fel anat, 62-64; NIH res grant, 64-72. Mem: Am Vet Med Asn; Am Asn Anat; Am Asn Vet Anat. Res: Electron microscopy; autoradiography; anatomy and cytology of skeletal muscles. Mailing Add: Col of Vet Med & Biomed Sci Colo State Univ Ft Collins CO 80521

VENABLE, PATRICIA LENGEL, b Elyria, Ohio, Aug 6, 30; m 65; c 2. BOTANY. Educ: Col Wooster, BA, 52; Ohio State Univ, MSc, 54, PhD(bot), 63. Prof Exp: Instr bot, Hanover Col, 54-55; instr biol, Muskingum Col, 55-56 & Col Wooster, 56-60; vis instr bot & zool, Ohio State Univ, Lakewood Br, 62; assoc prof biol, State Univ NY Col Buffalo, 63-65 & Rider Col, 66-70; MASTER BIOL, LAWRENCEVILLE SCH, 74- Mem: AAAS; Bot Soc Am; Fern Soc. Res: Morphological and biosystematic work with vascular plants, especially ferns and the Compositae. Mailing Add: 10 Monroe Ave Lawrenceville NJ 08648

VENABLE, WILLIAM HOWELL, JR, b Lakeland, Fla. OPTICAL PHYSICS. Educ: Univ Fla, BS, 55, MS, 56; Univ Ala, PhD(physics), 62. Prof Exp: Prof physics, Stillman Col, 62-64; asst prof, George Washington Univ, 64-66; PHYSICIST, NAT BUR STANDARDS, 66- Mem: Optical Soc Am. Res: Spectral reflectance and transmittance measurements; fluorometry; colorimetry; instrumentation. Mailing Add: Rm A317 Metrology Bldg Nat Bur of Standards Washington DC 20234

VENABLES, JOHN DUXBURY, b Cleveland, Ohio, Feb 6, 27; m 48; c 3. PHYSICS. Educ: Case Inst Technol, BS, 54; Univ Warwick, PhD, 71. Prof Exp: Physicist, Parma Res Ctr, Union Carbide Corp, 54-64; SR RES SCIENTIST, MARTIN MARIETTA LABS, MARTIN MARIETTA CORP, 64- Mem: AAAS; Am Phys Soc; Electron Micros Soc Am; Am Inst Mining, Metall & Petrol Engrs. Res: Defect structure of solids; radiation effects in solids; ordering effects in transition metal carbides; electron microscopy. Mailing Add: Res Inst Advan Studies Div of Martin Marietta Corp 1450 S Rolling Rd Baltimore MD 21227

VENARD, CARL ERNEST, b Marion, Ohio, Jan 10, 09; m 34; c 2. ZOOLOGY. Educ: Ohio State Univ, BA, 31, MSc, 32; NY Univ, PhD(helminth), 36. Prof Exp: Asst zool, Ohio State Univ, 32-34; asst biol, NY Univ, 34-36; from instr to prof zool & entom, 36-73, EMER PROF ZOOL & ENTOM, OHIO STATE UNIV, 73- Mem: Am Soc Parasitol; Entom Soc Am; Am Soc Zoologists. Res: Parasites of game birds and fishes; taxonomy and distribution of helminths; morphology of linguatulida; biology of fleas and mosquitoes. Mailing Add: 35 Indian Springs Dr Columbus OH 43214

VENDITTI, JOHN M, b Baltimore, Md, Feb 19, 27; m 51; c 3. PHARMACOLOGY, BIOCHEMISTRY. Educ: Univ Md, BS, 49, MS, 57; George Washington Univ, PhD(pharmacol), 65. Prof Exp: Biologist, 51-58, head screening sect, 63-66, CHIEF DRUG EVAL BR, NAT CANCER INST, 66-, PHARMACOLOGIST, 58- Mem: AAAS; Am Asn Cancer Res; Soc Exp Biol & Med; Am Soc Pharmacol & Exp Therapeut. Res: Experimental cancer chemotherapy; biochemical and pharmacological actions of potential antitumor agents. Mailing Add: Drug Eval Br Nat Cancer Inst Bethesda MD 20014

VENEMA, GERARD ALAN, b Grand Rapids, Mich, Jan 26, 49; m 69; c 2. TOPOLOGY. Educ: Calvin Col, AB, 71; Univ Utah, PhD(math), 75. Prof Exp: INSTR MATH, UNIV TEX, AUSTIN, 75- Mem: Am Math Soc. Res: Geometric topology and applications to shape theory. Mailing Add: Dept of Math Univ of Tex Austin TX 78712

VENER, RAYMOND EDWARD, chemical engineering, see 12th edition

VENERABLE, GRANT DELBERT, b Los Angeles, Calif, Aug 31, 42. CHEMISTRY, SCIENCE EDUCATION. Educ: Univ Calif, Los Angeles, SB, 65; Univ Chicago, SM, 67, PhD(phys chem), 70. Prof Exp: Resident assoc radiation chem, Chem Div, Argonne Nat Lab, 67-70; fel radiobiol, AEC Lab Nuclear Med, Sch Med, Univ Calif, Los Angeles, 70-71; instr, Duarte High Sch, Calif, 71-72; ASST PROF CHEM, CALIF POLYTECH STATE UNIV, 72- Concurrent Pos: Dir, Pilot Proj Innovation in Instr Process, Calif State Univ & Cols, 73-76. Mem: Am Chem Soc; AAAS. Res: Investigation of the nature of the learning process; theorizing on the deep structure of the nucleus. Mailing Add: Dept of Chem Calif Polytech State Univ San Luis Obispo CA 93407

VENERABLE, JAMES THOMAS, b Cobden, Ill, Aug 24, 23; m 49; c 4. ORGANIC CHEMISTRY. Educ: Univ Ill, AB, 44; Univ Wis, PhD(org chem), 49. Prof Exp: Res & develop chemist, Mallinckrodt Chem Works, 49-53, mgt res, 53-56, proj leader prod develop, 56-59; res supvr org chem, Morton Chem Co, 59-65, supvr org, photog & anal res & develop, Info Servs & Tech Personnel, Morton Int, Inc, 65-71; assoc prof chem, Wilbur Wright Col, 71; PRES, RIVERVIEW ENTERPRISES, 71- Mem: AAAS; Am Chem Soc; Sigma Xi. Res: Ketene acetals; pharmaceuticals; photographic chemicals; agricultural chemicals; thiocyanates and isothiocyanates; acridanes and polyphenols; plant tissue culture; water-proof writing materials. Mailing Add: 10425 Woodbine Lane Huntley IL 60142

VENERE, RALPH JOSEPH, SR, b Trenton, NJ, Oct 23, 42; m 64; c 2. PLANT PATHOLOGY. Educ: Drexel Univ, BS, 65; Mich State Univ, PhD(plant physiol), 71. Prof Exp: Dir microbiol lab, Qual Assurance Corp, Pa, 71-72; USDA grant via Dr R Gholson, Dept Biochem, Okla State Univ, 72-74; res assoc, 74-75, PRIN INVESTR, RES PROGS, LANGSTON UNIV, 75- Concurrent Pos: Coop State Res Serv res grant, 75. Mem: Am Phytopath Soc; Am Soc Plant Physiol. Res: Mechanism of resistance in bacterial blight of cotton; pectinase system of xanthomonas malvacearum. Mailing Add: Res Progs Langston Univ Langston OK 73050

VENEZIALE, CARLO MARCELLO, b Philadelphia, Pa, Oct 2, 32; m 59; c 2. BIOCHEMISTRY, ENDOCRINOLOGY. Educ: Haverford Col, BA, 54; Univ Pa, MD, 58; Univ Minn, MS, 64; Univ Wis, PhD(biochem), 69. Prof Exp: Intern, Grad Hosp, Univ Pa, 58-59; resident internal med, Mayo Clin, 61-64, staff asst metab & endocrinol, 64-65; instr biochem, 70-71, asst prof biochem & med, 71-74, ASSOC PROF BIOCHEM & MED, MAYO GRAD SCH MED, UNIV MINN, 74- Concurrent Pos: Consult, Mayo Clin, 65 & 69- Mem: Am Diabetes Asn; Endocrine Soc; Am Soc Biol Chemists; Am Soc Andrology. Res: Pathways and regulation of gluconeogenesis; mechanisms of action of androgens. Mailing Add: Dept of Molecular Med Mayo Grad Sch Med Univ Minn Rochester MN 55901

VENEZKY, DAVID LESTER, b Washington, DC, Sept 12, 24; m 50; c 2. INORGANIC CHEMISTRY. Educ: George Washington Univ, BS, 48; Univ NC, PhD(chem), 62. Prof Exp: Phys sci aide trace elements unit, US Geol Surv, 48-49; chemist, US Naval Res Lab, 49-55; instr chem, Univ NC, 58-60; asst prof inorg chem, Auburn Univ, 60-62; res chemist, 62-69, head reaction mechanism sect, Inorg Chem Div, 69-75, HEAD SOLUTION KINETICS SECT, US NAVAL RES LAB, 75- Mem: Am Chem Soc; Sigma Xi. Res: Coordination compounds and aggregation of inorganic substances in solutions; studies to elucidate the methods of preparation, structure and properties of inorganic polymers. Mailing Add: Code 6130 US Naval Res Lab Washington DC 20375

VENGRIS, JONAS, b Daglienai, Lithuania, Mar 26, 09; nat US; m 38; c 2. AGRONOMY. Educ: Dotnuva Agr Col, Lithuania, BS, 34, MS, 36; Univ Bonn, Dr agr sci, 39. Prof Exp: Asst field crops, Agr Col, Lithuania, 34-37, sr asst, 39-41, docent, 41-44; assoc prof, Baltic Univ, Ger, 46-49; from asst prof agron to assoc prof agron, 50-64, assoc prof plant & soil sci, 64-70, PROF PLANT & SOIL SCI, UNIV MASS, AMHERST, 70- Mem: Crop Sci Soc Am; Weed Sci Soc Am; Am Soc Agron. Res: Weed biology and control. Mailing Add: Dept of Plant & Soil Sci Univ of Mass Amherst MA 01002

VENHAM, LARRY LEE, b Akron, Ohio, June 24, 41; m 63; c 1. PEDODONTICS, PSYCHOLOGY. Educ: Ohio State Univ, DDS, 65, MS, 67, PhD(psychol), 72. Prof Exp: NIH fel, 67-69; ASST PROF DENT EDUC, HEALTH CTR SCH DENT MED, UNIV CONN, 70- Concurrent Pos: Am Inst Res Creative Talent Award, 72; Nat Inst Dent Res spec dent award, 75- Mem: Am Psychol Asn; Int Asn Dent Res; Soc Res Child Develop; Am Soc Dent Children. Res: Child development; situational stress, anxiety and coping behavior in response to dental stress; developmental factors in developing stress tolerance. Mailing Add: 390 Broad St Windsor CT 06095

VENIER, CLIFFORD GEORGE, b Trenton, Mich, June 17, 39; m 65; c 3. ORGANIC CHEMISTRY. Educ: Univ Mich, BS, 62; Ore State Univ, PhD(org chem), 66. Prof Exp: Res assoc chem, Univ Tex, 66-67; asst prof, 67-74, ASSOC PROF CHEM, TEX CHRISTIAN UNIV, 74- Concurrent Pos: Vis assoc prof, Univ Nijmegen, Neth, 75. Mem: AAAS; Am Chem Soc; The Chem Soc. Res: Organic sulfur chemistry; reaction mechanisms; quantum organic chemistry. Mailing Add: Dept of Chem Tex Christian Univ Ft Worth TX 76129

VENIT, STEWART MARK, b New York, NY, Apr 4, 46; m 72. MATHEMATICS. Educ: Queens Col, NY, BA, 66; Univ Calif, Berkeley, MA, 69, PhD(math), 71. Prof Exp: ASST PROF MATH, CALIF STATE UNIV, LOS ANGELES, 71- Mem: Am Math Soc. Res: Numerical solution of partial differential equations. Mailing Add: Dept of Math Calif State Univ Los Angeles CA 90032

VENKATACHALAM, TARACAD KRISHNAN, b Cochin, India, Apr 28, 37. ORGANIC CHEMISTRY. Educ: Univ Bombay, BSc, 58, MSc, 62; Univ Louisville, PhD(chem), 65. Prof Exp: Res chemist, 65-69, SR RES CHEMIST, E I DU PONT DE NEMOURS & CO, 69- Mem: Am Chem Soc; Indian Chem Soc; Royal Inst Chem. Res: Polymer technology; natural and synthetic resins; rubber chemistry; textile fibers. Mailing Add: Kevlar Spec Prod Res E I du Pont de Nemours & Co Wilmington DE 19898

VENKATARAGHAVAN, R, b Madras, India, June 29, 39. ANALYTICAL CHEMISTRY. Educ: Univ Madras, BSc, 58, MSc, 60; Indian Inst Sci, Bangalore, PhD(chem), 63. Prof Exp: Fel spectros, Nat Res Coun Can, 63-65; NIH res assoc mass spectros, Purdue Univ, Lafayette, 65-69; SR RES ASSOC CHEM, CORNELL UNIV, 69- Concurrent Pos: Consult, US Army Labs. Mem: AAAS; Am Chem Soc; Am Soc Mass Spectros. Res: Spectrochemistry; reaction kinetics; structural effects on infrared and ultraviolet spectra; instrumental effects on spectra; computer aided analytical techniques. Mailing Add: Dept of Chem Cornell Univ Ithaca NY 14850

VENKATARAMIAH, AMARANENI, b Atmakur, India, Aug 16, 28; m 49; c 5. COMPARATIVE PHYSIOLOGY, MARINE ZOOLOGY. Educ: Andhra Univ, India, BSc, 55; Sri Venkateswara Univ, India, MSc, 57, PhD(exp ecol), 65. Prof Exp: Asst prof zool, Andhra Loyola Col, India, 57-61; asst prof, Sri Venkateswara Univ, India, 65-66; PHYSIOLOGIST & HEAD SECT PHYSIOL, GULF COAST RES LAB, 69- Concurrent Pos: Prin investr salinity problems coastal water fauna, Environ Br, US Army Corps, Washington, DC, 70- Mem: AAAS; Am Fisheries Soc; Am Soc Zoologists; Fed Am Scientists; World Mariculture Soc. Res: Osmoregulatory, metabolic and nutritional problems of commercial shrimps and prawns in relation to salinity, temperature and dissolved oxygen parameters. Mailing Add: Gulf Coast Res Lab PO Drawer AG Ocean Springs MS 39564

VENKATASETTY, H V, electrochemistry, inorganic chemistry, see 12th edition

VENKATESAN, DORASWAMY, b Coimbatore, India. SPACE PHYSICS, ASTROPHYSICS. Educ: Loyola Col, Madras, India, BSc, 43; Benares Hindu Univ, MSc, 45; Gujarat Univ, India, PhD(cosmic rays), 55. Prof Exp: Lectr physics, K P Col, Allahabad, India, 47-48; lectr, Durbar Col, Rewa, 49; sr res asst cosmic rays, Phys Res Lab, Ahmedabad, 49-56; fel, Inst Electron Physics, Royal Inst Technol, Stockholm, 56-57; fel, Nat Res Coun Can, 57-60; res assoc space physics & astrophysics, Univ Iowa, 60-63, asst prof physics, 63-65, consult, High Altitude Balloon Prog, 65; assoc prof physics, Univ Alta, 65-66; assoc prof, 66-69, PROF

PHYSICS, UNIV CALGARY, 69- Mem: Am Geophys Union; Can Asn Physicists; Can Astron Soc; fel Brit Inst Physics. Res: Solar terrestrial relations; astrophysics involving studies of cosmic rays, radiation belts, ionospheric absorption, auroral x-rays, geomagnetism, solar activity, cosmic x-ray sources and interplanetary medium. Mailing Add: Dept of Physics Univ of Calgary Calgary AB Can

VENKETESWARAN, S, b Alleppey, Kerala, India, July 21, 31. CELL BIOLOGY, BOTANY. Educ: Univ Madras, BSc, 50; Univ Bombay, MSc, 53; Univ Pittsburgh, PhD(bot), 61. Prof Exp: Demonstr biol, Jai Hind Col, Univ Bombay, 52-57; ed asst, Coun Sci & Indust Res, Govt India, 57-58; teaching fel & asst, Univ Pittsburgh, 59-61, res assoc, 61-62, instr biol, 62-63, NASA res assoc, 63-65; asst prof, 65-68, ASSOC PROF BIOL, UNIV HOUSTON, 68- Concurrent Pos: Am Cancer Soc instnl grant, Univ Pittsburgh, 65; consult & NASA res grant, Lunar Sample Receiving Lab, 66-75; vis scientist, Argonne Nat Lab, 67. Mem: AAAS; Bot Soc Am; Soc Develop Biol; Tissue Cult Asn; Am Soc Cell Biol. Res: Plant morphogenesis using tissue culture techniques. Mailing Add: Dept of Biol Univ of Houston Houston TX 77004

VENNART, GEORGE PIERCY, b Boston, Mass, Apr 1, 26; m 51; c 3. PATHOLOGY. Educ: Wesleyan Univ, AB, 48; Univ Rochester, MD, 53. Prof Exp: Asst biol, Wesleyan Univ, 47-48; inetern intern & resident path, NC Mem Hosp, 53-56; asst prof, Col Physicians & Surgeons, Columbia Univ, 56-60; assoc prof path, Univ NC, 60-65; PROF PATH & CHMN DIV CLIN PATH, MED COL VA, 65- Concurrent Pos: Instr, Univ NC, 54-56; asst attend pathologist, Preby Hosp, New York, 56-60. Res: Experimental liver disease; platelet agglutination; pulmonary morphology and physiology. Mailing Add: Div of Clin Path Med Col of Va Richmond VA 23219

VENNEMAN, MARTIN RAY, b Cleveland, Ohio, Oct 10, 44; m 66; c 1. MICROBIOLOGY, IMMUNOLOGY. Educ: Ohio State Univ, BSc, 66, MSc, 67, PhD(microbiol), 69. Prof Exp: NIH res assoc, Bryn Mawr Col, 69-70; instr microbiol, 70, ASST PROF MICROBIOL, UNIV TEX, AUSTIN, 70- Mem: AAAS; Am Soc Microbiol; Reticuloendothelial Soc. Res: Induction and manifestation of cellular immune responses; mechanisms of cellular immunity. Mailing Add: Dept of Microbiol Univ of Tex Austin TX 78712

VENNES, JACK A, b Wheeler, Wis, June 12, 23. GASTROENTEROLOGY, INTERNAL MEDICINE. Educ: Univ Minn, BS, 47, MD, 51. Prof Exp: Intern med, Hennepin County Med Ctr, Minneapolis, 51-52; residency, Vet Admin Hosp, Minneapolis, 52-55; pvt pract, St Louis Park Med Ctr, Minneapolis, 57-63; staff physician gastroenterol, 64-67, asst chief med, 67-71, STAFF PHYSICIAN GASTROENTEROL, VET ADMIN HOSP, MINNEAPOLIS, 71- Concurrent Pos: Instr med, Univ Minn, 55-57, from asst prof to assoc prof, 65-76, prof med, 76- Mem: Am Gastroenterol Asn; Am Soc Gastrointestinal Endoscopy; Am Asn Study Liver Dis; Am Soc Clin Invest; Am Fedn Clin Res. Res: Development and applications of fiberoptic endoscopy to improved diagnosis in upper gastrointestinal tract, pancreas and biliary tree; treatment of gastrointestinal hemorrhage; non-surgical endoscopic removal of common duct gallstones; improved teaching methods of fiberoptic endoscopy. Mailing Add: Minneapolis Vet Admin Hosp 54th St & 48th Ave South Minneapolis MN 55417

VENNES, JOHN WESLEY, b Grenora, NDak, Aug 28, 24; m 48; c 3. BACTERIOLOGY. Educ: Univ NDak, BS, 51, MS, 52; Univ Mich, PhD(bact), 57. Prof Exp: Instr bact, Univ NDak, 52-54; asst, Univ Mich, 54-56; from instr to assoc prof, 56-66, PROF BACT, UNIV NDAK, 66- Mem: Am Soc Microbiol. Res: Bacterial physiology and industrial microbiology. Mailing Add: Dept of Microbiol Univ of NDak Grand Forks ND 58201

VENNESLAND, BIRGIT, b Kristiansand, Norway, Nov 17, 13; nat US. BIOCHEMISTRY, PLANT BIOCHEMISTRY. Educ: Univ Chicago, BS, 34, PhD(biochem), 38. Hon Degrees: DSc, Mt Holyoke Col, 60. Prof Exp: Asst biochem, Univ Chicago, 38-39; Am Asn Univ Women int fel, Harvard Med Sch, 39-41; from instr to prof, Univ Chicago, 41-68; dir, Max-Planck Inst Cell Biol, 68-70, DIR VENNESLAND RES DIV, MAX-PLANCK SOC, 70- Concurrent Pos: Mem study sect, Molecular Biol Panel, NSF, 54-63; mem physiol chem study sect, USPHS; mem biochem panel, Wooldridge Comt to Eval USPHS Prog, 64. Honors & Awards: Hales Award, Am Soc Plant Physiol, 50; Garvan Medal, Am Chem Soc, 64. Mem: AAAS; fel Am Soc Biol Chemists; Am Chem Soc; Am Soc Plant Physiol. Res: Mechanism of photosynthesis; enzyme mechanisms; carboxylation and oxidation-reduction reactions; nitrate reduction. Mailing Add: Germany, Federal Republic of

VENNOS, MARY SUSANNAH, b Oct 14, 31; Can citizen; m 58; c 4. CHEMISTRY. Educ: Univ London, BSc, 53; Univ NB, Fredericton, PhD(chem), 56. Prof Exp: Instr chem, Univ NB, 56-59; from asst prof to assoc prof, Russell Sage Col, 59-70; ASSOC PROF CHEM, ESSEX COMMUNITY COL, BALTIMORE COUNTY, MD, 70- Mem: Am Chem Soc. Res: Analytical instrumentation; polarography and chemical kinetics. Mailing Add: PO Box 244 Phoenix MD 21131

VENT, ROBERT JOSEPH, b Ford City, Pa, Feb 13, 40. UNDERWATER ACOUSTICS. Educ: San Diego State Univ, BS, 61, MS, 69. Prof Exp: Physicist underwater acoust, Navy Electronics Lab, 61-68; SUPV PHYSICIST UNDERWATER ACOUST, NAVAL UNDERSEA CTR, 68- Mem: Acoust Soc Am. Res: Underwater acoustics, especially attenuation, surface, bottom and volume scattering; applied ocean sciences. Mailing Add: Naval Undersea Ctr Code 409 Bldg 106 San Diego CA 92132

VENTRICE, CARL ALFRED, b York, Pa, Aug 7, 30; m 60; c 3. PLASMA PHYSICS, NUCLEAR PHYSICS. Educ: Pa State Univ, BS, 56, MS, 58, PhD(physics), 62. Prof Exp: Sr analyst, Anal Serv Inc, 63-64; assoc prof physics, Tenn Technol Univ, 64-66; assoc prof plasma physics, Auburn Univ, 66-68; PROF ELEC ENG, TENN TECHNOL UNIV, 68- Mem: AAAS; Am Phys Soc. Res: Interaction of electromagnetic waves in plasmas; plasma stability; lasers. Mailing Add: Dept of Elec Eng Tenn Technol Univ Cookeville TN 38501

VENTRIGLIA, ANTHONY E, b New York, NY, June 20, 22; m 53; c 2. ALGEBRA, APPLIED MATHEMATICS. Educ: Columbia Univ, AB, 42; Brown Univ, ScM, 43. Prof Exp: Instr math, Rutgers Univ, 47; from instr to asst prof, 47-61, ASSOC PROF MATH, MANHATTAN COL, 61- Concurrent Pos: Instr, Hunter Col, 61-63; adj asst prof, City Col New York; NSF fel, Inst Math Teachers, Univ Wyo, 59. Mem: AAAS; Am Math Soc; Math Asn Am; Am Acad Polit & Soc Sci; Am Asn Univ Prof. Res: Linear algebra and analysis. Mailing Add: 1 Georgia Ave Bronxville NY 10708

VENTURA, ARNOLDO K, b Kingston, Jamaica, Nov 16, 37; m 63; c 2. MICROBIOLOGY, VIROLOGY. Educ: Univ West Indies, BSc, 61, MS, 63, PhD(virol & microbiol), 67; Am Bd Med Microbiol, dipl, 73. Prof Exp: Resident microbiologist, Univ West Indies, Hosp, 67-69; asst prof virol, 70-75, DIR VIROL & ASST PROF PATH, SCH MED, UNIV MIAMI, 75- Concurrent Pos: Consult, WHO-Cornell Univ, 63-64; lab dir, State Fla, 70-; partic sci adv comt dengue, Pan-Am Health Orgn, 71, Venezuelan Equine Encephalitis, 72; dir virol, Cedars of Lebanon Hosp, 72-; consult epidemiol, Pan Am Health Orgn, 74; pres, Jamaican Cult & Civic Asn of Fla, Inc, 74-; consult virologist, Jamaican Govt, 76; adv, Dept Social Security, Dominican Repub, 75-; prin investr, Nat Inst Allergy & Infectious Dis grant, 75-78. Mem: AAAS; Am Soc Trop Med & Hyg; Am Soc Microbiol; Soc Epidemiol Res. Res: Ecology, epidemiology and pathogenesis of arbovirus diseases. Mailing Add: Dept of Path Univ Miami Sch Med Box 875 Miami FL 33152

VENTURA, JOAQUIN CALVO, b Cadiz, Spain, Mar 22, 29; Can citizen; m 61; c 2. PATHOLOGY. Educ: Univ Seville, MD, 52; Univ Montreal, PhD, 58; FRCP(C), 61. Prof Exp: SR LECTR PATH, UNIV MONTREAL, 58-; IN CHG LABS, SANTA CABRINI HOSP, 71- Concurrent Pos: Chief serv & dir res, St Joseph of Rosemont Hosp, Montreal, 62-69; consult pathologist, Sain-Jean de Dieu Hosp, Montreal, 69- & Maisonneuve Rosemont Hosp, Montreal, 70-; assoc dir labs, Santa Cabrini Hosp, 69-70. Honors & Awards: Miguel de Cervantes Award, Spain, 52. Mem: Can Asn Path; NY Acad Sci. Res: Chronic bronchitis, role of sensitization; pathogenesis of bronchiectasis; effects of pollution on chronic experimental bronchitis. Mailing Add: Santa Cabrini Hosp 5655 St Zotique E Montreal PQ Can

VENTURA, WILLIAM PAUL, b Braddock, Pa, Dec 1, 42; m 69; c 1. PHARMACOLOGY. Educ: Duquesne Univ, BS, 64, MS, 66; New York Med Col, PhD(pharmacol), 69. Prof Exp: Res assoc endocrinol, Duquesne Univ, 66; from instr to asst prof pharmacol, New York Med Col, 69-74; ASSOC PROF PHARMACOL & SCI COORDR, GRAD SCH NURSING, PACE UNIV, 74- Concurrent Pos: Lalor Found grant, 70. Mem: Am Physiol Soc; Soc Cryobiol; Int Fertil Asn; Animal Behav Soc. Res: Reproductive pharmacology, male and female reproductive studies. Mailing Add: Parent Rd Katonah NY 10536

VENUTO, CARMINE JOSEPH, b New York, NY, Aug 2, 23; m 56; c 3. ECONOMIC GEOLOGY, MINERALOGY. Educ: City Col New York, BS, 48; Harvard Univ, MA, 50. Prof Exp: Mine geologist, NJ Zinc Co, 51-53; res mineralogist, Foote Mineral Co, 53-56 & Int Minerals & Chem Corp, 56-58; asst prof geol, Villanova Univ, 58-59; RES SCIENTIST, TECHNOL CTR, ESB INC, 59- Mem: Mineral Soc Am; Am Crystallog Asn; Am Inst Mining, Metall & Petrol Eng; Electrochem Soc. Res: Battery active materials; electrochemical battery systems. Mailing Add: Technol Ctr ESB Inc 19 W College Ave Yardley PA 19067

VENUTO, PAUL B, b Flushing, NY, Feb 8, 33; m 58; c 3. PETROLEUM CHEMISTRY. Educ: Univ Pa, AB, 54, PhD(org chem), 62. Prof Exp: Res & develop chemist, Columbian Carbon Co, 57-59; sr res chemist, Mobil Oil Corp, 62-66, group leader heterogeneous catalysis, 66-67, group leader appl res & develop div, 67-69, res assoc, Paulsboro Res Lab, 69-75, ACTG MGR, ANAL & SPEC TECHNOL, CENT RES LAB, MOBIL RES & DEVELOP CORP, 75- Honors & Awards: Ipatieff Award, Am Chem Soc, 71. Mem: Am Chem Soc. Res: Organic heterogeneous catalysis; molecular sieve technology; new energy sources; catalysis in petroleum refining. Mailing Add: Mobil Res & Develop Corp Cent Res Div PO Box 1025 Princeton NJ 08540

VENZKE, WALTER GEORGE, b White Lake, SDak, June 18, 12; m 39; c 1. VETERINARY ANATOMY. Educ: Iowa State Col, DVM, 35, PhD(vet anat), 42; Univ Wis, MS, 37. Prof Exp: Asst genetics, Univ Wis, 35-37; instr vet anat, Iowa State Col, 37-41, asst prof, 41-42, vet physiol, 42; instr zool, 46, asst prof vet prev med, 46-48, assoc prof vet med, 48-53, PROF VET ANAT & HEAD DEPT, OHIO STATE UNIV, 54-, ASST DEAN & SECY, COL VET MED, 60- Mem: Am Vet Med Asn; Am Asn Anat; Conf Res Workers Animal Dis. Res: Endocrinology of the thymus and pineal gland. Mailing Add: Dept of Vet Anat Ohio State Univ 1900 Coffey Rd Columbus OH 43210

VERA, HARRIETTE DRYDEN, b Washington, Pa, Feb 22, 09. MEDICAL MICROBIOLOGY. Educ: Mt Holyoke Col, AB, 30; Yale Univ, PhD(bact), 38; Am Bd Microbiol, dipl. Prof Exp: Asst zool, Mt Holyoke Col, 30-31; teacher high sch, Conn, 31-37; from instr to asst prof physiol & hyg, Goucher Col, 38-43; res bacteriologist, Baltimore Biol Lab, 43-60; dir qual control lab prod, Becton, Dickinson & Co, 60-62, dir qual control, B-D Labs, Inc, 62-75; CONSULT, 75- Concurrent Pos: Vis lectr, Goucher Col, 46-51; consult, Becton, Dickinson & Co, 52-60 & US Dept Army, 56- Honors & Awards: Barnett L Cohen Award, Am Soc Microbiol, 63. Mem: Fel AAAS; fel Am Acad Microbiol; fel Am Pub Health Asn; Am Soc Microbiol; NY Acad Sci. Res: Bacterial morphology and physiology, especially nutrition. Mailing Add: 3501 Bellaire Dr N Apt 16 Ft Worth TX 76109

VERAGUTH, ARNOLD JOHN, b Winona, Minn, Mar 23, 15; m 43; c 3. ANALYTICAL CHEMISTRY. Educ: Wabash Col, AB, 37; Purdue Univ, MS, 39; Univ Ill, PhD(anal chem), 43. Prof Exp: Asst, State Geol Surv, Ill, 39-41; chemist, Hercules Powder Co, Del, 43-44; res engr, Battelle Mem Inst, 44-46; ANAL RES CHEMIST, LUBRIZOL CORP, 46- Mem: Am Chem Soc. Res: Infrared spectroscopy; polarography; organic synthesis of phthalocyanine dyes; analytical methods; problems in the electrolytic preparation of perchloric acid from sodium chloride. Mailing Add: 13101 W Geauga Trail Chesterland OH 44026

VERBANAC, FRANK, b Yugoslavia, Jan 12, 20; nat US; m 45; c 2. ORGANIC CHEMISTRY. Educ: Wayne State Univ, BS, 41; Univ Ill, PhD(chem), 49. Prof Exp: Chemist, Gelatin Prod Corp, 42-46 & Merck & Co, Inc, 49-57; sr res chemist, 57-60, group leader, 60-70, SR SCIENTIST, A E STALEY MFG CO, 70- Mem: AAAS; Am Chem Soc. Res: N-arylpyrazolines; antibiotics; natural products; carbohydrates; polymers; proteins. Mailing Add: 12 Dakota Dr Decatur IL 62526

VERBANC, JOHN JOSEPH, b New Brighton, Pa, Nov 30, 13; m 41; c 5. ORGANIC CHEMISTRY. Educ: Univ Notre Dame, BS, 35, MS, 36, PhD(org chem), 38. Prof Exp: Asst org chem, Univ Notre Dame, 35-37; org chemist, 38-46, group leader in chg res & develop chems for rubber & synthesis, Jackson Lab, 46-53, head, Elastomer Div, 53-73, DIV HEAD RES & DEVELOP, EXP STA, E I DU PONT DE NEMOURS & CO, INC, 57- Mem: AAAS; fel Am Inst Chemists; Am Chem Soc. Res: Development of accelerators; antioxidants; catalytic softening agents; colors; adhesives for rubber; synthetic rubbers; organic intermediates; isocyanate chemistry; polymerication of cyclic ethers. Mailing Add: E I du Pont de Nemours & Co Inc Exp Sta Wilmington DE 19898

VERBER, CARL MICHAEL, b New York, NY, May 20, 35; m 57; c 2. OPTICAL PHYSICS. Educ: Yale Univ, BS, 55; Univ Rochester, MA, 58; Univ Colo, PhD(physics), 61. Prof Exp: SR PHYSICIST, COLUMBUS LAB, BATTELLE MEM INST, 61- Mem: AAAS; Optical Soc Am; Inst Elec & Electronics Engrs. Res: Interaction of intense light with solids; optical properties of solids; integrated optics. Mailing Add: Dept of Physics Battelle Mem Inst 505 King Ave Columbus OH 43201

VERBER, JAMES LEONARD, b De Pere, Wis, Sept 2, 25; m 60; c 5. PHYSICAL OCEANOGRAPHY. Educ: Univ Wis, BS, 49, MS, 50. Prof Exp: Instr climat, Ohio State Univ, 50-53; chief hydrographer, Ohio State Div Shore Erosion, 53-60; chief oceanographer, Fed Water Pollution Control Admin, US Dept Interior, 60-67; chief oceanographer, Northeast Marine Health Sci Lab, 67-68, CHIEF, NORTHEAST TECH SERVS UNIT, FOOD & DRUG ADMIN, USPHS, 68- Mem: Am Geophys

Union; Am Soc Limnol & Oceanog; Marine Technol Soc; Nat Shellfisheries Asn; Int Soc Theoret & Appl Limnol. Res: Physical geography; climatology; physical limnology. Mailing Add: Northeast Tech Serv Bldg S-26 Food & Drug Admin USPHS Davisville RI 02854

VERBINSKI, VICTOR V, b Shickshinny, Pa, May 7, 22; m 58; c 5. NUCLEAR PHYSICS. Educ: Mass Inst Technol, SB, 48; Univ Pa, PhD(physics), 57. Prof Exp: Physicist, Gen Elec Co, 57-59, Oak Ridge Nat Lab, 59-67 & Gulf Gen Atomic, 67-74; MEM STAFF, IRT CORP, 74- Mem: Am Phys Soc; Am Nuclear Soc. Res: Low energy nuclear physics; neutron spectroscopy; nuclear structure physics; photonuclear reactions and fission studies. Mailing Add: IRT Corp 7650 Convay Court San Diego CA 92111

VERBISCAR, ANTHONY JAMES, b Chicago, Ill, Mar 22, 29; m 59; c 3. ORGANIC CHEMISTRY. Educ: DePaul Univ, BS, 51; Univ Notre Dame, PhD(org chem), 55. Prof Exp: Res chemist, Hercules Powder Co, 54 & Argonne Cancer Res Hosp, Ill, 56-57; vpres res, Regis Chem Co, 57-63; fel, Univ Calif, Los Angeles, 64; PRES, ANVER BIOSCI DESIGN, 64- Concurrent Pos: Chmn subcomt biogenic amines, Comt on Specifications & Criteria for Biochem Compounds, Nat Acad Sci-Nat Res Coun. Mem: Sigma Xi; Am Chem Soc; Am Inst Chemists; The Chem Soc; Drug Info Asn. Res: Organic synthesis; medicinal chemistry; nitrogen heterocyclics, latentiation, reference standards, biosynthesis, natural products; biogenic amines; metabolism of foreign compounds; biomedical literature analysis; radioactive tracer techniques. Mailing Add: Anver Biosci Design Inc 160 E Montecito Ave Sierra Madre CA 91024

VERBIT, LAWRENCE, b Philadelphia, Pa, Dec 14, 35; c 2. ORGANIC CHEMISTRY. Educ: Col William & Mary, BS, 59; Bryn Mawr Col, MA, 61, PhD(org chem), 63. Prof Exp: USPHS res fel, Univ Calif, Berkeley, 63-64; asst prof org chem, 64-68, ASSOC PROF ORG CHEM, STATE UNIV NY BINGHAMTON, 68- Concurrent Pos: NIH grant, 66-72. Mem: AAAS; Am Chem Soc; The Chem Soc; The Chem Soc. Res: Asymmetric synthetic reactions; determination of molecular geometry by means of optical rotatory dispersion and circular dichroism; synthesis and properties of liquid crystals. Mailing Add: Dept of Chem State Univ of NY Binghamton NY 13901

VERBRUGGE, CALVIN JAMES, b Sioux Falls, SDak, July 26, 37; m 61; c 2. ORGANIC CHEMISTRY. Educ: Calvin Col, BA, 59; Purdue Univ, PhD(org chem), 63. Prof Exp: SR RES CHEMIST, S C JOHNSON & SON, INC, 63- Mem: Sigma Xi; Am Chem Soc. Res: Emulsion and solution polymerization and coatings therefrom. Mailing Add: Chem Res Dept 1525 Howe St S C Johnson & Son Inc Racine WI 53403

VERBRUGGE, FRANK, b Chandler, Minn, Dec 22, 13; m 40; c 4. ACADEMIC ADMINISTRATION. Educ: Calvin Col, BA, 34; Univ Mo, MA, 40, PhD(physics), 42. Prof Exp: Instr physics, Univ Mo, 40-41; prof, Northeast Mo State Teachers Col, 41-43; from asst prof to prof, Carleton Col, 43-56; staff mem radiation lab, Mass Inst Technol, 44-46; assoc prof physics, 56-59, actg dean, 66-68, PROF PHYSICS, UNIV MINN, MINNEAPOLIS, 59-, ASSOC DEAN, INST TECHNOL, 59-, DIR, UNIV COMPUTER SERV, 68- Concurrent Pos: Consult, Ford Found Sci & Eng, Latin Am, 63-68. Mem: AAAS; Am Phys Soc; Am Asn Physics Teachers. Res: Absorption spectroscopy; enzyme and vitamin inactivation. Mailing Add: 1787 Shryer Ave W St Paul MN 55113

VERBURG, ROBERT MARTIN, b Muskegon, Mich, May 9, 19; m 44; c 3. INDUSTRIAL CHEMISTRY. Educ: Hope Col, AB, 41; Mass State Col, MS, 42. Prof Exp: From supvr carbonyl iron powder prod to mgr spec proj develop dept, Gen Aniline & Film Corp, 42-60, vpres & gen mgr photo div, 62-64; PRES, ANKEN INDUSTS, 64- Mem: Am Chem Soc. Mailing Add: Anken Indust 250 Madison Ave Morristown NJ 07960

VERBY, JOHN E, b St Paul, Minn, May 24, 23; m 46; c 4. FAMILY MEDICINE, COMMUNITY HEALTH. Educ: Carleton Col, BA, 44; Univ Minn, MB, BS, MD, 47. Prof Exp: Physician, pvt family pract, Minn, 49-68; PROF FAMILY PRACT & COMMUNITY HEALTH, MED SCH, UNIV MINN, MINNEAPOLIS, 69- Concurrent Pos: Sci assoc, Mayo Clin, Rochester, Minn, 67-68. Mem: Int Soc Gen Med; Am Asn Family Pract; AMA. Res: Thyroid disease. Mailing Add: 9609 Washburn Rd Bloomington MN 55431

VERCELLOTTI, JOHN R, b Joliet, Ill, May 2, 33; m; c 1. ORGANIC CHEMISTRY, BIOCHEMISTRY. Educ: St Bonaventure Univ, BA, 55; Marquette Univ, MS, 60; Ohio State Univ, PhD(chem), 63. Prof Exp: Asst chem, Marquette Univ, 58-60; fel Ohio State Univ, 60-63, lectr & vis res assoc, 63-64; asst prof, Marquette Univ, 64-67; asst prof, Univ Tenn, Knoxville, 67-70; assoc prof, 70-74, PROF BIOCHEM & NUTRIT, VA POLYTECH INST & STATE UNIV, 74- Concurrent Pos: Res chemist, Freeman Chem Corp, Wis, 59; USDA res grant, 64-73; consult, US Vet Hosp, Wood, Wis, 65-67 & Oak Ridge Nat Lab, 67-71; NSF res grant, 71-73. Mem: Am Chem Soc; The Chem Soc; Am Soc Biol Chem. Res: Synthesis and reactivity of glycoprotein model compounds; mucopolysaccharides; glycosidases. Mailing Add: Dept of Biochem & Nutrit Va Polytech Inst & State Univ Blacksburg VA 24601

VERCH, RICHARD LEE, b Wakefield, Mich, Feb 15, 37; m 66; c 1. AQUATIC BIOLOGY. Educ: Northland Col, BS, 62; Northern Mich Univ, MA, 66; Univ NDak, DA(biol), 71. Prof Exp: Asst prof biol, Bay de Noc Col, 66-69; asst prof, 71-75, ASSOC PROF BIOL & CHMN DIV NATURAL SCI, NORTHLAND COL, 75- Mem: Am Inst Biol Scientists; Nat Asn Biol Teachers; Nat Asn Sci Teachers. Res: Biology teaching, self study units. Mailing Add: Dept of Biol Northland Col Ashland WI 54806

VERDERBER, RUDOLPH RICHARD, physics, see 12th edition

VERDIER, PETER HOWARD, b Pasadena, Calif, Feb 16, 31; m 53; c 1. PHYSICAL CHEMISTRY. Educ: Calif Inst Technol, BS, 52; Harvard Univ, PhD(phys chem), 57. Prof Exp: Res assoc chem, Mass Inst Technol, 57-58; res fel, Harvard Univ, 58-59; res chemist, Union Carbide Res Inst, 59-64, staff consult, 64-65; chemist, 65-70, chief molecular characterization sect, Polymers Div, 70-75, CHEMIST, STRUCTURAL ANALYSIS & STAND SECT, NAT BUR STAND, 75- Mem: Am Phys Soc; The Chem Soc. Res: Chemical physics, especially molecular structure and dynamics; polymer solution properties. Mailing Add: Polymers Div Nat Bur Stand Washington DC 20234

VERDINA, JOSEPH, b Palermo, Italy, Dec 7, 21; US citizen; m 60; c 2. MATHEMATICS. Educ: Univ Palermo, PhD(math), 49. Prof Exp: Prof physics, Harbor Col, Calif, 58-59; PROF MATH, CALIF STATE UNIV, LONG BEACH, 59- Mem: Math Asn Am; Ital Math Union. Res: Geometrical transformations. Mailing Add: Dept of Math Calif State Univ Long Beach CA 90804

VERDOL, JOSEPH ARTHUR, organic chemistry, see 12th edition

VEREEN, LARRY EDWIN, b Loris, SC, Mar 24, 40. MICROBIOLOGY. Educ: Clemson Univ, BS, 63, MS, 64; Colo State Univ, PhD(microbiol), 68. Prof Exp: Asst prof food sci & biochem, Clemson Univ, 68-70; MEM FAC, DEPT BIOL, LANDER COL, 70- Mem: Am Soc Microbiol. Res: Behavior of Clostridium perfringens in vacuum-sealed foods. Mailing Add: Dept of Biol Lander Col Greenwood SC 29646

VERELL, RUTH ANN, b New York, NY, Mar 8, 35; m 66; c 2. ORGANIC CHEMISTRY. Educ: Allegheny Col, BS, 57; Univ Ill, MS, 58; Columbia Univ, PhD(chem), 62. Prof Exp: Res chemist, Nat Bur Standards, 62-64; prof asst, 64-65, asst prog dir, Instrnl Sci Equip Prog, 65-68, assoc prog dir, Col Sci Improv Progs, 68-69 & 71-73, PROJ MGR EXP PROJS & DEVELOP PROGS, NAT SCI FOUND, 73- Mem: AAAS; Am Chem Soc. Res: Cyclopropenones; alkaline conversions of labeled sugars.

VERGEER, TEUNIS, b Rotterdam, Neth, May 2, 01; nat US; m 26; c 3. BIOLOGY. Educ: Calvin Col, AB, 26; Univ Mich, MS, 28, PhD(zool), 32. Prof Exp: Asst, Univ Mich, 26-30; instr biol, Alma Col, 28; assoc prof, Hastings Col, 29-30; prof, Hope Col, 31-53, head dept, 41-53; asst prof physiol & pharmacol, Sch Med, Univ NDak, 53-57; chmn dept biol, Westminster Col, 57-60; prof biol, 60-71, EMER PROF BIOL, APPALACHIAN STATE UNIV, 71- Concurrent Pos: Examr, State Mich; NSF stipend, Ore State Univ & NC State Col; lectr, Univ Md, 61, prof, 62. Mem: Soc Exp Biol & Med; AMA. Res: Physiology; pharmacology; botany; parasitology. Mailing Add: 210 Watauga Dr Boone NC 28607

VERGHESE, MARGRITH WEHRLI, b Davos, Switz, May 12, 39; m 64; c 3. GENETICS, IMMUNOGENETICS. Educ: Iowa State Univ, BS, 61, PhD(poultry breeding), 64. Prof Exp: Res asst poultry breeding, Iowa State Univ, 62-64; res assoc quant genetics, NC State Univ, 65-68; researcher biostatist, Univ NC, Chapel Hill, 68-69; RES ASSOC, DUKE UNIV, 75- Honors & Awards: Nat Res Serv Award, NIH, 75. Mem: AAAS; Genetics Soc Am. Res: Linkage disequilibrium in major histocompatibility complex of mouse; quantitative genetics; selection theory for quantitative traits; interaction between artificial and natural selection in genetic populations; simulation of genetic populations. Mailing Add: 1228 Kingston Ridge Dr Cary NC 27511

VERGUIN, JACOB, b Orange City, Iowa, Nov 19, 13; m 42; c 5. PLANT PHYSIOLOGY. Educ: Iowa State Col, BS, 39, MS, 41, PhD(plant physiol), 47. Prof Exp: Instr bot, Iowa State Col, 41-42, instr plant physiol, 45-46; assoc prof bot & head dept, Univ SDak, 46-48; assoc prof hydrobiol, Franz Theodore Inst Hydrobiol, 48-55; prof biol, Bowling Green State Univ, 55-64; PROF BOT, SOUTHERN ILL UNIV, CARBONDALE, 64- Concurrent Pos: Consult, Commonwealth Edison, Chicago, 73- & Nat Environ Res Ctr, Environ Protection Agency, Nev, 75- Mem: AAAS; Ecol Soc Am; Am Soc Limnol & Oceanog; Am Fisheries Soc; Am Inst Biol Scientists. Res: Photosynthesis under natural conditions; respiration; diffusion problems; aquatic ecology. Mailing Add: Dept of Bot Southern Ill Univ Carbondale IL 62901

VERHALEN, LAVAL MATHIAS, b Knox City, Tex, May 8, 41; m 64; c 2. PLANT BREEDING, PLANT GENETICS. Educ: Tex Tech Col, BS, 63; Okla State Univ, PhD(plant breeding, genetics), 68. Prof Exp: From instr to asst prof agron, 67-71, ASSOC PROF AGRON, OKLA STATE UNIV, 71- Concurrent Pos: Chmn, 26th Cotton Improv Conf, 73-74. Mem: Sigma Xi; Am Soc Agron; Crop Sci Soc Am. Res: Cotton breeding; genetics, particularly population genetics; variety testing; row spacings; plant populations. Mailing Add: Dept of Agron Okla State Univ Stillwater OK 74074

VERHANOVITZ, RICHARD FRANK, b Walsenburg, Colo, Sept 20, 44; m 67; c 1. THEORETICAL NUCLEAR PHYSICS, NUMERICAL ANALYSIS. Educ: Wilkes Col, BS, 66; Lehigh Univ, MS, 69, PhD(physics), 74. Prof Exp: Mathematician, Defense Intel Agency, Dept Defense, 66; engr, Univac, Sperry-Rand Corp, 66-67; SR PROGRAMMER, SPACE DIV, GEN ELEC CO, 75- Mem: Am Phys Soc. Res: Low energy nuclear physics; two and three body interactions; charge symmetry and nuclear coulomb interactions; computer simulation and numerical analysis. Mailing Add: 1703 Meadowview Lane Mont Clare PA 19453

VERHEY, ROGER FRANK, b Grand Rapids, Mich, Sept 12, 38; m 60; c 4. MATHEMATICS. Educ: Calvin Col, AB, 60; Univ Mich, MA, 61, PhD(math), 66. Prof Exp: Lectr, 65-66, from asst prof to assoc prof, 66-72, PROF MATH, UNIV MICH-DEARBORN, 72-, CHMN DEPT MATH & STATIST, 71- Concurrent Pos: Fulbright lectr, Univ Ceylon, 69-70. Mem: Am Math Soc; Math Asn Am. Res: Immersions of the circle into the plane; extension of immersions to the disk from the point of view of the geometry of the image curve. Mailing Add: Dept of Math Univ of Mich Dearborn MI 48128

VERHEYDEN, JULIEN P H, b Brussels, Belg, May 22, 33; m 59; c 2. ORGANIC CHEMISTRY, MOLECULAR BIOLOGY. Educ: Free Univ Brussels, Lic en sci, 55, PhD(chem), 58. Prof Exp: Res assoc, Free Univ Brussels, 58-59; res assoc, Inst Sci Res Indust & Agr, Brussels, Belg, 60-61; fel, 61-63, res chemist, 63-72, HEAD BIO-ORG CEPT, SYNTEX INST MOLECULAR BIOL, 72- Mem: Am Chem Soc; Chem Soc Belg; The Chem Soc. Res: Carbohydrates; nucleosides; nucleotides. Mailing Add: Syntex Inst for Molecular Biol 3401 Hillview Ave Palo Alto CA 94304

VERHOEK, FRANK HENRY, b Grand Rapids, Mich, Feb 12, 09; m 40; c 3. PHYSICAL CHEMISTRY. Educ: Harvard Univ, SB, 29; Univ Wis, MS, 30, PhD(phys chem), 33; Oxford Univ, DPhil(phys chem), 35. Prof Exp: Asst chem, Univ Wis, 29-33; Rhodes scholar, Copenhagen Univ, 35-36; from instr to assoc prof chem, 36-53, supvr res found, 42-65, vchmn dept chem, 60-64 & 66-68, PROF CHEM, OHIO STATE UNIV, 53- Concurrent Pos: Res chemist, Gen Elec Co, 38; res assoc, Stanford Univ, 40; consult, Liberty Mirror Div, Libby-Owens-Ford Glass Co, 43-52; prin chemist, Argonne Nat Lab, 47; sr chemist, Olin Mathieson Chem Corp, 55; consult, US Naval Weapons Ctr, 57-62; vis prof, Univ Fla, 58-59; lectr chem bond approach proj, NSF, 59-68. Mem: Am Chem Soc. Res: Solution kinetics; gas kinetics; complex ion equilibria; strength of acids in nonaqueous solvents; solubility of electrolytes in nonaqueous solvents; hydrocarbon oxidation; oxidation of boron alkanes. Mailing Add: Dept of Chem Ohio State Univ 140 W 18th Ave Columbus OH 43210

VERHOEVEN, LEON A, b Calgary, Alta, Dec 6, 12; US citizen; m 42; c 4. FISH BIOLOGY. Educ: Univ Wash, BS, 42. Prof Exp: Apprentice, F W Kirsh Sheetmetal Works, Wash, 30-31; jr clerk, Richfield Oil Corp, 31-38; sci asst fisheries, US Fish & Wildlife Serv, 39, 40 & 41; asst scientist, Bristol Bay Res Comt, 46 & Int Pac Halibut Comn, 46-47; res assoc fisheries res inst, Univ Wash, 47-52, res asst prof, Col Fisheries, 52-57; proj supvr, Int Pac Salmon Fisheries Comn, BC, Can, 57-63; exec dir, 63-71, CONSULT & SPEC ASST, PAC MARINE FISHERIES COMN, 72- Concurrent Pos: Instr, Mt Hood Community Col, 72- Mem: Am Inst Fishery Res Biologists; Am Fisheries Soc. Res: Herring; halibut; Pacific salmon; environment; ecology. Mailing Add: 5320 SW Spruce Ave Beaverton OR 97005

VERHOOGEN, JOHN, b Brussels, Belg, Feb 1, 12; nat US; m 39; c 4. GEOPHYSICS.

Educ: Free Univ Brussels, MinEng, 33; Univ Liege, GeolEng, 34; Stanford Univ, PhD(volcanol), 36 Prof Exp: Asst geol, Free Univ Brussels, 36-39; Nat Sci Res Found fel, Belg, 39-40; chief prospecting serv, Kilo-Moto Gold Mines, Congo, 40-43; engr, Belg Congo Govt, 43-46; assoc prof geol, 47-52, PROF GEOL, UNIV CALIF BERKELEY, 52- Concurrent Pos: Guggenheim fel, 53-54, 61. Honors & Awards: Day Medal, Geol Soc Am, 58. Mem: Nat Acad Sci; fel Geol Soc Am; fel Am Acad Arts & Sci; fel Am Geophys Union. Res: Paleomagnetism; thermodynamics of geologic phenomena; volcanology. Mailing Add: Dept of Geol & Geophys Univ of Calif Berkeley CA 94720

VERITY, MAURICE ANTHONY, b Bradford, Eng, Apr 21, 31; c 3. PATHOLOGY, NEUROPATHOLOGY. Educ: Univ London, MB, BS, 56. Prof Exp: Intern surg, Paddington Gen Hosp, London, Eng & Portsmouth Group Hosps, 56; intern med, Royal Hosp, Wolverhampton, 57; clin pathologist, United Bristol Hosps, 58-59; vis asst prof pharmacol, 59-60, assoc resident path, Sch Med, 60-61, assoc mem, Brain Res Inst, 67, assoc prof, 68-74, PROF PATH, SCH MED, UNIV CALIF, LOS ANGELES, 74- Concurrent Pos: NIH travel award, Int Neuropath-Neurol Cong, Europe , 65 & 70; NIH fel biochem, Med Sch, Bristol Univ, 68-69; Milheim Found grant, Sch Med, Univ Calif, Los Angeles, 70-71, USPHS grant, 71-74. Mem: AAAS; Am Soc Exp Path; Am Asn Pathologists & Bacteriologists; for mem Royal Soc Med; Brit Biochem Soc; Histochem Soc. Res: Biochemical and histochemical studies of subcellular organelle function in pathologic states, including mercury intoxification, partial hepatectomy; investigations of neurogenic control of vascular smooth muscle. Mailing Add: Dept of Path Univ of Calif Med Ctr Los Angeles CA 90024

VERKADE, JOHN GEORGE, b Chicago, Ill, Jan 15, 35; m 59; c 3. BIOINORGANIC CHEMISTRY, ORGANOMETALLIC CHEMISTRY. Educ: Univ Ill, BS, 56, PhD(inorg chem), 60; Harvard Univ, AM, 57. Prof Exp: From instr to assoc prof, 60-70, PROF INORG CHEM, IOWA STATE UNIV, 70- Concurrent Pos: Grants, NSF, 61-, Petrol Res Found, 63-66 & NIH, 72-; Sloan fel, 66-68. Mem: Am Chem Soc; The Chem Soc. Res: Infrared and nuclear magnetic resonance spectroscopic studies of coordination compounds containing phosphorus ligands; solution and solid state conformational analysis of biologically active phosphorus compounds; stereospecific reactions of phosphorus compounds. Mailing Add: Dept of Chem Iowa State Univ Ames IA 50010

VERKHOVSKY, BORIS SAMUEL, b Odessa, USSR, Oct 8, 33; US citizen; m 72; c 2. SYSTEMS ANALYSIS, RESOURCE MANAGEMENT. Educ: State Univ Odessa, MS, 57; State Univ Latvia, PhD(opers res), 64. Prof Exp: Engr & researcher aerodynamics, Radio & Electronics Res Inst, USSR, 57-58; sr researcher appl math, Res Inst Test Equip, Moscow, 59; engr optimal control, Sci Res Inst Comput, 60; sci researcher transport & foreign trade, Cent Inst Econ & Math, Acad Sci USSR, Moscow, 63-65, head scientist large scale systs anal & resource mgt, 65-72; consult environ sci, IBM Thomas J Watson Res Lab, 74, vis scientist water resources, 74-75; vis prof civil eng, 74-75, ASSOC PROF, WATER RESOURCE PROG, DEPT CIVIL ENG, PRINCETON UNIV, 75- Honors & Awards: Alvin Johnson Award for Achievement, Am Coun for Emigres in the Prof, 75. Mem: Opers Res Soc Am; Int Water Resources Asn. Res: Management science; numerical methods; operations research; resources management; applied mathematics; systems science; water resources planning; developing systems control and management; environmental and energy problem modeling; mathematical programming; applied stochastic processes. Mailing Add: Dept of Civil Eng Eng Quadrangle Princeton Univ Princeton NJ 08540

VERKLER, ROBERT CHARLES, b Peoria, Ill, Sept 18, 15; m 46; c 3. OPERATIONS RESEARCH. Educ: Univ Ill, BS, 37, MS, 51. Prof Exp: Dir res, Indust Res Planning Coun, 47-53; tech instr, US Dept Air Force, 50-51; engr guided missile div, Gen Dynamics/Convair, 52-55; mem opers staff, Missiles & Space Div, Lockheed Aircraft Co, 56-61, eng mgt rep, Lockheed Aircraft Corp, Calif, 61-70; asst prof info systs, Calif State Univ, Los Angeles, 70-74; VPRES, PUB SYSTS CONSULT, INC, 74- Concurrent Pos: Consult, Paul J Fields Co, 49- & Govt Mex, 55; mem, Nat Defense Exec Reserve, US Dept Com, 66-; dean grad sch, Van Norman Univ, 67-; vpres, Uniform Safety Bd; adv, Int Conf Bldg Off, Calif; consult engr. Res: Space operations; planning satellite programs; computer research. Mailing Add: 1832 Stratford Pl Pomona CA 91766

VERLANGIERI, ANTHONY JOSEPH, b Newark, NJ, Aug 2, 45; m 67; c 3. BIOCHEMISTRY, TOXICOLOGY. Educ: Rutgers Univ, BS, 68; Pa State Univ, PhD(biochem), 73. Prof Exp: Asst biochem, Pa State Univ, 68-72; ASST PROF TOXICOL, COOK COL, RUTGERS UNIV, 72- Concurrent Pos: Consult toxicologist, Nutrit Int, Inc, 74- Mem: Am Col Vet Toxicol; Am Chem Soc. Res: Effects of sulfating agents on atherogenesis and the influence of lead intoxication on the central nervous system, behavior and learning. Mailing Add: Dept of Animal Sci Cook Col Rutgers Univ New Brunswick NJ 08903

VERLEUR, HANS WILLEM, physics, see 12th edition

VERLEY, FRANK A, b Kingston, Jamaica, Dec 18, 33; m 67. GENETICS. Educ: Univ Conn, BS, 59; Univ Ill, Urbana, MS, 60, PhD(genetics), 64. Prof Exp: Resident res assoc radiation genetics, Argonne Nat Lab, 64-67; from asst prof biol genetics to assoc prof biol genetics, 67-74, PROF BIOL GENETICS, NORTHERN MICH UNIV, 74- Mem: Genetics Soc Am; Am Inst Biol Sci; Am Genetics Asn; Biomet Soc Am. Res: Genetic effects of a recessive sex-linked lethal gene on prenatal development, steroid biosynthesis, electrophoretic patterns of nuclear and cytoplasmic proteins, growth rate and collagen and eleastin synthesis in mice. Mailing Add: 33 West Sci Bldg Dept of Biol Northern Mich Univ Marquette MI 49855

VERLY, WALTER G, b Peronnes, Belg, Apr 10, 23; m 47; c 3. BIOCHEMISTRY. Educ: Univ Liege, MD, 47. Prof Exp: Asst pharmacol, Univ Liege, 47-48; asst prof biochem, Cornell Univ, 51-52; asst, Univ Liege, 52-57, agrege, 57-63, assoc prof, 63-64; PROF BIOCHEM & HEAD DEPT, UNIV MONTREAL, 64- Concurrent Pos: Brit Coun fel biochem, Univ Edinburgh, 48-49; CRB fel, Med Col, Cornell Univ, 49-51; assoc, Nat Found Sci Res, Belg, 53-63; sci secy, Int Cong Biochem, Brussels, 55; consult, Belg Nuclear Ctr, 57-64 & Zool Sta, Naples, 58-59; mem & gen secy, Europ Group Cancer Chemother, 62-64. Mem: AAAS; Am Chem Soc; Am Soc Biol Chem; Can Biochem Soc; NY Acad Sci. Res: Intermediary metabolism; isotope effect; catecholamines; radiobiology; molecular genetics. Mailing Add: Dept of Biochem Univ of Montreal PO Box 6128 Montreal PQ Can

VERMA, DEVI C, b Barsalu-Haryana, India, Apr 30, 46. AGRICULTURAL BIOCHEMISTRY. Educ: Punjab Agr Univ, BS, 68; Univ Calif, Davis, MS, 70; State Univ NY Buffalo, PhD(plant biochem), 75. Prof Exp: FEL PLANT TISSUE CULT, W ALTON JONES CELL SCI CTR, 75- Mem: Am Soc Plant Physiologists; Int Asn Plant Tissue Cult. Res: Biochemistry of cellular differentiation; cell wall formation and development in plant tissue cultures. Mailing Add: W Alton Jones Cell Sci Ctr Old Barn Rd Lake Placid NY 12946

VERMA, GHASI RAM, b Sigari, India, Aug 1, 29; m 54; c 3. APPLIED MATHEMATICS. Educ: Birla Eng Col, India, BA, 50; Benaras Hindu Univ, MA, 54; Univ Rajasthan, India, PhD(math), 57. Prof Exp: Tutor math, Birla Eng Col, India, 54-57, lectr, 57-58; fel, Courant Inst Math Sci, NY Univ, 58-59; asst prof, Fordham Univ, 59-61; reader, Birla Inst Technol & Sci, India, 61-64; ASSOC PROF MATH, UNIV RI, 64- Concurrent Pos: Sr sci res fel, Coun Sci & Indust Res, New Delhi, India, 55-57. Mem: Am Math Soc; Math Asn Am; Soc Indust & Appl Math. Res: Elasticity; fluid mechanics. Mailing Add: Dept of Math Univ of RI Kingston RI 02881

VERMA, RAM D, b May 31, 29; Indian citizen; m 62; c 2. SPECTROSCOPY. Educ: Univ Agra, BSc, 52, MSc, 54, PhD(physics), 58. Prof Exp: Lectr physics, DAV Col, Aligarh, 54-55; res asst, Aligarh Muslim Univ, India, 55-57, sr res fel, 57-58; res assoc, Univ Chicago, 58-61; fel Nat Res Coun Can, 61-63; from asst prof to assoc prof, 63-71, PROF PHYSICS, UNIV NB, 71- Concurrent Pos: Can deleg, Int Conf Spectros, 67; vis prof, Univ Calif, Santa Barbara, Univ Stockholm & Bhabha Atomic Res Ctr, Bombay, India. Mem: Am Phys Soc; Can Asn Physicists. Res: Molecular structure and spectra of stable and free radicals; investigation covering region from near infrared to far ultraviolet. Mailing Add: Dept of Physics Univ of NB Fredericton NB Can

VERMA, SADANAND, b Muzaffarpur, India, Jan 24, 30. ALGEBRA, TOPOLOGY. Educ: Patna Univ, BSc, 50; Univ Bihar, MSc, 52; Wayne State Univ, MS & PhD(math), 58. Prof Exp: Hon lectr math, L S Col, Univ Bihar, 52-53, lectr univ, 53-55 & 58-60; asst prof, Univ Windsor, 60-65; assoc prof, Western Mich Univ, 65-67; PROF MATH, UNIV NEV, LAS VEGAS, 67-, CHMN DEPT, 68- Mem: Math Asn Am; Can Math Cong. Res: Algebraic topology; homotopy theory; elementary number theory; magnetohydrodynamics; general topology; elementary ordinary differential equations. Mailing Add: Dept of Math Univ of Nev Las Vegas NM 89109

VERMA, SUBHASH CHANDER, b Lyallpun, West Panjab, India, Mar 5, 41; m 68; c 1. PHARMACOLOGY. Educ: L M Col Pharm, India, BPharm, 64, MPharm, 67; Univ BC, PhD(pharmacol), 74. Prof Exp: Anal chemist, Pure Pharmaceut Works, India, 64-65; instr pharmaceut, L M Col Pharm, India, 65-67; instr pharm, GSVM Med Col, India, 68-69; teaching asst pharmaceut, 69-74, CAN HEART FOUND FEL & LECTR PHARMACOL, FAC PHARMACEUT SCI, UNIV BC, 74- Mem: Indian Hosp Asn; Indian Pharmaceut Mfr Asn. Res: Study of biochemical and mechanical effects of adrenergic and histaminergic drugs on the heart; other drugs affecting the cardiovascular system. Mailing Add: Fac of Pharmaceut Sci Univ of BC Vancouver BC Can

VERME, LOUIS JOSEPH, b New York, NY, Jan 24, 24; m 57; c 2. WILDLIFE BIOLOGY, ECOLOGY. Educ: Mich State Univ, BS, 50, MS, 53. Prof Exp: Fishery aide, US Fish & Wildlife Serv, 51; game biologist, 53-65, GAME RES BIOLOGIST, MICH DEPT NATURAL RESOURCES, 65- Mem: Wildlife Soc; Am Soc Mammal. Res: Relation of nutrition to reproduction, growth and physiology of deer; habitat quality and management; environmental resistances. Mailing Add: Cusino Wildlife Res Sta Shingleton MI 49884

VERMEER, DONALD E, b Oakland, Calif, Nov 9, 32; m 54. GEOGRAPHY. Educ: Univ Calif, Berkeley, BA, 54, MA, 59, PhD(geog), 64. Prof Exp: Asst prof geog, Univ Colo, Boulder, 62-68; assoc prof geog & anthrop, 68-73, chmn dept, 72-75, PROF GEOG & ANTHROP, LA STATE UNIV, BATON ROUGE, 73- Concurrent Pos: Fac fel, Univ Colo, Boulder, 66-67; vis lectr, Univ Cape Town, 67; La State Univ grad sch fac fel, 69 & 71; consult, World Bk Year Bk, 71 & 72; Am Coun Learned Soc res grant & Nat Geog Soc res grant, 75; hon vis prof, Univ Ibadan, 75-76. Mem: AAAS; Asn Am Geog; African Studies Asn. Res: Coastal geomorphology; troptropical agricultural and dietary problems; cross cultural comparison of the practice of earth eating (geophagy), Africa and the United States; place of compound crops in rural change and development. Mailing Add: Dept of Geog & Anthrop La State Univ Baton Rouge LA 70803

VERMEERSCH, JOYCE ANN, b Detroit, Mich, June 15, 45. NUTRITION, COMMUNITY HEALTH. Educ: Western Mich Univ, BS, 67; Univ Calif, Berkeley, MPH, 69, DrPH(nutrit), 75. Prof Exp: Asst prof nutrit, Sch Nursing, Med Col Ga, 69-71; ASST PROF NUTRIT, UNIV CALIF, DAVIS, 74- Concurrent Pos: Consult, Training Workshops, Nutrit Prog, Sch Pub Health, Univ Calif, Berkeley, 74-75. Mem: Am Pub Health Asn; Am Dietetic Asn; Soc Nutrit Educ. Res: Socioeconomic and cultural determinants of food consumption, expenditures, and nutritional status; evaluation of community nutrition programs. Mailing Add: Dept of Nutrit Univ of Calif Davis CA 95616

VERMEIJ, GEERAT JACOBUS, b Sappemeer, Neth, Sept 28, 46; m 72. EVOLUTIONARY BIOLOGY, BIOGEOGRAPHY. Educ: Princeton Univ, AB, 68; Yale Univ, MPhil, 70, PhD(biol), 71. Prof Exp: From instr to asst prof zool, 71-74, ASSOC PROF ZOOL, UNIV MD, COL PARK, 74- Concurrent Pos: J S Guggenheim Mem fel, 75. Mem: Soc Study Evolution; Ecol Soc Am; Soc Am Naturalists; Paleont Soc; Neth Malacol Soc. Res: Comparative ecology and history of shallow-water benthic marine communities; adaptive morphology, especially molluscs and decapods; temporal patterns of adaptation. Mailing Add: Dept of Zool Univ of Md College Park MD 20742

VERMEULEN, CARL WILLIAM, b Chicago, Ill, July 23, 39; m 61; c 2. MICROBIOLOGY, BIOCHEMISTRY. Educ: Hope Col, AB, 61; Univ Ill, Urbana, MS, 63, PhD(microbiol), 66. Prof Exp: Asst prof microbiol & biochem, 66-71, ASSOC PROF MICROBIOL & BIOCHEM, COL WILLIAM & MARY, 71- Mem: Am Inst Chemists. Res: Microbial genetics; molecular mechanisms of eubacterial acute infection; biochemical contributions to soil genesis. Mailing Add: Dept of Biol Col of William & Mary Williamsburg VA 23185

VERMEULEN, CORNELIUS WILLIAM, b Paterson, NJ, Aug 23, 12; m 36; c 1. UROLOGY. Educ: Calvin Col, AB, 33; Univ Chicago, MD, 37. Prof Exp: Instr surg, Univ Chicago, 41; assoc prof, Univ Ill, 46-53; PROF UROL & CHIEF SECT, PRITZKER SCH MED, UNIV CHICAGO, 53-, DEP DEAN DIV BIOL SCI, 66- Res: Urolithiasis. Mailing Add: Pritzker Sch of Med Univ of Chicago Chicago IL 60637

VERMILLION, FREDERICK J, JR, chemical engineering, paper chemistry, see 12th edition

VERMILLION, HERBERT EDWARD, chemistry, see 12th edition

VERMILLION, ROBERT EVERETT, b Kingsport, Tenn, Aug 17, 37; m 63; c 2. PHYSICS. Educ: King Col, AB, 59; Vanderbilt Univ, MS, 61, PhD(physics), 65. Prof Exp: Asst prof physics, 65-70, ASSOC PROF PHYSICS, UNIV NC, CHARLOTTE, 70- Mem: Am Phys Soc; Am Asn Physics Teachers. Res: Electric shock-tube production of plasmas; techniques and apparatus for teaching undergraduate physics; experimental plasma physics. Mailing Add: Dept of Physics Univ of NC at Charlotte Sta Charlotte NC 28223

VERMUND, HALVOR, b Norway, Aug 8, 16; nat US; m 43; c 2. RADIOLOGY. Educ: Univ Oslo, MD, 43; Univ Minn, PhD(radiol), 51. Prof Exp: Fel, Halden Munic & Vestfold County Hosps, Norway, 44-48; Picker Found fel radiol, Univ Minn, 51-53, res assoc radiol, 53, from asst prof to assoc prof, 54-57; prof radiol & dir radiation ther, Univ Wis-Madison Hosps, 57-68; PROF RADIOL & DIR RADIOTHER RES & DEVELOP, UNIV CALIF, IRVINE, 68- Concurrent Pos: Mem radiation study sect, NIH, 65-69; mem comt diag & ther of cancer, Am Cancer Soc. Mem: Fel Am Col Radiol; Soc Exp Biol & Med; Am Radium Soc; Radiol Soc NAm; Am Roentgen Ray Soc. Res: Radiation therapy; medical radiology; radioactive isotopes. Mailing Add: 101 City Dr S Orange CA 92668

VERNA, JOHN E, b Everett, Mass, Oct 8, 29; m 54; c 4. BIOLOGY. Educ: Northeastern Univ, BS, 52; Univ RI, MS, 54; Brown Univ, PhD(biol), 57. Prof Exp: From instr to asst prof microbiol, Med Sch, Univ Minn, 57-65; head, Cell Biol & Viral Oncol Br, Melpar, Inc, Va, 65-70; vpres biosci & bio prod div, 70-74, VPRES & GEN MGR, BIOSCI DIV, MELOY LABS, INC, 74- Mem: Am Soc Microbiol; NY Acad Sci. Res: Virology, bacteriology, immunology and cancer biology. Mailing Add: Meloy Labs Inc 6715 Electronic Dr Springfield VA 22151

VERNADAKIS, ANTONIA (MRS H L OCKERMAN), b Canea, Crete, Greece, May 11, 30; m 61. DEVELOPMENTAL NEUROBIOLOGY. Educ: Univ Utah, BA, 55, MS, 57, PhD(anat, pharmacol), 61. Prof Exp: Res assoc & res instr anat & pharmacol, Univ Utah Col Med, 61-64; interdisciplinary training prog fel pharmacol, Univ Calif Sch Med, San Francisco Med Ctr, 64-65; asst res physiologist, Univ Calif, Berkeley, 65-67; asst prof, 67-70, ASSOC PROF PSYCHIAT & PHARMACOL, UNIV COLO SCH MED, 70- Concurrent Pos: Res scientist develop award, NIMH, 69-79. Mem: Am Soc Pharmacol & Exp Therapeut; Am Physiol Soc; Am Neurochem Soc; Int Soc Neurochem; Int Soc Psychoneuroendocrinology. Res: Regulatory mechanisms in brain maturation; drugs and hormones; neurotransmission maturation; neural cell growth and differentiation using neural cell culture. Mailing Add: Univ of Colo Sch of Med 4200 E Ninth Ave Denver CO 80220

VERNARDAKIS, THEODORE GALACTION, b Limassol, Cyprus, Nov 14, 42; m 75. PHYSICAL CHEMISTRY, MATERIALS SCIENCE. Educ: Col Emporia, BS, 65; Okla State Univ, MS, 68, PhD(phys chem), 72. Prof Exp: RES ASSOC MAT SCI, UNIV CINCINNATI, 72- Mem: Am Chem Soc; Sigma Xi. Res: Surface chemistry, physical adsorption, infrared and Raman spectra of surfaces; mass spectrometric studies on the high temperature vaporization of graphite and various alloys; thermodynamic properties of metals, oxides and halides. Mailing Add: Dept of Mat Sci & Metall Eng Univ of Cincinnati Cincinnati OH 45221

VERNBERG, FRANK JOHN, b Fenton, Mich, Nov 6, 25; m 45; c 3. MARINE BIOLOGY, PHYSIOLOGICAL ECOLOGY. Educ: DePauw Univ, AB, 49, MA, 50; Purdue Univ, PhD(zool), 51. Prof Exp: From instr to prof zool, Duke Univ, 51-69, asst dir res, 58-63, asst dir marine lab, 63-69; BARUCH PROF MARINE BIOL & DIR BELLE W BARUCH COASTAL RES INST, UNIV SC, 69- Concurrent Pos: Guggenheim fel, 57-58; Fulbright-Hayes res award, Brazil, 65; lectr, Univs Kiel & Sao Paulo; mem comt manned orbital res lab, Am Inst Biol Sci, 66-; dir int biol prog-prog exp anal, Biogeog of the Sea, Nat Acad Sci, 67-69; consult, Environ Protection Agency, 74- Mem: AAAS; Am Physiol Soc; Am Soc Zoologists; Ecol Soc Am. Res: Physiological ecology of marine animals; distribution of decapod crustacea; tissue metabolism; mechanisms of temperature acclimation; physiological diversity of latitudinally separated populations. Mailing Add: Belle W Baruch Coastal Res Inst Univ of SC Columbia SC 29208

VERNBERG, WINONA B, b Pittsburg, Kans, Jan 9, 24; m 45; c 3. ENVIRONMENTAL PHYSIOLOGY. Educ: Kans State Col, AB, 44; DePauw Univ, MA, 47; Purdue Univ, PhD(zool), 51. Prof Exp: Instr zool, DePauw Univ, 47-49; res assoc marine lab, Duke Univ, 52-69; res prof, Belle W Baruch Coastal Res Inst, 69-75, PUB HEALTH PROF BIOL & PROG DIR ENVIRON HEALTH STAFF, SCH PUB HEALTH, UNIV SC, 75- Concurrent Pos: Nat Res Coun fel, Brazil & Fulbright travel award, 65; comt mem, Nat Adv Comt on Oceans & Atmosphere, 73-75. Mem: Sigma Xi; Am Soc Zoologists (treas, 74-77); Am Inst Biol Scientists. Res: Tissue metabolism and temperature acclimation in invertebrates; physiological diversity in latitudinally separated populations of animals; environmental physiology of marine animals. Mailing Add: Sch of Pub Health Univ of SC Columbia SC 29208

VERNEKAR, ANANDU DEVARAO, b Hosali, India, July 5, 32; m 59; c 3. METEOROLOGY. Educ: Univ Poona, BSc, 55, BSc, 56, MSc, 59; Univ Mich, MS, 63, PhD(meteorol), 66. Prof Exp: Sci asst meteorol, Upper Air Sect, India Meteorol Dept, 56-61; res scientist, Travelers Res Ctr, Inc, 67-69; asst prof meteorol, 69-73, ASSOC PROF METEOROL, UNIV MD, COLLEGE PARK, 73- Mem: Am Geophys Union; Am Meteorol Soc. Res: Dynamical meteorology; general circulation; theory of climate and statistical meteorology. Mailing Add: Inst for Fluid Dynamics Univ of Md College Park MD 20742

VERNER, JAMES HAMILTON, b Hitchin, Eng, Feb 22, 40; Can citizen; m 64; c 3. MATHEMATICS. Educ: Queen's Univ, Ont, BSc, 62, MSc, 65; Univ Edinburgh, PhD(comput sci), 69. Prof Exp: Lectr math, Royal Mil Col Can, 63-64; External Aids Off teaching adv, Umuahia, Eastern Nigeria, 64-66; RES ASSOC, QUEEN'S UNIV, ONT, 69-, ASST PROF MATH, 72- Concurrent Pos: Hon lectr math, Univ of Auckland, NZ, 75-76. Res: Numerical analysis, especially numerical solution of initial value problems for ordinary differential equations. Mailing Add: Dept of Math Queen's Univ Kingston ON Can

VERNER, JARED, b Baltimore, Md, Aug 16, 34; m 58; c 3. ANIMAL ECOLOGY. Educ: Wash State Univ, BS, 57; La State Univ, MS, 59; Univ Wash, PhD(zool), 63. Prof Exp: Res assoc zool, Univ Calif, Berkeley, 63-65; from asst prof to prof biol, Cent Wash State Col, 65-73; PROF BIOL, ILL STATE UNIV, 73- Concurrent Pos: NSF fel, 63-65, res grant, 66-71. Mem: Am Ornith Union; Cooper Ornith Soc; Wilson Ornith Soc; Soc Study Evolution; Ecol Soc Am. Res: Evolution and natural selection; avian social organization and communication systems; avian population ecology. Mailing Add: Dept of Biol Ill State Univ Normal IL 61761

VERNICK, SANFORD H, b New York, NY, Oct 29, 31; m 62; c 2. PATHOBIOLOGY. Educ: Univ Miami, BS, 53; Hofstra Univ, MA, 57; Fordham Univ, PhD(biol), 62. Prof Exp: Instr biol, Marymount Col, 59-62; assoc prof, Georgetown Univ, 62-67; staff fel viral biol br, Nat Cancer Inst, 67-71; assoc prof biol, C W Post Col, Long Island Univ, 71-75; ASSOC PROF PATH, EASTERN VA MED SCH, 75- Mem: Sigma Xi (secy, 67); Am Micros Soc; Am Soc Zoologists. Res: Cytological analysis of environmental pressures as pertaining to poikilothermic animals, especially fish; tumor induction in different aged skin grafts. Mailing Add: Dept of Path Eastern Va Med Sch Norfolk VA 23507

VERNIER, ROBERT L, b El Paso, Tex, July 29, 24; m 45; c 4. PEDIATRICS. Educ: Univ Dayton, BS, 48; Univ Cincinnati, MD, 52. Prof Exp: Clin fel pediat, Univ Ark, 52-54; clin fel pediat, Univ Minn, 54-55, USPHS res fel, 55-57, Am Heart Asn res fel, 57-59; asst prof pediat & Am Heart Asn estab investr, Med Sch, Univ Minn, 59-65; prof, Sch Med, Univ Calif, Los Angeles, 65-68; PROF PEDIAT, MED CTR, UNIV MINN, MINNEAPOLIS, 68- Mem: AAAS; Am Soc Clin Invest; Am Soc Exp Path; Soc Pediat Res; affil Royal Soc Med. Res: Clinical pediatrics; renal disease in childhood; electron microscopy in the kidney. Mailing Add: Dept of Pediat Univ of Minn Med Ctr Minneapolis MN 55455

VERNIER, VERNON GEORGE, b Norwalk, Conn, Nov 14, 24; m 55; c 4. PHARMACOLOGY. Educ: Univ Ill, BS, 47, MD, 49. Prof Exp: Intern, Res & Educ Hosp, Univ Ill, 49-50, res assoc pharmacol, Med Col, 50-51, instr, 51-52; res assoc, Sharpe & Dohme, Inc, 52-54; res assoc physiol, Merck Inst Therapeut Res, 56-63; mgr pharmacol sect, Stine Lab, 63-74, DIR PHARMACOL, E I DU PONT DE NEMOURS & CO, INC, 74- Concurrent Pos: Lectr, Sch Med, Temple Univ, 56-66, vis prof, 66- Mem: AAAS; Am Soc Pharmacol & Exp Therapeut; Am Col Neuropsychopharmacol; NY Acad Sci. Res: Neuropsychopharmacology; psychopharmacology; toxicology. Mailing Add: E I du Pont de Nemorus & Co Stine Lab PO Box 30 Newark DE 19711

VERNIKOS-DANELLIS, JOAN, b Alexandria, Egypt, May 9, 34; wid; c 2. PHARMACOLOGY, ENDOCRINOLOGY. Educ: Univ Alexandria, BPharm, 55; Univ London, PhD(pharmacol), 60. Prof Exp: Muelhaupt scholar, Ohio State Univ, 60-61, asst prof pharmacol, Col Med, 61-64; res assoc, 64-66, CHIEF HUMAN STUDIES BR, AMES RES CTR, NASA, 72- Honors & Awards: NASA Medal for Except Sci Achievement, 73. Mem: Endocrine Soc; Soc Exp Biol & Med; Am Physiol Soc; Am Soc Pharmacol & Exp Therapeut; Int Brain Res Orgn. Res: Endocrine pharmacology, particularly the use of drugs in the study of the mechanisms regulating pituitary adrenocorticotropic hormone secretion and the physiological response to stress. Mailing Add: Ames Res Ctr Human Studies Br NASA Moffett Field CA 94035

VERNON, EUGENE HAWORTH, b Mesilla Park, NMex, June 27, 01; m 41; c 4. ANIMAL BREEDING. Educ: Iowa State Col, BS, 23, MS, 48, PhD(animal breeding), 50. Prof Exp: Animal husbandman & supt-in-chg, Iberia Livestock Exp Farm, USDA, 50-58, animal husbandman, Animal Husb Res Div, 58-65; animal sci adv, NC State Air Mission to Peru, 65-68; dir Guyana Prog, 68-70, EMER PROF ANIMAL SCI, SCH AGR, TUCKEGEE INST, 70- Mem: Fel AAAS; Am Soc Animal Sci; Genetics Soc Am; Am Genetic Asn; Am Dairy Sci Asn. Res: Effects of inbreeding on sex ratios and on mortality in swine; crossbreeding of beef cattle. Mailing Add: 811 N Maple St Tuskegee AL 36083

VERNON, FRANK LEE, JR, b Dallas, Tex, Sept 16, 27; m 50; c 3. LOW TEMPERATURE PHYSICS, QUANTUM PHYSICS. Educ: Southern Methodist Univ, BS, 49; Univ Calif, Berkeley, MS, 52; Calif Inst Technol, PhD(elec eng, physics), 59. Prof Exp: Asst recorder geophys prospecting, Tex Co, 49-50; head, Sect Microwave Physics, Hughes Aircraft Co, 51-61; SR STAFF SCIENTIST LOW TEMPERATURE & QUANTUM PHYSICS, AEROSPACE CORP, 61- Concurrent Pos: Res fel physics, Calif Inst Technol, 59-60. Mem: Am Phys Soc; Inst Elec & Electronics Engrs; Sigma Xi. Res: Experimental and theoretical investigations in the fields of low temperature, microwave and quantum physics including superconductors, lasers and the interaction of electron tunneling mechanisms with high frequency radiation. Mailing Add: 1560 Knollwood Terr Pasadena CA 91103

VERNON, GREGORY ALLEN, b Akron, Ohio, July 27, 47. INORGANIC CHEMISTRY. Educ: Pa State Univ, BS, 69; Univ Ill, MS, 71, PhD(chem), 75. Prof Exp: RES ANAL CHEMIST, ATOMIC INT DIV, ROCKWELL INT, 75- Mem: Am Chem Soc. Mailing Add: Atomics Int Div 8900 DeSoto Ave Canoga Park CA 91304

VERNON, JACK ALLEN, b Kingsport, Tenn, Apr 6, 22; m 45; c 2. PSYCHOPHYSIOLOGY. Educ: Univ Va, BA, 48, MA, 50, PhD, 52. Prof Exp: From instr psychol to prof psychol, Princeton Univ, 52-66; PROF OTOLARYNGOL, MED SCH & DIR AUDITION LAB, UNIV ORE, 66- Concurrent Pos: Consult, Nat Acad Sci, 59- Mem: AAAS; Am Psychol Asn. Res: Sensory processes, especially audition and cutaneous mechanisms. Mailing Add: Kresge Hearing Res Lab Portland OR 97201

VERNON, JOHN ASHBRIDGE, b Camden, NJ, Jan 19, 40; m 62; c 2. ORGANIC CHEMISTRY. Educ: Rutgers Univ, BS, 61; Univ Md, PhD(org chem), 65. Prof Exp: Asst chem, Univ Md, 61-63, Gillette Harris res fel org chem, 63-64; res chemist, E I du Pont de Nemours & Co, Inc, 65-70; plant mgr, Pioneer Labs, Chesebrough-Ponds, Inc, 71-73; MGR LABS, VICK MFG DIV, RICHARDSON-MERRELL, INC, 73- Mem: Am Chem Soc. Res: Reactions of organic compounds over alumina; synthesis and reaction of azides; heterocyclic compounds; synthesis of liquid crystals. Mailing Add: Box V Hatboro PA 19040

VERNON, LEO PRESTON, b Roosevelt, Utah, Oct 10, 25; m 46; c 5. BIOCHEMISTRY. Educ: Brigham Young Univ, BA, 48; Iowa State Col, PhD, 51. Prof Exp: Fel, Enzyme Inst, Univ Wis, 51-52; res assoc, Washington Univ, 52-54; assoc prof chem, Brigham Young Univ, 54-61; dir, C F Kettering Res Lab, Kettering Found, Ohio, 61-70; dir res, 70-74, ASST ACAD VPRES RES, BRIGHAM YOUNG UNIV, 74- Concurrent Pos: Researcher, Nobel Inst, Stockholm, Sweden, 60-61. Mem: Am Soc Biol Chem; Am Soc Plant Physiol. Res: Photosynthesis; cytochrome chemistry; respiratory enzymes. Mailing Add: Res Div 673 Widtsoe Bldg Brigham Young Univ Provo UT 84601

VERNON, LONNIE WILLIAM, b Dallas, Tex, Mar 16, 22; m 49; c 3. PHYSICAL CHEMISTRY, FUEL SCIENCE. Educ: Rice Univ, BA, 48, MA, 50, PhD(chem), 52. Prof Exp: SR RES ASSOC, EXXON RES & ENG CO, 52- Mem: Am Chem Soc. Res: Coal conversion processes; surface chemistry; catalysis. Mailing Add: Baytown Res Facil Exxon Res & Eng Co PO Box 4255 Baytown TX 77520

VERNON, MINA LEE, b Winters, Tex, Apr 12, 27. ANATOMY. Educ: Tex State Col Women, BS, 49; Univ Okla, MS, 55, PhD(anat), 58. Prof Exp: Jr clerk, Dallas Chem Warfare Procurement Div, Eighth Serv Command, US Dept Army, 44-45; chief technician, Tissue Lab, Univ Hosp, Baylor Univ, 49-54, vis teaching fel, Col Dent, 53-54; asst, Univ Okla, 54-56; USPHS fel, Nat Cancer Inst, 56-58; chief electron micros lab, Hazelton Labs, Inc, 58-68; HEAD ELECTRON MICROS LAB, MICROBIOL ASSOCS, INC, BETHESDA, MD, 68-, PROJ DIR, 68- Mem: Am Soc Microbiol; Am Asn Cancer Res; AAAS; Electron Micros Soc Am; NY Acad Sci. Res: Histology; cytology; electron microscopy; ultrastructural studies related to viral-chemical oncology and slow latent virus diseases. Mailing Add: Electron Micros Lab Microbiol Assocs 5221 River Rd Bethesda MD 20016

VERNON, ROBERT CAREY, b Wilmington, Ohio, Feb 7, 23; m 49; c 2. SOLID STATE PHYSICS. Educ: Bates Col, BS, 47; Wesleyan Univ, MA, 49; Pa State Univ, PhD(physics), 52. Prof Exp: Asst physics, Pa State Univ, 49-52; instr, Williams Col, 52-54, lectr, 54-55, asst prof, 55-58; asst prof, Clarkson Col Technol, 58-60, assoc prof, 60-61; chmn dept physics, 61-72, PROF PHYSICS, SIMMONS COL, 61- Mem: Am Phys Soc; Am Asn Physics Teachers; Geol Soc Am. Res: Imperfections in nearly perfect crystals; optical properties of semiconductors. Mailing Add: Dept of Physics Simmons Col Boston MA 02115

VERNON, ROBERT ORION, geology, deceased

VERNON, RUSSEL, b Chicago, Ill, May 5, 30; m 52; c 4. PHYSICS, NUCLEAR ENGINEERING. Educ: Univ Chicago, PhB, 48, MS, 51; Univ Calif, Los Angeles, PhD(plasma physics), 60. Prof Exp: From physicist to sr tech specialist, Atomics Int Div, NAm Aviation, Inc, 54-67; mem tech staff, Bellcomm, Inc, 67-68, supvr flight exp, 68-72; MEM TECH STAFF NETWORK ANAL, ANAL STUDIES DEPT, BELL TEL LABS, 72- Concurrent Pos: Guest scientist, Swiss Fed Inst Reactor Res, 62-63. Mem: Am Phys Soc. Res: Applied mathematics; space applications; plasma physics; research administration. Mailing Add: Anal Studies Dept Bell Tel Labs Murray Hill NJ 07974

VERNON, WILLIAM W, b Concord, NH, Nov 1, 25; m 51; c 1. GEOLOGY. Educ: Univ NH, BA, 52; Lehigh Univ, MS, 55, PhD, 64. Prof Exp: Civil engr, Dept Pub Rds & Hwys, NH, 52-53; geologist, US Geol Surv, 56-57; from asst prof to assoc prof geol, 57-71, chmn dept, 65-74, PROF GEOL, DICKINSON COL, 71- Mem: AAAS; Geochem Soc; Nat Asn Geol Teachers; Archaeol Inst Am; Geol Soc Am. Res: Mineralogy, petrology and structure of igneous and metamorphic rocks in south-central New Hampshire; archaeological investigations of Early Man in New York State. Mailing Add: Dept of Geol Dickinson Col Carlisle PA 17013

VER NOOY, CHARLES DEPEW, III, organic chemistry, see 12th edition

VER NOOY, MALCOLM BENNETT, organic chemistry, see 12th edition

VERON, HARRY, physics, see 12th edition

VERONIS, GEORGE, b New Brunswick, NJ, June 3, 26; m 63; c 2. OCEANOGRAPHY. Educ: Lafayette Col, AB, 50; Brown Univ, PhD(appl math), 54. Hon Degrees: MA, Yale Univ, 66. Prof Exp: Staff meteorologist, Inst Advan Study, 53-56; staff mathematician, Woods Hole Oceanog Inst, 56-63; assoc prof oceanog, Mass Inst Technol, 61-63, res oceanogr, 64-66; PROF GEOPHYS & APPL SCI, YALE UNIV, 66- Concurrent Pos: Guggenheim fel, Stockholm, Sweden, 60-61 & 66-67. Mem: AAAS; fel Am Geophys Union; Am Acad Arts & Sci. Res: Ocean circulations; rotating and stratified fluids. Mailing Add: Dept of Geol & Geophys Yale Univ New Haven CT 06520

VEROSUB, KENNETH LEE, b New York, NY, July 10, 44; m 67; c 1. GEOPHYSICS. Educ: Univ Mich, BA, 66; Stanford Univ, MS, 71, PhD(physics), 73. Prof Exp: Asst prof geophysics, Amherst Col, 72-75; ASST PROF GEOPHYSICS, UNIV CALIF, DAVIS, 75- Mem: Am Geophys Union; Geol Soc Am. Res: History of the earth's magnetic field; quaternary seismicity of California and Alaska; geothermal resources in California. Mailing Add: Dept of Geol Univ of Calif Davis CA 95616

VERPLANCK, VINCENT, organic chemistry, see 12th edition

VER PLOEG, DAN ARTHUR, physical biochemistry, see 12th edition

VERPOORTE, JACOB A, b Utrecht, Neth, Oct 17, 36; m 61; c 2. BIOPHYSICAL CHEMISTRY. Educ: Univ Utrecht, MSc, 60; Univ Pretoria, PhD(biochem), 64. Prof Exp: Fel phys protein chem, Univ Alta, 64-65 & Harvard Univ, 65-67; asst prof, 67-72, ASSOC PROF PHYS PROTEIN CHEM, DALHOUSIE UNIV, 72- Concurrent Pos: Muscular Dystrophy Asn Can fel, 64-66. Res: Isolation and characterization of biologically active proteins; physico-chemical studies, including studies on the structure of proteins. Mailing Add: Dept of Biochem Dalhousie Univ Halifax NS Can

VERRALL, RONALD ERNEST, b Ottawa, Ont, Feb 26, 37; m 61; c 2. PHYSICAL CHEMISTRY. Educ: Univ Ottawa, Ont, BSc, 62, PhD(phys chem), 66. Prof Exp: Nat Res Coun Can-NATO fel, Mellon Inst, Carnegie-Mellon Univ, 66-68, vis fel, 68; asst prof chem, 68-72, ASSOC PROF CHEM, UNIV SASK, 72- Mem: AAAS. Res: Thermodynamic studies of electrolyte and non-electrolyte solutions; spectroscopic studies of the liquid phase; ultrasonic studies of liquid phase. Mailing Add: Dept of Chem Univ of Sask Saskatoon SK Can

VERRETT, MARY JACQUELINE, biochemistry, see 12th edition

VERRILLO, RONALD THOMAS, b Hartford, Conn, July 31, 27; m 50; c 3. PSYCHOPHYSICS, NEUROSCIENCES. Educ: Syracuse Univ, BA, 52; Univ Rochester, PhD(psychol), 58. Prof Exp: Asst prof spec educ, 57-62, res assoc, Bioacoust Lab, 59-63, res fel, Lab Sensory Commun, 63-67, assoc prof sensory commun, 67-74, PROF SENSORY SCI, SYRACUSE UNIV, 74- Concurrent Pos: NATO sr fel, Oxford Univ, 70-71. Honors & Awards: Res Award, Am Personnel & Guid Asn, 62. Mem: Acoust Soc Am; AAAS; Psychonomic Soc; Soc Neurosci. Res: Cutaneous sensitivity; effects of the physical parameters of vibratory stimuli on threshold and suprathreshold responses in humans. Mailing Add: Inst for Sensory Res Syracuse Univ Merrill Lane Syracuse NY 13210

VERSCHINGEL, ROGER H C, b Jan 19, 28; Can citizen; m 59. CHEMICAL INSTRUMENTATION. Educ: Sir George Williams Univ, BSc, 49; McGill Univ, PhD(chem), 55. Prof Exp: Lectr chem, Sir George Williams Univ, 54-56, from asst prof to assoc prof, 56-67, PROF CHEM, CONCORDIA UNIV, 67-, CHMN DEPT, 68- & DEAN, SIR GEORGE WILLIAMS FAC SCI, 73- Mem: Am Chem Soc; Chem Inst Can. Res: Chemical spectroscopy. Mailing Add: Dean Fac Sci Sir George Williams Concordia Univ 1455 de Maisonneuve Blvd W Montreal PQ Can

VERSCHUUR, GERRIT L, b Capetown, SAfrica, June 6, 37; m 66; c 1. ASTRONOMY. Educ: Rhodes Univ, SAfrica, BSc, 57, MSc, 60; Univ Manchester, PhD(radio astron), 65. Prof Exp: Jr lectr physics, Rhodes Univ, SAfrica, 60; lectr, Univ Manchester, 64-67; res assoc, Nat Radio Astron Observ, 67-69, asst scientist, 69-72, assoc scientist, 72-73; PROF ASTROGEOPHYS & DIR FISKE PLANETARIUM, UNIV COLO, BOULDER, 73- Mem: Am Astron Soc; Int Soc Planetarium Educr. Res: Interstellar neutral hydrogen studies; interstellar magnetic field measurements. Mailing Add: Fiske Planetarium Univ of Colo Boulder CO 80302

VERSEPUT, HERMAN WARD, b Grand Rapids, Mich, Sept 25, 21; m 51. PAPER TECHNOLOGY. Educ: Yale Univ, BE, 42; Lawrence Univ, MS, 48, PhD(pulp & paper technol), 51. Prof Exp: Res chemist, Robert Gair Co, Inc, 51-56, chief appl res sect, Gair Paper Prod Group, Continental Can Co, Inc, 56-61; dir res & develop, Folding Carton Div, Riegel Paper Corp, 61-66; sr develop engr, Beloit Corp, 67-71; RES MGR PAPERBOARD, BOXBOARD RES & DEVELOP ASN, 71- Mem: Tech Asn Pulp & Paper Indust; Am Chem Soc; Am Soc Naval Eng. Res: Product and process development in the manufacture of packaging materials from recycled fibers. Mailing Add: Boxboard Res & Develop Asn 350 S Burdick Mall Kalamazoo MI 49006

VERSES, CHRIST JAMES, b Stamford, Conn, Apr 12, 39; m 66; c 1. PHYSIOLOGY, MOLECULAR BIOLOGY. Educ: Valparaiso Univ, BS, 61; Univ Conn, PhD(bact), 67. Prof Exp: Fel microbiol, Univ Colo Med Ctr, 66-68; asst prof biol, Moorhead State Col, 68-69; res microbiologist, Pollution Control Industs Inc, 69-72; dir pub health lab, Water & Sewage Anal, State of Conn, 72-73; ASST PROF BIOL, SACRED HEART UNIV, 73- Concurrent Pos: Part-time lectr, Sacred Heart Univ, 72-73; owner anal lab. Mem: Fel Am Inst Chemists; NY Acad Sci; Am Soc Microbiologists; Am Inst Biol Sci. Res: Pollution survey of municipal harbor systems; mechanism of attachment of phage to a host cell; biochemical reactions; viral genetics; new sterilization device and indicators. Mailing Add: 295 Fairview Ave Fairfield CT 06430

VER STRATE, GARY WILLIAM, b Metuchen, NJ, Jan 29, 40; m 61; c 2. POLYMER CHEMISTRY, POLYMER PHYSICS. Educ: Hope Col, BS, 63; Univ Del, PhD(chem), 67. Prof Exp: Sr res chemist, 66-75, RES ASSOC, EXXON CHEM CO, LINDEN, 75- Mem: Am Phys Soc; Am Chem Soc. Res: Crystallinity; rheological properties; characterization by physical methods; light scattering; kinetics and molecular weight distribution; branching; cationic polymerization; chemical modification of polymers; polymer networks; liquid rubbers. Mailing Add: 25 Balmoral Ave Matawan NJ 07747

VERTER, HERBERT SIGMUND, b New York, NY, Jan 30, 36; m 60; c 1. ORGANIC CHEMISTRY. Educ: City Col New York, BS, 56; Harvard Univ, MA, 57, PhD(chem), 60. Prof Exp: NATO fel, Imp Col, Univ London, 60-61; asst prof chem, Cent Mich Univ, 61-66; assoc prof, 66-67, dean acad affairs, 73, PROF CHEM, INTER-AM UNIV PR, 67-, CHMN DEPT, 66-72 & 74- Mem: Am Chem Soc; Sigma Xi. Res: Chemistry of natural products; organic synthesis; mechanism; carbon oxides. Mailing Add: Dept of Chem Inter-Am Univ PR San German PR 00753

VERTES, VICTOR, b Cleveland, Ohio, Sept 10, 27. INTERNAL MEDICINE. Educ: Western Reserve Univ, BS, 49, MD, 53; Am Bd Internal Med, dipl, 62. Prof Exp: Intern med, Univ Hosps, Cleveland, 53-54; resident, Mt Sinai Hosp, Cleveland, 54-56; fel metab & endocrinol, New York Hosp-Cornell Med Ctr, 56-57; demonstr & instr med, 58-62 & 64, from asst clin prof to assoc clin prof, 64-72, PROF MED, SCH MED, CASE WESTERN RESERVE UNIV, 72- Concurrent Pos: Assoc vis physician, Mt Sinai Hosp, 58-64, dir metab & endocrine lab, 59-65, dir div med, 64-65, proj dir chronic dialysis ctr, 65-68; investr, USPHS, 65- Mem: Fel Am Col Physicians; Am Soc Artificial Internal Organs; dipl mem Pan-Am Med Asn; Europ Dialysis & Transplant Asn; Am Col Cardiol. Res: Treatment and mechanisms of uremia; role of the kidney in hypertension. Mailing Add: 1800 E 105th St Cleveland OH 44106

VERTREES, ROBERT LAYMAN, b Louisville, Ky, Nov 1, 39; m 66; c 3. AGRICULTURAL ECONOMICS, RESOURCE MANAGEMENT. Educ: Purdue Univ, BS, 61; Mich State Univ, MS, 67, PhD(resouce develop), 74. Prof Exp: Instr resource econ, Dept Agr & Food Econ, Univ Mass, 69-72; asst prof, Dept Econ, SDak State Univ, 73-76; ASST PROF RESOURCE DEVELOP, OHIO STATE UNIV, 76- Mem: Am Agr Econ Asn; Soil Conserv Soc Am. Res: Economic impacts of alternative water and land resource projects; public water and land resource projects, policies and programs. Mailing Add: Sch of Natural Resources Ohio State Univ Columbus OH 43210

VERVILLE, GEORGE JULIUS, b Chippewa Falls, Wis, May 4, 21; m 43; c 3. GEOLOGY. Educ: Univ Wis, PhB, 47, MS, 49, PhD, 51. Prof Exp: Paleontologist, Stanolind Oil & Gas Co, Okla, 58-59; sr paleontologist, Pan Am Petrol Corp, 59-65, res group supvr, 65-69, RES SECT MGR, AMOCO PROD CO, 69- Mem: Paleont Soc (treas); Am Asn Petrol Geologists. Res: Fusulinid research; stratigraphy. Mailing Add: Res Ctr Amoco Prod Co PO Box 591 Tulsa OK 74102

VERVOORT, GERARDUS, b Utrecht, Neth, July 17, 33; Can citizen; m 64; c 1. COMMUNICATIONS SCIENCE, MATHEMATICS EDUCATION. Educ: Loras Col, BA, 60; Univ Iowa, MSc, 64, PhD(math), 70. Prof Exp: Teacher elem sch, Neth, 53-55 & Indian Sch, Ont, Can, 55-58; itinerant teacher, Dept Northern Affairs, NW Territories, 59-60; teacher sec sch, Dept Indian Affairs, Alta, 60-62; ASSOC PROF MATH & EDUC, LAKEHEAD UNIV, 70- Concurrent Pos: IBM Corp grant, Lakehead Univ, 71-72; consult, Ont Educ Commun Authority, 74-75; dept commun res contract, 74-75 & 75-76. Mem: Math Asn Am; Can Asn Prof Educ. Res: Factors associated with instructor effectiveness in the teaching of university and college level mathematics courses; satellite communication as a means for delivery of higher education to people in remote areas. Mailing Add: Dept of Math Sci Lakehead Univ Thunder Bay ON Can

VERWEY, WILLARD FOSTER, b Walkill, NY, July 14, 13; m; c 2. MEDICAL MICROBIOLOGY. Educ: Rutgers Univ, BSc, 34; Johns Hopkins Univ, DSc(bact), 37. Prof Exp: Asst bact, Col Med, NY Univ, 37-38, instr, 38-40; dir bact res, Sharp & Dohme, Inc, 40-51 & Merck Inst Therapeut Res, 51-57; PROF MICROBIOL, UNIV TEX MED BR, GALVESTON, 57-, DIR CHOLERA RES LAB, BANGLADESH, 74- Mem: AAAS; Am Soc Microbiol; Soc Exp Biol & Med; fel Am Acad Microbiol. Res: Antibacterial chemotherapy; immunization against bacterial diseases, including development of dehydrated Brucella vaccine and purified somatic antigens for immunization against pertussis and cholera; experimental and field testing of purified cholera toxoid in humans. Mailing Add: Cholera Res Lab Dacca Dept of State Washington DC 20520

VERWOERDT, ADRIAN, b Voorburg, Neth, July 5, 27; US citizen; m 65; c 2. PSYCHIATRY, PSYCHOANALYSIS. Educ: Univ Amsterdam, MD, 52; Am Bd Psychiat & Neurol, dipl, 62. Prof Exp: Intern, Touro Infirmary, New Orleans, 53-54; resident, 54-55 & 58-60, chief resident, 59-60, from instr to assoc prof, 60-71, PROF PSYCHIAT, MED CTR, DUKE UNIV, 71-, DIR GERO-PSYCHIAT TRAINING, 66-; DIR RESIDENCY TRAINING PSYCHIAT, JOHN UMSTEAD HOSP, BUTNER, 68- Concurrent Pos: Consult, Vet Admin Regional Off, Winston-Salem, 61-63, Serv to Aging, NC Dept Pub Welfare, 66-68, Dorothea Dix Hosp, Raleigh, 69-72 & Cherry Hosp, Goldsboro, 72-; fel psychiat res, Duke Univ Med Ctr, 60-62, NIMH career teacher training award, 64-66; mem, NC Multiversity Comt, 68-; corresp ed, J Geriat Psychiat, 70-; instr psychoanal, Duke-Univ NC Psychoanal Inst, 72. Mem: Fel Am Psychiat Asn; Geront Soc; Am Psychosom Soc; Am Geriat Soc; dipl Pan-Am Med Asn. Res: Physical illness and depressive symptomatology; psychological reactions in fatal illness; depression in the aged; sexual behavior in senescence; training in geriatric psychiatry; psychiatric education. Mailing Add: Dept of Psychiat Duke Univ Med Ctr Durham NC 27710

VERZARIU, POMPILIU EMIL, physics, see 12th edition

VERZEANO, MARCEL, b Romania, Dec 1, 11; nat US. MEDICINE. Educ: Univ Pisa, MD, 36. Prof Exp: Res fel, Harvard Med Sch & Mass Gen Hosp, 47-50; asst res anatomist, Sch Med, Univ Calif, Los Angeles, 51-53, from asst prof biophys to prof, 53-66; PROF PSYCHOBIOL, SCH MED, UNIV CALIF, IRVINE, 67- Honors & Awards: Knight, Crown of Italy, 45. Mem: Fel Royal Soc Med; Am Physiol Soc; Am

EEG Soc. Res: Neurophysiology; brain and behavior. Mailing Add: Dept of Psychobiol Univ of Calif Sch Med Irvine CA 92717

VESCIAL, FREDERICK, lasers, see 12th edition

VESECKY, JOHN FENWICK, astronomy, space science, see 12th edition

VESELL, ELLIOT S, b New York, NY, Dec 24, 33. MOLECULAR PHARMACOLOGY, BIOCHEMICAL PHARMACOLOGY. Educ: Harvard Univ, MD, 59. Prof Exp: Intern pediat, Mass Gen Hosp, 59-60; res assoc human genet & asst physician, Rockefeller Inst, 60-62; asst resident med, Peter Bent Brigham Hosp, 62-63; clin assoc, Nat Inst Arthritis & Metab Dis, 63-65, head sect pharmacogenet, Lab Chem Pharmacol, Nat Heart Inst, 65-69; PROF PHARMACOL, GENETICS & MED & CHMN DEPT PHARMACOL, COL MED, HERSHEY MED CTR, PA STATE UNIV, 69-, ASST DEAN GRAD STUDIES, 74- Concurrent Pos: Clin asst prof, Sch Med, Georgetown Univ, 65-; vis prof, Med Col Ala, 66-67; physician, Johns Hopkins Hosp, 66-70; William N Creasy vis prof clin pharmacol, Sch Med, George Washington Univ, 75; Pfizer lectr clin pharmacol, Univ Iowa, 76. Honors & Awards: Samuel James Meltzer Award, 67; Am Soc Pharmacol & Exp Therapeut Award, 71. Mem: Harvey Soc; Soc Human Genetics; Soc Exp Biol & Med; Am Soc Clin Invest; Am Fedn Clin Res. Res: Multiple molecular forms of enzymes; effect of heredity on metabolism of drugs; influence of isozymes. Mailing Add: Hershey Med Ctr Pa State Univ Col Med Hershey PA 17033

VESELY, JOHN ANTHONY, b Prague, Czech, Apr 26, 25; Can citizen; m 59; c 2. ANIMAL BREEDING, GENETICS. Educ: Univ BC, BSA, 55, MSA, 57; NC State Univ, PhD(breeding, genetics), 70. Prof Exp: Res officer sheep breeding, Res Br, 57-66, RES SCIENTIST SHEEP & DAIRY CATTLE BREEDING, RES BR, CAN DEPT AGR, 69- Mem: Am Soc Animal Sci; Can Soc Animal Sci. Res: Improvement of livestock production through breeding; population genetics; evolution; statistics. Mailing Add: Res Sta Can Dept of Agr Lethbridge AB Can

VESLEY, DONALD, b Astoria, NY, Nov 7, 32; m 62; c 1. ENVIRONMENTAL HEALTH. Educ: Cornell Univ, BS, 55; Univ Minn, MS, 58, PhD(environ health), 68. Prof Exp: From instr to asst prof, 60-74, ASSOC PROF PUB HEALTH, UNIV MINN, MINNEAPOLIS, 74- Mem: Am Soc Microbiol; Am Pub Health Asn. Res: Environmental microbiology. Mailing Add: Univ Minn Sch Pub Health 1158 Mayo Minneapolis MN 55455

VESLEY, GEORGE F, chemistry, see 12th edition

VESLEY, RICHARD EUGENE, mathematics, see 12th edition

VESSEL, EUGENE DAVID, b Mt Olive, Ill, Dec 1, 27; m 46; c 2. ORGANIC POLYMER CHEMISTRY. Educ: Univ Ill, BS, 57; Univ Iowa, MS, 59, PhD(org chem), 60. Prof Exp: Res chemist, Chevron Res Corp, Standard Oil Calif, 60-61; sr res chemist, United Tech Ctr, United Aircraft Corp, 61-67; RES SPECIALIST NONMETALLIC MAT, COM AIRPLANE GROUP, BOEING CO, 67- Mem: Am Chem Soc. Res: Adhesives; elastomers; plastics; composite materials; materials research; fire research and technology. Mailing Add: 4122 135th Place SE Bellevue WA 98006

VESSEL, MATTHEW F, b Chisholm, Minn, Apr 21, 12; m 38; c 2. SCIENCE EDUCATION, BOTANY. Educ: St Cloud Teachers Col, BE, 36; Cornell Univ, PhD(sci educ limnol & bot), 40. Prof Exp: Teacher high sch, Minn, 36-38; PROF BIOL & SCI EDUC, SAN JOSE STATE UNIV, 40-43 & 46-, HEAD DEPT SCI EDUC, 59-, ASSOC DEAN SCH SCI, 69- Concurrent Pos: Field biologist, State Conserv Dept, Minn, 37; consult, San Mateo Schs, 46-49 & Palo Alto Schs, 58-59; dean sch sci, Calif State Univ, San Jose, 66-67, actg dean, 67-69. Mem: AAAS; Am Nature Study Soc; Nat Asn Res Sci Teaching; Nat Sci Teachers Asn. Mailing Add: Sch of Sci S-127 San Jose State Univ San Jose CA 95192

VESSELINOVITCH, STAN DUSHAN, b Zagreb, Yugoslavia, Feb 20, 22; Can citizen; m 47; c 3. PHYSIOLOGY, ONCOLOGY. Educ: Univ Belgrade, DVM, 49; Univ Toronto, MVSc, 57, DVSc, 58. Prof Exp: Lectr physiol, Univ Belgrade, 49-51; res assoc oncol, Univ Toronto, 52-53, asst prof, 54-58; res assoc, Michael Reese Hosp, Chicago, Ill, 59-61, asst dir oncol, 62-64; assoc prof, Chicago Med Sch, 64-69; assoc prof radiol, 69-72, PROF RADIOL & PATH, UNIV CHICAGO, 72- Concurrent Pos: Res assoc fel, Nat Cancer Inst Can, 55-57, prin investr res grant, 57-58; co-investr res grant, Nat Cancer Inst, 59-64, prin investr, 64-69, res contracts, 69- Mem: AAAS; Am Vet Med Asn; Am Asn Cancer Res; NY Acad Sci. Res: Animal physiology; environmental carcinogenesis; factors and mechanisms modifying carcinogenesis. Mailing Add: Dept of Radiol Univ of Chicago Chicago IL 60637

VESSEY, STEPHEN H, b Stamford, Conn, Mar 7, 39; m 62. ANIMAL BEHAVIOR, ECOLOGY. Educ: Swarthmore Col, BA, 61; Pa State Univ, MS, 63, PhD(zool), 65. Prof Exp: Biologist, NIH, 65-69; asst prof biol, 69-74, ASSOC PROF BIOL SCI, BOWLING GREEN STATE UNIV, 74- Mem: AAAS; Animal Behav Soc; Am Soc Mammal; Am Inst Biol Sci; Int Primatol Soc. Res: Population regulatory mechanisms in mammals; social behavior and population dynamics in primates. Mailing Add: Dept of Biol Bowling Green State Univ Bowling Green OH 43403

VESSEY, THEODORE ALAN, b St Paul, Minn, June 16, 38; m 65; c 1. MATHEMATICAL ANALYSIS. Educ: Univ Minn, BA, 60, PhD(math), 66. Prof Exp: Assoc res engr, Honeywell, Inc, 62-63; asst prof math, Univ Wis-Milwaukee, 66-70; asst prof physics, 70-72, ASSOC PROF MATH, ST OLAF COL, 72- Mem: Am Math Soc; Math Asn Am. Res: Cluster set theory in determination of boundary behavior of complex functions. Mailing Add: Dept of Math St Olaf Col Northfield MN 55057

VESSOT, ROBERT F C, b Montreal, Que, Apr 16, 30; m; c 3. PHYSICS. Educ: McGill Univ, BA, 51, MSc, 54, PhD(physics), 57. Prof Exp: Mem, Div Sponsored Res Staff, Mass Inst Technol, 56-60; mgr maser res & develop, Varian Assocs, 60-67 & Hewlett-Packard Co, 67-69; physicist, 69-72, PRIN INVESTR GRAVITATIONAL REDSHIFT EXP, SMITHSONIAN ASTROPHYS OBSERV, CAMBRIDGE, 72- Concurrent Pos: Mem, US Study Group VII, Int Radio Consultive Comt, 65- Mem: Am Phys Soc; Can Asn Physicists. Res: Physical electronics; noise in electron beams; atomic beams; atomic resonance physics; atomic hydrogen maser frequency standard. Mailing Add: 334 Ocean Ave Marblehead MA 01945

VEST, EDWIN DEAN, plant ecology, human ecology, see 12th edition

VEST, FLOYD RUSSELL, b Orland, Calif, Feb 12, 34; m 55; c 2. MATHEMATICS EDUCATION. Educ: ECent State Col, BSEd, 56; Univ Okla, MA, 59; NTex State Univ, EdD(math), 68. Prof Exp: Instr math, ETex State Univ, 59-61; asst prof, 61-74, ASSOC PROF MATH, N TEX STATE UNIV, 74- Mem: Math Asn Am. Res: Learning theory; curriculum. Mailing Add: Dept of Math NTex State Univ Denton TX 76203

VEST, HYRUM GRANT, JR, b Salt Lake City, Utah, Sept 23, 35; m 58; c 4. PLANT PATHOLOGY, PLANT GENETICS. Educ: Utah State Univ, BS, 60, MS, 65; Univ Minn, PhD(plant path), 67. Prof Exp: Res plant pathologist crops res div, Agr Res Serv, USDA, Md, 67-70; ASSOC PROF HORT, MICH STATE UNIV, 70- Mem: Am Soc Hort Sci; Am Soc Agron. Res: Genetics of nodulation and nitrogen fixation in soybean; breeding and genetics of onions, lettuce and asparagus. Mailing Add: Dept of Hort Mich State Univ East Lansing MI 48824

VEST, MARVIN LEWIS, b Elkins, WVa, May 17, 06; m 30; c 2. MATHEMATICAL ANALYSIS. Educ: Davis & Elkins Col, BS, 27; WVa Univ, MS, 32; Univ Mich, AM, 42, PhD(math), 48. Hon Degrees: ScD, Davis & Elkins Col, 70. Prof Exp: Prin high sch, WVa, 27-28, instr, 28-31; asst math, WVa Univ, 31-32, instr, 32-33; assoc prof, Davis & Elkins Col, 33-38; from instr to prof, 38-73, EMER PROF MATH, WVA UNIV, 73- Concurrent Pos: Consult, US Bur Mines, 51-56. Mem: Am Math Soc; Math Asn Am. Res: Algebraic geometry; mechanics; birational space transformations associated with congruences of lines. Mailing Add: Dept of Math WVa Univ Morgantown WV 26506

VEST, ROBERT D, organic chemistry, inorganic chemistry, see 12th edition

VESTAL, BEDFORD MATHER, b Gainesville, Tex, Mar 8, 43; m 65; c 2. ANIMAL BEHAVIOR, VERTEBRATE ZOOLOGY. Educ: Austin Col, BA, 65; Mich State Univ, MS, 67, PhD(zool), 70. Prof Exp: Instr biol, Univ Mo-St Louis, 69-70, asst prof, 70-73; ASST PROF ZOOL, UNIV OKLA & RES CUR, OKLAHOMA CITY ZOO, 73- Concurrent Pos: Res assoc, Mich State Univ, 70-71; NIMH res grant, 74. Mem: AAAS; Soc Study Evolution; Animal Behav Soc; Ecol Soc Am; Am Soc Mammal. Res: Comparative social behavior of mammals; behavioral ecology and behavior development of rodents; space related behavior of zoo animals. Mailing Add: Oklahoma City Zoo Rte One Box 478 Oklahoma City OK 73111

VESTAL, CLAUDE KENDRICK, b High Point, NC, Mar 11, 16; m 46. METEOROLOGY, CLIMATOLOGY. Educ: Guilford Col, AB, 46. Prof Exp: Observer, Weather Bur, NC, 37-39, observer & forecaster, DC, 39-42, proj head climat, NY, 42-43, sect head, DC, 43-50; foreign serv staff officer, Dept State, Monrovia, Liberia, 50-52; sect head, Weather Bur, DC, 52-56, regional climatologist, Tex, 57-71; CONSULT, 71- Res: Application of modern statistical methods to climatological data analysis, for design purposes and evaluation of operational risks. Mailing Add: 1720 Gun Wood Pl Crofton MD 21114

VESTAL, JAMES ROBIE, b Orlando, Fla, Oct 16, 42; m 67. MICROBIOLOGY. Educ: Hanover Col, BA, 64; Miami Univ, MS, 66; NC State Univ, PhD(microbiol), 69. Prof Exp: Fed Water Pollution Control Admin res assoc microbiol, Syracuse Univ, 69-71; ASST PROF MICROBIOL, UNIV CINCINNATI, 71- Mem: AAAS; Am Soc Microbiol. Res: Microbial degradation of hydrocarbons; bacterial cell wall structure and function; iron oxidizing bacteria; microbial metabolism. Mailing Add: Dept of Biol Sci Univ of Cincinnati Cincinnati OH 45221

VESTAL, PAUL ANTHONY, b Roosevelt, Okla, Nov 2, 08; m 36; c 3. BOTANY. Educ: Colo Col, BA, 30; Harvard Univ, MA, 33, PhD(biol), 35. Prof Exp: Instr bot, Harvard Univ, 35-42, res cur, 36-42; from asst prof to prof, 42-69, Bush prof sci, 69-74, EMER PROF BIOL, ROLLINS COL, 74- Concurrent Pos: Hays-Fulbright Act vis lectr, Repub Uruguay, 63; consult & writer, Biol Sci Curric Study, 60-61, 63-65; sci writer. Mem: AAAS; Bot Soc Am; Soc Econ Bot; Am Inst Biol Sci; Nat Audubon Soc; Nat Asn Biol Teachers. Res: Plant anatomy in relation to taxonomy; economic botany; ethnobotany of the Navajo Indians; prehistoric plant remains; lower plant groups. Mailing Add: Bush Sci Ctr Rollins Col Winter Park FL 32789

VESTER, JOHN WILLIAM, b Cincinnati, Ohio, June 5, 24. BIOCHEMISTRY, INTERNAL MEDICINE. Educ: Univ Cincinnati, MD, 47. Prof Exp: Porter fel res med, Hosp Univ Pa, 54-56; from asst prof to assoc prof biochem & nutrit, Grad Sch Pub Health, Univ Pittsburgh, 56-61, from asst prof to assoc prof med, Sch Med, 56-67, asst prof biochem, 61-67; assoc prof biochem, 67-74, assoc prof med, 67-71, PROF MED, COL MED, UNIV CINCINNATI, 71-, PROF BIOCHEM, 74-; DIR RES, GOOD SAMARITAN HOSP, 67- Concurrent Pos: Asst ward chief, Hosp Univ Pa, 54-56; chief sect isotopes & metab, Vet Admin Hosp, Pittsburgh, Pa, 61-67 & assoc chief of staff, 62-67. Mem: AMA; Am Fedn Clin Res; Am Diabetes Asn; fel Am Col Physicians; fel Am Col Cardiol. Res: Diabetes and mechanism of insulin actions; alcoholism; obesity; muscle diseases. Mailing Add: Dept of Med Res Good Samaritan Hosp Cincinnati OH 45220

VESTERGAARD, PER B, b Firenze, Italy, Aug 30, 20; m 49; c 3. PSYCHIATRY. Educ: Copenhagen Univ, MD, 47. Prof Exp: Intern, Herning County Hosp, Denmark, 47-48; resident med & psychiat, Aarhus Univ Clin, 48-49; dir res labs psychiat, St Hans Hosp, Rosklide, 49-53; sr res med biochemist, Res Facil, 53-58, assoc res scientist, 58-66, PRIN RES SCIENTIST, RES CTR, ROCKLAND PSYCHIAT CTR, 66- Mem: AAAS; Am Chem Soc; NY Acad Sci; for mem Danish Med Asn. Res: Endocrinology; automated chromatography; computerized analytical systems; periodic catatonia; schizophrenia. Mailing Add: Rockland Psychiat Ctr Res Ctr Orangeburg NY 10962

VESTLING, CARL SWENSSON, b Northfield, Minn, May 6, 13; m 38; c 3. BIOCHEMISTRY. Educ: Carleton Col, BA, 34; Johns Hopkins Univ, PhD(biochem), 38. Prof Exp: From instr to prof chem, Univ Ill, Urbana, 38-63; PROF BIOCHEM & HEAD DEPT, UNIV IOWA, 63- Concurrent Pos: Guggenheim fel, Nobel Inst, Sweden, 54. Mem: AAAS; Am Chem Soc; Am Soc Biol Chemists; Soc Exp Biol & Med; Brit Biochem Soc. Res: Isolation, structure, mechanism, lactate and malate dehydrogenases and certain other liver and hepatoma enzymes. Mailing Add: Dept of Biochem Univ of Iowa Iowa City IA 52242

VESTLING, MARTHA MEREDITH, b Urbana, Ill, Sept 4, 41. ORGANIC CHEMISTRY. Educ: Oberlin Col, AB, 62; Northwestern Univ, Evanston, PhD(chem), 67. Prof Exp: Fel, Univ Fla, 67-68; res assoc, Vanderbilt Univ, 68-69; asst prof chem, 70-75, ASSOC PROF CHEM, STATE UNIV NY COL BROCKPORT, 75- Mem: Am Chem Soc; The Chem Soc; Am Soc Mass Spectrometry. Res: Organosulfur and organonitrogen chemistry; reaction mechanisms; mass spectrometry. Mailing Add: Dept of Chem State Univ of NY Col Brockport NY 14420

VETHAMANY, VICTOR GLADSTONE, b Servaikaramadam, India, Feb 7, 35; Can citizen; m 62; c 1. BIOLOGY, ANATOMY. Educ: Univ Madras, BA, 54, MA, 57; Univ Toronto, PhD(zool), 65. Prof Exp: Asst lectr biol, Univ Madras, 57-59; high sch teacher, Ethiopia, 60-61; demonstr zool, Univ Toronto, 61-64; res assoc path, Isaac Albert Res Inst, Kingsbrook Med Ctr, 65-67; ASSOC PROF ANAT, MED SCH, DALHOUSIE UNIV, 67- Concurrent Pos: Fel, Isaac Albert Res Inst, Kingsbrook Med Ctr, NY, 65-67. Mem: AAAS; Histochem Soc; Can Asn Anat; NY Acad Sci; Can Soc Cell Biol. Res: Ultrastructure and histochemistry of blood cells and blood forming organs; ultrastructural studies on Niemann-Pick disease; comparative ultrastructural studies of blood in vertebrates and invertebrates; chemotaxis and cell injury. Mailing Add: Dept of Anat Dalhousie Univ Med Sch Halifax NS Can

VETRANO, JAMES BOND, b Hartford, Conn, Feb 14, 30; m 53; c 4. ENVIRONMENTAL MANAGEMENT. Educ: Bates Col, BS, 51; Univ Wash, MBA, 72. Prof Exp: Res chemist, Metal Hydrides, Inc, 51-55; proj leader, Battelle Mem Inst, 55-58; group leader, Atomics Int, 58-66; mgr, Battelle Northwest, 66-72, res mgr, 72-75; ENVIRON ENGR, WASH PUB POWER SUPPLY SYST, 75- Res: Socio-economic effects of electic power plant construction and operation; chemical and metallurgical hydrides; nuclear reactor fuels and moderators; metallurgy of superconductor materials; phase equilibria; nondestructive testing methods. Mailing Add: Wash Pub Power Supply Syst 3000 George Washington Way Richland WA 99352

VETTE, JAMES IRA, b Evanston, Ill, Mar 4, 27; m 51; c 4. PHYSICS. Educ: Rice Univ, BS, 52; Calif Inst Technol, PhD(physics), 58. Prof Exp: Jr geophysicist, Humble Oil Co, 52; asst physics, Calif Inst Technol, 52-54; staff scientist, Sci Res Lab, Convair Div, Gen Dynamics Corp, 58-62, sr staff scientist, Sci Res Lab, Astronaut Div, 62; mgr nuclear physics, Vela Satellite Prog, Aerospace Corp, 62-63, staff scientist, Space Physics Lab, 63-67; DIR, NAT SPACE SCI DATA CTR, NASA GODDARD SPACE FLIGHT CTR, 67- Mem: AAAS; Am Phys Soc; Am Geophys Union; Inst Elec & Electronics Engrs. Res: Synchrotron; meson physics; high altitude radiation with balloons; solar physics; high energy physics; cosmic rays; magnetospheric physics; satellite measurements; space physics. Mailing Add: Nat Space Sci Data Ctr Code 601 NASA Goddard Space Flight Ctr Greenbelt MD 20771

VETTER, JAMES LOUIS, b St Louis, Mo, Jan 26, 33; m 54; c 2. FOOD TECHNOLOGY. Educ: Washington Univ, AB, 54; Univ Ill, MS, 55, PhD(food technol), 58. Prof Exp: Food technologist, Monsanto Chem Co, 58-63; mgr res & develop labs, Keebler Co, Ill, 63-67, dir res & develop, 67-72; corp dir res & develop, Confectionery, Nut & Snack Prod Lab, 72-74, VPRES & TECH DIR RES & DEVELOP LAB, PLANTERS/CURTISS DIV, STANDARD BRANDS, INC, 74- Mem: Am Chem Soc; Am Asn Cereal Chem; Inst Food Technologists. Res: Development of new products and processes in various food areas with primary emphasis on cookies, crackers and snacks. Mailing Add: Res & Develop Lab Stand Brands 3401 Mt Prospect Rd Franklin Park IL 60131

VETTER, RICHARD C, b Homer, Mich, Apr 17, 23; m 51; c 3. OCEANOGRAPHY. Educ: Albion Col, BA, 49; Univ Calif, San Diego, MS, 51. Prof Exp: Instr col math, Far East Exten Div, Univ Calif, 51; oceanogr, Off Naval Res, 51-57; exec secy comt oceanog, 57-70, EXEC SECY OCEAN AFFAIRS BD, NAT ACAD SCI, 70- Concurrent Pos: Consult develop oceanog, State Univs, Fla, 64 & Univ Va, 67. Mem: AAAS; Marine Technol Soc (vpres, 64-67); Nat Oceanog Asn; Am Geophys Union; Am Soc Limnol & Oceanog. Res: Characteristics of surface ocean waves, particularly the two-dimensional ocean wave spectra. Mailing Add: Nat Acad Sci Ocean Affairs Bd 2101 Constitution Ave NW Washington DC 20418

VETTER, RICHARD J, b Castlewood, SDak, July 17, 43; m 65; c 3. HEALTH PHYSICS, RADIOBIOLOGY. Educ: SDak State Univ, BS, 66, MS, 68; Purdue Univ, PhD(bionucleonics), 70. Prof Exp: Asst prof biol, Point Park Col, Pittsburgh, 69-70; asst prof bionucleonics, 70-75, ASSOC PROF BIONUCLEONICS, PURDUE UNIV, 75- Concurrent Pos: Asst radiol control officer, Purdue Univ, 70- Mem: AAAS; Health Physics Soc; Int Radiation Protection Asn; Int Microwave Power Inst. Res: Biological effects of ionizing and nonionizing radiation including radioecology; applications of microwaves, especially as a modality in treatment of cancer. Mailing Add: Dept of Bionucleonics Purdue Univ West Lafayette IN 47907

VETTER, RICHARD L, b Henry Co, Ill, Dec 28, 30; m 50; c 2. ANIMAL SCIENCE, ENVIRONMENTAL SCIENCES. Educ: Univ Ill, BS, 53, MS, 57; Univ Wis, PhD(nutrit, biochem), 60. Prof Exp: Asst, Univ Ill, 55-57; asst biochem, Univ Wis, 57-60, fel, 60-61; res nutritionist, Hess & Clark Co Div, Richardson-Merrill, Inc, 61-62; from asst prof to assoc prof animal sci, 62-72, PROF ANIMAL SCI, IOWA STATE UNIV, 72- Concurrent Pos: Researcher, Inst Animal Physiol, Cambridge, Eng, 68-69. Mem: Am Soc Animal Sci; Am Soc Dairy Sci; Fedn Am Soc Exp Biol. Res: Nutrition and metabolic disorders; chemistry, nutrition and utilization of plant and animal wastes; energy and nutrient conservation. Mailing Add: Dept of Animal Sci Iowa State Univ Ames IA 50010

VETTER, ROBERT JOSEPH, chemistry, see 12th edition

VETTERLING, JOHN MARTIN, b Fitzsimons, Colo, July 21, 34; m 57; c 3. PROTOZOOLOGY, PARASITOLOGY. Educ: Colo State Univ, BS, 56, MS, 62; Univ Ill, PhD(vet med sci), 65. Prof Exp: SR RES PARASITOLOGIST, BELTSVILLE PARASITOL LAB, AGR RES SERV, USDA, 65- Mem: Soc Protozool; Am Soc Parasitol; Electron Micros Soc Am; World Asn Advan Vet Parasitol. Res: Biology and taxonomy of coccidia; cytology, cytochemistry and electron microscopy of intestinal protozoan parasites. Mailing Add: Beltsville Parasitol Lab Agr Res Serv USDA Beltsville MD 20705

VEUM, TRYGVE LAURITZ, b Virogua, Wis, Mar 16, 40; m 67; c 2. ANIMAL NUTRITION, VETERINARY PHYSIOLOGY. Educ: Univ Wis, BS, 62; Cornell Univ, MS, 65, PhD(animal nutrit, vet physiol & path), 68. Prof Exp: Asst prof, 67-74, ASSOC PROF ANIMAL NUTRIT, UNIV MO-COLUMBIA, 74- Mem: Am Soc Animal Sci. Res: Swine nutrition. Mailing Add: Animal Sci Res Ctr Univ of Mo-Columbia Columbia MO 65201

VEVERBRANTS, EGILS, b Melluzi, Latvia, July 16, 35; m 68; c 1. MEDICINE. Educ: Univ NH, BS, 53; George Washington Univ, MD, 64. Prof Exp: From intern to resident, Boston City Hosp, 64-66; resident, RI Hosp, 68-70; ASST PROF MED, SCH MED & DENT, UNIV ROCHESTER, 70- Concurrent Pos: Res fel, Thorndike Mem Lab, 66-68; fel, RI Hosp, 68-70; assoc physician, Strong Mem Hosp, Rochester, 70- Mem: Am Soc Nephrology. Res: Acid-base balance; renal gluconeogenesis; hemodialysis. Mailing Add: Rochester Gen Hosp 1425 Portland Ave Rochester NY 14621

VEVERKA, JOSEPH, b Pelrimov, Czech, June 8, 41; Can citizen. ASTRONOMY, PLANETARY SCIENCES. Educ: Queen's Univ, Ont, BSc, 64, MSc, 65; Harvard Univ, MA & PhD(astron), 70. Prof Exp: Res assoc, 70-71, SR RES ASSOC, PLANETARY SCI, LAB PLANETARY STUDIES, CTR RADIOPHYS & SPACE RES, CORNELL UNIV, 71-, ASST PROF ASTRON, 74- Mem: Am Astron Soc; Royal Astron Soc Can; Int Astron Union. Res: Physical properties of planetary surfaces and atmospheres; nature and origin of natural satellites. Mailing Add: Lab for Planetary Studies Ctr for Radiophys & Space Res Cornell Univ Ithaca NY 14850

VEZINA, CLAUDE, b Oka, Que, Feb 19, 26; m 52; c 3. MICROBIOLOGY. Educ: Univ Montreal, BA, 46, BSA, 50, MSc, 52; Univ Wis, PhD(bact), 56. Prof Exp: Prof bact & biochem, Oka Agr Inst, Univ Montreal, 50-60; head gen microbiol, Res Labs, Ayerst, McKenna & Harrison, Ltd, 60-64, ASSOC DIR RES-MICROBIOL, AYERST LABS, AM HOME PRODS CORP, 64- Concurrent Pos: Nat Res Coun Can res assoc, 57; asst prof, Fac Med, Univ Montreal. Mem: Am Soc Microbiol; Mycol Soc Am; Soc Indust Microbiol; Am Chem Soc; Soc Gen Microbiol. Res: Microbial genetics and physiology; transformation of steroids; antibiotics; heterokaryosis in fungi; recombination in streptomyces and nocardia; nocardiophages. Mailing Add: Dept Microbiol Ayerst Labs Am Home Prod Corp Box 6115 Montreal PQ Can

VEZZETTI, DAVID JOSEPH, b Hoboken, NJ, Mar 20, 38; m 63; c 2. STATISTICAL MECHANICS. Educ: Stevens Inst Technol, ME, 59, MS, 61, PhD(physics), 64. Prof Exp: Asst res scientist math, Courant Inst Math Sci, NY Univ, 63-65; from asst prof to assoc prof, 65-73, PROF PHYSICS, UNIV ILL, CHICAGO CIRCLE, 73- Mem: Am Phys Soc. Res: Equilibrium and non-equilibrium statistical mechanics; statistical theory of wave propagation in random media. Mailing Add: Dept of Physics Univ of Ill Chicago Circle Chicago IL 60680

VEZZOLI, GARY CHRISTOPHER, b Mt Vernon, NY, Dec 16, 42. SOLID STATE PHYSICS, CHEMICAL PHYSICS. Educ: Fordham Univ, BS, 64; Boston Col, MS, 66; Pa State Univ, PhD(solid state sci), 69. Prof Exp: Res physicist, Inst Explor Res, Ft Monmouth, NJ, 69-72; RES PHYSICIST, FELTMAN RES LAB, PICATINNY ARSENAL, 72- Concurrent Pos: Res adv, Nat Res Coun, 73-76. Honors & Awards: Res & Eng Awards for Basic Res & Develop, Picatinny Arsenal, 73, 74 & 75. Mem: Sigma Xi; Am Ceramic Soc. Res: Electro-optical effects in non-linear materials; effects of temperature, pressure and electric field on the structure and properties of ring and chain molecules and electronically active glasses; nonequilibrium effects during phase transitions. Mailing Add: Eng Sci Div Feltman Res Lab Picatinny Arsenal Dover NJ 07801

VIA, FRANCIS ANTHONY, b Frostburg, Md, Nov 30, 43; m 70. CHEMISTRY. Educ: WVa Univ, BS, 65; Ohio State Univ, MS, 67, PhD(phys org chem), 70. Prof Exp: Teaching asst chem, Ohio State Univ, 65-66, res fel phys org chem, 66-70; from res chemist to sr res chemist, 70-75, SUPVR ORG CHEM, STAUFFER CHEM CO, 75- Mem: Am Chem Soc. Res: Synthetic methods and structure-reactivity relationships in organophosphorus chemistry; organic and inorganic photochemistry applied to initiation of polymerization; catalysis of transesterification reactions. Mailing Add: Eastern Res Ctr Stauffer Chem Co Dobbs Ferry NY 10522

VIA, GIORGIO G, b Rome, Italy, Dec 6, 28; US citizen; m 55; c 3. PHYSICS. Educ: Univ Pisa, PhD(physics), 57. Prof Exp: Assoc physicist res lab, Switz, 58-60, staff physicist fed systs div, NY, 60-61, 63, adv physicist, 64, adv physicist space systs ctr advan progs, 64-68, sr physicist, 68-70, SR PHYSICIST, COMPONENTS DIV, MANASSAS, IBM CORP, 70- Res: Applied research in solid state, semiconductors and magnetics; electron diffraction studies; electronoptics devices and advanced storage techniques; scientific programs planning and implementation; large scale integration technology; product evaluation and testing; high resolution photolitographic systems. Mailing Add: 5954 Woodacre Ct McLean VA 22101

VIA, WILLIAM FREDRICK, JR, b Ironton, Ohio, Dec 27, 20; m 47; c 2. PEDODONTICS. Educ: Ohio State Univ, DDS, 45; Univ Mich, MS, 53. Prof Exp: Instr oper dent, Col Dent, Ohio State Univ, 48-51; instr, Col Dent, Univ Calif, 51-52; mem staff, Henry Ford Hosp, Detroit, Mich, 53-68; chmn dept oral radiol, Sch Dent Med, Univ Conn, 68-69; PROF ORAL DIAG & CHMN DEPT, SCH DENT, UNIV NC, CHAPEL HILL, 69- Mem: AAAS; fel Am Col Dentists; fel Am Acad Dent Radiol; Am Acad Pedodontics; Int Asn Dent Res. Res: Prenatal, neonatal and post-natal influences upon dental enamel development; healing of dental pulp following bacterial, chemical or mechanical trauma; oral roentgenographic technique. Mailing Add: Dept of Oral Diag Sch of Dent Univ of NC Chapel Hill NC 27514

VIAL, JAMES LESLIE, b Taft, Calif, Dec 19, 24. VETEBRATE BIOLOGY, POPULATION ECOLOGY. Educ: Calif State Univ, Long Beach, BA, 52, MA, 54; Univ Southern Calif, PhD, 65. Prof Exp: From instr to assoc prof biol, Los Angeles Valley Col, 55-61; vis prof zool, Univ Costa Rica, 61-62, Ford Found Prof ecol, 62-64; asst prof biol, Western Mich Univ, 64-66; from assoc prof to prof, Univ Mo-Kansas City, 66-75, asst dean res & dir res admin, 67-68; PROF BIOL & CHMN FAC NATURAL SCI, UNIV TULSA, 75- Concurrent Pos: Assoc dir, Orgn Trop Studies, Inst Trop Ecol, Costa Rica, 63, dir, 64; herpet ed, Am Soc Ichthyologists & Herpetologists, 72- Mem: AAAS; Ecol Soc Am; Am Soc Mammal; Soc Study Amphibians & Reptiles; Asn Trop Biol. Res: Vertebrate ecology and population dynamics, especially amphibians and reptiles. Mailing Add: Fac of Natural Sci Univ of Tulsa Tulsa OK 74104

VIAL, LESTER JOSEPH, JR, b New Orleans, La, Mar 19, 44; m 71. MEDICINE, PATHOLOGY. Educ: La State Univ, New Orleans, BS, 66; Med Sch, La State Univ, MD, 70; Am Bd Path, dipl, 74. Prof Exp: Intern path, Charity Hosp, New Orleans, 70-71, resident, 71-74; instr, 74-75, ASST PROF PATH, MED SCH, LA STATE UNIV, 75- Concurrent Pos: Vis staff, Charity Hosp, New Orleans, 74- Mem: Am Soc Clin Path; Am Soc Cytol. Mailing Add: Dept of Path Sch Med La State Univ 1542 Tulane Ave New Orleans LA 70112

VIAL THEODORE MERRIAM, b Ware, Iowa, Feb 27, 21; m 49; c 5. RUBBER CHEMISTRY. Educ: Univ Md, BS, 42; Univ Ill, PhD(chem), 49. Prof Exp: Res chemist, Chas Pfizer & Co, 48-50; res chemist, 50-51, tech rep, 51-58, tech mgr rubber chem dept, 58-60, com develop mgr, 60-63, sales develop mgr, 63-66, GROUP LEADER, RES DEPT, AM CYANAMID CO, 66- Mem: AAAS; Am Chem Soc; Com Develop Asn. Res: Rubber and elastomer compounding; theory and application of vulcanization and protective agents; polyacrylate and other specialty elastomers. Mailing Add: Am Cyanamid Co Res Dept Bound Brook NJ 08805

VIALE, RICHARD O, b Eureka, Calif, June 9, 40; m 62; c 2. BIOCHEMISTRY, BIOPHYSICS. Educ: Chico State Col, AB, 62; Univ Calif, Davis, PhD(biophys), 55. Prof Exp: NIH fel, Sch Med, Univ Pa, 66-67; assoc, 67-71, ASST PROF BIOCHEM, SCH MED, UNIV PA, 71- Mem: AAAS; Am Chem Soc. Res: Applied mathematics in biochemistry; methods of education in biochemistry and medicine; application of computers to biochemistry. Mailing Add: Dept of Biochem Univ of Pa Sch of Med Philadelphia PA 19143

VIAMONTE, MANUEL, JR, b Havana, Cuba, Mar 19, 30; US citizen; m 55; c 2. RADIOLOGY. Educ: Univ Havana, MD, 55. Prof Exp: Fel radiol, Hosp Univ Pa, 56-58 & Children's Hosp, Philadelphia, 58; James Picker Found scholar radiol res, 60-63; PROF RADIOL, SCH MED, UNIV MIAMI, 66-, CHMN DEPT, 69- Concurrent Pos: Am Cancer Soc fel, 57-58; fel registry radiol path, Armed Forces Inst Path, 58; consult, Vet Admin Hosp, Coral Gables, Fla, 66 & Armed Forces Inst Radiol, Washington, DC, 67, 68 & 72; dir dept radiol, Mt Sinai Hosp Greater Miami, 68-, dir postgrad med educ, 69-; assoc dean hosp affairs, Univ Miami, 69-72; dir dept radiol, Jackson Mem Hosp, Miami, 69- Honors & Awards: Swiss Univ Award, 67. Mem: Am Col Chest Physicians; Am Col Radiol; Am Roentgen Ray Soc; Asn Univ Radiol; Radiol Soc NAm. Res: Angiographic examinations; nonvascular special procedures; isotopic-angiographic correlations. Mailing Add: Dept Radiol Mt Sinai Med Ctr 4300 Alton Rd Miami Beach FL 33140

VIAN, RICHARD W, b Lebanon, Ohio, Sept 28, 35; m 55; c 4. GEOLOGY. Educ: Miami Univ, Ohio, BA, 57, MS, 59; Univ Mich, PhD(mineral), 65. Prof Exp: Geologist, Gulf Res & Develop Co, 62-64; res asst mineral, Univ Mich, 65-66; asst prof geol, Kans State Univ, 66-71; ASST PROF GEOL, MIAMI UNIV, MIDDLETOWN, 71- Mem: Mineral Soc Am; Geol Soc Am. Res: Mineralogy; igneous and metamorphic petrology; geochemistry. Mailing Add: 2904 Orlando Ave Middletown OH 45042

VIANNA, NICHOLAS JOSEPH, b New York, NY, Dec 20, 42; m 67; c 2. EPIDEMIOLOGY. Educ: St Peter's Col, BS, 63; Cornell Univ Med Col, 67; Albany Med Col, MSPH, 71. Prof Exp: Intern & resident med, Montefiore Hosp & Med Ctr, New York, 68-70; clin instr, Albany Med Col, 70-71; fel infectious dis, NJ Col Med & Sloan Kettering Mem Inst, 71-72; DIR LYMPHOMA RES, CANCER CONTROL BUR, NY STATE DEPT HEALTH, ALBANY, 72-; ASST PROF, DEPT MED & PREV MED, ALBANY MED COL, 72- Concurrent Pos: Asst to the dir, Bur Epidemiol & Cancer, NY State Health Dept, Albany, 69-71; officer epidemiol, Ctr Dis Control, Atlanta, 69-71; consult, Kettering Inst Cancer Res, NY, 72-73 & Cancer Epidemiol Sect, WHO, Geneva, Switz, 72-73; travel fel, WHO, Oxford Univ, 72; dir, Coeymans Med Clin, Albany, NY, 72-; consult infectious dis, Ellis Hosp, Schenectady, NY, 73- Res: Epidemiology of lymphoreticular malignancies with major emphasis on etiology. Mailing Add: NY State Dept Health Tower Bldg Empire State Plaza Albany NY 12237

VIAU, ANNA TEMLER, physiology, pharmacology, see 12th edition

VIAU, JEAN-PAUL, biochemistry, see 12th edition

VIAVANT, WILLIAM JOSEPH, b San Antonio, Tex, Jan 2, 22; m 50; c 4. COMPUTER SCIENCE. Educ: Univ Tex, BS, 44, PhD(physics), 54. Prof Exp: Res physicist, Univ Tex, 51-55; res scientist, Shell Develop Co, 55-57; dir sci comput, Univ Okla, 57-62; NSF sr fel, Cambridge & Regnecentralen, Denmark, 62-63; PROF COMPUT SCI, UNIV UTAH, 64- Concurrent Pos: Consult, Owens-Ill Co, 64-71, Gen Elec Co, 66 & Univ Jundi Shapur, Ahwaz, Iran, 74-75; Iran minister of higher educ, 74-75. Mem: Asn Comput Mach. Res: Computer organization and programming; simulation; person-to-machine communication; academic programs in computer science; small computer systems. Mailing Add: Dept of Comput Sci Univ of Utah Salt Lake City UT 84112

VICCIONE, DANIEL MICHAEL, b Providence, RI, Aug 3, 39; m 61; c 3. ACOUSTICS. Educ: Univ RI, BSEE, 61, PhD(elec eng), 71; NY Univ, MSEE, 63. Prof Exp: Mem tech staff commun, Bell Tel Labs, 61-63; instr comput sci, Univ RI, 67-69; sr engr acoustics res, Submarine Signal Div, Raytheon Co, 69-70, sr tech staff sonar systs, 70-74; MEM STAFF, NAVAL UNDERWATER SYSTS CTR, US NAVY, 74- Mem: Inst Elec & Electronics Engr; Am Phys Soc. Res: Statistical signal detection; statistical detection theory; acoustic propagation underwater; adaptive signal processing. Mailing Add: Naval Underwater Systs Ctr Ft Trumbull New London CT 06320

VICE, JOHN LEONARD, b Evergreen Park, Ill, Jan 12, 42; m 63. MICROBIOLOGY, BIOCHEMISTRY. Educ: Loyola Univ Chicago, BS, 63; Univ Ill, Chicago, MS, 65, PhD(microbiol), 69. Prof Exp: Res microbiologist, Nat Cancer Inst, 65-66; dir clin microbiol, Alexian Bros Med Ctr, Ill, 67-68; DIR CLIN MICROBIOL, MED CTR, LOYOLA UNIV CHICAGO, 68- Concurrent Pos: Consult, Community Gen Hosp, Sterling, Ill, 72- Mem: Am Soc Microbiol. Res: Immunogenetics of rabbit immunoglobulins; rabbit lymphocyte antigens; immunochemical characterization of gram negative nonfermentative bacteria. Mailing Add: Chicago IL

VICEPS-MADORE, DACE I, b Esslingen, Ger, Feb 22, 47; US citizen; m 75. CELL BIOLOGY. Educ: Univ Rochester, AB, 69; Temple Univ, MA, 71, PhD(biol), 74. Prof Exp: Res investr cell biol, Wistar, Inst, 73-75; FEL CELL BIOL, INST CANCER RES, 75- Mem: AAAS; Asn Women Sci; Tissue Cult Asn; Sigma Xi. Res: Understanding and control of the regulation of differentiation on the cellular level; use of mutagen application in tissue culture as a tool to obtain cells altered in genotypic expression. Mailing Add: Inst for Cancer Res 7701 Burholme Ave Philadelphia PA 19111

VICHER, EDWARD ERNEST, b Chicago, Ill, Nov 23, 14; m 50; c 2. BACTERIOLOGY. Educ: Univ Ill, BS, 35, MS, 37, PhD(bact), 42. Prof Exp: Asst bact, Col Pharm, 35-39, from instr to assoc prof, Col Med, 39-69, PROF BACT, COL MED, UNIV ILL, 69-, CONSULT, RES & EDUC HOSPS, 46- Concurrent Pos: Pharmacist, Ill, 35-; consult, US Vet Admin Hosps, Hines & Chicago, 46-56; consult, Toni Co, 52-72. Mem: Am Soc Microbiol; Med Mycol Soc of the Americas. Res: Medical microbiology; clinical diagnostic bacteriology; mycology and parasitology; hospital asepsis and disinfection; metabolism of the dermatophytes; cariogenic streptococci. Mailing Add: Dept of Microbiol Univ of Ill Col Med Chicago IL 60612

VICHICH, THOMAS E, b Calumet, Mich, Oct 5, 17; m 45; c 2. MATHEMATICS. Educ: Mich Col Mining & Technol, BS, 39; Univ Mich, MS, 52. Prof Exp: From instr to assoc prof, 41-61, PROF MATH, MICH TECHNOL UNIV, 61- Res: Differential and integral calculus; differential equations. Mailing Add: Dept of Math Mich Technol Univ Houghton MI 49931

VICIEDO, EUSEBIO, b Havana, Cuba, Aug 14, 01; m 35; c 1. MICROBIOLOGY, BIOCHEMISTRY. Educ: Inst Sci, Cuba, BS, 18; Univ Havana, MS, 20, PhD(biol), 22. Prof Exp: Asst biochemist, Iturrioz Labs, Cuba, 25-30; dir pharmaceut chem, Plasencia Labs, 32-45; yeast mfr, Zimotecnica Corp, 45-52; researcher microbiol, Cuban Res Inst, 57-61 & Arroyo Pharmaceut Corp, 61-64; microbiologist, Atlas Yeast Corp, 64-66; BIOCHEMIST, BACARDI CORP, SAN JUAN, 66- Concurrent Pos: Prof, Univ Oriente, Cuba, 53-59. Mem: Am Chem Soc; Soc Indust Microbiol; Brit Soc Appl Bact. Res: Industrial microbiology; yeast manufacture; alcohol fermentation; acetone-butanol fermentation. Mailing Add: Independencia 585 Hato Rey PR 00918

VICK, ALPHONSO ROSCOE, b Wilson Co, NC, Mar 12, 20; m 49; c 2. BOTANY, ZOOLOGY. Educ: J C Smith Univ, AB, 42; NC Col Durham, MS, 47; Univ Mich, AM, 56; Syracuse Univ, PhD(bot), 61. Prof Exp: Prof biol, Bennett Col, 61-62; prof, Winston-Salem State Col, 62-67, chemn sci dept, 63-67; PROF BIOL, NC A&T STATE UNIV, 67- Mem: AAAS; Nat Inst Sci; Inst Biol Sci Med; Sigma Xi. Res: A comparison of the accumulation rates of radioactive phosphorus in urban vital organs of fish; radioactive phosphorus is fed the specimens over a given period followed by dissecting out the organs and exposing them to a radioactive detecting machine to ascertain the concentration of the isotope in the respective organs. Mailing Add: 1601 S Benbow Rd Greensboro NC 27406

VICK, CHARLES BOOKER, b Seaboard, NC, Sept 15, 32; m 63; c 2. FOREST PRODUCTS. Educ: Duke Univ, AB, 54, MF, 58. Prof Exp: Forest prod technologist, 58-60, wood scientist, 60-74, PRIN WOOD SCIENTIST, FORESTRY SCI LAB, SOUTHEASTERN FOREST EXP STA, 74- Honors & Awards: Cert appreciation, USDA, 73, Cert Merit, 75. Mem: Am Soc Testing & Mat; Forest Prod Res Soc. Res: Housing research, especially development and application of structural adhesives and mechanical fasteners for more efficient structural design of light-frame constructions. Mailing Add: Forestry Sci Lab Carlton St Athens GA 30601

VICK, GEORGE R, b New Waverly, Tex, Dec 18, 20; m 43; c 2. ALGEBRA, GEOMETRY. Educ: Sam Houston State Teachers Col, BA, 41, MA, 42; Univ Tex, PhD(math), 64. Prof Exp: Teacher high sch, Tex, 41-42; from asst prof to assoc prof math, 46-51, assoc prof, 55-56, actg dir dept, 56-64, dir dept, 64-67, PROF MATH, SAM HOUSTON STATE UNIV, 64- Mem: Math Asn Am. Res: Foundations of geometry; mathematical pedagogy at all levels, especially training and re-training teachers. Mailing Add: Dept of Math Sam Houston State Univ Huntsville TX 77340

VICK, GERALD KIETH, b Dixon, Ill, Mar 6, 30; m 50; c 3. ORGANIC CHEMISTRY. Educ: Univ Ill, BS, 52; Univ Rochester, PhD(org chem), 56. Prof Exp: Res chemist, Esso Res & Eng Co, 55-58, proj leader engine oils, 58-62, sect head motor fuels & lubricants, 62-65, sr staff adv petrol fuels & lubricants, 65-67, dir lubricants & specialties lab, 67-75, SR PLANNING ADV, CORP PLANNING DEPT, EXXON CORP, 75- Mem: AAAS; Am Chem Soc; Soc Automotive Eng. Res: Product quality research in the field of petroleum fuels and lubricants. Mailing Add: Exxon Corp 1251 Avenue of the Americas New York NY 10020

VICK, JAMES WHITFIELD, b Hope, Ark, Mar 8, 42; m 64; c 2. MATHEMATICS. Educ: La State Univ, Baton Rouge, BS, 64; Univ Va, MA, 66, PhD(math), 68. Prof Exp: Instr math, Princeton Univ, 68-70; asst prof, 70-73, ASSOC PROF MATH, UNIV TEX, AUSTIN, 73- Mem: Am Math Soc. Res: Algebraic and differential topology, K-theory and transformation groups. Mailing Add: Dept of Math Univ of Tex Austin TX 78712

VICK, MAURICE M, b Russell, Ark, Nov 26, 09; m 35; c 2. ANALYTICAL CHEMISTRY. Educ: Ouachita Baptist Col, BA, 31; La State Univ, MS, 33, PhD(chem), 40. Prof Exp: Assoc prof chem, Ouachita Baptist Col, 33-36; from instr to prof, 40-74, EMER PROF CHEM, LA STATE UNIV, BATON ROUGE, 74- Mem: AAAS; Am Chem Soc. Res: Application of radioactive tracers in analytical chemistry; development of nonsulfide analyses for metals; ion exchange. Mailing Add: Dept of Chem La State Univ Baton Rouge LA 70803

VICK, ROBERT LORE, b Courtland, Miss, Sept 1, 29; m 53; c 1. PHARMACOLOGY, PHYSIOLOGY. Educ: Univ Miss, BS, 52, MS, 54; Univ Cincinnati, PhD(pharmacol), 57. Prof Exp: Actg asst prof pharm, Southwestern State Col, 53; instr, Univ Miss, 53-54; res assoc pharmacol, Univ Cincinnati, 57-58; instr physiol, State Univ NY Upstate Med Ctr, 58-61; from asst prof to assoc prof, 61-72, PROF PHYSIOL, BAYLOR COL MED, 72- Concurrent Pos: USPHS career develop award, 66-71. Mem: AAAS; Am Physiol Soc; Am Soc Pharmacol & Exp Therapeut; Soc Exp Biol & Med. Res: Heart and circulation; ion movements; electrophysiology; autonomic nervous system. Mailing Add: Dept of Physiol Baylor Col of Med Houston TX 77030

VICKERS, DAVID HYLE, b Sturgis, Miss, Jan 14, 40; m 70; c 2. ENTOMOLOGY, BIOCHEMISTRY. Educ: Miss State Univ, BS, 61, MS, 64; La State Univ, PhD(entom), 69. Prof Exp: Instr entom, Southeastern La Col, 64-66; asst prof physiol, 69-72, ASSOC PROF PHYSIOL, FLA TECHNOL UNIV, 72-, ACTG CHMN DEPT, 74- Mem: AAAS; Am Inst Biol Sci; Entom Soc Am. Res: Pollution of natural waters by solid wastes; insect lipids; metabolism of pesticides by insects, particularly chemosterilants; nitrogen metabolism in insects. Mailing Add: Dept of Biol Sci Fla Technol Univ Orlando FL 32816

VICKERS, JAMES HUDSON, b Columbus, Ohio, Apr 21, 30; m 64; c 1. VETERINARY MEDICINE, PATHOLOGY. Educ: Ohio State Univ, BSc, 52, DVM, 58; Univ Conn, MS, 66. Prof Exp: Vet, Columbus Zoo, Ohio, 58-60 & Lab Animal Colony, Lederle Labs, Am Cyanamid Co, NY, 60-64, pathologist, 64-66, head dept vet path, 66-68, head dept exp path, 68-70; vpres & dir res, Primelabs, Inc, 70-73; DIR PATH & PRIMATOL BR, BUR BIOL, FOOD & DRUG ADMIN, 73- Concurrent Pos: Lectureship, State Univ NY Downstate Med Sch; mem, Comp Path Colloquy; Food & Drug Admin rep, Dept Housing Educ & Welfare Primate Steering Comt, 74- Mem: Am Vet Med Asn; Am Asn Zool Vets; Am Asn Neuropath; Am Asn Avian Path; Indust Vet Asn. Res: Diseases and pathology of primates; laboratory animals and exotic zoological species; testing and quality control of vaccines, especially polio vaccine, toxicological testing and pathology of pharmaceuticals. Mailing Add: Box 142M24 RFD 1 Ijamsville MD 21754

VICKERS, STANLEY, b Blackpool, Eng, Sept 27, 39. DRUG METABOLISM. Educ: Univ London, BSc, 62; State Univ NY Buffalo, PhD(biochem pharm), 67. Prof Exp: Fel, Univ Kans, 66-69; RES FEL DRUG METAB, MERCK INST THERAPEUT RES, 69- Mem: Am Chem Soc; AAAS; The Chem Soc. Res: Detoxification mechanisms and metabolic transformations which control the fate of foreign compounds. Mailing Add: Merck Inst for Therapeut Res West Point PA 19486

VICKERS, THOMAS J, b Miami, Fla, Mar 29, 39; m 63; c 4. ANALYTICAL CHEMISTRY. Educ: Spring Hill Col, BS, 61; Univ Fla, PhD(chem), 64. Prof Exp: Asst prof, 66-71, ASSOC PROF ANAL CHEM, FLA STATE UNIV, 71- Mem: Am Chem Soc; Soc Appl Spectros. Res: Spectroscopic methods of trace element analysis, atomic emission, absorption and fluorescence flame spectrometry; new excitation sources for emission spectroscopy; non-flame atomization techniques. Mailing Add: Dept of Chem Fla State Univ Tallahassee FL 32306

VICKERS, WILLIAM W, b San Francisco, Calif, June 21, 23; m 54; c 3. ATMOSPHERIC PHYSICS. Educ: Univ Calif, BA, 54, MA, 56; McGill Univ, PhD(hydrol, meteorol), 65. Prof Exp: Res assoc, Inst Polar Studies, Ohio State Univ, 57-61; head geophys res group, Tech Opers, Inc, 61-66; SR SCI EXEC, EG&G, INC & ENVIRON SENSOR SYSTS DIV, MITRE CORP, BEDFORD, 66- Concurrent Pos: Mitre Corp grant, 69-70. Mem: Am Geophys Union; Am Meteorol Soc. Res: Weather modification; atmospheric pollution transport; atmospheric refraction; cloud dynamics. Mailing Add: Indian Hill Prides Crossing MA 01965

VICKERS, ZATA MARIE, b Salem, Ore, Oct 13, 50. FOOD SCIENCE. Educ: Ore State Univ, BS, 72; Cornell Univ, PhD(food sci), 75. Prof Exp: ASST PROF FOOD SCI, UNIV MINN, ST PAUL, 75- Mem: Inst Food Technol. Res: Relationships between the physical, acoustical and sensory properties of foods. Mailing Add: Dept of Food Sci & Nutrit Univ Minn St Paul MN 55108

VICKERY, HUBERT BRADFORD, b Yarmouth, NS, Feb 28, 93; nat US; m 36. BIOCHEMISTRY. Educ: Dalhousie Univ, BSc, 15, MSc, 18; Yale Univ, PhD(org chem), 22. Hon Degrees: DSc, Yale Univ, 48; LLD, Dalhousie Univ, 73. Prof Exp: Teacher high sch, NS, Can, 15-17; chemist, Imp Oil, Ltd, 17-19; teacher chem, Norwalk Col, NS, 19-20; asst biochemist, 22-28, biochemist in chg, 28-63, EMER BIOCHEMIST, CONN AGR EXP STA, 63- Concurrent Pos: Lectr, Yale Univ, 24-63; res assoc, Carnegie Inst, 29-38; observer, Bikini, 46. Honors & Awards: Hales Prize, Am Soc Plant Physiologists, 33. Mem: Nat Acad Sci; Am Chem Soc; Am Soc

Biol Chem (pres, 50); Soc Exp Biol & Med; Am Soc Plant Physiologists. Res: Amino acids; proteins; constituents of plant tissues; organic acids of plants; metabolism of plants; history of protein chemistry; nonemclature of amino acids. Mailing Add: Conn Agr Exp Sta New Haven CT 06504

VICKERY, LARRY EDWARD, b Atlanta, Ga, Nov 26, 45; c 2. BIOPHYSICAL CHEMISTRY. Educ: Univ Calif, Santa Barbara, BA, 67, PhD(biol), 71. Prof Exp: Res assoc biochem, Western Regional Res Lab, USDA, 71; res assoc biophys chem, Lawrence Berkeley Lab, 72; fel, 73-74, RES ASSOC BIOPHYS CHEM, DEPT CHEM, UNIV CALIF, BERKELEY, 75- Concurrent Pos: Res assoc, Nat Res Coun, Nat Acad Sci, Nat Acad Engr, 71; fel, Nat Inst Gen Med Sci, USPHS, 73-74. Mem: Biophys Soc. Res: Physical chemistry of proteins—kinetics and mechanisms of conformational changes, interactions with metals and heme groups, metal ion transport. Mailing Add: Dept of Chem Univ of Calif Berkeley CA 94720

VICKERY, ROBERT KINGSTON, JR, b Saratoga, Calif, Sept 18, 22; m 51; c 2. PLANT EVOLUTION. Educ: Stanford Univ, AB, 44, AM, 48, PhD(biol, bot), 52. Prof Exp: Instr bot, Pomona Col, 50-51; from instr to assoc prof biol, 52-64, head dept genetics & cytol, 62-65, PROF BIOL, UNIV UTAH, 64- Concurrent Pos: Researcher, Carnegie Inst Wash, 48-52; res fel, Calif Inst Technol, 55; mem gen biol & genetics fel panel, NIH, 65-69, 7th Int Bot Cong & 10th-12th Int Gen Cong; assoc ed, Evolution, Soc Study Evolution, 68-72; mem, Int Orgn Plant Biosysts Coun, 75- Honors & Awards: Distinguished Teaching Award, Univ Utah, 72. Mem: AAAS; Soc Study Evolution; Am Soc Nat; Ecol Soc Am; Genetics Soc Am. Res: Cytogenetics, ecologic, numerical and classical taxonomic, and biochemical approaches to problems of the evolutionary mechanisms and patterns of the genus Mimulus, particularly sections Simiolus and Erythranthe. Mailing Add: Dept of Biol Univ of Utah Salt Lake City UT 84112

VICKERY, VERNON RANDOLPH, b South Ohio, NS, June 6, 21; m 47; c 3. TAXONOMY, ENTOMOLOGY. Educ: McGill Univ, BSc, 49, MSc, 57, PhD(entom), 64. Prof Exp: Teacher pub sch, NS, 40-41; entomologist, NS Dept Agr, 49-60; assoc prof entom, 60-74, PROF ENTOM, MACDONALD COL, McGILL UNIV & CUR LYMAN ENTOM MUS & RES LAB, 74- Mem: AAAS; Entom Soc Am; Entom Soc Can; Royal Entom Soc London. Res: Economic entomology; insect behavior and ecology; taxonomy of Orthoptera. Mailing Add: Lyman Entom Mus & Res Lab Macdonald Col Ste Anne de Bellevue PQ Can

VICKREY, HERTA MILLER, b San Gregorio, Calif, Feb 10, 25; m 45; c 4. IMMUNOLOGY, MEDICAL MICROBIOLOGY. Educ: Jan Jose State Col, BA, 57; Univ Calif, Berkeley, MA, 63, PhD(bact & immunol), 70. Prof Exp: Microbiologist, Viral & Rickettsial Dis Lab, Calif State Dept Pub Health, 57-60, 61-62; res bacteriologist, Univ Calif, Berkeley, 63-64; asst prof immunol, virol & microbiol, Univ Victoria, BC, 70-72; RES ASSOC CANCER IMMUNOL, DEPT RES & EDUC, WAYNE COUNTY GEN HOSP, 72- Concurrent Pos: Bacteriologist, Children's Hosp Med Ctr Northern Calif, Oakland, 58-70; res grants, Univ Victoria, BC, 70-72; med staff res & educ grants, 73, 74 & 75. Mem: Am Soc Microbiol. Res: Cellular immunological resistance to tuberculosis and oncogenesis; autoimmunity relative to multiple sclerosis; application of in vitro cell culture assays to autoallergies, oncogenesis and hypersensitivities; tissue culture studies on immunotherapy of lymphatic leukemia; methods for studies on responses of rabbit liver cells to various stimuli. Mailing Add: Dept of Res & Educ Wayne County Gen Hosp Box 124 Eloise MI 48132

VICKROY, DAVID GILL, b San Antonio, Tex, July 5, 41; m 64; c 2. ANALYTICAL CHEMISTRY. Educ: Vanderbilt Univ, BA, 63; Rice Univ, MA, 66; Univ Tenn, PhD(inorg chem), 69. Prof Exp: Res chemist, 69-72, sr res chemist, 72-75, RES ASSOC ANAL CHEM, CELANESE FIBERS CO, 75- Mem: Sigma Xi; Am Chem Soc. Res: Analytical characterization of complex mixtures such as tobacco smoke and environmental samples; development of synthetic alternatives to natural products. Mailing Add: Celanese Fibers Co PO Box 1414 Charlotte NC 28232

VICKROY, VIRGIL VESTER, JR, b San Antonio, Tex, Aug 8, 31; m 55; c 2. POLYMER CHEMISTRY, PHYSICAL CHEMISTRY. Educ: Auburn Univ, BS, 52; Univ Akron, MS, 62, PhD(polymer sci), 65. Prof Exp: Jr chemist, B F Goodrich Co, 55-61; res chemist, Harrison-Morton Labs, Ohio, 61-63; res chemist, Univ Akron, 63-65; SR RES CHEMIST, MONSANTO CO, 65- Mem: Am Chem Soc. Res: Effect of thermal and thermo-oxidative history on morphological, mechanical and molecular properties of polyolefins; physical chemistry of Ziegler catalyst systems. Mailing Add: Petrochem & Polymers Dept Monsanto Co PO Box 1311 Texas City TX 77590

VICTOR, ANDREW C, b New York, NY, Nov 4, 34; m 55; c 3. ENGINEERING PHYSICS. Educ: Swarthmore Col, BA, 56; Univ Md, MS, 61. Prof Exp: Physicist, Nat Bur Stand, 56-62; head, Stand Lab, US Naval Ord Test Sta, 62-63, physicist, systs anal, 63-64 & appl propulsion res, 64-68, head, Anal Br, 68-73, HEAD, PROPULSION ANAL BR, US NAVAL WEAPONS CTR, 73- Concurrent Pos: Navy rep plume technology steering comt, Joint Army, Navy, NASA, Air Force Interagnecy Propulsion Comt, 70- Mem: Am Phys Soc; Am Chem Soc; assoc fel Am Inst Aeronaut & Astronaut; Sigma Xi. Res: Missile propulsion analysis; rocket exhaust plume; electromagnetic interactions and signature phenomena; ramjet engine cost analysis. Mailing Add: 712 N Peg St Ridgecrest CA 93555

VICTOR, LEONARD BAKER, b Schenectady, NY, Aug 3, 34; m 66; c 2. ANATOMIC PATHOLOGY, CLINICAL PATHOLOGY. Educ: NY Univ, AB, 53; Univ Brussels, MD, 60; Royal Col Trop Med, TMD, 60. Prof Exp: Intern & resident path, Strong Hosp, Univ Rochester, 61-65, sr instr, Univ, 65-67; assoc prof path & lab med, Meharry Med Col, 68-72, assoc prof path, Grad Sch & dir Meharry Multiphasic Lab, 68-72; PROF PATH, UNIV TENN, MEMPHIS & DIR CLIN LABS, CITY OF MEMPHIS HOSPS, 72- Concurrent Pos: Dep med examr, Monroe County, NY, 65-67; consult, State Hosp, Rochester, NY, 65-67; assoc prof, Sch Eng, Vanderbilt Univ, 69-72; chmn, Comt Health Fitness Sci, Nashville, 70-71; chmn, Nat Adv Task Force for Regional Med Prog Eval of Automated Multiphasic Health Testing, 72; med dir & dir, Mid-South Comprehensive Home Health Serv Agency, 73; mem bd dirs, Community Blood Plan of Memphis, 73- Mem: Fel Col Am Path; fel Am Soc Clin Path; fel Soc Advan Med Systs; fel Royal Soc Health; AMA. Res: Administrative and academic medicine, including curriculum development and interdisciplinary functioning; pathology and prospective medicine, including automation, computerization and management techniques. Mailing Add: 957 Green Oaks Dr Memphis TN 38117

VICTORIA, JANICE KRUTAK, b Alhambra, Calif, Oct 23, 40; m 66. ANIMAL BEHAVIOR. Educ: Univ Calif, Los Angeles, BA, 63, MA, 66, PhD, 69. Prof Exp: Asst prof biol, Prince George's Community Col, 71-73; asst prof, Univ San Diego, 74-75; CHMN EDUC ACTIV, SAN DIEGO NATURAL HIST MUS, 75- Mem: Animal Behav Soc; Sigma Xi. Res: Social behavior of weaverbirds; influence of various environmental stimuli on female reproductive behavior; social facilitation effects. Mailing Add: Natural Hist Mus PO Box 1390 San Diego CA 92112

VICTORICA, BENJAMIN (EDUARDO), b Mendoza, Arg, June 9, 36; m 63; c 3. PEDIATRICS. Educ: Nat Univ Cuyo, MD, 62; Educ Coun For Med Grad, cert, 63; Am Bd Pediat, dipl, 68, cert pediat cardiol, 71. Prof Exp: Intern, MedSch, Nat Univ Cuyo, 63-63 & St Benedict's Hosp, Ogden, Utah, 63-64; from resident pediat to chief resident, 64-66, instr & spec trainee pediat cardiol, 67-70, asst prof, 70-74, ASSOC PROF PEDIAT CARDIOL, COL MED, UNIV FLA, 74- Mem: Fel Am Acad Pediat; Am Col Cardiol. Res: Pediatric cardiology. Mailing Add: Dept of Pediat Univ of Fla Col of Med Gainesville FL 32601

VICTORIUS, CLAUS, b Hamburg, Ger, Aug 24, 23; nat US; m 52; c 2. ORGANIC POLYMER CHEMISTRY. Educ: Guilford Col, BS, 43; Univ NC, MA, 46. Prof Exp: RES ASSOC, EXP STA, E I DU PONT DE NEMOURS & CO, INC, WILMINGTON, DEL, 46- Mem: Am Chem Soc. Res: Organic coatings; automotive finishes; powder coatings. Mailing Add: 21 Paxon Hollow Rd Media PA 19063

VIDA, JULIUS ADALBERT, b Losonc, Czech, May 30, 28; US citizen; c 2. MEDICINAL CHEMISTRY. Educ: Pazmany Peter Univ, Budapest, Dipl, 50; Carnegie Inst Technol, MS, 59, PhD(org chem), 60. Prof Exp: Chemist, United Pharmaceut Co, Wander Co, Hungary, 50-56 & Merck & Co, Inc, NJ, 57-58; res fel, Harvard Univ, 61-62; chemist, Worcester Found Exp Biol, Mass, 62-67; group leader, T Clark Lab, Kendall Co, Lexington, Mass, 67-72, sect head, 72-75; ASST DIR RES PLANNING & LICENSING, BRISTOL LAB, 75- Concurrent Pos: Adj prof med chem, Grad Sch Pharmaceut Sci, northeastern Univ, 73-75. Mem: Am Chem Soc. Res: Drugs acting on the central nervous systems; heterocyclic compounds; steroids. Mailing Add: Bristol Lab PO Box 657 Syracuse NY 13201

VIDAL, FREDERICK, organic chemistry, physiological chemistry, see 12th edition

VIDALE, ROSEMARY J, b New Haven, Conn, Mar 27, 31; m 57; c 3. GEOLOGY. Educ: Oberlin Col, BA, 52; Univ Mich, MS, 54; Yale Univ, PhD(geol), 68. Prof Exp: ASST PROF GEOL, STATE UNIV NY BINGHAMTON, 68- Mem: AAAS; Am Geophys Union; Geol Soc Am. Res: Metamorphic geology; experimental petrology and geochemistry. Mailing Add: Dept of Geol State Univ of NY Binghamton NY 13901

VIDAURRETA, LUIS E, b Havana, Cuba, Dec 15, 20; US citizen; m 43; c 2. ANALYTICAL CHEMISTRY. Educ: Univ Havana, PhD(chem), 43. Prof Exp: Prof anal chem, Univ Havana, 43-65; ASSOC PROF CHEM, LA STATE UNIV, BATON ROUGE, 66- Mem: Am Chem Soc. Res: Instrumental analysis; gas chromatography; sugar and sugar by-products analysis. Mailing Add: Coates 167 La State Univ Baton Rouge LA 70803

VIDAVER, ANNE MARIE KOPECKY, b Vienna, Austria, Mar 29, 38; US citizen; m 62; c 2. BACTERIOLOGY. Educ: Russell Sage Col, BA, 60; Ind Univ, Bloomington, MA, 62, PhD(bact), 65. Prof Exp: Instr bact, 65-66, res assoc plant path, 66-72, asst prof, 72-74, ASSOC PROF PLANT PATH, UNIV NEBR, LINCOLN, 74- Mem: Am Soc Microbiol; Am Phytopath Soc. Res: Phytopathogenic bacteria; bacteriophages; bacteriocins. Mailing Add: Dept of Plant Path Univ of Nebr Lincoln NE 68503

VIDAVER, GEORGE ALEXANDER, b Detroit, Mich, Apr 17, 30; m 62; c 2. BIOCHEMISTRY. Educ: Univ Chicago, BA, 51, PhD(biochem), 55. Prof Exp: Res assoc biochem, Univ Chicago, 55-56, Inst Enzyme Res, Univ Wis, 56-57, Med Sch, Northwestern Univ, 57-58 & Ind Univ, 58-65; assoc prof, 66-73, PROF BIOCHEM, UNIV NEBR, LINCOLN, 73- Concurrent Pos: Nat Found Infantile Paralysis fel, 56-57; NSF fel, 59-60, res grant, 68-70; USPHS fel, 61-62, res grants, 66- Mem: AAAS; Am Chem Soc; Am Soc Biol Chem; Fedn Am Socs Exp Biol. Res: Active transport of amino acids; membrane structure. Mailing Add: Dept of Chem Univ of Nebr Lincoln NE 68508

VIDAVER, WILLIAM ELLIOTT, b San Francisco, Calif, Feb 2, 21; m 51; c 3. PLANT PHYSIOLOGY. Educ: San Francisco State Col, AB, 58; Stanford Univ, PhD(biol), 64. Prof Exp: Fel plant biol, Carnegie Inst Dept Plant Biol, 63-65; assoc prof biol, 65-69, PROF BIOL, SIMON FRASER UNIV, 69- Concurrent Pos: Nat Res Coun Can operating grants, 65-. Mem: AAAS; Am Soc Plant Physiol; Can Soc Plant Physiol. Res: Algal physiology; mechanisms of photosynthesis; mechanisms of phytochrome action; germination and dormancy. Mailing Add: Dept of Biol Sci Simon Fraser Univ Burnaby BC Can

VIDMAR, PAUL JOSEPH, b Vallejo, Calif, May 22, 44. PLASMA PHYSICS. Educ: Univ Notre Dame, BS, 66; Univ Calif, San Diego, MS, 72, PhD(physics), 75. Prof Exp: RES ASSOC PHYSICS, FUSION RES CTR, UNIV TEX, AUSTIN, 75- Mem: Am Phys Soc. Res: Experimental studies of turbulence and nonlinear wave phenomena in plasmas. Mailing Add: Dept of Physics Univ of Tex Austin TX 78712

VIDOLI, VIVIAN ANN, b Bridgeport, Conn, Nov 2, 41. PHYSIOLOGY, NEUROPHYSIOLOGY. Educ: Southern Conn State Col, BS, 63; Ariz State Univ, MS, 66, PhD(zool & physiol), 69. Prof Exp: Asst prof, 70-73, ASSOC PROF BIOL, CALIF STATE UNIV, FRESNO, 73-, ASST DIR DIV HEALTH PROFESSIONS, 75- Concurrent Pos: Consult, Area Health Educ Consortium, San Joaquin Valley, Calif, 74- Mem: Am Physiol Soc; AAAS. Res: Anatomical and physiological correlates of sensory mechanisms. Mailing Add: Dept of Biol Calif State Univ Shaw & Cedar Ave Fresno CA 93710

VIDONE, ROMEO ALBERT, b Greenwich, Conn, July 1, 30; m 55; c 3. PATHOLOGY. Educ: Davis & Elkins Col, BS, 52; Yale Univ, MD, 57. Prof Exp: From instr to assoc prof, 59-68, ASSOC CLIN PROF PATH, YALE UNIV, 68-; DIR LABS, CHARLOTTE HUNGERFORD HOSP, 68- Concurrent Pos: Asst clin prof, Health Ctr, Univ Conn, 72-; pvt pract. Mem: Am Soc Clin Path; Col Am Path; AMA; Int Acad Path; Am Cancer Soc. Res: Cardiopulmonary physiology and pathology; cancer, clinicopathologic correlation with emphasis on mucin production by tumors. Mailing Add: Dept of Path Charlotte Hungerford Hosp Torrington CT 06790

VIDULICH, GEORGE A, b New York, NY, Nov 20, 31. PHYSICAL CHEMISTRY. Educ: Univ Conn, BA, 59; Brown Univ, PhD(phys chem), 64. Prof Exp: NIH fel phys chem, Max Planck Inst Phys Chem, Ger, 63-64; fel chem, Mellon Inst, 65-66; ASST PROF CHEM, HOLY CROSS COL, 66- Concurrent Pos: Holy Cross Col fac & Off Saline Water sr fels, Calif State Univ, Los Angeles, 71-72. Mem: AAAS; Am Phys Soc; Biophys Soc; Am Chem Soc. Res: Properties and structures of aqueous and electrolyte solutions; mobilities of ions in aqueous and nonaqueous solutions. Mailing Add: Dept of Chem Holy Cross Col Worcester MA 01610

VIEBROCK, FREDERICK WILLIAM, b Staten Island, NY, Nov 23, 35; m 56; c 2. BIOCHEMISTRY. Educ: Wagner Col, BS, 57; Polytech Inst Brooklyn, MS, 68; Va Polytech Inst & State Univ, PhD(biochem), 74. Prof Exp: Res scientist enzymol, Wallerstein Lab, Div Travenol Lab, 58-69; SR SCIENTIST BIOCHEM, JOHNSON & JOHNSON RES CTR, 72- Mem: AAAS; Am Chem Soc; Sigma Xi. Res: Purification and kinetic analysis of enzymes of the purine metabolic pathways, wound healing, eczematous skin diseases, local anesthetic. Mailing Add: Johnson & Johnson Res Ctr US Rte 1 New Brunswick NJ 08903

VIEHLAND, LARRY ALAN, b St Louis, Mo, Apr 30, 47; m 69; c 1. CHEMICAL PHYSICS. Educ: Mass Inst Technol, BS, 69; Univ Wis-Madison, PhD(chem), 73. Prof Exp: RES ASSOC CHEM, BROWN UNIV, 73- Mem: Sigma Xi. Res: Theoretical chemistry and atomic physics, specifically kinetic theory and nonequilibrium statistical mechanics as a tool for understanding intermolecular potentials and other microscopic properties. Mailing Add: Dept of Chem Brown Univ Providence RI 02912

VIEIRA, ERNEST CHARLES, b Lawrence, Mass, Mar 3, 27; m 54; c 2. PHYSICAL CHEMISTRY. Educ: Northeastern Univ, BS, 52; Pa State Univ, PhD(phys chem), 56. Prof Exp: Chemist, 55-58, RES CHEMIST, ORG CHEM DIV, AM CYANAMID CO, BOUND BROOK, 58- Mem: Am Chem Soc. Res: Evaluation of thermal and chemical hazards. Mailing Add: 133 Edgewood Dr Somerville NJ 08876

VIELE, GEORGE WASHINGTON, b Wausau, Wis; m 58; c 2. GEOLOGY, TECTONICS. Educ: Yale Univ, BS, 51; Univ Utah, PhD(geol), 60. Prof Exp: Geologist, US Geol Surv, 51-56 & Stand Oil Co Calif, 57-59; from asst prof to assoc prof geol, 59-72, PROF GEOL, UNIV MO-COLUMBIA, 72-, CHMN DEPT, 74- Mem: Geol Soc Am; Am Asn Petrol Geol; Am Geophys Union. Res: Structural geology; regional tectonics; Northern Rocky and Ouachita Mountains. Mailing Add: Dept of Geol Univ of Mo Columbia MO 65201

VIER, DWAYNE TROWBRIDGE, b Washington, DC, Sept 17, 14; m 51; c 2. PHYSICAL CHEMISTRY. Educ: Univ NH, BS, 37, MS, 39; Columbia Univ, PhD(phys chem), 43. Prof Exp: Asst, Univ NH, 37-39, asst Columbia Univ, 39-40, asst chem, 40-42, res chemist, S A M Labs, 43-45; assoc scientist, Manhattan Dist, 45-46, group leader, Los Alamos Sci Lab, 46-70, STAFF MEM, LOS ALAMOS SCI LAB, UNIV CALIF, 70- Mem: Am Chem Soc; AAAS. Res: Fields in physical chemistry; inorganic chemistry of rare radioactive elements; high temperature chemistry; equation of state. Mailing Add: 764 43rd St Los Alamos NM 87544

VIERCK, CHARLES JOHN, JR, b Columbus, Ohio, July 6, 36; m 60; c 1. PSYCHOPHYSIOLOGY. Educ: Univ Fla, BS, 59, MS, 61, PhD(psychol), 63. Prof Exp: Fel physiol psychol, Inst Neurol Sci, Univ Pa, 63-65; asst prof, 65-71, ASSOC PROF PHYSIOL PSYCHOL, COL MED, UNIV FLA, 71- Concurrent Pos: Nat Inst Neurol Dis & Stroke res grant, 67-75, mem neurol B study sect. Mem: AAAS; Am Psychol Asn; Psychonomic Soc; Soc Neurosci; Int Neuropsychol Soc. Res: Central nervous system mechanisms relating to somesthetic discrimination; discrimination and perception of pain; recovery of function after nervous system damage. Mailing Add: Dept of Neurosci Ctr Neurobiol Sci Univ of Fla Col of Med Gainesville FL 32601

VIERECK, LESLIE A, b New Bedford, Mass, Feb 20, 30; m 55; c 3. PLANT ECOLOGY, PLANT TAXONOMY. Educ: Dartmouth Col, BA, 51; Univ Colo, MA, 57, PhD(plant ecol), 62. Prof Exp: Asst bot, McGill Subarctic Res Sta, 54-55; asst, Herbarium, Univ Colo, 55-57, actg cur, 56-57, res assoc ecol, Inst Arctic & Alpine Res, 55-59; res assoc ecol, Univ Alaska, 59-60, asst prof bot, 60-61; res biologist, Alaska Dept Fish & Game, 61-63; PRIN PLANT ECOLOGIST, INST NORTHERN FORESTRY, 63- Mem: Fel AAAS; fel Arctic Inst NAm; Ecol Soc Am; Am Bryol & Lichenological Soc. Res: Plant ecology and plant taxonomy of arctic, subarctic and alpine regions. Mailing Add: Inst of Northern Forestry Fairbanks AK 99701

VIERIMA, TERI L, b Long Island, NY, Nov 19, 48; m 74. CHEMICAL PHYSICS. Educ: St Olaf Col, BA, 70; Yale Univ, MPhil, 73, PhD(physics), 74. Prof Exp: ASST PROF PHYSICS, BATES COL, 75- Concurrent Pos: Res staff physicist, Yale Univ & vis asst prof physics, Trinity Col, Conn, 74-75. Mem: Am Phys Soc; Am Asn Physics Teachers. Res: Fine and hyperfine structure of diatomic and triatomic molecules. Mailing Add: Dept of Physics & Astron Bates Col Lewiston ME 04240

VIERNSTEIN, LAWRENCE J, b New York, NY, Feb 20, 19; m 69; c 2. BIOMEDICAL ENGINEERING. Educ: Okla State Univ, BS, 50, MS, 51; Johns Hopkins Univ, PhD, 70. Prof Exp: Mem assoc staff, Appl Physics Lab, Johns Hopkins Univ, 52-57, sr physicsist, 57-59, mem prof staff, Dept Physiol, 61-66, MEM PROF STAFF, WILMER INST, JOHNS HOPKINS UNIV, 66- & APPL PHYSICS LAB, 59- Res: Theoretical biology and biomedical engineering; neurophysiology; artificial intelligence in medicine. Mailing Add: Johns Hopkins Univ Appl Physics Lab Johns Hopkins Rd Laurel MD 20810

VIERS, JIMMY WAYNE, b Grundy, Va, Feb 26, 43; m 65; c 1. PHYSICAL CHEMISTRY. Educ: Berea Col, AB, 65; Wake Forest Univ, MA, 67; Stanford Univ, PhD(chem), 71. Prof Exp: ASST PROF CHEM, VA POLYTECH INST & STATE UNIV, 71- Mem: Am Chem Soc. Res: Quantum chemistry, electron-atom scattering. Mailing Add: Dept of Chem Va Polytech Inst & State Univ Blacksburg VA 24061

VIETH, JOACHIM, b Hamburg, Ger, Oct 26, 25; m 53; c 2. PLANT MORPHOLOGY. Educ: Univ Saarlandes, Lic natural sci, 53, dr rer nat, 57; Univ Dijon, DSc(bot), 65. Prof Exp: Asst bot, Univ Saarlandes, 53-57; res fel, Nat Ctr Sci Res, Univ Dijon, 57-65; vis prof, 66-67, ASSOC PROF BOT, UNIV MONTREAL, 67- Mem: Bot Soc France; Can Bot Asn. Res: Anatomy of flowers and inflorescences, both normal and anomalous; experimental morphology; relationship between vegetative and inflorescential regions, between normal and anomalous forms experimentally induced. Mailing Add: Dept of Bot Univ of Montreal Montreal PQ Can

VIETMEYER, NOEL DUNCAN, b Wellington, NZ, Nov 9, 40; m 65; c 2. ECONOMIC BIOLOGY, SCIENCE WRITING. Educ: Univ Otago, NZ, BSc, 63; Univ Calif, Berkeley, PhD, 67. Prof Exp: Lectr org chem, Univ Calif, Berkeley, 67-68; NIH fel, Stanford Univ, 68-69, fel, 68-70; PROF ASSOC, NAT ACAD SCI, 70- Res: Organic chemistry; utilization of aquatic weeds; economic development of Simmondsia californica; introduction of technology into developing countries; neglected tropical plants and animals with promising economic potential. Mailing Add: Nat Acad of Sci Washington DC 20418

VIETOR, DONALD MELVIN, b Urbana, Ill, Sept 29, 45; m 71; c 1. CROP PHYSIOLOGY. Educ: Univ Minn, BS, 67, MS, 69; Cornell Univ, PhD(crop sci), 75. Prof Exp: Biol sci asst soil sci, US Army Cold Regions Res & Engr Lab, 69-71; ASST PROF AGRON, DEPT PLANT & SOIL SCI, UNIV MASS, 74- Mem: Am Soc Agron; Crop Sci Soc Am; Am Soc Plant Physiologists. Res: Effect of genotype, environment, and management practices on photosynthesis, respiration, assimilate distribution, and dry matter production of perennial forages. Mailing Add: Dept of Plant & Soil Sci Stockbridge Hall Univ of Mass Amherst MA 01002

VIETS, FRANK GARFIELD, JR, b Stanberry, Mo, Apr 3, 16; m 38; c 3. SOIL SCIENCE. Educ: Colo Agr & Mech Col, BS, 37; Univ Calif, MS, 39, PhD(plant physiol), 42. Prof Exp: Agent div cereal crops, USDA, Calif, 37-39; asst div plant nutrit, Univ Calif, 39-42; supv chemist, Cutter Labs, Calif, 42-44; assoc agr chemist, Exp Sta, SDak State Col, 44-45; agronomist div soil mgt & irrig, USDA, 45-49, soil scientist, Wash, 49-53, soil & water conserv res div, Agr Res Serv, 53-74; CONSULT, US AID, 74- Concurrent Pos: Vis prof, Univ Ill, 59 & Iowa State Univ, 64; ed in chief, Soil Sci Soc Am, 63-65. Honors & Awards: Superior Serv Award, USDA, 55, Distinguished Serv Award, 72. Mem: Fel AAAS; Soil Sci Soc Am (vpres, 66, pres, 67); fel Am Soc Agron; Am Soc Sugar Beet Technol; Int Soc Soil Sci. Res: Mineral nutrition of plants; zinc deficiency in soils and plants; water pollution by animal wastes, fertilizers and agriculture; soil fertility and productivity; tropical soils. Mailing Add: 102 Yale Way Ft Collins CO 80521

VIETTE, MICHAEL ANTHONY, b Pittsburg, Kans, Mar 27, 41; m 63; c 1. PHYSICS. Educ: Kans State Col Pittsburg, BA, 64, MS, 66; Univ Mo-Rolla, PhD(physics), 72. Prof Exp: Res asst cloud physics, Grad Ctr Cloud Physics Res, Univ Mo-Rolla, 66-70; ASST PROF PHYSICS, UNIV MAINE, ORONO, 71- Concurrent Pos: NSF res grant, 72- Mem: Am Phys Soc; Am Geophys Union; Am Meteorol Soc. Res: Condensation and growth of micron sized water droplets. Mailing Add: Bennett Hall Univ of Maine Orono ME 04472

VIETTI, TERESA JANE, b Ft Worth, Tex, Nov 5, 27. PEDIATRICS, HEMATOLOGY. Educ: Rice Inst, AB, 49; Baylor Univ, MD, 53; Am Bd Pediat, dipl, 59. Prof Exp: Instr pediat, Wayne State Univ, 58 & Southwestern Med Sch, Univ Tex, 58-60; vis pediatrician, Hacettepe Children's Hosp, Ankara, Turkey, 60-61; from asst prof to assoc prof pediat, 61-72, PROF PEDIAT, SCH MED, WASH UNIV, 72-, ASSOC PROF PEDIAT IN RADIOL, 71- Concurrent Pos: Dir hemat labs, attend pediatrician & consult, Tex Children's Hosp, Dallas, 58-60; attend pediatrician & consult, Parkland Mem Hosp, 58-60; Am Cancer Soc fel, 58-59; USPHS trainee, 59-60, grant, 61-; mem, Southwest Oncol Group, 61-; asst pediatrician, St Louis Children's Hosp, 61-65, assoc pediatrician, 65-, dir div hemat & oncol, 70-; from asst pediatrician to assoc pediatrician, Barnes & Allied Hosps, 61-65; consult, Homer G Phillips Hosp & St Louis County Hosp; assoc in pediat, Mo Crippled Children's Serv; mem, Cancer Clin Invest Rev Comt, 74; pediat consult high risk maternity & child care prog, Mo Div Health, 75. Mem: Am Acad Pediat; Am Hemat Soc; Int Soc Hemat; Am Asn Cancer Res; Am Pediat Soc. Res: Cancer chemotherapy. Mailing Add: Dept of Pediat Wash Univ Sch of Med St Louis MO 63110

VIEWEG, HERMANN FREDERICK, physical chemistry, deceased

VIG, BALDEV K, b India, Oct 1, 35; m; c 2. CYTOGENETICS. Educ: Khalsa Col, India, BSAgr, 57; Panjab Univ, India, MS, 61; Ohio State Univ, PhD(genetics), 67. Prof Exp: Demonstr agr, Khalsa Col, India, 58-61; assoc prof bot, Rajasthan Col Agr, India, 61-64; res cytogeneticist, Dept Pediat, Children's Hosp, Ohio State Univ, 67-68, res award, 68; asst prof biol, 68-72, res adv grants, 69-79, ASSOC PROF BIOL, UNIV NEV, RENO, 72- Concurrent Pos: Consult, Western Environ Res Ctr, Environ Protection Agency. Mem: Am Genetics Asn; Genetic Soc Am; Genetics Soc Can; Environ Mutagen Soc. Res: Mode and mechanisms of somatic crossingover in Glycine max; chromosome structure and rejoining in human leukocytes in vitro; action of antileukemic drugs on chromosomes. Mailing Add: Dept of Biol Genetics Lab Univ of Nev Reno NV 89507

VIGDAHL, ROGER LEON, biochemistry, see 12th edition

VIGEE, GERALD S, b Crowley, La, Mar 4, 31; m; c 3. PHYSICAL INORGANIC CHEMISTRY. Educ: US Mil Acad, BS, 54, La State Univ, Baton Rouge, BS, 60, PhD(chem), 68. Prof Exp: Prof engr, NASA, Ala, 60-61; propulsion design engr, Chrysler Corp, 61-64; asst prof chem, Univ Miss, 68-69; ASSOC PROF CHEM, UNIV ALA, BIRMINGHAM, 69- Res: Synthesis of coordination complexes, investigation of the magneto chemistry and spectroscopic energy levels of these complexes. Mailing Add: Dept of Chem Univ of Ala Birmingham AL 35299

VIGFUSSON, NORMAN V, b Ashern, Man, July 1, 30; m 54; c 5. GENETICS, MEDICAL GENETICS. Educ: Univ Man, BSA, 51; Univ Alta, PhD(fungal genetics), 69. Prof Exp: Asst prof genetics, 69-72, ASSOC PROF BIOL, EASTERN WASH STATE COL, 72- Concurrent Pos: Genetic consult, Sacred Heart Med Ctr, Spokane, Wash, 75- Mem: Genetics Soc Am; Brit Med Asn; AAAS. Res: Sexuality in Neorospora crassa with respect to stages and control of the sexual cycle and attempts to arrive at elucidation of incompatibility control mechanism; human cytogenetics. Mailing Add: Dept of Biol Eastern Wash State Col Cheney WA 99004

VIGIL, EUGENE LEON, b Chicago, Ill, Mar 14, 41; m 63; c 3. CELL BIOLOGY. Educ: Loyola Univ Chicago, BS, 63; Univ Iowa, MS, 65, PhD(bot), 67. Prof Exp: NIH fel, Univ Wis-Madison, 67-69; trainee cell biol, Univ Chicago, 69-71; ASST PROF CELL BIOL, MARQUETTE UNIV, 71- Concurrent Pos: Distinguished vis scientist, Dept Physiol & Biophys, Colo State Univ, 74-75. Mem: AAAS; Am Soc Cell Biol; Am Soc Plant Physiol; Histochem Soc. Res: Structure and function of animal and plant microbodies; effects of plant growth regulations on barley aleurone cells; biogenesis of mitochondria in yeast; cytochemical localization of photosynthetic reactions in chloroplasts. Mailing Add: Dept of Biol Marquette Univ 530 N 15th St Milwaukee WI 53233

VIGLIERCHIO, DAVID RICHARD, b Madera, Calif, Nov 25, 25; m 67; c 1. NEMATOLOGY. Educ: Calif Inst Technol, BS, 50, PhD(bio-org chem), 55. Prof Exp: Jr res nematologist, 55-57, asst res nematologist, 57-63, assoc nematologist, 63-39, NEMATOLOGIST, UNIV CALIF, DAVIS, 69- Concurrent Pos: Fulbright fel, 64-65; J S Guggenheim fel, 65; partic, US Antarctic Prog, 69-70; Nat Acad Sci exchange USSR, 70-71. Mem: Am Chem Soc; Soc Nematol; Soc Europ Nematol. Res: Chemistry and physiology of plant parasitic and free-living nematodes; host-parasite relationships; physiological methods of nematode control. Mailing Add: Dept of Nematol Univ of Calif Davis CA 95616

VIGNALE, MICHAEL JOSEPH, b Methuen, Mass, Dec 18, 24; m 49; c 2. ORGANIC CHEMISTRY. Educ: Boston Univ, AB, 51, PhD(chem), 56. Prof Exp: Sr res chemist, Monsanto Chem Soc, 55-60, Foster Grant Co, 60-64 & Metallomer Labs, 64-66; ASSOC PROF CHEM, MASS STATE COL FITCHBURG, 66- Concurrent Pos: Instr, Western New Eng Col, 56- Mem: Am Chem Soc. Res: Organometallic chemistry. Mailing Add: Dept of Chem Mass State Col Fitchburg MA 01420

VIGNES, ROBERT PAUL, b New Orleans, La, Jan 17, 44; m 65; c 1. ORGANIC CHEMISTRY. Educ: Tulane Univ, BS, 66, PhD(chem), 74. Prof Exp: Chemist, E I du Pont de Nemours & Co, 66-69; qual control lab supvr anal, 71-74, div chemist, 74-75, RES CHEMIST ADHESIVES, E I DU PONT DE NEMOURS & CO, 75- Mem: Am Chem Soc. Res: Development of new adhesive systems. Mailing Add: Elaschem Dept E I du Pont de Nemours & Co Wilmington DE 19898

VIGNOS, JAMES HENRY, b Cleveland, Ohio, July 27, 33; m 62; c 2. PHYSICS. Educ: Case Inst, BS, 55; Yale Univ, MS, 57, PhD(physics), 62. Prof Exp: Vis res scientist, Low Temperature Inst, Bavarian Acad Sci, Germany, 62-64; resident res assoc, Chem Div, Argonne Nat Lab, 64-66; asst prof physics, Dartmouth Col, 66-72; MEM STAFF, RES CTR, FOXBORO CO, 72- Concurrent Pos: Fulbright res scholar, 62-63; von Humboldt fel, 63-64. Mem: AAAS; Am Asn Physics Teachers; Am Phys

Soc. Res: Low temperature physics; liquid and solid helium; superconductivity; ultrasonics. Mailing Add: Res Ctr Foxboro Co Foxboro MA 02035

VIGNOS, PAUL JOSEPH, JR, b Canton, Ohio, Nov 10, 19; m 46; c 3. BIOCHEMISTRY, RHEUMATOLOGY. Educ: Univ Notre Dame, BS, 41; Western Reserve Univ, MD, 44. Prof Exp: From intern to resident, Univ Hosps, Cleveland, 44-46; resident, Presby Hosp, New York, 48-49; Am Cancer Soc fel, Univ Hosps, Cleveland, 49-50; Rees fel med, Sch Med, 50-51, USPHS fel pharmacol, 51-52, from instr to asst prof, 52-66, ASSOC PROF MED, SCH MED, CASE WESTERN RESERVE UNIV, 66- Res: Bioclinical effect of disease of the locomotor system, particularly skeletal muscle and joints; biochemistry of normal and diseased muscle; metabolic effects of corticosteroid on skeletal muscle. Mailing Add: Dept of Med Case Western Reserve Univ Cleveland OH 44106

VIGO, TYRONE LAWRENCE, b New Orleans, La, Feb 1, 39; m 63; c 1. POLYMER CHEMISTRY, TEXTILE CHEMISTRY. Educ: Loyola Univ, La, BS, 60; Tulane Univ La, MS, 63, PhD(org chem), 69. Prof Exp: Res chemist, 63-66, PROJ LEADER TEXTILE & POLYMER CHEM, SOUTHERN REGIONAL RES CTR, AGR RES SERV, USDA, 68- Mem: Am Chem Soc; Am Asn Textile Chemists & Colorists. Res: Chemical modification of cellulose by application of new synthetic techniques; modification of textiles; synthetic organic chemistry; polymer chemistry and physics of natural polymers; industrial microbiology of polymeric materials. Mailing Add: Southern Reg Res Ctr USDA 1100 Robert E Lee Blvd New Orleans LA 70179

VIGRASS, LAURENCE WILLIAM, b Melfort, Sask, May 9, 29; m 54; c 3. GEOLOGY. Educ: Univ Sask, BE, 51, MSc, 52; Stanford Univ, PhD(geol), 61. Prof Exp: Geologist, Calif Stand Co, 52-55; res geologist, Imp Oil Ltd, 58-65; consult geologist, Western Resources Consult Ltd, 65-68; assoc prof geol, Univ Sask, Regina, 68-73, actg chem dept, 72-73, assoc prof geol, 68-73, PROF GEOL, UNIV REGINA, 73- Mem: Geol Asn Can; Can Inst Mining & Metall; Am Asn Petrol Geol. Res: Sedimentary geology; occurence of petroleum and natural gas; sedimentary geochemistry. Mailing Add: Dept of Geol Sci Univ of Regina Regina SK Can

VIJAYAN, VIJAYA KUMARI, b Trivandrum, India, Feb 25, 42; m 66; c 2. HUMAN ANATOMY, NEUROANATOMY. Educ: Univ Kerala, MBBS, 65; Univ Calif, Davis, PhD(anat), 72. Prof Exp: Tutor human anat, Med Col, Trivandrum, India, 65-68; ASST PROF HUMAN ANAT, MED SCH, UNIV CALIF, DAVIS, 73- Res: Biochemistry and ultrastructure of developing and aging nervous system; electrophoresis; neurotransmitters. Mailing Add: Dept of Human Anat Univ of Calif Sch of Med Davis CA 95616

VIJAYENDRAN, BHEEMA R, b Bangalore, India, 1941; m 70. PHYSICAL CHEMISTRY. Educ: Univ Madras, BTech, 63, MTech, 65; Univ Southern Calif, PhD(chem), 69. Prof Exp: Lectr, Cent Leather Res Inst, India, 65-66; indust fel surface chem, R J Reynolds Indust, NC, 69-70; mgr res, Copier Prod Div, 74-75, CHEMIST, COPIER PROD DIV, PITNEY BOWES, INC, 70- Concurrent Pos: Teaching asst, Univ Southern Calif, 66-68. Mem: Am Chem Soc; Am Oil Chem Soc; Brit Oil & Colour Chem Asn. Res: Physical chemistry of surfaces; colloidal systems; emulsions; biopolymers and synthetic polymers; adhesions and their application in graphic arts such as printing, photography, xero-graphy and other reprographic techniques. Mailing Add: Copier Prod Div Pitney Bowes Inc Danbury CT 06810

VIKIS, ANDREAS CHARALAMBOUS, b Moni, Cyprus, July 8, 42; Can citizen; m 71; c 1. PHYSICAL CHEMISTRY. Educ: Col Emporia, BSc, 64; Kans State Univ, PhD(phys chem), 69. Prof Exp: Fel, Univ Toronto, 69-70, lectr & res assoc, 70-74, asst prof & res assoc, 74-75; ASST RES OFFICER CHEM, NAT RES COUN CAN, 75- Mem: Am Chem Soc. Res: Energy transfer processes and chemical reactivity of electronically excited atoms and molecules; laser photochemistry relating to isotope enrichment. Mailing Add: Chem Div Nat Res Coun of Can Ottawa ON Can

VIKSNE, ANDY, b Jan 27, 34; m 64; c 3. GEOPHYSICS. Educ: Harvard Univ, AB, 56; Univ Utah, MS, 58. Prof Exp: Geophysicist, Texaco, Inc, 59-65 & Environ Res Corp, Systs Sci Corp, 65-66; scientist, Raytheon Co, Va, 67-68; geophysicist, US Bur Mines, 68-72; GEOPHYSICIST, US BUR RECLAMATION, 72- Mem: Soc Explor Geophys; Europ Asn Explor Geophysicists. Res: Application of remote sensing instruments in earth resouces surveys and engineering geology problems; application of geophysical exploration methods in solving geotechnical engineering problems concerned with water resources developments; in situ determination of elastic moduli for earth dams and foundation sites. Mailing Add: 7719 S Eaton Way Littleton CO 80123

VILA, SAMUEL CAMPDERROS, b Rubi, Spain, May 7, 30. ASTROPHYSICS. Educ: Univ Barcelona, Lic physics, 52; Univ Rochester, PhD(astron), 65. Prof Exp: Res assoc astron, Ind Univ, 65-67 & Inst Space Studies, NASA, NY, 67-69; asst prof, 69-74, ASSOC PROF ASTRON, UNIV PA, 74- Mem: Am Astron Soc; Int Astron Union. Res: Late stages of stellar evolution. Mailing Add: Dept of Astron Univ of Pa Philadelphia PA 19174

VILAKAZI, ABSOLOM, b Natal, SAfrica, Dec 27, 14; m 43; c 5. ANTHROPOLOGY. Educ: Univ Natal, BA, 49 & 51, PhD, 59; Hartford Sem Found, MA, 54; Trinity Col, Conn, MA, 55. Prof Exp: From asst prof to assoc prof anthrop & African studies, Hartford Sem Found, 57-62; chief soc res sect, UN Econ Comn Africa, 62-65; PROF AFRICAN STUDIES, SCH INT SERV, AM UNIV, 65-, PROF ANTHROP, 66- Mem: Fel Am Anthrop Asn; Soc Appl Anthrop; fel African Studies Asn; Am Ethnol Soc; Soc Sci Study Relig. Res: Separatist church movements and Messianism in Africa; Bantu magico-medical concepts and practices; social change in Africa; education in African traditions and transitional societies. Mailing Add: Dept of African Studies Sch of Int Serv Am Univ Washington DC 20016

VILBRANDT, CHARLES FRANK, b Oak Harbor, Ohio, Aug 12, 17; m 40; c 2. PHYSICAL CHEMISTRY. Educ: Univ NC, BS, 39; Univ Wis, PhD(phys chem), 42. Prof Exp: Chem engr, Eastman Kodak Co, 42-44, develop engr, 44-45, chem engr, 45-48, tech supvr, 48-59, asst supt, Film Emulsion Div, 59-75; RETIRED. Res: Ultra-centrifugal analysis; process design; various aspects of film manufacturing. Mailing Add: 210 Glenwood Trail Southern Pines NC 28387

VILCEK, JAN TOMAS, b Bratislava, Czech, June 17, 33; m 62. MICROBIOLOGY, VIROLOGY. Educ: Univ Bratislava, MD, 57; Czech Acad Sci, CSc(virol), 62. Prof Exp: Res assoc virol, Inst Virol Czech Acad Sci, Bratislava, 57-59, head lab, 62-64; from asst prof to assoc prof microbiol, 65-72, PROF MICROBIOL, SCH MED, NY UNIV, 72- Concurrent Pos: Am Cancer Soc grant, 65-66; USPHS grant, 65-79, career develop award, 68-73 & contract, 70-76; Irwin Strassburger Mem Med Found grant, 69-73; ed, Arch Virol 72-74, ed in chief, 75- Mem: AAAS; Am Soc Microbiol; Brit Soc Gen Microbiol. Res: Virus interference; interferon; antiviral substances. Mailing Add: Dept of Microbiol NY Univ Sch of Med 550 First Ave New York NY 10016

VILCHES, OSCAR EDGARDO, b Mercedes, Arg, Feb 20, 36. PHYSICS. Educ: Nat Univ Cuyo, lic physics, 59, Dr en Fisica, 66. Prof Exp: Investr physics, Physics Inst, Arg, 60-64; res asst, Univ Ill, Urbana, 64-65, res assoc, 65-67; res assoc, Univ Calif, San Diego, 67-68; asst prof, 68-73, ASSOC PROF PHYSICS, UNIV WASH, 73- Concurrent Pos: NSF res grant, 70- Mem: Am Phys Soc. Res: Properties of liquid and solid helium and helium films; very low temperature magnetism and superconductivity. Mailing Add: Dept of Physics Univ of Wash Seattle WA 98195

VILCINS, GUNARS, b Riga, Latvia, May 8, 30; m 61; c 2. ANALYTICAL CHEMISTRY, SPECTROSCOPY. Educ: Univ Richmond, BS, 54, MS, 62. Prof Exp: Assoc chemist, 57-63, RES CHEMIST, PHILIP MORRIS, INC, 63- Mem: Am Chem Soc; Soc Appl Spectros; Coblentz Soc. Res: Infrared and Raman spectroscopy; cigarette smoke; tobacco; low temperature studies; microanalysis. Mailing Add: 2504 Haviland Dr Richmond VA 23229

VILETTO, JOHN, JR, b Arnold, Pa, Dec 5, 26; m 68; c 2. GEOGRAPHY, GEOMORPHOLOGY. Educ: Pa State Univ, BS, 59, MS, 61, PhD(geol), 68. Prof Exp: Res geomorphol, geol & geog, US Army Natick Labs, 67-71; RES GEOL, GEOG & CARTOG, US ARMY ENG TOPOG LAB, 71- Concurrent Pos: Tech eval comt mem; US Army Air Mobility Res & Develop Lab, 74-75. Res: Design and development of maps and other graphics to depict selected natural environmental elements relevant to design, testing and issue of all types of military equipment. Mailing Add: US Army Eng Topog Lab Ft Belvoir VA 22060

VILKS, GUSTAVS, b Riga, Latvia, May 7, 29; Can citizen; m 55; c 2. MICROPALEONTOLOGY. Educ: McMaster Univ, BSc, 61; Dalhousie Univ, MSc, 66. Prof Exp: MICROPALEONTOLOGIST, BEDFORD INST OCEANOG, 62- Mem: Geol Asn Can. Res: Ecology and paleoecology of Recent Foraminifera in the Canadian Arctic and Gulf of St Lawrence; ecology of planktonic Foraminifera in the North Atlantic. Mailing Add: Atlantic Geosci Ctr Bedford Inst of Oceanog Dartmouth NS Can

VILL, JOHN JOSEPH, b Elizabeth, NJ, Oct 25, 36; m 59; c 2. PHYSICAL ORGANIC CHEMISTRY. Educ: Rutgers Univ, Newark, BA, 58; Rutgers Univ, NB, PhD(phys org chem), 61. Prof Exp: Res chemist, cent Res Div, W R Grace & Co, Md, 61-64, mgr res & develop, Hatco Chem Div, NJ, 64-69; PROCESS DEVELOP SUPVR, MED CHEM, GANE'S CHEM WORKS, INC, CARLSTADT, 69- Mem: Am Chem Soc. Res: Process development of fine and specialty organic chemicals. Mailing Add: 20 Broad St Denville NJ 07834

VILLA, JUAN FRANCISCO, b Matanzas, Cuba, Sept 23, 41; US citizen; m 67; c 4. INORGANIC CHEMISTRY. Educ: Univ Miami, BS, 65, MS, 67, PhD(inorg chem), 69. Prof Exp: Teaching asst chem, Univ Miami, 65-69; res assoc inorg chem, Univ NC, Chapel Hill, 69-71; asst prof chem, 71-73, ASSOC PROF CHEM, LEHMAN COL, 74- Concurrent Pos: George N Shuster fel, Lehman Col, 71-72 & 74-76; Petrol Res Fund fel, 71-73. Mem: AAAS; Am Chem Soc; The Chem Soc; NY Acad Sci; Am Inst Chemists. Res: Study of transition metal coordination compounds of biological importance including synthesis, electron paramagnetic resonance spectroscopy, magnetic susceptibility measurements, electronic and infrared spectra; ligand field and molecular orbital calculations. Mailing Add: 14 Clark Dr Spring Valley NY 10977

VILLA, VICENTE DOMINGO, b Laredo, Tex, Dec 1, 40; m 62; c 2. MICROBIAL PHYSIOLOGY. Educ: Univ Tex, Austin, BA, 64; Rice Univ, PhD(microbiol), 70. Prof Exp: Fel molecular biol, Molecular Biol Lab, Univ Wis, 69-71; res assoc, Rosenstiel Res Ctr, Brandeis Univ, 71-72; asst prof, 72-75, ASSOC PROF BIOL, N MEX STATE UNIV, 76- Concurrent Pos: NIH fel, 70-71; ad hoc consult, Minority Biomed Support Prog, NIH, 72-75, mem gen res support prog adv comt, Div Res Resources, 76-80; panelist, NSF Rev Panel-Res Initiation & Support Prog, 76. Mem: Am Soc Microbiol. Res: Regulation of fermentative and oxidative metabolism in fungi and its relation to the morphogenesis of the organism. Mailing Add: Dept of Biol Box 3AF NMex State Univ Las Cruces NM 88003

VILLABLANCA, JAIME ROLANDO, b Chillan, Chile, Feb 28, 29; m 55; c 5. NEUROPHYSIOLOGY, EXPERIMENTAL NEUROLOGY. Educ: Univ Chile, Bachelor, 46, Lic Med, 53, Dr(med), 54; Univ Calif, Los Angeles, cert neurophysiol, 68. Prof Exp: From instr to prof pathophysiol, Sch Med, Univ Chile, 54-71; assoc res anat & psychiat, 71-72, PROF PSYCHIAT, UNIV CALIF, LOS ANGELES, 72- Concurrent Pos: Rockefeller Found fel physiol, Johns Hopkins Univ, 59-61; fel, Neurol Unit, Harvard Med Sch, 61; US Air Force Off Sci res grant, 62-65; NIH int res fel anat, Univ Calif, Los Angeles, 66-68; Found Fund Res in Psychiat grant, 69-72; Nat Inst Child Health & Human Develop prog proj grant, 72-77. Mem: NY Acad Sci; Am Physiol Soc; Soc Neurosci; Asn Psychophysiol Study Sleep; Int Brain Res Orgn. Res: Neurophysiology of sleep-wakefulness; neurological, behavioral and electrophysiological effects of lesions upon the mature and upon the developing brain; physiology and pathophysiology of the basal ganglia. Mailing Add: Dept of Psychiat Univ of Calif Los Angeles CA 90024

VILLAFRANCA, JOSEPH JOHN, b Silver Creek, NY, Mar 23, 44; m 67; c 2. BIOCHEMISTRY, BIO-ORGANIC CHEMISTRY. Educ: State Univ NY Col Fredonia, BS, 65; Purdue Univ, Lafayette, PhD(biochem), 69. Prof Exp: USPHS fel, Inst Cancer Res, 69-71; ASST PROF CHEM, PA STATE UNIV, UNIVERSITY PARK, 71- Concurrent Pos: Res Corp grant, Pa State Univ, 72-73, NSF grant, 72-76 & USPHS grant, 74-77. Mem: Am Chem Soc; Biophys Soc. Res: Mechanism of enzyme action studied by magnetic resonance techniques. Mailing Add: 21 Chandlee Lab Dept of Chem Pa State Univ University Park PA 16802

VILLALON, BENIGNO, plant pathology, plant virology, see 12th edition

VILLANI, FRANK JOHN, b Brooklyn, NY, May 9, 21; m 51; c 4. CHEMISTRY. Educ: Brooklyn Col, BA, 41; Fordham Univ, MS, 43, PhD(chem), 46. Prof Exp: Org chemist, 46-64, FEL MED CHEM, SCHERING CORP, BLOOMFIELD, 64- Mem: AAAS; Am Chem Soc; Am Inst Chem; NY Acad Sci. Res: Aldehyde condensations; synthetic medicinals; heterocyclic chemistry. Mailing Add: 55 McKinley Ave West Caldwell NJ 07006

VILLANUEVA, GERMAN BAIT, b Davao City, Philippines, Apr 21, 41; m 69; c 1. BIOCHEMISTRY. Educ: Ateneo de Manila Univ, BA, 62; Fordham Univ, MS, 69, PhD(biochem), 71. Prof Exp: Instr chem, Mindanao Univ, 62-63 & Ateneo de Davao, 63-65; lab instr, Fordham Univ, 66-69; ASST PROF, COL WHITE PLAINS, 70- Concurrent Pos: Fulbright exchange fel, Inst Int Educ, NY, 66; Camille & Henry Dreyfus Found & NY State Heart Assembly res grant, Manhattan Col, 72- Mem: AAAS; Am Chem Soc. Res: Difference spectroscopy and chemical modifications of proteins, including conformational studies by optical rotatory dispersion and circular dichroism; the role of thrombin in blood clotting. Mailing Add: Dept of Sci & Math Col of White Plains Pleasantville NY 10570

VILLAREJO, DON, physics, see 12th edition

VILLAREJO, MERNA, b New York, NY, June 19, 39; m 59; c 2. BIOCHEMISTRY. Educ: Univ Chicago, BS, 59, PhD(biochem), 63. Prof Exp: Res assoc & asst prof biochem, Univ Chicago, 63-68; RES ASSOC BIOL CHEM, SCH MED, UNIV CALIF, LOS ANGELES, 68- Concurrent Pos: USPHS fel, Univ Chicago, 63-65. Res: Protein structure and function; enzyme complementation of beta-galactosidase. Mailing Add: Dept of Biol Chem Univ of Calif Sch of Med Los Angeles CA 90024

VILLAREJOS, VICTOR MOISES, b La Paz, Bolivia, Sept 4, 18; US citizen; m 41; c 4. EPIDEMIOLOGY, TROPICAL MEDICINE. Educ: Univ Heidelberg, MD, 41; Tulane Univ, MPH & TM, 59, DrPH, 61. Prof Exp: Chief serv med, Miraflores Gen Hosp, La Paz, 45-58; from assoc prof to prof trop med, Med Sch, Univ La Paz, 47-58; assoc prof trop med, Sch Med, 61-66, chief epidemiol sect, La State Univ Int Ctr Med Res & Training, San Jose, Costa Rica, 62-66, prog coordr, 66-69, PROF TROP MED, SCH MED, LA STATE UNIV, NEW ORLEANS, 66-, DIR, LA STATE UNIV-INT CTR MED RES & TRAINING, SAN JOSE, COSTA RICA, 69- Concurrent Pos: Alexander von Humboldt fel, Ger, 41-42; Pan Am Health Orgn fel, 58-59; USPHS & Armed Forces Epidemiol Bd res grants. Honors & Awards: Geiger Medal, Tulane Univ, 61. Mem: Am Soc Trop Med & Hyg; Am Soc Parasitol; Am Pub Health Asn. Res: Epidemiology of diseases prevalent in tropical areas; pathogenesis of E histolytica; diarrheal diseases; infectious hepatitis. Mailing Add: Dept of Trop Med Sch of Med La State Univ New Orleans LA 70112

VILLAR-PALASI, CARLOS, b Spain, Mar 3, 28; m 57; c 4. PHARMACOLOGY, BIOCHEMISTRY. Educ: Univ Valencia, MS, 51; Univ Madrid, PhD(biochem), 55; Univ Barcelona, MPharm, 62. Prof Exp: Res assoc biochem, Univ Madrid, 60-63; res assoc, Univ Minn, Minneapolis, 64-65, asst prof, 65-69; assoc prof pharmacol, 69-72, PROF PHARMACOL, UNIV VA, 72- Concurrent Pos: Span Res Coun fel, Univ Hamburg, 53-54; NIH fel, Western Reserve Univ, 57-60; NIH grant, Univ Va, 70- Honors & Awards: AAAS Res Award, 60; Span Soc Biochem Res Award, 72. Mem: AAAS; NY Acad Sci; Am Soc Biol Chemists; Span Soc Biochem; Am Soc Pharmacol & Exp Therapeut. Res: Glycogen metabolism and the mechanism of action of cyclic adenosine monophosphate; mechanisms of control, metabolic and hormonal; effects of insulin, epinephrine and glucagon and glycogen metabolism as well as the mechanisms by which epinephrine exerts its effect on muscle contraction. Mailing Add: Dept of Pharmacol Univ of Va Charlottesville VA 22903

VILLARREAL, JESSE JAMES, b San Antonio, Tex, Oct 22, 13; m 35; c 2. SPEECH PATHOLOGY. Educ: Univ Tex, BA, 35, MA, 37; Northwestern Univ, PhD(speech path, audiol), 47. Prof Exp: Dir speech & hearing clin, 39-62, prof speech, 52-65, chmn dept speech, 62-68, PROF SPEECH COMMUN & EDUC, UNIV TEX, AUSTIN, 65- Mem: Fel Am Speech & Hearing Asn; Speech Commun Asn. Res: English as a second language. Mailing Add: 5104 Crestway Dr Austin TX 78731

VILLARREAL, JULIAN ERNESTO, b Mexico, DF, Apr 10, 37; m 59; c 6. PHARMACOLOGY. Educ: Nat Univ Mex, BS, 53, MD, 60; Univ Mich, Ann Arbor, PhD(pharmacol), 69. Prof Exp: Lab asst pharmacol, Med Sch, Nat Univ Mex, 57, res asst exp path, 58-59, res asst pharmacol, 60; from instr to asst prof, 65-69, ASSOC PROF PHARMACOL, SCH MED, UNIV MICH, ANN ARBOR, 70-; DIR BEHAV PHARMACOL, MILES INST EXP THERAPEUT, MEX, 72- Concurrent Pos: NIMH grant, Univ Mich Ann Arbor, 71-; consult, NIMH, 71- Mem: AAAS; Am Soc Pharmacol & Exp Therapeut; Behav Pharmacol Soc. Res: Psychopharmacology; narcotic dependence; nracotic antagonists. Mailing Add: Miles Inst of Exp Therapeut Apartado Postal 22026 Mexico DF Mexico

VILLARROEL, FERNANDO, b Valparaiso, Chile, Aug 18, 35; US citizen; m 59. BIOMEDICAL ENGINEERING, CHEMICAL ENGINEERING. Educ: Mil Politech Acad, Chile, BS, 60; Univ Md, College Park, MS, 67, PhD(chem eng), 70. Prof Exp: Assoc engr, Nat Instrument Lab, 62-66; proj leader, Biomed Eng Group, Harry Diamond Labs, 67-72; PROG COORDR, ARTIFICIAL KIDNEY-CHRONIC UREMIA PROG, NAT INST ARTHRITIS, METAB & DIGESTIVE DIS, 73- Mem: Am Soc Artificial Internal Organs; Am Inst Chem Engrs. Res: Artificial internal organs. Mailing Add: Bldg 31 Rm 9A07 Nat Inst of Health Bethesda MD 20014

VILLARS, CHARLES EARL, b Tecumseh, Nebr, Oct 28, 24; m 47; c 6. ORGANIC POLYMER CHEMISTRY, INDUSTRIAL ORGANIC CHEMISTRY. Educ: Univ Nebr, BS, 50, MS, 51; Mich State Univ, PhD(org chem), 59. Prof Exp: Res chemist, Britton Res Lab, Dow Chem Co, 51-56 & Benger Res Lab, E I du Pont de Nemours & Co, Inc, Va, 59-64; sr sceintist res & develop lab, Pillsbury Co, 64-69; sr scientist, 69, TECH SPECIALIST, TECH DEPT, FED CARTRIDGE CORP, 69- Mem: Am Chem Soc; NY Acad Sci; Soc Plastic Engrs. Res: Degradeable plastics; water soluble gums. Mailing Add: 1130 Wills Place Minneapolis MN 55422

VILLARS, FELIX MARC HERMANN, b Biel, Switz, Jan 6, 21; nat US; m 49; c 4. THEORETICAL PHYSICS. Educ: Swiss Fed Inst Technol, Dipl, 45, DSc, 46. Prof Exp: Res asst physics, Swiss Fed Inst Technol, 46-49; vis mem, Inst Adv Study, 49-50; res assoc 50-52, from asst prof to assoc prof, 52-60, PROF PHYSICS, MASS INST TECHNOL, 60- Concurrent Pos: Consult, Lincoln Lab, 54-64, 68-70; Guggenheim fel, 56-57; lectr physics, Harvard Med Sch, 74- Mem: Am Phys Soc; Am Acad Arts & Sci. Res: Nuclear physics, mainly nuclear models and reactions; quantum field theory; physics of upper atmosphere; turbulence; plasma probes; biophysics. Mailing Add: Dept of Physics Rm 6-311 Mass Inst of Technol Cambridge MA 02139

VILLEE, CLAUDE ALVIN, JR, b Lancaster, Pa, Feb 9, 17; m 52; c 4. BIOCHEMISTRY. Educ: Franklin & Marshall Col, BS, 37; Univ Calif, PhD(physiol genetics), 41. Hon Degrees: AM, Harvard Univ, 57. Prof Exp: Res assoc zool, Univ Calif, 41-42; from instr to asst prof, Univ NC, 42-45; from instr to assoc prof biol chem, 46-63, PROF BIOL CHEM, HARVARD UNIV, 63-, ANDELOT PROF, 64-, TUTOR PRECLIN SCI, MED SCH, 47- Concurrent Pos: Asst prof, Armstrong Col, 41-42; tech aide, Come Growth, Nat Res Coun, 46; Lalor fel, Marine Biol Lab, Woods Hole, 47 & 48; Guggenheim fel, Denmark, 49-50; res assoc, Boston Lying-in-Hosp, 50-; consult, Mass Gen Hosp, 50- & NSF, 59-; consult, NIH, 58-, mem nat adv child health & human develop coun, 63-65; dir lab reproductive biol, Boston Hosp Women, 66-; mem sci adv comt, Ore Regional Primate Ctr, 70-; distinguished vis prof, Univ Belgrade & Mahidol Univ, Bangkok, 74. Honors & Awards: Ciba Award, Endocrine Soc, 56; Rubin Award, Am Soc Study Sterility, 57. Mem: Hon mem Soc Gynec Invest; hon fel Am Col Obstet & Gynec; hon fel Am Gynec Soc; Am Soc Biol Chemists; Genetic Soc Am. Res: Nucleic acid chemistry and metabolism; carbohydrate metabolism; effects of hormones on intermediary metabolism; function of the placenta; biochemical genetics; metabolism of fetal tissues. Mailing Add: Dept of Biol Chem Harvard Med Sch Boston MA 02115

VILLEE, DOROTHY, b Charleston, SC, Nov 25, 27; m 52; c 4. ENDOCRINOLOGY, BIOCHEMISTRY. Educ: Barnard Col, Columbia Univ, BA, 50; Harvard Med Sch, MD, 55. Prof Exp: Instr pediat, 56-60, res fel biochem, 60-61, assoc pediat, 62-74, ASST PROF PEDIAT, HARVARD MED SCH, 74- Concurrent Pos: Fel, Mass Gen Hosp, 57-59, USPHS res grant, 64- Mem: Endocrine Soc; Am Med Women's Asn. Res: Growth and differentiation of normal and abnormal endocrine tissue and in the factors controlling growth and differentiation. Mailing Add: Children's Hosp Med Ctr 300 Longwood Ave Boston MA 02115

VILLELLA, JOHN BAPTIST, b Walston, Pa, Mar 31, 15; m 59. ZOOLOGY. Educ: Gettysburg Col, AB, 42; Univ Mich, MS, 49, PhD(zool), 54. Prof Exp: Jr chemist, USDA, 46-47; asst radiation biol, AEC, 53-55; AEC res assoc parasitol, Univ Mich, 55-57, res assoc, Phoenix Mem Lab, 57-61; mem fac med zool, Sch Trop Med, Univ PR & PR Nuclear Ctr, 61-66; prof biol, Savannah State Col, 66-71; PROF BIOL, INTER-AM UNIV PR, 71- Mem: Am Soc Parasitol; Am Micros Soc; Am Inst Biol Sci; AAAS. Res: Host-parasite relationships. Mailing Add: Dept of Biol Inter-Am Univ of PR San German PR 00753

VILLEMEZ, CLARENCE LOUIS, JR, b Port Arthur, Tex, Sept 6, 38; m 64; c 4. BIOCHEMISTRY. Educ: Harvard Univ, AB, 58; Purdue Univ, MS, 61, PhD(biochem), 63. Prof Exp: Fel biochem, Purdue Univ, 63-65; asst res biochemist, Univ Calif, Berkeley, 65-66; res assoc biochem, Univ Colo, 66-67; from asst prof to assoc prof, Ohio Univ, 67-72; assoc prof, 72-74, PROF BIOCHEM, UNIV WYO, 74- Mem: Am Soc Biol Chemists; Am Soc Plant Physiologists. Res: Polysaccharide biosynthesis; cell wall formation. Mailing Add: Div of Biochem Univ of Wyo Laramie WY 82071

VILLEMURE, M PAUL JAMES, b Newberry, Mich, Nov 28, 28. MATHEMATICS. Educ: Siena Heights Col, BS, 50; Univ Notre Dame, PhD(math), 58. Prof Exp: Teacher high sch, 58-59; instr math & sci, Col San Antonio, 51-54; from instr math to asst prof math, 59-69, PROF MATH, BARRY COL, 69- Mem: Math Asn Am. Mailing Add: Dept of Math Barry Col Miami FL 33161

VILLENEUVE, PAUL YVON, b Montreal, Que, Oct 24, 43; m 68. URBAN GEOGRAPHY. Educ: Univ Ottawa, BA, 64; Laval Univ, lic geog, 67; Univ Wash, PhD(geog), 71. Prof Exp: ASST PROF GEOG, LAVAL UNIV, 71- Mem: Asn Am Geogrs; Can Asn Geogrs. Res: Minority groups behavior in cities; spatial diffusion and mobility theories; quantitative methods in social urban geography. Mailing Add: Dept of Geog Laval Univ Quebec PQ Can

VILLMOW, JACK R, b Milwaukee, Wis, Jan 20, 26; m 51. GEOGRAPHY. Educ: Univ Wis, PhD, 55. Prof Exp: Asst geog, Univ Wis, 48-51, integrated lib studies, 49-51; instr geog, Wellesley Col, 51-55, asst prof, 55-56; asst prof, Ohio State Univ, 56-59; vis lectr grad schs, Boston & Clark Univs, 59-60; assoc prof, Ohio State Univ, 60-65; assoc prof, Kenosha Ctr, Univ Wis, 65-68; PROF GEOG, NORTHERN ILL UNIV, 68- Concurrent Pos: Visitor, Harvard Russian Ctr, 59-60. Mem: Am Geog Soc; Asn Am Geogr; Nat Coun Geog Educ; Am Meteorol Soc. Res: Meteorology; geography of Europe and the Union of Soviet Socialist Republics. Mailing Add: Dept of Geog Northern Ill Univ De Kalb IL 60115

VILMS, JURI, solid state physics, see 12th edition

VILTER, RICHARD WILLIAM, b Cincinnati, Ohio, Mar 21, 11; m 35; c 1. MEDICINE. Educ: Harvard Univ, BA, 33, MD, 37; Am Bd Internal Med, dipl; Am Bd Nutrit, dipl. Prof Exp: Intern, Cincinnati Gen Hosp, Ohio, 37-38, sr asst resident, 40-41, chief resident, Med Serv, 41-42; from asst prof to assoc prof med, 42-56, asst to dean col med, 43-45, asst dean, 45-52, asst dir dept, 52-56, dir lab hemat & nutrit, 45-56, TAYLOR PROF MED & DIR DEPT INTERNAL MED, COL MED, UNIV CINCINNATI, 56- Concurrent Pos: Fel nutrit, Hillman Hosp, Ala, 39-40; attend physician, Cincinnati Gen Hosp, 45-; Musser lectr, Sch Med, Tulane Univ, 65; chmn hemat study sect, NIH, 65-69. Honors & Awards: Goldberger Award, AMA, 50. Mem: Am Soc Clin Invest; Am Clin & Climat Asn (vpres, 64-65); Asn Am Physicians; fel Am Col Physicians (secy-gen, 74-77); Am Soc Hemat. Res: Hematology; nutrition; refractory and aplastic anemias; megaloblastic anemias; nutritional anemias. Mailing Add: 6067 Col of Med Bldg 231 Bethesda Ave Cincinnati OH 45267

VIMMERSTEDT, JOHN P, b Jamestown, NY, June 5, 31; m 53; c 4. SOIL SCIENCE, FORESTRY. Educ: State Univ NY Col Forestry, Syracuse, BS, 53; Yale Univ, MS, 58, DF, 65. Prof Exp: Res forester, Southeastern Forest Exp Sta, 55-58; asst prof forest soils, 63-68, ASSOC PROF FOREST SOILS, OHIO AGR RES & DEVELOP CTR, 68- Mem: Soil Sci Soc Am; Soc Am Foresters; AAAS; Sigma Xi. Res: Cation exchange properties of plant roots; mineral nutrition of trees; reclamation of spoil banks from coal mining; ecological impacts of forest recreation; soil fauna. Mailing Add: Dept of Forestry Ohio Agr Res & Develop Ctr Wooster OH 44691

VINAL, RICHARD S, b Worcester, Mass, May 5, 38; m 61; c 2. INORGANIC CHEMISTRY. Educ: Bates Col, BS, 60; Cornell Univ, PhD(inorg chem), 65. Prof Exp: Sr res chemist, 65-71, RES ASSOC, EASTMAN KODAK CO, 71- Mem: Am Chem Soc; Soc Photog Scientists & Engrs. Res: Transition metal; coordination chemistry; structural studies; photographic science; redox reactions; non-silver photographic systems development. Mailing Add: Eastman Kodak Co Res Labs 343 State St Rochester NY 14650

VINCE, ROBERT, b Auburn, NY, Nov 20, 40; m 61; c 2. MEDICINAL CHEMISTRY. Educ: Univ Buffalo, BS, 62; State Univ NY Buffalo, PhD(med chem), 66. Prof Exp: Asst prof med chem, Col Pharm, Univ Miss, 66-67; from asst prof to assoc prof, 67-76, PROF MED CHEM, COL PHARM, UNIV MINN, 76- Concurrent Pos: Vis scientist, Roche Inst Molecular Biol, 74-75; Nat Cancer Inst res career develop award, 72-76. Honors & Awards: Lunsford Richardson Grad Res Award, Richarson-Merrell Inc, 66. Mem: Am Chem Soc; Am Pharmaceut Asn; Am Asn Cancer Res. Res: Design and synthesis of inhibitors of protein biosynthesis; nucleoside analogs as cancer chemotherapy agents. Mailing Add: Col of Pharm Univ of Minn Minneapolis MN 55455

VINCENT, DAVID NOEL, polymer chemistry, see 12th edition

VINCENT, DAYTON GEORGE, b Hornell, NY, Apr 23, 36; m 59, 75; c 3. METEOROLOGY. Educ: Univ Rochester, AB, 58; St Louis Univ, dipl meteorol, 59; Univ Okal, MS, 64; Mass Inst Technol, PhD(meteorol), 70. Prof Exp: Weather officer meteorol, Air Weather Serv, US Air Force, 59-62; res asst, Univ Okla Res Inst, 64; res meteorologist, Naval Weapons Lab, Dahlgren, Va, 64-65; res assoc, Mass Inst Technol, 69-70; asst prof atmospheric sci, 70-74, ASSOC PROF ATMOSPHERIC SCI, DEPT GEOSCI, PURDUE UNIV, WEST LAFAYETTE, 74- Concurrent Pos: Mem Trop Meteorol Group to study scale interactions, 73-; Int travel grant, Am Meteorol Soc, 74. Mem: Am Meteorol Soc; Sigma Xi. Res: Case study approach using observed data to dianose kinematic and thermodynamic properties and energy processes during stages in the life cycle of cyclone systems in the tropics and mid-latitudes. Mailing Add: Dept of Geosci Purdue Univ West Lafayette IN 47907

VINCENT, DONALD LESLIE, b St John, NB, CAn, July 23, 21; m 49; c 3. ORGANIC CHEMISTRY. Educ: Acadia Univ, BSc, 42; McGill Univ, PhD(chem), 53. Prof Exp: Instr chem, Acadia Univ, 46-49; asst res officer, Nat Res Coun Can, 53-60; org group leader res lab, Coal Tar Prod Div Dom Tar & Chem Co, Ltd, 60-63; ORGANIC CHEMICALS GROUP LEADER, DOMTAR RES CTR, 63- Mem: Chem Inst Can. Res: Wood chemistry; lignin; carbohydrates; pulp and paper; natural products; organic synthesis. Mailing Add: 153 Douglas Shand Ave Pointe Claire PQ Can

VINCENT, GEORGE PAUL, b Cleveland, Ohio, Oct 20, 01; m 28; c 1. CHEMISTRY. Educ: Hiram Col, AB, 23; Cornell Univ, MS, 24, PhD(phys chem), 27. Prof Exp: Asst chem, Cornell Univ, 23-25; res chemist, Eastman Kodak Co, 27-30; res chemist, Olin Mathieson Chem Corp, NY, 30-33, mgr res lab, 33-34, asst to res dir, 35-38, mgr sales develop, 49-52, mgr sales, Hydrocarbon Div, 52-53, govt serv, 53-67; CONSULT, 67- Concurrent Pos: Fed Govt liaison in food additive fields. Mem: Am Chem Soc; Am Inst Chem Eng. Res: Electrochemistry; sodium chlorite manufacturing methods and uses; inorganic chemistry in alkali-chlorine industry. Mailing Add: Chemists Club 52 E 41st St New York NY 10017

VINCENT, GERALD GLENN, b Winnipeg, Man, Apr 13, 34; m 56; c 4. RESEARCH ADMINISTRATION, POLYMER CHEMISTRY. Educ: Univ Man, BS, 58, MS, 60, PhD(phys chem), 63. Prof Exp: Res chemist adhesion, Dow Chem Co, 63-65, proj mgr adhesives, 65-67; sect leader resins, DeSoto Inc, Des Plaines, 67-69, tech mgr aerospace, 69-72, mgr resin res, 72-73; asst dir res, 73-74, DIR RES, CENT RES, CROWN ZELLERBACH, 74- Mem: Am Chem Soc; Sigma Xi; Tech Asn Pulp & Paper Indust; Indust Res Inst. Res: Adhesion chemistry. Mailing Add: Cent Res Crown Zellerbach Camas WA 98607

VINCENT, GORDON ROSS, b East Orange, NJ, June 2, 24; m 45, 63; c 5. DENTISTRY, DENTAL MATERIALS. Educ: Univ Pa, DDS, 46; Univ Ore, MS, 69. Prof Exp: Pvt pract, Glenside, Pa, 49-65; from asst prof to assoc prof oper dent, Dent Sch, Univ Ore, 67-71; assoc dean dent sch, 71-73, dir dent mat, 73-75, PROF OPER DENT, NJ DENT SCH, COL MED & DENT NJ, 71-, CHMN DENT BIOMAT, 75- Concurrent Pos: Mem, Am Asn Dent Schs, 67- Mem: Am Dent Asn; Int Asn Dent Res; Asn Am Med Cols; Am Asn Higher Educ. Res: Dental casting investments; clinical behavior of restorative materials; postural relations to temporomandibular joint problems; ease of plaque removal from restorative surfaces; intercoronal cavity design; articulation and occlusion by computer analysis. Mailing Add: NJ Dent Sch Col of Med & Dent NJ Newark NJ 07103

VINCENT, HAROLD ARTHUR, b Lake City, Iowa, Jan 19, 30; m 57; c 2. ANALYTICAL CHEMISTRY. Educ: Univ Iowa, BS, 53; Univ Nev, MS, 60; Univ Ariz, PhD(chem), 64. Prof Exp: Chemist, Mining Anal Lab, Univ Nev, Reno, 56-68; CHIEF CHEMIST, ANACONDA CO, 68- Mem: AAAS; Am Chem Soc; Soc Appl Spectros. Res: Electroanalytical chemistry; fast neurtron activation analysis; flame emission and absorption spectroscopy; x-ray emission spectroscopy; thermal methods of analysis. Mailing Add: Gen Mining Div Anaconda Co Tucson AZ 85726

VINCENT, JAMES SIDNEY, b Redlands, Calif, Sept 19, 35; m 69; c 1. PHYSICAL CHEMISTRY. Educ: Univ Redlands, BS, 57; Harvard Univ, PhD(chem), 63. Prof Exp: Fel, Harvard Univ, 63-64 & Calif Inst Technol, 64-65; asst prof chem, Univ Calif, Davis, 65-71; ASSOC PROF CHEM, UNIV MD, BALTIMORE COUNTY, 71- Mem: Am Phys Soc. Res: Electron paramagnetic resonance of triplet state molecules. Mailing Add: Dept of Chem Univ of Md Baltimore County Baltimore MD 21228

VINCENT, JERRY WILLIAM, b Chicago, Ill, June 24, 35; m 59; c 3. PALEONTOLOGY, PALEOECOLOGY. Educ: Univ Maine, Orono, BA, 58; Tex A&M Univ, MEd, 66, PhD(geol), 71. Prof Exp: Teacher geol & biol, N Yarmouth Acad, 58-61; teacher high sch, Maine, 61-67; ASSOC PROF GEOL, STEPHEN F AUSTIN STATE UNIV, 69- Mem: Soc Econ Paleont & Mineral; Nat Asn Geol Teachers. Res: Paleoecology of carbonate rocks, numerical taxonomy of various fossil taxa; biostratigraphy of lower Cretaceous rocks of Texas; invertebrate paleontology; earth science education in elementary and secondary schools. Mailing Add: Dept of Geol Stephen F Austin State Univ Nacogdoches TX 75961

VINCENT, JOSEPH FRANCIS, b Garnsey, Ala, Dec 12, 12; m 40; c 3. BIOCHEMISTRY. Educ: Ala Polytech Inst, BS, 36; Ohio State Univ, MA, 38, PhD(physiol chem), 40. Prof Exp: Asst physiol chem, Ohio State Univ, 38-40; res biochemist, Goodyear Tire & Rubber Co, 40-45; food technologist, Southern Res Inst, 45-46; assoc prof, 46-47, PROF CHEM, GA COL MILLEDGEVILLE, 47-, HEAD DEPT, 49- Mem: Am Chem Soc. Res: Packaging foods; metabolism of fresh foods; toxicity of industrial chemicals; peanut products; peptide synthesis; radiation chemistry. Mailing Add: 620 W Charlton St Milledgeville GA 31061

VINCENT, LLOYD DREXELL, b DeQuincy, La, Jan 7, 24; m 51; c 2. NUCLEAR PHYSICS. Educ: Univ Tex, BS, 52, MA, 53, PhD(physics), 60. Prof Exp: From asst prof to assoc prof, Univ Southwestern La, 53-58; instr, Tex A&M, 55-56; res scientist, Tex Nuclear Corp, 59-60, prof & dir physics dept, Sam Houston State Col, 60-65, asst to pres, 65-67; PRES, ANGELO STATE UNIV, 67- Mem: Am Phys Soc; Am Asn Physics Teachers. Mailing Add: President's Off Angelo State Univ San Angelo TX 76901

VINCENT, MONROE MORTIMER, b Cleveland, Ohio, July 28, 12; m 41; c 1. CELL BIOLOGY, IMMUNOBIOLOGY. Educ: Case Western Reserve Univ, BA, 34. Prof Exp: Res asst parasitol, Univ Chicago, 36-41; vpres, North-Strong Corp, 48-50; vpres, Microbiol Assocs, 51-74; CONSULT, 74- Concurrent Pos: Assoc ed, In Vitro, 70-; mem bd dir, Am Found Biol Res, 74-; exec ed, Tissue Cult manual, 74- Mem: Tissue Culture Asn; Soc Cryobiology (treas, 73-74); Am Soc Cell Biol; Am Soc Trop Med. Res: Cell tissue and organ culture, virology, cell-mediated immunity, cryobiology. Mailing Add: 3905 Jones Bridge Rd Chevy Chase MD 20015

VINCENT, MURIEL C, b Spokane, Wash, Sept 8, 22. PHARMACY, PHARMACEUTICAL CHEMISTRY. Educ: Ore State Col, BS, 44; Univ Wash, MS, 51, PhD, 55. Prof Exp: Instr pharm, Univ Wash, 53-54 & Ore State Col, 54-56; from asst prof to assoc prof, 56-58, chmn dept, 58-74, PROF PHARM, COL PHARM, N DAK STATE UNIV, 58-, ASST DEAN, 65- Concurrent Pos: Consult, Vet Admin Hosp, 60-74. Mem: AAAS; Am Chem Soc; Am Pharmaceut Asn; NY Acad Sci. Res: Application of ion exchange resins and chromatography to the analysis of pharmaceutical products; drug absorption. Mailing Add: NDak State Univ Col of Pharm Fargo ND 58102

VINCENT, PHILLIP G, b East Machias, Maine, July 18, 41. BIOCHEMISTRY, MICROBIOLOGY. Educ: Univ Md, BS, 64, PhD(fungus physiol), 67. Prof Exp: Res asst bot, Univ Md, 64, res fel fungus physiol & biochem, 65-67; res phytopathologist, 67-68, RES MICROBIOLOGIST, FIELD CROPS & ANIMAL PROD RES BR, MKT QUAL RES DIV, AGR RES SERV, USDA, 68- Mem: AAAS; Am Soc Microbiol; Am Phytopath Soc; Scand Soc Plant Physiol. Res: Fungus physiology and biochemistry; mechanism of action of toxicants; phytopathology; plant physiology; biochemical characteristics and mechanisms of quality deterioration of meat. Mailing Add: Agr Mkt Res Inst Bldg 006 Agr Res Ctr Beltsville MD 20705

VINCENT, ROBERT CORBIN, b Maine, NY, Sept 17, 12. ANALYTICAL CHEMISTRY. Educ: Cornell Univ, AB, 35, MA, 37, PhD(inorg chem), 40. Prof Exp: Asst chem, Cornell Univ, 36-40; instr, 40-41, asst prof anal chem, 46-48, assoc prof, 48-60, PROF CHEM, GEORGE WASHINGTON UNIV, 60- Mem: Am Chem Soc. Res: Sulfur dioxide as a volatile component in binary systems. Mailing Add: Dept of Chem George Washington Univ Washington DC 20006

VINCENT, WALTER SAMPSON, JR, b Veneta, Ore, Aug 6, 21; m 42; c 3. ZOOLOGY. Educ: Ore State Col, BS, 46, MS, 48; Univ Pa, PhD(zool), 52. Prof Exp: Asst zool, Ore State Col, 46; asst instr, Univ Pa, 48; res assoc genetics, Iowa State Col, 51-52; from instr to asst prof anat, Col Med, State Univ NY Upstate Med Ctr, 52-61; assoc prof, Sch Med, Univ Pittsburgh, 61-69; vis prof, Dept Biol, Brooklyn Col, 69-70; PROF BIOL SCI & CHMN DEPT, UNIV DEL, 71- Concurrent Pos: USPHS sr res fel, 59-61, career develop award, 61-63; Lalor fel, Marine Biol Lab, Woods Hole, 55, trustee, 66-, mem exec comt, 71-, mem corp; res prof, Univ Edinburgh, 64, Sci Res Coun sr vis res fel, 67; mem exec comt, Marine Biol Lab, 71-75. Mem: AAAS; Am Soc Cell Biol; Genetics Soc Am; Soc Gen Physiol; Int Inst Embryol. Res: Chemistry and function of oocytes; biosynthesis of ribosomes; evolution of ribosomal genes. Mailing Add: Dept of Biol Sci Univ of Del Newark DE 19711

VINCENT, WILLIAM FRANKLIN, clinical chemistry, laboratory quality control, see 12th edition

VINCENZ, STANISLAW ALEKSANDER, b Oskrzesince, Poland, Feb 4, 15; m 49; c 1. GEOPHYSICS. Educ: Univ London, ARCS & BSc, 37, DIC, 39, PhD(geophys), 52. Prof Exp: Demonstr geophys, Imp Col, London, 48-49, asst lectr, 49-51, res asst, 51-53; geophysicist & head geophys div, Jamaica Indust Develop, Corp, WI, 53-61; assoc prof geophys & geophys eng, 61-67, PROF GEOPHYS & GEOPHYS ENG, ST LOUIS UNIV, 67- Concurrent Pos: NSF prin investr grants, St Louis Univ, 62-67, 67-68, 69-70 & 71-73; prin investr, NSF, 74-75 & US Geol Surv, 73-76. Mem: AAAS; Soc Explor Geophys; Am Geophys Union; fel Royal Astron Soc; Europ Asn Explor Geophys. Res: Exploration geophysics; rock magnetism and paleomagnetism; geomagnetism. Mailing Add: Dept of Earth & Atmospheric Sci St Louis Univ PO Box 8099 Laclede Sta St Louis MO 63156

VINCENZI, FRANK FOSTER, b Seattle, Wash, Mar 14, 38; m 60; c 2. PHARMACOLOGY. Educ: Univ Wash, BS, 60, MS, 62, PhD(pharmacol), 65. Prof Exp: NSF fel, Berne, 65-67; asst prof, 67-72, ASSOC PROF PHARMACOL, SCHS MED & PHARM, UNIV WASH, 72- Mem: AAAS; NY Acad Sci; Am Soc Pharmacol & Exp Therapeut; Cardiac Muscle Soc. Res: Autonomic transmitters; mechanisms of cardioactive drugs; membrane transport; red blood cell physiology and pathology. Mailing Add: Dept of Pharmacol Univ of Wash Seattle WA 98195

VINCIGUERRA, MICHAEL JOSEPH, b New York, NY, Mar 19, 45; m 70; c 1. PHYSICAL CHEMISTRY. Educ: Iona Col, BS, 66; Adelphi Univ, MS, 69, PhD(phys chem), 71. Prof Exp: Asst prof, 70-75, ASSOC PROF CHEM, STATE UNIV NY AGR & TECH COL FARMINGDALE, 75- Concurrent Pos: Res asst, Adelphi Univ, 71-72; adj asst prof, St John's Univ, NY, 72; res consult, Unichem Res Assoc, 75- Mem: Am Chem Soc; NY Acad Sci. Res: Physical properties of bio-polymers; light scattering by biological gels; nuclear magnetic relaxation of polymer solutions. Mailing Add: Dept of Chem State Univ of NY Agr & Tech Col Farmingdale NY 11735

VINCOW, GERSHON, b New York, NY, Feb 27, 35; m 64; c 2. PHYSICAL CHEMISTRY. Educ: Columbia Univ, AB, 56, MA, 57, PhD(chem), 59. Prof Exp: Fel, Hebrew Univ, Israel, 60; NSF fel, Calif Inst Technol, 60-61; from asst prof to prof chem, Univ Wash, 61-71; CHMN DEPT CHEM, SYRACUSE UNIV, 71- Concurrent Pos: Sloan Found res fel, 64-67; NSF fel, Harvard Univ, 70-71. Mem: Am Chem Soc; Am Phys Soc; AAAS. Res: Electron paramagnetic resonance spectroscopy. Mailing Add: Dept of Chem Syracuse Univ 108 Bowne Hall Syracuse NY 13210

VINCZE, LAJOS, b Ghiorac, Rumania, Jan 26, 20; US citizen; m 59; c 2. ANTHROPOLOGY. Educ: Univ Hungary, Philos D, 44; Bolyae Univ, Rumania, Lycee prof dipl, 45. Hon Degrees: PhD, Univ Hungary, 44. Prof Exp: Asst lectr ethnog, Univ Hungary, 43-45; prof Lycee, State Lycee, Salonta, Rumania, 45-47; teacher Span & Fr, Millersburg Mil Inst, 64-66; asst prof Span, Fr & antr- op, Morehead State Univ, 66-68; asst prof anthrop, 68-72, ASSOC PROF ANTHROP, BOWLING GREEN STATE UNIV, 72- Concurrent Pos: Researcher, Transylvanian Sci Inst, 42-45; grant, Bolyae Univ, Rumania, 45-46; lectr, Free Univ, Hungarian Sci Acad, Buenos Aires, 52-58; exchange scientist, Nat Acad Sci, 71. Mem: Fel Am Anthrop Asn. Res: Peasant societies; ethnolinguistics. Mailing Add: Dept of Sociol Bowling Green State Univ Bowling Green OH 43403

VIND, HAROLD PENNINGTON, b Huron, SDak, May 5, 15; m 67; c 2. ENVIRONMENTAL CHEMISTRY. Educ: Univ Minn, BS, 39; Univ Southern Calif, MS, 50, PhD(biochem), 55. Prof Exp: Res chemist northern regional lab, USDA, Ill, 41-47; biochemist, Vet Admin Hosp, Long Beach, Calif, 49-52; SR PROJ SCIENTIST, US NAVY, CIVIL ENG LAB, 55- Mem: Fel AAAS; fel Am Inst Chem; Am Chem Soc; Marine Technol Soc. Res: Biodegradation of pesticides and construction materials in the ocean; marine corrosion; marine materials; environmental hazards of construction materials; detection and removal of hazardous lead-based paint; environmental chemistry. Mailing Add: Code L 52 Mat Sci Div Civil Eng Lab Naval Construct Port Hueneme CA 93043

VINEGAR, RALPH, b New York, NY, June 28, 24; m 48; c 2. PHARMACOLOGY. Educ: NY Univ, BA, 49, MS, 50; Cornell Univ, PhD(biol growth, pharm), 57. Prof Exp: Head dept cytol, Wallace Labs, Inc, 57-65; SR PHARMACOLOGIST, WELLCOME RES LABS, 65- Res: Chemical and phsyical effects on malignant, hematological and reticuloendothelial cells; development of edema and inflammation; mechanism of action of anti-inflammatory and analgesic drugs. Mailing Add: Lab 1231 Dept Pharmacol Wellcome Res Labs Rsearch Triangle Park NC 27709

VINES, HERBERT MAX, b Ala, Feb 4, 18; m 42; c 2. PLANT PHYSIOLOGY. Educ: Ala Polytech Inst, BS, 40; Univ Calif, MS, 49; Univ Calif, Los Angeles, PhD, 59. Prof Exp: Specialist postharvest physiol, Univ Calif, 49-53; tech rep nutrit, Shell Chem Co, 53-56; technician plant biochem, Univ Calif, 57-59, fel, 59-60; assoc biochemist, Citrus Exp Sta, Univ Fla, 61-67; PROF HORT, UNIV GA, 67- Mem: Am Soc Hort Sci; Am Soc Plant Physiol. Res: Post-harvest physiology; plant metabolism, especially metabolic blocks in electron transport system. Mailing Add: Dept of Hort Plant Sci Bldg Univ of Ga Athens GA 30601

VINEYARD, BILLY DALE, b Clarkton, Mo, Sept 7, 31; m 56; c 1. ORGANIC CHEMISTRY. Educ: Southeast Mo State Col, BS, 53; Univ Mo, PhD(org chem), 59. Prof Exp: Res chemist, Celanese Corp, 59-60; sr res chemist, Org Div, 60-64, res specialist, 64-68, SCI FEL, MONSANTO CO, 68- Mem: Am Chem Soc. Res: Catalytic homogeneous asymmetric hydrogenation. Mailing Add: Monsanto Co 800 N Lindbergh Blvd St Louis MO 63166

VINEYARD, GEORGE HOAGLAND, b St Joseph, Mo, Apr 28, 20; m 45; c 2. PHYSICS. Educ: Mass Inst Technol, BS, 41, PhD(physics), 43. Prof Exp: Mem staff, Radiation Lab, Mass Inst Technol, 42-45; from asst prof to prof physics, Univ Mo,

46-54; from physicist to sr physicist, 54-61, chmn dept physics, 61-66, assoc dir, 66-67, dep dir, 67-72, DIR, BROOKHAVEN NAT LAB, 73- Concurrent Pos: Mem solid state sci comt, Nat Res Coun, 65; mem, Advan Res Proj Agency, Mat Res Coun, 67-; mem math & phys sci adv comt, NSF, 68-71. Mem: Fel Am Phys Soc. Res: Crystal growth; phase transitions; magnetron, klystrons and other microwave devices; theory of x-ray and neutron scattering; structure of liquids; radiation effect; solid state physics. Mailing Add: Brookhaven Nat Lab Upton NY 11973

VINGE, CLARENCE L, b Ishpeming, Mich, Dec 30, 15; m 43; c 2. ECONOMIC GEOGRAPHY, HISTORICAL GEOGRAPHY. Educ: Northern Mich Univ, BA, 38; Univ Wis, PhD(geog), 46. Prof Exp: Bus economist geog, War Prod Bd, 41-42; assoc prof, Southern Ill Univ, 46-47; from asst prof to prof, 47-76, EMER PROF GEOG, MICH STATE UNIV, 76- Mem: Asn Am Geog; Am Geog Soc; Nat Coun Geog Educ. Res: Economic geography, particularly influence of government. Mailing Add: 4115 Hulett Rd Okemos MI 48864

VINGIELLO, FRANK ANTHONY, b New York, NY, Aug 20, 21; m 42; c 3. CHEMISTRY. Educ: Polytech Inst Brooklyn, BS, 42; Duke Univ, PhD(org chem), 47. Prof Exp: Lab instr org chem, Polytech Inst Brooklyn, 43-44; lab instr, Duke Univ, 44-47; instr, Univ Pittsburgh, 47; res assoc, Northwestern Univ, 47-48; from asst prof chem to assoc prof chem, Va Polytech Inst & State Univ, 48-57, prof org chem, 57-68; PROF ORG CHEM, NORTHEAST LA UNIV, 68- Concurrent Pos: Chemist, WVa Ord Works, 42-43; consult chem indust. Honors & Awards: J Shelton Horsley Award, Va Acad Sci, 66. Mem: Am Chem Soc; NY Acad Sci. Res: Organic synthesis in steroids and aromatic molecules; mechanisms of organic reactions; cyclization of o-benzylphenones; synthesis of aromatic hydrocarbons, research in air pollution. Mailing Add: Dept of Chem Northeast La Univ Monroe LA 71201

VINH, NGUYEN XUAN, b Yenbay, Viet Nam, Jan 3, 30; m 55; c 4. CELESTIAL MECHANICS. Educ: Air Inst, France, BS, 53; Univ Marseille, MS, 54; Univ Colo, MS, 63, PhD(aerospace), 65; Univ Paris, DSc(math), 72. Prof Exp: Asst prof aerospace eng, Univ Colo, 65-68; assoc prof, 68-72, PROF AEROSPACE ENG, UNIV MICH, ANN ARBOR, 72- Concurrent Pos: Guest lectr, US Air Force Acad, 66; vis lectr, US Air Force Acad, 66; vis lectr, Univ Calif, Berkeley, 67. Mem: Math Asn Am. Res: Ordinary differential equations; astrodynamics and optimization of space flight trajectories; theory of non linear oscillations. Mailing Add: Dept of Aerospace Eng Univ of Mich Ann Arbor MI 48104

VINICK, FREDRIC JAMES, b Amsterdam, NY, June 18, 47; m 70; c 1. SYNTHETIC ORGANIC CHEMISTRY. Educ: Williams Col, BA, 69; Yale Univ, PhD(chem), 73. Prof Exp: NIH res fel, Columbia Univ, 73-75; SR SCIENTIST ORG CHEM, PHARMACEUT DIV, CIBA-GEIGY CORP, 75- Mem: Am Chem Soc. Res: Synthesis of biologically and/or medicinally important compounds; development of new synthetic methods.

VINING, LEO CHARLES, b Whangarei, NZ, Mar 28, 25; m 53; c 4. BIO-ORGANIC CHEMISTRY. Educ: Univ NZ, BSc, 48, MSc, 49; Cambridge Univ, PhD, 51. Prof Exp: Scholar, Univ Kiel, Ger, 51-53; fel, Rutgers Univ, 53-54, instr, Inst Microbiol, 54-55; asst res off, Prairie Regional Lab, Nat Res Coun Can, 55-58, assoc res off, 58-62, sr res off, Atlantic Regional Lab, 62-69, prin res off, 69-71; PROF BIOL, DALHOUSIE UNIV, 71- Concurrent Pos: Merck, Sharpe & Dohme lectr, 65. Honors & Awards: Harrison Prize, Royal Soc Can, 72; Can Soc Microbiologists Award, 76. Mem: Fel The Chem Soc; fel Can Inst Chem; Can Soc Microbiol; Am Soc Microbiol; fel Royal Soc Can. Res: Chemistry of antibiotics; fungal metabolites; biosynthesis of natural products; control of secondary metabolism. Mailing Add: Dept of Biol Dalhousie Univ Halifax NS Can

VINJE, MARY M, b Madison, Wis, Sept 8, 13; m 38. BIOLOGY. Educ: Univ Wis, PhD(bot), 38. Prof Exp: Asst bot, Univ Wis, 35-38; teacher high schs, 45-48; from asst prof to assoc prof biol, 48-58, PROF BIOL, ST AMBROSE COL, 58-, HEAD DEPT, 74- Concurrent Pos: NSF vis scientist, Iowa High Schs, 60-66. Mem: AAAS; Nat Asn Biol Teachers; Nat Sci Teachers Asn; Bot Soc Am; Sigma Xi. Res: Aerobiology, particularly the investigation of microbiota in atmosphere; isolation of a fungus from an ozone meter; origin of blue rain. Mailing Add: Dept of Biol St Ambrose Col Davenport IA 52803

VINOCUR, MYRON, b Detroit, Mich, Feb 27, 31; m 58; c 2. MEDICINE, PATHOLOGY. Educ: Univ Mich, AB, 53, MD, 56. Prof Exp: Intern, Harper Hosp, 56-57; resident path, Wayne State Univ, 57-61; pathologist, Lederle Labs, Am Cyanamid Co, 63-66, head dept path, 66-68; head exp path, Geigy Pharmaceut, 68-72; DIR PATH & TOXICOL & ASSOC DIR CLIN RES, BOEHRINGER INGELHEIM LTD, ELMSFORD, 72- Mem: Soc Toxicol; Am Soc Clin Path; Int Acad Path. Res: Anatomical, clinical and experimental pathology; evaluation of drug induced morphological findings; development, preparation and monitoring clinical trials for new drug development, preclinical animal toxicology and pathology studies. Mailing Add: PO Box 356 Monsey NY 10952

VINOGRAD, JEROME, b Milwaukee, Wis, Feb 9, 13; m 37; c 2. MOLECULAR BIOLOGY, BIOLOCIAL CHEMISTRY. Educ: Univ Calif, Los Angeles, MA, 37; Stanford Univ, PhD(chem), 50. Prof Exp: Res assoc, Stanford Univ, 39-41; chemist, Shell Develop Co, 41-49; sr res fel phys chem, 51-57, res assoc chem, 57-65, PROF CHEM & BIOL, CALIF INST TECHNOL, 65- Concurrent Pos: Consult, Beckman Instruments-Spinco Div, 63-; Nat Cancer Inst working group for molecular control comt, 72-; Jesse W Beams lectr, Univ Va, 72; Falk-Plaut lectr, Columbia Univ, 72. Honors & Awards: Am Chem Soc-Kendall Co Award Colloid & Surface Chem, 70; T Duckett Jones Award, 72. Mem: Nat Acad Sci; AAAS; Am Chem Soc; The Chem Soc; Am Soc Biol Chem. Res: Physical biochemistry; chemistry of circular DNA; replication of mitochondrial DNA; theory and application of the ultracentrifuge; virology. Mailing Add: Norman W Church Lab Div of Chem & Chem Eng Calif Inst Tech 1201 E Calif Blvd Pasadena CA 91109

VINOGRADE, BERNARD, b Chicago, Ill, May 7, 15; m 42; c 4. MATHEMATICS. Educ: City Col, BS, 37; Univ Mich, MA, 40, PhD(math), 42. Prof Exp: Instr math, Univ Wis, 42-44 & Tulane Univ, 44-45; staff mem, Radiation Lab, Mass Inst Technol, 45; from asst prof to assoc prof math, 45-55, chmn dept, 61-64, actg head, 60, PROF MATH, IOWA STATE UNIV, 55- Concurrent Pos: Opers analyst, Standby Unit, US Air Force, 50-; vis prof, San Diego State Col, 59-60 & City Col New York, 64-65. Mem: Am Math Soc; Math Asn Am. Res: Abstract and linear algebra; field theory; forest resouce management. Mailing Add: Dept of Math Iowa State Univ Ames IA 50010

VINOGRADOV, SERGE, b Beirut, Lebanon, Aug 27, 33; US citizen; m; c 2. BIOCHEMISTRY. Educ: Am Univ Beirut, BA, 52, MA, 54; Ill Inst Technol, PhD(phys chem), 59. Prof Exp: Fel, Univ Alta, 59-62; res assoc, Yale Univ, 62-66; asst prof biochem, 66-68, assoc prof, 68-71, PROF BIOCHEM & ADJ ASSOC PROF BIOL, WAYNE STATE UNIV, 71- Mem: Am Soc Biol Chem; The Chem Soc; Am Chem Soc; Biophys Soc. Mailing Add: Dept of Biochem Wayne State Univ Sch of Med Detroit MI 48201

VINOPAL, ROBERT THOMAS, b Titusville, Pa. MICROBIAL PHYSIOLOGY. Educ: Harvard Univ, AB, 63; Univ Calif, Davis, PhD(microbiol), 72. Prof Exp: ASST PROF BIOL, UNIV CONN, STORRS, 73- Mem: Am Soc Microbiol. Res: Bacterial physiology and genetics; molecular basis of adaptation. Mailing Add: Univ of Conn Box U-44 Storrs CT 06268

VINSON, DAVID BERWICK, b Houston, Tex, Oct 7, 17; m 40; c 1. PSYCHOPHYSIOLOGY. Educ: Univ Calif, Los Angeles, BA, 41; Univ London, PhD, 52. Prof Exp: Asst psychol, Baylor Col Med, 48-50, asst prof, 52-54; res psychologist, Inst Psychiat, Univ London, 50-52; CONSULT LIFE SCI, 54-; DIR, TEX ACAD ADVAN LIFE SCI, 60-; PRES, ASSESSMENT SYSTS INC, 67- Concurrent Pos: Res psychologist, Med Br, Univ Tex, 48-50; clin psychologist, Vet Admin Regional Off, 48-50; consult, Neurosurg Unit, Netherne Hosp, London, Eng, 50-52; mem 5th Int Cong Geront & 13th & 16th Int Cong Psychol. Mem: Aerospace Med Asn; Am Inst Aeronaut & Astronaut; Am Psychol Asn; Inst Elec & Electronics Eng; Brit Psychol Soc. Res: Man-machine systems; biomedical sciences. Mailing Add: 1107 Fannin Bank Bldg Houston TX 77025

VINSON, JAMES S, b Chambersburg, Pa, May 17, 41; m 67. PHYSICS. Educ: Gettysburg Col, BA, 63; Univ Va, MS, 65, PhD(physics), 67. Prof Exp: Res asst low temperature physics, Univ Va, 64-67; asst prof physics, MacMurray Col, 69-71; assoc prof, 71-75, PROF PHYSICS, UNIV NC, ASHEVILLE, 75-, CHMN DEPT, 71-, DIR COMPUT CTR, 74- Mem: AAAS; Am Phys Soc; Am Asn Physics Teachers; Nat Sci Teachers Asn; World Future Soc. Res: Low temperature physics; scintillations in liquid helium; quantum mechanics; computer based instructions; future studies. Mailing Add: Dept of Physics Univ of NC Asheville NC 28804

VINSON, JOE ALLEN, b Ft Smith, Ark, Nov 16, 41; m 66; c 1. ANALYTICAL CHEMISTRY. Educ: Univ Calif, Berkeley, BS, 63; Iowa State Univ, MS, 66, PhD(org & anal chem), 67. Prof Exp: Res asst, Anal Chem Sect, Ames Lab, AEC, 66-67; asst prof chem, Shippensburg State Col, 67-68 & Washington & Jefferson Col, 68-72; mem staff, J T Baker Chem Co, 72, prod develop chemist, 72-74; ASST PROF CHEM, UNIV SCRANTON, 74- Concurrent Pos: Res Corp Cottrell grant, 69-70; Law Enforcement Assistance Admin grant, 71-72. Mem: Am Chem Soc; NY Acad Sci. Res: Use of dipolar aprotic solvents in organic and anlytical chemistry; clinical, drug and pollution analysis; thin layer chromatography; analysis of marijuana in biological fluids. Mailing Add: Dept of Chem Univ of Scranton Scranton PA 18072

VINSON, JOHN WILLIAM, b Tampa, Fla, Apr 15, 16. RICKETTSIAL DISEASES, VENEREAL DISEASES. Educ: Duke Univ, BS, 40; Harvard Univ, SDHyg, 58; Am Bd Microbiol, Dipl, 65. Prof Exp: Head virol lab, Chas Pfizer & Co, Inc, NY, 48-52; asst microbiol, 55-58, res fel, 58-61, res assoc, 61-67, sr res assoc, 67-69, ASSOC PROF MICROBIOL, SCH PUB HEALTH, HARVARD UNIV, 69- Honors & Awards: Hans Zinsser Mem Award, 64. Mem: AAAS; Am Soc Microbiol; Med Soc Study Venereal Dis; Am Venereal Dis Asn. Res: Trench fever, typhus fever and tsutsugamushi disease; sexually transmitted diseases; rickettsiae-host cell relationships; Rocky Mountain spotted fever. Mailing Add: Dept of Microbiol 665 Huntington Ave Harvard Univ Sch of Pub Health Boston MA 02115

VINSON, LEONARD J, b New York, NY, May 29, 15; m 42; c 5. BIOCHEMISTRY. Educ: Brooklyn Col, AB, 36; Fordham Univ, MS, 37, PhD(biochem), 43. Prof Exp: Nutrit Found fel, Fordham Univ, 44; supvr biochem res, Armour Res Found, 45-49; sect chief biochem, 49-65, res mgr, Biol Dept, 65-74, DIR BIOMED AFFAIRS, LEVER BROS CO, 74- Concurrent Pos: Lectr, NY Univ Med Ctr, 56- Mem: AAAS; Soc Toxicol; Am Chem Soc; Sigma Xi; Soc Invest Dermat. Res: Medical, regulatory and clinical developments dealing with human and environmental health and safety. Mailing Add: 179 Rodney St Glen Rock NJ 07452

VINSON, RICHARD G, b Prattville, Ala, Nov 18, 31; m 55; c 2. MATHEMATICS. Educ: Huntingdon Col, BA, 54; Fla State Univ, MA, 56; Univ Ala, PhD(math), 62. Prof Exp: Instr math, Fla State Univ, 55-56, Univ Tenn, 56-58 & Univ Ala, 58-61; prof, Huntingdon Col, 61-69; PROF MATH, UNIV S ALA, 69- Concurrent Pos: Instr state-wide educ TV network, 63-67; dir NSF two-yr col prog. Mem: AAAS; Am Math Soc; Math Asn Am. Res: Non-Euclidean geometry. Mailing Add: Dept of Math Univ of SAla Mobile AL 36688

VINSON, S BRADLEIGH, b Mansfield, Ohio, Apr 8, 38; m 60; c 2. ENTOMOLOGY. Educ: Ohio State Univ, BS, 61; Miss State Univ, MS, 63, PhD(entom), 65. Prof Exp: Res asst entom, Miss State Univ, 64-65, asst prof, 65-69; assoc prof, 69-75, PROF ENTOM, TEX A&M UNIV, 75- Mem: AAAS; Entom Soc Am; Am Inst Biol Sci; Am Soc Zool. Res: Vertebrate insecticide resistance; mechanisms of arthropod resistance; insect physiology; parasite-host and predator-host relationships. Mailing Add: Dept of Entom Tex A&M Univ College Station TX 77843

VINSON, WILLIAM ELLIS, b Greensboro, NC, Apr 4, 43; m 63; c 2. POPULATION GENETICS, DAIRY SCIENCE. Educ: NC State Univ, BS, 65, MS, 68; Iowa State Univ, PhD(pop genetics), 71. Prof Exp: ASST PROF DAIRY CATTLE GENETICS, DEPT DAIRY SCI, VA POLYTECH INST & STATE UNIV, 71- Mem: Am Dairy Sci Asn; Am Genetic Asn; Am Soc Animal Sci; Biomet Soc; Sigma Xi. Res: Direct and correlated responses to selection; pedigree evaluation of genetic merit; inheritance of discontinuous characters; genetic evaluations from field data; computer simulation of genetic populations. Mailing Add: Dept of Dairy Sci Va Polytech Inst & State Univ Blacksburg VA 24061

VINSONHALER, CHARLES I, b Winfield, Kans, Mar 29, 42. MATHEMATICS. Educ: Calif Inst Technol, BS, 64; Univ Wash, PhD(math), 68. Prof Exp: ASST PROF MATH, UNIV CONN, 68- Mem: Am Math Soc. Res: Ring theory and Abelian groups. Mailing Add: Dept of Math Univ of Conn Storrs CT 06268

VINT, LARRY FRANCIS, b Davenport, Iowa, May 12, 41; m 64; c 3. ANIMAL BREEDING, POULTRY BREEDING. Educ: Iowa State Univ, BS, 63, MS, 69, PhD(animal breeding), 71. Prof Exp: Data processing mgr biomet, Pilch-De Kalb, De Kalb AgRes, 71-72, dir res poultry breeding, 72-73, res investr corp develop, De Kalb AgRes, 73-74; geneticist animal breeding, USDA, 74; ASST DIR RES POULTRY BREEDING, DE KALB AgRES, 74- Mem: Am Soc Animal Sci; Poultry Sci Asn; Coun Agr Sci & Technol. Res: Genetic improvement in poultry populations. Mailing Add: De Kalb AgRes Inc Sycamore Rd De Kalb IL 60115

VINTI, JOHN PASCAL, b Newport, RI, Jan 16, 07. CELESTIAL MECHANICS. Educ: Mass Inst Technol, SB, 27, ScD(physics), 32. Prof Exp: Harrison res fel physics, Univ Pa, 32-34; res asst, Mass Inst Technol, 34-35; instr, Brown Univ, 36-37; asst prof, The Citadel, 37-38; instr, Worcester Polytech Inst, 39-41; physicist ballistic res labs, Aberdeen Proving Ground, 41-57 & Nat Bur Stand, 57-65; prof appl math, NC State Univ, 66; consult, Exp Astron Lab, 67-70, vis assoc prof aeronaut & astronaut, Inst, 69-70, LECTR AERONAUT & ASTRONAUT, MASS INST TECHNOL, 71-, CONSULT, MEASUREMENT SYSTS LAB, 70- Concurrent Pos: Lectr, Univ Del, 48-49 & 54-55 & Univ Md, 50-52; consult, Army Chem Ctr, 52; prof lectr, Georgetown Univ, 63-64; adj prof, Cath Univ Am, 65-66; mem, Comn Celestial

Mech, Int Astron Union. Honors & Awards: Cert Award, Nat Bur Stand, 61. Mem: Fel AAAS; fel Am Phys Soc; assoc fel Am Inst Aeronaut & Astronaut; Am Astron Soc; Am Geophys Union. Res: Possible effects of variation of the gravitational constant, both in dynamical astronomy and in cosmology. Mailing Add: Measurement Systs Lab Rm W91-202 Mass Inst Technol Cambridge MA 02139

VINTON, PAUL WESLEY, b Geneva, NY, Oct 8, 10; m 45; c 1. DENTISTRY. Educ: Syracuse Univ, BS, 34; Univ Ala, AB, 39, BS & MA, 40; Tufts Univ, DMD, 47. Prof Exp: Instr chem, Univ Ala, 37-40; prin high sch, NY, 40; asst prof biol, Siena Col, 40-44, asst registr, 41, asst dean, 41-44; physician pvt pract, 47-48; from instr to assoc prof prosthodontics, 48-52, dir preclin studies, 51-52; assoc prof, Univ NC, 52-55; from assoc prof to prof, Univ NC, 52-56, head dept, 55-56; PROF PROSTHODONTICS & HEAD DEPT, NJ DENT SCH, COL MED & DENT NJ, 56- Concurrent Pos: Consult, Regional Off, Vet Admin, NC, 55- & Vet Admin Hosp, Brooklyn, NY, 60- Mem: AAAS; Am Dent Asn; fel Am Col Dent; fel Am Geriat Soc; Int Asn Dent Res. Res: Dental prosthetics; gerontology; geriatrics. Mailing Add: Dept of Prosthodontics NJ Dent Sch Col Med & Dent NJ Jersey City NJ 07304

VINTON, WILLIAM HOWELLS, b Columbus, Ohio, Dec 26, 18; m 57; c 3. ORGANIC CHEMISTRY, INORGANIC CHEMISTRY. Educ: Yale Univ, BA, 40, PhD(org chem), 44. Prof Exp: Res chemist, E I du Pont de Nemours & Co, Del, 43-46, group leader org chem, photo prod dept, NJ, 46-50, res mgr, 50-54, chief supvr plant tech dept, 54-55, mgr new prod develop, 55-58, asst mgr graphic arts prod, 59-60, patent specialist, Int Dept, 60-64, sr chemist, Org Chem Dept, 64-67, patent specialist, Develop Dept, 67-69; PROF CHEM, CHEYNEY STATE COL, 69- Concurrent Pos: Pres, Trade Winds of Vero Beach Inc, Fla, 69-71. Mem: Soc Motion Picture & TV Eng; Am Chem Soc. Res: Amino acid synthesis; sulfur and photographic chemistry; decontamination of war gases; anti-malarials; color developers; dyestuffs and dyeing; patents. Mailing Add: Dept of Chem Cheyney State Col Cheyney PA 19319

VINYARD, WILLIAM CORWIN, b McArthur, Calif, Apr 30, 22; m 60. BOTANY. Educ: Chico State Col, BA, 42; Mich State Univ, MS, 51, PhD(bot), 58. Prof Exp: Instr bot, Univ Okla, 53-54; instr, Mich State Univ, 55-56; instr, Univ Mont, 56-57; asst prof, Univ Kans, 57-58; from asst prof to assoc prof, 58-72, PROF BOT, CALIF STATE UNIV, HUMBOLDT, 72- Concurrent Pos: Algological consult, Calif State Dept Water Resources, 62-64 & Klamath Basin Study, US Dept Interior, 67- Mem: AAAS; Am Micros Soc; Phycol Soc Am; Int Asn Plant Taxon; Int Phycol Soc. Res: Taxonomy and ecology of freshwater algae; synopsis of the desmids of North America; algae of western North America; relation of algae to water pollution; algal food of herbivorous fishes; eipzoic algae. Mailing Add: Dept of Biol Calif State Univ Humboldt Arcata CA 95521

VIOHL, PAUL, b Charleston, SC, Dec 12, 13; m 40; c 2. ORGANIC CHEMISTRY. Educ: Col Charleston, BS, 35; Johns Hopkins Univ, PhD(chem), 39. Prof Exp: Lab asst chem, Johns Hopkins Univ, 36-39; res chemist, US Rubber Co, 39-46, asst develop mgr, 46-50, sr res scientist, Uniroyal Inc, 50-70; CONSULT, 70- Mem: Am Chem Soc. Res: Synthetic and physical organic chemistry; rubber technology. Mailing Add: 134 N Muhlenberg St Woodstock VA 22664

VIOLA, ALFRED, b Vienna, Austria, July 8, 28; nat US; m 63. ORGANIC CHEMISTRY. Educ: Johns Hopkins Univ, BA, 49, MA, 50; Univ Md, PhD(chem), 55. Prof Exp: Asst instr chem, Johns Hopkins Univ, 49-50; asst, Univ Md, 50-54; res assoc, Boston Univ, 55-57; from asst prof to assoc prof chem, 57-68, PROF CHEM, NORTHEASTERN UNIV, 68- Mem: Am Chem Soc; Sigma Xi. Res: Preparation and properties of unsaturated organic compounds; thermal rearrangements; stereochemistry; pericyclic reactions of acetylenes. Mailing Add: Dept of Chem Northeastern Univ 360 Huntington Ave Boston MA 02115

VIOLA, HERMAN JOSEPH, b Chicago, Ill, Feb 24, 38; m 64; c 3. ANTHROPOLOGY. Educ: Marquette Univ, BS, 60, MA, 64; Ind Univ, PhD(hist, anthrop), 70. Prof Exp: Ed Prologue, Nat Archives & Records Serv, 68-72; DIR NAT ANTHROP ARCHIVES, SMITHSONIAN INST, 72- Concurrent Pos: Am Philos Soc grant, 71. Mem: Soc Am Arch; Orgn Am Historians; Am Hist Asn. Res: Federal American Indian policy, precivil war; portraiture of American Indians, primarily Charles Bird King and Henry Inman. Mailing Add: Nat Anthrop Archives Smithsonian Inst Washington DC 20560

VIOLA, JOHN THOMAS, b Haverhill, Mass, Mar 6, 38; m 60; c 2. PHYSICAL CHEMISTRY. Educ: Univ NH, BS, 60; Pa State Univ, MS, 61; Mass Inst Technol, PhD(chem), 67. Prof Exp: US Air Force, 61-, nuclear effects res officer thermodyn, Air Force Weapons Lab, Albuquerque, NMex, 61-64, lab proj officer, 67-71; instr & asst prof chem, US Air Force Acad, Colo, 71-74, assoc prof & dir advan courses chem, 74-75, PROG MGR CHEM, DIRECTORATE CHEM SCI, AIR FORCE OFF SCI RES, US AIR FORCE, 75- Mem: Am Chem Soc; Sigma Xi. Res: Identification, selection and management of research in chemical dynamics, chemical kinetics and atomic and molecular phenomena. Mailing Add: Air Force Off of Sci Res Dir of Chem Sci Bolling AFB Washington DC 20332

VIOLA, VICTOR E, JR, b Abilene, Kans, Apr 8, 35; m 62; c 3. NUCLEAR CHEMISTRY. Educ: Univ Kans, AB, 57; Univ Calif, Berkeley, PhD(nuclear chem), 61. Prof Exp: Instr & res fel nuclear chem, Univ Calif, Berkeley, 61-62; NSF fel, Europ Orgn Nuclear Res, 63, Ford fel, 64; res assoc chem, Argonne Nat Lab, 64-66; from asst prof to assoc prof, 66-74, PROF CHEM, UNIV MD, COLLEGE PARK, 74- Concurrent Pos: Consult, Argonne Nat Lab, 66-73; vis prof, Univ Calif, Berkeley, 73-74. Mem: AAAS; Am Chem Soc; Am Phys Soc; Sigma Xi. Res: Reaction mechanism studies in heavy ion collisions; synthesis of the elements in nature; nuclear fission at moderate excitation energies; systematics of heavy element lifetimes and energetics. Mailing Add: Dept of Chem Univ of Md College Park MD 20742

VIOLANTE, MICHAEL ROBERT, b Buffalo, NY, Mar 6, 44; m 67. INORGANIC CHEMISTRY, PHYSICAL CHEMISTRY. Educ: State Univ NY Buffalo, BA, 66; Fla State Univ, PhD(inorg chem), 70. Prof Exp: RES SCIENTIST PULP & PAPER, RES & DEVELOP DIV, UNION-CAMP CORP, 70- Mem: Am Chem Soc; Tech Asn Pulp & Paper Indust. Res: Wood pulping; solvent effects and aqueous solution chemistry; membrane transport phenomena. Mailing Add: Res & Develop Div Union-Camp Corp Box J12 Princeton NJ 08540

VIOLET, CHARLES EARL, b Des Moines, Iowa, May 1, 24; m 51; c 5. PHYSICS. Educ: Univ Chicago, BS, 48; Univ Calif, AB, 49, PhD(physics), 53. Prof Exp: Asst physics, 49-50, physicist, 50-57, testing group dir, Oper Plumbbob, 57-58, dep test mgr, Oper Hardtack, 58-59, test div leader, 59-61, PHYSICIST, LAWRENCE LIVERMORE LAB, UNIV CALIF, 61- Mem: Am Phys Soc. Res: Mössbauer resonance spectroscopy; dilute alloy magnetism; nuclear moments; laser plasma interactions. Mailing Add: Lawrence Livermore Lab Univ of Calif Box 808 Livermore CA 94550

VIOLETT, THEODORE DEAN, b Great Bend, Kans, Apr, 27, 32; m 53; c 3. PHYSICS. Educ: Univ Mo, BS, 53, MA, 54; Univ Colo, PhD(physics), 59. Prof Exp: PROF PHYSICS, WESTERN STATE COL COLO, 59- Mem: Am Phys Soc; Am Asn Physics Teachers. Res: Vacuum ultraviolet radiation and solar spectroscopy. Mailing Add: Dept of Physics Western State Col Gunnison CO 81230

VIPPERMAN, POSEY ELMER, JR, b Stuart, Va, Apr 29, 31; m 55; c 4. NUTRITION. Educ: Va Polytech Inst, BS, 55, MS, 62; Univ Mo-Columbia, PhD(animal nutrit), 67. Prof Exp: Instr animal sci, Va Polytech Inst, 62; asst prof, Univ Nebr, 66-70; ASST PROF ANIMAL SCI, UNIV FLA, 70- Mem: Am Soc Animal Sci. Res: Animal Nutrition. Mailing Add: Agr Res Ctr Univ of Fla Box 878 Marianna FL 32446

VIRGIN, WALTER JAMES, plant pathology, see 12th edition

VIRGO, BRUCE BARTON, b Vancouver, BC, Mar 18, 43; Can citizen; m 69. TOXICOLOGY, REPRODUCTIVE PHYSIOLOGY. Educ: Univ BC, BSc, 65, MSc, 70, PhD(pharmacol, toxicol), 74. Prof Exp: Res biologist econ ornith, Can Wildlife Serv, Govt Can, 65-66, contractee ecol, 67-68; Nat Res Coun fel toxicol, McGill Univ, 74-75; Nat Res Coun fel pharmacol, Univ Montreal, 75; ASST PROF PHYSIOL, UNIV WINDSOR, 75- Mem: Soc Toxicol; Pharmacol Soc Can; Can Asn for Res Toxicol. Res: Physiology, pharmacology and toxicology of reproduction processes at all organizational levels; toxicology of environmental chemicals with emphasis on the effects of chronic, low-dose exposure. Mailing Add: Dept of Biol Univ of Windsor Windsor ON Can

VIRKAR, RAGHUNATH ATMARAM, b Vir, Maharashtra, India, Dec 12, 30; m 53; c 1. PHYSIOLOGY: INVERTEBRATE ZOOLOGY. Educ: Univ Bombay, BS, 50, MS, 52; Univ Minn, PhD(zool), 64. Prof Exp: Demonstr zool, Wilson Col, Bombay, 52-53; lectr, Vithalbhai Patel Col, 53-61; instr, Univ Minn, 64; asst res biologist, Univ Calif, Irvine, 65-66 & Univ Calif, Riverside, 66; asst prof biol, Wis State Univ-Superior, 66-68; assoc prof, Newark State Col, 68-73, PROF BIOL, KEAN COL NJ, 73- Mem: AAAS; Am Soc Zool. Res: Nutritional role of dissolved organic matter in invertebrates; role of free amino acids in osmoregulation. Mailing Add: Dept of Biol Kean Col of NJ Union NJ 07083

VIRKKI, NILO, b Vuoksela, Finland, May 7, 24; m 49; c 2. INSECT CYTOGENETICS. Educ: Univ Helsinki, Lic phil, 51, PhD(genetics), 52. Prof Exp: Inspector pest animals, Vet Sect, Health Bd, Helsinki, Finland, 49-53; lab keeper, Univ Helsinki, 53-61, asst prof genetics, 55-61; from asst cytogeneticist to assoc cytogeneticist, 61-64, CYTOGENETICIST, AGR EXP STA, UNIV PR, 64- Concurrent Pos: Fel fores insect lab, Can Dept Agr, Ont, 55-56; NSF grants, 65-68. Res: Problems concerning evoultion of karyotypes in the beetle family Alticidae, especially formation of giant asynaptic sex chromosomes and their mode of orientation and segregation in meiosis. Mailing Add: Genetics Dept Agr Exp Sta Rio Piedras PR 00928

VIRMANI, YASH PAUL, b Lahore, India, Sept 10, 35; US citizen; m 63; c 2. PHYSICAL CHEMISTRY. Educ: Univ Agra, BSc, 55; Univ Rajasthan, MSc, 58; Bhaba Atomic Res Ctr, Trombay, India, dipl, 59; Univ Bombay, PhD(phys chem), 66. Prof Exp: Sci officer radiation chem, Bhaba Atomic Res Ctr, Trombay, India, 59-67; fel, Univ Fla, 67-69; sr res assoc solid state chem, Univ Kans, 69-73; RES CHEMIST, CONCRETE TECHNOL, DEPT TRANSP, KANS, 73- Concurrent Pos: Consult, Gulf Oil, Kans, 71. Res: Electroosmotic technique for removal of chloride from concrete and for emplacement of sealents; rapid in situ determination of chloride ion in Portland Cement concrete bidge decks. Mailing Add: Dept of Transp 2300 Van Buren Topeka KS 66611

VIRNSTEIN, ROBERT W, b Washington, DC, Mar 19, 43; m 69. MARINE ECOLOGY. Educ: Johns Hopkins Univ, BA, 66; Univ SFla, MA, 72; Col William & Mary, PhD(marine sci), 76. Prof Exp: Teacher high schs, Md, 66-69; MARINE SCIENTIST, VA INST MARINE SCI, 75- Mem: Ecol Soc Am; Am Soc Limnol & Oceanog; Int Asn Meiobenthologists. Res: Estuarine benthic ecology; trophic relationships; experimental field ecology; benthic invertebrate taxonomy. Mailing Add: Va Inst of Marine Sci Gloucester Point VA 23062

VISCELLI, THOMAS ALFONSE, b Brooklyn, NY, Jan 27, 24; m 50; c 3. BIOCHEMISTRY. Educ: St John's Col, NY, BS, 50; Fordham Univ, MS, 53; Columbia Univ, PhD(biochem), 59. Prof Exp: Asst urol, Col Physicians & Surgeons, Columbia Univ, 56-60; exec dir labs, High Tor Found, Inc, Palisades, NY, 60-73; CONSULT, 73- Mem: AAAS; fel Am Inst Chem; Am Chem Soc. Res: Smoking and lung cancer and effect on vascular system; reduction of carcinogens in tobacco smoke; role of radical scavengers and deuterium isotope effects in reduction of tobacco smoke carcinogenicity. Mailing Add: 48 Hamilton Ave Hasbrouck Heights NJ 07604

VISCO, EUGENE PAUL, b Boston, Mass, Apr 20, 27. OPERATIONS RESEARCH, STATISTICS. Educ: Univ Miami, BS, 50. Prof Exp: Sci aide field experimentation, Dugway Proving Ground, Utah, 51-53, mathematician, 53-56; opers analyst, Opers Res Off, Johns Hopkins Univ, 56-61 & Res Anal Corp, Va, 61-66; sr staff analyst, Tech Opers, Inc, 66-68; sr opers analyst, Geomet, Inc, 68-74; DIR HEALTH & SOCIAL STUDIES, NAT INST COMMUNITY DEVELOP, WASHINGTON, DC, 74- Concurrent Pos: Lectr, Sch Nursing, Univ Md, 74- Mem: Opers Res Soc Am; Am Pub Health Asn; Int Inst Strategic Studies. Res: Social systems operational analysis and evaluation with emphasis on public health services. Mailing Add: Nat Inst Community Develop 2021 K St NW Washington DC 20006

VISCO, RAYMOND JOHN, parasitology, bacteriology, see 12th edition

VISCO, ROBERT EDWARD, electrochemistry, see 12th edition

VISCONTI, JAMES ANDREW, b St Louis, Mo, Apr 13, 39; m 63; c 2. PHARMACY. Educ: St Louis Col Pharm, BS, 61, MS, 63; Univ Miss, PhD(pharm), 69. Prof Exp: Resident, John Cochran Vet Admin Hosp, St Louis, Mo, 63, staff pharmacist, 64-66; res pharmacist, Vet Admin Hosp, Long Beach, Calif, 63-64; asst prof pharm, 68-72, ASSOC PROF PHARM, COL PHARM, OHIO STATE UNIV, 72-, DIR DRUG INFO CTR, UNIV HOSPS, 68- Concurrent Pos: Lehn & Fink Pharm gold medal award, 61-62; Robert Lincoln McNeil citation fel award, 66-67. Mem: Am Soc Hosp Pharmacists; Am Pharmaceut Asn; assoc mem AMA; Drug Info Asn. Res: Epidemiology and econmics of adverse drug reactions; pharmacology of drug-drug drug-laboratory tests and drug-food interactions; computerized drug information services; health and disease economics. Mailing Add: Drug Info Ctr Ohio State Univ Hosp Dept of Pharm 410 W Tenth Ave Columbus OH 43210

VISEK, KENNETH EDWARD, organic chemistry, see 12th edition

VISEK, WILLARD JAMES, b Sargent, Nebr, Sept 19, 22; m 49; c 3. NUTRITION, TOXICOLOGY. Educ: Univ Nebr, BSc, 47; Cornell Univ, MSc, 49, PhD(nutrit), 51; Univ Chicago, MD, 57. Prof Exp: Asst animal husb, Cornell Univ, 48-51; AEC fel, Univ-Atomic Energy Agr Res Prog, Tenn, 51-52, res assoc, 52-53; res asst pharmacol,

Univ Chicago, 53-57, from asst prof to assoc prof, 57-64; prof nutrit & comp metab, Cornell Univ, 64-75; PROF CLIN SCI, SCH BASIC MED SCI, UNIV ILL, URBANA-CHAMPAIGN, 75- Concurrent Pos: Intern univ hosp & clins, Univ Chicago, 57-59; mem teratol subcomt, Comn Drug Safety, 63 & subcomt animal nutrit, Nat Res Coun-Nat Acad Sci, 65-72, adv coun, Inst Lab Animal Resources, 66-69, subcomt animal care facilities surv, 68-70; consult sect health related facilities, USPHS, 67; George Henry Durgin lectr, Bridgewater State Col, 70; Nat Cancer Inst-USPHS spec fel, Mass Inst Technol, 70-71; res fel, Mass Gen Hosp, 70-71; grad fac rep nutrit, Cornell Univ, 74- Mem: AAAS; Am Soc Pharmacol & Exp Therapeut; Soc Exp Biol & Med; Am Soc Animal Sci; Am Soc Clin Nutrit. Res: Effects of ammonia on energy metabolism, area cycle activity and nucleic acid synthesis; interactions of amino acids; enzyme immunity; influence of diet on cancer incidence and toxicity of environmental pollutants. Mailing Add: Sch Basic Med Sci Univ of Ill Urbana IL 61801

VISELTEAR, ARTHUR JACK, b New York, NY, Mar 19, 38; m 66; c 2. PUBLIC HEALTH, HISTORY OF MEDICINE. Educ: Tulane Univ, BA, 59; Univ Calif, Los Angeles, MPH, 63, PhD(hist), 65. Prof Exp: Lectr pub health, Univ Calif, Los Angeles, 65-69; asst prof pub health, 69-74, ASSOC PROF PUB HEALTH & RES ASSOC HIST OF SCI & MED, SCH MED, YALE UNIV, 74- Concurrent Pos: Robert Wood Johnson Health Policy fel, Inst Med, Nat Acad Sci, 74-75. Mem: AAAS; Am Asn Hist Med; Am Pub Health Asn. Res: History and health policy; social medicine; health services research. Mailing Add: Dept Epidemiol & Pub Health Yale Univ Sch of Med New Haven CT 06510

VISHER, FRANK N, b Twin Falls, Idaho, Mar 10, 23; m 48; c 4. HYDROLOGY, GEOLOGY. Educ: Tex Tech Col, BS, 46 & 47. Prof Exp: Apprentice engr, Tex Hwy Dept, 47-48; geologist, 48-56, engr, 56-66, hydrologist, Fla, 66-67, RES HYDROLOGIST, US GEOL SURV, 67- Mem: AAAS; Geol Soc Am. Res: Hydrologic studies, especially the principals of occurrence of ground water, water budget studies, fresh-salt water interrelationships, geochemistry of water and relation of geomorphology to ground water. Mailing Add: US Geol Surv Denver Fed Ctr Bldg 25 Lakewood CO 80225

VISHER, GLENN S, b May 20, 30; US citizen; m 53; c 3. GEOLOGY. Educ: Univ Cincinnati, BS, 52; Northwestern Univ, MS, 56, PhD(geol), 60. Prof Exp: Explorationist, Shell Oil Co, 58-60; res geologist, Sinclair Res, Inc, 60-66; lectr, 64-66, assoc prof geol, 66-71, PROF GEOL, UNIV TULSA, 71- Concurrent Pos: Lectr training courses, Domestic & Int Petrol Co Personnel. Mem: Am Asn Petrol Geol; fel Geol Soc Am; Soc Econ Paleont & Mineral; Int Asn Sedimentol. Res: Stratigraphic models; depostional processes; physical characteristics of sandstone units; texture of sandstones; petrology of shales and sandstones. Mailing Add: Dept of Earth Sci Univ of Tulsa 600 S College Tulsa OK 74104

VISHER, HALENE HATCHER, b Murray, Ky, June 18, 09; c 1. ECONOMIC GEOGRAPHY, RESOURCE GEOGRAPHY. Educ: Murray State Univ, AB, 30; George Peabody Col, MA, 32; Ind Univ, MA, 55, PhD(conserv educ, regional geog), 60. Prof Exp: Teacher geog & Eng, Paducah City Schs, Ky, 30-42; instr geog, George Peabody Col, 42-44; cartographer, Off Strategic Serv, Washington, DC, 44-45; asst prof geog, Murray State Univ, 45-48; specialist geog & conserv, US Off Educ, 48-51; AUTH, LECTR & CONSULT, PUB SCH SYSTS & PUBL CO, 51- Concurrent Pos: US Off Educ rep, Inter-Am Conserv Conf, 48; mem US organizing comt, Int Tech Conf Protection Nature, UNESCO, Lake Success, 49; US Off Educ rep, UN Sci Conf Conserv & Utilization of Resources, 49; secy, Conserv Educ Comn, Int Union Protection of Nature, 49-51; lectr conserv, Purdue Univ, 51; mem, Pop Coun. Mem: AAAS; Asn Am Geog; Nat Coun Geog Educ; Conserv Educ Asn; Am Soc Ecol Educ. Res: Conservation, especially our public lands, environmental quality and education, population growth and resources; regional geography, especially of the South, Indiana and Latin America. Mailing Add: 1221 Dogwood Dr Murray KY 42071

VISHNIAC, HELEN SIMPSON, b New Haven, Conn, Dec 22, 23; m 51; c 3. MICROBIAL ECOLOGY, MYCOLOGY. Educ: Univ Mich, BA, 45; Radcliffe Col, MA, 47; Columbia Univ, PhD(bot), 50. Prof Exp: Tutor biol, Queens Col, NY, 48-51, instr, 51-52; lectr microbiol, Sch Med, Yale Univ, 53-61; RES ASSOC BIOL, UNIV ROCHESTER, 74- Concurrent Pos: Lectr, Nazareth Col Rochester, 75-76; fel, Am Asn Univ Women, 75-76. Mem: AAAS; Mycol Soc Am; Soc Gen Microbiol; Am Soc Microbiol. Res: Antarctic yeasts as members of a minimal community; ecology and physiology of Rhizophlyctis rosea as a model soil-inhabiting chytridiaceous fungus. Mailing Add: Dept of Biol Univ of Rochester Rochester NY 14627

VISHNIAC, WOLF VLADIMIR, microbiology, biochemistry, deceased

VISNER, SIDNEY, b New York, NY, Dec 10, 17; m 45; c 2. PHYSICS. Educ: City Col New York, BS, 37, MS, 38; Univ Tenn, PhD(physics), 51. Prof Exp: Lab asst physics, Columbia Univ, 41-43, res scientist & sect leader, S A M Labs, 43-45; sr res physicist & dept head, Gaseous Diffusion Plant, Carbide & Carbon Chems Corp, 45-50, head physicist, 50-55; mgr physics dept, Nuclear Div, 55-69, DIR PHYSICS & COMPUT ANAL, NUCLEAR POWER DEPT, COMBUSTION ENG, INC, 69- Mem: AAAS; Am Phys Soc; fel Am Nuclear Soc. Res: Hydrodynamic studies in isotope separation by gaseous diffusion; transport properties of gases at low pressures; neutron reactor physics and analysis; experimental physics; nuclear reactor development. Mailing Add: Nuclear Power Dept Combustion Eng Inc Windsor CT 06095

VISOTSKY, HAROLD M, b Chicago, Ill, May 25, 24; m 56; c 2. MEDICINE, PSYCHIATRY. Educ: Univ Ill, BS, 48, MD, 51. Prof Exp: Coordr psychiat residency training, Univ Ill, Chicago Circle, 55-59, from asst prof to assoc prof psychiat, 59-69; PROF PSYCHIAT & CHMN DEPT, MED SCH, NORTHWESTERN UNIV, 69-, DIR INST PSYCHIAT, UNIV & MEM HOSP, 75- Concurrent Pos: Nat Found Infantile Paralysis res fel, 55-56; mem psychiat educ prog comn, State Dept Welfare, Ill, 55-59; chief of serv, Chicago State Psychiat Hosp, 57-59; dir ment health sect, City Bd Health, Chicago, 59-62; dir, Ill Dept Ment Health, 62-69; chmn task force, Joint Comn Ment Health of Children, 66-; mem adv comt, Secy Dept Health, Educ & Welfare, 66-67. Honors & Awards: Edward A Strecker Award, Inst of Pa Hosp, 69. Mem: Am Psychosom Soc; AMA; Am Psychiat Asn (vpres, 73-74); NY Acad Sci; Am Orthopsychiat Asn (pres, 76-77). Res: Social and milieu psychiatry; effects of hallucinogenic drugs in understanding mental illness. Mailing Add: Dept of Psychiat Northwestern Univ Med Sch Chicago IL 60611

VISSAT, PETER LOUISA, b Frisanco, Italy, Aug 9, 13; m 40; c 4. PHYSICS. Educ: Univ Pittsburgh, BS, 35, PhD(physics), 41. Prof Exp: Asst physics, Univ Pittsburgh, 35-39, lectr, 42; asst prof, Seton Hill Col, 39-42; instr, US Navy Officers & Exten Courses, Cornell Univ, 42-43; res engr, Tube Turns, Inc, 43-48; mgr prod eng dept, Taylor Forge & Pipe Works, 48-62, mgr eng, Int Div, 62-70, mgr eng develop, Taylor Forge Inc, 70; V PRES MFG, INT DIV, ENERGY PROD GROUP, GULF & WESTERN, 70- Concurrent Pos: Supvr, Pa State Col, 41-42; exten fac mem, Purdue Univ, 44 & Univ Louisville, 47-48. Mem: AAAS; Am Phys Soc; Am Soc Mech Eng; Soc Exp Stress Anal; Am Soc Nondestructive Testing. Res: Properties of metals; induction heating; high electrostatic fields; fluid-flow; high vacuum techniques; stress analsyis. Mailing Add: 615 N La Grange Rd La Grange Park IL 60525

VISSCHER, MAURICE B, b Holland, Mich, Aug 25, 01; m 25; c 4. PHYSIOLOGY. Educ: Hope Col, BS, 22; Univ Minn, MS, 24, PhD(physiol), 25, MD, 31. Prof Exp: From fel to asst prof, Univ Minn, 22-25; prof physiol & head dept, Col Med, Univ Tenn, 27-29; prof physiol & pharmacol, Univ Southern Calif, 29-31; prof physiol, Col Med, Univ Ill, 31-36; prof & head dept, 36-60, distinguished serv prof, 60-68, regents prof, 68-71, EMER REGENTS PROF PHYSIOL, UNIV MINN, MINNEAPOLIS, 71- Concurrent Pos: Mem subcomt clin invest, Div Med Sci & Comt on UNESCO, Nat Res Coun; secy coun, Int Union Physiol Sci, 53-; sci co-dir, Italian Med Nutrit Mission, UNRRA & Unitarian Serv Comt, chmn Austrian med teaching. Honors & Awards: Res Achievement Award, Am Heart Asn, 62; Distinguished Serv Physiol, Am Physiol Soc, 75. Mem: Nat Acad Sci; AAAS; Nat Soc Med Res (pres, 65-); Am Physiol Soc (secy, 46-48, pres, 48-49); Soc Exp Biol & Med. Res: Circulation; respiration; physiology of secretion and absorption; cardiovascular physiology; biological transport of materials; ionic factors in cardiac function. Mailing Add: Dept of Physiol Stone Lab Univ of Minn 421 29th Ave SE Minneapolis MN 55414

VISSCHER, SARALEE NEUMANN, b Lewistown, Mont, Jan 9, 29; m 69; c 4. ENTOMOLOGY. Educ: Univ Mont, BA, 49; Mont State Univ, MS, 58, PhD(entom), 63. Prof Exp: Asst prof entom, Mont State Univ, 62-65; NIH res fel insect develop, Univ Va, 65-66; assoc prof, 67-71, PROF ENTOM, MONT STATE UNIV, 72- Concurrent Pos: Co-investr, NIH res grant, 65-69. Mem: Fel AAAS; Entom Soc Am; Am Soc Zoologists; Soc Develop Biol; Sigma Xi. Res: Neuroendocrine involvement in the regulation of embryonic developmental physiology, especially diapause of insects; the importance of maternal physiology on embryonic development of the progeny as related to grasshopper population dynamics. Mailing Add: Dept of Biol Mont State Univ Bozeman MT 59715

VISSCHER, WILLIAM M, b Memphis, Tenn, May 16, 28; m 51; c 4. THEORETICAL PHYSICS. Educ: Univ Minn, BA, 49; Cornell Univ, PhD(theoret physics), 53. Prof Exp: Res assoc physics, Univ Md, 53-56; STAFF MEM THEORET PHYSICS, LOS ALAMOS SCI LAB, 56- Concurrent Pos: Vis prof, Univ Wash, 67. Mem: Fel Am Phys Soc. Res: Meson theory; theory of nuclear structure and spectra; lattice dynamics and Mössbauer effect; particle accelerator physics; solid state physics; statistical mechanics; random packing. Mailing Add: Theoret Div Los Alamos Sci Lab Los Alamos NM 87544

VISSER, DONALD WILLIS, b Holland, Mich, June 8, 15; m 44; c 1. BIOCHEMISTRY. Educ: Hope Col, AB, 37; Syracuse Univ, MS, 39; Univ Colo, PhD(biochem), 47. Prof Exp: Mem staff chem res, Agfa Ansco, NY, 39-40 & Dow Chem Co, Mich, 40-42; from instr to assoc prof biochem, 47-56, PROF BIOCHEM, SCH MED, UNIV SOUTHERN CALIF, 56- Mem: AAAS; Am Soc Biol Chemists; Soc Exp Biol & Med; Am Chem Soc. Res: Synthesis and biological properties of nucleoside analogs; transport of purines, pyrimidines and nucleosides in bacteria and animal cells. Mailing Add: Dept of Biochem Univ of Southern Calif Sch Med Los Angeles CA 90033

VISTE, ARLEN E, b Austin, Minn, Aug 13, 36; m 59; c 3. INORGANIC CHEMISTRY. Educ: St Olaf Col, BA, 58; Univ Chicago, PhD(inorg chem), 62. Prof Exp: Asst prof chem, St Olaf Col, 62-63; NSF fel, Columbia Univ, 63-64; assoc prof, 64-72, PROF CHEM, AUGUSTANA COL, SDAK, 72- Concurrent Pos: Partic fac res participation prog, Argonne Nat Lab, Ill, 70-71. Mem: Am Chem Soc; The Chem Soc. Res: Reaction mechanisms; spectroscopy. Mailing Add: 1500 W 30th St Sioux Falls SD 57105

VISTE, KENNETH LYLE, plant physiology, agronomy, see 12th edition

VISWANATHA, THAMMAIAH, b Channapatna, India, Sept 22, 26. BIOCHEMISTRY. Educ: Univ Mysore, PhD, 55. Prof Exp: Rask-Orsted fel, Carlsberg Lab, Denmark, 56-57; res assoc, Univ Minn, 57-58; vis scientist, Nat Inst Arthritis & Metab Dis, 58-62 & Inst Molecular Biol & Dept Chem, Univ Ore, 62-64; PROF CHEM, UNIV WATERLOO, 64- Res: Enzymes; proteins; nucleic acids. Mailing Add: Dept of Chem Univ of Waterloo Waterloo ON Can

VISWANATHAN, KADAYAM SANKARAN, b Madras, India, Apr 25, 37; m 67. STATISTICAL MECHANICS. Educ: Univ Madras, BSc, 57; Univ Calif, Riverside, MA, 64, PhD(physics), 65. Prof Exp: Res officer crystallog, Atomic Energy Estab, India, 57-60; asst prof, 65-70, ASSOC PROF THEORET PHYSICS, SIMON FRASER UNIV, 70- Mem: Am Phys Soc; Can Asn Physicists. Res: Phase transitions; statistical mechanics of equilibrium and non-equilibrium; non-linear waves and the theory of solitons. Mailing Add: Dept of Physics Simon Fraser Univ Burnaby BC Can

VISWANATHAN, MURI A, b Kozhikode, India, Oct 26, 28; Can citizen; m 59; c 2. PLANT PATHOLOGY, PHYSIOLOGY. Educ: Univ Madras, BSc, 49; Indian Agr Res Inst, New Delhi, AIARI, 52; McGill Univ, PhD(plant path), 64. Prof Exp: Res asst plant breeding, Cotton Breeding Sta, Madras Agr Dept, Coimbatore, India, 49-50; asst plant path, Indian Agr Res Inst, New Delhi, 52-60; Swiss Nat Res Found fel, Univ Geneva, 65-66; asst prof biol, Prince Wales Col, Univ PEI, 66-67; prof assoc, 67-70, ASST PROF PLANT PATH, MACDONALD COL, McGILL UNIV, 70- Mem: Am Phytopath Soc; Can Bot Asn; Can Phytopath Soc. Res: Physiology of disease; use of histochemical techniques; role of growth substances in symptom expression. Mailing Add: Dept of Plant Path Macdonald Col McGill Univ Montreal PQ Can

VITALE, AMERICUS CHRISTOPHER, organic chemistry, see 12th edition

VITALE, JOSEPH JOHN, b Boston, Mass, Dec 14, 24; m 49; c 1. NUTRITION, BIOCHEMISTRY. Educ: Northeastern Univ, BS, 47; NY Univ, MS, 49; Harvard Univ, DSc(nutrit biochem), 51; Antioquia Univ, Colombia, MD, 66. Prof Exp: Res assoc nutrit, Sch Pub Health, Harvard Univ, 51-54, assoc, 54-55, asst prof, 55-66; prof food, nutrit & med, Univ Wis, 66-67; dir nutrit progs, Sch Med, Tufts Univ, 67-72; PROF PATH & COMMUNITY MED, SCH MED, BOSTON UNIV, 72- Concurrent Pos: Res assoc path, Sch Med, Boston Univ, 52-66; spec consult, Interdept Comt Nutrit for Nat Defense, 59-; Claude Bernard prof, Med Sch, Univ Montreal, 60; vis prof, Univ del Valle, Colombia, 60-62. Mem: Am Inst Nutrit; Am Soc Clin Nutrit; Brit Nutrit Soc. Res: Atherosclerosis, gastrointestinal metabolism, nutritional anemias and public health. Mailing Add: Mallory Inst of Path Boston Univ Sch of Med Boston MA 02118

VITALE, RICHARD ALBERT, b New Haven, Conn, Sept 7, 44. MATHEMATICS, STATISTICS. Educ: Harvard Univ, AB, 66; Brown Univ, PhD(appl math), 70. Prof Exp: ASST PROF APPL MATH, BROWN UNIV, 70- Mem: Am Math Soc; Asn Comput Math; Inst Math Statist. Res: Time series and pattern analysis. Mailing Add: Div of Appl Math Brown Univ Providence RI 02912

VITALE, WILLIAM RICHARD, chemistry, see 12th edition

VITALIANO, CHARLES JOSEPH, b New York, NY, Apr 2, 10; m 40; c 2. GEOLOGY. Educ: City Col New York, BS, 36; Columbia Univ, AM, 38, PhD(mineral), 44. Prof Exp: Asst mineral, Columbia Univ, 37-39, lab instr gems & precious stones, Exten, 39-40; instr ceramic petrog, Rutgers Univ, 40-42; from asst geologist to assoc geologist, US Geol Surv, 42-46; assoc prof, 47-57, PROF GEOL, IND UNIV, BLOOMINGTON, 57- Concurrent Pos: Geologist, US Geol Surv, 46-59; Fulbright scholar, Univ NZ, 54-55; NSF grant, 57-60. Mem: Fel Mineral Soc Am; Soc Econ Geologists; fel Geol Soc Am; Geochem Soc. Res: Geology and ore deposits of the Paradise Peak Quadrangle, Nevada; igneous and metamorphic petrography of western Nevada, southern New Zealand and southwest Montana; volcanic rocks of western United States; archaeological geology of Mediterranean regions. Mailing Add: Dept of Geol Ind Univ Bloomington IN 47401

VITALIANO, DOROTHY BRAUNECK, b New York, NY, Feb 10, 16; m 40; c 2. GEOLOGY. Educ: Barnard Col, AB, 36; Columbia Univ, AM, 38, MPhil, 73. Prof Exp: Teaching asst geol, Barnard Col, 36-39; field asst, 42-43, GEOLOGIST, US GEOL SURV, 53- Mem: Geol Soc Am; Am Geophys Union; Int Planetological Asn; Geosci Info Soc; AAAS. Res: Dating of Bronze Age eruption of Santorini volcano relative to stages of Minoan culture; scientific basis of Atlantis; geomythology; technical translation; tephrochronology. Mailing Add: Geol Bldg Rm 227 Geol Surv 1005 E Tenth St Bloomington IN 47401

VITCHA, JAMES F, b Cleveland, Ohio, Mar 5, 10; m 36; c 1. ORGANIC CHEMISTRY. Educ: Ohio Wesleyan Univ, BA, 32; Univ Ill, MS, 33. Prof Exp: Res chemist, Solvay Process Corp, NY, 33-43; chief chemist, Ohio Chem & Mfg Co, 43-50 & Puritan Compressed Gas Corp, Mo, 50-55; sect head, Cent Res Labs, 55-65, mgr chem prod develop, Ohio Med Prod Div, 65-75, CONSULT, OHIO MED PROD DIV, AIRCO INC, 75- Mem: Am Chem Soc. Res: Research in anesthetics and anesthetic gases; medical equipment; clinical evaluation of new anesthetics; drug regulatory affairs. Mailing Add: Ohio Med Prod Div Airco Inc Murray Hill NJ 07974

VITEK, JIRI JAKUB, b Praha, Czech, Mar 29, 35; m 66; c 2. RADIOLOGY. Educ: Charles Univ, Praha, MD, 59, PhD(neuroradiol), 68. Prof Exp: ASSOC PROF RADIOL & ASST PROF NEUROL, MED CTR, UNIV ALA, BIRMINGHAM, 70- Mem: AMA. Res: Neuroradiology; stroke. Mailing Add: Dept of Radiol Univ of Ala Med Ctr Birmingham AL 35233

VITOLS, EBERHARDS, biochemistry, deceased

VITOSH, MAURICE LEE, b Odell, Nebr, Jan 16, 39; m 63; c 1. AGRONOMY, SOIL SCIENCE. Educ: Univ Nebr, BS, 62, MS, 64; NC State Univ, PhD(soils), 68. Prof Exp: Agronomist, NC Dept Agr, 65-68; EXTEN SPECIALIST SOIL FERTILITY & ASSOC PROF CROP & SOIL SCI, MICH STATE UNIV, 68- Mem: Am Soc Agron; Soil Sci Soc Am; Potato Asn Am. Res: Soil fertility with potato, corn, soybeans and field beans. Mailing Add: Dept Crop & Soil Sci 107 Soil Sci Bldg Mich State Univ East Lansing MI 48823

VITOUSEK, MARTIN J, b Honolulu, Hawaii, July 30, 24; m 65; c 4. GEOPHYSICS. Educ: Stanford Univ, BS, 49, PhD(math), 54. Prof Exp: Radar lab worker, Pearl Harbor, 43-46; asst prof math, Univ Hawaii, 53-55; sr engr, Scripps Inst, Calif, 56-59; from assoc geophysicist to geophysicist, 61-74, SPECIALIST OCEANOG INSTRUMENT, HAWAII INST GEOPHYS, UNIV HAWAII AT MANOA, 74- Mem: Am Geophys Union; Instrument Soc Am; Marine Technol Soc; Solar Energy Soc; Inst Elec & Electronics Engrs. Res: Applied mathematics; solid earth geophysics and oceanography; long period ocean waves, instrumentation and analysis. Mailing Add: Hawaii Inst of Geophys Univ of Hawaii at Manoa Honolulu HI 96822

VITOUSEK, PETER MORRISON, b Honolulu, Hawaii, Jan 24, 29. 49. ECOLOGY. Educ: Amherst Col, BA, 71; Dartmouth Col, PhD(biol sci), 75. Prof Exp: ASST PROF ZOOL, IND UNIV, BLOOMINGTIN, 75- Mem: Ecol Soc Am; AAAS. Res: Regulation of nutrient cycling in terrestrial ecosystems; land-water interactions. Mailing Add: Dept of Zool Ind Univ Bloomington IN 47401

VITROGAN, DAVID, b Russia, May 8, 12; nat US; m 38; c 1. SCIENCE EDUCATION. Educ: Brooklyn Col, BS, 33; City Col New York, MS, 41; Polytech Inst Brooklyn, MEE, 48; NY Univ, PhD, 65. Prof Exp: Res & develop engr, Int Tel & Tel Corp, 40-45; asst prof elec eng, Polytech Inst Brooklyn, 45-49, res assoc, Microwave Res Inst, 45-47; assoc prof, Pratt Inst, 49-60, chmn elec tech, 50-60; dir sci & tech div, Sch Continuing Studies, NY Univ, 60-64; adminr tech training, Bendix Corp, 64-65; assoc prof educ, Ferkauf Grad Sch Humanities & Soc Sci, Yeshiva Univ, 65-69; prof eng technol & assoc dean instr, Luzerne County Col, Pa, 69-70; dir, Div Math Phys Sci & Eng, Community Col, Philadelphia, 70-73; PROF EDUC, MEDGAR EVERS COL, CITY UNIV NEW YORK, 73- Concurrent Pos: Indust consult, 49-; deleg, Int Conf on Environ Americas, Off Am States, 71; pres, Tech Socs Coun NY; dir NSF grant, Circuit Lab, Medgar Evers Col, City Univ New York, 75-76. Mem: Fel AAAS; sr mem Inst Elec & Electronics Engrs. Res: Solid state devices; microwave electronics; direction and supervision of science in colleges and technical colleges; development of learning activity packets integrating language arts and science experiences as a means for integrating a competency-based teacher education program. Mailing Add: 130 St Edward St Brooklyn NY 11201

VITT, DALE HADLEY, b Washington, Mo, Feb 9, 44; c 2. BOTANY, BRYOLOGY. Educ: Southeast Mo State Col, BS, 67; Univ Mich, Ann Arbor, MS, 68, PhD(bot), 70. Prof Exp: Asst prof bot, 70-74, ASSOC PROF BOT, UNIV ALTA, 75- Mem: Am Soc Plant Taxon; Am Bryol & Lichenological Soc; Can Bot Soc; Danish Bryol Soc. Res: Taxonomic, phylogenetic and ecological studies of bryophytes; monographic treatment of arctic, antarctic and temperate mosses; ecological analyses and productivity of bryophytes in arctic and alpine tundras and in boreal peatlands. Mailing Add: Dept of Bot Univ of Alta Edmonton AB Can

VITTI, TRIESTE GUIDO, b Detroit, Mich, May 22, 25; m 53; c 4. BIOPHARMACEUTICS. Educ: Univ Detroit, BS, 49, MS, 51; Wayne State Univ, PhD(biochem), 61. Prof Exp: Lectr pharmacol, Fac Med, Univ Man, 64-67; chief bioavailability, Upjohn Co, Mich, 67-71; dir clin res, Bur Drugs, Food & Drug Admin, 71-72; PROF PHARMACEUT, FAC PHARM, UNIV MAN, 72- Concurrent Pos: USPHS fel, Univ Man, 64-67; consult, Biodecision Labs, Pittsburgh, Pa, 72-; mem permanent adv expert comt bioavailability, Health & Welfare, Health Progs Bd, Can, 74- Mem: Am Chem Soc; Pharmacol Soc Can; Am Pharmaceut Asn. Res: Bioavailability; clinical pharmacology. Mailing Add: Fac of Pharm Univ of Man Winnipeg MB Can

VITTITOE, CHARLES NORMAN, b Louisville, Ky, Oct 3, 34; m 58; c 2. ELECTRODYNAMICS. Educ: Univ Ky, BS, 56; Univ Wis, MS, 58; Univ Ky, PhD(physics), 63. Prof Exp: Instr physics, Univ Ky, 59-60, from res asst to res assoc, 62-63; asst prof, Univ Ohio, 63-66, univ res comt summer grant, 66; staff mem, Radiation Phenomena Div, 66-70, STAFF MEM, THEORET DIV, SANDIA LABS, 70- Mem: Am Phys Soc. Res: Pion-nucleon interactions; reconstruction and identification of bubble chamber tracks; elementary particle physics; electromagnetic theory; nuclear weapon effects; electrodynamics; optical transport theory. Mailing Add: Sandia Labs Sandia Base Br Box 5800 Albuquerque NM 87115

VITTOR, BARRY ADOLPH, b Pittsburgh, Pa, June 25, 44; m 72. ECOLOGY. Educ: Univ Calif, Riverside, BA, 66; San Diego State Col, MA, 68; Univ Ore, PhD(ecol), 71. Prof Exp: ASST PROF MARINE SCI, UNIV ALA, BIRMINGHAM, 71- Concurrent Pos: Adj asst prof, Univ Ala, Tuscaloosa, 71- Mem: Ecol Soc Am. Res: Evolution of life history characters; ecological consequences of dredging in bays and estuaries. Mailing Add: Dept of Biol Univ of Ala Birmingham AL 35294

VITTORIA, CARMINE, b Avella, Italy, May 15, 41; US citizen; m 67; c 3. MAGNETISM. Educ: Toledo Univ, BS, 62; Yale Univ, MS, 67, PhD(physics), 70. Prof Exp: Elec engr bionics, Naval Ord Lab, 62-63; teacher elec eng, Toledo Univ, 63-64; PHYSICIST, NAVAL RES LAB, 70- Concurrent Pos: Naval Res Coun adv, Naval Res Lab, 74. Honors & Awards: Outstanding Achievement Award, Naval Res Lab, 72 & 74. Mem: Am Phys Soc; Inst Elec & Electronics Engrs; Sigma Xi. Res: Electromagnetic wave propagation in magnetic materials. Mailing Add: Naval Res Lab Code 6453 V Washington DC 20375

VITTORIO, PAUL VINCENT, biochemistry, see 12th edition

VITTUM, MORRILL THAYER, b Haverhill, Mass, May 4, 19; m 41; c 3. HORTICULTURE, AGRONOMY. Educ: Univ Mass, BS, 39; Univ Conn, MS, 41; Purdue Univ, PhD(soil sci), 44. Prof Exp: Asst agron, Univ Conn, 39-41; asst agron, Purdue Univ, 41-42, tech asst soils, 42-45; from asst prof to assoc prof, 46-59, PROF VEG CROPS, NY STATE AGR EXP STA, CORNELL UNIV, 59-, HEAD DEPT, 60-69 & 71- Concurrent Pos: Actg asst olericulturist, Univ Calif, Davis, 56-57; vis prof hort, Ore State Univ, 64; proj leader, Univ Philippines-Cornell Grad Educ Prog, Col Agr, Univ Philippines, 69-71; actg horticulturist, Coop State Res Serv, USDA, 73-74. Mem: Am Soc Agron; Soil Sci Soc Am; Am Soc Hort Sci; Int Soc Hort Sci; Int Soc Soil Sci. Res: Effects of fertilizers, irrigation, rotation and cultural practices on the yield and quality of processing vegetables; evapotranspiration and soil-plant-water relationships. Mailing Add: NY State Agr Exp Sta Cornell Univ Geneva NY 14456

VITUCCI, JAMES CHARLES, biochemistry, see 12th edition

VITULLO, VICTOR PATRICK, b Chicago, Ill, Oct 18, 39; m 62; c 1. ORGANIC CHEMISTRY. Educ: Loyola Univ, Chicago, BS, 61; Ill Inst Technol, PhD(chem), 65. Prof Exp: NSF fel, Mass Inst Technol, 65-66; res chemist, E I du Pont de Nemours & Co, Inc, 66-68; instr, Univ Kans, 68-69; asst prof, 69-74, ASSOC PROF CHEM, UNIV MD, BALTIMORE COUNTY, 69- Mem: AAAS; Am Chem Soc. Res: Physical organic chemistry; organic reaction mechanisms; acid-base catalysis; transition state structure; isotope effects. Mailing Add: Dept of Chem Univ of Md Baltimore County Baltimore MD 21228

VIVIAN, DONALD LINDSAY, organic chemistry, deceased

VIVIAN, VIRGINIA M, b Barneveld, Wis, July 1, 23. NUTRITION. Educ: Univ Wis, BS, 45; Columbia Univ, MS, 47; Univ Wis, PhD(home econ, biochem), 59. Prof Exp: Instr foods, nutrit & dietetics, Sch Nursing, Presby Hosp, 48-49; instr foods, nutrit & dietetics, Sch Nursing, Univ Mich, 49-51, asst dir dietary dept, Univ Hosp, 51-55; from asst prof to assoc prof, 59-68, PROF HOME ECON, OHIO AGR RES & DEVELOP CTR, 68-, PROF, OHIO STATE UNIV, 68- Mem: AAAS; Am Home Econ Asn; Am Dietetic Asn; Am Inst Nutrit; Am Inst Chemists. Res: Amino acid-lipid metabolism with humans, preschool dietary adequacy and nutrition status studies. Mailing Add: Sch of Home Econ Ohio State Univ Columbus OH 43210

VIVILECCHIA, RICHARD, b Everett, Mass, Jan 2, 42; m 69; c 1. ANALYTICAL CHEMISTRY. Educ: Univ Mass, Amherst, BS, 64; Northeast Univ, MS, 68, PhD(chem), 70. Prof Exp: Nat Cancer Inst Can fel, Dalhousie Univ, 70-71; res scientist, Sandoz-Wander, Inc, 71-72; SR RES SCIENTIST, WATERS ASSOCS INC, 72- Mem: Am Chem Soc; Brit Inst Petrol. Res: Chromatographic methods of analysis, silica chemistry. Mailing Add: Waters Assocs Inc Maple St Milford MA 01757

VIVONA, STEFANO, b St Louis, Mo, Mar 25, 19; m 50; c 5. PREVENTIVE MEDICINE, PUBLIC HEALTH. Educ: St Louis Univ, MD, 43; Harvard Univ, MPH, 52. Prof Exp: Chief prev med, 5th Army Corps, US Army, Ger, 52-54, chief prev med, 7th Army, 54-55, actg chief biostatist, Walter Reed Army Inst Res, DC, 55-58, chief prev med res admin, Med Res & Develop Command, 58-60, prev med, 8th Army, Korea, 60-61, dir div commun dis & immunol, 62-64, chief med res team, Vietnam, 64-65, dir med component, SEATO, Thailand, 65-67, dir div prev med, Walter Reed Army Inst Res, 67-69; proj officer, 69-70, vpres res grant awards, 70-74, VPRES RES, RES DEPT, AM CANCER SOC, INC, 74- Concurrent Pos: Fel biologics res, Walter Reed Army Inst Res, 61-62. Mem: AMA; Am Pub Health Asn; Am Asn Cancer Res; Am Col Prev Med. Res: Infectious diseases; epidemiologic aspects of biostatistics; cancer. Mailing Add: Res Dept Am Cancer Soc Inc 777 Third Ave New York NY 10017

VIZARD, DOUGLAS LINCOLN, b Worcester, Mass, Sept 13, 44; m 62; c 3. BIOPHYSICS, MOLECULAR BIOLOGY. Educ: Worcester Polytech Inst, BS, 66; Pa State Univ, MS, 69, PhD(biophys), 71. Prof Exp: NIH fel biophys, 71-73, Robert A Welsch Found fel, 73-74, fel biochem, 75, NIH FEL BIOPHYS, M D ANDERSON HOSP & TUMOR INST, 75- Concurrent Pos: NIH fel, Nat Cancer Inst, 71-75. Mem: Biophys Soc; Sigma Xi. Res: Organization of duplex viral DNA and RNA genomes; organization of chromatin; high resolution thermal denaturation. Mailing Add: Dept of Physics M D Anderson Hosp & Tumor Inst Houston TX 77035

VLAD, PETER, b Cluj, Romania, Mar 19, 22; m 54; c 3. CARDIOLOGY. Educ: Univ Paris, DPH, 49; Univ Buffalo, MD, 58. Prof Exp: Mem staff cardiol, Hosp Broussaisla-Charite Paris, 49-51; instr pediat, Med Sch, Univ Toronto, 51-55; jr intern, Mt Sinai Hosp, Toronto, 55-56; asst, Chronic Dis Res Inst & instr pediat, Sch Med, Univ Buffalo, 56-58, assoc, 58-59, asst prof, 59-63; from assoc prof to prof, Univ Iowa, 63-69; PROF PEDIAT, SCH MED, STATE UNIV NY BUFFALO, 69- Concurrent Pos: Chief, Div Cardiol, Children's Hosp Buffalo, 74- Mem: Am Pediat Soc; Soc Pediat Res; Am Heart Asn; Can Cardiovasc Soc. Res: Congenital heart disease. Mailing Add: Dept of Pediat State Univ of NY Sch of Med Buffalo NY 14214

VLADIMIROFF, THEODORE, theoretical chemistry, see 12th edition

VLADYKOV, VADIM DMITRIJ, b Kharkov, Russia, Mar 18, 98; Can citizen; m 38; c 2. TAXONOMY, ICHTHYOLOGY. Educ: Charles Univ, Prague, Dr Rer Nat (zool), 25. Prof Exp: Biologist, Fishery Invests of Subcarpathia under auspices of Charles Univ, Natural Hist Mus Prague & Ministry Educ & Agr Czech, 23-27; investr ichthyol, Nat Mus Natural Hist, Paris, 28-30; biologist, Biol Bd Can, 30-36; in-charge of invests on shad & striped bass, State of Md, 36-37; prof ichthyol, Univ Montreal,

38-42; dir biol lab, Dept Fisheries, Que, 43-58; prof biol, Univ Ottawa, 58-61; fishery expert in Iran for Caspian Sea, Food & Agr Orgn, UN, Rome, 61-62; PROF BIOL, UNIV OTTAWA, 63- Concurrent Pos: Correspondent, Nat Mus Natural Hist, Paris, 29. Mem: Int Asn Theoret & Appl Limnol; Can Soc Zool; Am Fisheries Soc; Am Soc Ichthyol & Herpet; Am Soc Limnol & Oceanog. Res: Taxonomy of Holarctic lampreys; taxonomy of Salmonidae; taxonomy of sturgeons; biology of eels; fishes of Iran. Mailing Add: Dept of Biol Univ of Ottawa Ottawa ON Can

VLAHAKIS, GEORGE, b New York, NY, Oct 12, 23; m 49; c 2. GENETICS. Educ: Johns Hopkins Univ, AB, 51; Univ Tex, MA, 53. Prof Exp: Biologist, NIH, 52-53; chemist, US Testing Co, 54-55; BIOLOGIST, NAT CANCER INST, 55- Mem: AAAS; Am Genetic Asn; Am Inst Biol Sci. Res: Role of genes and their relationship to non-genetic factors in the development of tumors in mice. Mailing Add: 1720 Evelyn Dr Rockville MD 20852

VLAHCEVIC, ZDRAVKO RENO, b Yugoslavia, June 30, 31; US citizen; m 61; c 2. GASTROENTEROLOGY. Educ: Univ Zagreb, MD, 67. Prof Exp: Instr med, Sch Med, Tufts Univ, 62-64; jr instr, Western Reserve Univ, 65-66; from asst prof to assoc prof, 66-74, PROF MED, MED COL VA, 74- Concurrent Pos: Mem staff, Vet Admin Hosp, Richmond, Va. Mem: AAAS; AMA; Am Fedn Clin Res; Am Asn Study Liver Dis; Am Gastroenterol Asn; NY Acad Sci. Res: Cholesterol and bile acid metabolism in cholesterol gallstone and cirrhotic patients; mechanism of feedback regulation of bile acid and cholesterol synthesis in man. Mailing Add: Dept of Gastroenterol Vet Admin Hosp Richmond VA 23249

VLAMIS, JAMES, b US, June 8, 14; m 41; c 4. PLANT NUTRITION. Educ: Univ Calif, BS, 35, PhD(plant physiol), 41. Prof Exp: Asst bot, Univ Calif, 37-41 & Harvard Univ, 41-43; from jr soil chemist to assoc soil chemist, 46-60, assoc plant physiologist, 60-66, PLANT PHYSIOLOGIST, UNIV CALIF, BERKELEY, 66- Concurrent Pos: Fulbright fel, Portugal, 62-63. Mem: Am Soc Plant Physiol. Res: Soil-plant relations; mineral nutrition of plants; soil chemistry and fertility in the growth of plants; availability of soil nutrients as a factor in agriculture and in plant ecology. Mailing Add: Dept of Soils & Plant Nutrit Univ of Calif Berkeley CA 94720

VLAOVIC, MILAN STEPHEN, b Novi Sad, Yugoslavia, Feb 1, 36; m 69; c 2. VETERINARY PATHOLOGY. Educ: Univ Belgrade, DVM, 61; Univ Sask, MSc, 70; Univ Mo-Columbia, PhD(vet med), 74. Prof Exp: Gen practr, WGer, 65-67; tech officer, Can Dept Agr, 67-68; res asst vet microbiol, Univ Sask, 68-70; res asst, Wash State Univ, 70-71; res assoc, Univ Mo, 71-74; VET PATHOLOGIST, FREDERICK CANCER RES CTR, 74- Mem: Am Vet Med Asn; Am Asn Lab Animal Sci; Soc Pharmacol & Environ Pathologists; Tissue Cult Asn. Res: Genetic control of endogenous C-type virus expression in mice as monitored by in vitro and in vivo parameters. Mailing Add: Frederick Cancer Res Ctr PO Box B Frederick MD 21701

VLASES, GEORGE CHARPENTIER, b New York, NY, Oct 22, 36; m 58; c 3. PLASMA PHYSICS. Educ: Johns Hopkins Univ, BES, 58; Calif Inst Technol, MS, 59, PhD(aeronaut), 63. Prof Exp: Res fel aeronaut, Calif Inst Technol, 63; from asst prof to assoc prof aerospace eng sci, Univ Colo, Boulder, 63-69; res assoc prof, 69-73, PROF NUCLEAR ENG, AEROSPACE RES LAB, UNIV WASH, 73- Concurrent Pos: Consult, Aerospace Corp, Calif, 63 & Math Sci Northwest, Inc. Mem: Am Phys Soc. Mailing Add: Aerospace Res Lab FL-10 Univ of Wash Seattle WA 98195

VLASSIS, CONSTANTINE G, inorganic chemistry, see 12th edition

VLASTARAS, ATHANASIOS S, polymer chemistry, physical chemistry, see 12th edition

VLCEK, DONALD HENRY, b Holyrood, Kans, Nov 17, 18; m 44; c 4. ELECTRONICS, RESEARCH ADMINISTRATION. Educ: US Mil Acad, BA, 43; Stanford Univ, ME, 49. Prof Exp: Chief electronic requirements br, HQ, Air Defense Command, US Air Force, 49-51, electronics staff officer to Asst Secy Defense for Res & Develop, 52-54, semi-atomic ground environ proj officer, Air Defense Systs Proj Off, Hq, Air Res & Develop Command, 55-56, chief track test div, Missile Develop Ctr, 56-61, test instrumentation develop div, Hq, Air Force Systs Command, 61-64, dir plans & requirements, Hq, Nat Range Div, 64-67, dir eng, Hq, Ground Electronics Eng Installation Agency, 67-69, Comdr, Ballistic Missile Early Warning Site, Eng, 69-72; COORDR, CANCER RES CTR, MED SCH, OHIO STATE UNIV, 72- Mem: Am Inst Aeronaut & Astronaut; Inst Elec & Electronics Engrs. Res: Captive testing and the influence of the rate of change of acceleration upon the behavior of components and systems. Mailing Add: 2588 Edgevale Upper Arlington Columbus OH 43221

VLIEKS, ARNOLD EVALD, b Staten Island, NY, Feb 12, 45; m 70; c 2. EXPERIMENTAL NUCLEAR PHYSICS, ASTROPHYSICS. Educ: Rensselaer Polytech Inst, BS, 66; Ohio State Univ, MS, 71, PhD(physics), 73. Prof Exp: Systs test engr, Boeing Co, 66-67; control systs engr, NASA, 67-68; LECTR & RES FEL PHYSICS, SCARBOROUGH COL, UNIV TORONTO, 73- Mem: Am Phys Soc. Res: Experimental nuclear physics involving the evaluation of reaction cross sections of nuclei of interest in late-stage stellar nucleosynthesis; evaluation of stellar reaction rates at appropriate stellar temperatures. Mailing Add: Dept of Physics Univ of Toronto Toronto ON Can

VLITOS, AUGUST JOHN, b Vandergrift, Pa, Mar 30, 23; m 48; c 3. PLANT PHYSIOLOGY. Educ: Okla Agr & Mech Col, BS, 48; Iowa State Col, MS, 50; Columbia Univ, PhD(plant physiol, biochem), 53. Prof Exp: Asst plant path, Okla Agr & Mech Col, 41-42 & 46-48; asst plant path, Iowa State Col, 49-50; asst plant pathologist, Boyce Thompson Inst Plant Res, 50-51, sr fel plant physiol, 51-59; dir res, Caroni, Ltd & St Madeleine Sugar Co, Ltd, Trinidad, 59-67; CHIEF EXEC, GROUP RES & DEVELOP, PHILIP LYLE MEM RES LAB, TATE & LYLE, LTD, 67- Concurrent Pos: Plant physiologist, Carbide & Carbon Chem Co, 51-59; vis prof, Univ Reading, 72- Mem: AAAS; Am Phytopath Soc; fel Inst Biol UK. Res: Plant growth regulators; biochemistry; chemistry of sugar refining; bioanalysis; chemurgy. Mailing Add: Philip Lyle Mem Res Lab Tate & Lyle Ltd Univ Reading Reading England

VNEK, JOHN, b Kosice, Czech, May 6, 32; US citizen; m 63; c 2. PROTEIN CHEMISTRY, ENZYMOLOGY. Educ: Bratislava Tech Univ, BS, 58, MS, 59; Czech Acad Sci, PhD(biochem), 67. Prof Exp: Asst prof biochem, Fac Med, Safarika Univ, Czech, 59-65, assoc prof, Fac Natural Sci, 65-67; Am Heart Asn res fel, Columbia Univ, 68-69; instr, Mt Sinai Sch Med, 70-71; RES ASSOC, NEW YORK BLOOD CTR, 72- Mem: Am Chem Soc. Res: Biochemistry; immunochemistry; metabolism. Mailing Add: New York Blood Ctr 310 E 67th St New York NY 10021

VOADEN, DENYS J, organic chemistry, see 12th edition

VOBACH, ARNOLD R, b Chicago, Ill, Nov 20, 32; m 57; c 2. MATHEMATICS. Educ: Harvard Univ, AB, 54, SB, 56; Ill Inst Technol, MS, 59; La State Univ, PhD(math), 63. Prof Exp: Instr math, Univ Ga, 62-63, asst prof, 63-68; ASSOC PROF MATH, UNIV HOUSTON, 68- Mem: Am Math Soc; Math Asn Am. Res: Topology. Mailing Add: Dept of Math Univ of Houston Cullen Blvd Houston TX 77004

VOBECKY, JOSEF, b Brno, Czech, Sept 29, 23; m 55; c 3. EPIDEMIOLOGY. Educ: Masaryk Univ, Brno, MD, 50; Postgrad Med Sch, Prague, DPH, 56, dipl epidemiol, 60. Prof Exp: Epidemiologist, Czech Pub Health Serv, Prague, 50-55, head dept epidemiol, Brno, 56-62; dir dept epidemiol, Inst Epidemiol & Microbiol, Prague, 63-68; asst prof epidemiol, 69-70, ASSOC PROF EPIDEMIOL, MED FAC, UNIV SHERBROOKE, 71- Concurrent Pos: Vis prof, Fac Med, Charles Univ, Prague, 60-62; consult field proj, WHO, Mongolia, 63-65, Iraq, 66, lectr, 66-69; sr lectr, Postgrad Med Sch, Prague, 66-69. Mem: Int Epidemiol Asn; Am Col Prev Med; AAAS; Am Pub Health Asn; NY Acad Sci. Res: Epidemiology of infectious diseases; epidemiological surveillance; chronic disease epidemiology; environmental factors. Mailing Add: Dept of Epidemiol Univ of Sherbrooke Fac of Med Sherbrooke PQ Can

VOCCI, FRANK JOSEPH, b Baltimore, Md, Aug 13, 24; m 48; c 8. BIOCHEMISTRY. Educ: Loyola Col, Md, BS, 49. Prof Exp: Group leader & gen chemist, Aerosol Br, 49-56, gen chemist & asst chief, Basic Toxicol Br, 56-61, CHIEF BASIC TOXICOL BR, TOXICOL DIV, EDGEWOOD ARSENAL, 61- Mem: AAAS; Sigma Xi; Am Chem Soc; Am Indust Hyg Asn. Res: Toxicological effects of poisonous substances; analysis of biological fluids and tissue extracts; kinetics of enzyme reacions; permeability of biological membranes; mode of action of alkylating and riot control agents. Mailing Add: Basic Toxicol Br Biomed Lab Bldg 3220 Edgewood Arsenal MD 21010

VODKIN, MICHAEL HAROLD, b Boston, Mass, Dec 4, 42; m 75. GENETICS. Educ: Boston Col, BS, 64, MS, 66; Univ Ariz, PhD(genetics), 71. Prof Exp: Fel genetics, Cornell Univ, 71-73; ASST PROF BIOL, UNIV SC, 73- Mem: Sigma Xi; Genetics Soc Am; Am Soc Microbiol. Res: Genetics and biochemistry of the killer factor, Saccharomyces cerevisiae, in yeast. Mailing Add: Dept of Biol Univ of SC Columbia SC 29208

VOEDISCH, ROBERT W, b Ft Eustis, Va, Nov 5, 24; m 53; c 3. ORGANIC CHEMISTRY. Educ: Beloit Col, BS, 48. Prof Exp: Res & develop chemist, 50-55, group leader, 55-56, chief chemist, 56-60, tech dir, 60-67, VPRES RES & DEVELOP, LAWTER CHEM, INC, CHICAGO, 67- Mem: AAAS; Am Chem Soc; Fedn Socs Paint Technol; Soc Appl Spectros; Am Asn Textile Chem & Colorists. Res: Luminescent compounds; ink vehicles; synthetic resins; alkyds, phenolics, maleics, ketone and polyamide resins. Mailing Add: 9400 Normandy Ave Morton Grove IL 60053

VOEGELIN, CHARLES FREDERICK, b New York, NY, Jan 14, 06; m 31, 54; c 1. ANTHROPOLOGY. Educ: Stanford Univ, BA, 27; Univ Calif, PhD(anthrop), 32. Prof Exp: Fel, Yale Univ, 33-34, Am Coun Learned Socs fel, 34-35, Nat Res Coun fel, 35-36; from asst prof to assoc prof anthrop, DePauw Univ, 36-41; assoc prof, 41-47, prof & chmn dept, 47-66, dir ling inst, 52-53, DISTINGUISHED PROF ANTHROP, IND UNIV, BLOOMINGTON, 67-, PROF LING, 67-, ED ANTHROP & LING PUBL, 48- Concurrent Pos: Lectr, Univ Mich, Univ NC & Ling Inst, Ind Univ, 38-41; ed, Int J Am Ling, 44-; lectr, Northwestern Univ, 46; Guggenheim fel, Univ Calif, 48. Mem: Fel Anthrop Am Asn; Ling Soc Am (pres, 54); Am Acad Arts & Sci. Res: Structures of American Indian languages; ethno-linguistics. Mailing Add: Dept of Anthrop Ind Univ Bloomington IN 47401

VOEKS, JOHN FORREST, b Seattle, Wash, Aug 16, 22; m. PHYSICAL CHEMISTRY. Educ: Univ Wash, PhD(chem), 51. Prof Exp: ASSOC SCIENTIST, DOW CHEM CO, 51- Res: Physical chemistry of polymers. Mailing Add: Res Dept Dow Chem Co Walnut Creek CA 94598

VOELKER, ALAN MORRIS, b Eau Claire, Wis, Aug 12, 38; m 60; c 2. SCIENCE EDUCATION. Educ: Wis State Univ-River Falls, BS, 59; Syracuse Univ, MS, 63; Univ Wis-Madison, PhD(sci educ), 67. Prof Exp: Teacher chem, physics, gen sci & math & chmn dept, High Schs, Wis, 59-64; asst prof sci educ, Ohio State Univ, 67-69; asst prof sci educ, Univ Wis-Madison & prin investr cognitive learning, Res & Develop Ctr, 69-73; ASSOC PROF SCI EDUC, COL EDUC, NORTHERN ILL UNIV, 73- Concurrent Pos: Mem adv bd sci educ sect, Educ Resources Info Ctr, Info Anal Ctr Sci, Math & Environ Educ, 72-76. Mem: Fel AAAS; Nat Sci Teachers Asn; Nat Asn Res Sci Teaching; Am Educ Res Asn. Res: Science concept learning; science teacher education; attitudes toward science. Mailing Add: Gabel Hall Northern Ill Univ De Kalb IL 60115

VOELKER, HOWARD H, dairy science, see 12th edition

VOELKER, RICHARD WILLIAM, b Stanton, Nebr, July 16, 36; m 61; c 4. VETERINARY PATHOLOGY. Educ: Kans State Univ, BS & DVM, 59; Purdue Univ, MS, 64, PhD(vet path), 69; Am Col Vet Pathologists, cert, 70. Prof Exp: Vet food inspector, US Army Vet Corps, 59-61; med lab officer, US Armed Forces, Europe, 61-64; instr vet path, Purdue Univ, West Lafayette, 64-68; staff pathologist, Hazleton Labs, 68-71; sect head toxicol, William S Merrell Co, 71-73; DIR PATH, HAZLETON LABS AM, 73- Concurrent Pos: Adj asst prof path, Med Sch, Univ Cincinnati, 71-73. Mem: Am Vet Med Asn; Int Acad Path; Am Col Vet Pathologists; Soc Pharmacol & Environ Pathologists; Indust Vet Asn. Res: Toxocologic pathology in the investigation and description of various tissue responses caused by a wide variety of chemical and pharmaceutical compounds; tumor induction in laboratory animals by a wide variety of environmental chemicals given by various routes of administration. Mailing Add: Hazleton Labs Am 9200 Leesburg Turnpike Vienna VA 22180

VOELKER, ROBERT ALLEN, b Palmer, Kans, Jan 24, 43; m 65. GENETICS. Educ: Concordia Teachers Col, Nebr, BSEd, 65; Univ Nebr, Lincoln, MS, 67; Univ Tex, Austin, PhD(zool), 70. Prof Exp: Asst, Univ Nebr, Lincoln, 67; NSF fel, Univ Ore, 70-71; RES ASSOC GENETICS, NC STATE UNIV, 71- Concurrent Pos: Vis asst prof, Dept Genetics, NC State Univ, Raleigh, 73- Mem: Genetics Soc Am; Soc Study Evolution; Sigma Xi. Res: Drosophila population genetics; drosophila salivary gland chromosome cytogenetics; meiotic drive in drosophila. Mailing Add: Dept of Genetics NC State Univ Raleigh NC 27607

VOELLER, BRUCE RAYMOND, developmental biology, plant physiology, see 12th edition

VOELZ, FREDERICK, b Wheaton, Ill, May 22, 27; m 50; c 2. CHEMICAL PHYSICS. Educ: Ill Inst Technol, BS, 51, MS, 53, PhD(physics, math), 55. Prof Exp: Asst, Ill Inst Technol, 51-55; proj chemist, Sinclair Res, Inc, 55-62, sr res physicist, 62-69, SR PROJ ENGR SPEC PROJS, ATLANTIC RICHFIELD CO, 69- Mem: Am Chem Soc; Am Phys Soc; Soc Automotive Eng; Air Pollution Control Asn. Res: Raman and infrared spectroscopy; molecular structure; automotive exhaust emissions instrumentation and testing. Mailing Add: Harvey Tech Ctr Atlantic Richfield Co Harvey IL 60426

VOELZ, GEORGE LEO, b Wittenberg, Wis, Oct 13, 26; m 50; c 4. OCCUPATIONAL MEDICINE. Educ: Univ Wis, BS, 48, MD, 50. Prof Exp: AEC fel indust med, 51-52; indust physician, Los Alamos Sci Lab, Univ Calif, 52-57; chief med br, Idaho Opers Off, AEC, 57-63, asst dir, Health Serv Lab, 63-67, dir, 67-70; HEALTH DIV LEADER, LOS ALAMOS SCI LAB, 70- Mem: Indust Med Asn; Am Acad Occup Med; Am Col Prev Med; Am Indust Hyg Asn; Health Physics Soc. Res: Occupational health problems, especially in the atomic energy industries; radiological health problems; radiobiological research and radiation dosimetry. Mailing Add: Health Div Los Alamos Sci Lab PO Box 1663 Los Alamos NM 87544

VOELZ, HERBERT GUSTAV, b Kartzig, Ger, Sept 15, 20; US citizen; m 49. MICROBIOLOGY. Educ: Univ Greifswald, dipl biol, 56, Dr rer nat(microbiol), 59. Prof Exp: Sci asst, Inst Microbiol, Univ Greifswald, 56-59; sci assoc, Inst Hyg & Bact, Wernigerode, Ger, 59-60; fel microbiol, Sch Med, Univ Ind, 60-62, asst prof, 62-64; from asst prof to assoc prof, 64-72, PROF MICROBIOL, SCH MED, WVA UNIV, 72- Concurrent Pos: USPHS res grant, 63- Mem: Am Soc Microbiol; Electron Micros Soc Am. Res: Cytology of microorganisms and structure and function of their cell organelles. Mailing Add: Dept of Microbiol WVa Univ Sch of Med Morgantown WV 26505

VOET, ANDRIES, b Amsterdam, Holland, Nov 21, 07; nat US; m 32; c 4. CHEMISTRY. Educ: Univ Amsterdam, PhD(chem), 35. Prof Exp: Instr chem, Univ Amsterdam, 31-35; Neth-Am Found fel, 35; chemist, Van Son's Ink Works, 35-40; res chemist, Gen Printing Ink Corp, 40-43 & Neville Co, 43; dir phys res, J M Huber Corp, 43-70; ASSOC PROF CHEM, UNIV STRASBOURG, 71-; ASSOC PROF, CTR PHYS CHEM OF SOLID SURFACES, NAT CTR SCI RES, 71- Concurrent Pos: Consult, Tech Indust Intel Agency, US Dept Com. Mem: Am Chem Soc; Soc Rheol; Am Phys Soc; fel NY Acad Sci. Res: Printing inks; pigments; resins; carbon blacks; colloidal systems; rheology; surface phenomena; elastomer reinforcement. Mailing Add: Ctr Phys Chem Solid Surfaces 24 President Kennedy Ave 68 Mulhouse France

VOET, DONALD HERMAN, b Amsterdam, Neth, Nov 29, 38; US citizen; m 65; c 2. CRYSTALLOGRAPHY, BIOCHEMISTRY. Educ: Calif Inst Technol, BS, 60; Harvard Univ, PhD(chem), 67. Prof Exp: Res assoc biol, Mass Inst Technol, 66-69; asst prof, 69-74, ASSOC PROF CHEM, UNIV PA, 74- Mem: Am Chem Soc; Am Crystallog Asn. Res: X-ray structural determination of molecules of biological interest such as proteins, nucleic acid model compounds, coenzymes and drugs. Mailing Add: Dept of Chem Univ of Pa Philadelphia PA 19104

VOGAN, ERIC LLOYD, b London, Ont, Sept 3, 24; m 51; c 3. PHYSICS. Educ: Univ Western Ont, BSc, 46, MSc, 47; McGill Univ, PhD(physics), 52. Prof Exp: Sci officer, Defense Res Telecommun Estab, Can, 52-57; Can liaison officer, Lincoln Lab, Mass Inst Technol & Air Force Cambridge Res Ctr, 57-60; sci officer, Defense Res Telecommun Estab, Can, 60-64; assoc prof physics, 64-70, PROF PHYSICS, UNIV WESTERN ONT, 70- Mem: Am Asn Physics Teachers; Am Geophys Union; Can Asn Physicists. Res: Aeronomy; physics of the upper atmosphere. Mailing Add: Dept of Physics Univ of Western Ont London ON Can

VOGE, HERVEY HARPER, chemistry, see 12th edition

VOGE, MARIETTA, b Yugoslavia, July 7, 19; nat US; m 42; c 1. PARASITOLOGY. Educ: Univ Calif, AB, 44, MA, 46, PhD(zool), 50. Prof Exp: From asst to assoc zool, Univ Calif, 44-51; assoc med microbiol, 52-54, from instr to assoc prof, 54-68, PROF MED MICROBIOL, SCH MED, UNIV CALIF, LOS ANGELES, 68- Mem: Am Soc Parasitol; Brit Soc Parasitol; Am Soc Trop Med & Hyg; Brazilian Soc Trop Med. Res: Helminthology, particularly cestodes; systematics; distribution; ecology; development in vitro; anatomy of digestive tract of mammals. Mailing Add: Univ of Calif Sch of Med Los Angeles CA 90024

VOGEL, ALFRED MORRIS, b New York, NY, Mar 11, 15; m 40; c 2. CHEMISTRY, INSTRUMENTATION. Educ: City Col New York, BS, 34; NY Univ, MS, 48, PhD(chem), 50. Prof Exp: Jr chemist, US Dept Navy, 36-41; chemist, New York City Bd Transp, 41-47; asst, NY Univ, 47-49; instr, Sch Indust Technol, 49-51; from asst prof chem to prof chem, Adelphi Univ, 50-67, chmn dept, 53-67; PROF CHEM, C W POST COL, LONG ISLAND UNIV, 67-, CHMN DEPT, 74- Mem: Am Chem Soc. Res: Analytical and chemical instrumentation; methods of assay of pharmaceutical products; coordination compounds. Mailing Add: 77 Cedar Rd Malverne NY 11565

VOGEL, BEATRICE ROESLE, biology, see 12th edition

VOGEL, CALVIN, organic chemistry, organometallic chemistry, see 12th edition

VOGEL, FRANCIS STEPHEN, b Middletown, Del, Sept 29, 19; m 49; c 5. PATHOLOGY. Educ: Villanova Col, AB, 41; Western Reserve Univ, MD, 44; Am Bd Path, dipl, 51. Prof Exp: From asst prof to assoc prof path, Med Col, Cornell Univ, 50-61, asst prof path in surg, 50-61; PROF PATH, MED CTR, DUKE UNIV, 61- Concurrent Pos: Consult, Vet Admin Hosp, New York. Mem: Am Soc Exp Path; Am Asn Pathologists & Bacteriologists. Res: Neuropathology; metabolic function of mitochondrial nucleic acids. Mailing Add: Dept of Path Duke Univ Med Ctr Durham NC 27706

VOGEL, GEORGE, b Prague, Czech, May 28, 24; nat US; m 50; c 2. ORGANIC CHEMISTRY. Educ: Prague Inst Technol, DSc(chem), 50. Prof Exp: Lectr chem, Univ Col, Ethiopia, 51-54; res chemist, Monsanto Chem, Ltd, Eng, 54-55; res fel, Ohio State Univ, 55-56; from asst prof to assoc prof, 56-69, PROF CHEM, BOSTON COL, 69- Mem: Am Chem Soc. Res: Heterocyclic chemistry; steric effects in conjugated systems. Mailing Add: Dept of Chem Boston Col Chestnut Hill MA 02167

VOGEL, GLENN CHARLES, b Columbia, Pa, Mar 7, 43; m 69. INORGANIC CHEMISTRY. Educ: Pa State Univ, University Park, BS, 65; Univ Ill, Urbana, MS, 67, PhD(inorg chem), 70. Prof Exp: Asst prof, 70-74, ASSOC PROF CHEM, ITHACA COL, 74- Concurrent Pos: Am Chem Soc-Petrol Res Fund grant, 73. Mem: Am Chem Soc. Res: Complex formation of metalloporphyrins; hydrogen bonding; ternary copper catechol complexes. Mailing Add: Dept of Chem Ithaca Col Ithaca NY 14850

VOGEL, HENRY, b New York, NY, Sept 2, 16; m 47; c 1. MICROBIOLOGY, SEROLOGY. Educ: La State Univ, BS, 40; NY Univ, MS, 49, PhD(biol), 56. Prof Exp: Bacteriologist, Jewish Mem Hosp, New York, 45-49; sr bacteriologist, Willard Parker Hosp, New York, 49-52; sr bacteriologist, 52-66, SR RES SCIENTIST, BUR LABS, NEW YORK CITY DEPT HEALTH, 66- Mem: NY Acad Sci; Brit Soc Appl Bact; Brit Soc Gen Microbiol. Res: Enteric microbiology; metabolism of cold-blooded acid fast organisms; metabolism of Leptospira; serologic studies of genetic relationships and diagnosis. Mailing Add: City of New York Dept of Health Bur Labs 455 First Ave New York NY 10016

VOGEL, HENRY A, organic chemistry, see 12th edition

VOGEL, HENRY ELLIOTT, b Greenville, SC, Sept 16, 25; m 53; c 4. SOLID STATE PHYSICS. Educ: Furman Univ, BS, 48; Univ NC, MS, 50, PhD(physics), 62. Prof Exp: From instr to assoc prof, 50-65, head dept physics, 67-71, PROF PHYSICS, CLEMSON UNIV, 65-, DEAN, COL SCI, 71- Mem: Am Phys Soc; Am Asn Physics Teachers. Res: Superconductivity; thin vacuum-deposited films; tunneling between films. Mailing Add: Kinard Lab Physics Rm 119 Clemson Univ Clemson SC 29631

VOGEL, HERWARD A, organic chemistry, polymer chemistry, see 12th edition

VOGEL, HOWARD H, JR, b New York, NY, Nov 30, 14; m 40; c 4. ZOOLOGY. Educ: Bowdoin Col, AB, 36; Harvard Univ, MA, 37, PhD(biol), 40. Prof Exp: From asst prof to assoc prof, Wabash Col, 41-48, actg head dept, 46; assoc prof biol sci, Univ Chicago, 48-59; assoc biol & group leader neutron radiobiol, Argonne Nat Lab, 49-67; PROF RADIOL, PHYSIOL & BIOPHYS & HEAD RADIATION BIOL, DEPT RADIOL, MED UNITS, UNIV TENN, MEMPHIS, 67-, PROF RADIATION ONCOL, 73- Concurrent Pos: Ornithologist, Bowdoin-MacMillan Arctic Exped, 34; assoc, Roscoe B Jackson Mem Lab. Mem: AAAS; Am Soc Zool; assoc Am Ornith Union; assoc Arctic Int NAm; Transplantation Soc. Res: History of arctic aviation; social behavior of birds and mammals; skin transplantation; radiobiology; biological effects of neutrons; radiation carcinogenesis. Mailing Add: Dept of Radiol Univ of Tenn Med Units Memphis TN 38103

VOGEL, JAMES ALAN, b Snohomish, Wash, Dec 22, 35; m 59; c 3. EXERCISE PHYSIOLOGY, ENVIRONMENTAL MEDICINE. Educ: Wash State Univ, BS, 57; Rutgers Univ, PhD(physiol), 61. Prof Exp: Res physiologist, US Army Med Res & Nutrit Lab, Fitzsimons Gen Hosp, 61-67, RES PHYSIOLOGIST, US ARMY RES INST ENVIRON MED, 67-, DIR DEPT PHYSIOL, 73- Honors & Awards: Civilian Outstanding Performance Award, US Dept Army, 65. Mem: AAAS; Am Physiol Soc. Res: Cardiac output physiology; exercise and physical fitness training; cardiopulmonary physiology of high altitude. Mailing Add: US Army Res Inst of Environ Med Natick MA 01760

VOGEL, JAMES GARRETT, inorganic chemistry, analytical chemistry, see 12th edition

VOGEL, JAMES JOHN, b Longmont, Colo, June 16, 35; m 60; c 2. BIOCHEMISTRY, NUTRITION. Educ: William Jewell Col, AB, 57; Univ Wis, MS, 59, PhD(biochem), 61. Prof Exp: Res fel dent biochem, Forsyth Dent Ctr, Harvard Univ, 61-63; res assoc biochem, Med Sch, Univ Minn, 63-67; asst mem, 67-71, ASSOC PROF, UNIV TEX DENT SCI INST, HOUSTON, 71- Mem: Int Asn Dent Res; Am Chem Soc. Res: Dietary factors involved in dental caries; nutritional and metabolic aspects of magnesium, phosphorus and fluorine with respect to skeletal tissues; microbiologic calcification; phospholipid-protein interactions in calcification. Mailing Add: Univ of Tex Dent Sci Inst PO Box 20068 Houston TX 77025

VOGEL, KENNETH HENRY, physical chemistry, chemical engineering, see 12th edition

VOGEL, MARTIN, b Los Angeles, Calif, Mar 7, 35; m 63; c 2. ORGANIC CHEMISTRY, POLYMER CHEMISTRY. Educ: Calif Inst Technol, BS, 55, PhD(org chem), 61. Prof Exp: Asst prof org chem, Rutgers Univ, 60-65; SR RES CHEMIST, ROHM AND HAAS CO, 65- Mem: Am Chem Soc. Res: Organic coatings. Mailing Add: 550 Pine Tree Rd Jenkintown PA 19046

VOGEL, NORMAN WILLIAM, b Brooklyn, NY,May 17, 17; m 47; c 4. ZOOLOGY. Educ: Univ Mich, AB, 40, MS, 43; Univ Ind, PhD(zool), 56. Prof Exp: Asst zool, Univ Mich, 42-43; asst physiol, Vanderbilt Univ, 43-44; Lawrason Brown res fel, Saranac Lab, NY, 48-49; instr biol, Champlain Col, 49-52; asst zool, Univ Ind, 52-55; from asst prof biol to assoc prof biol, 56-66, PROF BIOL, WASHINGTON & JEFFERSON COL, 66- Concurrent Pos: USPHS res grant, Nat Cancer Inst, 58-59; Fulbright-Hays lectr physiol, Fac Med, Univ Nangrahar, Afghanistan, 68-69. Mem: AAAS; Am Inst Biol Sci. Res: Role of the pituitary gland in chick growth and development, prior to hatching by accomplishing hypophysectomy through ablation of the free-head region of the early embryo. Mailing Add: Dept of Biol Washington & Jefferson Col Washington PA 15301

VOGEL, PAUL WILLIAM, b Swayzee, Ind, Oct 5, 19; m 48; c 2. ORGANIC CHEMISTRY. Educ: DePauw Univ, AB, 41; Ind Univ, PhD(org chem), 46. Prof Exp: Asst, Ind Univ, 41-44; res chemist, 45-46, res supvr, 47-50, tech asst to dir res & develop, 50-53, supvr org & anal res, 53-59, dir testing, 59-62, DIR CHEM RES, LUBRIZOL CORP, 62- Mem: AAAS; Am Chem Soc. Res: Synthetic organic chemistry; pyroxonium and pyrylium salts; organic phosphorous compounds; lubricant additives. Mailing Add: Lubrizol Corp Wickliffe OH 44092

VOGEL, PETER, b Prague, Czech, Aug 12, 37. NUCLEAR PHYSICS. Educ: Czech Inst Technol, Prague, EngrTechPhysics, 60; Acad Sci USSR, CandSci(physics), 66. Prof Exp: Res fel, Joint Inst Nuclear Res, Dubna, USSR, 62-66, Nuclear Res Inst, Rez, Czech, 66-68 & Niels Bohr Inst, Copenhagen, Denmark, 68-69; res assoc, Nordic Inst Theoret Atomic Physics, Univ Bergen, 69-70; sr res fel physics, 70-75, RES ASSOC PHYSICS, CALIF INST TECHNOL, 75- Res: Nuclear structure theory; vibrations, rotations and deformations of nuclei; intermediate energy physics, mesonic atoms. Mailing Add: Phys 34 Norman Bridge Lab Physics Calif Inst of Technol Pasadena CA 91125

VOGEL, PHILIP CHRISTIAN, b Fargo, NDak, Nov 28, 41; m 66. PHYSICAL ORGANIC CHEMISTRY. Educ: Lawrence Univ, AB & BS, 63; Ind Univ, PhD(chem), 67. Prof Exp: From res assoc to sr res assoc chem, Yeshiva Univ, 67-70; asst prof chem, Col Pharmaceut Sci, Columbia Univ, 70-73; guest scientist, Max Planck Inst Chem, 73-75; RES STAFF SCIENTIST, BASF WYANDOTTE CORP, 75- Mem: Am Chem Soc. Res: Theory of isotope effects in organic reaction mechanisms; theory of structure of aqueous solutions. Mailing Add: BASF Wyandotte Corp Wyandotte MI 48192

VOGEL, PHILIP E, b Climax Springs, Mo, Sept 8, 26; m 53; c 2. GEOGRAPHY. Educ: Cent Mo Univ, BS, 53; Univ Nebr, MA, 56, PhD(geog), 60. Prof Exp: Asst prof geog, Southern Ill Univ, 59-64; assoc prof, Univ Omaha, 64-66; assoc prof, Ore Col Educ, 66-68; PROF GEOG, UNIV NEBR, OMAHA, 68- Mem: Asn Am Geog; Am Geog Soc. Res: Agricultural geography. Mailing Add: Dept of Geog Univ of Nebr Omaha NE 68101

VOGEL, PHILIP H, comparative anatomy, radiation biology, see 12th edition

VOGEL, PHILIP JAMES, b Melrose, Minn, Feb 21, 06; m 36; c 2. MEDICINE. Educ: Col Med Evangelists, MD, 34; Am Bd Neurol Surg, dipl. Prof Exp: Intern, Orange County Hosp, 33-34; resident neurosurg, White Mem Hosp, Los Angeles, 40-41, Lahey Clin, Boston, 41-42 & Boston City Hosp, 42-43; head dept neurosurg, Univ Calif, Irvine-Calif Med Col, 63-69; PROF NEUROSURG, LOMA LINDA UNIV, 69- Concurrent Pos: Pvt practr, 34-40; chmn dept neurosurg, White Mem Med Ctr, 46- Mem: Am Asn Neurol Surgeons; AMA; fel Am Col Surgeons. Res:

Hydrocephalus; multiple sclerosis. Mailing Add: Dept of Neurosurg Loma Linda Univ Loma Linda CA 92354

VOGEL, RALPH A, b Brooklyn, NY, June 13, 23; m 52; c 3. MICROBIOLOGY. Educ: Wagner Col, BS, 46; Univ Buffalo, MS, 49; Duke Univ, PhD(microbiol), 52. Prof Exp: Instr bact, Sch Med, Duke Univ, 52-53; instr, 54-70, ASST PROF MICROBIOL, SCH MED, EMORY UNIV, 70-; MICROBIOLOGIST, VET ADMIN HOSP, 54- Concurrent Pos: Fel, Yale Univ, 52; instr, Ga State Col, 53- Mem: Fel Am Acad Microbiol; Am Soc Microbiol; Sigma Xi; NY Acad Med. Res: Immunology of the mycosis; pathogenesis of bacterial and mycotic diseases. Mailing Add: Dept of Microbiol Emory Univ Sch of Med Atlanta GA 30322

VOGEL, RICHARD CLARK, b Ames, Iowa, Jan 28, 18; m 44; c 3. PHYSICAL CHEMISTRY. Educ: Iowa State Univ, BS, 39; Pa State Univ, MS, 41; Harvard Univ, AM, 43, PhD(chem), 46. Prof Exp: Asst prof chem, Ill Inst Technol, 46-49; sr chemist, Argonne Nat Lab, 49-54, assoc dir, Chem Eng Div, 54-63, div dir, 63-73; MGR, CHEM PROCESS RES & TECHNOL CTR, EXXON NUCLEAR CO, INC, 73- Mem: Am Chem Soc; fel Am Nuclear Soc; Am Inst Chem Engrs. Res: Physical inorganic chemistry; fluorine chemistry; pyrochemical processes; separations processes; chemical problems in nuclear reactor safety; direct conversion of heat to electricity. Mailing Add: Exxon Nuclear Co Inc 2955 George Washington Way Richland WA 99352

VOGEL, RICHARD E, b Chicago, Ill, July 7, 30; m 53; c 2. COMPUTER SCIENCE. Educ: Colo State Univ, BS, 53; Univ NMex, MS, 60. Prof Exp: Staff scientist, Los Alamos Sci Lab, 57-59; consult statist & comput, Corp Econ & Indust Res, 59-60; dir comput facil, Kaman Nuclear Div, Kaman Aircraft Corp, 60-69; PRES, DATA MGT ASSOCS INC, 69- Concurrent Pos: Lectr, Univ Colo, 61-; adj prof, Colo Col, 65- Mem: Am Meteorol Soc. Res: Statistics; meteorology; mathematics. Mailing Add: 809 Aurora Dr Colorado Springs CO 80906

VOGEL, RONALD FRANK, b Maryville, Mo, Dec 27, 41. ACOUSTICS. Educ: Iowa State Univ, BS, 66, MS, 68, PhD(elec eng), 71. Prof Exp: Technician quartz crystal oscillators, Pioneer Cent Div, Bendix Corp, 62-63; fel acoust & Themis prog mat sci, Iowa State Univ, 71-72; scientist, Agridustrial electronics Inc, 72-75; CONSULT, INSTRUMENTS & LIFE SUPPORT DIV, BENDIX CORP, 75- Concurrent Pos: Consult, Iowa State Univ, 72- Res: Properties of dielectrics, piezoelectric acoustics and related mathematical research. Mailing Add: 818 26th St Bettendorf IA 52722

VOGEL, STEVEN, b Beacon, NY, Apr 7, 40; m 63 & 74; c 1. ZOOLOGY, PHYSIOLOGY. Educ: Tufts Univ, BS, 61; Harvard Univ, PhD(biol), 66. Prof Exp: Asst prof, 66-71, ASSOC PROF ZOOL, DUKE UNIV, 71- Res: Fluid flow through and around organisms; induced flow and ventilatory processes in sponges, leaves, burrows and mounds; convective cooling. Mailing Add: Dept of Zool Duke Univ Durham NC 27706

VOGEL, THOMAS A, b Janesville, Wis, July 5, 37; m 60; c 3. GEOLOGY. Educ: Univ Wis, BS, 59, MS, 61, PhD(geol), 63. Prof Exp: Asst prof geol, Rutgers Univ, New Brunswick, 63-68; assoc prof, 68-74, PROF GEOL, MICH STATE UNIV, 74- Concurrent Pos: Vis prof, Univ SC, 74-75. Mem: Geol Soc Am; Am Geophys Union. Res: Feldspars as petrogenetic indicators; textural analysis of rock processes; emplacement of batholiths; petrology of the Grenville rocks in southeast Ontario; igneous rocks of Northwest Africa; coexisting basic and acid melts. Mailing Add: Dept of Geol Mich State Univ East Lansing MI 48824

VOGEL, WILLIS GENE, b Seward, Nebr, Nov 27, 30; m 54; c 4. RANGE SCIENCE. Educ: Univ Nebr, BS, 52; Mont State Col, MS, 61. Prof Exp: Range conservationist, Soil Conserv Serv, USDA, Idaho, 59-60, range conservationist, Forest Serv, Mo, 60-63, RANGE SCIENTIST, FOREST SERV, NORTHEASTERN FOREST EXP STA, USDA, 63- Mem: Am Soc Agron; Soc Range Mgt; Soil Conserv Soc Am. Res: Herbaceous revegetation of coal strip mine spoils. Mailing Add: Northeastern Forest Exp Sta USDA 204 Center St Berea KY 40403

VOGEL, WOLFGANG HELLMUT, b Dresden, Ger, Aug 4, 30; m 61; c 1. BIOCHEMISTRY, PHARMACOLOGY. Educ: Dresden Tech Univ, BS, 49; Stuttgart Tech Univ, MS, 56, PhD(chem), 58. Prof Exp: Fel med, State Univ NY Upstate Med Ctr, 58-59; chemist, Farbwerke Hoechst, Ger, 59-61; res assoc biochem pharmacol, Col Med, Univ Ill, 61-63, instr, 63-64; vis scientist, NIH, 64-65; asst prof pharmacol, Col Med, Univ Ill, 65-67; assoc prof pharmacol, 67-74, PROF PHARMACOL, JEFFERSON MED COL, 74- Concurrent Pos: Med res assoc, L B Mendel Res Lab, Elgin State Hosp, 65-67. Res: Biochemistry of mental disorders; biochemical pharmacology; drug-enzyme-interactions; development of drug assays; neurochemical correlates of behavior. Mailing Add: Dept of Pharmacol Jefferson Med Col Philadelphia PA 19107

VOGELEY, CLYDE EICHER, JR, b Pittsburgh, Pa, Oct 19, 17; m 47; c 2. APPLIED MATHEMATICS. Educ: Carnegie Inst Technol, BFA, 40; Univ Pittsburgh, BS, 44, MS, 46, PhD(appl math), 49. Prof Exp: Teacher pub schs, Pa, 40-41; instr air crew physics, Univ Pittsburgh, 43-44; res engr, Res Labs, Westinghouse Elec Corp, 44-54; asst prof math, Univ Pittsburgh, 49-56; sr scientist, Bettis Atomic Power Lab & adminstr Bettis fel & doctoral progs, Westinghouse Elec Corp, 56-57, adminstr, Bettis Reactor Eng Sch, 57-59, supvr tech training & dir sch, 59-61, mgr Bettis Tech Training, 61-71, MGR, BETTIS REACTOR ENG SCH, BETTIS ATOMIC POWER LAB, 71- Concurrent Pos: Adj prof, Univ Pittsburgh, 56-; Consult math sect, Bettis Shielding Sch, Westinghouse Elec Corp, 56, mem, eng & scientist adv comt, Bettis Atomic Power Lab, 69-,mem statist adv comt, Westinghouse Learning Corp, 70-71. Mem: Am Phys Soc; Inst Elec & Electronics Engrs; AAAS. Res: Microwave component designs; transmitting and receiving systems; crystal mixers; mathematics related to reactor engineering; design and presentation of professional level training programs in field of reactor engineering. Mailing Add: 185 Peach Dr Pittsburgh PA 15236

VOGELFANGER, ELLIOT AARON, b New York, NY, Apr 5, 37; m 58; c 2. POLYMER SCIENCE, ORGANIC CHEMISTRY. Educ: Columbia Univ, BA, 58; Univ Calif, Los Angeles, PhD(phys org chem), 63. Prof Exp: Res chemist, Esso Res & Eng Co, 63-66; group leader polymer sci, Celanese Corp, 66-74; MGR RES & DEVELOP, SOLTEX POLYMER CORP, 74- Concurrent Pos: Lectr, Hunter Col, 65-71. Mem: Am Chem Soc; Soc Plastics Engrs; Am Soc Testing & Mat. Res: Polymer rheology; high temperature polymers; physical organic chemistry; polyolefins research, application, development, and catalysis. Mailing Add: 27 Hibury Dr Houston TX 77024

VOGELFANGER, ISAAC JOEL, b Gliniany, Austria, Sept 8, 09; Can citizen; m 51; c 2. EXPERIMENTAL SURGERY. Educ: John Casimir Univ, Poland, MD, 34, MScD, 38; Royal Col Physicians & Surgeons Can, cert specialist gen surg, 54; FRCS(C), 72. Prof Exp: Assoc biochem, John Casimir Univ, 34-36, asst prof surg, 39-41; chief surgeon, Mil Hosp, 41-43, chief surgeon, Mil Hosp & Labor Camp, 43-49; head dept surg, Govt Hosp, Israel, 49-51; mem surg dept, Mt Sinai Hosp, New York, 51-52; from intern to attend surgeon, 52-62, CHIEF EXP SURG, OTTAWA CIVIC HOSP, 62-; PROF SURG, UNIV OTTAWA, 72- Concurrent Pos: Nat Res Coun awards, 57-; sr res assoc, Ont Heart Found, 64- Mem: Fel Am Col Surgeons; NY Acad Sci; Transplantation Soc; Can Med Asn. Res: Development of the Canadian Vogelfanger stapling instruments, immunological differences in transplantation of specific organs studied in kidney isotransplantation; antral vagal relationship in hydrochloric acid secretion studied on completely denervated fundic pouches; microvascular surgery and physiology of vessels; immunobiology and heart and liver transplantation research. Mailing Add: Dept of Exp Surg Ottawa Civic Hosp Ottawa ON Can

VOGELHUT, PAUL OTTO, b Vienna, Austria, Dec 2, 35; m 60; c 3. BIOPHYSICS. Educ: Univ Calif, Berkeley, AB, 57, PhD(biophys), 62. Prof Exp: Asst prof bioelectronics, Univ Calif, Berkeley, 62-70; SR RES PHYSICIST, AMES CO DIV, MILES LABS INC, 70- Concurrent Pos: Consult, Electro-Neutronics, 65-68; Ames Co div, Miles Labs Inc, 69-70. Mem: AAAS. Res: Analysis of constituents of body fluids for clinical diagnosis by automated techniques. Mailing Add: Ames Co Div Miles Labs Inc 1127 Myrtle St Elkhart IN 46514

VOGELI, BRUCE R, b Alliance, Ohio, Nov 25, 29; m 56; c 2. MATHEMATICS. Educ: Mt Union Col, BS, 51; Kent State Univ, MA, 57; Univ Mich, PhD(math educ), 60. Prof Exp: Assoc prof math, Bowling Green State Univ, 59-65; PROF MATH, TEACHERS COL, COLUMBIA UNIV, 65- Concurrent Pos: Vis prof, Lenin Inst, Moscow, USSR, 64 & Kurukshetra Univ, India, 65; consult, Ministry Educ, Chile, 66-67 & Gen Learning Corp, 66-67. Mem: Math Asn Am. Res: Mathematics education; international mathematical activities. Mailing Add: Dept of Math Teachers Col Columbia Univ New York NY 10027

VOGELMANN, HUBERT WALTER, b Buffalo, NY, Nov 13, 28; m 51; c 2. BOTANY. Educ: Heidelberg Col, BS, 51; Univ Mich, MA, 52, PhD(bot), 55. Prof Exp: Asst prof taxon bot, 59-62, assoc prof, 62-70, PROF BOT, UNIV VT, 70- Res: Ecology of alpine-arctic plants; taxonomy of higher plants. Mailing Add: Dept of Bot Univ of Vt Burlington VT 05401

VOGELSONG, DONALD CLAIR, b York, Pa, Oct 11, 31; m 54; c 1. PHYSICAL CHEMISTRY, POLYMER CHEMISTRY. Educ: Franklin & Marshall Col, BS, 53; Northwestern Univ, PhD(phys chem), 56. Prof Exp: NSF fel, Oxford Univ, 56-57; res chemist, 57-76, SPECIALIST, TEXTILE FIBERS DEPT, E I DU PONT DE NEMOURS & CO, INC, 76- Mem: Am Chem Soc. Res: Crystal structure of polymers by x-ray diffraction; mechanical properties of synthetic polymer fibers. Mailing Add: Textile Fibers Dept E I du Pont de Nemours & Co Martinsville VA 24112

VOGET, FRED W, b Salem, Ore, Feb 12, 13; m 42; c 3. CULTURAL ANTHROPOLOGY, ETHNOLOGY. Educ: Univ Ore, BA, 36; Yale Univ, PhD(anthrop), 48. Prof Exp: Instr anthrop, Univ Nebr, Lincoln, 47-48; asst prof, McGill Univ, 48-52; prof, Univ Ark, Fayetteville, 52-59; prof, Univ Toronto, 59-65; PROF ANTHROP, SOUTHERN ILL UNIV, EDWARDSVILLE, 65- Concurrent Pos: Can Coun fel hist ethnol theory, 64-65; Fulbright-Hays res fel, Ger, 71; mem exec comt, Soc Appl Anthrop, 74-76. Mem: Fel Am Anthrop Asn; Am Ethnol Soc; Soc Appl Anthrop; Soc Am Archaeol. Res: Culture change, especially acculturation; nativistic movements; North American ethnology; history of ethnological theory. Mailing Add: Dept of Anthrop Southern Ill Univ Edwardsville IL 62025

VOGH, BETTY POHL, b Georgetown, Ohio, Apr 19, 27; m 47; c 4. PHARMACOLOGY, PHYSIOLOGY. Educ: Tex Woman's Univ, BA, 46; Univ Fla, PhD(physiol), 64. Prof Exp: Res assoc physiol, 65-66, res assoc pharmacol, 66-67, asst prof, 68-74, ASSOC PROF PHARMACOL, COL MED, UNIV FLA, 74- Res: Physiology and pharmacology of body fluids; pH regulation of cerebrospinal fluid. Mailing Add: Dept of Pharmacol Univ of Fla Col of Med Gainesville FL 32610

VOGL, ALFRED, internal medicine, deceased

VOGL, OTTO, b Traiskirchen, Austria, Nov 6, 27; nat US; m 55; c 2. POLYMER CHEMISTRY. Educ: Univ Vienna, PhD, 50. Prof Exp: Instr chem, Univ Vienna, 48-53; res assoc, Univ Mich, 53-55 & Princeton Univ, 55-56; chemist, E I du Pont de Nemours & Co, Del, 56-70; PROF POLYMER SCI ENG, UNIV MASS, AMHERST, 70- Concurrent Pos: Vis prof, Kyoto Univ, Osaka Univ, 68, Royal Inst Technol, Stockholm, 71 & Univ Freiburg, 73; comt mem macromol chem, Nat Res Coun-Nat Acad Sci, 75- Mem: AAAS; Am Chem Soc (treas div polymer chem, 69-72); Austrian Chem Soc; Japanese Soc Polymer Sci. Res: Ionic and stereoselective polymerization; polyaldehydes; ring opening polymerization; regular copolyamides; reactions on polymers; functional polymers. Mailing Add: Dept of Polymer Sci & Eng Univ of Mass Amherst MA 01002

VOGL, RICHARD J, b Milwaukee, Wis, Jan 19, 32; m 61; c 3. BOTANY, ECOLOGY. Educ: Marquette Univ, BS, 53, MS, 55; Univ Wis, PhD(ecol), 61. Prof Exp: Instr bot, Marquette Univ, 55-56; res asst, Univ Wis, 58-61; PROF BOT, CALIF STATE UNIV, LOS ANGELES, 61- Concurrent Pos: Ed, Ecol Soc Am, 72-75. Mem: Ecol Soc Am; Wildlife Soc. Res: Plant and fire ecology. Mailing Add: Dept of Biol Calif State Univ Los Angeles CA 90032

VOGL, THOMAS PAUL, b Vienna, Austria, July 10, 29; nat US; c 3. BIOMEDICAL ENGINEERING. Educ: Columbia Univ, BA, 52; Univ Pittsburgh, MS, 57; Carnegie-Mellon Univ, PhD(elec eng), 69. Prof Exp: Sr res physicist, Res Lab, Westinghouse Elec Corp, 52-60; head infrared sect, Res Labs, Hughes Aircraft Co, 60-61; mgr optical physics, Res Labs, Westinghouse Elec Corp, 61-69, mgr optics, 69-74; PRIN STAFF OFFICER, ASSEMBLY LIFE SCI, NAT ACAD SCI, 74- Concurrent Pos: Lectr, Univ Calif, Los Angeles, 59-74; mem comt photother in newborn & subcomt bioeng aspects, Div Med, Nat Acad Sci-Nat Res Coun; adj prof radiation physics, Dept Pediat, Col Physicians & Surgeons, Columbia Univ, 73-, mem, Bioeng Inst, 75- Mem: Am Phys Soc; Optical Soc Am; Environ Mutagen Soc; Am Soc Photobiol. Res: Infrared detection and imaging; photoconductive and photoelectric effects and materials; optical imaging and illuminating systems; non-linear optimization; clinical applications of light; phototherapy of hyperbilirubinemia. Mailing Add: 2101 Constitution Ave NW Washington DC 20418

VOGLESONG, WILLIAM FREDERICK, b Saginaw, Mich, May 14, 22; m 46; c 4. PHOTOMETRY, PHOTOGRAMMETRY. Educ: Albion Col, AB, 44; Univ Mich, Ann Arbor, MS, 48. Prof Exp: Develop engr, Manhattan Proj, Tenn Eastman Corp, 44-45; staff physicist, Sharples Chem, 45-46; fel physics, Univ Mich, Ann Arbor, 46-48; asst prof, Eastern Mich Univ, 48-50; develop engr, 50-52, photog engr, 52-65, SUPVR PHYS LAB, PHOTOG TECHNOL DIV, EASTMAN KODAK CO, 65- Concurrent Pos: Lectr, Rochester Inst Technol, 56-; rep comt photog sensitometry, Am Nat Stand Asn, 74- Mem: Soc Photog Sci & Eng. Res: Photographic sensitometry of color materials; application of computers to laboratory measurement and data systems; exposure determination for color printing systems. Mailing Add: Phys Lab Photog Tech Div Eastman Kodak Co Kodak Park Rochester NY 14650

VOGT, ALBERT R, b St Louis, Mo, Apr 6, 38; m 63; c 1. PLANT PHYSIOLOGY. Educ: Univ Mo, BS, 61, MS, 62, PhD(forestry), 66. Prof Exp: Instr forestry, Univ Mo, 65-66; asst prof tree physiol, 66-70, from actg assoc chmn to assoc chmn, Div Forestry, 69-75, ASSOC PROF TREE PHYSIOL, OHIO AGR RES & DEVELOP CTR, OHIO STATE UNIV, 70-, ACTG CHMN, DIV FORESTRY, 75- Mem: Am Soc Plant Physiol; Soc Am Foresters. Res: Physiology of tree growth and development; bud dormancy in oak; flowering of trees. Mailing Add: Dept of Forestry Ohio Agr R&D Ctr Ohio State Univ Wooster OH 44691

VOGT, BERTHOLD RICHARD, organic chemistry, see 12th edition

VOGT, CLIFFORD MARSHALL, b Chicago, Ill, May 19, 27; m 55; c 1. ORGANIC CHEMISTRY, POLYMER CHEMISTRY. Educ: Univ Ill, BS, 50; Purdue Univ, MS, 56, PhD(org chem), 57. Prof Exp: Chemist, Sherwin-Williams Co, Ill, 50-51; lab technician, Argonne Nat Labs, 51-52; Purdue Res Found fel, 57-58; res chemist, Am Cyanamid Co, 58-64; sr res chemist, Celanese Res Co, 64-72; SR RES SCIENTIST, KIMBERLY CLARK CO, 72- Mem: Am Chem Soc. Res: Monomer and polymer synthesis and characterization; wet, dry and melt spinning systems; thermally stable fibers; non-flammable fibers; non-wovens; film system modifications; filtration systems; textile evaluations; fiber color, dyeability and structure modification. Mailing Add: Res & Eng Ctr Kimberly Clark Co Neenah WI 54956

VOGT, DALE WILLIAM, b Collinsville, Ill, Dec 22, 31; m 54; c 1. ANIMAL BREEDING, ANIMAL GENETICS. Educ: Southern Ill Univ, Carbondale, BS, 56; Univ Minn, St Paul, MS, 57, PhD(quant genetics), 61. Prof Exp: Res assoc quant genetics, Univ Tex M D Anderson Hosp & Tumor Inst, 61-62; asst prof, Va Polytech Inst, 62-64, Iowa State Univ, 64-65 & Va Polytech Inst, 65-68; animal prod off, Food & Agr Orgn, UN, Iran, 68-70; ASSOC PROF QUANT GENETICS & ANIMAL CYTOGENETICS, UNIV HAWAII, 70- Concurrent Pos: Nat Inst Child Health & Human Develop res grant, Va Polytech Inst, 67-68. Mem: Am Soc Animal Sci. Res: Quantitative animal genetics; animal cytogenetics. Mailing Add: Dept of Animal Sci Univ of Hawaii Honolulu HI 96822

VOGT, ERICH W, b Steinbach, Can, Nov 12, 29; m 52; c 5. NUCLEAR PHYSICS, THEORETICAL PHYSICS. Educ: Univ Man, BSc, 51, MSc, 52; Princeton Univ, PhD(physics), 55. Prof Exp: Nat Res Coun Can fel, Univ Birmingham, 55-56; from asst res officer to sr res officer, Atomic Energy Can Ltd, 56-65; PROF PHYSICS, UNIV BC, 65- Concurrent Pos: Vis assoc prof, Univ Rochester, 58-59; Nat Res Coun Can sr travelling fel, Oxford Univ, 71-72. Honors & Awards: Centennial Medal Can, 67. Mem: Fel Am Phys Soc; fel Royal Soc Can; Can Asn Physicists (pres, 70-71). Res: Theory of nuclear reactions, nuclear structure and intermediate energy physics. Mailing Add: Dept of Physics Univ of BC Vancouver BC Can

VOGT, EVON ZARTMAN, b Gallup, NMex, Aug 20, 18; m 41; c 4. ANTHROPOLOGY. Educ: Univ Chicago, AB, 41, MA, 46, PhD, 48. Prof Exp: From instr to assoc prof anthrop, Harvard Univ, 48-59, asst cur Am ethnol, Peabody Mus, 50-59, cur Mid Am ethnol, 60-74, chmn dept anthrop, 69-73, PROF SOCIAL ANTHROP, HARVARD UNIV, 59- Concurrent Pos: Fel, Ctr Advan Study Behav Sci, 56-57; mem adv panel anthrop, NSF, 64-66; bd mem div anthrop & psychol, Nat Res Coun, 55-57. Honors & Awards: Bernardino de Sahagun Prize, Repub Mex, 69. Mem: Am Anthrop Asn; Soc Am Archaeol; Am Acad Arts & Sci; Royal Anthrop Soc Gt Brit & Ireland. Res: Social anthropology; symbolic analysis of ritual; middle American ethnology. Mailing Add: Dept of Anthrop Harvard Univ Cambridge MA 02138

VOGT, FRANK CONRAD, b Brooklyn, NY, Apr 24, 20; m 44; c 3. PEDIATRICS. Educ: Wesleyan Univ, AB, 42; Cornell Univ, MD, 45; Am Bd Pediat, dipl. Prof Exp: Intern, Brooklyn Hosp, NY, 45-46, resident pediat, 48-49; asst resident, Long Island Col Hosp & resident, House of St Giles the Cripple, Brooklyn, NY, 49-50; attend pediatrician, Brooklyn Hosp & House of St Giles the Cripple, 50-53; attend physician, Children's Hosp, Columbus, Ohio, 53-65; physician, Dept Clin Invest, Med Res Div, Schering Corp, 65-68, assoc dir clin res, 68-69; CLIN ASST PROF, DEPT PEDIAT, COL MED & DENT NJ, 66-; SR ASSOC DIR CLIN RES, DEPT CLIN INVEST, MED RES DIV, SCHERING CORP, 69- Concurrent Pos: Pvt pract, Brooklyn, NY, 50-53 & Marion, Ohio, 53-65; instr, Col Med, Ohio State Univ, 53-66; mem bd dirs, Marion Ment Health Clin, Family Serv & Frederick C Smith Med Found, Ohio, 53-66; chief pediat, Marion Gen Hosp, Ohio, 54-65; chief staff & pediat, Commun Mem Hosp, Ohio, 64-65. Mem: Fel Am Acad Pediat; fel Am Acad Allergy; Am Acad Clin Toxicol; AMA; NY Acad Sci. Res: Pediatric allergy; clinical research. Mailing Add: Dept of Clin Invest Med Res Div Schering Corp Bloomfield NJ 07003

VOGT, GEORGE BRITTON, b Baltimore, Md, Apr 10, 20. BIOSYSTEMATICS. Educ: Univ Md, BS, 41, MS, 49. Prof Exp: Entomologist malaria control in war areas, USPHS, 42-45 & dysentery & diarrhea control proj, 45-47; asst prof entom, Univ Md, 47-49; entomologist, 49-72, RES ENTOMOLOGIST, SOUTHERN WEED SCI LAB, AGR RES SERV, USDA, 72- Concurrent Pos: Entomologist, Tech & Econ Mission, Burma, 51-53; explor, Spain & Southwest Asia, 56 & SAm, 60-62. Mem: Entom Soc Am; Ecol Soc Am; AAAS; Am Inst Biol Sci; Soc Syst Zool. Res: Host plant specificity and biosystematics of alticine Chrysomelidae and leaf-mining Buprestidae; biological control of alligator weed. Mailing Add: Southern Weed Sci Lab USDA PO Box 225 Stoneville MS 38776

VOGT, HERWART CURT, b Elizabeth, NJ, Sept 14, 29; m 58; c 2. POLYMER CHEMISTRY. Educ: Northwestern Univ, 52; Univ Del, MS, 54, PhD(chem), 57. Prof Exp: Res chemist, Hercules Inc, 57-59; sr res chemist, 59-66, res assoc, 67-73, SUPVR, BASF WYANDOTTE CORP, 73- Concurrent Pos: Vis lectr, Oakland Univ, 66-69 & Wayne State Univ, 69-71; exchange chemist, BASF-AG, WGer, 71-73. Mem: Am Chem Soc; Soc Plastics Eng; Sigma Xi; The Chem Soc. Res: Organic phosphorus compounds pertaining to polymers; novel halogen containing unsaturated polyesters; isocyanate and urethane chemistry; chlorine containing elastomers; research and development in noncellular urethane plastics. Mailing Add: Cent Res BASF Wyandotte Corp Wyandotte MI 48192

VOGT, JAMES ROBERT, nuclear chemistry, see 12th edition

VOGT, LESTER HERBERT, JR, b New York, NY, Nov 9, 31. INORGANIC CHEMISTRY. Educ: Hofstra Univ, BA, 55; Rensselaer Polytech Inst, MS, 61, PhD(chem), 64. Prof Exp: Chemist, US testing Co, 55-56; res chemist, Gen Elec Res Lab, 56-62 & Res & Develop Ctr, 65-70, supvr, Advan Process Lab, Chem Prod Sect, 70-73, MGR LAMP CHEM ENG, QUARTZ & CHEM PROD SECT, GEN ELEC CO, NY, 73- Concurrent Pos: NIH spec fel, Univ Calif, Berkeley, 64-65. Mem: Am Chem Soc. Res: Silicones; organic semiconductors; coordination compounds; x-ray crystallography; synthetic oxygen-carrying chelates; organic and inorganic phosphors. Mailing Add: Quartz & Chem Prod Sect Gen Elec Co 1099 Ivanhoe Rd Cleveland OH 44110

VOGT, MARGUERITE MARIA PAULETTE, b Berlin, Ger, Feb 19, 13; nat US. MEDICINE. Educ: Univ Berlin, MD, 37. Prof Exp: Sr res fel animal virol, Calif Inst Technol, 50-63; res assoc, 63-72, SR RES ASSOC ANIMAL VIROL, SALK INST BIOL STUDIES, 72- Res: Genetics; virology. Mailing Add: Salk Inst for Biol Studies PO Box 1809 San Diego CA 92112

VOGT, MOLLY THOMAS, b Lyndhurst, Eng, Apr 15, 39; m 64; c 2. BIOCHEMISTRY. Educ: Bristol Univ, BSc, 60; Univ Pittsburgh, PhD(biochem), 67. Prof Exp: Jr res officer, Toxicol Unit, Med Res Coun, Eng, 60-62; res asst biochem path, Sch Med, 62-63, asst prof biochem, 70-72, chmn div health related prof interdisciplinary progs, 72-74, ASSOC PROF BIOCHEM, SCH HEALTH RELATED PROF, UNIV PITTSBURGH, 72- Concurrent Pos: NIH fel biochem, Sch Med, Univ Pittsburgh, 67-70, Health Res Serv Found grant, 70-71; Am Coun Educ Admin intern, 74-75. Mem: AAAS; Am Inst Biol Scientists. Res: Cell metabolism in health and disease; phagocytic process; mitochondrial metabolism; biochemical effects of typical air pollutants. Mailing Add: Sch Health Related Professions Univ of Pittsburgh Pittsburgh PA 15261

VOGT, PETER KLAUS, b Braunau, Ger, Mar 10, 32; US citizen; m 61; c 1. BIOLOGY, VIROLOGY. Educ: Univ Tübingen, PhD, 59. Prof Exp: From asst prof to assoc prof path, Sch Med, Univ Colo, 62-67; from assoc prof to prof microbiol, Sch Med, Univ Wash, 67-71; HASTINGS PROF MICROBIOL, SCH MED, UNIV SOUTHERN CALIF, 71- Concurrent Pos: Damon Runyon cancer fel, Virus Lab, Univ Calif, Berkeley, 59-62; res grants, USPHS, 62- & Am Cancer Soc, 63-68; mem virol study sect, NIH, 67-71; mem cell biol & virol adv comt, Am Cancer Soc, 72-76. Mem: AAAS; Am Soc Microbiol. Res: Mechanism of neoplastic cellular transformation induced by viruses. Mailing Add: Dept of Microbiol Univ of Southern Calif Sch Med Los Angeles CA 90033

VOGT, PETER RICHARD, b Hamburg, Ger, June 8, 39; US citizen; m 67; c 1. MARINE GEOPHYSICS. Educ: Calif Inst Technol, BS, 61; Univ Wis, MA, 65, PhD(oceanog), 68. Prof Exp: GEOPHYSICIST, US NAVAL OCEANOG OFF, 67- Honors & Awards: Henry A Kaminski Award, Sci Res Soc Am. Mem: Am Geophys Union; fel Geol Soc Am. Res: Geophysical research on the constitution and history of the crust beneath the sea, especially the analysis of marine magnetic field anomalies and mantle hot spot phenomena as related to ocean floor movement, crustal composition and continental drift. Mailing Add: Code 6120 US Naval Oceanog Off Chesapeake Beach MD 20732

VOGT, ROCHUS E, b Neckarelz, Ger, Dec 21, 29; m 58; c 2. PHYSICS. Educ: Univ Chicago, SM, 57, PhD(physics), 61. Prof Exp: Res assoc cosmic rays, Univ Chicago, 61-62; from asst prof to assoc prof, 62-70, PROF PHYSICS, CALIF INST TECHNOL, 70- Mem: Am Phys Soc; Am Asn Physics Teachers; Am Geophys Union; AAAS. Res: Cosmic rays; astrophysics. Mailing Add: George W Downs Lab 220-47 Calif Inst of Technol Pasadena CA 91125

VOGT, THOMAS CLARENCE, JR, b San Antonio, Tex, Sept 21, 32; m 63; c 6. PHYSICAL CHEMISTRY. Educ: St Mary's Univ, BS, 54, PhD(phys chem), 61. Prof Exp: Res chemist, Mobil Oil Corp, 57; asst, Univ Notre Dame, 57-58, fel diffusion kinetics, Radiation Proj, 58-61; ASSOC CHEMIST, FIELD RES LAB, MOBIL RES & DEVELOP CORP, 61- Mem: Sigma Xi; Am Chem Soc; Soc Petrol Eng. Res: Reaction kinetics of heterogeneous systems; diffusion and recombination of free radicals in liquid systems; effects of high pressure on reaction rates; chemical stimulation of petroleum production wells; hydraulic fracturing of subsurface formations. Mailing Add: Mobil Res & Develop Corp PO Box 900 Dallas TX 75221

VOHRA, PRAN NATH, b Gwaliar, India, June 11, 19. NUTRITION, BIOCHEMISTRY. Educ: Univ Panjab, India, MSc, 42; Wash State Univ, MS, 54; Univ Calif, Davis, PhD(nutrit), 59. Prof Exp: Res asst chem, Sci & Indust Res Orgn, India, 42-49; int trainee fermentations, Joseph E Seagram & Sons, Ky, 49-50; specialist nutrit, Dept Poultry Husb, Univ Calif, Davis, 58-59; asst, BO&C Mills, Eng, 59-60; specialist poultry, US AID India, 61-62; asst res nutritionist, Dept Poultry Husb, 62-70, assoc prof avian sci, 70-74, PROF AVIAN SCI, UNIV CALIF, DAVIS, 74- Honors & Awards: Award, Am Feed Mfrs Asn. Mem: Poultry Sci Asn; Am Inst Nutrit; Brit Biochem Soc; Brit Nutrit Soc. Res: Trace elements in nutrition; improvement of nutrition in developing countries; comparative nutrition of avian species. Mailing Add: Dept of Avian Sci Univ of Calif Davis CA 95616

VOHS, PAUL ANTHONY, JR, b Kansas City, Kans, Jan 19, 31; m 53; c 5. ZOOLOGY. Educ: Kans State Univ, BS, 55; Univ Southern Ill, MA, 58; Iowa State Univ, PhD, 64. Prof Exp: Proj leader, Coop Proj, Ill Dept Conserv, Ill Natural Hist Surv & Southern Ill Univ, 55-58; res assoc, Coop Wildlife Res Proj, Univ Southern Ill, 59-61; res asst, Coop Wildlife Res Unit, Iowa State Univ, 61-62, instr zool & entom, 63-64, asst prof zool, 64-68, assoc prof wildlife biol, 68; assoc prof wildlife ecol, Ore State Univ, 68-73, prof & exten wildlife specialist, Coop Exten Serv & Dept Fisheries & Wildlife, 73-74; PROF WILDLIFE & FISHERIES SCI & HEAD DEPT, S DAK STATE UNIV, BROOKINGS, 74- Mem: AAAS; Wilson Ornith Soc; Wildlife Soc; Am Soc Mammal; Ecol Soc Am. Res: Vertebrate ecology; response of birds and mammals to manipulations of habitat; serology; genetics, biology and ecology of wild ungulates. Mailing Add: Dept Wildlife & Fisheries SDak State Univ Brookings SD 57006

VOICHICK, MICHAEL, b Yonkers, NY, May 28, 34; m 60; c 3. MATHEMATICS. Educ: Oberlin Col, BA, 57; Brown Univ, PhD(math), 62. Prof Exp: Res instr math, Dartmouth Col, 62-64; asst prof, 64-68, ASSOC PROF MATH, UNIV WIS-MADISON, 68- Mem: Am Math Soc; Math Asn Am. Res: Function theory. Mailing Add: Dept of Math Univ of Wis Madison WI 58706

VOIGHT, BARRY, b Yonkers, NY, Dec 17, 37; m 59; c 2. GEOLOGY. Educ: Univ Notre Dame, BS, 59 & 60, MS, 61; Columbia Univ, PhD(struct geol), 65. Prof Exp: Asst prof eng geol, 64-70, ASSOC PROF GEOL, PA STATE UNIV, 70- Concurrent Pos: Res grants, Pa State Mineral Conserv Fund, 66-67 & NSF, 67-; with Dept Mineral Eng, 68-; vis prof, Delft Technol Inst, Univ Toronto, 72-73. Mem: Am Soc Civil Eng; Asn Eng Geol; Geol Soc Am; Int Soc Rock Mech; Am Acad Mech. Res: Geology of eastern North America; stress measurement instrumentation and interpretation; residual stresses in rocks; boudinage, fault, fold mechanics; engineering geology; rock mechanics; mechanics of landslides; geology of Yellowstone Park area. Mailing Add: Dept Geol Col Earth & Mineral Sci Pa State Univ University Park PA 16802

VOIGT, ADOLF FRANK, b Upland, Calif, Jan 31, 14; m 41; c 2. RADIOCHEMISTRY. Educ: Pomona Col, BA, 35; Claremont Cols, MA, 36; Univ Mich, PhD(phys chem), 41. Prof Exp: Instr chem, Smith Col, 41-42; chemist, Off Sci Res & Develop Contract, 42-43, chemist & group leader, Manhattan Proj, 43-46, from asst prof to assoc prof, 46-55, PROF CHEM, IOWA STATE UNIV, 55-, SR CHEMIST, INST ATOMIC RES, UNIV & AMES LAB, US ENERGY RES & DEVELOP ADMIN, 55-, ASST DIR, 59-, CHIEF REACTOR DIV, 68- Concurrent Pos: Asst to dir, Inst Atomic Res, Iowa State Univ & Ames Lab, 50-59. Mem: AAAS; Am Phys Soc; Am Chem Soc; Am Nuclear Soc; Am Inst Chemists. Res: Radiochemistry; activation analysis, chemical results of nuclear transformations;

nuclear chemistry; separation of short-lived fission products and study of their decay. Mailing Add: Ames Lab ERDA Iowa State Univ Ames IA 50011

VOIGT, CHARLES FREDERICK, b Woodside, NY, Dec 17, 42; m 62; c 3. ORGANIC CHEMISTRY. Educ: Univ SFla, Tampa, BA, 65; Duke Univ, PhD(org chem), 70. Prof Exp: Assoc indexer org chem, 70-71, sr assoc ed macromolecular chem, 71-73, SR ED APPL CHEM, CHEM ABSTR SERV, OHIO STATE UNIV, 73- Mem: Am Chem Soc. Res: Heterocyclic chemistry; polymers; applied chemistry. Mailing Add: Chem Abstr Serv Ohio State Univ Columbus OH 43210

VOIGT, DAVID QUENTIN, b Reading, Pa, Aug 9, 26; m 51; c 2. ANTHROPOLOGY, SOCIOLOGY. Educ: Albright Col, BS, 48; Columbia Univ, MA, 49; Syracuse Univ, PhD(sociol, anthrop), 62. Prof Exp: PROF SOCIOL & ANTHROP, ALBRIGHT COL, 64- Concurrent Pos: Adj prof, Franklin & Marshall Col, 70- Mem: Am Sociol Asn; Am Anthrop Asn. Res: Sociology of leisure and sports; American baseball. Mailing Add: Dept of Sociol Albright Col Reading PA 19604

VOIGT, EVA-MARIA, b Dortmund, WGer, Feb 2, 28; Can citizen. PHYSICAL CHEMISTRY. Educ: McMaster Univ, BSc, 53, MSc, 54; Univ BC, PhD(phys chem), 63. Prof Exp: Head res sect, Aylmer Foods, Inc, 55-56; lectr chem, Mt Allison Univ, 56-57; fel phys chem, Univ Calif, Berkeley, 63-65; asst prof, 66-69, ASSOC PROF CHEM, SIMON FRASER UNIV, 69- Mem: AAAS; Am Chem Soc; Am Phys Soc; Chem Inst Can; Can Inst Phys. Res: Molecular spectroscopy; charge-transfer interactions; energy transfer. Mailing Add: Dept of Chem Simon Fraser Univ Burnaby BC Can

VOIGT, GARTH KENNETH, b Merrill, Wis, Jan 17, 23; m 46; c 3. SOILS, PLANT NUTRITION. Educ: Univ Wis, BS, 48, MS, 49, PhD(soils), 51. Prof Exp: From instr to asst prof soils, Univ Wis, 51-55; from asst prof to prof, 55-67, actg dean, Sch Forestry, 70-71 & 75-76, dir admis, 70-75, dir grad studies, Dept Forestry, 71-75, MARGARET K MUSSER PROF FOREST SOILS, YALE UNIV, 67- Concurrent Pos: Collabr, Lake States Forest Exp Sta, US Forest Serv, 54-60. Mem: Am Soc Plant Physiol; Soil Sci Soc Am; Am Soc Agron. Res: Relationships between soil and the growth of plants. Mailing Add: Yale Univ Sch of Forestry New Haven CT 06511

VOIGT, JOHN L, pharmacy, see 12th edition

VOIGT, JOHN WILBUR, b Sullivan, Ind, July 6, 20; m 43; c 2. PLANT ECOLOGY. Educ: Univ Nebr, PhD, 50. Prof Exp: From asst prof to assoc prof, 50-60, dean gen studies div, 62-73, PROF BOT, SOUTHERN ILL UNIV, CARBONDALE, 60- Mem: Ecol Soc Am; Soc Range Mgt; Sigma Xi. Res: Geography of southern Illinois vascular plants; vegetation of southern Illinois; prairie and pasture research. Mailing Add: Rte 4 Carbondale IL 62901

VOIGT, PAUL WARREN, b Ann Arbor, Mich, Mar 20, 40; m 63; c 3. PLANT BREEDING. Educ: Iowa State Univ, BS, 62; Univ Wis, MS, 64, PhD(agron), 67. Prof Exp: Res geneticist, Southern Great Plains Field Sta, 67-74, RES GENETICIST, GRASSLAND-FORAGE RES CTR, AGR RES SERV, USDA, 74- Mem: Am Soc Agron; Crop Sci Soc Am; Soc Range Mgt; AAAS. Res: Forage grass breeding and genetics. Mailing Add: Grassland-Forage Res Ctr PO Box 748 Temple TX 76501

VOIGT, ROBERT GARY, b Olney, Ill, Dec 21, 39; m 62; c 1. NUMERICAL ANALYSIS. Educ: Wabash Col, BA, 61; Purdue Univ, West Lafayette, MS, 63; Univ Md, College Park, PhD(math), 69. Prof Exp: Res assoc, Comput Sci Dept, Univ Md, 69-70, vis asst prof, 70-71; mathematician, Naval Ship Res & Develop Ctr, Washington, DC, 71-73; ASST DIR, INST COMPUT APPLNS IN SCI & ENG, 73- Mem: Soc Indust & Appl Math; Asn Comput Mach; Am Math Soc; AAAS. Res: Numerical analysis for parallel and vector computers and the application of micro processors to scientific computing. Mailing Add: ICASE MS-132 C NASA-Langley Res Ctr Hampton VA 23665

VOIGT, ROBERT LEE, b Hebron, Nebr, Nov 23, 24; m 51; c 4. PLANT BREEDING. Educ: Univ Nebr, BS, 49, MS, 55; Iowa State Univ, PhD(crop breeding), 59. Prof Exp: Instr soybeans, Iowa State Univ, 55-59; from asst prof & asst plant breeder to assoc prof & assoc plant breeder, 59-69, PROF PLANT BREEDING & PLANT BREEDER, AGR EXP STA, UNIV ARIZ, 69- Concurrent Pos: Ed, Sorghum Newsletter, Sorghum Improvement Conf NAm, 72- Mem: Am Soc Agron. Res: Crop breeding; forage and grain sorghum; soybeans. Mailing Add: Dept of Plant Sci Univ of Ariz Tucson AZ 85721

VOIGT, WALTER, b Havana, Cuba, Feb 26, 38; US citizen; m 61; c 3. BIOCHEMISTRY, ENDOCRINOLOGY. Educ: Univ Villanueva, Cuba, MS, 60; Univ Miami, PhD(biochem), 68. Prof Exp: Res assoc skin biochem, 69-70, asst prof dermat, 70-74, ASSOC PROF DERMAT & ONCOL, SCH MED, UNIV MIAMI, 74- Concurrent Pos: Fel bile acid metab, Sch Med, Univ Miami, 68-69, Am Cancer Soc grant, 70-71, Nat Cancer Inst grant, 72-75. Mem: AAAS; Brit Biochem Soc; Am Chem Soc; Endocrine Soc; Am Fedn Clin Res. Res: Mechanism of androgen action and prostatic neoplasia; enzymes of bile acids and steroid metabolism; biochemistry of the skin; membrane electron transport; cancer. Mailing Add: Dept of Dermat Univ of Miami Sch of Med Miami FL 33152

VOISINET, DONALD LOUIS, b Buffalo, NY, May 5, 19; m 48; c 8. CHEMISTRY. Educ: Canisius Col, BS, 41, MS, 48. Prof Exp: Anal chemist res lab, Linde Prod Co Div, 41-43, anal chemist, Atomic Energy Comn Proj, Ceramics & Chandler Plant, 43-48, assoc chemist, Silicone Chems Dept, 49-57, CHIEF CHEMIST, SILICONES DIV, UNION CARBIDE CORP, 57- Mem: Am Chem Soc. Res: Extraction of gold from ores; rare gas purification; silicone chemistry. Mailing Add: Union Carbide Corp Sisterville WV 26175

VOITLE, ROBERT ALLEN, b Parkersburg, WVa, May 12, 38; m 59; c 2. POULTRY PHYSIOLOGY. Educ: Univ WVa, BS, 62, MS, 64; Univ Tenn, PhD(physiol), 69. Prof Exp: Asst prof physiol & asst poultry physiologist, 69-74, ASSOC PROF PHYSIOL & ASSOC POULTRY PHYSIOLOGIST, UNIV FLA, 74- Mem: Poultry Sci Asn. Res: Environmental and reproductive physiology with special emphasis on the effect of nutrition and photoperiod; breeding and genetics, especially radiation effects. Mailing Add: Dept of Poultry Sci Univ of Fla Gainesville FL 32611

VOKES, EMILY HOSKINS, b Monroe, La, May 21, 30; m 59. INVERTEBRATE PALEONTOLOGY, MALACOLOGY. Educ: Tulane Univ, La, BS, 60, MS, 62, PhD(paleont), 67. Prof Exp: Cur paleont, Dept Geol, 57-74, ASSOC PROF EARTH SCI, TULANE UNIV, 73-, CHMN DEPT, 74- Concurrent Pos: Lectr geog, Tulane Univ, 69-, assoc ed, Tulane Studies Geol & Paleont, 70-; vis prof, Univ Rio Grande do Sul, Brazil, 71- Mem: Am Malacol Union; Paleont Soc; Malacol Soc London; Paleont Res Inst. Res: Systematic paleontology and zoology of Cenozoic Gastropoda, including both fossil and recent members. Mailing Add: Dept of Earth Sci Tulane Univ New Orleans LA 70118

VOKES, HAROLD ERNEST, b Windsor, Ont, June 27, 08; nat US; m 32 & 59; c 4. STRATIGRAPHY, INVERTEBRATE PALEONTOLOGY. Educ: Occidental Col, BA, 31; Univ Calif, PhD(paleont), 35. Prof Exp: Hon fel paleont, Yale Univ, 35-36; asst geologist, State Geol Surv, Ill, 37; from asst cur to assoc cur invert paleont, Am Mus Natural Hist, 37-43, actg chmn dept invert, 43; geologist, US Geol Surv, 43-45; from assoc prof to prof geol, Johns Hopkins Univ, 45-56; prof, 56-72, chmn dept, 57-67 & 70-71, W R IRBY PROF GEOL, TULANE UNIV, 72- Concurrent Pos: Guggenheim fel, Am Univ Beirut, 40; geologist, US Geol Surv, Pl, 52-53; vis prof, Univ Rio Grande do Sul, Brazil, 71; mem, Int Comn Zool Nomenclature; trustee, Paleont Res Inst, 72- Mem: Fel Geol Soc Am (vpres, 52); Soc Study Evolution; Paleont Soc (secy, 40-49, pres, 51). Res: Cretaceous and Tertiary stratigraphy and molluscan paleontology; fossil and recent pelecypoda. Mailing Add: Dept of Geol Tulane Univ New Orleans LA 70118

VOLAVKA, JAN, b Prague, Czech, Dec 29, 34; m 64. PSYCHIATRY, ELECTROPHYSIOLOGY. Educ: Charles Univ, Prague, BA & MD, 59; Czech Acad Sci, PhD(med sci), 65. Prof Exp: Intern internal med, Psychiat Hosp, Horni Berkovice, Czech, 59-60, resident psychiat, 60-63; resident psychiatrist, Psychiat Res Inst, Prague, 63-66; electroencephalographer, London Hosp, Eng, 66-67; resident psychiatrist, Psychiat Res Inst, Prague, 67-68; fel neurophysiol, Max Planck Inst Psychiat, 68-69; asst prof, 69-73, ASSOC PROF PSYCHIAT, NEW YORK MED COL, 73- Concurrent Pos: Prin investr, Nat Inst Drug Abuse grant, 73-76. Mem: Soc Biol Psychiat; Am Electroencephalog Soc. Res: Psychopharmacology; EEG; drug addiction; experimental design; statistics. Mailing Add: 400 Central Park West New York NY 10025

VOLBORTH, ALEXIS, b Viipuri, Finland, July 11, 24; nat US; m 47; c 7. GEOCHEMISTRY, ANALYTICAL CHEMISTRY. Educ: Univ Helsinki, PhC, 50, PhLic & PhD(geol, mineral), 54. Prof Exp: Res asst, Geol Surv, Finland, 50, field asst, 52; asst, Inst Technol, Finland, 50-51; field geologist, Finnish Mineral Co, 53; sr asst geol, Univ Helsinki, 53-54; traveling res fel, Outokumpu Found, 54-55 & Calif Inst Technol, 55-56; from asst mineralogist to mineralogist, Nev Mining Anal Lab, Univ Nev, Reno, 56-68, res assoc & consult, Desert Res Inst, 61-62, assoc prof, Univ, 63, prof, 64-68, mem radioactivity safety bd, 64-66; Killam vis prof geol, Dalhousie Univ, 68-71, Killam res prof, 71-72; vis prof, Lunar Sci Inst, Univ Houston, 72-73; vis res chemist, Univ Calif, Irvine, 73-75; PROF GEOL & CHEM, NDAK STATE UNIV, 75- Concurrent Pos: Australian Acad Sci sr fel, 65; J S Guggenheim Mem Found fel, 65-66; adj prof geol, Mackay Sch Mines, Univ Nev, Reno, 69-73; prin investr, Stoichiometry Study of Lunar Rocks, NASA, 72-73; vis prof, Univ Calif, Irvine, 75- Consult, US AEC, 61-63, NASA, 65-73 & Anaconda Co, 68. Honors & Awards: White Cross, Finnish Chem Soc, 55. Mem: Fel Mineral Soc Am; fel Am Inst Chemists; Am Chem Soc; Am Nuclear Soc; Soc Econ Geologists. Res: Geochemistry and analytical chemistry of complex systems, mainly nondestructive instrumental neutron activation and x-ray, fluorescence analysis of major and trace elements; oxygen stoichiometry in rocks, minerals, chemicals and industrial products; mineralogy of and deficiency of oxygen in lunar rocks and fines; nondestructive analysis of coal and lignite. Mailing Add: Dept of Geol NDak State Univ Fargo ND 58102

VOLBRECHT, STANLEY GORDON, b Lodi, Calif, Sept 12, 23; m 45; c 4. MINING GEOLOGY. Educ: Col of the Pac, BA, 53; Stanford Univ, MS, 62. Prof Exp: Explor geologist, Am Copper Co, 54-55; instr geol, Stockton Col, 56-61; from asst prof to assoc prof, 61-70, PROF GEOL, UNIV OF THE PAC, 70-, CHMN DEPT GEOL & GEOG, 66- Mem: Nat Asn Geol Teachers; Geol Soc Am. Res: Economics. Mailing Add: Dept of Geol & Geog Univ of the Pac Stockton CA 95204

VOLCANI, BENJAMIN ELAZARI, b Ben-Shemen, Israel, Jan 4, 15; m 48; c 1. MICROBIOLOGY, BIOCHEMISTRY. Educ: Hebrew Univ, MSc, 36, PhD, 41. Prof Exp: Vis scientist microbiol, Inst Tech, Delft Univ, 37-38 & chem, State Univ Utrecht, 38-39; mem staff, Sieff Res Inst, Weizmann Inst, 39-58; PROF MICROBIOL, SCRIPPS INST OCEANOG, UNIV CALIF, 69- Concurrent Pos: Res fel, Univ Calif, Berkeley, 45-46, res assoc biochem, 56-59; res fel microbiol, Hopkins Marine Sta, Stanford Univ, 46-47 & Calif Inst Technol, 47; res fel biochem, Univ Wis, 48; res assoc, Pasteur Inst, Paris, 51; vis prof, Univ Col, Welsh Nat Sch Med, Cardiff, Wales, UK, 73-74. Mem: Am Soc Cell Biol; Am Soc Microbiol; Soc Gen Microbiol; Brit Biochem Soc. Res: Microbial metabolism and ecology; antimetabolites; bacterial pigments; halophilic microorganisms; biochemistry and ultra-fine structure of the diatoms; siliceous organisms and dinoflagellates; silicon metabolism; mineralization in biological systems; role of silicon in life processes and pathogenicity. Mailing Add: Scripps Inst of Oceanog Univ of Calif La Jolla CA 92093

VOLCHECK, EMIL JOHN, JR, organic chemistry, see 12th edition

VOLCHOK, HERBERT LEE, b New York, NY, Oct 29, 26; m 51; c 5. GEOCHEMISTRY. Educ: Utica Col, AB, 49; Columbia Univ, MA, 51, PhD(geochem), 55. Prof Exp: Asst geochem, Columbia Univ, 49-52, res assoc, 52-55; off & tech dir, Isotopes, Inc, 55-64; PHYS SCIENTIST, HEALTH & SAFETY LAB, US ENERGY RES & DEVELOP ADMIN, 64- Mem: AAAS; Am Geophys Union; Am Geog Soc. Res: Isotope geochemistry; oceanography; distribution of debris from nuclear weapons tests; studies of energy related environmental pollution. Mailing Add: US Energy Res & Dev Admin Health & Safety Lab 376 Hudson St New York NY 10014

VOLCKMANN, RICHARD PETER, b Mt Vernon, NY, Oct 14, 38; m 60; c 1. GEOLOGY. Educ: Colgate Univ, AB, 60; Univ Mich, Ann Arbor, MS, 61, PhD(geol), 65. Prof Exp: Horace Rackham Sch Grad Studies, Univ Mich fel, State Univ Utrecht & Tarragona, Spain, 65-66; GEOLOGIST, US GEOL SURV, 67- Mem: AAAS; Geol Soc Am. Res: Aereal geology; structural geology and tectonics; petrology-petrography; glacial geology and chronology. Mailing Add: US Geol Surv 928 Nat Ctr Reston VA 22092

VOLD, BARBARA SCHNEIDER, b Oakland, Calif, Jan 3, 42. BIOCHEMISTRY, MICROBIOLOGY. Educ: Univ Calif, Berkeley, BA, 63; Univ Ill, MS, 64, PhD(cell biol), 67. Prof Exp: NIH fel biol, Mass Inst Technol, 67-69; ASSOC MICROBIOL, SCRIPPS CLIN & RES FOUND, 69- Concurrent Pos: Nat Inst Gen Med Sci career develop award, 71-76; consult, Physiol Chem Study Sect, NIH, 73-77. Mem: Am Soc Microbiol; Am Soc Biol Chem. Res: Structure and function of transfer ribonucleic acids; changes in nucleic acids during development. Mailing Add: Dept of Microbiol Scripps Clin & Res Found La Jolla CA 92037

VOLD, MARJORIE JEAN, b Ottawa, Ont, Oct 25, 13; US citizen; m 36; c 3. COLLOID CHEMISTRY. Educ: Univ Calif, BS, 34, PhD(chem), 36. Prof Exp: Jr res assoc chem, Stanford Univ, 37-41; res assoc & lectr, 41-58, ADJ PROF CHEM, UNIV SOUTHERN CALIF, 58- Concurrent Pos: Res chemist, Union Oil Co Calif, 42-46; Guggenheim Mem fel, State Univ Utrecht, 53-54. Honors & Awards: Garvan Medal, Am Chem Soc, 67. Mem: Am Chem Soc. Res: Association colloids; mesomorphic phases; gels and other colloidal solids; stability of emulsions, foams, films and suspensions; adsorption; rheology; computer simulation of colloidal processes. Mailing Add: 17465 Plaza Animado 144 San Diego CA 92128

VOLD, REGITZE ROSENØRN, b Copenhagen, Denmark, July 2, 37; US citizen; m 72. NUCLEAR MAGNETIC RESONANCE. Educ: Tech Univ Denmark, MS, 60, Lic Techn(org chem), 62. Prof Exp: Lectr org chem, Tech Univ Denmark, 62; fel chem, Univ NMex, 62-64; staff fel magnetic resonance, NIH, 65-71; RES ASSOC MAGNETIC RESONANCE, UNIV CALIF, SAN DIEGO, 71- Concurrent Pos: Fel, NIH, 68-69; guest worker, Nat Bur Standards, 69-71; lectr chem, Univ Calif, San Diego, 72-74; mem bd trustees, Exp Nuclear Magnetic Resonance Conf, 74-, treas, 75- Mem: Am Chem Soc; AAAS. Res: Nuclear magnetic resonance as used in study of molecular dynamics and liquid structure by means of relaxation in complex spin systems. Mailing Add: Dept of Chem Univ of Calif at San Diego La Jolla CA 92093

VOLD, ROBERT DONALD, b Boston, Mass, Dec 11, 10; m 36; c 3. COLLOID CHEMISTRY. Educ: Univ Nebr, AB, 31, MS, 32; Univ Calif, Berkeley, PhD(chem), 35. Prof Exp: Res chemist, Procter & Gamble Co, 35-37; res assoc chem, Stanford Univ, 37-41; from asst prof to prof, 41-74, head dept, 50-53, EMER PROF CHEM, UNIV SOUTHERN CALIF, 74- Concurrent Pos: Fulbright scholar, State Univ Utrecht, 53-54; vis prof, Indian Inst Sci, 55-57; vis assoc, Calif Inst Technol, 64. Honors & Awards: Tolman Medal, 70. Mem: AAAS; Am Chem Soc; The Chem Soc. Res: Phase behavior of colloidal systems; structure of gels; stability of aqueous and nonaqueous suspensions and emulsions; kinetics of flocculation; properties of surface films; characterization of macromolecules by buoyant density, sedimentation constants and light scattering. Mailing Add: 17465 Plaza Animado 144 San Diego CA 92128

VOLD, ROBERT LAWRENCE, b Los Angeles, Calif, Sept 20, 42; m 63. Educ: Univ Calif, Berkeley, BS, 63; Univ Ill, Urbana, MS, 65, PhD(chem), 66. Prof Exp: Asst prof, 68-74, ASSOC PROF CHEM, UNIV CALIF, SAN DIEGO, 74- Concurrent Pos: A P Sloan fel, 72-74. Mem: Am Inst Physics. Res: Nuclear magnetic resonance; relaxation mechanisms; theory and applications of pulsed nuclear magnetic resonance techniques. Mailing Add: 14092 Rue San Remo Del Mar CA 92014

VOLDENG, ALBERT NELSON, b Wellington, Kans, Nov 25, 38; m 61; c 2. MEDICINAL CHEMISTRY. Educ: Univ Kans, BS, 60, PhD(med chem), 64. Prof Exp: Asst prof pharmaceut chem, 64-68, assoc prof med chem, 68-73, PROF MED CHEM & CHMN DEPT, SCH PHARM, UNIV ARK FOR MED SCI, 73- Mem: Am Chem Soc; Am Acad Clin Toxicol; Soc Toxicol; Am Asn Cols Pharm. Mailing Add: Col of Pharm Univ Ark Little Rock AR 72201

VOLENEC, FRANK JERRY, b Omaha, Nebr, Mar 23, 40; m 60; c 1. VIROLOGY, BIOCHEMISTRY. Educ: Univ Nebr, Lincol, BS, 62; Cornell Univ, PhD(virol), 69. Prof Exp: Res assoc vet virol, Jensen Salsbury Labs Div, Richardson-Merrell, 62-65, sr res virologist, 69-71; dir res & prod, Vet Biol, Inc, Ralston Purina Co, 71-73; WITH NAT LABS CORP, 73- Mem: Am Soc Microbiol. Res: Diseases of domestic animals, especially viral; antigenic analysis of pathogens; immunological problems and defense mechanisms of host animals; bovine virus diarrhea virus. Mailing Add: Nat Labs Corp 12300 Santa Fe Dr Lenexa KS 66215

VOLICER, LADISLAV, b Prague, Czech, May 21, 35; m 72; c 1. PHARMACOLOGY, CARDIOVASCULAR DISEASES. Educ: Charles Univ, Prague, MD, 59; Czech Acad Sci, PhD(pharmacol), 64. Prof Exp: Resident med, Hosp Jindr Hradec, 59-61; instr pharmacol, Sch Pediat, Charles Univ, Prague, 61-65; vis assoc, Nat Heart Inst, Md, 65-66; res assoc & lectr, Inst Pharmacol, Czech Acad Sci, 66-68; res asst prof, Sch Med, Univ Munich, 68-69; asst prof, 69-72, ASSOC PROF PHARMACOL, SCH MED & GRAD SCH, BOSTON UNIV, 72-, ASST PROF MED, 75- Concurrent Pos: Asst vis physician, Boston City Hosp, 75- Mem: AAAS; Am Soc Pharmacol & Exp Therapeut; NY Acad Sci; Am Col Clin Pharmacol. Res: Pharmacology of hypertension; relationship between angiotensin and norepinephrine; cyclic nucleotides in blood vessels; role of cyclic nucleotides in drug abuse and addiction. Mailing Add: Dept of Pharmacol Boston Univ Boston MA 02118

VOLIN, RAYMOND BRADFORD, b Kalispell, Mont, May 22, 43; m 64. PLANT PATHOLOGY, PLANT BREEDING. Educ: Mont State Univ, BS, 66, MS, 68, PhD(plant path), 71. Prof Exp: Trainee agron, Mont State Univ, 66-68, res asst plant path, 68-71; ASST PROF PLANT PATH, AGR RES & EDUC CTR, UNIV FLA, 71- Mem: Am Phytopath Soc; Am Soc Crop Sci. Res: Genetic improvement of field and vegetable crops; physiological relationship and genetic interaction between plant hosts and plant disease organisms. Mailing Add: 55 NE 18th St Homestead FL 33030

VOLK, BOB GARTH, b Auburn, Ala, July 13, 43; m 66; c 1. SOIL CHEMISTRY. Educ: Ohio State Univ, BA, 65, MS, 67; Mich State Univ, PhD(soil chem), 70. Prof Exp: Asst prof soil chem & asst soil chemist, 70-73, ASSOC PROF SOIL SCI, UNIV FLA & ASSOC SOIL CHEMIST, AGR RES & EDUC CTR, BELLE GLADE, 73- Mem: Am Soc Agron. Res: Organic soil chemistry; carbon dioxide evolution from organic soils; subsidence. Mailing Add: Dept Soil Sci 2169 McCarty Hall Univ of Fla Gainesville FL 32611

VOLK, GARTH WILLIAM, b Maple Valley, Wis, June 12, 05; m 35; c 2. SOIL CHEMISTRY. Educ: Univ Wis, BS, 34, MS, 35, PhD(soil chem), 36. Prof Exp: Soil chemist, United Fruit Co, 28-33; asst soil chemist, Okla Agr & Mech Col, 36-38; assoc soil chemist, Ala Polytech Inst, 38-43, soil chemist, 43-44; assoc agronomist, Ohio Exp Sta, 44-47, PROF AGRON & CHMN DEPT, OHIO STATE UNIV, 47- Mem: Am Soc Agron; Soil Sci Soc Am; Int Soc Soil Sci. Res: Fertility. Mailing Add: Dept of Agron Ohio State Univ 1885 Neil Ave Columbus OH 43210

VOLK, GAYLORD MONROE, b Maple Valley Township, Wis, Aug 5, 08; m 36; c 2. SOIL CHEMISTRY. Educ: Univ Wis, MS, 33, PhD(soil chem), 46. Prof Exp: Soil chemist, United Fruit Co, Honduras, 34; assoc soil scientist soil conserv serv, USDA, NMex, 35-39; PROF SOILS, INST FOOD & AGR SCI, UNIV FLA, 39- Concurrent Pos: Mem Univ Fla Mission to Costa Rica, 56-57. Mem: Fel Am Soc Agron; Soil Sci Soc Am. Res: Soil fertility; turf; significance of moisture translocation from soil zones of low moisture tension to zones of high moisture tension by plant roots. Mailing Add: Inst of Food & Agr Sci 106 Newall Hall Univ of Fla Gainesville FL 32601

VOLK, HERBERT F, physical chemistry, inorganic chemistry, see 12th edition

VOLK, MURRAY EDWARD, b Cleveland, Ohio, Aug 23, 22; m 49; c 3. ORGANIC CHEMISTRY, RADIOCHEMISTRY. Educ: Oberlin Col, BA, 43; Univ Chicago, MS, 48; Temple Univ, PhD(chem), 53. Prof Exp: Assoc chemist, Nuclear Instrument & Chem Corp, 53-55; pres, Volk Radiochem Co, 55-65; mkt mgr res prod, Miles Labs, 66-69; PRES, ISOLAB INC, 69- Mem: AAAS; Am Chem Soc; Am Asn Clin Chem; Soc Nuclear Med. Res: Application of isotopes to biological and chemical research; preparation of radioactive pharmaceuticals. Mailing Add: Isolab Inc Drawer 4350 Akron OH 44321

VOLK, RICHARD JAMES, b Tela, Honduras, Nov 5, 28; m 51; c 3. PLANT NUTRITION. Educ: Purdue Univ, BS, 50, MS, 51; NC State Univ, PhD(soil chem), 54. Prof Exp: Res specialist, Crops Div, Biol Warfare Labs, Ft Detrick, Md, 54-56; from asst prof to assoc prof, 56-66, PROF SOIL SCI, NC STATE UNIV, 66- Concurrent Pos: Grants, NSF, 62-64, Am Potash Inst, 63-66 & res contract, USDA, 65-68. Honors & Awards: Co-recipient Campbell Award, Am Inst Biol Sci, 65. Mem: Am Soc Plant Physiol; Soil Sci Soc Am; Crop Sci Soc Am; Am Soc Agron. Res: Application of mass spectrometry and stable isotopes to plant nutrition and biochemistry; absorption and metabolism of ammonium and nitrate nitrogen by plants; regulatory role of mineral nutrition in photosynthesis and respiration. Mailing Add: Dept of Soil Sci NC State Univ Raleigh NC 27607

VOLK, THOMAS LEWIS, b Dayton, Ohio, Nov 4, 33; m 61; c 4. HUMAN PATHOLOGY, ENDOCRINOLOGY. Educ: Univ Dayton, BS, 55; Marquette Univ, MD, 59. Prof Exp: Instr path, Ohio State Univ, 65-66; asst prof, Univ Kans, 66-67; asst prof path, Univ Calif, Davis, 68-72; DIR LABS, KAWEAH DELTA DIST HOSP, 72- Concurrent Pos: NIH path training grant, 63-65; Am Cancer Soc adv clin fel, 65-66; consult, Vet Admin Hosp, Kansas City, Mo, 67-68 & Sacramento County Hosp, Calif, 68-72. Mem: AMA; Int Acad Path; Am Asn Path & Bact. Res: Ultrastructural-functional relationships of steroidogenesis in the adrenal cortex and placenta, ovary and testis; ultrastructural changes in the adrenal cortex and placenta, produced by drugs inhibiting steroidogenesis. Mailing Add: Kaweah Delta Dist Hosp 400 W Mineral King Visalia CA 93277

VOLK, VERIL VAN, b Montgomery, Ala, Nov 18, 38. SOIL CHEMISTRY. Educ: Ohio State Univ, BS, 60, MS, 61; Univ Wis, PhD(soils), 66. Prof Exp: Proj assoc soils, Univ Wis, 66; asst prof, 66-74, ASSOC PROF SOILS, ORE STATE UNIV, 74- Mem: Am Soc Agron; Weed Sci Soc Am; Mineral Soc Am. Res: Ion exchange and soil acidity interactions; clay mineral surface morphology; detergent contamination and their adsorption on soil colloids; adsorption and movement of pesticides in soils; agricultural waste disposal on soils. Mailing Add: Dept of Soil Sci Ore State Univ Corvallis OR 97330

VOLK, WESLEY AARON, b Mankato, Minn, Nov 23, 24; m 45; c 2. MICROBIOLOGY. Educ: Univ Wash, BS & BS(food technol), 48, MS, 49, PhD, 51. Prof Exp: From asst prof to assoc prof, 51-64, PROF MICROBIOL, SCH MED, UNIV VA, 64- Concurrent Pos: NIH spec fel, 62-63; spec fel, Max Planck Inst Immunobiol, 69-70. Mem: Am Soc Microbiol; Am Soc Biol Chemists. Res: Carbohydrate metabolism; structure and function of endotoxins. Mailing Add: Dept of Microbiol Univ of Va Sch of Med Charlottesville VA 22901

VOLKAN, VAMIK, b Nicosia, Cyprus, Dec 13, 32; US citizen; m; c 4. PSYCHIATRY. Educ: Univ Ankara, MD, 56; Wash Psychoanal Inst, grad, 71. Prof Exp: Staff physician, NC State Hosp, 61-63; from instr to assoc prof, 63-72, PROF PSYCHIAT, SCH MED, UNIV VA, 72- Concurrent Pos: Clin instr, Sch Med, Univ NC, 61-63; consult, Inst Criminal Law & Procedures, Georgetown Univ, 66- Mem: AAAS; fel Am Psychiat Asn; fel Royal Soc Med; AMA; NY Acad Sci. Res: Psychotherapy of schizophrenia and pathological grief reactions. Mailing Add: Sch of Med Univ of Va Charlottesville VA 22901

VOLKER, EUGENE JENO, b Sopron, Hungary, May 13, 42; US citizen. ORGANIC CHEMISTRY. Educ: Univ Md, BS, 64; Mass Inst Technol, MS, 67; Univ Del, PhD(chem), 70. Prof Exp: Asst prof, 69-75, ASSOC PROF CHEM, SHEPHERD COL, WVA, 75- Mem: Am Chem Soc. Res: Photochemistry; synthesis of heterocyclic compounds. Mailing Add: Dept of Chem Shepherd Col Shepherdstown WV 25443

VOLKER, JOSEPH FRANCIS, b Elizabeth, NJ, Mar 9, 13; m 37; c 3. BIOCHEMISTRY, DENTISTRY. Educ: Ind Univ, DDS, 36; Univ Rochester, AB, 38, MS, 39, PhD(biochem), 41; FRCS, 61; FRCS(I), 73. Hon Degrees: DSc, Univ Med Sci, Bangkok, 67; Dr Odontol, Univ Lund, 68; DSc, Ind Univ, 70; DSc, Univ Ala, 70; Dr, Louis Pasteur Univ, 72; DS, Univ Rochester, 75. Prof Exp: Dent res, Mountainside Hosp, NJ, 36-37; asst prof dent, Sch Med & Dent, Univ Rochester, 41-42; prof clin dent, Sch Dent, Tufts Col, 42-47, dean, 47-49; dean, Sch Dent, 48-62, dir res & grad study, 55-65, vpres health affairs, 62-66, vpres, Birmingham Affairs & dir med ctr, 66-68, exec vpres univ, 68-69, PRES, UNIV ALA, BIRMINGHAM, 69- Concurrent Pos: Mem, Unitarian Med Teaching Mission, Czech, 46, Ger, 48; specialist, US Dept State, Thailand, 51-; dir, Ariz Med Sch Study, 60-61; mem, Inst Med, Nat Acad Sci, 71. Honors & Awards: Mem, Order of White Lion, Czech, 46; Comdr, Order of Crown, Thailand, 59; Comdr, Order of the Falcon, Repub Iceland, 69. Mem: AAAS; Soc Exp Biol & Med; Int Asn Dent Res; Sigma Xi; hon mem Stomatol Soc Czech. Res: Dental caries; mineral metabolism; oral physiology of carbohydrates. Mailing Add: Off of the Pres Univ of Ala Univ Sta Birmingham AL 35294

VOLKERT, WYNN ARTHUR, b St Louis, Mo, Apr 6, 41; m 67; c 2. RADIOCHEMISTRY, RADIOBIOLOGY. Educ: St Louis Univ, BS, 63; Univ Mo-Columbia, PhD(chem), 68. Prof Exp: NASA fel, 67-69, asst prof, 69-72, ASSOC PROF RADIOL SCI, UNIV MO-COLUMBIA, 72- Concurrent Pos: NSF grant, Univ Mo-Columbia, 73-; consult, Vet Admin Hosp, Columbia, 73- Mem: Radiation Res Soc; Biophys Soc; Soc Exp Biol & Med; Am Asn Physicists in Med; Am Chem Soc. Res: Radiation and photochemistry of amino acids. Mailing Add: Dept of Radiol Univ of Mo Med Ctr Columbia MO 65201

VOLKIN, ELLIOT, b Mt Pleasant, Pa, Apr 23, 19; m 47; c 2. BIOCHEMISTRY. Educ: Pa State Col, BS, 42; Duke Univ, MA, 45, PhD(biochem), 47. Prof Exp: Res assoc biochem, Duke Univ, 47-48; SCI DIR BIOCHEM, OAK RIDGE NAT LAB, 48- Mem: Fel AAAS; Am Chem Soc; Am Soc Biol Chem; Am Soc Microbiol; NY Acad Sci. Res: Biochemical and biophysical studies of nucleic acids and nucleoproteins. Mailing Add: Biol Div Oak Ridge Nat Lab Oak Ridge TN 37830

VOLKMAN, ALVIN, b Brooklyn, NY, June 10, 26; m 47; c 5. PATHOLOGY, IMMUNOLOGY. Educ: Union Col, BS, 47; Univ Buffalo, MD, 51; Oxford Univ, DPhil, 63. Prof Exp: Asst prof path, Columbia Univ, 60-66; asst mem, 66-68, ASSOC MEM, TRUDEAU INST, INC, 68- Concurrent Pos: Arthritis & Rheumatism Found fel, 52-54; Am Cancer Soc scholar, 61-62; adj assoc prof path, Sch Med, NY Univ, 69- Mem: AAAS; Am Soc Hemat; Am Thoracic Soc; NY Acad Sci. Res: Experimental pathology related to the origin and production of white blood cells; their participation in inflammation and immunological processes. Mailing Add: Trudeau Inst Inc Saranac Lake NY 12983

VOLKOFF, GEORGE MICHAEL, b Moscow, Russia, Feb 23, 14; Can Citizen; m 40; c 3. THEORETICAL PHYSICS. Educ: Univ BC, BA, 34, MA, 36; Univ Calif, PhD(theoret physics), 40. Hon Degrees: DSc, Univ BC, 45. Prof Exp: Asst prof physics, Univ BC, 40-43; assoc res physicist, Montreal Lab, Nat Res Coun, Can, 43-45, res physicist & head theoret physics br, Atomic Energy Proj, Que & Ont, 45-46; head dept, 61-72, PROF PHYSICS, UNIV BC, 46-, DEAN FAC SCI, 72- Concurrent Pos: Ed, Can J Physics, 50-56; mem, Nat Res Coun Can, 69-75. Honors & Awards: Mem, Order of the Brit Empire, 46. Mem: Fel AAAS; fel Am Phys Soc; Am Asn Physics Teachers; fel Royal Soc Can; Can Asn Physicists (vpres, 61-62, pres, 62-63).

Res: Theoretical nuclear physics; neutron diffusion; nuclear magnetic and quadrupole resonance. Mailing Add: Fac of Sci Univ of BC Vancouver BC Can

VOLKOV, ANATOLE BORIS, b San Francisco, Calif, Oct 29, 24; m 50; c 2. NUCLEAR PHYSICS. Educ: Univ NC, BS, 48; Univ Wis, MS, 50, PhD(physics), 53. Prof Exp: Longwood fel, Univ Del, 53-55; asst prof physics, Univ Miami, 58-59; sr lectr, Israel Inst Technol, 59-62; res intermediate scientist, Weizmann Inst, 62-63; Ford Found fel, Niels Bohr Inst, Copenhagen, Denmark, 63-64; from asst prof to assoc prof, 64-68, PROF PHYSICS, McMASTER UNIV, 68- Mem: Fel Am Phys Soc; Can Asn Physicists. Res: Theoretical physics, especially low energy nuclear physics and nuclear deformations. Mailing Add: Dept of Physics McMaster Univ Hamilton ON Can

VOLL, MARY JANE, b Baltimore, Md, June 29, 33. MICROBIAL GENETICS. Educ: Loyola Col, BA, 55; Johns Hopkins Univ, MSc, 61; Univ Pa, PhD(microbiol), 64. Prof Exp: Staff fel microbiol, NIH, 64-66, USPHS fel, 66-67, microbiologist, 67-69; res assoc biol, Johns Hopkins Univ, 69-71; ASST PROF MICROBIOL, UNIV MD, COLLEGE PARK, 71- Mem: Am Soc Microbiol; Am Inst Biol Sci. Res: Homospecific and heterospecific gene transfer in enteric bacteria; environmental mutagenesis. Mailing Add: Dept of Microbiol Univ of Md College Park MD 20742

VOLLAND, LEONARD ALLAN, b Cleveland, Ohio, Apr 26, 37; m 63; c 2. PLANT ECOLOGY. Educ: Univ Idaho, BS, 59; Ore State Univ, MS, 63; Colo State Univ, PhD(quant ecol), 74. Prof Exp: Forester natural resources, 59-66, plant ecologist, 66-73, QUANT ECOLOGIST, US FOREST SERV, 73- Mem: Soc Am Foresters; Soc Range Mgt. Res: Plant community ecology and its application to natural resource management. Mailing Add: US Forest Serv PO Box 3623 Portland OR 97208

VOLLE, ROBERT LEON, b Houston, Pa, June 2, 30; m 52; c 5. PHARMACOLOGY. Educ: WVa Wesleyan Col, BS, 53; Univ Kans, PhD(pharmacol), 59. Prof Exp: From instr to assoc prof pharmacol, Sch Med, Univ Pa, 60-65; prof, Sch Med, Tulane Univ, 65-68; PROF PHARMACOL, SCH MED, UNIV CONN, 68- Concurrent Pos: Marsh fel pharmacol, Sch Med, Univ Pa, 59-60, Pa Plan scholar, 60-63; USPHS career develop award, 63-65. Mem: Fel AAAS; Am Soc Pharmacol & Exp Therapeut; NY Acad Sci. Res: Neuropharmacology. Mailing Add: Dept of Pharmacol Univ of Conn Sch of Med Farmington CT 06032

VOLLENWEIDER, RICHARD A, b Zurich, Switz, June 27, 22; m 65. LIMNOLOGY. Educ: Univ Zurich, dipl biol, 46, PhD(biol), 51. Prof Exp: Teacher undergrad schs, Lucern, Switz, 49-54; fel limnol, Ital Hydrobiol Inst, Palanza, Italy, 54-55 & Swiss Swed Res Coun, Uppsala, 55-56; field expert limnol & fisheries, UNESCO Univ Agr, Egypt, 57-59; res assoc limnol, Ital Hydrobiol Inst, Pallanza, 59-66; consult water pollution, Orgn Econ Coop Develop, Paris, France, 66-68; chief limnologist & head fisheries res bd, 68-70, chief, Lakes Res Div, 70-73, SR SCIENTIST, CAN CENTRE INLAND WATERS, 73- Mem: Swiss Soc Microbiol; Int Asn Theoret & Appl Limnol; Int Asn Great Lakes Res. Res: Inland water research; biological communities; water chemistry and physics; eutrophication; water pollution. Mailing Add: Can Ctr Inland Waters Box 5050 Burlington ON Can

VOLLMAN, RUDOLF F, b Ger, Mar 17, 12; nat US; m 38; c 1. OBSTETRICS & GYNECOLOGY. Educ: Univ Geneva, MD, 50. Prof Exp: Res assoc, Inst Vet Path, Univ Zurich, 42-43; intern, Jewish Mem Hosp, New York, 52-53, resident obstet & gynec, 53-54; med officer, Sect Develop & Regeneration, Lab Neuroanat Sci, PR, 59-60; HEAD OBSTET SECT, PERINATAL RES BR, NAT INST NEUROL DIS & STROKE, 60-, CONSULT, COLLAB PROJ, 59- Mem: AAAS; Am Asn Anatomists; NY Acad Sci; Int Fertil Asn. Res: Adolescent development; menstrual cycle; physiology and pathology of pregnancy; sterility and fertility; pregnancy wastage; duration of pregnancy; onset and mechanism of labor; menopause. Mailing Add: Perinatal Res Br Nat Inst of Neurol Dis & Stroke Bethesda MD 20014

VOLLMAR, ARNULF R, b Pluderhausen, Ger, Apr 15, 28. ORGANIC CHEMISTRY. Educ: Univ Heidelberg, dipl chem, 55, PhD(org chem), 57. Prof Exp: Res assoc, Univ Heidelberg, 57-58; fel, Univ Calif, Los Angeles, 58-60; res chemist, Chevron Res Corp, Calif, 60-64; assoc prof, 65-74, PROF CHEM, CALIF STATE POLYTECH UNIV, POMONA, 74- Concurrent Pos: NSF res grant, 69. Mem: Am Chem Soc. Res: Chemistry of tetrazole ethers and isocyanide. Mailing Add: Dept of Chem Calif State Polytech Univ Pomona CA 91766

VOLLMER, ERWIN PAUL, b New York, NY, Jan 16, 06; m 34; c 2. PHYSIOLOGY. Educ: Dartmouth Col, AB, 29; NY Univ, MS, 39, PhD(physiol), 41. Prof Exp: Bacteriologist, Calco Chem Co Div, Am Cyanamid Co, 37; asst instr biol, NY Univ, 41-42; tutor, Brooklyn Col, 42-43; physiologist, US Naval Med Inst, 47-56; chief endocrinol, Cancer Chemother Nat Serv Ctr, 56-66; CHIEF ENDOCRINE EVAL BR, GEN LABS & CLINS & EXEC SECY BREAST CANCER TASK FORCE, NAT CANCER INST, BETHESDA, 66- Mem: AAAS; Endocrine Soc; NY Acad Sci. Res: Physiology of resistance to infection; endocrine factors in hemopoiesis; endocrine etiology and chemotherapy in cancer. Mailing Add: 7202 44th St Chevy Chase MD 20015

VOLLMER, JAMES, b Philadelphia, Pa, Apr 19, 24; m 46; c 3. PHYSICS. Educ: Union Col, BS, 45; Temple Univ, MA, 51, PhD(physics), 56; Harvard Univ, advan mgt prog, 71. Prof Exp: Instr physics, Temple Univ, 46-51; res engr, Indust Div, Honeywell Inc, 51-59; engr appl res, Radio Corp Am, 59, group leader appl plasma physics, 59-63, mgr appl physics, 63-66, mgr appl res, 66-68, dir, Advan Technol Labs, 68-72, gen mgr, Palm Beach Div, 72-74, div vpres & gen mgr, 74-75, DIV VPRES & GEN MGR, GOVT COMMUN SYSTS DIV, RCA CORP, 75- Concurrent Pos: Lectr, Temple Univ, 57-59; adj prof, Drexel Inst Technol, 64-66; chmn session on low noise technol, Int Conf Commun, 66. Mem: Fel AAAS; fel Inst Elec & Electronics Engrs; Am Phys Soc. Res: Infrared properties of materials; plasma physics; quantum electronics; microsonics; lasers; photosensors; radiometry. Mailing Add: Govt Commun Systs Div RCA Del & Cooper Camden NJ 08102

VOLMAN, DAVID H, b Los Angeles, Calif, July 10, 16; m 44; c 3. PHYSICAL CHEMISTRY. Educ: Univ Calif, Los Angeles, AB, 37, AM, 38; Stanford Univ, PhD(chem), 40. Prof Exp: Asst chem, Univ Calif, Los Angeles, 37-38, res chemist, Nat Defense Res Comt Proj, 41-42; asst chem, Stanford Univ, 38-39; instr chem & jr chemist, Exp Sta, Univ Calif, 40-41; res chemist, Off Sci Res & Develop, Northwestern Univ, 41-45 & Univ Ill, 45-46; from asst prof & asst chemist to assoc prof & assoc chemist, Exp Sta, 46-56, PROF CHEM, UNIV CALIF, DAVIS, 56-, CHMN DEPT, 74- Concurrent Pos: Guggenheim fel, Harvard Univ, 49-50. Mem: Am Chem Soc. Res: Photochemistry; kinetics; electron spin resonance. Mailing Add: Dept of Chem Univ of Calif Davis CA 95616

VOLPE, ANGELO ANTHONY, b New York, NY, Nov 8, 38; m 65. ORGANIC CHEMISTRY, POLYMER CHEMISTRY. Educ: Brooklyn Col, BS, 59; Univ Md, MS, 62, PhD(org chem), 66. Hon Degrees: ME, Stevens Inst Technol, 75. Prof Exp: Res chemist, US Naval Ord Lab, 61-66; from asst prof to assoc prof, 66-74, actg head dept chem & chem eng, 74-75, PROF CHEM, STEVENS INST TECHNOL, 74- Mem: Am Chem Soc. Res: Correlation of polymer properties to molecular structure; synthesis and mechanisms of formation and degradation of thermally stable polymers; monomer synthesis; synthesis and study of biopolymers. Mailing Add: Dept Chem & Chem Eng Stevens Inst of Technol Hoboken NJ 07030

VOLPE, ERMINIO PETER, b New York, NY, Apr 7, 27; m 55; c 3. ZOOLOGY. Educ: City Col New York, BS, 48; Columbia Univ, MA, 49, PhD(zool), 52. Prof Exp: Asst zool, Columbia Univ, 48-51; instr biol, City Col New York, 51-52; from asst prof to assoc prof, 52-60, chmn dept, Col, 54-64, chmn dept, Univ, 64-66, assoc dean grad sch, 67-69, PROF ZOOL, NEWCOMB COL, TULANE UNIV, 60-, CHMN DEPT, UNIV, 69- Concurrent Pos: Mem steering comt, Biol Sci Curric Study, 66-69; consult comn undergrad educ biol sci, NSF, 67-70; US Nat comnr, UNESCO, 68-; mem exam comt, Col Entrance Exam Bd, Princeton Univ, 69-; ed, Am Zoologist, Am Soc Zoologists, 76-81. Mem: AAAS; Genetics Soc Am; Soc Study Evolution; Am Soc Zoologists; Soc Syst Zool. Res: Embryology, genetics and evolution of amphibians; transplantation immunity and tolerance in anurans. Mailing Add: Dept of Biol Tulane Univ New Orleans LA 70118

VOLPE, ROBERT, b Toronto, Ont, Mar 6, 26; m 49; c 5. ENDOCRINOLOGY. Educ: Univ Toronto, MD, 50; FRCP(C), 56. Prof Exp: Dept Vet Affairs med res fel, Clin Invest Unit, Sunnybrook Hosp, Toronto, 52-53; Med Res Coun Can fel, Toronto Gen Hosp, 55-57; sr res fel endocrinol, Fac Med, Univ Toronto, 57-65; from clin teacher to assoc prof, 65-71, PROF ENDOCRINOL, FAC MED, UNIV TORONTO, 71-; DIR ENDOCRINE RES LAB, WELLESLEY HOSP, 67- Concurrent Pos: Physician-in-chief, Dept Med, Wellesley Hosp, Toronto, 67- Mem: Fel Am Col Physicians; Am Thyroid Asn (1st vpres, 75); Am Fedn Clin Res; Can Soc Clin Invest; Can Soc Endocrinol & Metab (pres, 72). Res: Immune mechanisms in thyroid disease; thyroid hormone metabolism and kinetics. Mailing Add: 3 Daleberry Pl Don Mills ON Can

VOLPERT, EUGENE M, b Koenigsberg, Ger, Feb 15, 25; US citizen. BIOCHEMISTRY, ENDOCRINOLOGY. Educ: NY Univ, BA, 45; Univ Wis, MS, 47; Univ Paris, PhD(biochem), 59. Prof Exp: Res assoc biochem, Col Physicians & Surgeons, Columbia Univ, 59-65; biochemist, Endocrine Serv, Montefiore Hosp, Bronx, 65-67; assoc scientist, Med Found Buffalo, 67-75; BIOLOGIST, NUCLEAR RES, VET ADMIN HOSP, BROOKLYN, 75- Mem: Am Chem Soc; Am Thyroid Asn; Endocrine Soc. Res: Biochemistry and binding of thyroid hormones; relation with pituitary gland and pituitary tumors; thyroid gland mechanisms. Mailing Add: Nuclear Res Vet Admin Hosp Brooklyn NY 11209

VOLPITTO, PERRY PAUL, b Italy, July 9, 05; nat US; m 37; c 3. ANESTHESIOLOGY. Educ: Washington & Jefferson Col, BS, 28; Western Reserve Univ, MD, 33; Am Bd Anesthesiol, dipl. Prof Exp: From assoc prof to prof anesthesiol, 37-73, chmn dept, 38-72, EMER PROF ANESTHESIOL, MED COL GA, 73- Concurrent Pos: Area consult, US Vet Admin, 48-68; consult, US Army Hosp, Ft Gordon, 50-74. Honors & Awards: Hardeman Award, 66 & Distinguished Serv Award, 75, Med Asn Ga; Distinguished Serv Award, Am Soc Anesthesiol, 74. Mem: Am Soc Anesthesiol (pres, 65); AMA. Res: Treatment of barbiturate poisoning; amnesia in labor; treatment of apnea of the newborn; intra-arterial blood pressure in children and during anesthesia; use of stellate ganglion block in cerebral vascular accidents; intravenous barbiturates and muscle relaxants for endotracheal intubation; electronarcosis utilizing a combination of alternating and direct currents. Mailing Add: Dept of Anesthesiol Med Col of Ga Augusta GA 30902

VOLPP, GERT PAUL JUSTUS, b Loerrach, Ger, July 30, 30; nat US; m 62; c 4. ORGANIC CHEMISTRY. Educ: Univ Basel, PhD(chem), 58. Prof Exp: Res fel org chem, Harvard Univ, 58-63; interdisciplinary scientist, 63-65, mgr explor org res, 65-72, mgr prod res, 72-73, TECH DIR ALKALI CHEM, FMC CORP, 73- Mem: Am Chem Soc; Ger Chem Soc. Res: Synthetic organic chemistry; intermediates for dyestuffs; additives for plastics; antioxidants; detergent chemistry; agricultural chemistry; manufacturing technology for soda ash, caustic, chlorine, glycerine, allyl alcohol, barium and strontium chemicals. Mailing Add: Chem Res & Develop Ctr FMC Corp PO Box 8 Princeton NJ 08540

VOLTZ, STERLING ERNEST, b Philadelphia, Pa, Apr 17, 21; m 43; c 2. PHYSICAL CHEMISTRY. Educ: Temple Univ, AB, 43, MA, 47, PhD(phys chem), 52. Prof Exp: Lab asst, Temple Univ, 46-47; res fel, Univ Pa, 47-48; instr, Temple Univ, 48-51; res chemist, Houdry Process Corp, 51-58; group leader, Sun Oil Co, 58-60; supv chemist, Missile & Space Div, Gen Elec Co, 60-62, consult liaison scientist, 62-68; RES ASSOC, MOBIL RES & DEVELOP CORP, PAULSBORO, NJ, 68- Mem: AAAS; Am Chem Soc; Catalysis Soc. Res: Catalysis; surface and solid state chemistry; chemical kinetics; electrochemistry; fuel cells; petroleum and petrochemical processes; synthetic fuels; automotive emission control systems; program management; research administration and planning. Mailing Add: 6 E Glen Circle Media PA 19063

VOLWILER, WADE, b Grand Forks, NDak, Sept 16, 17; m 43; c 3. MEDICINE. Educ: Oberlin Col, AB, 39; Harvard Med Sch, MD, 43; Am Bd Internal Med, dipl, 50; Am Bd Gastroenterol, dipl, 54. Prof Exp: From intern to resident med, Mass Gen Hosp, Boston, 43-45, asst, 45-48; asst, Harvard Med Sch, 46-48; from instr to assoc prof, 49-59, PROF MED, SCH MED, UNIV WASH, 59-, HEAD DIV GASTROENTEROL, 50- Concurrent Pos: Teaching fel med, Harvard Med Sch, 45-46; res fel gastroenterol, Mass Gen Hosp, Boston, 45-48; Am Gastroenterol Asn res fel, 47; Nat Res Coun fel, 48-49; Markle scholar, 50-55; res assoc, Mayo Found, Univ Minn, 48-49; attend physician, King County Hosp Syst, Seattle, 50- & Vet Admin Hosp, 51-; consult, USPHS Hosp, 55- & Univ Wash Hosp, 60-; mem subspecialty bd gastroenterol, Am Bd Internal Med, 70-76. Honors & Awards: Distinguished Achievement Award, Univ Minn, 64. Mem: Am Soc Clin Invest; Am Gastroenterol Asn (secy, 59-62, pres, 67); Asn Am Physicians; Am Asn Study Liver Dis (pres, 56). Res: Liver diseases; gastroenterology; plasma proteins. Mailing Add: Dept of Med Univ of Wash Sch of Med Seattle WA 98195

VOLZ, EMIL CONRAD, b Saginaw, Mich, Sept 22, 91; m 20. HORTICULTURE. Educ: Mich State Col, 14; Cornell Univ, MS, 18. Prof Exp: Instr floricult & veg crops, Iowa State Col, 14-15; instr floricult, Cornell Univ, 15-18; asst prof hort, Univ Ill, 18-21; prof hort, Iowa State Univ, 21-75; RETIRED. Concurrent Pos: Lectr, Mich State Col, 17. Res: Greenhouse soil; soilless culture in greenhouses; sub-irrigation; acid soil behavior in greenhouses; breeding geraniums. Mailing Add: 619 Ash Ave Ames IA 50010

VOLZ, FREDERIC ERNST, b Singen, Ger, Oct 29, 22; m 57; c 3. ATMOSPHERIC PHYSICS. Educ: Univ Frankfurt, dipl, 50, PhD(meteorol), 54. Prof Exp: Res asst, Lichtklimat Observ Arusa, 50-52; res asst meteorol, Univ Mainz, 52-57; res fel atmospheric physics, Harvard Univ, 57-61 & Astron Inst, Univ Tübingen, 62-67; RES PHYSICIST, AIR FORCE CAMBRIDGE RES LABS, 67- Mem: Am Meteorol Soc; Am Geophys Union; Optical Soc Am; Ger Meteorol Soc. Res: Atmospheric optics, optical constants of aerosol, twilight, stratospheric aerosol. Mailing Add: Air Force Cambridge Res Labs Bedford MA 01731

VOLZ, MICHAEL GEORGE, b Long Beach, Calif, Nov 30, 45; m 68; c 2.

AGRICULTURAL MICROBIOLOGY, PLANT PHYSIOLOGY. Educ: Univ Calif, Berkeley, BS, 67, PhD(soil sci, plant physiol), 72. Prof Exp: Res biochemist, Univ Calif, Berkeley, 72-74, asst res biochemist, 74-75; ASST PLANT PHYSIOLOGIST, CONN AGR EXP STA, 75- Mem: Am Soc Agron; Soil Sci Soc Am; Crop Sci Soc Am. Res: Investigations into the interactions of plant roots and soil microbes as they relate to soil nutrient transformations. Mailing Add: Dept Ecol & Climatol Conn Agr Exp Sta New Haven CT 06504

VOLZ, PAUL ALBERT, b Ann Arbor, Mich, Mar 26, 36. MYCOLOGY, BOTANY. Educ: Heidelberg Col, BA, 58; Mich State Univ, MS, 62, PhD(mycol), 66. Prof Exp: Instr bot, Univ Wis-Milwaukee, 62-63; USPHS res grant, Med Ctr, Ind Univ, 67-68; assoc prof bot & mycol, Purdue Univ, 68-69; asst prof & mycologist, 69-72, ASSOC PROF BOT & MYCOL, EASTERN MICH UNIV, 72- Concurrent Pos: Sr res assoc, Nat Res Coun, 71-73; res contractor, NASA Manned Spacecraft Ctr, 71-74; vis prof mycol, Nat Taiwan Univ, 74-75. Mem: AAAS; Am Inst Biol Sci; Electron Micros Soc Am; Am Fern Soc; Asn Trop Biol. Res: Fern anatomy; marine and soil fungi of the Bahamas and The Republic of China; keratinophilic fungi; drug sensitivity and nutritional requirements of fungi; effects of space flight parameters on select fungal species; fungal cytogenetics and morphology. Mailing Add: Dept of Biol Eastern Mich Univ Ypsilanti MI 48197

VOLZ, WILLIAM BECKHAM, b Chickasha, Okla, July 9, 42; m 66; c 2. ELECTROOPTICS. Educ: Okla State Univ, BS, 63, PhD(chem), 70. Prof Exp: Assoc scientist, 70-76, SR SCIENTIST MAT SCI, TEX DIV, VARO, INC, 76- Mem: Am Chem Soc. Res: Thermal imaging based on the quantum counter principle; photocathode sensitivity at 1.06 microns; laser damage; applications of low-light level image intensifiers in medicine, science and military technology. Mailing Add: Varo Inc Tex Div 2201 Walnut PO Box 828 Garland TX 75040

VOMHOF, DANIEL WILLIAM, b Grant, Nebr, Apr 19, 38; m 60; c 3. FORENSIC SCIENCE, CHEMISTRY. Educ: Augsburg Col, BA, 62; Univ Ariz, MS, 66, PhD(plant physiol), 67; Am Inst Chemists, cert, 69. Prof Exp: Chemist, Ariz Agr Exp Sta, Univ Ariz, 63-67; res chemist, Corn Refiners Asn, 67-69; dir, Region IX Lab, US Bur Customs, Ill, 69-72, forensic scientist, Region VII, 72-74, dir, Lab Div, US Customs Serv, Region IX, Chicago, 74; PRES, EXPERT WITNESS SERV, 74- Concurrent Pos: Res assoc, Nat Bur Stand, 67-69. Honors & Awards: Citation, Nat Bur Stand, 69; Spec Achievement Award, US Treas Dept, 71, 73 & 74. Mem: AAAS; Am Chem Soc; fel Am Inst Chemists; Independent Asn Questioned Doc Examrs; Am Soc Testing & Mat. Res: Accident dynamics; document identification; biomechanics; analytical chemistry; driver behavior. Mailing Add: Expert Witness Serv 5240 Wood St La Mesa CA 92041

VOMOCIL, JAMES ARTHUR, b Jacumba, Calif, Sept 12, 26; m 46; c 3. SOIL SCIENCE, AGRONOMY. Educ: Univ Ariz, BS, 50; Mich State Univ, MS, 52; Rutgers Univ, PhD, 55. Prof Exp: Asst soil sci, Mich State Univ, 50-52, instr, 52; asst, Rutgers Univ, 52-55; instr soil physics, Univ Calif, Davis, 55, from asst prof to assoc prof & assoc exp sta, 55-67; EXTEN SOILS SPECIALIST & PROF SOILS, ORE STATE UNIV, 67- Mem: Am Soc Agron; Soil Sci Soc Am; Int Soc Soil Sci. Res: Soil physical condition and plant growth; soil strength and deformation. Mailing Add: Dept of Soils Ore State Univ Corvallis OR 97331

VON DAVID LEE, organic chemistry, biochemistry, see 12th edition

VON, ISAIAH, b Philadelphia, Pa, Dec 28, 18; m 45; c 3. INDUSTRIAL ORGANIC CHEMISTRY. Educ: Univ Buffalo, BA, 40; Univ Pa, MS, 41, PhD(org chem), 43. Prof Exp: Res assoc, Nat Defense Res Comt Proj, Univ Pa, 43-45, mem comt on med res proj, 45-46; res chemist, 46-53, develop chemist, 53-54, group leader, 54-56, sect chief chemist, 56-64, DEP CHIEF CHEMIST, AM CYANAMID CO, 65- Mem: Am Chem Soc; Am Asn Textile Chemists & Colorists. Res: Dyestuffs; pigments; organic intermediates. Mailing Add: 1005 W 8th St Plainfield NJ 07063

VONA, JOSEPH ALBERT, b Brooklyn, NY, Aug 15, 20; m 46; c 2. ORGANIC CHEMISTRY. Educ: Brooklyn Col, BA, 41, MA, 44; Polytech Inst Brooklyn, PhD, 54. Prof Exp: Head lab sect plastics res, Barrett Chem Co, 45-46; res & develop chemist, Nat Lead Co, 46-50; asst to tech dir, Baker Castor Oil Co, 50-55; mgr, Tech Serv Lab, 55-69, DIR MTD LAB, CELANESE CHEM CO, 69- Mem: Am Chem Soc; Com Develop Asn. Res: Research and development in radiation technology; new types of coatings; emulsion solution and bulk polymerization of monomers; new compounds which can produce durable coatings. Mailing Add: Celanese Chem Co Box 1000 Summit NJ 07901

VON ALMEN, WILLIAM FREDERICK I, b Olney, Ill, May 6, 28; m 50; c 4. PALYNOLOGY, GEOLOGY. Educ: Southern Ill Univ, Carbondale, BA, 57; Univ Mo-Columbia, MA, 59; Mich State Univ, PhD(geol), 70. Prof Exp: Geologist, Pure Oil Co, 59-60; geologist, Stand Oil Co Tex, 60-64, geologist-palynologist, 66-68, palynologist, Chevron Oil Field Res Co, 68-69, from lead palynologist to div paleontologist, 69-71, SR PALEONTOLOGIST, CALIF CO DIV, CHEVRON OIL CO, 71- Mem: Am Asn Stratig Palynol. Res: Palynology of Devonian-Mississippian Boundary; Mesozoic palynostratigraphy. Mailing Add: Chevron Oil Co 1111 Tulane Ave New Orleans LA 70112

VON ARX, WILLIAM STELLING, b Highland Mills, NY, Sept 27, 16; wid; c 2. OCEANOGRAPHY. Educ: Brown Univ, AB, 42; Yale Univ, ScM, 43; Mass Inst Technol, ScD, 55. Prof Exp: Instr physics, Yale Univ, 43-45; phys oceanogr, 45-68, SR SCIENTIST, WOODS HOLE OCEANOG INST, 68- Concurrent Pos: Lectr Harvard Univ, 48; consult, Nat Acad Sci, 51-, 56- 56 & President's Sci Adv Comt, 60-66; del, Int Asn Phys Oceanog, 54-; from assoc prof to prof oceanog, Mass Inst Technol, 56-70; physicist, Off Sci Res & Develop; mem adv comt, US Coast & Geod Surv, 62-66; mem geophys inst, Univ Alaska, 63-71; mem space sci bd, Nat Acad Sci, 64-; mem spec comn weather modification, NSF, 65-66; mem exec comt, Earth Sci Div, Nat Res Coun, 65-68; mem comt space sci & appln, NASA, 67-; mem council, Smithsonian Inst, Washington, DC, 70-74, hon mem, 74- Mem: Am Soc Limnol & Oceanog; fel Am Meteorol Soc; fel Am Geophys Union; fel Am Acad Arts & Sci. Res: Methods for measuring ocean currents; laboratory studies of the ocean circulation; studies of the short-period variations in the structure of the Gulf Stream; ultrawide field optics; stabilized optical systems for geophysical measurements at sea; marine geodesy; energy of processes in the solar-terrestrial heat balance. Mailing Add: Woods Hole Oceanog Inst Woods Hole MA 02543

VON BACHO, PAUL STEPHAN, JR, b Rochester, NY, Dec 28, 37; m 60; c 2. PHOTOGRAPHIC CHEMISTRY. Educ: Univ Rochester, BS, 65, MS, 71, PhD(mat sci), 76. Prof Exp: Res chemist, 65-74, SR RES CHEMIST COLOR PHOTOG, EASTMAN KODAK CO RES LABS, 74- Mem: Am Chem Soc; Am Vacuum Soc; Soc Photog Scientists & Engrs. Res: Use of analytical, radiotracer and instrumental techniques to study color photographic films and process solutions. Mailing Add: Eastman Kodak Co Res Labs 1669 Lake Ave Rochester NY 14650

VON BAEYER, HANS CHRISTIAN, b Berlin, Ger, Apr 6, 38; Can citizen; m 61. c 2. THEORETICAL PHYSICS. Educ: Columbia Univ, AB, 58; Univ Miami, MSc, 61; Vanderbilt Univ, PhD(physics), 64. Prof Exp: Res assoc physics, McGill Univ, 64-65, asst prof, 65-68; from asst prof to assoc prof, 68-75, PROF PHYSICS, COL WILLIAM & MARY, 75-, CHMN DEPT, 72- Mem: Am Phys Soc; Fedn Am Sci; Am Asn Univ Prof. Res: Theory of elementary particles. Mailing Add: Dept of Physics Col of William & Mary Williamsburg VA 23185

VON BAUMGARTEN, RUDOLF JURY, b Freiburg, Ger, Oct 18, 22; m 51; c 1. NEUROPHYSIOLOGY. Educ: Univ Freiburg, MD, 50. Prof Exp: Asst neurol, Univ Freiburg, 50-51, asst neurosurg, 51-54; asst gen surg, Miners Hosp Aachen, 54-55; assoc prof neurophysiol, Univ Göttingen, 55-62, prof physiol, 62-66; res neurophysiologist, Ment Health Res Inst, Univ Mich, Ann Arbor, 66-70, prof physiol, 67-70; PROF PHYSIOL & HEAD DEPT, UNIV MAINZ, 70- Concurrent Pos: Res fels, Dept Physiol, Univ Pisa, 51, Dept Anat, Univ Calif, Los Angeles, 60-61, All India Inst Med Res, New Delhi, 64 & Dept Physiol, Med Sch, Univ Southern Calif, 64 & 65; NIH consult, Dept Clin Neuropharmacol, St Elizabeth Hosp, DC, 60-63; consult, Dept Physiol, Univ Southern Calif, 64 & 65. Mem: NY Acad Sci; Ger Physiol Soc; Ger Electroencephalog Soc; Am Physiol Soc. Res: Neurophysiology of information-storage, optic cortex, reticular formation, respiratory and cardiovascular centers; aerospace medicine; micro-recordings in these structures; intracellular recording in mollusks; olfactory system in mammals and fish. Mailing Add: Physiol Inst Johannnes Gutenberg Univ Mainz Mainz Germany

VON BECKH, HARALD JOHANNES, b Vienna, Austria, Nov 17, 17; nat US; m 49; c 6. AEROSPACE MEDICINE. Educ: Univ Vienna, MD, 40; Nat Bd Cert, Buenos Aires, Arg, cert specialist aviation med, 56. Prof Exp: Staff mem & lectr aviation med, Aeromed Acad, Berlin, Ger, 41-43; prof, Nat Inst Aviation Med, Buenos Aires, 47-56; mem sci staff, Aeromed Res Lab, Holloman AFB, NMex, 57-64, chief scientist, 64-70; DIR MED RES, CREW SYSTS DEPT, NAVAL AIR DEVELOP CTR, 70- Concurrent Pos: Mem comt bioastronaut, Armed Forces-Nat Res Coun, 58-61; mem bioastronaut comt, Int Astronaut Fedn, 61-; hon mem, Ctr Astronaut Studies Portugal, 61- Honors & Awards: Melbourne W Boynton Award, Am Astronaut Soc, 72; Arnold D Tuttle Award, Aerospace Med Asn, 72; Claude Bernard Medal, Fr Asn Astronaut Res, 72. Mem: Assoc fel Am Inst Aeronaut & Astronaut; fel Aerospace Med Asn; Asn Mil Surgeons US; hon mem Herman Oberth Soc; hon mem Span Soc Aerospace Med. Res: Neurophysiology, hemodynamics and deconditioning effect of weightlessness; effects of accelerations on humans and test animals; protective devices against accelerations of space and atmospheric flight. Mailing Add: PO Box 1220 Warminster PA 18974

VON BLOEKER, JACK CHRISTIAN, JR, b Sacramento, Calif, July 11, 09; m 37; c 3. VERTEBRATE ZOOLOGY. Educ: Univ Calif, BA, 37, MA, 38. Prof Exp: Vert zoologist, Silliman Collections, 36-38; field zoologist, Los Angeles Mus, 38-39, cur ornith & mammal, 39-41; asst vert zool, Allan Hancock Found, Univ Southern Calif, 41-50; PROF BIOL, LOS ANGELES CITY COL, 50- Concurrent Pos: Allan Hancock scholarship, 49-52; ed, Western Found Vert Zool, 65- Mem: Am Soc Mammal; AAAS; Cooper Ornith Soc (treas, 52-64); Soc Syst Zool; Am Ornith Union. Res: Recent mammal fauna of islands off the coast of California and Lower California; habits and distribution and ecology of American mammals, bird, reptiles and amphibians; taxonomy and distribution of scarab beetles. Mailing Add: Dept of Life Sci Los Angeles City Col Los Angeles CA 90029

VON BODUNGEN, GEORGE ANTHONY, b New Orleans, La, Oct 12, 40; m 69; c 2. PHYSICAL CHEMISTRY. Educ: Loyola Univ, New Orleans, 62; Tulane Univ La, PhD(phys chem), 66. Prof Exp: SR RES CHEMIST, COPOLYMER RUBBER & CHEM CORP, 66- Mem: Am Chem Soc; Soc Plastics Engrs. Res: Synthesis and rheology of impact resistant plastics and thermo plastic elastomers; computer simulations and mathematical models of process and products; applied mathematics. Mailing Add: Res Dgpt PO Box 2591 Copolymer Rubber & Chem Corp Baton Rouge LA 70821

VON BORSTEL, ROBERT CARSTEN, b Kent, Ore, Jan 24, 25; m 48; GENETICS. Educ: Ore State Col, BA, 47, MS, 49; Univ Pa, PhD(zool), 53. Prof Exp: Fel, Carnegie Inst, NY, 52-53; biologist, Oak Ridge Nat Lab, 53-71; PROF GENETICS & CHMN DEPT, UNIV ALTA, 71- Concurrent Pos: NSF fel, Univ Pavia, 59-60. Mem: Genetics Soc Am; Am Soc Nat; Genetic Soc Can; Int Asn Environ Mutagen Socs (secy, 73-); Am Environ Mutagen Soc. Res: Dominant lethality and cell-killing by radiation; microorganism genetics; spontaneous mutation rates; mutator genes in yeast. Mailing Add: 12312 Grandview Dr Edmonton AB Can

VON BRAMER, PAUL THOMAS, organic chemistry, see 12th edition

VON BRAUN, WERNHER, b Wirsitz, Ger, Mar 23, 12; nat US; m 47; c 3. PHYSICS, AEROSPACE ENGINEERING. Educ: Berlin Inst Technol, BS, 32; Univ Berlin, PhD(physics), 34. Hon Degrees: Twenty from US & foreign univs & cols, 58-75. Prof Exp: Chief testing sta, Ger Army Proving Ground, Kummersdorf, 32-37, tech dir, Rocket Ctr, Peenemünde, 37-45; proj dir res & develop serv, US Army Ord Corps, Tex, 45-50, chief guided missile develop div, Redstone Arsenal, 50-56, phys scientist & dir develop opers div, 56-60; dir Marshall Space Flight Ctr, NASA, 60-70, dep assoc adminr for planning, NASA Hq, Washington, DC, 70-72; VPRES ENG & DEVELOP, FAIRCHILD INDUSTS, INC, 72-; PRES, NAT SPACE INST, WASHINGTON, DC, 75- Concurrent Pos: Fel, Ind Inst Technol; pres, Rocket City Astron Asn, 55-; chmn int sponsors comt, Robert Hutchings Goddard Libr Prog, Clark Univ. Honors & Awards: Astronaut Award, Am Inst Aeronaut & Astronaut, 56, Oberth Award, 61 & Hill Award, 65; Space Flight Award, Am Astronaut Soc, 57; Distinguished Civilian Serv Award, Dept Defense & Except Civilian Serv Decoration, Dept Army, 57; Goddard Mem Trophy, 58; Fed Civilian Serv Award, 59; Harrison Award & Crowell Gold Medal, Am Ord Asn, Holt Gold Medal, Rollins Col & Distinguished Serv Award, Southern Asn Sci & Indust, 59; Gold Medal, Brit Interplanetary Soc, 61; Order of Merit for Res & Invention, France & Cresson Award, Franklin Inst, 62; Sci & Eng Award, Drexel Inst, 63; NASA Medal, 64; Award, Aerospace Elec Soc & Am Astronaut Soc, 65; Diesel Gold Medal, Ger Soc Inventions, 65; Galabert Int Astronaut Prize, France, 65; Award, Ger Soc Aviation & Space Med, 66; Greek Fel Award, Hellenic Astronaut Soc. Mem: Nat Acad Eng; hon fel Am Inst Aeronaut & Astronaut; hon fel Brit Interplanetary Soc; hon fel Norweg Interplanetary Soc. Res: Rocket and space technology. Mailing Add: Fairchild Industs Inc Fairchild Dr Germantown MD 20767

VONBUN, FRIEDRICH OTTO, b Vienna, Austria, June 22, 25; US citizen; m 52; c 2. PHYSICS, MATHEMATICS. Educ: Vienna Tech Univ, MS, 52; Graz Tech Univ, PhD(physics, math), 56. Prof Exp: Physicist, US Army Signal Corps, 53-57, chief molecular beam sect, Atomic Resonance Br, 57-59, sr scientist & dir frequency control div, 59-60; consult, Tracking & Data Systs Directorate, 60-61, head plans off, 61-63, chief systs anal off, 63-65, chief mission anal off, 65-67, chief mission & trajectory anal div, 67-71, chief trajectory anal & geodyn div, Mission & Data Opers Directorate, 71-72, chief geodyn prog div, 72-74, ASST DIR APPLICATIONS SCI, APPLICATIONS DIRECTORATE, GODDARD SPACE FLIGHT CTR, NASA, 74- Concurrent Pos: Mem panel tracking & data anal, Nat Acad Sci, 63-65; spec adv,

Range Tech Adv Group, 66-; mem, Steering Comt, Working Group 1, COSPAR, 72-; chmn, Working Group Earth & Ocean Dynamics, Int Astronaut Fedn, 75 & Working Group Global Data Collection, 76. Mem: Assoc fel Am Inst Aeronaut & Astronaut; Am Geophys Union; Int Astronaut Fedn. Res: Space systems analysis; navigation; geodynamics; ocean dynamics; gravity and magnetic field studies; active and passive microwave observations of the Earth's surface from space; application of space science and technology toward solutions of practical problems. Mailing Add: Applications Directorate NASA Goddard Space Flight Ctr Code 900 Greenbelt MD 20771

VONDELL, RICHARD M, b Amherst, Mass, Apr 6, 30; m 59; c 4. FOOD SCIENCE. Educ: Univ Mass, BS, 52, PhD(food sci), 63; Univ NH, MS, 56. Prof Exp: Res assoc animal nutrit, Univ NH, 56-57; asst instr food sci, Univ Mass, 59-62; food technologist, 63-65, DIR PILOT PLANT OPERS, KELLOGG CO, 65- Mem: Inst Food Technologists. Res: Cereal technology; vapor-phase destruction of microorganisms. Mailing Add: Kellogg Co Battle Creek MI 49016

VONDERBRINK, SALLY ANN, b Cincinnati, Ohio, Dec 1, 34. ANALYTICAL CHEMISTRY. Educ: Col Mt St Joseph, AB, 56; Univ Cincinnati, PhD(chem), 66. Prof Exp: Documentalist, Procter & Gamble Co, 56-62; from instr to assoc prof anal chem, Col Mt St Joseph, 65-71; CHMN SIC DEPT, ST XAVIER HIGH SCH, 71- Concurrent Pos: Lectr, Univ Cincinnati, 66- Mem: AAAS; Am Chem Soc; Nat Sci Teachers Asn. Res: Chemical literature and documentation; spectrophotometric analysis. Mailing Add: St Xavier High Sch 600 North Bend Rd Cincinnati OH 45224

VONDER HAAR, THOMAS HENRY, b Quincy, Ill, Dec 28, 42; m 61; c 3. METEOROLOGY, SPACE SCIENCE. Educ: St Louis Univ, BS, 63; Univ Wis-Madison, MS, 64, PhD(meteorol), 68. Prof Exp: Assoc scientist meteorology, Space Sci & Eng Ctr, Univ Wis, 68-70; ASSOC PROF ATMOSPHERIC SCI, COLO STATE UNIV, 70-, HEAD DEPT, 74- Concurrent Pos: Consult, US Army, McDonnell-Douglas Corp, Ball Bros & Res Corp, 69-; mem int radiation comn, Int Union Geod & Geophys, 75. Mem: Am Meteorol Soc. Res: Application of measurements from meteorological satellites to problems of atmospheric and environmental science; radiation measurement; air pollution; weather forecasting. Mailing Add: Dept of Atmospheric Sci Colo State Univ Ft Collins CO 80521

VON DOHLEN, WERNER CLAUS, b Langen, Ger, Nov 7, 24; nat US; m 47; c 3. SURFACE CHEMISTRY. Educ: Rutgers Univ, BSc, 50; Brown Univ, PhD(phys chem), 54. Prof Exp: Res chemist, 54-64, PROJ LEADER, UNION CARBIDE CORP, 64- Concurrent Pos: Instr, WVa State Col, 59-61, asst prof, 61-62. Mem: Am Chem Soc; Am Vacuum Soc. Res: Surface science, electron spectroscopy; x-ray crystallography of organic compounds; heterogeneous and homogeneous catalysis; olefin polymerization; kinetics and mechanism of polymerization; polymer morphology and physical properties; structure of catalysts; electron microscopy. Mailing Add: Charleston Tech Ctr Res & Develop Union Carbide Corp PO Box 8361 South Charleston WV 25303

VONDRA, BENEDICT LORENZ, JR, chemistry, see 12th edition

VONDRA, CARL FRANK, b Seward, Nebr, June 3, 34; m 55; c 4. GEOLOGY. Educ: Univ Nebr, BS, 56, MS, 58, PhD(geol), 63. Prof Exp: Develop geologist, Calif Co, 61-62; geologist, Pan Am Petrol Corp, 62-63; from asst prof to assoc prof, 63-71, PROF GEOL, IOWA STATE UNIV, 71- Mem: Geol Soc Am; Am Asn Petrol Geol; Paleont Soc; Soc Vert Paleont; Soc Econ Paleont & Mineral. Res: Stratigraphy of the Eocene deposits of the Big Horn Basin, Wyoming; stratigraphy of the upper Eocene and Oligocene deposits of Egypt; stratigraphy of the Siwalik deposits in northern India; stratigraphy and sedimentation of the Plio-Pleistocene deposits in the East Rudolf Basin, Kenya. Mailing Add: Dept of Earth Sci Iowa State Univ Ames IA 50010

VONDRAK, EDWARD ANDREW, b Chicago, Ill, Nov 12, 38; m 61; c 3. PHYSICS. Educ: Knox Col, AB, 60; Vanderbilt Univ, MA, 63, PhD(physics), 65. Prof Exp: Teaching fel physics, Vanderbilt Univ, 61-64; from asst prof to assoc prof, 67-72, PROF PHYSICS, IND CENT UNIV, 72- Mem: Am Asn Physics Teachers; Am Phys Soc; Soc Appl Spectros. Res: Infrared spectroscopy; molecular structure. Mailing Add: Dept of Math & Physics Ind Cent Univ 1400 E Hanna Ave Indianapolis IN 46227

VON DREELE, PATRICIA HAGAN, biophysical chemistry, see 12th edition

VON DREELE, ROBERT BRUCE, b Minneapolis, Minn, Dec 10, 43; m 68. SOLID STATE CHEMISTRY, CRYSTALLOGRAPHY. Educ: Cornell Univ, BS, 66, PhD(chem), 71. Prof Exp: Asst prof, 71-75, ASSOC PROF CHEM, ARIZ STATE UNIV, 75- Concurrent Pos: NSF fel dept inorg chem, Oxford Univ, 72-73. Mem: Am Chem Soc; Am Crystallog Asn; AAAS. Res: X-ray crystal structure analysis; solid state chemistry of metal oxides; conformation of biologically significant molecules. Mailing Add: Dept of Chem Ariz State Univ Tempe AZ 85281

VONEIDA, THEODORE J, b Auburn, NY, Aug 26, 30; m 56; c 3. NEUROBIOLOGY. Educ: Ithaca Col, BS, 53; Cornell Univ, MEd, 54, PhD(zool), 60. Prof Exp: Res assoc neuroanat, Walter Reed Army Inst Res, 54-56; asst comp neurol, Cornell Univ, 56-59; assoc prof, 62-75, PROF ANAT & BIOL, SCH MED, CASE WESTERN RESERVE UNIV, 75- Concurrent Pos: USPHS res fel neurobiol, Calif Inst Technol, 60-62. Mem: AAAS; Am Asn Anatomists. Res: Utilization of neuroanatomical and behavioral techniques to investigate the central nervous system. Mailing Add: Dept of Anat Case Western Reserve Univ Sch Med Cleveland OH 44106

VON ELBE, GUENTHER, chemistry, see 12th edition

VON ELBE JOACHIM HERMAN, food science, see 12th edition

VON ESSEN, CARL FRANCOIS, b Tokyo, Japan, May 17, 26; nat US; m 50; c 3. RADIOTHERAPY. Educ: Stanford Univ, AB, 48, MD, 52; Am Bd Radiol, dipl, 58. Prof Exp: Res fel cancer, Stanford Univ, 57-59; from instr to assoc prof radiol, Yale Univ, 59-69; PROF RADIOL & ONCOL & DIR RADIATION THER, UNIV CALIF, SAN DIEGO, 69- Concurrent Pos: Vis prof, Christian Med Col, Vellore, India, 65-66; mem rev comt, Radiation Study Sect, NIH, 67-69 & Cancer Res Ctr, 70-74; mem staff, Ludwig Inst Cancer Res, 75-76. Mem: Am Soc Ther Radiologists; Radiation Res Soc; Am Asn Cancer Res; Radiol Soc NAm. Res: Biological effects of radiation; clinical and experimental time-dose relationships; epidemiology of oral cancer. Mailing Add: Univ Hosp 225 W Dickinson St San Diego CA 92103

VON EULER, LEO HANS, b Stockholm, Sweden, Jan 31, 31; US citizen; m 55; c 2. PATHOLOGY. Educ: Williams Col, BA, 52; Yale Univ, MD, 59. Prof Exp: Trainee path, Sch Med, Yale Univ, 59-61, trainee pharmacol, 61-63; fel hematol, Dept Clin Path, NIH, 65-66; scientist, Sect Nutrit Biochem, Nat Inst Arthritis & Metab Dis, 66-67, prog adminr path res training progs, Nat Inst Gen Med Sci, 67-72, spec asst to dir, 72-74, DEP DIR, NAT INST GEN MED SCI, 74- Mem: Am Asn Path & Bact; Am Soc Exp Path. Res: Purine and pyrimidine metabolism; orotic acid induced fatty liver in the rat; biochemical and histological changes. Mailing Add: Westwood Bldg Rm 904 Nat Inst of Gen Med Sci Bethesda MD 20014

VON FISCHER, WILLIAM, b St Paul, Minn, Mar 3, 10; m 37; c 3. CHEMISTRY. Educ: Univ Minn, BChem, 32, MS, 33, PhD(anal & phys chem), 37. Prof Exp: Asst, Inst Technol, Univ Minn, 33-37; from instr to prof chem, Case Inst Technol, 37-56, asst head dept chem & chem eng, 48, head, 48-56; coordr res & develop, Glidden Co, 56-58, vpres res, 58-60; consult, 60-63; vpres, Day-Glo Color Corp, Ohio, 63-71; res adminr, Inst Environ Studies, Univ Ill, Urbana, 71-74; CONSULTANT, 74- Mem: AAAS; Am Chem Soc; Am Inst Chem Eng; Fedn Socs Paint Technol. Res: Analytical methods; organic protective coatings; co-precipitation and aging; use of radioactive indicators; preparation of organic indicators; synthetic rubber; environmental studies. Mailing Add: Rte 3 Box 437C Melbourne Beach FL 32951

VON FRANKENBERG, CARL ALEXANDER, b Gera, Germany, Nov 22, 32; US citizen; m 57; c 2. PHYSICAL CHEMISTRY. Educ: Swarthmore Col, BA, 56; Univ Pa, PhD(chem), 61. Prof Exp: Asst prof, 61-69, ASSOC PROF CHEM, UNIV DEL, 69- Concurrent Pos: NSF fel, Cornell Univ, 61. Mem: Am Chem Soc. Res: Polymer solution theory; statistical mechanics. Mailing Add: 409 Apple Rd Newark DE 19711

VON GIERKE, HENNING EDGAR, b Karlsruhe, Ger, May 22, 17; m 50; c 2. BIOACOUSTICS, BIOMECHANICS. Educ: Karlsruhe Tech, Dipl Ing, 43, DrEng, 44. Prof Exp: Asst acoust, Inst Theoret Elec Eng & commun techniques, Karlsruhe Tech, 44-47, lectr, 46; consult, 47-54, chief bioacoust br, 54-63, DIR BIODYNAMICS & BIONICS DIV, AEROSPACE MED RES LABS, WRIGHT-PATTERSON AFB, 63- Concurrent Pos: Mem comt hearing & bioacoust, Armed Forces Nat Res Coun, 53-, mem bioastronaut comt, 59-61; mem adv comt flight med & biol, NASA, 60-61; assoc prof, Ohio State Univ, 63-; mem, White House Ad Hoc Panel Jet Aircraft Noise, 66. Honors & Awards: Distinguished Civilian Serv Award, Dept Defense, 63; Eric LiljenKrantz Award, Aerospace Med Asn, 66 & Arnold D Tuttle Award, 74. Mem: Fel Acoust Soc Am; fel Aerospace Med Asn (vpres, 66-67); hon fel Inst Environ Sci; cor mem Int Acad Astronaut; Int Acad Aviation & Space Med. Res: Physical, physiological and psychological acoustics; biodynamics; effects of noise, vibration and impact on man; communication biophysics; bionics. Mailing Add: 1325 Meadow Lane Yellow Springs OH 45387

VON GOELER, EBERHARD, b Berlin, Ger, Feb 22, 30; m 60; c 3. HIGH ENERGY PHYSICS. Educ: Univ Ill, MS, 55, PhD(physics), 61. Prof Exp: Res assoc physics, Univ Ill, 60-61; res scientist, Deutsches Elektronen Synchrotron, Hamburg, Ger, 61-63; from asst prof to assoc prof, 63-73, PROF PHYSICS, NORTHEASTERN UNIV, 73- Concurrent Pos: Vis prof, Univ Hamburg, 67-68; vis scientist, Nat Accelerator Lab, Ill, 71-72. Mem: Am Phys Soc. Res: Surface physics; tests of quantum electrodynamics; photoproduction of vector mesons, antibaryons; meson spectrometry; high mass bosons; counter techniques in high energy physics. Mailing Add: Dept of Physics Northeastern Univ Boston MA 02115

VON GRAEVENITZ, ALEXANDER W C, b Leipzig, Ger, Nov 8, 32; US citizen; m 60; c 3. MEDICAL MICROBIOLOGY. Educ: Univ Tübingen, BS, 51; Univ Bonn, MD, 56. Prof Exp: Res fel pharmacol, Univ Bonn, 56; Fulbright travel grant, 57; res fel microbiol, Univ Mainz, 58-60 & Yale Univ, 61-63; asst prof microbiol & lab med, 63-69, assoc prof, 69-73, PROF LAB MED, YALE UNIV, 73-; DIR CLIN MICROBIOL LABS, YALE-NEW HAVEN HOSP, 63- Concurrent Pos: Mem, Conf State Pub Health Dirs, 67-; mem standards & exam comt, Am Bd Med Microbiol, 73- Mem: Am Acad Microbiol; Am Soc Microbiol; Ger Soc Hyg & Microbiol; Asn Clin Sci. Res: Pathogenicity and diagnosis of gram-negative rods. Mailing Add: 789 Howard Ave New Haven CT 06504

VON GUTFELD, ROBERT J, b Berlin, Ger, Mar 5, 34; US citizen. SOLID STATE PHYSICS. Educ: Queens Col, BS, 54; Columbia Univ, MA, 57; NY Univ, PhD(physics), 65. Prof Exp: Substitute instr physics, Queens Col, 54-55; engr, Sperry Gyroscope Co, 57-60; RES STAFF MEM, T J WATSON RES CTR, IBM CORP, 60- Mem: Am Phys Soc. Res: Thermal transport in solids using heat pulse techniques; amorphous semiconductor and dye laser research and applications to optical memories; transverse thermoelectric effects in metallic thin films. Mailing Add: T J Watson Res Ctr IBM Corp Yorktown Heights NY 10598

VON HAAM, EMMERICH, b Vienna, Austria, Aug 24, 03; nat US; m 31; c 5. PATHOLOGY. Educ: Univ Vienna, MD, 26. Prof Exp: Instr clin path & internal med, Univ Vienna, 26-28; fel, Physiol Inst, Morristown, NJ, 28-29; instr internal med, Univ Vienna, 29-30; assoc prof path, Univ Ark, 30-31; asst prof path & bact, Sch Med, La State Univ, 32-37; chmn dept, 37-70, PROF PATH, COL MED, OHIO STATE UNIV, 37- Honors & Awards: Am Soc Clin Path Gold Medal, 36. Mem: Soc Exp Biol & Med; Am Soc Clin Path; Am Soc Trop Med & Hyg; Am Asn Cancer Res; Am Asn Pathologists & Bacteriologists. Res: Physiology and pathology of the spleen; iron metabolism; lymphogranuloma inguinale; cancer; cytology; pathology of heart failure. Mailing Add: Dept of Path Ohio State Univ Columbus OH 43210

VON HAGEN, D STANLEY, b Nashville, Tenn, Dec 21, 37; m 59; c 2. PHARMACOLOGY. Educ: Carson-Newman Col, BS, 59; Vanderbilt Univ, PhD(pharmacol), 65. Prof Exp: Res assoc pharmacol, Vanderbilt Univ, 65-66, instr, 66; instr, 67-69, ASST PROF PHARMACOL, NJ MED SCH, COL MED & DENT, NJ, 69- Mem: AAAS; NY Acad Sci. Res: Smooth muscle physiology and pharmacology; physiological role of calcium ion in smooth muscle function. Mailing Add: Dept of Pharmacol New Jersey Med Sch Newark NJ 07103

VON HERRMANN, PIETER, nuclear physics, see 12th edition

VON HERZEN, RICHARD P, b Los Angeles, Calif, May 21, 30; m 58; c 2. MARINE GEOPHYSICS. Educ: Calif Inst Technol, BS, 52; Harvard Univ, AM, 56; Univ Calif, PhD(oceanog), 60. Prof Exp: Lab asst oceanog, Scripps Inst Oceanog, 52-53, geophysicist, 58-60, asst res geophysicist, 60-64; dep dir off oceanog, UNESCO, 64-66; assoc scientist, 66-73, SR SCIENTIST, WOODS HOLE OCEANOG INST, 73- Concurrent Pos: Assoc ed J Geophys Res, Am Geophys Union, 69-71; vis res geophysicist & lectr, Scripps Inst Oceanog, 74-75. Mem: Am Geophys Union. Res: Structure and dynamics of the earth beneath the ocean floor. Mailing Add: Woods Hole Oceanog Inst Woods Hole MA 02543

VON HIPPEL, FRANK, b Cambridge, Mass, Dec 26, 37; div. THEORETICAL PHYSICS. Educ: Mass Inst Technol, SB, 59; Oxford Univ, DPhil(physics), 62. Prof Exp: Res assoc physics, Univ Chicago, 62-64; res assoc, Cornell Univ, 64-66; asst prof, Stanford Univ, 66-70; mem staff theory group, High Energy Physics Div, Argonne Nat Lab, 70-73; resident fel, Nat Acad Sci, 73-74; RES SCIENTIST CTR ENVIRON STUDIES, PRINCETON UNIV, 74- Concurrent Pos: Sloan Found fel, 67-70; consult nuclear energy policy, Off Technol Assessment, Gen Acct Off, House Interior Comt, US Congress, 75. Mem: Am Phys Soc; AAAS. Res: Nuclear energy policy and energy policy more generally. Mailing Add: Ctr Environ Studies Princeton Univ Eng Quadrangle Princeton NJ 08540

VON HIPPEL, PETER HANS, b Göttingen, Ger, Mar 13, 31; nat US; m 54; c 3. BIOPHYSICAL CHEMISTRY. Educ: Mass Inst Technol, BS, 52, MS, 53, PhD(biophys), 55. Prof Exp: Asst phys biochem, Mass Inst Technol, 53, NIH fel, 55-56; phys biochemist, US Naval Med Res Inst, Md, 56-59; asst prof biophys, Dartmouth Med Sch, 59-61, assoc prof biochem, 61-67; res assoc, 67-69, PROF CHEM, UNIV ORE, 67-, DIR INST MOLECULAR BIOL, 69- Concurrent Pos: Sr fel, USPHS, 59-67, mem study sect biophys & biophys chem, 63-67; chmn, Gordon Res Conf Physics & Phys Chem of Biopolymers, 68; corp vis comn, Dept Biol, Mass Inst Technol, 73-; Guggenheim found fel, 73-74; mem bd sci counr, Nat Inst Arthritis, Metab & Digestive Dis, NIH, 74- Mem: AAAS; Biophys Soc (pres-elect, 72-73, pres, 73-74); Am Soc Biol Chem; Am Chem Soc; Soc Gen Physiol. Res: Physical biochemistry of macromolecules; structure, function and interactions of proteins and nucleic acids; molecular aspects of control of genetic expression. Mailing Add: Inst of Molecular Biol Univ of Ore Eugene OR 97403

VON HOLDT, RICHARD ELTON, b Chicago, Ill, Jan 9, 17; m 45; c 2. MATHEMATICS. Educ: De Paul Univ, BPh, 40, MS, 42; Northwestern Univ, PhD(math), 55. Prof Exp: Instr math, Northwestern Univ, 45-50; mem staff, Los Alamos Sci Lab, 50-53; comput engr, Douglas Aircraft Co, 53-54; MATHEMATICIAN, LAWRENCE LIVERMORE LAB, UNIV CALIF, 54- Res: Numerical analysis; matrix theory and applications; applied mathematics; programming; compound matrices. Mailing Add: Lawrence Livermore Lab Univ of Calif Livermore CA 94550

VON HUENE, ROLAND, b Los Angeles, Calif, Jan 30, 29; m 53; c 3. GEOLOGY. Educ: Univ Calif, Los Angeles, AB, 53, PhD, 60. Prof Exp: Gen geologist, US Naval Ord Test Sta, 53-67; GEOPHYSICIST, US GEOL SURV, 67- Concurrent Pos: Asst, Univ Calif, Los Angeles, 55-57; Fulbright grant, Innsbruck, 57-58; mem comt Alaska earthquake, Nat Acad Sci; Pac site panel, Joint Ocean Insts for Deep Earth Sampling; dep chief off marine geol, US Geol Surv, 73-75; panel mem active margin & site surv, Int Prog Ocean Drilling, 74-; mem nat comt, US Geodynamics Comt, 75- Mem: AAAS; Am Geophys Union; Sigma Xi; Geol Soc Am; Ger Geol Asn. Res: Marine gravity; magnetics; seismic profiling; tectonics; structural geology; gravimetry. Mailing Add: US Geol Surv 345 Middlefield Rd Menlo Park CA 94025

VON KAULLA, KURT NIKOLAJ, b Darmstadt, Ger, July 17, 12; nat US; m 41. EXPERIMENTAL MEDICINE. Educ: Univ Freiburg, MD, 38. Prof Exp: Head res lab, Dept Surg, Univ Freiburg, 43-46; mem staff, Res Dept, Geigy Pharmaceut, Basle, Switz, 46-51; investr, Res Lab, Dept Obstet & Gynec, Basle Univ, 51-52; from asst prof to prof med, Sch Med, Univ Colo, Denver, 53-76, dir coagulation lab, Med Ctr, 61-74. Concurrent Pos: Res fel, Res Lab, Dept Surg, Univ Freiburg, 40-45; res fel, Med Div, Goldwater Mem Hosp, NY Univ, 52-53; Am Heart Asn fel, 56-59; mem comt thrombolytic agents, Nat Heart Inst, 64-68; mem, Ger Study Group Coagulation Res. Mem: Am Chem Soc; Soc Exp Biol & Med; Am Physiol Soc; Am Soc Hemat; hon mem Mex Soc Hemat. Res: Anticoagulants; fibrinolytic enzymes; synthetic fibrinolytic agents; clotting factors in body fluids; blood coagulation in experimental and pathological conditions and related fields. Mailing Add: Stechertweg 2 7800 Freiburg/Brsg West Germany

VON KESZYCKI, CARL HEINRICH, b Utrecht, Holland, Sept 4, 22; US citizen; m 56. PHYSICS. Educ: Univ Munich, BS, 52, PhD(physics), 59. Prof Exp: Res engr, Lockheed-Calif Co, 60-63, scientist, Lockheed-Ga Co, 63-64, assoc dir res, 64-67, sr res & develop engr, 67-70; pres, Sylesia Corp, 70-72; sr fel, Mellon Inst Sci, 72-74; DIR PHYS SCI TECHNOL, ESSEX INT, INC, 72-; DIR SPEC PROJS, GRUMMAN AEROSPACE CORP, 74- Mem: Am Phys Soc; Am Inst Aeronaut & Astronaut. Res: Nuclear physics; aeronautical engineering. Mailing Add: 22 West Mall Dr Huntington NY 11743

VON KOEPPEN, ANDREAS, b Petrograd, Russia, Aug 10, 17; US citizen; m 48; c 2. PULP TECHNOLOGY, PAPER TECHNOLOGY. Educ: Tech Acad Forest Utilization, Leningrad, Chem Engr, 39; Darmstadt Tech Univ, Dr Eng, 48. Prof Exp: Asst mgr pulp mill, Waldhof Zellstoffabriken, WGer, 43-45; res officer grade III, Div Forest Prod, Commonwealth Sci & Indust Res Orgn, Melbourne, Australia, 50-56; res group leader, WVa Pulp & Paper Co, 56-58; res group leader, Packaging Corp Am; asst res prof, Western Mich Univ, 59-61; asst res dir, Packaging Corp Am, 61-66; tech dir pulp & paper eng, Technopulp, Inc, 66-67; sr res engr, Copeland Process Corp, 67-68; VPRES, WRIGHT CHEM CORP, CHICAGO, 68- Concurrent Pos: Consult, Am Box-Board Co, 59-61; del, Ger Pulp & Paper Engrs & Chemists, Baden-Baden, 63; Int Water Conf, Pittsburgh, Pa, 71. Mem: Am Chem Soc; fel Am Inst Chem; Tech Asn Pulp & Paper Indust; Nat Asn Corrosion Engr. Res: Structural and chemical differences between sulfite and sulfate pulps; investigation of tropical woods from New Guinea and other Pacific islands for pulp and paper making; new processes for production of ultra high yield pulp from hardwoods; corrosion inhibitors; paper chemicals for paper recycling. Mailing Add: Wright Chem Corp 1319 Wabansia Ave Chicago IL 60622

VON KORFF, RICHARD WALTER, b Davenport, Iowa, Jan 6, 16; m 43; c 3. BIOCHEMISTRY, ENZYMOLOGY. Educ: Univ Minn, BA, 47, PhD(physiol chem), 51; Am Bd Clin Chem, dipl, 74. Prof Exp: Anal chemist, Testing & Res Lab, Deere & Co, Ill, 37-41; asst & sr sci aide, Anal & Phys Chem Div, Northern Regional Res Lab, USDA, 41-43, jr chemist, Agr Residues Div, 43-45; asst prof pediat & physiol chem, Univ Minn, 55-66; dir biochem res, Friends of Psychiat Res, Spring Grove State Hosp, Baltimore, 66-68; DIR BIOCHEM RES, MD PSYCHIAT RES CTR, 68- Concurrent Pos: Fel, Inst Enzyme Res, Univ Wis, 51-52; Whitney Found fel biochem, Dept Pediat, Heart Hosp, Univ Minn, 52-53, Am Heart Asn fel, 53-55, USPHS sr res fel biochem, 60-66; chmn subcomt enzymes, Comt Biol Chem, Nat Acad Sci-Nat Res Coun, 61-67 & 69-; adj prof, Dept Med Chem, Sch Pharm, 71-73. Mem: Am Soc Biol Chemists; Am Chem Soc; Am Soc Cell Biol; NY Acad Sci; Am Soc Neurochem. Res: Enzymic control mechanisms; interactions of intracellular enzyme systems; monoamine oxidase. Mailing Add: Dept of Biochem Md Psychiat Res Ctr Baltimore MD 21228

VON LEDEN, HANS VICTOR, b Ger, Nov 20, 18; nat US; m 48; c 2. OTOLARYNGOLOGY. Educ: Loyola Univ, Ill, MD, 42; Am Bd Otolaryngol, dipl, 45. Prof Exp: Intern, Mercy Hosp-Loyola Univ Clins, 41-42; resident, Presby Hosp, Chicago, 42-43; fel otolaryngol & plastic surg, Mayo Found, Univ Minn, 43-45, first asst, Mayo Clin, 45; clin assoc otolaryngol, Stritch Sch Med, Loyola Univ, Ill, 47-51; from asst prof to assoc prof, Med Sch, Northwestern Univ, 52-61; assoc prof surg, Sch Med, Univ Calif, Los Angeles, 61-66; pres, 59-65, MED DIR, INST LARYNGOL & VOICE DIS, 65-; PROF BIO-COMMUN, UNIV SOUTHERN CALIF, 66- Concurrent Pos: Consult, US Navy, 47-; assoc prof, Cook County Grad Sch Med, 49-58; med dir, William & Harriet Gould Found, 55-59; pres, Inst Laryngol & Voice Dis, 59-65, med dir, 65-; vis prof, US & 25 for countries. Honors & Awards: Bucranio, Univ Padua, 58; Gold Medal, Ital Res Cross, 59; Award Honor, Am Acad Ophthal & Otolaryngol, 59; Hektoen Medal, AMA, 60; Sci Awards, Am Speech & Hearing Asn, 60, 62 & 65; Gold Medal, Ill State Med Soc, 60; Casselberry Award, Am Laryngol Asn, 62; Manuel Garcia Prize, Int Asn Logoped & Phoniatrics, 68. Mem: Fel AAAS; fel Am Acad Ophthal & Otolaryngol; fel Am Col Surgeons; fel Int Col Surgeons (pres, 72); fel Am Speech & Hearing Asn. Res: Voice and speech; laryngology. Mailing Add: Inst Laryngol & Voice Dis 10921 Wilshire Blvd Los Angeles CA 90024

VON LICHTENBERG, FRANZ, b Miskole, Hungary, Nov 29, 19; US citizen; m 49; c 6. PATHOLOGY, TROPICAL MEDICINE. Educ: Nat Univ Mex, MD, 45. Hon Degrees: Nicaragua Univ, Hon Dr, 59. Prof Exp: Asst prof path, Nat Univ Mex, 47-48, ord prof, 48-53; from asst prof to assoc prof, Univ PR, 53-58; from instr to assoc prof, 58-72, PROF PATH, HARVARD MED SCH, 73-; PATHOLOGIST, PETER BENT BRIGHAM HOSP, 69- Concurrent Pos: W K Kellog Latin Am fel, 50-51; chief consult, Hosp das Clinicas, Bahia, Brazil, 51-52; Smith-Mundt lectr, Nicaragua, 59; mem, Comn Pediat Dis, Armed Forces Epidemiol Bd, 60-69; sr assoc, Peter Bent Brigham Hosp, 61-68; consult, Expert Group Parasitic Dis, WHO, 65; vis res scientist, Univ Col Ibadan, 66; chmn, Panel Parasitic Dis, US-Japan Coop Med Sci Prog, 73- Mem: Am Asn Pathologists & Bacteriologists; Am Soc Exp Path; Am Soc Trop Med & Hyg. Res: Pathology of tropical diseases; immunopathology; general pathology; schistosomiasis. Mailing Add: Dept of Path Peter Bent Brigham Hosp Boston MA 02115

VON MALTZAHN, KRAFT EBERHARD, b Rostock, Ger, Dec 17, 25; Can citizen; m 49; c 3. BOTANY. Educ: Yale Univ, MS, 50, PhD(bot), 54. Prof Exp: From asst prof to assoc prof, 54-62, PROF BIOL, DALHOUSIE UNIV, 62-, GEORGE S CAMPBELL PROF & HEAD DEPT, 63- Mem: Can Bot Asn; Can Soc Plant Physiol. Res: Plant morphogenesis. Mailing Add: Dept of Biol Dalhousie Univ Halifax NS Can

VON MEERWALL, ERNST DIETER, b Vienna, Austria, Dec 29, 40. PHYSICS. Educ: Northern Ill Univ, BS, 63, MS, 65; Northwestern Univ, Evanston, PhD(physics), 69. Prof Exp: Res assoc, Dept Metall & Mat Res Lab, Univ Ill, Urbana, 69-71; asst prof, 71-74, ASSOC PROF PHYSICS, UNIV AKRON, 74- Mem: Am Phys Soc; Sigma Xi. Res: Solid state experiment; nuclear magnetic resonance, Mössbauer effect and magnetic susceptibility; alloys; nuclear quadrupole effect; polymers; numerical methods. Mailing Add: Dept of Physics Univ of Akron Akron OH 44325

VON MERING, OTTO O, b Oct 21, 22; US citizen; m 54; c 3. ANTHROPOLOGY. Educ: Williams Col, BA, 43; Harvard Univ, PhD(social anthrop), 56. Prof Exp: Libr asst, New York Pub Libr, 44-45; instr, Belmont Hill Sch, Mass, 45-47; instr sociol & anthrop, Boston Univ, 47-48; tutor & counsr emotionally & intellectually retarded young adults, Mass, 48-50; res asst, Lab Social Rels, Harvard Univ, 50-51; res asst, Mass Ment Health Ctr, 51-53; asst prof social anthrop, Sch Med, Univ Pittsburgh, 55-60, assoc prof, Dept Psychiat & Anthrop, 60-71, prof anthrop, Sch Med & Col Arts & Sci, 65-71; PROF ANTHROP, COL ARTS & SCI & SCH MED, UNIV FLA, 71- Concurrent Pos: Instr, Cambridge Jr Col, 48-49 & 51-52; Russell Sage Found soc sci fel, 53-55; Richard-Merton guest prof, Univ Heidelberg, 62-63; guest prof, Inst Psychoanal & Psychosom Med, Frankfort, Ger, 62-63; Fulbright vis lectr, 62-63; res lectr, Sigmund Freud Inst, 64; Maurice Falk Med Fund training prog develop grant, 66-69; mem continuing conf psychiat & med educ, Am Psychiat Asn-Am Asn Med Cols, 66-68; mem, Joint Comn Ment Health of Children, Inc, Md, 67-69; vis prof anthrop, Dartmouth Col, 70-71; NIMH spec fel, 71-72; corresp ed, J Geriat Psychiat, 73-; commentary ed, J Human Orgn, 74- Mem: Fel AAAS; fel Am Acad Psychosom Med; fel Am Anthrop Asn; fel Am Ethnol Soc; fel Royal Anthrop Inst Gt Brit & Ireland. Res: Anthropomedical perspectives on aging and child development; patterns of affective disorders in ethnic families; culture context and value in clinical diagnosis and treatment; consumer health education; impact on care delivery systems. Mailing Add: 818 NW 21st St Gainesville FL 32603

VON MOLNAR, STEPHAN, b Leipzig, Ger, June 26, 35; US citizen; m 56; c 2. SOLID STATE PHYSICS. Educ: Trinity Col (Conn), BS, 57; Univ Maine, MS, 59; Univ Calif, Riverside, PhD(physics), 65. Prof Exp: Mem res staff physics, Polychem Div, Exp Sta, E I du Pont de Nemours & Co, 59-60; MEM RES STAFF PHYSICS, THOMAS J WATSON RES CTR, IBM CORP, 65- Concurrent Pos: Sr res fel, Imp Col, Univ London, 73-74. Mem: Am Phys Soc. Res: Paramagnetic and ferromagnetic resonance; transport properties of ferromagnetic semiconductors; tunneling spectroscopy of superconductors and semiconductors; low temperature specific heat. Mailing Add: Thomas J Watson Res Ctr IBM Corp PO Box 218 Yorktown Heights NY 10598

VONNEGUT, BERNARD, b Indianapolis, Ind, Aug 29, 14; wid; c 5. PHYSICAL CHEMISTRY. Educ: Mass Inst Technol, BS, 36, PhD(phys chem), 39. Prof Exp: Res assoc, Preston Labs, Pa, 39-40 & Hartford Empire Co, Conn, 40-41; chem eng, Mass Inst Technol, 41-42, meteorol, 42-45; res labs, Gen Elec Co, 45-52; mem staff, Arthur D Little Inc, 52-67; PROF ATMOSPHERIC SCI, STATE UNIV NY ALBANY, 67-, SR RES SCIENTIST ATMOSPHERIC SCI RES CTR, 67- Mem: AAAS; Am Meteorol Soc; Am Geophys Union; Meteorol Soc Japan; Royal Meteorol Soc. Res: Nucleation phenomena; cloud seeding; surface chemistry; aerosols; atmospheric electricity. Mailing Add: Dept of Atmospheric Sci State Univ of NY Albany NY 12222

VON NOORDEN, GUNTER KONSTANTIN, b Frankfurt, Ger, Mar 19, 28; US citizen; m; c 1. OPHTHALMOLOGY. Educ: Univ Frankfurt, MD, 54; State Univ Iowa, MS, 60. Prof Exp: Rotating intern, St Vincent Infirmary, Little Rock, Ark, 54-56; fel ophthal, Cleveland Clin, 56-57; resident, Med Ctr, State Univ Iowa, 57-60, asst prof, Sch Med, 61-63; from assoc prof to prof, Johns Hopkins Univ, 63-72; PROF OPHTHAL, BAYLOR COL MED, 72- Concurrent Pos: Nat Inst Neurol Dis & Blindness spec trainee, Univ Tübingen, 60-61; Nat Inst Neurol Dis & Blindness spec fel, Univ Iowa, 61-62; mem, Armed Forces Nat Res Coun Vision, 64-68; Int Strabismological Asn Bielschowsky lectr, 70; adj prof neurol sci, Sch Biol Sci, Univ Tex, 72-; pres, Am Orthoptic Coun, 73-74. Honors & Awards: Hectoen Gold Medal, AMA, 60; Honor Award, Am Acad Ophthal & Otolaryngol, 70. Mem: Am Ophthal Soc; fel Am Acad Ophthal & Otolaryngol; Asn Res Strabismus (secy, 72-73); Int Strabismological Asn (secy-treas, 68-74); Pan Am Ophthal Asn. Res: Investigation of clinical and laboratory aspects of neuromuscular anomalies of the eyes, especially amblyopia; improvement of our knowledge of strabismus and assistance in the understanding of the basic morphological and neurophysiological aspects of different forms of amblyopia. Mailing Add: Tex Children's Hosp 6621 Fannin Houston TX 77025

VON OSTWALDEN, PETER WEBER, b Reichenberg, Czech, June 1, 23; m 46; c 1. ORGANIC CHEMISTRY. Educ: Univ Graz, Doctorandum, 50; Columbia Univ, MA, 54, PhD(pyridine chem), 58. Prof Exp: Process develop chemist, Merck & Co, Inc, Cherokee Plant, Pa, 57-63; ASSOC PROF CHEM, YOUNGSTOWN STATE UNIV, 63- Concurrent Pos: Vis assoc, Calif Inst Technol, 70-71. Mem: Am Chem Soc. Res: Steroid and pyridine chemistry; heterocyclic nitrogen oxides; chemistry of heterocyclic compounds; spectroscopy; organic applications; steroid chemistry. Mailing Add: Dept of Chem Youngstown State Univ Youngstown OH 44503

VON RIESEN, DANIEL DEAN, b Beatrice, Nebr, Nov 20, 43; m 68. ORGANIC CHEMISTRY. Educ: Hastings Col, BA, 65; Univ Nebr, PhD(chem), 71. Prof Exp: Instr chem, Hastings Col, 70-71; asst prof, Hamilton Col, 71-72; ASST PROF CHEM,

ROGER WILLIAMS COL, 72- Mem: Am Chem Soc. Res: Cycloaddition reactions of heterocumulenes; chemical education. Mailing Add: Dept of Chem Roger Williams Col Bristol RI 02809

VON RIESEN, VICTOR LYLE, b Marysville, Kans, Jan 21, 23; m 45; c 2. MEDICAL MICROBIOLOGY. Educ: Univ Kans, BA, 48, MA, 50, PhD, 55. Prof Exp: Asst instr bact, Univ Kans, 48-49, instr, 49-55, instr med microbiol, Sch Med, 55-56, asst prof, 56-59; from asst prof to assoc prof, 59-69, PROF MED MICROBIOL, UNIV NEBR MED CTR, OMAHA, 69- Mem: Am Soc Microbiol; Soc Appl Bact; spec affil AMA. Res: Medical bacteriology; microbial physiology. Mailing Add: Dept of Med Microbiol Univ of Nebr Med Ctr Omaha NE 68105

VON ROHR, BEATRICE LOUISE, b St Louis, Mo, Apr 7, 25; m 47. APPLIED MATHEMATICS. Educ: Wash Univ, AB, 46; Gonzaga Univ, MS, 64. Prof Exp: Asst math, Wash Univ, 46-47, instr, 47-51, res mathematician, Barnes Hosp Div, 51-52, asst math, univ, 52-59; chmn dept math high sch, Mo, 59-60; prof, Hannibal-La Grange Col-St Louis, 61-62; assoc prof, Shelton Col, 67-68; prof, Hannibal-La Grange Campus, 68-69, head dept, St Louis Campus, 69-71, PROF MATH, MO BAPTIST COL, ST LOUIS CAMPUS, 69- Concurrent Pos: Vpres, Electrovision Co, 68-; math consult parochial schs, E US, 71- & Electrovision Co, 74-; judge high sch St Louis Post-Dispatch Sci Fair, 71-; lectr high schs, Mo, 72. Mem: Math Asn Am; Soc Indust & Appl Math. Res: Research for advanced calculus and algebra texts. Mailing Add: 1119 Sanford St Louis MO 63139

VON ROOS, OLDWIG, b Koblenz, Ger, Jan 23, 25; m 53; c 4. THEORETICAL PHYSICS. Educ: Univ Marburg, BS, 50, MS, 54, PhD(physics), 56. Prof Exp: Staff scientist, Jet Propulsion Lab, 57-61, group supvr physics, 61-65; sr staff scientist, Astrophys Res Co, Calif, 65-72; MEM TECH STAFF, JET PROPULSION LAB, 72- Concurrent Pos: Vis prof, Univ Southern Calif, 63-65. Mem: AAAS; fel Am Phys Soc. Res: Atomic physics; quantum plasma theory; electromagnetic theory. Mailing Add: Jet Propulsion Lab 4800 Oak Grove Dr Pasadena CA 91109

VON ROSENBERG, JOSEPH LESLIE, JR, b Lockhart, Tex, Aug 22, 32; m 58; c 3. ORGANIC CHEMISTRY. Educ: Univ Tex, Austin, BA, 54, PhD(chem), 63. Prof Exp: Res chemist, Ethyl Corp, 57-58; res chemist, Celanese Chem Co, 62-64; Robert A Welch fel, Univ Tex, Austin, 64-65; asst prof, 65-69, ASSOC PROF CHEM, CLEMSON UNIV, 69- Mem: Am Chem Soc; The Chem Soc. Res: Physical organic and organometallic chemistry. Mailing Add: Dept of Chem Clemson Univ Clemson SC 29631

VON RUDLOFF, ERNST MAX, b Munich, Ger, May 27, 23; nat Can; m 53; c 2. ORGANIC CHEMISTRY. Educ: Univ Pretoria, BS, 48, MSc, 50, DSc(org chem), 53. Prof Exp: Tech asst org chem, Univ Pretoria, 47-50; fel, 53-55, res officer, 55-74, SR RES OFFICER, PRAIRIE REGIONAL LAB, NAT RES COUN CAN, 74- Concurrent Pos: Mem wood chem working group, Int Union Forest Res Orgn. Mem: Phytochem Soc NAm. Res: Chemistry of natural products; chemosystematic studies on North American conifer species; wood and plant extractives; terpenes; plant phenolics; carbohydrates, fats and waxes; wool wax; dehydration of alcohols, oxidation of olefinic double bonds. Mailing Add: Prairie Regional Lab Nat Res Coun of Can Saskatoon SK Can

VON RÜMKER, ROSMARIE, b Halberstadt, Ger, July 30, 26; nat US. AGRICULTURAL CHEMISTRY. Educ: Univ Bonn, Dipl & DAgr(plant path, entom, agr econ), 50. Prof Exp: Farm adminr seed breeding, Ger, 50-51; agr res biologist, Farbenfabriken Bayer, Inc, 51-54; dir res, Chemagro Corp, NY, 54-58, vpres res & develop, Kansas City, Mo, 59-71; MANAGING PARTNER, RvR CONSULTS, 71- Concurrent Pos: Mem, Agr Res Inst. Mem: Entom Soc Am; Am Chem Soc; Am Soc Agr Eng; Weed Sci Soc Am. Res: Crop protection; environmental effects of pesticides; pest control problems and opportunities; development of new pesticides and animal health products; economics of pest control and pesticide development. Mailing Add: PO Box 553 Shawnee Mission KS 66201

VON SALLMANN, LUDWIG, ophthalmology, deceased

VON SALTZA, MALCOLM H, biochemistry, organic chemistry, see 12th edition

VON SCHMELING, BOGISLAV G, b Guedenhagen, Ger, Mar 11, 24; nat US; m 55; c 2. AGRICULTURE. Educ: Munich Tech Univ, dipl, 51, PhD(agr chem), 53. Prof Exp: Res biologist, Va-Carolina Chem Corp, 54-59; res biologist, Uniroyal, Inc, 59-70, dir develop & mkt agr chem, 70-73, MKT MGR, AGR CHEM-OVERSEAS, UNIROYAL CHEM, 73- Mem: Am Phytopath Soc. Res: Agricultural chemicals, especially pesticides, herbicides and plant growth regulators. Mailing Add: Uniroyal Chem EMIC Bldg Spencer St Naugatuck CT 06770

VON SCHRILTZ, DON MORRIS, b Salina, Kans, June 8, 41; m 65; c 2. CHEMISTRY. Educ: Rice Univ, BA, 63; Duke Univ, PhD(chem), 66. Prof Exp: NIH fel, Univ Calif, Los Angeles, 66-67; sr res chemist, 67-73, ASST CHIEF CHEMIST, PLASTICS DEPT, E I DU PONT DE NEMOURS & CO, INC, 73- Mem: Am Chem Soc. Res: Carbanions; sulfur chemistry; polyamides. Mailing Add: 109 Crestwood Dr Parkersburg WV 26101

VON SCHUCHING, SUSANNE, b Berlin, Ger; nat US; m 31; c 1. ORGANIC CHEMISTRY, BIOCHEMISTRY. Educ: Univ Berlin, PhD(org chem), 37. Prof Exp: Res asst, Sch Med, Tufts Univ, 42-44; res asst, Manhattan Proj, 44-45; chemist, Mercy Hosp, Baltimore, Md, 45-47; chemist, Isotope Ctr, Sch Med, Johns Hopkins Univ, 47-54; chief biochemist, Radioisotope Serv, Vet Admin Ctr, Ft Howard, Md, 54-55, chief biochemist & asst dir res, Martinsburg, WVa, 55-71; ASST PROF NUCLEAR MED, MED CTR, IND UNIV, INDIANAPOLIS, 71- Mem: Am Chem Soc. Res: Synthesis of organic compounds labeled with radioisotopes; metabolism of labeled compounds; application of radioisotopes to medical problems. Mailing Add: Dept of Radiol Ind Univ Med Ctr Indianapolis IN 46202

VON STRANDTMANN, MAX, b Grodno, Bielorussia, Apr 17, 27; nat US; m 50; c 4. MEDICINAL CHEMISTRY. Educ: Hochsch Bamberg, Ger, BS, 52, MS, 53; Univ Erlangen, PhD, 55. Prof Exp: Scientist, Chem Fabrik, Bamberg, 55-57; scientist, 57-61, sr scientist, 61-65, sr res assoc, 65-75, ASSOC DIR, WARNER-LAMBERT RES INST, 75- Mem: Am Chem Soc. Res: Development of new synthetic methods in heterocyclic chemistry; medicinal chemistry research in allergy, central nervous system and antimicrobial areas; structure elucidation and structure modification of natural products. Mailing Add: Warner-Lambert Res Inst 170 Tabor Rd Morris Plains NJ 07950

VON STRYK, FREDERICK GEORGE b Pollenhof, Estonia, Sept 6, 12; Can citizen; m 44; c 1. ORGANIC CHEMISTRY. Educ: Tartu State Univ, Chem, 34; Univ Leipzig, dipl chem, 38, Dr rer nat(chem), 40. Prof Exp: Res asst org chem, Univ Leipzig, 40-41; res chemist, Badische Soda & Anilin Fabrik, Ger, 41-48, Tex Co, Que, 48-51 & Dominion Rubber Co, Ont, 51-65; RES SCIENTIST, CAN DEPT AGR, 65- Mem: Chem Inst Can. Res: Plant protection by chemicals, such as insecticides, fungicides and herbicides; translocation of these chemicals in plants and changes in their structure due to plant metabolism. Mailing Add: Can Dept of Agr Res Sta Harrow ON Can

VON VOIGTLANDER, PHILIP FRIEDRICH, b Jackson, Mich, Feb 3, 46; m 68; c 1. NEUROPHARMACOLOGY. Educ: Mich State Univ, BS, 68, DVM, 69, MS, 71, PhD(pharmacol), 72. Prof Exp: NIH trainee cent nerv syst pharmacol, Mich State Univ, 69-72; RES SCIENTIST, CENT NERV SYST PHARMACOL, UPJOHN CO, 72- Mem: AAAS; Soc Neurosci; Am Soc Pharmacol & Exp Therapeut. Res: Development of animal models of psychiatric and neurological diseases for the purpose of studying the mechanisms of action of centrally acting drugs and identification of new therapeutic agents. Mailing Add: Upjohn Co Kalamazoo MI 49001

VON WEYSSENHOFF, HANNS, b Hannover, Ger, June 8, 29; m 61. PHYSICAL CHEMISTRY, PHYSICS. Educ: Univ Göttingen, Dipl, 55; Univ Bonn, Dr rer nat, 59. Prof Exp: Res assoc phys chem, Rensselaer Polytech Inst, 59-61; mem res staff, Jet Propulsion Lab, Calif Inst Technol, 61-63, res assoc, Northwestern Univ, 63-66; asst prof chem, Ill Inst Technol, 66-71; PROF PHYS CHEM, HANNOVER TECH UNIV, 71- Mem: Am Phys Soc. Res: Photochemistry; chemical kinetics of gases; fluorescence and energy transfer of excited molecules; radiationless processes. Mailing Add: Inst of Phys Chem Hannover Tech Univ Hannover West Germany

VON WICKLEN, FREDERICK CHARLES, b Mt Vernon, Ohio, Sept 3, 00; m 41; c 1. CHEMISTRY. Educ: Univ Louisville, BS, 22, MS, 23; Columbia Univ, MA, 31, PhD(phys chem), 34. Prof Exp: Res chemist, Acme White Lead Co, Mich, 23-25; asst chief chemist, Graham-Paige Motors, 25-29; res chemist, Titanium Pigment Co, NY, 34-36; asst, Columbia Univ, 36-37; prof chem, Col Ozarks, 37-39; res chemist, Takamine Lab, NJ, 39-40; prof chem, York Col, 41-46; assoc prof, Univ Omaha, 46-48; prof, Lambuth Col, 49-52; assoc prof, Westminster Col (Mo), 52-53; prof, Panhandle Agr & Mech Col, 54-61; prof, 61-70, EMER PROF CHEM, SOUTHWESTERN STATE COL, 70- Concurrent Pos: Instr, Univ Nebr, 44-46. Mem: AAAS; fel Am Inst Chem; Am Chem Soc. Res: Chemistry of colloidal solutions of chromic chloride; preparation and testing of starch-splitting enzymes. Mailing Add: Dept of Chem Southwestern State Col Weatherford OK 73096

VON WINBUSH, SAMUEL, b Henderson, NC, Aug 2, 32; m 62; c 1. INORGANIC CHEMISTRY, PHYSICAL CHEMISTRY. Educ: Tenn State Univ, AB, 53; Iowa State Univ, MS, 56; Univ Kans, PhD(inorg chem), 60. Prof Exp: Asst prof chem & chmn dept, Tenn State Univ, 60-62; prof, NC A&T State Univ, 62-65; prof, Fisk Univ, 65-71; PROF CHEM, STATE UNIV NY COL OLD WESTBURY, 71- Concurrent Pos: Consult metals & ceramics div, Oak Ridge Nat Lab, 66-; vis prof, Wesleyan Univ, 69-70; consult, State Univ NY Col Old Westbury, 70-71. Mem: AAAS; Am Chem Soc. Res: Coordination chemistry; ligand field and charge transfer spectra; inorganic polymers; unfamiliar oxidation states of metals in molten salts and other nonaqueous solvents. Mailing Add: Dept of Chem State Univ NY Old Westbury NY 11568

VON ZELLEN, BRUCE WALFRED, b Ann Arbor, Mich, Feb 14, 22; m 49; c 1. PARASITOLOGY, PROTOZOOLOGY. Educ: Northern Mich Col, AB, 47; Univ Mich, MS, 49; Duke Univ, PhD(zool), 59. Prof Exp: Lab asst physiol & zool, Univ Mich, 49-50; assoc prof biol sci, Ky Wesleyan Col, 50-57; instr zool, Duke Univ, 57-59; ASSOC PROF BIOL SCI, NORTHERN ILL UNIV, 59- Concurrent Pos: Consult, panels on equip grants, NSF, Washington, DC & Chicago. Mem: AAAS; Am Soc Trop Med & Hyg; Soc Protozool; Am Soc Parasitol. Res: Parasitic protozoa; life history of coccidiosis; blood parasite infection; cell culture. Mailing Add: Dept of Biol Sci Northern Ill Univ De Kalb IL 60115

VOOGT, JAMES LEONARD, b Grand Rapids, Mich, Feb 8, 44; m 66; c 2. PHYSIOLOGY, NEUROENDOCRINOLOGY. Educ: Mich Technol Univ, BS, 66; Mich State Univ, MS, 68, PhD(physiol), 70. Prof Exp: NIH fel, Med Ctr, Univ Calif, San Francisco, 70-71; ASST PROF PHYSIOL, UNIV LOUISVILLE, 71- NIH RES GRANT, 72- Mem: Int Soc Neuroendocrinol; Endocrine Soc; Soc Study Reproduction. Res: Control of anterior pituitary function by hypothalamus; feedback systems; hormone analysis; reproduction control. Mailing Add: Dept of Physiol & Biophys Univ of Louisville Louisville KY 40201

VOOK, FREDERICK LUDWIG, b Milwaukee, Wis, Jan 17, 31; m 58; c 2. PHYSICS. Educ: Univ Chicago, BA, 51, BS, 52; Univ Ill, MS, 54, PhD(physics), 58. Prof Exp: Mem staff, 58-62, div supvr, 62-71, DEPT MGR, SANDIA LABS, 71- Mem: Fel Am Phys Soc. Res: Defects in solids, primarily semiconductors; defects investigated by means of radiation damage at low temperatures; infrared absorption; ion implantation in semiconductors; ion backscattering and channeling studies of solids. Mailing Add: Dept 5110 Sandia Labs Albuquerque NM 87115

VOOK, RICHARD WERNER, b Milwaukee, Wis, Aug 2, 29; m 57; c 4. SOLID STATE PHYSICS. Educ: Carleton Col, BA, 51; Univ Ill, MS, 52, PhD(physics), 57. Prof Exp: Mem res staff, Res Ctr, Int Bus Mach Corp, NY, 57-61; res labs, Franklin Inst, Pa, 61-65; assoc prof metall, 65-70, PROF MAT SCI, SYRACUSE UNIV, 70-, DIR ELECTRON MICROS LAB, 68- Concurrent Pos: Consult, Amperex Corp, 72. Mem: Am Phys Soc; Am Vacuum Soc; Electron Micros Soc Am; Metal Soc of the Am Inst Metal Engrs. Res: Electron microscopy and diffraction; x-ray diffraction; thin films; epitaxial growth; surface physics; imperfections in solids. Mailing Add: Dept of Chem Eng & Mat Sci Syracuse Univ 409 Link Hall Syracuse NY 13210

VOORHEES, BURTON HAMILTON, b Tucson, Ariz, Dec 3, 42. BIOMATHEMATICS. Educ: Univ Calif, Berkeley, AB, 64; Univ Ariz, MS, 66; Univ Tex, Austin, PhD(physics), 71. Prof Exp: Asst prof math & physics, Pars Col, Iran, 71-73; RES ASSOC MATH, UNIV ALTA, 73- Res: Stochastic geometry; quantization of gravitation; black hole physics; mathematical models of evolutive processes. Mailing Add: Dept of Math Univ of Alta Edmonton AB Can

VOORHEES, JOHN JAMES, b Cleveland, Ohio, Dec 5, 38; m 61; c 4. DERMATOLOGY, MEDICAL RESEARCH. Educ: Bowling Green State Univ, BS, 60; Univ Mich, Ann Arbor, MD, 63. Prof Exp: Intern internal med, 63-64, trainee clin internal med, 66, NIH trainee biochem, 67-68, trainee clin dermat, 67-69, Carl Herzog scholar biochem, 68-70, from instr to assoc prof dermat, 69-74, PROF DERMAT, MED SCH, UNIV MICH, ANN ARBOR, 74-, CHMN DEPT, 75-, CHIEF DERMAT SERV, UNIV HOSP, 75- Concurrent Pos: Consult dermat, Dept Med, Wayne County Gen Hosp, 69- & Vet Admin Hosp, Ann Arbor, 71-; assoc ed, J Cutaneous Path, 72-; mem med & sci adv bd, Nat Psoriasis Found, 71-; mem revision panel, 1980 Ed, US Pharmacopeia, 75-80. Honors & Awards: Taub Int Mem Award Psoriasis Res, 73; Henry Russell Award Distinguished Res, Univ Mich, 73; Outstanding Serv Award, Nat Psoriasis Found, 73. Mem: Am Soc Clin Invest; Am Soc Pharmacol & Exp Therapeut; Am Soc Exp Path; Am Soc Cell Biol; Soc Exp Biol & Med. Res: Role of cyclic nucleotides, glucocorticoids, the arachidonate, HETE, thromboxane, prostaglandin cascade and immunology in the molecular pathophysiology and pharmacology of skin diseases with inflammation, induced proliferation and reduced differentiation. Mailing Add: Dept of Dermat Med Sch Univ of Mich Ann Arbor MI 48109

VOORHEES, KENT JAY, b Provo, Utah, Sept 7, 43; m 66; c 2. PHYSICAL ORGANIC CHEMISTRY, ANALYTICAL CHEMISTRY. Educ: Utah State Univ, BS, 65, MS, 68, PhD(org chem), 70. Prof Exp: Res fel phys chem, Mich State Univ, 70-71; instr org chem, 71-73, ASST RES PROF ANAL POLYMER CHEM, UNIV UTAH, 73- Mem: Am Chem Soc. Res: Formulation of complex structures and/or degradation mechanisms by the application of gas chromatography and mass spectrometry in the analysis of thermal decomposition products of oil shale, petroleum and man made polymers. Mailing Add: 3343 S Davis Blvd Bountiful UT 84010

VOORHEES, WARD BYRON, b Danvers, Minn, Apr 4, 36; m 62; c 3. SOIL CONSERVATION. Educ: Univ Minn, BS, 59; Iowa State Univ, MS, 69. Prof Exp: Soil scientist, 59-66, RES SOIL SCIENTIST, AGR RES SERV, USDA, 66- Mem: Am Soc Agron; Soil Sci Soc Am; Int Soc Soil Sci; Soil Conserv Soc Am; Am Soc Agr Eng. Res: Basic and applied research on soil and water conservation with emphasis on tillage, soil compaction, root growth and plant response. Mailing Add: USDA Agr Res Serv Soil & Water Conserv Res Div Morris MN 56267

VOORHESS, MARY LOUISE, b Livingston Manor, NY, June 2, 26. PEDIATRICS, ENDOCRINOLOGY. Educ: Univ Tex, BA, 52; Baylor Univ, MD, 56. Prof Exp: From intern to resident pediat, Albany Med Ctr, New York, 46-59; from asst prof to prof pediat, State Univ NY Upstate Med Ctr, 61-76; PROF PEDIAT, STATE UNIV NY BUFFALO, 76- Concurrent Pos: Res fel pediat endocrinol & genetics, State Univ NY Upstate Med Ctr, 59-61; Nat Cancer Inst res grant, 62-69, career develop award, 61-71. Mem: Endocrine Soc; Am Fedn Clin Res; Am Acad Pediat; Lawson Wilkins Pediat Endocrine Soc; Am Pediat Soc. Res: Pediatric endocrinology; catecholamine metabolism in children; biochemistry in tumors of neural crest origin. Mailing Add: Children's Hosp 219 Bryant St Buffalo NY 14222

VOORHIES, BARBARA, b New York, NY, Mar 10, 39. ANTHROPOLOGY, ARCHAEOLOGY. Educ: Tufts Univ, BS, 61; Yale Univ, PhD(anthrop), 69. Prof Exp: Asst prof anthrop, San Diego State Col, 69-70; vis asst prof, 70-71, ASST PROF ANTHROP, UNIV CALIF, SANTA BARBARA, 71- Concurrent Pos: NSF fel, 72-73. Mem: Am Anthrop Asn; Soc Am Archaeol. Res: Maya prehistory; early coastal village life in Mesoamerica. Mailing Add: Dept of Anthrop Univ of Calif Santa Barbara CA 93107

VOORHIES, JOHN DAVIDSON, b Hartford, Conn, Nov 26, 33; m 59. SURFACE CHEMISTRY. Educ: Princeton Univ, AB, 55, MA, 57, PhD(chem), 58. Prof Exp: Res chemist, 58-62, sr res chemist, 62-64, group leader, 64-71, SR RES SCIENTIST, AM CYANAMID CO, 71- Concurrent Pos: Vis scholar, Dept Chem Eng, Stanford Univ, 70-71. Mem: AAAS; Am Chem Soc; Electrochem Soc. Res: Electrochemistry; electrochemical power sources; electrochemistry of organic compounds; electroanalytical techniques; chronopotentiometry; polarography; coulometry; interfacial chemistry; heterogeneous catalysis; hydrotreating catalysts. Mailing Add: 14 Harrison Ave New Canaan CT 06840

VOORHIES, MICHAEL REGINALD, b Orchard, Nebr, June 17, 41; m 68. VERTEBRATE PALEONTOLOGY. Educ: Univ Nebr, BS, 62; Univ Wyo, PhD(geol), 66. Prof Exp: From asst prof to assoc prof geol, Univ Ga, 66-75; ASSOC CUR FOSSIL VERT, UNIV NEBR STATE MUS, 75- Mem: Paleont Soc; Soc Vert Paleont; Soc Study Evolution; Soc Syst Zool. Res: Taphonomy and population dynamics of Cenozoic mammals; community evolution; neogene stratigraphy of the Great Plains. Mailing Add: State Mus Morrill Hall Univ of Nebr Lincoln NE 68508

VOORHIS, ARTHUR DAVID, physics, see 12th edition

VOORHOEVE, RUDOLF JOHANNES HERMAN, b Sentang, Sumatra, Indonesia, Oct 4, 38; m 62, 68, 75; c 5. SURFACE CHEMISTRY, PHYSICAL INORGANIC CHEMISTRY. Educ: Delft Univ Technol, Ing, 61, Dr(organosilicon & catalytic chem), 64. Prof Exp: Instr organosilicon & catalysis chem, Delft Univ Technol, 61-64; res chemist, Nat Defense Res Orgn, 64-66; res chemist, Koninklyke/Shell Lab, Amsterdam, 66-68; MEM TECH STAFF, BELL LABS, 68- Mem: Am Chem Soc; Am Phys Soc. Res: Organosilicon chemistry, especially direct synthesis of organohalosilanes; heterogeneous catalysis, studies by gas-solid kinetics, mechanistic studies; gas-solid reactions for chemical vapors deposition, molecular beams. Mailing Add: 1E-334 Bell Labs Murray Hill NJ 07974

VOORS, ANTONIE WOUTER, b Opperdoes, Neth, Dec 11, 24; m 56; c 2. EPIDEMIOLOGY. Educ: State Univ Utrecht, MD, 51; Univ Amsterdam, cert trop med & hyg, 53; Univ NC, MPH, 56, DrPH, 65. Prof Exp: Resident med, Deaconess Hosp, Hilversum, Neth, 51-53; govt physician, Govt Neth, New Guinea, 53-55, chief, Div Health Educ, 56-59, sr govt physician, 60-63; fel epidemiol, Sch Pub Health, Univ NC, Chapel Hill, 63-65; sr resident epidemiol, NY State Dept Health, 65-66; from instr to assoc prof, Sch Pub Health, Univ NC, Chapel Hill, 66-73; ASSOC PROF PREV MED, LA STATE UNIV SCH MED, NEW ORLEANS, 73- Mem: Am Pub Health Asn; Soc Epidemiol Res; Soc Environ Geochem & Health. Res: Determinants of diseases using mathematical models, especially cost effectiveness in detection and control. Mailing Add: Dept of Prev Med La State Univ Sch of Med New Orleans LA 70112

VOOS, JANE RHEIN, b Neuremberg, Ger, Oct 2, 27; m 50; c 3. MICROBIOLOGY. Educ: Hunter Col, BA, 52; Columbia Univ, PhD(biol sci), 68. Prof Exp: Res bacteriologist, Bellevue Hosp, New York, 53-54; instr biol, Stern Col Women, 56-58; from asst prof to assoc prof, 68-75, asst to dean, 71-72, PROF BIOL SCI, WILLIAM PATERSON COL, NJ, 75-, CHMN DEPT, 72- Mem: AAAS; Bot Soc Am; Mycol Soc Am. Res: Electron microscopy of spores and modern pollens. Mailing Add: Dept of Biol Sci William Paterson Col Wayne NJ 07470

VOOTS, RICHARD JOSEPH, b Quincy, Ill, Feb 18, 21; m 45; c 3. ACOUSTICS. Educ: NY Univ, BS, 44; Univ Iowa, PhD, 55. Prof Exp: Engr, Boeing Aircraft Co, Kans, 46; instr, Burrton High Sch, 47; instr, Benton High Sch, 47-48; physics lab, Univ Wichita, 48; audio technician, Bennett Music House, 49-51; asst psychol, 51-52, res assoc speech path & audiol, 51-61, RES ASST PROF SPEECH PATH, AUDIOL & MUSIC, COL MED, UNIV IOWA, 61-, RES ASSOC OTOLARYNGOL, UNIV HOSPS, 55- Mem: AAAS; Acoust Soc Am. Res: Otological and musical acoustics; auditory stimulus-response relationship, particularly the frequency-pitch dimension; automated apparatus for sensitivity threshold and differential pitch threshold audiometry; physical acoustics of musical instruments. Mailing Add: Rm 4 Otolaryngol Res Lab Univ of Iowa Med Res Ctr Iowa City IA 52240

VOPICKA, EDWARD, b Cleveland, Ohio, Oct 15, 11; m 35; c 2. INDUSTRIAL ORGANIC CHEMISTRY. Educ: Case Univ, BS, 31, MS, 32; Western Reserve Univ, PhD(org chem), 35. Prof Exp: Lab technician, Standard Oil Co Ohio, 34; res chemist, E I du Pont de Nemours & Co, 35; Am Czechoslovakian student exchange fel, Charles Univ, Prague, 35-36; res chemist, Givaudan-Delawanna, Inc, NJ, 37-44; asst dir res & plant supt, S B Penick & Co, 44-50; TECH DIR, INT FLAVORS & FRAGRANCES, INC, UNION BEACH, 50- Mem: Am Chem Soc. Res: Electrochemistry; organic chemistry synthesis; interaction of 2, 4-dichloroquinazoline in alcohol with ammonia and methyl amine; saponification and thermal decomposition of mixed diacyl derivatives of 2-amino-4-chlorothiophenol. Mailing Add: 5 Circle Dr Rumson NJ 07760

VOPICKA, ELLEN VANDERSEE, b Passaic, NJ, Dec 6, 41. DEVELOPMENTAL BIOLOGY, MICROBIOLOGY. Educ: Cedar Crest Col, BA, 63; Wake Forest Col, MA, 65; Univ Vt, PhD(zool), 68. Prof Exp: From asst prof to assoc prof biol, Mary Baldwin Col, 68-74, ASSOC PROF BIOL, MERCY COL, 74- Mem: AAAS; Am Inst Biol Sci. Res: Developmental biology of amphibian eggs; regeneration of amphibian limbs; bacterial ecology, especially with respect to myxobacteria. Mailing Add: Dept of Natural Sci Mercy Col Dobbs Ferry NY 10522

VORACHEK, JAMES HAROLD, radiation chemistry, see 12th edition

VORBECK, MARIE L, b Rochester, NY, June 24, 33. BIOCHEMISTRY, MICROBIOLOGY. Educ: Cornell Univ, BS, 55, PhD(bact biochem), 62; Pa State Univ, MS, 58. Prof Exp: Fel biochem, Sch Med & Dent, Univ Rochester, 62-64; asst prof microbiol, Med Sch, Temple Univ, 64-66; asst prof biochem, Jefferson Med Col, 67-68; assoc prof, 68-74, PROF PATH & BIOCHEM, SCH MED, UNIV MO-COLUMBIA, 74- Mem: Am Chem Soc; Am Soc Microbiol; NY Acad Sci. Res: Structure, function biosynthesis of membrane lipids; lipids in microorganisms; membrane process; development of analytical methods in lipid biochemistry. Mailing Add: Dept of Path Univ of Mo Columbia MO 65201

VORCHHEIMER, NORMAN, b Thungen, Ger, Sept 10, 35; US citizen; m 67; c 2. POLYMER CHEMISTRY. Educ: Brooklyn Col, BS, 57; Polytech Inst Brooklyn, PhD(org chem), 62. Prof Exp: Res chemist textile fibers, E I du Pont de Nemours & Co, 62-67; sr res chemist polymer synthesis, Betz Labs Inc, 67-70; exec vpres, Shasta Fund Inc, 70-71; sr res chemist, 71-73, GROUP LEADER POLYMER SYNTHESIS, BETZ LABS INC, TREVOSE, 73- Mem: Am Chem Soc; The Chem Soc. Res: Water-soluble polymers. Mailing Add: PO Box 403 Buckingham PA 18912

VORE, MARY EDITH, b Guatemala City, Guatemala, June 27, 47; US citizen. TOXICOLOGY, PHARMACOLOGY. Educ: Asbury Col, BA, 68; Vanderbilt Univ, PhD(pharmacol), 72. Prof Exp: Fel, Dept Biochem & Drug Metab, Hoffmann-LaRoche Inc, 72-74; ASST PROF TOXICOL, DEPT PHARMACOL, UNIV CALIF MED CTR, SAN FRANCISCO, 74- Mem: Am Soc Pharmacol & Exp Therapeut. Res: The biochemical properties of liver and lung microsomal mixed-function oxidases and their role in the metabolism of drugs and xenobiotics to toxic reactive intermediates. Mailing Add: Dept of Pharmacol & Toxicol Univ of Calif Med Ctr San Francisco CA 94143

VORHERR, HELMUTH WILHELM, b Alzey, WGer, Feb 6, 28; m 55; c 2. OBSTETRICS & GYNECOLOGY. Educ: Univ Mainz, MD, 55, specialist obstet & gynec, 62. Prof Exp: Mem staff obstet & gynec, Univ Frankfurt, 62-65; res pharmacologist, Cedars-Sinai Med Ctr, Los Angeles, Calif, 65-68; assoc prof, 68-71, PROF OBSTET, GYNEC & PHARMACOL, SCH MED, UNIV NMEX, 71- Concurrent Pos: Damon Runyon Mem Fund res fel, Cedars-Sinai Med Ctr & Univ Calif, Los Angeles, 65-66, NIH spec fel, 66-67; asst prof, Sch Med, Univ Calif, Los Angeles, 66-68. Mem: Am Fedn Clin Res; Am Soc Pharmacol & Exp Therapeut; Soc Gynec Invest. Res: Physiology, bioassay and pharmacology of neurohypophysial hormones; role of oxytocin for onset of premature labor; labor; puerperium and lactation; uterine relaxation; prevention of premature labor. Mailing Add: Dept of Obstet & Gynec Univ of NMex Albuquerque NM 87131

VORIS, AARON LEROY, b Burgin, Ky, July 1, 06; m 34, 56; c 3. NUTRITION. Educ: Georgetown Col, Ky, BA, 27; Pa State Col, MS, 36; Cornell Univ, PhD(nutrit), 39; Am Bd Nutrit, dipl. Prof Exp: Jr chemist, US Bur Internal Revenue, 29; from asst to asst prof nutrit, Pa State Col, 29-37; asst, Cornell Univ, 37-38; asst prof, Pa State Col, 39-42; exec secy, Food & Nutrit Bd, Nat Acad Sci-Nat Res Coun, 46-71; RETIRED. Concurrent Pos: Consult, USPHS Nutrit Surv, Chile, 60. Mem: AAAS; Am Pub Health Asn; Am Inst Nutrit; Chilean Soc Nutrit. Mailing Add: Oak Leaf Dr Pine Knoll Shores Rte 1 Morehead City NC 28557

VORIS, HAROLD CORNELIUS, b Pleasant, Ind, Apr 18, 02; m 31; c 4. NEUROSURGERY. Educ: Hanover Col, AB, 23; Univ Chicago, PhD, 29, Rush Med Sch, MD, 30; Univ Minn, MS, 34. Hon Degrees: DSc, Hanover Col, 45. Prof Exp: Instr biol & physics, Morris Harvey Col, 23-24; instr anat, Univ Chicago, 26-27; fel neurosurg, Mayo Clin, Minn, 31-34; assoc clin prof, Stritch Sch Med, Loyola Univ Chicago, 34-35, clin prof neurol surg, 36-64; prof neurol surg, 64-75, EMER PROF NEUROL SURG, UNIV ILL COL MED, 75-; EMER PROF, RUSH MED SCH, 71- Concurrent Pos: Mem staff, Mercy Hosp & Med Ctr, 34-75; consult, Hines Vet Admin Hosp, 55-; consult, Presby-St Lukes Med Ctr, 70-75; pvt pract. Mem: Am Asn Neurol Surg; fel Am Asn Surg of Trauma; AMA; fel Am Col Surg; fel Int Col Surg. Res: Neuroanatomy. Mailing Add: 1550 Lake Shore Drive Chicago IL 60610

VORIS, HAROLD K, b Chicago, Ill, Oct 5, 40. HERPETOLOGY. Educ: Hanover Col, BA, 62; Univ Chicago, PhD(biol), 69. Prof Exp: Instr biol, Yale Univ, 67-69; asst prof, Dickinson Col, 69-73; ASST CUR, FIELD MUS NATURAL HIST, 73- Mem: Am Soc Ichthyol & Herpet; Soc Syst Zool; Ecol Soc Am; Soc Study Evolution. Res: Evolution and systematics; ecology; sea snake ecology and systematics; rain forest ecosystems; numerical taxonomy. Mailing Add: Div of Reptiles Field Mus of Natural Hist Chicago IL 60605

VORIS, ROBERT SITES, chemistry, see 12th edition

VORRES, KARL S, b Chicago, Ill, Mar 14, 27; m 52; c 4. PHYSICAL CHEMISTRY. Educ: Mich State Univ, BS, 52, MS, 53; Univ Iowa, PhD(phys chem), 58. Prof Exp: Org chemist, Isotopes Specialties Co, 55-56; instr phys chem, Univ Iowa, 58-60; asst prof phys & inorg chem, Univ Miami, 60-62; asst prof inorg chem, Purdue Univ, 62-65; chief chem sect, Res Ctr, Babcock & Wilcox Co, 65-69, mgr chem & combustion sect, 69-75; ASST DIR PLANNING & DEVELOP, INST GAS TECHNOL, 75- Concurrent Pos: Consult, Inst Gas Technol, 59-61 & Babcock & Wilcox Co, 64-65. Mem: Am Chem Soc; Am Ceramic Soc. Res: Physical chemistry of solid state; metal oxidation; phase studies; radiochemistry; supervision of research; air pollution; coal gasification; combustion. Mailing Add: Inst Gas Technol 3424 S State St Chicago IL 60616

VORST, JAMES J, b Cloverdale, Ohio, Mar 20, 42; m 66; c 3. AGRONOMY. Educ: Ohio State Univ, BS, 64, MS, 66; Univ Nebr, PhD(agron), 69. Prof Exp: Teaching asst agron, Ohio State Univ, 64-66; instr, Univ Nebr, 66-69; MEM FAC, DEPT AGRON, PURDUE UNIV, LAFAYETTE, 69- Mem: Am Soc Agron. Res: Crop production and physiology; simulated hail damage to agronomic crops; teaching methods in agronomy. Mailing Add: Dept of Agron Purdue Univ West Lafayette IN 47906

VORWALD, ARTHUR JOHN, pathology, industrial medicine, deceased

VOS, KENNETH DEAN, b Oskaloosa, Iowa, Nov 13, 35; m 60; c 2. PHYSICAL CHEMISTRY. Educ: Cent Col, Iowa, BA, 57; Mich State Univ, PhD(phys chem), 63. Prof Exp: Res asst chem, Los Alamos Sci Lab, 60; staff assoc, John Jay Hopkins Lab, Gen Atomic Div, Gen Dynamics Corp, Calif, 63-68; sr res chemist, 68-72, SUPVR PRESSURIZED PROD RES SECT, S C JOHNSON & SON, INC, 72- Mem: AAAS; Am Chem Soc; Am Phys Soc; fel Am Inst Chemists. Res: Electron paramagnetic presonance of metalamines and free radicals; semiconductor and high polymer chemistry biomedical research; fine particle research. Mailing Add: Pressurized Prod Res Sect S C Johnson & Son Inc Racine WI 53403

VOSBURG, DAVID LEE, b Enid, Okla, Dec 24, 30; m 60; c 2. STRATIGRAPHY. Educ: Phillips Univ, BS, 52; Univ Okla, MS, 54, PhD(geol), 63. Prof Exp: From instr to asst prof geol, Univ RI, 60-65; asst prof & chmn dept, Phillips Univ, 65-66; asst prof, 66-67, ASSOC PROF GEOL, ARK STATE UNIV, 67- Mem: Am Asn Petrol Geol. Res: Occurrence and distribution of subsurface evaporites within shelf sediments related to the Permian Basin, especially economic potential and stratigraphic relations. Mailing Add: Dept of Geol Ark State Univ State University AR 72467

VOSBURGH, KIRBY GANNETT, b Pasadena, Calif, May 27, 44; m 67; c 2. APPLIED PHYSICS, EXPERIMENTAL PHYSICS. Educ: Cornell Univ, BS, 65, MS, 67; Rutgers Univ, PhD(physics), 71. Prof Exp: Res asst applied physics, Cornell Univ, 65-67; mem tech staff accelerator physics, Princeton-Penn Accelerator, Princeton Univ, 67-68; res fel physics, Rutgers Univ, 68-71; mem tech staff & asst to dir particle physics, Princeton Particle Accelerator, Princeton Univ, 71-72; PHYSICIST, CORP RES & DEVELOP, GEN ELEC, 72- Mem: Am Phys Soc; Am Asn Physicists Med. Res: Power generation and transmission; energy conversion; medical x-ray imaging; radiation therapy; electron optics; accelerator design; high energy heavy ion physics; experimental particle physics, non-destructive testing. Mailing Add: Corp Res & Develop Gen Elec Co PO Box 8 Schenectady NY 12301

VOSBURGH, WILLIAM GEORGE, b Flint, Mich, May 30, 25; m 46; c 6. ORGANIC CHEMISTRY. Educ: Mich State Col, BS, 49, MS, 50; Univ Del, PhD, 56. Prof Exp: Chemist, 50-51, res chemist, 51-75, SR RES CHEMIST, TEXTILE FIBERS DEPT, E I DU PONT DE NEMOURS & CO, INC, WILMINGTON, DEL, 75- Mem: Am Chem Soc; Sigma Xi. Res: Organic synthesis; vinyl and condensation polymers; flame proofing; spunbonded nonwoven fabrics; latex development. Mailing Add: 1151 Lake Dr West Chester PA 19380

VOSE, GEORGE PARLIN, b Machias, Maine, May 3, 22. BIOMEDICAL ENGINEERING. Educ: Pa State Univ, BA, 52; Southern Methodist Univ, MS, 56. Prof Exp: Res assoc, 52-53, from asst prof to assoc prof human nutrit, 54-65, PROF RADIOGRAPHIC RES, TEX WOMAN'S UNIV, 65- Mem: Am Nuclear Soc. Res: X-ray densitometry of biologic tissues; bone strength; electron microscopy; treatment of bone demineralization. Mailing Add: Bone Metab Lab Box 23546 TWU Sta Tex Woman's Univ Denton TX 76204

VOSE, JOHN RANDAL, b Cheshire, Eng, June 21, 41; m 67; c 3. AGRICULTURAL BIOCHEMISTRY. Educ: Univ Wales, BSc, 64; Univ Alta, PhD(plant PhD(plant biochem), 68. Prof Exp: Nat Res Coun fel biochem, Univ Liverpool, 68-69; fel bot, Univ BC, 69-70; res mgr plant enzymes, Reckitt & Colman Ltd Can, 70-72; assoc res chemist, R T French Co, 72-74, mgr biochem res, 73-74; ASSOC RES OFFICER, PRAIRIE REGIONAL LAB, NAT RES COUN CAN, 75- Res: Starch chemistry; process development; industrial applications of enzymes; biochemistry of oilseeds and grain legumes. Mailing Add: Prairie Regional Lab Nat Res Coun of Can Saskatoon SK Can

VOSKO, SEYMOUR H, b Montreal, Que, Sept 9, 29; m 55; c 2. THEORETICAL PHYSICS. Educ: McGill Univ, BEng Phys, 51, MSc, 52; Carnegie Inst Technol, PhD(theoret physics), 57. Prof Exp: Instr physics, Carnegie Inst Technol, 56-58, vis asst prof, 58-60; from asst prof to assoc prof, McMaster Univ, 60-64; fel scientist, Westinghouse Res Labs, Pa, 64-70; PROF PHYSICS, UNIV TORONTO, 70- Mem: Am Phys Soc; Can Asn Physicists. Res: Theoretical solid state physics; applications of many-body perturbation theory; nuclear magnetic resonance in metals and alloys; electron-phonon interaction; phonons in metals; magnetism in metals. Mailing Add: Dept of Physics Univ of Toronto Toronto ON Can

VOSKUIL, WALTER HENRY, b Cedar Grove, Wis, Aug 22, 92; m 29; c 1. GEOLOGY. Educ: Univ Wis, BS, 21, MS, 22, PhD(geog), 24. Prof Exp: Asst geog, Univ Wis, 22-24; from instr to asst prof indust, Wharton Sch, Univ Pa, 24-30; chief mineral economist, Nat Indust Conf Bd, 30-31; State Geol Surv, Ill, 31-60; distinguished vis prof, 60-72, EMER DISTINGUISHED VIS PROF MINERAL ECON, MACKAY SCH MINES, UNIV NEV, 72- Concurrent Pos: Mem Coun Foreign Rels, 43; prof, Univ Ill, 49-60. Honors & Awards: Mineral Econ Award, Am Inst Mining, Metall & Petrol Eng, 74. Mem: Geol Soc Am; Asn Am Geog; Am Inst Mining, Metall & Petrol Eng. Res: Minerals in modern industry; international aspects of the mineral industry. Mailing Add: 2173 Vale St Reno NV 89502

VOSS, EDWARD GROESBECK, b Delaware, Ohio, Feb 22, 29. TAXONOMIC BOTANY. Educ: Denison Univ, BA, 50; Univ Mich, MA, 51, PhD(bot), 54. Prof Exp: Asst syst bot, Biol Sta, 49, bot, univ, 50-51, biol sta, 51-53, res assoc, bot gardens, 54, res asst, Metab Res Lab, Univ Hosp, 54-56, res assoc herbarium, 56-61, from asst prof to assoc prof bot, 60-69, PROF BOT, UNIV MICH, ANN ARBOR, 69- CUR VASCULAR PLANTS, HERBARIUM, 61- Concurrent Pos: Secy gen comt on bot nomenclature & ed comn, Int Code of Bot Nomenclature, 69-; vice rapporteur, Bur of Nomenclature, Int Bot Congresses, 69 & 75. Mem: Am Soc Plant Taxon; Soc Syst Zool; Lepidop Soc; Int Asn Plant Taxon; Soc Bibliog Natural Hist. Res: Floristics; vascular flora and vegetational history of Great Lakes region; history of biology; nomenclature; Lepidoptera of Michigan; natural areas in Michigan. Mailing Add: Herbarium North Univ Bldg Univ of Mich Ann Arbor MI 48104

VOSS, EDWARD WILLIAM, JR, b Chicago, Ill, Dec 2, 33; m 58; c 2. IMMUNOCHEMISTRY, MICROBIOLOGY. Educ: Cornell Col, AB, 55; Univ Ind, Indianapolis, MS, 64, PhD(immunol), 66. Prof Exp: USPHS fel, Sch Med, Wash Univ, 66-67; from asst prof to assoc prof immunochem, 67-74, PROF MICROBIOL, UNIV ILL, URBANA, 74- Concurrent Pos: Fac fel sci, NSF, 75. Mem: AAAS; Am Asn Immunol; Sigma Xi; Reticuloendothelial Soc; Fedn Am Scientists.Res: Structure and biosynthesis of immunoglobins and effects of drugs on lymphocytes and antibody synthesis. Mailing Add: Dept of Microbiol Univ of Ill 217 Burrill Hall Urbana IL 61801

VOSS, ELBERT, pharmacognosy, see 12th edition

VOSS, GILBERT LINCOLN, b Hypoluxo, Fla, Feb 12, 18; m 52; c 2. BIOLOGICAL OCEANOGRAPHY, SYSTEMATIC ZOOLOGY. Educ: Univ Miami, BS, 51, MS, 52; George Washington Univ, PhD(biol), 55. Prof Exp: Asst marine biol, 51-55, from res instr to res assoc prof, 58-61, chmn div biol, 61-73, PROF MARINE SCI, UNIV MIAMI, 61- Mem: AAAS; Soc Syst Zool; Am Soc Limnol & Oceanog; Am Soc Zool. Res: Biological oceanography; marine ecology; systematics of marine invertebrates, especially biology, systematics and Cephalopod fisheries; deepsea biology. Mailing Add: Rosenstiel Sch Marine & Atmos Sci Univ of Miami Miami FL 33149

VOSS, GORDON D, b Elmhurst, Ill, June 25, 34; m 64. FOOD TECHNOLOGY. Educ: Univ Wis, BS, 58, MS, 59. Prof Exp: Food technologist, Armour & Co, Ill, 59-69; dir airline develop, Marriot Corp, 69-70; MGR TECH SERV, MORTON DIV, ITT CONTINENTAL BAKING CO, 71- Mem: Inst Food Technologists. Res: Product and process development of frozen prepared food for institutional use. Mailing Add: ITT Continental Baking Co Morton Div Box 7445 Charlottesville VA 22906

VOSS, JACK GODDARD, b Chicago, Ill, Mar 23, 18; m 44; c 2. MICROBIOLOGY. Educ: Univ Calif, Los Angeles, AB, 39, MA, 41; Univ Wis, PhD(agr bact), 43. Prof Exp: Res bacteriologist, Golden State Co, Ltd, Calif, 43-46; res bacteriologist, Nat Distillers & Chem Corp, Ohio, 46-50; RES BACTERIOLOGIST, PROCTER & GAMBLE CO, 50- Mem: Soc Invest Dermat; Am Soc Microbiol. Res: Bacterial cell walls; skin microbiology; allergy. Mailing Add: 149 Fleming Rd Cincinnati OH 45215

VOSS, RAYMOND OLSON, b Minn, June 23, 18; m 44; c 2. CHEMISTRY. Educ: Stetson Univ, BS, 41. Prof Exp: Chemist, 42-44, res chemist, 44-50, res supvr, 50-55, mgr ceramic res, 56, mgr consumer prod develop, 56-63, DIR PROD DEVELOP, CORNING GLASS WORKS, 63- Concurrent Pos: Advan mgt prog, Harvard Bus Sch, 67. Mem: Am Chem Soc; Am Ceramic Soc. Res: Glass ceramics; strength of brittle materials; laminate technology. Mailing Add: 310 Steuben St Corning NY 14830

VOSS, REGIS D, b Cedar Rapids, Iowa, Jan 4, 31; m 56; c 3. SOIL FERTILITY. Educ: Iowa State Univ, BS, 52, MS, 60, PhD(soil fertil), 62. Prof Exp: Res asst soil fertil, Iowa State Univ, 57-62; agriculturist, Test Demonstration Br, Tenn Valley Auth, 62-64; from asst prof to assoc prof, 64-69, PROF AGRON, IOWA STATE UNIV, 69- Concurrent Pos: Vis prof, Univ Ill, 70-71; consult, Int Maize & Wheat Improvement Ctr, Arg, 71- Mem: AAAS; Am Soc Agron; Soil Sci Soc Am. Res: Effect of uncontrolled factors on the response of field crops to applied fertilizers by using biological statistical methods. Mailing Add: Dept of Agron Iowa State Univ Ames IA 50010

VOSSEN, JOHN LOUIS, b Philadelphia, Pa, Apr 4, 37; m 63; c 1. PHYSICS. Educ: St Joseph's Col, Pa, 58. Prof Exp: Engr, RCA Semiconductor & Mat Div, 58-62, group leader thin-film physics, RCA Advan Commun Lab, 62-65, MEM TECH STAFF, PROCESS RES LAB, DAVID SARNOFF RES CTR, RCA LABS, 65- Honors & Awards: Achievement Awards, RCA Labs, 68, 69 & 71. Mem: Am Vacuum Soc; Am Phys Soc; Electrochem Soc; Inst Elec & Electronic Engrs. Res: Study of the methods by which the properties of thin films may be controlled, principally, sputtering, ion plating, plasma anodization and evaporation. Mailing Add: Process Res Lab David Sarnoff Res Ctr RCA Labs Princeton NJ 08540

VOSTAL, JAROSLAV JOSEPH, b Prague, Czech, Mar 17, 27; m 52; c 4. PHARMACOLOGY, TOXICOLOGY. Educ: Charles Univ, Prague, MD, 51; Czech Acad Sci, PhD(med sci), 61. Prof Exp: Physician, Regional Inst Nat Health, Jihlava, Czech, 51-55; vis scientist, Nat Inst Pub Health, Stockholm, Sweden, 67-68; ASSOC PROF PHARMACOL & TOXICOL, SCH MED, UNIV ROCHESTER, 68-, ASSOC PROF PREV MED & COMMUN HEALTH, 69- Concurrent Pos: Mem, Permanent Comn & Int Asn Occup Health, 66-, mem, Int Subcomt Toxicol Metals, 69-; chmn, Panel Fluorides, Nat Acad Sci-Nat Res Coun, 70-71, mem, Comt Biol Effects Atmospheric Pollutants, 70- Mem: AAAS; Am Soc Pharmacol & Exp Therapeut; Soc Toxicol. Res: Pharmacology of organomercurial compounds; toxicology of heavy metals and inorganic poisons, inter-species differences in pharmacokinetics and biotransformation of toxic substances. Mailing Add: Dept of Pharmacol & Toxicol Univ of Rochester Sch of Med Rochester NY 14642

VOSTI, DONALD CURTIS, b Modesto, Calif, Aug 26, 27; m 51; c 6. ORGANIC CHEMISTRY, FOOD TECHNOLOGY. Educ: Univ Calif, BS, 47, PhD, 52. Prof Exp: Chemist, E & J Gallo Winery, Calif, 47-52; sr chemist, 52-61, supvr, Container Specif Group, Western Area Lab, 61-62, supvr, Eval & Inspection Group, 62-64, supvr, Packaging Technol Group, 64-66, mgr, Customer Rels Sect, 66-69, proj mgr, Prod Technol Sect, 69-72, assoc dir, Gen Technol Sect, 72-73, PROJ MGR PLASTIC HOT FILL FOODS, AM CAN CO, 73- Mem: Inst Food Technologists. Res: Commercialization of new rigid containers developed for foods, non-foods and beverages; container utilization technology involving food processing, microbiological spoilage of food, container corrosion and related problems; container materials and manufacturing process improvments; development of plastic container for hot filled foods. Mailing Add: Am Can Co Res & Develop 433 N Northwest Hwy Barrington IL 60010

VOTAW, CHARLES ISAC, b Farris, Okla, Jan 3, 33; m 56; c 2. MATHEMATICS. Educ: Okla State Univ, BS, 57; NTex State Univ, MS, 67; Univ Kans, PhD(math), 71. Prof Exp: Staff mem, Sandia Corp, 57-60; mkt engr, Raytheon Co, 60-62; prod engr, Tex Instruments, 62-65; prod & test engr, Hunt Electronics, 65-66; ASST PROF MATH, FT HAYS KANS STATE COL, 71- Mem: Math Asn Am; Am Math Soc. Res: Topology; applied mathematics. Mailing Add: Dept of Math Ft Hays Kans State Col Hays KS 67601

VOTAW, CHARLES LESLEY, b Chicago, Ill, Oct 11, 29; m 53; c 3. NEUROANATOMY. Educ: Hope Col, AB, 51; Univ Mich, MD, 55, PhD(anat), 58. Prof Exp: Intern, St Joseph Mercy Hosp, Ann Arbor, 55-56; from instr to assoc prof, 56-70, asst dean curric, 71-75, PROF ANAT, MED SCH, UNIV MICH, ANN ARBOR, 70-, ASSOC DEAN CURRIC, 76- Concurrent Pos: NIH spec res fel, Univ Calif, Los Angeles, 62-63. Mem: AAAS; Am Asn Anatomists; Am Acad Neurol; Soc Neurosci. Res: Physiology and anatomy of the hippocampus and other rhinenciphalic structures; medical education. Mailing Add: 5704 Med Sci II Univ of Mich Med Sch Ann Arbor MI 48104

VOTAW, DAVID FREEMAN, JR, mathematics, see 12th edition

VOTAW, ROBERT GRIMM, b St Louis, Mo, Sept 13, 38; m 61; c 3. BIOCHEMISTRY, MEDICAL EDUCATION. Educ: Wesleyan Univ, BA, 60; Case Western Reserve Univ, PhD(microbiol), 66. Prof Exp: Instr microbiol, Sch Med, Case Western Reserve Univ, 66-67; instr, 67-69, ASST PROF BIOCHEM, SCH MED, UNIV CONN, 69-, ASST DEAN EDUC, 74- Mem: AAAS; Am Soc Microbiol; Am Chem Soc; Asn Am Med Cols. Res: Structure and function of enzymes and other macromolecules which bind thiamin diphosphate; role of thiamin compounds in metabolism; medical education and innovative approaches to self-paced instruction. Mailing Add: Dept of Biochem Univ of Conn Health Ctr Farmington CT 06032

VOTER, ROGER CONANT, b Boston, Mass, July 10, 22; m 45; c 5. CHEMISTRY. Educ: Wesleyan Univ, AB, 44; Iowa State Col, PhD(chem), 51. Prof Exp: Mem mgt staff, 51-64, MEM PATENTS & CONTRACTS STAFF, E I DU PONT DE NEMOURS & CO, INC, 64- Mem: Am Chem Soc. Res: Organic analytical reagents; instrumental analysis; physical measurements of polymers; polymer product development; patents; licensing and contracts. Mailing Add: Patents & Contracts E I du Pont de Nemours & Co Inc 1007 Market St Wilmington DE 19898

VOTH, HAROLD MOSER, b Newton, Kans, Dec 29, 22; m 46; c 3. PSYCHIATRY. Educ: Washburn Univ, BS, 43; Univ Kans, MD, 47; Menninger Sch Psychiat, psychiatrist, 52; Topeka Inst Psychoanal, psychoanalyst, 62. Prof Exp: Asst sect chief, Vet Admin Hosp, Topeka, Kans, 52-53, asst chief acute intensive treatment sect, 53-54, chief women's neuropsychiat serv, 54-57; STAFF PSYCHIATRIST, MENNINGER FOUND, 57- Concurrent Pos: Mem fac, Menninger Sch Psychiat, 55-; NIMH res grant, Menninger Found, 63-72; examr, Am Bd Psychiat & Neurol, 70-71; consult, Walter Reed Army Med Ctr, 72-; assoc chief psychiat for educ, Vet Admin Hosp, Topeka, Kans, 75- Mem: Fel AAAS; fel Am Psychiat Asn; Am Psychoanal Asn; Am Col Psychoanal; AMA. Res: Personality organization; psychotherapy; the study of autokinesis as a research and clinical instrument. Mailing Add: Res Dept Menninger Found Box 829 Topeka KS 66604

VOTH, ORVILLE LESTER, b Can, Jan 4, 24; US citizen; m 45; c 4. BIOCHEMISTRY. Educ: Bethel Col (Kans), AB, 48; Okla State Univ, MS, 50; Pa State Univ, PhD(biochem), 57. Prof Exp: Biochemist, Biochem Res Lab, Ill, 50-52; asst biochemist, Univ WVa, 52-55; assoc prof biochem, Kans Wesleyan Univ, 57-66; pres & prof chem, Bethel Col (Kans), 66-71; VPRES ACAD AFFAIRS, KANS WESLEYAN UNIV, 71- Mem: AAAS; Am Chem Soc. Res: Protein and amino acid interactions with tocopherol. Mailing Add: Kans Wesleyan Univ Salina KS 67401

VOTH, PAUL DIRKS, b Gotebo, Okla, June 12, 05; m 30; c 2. BOTANY. Educ: Bethel Col, Kans, AB, 29; Univ Chicago, SM, 30, PhD(bot), 33. Prof Exp: Instr biol, Tex Tech Col, 30-32; from instr to assoc prof bot, 33-48, prof & secy dept, 48-68, prof biol, 68-70, chmn div biol sequence, 60-68, EMER PROF BIOL, UNIV CHICAGO, 70- Concurrent Pos: Exchange instr, Cornell Univ, 38; consult, World Book Encycl, 67-71; prof biol sci, Northern Ill Univ, 70-75. Mem: Fel AAAS; Am Soc Plant Physiol; Bot Soc Am; Phycol Soc Am; Am Bryol & Lichenological Soc (secy-treas, 39-41). Res: Anatomy of ferns and gymnosperms; mineral nutrition, culture, speciation and microhabitat of Marchantia and other bryophytes; fertility and seed germination in Hemerocallis. Mailing Add: 620 Normal Rd De Kalb IL 60115

VOUGHT, ELDON JON, b Chicago, Ill, May 21, 35; m 59; c 4. TOPOLOGY. Educ: Manchester Col, AB, 57; Univ Mich, Ann Arbor, MA, 58; Univ Calif, Riverside, PhD(math), 67. Prof Exp: Instr math, Pomona Col, 60-61; assoc prof math, Calif State Polytech Univ, Pomona, 61-70; PROF MATH, CALIF STATE UNIV, CHICO, 70- Concurrent Pos: NSF res fel, 71-73; vis prof math, Ariz State Univ, 75-76. Mem: Am Math Soc; Math Asn Am; Sigma Xi. Res: The study of the invariance of various topological properties of compact, connected metric spaces under certain types of continuous functions, for example, local homeomorphisms and confluent functions. Mailing Add: Dept of Math Calif State Univ Chico CA 95926

VOUGHT, ROBERT HOWARD, b Ridgway, Pa, Jan 30, 20; m 49; c 3. PHYSICS. Educ: Allegheny Col, BA, 41; Univ Pa, PhD(physics), 46. Prof Exp: Res physicist, Univ Pa, 42-46, instr phsyics, 44-46; res assoc, Gen Elec Co, 46-48; asst prof, Union Univ, NY, 48-53, assoc prof, 53-56; SOLID STATE PHYSICIST, GEN ELEC CO, 56- Concurrent Pos: Exchange prof, St Andrews Univ, 53-54. Mem: Am Phys Soc; Am Asn Physics Teachers; AAAS; Am Nuclear Soc. Res: Semiconductors; mass spectrometry; surface physics; corrosion; energy conversion materials and devices; optoelectron devices; reactor technology; radiation damage. Mailing Add: 1465 Myron St Schenectady NY 12309

VOUGHT, ROBERT LOUIS, b Bradford Co, Pa, July 17, 08; m 32; c 3. EPIDEMIOLOGY, PREVENTIVE MEDICINE. Educ: Syracuse Univ, MD, 32; Johns Hopkins Univ, MPH, 37. Prof Exp: Intern, Samaritan Hosp, Troy, NY, 32-33; pvt pract, 33-36; dir syphilis control, Buffalo & dist health officer, Jamestown, State Dept Health, NY, 36-42, regional health dir, White Plains, 48-52; chief field party, Inst Inter-Am Affairs, Guatemala & Ecuador, 42-47; from instr to assoc prof epidemiol, Sch Pub Health, Columbia Univ, 51-55; dir med coordination, Bristol Labs, Inc, 55-61; CHIEF METAB DIS EPIDEMIOL UNIT, EPIDEMIOL & FIELD STUDIES BR, NAT INST ARTHRITIS, METAB & DIGESTIVE DIS, 61- Concurrent Pos: Mem, Found Advan Educ in Sci. Mem: AAAS; fel Am Pub Health Asn; Am Thyroid Asn; Soc Environ Geochem & Health; Soc Epidemiol Res. Res: Epidemiology of metabolic diseases; iodine metabolism. Mailing Add: Nat Inst of Arthritis Metab & Digestive Dis Bethesda MD 20014

VOULGAROPOULOS, EMMANUEL, b Lowell, Mass, Apr 16, 31; m 59; c 2. PUBLIC HEALTH. Educ: Tufts Col, BSc, 52; Cath Univ Louvain, MD, 57; Johns Hopkins Univ, MPH, 62. Prof Exp: Med dir, Med Int Corp, Cambodia, 58-60, exec field dir for Asia, Cent Am & Africa, 60-61; dep, Health Serv Develop Proj, Vietnam AID, Saigon, 62-64, chief pub health div, 64-65; assoc prof pub health, 65-68, head, Int Health Prog, 65-70, head, Int Health-Pop & Family Planning Studies Prog, 70-71, PROF PUB HEALTH, SCH PUB HEALTH, UNIV HAWAII, MANOA, 68-, ASSOC DEAN, 74- Concurrent Pos: Consult, AID, 62, US Civil Admin Ryuku Islands, 65-67, US Trust Territory Pac Islands, 65-71, Govt of Guam & Govt of Am Samoa, 65-71; consult, Peace Corps, 66-69, SPac Comn, Noumea New Caledonia, 67 & USPHS Global Community, Arlington, Va, 69; vis prof pub health & adv, Udayana Community Health Prog, Udayana State Univ, 71-72; vis prof pub health & adv to fac pub health, Univ Indonesia, 72-73; consult, Ministry Health, Govt Indonesia, 71-72, Ministry Educ & Culture, 71-73 & Ministry Social Welfare, 73; consult, Int Asn Schs Social Work, NY, 73. Res: Health and medical education systems; health delivery systems; international health. Mailing Add: Sch of Pub Health Univ of Hawaii at Manoa Honolulu HI 96822

VOURAS, PAUL PETER, b Vamvakou, Greece, Nov 27, 21; US citizen. GEOGRAPHY. Educ: Cent Conn State Col, BA, 49; Clark Univ, MA, 51; Ohio State Univ, PhD(geog), 56. Prof Exp: PROF GEOG, WILLIAM PATERSON COL, NJ, 56- Concurrent Pos: Nat Acad Sci-Off Naval Res field work res grant, Northern Greece, 59-60; Am Philos Soc field res travel grant, Greece, 67. Mem: Am Geog Soc; Asn Am Geog; Nat Coun Geog Educ. Mailing Add: 740 Stevens Ct Paramus NJ 07652

VOURNAKIS, JOHN NICHOLAS, b Cambridge, Ohio, Dec 1, 39; m 61; c 1. BIOPHYSICS, MOLECULAR BIOLOGY. Educ: Albion Col, BA, 61; Cornell Univ, PhD(chem), 68. Prof Exp: Nat Acad Sci exchange fel, Inst Org Chem & Biochem, Prague, Czech, fall 68; NIH fel biol, Mass Inst Technol, 69-71, res assoc, 71-72; res assoc, Harvard Univ, 72-73; ASST PROF BIOL, SYRACUSE UNIV, 73- Concurrent Pos: Vis assoc prof, Amherst Col, 70-71; NIH res grant gen med, 75. Mem: Biophys Soc; Soc Develop Biol. Res: Secondary and tertiary structure of eukaryotic mRNA; developmental regulation of transcription. Mailing Add: Dept of Biol Res Labs Syracuse Univ 130 College Place Syracuse NY 13210

VOUROS, PAUL, b Thessaloniki, Greece, Apr 1, 38; US citizen; m 65; c 2. ANALYTICAL CHEMISTRY, ORGANIC CHEMISTRY. Educ: Wesleyan Univ, BA, 61; Mass Inst Technol, PhD(chem), 65. Prof Exp: Staff scientist, Tech Opers, Inc, 66-67, proj mgr, 67-68; asst prof chem, Inst Lipid Res, Col Med, Baylor Univ, 68-74; SR SCIENTIST INST CHEM ANAL, APPLN & FORENSIC SCI, NORTHEASTERN UNIV, 74- Mem: Am Chem Soc; Sigma Xi; Am Soc Mass Spectrometry. Res: Organic mass spectrometry; mass spectrometry of biological compounds; gas chromatography-mass spectrometry; photographic ion detection; applications of chromatography and mass spectrometry to forensic problems. Mailing Add: Inst Chem Anal Appln Forensic Sci Northeastern Univ Boston MA 02115

VOURVOPOULOS, GEORGE, b June 11, 36; US citizen; m 63; c 3. NUCLEAR PHYSICS. Educ: Nat Univ Athens, BS, 58; Fla State Univ, MS, 65, PhD(physics), 67. Prof Exp: Assoc prof, 67-75, PROF PHYSICS, FLA A&M UNIV, 75-, CHMN DEPT, 74- Concurrent Pos: Res fel, Israel Inst Technol, 69-70; Res Corp grant, Fla A&M Univ, 71-, NSF grant, 72-74. Mem: Am Phys Soc. Res: Nuclear reactions; nuclear spectroscopy. Mailing Add: Dept of Physics Fla A&M Univ Tallahassee FL 32307

VOXMAN, WILLIAM L, b Iowa City, Iowa, Feb 1, 39; m 63; c 2. TOPOLOGY. Educ: Univ Iowa, BA, 60, MS, 63, PhD(math), 68. Prof Exp: Latin Am teaching fel & prof math, Univ Chile, 68-69; prof, Concepcion Univ, 69; Fulbright travel grant & prof, State Tech Univ, Chile, 69-70; asst prof, 70-72, ASSOC PROF MATH, UNIV IDAHO, 72- Concurrent Pos: Latin Am teaching fel & prof math, Nat Polytech Sch, Quito, Ecuador, 74-76. Mem: Am Math Soc. Res: General topology, especially upper semicontinuous decompositions of topological spaces. Mailing Add: Dept of Math Univ of Idaho Moscow ID 83843

VOYVODIC, LOUIS, b Yugoslavia, Oct 22, 21; US citizen; m 51; c 5. HIGH ENERGY PHYSICS. Educ: McGill Univ, BSc, 43, PhD(physics), 48. Prof Exp: Physicist cosmic rays, Nat Res Coun Can, 48-56; tech proj dir radiation physics, Isotope Prod Ltd Can, 56-58; physicist, Armour Res Found, Ill Inst Technol, 58-60; physicist, Argonne Nat Lab, 62-72; PHYSICIST, FERMI NAT ACCELERATOR LAB, 72- Mem: Am Inst Physics; Am Phys Soc; Fedn Am Scientists. Res: Interactions of elementary particles at high energies, particularly as studied by optical track chamber techniques, and development of improved detection techniques. Mailing Add: Fermi Nat Accelerator Lab PO Box 500 Batavia IL 60510

VOZOFF, KEEVA, b Minneapolis, Minn, Jan 26, 28; m 57; c 4. GEOPHYSICS. Educ: Univ Minn, BPhys, 49; Pa State Univ, MSc, 51; Mass Inst Technol, PhD(geophys), 56. Prof Exp: Geophysicist, Nucom-McPhar Geophys, Ltd, 55-58; assoc prof geophys, Univ Alta, 58-64; vpres, Geosci Inc, Mass, 64-69; consult, 69-72; PROF GEOPHYS, MACQUARIE UNIV, AUSTRALIA, 72- Concurrent Pos: Mem earth sci ad hoc comt on Soviet-Australian coop, 74-75. Mem: Am Soc Explor Geophys; Am Geophys Union; Europ Asn Explor Geophys; Petrol Explor Soc Australia; Soc Explor Geophys. Res: Electrical and electromagnetic methods of determining earth structure; natural electromagnetic fields. Mailing Add: Sch of Earth Sci Macquarie Univ Sydney 2113 Australia

VOZZA, JOHN F, b Morenci, Ariz, Oct 19, 16; m 50. ORGANIC CHEMISTRY. Educ: Univ Ariz, BS, 38, MS, 39; Univ Wis, PhD(chem), 48. Prof Exp: Instr chem, Univ Ariz, 39-43, Navy-War Training Serv, 43-45; asst instr, 45-48, assoc prof exten div, 48-63, prof ctr syst, 63-68, PROF CHEM, UNIV WIS-PARKSIDE, 68- Concurrent Pos: Indust chem consult, 64- Mem: AAAS; Am Chem Soc; NY Acad Sci. Res: Grignard reaction involving gamma-butyrolactone; mercury sulfides formed with thioacetamide; reactions of 2-methyl-pyridine-1-oxide; reduction and halogenation of azoxy benzene. Mailing Add: Dept of Chem Univ of Wis-Parkside Kenosha WI 53140

VOZZO, JOHN ANTHONY, plant physiology, see 12th edition

VRANIC, MLADEN, b Zagreb, Yugoslavia, Apr 3, 30; Can citizen; m 57; c 2. PHYSIOLOGY, ENDOCRINOLOGY. Educ: Univ Zagreb, MD, 55, DSc(physiol), 62. Prof Exp: Fel, Fac Med, Univ Toronto, 63-65, from asst prof to assoc prof, 65-72, PROF PHYSIOL, FAC MED, UNIV TORONTO, 72- Concurrent Pos: Mem, Inst Biomed Electronics, Univ Toronto, 71-, Inst Med Sci, 73- & dir, C H Best Found. Mem: Endocrine Soc; Can Diabetes Asn; Am Diabetes Asn; Can Physiol Soc; Am Physiol Soc. Res: Metabolic roles and interactions of insulin and glucagon in health and disease—diabetes, hyperlipemia, obesity; origin, structure and secretion of nonpancreatic glucagon; endocrine responses and effects during exercise; tracer methodology. Mailing Add: Dept of Physiol Med Sci Bldg Univ of Toronto Fac of Med Toronto ON Can

VRANKA, ROBERT G, physical chemistry, see 12th edition

VRATNY, FREDERICK, b Detroit, Mich, Mar 23, 31; c 2. PHYSICAL CHEMISTRY, ANALYTICAL CHEMISTRY. Educ: Univ Mich, BS, 53; Ind Univ, PhD(chem), 57. Prof Exp: Asst chem, Ind Univ, 53-56; instr, Purdue Univ, 56-60; MEM STAFF, BELL LABS, INC, 60- Mem: Am Phys Soc; Electrochem Soc; Am Vacuum Soc; Am Soc Testing & Mat; NY Acad Sci. Res: Solid state reactions; Raman and infrared spectra of solids and adsorbates; thin film; plasma; dielectrics; active thin film devices; electrochemical processes. Mailing Add: Bell Labs Murray Hill NJ 07971

VRATSANOS, SPYROS M, b Athens, Greece, Apr 10, 20; US citizen; m 58; c 2. BIOCHEMISTRY, ORGANIC CHEMISTRY. Educ: Univ Athens, dipl chem, 50; Adelphi Univ, MS, 56; Fordham Univ, PhD(enzymol, org chem), 61. Prof Exp: Asst prof biochem, Adelphi Univ, 61-63; res assoc, 63-65, ASST PROF MICROBIOL, COL PHYSICIANS & SURGEONS, COLUMBIA UNIV, 65- Mem: Am Chem Soc; Neuberg Socl Harvey Soc. Res: Organophosphorous compounds; origin of life on the earth; proteins; active sites of enzymes; conversion of light energy to chemical signals; chemistry of vision; immunochemistry. Mailing Add: Dept of Microbiol Columbia Univ New York NY 10032

VRBA, FREDERICK JOHN, b Cedar Rapids, Iowa, May 25, 49; m 71. ASTRONOMY. Educ: Univ Iowa, BA, 71; Univ Ariz, PhD(astron), 76. Prof Exp: STAFF ASTRONR, FLAGSTAFF STA, NAVAL OBSERV, 76- Mem: Am Astron Soc. Res: Infrared and polarimetric observations of young stars, dark nebulae, and the general interstellar medium; electronic camera photometry and polarimetry; trigonometric parallaxes of nearby stars. Mailing Add: Naval Observ Flagstaff Sta Box 1149 Flagstaff AZ 86001

VRBA, RUDOLF, b Topolcany, Czech, Sept 11, 24; m; c 2. NEUROCHEMISTRY, IMMUNOCHEMISTRY. Educ: Prague Tech Univ, IngChem, 49, DrTechnSc, 51; Czech Acad Sci, CSc(chem), 56. Prof Exp: Staff mem, Penicillin Factory, Czech, 52-53; mem sci staff, Inst Indust Hyg & Occup Dis, Ministry Health, Czech, 53-58; staff mem, Vet Res Inst, Ministry Agr, Israel, 58-60; mem sci staff, Neuropsychiat Res Unit, Brit Med Res Coun, 60-67; ASSOC PROF PHARMACOL, FAC MED, UNIV BC, 67- Concurrent Pos: Rockefeller grant, Psychiat Res Unit, Brit Med Res Coun,

60-62; assoc, Med Res Coun Can, 68-75; vis lectr pharmacol, Harvard Med Sch, 73-75. Mem: Brit Biochem Soc; Int Neurochem Soc; Can Biochem Soc. Res: Physiological chemistry; biochemical aspects of cancer, diabetes and immunology. Mailing Add: Dept of Pharmacol Univ of BC Fac of Med Vancouver BC Can

VRBASKI, THEODORE, organic chemistry, physical chemistry, see 12th edition

VREBALOVICH, THOMAS, b Los Angeles, Calif, July 10, 26; m 51; c 2. SPACE PHYSICS, FLUID MECHANICS. Educ: Calif Inst Technol, BS, 48, MS, 49, PhD(aeronaut eng), 54. Prof Exp: From res scientist to sr res scientist, Jet Propulsion Lab, Calif Inst Technol, 52-61, res specialist, 61-62, Ranger proj scientist, 63-65, group supvr photosci, 65-66, Surveyor assoc proj scientist, 65-67, div rep space sci, 66-67, Voyager landed capsule syst scientist, 67-68, on leave, 68-70, asst proj scientist, Mariner Mars 1971 proj, 70-73, mission sci coordr, Mariner Jupiter Saturn Proj, 73-74, mgr res, 74-75; COUN SCI & TECHNOL AFFAIRS, AM EMBASSY, DEPT OF STATE, NEW DELHI, INDIA, 75- Concurrent Pos: Instr, Univ Southern Calif, 56 & Univ Calif, Los Angeles, 57; consult, Flow Corp, 66-; vis prof aeronaut, Indian Inst Technol, Kanpur, 68-70; mem bd gov, Photog Art & Sci Found, 72- Honors & Awards: Fairbanks Mem Award, Soc Photog Instrumentation Engrs, 66. Mem: Am Inst Aeronaut & Astronaut; Am Phys Soc. Res: Supersonic aerodynamics; space photography and science. Mailing Add: Am Embassy New Delhi Dept of State Washington DC 20521

VREDEVELD, NICHOLAS GENE, b Hudsonville, Mich, May 5, 29; m 53; c 2. PLANT PATHOLOGY. Educ: Calvin Col, AB, 51; Mich State Univ, MS, 55, PhD(plant path), 65. Prof Exp: Biochemist, St Lawrence Hosp, Lansing, Mich, 62-64; asst prof, 64-67, ASSOC PROF BIOL, UNIV TENN, CHATTANOOGA, 67- Mem: Am Phytopath Soc. Res: Fungicides and fungus physiology, environmental effect on plant disease distribution. Mailing Add: 3007 Ozark Circle Chattanooga TN 37415

VREDEVOE, DONNA LOU, b Ann Arbor, Mich, Jan 11, 38; c 1. IMMUNOLOGY, MICROBIOLOGY. Educ: Univ Calif, Los Angeles, BA, 59, PhD(microbiol), 63. Prof Exp: Instr bact, 63, asst res immunologist, 64-67, asst prof nursing res, 67-70, ASSOC PROF NURSING RES, UNIV CALIF, LOS ANGELES, 70-, CONSULT, LAB NUCLEAR MED & RADIATION BIOL, 67-, ASST DIR SPACE PLANNING, CANCER CTR, 74- Concurrent Pos: USPHS fel microbiol, Stanford Univ, 63-64; res grants, Calif Inst Cancer Res, Univ Calif, Cancer Res Coord Comt, Am Cancer Soc, Calif Div & USPHS. Mem: Am Asn Immunol; Am Soc Microbiol; Am Asn Cancer Res. Res: Stereotyping for human kidney transplantation; delayed hypersensitivity; immunosuppression; immunotherapy; tumor immunology; mouse lymphoma. Mailing Add: Sch of Nursing Univ of Calif Ctr Health Sci Los Angeles CA 90024

VREDEVOE, LAWRENCE A, b Ann Arbor, Mich, Aug 2, 40; m 66; c 1. SURGERY. Educ: Univ Calif, Los Angeles, BA, 62, MA, 64, PhD(physics), 66; Univ Calif, San Francisco, MD, 75. Prof Exp: Mem tech staff theoret physics, Sci Ctr, NAm Rockwell Corp, Calif, 66-70; assoc prof physics, Ind Univ Bloomington, 70-72; mem staff med training, Sch Med, Univ Calif, San Francisco, 72-75, SURG RESIDENT, CTR HEALTH SCI, UNIV CALIF, LOS ANGELES, 75- Mem: Am Phys Soc. Res: Anharmonic phonon interactions; vibronic spectra of solids; paraelectric resonance and relaxation effects of electric dipole impurities in solids; optical properties of rare-earth ions; magnetic scattering of electrons. Mailing Add: 428 21st St Santa Monica CA 90402

VREELAND, HERBERT HAROLD, III, b New York, NY, May 21, 20; m 59; c 3. ANTHROPOLOGY. Educ: Yale Univ, BA, 41, PhD(anthrop), 53. Prof Exp: Lectr anthrop, Dept Social Rels, Harvard Univ, 54-55; dep dir, For Area Studies, Am Univ, 57-67; sr res scientist, Human Sci Res, Inc, 67-75; SOCIAL SCI ANALYST, CTR STUDIES METROP PROBS, NIMH, DEPT HEALTH, EDUC & WELFARE, 75- Concurrent Pos: Adj prof, Am Univ, 68-69. Mem: Fel Am Anthrop Asn. Res: Social impact assessment; energy conservation; urban social problems related to mental health. Mailing Add: 7004 Braeburn Ct Bethesda MD 20034

VREELAND, JAY HENRY, physical chemistry, see 12th edition

VREELAND, JOHN ALLEN, b Orlando, Fla, Jan 6, 25; m 52; c 2. PHYSICS. Educ: Presby Col, BS, 49; Univ Wis, MA, 51, PhD(physics), 55. Prof Exp: Asst physicist, Univ Wis, 50-55; scientist, Atomic Power Div, Westinghouse Elec Corp, 55-60; sr nuclear specialist, Rocketdyne Div, NAm Aviation, 60-62; mgr nuclear anal dept, nuclear rocket opers, Aerojet-Gen Corp, 62-69; PROF NUCLEAR ENG, SCH ENG, CALIF STATE UNIV, SACRAMENTO, 69- Concurrent Pos: Lectr, Univ Calif, 60-; consult, Univ Fla, 62; mem Atomic Indust Forum. Mem: Am Physics Soc; Am Nuclear Soc; Am Inst Aeronaut & Astronaut. Res: Reactor physics; nuclear structure; analysis and detection of nuclear transport phenomena; systems analysis related to nuclear power plant design. Mailing Add: Sch of Eng Calif State Univ Sacramento CA 95819

VREMAN, HENDRIK JAN, b Soest, Netherlands, Jan 22, 39; US citizen; m 64; c 2. BIOLOGICAL CHEMISTRY. Educ: Univ NC, Chapel Hill, BA, 68; Univ Wis-Madison, PhD(bot), 73. Prof Exp: Lab asst dairy chem & bact, United Gooi Dairies, Hilversum, Netherlands, 56-57; anal chemist vet med, Lab Medical Vet Medicine, Univ Utrecht, 57-60; anal chemist pub health, Pharmaceut & Toxicol Lab, Nat Inst Pub Health, Utrecht, Netherlands, 60-62; res technician biochem, E R Johnson Found, Univ Pa, 62-64; res technician physiol & pharmacol, Duke Univ, 64-68; res asst bot, Inst Plant Develop, Univ Wis-Madison, 68-73; Nat Res Coun res assoc, Western Regional Res Ctr, Agr Res Serv, USDA, Calif, 73-75; RES ASSOC MED, STANFORD UNIV SERVS, VET ADMIN HOSP, 75- Mem: AAAS; Am Soc Plant Physiologists. Res: Role of cytokinins in plant growth and development; isolation, separation, identification of naturally occurring cytokinins; plant cell and tissue cultures; acetate metabolism in humans with chronic renal failure. Mailing Add: Dialysis Unit Vet Admin Hosp 3801 Miranda Ave Palo Alto CA 94304

VRIELAND, GAIL EDWIN, b Grand Rapids, Mich, Jan 4, 38; m 63; c 3. INDUSTRIAL CHEMISTRY. Educ: Calvin Col, AB, 59; Northwestern Univ, PhD(chem), 63. Prof Exp: RES SPECIALIST CHEM, CENT RES LAB, DOW CHEM CO, 63- Res: Heterogeneous catalysis and high temperature vapor phase reaction including oxidations and hydrocyanation. Mailing Add: Cent Res Lab 438 Bldg Dow Chem Co Midland MI 48640

VRIESEN, CALVIN W, b Elkhart Lake, Wis, Aug 31, 16; m 41; c 2. ORGANIC POLYMER CHEMISTRY, SYNTHETIC ORGANIC CHEMISTRY. Educ: Univ Minn, BS, 39, MS, 47; Purdue Univ, PhD(org chem), 52. Prof Exp: Assoc prof chem, Ill Col, 47-49; res chemist, Chattanooga Nylon Plant, E I du Pont de Nemours & Co, 52-56, Chambers Works, 56-58; res chemist, 58-62, staff chemist, 62-68, sr scientist, 68-74, GROUP SUPVR, THIOKOL CORP, ELKTON, MD, 74- Honors & Awards: Aerospace Scientist of Year Award, Am Inst Aeronaut & Astronaut, 67. Mem: Am Inst Aeronaut & Astronaut; Am Chem Soc. Res: Condensation, cationic, anionic polymerization; synthesis of new binders, oxidizers, coolants for solid rocket propellants. Mailing Add: 12 Mitchell Circle Newark DE 19713

VRIJENHOEK, ROBERT CHARLES, b Rotterdam, Netherlands, Mar 13, 46; US citizen; m 68; c 2. EVOLUTIONARY BIOLOGY. Educ: Univ Mass, BA, 68; Univ Conn, PhD(zool), 72. Prof Exp: Asst prof biol, Southern Methodist Univ, 72-74; ASST PROF ZOOL, RUTGERS UNIV, NEW BRUNSWICK, 74- Concurrent Pos: NSF grant, 74. Mem: Soc Study Evolution; Am Soc Ichthyol & Herpetol; Genetics Soc Am; Am Genetic Asn. Res: Population genetic studies of evolutionary relationships and genetic variation in fishes; the effects of various mating systems on the genetic structure and evolutionary potential of populations. Mailing Add: Dept of Zool Rutgers Univ Busch Campus New Brunswick NJ 08903

VROMAN, HUGH EGMONT, b Detroit, Mich, Apr 18, 28; m 59. BIOCHEMISTRY. Educ: Univ Md, BS, 50, PhD(zool), 62. Prof Exp: Biologist, Nat Heart Inst, 57-58, biochemist, 58-61; res biologist, Insect Physiol Lab, Agr Res Serv, USDA, 61-66; biochemist, Dept Dermat, Sch Med, Univ Miami, 66-69; chmn dept biol, 71-73, PROF BIOL, CLAFLIN COL, 69- Mem: AAAS; Entom Soc Am; Am Soc Zool; Brit Biochem Soc; Am Inst Biol Sci. Res: Cholesterol metabolism; lipid biosynthesis by insects; insect hormones; sterol metabolism by insects; lipid biosynthesis and metabolism in skin. Mailing Add: Dept of Biol Claflin Col Orangeburg SC 29115

VROMAN, LEO, b Gouda, Holland, Apr 10, 15; nat; m 47; c 2. PHYSIOLOGY. Educ: Jakarta Med Col, Indonesia, Drs, 41; Univ Utrecht, PhD(animal physiol), 58. Prof Exp: Asst zool, anat & physiol, Jakarta Med Col, 41; res assoc, St Peter's Gen Hosp, New Brunswick, NJ, 46-55; asst, Mt Sinai Hosp, New York, 56-58; sr physiologist, Stress-Tension Proj, Dept Animal Behav, Am Mus Natural Hist, 58-61; BIOCHEMIST, VET ADMIN HOSP, BROOKLYN, 61- Res: Blood clotting; behavior of proteins at interfaces; biomaterials. Mailing Add: 2365 E 13th St Brooklyn NY 11229

VROMEN, BENJAMIN H, b Leeuwarden, Netherlands, June 6, 21; US citizen; m 53; c 2. PHYSICAL CHEMISTRY. Educ: Hebrew Univ, MS, 46, PhD(phys chem), 51. Prof Exp: Mem staff, Weizman Inst, 49-54; dir lab, Kadimah Chem Co, Israel, 54-58; group leader res & develop, Graver Water Conditioning Co, 58-60; mem tech staff, Bell Tel Labs, NJ, 61-68; ADV CHEMIST, COMPONENTS DIV, IBM CORP, 68- Concurrent Pos: Vis lectr, Polytech Inst Brooklyn, 67-68. Mem: Electrochem Soc; Am Vacuum Soc; Royal Netherlands Chem Soc. Res: Properties of sputtered and evaporated films; anodic reactions; separation processes; diffusion; thin film dielectrics. Mailing Add: Dept 265 Bldg 300-100 IBM Corp Hopewell Junction NY 12533

VROOM, ALAN HEARD, b Montreal, Que, Can, Oct 5, 20; m 43; c 2. INSTRUMENTATION, MATERIALS SCIENCE. Educ: McGill Univ, BSc, 42, m, Nat Res Coun Can, 44-46; asst dir res pulp & paper, Fraser Co, Ltd, 46-49; Hibbert Mem fel & hon lectr, McGill Univ, 50; res fel bark chem, Pulp & Paper Res Inst Can, 50-51; asst chief appl chem sect, Weyerhaeuser Timber Co, 51-52, chief appl physics sect, 52-54, chief appl chem sect, 54; asst dir res, Consol Paper Corp, Ltd, 55-56, dir res, 56-67, dir res & develop Consol-Bathurst Ltd, 67-71; spec consult, Nat Res Coun Can, 71-73; PRES, SULPHUR INNOVATIONS, LTD, 73- Concurrent Pos: Consult, Pulp & Paper Res Inst, Can, 51-54. Mem: Chem Inst Can; Can Soc Chem Eng. Res: Bark chemistry; pulp and paper; wood and fiber technology; sulfur utilization; development of new sulfur-based construction materials, primarily sulfur concrete. Mailing Add: 3015 58th Ave SE Calgary AB Can

VROOM, DAVID ARCHIE, b Vancouver, BC, Sept 12, 41; m 69; c 1. CHEMICAL PHYSICS. Educ: Univ BC, BSc, 63, PhD(phys chem), 67. Prof Exp: Nat Res Coun Can overseas fel, 67-68; staff chemist, Atomic Physics Br, Gulf Radiation Technol Div, Gulf Energy & Environ Systs, 68-73; PRIN SCIENTIST, IRT CORP, 73- Mem: Am Phys Soc; Radiation Res Soc. Res: Photoionization and photoelectron spectroscopy; electron impact studies of excitation; dissociation and ionization; low energy ion neutral reactions and pulse radiolysis studies. Mailing Add: PO Box 80817 San Diego CA 92138

VROOM, KENNETH EDWIN, b Montreal, Can, May 28, 27. MATHEMATICS. Educ: McGill Univ, BSc, 48. Prof Exp: Supvr statist, Bell Tel Co Can, 48-53; supvr statist methods, 53-60, chmn tech serv dept, 60-69, SECY, PULP & PAPER RES INST CAN, 67-, DIR INFO SERV DIV, 69-, DIR SCI & BUS SERV, 71- Mem: Tech Asn Pulp & Paper Indust; Am Statist Asn; Biomet Soc; Can Pulp & Paper Asn. Res: Statistical and mathematical analysis; operations and pulping research; applications of computers to research; information science; research administration. Mailing Add: Pulp & Paper Res Inst Can 570 St John's Blvd Pointe Claire PQ Can

VUCKOVIC, VLADETA, b Aleksinac, Yugoslavia, Mar 30, 23; m 54; c 1. MATHEMATICS. Educ: Univ Belgrade, MS, 49; Serbian Acad Sci, 53. Prof Exp: Instr math, Univ Belgrade, 49-52; sci collabr, Math Inst, Serbian Acad Sci, 52-54; prof math, Teacher Inst, Zrenjanin, Yugoslavia, 54-60; from asst prof to assoc prof, Univ Belgrade, 60-63; asst prof, 63-66, ASSOC PROF MATH, UNIV NOTRE DAME, 66- Mem: Math Asn Am; Asn Symbolic Logic. Res: Foundations of mathematics; mathematical analysis; summability of divergent series and integrals; theory of recursive functions. Mailing Add: Dept of Math Univ of Notre Dame Notre Dame IN 46556

VUICICH, GEORGE, b Chisholm, Ninn, Nov 12, 25; m 47; c 3. URBAN GEOGRAPHY. Educ: Univ Iowa, BA, 50, MA, 55, PhD, 60. Prof Exp: Instr geog, Univ Iowa, 52-55; asst prof, Western Mich Univ, 55-58; instr, Univ Iowa, 58-60; asst prof, Wis State Univ, Eau Claire, 60-65; assoc dir high sch geog proj, 65-68; PROF GEOG, WESTERN MICH UNIV, 68- Mem: Asn Am Geog. Res: Application of statistical techniques to urban geography; geography of the Soviet Union; geography education. Mailing Add: Dept of Geog Western Mich Univ Wood Hall Kalamazoo MI 49008

VUILLEMIN, JOSEPH J, b Waco, Tex, July 22, 34; m 57; c 3. PHYSICS. Educ: Univ Tex, BS, 56; Baylor Univ, MS, 57; Univ Chicago, PhD(physics), 65. Prof Exp: NSF fel, 65-66; asst prof, 66-70, ASSOC PROF PHYSICS, UNIV ARIZ, 70- Concurrent Pos: Sci res coun fel, Univ Bristol, 74-75. Mem: Am Phys Soc. Res: Electronic structure of metals and low temperature physics. Mailing Add: Dept of Physics Univ of Ariz Tucson AZ 85721

VUKOVICH, FRED MATTHEW, b Chicago, Ill, July 13, 39; m 66; c 3. DYNAMIC METEOROLOGY, PHYSICAL OCEANOGRAPHY. Educ: Parks Col Aeronaut Technol, St Louis, BS, 60; St Louis Univ, MS, 63, PhD(meteorol), 66. Prof Exp: Res meteorologist, Meteorol Res Inc, 63-64; res meteorologist, 66-68, supvr, 68-71, sr scientist, 71-74, MGR GEOSCI DEPT, RES TRIANGLE INST, 74- Concurrent Pos: Assoc prof, Duke Univ, 67- Mem: Am Meteorol Soc. Res: Dynamic meteorology of urban atmosphere; physical oceanographic studies on continental shelf and gulf stream; satellite oceanography; atmospheric gravity waves and synoptic scale energetics. Mailing Add: Res Triangle Inst PO Box 12194 Research Triangle Park NC 27709

VULLIET, WILLIAM GEORGE, b Pueblo, Colo, July 11, 28; m 51; c 2. ATOMIC PHYSICS, MOLECULAR PHYSICS. Educ: San Diego State Col, BS, 49; Univ Iowa,

MA, 54. Prof Exp: Engr, Collins Radio Co, 53-54; res engr, Gen Dynamics/Convair, 54-59; SR SCI ADV, IRT CORP, 59- Concurrent Pos: Consult, Los Alamos Sci Lab, 73- Res: Equation of state and radiative opacities; radiative transfer under non-local thermodynamic equilibrium; nuclear weapons effects; pollutant formation kinetics during combustion. Mailing Add: 13935 Putney Rd Poway CA 92064

VULLO, WILLIAM JOSEPH, b Buffalo, NY, June 14, 33; m 59; c 2. ORGANIC CHEMISTRY. Educ: Univ Buffalo, BA, 55; Northwestern Univ, PhD(org chem), 59. Prof Exp: Instr gen chem, Northwestern Univ, 59; sr chemist, Hooker Chem Corp, 59-69; MGR TEXTILE RES, MOHASCO INDUSTS, 69- Mem: Am Chem Soc; Am Asn Textile Chem & Colorists. Res: Organometallic chemistry; organic reaction mechanisms; metal conversion coatings and treatments; cellulose reactive chemicals; organic phosphorus and fluorine chemistry; fire retardants; textile chemicals and finishes. Mailing Add: Tech Res & Serv Dept Mohasco Industs 57 Lyon St Amsterdam NY 12010

VURAL, BAYRAM, solid state physics, electromagnetics, see 12th edition

VUYLSTEKE, ARTHUR ADOLPH, physics, see 12th edition

VYAS, GIRISH NARMADASHANKAR, b Aglod, India, June 11, 33; m 62; c 2. IMMUNOLOGY, GENETICS. Educ: Univ Bombay, BSc, 54, MSc, 57, PhD(microbiol), 64. Prof Exp: Asst res officer, Blood Group Ref Ctr, Indian Coun Med Res, Bombay, 57-64; officer-in-chg, Bombay Munic Blood Ctr, King Edward Mem Hosp, India, 64-65; lectr immunol, 67-69, asst prof path, 69-73, ASSOC PROF LAB MED & DIR BLOOD BANK, SCH MED, UNIV CALIF, SAN FRANCISCO, 73- Concurrent Pos: Jr res fel hemat, J J Hosp, Bombay, India, 56-57; fel genetics, Western Reserve Univ, 65-67. Mem: AAAS; Am Soc Human Genetics; Am Asn Immunol. Res: Microbiology; blood group serology; immunogenetics; blood banking; transfusion and circulatory physiology; genetics of gamma globulin and its structure. Mailing Add: Univ of Calif Med Ctr San Francisco CA 94143

VYE, MALCOLM VINCENT, b Gary, Ind, Feb 17, 36; m 60; c 1. HEMATOLOGY, PATHOLOGY. Educ: Marquette Univ, MD, 61. Prof Exp: Asst instr, 62-64, from instr to asst prof path, Univ Ill Col Med, 64-66 & 68-71; ASST PROF PATH, MED SCH, NORTHWESTERN UNIV, EVANSTON, 71-; ASSOC PATHOLOGIST, EVANSTON HOSP, 71- Concurrent Pos: Mem staff, Armed Forces Inst Path. Mem: AAAS; Am Soc Clin Path; Am Asn Path & Bact; Col Am Pathologists. Res: Cell differentiation embryonic muscle; ultrastructure of glycogen; hematology laboratory methodology. Mailing Add: Dept of Path Northwestern Univ Med Sch Evanston IL 60201

VYGANTAS, AUSTE MARIJA, b Kaunas, Lithuania, Jan 15, 42; US citizen; m 69; c 1. BIOCHEMISTRY, ORGANIC CHEMISTRY. Educ: St Xavier's Col, BS, 63; Univ Ill, Urbana, MS, 65, PhD(bio-org chem), 67. Prof Exp: Fel, Univ Chicago, 67-70; ASST PROF BIOCHEM, MED SCH, NORTHWESTERN UNIV, CHICAGO, 70- Mem: AAAS; Am Chem Soc. Res: Biosynthesis of antibiotics. Mailing Add: Dept of Biochem Northwestern Univ Med Sch Chicago IL 60611

W

WAACK, RICHARD, b Syracuse, NY, May 18, 31; m 60. PHYSICAL CHEMISTRY, POLYMER CHEMISTRY. Educ: State Univ NY, BS, 53, MS, 54, PhD, 58. Prof Exp: Tech serv rep, Dow Chem Co, Mich, 54-56; chemist, Solvay Process Div, Allied Chem Co, NY, 58-59; res chemist, Eastern Res Lab, Dow Chem Co, Mich, 59-60, res chemist, Phys Res Lab, 67-69; MGR DEVELOP LAB, POLAROID CORP, WALTHAM, 69- Mem: Am Chem Soc; Soc Photog Sci & Eng. Res: Polymer synthesis; spectroscopy; ionic polymerization mechanisms; photographic science; silver halide emulsion technology; colloidal processes; water soluble polymers; diffusion processes. Mailing Add: 19 Morrill Dr Wayland MA 01778

WAAG, CHARLES JOSEPH, b Oct 25, 31; US citizen; m 56; c 1. GEOLOGY. Educ: Univ Pittsburgh, BS, 56, MS, 58; Univ Ariz, PhD(geol), 68. Prof Exp: Sr geologist, Orinoco Mining Co, US Steel Corp, 58-63; sr geologist, Va Div Mineral Resources, 63-64; asst & lectr, Univ Ariz, 64-68; asst prof, 68-71, ASSOC PROF GEOL, GA STATE UNIV, 71- Concurrent Pos: Consult hydrogeol, 69- Mem: Geol Soc Am. Res: Glaciers as models in structural geology; gravity tectonics attendant to mantled gneiss domes in the Basin and Range Province. Mailing Add: Dept of Geol Ga State Univ Atlanta GA 30303

WAAGE, EDWARD VERN, b Oakland, Calif, Oct 27, 42; m 68. PHYSICAL CHEMISTRY. Educ: Reed Col, BA, 64; Univ Wash, PhD(chem), 70. Prof Exp: ASST PROF CHEM, ILL STATE UNIV, 70- Mem: Am Chem Soc; Am Phys Soc. Res: Gas phase kinetics, photochemistry; crystal growth. Mailing Add: Dept of Chem Ill State Univ Normal IL 61761

WAAGE, KARL MENSCH, b Philadelphia, Pa, Dec 17, 15; m 42; c 2. GEOLOGY. Educ: Princeton Univ, AB, 39, MA, 42, PhD(geol), 46. Prof Exp: From instr to assoc prof, 46-67, PROF GEOL, YALE UNIV, 67-, CUR INVERT PALEONT, PEABODY MUC, 59- Concurrent Pos: Geologist, US Geol Surv, Washington, DC, 42- Mem: Fel Geol Soc Am; assoc Paleont Soc. Res: Field exploration for non-metalliferous deposits; stratigraphic geology and paleontology of cretaceous of western interior. Mailing Add: Peabody Mus Yale Univ New Haven CT 06520

WAALAND, JOSEPH ROBERT, b San Mateo, Calif, Feb 22, 43; m 69. ALGOLOGY, CYTOLOGY. Educ: Univ Calif, Berkeley, BA, 66, PhD(bot), 69. Prof Exp: Asst prof, 69-75, ASSOC PROF BOT, UNIV WASH, 75- Mem: Bot Soc Am; Phycol Soc Am; Int Phycol Soc; Marine Biol Asn UK. Res: Development, cytology and ecology of algae; aquaculture of marine algae. Mailing Add: Dept of Bot Univ of Wash Seattle WA 98195

WAALKES, T PHILLIP, b Belmond, Iowa, Oct 30, 19; m 45; c 6. PUBLIC HEALTH. Educ: Hope Col, AB, 41; Ohio State Univ, PhD(org chem), 45; George Washington Univ, MD, 51. Prof Exp: Asst chem, Ohio State Univ, 41-43, res assoc, Res Found, 44, Am Petrol Inst res assoc, Univ, 45, instr chem, 46-47; intern, USPHS Hosp, 51-52, res med, 52-55, Nat Heart Inst, 55-58, asst chief in chg clin activ, Cancer Chemother Nat Serv Ctr, 58-63, ASSOC DIR, LAB CHEM PHARMACOL, NAT CANCER INST, 63- Mem: AAAS. Res: Organic fluorine compounds; biochemistry; amino acids; fluorinated derivatives of propane and propylene; addition of fluorine to double bonds; cancer chemotherapy. Mailing Add: Lab Chem Pharmacol Nat Cancer Inst NIH Bethesda MD 20014

WABECK, CHARLES J, b Montague, Mass, July 16, 38; m 64; c 2. FOOD SCIENCE. Educ: Univ Mass, BS, 62; Univ NH, MS, 64; Purdue Univ, PhD(food sci), 66. Prof Exp: Res assoc poultry & frozen foods, Armour & Co, 66-69; dir res frozen foods, Ocoma Foods Co, Nebr, 69; ASST PROF POULTRY PROD, UNIV MD, COLLEGE PARK, 69- Mem: Inst Food Technologists; Poultry Sci Asn. Res: Research and development; quality control; frozen foods; poultry and meat products. Mailing Add: Dept of Poultry Sci Univ of Md College Park MD 20742

WABER, JAMES THOMAS, b Chicago, Ill, Apr 8, 20; m 51; c 3. ATOMIC PHYSICS, SOLID STATE PHYSICS. Educ: Ill Inst Technol, BS, 41, MS, 43, PhD(metall), 46. Prof Exp: Res assoc, Ill Inst Technol, 46, asst prof chem, 46-47; assoc metallurgist, Los Alamos Sci Lab, 47-49, staff mem, 49-66; PROF MAT SCI, NORTHWESTERN UNIV, EVANSTON, 67- Concurrent Pos: NSF sr fel, Univ Birmingham, 60-61; chmn comt alloy phases, past chmn nuclear metall comt & mem exec comt, Inst Metals Div, NY; partic, Robert A Welch Found Conf on Chem Res XIII Mendeleef Centennial-The Transuranium Elements. Honors & Awards: Turner Prize, Electrochem Soc, 47; Whitney Prize, Nat Asn Corrosion Eng, 63. Mem: Am Soc Metals; Electrochem Soc; Am Inst Mining, Metall & Petrol Engrs; Nat Asn Corrosion Engrs. Res: Corrosion and oxidation of metals; relativistic self-consistent field Dirac-Slater and Hartree-Fock calculations for atoms and ions; energy band calculations; chemistry and physics of superheavy elements. Mailing Add: 2324 Hartzell St Evanston IL 60201

WACASEY, JERVIS WINN, b Clarksville, Tex, Nov 1, 29; m 56; c 2. BIOLOGICAL OCEANOGRAPHY. Educ: Tex Technol Col, BS, 54, MS, 55; Mich State Univ, PhD(zool), 61. Prof Exp: Instr biol, Eastern Ill Univ, 61-64; NIH fel, Inst Marine Sci, Univ Miami, 64-66; RES SCIENTIST II BIOL OCEANOG, ARCTIC BIOL STA, FISHERIES RES BD CAN, 67- Mem: Am Soc Limnol & Oceanog; Soc Syst Zool. Res: Ecology of benthic invertebrates in the arctic marine ecosystem. Mailing Add: Dept of Environ Arctic Biol Sta Box 400 Ste-Anne-de-Bellevue PQ Can

WACHHOLZ, BRUCE WILLIAM, b Chicago, Ill, Aug 16, 36; m 63; c 1. RADIATION BIOLOGY. Educ: Valparaiso Univ, BA, 58; Univ Rochester, MS, 59, PhD(radiation biol), 67. Prof Exp: Sr res scientist, Pac Northwest Labs, Battelle Mem Inst, 66-71; RADIATION BIOLOGIST, MED BR, DIV BIOL & MED, ENERGY RES & DEVELOP ADMIN, 71- Mem: AAAS; Am Phys Soc; Radiation Res Soc; NY Acad Sci; Geront Soc. Res: Pathological, physiological and endocrinological effects of radiation; metabolism and toxicity of radionuclides; gerontology. Mailing Add: Med Br Div Biol & Med Energy Res & Develop Admin Washington DC 20545

WACHMAN, HAROLD YEHUDA, b Tel Aviv, Israel, Dec 2, 27; nat US; m 54; c 1. SURFACE PHYSICS. Educ: City Col New York, BS, 49; Univ Mo, MA, 52, PhD(phys chem), 57. Prof Exp: Specialist chem physics, Aerosci Lab, Gen Elec Co, 57-63; vis prof, 63-64, assoc prof, 64-69, PROF AERONAUT & ASTRONAUT, MASS INST TECHNOL, 69- Concurrent Pos: Consult, Space Sci Lab, Gen Elec Co; vis sr res fel, Jesus Col, Oxford Univ, 72-73. Mem: Am Chem Soc. Res: Rarefied gas phenomena; adsorption; high temperature chemical equilibrium studies; gas surface interactions; nucleation phenomena. Mailing Add: Dept of Aeronaut & Astronaut Mass Inst of Technol Cambridge MA 02139

WACHMAN, MURRAY, b Tel Aviv, Israel, Feb 1, 31; US citizen; m 58; c 3. MATHEMATICS. Educ: Brooklyn Col, BA, 53; NY Univ, MS, 56, PhD(math), 61. Prof Exp: Math analyst, Repub Aviation Corp, 57-59; appl mathematician, Gen Elec Co, 59-63, consult mathematician, 63-65, group leader appl math, 65-67; assoc prof, 67-73, PROF MATH, UNIV CONN, 73- Concurrent Pos: Consult, Missile & Space Div, Gen Elec Co, 67-69. Mem: Soc Indust & Appl Math; Am Math Soc. Res: Representation of functions; Boltzmann equation; elliptic partial differential equation; two point boundary value problem; biological and economic models. Mailing Add: Dept of Math Univ of Conn Storrs CT 06268

WACHS, GERALD N, b Chicago, Ill, Nov 5, 37; m 62; c 4. DERMATOLOGY. Educ: Univ Ill, BS, 58, MD, 62; Am Bd Dermat, dipl, 68. Prof Exp: Intern med, Michael Reese Hosp, Ill, 62-63; resident dermat, Univ Calif, 63-65, chief resident, 65-66; mem dept clin invest, 66-67, asst med dir, 67-69, assoc med dir, 69-74, DIR NEW PROD PLANNING, SCHERING LABS, KENILWORTH, NJ, 74- Concurrent Pos: Clin asst dermatologist, St Vincent's Hosp, NY, 67-; attend staff dermatologist, Mary Manning Walsh Home, New York; attend staff, St Barnabus Med Ctr, Livingston, NJ. Mem: Fel Am Col Physicians; fel Am Acad Dermat; fel Am Col Allergists; Int Soc Trop Dermat; Am Acad Allergy. Res: Clinical investigation of drugs in dermatology and allergy; development of new concepts in approaching the therapy of difficult diseases in dermatology and allergy. Mailing Add: 459 Long Hill Dr Short Hills NJ 07078

WACHSBERGER, PHYLLIS RACHELLE, b New York, NY; m 67; c 1. CELL BIOLOGY. Educ: City Univ New York, BS, 64; Med Col Pa, PhD(physiol & biophys), 71. Prof Exp: Res instr muscle biophys, Dept Physiol & Biophys, Med Col Pa, 70-71; RES ASSOC MUSCLE BIOPHYS, DEPT ANAT, UNIV PA, 73- Concurrent Pos: NIH fel, Dept Anat, Univ Pa, 71-73. Mem: Biophys Soc; AAAS; Am Inst Biol Sci; Sigma Xi. Res: Studies of the self assembly of synthetic vertebrate smooth muscle myosin filaments; comparative studies of molecular substructure of myosin filaments from various muscle types. Mailing Add: 433 School Lane Strafford Wayne PA 19087

WACHSMAN, JOSEPH T, b New York, NY, July 25, 27; m 60; c 1. MICROBIOLOGY, BIOCHEMISTRY. Educ: NY Univ, AB, 48; Univ Calif, PhD(microbiol), 55. Prof Exp: USPHS fel, Brussels, Belg, 55-57; asst prof, 57-63, ASSOC PROF MICROBIOL, UNIV ILL, URBANA-CHAMPAIGN, 63- Concurrent Pos: USPHS res career develop award, 62- Mem: Am Soc Microbiol. Res: Enzyme localization in the bacterial cell, especially on structure and properties of cytoplasmic membrane. Mailing Add: Dept of Microbiol Univ of Ill Urbana IL 61801

WACHTEL, ALLEN W, b New York, NY, Aug 13, 25; m 46, 61; c 2. CELL BIOLOGY, CYTOLOGY. Educ: Columbia Univ, BS, 53, MA, 54, PhD(zool), 62. Prof Exp: Res asst cell biol, Cell Res Lab, Mt Sinai Hosp, New York, 56-63; from asst prof zool to assoc prof zool, 63-72, PROF ZOOL, UNIV CONN, 72- Mem: Am Soc Cell Biol; Electron Micros Soc Am; Histochem Soc. Res: Histochemistry; cytology of electric organs; receptor structure. Mailing Add: Sect of Genetics & Cell Biol Univ of Conn U-131 Storrs CT 06268

WACHTEL, ANSELM, b Vienna, Austria, Aug 20, 20; US citizen; m 44; c 2. SOLID STATE CHEMISTRY. Educ: Bundeslehr-und Versuchsanstalt Chem Indust, Vienna, Austria, BS, 38. Prof Exp: Chief chemist, VCA Labs, NJ, 47-54; from asst res engr to res engr, 54-62, SR RES ENGR, LAMP DIV, WESTINGHOUSE ELEC CORP, BLOOMFIELD, 62- Mem: Electrochem Soc; Am Chem Soc; fel Am Inst Chemists. Res: Luminescence; development of new phosphors useful in electroluminescence or in fluorescent lamps; pertinent fields of solid state chemistry and physics. Mailing Add: Westinghouse Elec Corp Lamp Div 1 Westinghouse Plaza Bloomfield NJ 07003

WACHTEL, HOWARD, b New York, NY, July 5, 39; m 67; c 1. BIOMEDICAL ENGINEERING, PHYSIOLOGY. Educ: Cooper Union, BSEE, 60; Drexel Inst Technol, MS, 61; NY Univ, PhD(physiol), 67. Prof Exp: Res asst biomed eng, NY Univ, 61-67; NSF grant, 68-70, asst prof biomed eng, 68-71, ASSOC PROF

BIOMED ENG, DUKE UNIV, 71-, ASST PROF PHYSIOL, 68- Concurrent Pos: NIH grant, Duke Univ, 69- & NIMH grant, 71- Mem: Am Physiol Soc; Soc Gen Physiol; Biomed Eng Soc. Res: Neurophysiology; neuronal basis of behavior; neuronal interactions; slow wave generation in neurons; prolonged synaptic events; organization of neural nets; microwave effects on neurons. Mailing Add: Div of Physiol Duke Univ Med Ctr Durham NC 27706

WACHTEL, JACQUES LOUIS, chemistry, see 12th edition

WACHTEL, JONATHAN MARK, b Bronx, NY, Feb 3, 42; m 64; c 2. PLASMA PHYSICS. Educ: Mass Inst Technol, SB, 63; Yale Univ, MS, 65, PhD(physics), 67. Prof Exp: Res asst physics, Yale Univ, 63-67, res staff appl scientist, 67 & 69; asst prof, 69-74, ASSOC PROF PHYSICS, BELFER GRAD SCH SCI, YESHIVA UNIV, 74- Concurrent Pos: Consult, Defense Nuclear Agency, 73- Res: Experimental plasma physics. Mailing Add: Belfer Grad Sch of Sci Yeshiva Univ New York NY 10033

WACHTEL, LOUIS WILLIAM, b Brooklyn, NY, Mar 22, 20; m 47, 65; c 2. BIOCHEMISTRY. Educ: Brooklyn Col, BA, 39; Univ Wis, MS, 41, PhD(biochem), 43. Prof Exp: Asst, Univ Wis, 40-43; res scientist, Upjohn Co, 46-52; res scientist, Med Serv Corps, US Navy, Md, 52-70, head, Res Dept, US Naval Dent Sch, 69-70; chief, Caries Contract Progs Br, Nat Inst Dent Res, 70-72, chief, Biomat Prog Br, 72-74. Mem: Am Chem Soc; Sigma Xi; Int Asn Dent Res; fel Am Col Dent; Asn Mil Surgeons US. Res: In-vitro dental caries; x-radiation effects; physiology of trace elements in the animal body; caries prevention. Mailing Add: PO Box 88 Lake of the Ozarks Kaiser MO 65047

WACHTELL, GEORGE PETER, b New York, NY, Mar 18, 23; m 50; c 2. PHYSICS. Educ: Princeton Univ, PhD(physics), 51. Prof Exp: Mem staff, Radiation Lab, Mass Inst Technol, 43-45; asst, Princeton Univ, 45-51; PRIN SCIENTIST, ENERGY SYSTS LAB, RES LABS, FRANKLIN INST, PA, 51- Res: Optics; supersonics; heat transfer; fluid dynamics. Mailing Add: Energy Systs Lab Franklin Inst Res Labs Philadelphia PA 19103

WACHTER, RALPH FRANKLIN, b Frederick, Md, Mar 6, 18; m 47; c 3. BIOCHEMISTRY, VIROLOGY. Educ: Univ Notre Dame, BS, 39; Catholic Univ, MS, 41; Purdue Univ, PhD(biochem), 50. Prof Exp: City chemist, Frederick, Md, 41-43; biochemist, US Army Biol Labs, 50-72; RES CHEMIST, RICKETTSIOL DIV, US ARMY MED RES INST INFECTIOUS DIS, 72- Mem: Am Chem Soc; Am Soc Microbiol. Res: Biochemical aspects of virology; virus stabilization and inactivation; biochemical and biological characterization of rickettsiae; rickettsial vaccines. Mailing Add: Rickettsiol Div US Army Med Res Inst Infect Dis Frederick MD 21701

WACHTL, CARL, b Vienna, Austria, Oct 3, 06; nat US. BIOCHEMISTRY. Educ: Univ Tex, BS, 48; Northwestern Univ, PhD(biochem), 53. Prof Exp: Res assoc, Lithographic Tech Found, 48-49; biochemist, Kresge Eye Inst, 53-60; head phys & inorg chem ed, Chem Abstr Serv, Ohio, 60-71; VIS SCHOLAR CHEM,TECHNOL INST, NORTHWESTERN UNIV, EVANSTON, 71- Concurrent Pos: Asst prof, Wayne State Univ. Mem: Fel AAAS; Am Chem Soc; Asn Res Vision & Ophthal. Res: Synthesis of inhibitors of dental caries; permeability of teeth; lens metabolism; culture of the lens of the eye in natural and synthetic media; radiation cataract; amino acid and protein metabolism of ocular lens. Mailing Add: PO Box 1549 Evanston IL 60204

WACHTMAN, JOHN BRYAN, JR, b Conway, SC, Feb 6, 28; m 55. SOLID STATE SCIENCE. Educ: Carnegie Inst Technol, BS, 48, MS, 49; Univ Md, PhD(physics), 61. Prof Exp: Physicist, 51-62, chief phys properties sect, 62-68, CHIEF INORG MAT DIV, NAT BUR STANDARDS, 68- Concurrent Pos: Ed, Ceramics & Glass, 68- & Sci & Technol, 68-; trustee, Edward Orton, Jr Ceramic Found, 70-; mem ceramic eng adv bd, Univ Ill Urbana, 73-76; prog mgr mat, Off Technol Assessment, US Congress, 74-75; mem adv coun ceramics, Univ NY Alfred, 74-77. Honors & Awards: Gold Medal, Dept of Commerce, 71; Sosman Mem Lectr Award, Am Ceramic Soc, 74. Mem: Am Phys Soc; Am Ceramic Soc; Am Soc Testing & Mat; Nat Inst Ceramic Eng; Fedn Mat Socs (secy/treas, 73, pres-elect, 74, pres, 75). Res: Mechanical properties and effective utilization of inorganic materials. Mailing Add: Nat Bur of Standards Washington DC 20234

WACK, JOSEPH PIERRE, anatomy, see 12th edition

WACK, PAUL EDWARD, b Council Bluffs, Iowa, Apr 28, 19; m 52; c 4. NUCLEAR PHYSICS. Educ: Creighton Univ, AB, 41; Univ Notre Dame, MS, 42, PhD(physics), 47. Prof Exp: Asst physics, Univ Notre Dame, 41-43, instr, 43-46; res assoc, Off Naval Res, 46-47; dir dept physics, Creighton Univ, 47-49; from asst prof to assoc prof physics, 49-68, head dept, 66-73, PROF PHYSICS, UNIV PORTLAND, 68- Concurrent Pos: Res assoc, Off Rubber Reserve, 43-45; res assoc, Gen Tire Co, 45-46. Mem: Am Phys Soc; Am Asn Physics Teachers. Res: Electron optics; stress relaxation, low temperature behavior, equation of state and electrical conductivity of natural and synthetic rubbers; nuclear spectroscopy. Mailing Add: Dept of Physics Univ of Portland Portland OR 97203

WACKER, PAUL FREDERICK, b Lancaster, Ohio, May 25, 14; m 39; c 1. ELECTROMAGNETICS. Educ: Ohio State Univ, BA, 36, MA, 39; Catholic Univ, PhD, 54. Prof Exp: Res engr, Battelle Mem Inst, 39-40; tech abstr, Standard Oil Develop Co, 40-42; chemist, Petrol Conversion Corp, 42-44; chemist, 44-51, physicist, 51-58, CONSULT THEORET PHYSICS & APPL MATH, ELECTROMAGNETICS DIV, NAT BUR STANDARDS, 58- Concurrent Pos: Asst, Catholic Univ, 46-47. Honors & Awards: Silver Medal, US Dept of Commerce, 74. Mem: Inst Elec & Electronics Engrs; Antennas & Propagation Soc; Soc Indust & Appl Math; Am Math Soc; Sigma Xi. Res: Theory of near-field measurements of antennas, electroacoustic transducers; spherical scanning and extrapolation techniques; application of representation theory of continuous groups. Mailing Add: Nat Bur of Standards Boulder CO 80302

WACKER, PETER OSCAR, b Orange, NJ, Aug 7, 36; m 62; c 2. CULTURAL GOEGRAPHY, HISTORICAL GEOGRAPHY. Educ: Montclair State Col, BA, 59; La State Univ, MA, 61, PhD(cult geog), 66. Prof Exp: Instr geog, La State Univ, New Orleans, 62-64; from instr to assoc prof geog, 64-72, chmn dept, 73-76, PROF GEOG, RUTGERS UNIV, 72- Concurrent Pos: Guggenheim fel, Rutgers Univ, 71-72, fac fel, 71-72; NJ Hist Soc fel, 72. Honors & Awards: Cert Merit, Am Coun State & Local Hist, 68. Mem: Am Geog Soc; Asn Am Geog. Res: Historical cultural geography of transatlantic movements of populations and culture traits. Mailing Add: Dept of Geog Rutgers Univ New Brunswick NJ 08903

WACKER, WALDON BURDETTE, b Garrison, NDak, Aug 13, 23; m 55; c 4. IMMUNOLOGY, MICROBIOLOGY. Educ: Washington Univ, AB, 49; Univ Mich, MS, 51; Ohio State Univ, PhD(bact), 57. Prof Exp: Res assoc virol, Ohio State Univ, 58-59; asst prof, 59-67, ASSOC PROF MICROBIOL, UNIV LOUISVILLE, 67- Concurrent Pos: NIH career develop award, Univ Louisville, 62-69, NIH res grant, 62- Mem: Asn Res Vision & Ophthal; Sigma Xi. Res: Autoimmune disease; immunopathology; uveitis. Mailing Add: Eye Res Inst Dept Ophthal Univ of Louisville Sch of Med Louisville KY 40202

WACKER, WARREN ERNEST CLYDE, biochemistry, see 12th edition

WACKERLE, JERRY (DONALD), physics, fluid dynamics, see 12th edition

WACKERNAGEL, HANS BEAT, b Basel, Switz, Aug 31, 31; US citizen; div; c 4. ASTRODYNAMICS, DATA PROCESSING. Educ: Univ Basel, PhD(astron), 58. Prof Exp: Observer, Observ Neuchatel, 53-54; res asst, Observ Basel, 55-58; astronr, Proj Spacetrack, Air Force Cambridge Res Labs, Mass, 58-59, 496L Syst Proj Off, 60-61, First Aeorspace Control Squadron, Ent AFB, Colo, 61-62, Ninth Aerospace Defense Div, 62-68 & Fourteenth Aerospace Force, 68-73, comput specialist, Hq NAm Air Defense, 73-75, opers res analyst, Second Commun Squadron, Buckley Air Nat Guard Base, 75, MATHEMATICIAN, HQ N AM AIR DEFENSE, ENT AFB, COLO, 75- Concurrent Pos: Lectr, Univ Colo, Colorado Springs Ctr, 64-70. Mem: Am Astron Soc; fel Brit Interplanetary Soc; Swiss Astron Soc. Res: Design and evaluation of advanced space defense systems; applied celestial mechanics. Mailing Add: 2939 Country Club Dr Colorado Springs CO 80909

WACKMAN, PETER HUSTING, b Cleveland, Ohio, June 16, 28; m 51; c 3. THEORETICAL PHYSICS. Educ: Univ Wis, BS, 51, MS, 53; Univ Pittsburgh, PhD(physics), 60. Prof Exp: Sr scientist, Bettis Atomic Power Lab, Westinghouse Elec Corp, 53-60; staff scientist, A-C Spark Plug Div, Gen Motors Corp, 60-61; ASSOC PROF MECH ENG, MARQUETTE UNIV, 61- Concurrent Pos: Mem staff, McGraw Edison Power Syst, 69- Mem: Am Phys Soc; Am Nuclear Soc. Res: Nuclear structure; low energy nuclear physics; nuclear reactor physics; inertial guidance systems; radiation effects; materials science. Mailing Add: Dept of Mech Eng Marquette Univ Milwaukee WI 53233

WADA, JAMES YASUO, b Lomita, Calif, May 15, 34; m 57; c 4. LASERS, PLASMA PHYSICS. Educ: Univ Calif, Los Angeles, BS, 56; Univ Southern Calif, MS, 58, PhD(elec eng), 63. Prof Exp: Sect head, Elec-Gasdyn Lasers, Hughes Res Labs, Malibu, 56-74, MEM STAFF, HUGHES SPACE-COMMUN GROUP, LOS ANGELES, 74- Concurrent Pos: Lectr, Univ Southern Calif, 64, asst prof, 64-66. Mem: Am Phys Soc; Inst Elec & Electronics Eng. Res: High power lasers and optics; physical optics; microwave tubes; electromagnetic theory. Mailing Add: Hughes Space-Comm GP Box 92919 Bldg 366 MS 720 Los Angeles CA 90009

WADA, JUHN A, b Tokyo, Japan, Mar 28, 24; nat Can; m 56. MEDICINE. Educ: Hokkaido Imp Univ, Japan, MD, 45, DMedSci, 51; FRCPS(C), 72. Prof Exp: Asst prof neurol & psychiat, dir labs exp neurol & brain surgeon-in-chief, Univ Hosps, Hokkaido Imp Univ, Japan, 52-57; res assoc neurol, 57-59, asst prof neurol res & psychiat & chief labs EEG & neurophysiol, 60-63, assoc prof med neurol & dir EEG labs, 63-70, PROF NEUROL SCI, UNIV BC & DIR EEG DEPT, HEALTH SCI CTR HOSP, 70- Concurrent Pos: Fel, Univ Minn, 54-55 & Montreal Neurol Inst, McGill Univ, 55-56; Can Med Res Coun assoc, 66; attend neurologist & assoc dir EEG dept, Vancouver Gen Hosp. Mem: Am Electroencephalog Soc; Am Epilepsy Soc; fel Am Acad Neurol; Can Neurol Soc; Can Soc Electroencephalog. Res: Neurological mechanism of human behavior; epilepsy; electrical activity of brain; cerebral speech function. Mailing Add: Health Sci Ctr Hosp Univ of BC Vancouver BC Can

WADA, WALTER W, b Loomis, Calif, Feb 26, 19; m 46; c 4. PHYSICS. Educ: Univ Utah, BA, 43; Univ Mich, MA, 46, PhD, 51. Prof Exp: Physicist nucleonics div, US Naval Res Lab, 51-66; PROF PHYSICS, OHIO STATE UNIV, 66- Concurrent Pos: Lectr, Univ Md, 51-62; vis prof, Northwestern Univ, 62-64. Mem: Am Phys Soc. Res: Quantum theory of fields and applications in electrodynamics; theoretical high energy physics. Mailing Add: Dept of Physics Ohio State Univ Columbus OH 43210

WADDELL, CHARLES NOEL, b Omaha, Nebr, Nov 11, 22; m 45; c 6. PHYSICS. Educ: Univ Calif, BA, 50, PhD(physics), 58. Prof Exp: Assoc physics, Univ Calif, 50-52, physicist, Radiation Lab, 52-58; asst prof, 58-65, ASSOC PROF PHYSICS, UNIV SOUTHERN CALIF, 65- Mem: Am Phys Soc. Res: Nuclear reaction mechanisms and nuclear structure; nucleon-nucleon interactions; (p, 2p) and knock-out reactions. Mailing Add: Dept of Physics Univ of Southern Calif University Park Los Angeles CA 90007

WADDELL, ERIC WILSON, b Baltimore, Md, July 8, 39; Can citizen; m 63; c 3. CULTURAL GEOGRAPHY, ANTHROPOLOGY. Educ: Oxford Univ, BA, 61; McGill Univ, MA, 63; Australian Nat Univ, PhD(human geog), 69. Prof Exp: Res asst New Guinea res unit, Australian Nat Univ, 63-65; asst prof, 69-74, ASSOC PROF GEOG, MCGILL UNIV, 74- Concurrent Pos: Vis assoc prof geog, Univ Hawaii, 73; vis fel, Australian Nat Univ, 73. Mem: Can Asn Geog; fel Am Anthrop Asn. Res: Cultural ecology of tropical agricultural systems, particularly in Melanesia; maritime communities in Eastern Canada; problems of ethnic boundaries; inter-ethnic relations; minority groups. Mailing Add: Dept of Geog McGill Univ PO Box 6070 Sta A Montreal PQ Can

WADDELL, HENRY THOMAS, b Wilson, Ark, Apr 19, 18; m 45; c 2. BOTANY. Educ: Peabody Col, BS, 49, MA, 51; Univ Fla, PhD(plant path), 59. Prof Exp: Asst prof biol, Martin Br, Univ Tenn, 49-56; assoc prof, Peabody Col, 59-63; PROF BIOL, LAMAR UNIV, 63- Mem: AAAS; Am Phytopath Soc; Mycol Soc Am. Res: Plant Pathology; mycology. Mailing Add: Dept of Biol Lamar Univ Beaumont TX 77710

WADDELL, JAMES, b Hamilton, Scotland, Aug 29, 98; nat US; m 26; c 3. NUTRITION. Educ: Univ Sask, BSA, 20; Iowa State Col, MS, 22; Univ Wis, PhD(biochem nutrit), 26. Prof Exp: Exten specialist, Univ Sask, 20-21; asst chief dairy farm sect & instr dairy husb, Iowa State Col, 22-23; res assoc, Univ Wis, 26-30; dir biol lab, Acetol Prod, Inc, E I du Pont de Nemours & Co, Del, 30-52, mgr nutrit sect, Stine Lab, 52-53, mgr tech develop, animal indust & nutrit sect, indust & biochem dept, 53-63; exec secy, Am Inst Nutrit, Md, 65-71; consult, 71-75; RETIRED. Mem: Fel, Am Inst Nutrit. Res: Calcium and phosphorus metabolism; unidentified growth factors for chicks; copper as an anti-anemic substance; synthetic amino acids as feed supplements; identification of pro-vitamin D steroids. Mailing Add: 907 Centre Rd Wilmington DE 19807

WADDELL, MATHEWS CARY, b St Paul, Minn, Aug 16, 16; m 48. MATHEMATICS. Educ: Hamilton Col, BS, 38; Univ Minn, MA, 40; Johns Hopkins Univ, PhD(math), 50. Prof Exp: Asst prof, Western Reserve Univ, 50-51; MEM PRIN PROF STAFF, APPL PHYSICS LAB, JOHNS HOPKINS UNIV, 51- Mem: Am Math Soc; Opers Res Soc Am. Mailing Add: Appl Physics Lab Johns Hopkins Univ 8621 Georgia Ave Silver Spring MD 20910

WADDELL, ROBERT CLINTON, b Mattoon, Ill, Aug 15, 21; m 60; c 3. PHYSICS. Educ: Eastern Ill Univ, BS, 47; Univ Ill, MS, 48; Iowa State Univ, PhD(physics), 55. Prof Exp: From instr to assoc prof, 48-60, PROF PHYSICS, EASTERN ILL UNIV, 60- Concurrent Pos: Res asst, Iowa State Univ, 53-55. Mem: Am Asn Physics

Teachers; Am Phys Soc. Res: Low-energy nuclear physics. Mailing Add: Dept of Physics Eastern Ill Univ Charleston IL 61920

WADDELL, THOMAS GROTH, b Madison, Wis, July 29, 44; m 67. BIO-ORGANIC CHEMISTRY. Educ: Univ Wis-Madison, BS, 66; Univ Calif, Los Angeles, PhD(org chem), 69. Prof Exp: Scholar org chem, Univ Calif, Los Angeles, 69; NIH res fel, Univ Calif, Berkeley, 70-71; ASST PROF CHEM, UNIV TENN, CHATTANOOGA, 71- Mem: Am Chem Soc. Res: Chemical plant taxonomy; chemical constituents of medicinal plants; synthesis of naturally occurring drugs. Mailing Add: Dept of Chem Univ of Tenn Chattanooga TN 37401

WADDELL, WALTER HARVEY, b Chicago, Ill, Sept 26, 47; m 73. PHOTOCHEMISTRY. Educ: Univ Ill, Chicago, BS, 69; Univ Houston, PhD(chem), 73. Prof Exp: Res assoc chem, Columbia Univ, 73-75; ASST PROF CHEM, CARNEGIE-MELLON UNIV, 75- Concurrent Pos: NIH res fel, Nat Eye Inst, 75. Mem: Am Chem Soc; Am Soc Photobiol. Res: Spectroscopic and photochemical investigations of the protein-chromophore interactions in the visual protein rhodopsin and the photo-oxidation mechanism of polymers using magnetic resonant and pulsed laser excitation techniques. Mailing Add: Dept of Chem Mellon Inst Sci Carnegie-Mellon Univ Pittsburgh PA 15213

WADDELL, WILLIAM JOSEPH, b Commerce, Ga, Mar 16, 29; m 51; c 4. PHARMACOLOGY. Educ: Univ NC, AB, 51, MD, 55. Prof Exp: From asst prof to assoc prof pharmacol, Univ NC, Chapel Hill, 58-71, assoc prof oral biol, 67-69, prof oral biol & assoc dir dent res ctr, 69-72, assoc div dir, Ctr Res Pharmacol & Toxicol, 66-67; PROF PHARMACOL, UNIV KY, 72- Concurrent Pos: USPHS res fel, Univ NC, Chapel Hill, 55-58; NIH spec fel, Royal Vet Col, Sweden, 65-66. Mem: Teratology Soc; Int Soc Quantum Biol; Am Soc Pharmacol & Exp Therapeut; Soc Exp Biol & Med; Am Physiol Soc. Res: Intracellular pH; teratogenic agents. Mailing Add: Dept of Pharmacol Univ of Ky Lexington KY 40506

WADDELL, WILLIAM RHOADS, b Ft Smith, Ark, Oct 12, 18; m 44; c 4. SURGERY. Educ: Univ Ariz, BS, 40; Harvard Univ, MD, 43; Am Bd Surg, dipl, 54; Am Bd Thoracic Surg, dipl, 55. Prof Exp: From instr to asst clin prof surg, Harvard Med Sch, 52-61; chmn dept, 61-72, PROF SURG, MED CTR, UNIV COLO, DENVER, 61- Concurrent Pos: USPHS res grant, 64-72; asst, Mass Gen Hosp, 52-54, asst surgeon, 55-58, assoc vis surgeon, 58-61; mem courtesy staff, Faulkner Hosp, 53-54, assoc staff, Surg Serv, 55-61; consult, Mass Eye & Ear Infirmary, 56-61 & Vet Admin Hosp, Grand Junction & Denver, Colo, 61-; mem active staff, Denver Gen Hosp, 61-72; consult, Mercy Hosp, 68- & Gen Rose Mem Hosp, 70-; Glover H Copher vis lectr, Sch Med, Washington Univ, 72. Mem: AAAS; Am Asn Thoracic Surg; Am Thoracic Soc; fel Am Col Surgeons; Am Surg Asn. Res: Surgical physiology; physiological and clinical research on lipid metabolism, gastric physiology, transplantation and clinical surgery; cancer research surgery; enzyme inhibition synthesis. Mailing Add: 5745 E Sixth Ave Denver CO 80220

WADDEY, WALTER EDWIN, organic chemistry, see 12th edition

WADDILL, VAN HULEN, b Brady, Tex, Aug 24, 47; m 69. ENTOMOLOGY. Educ: Tex A&M Univ, BS, 70, MS, 71; Clemson Univ, PhD(entom), 74. Prof Exp: ASST PROF ENTOM, AGR RES & EDUC CTR, INST FOOD & AGR SCI, UNIV FLA, 75- Mem: Sigma Xi; Entom Soc Am; Int Orgn Biol Control. Res: Management of insect pests of vegetables. Mailing Add: Univ of Fla Agr Res & Educ Ctr 18905 SW 280th St Homestead FL 33030

WADDINGTON, CECIL JACOB, b Cambridge, Eng, July 6, 29; m 56. PHYSICS, ASTROPHYSICS. Educ: Bristol Univ, BSc, 52, PhD(physics), 55. Prof Exp: Royal Soc McKinnon res studentship physics, Bristol Univ, 56-59, lectr, 59-62; assoc prof, 62-68, PROF, SCH PHYSICS & ASTRON, UNIV MINN, MINNEAPOLIS, 68- Concurrent Pos: Res assoc & lectr, Univ Minn, 57-58; Nat Acad Sci sr fel, Goddard Space Flight Ctr, Md, 61; sr vis fel, Imp Col, Univ London, 72-73; mem, Cosmic Ray Comn, Int Union Pure & Appl Physics, 72- Mem: AAAS; fel Am Phys Soc; Am Geophys Union; Am Astron Soc. Res: Nature and properties of the primary cosmic radiation, particularly electronic and nuclear emulsion detectors, charge and mass composition. Mailing Add: Sch of Physics & Astron Univ of Minn Minneapolis MN 55455

WADDINGTON, DONALD VAN PELT, b Norristown, Pa, Dec 31, 31; m 55; c 6. SOIL FERTILITY, PLANT SCIENCE. Educ: Pa State Univ, BS, 53; Rutgers Univ, MS, 60; Univ Mass, PhD(agron), 64. Prof Exp: Asst chemist, Eastern States Farmer's Exchange, Inc, 56-57; instr agron, Univ Mass, 60-65; asst prof soil technol, 65-68, assoc prof soil sci, 68-75, PROF SOIL SCI, PA STATE UNIV, UNIVERSITY PARK, 75- Mem: Am Soc Agron; Soil Sci Soc Am; Int Soil Sci Soc; Soil Conserv Soc; Int Turfgrass Soc. Res: Soil physical properties, especially soil modification for turfgrass; turfgrass nutrition. Mailing Add: RD 1 Box 231 Boalsburg PA 16827

WADDINGTON, JOHN T, plant physiology, horticulture, see 12th edition

WADDLE, BILLY MACK, plant breeding, see 12th edition

WADDLE, BRADFORD AVON, b Tex, Jan 26, 20; m 45. AGRONOMY, PLANT BREEDING. Educ: Agr & Mech Col, Tex, BS, 42, MS, 50; Purdue Univ, PhD(plant breeding), 54. Prof Exp: Instr, Hunt County Voc Schs, Tex, 46-47; jr agronomist, Greenville Cotton Sta, USDA, 48; instr cotton breeding, Agr Exp Sta, Univ Tex, 50; asst agron, Agr Exp Sta, 51-56, assoc, 56-59, PROF AGRON & ALTHEIMER CHAIR COTTON RES, UNIV ARK, 59- Mem: Am Soc Agron; Am Genetic Asn. Res: Cotton breeding and genetics, especially breeding for resistance to disease and insects. Mailing Add: Dept of Agron Univ of Ark Fayetteville AR 72701

WADDLE, HOWARD MEFFERT, b Americus, Kans, Sept 13, 05; m; c 2. ORGANIC CHEMISTRY. Educ: Baker Univ, AB, 28; Univ Colo, AM, 30; Univ Wis, PhD(org chem), 40. Prof Exp: Asst chem, Univ Colo, 28-30; from instr to prof, Ga Tech Col, 30-44; head chem dept, 44-66, ACTG DIR RES, RES CTR, WEST POINT-PEPPERELL, INC, 67- Concurrent Pos: Asst, Univ Wis, 37-38. Mem: Am Chem Soc; Am Asn Textile Chemists & Colorists. Res: Natural products; abietic acid; chlorination of esters; chemistry as applied to textile fiber finishes; cellulose degradation; modification of cellulose; polymerization; resins; cross-linking reactions. Mailing Add: 410 E Third St West Point GA 31833

WADE, ADELBERT ELTON, b Hilliard, Fla, Apr 29, 26; m 50; c 2. PHARMACOLOGY, BIOCHEMISTRY. Educ: Univ Fla, BS, 54, MS, 56, PhD(pharmacol), 59. Prof Exp: Asst chemother, Univ Fla, 54-56, asst biochem, 56-57, asst chemother, 57-59; from asst prof to assoc prof, 59-67, PROF PHARMACOL, UNIV GA, 67-, HEAD DEPT, 68- Mem: Am Asn Cols Pharm; Soc Exp Biol & Med; Am Soc Pharmacol & Exp Therapeut; Int Soc Biochem Pharmacol. Res: Immunology and chemotherapy of Dictyocaulus viviparus; mechanism of drug metabolism; effects of diet on drug metabolism. Mailing Add: Dept of Pharmacol Univ of Ga Sch of Pharm Athens GA 30602

WADE, CAMPBELL MARION, b Elizabethtown, Ky, Nov 25, 30; m 56; c 4. ASTRONOMY. Educ: Harvard Univ, AB, 54, AM, 55, PhD(astron), 57. Prof Exp: Res officer, Div Radiophysics, Commonwealth Sci & Indust Res Orgn, Australia, 57-59; res assoc, 60-66, SCIENTIST, NAT RADIO ASTRON OBSERV, 66- Concurrent Pos: Adv ed, Soviet Astron, Am Inst Physics, 69- Mem: Am Astron Soc; Int Astron Union; Int Union Radio Sci. Res: Galactic and extragalactic radio astronomy. Mailing Add: Nat Radio Astron Observ Edgemont Rd Charlottesville VA 22901

WADE, CHARLES GORDON, b Griggsville, Ill, Apr 5, 37; m 64. PHYSICAL CHEMISTRY. Educ: Southern Ill Univ, BA, 60; Mass Inst Technol, PhD(phys chem), 65. Prof Exp: Res assoc chem, Enrico Fermi Inst Nuclear Studies, Univ Chicago, 65-67; asst prof, 67-73, ASSOC PROF CHEM, UNIV TEX, AUSTIN, 73- Mem: Am Phys Soc; Am Chem Soc; Sigma Xi; AAAS. Res: Nuclear magnetic resonance relaxation; electron spin resonance of photo-excited triplet states of aromatic molecules; transport properties of fluids; properties of liquid crystals; structure of membrane systems. Mailing Add: Dept of Chem Univ of Tex Austin TX 78712

WADE, CLARENCE W R, b Laurinburg, NC, Mar 31, 27; m 55; c 1. ORGANIC CHEMISTRY. Educ: J C Smith Univ, BS, 48; Tuskegee Inst, MS, 50; Georgetown Univ, PhD(org chem), 65. Prof Exp: From instr to asst prof chem, St Augustine's Col, 50-57; from chemist to res chemist, Nat Bur Stand, 57-66; res chemist, US Army Med & Biomech Lab, Walter Reed Army Med Ctr, 66-68, chief, Synthesis Br, 68-70, chief mat & applns div, 70-72, CHIEF, MAT & APPLNS DIV, US ARMY MED & BIOMECH RES LAB, FT DETRICK, MD, 72- Concurrent Pos: Consult, Nat Heart Inst, 68-70. Mem: AAAS; Am Chem Soc. Res: Development of inert or degradable implant materials, tissue and bone adhesives, sutures, tendons, vascular tubes, wound and burn dressings, bone repair polymers; mechanisms of implant degradation. Mailing Add: US Army Med & Biomech Res Lab Ft Detrick Frederick MD 21701

WADE, DALE A, b Buffalo, SDak, May 23, 28; m 53; c 5. WILDLIFE MANAGEMENT. Educ: SDak State Univ, BS, 69, PhD(animal sci), 72. Prof Exp: Mem staff mammal control, US Fish & Wildlife Serv, 62-65; wildlife specialist, Colo State Univ, 72-74; WILDLIFE SPECIALIST, UNIV CALIF, DAVIS, 74- Concurrent Pos: Consult, US Environ Protection Agency, 70-75; mem, Eisenhower Consortium, 73-74. Mem: Sigma Xi; AAAS; Am Inst Biol Sci; Soc Range Mgt; Wildlife Soc. Res: Evaluation of biological, economic conflicts and possible solutions in human, wildlife and agricultural relationships. Mailing Add: Dept of Animal Physiol Univ of Calif Davis CA 95616

WADE, DAVID ROBERT, b London, Eng, May 25, 39; m 62; c 3. BIOCHEMISTRY. Educ: Univ Cambridge, BA & MA, 63, PhD(biochem), 67. Prof Exp: Res assoc physiol, Col Med, Pa State Univ, 67-69; Bank Am Giannini fel biochem, Sch Med, Univ Calif, Davis, 69-71; USPHS grant metab regulation & asst prof physiol, Col Med, Pa State Univ, 71-74; ASSOC PROF PHYSIOL, SCH MED, SOUTHERN ILL UNIV, 74- Res: Regulation of melanin synthesis. Mailing Add: Dept of Physiol Southern Ill Univ Sch of Med Carbondale IL 62901

WADE, EARL KENNETH, b Toledo, Iowa, July 13, 14; m 47; c 3. PLANT PATHOLOGY. Educ: Univ Wis, BS, 38, MS, 50. Prof Exp: Instr high sch, Wis, 38-42; asst potato cert serv, 46-50, EXTEN PLANT PATHOLOGIST, UNIV WIS-MADISON, 50-, PROF PLANT PATH, 69- Mem: Am Phytopath Soc; Am Potato Asn. Res: Vegetable and fruit diseases. Mailing Add: Dept of Plant Path Univ of Wis Madison WI 53706

WADE, FRANKLIN ALTON, b Akron, Ohio, Feb 5, 03; m 38; c 1. EXPLORATION GEOLOGY. Educ: Kenyon Col, BS, 25, AM, 26; Johns Hopkins Univ, PhD(geol), 37. Hon Degrees: DSc, Kenyon Col, 62. Prof Exp: Instr geol & chem, Univ Del, 29-31; geologist, Byrd Antarctic Exped II, 33-35; asst prof geol, Miami Univ, 36-39; sr scientist, US Antarctica Serv, 39-41; from asst prof to prof geol, Tex Tech Univ, 46-73, head dept, 54-64, RES ASSOC & DIR, ANTARCTICA RES CTR, MUS, TEX TECH UNIV, 73- Concurrent Pos: Dir studies, Camp Norton, Wyo, 47-54; chief opers anal off, US Dept Air Force, Japan & Korea, 50-51; party leader, Antarctic geol explor, 62-63 & 64-65; sr scientist, Byrd Coast Surv Party, Antarctica, 66-68. Honors & Awards: Congressional Medal, 37 & 45; Meritorious Civilian Serv Award, US Dept Air Force, 51. Mem: Fel Geol Soc Am; Mineral Soc Am; Nat Asn Geol Teachers; Am Geophys Union; Am Polar Soc (pres, 68-). Res: Antarctic geology; igneous and metamorphic petrology; Circum-Pacific Map Project; geologic, tectonic, mineral resources, energy resources. Mailing Add: The Museum Texas Tech Univ Lubbock TX 79409

WADE, GEORGE WESLEY, b Nashville, Tenn, Sept 3, 10; m 46; c 2. ORAL SURGERY. Educ: Ind Univ, AB, 35; Butler Univ, MA, 40; Howard Univ, DDS, 47. Prof Exp: Teacher high sch, Ind, 35-42; asst prof endodontia & oral diag, Sch Dent, Howard Univ, 47-59, from asst prof to prof oral med, 59-68; DENT OFFICER, FOOD & DRUG ADMIN, ROCKVILLE, MD, 68- Mem: AAAS; Am Asn Endodont; NY Acad Sci. Res: Endodontia; research in areas of wound healing and apical pathology following traumatic pulpal injuries. Mailing Add: 5029 Illinois Ave NW Washington DC 20011

WADE, JAMES JOSEPH, b St Paul, Minn, Jan 7, 46; m 70; c 1. MEDICINAL CHEMISTRY. Educ: Col St Thomas, BA, 68; Univ Minn, PhD(org chem), 72. Prof Exp: NIH fel org chem, Univ Rochester, 72-73; SR MED CHEMIST, RIKER LABS, 3M CO, 73- Mem: Am Chem Soc. Res: Design and synthesis of organic compounds for possible medicinal use, particularly in the antiallergy and antithrombotic areas. Mailing Add: Riker Labs 3M Co 3M Ctr Bldg 218-1 St Paul MN 55101

WADE, LEO, JR, radiation biophysics, see 12th edition

WADE, LEO J, internal medicine, oncology, deceased

WADE, LUTHER IRWIN, b Dallas, Tex, Nov 27, 16; m 36; c 6. MATHEMATICS. Educ: Duke Univ, AB, 38, PhD(math), 41. Prof Exp: Instr math, Johns Hopkins Univ, 41-42; Nat Res Coun fel, Inst Advan Study, 42-43; from instr to asst prof, Duke Univ, 43-48; PROF MATH & HEAD DEPT, LA STATE UNIV, BATON ROUGE, 48- Mem: Am Math Soc; Math Asn Am. Res: Number theory of polynomials in finite fields; abstract algebra. Mailing Add: Dept of Math La State Univ Baton Rouge LA 70803

WADE, NELSON JOHN, biology, deceased

WADE, PETER CAWTHORN, b Washington, DC, Feb 15, 44; m 66; c 1. MEDICINAL CHEMISTRY, ORGANIC CHEMISTRY. Educ: Middlebury Col, AB, 66; Univ Wash, PhD(org chem), 71. Prof Exp: RES SCIENTIST, SQUIBB INST MED RES, 71- Mem: Am Chem Soc; AAAS. Res: Anxiolytic, antidepressive, neuroleptic and anti-inflammatory agents; heterocyclic chemistry. Mailing Add: Squibb Inst for Med Res PO Box 4000 Princeton NJ 08540

WADE, RICHARD ARCHER, b Fitchburg, Mass, Aug 16, 30; m 70; c 2.

BIOLOGICAL OCEANOGRAPHY. Educ: Univ Miami, BS, 56, MS, 62, PhD(biol oceanog), 68. Prof Exp: Marine scientist, Ayerst Labs, Div Am Home Prod Corp, 66-68; head dept ecol & pollution, Va Inst Marine Sci, 68-69; chief lab, Environ Protection Agency, Fed Water Qual Admin, 68-69 & 70-71; exec secy, Sport Fishing Inst, 71-72; exec dir, Am Fisheries Soc, 72-75; MARINE ECOLOGIST, US FISH & WILDLIFE SERV, 75- Concurrent Pos: Consult, NIH Pesticide Proj, Univ Miami, 66-68; mem, Water Qual Mgt Comt, US Govt Interagency Group, 68-69; mem res subcomt, Fed Comt Pest Control, 68; clin res assoc, Med Univ SC, 70-71; mem, Subcomt Marine Water Qual Criteria, Nat Acad Sci, 71; treas, Sport Fishing Res Found, 71-72. Mem: Am Fisheries Soc; Am Soc Ichthyol & Herpet; Gulf & Caribbean Fisheries Inst; Marine Technol Soc. Res: Coastal ecosystems of the United States, including dredge disposal, offshore oil and gas development, development of deepwater ports, power plant construction and operation; marine and estuarine water quality problems. Mailing Add: Nat Space Technol Labs US Fish & Wildlife Serv Bay St Louis MS 39520

WADE, ROBERT CHARLES, b Lakewood, Ohio, May 18, 20; m 44; c 8. INORGANIC CHEMISTRY. Educ: Oberlin Col, AB, 42. Prof Exp: Res chemist, E I du Pont de Nemours & Co, 42-44 & 46-49; res chemist & group leader, Nat Distillers Prod Corp, 49-53; asst dir res, Metal Hydrides, Inc, 53-59, mgr tech mkt, 59-67, mgr explor develop, 67-68; mgr explor res, 68-74, SR SCIENTIST, VENTRON CORP, 68- Mem: Am Chem Soc; Am Inst Chemists; Catalysis Soc; Tech Asn Pulp & Paper Indust. Res: Inorganic and organic investigations on sodium metal; reduction of metal chlorides; metal-hydrogen systems; sodium and alumino hydrides; borohydride chemistry; organometallic chemistry; pulp and pulp bleaching; catalysts. Mailing Add: 61 S Village Green Ipswich MA 01938

WADE, ROBERT HAROLD, b Opportunity, Wash, Sept 16, 20; m 44; c 2. ORGANIC CHEMISTRY, POLYMER CHEMISTRY. Educ: Univ Wash, BS, 46, PhD(chem), 51. Prof Exp: Res chemist, M W Kellog Co div, Pullman, Inc, 51-57; org chemist & proj leader, Stanford Res Inst, 57-63; SR RES SCIENTIST, US NAVAL UNDERSEA CTR, 63- Mem: Am Chem Soc. Res: Synthesis of polynuclear aromatic compounds; high temperature metal-chelate polymers; physical and chemical fate of fluoride in plants; synthesis and properties of water soluble and friction reducing polymers; marine natural products. Mailing Add: US Naval Undersea Ctr San Diego CA 92132

WADE, ROBERT SIMSON, b Gorrie, Ont, Aug 20, 20; m 43; c 3. CHEMISTRY, RESEARCH ADMINISTRATION. Educ: Univ Western Ont, BA, 42, MA, 43. Prof Exp: Res chemist, Imp Oil, Ltd, 43-47; chief chemist, Imp Tobacco Co Can, Ltd, 47-53, mgr lab, 53-57, mgr res develop & tech serv, 57-69, MGR RES & DEVELOP, IMP TOBACCO LTD, 69- Mem: Chem Inst Can; Can Res Mgt Asn. Res: Growing, processing and manufacturing of tobacco and tobacco products; development of processes and products; technology of tobacco and tobacco smoke. Mailing Add: Res & Develop Dept Imp Tobacco Ltd 3810 Antoine St Montreal PQ Can

WADE, THOMAS LEONARD, JR, b Ridgeway, Va, Apr 21, 05; m 31; c 4. MATHEMATICS. Educ: Univ Va, BS, 29, MS, 30, PhD(math), 33. Prof Exp: Asst instr math, Univ Va, 29-34; prof, Mercer Univ, 34-39; asst prof, Univ Ala, 39-43; prof, 43-75, head dept, 43-64, EMER PROF MATH, FLA STATE UNIV, 75- Mem: Am Math Soc; Math Asn Am. Res: Algebraic invariants; tensor algebra. Mailing Add: 1003 Washington St Tallahassee FL 32303

WADE, WILBERT ERNEST, natural science, botany, see 12th edition

WADE, WILLIAM FRANK, b Mansfield, Ark, Feb 25, 38. VERTEBRATE ZOOLOGY, ICHTHYOLOGY. Educ: Okla State Univ, BS, 61, PhD(zool), 68; Univ Okla, MS, 64. Prof Exp: Res biologist & adj asst prof, Univ Okla, 68-69; ASSOC PROF GEN BIOL, WILDLIFE CONSERV, MAMMAL, ICHTHYOL & HERPET & DIR FISHERIES & APPL AQUACULT, SOUTHEASTERN STATE COL, 69- Mem: Am Fisheries Soc. Res: Wildlife conservation; fisheries management; pollution investigation. Mailing Add: Dept of Biol Southeastern State Col Durant OK 74701

WADE, WILLIAM H, b San Antonio, Tex, Nov 3, 30; m 51. PHYSICAL CHEMISTRY. Educ: St Mary's Univ, Tex, BS, 51; Univ Tex, PhD(chem), 55. Prof Exp: Res scientist, Univ Calif, Berkeley, 55-58; res scientist, 58-61, from asst prof to assoc prof, 61-72, PROF CHEM, UNIV TEX, AUSTIN, 72- Mem: Am Chem Soc. Res: Surface chemistry; electrode kinetics; catalysis; nuclear scattering processes. Mailing Add: Dept of Chem Univ of Tex Austin TX 78712

WADE, WILLIAM HOWARD, b Stoughton, Wis, Apr 18, 23; m 43; c 1. ENTOMOLOGY. Educ: Univ Calif, BS, 50, PhD(entom), 56. Prof Exp: Res asst, Univ Calif, 50-53; mgr tech serv & prod promotion, Agr Chem Div, 53-72, mgr develop, 72-75, MGR TECH SERV, AGR CHEM DIV, FMC CORP, 75- Mem: Entom Soc Am; Sigma Xi. Res: Insect biology; field evaluation of pesticides. Mailing Add: 214 W Andrews Fresno CA 93705

WADE, WILLIAM RAYMOND, II, b Los Angeles, Calif, Oct 28, 43; m 65; c 2. MATHEMATICS. Educ: Univ Calif, Riverside, BA, 65, MA, 66, PhD(math), 68. Prof Exp: Asst prof, 68-72, ASSOC PROF MATH, UNIV TENN, KNOXVILLE, 72- Concurrent Pos: Consult, Oak Ridge Nat Lab, 69- Mem: Am Math Soc; Math Asn Am. Res: Fourier analysis on groups; Haar and Walsh series; sets of uniqueness; transform theory. Mailing Add: Dept of Math Univ of Tenn Knoxville TN 37916

WADEHRA, INDERJIT LAL, polymer chemistry, physical chemistry, see 12th edition

WADELIN, COE WILLIAM, b Dover, Ohio, Aug 18, 27; m 50; c 1. ANALYTICAL CHEMISTRY. Educ: Mt Union Col, BS, 50; Purdue Univ, MS, 51, PhD, 53. Prof Exp: Res chemist, 53-65, sect head anal chem, Res Div, 65-75, SECT HEAD SPECTROS, RES DIV, GOODYEAR TIRE & RUBBER CO, AKRON, OHIO, 75- Concurrent Pos: Fel, Ctr Advan Eng Study, Mass Inst Technol, 68-69. Mem: Am Chem Soc. Res: Analysis of polymers and organic chemicals; absorption spectroscopy. Mailing Add: 2365 17th St Cuyahoga Falls OH 44223

WADELL, LYLE H, b Elsie, Mich, Mar 7, 34; m 57; c 4. ANIMAL BREEDING. Educ: Mich State Univ, BS, 55, MS, 57; Iowa State Univ, PhD(animal breeding, statist, genetics), 59. Prof Exp: Res assoc animal breeding res, 59-60, res animal geneticist, 60-61, admin supvr comput ctr mgt, 61-66, DIR COMPUT CTR MGT, CORNELL UNIV, 66- Mem: Am Dairy Sci Asn; Am Soc Animal Sci. Res: Computing center management; data processing techniques; statistics. Mailing Add: 468 Auburn Rd Groton NY 13073

WADEY, WALTER GEOFFREY, b Whangarei, NZ, Sept 9, 18; nat US; m 45; c 3. PHYSICS. Educ: Univ Mich, BSc, 41, MA, 42, PhD(physics), 47. Prof Exp: Res assoc, Radio Res Lab, Harvard Univ, 43-45; instr physics, Yale Univ, 47-50, asst prof, 50-56; prof, Southern Ill Univ, 56-57; mgr sci prog, Remington Rand Univac Div, Sperry Rand Corp, 57-58, tech coordr, 59; physicist, Hughes Aircraft Co, 59-60; mgr advan electromech develop dept, Univac Div, Sperry Rand Corp, 60-62; chief scientist, Bowles Eng Corp, 62-63 & Wash Tech Assocs, 63-64; SR SCIENTIST, OPERS RES, INC, SILVER SPRING, MD, 64- Mem: AAAS; Am Phys Soc; Asn Comput Mach; Opers Res Soc Am; Marine Technol Soc. Res: Experimental nuclear physics; nuclear spectroscopy; linear electron accelerators; electronics; computer programming and arithmetics; fluid mechanics; electromechanical design; fluid-amplifier technology; operations research; systems analysis; anti-submarine warfare; information systems. Mailing Add: 7505 Holiday Terrace Bethesda MD 20034

WADKE, DEODATT ANANT, b July 7, 38; Indian citizen; m 67; c 1. PHYSICAL PHARMACY. Educ: Banaras Hindu Univ, BPharm, 61; Ohio State Univ, MS, 63; State Univ NY Buffalo, PhD(pharmaceut), 67. Prof Exp: Res assoc formulations, Merck Sharpe & Dohme Res Labs, 66-69; res investr pharmaceut res & develop, 69-71, sr res investr, 72-73, HEAD, PREFORMULATION STUDIES SECT, SQUIBB INST MED RES, 73- Mem: Am Pharmaceut Asn; Acad Pharmaceut Sci. Res: Thermodynamics of dissolution, solubilization and absorption; dissolution of polyphase systems; drug stability; pharmacokinetics. Mailing Add: Squibb Inst for Med Res New Brunswick NJ 08903

WADKINS, CHARLES LEROY, b Joplin, Mo, May 8, 29; m 52; c 2. BIOCHEMISTRY. Educ: Univ Kans, AB, 51, PhD(biochem), 56. Prof Exp: Instr biochem, Univ Kans, 55-56; from instr to assoc prof, Sch Med, Johns Hopkins Univ, 57-66; PROF BIOCHEM & CHMN DEPT, MED CTR, UNIV ARK, LITTLE ROCK, 66- Concurrent Pos: Fel biochem, Sch Med, Johns Hopkins Univ, 56-57; USPHS sr res fel, 59-64; mem adv panel, NSF. Mem: Am Chem Soc; Am Soc Biol Chemists; Brit Biochem Soc. Res: Biological oxidation reactions; oxidative phosphorylation; mechanism and control of biological calcification reactions. Mailing Add: Dept of Biochem Univ of Ark Med Ctr Little Rock AR 72201

WADLEIGH, CECIL HERBERT, b Gilbertville, Mass, Oct 1, 07; m 30; c 4. PLANT PHYSIOLOGY. Educ: Mass Col, BS, 30; Ohio State Univ, MS, 32; Rutgers Univ, PhD(plant physiol), 35. Prof Exp: Asst, Rutgers Univ, 33-36; asst prof agron, Univ Ark, 36-41; sr chemist, Regional Salinity Lab, Bur Plant Indust, USDA, 41-42 & Bur Plant Indust, Soils & Agr Eng, 42-48, prin plant physiologist, 48-51, head physiologist in chg div sugar plant invests, 51-54, head sect soils & plant relationships, Soil & Water Conserv Res Br, 54-55, dir, Soil & Water Conserv Res Div, 55-74; RETIRED. Mem: Nat Acad Sci; Soil Conserv Soc Am; Am Soc Agron; Soil Sci Soc Am; Am Soc Sugar Beet Technol. Res: Mineral nutrition of plants; carbohydrate and nitrogen metabolism of plants; salt tolerance of plants; soil moisture stress. Mailing Add: 5621 Whitfield Chapel Rd Lanham MD 20801

WADLEY, BRYCE NEPHI, b Pleasant Grove, Utah, Dec 8, 09; m 42; c 5. PLANT PATHOLOGY. Educ: Brigham Young Univ, BS, 36; Iowa State Col, MS, 42, PhD(plant path), 47. Prof Exp: Agr inspector, State Dept Agr, Utah, 36-40; assoc plant pathologist, Bur Plant Indust, Soils & Agr Eng, 48-53, plant pathologist, 53-73, RES PLANT PATHOLOGIST, AGR RES SERV, USDA, UTAH STATE UNIV, 73- Mem: Am Phytopath Soc; Am Soc Hort Sci. Res: Virus diseases of stone and pome fruits. Mailing Add: USDA Agr Res Serv Utah State Univ UMC-45 Logan UT 84322

WADLEY, MARGIL WARREN, b Cisco, Tex, Dec 4, 31; m 66; c 1. INORGANIC CHEMISTRY. Educ: Bethany Nazarene Col, BS, 53; Okla State Univ, MS, 60; Purdue Univ, PhD(inorg chem), 63. Prof Exp: Sr chemist, Autonetics Div, NAm Aviation, Inc, 63-64 & Korad Dept, Union Carbide Corp, 64-65; sr res engr, Autonetics Div, NAm Aviation, Inc, 65 & 66-69, Space Systems Div, 65-66; environ scientist, 69-71, prin chemist, 71-75, SUPV CHEMIST, SOUTHERN CALIF AIR POLLUTION CONTROL DIST, 75- Concurrent Pos: Adj lectr, Univ Southern Calif; mem bd, Los Angeles, Henry George Sch Soc Sci. Mem: Am Chem Soc; Sigma Xi; Am Inst Chemists. Res: Electrochemistry of nonaqueous solvent systems; improved materials for medical and semiconductor applications; political economics; size and mass distribution of airborne particulate matter and associated visibility relationships. Mailing Add: 520 E Riverdale Ave Orange CA 92665

WADLINGER, ROBERT LOUIS PETER, b Philadelphia, Pa, Mar 20, 32; m 55; c 3. CHEMISTRY. Educ: LaSalle Col, AB, 53; Cath Univ, PhD(photochem kinetics, phys chem), 61. Prof Exp: Res asst, Benjamin Franklin Inst Labs, Pa, 53-55; sr res chemist, Mobil Oil Co, Inc, Labs, NJ, 60-62; assoc prof chem, State Univ NY Col Oneonta, 62-65; ASSOC PROF CHEM, NIAGARA UNIV, 65- Mem: Am Chem Soc. Res: Zeolite syntheses. Mailing Add: 4963 Creek Rd Lewiston NY 14092

WADMAN, W HUGH, b Marlborough, Eng, Sept 18, 26; m 52; c 2. CHEMISTRY, BIOCHEMISTRY. Educ: Bristol Univ, BSc, 47, PhD, 51. Prof Exp: Res assoc plant biochem, Univ Calif, 51-53; group leader cellulose chem, Rayonier Inc, Wash, 53-55; PROF ORG CHEM, UNIV OF THE PAC, 55- Mem: Am Chem Soc. Res: Carbohydrate chemistry and chromatographic techniques; origin of life; primitive biochemical systems. Mailing Add: Raymond Col Univ of the Pac Stockton CA 95204

WADSWORTH, DALLAS FREMONT, b Arcadia, Okla, Mar 2, 22; m; c 1. PLANT PATHOLOGY. Educ: Okla State Univ, BS, 48, MS, 49; Univ Calif, PhD(plant path), 66. Prof Exp: From asst prof to assoc prof, 71-74, PROF BOT & PLANT PATH, OKLA STATE UNIV, 74- Mem: Am Phytopath Soc. Res: Diseases of peanuts; plant virology. Mailing Add: Dept of Plant Path Okla State Univ Stillwater OK 74074

WADSWORTH, DONALD VAN ZELM, b Mamaroneck, NY, July 14, 31. GEOPHYSICS, TELECOMMUNICATIONS. Educ: Williams Col, Mass, BA, 53; Mass Inst Technol, PhD(geophys), 58. Prof Exp: Dept head, Bell Labs, 58-75; PRES, WADSWORTH ENG, 75- Concurrent Pos: Consult, Comt Undersea Warfare, Nat Acad Sci, 66-71. Mem: Sr mem Inst Elec & Electronics Eng. Res: Telecommunications; celestial mechanics. Mailing Add: 1211 Via Granate Sierra Madre CA 91024

WADSWORTH, FRANCIS THOMAS, organic chemistry, see 12th edition

WADSWORTH, FRANK H, b Chicago, Ill, Nov 26, 15; m 41; c 2. FORESTRY. Educ: Univ Mich, BSF & MF, 37, PhD(forestry), 50. Prof Exp: DIR INST TROP FORESTRY, US FOREST SERV, 55- Honors & Awards: Fernow Award, Am Forestry Asn, 73. Mem: Soc Am Foresters. Mailing Add: Sacarello 1016 Rio Piedras PR 00924

WADSWORTH, GEORGE PROCTOR, b Methuen, Mass, Apr 6, 09; m 35; c 2. MATHEMATICS. Educ: Mass Inst Technol, BS, 30, MS, 31, PhD(math), 33. Prof Exp: From instr to prof, 31-74, EMER PROF MATH, MASS INST TECHNOL, 74- Res: Systems of partial differential equations; geometry of algebraic Pfaffians; mathematical statistics; application of time series analysis to economic and geophysical data; engineering applications of mathematics; use of statistical methods in industry; probability theory; operations research and applications. Mailing Add: Dept of Math Rm 2-108 Mass Inst Technol Cambridge MA 02139

WADSWORTH, GLADYS ELIZABETH, b Reading, Pa, Oct 6, 14. ANATOMY. Educ: EStroudsburg State Col, BS, 36; Columbia Univ, MA, 42; Univ Md, PhD(anat), 55. Prof Exp: High sch teacher, Pa, 37-42; chief phys therapist, Georgetown Univ

Hosp, 47-48; phys therapist, Children's Hosp Sch, Baltimore, 49-50; instr anat, Sch Med, Univ Md, 50-56, assoc phys ther & head dept, 56-62; consult phys ther, Div Educ, Am Phys Ther Asn, NY, 62-64; asst prof, 64-67, ASSOC PROF ANAT, SCH MED, UNIV MD, BALTIMORE CITY, 68- Mem: AAAS; Am Asn Anatomists; Am Phys Ther Asn; Int Soc Electromyographic Kinesiology. Res: Physical therapy; biomechanics of human movement; electromyographic kinesiology; muscle function and testing in man. Mailing Add: Dept of Anat Univ of Md Sch of Med Baltimore MD 21201

WADSWORTH, JOSEPH ALLISON CANNON, b Charlotte, NC, Mar 22, 13; m 42; c 3. OPHTHALMOLOGY, MEDICINE. Educ: Davidson Col, BS, 35; Duke Univ, MD, 39. Prof Exp: Assoc clin prof ophthal, Columbia Univ, 59-65; attend surgeon, Roosevelt Hosp, 61-65; PROF OPHTHAL & CHMN DEPT, MED CTR, DUKE UNIV, 65- Concurrent Pos: Consult ophthal, New Rochelle Hosp, NY, 55-65; attend ophthalmologist, Presby Hosp & Vanderbilt Clin, New York, 59-65; consult ophthal, Watts & Vet Admin Hosps, Durham, NC, 65-; pres, Am Bd Ophthal, 73-74; mem, NC State Comn for Blind. Mem: Am Acad Ophthal & Otolaryngol (first vpres, 74); fel Am Col Surgeons; Pan-Am Asn Ophthal; Am Asn Ophthal; Am Ophthal Soc (pres, 75-76). Res: Ophthalmic pathology; ocular tumors. Mailing Add: 1532 Pinecrest Rd Durham NC 27705

WADSWORTH, WILLIAM BINGHAM, b Cortland, NY, Dec 4, 34; m 62; c 2. PERTOLOGY, MINERALOGY. Educ: Brown Univ, AB, 57; Northwestern Univ, MS, 62, PhD(igneous petrol), 66. Prof Exp: Asst prof geol, Univ SDak, 63-66; from asst prof to assoc prof, Idaho State Univ, 66-72; ASSOC PROF GEOL, WHITTIER COL, 72-, CHMN DEPT, 73- Concurrent Pos: NSF sci fac fel, Pomona Col, 71-72. Mem: AAAS; Geol Soc Am; Soc Econ Paleont & Mineral; Am Geophys Union; Mineral Soc Am. Res: Quantitative petrography and geochemistry of igneous rocks; petrology of granitic plutons, especially copper-bearing porphyries; structural petrology and fabric analysis; statistical design in geology. Mailing Add: Dept of Geol Sci Whittier Col Whittier CA 90608

WADSWORTH, WILLIAM STEELE, JR, b Hartford, Conn, May 6, 27; m 56; c 3. ORGANIC CHEMISTRY. Educ: Trinity Col, Conn, BS, 50, MS, 52; Pa State Univ, PhD(chem), 56. Prof Exp: Res chemist, Rohm & Haas Co, 56-63; assoc prof, 63-68, PROF CHEM, SDAK STATE UNIV, 68- Mem: Am Chem Soc. Res: New reactions and mechanisms in organic chemistry; heterocyclic and organophosphorus chemistry. Mailing Add: Dept of Chem S Dak State Univ Brookings SD 57007

WAECH, THEODORE G, b Chicago, Ill, June 20, 41; m 67; c 1. PHYSICAL CHEMISTRY. Educ: Univ Ill, Urbana, BS, 63; Univ Wis-Madison, PhD(phys chem), 68. Prof Exp: Res assoc chem, McGill Univ, 68-70; asst prof chem, St Norbert Col, 70-71; ASST PROF CHEM, CONCORDIA COL, WIS, 71- Mem: Am Chem Soc. Res: Molecular beams. Mailing Add: Dept of Chem Concordia Col Milwaukee WI 53208

WAEHNER, KENNETH ARTHUR, b Milwaukee, Wis, Apr 12, 26; m 52; c 3. ANALYTICAL CHEMISTRY. Educ: Marquette Univ, BS, 52; Univ Wis-Madison, MS, 53. Prof Exp: Lab supvr anal chem, Fansteel Metall Corp, Ill, 53-66; MGR SPECTROG LAB, XEROX CORP, 66- Mem: Soc Appl Spectros; Am Chem Soc. Res: Inorganic analyses; application of emission spectrography and classical chemical analyses to the characterization of semiconductors and refractory metals. Mailing Add: Mat Analyses Labs Xerox Corp Webster NY 14580

WAELSCH, SALOME GLUECKSOHN, b Ger, Oct 6, 07; nat US; m 43; c 2. GENETICS, DEVELOPMENTAL BIOLOGY. Educ: Univ Freiburg, PhD(zool), 32. Prof Exp: Res assoc & lectr zool, Columbia Univ, 36-55; assoc prof anat, 55-58, PROF GENETICS, ALBERT EINSTEIN COL MED, 58-, CHMN DEPT, 63- Mem: Am Soc Zoologists; Am Asn Anatomists; Genetics Soc Am; Soc Develop Biol. Res: Developmental and mammalian genetics; role of genes in differentiation. Mailing Add: Dept of Genetics Albert Einstein Col of Med Bronx NY 10461

WAFF, HARVE S, b Norfolk, Va, Oct 5, 40; m 61; c 2. GEOPHYSICS. Educ: Col William & Mary, BS, 62; Univ Ore, MS, 66, PhD(physics), 70. Prof Exp: Engr autonetics div, NAm Rockwell Corp, 62-64; NASA fel, Ctr Volcanology, Univ Ore, 70-72; RES FEL, HOFFMAN LAB, HARVARD UNIV, 72- Res: Electronic properties of minerals; chemistry of silicate liquids; optical properties of solids; solid earth geophysics; group theory. Mailing Add: Hoffman Lab Harvard Univ Cambridge MA 02138

WAFFLE, ELIZABETH LENORA, b Marion, Iowa, Feb 14, 38. PARASITOLOGY, INVERTEBRATE ZOOLOGY. Educ: Cornell Col, BA, 60; Univ Iowa, MS, 63; Iowa State Univ, PhD(parasitol), 67. Prof Exp: Assoc prof biol, Armstrong State Col, 66-67; asst prof, Iowa Wesleyan Col, 67-68; ASST PROF BIOL, EASTERN MICH UNIV, 68- Concurrent Pos: Consult, Parasitol Prog, Ann Arbor Biol Ctr, 70- Mem: Am Soc Parasitol; Am Soc Zoologists; Wildlife Dis Asn; Am Inst Biol Sci. Res: Dog heartworm and other parasites of dogs; parasites of fish; entomology; marine biology. Mailing Add: Dept of Biol Eastern Mich Univ Ypsilanti MI 48197

WAGENAAR, EMILE B, b Poerwokerto, Indonesia, Apr 7, 23; Can citizen; m 54; c 3. CELL BIOLOGY, GENETICS. Educ: St Agr Univ, Wageningen, Ir, 54; Univ Alta, PhD(genetics), 58. Prof Exp: Nat Res Coun Can res fel, 58-60; res scientist, Genetics & Plant Breeding Res Inst, Can Dept Agr, 60-65; res zoologist, Univ Calif, Berkeley, 65-67; assoc prof, 67-69, PROF BIOL, UNIV LETHBRIDGE, 69- Concurrent Pos: Vis assoc & sessional lectr, Ottawa, Ont, 64-65. Mem: Genetics Soc Am; Genetics Soc Can. Res: Cytogenetics and evolution of species in Triticum and Hordeum; chemistry, ultrastructure and behavior of chromosomes in nucleus and during cell division. Mailing Add: Dept of Biol Sci Univ of Lethbridge Lethbridge AB Can

WAGENAAR, RAPHAEL OMER, b Spokane, Wash, Jan 9, 16; m 51; c 2. DAIRY BACTERIOLOGY. Educ: Wash State Univ, BS, 42, MS, 47; Univ Minn, PhD(dairy bact), 51. Prof Exp: Asst dairy bact, Univ Minn, 49-51; res assoc, Food Res Inst, Univ Chicago, 51-56; sect leader microbiol, Food Develop Dept, 56-62, RES ASSOC MICROBIOL, FOOD DEVELOP ACTIV, TECHNOL CTR, GEN MILLS, INC, 62- Mem: Am Soc Microbiol; Am Dairy Sci Asn; Inst Food Technologists. Res: Bacterial food poisoning; lactic acid bacteria; effect of irradiation on bacterial spores and toxins; psychrophilic bacteria causing food spoilage. Mailing Add: Tech Ctr Gen Mills Inc 9000 Plymouth Ave N Minneapolis MN 55427

WAGENBACH, GARY EDWARD, b Barron, Wis, Mar 24, 40; m 60; c 3. PARASITOLOGY, ZOOLOGY. Educ: Univ Wis-River Falls, BS, 62; Univ Wis-Madison, MS, 64, PhD(zool), 68. Prof Exp: NIH proj assoc, Univ Wis-Madison, 68-69; asst prof, 69-74, ASSOC PROF BIOL, CARLETON COL, 74- Mem: AAAS; Am Soc Parasitol; Soc Protozoologists. Res: Biology of parasites especially coccidia and digenetic trematodes. Mailing Add: Dept of Biol Carleton Col Northfield MN 55057

WAGENER, JOHANN CHRISTIAN SIEGFRIED, b Hamburg, Ger, Aug 14, 08; m 38; c 1. ELECTRONICS. Educ: Univ Berlin, PhD(physics), 35. Prof Exp: Physicist, Telefunken Labs, Ger, 35-45; owner, Electronic Tubes Lab, 45-47; prin sci officer, Post Off Res Sta, London, Eng, 47-53; head lab, Kemet Co, 54-57, mgr res & develop, Kemet Co Div, Union Carbide Corp, 57-65, gen mgr, Electronics Div, London, 65-67, dir electronics, Union Carbide Europe, Switz, 67-71; consult, 71-75; RETIRED. Mem: Sr mem Inst Elec & Electronics Engrs. Res: Vacuum and solid state physics. Mailing Add: Bodenschneidenstrasse 11 Tegernsee West Germany

WAGENKNECHT, AUSTIN CLAYTON, b Longmont, Colo, June 1, 19; m 43; c 2. BIOCHEMISTRY. Educ: Univ Wis, BS, 42, MS, 48, PhD(biochem), 50. Prof Exp: Asst biochem & plant path, Univ Wis, 41-42; from asst prof to assoc prof biochem, Agr Exp Sta, Cornell Univ, 50-60; prin scientist, 60-64, head fundamental food res, 65-73, HEAD DEVELOP FOOD RES DEPT, GEN MILLS, INC, 73- Concurrent Pos: Gerber Found fel, Univ Ill, 56. Mem: AAAS; Asn Cereal Chemists; Am Chem Soc; Am Soc Biol Chem; Inst Food Technologists. Res: Lipids; enzymes; food flavors and off-flavors; food preservation; chemistry of natural products. Mailing Add: Gen Mills Bell Tech Ctr 9000 Plymouth Ave N Minneapolis MN 55427

WAGENKNECHT, BURDETTE LEWIS, b Cotter, Iowa, Sept 9, 25; m 51; c 4. BOTANY. Educ: Univ Iowa, BA, 48, MS, 54; Univ Kans, PhD(bot), 58. Prof Exp: Instr biol & phys sci, Franklin Col, 54-55; asst bot, Univ Kans, 57-58; hort taxonomist, Arnold Arboretum, 58-61; from asst prof to assoc prof biol, Norwich Univ, 61-68; PROF BIOL & HEAD DEPT, WILLIAM JEWELL COL, 68- Mem: AAAS; Am Inst Biol Sci. Res: Floristics of Washington County, Iowa; Heterotheca; taxonomy of cultivated wood plants; registration of cultivars in the genus Buxus. Mailing Add: Dept of Biol William Jewell Col Liberty MO 64068

WAGENKNECHT, JOHN HENRY, b Washington, Iowa, Jan 30, 39; m 60; c 3. ORGANIC CHEMISTRY, ELECTROCHEMISTRY. Educ: Monmouth Col, AB, 60; Univ Iowa, PhD(chem), 64. Prof Exp: Sr res chemist, 64-70, res specialist, 70-74, SCI FEL, MONSANTO CO, 74- Mem: Electrochem Soc; Int Soc Electrochem; Am Chem Soc. Res: Synthesis of organic chemicals by electrochemistry; electroanalytical chemistry; electrical discharge chemistry; scale-up of electro-organic processes. Mailing Add: Monsanto Co 800 N Lindbergh Blvd St Louis MO 63166

WAGGENER, DONALD TODD, b Dawson, Nebr, July 22, 13; m 36. DENTISTRY. Educ: Univ Nebr, DDS, 36. Prof Exp: Instr oral surg, 42, asst prof oral surg & path, 48-49, from asst prof to prof oral path, 49-70, chmn dept oral path, 49-70, assoc prof surg, Col Med, 57-70, PROF DENT & RADIOL, COL DENT, UNIV NEBR, LINCOLN, 70- Concurrent Pos: Fel dent surg, Mayo Found, Univ Minn, 38-41. Mem: Am Dent Asn; Int Asn Dent Res; fel Am Col Dent; fel Am Acad Dent Radiol; Am Acad Oral Path. Res: Oral diagnosis, pathology and roentgenology. Mailing Add: Dept of Radiol Univ of Nebr Col of Dent Lincoln NE 68508

WAGGENER, ROBERT GLENN, b Benton, Ky, June 12, 32; m 59; c 2. MEDICAL PHYSICS, BIOPHYSICS. Educ: Univ Tex, Austin, BA, 54, MA, 63; Univ Tex M D Anderson Hosp & Tumor Inst Houston, PhD(biophys), 67; Am Bd Radiol, cert, 72. Prof Exp: Res asst physics, Nuclear Physics Lab, Balcones Res Ctr, Univ Tex, Austin, 60-61; pres, Nucleonics Res & Develop Corp, Tex, 61-63; radiol health specialist, Tex State Health Dept, 63-64; Nat Cancer Inst fel physics, Univ Tex M D Anderson Hosp & Tumor Inst Houston, 67-68; asst prof, 68-72, ASSOC PROF RADIOL, UNIV TEX MED SCH SAN ANTONIO, 72- Concurrent Pos: Consult, Brooke Army Med Ctr, Ft Sam Houston, Tex, 71- Mem: Am Asn Physicists in Med; Biophys Soc; Am Col Radiol; Soc Nuclear Med; Radiol Soc NAm. Res: Measurement of x-ray spectra; calculation of information content in diagnostic x-rays; dosimetry and measurement of ionizing radiation; computerized tomography. Mailing Add: Dept of Radiol Univ of Tex Med Sch San Antonio TX 78284

WAGGENER, RONALD E, b Green River, Wyo, Oct 6, 26; m 48; c 4. RADIOLOGY. Educ: Univ Nebr, BS, 49, MS, 53, MD, 54, PhD, 57; Am Bd Radiol, dipl, 59. Prof Exp: ASSOC PROF RADIOL, UNIV NEBR MED CTR, OMAHA, 58- Concurrent Pos: Radiotherapist, Methodist Hosp, 59- Mem: Am Asn Cancer Res; Radiol Soc NAm; Royal Soc Med; Brit Inst Radiol; fel Am Col Radiol. Res: Biological effects of radiation stressing the hematological effects of ionizing rays. Mailing Add: Nebr Methodist Hosp 8303 Dodge St Omaha NE 68114

WAGGENER, THOMAS RUNYAN, b Indianapolis, Ind, July 20, 38; m 64; c 2. FOREST ECONOMICS. Educ: Purdue Univ, BSF, 62; Univ Wash, MF, 63, MA, 65, PhD(forest econ), 66. Prof Exp: Asst prof, 67-71, chmn mgt & soc sci div, 72-75, ASSOC PROF FOREST ECON, COL FOREST RESOURCES, UNIV WASH, 71- Concurrent Pos: Assoc coordr course in trop forestry, Orgn Trop Studies, Inc, Costa Rica & Honduras, 68 & 71; economist & analyst, Pub Land Law Rev Comn, DC, 68-69. Mem: Am Econ Asn; Soc Am Foresters. Res: Natural resources economics and analysis of economic impact of resource management policies; regional economic analysis; industrial organization and market structure. Mailing Add: Col Forest Resources AR-10 Univ of Wash Seattle WA 98195

WAGGENER, WILLIAM COLE, b Princeton, Ky, Feb 5, 17; m 51; c 2. INORGANIC CHEMISTRY. Educ: Centre Col, Ky, AB, 39; Univ Buffalo, PhD(inorg chem), 49. Prof Exp: Asst chem, Univ Buffalo, 39-41; supvr, Ammonia Oxidation Plant, Lake Ont Ord Works, 41-42; chemist, Res Lab, Nat Carbon Co, 43-45; MEM STAFF, OAK RIDGE NAT LAB, 49- Mem: AAAS; Am Chem Soc; Sigma Xi. Res: Chemistry of thorium; inorganic complexes; coordination properties of thiocyanates; solution spectrophotometry over wide ranges of temperature and pressure; near infrared spectroscopy of pure substances in condensed states; liquid effluents from light water-cooled nuclear reactors.

WAGGLE, DOYLE H, b Osborne, Kans, Aug 11, 39; m 60; c 2. CEREAL CHEMISTRY. Educ: Ft Hays Kans State Col, BS, 61; Kans State Univ, MS, 63, PhD(milling indust), 66. Prof Exp: Res asst feed technol, Kans State Univ, 65-66, res assoc, 66-67; process res chemist, 67-68, mgr process res, 68-72, DIR RES & DEVELOP, VENTURE MGT, RALSTON PURINA CO, 72- Mem: Am Asn Cereal Chemists; Inst Food Technologists; Am Chem Soc. Res: Chemistry of processes related to foods and feeds. Mailing Add: Ralston Purina Co 835 S Eighth St St Louis MO 63188

WAGGONER, ALAN STUART, b Los Angeles, Calif, Jan 8, 42; m 65. BIOPHYSICAL CHEMISTRY. Educ: Univ Colo, BA, 65; Univ Ore, PhD(chem), 69. Prof Exp: NIH fel, Yale Univ, 69-71; ASST PROF CHEM, AMHERST COL, 71- Mem: Am Chem Soc. Res: Spectroscopic studies of biological membranes. Mailing Add: Dept of Chem Amherst Col Amherst MA 01002

WAGGONER, JACK HOLMES, JR, b Pittsburgh, Pa, Sept 4, 27; m 61. THEORETICAL PHYSICS. Educ: Ohio State Univ, BS, 49, PhD(physics), 57. Prof Exp: Asst photochem, Res Found, Ohio State Univ, 49-53, res assoc, 58-59, proj assoc supvr, 59, proj supvr, 59, asst physics, Ohio State Univ, 53-55, instr, 57-58, asst prof, 58-59; asst prof, Univ Calif, Riverside, 59-61; asst prof, 61-65, ASSOC PROF PHYSICS, HARVEY MUDD COL, 65- Concurrent Pos: Vis assoc, Calif Inst Technol, 67-68. Mem: AAAS; Am Phys Soc; Am Inst Physics; Am Asn Physics

Teachers. Res: Methods of theoretical physics; theory of molecular spectroscopy. Mailing Add: Dept of Physics Harvey Mudd Col Claremont CA 91711

WAGGONER, JAMES ARTHUR, b West Lafayette, Ind, Dec 31, 31; m 53; c 4. NUCLEAR PHYSICS. Educ: Univ Ill, BS, 53; Cornell Univ, PhD(exp physics), 60. Prof Exp: Physicist, Lawrence Radiation Lab, Univ Calif, 60-70, Physics Int Co, Calif, 71 & Maxwell Labs, 71; PHYSICIST, SCHLUMBERGER WELL SERVS, 71- Concurrent Pos: Consult to NASA. Mem: Am Phys Soc. Res: Lunar and planetary surface composition analysis using neutron inelastic scattering; Van Allen zone charged particles; geophysical instrumentation. Mailing Add: Schlumberger Well Servs PO Box 2175 Houston TX 77001

WAGGONER, JAMES NORMAN, b Elgin, Ill, Nov 10, 25; m; c 2. AEROSPACE MEDICINE. Educ: Ind Univ, BS, 46, MD, 49; Am Bd Prev Med, dipl & cert aviation med, 57. Prof Exp: Flight surgeon, United Air Lines, Inc, 52-59; CORP DIR MED SAFETY & LIFE SCI, GARRETT CORP, 59- Concurrent Pos: Adj prof, Univ Southern Calif, 58-; Aerospace Med Asn fel, 60; Indust Med Asn fel, 64; sr med consult, Western US, Air France, 72- Honors & Awards: Spec Aerospace Med Honor Citation, AMA, 62; Louis H Bauer Founders Award, Eaton Lab-Norwich Pharmacal Co, 68. Mem: Aerospace Med Asn (pres, 67-68); Airline Med Dir Asn (pres, 63-64); Am Col Prev Med; Am Inst Aeronaut & Astronaut; Int Astronaut Fedn. Res: Effects of microbiological contamination of closed ecological systems; aeroembolism; combined stressors of spaceflight on human and subhuman subjects; effects of trace contaminants on humans and animals. Mailing Add: Garrett Corp 9851 Sepulveda Blvd Los Angeles CA 90009

WAGGONER, MARGARET ANN, b Centerville, Iowa, Aug 27, 26. NUCLEAR PHYSICS, HISTORY OF SCIENCE. Educ: Univ Iowa, BA, 46, MA, 48, PhD, 50. Prof Exp: From instr to asst prof physics, Vassar Col, 50-58; asst prof, Stanford Univ, 58-61; assoc prof, Univ Md, 61-65 & Univ Iowa, 65-70; prof, Smith Col, 70-75; PROF PHYSICS & HIST & PHILOS OF SCI & PRES, WILSON COL, 75- Mem: Am Phys Soc; Am Asn Physics Teachers. Res: Nuclear physics and reactions; beta and gamma ray spectroscopy; history and philosophy of science. Mailing Add: Off of Pres Wilson Col Chambersburg PA 17201

WAGGONER, PAUL EDWARD, b Appanoose Co, Iowa, Mar 29, 23; m 45; c 2. CLIMATOLOGY. Educ: Univ Chicago, SB, 46; Iowa State Col, MS, 49, PhD, 51. Prof Exp: From asst to assoc plant pathologist, 51-56, chief dept soils & climat, 56-69, vdir, 69-71, DIR, CONN AGR EXP STA, 72- Concurrent Pos: Guggenheim fel, 63; lectr, Yale Univ, 62. Honors & Awards: Am Meteorol Soc Award, 67. Mem: Fel Am Soc Agron; fel Am Phytopath Soc; Am Meteorol Soc; Ecol Soc Am; Am Soc Plant Physiologists. Res: Agriculture; plant pathology; effect of environment on plants, especially plant diseases. Mailing Add: Conn Agr Exp Sta PO Box 1106 New Haven CT 06504

WAGGONER, PHILLIP RAY, b Parkersburg, WVa, Apr 4, 43; m 67; c 2. DEVELOPMENTAL BIOLOGY. Educ: WVa Univ, BS, 65, MS, 68, PhD(genetics & develop biol), 72. Prof Exp: Instr biol, Fairmont State Col, 68 & WVa Univ, 68-69; ASST PROF ANAT, WAYNE STATE UNIV, 72- Mem: Am Asn Anat; AAAS; Sigma Xi. Res: Development of the vertebrate eye. Mailing Add: Dept of Anat Sch of Med Wayne State Univ Detroit MI 48201

WAGGONER, RAYMOND WALTER, b Carson City, Mich, Aug 2, 01; m 30; c 2. PSYCHIATRY, NEUROLOGY. Educ: Univ Mich, MD, 24; Univ Pa, ScD, 30. Prof Exp: Intern, Harper Hosp, Detroit, 24-25; resident, Philadelphia Orthop Hosp & Infirmary Nerv Dis, 25-26; lab intern, Pa Hosp, Philadelphia, 26; from asst prof to assoc prof neurol, Med Sch, 29-36, from asst neurologist to neurologist, Univ Hosp, 29-36, chmn dept psychiat, Med Sch, 37-71, prof psychiat & dir neuropsychiat inst, 37-72, EMER PROF PSYCHIAT & EMER DIR NEUROPSYCHIAT INST, MED SCH, UNIV MICH, ANN ARBOR, 72- Concurrent Pos: Consult, spec comt rights ment ill, Am Bar Found, 59-66, indust personnel security, Dept Defense, 66-, Surgeon-Gen, US Army, Selective Serv Syst, Peace Corps, Vet Admin & Social Security Admin; mem med adv bd, Social Security Admin, 65-; adv comt, Nat Paraplegic Found; mem test comt psychiat, Nat Bd Med Examr, 65-70; mem, Res Socs Coun, 70-73; vpres bd trustees, Mich Inst Pastoral Care; consult & bd mem, Reproductive Biol Res Found, 70-; consult, Mich State Dept Ment Health, 74-; distinguished vis prof psychiat, Univ Louisville, 74- Honors & Awards: E B Bowis Award, Am Col Psychiat, 68. Mem: Fel AAAS; Am Acad Psychoanal; fel Am Col Psychiat (vpres, 64-65, pres elect, 65-66, pres, 66-67); Am Geriat Soc; fel Am Psychiat Asn (vpres, 60-61, pres, 69-70). Res: Personality studies in chorea; the convulsive state; myopathies; psychotherapy. Mailing Add: 3333 Geddes Rd Ann Arbor MI 48105

WAGGONER, TERRY BILL, b Stillwater, Okla, Oct 25, 34; m 66; c 1. ORGANIC CHEMISTRY, BIOCHEMISTRY. Educ: Okla State Univ, BS, 57, MS, 59; Mich State Univ, PhD(chem), 64. Prof Exp: Instr chem, Okla State Univ, 59-60; sr res chemist, M & T Chem, Inc, 63-66; res chemist, 66-69, MGR BIOCHEM RES, CHEMAGRO DIV, MOBAY CHEM CORP, 69- Mem: Am Chem Soc; Am Inst Chemists. Res: Pesticide residues in plants, animals, water, soil; analytical residue methods; metabolism of pesticides in plant and animal systems; radioactive tracer methods. Mailing Add: Biochem Sect Chemagro Div Mobay Chem Corp PO Box 4913 Kansas City MO 64120

WAGGONER, WILBUR J, b Sutherland, Iowa, May 28, 24; m 46; c 3. MATHEMATICS. Educ: Buena Vista Col, BA, 47; Drake Univ, MSE, 50; Univ Wyo, EdD, 56. Prof Exp: Prin, coach & teacher high sch, 47-51; supt twp sch, 51-55; asst, Univ Wyo, 55-56; from asst prof to assoc prof, 56-62, PROF MATH, CENT MICH UNIV, 62-, ACTG DEAN, SCH GRAD STUDIES, 73- Mem: Am Statist Asn; Nat Coun Teachers Math. Res: Statistics. Mailing Add: Dept of Math Cent Mich Univ Mt Pleasant MI 48858

WAGGONER, WILLIAM CHARLES, b Alma, Mich, Jan 18, 36; m 58; c 4. PHYSIOLOGY, TOXICOLOGY. Educ: Hope Col, AB, 58; Mich State Univ, MS, 61, PhD(physiol), 63. Prof Exp: Teaching fel physiol & pharmacol, Med Sch, Mich State Univ, 60-63; res physiologist, Colgate-Palmolive Res Ctr, 63-64, res projs coordr oral health, 64-66; asst dir med res, Unimed, Inc, 66-70; assoc dir med serv & govt affairs, Wallace Pharmaceut, NJ, 70-74; ADMINR, MED & REGULATORY AFFAIRS, BABY PROD CO, JOHNSON & JOHNSON, 74- Concurrent Pos: Consult fac mem, Inst Clin Toxicol, 74- Mem: Am Physiol Soc; Am Acad Clin Toxicol; Soc Toxicol. Res: Clinical toxicology and pharmacology. Mailing Add: 24 Berkshire Dr Warren NJ 07060

WAGGONER, WILLIAM HORACE, b Ravenna, Ohio, June 8, 24; m 46; c 1. INORGANIC CHEMISTRY. Educ: Hiram Col, AB, 49; Western Reserve Univ, MS, 51, PhD(chem), 53. Prof Exp: Asst chem, Hiram Col, 48-49; asst prof, 52-59, ASSOC PROF INORG CHEM, UNIV GA, 59- Mem: Am Chem Soc. Res: Solubilities in non-aqueous systems; history of chemistry; spectral properties of inorganic materials. Mailing Add: Dept of Chem Univ of Ga Athens GA 30601

WAGH, PREMANAND VINAYAK, b Sadashivgad, India, July 9, 34; US citizen; m 65; c 3. BIOCHEMISTRY. Educ: Univ Bombay, BSc, 54, MSc, 56; Southern Ill Univ, Carbondale, MS, 62; Univ Minn, Minneapolis, PhD(animal sci, biochem), 65. Prof Exp: Res assoc biochem, Univ Ill Med Ctr, 65; res assoc, State Univ NY Buffalo, 65-68, instr, 68-69; res biochemist, 69-73, CHIEF CONNECTIVE TISSUE RES, VET ADMIN HOSP, 74-; ASST PROF BIOCHEM, MED CTR, UNIV ARK, LITTLE ROCK, 69- Mem: Am Chem Soc; Am Fedn Clin Res; Geront Soc; Am Soc Biol Chemists. Res: Cardiovascular glycoproteins; structure and function of connective tissue glycoproteins; glycoproteins and mucopolysaccharides in aging. Mailing Add: Connective Tissue Lab Vet Admin Hosp 300 E Roosevelt Rd Little Rock AR 72206

WAGHORNE, DICK, biochemistry, see 12th edition

WAGLE, GILMOUR LAWRENCE, b Staten Island, NY, Nov 17, 22; m 49; c 3. PHARMACOLOGY. Educ: Wagner Col, BS, 50; Rutgers Univ, MS, 56; Princeton Univ, MA, 59, PhD(biol), 60. Prof Exp: Pharmacologist, Res Dept, Ciba Pharmaceut Prod, Inc, NJ, 48-61; sr pharmacologist, Chas Pfizer & Co, 61-64; asst to dir res admin, Toxicol Res Sect, 64-72, ASST DIR, MED CONTROLS BR, OFF GOVT CONTROLS, LEDERLE LABS, AM CYANAMID CO, 72- Mem: AAAS; Soc Toxicol. Res: Cardiovascular, renal and central nervous system pharmacology; government regulatory affairs; toxicology. Mailing Add: Am Cyanamid Co Lederle Labs Pearl River NY 10965

WAGLE, ROBERT FAY, b Jamestown, NDak, Sept 3, 16. FORESTRY, BOTANY. Educ: Univ Minn, BS, 40; Univ Wash, MF, 55; Univ Calif, PhD(bot), 58. Prof Exp: Asst forestry, Univ Wash, 47-48; logging engr, Shasta Plywood Co, Calif, 48-49; sr lab asst, Univ Calif, 49-54; asst, Calif Forest & Range Exp Sta, US Forest Serv, 54-57; assoc prof watershed mgt, 57-69, PROF WATERSHED MGT & WATERSHED SPECIALIST, UNIV ARIZ, 69- Concurrent Pos: Consult fire & silvicult, Univ Nev, 64. Mem: Fel AAAS; Soc Am Foresters; Ecol Soc Am. Res: Silvics, genetics, ecology and silviculture; plant variation and its relationship to environment; nutrient and water relationships of wildland plants; effects of fire on plants and their environments. Mailing Add: Dept of Renewable Water Resources Univ of Ariz Col of Agr Tucson AZ 85721

WAGLE, SHREEPAD R, b Bombay, India, Jan 1, 31; nat US; m 62; c 2. PHARMACOLOGY. Educ: Univ Bombay, BS, 52, MS, 55; Univ Ill, PhD(biochem, nutrit), 59. Prof Exp: Res chemist, Haffkine Inst, India, 52-55; asst, Univ Ill, 55-59; res chemist, Sigma Lab, India, 60; from res assoc to asst prof, 60-65, assoc prof, 65-68, PROF PHARMACOL & DIR GRAD PROGS, SCH MED, IND UNIV, INDIANAPOLIS, 68- Mem: Am Cancer Soc; Am Inst Nutrit; Soc Exp Biol & Med; Am Diabetes Asn; Am Soc Pharmacol & Exp Therapeut. Res: Cofactors in protein and RNA biosynthesis; role of hormones and nutritional factors in protein synthesis and cancer cells; protein kinases; lipases and cyclic adenosine monophosphates; metabolism of isolated liver cells. Mailing Add: 350 Med Sci Bldg Ind Univ Sch of Med Indianapolis IN 46202

WAGLEY, CHARLES W, b Clarksville, Tex, Nov 9, 13; m 42; c 1. ANTHROPOLOGY, ETHNOLOGY. Educ: Columbia Col, AB, 36; Columbia Univ, PhD(anthrop), 41. Hon Degrees: Dr, Univ Bahia, 62; LLD, Univ Notre Dame, 64. Prof Exp: Res assoc anthrop, Nat Mus, Brazil, 41-42; mem field staff, Inst Inter-Am Affairs-USPHS, 42-45; mem staff, Guggenheim Found, 45-47; mem staff, Soc Sci Res Coun, 48-49; from assoc prof to prof anthrop, Columbia Univ, 49-65, dir, Inst Latin Am Studies, 61-69, Franz Boas prof anthrop, 65-71; GRAD RES PROF ANTHROP & LATIN AM STUDIES, UNIV FLA, 71- Concurrent Pos: Asst prof anthrop, Columbia Univ, 46-49; consult & lectr, For Serv Inst, US Dept State, 49-53, consult & lectr Latin Am studies, 65-69; dir, Bahia State-Columbia Univ Community Study Proj, 51-52; consult & contribr, Div Social Affairs, UN, 51-52; consult, Brazilian Ministry Educ, 53; mem, UNESCO Mission, Brazil, 55; fel, Ctr Advan Study Behav Sci, 57-58; mem, Joint Comt Latin Am Studies, Soc Sci Res Coun & Am Coun Learned Socs, 60-66; mem, Adv Bd, Brazilian Am Cult Inst, Brazilian Embassy, 64-68; fel, Ctr Inter-Am Rels, 67-; mem, Soc Sci Rev Comt, Nat Inst Ment Health, 69- Honors & Awards: Off, Nat Order Southern Cross, Brazil, 45, Comendador, 64; Medalha da Guerra, Brazil, 46. Mem: Fel Am Ethnol Soc (pres, 57-58); fel Am Anthrop Asn (pres, 70-71); fel Am Acad Arts & Sci; fel Am Philos Soc; Coun For Rels. Res: Social anthropology; social change; minority groups and race relations; community studies; aboriginal and contemporary Latin American culture and society. Mailing Add: Dept of Anthrop Univ of Fla Gainesville FL 32610

WAGLEY, PHILIP FRANKLIN, b Mineral Wells, Tex, Feb 5, 17; m 53. MEDICINE. Educ: Southern Methodist Univ, BS, 38; Johns Hopkins Univ, MD, 43; Am Bd Internal Med, dipl. Prof Exp: Instr med, 45-47 & 49-64, asst prof, 64-74, ASSOC PROF MED, JOHNS HOPKINS UNIV, 74- Concurrent Pos: Am Col Physicians res fel, Harvard Med Sch, 47-48; Nat Res Coun fel med sci, Mass Inst Technol, 48-49. Mem: Am Clin & Climat Asn; fel Am Col Physicians; Am Thoracic Soc; Am Soc Hemat. Res: Chest disease. Mailing Add: 9 E Chase St Baltimore MD 21212

WAGMAN, DONALD DAVID, b Detroit, Mich, July 21, 16; m 46; c 4. CHEMISTRY. Educ: George Washington Univ, BS, 36, MA, 41. Prof Exp: Clerk & messenger, Interstate Com Comn, DC, 34-36; lab asst, 36-37, jr chemist, 37-42, from asst chemist to chemist, 42-63, chief thermochem sect, 63-68, CHIEF CHEM THERMODYN DATA CTR, NAT BUR STANDARDS, 68- Mem: Am Chem Soc. Res: Thermochemistry; thermodynamic correlations; preparation of tables of selected values of chemical thermodynamic properties. Mailing Add: Chem Process Data Eval Sect Nat Bur of Standards Washington DC 20234

WAGMAN, GERALD HOWARD, b Newark, NJ, Mar 4, 26; m 48; c 2. MICROBIAL BIOCHEMISTRY. Educ: Lehigh Univ, BS, 46; Va Polytech Inst, MS, 47. Prof Exp: Tech asst antibiotics, Squibb Inst Med Res, 47-49, electronics, 49-54, microbial biochemist, 54-57; from assoc biochemist to sr biochemist, 57-69, sect leader, 69-70, mgr antibiotics dept, 70-74, ASSOC DIR MICROBIAL SCI/ANTIBIOTICS, SCHERING CORP, 74- Mem: AAAS; Am Chem Soc; Am Soc Microbiol; fel Am Inst Chemists; Sigma Xi. Res: Antibiotics, especially isolation, identification and evaluation; strain development; fermentation biosynthesis; isolation of natural products. Mailing Add: Antibiotics Res Schering Corp 60 Orange St Bloomfield NJ 07003

WAGMAN, IRVING HENRY, b New York, NY, Aug 15, 16; m 42; c 4. NEUROPHYSIOLOGY, PHYSIOLOGY. Educ: City Col New York, BS, 36; Univ Calif, MA, 37, PhD(physiol), 41. Prof Exp: Asst physiol, Sch Med, Univ Calif, 38-41, instr, 42, res assoc physiol & med, 41-43; asst physiologist, NIH, 43; assoc physiol, Jefferson Med Col, 46-48, from asst prof to assoc prof, 48-54; res assoc neurol, Mt Sinai Hosp, NY, 54-61; res physiologist, Biomech Lab, Med Ctr, Univ Calif, San Francisco, 61-65; PROF PHYSIOL & RES PHYSIOLOGIST, UNIV CALIF, DAVIS, 65- Concurrent Pos: John Found fel, Univ Pa, 43-46. Mem: Fel AAAS; Am Neurol Asn; Am Physiol Soc; Harvey Soc; Soc Neurosci. Res: Physiology of visual sense organ and pupil of the eye; electrical activity of single visual receptors; electrical characteristics of peripheral nerves; physiology of peripheral vascular, peripheral neuro-muscular and ocular motor systems; integration of cutaneous input. Mailing Add: Dept of Animal Physiol Univ of Calif Davis CA 95616

WAGMAN, JACK, physical chemistry, see 12th edition

WAGMAN, NICHOLAS EMORY, b Saratoga Springs, NY, Feb 22, 05; m 28; c 4. ASTRONOMY. Educ: Wesleyan Univ, BA, 27, MA, 28; Univ Pittsburgh, PhD(astron), 37. Prof Exp: Jr astronr, US Naval Observ, 28-30; astronr, Allegheny Observ, 30-41, from assoc prof to prof, 41-74, actg dir, 41-47, dir, 47-70, mem staff, 70-74, EMER PROF ASTRON, UNIV PITTSBURGH & EMER DIR, ALLEGHENY OBSERV, 74- Concurrent Pos: Lectr, Buhl Planetarium. Mem: AAAS; Am Astron Soc; Int Astron Union. Res: Stellar parallax; astrometric binaries. Mailing Add: 3726 Perrysville Ave Pittsburgh PA 15214

WAGNER, ALAN R, b Columbus, Ohio, Aug 20, 23; m 52; c 3. PATHOLOGY. Educ: Ohio State Univ, DVM, 46, MSc, 52. Prof Exp: Sr veterinarian, UNRRA, 46-47; field veterinarian, USDA, 47; veterinarian, Columbus Health Dept, 47-48; pathologist, Path Serv Labs, Ohio State Dept Agr, 48-56 & Lederle Labs Div, Am Cyanamid Co, 56-58; dir vet med res, Warren-Teed Pharmaceut Inc, 58-69; pres, Arlington Res Labs, Inc, 69-72; ASST DIR & PATHOLOGIST, LAB ANIMAL CTR, OHIO STATE UNIV, 72- Mem: Am Vet Med Asn; Am Asn Lab Animal Sci; NY Acad Sci; Soc Toxicol. Res: Pathology in fields of human and animal investigations and clinical as well as histopathological and bacteriological determination; new pharmaceutical and biological products. Mailing Add: Lab Animal Ctr Ohio State Univ 5769 Godown Rd Columbus OH 43220

WAGNER, ALBERT FORDYCE, b Rochester, NY, Feb 3, 45; m 69; c 1. CHEMICAL PHYSICS. Educ: Boston Col, BS, 66; Calif Inst Technol, PhD(chem), 72. Prof Exp: Presidential intern, 72-74, ASST CHEMIST, CHEM DIV, ARGONNE NAT LAB, 74- Mem: Am Inst Physics. Res: Theory and modeling of chemical reactions occurring at energies of less than ten electron volts. Mailing Add: Chem Div Argonne Nat Lab Argonne IL 60439

WAGNER, ALVIN, parasitology, see 12th edition

WAGNER, ARTHUR FRANKLIN, b Jersey City, NJ, Oct 25, 22; m 45; c 3. ORGANIC CHEMISTRY. Educ: Princeton Univ, AB, 48, MA, 49, PhD(org chem), 51. Prof Exp: Sr chemist, 51-61, res assoc org chem, 61-65, SR RES FEL ORG CHEM, MERCK SHARP & DOHME RES LABS, 65- Mem: Am Chem Soc; AAAS. Res: Synthetic organic chemistry in natural products, isolation, structure determination and synthesis of vitamins and cofactors; synthesis of benzimidazoles; synthesis of peptides, biopolymers and immobilized biopolymers. Mailing Add: Synthetic Chem Res Merck Sharpe & Dohme Res Labs Rahway NJ 07065

WAGNER, BERNARD MEYER, b Philadelphia, Pa, Jan 17, 28; m 51; c 2. PATHOLOGY. Educ: Hahnemann Med Col, MD. Prof Exp: Dir exp path labs, Hahnemann Med Col, 54-55; asst prof path, Med Sch & Grad Sch Med, Univ Pa, 56-58; assoc prof path & Robert L King chair cardiovasc res, Sch Med, Univ Wash, 58-60; prof path & chmn dept, New York Med Col, 60-67, clin prof, 67-68; CLIN PROF PATH, COLUMBIA UNIV, 68-; DIR LABS, BEEKMAN DOWNTOWN HOSP, NEW YORK, 71- Concurrent Pos: Dazian Found Med Res fel, Hahnemann Med Col, 53, Am Heart Asn Southeast Pa fels, 54-55; asst vis chief serv, Philadelphia Gen Hosp, 54-58; pathologist & dir path, Children's Hosp, Philadelphia, 55-58; lectr, Philadelphia Col Pharm, 56-58; attend pathologist, Vet Admin Hosp, 58-; Burroughs Wellcome Fund travel grant & spec investr, Hosp for Sick Children, London, Eng, 59; vpres, Warner Lambert Res Inst, 67-; dir labs, Francis Delafield Hosp, 68- Mem: Am Soc Exp Path; Soc Pediat Res; Histochem Soc; Am Asn Path & Bact; Am Rheumatism Asn. Res: Diseases of connective tissue; rheumatic heart disease. Mailing Add: Beekman Downtown Hosp 170 William St New York NY 10038

WAGNER, CARL GEORGE, b Newark, NJ, Sept 26, 43; m 69. MATHEMATICS. Educ: Princeton Univ, AB, 65; Duke Univ, PhD(math), 69. Prof Exp: ASST PROF MATH, UNIV TENN, KNOXVILLE, 69- Mem: Am Math Soc. Res: Function theory in fields with non-archimedean absolute value; combinatorics; arithmetic functions. Mailing Add: Dept of Math Univ of Tenn Knoxville TN 37916

WAGNER, CHARLES DANIEL, chemistry, see 12th edition

WAGNER, CHARLES EUGENE, b Memphis, Tenn, June 21, 23; m 49; c 3. MORPHOLOGY. Educ: Princeton Univ, AB, 47; Ind Univ, PhD(zool), 54. Prof Exp: Asst zool, Ind Univ, 48-50; from instr to assoc prof, 52-71, from asst dean to assoc dean, 61-74, PROF ANAT, SCH MED, UNIV LOUISVILLE, 71- Res: Experimental morphology; regeneration; movements at synovial joints. Mailing Add: Dept of Anat Univ of Louisville Sch of Med Louisville KY 40201

WAGNER, CHARLES ROE, b Olivet, SDak, Dec 3, 25; m 50; c 2. ORGANIC CHEMISTRY. Educ: SDak Sch Mines & Technol, BSc, 50; Mich State Col, PhD, 55. Prof Exp: Res chemist, 55-62, res supvr, Specialty Chem Div, 62-69, mgr process res, 69-71, dir develop, 71-74, DIR, BUFFALO RES LABS, SPECIALTY CHEM DIV, ALLIED CHEM CORP, 74- Mem: Am Chem Soc; Am Inst Chem Eng. Res: Organic isocyanates; organic acids. Mailing Add: Allied Chem Corp Buffalo Res Labs PO Box 1069 Buffalo NY 14240

WAGNER, CONRAD, b Brooklyn, NY, Nov 1, 29; m 53; c 2. BIOCHEMISTRY, MICROBIOLOGY. Educ: City Col New York, BA, 51; Univ Mich, MS, 52, PhD(biochem), 56. Prof Exp: USPHS fel biochem, NIH, 59-61; from asst prof to assoc prof biochem, 61-75, PROF BIOCHEM, SCH MED, VANDERBILT UNIV, 75- Concurrent Pos: Res biochemist, Vet Admin Hosp, 61-68, chief biochem res, 68-, assoc chief of staff res, 74- Mem: Am Soc Microbiol; Am Soc Biol Chemists; Am Inst Nutrit. Res: Gluconeogenesis from lipid in Tetrahymena pyriformis; sulfonium compounds and one carbon metabolism in bacteria; role and function of natural folate coenzymes; regulation of tryptophan-niacin relation in animals and microorganisms. Mailing Add: Dept of Biochem Vanderbilt Univ Sch of Med Nashville TN 37232

WAGNER, DANIEL HOBSON, b Jersey Shore, Pa, Aug 24, 25; m 49; c 4. MATHEMATICS. Educ: Haverford Col, BS, 47; Brown Univ, PhD(math), 51. Prof Exp: Staff mem sci, Opers Eval Group, Mass Inst Technol, 51-56; supvr math anal, Burroughs Corp, 56-58; partner, Kettelle & Wagner, 58-63; PRES, DANIEL H WAGNER ASSOCS, 63- Concurrent Pos: Chmn reliability task group, Off Asst Secy Defense Res & Eng, 56; lectr, Swarthmore Col, 58 & Univ Pa, 58 & 61-62. Mem: Am Math Soc; Opers Res Soc Am; Soc Indust & Appl Math; Math Asn Am. Res: Operations research; constrained optimization; measurable set-valued functions. Mailing Add: Station Square One Paoli PA 19301

WAGNER, DAVID KENDALL, b Berkeley, Calif, Aug 7, 45; m 67. EXPERIMENTAL SOLID STATE PHYSICS. Educ: Pomona Col, BA, 67; Cornell Univ, PhD(physics), 72. Prof Exp: RES ASSOC PHYSICS, CORNELL UNIV, 72- Mem: Am Phys Soc. Res: Investigation of electronic scattering mechanisms in metals. Mailing Add: Lab of Atomic & Solid State Phys Cornell Univ Ithaca NY 14853

WAGNER, DAVID LOREN, b Erie, Pa, Nov 19, 42; m 64; c 1. LOW TEMPERATURE PHYSICS, SOLID STATE PHYSICS. Educ: Case Western Reserve Univ, BS, 64, MS, 66, PhD(physics), 70. Prof Exp: From asst prof to assoc prof, 70-75, PROF PHYSICS, EDINBORO STATE COL, 75-, CHMN DEPT, 72- Concurrent Pos: NSF acad year exten grant, 71-73, student sci training grant, 75. Mem: Am Asn Physics Teachers; AAAS. Res: Fermi surface of metals and semimetals. Mailing Add: Dept of Physics Edinboro State Col Edinboro PA 16444

WAGNER, EDWARD D, b Eureka, SDak, June 28, 19; m 42; c 3. PARASITOLOGY. Educ: Walla Walla Col, BA, 42; Wash State Univ, MS, 45; Univ Southern Calif, PhD, 53. Prof Exp: Instr biol, Atlantic Union Col, 45-47 & Andrews Univ, 47-49; assoc, Univ Southern Calif, 50-52; head dept parasitol, Sch Trop & Prev Med, 50-59, instr microbiol, Sch Med, 53-56, from asst prof to assoc prof, 56-68, PROF MICROBIOL, SCH MED, LOMA LINDA UNIV, 68- Mem: Am Soc Trop Med & Hyg; Am Soc Parasitol; Am Micros Soc; Soc Protozool; Japanese Soc Parasitol. Res: Schistosomiasis; parasite therapy; helminths. Mailing Add: Dept of Microbiol Loma Linda Univ Sch of Med Loma Linda CA 92354

WAGNER, EDWARD KNAPP, b Akron, Ohio, May 4, 40; m 61; c 2. ANIMAL VIROLOGY, BIOCHEMISTRY. Educ: Univ Calif, Berkeley, BA, 62; Mass Inst Technol, PhD(biochem), 67. Prof Exp: Helen Hay Whitney Found fel, Univ Chicago, 67-70; asst prof, 70-75, ASSOC PROF VIROL, UNIV CALIF, IRVINE, 75- Concurrent Pos: Nat Cancer Inst res grant, 70- Mem: AAAS; Am Soc Microbiol; Tissue Cult Asn; Am Soc Biol Chemists; Am Soc Cell Biol. Res: Control of gene action in animal virus infection; mechanism of viral carcinogenesis; control of information transfer between nucleus and cytoplasm in eucaryotic cells. Mailing Add: Dept of Molecular Biol & Biochem Univ of Calif Irvine CA 92664

WAGNER, EDWARD LEWIS, chemical physics, deceased

WAGNER, ERIC G, b Ossining, NY, Oct 1, 31; m 60; c 2. MATHEMATICS. Educ: Harvard Univ, BA, 53; Columbia Univ, MA, 59, PhD(math), 63. Prof Exp: Tech engr, IBM Corp, 53-54, assoc engr switching theory, 56-58, RES STAFF MEM, T J WATSON RES CTR, IBM CORP, 58- Concurrent Pos: Lectr, NY Univ, 64-65, adj asst prof, 65-66; sr vis res fel, Queen Mary Col, Univ London, 73-74. Mem: Am Math Soc; Asn Comput Mach; Asn Symbolic Logic. Res: Automata theory, computability theory and category theory with emphasis on their relationship to computer science; theory of programming languages and logical design. Mailing Add: IBM Watson Res Ctr PO Box 218 Yorktown Heights NY 10598

WAGNER, EUGENE ROSS, b Monroe, Wis, Nov 21, 37; m 58; c 2. MEDICINAL CHEMISTRY. Educ: Univ Wis, BS, 59, PhD(org chem), 64. Prof Exp: Chemist, Spec Assignment Prog, 63-64, Dow Human Res & Develop Labs, Pitman-Moore Div, 64-67, sr res chemist, 67-72, res specialist, Chem Biol Res, 72-75, SR RES SPECIALIST, PHARMACEUT CHEM, DOW CHEM CO, 75- Mem: Am Chem Soc. Res: Organic synthesis of biologically active compounds; synthesis of hypocholesteremic agents and radiolabeled compounds. Mailing Add: 438 Bldg Dow Chem Co Midland MI 48640

WAGNER, EUGENE STEPHEN, b Gary, Ind, Mar 30, 34; m 62; c 3. BIOLOGICAL CHEMISTRY, PHYSICAL CHEMISTRY. Educ: Ind Univ, BS, 59; Purdue Univ, PhD(chem), 64. Prof Exp: Instr chem, Purdue Univ, 62-64; sr phys chemist, Eli Lily & Co, Ind, 64-71; asst prof, 71-75, ASSOC PROF CHEM, BALL STATE UNIV & MUNCIE CTR MED EDUC, 75- Mem: Am Chem Soc. Res: Chemical kinetics; reaction mechanisms; mechanism of penicillin hypersensitivity. Mailing Add: Dept of Chem Ball State Univ Muncie IN 47306

WAGNER, FLORENCE SIGNAIGO, b Birmingham, Mich; m 48; c 2. BOTANY. Educ: Univ Mich, Ann Arbor, AB, 41, MA, 43; Univ Calif, Berkeley, PhD(bot), 52. Prof Exp: Res asst soc sci, Off Coordr Inter-Am Affairs, 43-45 & Off Strategic Serv, 45; from res asst to res assoc bot, 61-73, SR RES ASSOC BOT, UNIV MICH, ANN ARBOR, 73- Concurrent Pos: Lectr, Univ Ctr Adult Educ, Ann Arbor, 71- Mem: Am Fern Soc; Brit Pteridological Soc; Bot Soc Am; Am Soc Plant Taxonomists. Res: Analysis of chromosomal behavior in ferns and fern hybrids and comparative studies of their morphology. Mailing Add: Dept of Bot Univ of Mich Ann Arbor MI 48104

WAGNER, FRANCES JOAN ESTELLE, paleontology, see 12th edition

WAGNER, FRANK A, JR, b New Haven, Conn, Apr 19, 32; m 61; c 4. ORGANIC CHEMISTRY. Educ: Yale Univ, BS, 58; Rutgers Univ, New Brunswick, PhD(org chem), 68. Prof Exp: Chemist, 58-65, res chemist, 65-68, SR RES CHEMIST, AM CYANAMID CO, 68- Mem: Am Chem Soc. Res: Synthesis of compounds as possible pesticidal agents; design of procedures for large-scale syntheses. Mailing Add: Am Cyanamid Co PO Box 400 Princeton NJ 08540

WAGNER, FRANK JOSEPH, mathematics, see 12th edition

WAGNER, FRANK S, JR, b Temple, Tex, Aug 26, 25; m 53; c 5. ORGANIC CHEMISTRY. Educ: Southwest Tex State Col, BA & MA, 47. Prof Exp: Assoc prof chem, Schreiner Inst, Tex, 48-50; analyst, Celanese Corp, 50-52, group leader, 52-53, librn, 53-65, HEAD INFO CTR, TECH CTR, CELANESE CHEM CO, 65- Mem: AAAS; Am Chem Soc; Spec Libr Asn; Egypt Explor Soc. Res: Application of machine methods to critical literature reviews and commerical intelligence activities; writing of encyclopedic reviews. Mailing Add: Celanese Chem Co PO Box 9077 Corpus Christi TX 78408

WAGNER, FREDERIC HAMILTON, b Corpus Christi, Tex, Sept 26, 26; m 49; c 2. BIOLOGY. Educ: Southern Methodist Univ, BS, 49; Univ Wis, MS, 53, PhD(wildlife mgt, zool), 61. Prof Exp: Refuge asst, US Fish & Wildlife Serv, 45; asst zool & bot, Southern Methodist Univ, 46-49; asst wildlife mgt, Univ Wis, 49-51; res fel, Wildlife Mgt Inst, 51; res biologist, Wis Conserv Dept, 52-58; asst prof wildlife resources, Utah State Univ, 58-59; res biologist, Wis Conserv Dept, 59-61; assoc prof, 61-66, PROF WILDLIFE RESOURCES, UTAH STATE UNIV, 66-, ASSOC DEAN, COL NATURAL RESOURCES, 70- Concurrent Pos: Dir Desert Biome, US-Int Biol Prog & mem US Exec Comt, Int Biol Prog, 71-74; mem comt predator control, President's Coun Environ Qual, 71. Honors & Awards: Award, Wildlife Soc, 68. Mem: Ecol Soc Am; Am Soc Mammal; Wildlife Soc; Am Inst Biol Sci; Cooper Ornith Soc. Res: Vertebrate population ecology, especially population dynamics, limiting factors and homeostatic mechanisms; wildlife management; conservation of natural resources; systems ecology. Mailing Add: 1066 N 1730 E Logan UT 84321

WAGNER, FREDERICK WILLIAM, b Erie, Pa, Feb 4, 40; m 65. BIOCHEMISTRY. Educ: Southwest Tex State Col, BSc, 62; Tex A&M Univ, PhD(biochem), 66. Prof Exp: US Air Force res assoc, Brooks AFB, 66-67; res fel biochem & biophys, Tex A&M Univ, 67-68; asst prof biochem & nutrit, 68-73, ASSOC PROF, LAB AGR BIOCHEM, UNIV NEBR, LINCOLN, 73- Res: Structure and function of proteins with special emphasis on proteolytic enzymes. Mailing Add: Dept of Biochem & Nutrit Univ of Nebr Lincoln NE 68503

WAGNER, GEORGE HENRY, soil microbiology, biochemistry, see 12th edition

WAGNER, GEORGE HOYT, b Mulberry, Ark, Dec 28, 14; m 39; c 3. GEOLOGY. Educ: Univ Ark, BS, 37; Univ Iowa, MS, 39, PhD(phys chem), 41. Prof Exp: Chemist, Univ Ark, 35-37; asst chem, Univ Iowa, 37-41; res chemist, Linde Div, Union Carbide Corp, 41-47, head div phys chem, 47-51, asst to supt, 51-53, res supvr, 53-55, mgr res, 55-59, dir, 59-64, mgr develop, 64-65, dir res, mining & metals div, 65-66, vpres, 66-70, ferroalloy div, 70-71; CONSULT RES & DEVELOP, 71- Concurrent Pos: Adj prof, Univ Ark, 74- Honors & Awards: Schoelkopf Medal, Am Chem Soc, 60. Mem: Am Chem Soc; Am Phys Soc; Welding Soc; Am Inst Mining, Metall & Petrol Engrs; Soc Chem Indust. Res: Synthetic lubricants; corrosion inhibition; organometallics; geochemistry and economic geology. Mailing Add: Box 144 Fayetteville AR 72701

WAGNER, GEORGE JOSEPH, b Buffalo, NY, Sept 15, 43; m 70; c 1. PLANT PHYSIOLOGY. Educ: State Univ NY Buffalo, BA, 70, MA, 71, PhD(biol), 74. Prof Exp: RES ASSOC PLANT BIOCHEM, BROOKHAVEN NAT LAB, 74- Mem: Am Soc Plant Physiologists; AAAS; Phytochem Soc NAm. Res: Study of the physiology and biochemistry of the mature plant cell vacuole and the metabolism of vacuolar constituents. Mailing Add: Biol Div Brookhaven Nat Lab Upton NY 11973

WAGNER, GEORGE RICHARD, b Chicago, Ill, Nov 12, 33; m 54; c 3. SOLID STATE PHYSICS. Educ: Univ Ill, Urbana, BS, 60; Carnegie-Mellon Univ, MS, 62, PhD(physics), 65. Prof Exp: Sr engr, 65-73, FEL SCIENTIST PHYSICS, RES & DEVELOP CTR, WESTINGHOUSE ELEC CORP, 73- Mem: Am Phys Soc. Res: Alternating current losses, stability and properties which characterize superconducting wires for use in magnets and machines. Mailing Add: Res & Develop Ctr Westinghouse Elec Corp Pittsburgh PA 15235

WAGNER, GERALD GALE, b Plainview, Tex, June 3, 41; m 62; c 2. IMMUNOLOGY. Educ: Tex Tech Col, BS, 63; Univ Kans, MA, 65, PhD, 68. Prof Exp: Microbiologist, Immunol Div, Plum Island Animal Dis Lab, Animal Dis & Parasite Div, 68-70, MICROBIOLOGIST, COOP RES DIV, E AFRICAN VET RES ORGN, USDA, 71- Concurrent Pos: Nat Acad Sci-Agr Res Serv res fel, Plum Island Animal Dis Lab, 68-70. Mem: Am Soc Microbiol; Am Asn Immunol. Res: Immunochemistry; viral and protozoal immunology. Mailing Add: USDA Coop Res Div EAfrican Vet Res Orgn PO Kabete Kenya East Africa

WAGNER, GERALD RICHARD, b Norfolk, Nebr, Sept 2, 34; m 59; c 3. STATISTICS, OPERATIONS RESEARCH. Educ: Univ Nebr, Lincoln, BS, 58, MS, 60; Iowa State Univ, PhD(nutrit & statist), 64. Prof Exp: Res statistician, Res & Develop, Swift & Co, 64-65; head exp statist, 65-67, chief corp statist & mgr opers res, 67-68; vpres consult, MRI Systs Corp, 68-69; HEAD, OPERS RES GROUP, UNIV TEX, AUSTIN, 69- Concurrent Pos: Lectr & consult, Univ Tex, Austin, 68-69 & Univ Wis, Honeywell, Inc, Gen Mills, Ling-Temco-Vought & MRI Systs Corp, 69- Mem: Am Statist Asn; Inst Mgt Sci. Res: Application of computers in learning and innovative teaching methods; policy capturing and judgment analysis; on-line interactive information and decision systems; experimental applied statistics; computer assisted decision making. Mailing Add: Dept of Mech Eng Univ of Tex Austin TX 78712

WAGNER, GERALD ROY, b Evansville, Ind, Feb 14, 28; m 50; c 2. ORGANIC CHEMISTRY, GEOLOGY. Educ: Mt Union Col, BS, 50; Univ Ark, MS, 52. Prof Exp: Res chemist, Com Solvents Corp, NY, 52-53, Olin Mathieson Chem Corp, 53-55 & Nat Aniline Div, Allied Chem Corp, 55-61; res chemist, 61-66, asst prof, 66, head dept, 66-72, PROF CHEM, ERIE COMMUNITY COL, 66- Mem: Am Chem Soc; Am Soc Eng Educ; Nat Asn Geol Teachers; Am Inst Chemists. Res: Geological education; chemical education. Mailing Add: Dept of Chem Erie Community Col Buffalo NY 14221

WAGNER, HANS, b July 19, 32; US citizen. ORGANIC CHEMISTRY. Educ: Univ Iowa, BS, 55; Pa State Univ, PhD(chem), 59. Prof Exp: Sr res investr, 59-70, GROUP LEADER CHEM, G D SEARLE & CO, 70- Mem: Am Chem Soc. Res: Dipolar cycloadditions; mesoionic compounds; heterocyclic azido compounds. Mailing Add: G D Searle & Co Box 5110 Chicago IL 60680

WAGNER, HARRY HENRY, b San Diego, Calif, Jan 10, 33; m 56; c 3. FISH BIOLOGY, ECOLOGY. Educ: Humboldt State Col, BS, 55; Ore State Univ, MS, 59, PhD(fisheries), 70. Prof Exp: Fishery res biologist & physiol ecologist, 59-68, fishery res coordr, 69-73, CHIEF DIV WILDLIFE RES, ORE WILDLIFE COMN, 73- Concurrent Pos: Courtesy assoc prof, Ore State Univ, 59- Mem: Am Fisheries Soc; Am Inst Fishery Res Biol. Res: Parr-smolt transformation of anadromous salmonids. Mailing Add: Div Wildlife Res Ore Wldlf Comn 303 Exten Hall Ore State Univ Corvallis OR 97331

WAGNER, HARRY MAHLON, b Iola, Kans, June 1, 24; m 44; c 5. MATHEMATICS. Educ: Naval Postgrad Sch, BS, 54; Kans State Teachers Col, MS, 64; Univ Ark, EdD(higher educ), 69. Prof Exp: Instr math, John Brown Univ, 64-67; res grad asst psychol, Univ Ark, 67-69; asst prof, 69-72, ASSOC PROF MATH, CAMERON UNIV, 72- Mem: Math Asn Am. Mailing Add: Dept of Math Cameron Univ Lawton OK 73501

WAGNER, HENRY GEORGE, b Washington, DC, Sept 13, 17; m 45; c 3. NEUROSCIENCES, AEROSPACE MEDICINE. Educ: George Washington Univ, AB, 39, MD, 42; Univ Pa, cert ophthal, 49. Prof Exp: Intern, Naval Hosp, Brooklyn, NY, 42-43, med officer, Naval Med Sch, Md, 43, flight surgeon, Sch Aviation Med, Naval Air Sta, Fla, 43-44, flight surgeon, Naval Air Base, Guam, 44-45, flight surgeon, Naval Air Sta, Tex, 45-46, asst supt aeromed equip lab, Naval Air Exp Sta, Pa, 46-48, med res investr, Naval Med Res Inst, Md, 51-54, sr med officer, USS Valley Forge, 54-56, head physiol div, Naval Med Res Inst, 56-60, cmndg officer, 60-61, exec officer, 61-64, actg dir physiol sci dept, 61-64, dir aerospace crew equip lab, Naval Air Eng Ctr, Pa, 64-66; dir intramural res, 66-74, HEAD SECT NEUROANAL INTERACTIONS, LAB NEUROPHYSIOL, NAT INST NEUROL, COMMUNICATIVE DIS & STROKE, 74- Concurrent Pos: Fel biophys, Johns Hopkins Univ, 49-51, hon prof, 58-64; mem comt vision, hearing & bioacoust, Nat Acad Sci-Nat Res Coun. Mem: AAAS; Am Physiol Soc; Soc Neurosci; Am Col Prev Med; AMA. Res: Neurophysiology of the visual system. Mailing Add: Nat Inst of Neurol Commun Dis & Stroke Bethesda MD 20014

WAGNER, HENRY N, JR, b Baltimore, Md, May 12, 27; m 51; c 4. INTERNAL MEDICINE, NUCLEAR MEDICINE. Educ: Johns Hopkins Univ, AB, 48, MD, 52. Prof Exp: From asst prof med & radiol, sch med to assoc prof med, radiol & radiol sci, 59-67, assoc prof med, 67-68, PROF RADIOL SCI & RADIOL, SCH MED & SCH HYG & PUB HEALTH, JOHNS HOPKINS UNIV, 67-, PROF MED, 68-, ACTG CHMN RADIOL SCI, 72- Mem: Am Fedn Clin Res; Asn Am Physicians; Am Soc Clin Invest; Soc Nuclear Med (past pres); World Fedn Nuclear Med & Biol (pres). Mailing Add: Johns Hopkins Univ Sch of Hyg & Pub Health Baltimore MD 21205

WAGNER, HERMAN BLOCK, b Baltimore, Md, Dec 25, 23; m 46; c 5. PHYSICAL CHEMISTRY. Educ: Johns Hopkins Univ, BE, 44, MA, 45, PhD(phys chem), 48. Prof Exp: Chemist, Johns Hopkins Univ, 44-45; instr phys & inorg chem, Loyola Col, Md, 46-48, asst prof, 48-49; res chemist, E I du Pont de Nemours & Co, 49-55; assoc res prof, Rutgers Univ, 55-57; dir chem res, Tile Coun Am Res Ctr, 57-61; PROF CHEM, DREXEL UNIV, 61- Concurrent Pos: Consult, Am Cyanamid Co, Tile Coun Res Ctr & Dow Chem Co. Mem: Am Chem Soc; Sigma Xi. Res: Corrosion; electrochemistry; polymers; photochemistry; hydraulic cements. Mailing Add: Dept of Chem Drexel Univ Philadelphia PA 19104

WAGNER, HERMAN LEON, b New York, NY, Mar 21, 21; m 54; c 1. POLYMER CHEMISTRY. Educ: City Col, New York, BS, 42; Polytech Inst Brooklyn, MS, 46; Cornell Univ, PhD(chem), 50. Concurrent Pos: Chemist, SAM Labs, Manhattan Proj, Columbia Univ, 42-46; res assoc, Cornell Univ, 50-51; phys chemist, E I du Pont de Nemours & Co, 51-55, M W Kellog Co, 55-57 & Celanese Corp Am, 57-68; RES CHEMIST, NAT BUR STANDARDS, 68- Mem: Am Chem Soc. Res: Physical chemistry of high polymers; dilute solution properties; characterization of high polymers; thermal analysis; composites; melt rheology; fibers; correlation of molecular structure with physical properties. Mailing Add: Nat Bur of Standards Washington DC 20234

WAGNER, JAMES BRUCE, JR, b Hampton, Va, July 28, 27; m 51; c 3. PHYSICAL CHEMISTRY. Educ: Univ Va, BS, 50, PhD(chem), 55. Prof Exp: Fel metall, Mass Inst Technol, 54-56; asst prof, Pa State Univ, 56-58 & Yale Univ, 58-62; assoc prof, 62-65, PROF MAT SCI, NORTHWESTERN UNIV, 65-, DIR, MAT RES CTR, 72- Concurrent Pos: Ford Found resident engr pract, Semiconductor Prod Div, Motorola, Inc, Ariz, 68-69. Mem: Am Phys Soc; Am Inst Mining, Metall & Petrol Eng; Electrochem Soc. Res: Oxidation of metals; thermodynamics and transport properties of compound semiconductors. Mailing Add: Dept of Mat Sci Northwestern Univ Evanston IL 60201

WAGNER, JOHN A, biology, deceased

WAGNER, JOHN ALEXANDER, b Kansas City, Mo, Feb 9, 35; m 63; c 2. ENTOMOLOGY. Educ: Northwestern Univ, BS, 57, MS, 59, PhD(biol), 62. Prof Exp: MEM FAC BIOL & SCI, KENDALL COL, 62- Concurrent Pos: Collabr & consult, Encycl Britannica Films, 62-75; res assoc, Dept Biol Sci, Northwestern Univ, 63-76; mem, Environ Studies Group, Alfred Benesch & Co, Consult Engrs, 74-76. Mem: Am Inst Biol Sci; Soc Syst Zool; Coleopterists Soc; Nature Conservancy. Res: Coleoptera, family Pselaphidae, especially nearctic and neotropics. Mailing Add: Dept of Biol Kendall Col Evanston IL 60204

WAGNER, JOHN ALFRED, b Baltimore, Md, Feb 9, 11; m 47; c 3. NEUROPATHOLOGY. Educ: Washington Col, Md, BS, 34; Univ Md, MD, 38; Am Bd Path, dipl, 47 & cert anat path & neuropath. Prof Exp: Intern, Univ Hosp, 38-40, asst resident med, Mercy Hosp, 40-41, from assoc prof to prof & head dept neuropath, 47-72, EMER PROF NEUROPATH, UNIV MD, BALTIMORE CITY, 72-, ATTEND PATHOLOGIST, UNIV HOSP, 46- Concurrent Pos: Weaver fel path, Sch Med, Univ Md, Baltimore City, 41-42, Hitchcock fel, 42-46; pathologist, St Agnes Hosp, 43-54 & Lutheran Hosp Md, 45-75; consult, Hosps, 48-; ed, Bull Univ Md Sch Med. Mem: Am Asn Neuropath; Am Med Writers' Asn; Am Acad Neurol; Int Acad Path. Res: Diseases of the central nervous system. Mailing Add: 115 Overhill Rd Baltimore MD 21210

WAGNER, JOHN CLEAVER, virology, immunology, see 12th edition

WAGNER, JOHN GARNET, b Weston, Ont, Mar 28, 21; m 46. PHARMACEUTICAL CHEMISTRY, ORGANIC CHEMISTRY. Educ: Univ Toronto, PhmB, 47; Univ Sask, BSP, 48, BA, 49; Ohio State Univ, PhD(pharmaceut chem), 52. Prof Exp: Instr pharmaceut chem, Ohio State Univ, 51-52, asst prof, 52-53; res scientist, Upjohn Co, Mich, 53-56, sect head, 56-63, sr res scientist, 63-68; asst dir res & develop, Pharm Serv, Univ Hosp, 68-72, PROF PHARM, COL PHARM, UNIV MICH, ANN ARBOR, 68-, MEM STAFF, UPJOHN CTR CLIN PHARMACOL, 73- Honors & Awards: Ebert Prize, Am Pharmaceut Asn, 61; Hoest-Madsen Medal, Int Pharmaceut Fedn, 72; Propter Merita Medal, Czech Med Soc, 74. Mem: AAAS; Am Pharmaceut Asn; Am Fedn Clin Res; Am Soc Clin Pharmacol & Therapeut; Am Soc Pharmacol & Exp Therapeut. Res: Pharmacokinetics and biopharmaceutics; absorption, metabolism and excretion of drugs in man. Mailing Add: Upjohn Ctr for Clin Pharmacol Univ of Mich Ann Arbor MI 48104

WAGNER, JOSEPH RICHARD, biochemistry, see 12th edition

WAGNER, KENNETH A, b Union City, Ind, Nov 30, 19; m 45. BOTANY. Educ: DePauw Univ, AB, 41, AM, 46; Univ Mich, PhD, 51. Prof Exp: Instr, Univ Tenn, 47-49; asst prof, Fla State Univ, 49-54; prof & head dept biol, Norfolk Div, Col William & Mary, 54-59; botanist & sci coordr, Powell Lab Div, Carolina Biol Supply Co, 59-66; prof biol, NC Wesleyan Col, 66-69; PROF BIOL, FERRIS STATE COL, 69- Concurrent Pos: Consult, Army Air Force Trop Test Detachment, Panama, 44; partic, Fla State Archaeol Exped, Cuba, 53; mem, Nat Wildlife Fedn. Mem: Nat Asn Biol Teachers; Am Inst Biol Sci. Res: North American liverworts; cypress; east coast salt marshes; desert ecology. Mailing Add: Dept of Biol Ferris State Col Big Rapids MI 49307

WAGNER, KIT KERN, b Chickasha, Okla, Nov 13, 47; m 70. DYNAMIC METEOROLOGY. Educ: Univ Okla, BS, 70, MS, 71, PhD(meteorol), 75. Prof Exp: Res asst meteorol, Univ Okla Res Inst, 70-75; meteorologist, Nat Severe Storm Lab, Nat Oceanic & Atmospheric Admin, 75; ASST PROF METEOROL, UNIV CALIF, DAVIS, 75- Mem: Am Meteorol Soc. Res: Mesoscale dynamic meteorology and boundary layer hydrodynamic instabilities. Mailing Add: Dept Land Air & Water Resources Univ of Calif Davis CA 95616

WAGNER, KLAUS PETER, b Berlin, Ger, Aug 23, 42; m 71. ORGANIC POLYMER CHEMISTRY, ORGANOMETALLIC CHEMISTRY. Educ: Univ Witwatersrand, BSc, 68; Univ Wis-Madison, PhD(inorg chem), 74. Prof Exp: Res technician, African Explosives & Chem Indust, 61-65; sr bursar inorg chem, Univ Witwatersrand, 67-69; res asst inorg chem, Univ Wis-Madison, 71-74; RES CHEMIST, RES CTR, HERCULES, INC, 74- Mem: Am Chem Soc; The Chem Soc. Res: Transition metal organometallic chemistry; synthesis and properties of compounds as related to structure and uses as reagents in organic synthesis and in catalysis; inorganic and organic polymer chemistry and applications; paper and paper additive chemistry. Mailing Add: Hercules Res Ctr Hercules Inc Wilmington DE 19899

WAGNER, LAWRENCE CARL, b Campbellsport, Wis, Dec 28, 46. HIGH TEMPERATURE CHEMISTRY. Educ: Marquette Univ, BS, 68; Purdue Univ,

PhD(chem), 74. Prof Exp: APPOINTEE CHEM DIV, ARGONNE NAT LAB, 74- Mem: Am Chem Soc; Am Soc Mass Spectrometry; Sigma Xi. Res: High temperature mass spectrometry; photoelectron spectrometry; diffusion controlled processes. Mailing Add: Chem Div Argonne Nat Lab 9700 Cass Ave Argonne IL 60439

WAGNER, MARTIN JAMES, b Independence, Kans, Oct 4, 31; m 53; c 3. BIOCHEMISTRY. Educ: Kans State Col Pittsburg, BS, 54; Ind Univ, PhD(biochem), 58. Prof Exp: Asst chem, Ind Univ, 54-58; from asst prof to assoc prof, 58-63, PROF BIOCHEM, BAYLOR COL DENT, 63-, CHMN DEPT, 58- Mem: Am Chem Soc; Int Asn Dent Res. Res: Intermediary metabolism of inorganic fluoride ion, following its ingestion in trace quantities in food and drinking water. Mailing Add: Dept of Biochem Baylor Col of Dent Dallas TX 75226

WAGNER, MELVIN PETER, b Nebr, Nov 16, 26; m 53; c 5. ORGANIC CHEMISTRY, POLYMER CHEMISTRY. Educ: Creighton Univ, BS, 49, MS, 52; Univ Akron, PhD(polymer chem), 60. Prof Exp: Res chemist, 52-60, sr res chemist, 60-64, SUPVR RUBBER CHEM RES, BARBERTON LAB, CHEM DIV, PPG INDUSTS, 64- Mem: Am Chem Soc. Res: High polymers; rubber reinforcement; vulcanization. Mailing Add: PPG Industs PO Box 31 Barberton OH 44203

WAGNER, MORRIS, b Chicago, Ill, Aug 6, 17; m 47; c 4. BACTERIOLOGY, IMMUNOLOGY. Educ: Cornell Univ, BS, 41; Univ Notre Dame, MS, 46; Purdue Univ, PhD, 66. Prof Exp: Asst, 41-43, bacteriologist, Lobund Labs, 43-46, from instr to assoc prof, Univ, 46-69, PROF MICROBIOL & ASST CHMN DEPT, UNIV NOTRE DAME, 69-, RES SCIENTIST, LOBUND LAB, 41- Concurrent Pos: NSF sci fac fel, 63; mem Subcomt Stand Gnotobiotics, Nat Acad Sci. Mem: Am Soc Microbiol; Asn Gnotobiotics. Res: Gnotobiotics; experimental dental caries; defined intestinal flores. Mailing Add: Dept of Microbiol Univ of Notre Dame Notre Dame IN 46556

WAGNER, MYRON L, b Oak Park, Ill, Nov 11, 23; m 47; c 3. CHEMISTRY. Educ: Carthage Col, AB, 48; Univ Iowa, MS, 52, PhD(chem), 54. Prof Exp: Res assoc, Cornell Univ, 54-56; instr chem, Univ Iowa, 56-57; from asst prof to assoc prof, 57-63, PROF CHEM, UNIV MO-KANSAS CITY, 63- Mem: Am Chem Soc. Res: Physical chemistry of systems of biological importance. Mailing Add: Dept of Chem Univ of Mo Kansas City MO 64110

WAGNER, NEAL RICHARD, b Topeka, Kans, May 4, 40; m 71. TOPOLOGY. Educ: Univ Kans, AB, 62; Univ Ill, Urbana-Champaign, AM, 64, PhD(math), 70. Prof Exp: ASST PROF MATH, UNIV TEX, EL PASO, 69- Mem: Am Math Soc; Math Asn Am. Res: Topology of function spaces; conformal mapping theory. Mailing Add: Dept of Math Univ of Tex El Paso TX 79968

WAGNER, NORMAN KEITH, b Longview, Wash, Oct 3, 32; m 54; c 2. MICROMETEOROLOGY. Educ: Univ Wash, BS, 54, MS, 56; Univ Hawaii, PhD(meteorol), 66. Prof Exp: Instr meteorol, Univ Tex, 56-57 & 58-63, res meteorologist, 57-58; asst prof meteorol, Univ Hawaii, 65, asst researcher, 65-66; asst prof, 66-70, ASSOC PROF METEOROL, UNIV TEX, AUSTIN, 70-, DIR, ATMOSPHERIC SCI GROUP, 72- Mem: Am Meteorol Soc. Res: Meteorological instrumentation; micrometeorology; atmospheric boundary layer. Mailing Add: Atmospheric Sci Group Univ of Tex Austin TX 78712

WAGNER, ORVIN EDSON, physics, see 12th edition

WAGNER, PAUL, b Troy, NY, Dec 27, 23; m 62; c 2. PHYSICAL CHEMISTRY. Educ: State Univ NY, AB, 47, AM, 48; Univ Rochester, PhD(phys chem), 52. Prof Exp: Asst, Cornell Univ, 48 & Univ Rochester, 49-50; res scientist, Lewis Flight Propulsion Lab, Nat Adv Comt Aeronaut, 52-56; MEM STAFF, LOS ALAMOS SCI LAB, UNIV CALIF, 56- Concurrent Pos: Mem, Int Thermal Expansion Comt & Bd Govrs, Int Thermal Conductivity Conf. Mem: Am Soc Metals; Am Soc Testing & Mat; Sigma Xi. Res: Spectroscopy; combustion; high temperature properties of materials; diffusion at high temperatures; thermal, electrical, expansion properties at high temperatures; structural interpretations; all aspects of materials research and development for high temperature reactor application, including advanced high temperature nuclear fuels. Mailing Add: Los Alamos Sci Lab MS-734 Univ Calif Los Alamos NM 87545

WAGNER, PETER J, b Chicago, Ill, Dec 25, 38; m 63; c 5. PHYSICAL ORGANIC CHEMISTRY, PHOTOCHEMISTRY. Educ: Loyola Univ, Ill, BS, 60; Columbia Univ, MA, 61, PhD(chem), 63. Prof Exp: Res assoc chem, Columbia Univ, 63-64; NSF fel, Calif Inst Technol, 64-65; from asst prof to assoc prof, 65-70, PROF CHEM, MICH STATE UNIV, 70- Concurrent Pos: Sloan fel, 68-70; NSF sr fel, Univ Calif, Los Angeles, 71-72; consult, Hercules, Inc, 73-; assoc ed, J Am Chem Soc, 75- Mem: AAAS; Am Chem Soc. Res: Mechanisms of free radical and photochemical reactions; electronic energy transfer; spectroscopy of excited states. Mailing Add: Dept of Chem Col of Natural Sci Mich State Univ East Lansing MI 48823

WAGNER, PHILIP LAURENCE, b San Jose, Calif, Oct 7, 21; m 65; c 1. CULTURAL GEOGRAPHY. Educ: Univ Calif, Berkeley, AB, 47, MA, 50, PhD(geog), 53. Prof Exp: Teacher geog, Univ Calif Exten, Far E Prog, 53-54; res assoc, Soc Sci Div, Univ Chicago, 54-55, asst prof, 55-61; from asst prof to prof, Univ Calif, Davis, 61-67; PROF GEOG, SIMON FRASER UNIV, 67- Mem: Asn Am Geog; Am Geog Asn; Am Anthrop Asn. Res: Geographical perspective as human nature and culture; geographical aspects of language, religion, social organization and the arts; cultural groups and differences in relation to environments. Mailing Add: Dept of Geog Simon Fraser Univ Burnaby BC Can

WAGNER, RAPHAEL DARREL, b Charlestown, Wis, May 24, 13; m 39; c 2. MATHEMATICS. Educ: Univ Wis, BS, 39, PhM, 40, PhD(algebra), 46. Prof Exp: Instr, 42-47, from asst prof to assoc prof, 47-68, PROF MATH, EXTEN DIV, UNIV WIS-MADISON, 68- Concurrent Pos: Ed, Wis Teacher Math, Wis Math Coun. Mem: Math Asn Am. Res: Abstract algebra. Mailing Add: 701 Extension Bldg Univ of Wis-Madison Madison WI 53706

WAGNER, RAYMOND LEE, b Kansas City, Mo, Aug 21, 46; m 69; c 1. THEORETICAL ASTROPHYSICS, ASTRONOMY. Educ: Rice Univ, BA, 68; Univ Tex, Austin, PhD(astron), 72. Prof Exp: Asst prof astron, Univ Wash, 72-74, res assoc, 73-74; ASST PROF ASTRON & PHYSICS, LA STATE UNIV, BATON ROUGE, 74- Concurrent Pos: NSF res grant, 75- Mem: Am Astron Soc; AAAS; Sigma Xi; Am Asn Univ Profs. Res: Stellar structure and evolution; stellar stability; nucleosynthesis; peculiar stars. Mailing Add: Dept of Physics & Astron La State Univ Baton Rouge LA 70803

WAGNER, RAYMOND PETER, physical inorganic chemistry, see 12th edition

WAGNER, RICHARD CARL, b Orange, NJ, Sept 25, 41. MATHEMATICS. Educ: Rutgers Univ, AB, 63; Univ Chicago, MS, 64, PhD(math), 68. Prof Exp: Asst prof, 68-74, ASSOC PROF MATH, FAIRLEIGH DICKINSON UNIV, 74- Mem: AAAS; Am Math Soc; Math Asn Am; London Math Soc. Res: Quadratic forms; algebraic k-theory. Mailing Add: Dept of Math Fairleigh Dickinson Univ Madison NJ 07940

WAGNER, RICHARD H, b West Haven, Conn, Oct 31, 34. PLANT ECOLOGY. Educ: Univ Conn, BA, 56; Duke Univ, MA, 58, PhD(bot), 63. Prof Exp: Asst prof biol, Greensboro Col, 59-63; res assoc & fel, Brookhaven Nat Lab, 63-65; asst prof bot, Pa State Univ, University Park, 65-74; CONSULT & WRITER, 74- Mem: AAAS; Asn Trop Biol; Ecol Soc Am; Brit Ecol Soc. Res: Autoecology of stress environment species; factor interaction as affecting phenology and life history; experimental ecology of weedy species. Mailing Add: c/o W W Norton Co 500 Fifth Ave New York NY 10024

WAGNER, RICHARD JOHN, b Barnesville, Minn, Jan 13, 36; m 58; c 2. SOLID STATE PHYSICS. Educ: St John's Univ, Minn, BS, 58; Univ Calif, Los Angeles, MS, 60, PhD(physics), 66. Prof Exp: Mem tech staff eng, 58-68, staff physicist, 68-69, sr tech staff asst physics, 69-71, sr staff physicist, 71-75, SR SCIENTIST, HUGHES AIRCRAFT CO, 75- Concurrent Pos: Teaching asst, Univ Calif, Los Angeles, 61-63, res asst, 63-66, asst res physicist, 66-67. Mem: Am Phys Soc. Res: Solid state physics, particularly as applicable to solid state microwave devices. Mailing Add: Radar Microwave Lab Hughes Aircraft Co Culver City CA 90230

WAGNER, RICHARD JOHN, b New Ulm, Minn, Dec 3, 32; m 53; c 4. MATHEMATICAL PHYSICS. Educ: Univ Minn, BA, 53, MS, 55; Rice Univ, PhD(physics), 58. Prof Exp: Mem tech staff theoret physics, Ramo-Wooldridge Corp, 58-59 & TRW Space Tech Labs, 59-61, sect head quantum theory, 62-65, sect head, TRW Systs, 65-66, SECT HEAD WAVE PROPAGATION, TRW SYSTS, 66- Concurrent Pos: Cedric K Ferguson Medal, Am Inst Mining Metall & Petrol Engrs, 58. Mem: Am Phys Soc. Res: Scattering theory; missile and space vehicle re-entry physics; radio propagation; electromagnetic diffraction theory; radar concealment; underwater acoustics; statistical scattering theory; remote sensing; laser propagation. Mailing Add: Res Group Bldg R1-1196 TRW Systs One Space Park Redondo Beach CA 90278

WAGNER, RICHARD LLOYD, b Manitowoc, Wis, May 30, 34; m 56; c 5. POLYMER CHEMISTRY, PHOTOCHEMISTRY. Educ: Univ Wis-Madison, BS, 60. Prof Exp: Chemist, Hercules Res Ctr, 60-65, res chemist & proj leader mat sci & appl res, 66-71, sr venture analyst, New Enterprise Dept, 71-73, supvr mkt serv, 73-75, MGR ENG & DEVELOP, GRAPHIC SYSTS DIV, ORG DEPT, HERCULES INC, 75- Mem: Am Chem Soc. Res: Applications research and product development work related to uses of company products in graphic arts areas and other commercially important areas; product and market development with photechnical systems in graphic arts uses; chemical and equipment systems for graphic arts products involving photopolymers. Mailing Add: 2319 Wynnwood Rd Wilmington DE 19810

WAGNER, RICHARD VERNON, b San Bruno, Calif, Dec 10, 21; m 60. ANTHROPOLOGY. Educ: Univ Calif, Berkeley, BA, 51, MA, 62, PhD(anthrop), 69. Prof Exp: Res assoc anthrop, Am Univ, 67-68; asst prof, Calif State Univ, San Diego, 69-70; maitre asst Am studies, Univ Provence, 70-71; ASST PROF ANTHROP, CALIF STATE UNIV, SAN DIEGO, 71- Mem: Am Anthrop Asn. Res: European cultures, particularly France; anthropology of rural communities in complex society; historical anthropology; social anthropology, with a focus on status structure and interaction. Mailing Add: Dept of Anthrop Calif State Univ San Diego CA 92115

WAGNER, ROBERT EARL, b Garden City, Kans, Mar 6, 21; m 48; c 3. AGRONOMY. Educ: Kans State Col, BS, 42; Univ Wis, MS, 43, PhD(agron, bot), 50. Prof Exp: Forage crops specialist, Ft Hays Exp Sta, USDA, 43-45, res agronomist, Forage & Range Sect, Plant Indust Sta, 45-54, proj leader pasture & range invests, 54-56; head dept agron, Univ Md, 56-59; eastern dir, Am Potash Inst, Washington, DC, 59-67, vpres, 67; dir coop exten serv, Univ Md, College Park, 67-75; PRES, POTASH INST, 75- Concurrent Pos: Pres, Am Forage & Grassland Coun, 62-64, chmn, Bd Dirs, 65-67; chmn, Nat Exten Comt Orgn & Policy, 72-73. Honors & Awards: Medallion Award, Am Forage & Grassland Coun. Mem: Fel AAAS; fel Am Soc Agron; Soc Range Mgt; Crop Sci Soc Am; Soil Conserv Soc Am. Res: Forage and range culture, production and management; reseeding; fertilization; grazing management and physiology of pasture and range plants. Mailing Add: Potash Inst 1649 Tullie Circle NE Atlanta GA 30329

WAGNER, ROBERT EDWIN, b Akron, Ohio, May 5, 20; m 52; c 2. ORGANIC CHEMISTRY. Educ: Mass Inst Technol, SB, 42; Princeton Univ, MA, 49, PhD(chem), 51. Prof Exp: Jr technologist, Shell Oil Co, 43-47; res chemist, 53-60, TECH SERV SUPVR, EXP STA LAB, E I DU PONT DE NEMOURS & CO, INC, 60- Concurrent Pos: Mem, Nat Defense Res Comt, 42. Res: Cellulose and polymer chemistry. Mailing Add: Exp Sta Lab E I du Pont de Nemours & Co Inc Wilmington DE 19898

WAGNER, ROBERT G, b Kansas City, Mo, Apr 2, 34; m 57; c 3. SOLID STATE PHYSICS. Educ: Grinnell Col, AB, 56; Univ Mo, MS, 60, PhD(physics), 66. Prof Exp: Res engr, NAm Aviation, Inc, 56-58; res asst solid state physics, 60-66, from res scientist to assoc scientist, 66-71, SR GROUP ENGR ELECTRONICS, McDONNELL DOUGLAS CORP, 71- Concurrent Pos: Asst prof, Univ Mo-St Louis, 71- Mem: AAAS; Inst Elec & Electronics Engrs; Am Phys Soc. Res: Thins films; device physics. Mailing Add: 7052 Kingsbury St Louis MO 63130

WAGNER, ROBERT H, b Peru, Ind, Aug 11, 21; m 45; c 6. BIOCHEMISTRY. Educ: DePauw Univ, AB, 43; Univ Cincinnati, PhD(biochem), 50. Prof Exp: Asst, DePauw Univ, 43-44; asst, Res Found, Children's Hosp, 46-50; res assoc path & instr biochem, 53-56, asst prof path & biochem, 57-61, assoc prof, 61-67, PROF PATH, SCH MED, UNIV NC, CHAPEL HILL, 67-, PROF BIOCHEM, 72- Concurrent Pos: USPHS sr fel, 59-63, res career develop fel, Univ NC, Chapel Hill, 64-69. Honors & Awards: Muray Thelin Hemophilia Award. Mem: AAAS; Am Chem Soc; Soc Exp Biol & Med; Am Inst Chemists; Am Soc Exp Path. Res: Plasma proteins; enzymes; blood clotting; antihemophilic factors. Mailing Add: Dept of Path Univ of NC Sch of Med Chapel Hill NC 27514

WAGNER, ROBERT MARVIN, b Fresno, Calif, Sept 5, 20; c 2. PHYSICAL ORGANIC CHEMISTRY. Educ: Fresno State Col, BA, 41; Stanford Univ, MSc, 47, PhD(phys org chem), 49. Prof Exp: Teaching asst, Fresno State Col, 38-41; asst chemist, US Bur of Mines, Nev, 41-43; asst, Stanford Univ, 46-48; res org chemist, Hanford Atomic Prod Opers, Gen Elec Co, Wash, 48-53; group leader, Saran Yarns Co, 54-55, head chem res sect, 55-57; group leader radiation chem, Stanford Res Inst, 57-60; head chem dept, Hazelton Nuclear Sci Corp, 61-62; head spec proj br, Test Develop Div, Qual Eval & Eng Lab, US Naval Weapons Sta, 62-71, SUPVRY GEN PHYS SCIENTIST, RADIATION PROTECTION OFF, QUAL EVAL & ENG LAB, US NAVAL WEAPONS STA, 71- Concurrent Pos: Head, Radiation Physics Div, Defense Nuclear Agency, DC, 67-70. Mem: Am Chem Soc; fel Am Inst Chemists. Res: Organosulfur and organophosphorus compounds; polyvinyl-vinylidene derivatives and plastics research; ultraviolet and infrared spectrophotometry; thermo-osmometry, instrumentation; radiation health physics; nuclear magnetic resonance;

gas-liquid-partition-chromatography techniques. Mailing Add: PO Box 814 Concord CT 94522

WAGNER, ROBERT PHILIP, b New York, NY, May 11, 18; m 47; c 3. GENETICS. Educ: City Col New York, BS, 40; Univ Tex, PhD(genetics), 43. Prof Exp: Instr zool, Univ Tex, 43-44; res biologist, Nat Cotton Coun, Dallas, 44-45; from asst prof to assoc prof, 45-56, PROF ZOOL, UNIV TEX, AUSTIN, 56- Concurrent Pos: Nat Res Coun fel, Calif Inst Technol, 46; Guggenheim fel, 57; mem genetics panel, NSF, 61-64 & prog projs comt, 64-68 & Genetics Training Grant Comt, Nat INst Gen Med Sci, 70-73. Mem: AAAS; Soc Study Evolution; Genetics Soc Am (secy, 65-66, vpres, 70, pres, 71); Am Soc Naturalists; Am Soc Biol Chemists. Res: Physiological and biochemical aspects of Neurospora; genetics of mitochondria; mammalian cell genetics. Mailing Add: Dept of Zool Univ of Tex Austin TX 78712

WAGNER, ROBERT RODERICK, b New York, NY, Jan 5, 23. VIROLOGY, MICROBIOLOGY. Educ: Yale Univ, MD, 46. Prof Exp: Intern med, New Haven Hosp, 46-47, asst resident, 49-50; instr med, Yale Univ, 51-53, asst prof, 53-55; from asst prof to prof microbiol, Johns Hopkins Univ, 56-67, from asst dean to assoc dean med fac, 57-63; PROF MICROBIOL & CHMN DEPT, UNIV VA, 67- Concurrent Pos: USPHS fel, Nat Inst Med Res, London, 50-51; vis fel, Dept Path, All Souls Col, Oxford Univ, 67; vis scientist, Dept Path, All Souls Col, Oxford Univ, 67; ed-in-chief, J Virol, 66-; consult, NIH, NSF & Am Cancer Soc; Josiah Macy Jr Found Fac Scholar, Oxford Univ, 75-76. Mem: Soc Exp Biol & Med; Am Soc Microbiol; Am Soc Clin Invest; Asn Am Med Cols; Asn Am Physicians. Res: Host susceptibility to infection. Mailing Add: Dept of Microbiol Univ of Va Charlottesville VA 22901

WAGNER, ROBERT THOMAS, b Winona, Minn, July 15, 23; m; c 3. NUCLEAR PHYSICS. Educ: US Mil Acad, BS, 46; Univ Va, PhD(nuclear physics), 55. Prof Exp: Staff mem, Los Alamos Sci Lab, 55-64; prof physics, St Mary's Col, Minn, 64-66; chief tech develop div, Nike-X Syst Off, US Army, 66-67; PROF & HEAD PHYSICS DEPT, NORTHERN MICH UNIV, 67- Mem: AAAS; Am Phys Soc. Res: Ferroelectric ceramics; particle accelerators; fission physics; nuclear decay schemes; neutron and x-ray transport and diffusion; theoretical mechanics; scintillation radiation detectors. Mailing Add: Dept of Physics Northern Mich Univ Marquette MI 49885

WAGNER, ROBERT WANNER, b Nesquehoning, Pa, May 5, 13; m 42; c 3. MATHEMATICS. Educ: Ohio Univ, AB, 34; Univ Mich, AM, 35, PhD(math), 37. Prof Exp: Instr math, Univ Wis, 37-39; from instr to assoc prof, Oberlin Col, 39-50; assoc dean, Col Arts & Sci, 61-70, actg dir, Off Instnl Studies, 70-72, PROF MATH, UNIV MASS, AMHERST, 50- Mem: Am Math Soc; Math Asn Am. Res: Differential equations; function theory of linear algebras. Mailing Add: Dept of Math & Statist Univ of Mass Amherst MA 01002

WAGNER, ROGER CURTIS, b Aitkin, Minn, May 4, 43; m 68; c 1. CELL BIOLOGY. Educ: Hamline Univ, BS, 65; Ohio Univ, MS, 67; Univ Minn, Minneapolis, PhD(cell biol), 71. Prof Exp: Teaching asst zool, anat & hist, Ohio Univ, 65-67; teaching asst biol & physiol, Univ Minn, Minneapolis, 67-68, res asst electrophysiol, St Paul, 68-69; teaching asst cell biol, Med Sch, Yale Univ, 71-74; ASST PROF BIOL SCI, UNIV DEL, 74- Concurrent Pos: Nat Cancer Inst fel cell biol, Sch Med, Yale Univ, 71- Mem: AAAS; Am Soc Cell Biol; Biophys Soc; Am Inst Biol Sci. Res: Mechanism and function of macropinocytosis and micropinocytosis in mammalian cells; biomembranes; cell and molecular biology; histology. Mailing Add: Dept of Biol Sci Univ of Del Newark DE 19711

WAGNER, ROMEO BARRICK, b Hopewell, Va, Dec 14, 17; m 48; c 1. ORGANIC CHEMISTRY. Educ: Gettysburg Col, AB, 38; Pa State Univ, MS, 40, PhD(chem), 41. Prof Exp: Asst chem, Pa State Univ, 41-42, instr, 42-45, asst prof, 45-51; res chemist, Res Ctr, 51-56, res supvr, 56-59, mgr, Naval Stores Res Div, 60-61, tech asst to dir res, 61-64, TECH ASST TO DIR, RES CTR, HERCULES, INC, 64- Mem: Am Chem Soc. Res: Synthetic organic chemistry; steroids; penicillin; insecticides; cellulose; resin acids; terpenes; phenols. Mailing Add: Res Ctr Hercules Inc Wilmington DE 19899

WAGNER, RONALD L, biochemistry, see 12th edition

WAGNER, ROSS IRVING, b Los Angeles, Calif, Apr 8, 25; m 49; c 3. INORGANIC CHEMISTRY. Educ: Univ Calif, Los Angeles, BS, 47; Univ Southern Calif, MS, 50, PhD(chem), 53. Prof Exp: Asst chem, Univ Calif, 49-53; sr res chemist, Am Potash & Chem Corp, 53-63; MEM TECH STAFF, ROCKETDYNE DIV, ROCKWELL INT, 63- Mem: Am Chem Soc. Res: Chemistry of boron and phosphorus sulfur and fluorine. Mailing Add: Rocketdyne Div Rockwell Int D/522-198-6633 Canoga Ave Canoga Park CA 91304

WAGNER, ROY, b Cleveland, Ohio, Oct 2, 38; m 68; c 2. ANTHROPOLOGY. Educ: Harvard Col, AB, 61; Univ Chicago, AM, 62, PhD(anthrop), 66. Prof Exp: Asst prof anthrop, Southern Ill Univ, 66-68; assoc prof, 68-74, PROF ANTHROP & CHMN DEPT, UNIV VA, 74- Concurrent Pos: Soc Sci Res Coun fac res grant, 68-69. Mem: Fel Am Anthrop Asn; Royal Anthrop Inst Gt Brit & Ireland. Res: Social, religious and symbolic anthropology; ideology and the study of figurative expression; New Guinea highlands and southern foothills; modern American society. Mailing Add: Dept of Anthrop Univ of Va Charlottesville VA 22903

WAGNER, RUSSEL OLSON, b Racine, Wis, Feb 18, 18; m 40; c 2. ECOLOGY, BOTANY. Educ: Univ Wis, BS, 40, MS, 47, PhD(bot), 60. Prof Exp: Teacher high schs, Wis, 40-42; instr biol, 47-53, asst prof biol sci, 53-57, assoc prof, 57-60, head dept life sci, 63-68, PROF BIOL SCI, UNIV WIS-PLATTEVILLE, 60- Concurrent Pos: Dir Wis univ biol sessions, Pigeon Lake Field Sta, 70-73. Mem: Ecol Soc Am; Entom Soc Am. Res: Influence of reproduction on distribution pattersn of prairie plants; ecology of a mound building ant found in prairie remnants, Formica cinerea. Mailing Add: Dept of Biol Sci Univ of Wis Platteville WI 53818

WAGNER, THOMAS EDWARDS, b Cleveland, Ohio, Nov 29, 42; m 66; c 1. BIOCHEMISTRY, ENDOCRINOLOGY. Educ: Princeton Univ, BS, 64; Northwestern Univ, PhD(biochem), 66. Prof Exp: Asst prof chem, Wellesley Col, 66-68; asst prof biochem, Med Col, Cornell Univ, 68-70; ASST PROF CHEM, OHIO UNIV, 70- Concurrent Pos: Petrol Res Fund grant, 67-70; assoc endocrinol, Sloan-Kettering Inst Cancer Res, 68-70. Mem: Am Chem Soc. Res: Study of weak interactions due to hydrophobic groups and their role in enzyme and hormone mechanism. Mailing Add: Dept of Chem Ohio Univ Athens OH 45701

WAGNER, TIMOTHY KNIGHT, b Pearl River, NY, July 5, 39; m 65; c 1. SOLID STATE PHYSICS. Educ: Univ Rochester, BS, 61; Univ Md, PhD(physics), 68. Prof Exp: Assoc solid state physic, Ames Lab, AEC, Iowa, 67-68, asst physicist, 68-70; from instr to asst prof, Iowa State Univ, 68-70; assoc prof, 70-74, PROF PHYSICS & CHMN DEPT, EAST STROUDSBURG STATE COL, 74- Mem: Am Phys Soc; Am Asn Physics Teachers. Res: Fermi surface measurements using radio-frequency size effect; ferromagnetic resonance in rare earth metals. Mailing Add: Dept of Physics East Stroudsburg State Col East Stroudsburg PA 18301

WAGNER, VAUGHN EDWIN, b Sharon, Pa, Sept 5, 42; m 68; c 1. ENVIRONMENTAL HEALTH, MEDICAL ENTOMOLOGY. Educ: Grove City Col, BS, 64; Pa State Univ, MS, 67; Mich State Univ, PhD(med entom), 75; Am Registry Prof Entom, cert. Prof Exp: Sanitarian, Allegheny Co Health Dept, Pa, 64-65; res assoc, USDA, 67-68; med entomologist, Dutchess Co Health Dept, NY, 68-71; res assoc environ health, Mich State Univ, 74-75; ASST PROF ENVIRON HEALTH, YORK COL, CITY UNIV NY, 75- Mem: Sigma Xi; AAAS; Entom Soc Am; Am Mosquito Control Asn. Res: Epidemiology of arthropod-borne diseases; ecology and population dynamics of invertebrate disease vectors. Mailing Add: Health Professions York Col City Univ NY Jamaica NY 11451

WAGNER, WARREN HERBERT, JR, b Washington, DC, Aug 29, 20; m 48; c 2. BOTANY. Educ: Univ Pa, AB, 42; Univ Calif, PhD, 50. Prof Exp: Res fel, Harvard Univ, 50-51; from instr to assoc prof, 51-61, dir bot gardens, 66-71, PROF BOT, UNIV MICH, ANN ARBOR, 61-, CUR HERBARIUM, 62-, CHMN DEPT, 75- Concurrent Pos: Mem, Ad Hoc Comt Plant Taxon, Nat Acad Sci, 56-57, Plant Sci Planning Comt, 64-65, dep for bot, Subcomt Syst Biogeog, US Nat Comt, Int Biol Prog, 65-68; trustee, Cranbrook Inst Sci, 63-; mem, Fairchild Trop Garden Res Comt, 66-69; mem, Smithsonian Inst Coun, 67-72 & Panel Syst Biol, NSF; consult mem, Int Union Conserv Nature & Natural Resources, 72- Mem: Fel AAAS (secy, 63-67, vpres bot sci sect, 68); Soc Study Evolution (vpres, 66, pres, 72); Am Soc Plant Taxon (pres, 66); Am Fern Soc (secy, 51-53, cur, 57-, pres, 70-71); Bot Soc Am (vpres, 72). Res: Morphology, life cycles, evolution and systematics of vascular plants, especially pteridophytes; science education; biology of higher plants, especially ferns. Mailing Add: Dept of Bot Univ of Mich Ann Arbor MI 48104

WAGNER, WARREN RICHARD, geology, see 12th edition

WAGNER, WILLIAM CHARLES, b Elma, NY, Nov 12, 32; m 54; c 4. REPRODUCTIVE ENDOCRINOLOGY. Educ: Cornell Univ, DVM, 56, PhD(physiol), 68. Prof Exp: Asst vet med in pract with H K Fuller, DVM, Interlaken, NY, 56-57; field vet, Cornell Univ, 57-60, res vet, 60-65, NIH fel animal physiol, 65-68; from asst prof to assoc prof, 68-74, PROF PHYSIOL, IOWA STATE UNIV, 74- Concurrent Pos: Von Humboldt sr vis US scientist fel, Inst Physiol, Munich Tech Univ, 73-74. Mem: Brit Soc Study Fertility; Am Col Theriogenologists; Am Physiol Soc; Soc Study Reproduction; Am Vet Med Asn. Res: Agalactia syndromes; parturition mechanisms; effect of stress on reproduction in female; physio-pathology of animal fertility; fetal endocrinology; pituitary-adrenal-gonad interactions. Mailing Add: Vet Med Res Inst Iowa State Univ Col of Vet Med Ames IA 50011

WAGNER, WILLIAM EDWARD, JR, b New York, NY, June 17, 25; m 63; c 1. CLINICAL PHARMACOLOGY. Educ: Princeton Univ, BA, 45; Columbia Univ, MD, 50. Prof Exp: SR FEL CLIN PHARMACOL, PHARMACEUT DIV, CIBA-GEIGY CORP, 51-; INSTR CLIN MED, MED SCH, NY UNIV, 60- Mem: AMA; Am Soc Clin Pharmacol & Therapeut; Am Acad Family Pract. Res: Drug metabolism; pharmacokinetics; biopharmaceuticals. Mailing Add: 3301 Valley Rd Millington NJ 07946

WAGNER, WILLIAM FREDERICK, b Canton, Mo, Sept 13, 16; m. CHEMISTRY. Educ: Culver-Stockton Col, AB, 38; Univ Chicago, SM, 40; Univ Ill, PhD(anal chem), 47. Prof Exp: Asst chemist, State Geol Surv, Ill, 40-45; asst chem, Univ Ill, 45-47; asst prof, Hanover Col, 47-49; from instr to assoc prof, 49-58, chmn dept, 65-68, PROF CHEM, UNIV KY, 58- Mem: AAAS; Am Chem Soc. Res: Spectrographic analysis emission; x-ray applied to chemical analysis; solvent extraction of metal chelates; thermal methods of analysis. Mailing Add: Dept of Chem Univ of Ky Lexington KY 40506

WAGNER, WILLIAM GERARD, b St Cloud, Minn, Aug 22, 36; m 68; c 3. THEORETICAL PHYSICS, QUANTUM ELECTRONICS. Educ: Calif Inst Technol, BS, 58, PhD(physics), 62. Prof Exp: Mem tech staff, Res Labs, Hughes Aircraft Co, 62-65, sr staff physicist, 65-70; assoc prof, 66-69, PROF PHYSICS & ELEC ENG, UNIV SOUTHERN CALIF, 69- Concurrent Pos: Consult, Rand Corp, 60-65; Tolman res fel theoret physics, Calif Inst Technol, 62-65; asst prof, Univ Calif, Irvine, 65-66; consult, Janus Mgt Corp, 70-71 & Croesus Capital Corp, 71-74; dean, Div Natural Sci & Math, Col Letters, Arts & Sci, Univ Southern Calif, 73-, spec asst, Acad Record Serv, 75- Mem: Am Phys Soc; Financial Mgt Asn. Res: Investment analysis; lasers; nonlinear optics; computer applications; propagation of high intensity optical beams. Mailing Add: Div Natural Sci & Math LAS Univ Southern Calif Los Angeles CA 90007

WAGNER, WILLIAM JOHN, chemistry, see 12th edition

WAGNER, WILLIAM JOHN, b Gary, Ind, Mar 29, 38; m 62; c 5. ASTROPHYSICS. Educ: John Carroll Univ, BS, 60, MS, 62; Univ Colo, PhD(astro-geophys), 69. Prof Exp: Sr physicist, Rocketdyne Div, NAm Aviation, Inc, 62-64, sci eng fel, 64-69; ASTROPHYSICIST & BIG DOME FACIL SECT CHIEF, SACRAMENTO PEAK OBSERV, AIR FORCE GEOPHYS LABS, 69- Mem: Fel AAAS; Int Astron Union; Am Astron Soc; Am Phys Soc; Am Geophys Union. Res: Observational research concerning solar physics, solar activity, the corona and solar wind; solar-terrestrial physics; spectroscopy. Mailing Add: Sacramento Peak Observ Sunspot NM 88349

WAGNER, WILLIAM SHERWOOD, b Mora, Minn, Sept 21, 28; m 62. ORGANIC CHEMISTRY. Educ: Univ Minn, BChem, 49; Univ Mo, PhD(org chem), 52. Prof Exp: Sr res chemist, Chemstrand Corp, Ala, 52-59; mgr org res paper prod, Fiber Prod Res Ctr, Inc, 59-62, asst dir paper prod, 62-63; res assoc, Celanese Res Co, 63-65, head spinning res sect, 65-70; group mgr, Hoechst Fibers, Inc, 70-71, mgr res & labs, 71-72, DIR DEVELOP, HOECHST FIBERS INDUSTS, 72- Mem: Am Chem Soc; Am Inst Chemists. Res: Synthetic fibers; polymers; organic synthesis. Mailing Add: 696 Perrin Dr Spartanburg SC 29302

WAGNER, WILTZ WALKER, JR, b New Orleans, La, July 7, 39; m 67; c 1. PULMONARY PHYSIOLOGY. Educ: Colo State Univ, PhD(physiol), 74. Prof Exp: Res fel physiol, 60-67, res assoc, 67-74, INSTR MED, MED CTR, UNIV COLO, DENVER, 74- Concurrent Pos: Site vis, NIH, 74; consult, Med Sch, Univ Calif, Los Angeles, 74 & Univ Calif, La Jolla, 75. Honors & Awards: Honors Achievement Award, Angiol Res Found, 65. Mem: Fel Royal Micros Soc; Sigma Xi; Am Physiol Soc; Microcirc Soc; Biol Photog Asn. Res: Pulmonary microcirculation using methods for direct visualization of capillary perfusion in vivo; capillary control mechanisms and functional implications in health and disease. Mailing Add: CVP Lab Med Ctr Univ Colo Denver CO 80220

WAGNER-MERNER, DIANE TESTRAKE, biology, mycology, see 12th edition

WAGNON, HARVEY KEITH, b Douglas, Ariz, Mar 1, 16; m 56; c 1. PLANT PATHOLOGY. Educ: Univ Calif, BA, 42; Univ Mich, MS, 48, PhD(bot), 51. Prof Exp: PROG SUPVR & PLANT PATHOLOGIST, CALIF DEPT FOOD & AGR, 51- Mem: Am Phytopath Soc. Res: Virus diseases of deciduous fruit and nut trees, brambles, grapevines and roses; registration and certification procedures for virus-free nursery stocks; control and eradication procedures for plant pests. Mailing Add: 6925 Southhampton Way Sacramento CA 95823

WAGONER, DALE E, b Niagra Falls, NY, Oct 12, 36; m 65; c 4. GENETICS, BIOLOGY. Educ: Ind Univ, AB(music) & AB(zool), 59, MA, 64, PhD(genetics), 65. Prof Exp: Res asst Drosophila genetics, H J Muller Lab, Ind Univ, 60; from asst prof to assoc prof entom, Grad Fac, NDak State Univ, 68-75; PROF BIOL, MAHARISHI INT UNIV, 75- Concurrent Pos: Res geneticist, Metab & Radiation Res Lab, USDA, 64-75. Mem: AAAS; Genetics Soc Am; Am Inst Biol Sci; Am Genetic Asn; Entom Soc Am. Res: Basic formal genetics of house flies; karyotype-linkage group relationship; insect control by the use of genetic mechanisms, such as chromosomal translocation, meiotic drive, hybrid sterility, cytoplasmic incompatibility, compound chromosomes and conditional lethal mutations. Mailing Add: Dept of Biol Maharishi Int Univ Fairfield IA 52556

WAGONER, EARL V, JR, b Joplin, Mo, Apr 17, 26; m 50; c 5. BIOPHYSICS, ENGINEERING PHYSICS. Educ: Univ Calif, Los Angeles, BS, 50. Prof Exp: Sr res analyst, 50-56, proj engr, 56-57, sr design engr, 57-58, supvr advan projs, 58-60, chief res, 60-64, mgr res, 64-68, PROG MGR RES, N AM ROCKWELL CORP, 68- Concurrent Pos: Med systs & pollution abatement consult. Mem: Inst Elec & Electronics Engrs; Am Phys Soc; Am Inst Mgt. Res: Military electronic systems; medical electronic systems and devices; biological monitoring systems; waste recovery and pollution control; industrial process synthesis; space systems; insulation development. Mailing Add: 2026 Delasonde Dr San Pedro CA 90732

WAGONER, GLEN, b Terreton, Idaho, July 28, 27; m 52; c 1. PHYSICS. Educ: Idaho State Col, BS, 49; Univ Chicago, MS, 52; Univ Calif, PhD(physics), 57. Prof Exp: PHYSICIST, RES LABS, UNION CARBIDE CORP, 57- Concurrent Pos: Prof lectr, Case Inst Technol, 59-61. Mem: Am Phys Soc. Res: Solid state physics; magnetic resonance; electronics; electric arcs. Mailing Add: Parma Tech Ctr Union Carbide Corp PO Box 6116 Cleveland OH 44101

WAGONER, JOHN ALLEN, b Red Cloud, Nebr, Jan 27, 13; m 44; c 2. CARBOHYDRATE CHEMISTRY. Educ: Kans State Teachers Col, BS, 39; Kans State Col, MS, 41, PhD(chem), 43. Prof Exp: Teacher schs, Kans, 31-37 & high sch, Mo, 39-40; asst, Kans State Col, 40-43, Kans Indust Develop Comn fel & res chemist, 43-46; res chemist, A E Staley Mfg Co, 46-59, sr res chemist, 59-60, group leader phys chem res, 61-72; CHEM CONSULT, 72- Mem: Am Chem Soc. Res: Starch properties and sources; protein derivatives; carbohydrate oxidation; physical properties of polymers. Mailing Add: 605 Van Buren Hugoton KS 67951

WAGONER, ROBERT VERNON, JR, b Teaneck, NJ, Aug 6, 38; m 63; c 2. THEORETICAL ASTROPHYSICS. Educ: Cornell Univ, BME, 61; Stanford Univ, MS, 62, PhD(physics), 65. Prof Exp: Res fel physics, Calif Inst Technol, 65-68; from asst prof to assoc prof astron, 68-73, ASSOC PROF PHYSICS, STANFORD UNIV, 73- Concurrent Pos: Sloan res fel, 69-71. Mem: Am Phys Soc; Am Astron Soc. Res: Relativistic astrophysics; cosmology; gravitation theory; nucleosynthesis. Mailing Add: Dept of Physics Stanford Univ Stanford CA 94305

WAGONER, RONALD LEWIS, b Fairfield, Calif, Aug 4, 42; m 60; c 2. MATHEMATICS. Educ: Fresno State Col, BA, 65, MA, 66; Univ Ore, PhD(math), 69. Prof Exp: Asst prof, 69-74, ASSOC PROF MATH, CALIF STATE UNIV, FRESNO, 74- Concurrent Pos: Math specialist, Fresno City Unified Sch Dist, 70-71. Res: Ring theory: associative rings with identity. Mailing Add: Dept of Math Calif State Univ Fresno CA 93710

WAGREICH, HARRY, b New York, NY, Nov 29, 07; m 36; c 1. BIOCHEMISTRY. Educ: City Col New York, BS, 28; Columbia Univ, MA, 30, PhD(chem), 38. Prof Exp: Tutor chem, 32-36, from instr to prof, 37-74, EMER PROF CHEM, CITY COL NEW YORK, 74- Concurrent Pos: Greenbaum Fund res grant, 40; Plotz Found grant, 42; res grant, City Col New York, 48. Mem: Am Chem Soc; NY Acad Sci. Res: Enzymes; human detoxication of phenyl-acetic acid, benzoic acid and borneol; hormones; tensile strength of coagulated plasma and fibrinogen; tyrosinase; invertase; amine oxidase; coupled oxidation products; clinical chemistry. Mailing Add: Dept of Chem City Col New York 139th St & Convent Ave New York NY 10031

WAGREICH, PHILIP DONALD, b New York, NY, July 25, 41; m 62; c 2. PURE MATHEMATICS. Educ: Brandeis Univ, BA, 62; Columbia Univ, PhD(math), 66. Prof Exp: Lectr math, Brandeis Univ, 66-68; asst prof, Univ Pa, 68-74; ASSOC PROF, UNIV Ill, Chicago Circle, 74- Concurrent Pos: Off Naval Res fel, 68-69; mem, Inst Advan Study, 68-70; NSF res grants, 74- Mem: Am Math Soc. Res: Algebraic geometry; topology; transformation groups. Mailing Add: Dept of Math Box 4348 Univ of Ill at Chicago Circle Chicago IL 60680

WAGSTAFF, DAVID JESSE, b Lehi, Utah, Feb 22, 35; m 63; c 3. TOXICOLOGY. Educ: Utah State Univ, BS, 59, PhD(toxicol), 70; Cornell Univ, DVM, 62. Prof Exp: Vet epidemiologist, USPHS, 62-64; vet meat inspector, USDA, 64-65; vet epidemiologist, USPHS, 65-66; asst prof toxicol, Univ Mo-Columbia, 69-73; TOXICOLOGIST, FOOD & DRUG ADMIN, 73- Concurrent Pos: NIH fel toxicol, Utah State Univ, 66-69; mem, Am Bd Vet Toxicol. Mem: Soc Toxicol; Am Col Vet Toxicol; Am Vet Med Asn. Res: Induction of liver microsomal enzymes; drug toxicity; environmental contaminants; toxicants in natural foods; interaction of toxicology with other fields; toxicant interactions; poisonous plants; epidemiology. Mailing Add: Agr Res Ctr Bldg 339-D Food & Drug Admin Beltsville MD 20705

WAGSTAFF, H REID, b Newton, Mass; m 62; c 3. ECONOMIC GEOGRAPHY. Educ: Univ Mich, AB, 56, AM, 63, PhD(geog), 67. Prof Exp: Teaching fel geog, Univ Mich, 61-63; instr, Eastern Mich Univ, 63-65, asst prof, 65-69; ASSOC PROF GEOG, ARIZ STATE UNIV, 69- Mem: Asn Am Geogrs. Res: Geography of energy, especially its production, exchange and consumption, as well as its effect on the environment; potential coal development in Arizona; energy and environmental impacts of desalinization plants. Mailing Add: Dept of Geog Ariz State Univ Tempe AZ 85281

WAGSTAFF, PAUL ARLEN, b Van Wert, Ohio, Sept 17, 34; m 66; c 2. MICROBIOLOGY. Educ: Ohio State Univ, BA, 56, MSc, 64, PhD(microbiol), 65. Prof Exp: Immunologist, Human Health Res & Develop Ctr, Dow Chem Co, Ind, 65-67; assoc scientist, Ortho Res Found, 67-68, scientist, 68-75, SR SCIENTIST, ORTHO PHARMACEUT CORP, 75- Mem: Am Soc Microbiol. Res: Detection of autoantigens and autoantibodies by immunofluorescence; immunology of parasite infections and reproduction. Mailing Add: Ortho Pharmaceut Corp Raritan NJ 08869

WAGSTAFF, SAMUEL STANDFIELD, JR, b New Bedford, Mass, Feb 21, 45. MATHEMATICS. Educ: Mass Inst Technol, BS, 66; Cornell Univ, PhD(math), 70. Prof Exp: Instr math, Univ Rochester, 70-71; vis mem, Inst Advan Study, 71-72; vis lectr, 72-75, ASST PROF MATH, UNIV ILL, 75- Mem: Am Math Soc; Math Asn Am. Res: Number theory. Mailing Add: Dept of Math Univ of Ill Urbana IL 61801

WAHAB, JAMES HATTON, b Bridgeton, NC, Aug 29, 20; m 47; c 2. MATHEMATICS. Educ: Col William & Mary, BS, 40; Univ NC, AM, 50, PhD(math), 51. Prof Exp: Instr math & eng, Norfolk Div, Col William & Mary, 40-42 & 46-47, asst prof, 47-48; instr math, Univ NC, 50-51; from asst prof to assoc prof, Ga Inst Technol, 51-58; prof, La State Univ, 58-61, chmn dept, 60-61; prof, NC State Col, 61-63; prof & chmn dept, Univ NC, Charlotte, 63-68, actg acad dean, 64-68; head dept, 68-73, PROF MATH, UNIV SC, 68- Mem: Am Math Soc; Math Asn Am. Res: Irreducibility of legendre polynomials; algebra; statistics; numerical analysis. Mailing Add: Dept of Math Univ of SC Columbia SC 29208

WAHBA, GRACE, b Washington, DC. MATHEMATICAL STATISTICS. Educ: Cornell Univ, BA, 56; Univ Md, College Park, MA, 62; Stanford Univ, PhD(math statist), 66. Prof Exp: Res mathematician, Opers Res, Inc, 57-61; systs analyst, Int Bus Mach Corp, 61-66; res assoc math statist, Stanford Univ, 66-67; asst prof, 67-70, ASSOC PROF STATIST, UNIV WIS-MADISON, 71- Mem: Inst Math Statist; Am Math Soc; Soc Indust & Appl Math. Res: Stochastic processes; approximation theory. Mailing Add: Dept of Statist Univ of Wis Madison WI 53705

WAHBA, ISAAC JACK, b Alexandria, Egypt, Nov 12, 29; US citizen; m 60; c 2. FOOD CHEMISTRY. Educ: Univ Calif, AB, 49, BS, 51; Univ Mo, MS, 53, PhD(hort), 54. Prof Exp: Chemist, Inter-Am Serv Agr Coop, Panama, 54-55; pvt consult food processing, Repub Panama, 55-59; res fel food sci, Rutgers Univ, 59-60, instr foods, 60-61; res specialist radiation chem, Cornell Univ, 61-64; res food scientist, Gen Mills, Inc, 64-65, sr res chemist, James Ford Bell Tech Ctr, 65-71, RES ASSOC, JAMES FORD BELL TECH CTR, GEN MILLS, INC, 71- Mem: Inst Food Technologists; Am Soc Hort Sci; Am Hort Soc. Res: Food science and technology; agricultural chemistry and biochemistry; soil and plant chemistry, horticulture; human nutrition; controlled environment agriculture, hydroponics; plant physiology. Mailing Add: 8345 Julianne Terrace Minneapolis MN 55427

WAHL, ARNOLD C, chemical physics, quantum mechanics, see 12th edition

WAHL, ARTHUR CHARLES, b Des Moines, Iowa, Sept 8, 17; m 43; c 1. NUCLEAR CHEMISTRY. Educ: Iowa State Univ, BS, 39; Univ Calif, PhD(chem), 42. Prof Exp: Res assoc, Manhattan Proj, Univ Calif, 42-43, group leader, Los Alamos Sci Lab, 43-46; assoc prof chem, 46-53, FARR PROF RADIOCHEM, WASHINGTON UNIV, 53- Concurrent Pos: Consult, Los Alamos Sci Lab, 50-; NSF fel, 67. Honors & Awards: Am Chem Soc Award, 66. Mem: Am Chem Soc. Res: Nuclear-charge distribution in fission; rapid electron-transfer reactions. Mailing Add: Dept of Chem Washington Univ St Louis MO 63130

WAHL, EBERHARD WILHELM, b Berlin, Ger, May 24, 14; US citizen; m 46; c 2. METEOROLOGY, SPACE SCIENCES. Educ: Univ Berlin, PhD(astron), 37. Prof Exp: Asst prof meteorol, Meteorol Inst, Univ Berlin, 37-45; scientist, N W Ger Weather Serv, 46-49; proj scientist, Geophys Res Directorate, US Air Force, 49-58, astronr, Proj Spacetrack, 58-61, tech dir, 61-63; chmn dept, 70-73, PROF METEOROL, UNIV WIS-MADISON, 63- Concurrent Pos: Fel astron, Observ, Univ Berlin, 37-41; vis prof, Univ Wis, 62-63; trustee, Univ Corp Atmospheric Res, Boulder, Colo, 75- Mem: Fel AAAS; Am Meteorol Soc; Am Geophys Union; Ger Meteorol Soc. Res: Meteorology of large scale circulation; dynamic climatology; satellite meteorology. Mailing Add: Dept of Meteorol Univ of Wis Madison WI 53706

WAHL, FLOYD MICHAEL, b Hebron, Ind, July 7, 31; m 53; c 4. MINERALOGY. Educ: DePauw Univ, AB, 53; Univ Ill, MA, 57, PhD(mineral & geochem), 58. Prof Exp: Instr geol, Univ Ill, Urbana, 58-59, res asst prof, 59-60, from asst prof to assoc prof, 60-69; prof & chmn dept, 69-73, dir div phys sci & math, 71-73, ASSOC DEAN GRAD SCH, UNIV FLA, 73- Mem: Fel Geol Soc Am; Mineral Soc Am; Geochem Soc; Clay Minerals Soc; Am Inst Prof Geologists. Res: Clay mineralogy and sedimentary geochemistry; development of mineral resources; chemical alteration and those factors that lead to and control element concentration; phase changes in minerals at elevated temperatures. Mailing Add: 237 Grinter Hall Univ of Fla Gainesville FL 32611

WAHL, GEOFFREY MYLES, b Los Angeles, Calif, Apr 6, 48. MOLECULAR BIOLOGY. Educ: Univ Calif, Los Angeles, BA, 70; Harvard Univ, PhD(biochem), 76. Prof Exp: Res asst prof biol, Univ Utah, 75-76; FEL BIOCHEM, MED SCH, STANFORD UNIV, 76- Res: Isolation of mammalian nonsense suppressors; control of gene expression in eukaryotes. Mailing Add: Dept of Biochem Stanford Univ Med Sch Stanford CA 94305

WAHL, GEORGE HENRY, JR, b New York, NY, Sept 17, 36; m 58; c 3. STRUCTURAL CHEMISTRY. Educ: Fordham Univ, BS, 58; NY Univ, MS, 61, PhD(org chem), 63. Prof Exp: Res chemist, Pittsburgh Plate Glass Chem Co, Ohio, 58-59; NIH res fel org chem, Cornell Univ, 63-64; from asst prof to assoc prof, 64-75, PROF ORG CHEM, NC STATE UNIV, 75- Concurrent Pos: Guest prof, Swiss Fed Inst, Zurich, 73-74. Honors & Awards: Sigma Xi Res Award, 74. Mem: Am Chem Soc; The Chem Soc. Res: Organic stereochemistry; nuclear magnetic resonance spectroscopy; mass spectrometry; synthesis of unusual structures for physical investigation; synthesis and structure of adamantane and biphenyl derivatives. Mailing Add: Dept of Chem NC State Univ Raleigh NC 27607

WAHL, JOHN SCHEMPP, b Tungjen, China, Aug 8, 20; US citizen; m 43; c 3. NUCLEAR PHYSICS. Educ: Iowa State Teachers Col, BA, 41; Univ Iowa, MS, 44, PhD(physics), 52. Prof Exp: Asst, Sylvania Elec Prod Co, Pa, 41; res assoc, Physics Eng, Develop Proj, Iowa, 44-45; instr physics, Univ Iowa, 47-48; mem res staff, Los Alamos Sci Lab, Univ Calif, 49-54; MEM RES STAFF, SCHLUMBERGER-DOLL RES CTR, 54- Mem: Am Phys Soc. Res: Nuclear physics; neutron and gamma ray interactions in extended media. Mailing Add: Schlumberger-Doll Res Ctr Box 307 Ridgefield CT 06877

WAHL, JONATHAN MICHAEL, b Washington, DC, Jan 29, 45; m 70. MATHEMATICS. Educ: Yale Univ, BS & MA, 65; Harvard Univ, PhD(math), 71. Prof Exp: Instr math, Univ Calif, Berkeley, 70-72; vis, Inst Advan Study, Princeton Univ, 72-73; asst prof, 73-75, ASSOC PROF MATH, UNIV NC, CHAPEL HILL, 75- Concurrent Pos: NSF res grant, 71- Mem: Am Math Soc. Res: Singularities; deformation theory; algebraic geometry. Mailing Add: Dept of Math Univ of NC Chapel Hill NC 27514

WAHL, MILTON HEINS, b Emden, Ill, Sept 17, 08; m 33; c 3. CHEMISTRY, SCIENCE ADMINISTRATION. Educ: Cent Wesleyan Col, AB, 28; Univ Mo, AM, 30; Univ Ill, PhD(phys chem), 33. Prof Exp: Asst, Columbia Univ, 33-35; res chemist, E I du Pont de Nemours & Co, 35-42, tech asst, 42-43, tech specialist, 43-44, process mgr, 44-45, head develop sect, 45-50, mgr liaison off, 50-52, dir, Savannah River Lab, 52-61, asst tech dir, 55-61, mgr, Atomic Energy Div, 61-69, asst gen mgr, Explosives Dept, 69-72, Polymer Intermediates Dept, 72-73; MANAGING DIR, ACAD NATURAL SCI, PHILADELPHIA, 74- Mem: AAAS; Am Chem Soc; Am Nuclear Soc; Am Inst Chemists; Asn Mus Dirs. Res: Reaction kinetics; molecular rays;

explosives; separation of isotopes; development of organic chemicals; atomic energy. Mailing Add: 304 Country Club Dr Wilmington DE 19803

WAHL, PATRICIA WALKER, b La Grande, Ore, Dec 6, 38; m 63; c 1. BIOSTATISTICS. Educ: San Jose State Col, BA, 60; Univ Wash, PhD(biostatist), 71. Prof Exp: Res analyst comput programming, Lockheed Missiles, Lockheed Aircraft Corp, 60-62; systs analyst, Control Data Corp, 63-64; head programmer, 64-66, instr biostatist, 71-73, ASST PROF BIOSTATIST, UNIV WASH, 74- Mem: Am Statist Asn; Biomet Soc. Res: Use of regression analysis and other multivariate statistical techniques for exploratory data analysis; effect on classification by discriminant analysis when model assumptions fail. Mailing Add: 1509 Killarney Way Bellvue WA 98004

WAHL, SHARON KNUDSON, b Mt Vernon, Wash, Mar 16, 45; m 71. IMMUNOLOGY. Educ: Pac Lutheran Univ, BS, 67; Univ Wash, PhD(biol struct), 71. Prof Exp: Fel path, Sch Med, Univ Wash, 71-72; fel cellular immunol, 72-74, staff fel humoral immunity, 74-75, SR STAFF FEL HUMORAL IMMUNITY, NAT INST DENT RES, 75- Mem: Am Asn Immunol; Sigma Xi; Reticuloendothelial Soc. Res: Mechanisms of activation and characterization of T and B lymphocyte participation in cellular immune reactions and effect of immunosuppressive agents on these responses; influence of immune system on connective tissue metabolism. Mailing Add: Nat Inst Dent Res Bldg 30 Rm 332 9000 Rockville Pike Bethesda MD 20014

WAHL, WERNER HENRY, b Buffalo, NY, Oct 1, 30; m 51; c 2. NUCLEAR CHEMISTRY, RADIOCHEMISTRY. Educ: Univ Buffalo, BA, 54; Purdue Univ, MS, 56, PhD(phys inorg chem), 57. Prof Exp: Chem operator, Pathfinder Chem Corp, 49; asst, Linde Co Div, Union Carbide Co, 51-53; asst, Durez Plastics, Inc, 53; asst chem, Univ Buffalo, 53; asst, Purdue Univ, 54-57; res chemist, Univ Carbide Nuclear Corp, 57-61, group leader, 61-65, asst mgr res, 65-66, dir radiopharmceut, Neisler Labs, Inc, Union Carbide Corp, NY, 66-69; dir opers, Mallinckrodt/Nuclear, Mo, 69-70; vpres & gen mgr, 70, exec vpres, 71, PRES, AMERSHAM-SEARLE COPR, 71- Mem: Fel AAAS; Am Chem Soc; Am Nuclear Soc; Soc Nuclear Med; fel Am Inst Chem. Res: Natural radioactivity; radiation chemistry of gases; activation analysis; radiopharmaceuticals; nuclear medicine. Mailing Add: Amersham-Searle Corp 2636 S Clearbrook Dr Arlington Heights IL 60005

WAHL, WILLIAM GEORGE, b Crosby, Minn, June 16, 18; Can citizen; m 24; c 3. GEOLOGY, GEOPHYSICS. Educ: Mich State Univ, BSc, 40; McGill Univ, MSc, 41, PhD(geol), 47. Prof Exp: Geologist, Que Dept Mines, 46-48; asst prof geol, Mt Allison Univ, 47-48; geologist, Bethlehem Steel Co, 48-50 & Steel Co Can, 50-55; PRES, W G WAHL LTD, 55- Mem: Am Geophys Union; Geol Soc Am; Am Inst Mining, Metall & Petrol Eng; Soc Econ Geologists; Geol Asn Can. Res: Geophysical and geological exploration. Mailing Add: W G Wahl Ltd Ste 1101 302 Bay St Toronto ON Can

WAHLBECK, PHILLIP GLENN, b Kankakee, Ill, Mar 29, 33; m 56; c 3. HIGH TEMPERATURE CHEMISTRY, SURFACE CHEMISTRY. Educ: Univ Ill, BS, 54, PhD(chem), 58. Prof Exp: Asst chem, Univ Ill, 54-58; res assoc, Univ Kans, 58-60; from instr to assoc prof, 60-72, PROF CHEM & CHMN DEPT, WICHITA STATE UNIV, 72- Concurrent Pos: Vis prof, Tech Univ Norway, 70. Mem: Am Chem Soc. Res: Molecular beams; thermodynamics; transition metal hydrides, oxides, selenides and tellurides; gas-surface interactions; mean residence times; surface diffusion; spatial distributions of restituted molecules. Mailing Add: Dept of Chem Wichita State Univ Wichita KS 67208

WAHLERT, HOWARD ELMER, mathematics, deceased

WAHLERT, JOHN HOWARD, b New York, NY, May 12, 43; m 69. VERTEBRATE PALEONTOLOGY. Educ: Amherst Col, BA, 65; Harvard Univ, MA, 66, PhD(geol), 72. Prof Exp: CURATORIAL ASST VERT PALEONT, AM MUS NATURAL HIST, 72- Mem: Soc Vert Paleont; Soc Study Evolution; Am Soc Mammal. Res: Cenozoic rodents and their anatomy, taxonomy and phylogeny. Mailing Add: Dept of Vert Paleont Am Mus of Natural Hist New York NY 10024

WAHLGREN, HAROLD EMIL, b Chicago, Ill, Mar 7, 29. FOREST PRODUCTS. Educ: Iowa State Univ, BS, 51, MS, 56; Duke Univ, DF, 68. Prof Exp: Photogram engr, Hydrographic Off, US Dept Navy, 51-52; asst wood technol, Iowa State Univ, 55-56; forest prod technologist, 56-67, PROJ LEADER TIMBER QUAL & PROD POTENTIAL RES, FOREST PROD LAB, US FOREST SERV, 67- Mem: Soc Am Foresters; Forest Prod Res Soc; Soc Wood Sci & Technol. Res: Improving wood quality and evaluating product potential of nation's lumber resource; efficient and effective utilization of forest residues. Mailing Add: US Forest Prod Lab PO Box 5130 Madison WI 53705

WAHLGREN, MORRIS A, b Wildrose, NDak, May 31, 29; m 55; c 3. ENVIRONMENTAL CHEMISTRY. Educ: Jamestown Col, BS, 51; Univ Mich, PhD(chem), 61. Prof Exp: Radiochemist, Atomic Energy Div, Phillips Petrol Co, Idaho, 53-56; asst chemist, Chem Div, 61-66, assoc chemist, Radiol & Environ Res Div, 66-72, CHEMIST, RADIOL & ENVIRON RES DIV, ARGONNE NAT LAB, 72- Mem: AAAS; Am Chem Soc; fel Am Inst Chemists; Int Soc Limnol. Res: Nuclear and analytical chemistry; radiochemical separations; chemical limnology; behavior of artificial radionuclides in the Great Lakes. Mailing Add: Dept 203 E161 Argonne Nat Lab 9700 S Cass Ave Argonne IL 60439

WAHLIG, CHARLES F, b New York, NY, Aug 11, 24; m 51; c 4. PHYSICS. Educ: Queens Col, NY, BS, 43; Univ Pa, PhD(physics), 53. Prof Exp: Lab asst, Los Alamos Sci Lab, 44-46; instr physics, Univ Pa, 47-50; asst prof, Drexel Inst Technol, 50-53; physicist, Photo Prod Dept, 53-58, PHYSICIST, ORG CHEM DEPT, EXP STA, E I DU PONT DE NEMOURS & CO, INC, 58- Res: Solid state physics; luminescence; Hall effect; organic materials. Mailing Add: Org Chem Dept Res & Develop Div E I du Pont de Nemours & Co Inc Wilmington DE 19898

WAHLIG, MICHAEL ALEXANDER, b New York, NY, Oct 21, 34; m 56; c 4. ENERGY CONVERSION. Educ: Manhattan Col, BS, 55; Mass Inst Technol, PhD(physics), 62. Prof Exp: Res assoc physics, Mass Inst Technol, 62-66; res staff physics, 66-72, MEM RES STAFF SOLAR ENERGY, ENERGY & ENVIRON DIV, LAWRENCE BERKELEY LAB, UNIV CALIF, BERKELEY, 72- Mem: Am Phys Soc; Int Solar Energy Soc; AAAS. Res: Research, development and analysis of the availability, conversion, and use of solar energy for providing heating, cooling and electric power. Mailing Add: Lawrence Berkeley Lab Univ of Calif Berkeley CA 94720

WAHLSTROM, ERNEST EUGENE, b Boulder, Colo, Dec 30, 09; m 31; c 2. PETROLOGY. Educ: Univ Colo, BA, 31, MA, 33; Harvard Univ, MA, 36, PhD(geol), 39. Prof Exp: Instr geol, 36-38, asst prof geol & mineral, 38-42, assoc prof geol, 42-47, dean fac, 68-71, PROF GEOL, UNIV COLO, BOULDER, 47- Concurrent Pos: Assoc geologist, US Geol Surv, 43-44, geologist, 60-65; consult geologist, 43- Mem: AAAS; fel Geol Soc Am; fel Mineral Soc Am; Am Inst Prof Geologists; Soc Econ Geologists. Res: Igneous petrology; mining and engineering geology; dams and tunnels. Mailing Add: Dept of Geol Sci Univ of Colo Boulder CO 80302

WAHLSTROM, LAWRENCE F, b Aurora, Wis, Feb 4, 15; m 38; c 2. MATHEMATICS. Educ: Lawrence Col, BA, 36; Univ Wis, MA, 37, PhD(math educ), 50. Prof Exp: Pub sch teacher, Ill, 37-41, chmn dept math, jr high sch, 41-45; chmn dept, Elgin Acad, 45-47; asst, Univ Wis, 47-48; PROF MATH & CHMN DEPT, UNIV WIS-EAU CLAIRE, 48- Concurrent Pos: NSF fac sci grant, 57-58. Mem: Math Asn Am. Res: Geometry. Mailing Add: Dept of Math Univ of Wis Eau Claire WI 54701

WAHLSTROM, RICHARD CARL, b Craig, Nebr, Feb 13, 23; m 47; c 3. ANIMAL SCIENCE. Educ: Univ Nebr, BS, 48; Univ Ill, MS, 50, PhD(animal nutrit), 52. Prof Exp: Asst animal husb, Univ Ill, 48-51; res assoc nutrit, Merck Inst Therapeut Res, 51-52; assoc prof animal husb, 52-59, head dept, 60-67, PROF ANIMAL HUSB, SDAK STATE UNIV, 59- Mem: Am Soc Animal Sci; Am Inst Nutrit. Res: Swine nutrition; antibiotics; selenium poisoning; protein levels and amino acid requirements; high protein cereals; mineral nutrition; by-product feeds. Mailing Add: Dept of Animal Sci SDak State Univ Brookings SD 57006

WAHNSIEDLER, WALTER EDWARD, b Ind, Jan 23, 47; m 69; c 1. CHEMICAL PHYSICS. Educ: Purdue Univ, BS, 67, PhD(chem physics), 75. Prof Exp: Vis scholar mat sci, Northwestern Univ, 74; SCIENTIST PHYS CHEM, ALUMINUM CO AM, 75- Mem: Am Chem Soc. Res: Theoretical solid state studies; applications of computers to chemistry; properties of oxides; aluminum smelting; environmental impact of industry. Mailing Add: 2732 Kingston Dr Natrona Heights PA 15065

WAHR, JOHN CANNON, b Ann Arbor, Mich, Apr 2, 26; m 49; c 2. PHYSICS. Educ: Univ Mich, BSE, 48, MS, 49, PhD(physics), 53. Prof Exp: Asst, Univ Mich, 48-49; PHYSICIST, CENT RES, DOW CHEM CO, 53- Mem: AAAS; Am Phys Soc; Optical Soc Am. Res: Quantum electronics; holography; atomic and molecular physics; surface physics. Mailing Add: 705 Crescent Dr Midland MI 48640

WAHRHAFTIG, AUSTIN LEVY, b Sacramento, Calif, May 5, 17; m 57. PHYSICAL CHEMISTRY. Educ: Univ Calif, AB, 38; Calif Inst Technol, PhD(phys chem), 41. Prof Exp: Fel, Calif Inst Technol, 41-45; res chemist, Dr W E Williams, 45-46; univ fel, Ohio State Univ, 46-47; from asst prof to assoc prof chem, 47-59, PROF CHEM, UNIV UTAH, 59- Mem: AAAS; Am Chem Soc; Am Phys Soc. Res: Molecular spectra; mass spectrometry; kinetics of gas-phase ion reactions. Mailing Add: Dept of Chem Univ of Utah Salt Lake City UT 84112

WAHRHAFTIG, CLYDE (ADOLPH), b Fresno, Calif, Dec 1, 19. GEOLOGY. Educ: Calif Inst Technol, BS, 41; Harvard Univ, MA, 47, PhD, 53. Prof Exp: Jr geologist, US Geol Surv, 41 & 42-43, asst geologist, 43-45; assoc prof, 60-67, PROF GEOL, UNIV CALIF, BERKELEY, 67-; GEOLOGIST, US GEOL SURV, 45- Concurrent Pos: Am Geol Inst vis geoscientist, 69; mem comt geol sci, Nat Acad Sci, 70-72; consult, Conserv Found, 71-72. Honors & Awards: Kirk Bryan Award, Geol Soc Am, 67. Mem: AAAS; fel Geol Soc Am; Am Geophys Union. Res: Geomorphology; igneous petrology; stratigraphy and sedimentation; geology applied to land use; geology of California and Alaska. Mailing Add: Dept of Geol & Geophys Univ of Calif Berkeley CA 94720

WAI, CHIEN MOO, b China, Aug 8, 37; m 65; c 2. NUCLEAR CHEMISTRY, GEOCHEMISTRY. Educ: Nat Taiwan Univ, BS, 60; Univ Calif, Irvine, PhD(chem), 67. Prof Exp: Fel, Univ Calif, Los Angeles, 66-69; asst prof chem & geol, 69-73, ASSOC PROF CHEM & GEOL, UNIV IDAHO, 73- Concurrent Pos: Vis assoc prof, Inst Geophys & Planetary Physics, Univ Calif, Los Angeles, 75-76. Mem: AAAS; Am Chem Soc; Geochem Soc. Res: Chemical effects of nuclear transformation; origin of meteorites; heavy metal pollution. Mailing Add: Dept of Chem Univ of Idaho Moscow ID 83843

WAIBEL, PAUL EDWARD, b Hawthorne, NJ, June 22, 27; m 71; c 3. POULTRY NUTRITION. Educ: Rutgers Univ, BS, 48; Univ Wis, MS, 51, PhD(poultry nutrit, biochem), 53. Prof Exp: Teaching asst poultry husb, Univ Wis, 49-53; res assoc poultry nutrit, Cornell Univ, 53-54; res assoc, 54-55, from asst prof to assoc prof, 55-64, PROF POULTRY NUTRIT, UNIV MINN, ST PAUL, 64- Honors & Awards: Poultry Sci Asn Res Award. Mem: AAAS; Am Inst Nutrit; Am Chem Soc; Poultry Sci Asn; NY Acad Sci. Res: Nutrition of turkeys. Mailing Add: Dept of Animal Sci Univ of Minn St Paul MN 55108

WAID, MARGARET COWSAR, b Baton Rouge, La, Feb 21, 41; m 63; c 2. APPLIED MATHEMATICS. Educ: La State Univ, Baton Rouge, BS, 61, MS, 63; Tex Tech Univ, PhD(math), 71. Prof Exp: Teacher pub schs, La, 65-67; instr math, Tex Tech Univ, 67-71; asst prof math, DC Teachers Col, 71-72; ASST PROF MATH, UNIV DEL, 72- Mem: Am Math Soc; Soc Indust & Appl Math; London Math Soc; Asn Women in Math; Am Soc Eng Educ. Res: Partial differential equations, especially those of degenerate parabolic type, including applications to fluid flow through porous media, cardiovascular mechanics; analysis, especially nonlinear functional analysis and numerical analysis. Mailing Add: Dept of Math Univ of Del Newark DE 19711

WAIDE, JACK BOID, b El Paso, Tex, Aug 29, 47; m 67; c 1. ECOLOGY. Educ: Univ Tex, Austin, BA, 70; Univ Ga, PhD(ecol), 76. Prof Exp: Asst biol & zool, Univ Tex, Austin, 68-70; res asst ecol, Univ Ga, 73-76; ASST PROF ZOOL, CLEMSON UNIV, 76- Mem: Ecol Soc Am; AAAS; Am Inst Biol Sci; Soc Comput Simulation; Soc Gen Systs Res. Res: Nutrient cycles in terrestrial ecosystems and their response to perturbation; soil biota and substrate quality as regulators of terrestrial decomposition processes; systems and theoretical ecology; lysimeter studies of soil chemistry. Mailing Add: Dept of Zool Clemson Univ Clemson SC 29631

WAIFE, SHOLOM OMI, b New York, NY, Feb 20, 19; m 42; c 2. INTERNAL MEDICINE, MEDICAL EDUCATION. Educ: Johns Hopkins Univ, AB, 40; NY Univ, MD, 43; Am Bd Internal Med, dipl, 51. Prof Exp: Res assoc med, Sch Med, Yale Univ, 45; resen resident physician, Long Island Col Hosp, 45-46; asst med, Med Sch, Johns Hopkins Univ, 46-48; instr, Sch Med, Univ Pa, 48-52; assoc, 52-60, asst prof, 60-68, ASSOC PROF MED, SCH MED, IND UNIV, INDIANAPOLIS, 68-; DIR MED SERV DIV, RES LAB, ELI LILLY & CO, 64- Concurrent Pos: Ed-in-chief, Am J Clin Nutrit, 52-62; head med educ dept, Eli Lilly & Co, 52-64; co-ed, Perspectives Biol & Med, 58-64. Mem: Am Diabetes Asn; Am Med Writers' Asn; fel Am Col Physicians; Am Fedn Clin Res. Res: Metabolism; diabetes; obesity; vitamins. Mailing Add: Lilly Res Labs 307 E McCarty St Indianapolis IN 46206

WAILES, JOHN LEONARD, b Loveland, Colo, Oct 9, 23; m 47; c 3. PHARMACY. Educ: Univ Colo, BS, 47, MS, 50, PhD(pharm), 54. Prof Exp: Chemist, US Food & Drug Admin, 47-48; pharmacist, Park-Hill Drug Co, Colo, 48-50; instr pharm, Univ Colo, 50-54; from asst prof to assoc prof, 43-61, PROF PHARM, UNIV MONT, 61- Concurrent Pos: USPHS grant, 59; with Merck Sharp & Dohme Div, Merck & Co, Colo, 51-54. Mem: Am Pharmaceut Asn; Asn Cols Pharm. Res: Respiration of mold and yeast in the presence and absence of inhibitors and antagonists using the Warburg apparatus; preservation of pharmaceutical products; synergism and antagonism of various preservatives and their possible inactivation; complexing of macromolecules. Mailing Add: Dept of Pharm Univ of Mont Missoula MT 59801

WAINBERG, MARK ARNOLD, b Montreal, Que, Apr 21, 45; m 69; c 2. CANCER. Educ: McGill Univ, BSc, 66; Columbia Univ, PhD(microbiol), 72. Prof Exp: Lectr immunol, Hebrew Univ-Hadassah Med Sch, 72-74; STAFF INVESTR TUMOR IMMUNOL, LADY DAVIS INST MED RES, JEWISH GEN HOSP, MONTREAL, 74- Concurrent Pos: Europ Molecular Biol Orgn res fel, 72-74; Que Med Res Coun res scholar, 75-; Nat Cancer Inst Can res grant, 75-; researcher, Dept Microbiol & Immunol, Univ Montreal, 75- Mem: Am Soc Microbiol; Sigma Xi; Can Soc Immunol; NY Acad Sci; Can Oncol Soc. Res: Cellular and humoral anti-tumor immunity of chickens bearing tumors induced by Rous sarcoma virus; how immunoprophylaxis and immunotherapy may serve to render such immunity more efficient. Mailing Add: Lady Davis Inst for Med Res Jewish Gen Hosp 3755 Cote Ste Catherine Rd Montreal PQ Can

WAINE, MARTIN, b Berlin, Ger, Apr 8, 33; US citizen; m 63; c 2. PHYSICS. Educ: Columbia Univ, BS, 58; Yale Univ, MS, 59, PhD(physics), 65. Prof Exp: Asst prof physics, Mt Holyoke Col, 64-70; prin engr, MRC Corp, Md, 71-72; chief engr, Diamondex Enterprises Inc, 72-74; vpres mfg, Evershield Prod Inc, 74-75; PRIN ENGR, MRC CORP, 75- Mem: Am Phys Soc. Res: Dynamic nuclear orientation; nuclear magnetic resonance; instrumentation and control theory. Mailing Add: 616 Wheel Rd Bel Air MD 21014

WAINER, ARTHUR, b Cincinnati, Ohio, Jan 28, 38; m 57; c 3. BIOCHEMISTRY. Educ: Univ Miami, BS, 57; Univ Fla, PhD(biochem), 61. Prof Exp: Instr biochem, Univ Fla, 61-62; from instr to assoc prof, Bowman Gray Sch Med, 62-70; PROF CHEM, EDINBORO STATE COL, 70- Mem: AAAS; Am Chem Soc; Am Soc Biol Chemists; Am Asn Clin Chemists. Res: Sulfur amino acid metabolism, ion exchange column chromatography. Mailing Add: Dept of Chem Edinboro State Col Edinboro PA 16412

WAINER, EUGENE, b Pittsburgh, Pa, Feb 14, 08; m 32; c 2. INORGANIC CHEMISTRY. Educ: Univ Akron, BS, 29; Cornell Univ, PhD(inorg chem), 33. Prof Exp: Asst dir res, Titanium Alloy Mfg Co, 34-46; dir res, 46-68, pres, 68-74, CHMN BD, HORIZONS RES INC, 74- Concurrent Pos: Res adv grad sch, Rutgers Univ, 46-47. Honors & Awards: Kosar Mem Award, Soc Photog Sci & Eng, 69. Mem: Am Chem Soc; fel Am Ceramic Soc; fel Am Inst Chemists; Soc Photog Sci & Eng; Am Electrochem Soc. Res: Ceramics; physics; metallurgy of zirconium and titanium; dielectrics; semiconductors; extractive metallurgy of the transition elements; organic and inorganic photochemistry; photoscience; free radical chemistry. Mailing Add: Horizons Res Inc 23800 Mercantile Rd Cleveland OH 44122

WAINFAN, ELSIE, b New York, NY, Aug 2, 26; m 47; c 2. BIOCHEMISTRY. Educ: City Col New York, BS, 47; Univ Southern Calif, PhD(biochem), 54. Prof Exp: Res technician, NY Psychiat Inst, 47-49; USPHS fel, Med Sch, Univ Ore, 55-56; res assoc biochem, Cornell Univ, 56-59; res assoc, Col Physicians & Surgeons, Columbia Univ, 59-67; asst prof, Univ Southern Calif, 67-68; ASSOC INVESTR, NY BLOOD CTR, 68- Mem: Am Chem Soc; Am Soc Biol Chemists; Am Soc Microbiologists; Am Asn Cancer Res. Res: Bacteriophage, nucleic acids, enzymes; metabolic inhibitors. Mailing Add: New York Blood Ctr New York NY 10021

WAINIO, WALTER W, b Astoria, Ore, Sept 8, 14; m 49; c 1. ENZYMOLOGY, BIOCHEMISTRY. Educ: Univ Mass, BS, 36; Pa State Univ, MS, 40; Cornell Univ, PhD(physiol), 43. Prof Exp: Asst animal nutrit, Pa State Univ, 36-38, instr, 38-41; asst prof physiol, Med Col, Cornell Univ, 41-43; asst prof, Col Dent, NY Univ, 43-48; assoc res specialist, 48-50, assoc prof biochem, 50-59, chmn dept physiol & biochem, 60-63 & 66-67, chmn dept biochem, 67-72, PROF BIOCHEM, RUTGERS UNIV, 59-, CHMN DEPT, 75- Concurrent Pos: Mem, Marine Biol Labs, Woods Hole. Mem: AAAS; Am Chem Soc; Am Soc Biol Chemists; fel NY Acad Sci; Brit Biochem Soc. Res: Cytochromes. Mailing Add: Nelson Biol Labs Rutgers Univ Dept of Biochem New Brunswick NJ 08903

WAINWRIGHT, LILLIAN K (SCHNEIDER), b Brooklyn, NY, June 30, 23; m 52; c 2. GENETICS. Educ: Brooklyn Col, BA, 43; Columbia Univ, MA, 51, PhD(zool), 56. Prof Exp: Res asst zool, Columbia Univ, 43-52; from asst prof to assoc prof biol, 61-70, PROF BIOL, MT ST VINCENT UNIV, 70- Mem: Genetics Soc Am; Can Soc Cell Biol. Res: Genetics of microorganisms; regulation of the initiation of hemoglobin synthesis in the blood island cells of chick embryos. Mailing Add: Dept of Biol Mt St Vincent Univ Halifax NS Can

WAINWRIGHT, STANLEY D, b Hull, Eng, Apr 15, 27; Can citizen; m 52; c 2. CHEMICAL EMBRYOLOGY. Educ: Cambridge Univ, BA, 47; Univ London, PhD(biochem), 50. Prof Exp: Brit Med Res Coun exchange scholar biochem, Physiol Microbiol Serv, Pasteur Inst, Paris, 50-51; res assoc microbial genetics, Columbia Univ, 51-52; Nat Res Coun Can & Atomic Energy Can, Ltd fel, Biol Div, Atomic Energy Can, Ltd, 52-55; res assoc microbial physiol, Yale, 55-56; res asst prof & Med Res Coun assoc biochem, 56-58, res assoc prof & Med Res Coun assoc, 58-64, PROF BIOCHEM, DALHOUSIE UNIV, 65- Mem: AAAS; Genetics Soc Am; Am Soc Cell Biologists; Can Biochem Soc. Can Soc Cell Biol (pres, 75-76). Res: Regulation of protein biosynthesis; regulation of onset of hemoglobin synthesis in developing chick blastodiscs; biochemical neuroendocrinology of the developing chick pineal gland. Mailing Add: Dept of Biochem Dalhousie Univ Fac of Med Halifax NS Can

WAINWRIGHT, STEPHEN ANDREW, b Indianapolis, Ind, Oct 9, 31; m 56; c 4. INVERTEBRATE ZOOLOGY. Educ: Duke Univ, BS, 53; Univ Cambridge, BA, 58, MA, 63; Univ Calif, Berkeley, PhD(zool), 62. Prof Exp: NSF fel med physics, Karolinska Inst, Sweden, 62-63; NSF fel biol, Woods Hole Oceanog Inst, 63-64; ASSOC PROF ZOOL, DUKE UNIV, 64- Mem: Soc Exp Biol & Med; Am Soc Enol; Marine Biol Asn UK. Res: Functional morphology of supportive systems of animals and plants from the macromolecular through the organism levels of organization. Mailing Add: Dept of Zool Duke Univ Durham NC 27706

WAISBROT, SAMUEL WILLIAM, b New York, NY, Jan 12, 16; m 49; c 2. ORGANIC CHEMISTRY. Educ: Ohio State Univ, AB, 38, PhD(org chem), 41. Prof Exp: Asst anal chem, Ohio State Univ, 38-41; asst chemist, Western Regional Res Lab, Bur Agr Chem & Eng, USDA, 41-43; res chemist, Celanese Corp Am, 43-46; sr res chemist, Permutit Co, 46-48; sr res chemist, 50-53, org proj engr, 53-65, RES SCIENTIST, GOODYEAR TIRE & RUBBER CO, 65- Mem: Am Chem Soc. Res: Carbohydrates; preparation of resins and ion exchange resins; preparation of chemicals for use in manufacture of rubber and plastic products; film forming cellulose derivatives; radioactive isotope dating; information research. Mailing Add: 2036 Thornhill Dr Akron OH 44313

WAISMAN, JERRY, b Borger, Tex, Sept 14, 34; m 58; c 3. PATHOLOGY. Educ: Univ Tex, BA, 56, MD, 60. Prof Exp: Pathologist & chief lab div path, US Air Force Hosp, Sheppard AFB, Tex, 62-64; fel path, Univ Utah, 64-65, instr path, 65-68; asst prof, 68-72, ASSOC PROF PATH, UNIV CALIF, LOS ANGELES, 72- Concurrent Pos: Attend physician, Ft Douglas Vet Admin Hosp, Salt Lake City, 67, part-time sr physician, 67-68. Mem: Int Acad Path; Am Asn Path & Bact. Res: Ultrastructure of benign and malignant neoplasms. Mailing Add: Dept of Path Univ of Calif Ctr Health Sci Los Angeles CA 90024

WAISS, ANTHONY CHAN, JR, organic chemistry, see 12th edition

WAIT, DAVID FRANCIS, b Sidney, Nebr, Sept 28, 33; m 56; c 4. METROLOGY. Educ: Colo State Univ, BS, 55, MS, 57; Univ Mich, PhD(physics), 63. Prof Exp: Instr & res asst physics, Univ Mich, 62-63; sr scientist, Laser Systs Ctr, Lear Siegler, Inc, 63; PHYSICIST, NAT BUR STAND, 63- Mem: Sigma Xi; Inst Elec & Electronics Engrs; Microwave Theory & Tech Soc. Res: Noise in communications; radiometers; microwave cryogenic noise standards. Mailing Add: Noise & Interference Sect Nat Bur of Standards Boulder CO 80302

WAIT, JAMES RICHARD, b Ottawa, Can; US citizen. GEOENVIRONMENTAL SCIENCE. Educ: Univ Toronto, BASc, 48, MASc, 49, PhD(elec eng), 51. Prof Exp: CONSULT APPL PHYSICS, US DEPT COMMERCE, BOULDER, COLO, 65- Concurrent Pos: Mem nat comt, Int Union Radio Sci, 58-61 & 65-68, secy, US Nat Comt, 75-; adj prof elec eng, Univ Colo, Boulder, 61-, fel, Coop Inst Res Environ Sci, 68- Honors & Awards: Gold Medal, US Dept Commerce, 58; Flemming Award, US Chamber Commerce, 64; Harry Diamond Award, Inst Elec & Electronics Eng, 64; NOAA Res & Achievement Award, Nat Oceanic & Atmospheric Admin, 73. Mem: Fel Inst Elec & Electronics Eng; Int Union Radio Sci. Res: Applications of electromagnetic theory to problems in geophysics and telecommunications. Mailing Add: RB 1 Rm 242 US Dept Commerce Boulder CO 80302

WAIT, SAMUEL CHARLES, JR, b Albany, NY, Jan 26, 32; m 57; c 2. PHYSICAL CHEMISTRY. Educ: Rensselaer Polytech Inst, BS, 53, MS, 55, PhD(chem), 56. Prof Exp: Fulbright fel, Univ Col, London, 56-57, asst lectr chem, 57-58; res fel, Univ Minn, 58-59; asst prof, Carnegie Inst Technol, 59-60; chemist, Nat Bur Stand, 60-61; from asst prof to assoc prof chem, 61-71, asst dean sch sci, 72-74, PROF CHEM, RENSSELAER POLYTECH INST, 71-, ASSOC DEAN SCI, 74- Concurrent Pos: Mem adv coun sci & math, Schenectady County Community Col, 75-78. Mem: Am Chem Soc; Optical Soc Am; Coblentz Soc. Res: High resolution ultraviolet, infrared and Raman spectroscopy; asymmetric rotor theory and calculation; quantum biology; molecular orbital theory; vibrational and fine structural analyses; theoretical methods; simple and polyatomic systems. Mailing Add: Dept of Chem Rensselaer Polytech Inst Troy NY 12181

WAITE, ALAN C, b Springfield, Mass, Aug 6, 33; m 66; c 2. MATHEMATICS. Educ: Bob Jones Univ, BS, 56; Univ Hartford, BSME, 58; Univ Ala, MA, 63. Concurrent Pos: Develop engr, United Aircraft Corp, 56-58; from instr to assoc prof math, Bob Jones Univ, 58-69, chmn dept, 64-69; MGR COMPUT CTR, J E SIRRINE CO, GREENVILLE, 69- Concurrent Pos: Consult, J E Sirrine Co, 66-69. Mem: Math Asn Am. Res: Number theory; primitive roots; p-adic numbers; numerical analysis and methods of polynomials; difference equations. Mailing Add: Groveland Dr Rte 3 Taylors SC 29687

WAITE, ALBERT B, b Holbrook, Ariz, Mar 4, 36; m 60; c 4. REPRODUCTIVE PHYSIOLOGY, GENETICS. Educ: Utah State Univ, BS, 61, MS, 62; Univ Mo, PhD(reprod physiol), 66. Prof Exp: Instr reprod physiol, Univ Mo, 66-67; asst prof animal husb, 67-71, assoc prof agr, 71-74, PROF AGR, CENT MO STATE UNIV, 74- Mem: Am Soc Animal Sci; Am Fertil Soc; Brit Soc Study Fertil. Res: Estrus synchronization in sheep and swine; embryonic mortality in domestic animals. Mailing Add: Dept of Agr Cent Mo State Univ Warrensburg MO 64093

WAITE, DANIEL ELMER, b Grand Rapids, Mich, Feb 19, 26; m 48; c 4. ORAL SURGERY. Educ: Univ Iowa, DDS, 53, MS, 55; Am Bd Oral Surg, dipl, 59. Prof Exp: Resident oral surg, Univ Hosp, Univ Iowa, 53-55, from instr to prof & head dept, Col Dent, 55-59, assoc prof, Hosp Dent Dept, Univ Hosps, 57-63; asst prof dent, Mayo Grad Sch Med, 63-68, PROF ORAL SURG & CHMN DIV, SCH DENT, UNIV MINN, MINNEAPOLIS, 68- Concurrent Pos: Mem staff, Proj Hope, Peru, Ceylon & Haiti; trustee, Park Col, Mo. Mem: Am Soc Oral Surg; Am Dent Asn; Am Col Dent; Int Asn Dent Res. Mailing Add: Div of Oral Surg Univ of Minn Sch of Dent Minneapolis MN 55455

WAITE, JOHN HENRY, b St Cloud, Minn, Aug 6, 33; m 42; c 5. SURGERY. Educ: State Univ NY Buffalo, MD, 47; Ohio State Univ, MMSc, 53. Prof Exp: Instr surg, Ohio State Univ, 50-53; investr, NIH, 54-57; clin instr surg, Univ Wash, 59-61; clin asst prof, Tulane Univ, 61-64, clin assoc prof, 64-67; assoc prof, 67-72, PROF SURG, LA STATE UNIV, BATON ROUGE, 73- Mem: Am Col Surg. Res: Clinical cancer therapy. Mailing Add: Earl K Long Mem Hosp 5825 Airline Hwy Baton Rouge LA 70805

WAITE, LEONARD CHARLES, b Reynoldsville, Pa, Sept 10, 41; m 60; c 3. PHARMACOLOGY. Educ: Alderson-Braddus Col, BS, 65; WVa Univ, MS, 67; Univ Mo-Columbia, PhD(pharmacol), 69. Prof Exp: Asst prof, 70-74, ASSOC PROF PHARMACOL, SCH MED, UNIV LOUISVILLE, 74- Mem: Endocrine Soc; Am Soc Pharmacol & Exp Therapeut. Res: Endocrinology; physiology; calcium metabolism. Mailing Add: Dept of Pharmacol Univ of Louisville Sch of Med Louisville KY 40201

WAITE, MARILYNN RANSOM FAIRFAX, b Mt Vernon, NY, May 1, 42. VIROLOGY. Educ: Bryn Mawr Col, BA, 63; Dartmouth Col, PhD(molecular biol), 71. Prof Exp: Instr microbiol, Dartmouth Med Sch, 73; ASST PROF MICROBIOL, UNIV TEX, AUSTIN, 73- Concurrent Pos: Anna Fuller Fund fel, Med Sch, Dartmouth Col, 71-73; NIH fel, 73. Mem: AAAS; Am Soc Microbiol. Res: Replication of RNA tumor viruses and Sindbis virus; interaction of viral proteins with membranes of infected cells. Mailing Add: Dept of Microbiol Univ of Tex Austin TX 78712

WAITE, MOSELEY, b Durham, NC, Oct 22, 36; m 59; c 3. BIOCHEMISTRY, ORGANIC CHEMISTRY. Educ: Rollins Col, BS, 58; Duke Univ, PhD(biochem), 63. Prof Exp: Asst prof, 67-71, ASSOC PROF BIOCHEM, BOWMAN GRAY SCH MED, 71- Concurrent Pos: Am Cancer Soc fel biochem, Duke Univ, 62-65; Am Heart Asn advan fel, Univ Utrecht, 65-67; grants, Am Heart Asn, 66-69, USPHS, 67-74 & NC Heart Asn, 70-73; USPHS res career develop award, 73-78. Mem: AAAS; Am Chem Soc; Am Soc Biol Chemists. Res: Phospholipid and fatty acid metabolism; enzyme purification and characterization; relation of metabolism of lipids to certain morphological changes, especially mitochondrion. Mailing Add: Dept of Biochem Bowman Gray Sch of Med Winston-Salem NC 27103

WAITES, ROBERT ELLSWORTH, b Middletown, Ohio, Apr 28, 16; m 45; c 2. ENTOMOLOGY. Educ: Otterbein Col, BS, 41; Ohio State Univ, MS, 46, PhD(entom), 49. Prof Exp: Asst entomologist, 51-68, ASSOC ENTOMOLOGIST, AGR EXP STA, UNIV FLA, 68-, ASSOC PROF ENTOM, 74- Mem: AAAS; Int

Orgn Biol Control; Entom Soc Am. Res: Chemical control of insects on vegetable crops; systematics and biology of the Coccinellidae; arthropod pests of poultry. Mailing Add: Dept of Entom Agr Exp Sta Univ of Fla Gainesville FL 32601

WAITHE, WILLIAM IRWIN, b New York, NY, May 3, 37; c 4. CELL PHYSIOLOGY. Educ: St Francis Col, BS, 58; NY Univ, MS, 63, PhD(cell biol), 69. Prof Exp: Res asst med genetics, Sch Med, NY Univ, 58-66; res assoc immunol & cell biol, Mt Sinai Sch Med, City Univ NY, 66-68, instr genetics, 68-71; ASST PROF IMMUNOL & CELL BIOL, FAC MED, UNIV LAVAL, 71- Concurrent Pos: Scholar, Career Develop Award, Med Res Coun Can, 71-76. Mem: AAAS; Can Soc Cell Biol. Res: Regulation of growth by nuclear and cytoplasmic proteins; biochemical mechanisms controlling lymphocyte activation in vitro; the role of serum factors in controlling the in vitro immune response. Mailing Add: Ctr Hemat & Immunol Hosp St Sacrament Quebec PQ Can

WAITKINS, GEORGE RAYMOND, b Glasgow, Scotland, Feb 28, 11; nat US; m 37; c 2. PHYSICAL CHEMISTRY. Educ: Syracuse Univ, 33, MS, 34, PhD(chem), 38. Prof Exp: Res engr, Battelle Mem Inst, 38-43; chemist, Mutual Chem Co Am, 43; chem supvr, Can Copper Refiners, Ltd, Que, 44-45; res chemist, Calco Div, Am Cyanamid Co, 45-52; asst mgr res dept, Am Zinc Lead & Smelting Co, 52-62; PHYS CHEMIST, MATTIN LABS, MEARL CORP, OSSINING, 62- Mem: Am Chem Soc; Am Inst Chemists; AAAS. Res: Inorganic, organic and nacreous pigments; crystal growth. Mailing Add: 1 Hughes St Croton-on-Hudson NY 10520

WAITKUS, PHILLIP ANTHONY, polymer chemistry, see 12th edition

WAITMAN, REUBEN HOMER, b Evansville, Ind, Apr 14, 14; m 47; c 5. FOOD CHEMISTRY. Educ: Univ Evansville, BA, 51. Prof Exp: Proj leader, Igleheart Bros, Ind, 53-57; proj leader, Jell-o Div, Gen Foods Corp, 57-59, sect head, 59-64, res specialist, 64-67, RES SPECIALIST, CORP PROD DEVELOP, GEN FOODS CORP, 67- Concurrent Pos: Ed cereal chem & milling tech sect, Biol Abstracts, 52- Mem: Am Chem Soc; Am Asn Cereal Chem; Inst Food Technologists. Res: Wheat flour, colloid and analytical chemistry; food technology; market research. Mailing Add: 29 Lark St Pearl River NY 10965

WAITS, BERT KERR, b New Orleans, La, Dec 21, 40; m 63; c 2. MATHEMATICS. Educ: Ohio State Univ, BSc, 62, MSc, 64, PhD(math educ), 69. Prof Exp: Asst to chmn dept math, 65-69, asst prof math, 69-75, ASSOC PROF MATH, OHIO STATE UNIV, 75- Mem: Math Asn Am; Nat Coun Teachers Math. Res: Mathematics education; individualized instruction at the college level; computer assisted instruction; audiovisual pedagogical experimentation. Mailing Add: Dept of Math Ohio State Univ Columbus OH 43210

WAITS, EWELL DOUGLAS, botany, see 12th edition

WAITZ, JAY ALLAN, b Elizabeth, NJ, Nov 26, 35; m 60; c 2. CHEMOTHERAPY. Educ: Univ Idaho, BS, 57, MS, 59; Univ Ill, PhD(parasitol), 62. Prof Exp: Assoc res parasitologist, Parke, Davis & Co, 62-65, res parasitologist, 65-66; sr microbiologist, 66-68, sect head, 69-70, mgr chemother dept, 70-73, ASSOC DIR MICROBIOL, SCHERING CORP, 73- Mem: Am Soc Parasitol; Am Micros Soc; Am Soc Microbiol; Am Soc Trop Med & Hyg; Soc Protozool. Res: Chemotherapy of parasitic, bacterial and fungal diseases; parasite physiology and histochemistry. Mailing Add: Dept of Chemother Schering Corp Bloomfield NJ 07003

WAITZMAN, MORTON BENJAMIN, b Chicago, Ill, Nov 8, 23; m 49; c 3. PHYSIOLOGY, BIOCHEMISTRY. Educ: Univ Miami, BS, 48; Univ Ill, MS, 50, PhD(physiol), 53. Prof Exp: Res asst physiol, Univ Ill, 51-54; res assoc, Dept Pharmacol & Lab Res Ophthal, Sch Med, Western Reserve Univ, 54-56, instr, 56-59, asst prof ophthalmic res & pharmacol & dir lab res ophthal, 59-62; assoc prof, 62-68, PROF OPHTHAL, SCH MED, EMORY UNIV, 68-, DIR LAB OPHTHALMIC RES, 62- Mem: AAAS; Asn Res Vision & Ophthal; Am Physiol Soc; NY Acad Sci. Res: Ophthalmic research; metabolic and hormonal aspects of aqueous humor and cerebrospinal fluid production; glaucoma; neuro-chemistry; autonomic nature of ocular extracts. Mailing Add: Lab for Ophthalmic Res Emory Univ Sch of Med Atlanta GA 30322

WAJDA, EDWARD STANLEY, b Schenectady, NY, Oct 31, 24; m 50; c 2. PHYSICS. Educ: Union Col, NY, BS, 45; Cornell Univ, MS, 48; Rensselaer Polytech Inst, PhD(physics), 53. Prof Exp: Instr physics, Amherst Col, 46-47; Col, instr, Union Col, NY, 48-49, asst prof, 53-55; SR PHYSICIST, IBM CORP, 55- Mem: Inst Elec & Electronics Engrs. Res: Solid state physics; semiconductors. Mailing Add: 39 Spy Hill Poughkeepsie NY 12603

WAJDA, ISABEL, b Cracow, Poland, Apr 3, 13; US citizen; m 34. PHARMACOLOGY, NEUROCHEMISTRY. Educ: Jagiellonian Univ, BSc, 36; Univ Birmingham, PhD(pharmacol), 51. Prof Exp: Res pharmacologist, Oxford Univ, 43-46; res pharmacologist, Med Sch, Univ Birmingham, 46-51; head dept pharmacol, Med Sch, Univ Mendoza, Arg, 53-55; instr, New York Med Col, Flower & Fifth Ave Hosps, 55, asst prof, 56-59; sr res scientist, NY State Psychiat Inst, 59-66, assoc res scientist, 66-68; ASSOC RES SCIENTIST, NY STATE RES INST NEUROCHEM & DRUG ADDICTION, 68- Concurrent Pos: Assoc prof pharmacol & chmn dept, Sch Dent, Fairleigh Dickinson Univ, 68-74. Mem: Am Soc Pharmacol & Exp Therapeut; Am Soc Neurochem; Brit Pharmacol Soc; Int Soc Neurochem; Int Soc Biochem Pharmacol. Res: Biological standardization; biochemistry and pharmacology of the central nervous system related to neurochemical transmission; enzyme metabolism in experimental autoimmune diseases; metabolism of neurotransmitters in drug addiction. Mailing Add: NY State Res Inst Neurochem & Drug Addiction Ward's Island NY 10035

WAJDA, STANISLAW HENRY, anatomy, histology, deceased

WAKADE, ARUN RAMCHANDRA, b Wai, India, Dec 3, 40. PHARMACOLOGY. Educ: Univ Bombay, BS, 61; State Univ NY Downstate Med Ctr, MS, 64, PhD(pharmacol), 67. Prof Exp: Govt India sr sci fel pharmacol, Med Col, Univ Baroda, 68-70; from instr to asst prof pharmacol, 70-75, ASSOC PROF PHARMACOL, STATE UNIV NY DOWNSTATE MED CTR, 75- Mem: Am Soc Pharmacol & Exp Therapeut; Indian Asn; Physiol & Pharmacol. Res: Adrenergic mechanism in peripheral autonomic nervous system. Mailing Add: Dept of Pharmacol State Univ of NY Downstate Med Ctr Brooklyn NY 11203

WAKE, DAVID BURTON, b Webster, SDak, June 8, 36; m 62; c 1. EVOLUTIONARY BIOLOGY. Educ: Pac Lutheran Univ, BA, 58; Univ Southern Calif, MSc, 60, PhD(biol), 64. Prof Exp: Asst biol, Univ Southern Calif, 58-59, head lab assoc, 62-63, instr, 63-64; instr anat & biol, Univ Chicago, 64-66, asst prof, 66-69; assoc prof zool, 69-73, assoc cur, Mus Vert Zool, 69-71, PROF ZOOL, UNIV CALIF, BERKELEY, 73-, DIR, MUS VERT ZOOL, 71-, CUR HERPET, 73- Honors & Awards: Quantrell Award, Univ Chicago, 67. Mem: AAAS; Am Soc Ichthyologists & Herpetologists; Am Asn Anat; Am Soc Zoologists; Soc Study Evolution. Res: Functional and evolutionary morphology of lower vertebrates; evolution, systematics and zoogeography of modern Amphibia, with emphasis on salamanders; evolutionary theory. Mailing Add: Mus of Vert Zool Univ of Calif Berkeley CA 94720

WAKE, MARVALEE H, b Orange, Calif, July 31, 39; m 62; c 1. VERTEBRATE BIOLOGY. Educ: Univ Southern Calif, BA, 61, MS, 64, PhD(biol), 68. Prof Exp: Teaching asst biol, Univ Ill, Chicago, 64-66, instr, 66-68, asst prof, 68-69; lectr, 69-73, ASST PROF BIOL, UNIV CALIF, BERKELEY, 73- Mem: AAAS; Am Soc Ichthyologists & Herpetologists; Soc Study Evolution. Res: Evolution of vertebrates; morphology; reproductive biology. Mailing Add: Dept of Zool 4079 LSB Univ of Calif Berkeley CA 94720

WAKE, WILLIAM HENRY, b Chicago, Ill, June 11, 22; m 44; c 1. GEOGRAPHY. Educ: Stanford Univ, AB, 49; Columbia Univ, AM, 49; Univ Calif, Los Angeles, PhD(geog), 61. Prof Exp: Instr geog, Wash State Col, 49-51; teaching asst, Univ Calif, Los Angeles, 51-54; asst prof, Univ Southern Calif, 55-64, chmn dept, 63-64; from asst prof to assoc prof, Fresno State Col, Bakersfield Ctr, 64-70; assoc prof, 70-74, PROF GEOG, CALIF STATE COL, BAKERSFIELD, 74- Concurrent Pos: Southwest rep, Theodore Roosevelt Centennial Comn, 56-58; mem, Calif State Adv Bd, US Bur Land Mgt, 65- Geog Adv Panel, Statewide Soc Sci Study Comt, 65-66 & Conserv Adv Comt, Calif State Bd Educ, 67-70; consult, NSF, 67-68, US Off Educ, 68-70, US Bur Land Mgt Planning Staff, 70, Col Coop Sci Serv, NSF, 70-71 & Col Sci Improv Proj, 72-; state coordr, Nat Coun Geog Educ, 68- Honors & Awards: Merit Award, Calif Conserv Coun, 64. Mem: Fel AAAS; fel Nat Coun Geog Educ; Asn Am Geogr; Am Soc Photogram; Royal Geog Soc. Res: Geographic and conservation education; historical geography; India-South Asia; cartography and educational media. Mailing Add: Dept of Earth Sci Calif State Col Bakersfield CA 93309

WAKEFIELD, GENE F, b Albuquerque, NMex, Dec 22, 33; m 57; c 1. PHYSICAL CHEMISTRY, RESEARCH ADMINISTRATION. Educ: Colo State Univ, BS, 55; Iowa State Univ, PhD(phys chem), 61; Southern Methodist Univ, MBA, 75. Prof Exp: Jr chemist, Ames Lab, 55-56, res chemist, 56-57; asst, Iowa State Univ, 57-61; mem tech staff, 61-71, TECH MGR, TEX INSTRUMENTS INC, 71- Mem: Metall Soc; Electrochem Soc; Am Ceramic Soc; Int Solar Energy Soc; Am Soc Testing & Mat. Res: Preparation and properties of metals, refractory compounds; high temperature chemistry and equipment; diffusion of solids; vapor deposition of materials; preparation, crystal growth and processing of electronic-grade silicon; photovoltaic devices; technoeconomic studies. Mailing Add: 406 Tyler Richardson TX 75080

WAKEFIELD, LUCILLE MARION, b Dayville, Conn, June 13, 25. NUTRITION. Educ: Univ Conn, BS, 49, MS, 56; Ohio State Univ, PhD(nutrit), 65. Prof Exp: Intern dietetics, Mt Auburn Hosp, 49-50; therapeut dietitian, New Brit Gen Hosp, 50-52, admin dietitian, 52-53; dir dietetics, Auburn Mem Hosp, 53-57; asst prof food & nutrit, Univ Vt, 57-65, head dept nutrit & inst mgt, 62-65; prof foods & nutrit & head dept, Kans State Univ, 65-75; PROF FOOD & NUTRIT & HEAD DEPT, FLA STATE UNIV, 75- Mem: AAAS; fel Am Inst Chemists; Am Dietetic Asn; Am Pub Health Asn; Am Inst Nutrit. Res: Nutritional, sociological, psychological aspects of humans and their body composition as it relates to population groups and nutritional status; clinical nutrition and community health problems; nutrition in aging; food patterning. Mailing Add: Dept of Foods & Nutrit Fla State Univ Sandel Hall Tallahassee FL 33203

WAKEFIELD, LYNN BURRITT, organic chemistry, see 12th edition

WAKEFIELD, ROBERT CHESTER, b Providence, RI, Sept 14, 25; m 49; c 3. AGRONOMY. Educ: Univ RI, BS, 50; Rutgers Univ, MS, 51, PhD, 54. Prof Exp: Res assoc farm crops, Rutgers Univ, 51-54; from asst prof to assoc prof agron, 54-65, chmn dept, 61-70, PROF AGRON, UNIV RI, 65- Res: Crop ecology; landscape ecology. Mailing Add: Dept of Plant & Soil Sci Univ of RI Kingston RI 02881

WAKEFIELD, SHIRLEY LORRAINE, b Milwaukee, Wis, Nov 20, 34; m 57; c 2. SOLID STATE CHEMISTRY. Educ: Univ Wis-Madison, BS, 57; Univ Cincinnati, PhD(phys chem), 69; Xavier Univ, MBA, 76. Prof Exp: Chemist, Forest Prod Lab, US Forest Serv, 57-58; lab technician, Pulmonary Dis Res Lab, Vet Admin Hosp, 59-62; staff engr, Avco Electronics, 69-73; spec progs mgr, Measurements & Sensors Develop, 73-75, MGR COMPOSITES, MAT & PROCESSES TECHNOL PROGS, ADVAN ENG & TECHNOL PROG DEPT, AIRCRAFT ENGINE GROUP, GEN ELEC CO, 75- Mem: Am Chem Soc; Am Phys Soc; Am Soc Nondestructive Testing; Soc Advan Mat & Process Eng. Res: Infrared detectors; nuclear quadrupole resonance spectroscopy; pulmonary disease; gas chromatography; high energy x-ray, Raman spectroscopy; ceramics; high temperature alloy development. Mailing Add: Gen Elec Aircraft Engine Group MD H-9 175 & Bypass 50 Cincinnati OH 45215

WAKEHAM, HELMUT, b Hamburg, Ger, Apr 15, 16; US citizen; m 39; c 3. PHYSICAL CHEMISTRY. Educ: Univ Nebr, BA, 36, MA, 37; Univ Calif, PhD(phys chem), 39. Prof Exp: Asst chem, Univ Nebr, 36-37; asst, Univ Calif, 37-39; res chemist, Stand Oil Co, Calif, 39-41; res chemist, Southern Regional Res Lab, USDA, 41-47; from res assoc to proj head & dir res, Chem Physics Sect, Textile Res Inst, NJ, 49-56; dir, Ahmedabad Textile Industs Res Asn, India, 56-58; staff asst to vpres & chief opers & subsidiaries, 58-60, dir res ctr, 60-66, VPRES RES & DEVELOP, RES CTR PHILIP MORRIS, INC, 61- Concurrent Pos: Mem, Tobacco Working Group, 68-; mem, Nat Cancer Plan, 71; mem bd dir, Indust Res Inst, Inc, 71-72. Mem: AAAS; Am Chem Soc; Fiber Soc; Soc Rheol; Am Crystallog Asn; fel Am Inst Chemists. Res: Surface physics and chemistry; colloids; hydrocarbon and vegetable oils; cellulose chemistry; physical and mechanical properties of textile fibers; textile fiber and filament processing; determination of melting and freezing temperatures; calorimetry and thermometry. Mailing Add: Philip Morris Res Ctr PO Box 26583 Richmond VA 23261

WAKELEY, JAY TOWNSEND, b Honesdale, Pa, July 19, 19; m 41; c 2. OPERATIONS RESEARCH, APPLIED STATISTICS. Educ: Iowa State Univ, BS, 41; NC State Univ, MS, 46, PhD(exp statist), 50. Prof Exp: Asst statistician, NC State Univ, 46-48, asst prof statist, 48-51; oper analyst, A F Spec Weapons Ctr, 51-54; mathematician, Rand Corp, 54-63; mgr opers anal, NAm Aviation, Inc, 63-66; dir new progs, Gen Res Corp, 66-69; dir opers res & econ div, 69-71, DIR CTR HEALTH STUDIES, RES TRIANGLE INST, 71- Concurrent Pos: Adj prof, NC State Univ, 70-, consult, Ctr Alcohol Studies, 74- Mem: AAAS; Opers Res Soc Am. Res: Operations research as applied to health economics, health services delivery and health systems research. Mailing Add: Res Triangle Inst Research Triangle Park NC 27709

WAKELIN, JAMES HENRY, JR, b Holyoke, Mass, May 6, 11; m 38; c 3. PHYSICS. Educ: Dartmouth Col, AB, 32; Cambridge Univ, BA, 34, MA, 39; Yale Univ, PhD(physics), 40. Prof Exp: Sr physicist, B F Goodrich Co, Ohio, 39-43; dir res, Eng Res Assocs, Inc, DC & Minn, 46-48; assoc dir res, Textile Res Inst, NJ, 48-51, dir res, 51-54; founding dir & vpres, Chesapeake Instrument Corp, Md, 54-59; asst secy res & develop, Dept Navy, 59-64; chmn adv bd, Teledyne Ryan Aeronaut, Calif, 64-68, mem adv bd, 68-71; asst secy sci & technol, Dept Com, 71-72; pres, Res Anal Corp, McLean, Va, 72-75, chmn bd dirs, 65-71. Concurrent Pos: Independent consult, 54-59; dir, Nassau Fund, Princeton, NJ, 55-59; pres, Sci Eng Inst, Waltham, Mass, 64-67; consult, United Aircraft Corp, Conn, 67-69; dir, Greyrad Corp, NJ, 67-69, Oceans Gen, Inc, Fla, 68-71, Aqua Int, Calif, 69-70 & Wellington & Assoc Mutual Funds, Del, 70-71; spec asst to Gov Del for Marine & Coastal Affairs, 70-71; head US deleg, Intergovt Conf Oceanog, Copenhagen, 60 & Intergovt Oceanog Comn, UNESCO, Paris, 61; chmn, Interagency Comt Oceanog, 60-64; mem, Naval Res Adv Comt, Dept Navy, 64-71; chmn bd, Oceanic Found, Honolulu, Hawaii, 66-71; mem corp, Woods Hole Oceanog Inst, 66-71; chmn, President's Task Force Oceanog, 69 & Com Tech Adv Bd, 71-72; mem, Fed Coun Sci & Technol, Off of the President, 71-72; chmn, Interagency Comt Atmospheric Sci, 71-72 & Comt Govt Patent Policy, 71-72; mem, Comt Res Appl to Nat Needs, 71-72; mem bd trustees, Comt Res & Explor, Nat Geog Soc, 63-; overseers' vis comt, Dept Astron, Harvard Univ, 64-69 & 70-; chmn bd overseers, Thayer Sch Eng, Dartmouth Col, 67-70; mem bd adv to President, Naval War Col, 70-74; mem vis comt, Dept Ocean Eng, Mass Inst Technol, 71-; pres, Grad Sch Asn & mem, Alumni Bd, Yale Univ, 71-; mem, Textile Res Inst. Honors & Awards: Distinguished Pub Serv Awards, Dept Navy, 61 & 64; Rear Admiral William S Parsons Award, Navy League of US, 65. Mem: Hon mem Nat Security Indust Asn; hon mem Marine Technol Soc (pres, 66-68); Am Phys Soc; Fiber Soc; Rheol Soc. Res: Magnetization near boundaries; x-ray diffraction of natural rubber and synthetic polymers; mathematical computing devices; physics of textile fibers; oceanography. Mailing Add: 1809 45th St NW Washington DC 20007

WAKELYN, PHILLIP JEFFREY, b Akron, Ohio, Apr 29, 40. TEXTILE CHEMISTRY. Educ: Emory Univ, BS, 63; Ga Inst Technol, MS, 68; Univ Leeds, PhD(textile chem), 71. Prof Exp: Res chemist, Fibers Div, Dow Chem Co, 63-66 & Dow-Badische Co, 66-67; res assoc textile chem, Textile Res Ctr, Tex Tech Univ, 71-73, head chem res, 73; MGR ENVIRON & SAFETY TECHNOL, NAT COTTON COUN, 73- Concurrent Pos: Lectr textile chem, Tex Tech Univ, 72-74, adj prof chem eng, 74- Mem: Am Chem Soc; Sigma Xi; sr mem Am Asn Textile Chemists & Colorists; assoc Brit Textile Inst. Res: Physical and chemical properties of textile fibers; additives to fibers; antistats; lubricants; fire retardants; chemistry of wool and cotton; environmental, health and safety problems; cotton dust and occupational diseases. Mailing Add: Nat Cotton Coun Box 12285 Memphis TN 38112

WAKEMAN, DONALD LEE, b Lebanon, Mo, Nov 17, 29; m 50; c 3. ANIMAL HUSBANDRY. Educ: Okla State Univ, BSA, 51; Univ Fla, MSA, 55. Prof Exp: Instr animal sci, Univ Tenn, 51; instr, 55-57, asst prof, Univ & asst animal husbandman, Agr Exp Sta, 57-67, assoc prof, Univ, 67-75, PROF ANIMAL SCI, UNIV FLA, 75-, ASSOC ANIMAL HUSBANDMAN, AGR EXP STA, 67- Mem: Am Soc Animal Sci. Res: Animal production and nutrition; beef cattle production. Mailing Add: Dept of Animal Sci Univ of Fla Gainesville FL 32601

WAKEMAN, IRVING B, b Fairfield, Conn, June 20, 22; m 51; c 4. ANALYTICAL CHEMISTRY. Educ: Middlebury Col, BA, 44. Prof Exp: Chemist, Heyden Chem Corp, 44-63; chemist, 63-68, proj leader instrumental anal, 68-70, GROUP LEADER INSTRUMENTAL ANAL, TENNECO CHEM INC, 70- Mem: Am Chem Soc; Soc Appl Spectros. Res: Gas chromatography; infrared spectroscopy; pollution control. Mailing Add: Tenneco Chem Inc PO Box 365 Piscataway NJ 08854

WAKERLIN, GEORGE EARLE, b Chicago, Ill, July 1, 01; m 52; c 2. MEDICINE. Educ: Univ Chicago, BS, 23, PhD(physiol), 26; Univ Wis, SM, 24; Rush Med Col, MD, 29. Prof Exp: Instr pharmacol, Univ Wis, 23-25; assoc instr physiol, Univ Chicago, 25-26, asst med, 30-31; surg house officer, Hopkins Hosp, 28-29, med house officer, 29-30; from asst prof to prof physiol & pharmacol, Sch Med, Univ Louisville, 31-37; prof & head dept, Col Med, Univ Ill, 37-58, asst dean, Rush-Presby Div, 45-46; med dir, Am Heart Asn, 58-66; prof med, 66-71, EMER PROF MED, SCH MED, UNIV MO-COLUMBIA, 71-; MED WRITING, 73- Concurrent Pos: Clin prof, Col Physicians & Surgeons, Columbia Univ, 59-60, adj prof, 60-66; mem coun arteriosclerosis, Am Heart Asn; dir, Mo Regional Med Prog, 66-67, dir planning, 67-71, cent dist consult, 71-73. Honors & Awards: Am Cancer Soc Medal, 57; Gold Heart Award, Am Heart Asn, 66. Mem: AAAS; Geront Soc; Am Cancer Soc; Am Physiol Soc; Am Soc Pharmacol & Exp Therapeut. Res: Chemotherapy, immunity and serology of experimental and clinical syphilis; hematopoiesis and pathogenesis of pernicious anemia; antihormones; pathogenesis and treatment of experimental and clinical hypertensions; pathogenesis of experimental atherosclerosis; medical administration. Mailing Add: 2120 Pacific Ave San Francisco CA 94115

WAKERLING, RAYMOND KORNELIOUS, b Oakland, Calif, June 16, 14; m 40. MATHEMATICS. Educ: Univ Calif, AB, 36, PhD(math), 39. Prof Exp: Asst math, Univ Calif, 36-39; instr, Tex Tech Col, 39-42; asst prof, Fresno State Col, 42-43; physicist, 43-46, dir tech info div, 46-74, ASSOC DIR EMPLOYEE & INFO SERVS, LAWRENCE BERKELEY LAB, UNIV CALIF, 74- Mem: Am Soc Info Sci. Res: Algebraic geometry; scientific information problems. Mailing Add: Lawrence Berkeley Lab Univ of Calif Berkeley CA 94720

WAKIL, SALIH J, b Karballa, Iraq; nat US; m 52; c 4. BIOCHEMISTRY. Educ: Am Univ Beirut, BA, 48; Univ Wash, PhD(biochem), 52. Prof Exp: Res assoc, Inst Enzyme Res, Univ Wis, 52-56, asst prof, Univ, 56-59; from asst prof to prof biochem, Sch Med, Duke Univ, 59-71; PROF BIOCHEM & CHMN DEPT, BAYLOR COL MED, 71- Honors & Awards: Am Chem Soc Award, 67. Mem: Am Chem Soc. Res: Genetic and metabolic control of fatty acid metabolism. Mailing Add: Dept of Biochem Baylor Col of Med Houston TX 77025

WAKIM, KHALIL GEORGES, b Sidon, Lebanon, July 17, 07; nat US; m 36. INTERNAL MEDICINE. Educ: Am Univ Beirut, BA, 29, MD, 33; Univ Minn, PhD(physiol), 41. Prof Exp: Instr physiol, Am Univ Beirut, 33-38; actg prof, Col Med, Univ Iowa, 40; from assoc prof to prof, Sch Med, Ind Univ, 41-46; prof, 46-71, EMER PROF PHYSIOL, MAYO GRAD SCH MED, UNIV MINN, 71-, EMER CONSULT, MAYO CLIN & MAYO FOUND, ROCHESTER, MINN, 71-; PROF CLIN PHYSIOL, SCH MED, IND UNIV, INDIANAPOLIS, 71-; COORDR MED EDUC, TERRE HAUTE MED EDUC FOUND, 71- Concurrent Pos: Consult, Mayo Clin, 46-71; consult, Off Surgeon Gen, US Army. Honors & Awards: First Order Merit, Repub Syria, 50; Highest Order Merit, Repub Lebanon, 50, Knight, Order of the Cedar, Lebanon, 67. Mem: AAAS; Am Physiol Soc; Soc Exp Biol & Med; Am Soc Trop Med & Hyg; Am Soc Pharmacol & Exp Therapeut. Res: Liver and gastrointestinal tract; heart and circulation; effects of physical agents; kidney; neurophysiology; muscle physiology. Mailing Add: Terre Haute Med Educ Found PO Box 546 Terre Haute IN 47808

WAKSBERG, ARMAND L, b Paris, France, Oct 11, 34; Can citizen; m 60; c 1. MATHEMATICS, PHYSICS. Educ: McGill Univ, BS, 56, MS, 60. Prof Exp: Scientist, Can, Ltd, 56-58 & Can Aviation Electronics, 60-63; SR MEM SCI STAFF, RES DEPT, RCA LTD, 63- Mem: Inst Elec & Electronics Engrs. Res: Lasers, including sidelight spectroscopy, laser noise and phase locking phenomena; laser communications. Mailing Add: Res Dept RCA Ltd Ste-Anne-de-Bellevue PQ Can

WAKSBERG, JOSEPH, b Kielce, Poland, Sept 20, 15; US citizen; m 41; c 2. APPLIED STATISTICS. Educ: City Col New York, BS, 36. Prof Exp: Jr mathematician, US Navy Dept, 37-38; asst proj dir math, US Works Proj Admin, 38-40; asst proj dir, US Bur Census, 40-59, asst chief construction statist div, 59-63, chief statist methods div, 63-71, assoc dir statist, Bur, 72-73; VPRES, WESTAT INC, 73- Concurrent Pos: Instr statist, USDA Grad Sch, 63-; consult, CBS News, 66- Honors & Awards: Gold Medal for Distinguished Contrib, US Dept Com, 68. Mem: Fel Am Statist Asn; Inst Math Statist; Pop Asn Am; Int Asn Surv Statist. Res: Sample design for surveys; research in survey methodology, especially sampling and response errors. Mailing Add: 6302 Tone Dr Bethesda MD 20034

WAKSMAN, BYRON HALSTEAD, b New York, NY, Sept 15, 19; m 44; c 2. IMMUNOLOGY. Educ: Swarthmore Col, BA, 40; Univ Pa, MD, 43. Prof Exp: Intern, Michael Reese Hosp, Ill, 44; res assoc neuropath, Harvard Med Sch, 49-52, assoc bact & immunol, 52-57, asst prof, 57-63; prof microbiol, 63-74, PROF PATH, YALE UNIV, 74- Concurrent Pos: Fel, Mayo Clin, 46-48; NIH fel, Columbia Univ, 48-49; res fel neuropath, Mass Gen Hosp, 49-52; assoc bacteriologist, Mass Gen Hosp, 52-63; consult assoc bacteriologist, Mass Eye & Ear Infirmary, 57-63; mem microbiol fels panel, NIH, 61-64, mem study sect B on allergy & Immunol, 65-69; mem res rev panel, Nat Multiple Sclerosis Soc, 61-66; mem expert adv panel immunol, WHO, 63-68; chmn dept microbiol, Yale Univ, 64-70 & 72-74. Mem: Fel AAAS; Am Asn Immunol (secy-treas, 61-64, pres, 70-71); Am Soc Microbiol; Soc Exp Biol & Med; Am Rheumatism Asn. Res: Role of thymus and lymphocytes in immune responses; immunologic tolerance; delayed hypertensitivity; immunologic and pathologic character of experimental autoallergic diseases. Mailing Add: Dept of Path Yale Univ New Haven CT 06510

WAKSMAN, SELMAN A, microbiology, deceased

WALASZEK, EDWARD JOSEPH, b Chicago, Ill, July 4, 27; m 55; c 2. PHARMACOLOGY. Educ: Univ Ill, BSc, 49; Univ Chicago, PhD(pharmacol), 53. Prof Exp: Asst prof neurophysiol & biochem, Univ Ill, 55-56; asst prof, 57-59, assoc prof, 59-62, PROF PHARMACOL, UNIV KANS MED CTR, KANSAS CITY, 62-, CHMN DEPT, 64- Concurrent Pos: Res fel, Univ Edinburgh, 53-55; USPHS spec res fel, 56-61, res career develop award, 61-63, res career award, 63-64; mem health study sect med chem, NIH, 62-66, mem res career award study sect, Nat Inst Gen Med Sci, 66-70; mem adv coun, Int Union Pharmacol, 72-76; mem, Health Study Sect Pharmacol-Toxicol, 74-78; mem comt teaching of sci, Int Coun Sci Unions. Mem: AAAS; Am Chem Soc; Soc Neurosci; Am Soc Pharmacol & Exp Therapeut; fel Am Soc Clin Pharmacol & Therapeut. Res: Pharmacologically active polypeptides; neurohumoral substances; naturally-occurring biogenic amines; pharmacology and physiology of the central nervous system. Mailing Add: Dept of Pharmacol Univ of Kans Med Ctr Kansas City KS 66103

WALAWENDER, MICHAEL JOHN, b Auburn, NY, Dec 16, 39; m 67. PETROLOGY, GEOLOGY. Educ: Syracuse Univ, BS, 65; SDak Sch Mines & Technol, MS, 67; Pa State Univ, University Park, PhD(petrol), 72. Prof Exp: Res asst mineral, SDak Sch Mines & Technol, 65-67; asst petrol, Pa State Univ, University Park, 67-72; ASST PROF GEOL, SAN DIEGO STATE UNIV, 72- Mem: Geol Soc Am. Res: Igenous and metamorphic petrology; mineralogy; planetology. Mailing Add: Dept of Geol San Diego State Univ San Diego CA 92182

WALBA, HAROLD, b Chelsea, Mass, Mar 10, 21; m 46; c 2. ORGANIC CHEMISTRY. Educ: Univ Mass, BS, 46; Univ Calif, PhD(chem), 49. Prof Exp: From instr to assoc prof, 49-58, chmn dept, 61-64, PROF CHEM, SAN DIEGO STATE UNIV, 58- Mem: AAAS; Am Chem Soc; The Chem Soc. Res: Substituent effects and their transmission in organic molecules; tautomerism; acid-base strengths. Mailing Add: Dept of Chem San Diego State Univ San Diego CA 92182

WALBERG, CLIFFORD BENNETT, b Watkins, Minn, Feb 24, 15; m 46; c 4. CLINICAL CHEMISTRY. Educ: Univ Sask, BS, 39; Univ Southern Calif, AB, 43, MS, 45, PhD(biochem), 57. Prof Exp: CLIN CHEMIST, TOXICOL LAB, LOS ANGELES COUNTY-UNIV SOUTHERN CALIF MED CTR, 57-, ASST PROF PATH, SCH MED, UNIV SOUTHERN CALIF, 69- Mem: Am Asn Clin Chem. Res: Clinical biochemistry; toxicology. Mailing Add: Toxicol Lab Los Angeles Co-Univ Southern Calif Med Ctr 1200 N State St Los Angeles CA 90033

WALBORG, EARL FREDRICK, JR, b Chicago, Ill, Nov 13, 35; m 58; c 3. BIOCHEMISTRY. Educ: Austin Col, BA, 58; Baylor Univ, PhD(biochem), 62. Prof Exp: Asst biochemist & asst prof biochem, 65-70, assoc prof biochem, 70-73, ASSOC BIOCHEMIST, UNIV TEX SYST CANCER CTR, M D ANDERSON HOSP & TUMOR INST, HOUSTON, 70-, CHIEF SECT PROTEIN STRUCTURE, DEPT BIOCHEM, 70- PROF BIOCHEM, 73-; MEM GRAD FAC, UNIV TEX GRAD SCH BIOMED SCI, HOUSTON, 70- Concurrent Pos: USPHS res fel physiol chem, Univ Lund, 62-65; Eleanor Roosevelt Int Cancer fel biochem, Neth Cancer Inst, Amsterdam, 74. Mem: AAAS; Am Chem Soc; fel Am Inst Chem; Am Asn Cancer Res; Am Soc Biol Chem. Res: Chemistry of the cell-surface, glyoproteins. Mailing Add: Dept of Biochem Syst Cancer Ctr M D Anderson Hosp & Tumor Inst Houston TX 77025

WALBORSKY, HARRY M, b Lodz, Poland, Dec 25, 23; nat US; m 53; c 4. ORGANIC CHEMISTRY. Educ: City Col New York, BS, 45; Ohio State Univ, PhD(chem), 49. Prof Exp: Res assoc, Calif Inst Technol, 48; res assoc, Atomic Energy Proj, Univ Calif, Los Angeles, 49-50; from asst prof to assoc prof chem, 50-59, PROF CHEM, FLA STATE UNIV, 59- Concurrent Pos: USPHS fel, Basel, Switz, 52-53. Mem: Am Chem Soc; The Chem Soc. Res: Small ring compounds; organometallics; asymmetric synthesis; electrolytic and dissolving metal reductions; synthetic methods. Mailing Add: Dept of Chem Fla State Univ Tallahassee FL 32306

WALBOT, VIRGINIA ELIZABETH, US citizen. DEVELOPMENTAL BIOCHEMISTRY. Educ: Stanford Univ, AB, 67; Yale Univ, MPhil, 69, PhD(biol), 72. Prof Exp: NIH fel, Univ Ga, 72-75; ASST PROF BIOL, WASHINGTON UNIV, 75- Mem: AAAS; Bot Soc Am; Soc Develop Biol; Am Soc Plant Physiol; Am Soc Cell Biol. Res: Plant biochemistry and development; cell biology; botany. Mailing Add: Dept of Biol Washington Univ St Louis MO 63130

WALBRICK, JOHNNY MAC, b Wichita Falls, Tex, Sept 14, 41; m 64; c 2. INDUSTRIAL ORGANIC CHEMISTRY. Educ: Midwestern Univ, BS, 63; Univ Fla, PhD(chem), 67. Prof Exp: NSF fel, Univ Fla, 67-68; res chemist, Res Labs, 70-74, MGR RES, MERICHEM CO, 73- Mem: Am Chem Soc; The Chem Soc; Electrochem Soc. Res: Mechanism of electroorganic reaction processes; new industrial processes for organic chemicals. Mailing Add: 14431 Woodforest Blvd Houston TX 77015

WALBURG, CHARLES HERMAN, b Mankato, Minn, June 9, 26; m 49; c 2. FISH BIOLOGY. Educ: Univ Minn, BS, 49. Prof Exp: Fishery res biologist, Atlanta Coast Shad & Blue Crab Prog, Biol Lab, US Bur Com Fisheries, NC, 50-61, fishery res biologist, NCent Reservoir Invests, US Bur Sport Fisheries & Wildlife, 62-75, CHIEF

NCENT RESERVOIR INVESTS, US FISH & WILDLIFE SERV, 75- Mem: Am Fisheries Soc; Am Inst Fishery Res Biologists; Sigma Xi. Res: Population dynamics; estuarine and freshwater ecology; biology of reservoirs; effect of industrial development on aquatic biota. Mailing Add: NCent Reservoir Invests PO Box 139 Yankton SD 57078

WALBURG, HARRY E, JR, b Newark, NJ, Feb 6, 32; m 54; c 4. VETERINARY MEDICINE. Educ: Dartmouth Col, AB, 53; Va Polytech Inst, MS, 58; Univ Ga, DVM, 58; Univ Ill, PhD(radiobiol), 61. Prof Exp: Biologist, 61-73, DIR COMP ANIMAL RES LAB, OAK RIDGE NAT LAB, 73- Honors & Awards: Animal Care Panel Res Award, 65. Mem: Radiation Res Soc; Geront Soc; Soc Exp Biol & Med; Am Asn Cancer Res; Am Inst Biol Sci. Res: Radiation carcinogenesis and radiation induced life-shortening and aging. Mailing Add: 121 Morningside Dr Oak Ridge TN 37830

WALCH, HENRY ANDREW, JR, b Minneapolis, Minn, June 3, 22; m 53; c 2. MYCOLOGY. Educ: Univ Calif, Los Angeles, BA, 50, PhD(microbiol), 54. Prof Exp: Mycol technician, Univ Calif, Los Angeles, 50-54, res asst, 54-55; from instr to assoc prof microbiol, 55-64, chmn dept, 60-64 & 72-75, PROF MICROBIOL, SAN DIEGO STATE UNIV, 64- Concurrent Pos: Res grants, San Diego Imp Counties Tuberc & Respiratory Health Asn, 57-72 & Respiratory Dis Asn Calif, 72-73; consult & lectr, Sharp Mem Hosp, San Diego, 58- & US Naval Hosp, 64-; consult, Palomar Mem Hosp, Escondido, Calif; NIH spec fel, Mycol Unit, Commun Dis Ctr, Atlanta, Ga. Mem: Am Soc Microbiol; Mycol Soc Am; Am Inst Biol Sci; AAAS. Res: Human and animal pathogenic fungi, particularly virulence factors, immunology and ecology. Mailing Add: Dept of Microbiol San Diego State Univ San Diego CA 92182

WALCH, HERBERT NICKOLAS, agricultural economics, see 12th edition

WALCHER, DWAIN N, b Ill, Apr 7, 15; m 39; c 2. PEDIATRICS, HUMAN DEVELOPMENT. Educ: Univ Chicago, BS, 38, MD, 40. Prof Exp: Instr pediat, Sch Med, Yale Univ, 43-46; asst prof, Med Ctr, Ind Univ, 47-52, assoc prof, 52-62, prof, 62-63; prog dir growth & develop, Nat Inst Child Health & Human Develop, 63-66, assoc dir prog planning & eval, 66-69; PROF HUMAN DEVELOP & DIR INST STUDY HUMAN DEVELOP, PA STATE UNIV, 69- Mem: Soc Pediat Res; Am Acad Pediat; Infectious Dis Soc Am; Int Orgn Study Human Develop (exec secy-treas). Res: Infectious diseases in children. Mailing Add: Col of Human Develop Pa State Univ University Park PA 16802

WALCK, ROBERT E, JR, physical chemistry, see 12th edition

WALCOTT, BENJAMIN, b Boston, Mass, May 31, 41; m 72. COMPARATIVE PHYSIOLOGY. Educ: Harvard Univ, BA, 63; Univ Ore, PhD(biol), 68. Prof Exp: USPHS physiol trainee, Univ Ore, 64-67, instr biol, 67-68; vis res fel biol, Res Sch Biol Sci, Australian Nat Univ, 69-71, fel biol, 71-72; ASST PROF ANAT, HEALTH SCI CTR, STATE UNIV NY STONY BROOK, 72- Mem: Am Soc Cell Biol; foreign mem Brit Soc Exp Biol; Soc Gen Physiol; Biophys Soc; Soc Neurosci. Res: The role of visual interneurons in the regulation of arthropod behavior using intracellular recording techniques; correlation between molecular structure and mechanical responses of muscles from different vertebrates and invertebrates. Mailing Add: Dept Anat Sci Sch Basic Health Sci Health Sci Ctr State Univ NY Stony Brook NY 11794

WALCOTT, CHARLES, b Boston, Mass, July 19, 34. BEHAVIORAL PHYSIOLOGY. Educ: Harvard Univ, AB, 56; Cornell Univ, PhD, 59. Prof Exp: Asst, Cornell Univ, 56-58; res fel biol, Harvard Univ, 59-60, asst prof appl biol, Div Eng & Appl Physics, 60-65; asst prof biol, Tufts Univ, 65-67; assoc prof biol, 67-74, actg dir, Ctr Curriculum Develop, 67-71, PROF BIOL, STATE UNIV NY STONY BROOK, 74-, CHMN DEPT CELLULAR & COMP BIOL, 71- Concurrent Pos: Dir, Natural Sci TV Proj, 59-60; dir elem sci study, Educ Develop Ctr, Inc, 65-67. Mem: AAAS; Am Soc Zoologists; Nat Sci Teachers Asn; Nat Asn Biol Teachers; Animal Behav Soc. Res: Neurophysiology; animal behavior and orientation. Mailing Add: Dept of Biol State Univ of NY Stony Brook NY 11790

WALCZAK, HUBERT R, b South Saint Paul, Minn, Jan 21, 34; m 61; c 3. MATHEMATICS. Educ: Col St Thomas, BA, 55; Univ Minn, PhD(math), 63. Prof Exp: Asst prof, 63-72, ASSOC PROF MATH, COL ST THOMAS, 72- Mem: Math Asn Am. Res: Analysis and quasiconformal mappings. Mailing Add: Dept of Math Col of St Thomas St Paul MN 55105

WALD, FRANCINE JOY, b Brooklyn, NY, Jan 13, 38; m 64; c 2. SOLID STATE PHYSICS, SCIENCE EDUCATION. Educ: City Col New York, BEE, 60; Polytech Inst Brooklyn, MS, 62, PhD(chem physics), 69. Prof Exp: Engr solid state physics, Remington Rand Univac Div, 60; instr physics, Polytech Inst Brooklyn, 62-64, adj res assoc, 69-70; SCI CONSULT PHYSICS & BIOL, FRIENDS SEM, 72- Concurrent Pos: Lectr phys sci, New York Community Col, 69 & 70. Mem: Am Phys Soc; Sigma Xi. Res: Investigating how children of various ages respond to science, particularly physics, how they assimilate the concepts and language encountered. Mailing Add: 520 La Guardia Pl New York NY 10012

WALD, GEORGE, b New York, NY, Nov 18, 06; m 31, 58; c 4. BIOCHEMISTRY. Educ: NY Univ, BS, 27; Columbia Univ, AM, 28, PhD(zool), 32. Hon Degrees: AM, Harvard Univ, 44; MD, Univ Berne, 57; DSc, Yale Univ, 58, Wesleyan Univ, 62, NY Univ, 65, McGill Univ, 66 & Univ Rennes, 70. Prof Exp: Asst instr biol, NY Univ, 27-28; asst biophys, Columbia Univ, 28-32, Nat Res Coun fel, 32-34; instr & tutor biochem sci, 34-35, instr & tutor biol, 35-39, fac instr & tutor, 39-44, assoc prof, 44-48, PROF BIOL, HARVARD UNIV, 48- Concurrent Pos: Chmn div comt biol & med sci, NSF, 54-55; vis prof, Univ Calif, 56. Honors & Awards: Nobel Prize in Physiol & Med, 67; Lilly Award, Am Chem Soc, 39; Lasker Award, Am Pub Health Asn, 53; Proctor Medal, Asn Res Vision & Ophthal, 55; Rumford Medal, Am Acad Arts & Sci, 59; Ives Medal, Optical Soc Am, 66; Paul Karrer Medal, Univ Zurich, 67; T Duckett Jones Mem Award, Whitney Found, 67; Bradford Washburn Medal, Boston Mus Sci, 68; Max Berg Award, 69; Joseph Priestley Award, Dickinson Col, 70; Albert Einstein Award, Yeshiva Univ, 67; Jaffe Award, Mt Sinai Sch Med, 72. Mem: Nat Acad Sci; AAAS; Am Physiol Soc; fel Am Philos Soc; Am Chem Soc. Res: Chemistry and physiology of vision; biochemical evolution. Mailing Add: Biol Labs A-302 Harvard Univ Cambridge MA 02138

WALD, MILTON M, b San Francisco, Calif, Oct 29, 25; m 56; c 2. CHEMISTRY. Educ: Univ Calif, Los Angeles, BS, 49; Univ Southern Calif, PhD(chem), 54. Prof Exp: Res assoc, Brookhaven Nat Lab, 54-56; CHEMIST, SHELL DEVELOP CO, 56- Mem: Am Chem Soc. Res: Organic and catalytic chemistry; petroleum chemistry. Mailing Add: Shell Develop Co PO Box 481 Houston TX 77001

WALD, NIEL, b New York, NY, Oct 1, 25; m 53; c 2. PUBLIC HEALTH, RADIATION MEDICINE. Educ: Columbia Univ, AB, 45; NY Univ, MD, 48. Prof Exp: Intern & resident med affiliated hosps, NY Univ, 48-49, 50-52; sr hematologist & head radioisotope lab, Atomic Bomb Casualty Comn, Japan, 54-57; head biologist, Health Physics Div, Oak Ridge Nat Lab, 57-58; assoc res prof, 58-60, assoc prof, 60-62, PROF RADIATION HEALTH, GRAD SCH PUB HEALTH, UNIV PITTSBURGH, 62-, CHMN DEPT, 69-, PROF RADIOL, SCH MED, 65- Concurrent Pos: Fel immunohemat, NY Univ affiliated hosps, 49-50; asst prof med, Grad Sch Pub Health, Univ Pittsburgh, 58-65; mem Pa Governor's Adv Comt Atomic Energy Develop & Radiation Contro, 66-, chmn, 74-; consult, Div Oper Safety, US AEC, 68-75, Div Compliance, 69-75, US Navy Submarine & Radiation Med Div, 73-, US Energy Res & Develop Admin & US Nuclear Regulatory Comn, 75-; mem, Nat Coun Radiation Protection, 70-. Mem: Health Physics Soc (pres, 73-74); Radiation Res Soc; Soc Nuclear Med; Soc Human Genetics; AMA. Res: Diagnosis and treatment of radiation injury; health physics; cytogenetics; automatic biomedical image processing. Mailing Add: Grad Sch of Pub Health Univ of Pittsburgh Pittsburgh PA 15261

WALD, ROBERT MANUEL, b New York, NY, June 29, 47. THEORETICAL PHYSICS. Educ: Columbia Univ, AB, 68; Princeton Univ, PhD(physics), 72. Prof Exp: Res assoc physics, Univ Md, 72-74; res assoc, 74-76, ASST PROF PHYSICS, UNIV CHICAGO, 76- Mem: Am Phys Soc. Res: General relativity and gravitation; black holes; quantum field theory in curved spacetime. Mailing Add: Enrico Fermi Inst Univ of Chicago Chicago IL 60637

WALD, SAMUEL STANLEY, b New York, NY, Feb 5, 06; m 32; c 2. ROENTGENOLOGY. Educ: NY Univ, DDS, 28. Prof Exp: CLIN PROF RADIOL, DIAG & ORAL CANCER, COL DENT, NY UNIV, 28-, ASSOC PROF ROENTGENOL, 54-, CLIN PROF RADIOL, SCH MED & POSTGRAD MED SCH, 30- Concurrent Pos: Head dept roentgenol & diag, Guggenheim Dent Clin & Sch Dent Hyg, 30-42; consult, Dent Clins, Community Serv Soc, NY, 30-; spec lectr, US Naval Hosp, St Albans, 48- & Vet Admin, Brooklyn, 51-; chmn adv comt radiol, State Dept Health, NY & mem, Mayor's Adv Comt Radiation & Dent, New Yor, 56-; vis prof, NJ Col Med & Dent, 58-; consult radiol, diag & oral med, Mem Hosp Cancer & Allied Dis, New York. Mem: Fel Royal Soc Health; Sci Res Soc Am; Am Dent Asn; Asn Mil Surg US; fel Am Col Dent. Res: Radiology and diagnosis; oral surgery and oral cancer; dental medicine. Mailing Add: 420 E 72nd St New York NY 10021

WALD, WILBUR J, b Kansas City, Mo, Mar 16, 14. POLYMER CHEMISTRY. Educ: Rockhurst Col, BS, 36; Mass Inst Technol, MS, 38, PhD(org chem), 42. Prof Exp: Mat engr, Westinghouse Elec & Mfg Co, 40-43; mat engr, Marbon Corp, 43-47; proj chemist, Stand Oil Co, Ind, 47-49; mgr tech serv dept, Neville Chem Co, 49-59, sr scientist & staff specialist, 59-62; mgr tech liaison & tech serv, Fiberite Corp, Minn, 63-69; RETIRED. Mem: Am Chem Soc. Res: Wood and lignin chemistry; natural and synthetic resins; elastomers; adhesives; plasticizers; raw materials and products of potential use in the missile and space fields. Mailing Add: 227 W Fourth St Winona MN 55987

WALDBAUER, EUGENE CHARLES, b Philadelphia, Pa, July 4, 26; m 56; c 5. NATURAL HISTORY. Educ: East Stroudesburg State Col, BS, 52; Pa State Univ, MS, 56; Cornell Univ, PhD(wildlife biol, natural hist & parasitol), 66. Prof Exp: From asst prof to assoc prof, 56-69, PROF BIOL, STATE UNIV NY COL CORTLAND, 69- Mem: Wilderness Soc; Ecol Soc Am; Wildlife Soc; Am Nature Study Soc; Sigma Xi. Res: Flora of Cortland County, New York; pollen analysis of central New York bogs. Mailing Add: Dept of Biol State Univ NY Col Cortland NY 13045

WALDBAUER, GILBERT PETER, b Bridgeport, Conn, Apr 18, 28; m 55; c 2. ENTOMOLOGY. Educ: Univ Mass, BS, 53; Univ Ill, MS, 56, PhD, 60. Prof Exp: Asst, 53-58, from instr to assoc prof, 58-71, PROF ENTOM, UNIV ILL, URBANA, 71- Mem: AAAS; Entom Soc Am; Animal Behav Soc. Res: Biology, behavior and physiology of insects. Mailing Add: Dept of Entom Univ of Ill Urbana IL 61803

WALDBAUM, DAVID ROBERT, geology, physical chemistry, see 12th edition

WALDBILLIG, JAMES OLIVER, organic chemistry, see 12th edition

WALDE, RALPH ELDON, b Perham, Minn, Mar 8, 43; m 72; c 1. MATHEMATICS. Educ: Univ Minn, Minneapolis, BA, 64; Univ Calif, Berkeley, PhD(math), 67. Prof Exp: Asst prof math, Univ Minn, Minneapolis, 67-72; ASST PROF MATH, TRINITY COL, CONN, 72- Res: Lie algebras; non-associative algebras. Mailing Add: Dept of Math Trinity Col Hartford CT 06106

WALDEN, CLYDE HARRISON, b Kansas City, Mo, Dec 19, 21; m 46; c 3. SCIENCE ADMINISTRATION. Educ: William Jewell Col, BA, 42; Univ Colo, MS, 46, PhD(phys chem), 49. Prof Exp: Mgr uranium accountability, Mallinckrodt Chem Co, 48-51; sr scientist sec recovery oil, Phillips Petrol Co, 51; mgr qual control, Nat Lead Co, Ohio, 51-57; dir qual control, Gen Tire & Rubber Co, 57-64; DIR PROCESS TECHNOL, KAISER ALUMINUM & CHEM CORP, 64- Mem: Am Chem Soc; fel Am Soc Qual Control. Res: Administration of technical and engineering functions; heats of chemical reactions. Mailing Add: 50 Bellevue Ave Piedmont CA 94611

WALDEN, DAVID BURTON, b New Haven, Conn, Mar 29, 32; m; c 2. PLANT GENETICS, CYTOGENETICS. Educ: Wesleyan Univ, BA, 54; Cornell Univ, MSc, 58, PhD(genetics), 59. Prof Exp: Fel bot, Ind Univ, 59-61; from asst prof to assoc prof, 61-71, actg chmn dept, 71-73, PROF BOT, UNIV WESTERN ONT, 71- Concurrent Pos: Vis prof, Dept Genetics, Univ Birmingham, 73-74 & Dept Genetics & Develop & Dept Agron, Univ Ill, 74; pres, Biol Coun Can, 74-75. Mem: AAAS; Crop Sci Soc Am; Genetics Soc Am; Genetics Soc Can; Am Inst Biol Sci. Res: Pollen biology; corn genetics; plant and human cytogenetics; somatic cell genetics. Mailing Add: Dept of Plant Sci Univ of Western Ont London ON Can

WALDEN, GEORGE ELLIS, b Tuscumbia, Ala, May 30, 30; m 50; c 3. PHYSICAL CHEMISTRY, ANALYTICAL CHEMISTRY. Educ: Florence State Col, BS, 56; Univ Ala, MS, 58, PhD(phys chem), 59. Prof Exp: Chemist, US Bur Mines, Ala, 57-59; res chemist, Chemstrand Corp, 59-60; develop chemist, Union Carbide Nuclear Corp, 60-66; ASSOC PROF CHEM, DAVID LIPSCOMB COL, 66- Mem: Am Chem Soc. Res: X-ray emission; x-ray diffraction; gas chromatography; application of x-ray emission techniques to analytical chemistry and light element analysis. Mailing Add: Dept of Chem David Lipscomb Col Nashville TN 37203

WALDEN, RICHARD TRUSSELL, b Casper, Wyo, Mar 13, 21; m 42; c 6. MEDICINE. Educ: Wash State Univ, BS, 42, DVM, 44; Loma Linda Univ, MD, 52; Temple Univ, MS, 60; Am Bd Internal Med, dipl, 64; Am Bd Prev Med, dipl, 66. Prof Exp: From instr to assoc prof prev med, Loma Linda Univ, 55-67, from co-chmn to chmn dept, 64-72, chmn dept epidemiol, 67-69, prof prev med & epidemiol & asst dean Sch Pub Health, 67-75. Concurrent Pos: Dir, Exec Health Serv, 60-75; coordr, Area VI, Calif Regional Med Progs, 69-72; fel, Coun Atherosclerosis & Epidemiol & Heart Dis, Am Heart Asn. Mem: Fel Am Col Physicians; fel Am Pub Health Asn; Asn Teachers Prev Med; AMA; Pan-Am Med Asn. Mailing Add: 1810 Sherman Ave Stevens Point WI 54481

WALDEN, WILLIAM EARL, b Paragould, Ark, Oct 8, 29; m 51; c 4. COMPUTER SCIENCES, MATHEMATICS. Educ: NMex State Univ, AB, 50, MS, 59, PhD(math), 64. Prof Exp: Mathematician-programmer, Land-Air, Inc, 51-54; staff mem, Los Alamos Sci Lab, Univ Calif, 54-64; assoc prof math & dir comput ctr, Univ Nebr, Omaha, 64-67; assoc prof info sci & math & dir comput ctr, 67-70, assoc prof comput sci, 70-72, PROF COMPUT SCI, WASH STATE UNIV, 72-, DIR SYSTS & COMPUT, 70- Mem: Asn Comput Mach; Am Soc Info Sci. Res: Mathematical applications of digital computers; theory of information storage and retrieval. Mailing Add: Dept of Comput Sci Wash State Univ Pullman WA 99163

WALDER, ORLIN E, b Hayti, SDak, Sept 21, 05; m 31; c 1. MATHEMATICS. Educ: Huron Col, BSc, 28, LHD, 57; Univ Nebr, AM, 30. Prof Exp: Asst, Univ Nebr, 28-30 & Univ Minn, 34; instr, 30-33, from asst prof to assoc prof, 34-36, PROF MATH, S DAK STATE UNIV, 46-, DIR STUDENT AFFAIRS, 50-, DEAN MEN, 59- Mem: Math Asn Am. Mailing Add: Dept of Math SDak State Univ Univ Sta Brookings SD 57006

WALDERN, DONALD E, b Lacombe, Alta, June 8, 28; m 53; c 4. ANIMAL NUTRITION, BIOCHEMISTRY. Educ: Univ BC, BSA, 51, MSA, 54; Wash State Univ, PhD(nutrit, biochem), 62. Prof Exp: Res scientist, Exp Sta, Can Dept Agr, 53-57 & 61-62; assoc prof dairy nutrit, Wash State Univ, 62-67; res scientist, 67-73, DIR RANGE RES STA, AGR CAN, 73- Mem: Am Dairy Sci Asn; Am Soc Animal Sci; Agr Inst Can. Res: Nutritive value of forages and cereal grains for dairy and beef cattle; complete feeds for dairy cows and early weaned calves; relationship of blood biochemical parameters to performance factors in dairy and beef cows; nutritional management systems for beef cows grazing grassland and forested rangeland. Mailing Add: Range Res Sta Agr Can Kamloops BC Can

WALDHALM, DONALD GEORGE, b Enderlin, NDak, Jan 6, 24; m 45; c 4. VETERINARY MICROBIOLOGY. Educ: Univ Minn, BA, 48, MS, 50; Univ Ill, PhD, 53. Prof Exp: Res scientist microbiol, Res Labs, Swift & Co, Ill, 53-54; med bacteriologist, Rocky Mountain Lab, USPHS, Mont, 54-56; asst prof biol, Carroll Col, Mont, 56-60; asst prof bact, 60-66, ASSOC VET MICROBIOLOGIST, UNIV IDAHO, 66-, ASSOC RES PROF VET SCI, 68- Mem: Am Soc Microbiologists. Res: Microbial physiology; pathogenic microorganisms; abortion diseases in livestock; diseases of neonatal calves and lambs. Mailing Add: Univ of Idaho Exp Sta Rte 8 Caldwell ID 83605

WALDHAUSEN, JOHN ANTON, b New York, NY, May 22, 29; m 57; c 3. SURGERY. Educ: Col Great Falls, BS, 50; St Louis Univ, MD, 54. Prof Exp: Intern, Johns Hopkins Hosp, Baltimore, 54-55, resident, 56-57; surgeon, Nat Heart Inst, Md, 57-59; resident, Hosp, Univ Pa, 59-60; resident, Med Ctr, Ind Univ, Indianapolis, 60-62, instr surg, Sch Med, 62-63, asst prof, 63-66; assoc prof, Sch Med, Univ Pa, 66-70; interim provost & dean, Col Med, 72-73, PROF SURG & CHMN DEPT, HERSHEY MED CTR, PA STATE UNIV, 70- Concurrent Pos: Surg fel, Johns Hopkins Hosp, Baltimore, Md, 55-56; NIH career develop award, Sch Med, Ind Univ, Indianapolis, 63-66; assoc surgeon, Children's Hosp, Philadelphia, 66-70 & Hosp Univ Pa, 66-70; mem surg study sect B, NIH, 74- Mem: Soc Univ Surg; Am Col Cardiol; Am Surg Asn; Am Asn Thoracic Surg; Am Col Surg. Res: Effects of operative repair of congenital heart defects on pulmonary circulation; newer methods in repair of congenital heart defects; effects of cardiac surgery on ventricular function. Mailing Add: Milton S Hershey Med Ctr Pa State Univ Hershey PA 17033

WALDICHUK, MICHAEL, b Mitkau, Roumania, Oct 23, 23; nat Can; m 55; c 3. OCEANOGRAPHY. Educ: Univ BC, BA, 48, MA, 50; Univ Wash, PhD, 55. Prof Exp: Assoc scientist, Pac Biol Sta, Fisheries Res Bd Can, Nanaimo, BC, 54-58, sr scientist, 58-63, prin scientist, 63-66, oceanogr-in-chg, 66-69; oceanog consult & secy, Can Comt Oceanog, Ottawa, 69-70; PROG HEAD, PAC ENVIRON INST, FISHERIES RES BD CAN, 70- Concurrent Pos: Mem, Intergovt Maritime Consult Orgn-Food & Agr Orgn-UNESCO-World Meteorol Orgn-Int Atomic Energy Agency-UN Joint Group Experts Sci Aspects of Marine Pollution, 69-, chmn, 70-; partic coastal wastes mgt study session, Nat Acad Sci, Wyo, 69, workshop on marine environ qual, 71; mem panel marine aquatic life & wildlife, Comt Water Qual Criteria, Washington, DC, 71- Mem: AAAS; Am Chem Soc; Am Geophys Union; Am Soc Limnol & Oceanog; Chem Inst Can. Res: Chemical and physical oceanography; industrial wastes and water pollution. Mailing Add: Fisheries Res Bd Can Pac Environ Inst 4160 Marine Dr Vancouver BC Can

WALDINGER, HERMANN V, b Vienna, Austria, June 17, 23; US citizen; m 48; c 2. MATHEMATICS. Educ: Pomona Col, BA, 43; Brown Univ, MS, 44; Columbia Univ, PhD(math), 51. Prof Exp: Res engr, Repub Aviation Corp, 45-46; appl mathematician, M W Kellogg Co, 46-53; sr mathematician, Nuclear Develop Corp Am, 53-59; sr scientist, Repub Aviation Corp, 59-61; ASSOC PROF MATH, POLYTECH INST NEW YORK, 61- Mem: AAAS; Am Math Soc; Math Asn Am. Res: Group theory. Mailing Add: Dept of Math Polytech Inst of New York Brooklyn NY 11201

WALDINGER, RICHARD J, b Brooklyn, NY, Mar 1, 44. COMPUTER SCIENCE. Educ: Columbia Univ, AB, 64; Carnegie-Mellon Univ, PhD(comput sci), 69. Prof Exp: RES MATHEMATICIAN, ARTIFICIAL INTEL CTR, STANFORD RES INST, 69- Res: Artificial intelligence; automatic program synthesis and verification; mechanical theorem proving; robotics. Mailing Add: Artificial Intel Ctr Stanford Res Inst Menlo Park CA 94025

WALDMAN, BERNARD, b New York, NY, Oct 12, 13; m 42, 64; c 4. NUCLEAR PHYSICS. Educ: NY Univ, AB, 34, PhD(physics), 39. Prof Exp: Asst physics, NY Univ, 35-38; res assoc, Univ Notre Dame, 38-40, instr, 40-42, asst prof, 42-43; staff mem & group leader, Los Alamos Sci Lab, 43-45; assoc prof physics, 45-51, assoc dean col sci, 64-67, PROF PHYSICS, UNIV NOTRE DAME, 51-, DEAN COL SCI, 67- Concurrent Pos: Mem staff, Midwestern Univs Res Asn, 58-59, vpres, 59-60, 65-, lab dir, 60-65; mem physics adv panel, NSF, 65-68, chmn, 68; trustee, Univs Res Asn, Inc, 65-71. Mem: Fel Am Phys Soc; Am Asn Physics Teachers. Res: Medium energy electrons and x-rays; electrostatic and high energy accelerators. Mailing Add: Col of Sci Univ of Notre Dame Notre Dame IN 46556

WALDMAN, JOSEPH, b Philadelphia, Pa, May 12, 06; m; c 3. OPHTHALMOLOGY. Educ: Jefferson Med Col, MD, 30; Am Bd Ophthal, dipl, 35. Prof Exp: PROF OPHTHAL, JEFFERSON MED COL, 30- Mem: AMA. Mailing Add: Dept of Ophthal Jefferson Med Col Philadelphia PA 19107

WALDMAN, ROBERT H, b Dallas, Tex, Dec 21, 38; m 63; c 3. IMMUNOLOGY. Educ: Rice Inst, BA, 59; Wash Univ, MD, 63. Prof Exp: Intern, Johns Hopkins Hosp, 64, resident, 65; clin assoc, Nat Inst Allergy & Infectious Dis, 65-67; PROF & ACTG CHMN DEPT MED, COL MED, UNIV FLA, 67- Concurrent Pos: Attend physician, Vet Admin Hosp, Gainesville, Fla, 68; consult cholera, WHO, 69, vis scientist, Int Res & Training Ctr Immunol, 69-70; mem cholera adv comt, NIH. Mem: Am Soc Clin Invest; Am Fedn Clin Res; Am Asn Immunol; Am Soc Microbiol; fel Am Col Physicians. Res: Immunology of viral respiratory infections; study of the secretory immunologic system; immunization by application of antigen to mucous surfaces; host defense mechanisms resident on mucosal surfaces. Mailing Add: Dept of Med Univ of Fla Col of Med Gainesville FL 32610

WALDMANN, THOMAS A, b New York, NY, Sept 21, 30; m 58; c 3. MEDICINE, IMMUNOLOGY. Educ: Univ Chicago, AB, 51; Harvard Univ, MD, 55. Prof Exp: Intern, Mass Gen Hosp, 55-56; clin assoc, 56-58, sr investr, Metab Br, 59-65, HEAD IMMUNOPHYSIOL SECT, NAT CANCER INST, 65-, CHIEF METAB BR, 71- Concurrent Pos: Am Heart Asn fel, Nat Cancer Inst, 58-59; mem, Nat Cancer Plan Comt, 72. Honors & Awards: Superior Serv Award, Dept Health, Educ & Welfare. Mem: AAAS; Am Fedn Clin Res; Am Soc Clin Invest; Asn Am Physicians; Am Physiol Soc. Res: Factors controlling plasma protein synthesis, transport and catbolism; immunologic deficiency syndromes; cancer immunology. Mailing Add: Metab Br Nat Cancer Inst Bethesda MD 20014

WALDO, WILLIS HENRY, b Detroit, Mich, Sept 27, 20; m 49; c 5. INORGANIC CHEMISTRY. Educ: Washington & Jefferson Col, BS, 42; Univ Md, MS, 50. Prof Exp: Chemist, E I du Pont de Nemours & Co, 42-45; chemist, Socony Vacuum Oil Co, 45-46; asst chem, Univ Md, 46-49; tech ed, Monsanto Chem Co, 49-60, ADMIN MGR AGR DIV, MONSANTO CO, 60-, ED, MONSANTO TECH REV, 56- Concurrent Pos: Mem, Nat Acad Sci-Nat Res Coun Comt on Mod Methods Handling Chem Info, 62-66; lectr, Southern Ill Univ, Edwardsville, 72-74. Mem: Am Chem Soc; Sigma Xi; Am Soc Info Sci. Res: Chromium complexes; sulfur; machine documentation. Mailing Add: Agr Dept Monsanto Co St Louis MO 63166

WALDON, EDGAR F, b Meerut, India, Feb 26, 26; m 50; c 3. AUDIOLOGY. Educ: Univ Delhi, BA, 55; Syracuse Univ, MS, 58; Ohio State Univ, PhD(audiol, speech path), 63. Prof Exp: Asst teacher, Govt Lady Noyce Sch Deaf, 45-57; asst prof audiol & audiologist, Cath Univ, 63-67; assoc prof & coordr audiol, Bradley Univ, 67-72; US GOVT GRANT, TRAINING TEACHERS DEAF & BLIND, FED CITY COL, 72- Concurrent Pos: Nat Inst Child Health & Human Develop grant, 64-67; Dept Health, Educ & Welfare grant, 65; consult, St Francis Hosp, Washington, DC, 64; consult, Dept Spec Educ Proj Deaf-Retarded, Cath Univ, 67. Mem: Am Speech & Hearing Asn; Coun Except Children; Am Asn Ment Deficiency; Int Asn Logopedics & Phoniatrics. Res: Testing hearing of infants, children and subnormal children; development of the baby cry test, a new audiometric test; study of spoken and written language of children with impaired hearing. Mailing Add: Dept of Commun Sci Fed City Col 724 Ninth St NW Washington DC 20002

WALDREN, CHARLES ALLEN, b Syracuse, Kans, June 2, 34; m 61; c 3. BIOPHYSICS. Educ: Univ Colo, Boulder, BA, 59; Univ Colo Med Ctr, Denver, MS, 65, PhD(biophys), 72. Prof Exp: Chemist, 60-61, fel, 61-65, res tech III, 65-67, instr, 67-75, ASST PROF BIOPHYS & GENETICS, UNIV COLO MED CTR, 75-; ASST DIR, ELEANOR ROOSEVELT INST CANCER RES, 74- Concurrent Pos: Vis scientist & fel, Dept Zool, Div Cell Biol, Univ Cambridge, 72-73. Mem: Sigma Xi; Am Soc Cell Biol; Human Genetics Soc; Radiation Res Soc. Res: Genetic-biochemical analysis of mutagenesis and genome repair mechanisms in somatic mammalian cells. Mailing Add: Eleanor Roosevelt Inst Cancer Res Box B129 4200 E Ninth Ave Denver CO 80220

WALDREP, ALFRED CARSON, JR, b Orange, Tex, Apr 17, 23; m; c 4. ORAL SURGERY, DENTISTRY. Educ: Loyola Univ, La, DDS, 46; Baylor Univ, BS, 59, MS, 61; Am Bd Oral Surg, dipl, 63. Prof Exp: Oral surgeon, Valley Forge Gen Hosp, Valley Forge Pa Hq, US Army, San Antonio, Tex, 46-54, dent surgeon, Task Force 7, Cent Pac, 54-55, chief hosp surg dent, US Army Hosp, Ft Polk, La, 55-58, resident oral surg, Hosp, Baylor Univ, 58-59, resident, Brooks Army Med Ctr, 59-61, consult, US Army Med Area, Stuttgart, Ger, 61-64, chief host dent, US Army Hosp, Ft Polk, La, 64-68; PROF ORAL SURG, MED UNIV SC, 68- Concurrent Pos: Fel, Hosp, Baylor Univ, 58-59 & Brooks Army Med Ctr, 59-61; consult, Vet Admin Hosp, Charleston, SC, 68-, US Navy Hosp, 69- & SC Dept Corrections, 69- Mem: Am Dent Asn; Am Soc Oral Surg; assoc Brit Asn Oral Surg. Res: Dental education; precautions for patients on drug therapy; oral surgery for patients on anticoagulant therapy. Mailing Add: Dept of Oral Surg Med Univ of SC Charleston SC 29401

WALDREP, THOMAS WILLIAM, b Madison, Fla, Feb 14, 34; m 55; c 2. PLANT PHYSIOLOGY. Educ: Univ Fla, BSA, 61; Univ Ky, MSA, 63; NC State Univ, PhD(crop sci), 67. Prof Exp: Instr crop sci, NC State Univ, 66-67, asst prof, 67-69; RES SCIENTIST, LILLY RES LABS, ELI LILLY & CO, 69- Mem: Weed Sci Soc Am. Res: Basic and applied research in plant physiology; growth regulators and herbicides; herbicide research. Mailing Add: Lilly Res Labs Eli Lilly & Co Greenfield IN 46140

WALDRON, ACIE CHANDLER, b Malad, Idaho, Feb 4, 30; m 57; c 5. AGRONOMY, ENTOMOLOGY. Educ: Brigham Young Univ, BSc, 57; Ohio State Univ, MSc, 59, PhD(agron), 61. Prof Exp: Res asst agron, Agr Exp Sta, Ohio State Univ, 57-61; develop chemist, Agr Div, Am Cyanamid Co, NJ, 61-66; EXTEN SPECIALIST PESTICIDE CHEM & STATE COORDR AGR CHEM, OHIO COOP EXTEN SERV, OHIO STATE UNIV, 66- Concurrent Pos: Rep, Ohio State Univ & Ohio Agr Res & Develop Ctr, 73-; training coordr, Ohio Coop Exten Serv, 75-; mem, Coun Agr Sci & Technol. Mem: Am Chem Soc; Am Soc Agron; Asn Off Anal Chem; Soil Sci Soc Am. Res: Pesticide residue chemistry; pesticide residues in plant and animal crops, in soil and water; pesticide safety; chemistry of organic nitrogen and phosphorous in soil organic matter. Mailing Add: Ohio State Univ Coop Exten Serv 1735 Neil Ave Columbus OH 43210

WALDRON, CHARLES A, b Minneapolis, Minn, July 16, 22; m 43; c 2. ORAL PATHOLOGY. Educ: Univ Minn, DDS, 45, MSD, 51; Am Bd Oral Path, dipl, 59. Prof Exp: From asst prof to prof path, Sch Dent Wash Univ, 50-57; PROF PATH, SCH DENT, EMORY UNIV, 57- Concurrent Pos: Consult, Vet Admin Hosp, Ga, 57 & Dent Intern Prog, Ft Benning, 59; mem med bd dirs, Am Bd Oral Path, 59-70; sci adv bd consult, Armed Forces Inst Path, 70- Mem: Am Dent Asn; fel Am Col Dent; Am Acad Oral Path (pres, 59, ed, 70). Res: Oral tumors; diagnostic oral pathology. Mailing Add: Sch of Dent Emory Univ Atlanta GA 30322

WALDRON, HAROLD FRANCIS, b Manchester, Ohio, Sept 24, 29; m 49; c 1. APPLIED CHEMISTRY. Educ: Capital Univ, BS, 52; Purdue Univ, West Lafayette, MS, 54. Prof Exp: Chemist, Uranium Div, Mallinckrodt Chem Works, 54-62, supvr anal methods develop, 62-66, supvr anal res, Opers Div, 66-67, res group leader, Indust Chem Div, 67-69, res mgr chem group, 69-73, RES & DEVELOP MGR, FOOD PROD DIV, MALLINCKRODT, INC, 73- Mem: Am Chem Soc; Inst Food Technologists. Res: Chemical analytical methods, especially the use of vacuum techniques and application of complex ion formation; design and application of electronic instrumentation; general inorganic and uranium chemistry; process development. Mailing Add: 6 Garden Lane Kirkwood MO 63122

WALDRON, HOWARD HAMILTON, b Nampa, Idaho, Nov 6, 17; m 43; c 3. GEOLOGY. Educ: Univ Wash, BS, 40. Prof Exp: Photogrammetrist, US Hydrographic Off, Washington, DC, 42-46; geologist, US Geol Surv, 46-73; STAFF CONSULT GEOL, SHANNON & WILSON, INC, 73- Concurrent Pos: Tech adv,

Geol Surv Indonesia, 60-62, Costa Rica, 64 & Colo Eng Coun, 65-73; eng geol consult & adv, AEC, 67-72 & Vet Admin, 71-73; mem, Earthquake Eng Res Inst. Mem: Fel AAAS; Asn Eng Geol; fel Geol Soc Am; Am Soc Civil Eng. Res: Engineering geology; areal and glacial geology of Pacific Northwest; urban and environmental geology; geologic and earthquake hazards evaluations. Mailing Add: 1105 N 38th St Seattle WA 98105

WALDRON, INGRID LORE, b Nyack, NY, Dec 8, 39. PSYCHOSOMATIC MEDICINE. Educ: Radcliffe Col, AB, 61; Univ Calif, Berkeley, PhD(biol), 67. Prof Exp: NSF fel, Univ Cambridge, 67-68; ASSOC PROF BIOL, UNIV PA, 68- Concurrent Pos: NIH grant, Univ Pa, 69-72. Res: Human biology; sex differences; social and psychological origins of disease. Mailing Add: Dept of Biol Univ of Pa Philadelphia PA 19174

WALDRON, ROBERT DOUGLAS, physical chemistry, see 12th edition

WALDRON, STEPHEN, b Taunton, Mass, Feb 7, 25; m 44; c 4. PHYSICS. Educ: Lafayette Col, AB, 49; Mass Inst Technol, PhD, 53. Prof Exp: Instr physics, Lafayette Col, 49-50; sci analyst, Opers Eval Group, US Dept Navy, Mass Inst Technol, 52-63; sci analyst, Inst Naval Studies, Ctr Naval Anal, 63-66; SR STAFF, A D LITTLE, INC, 66- Mem: AAAS; Am Phys Soc; Opers Res Soc Am. Res: Microwave electronics; cryogenics; solid state physics; aircraft and ship preliminary design; ship dynamics; surface oceanography; analysis of military, government and industrial systems and policy problems; police management and operations; law enforcement; addict rehabilitation. Mailing Add: PO Box 198 Durham NH 03824

WALDROP, ANN LYNEVE, b Winters, Tex, Oct 9, 39; m 64. CRYSTALLOGRAPHY. Educ: McMurry Col, BA, 61; Vanderbilt Univ, MA, 65; Mass Inst Technol, PhD(crystallog), 70. Prof Exp: Teacher high sch, Mass, 64-65; ASST PROF CHEM, SIENA COL, NY, 70- Mem: Am Crystallog Asn; Am Chem Soc. Res: Determination of crystal and molecular structure by x-ray crystallography; relationship between structure and properties; polymorphism. Mailing Add: Dept of Chem Siena Col Londonville NY 12211

WALDROP, FRANCIS N, b Asheville, NC, Oct 5, 26; m 50; c 2. MEDICINE. Educ: Univ Minn, AB, 46; George Washington Univ, MD, 50. Prof Exp: Intern med, Univ Hosp, George Washington Univ, 50-51; resident psychiat, 51-54, med officer, 54-59, assoc dir res, 59-65, DIR CLIN & BEHAV STUDIES RES CTR, ST ELIZABETH'S HOSP, US DEPT HEALTH, EDUC & WELFARE, 65-, DIR PROF TRAINING PSYCHIAT, 63-, ACTG DEP ADMINR, ALCOHOL, DRUG ABUSE & MENT HEALTH ADMIN, 75- Concurrent Pos: Clin asst prof, George Washington Univ, 62-65, clin assoc prof, 65-; spec asst res & training, Nat Inst Ment Health, 66-68, dep dir, Nat Ctr Ment Health Serv, Training & Res, 68-71, from assoc dir to dir, Div Manpower & Training Progs, 71-75. Honors & Awards: Superior Serv Award, US Dept Health, Educ & Welfare, 62, Distinguished Serv Award, 64. Mem: AAAS; AMA; fel Am Psychiat Asn. Res: Clinical psychiatry; psychopharmacology; drug dependence; basic biological sciences in relation to psychiatric disorders. Mailing Add: ADAMHA 12-105 Parklawn Bldg 5600 Fishers Lane Rockville MD 20852

WALDROP, MORGAN A, b Ft Worth, Tex, Jan 8, 37. SOLID STATE PHYSICS, ATOMIC PHYSICS. Educ: Rice Inst, BA, 59, MA, 62, PhD(physics), 64. Prof Exp: SR RES PHYSICIST, PHILLIPS PETROL CO, 63- Mem: Am Phys Soc. Res: Nuclear magnetic and electron paramagnetic resonance in solids; atomic aspects of heterogeneous catalysis. Mailing Add: 1330 Melmart Bartlesville OK 74003

WALDROUP, PARK WILLIAM, b Maryville, Tenn, Oct 17, 37; m 61; c 4. NUTRITION, BIOCHEMISTRY. Educ: Univ Tenn, Knoxville, BS, 59; Univ Fla, MS, 62, PhD(nutrit, biochem), 65. Prof Exp: Res assoc poultry nutrit, Univ Fla, 64-65, asst prof, 65-66; asst prof, 66-71, ASSOC PROF POULTRY NUTRIT, UNIV ARK, FAYETTEVILLE, 71- Mem: Poultry Sci Asn; Am Inst Nutrit; Animal Nutrit Res Coun. Res: Studies concerned with nutrient requirements of poultry in terms of nutrient balance and interrelationships of nutrients; effects of processing on nutritive value of feeds. Mailing Add: Dept of Animal Sci Univ of Ark Fayetteville AR 72701

WALDSTEIN, SHELDON SAUL, b Chicago, Ill, June 23, 24; m 52; c 3. INTERNAL MEDICINE. Educ: Northwestern Univ, BS, 46, MD, 47, MS, 51; Am Bd Internal Med, dipl. Prof Exp: Intern & resident internal med, Cook County Hosp, 47-51; from clin asst to assoc prof, 51-66, PROF INTERNAL MED, MED SCH, NORTHWESTERN UNIV, CHICAGO, 66- Concurrent Pos: Res assoc, Hektoen Inst & assoc attend physician, Cook County Hosp, 54-57, attend physician, 57-, chief, Northwestern Med Div, 59-62, exec dir dept med, 62-64, chmn dept med, 64-69; exec dir, NSuburban Asn Health Resources, Northbrook, 69- Mem: AAAS; Endocrine Soc; AMA; Am Col Physicians; Am Fedn Clin Res. Res: Endocrinology. Mailing Add: 265 Walden Dr Glencoe IL 60022

WALECKA, JERROLD ALBERTS, b Highland Park, Ill, Aug 30, 30; m 53; c 2. ORGANIC CHEMISTRY. Educ: Lawrence Col, BS, 51, MS, 53, PhD(pulp & paper), 56. Prof Exp: Asst tech dir, WVa Pulp & Paper Co, 55-59; res chemist, Mead Corp, 59-63, assoc res dir new prod develop, 63-70; GEN MGR RES & DEVELOP, FOREST PROD GROUP, CONTINENTAL CAN CO, 70- Mem: Tech Asn Pulp & Paper Indust. Res: Pulp and paper industry; new products; cellulose and plastic chemistry; photochemistry; reproduction processes; graphic arts; information handling; electronic data processing; packaging. Mailing Add: Continental Can Co PO Box 1425 Augusta GA 30903

WALECKA, JOHN DIRK, b Milwaukee, Wis, Mar 11, 32; m 54; c 3. THEORETICAL PHYSICS. Educ: Harvard Col, BA, 54; Mass Inst Technol, PhD(physics), 58. Prof Exp: NSF fel, Europ Orgn Nuclear Res, Switz, 58-59; NSF fel, 59-60, from asst prof to assoc prof physics, 60-66, assoc dean humanities & sci, 70-72, chmn acad senate, 73-74, mem adv bd, 73-76, chmn, 75-76, PROF PHYSICS, STANFORD UNIV, 66- Concurrent Pos: A P Sloan Found fel, 62-66; mem sci prog adv comts, Bates Linac, Mass Inst Technol, 71-, Nevis Cyclotron, Columbia Univ, 71- & Los Alamos Meson Physics Facil, 74-76. Mem: Fel Am Phys Soc. Res: Nuclear structure; high energy physics. Mailing Add: Dept of Physics Stanford Univ Stanford CA 94305

WALES, BRUCE ALFRED, plant ecology, meteorology, see 12th edition

WALES, DAVID BERTRAM, b Vancouver, BC, July 31, 39; m 61; c 2. MATHEMATICS. Educ: Univ BC, BSc, 61, MA, 62; Harvard Univ, PhD(math), 67. Prof Exp: Bateman res fel math, 67-68, asst prof, 68-71, ASSOC PROF MATH, CALIF INST TECHNOL, 71- Mem: Am Math Soc. Res: Representation theory of finite groups. Mailing Add: Dept of Math Calif Inst of Technol Pasadena CA 91109

WALES, MICHAEL, physical chemistry, see 12th edition

WALES, ROBERT DAVID, physical chemistry, chemical engineering, see 12th edition

WALES, WALTER D, b Oneonta, NY, Aug 2, 33; m 55; c 2. PHYSICS. Educ: Carleton Col, BA, 54; Calif Inst Technol, MS, 55, PhD(physics), 60. Prof Exp: From instr to assoc prof, 59-72, PROF PHYSICS, UNIV PA, 72-, CHMN DEPT, 73- Concurrent Pos: Assoc dir, Princeton Pa Accelerator, 68-71; physicist, AEC, 72-73. Mem: Am Phys Soc. Res: Particle physics. Mailing Add: Dept of Physics Univ of Pa Philadelphia PA 19174

WALETZKY, EMANUEL, zoology, see 12th edition

WALFORD, GORDON LYN, organic chemistry, see 12th edition

WALFORD, LIONEL ALBERT, b San Francisco, Calif, May 29, 05; m 37; c 3. FISH BIOLOGY. Educ: Stanford Univ, AB, 29; Harvard Univ, MA, 32, PhD(biol), 35. Hon Degrees: DSc, Hartwick Col, 64 & Monmouth Col, 69. Prof Exp: Fishery biologist, Calif Dept Fish & Game, 26-31; asst biol, Harvard Univ, 34-35; instr zool, Santa Barbara State Col, 35-36; asst aquatic biologist, Bur Fisheries, US Fish & Wildlife Serv, 36-37, assoc aquatic biologist, 37-40, aquatic biologist, 40-45, asst chief div info, 45-47, chief sect marine fisheries, 47-48, chief br fishery biol, 48-58, dir, Washington Biol Lab, 58-60, dir, Sandy Hook Marine Lab, 60-71, sr scientist, 71-74; EXEC DIR, NJ MARINE SCI CONSORTIUM, 74- Concurrent Pos: Ichthyol ed, Copeia, 37-47; lectr, Stanford Univ, 40-45; chmn comt res & statist, Int Comn Northwest Atlantic Fisheries, 55-58; mem div biol & agr, Nat Res Coun, 55-60; mem comt effects atomic radiation on oceanog & fisheries, Nat Acad Sci Adv US sect, Int NPac Fisheries Comn, 53-55, mem panel natural resources, Comt Oceanog, 58; sr res assoc, Lamont Geol Observ, 62-; adj prof, Columbia Univ, 64- Honors & Awards: US Dept Interior Distinguished Serv Award, 70. Mem: AAAS; Am Soc Ichthyologists & Herpetologists; Am Soc Limnol & Oceanog. Res: Marine ecology; ichthyology; biogeography; biogeography of Atlantic fishes. Mailing Add: 37 Navesink Ave Rumson NJ 07760

WALFORD, LIONEL K, b Cardiff, UK, May 19, 39; m 63. SOLID STATE PHYSICS. Educ: Univ Wales, BSc, 60; Univ Cambridge, PhD(physics), 63. Prof Exp: PROF PHYSICS, SOUTHERN ILL UNIV, EDWARDSVILLE, 63-, CHMN DEPT, 75- Concurrent Pos: Consult res div, McDonnell Co, Mo, 64-71. Mem: Am Crystallog Asn; fel Brit Inst Physics. Res: X-ray crystallography; defects in semiconductors; structure of intermetallic compounds; charge transfer in alloys; amorphous phases; general properties of material. Mailing Add: Dept of Physics Southern Ill Univ Sch Sci & Tech Edwardsville IL 62026

WALFORD, ROY LEE, JR, b San Diego, Calif, June 29, 24; m 50; c 3. PATHOLOGY. Educ: Univ Chicago, BS, 46, MD, 48. Prof Exp: Intern, Gorgas Hosp, CZ, 50-51; resident path, Vet Admin Hosp, Los Angeles, 51-52; chief lab, Chanute AFB Hosp, Ill, 52-54; from asst prof to prof path, 54-70, PROF PATH & HEMATOPATH, SCH MED, UNIV CALIF, LOS ANGELES, 70- Concurrent Pos: Attend physician, Brentwood Vet Admin Hosp, 55-56; consult, Los Angeles Harbor Gen Hosp, 59- Mem: Am Soc Exp Path; Am Asn Path & Bact; Col Am Path. Res: Hematologic pathology; immunology of the white blood cell; structure, metabolism and diseases of elastic tissue; homograft immunity; gerontology. Mailing Add: 13-267 Ctr for Health Sci Univ of Calif Sch of Med Los Angeles CA 90024

WALGENBACH, DAVID D, b Marshall, Minn, Sept 4, 37; m 61; c 4. ENTOMOLOGY, AGRONOMY. Educ: Iowa State Univ, BS, 59; Univ Wis-Madison, MS, 62, PhD(entom), 65. Prof Exp: Asst prof biol, Stout State Univ, 64-65; field tech specialist entom, Chevron Chem Co, 65-66, crop specialist, Agr Chem Res, 66-67, sr res specialist, 67-73; MEM FAC, DEPT OF ENTOM & ZOOL, S DAK STATE UNIV, 73- Mem: AAAS; Entom Soc Am; Am Soc Agron. Res: Foreward agricultural chemical research; plant physiology. Mailing Add: Dept of Entom & Zool SDak State Univ Brookings SD 57006

WALHOOD, VILAS TRUMAN, plant physiology, see 12th edition

WALI, KAMESHWAR C, b Bijapur, India, Oct 15, 27; m 52; c 3. THEORETICAL PHYSICS. Educ: Univ Bombay, BSc, 48; Benares Hindu Univ, MSc, 52, MA, 54; Univ Wis, PhD(theoret physics), 59. Prof Exp: Res asocxthassoc theoret physics, Univ Wis, 59-60 & Johns Hopkins Univ, 60-62; from asst physicist to sr physicist, Argonne Nat Lab, 62-69; PROF PHYSICS, SYRACUSE UNIV, 69- Concurrent Pos: Co-ed, Int Conf Weak Interactions, Argonne Nat Lab, 65; vis mem, Inst Advan Sci Study, Bures-sur-Yvette, France, 71-72. Mem: Am Phys Soc; Sigma Xi. Res: Elementary particles; high energy physics; higher symmetries; dispersion theory. Mailing Add: Dept of Physics Syracuse Univ Syracuse NY 13210

WALI, MOHAN KISHEN, b Srinagar, India, Mar 1, 37; m 60; c 2. PLANT ECOLOGY, ENVIRONMENTAL BIOLOGY. Educ: Univ Jammu & Kashmir, India, BSc, 57; Allahabad Univ, MSc, 60; Univ BC, PhD(plant ecol), 69. Prof Exp: Demonstr bot, S P Col, Srinagar, India, 61-63, lectr, 63-65; teaching asst biol, Univ BC, 66-67, teaching asst plant ecol, 68-69; asst prof, 69-73, ASSOC PROF BIOL, UNIV NDAK, 73- Concurrent Pos: Res asst, Nat Res Coun Can, 67-69; prin investr, Proj Reclamation, 75- Mem: Ecol Soc Am; Can Bot Asn; Am Inst Biol Sci; Brit Ecol Soc; Int Asn Ecol. Res: Environmental ecology; influence of water, nutrients, temperature and light on plant populations and communities; nutrient cycling; ecosystem model building; soil-plant relationship; pollution; phytosociology; systems approach to the reclamation of strip mined areas. Mailing Add: Dept of Biol Univ of NDak Grand Forks ND 58201

WALIA, JASJIT SINGH, b Lahore, India, Mar 19, 34; m 66; c 1. ORGANIC CHEMISTRY. Educ: Univ Punjab, India, BSc(hons), 55, MS(hons), 56; Univ Southern Calif, PhD(org chem), 60. Prof Exp: Res assoc org chem, Univ Southern Calif, 60; res assoc, Mass Inst Technol, 60-61; lectr, Benaras Hindu Univ, 62-66; assoc prof, 66-73, PROF ORG CHEM, LOYOLA UNIV, LA, 73- Mem: Am Chem Soc; The Chem Soc; Int Soc Heterocyclic Chem. Res: New reactions of organic nitrogen compounds; new synthetic reactions; heterocumulene intermediates; carbanion chemistry; electronic interaction through N—H...N bond; cyanide ion-catalyzed reactions; enamino carbonyl compounds; mechanism of reactions; new compounds of potential therapeutic and/or other commercial importance. Mailing Add: Dept of Chem Loyola Univ New Orleans LA 70118

WALKENSTEIN, SIDNEY S, b Philadelphia, Pa, Dec 21, 20; m 46; c 1. DRUG METABOLISM. Educ: Temple Univ, BS, 42, AM, 50, PhD(biochem), 53. Prof Exp: Biochemist, Mold Metab, Pitman-Dunn Labs, 52-53; chief radiochemist drug metab, Wyeth Inst Med Res, 53-58; pharmaceut specialist, Union of Burma Appl Res Inst, 58-60; sr res scientist, Wyeth Labs, Inc, 60-62, mgr radiochem sect, 62-67; ASSOC DIR BIOCHEM, SMITH KLINE & FRENCH LABS, 67- Concurrent Pos: Consult, Burma pharmaceut indust, 58-60. Mem: Am Soc Pharmacol & Exp Therapeut; Am Pharmaceut Asn; Acad Pharmaceut Sci; Int Soc Biochem Pharmacol; NY Acad Sci. Res: Metabolism of aldehydes and fatty acids; effects of toxins on yeast respiration; panto-thenate-deficient yeast metabolism; utilization of hydrocarbons by molds; biotransformation and physiological disposition of isotopically-labeled drugs; medicinal plants; mechanism of drug action; trace drug analysis; pharmacokinetics; biopharmaceutics. Mailing Add: Biochem Dept Smith Kline & French Labs Philadelphia PA 19101

WALKER, ALAN, b Bridlington, Eng, Apr 30, 37; m 62. INORGANIC CHEMISTRY. Educ: Univ Nottingham, BSc, 59, PhD(inorg chem), 62. Prof Exp: Dept Indust & Sci Res res fel inorg chem, Univ Nottingham, 62-63; resident res assoc fel chem, Argonne Nat Lab, 63-65; ASSOC PROF CHEM & ASSOC DEAN, UNIV TORONTO, 65- Concurrent Pos: Nat Res Coun Can res grant, 65- Mem: Fel The Chem Soc. Res: Nonaqueous solvent chemistry, particularly of nitrates; inorganic spectroscopy; reactions of coordinated ligands to platinum group metals. Mailing Add: Dept of Chem Univ of Toronto Toronto ON Can

WALKER, ALMA TOEVS, b Charlson, NDak, Aug 6, 11; m 41. BOTANY. Educ: Iowa State Col, BS, 40, PhD, 52; Tex State Col Women, MA, 43. Prof Exp: Asst prof home econ, Col Idaho, 40-42; anal chemist, Armour & Co, Inc, Tex, 44; asst prof home econ, Utah, 45-46; asst, Ohio State Univ, 46-48; microbiologist, Tuberc Res Lab, Vet Admin Hosp, Atlanta, Ga, 57-58; assoc investr zool, 59-62, RES PHYSIOLOGIST, UNIV GA, 62- Concurrent Pos: Grants in aid, Am Acad Arts & Sci, 62 & Sigma Xi, 65. Mem: AAAS; Bot Soc Am; Mycol Soc Am; Am Bryol & Lichenological Soc. Res: Fungus physiology; catalase of mycobacteria; lipid deposition in migratory birds; lichens; medicinal plants. Mailing Add: 1095 Ivywood Dr Athens GA 30601

WALKER, ARTHUR BERTRAM CUTHBERT, JR, b Cleveland, Ohio, Aug 24, 36; m 59; c 1. SPACE PHYSICS, ASTRONOMY. Educ: Case Inst Technol, BS, 57; Univ Ill, Urbana, MS, 58, PhD(physics), 62. Prof Exp: Mem tech staff, Space Physics Lab, Aerospace Corp, 65-68, staff scientist, 68-70, sr staff scientist, 70-72, 72-75; ASSOC PROF APPL PHYSICS, STANFORD UNIV, 75-, ASSOC DEAN GRAD STUDIES, 75- Concurrent Pos: Mem exec comt, Inst Plasma Res, Stanford Univ, 75-; consult, Aerospace Corp, Rand Corp & R & D Assocs, Los Angeles, 75- Mem: Am Phys Soc; Am Geophys Union; Am Astron Soc; Int Scientific Union. Res: Solar physics; solar coronal structure; solar x-rays; solar abundances; high energy astrophysics; stellar x-ray sources; interstellar medium; physics of the upper atmosphere. Mailing Add: Inst for Plasma Res Stanford Univ Via Crespi Stanford CA 94305

WALKER, ARTHUR EARL, b Winnipeg, Man, Mar 12, 07; nat US. MEDICINE. Educ: Univ Alta, BA, 26, MD, 30. Hon Degrees: LLD, Univ Alta, 52. Prof Exp: Intern, Toronto Western Hosp, 30-31; Smith fel, Univ Chicago, 31, res neurol & neurosurg, 31-34, instr neurol, 37-38, from instr to assoc prof neurol surg, 38-45, prof & chief div, 46-47; instr neurosurg, Univ Iowa, 34; Rockefeller fel, Yale Univ, Univ Amsterdam & Brussels, 35-37; prof neurol surg, 47-72, EMER PROF NEUROL SURG, JOHNS HOPKINS UNIV, 72- Concurrent Pos: Vis prof, Sch Med, Univ NMex, 71- Mem: Soc Neurol Surg; Am Asn Neurol Surg; Am Electroencephalog Soc (pres, 54); Am Neurol Asn (pres, 65-); fel Am Col Surg. Res: Neurophysiological basis of epilepsy; anatomy and physiology of thalamus; experimental physiology of cerebral cortex; cerebello-cerebral relationships; visual mechanisms; neurosurgical therapy of pain; physiology of cerebral injuries; cerebral death; intracranial pressure. Mailing Add: 2211 Lomas Blvd NE Albuquerque NM 87131

WALKER, AUGUSTUS CHAPMAN, b Brooklyn, NY, Oct 2, 23; m 47; c 4. RESEARCH ADMINISTRATION, ACADEMIC ADMINISTRATION. Educ: Harvard Univ, BS, 48. Prof Exp: Asst biochemist, Thanhauser Lab, New Eng Med Ctr, 48-51; res chemist high polymers, Cryovac Div, W R Grace Co, 52-57; res assoc, Plastics Lab, Mass Inst Technol, 57-58; lectr, Lowell Technol Inst, 57-59, asst prof chem, 58-59; consult, group leader, sect chief & asst to gen mgr, Res & Adv Develop Div, Avco Corp, 59-65; dir res, Polymer Corp, Pa, 65-70; dir res, Resin Products Div, PPG Indust, Inc, 70-73; PRES, EFFECTIVE RES, PITTSBURGH, 73-; SR LECTR ENG MGT, CARNEGIE-MELLON UNIV, 74-, CONSULT & ACTG DIR, OFF POST-COL PROF EDUC, CARNEGIE INST TECHNOL, 75- Honors & Awards: Award, Am Inst Chem Engrs, 63. Mem: Am Chem Soc; Soc Plastic Engrs; Am Soc Eng Educ. Res: Methods of managing technical activities. Mailing Add: Effective Res 141 Westland Dr Pittsburgh PA 15217

WALKER, BENNIE FRANK, b Mt Pleasant, Tex, Sept 19, 37. PHYSICAL CHEMISTRY. Educ: Sam Houston State Col, BS, 59, MA, 62; Univ Tex, Austin, PhD(phys chem), 70. Prof Exp: Instr chem, Sam Houston State Col, 59-62; asst prof, 68-72, ASSOC PROF CHEM, STEPHEN F AUSTIN STATE UNIV, 72- Mem: Am Chem Soc. Res: Kinetics; hydrogen bonding; applications of computers in chemistry. Mailing Add: Dept of Chem Stephen F Austin State Univ Nacogdoches TX 75961

WALKER, BERNARD FORESTIER, b Victoria, BC, Oct 15, 13; m 41; c 1. PHYSICAL CHEMISTRY. Educ: Univ BC, BA, 34; McGill Univ, MSc, 35, PhD(chem), 37. Prof Exp: Res chemist, Int Paper Co, Can, 37-42; tech asst, Can Govt, 42-44; dir res, Lake St John Power & Paper Co, 44-46; dir res, Fraser Co, Ltd, 46-53; asst dir res, Am Enka Corp, NC, 53-68; dir res, Huyck Res Ctr, NY, 68-71; dir res, Barrow Res Lab, 71-74; LAB DIR, HUYCK RES, 74- Mem: Am Chem Soc; Tech Asn Pulp & Paper Indust; Can Pulp & Paper Asn. Res: Synthetic and natural fiber technology; cellulose chemistry; pulp and paper; dissolving pulps. Mailing Add: Huyck Res 100 Washington St Rensselaer NY 12144

WALKER, BOYD WALLACE, b Manhattan, Kans, May 26, 17; m 43; c 4. ICHTHYOLOGY. Educ: Univ Mich, AB, 40, MS, 42; Univ Calif, Los Angeles, PhD, 49. Prof Exp: From asst prof to assoc prof, 48-60, PROF ZOOL, UNIV CALIF, LOS ANGELES, 60- Mem: AAAS; Am Soc Ichthyologists & Herpetologists; Soc Study Evolution; Am Fisheries Soc; Soc Syst Zool. Res: Systematics, zoogeography, ecology and behavior of fishes, especially the fauna of eastern tropical Pacific. Mailing Add: Dept of Biol 1330 Life Sci Bldg Univ of Calif Los Angeles CA 90024

WALKER, BRIAN LAWRENCE, b Liverpool, Eng, Jan 4, 37; m 64; c 2. BIOCHEMISTRY, NUTRITION. Educ: Univ Ill, PhD(food sci), 62. Prof Exp: Lab asst chem, Albright & Wilson, Mfg Chemists, 54-57; res assoc lipid chem, Univ Ill, 62-64; from asst prof to assoc prof nutrit, 64-75, PROF NUTRIT, UNIV GUELPH, 75- Concurrent Pos: Grants, Nat Res Coun Can, 65- & Med Res Coun Can, 66-68. Mem: Assoc Royal Inst Chem; Am Inst Nutrit; Am Oil Chem Soc; Nutrit Soc Can; Nutrit Today Soc. Res: Lipid chemistry and biochemistry; essential fatty acids; lipids and cell membranes; perinatal nutrition and development. Mailing Add: Dept of Nutrit Univ of Guelph Guelph ON Can

WALKER, BRUCE EDWARD, b Montreal, Que, June 17, 26; m 48; c 4. ANATOMY. Educ: McGill Univ BSc, 47, MSc, 52, PhD(genetics), 54; Univ Tex, MD, 66. Prof Exp: Lectr anat, McGill Univ, 54-57; from asst prof to assoc prof, Univ Tex Med Bd, Galveston, 57-67; prof anat & chmn dept, 67-75; PROF ANAT, MICH STATE UNIV, 75- Mem: Am Asn Anat. Res: Experimental teratology; muscle disease; cell differentiation. Mailing Add: Dept of Anat Mich State Univ East Lansing MI 48823

WALKER, BRYAN D, nuclear physics, see 12th edition

WALKER, CAROL L, b Martinez, Calif, Aug 19, 35; m 62; c 4. MATHEMATICS. Educ: Univ Colo, BME, 57; NMex State Univ, MS, 61, PhD(math), 63. Prof Exp: Mem math, Inst Advan Study, 63-64; from asst prof to assoc prof, 64-72, PROF MATH, N MEX STATE UNIV, 72- Concurrent Pos: NSF fel, 63-64, NSF grants, 64-72. Mem: Am Math Soc. Res: Algebra, primarily Abelian group theory and homological algebra. Mailing Add: Dept of Math NMex State Univ Las Cruces NM 88001

WALKER, CAROLYN CRANE, cell biology, developmental genetics, see 12th edition

WALKER, CHARLES A, b Foreman, Ark, Dec 14, 35; m 57; c 3. NEUROPHARMACOLOGY. Educ: Ark Agr, Mech & Normal Col, BS, 57; Wash State Univ, MS, 59; Loyola Univ, PhD(pharmacol), 69. Prof Exp: Res asst, Wash State Univ, 57-59; asst prof biol, Ft Valley State Col, 59-63; asst prof physiol, Tuskegee Inst, 63-65, assoc prof pharmacol, 68-71, prof vet pharmacol & chmn dept pharmacol, 71-74; PROF PHARM & DEAN SCH PHARM, FLA A&M UNIV, 74- Mem: AAAS; Am Asn Clin Chem; NY Acad Sci; Am Soc Pharmacol & Exp Therapeut; Int Soc Chronobiol. Res: Circadian rhythms of biogenic amines in the central nervous system. Mailing Add: Apt Q-8 1111 E Lafayette St Tallahassee FL 32307

WALKER, CHARLES CAREY, b Bellvue, Pa, Sept 30, 36; m 64. ORGANIC POLYMER CHEMISTRY. Educ: John Carroll Univ, BS, 58, MS, 60; Ohio State Univ, PhD(org chem), 65. Prof Exp: Asst prof chem, Utica Col, Syracuse Univ, 65-67; RES CHEMIST, FILM DEPT, E I DU PONT DE NEMOURS & CO, INC, 67- Mem: Am Chem Soc. Res: Chemistry of the preparation, catalysis, degradation and recovery of polyesters. Mailing Add: Film Dept E I du Pont de Nemours & Co Inc Circleville OH 43113

WALKER, CHARLES EDWARD, JR, b Kosciusko, Miss, June 16, 39; m 67; c 1. PHYSICS. Educ: Univ Calif, Berkeley, BA, 61; Univ Ill, Urbana, MS, 63; Univ Wyo, PhD(physics), 68. Prof Exp: Asst prof physics, Univ Wyo, 68; asst prof, 68-75, ASSOC PROF PHYSICS, UNIV WIS-RIVER FALLS, 75- Mem: Am Phys Soc. Res: Non-mesic decay of heavy hypernuclei. Mailing Add: Dept of Physics Univ of Wis River Falls WI 54022

WALKER, CHARLES EUGENE, b Winterset, Iowa, Dec 17, 36; m 58; c 3. CEREAL CHEMISTRY. Educ: Iowa State Univ, BS, 59; NDak State Univ, PhD(cereal chem), 66. Prof Exp: Milling res engr, J F Bell Tech Ctr, Gen Mills Co, 59-62; prof chem & chmn div sci & math, Valley City State Col, 65-74; PROD DEVELOP MGR, FAIRMONT FOODS CENT RES LABS, 74- Mem: Am Asn Cereal Chem. Res: Development of new and improved bakery and snack products; nutritional fortification and enrichment. Mailing Add: Fairmont Foods Res & Develop Labs 8625 I St Omaha NE 68127

WALKER, CHARLES FREDERIC, b Columbus, Ohio, Dec 27, 04. ZOOLOGY. Educ: Ohio State Univ, AB, 30, MS, 31; Univ Mich, PhD(zool), 35. Prof Exp: Asst, Dept Natural Hist, State Mus, Ohio, 27-33; biologist, Soil Conserv Serv, USDA, 35-38; assoc prof zool, Stone Lab, Ohio State Univ, 38-47; from assoc cur to cur reptiles & amphibians, Mus Zool, 47-75, prof zool, Univ, 62-75, EMER PROF ZOOL, UNIV MICH, ANN ARBOR, 75- Mem: Am Soc Ichthyologists & Herpetologists (vpres, 48, 62, pres, 68); Soc Syst Zool; Am Ornith Union. Res: Herpetology. Mailing Add: Mus Bldg Univ of Mich Ann Arbor MI 48104

WALKER, CHARLES R, b Chicago, Ill, Dec 18, 28; m 50; c 4. BIOCHEMISTRY, FISH BIOLOGY. Educ: Southern Ill Univ, BA, 51, MA, 52. Prof Exp: Biochemist & fishery biologist, Mo Conserv Comn, 52-61; biochemist, Fish Control Lab, 61-67, chief br pest control res, Div Fishery Res, 67-72, CHIEF OFF ENVIRON ASSISTANCE, DIV FISHERY RES, US FISH & WILDLIFE SERV, 72- Concurrent Pos: Consult fishery biol, pesticides & pond cult, 54-61; lectr, Viterbo Col, 64-65; instr, USDA Grad Sch, 69-; mem adj fac environ systs mgt, Am Univ, 75- Mem: Am Chem Soc; Am Fisheries Soc; Weed Sci Soc Am; Am Soc Testing & Mat; Am Soc Limnol & Oceanog. Res: Fishery research; aquatic ecology, fish-pesticide relationships; pollution biology; pond culture; aquatic herbicides; toxicity; efficacy residues of drugs and pest control agents for fisheries; analytical chemistry; limnology; soil science; environment impact statements control; environmental impact assessments; monitoring environmental contaminants. Mailing Add: US Fish & Wildlife Serv Dept of the Interior Washington DC 20240

WALKER, CHARLES ROBERT, b Yreka, Calif, Feb 16, 46; m 75. DEVELOPMENTAL BIOLOGY. Educ: Univ Calif, Davis, BS, 67, PhD(genetics), 75; Ore State Univ, MS, 69. Prof Exp: Staff researcher biochem, Univ Calif, Davis, 69-71 & develop biol, 71-73; FEL DEVELOP BIOL, UNIV CALIF, BERKELEY, 75- Res: Regulation of skeletal muscle differentiation and maturation; role of muscular contraction in growth and maturation; embryonic muscle cell culture. Mailing Add: Dept of Zool Univ of Calif Berkeley CA 94720

WALKER, CHARLES THOMAS, b Chicago, Ill, Sept 5, 32; m 53; c 3. SOLID STATE PHYSICS. Educ: Univ Louisville, AB, 56, MS, 58; Brown Univ, PhD(physics), 61. Prof Exp: Res asst physics, Brown Univ, 58-60; res assoc, Cornell Univ, 61-63; asst prof, Northwestern Univ, Evanston, 63-67, assoc prof, 67-71; PROF PHYSICS, ARIZ STATE UNIV, 71- Concurrent Pos: Guggenheim fel, Oxford Univ, 67-68; vis prof, Munich Tech Univ, 71; consult, Motorola, Inc, 74- Mem: AAAS; fel Am Phys Soc; Am Asn Physics Teachers. Res: Light scattering; lattice dynamics and impurity studies in solids; magnetism. Mailing Add: Dept of Physics Ariz State Univ Tempe AZ 85281

WALKER, CHRISTOPHER BLAND, b Lakeland, Fla, July 25, 25; m 61; c 3. PHYSICS. Educ: Davidson Col, BS, 48; Mass Inst Technol, PhD(physics), 51. Prof Exp: Fulbright scholar, France, 51-52; instr physics, Mass Inst Technol, 52-53; from asst prof to assoc prof, Inst Metals, Chicago, 53-63; RES PHYSICIST, ARMY MAT & MECH RES CTR, 63- Concurrent Pos: Guggenheim fel, 63-64. Mem: Am Phys Soc; Am Crystallog Asn; Fr Soc Mineral & Crystallog. Res: X-ray diffraction; imperfections in crystals; thermal vibrations; neutron inelastic scattering. Mailing Add: Army Mat & Mech Res Ctr Watertown MA 02172

WALKER, COURTNEY EMERY, b Fallis, Okla, Dec 20, 11; m 45; c 1. AGRICULTURE. Educ: Langston Univ, BS, 36; Univ Mass, MS, 45, PhD, 48. Prof Exp: Teacher, Okla Pub Sch, 36-37; Nat Youth Admin dir, Langston Univ, 37-38; county agent, Agr Exten Serv, Okla, 38-44; prof agr & dean sch agr & home econ, 48-70, PROF ANIMAL SCI, FLA A&M UNIV, 70-, DIR SPONSORED RES, 71- Mailing Add: Dept of Animal Sci Fla A&M Univ Tallahassee FL 32307

WALKER, DAN B, b Connersville, Ind, Apr 18, 45; m 69. PLANT ANATOMY. Educ: Ind Univ, Bloomington, AB, 68; Univ Calif, Berkeley, PhD(bot), 74. Prof Exp: Lectr bot, Univ Calif, Berkeley, 73-74; ASST PROF BOT, UNIV GA, 74- Mem: Sigma Xi; AAAS; Bot Soc Am; Am Soc Plant Physiologists. Res: Investigations of structure-function and developmental problems at the cellular level in higher plants, especially on the mechanisms of intercellular transport and communication in plants. Mailing Add: Dept of Bot Univ of Ga Athens GA 30602

WALKER, DANIEL ALVIN, b Cleveland, Ohio, Dec 18, 40; m 70; c 1.

SEISMOLOGY. Educ: John Carroll Univ, BS, 63; Univ Hawaii, MS, 65, PhD(geophys), 71. Prof Exp: Res asst seismol, 63-68, jr seismologist, 69-71, ASST GEOPHYSICIST, HAWAII INST GEOPHYS, UNIV HAWAII, 72- Mem: Seismol Soc Am; Am Geophys Union; AAAS. Mailing Add: Hawaii Inst Geophys Univ of Hawaii Honolulu HI 96822

WALKER, DARRELL E, plant breeding, deceased

WALKER, DAVID, b Troy, NY, NY, Aug 9, 46; m 68. PETROLOGY. Educ: Oberlin Col, AB, 68; Harvard Univ, AM, 70, PhD(geol), 72. Prof Exp: Lectr geol, 73-74, RES FEL GEOPHYS, HARVARD UNIV, 72- Honors & Awards: F W Clarke Medal, Geochem Soc, 75. Mem: Geol Soc Am; Am Geophys Union; Geochem Soc; Mineral Soc Am; AAAS. Res: General geology with specialty in petrology, particularly experimental petrology studies of lunar basaltic samples. Mailing Add: Hoffman Lab Harvard Univ 20 Oxford St Cambridge MA 02138

WALKER, DAVID CROSBY, b York, Eng, June 16, 34; m 61; c 2. RADIATION CHEMISTRY. Educ: Univ St Andrews, BSc, 55, Hons, 56; Univ Leeds, PhD(chem), 59. Hon Degrees: DSc, Univ St Andrews, 74. Prof Exp: Fel chem, Nat Res Coun Can, 59-61; res lectr, Univ Leeds, 61-64; from asst prof to assoc prof, 64-75, PROF CHEM, UNIV BC, 75- Mem: Chem Inst Can; fel The Chem Soc; Am Chem Soc; Am Inst Physics; Radiation Res Soc. Res: Radiation chemistry of water and organic liquids; solvated electron studies in polar liquids. Mailing Add: Dept of Chem Univ of BC Vancouver BC Can

WALKER, DAVID GEORGE, physical chemistry, see 12th edition

WALKER, DAVID KENNETH, b Youngstown, Ohio, Apr 4, 43; m 67. PHYSICS. Educ: Pa State Univ, University Park, BS, 65; WVa Univ, MS, 68, PhD(physics), 71. Prof Exp: Instr physics, WVa Univ, 69-71; asst prof, 71-75, ASSOC PROF PHYSICS, WAYNESBURG COL, 75- Mem: Electrostatics Soc Am; Am Phys Soc; Am Asn Physics Teachers. Res: Electrophysics, electrostatics; atmospheric electric field; operation of motors from atmospheric field; electrets; electromechanical devices; windmills. Mailing Add: Dept of Chem & Physics Waynesburg Col Waynesburg PA 15370

WALKER, DAVID RUDGER, b Ames, Iowa, Sept 15, 29; m 48; c 10. POMOLOGY. Educ: Utah State Univ, BS, 51, MS, 52; Cornell Univ, PhD, 55. Prof Exp: From asst prof to assoc prof hort, NC State Col, 55-60; assoc prof, 60-65, PROF PLANT SCI, UTAH STATE UNIV, 65- Mem: Fel Am Soc Hort Sci; Am Pomol Soc. Res: Plant hardiness; mineral nutrition; growth substances. Mailing Add: Dept of Plant Sci Utah State Univ Logan UT 84321

WALKER, DAVID TUTHERLY, b Huntington, WVa, July 10, 22; m 57; c 1. MATHEMATICS. Educ: Wofford Col, BS, 49; Univ Ga, MS, 51, PhD, 55. Prof Exp: Asst math, Univ Ga, 51-53; instr, Univ SC, 53-54; asst, Univ Ga, 54-55; from asst prof to assoc prof, 55-67, PROF MATH, MEMPHIS STATE UNIV, 67- Res: Mathematical analysis; modern algebra; theory of numbers; geometry. Mailing Add: Dept of Math Memphis State Univ Memphis TN 38111

WALKER, DAVID WHITMAN, b Seattle, Wash, Sept 20, 23; m 47; c 3. ENTOMOLOGY. Educ: Univ Wash, BS, 50; Wash State Univ, MS, 51, PhD(entom), 59. Prof Exp: Chief sanitarian & malaria officer, UN Relief for Palestinian Refugees Comn, 48-49; asst, Wash State Univ, 50-51, proj leader stored prod invests, 53-59; entomologist, USPHS Malaria Prog, Philippines, 51-53; assoc prof biol, Univ, 59-63, sr scientist, Nuclear Ctr, 63-75, PROF BIOL, UNIV PR, MAYAGUEZ, 74-, SCIENTIST, PR NUCLEAR CTR, 75- Concurrent Pos: Consult, Commodity Stabilization Serv, USDA, 56-60, indust, 57-64 & Commonwealth Exp Sta, 61- Mem: Entom Soc Am. Res: Sterility of Lepodoptera and Hemiptera; insect behavior; chemical and biological control of insects; toxicology; induced sterility in insects; rice insects; banana insects. Mailing Add: Entom Lab Univ of PR Mayaguez PR 00708

WALKER, DENNIS KENDON, b Sacramento, Calif, Aug 1, 38; m 60; c 4. BOTANY. Educ: Humboldt State Col, BA, 60; Univ Calif, Davis, MS, 64, PhD(bot), 66. Prof Exp: Asst prof, 65-72, ASSOC PROF BOT, CALIF STATE UNIV, HUMBOLDT, 72- Mem: Electron Micros Soc Am; Bot Soc Am. Res: Plant morphology; developmental plant anatomy and plant ultrastructure, specifically the ultrastructure of differentiating elements of vascular tissues. Mailing Add: Dept of Biol Calif State Univ Humboldt Arcata CA 95521

WALKER, DEWARD EDGAR, JR, b Johnson City, Tenn, Aug. 3, 35; m 59; c 4. ANTHROPOLOGY. Educ: Univ Ore, BA, 61, PhD(anthrop), 64. Prof Exp: Asst prof anthrop, George Washington Univ, 64-65 & Wash State Univ, 65-67; assoc prof anthrop & head dept, Univ Idaho, 67-69; assoc prof, 69-72, PROF ANTHROP, UNIV COLO, BOULDER, 72- Concurrent Pos: Assoc dean grad sch & res assoc, Inst Behav Sci, Univ Colo; affil fac, Univ Idaho; appl anthropologist, Am Indian Civil Liberties Trust, Nez Perce Tribe of Idaho, Yakima Tribe of Wash & Native Am Rights Fund; workshop dir & acad adv, Am Indian Develop, Inc; NSF res grants, 64-70; Nat Endowment for Humanities res grants, 70-71; USPHS res grants, 71-75; ed, Human Orgn, 70-76; assoc ed, Am Anthropologist, 72-75. Mem: AAAS; fel Am Anthrop Asn; fel Soc Appl Anthrop; Am Ethnol Soc; Am Acad Polit & Soc Sci. Res: Acculturation, religion, political, economical and social organization of American Indians of Northwestern North America. Mailing Add: Dept of Anthrop Univ of Colo Boulder CO 80302

WALKER, DON WESLEY, b Ft Worth, Tex, July 30, 42; m 64; c 2. NEUROSCIENCE, NEUROPHARMACOLOGY. Educ: Univ Tex, Arlington, BA, 64; Tex Christian Univ, MA & PhD(psychol), 68. Prof Exp: Asst prof, 70-75, ASSOC PROF NEUROSCI & PSYCHOL, UNIV FLA, 75-; RES INVESTR, VET ADMIN HOSP, GAINESVILLE, 70- Concurrent Pos: Nat Inst Ment Health training grant, Col Med, Univ Fla, 68-70, NIH res grant neurosci, 72-; Vet Admin res fund grant, Vet Admin Hosp, Gainesville, 70- Mem: AAAS; Soc Neurosci; Am Psychol Asn. Res: Neurobiology of drug dependence; chronic effects of ethanol on the brain; neural mechanisms of feeding behavior and weight regulation. Mailing Add: Dept of Neurosci Univ of Fla Col of Med Gainesville FL 32601

WALKER, DONALD F, b Brush, Colo, July 16, 23; m 44; c 1. VETERINARY MEDICINE. Educ: Colo State Univ, DVM, 44. Prof Exp: Pvt pract, Grassland Hosp, 45-58; assoc prof, 58-66, PROF LARGE ANIMAL SURG & MED, AUBURN UNIV, 66- Res: Obstetrical surgery; ultrasonic therapy. Mailing Add: Dept of Large Animal Surg & Med Auburn Univ Auburn AL 36830

WALKER, DONALD GREGORY, b East Rochester, NY, Feb 17, 25; m 49; c 6. ANATOMY. Educ: Univ Calif, PhD(anat), 51. Prof Exp: Asst anat, Univ Calif, 48-51; from instr to asst prof, Med Sch, Univ Ore, 51-54; from asst prof to assoc prof, 54-66, PROF ANAT, MED SCH, JOHNS HOPKINS UNIV, 66- Concurrent Pos: USPHS spec fel res, 57, res career develop award, 62-67; Lederle Award, 57-62. Mem: Histochem Soc; Am Asn Anatomists. Res: Pituitary and thyroid glands in early skeletal development; elastic tissue in lathyrism; oxidative enzymes in muscle; developing bone; development of cytochemical methods for mitochondrial enzymes; osteoclast cytochemistry; osteopetrosis; bone collagenase; thyroid C cell. Mailing Add: Dept of Anat Johns Hopkins Univ Med Sch Baltimore MD 21205

WALKER, DONALD I, b Lombard, Ill, Jan 13, 22; m 44; c 3. ANALYTICAL CHEMISTRY, CHEMICAL MICROSCOPY. Educ: Univ Ill, BS, 48; Univ Colo, PhD(chem), 56. Prof Exp: Asst chem, Univ Colo, 48-50, asst instr, 53-56; res chemist, Los Alamos Sci Lab, Univ Calif, 50-53; dep dir health & safety div, Idaho Opers Off, AEC, 56-57, dir licensee compliance div, 57-60, dir region VIII, Div Compliance, 60-62, dir region IV, 62-70, DIR HEALTH SERV LAB, IDAHO OPERS OFF, ENERGY RES & DEVELOP ADMIN, 70- Concurrent Pos: Consult, Rocky Flats Div, Dow Chem Co, 54. Mem: AAAS; Am Chem Soc; Health Physics Soc; Am Inst Chemists. Res: Administration of environmental monitoring, radiation dosimetry, ecology. Mailing Add: Energy Res & Develop Admin 550 Second St Idaho Falls ID 83401

WALKER, DUARD LEE, b Bishop, Calif, June 2, 21; m 45; c 4. VIROLOGY, MICROBIOLOGY. Educ: Univ Calif, Berkeley, AB, 43, MA, 47; Univ Calif, San Francisco, MD, 45; Am Bd Med Microbiol, dipl. Prof Exp: Asst resident physician internal med, Stanford Univ Serv, San Francisco Hosp, 50-52; assoc prof, 52-59, PROF MED MICROBIOL, MED SCH, UNIV WIS-MADISON, 59-, CHMN DEPT, 70- Concurrent Pos: Nat Res Coun fel, Rockefeller Inst, New York, 47-49; USPHS fel, George Williams Hooper Found, Univ Calif, San Francisco, 49-50, res assoc, 50-51; consult, Naval Med Res Unit 4, Great Lakes, Ill, 58-74; mem microbiol training comt, Nat Inst Gen Med Sci, 66-70; mem, Nat Adv Allergy & Infectious Dis Coun, 70-74. Mem: Am Asn Immunol; Am Soc Microbiol; fel Am Acad Microbiol; Reticuloendothelial Soc; Soc Exp Biol & Med. Res: Persistent and chronic viral infections; host response to viral infection. Mailing Add: Dept of Med Microbiol Univ of Wis Med Sch Madison WI 53706

WALKER, EARNEST ARTMAN, plant physiology, plant pathology, see 12th edition

WALKER, EDWARD JOHN, b Detroit, Mich, Apr 16, 27; m 60; c 1. SOLID STATE PHYSICS. Educ: Univ Mich, BSE, 49; Yale Univ, PhD(physics), 60. Prof Exp: Asst electronics, Tube Lab, Nat Bur Stand, 49-53; PHYSICIST, RES CTR, IBM CORP, 60- Mem: Am Phys Soc. Res: Semiconductor physics. Mailing Add: Spring Valley Rd Ossining NY 10562

WALKER, EDWARD ROBERT, b Winnipeg, Man, July 29, 22; m 54; c 3. METEOROLOGY. Educ: Univ Manitoba, BSc, 43; Univ Toronto, MA, 49; McGill Univ, PhD(meteorol), 61. Prof Exp: Meteorologist, Meteorol Serv Can, 43-59; res asst meteorol, McGill Univ, 59-60; res micrometeorologist, Defence Res Bd, Can, 61-67; RES ARCTIC METEOROLOGIST, CAN DEPT ENVIRON, 67- Mem: Am Meteorol Soc; Royal Astron Soc Can; fel Royal Meteorol Soc. Res: Arctic meteorology; oceanography. Mailing Add: 3350 Woodburn Ave Victoria BC Can

WALKER, ELBERT ABNER, b Huntsville, Tex, Mar 11, 30; m 51; c 3. MATHEMATICS. Educ: Sam Houston State Col, BA, 50, MA, 52; Univ Kans, PhD(math), 55. Prof Exp: High sch teacher, Tex, 50-52; mathematician, US Dept Defense, Washington, DC, 55-56; asst prof math, Univ Kans, 56-57; from asst prof to assoc prof, 57-63, PROF MATH, N MEX STATE UNIV, 63- Mem: Am Math Soc; Math Asn Am. Res: Abelian group theory; category theory; ring theory. Mailing Add: Dept of Math Sci NMex State Univ Las Cruces NM 88003

WALKER, ELIZABETH REED, b Rochester, Pa, July 2, 41; m 67; c 1. HUMAN ANATOMY. Educ: Mich State Univ, BA, 63; WVa Univ, MS, 71, PhD(human anat), 75. Prof Exp: Res asst microbiol, Rockefeller Univ, 64-66; technologist electron micros, 67-71, lectr human anat, 74-75, INSTR HUMAN ANAT, W VA UNIV, 75- Res: Investigation of rheumatology by transmission and scanning electron microscopy, particularly pathogenesis of rheumatoid arthritis and other connective tissue diseases, and pulmonary research, with emphasis on macrophage uptake of respirable mineral particulates. Mailing Add: Dept of Anat WVa Univ Sch of Med Morgantown WV 26506

WALKER, ETTA FRANCES, b Houston, Tex, Jan 11, 44. BIOPHYSICS. Educ: NMex Inst Mining & Technol, BS, 65; Stanford Univ, PhD(biophys), 71. Prof Exp: Asst prof physics, Prairie View Agr & Mech Col, 71-72; CHMN DIV MATH & NATURAL SCI, HOUSTON COMMUNITY COL, 72- Concurrent Pos: Adj instr, Baylor Col Med, 71- Res: Viral carcinogenesis; biochemical virology. Mailing Add: Dept of Math Sci Houston Community Col Houston TX 77027

WALKER, EUGENE HOFFMAN, b New York, NY, Mar 28, 15; m 47; c 3. ECONOMIC GEOLOGY. Educ: Harvard Univ, BA, 37, MA, 42, PhD(geol), 47. Prof Exp: Geologist, Shell Oil Co, Tex, 37-39; geologist, Patino Mines, Bolivia, 42-45; instr geol, Univ Mich, 45-49; GEOLOGIST, US GEOL SURV, 49- Mem: Geol Soc Am. Res: Ground water geology and hydrology; geomorphology; glacial geology. Mailing Add: US Geol Surv 150 Causeway St Boston MA 02114

WALKER, FRANCIS EDWIN, b Morris, Ill, Nov 29, 31; m 51; c 3. AGRICULTURAL ECONOMICS. Educ: Univ Ill, BS, 54, MS, 58, PhD(agr econ), 60. Prof Exp: Asst prof agr econ, Purdue Univ, 60-61; from asst prof to assoc prof, 61-68, PROF AGR ECON, OHIO STATE UNIV, 68- Concurrent Pos: Bk rev ed, Am Agr Econ Asn, 75-77. Mem: Am Statist Asn; Am Agr Econ Asn. Res: International trade policy; interregional competition. Mailing Add: Dept of Agr Econ Ohio State Univ 2120 Fyffe Rd Columbus OH 43210

WALKER, FRANCIS H, b San Francisco, Calif, Jan 15, 36; m 66; c 2. ORGANIC CHEMISTRY. Educ: Stanford Univ, BS, 58, MS, 60. Prof Exp: CHEMIST, STAUFFER CHEM CO, 60- Mem: AAAS; Am Chem Soc. Res: Organic synthesis of agricultural chemicals. Mailing Add: Stauffer Chem Co 1200 S 47th St Richmond CA 94804

WALKER, FRANKLIN EARL, organic chemistry, biochemistry, see 12th edition

WALKER, GEORGE BERNARD, JR, organic chemistry, see 12th edition

WALKER, GEORGE EDWARD, b Chillicothe, Ohio, Nov 5, 40; m 64; c 3. THEORETICAL NUCLEAR PHYSICS. Educ: Wesleyan Univ, BA, 62; Case Western Reserve Univ, MS, 64, PhD(physics), 66. Prof Exp: Res assoc physics, Los Alamos Sci Lab, 66-68; res assoc, Stanford Univ, 68-70; asst prof, 70-73, ASSOC PROF PHYSICS, IND UNIV, BLOOMINGTON, 73- Concurrent Pos: Vis staff mem, Los Alamos Sci Lab, 68- Mem: Am Phys Soc; Am Asn Physics Teachers. Res: Nuclear theory; electron scattering; pion-nucleus interactions; nucleon-nucleus interactions; heavy ion scattering. Mailing Add: Dept of Physics Ind Univ Bloomington IN 47401

WALKER, GEORGE WILLIAM RUTHERFORD, cytogenetics, see 12th edition

WALKER, GLENN ANTHONY, b Louisville, Ky, July 16, 36; m 57; c 4. BIOCHEMISTRY. Educ: Bellarmine Col, Ky, BA, 58; Mich State Univ, PhD(biochem, 63). Prof Exp: Fel biochem, Scripps Clin & Res Labs, Univ Calif, San Diego, 63-66, sr fel, 66-69; ASSOC PROF BIOCHEM, MED SCH, GEORGE WASHINGTON UNIV, 69- Mem: AAAS. Res: Chemistry and biochemistry of B-twelve enzymes; chemotherapeutic effect of B-twelve analogues; mechanism of cell replication. Mailing Add: Med Sch Dept of Biochem George Washington Univ Washington DC 20005

WALKER, GLENN KENNETH, b South Weymouth, Mass, May 15, 48; m 74. CELL BIOLOGY, PROTOZOOLOGY. Educ: Univ Mass, Amherst, BS, 70; Northern Ariz Univ, MS, 72; Univ Md, College Park, PhD(cell biol), 75. Prof Exp: Teaching asst biol, Northern Ariz Univ, 70-72; teaching asst zool, cell biol & protozool, Univ Md, College Park, 72-75; SCHOLAR CELL CHEM, CELL CHEM LAB, UNIV MICH, ANN ARBOR, 75- Mem: Am Micros Soc; Soc Protozoologists. Res: Examination of the molecular mechanisms associated with pathologies which may represent abnormalities in epidermal differentiation; studies include in situ DNA hybridization and amino acid localization by autoradiography. Mailing Add: Dept Indust & Environ Health Univ Mich Cell Chem Lab SPH I Ann Arbor MI 48104

WALKER, GORDON ARTHUR HUNTER, b Kinghorn, Scotland, Jan 30, 36; m 62; c 2. ASTROPHYSICS. Educ: Univ Edinburgh, BSc, 58; Univ Cambridge, PhD(astrophys), 62. Prof Exp: Nat Res Coun fel astrophys, Dept Mines & Technol Surv, Dom Astrophys Observ, 62-63, res scientist II, 63-69; assoc prof, 69-74, PROF, UNIV BC, 74-, DIR INST ASTRON & SPACE SCI, 72- Mem: AAAS; Am Astron Soc. Res: Interstellar materials, particularly interstellar dust; early type stars, their distance, luminosity and rotational velocities; photoelectric photometry; telescope auxilliary instrumentation; low light level multichannel detection systems. Mailing Add: Inst of Astron & Space Sci Univ of BC Vancouver BC Can

WALKER, GORDON LOFTIS, b Salt Lake City, Utah, Oct 29, 12; m 38; c 2. MATHEMATICS. Educ: La State Univ, BS, 37, MA, 38; Cornell Univ, PhD(math), 42. Prof Exp: Instr math, Cornell Univ, 38-42; asst prof, Univ Del, 42-45, Temple Univ, 45-47 & Purdue Univ, 47-54; head math sect, Res Ctr, Am Optical Co, 54-59; EXEC DIR, AM MATH SOC, 59- Concurrent Pos: Prof, Cornell Univ, 44; consult, Pratt Whitney Aircraft Co, Del, 44-45. Mem: Am Math Soc; Math Asn Am; Math Union, Arg & Italy; Ger Soc Appl Math & Mech. Res: Geometry; algebra. Mailing Add: Am Math Soc PO Box 6248 Providence RI 02906

WALKER, GORDON NORTHROP, organic chemistry, see 12th edition

WALKER, GRAYSON HOWARD, b North Wilkesboro, NC, Dec 9, 38; m 71. STATISTICAL MECHANICS, CHEMICAL PHYSICS. Educ: Univ NC, BS, 61; Univ Ill, MS, 62; Ga Inst Technol, PhD(physics), 69. Prof Exp: From instr to assoc prof, 67-74, PROF PHYSICS, CLARK COL, 74- Mem: Am Phys Soc; Am Meteorol Soc; Am Asn Physics Teachers. Res: Applications of the methods of statistical physics to problems in chemical physics, atomospheric research, planetary atmospheres; experimental studies in planetary atmospheres involving light scattering techniques. Mailing Add: 772 Willivee Dr Decatur GA 30033

WALKER, HARLEY JESSE, b Bushnell, Mich, July 4, 21; m 53; c 3. GEOGRAPHY. Educ: Univ Calif, AB, 47, MA, 54; La State Univ, PhD, 60. Prof Exp: Asst prof geog, Ga State Col, 51-56, assoc prof geog & head dept, 57-59; geogr, Off Naval Res, 59-60; from asst prof to assoc prof geog, 60-67, chmn dept, 63-71, PROF GEOG, LA STATE UNIV, BATON ROUGE, 67- Concurrent Pos: Liaison scientist, Off Naval Res, London, 68-69. Mem: Asn Am Geogr; Am Geog Soc; Arctic Inst NAm. Res: Arctic and tropics; hydrology, climatology and coastal morphology. Mailing Add: Dept of Geog & Anthrop La State Univ Baton Rouge LA 70803

WALKER, HARRY MATTHEWS, organic chemistry, chemical engineering, see 12th edition

WALKER, HOMER FRANKLIN, b Beaumont, Tex, Sept 7, 43; m 70. MATHEMATICS. Educ: Rice Univ, BA, 66; NY Univ, MS, 68, PhD(math), 70. ASST PROF MATH, TEX TECH UNIV, 70- Mem: Am Math Soc; Math Asn. Res: Partial differential equations; scattering theory. Mailing Add: Dept of Math Tex Tech Univ Lubbock TX 79409

WALKER, HOMER WAYNE, b Saxonburg, Pa, May 22, 25; m 63; c 1. FOOD MICROBIOLOGY. Educ: Pa State Univ, BS, 51; Univ Wis, MS, 53, PhD(bact), 55. Prof Exp: Asst bact, Univ Wis, 51-55; from asst prof to assoc prof, 55-66, PROF FOOD TECHNOL, IOWA STATE UNIV, 66- Mem: Am Soc Microbiol; Inst Food Technologists; Soc Indust Microbiol; Int Asn Milk, Food & Environ Sanit; Brit Soc Appl Bact. Res: Resistance of bacterial spores to heat and chemicals; bacterial toxins; antibiotics in foods and use as preservatives; microbiology of processed poultry and meats; sanitary bacteriology of food and water. Mailing Add: Dept of Food Technol Iowa State Univ Ames IA 50011

WALKER, HOWARD DAVID, b New York, NY, May 7, 25; m 47; c 3. BIOCHEMISTRY. Educ: NY Univ, BA, 47, MS, 48; Univ Calif, Los Angeles, PhD(biochem), 55. Prof Exp: USPHS fel biochem, Am Meat Inst Found, Chicago, 55-56; instr, Northwestern Univ, 56-57; from asst prof to assoc prof chem, 57-68, PROF CHEM, CALIF POLYTECH STATE UNIV, SAN LUIS OBISPO, 68- Concurrent Pos: Group leader, Vet Admin Hosp, Downey, Ill, 56-57. Mem: Am Chem Soc. Res: Chemistry of pesticides and foods. Mailing Add: Dept of Chem Calif Polytech State Univ San Luis Obispo CA 93407

WALKER, HOWARD GEORGE, JR, b Oak Park, Ill, Aug 1, 21; m 48; c 4. AGRICULTURAL CHEMISTRY. Educ: Duke Univ, BS, 43, PhD(chem), 47. Prof Exp: Res assoc allylic systs, Univ Calif, Los Angeles, 47-48; CHEMIST, WESTERN UTILIZATION RES & DEVELOP DIV, AGR RES SERV, USDA, 48- Mem: Am Asn Cereal Chem; Am Chem Soc. Res: Sugar beet, synthetic organic, physical and cereal chemistry. Mailing Add: Western Regional Res Lab USDA Agr Res Serv Berkeley CA 94710

WALKER, IAN GARDNER, b Saskatoon, Sask, Apr 20, 28; m 52; c 4. BIOCHEMISTRY, CELL BIOLOGY. Educ: Univ Sask, BA, 48; Univ Toronto, MA, 51, PhD(biochem), 54. Prof Exp: Defense sci serv officer, Defence Res Med Labs, Univ Toronto, 54-60, spec lectr, Fac Pharm, 54-62; from asst prof to assoc prof, 66-74, PROF BIOCHEM, CANCER RES LAB & DEPT BIOCHEM, UNIV WESTERN ONT, 74- Concurrent Pos: Nat Cancer Inst Can fel, Ont Cancer Inst, 60-62; Eleanor Roosevelt Int Cancer fel, 68-69. Mem: Am Asn Cancer Res; Can Biochem Soc; Can Soc Cell Biol. Res: Biochemistry of nucleic acids, cell division, anticancer agents; biochemistry and toxicology of omega-fluorinated compounds; repair synthesis of DNA; toxicity of oxygen at high pressures. Mailing Add: Dept of Biochem Univ of Western Ont London ON Can

WALKER, IAN MUNRO, b Toronto, Ont, Aug 18, 40; US citizen; m 66. INORGANIC CHEMISTRY. Educ: Bowdoin Col, BA, 62; Brown Univ, PhD(chem), 67. Prof Exp: NIH fel chem, Univ Ill, Urbana, 67-68; asst prof, 68-73, ASSOC PROF CHEM, YORK UNIV, 73- Concurrent Pos: Consult, Ashland Oil (Can), Ltd, 72- Mem: Am Chem Soc; The Chem Soc. Res: Structure of ion-aggregates in solution; low symmetry fields in transition-metal complexes. Mailing Add: Dept of Chem York Univ Downsview ON Can

WALKER, ISAAC FLETCHER, chemistry, see 12th edition

WALKER, J CALVIN, b Mooresville, NC, Jan 16, 35; m 58; c 3. NUCLEAR PHYSICS, SOLID STATE PHYSICS. Educ: Harvard Univ, AB, 56; Princeton Univ, PhD(physics), 61. Prof Exp: Instr physics, Princeton Univ, 61-62; fel, Atomic Energy Res Estab, Harwell, Eng, 62-63; from asst prof to assoc prof, 63-70, PROF PHYSICS, JOHNS HOPKINS UNIV, 70- Concurrent Pos: Alfred P Sloan Found fel, 66-68. Mem: Am Phys Soc. Res: Atomic beam studies of radioactive nuclei; solid state and nuclear studies using gamma resonance techniques. Mailing Add: Dept of Physics Johns Hopkins Univ Baltimore MD 21218

WALKER, JACK, b Derbyshire, Eng, June 30, 29; Can citizen. PHYSICAL CHEMISTRY, CHEMICAL ENGINEERING. Educ: Univ Birmingham, BSc, 50, PhD(chem eng), 56. Prof Exp: Lectr chem eng, Univ Toronto, 54-55; chemist, Husky Oil & Ref Ltd, Can, 55-57; from chemist to sr res chemist, 57-70, res adv lube process, 70-74, mgr lube process, 74-75, MGR PROCESS RES, IMP OIL ENTERPRISES LTD, 75- Mem: Fel Chem Inst Can; assoc Royal Inst Chemists; Can Soc Chem Eng. Res: Petroleum processing technology; lubricating oil processing. Mailing Add: Res Dept Imp Oil Enterprises Ltd Sarnia ON Can

WALKER, JAMES BENJAMIN, b Dallas, Tex, May 15, 22; m 56; c 3. BIOCHEMISTRY. Educ: Rice Inst, BS, 43; Univ Tex, MA, 49, PhD(biochem), 52. Prof Exp: Res scientist, Biochem Inst, Univ Tex, 52-55; Nat Cancer Inst fel biochem, Univ Wis, 55-56; from asst prof to assoc prof, Baylor Col Med, 56-64; PROF BIOCHEM, RICE UNIV, 64- Concurrent Pos: USPHS sr res fel, 57-64. Mem: Am Soc Biol Chemists; Am Chem Soc; Am Soc Microbiol. Res: Feedback and hormonal controls in biosynthetic reactions; biosynthesis of creatine and streptomycin; arginine metabolism; mechanism of enzyme action; differentiation; feedback repression during embryonic dev,elopment. Mailing Add: Dept of Biochem Rice Univ Houston TX 77001

WALKER, JAMES CALLAN GRAY, b Johannesburg, SAfrica, Jan 31, 39; m 59; c 1. GEOPHYSICS. Educ: Yale Univ, BS, 60; Columbia Univ, PhD(geophys), 64. Prof Exp: Res assoc aeronomy, Inst Space Studies, New York, 64-65; res fel, Queen's Univ, Belfast, 65-66; res assoc, Goddard Space Flight Ctr, NASA, 66-67; asst prof geol, Yale Univ, 67-70, assoc prof geophys, 70-74; SR RES ASSOC, NAT ASTRON & IONOSPHERE CTR, 74- Concurrent Pos: Adj asst prof, NY Univ, 64-65; mem comt solar terrestrial res, Geophys Res Bd, Nat Acad Sci, 70-74; assoc ed, J Geophys Res, 72-75. Mem: AAAS; Am Geophys Union. Res: Aeronomy; atmospheric physics; ionospheric physics; evolution of the atmosphere. Mailing Add: Arecibo Observ Box 995 Arecibo PR 00612

WALKER, JAMES ELLIOT CABOT, b Bryn Mawr, Pa, Sept 28, 26; m 65; c 1. INTERNAL MEDICINE. Educ: Williams Col, BA, 49; Univ Pa, MD, 53; Harvard Univ, MS, 66. Prof Exp: Intern, Univ Wis Hosp, 53-54; resident med, Univ Mich Hosp, 54-55; res fel, Harvard Med Sch, 57-60; sr resident, Peter Bent Brigham Hosp, 59-60, asst to dir & dir ambulatory serv, 60-65; prof med & sociol, 65-67, prof clin med & health care & chmn dept, 67-71, PROF MED & CHMN DEPT COMMUNITY MED & HEALTH CARE, SCH MED, UNIV CONN, 71- Concurrent Pos: Mass Heart Asn fel, 58-59; Commonwealth Fund traveling fel, 65-66; from instr to lectr, Harvard Med Sch, 60-66, dir div med care res, Dept Med, 63-66; consult, Univ Wis Hosp & Univ NB Hosp, 65; chief med serv, Univ Conn Health Ctr, McCook Div & actg chief med serv, Vet Admin Hosps, Newington, Univ Conn, 69-71. Mem: AAAS; Asn Am Med Cols; fel Am Col Physicians; Am Fedn Clin Res; AMA. Res: Pulmonary physiology; airway temperatures; delivery of health care services; responsibilities of medical education and the university to medical care and society. Mailing Add: Univ of Conn Sch of Med Farmington CT 06032

WALKER, JAMES FREDERICK, b Riverton, Ala, July 30, 04; m 27; c 1. PHYSIOLOGY, HISTOLOGY. Educ: Univ Miss, AB, 27, MS, 31; Univ Iowa, PhD(zool), 35. Prof Exp: Instr sci, Miss Southern Col, 26-30, actg head dept, 29-30, assoc prof, 30-31; asst zool, Univ Iowa, 31-32; assoc prof sci, Miss Southern Col, 32-33 & 35-41; instr naval aviation ground sch, Boston Univ, 41-43; instr naval preflight sch, Univ Iowa, 43-44; from assoc prof to prof anat, Univ Calif-Calif Col Med, 44-45; prof biol, 45-72, head div biol sci, 46-57, chmn dept biol, 57-68, assoc dean arts & sci, 68-70, DISTINGUISHED EMER UNIV PROF BIOL, UNIV SOUTHERN MISS, 72- Concurrent Pos: Researcher zool, Univ Calif, Los Angeles, 62-63. Mem: Fel AAAS; assoc Am Physiol Soc. Res: Experimental histology; marine biology; histology and cytology of invertebrates; physiology of vertebrates and invertebrates; histology and histochemistry of the commercial shrimp integument and the phenomena of black spotting in shrimp integument; transverse fission of two types in hydra; correlation of excess ive bud formation and excessive tentacles in individual hydra. Mailing Add: Dept of Biol Univ of Southern Miss Box 145 Hattiesburg MS 39401

WALKER, JAMES FREDERICK, JR, b Minneapolis, Minn, July 22, 37; m 59; c 3. THEORETICAL PHYSICS. Educ: Univ Minn, BPhys, 59, MS, 61, PhD(physicsP 64. Prof Exp: Asst res scientist, NY Univ, 64-66; mem res staff, Mass Inst Technol, 66-68; ASST PROF PHYSICS, UNIV MASS, AMHERST, 68- Mem: AAAS; Am Phys Soc. Res: Atomic collision processes; three nucleon problem; parity violations in nuclear physics; nuclear reaction theories; nuclear structure theory. Mailing Add: Dept of Physics & Astron Univ of Mass Amherst MA 01002

WALKER, JAMES JOSEPH, b Philadelphia, Pa, Dec 29, 33; m 57; c 3. THEORETICAL PHYSICS. Educ: Univ NMex, BS, 59; Univ SC, PhD(physics), 65. Prof Exp: Gen mgr, EG&G, Inc, NMex, 65-75; GROUP LEADER NEUTRON PHYSICS, LOS ALAMOS SCI LAB, J-16, 75- Mem: Sigma Xi. Res: Integral equations; holography; nuclear physics. Mailing Add: 829 Gonzales Rd Santa Fe NM 87501

WALKER, JAMES KING, b Greenock, Scotland, Oct 9, 35; m 60; c 2. PARTICLE PHYSICS. Educ: Glasgow Univ, BSc, 57, PhD(physics), 60. Prof Exp: Res scientist, Ecole Normale Superieure, Paris, 60-62; res assoc physics, Harvard Univ, 62-64, from asst prof to assoc prof, 64-69; SCIENTIST, NAT ACCELERATOR LAB, 69- Concurrent Pos: Consult, Pilot Chem Co, 67-69. Res: Electromagnetic properties and structure of elementary particles; elementary particle physics. Mailing Add: Physics Dept Nat Accelerator Lab PO Box 500 Batavia IL 60510

WALKER, JAMES MARTIN, b Jonesboro, La, Oct 21, 38; m 61; c 2. HERPETOLOGY. Educ: La Polytech Inst, BS, 60, MS, 61; Univ Colo, PhD(zool), 66. Prof Exp: ASSOC PROF ZOOL, UNIV ARK, FAYETTEVILLE, 65- Mem: Am Soc Ichthyol & Herpet; Herpetologists League. Res: Reptiles and amphibians of North

America, with special interest in the ecology and systematics of lizards of the genus Cnemidophorus of the family Teiidae. Mailing Add: Dept of Zool Univ of Ark Fayetteville AR 72701

WALKER, JAMES RICHARD, b Boise, Idaho, Feb 26, 33; m 61; c 3. PHYSIOLOGY. Educ: Ariz State Univ, BS, 56; Univ Miss, PhD(physiol), 65. Prof Exp: Instr, 65-66, ASST PROF PHYSIOL, UNIV TEX MED BR, GALVESTON, 66-, ASST DIR INTEGRATED FUNCTIONAL LAB, 74- Concurrent Pos: Mem staff, Commun Sci Lab, Univ Fla, 72-73. Mem: Acoust Soc Am; Am Inst Physics. Res: Auditory neurophysiology, biological control systems. Mailing Add: Dept of Physiol Univ of Tex Med Br Galveston TX 77550

WALKER, JAMES ROY, b Chestnut, La, Nov 8, 37; m 59; c 2. MICROBIOLOGY. Educ: Northwestern State Col, La, BS, 60; Univ Tex, PhD(microbiol), 63. Prof Exp: Nat Cancer Inst fel biochem sci, Princeton Univ, 65-67; asst prof microbiol, 67-71, ASSOC PROF MICROBIOL, UNIV TEX, AUSTIN, 71- Concurrent Pos: Res assoc dept chem, Harvard Univ, 72-73. Mem: Genetics Soc Am; Am Soc Microbiol. Res: Microbial genetics; regulation of cell division. Mailing Add: Dept of Microbiol Univ of Tex Austin TX 78712

WALKER, JAMES WILLARD, b Taylor, Tex, Mar 23, 43; m 73. EVOLUTIONARY BIOLOGY. Educ: Univ Tex, Austin, BA, 64; Harvard Univ, PhD(biol), 70. Prof Exp: ASST PROF BOT, UNIV MASS, AMHERST, 69- Honors & Awards: George R Cooley Award, Am Soc Plant Taxon, 72. Mem: AAAS; Bot Soc Am; Am Soc Plant Taxon; Am Inst Biol Sci; Linnean Soc London. Res: Angiosperm systematics; morphology, phylogeny and evolution of primitive angiosperms; pollen morphology of ranalian dicots. Mailing Add: Dept of Bot Univ of Mass Amherst MA 01002

WALKER, JAMES WILSON, b NC, July 17, 22; m 45; c 2. MATHEMATICAL STATISTICS. Educ: Univ NC, PhD(math statist), 57. Prof Exp: Intel specialist, US Dept Air Force, 50-55; asst statist, Univ NC, 55-56; from asst prof to assoc prof math, 56-64, res assoc eng exp sta, 58-65, PROF MATH, GA INST TECHNOL, 64- Concurrent Pos: Consult, WVa Pulp & Paper Co, 66-68 & Union Camp, Inc, 74-75. Mem: Am Statist Asn; Math Asn Am. Res: Statistical inference from grouped data; optimal grouping of statistical data; inefficiency of certain estimates based on grouped data. Mailing Add: Dept of Math Ga Inst of Technol Atlanta GA 30332

WALKER, JEARL DALTON, b Pensacola, Fla, Jan 20, 45; m 67; c 3. OPTICS. Educ: Mass Inst Technol, BS, 67; Univ Md, PhD(physics), 73. Prof Exp: ASST PROF PHYSICS, CLEVELAND STATE UNIV, 73- Res: Scattering of light from particles. Mailing Add: Dept of Physics Cleveland State Univ Cleveland OH 44115

WALKER, JERRY, b Michigantown, Ind, Mar 6, 37; m 57; c 2. VETERINARY MEDICINE, MICROBIOLOGY. Educ: Mich State Univ, BS, 59, DVM, 61; Univ Mich, MS, 65. Prof Exp: Vet practitioner, Bardens Small Animal Hosp, 61-62; vet corps, US Army, 62-, vet process develop div, Ft Detrick, 62-63, prin investr, 65-68, chief dept vet med, 9th Med Lab, Vietnam, 68-69, asst chief virol div, US Army Med Res Inst Infectious Dis, 69-70, chief dept rickettsial dis, US Army Med Res Univ, Inst Med Res, Kuala Lumpur, Malaysia, 70-72, CHIEF DIV AEROBIOL, US ARMY MED RES INST INFECTIOUS DIS, VET CORPS, US ARMY, 72- Honors & Awards: Outstanding Res Award, Sci Res Soc Am, 66. Mem: Tissue Cult Asn; Am Vet Med Asn; Sigma Xi. Res: Pathogenesis and therapy of infectious diseases, both bacterial and viral; tissue culture. Mailing Add: US Army Res Inst Infectious Dis Ft Detrick Frederick MD 21701

WALKER, JERRY ARNOLD, b Olney, Ill, Mar 4, 48. SYNTHETIC ORGANIC CHEMISTRY. Educ: Univ Ill, BS, 69; Mass Inst Technol, PhD(org chem), 73. Prof Exp: Fel org chem, Univ Calif, Los Angeles, 73-74 & Calif Inst Technol, 74-75; RES CHEMIST, UPJOHN CO, 75- Mem: Am Chem Soc; The Chem Soc. Res: Research and development of methods for the synthesis of biologically active compounds. Mailing Add: Upjohn Co Portage Rd Kalamazoo MI 49001

WALKER, JERRY TYLER, b Cincinnati, Ohio, Sept 7, 30; m 53; c 2. PLANT PATHOLOGY. Educ: Univ Miami, Ohio, BA, 52; Ohio State Univ, MSc, 57, PhD, 60. Prof Exp: Asst, Ohio State Univ, 55-59, asst agr exp sta, 59-61; plant pathologist, Brooklyn Bot Garden, 61-69; ASSOC PROF PLANT PATH & HEAD DEPT, AGR EXP STA, UNIV GA, 69- Concurrent Pos: NSF grant, 63-66; actg chmn res, Kitchawan Lab, NY, 67-69. Mem: Am Phytopath Soc; Soc Nematol. Res: Phytonematology, including control; diseases of ornamentals; fungicides. Mailing Add: Dept of Plant Path Agr Exp Sta Univ of Ga Experiment GA 30212

WALKER, JOANNE GILLESPIE, b Rockford, Ill, July 20, 31. PHYSIOLOGY. Educ: Univ Ill, BS, 53, MS, 55, PhD(physiol), 59. Prof Exp: Asst prof biol, Muskingum Col, 58-61; res assoc microbiol, 61-66, res assoc vet path, 66-70, RES ASSOC PHARMACOL, OHIO STATE UNIV, 70- Mem: AAAS; Am Soc Zoologists. Res: Oxygen toxicity; respiration. Mailing Add: Dept of Pharmacol Ohio State Univ Columbus OH 43210

WALKER, JOE AARON, b San Augustine, Tex, Sept 21, 21; m 46; c 3. MEDICINE, ANESTHESIOLOGY. Educ: Stephen F Austin State Col, BS, 43; Univ Tex, MD, 46. Prof Exp: From instr to assoc prof, 59-72, PROF ANESTHESIOL, UNIV TEX MED BR, GALVESTON, 72- Mem: Am Soc Anesthesiol. Res: Cardiovascular effects of anesthetic drugs and methods of treating different types of peripheral myoneural blocks. Mailing Add: Dept of Anesthesiol Univ of Tex Med Br Galveston TX 77550

WALKER, JOE M, b Baxter Springs, Kans, Mar 8, 30; m 53; c 2. ANALYTICAL CHEMISTRY. Educ: Kans State Col Pittsburg, BS, 51, MS, 53; Kans State Univ, PhD(chem), 58. Prof Exp: Instr sci, Kans State Col Pittsburg, 50-53; asst prof chem, Univ Ottawa, 54-56; from asst prof to assoc prof, 57-61, PROF CHEM, KANS STATE COL PITTSBURG, 61- Mem: Am Chem Soc. Res: Instrumental methods; high frequency methods in analytical chemistry; trace carbon analysis; gas chromatography; analytical chemistry of boron. Mailing Add: Dept of Chem Kans State Col Pittsburg KS 66762

WALKER, JOHN FREDERICK, organic chemistry, see 12th edition

WALKER, JOHN J, b Alma, Nebr, July 4, 35; m 60. ORGANIC CHEMISTRY. Educ: Univ Nebr, Lincoln, BS, 58; Atlanta Univ, MS, 68; Ga Inst Technol, PhD(org chem), 74. Prof Exp: Instr, Ga Inst Technol, 70-73; TECH DIR, DETTELBACH CHEM CORP, 73- Concurrent Pos: Adj prof, DeKalb Community Col, 73- Mem: Am Chem Soc; Am Inst Chemists. Res: Synthesis of physiologically active barbiturates. Mailing Add: 2154 Drew Valley Rd NE Atlanta GA 30319

WALKER, JOHN LAWRENCE, JR, b Whitewater, Wis, Dec 12, 31; m 56; c 2. PHYSIOLOGY. Educ: Univ Wis, BS, 56; Duke Univ, MA, 58; Univ Minn, Minneapolis, PhD(physiol), 63. Prof Exp: Instr physiol, Univ Minn, Minneapolis, 62-64, asst prof, 64-65; asst prof, 66-71, ASSOC PROF PHYSIOL, UNIV UTAH, 71- Prof Exp: USPHS fel, 64-66, res grant, 66. Res: Mechanism of movement of ions and molecules through membranes, especially permeation of ions and electrical properties of membranes; ion selective microelectrodes. Mailing Add: Dept of Physiol Univ of Utah Salt Lake City UT 84112

WALKER, JOHN MARTIN, b Norfolk, Va, July 6, 35; m 55; c 2. WATER POLLUTION, SOIL SCIENCE. Educ: Rutgers Univ, BS, 57, MS, 59, PhD(agron), 61. Prof Exp: Asst soil fertility and plant nutrit, Purdue Univ, 57-60; NATO fel soil chem, Rothamsted Exp Sta, Eng, 61-62; res soil scientist soils lab plant indust sta, Soil & Water Conserv Res Div, Agr Res Serv, USDA, 63-72, soil scientist biol waste mgt lab, 72-74, actg chief, 75; REGIONAL SCI ADV WASTEWATER, SLUDGE & SOIL, OFF RES & DEVELOP, US ENVIRON PROTECTION AGENCY, 75- Concurrent Pos: Adj prof dept crop & soil sci, Mich State Univ, 75- Mem: Fel AAAS; Am Soc Agron; Soil Sci Soc Am; Int Soc Soil Sci; Water Pollution Control Fedn. Res: Utilization of sewage sludge and wastewater treatment and use on land; soil temperature effects on movement and uptake of water and ions; plant response to controlled environments. Mailing Add: Dept of Crop & Soil Sci Mich State Univ East Lansing MI 48824

WALKER, JOHN ROBERT, b Newbern, Tenn, Nov 27, 31; m 55; c 1. ENTOMOLOGY, RESEARCH ADMINISTRATION. Educ: La State Univ, BS, 55, MS, 59; Iowa State Univ, PhD(entom), 62. Prof Exp: Asst prof, Univ, 62-65, asst to vpres res, Univ Syst, 66-68, ASSOC PROF ENTOM, LA STATE UNIV, BATON ROUGE, 65-, ASST TO V PRES INSTR & RES, LA STATE UNIV SYST, 68- Mem: AAAS; Entom Soc Am. Res: Effects of ionizing radiations on reproductive system of insects. Mailing Add: Off of VPres for Instr & Res La State Univ Syst Baton Rouge LA 70803

WALKER, JOSEPH, b Rockford, Ill, Dec 28, 22; m 44; c 4. ANALYTICAL CHEMISTRY, ORGANIC CHEMISTRY. Educ: Beloit Col, BS, 43; Univ Wis, MS, 48, PhD(chem), 50. Prof Exp: Sr res chemist res ctr, Pure Oil Co, Ill, 50-51, proj technologist, 51-56, sect supvr phys chem, 56-58, dir anal res & serv div, 58-64, res coordr, 64-65, dir res, 65-66, ASSOC DIR RES, RES CTR, UNION OIL CO CALIF, BREA, 66- Mem: Am Chem Soc. Res: Petroleum technology; analysis of petroleum products; petrochemicals research; fuels development; petroleum research administration. Mailing Add: 3010 Anacapa Place Fullerton CA 92635

WALKER, KEITH GERALD, b Carthage, Mo, Aug 22, 41; m 63; c 2. ATOMIC PHYSICS, MOLECULAR PHYSICS. Educ: Bethany Nazarene Col, BS, 63; Ohio State Univ, MS, 66; Univ Okla, PhD(physics), 71. Prof Exp: Technician, State of Ohio, summer 64; instr physics, 65-67, from asst prof to assoc prof, 67-72, PROF PHYSICS, BETHANY NAZARENE COL, 72- Mem: Am Asn Physics Teachers; Optical Soc Am. Res: Electron-atom impact and resulting cross-sections. Mailing Add: Dept of Physics Bethany Nazarene Col Box 339 Bethany OK 73008

WALKER, KELSEY, JR, b Columbus, Tex, Nov 16, 25; m 45; c 5. THEORETICAL GAS DYNAMICS, APPLIED MATHEMATICS. Educ: Rensselaer Polytech Inst, BAE, 50; Mass Inst Technol, SM, 52. Prof Exp: Res scientist, Douglas Aircraft Co, 52-54; sr scientist, Lockheed Missile Systs Div, 54-55 & Aeronutronic Systs, Inc, 56-57; sect head syst anal, Space Tech Labs, 57-61; dept mgr systs eng, Aerospace Corp, 61-66; mgr sr staff, Los Angeles Opers, 66-72, asst prog mgr site defense, Ballistic Missile Defense Bus Area, 72-75, ASST MGR ADV DEFENSE SYSTS, BALLISTIC MISSILE DEFENSE BUS AREA, TRW SYSTS, REDONDO BEACH, 75- Mem: Am Inst Aeronaut & Astronaut; Sigma Xi. Res: Transonic gas dynamics; supersonic wing body interference; hypersonic gas dynamics; reentry body ablation theory; flight mechanics of reentry vehicles and ballistic missiles; systems analysis and design of ballistic missiles and space craft. Mailing Add: 5011 Casa Dr Tarzana CA 91356

WALKER, KENNETH MERRIAM, b Blaine, Ore, Mar 30, 21; m 41; c 4. BIOLOGY. Educ: Ore State Col, BS, 42, MS, 49, PhD(zool), 55. Prof Exp: From instr to asst prof biol, Univ Puget Sound, 51-57; from asst prof to assoc prof, 57-69, PROF BIOL, ORE COL EDUC, 69- Mem: Am Soc Mammalogists. Res: Vertebrate taxonomy and ecology. Mailing Add: Dept of Biol Ore Col of Educ Monmouth OR 97361

WALKER, KENNETH RUSSELL, b Spartanburg, SC, June 21, 37; div; c 3. STRATIGRAPHY, PALEOECOLOGY. Educ: Univ NC, Chapel Hill, BS, 59, MS, 64; Yale Univ, MPh, 67, PhD(paleoecol), 69. Prof Exp: Asst prof geol, 68-72, ASSOC PROF GEOL, UNIV TENN, KNOXVILLE, 72- Mem: Paleont Soc; Am Asn Petrol Geol; Soc Econ Paleont & Mineral; Geol Soc Am. Res: Ancient marine organic communities; lower paleozoic paleoenvironments; trophic relationships in organic communities; Ordovician problems; Holocene and ancient carbonate environments; invertebrate paleontology. Mailing Add: Dept of Geol Sci Univ of Tenn Knoxville TN 37916

WALKER, LAURENCE COLTON, b Washington, DC, Sept 8, 24; m 48; c 4. SILVICULTURE, SOILS. Educ: Pa State Univ, BS, 48; Yale Univ, MF, 49; State Univ NY, PhD(silvicult, soils), 53. Prof Exp: Forester, US Forest Serv, 58-51; asst, State Univ NY Col Forestry, Syracuse, 51-53; res forester, US Forest Serv, 53-54; prof silvicult res, Univ Ga, 54-63; PROF FORESTRY & DEAN SCH, STEPHEN F AUSTIN STATE UNIV, 63- Concurrent Pos: Consult, Nat Plant Food Inst. Mem: Fel AAAS; fel Soc Am Foresters; Soil Sci Soc Am. Res: Silvicides for hardwood control; soil-water relationships in forests; forest fertilization. Mailing Add: Sch of Forestry Stephen F Austin State Univ Nacogdoches TX 75961

WALKER, LAURENCE GRAVES, b Houston, Tex, Aug 28, 37; m 59; c 2. GEOLOGY. Educ: Univ Tex, BS, 60; Univ Calif, Los Angeles, MA, 62; Harvard Univ, PhD(geol), 67. Prof Exp: Geologist, Humble Oil & RegiRefining Co, 65-66; asst prof, 66-70, ASSOC PROF GEOL, MEMPHIS STATE UNIV, 70- Mem: Geol Soc Am. Res: Paleontology; paleoecology; stratigraphy. Mailing Add: Dept of Geol Memphis State Univ Memphis TN 38152

WALKER, LAURENCE RICHARD, physics, see 12th edition

WALKER, LEIGH E, organic chemistry, see 12th edition

WALKER, LEON BRYAN, JR, b Gulfport, Miss, June 9, 25; m 47; c 1. ANATOMY. Educ: Univ Houston, BS, 50, MS, 52; Duke Univ, PhD(anat), 55. Prof Exp: Asst biol, Univ Houston, 50-52; asst anat, Duke Univ, 53-55; instr, Sch Med, Temple Univ, 55-59; from asst prof to assoc prof, 59-71, PROF ANAT, SCH MED, TULANE UNIV, 71- Mem: Am Asn Anatomists. Res: Muscle innervation; morphology of neuromuscular spindles; stress-strain studies in tendon. Mailing Add: Dept of Anat Tulane Univ New Orleans LA 70112

WALKER, LEROY HAROLD, b Union, Utah, Sept 24, 33; m 63; c 2. MATHEMATICS, ELECTRICAL ENGINEERING. Educ: Univ Utah, BS, 55; Mass Inst Technol, SM, 57, EE, 58; Univ Calif, Los Angeles, PhD(math), 68. Prof Exp: Res assoc opers res ctr, Mass Inst Technol, 58-60; ASST PROF MATH, BRIGHAM YOUNG UNIV, 68- Concurrent Pos: Fac res fel, Brigham Young Univ, 69-70. Mem: Inst Math Statist; Math Asn Am; Am Math Soc; Asn Comput Mach. Res: Stopping rules for stochastic processes. Mailing Add: Dept of Math Brigham Young Univ Provo UT 84601

WALKER, LESTER EUGENE, physiology, zoology, see 12th edition

WALKER, LILLIE CUTLAR, b Wilmington, NC, June 9, 02; m 50. PREVENTIVE MEDICINE, PEDIATRICS. Educ: Univ NC, BS, 23, PhD(chem), 27; Univ Chicago, MD, 42. Prof Exp: Instr bact, Sch Med, Tufts Univ, 29-30 & Sch Med, Duke Univ, 30-31; instr pediat, Univ Pa, 44-45; assoc prof prev med, Univ Tenn, Memphis, 53-72; RETIRED. Mem: AMA; Am Med Women's Asn; Asn Teachers Prev Med; Am Pub Health Asn; Am Acad Cerebral Palsy. Mailing Add: Box 268 Little Switzerland NC 28749

WALKER, MARSHALL JOHN, b Bath, NY, Jan 23, 12; m 37; c 2. OPTICS. Educ: Cornell Univ, BChem, 33, AM, 36; Pa State Univ, PhD(physics), 50. Prof Exp: Asst physicist photo prods dept, E I du Pont de Nemours & Co, 36-39, asst chemist, 39-42; assoc physicst, Nat Bur Standards, 42-44; assoc physicist, Allegany Ballistics Lab, 44-45; asst, Pa State Univ, 46-49; from asst prof to assoc prof physics, 49-63, PROF PHYSICS, UNIV CONN, 63- Mem: Am Phys Soc; Optical Soc Am; Am Asn Physics Teachers. Res: Philosophy of science. Mailing Add: Star Route Chaplin CT 06235

WALKER, MATTHEW, b Water Proof, La, Dec 7, 06; m; c 4. SURGERY. Educ: La State Univ, AB, 29; Meharry Med Col, MD, 34. Prof Exp: Instr physiol & asst prof surg & gynec, 39-42, instr path, 41-43, assoc prof surg & gynec, 42-44, asst dean, 52, PROF SURG & GYNEC, MEHARRY MED COL, 44- Concurrent Pos: Gen Educ Bd fel, Howard Univ, 38-39; dir surg, Taborian Hosp, Miss; sr consult, Riverside Sanitarium & Hosp; mem med staff, Nashville Mem & Metrop Gen Hosps. Mem: Nat Med Asn (pres, 54); fel Am Col Surgeons; fel Int Col Surgeons. Res: Experimental peritonitis and penicillin; wound healing; streptomycin; massive intestinal resections; nutrition and survival; radioactive gold injection in inoperable cancer of gastrointestinal tract; anti-cancer drugs in treatment of cancer; prevention and treatment of metastases in mice and humans. Mailing Add: Dept of Surg Meharry Med Col Nashville TN 37208

WALKER, MERLE F, b Pasadena, Calif, Mar 3, 26; m 59; c 1. ASTRONOMY. Educ: Univ Calif, AB, 49, PhD(astron), 52. Prof Exp: Asst astron, Univ Calif, 49-52, jr res astronr, 55-56; Carnegie fel, Mt Wilson & Palomar Observ, 52-54; res assoc, Yerkes Observ, Univ Chicago, 54-55; instr, Warner & Swasey Observ, Case Inst Technol, 56-57; from asst astronr to assoc astronr, 57-71, PROF ASTRON & ASTRONOMER, LICK OBSERV, UNIV CALIF, SANTA CRUZ, 71- Concurrent Pos: Sr resident astronr, Cerro Tololo Interam Observ, 68-69. Mem: Int Astron Union; Am Astron Soc. Res: Photoelectric photometry of short period variable stars; photoelectric magnitudes and colors of stars; stellar spectra and radial velocities; electronic image intensification; astronomical seeing and observatory sites. Mailing Add: Lick Observ Univ of Calif Santa Cruz CA 95060

WALKER, MICHAEL BARRY, b Regina, Sask. PHYSICS. Educ: McGill Univ, BEng, 61; Oxford Univ, PhD, 65. Prof Exp: Fel, 66-68, ASSOC PROF PHYSICS, UNIV TORONTO, 68- Mem: Can Asn Physicists. Res: Solid state physics, especially magnetism. Mailing Add: Dept of Physics Univ of Toronto Toronto ON Can

WALKER, MICHAEL DIRCK, b New York, NY, Jan 24, 31; m 53; c 3. NEUROSURGERY. Educ: Yale Univ, BA, 56; Boston Univ, MD, 60. Prof Exp: Intern surg, Mass Mem Hosps, 60-61; resident neurosurg, Boston City Hosp & Lahey Clin, 61-65; sr investr pharmacol, Nat Cancer Inst, 65-67; CHIEF SECT NEUROSURG, BALTIMORE CANCER RES CTR, 67-, CHIEF CTR, 71- Concurrent Pos: Chmn, Brain Tumor Study Group, 67; assoc dir, Div Cancer Treatment, Nat Cancer Inst, 73-; assoc prof neurosurg, Sch Med, Univ Md, 73-; asst prof neurol surg, Sch Med, Johns Hopkins Univ, 74- Mem: Am Asn Cancer Res; Am Acad Neruol; Am Soc Clin Oncol; Cong Neurol Surg; NY Acad Sci. Res: Neurological surgery; analysis and treatment of human brain tumors with cytotoxic agents able to penetrate the blood-brain barrier. Mailing Add: Baltimore Cancer Res Ctr 22 S Greene St Baltimore MD 21201

WALKER, MICHAEL SIDNEY, b Hull, Eng, Sept 13, 40; m 66; c 2. PHOTOPHYSICS. Educ: Univ Sheffield, BS, 62, PhD(chem), 65. Prof Exp: AEC fel dept chem, Univ Minn, Minneapolis, 65-67; scientist res labs, 67-71, MGR LABS, XEROX CORP, 71- Res: Photophysics of organic molecules including semiconductors and photoconductors; materials characterization. Mailing Add: 159 New Wickham Dr Penfield NY 14526

WALKER, MICHAEL STEPHEN, b Detroit, Mich, Dec 16, 39; m 70; c 2. ENGINEERING PHYSICS, EXPERIMENTAL SOLID STATE PHYSICS. Educ: Mass Inst Technol, BS, 61; Carnegie Inst Technol, MS, 64; Carnegie Mellon Univ, PhD(physics), 71. Prof Exp: Sr engr, Westinghouse Elec Corp, 61-75; MGR MAT & DEVICE DEVELOP, INTERMAGNETICS GEN CORP, 76- Mem: Am Phys Soc; Inst Elec & Electronics Engrs; Sigma Xi. Res: Research and development on superconducting materials and devices, including the development of niobium-tin multifilament conductors and superconducting magnets for machinery, energy storage and for plasma confinement for fusion devices. Mailing Add: Intermagnetics Gen Corp PO Box 566 Guilderland NY 12084

WALKER, NATHANIEL, b Cincinnati, Ohio, Apr 9, 09; m 34; c 2. FOREST MANAGEMENT, FOREST ECONOMICS. Educ: Colo Col, BS, 33; Pa State Univ, MS, 55; NC State Univ, PhD, 70. Prof Exp: Dist forest ranger, US Forest Serv, 33-44; asst exten forester, Exten Serv, 44-47, from assoc prof to prof forestry, 47-74, EMER PROF FORESTRY, OKLA STATE UNIV, 74- Mem: Soc Am Foresters. Res: Frequency analyses and measurements; development and yield of cottonwood and red cedar; stand structures and valuation in several species of forest trees; forest soil-site evaluations. Mailing Add: 1906 W Admiral Stillwater OK 74074

WALKER, NEIL ALLAN, b Flint, Mich, May 4, 24; m 53; c 2. ACAROLOGY. Educ: Southern Methodist Univ, BS, 47; Univ Mich, MA, 48; Univ Calif, Berkeley, PhD(entom), 64. Prof Exp: Fisheries res technician inst fisheries res, Mich Dept Conserv, 49-50; from asst prof to assoc prof zool, 58-67, chmn dept biol sci & agr, 70-73, PROF ZOOL, FT HAYS KANS STATE COL, 67- Concurrent Pos: Consult, Dept Sci & Indust Res, NZ, 65 & 72. Mem: Entom Soc Am; Soc Syst Zool; Acarol Soc Am. Res: Taxonomy; biogeography; ecology of ptyctimous Oribatei; biology of Opiliones. Mailing Add: Dept of Biol Sci Ft Hays Kans State Col Hays KS 67601

WALKER, PHILIP CALEB, b Pittsburgh, Pa, Nov 26, 11; m 40; c 4. BIOLOGY, PALYNOLOGY. Educ: Univ Pittsburgh, BS, 34, PhD, 58. Prof Exp: Asst bot & biol, Univ Pittsburgh, 34-40; park naturalist, Bur Parks, Pittsburgh, Pa, 40-43; asst prof biol, WVa Inst Technol, 46-51; PROF BIOL, STATE UNIV NY COL PLATTSBURGH, 51- Concurrent Pos: Instr bot, Geneva Col, 37-38; panelist instrnl sci equip prog, NSF. Mem: AAAS. Res: Forest ecology and sequence; palynology and forest sequence studies in pleistocene and post-Wisconsin glacial bogs in New York, Pennsylvania and New Jersey; forest sequence of Hartstown bog area in Pennsylvania; palynology of Adirondack bogs and anemophilous and zoophilous pollen transport in the atmosphere. Mailing Add: Dept of Biol State Univ of NY Col Plattsburgh NY 12901

WALKER, RAYMOND JOHN, b Los Angeles, Calif, Oct 26, 42. SPACE PHYSICS. Educ: San Diego State Univ, BA, 64; Univ Calif, Los Angeles, MS, 69, PhD(planetary & space physics), 73. Prof Exp: RES ASSOC PHYSICS, UNIV MINN, MINNEAPOLIS, 73- Mem: Am Geophys Union. Res: Magnetospheric physics; the dynamics of charged particles in the magnetosphere; numerical studies of magnetospheric convection; organization and analysis of multi-parameter satellite data sets. Mailing Add: Sch of Physics & Astron Univ of Minn Minneapolis MN 55455

WALKER, RAYMOND LLOYD, b Atwood, Tenn, July 16, 27; m 53; c 4. ANALYTICAL CHEMISTRY. Educ: Lambuth Col, BA, 50. Prof Exp: Anal chemist, 51-71, GROUP LEADER MASS SPECTROMETRY, NUCLEAR DIV, UNION CARBIDE CORP, 71- Mem: Am Chem Soc. Res: Isotopic analysis by surface ionization of actinides and isotope dilution techniques; burn up analysis and accurate minor isotope measurements for safeguards applications. Mailing Add: 106 Pacific Rd Oak Ridge TN 37830

WALKER, RICHARD ALDEN, b Chicago, Ill, Jan 15, 21; m 54; c 4. UNDERWATER ACOUSTICS, OCEAN ENGINEERING. Educ: Iowa State Col, PhD(physics), 52. Prof Exp: Res assoc, Iowa State Col, 50-52; MEM TECH STAFF, BELL TEL LABS, INC, 52- Mem: AAAS. Res: Underwater acoustics; seismology; communications engineering; ultrasonics; re-entry physics. Mailing Add: Bell Tel Labs Whippany NJ 07981

WALKER, RICHARD BATTSON, b Tennessee, Ill, Oct 24, 16; m; c 3. BOTANY. Educ: Univ Ill, BS, 38; Univ Calif, PhD(bot), 48. Prof Exp: From instr to assoc prof, 48-60, chmn dept, 62-71, PROF BOT, UNIV WASH, 60- Mem: AAAS; Ecol Soc Am; Bot Soc Am; Am Soc Plant Physiologists. Res: Mineral nutrition and water relations of conifers; comparative calcium-magnesium nutrition; iron nutrition. Mailing Add: Dept of Bot Univ of Wash Seattle WA 98195

WALKER, RICHARD FRANCIS, b Amboy, NJ, Dec 14, 39; m 70. ENDOCRINOLOGY, PHYSIOLOGY. Educ: Rutgers Univ, Newark, BS, 61; NMex State Univ, MS, 69; Rutgers Univ, New Brunswick, PhD(zool), 71. Prof Exp: ASST PROF ZOOL, CLEMSON UNIV, 71- Mem: AAAS; Am Soc Zool. Res: Endocrine regulatory mechanisms; differential effects of light on pineal function. Mailing Add: Dept of Zool Clemson Univ Clemson SC 29631

WALKER, RICHARD ISLEY, b Winston-Salem, NC, Aug 9, 29; m 56; c 2. MEDICINE. Educ: Univ NC, BS, 50; Harvard Univ, MD, 54. Prof Exp: From instr to assoc prof, 62-70, PROF MED, SCH MED, UNIV NC, CHAPEL HILL, 70- Concurrent Pos: NIH res fel, 59-62; Leukemia Soc scholar, 62-66; NIH res career develop award, 66- Mem: AAAS; Am Fedn Clin Res; Am Soc Hemat. Res: Physiology of leucocyte functions. Mailing Add: Dept of Med S013 Clin Bldg Univ of NC Sch of Med Chapel Hill NC 27514

WALKER, RICHARD V, b Pueblo, Colo, Mar 8, 18; m 45; c 3. MEDICAL BACTERIOLOGY, IMMUNOLOGY. Educ: Univ Calif, Berkeley, BS, 49, MPH, 52, PhD(bact), 60. Prof Exp: Assoc & instr, Pub Health Lab, Sch Pub Health, Univ Calif, Berkeley, 49-57; from grad res immunologist to asst res immunologist, George Williams Hooper Found, Med Ctr, Univ Calif, San Francisco, 57-65; asst res immunologist, Nat Ctr Primate Biol, Univ Calif, Davis, 65-67; ASSOC PROF ZOOL, OHIO UNIV, 67- Mem: Am Soc Microbiol. Res: Bacterial toxins; immunochemistry; fluorescent antibody; Salmonella-Shigella diagnosis. Mailing Add: Dept of Zool & Microbiol Ohio Univ Athens OH 45701

WALKER, ROBERT EDGAR, animal ecology, animal behavior, see 12th edition

WALKER, ROBERT HUGH, b O'Donnell, Tex, Sept 8, 35; m 55; c 4. PHYSICS. Educ: Tex Christian Univ, BS, 57, MS, 59; Mass Inst Technol, PhD(physics), 62. Prof Exp: Asst prof physics, 64-67, assoc dean col arts & sci, 73-74, ASSOC PROF PHYSICS, UNIV HOUSTON, 67-, DEAN COL NATURAL SCI & MATH, 74- Concurrent Pos: Lectr, Baylor Col Med, 66-; consult, Int Inst Educ, 67-69. Mem: Am Phys Soc; Am Asn Physics Teachers. Res: Theoretical physics; solid state physics; atomic physics. Mailing Add: Col of Natural Sci & Math Univ of Houston Houston TX 77004

WALKER, ROBERT JOHN, b Pittsburgh, Pa, May 5, 09. MATHEMATICS. Educ: Carnegie Inst Technol, BS, 30; Princeton Univ, PhD(math), 34. Prof Exp: Instr math, Princeton Univ, 34-35; instr, 35-38, from asst prof to assoc prof, 38-48, prof, 48-74, chmn dept, 50-61, EMER PROF MATH, CORNELL UNIV, 74- Concurrent Pos: Lectr, Princeton Univ, 40-41; mathematician, Aberdeen Proving Ground, Md, 43-45; vis prof, Fla State Univ. Mem: AAAS; Am Math Soc; Math Asn Am. Res: Singularities of algebraic manifolds; magic squares; artillery rockets; numerical and combinatorial analysis. Mailing Add: 201 Adeline Ave Pittsburgh PA 15228

WALKER, ROBERT LEE, b St Louis, Mo, June 29, 19; m 46. PHYSICS. Educ: Univ Chicago, BS, 41; Cornell Univ, PhD(exp physics), 48. Prof Exp: Asst metall lab, Univ Chicago, 42-43; scientist, Los Alamos Sci Lab, 43-46; res assoc, Cornell Univ, 48-49; from asst prof to assoc prof, 49-59, PROF PHYSICS, CALIF INST TECHNOL, 59- Mem: Am Phys Soc. Res: Photoproduction experiments and analyses; interaction of gamma rays with matter; high energy physics. Mailing Add: Lauritsen Lab 356-48 Calif Inst of Technol Pasadena CA 91109

WALKER, ROBERT MOWBRAY, b Philadelphia, Pa, Feb 6, 29; m 51; c 2. SPACE PHYSICS. Educ: Union Univ, NY, BS, 50; Yale Univ, MS, 51, PhD(physics), 54. Hon Degrees: DSc, Union Univ, NY, 67. Prof Exp: Res assoc, Gen Elec Co, 54-66; dir lab space physics, 66-75, McDONNELL PROF PHYSICS, WASH UNIV, 66-, DIR, McDONNELL CTR SPACE SCI, 75- Concurrent Pos: NSF sr fel, 62; vis prof, Univ Paris, 62-63; adj prof, Rensselaer Polytech Inst, 65-66; mem, Lunar Sample Anal Planning Team, 68-70; mem, Lunar Sample Rev Bd, 70-72; vis prof, Calif Inst Technol, 72; mem bd dirs, Vols Tech Assistance; mem bd dirs, Univs Space Res Asn, 69-71; mem, Lunar Sci Inst Adv Comt, 72-, mem bd sci & technol for int develop, Nat Acad Sci, 74-77. Honors & Awards: Am Nuclear Soc Award, 64; Yale Eng Asn Award, 66; NASA Medal Except Sci Achievement, 70; E O Lawrence Award, AEC, 71. Mem: Fel AAAS; fel Am Phys Soc; fel Meteoritical Soc; Am Geophys Union; Am Astron Soc. Res: Radiation effects in solids; development of dielectric nuclear track detectors and their application to nuclear science; geochronology; space science; cosmic rays; meteorites; astrophysics; planetary surfaces; archeometry. Mailing Add: Dept of Physics Wash Univ St Louis MO 63130

WALKER, ROBERT PAUL, b Washington, DC, Mar 15, 43; m 65; c 1. MATHEMATICS. Educ: Univ Md, BS, 65; Mass Inst Technol, PhD(math), 68. Prof Exp: Asst prof math, Univ NC, Chapel Hill, 68-75; MEM FAC, TALLADEGA COL, 75- Mem: Am Math Soc; Math Asn Am; Nat Asn Mathematicians. Res: Algebraic topology; mathematics education. Mailing Add: Dept of Math Talladega Col Talladega AL 35160

WALKER, ROBERT W, b Arlington, Mass, Mar 15, 33; m 59; c 1. MICROBIOLOGY. Educ: Univ Mass, BS, 55, MS, 59; Mich State Univ, PhD(microbiol), 63. Prof Exp: Hatch fel, 63-64; instr microbiol, 64-65, asst prof, 65-74, ASSOC PROF MICROBIOL, UNIV MASS, AMHERST, 74- Res: Structure and function of microbial lipids; biosynthesis of mycolic acids and other mycobacterial lipids. Mailing Add: Dept of Environ Sci Marshall Hall Univ of Mass Amherst MA 01002

WALKER, ROBERT WINN, b Montgomery, Ala, Jan 5, 25; m 49; c 3. PHYSICAL CHEMISTRY. Educ: Auburn Univ, BS, 48; Mass Inst Technol, PhD(phys chem), 52. Prof Exp: Res chemist, Redstone Labs, 52-53, group leader propellant res, 53-59, sect head phys & polymer chem, 59-65, lab head ion exchange appln res div, 65-73, MGR ION EXCHANGE RES DEPT, ROHM AND HAAS CO, 73- Mem: Am Chem Soc. Res: Molecular structure; chemical thermodynamics; rocket propulsion; ion exchange resins; adsorbents; flocculants. Mailing Add: 6008 Cannon Hill Rd Ft Washington PA 19034

WALKER, ROGER GEOFFREY, b London, Eng, Mar 26, 39; m 65; c 2. SEDIMENTOLOGY. Educ: Oxford Univ, BA, 61, DPhil(geol), 64. Prof Exp: NATO fel geol, Johns Hopkins Univ, 64-66; from asst prof to assoc prof, 66-73, PROF GEOL, McMASTER UNIV, 73- Honors & Awards: Past Pres' Medal, Geol Asn Can, 75. Mem: Geol Soc Am; Soc Econ Paleont & Mineral; Int Asn Sedimentol; Am Asn Petrol Geologists; Geol Asn Can. Res: Sedimentary facies analysis; sedimentology of turbidites; quantitative basin analysis; sedimentology of Archaean greenstone belts. Mailing Add: Dept of Geol McMaster Univ Hamilton ON Can

WALKER, ROLAND, b Stellenbosch, SAfrica, Feb 8, 07; US citizen; m 42; c 2. FISH PATHOLOGY. Educ: Oberlin Col, AB, 28, AM, 29; Yale Univ, PhD(zool), 34. Prof Exp: Asst biol, Yale Univ, 29-31; instr physiol, Oberlin Col, 31-32; instr biol, 34-42, from asst prof to assoc prof, 42-54, prof, 54-72, EMER PROF BIOL, RENSSELAER POLYTECH INST, 72- Mem: AAAS; Am Soc Zool; Wildlife Dis Asn. Res: Neurology of fish and crustacea; fish parasitology; ultrastructure of fish blood and tumors cells with virus. Mailing Add: Dept of Biol Rensselaer Polytech Inst Troy NY 12181

WALKER, RONALD ELLIOT, b Avon, SDak, Mar 14, 29; m 53; c 2. PHYSICS, PHYSICAL CHEMISTRY. Educ: SDak Sch Mines & Technol, BS, 50; Univ Md, MS, 55, PhD(physics), 58. Prof Exp: Assoc physicist, 51-55, sr physicist, 55-60, PRIN STAFF PHYSICIST, APPL PHYSICS LAB, JOHNS HOPKINS UNIV, 60- Concurrent Pos: Instr eve col, Johns Hopkins Univ, 66- Mem: Am Phys Soc. Res: Atomic and molecular physics; laser physics; bioengineering. Mailing Add: Physics Lab Johns Hopkins Univ 8621 Georgia Ave Silver Spring MD 20910

WALKER, RUDGER HARPER, b Rexburg, Idaho, Aug 20, 02; m 23; c 5. SOIL SCIENCE. Educ: Brigham Young Univ, BS, 23; Iowa State Univ, MS, 25, PhD(soils), 27. Prof Exp: Teacher high sch, Utah, 23-24; asst, Iowa State Univ, 24-27, asst chief soil bact exp sta, 28-31, res assoc prof soils, 31-36; asst prof agron, Colo State Univ & asst agronomist exp sta, 27-28; conservationist intermountain forest & range exp sta, US Forest Serv, 36-38; dean sch agr & dir exp sta, Utah State Univ, 38-55, dean col agr, 55-60; prof agron, 60-72, dean col biol & agr sci, 60-68, chmn dept agron & hort, 68-72, EMER PROF AGRON & EMER DEAN COL BIOL & AGR SCI, BRIGHAM YOUNG UNIV, 72- Concurrent Pos: Dir US salinity lab, Agr Res Serv, USDA, 40, asst dir, For Agr Serv, 52-53, mem nat sugar res comt, 54-59; chmn, Food & Agr Orgn, UN Mission, Thailand, 48, mem US nat comn, UNESCO, 53-57; mem comt instnl projs abroad, Am Coun Educ, 54-57; trustee, Am Univ Beirut, 55-; supvr, Utah State Univ & Int Coop Admin Contracts, Iran, 58-60. Mem: Fel AAAS; fel Am Soc Agron; Soil Sci Soc Am; Am Soc Hort Sci. Res: Soil microbiology; range reseeding and conservation. Mailing Add: 1457 N Cherry Lane Provo UT 84601

WALKER, RUFUS FLOYD, JR, b Paris, Ark, Sept 13, 35; m 60; c 1. BIOMETRICS, BIOPHYSICS. Educ: Harvard Univ, AB, 57, PhD(physics), 65. Prof Exp: Physicist, SPERT Proj, Phillips Petrol Co, 58 & 59; asst prof physics, Tufts Univ, 63-68; assoc prof, Centenary Col La, 68-72, dir comput ctr & chmn dept physics, 71-72; prof biophys & head comput sci prog, 72-75, clin assoc prof, 70-72, PROF BIOMET & HEAD DEPT, MED CTR, SCH MED, LA STATE UNIV, SHREVEPORT, 75- Concurrent Pos: Consult nuclear med, Vet Admin Hosp, Shreveport, 71- Mem: Am Asn Physics Teachers; Asn Comput Mach; Am Phys Soc; Am Statist Asn; Health Physics Soc. Res: Biometrics; computer application in medical education and research; physics in medicine; electronic instrumentation. Mailing Add: 1414 Captain Shreve Dr Shreveport LA 71104

WALKER, RUSSELL GLENN, astronomy, see 12th edition

WALKER, RUSSELL WAGNER, b Fredericktown, Ohio, Nov 14, 24; m 46; c 3. ORGANIC CHEMISTRY, PHYSICAL CHEMISTRY. Educ: Ohio Wesleyan Univ, BS, 47; Ohio State Univ, PhD(org chem), 52. Prof Exp: Res assoc, Am Petrol Inst, 48-52; res chemist, Sinclair Res, Inc, 53-57, group leader, 57-60, div dir, 60-66, res dir, Sinclair Petrochem, Inc, 66-67, tech mgr, Sinclair Res, Inc, 67-68, vpres & dir assoc opers, Sinclair Petrochem, 68-69, mgr res & develop, Sinclair-Koppers Co, 69-74; DIR RES & DEVELOP, ARCO POLYMERS INC, 74- Mem: AAAS; Am Chem Soc; Indust Res Inst. Res: Relation of hydrocarbon structure to combustion characteristics; biodegradation and environmental pollution. Mailing Add: ARCO Polymers Inc 1500 Market St Philadelphia PA 19101

WALKER, RUTH ANGELINA, b New York, NY, July 11, 20. ORGANIC CHEMISTRY. Educ: Vassar Col, BA, 42; Yale Univ, PhD(org chem), 45. Prof Exp: Asst chem, Chas Pfizer & Co, NY, 45; chemist col med, NY Univ, 45-50; sr res chemist, Celanese Corp Am, 50-56; sr res chemist, Johnson & Johnson, 57; instr chem, Hunter Col, Bronx, 57-60, from asst prof to assoc prof, 61-68, assoc prof, Lehman Col, 69-71, PROF CHEM, LEHMAN COL, CITY UNIV NEW YORK, 71- Concurrent Pos: Sigma Delta Epsilon grant-in-aid metal complexes hydroxanthraquinones, Hunter Col, 63, Sigma Xi grant-in-aid res, 65, George N Shuster fel grant, 65, City Univ New York res grant, 65; chmn, Lehman Col Comt Curric, 68- Honors & Awards: Award, Am Asn Textile Chem & Colorists, 60. Mem: AAAS; Sigma Xi; Am Chem Soc; fel NY Acad Sci; fel Am Inst Chem. Res: Synthesis of medicinal products; dyestuff synthesis; organometallic complexes of 1,4-dihydroxyanthraquinone; methods of teaching. Mailing Add: Lehman Col City Univ of New York Park Blvd W Bronx NY 10468

WALKER, SHEPPARD MATTHEW, b Perkinston, Miss, Feb 2, 09; m 32; c 1. PHYSIOLOGY, BIOPHYSICS. Educ: Western Ky Univ, BS, 32, AM, 33; La State Univ, PhD(physiol), 41. Prof Exp: Instr biol, Perkinston Jr Col, 33-38; asst prof sci, Delta State Teachers Col, Miss, 41-42; from instr to asst prof physiol, Sch Med, Wash Univ, 42-49; assoc prof, 49-62, PROF PHYSIOL, SCH MED, UNIV LOUISVILLE, 62- Concurrent Pos: Mem, Spec Rev Muscle Contraction, 60 & 67; actg chmn dept physiol & biophys, Sch Med, Univ Louisville, 65-67. Mem: Soc Exp Biol & Med; Am Physiol Soc; Biophys Soc. Res: Muscle structure and function; development of fine structures in muscle fibers; neurophysiology. Mailing Add: Health Sci Ctr Dept of Physiol Univ of Louisville Sch of Med Louisville KY 40201

WALKER, STROTHER HOLLAND, b Denver, Colo, Jan 19, 14; m 40; c 3. MATHEMATICS, STATISTICS. Educ: Harvard Col, SB, 34; Georgetown Univ, MA, 62; Johns Hopkins Univ, PhD(biostatist), 65. Prof Exp: Res group chmn, Opers Res Off, Johns Hopkins Univ, 54-61, res group chmn, Res Anal Corp, 61-66; div head biostatist, 66-74, PROF BIOMET & CHMN DEPT, UNIV COLO MED CTR, DENVER, 74- Concurrent Pos: NIH spec fel biostatist sch hyg & pub health, Johns Hopkins Univ, 62-65. Mem: Am Math Soc; Math Asn Am; Am Statist Asn; Opers Res Soc Am. Res: Development of mathematical and statistical models of biomedical processes and of methods of estimating risk in cardiovascular disease. Mailing Add: Dept Biomet Univ Colo Med Ctr 4200 E Ninth Ave Denver CO 80220

WALKER, TERRY M, b Chicago, Ill, Dec 15, 38; m 60; c 2. OPERATIONS RESEARCH, COMPUTER SCIENCE. Educ: Fla State Univ, BS, 61; Univ Ala, PhD(statist), 66. Prof Exp: Asst prof comput sci & economet, Ga State Col, 65-67; assoc prof quant mgt sci, Univ Houston, 67-75; PROF COMPUT SCI & HEAD DEPT, UNIV SOUTHWESTERN LA, 74- Concurrent Pos: lectr & Ford Found consult, Atlanta, 66-67. Mem: Asn Comput Mach; Am Statist Asn. Res: Simulation of industrial processes; distribution sampling of statistical populations. Mailing Add: Dept of Comput Sci USL Box 1210 Univ of Southwestern La Lafayette LA 70501

WALKER, THEODORE ROSCOE, b Madison, Wis, Feb 8, 21; m 49; c 4. GEOLOGY. Educ: Univ Wis, PhB, 47, PhD, 52. Prof Exp: Asst geologist, State Geol Surv, Ill, 52-53; from asst prof to assoc prof geol, 53-65, fac res lectr, 72-73, PROF GEOL, UNIV COLO, BOULDER, 65-, CHMN DEPT GEOL SCI, 72- Concurrent Pos: NSF sr fel, 62-63, grants, 65-; Am Asn Petrol Geol distinguished lectr, 65. Mem: AAAS; Geol Soc Am; Soc Econ Paleont & Mineral; Am Asn Petrol Geol. Res: Sedimentation; sedimentary petrology. Mailing Add: Dept of Geol Sci Univ of Colo Boulder CO 80302

WALKER, THOMAS CARL, b Teaneck, NJ, Aug 13, 44; m 70; c 1. BIOMATERIALS, INSTRUMENTATION. Educ: Fairleigh Dickinson Univ, BS, 70; Univ Va, MS, 72, PhD(mat sci), 75. Prof Exp: Res technician seismol, Lamont-Doherty Geol Observ, Columbia Univ, 66-70; RES ASSOC PHYSICS, UNIV VA, 75- Mem: Am Asn Physics Teachers. Res: Improvements of the ultra sensitive magnetic susceptometer; using this susceptometer to conduct measurements on heme proteins. Mailing Add: 814 Cabell Ave Charlottesville VA 22903

WALKER, THOMAS JEFFERSON, b Dyer Co, Tenn, July 24, 31; m 59; c 2. ENTOMOLOGY. Educ: Univ Tenn, BA, 53; Ohio State Univ, MSc, 54, PhD(entom), 57. Prof Exp: From asst prof to assoc prof biol sci & entom, 57-68, PROF BIOL SCI & ENTOM, UNIV FLA, 68- Concurrent Pos: Res assoc dept tropic res, NY Zool Soc, 66; res assoc, Fla State Collection Arthropods, 63-; ed, Fla Entomologist, 64-66. Mem: Fel AAAS; Entom Soc Am; Soc Study Evolution; Soc Syst Zool. Res: Acoustical behavior of insects; systematics, behavior, ecology and evolution of Gryllidae and Tettigoniidae. Mailing Add: Dept of Entom & Nematol Univ of Fla Gainesville FL 32601

WALKER, VICTOR CHARLES ROWAN, veterinary medicine, see 12th edition

WALKER, WALDO SYLVESTER, b Fayette, Iowa, June 12, 31; m 52; c 2. BOTANY. Educ: Upper Iowa Univ, BS, 53; Univ Iowa, MS, 57, PhD(bot), 59. Prof Exp: From asst prof to assoc prof biol, 58-68, assoc dean, 63-65, dean admin, 69-73, PROF BIOL, GRINNEL COL, 68-, DEAN COL, 73- Mem: AAAS; Bot Soc Am. Res: Ultrastructure of plants; experimental morphology; plant physiology. Mailing Add: Dept of Biol Grinnell Col Grinnell IA 50112

WALKER, WARREN ELLIOTT, b New York, NY, Apr 7, 42; m 70; c 2. OPERATIONS RESEARCH, URBAN RESEARCH AND DEVELOPMENT. Educ: Cornell Univ, BA, 63, MS, 64, PhD(opers res), 68. Prof Exp: Pres, Compuvisor, Inc, 68-70; sr opers res analyst & proj dir, NY City-Rand Inst, Rand Corp, 70-75; ASST V PRES, CHEM BANK, 75- Concurrent Pos: Consult, US Environ Protection Agency Off Solid Waste Mgt Progs, 68-72; adj prof opers res, Columbia Univ, 71-; pres, Urbatronics, Inc, 75- Honors & Awards: Lanchester Prize, Opers Res Soc Am, 74. Mem: Inst Mgt Sci; Opers Res Soc Am. Res: Development and use of new tools for public policy analysis and commercial bank planning. Mailing Add: 713 Salem St Teaneck NJ 07666

WALKER, WARREN FRANKLIN, JR, b Malden, Mass, Sept 27, 18; m 44; c 4. ZOOLOGY. Educ: Harvard Univ, SB, 41, PhD(zool), 46. Prof Exp: Instr anat sch med, Boston Univ, 45-47; instr zool, 47-48, from asst prof to assoc prof, 49-57, actg provost, 74-75, chmn dept biol, 67-74, PROF ZOOL, OBERLIN COL, 57- Mem: AAAS; Am Asn Anat; Am Soc Zool; Soc Syst Zool; Am Soc Ichthyol & Herpet. Res: Herpetology of South America; vertebrate anatomy and evolution; myology; vertebrate locomotion. Mailing Add: Dept of Biol Oberlin Col Oberlin OH 44074

WALKER, WELLINGTON EPLER, b Owenton, Ky, July 20, 31; m 64; c 2. INDUSTRIAL CHEMISTRY, ORGANIC CHEMISTRY. Educ: Univ Ky, BS, 53; Marshall Univ, MS, 58. Prof Exp: Chemist, 53-65, PROF SCIENTIST, TECH CTR, UNION CARBIDE CORP, 65- Res: Homogeneous catalysis via organometallic complexes of transition metals, especially group VIII complexes. Mailing Add: Union Carbide Corp Tech Ctr Res & Develop Box 8361 South Charleston WV 25303

WALKER, WILBUR GORDON, b Lena, La, Sept 18, 26; m 47; c 4. INTERNAL MEDICINE. Educ: Tulane Univ, MD, 51. Prof Exp: Intern med, Osler Med Serv, Johns Hopkins Hosp, 51-52; resident, Tulane Serv, Charity Hosp, La, 52-53; asst resident, Osler Med Serv, Hosp, 53-54, Am Heart Asn fel, 54-56, resident physician, 56-57, from asst prof to assoc prof, Univ, 58-68, PROF MED, SCH MED, JOHNS HOPKINS UNIV, 68-, PHYSICIAN, JOHNS HOPKINS HOSP, 57-, DIR CLIN RES CTR, 60- Concurrent Pos: Estab investr, Am Heart Asn, 57-; consult, Vet Admin Hosp, Loch Raven; dir renal div, Johns Hopkins & Good Samaritan Hosps; physician-in-chg, Dept Res Med, Good Samaritan Hosp. Mem: Am Physiol Soc; Am Soc Nephrology; Am Soc Clin Invest; Am Fedn Clin Res. Res: Renal function and renal diseases; electrolyte metabolism; renal ion exchange mechanisms; metabolic investigations in renal disease; energy metabolism and protein synthesis in isolated glomeruli; capillary and glomerular permeability to protein; renin-angiotensin-aldosterone physiology. Mailing Add: Johns Hopkins Hosp Baltimore MD 21205

WALKER, WILLARD BREWER, b Boston, Mass, July 29, 26; m 52; c 2. ETHNOLOGY, APPLIED ANTHROPOLOGY. Educ: Harvard Univ, AB, 50; Cornell Univ, PhD(gen ling), 64. Prof Exp: Summer instr, Cornell Univ, 61; res assoc Cherokee lit & lang, Univ Chicago, 64-66; asst prof anthrop, 66-70, ASSOC PROF ANTHROP, WESLEYAN UNIV, 70- Mem: Fel Am Anthrop Asn; Soc Appl Anthrop; Ling Soc Am; Am Ethnol Soc. Res: American Indian cultures and languages; application of linguistic analysis to current social problems; language as a key to world-view; native American Indian writing systems; Indian education; action anthropology. Mailing Add: Dept of Anthrop Wesleyan Univ Middletown CT 06457

WALKER, WILLIAM CHARLES, b Santa Barbara, Calif, Aug 22, 28; m 51; c 3. SOLID STATE PHYSICS. Educ: Univ Calif, AB, 50; Univ Southern Calif, MS, 53, PhD(physics), 55. Prof Exp: Asst physics, Univ Southern Calif, 50-52, res assoc, 52-55; instr, 55-57, from asst prof to assoc prof, 57-68, chmn dept, 72-75, PROF PHYSICS, UNIV CALIF, SANTA BARBARA, 68- Concurrent Pos: Consult, Servomechanisms, Inc, 59-60; Nat Acad Sci-NASA fel, Goddard Space Flight Ctr, 63-64; consult, Sloan Technol, 68-74. Mem: AAAS; Am Phys Soc; Optical Soc Am. Res: Solid state and ultraviolet spectroscopy; optical properties of solids. Mailing Add: Dept of Physics Univ of Calif Santa Barbara CA 93106

WALKER, WILLIAM COMSTOCK, b Milwaukee, Wis, July 6, 21; m 45; c 2. PAPER CHEMISTRY, PHYSICAL CHEMISTRY. Educ: Lehigh Univ, BS, 43, MS, 44, PhD(phys chem), 46. Prof Exp: Inst res fel, Lehigh Univ, 46-47; dir, Nat Printing Ink Res Inst, 47-55, res asst prof chem, 53-55; res dir, 55-65, TECH ASST TO CORP VPRES RES, WESTVACO CORP, NEW YORK, NY, 64- Concurrent Pos: Instr, Muhlenberg Col, 46-47. Mem: Am Chem Soc; Tech Asn Graphic Arts; Tech Asn Pulp & Paper Indust; Can Pulp & Paper Asn. Res: Printability of paper; electrostatic copying; paper production technology; printing inks; adsorption of gases on solids; removal of sulfur oxides from flue gases.

WALKER, WILLIAM DELANY, b Dallas, Tex, Nov 23, 23; m 46; c 3. NUCLEAR PHYSICS. Educ: Rice Inst, BA, 44; Cornell Univ, PhD(cosmic ray physics), 49. Prof Exp: Physicist, US Naval Res Lab, 44-45; asst prof physics, Rice Inst, 49-51; lectr, Univ Calif, 51-52; asst prof, Univ Rochester, 52-54; asst prof, Univ Wis, Madison, 54-57, from assoc prof to prof, 57-67, Max Mason prof, 67-71, chmn dept, 64-66; PROF PHYSICS, DUKE UNIV, 71-, CHMN DEP, 75- Concurrent Pos: Mem physics panel, NSF, 64-67; mem high energy surv comt, Nat Acad Sci, 64-65; chmn, Argonne User's Group, 64-66; mem user's exec comt, Nat Accelerator Lab, 72; chmn, Fermilab User's Exec Comt, 73-74. Mem: Fel Am Phys Soc. Res: Strong interaction physics; technology of bubble chambers; hadronic strong interactions. Mailing Add: Dept of Physics Duke Univ Durham NC 27706

WALKER, WILLIAM EVERETT, II, physics, see 12th edition

WALKER, WILLIAM HERBERT, reactor physics, see 12th edition

WALKER, WILLIAM HOWARD, b Chicago, Ill, Jan 28, 43; c 2. AQUATIC ECOLOGY. Educ: Elmhurst Col, BA, 65; La State Univ, MS, 67; Univ Pittsburgh, PhD(biol), 75. Prof Exp: From instr to asst prof, 68-74, ASSOC PROF BIOL, SETON HILL COL, 74-, CHMN DEPT, 73-; OWNER, AQUATIC FEEDING SPECIALISTS, 73- Concurrent Pos: Environ consult, Environ Systs Dept, Westinghouse Elec Corp, 70-; aquatic consult, Aquatic Ecol Assocs, Pittsburgh, 74-; aquatic feeding consult, QLM Labs, Oswego, NY, 75- Mem: Am Fisheries Soc; Am Soc Ichthyologists & Herpetologists; Ecol Soc Am; Freshwater Biol Asn. Res: Qualification and quantification of the feeding of fishes to include aquatic ecological and pollution associations. Mailing Add: Aquatic Feeding Specialists 403 N Maple Ave Greensburg PA 15601

WALKER, WILLIAM M, b Savannah, Tenn, Sept 17, 28; m 51; c 4. SOIL FERTILITY, BIOMETRICS. Educ: Florence State Col, BS, 50; Univ Tenn, MS, 57; Iowa State Univ, PhD(soil fertil), 61. Prof Exp: Res assoc agron, Iowa State Univ, 57-61; asst agronomist, Univ Tenn, 61-66; asst prof, 66-69, ASSOC PROF BIOMET, UNIV ILL, URBANA, 69- Mem: Am Soc Agron; Soil Sci Soc Am; Am Statist Asn. Res: Application of biomathematics to soil-plant relationships. Mailing Add: Dept of Agron 501C Turner Hall Univ of Ill Urbana IL 61801

WALKER, WILLIAM STANLEY, b Glendale, Calif, Apr 21, 40; m 64; c 2. IMMUNOLOGY, MICROBIOLOGY. Educ: Univ Southern Calif, AB, 63, PhD(microbiol), 68. Prof Exp: Lectr bact, Univ Southern Calif, 64-65; sci officer first class, Dept Immunohaemat, Acad Hosp, State Univ Leiden, 71; asst mem labs virol & immunol, 72-74, ASSOC MEM DIV IMMUNOL, ST JUDE CHILDREN'S RES HOSP, 75- Concurrent Pos: Fel immunol, Pub Health Res Inst City New York, Inc, 68-71. Mem: Reticuloendothelial Soc; Am Asn Immunol. Res: Polymorphonuclear leukocyte chemotaxis; immunobiology of macrophages. Mailing Add: St Jude Children's Res Hosp 332 N Lauderdale Memphis TN 38101

WALKER, WILLIAM WALDRUM, b Alexander City, Ala, Jan 16, 33; m 58; c 3. PHYSICS. Educ: Auburn Univ, BS, 55; Univ Va, MA, 57, PhD(physics), 59. Prof Exp: Asst prof physics, Col William & Mary, 59-60; physicist, Signal Res & Develop Lab, US Army, NJ, 60; from asst prof to assoc prof physics, 61-71, PROF PHYSICS, UNIV ALA, 71- Mem: Am Phys Soc; Am Asn Physics Teachers. Res: Positron lifetimes; low energy nuclear physics. Mailing Add: Dept of Physics Univ of Ala University AL 35486

WALKER-NASIR, EVELYNE, b Geneva, Switz, Mar 16, 36; m 74; c 1. CARBOHYDRATE CHEMISTRY. Educ: Col Geneva, Switz, BA, 55; Univ Geneva, BSc, 57, Fed Dipl, 61, PhD(pharmaceut sci), 66. Prof Exp: Asst pharmacist, Pharm Coop & Pharm Pervenches, Switz, 58-59; pharmacist, Pharm Pervenches, 61-62; res asst carbohydrate chem, Lab Carbohydrate Res, Mass Gen Hosp, Boston, 62-64; chemist pharmaceut & med chem, Labs, Vifor SA, Switz, 66-67, head sci dept, 67-70; res fel biol chem, Harvard Med Sch, 71-73; res fel biochem, 71-73, ASST BIOCHEM, MASS GEN HOSP, BOSTON, 73-; ASSOC BIOL CHEM, HARVARD MED SCH, 73- Res: Chemistry of carbohydrates with biological interest; synthesis of muramic acid derivatives, of glycopeptides and of oligosaccharide-lipid derivatives. Mailing Add: Lab for Carbohydrate Res Harvard Med Sch at Mass Gen Hosp Boston MA 02114

WALKIEWICZ, THOMAS ADAM, b Erie, Pa, Dec 25, 39; m 62; c 3. NUCLEAR PHYSICS, SOLID STATE PHYSICS. Educ: Xavier Univ, Ohio, BS, 62; Pa State Univ, PhD(physics), 69. Prof Exp: Asst prof physics, East Stroudsburg State Col, 69; physicist, Picatinny Arsenal, NJ, 69-70; assoc prof physics, 70-73, PROF PHYSICS, EDINBORO STATE COL, 73- Concurrent Pos: AEC collab researcher, Oak Ridge Nat Lab, 71-77, consult, 75-76. Mem: Am Phys Soc; Am Asn Physics Teachers. Res: Experimental low-energy nuclear physics, especially gamma ray spectroscopy; neutron and charged particle reactions; experimental solid state physics, especially electrooptical and radiation effects in amorphous semiconductors. Mailing Add: Dept of Physics Edinboro State Col Edinboro PA 16444

WALKINGTON, DAVID L, b Waukegan, Ill, July 20, 30; m 55; c 1. BOTANY. Educ: Ariz State Univ, BA, 57, MS, 59; Claremont Grad Sch, PhD(bot), 65. Prof Exp: Instr bot, Ariz State Univ, 59-60; res assoc, 60-63, lectr, 62-63, from asst prof to assoc prof biol, 63-72, actg assoc dean sch math, sci & eng, 73-74, actg dean, 74-75, assoc dean, 75-76, PROF BIOL, CALIF STATE UNIV, FULLERTON, 72- Concurrent Pos: NSF res grant, 65-66; pres bd trustees, Mus Asn NOrange County, 75- Mem: AAAS; Bot Soc Am; Am Inst Biol Sci. Res: Morphology and chemistry of pollen; chemotaxonomy of cacti; naturally occurring antibiotics in plants especially bryophytes and cacti. Mailing Add: Sch of Math Sci & Eng Calif State Univ Fullerton CA 92631

WALKINSHAW, CHARLES HOWARD, JR, b Blairsville, Pa, Nov 14, 35; m 57; c 3. PHYTOPATHOLOGY. Educ: Univ Fla, BSA, 57; Univ Wis, PhD(plant path), 60. Prof Exp: Asst plant path, Univ Wis, 57-60, proj assoc, 60-61, trainee biochem & path sch med, 61-63; plant pathologist, Southern Forest Exp Sta, US Forest Serv, Miss, 63-65; asst prof microbiol sch med, Univ Miss, 65-68; plant pathologist med support br, NASA-Manned Spacecraft Ctr, 68-73; PRIN PLANT PATHOLOGIST, SOUTHERN FOREST EXP STA, US FOREST SERV, 73- Honors & Awards: Super Serv Award, USDA, 72. Mem: AAAS; Am Phytopath Soc; Am Soc Cell Biol; Soc Am Foresters; Tissue Cult Asn. Res: Plant diseases; plant tissue culture; fungus diseases of pines; biochemistry of plant diseases. Mailing Add: Forest & Wood Prod Dis Lab Box 2008 Evergreen Sta Gulfport MS 39501

WALKLING, ROBERT ADOLPH, b Philadelphia, Pa, Sept 11, 31; m 59; c 2. ACOUSTICS. Educ: Swarthmore Col, BA, 53; Harvard Univ, SM, 54, PhD(acoustics), 62. Prof Exp: Res fel appl physics, Harvard Univ, 62-63; asst prof physics, Bowdoin Col, 63-69; assoc prof, Univ Maine, Portland, 69-70, ASSOC PROF PHYSICS, UNIV MAINE, PORTLAND-GORHAM, 70- Mem: AAAS; Acoust Soc Am; Audio Eng Soc; Am Sci Affiliation; Am Asn Physics Teachers. Res: Electroacoustics; noise and vibration; architectural and musical acoustics. Mailing Add: Dept of Phys Sci & Eng Univ of Maine at Portland-Gorham Portland ME 04103

WALKLING, WALTER DOUGLAS, b Baltimore, Md, Feb 27, 39; m 61; c 2. PHARMACY. Educ: Univ Md, BS, 61, MS, 63, PhD(pharm), 66. Prof Exp: Pharmacist, Yager Drug Co, 64-66; Nat Inst Gen Med Sci fel, Swiss Fed Inst Technol, 66-67; sr pharmaceut chemist, Eli Lilly & Co, 67-70; sr scientist, 70-72, GROUP LEADER, PHARM RES DEPT, McNEIL LABS, 72- Mem: Acad Pharmaceut Sci; Am Pharmaceut Asn; Am Chem Soc. Res: Design and evaluation of pharmaceutical dosage forms. Mailing Add: Pharm Res Dept McNeil Labs Ft Washington PA 19034

WALKOF, CHARLES, horticulture, see 12th edition

WALKOWIAK, EDMUND FRANCIS, b Webster, Mass, June 12, 35; m 60; c 2. PHYSIOLOGY. Educ: Boston Univ, AB, 58; Univ Conn, PhD(zool), 67. Prof Exp: Instr biol, Salem State Col, 59-62; from asst prof to assoc prof, Augusta Col, 66-69; assoc prof, 70-74, PROF PHYSIOL, MASS COL OPTOM, 74- Concurrent Pos: Mem biol curric comt, Univ Syst Ga, 69. Mem: Am Inst Biol Sci; Am Soc Mammal. Res: Application of serological and electrophoretic techniques to the study of tear films. Mailing Add: Dept of Physiol Mass Col of Optom Boston MA 02115

WALKUP, DAVID WILLIAM, mathematics, see 12th edition

WALKUP, JOHN HARPER, b Jackson, Miss, Jan 20, 15; m 43; c 1. PHYSICAL CHEMISTRY. Educ: Westminster Col, Mo, BA, 37; Western Reserve Univ, PhD(chem), 43. Prof Exp: Teacher high sch, Miss, 37-40; sr res chemist, Pa Salt Mfg Co, 43-45; from asst prof to assoc prof chem, 45-53, PROF CHEM, CENTRE COL KY, 53-, CHMN DEPT, 60- Concurrent Pos: Mem res staff, WVa Pulp & Paper Co, 51. Mem: Am Chem Soc; Electrochem Soc; The Chem Soc. Res: Wood chemistry; vapor-liquid equilibrium; hydrogen bonding studies; specific ion electrodes. Mailing Add: Dept of Chem Centre Col of Ky Danville KY 40422

WALL, CHARLES EPHRAIM, b Chandler, Okla, Jan 21, 14; m 35; c 5. CHEMISTRY. Educ: Okla State Univ, BS, 35, MS, 58, EdD, 67. Prof Exp: Instr pub schs, Okla, 35-39; mgr, Wall & Sons' Farms, 39-56; instr pub schs, Okla, 56-60; NSF traveling sci teacher, Okla State Univ, 60-61, teaching asst chem, 61-63, instr, 64-66; assoc prof, 66-68, PROF PHYS SCI, LANGSTON UNIV, 68- Mem: Am Chem Soc; Am Asn Physics Teachers; Nat Sci Teachers Asn. Res: Course content and method of presentation for students in beginning chemistry and physics. Mailing Add: Dept of Phys Sci Langston Univ Langston OK 73050

WALL, CHARLES ROBERT, b Palestine, Tex, Mar 16, 41; div; c 2. MATHEMATICS. Educ: Tex Christian Univ, BA, 63, MA, 64; Univ Tenn, PhD(math), 70. Prof Exp: Lectr math, Univ Tenn, 70; asst prof, ETex State Univ, 70-72; vis asst prof, 72-74, ASST PROF MATH, UNIV SC, 74- Mem: Am Math Soc; Math Asn Am. Res: Arithmetic function theory; bounding density functions associated with arithmetic functions. Mailing Add: Dept of Math Univ of SC Columbia SC 29208

WALL, CONRAD, III, b Boston, Mass, June 13, 39; m 61; c 2. BIOENGINEERING. Educ: Tulane Univ, BS, 62, MS, 68; Carnegie-Mellon Univ, PhD(bioeng), 75. Prof Exp: Proj officer elec eng, US Army AV Labs, 63-65; mem tech staff appl physics, Boeing Co, 65-70; NIH RES ASSOC SENSORY PHYSIOL, DEPT OTOLARYNGOL, MED SCH, UNIV PITTSBURGH, 75- Mem: Soc Neurosci; Inst Elec & Electronics Engrs. Res: Information processing in the vestibular and visual systems; digital signal processing of clinical and experimental sensory systems data. Mailing Add: Eye & Ear Hosp of Pittsburgh 230 Lothrop St Pittsburgh PA 15213

WALL, DONALD DINES, b Kansas City, Mo, Aug 13, 21; m 43; c 3. MATHEMATICS. Educ: Univ Calif, PhD(math), 49. Prof Exp: Instr, Santa Barbara Col, 49-51; appl sci rep, IBM Corp, NY, 51-74, MEM STAFF, DATA PROCESSING BR, IBM CORP, GA, 74- Mem: Am Math Soc; Math Asn Am; Asn Comput Mach. Res: Number theory; computing machines. Mailing Add: Data Processing Br IBM Corp Atlanta Com Life Ga Bldg Atlanta GA 30308

WALL, FRANCIS JOSEPH, b Moss Point, Miss, Mar 22, 27; m 50; c 3. BIOSTATISTICS, APPLIED STATISTICS. Educ: Sul Ross State Col, BS, 47; Univ Colo, MS, 56; Univ Minn, PhD, 61. Prof Exp: Sr statistician, Dow Chem Co, 52-57; statistician, Remington Rand Univac Div, Sperry Rand Corp, 57-61; sr res mathematician, Dikewood Corp, 61-69; biostatistician, Lovelace Found Med Educ & Res, 69-71; CONSULT BIOSTATIST & STATIST ANAL, 72- Concurrent Pos: Consult, Shell Oil Co, Colo, 56; vis lectr, Univ NMex, 67, adj asst prof sch med, 70-74, clin assoc, 68-69. Mem: Biomet Soc; Am Statist Asn; Sigma Xi. Res: Application of statistical methods to biomedical research; design and analysis of clinical trials; applications of statistical methods especially experimental design to the pharmaceutical, chemical and electronic industries. Mailing Add: 290 Alamosa Rd NW Albuquerque NM 87107

WALL, FREDERICK THEODORE, b Chisholm, Minn, Dec 14, 12; m 40; c 2. PHYSICAL CHEMISTRY. Educ: Univ Minn, BSh, 33, PhD(chem), 37. Prof Exp: Instr chem, Univ Ill, Urbana, 37-39, assoc, 39-41, from asst prof to assoc prof, 41-46, prof, 46-64, dean grad col, 55-63; prof chem & chmn dept, Univ Calif, Santa Barbara, 64-66, vchancellor res, 65-66; vchancellor grad studies & res & prof chem, Univ Calif, San Diego, 66-69; exec dir, Am Chem Soc, 69-72; PROF CHEM, RICE UNIV, 72- Honors & Awards: Award, Am Chem Soc, 45. Mem: Nat Acad Sci; AAAS; Am Acad

Arts & Sci; Am Chem Soc (ed, J Phys Chem, 65-69); Am Phys Soc. Res: Physical chemistry of macromolecular configurations; theory of reaction rates. Mailing Add: Dept of Chem Rice Univ Houston TX 77001

WALL, GREGORY JOHN, b Toronto, Ont, Aug 16, 44; m 68; c 1. SOIL SCIENCE. Educ: Univ Guelph, BSA, 67, MSc, 69; Ohio State Univ, PhD(soil sci), 73. Prof Exp: Res officer soil sci, Agr Can, 67-70; teaching asst agron, Ohio State Univ, 70-71, res assoc soil mineral, 71-73; SOIL SCIENTIST, AGR CAN, 73- Mem: Can Soc Soil Sci; Am Soc Agron; Int Asn Great Lakes Res; Int Soc Soil Sci; Soil Conserv Soc Am. Res: Sources and magnitude of water pollution by sediment in agricultural regions; mineralogy and exchange properties of fluvial sediments; variability of soil physical and engineering properties in the mineralogy of soils. Mailing Add: Dept Land Resource Sci Agr Can Univ Guelph Guelph ON Can

WALL, JAMES GRAHAM, b Benson, NC, Jan 21, 09; m 42; c 1. MATHEMATICS. Educ: Univ NC, AB, 32, MA, 33; Univ Ga, EdD, 54. Prof Exp: Instr math, NC State Col, 46-48; instr, Ga Inst Technol, 48-50; asst, Univ Ga, 50-53; from asst prof to assoc prof, 53-73, PROF MATH, VALDOSTA STATE COL, 73- Mem: Am Math Soc; Math Asn Am. Res: Analytical study of critical thinking. Mailing Add: Dept of Math Valdosta State Col Valdosta GA 31601

WALL, JAMES ROBERT, b Lenoir, NC, Dec 7, 29; m 57; c 1. PLANT GENETICS. Educ: Va Polytech Inst, BS, 51; Cornell Univ, PhD, 55. Prof Exp: Geneticist, USDA, 56-62; asst prof biol, La State Univ, New Orleans, 62-66; from asst prof to assoc prof, Tex Tech Univ, 66-69; assoc prof, 69-74, PROF BIOL, GEORGE MASON UNIV, 74- Mem: Bot Soc Am; Genetics Soc Am; Soc Study Evolution. Res: Interspecific hybridization in Cucurbita and mechanisms of speciation in Phaseolus; experimental evolution. Mailing Add: Dept of Biol George Mason Univ Fairfax VA 22030

WALL, JOHN HALLETT, b St Stephen, NB, Aug 10, 24; m 61; c 1. MICROPALEONTOLOGY. Educ: Univ NB, BSc, 45; Univ Alta, MSc, 51; Univ Mo, PhD(geol), 58. Prof Exp: Asst geologist, NB Dept Mines, 43; asst geologist, Geol Surv Can, 44; jr geologist, Imp Oil Ltd, 45-46, subsurface geologist & micropaleonto micropaleontologist, 47-51; subsurface geologist, J C Sproule & Assocs, Explor Consults, 52; micropaleontologist, Calif Standard Co, 53; res officer, Res Coun Alta, 57-74; RES SCIENTIST, GEOL SURV CAN, 74- Concurrent Pos: Lectr geol, Univ Alta, 60-72. Mem: AAAS; fel Geol Soc Am; Soc Econ Paleont & Mineral; Paleont Soc; fel Geol Asn Can. Res: Mesozoic microfossils of western and arctic Canada. Mailing Add: 3303 33rd St NW Calgary AB Can

WALL, JOSEPH S, b Madison, Wis, Nov 17, 42. BIOPHYSICS. Educ: Univ Wis-Madison, BS, 64; Univ Chicago, PhD(biophys), 71. Prof Exp: Fel biophys, Univ Chicago, 71-73; ASSOC BIOPHYSICIST DEPT BIOL, BROOKHAVEN NAT LAB, 73- Res: Development and biological application of the high resolution scanning transmission electron microscope. Mailing Add: Dept of Biol Brookhaven Nat Lab Upton NY 11973

WALL, JOSEPH SENNEN, b Chicago, Ill, June 2, 23; m 50. BIOCHEMISTRY. Educ: Univ Chicago, BS, 46, MS, 49; Univ Wis, PhD(biochem), 52. Prof Exp: Instr chem, Lincoln Col, 47-49; asst biochem, Univ Wis, 50-52; instr pharmacol sch med, NY Univ, 52-56; head chem reactions & structure invests, 56-72, RES LEADER PROTEIN REACTIONS & STRUCTURES UNIT, CEREAL PROPERTIES LAB, NORTHERN REGIONAL RES & DEVELOP DIV, USDA, 72- Mem: AAAS; Am Chem Soc; Am Soc Biol Chem; Am Asn Cereal Chem; Inst Food Technologists. Res: Mechanism of nitrogen fixation by bacteria; isolation of natural products; hormonal regulation of carbohydrate metabolism in mammals; protein chemistry; cereal chemistry; nutrition; enzymology. Mailing Add: Northern Regional Res Lab USDA Peoria IL 61604

WALL, LAWRENCE SCOTT, physics, see 12th edition

WALL, MALCOLM JEFFERSON, JR, b Meridian, Miss, Jan 25, 41; m 65; c 2. PHYSIOLOGY. Educ: Lamar Univ, BS, 63; Univ Tex Med Br, Galveston, PhD(physiol), 70; NTex State Univ, MA, 67. Prof Exp: Asst biol, NTex State Univ, 63-66; from instr to asst prof physiol & med, Med Col Wis, 70-74; RES ASSOC PHYSIOL & BIOPHYS, UNIV TEX MED BR, GALVESTON, 74- Concurrent Pos: USPHS grant, Med Col Wis, 72-74; lectr, Univ Wis-Milwaukee, 72. Mem: AAAS; Am Inst Biol Sci; Biophys Soc; Assoc Am Physiol Soc. Res: Intestinal transport mechanisms for electrolytes, orgainc solutes and water. Mailing Add: Dept of Physiol & Biophys Univ of Tex Med Br Galveston TX 77550

WALL, MONROE ELIOT, b Newark, NJ, July 25, 16; m 41; c 2. BIOCHEMISTRY. Educ: Rutgers Univ, BS, 36, MS, 38, PhD(biochem), 39. Prof Exp: Chemist, NJ Exp Sta, 39-40; res chemist, Wallerstein Labs, 40; res assoc, Barrett Co, 41; from asst chemist to supvr plant steroid units eastern regional res lab, Bur Agr & Indust Chem, USDA, 41-53, supvr eastern utilization res br, Agr Res Serv, 53-60; head natural prod lab, 60-66, dir chem & life sci lab, 66-71, VPRES PHYS & LIFE SCI DIV, RES TRIANGLE INST, 71- Concurrent Pos: Adj prof, NC State Univ, 62- & Univ NC, Chapel Hill, 66-; NSF vis scientist, Am Col Pharmacists, 64-; consult, NIH, 62- Honors & Awards: Merck, Sharp & Dohme Res Achievement Award Natural Prod, Am Pharmaceut Asn, 70. Mem: AAAS; Am Chem Soc; fel Acad Pharmaceut Sci; NY Acad Sci; Soc Econ Bot (pres, 75). Res: Plant chemistry; steroids; cancer chemotherapy; drug metabolism; chemistry and metabolism of cannabinoids. Mailing Add: Life Sci Div Res Triangle Inst PO Box 12194 Res Triangle Park NC 27709

WALL, NATHAN SANDERS, b Chicago, Ill, May 25, 25; m m 51; c 2. NUCLEAR PHYSICS. Educ: Rensselaer Polytech Inst, BS, 49; Mass Inst Technol, PhD(physics), 54. Prof Exp: Asst, Mass Inst Technol, 49-53; res assoc, Ind Univ, 53-54; res assoc & asst prof, Univ Rochester, 54-55; asst prof physics, Mass Inst Technol, 55-64; assoc prof, 64-67, PROF PHYSICS, UNIV MD, COLLEGE PARK, 67- Concurrent Pos: NSF sr fel, 61; NATO fel, 71; mem subcomt nuclear structure, Nat Acad Sci-Nat Res Coun; mem sci adv comt, Space Radiation Effects Lab; consult, Qm Corps, US Dept Army, AEC & Inst Defense Anal. Mem: Fel Am Phys Soc. Res: Nuclear structure. Mailing Add: Dept of Physics Univ of Md College Park MD 20740

WALL, ROBERT ALLEN, b Melrose, Mass, Feb 17, 33; m 57; c 3. PHYSICAL CHEMISTRY. Educ: Northeastern Univ, BS, 56; Pa State Univ, PhD(chem), 62. Prof Exp: Staff scientist, Orion Res Inc, 62-63; group leader phys chem, Clairol Res Labs, 63-66, res mgr, 66-68, assoc dir prod develop, Res & Develop Lab, Clairol Inc, 68, dir prod develop, 68-72, VPRES PROD & TECH DEVELOP, RES & DEVELOP LAB, CLAIROL INC, 72- Mem: Am Inst Chem; Am Phys Soc; Soc Rheol; Am Chem Soc; Asn Res Dirs. Res: Dynamic mechanical properties and transitions in polymers; ion specific electrodes instrumentation; rheology; keratin physics; colloid and dyeing chemistry; hair care products development; hair coloring science; process development for cosmetics. Mailing Add: 17 Roland Dr Darien CT 06820

WALL, ROBERT C, biology, conservation, see 12th edition

WALL, ROBERT ECKI, b Aurora, Ill, Aug 1, 35; m 63; c 3. GEOPHYSICS. Educ: Carleton Col, AB, 57; Columbia Univ, PhD(geophys), 65. Prof Exp: Res assoc marine geophys, Lamont Geol Observ, Columbia Univ, 65-66; sci officer marine geol & geophys, Off Naval Res, 66-70; prog dir submarine geol & geophys, 70-75, HEAD OCEANOG SECT, NSF, 75- Mem: AAAS; Am Geophys Union; Geol Soc Am; Soc Explor Geophys. Res: Marine geophysics; geological oceanography. Mailing Add: Nat Sci Found Washington DC 20550

WALL, ROBERT GENE, b Mo, Nov 17, 37; m 64; c 3. ORGANIC CHEMISTRY. Educ: Ore State Univ, BS, 61, MS, 63, PhD(org chem), 66. Prof Exp: NIH fel, Univ Mich, Ann Arbor, 66-67; res chemist, 67-72, SR RES CHEMIST, CHEVRON RES CO, RICHMOND, CALIF, 72- Mem: AAAS; Am Chem Soc. Res: Physical organic chemistry; heterogeneous catalysis. Mailing Add: 2826 Wright Ave Pinole CA 94564

WALL, ROBERT LEROY, b Doylestown, Pa, July 7, 21; m 52; c 4. MEDICINE. Educ: Oberlin Col, AB, 43; Temple Univ, MD, 46; Am Bd Internal Med, dipl, 55. Prof Exp: Intern, Bryn Mawr Hosp, 46-47, resident clin path, 47-48; mem staff med res, Univ Hosp, 50-52, asst prof med & dir lymphoma clin, Col Med, 52-57, assoc prof med, 57-65, asst dean res & secy col med, 71-73, PROF MED, COL MED, OHIO STATE UNIV, 65- Concurrent Pos: Consult, Dayton Vet Hosp, 54-; mem plasma proteins comt, Nat Res Coun, 63-71. Mem: AAAS; Am Soc Hemat; AMA; Am Asn Cancer Res; Am Fedn Clin Res. Res: Blood preservation; plasma protein fractionation; qualitative changes in globulins of diseased plasma. Mailing Add: Univ Hosp Rm N1016 410 Tenth Ave Columbus OH 43210

WALL, RONALD EUGENE, b Dryden, Ont, Feb 19, 36; m 57; c 3. PLANT PATHOLOGY. Educ: Ont Agr Col, BSA, 58; Univ Wis, PhD(plant path), 62. Prof Exp: Plant pathologist, Can Dept Agr, 62-66; plant pathologist, 66-73, RES SCIENTIST, DEPT ENVIRON, CAN DEPT FORESTRY, 73- Mem: Am Phytopath Soc; Can Phytopath Soc. Res: Decays of maturing plants; trunk rots of forest trees; diseases of conifer seedlings. Mailing Add: Maritimes Forest Res Ctr Fredericton NB Can

WALL, THOMAS RANDOLPH, b Lakeland, Fla, Mar 23, 43. MOLECULAR BIOLOGY, IMMUNOLOGY. Educ: Univ SFla, AB, 65; Ind Univ, PhD(microbiol), 70. Prof Exp: Res assoc cell & molecular biol, Columbia Univ, 70-72; ASST PROF MICROBIOL & IMMUNOL, SCH MED, UNIV CALIF, LOS ANGELES, 72- Concurrent Pos: Damon Runyon Mem Fund Cancer Res fel, Columbia Univ, 70-72; assoc mem, Molecular Biol Inst, Univ Calif, Los Angeles, 72- Mem: Am Soc Microbiol. Res: Organization, expression and regulation of viral and cellular genes in eukaryotic cells; molecular analysis of the nature and expression of immunoglobulin genes; developmental aspects of the immune response. Mailing Add: Dept of Microbiol & Immunol Univ of Calif Sch of Med Los Angeles CA 90024

WALL, WILLIAM JAMES, b Northampton, Mass, June 25, 21; m 46; c 1. ENTOMOLOGY. Educ: Univ Mass, BS, 42, MS, 49; Univ Calif, PhD(entom), 52. Prof Exp: Instr biol, Univ Mass, 46-49; asst, Univ Calif, 49-52; instr biol, State Univ NY Albany, 52-54, asst prof, 54-56; PROF BIOL, BRIDGEWATER STATE COL, 56- Concurrent Pos: NIH-USPHS-Nat Commun Dis Ctr res grant, Cape Cod, Mass, 67-70; entomologist, Cape Cod Mosquito Control Proj, 58-; mem sci adv panel, Volta River Basin Area, WAfrica, WHO, 74-79. Mem: AAAS; Entom Soc Am; Am Mosquito Control Asn. Res: Biology, life history, control and taxonomy of Thysanura, Tabanidae, Ceratopogonidae and Culicidae. Mailing Add: Dept of Biol Bridgewater State Col Bridgewater MA 02324

WALLACE, ALEXANDER CAMERON, b St Thomas, Ont, Aug 27, 21; m 48; c 4. MEDICINE, PATHOLOGY. Educ: Univ Western Ont, BA, 47, MD, 48; FRCP(C), 72. Prof Exp: Clin intern, Victoria Hosp, London, Ont, 48-49; intern path, New Haven Hosp, 49-51; resident, 51-52; lectr med res, Univ Western Ont, 52-53, asst prof, 53-55; assoc prof path, Univ Man, 55-61; dir cancer res lab, 61-65, head dept path, 65-74, PROF PATH, FAC MED, UNIV WESTERN ONT, 65- Concurrent Pos: Markle Found scholar, 52-57; pathologist, Winnipeg Munic Hosp, 55-61; mem res adv comt, Nat Cancer Inst Can, 60-68, dir, 69-; consult, Westminster Hosp, London, 63-; pathologist, Univ Hosp, London, 72- Mem: Fel Am Col Physicians; Am Asn Cancer Res; Can Asn Path; Int Acad Path. Res: Cancer research; biology of neoplasia; metastases; renal pathology. Mailing Add: Dept of Path Univ of Western Ont Fac of Med London ON Can

WALLACE, ALEXANDER DONIPHAN, b Hampton, Va, Aug 21, 05; m 31; c 1. MATHEMATICS. Educ: Univ Va, BS, 35, MS, 36, PhD(math), 39. Prof Exp: Instr math, Univ Va, 38-40; instr, Princeton Univ, 40-41; instr, Univ Pa, 41-43, asst prof, 43-47; prof, Tulane Univ, 47-63; prof, Univ Fla, 63-66; prof, Univ Miami, Fla, 66-67; prof, Univ Fla, 67-73; RETIRED. Mem: Am Math Soc; Math Asn Am. Res: Set-theoretic and algebraic topology; topological algebra. Mailing Add: 306 E Gatehouse Dr Apt H Metairie LA 70001

WALLACE, ANDREW GROVER, b Columbus, Ohio, Mar 22, 35; m 57; c 3. INTERNAL MEDICINE, CARDIOVASCULAR PHYSIOLOGY. Educ: Duke Univ, BS, 58, MD, 59. Prof Exp: Intern med, Med Ctr, Duke Univ, 59-60, asst resident, 60-61; investr cardiovasc physiol, Nat Heart Inst, 61-63; chief resident med, 63-64, assoc, 64-65, from asst prof to assoc prof, 65-70, dir cardiac intensive care unit, 65-70, PROF MED, ASST PROF PHYSIOL, ASST DIR GRAD MED EDUC & CHIEF CARDIOL DIV, MED CTR, DUKE UNIV, 70- Concurrent Pos: Fel cardiol, Duke Univ, 61; Markle scholar acad med, 65-70; USPHS career develop award, 65-70. Mem: Am Fedn Clin Res; Am Heart Asn. Res: Cardiology; electrocardiology; electrophysiology of the heart. Mailing Add: 3413 Rugby Rd Durham NC 27707

WALLACE, ANDREW HUGH, b Glasgow, Scotland, June 14, 26. MATHEMATICS. Educ: Univ Edinburgh, MA, 46; St Andrews Univ, PhD(math), 49. Prof Exp: Asst lectr math, Univ Col, Dundee, 46-49, lectr, 49-50; Commonwealth Fund fel, Univ Chicago, 50-52; lectr, Univ Col, Dundee, 52-53; from lectr math to sr lectr, Univ Col, NStaffordshire, 53-57; asst prof, Univ Toronto, 57-59; asst prof, Ind Univ, 59-61, prof, 61-64; grant in aid, Inst Adv Study, 64-65; PROF MATH, UNIV PA, 65- Concurrent Pos: Vis prof & assoc dir, Univ Pa Group, Pahlavi Univ, Iran, 71-72. Mem: Am Math Soc; Can Math Cong; London Math Soc. Res: Algebra; algebraic geometry and topology. Mailing Add: Dept of Math Univ of Pa Philadelphia PA 19104

WALLACE, ANTHONY FRANCIS CLARKE, b Toronto, Ont, Apr 15, 23; US citizen; m 42; c 4. CULTURAL ANTHROPOLOGY. Educ: Univ Pa, BA, 47; Univ Pa, MA, 49, PhD(anthrop), 50. Prof Exp: Instr anthrop, Bryn Mawr Col, 48-50; asst instr anthrop, Univ Pa, 48-49, instr sociol, 50-52, res asst prof anthrop, 52-55, vis assoc prof, 55-61, chmn dept, 61-71; res assoc, 55-60, dir clin res, 60-61, MED RES SCIENTIST, EASTERN PA PSYCHIAT INST, 61-; PROF ANTHROP, UNIV PA, 61- Concurrent Pos: Consult, Philadelphia Housing Authority, 51-52; Soc Sci Res Coun fac res fel, 51-54; consult comt disaster studies, Nat Res Coun, 53-56, mem, 56-57, mem div behav sci, 63-66; prin investr, NIMH, 54-57, mem fel rev panel behav sci, 61-64, mem behav sci study sect, 64-68; mem res adv comt, Commonwealth Ment Health Res Found, 57-; consult, Vet Admin, Perry Point, Md, 58-60; mem tech adv comt, NJ Neuro-Psychiat Inst, 58-61; mem environ panel, Coop Res Prog, US Off Educ, 62, mem res adv coun, 65-68; mem soc sci res adv comt, NSF, 68-69, chmn, 71, grant sci res, 72-74; mem, Surgeon Gen Sci Adv Comt TV & Social Behav, 69-71; mem bd dirs, Found Fund Res in Psychiat, 69-71. Mem: Nat Acad Sci; Am Philos Soc; Am Anthrop Asn (pres, 71-72); Am Philos Soc; Am Acad Arts & Sci. Res: Culture change; culture and cognition; culture and personality. Mailing Add: Dept of Anthrop Univ of Pa Philadelphia PA 19104

WALLACE, ARTHUR, b Bear River, Utah, Jan 4, 19; m 43; c 4. PLANT NUTRITION, PLANT PHYSIOLOGY. Educ: Utah State Univ, BS, 43; Rutgers Univ, PhD, 49. Prof Exp: Chief div environ biol, 66-72, PROF PLANT NUTRIT, UNIV CALIF, LOS ANGELES, 49-, ASST CHIEF DIV ENVIRON BIOL, 72- Mem: Am Soc Plant Physiol; Am Chem Soc; Am Soc Hort Sci; fel Am Soc Agron. Res: Inorganic plant nutrition and related physiology; major cations, nitrogen and micronutrient elements including their supply to plants by synthetic chelating agents; comparative mineral nutrition of plants; ecophysiology. Mailing Add: Lab Nuclear Med Radiation Biol Univ of Calif Los Angeles CA 90024

WALLACE, ATWELL MILTON, plant physiology, see 12th edition

WALLACE, BEN J, b Ft Smith, Ark, Aug 21, 37; m 57; c 2. ETHNOLOGY, APPLIED ANTHROPOLOGY. Educ: Univ Okla, BA, 61, MA, 62; Univ Wis, PhD(anthrop), 66. Prof Exp: Instr anthrop, Univ Wis, 67; asst prof, Univ Calif, Santa Barbara, 67-69; assoc prof, 69-72, PROF ANTHROP, SOUTHERN METHODIST UNIV, 72- Concurrent Pos: Consult, Int Sch Bus, Univ Dallas, 70- Mem: Am Anthrop Asn; Asian Soc. Res: Agricultural development; Southeast Asian peoples and cultures; American Indians; youth and hippie culture; culture change. Mailing Add: Dept of Anthrop Southern Methodist Univ Dallas TX 75222

WALLACE, BRUCE, b McKean, Pa, May 18, 20; m 45. GENETICS. Educ: Columbia Univ, AB, 41, PhD(genetics), 49. Prof Exp: Res assoc dept genetics, Carnegie Inst, 47-49; res assoc, LI Biol Asn, 49-58; assoc prof genetics, 58-61, PROF GENETICS, CORNELL UNIV, 61- Mem: Soc Study Evolution (pres, 74); Genetics Soc Am (secy, 68-70, pres, 74); Am Acad Arts & Sci; Am Soc Naturalists (secy, 56-58, pres, 70). Res: Physiology, population dynamics and speciation of Drosophila. Mailing Add: Sect of Genetics Develop Physiol Cornell Univ Ithaca NY 14850

WALLACE, CHARLES RAY, b Dec 20, 38; US citizen; m 67; c 2. ICHTHYOLOGY, ECOLOGY. Educ: Univ Ark, BA, 64, MS, 65; Univ Nebr, PhD(zool), 69. Prof Exp: Asst prof biol, Kearney State Col, 69-71; asst prof, Pan Am Univ, 71-74; MEM STAFF, GAME PARKS COMN, LINCOLN, NEBR, 74- Concurrent Pos: NSF grant, 71-73. Mem: Am Fisheries Soc. Res: Ecology of fishes. Mailing Add: Dept of Biol Pan Am Univ Edinburg TX 78539

WALLACE, CHESTER ALAN, b Rockville Centre, NY, Oct 1, 42; m 68; c 2. STRATIGRAPHY, SEDIMENTARY PETROLOGY. Educ: Antioch Col, BA, 65; Univ Calif, Santa Barbara, PhD(geol), 72. Prof Exp: Asst prof geol, Ga Southwestern Col, 69-74; GEOLOGIST, US GEOL SURV, 74- Mem: Geol Soc Am; Soc Econ Paleontologists & Mineralogists. Res: Stratigraphy, sedimentology, paleocurrent analysis, and diagenesis of clastic sedimentary rocks; tectonic history and sedimentary basin analysis of Precambrian sedimentary rocks in the western United States. Mailing Add: US Geol Surv Cent Environ Geol Denver Fed Ctr MS 913 Denver CO 80225

WALLACE, CRAIG KESTING, b Woodbury, NJ, Dec 4, 28; m 60; c 2. MEDICINE. Educ: Princeton Univ, AB, 50; NY Med Col, MD, 55. Prof Exp: Instr med, Jefferson Med Col Hosp, 60-64; from asst prof to assoc prof, Sch Med & Sch Hyg & Pub Health, Johns Hopkins Univ, 64-72, assoc dir, Ctr Med Res & Training, 67-72; CMNDG OFFICER, US NAVAL MED RES UNIT, ETHIOPIA, 72- Concurrent Pos: Clin investr, US Naval Med Res Unit, Taiwan, 60-63; adv & consult, WHO, 62-; consult, Magee Mem Hosp, Philadelphia, Pa, 63-64; resident coordr & head med prog, Ctr Med Res & Training, Calcutta, India, 64-66; asst physician, Johns Hopkins Hosp, 66-69, physician, 69-72; physician & consult, Vet Admin Hosp, Perry Point, Md, 67-72; mem bact & mycol study sect, NIH, 68-72, chmn, 71-72; physician, Good Samaritan Hosp, Baltimore, Md, 69-72. Mem: AAAS; fel Am Col Physicians; Infectious Dis Soc Am; Am Soc Microbiol; Am Soc Trop Med & Hyg. Res: Pathophysiology and treatment of infectious diseases, especially in the areas of tropical enteric infections; physician training and research in international medicine. Mailing Add: US Naval Med Res Unit 5 APO New York NY 09319

WALLACE, DAVID JEFFERSON, organic chemistry, see 12th edition

WALLACE, DAVID LEE, b Homestead, Pa, Dec 24, 28; m 55; c 3. STATISTICS. Educ: Carnegie Inst Technol, BS, 48, MS, 49; Princeton Univ, PhD(math), 53. Prof Exp: Moore instr math, Mass Inst Technol, 53-54; from asst prof to assoc prof statist, 54-67, PROF STATIST, UNIV CHICAGO, 67- Concurrent Pos: Fel, Ctr Advan Study in Behav Sci, 60-61; mem comput & biomath sci study sect, NIH, 70-74. Mem: Fel Am Statist Asn; Inst Math Statist; Psychomet Soc; Asn Comput Mach; fel Royal Statist Soc. Res: Theoretical statistics; computer methods. Mailing Add: Dept of Statist Univ of Chicago Chicago IL 60637

WALLACE, DONALD ALBIN, b Dallas Ctr, Iowa, Apr 16, 05; m 28; c 1. PHARMACOLOGY, CHEMISTRY. Educ: Univ Iowa, AB, 29; Carnegie Inst Technol, MS, 30; Univ Chicago, PhD(phys chem), 32. Prof Exp: Chemist, Dearborn Chem Co, 32-33; res chemist, Kraft-Phenix Cheese Corp, 33-35; chief chemist, Petrolagar Labs, Inc, 35-38; assoc chemist, Am Dent Asn, 38-42, secy coun dent therapeut, 42-49, dir bur chem, 42-48, dir sci exhibits, 48-49; asst mgr res, Pepsodent Div, Lever Bros Co, 49-51; from assoc prof to prof appl mat med & therapeut, 51-65, PROF RADIOL, COL DENT, UNIV ILL MED CTR, 65- Mem: AAAS; affil AMA; Int Asn Dent Res. Res: Water fluoridation; dental medicines and dentifrices. Mailing Add: Univ of Ill at the Med Ctr Box 6998 Chicago IL 60612

WALLACE, DONALD HOWARD, b Driggs, Idaho, June 27, 26; m 49; c 5. PLANT GENETICS. Educ: Utah State Agr Col, BS, 53; Cornell Univ, PhD(plant breeding), 58. Prof Exp: Asst plant breeding, 53-55 & 57, actg asst prof plant breeding & veg crops, 55-57, from asst prof to assoc prof, 58-71, PROF PLANT BREEDING & VEG CROPS, CORNELL UNIV, 71- Mem: Am Soc Hort Sci; Crop Sci Soc Am; Am Soc Plant Physiol. Res: Development of hybrid and improved varieties of vegetables; physiological genetics of crop plants. Mailing Add: Dept of Plant Breeding Cornell Univ Ithaca NY 14850

WALLACE, DORIS DAVID, b Lentner, Mo, July 10, 15; m 66. SPEECH PATHOLOGY, AUDIOLOGY. Educ: Northeastern Mo State Teachers Col, BS, 37; Univ Mo, MA, 45; NY Univ, PhD(speech educ), 64. Prof Exp: Teacher high schs, Mo, 37-45; supvr speech correction, Pub Schs, Ill, 45-47; instr speech, Col Emporia, 47-48; asst prof, Culver-Stockton Col, 48-53; instr speech & speech correction, Mankato State Col, 53-56; lectr speech, Queen's Col, NY, 58; asst prof except children educ, State Univ Ny Col, Buffalo, 58-61; instr speech correction, Univ Nebr, 61-62; asst prof speech path, Northern Ill Univ, 62-67; PVT PRACT, 66- Concurrent Pos: Exten instr, Northeastern Mo State Col, 67-69; consult speech pathologist, Regional Diag Clin, Hannibal, 67-73. Mem: Am Speech & Hearing Asn; Speech Commun Asn. Res: History of public address; therapy for stutterers; training of speech correctionists; learning disabilities; language of the mentally retarded. Mailing Add: 208 N Macon Clarence MO 63437

WALLACE, DOUGLAS CECIL, b Cumberland, Md, Nov 6, 46. GENETICS. Educ: Cornell Univ, BS, 68; Yale Univ, MPh, 72, PhD(somatic cell genetics), 75. Prof Exp: Res microbiologist, USPHS Northwestern Water Hyg Lab, Gig Harbor, Wash, 68-70; FEL SOMATIC CELL GENETICS, SCH MED, YALE UNIV, 75- Mem: Sigma Xi; Am Soc Microbiol; AAAS. Res: Study of the genetics, biogenesis and phylogenetic relationships of mammalian mitochondria using the techniques of somatic cell genetics and molecular biology. Mailing Add: Dept of Human Genetics Yale Univ Sch of Med New Haven CT 06510

WALLACE, DWIGHT TOUSCH, b Oakland, Calif, May 25, 27; m 50; c 2. ANTHROPOLOGY, ARCHAEOLOGY. Educ: Univ Calif, AB, 50, PhD, 57. Prof Exp: Researcher & proj coordr, Fulbright Comn, Lima, 57-59; asst prof anthrop, Univ NC, 59-60, Univ Ga, 60-61 & Univ Ore, 61-68; ASSOC PROF ANTHROP, STATE UNIV NY ALBANY, 68- Mem: Am Anthrop Soc; Soc Am Archaeol. Res: Peruvian archaeology; South and Middle American culture history; technology and culture change. Mailing Add: Dept of Anthrop State Univ of NY Albany NY 12222

WALLACE, EDWARD HAMILTON, chemistry, see 12th edition

WALLACE, EDWIN GARFIELD, b Akron, Ohio, Jan 27, 17; m 44; c 6. ORGANIC CHEMISTRY. Educ: Univ Miami, Ohio, AB, 38; Ohio State Univ, PhD(chem), 42. Prof Exp: Res chemist, Eastman Kodak Co, NY, 42-47; res chemist, US Naval Ord Test Sta, Calif, 47-48; res chemist & group leader, Shell Chem Corp, 48-54; from asst to vpres res to mgr chem res, 54-65, lab dir, 65-67, SR RES ASSOC, WESTERN RES CTR, STAUFFER CHEM CO, 67- Mem: Am Chem Soc. Res: Agricultural chemicals; industrial organic chemicals and products; polymers; fluorine chemicals. Mailing Add: Western Res Ctr Stauffer Chem Co 1200 S 47th St Richmond CA 94804

WALLACE, FRANKLIN GERHARD, b Deer River, Minn, Apr 3, 09; m 35, 75; c 4. PARASITOLOGY. Educ: Carleton Col, BA, 28; Univ Minn, MA, 30, PhD(zool), 33. Prof Exp: Asst prof biol, Lingnam Univ, 33-37; instr zool, 37-42, assoc prof, 46-63, PROF ZOOL, UNIV MINN, MINNEAPOLIS, 63- Concurrent Pos: Consult, Vet Admin Hosp, 46-73; mem trop med & parasitol study asst, NIH, 70-74. Mem: Am Micros Soc (vpres, 66); Am Soc Parasitol (vpres, 74); Am Soc Trop Med & Hyg; Soc Protozool (pres-elect, 76). Res: Trypanosomatid flagellates and parasites of insects. Mailing Add: Dept of Zool Univ of Minn Minneapolis MN 55455

WALLACE, FREDERIC ANDREW, b Boston, Mass, Apr 28, 33; m 54; c 2. PHYSICAL CHEMISTRY, ANALYTICAL CHEMISTRY. Educ: Harvard Univ, AB, 58; Calif Inst Technol, MS, 60; Tufts Univ, PhD(chem), 67. Prof Exp: GROUP LEADER PHYS & ANAL CHEM RES, CENT ANAL LAB, RES DIV, POLAROID CORP, 67- Concurrent Pos: Chemist, Monsanto Chem Co, 70-74. Mem: Am Chem Soc. Res: Investigations of structure and bonding of organo-silver complexes and salts; measurement of stability constants, and solubility products of same; chemical analysis and method development for photographic chemicals. Mailing Add: Polaroid Corp 750 Main St 3-D Cambridge MA 02139

WALLACE, GARY OREN, b Stewart Co, Tenn, Apr 2, 40; m 62; c 2. ECOLOGY. Educ: Austin Peay State Univ, BS, 62; Univ Tenn, MS, 64, PhD(zool), 70. Prof Exp: Teacher biol, Maryville Col, 64-65; ASST PROF BIOL, MILLIGAN COL, 67-68 & 71- Concurrent Pos: Ed, The Migrant, 71- Mem: Wilson Ornith Soc; Nat Audubon Soc; Sigma Xi. Res: Abundance and distribution of certain bird populations. Mailing Add: Dept of Biol Milligan Col Milligan College TN 37682

WALLACE, GEORGE EGBERT, b Ohio, Ill, Mar 24, 12; m 46; c 4. ENTOMOLOGY. Educ: Univ Pittsburgh, BS, 33, MS, 36, PhD(zool), 40. Prof Exp: Asst zool & biol, Univ Pittsburgh, 34-37; asst entom, 37-41, from asst cur to assoc cur sect insects & spiders, 41-51, CUR SECT INSECTS & SPIDERS, CARNEGIE MUS NATURAL HIST, 51- Mem: Entom Soc Am. Res: Taxonomy of Pteromalidae. Mailing Add: Sect of Insects & Spiders Carnegie Mus of Natural Hist Pittsburgh PA 15213

WALLACE, GEORGE JOHN, b Waterbury, Vt, Dec 9, 06; m 34; c 2. ORNITHOLOGY. Educ: Univ Mich, AB, 32, MA, 33, PhD(zool), 36. Hon Degrees: ScD, Cent Mich Univ, 70. Prof Exp: Biologist, Vt Fish & Game Serv, 36-37; dir, Pleasant Valley Sanctuary, Mass, 37-42; game ecologist game div, Mich Dept Conserv, 42; instr zool, 42-45, from asst prof to assoc prof, 45-54, prof, 54-72, EMER PROF ZOOL, MICH STATE UNIV, 72- Concurrent Pos: Consult AEC, Argonne Nat Lab, 74-75. Mem: Wilson Ornith Soc; Cooper Ornith Soc; Nat Audubon Soc; Am Ornith Union; Nat Wildlife Fedn. Res: Predatory birds; effects of insecticides on birds; neotropical thrushes; conservation of natural resources; books and encyclopedia articles on birds. Mailing Add: Route 1 Box 1638 Grayling MI 49738

WALLACE, GERALD WAYNE, b Sault Ste Marie, Mich, July 20, 33; m 54; c 3. ANALYTICAL CHEMISTRY, PHYSICAL CHEMISTRY. Educ: Univ Mich, BS, 55; Purdue Univ, MS, 57, PhD(anal chem), 59. Prof Exp: Sr chemist, Esso Res & Eng Co, 59-60; sr anal chemist, 60-65, res scientist, 65-67, head anal develop phys, 67-71, DIR PHYS CHEM RES DIV, ELI LILLY & CO, 71- Mem: Am Chem Soc. Res: Spectroscopy; fluorescence; photochemistry. Mailing Add: Eli Lilly & Co Indianapolis IN 46206

WALLACE, GORDON DEAN, b Los Angeles, Calif, Dec 17, 27; m 52; c 2. EPIDEMIOLOGY. Educ: Colo State Univ, BS, 52, DVM, 54; Univ Calif, MPH, 62. Prof Exp: Epidemiologist, Commun Dis Ctr, USPHS, 54-61, MED RESEARCHER, NAT INST ALLERGY & INFECTIOUS DIS, 61- Concurrent Pos: Clin assoc prof trop med & med microbiol, Sch Med, Univ Hawaii, 67. Mem: AAAS; Am Soc Trop Med & Hyg. Res: Epidemiology of infectious diseases, including eosinophilic meningitis, influenza and toxoplasmosis. Mailing Add: Pac Res Sect NIH PO Box 1680 Honolulu HI 96806

WALLACE, GRAHAM FRANKLIN, b Santa Rosa, Calif, Mar 27, 35; m 59; c 2. COMPUTER SCIENCE. Educ: Pomona Col, BA, 57; Univ Calif, Berkeley, MA, 60. Prof Exp: Mathematician math sci dept, 60-64, res mathematician, 64-66, asst mgr admin, 66-70, sr systs programmer info sci lab, 70-74, SR RES ENGR INFO SCI LAB, STANFORD RES INST, 74- Concurrent Pos: Chmn seventh symposium gaming & chmn steering comt, Nat Gaming Coun, 68; vchmn tech prog eight int conf, Inst Elec & Electronics Engrs Comput Soc, 74. Mem: AAAS; Sigma Xi. Res: Theory and application of simulation in the study of systems; design, analysis and evaluation of computer-based information systems. Mailing Add: 1331 Hillview Dr Menlo Park CA 94025

WALLACE, HAROLD DEAN, b Walnut, Ill, June 8, 22; m 45; c 4. ANIMAL

NUTRITION. Educ: Univ Ill, BS, 45, MS, 47; Cornell Univ, PhD(animal nutrit), 50. Prof Exp: From asst prof to assoc prof animal nutrit, 50-61, PROF ANIMAL NUTRIT, UNIV FLA, 61- Honors & Awards: Am Feed Mfrs Award, 62. Mem: AAAS; Am Soc Animal Sci; Am Dairy Sci Asn; Am Inst Nutrit. Res: Swine nutrition; vitamins; antibiotics; amino acids; carcass quality. Mailing Add: Dept of Animal Sci Univ of Fla Gainesville FL 32601

WALLACE, HELEN M, b Hoosick Falls, NY, Feb 18, 13. PUBLIC HEALTH. Educ: Wellesley Col, AB, 33; Columbia Univ, MD, 37; Harvard Univ, MPH, 43. Prof Exp: With New York Dept Health, 43-55; prof prev med, New York Med Col, 55-56; prof maternal & child health, Univ Minn, 56-59; chief child health studies, US Children's Bur, 59-62; PROF MATERNAL, CHILD & FAMILY HEALTH, SCH PUB HEALTH, UNIV CALIF, BERKELEY, 62- Concurrent Pos: WHO traveling fel, 57; consult, WHO, Uganda, 61, Philippines, 66, India, 68 & 69, Turkey, 69, Geneva, 70 & 74 & Iran, 72; Ford Found consult, Sch Pub Health, Univ Antioquaia, Colombia, 71; consult, Health Bur, Panama Canal Co, 72, India, Thailand, Burma & Ceylon, 75; dir Uganda Prog, Univ Calif. Mem: Asn Teachers Maternal & Child Health (pres); Am Pub Health Asn. Res: Maternal and child health. Mailing Add: Univ of Calif Sch of Pub Health Berkeley CA 94708

WALLACE, HERBERT STEPHEN, zoology, see 12th edition

WALLACE, HERBERT WILLIAM, b Brooklyn, NY, Dec 11, 30; m 54; c 3. BIOCHEMISTRY, SURGERY. Educ: Harvard Univ, AB, 52; Tufts Univ, MD, 56, MS, 60. Prof Exp: Asst attend surgeon, Elmhurst Gen Hosp Div, Mt Sinai Hosp, 65-66; assoc, 66-70, asst prof, 70-72, ASSOC PROF SURG, SCH MED, UNIV PA, 72-, ASSOC PHYSIOL, 70-, ASSOC PROF BIOENG, COL ENG & APPL SCI,74-; RES ASSOC, DIV CARDIOL, PHILADELPHIA GEN HOSP, 71- Concurrent Pos: Res fel, Nat Heart Inst, 58-59; teaching fel surg, Sch Med, Univ Pittsburgh, 61-62; Am Thoracic Soc fel, 62-65; John Polachek Found Med Res fel, Div Cardio-Thoracic Surg, Mt Sinai Hosp, New York, 65-66; asst surgeon, Grad Hosp, Univ Pa, 66-74, assoc surg, 74-, assoc dir, Gen Clin Res Ctr, 67-70, actg dir, 70-73. Mem: AAAS; Asn Acad Surg; Am Chem Soc; Am Col Surg; Am Fedn Clin Res. Res: Biochemistry and physiology of extracorporeal circulation and respiration; lung metabolism. Mailing Add: Dept of Surg Univ of Pa Grad Hosp Philadelphia PA 19146

WALLACE, JACK E, b Harrisburg, Ill, Jan 5, 34; m 55; c 2. ANALYTICAL BIOCHEMISTRY. Educ: Univ Southern Ill, BA, 55, MA, 57; Purdue Univ, PhD(biochem), 61. Prof Exp: Instr chem, Univ Southern Ill, 55-57; instr biochem, Purdue Univ, 59-61; chemist anal lab, State Chemist's Lab, Ind, 57-59; chief forensic toxicol br, US Air Force Sch Aerospace Med, 61-72; ASSOC PROF PATH, UNIV TEX HEALTH SCI CTR, SAN ANTONIO, 72- Concurrent Pos: Consult, Army Med Lab, Ft Sam Houston, Tex, Audie Murphy Vet Admin Hosp, San Antonio, Tex, Wilford Hall US Air Force Med Ctr, Lackland AFB, Tex & Southwest Bioclin Lab, San Antonio, Tex. Mem: Sr mem Am Chem Soc; fel Am Inst Chem; fel Am Acad Forensic Sci; Am Acad Clin Toxicol; Am Asn Clin Chemists. Res: Forensic toxicology; enzymology; drug metabolism; drug analysis in biological specimens; pharmacology; clinical chemistry; biochemical pathology. Mailing Add: Dept of Path Univ of Tex Health Sci Ctr San Antonio TX 78284

WALLACE, JAMES, b New Brunswick, NJ, Oct 6, 32; m 58; c 5. OPTICS, FLUID MECHANICS. Educ: US Merchant Marine Acad, BS, 54; Univ Notre Dame, MS, 59; Brown Univ, PhD(fluid mech), 63. Prof Exp: Res engr, Bendix Corp, 58-59; prin res scientist optics, Fluid Mech, Avco Everett Res Lab & Avco Systs Div, 63-73; PRES, FAR FIELD, INC, 73- Mem: Optical Soc Am; Inst Elec & Electronics Engrs. Res: Theoretical optics; lasers; fluid mechanics. Mailing Add: 6 Thoreau Way Sudbury MA 01776

WALLACE, JAMES BRUCE, b Williamsburg Co, SC, Mar 2, 39; m 62; c 1. ENTOMOLOGY, HYDROBIOLOGY. Educ: Clemson Univ, BS, 61; Va Polytech Inst, MS, 63, PhD(entom), 67. Prof Exp: Res asst entom, Va Polytech Inst, 61-66; asst prof, 67-71, ASSOC PROF ENTOM, UNIV GA, 71-, MEM STAFF, INST ECOL, 69- Concurrent Pos: Environ Protection Agency res grant, Univ Ga, 68-72; NSF res grant, 74-76. Mem: AAAS; Royal Entom Soc London; Am Entom Soc; Entom Soc Am; NAm Benthol Soc. Res: Immature insects; aquatic entomology; taxonomy and biology; ecology and biology of filter feeding insects. Mailing Add: Dept of Entom Univ of Ga Athens GA 30602

WALLACE, JAMES MERRILL, b Ripley, Miss, Oct 13, 02; m 29; c 1. PLANT PATHOLOGY. Educ: Miss State Col, BS, 23; Univ Minn, MS, 27, PhD(plant path), 29. Prof Exp: Asst pathologist, State Plant Bd, Miss, 24-25; instr plant path, Univ Minn, 25-28; assoc botanist & plant pathologist exp sta, Clemson Col, 28-29; assoc plant pathologist div sugar plant invests bur plant indust, USDA, 29-41; assoc plant pathologist citrus exp sta, 42-48, plant pathologist, 48-70, prof plant path, 61-70, EMER PROF PLANT PATH, UNIV CALIF, RIVERSIDE, 70- Concurrent Pos: Rockefeller travel grant, SAm, 59 & Japan, 63; consult, US Opers Mission, Israel, 55, Cyprus Palestine Plantations Co, Cyprus, 55 & 66, SAfrican Coop Citrus Exchange, 59, Bur Plant Indust, Philippines, 63, Joint Comn Rural Reconstruct, Taiwan, 63, Farmers Asn, Peru, 64, USAID, India, 67 & Morocco, 70. Mem: Int Orgn Citrus Virol (1st pres, 57-60); fel Am Phytopath Soc; Indian Phytopath Soc. Res: Plant virology; virus diseases of citrus and avocado; immunological and defense reactions in plants. Mailing Add: Dept of Plant Path Univ of Calif Riverside CA 92502

WALLACE, JAMES WILLIAM, JR, b Cincinnati, Ohio, July 31, 40; m 62; c 2. PLANT BIOCHEMISTRY, PLANT PHYSIOLOGY. Educ: Miami Univ, BS, 62, MS, 64; Univ Tex, Austin, PhD(plant biochem), 67. Prof Exp: Technician, Kimberly-Clarke Corp, 62; asst prof, 67-72, ASSOC PROF BIOL, WESTERN CAROLINA UNIV, 72- Mem: Am Chem Soc; Bot Soc Am; Phytochem Soc NAm. Res: Flavanoids, biosynthesis and physiology. Mailing Add: Dept of Biol Western Carolina Univ Cullowhee NC 28723

WALLACE, JOAN M, b Rochester, NY, Mar 7, 28. PLANT BIOCHEMISTRY. Educ: Cornell Univ, BS, 51; Rutgers Univ, MS, 54, PhD(plant physiol), 57. Prof Exp: Res fel biol, Calif Inst Technol, 58-63; RES CHEMIST, WESTERN REGIONAL RES LAB, USDA, 63- Mem: AAAS; Am Soc Plant Physiol; Am Inst Biol Sci; Sigma Xi. Res: Phytochemistry of plant-predator relationships. Mailing Add: Western Regional Res Lab USDA Berkeley CA 94710

WALLACE, JOHN DOYLE, acoustics, deceased

WALLACE, JOHN GEORGE, organic chemistry, see 12th edition

WALLACE, JOHN HOWARD, b Cincinnati, Ohio, Mar 8, 25; m 45; c 2. MICROBIOLOGY, IMMUNOLOGY. Educ: Howard Univ, BS, 47; Ohio State Univ, MS, 49, PhD(bact), 53. Prof Exp: Asst virol, Children's Hosp Res Found, Cincinnati, 47 & 49-51; asst bact, Ohio State Univ, 51-53, asst instr, 53; asst bacteriologist, Leonard Wood Mem Lab & res assoc bact & immunol, Harvard Univ Med Sch, 55-59; from asst prof to prof microbiol, Meharry Med Col, 59-66; from assoc prof to prof, Sch Med, Tulane Univ, 66-70; prof, Ohio State Univ, 70-72; PROF MICROBIOL & IMMUNOL & CHMN DEPT MICROBIOL, SCH MED, UNIV LOUISVILLE, 72- Concurrent Pos: USPHS fel, Ohio State Univ, 54-55; USPHS sr res fel, 59-61; NIH career res develop award, 61-66; mem bd sci counr, Nat Inst Allergy & Infectious Dis, 72-75, chmn, 75-76. Mem: Am Asn Immunol; Transplantation Soc; fel Am Acad Microbiol; Am Asn Cancer Res; Soc Exp Biol & Med. Res: Virus modified erythrocytes; in vitro cultivation of murine leprosy bacilli; tissue culture in agar systems; delayed hypersensitivity; immunology of leprosy; tissue transplantation; tumor immunology; effects of tobacco smoke on immune responses. Mailing Add: Sch of Med Dept of Microbiol & Immunol Univ Louisville Health Sci Ctr Louisville KY 40201

WALLACE, JOHN LONGSTREET, b Philadelphia, Pa, Dec 14, 45; m 64; c 2. EXPERIMENTAL SOLID STATE PHYSICS. Educ: Temple Univ, AB, 64; Calif Inst Technol, MS, 66, PhD(physics), 71. Prof Exp: RES SCIENTIST, AIL, DIV CUTLER HAMMER INC, 71- Res: The development of gallium arsenide microwave semiconductor devices. Mailing Add: 43 Reynolds St Huntington Station NY 11746

WALLACE, JOHN M, b St Louis, Mo, Aug 4, 25; m 52; c 5. MEDICINE. Educ: Wash Univ, AB, 46, MD, 50. Prof Exp: Assoc med & chief cardiol, Durham Vet Admin Hosp, NC, 58-62; from asst prof to assoc prof med, 62-71, PROF INTERNAL MED, UNIV TEX MED BR GALVESTON, 71- Concurrent Pos: Fel, Duke Univ, 55-58. Res: Cardiology; hypertension. Mailing Add: Div of Cardiol Univ of Tex Med Br Galveston TX 77550

WALLACE, JOHN MICHAEL, b Flushing, NY, Oct 28, 40; c 2. METEOROLOGY. Educ: Webb Inst Naval Archit, BS, 62; Mass Inst Technol, PhD(meteorol), 66. Prof Exp: Asst prof, 66-70, ASSOC PROF ATMOSPHERIC SCI, UNIV WASH, 70-, ADJ ASSOC PROF ENVIRON STUDIES, 73- Honors & Awards: Macelwane Award, Am Geophys Union, 72. Mem: Am Meteorol Soc; Am Geophys Union; Meteorol Soc Japan. Res: General circulation; tropical meteorology. Mailing Add: Dept of Atmospheric Sci Univ of Wash Seattle WA 98195

WALLACE, KYLE DAVID, b Nancy, Ky, Apr 3, 43; m 64; c 2. MATHEMATICS. Educ: Eastern Ky State Col, BS, 63; Vanderbilt Univ, MS, 65, PhD(math), 70. Prof Exp: Instr math, Easten Ky Univ, 65-67; asst prof, 70-75, ASSOC PROF MATH, WESTERN KY UNIV, 75- Mem: Am Math Soc; Math Asn Am. Res: Infinite Abelian groups; structure and classification of groups. Mailing Add: Dept of Math Western Ky Univ Bowling Green KY 42101

WALLACE, OLIVER P, SR, b Claremont, NH, June 30, 14; m 40; c 4. FORESTRY. Educ: Univ NH, BS, 37; Univ Mich, BSF, 38, MF, 47, PhD, 53. Prof Exp: Forester, Northeastern Timber Salvage Admin, US Forest Serv, 38-43; head dept forestry, Paul Smith's Col, 47-51; ASSOC PROF FORESTRY, UNIV NH, 53- Mem: Soc Am Foresters. Res: Forest marketing; forest resource and forest industry economics. Mailing Add: Pettee Hall Univ of NH Durham NH 03824

WALLACE, PAUL FRANCIS, b Tyrone, Pa, June 11, 27; m 51; c 3. SURFACE CHEMISTRY. Educ: Pa State Univ, BS, 50, MS, 52. Prof Exp: Res asst ceramics, Pa State Univ, 50-52; res engr, Res Labs, Carborundum Co, 52-54; res engr, 54-70, SR SCIENTIST, CHEM METAL DIV, ALCOA RES LABS, ALUMINUM CO AM, 70- Mem: The Electrochem Soc. Res: Surface chemistry of aluminum and aluminum alloys as related to fabricating processes, finishes and mechanisms of corrosion. Mailing Add: Alcoa Labs Alcoa Tech Ctr Alcoa Ctr PA 15069

WALLACE, PHILIP RUSSELL, b Toronto, Ont, Apr 19, 15; m 40; c 3. THEORETICAL PHYSICS. Educ: Univ Toronto, BA, 37, MA, 38, PhD, 40. Prof Exp: Instr math, Univ Cincinnati, 40-42; instr, Mass Inst Technol, 42; assoc res physicist, Atomic Energy Div, Nat Res Coun, 43-46; from assoc prof to prof appl math, 46-63, dir inst theoret physics, 66-70, PROF PHYSICS, PHYS SCI CTR, McGILL UNIV, 63- Concurrent Pos: Mem comn higher educ, Super Coun Educ, Que, 70-73; mem grant selection comt physics, Nat Res Coun Can, 71-74; ed, Can J Physics, 73- Mem: Am Phys Soc; Am Asn Physics Teachers; Can Asn Physicists; Europ Phys Soc; Royal Soc Can. Res: Theoretical and solid state physics; physics of semiconductors and semimetals in intense magnetic fields. Mailing Add: Dept of Physics McGill Univ Montreal PQ Can

WALLACE, RAPHAEL HERMAN, b NS, Aug 24, 15; m 45; c 4. MICROBIOLOGY. Educ: Dalhousie Univ, BSc, 39; McGill Univ, MSc, 46, PhD(agr bact), 48. Prof Exp: Agr scientist cent exp farm, Can Dept Agr, 46-50; from asst prof to assoc prof agr bact, Macdonald Col, McGill Univ, 50-58; lab dir, Dow Brewery, Ltd, 58-66; qual control mgr, Can Breweries Que Ltd, 66-72, QUAL CONTROL MGR, O'KEEFE BREWING CO LTD, 72- Concurrent Pos: Nuffield Found travel grant, 55. Mem: Am Soc Microbiol; Soc Indust Microbiol; Am Soc Brewing Chem; Master Brewers Asn Am; Can Soc Microbiol. Res: Soil, food and brewing microbiology; qualitative studies; rhizosphere of plants; fruits and fruit products. Mailing Add: O'Keefe Brewing Co Ltd 990 Notre Dame St W Montreal PQ Can

WALLACE, REUBEN HENRY, b Coolidge, Tex, Sept 13, 22; m 51; c 3. PHYSICS. Educ: Univ Tex, BS, 51. Prof Exp: Mem electronics group, Underwater Sound Lab, Harvard Univ, 42-45; mem staff, US Naval Underwater Sound Lab, 45; from mem staff to res physicist & head acoust div, 51-67, asst dir, 67-70, ASSOC DIR APPL RES LAB, UNIV TEX, AUSTIN, 70- Mem: Acoust Soc Am. Res: Sonar mechanisms; electronics; acoustics. Mailing Add: Appl Res Lab Univ of Tex Box 8029 Austin TX 78712

WALLACE, RICHARD MAITHA, b Ellwood City, Pa, Sept 29, 21; m 42; c 5. PHYSICAL CHEMISTRY. Educ: Bethany Col, WVa, BS, 43; Univ Kans, PhD(chem), 53. Prof Exp: Chemist, Hercules Powder Co, 43-45; instr chem, ETenn State Col, 48-50; chemist, 52-62, RES ASSOC, SAVANNAH RIVER LAB, E I DU PONT DE NEMOURS & CO, INC, 62- Mem: Am Chem Soc. Res: Membrane equilibrium; ruthenium chemistry; nuclear waste disposal; actinide chemistry; uranium geochemistry. Mailing Add: Savannah River Lab E I du Pont de Nemours & Co Aiken SC 29801

WALLACE, ROBERT ALLAN, b Chicago, Ill, Aug 23, 30. BIOCHEMISTRY, ORGANIC CHEMISTRY. Educ: Northern Ill Univ, BS, 53; Univ Bonn, MS, 57, PhD(org chem), 59. Prof Exp: Fel, Calif Inst Technol, 60-62; assoc prof biochem, 62-71, PROF BIOCHEM, HUMBOLDT STATE UNIV, 71- Mem: AAAS; Am Chem Soc; Soc Ger Chem; Ger Soc Nature Study. Res: Enzyme and pteridine chemistry. Mailing Add: Dept of Chem Humboldt State Univ Arcata CA 95521

WALLACE, ROBERT B, b Stoneham, Mass, Jan 16, 37. PSYCHOBIOLOGY. Educ: Boston Univ, AB, 60, AM, 61, PhD(psychol), 66. Prof Exp: Instr psychol, 66-67, lectr, 67-68; asst prof, 68-72, ASSOC PROF PSYCHOL, UNIV HARTFORD, 72- Concurrent Pos: Res assoc, Mass Inst Technol, 66-68 & Inst Living, 71-; vis assoc prof, Univ Conn Health Ctr, 74- Mem: AAAS; Am Psychol Asn; Soc Neurosci; Psychonomic Soc; NY Acad Sci. Res: Relation of central nervous system structure to behavior; plasticity of mammalian nervous system; postnatal neurogenesis. Mailing Add: Dept of Psychol Univ of Hartford 200 Bloomfield Ave West Hartford CT 06117

WALLACE, ROBERT BRUCE, b Washington, DC, Apr 12, 31; m 55; c 3. THORACIC SURGERY, CARDIOVASCULAR SURGERY. Educ: Columbia Col, BA, 53; Columbia Univ, MD, 57. Prof Exp: From intern to resident surg, St Vincents Hosp, New York, 57-62; resident thoracic surg, Baylor Univ, 72-73; CONSULT GEN, THORACIC & CARDIOVASC SURG, MAYO CLIN, ROCHESTER, MINN,64-, PROF SURG & CHMN DEPT, 68- Concurrent Pos: Spec fel cardiovasc surg, Mayo Grad Sch Med, Univ Minn, 63-64. Mem: Int Cardiovasc Soc; Am Col Surg; Am Asn Thoracic Surg; Soc Clin Surg; Soc Vasc Surg (treas, 73-). Mailing Add: 200 First St SW Rochester NY 55901

WALLACE, ROBERT EARL, b New York, NY, July 16, 16; m 45; c 1. GEOLOGY. Educ: Northwestern Univ, BS, 38; Calif Inst Technol, MS, 40, PhD(struct geol, vert paleont), 46. Prof Exp: Geologist, 42-70, chief southwestern br, 60-65, chief, Nat Ctr Earthquake Res, 72-73, REGIONAL GEOLOGIST, US GEOL SURV, 70-, CHIEF SCIENTIST, OFF EARTHQUAKE STUDIES, 73- Concurrent Pos: Prof, Wash State Univ, 46-51; vis lectr, Stanford Univ, 60; mem comt seismol, Nat Acad Sci-Nat Res Coun; chmn earthquake eng res inst, US/USSR Environ Agreement, US Working Group Earthquake Prediction. Mem: Fel Geol Soc Am; Soc Econ Geol; Seismol Soc Am. Res: Active faults; tectonics; earthquakes; engineering geology and environment; mineral deposits. Mailing Add: US Geol Surv Menlo Park CA 94025

WALLACE, ROBERT HENRY, b Cleveland, Ohio. MEDICAL PHYSICS. Educ: Transylvania Col, BA, 64; Purdue Univ, West Lafayette, MS, 68, PhD(bionucleonics), 70. Prof Exp: Assoc radiol health physicist, Ky State Dept Health, 65-66; ASST PROF RADIOL, ALBANY MED COL, 70-, RADIATION PHYSICIST, ALBANY MED CTR HOSP, 70- Mem: Am Asn Physicists in Med; Soc Nuclear Med; Health Physics Soc. Res: Radiation dosimetry and protection; nuclear medicine instrumentation. Mailing Add: Radiol Dept Med Physics Div Albany Med Ctr Hosp Albany NY 12208

WALLACE, ROBERT WILLIAM, b Central Falls, RI, Jan 1, 43; m 65; c 2. SCIENCE EDUCATION, ENVIRONMENTAL CHEMISTRY. Educ: Providence Col, BS, 64; Niagara Univ, MS, 66; Boston Univ, PhD(chem), 73. Prof Exp: ASST PROF CHEM, BENTLEY COL, 72- Mem: Am Chem Soc; AAAS. Res: Air and water quality as well as the quality of consumer products, including chemical composition versus the list of ingredients. Mailing Add: Dept of Sci Bentley Col Beaver & Forest St Waltham MA 02154

WALLACE, ROBERTS MANNING, b Chicago, Ill, Feb 21, 15; m 47; c 2. ECONOMIC GEOLOGY. Educ: Beloit Col, AB, 38; Univ Ariz, BS, 49, MS, 51, PhD(geol), 55. Prof Exp: GEOLOGIST, US GEOL SURV, 54- Honors & Awards: Meritorious Serv Award, Int Coop Admin, 59. Mem: Soc Econ Geol; fel Geol Soc Am. Res: Aerial photogeology with special interest in interpretation of geologic structures as seen from the air. Mailing Add: US Geol Surv Cent Environ Br 2035 Regency Rd Lexington KY 40503

WALLACE, ROBIN A, b Chicago, Ill, Nov 11, 33; m 55; c 2. REPRODUCTIVE BIOLOGY. Educ: Columbia Univ, BA, 55, PhD(zool), 61. Prof Exp: Res assoc, Sloan-Kettering Inst, NY, 57; consult, 60-61, USPHS fel, 61-63, STAFF MEM BIOL DIV, OAK RIDGE NAT LAB, 63- Concurrent Pos: Vis investr, Nat Res Coun Can, 63-64; mem, Marine Biol Lab Corp; dir reprod biol prog, Marine Biol Lab, Woods Hole, 74 & 75; co-ed, Develop Biol, 72-74; mem develop biol panel, NSF, 75-76. Mem: AAAS; Am Soc Cell Biol; Soc Develop Biol; Fedn Am Sci; Soc Gen Physiol. Res: Comparative biochemical studies on yolk proteins and the mechanisms of oocyte growth. Mailing Add: Biol Div Oak Ridge Nat Lab PO Box Y Oak Ridge TN 37830

WALLACE, ROGER WAYNE, JR, nuclear physics, see 12th edition

WALLACE, RONALD GARY, b Cadiz, Ohio, July 6, 38; m 65; c 1. GEOMORPHOLOGY. Educ: Kent State Univ, BS, 61; Ohio State Univ, MS, 64, PhD(geol), 67. Prof Exp: Petroleum geologist, Standard Oil Co, Tex, 67-69; asst prof geol, Ohio State Univ, 69-70; ASST PROF GEOL, EASTERN ILL UNIV, 70- Mem: Geol Soc Am. Res: Alpine mass movement; strip mine erosion. Mailing Add: Dept of Geog & Geol Eastern Ill Univ Charleston IL 61920

WALLACE, RONALD RICHARD, b Sask, Mar 23, 46. FRESH WATER ECOLOGY. Educ: Univ Sask, BA, 68; Queen's Univ, Ont, MSc, 70; Univ Waterloo, PhD(freshwater ecol), 73. Prof Exp: Sr biologist, Northwest Territories Dist Off, Environ Protection Serv, 73-75; CHMN & PROJ LEADER AQUATIC FAUNA, ALTA OIL SANDS ENVIRON RES PROG & RES SCIENTIST, FRESHWATER INST, ENVIRON CAN, 75- Concurrent Pos: Spec consult, Onchocerciasis Prog, WHO, 75-76 & Northern Assessment Group, Berger Pipeline Enquiry, 75; panelist & author, Environ Secretariat on Phenoxy-herbicides, Nat Res Coun Can, 75; panelist, Northern Develop Enquiry, Sci Coun Can, 75-76. Mem: Benthological Soc NAm; Brit Freshwater Biol Asn. Res: Effects of chemical pollutants, particularly pesticides, on the invertebrata of running waters. Mailing Add: Freshwater Inst Environ Can 501 University Crescent Winnipeg MB Can

WALLACE, SIDNEY, b Philadelphia, Pa, Feb 26, 29; m; c 3. RADIOLOGY. Educ: Temple Univ, BA, 49, MD, 54; Am Bd Radiol, dipl, 62. Prof Exp: Intern, Philadelphia Gen Hosp, 54-55; resident radiol, Hosp, Jefferson Med Col, 59-62, from instr to asst prof, Col, 62-66; assoc prof, 66-69, PROF RADIOL, UNIV TEX M D ANDERSON HOSP & TUMOR INST HOUSTON, 69- Concurrent Pos: Fel radiol, Univ Lund, 63-64. Mem: AMA; Am Col Radiol; Int Soc Lymphology. Res: Lymphangiography and angiography. Mailing Add: M D Anderson Hosp & Tumor Inst 6723 Bertner Ave Houston TX 77025

WALLACE, STEPHEN CHARLES, chemical physics, see 12th edition

WALLACE, STEPHEN JOSEPH, b Youngstown, Ohio, May 10, 39; m 61; c 2. THEORETICAL NUCLEAR PHYSICS. Educ: Case Inst Technol, BS, 61; Univ Wash, MS, 69, PhD(physics), 71. Prof Exp: Res engr, Boeing Co, 61-68; res asst physics, Univ Wash, 68-71; res assoc, Univ Fla, 71-72 & Harvard Univ, 72-74; ASST PROF PHYSICS, UNIV MD, COLLEGE PARK, 74- Mem: Am Phys Soc. Res: Intermediate to high energy nuclear multiple scattering theory; Eikonal expansion methods. Mailing Add: Dept of Physics & Astron Univ of Md College Park MD 20742

WALLACE, STEWART RAYNOR, b Freeport, NY, Mar 31, 19; m 46; c 2. GEOLOGY. Educ: Dartmouth Col, BA, 41; Univ Mich, MS, 48, PhD(geol), 53. Prof Exp: Geologist, US Geol Surv, 48-55; resident geologist, Climax Molybdenum Co, 55-58, chief geologist, 58-64, chief geol & explor, 64-69; PRES & DIR EXPLOR, MINE FINDERS, INC, 70- Honors & Awards: D C Jackling Award, Am Inst Mining, Metall & Petrol Engrs, 74. Mem: Geol Soc Am; Geochem Soc; Am Inst Mining, Metall & Petrol Engrs; Soc Econ Geol; Asn Explor Geochemists. Res: Genesis of metallic ore deposits; igneous and metamorphic structure and petrology related to ore deposits. Mailing Add: Route 3 Box 462 Golden CO 80401

WALLACE, SUSAN SCHOLES, b Brooklyn, NY, Jan 10, 38; m 60; c 3. MOLECULAR BIOLOGY, BIOPHYSICS. Educ: Marymount Col, NY, BS, 59; Univ Calif, Berkeley, MS, 61; Cornell Univ, PhD(biophys), 65. Prof Exp: USPHS fel, Columbia Univ, 65-67; instr biol sci, 67-68, asst prof, 69-73, ASSOC PROF BIOL SCI, LEHMAN COL, 73- Concurrent Pos: City Univ New York res grant, Lehman Col, 68-72, NIH res grant, 72-; vis prof, Albert Einstein Col Med, 74-75; scholar, Am Cancer Soc, 74-75. Mem: AAAS; Biophys Soc; Radiation Res Soc; Am Soc Microbiologists. Res: Radiation repair processes in bacteria and bacteriophages; in vivo and in vitro enzymatic repair of ionizing radiation damage. Mailing Add: Dept of Biol Lehman Col Bronx NY 10468

WALLACE, TERRY CHARLES, b Phoenix, Ariz, May 18, 33; m 55; c 5. PHYSICAL CHEMISTRY. Educ: Ariz State Univ, BS, 55; Iowa State Univ, PhD(phys chem), 58. Prof Exp: Mem staff, 58-70, ALT GROUP LEADER, LOS ALAMOS SCI LAB, UNIV CALIF, 70- Concurrent Pos: Chief tech projs br, Environ Test Div, Dugway Proving Ground. Mem: AAAS; Am Chem Soc; Am Inst Chemists. Res: Structural properties and mass transport of materials at high temperature; high temperature thermodynamics; chemical vapor deposition. Mailing Add: Los Alamos Sci Labs PO Box 1663 Los Alamos NM 87544

WALLACE, THOMAS HOMKOWYCZ, b Boston, Mass, Aug 27, 12; m 47; c 1. SPECTROSCOPY. Educ: Boston Univ, BS, 33, PhD(physics), 39; Harvard Univ, AM, 36. Prof Exp: Asst physics, Boston Univ, 34-35; asst, Williams Col, 35-38; from instr math & physics to asst prof physics, Northeastern Univ, 39-42; asst prof, Bowdoin Col, 43-44; asst prof, Williams Col, 44-45; tech asst to historian, Off Sci Res & Develop, 45; from asst prof to assoc prof, 46-60, PROF PHYSICS, NORTHEASTERN UNIV, 60-, CHMN DEPT, LINCOLN COL, 41- Mem: Am Phys Soc; Am Asn Physics Teachers. Res: Separations of the components of the H-alpha complex of hydrogen. Mailing Add: Dept of Physics Northeastern Univ Boston MA 02043

WALLACE, THOMAS J, physical organic chemistry, see 12th edition

WALLACE, THOMAS PATRICK, b Washington, DC, Apr 11, 35; m 59; c 3. PHYSICAL CHEMISTRY, POLYMER SCIENCE. Educ: State Univ NY Col Potsdam, BS, 58; Syracuse Univ, MS, 61; St Lawrence Univ, MS, 64; Clarkson Col Technol, PhD(phys chem), 68. Prof Exp: Asst prof chem, State Univ NY Col Potsdam, 61-67; asst prof, 68-70, head dept chem, 70-72, assoc dean col sci, 72-73, ASSOC PROF CHEM, ROCHESTER INST TECHNOL, 70-, DEAN COL SCI, 73- Mem: AAAS; Am Chem Soc. Res: Light scattering; polymer latex systems and characterization; size distribution analysis; statistical thermodynamics of dilute polymer solutions. Mailing Add: Col of Sci Rochester Inst of Technol Rochester NY 14623

WALLACE, TRACY I, b Irvine, Ky, Nov 20, 24; m 51; c 3. INTERNAL MEDICINE. Educ: Univ Ky, BS, 46; Univ Cincinnati, MD, 49. Prof Exp: Staff physician, Vet Admin Hosp, McKinney, Tex, 55-57, asst chief med serv, 57-59, CHIEF MED SERV, TEMPLE VET ADMIN HOSP, 61- Mem: AMA. Res: Adrenal function in patients with pulmonary emphysema and other hypoxic states. Mailing Add: Vet Admin Ctr Temple TX 76501

WALLACE, VICTOR LEW, b Brooklyn, NY, Mar 20, 33; m 62. COMPUTER SCIENCE. Educ: Polytech Inst Brooklyn, BS, 55; Univ Mich, PhD(elec eng), 69. Prof Exp: Mem tech staff syst eng, Bell Tel Labs, 55-56; mathematician & programmer, IBM Corp, 56-57; instr elec eng, Univ Mich, 57-62; assoc res scientist, 62-69; assoc prof comput sci, Univ NC, 69-76; PROF & CHMN COMPUT SCI, UNIV KANS, 76- Concurrent Pos: Vis scientist, Imp Col, Univ London, 70. Mem: Asn Comput Mach; Inst Elec & Electronics Engrs; Inst Mgt Sci; Am Asn Univ Profs. Res: Computer system modeling; operating system theory; computer graphics software; man-machine interface in computer-aided design. Mailing Add: Dept of Comput Sci Strong Hall Univ of Kans Lawrence KS 66044

WALLACE, VOLNEY, b Idaho Falls, Idaho, Oct 9, 25; m 44; c 5. AGRICULTURAL CHEMISTRY. Educ: Univ Idaho, BS, 46; Purdue Univ, MS, 49, PhD(agr chem), 53. Prof Exp: Asst agr chem, Purdue Univ, 47-53; res assoc, Wash State Univ, 53-55; asst prof biochem, SDak State Col, 55-61; RES CHEMIST, DESERET TEST CTR, 61- Mem: Am Chem Soc. Res: Carotenes; ruminant bloat; vegetable silica; enzymatic analysis of anticholinesterase agents; analysis of alkaloids and fatty ions by liquid-liquid extraction of colored salts; solar energy; self-sufficiency gardening. Mailing Add: Box 534 Dugway UT 84022

WALLACE, WAYNE ALEXANDER, b Los Angeles, Calif, Nov 10, 11; m 45; c 2. RESOURCE GEOGRAPHY, GEOGRAPHY OF THE SOVIET UNION. Educ: Univ Calif, Los Angeles, BA, 33, MA, 42. Prof Exp: From asst prof to assoc prof geog, 46-70, ASSOC PROF ENVIRON SCI, UNIV VA, 70-, ASST DEAN COL ARTS & SCI, 65- Concurrent Pos: Expert consult, Amoco Int, 59- Mem: Asn Am Geog; Am Geog Soc; Royal Geog Soc. Res: Territorial claims and conflicts on the Continental Shelf. Mailing Add: Col of Arts & Sci Univ of Va Charlottesville VA 22903

WALLACE, WILLIAM DONALD, b Detroit, Mich, Sept 19, 33. SOLID STATE PHYSICS. Educ: Eastern Mich Univ, BA, 55; Univ Md, College Park, MS, 60; Wayne State Univ, PhD(physics), 66. Prof Exp: Asst prof physics, Eastern Mich Univ, 59-62; res asst, Cornell Univ, 67-70; asst prof, 70-74, ASSOC PROF PHYSICS, OAKLAND UNIV, 74- Concurrent Pos: Leverhulme vis fel, Univ Essex, 66-67. Mem: AAAS; Am Phys Soc; Am Asn Physics Teachers. Res: Ultrasonic properties and magnetic properties of solids; metals and low temperature physics. Mailing Add: Dept of Physics Oakland Univ Rochester MI 48063

WALLACE, WILLIAM EDWARD, b Fayette, Miss, Mar 11, 17; m 47; c 3. SOLID STATE CHEMISTRY. Educ: Miss Col, BA, 36; Univ Pittsburgh, PhD(chem), 41. Prof Exp: Asst, Univ Pittsburgh, 36-40; Carnegie Found fel, 40-42, sr res fel, 42-44; res assoc, Ohio State Univ, 44-45; from asst res prof to assoc res prof chem, 45-53, PROF CHEM, UNIV PITTSBURGH, 53-, CHMN DEPT,. 63- Mem: AAAS; Am Chem Soc. Res: Magnetic and low temperature heat capacities of intermetallic compounds containing lanthanides; lanthanide hydrides; heterogeneous catalysis and surface science. Mailing Add: Dept of Chem Univ of Pittsburgh Pittsburgh PA 15213

WALLACE, WILLIAM EDWARD, JR, b Charleston, WVa, Sept 25, 42; m 70; c 1. CHEMICAL PHYSICS, BIOPHYSICS. Educ: WVa Univ, BS, 63, MS, 67, PhD(physics), 69. Prof Exp: Nat Res Coun-US Bur Mines res assoc, Morgantown Energy Res Ctr, US Bur Mines, 70-71, res physicist, 71-74, SUPVRY RES PHYSICIST, MORGANTOWN ENERGY RES CTR, US ENERGY RES & DEVELOP ADMIN, 74- Res: Chemical and surface physics studies of solids removal from coal derived liquids; biophysics studies of pulmonary surfactant interactions with solids. Mailing Add: Morgantown Energy Res Ctr US Energy Res & Develop Admin Morgantown WV 26505

WALLACE, WILLIAM ELDRED, b Wichita, Kans, Feb 8, 17; m 41; c 3. CHEMISTRY. Educ: Univ Wichita, BS, 50; Univ Ill, MS, 42, PhD(org chem), 44. Prof Exp: RES CHEMIST, CHEM DIV, GAF CORP, 44- Mem: Am Chem Soc; Am Asn Textile Chemists & Colorists. Res: Grignard reaction; Friedel-Crafts reaction; synthetic organic chemicals; azoic dyes; synthetic dyestuffs. Mailing Add: RD 1 Box 190A Rensselaer NY 12144

WALLACE, WILLIAM HUSTON, b Chicago, Ill, Dec 10, 24; m 50; c 2. ECONOMIC GEOGRAPHY, GEOGRAPHY OF NEW ENGLAND. Educ: Beloit Col, BS, 48; Univ Wis, MA, 50, PhD(geog), 56. Prof Exp: Asst geog, Univ Wis, 48-52; lectr, Univ Auckland, 52-54; instr, Rutgers Univ, 54-56, asst prof, 56-57; from asst prof to assoc prof, 57-65, PROF GEOG, UNIV N H, 65- Concurrent Pos: John Simon Guggenheim fel, Univ NH, 63-64; mem & chmn, Grad Record Exam Comt Geog, 65-72; Fulbright lectr, Univ Oslo, 70-71. Mem: Asn Am Geog; Am Geog Soc; Can Asn Geog. Res: Geography of transportation, especially rail transport; geography of manufacturing; historical geography. Mailing Add: Dept of Geog James Hall Univ of NH Durham NH 03824

WALLACE, WILLIAM J, b Knoxville, Tenn, July 27, 35; m 58; c 2. INORGANIC CHEMISTRY. Educ: Carson-Newman Col, BS, 56; Purdue Univ, PhD(inorg chem), 61. Prof Exp: Asst prof inorg chem, Univ Miss, 60-63; from asst prof to assoc prof inorg & phys chem, 63-72, actg chmn dep, 68-69, PROF INORG & PHYS CHEM & COORDR SCI DIV, MUSKINGUM COL, 72- Concurrent Pos: Res grants, Res Corp, 61-63; res grants, Am Acad Arts & Sci, 62-63; dir, NSF Res Partic High Sch Teachers, 66, 67 & 68; res fel with P S Braterman, Univ Glasgow, 70-71. Mem: AAAS; Am Chem Soc. Res: Study of systems using polyether solvents with inorganic compounds, especially solubility, reaction and spectral phenomena; iron carbonyl reactions with Lewis bases. Mailing Add: Dept of Chem Muskingum Col New Concord OH 43762

WALLACE, WILLIAM JAMES LORD, b Salisbury, NC, Jan 13, 08; m 29; c 1. PHYSICAL CHEMISTRY. Educ: Univ Pittsburgh, BS, 27; Columbia Univ, AM, 31; Cornell Univ, PhD(phys chem), 37; Livingstone Col, LLD, 59; Concord Col, LHD, 70; Alderson Broaddus Col, DSc, 71. Prof Exp: Instr chem, Livingston Col, 27-32; instr, Lincoln Univ, Mo, 32-33; instr, 33-34, from asst prof to assoc prof, 34-43, prof, 43-75, actg admin asst to pres, 44-45, admin asst, 45-50, actg pres, 52-53, pres, 53-75, EMER PROF CHEM & EMER PRES, WVA STATE COL, 75- Honors & Awards: Outstanding Civilian Serv Medal, Dept Army, 72; Annual Award, Educ Comn States, 75. Mem: Am Chem Soc. Res: Freezing points of aqueous solutions of alpha amino acids; teaching problems in general chemistry. Mailing Add: WVa State Col PO Box 416 Institute WV 25112

WALLACE, WILLIAM JOHN, inorganic chemistry, see 12th edition

WALLACE, WILLIE ROBERT, b New Orleans, La, Dec 5, 38; m 60; c 2. MICROBIOLOGY, DAIRY SCIENCE. Educ: Southeastern La Col, BS, 60; La State Univ, MS, 61, PhD(dairy bact), 63. Prof Exp: Asst prof vet sci, La State Univ, 63-66; asst prof microbiol, Northwestern State Col, La, 66-67; asst prof, Southeastern La Univ, 67-74; GEN MGR, GULF DAIRY ASN INC, 74- Mem: AAAS; Am Soc Microbiol; Conf Res Workers Animal Dis; Am Dairy Sci Asn; Inst Food Technol. Res: Molecular action of aflatoxin; Streptococcus agalactiae cell wall characterization and antigenic response to various fractions. Mailing Add: 1 Silman Ave Hammond LA 70401

WALLACE-HAAGENS, MARY JEAN, b Gary, Ind, Nov 10, 30. ENDOCRINOLOGY, REPRODUCTIVE PHYSIOLOGY. Educ: St Mary's Col, BS, 57; Univ Notre Dame, PhD(biol), 64. Prof Exp: Instr biol, Cardinal Cushing Col, 64-65; asst prof, St Mary's Col, Ind, 65-69; res assoc ctr pop studies, Harvard Sch Pub Health, 69-72; res fel gynec, Vincent Mem Hosp, Mass Gen Hosp, 72-73; ASST PROF BIOL, RI COL, 73- Concurrent Pos: Res assoc ctr pop studies, Harvard Med Sch, 66-67. Mem: AAAS; Endocrine Soc; Am Soc Zool. Res: Detection of ovulation; study of the menstrual cycle; radioimmunoassay and bioassay; sperm survival; effects of aging on ovarian function. Mailing Add: Dept of Biol RI Col Providence RI 02908

WALLACH, DONALD P, b New York, NY, Sept 16, 27; m 51. BIOCHEMISTRY, PHARMACOLOGY. Educ: Mich State Univ, BS, 47, MS, 48; Univ Wis, PhD(pharmacol), 53. Prof Exp: RES ASSOC PHARMACOL, UPJOHN CO, 56- Concurrent Pos: Fel pharmacol, Alumni Res Found, Univ Wis, 53-54; USPHS fel enzyme chem, Univ Kans Med Ctr, Kansas City, 54-56. Mem: Am Soc Pharmacol & Exp Therapeut; Am Soc Biol Chem. Res: Applications of enzymology to problems in pharmacology. Mailing Add: Upjohn Co Dept of Exp Biol 301 Henrietta St Kalamazoo MI 49001

WALLACH, EDWARD E, b Brooklyn, NY, Oct 8, 33; m 56; c 2. OBSTETRICS & GYNECOLOGY. Educ: Swarthmore Col, BA, 54; Cornell Univ, MD, 58. Prof Exp: Intern internal med, Cornell Med Div, Bellevue Hosp, New York, 58-59; resident obstet & gynec, Kings County Hosp, 59-63; assoc, 65-66, from asst prof to assoc prof, 66-71, PROF OBSTET & GYNEC, SCH MED, UNIV PA, 71-; DIR OBSTET & GYNEC, PA HOSP, 71- Concurrent Pos: Res fel reprod physiol, Worcester Found Exp Biol, 61-62; Josiah Macy, Jr Found fel, 65-; Lalor Found awards, 67-69. Mem: Am Col Obstetricians & Gynecologists; Soc Gynec Invest; Am Fertil Soc; Soc Study Reprod; Endocrine Soc. Res: Reproductive biology; ovarian physiology; gynecologic endocrinology; infertility; family planning. Mailing Add: Dept of Obstet & Gynec Pa Hosp Philadelphia PA 19107

WALLACH, JACQUES BURTON, b New York, NY, Jan 25, 26; m 53; c 3. MEDICINE, PATHOLOGY. Educ: Long Island Col Med, MD, 47; Am Bd Path, dipl, 55. Prof Exp: From instr to asst prof, 54-59, VIS ASST PROF PATH, ALBERT EINSTEIN COL MED, 59- Concurrent Pos: Resident fel & asst pathologist, Queens Gen Hosp Ctr, New York, 48-55; asst vis pathologist, Bronx Munic Hosp Ctr, 54-59; consult, NY Zool Park, 54-; vis asst prof, Rutgers Med Sch, Col Med & Dent NJ, 69-72, vis assoc prof, 72- Mem: Fel Am Soc Clin Path; fel Col Am Path; Am Col Cardiol; fel Am Col Physicians; NY Acad Sci. Res: Rheumatic heart disease; clinical pathology; comparative pathology; histochemistry; cardiology; cancer. Mailing Add: 18 Dartmouth Rd Cranford NJ 07016

WALLACH, MARSHALL BEN, b Buffalo, NY, June 15, 40; m 63; c 2. NEUROPHARMACOLOGY. Educ: Columbia Univ, BS, 62; Univ Minn, PhD(pharmacol), 67. Prof Exp: Assoc res scientist neuropharmacol, Dept Psychiat, Med Ctr, NY Univ, 68-70, res scientist, 70-71, instr, 71-72; NEUROPHARMACOLOGIST, SYNTEX RES, 73- Concurrent Pos: Consult, New York City Rand Inst, 71-72. Mem: AAAS; Am Soc Exp Pharmacol & Therapeut. Res: Models of psychiatric and neurological diseases; electrophysiology of sleep; neurohumoral mechanisms; anorexigens; antidepressants; neuroleptics; analgesics; antiparkinsonian agents; anticonvulsants; antitussives and psychotomimetic agents. Mailing Add: Dept Exp Pharmacol Syntex Res 3401 Hillview Ave Palo Alto CA 94304

WALLACH, MORTON LAWRENCE, physical chemistry, see 12th edition

WALLACH, STANLEY, b Brooklyn, NY, Dec 10, 28; m 54, 73; c 3. MEDICINE. Educ: Cornell Univ, AB, 48; Columbia Univ, MA, 49; State Univ NY Downstate Med Ctr, MD, 53. Prof Exp: Intern med, Univ Med Serv, Kings County Hosp, Brooklyn, 53-54; resident, Col Med, Univ Utah, 54-56; clin & res fel, Mass Gen Hosp, 56-57; from instr to prof, State Univ NY Downstate Med Ctr, 57-73, prog dir, USPHS Clin Res Ctr, 66-73; PROF MED & ASST CHMN DEPT, ALBANY MED COL, 73-; CHIEF MED, VET ADMIN HOSP, 73- Concurrent Pos: Vis physician, Kings County Hosp, Brooklyn, 57-73; career scientist, Health Res Coun City of New York, 61-71; consult, St Johns Episcopal Hosp, Brooklyn, 65-73; attend physician, State Univ Hosp, Brooklyn, 66-73; consult & lectr, US Naval Hosp, St Albans, NY, 66-73; attend, physician, Albany Med Ctr, 73- Honors & Awards: Hektoen Silver Award, AMA, 59. Mem: Endocrine Soc; Am Soc Clin Invest; Am Fedn Clin Res; Harvey Soc; fel Am Col Physicians. Res: Endocrine and metabolic diseases; radioisotopes; calcium, magnesium and mineral metabolism; metabolic bone disease. Mailing Add: Vet Admin Hosp Albany NY 12208

WALLACH, SYLVAN, b San Antonio, Tex, Jan 9, 14; m 38; c 5. MATHEMATICS. Educ: Rutgers Univ, BS, 34; Johns Hopkins Univ, PhD(math), 48. Prof Exp: Chemist Mass & Waldstein Co, NJ, 36-38; patent examr, US Patent Off, Washington, DC, 38-42; indust analyst, War Prod Bd, 42-45; instr math, Johns Hopkins Univ, 48-49; sr scientist atomic power div, Westinghouse Elec Corp, 49-52; sr scientist, Walter Kidde Nuclear Labs, Inc, 52-57; asst proj mgr, Gibbs & Cox, Inc, 57-61; chmn dept math, 62-72, PROF MATH, C W POST COL, LI UNIV, 62- Mem: AAAS; Am Math Soc; Math Asn Am. Mailing Add: 110 Crabapple Rd Manhasset NY 11030

WALLANDER, JEROME F, b Cato, Wis, Aug 29, 39; m 65; c 2. FOOD SCIENCE, BIOCHEMISTRY. Educ: Univ Wis, BS, 62, MS, 65, PhD(food sci), 68. Prof Exp: Sr scientist, Mead Johnson Nutritionals, 67-71, ASSOC DIR FOOD PROD DEVELOP, MEAD JOHNSON & CO, 71- Mem: Sigma Xi; Am Dairy Sci Asn; Inst Food Technol. Res: Milk lipase and milk protein studies. Mailing Add: Food Prod Develop Mead Johnson & Co Evansville IN 47721

WALLBANK, ALFRED MILLS, b Farmington, Mich, May 13, 25; m 49; c 3. VIROLOGY. Educ: Mich State Univ, BS, 48, MS, 53, PhD, 57. Prof Exp: Bacteriologist, Barry Labs, Inc, Mich, 48-50; bacteriologist, Blood Sterilization Proj, Henry Ford Hosp, Detroit, 50-52; asst microbiol & pub health, Mich State Univ, 53-56; res assoc & microbiologist, Dept Surg, Duke Univ, 56-59; res asst prof microbiol & virologist, Sch Vet Med, Univ Pa, 59-66; head dept virol, Va Inst Sci Res, 66-67; ASSOC PROF MICROBIOL, MED COL, UNIV MAN, 67- Mem: Am Soc Microbiol; Am Asn Cancer Res; Soc Exp Biol & Med; Tissue Cult Asn; Am Acad Microbiol. Res: Viruses. Mailing Add: Dept of Med Microbiol Univ of Man Med Col Winnipeg MB Can

WALLBRUNN, HENRY MAURICE, b Chicago, Ill, Apr 30, 18; m 53; c 4. ZOOLOGY. Educ: Univ Chicago, BS, 40, PhD, 51. Prof Exp: Instr zool, Univ Chicago, 50; asst prof biol, 51-62, ASSOC PROF ZOOL, UNIV FLA, 63- Concurrent Pos: Fel statist & zool, Univ Chicago, 53-54. Mem: Genetics Soc Am; Soc Study Evolution. Res: Genetics; genetics and evolution of orchids. Mailing Add: Dept of Zool Univ of Fla Gainesville FL 32603

WALLCAVE, LAWRENCE, b Schenectady, NY, Apr 21, 26; m 65. BIOCHEMISTRY. Educ: Univ Calif, Berkeley, BS, 48; Calif Inst Technol, PhD, 53. Prof Exp: Assoc prof biochem, 68-74, PROF BIOCHEM, EPPLEY INST MED CTR, UNIV NEBR, OMAHA, 74- Res: Chemical carcinogenesis; analytical biochemistry. Mailing Add: Eppley Inst Univ of Nebr Med Ctr Omaha NE 68105

WALLE, OSCAR THEODORE, b Utica, NY, Sept 3, 11; m 37; c 4. ZOOLOGY. Educ: Univ Tulsa, BS, 40; Univ Tex, MA, 47. Hon Degrees: DD, Concordia Sem, 61. Prof Exp: Instr phys sci, Concordia Col, Tex, 34-36; instr, Concordia Col Inst, 36-37; assoc prof natural sci, St Paul's Col, 40-52; pres, Calif Concordia Col, 52-56; PROF NATURAL SCI, CONCORDIA SR COL, 56- Mem: Soc Study Evolution; Hist Sci Soc. Res: History of biology, particuarly the evolution concept as related to the philosophy of science. Mailing Add: 1723 St Joseph Ctr Rd Ft Wayne IN 46825

WALLEN, CLARENCE JOSEPH, b Phoenix, Ariz, Oct 17, 16. MATHEMATICS. Educ: St Louis Univ, BA, 45, MS, 46, PhD(math), 56; Alma Col, STL, 52. Prof Exp: Instr math, 46-48 & 56-57, from asst prof to assoc prof, 57-70, chmn dept, 70-73, PROF MATH, LOYOLA MARYMOUNT UNIV, LOS ANGELES, 70-, MEM BD TRUSTEES, 67- Mem: AAAS; Am Math Soc; Math Asn Am. Res: Limit theory in mathematics. Mailing Add: Dept of Math Loyola Marymount Univ 7101 W 80th Los Angeles CA 90045

WALLEN, DONALD GEORGE, b Ont, Can, July 6, 33; m 65; c 1. ECOLOGY, CELL PHYSIOLOGY. Educ: Dalhousie Univ, BS, 61, MEd, 62; Simon Fraser Univ, MS, 67, PhD(algal physiol & ecol), 70. Prof Exp: ASST PROF BIOL, UNIV WINDSOR, 70- Mem: Am Soc Limnol & Oceanog; Phycol Soc Am; Can Soc Plant Physiol. Res: Interrelationship between light quality, intensity and temperature on growth, photosynthesis and metabolism of algae; influence of fertilizers from farm land drainage on photosynthesis and eutrophication in lakes. Mailing Add: Dept of Biol Univ of Windsor Windsor ON Can

WALLEN, IRVIN EUGENE, b Afton, Okla, Oct 4, 21; m 45; c 5. OCEANOGRAPHY. Educ: Okla State Univ, BS, 41, MS, 46; Univ Mich, PhD(zool, limnol), 50. Prof Exp: Mem fac zool, Okla State Univ, 48-49, from asst prof to assoc prof, 49-56; asst dir sci teaching improvement prog, AAAS, Washington, DC, 56-57; sr training officer, AEC, 57-59, marine biologist, 59-62; asst dir oceanog, Mus Natural Hist, 62-66; head, Off Oceanog & Limnol, Washington, DC, 66-69, dir, Off Environ Sci, 69-71, DIR, FT PIERCE BUR, SMITHSONIAN INST, 71-, SPEC ASST MARINE PROGS, 71-, DIR, HARBOR BR FOUND, 71- Concurrent Pos: Grantee numerous orgns, 50-; founder, Okla Petrol Ref Waste Control Coun, 55; dir, Eniwetok Marine Biol Lab, Marshall Islands, 59-62; US rep to consult comn, Indian Ocean Biol Ctr, Cochin, India, 63-64 & Int Coun Sci Explor, Mediterranean, Rome, Italy, 65; mem tech asst comt, Intergovt Oceanog Comn, UNESCO, 64-69; adj instr oceanog, Univ Md, 65; consult, Marine Biol Lab, Woods Hole Inst Oceanog, 66-67, Cape Haze Marine Lab, Fla, 67 & Am Embassies, Iran, 66, 68 & Chile, 70; corp mem, Woods Oceanog Inst, 67-; consult cols & univs; chmn, US Del, Int Comn Sci Explor, Mediterranean, Monaco, 68-69; nat cor, US Coop Invests Mediterranean, 69-71; mem tech assistance bd, Link Found, 69-71, trustee, 71-; mem, Sci Adv Comt Int Atlantic Salmon Found, 70-; mem bd dirs, Iran Found; trustee, Atlantic Found, Tai Ping Found & Harbor Br Found. Honors & Awards: Outstanding Performance Award, Smithsonian Inst, 67. Mem: Fel AAAS; Am Inst Biol Sci; Am Soc Limnol & Oceanog (pres, 71-); Marine Technol Soc; Soc Syst Zool. Mailing Add: RR 1 Box 194-C Fort Pierce FL 33450

WALLEN, LOWELL LAWRENCE, b Rockford, Ill, May 22, 21; m 49; c 2. BIO-ORGANIC CHEMISTRY. Educ: Wheaton Col, Ill, BS, 44; Univ Ark, MS, 48; Iowa State Col, PhD, 54. Prof Exp: Asst chemist, Goodyear Tire & Rubber Co, Ohio, 44-46; asst, Univ Ark, 46-48; asst, Iowa State Col, 51-54; BIOCHEMIST, NORTHERN REGIONAL RES LAB, USDA, 54- Mem: Am Chem Soc; Coblentz Soc. Res: Chemistry of fermentation products; microbiological type reactions; oxidation; reduction; fermentations; chemical structure elucidation; bio-organic chemistry; infrared spectroscopy; microbial degradation of lignin. Mailing Add: 2929 N Gale Ave Peoria IL 61604

WALLEN, VICTOR REID, b Kenmore, Ont, June 16, 23; m 52; c 2. PLANT PATHOLOGY. Educ: McGill Univ, BSc, 46, MSc, 49, PhD(plant path), 54. Prof Exp: Asst, McGill Univ, 45; agr res officer, Div Bot & Plant Path, Sci Serv, 46-59, agr res officer, Cell Biol Res Inst, Res Br, 59-62, head phytopath sect, 62-67, chief sect, 67-70, chief crop loss sect, 70-73, HEAD CROP DIS LOSS SECT, OTTAWA RES STA, 73-, NAT COORDR CROP DIS LOSS PROG, 70- Concurrent Pos: Asst, Cornell Univ, 46-47 & McGill Univ, 48-49. Mem: Can Phytopath Soc (secy-treas, 59-61); Agr Inst Can; affil Royal Med Soc. Res: Seedborne diseases; fungicides; remote sensing for plant diseases; crop disease loss evaluation. Mailing Add: Ottawa Res Sta Can Dept of Agr Ottawa ON Can

WALLENBERGER, FREDERICK THEODORE, b St Peter, Austria, Aug 28, 30; nat US. ORGANIC POLYMER CHEMISTRY. Educ: Fordham Univ, MS, 56, PhD(chem), 58. Prof Exp: Instr chem, Fordham Univ, 57-58; res fel, Harvard Univ, 58-59; res chemist pioneering res div, 59-63, res supvr, Carothers Res Lab, Nylon Tech Div, 63-68, supvr res & develop carpet fiber & fiberfill div, Textile Res Lab, 68-73, PROD DEVELOP SUPVR TEXTILE FIBERS DEPT, NEW VENTURES DIV, E I DU PONT DE NEMOURS & CO, INC, 73- Concurrent Pos: Gen chmn 5th mid Atlantic regional meeting, Am Chem Soc, 70, chmn 1st conf chem & environ, 70. Mem: AAAS; Am Chem Soc; NY Acad Sci. Res: Organic polymer synthesis; high temperature fibers; textile technology; kinetics; light absorption; ozonization; high performance fibers; foam and pneumacel technology. Mailing Add: Textile Fibers Dept E I du Pont de Nemours & Co Inc Wilmington DE 19898

WALLENFELDT, EVERT, b Stanton, Iowa, June 26, 04; m 28; c 2. FOOD SCIENCE. Educ: Iowa State Col, BS, 26; Cornell Univ, MS, 29. Prof Exp: Instr high sch, Wis, 26-28; dairy fieldman, Borden Farm Prod Co, Ill, 29-31, supvr spec prod & tech probs, 31-34; supvr, Borden-Wieland Co, 34-37; res bacteriologist, Borden Co, 37-38; prof dairy indust, 38-72, EMER PROF FOOD SCI, UNIV WIS-MADISON, 72- Mem: AAAS; Am Dairy Sci Asn. Res: Market milk; butter manufacturing; concentrated milks and related products. Mailing Add: Dept of Food Sci Univ of Wis Madison WI 53706

WALLENIUS, ROGER WYNN, b Baltimore, Md, Oct 21, 41; m 68; c 2. ANIMAL NUTRITION. Educ: Pa State Univ, BS, 63; Ohio State Univ, MS, 65, PhD(animal nutrit), 69. Prof Exp: ASST PROF ANIMAL SCI, WASH STATE UNIV, 70- Mem: Am Dairy Sci Asn; Am Soc Animal Sci. Res: Ruminant nutrition; nutritional aspects of maximizing production in dairy cattle. Mailing Add: Dept of Animal Sci 263 Clark Wash State Univ Pullman WA 99163

WALLENMEYER, WILLIAM ANTON, b Evansville, Ind, Feb 3, 26; m 52; c 4. HIGH ENERGY PHYSICS. Educ: Purdue Univ, BS, 50, MS, 54, PhD(physics), 57. Prof Exp: Asst physics, Purdue Univ, 50-54 & 56; jr res assoc high energy particle interactions, Brookhaven Nat Lab, 54-55; asst prof physics, Wabash Col, 55-56; physicist, Midwestern Univs Res Asn, Wis, 56-60, div dir particle accelerators, 60-62; physicist res div, US AEC, 62-66, asst dir high energy physics, 66-75, ASST DIR HIGH ENERGY PHYSICS, DIV PHYS RES, US ENERGY RES & DEVELOP ADMIN, 75- Mem: AAAS; fel Am Phys Soc. Res: Cloud chambers; elementary particle physics; accelerator physics. Mailing Add: Div of Phys Res Energy Res & Develop Admin Washington DC 20545

WALLENTINE, MAX V, b Paris, Idaho, Apr 19, 31; m 53; c 8. ANIMAL SCIENCE. Educ: Utah State Univ, BS, 55; Cornell Univ, MS, 56, PhD(animal sci & physiol), 60. Prof Exp: Asst animal sci, Cornell Univ, 55-56 & 58-60, res assoc meat sci, 60-61; asst prof meat & animal sci, Purdue Univ, 61-62, from asst prof to assoc prof, 62-71, PROF MEAT & ANIMAL SCI, BRIGHAM YOUNG UNIV, 71-, DIR AGR, 68-, ASST DEAN COL BIOL & AGR, 71- Mem: Am Soc Animal Sci; Am Meat Sci Asn. Res: Ultrasonic evaluation of live meat animals; carcass effects from stilbestrol and pelleted roughages; early breeding of ewe lambs; effects of nutritional level. Mailing Add: 301 Widstoe Bldg Brigham Young Univ Provo UT 84602

WALLER, ADOLPH EDWARD, botany, deceased

WALLER, COY WEBSTER, b Dover, NC, Feb 25, 14; m; c 4. PHARMACEUTICAL CHEMISTRY. Educ: Univ NC, BS, 37; Univ Buffalo, MS, 39; Univ Minn, PhD(pharmaceut chem), 42. Prof Exp: Instr pharm, State Col Wash, 42-44; dir org chem, Lederle Labs, Am Cyanamid Co, 44-57; dir chem res, Mead Johnson & Co, 57-61; vpres res, Res Ctr, 61-68; assoc dir, Res Inst Pharmaceut Sci, 68-69, PROF PHARM, UNIV MISS, 68-, DIR, RES INST PHARMACEUT SCI, 69- Mem: Am Chem Soc; Am Pharmaceut Asn; The Chem Soc. Res: Structural determination of natural organic compounds. Mailing Add: Res Inst of Pharmaceut Sci Univ of Miss Sch of Pharm University MS 38677

WALLER, DAVID PERCIVAL, b Buffalo, NY, Jan 18, 43. ORGANIC CHEMISTRY, PHOTOGRAPHIC CHEMISTRY. Educ: Tex Christian Univ, BS, 64; Northeastern Univ, MS, 75. Prof Exp: Res asst chem, Res Found, Tex Christian Univ, 64-65; from asst scientist to assoc scientist chem, 65-74, SCIENTIST CHEM, POLAROID CORP, 74- Mem: Am Chem Soc; The Chem Soc. Res: Organic synthesis; novel photographic systems; dye chemistry. Mailing Add: Polaroid Corp 730 Main St Cambridge MA 02139

WALLER, ERNEST FREDERICK, b Stewartsville, Mo, Aug 17, 08; m 33; c 2. VETERINARY PATHOLOGY. Educ: Iowa State Col, DVM, 31, MS, 38. Prof Exp: Diagnostician, Dept Vet Med, Univ Minn, 31-34; asst prof vet path, Iowa State Col, 34-40, jr vet in-chg Bangs Dis Lab, Bur Animal Indust, USDA, 35-40; asst prof poultry husb, Univ NH & poultry pathologist, Exp Sta, 40-45; prof animal path, Univ Vt, 45-50; prof animal & poultry indust & head dept, Univ Del, 50-59; dir extramural res animal husb, Sterling-Winthrop Res Inst, NY, 60-68; tech dir, Sterwin Labs Inc, 68-70; DIR, GA POULTRY LAB 6, 70- Mem: Am Vet Med Asn; Poultry Sci Asn; Am Asn Avian Path; Indust Vet Asn. Res: Virology of poultry diseases; veterinary pharmacology. Mailing Add: Ga Poultry Lab 6 Coastal Exp Sta Tifton GA 31794

WALLER, FRANCIS JOSEPH, b Rome, NY, Mar 12, 43; m 69. ORGANIC CHEMISTRY. Educ: Niagara Univ, BS, 65; Univ Vt, PhD(org chem), 70. Prof Exp: Res investr low temperature kinetics, St Louis Univ, 70-71; vis asst prof org chem, St Lawrence Univ, 71-72; asst prof, Simmons Col, 72-74; RES CHEMIST, E I DU PONT DE NEMOURS & CO, INC, 74- Mem: AAAS; Am Chem Soc; The Chem Soc. Res: Homogeneous catalysis; process development. Mailing Add: Plastic Prod & Resins Dept E I du Pont de Nemours & Co Inc Wilmington DE 19898

WALLER, GEORGE ROZIER, JR, b Clinton, NC, July 14, 27; m 47; c 3. BIOCHEMISTRY. Educ: NC State Univ, BS, 50; Univ Del, MS, 52; Okla State Univ, PhD(biochem), 61. Prof Exp: Instr agr & biol chem, Univ Del, 50-53; res chemist, Imp Paper & Color Div, Hercules Powder Co, NY, 53-56; from asst prof to assoc prof, 56-67, PROF BIOCHEM, OKLA STATE UNIV, 67-, ASST DIR AGR EXP STA, 69- Concurrent Pos: NIH fel, Nobel Med Inst & Karolinska Inst, Sweden, 63-64; Seydell-Woolley lectr, Ga Inst Technol, 68; mem, Okla Environ Qual Task Force & Gov Task Force on Recommending Sci Policy Struct for State of Okla, 70; pres, Midcontinent Environ Ctr Asn, 70-75. Mem: AAAS; Am Soc Biol Chemists; Am Soc Mass Spectrometry; Am Chem Soc; Phytochem Soc NAm. Res: Plant biochemistry, especially the metabolism of pyridine and isoprenoid compounds; biochemical applications of mass spectrometry; alkaloid metabolism; biochemical applications of mass spectrometry; urinary metabolites of livestock. Mailing Add: Dept of Biochem Okla State Univ Stillwater OK 74074

WALLER, GORDON DAVID, b Gale, Wis, July 19, 35; m 59; c 2. APICULTURE. Educ: Wis State Univ-River Falls, BS, 59; Utah State Univ, MS, 67, PhD(entom, statist), 73. Prof Exp: Sci teacher, Pub Schs, Wis, 62-64; RES ENTOMOLOGIST, BEE RES LAB, AGR RES SERV, USDA, 67-; ASST PROF ENTOM, UNIV ARIZ, 69- Mem: Bee Res Asn; Entom Soc Am; Int Comn Bee Bot. Res: Applied pollination ecology with special emphasis on honey bee responses to olfactory and gustatory stimuli and the evaluation of foraging behavior of honey bees subjected to different management practices. Mailing Add: USDA Agr Res Serv Bee Res Lab 2000 E Allen Rd Tucson AZ 85719

WALLER, HARDRESS JOCELYN, b San Diego, Calif, Aug 27, 28; m 53; c 3. PHYSIOLOGY. Educ: San Diego State Col, AB, 50; Univ Wash, PhD(physiol), 57. Prof Exp: From instr to asst prof, 58-68, ASSOC PROF PHYSIOL, ALBERT EINSTEIN COL MED, YESHIVA UNIV, 68- Mem: AAAS; Am Physiol Soc. Res: Electrical activity of central nervous system. Mailing Add: Dept of Physiol Albert Einstein Col of Med New York NY 10461

WALLER, JAMES R, b Eureka, Mont, Dec 8, 31; m 55; c 8. MICROBIOLOGY, BIOCHEMISTRY. Educ: St John's Univ, Minn, BA, 57; Univ Minn, MS, 60, PhD(microbiol), 64. Prof Exp: Res assoc microbial physiol, Univ Cincinnati, 63-66; asst prof, 66-70, ASSOC PROF MICROBIOL, UNIV NDAK, 70- Mem: Am Soc Microbiol. Res: Physiology and mechanisms of vitamin transport in microorganisms and tissue cultures; water quality of impounded waters used for recreation. Mailing Add: Dept of Microbiol Univ of NDak Grand Forks ND 58202

WALLER, JEROME HOWARD, human genetics, behavioral genetics, see 12th edition

WALLER, JOHN D, mathematics, see 12th edition

WALLER, JULIAN ARNOLD, b New York, NY, Apr 17, 32; m 56; c 2. MEDICINE, PUBLIC HEALTH. Educ: Columbia Univ, AB, 53, Boston Univ, MD, 57; Harvard Univ, MPH, 60; Am Bd Prev Med, dipl, 65. Prof Exp: Intern, Mary Fletcher Hosp, Burlington, Vt, 57-58; resident, Contra Costa Health Dept, Calif, 58-59; regional consult chronic dis & voc rehab, USPHS, 60-62; resident, Calif Dept Pub Health, 62, coordr accident prev, 62-63, med officer occup health, 64-66, med officer chronic dis, 66-67, chief emergency health serv, 68; prof community med, 68-72, PROF & CHMN DEPT EPIDEMIOL & ENVIRON MED, UNIV VT, 72- Concurrent Pos: Consult, US Dept Transp, 67-, Vt State Health Dept, 68- & Vt Dept Ment Health, 70-; mem safety & occup health study sect, Dept Health, Educ & Welfare, 68-71; mem, Nat Hwy Safety Adv Comt, 69-72. Honors & Awards: Metrop Life Award of Merit Accident Res, 72. Mem: AAAS; fel Am Pub Health Asn; Am Asn Automotive Med (pres elect, pres, 73). Res: Epidemiology and control of highway and non-highway injury; emergency health services; epidemiology and control of problem drinking. Mailing Add: Dept Epidemiol & Environ Med Given Bldg Univ of Vt Burlington VT 05401

WALLER, MARION VAN NOSTRAND, b Flushing, NY, Jan 6, 19; m 64; c 2. SEROLOGY, IMMUNOLOGY. Educ: Hunter Col, BA, 40; Med Col Va, MS, 53, PhD(immunol), 59. Prof Exp: Dir blood grouping res lab, 52-61, dir blood bank, 57-61, from asst prof to assoc prof med, 61-73, PROF MED, MED COL VA, 73-, DIR SEROL STUDIES, CONNECTIVE TISSUE DIV, 61- Concurrent Pos: Arthritis & Rheumatism Asn fel, Lister Inst Prev Med, London, 61-62; dir serol studies for child develop study, Connective Tissue Div, Med Col, Va, 57-66, asst prof clin path, 60-61. Mem: Am Rheumatism Asn; Int Soc Hemat; Am Soc Hemat. Res: Antiglobulin antibodies in human sera; humoral antibodies in patients with renal transplants, influence of infection on antiglobulin antibodies, serum agglutinators. Mailing Add: Dept of Med Med Col of Va Richmond VA 23298

WALLER, MARY CONCETTA, b Louisville, Ky, June 25, 14. INORGANIC CHEMISTRY. Educ: Creighton Univ, BS, 44; Marquette Univ, MS, 50; St Louis Univ, PhD(chem), 61. Prof Exp: Teacher, St Martin Sch, Ky, 35-37 & Blessed Sacrament Sch, Nebr, 37-44; from instr to prof chem, 44-62, pres, 62-64, PROF CHEM, URSULINE COL, KY, 65-; SECOND VPRES, URSULINE SOC & ACAD EDUC, 72- Concurrent Pos: Prof chem, Bellarmine Col, Ky, 66-72; gen councilor, Ursuline Soc & Acad Educ, Ky, 68-72. Mem: Am Chem Soc. Res: Reduction reactions in aqueous sodium borohydride; coordination chemistry. Mailing Add: Ursuline Soc & Acad of Educ 3105 Lexington Rd Louisville KY 40206

WALLER, RAY ALBERT, b Grenola, Kans, Mar 4, 37; m 60; c 2. MATHEMATICAL STATISTICS. Educ: Southwestern Col, BA, 59; Kans State Univ, MS, 63; Johns Hopkins Univ, PhD(math statist), 67. Prof Exp: Instr math, St John's Sch, PR, 60-61; asst prof math, Towson State Col, 66-67; from asst prof to assoc prof statist, Kans State Univ, 67-74; STAFF MEM, LOS ALAMOS SCI LAB, 74- Concurrent Pos: Consult, White Sands Missile Range, 68-72. Mem: Am Statist Asn; Inst Math Statist. Res: Bayesian inference and reliability estimation. Mailing Add: Group C-5 MS 254 Los Alamos Sci Lab Los Alamos NM 87545

WALLER, ROGER MILTON, b Taylor, Wis, Dec 30, 26; m 55; c 3. GROUNDWATER GEOLOGY. Educ: Univ Wis, BS, 50; Univ Ariz, MS, 69. Prof Exp: Jr computer seismic explor, Nat Geophys Co, 50-51; groundwater geologist, 51-55, admin hydrologist, Alaska Dist, 56-63, res hydrologist, Alaska Earthquake Effects, 64-65, hydrologist, NY, 66-68, Wis, 68-71, assoc dist chief, Ohio Dist, 72-73, HYDROLOGIST, US GEOL SURV, NY, 73- Concurrent Pos: Hydrol panel mem, Nat Acad Sci Comt Alaska Earthquake, 64-74. Mem: Geol Soc Am; Am Geophys Union. Res: Groundwater resources of glacial deposits. Mailing Add: US Geol Surv PO Box 1350 Albany NY 12201

WALLER, THOMAS RICHARD, b Chicago, Ill, July 18, 37; m 59; c 2. INVERTEBRATE PALEONTOLOGY, GEOLOGY. Educ: Univ Wis, BS, 59, MS, 61; Columbia Univ, PhD(geol), 66. Prof Exp: ASSOC CUR TERTIARY

MOLLUSCA, SMITHSONIAN INST, 66- Mem: AAAS; Paleont Soc; Geol Soc Am; Soc Syst Zool. Res: Cenozoic mollusca; evolution; zoogeography; bivalve morphology; upper Cenozoic biostratigraphy of eastern North America and Caribbean. Mailing Add: Dept of Paleobiol Smithsonian Inst Washington DC 20560

WALLER, WILLIAM HENRY, b Frontier, Mich, Aug 2, 09; m 32; c 2. PSYCHIATRY. Educ: DePauw Univ, AB, 30; Cornell Univ, PhD(exp neurol), 33; Univ Ga, MD, 48; Am Bd Psychiat & Neurol, dipl, 63. Prof Exp: Asst lab instr histol-embryol, Cornell Univ, 30-31, asst lab instr anat, Med Col, 31-32, instr, 32-33; instr, Med Sch, George Washington Univ, 33-37; prof, Med Sch, Univ SDak, 37-44; assoc prof, Med Sch, Univ Ga, 44-48; assoc prof, Med Sch, Boston Univ, 48-53; pvt pract, Mass, 53-57; resident psychiat, Metrop State Hosp, Waltham, Mass, 57-59; resident psychiat, US Vet Admin Hosp, Brockton, Mass, 59-60; resident child psychiat, S Shore Guid Ctr, Quincy, Mass, 60-62; supt protem, Dever Sch, Taunton, Mass, 62-63; dir psychiat, Foxboro State Hosp, Mass, 63-64; psychiatrist, Brockton Ment Health Clin, 64-69, dir, 69-73. Concurrent Pos: Mem, S Dak Bd Basic Sci Examrs, 39-44; consult-instr, US Vet Admin Hosp, Lenwood, Augusta, Ga, 46-48; consult, Work Exp Prog, Taunton Mass Bd Pub Welfare, 66-68. Res: Mammalian thalamus and its cortical connections and functions. Mailing Add: 1011 Washington St Abington MA 02351

WALLER, WILLIAM T, b Girard, Kans, Apr 29, 41; m 61; c 2. AQUATIC BIOLOGY, LIMNOLOGY. Educ: Kans State Col, BS, 65, MS, 67; Va Polytech Inst & State Univ, PhD(biol), 71. Prof Exp: Res asst, Kans State Col, 65-67, Univ Kans, 68, Va Polytech Inst & State Univ, 68-71 & Region VII, Environ Protection Agency, 72; assoc res scientist environ med, Med Ctr, NY Univ, 72-75; ASSOC PROF, GRAD SCH ENVIRON SCI, UNIV TEX, DALLAS, 75- Mem: Am Fisheries Soc. Res: Organismic, population and community structure and function as it relates to acute and chronic stresses. Mailing Add: Grad Sch of Environ Sci Univ of Tex at Dallas PO Box 688 Richardson TX 75080

WALLERSTEIN, GEORGE, b New York, NY, Jan 13, 30. ASTRONOMY. Educ: Brown Univ, AB, 51; Calif Inst Technol, PhD, 58. Prof Exp: From instr to assoc prof astron, Univ Calif, Berkeley, 58-64; PROF ASTRON, UNIV WASH, 65- Concurrent Pos: Bd trustees, Brown Univ, 75-80. Mem: Am Astron Soc; Royal Astron Soc; Arctic Inst NAm. Res: Spectra of variable stars; abundances of the elements in stellar atmospheres; colormagnitude diagrams; interstellar absorption lines. Mailing Add: Dept of Astron Univ of Wash Seattle WA 98195

WALLERSTEIN, HARRY, b New York, NY, Dec 11, 06; m 26; c 2. MEDICINE, HEMATOLOGY. Educ: George Washington Univ, AB, 27, MD, 30. Prof Exp: Assoc pathologist, Sydenham Hosp, 31-33; assoc pathologist, Sea View Hosp, 33; pathologist, Hosp, 34-37, assoc hematologist, 37-50, DIR MARCIA SLATER LEUKEMIA RES LAB, JEWISH MEM HOSP, 50-, DIR BLOOD BANK, 38- Concurrent Pos: From asst vis pathologist to assoc vis pathologist, Queens Gen Hosp, 42-55; consult hematologist, Morrisania City Hosp, 48 & Fordham City & St Elizabeth's Hosps, 51-; attend hematologist, Jewish Mem Hosp, 50-; vis pathologist & dir blood bank, Bronx Muniv Hosp Ctr, 56-; vis asst prof, Albert Einstein Col Med, 56-; mem adv comt blood banks & transfusion, New York Dept Health, 66- Mem: AAAS; Am Soc Human Genetics; fel Am Soc Clin Path; fel NY Acad Med; fel NY Acad Sci. Res: Total blood substitution in treatment of erythroblastosis fetalis; significance of erythrocyte antigens in erythroblastosis fetalis; nature and treatment of leukemia. Mailing Add: 720 Gramatan Ave Mt Vernon NY 10552

WALLERSTEIN, ROBERT SOLOMON, b Berlin, Ger, Jan 28, 21; US citizen; m 47; c 3. PSYCHIATRY, PSYCHOANALYSIS. Educ: Columbia Univ, BA, 41, MD, 44. Prof Exp: Intern med, Mt Sinai Hosp, New York, 44-45, asst resident, 45-46, resident, 48, resident psychiat, 49; resident, Vet Admin Hosp, Topeka, Kans, 49-51, chief psychosom sect, 51-53; assoc dir dept res, Menninger Found, Kans, 54-65, dir dept, 65-66; clin prof psychiat, Sch Med & Langley-Porter Neuropsychiat Inst, 67-75, PROF PSYCHIAT & CHMN DEPT, SCH MED & DIR, LANGLEY PORTER INST, UNIV CALIF, SAN FRANCISCO, 75-; CHIEF DEPT PSYCHIAT, MT ZION HOSP, SAN FRANCISCO, 66- Concurrent Pos: Lectr psychol, Menninger Sch Psychiat, Kans, 51-66; lectr, Topeka Inst Psychoanal, 59-66, training & supv analyst, 65-66; fel, Ctr Advan Studies Behav Sci, Stanford, Calif, 64-65; training & supv analyst, San Francisco Psychoanal Inst, 66-; mem res sci career develop comn, NIMH, 66-70, chmn, 68-70; vis prof psychiat, Sch Med, La State Univ & New Orleans Psychoanal Inst, 72-73. Honors & Awards: Heinz Hartmann Award, New York Psychoanal Inst, 68. Mem: Am Psychoanal Asn (pres, 71-72); Int Psychoanal Asn; fel Am Psychiat Asn; fel Am Col Physicians; fel Am Orthopsychiat Asn. Res: Psychotherapy research, especially the processes and outcomes of psychoanalytic therapy; supervision processes; alcoholism; psychosomatic medicine. Mailing Add: Langley-Porter Inst 401 Parnassus San Francisco CA 94143

WALLES, WILHELM EGBERT, b Enschede, Neth, May 25, 25; nat US; m 51; c 5. ORGANIC CHEMISTRY, POLYMER CHEMISTRY. Educ: Univ Amsterdam, MSc, 48, PhD(org chem, physics), 51, DSc, 53. Prof Exp: Asst indust res, Univ Amsterdam, 49-51; Nat Res Coun Can fel, 51-53; dir res, N V Neth Refining Co, 53-55; assoc scientist, 55-67, ASSOC RES SCIENTIST PLASTICS, CENT RES LAB, DOW CHEM CO, 67- Honors & Awards: IR 100 Award, 74. Mem: Am Chem Soc. Res: Synthesis and physical properties of high polymers; surface reactions of polymers; polymolecular complexes; magneto-organic chemistry; aerosol physics; surface chemistry of plastic films, fibers and articles. Mailing Add: Cent Res Lab 1702 Bldg Dow Chem Co Midland MI 48640

WALLEY, WILLIS WAYNE, b Brooklyn, Miss, July 26, 34; m 66. ZOOLOGY, ECOLOGY. Educ: Southern Miss Univ, BS, 56; Miss State Univ, MS, 61, PhD(zool)j 65. Prof Exp: Pub sch teacher, Miss, 56-61; asst prof biol, Southeastern La Col, 64-68; PROF BIOL & CHMN DEPT & CHMN DIV III, BELHAVEN COL, 68- Mem: AAAS; Am Inst Biol Sci; Nat Audubon Soc; Sigma Xi. Res: Absorption, metabolism and excretion of chlorinated hydrocarbons in birds; in vitro metabolism of dichloro-diphenyl-trichloro-ethane by various tissues of the common grackle; biogeochemical cycling of boron; site fertility and primary production in hydrosoils. Mailing Add: Dept of Biol Belhaven Col Jackson MS 39202

WALLICK, EARL TAYLOR, b Monticello, Ark, Jan 11, 38; m 62; c 2. BIOLOGICAL CHEMISTRY. Educ: Miss State Univ, BS, 60, MS, 62; Rice Univ, PhD(org chem), 66. Prof Exp: Res chemist, Dacron Res Lab, E I du Pont de Nemours & Co, Inc, 66-67; assoc prof chem, King Col, 67-71; trainee myocardial biol, 71-73, INSTR CELL BIOPHYS, BAYLOR COL MED, 73- Mem: Am Chem Soc. Res: Isotope effects; cardiac glycosides, adenosine triphosphase. Mailing Add: Dept of Cell Biophys Baylor Col of Med Houston TX 77025

WALLICK, GEORGE CASTOR, b Grand Rapids, Mich, July 2, 23; m 45; c 2. PHYSICS. Educ: Univ Mich, BS, 43, MS, 46, PhD(physics), 52. Prof Exp: Radio engr radar res & develop, US Naval Res Lab, 44-45; sr res technologist appl math, 51-59, RES ASSOC, FIELD RES LAB, MOBIL RES & DEVELOP CORP, DALLAS, 59- Mem: Am Phys Soc; Am Asn Physics Teachers; Asn Comput Mach. Res: Petroleum production; flow of fluids through porous media; non-Newtonian flow; heat transfer; numerical analysis; computer programming and utilization. Mailing Add: 518 Towne Pl Duncanville TX 75116

WALLICK, ROLLIN HERBERT, analytical chemistry, see 12th edition

WALLIHAN, ELLIS FLOWER, b Ontario, Calif, Sept 20, 11; m 36; c 2. PLANT NUTRITION. Educ: Univ Calif, BS, 34, MS, 36; Cornell Univ, PhD(forest soils), 38. Prof Exp: From asst prof to assoc prof forestry, Cornell Univ, 38-48; from asst chemist to assoc prof, 48-69, PROF SOIL SCI, UNIV CALIF, RIVERSIDE, 69- Concurrent Pos: Field engr, Western Elec Co, NY, 42-45. Mem: AAAS; Soil Sci Soc Am; Am Soc Plant Physiologists; Am Soc Hort Sci. Res: Iron nutrition. Mailing Add: Citrus Res Ctr & Agr Exp Sta Univ of Calif Riverside CA 92502

WALLIN, JACK ROBB, b Omaha, Nebr, Nov 21, 15; m; c 2. PLANT PATHOLOGY. Educ: Iowa State Col, BS, 39, PhD(plant path), 44. Prof Exp: Asst bot, Univ Mo, 39-41 & N Iowa Agr Exp Asn, 41-44; collabr, E May Seed Co, 44-45; res asst prof, Iowa State Col, 45-47; plant pathologist, Agr Res Serv, USDA, 47-58, sr plant pathologist, Iowa State Univ, 59-75; PROF PLANT PATH, UNIV MO-COLUMBIA, 75- Concurrent Pos: Mem working group, Comn Instruments & Methods Observation, World Meteorol Orgn. Honors & Awards: William F Peterson Award, Int Soc Biometeorol, 67. Mem: Am Phytopath Soc; Am Meteorol Soc; Int Soc Biometoerol. Res: Investigating the cause of epidemics of the southern corn leaf blight fungus and experimental forecasts of this disease and other diseases that lend themselves to forecasting from weather data; epidemiology; disease forecasting; aerobiology. Mailing Add: Dept of Plant Path Univ of Mo Columbia MO 65201

WALLIN, RICHARD FRANKLIN, b Chicago, Ill, Jan 31, 39; m 61; c 2. TOXICOLOGY, PHARMACOLOGY. Educ: Univ Ill, BS, 61, DVM, 63, MS, 64, PhD(vet med sci), 66. Prof Exp: Physiologist, McDonnell Aircraft Corp, 66-67; sr res pharmacologist, 67-70, dir res admin, 70-72, actg dir pharm & microbiol res, 72-73, ASSOC DIR PHARMACOL RES, BAXTER LABS, INC, MORTON GROVE, 73- Concurrent Pos: NIH fel, 63-66. Mem: Am Vet Med Asn; Am Soc Vet Physiol & Pharmacol; Am Asn Lab Animal Sci; Am Soc Pharmacol & Exp Therapeut; Soc Biomat. Res: Anesthesiology; biomaterials; laboratory animal medicine. Mailing Add: 2037 Chestnut Ave Wilmette IL 60091

WALLING, CHEVES, b Evanston, Ill, Feb 28, 16; m 40; c 5. PHYSICAL ORGANIC CHEMISTRY. Educ: Harvard Univ, BA, 37; Univ Chicago, PhD(org chem), 39. Prof Exp: Res chemist, Jackson Lab, E I du Pont de Nemours & Co, 39-42 & Gen Labs, US Rubber Co, 43-49; res assoc, Lever Bros Co, 49-52; prof chem, Columbia Univ, 52-70, chmn dept, 63-66; DISTINGUISHED PROF, UNIV UTAH, 70- Concurrent Pos: Tech aide, Comt Med Res, Off Sci Res & Develop, 45-46; chmn div chem & chem technol, Nat Res Coun, 72-; ed, J Am Chem Soc, 75- Mem: Nat Acad Sci; Am Acad Arts & Sci; AAAS; Am Chem Soc. Res: Organic reaction mechanisms; free radical reactions; polymerization; peroxides and autoxidation. Mailing Add: Dept of Chem Univ of Utah Salt Lake City UT 84112

WALLING, DERALD DEE, b Granger, Iowa, Feb 14, 37; m 58; c 3. MATHEMATICS. Educ: Iowa State Univ, BS, 58, MS, 61, PhD(math), 63. Prof Exp: Mathematician, Ames Lab, US AEC, 62-63; asst prof math, Univ Ariz, 63-66; ASSOC PROF MATH, TEX TECH UNIV, 66- Mem: Am Math Soc. Res: Numerical analysis; least squares; statistics; probability. Mailing Add: Dept of Math Tex Tech Univ Lubbock TX 79409

WALLING, MATHEW T, JR, inorganic chemistry, see 12th edition

WALLINGFORD, JOHN STUART, b El Paso, Tex, Apr 13, 35; m 70; c 2. PHYSICS. Educ: Univ Minn, Minneapolis, BPhys, 61; Fla State Univ, MS, 66, PhD(physics), 67. Prof Exp: Instr physics, Fullerton Jr Col, 61-62 & Cerritos Col, 62-63; asst, Fla State Univ, 63-66; from asst prof to assoc prof, Fla A&M Univ, 66-69; vis assoc prof, Temple Univ, 69-70; assoc prof, 70-75, PROF PHYSICS, PEMBROKE STATE UNIV, 75- Mem: Am Asn Physics Teachers; Am Inst Physics. Res: Consequences of general relativity theory; making science interesting and accessible to all; hydrolysis of organic wastes. Mailing Add: Dept of Physical Sci Pembroke State Univ Pembroke NC 28372

WALLIS, ANTHONY, b London, Eng, May 11, 42; Can citizen; m 68. COMPUTER SCIENCE, MATHEMATICS. Educ: Univ Manchester, BSc, 63, MSc, 64, PhD(chem), 68. Prof Exp: Res assoc math, Queen's Univ, Ont, 68-70; ASST PROF COMPUT SCI, YORK UNIV, 70- Mem: Asn Comput Mach. Res: Numerical techniques and applications to physical sciences; computer hardware; microprogramming. Mailing Add: Dept of Comput Sci York Univ Downsview ON Can

WALLIS, DONALD DOUGLAS JAMES H, b Brandon, Man, Apr 20, 43. IONOSPHERIC PHYSICS. Educ: Univ Alta, Calgary, BSc, 65; Univ Calgary, MSc, 68; Univ Alaska, PhD(geophys), 74. Prof Exp: Fel geophys, Univ Alta, 73-75; RES ASSOC GEOPHYS, UNIV CALGARY, 75- Mem: Can Asn Physicists; Am Geophys Union. Res: Auroral spectroscopy; ionospheric winds and currents; atmospheric changes; magnetospheric physics. Mailing Add: Dept of Physics Univ of Calgary Calgary AB Can

WALLIS, GEORGE, solid state physics, see 12th edition

WALLIS, JAMES RICHARD, hydrology, forestry, see 12th edition

WALLIS, ORTHELLO LANGWORTHY, b Hamilton, NY, May 1, 21; m 46; c 3. ECOLOGY, NATURAL SCIENCE. Educ: Univ Redlands, BA, 47; Ore State Col, MS, 48. Prof Exp: Park ranger, Lake Mead Nat Recreational Area Nat Park Serv, 48-49; marine fishery res biologist, US Fish & Wildlife Serv, 49-50; park ranger, Yosemite Nat Park, Nat Park Serv, 50-54, asst park naturalist, Lake Mead Recreational Area, 54-55, chief park naturalist, 55-57, chief aquatic biologist, Natural Hist Br, 57-65, chief aquatic res biologist, Off Natural Sci Studies, Washington, DC, 66-72, REGIONAL CHIEF SCIENTIST, WESTERN REGION, NAT PARK SERV, 72- Concurrent Pos: Mem UNESCO mission, Costa Rica, 72. Mem: Am Fisheries Soc; Am Soc Ichthyologists & Herpetologists. Res: Fisheries and aquatic resources of national parks; interpretation of marine and fresh water resources. Mailing Add: Western Region Nat Park Serv 450 Golden Gate Ave Box 36063 San Francisco CA 94102

WALLIS, RICHARD FISHER, b Washington, DC, May 14, 24; m 55. SOLID STATE PHYSICS. Educ: George Washington Univ, BS, 45, MS, 48; Cath Univ, PhD(chem), 52. Prof Exp: Fel, Inst Fluid Dynamics, Univ Md, 51-53; chemist, Appl Physics Lab, Johns Hopkins Univ, 53-56; physicist, US Naval Res Lab, 56-58, from actg head to head semiconductors br, 58-66; prof physics, Univ Calif, Irvine, 66-67; head semiconductors br, US Naval Res Lab, 67-69; PROF PHYSICS, UNIV CALIF, IRVINE, 69-, CHMN DEPT, 72- Concurrent Pos: Consult, Res Labs, Gen Motors

Corp, 58- Mem: Fel Am Phys Soc. Res: Quantum and statistical mechanics. Mailing Add: Dept of Physics Univ of Calif Irvine CA 92650

WALLIS, ROBERT CHARLES, b West Burlington, Iowa, Oct 30, 21; m 53; c 2. MEDICAL ENTOMOLOGY, PARASITOLOGY. Educ: Ohio Univ, BS, 48, MS, 50; Johns Hopkins Univ, DSc, 53. Prof Exp: Asst zool, Ohio Univ, 48-49, asst parasitol, Sch Hyg & Pub Health, Johns Hopkins Univ, 50-52; from asst prof to prof entom, Agr Exp Sta, Univ Conn, 53-63; ASSOC PROF EPIDEMIOL & CHIEF SECT MED ENTOM, SCH MED, YALE UNIV, 63- Concurrent Pos: Consult, US Army, 62-72. Mem: Am Soc Trop Med & Hyg; Am Soc Parasitol; Entom Soc Am; Am Mosquito Control Asn. Res: Epidemiology and natural history of eastern encephalitis; ecology and physiology of arthropod vectors of disease; mosquito biology and control. Mailing Add: Dept of Epidemiol & Pub Health Yale Univ Sch of Med New Haven CT 06520

WALLIS, ROBERT L, b Sheridan, Wyo, Sept 22, 34; m 60; c 2. PHYSICS. Educ: Univ Colo, BA, 56, MA, 58, PhD(physics), 62. Prof Exp: From asst prof to assoc prof, 62-75, PROF PHYSICS, BALDWIN-WALLACE COL, 75-, CHMN DEPT, 70- Mem: AAAS; Am Phys Soc; Am Asn Physics Teachers. Res: Educational techniques. Mailing Add: Dept of Physics Baldwin-Wallace Col Berea OH 44017

WALLMAN, HENRY, mathematics, see 12th edition

WALLMAN, SONIA RUDE, neurobiology, cell biology, see 12th edition

WALLMO, OLOF CHARLES, b Mason City, Iowa, July 1, 19; m 45; c 3. VERTEBRATE ECOLOGY. Educ: Utah State Agr Col, BS, 47; Univ Wis, MS, 49; Agr & Mech Col, Tex, PhD(wildlife mgt), 57. Prof Exp: Res biologist, US Fish & Wildlife Serv, 47; biologist, Ariz Game & Fish Dept, 48-51 & Tex Game & Fish Comn, 53-54; from asst prof to assoc prof wildlife mgt, Agr & Mech Col Tex, 55-62; res biologist, Ariz Game & Fish Dept, 62-64; RES WILDLIFE BIOLOGIST, ROCKY MOUNTAIN FOREST & RANGE EXP STA, US FOREST SERV, 64- Concurrent Pos: Fac affil, Grad Sch, Colo State Univ. Mem: AAAS; Wildlife Soc; Sigma Xi. Res: Wildlife habitat management research. Mailing Add: Rocky Mountain Forest & Range Exp Sta US Forest Serv Ft Collins CO 80521

WALLNER, JULIUS MICHAEL, b New Brunswick, NJ, Jan 9, 10; m 35; c 4. PSYCHIATRY. Educ: Rutgers Univ, BSc, 31; Univ Pa, MD, 35. Prof Exp: Intern, Colo Gen Hosp, Denver, 35-36; instr, Tulane Univ Outpatient Dept & Tulane Serv, Charity Hosp, 39-40; instr, Neuropsychiat Inst, Univ Mich, 40-43, asst prof, Med Ctr, 43-46; chief psychoneurosis sect, Winter Vet Admin Hosp, Topeka, Kans, 46-47; chief neuropsychiat serv, Vet Admin Hosp, Palo Alto, Calif, 47-53; dir psychiat residency training prog, Univ Hosp, 53-64, asst dir, Neuropsychiat Inst, 54-57, ASSOC PROF PSYCHIAT, MED CTR, UNIV MICH, ANN ARBOR, 53- Concurrent Pos: Commonwealth Fund fel psychiat, Colo Psychopathic Hosp, 36-39; consult, Vet Admin Hosp, Ann Arbor, Mich, 54-, Ypsilanti State Hosp, 57- & Lafayette Clin, 60- Mem: Am Psychiat Asn; Asn Res Nerv & Ment Dis; AMA; Asn Am Med Cols, Mailing Add: Dept of Psychiat Univ of Mich Med Ctr Ann Arbor MI 48104

WALLNER, STEPHEN JOHN, b Sioux Falls, SDak, Mar 22, 45; m 69; c 2. PLANT PHYSIOLOGY. Educ: SDak State Univ, BS, 67, MS, 69; Iowa State Univ, PhD(plant physiol), 73. Prof Exp: Plant physiologist, US Army Natick Labs, 73-75; ASST PROF HORT PHYSIOL, PA STATE UNIV, UNIVERSITY PARK, 75- Mem: Am Soc Plant Physiologists; Am Inst Biol Sci; Sigma Xi; AAAS; Int Asn Plant Tissue Cult. Res: Postharvest physiology, especially involving cell wall changes during fruit ripening. Mailing Add: Dept of Hort Pa State Univ University Park PA 16802

WALLNER, WILLIAM E, b Greenfield, Mass, Nov 25, 36; m 64; c 2. ENTOMOLOGY, PLANT PATHOLOGY. Educ: Univ Conn, BS, 59; Cornell Univ, PhD, 65. Prof Exp: Res asst entom, Cornell Univ, 59-65; assoc prof entom, 69-74, EXTEN ENTOMOLOGIST, MICH STATE UNIV, 65-, PROF ENTOM, 74- Mem: Entom Soc Am. Res: Biology and control of forest; ornamental and plantation insects with emphasis on population suppression and pest management of insects for forest-recreational areas. Mailing Add: Dept of Entom Mich State Univ East Lansing MI 48823

WALLOCH, RICHARD ARTHUR, b Chicago, Ill, July 9, 43; m 69; c 1. PHYSIOLOGY. Educ: Loyola Univ Chicago, BS, 65; Univ Ore, MS, 69, PhD(psychol), 71. Prof Exp: ASST PROF OTORHINOLARYNGOL, UNIV OKLA HEALTH SCI CTR, 71- Mem: Acoust Soc Am. Res: Development of electrical prosthesis for sensorineural hearing loss; relationship of various cochlear potentials. Mailing Add: Univ of Okla Health Sci Ctr PO Box 26901 Oklahoma City OK 73190

WALLRAFF, EVELYN BARTELS, b Chicago, Ill, Oct 21, 20; m 46; c 2. MICROBIOLOGY, IMMUNOLOGY. Educ: Rosary Col, BS, 40; Univ Chicago, MS, 42; Univ Ariz, PhD(virol, immunol), 61. Prof Exp: Res technician bact & immunol, Univ Chicago & Zoller Dent Clin, 41-43; from instr to asst prof bact, Univ Ariz, 43-47, res assoc med res, Univ Ariz & Southwestern Clin & Res Inst, 47-51; res microbiologist, Vet Admin Hosp, 61-71; RES ASSOC MICROBIOL & MED TECHNOL, UNIV ARIZ, 70-; PROF MICROBIOL & LIFE SCI, PIMA COL, 72- Concurrent Pos: Consult, Vet Admin Hosp, Tucson, 71- Mem: AAAS; Am Soc Microbiol; Am Asn Immunol; Reticuloendothelial Soc; Am Thoracic Soc. Res: Coccidioidin hypersensitivity used for study of mechanisms of delayed hypersensitivity in man; cellular and transplantation immunology; anti-macrophage serum. Mailing Add: 2708 E Mabel Tucson AZ 85716

WALLS, JAMES GRAY, geology, see 12th edition

WALLS, KENNETH W, b Ft Lauderdale, Fla, Dec 4, 28; m 56. IMMUNOLOGY, PARASITOLOGY. Educ: Ind Univ, AB, 49 & 50; Univ Mich, MS, 52, PhD(bact), 55. Prof Exp: Lab chief parasitol & mycol serol, 55-59, lab chief toxoplasmosis, 59-66, CHIEF PARASITOL SEROL UNIT, CTR DIS CONTROL, 66- Mem: AAAS; Am Soc Trop Med & Hyg. Res: Immunology, serology and epidemiology of parasitic diseases with special emphasis on toxoplasmosis and related diseases. Mailing Add: Parasitol Sect Ctr Dis Control 1600 Clifton Rd Atlanta GA 30333

WALLS, NANCY WILLIAMS, b Johnstown, Pa, Sept 19, 30; m 56. BACTERIOLOGY. Educ: Univ Mich, BS, 52, MS, 53, PhD(bact), 59. Prof Exp: Instr bact, Emory Univ, 58-59; asst res biologist, 59-61, res asst prof, 62-67, sr res biologist, 67-69, actg dir sch biol, 69-70, ASSOC PROF BIOL, GA INST TECHNOL, 69- Mem: AAAS; Am Soc Microbiol; Radiation Res Soc; Am Inst Biol Sci; NY Acad Sci. Res: Physiology of Clostridium botulinum; marine microbial ecology; anaerobic bacterial spores; mechanisms of bacterial toxin formation. Mailing Add: Sch of Biol Ga Inst of Technol Atlanta GA 30332

WALLS, ROBERT CLARENCE, b Batesville, Ark, Mar 9, 34; m 66; c 3. STATISTICS, MATHEMATICS. Educ: Harding Col, BS, 59; Univ Ark, MS, 61; Okla State Univ, PhD(statist), 67. Prof Exp: Mathematician, Res & Technol Dept, Texaco Inc, 61-64; asst prof, 67-72, ASSOC PROF BIOMET, SCH MED, UNIV ARK, LITTLE ROCK, 72- Mem: Am Statist Asn; Math Asn Am. Res: Nonparametric statistics; graphical methods in statistics; mathematical models in biology and medicine. Mailing Add: Div of Biomet Univ of Ark Med Ctr Little Rock AR 72201

WALMSLEY, FRANK, b New Bedford, Mass, June 26, 35; m 59; c 2. INORGANIC CHEMISTRY. Educ: Univ NH, BS, 57; Univ NC, Chapel Hill, PhD(chem), 62. Prof Exp: From asst prof to assoc prof, 62-75, PROF CHEM, UNIV TOLEDO, 75- Mem: Am Chem Soc; The Chem Soc; Am Sci Affil. Res: Spectral and magnetic properties of coordination compounds. Mailing Add: Dept of Chem Univ of Toledo Toledo OH 43606

WALMSLEY, JUDITH ABRAMS, b Oak Park, Ill, Feb 6, 36. INORGANIC CHEMISTRY. Educ: Fla State Univ, BA, 58; Univ NC, Chapel Hill, PhD(chem), 62. Prof Exp: Res scientist chem, Owens-Ill, Inc, 63-66; vis res assoc, 73-75, SR RES ASSOC CHEM, UNIV TOLEDO, 74- Mem: Am Chem Soc; AAAS; Sigma Xi. Res: Chemistry of metal complexes of biological significance; transition metal complexes of organophosphorus ligands; self-association of lactams in nonpolar solvents. Mailing Add: Dept of Chem Univ of Toledo Toledo OH 43606

WALMSLEY, MILDRED MARIE, b Briceton, Ohio, Sept 18, 08. GEOGRAPHY. Educ: Bowling Green State Univ, BS; Clark Univ, MA & PhD(geog). Prof Exp: Instr, 43-47 & 49-52, asst prof, 52-71, ASSOC PROF GEOG & GEOL, CASE WESTERN RESERVE UNIV, 71- Mem: Asn Am Geogr; Am Geog Soc; Coun Geog Educ. Res: West Indies, especially Jamaica. Mailing Add: Dept of Geog 4040 Crawford Hall Case Western Reserve Univ Cleveland OH 44106

WALNE, PATRICIA LEE, b Newark, NJ, Nov 27, 32. BOTANY. Educ: Hanover Col, BS, 54; Ind Univ, MS, 59; Univ Tex, PhD(phycol, cell biol), 65. Prof Exp: Res fel, Cell Res Inst, Univ Tex, 65-66; from asst prof to assoc prof, 66-73, PROF BOT, UNIV TENN, KNOXVILLE, 73- Concurrent Pos: Ed, Phycol Soc Am Newslett, 66-69; consult, Biol Div, Oak Ridge Nat Lab, 66-74; Fulbright-Hays res scholar, Comt Int Exchange Scholars, 74-75; Am Asn Univ Women sr fel, 74-75. Mem: Am Soc Cell Biol; Bot Soc Am; Phycol Soc Am (secy, 69-72, vpres, 73, pres, 74); Am Soc Plant Physiologists; Int Phycol Soc. Res: Cell biology, experimental phycology and cytology; ultrastructure and development of algae; photoreceptors in algal flagellates; correlation of structure and function. Mailing Add: Dept of Bot Univ of Tenn Knoxville TN 37916

WALNER, ARTHUR H, b New York, NY, Nov 25, 28; m 52; c 3. MATHEMATICAL STATISTICS. Educ: City Col New York, BS, 51. Prof Exp: Anal statistician, Picatinny Arsenal, 51-55; consult statistician, Naval Appl Sci Lab, NY, 55-67; dir opers planning div, Metrop Police Dept, 67-68; mem prog & policy planning, Gen Serv Admin, Washington, DC, 68-71; asst dir opers planning, Cost of Living Coun, 71-72; SR PLANNER, ENVIRON PROTECTION AGENCY, 72- Concurrent Pos: Lectr, Wayne State Univ, 56. Mem: Am Soc Qual Control; Biomet Soc; Opers Res Soc Am. Res: Design of experiments and analysis and interpretation for backgrounds in the physical sciences and engineering; management and statistical planning in crime and traffic analysis. Mailing Add: 11215 Oak Leaf Dr Silver Spring MD 20901

WALNUT, THOMAS HENRY, JR, b Philadelphia, Pa, May 22, 24; m 70; c 1. QUANTUM CHEMISTRY. Educ: Harvard Univ, AB, 47; Brown Univ, PhD(chem), 51. Prof Exp: Instr, Inst Study Metals, Univ Chicago, 50-52; from asst prof to assoc prof, 52-63, PROF CHEM, SYRACUSE UNIV, 63- Mem: Am Chem Soc; Am Phys Soc. Res: Magnetic susceptibility of molecules; quantum chemistry; development of shapes in biological systems. Mailing Add: Dept of Chem Syracuse Univ Syracuse NY 13210

WALP, RUSSELL LEE, b Youngstown, Ohio, Oct 23, 06; m; c 2. BOTANY. Educ: Univ Mich, BS, 30, MS, 31. Prof Exp: From instr to assoc prof, 31-61, asst prof geol, Pre-Flight Training, 42-44, PROF BIOL, MARIETTA COL, 61- Concurrent Pos: Owner & dir, Sea Pine Camp Girls, 34-; Fund Adv Educ fac fel, Stanford Univ, 51-52. Mem: Bot Soc Am. Res: Taxonomy of local flora; shrubs; ecology; land use survey; early history of science of Ohio. Mailing Add: Dept of Biol Marietta Col Marietta OH 45750

WALPER, JACK LOUIS, b Excel, Alta, Nov 29, 16; nat US; m 43. GEOLOGY. Educ: Univ Okla, BS, 47, MS, 49; Univ Tex, PhD(geol), 58. Prof Exp: Asst geol, Univ Okla, 47-48; asst prof, Univ Tulsa, 48-54; instr, Univ Tex, 55-58; assoc prof, Univ Tulsa, 58-63; PROF GEOL, TEX CHRISTIAN UNIV, 63- Concurrent Pos: Consult & mem bd dirs, Tex Archit Aggregate Co & Empresa Centro Americana, 74- Mem: Am Geophys Union; Asn Eng Geologists; Am Asn Petrol Geologists; Geol Soc Am; Nat Asn Geol Teachers. Res: Field exploration; tectonics, especially Central American tectonics. Mailing Add: Dept of Geol Tex Christian Univ Ft Worth TX 76129

WALPOLE, RONALD EDGAR, b Wiarton, Ont, June 19, 31; m 64; c 1. MATHEMATICS, STATISTICS. Educ: McMaster Univ, BA, 54, MSc, 55; Va Polytech Inst, PhD, 58. Prof Exp: From asst prof to assoc prof, 57-61, PROF MATH & STATIST, ROANOKE COL, 61- Concurrent Pos: Consult, White Sands Missile Range, 50-63. Mem: Inst Math Statist; Am Statist Asn; Biomet Soc. Res: Statistical design of experiments. Mailing Add: Dept of Math & Statist Roanoke Col Salem VA 24153

WALRADT, JOHN PIERCE, b Caldwell, Idaho, Feb 12, 42; m 64. FOOD SCIENCE, ANALYTICAL CHEMISTRY. Educ: Univ Idaho, BS, 65; Ore State Univ, MS, 67, PhD(food sci), 69. Prof Exp: Sr chemist, 69-71, proj leader, 71-73, GROUP LEADER, INT FLAVORS & FRAGRANCES, INC, 73- Mem: Am Chem Soc; Inst Food Technologists. Res: Flavor chemistry, natural component identification and associated analytical techniques. Mailing Add: Int Flavors & Fragrances Inc 1515 Hwy 36 Union Beach NJ 07735

WALRAFEN, GEORGE EDOUARD, chemical physics, see 12th edition

WALSER, ARMIN, b Nalzenhausen, Switz, Apr 6, 37; m 57; c 3. MEDICINAL CHEMISTRY. Educ: Swiss Fed Inst Technol, Dipl Ing Chem, 60, PhD(org chem), 63. Prof Exp: Fel org chem, Stanford Univ, 63-64; sr chemist, Nutley, NJ, 64-66, Basel, Switz, 66-69 & Nutley, NJ, 69-72, res fel med chem, 72-74, GROUP CHIEF MED CHEM, HOFFMANN-LA ROCHE, INC, 74- Mem: Am Chem Soc. Res: Synthesis of new compounds of potential pharmaceutical interest, in particular compounds acting on central nervous system. Mailing Add: Hoffmann-La Roche Inc Nutley NJ 07110

WALSER, MACKENZIE, b New York, NY, Sept 19, 24; m 65; c 4. MEDICINE. Educ: Yale Univ, BA, 44; Columbia Univ, MD, 48; Am Bd Internal Med, dipl, 56. Prof Exp: Intern med, Mass Gen Hosp, 48-49, asst resident, 49-50; from instr to asst prof, Univ Tex Southwestern Med Sch, Dallas, 50-52; investr, Nat Heart Inst, 54-57; from asst prof to assoc prof med & pharmacol, 57-70, PROF MED & PHARMACOL, SCH MED, JOHNS HOPKINS UNIV, 70- Concurrent Pos: Resident, City-County Hosp, Dallas, Tex, 50-52; physician, Johns Hopkins Hosp, 60-

Mem: Am Physiol Soc; Am Soc Clin Invest; Am Soc Pharmacol & Exp Therapeut; Biophys Soc; Asn Am Physicians. Res: Medical, Physiological, and pharmacological aspects of electrolyte metabolism and renal function. Mailing Add: Dept of Pharmacol Johns Hopkins Univ Sch of Med Baltimore MD 21205

WALSH, ALEXANDER HAMILTON, b Montclair, NJ, June 2, 31; m 57; c 2. VETERINARY PATHOLOGY. Educ: Cornell Univ, DVM, 57; Univ Wis-Madison, PhD(path), 74. Prof Exp: Veterinarian, Del-Tor Clin, Lexington, Ky, 57-59, Animal Rescue League of Boston, 59-61 & Green Mountain Animal Hosp, South Burlington, Vt, 61-69; res asst path, Univ Wis-Madison, 69-72; PATHOLOGIST, PFIZER, INC, 72- Concurrent Pos: Lectr fish path, NY State Vet Col, 72-; consult pathobiol, Univ Conn, 73-75; discussant, Charles Louis Davis Found Advan Vet Path, 73-, mem directorate, 75-; adj prof animal path, Univ RI, 76. Mem: Am Col Vet Path; AAAS; Wildlife Dis Asn; Soc Pharmaceut & Environ Path. Res: Evaluation of the toxic and carcinogenic potential of environmental chemicals on laboratory animals, particularly neurocarcinogens and the carcinogenic potential of that group of compounds upon fish. Mailing Add: 70 College St Old Saybrook CT 06475

WALSH, BERTRAM (JOHN), b Lansing, Mich, May 7, 38; m 60. MATHEMATICAL ANALYSIS. Educ: Univ Mich, PhD(math), 63. Prof Exp: Lectr math, Univ Mich, 63; from asst prof to assoc prof, Univ Calif, Los Angeles, 63-70; assoc prof, 70-71, PROF MATH, RUTGERS UNIV, NEW BRUNSWICK, 71- Concurrent Pos: Vis asst prof, Univ Wash, 66-67. Mem: Am Math Soc. Res: Functional analysis, locally convex spaces, linear transformations and spectral theory; measure theory; potential theory. Mailing Add: Dept of Math Rutgers Univ New Brunswick NJ 08903

WALSH, CHRISTOPHER THOMAS, b Boston, Mass, Feb 16, 44; m 66. BIOCHEMISTRY. Educ: Harvard Univ, BA, 65; Rockefeller Univ, PhD(life sci), 70. Prof Exp: Helen Hay Whitney Found fel, Brandeis Univ, 70-72; ASST PROF CHEM & BIOL, MASS INST TECHNOL, 72- Res: Enzymatic reaction mechanisms, phosphoryl and pyrophosphoryl transfers; flavin-dependent enzymes; membrane biochemistry and mechanism of active transport. Mailing Add: Dept of Chem Rm 18-025 Mass Inst of Technol Cambridge MA 02139

WALSH, DAVID ERVIN, b DeGraff, Minn, Aug 7, 39; m 63; c 2. CEREAL CHEMISTRY, BIOCHEMISTRY. Educ: St Cloud State Col, BS, 61; NDak State Univ, MS, 63, PhD(cereal chem), 69. Prof Exp: Proj leader cereal prod, Squibb Beech-Nut Corp, 63-65; instr cereal chem, 65-69, ASSOC PROF CEREAL CHEM & TECHNOL, N DAK STATE UNIV, 69- Mem: Am Asn Cereal Chemists; Inst Food Technologists. Res: Macaroni products; industrial engineering; computer applications to food processes; protein compositional studies of wheat. Mailing Add: Dept of Cereal Chem & Technol NDak State Univ Fargo ND 58102

WALSH, DON, b Berkeley, Calif, Nov 2, 31; m 62; c 2. PHYSICAL OCEANOGRAPHY, OCEAN ENGINEERING. Educ: US Naval Acad, BS, 54; Tex A&M Univ, MS, 67, PhD(oceanog), 68; San Diego State Col, MA, 68. Prof Exp: Officer-in-chg bathyscaphe Trieste, Navy Electronics Lab, US Navy, San Diego, Calif, 58-62, prin investr remote sensor oceanog proj, Tex A&M Univ, 65-68, sci liaison officer ocean eng, Submarine Develop Group One, San Diego, 69-70, spec asst to Asst Secy Navy Res & Develop, Navy Dept, Washington, DC, 70-73, dep dir, Navy Labs, Hq Naval Mat Command, 74-75; PROF OCEAN ENG & DIR INST MARINE & COASTAL STUDIES, UNIV SOUTHERN CALIF, 75- Concurrent Pos: Partic, Deep Freeze, Antarctic, 71; mem sea grant adv bd, Univ Southern Calif, 71-; fel, Woodrow Wilson Int Ctr Scholars, Smithsonian Inst, 72-74; mem US adv comt, Eng Comt Ocean Resources, Nat Acad Eng, 72- Honors & Awards: Gold Medal, City of Trieste, Italy, 60; Rear Adm W S Parsons Award, Navy League US, 72. Mem: AAAS; Am Geophys Union; Marine Technol Soc (vpres, 75-); US Naval Inst; hon mem Am Soc Oceanog (vpres, 69). Res: Application of deep submersibles to ocean sciences; deep ocean engineering research and development; application of remote sensors to oceanography. Mailing Add: 2425 Via Campesina Apt 4 Palos Verdes Estates CA 90274

WALSH, EDWARD JOHN, b Brooklyn, NY, Aug 29, 42; m 64; c 2. INORGANIC CHEMISTRY, POLYMER CHEMISTRY. Educ: Franklin & Marshall Col, BA, 64; Middlebury Col, MS, 66; Pa State Univ, University Park, PhD(inorg chem), 70. Prof Exp: ASST PROF CHEM, PA STATE UNIV, SHENANGO VALLEY CAMPUS, 70- Mem: Am Chem Soc; The Chem Soc. Res: Phosphazene derivatives; germazanes; trace elements in water aseneazenes. Mailing Add: Pa State Univ Dept Chem Shenango Val Camp Sharon PA 16146

WALSH, EDWARD JOSEPH, b Waterbury, Conn, May 5, 41; m 63; c 3. ORGANIC CHEMISTRY, PHOTOGRAPHIC SCIENCE. Educ: Yale Univ, BS, 63; Univ Wis-Madison, PhD(org chem), 68. Prof Exp: NIH fel, Cornell Univ, 68-69; SR RES CHEMIST, EASTMAN KODAK CO, 69- Concurrent Pos: Asst lectr, Univ Rochester, 69- Mem: Am Chem Soc; Soc Photog Scientists & Engrs. Res: Physical organic chemistry; mechanisms of thermal and photolytic processes. Mailing Add: Eastman Kodak Co Res Lab 1669 Lake Ave Rochester NY 14650

WALSH, EDWARD JOSEPH, JR, b Philadelphia, Pa, Aug 2, 35; m 59; c 2. ORGANIC CHEMISTRY. Educ: State Univ NY Albany, BS, 60; Univ NH, PhD(chem), 64. Prof Exp: Teaching asst, State Univ NY Albany, 60-61; asst prof, 64-68, ASSOC PROF CHEM, ALLEGHENY COL, 68- Concurrent Pos: NSF sci fac grant, Mass Inst Technol, 69-70. Mem: Am Chem Soc. Res: Free radical reactions involving the acyl radical; free radical reactions of certain organotin hydrides. Mailing Add: Dept of Chem Allegheny Col Meadville PA 16335

WALSH, EDWARD NELSON, b Chicago, Ill, Nov 22, 25; m 50; c 2. ORGANIC CHEMISTRY. Educ: Ill Inst Technol, BS, 48, PhD, 65; DePaul Univ, MS, 52. Prof Exp: Chemist, Swift & Co, Ill, 48-51 & Victor Chem Works, 51-59; supvr org res, 59-63, mgr chem res, 63-65, mgr chem prod develop sect, Dobbs Ferry, 65-69, SR SECT MGR ORG RES, STAUFFER CHEM CO, DOBBS FERRY, 69- Mem: AAAS; Am Chem Soc. Res: Organophosphorus compounds; agricultural chemicals; solvents; surfactants; flame retardants; synthetic lubricants; pharmaceutical intermediates; organometallics; photochemistry. Mailing Add: 33 Concord Dr New City NY 10956

WALSH, GARY LYNN, b Fremont, Nebr, June 30, 40; m 64. AIR POLLUTION, ENVIRONMENTAL SCIENCES. Educ: Midland Lutheran Col, BS, 62; Univ Nebr, MS, 65; Univ SDak, PhD(zool), 69. Prof Exp: Asst prof zool, Ind Univ Northwest, 69-70; air pollution control chief, Michigan City, Ind, 70-72; admin asst to air pollution control officer, 72-74, SUPVR, AIR POLLUTION CONTROL SECT, LINCOLN-LANCASTER COUNTY HEALTH DEPT, 74- Mem: Air Pollution Control Asn. Res: Role of vitamin B12, biotin and thiamine on seasonal fluctuations of euglenophyte populations; taxonomy of Antarctic freshwater and soil amoeba. Mailing Add: Div of Environ Health Lincoln-Lancaster Co Health Dept Lincoln NE 68502

WALSH, GERALD EDWARD, zoology, ecology, see 12th edition

WALSH, GERALD MICHAEL, b Portland, Ore, Sept, 1, 44; m 70; c 2. PHARMACOLOGY. Educ: Univ Santa Clara, BS, 66; Ore State Univ, PhD(pharmacol), 71. Prof Exp: Res asst pharmacol, Ore State Univ, 67-69; asst prof, Univ Ga, 70-74; ASST PROF RES MED, SCH MED, UNIV OKLA, 74- Concurrent Pos: NSF instnl res grant, Univ Ga, 72-73. Res: Cardiovascular pharmacology and toxicology; hypertension. Mailing Add: Health Sci Ctr Dept of Med Univ of Okla PO Box 26901 Oklahoma City OK 73190

WALSH, JAMES ALOYSIUS, b Brooklyn, NY, Dec 15, 33; m 60. PHYSICAL ORGANIC CHEMISTRY. Educ: Fordham Univ, BS, 55; Purdue Univ, MS, 58, PhD(org chem), 63. Prof Exp: Instr chem, Purdue Univ, 60-63; from asst prof to assoc prof, 63-73, chmn dept, 69-72, PROF CHEM, JOHN CARROLL UNIV, 73- Mem: Am Chem Soc; Sigma Xi. Res: Chemistry of organo-sulfur compounds, especially sulfoxides and derivatives of sulfurtrioxide; phthalocyanines. Mailing Add: Dept of Chem John Carroll Univ Cleveland OH 44118

WALSH, JOHN EDMOND, b New York, NY, Aug 20, 39; m 66; c 2. PLASMA PHYSICS. Educ: NS Tech Col, BSc, 62; Columbia Univ, MSc, 65, DSc, 68. Prof Exp: Res engr, US Army Signal Res & Develop Lab, Ft Monmouth, 62-65; asst prof, 68-74, ASSOC PROF PHYSICS, DARTMOUTH COL, 74- Mem: Am Phys Soc; Sigma Xi. Res: Diffusion of turbulent plasmas; electron scattering in turbulent plasmas; nonlinear interactions in plasmas. Mailing Add: Dept of Physics Dartmouth Col Hanover NH 03755

WALSH, JOHN JOSEPH, b Cambridge, Mass, Sept 11, 42; m 69. ECOLOGY, OCEANOGRAPHY. Educ: Harvard Univ, AB, 64; Univ Miami, MS, 68, PHD(marine sci), 69. Prof Exp: Fel, 69-70, RES ASST PROF OCEANOG, UNIV WASH, 70- Mem: AAAS; Am Soc Limnol & Oceanog; Ecol Soc Am. Res: Upwelling ecosystems; systems analysis; statistics; phytoplankton ecology; mathematical models; theoretical ecology. Mailing Add: Dept of Oceanog Univ of Wash Seattle WA 98195

WALSH, JOHN JOSEPH, b New York, NY, July 31, 24; wid; c 3. CARDIOLOGY. Educ: Long Island Col Med, MD, 48; Am Bd Internal Med, dipl, 58. Prof Exp: Intern USPHS Hosp, NY, 48-49, resident med, Seattle, Wash, 51-54, asst chief med, New Orleans, La, 54-56, dep chief, 56; from instr med to asst prof clin med, 57-60, dean sch med & coordr health serv, 68-69, PROF MED, SCH MED, TULANE UNIV, 60-, VPRES HEALTH AFFAIRS, 69-, CHANCELLOR MED CTR, 72- Concurrent Pos: Fel cardiol, Sch Med, Tulane Univ, 57-58, instr, 55-; vis physician, Charity Hosp, 55-; chief res activities, USPHS, 58-64, chief med, 63-64, med officer in charge, 64-66, dir div direct health serv, 66-68. Mem: Am Thoracic Soc; AMA; fel Am Col Cardiol; fel Am Col Physicians; fel Am Col Chest Physicians. Res: Cardiopulmonary diseases. Mailing Add: Tulane Univ Med Ctr 1430 Tulane Ave New Orleans LA 70112

WALSH, JOHN LAWRENCE, b San Francisco, Calif, Apr 5, 28. SOLID STATE PHYSICS. Educ: Univ Calif, BS, 50, MA, 52, PhD(physics), 65. Prof Exp: Physicist, Los Alamos Sci Lab, 52 & US Naval Electronics Lab, 53; mem tech staff, Hughes Aircraft Co, 56-60, mem tech staff, Res Labs, 60-62; mem tech staff, Sci & Tech Div, Inst Defense Anal, 62-70; MEM TECH STAFF, NAVAL RES LAB, 70- Concurrent Pos: Lectr, Univ Southern Calif, 61-62. Mem: Am Phys Soc; Sigma Xi; Optical Soc Am; Inst Elec & Electronics Engrs. Res: Solid state and gaseous laser physics; solid state physics research, using nuclear magnetic resonance as an experimental tool. Mailing Add: Naval Res Lab Code 6505 Washington DC 20390

WALSH, JOHN M, b Wichita Falls, Tex, Nov 6, 23; m 52; c 3. PHYSICS. Educ: Univ Tex, PhD(physics), 50. Prof Exp: Staff mem, Los Alamos Sci Lab, 50-60; staff mem, Gen Atomic Div, Gen Dynamics Corp, 60-67; mgr continuum mech div, Systs Sci & Software, 67-74; MEM STAFF, LOS ALAMOS SPACE LAB, 74- Mem: Am Phys Soc. Res: Shock hydrodynamics; fluid dynamics. Mailing Add: Los Alamos Space Lab MS 682 PO Box 1620 Los Alamos NM 87545

WALSH, JOHN PAUL, b Rochester, NY, Dec 29, 42; m 67; c 2. ORGANIC CHEMISTRY. Educ: Purdue Univ, Lafayette, BS, 64; Univ Wis-Madison, MS, 66; Univ Tex, Austin, PhD(org chem), 70. Prof Exp: Sr res chemist, Org Res Dept, Pennwalt Corp, Pa, 69-74; RES CHEMIST, PARA-CHEM INC, PHILADELPHIA, 74- Mem: Am Chem Soc. Res: Organic synthesis; organosulfur, nitrogen and phosphorus.

WALSH, JOHN RICHARD, b San Francisco, Calif, Aug 22, 20; m 44; c 5. INTERNAL MEDICINE. Educ: Creighton Univ, BS, 43, MD, 45, MSc, 51; Am Bd Internal Med, dipl, 53. Prof Exp: Actg asst med, Sch Med, Creighton Univ, 51-52, asst, 52-53, from instr to assoc prof, 53-57, prof & dir dept, 57-60; PROF MED, MED SCH, UNIV ORE, 60- Concurrent Pos: Ward physician, Vet Admin Hosp, Omaha, Nebr, 51-52, asst chief med serv, 52-53, actg chief, 53-54, chief, 54-56, chief, Portland, Ore, 60-, actg chief radioisotope serv, 62-70; assoc, Col Med, Univ Nebr, 53-55, asst prof, 55-56. Mem: Am Soc Hemat; Am Fedn Clin Res; fel Am Col Physicians; Int Soc Hemat. Res: Hematology. Mailing Add: Vet Admin Hosp Portland OR 97207

WALSH, JOHN THOMAS, b Lincoln, RI, Dec 23, 27; m 52; c 3. ANALYTICAL CHEMISTRY. Educ: Providence Col, BS, 50; Univ RI, MS, 52, PhD(chem), 65. Prof Exp: Res chemist, Rumford Chem Works, RI, 52-54; RES CHEMIST, US ARMY NATICK LABS, 54- Mem: Am Chem Soc; Am Soc Testing & Mat. Res: Analytical instrumentation research in the areas of gas chromatography and mass spectrometry; applications have been the composition study of natural products such as foods and biologicals; investigation of toxic pollutants in air, water and solid wastes by analytical methods of gas and liquid chromatography and mass spectrometry. Mailing Add: 15 Ridgeland Dr Cumberland RI 02864

WALSH, JOSEPH BROUGHTON, b Utica, NY, Sept 5, 30; m 62; c 1. GEOLOGY, GEOPHYSICS. Educ: Mass Inst Technol, SB, 52, SM, 54, ME, 56, ScD(mech eng), 58. Prof Exp: Engr, Foster Miller Assocs, Inc, 57-59, A-b DeLaval Ljungstrom Angturbin, 59-60 & Woods Hole Oceanog Inst, 60-63; part-time vis prof, 63-64; res assoc geol & geophys, 63-72, SR RES SCIENTIST, DEPT EARTH & PLANETARY SCI, MASS INST TECHNOL, 72- Res: Theoretical analysis of various properties of rock, especially strength, elastic moduli and seismic attenuation, and analysis of how these properties should affect behavior in situ. Mailing Add: Dept of Earth & Planetary Sci Mass Inst of Technol Cambridge MA 02139

WALSH, JOSEPH LEONARD, mathematics, deceased

WALSH, JOSEPH MATTHEW, b Ciudad Guzman, Jalisco, Mex, Oct 19, 20; US citizen. BIOCHEMISTRY, PHYSICAL CHEMISTRY. Educ: De La Salle Col, BA, 42; Univ Md, PhD(chem), 63. Prof Exp: Teacher, Inst Regiomontano, Monterrey, Mex, 46-51; from asst prin to prin parochial high schs, La, instr, 58-59, from asst prof to assoc prof, 63-71, chmn dept phys sci & math, 63-71, PROF CHEM, COL SANTA FE, 71- Concurrent Pos: NSF grant, 64-66. Mem: Am Chem Soc. Res:

Enzymology; flavonoids. Mailing Add: Dept of Phys Sci & Math Col of Santa Fe Santa Fe NM 87501

WALSH, KENNETH ALBERT, b Yankton, SDak, May 23, 22; m 44; c 5. CHEMICAL METALLURGY, CERAMICS. Educ: Yankton Col, BA, 42; Iowa State Univ, PhD(chem), 50. Prof Exp: Asst prof chem, Iowa State Univ, 50-51; mem staff, Los Alamos Sci Lab, 51-57; supvr inorg chem res, Int Minerals & Chem Corp, 57-60; ASSOC DIR CORP TECHNOL, BRUSH WELLMAN INC, ELMORE, 60- Mem: AAAS; Am Chem Soc; Am Soc Metals; Am Ceramic Soc. Res: Beryllium metal extraction; role of trace elements in properties of beryllium; beryllium chemicals, ecology, and electronic materials. Mailing Add: 2624 Fangboner Rd Fremont OH 43420

WALSH, KENNETH ANDREW, b Hemmingford, Que, Aug 7, 31; m 53; c 3. BIOCHEMISTRY. Educ: McGill Univ, BSc, 51; Purdue Univ, MS, 53; Univ Toronto, PhD, 59. Prof Exp: Jr res officer, Nat Res Coun Can, 53-55; res instr, 59-62, asst prof, 62-65, assoc prof biochem, 65-68, PROF BIOCHEM, UNIV WASH, 68- Mem: Am Soc Biol Chemists. Res: Structure and function of proteins; mechanisms of zymogen activation and protease action; amino acid sequence and protein conformation; molecular evolution; biochemistry of fertilization. Mailing Add: Dept of Biochem Univ of Wash Seattle WA 98195

WALSH, LEO MARCELLUS, b Moorland, Iowa, Jan 16, 31; m 58. SOIL FERTILITY, SOIL SCIENCE. Educ: Iowa State Univ, BS, 52; Univ Wis, MS, 57, PhD(soils), 59. Prof Exp: Asst prof & exten specialist, 59-64, assoc prof, 64-68, assoc chmn dept, 71, chmn, 72, PROF SOILS, UNIV WIS-MADISON, 68- Mem: Fel Am Soc Agron; Soil Conserv Soc Am. Res: Use of nitrogen and sulfur fertilizers; use of zinc, manganese and other micronutrients; soil fertility, especially for corn and other cash crops; disposal of wastes on agricultural land. Mailing Add: 263 Soils Bldg Univ of Wis Madison WI 53706

WALSH, MARY IGNACIO, plant morphology, comparative anatomy, see 12th edition

WALSH, MICHAEL JOSEPH, b York, Pa, Apr, 18, 42; m 66; c 1. BIOCHEMISTRY, PHARMACOLOGY. Educ: Univ Md, Baltimore, BS, 65; Ohio State Univ, PhD(pharmacol), 69. Prof Exp: Instr biochem, Baylor Col Med, Tex Med Ctr, 69-71; asst prof pharmacol, Bowman Gray Sch Med, Wake Forest Univ, 71-74; ASSOC PROF PHARMACOL, EASTERN VA MED SCH, 74- Concurrent Pos: Nat Inst Ment Health fel biochem, Baylor Col Med, 69-71; res assoc lab neurochem, Vet Admin Hosp, Houston, 69-71. Mem: AAAS; Soc Neurosci; Am Soc Neurochem. Res: Neuropharmacology; neurochemistry; psychopharmacology; transmitter function; neuroamine metabolism; mechanism of action of psychotropic drugs; role of biogenic amines in addiction; molecular mechanisms and animal models of drug dependence. Mailing Add: Dept of Pharmacol Eastern Va Med Sch PO Box 1980 Norfolk VA 23501

WALSH, MICHAEL PATRICK, b Boston, Mass, Feb 28, 12. CYTOLOGY. Educ: Boston Col, AB, 34, AM, 35; Fordham Univ, MS, 38, PhD(cytol), 48; Weston Col, STL, 42. Hon Degrees: Sixteen from US cols & univs, 61-75. Prof Exp: Actg prin, Fairfield Univ Prep, 42; instr biol, Boston Col, 43-45, assoc prof, 48-58, pres, 58-69; pres, Fordham Univ, 69-72; ACAD ADV TO PRES, UNIV MASS, BOSTON, 72- Concurrent Pos: Trustee for many US cols & univs. Mem: AAAS; Genetics Soc Am; Am Micros Soc; Am Soc Zoologists; Bot Soc Am. Res: Cytology of earthworm; cytology with application of chemicals to plants. Mailing Add: Univ of Mass One Washington Mall Boston MA 02108

WALSH, PATRICK NOEL, b New York, NY, Dec 7, 30; m 62; c 5. HIGH TEMPERATURE CHEMISTRY. Educ: Fordham Univ, BS, 51, MS, 52, PhD(chem), 56. Prof Exp: Res assoc high temperature chem, Ohio State Univ, 56-60; mem staff, Union Carbide Res Inst, NY, 60-66 & Space Sci & Eng Lab, Union Carbide Corp, 66-68, RES ASSOC, MAT SYSTS DIV, UNION CARBIDE CORP, 68- Mem: Am Soc Metals; Electrochem Soc; Am Chem Soc. Res: Thermodynamics and kinetics of high temperature chemical processes. Mailing Add: Union Carbide Corp 1500 Polco St Indianapolis IN 46224

WALSH, PETER, b New York, NY, Aug 21, 29; m 52; c 4. PHYSICS. Educ: Fordham Univ, BS, 51; NY Univ, MS, 53, PhD(physics), 60. Prof Exp: Sr scientist, Westinghouse Lamp Div, NY, 51-61; instr physics & math, Eve Sch, Wagner Col, 57-63; supvr physics res, Am Stand Res Lab, NJ, 61-63; PROF PHYSICS & DIR QUANTUM PHYSICS LAB, FAIRLEIGH DICKINSON UNIV, 63- Concurrent Pos: Dir, NSF Undergrad Partic Prog, Fairleigh Dickinson Univ, 63-; consult, S-F-D Labs, Nuclear Res Asn, Curtiss Wright Corp & Belock Instrument Corp, 63- & US Army Res Off for Picatinny Arsenal, 65- Mem: AAAS; Am Phys Soc; Optical Soc Am. Res: Amorphous semiconductors; lasers; quantum physics; optics; plasmas. Mailing Add: Dept of Physics Fairleigh Dickinson Univ Teaneck NJ 07666

WALSH, PETER NEWTON, b Chicago, Ill, Apr 16, 35; m 58; c 3. HEMATOLOGY. Educ: Amherst Col, BA, 57; Washington Univ, MD, 61; Oxford Univ, DPhil(med), 72; Am Bd Internal Med, dipl, 68. Prof Exp: Intern internal med, Barnes Hosp, St Louis, 61-62, resident, 62-63; fel hemat, Sch Med, Washington Univ, 63-69; res fel blood coagulation, Oxford Haemophilia Ctr, Churchill Hosp, 69-72; asst prof, 72-74, ASSOC PROF INTERNAL MED, HEALTH SCI CTR, TEMPLE UNIV, 75- Concurrent Pos: Sr resident, Palo Alto Stanford Hosp, 64-65; chief resident, Sch Med, Washington Univ, 65-66; asst physician, Barnes Hosp, 65-66; med liaison officer, Nat Heart Inst, 66-69; NIH res fel, 69-72; hon sr registr med, United Oxford Hosps, 69-72; assoc ed, Thrombosis et Diathesis Haemorrhagica, 76-; mem coun thrombosis, Am Heart Asn. Honors & Awards: First Int Prize, Viviana Luckhaus Found, Arg, 72; Res Career Develop Award, Nat Heart & Lung Inst, 72; Jane Nugent Cochems Prize, Univ Colo Sch Med, 74. Mem: Int Soc Thrombosis & Haemostasis; Am Physiol Soc; Am Soc Clin Invest. Res: Role of blood platelets in blood coagulation, hemostasis and thrombosis; coagulation factor purification; mechanisms of binding of coagulation factors to platelets; role of platelet coagulant activities in thrombosis. Mailing Add: Spec Ctr Thrombosis Res Rm 421-OMS Temple Univ Health Sci Ctr Philadelphia PA 19140

WALSH, RALPH THOMAS, b Grand Rapids, Mich, Apr 14, 40. PHYSIOLOGY. Educ: Ferris State Col, BS, 62; Wayne State Univ, MS, 65, PhD(physiol, pharmacol), 68. Prof Exp: Res assoc physiol & pharmacol, 68-71, ASST PROF PHYSIOL, SCH MED, WAYNE STATE UNIV, 71- Mem: Reticuloendothelial Soc; Int Soc Thrombosis & Haemostasis; assoc Am Physiol Soc. Res: Blood platelet kinetics and regulatory mechanisms; antithrombotic therapy in cerebrovascular disease; cellular morphology by electron microscopy, changes by disease and drugs. Mailing Add: Dept of Physiol Wayne State Univ Sch of Med Detroit MI 48201

WALSH, RAYMOND ANTHONY, b Brooklyn, NY, Aug 9, 39; m 65; c 2. CHEMISTRY. Educ: Univ San Francisco, BS, 61; Univ Calif, Davis, PhD(chem), 69. Prof Exp: Res chemist, Stanford Res Inst, 61-64; RES CHEMIST, ORG CHEM DEPT, E I DU PONT DE NEMOURS & CO, INC, 69- Mem: AAAS; Am Chem Soc. Res: Organic chemical synthesis and reaction mechanisms; carbene chemistry; polymers; colloid stability; textile chemistry. Mailing Add: Org Chem Dept E I du Pont de Nemours & Co Inc Wilmington DE 19898

WALSH, RAYMOND ROBERT, b Denver, Colo, Apr 9, 25; m 52; c 6. PHYSIOLOGY. Educ: Cornell Univ, AB, 50, PhD(zool), 53. Prof Exp: Assoc res physiologist, Brookhaven Nat Lab, 53-55; from instr to assoc prof physiol, Sch Med, Univ Colo, Denver, 55-71; prof, Sch Dent, Southern Ill Univ, 71-72; PROF BIOL & CHMN DEPT, ST LOUIS UNIV, 72- Mem: Am Physiol Soc; Am Soc Zoologists; Soc Exp Biol & Med. Res: Neurobiology; comparative physiology. Mailing Add: Dept of Biol St Louis Univ St Louis MO 63103

WALSH, ROBERT A, b Lawrence, Mass, Dec 6, 10; m 35; c 5. PHARMACY, BIOLOGY. Educ: Boston Univ, BS, 42, AM, 48. Prof Exp: Asst prof biol, Mass Col Pharm, 36-53; vpres & dir res, E L Patch Co, 53-67; assoc prof pub health & dir pub rels, 67-72, PROF BIOL & DIR CONTINUING EDUC, MASS COL PHARM, 72- Concurrent Pos: Prof lectr, Mass Col Pharm, 52-67; chmn, City Bd Health, Andover, 57- Mem: Am Soc Microbiol; Am Pharmaceut Asn. Res: Pharmaceutical development and cooperative clinical investigation; public health. Mailing Add: Dept of Biol Mass Col Pharm Boston MA 02115

WALSH, ROBERT AUBREY, applied mathematics, see 12th edition

WALSH, ROBERT MICHAEL, b Wilmington, Del, Jan 28, 38; m 61; c 4. PHYSICAL CHEMISTRY. Educ: Univ Del, BS, 60; Univ Calif, Berkeley, PhD(chem), 65. Prof Exp: Res chemist, 65-72, SR RES CHEMIST, HERCULES INC, 72- Mem: Soc Photog Scientists & Engrs; Am Chem Soc. Res: Photoprocesses in materials; photographic systems. Mailing Add: 1800 Mt Salem Lane Wilmington DE 19806

WALSH, ROGER NUGENT, b Brisbane, Australia, July 3, 46. PSYCHIATRY, PSYCHOBIOLOGY. Educ: Univ Queensland, BMedSci, 68, dipl psychol, 69, MB, BS, 70, PhD(neurophysiol), 74. Prof Exp: Intern med, Repatriation Hosp, Australia, 71; res officer physiol, Univ Queensland, 72; RESIDENT PSYCHIAT, STANFORD UNIV, 72- Concurrent Pos: Fel psychiat, Stanford Univ, 72-75, Found Fund Psychiat Res fel, 75-76; Fulbright scholarship, 72. Honors & Awards: Mead Johnson Award, 73; Roche Neurosci Award, 73 & 74; William C Menninger Award, Cent Neuropsychiat Asn, 74 & 75. Mem: Int Soc Develop Psychobiol; Neurosci Asn; Asn Transpersonal Psychol; Asn Humanistic Psychol; Asn Advan Behav Ther. Res: Effects of sensory environments on brain structure, function and behavior; behavior therapy; humanistic and transpersonal psychology; clinical psychiatry; menstrual cycle effects on performance; hormonal influences on imprinting. Mailing Add: Dept of Psychiat Stanford Univ Med Sch Stanford CA 94305

WALSH, SCOTT WESLEY, b Wauwatosa, Wis, July 23, 47. ENDOCRINOLOGY, REPRODUCTIVE PHYSIOLOGY. Educ: Univ Wis-Milwaukee, BS, 70; Univ Wis-Madison, MS, 72, PhD(endocrinol, reproductive physiol), 75. Prof Exp: ASST PROF PHYSIOL, SCH MED, UNIV N DAK, 75- Mem: Soc Study Reproduction; Sigma Xi. Res: Female reproductive endocrinology and physiology. Mailing Add: Dept of Physiol & Pharmacol Univ of NDak Sch Med Grand Forks ND 58202

WALSH, THOMAS DAVID, b Chicago, Ill, Oct 30, 36; m 58; c 2. CHEMISTRY. Educ: Univ Notre Dame, AB, 58; Univ Calif, PhD(chem), 62. Prof Exp: Asst prof chem, Univ Ga, 62-67; vis asst prof, Ohio State Univ, 67-68; assoc prof, Univ SDak, 68-70; asst prof, 70-72, ASSOC PROF CHEM, UNIV NC, CHARLOTTE, 72- Concurrent Pos: Danforth Found assoc, 64-; res assoc, Univ NC, Chapel Hill, 69-70. Mem: Am Chem Soc. Res: Organic reaction mechanisms; stereochemistry; organometallic chemistry; photochemistry; electrochemistry. Mailing Add: Dept of Chem Univ of NC Charlotte NC 28223

WALSH, THOMAS EDWARD, physics, electrical engineering, see 12th edition

WALSH, WALTER MICHAEL, JR, b Los Angeles, Calif, July 28, 31; m 56; c 2. SOLID STATE PHYSICS. Educ: Harvard Univ, AB, 54, AM, 55, PhD(physics), 58. Prof Exp: Res fel solid state physics, Harvard Univ, 58-59; mem tech staff, 59-67, HEAD SOLID STATE & PHYSICS OF METALS RES DEPT, BELL LABS, 67- Mem: Fel Am Phys Soc. Res: Experimental physics of solids using microwave resonance techniques; effects of pressure and temperature on solids; resonance and wave propagation phenomena in metals. Mailing Add: Bell Labs Murray Hill NJ 07974

WALSH, WILLIAM J, b Hamilton, Can, May 14, 24; m 50; c 3. INTERNAL MEDICINE. Educ: Univ Western Ont, BA, 47, MD, 48; FRCPS. Prof Exp: Asst dean med, 65-71, ASSOC DEAN MED, McMASTER UNIV, 71- Concurrent Pos: Consult, St Joseph's Hosp & Hamilton Civic Hosp, Hamilton, Can. Mem: Fel Am Col Physicians; Can Med Asn. Res: Medical education. Mailing Add: McMaster Univ Med Ctr Hamilton ON Can

WALSH, WILLIAM K, b Columbus, Ohio, Sept 29, 32. TEXTILE CHEMISTRY, CHEMISTRY ENGINEERING. Educ: Univ SC, BS, 54; NC State Univ, PhD(chem eng), 67. Prof Exp: Engr, Celanese Corp Am, SC, 59-60; res asst, 60-67, asst prof, 67-72, ASSOC PROF TEXTILE CHEM, NC STATE UNIV, 72- Mem: AAAS; Am Chem Soc; Am Asn Textile Chemists & Colorists; Fiber Soc. Res: Applications of ionizing radiation to textile chemistry; radiation graft copolymerization, cross-linking, mechanical properties of textiles; kinetics and diffusion during grafting; physical chemistry of polymers. Mailing Add: Dept of Textile Chem NC State Univ Raleigh NC 27607

WALSH, WILLIAM LOUIS, b Pittsburgh, Pa, Apr 20, 26; m 54; c 3. ORGANIC CHEMISTRY. Educ: Univ Pittsburgh, BS, 49, MEd, 75; Carnegie Inst Technol, MS, 52, PhD(chem), 53. Prof Exp: Res chemist, Gulf Res & Develop Co, Pa, 53-66, sr res chemist, 66-74; INSTR CHEM, SOUTH BUTLER SCHS, 74- Concurrent Pos: Fel coal res, Univ Pittsburgh, 74. Mem: Am Chem Soc. Res: Friedel-Crafts; mercaptans; alcohols; olefins; organometallics; alkyl halides; hydrocarbons; dehydrogenations; catalysis; catalyst preparation; halogenation; hydrocarbon oxidation; free radicals. Mailing Add: 2825 Phillips Ave Glenshaw PA 15116

WALSH, WILLIAM PATRICK, geophysics, computer science, see 12th edition

WALSKE, MAX CARL, b Seattle, Wash, June 2, 22; m 46; c 3. PHYSICS. Educ: Univ Wash, BS, 44; Cornell Univ, PhD(physics), 51. Prof Exp: Mem staff, Los Alamos Sci Lab, 51-55, asst leader theoret div, 55-56; dep res dir, Atomics Int Div, NAm Aviation, Inc, 56-59; mem US deleg, Conf Suspension Nuclear Tests, Geneva, Switz, 59-61; sci rep, AEC, London, Eng, 61-62; theoret physicist, Rand Corp, 62-63; sci attache, US Missions to NATO & Orgn Econ Coop & Develop, Paris, France, 63-65; staff mem, Los Alamos Sci Lab, 65-66; asst to secy defense & chmn mil liaison comt, US Dept Defense, 66-73; PRES, ATOMIC INDUST FORUM, INC, 73- Concurrent Pos: Consult, Los Alamos Sci Lab, 56-59 & 62-63. Mem: Am Phys Soc; Am Nuclear Soc. Res: Reactor physics. Mailing Add: 7005 Buxton Terr Bethesda MD 20034

WALSTEDT, RUSSELL E, b Minneapolis, Minn, June 12, 36; m 64; c 2. SOLID STATE PHYSICS. Educ: Mass Inst Technol, BS, 58; Univ Calif, Berkeley, PhD(physics), 62. Prof Exp: NSF fel, Clarendon Lab, Eng, 61-62; asst res physicist, Univ Calif, Berkeley, 62-65; MEM TECH STAFF SOLID STATE MAGNETISM, BELL LABS, 65- Mem: Am Phys Soc. Res: Nuclear magnetic resonance and its application to the study of magnetism in the solid state, particularly metals and alloys. Mailing Add: Dept 1d362 Box 261 Bell Labs Murray Hill NJ 07974

WALSTON, DALE EDOUARD, b Woodsboro, Tex, Dec 1, 30. MATHEMATICS. Educ: Tex A&M Univ, BA, 52; Univ Tex, MA, 59, PhD(math), 61. Prof Exp: Asst prof, 61-72, ASSOC PROF MATH, UNIV TEX, AUSTIN, 72- Concurrent Pos: Consult, Manned Space Ctr, NASA, 66. Mem: Am Math Soc; Math Asn Am. Res: Numerical solution of differential equations. Mailing Add: Dept of Math Univ of Tex Austin TX 78712

WALSTROM, ROBERT JOHN, b Omaha, Nebr, Apr 24, 22; m 44; c 2. ENTOMOLOGY. Educ: Univ Nebr, BS, 47, MS, 49; Iowa State Univ, PhD, 55. Prof Exp: State entomologist, State Dept Agr & Inspection, Nebr, 48-50; exten entomologist, Iowa State Univ, 50-55; PROF ENTOM, SDAK STATE UNIV, 55- Mem: AAAS; Entom Soc Am. Res: Control of beneficial and injurious legume insects; apiculture. Mailing Add: 1409 First St Brookings SD 57006

WALT, ALEXANDER JEFFREY, b Cape Town, SAfrica, June 13, 23; US citizen; c 3. SURGERY. Educ: Univ Cape Town, MB & ChB, 48; FRCS(C), 55; Univ Minn, MS, 56; FRCS, 56; Am Bd Surg, dipl, 62. Prof Exp: Lectr path, Univ Cape Town, 49-50; registr surg, St Martin's Hosp, Bath, Eng, 56-57; asst surgeon, Groote Schuur Hosp, Cape Town, 57-61; asst chief surg, Vet Admin Hosp, Allen Park, Mich, 61-62; from asst prof to assoc prof, 61-66, from asst dean to assoc dean med, 64-70, PROF SURG & CHMN DEPT, SCH MED, WAYNE STATE UNIV, 66- Concurrent Pos: Consult div physician manpower & clin cancer training comt, NIH, 72-73. Mem: Am Surg Asn; Int Soc Surg; Soc Surg Alimentary Tract; Am Asn Surg of Trauma. Res: Studies of the effects and clinical management of severe trauma of the liver and stomach of humans. Mailing Add: Dept of Surg Wayne State Univ Sch of Med Detroit MI 48201

WALT, MARTIN, b West Plains, Mo, June 1, 26; m 50; c 4. MAGNETOSPHERIC PHYSICS, SPACE PHYSICS. Educ: Calif Inst Technol, BS, 50; Univ Wis, MS, 51, PhD(physics), 53. Prof Exp: Mem staff, Los Alamos Sci Lab, 53-56; MGR PHYS SCI, LOCKHEED MISSILES & SPACE CO, 56- Mem: Am Phys Soc; Am Geophys Union. Res: Fast neutron physics; Van Allen radiation; aurora; cosmic rays. Mailing Add: 12650 Viscaino Ct Los Altos Hills CA 94022

WALTCHER, AZELLE BROWN, b New York, NY, Mar 27, 25; m 55; c 2. MATHEMATICS. Educ: Columbia Univ, BA, 45, MA, 46; NY Univ, PhD(math, educ), 54. Prof Exp: Asst math, Barnard Col, Columbia Univ, 45-46; instr, Hollins Col, 46-48; teacher, Calhoun Sch, NY, 48-52; from instr to assoc prof, Univ, 52-72, teaching fel, New Col, 61-72, PROF MATH, HOFSTRA UNIV, 72- Concurrent Pos: Mem fac, Sarah Laurence Col, 53-55. Mem: Am Math Soc; Math Asn Am. Res: Logic and foundations of mathematics; number theory; group theory. Mailing Add: Dept of Math Hofstra Univ Hempstead NY 11550

WALTCHER, IRVING, b Newport, RI, Mar 6, 17. POLYMER CHEMISTRY. Educ: Univ RI, BS, 38; Duke Univ, MA, 40; Ohio State Univ, PhD(org chem), 47. Prof Exp: Chemist, War Ord Dept, Ala, 41-42; res chemist, B F Goodrich Co, Ohio, 42-44; asst org chem, Res Found, Ohio State Univ, 44-47; res assoc chem, Polytech Inst Brooklyn, 47-48; assoc prof, State Univ NY Col Forestry, Syracuse, 48-55; asst prof, 55-58, ASSOC PROF CHEM, CITY COL NEW YORK, 58, DEP CHMN DEPT, 74- Mem: Am Chem Soc. Res: Selective hydrogenation of acetylenes; preparation and characterization of graft copolymers. Mailing Add: Dept of Chem City Col of New York New York NY 10031

WALTER, CHARLES FRANK, b Sarasota, Fla, June 19, 36. BIOCHEMISTRY, BIOMATHEMATICS. Educ: Ga Inst Technol, BS, 57; Fla State Univ, MS, 59, PhD(chem), 62. Prof Exp: NIH fel, Med Sch, Univ Calif, San Francisco, 62-64; from asst prof to assoc prof biochem, Med Sch, Univ Tenn, Memphis, 64-70; assoc prof biomath & biochem, M D Anderson Hosp & Tumor Inst, Univ Tex, Houston, 70-74; PROF CHEM ENG, UNIV HOUSTON, 74- Concurrent Pos: NIH career develop award, Med Sch, Univ Tenn, Memphis, 65-70. Mem: Biophys Soc; Am Soc Eng Educ; Soc Math Biol. Res: Solar energy conversion; biological control; enzyme mechanisms and kinetics; biochemical engineering; photoproduction of hydrogen coenzyme conformation; real-time computer applications. Mailing Add: Dept of Chem Eng Univ of Houston Houston TX 77004

WALTER, CHARLES ROBERT, JR, b Charlottesville, Va, Oct 31, 22; m 50; c 3. ORGANIC CHEMISTRY. Educ: Univ Va, BA, 43, PhD(chem), 49. Prof Exp: Res asst chem, Univ Ill, 49-50; asst prof, Univ NC, 50-52; sr res chemist, Nitrogen Div, Allied Chem Corp, 52-58, supvry res chemist, 58-60, mgr res, 60-66; PROF CHEM & CHMN DEPT, GEORGE MASON UNIV, 66- Mem: Am Chem Soc. Res: Synthetic organic chemistry; Diels-Alder reaction of quinoneimides; industrial process development; organic nitrogen chemicals; vapor phase catalysis; chlorination of olefins. Mailing Add: 6120 Sherborn Lane Springfield VA 22152

WALTER, CHARLTON M, b Altoona, Pa, July 1, 23; m 47; c 2. INFORMATION SCIENCE. Educ: Columbia Univ, BA, 49; Harvard Univ, MA, 51. Prof Exp: Mathematician, Commun Lab, Air Force Cambridge Res Ctr, 51-54, chief simulation & eval br, Comput & Math Sci Lab, 54-63, chief dynamic processes br, Data Sci Lab, 63-70, CHIEF MULTISENSOR SIGNAL PROCESSING BR, DATA SCI LAB, AIR FORCE CAMBRIDGE RES LABS, 70- Mem: AAAS; Inst Elec & Electronics Engrs; Soc Gen Syst Res. Res: Development of interactive, computer-based, display-oriented signal processing systems, with applications to environmental sensor data collection; statistical data reduction; dynamic modelling; simulation and systems evaluation. Mailing Add: Data Sci Lab L G Hanscom Field Air Force Cambridge Res Labs Bedford MA 01730

WALTER, DEAN IRVING, b Hollidaysburg, Pa, Aug 11, 20; m 59; c 2. CHEMISTRY. Educ: Juniata Col, BS, 40. Prof Exp: Chemist, Gen Chem Co, Del, 40; anal chemist, Off Inspector Naval Mat, Pa, 41-44; res chemist, 46-54, head anal chem br, 54-74, CONSULT ANAL CHEM, US NAVAL RES LAB, 74- Honors & Awards: Superior Civilian Serv Award, US Navy, 65 & Meritorious Civilian Serv Award, 71. Mem: Am Chem Soc; Sigma Xi; Am Sci Affil. Res: High vacuum methods of analysis for the determination of gases in metals; laboratory administration; analytical chemistry. Mailing Add: 9811 Caltor Lane Oxon Hill MD 20022

WALTER, DONALD OLIVER, b Los Angeles, Calif, Jan 19, 29; m 53; c 2. NEUROPHYSIOLOGY. Educ: Pomona Col, BA, 49; Univ Calif, Los Angeles, PhD(neurophysiol), 62. Prof Exp: Asst res anatomist, Dept Anat & Physiol, Brain Res Inst, 62-64, asst prof in residence, Univ, 64-67, ASSOC PROF IN RESIDENCE ANAT & PHYSIOL, UNIV CALIF, LOS ANGELES, 67- Mem: Am EEG Soc; Asn Comput Mach; Am Physiol Soc; Biomet Soc. Res: Neurophysiology of integrative brain functions; mathematical and statistical analysis, using digital computers of ongoing and evoked electrical activity of the brain. Mailing Add: C3-384 Ctr for the Health Sci Univ of Calif Los Angeles CA 90024

WALTER, EDWARD JOSEPH, b St Louis, Mo, Dec 6, 14; m 39; c 8. GEOPHYSICS. Educ: St Louis Univ, BS, 37, MS, 40, PhD(geophys), 44. Prof Exp: Seismic computer, Root Petrol Co, Ark, 37; asst seismologist, Shell Oil Co, 38, seismologist, 45-46; assoc prof math & asst dir seismol observ, 46-50, dir dept, 57-59, dir seismol observ, 62, PROF MATH, JOHN CARROLL UNIV, 50- Mem: AAAS; Seismol Soc Am; Soc Explor Geophys; Am Geophys Union. Res: Seismology; local earthquakes; crustal structure; wave motion; attenuation coefficients; volcanology; engineering seismology; air pollution. Mailing Add: Seismol Observ John Carroll Univ Cleveland OH 44118

WALTER, EUGENE LEROY, JR, medical microbiology, see 12th edition

WALTER, EVERETT L, b Rensselaer, Ind, July 1, 29; m 51; c 4. MATHEMATICS. Educ: Ariz State Univ, BS, 51; NMex State Univ, MS, 57, PhD(math), 61. Prof Exp: Dir comput, Army Field Forces, Ft Bliss, Tex, 54-56; instr math, NMex State Univ, 56-60; res mathematician, White Sands Missile Range, 61-62; assoc prof, 62-68, PROF MATH, NORTHERN ARIZ UNIV, 68- Mem: Math Asn Am; Am Math Soc. Res: Functional analysis, particularly in the study of generalized conjugate spaces. Mailing Add: Dept of Math Northern Ariz Univ Flagstaff AZ 86001

WALTER, FRED JOHN, b Topeka, Kans, Mar 31, 31; m 51; c 4. SOLID STATE PHYSICS, SCIENCE ADMINISTRATION. Educ: Kans State Univ, BS, 53; Univ Tenn, MS, 58. Prof Exp: Develop engr, Oak Ridge Nat Lab, 53-57, res assoc, Physics Div, 58-63; chief physicist, Radiation Instrument Develop Lab, Nuclear-Chicago Corp, Ill, 63-65; mgr semiconductor res & develop, 65-71, tech dir, 71-73, VPRES & GEN MGR, PHYS SCI DIV, ORTEC INC DIV, EG&G, INC, 73- Concurrent Pos: Consult, 60-63; guest lectr, Oak Ridge Assoc Univs, 67-68, Univ Tenn, 68 & London Trade Conf, US Dept Com, 68. Mem: Am Phys Soc; sr mem Inst Elec & Electronics Engrs; Sigma Xi. Res: Nuclear structure physics; low temperature nuclear alignment and cryogenic design; semiconductor radiation spectrometers; x-ray spectroscopy; semiconductor technology and surface physics. Mailing Add: 119 Caldwell Dr Oak Ridge TN 37830

WALTER, GERALD JOSEPH, b Jan 31, 38; US citizen; m 63; c 3. ORGANIC CHEMISTRY. Educ: Iona Col, BS, 59; St Joseph's Col, Pa, MS, 61; Villanova Univ, PhD(org chem), 70. Prof Exp: Res chemist, Rohm and Haas Co, 61-67; SR RES CHEMIST, PENNWALT CORP, 69- Mailing Add: Pennwalt Corp Three Parkway Philadelphia PA 19102

WALTER, HARTMUT, b Stettin, Ger, July 13, 40; m 69. BIOGEOGRAPHY, WILDLIFE MANAGEMENT. Educ: Univ Bonn, Dr rer nat(bird ecol), 67. Prof Exp: Harkness fel geog, Univ Calif, Berkeley, 67-68 & Univ Chicago, 68; assoc regional expert ecol & conserv for Africa, UNESCO Field Sci Off, Nairobi, Kenya, 70-72; actg asst prof, 72-73, asst prof, 73-74, ASSOC PROF BIOGEOG, UNIV CALIF, LOS ANGELES, 74- Honors & Awards: Hoerlein Prize, Ger Biol Asn, 60. Mem: AAAS; Ecol Soc Am; Asn Am Geogr; Int Asn Ecol; Brit Ecol Soc. Res: Mediterranean and tropical ecosystem dynamics; evolutionary ecology; wildlife in urban and other man-modified environments. Mailing Add: Dept of Geog Univ of Calif Los Angeles CA 90024

WALTER, HENRY ALEXANDER, b Muehlhausen, Ger, Jan 8, 12; nat US; m 39; c 4. ORGANIC CHEMISTRY. Educ: Univ Heidelberg, dipl, 39. Prof Exp: Res chemist, Plaskon Co, 39-42; asst prof chem, Univ Mo, 42-44; res specialist, Monsanto Chem Co, 44-62; sr scientist, Plastic Coating Corp, Scott Paper Co, 62-71, CONSULT, SCOTT GRAPHICS, INC, 71- Mem: AAAS; Am Chem Soc. Res: Polymer chemistry; technical information services; patent liaison. Mailing Add: Samuel B Sutphin Res Ctr Scott Graphics Inc Holyoke MA 01040

WALTER, HENRY CLEMENT, b Boston, Mass, Sept 12, 19; m 54; c 6. ORGANIC CHEMISTRY. Educ: Mass Inst Technol, SB, 41, PhD(org chem), 46. Prof Exp: Asst, Mass Inst Technol, 42-43 & 44-45; RES CHEMIST, EXP STA, E I DU PONT DE NEMOURS & CO, INC, 46- Mem: AAAS; Am Chem Soc. Res: Elastomers; adhesives. Mailing Add: 310 Hampton Rd Sharpley Wilmington DE 19803

WALTER, JOHN FITLER, b Philadelphia, Pa, Mar 19, 43; m 68. APPLIED PHYSICS. Educ: Drexel Univ, BSEE, 66, MS, 68, PhD(physics), 70. Prof Exp: PHYSICIST, APPL PHYSICS LAB, JOHNS HOPKINS UNIV, 70- Mailing Add: Johns Hopkins Rd Laurel MD 20810

WALTER, JOHN HARRIS, b Los Angeles, Calif, Dec 14, 27; m 55; c 3. ALGEBRA. Educ: Calif Inst Technol, BS, 51; Univ Mich, MS, 53, PhD, 54. Prof Exp: From instr to asst prof math, Univ Wash, 54-61; assoc prof, 61-66, PROF MATH, UNIV ILL, URBANA, 66- Concurrent Pos: NSF fel, 57-58; vis asst prof, Univ Chicago, 60-61, vis assoc prof, 65-66; res assoc, Harvard Univ, 67-68. Mem: Am Math Soc. Res: Finite groups; classical groups; representation theory. Mailing Add: Dept of Math Univ of Ill Urbana IL 61801

WALTER, JOSEPH L, b Braddock, Pa, Jan 23, 30. INORGANIC CHEMISTRY. Educ: Duquesne Univ, BS, 51; Univ Pittsburgh, PhD(chem), 55. Prof Exp: ASSOC PROF INORG CHEM, UNIV NOTRE DAME, 60- Concurrent Pos: NIH fels, 62-70; AEC fel, 63-67. Mem: Am Chem Soc; Soc Appl Spectros. Res: Normal coordinate analysis of inorganic coordination compounds using the Urey-Bradley Force Field calculations and the thermodynamic studies of metal chelate formation. Mailing Add: Dept of Chem Univ of Notre Dame Notre Dame IN 46556

WALTER, LOUIS S, b New York, NY, Aug 11, 33; m 57; c 2. GEOCHEMISTRY. Educ: City Col New York, BS, 54; Univ Tenn, MS, 55; Pa State Univ, PhD(geochem), 60. Prof Exp: Res fel geochem, Pa State Univ, 60-62; res assoc, Nat Acad Sci, 62-63; GEOCHEMIST, GODDARD SPACE FLIGHT CTR, NASA, 63- Mem: Am Geophys Union; Geochem Soc; Mineral Soc Am; Meteoritical Soc. Res: Experimental petrology and mineralogy; crystal chemistry; phase equilibria; petrography; electron microprobe analyses; planetology; theoretical petrology. Mailing Add: Goddard Space Flight Ctr NASA Code 920 Greenbelt MD 20771

WALTER, MABLE RUTH, b Groveland, Ill, 1904. GENETICS. Educ: NCent Col, Ill, AB, 29; Univ Ill, AM, 30, PhD(genetics), 34. Prof Exp: Asst, Univ Ill, 29-34; assoc prof biol, MacMurray Col, 35-42; mem staff foreign serv, Am Red Cross, 42-46, dir col activities, 46-52; teacher biol, Hagerstown Jr Col, 52-67, prof biol & chmn sci div, 67-75; RETIRED. Concurrent Pos: Secy, Jr Col Coun, Med Atlantic States, 67-70. Res: Comparison of human twins; inheritance of multiple births in cattle; inheritance of human diseases. Mailing Add: 8761 Ft Foote Rd SE Washington DC 20022

WALTER, PAUL HERMANN LAWRENCE, b Jersey City, NJ, Sept 22, 34; m 56; c 2. INORGANIC CHEMISTRY. Educ: Mass Inst Technol, SB, 56; Univ Kans, PhD(inorg chem), 60. Prof Exp: Res chemist, Cent Res Dept, E I du Pont de

Nemours & Co, 60-66, col rels rep, Employee Rels Dept, 66-67; asst prof, 67-70, ASSOC PROF CHEM, SKIDMORE COL, 70-, CHMN DEPT, 75- Concurrent Pos: Guest, Univ Stuttgart, 64-65. Mem: AAAS; Am Chem Soc; Am Meteorol Soc; fel Am Inst Chemists. Res: Solid state inorganic chemistry; chemical education; rhenium chemistry; inorganic analytical chemistry; environmental chemistry. Mailing Add: Dept of Chem Skidmore Col Saratoga Springs NY 12866

WALTER, REGINALD HENRY, b St John's, Antigua, Feb 13, 33; m 65. FOOD CHEMISTRY. Educ: Tuskegee Inst Technol, BS, 60, MS, 63; Univ Mass, PhD(food sci), 67. Prof Exp: Asst food technologist, Univ PR, 67-68; sr chemist, Int Flavors & Fragrances, Inc, 68-71; ASST PROF FOOD SCI, CORNELL UNIV, 71- Mem: Am Chem Soc. Res: Food waste pollution; pesticide residues. Mailing Add: Food Res Lab Cornell Univ Geneva NY 14456

WALTER, RICHARD D, b Alameda, Calif, Aug 16, 21; m 47. MEDICINE. Educ: St Louis Univ, MD, 46; Am Bd Psychiat & Neurol, cert psychiat, 55, cert neurol, 60. Prof Exp: From instr to assoc prof, 55-70, PROF NEUROL & CHMN DEPT, MED CTR, UNIV CALIF, LOS ANGELES, 70- Concurrent Pos: Res fel psychiat, Langley Porter Clin, Univ Calif, Los Angeles, 53-55; dir, Reed Neurol Res Ctr. Mem: Am Acad Neurol; fel Am EEG Soc; Am Psychiat Asn; Am Neural Asn. Res: Clinical neurophysiology and electroencephalography as it relates to the convulsive disorders. Mailing Add: Neuropsychiat Inst Univ of Calif Med Ctr Los Angeles CA 90024

WALTER, RICHARD L, b Chicago, Ill, Nov 1, 33; m 58; c 2. NUCLEAR PHYSICS. Educ: St Procopius Col, BS, 55; Univ Notre Dame - PhD(physics), 60. Prof Exp: Res assoc nuclear physics, Univ Wis, 59-61, instr physics, 61-62; from asst prof to assoc prof, 62-74, PROF PHYSICS, DUKE UNIV, 74- Concurrent Pos: Vis prof, Max Planck Inst Nuclear Physics, Heidelberg, Ger, 70-71; Fulbright res fel, 70-71; vis scientist, Los Alamos Sci Lab, 75. Mem: Am Phys Soc. Res: Neutron physics; polarization of nucleons produced in reactions; scattering of plarized nucleons; low energy accelerator physics; studies involving trace metals in the environment. Mailing Add: Dept of Physics Duke Univ Durham NC 27706

WALTER, RICHARD WEBB, JR, b West Chester, Pa, Oct 5, 44; m 67; c 1. MICROBIAL BIOCHEMISTRY. Educ: Pa State Univ, BS, 66; Mich State Univ, PhD(biochem), 72. Prof Exp: Fel biochem, Univ Colo Med Ctr, Denver, 72-74; SR RES BIOCHEMIST, DOW CHEM CO, 74- Mem: Am Soc Microbiol; Am Chem Soc; AAAS. Res: Biochemical transformation and synthesis of sterospecific molecules that are of interest to both the chemical and pharmaceutical industries and are difficult to synthesize by normal chemical measn. Mailing Add: 1701 Bldg Dow Chem Co Midland MI 48640

WALTER, ROBERT IRVING, b Johnstown, Pa, Mar 12, 20. PHYSICAL ORGANIC CHEMISTRY. Educ: Swarthmore Col, AB, 41; Johns Hopkins Univ, MA, 42; Univ Chicago, PhD(chem), 49. Prof Exp: Asst, Swarthmore Col, 40-41 & Johns Hopkins Univ, 41-42; res chemist, Wyeth, Inc, 42-44; instr chem, Univ Colo, 49-51; from res asst prof to res assoc prof, Rutgers Univ, 51-53; instr chem, Univ Conn, 53-55; assoc physicist, Brookhaven Nat Lab, 55-56; from asst prof to prof chem, Haverford Col, 56-68; PROF CHEM, UNIV ILL, CHICAGO CIRCLE, 68- Concurrent Pos: NSF fac fel, 60-61; mem, Adv Coun Col Chem, 66-70, sr staff assoc, 67; vis prof, Stanford Univ, 67; acad guest, Inst Phys Chem, Univ Zurich, 75-76. Mem: AAAS; Am Chem Soc. Res: Equilibria in porphyrin systems; preparation and properties of stable organic free radicals. Mailing Add: Dept of Chem Univ of Ill Box 4348 Chicago IL 60680

WALTER, RODERICH, b Darmstadt, Ger, July 16, 37; m 74; c 2. CHEMISTRY, PHYSIOLOGY. Educ: Univ Giessen, Vordiplom, 61; Univ Cincinnati, PhD(chem, physiol), 64. Prof Exp: Res assoc biochem, Med Col, Cornell Univ, 64-65; res assoc physiol, Mt Sinai Hosp, 65-66; from assoc prof to prof physiol & biophys, Mt Sinai Sch Med, 66-74, co-dir dept, Grad Sch, 66-74; PROF PHYSIOL & HEAD DEPT, UNIV ILL MED CTR, 74- Mem: AAAS; Am Chem Soc; Soc Exp Biol & Med; Biophys Soc; Soc Appl Spectros. Res: Membrane and transport phenomena; mechanism of hormone action; peptide and orgainc synthesis; sequence and conformational analysis of peptides and proteins; endocrinol. Mailing Add: Dept of Physiol Univ of Ill at Med Ctr Chicago IL 60612

WALTER, THOMAS JAMES, b Dodgeville, Wis, Aug 20, 39; m 65; c 3. ORGANIC CHEMISTRY. Educ: Univ Wis-Madison, BS, 61; Cornell Univ, MST, 69; Univ Ga, PhD(chem), 73. Prof Exp: Pub sch teacher, Wis, 65-68; RES CHEMIST, ETHYL CORP, 73- Mem: Am Chem Soc. Res: Paracyclophanes, fatty acids, esters and alcohols; hydrocarbon chemistry. Mailing Add: Ethyl Corp Res & Develop Lab PO Box 341 Baton Rouge LA 70821

WALTER, WALDEMAR MELCHERT, b Cambridge, Mass, Sept 20, 20; m 48; c 4. INVERTEBRATE ZOOLOGY, ECOLOGY. Educ: Harvard Univ, AB, 42; Duke Univ, PhD(zool), 54. Prof Exp: Asst proj leader, Mo Conserv Comn, 48-49; asst zool, Duke Univ, 49-52; asst prof biol, Col Charleston, 52-55; USPHS fel & res assoc ecol for Dr C W McNeil, Wash State Univ, 55-58; asst prof biol, Eastern Wash State Univ, 58-59; assoc prof, Tex Woman's Univ, 59-64; ASSOC PROF BIOL SCI, WESTERN ILL UNIV, 64- Concurrent Pos: NSF undergrad res partic grants, 60-64; res partic, Ecol Div, Oak Ridge Nat Lab, 66 & 67. Mem: Am Malacol Union; Am Soc Limnol & Oceanog; Brit Freshwater Biol Asn; Int Soc Limnol; Ecol Soc Am. Res: Ecology of freshwater snails; freshwater biology as affected by various contaminants. Mailing Add: Dept of Biol Sci Western Ill Univ Macomb IL 61455

WALTER, WILBERT GEORGE, b Lingle, Wyo, Nov 16, 33; m 55; c 2. PHARMACOGNOSY, MEDICINAL CHEMISTRY. Educ: Univ Colo, BA, 55, BS & MS, 58; Univ Conn, PhD(pharmaceut chem), 62. Prof Exp: Asst, Univ Colo, 55-58; asst, Univ Conn, 58-61, spec res technologist, 60; res assoc pharmacog & asst prof pharmaceut chem, Sch Pharm, Univ Tenn, 61-63; assoc prof, Sch Pharm, Univ Miss, 63-68; PROF PHARMACOG & BIOL & CHMN DEPT, COL PHARM, MED UNIV SC, 68- Mem: Am Pharmaceut Asn; Am Soc Pharmacog; Am Chem Soc; Soc Econ Bot. Res: Organic medicinal chemistry; natural product chemistry. Mailing Add: Dept of Biol Med Univ of SC Charleston SC 29401

WALTER, WILLIAM ARNOLD, JR, b Pittsburgh, Pa, May 17, 22; m 46; c 2. EPIDEMIOLOGY. Educ: Ind Univ, AB, 43, MD, 45; Johns Hopkins Univ, MPH, 51. Prof Exp: Med officer, Ky State Bd Health, Louisville, 48-50; dir, Venereal Dis Control Div, Fla State Bd Health, Jacksonville, 51-53, assoc dir, Bur Prev Dis, 53-55; epidemiologist, Epidemiol Sect, Nat Cancer Inst, 55-57; med officer in-chg, Houston Pulmonary Cytol Proj, Univ Tex M D Anderson Hosp & Tumor Inst, 57-59; med officer in-chg uterine cancer cytol proj, Women's Med Col Pa, 59-60; res grants adminr, Grants & Training, 60-66, CHIEF, SPEC PROG BR, EXTRAMURAL ACTIV, NAT CANCER INST, 66- Mem: Am Pub Health Asn; AMA. Res: Chronic disease epidemiology with special interest in geographic pathology of leukemia. Mailing Add: 6310 Wilson Lane Bethesda MD 20014

WALTER, WILLIAM GOFF, b Lake Placid, NY, Nov 1, 14; m 41; c 4. MICROBIOLOGY. Educ: Cornell Univ, BS, 38, MS, 41; Mich State Univ, PhD, 52; Am Bd Med Microbiol, dipl. Prof Exp: Spec investr, NY Exp Sta, Geneva, 40-41, asst, 41-42; from asst prof to assoc prof, 42-61, PROF MICROBIOL, MONT STATE UNIV, 51-, HEAD DEPT MICROBIOL, 68- Honors & Awards: Walter S Mangold Award, Nat Environ Health Asn, 72. Mem: Am Soc Microbiol; fel Am Pub Health Asn; fel Am Acad Microbiol; Nat Environ Health Asn (pres, Nat Asn Sanitarians, 62); Am Acad Environ Engrs. Res: Environmental and public health; surface contamination. Mailing Add: Dept of Microbiol Mont State Univ Bozeman MT 59715

WALTER, WILLIAM MOOD, JR, b Sumter, SC, Nov 20, 36; m 59; c 1. FOOD SCIENCE. Educ: The Citadel, BS, 58; Univ Ga, PhD(org chem), 63. Prof Exp: Res asst, Univ Ga, 60-63; asst prof, 65-70, ASSOC PROF FOOD SCI, NC STATE UNIV, 70-; RES CHEMIST, AGR RES SERV, USDA, 65- Mem: Am Chem Soc; Inst Food Technologists. Res: Effect of processing and storage on organic constituents of food; emphasis on quality and nutritional value of processed foods. Mailing Add: Dept of Food Sci NC State Univ Raleigh NC 27607

WALTER, WILLIAM TRUMP, b Jamaica, NY, Dec 28, 31; m 60; c 4. ELECTROPHYSICS. Educ: Middlebury Col, AB, 53; Mass Inst Technol, PhD(physics), 62. Prof Exp: Res asst, Res Lab Electronics, Mass Inst Technol, 59-62; sr scientist, TRG, Inc, 62-67; RES SCIENTIST, POLYTECH INST NEW YORK, 67- Concurrent Pos: Guest, Res Lab Electronics, Mass Inst Technol, 62-63; pres, Laser Consults, Inc, 68- Mem: AAAS; Am Phys Soc; Optical Soc Am; Am Asn Physics Teachers; Fedn Am Scientists. Res: Gas laser research; metal vapor lasers; gas discharges; atomic physics; resonance phenomena in dilute gases, including optical pumping, orientation and nuclear magnetic resonance; optical and radiofrequency spectroscopy. Mailing Add: Dept of Elec Eng & Electrophys Polytech Inst of New York Rte 110 Farmingdale NY 11735

WALTERS, ARTHUR E, physics, see 12th edition

WALTERS, CARL JOHN, b Albuquerque, NMex, Sept 14, 44. SYSTEMS ECOLOGY. Educ: Humboldt State Col, BS, 65; Colo State Univ, MS, 67, PhD(fisheries), 69. Prof Exp: Res asst fisheries biol, 66-67, NSF fel, 67-69, CONSULT COLO COOP WILDLIFE UNIT, COLO STATE UNIV, 68-; ASST PROF ZOOL & ANIMAL RESOURCE ECOL, UNIV BC, 69- Concurrent Pos: Can Dept Environ & Natural Resources consult, 71-; res scholar, Int Inst Appl Systs Anal, Austria, 74-75. Mem: Am Fisheries Soc. Res: Dynamics of ecological communities; application of mathematical models and computer simulation techniques to problems in resource ecology. Mailing Add: Inst of Animal Resource Ecol Univ of BC Hut B8 Vancouver BC Can

WALTERS, CHARLES PHILIP, b Kansas City, Mo, May 1, 15; m 36; c 3. ASTROGEOLOGY. Educ: Kans State Univ, BS, 36, MS, 38; Cornell Univ, PhD, 57. Prof Exp: Asst geologist, Continental Oil Co, 44-48; from asst prof to assoc prof, 45-72, PROF GEOL, KANS STATE UNIV, 72- Concurrent Pos: Ford Found fac fel, 51-52; mem comt exam natural sci test & geol subj matter test, Educ Testing Serv, 62-70. Mem: AAAS; Am Asn Petrol Geologists; Am Quaternary Asn; Meteoritical Soc. Res: Structural and tectonic geology; geophysics; planetology; environmental geology; deterioration of Kansas salt beds; tektites from cryptovolcanic eruptions. Mailing Add: Dept of Geol Kans State Univ Manhattan KS 66506

WALTERS, CHARLES SEBASTIAN, b Detroit, Mich, Aug 18, 13; m 39; c 2. FORESTRY. Educ: Purdue Univ, BS, 38; Yale Univ, DFor, 57. Prof Exp: Asst, Tenn Valley Authority, 40 & Univ Ill, 40-41; proj forester, Timber Prod War Proj, Ill, 41-45; asst chief wood technol & utilization in forestry, 45-47, from asst prof to assoc prof, 47-57, PROF WOOD TECHNOL & UTILIZATION, UNIV ILL, URBANA, 57- Concurrent Pos: Consult, Indonesia, 71; adv, Nat Bur Standards, 71-74; mem standing comt hardboard, US Bur Standards, 74-77. Honors & Awards: Wood Salute, Wood & Wood Prod, 70. Mem: AAAS; Sigma Xi; Forest Prod Res Soc; Soc Wood Sci & Technol; Am Soc Testing & Mat. Res: Technology of wood and its use; wood preservation. Mailing Add: Dept of Forestry 219 Mumford Hall Univ of Ill Urbana IL 61801

WALTERS, CORA ETTA, b Wesley, Iowa, Oct 18, 14. PHYSIOLOGY. Educ: Univ Wis, BS & MS, 38; Univ Iowa, PhD(phys educ, physiol), 51. Prof Exp: Instr phys educ, Univ Chicago, 44-49; asst, Univ Iowa, 49-50, asst physiol, 50-51; asst prof phys educ res, Univ Mich, 51-54; res assoc clin physiol, Res & Educ Hosps, Univ Ill, 54-55 & Rehab Inst Chicago, Ill, 55-56; assoc prof phys educ, Fla State Univ, 56-60, assoc prof human develop, 60-69, prof home & family life, 69-75; RETIRED. Mem: Soc Res Child Develop; NY Acad Sci; Soc Psychophysiol Res. Res: Neurophysiological bases of learning; neurological and physiological development of children; human fetal activity; physical and psychological correlates of postnatal behavior; human and child development; fetal activity and aggression and apprehension in the two-year-old; antecedents of achievement motive and cognitive style in preschoolers. Mailing Add: 225 Atkinson Dr Tallahassee FL 32304

WALTERS, CRAIG THOMPSON, b Columbus, Ohio, July 23, 40; m 62; c 1. PLASMA PHYSICS, LASERS. Educ: Ohio State Univ, BS & MS, 63, PhD(physics), 71. Prof Exp: From res physicist to sr physicist, 63-71, assoc fel plasma physics, 71-73, sr researcher laser effects, 73-75, ASSOC SECT MGR LASER EFFECTS & ELECTROMAGNETICS, BATTELLE COLUMBUS LABS, BATTELLE MEM INST, 75- Concurrent Pos: Mem adv panel laser-supported absorption waves, Defense Advan Proj Agency, 74- Honors & Awards: NASA Tech Brief Award, 69; Distinguished Inventor Awards, Battelle Mem Inst, 73 & 74. Mem: Am Phys Soc. Res: Generation, propagation and material interaction of intense energy beams consisting of electromagnetic radiation or fundamental particles; study of plasma produced in laser beam interaction with condensed matter. Mailing Add: Battelle Columbus Labs 505 King Ave Columbus OH 43201

WALTERS, CURLA SYBIL, b Jamaica, June 3, 29. IMMUNOLOGY, MICROBIOLOGY. Educ: Andrews Univ, BA, 61; Howard Univ, MSc, 64; Georgetown Univ, PhD(microbiol & immunol), 69. Prof Exp: From instr to asst prof immunol, Med Ctr, Univ Colo, Denver, 71-74; ASSOC PROF DEPT MED, MED CTR, HOWARD UNIV, 74- Concurrent Pos: Am Asn Univ Women fel, Karolinska Inst, Sweden, 69-70; NIH training grant, Med Ctr, Univ Colo, 71. Mem: AAAS; Am Asn Immunol. Res: Basic and tumor immunology. Mailing Add: Dept of Immunol Univ of Colo Med Ctr Denver CO 80220

WALTERS, DAVID ROBERT, organic chemistry, see 12th edition

WALTERS, DAVID ROYAL, b Battle Creek, Mich, Apr 9, 41. CELL BIOLOGY, DEVELOPMENTAL BIOLOGY. Educ: Univ Mich, Ann Arbor, BS, 63; Harvard Univ, AM, 64, PhD(biol), 67. Prof Exp: Instr, Harvard Univ, 67-69, lectr, 69-70; ASST PROF BIOL, BOSTON UNIV, 70- Concurrent Pos: NSF res grant, Boston Univ, 72-74. Mem: AAAS; Am Inst Biol Sci; Am Soc Zoologists. Res: Cellular and developmental biology of insects; mechanism of intercellular adhesion and its relation to tissue metamorphosis; physiology of insect blood cells. Mailing Add: Dept of Biol Boston Univ Boston MA 02215

WALTERS, DOUGLAS BRUCE, b Brooklyn, NY, Apr 6, 42; m 69. ANALYTICAL CHEMISTRY. Educ: Long Island Univ, BS, 63, MS, 65; Univ Ga, PhD(chem), 71. Prof Exp: Res chemist, Farbewerke Hoechst A G, Frankfurt, Ger, 65 & Southeast Water Lab, Environ Protection Agency, Ga, 69-71; RES CHEMIST, RUSSELL RES LAB, US DEPT AGR, 71- Mem: Am Chem Soc; The Chem Soc. Res: Organophosphorus chemistry; nuclear magnetic resonance; inorganic chemistry; chelation; food and flavor chemistry; development of analytical methods; pollution chemistry; phase diagrams; tobacco chemistry; separation techniques. Mailing Add: Russell Res Lab USDA PO Box 5677 Athens GA 30604

WALTERS, EDWARD ALBERT, b Whitefish, Mont, Jan 2, 40; m 64; c 2. PHYSICAL ORGANIC CHEMISTRY. Educ: Pac Lutheran Univ, BS, 62; Univ Minn, Minneapolis, PhD(org chem), 66. Prof Exp: Res assoc chem, Cornell Univ, 66-68; from asst prof to assoc prof, 68-74, PROF CHEM, UNIV N MEX, 74- Concurrent Pos: Res Corp grant, Univ NMex, 69-70, NSF grant, 69-72. Mem: AAAS; Am Chem Soc. Res: Fast proton transfer reactions in solution; kinetic isotope effects; potential energy surfaces for reactive collisions. Mailing Add: Dept of Chem Univ of NMex Albuquerque NM 87131

WALTERS, ELEANOR BOYD, b Gunnison, Miss, Mar 28, 14. MATHEMATICS. Educ: Delta State Col, BS, 34; Duke Univ, MA, 39; Columbia Univ, EdD(math), 56. Prof Exp: High sch teacher, Miss, 34-38; asst math, Duke Univ, 38-39; high sch teacher, Miss, 39-41 & Fla, 41-43; from asst prof to assoc prof, 43-55, PROF MATH & HEAD DEPT, DELTA STATE UNIV, 56- Mem: Fel AAAS; Am Math Soc; Math Asn Am. Mailing Add: Dept of Math Delta State Univ Cleveland MS 38732

WALTERS, FRED HENRY, b Owen Sound, Ont, Aug 8, 47. ANALYTICAL CHEMISTRY. Educ: Univ Waterloo, BSc, 71; Univ Mass, PhD(chem), 75. Prof Exp: Res asst, Toronto Gen Hosp, Kitchener Waterloo Hosp & Ashland Oil Can, 66-71; teaching asst chem, Univ Mass, 71-75; RES ASSOC CHEM, UNIV WINDSOR, 75- Mem: The Chem Soc; Am Chem Soc; Can Inst Chem. Res: Electrochemical synthesis of inorganic compounds; high pressure liquid chromatography; organometallic chemistry; metals in biological systems. Mailing Add: Dept of Chem Univ of Windsor Windsor ON Can

WALTERS, GEOFFREY KING, b Baton Rouge, La, Aug 23, 31; m 54; c 3. ATOMIC PHYSICS. Educ: Rice Univ, BA, 53, PhD(physics), 56. Prof Exp: NSF fel, Duke, 56-57; br mgr & physicist, Tex Instruments, Inc, 57-62, corp res assoc, 62-63; prof physics, 63-64, actg dean sci & eng, 68-69 & 72-73, PROF PHYSICS & SPACE SCI, RICE UNIV, 64-, CHMN DEPT PHYSICS, 73- Concurrent Pos: Actg chief fire technol div, Nat Bur Standards, 71-72. Mem: AAAS; Am Phys Soc; Am Geophys Union. Res: Magnetic resonance; low temperature, solid state atomic collisions and reactions; optical pumping and dynamic nuclear orientation; solar-terrestrial relationships; radio-astronomy; fire research and safety. Mailing Add: Dept of Physics Rice Univ Houston TX 77001

WALTERS, HUBERT JACK, b Spring, Tex, Nov 2, 15; m 47; c 3. PLANT PATHOLOGY. Educ: NMex Col, BS, 41; Univ Ill, MS, 48; Univ Nebr, PhD(plant path), 51. Prof Exp: Asst prof plant path & asst plant pathologist, Univ Wyo, 50-54; from asst prof plant path & asst plant pathologist to assoc prof plant path & assoc plant pathologist, 54-62, PROF PLANT PATH & PLANT PATHOLOGIST, UNIV ARK, FAYETTEVILLE, 62- Mem: Am Phytopath Soc. Res: Virus diseases; beetle transmission of plant viruses; diseases of soybeans. Mailing Add: Dept of Plant Path Univ of Ark Fayetteville AR 72701

WALTERS, JACK HENRY, b Toronto, Ont, Apr 2, 25; m 49; c 3. OBSTETRICS, GYNECOLOGY. Educ: Univ Western Ont, BA, 46, MD, 51; FRCPS(C), 57; FRCOG, 67. Prof Exp: Chief dept obstet & gynec & dir cytol, St Joseph's Hosp, London, Ont, 58-73; prof obstet & gynec, Univ Western Ont, 66-73; PROF OBSTET & GYNEC & CHMN DEPT, MED COL OHIO, 73- Concurrent Pos: Can Cancer Soc McEachern traveling fel, 56-57. Mem: Can Med Asn; Soc Obstet & Gynaec Can; Am Soc Cytol; fel Am Col Obstet & Gynec. Res: Gynecological, particularly hormonal cytology; screening programs; perinatal mortality, particularly statistical research in computer programs; manpower studies; obstetrics-gynecology health care; delivery systems. Mailing Add: Dept of Obstet & Gynec Med Col of Ohio Toledo OH 43614

WALTERS, JAMES LEE, b Chicago, Ill, Oct 30, 16; m 41. CYTOGENETICS. Educ: Univ Chicago, BS, 37; Univ Calif, Berkeley, PhD(genetics), 48. Prof Exp: Asst bot, Univ Calif, Berkeley, 38-42, lectr, 42, asst genetics, 46-47; instr bot, 47-50, from asst prof to assoc prof biol, 50-72, acad asst to chancellor, 74-75, PROF BIOL, UNIV CALIF, SANTA BARBARA, 72- Mem: AAAS; Bot Soc Am; Soc Study Evolution; Genetics Soc Am. Res: Cytogenetics of Paeonia; cytology of structural hybrids; cytology of grasshoppers. Mailing Add: Dept of Biol Sci Univ of Calif Santa Barbara CA 93106

WALTERS, JOHN P, b Elgin, Ill, July 4, 38; m 61; c 2. ANALYTICAL CHEMISTRY, SPECTROSCOPY. Educ: Purdue Univ, BS, 60; Univ Ill, Urbana, PhD(chem), 64. Prof Exp: Res assoc spectros, Univ Ill, Urbana, 64-65; asst prof, 65-72, PROF ANAL CHEM, UNIV WIS-MADISON, 72- Mem: Soc Appl Spectros. Res: Time-resolved emission spectroscopy; mechanisms of spectroscopic discharges; spectrochemical methods and instrumentation. Mailing Add: Dept of Chem Univ of Wis Madison WI 53706

WALTERS, JOHN PHILIP, b Manhattan, Kans, Sept 26, 41; m 63; c 4. PHYSICAL CHEMISTRY. Educ: Kans State Univ, BS, 63; Iowa State Univ, PhD(phys chem), 68. Prof Exp: Sr res chemist, Phillips Petrol Co, Okla, 68-70, RES CHEMIST, PHILLIPS FIBER CORP, 70- Mem: AAAS; Am Phys Soc. Res: Polymer physics. Mailing Add: Res Dept Phillips Fibers Corp Box 66 Greenville SC 29602

WALTERS, LEE RUDYARD, b New York, NY, Jan 20, 28; m 50; c 2. ORGANIC CHEMISTRY. Educ: Bucknell Univ, BS, 54; Univ Kans, PhD(chem), 58. Prof Exp: Asst org chem, Univ Kans, 55-58; res chemist, Atlas Powder Co, Del, 58-59; ASST PROF CHEM, LAFAYETTE COL, 59- Mem: Am Chem Soc. Res: Nitrogen heterocyclic and organometallic compounds; chemistry of natural products. Mailing Add: Dept of Chem Lafayette Col Easton PA 18042

WALTERS, LESTER JAMES, JR, b Tulsa, Okla, June 3, 40; m 67. GEOCHEMISTRY. Educ: Univ Tulsa, BS, 62; Mass Inst Technol, PhD(geochem), 67. Prof Exp: Res scientist, Marathon Oil Co, 67-69; tech assistance expert, Int Atomic Energy Agency, 69; asst prof, 70-74, ASSOC PROF GEOL, BOWLING GREEN STATE UNIV, 74- Mem: Am Chem Soc; Am Geophys Union; Am Asn Petrol Geologists; Geochem Soc; Soc Explor Paleontologists & Mineralogists. Res: Geochemical cycle of iodine, bromine, and chlorine in sediments; neutron activation analysis; stable isotope geochemistry; heavy metal pollution in Lake Erie. Mailing Add: 907 Klotz Rd Bowling Green OH 43402

WALTERS, LOWELL EUGENE, b Freedom, Okla, Jan 13, 19; m 42; c 2. ANIMAL SCIENCE. Educ: Okla State Univ, BS, 40; Univ Mass, MS, 42. Prof Exp: Instr animal husb, La State Univ, 42-44; asst prof, Univ Mass, 44-46; from asst prof to assoc prof, 46-57, PROF ANIMAL HUSB, OKLA STATE UNIV, 58- Res: Meats; beef quality; carcass composition of beef, pork and lamb; potassium 40 techniques in live animal and carcass evaluation. Mailing Add: Dept of Animal Sci Okla State Univ Stillwater OK 74074

WALTERS, MARTA SHERMAN, b Los Angeles, Calif, Aug 11, 14; m 41. CYTOGENETICS. Educ: Univ Calif, Berkeley, BA, 35, MS, 37, PhD(genetics), 44. Prof Exp: Biologist, Radiation Lab, Univ Calif, 47; cytol asst, Div Genetics, Univ Calif, Berkeley, 48, Guggenheim fel & univ res fel, 49-50; RES ASSOC BIOL SCI, UNIV CALIF, SANTA BARBARA, 49-; RES ASSOC, SANTA BARBARA BOT GARDEN, 49- Concurrent Pos: USPHS res career develop award, 63-69. Mem: Am Soc Cell Biol; Bot Soc Am; Soc Study Evolution; Genetics Soc Am. Res: Chromosome cytology. Mailing Add: Santa Barbara Bot Garden 1212 Mission Canyon Rd Santa Barbara CA 93105

WALTERS, MARTHA I, b Logan, Ohio, Oct 21, 25. CLINICAL CHEMISTRY. Educ: Ohio State Univ, BSc, 47, MSc, 59, PhD(clin path), 70. Prof Exp: Supvr biochem labs, Ohio State Univ Hosp, 49-59, instr clin path, Univ, 55-59; sr res asst, Bellevue Med Ctr, NY Univ, 59-62; dir biochem, St Francis Hosp, Bronx, NY, 63-67 & Ohio Valley Hosp, Steubenville, 71-73; SUPVR ENDOCRINOL & TOXICOL, CONSOL BIOMED LABS, 73- Mem: Am Chem Soc; Am Asn Clin Chemists; Sigma Xi. Res: Relationship between metabolism and function in erythrocytes and leukocytes. Mailing Add: 952 Chelsea Ave Columbus OH 43209

WALTERS, ORVILLE SELKIRK, psychiatry, deceased

WALTERS, PHILIP MARION, b Oblong, Ill, Feb 10, 16; m 40; c 2. TEXTILE CHEMISTRY. Educ: Univ Ill, BS, 38; Univ Wis, PhD(org chem), 42. Prof Exp: Asst chem, Univ Wis, 38-41; tech investr, Textile Fibers Dept, 42-53, indust sales develop mgr, 53-55, merchandising mgr tire yarn, 55-60, indust tech mgr, 60-70, PROF MGR, INDUST FIBERS DIV, E I DU PONT DE NEMOURS & CO, INC, 70- Res: Textiles. Mailing Add: Indust Fibers Div E I du Pont de Nemours & Co Inc Wilmington DE 19898

WALTERS, REGINALD MICHAEL, chemistry, see 12th edition

WALTERS, RICHARD FRANCIS, b Teleajen, Romania, Aug 30, 30; US citizen; m 52; c 2. INFORMATION SCIENCE, MEDICAL EDUCATION. Educ: Williams Col, BA, 52, Univ Wyo, MA, 53; Univ Bordeaux, dipl natural sci, 55; Stanford Univ, PhD(geol), 57. Prof Exp: Res geologist, Humble Oil & Ref Co, 56-67; asst prof, 67-73, ASSOC PROF MED EDUC & BIOMED ENG, SCH MED, UNIV CALIF, DAVIS, 73- Mem: AAAS; Biomed Eng Soc; Asn Comput Mach; Simulation Coun. Res: Interactive information systems; instructional use of computers; simulation of human physical performance under environmental stress; computerized problem-oriented medical record systems. Mailing Add: Univ of Calif Sch of Med Davis CA 95616

WALTERS, RONALD ARLEN, b Greeley, Colo, Apr 25, 40. BIOCHEMISTRY, RADIOBIOLOGY. Educ: Colo State Univ, BS, 62, MS, 64, PhD(radiation biol), 67. Prof Exp: Engr radiation biol, Gen Elec Co, 62-63; STAFF MEM BIOCHEM, LOS ALAMOS SCI LAB, UNIV CALIF, 67- Mem: AAAS; Am Chem Soc; Biophys Soc; Radiation Res Soc; Am Soc Cell Biol. Res: Cellular biology. Mailing Add: 1559 41st St Los Alamos NM 87544

WALTERS, STANLEY SIMON, mathematics, see 12th edition

WALTERS, THOMAS RICHARD, b Milwaukee, Wis, May 9, 29. PEDIATRICS, HEMATOLOGY. Educ: Marquette Univ, MD, 54. Prof Exp: Instr pediat, Stanford Univ, 61-62; asst prof, Univ Kans Med Ctr, Kansas City, 62-67; assoc prof, Univ Tenn, Memphis, 67-71; ASSOC PROF PEDIAT & DIR DIV HEMAT-ONCOL, NJ MED SCH, 71- Concurrent Pos: Assoc mem hemat, St Jude Children's Res Hosp, Memphis, 67-71. Mem: Am Soc Hemat; Am Acad Pediat; Am Asn Cancer Res. Res: Multi-disciplinary chemotherapy programs. Mailing Add: Dept of Pediat NJ Med Sch Newark NJ 07103

WALTERS, VIRGINIA F, b New York, NY, May 26, 25; m 45; c 2. PHYSICS. Educ: Smith Col, AB, 47; Western Reserve Univ, MA, 58, PhD(physics), 65. Prof Exp: Physicist, DeMornay Budd, Inc, 47-49; res asst microwave components, Radiation Lab, Columbia Univ, 49-50; lectr elem physics, Adelphi Col, 50-51; physicist, Servo Corp Am, 51; asst physics, Western Reserve Univ, 54-65, fel, 65-66; asst prof, Cleveland State Univ, 66-67; res physicist, Carnegie-Mellon Univ, 67-68; asst prof phys sci, Point Park Col, 68-69; teacher physics & math, Western Reserve Acad, 69-74; LECTR PHYSICS, CLEVELAND STATE UNIV, 75- Mem: Am Phys Soc; Am Asn Physics Teachers. Res: Positron annihilation; nuclear instrumentation. Mailing Add: 20019 Sussex Rd Cleveland OH 44122

WALTERS, VLADIMIR, b New York, NY, Dec 18, 27; m 60. ICHTHYOLOGY. Educ: Cornell Univ, BS, 47, MS, 48; NY Univ, PhD(zool), 54. Prof Exp: Asst physiol, US Naval Arctic Res Lab, Alaska, 48-49; res assoc fishes, Am Mus Natural Hist, 55-56, asst cur, 56-61; from asst prof to prof, 61-74, EMER PROF ZOOL, UNIV CALIF, LOS ANGELES, 74- Mem: AAAS; Am Soc Naturalists; Am Soc Zoologists; Am Soc Ichthyologists & Herpetologists; Arctic Inst NAm. Res: Systematics and physiology of fishes; arctic biology; zoogeography; hydrodynamics of swimming. Mailing Add: Dept of Zool 2203 Life Sci Bldg Univ of Calif Los Angeles CA 90024

WALTERS, WILLIAM BEN, b Highland, Kans, Apr 26, 38; m 62; c 2. NUCLEAR CHEMISTRY, PHYSICAL CHEMISTRY. Educ: Kans State Univ, BS, 60; Univ Ill, PhD(chem), 64. Prof Exp: Res assoc chem, Mass Inst Technol, 64-65, asst prof, 65-70; ASSOC PROF CHEM, UNIV MD, COLLEGE PARK, 70- Mem: Am Chem Soc; Am Phys Soc. Res: Radioactive decay; nuclear spectroscopy; new isotopes and isomers; nuclear reactions; positron annihilation; nuclear structure; neutron-capture gamma-ray spectroscopy. Mailing Add: Dept of Chem Univ of Md College Park MD 20742

WALTERS, WILLIAM LE ROY, b Racine, Wis, Mar 30, 32; m 55; c 4. PHYSICS. Educ: Univ Wis, BS, 54, MS, 58, PhD(physics), 61. Prof Exp: From asst prof to prof physics, 61-68, assoc dean sci, Col Lett & Sci, 65-68, exec asst chancellor, 68-70, actg dean col appl sci & eng, 70, VCHANCELLOR, UNIV WIS-MILWAUKEE, 71- Mem: AAAS; Am Asn Physics Teachers; Am Phys Soc; Am Vacuum Soc. Res: Low energy nuclear, vacuum and surface physics. Mailing Add: Univ of Wis Milwaukee WI 53201

WALTHER, ADRIAAN, b The Hague, Holland, Apr 22, 34; m 60; c 2. OPTICS. Educ: Delft Univ Technol, PhD(physics), 59. Prof Exp: Mem res staff, Diffraction Ltd Inc, 60-72; PROF ENG & SCI, WORCESTER POLYTECH INST, 72- Concurrent Pos: Am Optical Soc vis prof, Worcester Polytech Inst, 68- Mem: Optical Soc Am; Neth

Phys Soc. Res: Geometrical and physical optics. Mailing Add: 20 Whittier Dr Acton MA 01720

WALTHER, ALINA, b Rosenthal, USSR, Aug 15, 23; Can citizen. PLANT PHYSIOLOGY, PLANT ECOLOGY. Educ: Sir George Williams Univ, BCom, 53, BSc, 61; McGill Univ, MSc, 63; Univ Toronto, PhD(plant physiol), 68. Prof Exp: Spec lectr, 67-68, asst prof, 68-72, ASSOC PROF BIOL, UNIV REGINA, 72- Concurrent Pos: Nat Res Coun res grant, 69-70. Mem: AAAS; Am Inst Biol Sci; Am Soc Plant Physiologists; Can Soc Plant Physiologists; Ecol Soc Am. Res: Plant senescence, especially metabolic changes in developing and senescing sunflower cotyledons and leaves. Mailing Add: Dept of Biol Univ of Regina Regina SK Can

WALTHER, FRANK H, b Williamsport, Pa, Aug 4, 30; m 54; c 3. MINERALOGY. Educ: Franklin & Marshall Col, BSc, 52. Prof Exp: Mgr mineral res, 56-72, MGR RES, HARBISON-WALKER REFRACTORIES, 72- Mem: Fel Am Ceramic Soc. Res: Mineralogical aspects of refractory technology. Mailing Add: Harbison-Walker Refractories Garber Res Ctr West Mifflin PA 15122

WALTHER, FRITZ R, b Chemnitz, Ger, Sept 8, 21. ANIMAL BEHAVIOR. Educ: Univ Frankfurt, BS, 44, MS, 56, PhD(zool), 63. Prof Exp: Teacher, Fed Ministry Educ, Ger, 51-59; sci dir res & admin, Opel Zoo, 63-63; res scientist, Zurich Zool, Switz, 64 & Serengeti Nat Park, Tanzania, 65-67; assoc prof zool, Univ Mo-Columbia, 67-70; PROF WILDLIFE, TEX A&M UNIV, 70- Concurrent Pos: Grants, Ger Res Soc, 63, Gertrud Rüegg Found, 64 & 67, Fritz Thyssen Found, 65-67 & 74-75, Res Coun Univ Mo, 69-70, Smithsonian Foreign Currency, 70-72 & Caesar Vleberg Found, 74-75. Mem: Animal Behav Soc. Res: Ethology of game animals, especially horned ungulates. Mailing Add: Dept of Wildlife & Fisheries Sci Tex A&M Univ College Station TX 77843

WALTKING, ARTHUR ERNEST, b New York, NY, Nov 7, 37; m 61; c 2. ANALYTICAL CHEMISTRY, FOOD CHEMISTRY. Educ: Lehigh Univ, BA, 59. Prof Exp: Asst chemist, 59-64, chemist, 64-66, group leader anal res, 66-67, sect head anal serv, 67-72, MGR ANAL SERV, BEST FOODS DIV, CPC INT INC, 72- Concurrent Pos: Assoc referee oxidized oils, Asn Off Anal Chemists, 71-; rep, Am Oil Chemists Soc, Joint Am Oil Chemists Soc-Asn Off Anal Chemists-Am Asn Cereal Chemists Comt Mycotoxins, 75- Honors & Awards: Golden Peanut Award, Nat Peanut Coun, 71. Mem: Am Oil Chemists Soc; Am Chem Soc. Res: The development of analytical methodology for analysis of food products for mycotoxins, essential fatty acids and polymers derived from oxidation of heat abuse of vegetable oils. Mailing Add: Best Foods Res Ctr 1120 Commerce Ave Box 1534 Union NJ 07083

WALTMAN, PAUL E, b St Louis, Mo, Oct 17, 31; m 53; c 3. MATHEMATICAL ANALYSIS, BIOMATHEMATICS. Educ: St Louis Univ, BS, 52; Baylor Univ, MA, 54; Univ Mo, MA, 60, PhD(math), 62. Prof Exp: Staff mem, Lincoln Lab, Mass Inst Technol, 57-58, Mitre Corp, 62-63 & Sandia Corp, 63-65; from asst prof to assoc prof, 65-68, PROF MATH, UNIV IOWA, 68- Mem: AAAS; Soc Indust & Appl Math; Am Math Soc. Res: Ordinary differential equations. Mailing Add: Dept of Math Univ of Iowa Iowa City IA 52240

WALTMANN, WILLIAM LEE, b Cedar Falls, Iowa, July 5, 34; m 58; c 3. MATHEMATICS. Educ: Wartburg Col, BA, 56; Iowa State Univ, MS, 58, PhD(math), 64. Prof Exp: Instr math, Wartburg Col, 58-61 & Iowa State Univ, 63-64; assoc prof, 64-72, PROF MATH, WARTBURG COL, 72-, CHMN DEPT, 71- Mem: Am Math Soc; Math Asn Am; Soc Indust & Appl Math. Res: Inversion of matrices; non-associative rings; tridiagonalization of matrices. Mailing Add: Dept of Math Wartburg Col Waverly IA 50677

WALTNER, ARTHUR, b Moundridge, Kans, Nov 28, 14; m 41; c 2. NUCLEAR PHYSICS. Educ: Bethel Col, Kans, AB, 38; Kans State Col, MS, 43; Univ NC, PhD, 49. Prof Exp: High sch teacher, 38-41; instr elec theory, Naval Training Sch, Ky State Teachers Col, Morehead, 42-43; lectr physics, Univ NC, 43-46, instr, 46-48; from asst prof to assoc prof, 48-56, PROF PHYSICS, NC STATE UNIV, 56- Concurrent Pos: Exchange physicist, A B Atomenergie, Stockholm, Sweden, 52-53. Mem: AAAS; fel Am Phys Soc. Res: Nuclear instrumentation; nuclear reactions due to heavy ions; x-ray production cross sections. Mailing Add: Dept of Physics NC State Univ Raleigh NC 27607

WALTON, ALAN, b Durham, Eng, May 29, 32; m 56; c 3. CHEMICAL OCEANOGRAPHY, GEOCHEMISTRY. Educ: Univ Durham, BSc, 53, PhD(chem), 56; Glasgow Univ, DSc, 71. Prof Exp: Res scientist nuclear chem, Isotopes Inc, NJ, 59-65; lectr, Glasgow Univ, 65-67, sr lectr, 67-70; HEAD CHEM OCEANOG DIV, ATLANTIC OCEANOG LAB, BEDFORD INST OCEANOG, 70- Mem: Fel Royal Inst Chem; Am Geophys Union; Geochem Soc. Res: Air-sea exchange; isotope applications in marine environment; carbon 14 dating; marine and atmospheric pollution. Mailing Add: Atlantic Oceanog Lab Bedford Inst of Oceanog Dartmouth NS Can

WALTON, ALAN GEORGE, b Birmingham, Eng, Apr 3, 36; wid; c 2. BIOPHYSICAL CHEMISTRY. Educ: Univ Nottingham, BSc, 57, PhD(chem), 60, DSc(biophys chem), 73. Prof Exp: Res assoc chem, Ind Univ, 60-62; from asst prof to assoc prof, 62-71, PROF MACROMOLECULAR SCI, CASE WESTERN RESERVE UNIV, 71- Concurrent Pos: Vis lectr, Harvard Med Sch, 71. Mem: Assoc Royal Inst Chem; Am Chem Soc; Biophys Soc. Res: Conformation and structure of synthetic biopolymers and fibrous proteins; molecular hematology; cell adhesion. Mailing Add: Dept of Macromolecular Sci Case Western Reserve Univ Cleveland OH 44106

WALTON, BARBARA ANN, b Baltimore, Md, Mar 30, 40; m 70. VERTEBRATE EMBRYOLOGY, HISTOLOGY. Educ: Ind Univ Pa, BSEd, 62; Univ Okla, MNatSci, 66, PhD(zool), 70. Prof Exp: Teacher, Franklin Twp Sch Dist, Pa, 62-63 & Bethel Park Jr High Sch, 63-65; ASST PROF BIOL, UNIV TENN, CHATTANOOGA, 70- Mem: Am Inst Biol Sci; Am Soc Zoologists. Res: Development of the chicken embryo following x-irradiation and application of other teratogens; histology and histochemistry of embryonic development. Mailing Add: Dept of Biol Univ of Tenn Chattanooga TN 37401

WALTON, BARRY LOUIS, b Buffalo, NY, Sept 4, 46; m 74. SOLID STATE PHYSICS, OPERATIONS ANALYSIS. Educ: Univ Calif, Santa Cruz, BA, 68, MS, 71, PhD(physics), 74. Prof Exp: Fel solid state physics, Ames Lab, Iowa State Univ & Energy Res & Develop Admin, 73-74; OPERS ANALYST, STANFORD RES INST, 74- Mem: Am Phys Soc; AAAS. Res: Energy and environmental impacts of new technologies and their policy implications, including the impacts of synthetic fuels technolgies; vortex nucleation in superconductors. Mailing Add: Stanford Res Inst 333 Ravenswood Ave Menlo Park CA 94025

WALTON, BRYCE CALVIN, b Lead, SDak, June 5, 23; m 46; c 4. PARASITOLOGY. Educ: Univ Southern Calif, MS, 50; Univ Md, PhD(zool, parasitol), 56. Prof Exp: US Army, 50-, res parasitologist, Walter Reed Army Inst Res, 52-56, chief dept med zool, 406th Med Gen Lab, 56-59, parasitologist, Third Army Med Lab, 59-62, chief parasitic dis sect, Middle Am Res Unit, 62-65, cmndg officer, Med Res Unit, Panama, CZ, 65-69 & US Army Res & Develop Group Far East, 69-72, CMNDG OFFICER, ARMY MED RES UNIT-PANAMA, 72- Concurrent Pos: Consult, Western Pac regional off, WHO, 58-59, mem study group serol Chagas' dis, Pan Am Health Orgn-WHO, 66-67. Mem: Am Soc Parasitol; Am Soc Trop Med & Hyg; Am Micros Soc. Res: American trypanosomiasis and leishmaniasis; toxoplasmosis; immuno-diagnosis of parasitic disease; systematics. Mailing Add: US Army Med Res Unit Panama Box 1809 APO New York NY 09826

WALTON, CHARLES ANTHONY, b Auburn, Ala, Apr 3, 26; m 46; c 2. PHARMACY, PHARMACOLOGY. Educ: Ala Polytech Inst, BS, 49; Purdue Univ, MS, 50, PhD, 56. Prof Exp: Instr, Ala Polytech Inst, 49; from asst prof to prof mat med, Col Pharm, 50-72, head dept, 56-66, dir drug info ctr, 66-72, PROF ORAL BIOL, COL DENT, UNIV KY, 72- Concurrent Pos: Fulbright prof, Univ Cairo, 64-65. Mem: Am Pharmaceut Asn; Am Soc Hosp Pharmacists; fel Am Soc Clin Pharmacol & Therapeut; Am Asn Poison Control Ctrs. Res: Clinical pharmacy and drug information services. Mailing Add: Dept of Oral Biol Univ of Ky Col of Dent Lexington KY 40506

WALTON, CHARLES HUTCHINSON ACOURT, b Mather, Man, May 24, 06; m 35; c 2. ALLERGY. Educ: Univ Man, BSc, 26, MSc, 28, MD, 32; FRCP(C). Prof Exp: Demonstr biochem, Univ Man, 26-27, lectr, 27-28, demonstr med, 33-35, lectr, 35-45, from asst prof to assoc prof, 45-72; MEM, WINNIPEG CLIN, 67- Mem: Fel Am Col Physicians; fel Am Acad Allergy; Can Med Asn; Can Acad Allergy (pres, 44, 48 & 53); fel Royal Soc Med. Res: Biochemistry. Mailing Add: Winnipeg Clin 425 St Mary Ave Winnipeg MB Can

WALTON, CHARLES WILLIAM, b Carlinville, Ill, Apr 3, 08; m 33; c 3. CHEMISTRY. Educ: Univ Ill, BS, 30; Univ Mich, MS, 31, PhD(chem), 33. Hon Degrees: DSc, Blackburn Col, 65. Prof Exp: Lectr & lab asst, Univ Mich, 30-31; res chemist, Goodyear Tire & Rubber Co, 33-39, supvr phys chem, 40-41, mgr & tech coordr, Synthetic Rubber Div, 41-44, mgr, Chem Prod Develop Div, 44-46; asst to exec vpres, 47, mgr, New Prod Div, 48-51, gen mgr, Adhesives & Coatings Div, 52-59, div vpres & gen mgr, Adhesives, Coatings & Sealers Div, 59-61, vpres res, 61-62, vpres res & develop, 62-69, CONSULT, MINN MINING & MFG CO, 69- Concurrent Pos: Trustee, Blackburg Col, 67- Mem: Am Chem Soc; Soc Chem Indust; Commercial Develop Asn. Res: Synthetic rubber; high polymers; surface chemistry and solids. Mailing Add: PO Box 136 Sugarloaf Shores FL 33044

WALTON, CYPRIAN JAMES, b Brooklyn, NY, July 1, 09. BIOLOGY, ENTOMOLOGY. Educ: Manhattan Col, BA, 32, MA, 36; Fordham Univ, MS, 39, PhD(biol), 44. Prof Exp: Teacher, St James High Sch, NY, 32-33 & Bishop Loughlin High Sch, 33-34; prin, Hillside Sch, 34-35; asst prin, St Joseph's High Sch, NH, 35-36; from instr to prof, 36-75, head dept, 47-73, PROF LECTR BIOL, MANHATTAN COL, 75- Concurrent Pos: Am Cancer Soc New York City Div res grant, 57-64. Mem: AAAS; Entom Soc Am; Asn Am Med Cols; NY Acad Sci. Res: Life cycles of plant lice causing tumors on shade trees in the northeastern United States. Mailing Add: Dept of Biol Manhattan Col Bronx NY 10471

WALTON, DAN WELLS, mammalogy, anatomy, see 12th edition

WALTON, DANIEL C, b Philadelphia, Pa, May 16, 34; m 60; c 2. PLANT BIOCHEMISTRY, PLANT PHYSIOLOGY. Educ: Univ Del, BS, 55; State Univ NY Col Forestry, Syracuse Univ, PhD(plant physiol), 62. Prof Exp: Chem engr, E I du Pont de Nemours & Co, 55-58; fel, Univ Tex, 62-63; from asst prof to assoc prof, 63-74, PROF BIOCHEM, STATE UNIV NY COL FORESTRY, SYRACUSE UNIV, 74- Concurrent Pos: Vis mem, Dept Bot, Univ Col Wales, 71-72. Mem: AAAS; Am Chem Soc; Am Soc Plant Physiologists. Res: Seed germination; plant growth regulation. Mailing Add: Dept of Chem State Univ of NY Syracuse NY 13210

WALTON, DEAN KIRKLAND, physical chemistry, see 12th edition

WALTON, DEREK, b Sao Paulo, Brazil, Mar 1, 31; nat US; m 54; c 4. PHYSICS. Educ: Univ Toronto, MSc, 54; Harvard Univ, PhD(appl physics), 58. Prof Exp: Asst res engr, Univ Calif, 57-58; sr physicist, Convair Div, Gen Dynamics Corp, 58-60; res assoc eng physics, Cornell Univ, 60-62; physicist, Oak Ridge Nat Lab, 62-68; assoc prof, 68-74, PROF PHYSICS, McMASTER UNIV, 74- Mem: Am Phys Soc. Res: Phonon-defect interactions; thermal conductivity; spin-phonon interactions; phase transformations. Mailing Add: Dept of Physics McMaster Univ Hamilton ON Can

WALTON, EDWARD, b Mt Carmel, Pa, May 11, 21; m 46; c 2. ORGANIC CHEMISTRY. Educ: Univ Md, BS, 42, PhD(org chem), 48. Prof Exp: Res chemist, Off Emergency Mgt & Comt Med Res Contract, Univ Md, 43-46; RES CHEMIST, MERCK SHARPE & DOHME RES LABS, 47- Concurrent Pos: Vis scientist, Univ Basel, 59-60; mem subcomt carbohydrates & related compounds, Nat Res Coun, 69- Mem: AAAS; Am Chem Soc. Res: Synthetic antimalarials; vitamins; antibiotics; peptides; nucleosides; carbohydrates. Mailing Add: Merck Sharp & Dohme Res Labs Rahway NJ 07065

WALTON, GEORGE, b Edmunton, Eng, Aug 14, 14; nat US; m 49; c 1. PHYSICAL CHEMISTRY, FORENSIC SCIENCE. Educ: San Diego State Col, AB, 36; Columbia Univ, MA, 39, PhD(phys chem), 41. Prof Exp: Asst, Columbia Univ, 40-41; asst prof chem, Col Pharm, Univ Cincinnati, 41-44, assoc prof, 46-47; consult chemist, 47-50; sr res chemist, Drackett Co, 50-52, tech adminr, 52-57, sr scientist, 57-62; geochem prospecting, 62-66; asst dir, Southwestern NMex Media Ctr, 66-68, asst prof, 68-74, ASSOC PROF PHYS SCI, WESTERN NMEX UNIV, 75- Mem: Am Chem Soc. Res: X-ray crystallography applied to chemical problems; paper chromatography of metal ions; silver in ores by atomic absorption spectroscopy. Mailing Add: 1312 S Silver Ave Deming NM 88030

WALTON, GERALD STEVEN, b Kansas City, Kans, July 23, 35; m 56; c 5. PLANT PATHOLOGY. Educ: Wabash Col, AB, 57; Rutgers Univ, PhD(plant path), 61. Prof Exp: Asst plant pathologist, 61-69, ASSOC PLANT PATHOLOGIST, CONN AGR EXP STA, 69- Mem: Am Phytopath Soc. Res: Nature of resistance to plant diseases, methods of disease control and determination of causal factor of a disease when unknown. Mailing Add: Dept of Plant Path Conn Agr Exp Sta PO Box 1106 New Haven CT 06504

WALTON, GRANT FONTAIN, b Philadelphia, Pa, Nov 19, 24; m 50; c 3. SOIL SCIENCE, RESOURCE ADMINISTRATION. Educ: Rutgers Univ, New Brunswick, BS, 55, MS, 70, PhD(soil sci, resource planning), 72. Prof Exp: Res asst wildlife biol, Rutgers Univ, New Brunswick, 51-55; res biologist, US Fish & Wildlife Serv, 55-56; res asst wildlife biol & forestry, Rutgers Univ, New Brunswick, 56-57; conserv dir, Stony Brook-Millstone Watershed Asn, NJ, 57-59; exec secy, State Soil Conserv Comn, NJ Dept Agr, 59-65; instr resource develop, 65-66, lectr, 66-70, assoc prof environ resources, 70-73, PROF ENVIRON RESOURCES, RUTGERS UNIV, NEW BRUNSWICK, 73-, CHMN DEPT, 70-, ACTG DEAN COOK COL & DIR AGR EXP STA, 75- Concurrent Pos: Environ consult, 55-71; mem land use policy adv comn, Nat Asn Conserv Dist, 68-69; dir div environ qual, NJ Dept Environ Protection, 71-73; assoc prof water pollution, Columbia Univ, 71- Honors & Awards: Cert of Merit, Soil Conserv Soc Am, 63. Mem: AAAS; Arctic Inst NAm; Soil Conserv Soc Am; Air Pollution Control Asn. Res: Enivornmental effects of land use changes; environmental resource management systems analysis; pedological and environmental studies of the high Arctic. Mailing Add: 105 Wynnwood Ave Piscataway NJ 08854

WALTON, HAROLD FREDERIC, b Tregony, Eng, Aug 25, 12; nat US; m 38; c 3. ANALYTICAL CHEMISTRY. Educ: Oxford Univ, BA, 34, PhD(chem), 37. Prof Exp: Procter vis fel, Princeton Univ, 37-38; res chemist, Permutit Co, 38-40; from instr to asst prof chem, Northwestern Univ, 40-46; from asst prof to assoc prof, 47-56, chmn dept, 62-66, PROF CHEM, UNIV COLO, BOULDER, 56- Concurrent Pos: Vis Fulbright-Hays lectr, Trujillo, 66-67 & 70 & Lima, 66-67; vis prof, Pedag Inst, Caracas, Venezuela, 72. Mem: AAAS; Am Chem Soc; The Chem Soc; corresp mem Chem Soc Peru. Res: Ion exchange; chromatography. Mailing Add: Dept of Chem Univ of Colo Boulder CO 80302

WALTON, HENRY MILLER, b Frankfurt am Main, Ger, May 7, 12; nat US; m 50; c 2. ORGANIC CHEMISTRY. Educ: Univ Frankfurt, PhD(philos), 34; Univ Chicago, PhD(org chem), 38. Prof Exp: Res chemist, Continental Carbon Co, NY, 39; Chas Pfizer & Co fel, Columbia Univ, 40-42; sr res chemist, Warner Inst Therapeut Res, 43-47; group leader fundamental res lab, Nat Dairy Res Labs, Inc, 47-57; sr res chemist, A E Staley Mfg Co, 57-61, res assoc, 61-69, patent chemist, 69-72; patent consult, 72-74; PROF PHILOS, RICHLAND COMMUNITY COL, 74- Mem: Am Chem Soc; Am Philos Asn; Philos Sci Asn. Res: Vitamins A and E; unsaturated aliphatics; condensation polymers; free radical polymerization reactions; starches; sugars; enzymes; immobilized enzymes; philosophy of science. Mailing Add: 115 S Stevens Ave Decatur IL 62522

WALTON, JOHN JOSEPH, b Sterling, Ill, Aug 25, 34; m 59; c 2. ATMOSPHERIC PHYSICS. Educ: Northwestern Univ, BS, 56; Univ Kans, PhD(physics), 61. Prof Exp: Physicist, Lawrence Radiation Lab, 61-72, PHYSICIST, LAWRENCE LIVERMORE LAB, 72- Concurrent Pos: Lectr appl sci, Univ Calif, 67-71. Res: Regional and global atmospheric modeling with emphasis on transport processes. Mailing Add: Atmosperhic & Geophys Sci Div Lawrence Livermore Lab L-142 Livermore CA 94550

WALTON, JOHN RICHARD, nuclear chemistry, nuclear physics, see 12th edition

WALTON, KENNETH NELSON, b Winnipeg, Man, May 1, 35; US citizen; m; c 4. MEDICINE. Educ: Univ Man, MD, 59; Am Bd Urol, dipl, 68. Prof Exp: Intern, Winnipeg Gen Hosp, 59-60, resident surg & path, 60-61; jr asst resident urol, Johns Hopkins Hosp, 61-62, res fel, 62-63, sr asst resident, 63-64, co-head resident, 64-65; from instr to asst prof urol, Dept Surg, Med Ctr, Univ Ky, 65-69, chmn div urol, 68-69; LOUIS McDONALD ORR PROF SURG & CHMN DEPT UROL, SCH MED, EMORY UNIV, 69- Concurrent Pos: Am Cancer Soc res fel, 63-64. Mem: Am Col Surgeons; AMA; Am Soc Nephrology; Am Urol Asn; Pan-Am Med Asn. Res: Kidney transplants; factors influencing renai oxygen consumption; effect of hypothermia and the inhibition of the tubular transport of sodium; difference in carbohydrate metabolism between prostatic cancers which are endocrine sensitive and those which are not. Mailing Add: 867 Castle Falls Dr NE Atlanta GA 30329

WALTON, LEWIS F, b Augusta, Ga, Jan 29, 07; m 39; c 2. MATHEMATICS. Educ: Emory Univ, BS, 29, MS, 31; Univ Calif, PhD(math), 40. Prof Exp: Instr math, Emory Univ, 31-35 & San Diego State Col, 39-43; lectr, 43-46, from asst prof to prof, 46-74, chmn dept, 54-55, actg dean, 55-56, EMER PROF MATH, UNIV CALIF, SANTA BARBARA, 74- Mem: Am Math Soc. Res: Algebra; mathematical logic. Mailing Add: Dept of Math Univ of Calif Santa Barbara CA 93106

WALTON, PETER DAWSON, b Leeds, Eng, Oct 18, 24; Can citizen; m 49; c 4. PLANT BREEDING. Educ: Univ Durham, BSc, 49, MSc, 53; Univ Lancaster, PhD(pop genetics), 61. Prof Exp: Plant breeder, Res Div, Ministry of Agr, Sudan, 50-55; sr plant breeder, Empire Cotton Growing Corp, 55-63; lectr agr & bot, Ahmadu Bello Univ, Nigeria, 63-67; assoc prof plant sci, Univ Sask, 67-69; PROF PLANT SCI, UNIV ALTA, 69- Mem: Brit Inst Biol; Am Soc Agron; Genetics Soc Can; Am Forage & Grassland Coun. Res: Forage crop breeding with special reference to the study of winter hardiness and the use of heterosis. Mailing Add: Dept of Plant Sci Univ of Alta Edmonton AB Can

WALTON, PHILIP WILSON, b Dublin, Ireland, May 22, 40; m 68; c 2. MEDICAL PHYSICS. Educ: Trinity Col, Dublin, BA, 62, PhD(physics), 67. Prof Exp: Physicist, Regional Physics Dept, Western Regional Hosp Bd, Scotland, 66-67, sr physicist, 67-68; asst prof, 68-70, ASSOC PROF MED PHYSICS, RADIATION PHYSICS DIV, MED COL VA, 70- Concurrent Pos: Radiation safety officer & physics consult, McGuire Vet Admin Hosp, Richmond, Va, 70- Mem: Am Asn Physicists in Med; Brit Hosp Physicists Asn. Res: Application of radioisotopes in medicine; instrumentation, notably the imaging of in vivo isotope distributions. Mailing Add: Radiation Physics Div Box 72 Med Col of Va Richmond VA 23298

WALTON, RICHARD BRUCE, b Fargo, NDak, Apr 1, 19; m 40; c 3. PHYSICS. Educ: Ore State Col, BA, 43, MS, 55. Prof Exp: From instr to asst prof, Vanport Exten Ctr, 46-52, asst prof, Univ, 52-75, ASSOC PROF PHYSICS, PORTLAND STATE UNIV, 75- Prof Exp: Instr, El Camino Col, 59-60; consult, US Naval Ord Test Sta, China Lake, Calif, 59-65. Mem: AAAS; Am Phys Soc; Am Asn Physics Teachers; Am Geophys Union. Res: Optics; solid state physics; photography; physical oceanography. Mailing Add: 7618 Fowler Ave Portland OR 97217

WALTON, RICHARD JOSEPH, b Auckland, NZ, Aug 30, 14; m 46; c 3. MEDICINE. Educ: Univ Otago, NZ, MB & Chb, 39. Prof Exp: PROF RADIOL, FAC MED, UNIV MAN, 52-; DIR RADIOTHERAPY, MAN CANCER TREAT & RES FOUND, 73- Concurrent Pos: Brit Empire Cancer Campaign fel, Royal Marsden Hosp, London, Eng, 51-52; consult radiotherapy, St Boniface Gen Hosp, 56-; dir, Man Cancer Treat & Res Found, 57-73; vpres med, Health Sci Ctr, 73- Mem: Nat Cancer Inst Can (vpres, 60, pres, 63-65). Res: Cancer. Mailing Add: Health Sci Ctr 700 William Ave Winnipeg MB Can

WALTON, ROBERT BRUCE, b Jersey City, NJ, Nov 30, 15; m 39; c 4. MICROBIOLOGY. Educ: Rutgers Univ, BSc, 48, PhD(microbiol), 53. Prof Exp: From asst res microbiologist to assoc res microbiologist, 40-53, SR RES MICROBIOLOGIST, MERCK SHARP & DOHME RES LABS, 53- Mem: AAAS; Am Soc Microbiol; Am Chem Soc; Soc Indust Microbiol. Res: Antibiotics; immunology; microbial nutrition and physiology; fermentations; actinophage. Mailing Add: 798 Central Ave Rahway NJ 07065

WALTON, ROBERT EUGENE, b Shattuck, Okla, Jan 15, 31; m 59. ANIMAL BREEDING, ANIMAL GENETICS. Educ: Okla State Univ, BS, 52, MS, 56; Iowa State Univ, PhD, 61. Prof Exp: Farm mgr, Westhide Farms, Eng, 53-54; asst prof dairy sci, Univ Ky, 58-62; geneticist, 68, PRES & CHIEF EXEC OFFICER, AM BREEDERS SERV, 68- Concurrent Pos: Mem prog mgt develop, Harvard Univ, 70. Mem: Am Soc Animal Sci; Am Dairy Sci Asn; Biomet Soc; Nat Asn Animal Breeders (pres, 72-74). Res: Application of genetical and statistical tools to problems of animal breeding and improvement; estimation of parameters of domestic large animal populations; genetic evaluation of dairy sires. Mailing Add: Rte 2 De Forest WI 53532

WALTON, ROBERT RALPH, entomology, see 12th edition

WALTON, RODDY BURKE, b Goldthwaite, Tex, Dec 9, 31; m 61; c 1. NUCLEAR PHYSICS. Educ: Tex A&M Col, BS, 52; Univ Wis, MS, 54, PhD(nuclear physics), 57. Prof Exp: Nuclear res officer, Air Force Weapons Lab, 57-59; staff physicist, Gen Atomic Div, Gen Dynamics Corp, 59-67; STAFF MEM, LOS ALAMOS SCI LAB, 67- Mem: Am Phys Soc; Am Nuclear Soc; Inst Nuclear Mat Mgt. Res: Neutron physics; photonuclear research; positron production with an electron linear accelerator; delayed gamma rays and delayed neutrons from fission; non-destructive assay applications. Mailing Add: A-1 Los Alamos Sci Lab Los Alamos NM 87545

WALTON, THEODORE ROSS, b Takoma Park, Md, Feb 26, 31; m 56; c 4. ORGANIC CHEMISTRY. Educ: Univ Md, BS, 55; Ohio State Univ, PhD(org chem), 60. Prof Exp: Proj dir chem, Atlantic Res Corp, 60-63; res chemist, Author W Sloan Found Va, 63-64; RES CHEMIST, US NAVAL RES LAB, WASHINGTON, DC, 64- Mem: Am Chem Soc. Res: Organic and polymer synthesis and chemistry; rocket motor case thermal insulation and material compatibility; fire retardant coating systems; water based paints; adhesive bonding. Mailing Add: US Naval Res Lab 4555 Overlook Ave SW Washington DC 20032

WALTON, THOMAS PEYTON, III, b Archer, Fla, Dec 7, 22; m 64; c 3. SURGERY. Educ: Tulane Univ, MD, 50; Am Bd Surg, dipl, 64. Prof Exp: Intern, Charity Hosp La, New Orleans, 50-51; physician for Seminole Indian Nation, Fla, 51-52; resident surg, Baptist Hosp, Nashville, Tenn, 52-55; chief resident, Nashville Gen Hosp, 55-56; pvt pract surg, Tampa, Fla, 57-67; asst prof, La State Univ Med Ctr New Orleans, 68-69; vchief surg, Vet Admin Hosp, Big Spring, Tex, 69-70; CLIN DIR & CHIEF SURG, LAFAYETTE CHARITY HOSP, 70- Concurrent Pos: VChief surg, Tampa Gen Hosp, Fla, 61-62; chief of staff, Clara Frye Hosp, Tampa, 62-63; asst prof surg, Med Sch, Univ New Orleans, 70-73; guest examr, Am Bd Surg, 71. Mem: Fel Am Col Surgeons. Res: Emergency treatment of upper gastrointestinal bleeding by comparison of vagotomy and drainage with vagotomy and resection. Mailing Add: Lafayette Charity Hosp Lafayette LA 70501

WALTON, WARREN LEWIS, b La, Dec 13, 14; m 43; c 3. ORGANIC CHEMISTRY. Educ: Millsaps Col, BS, 35; La State Univ, MS, 37; Univ Ill, PhD, 41. Prof Exp: Br chemist, The Coca-Cola Co, 37-38; res & develop chemist, Hercules Co, 41-46; res & develop chemist silicone synthesis, Gen Elec Co, NY, 46-50, head anal unit, Insulating Mat Dept, 50-69, instrumental anal chemist, 50-71; SELF-EMPLOYED, 71- Res: Infrared spectroscopy; microscopy, gas and gelpermeation chromatography, nuclear magnetic resonance spectroscopy. Mailing Add: 3017 Sunset Lane Schenectady NY 12303

WALTON, WILLIAM RALPH, b Ft Worth, Tex, Apr 11, 23; m 49; c 2. GEOLOGICAL OCEANOGRAPHY. Educ: Amherst Col, BA, 49; Univ Calif, MS, 52, PhD(oceanog), 54. Prof Exp: Paleoecologist, Gulf Res & Develop Co, 53-57; staff geologist, Pan Am Petrol Corp, 57-60, div consult geologist, 60-63, geol res dir, Amoco Prod Co, Okla, 63-75, CHIEF GEOLOGIST, AMOCO PROD CO, ILL, 75- Honors & Awards: Pres Award, Am Asn Petrol Geol, 57. Mem: Soc Econ Paleont & Mineral; Am Asn Petrol Geol. Res: Biostratigraphy of gulf coast tertiary; paleoecology, marine geology and sedimentology. Mailing Add: Amoco Prod Co 200 E Randolph Code 5105 Chicago IL 60680

WALTON, WILLIAM UPTON, b Salt Lake City, Utah, Mar 2, 26; m 49; c 3. PHYSICS, SCIENCE EDUCATION. Educ: Univ Mich, BS, 49, MS, 50. Prof Exp: Instr physics, Col Wooster, 51-52 & Rochester State Jr Col, 52-61; assoc prof, Webster Col, 61-66; sci educator, African Educ Prog, Educ Develop Ctr, Mass, 66-69; SR RES SCIENTIST PHYSICS, EDUC RES CTR, MASS INST TECHNOL, 69- Concurrent Pos: Consult elem sci prog, Educ Develop Ctr, Mass, 62-70, African Sci Prog, 63-, dir proj CALC, 73-; consult, Madison Proj, NY, 62-66; Ford Found consult, Inst Educ, Univ Nigeria, 63-64; dir, NSF Summer Inst Teachers, Webster Col, 64-65; consult dimensional communications; consult, Hayden Planetarium Proj, 70-71; dir, NSF Instructional Equip grant, Mass Inst Technol, 71-72. Mem: Math Asn Am; Am Asn Physics Teachers; Am Inst Physics. Res: Teaching in physics and mathematics. Mailing Add: Educ Develop Ctr-Proj CALC Newton MA 02160

WALTS, CHARLES C, dairy industry, deceased

WALTZ, ARTHUR G, b Irwin, Pa, Feb 14, 32; div. NEUROLOGY, CARDIOVASCULAR DISEASES. Educ: Univ Mich, BS, 52, MD, 55; Am Bd Psychiat & Neurol, dipl neurol, 62. Prof Exp: Rotating intern, Hosp Univ Pa, 55-56; asst resident neurol, Neurol Unit, Boston City Hosp, Mass, 56-57, sr resident, 57-58; res assoc, Sch Med, Wayne State Univ, 58-59, instr, 59; from instr to assoc prof, Mayo Grad Sch Med, Univ Minn, 62-71; prof neurol, Med Sch, Univ Minn, Minneapolis, 71-74; CHMN DEPT NEUROL, PAC MED CTR, 75- Concurrent Pos: Teaching fel neurol, Harvard Med Sch, 56-58; asst to staff neurol, Mayo Clin, 61, consult, 62-71; adj prof neurol, Univ of the Pac, 75-; fel stroke coun, Am Heart Asn. Mem: Fel Am Acad Neurol; Am Neurol Asn; Asn Res Nerv & Ment Dis. Res: Cerebral circulation, including blood flow, microcirculation and fluid balance, in normal and ischemic brain. Mailing Add: Pac Med Ctr Dept of Neurol PO Box 7999 San Francisco CA 94120

WALTZ, JOSEPH ELMER, physical chemistry, see 12th edition

WALTZ, MAYNARD CARLETON, b Damariscotta, Maine, Aug 19, 16; m 41; c 2. SEMICONDUCTORS. Educ: Colby Col, BA, 38; Wesleyan Univ, MA, 40. Prof Exp: Mem staff, Radiation Lab, Mass Inst Technol, 42-45; mem tech staff, Pa, 45-69, DEPT HEAD, BELL TEL LABS, INC, NJ, 69- Mem: Inst Elec & Electronics Eng. Res: Measurement of log decrement of crystal quartz; measurement and measuring techniques for microwave noise; noise theory of reflex klystrons; varistors and transistors; integrated circuits; reliability of solid-state devices; technical relations information. Mailing Add: Bell Tel Labs Mountain Ave Murray Hill NJ 07974

WALTZ, WILLIAM LEE, b Berkeley, Calif, June 3, 40; m 63; c 2. PHOTOCHEMISTRY, RADIATION CHEMISTRY. Educ: Miami Univ, BS, 62; Northwestern Univ, PhD(phys inorg chem), 67. Prof Exp: Res assoc chem, Univ Southern Calif, 66-68; sr chemist, Cent Res Lab, 3M Co, Minn, 68-69; asst prof, 69-73, ASSOC PROF CHEM, UNIV SASK, 69- Mem: AAAS; Am Chem Soc; The Chem Soc; Can Inst Chem. Res: Inorganic materials. Mailing Add: Dept of Chem & Chem Eng Univ of Sask Saskatoon SK Can

WALUM, HERBERT, b Bremerton, Wash, Aug 14, 36; m 59;c 2. MATHEMATICS. Educ: Reed Col, BA, 58; Univ Colo, PhD(math), 62. Prof Exp: Asst prof math, Harvey Mudd Col, 62-64; asst prof, 64-71, ASSOC PROF MATH, OHIO STATE UNIV, 71- Mem: Am Math Soc; Math Asn Am. Res: Number theory. Mailing Add: Dept of Math Ohio State Univ Columbus OH 43210

WALWICK, EARLE RICHARD, b San Diego, Calif, Oct 9, 29; m 51; c 2. CLINICAL CHEMISTRY. Educ: Univ Calif, Berkeley, AB, 52, PhD(biochem), 58. Prof Exp: Biochemist, US Naval Sch Aviation Med, Fla, 52-53; aviation physiologist, Naval Air Sta, 53-55; biochemist, Naval Hosp, Oakland, Calif, 57-58, biochemist, Naval Radiol Defense Lab, San Francisco, 58-61; supvr res, Aeronutronic Div, Philco-Ford Corp, 61-71; mem staff, US Govt, 71-72; CLIN CHEMIST & HOSP-PATHOLOGIST, CENT LAB OF ORANGE COUNTY, INC, 74- Concurrent Pos: NIH fel path, Univ Calif, Irvine, 72-74. Mem: Am Asn Clin Chem. Mailing Add: Satinwood Way Irvine CA 92715

WALZ, ALVIN EUGENE, b Hot Springs, SDak, Jan 12, 19. PHYSICAL CHEMISTRY, ANALYTICAL CHEMISTRY. Educ: Northern State Teachers Col, BS, 43; Univ Iowa, MS, 45, PhD(phys chem), 50. Prof Exp: Teacher, High Sch, Iowa, 43-48; from asst prof to prof chem, Mankato State Col, 50-63; PROF CHEM, CALIF LUTHERAN COL, 63-, CHMN DEPT, 66- Mem: AAAS; Am Chem Soc. Res: Reaction rates; electron affinity; methyl stibine. Mailing Add: 119 Sirius Circle Thousand Oaks CA 91360

WALZ, ARTHUR JOSEPH, b Calif, June 28, 18; m 43; c 5. ECONOMIC ENTOMOLOGY. Educ: Univ Calif, BS, 42, MS, 48. Prof Exp: Asst entomologist, Br Exp Sta, Univ Idaho, 48-56; mem staff, Asgrow Seed Co, 56-67; EXTEN POTATO SPECIALIST, PARMA RES & EXTEN CTR, UNIV IDAHO, 67- Mem: Entom Soc Am; Potato Asn Am. Res: Insects affecting onions and potatoes. Mailing Add: Res & Exten Ctr Univ of Idaho Parma ID 83660

WALZ, DONALD THOMAS, b Newark, NJ, Oct 25, 24; m 59; c 3. PHARMACOLOGY. Educ: Upsala Col, BS, 50; Rutgers Univ, MS, 51; Georgetown Univ, PhD(pharmacol), 59. Prof Exp: Jr pharmacologist, Hoffmann-La Roche, Inc, 51-55; sr investr pharmacol, 60-68, asst dir pharmacol, 68-74, ASSOC DIR PHARMACOL, SMITH, KLINE & FRENCH LABS, 74- Concurrent Pos: Nat Inst Neurol Dis & Blindness fel, Sch Med, Georgetown Univ, 59-60. Mem: AAAS; Am Diabetes Asn; Am Soc Pharmacol & Exp Therapeut; Acad Pharmaceut Sci; Am Rheumatism Asn. Res: Biochemical neuropharmacology metabolism; carbohydrate metabolism; cardiovascular-autonomic pharmacology. Mailing Add: Smith Kline & French Labs 1500 Spring Garden St Philadelphia PA 19101

WALZ, FREDERICK GEORGE, b Brooklyn, NY, May 11, 40; m 62; c 6. BIOCHEMISTRY. Educ: Manhattan Col, BS, 62; State Univ NY Downstate Med Ctr, PhD(biochem), 66. Prof Exp: NIH res fel biochem, Cornell Univ, 66-68; asst prof, State Univ NY Albany, 68-75; ASST PROF CHEM, KENT STATE UNIV, 75- Concurrent Pos: NSF res grant, State Univ NY Albany, 69-75. Mem: Am Chem Soc; Biophys Soc. Res: Rapid kinetic studies on enzymes; conformation of polynucleotides. Mailing Add: Dept of Chem Kent State Univ Kent OH 44242

WAMBACH, ROBERT F, b Detroit, Mich, Sept 29, 30; m 52; c 3. FOREST ECONOMICS, RESOURCE ADMINISTRATION. Educ: Univ Mont, BSF, 57; Univ Mich, MF, 59, PhD(forest econ), 67. Prof Exp: Forester, Forest Inventory, Intermountain Forest Exp Sta, US Forest Serv, 57-58, res forester silvicult northern conifers, Lake State Forest Exp Sta, 59-62, res proj leader plantation mgt, 62-64, res proj leader silvicult northern conifers, NCent Forest Exp Sta, 64-67; assoc prof forest econ, 67-71, PROF FORESTRY, UNIV MONT, 71-, DEAN FORESTRY SCH & DIR FOREST & CONSERV EXP STA, 72- Mem: AAAS; Soc Am Foresters; Am Econ Asn. Res: Growth and yield studies in northern conifers; silviculture of northern conifers; economic aspects of forest management, especially forest plantation management; development of methodology for decision-making in forestry. Mailing Add: Univ of Mont Sch of Forestry Missoula MT 59801

WAMPFLER, GENE LEROY, chemistry, see 12th edition

WAMPLER, DALE LEE, b Boones Mill, Va, June 25, 36; m 62; c 2. STRUCTURAL CHEMISTRY, INORGANIC CHEMISTRY. Educ: Bridgewater Col, BA, 57; Univ Wis, PhD(chem), 62. Prof Exp: From instr to assoc prof, 61-69, chmn dept, 65-68, PROF CHEM, JUNIATA COL, 69- Concurrent Pos: Sr vis fel, Univ Cambridge, 69-70. Mem: Am Chem Soc; Am Crystallog Asn. Res: Structural chemistry of transition metal complexes; single crystal x-ray diffraction techniques. Mailing Add: Dept of Chem Juniata Col Huntingdon PA 16653

WAMPLER, DONALD EUGENE, b Luanfu, China, July 24, 35; US citizen; m 61; c 3. BIOCHEMISTRY. Educ: Bridgewater Col, BA, 59; Mich State Univ, MS, 61, PhD(biochem), 65. Prof Exp: Res assoc biochem, Mich State Univ, 61-62; res assoc, Dartmouth Med Sch, 65-66; asst prof biochem, Univ Conn, 68-75; BIOCHEMIST, BIO-GANT CORP, 75- Concurrent Pos: NIH fel, Univ Mass, Amherst, 66-68. Mem: AAAS; Am Soc Biol Chem; Am Chem Soc. Res: Plasma fractionation, Factor VIII structure. Mailing Add: Orchard Rd Farmington CT 06032

WAMPLER, E JOSEPH, b Taiku, China, Jan 27, 33; US citizen; m 56; c 3. ASTRONOMY. Educ: Univ Va, BA, 58; Univ Chicago, MS, 59, PhD(astron), 62. Prof Exp: Res asst astron, Univ Chicago, 62-63; fel in residence, Miller Inst Basic Res in Sci, 63-65, actg asst astronomer, Observ, 65-66, asst astronomer, 66-70, assoc astronomer, 70-73, ASTRONOMER, LICK OBSERV, UNIV CALIF, 73- Mem: Am Astron Soc; Int Astron Union. Res: Quasi stellar sources; extra-galactic astronomy; instrumentation. Mailing Add: Lick Observ Univ of Calif Santa Cruz CA 95064

WAMPLER, FRED BENNY, b Kingsport, Tenn, Apr 2, 43; m 63; c 3. PHYSICAL CHEMISTRY, PHOTOCHEMISTRY. Educ: Univ Tenn, Knoxville, BS, 65; Univ Mo-Columbia, PhD(phys chem), 70. Prof Exp: Fel phys chem, Ohio State Univ, 70-72; sr scientist, Allison Div, Gen Motors Corp, 72-74; STAFF MEM, LOS ALAMOS SCI LAB, UNIV CALIF, 74- Mem: Am Chem Soc. Res: Kinetics; application of lasers to chemical problems; air pollution; laser spectroscopy; atmospheric chemistry. Mailing Add: Los Alamos Sci Lab PO Box 1663 Los Alamos NM 87544

WAMPLER, JESSE MARION, b Harrisonburg, Va, Oct 31, 36; m 62; c 2. GEOCHEMISTRY, GEOCHRONOLOGY. Educ: Bridgewater Col, BA, 57; Columbia Univ, PhD(geochem), 63. Prof Exp: Res asst geochem, Lamont Geol Observ, NY, 60-63; res assoc, Brookhaven Nat Lab, 63-65; asst prof, 65-69, ASSOC PROF GEOPHYS SCI, GA INST TECHNOL, 69- Mem: Geochem Soc; Am Geophys Union. Res: Nuclear geochemistry; geochemistry of argon and potassium; potassium-argon geochronology; lead isotope geochemistry. Mailing Add: Sch of Geophys Sci Ga Inst of Technol Atlanta GA 30332

WAMPLER, JOE FORREST, b Chanute, Kans, Dec 13, 26; m 49; c 2. MATHEMATICS. Educ: Univ Kans, AB, 50, MA, 52; Univ Nebr, PhD, 67. Prof Exp: Head dept math & physics, York Col, 51-53, head dept math & registr, 53-54; assoc prof, 54-66, PROF MATH, NEBR WESLEYAN UNIV, 66-, HEAD DEPT, 54-, CHMN DIV NATURAL SCI, 73- Concurrent Pos: Woods Found grant, 60-61; NSF coop teacher develop grant, 66. Mem: Math Asn Am. Res: Liouville's methods; use of various measures of aptitude to predict achievement in college mathematics. Mailing Add: Dept of Math Nebr Wesleyan Univ Lincoln NE 68504

WAMPLER, STANLEY NORMAN, b Irwin, Pa, Dec 9, 31. VETERINARY MEDICINE, RADIOBIOLOGY. Educ: Pa State Univ, BS, 53; Univ Pa, VMD, 56; Univ Rochester, MS, 61. Prof Exp: Vet, US Army, 56-60, radiobiologist, Walter Reed Inst Res, 61-65; res sr vet, Pesticide Res Lab, USPHS, 65-67; ASST PROF LAB ANIMAL MED, SCH VET MED, UNIV PA, 67- Concurrent Pos: Lectr, Col Med, Univ Miami, 65-67; from asst prof to assoc prof, Hahnemann Med Col, 67-; asst prof, Sch Med, Temple Univ & Med Col Pa, 67-; dir lab animal med, Fed Med Resources, 67- Mem: Sigma Xi. Res: Veterinary laboratory animal medicine. Mailing Add: RD 2 Honeybrook PA 19344

WAMSER, CARL CHRISTIAN, b New York, NY, Aug 10, 44; m 67; c 2. ORGANIC CHEMISTRY, PHOTOCHEMISTRY. Educ: Brown Univ, ScB, 66; Calif Inst Technol, PhD(chem), 69. Prof Exp: US Air Force Off Sci Res-Nat Res Coun fel, Harvard Univ, 69-70; ASSOC PROF CHEM, CALIF STATE UNIV, FULLERTON, 70- Concurrent Pos: Vis assoc prof, Univ Southern Calif, 75-76. Mem: Am Chem Soc; The Chem Soc; Nat Audubon Soc. Mailing Add: Dept of Chem Calif State Univ Fullerton CA 92634

WAMSER, CHRISTIAN ALBERT, b Long Island City, NY, July 15, 13; m 40; c 2. INORGANIC CHEMISTRY. Educ: Cooper Union, BS, 34. Prof Exp: Anal chemist, J F Jelenko & Co, Inc, NY, 34-41; supvr anal group, Gen Chem Div, Allied Chem Corp, 41-48; res chemist, Vitro Corp Am, NJ, 53-62; res chemist, Gen Chem Div, 62-65, res scientist, 65-69, res scientist, Syracuse Tech Ctr, 69-74, RES ASSOC, INDUST CHEM DIV, SYRACUSE TECH CTR, ALLIED CHEM CORP, NY, 74- Mem: Am Chem Soc. Res: Industrial inorganic chemistry, uranium, fluorine, chromium and aluminum compounds. Mailing Add: 207 Rebhahn Dr Camillus NY 13031

WAMSLEY, ROBERT ALAN, b Cameron, Mo, Aug 8, 27; m 54; c 2. PAPER CHEMISTRY. Educ: Cent Mo State Col, BS, 51; Prof Exp: Res chemist, Crossett Co, Ark, 55-61; res assoc, Int Paper, 61-65; mgr tech serv paper, Celanese Coatings Co, 65-73, PROD MGR RESINS, STEINHALL DIV, CELANESE CORP, 73- Mem: Tech Asn Pulp & Paper Indust. Res: Pulp and paper; paper coatings; adhesives; electrophotographic coating; colloid chemistry. Mailing Add: 809 Vannah Anchorage KY 40001

WAN, ABRAHAM TAI-HSIN, b Tsingtao, China, Oct 14, 28; US citizen; m 59; c 2. CLINICAL CHEMISTRY, ENDOCRINOLOGY. Educ: Nat Taiwan Univ, BS, 54; Univ Minn, Minneapolis, MS, 60; Univ Nebr, Omaha, PhD(med biochem), 64. Prof Exp: Biochemist, Dept Pub Health, Sask, Can, 63-64 & Sunland Hosp, Orlando, 64-67; ASSOC DIR CLIN CHEM, CLIN LAB, DEPT PATH, NORFOLK GEN HOSP, 67-; ASST PROF, EASTERN VA MED SCH, 74- Concurrent Pos: NIH grant, Sunland Hosp, Orlando, Fla, 64-67 & Norfolk Gen Hosp, 72-; adj prof, Old Dom Univ, 69- Mem: Am Chem Soc; Am Asn Clin Chemists; Am Soc Clin Pathologists. Res: Measurement of Renin levels in various parts of human brain, lung, ovary and kidney tissues and their relations to normal and hypertensive conditions. Mailing Add: Clin Lab Norfolk Gen Hosp Norfolk VA 23507

WAN, FREDERIC YUI-MING, b Shanghai, China, Jan 7, 36; US citizen; m 60. APPLIED MATHEMATICS, SOLID MECHANICS. Educ: Mass Inst Technol, SB, 59, SM, 63, PhD(math), 65. Prof Exp: Staff mem struct mech, Lincoln Lab, Mass Inst Technol, 59-62, staff assoc, 62-65, from instr to assoc prof appl math, 65-74; PROF MATH & DIR INST APPL MATH & STATIST, UNIV BC, 74- Concurrent Pos: Sloan Found fel, 73; mem comt appl math, Can Math Cong, 75- Mem: Am Soc Mech Eng; Soc Indust & Appl Math; Can Math Cong. Res: Classical elasticity; shell theory; random vibrations; stochastic ordinary and partial differential equations; economics of exhaustible resources; fisheries management. Mailing Add: Dept of Math Univ of BC Vancouver BC Can

WAN, JEFFREY KWOK-SING, b Hong Kong, June 4, 34; m 62; c 2. PHYSICAL CHEMISTRY. Educ: McGill Univ, BSc, 58; Univ Alta, PhD(phys chem), 62. Prof Exp: Fel, Univ Alta, 62-63; asst res chemist, Univ Calif, 63-65; Nat Res Coun Can fel chem, 65-66; from asst prof to assoc prof, 66-74, PROF CHEM, QUEEN'S UNIV, 74- Mem: Chem Inst Can. Res: Photochemistry and electron paramagnetic resonance spectroscopy. Mailing Add: Dept of Chem Queen's Univ Kingston ON Can

WAN, KWOK MING, b Sept 15, 37; US citizen. ORGANIC BIOCHEMISTRY, BUSINESS. Educ: Chung Hsing Univ, Taiwan, BSc, 58; Univ Sask, MSc, 61, PhD(chem), 69; Northwestern Univ, MBA, 74. Prof Exp: Instr chem, Hong Kong Baptist Col, 58-59; res chemist, Dearborn Chem Div, W R Grace Co, 62-63; res assoc biochem, Med Sch, Northwestern Univ, Chicago, 63-66; sr res chemist, Armour Indust Chem Div, Armour & Co, 69-71; SR RES CHEMIST, ARMAK CHEM DIV, AKZONA CO, 72- Mem: Am Chem Soc; Am Oil Chem Soc. Res: Organic synthesis; radioactive tracers in studies on organic and biochemical reaction mechanisms, fatty acid and food research. Mailing Add: Armak Co 8401 W 47th St McCook IL 60525

WAN, SUK HAN, b Malaysia. PHARMACODYNAMICS, CLINICAL PHARMACOLOGY. Educ: Univ Singapore, BPharm(Hons), 65; Univ Calif, San Francisco, PhD(pharm chem), 71. Prof Exp: Res assoc pharmacokinetics, Univ Kans Med Ctr, Kansas City, 71-74; ASST PROF PHARM, UNIV SOUTHERN CALIF, 74- Mem: Am Pharmaceut Asn; Acad Pharmaceut Sci. Res: The distribution, metabolism, excretion and absorption of drugs in man and animals, and factors affecting the above processes. Mailing Add: Sch of Pharm Univ Southern Calif 1985 Zonal Ave Los Angeles CA 90033

WAN, YIEH-HEI, b China, Feb 17, 47; m 75. TOPOLOGY. Educ: Nat Taiwan Univ, BS, 68; Univ Calif, Berkeley, PhD(math), 73. Prof Exp: Asst prof, 73-74, GEORGE WILLIAM HILL INSTR MATH, STATE UNIV NY BUFFALO, 74- Res: Application of global analysis to the study of general competitive equilibrium theory in mathematical economics; bifurcation theory for dynamical systems. Mailing Add: Dept of Math State Univ NY at Buffalo Amherst NY 14226

WAND, RONALD HERBERT, b Gloucester, NSW, Australia, July 20, 37; m 72; c 3. IONOSPHERIC PHYSICS. Educ: Univ Sydney, BSc, 58, PhD(physics), 65. Prof Exp: Res assoc ionospheric physics, Cornell Univ, 65-69; sr res fel, Univ Sydney, 70-71; STAFF SCIENTIST RADIOPHYSICS, LINCOLN LAB, MASS INST TECHNOL, 72- Concurrent Pos: Mem Comn III, Int Union Radio Sci. Mem: Am Geophys Union. Res: Ionospheric research using incoherent scatter radars and radar propagation studies. Mailing Add: Lincoln Lab Mass Inst Technol Lexington MA 02173

WANDER, IRVIN WOODROW, b Spencer, Ohio, Jan 14, 14; m 37; c 4.

HORTICULTURE. Educ: Col Wooster, BA, 35; Ohio State Univ, MS, 37, PhD(hort), 46. Prof Exp: Potash fel, Ohio Exp Sta, 37-39, asst horticulturist, 39-47, soil chemist, Citrus Exp Sta, Fla, 47-57; secy & gen mgr, 57-74, PRES, GROWERS FERTILIZER COOP, 74- Mem: Am Chem Soc; Am Soc Hort Sci; Am Soc Agron. Res: Spectroscopy; plant nutrition; soil chemistry; effect of cultural and fertilizer treatments on the potassium content of soil leaves and fruit of apple trees; methods of potassium analysis; nitrogen sources and effects on soil reaction; nutrition of citrus. Mailing Add: Growers Fertilizer Coop 312 N Buena Vista Dr Lake Alfred FL 33850

WANDER, JOSEPH DAY, b Columbus, Ohio, July 20, 41; m 67. NEUROCHEMISTRY. Educ: Case Inst Technol, BS, 63; Ohio State Univ, PhD(chem), 70. Prof Exp: Res fel chem, Tulane Univ, 70-71, La State Univ, Baton Rouge, 71-72 & Ohio State Univ, 72-74; DIR CHARLES B STOUT NEUROSCI LAB, UNIV TENN CTR HEALTH SCI, MEMPHIS, 74- Mem: AAAS; Am Chem Soc; The Chem Soc; assoc Royal Inst Chem; Soc Neurosci. Res: Applications of analytical spectroscopic methods to problems of molecular structure and sterochemistry in the neurosciences. Mailing Add: Charles B Stout Neurosci Lab Univ Tenn Ctr Health Sci Memphis TN 38163

WANDERER, PETER JOHN, JR, b Monroe, La, Aug 5, 43; m 72; c 1. HIGH ENERGY PHYSICS. Educ: Univ Notre Dame, BS, 65; Yale Univ, PhD(physics), 70. Prof Exp: Res assoc high energy physics, Lab Nuclear Studies, Cornell Univ, 70-73 & Univ Wis-Madison, 73-75; ASSOC SCIENTIST HIGH ENERGY PHYSICS, BROOKHAVEN NAT LAB, 75- Mem: Am Phys Soc. Res: Experimental research in neutrino-nucleon interactions. Mailing Add: Bldg 510-A Brookhaven Nat Lab Upton NY 11973

WANDS, RALPH CLINTON, b Norwich, NY, May 12, 19; m 42; c 3. TOXICOLOGY. Educ: Kent State Univ, BS, 41; Univ Minn, MS, 48; Am Bd Indust Hyg, dipl. Prof Exp: Chemist, Firestone Tire & Rubber Co, 41-42, develop engr, 42-43; res chemist, Minn Mining & Mfg Co, 51-61, res chemist, Indust Hyg & Toxicol, 61-64; prof assoc, 64-66, DIR, ADV CTR TOXICOL, NAT ACAD SCI, 66- Mem: AAAS; Am Chem Soc; Am Indust Hyg Asn; Soc Toxicol. Res: Industrial toxicology and hygiene; chemistry. Mailing Add: Nat Acad of Sci 2101 Constitution Ave Washington DC 20418

WANEK, ALEXANDER ANDREW, b Chicago, Ill, June 7, 16; m 43. GEOLOGY. Educ: Univ NMex, BS, 48. Prof Exp: Geologist, Mineral Fuels Invests, Rocky Mountain Region, US Geol Surv, 48-63, area geologist, Alaska Area, Conserv Div, 63-75; RETIRED. Mem: Geol Soc Am; Am Asn Petrol Geol. Res: Mineral land classification and evaluation and petroleum resource evaluation of the Alaskan Outer Continental Shelf. Mailing Add: PO Box 1402 Sequim WA 98382

WANER, JOSEPH LLOYD, b Detroit, Mich, Feb 4, 42; m 71. MICROBIOLOGY. Educ: Loyola Univ Chicago, BS, 64, MS, 66, PhD(microbiol), 69. Prof Exp: Fel, 69-71, res assoc, 71-74, ASST PROF TROP PUB HEALTH, SCH PUB HEALTH, HARVARD UNIV, 74- Mem: Am Soc Microbiol; Am Soc Trop Med & Hyg. Res: Virology; antigencity of the human herpes viruses with special attention to the human cytomegalouviruses. Mailing Add: Dept Trop Pub Health Harvard Univ Sch Pub Health Boston MA 02115

WANG, AN-CHUAN, b Tsing Tao City, China, Dec 28, 36; m 65; c 2. IMMUNOGENETICS. Educ: Nat Taiwan Univ, BS, 59; Univ Tex, PhD(genetics), 66. Prof Exp: Sec teacher, Taiwan, 61; res assoc genetics, Univ Tex, 66-67; asst prof genetics, Med Ctr, Univ Calif, San Francisco, 70-72, assoc prof microbiol & assoc researcher med, 72-75; ASSOC PROF IMMUNOL, MED UNIV SC, 75- Concurrent Pos: Fel, Med Ctr, Univ Calif, San Francisco, 67-70; USPHS & NSF res grants, 70-; USPHS career development award, 74; Am Cancer Soc fac res award, 74-79. Mem: AAAS; Am Soc Human Genetics; Genetics Soc Am; Am Asn Immunol. Res: Biochemical and genetical analyses of human plasma proteins, with special emphasis on immunoglobulins; immunology. Mailing Add: Dept of Basic & Clin Immunol & Microbiol Med Univ of SC Charleston SC 29401

WANG, AUGUSTINE WEISHENG, b China, Jan 5, 37; US citizen; m 63; c 3. MICROBIOLOGY. Educ: Nat Taiwan Univ, BS, 60; Ore State Univ, MS, 65; Syracuse Univ, PhD(microbiol), 69. Prof Exp: NIH fel, Case Western Reserv Univ, 69-70; asst prof microbiol, 70-73, ASSOC PROF MICROBIOL, MISS STATE UNIV, 73- Mem: Am Soc Microbiol. Res: Physiology of blue-green algae and other autotrophs. Mailing Add: Dept of Microbiol Miss State Univ State College MS 39762

WANG, BEN SHIH-PIN, b Huang-hsien, China, Sept 21, 27; Can citizen; m 60; c 3. FOREST PHYSIOLOGY. Educ: Nat Chung-hsing Univ, Taiwan, BScF, 52; Univ BC, MF, 60. Prof Exp: Res officer silvicult, 60-66, FORESTRY OFFICER, DEPT ENVIRON, CAN FORESTRY SERV, 66- Honors & Awards: Merit Award, Fed Govt Can, 75. Mem: Can Inst Forestry; Soc Am Foresters; Int Seed Testing Asn; Ont Prof Foresters Asn; Can Tree Improv Asn. Res: Seed certification standards; methodology of forest gene conservation; management of germplasm banks; methods of collection, processing, treatment and storage of forest seed, and testing of seed quality; seed dormancy; seed x-radiography. Mailing Add: Nat Forest Tree Seed Ctr Petawawa Forest Exp Sta Chalk River ON Can

WANG, CHANG-YI, b Kweichow, China, Aug 26, 39; US citizen; m 66; c 3. APPLIED MATHEMATICS. Educ: Nat Taiwan Univ, BS, 60; Mass Inst Technol, MS, 63, PhD, 66. Prof Exp: Fel appl math, Calif Inst Technol, 66-67, assoc appl math, Jet ProPropulsion Lab, 67-68; asst prof math, Univ Calif, Los Angeles, 68-69; asst prof, 69-70, ASSOC PROF MATH, MICH STATE UNIV, 70- Concurrent Pos: Vis prof, Nat Taiwan Univ, 71-72. Mem: Soc Indust & Appl Math. Res: Fluid mechanics. Mailing Add: Dept of Math Mich State Univ East Lansing MI 48824

WANG, CHAO-CHENG, b Rep. of China, July 20, 38; m 63; c 2. CONTINUUM MECHANICS. Educ: Nat Taiwan Univ, BS, 59; Johns Hopkins Univ, PhD(mech), 65. Prof Exp: Res asst math, Chinese Acad Sci, 60-61; instr mech, Johns Hopkins Univ, 63-64, lectr & fel, 65-66, from asst prof to assoc prof, 66-68; PROF MECH, RICE UNIV, 68- Concurrent Pos: NSF res grant, 66-75. Mem: Soc Natural Philos. Mailing Add: 5306 Grape Houston TX 77035

WANG, CHARLES T P, b Shantung, China, May 6, 30; US citizen; m 60; c 2. PHYSICS. Educ: Taiwan Norm Univ, BS, 55; Southern Ill Univ, Carbondale, MS, 59; Wash Univ, PhD(physics), 66. Prof Exp: Lectr physics, Southern Ill Univ, Edwardsville, 59-62; from asst prof to assoc prof, Parks Col, St Louis Univ, 65-67; PROF PHYSICS, STATE UNIV NY COL ONEONTA, 67- Concurrent Pos: NSF res participation fel, La State Univ, Baton Rouge, 69 & 71. Mem: Am Asn Physics Teachers. Res: Strong dynamical correlations in nuclear matter. Mailing Add: Dept of Physics State Univ of NY Col Oneonta NY 13820

WANG, CHIH CHUN, b Peking, China, Oct 9, 32; m 59; c 3. MATERIALS SCIENCE, PHYSICAL CHEMISTRY. Educ: Nat Taiwan Univ, BSc, 55; Kans State Univ, MSc, 59; Colo State Univ, PhD(phys chem), 62. Prof Exp: Res assoc, High Temp Phys Chem Res Lab, Univ Kans, 62-63; mem tech staff solid state mat res, 63-73, FEL TECH STAFF, RCA LABS, DAVID SARNOFF RES CTR, RCA CORP, PRINCETON, 73- Honors & Awards: Achievement Award, RCA Labs, 66, 68 & 71. Mem: Am Chem Soc; Electrochem Soc. Res: Electronic materials; thin films; crystal growth; chemical vapor deposition; high pressure and high temperautre chemistry; thermodynamics; x-ray crystallography; vidicon materials and devices. Mailing Add: 41 Maple Stream Rd Highstown NJ 08520

WANG, CHIH HSING, b Shanghai, China, Sept 20, 17; nat US; m 58; c 1. RADIOCHEMISTRY. Educ: Shantung Univ, China, BS, 37; Ore State Univ, MS, 47, PhD(chem), 50. Prof Exp: From asst prof to assoc prof, 51-58, PROF CHEM, ORE STATE UNIV, 58-, DIR RADIATION CTR, 62-, DIR INST NUCLEAR SCI & ENG, 64-, HEAD DEPT NUCLEAR ENG, 74- Concurrent Pos: Consult, NSF, 65-69; chmn, Ore Nuclear & Thermal Energy Coun, 72-73. Mem: Fel AAAS; Am Chem Soc; Am Soc Biol Chem; Am Soc Plant Physiol; Am Nuclear Soc. Res: Nuclear education; radiotracer methodology. Mailing Add: Radiation Ctr Ore State Univ Corvallis OR 97331

WANG, CHI-HUA, b Peiping, China, Apr 18, 23; m 49; c 1. ORGANIC CHEMISTRY. Educ: St John's Univ, China, BS, 45; Cath Univ, China, MS, 47; St Louis Univ, PhD(org chem), 51. Prof Exp: Fel, Brandeis Univ, 51-53, from instr to assoc prof chem, 53-62; sr chemist, Arthur D Little, Inc, 62-64; assoc prof chem, Wellesley Col, 64-68; assoc prof, 68-70, PROF CHEM, UNIV MASS, HARBOR CAMPUS, 70- Mem: Am Chem Soc; Sigma Xi. Res: Chemistry of free radicals in solution; mechanism of organic reactions. Mailing Add: Dept of Chem Univ of Mass Boston MA 02125

WANG, CHIN HSIEN, b Taiwan, Sept 4, 39; US citizen; m 63; c 4. CHEMISTRY. Educ: Nat Taiwan Univ, BS, 61; Utah State Univ, MS, 64; Mass Inst Technol, PhD(phys chem), 67. Prof Exp: Mem tech staff, Bell Tel Labs, 67-69; asst prof, 69-73, ASSOC PROF PHYS CHEM, UNIV UTAH, 73- Concurrent Pos: Petrol Res Fund grant, 70-72; Res Corp grant, 72-73; adj assoc prof elec eng, Univ Utah, 73-; Alfred P Sloan Found fel, 73. Mem: Am Phys Soc; Am Chem Soc. Res: Light scattering and Raman spectroscopy; nuclear magnetic resonance. Mailing Add: Dept of Chem Univ of Utah Salt Lake City UT 84112

WANG, CHIN SAN, b Taipei, Taiwan, Dec 18, 39; m 66; c 3. TOPOLOGY, SYSTEMS SCIENCE. Educ: Northeast Mo State Univ, BS, 68; Ohio State Univ, MS & PhD(math), 75. Prof Exp: ASST PROF COMPUT SCI, CHRISTOPHER NEWPORT COL, 75- Mem: Am Math Soc. Res: A generalization of Kuratowski's theorem on projective plane. Mailing Add: Dept of Comput Sci Christopher Newport Col Newport News VA 23606

WANG, CHING CHUNG, b Peking, China, Feb 10, 36; m 63; c 2. BIOCHEMISTRY, PARASITOLOGY. Educ: Nat Taiwan Univ, BS, 58; Univ Calif, Berkeley, PhD(biochem), 66. Prof Exp: Fel biochem, Col Physicians & Surgeons, Columbia Univ, 66-67; res assoc, Princeton Univ, 67-69; res fel, 69-75, SR RES FEL, MERCK INST THERAPEUT RES, 75- Res: Enzymes and active transport in microorganisms; development of protozoan parasites, including macromolecular biosynthesis and enzyme regulation.

WANG, CHING-PING SHIH, b Shanghai, China, Feb 16, 47; m 71. THEORETICAL SOLID STATE PHYSICS. Educ: Tung-Hai Univ, Taiwan, BS, 69; La State Univ, Baton Rouge, MS, 71, PhD(physics), 74. Prof Exp: RES ASSOC PHYSICS, DEPT PHYSICS & ASTRON, LA STATE UNIV, BATON ROUGE, 74- Mem: Am Phys Soc. Res: Energy band calculations; investigation of the magnetic and electronic properties of transition metals using the tight binding method. Mailing Add: Dept of Physics & Astron La State Univ Baton Rouge LA 70803

WANG, CHI-SUN, b Shanghai, China, Oct 8, 42; m 73; c 1. BIOCHEMISTRY. Educ: Nat Taiwan Univ, BS, 66; Univ Okla, PhD(biochem), 71. Prof Exp: Staff scientist, 74-75, ASST MEM MEMBRANE BIOCHEM, OKLA MED RES FOUND, 75- Res: The biochemical nature and function of human erythrocyte membrane protein. Mailing Add: 825 NE 13th St Oklahoma City OK 73104

WANG, CHIU-CHEN, b Canton, China, Nov 5, 22; US citizen; m 55; c 1. RADIOTHERAPY. Educ: Nat Kwei-Yang Med Col, China, MD, 48; Am Bd Radiol, dipl, 53. Prof Exp: Rotation intern, Canton Hosp, China, 47-48, asst resident med, 48-49; intern, Univ Hosp, Syracuse, NY, 49-50; asst resident radiol, Mass Gen Hosp, Boston, 50-51, resident, 52, clin fel, 53-56; asst radiol, 58-60, instr, 60-61, clin assoc, 62-67, asst clin prof, 68-69, asst prof, 69-70, from asst prof to assoc prof radiation ther, 70-75, PROF RADIATION THER, HARVARD MED SCH, 75-; RADIATION THERAPIST & HEAD CLIN SERV, MASS GEN HOSP, 73- Concurrent Pos: Teaching fel radiol, Harvard Sch Med, 53-56; Damon Runyon res grant, Donner Lab, Univ Calif, Berkeley, 61-62; from asst radiologist to radiologist, Mass Gen Hosp, Boston, 58-71; consult radiologist, Lawrence Berkeley Lab, Univ Calif, Berkeley, 62-65 & Mass Eye & Ear Infirmary, Boston, Waltham Hosp, Emerson Hosp & Malden Hosp, Mass, 62-; mem cont disaster planning, Mass Gen Hosp, 62; guest examr, Am Bd Radiol, 66. Mem: Fel Am Col Radiol; Am Radium Soc; Radiol Soc NAm; Am Soc Therapeut Radiol. Res: Clinical radiation and oncology; radiobiology. Mailing Add: Dept of Radiation Med Mass Gen Hosp Boston MA 02114

WANG, CHI-WU, b Tientsin, China, May 4, 13; nat US; m 56; c 3. FOREST GENETICS. Educ: Nat Tsing-Hua Univ, China, BS, 33; Yale Univ, MS, 47; Harvard Univ, PhD(biol), 53. Prof Exp: Res asst, Fan Mem Inst Biol, China, 33-43; assoc prof, Nat Kwangsi Univ, 43-46; res assoc, Harvard Forest, Harvard Univ, 53-54; from asst prof to assoc prof forestry, Univ Fla, 54-60; PROF FOREST GENETICS, UNIV IDAHO, 60- Concurrent Pos: Res assoc, Sch Forestry, Univ Minn, 58-59. Mem: AAAS; Ecol Soc Am; Soc Am Foresters; Soc Study Evolution; Am Genetic Asn. Res: Forest genetics and tree improvement; silvics; ecology; silviculture. Mailing Add: Univ of Idaho Col of Forestry Moscow ID 83843

WANG, CHUN SHAN, b Tainan, Taiwan, Dec 28, 32; US citizen; m 59; c 3. ORGANIC CHEMISTRY. Educ: Cheng Kung Univ, Taiwan, BS, 57; Ill Inst Technol, PhD(chem), 66. Prof Exp: Chem engr, Chinese Petrol Corp, 59-62; chemist, 66, res chemist, 66-69, sr res chemist, 69-72, sr res specialist, 72-75, ASSOC SCIENTIST, DOW CHEM CO, 75- Mem: Am Chem Soc; Sigma Xi. Res: Synthesis and process development of pesticides, organometallic compounds and animal health products; aqueous wastes treatment. Mailing Add: OCR Dow Chem Co Midland MI 48640

WANG, CHUNG CHIAN, b China, Dec 22, 39; US citizen; m 66. COMPUTER SCIENCE. Educ: Cheng-Kung Univ, Taiwan, BS, 61; Univ Rochester, MS, 64; State Univ NY Buffalo, PhD(comput sci), 74. Prof Exp: Res staff mem, T J Watson Res Ctr, IBM Corp, 64-65, sr assoc engr, Syst Develop Lab, 65-66; sr programmer data processing, Soc Sci Res Coun, 67; res staff mem, San Jose Res Lab, IBM Corp, 67-68; sr systs analyst, Moore Assocs, 69; instr & res asst comput sci, State Univ NY Buffalo, 70-72; ASST PROF COMPUT SCI, UNIV KY, 73- Mem: Asn Comput Mach; Soc Indust & Appl Math. Res: Computer operating systems; data base management; combinatorial algorithms; graph theory. Mailing Add: Dept of Comput Sci Univ of Ky Lexington KY 40506

WANG, CHUNG SHAN, b Fukien, China, Dec 16, 37; m 69. PHYSICS. Educ: Cheng Kung Univ, Taiwan, BSc, 62; Univ Idaho, MS, 66, PhD(physics), 69. Prof Exp: ASST PROF ATMOSPHERIC SCI, STATE UNIV NY ALBANY, 69- Mem: Am Geophys Union; Am Asn Physics Teachers. Res: Aeronomy; magnetospheric physics; solar-terrestrial relationship. Mailing Add: Dept of Atmospheric Sci State Univ NY Albany NY 12222

WANG, CHUN-JUAN KAO, b Mukden, China, Jan 10, 28; m 55; c 3. MYCOLOGY. Educ: Nat Taiwan Univ, BS, 50; Vassar Col, MS, 52; Univ Iowa, PhD(mycol), 55. Prof Exp: Asst, Univ Iowa, 52-55; res assoc, Clin Labs, Jewish Hosp, Cincinnati, Ohio, 55-58; from asst prof to assoc prof, 59-72, PROF FOREST BOT & PATH, NY STATE COL ENVIRON SCI & FORESTRY, SYRACUSE UNIV, 72- Concurrent Pos: Instr, Sch Med, Univ Cincinnati, 57-58. Mem: Mycol Soc Am; Brit Mycol Soc. Res: Medical mycology; soil fungi; pulp and paper fungi; imperfect fungi. Mailing Add: Dept of Forest Bot & Path NY State Col Envrn Sci & Forest Syracuse NY 13210

WANG, DALTON TA TUNG, b Peking, China, Nov 11, 25; m 46. PLANT BIOCHEMISTRY. Educ: Fu-jen Univ, China, BA, 47; Univ Sask, BSA, 52, MSc, 54; McGill Univ, PhD, 57. Prof Exp: Res fel bot, Univ Man, 57-61; fel biochem, Univ Wis, 61-62; biochemist, Boyce Thompson Res Inst, 62-70; res assoc chem, Purdue Univ, Lafayette, 71-74; ASST PROF BIOCHEM, ROCKEFELLER UNIV, 74- Mem: Am Soc Plant Physiol; Am Chem Soc. Res: Mechanism of protease and protein-protease inhibitor interaction. Mailing Add: Rockefeller Univ York Ave & 66th St New York NY 10021

WANG, FRANCIS WEI-YU, b Peikang, Taiwan, July 21, 36; US citizen; m 66; c 3. POLYMER SCIENCE. Educ: Calif Inst Technol, BS, 61, MS, 62; Univ Calif, San Diego, PhD(chem), 71. Prof Exp: Chemist, Pac Soap Co, 62-66; USPHS fel, 71-72; RES CHEMIST, POLYMERS DIV, NAT BUR STANDARDS, 72- Mem: Am Chem Soc; Am Phys Soc. Res: Thermodynamic and frictional properties of polymer solutions; photophysical processes in polymer molecules; ultracentrifugal analysis of macromolecules. Mailing Add: Rm A209 Bldg 224 Polymers Div Nat Bur Standards Washington DC 20234

WANG, FRANK FENG HUI, b Hopeh, China, Mar 21, 24; nat US; m 58; c 3. MARINE GEOLOGY. Educ: Nat Southwestern Assoc Univ, China, BS, 45; Univ Wash, PhD(geol), 55. Prof Exp: Asst geol, Nat Southwestern Assoc Univ, China, 45-46; instr, Nat Peking Univ, 46-48; asst, Univ Wash, 50-54; sedimentalogist-stratigrapher, Western Gulf Oil Co, 54-57; res geologist, Gulf Res & Develop Co, 57-63; marine geologist, Int Minerals & Chem Corp, 64-67; MARINE GEOLOGIST, US GEOL SURV, 67- Concurrent Pos: Spec consult, Chinese Petrol Corp, 58; vis scholar, Northwestern Univ, 64; tech adv, UN, 67-, spec adv & sr marine geologist, UN Develop Prog on Regional Offshore Prospecting in E Asia, 72-74, prin marine geologist, 74-; vchmn marine geol panel, US-Japan Coop Prog Natural Resources, 70- Mem: Geochem Soc; Am Asn Petrol Geol; Am Geophys Union; Marine Technol Soc. Res: Marine mineral resources; ocean mining; oceanographic data processing; mathematical models in marine geology. Mailing Add: US Geol Surv Off Marine Geol 345 Middlefield Rd Menlo Park CA 94025

WANG, FREDERICK E, b She-Tou, Formosa, Aug 1, 32; US citizen; m 61; c 2. CHEMISTRY, PHYSICS. Educ: Memphis State Univ, BS, 56; Univ Ill, MS, 57; Syracuse Univ, PhD(phys chem), 60. Prof Exp: Fel, Harvard Univ, 60-61; res assoc metal alloys, Syracuse Univ, 61-63; CHEMIST, US NAVAL SURFACE WEAPONS CTR, 63- Concurrent Pos: Fulbright exchange lectr, 67-68. Honors & Awards: Meritorious Civilian Serv Award, 65 & 73. Mem: Am Phys Soc; Am Crystallog Asn. Res: Metal and alloy physicsl order-disorder phenomena; superconductivity. Mailing Add: US Naval Surface Weapons Ctr White Oak Lab WR-32 Silver Spring MD 20910

WANG, GUANG TSAN, b Taiwan, China, Mar 6, 35; US citizen; m 62; c 2. VETERINARY MEDICINE, PARASITOLOGY. Educ: Nat Taiwan Univ, DVM, 58; Univ Ill, Urbana, MS, 64, PhD(vet med sci), 68. Prof Exp: Vet, Taiwan Serum Vaccine Labs, 60-62; res asst parasitol, Univ Ill, 62-68; res vet, 68-74, GROUP LEADER PARASITOL DISCOVERY, AGR DIV, AM CYANAMID CO, PRINCETON, 74- Mem: Am Soc Parasitol; Am Vet Med Asn. Res: Toxicity and efficacy of anthelmintics, anticoccidials and antibiotics in domestic animals; industrial parasitic chemotherapy. Mailing Add: 41 Slayback Dr Princeton Junction NJ 08550

WANG, H E FRANK, b China, Oct 23, 29; US citizen; m 55; c 2. FLUID PHYSICS, SYSTEMS ENGINEERING. Educ: Nat Taiwan Univ, BS, 52; Bucknell Univ, MS, 54; Brown Univ, PhD(gas dynamics), 59. Prof Exp: Res engr gas dynamics, Boeing Co, 58-60; MEM TECH STAFF & PROG MGR AEROPHYS & SYSTS ENG, AEROSPACE CORP, 60- Mem: Assoc fel Am Inst Aeronaut & Astronaut.

WANG, HAO, b Tsinan, China, May 20, 21; m 48; c 3. MATHEMATICS. Educ: Nat Southwestern Assoc Univ, China, BS, 43; Tsing Hua Univ, MA, 45; Harvard Univ, PhD, 48; Oxford Univ, MA, 56. Prof Exp: Soc Fels fel, Harvard Univ, 48-51, asst prof philos, 51-56; reader philos of math, Oxford Univ, 56-61; Gordon McKay prof math logic & appl math, Harvard Univ, 61-67; PROF MATH, ROCKEFELLER UNIV, 67- Concurrent Pos: Res engr, Burroughs Corp, 53-54; fel, Rockefeller Found, 54-55; John Locke lectr philos, Oxford Univ, 55; mem tech staff, Bell Tel Labs, 59-60; res scientist, IBM Res Ctr, 73-74; vis, Inst Advan Study, Princeton, 75-76. Mem: Asn Symbolic Logic; fel Am Acad Arts & Sci; foreign fel Brit Acad. Res: Mathematical logic; epistemology; philosophy of mathematics; general philosophy; contemporary China. Mailing Add: Rockefeller Univ New York NY 10021

WANG, HARRY (HSI), b Kashan, China, Apr 23, 07; nat US; m 32; c 2. ZOOLOGY. Educ: Shanghai Univ Sci & Technol, China, SB, 30, SM, 31; Univ Chicago, PhD(zool), 43. Prof Exp: Asst biol, Shanghai Univ Sci & Technol, 29-31, instr, 31-33, asst prof, 36-38; asst prof, Hangchow Christian Col, 33-35; asst zool, Univ Chicago, 38-45; assoc prof biol, Berea Col, 45-46; instr zool, Univ Wis, 46-47; asst prof biol sci, Univ Col, Univ Chicago, 47-49; chief histol div, Am Meat Inst Found, 49-57; from assoc prof to prof anat, Stritch Sch Med, Loyola Univ Chicago, 57-74; RETIRED. Prof Exp: Vis prof, Nat Taiwan Univ, 61; Nat Sci Coun Can spec chair & vis prof, Dept Biomorphics, Nat Defense Med Ctr, Taipei, Taiwan, 68; external examr zool, Chinese Univ Hong Kong, 68-70; Fulbright lectr biol, Taiwan Norm Univ, 69-70. Mem: Soc Exp Biol & Med; Am Asn Anatomists. Res: Experimental morphogenesis of feather development; correlation of structure and organoleptic characteristics of meat; enzymatic tenderization of beef; proteolysis of living protoplasm; germinal epithelium of mammalian ovaries; analysis of cigarette toxicity using Paramecium as test organism; human embryology. Mailing Add: 1930 18th Ave San Francisco CA 94116

WANG, HERBERT FAN, b Shanghai, China, Sept 14, 46; US citizen; m 68; c 2. GEOPHYSICS. Educ: Univ Wis-Madison, BA, 66; Harvard Univ, AM, 68; Mass Inst Technol, PhD(geophys), 71. Prof Exp: Res assoc geophys, Mass Inst Technol, 71-72; ASST PROF GEOPHYS, UNIV WIS-MADISON, 72- Mem: Am Geophys Union. Res: Elasticity of minerals and rocks; rock mechanics applied to earthquake mechanism. Mailing Add: Dept Geol & Geophys Univ Wis 1215 W Dayton St Madison WI 53706

WANG, HOWARD HAO, b Shanghai, China, Jan 24, 42; m 63; c 2. NEUROSCIENCES, BIOCHEMICAL PHARMACOLOGY. Educ: Calif Inst Technol, BS, 63; Univ Calif, Los Angeles, PhD(neurophysiol), 68. Prof. Exp: USPHS fel, Univ Calif, Berkeley, 68-69; resident scientist, Neurosci Res Prog, Mass Inst Technol, 69-70; ASST PROF BIOL & FEL, STEVENSON COL, UNIV CALIF, SANTA CRUZ, 70- Mem: AAAS; Am Asn Anat; Biophys Soc; Soc Neurosci. Res: Mechanism of local anesthetic action and drug-membrane interaction; effect of environmental chemicals on membrane structure and function; molecular and cellular mechanisms of brain function. Mailing Add: Div of Natural Sci Univ of Calif Santa Cruz CA 95064

WANG, HSI-CHANG, cytogenetics, see 12th edition

WANG, HSIEN CHUNG, b Peking, China, Apr 18, 19; m 56; c 3. MATHEMATICS. Educ: Northwest Assoc Univ, China, BS, 41; Univ Manchester, PhD(math), 48. Prof Exp: Lectr math, La State Univ, 49-51; mem, Inst Advan Study, 51-52; res prof, Ala Polytech Inst, 52-55; lectr, Univ Wash, 55-56; vis assoc prof, Columbia Univ, 56-57; from assoc prof to prof, Northwestern Univ, 57-66; PROF MATH, CORNELL UNIV, 66- Mem: Am Math Soc. Res: Lie groups and differential geometry. Mailing Add: Dept of Math Cornell Univ Ithaca NY 14850

WANG, HSUEH-HWA, b Peiping, China, July 10, 23; US citizen; m 48; c 3. PHARMACOLOGY, PHYSIOLOGY. Educ: Nat Cent Univ, Nanking, China, MB, 46. Prof Exp: ASSOC PROF PHARMACOL, COL PHYSICIANS & SURGEONS, COLUMBIA UNIV, 70- Concurrent Pos: NY Heart fel, 53-54. Mem: AAAS; Am Physiol Soc; Am Soc Pharmacol & Exp Therapeut. Res: Coronary circulation; central nervous system control of circulation and the effects of pharmacological agents. Mailing Add: Col of Physicians & Surgeons Columbia Univ Dept of Pharmacol New York NY 10032

WANG, HWA LIH, b Chekiang, China, Nov 29, 21; m 49; c 2. BIOCHEMISTRY. Educ: Nat Cent Univ, China, BS, 45; Univ Wis, MS, 50, PhD(biochem), 52. Prof Exp: Asst biochem, Med Sch, Nat Cent Univ, China, 45; res assoc, Med Sch, Marquette Univ, 53-55, Med Sch, Univ Wis, 56-61 & Wash Univ, 61-62; RES CHEMIST, NORTHERN REGIONAL LAB, AGR RES SERV, USDA, 63- Mem: Am Inst Nutrit; Am Chem Soc. Res: Biochemistry and physiology of molds used in food fermentation; nutritional value of fermented food products. Mailing Add: Fermentation Lab USDA Northern Regional Res Lab Peoria IL 61604

WANG, I-SHOU, b Taipei, Taiwan, Nov 30, 37; m 67; c 2. POPULATION GEOGRAPHY, URBAN GEOGRAPHY. Educ: Nat Taiwan Univ, BS, 61; Univ Minn, MA, 66, PhD(geog), 71. Prof Exp: Asst prof, 68-72, ASSOC PROF GEOG, CALIF STATE UNIV, NORTHRIDGE, 72- Mem: Asn Am Geogr; Pop Asn Am; Asn Asian Studies. Res: Population mobility in general; Chinese migration, particularly historical and contemporary. Mailing Add: Dept of Geog Calif State Univ Northridge CA 91324

WANG, JAMES C, b China, Nov 18, 36; m 61; c 2. BIOPHYSICAL CHEMISTRY. Educ: Nat Taiwan Univ, BS, 59; Univ SDak, MA, 61; Univ Mo, PhD(chem), 64. Prof Exp: Res fel chem, Calif Inst Technol, 64-66; from asst prof to assoc prof, 66-74, PROF CHEM, UNIV CALIF, BERKELEY, 74- Concurrent Pos: Mem biophys & biochem study sect, NIH, 72-76. Mem: Am Soc Biol Chem; Biophys Soc. Res: Physico-chemical studies of nucleic acids. Mailing Add: Dept of Chem Univ of Calif Berkeley CA 94720

WANG, JEN YU, b Foochow, China, Mar 3, 15; c 1. METEOROLOGY. Educ: Fukien Christian Univ, China, BS, 38; Univ Chicago, cert, 54; Univ Wis, MS, 55, PhD(meteorol), 58. Prof Exp: Instr math & physics, Cols, China & Hong Kong, 38-42; prin meteorologist, Weather Bur, China, 42-47; assoc prof physics, Fukien Christian Univ, 47-50; asst meteorol, Weather Forecasting Res Ctr, Univ Chicago, 53-54; asst, Univ Wis, 54-57, res assoc, 57-60, asst prof, 60-64; assoc prof, 64-68, PROF METEOROL & DIR ENVIRON SCI INST, SAN JOSE STATE UNIV, 68- Concurrent Pos: Fel, United Bd Higher Educ in Asia, US, 50-54; consult, 10th Weather Squadron, US Air Force, China, 45, US Weather Bur, Washington, DC, 58, AEC Proj, 65 & Stanford Res Inst, 66-67; pres, Milieu Info Serv, 71- Mem: Am Meteorol Soc; Am Soc Agron; Am Geophys Union; Int Soc Biometeorol; fel Am Geog Soc. Res: New techniques in the investigation of environmental relationships between animals and plants; agricultural meteorology; ecology; phenology; phytoclimatology; environmental assessment studies. Mailing Add: Milieu Info Serv 33 E San Fernando St San Jose CA 95113

WANG, JERRY HSUEH-CHING, b Nanking, China, Mar 12, 37; m 62; c 2. BIOCHEMISTRY. Educ: Nat Taiwan Univ, BSc, 58; Iowa State Univ, PhD(biochem), 65. Prof Exp: Asst prof, 66-71, ASSOC PROF BIOCHEM, FAC MED, UNIV MAN, 71- Concurrent Pos: Nat Res Coun Can fel biochem, 65-66; Med Res Coun scholar, 66- Mem: Can Biochem Soc; Am Soc Biol Chemists. Res: Quaternary structure and regulatory property of enzymes; structure and function of ribosomes. Mailing Add: Dept of Biochem Univ of Man Fac of Med Winnipeg MB Can

WANG, JIN TSAI, b Inchon, Korea, Apr 7, 31; US citizen; m 58; c 2. INORGANIC CHEMISTRY, ANALYTICAL CHEMISTRY. Educ: Ore State Univ, BS, 57; Carnegie Inst Technol, PhD(chem), 68. Prof Exp: Instr, Pa State Univ, 68; ASST PROF CHEM, DUQUESNE UNIV, 68- Mem: Am Chem Soc. Res: Infrared and polarographic studies of metal complexes. Mailing Add: Dept of Chem Duquesne Univ Pittsburgh PA 15219

WANG, JIN-LIANG, b Chu-Nan, Taiwan, Aug 18, 37; US citizen; m 62; c 2. POLYMER CHEMISTRY. Educ: Taipei Inst Technol, dipl, 58; Kent State Univ, MS, 66; Univ Akron, PhD(polymer chem), 71. Prof Exp: Sr chem engr, Hwa Ming Paper Mill, Taiwan, 60-61; res chem engr, Taiwan Prov Tobacco & Wine Monopoly Bur, 61-63; SR RES CHEMIST, RES DIV, GOODYEAR TIRE & RUBBER CO, 66- Mem: Am Chem Soc; Sigma Xi. Res: Polymer synthesis, characterization and applications; polymerization mechanism and kinetics. Mailing Add: Res Div Goodyear Tire & Rubber 142 Goodyear Blvd Akron OH 44316

WANG, JUI HSIN, b Peking, China, Mar 16, 21; nat US; m 49; c 2. BIOCHEMISTRY. Educ: Nat Southwestern Assoc Univ, China, BSc, 45; Wash Univ, PhD(chem), 49. Hon Degrees: MA, Yale Univ, 60. Prof Exp: Fel radiochem, Wash Univ, 49-51; res fel chem, Yale Univ, 51-52, res asst, 52-53, from instr to prof, 53-62, Eugene Higgins Prof, 62-72; EINSTEIN PROF CHEM, STATE UNIV NY

BUFFALO, 72- Concurrent Pos: Guggenheim fel, Cambridge Univ, 60-61; mem biophys & biochem study sect, NIH, 65-69; Kennedy lectr, Wash Univ, 72. Mem: AAAS; Am Chem Soc; Am Soc Biol Chem; Biophys Soc; Am Inst Chem. Res: Diffusion in liquids; hemoglobin; mechanisms of enzyme action, particularly those related to oxidative phosphorylation, photosynthesis and ion-transport through biological membranes. Mailing Add: Bioenergetics Lab State Univ NY Buffalo NY 14214

WANG, JU-KWEI, b Peiping, China, Feb 15, 34. MATHEMATICS. Educ: Nat Taiwan Univ, BS, 55; Stanford Univ, PhD(math), 60. Prof Exp: Assoc researcher math, Acad Sinica, 60-62; vis assoc prof, Univ Calif, Berkeley, 62-63; vis lectr, Yale Univ, 63-65; from assoc prof to prof, Nat Taiwan Univ, 65-68; assoc prof, 68-70, PROF MATH, UNIV MASS, AMHERST, 70- Mem: Am Math Soc. Res: Functional analysis, including harmonic analysis; Banach algebras and function algebras. Mailing Add: Dept of Math & Statist Univ of Mass Amherst MA 01002

WANG, KE-CHIN, b Chekiang, China, Dec 8, 30; US citizen; m 55; c 2. HIGH TEMPERATURE CHEMISTRY, CERAMICS. Educ: Univ Wis-Madison, BA, 58, MS, 60; Ill Inst Technol, PhD(chem), 66. Prof Exp: Res assoc high temperature chem, Okla State Univ, 67-69 & Ill Inst Technol, 69-70; dir res, Ceramtec Industs, Inc, 70-71, vpres, 71-72; RES MGR INDUST CERAMICS, HARBISON-WALKER REFRACTORIES CO, DIV DRESSER INDUSTS, INC, 73- Mem: Am Chem Soc; Am Ceramic Soc. Res: Advanced ceramic materials and forming processes. Mailing Add: Garber Res Ctr Harbison-Walker Refractories Co Pittsburgh PA 15227

WANG, KEN HSI, b Shanghai, China, July 4, 34; m 62; c 3. NUCLEAR PHYSICS. Educ: Int Christian Univ, Tokyo, BA, 58; Yale Univ, PhD(physics), 63. Prof Exp: Res fel physics, Harvard Univ, 63-66; asst prof, 66-69, ASSOC PROF PHYSICS, BAYLOR UNIV, 69- Mem: Am Phys Soc; Am Asn Physics Teachers. Res: Nuclear reaction and scattering. Mailing Add: Dept of Physics Baylor Univ Waco TX 76706

WANG, KIA K, b Soochow, China, Oct 10, 24; nat US; m 43; c 3. GEOLOGY, STRATIGRAPHY. Educ: Nat Southwestern Assoc Univ, China, BS, 43; La State Univ, MS, 47, PhD, 51. Prof Exp: Geologist, Calif Oil Co, La, 46; res geologist, State Geol Surv, La, 47-50; asst prof, 50-70, ASSOC PROF GEOL, BROOKLYN COL, 70- Concurrent Pos: Asst, La State Univ, 46-50. Mem: Geol Soc Am; Soc Econ Paleontologists & Mineralogists; Am Asn Petrol Geologists. Res: Petroleum geology; micropaleontology. Mailing Add: Dept of Geol Brooklyn Col Brooklyn NY 11210

WANG, LAWRENCE CHIA-HUANG, b Wusih, China, Apr 5, 40; m 66. PHYSIOLOGY, ZOOLOGY. Educ: Taiwan Norm Univ, BSc, 63; Rice Univ, MA, 67; Cornell Univ, PhD(physiol), 70. Prof Exp: Vis asst prof biol, Univ Ore, 69-70; asst prof zool, 70-74, ASSOC PROF ZOOL, UNIV ALTA, 74- Concurrent Pos: Mem, Hibernation Info Exchange, Off Naval Res, 71-; operating grant, Nat Res Coun Can, 71- & Nat Defence Res Bd Can, 73-76. Mem: AAAS; Can Soc Zool; Can Physiol Soc; Soc Cryobiol. Res: Physiology of temperature regulation; hypothermia and hibernation in mammals. Mailing Add: Dept of Zool Univ of Alta Edmonton AB Can

WANG, LI CHUAN, b Kaiyuan, China; US citizen; m; c 2. BIOCHEMISTRY, SOIL FERTILITY. Educ: Nat Cent Univ, Nanking, China, BS, 45; Univ Wis-Madison, PhD(soils), 52, MS, 58. Prof Exp: Teaching asst, Nat Cent Univ, Nanking, 45-48; fel biochem, Univ Wis-Madison, 58-60; res chemist, Monsanto Chem Co, 60-62; assoc prof chem, Univ Alaska, 62-63; PRIN CHEMIST, NORTHERN REGIONAL RES LAB, USDA, 63- Mem: Am Chem Soc; Am Soc Plant Physiologists; Inst Food Technologists. Res: Soybean proteins, their utilization, flavor and functionalities. Mailing Add: 7013 N Teton Dr Peoria IL 61614

WANG, LING-LIE, b Hunan, China, Jan 18, 39; US citizen; m 64; c 1. HIGH ENERGY PHYSICS. Educ: Nat Taiwan Univ, BS, 61; Univ Calif, Berkeley, PhD(physics), 66. Prof Exp: Fel physics, Lawrence Radiation Lab, Berkeley, 67; mem, Inst Advan Study, 67-69; vis fel, Europ Ctr High Energy Physics, 69; from asst physicist to assoc physicist, 69-74, PHYSICIST, BROOKHAVEN NAT LAB, 74- Mailing Add: Dept of Physics Brookhaven Nat Lab Upton NY 11973

WANG, MAW SHIU, b Chang Hwa, Formosa, Nov 1, 25; m 51; c 3. AGRONOMY, SPECTROCHEMISTRY. Educ: Prov Agr Col, Formosa, BS, 51; Okla State Univ, MS, 56; Univ Ill, PhD, 59. Prof Exp: Jr engr, Chem Lab, Pingtong Sugar Exp Sta, Formosa, 51-54; asst soil chem, Univ Ill, Urbana, 55-59, res assoc, 59-61; sr res chemist, Cent Res Dept, 61-68, RES SPECIALIST, MONSANTO CO, 68- Concurrent Pos: Adj prof, St Louis Univ, 70- Honors & Awards: Megger's Award, Soc Appl Spectros, 73. Mem: Am Soc Testing & Mat; Soc Appl Spectros. Res: Emission spectroscopy; spectrochemical analysis of major and minor elements in agricultural material; traces in semiconductors and related materials. Mailing Add: Monsanto Co 800 N Lindbergh St Louis MO 63166

WANG, NAI-SAN, b Changhua, Taiwan, Jan 20, 36; Can citizen; m 63; c 4. PATHOLOGY. Educ: Nat Taiwan Univ, MD, 60; McGill Univ, MS, 69, PhD(path), 71. Prof Exp: Teaching & res fel, 67-71, asst prof, 71-74, ASSOC PROF PATH, McGILL UNIV, 74- Concurrent Pos: Consult, Mesothelum Ref Panel Can, Tumor Ref Ctr, 74- Mem: Am Asn Pathologists & Bacteriologists; Am Thoracic Soc; Can Thoracic Soc. Res: Ultrastructural studies of the lung and pleura in normal and diseased conditions. Mailing Add: Dept of Path McGill Univ 3775 University Ave Montreal PQ Can

WANG, NANCY YANG, b Peiping, China, Jan 20, 26; m 49; c 1. ORGANIC CHEMISTRY. Educ: Cath Univ, Peiping, BS, 45; St Louis Univ, MS, 51; Boston Univ, PhD(org chem), 65. Prof Exp: Fel biol, Mass Inst Technol, 64-65, fel nutrit, 65-66, res assoc chem, 66-68; res assoc, Retina Found, Boston Univ, 68-71; res assoc, 71-73, lectr, 73-75, RES ASSOC CHEM, UNIV MASS, BOSTON, 75- Mem: Sigma Xi. Res: Biochemistry. Mailing Add: Dept of Chem Univ of Mass Boston MA 02125

WANG, PETER CHENG-CHAO, b Shantung, China, Jan 11, 37; US citizen; m 60; c 2. STATISTICS. Educ: Pac Lutheran Univ, BA, 60; Wayne State Univ, MA, 62, PhD(math), 66. Prof Exp: Instr math, Wayne State Univ, 64-66; asst prof statist, Mich State Univ, 66-67; from asst prof to assoc prof, Univ Iowa, 67-70; ASSOC PROF STATIST, NAVAL POSTGRAD SCH, 70- Concurrent Pos: Vis assoc prof statist, Stanford Univ, 69-70; consult, B D M Serv Co, Tex, 71- Mem: Am Math Soc; Inst Math Statist; London Math Soc. Res: Stochastic models; combinatorics; forecast models for military systems; sound propagation models. Mailing Add: Code 53 Naval Postgrad Sch Monterey CA 93940

WANG, RICHARD I H, b Shanghai, China, Oct 12, 24; US citizen; m 58; c 2. PHARMACOLOGY, INTERNAL MEDICINE. Educ: St John's Univ, BS, 45; Utah State Univ, MS, 49; Univ Ill, PhD(pharmacol), 52; Northwestern Univ, MD, 55. Prof Exp: Intern, Presby Hosp, Chicago, Ill, 55-56; resident, Indianapolis Gen Hosp, Ind, 56-58; prin scientist, Roswell Park Mem Inst, 61-63; assoc prof pharmacol & med, 63-70, PROF CLIN PHARMACOL, MED COL WIS, 70-; CHIEF CLIN PHARMACOL SERV, WOOD VET ADMIN HOSP, 63-, DIR DRUG ABUSE TREATMENT & REHAB PROG, 71- Concurrent Pos: Consult physician, Milwaukee County Gen Hosp, 64-; attend physician, Milwaukee County Ment Health Ctr, 65- Mem: Am Soc Pharmacol & Exp Therapeut; Radiation Res Soc; Am Soc Clin Pharmacol & Therapeut. Res: Clinical pharmacology; radiation biology. Mailing Add: Vet Admin Ctr W National Ave at 54th St Wood WI 53193

WANG, RICHARD J, b Chungking, China, Oct 23, 41; US citizen; m 66; c 2. CELL BIOLOGY, GENETICS. Educ: Harvard Univ, BA, 64; Univ Colo, PhD(biophys), 68. Prof Exp: Fel cell biol, NY Univ, 68-70; res assoc biol, Mass Inst Technol, 70-71; asst prof, 71-76, ASSOC PROF BIOL, UNIV MO, COLUMBIA, 76- Concurrent Pos: NIH res career develop award, 72-77; biochem consult, Cancer Res Ctr, Columbia, Mo, 73-; investr, Dalton Res Ctr, Univ Mo, Columbia, 74- Mem: Am Soc Cell Biol; Tissue Cult Soc; AAAS. Res: Biochemical genetics of human and mammalian cells in culture; effect of near-ultraviolet and visible light on human cells. Mailing Add: Dalton Res Ctr Univ of Mo Columbia MO 65201

WANG, SAMUEL SHAN-NING, organic chemistry, polymer chemistry, see 12th edition

WANG, SAMUEL S M, b Chekiang, China, Sept 21, 17; US citizen; m 47; c 1. MEDICINAL CHEMISTRY. Educ: Nat Col Pharm, China, BSc, 40; Univ London, BPharm, 47; Wash State Univ, MSc, 49; Univ Wis, PhD(med chem), 52. Prof Exp: Sr chemist, Maltible Labs, Wallace & Tiernan Inc, 51-54; res chemist, Chas Pfizer & Co, 54-55; from assoc prof to prof pharmaceut chem, Mercer Univ, 55-63, head dept, 57-63; researcher med chem, Univ Va, 63-64; sr res chemist, 64-72, RES SPECIALIST, HUMAN HEALTH RES CTR, DOW CHEM CO, USA, 72- Concurrent Pos: Consult, Nat Res & Develop, Ind, 56-60 & L H Studebaker Labs, Inc, 57-63. Mem: Am Chem Soc; Am Pharmaceut Asn. Res: Synthetic medicinals; central nervous system; anti-inflammatory agents; antacids. Mailing Add: Human Health Res Ctr Dow Chem Co Zionsville IN 46077

WANG, SAN-PIN, b Taiwan, Nov 7, 20; m 46; c 5. MEDICAL MICROBIOLOGY. Educ: Keio Univ, Japan, MD, 44; Univ Mich, MPH, 52. Hon Degrees: Dr Med Sci, Keio Univ, Japan, 59. Prof Exp: Asst bact, Sch Med, Keio Univ, Kapan, 44-46; chief dept bact, Taiwan Prov Hyg Lab, 46-51; chief dept virol, Taiwan Serum Vaccine Lab, 52-58; med officer virus immunol, US Naval Med Res Unit, 58-64; vis assoc prof prev med, Sch Med, 64-66, assoc prof, 66-70, PROF PATHOBIOL, SCH PUB HEALTH & COMMUNITY MED, UNIV WASH, 70- Res: Biological products, rabies and smallpox vaccines; research on tropical diseases, rabies, influenza, encephalitis and trachoma. Mailing Add: Dept of Pathobiol SC-38 Univ of Wash Seattle WA 98195

WANG, SHAO-FU, b Dairen, China, Aug 26, 22; m 44; c 3. THEORETICAL PHYSICS, SOLID STATE PHYSICS. Educ: Nagoya Univ, PhD(physics), 64. Prof Exp: Lectr physics, Ching Kung Univ, Taiwan, 48-52, assoc prof, 52-56; from assoc prof to prof, Tunghai Univ, Taiwan, 56-65; assoc prof, 65-68, PROF PHYSICS, UNIV WATERLOO, 68- Concurrent Pos: Res assoc, Univ Ill, 61-63, vis assoc prof, 67. Mem: Am Phys Soc; Can Asn Physicists. Res: Theoretical study of electronic processes in crystals. Mailing Add: Dept of Physics Univ of Waterloo Waterloo ON Can

WANG, SHIH CHUN, b Tientsin, China, Jan 25, 10; nat US; m 39; c 2. PHYSIOLOGY, PHARMACOLOGY. Educ: Yenching Univ, China, BSc, 31; Peiping Union Med Col, MD, 35; Northwestern Univ, PhD(neurol), 40. Prof Exp: Asst physiol, Peiping Union Med Col, China, 35-37; instr neurol, Northwestern Univ, 38-40; from instr to prof neurol, 41-56, PROF PHARMACOL, COL PHYSICIANS & SURGEONS, COLUMBIA UNIV, 56-, PFEIFFER PROF, 76- Concurrent Pos: Guggenheim Found fel, 51-52; Commonwealth Fund travel fel, 66; China Med Bd vis prof, Nat Taiwan Univ, 58-59. Mem: AAAS; Am Soc Pharmacol & Exp Therapeut; Am Physiol Soc; Soc Exp Biol & Med; Harvey Soc; Asn Res Nerv & Ment Dis. Res: Physiology of automonic nervous system; afferent factor in traumatic shock; carotid sinus reflex; gastroenterology; respiratory physiology; central nervous system control of vomiting; motion sickness. Mailing Add: Col of Physicians & Surgeons Columbia Univ New York NY 10032

WANG, SHIH YI, b Peiping, China, June 15, 23; nat US; m 47; c 2. ORGANIC CHEMISTRY. Educ: Nat Peking Univ, China, BS, 44; Univ Wash, PhD(org chem), 52. Prof Exp: Fel org chem, Boston Univ, 52-54; res fel, Harvard Univ, 54-55; res assoc, Sch Med, Tufts Univ, 55-56, asst prof physiol, 56-61; from asst prof to assoc prof biochem, 61-66, PROF BIOCHEM, JOHNS HOPKINS UNIV, 66- Mem: AAAS; Am Chem Soc; The Chem Soc. Res: Photochemistry of nucleic acids and related compounds; chemistry of heterocyclic nitrogen containing compounds; synthesis and structure of natural products; polynuclear aromatic compounds. Mailing Add: Sch Hyg & Pub Health Dept Biochem Johns Hopkins Univ Baltimore MD 21205

WANG, SHU LUNG, b Szechwan, China, May 2, 25; m 47; c 2. DATA PROCESSING. Educ: Wash Univ, St Louis, BS, 49, MS, 50, DSc(chem eng), 53. Prof Exp: From asst prof to assoc prof chem eng, Kans State Univ, 52-57; sect head eng systs develop & dir comput lab, Linde Div, 57-66, mgr, Niagara Frontier Regional Comput Ctr, 66-69, MGR COMPUT APPLNS, UNION CARBIDE CORP, 69- Mem: Am Inst Chem Eng; Am Chem Soc; Am Soc Eng Educ; Asn Comput Mach. Res: Information system development and data processing management; process control systems engineering; adsorption and cryogenic gas separation processes; vapor-liquid equilibrium data. Mailing Add: Tarrytown Tech Ctr Union Carbide Corp Tarrytown NY 10591

WANG, SU-SUN, b Taipei, China, Jan 11, 34; m 64; c 2. BIOCHEMISTRY. Educ: Nat Taiwan Univ, BS, 57, MS, 59; Univ Calif, Berkeley, PhD(biochem), 66. Prof Exp: Res biochemist, Univ Calif, Berkeley, 66-68; USPHS fel peptide chem, Rockefeller Univ, 68-70; SR CHEMIST, HOFFMANN-LA ROCHE, INC, 70- Mem: AAAS; Am Chem Soc. Res: Protein and enzyme chemistry; peptide synthesis, classical and solid phase. Mailing Add: Chem Res Dept Hoffmann-La Roche Inc Nutley NJ 07110

WANG, TAITZER, b Taiwan, Feb 2, 39; m 68; c 1. BIOCHEMISTRY. Educ: Nat Univ Taiwan, BS, 61; Rice Univ, PhD(org chem), 67. Prof Exp: RES ASST PROF CELL BIOPHYS, BAYLOR COL MED, 75- Concurrent Pos: Nat Inst Arthritis & Metab Dis res fel org chem, Fla State Univ, 67-69; US Dept Defense res fel inorg chem, Univ Ky, 69-71; Welch fel biochem, Baylor Col Med, 71-73; Welch fel, Rice Univ, 73-74; NIH spec res fel cell biophys, Baylor Col Med, 74-75. Mem: Am Chem Soc. Res: Enzyme kinetics and synthetic chemistry; cell biophysics. Mailing Add: Dept of Cell Biophys Baylor Col of Med Houston TX 77025

WANG, THEODORE JOSEPH, b Chicago, Ill, Dec 8, 06; m 36; c 2. PHYSICS. Educ: Univ Ill, BS, 32, PhD(physics), 39. Prof Exp: Asst physics, Univ Ill, 35-39; physicist, Oakes Prod Corp, Ill, 39-40; fel, Univ Minn, 41; from instr to asst prof elec eng, Ohio State Univ, 42-48; physicist, Nat Bur Standards, 48-49; biophysicist, Nat Cancer Inst, 49-50; physicist, George Washington Univ, 50-52; asst prof physics, Univ Mass, 52-55; prof & head dept, SDak Sch Mines & Technol, 55-56; prof, Howard Univ, 56-59;

analyst, Opers Res Off, Res Anal Corp, 59-62; prin scientist, Booz, Allen Appl Res, 62-66; DIR, INST CREATIVE STUDIES, 67- Concurrent Pos: Consult physicist, 50-52; analyst, Opers Res Off, Johns Hopkins Univ, 57; analyst, Res Anal Corp, 57-58; prof, Howard Univ, 60-64, lectr, 64- Res: Radiation physics. Mailing Add: Inst Creative Studies 4700 Essex Ave Chevy Chase MD 20015

WANG, VICTOR KAI-KUO, b Quei-Chow, China, Mar 18, 44; m 69; c 1. PHYSICAL CHEMISTRY, INDUSTRIAL CHEMISTRY. Educ: Chung Yuan Col Sci & Eng, Taiwan, BS, 65; State Univ NY Binghamton, MS, 68; Univ Minn, PhD(phys chem), 73. Prof Exp: Res asst phys chem, State Univ NY Binghamton, 66-68; teaching assoc, Univ Minn, 68-73; RES CHEMIST, E I DU PONT DE NEMOURS & CO, INC, 73- Mem: Am Chem Soc. Res: Process research, kinetics and catalysis. Mailing Add: Indust Chem Dept E I du Pont de Nemours & Co Inc Wilmington DE 19808

WANG, VIRGINIA LI, b Canton, China, Apr 2, 33; US citizen; c 3. PUBLIC HEALTH. Educ: NY Univ, MA, 56; Univ NC, MPH, 65, PhD, 68. Prof Exp: Res dietitian, Montefiore Hosp, 56; instr nutrit, Univ NC, 62-64; health educ specialist, Univ Md, 68-74; ASSOC PROF PUB HEALTH EDUC, JOHNS HOPKINS UNIV, 74- Concurrent Pos: Temp consult, Pan Am Health Orgn, 74; consult, WHO, UN, 74 & 76, Nat Cancer Inst, NIH, Dept Health, Educ & Welfare, 74-77. Mem: Am Dietetic Asn; Am Pub Health Asn; Soc Pub Health Educ; Soc Appl Anthrop. Res: Evaluation of nutrition and health education programs in developed and developing societies; health care delivery and health education in the Peoples Republic of China; planning, development and evaluation of continuing education for the health professions. Mailing Add: Sch of Hyg & Pub HeaLth Johns Hopkins Univ Baltimore MD 21205

WANG, WUN-CHENG, b Taichung, China, Mar 10, 36; m 62; c 3. WATER POLLUTION. Educ: Nat Taiwan Univ, BS, 58, MS, 61; Univ Wis-Madison, PhD(water chem), 68. Prof Exp: Asst prof scientist, 67-70, ASSOC PROF SCIENTIST, WATER QUAL SECT, ILL STATE WATER SURV, 70- Mem: Am Soc Limnol & Oceanog; Am Chem Soc; Soc Int Limnologists; Am Soc Agron. Res: Water qualtiy; eutrophication; algal growth; suspended solid, natural particulate matter. Mailing Add: Ill State Water Surv PO Box 717 Peoria IL 61601

WANG, YANG, b Tangshan, China, May 12, 23; US citizen; m 66; c 4. CARDIOLOGY. Educ: Nat Med Col Shanghai, MB; Harvard Univ, MD, 52; Am Bd Internal Med, dipl, Am Bd Cardiovasc Dis, dipl. Prof Exp: Intern & resident med, Mass Gen Hosp, 52-54 & 56-57; from instr to assoc prof, 59-64, PROF MED, MED SCH, UNIV MINN, MINNEAPOLIS, 64-, DIR CARDIAC CATHETERIZATION LABS & DIR CARDIAC CLINS, UNIV HOSPS, 60 Concurrent Pos: P D White fel cardiol, Mass Gen Hosp, Boston, 57-58; fel physiol, Mayo Grad Sch Med, Univ Minn, 58-59; consult, Vet Admin Hosp, Minneapolis, 67; fel coun clin cardiol & circulation, Am Heart Asn; attend physician, Univ Minn Hosps, Minneapolis, 59- Mem: Fel AAAS; Am Fedn Clin Res; fel Am Col Physicians; Soc Exp Biol & Med; Asn Univ Cardiol. Res: Cardiovascular and exercise physiology; cardiac catherterization in humans. Mailing Add: Dept of Med Univ of Minn Hosps Minneapolis MN 55455

WANG, YA-YEN LEE, b Peking, China, Mar 1, 30; US citizen; m 56; c 3. MATHEMATICS, COMPUTER SCIENCE. Educ: Villa Maria Col, BS, 56; Univ Fla, MS, 58; Univ Idaho, PhD(math), 65. Prof Exp: Instr math, 60-62, asst prof, 65-72, acting dir comput ctr, 67, ASSOC PROF MATH, UNIV IDAHO, 72- Mem: Am Math Soc; Math Asn Am; Asn Comput Mach; Sigma Xi; Am Asn Univ Prof. Res: Differential geometry and equations; calculus of variations; numerical analysis; computer languages. Mailing Add: Dept of Math Univ of Idaho Moscow ID 83843

WANG, YEN, b Dairen, China, Oct 21, 28; m 62. RADIOLOGY, NUCLEAR MEDICINE. Educ: Nat Taiwan Univ, MD, 53; Univ Pa, DSc(med), 62. Prof Exp: Asst prof, 63-65, ASSOC PROF RADIOL, UNIV PITTSBURGH, 65-; DIR RADIOL, HOMESTEAD HOSP, 66-; DIR NUCLEAR MED, MAGEE-WOMENS HOSP, PITTSBURGH, 67- Concurrent Pos: Res fel, Picker Found Acad Sci, 61-63; vis scientist, Protein Found, 62-64; ed, Critical Rev Clin Radiol & Nuclear Med. Mem: AMA; Soc Nuclear Med; Am Roentgen Ray Soc; Am Physiol Soc; Am Radium Soc; Radiol Soc NAm. Res: Physiology; protein chemistry. Mailing Add: South Hill Health Syst Homestead PA 15120

WANG, YEU-MING ALEXANDER, b Chungking, China, May 2, 42; m 73. BIOCHEMISTRY, HEMATOLOGY. Educ: Taiwan Chung-Hsing Univ, BS, 65; Vanderbilt Univ, PhD(biochem), 71. Prof Exp: Res asst biochem, Med Sch, Vanderbilt Univ, 66-71, res assoc, 71-73, instr pediat & hemat, 73; ASST PROF PEDIAT & ASST BIOCHEMIST, M D ANDERSON HOSP & TUMOR INST, UNIV TEX SYST CANCER CTR, 73- Mem: AAAS. Res: Biochemistry of erythrocytes metabolism and function; biochemistry and pharmacology of cancer drugs. Mailing Add: Dept of Pediat M D Anderson Hosp & Tumor Inst Houston TX 77025

WANG, YING YAO, statistics, see 12th edition

WANG, YUNG-LI, b Canton, China, Jan 8, 37; m 63; c 1. SOLID STATE PHYSICS. Educ: Nat Taiwan Univ, BS, 59; Nat Tsing Hua Univ, Taiwan, MS, 61; Univ Pa, PhD(physics), 66. Prof Exp: Res assoc physics, Univ Pa, 66-67 & Univ Pittsburgh, 67-68; asst prof, 68-72, ASSOC PROF PHYSICS, FLA STATE UNIV, 72- Mem: Am Phys Soc. Res: Many body theory of spin systems, magnetic phase transitions; crystal-field effects and impurities in magnetic systems. Mailing Add: Dept of Physics Fla State Univ Tallahassee FL 32306

WANGAARD, FREDERICK FIELD, b Minneapolis, Minn, Jan 3, 11; m 36; c 3. FOREST PRODUCTS. Educ: Univ Minn, BS, 33; State Univ NY, MS, 35, PhD(wood technol), 39. Hon Degrees: MA, Yale Univ, 52. Prof Exp: Instr forestry, Univ Wash, 36-39, asst prof wood technol, 39-42; technologist, Forest Prod Lab, US Forest Serv, 42-45; from asst prof to prof forest prod, Yale Univ, 45-67; HEAD DEPT FOREST & WOOD SCI, COLO STATE UNIV, 68- Concurrent Pos: Adv, Food & Agr Orgn, Philippines, 57; Fulbright res scholar, Norway, 58. Mem: Forest Prod Res Soc (pres, 75); Soc Am Foresters; Tech Asn Pulp & Paper Indust; Int Acad Wood Sci. Res: Thermal conductivity of wood; properties of wood in relation to growth; peoperties of tropical woods; plywood and laminated wood; technology of wood fibers. Mailing Add: 1609 Hillside Dr Ft Collins CO 80521

WANGEMANN, ROBERT THEODORE, b Rhinelander, Wis, Apr 27, 33; m 54; c 2. BIOPHYSICS. Educ: Univ Wis-Madison, BS, 55; Univ Rochester, MS, 64; Med Col Va, PhD(biophys), 74. Prof Exp: Pharmacist, Northgate Drugs & Southside Drugs, 55-57; med supply officer, 57-62, health physicist, 64-67, instr nuclear sci, 67-70, CHIEF LASER-MICROWAVE DIV, ENVIRON HYG AGENCY, US ARMY, 73- Concurrent Pos: Mem C95 comt, Am Nat Standards Inst, 73- Mem: AAAS; Am Soc Photobiol; Health Physics Soc; Am Conf Govt Indust Hygienists; Am Asn Physicists in Med. Res: Biological effects of electromagnetic radiation; interactions of microwave and radio frequency energy at the biomole- cular level and the photochemical aspects of vision and optical radiation effects. Mailing Add: Laser-Microwave Div Environ Hyg Agency US Army Aberdeen Proving Ground MD 21010

WANGENSTEEN, OVE DOUGLAS, b St Paul, Minn, Mar 15, 42; m 65; c 2. PHYSIOLOGY. Educ: Univ Minn, Minneapolis, BS, 64, PhD(physiol), 68. Prof Exp: Res assoc physiol, Sch Med & Dent, State Univ NY, Buffalo, 68-70; ASST PROF PHYSIOL, MED SCH, UNIV MINN, MINNEAPOLIS, 70- Mem: Am Physiol Soc. Res: Respiratory and cardiovascular physiology. Mailing Add: Dept of Physiol Univ of Minn Med Sch Minneapolis MN 55455

WANGENSTEEN, OWEN HARDING, b Lake Park, Minn, Sept 21, 98; m 23, 54; c 3. SURGERY. Educ: Univ Minn, AB, 19, MB, 21, MD, 22, PhD(surg), 25; FRCS, 61. Hon Degrees: LLD, Univ Buffalo, 46; DSc, Univ Chicago, 56, St Olaf Col, 58, Temple Univ, 61 & Hamline Univ, 63; hon Dr, Univ Paris, 62. Prof Exp: Intern, Univ Hosp, Univ Minn, 22, resident surgeon, 25, instr, Univ, 26, asst prof, 27; asst, Prof F de Quervains' Surg Clin, Berne, Switz, 27-28; from assoc prof to regents' prof, 28-67, distinguished serv prof, 60, EMER REGENTS' PROF SURG, MED CTR, UNIV MINN, MINNEAPOLIS, 67-, DIR DEPT SURG, UNIV HOSP, 30- Concurrent Pos: Asst, Leon Asher Physiol Inst, Switz, 27-28; mem, Surg Study Sect, NIH, 49-53; mem, Heart Coun, 54-58 & Res Fac Coun, 59-63; consult, US Army, 59; USPHS grants. Honors & Awards: Gross Award, 35; Scott Award, 41; Alvarenza Prize, Philadelphia Col Med, 49; Am Cancer Soc Award, 49, Spec Citation, 62; Passano Award, 61. Mem: Nat Acad Sci; Am Soc Exp Path; Am Physiol Soc; Am Asn Thoracic Surgeons; Am Surg Asn. Res: Intestinal obstruction; etiology of appendicitis; peptic ulcer problem; precursors of visceral cancer; cancer detection. Mailing Add: 511 Diehl Hall Univ Minn Med Ctr PO Box 610 Minneapolis MN 55455

WANGENSTEEN STEPHEN LIGHTNER, b Minneapolis, Minn, Aug 30, 33; m 56; c 4. SURGERY. Educ: Univ Minn, BA, 54, BS, 55; Harvard Univ, MD, 58. Prof Exp: Instr surg, Columbia-Presby Med Ctr, 64-65; from asst prof to assoc prof, 67-70, PROF SURG, UNIV VA, 70- Concurrent Pos: Vascular fel, Columbia-Presby Med Ctr, 58-59, USPHS fel, 60-63. Mem: Asn Acad Surg (vpres, 67-68); Soc Univ Surg; Am Surg Asn. Res: Gastrointestinal pathophysiology; circulatory shock; vascular surgery. Mailing Add: Dept of Surg Univ of Va Med Ctr Charlottesville VA 22901

WANGERSKY, PETER JOHN, b Woonsocket, RI, Aug 26, 27; m 59. OCEANOGRAPHY. Educ: Brown Univ, ScB, 49; Yale Univ, PhD(zool), 58. Prof Exp: Marine chem technician, Scripps Inst, Univ Calif, 49-50; chemist, Chem Corps, US Dept Army, 50-51; chem oceanogr, US Fish & Wildlife Serv, 51-54; res asst prof marine sci, Marine Lab, Univ Miami, 58-61; res assoc, Bingham Oceanog Lab, Yale Univ, 61-65; assoc prof chem, 65-68, PROF OCEANOG, DALHOUSIE UNIV, 68- Concurrent Pos: Guggenheim fel, John Simon Guggenheim Found, 71-72; ed in chief, Marine Chem, 74- Mem: AAAS; Am Soc Limnog & Oceanog; Am Chem Soc; Ecol Soc Am. Res: Mechanisms of marine sedimentation; chemical oceanography; organic metabolites in sea water; population dynamics. Mailing Add: Dept of Oceanog Dalhousie Univ Halifax NS Can

WANGLER, THOMAS P, b Bay City, Mich, Aug 2, 37. PHYSICS. Educ: Mich State Univ, BS, 58; Univ Wis, PhD(physics), 64. Prof Exp: Res assoc phyics, Univ Wis, 64-65 & Brookhaven Nat Lab, 65-66; ASST PHYSICIST, ARGONNE NAT LAB, 66- Mem: Am Phys Soc. Res: Experimental high energy physics; cosmic rays; environmental science; nuclear physics. Mailing Add: Physics Div Bldg 203 Argonne Nat Lab Argonne IL 60439

WANGSNESS, PAUL JEROME, b Madison, Wis, Mar 27, 44; m 67; c 2. ANIMAL NUTRITION. Educ: Univ Wis-Madison, BS, 66; Iowa State Univ, PhD(nutrit & physiol), 71. Prof Exp: NDEA fel nutrit, Iowa State Univ, 66-69, NSF fel, 69-71; asst prof, 72-75, ASSOC PROF NUTRIT, PA STATE UNIV, 75- Mem: Am Dairy Sci Asn; Am Soc Animal Sci. Res: Regulatory mechanisms involved in the control of food intake and the regulation of energy balance in lean and obese animals. Mailing Add: 303 Animal Indust Bldg Pa State Univ University Park PA 16802

WANGSNESS, ROALD KLINKENBERG, b Sleepy Eye, Minn, July 24, 22; m 44; c 2. PHYSICS. Educ: Univ Minn, BA, 44; Stanford Univ, PhD(physics), 50. Prof Exp: Asst physics, Univ Minn, 42-44; jr scientist, Los Alamos Sci Lab, 44-45; asst physics, Univ Minn, 45-46; asst physics, Stanford Univ, 46-48; asst prof, Univ Md, 50-51, prof, 53-59; PROF PHYSICS, UNIV ARIZ, 59- Concurrent Pos: Physicist, Naval Ord Lab, Md, 51-59. Mem: Fel AAAS; fel Am Phys Soc; Am Asn Physics Teachers. Res: Nuclear induction; nuclear moments; ferrimagnetic resonance; anti-ferromagnetism; atomic spectra. Mailing Add: Dept of Physics Univ of Ariz Tucson AZ 85721

WANI, JAGANNATH K, b Maharashtra, India, Sept 10, 34; m 59; c 3. STATISTICS. Educ: Univ Poona, BSc, 58, Hons, 59, MSc, 60; McGill Univ, PhD(math statist), 67. Prof Exp: Lectr math, Col Agr, Dhulia, India, 60-61; res asst statist, Gokhale Inst Econ, Poona, India, 61-62; res asst math statist, McGill Univ, 62-65; asst prof math, Univ Lethbridge, 65-66; from asst prof to assoc prof, St Mary's Univ, NS, 66-69; ASSOC PROF STATIST, UNIV CALGARY, 69- Concurrent Pos: Nat Res Coun Can fel, St Mary's Univ & Univ Calgary, 67-72; Can Math Cong fel, McGill Univ, Queen's Univ & Univ Alta, 69-71. Mem: Inst Math Statist; Can Math Cong; Am Statist Asn; Statist Sci Asn Can. Res: Distribution theory and statistical inference. Mailing Add: Dept of Statist Univ of Calgary Calgary AB Can

WANIEK, RALPH WALTER, b Milan, Italy, June 1, 25; nat US; m 53; c 1. PHYSICS. Educ: Univ Vienna, PhD(physics), 50. Prof Exp: Res assoc nuclear physics, Inst Radium Res, Univ Vienna, 48-50; asst prof physics & math, Newton Col, 50-56; sr physicist, Cambridge Electron Accelerator, Harvard Univ & Mass Inst Technol, 55-58; dir res, Plasmadyne Corp, 58-60; PRES & DIR RES, ADVAN KINETICS, INC, 60- Concurrent Pos: Res fel, Synchrocyclotron Lab, Harvard Univ, 50-55; consult, Transistor Prod, Inc, 52-55 & Allied Res Assocs, 56-57; lectr, Boston Col, 56-58 & Exten, Univ Calif, Los Angeles, 59- Mem: Am Phys Soc; Am Inst Aeronaut & Astronaut; Inst Elec & Electronics Eng. Res: Plasma, laser, space and nuclear physics; solid state; production of very intense magnetic fields; problems of space propulsion. Mailing Add: Advan Kinetics Inc 1231 Victoria St Costa Mesa CA 92627

WANKEL, RONALD ALDEN, organic chemistry, see 12th edition

WANLASS, SYLVAN DEAN, physics, mathematics, see 12th edition

WANLESS, HAROLD ROGERS, b Champaign, Ill, Feb 14, 42; m 65; c 2. SEDIMENTOLOGY, MARINE GEOLOGY. Educ: Princeton Univ, AB, 64; Univ Miami, MS, 68; Johns Hopkins Univ, PhD(geol), 73. Prof Exp: Res scientist, 71-73, res asst prof, 73-75, ASST PROF MARINE GEOL, SCH MARINE & ATMOSPHERIC SCI, UNIV MIAMI, 75- Mem: Soc Econ Paleontologists & Mineralogists. Res: Environments and processes of modern coastal and shelf sediments; petrology and paleo-environmental reconstruction of ancient sedimentary rocks; fine-grained sediment dynamics; biotic influences on sediments; economic and environmental application. Mailing Add: Rosenstiel Sch of Marine & Atmos Sci Univ of Miami Miami FL 33149

WANLESS, ROBERT KENNETH, physics, see 12th edition

WANN, ELBERT VAN, b Grange, Ark, Dec 29, 30; m 50; c 1. GENETICS, PLANT BREEDING. Educ: Univ Ark, BS, 59, MS, 60; Purdue Univ, PhD(genetics), 62. Prof Exp: Res assoc veg crops, Univ Ill, 62-63; GENETICIST, USDA, 63- Honors & Awards: Asgrow Award, Am Soc Hort Sci, 72. Mem: Am Soc Hort Sci. Res: Genetics and breeding of sweet corn and tomatoes, as related to disease and insect resistance and the improvement of consumer quality. Mailing Add: US Veg Breeding Lab Box 3348 USDA Charleston SC 29407

WANNAMAKER, LEWIS WILLIAM, b Matthews, SC, May 19, 23; m 48; c 4. PEDIATRICS, MICROBIOLOGY. Educ: Duke Univ, MD, 46. Prof Exp: Pediat intern & resident, Duke Univ & Willard Parker Hosps, 46-48; res asst prev med, Western Reserve Univ, 48-50, res assoc, 50-52; from instr to assoc prof pediat, 52-58, PROF PEDIAT, MED SCH, UNIV MINN, MINNEAPOLIS, 58-, PROF MICROBIOL, 65-; CAREER INVESTR, AM HEART ASN, 58- Concurrent Pos: Mem, Streptococcal Dis Lab, Wyo, 48-50, asst dir, 50-52; vis investr, Rockefeller Inst, 55-57; guest investr, Cent Pub Health Lab & Nat Inst Med Res, Eng, 66-67 & Hyg Inst, Univ Kölm, 73-74; mem streptococcal & staphylococcal comn, Armed Forces Epidemiol Bd & dir, 67-72. Mem: Am Soc Clin Invest; Am Soc Microbiol; Soc Pediat Res; Am Asn Immunol; Am Fedn Clin Res. Res: Streptococcal and other infectious diseases; rheumatic fever; nephritis; biology of streptococci and staphylococci; bacterial nucleases and other extracellular enzymes. Mailing Add: Dept of Pediat Box 296 Univ of Minn Med Sch Minneapolis MN 55455

WANNEMACHER, ROBERT, JR, b Hackensack, NJ, Jan 12, 29; m 71. BIOCHEMISTRY, NUTRITION. Educ: Wagner Col, BS, 50; Rutgers Univ, MS, 51, PhD(biochem, nutrit), 60. Prof Exp: From res asst to res assoc nutrit & biochem, Bur Biol Res, Rutgers Univ, 51-60, from asst res prof to assoc res prof, 60-69; SR BIOCHEMIST, PHYS SCI DIV, US ARMY MED RES INST INFECTIOUS DIS, 69- Mem: AAAS; Am Inst Nutrit; Am Chem Soc; Biophys Soc; Soc Exp Biol & Med; Am Soc Biol Chem. Res: Protein and RNA metabolism; infectious diseases; regulatory mechanisms; endocrinology, cancer and radiation. Mailing Add: US Army Med Res Inst for Infectious Dis Ft Detrick Frederick MD 21701

WANNER, ROBERT LOUIS, animal nutrition, see 12th edition

WANNIER, GREGORY HUGH, b Basel, Switz, Dec 30, 11; nat US; m 39; c 2. THEORETICAL PHYSICS. Educ: Univ Basel, PhD(math physics), 35. Prof Exp: Asst, Univ Geneva, 35-36; Swiss-Am exchange fel, Princeton Univ, 36-37; instr, Univ Pittsburgh, 37-38; lectr, Bristol Univ, 38-39; instr, Univ Tex, 39-41; lectr physics, Univ Iowa, 41-46; res assoc, Theoret Physics Lab, Socony-Vacuum Oil Co, 46-49; mem tech staff, Bell Labs, 49-60; PROF PHYSICS, UNIV ORE, 60- Concurrent Pos: Prof, Univ Geneva, 55-56. Mem: Fel Am Phys Soc; Swiss Phys Soc. Res: Theoretical molecular and crystal structure; magnetism; statistical mechanics; gas discharges; electron and solid state physics. Mailing Add: Dept of Physics Univ of Ore Eugene OR 97403

WANTA, RAYMOND CASIMIR, b Milwaukee, Wis, Mar 3, 21; div; c 4. METEOROLOGY, ENVIRONEMTNAL SCIENCES. Educ: Univ Chicago, BS, 43; NY Univ, MS, 52. Prof Exp: Meteorol aide, US Weather Bur, Wis & NY, 39-42, res meteorologist, NY, Ala, Ohio & DC, 47-59; meteorologist, Booz-Allen Appl Res, Inc, 59-60; sr meteorologist, Allied Res Assocs, Inc, 60-63; ENVIRON CONSULT, 63- Concurrent Pos: Consult, Atomic Energy Comn, 47-53; guest meteorologist, Brookhaven Nat Lab, 47-54; consult, Tenn Valley Authority, 52-55, res meteorologist, 54-55; chief meteorologist, Community Air Pollution Prog, USPHS, 55-58 & Singco Inc, 65-68; assoc ed, Atmospheric Environ, 67-; sr meteorologist, Bolt Beranek & Newman Inc, 68-69; mem subcomt iron, Comt Med & Biol Effects Environ Pollutants, Nat Res Coun, 75- Mem: Affil mem Am Soc Testing & Mat; Am Indust Hyg Asn; Air Pollution Control Asn; Soc Natural Philos. Res: Atmospheric structure and turbulence; air pollution and atmospheric submodels; aeronautical meteorology; weather modification; climatology; micrometeorology; industrial hygiene; applied statistics. Mailing Add: PO Box 98 Bedford MA 01730

WANTLAND, EVELYN KENDRICK, b Suffolk, Va, June 22, 17; m 39, 64; c 1. MATHEMATICS. Educ: Univ Ill, BA, 48, MA, 49, PhD(math), 58. Prof Exp: Asst, Univ Ill, 48-49; prof math, Ferrum Jr Col, 49-51; asst prof, Ill Wesleyan Univ, 51-57; asst prof, Kans State Univ, 57-62; assoc prof, Univ Miss, 62-64; PROF MATH & HEAD DEPT, ILL WESLEYAN UNIV, 64- Mem: AAAS; Am Math Soc; Math Asn Am. Res: Complex variables. Mailing Add: 110 E Beecher St Bloomington IL 61701

WAPLES, DOUGLAS WENDLE, b Oklahoma City, Okla, July 29, 45; m 68. ORGANIC GEOCHEMISTRY. Educ: DePauw Univ, AB, 67; Stanford Univ, PhD(org chem), 71. Prof Exp: Alexander von Humboldt fel, Geochem Inst, Univ Göttingen, 71-72; org geochemist, Empresa Nacional del Petroleo, Chile, 72-73; RES CHEMIST, CHEVRON OIL FIELD RES CO, STANDARD OIL CALIF, LA HABRA, 73- Concurrent Pos: Fulbright travel grant, 72; Latin Am teaching fel & prof chem, Cath Univ Valparaiso, Chile, 72-73. Res: Application of chemical methods to petroleum exploration; development of theoretical models for the process of petroleum formation. Mailing Add: 1036 N Glenhaven Fullerton CA 92635

WAPNIR, RAUL A, biochemistry, analytical chemistry, see 12th edition

WAPPNER, REBECCA SUE, b Mansfield, Ohio, Feb 25, 44. PEDIATRICS, BIOCHEMICAL GENETICS. Educ: Ohio Univ, BS, 66; Ohio State Univ, MD, 70; Am Bd Pediat, dipl, 75. Prof Exp: Intern pediat, Children's Hosp, Columbus, Ohio, 70-71, resident, 71-72, asst chief resident, 72-73; fel pediat metab & genetics, 73-75, ASST PROF PEDIAT, SCH MED, IND UNIV, INDIANAPOLIS, 75- Mem: Am Acad Pediat; Am Soc Human Genetics; Am Med Women's Asn. Res: Inborn errors of metabolism; lysosomal storage disorders. Mailing Add: Dept of Pediat Ind Univ Sch Med Indianapolis IN 46202

WARAVDEKAR, VAMAN SHIVRAM, b Varavda, India, May 11, 14; nat US; m 60; c 2. BIOCHEMISTRY. Educ: Univ Bombay, MSc, 40, PhD(chem), 42. Prof Exp: Res assoc, V J Tech Inst, India, 42-45; res officer carcinogenesis, Tata Mem Hosp, Bombay, 45-48; prin investr, Georgetown Univ, 50-52; vis scientist, Nat Cancer Inst, 52-57; prof biochem, All-India Inst Med Sci, 57-58; chief biochem br, Armed Forces Inst Path, 58-65; dir cancer chemother dept, Microbial Assocs, Inc, DC, 65-72; res chemist, Off of Assoc Dir Drug Res & Develop, Div Cancer Treatment, 72-73, RES PLANNING OFFICER, NAT CANCER INST, 73- Concurrent Pos: Res fel chemother of cancer, Nat Cancer Inst, 48-50; lectr grad sch, Georgetown Univ, 50-55. Mem: AAAS; Soc Exp Biol & Med; Am Soc Pharmacol & Exp Therapeut; Am Chem Soc; Am Asn Cancer Res. Res: Cellular chemistry; metabolism; enzymology; fatty acid oxidation; protein synthesis; chemotherapy; cancer research. Mailing Add: Res Planning Off Nat Cancer Inst Bethesda MD 20014

WARBURTON, DAVID LEWIS, b Hackensack, NJ, Aug 10, 47; m 75. GEOCHEMISTRY. Educ: Univ Calif, San Diego, BA, 69; Univ Chicago, PhD(geochem), 76. Prof Exp: ASST PROF GEOL, FLA ATLANTIC UNIV, 75- Mem: Am Geophys Union; AAAS; Mineral Soc Am. Res: Synthetic samples of olivine prepared for examination by Mossbauer technique to determine iron ion distribution at the two nonequivalent distorted octahedral sites. Mailing Add: Dept of Geol Fla Atlantic Univ Boca Raton FL 33431

WARBURTON, DOROTHY, b Toronto, Ont, Jan 12, 36; m 57; c 4. HUMAN GENETICS. Educ: McGill Univ, BSc, 57, PhD(genetics), 61. Prof Exp: From res asst to res assoc human genetics, Montreal Children's Hosp & McGill Univ, 58-63; res assoc obstet & gynec, Col Physicians & Surgeons, Columbia Univ, 64-67; dir genetics serv, St Luke's Hosp Ctr, 67-68; instr obstet & gynec, 68-69, assoc human genetics & develop, 69-71, asst prof human genetics & develop, 71-75, asst prof pediat, 74-75, ASSOC PROF CLIN PEDIAT, COL PHYSICIANS & SURGEONS, COLUMBIA UNIV, 75-; DIR GENETICS DIAG LAB, PRESBY HOSP, 69- Mem: Am Soc Human Genetics. Res: Cytogenetics; congenital malformations; genetics of obstetrical variables; dermatoglyphics. Mailing Add: Col of Physicians & Surgeons Columbia Univ New York NY 10032

WARBURTON, ERNEST KEELING, b Worcester, Mass, Apr 26, 28; m 47; c 3. NUCLEAR PHYSICS. Educ: Miami Univ, BA, 49; Mass Inst Technol, SB, 51; Univ Pittsburgh, PhD(physics), 57. Prof Exp: From instr to asst prof physics, Princeton Univ, 58-61; assoc physicist, 61-63, physicist, 63-68, SR PHYSICIST, BROOKHAVEN NAT LAB, 68- Concurrent Pos: NSF fel, Oxford Univ, 63-64 & 68-69. Mem: Fel Am Phys Soc. Res: Low energy nuclear physics; nuclear reactions initiated by electrostatic generators or cyclotrons; theoretical reaction mechanisms and properties of atomic nuclei. Mailing Add: Physics Dept Brookhaven Nat Lab Upton NY 11973

WARBURTON, FREDERICK E, b Kingston, Ont, July 19, 28; m 57; c 3. EVOLUTION, POPULATION GENETICS. Educ: McGill Univ, BSc, 58, MSc, 61, PhD(genetics), 63. Prof Exp: Asst prof, 63-69, ASSOC PROF ZOOL, BARNARD COL, COLUMBIA UNIV, 69- Mem: AAAS; Ecol Soc Am; Lepidop Soc. Res: Sponges, embryology and larvae; population genetics and mathematical models of evolutionary processes; evolutionary ecology. Mailing Add: Dept of Biol Barnard Col Columbia Univ New York NY 10027

WARD, ALFORD L, b Rockville, Md, Aug 14, 19. PHYSICS. Educ: Univ Md, BS, 49, PhD, 54. Prof Exp: PHYSICIST, HARRY DIAMOND LABS, US ARMY MATERIEL COMMAND, ADELPHI, 54- Mem: Fel AAAS; fel Am Phys Soc. Res: Gaseous electronics; semiconductors. Mailing Add: 3804 Underwood St Chevy Chase MD 20015

WARD, ANDERSON JAY, microbiology, see 12th edition

WARD, ANGUS LORIN, b Willard, Utah, Oct 26, 23; m 47; c 2. WILDLIFE RESEARCH. Educ: Utah State Univ, BS, 50. Prof Exp: Asst, Bear River Nat Wildlife Refuge, 51-52, wildlife res biologist commensal rodent control, 52-53, wildlife res biologist mammal repellents, 53-54, wildlife res biologist & proj leader agr rodent damage control, 54-66, ELK HABITAT RES PROJ LEADER, US FISH & WILDLIFE SERV, 66- Mem: Wildlife Soc. Res: Wildlife food habits; ecology of animals and plants; habitat management research; nutritional studies of range plants; animal behavior. Mailing Add: Rocky Mt Forest & Range Exp Sta US Forest Serv PO Box 3313 Laramie WY 82070

WARD, ANTHONY THOMAS, b London, Eng, Mar 9, 41; m 66; c 2. PHYSICAL CHEMISTRY. Educ: Univ London, BSc, 62; Rensselaer Polytech Inst, MS, 64, PhD(phys chem), 66. Prof Exp: MEM TECH STAFF, XEROX CORP, 66- Mem: Electrochem Soc; Am Phys Soc. Mailing Add: Joseph C Wilson Ctr for Technol W-114 Xerox Corp Webster NY 14580

WARD, ARLIN BRUCE, b Fargo, NDak, Sept 19, 19; m 43; c 4. FOOD SCIENCE. Educ: Kans State Col, BS, 43, Kans State Univ, MS, 51. Prof Exp: Milling asst, Shellabarger Mill & Elevator Co, Salina, Kans, 46; from instr to asst prof, Kans State Col, 47-51; process engr, Pillsbury Co, Minneapolis, Minn, 51-53, milling dept mgr, Springfield, Ill, 54-59, mgr milling process develop, Minneapolis, Minn, 59-61; assoc prof, 61-65, PROF GRAIN SCI & INDUST, KANS STATE UNIV, 66- Honors & Awards: Gold Medal Award, Asn Oper Millers, 69; Ernst Amme Award, Ger Millers Asn, 74. Mem: Asn Oper Millers; Am Asn Cereal Chemists; Int Asn Cereal Chemists; Inst Food Technologists. Res: Wheat quality improvement testing methods for milling and baking characteristics of grains with emphasis on wheat; fine grinding and air classification of cereal flours. Mailing Add: Dept of Grain Sci & Indust Kans State Univ Manhattan KS 66506

WARD, ARTHUR ALLEN, JR, b Manipay, Ceylon, Feb 4, 16; US citizen; m 41. NEUROSURGERY. Educ: Yale Univ, BA, 38, MD, 42. Prof Exp: Demonstr path, McGill Univ, 43, demonstr neruol & neruosurg, 44-45; asst physiol, Yale Univ, 45; asst, Ill Neuropsychiat Inst, 46; instr neurosurg, Univ Louisville, 46-48; from asst prof to prof surg, 48-65, CHMN DEPT NEUROSURG, MED SCH, UNIV WASH, 65- Concurrent Pos: Fel, McGill Univ, 43. Mem: AAAS; Am Asn Neurol Surg; Am Epilepsy Soc (vpres, 49, pres, 72); Am Physiol Soc; Am EEG Soc (pres, 59-60). Res: Epilepsy; function of animal and human cerebral cortex; reticular formation of the midbrain. Mailing Add: Dept of Neurosurg Univ of Wash Med Sch Seattle WA 98195

WARD, ARTHUR GOWSELL, physics, see 12th edition

WARD, BENJAMIN FRANKLIN, JR, organic chemistry, see 12th edition

WARD, BENJAMIN QUINN, bacteriology, see 12th edition

WARD, BENNIE FRANKLIN LEON, b Millen, Ga, Oct 19, 48. THEORETICAL HIGH ENERGY PHYSICS. Educ: Mass Inst Technol, BS(physics) & BS(math), 70; Princeton Univ, MA, 71, PhD(physics), 73. Prof Exp: Instr physics, Princeton Univ, 73; res assoc physics, Stanford Linear Accelerator Ctr, Stanford Univ, 73-75; ASST PROF PHYSICS, PURDUE UNIV, WEST LAFAYETTE, 75- Mem: Am Phys Soc. Res: Continued development of the theory of previously introduced sources in the renormalization group equations; the applications of a previously introduced Lorentz invariant formulation of strong coupling theory. Mailing Add: Dept of Physics Purdue Univ West Lafayette IN 47907

WARD, CALVIN B, particle physics, see 12th edition

WARD, CALVIN HERBERT, b Strawberry, Ark, Mar 1, 33; m 54; c 3. PLANT PATHOLOGY, PHYSIOLOGY. Educ: NMex State Univ, BS, 55; Cornell Univ, MS, 58, PhD(plant path), 60. Prof Exp: Res biologist, US Air Force Sch Aerospace Med, 60-63, plant physiologist, 63-65; assoc prof biol, 66-70, PROF BIOL & ENVIRON SCI & CHMN DEPT ENVIRON SCI & ENG, RICE UNIV, 70- Concurrent Pos: Grants, NASA, 63-66 & 70-, Environ Protection Agency, 66- & US Dept Air Force, 68-70; mem environ biol adv panel, Am Inst Biol Sci, 66-71, chmn, 69-71, mem comt space shuttle impact eval, 74-; mem bd dirs, Southwest Ctr Urban Res, 69-; mem life sci comt, NASA, 71-; vis prof, Univ Tex Sch Pub Health Houston, 73-74, adj prof environ health, 74- Mem: AAAS; Am Phytopath Soc; Soc Nematol; Soc Indust Microbiol; Phycol Soc Am. Res: Algal and plant physiology; bioregeneration; effects of environment on plant growth; biology of pollution and pollution control. Mailing Add: Dept of Environ Sci & Eng Rice Univ Houston TX 77001

WARD, CALVIN LUCIAN, b Yancey, Tex, Jan 30, 28; m 66; c 2. GENETICS. Educ: Univ Tex, BA, 47, MA, 49, PhD(zool), 51. Prof Exp: AEC fel, Oak Ridge Nat Lab, 51-52; from instr to asst prof zool, 52-60, ASSOC PROF ZOOL, DUKE UNIV, 60- Mem: Genetics Soc Am; Soc Study Evolution. Res: Cytology and genetics of Drosophila; speciation; radiation genetics. Mailing Add: Dept of Zool Duke Univ Durham NC 27706

WARD, CHARLES BRADLEY, JR, b Pittsburgh, Pa, Apr 1, 20; m 49; c 2. INDUSTRIAL MICROBIOLOGY. Educ: Univ Southern Calif, BS, 49; Okla State Univ, MS, 50; Iowa State Univ, PhD(food technol), 56. Prof Exp: Bacteriologist, Arden Farms Co, Calif, 56-59, Dugway Proving Ground, Chem Corps, US Dept Army, 59-61 & Truesdail Labs, Calif, 61-64; BIOL CHEMIST, ARROWHEAD PURITAS WATERS, 65- Mem: Am Chem Soc; Inst Food Technol; Am Soc Microbiol. Res: Water, dairy and general industrial bacteriology. Mailing Add: 4929 Lauderdale Ave La Crescenta CA 91214

WARD, CHARLES O, b West Carthage, NY, Jan 21, 42; m 63; c 3. PHARMACOLOGY, TOXICOLOGY. Educ: Union Univ, NY, BS, 63; Temple Univ, MS, 65, PhD(pharmacol), 68. Prof Exp: Asst pharmacol, Temple Univ, 63-67; asst prof, 68-71, ASSOC PROF PHARMACOL, COL PHARM, ST JOHN'S UNIV, NY, 72- Mem: AAAS; Am Pharmaceut Asn; Soc Toxicol. Res: Teratology; autonomic pharmacology of the human placenta; effect of diet on ethanol toxicity; toxicology of drug and cosmetic aerosols. Mailing Add: Dept of Pharmacol St John's Univ Jamaica NY 11439

WARD, CHARLES RICHARD, b Tahoka, Tex, Mar 25, 40; m 61; c 2. ENTOMOLOGY. Educ: Tex Tech Col, BS, 62, MS, 64; Cornell Univ, PhD(med entom), 68. Prof Exp: Asst prof, 67-72, ASSOC PROF ENTOM, TEX TECH UNIV, 72- Mem: AAAS; Am Mosquito Control Asn; Entom Soc Am. Res: Pest management research; biology and non-insecticidal control of mosquitoes and ecology of desert and grasslands insects. Mailing Add: Entom Sect Agr Bldg Tex Tech Univ Lubbock TX 79409

WARD, CHARLOTTE REED, b Lexington, Ky, Feb 19, 29; m 51; c 4. PHYSICAL CHEMISTRY. Educ: Univ Ky, BS, 49; Purdue Univ, MS, 51, PhD(phys chem), 56. Prof Exp: Instr gen sci, Ala Educ TV, 58-60, 61-62 & 63-72, from instr to asst prof physics, Univ, 61-75, ASSOC PROF PHYSICS, AUBURN UNIV, 75- Concurrent Pos: Abstractor, Che Abstr, 58- Mem: Am Asn Physics Teachers. Res: Molecular spectroscopy; development of physical science courses. Mailing Add: Dept of Physics Auburn Univ Auburn AL 36830

WARD, COLEMAN YOUNGER, b Millican, Tex, Sept 20, 28; m 47; c 3. AGRONOMY, PHYSIOLOGY. Educ: Tex Tech Univ, BS, 50, MS, 54; Va Polytech Inst & State Univ, PhD(agron), 62. Prof Exp: Instr agricult, Eastern NMex Univ, 50-51; soil scientist, Soil Conserv Serv, USDA, 51-52; instr agron, Tex Tech Univ, 52-54; asst agronomist, Univ Fla, 54-55 & Va Agr Exp Sta, 55-61; from assoc prof to prof crop sci, Miss State Univ, 61-74; PROF AGRON & CHMN DEPT, UNIV FLA, 74- Honors & Awards: Merit Award, Am Forage & Grassland Coun, 70. Mem: Fel Am Soc Agron; Sigma Xi; Crop Sci Soc Am. Res: Physiology and ecology of turfgrasses and forage crops. Mailing Add: Dept of Agron Univ of Fla Gainesville FL 32611

WARD, CURTIS HOWARD, b Round Bottom, Ohio, June 21, 27; m 51; c 4. PHYSICAL CHEMISTRY. Educ: Ind State Teachers Col, BS, 47; Univ Ky, MS, 50; Purdue Univ, PhD(phys chem), 54. Prof Exp: Res chemist, Linde Co Div, Union Carbide Corp, 53-57; assoc prof chem, Auburn Univ, 57-60; sr staff scientist, Avco Corp, 60-61; assoc prof, 61-65, PROF CHEM, AUBURN UNIV, 65- Mem: Am Chem Soc. Res: Thermodynamics; molecular spectroscopy; organometallic chemistry. Mailing Add: Dept of Chem Auburn Univ Auburn AL 36830

WARD, DANIEL BERTRAM, b Crawfordsville, Ind, Mar 20, 28; m 56; c 4. PLANT TAXONOMY. Educ: Wabash Col, AB, 50; Cornell Univ, MS, 53, PhD(plant taxon), 59. Prof Exp: From asst prof to assoc prof, 58-75, PROF BOT, UNIV FLA, 75- Mem: Int Asn Plant Taxon; Am Soc Plant Taxonomists. Res: Vascular flora of Florida; methods of population analysis; preservation of endangered species. Mailing Add: Dept of Bot Univ of Fla Gainesville FL 32611

WARD, DARRELL N, b Logan, Utah, Jan 22, 24; m 46; c 7. BIOCHEMISTRY. Educ: Utah State Univ, BS, 49; Stanford Univ, MS, 51; PhD(biochem), 53. Prof Exp: Res assoc & instr biochem, Med Col, Cornell Univ, 52-55; from asst biochemist to assoc biochemist, 55-61, PROF BIOCHEM, HEAD DEPT & BIOCHEMIST, UNIV TEX M D ANDERSON HOSP & TUMOR INST, HOUSTON, 61-; MEM GRAD FAC, UNIV TEX GRAD SCH BIOMED SCI, HOUSTON, 61- Concurrent Pos: Asst prof, Univ Tex Dent Br, Houston, 56-60; from asst clin prof to assoc clin prof, Baylor Col Med, 56-62; mem reproductive biol study sect, NIH, 67-71, consult, Ctr Pop Res, 69-71. Mem: AAAS; Am Chem Soc; Am Asn Cancer Res; Am Soc Biol Chem; Endocrine Soc. Res: Protein purification; protein and peptide hormones. Mailing Add: Dept of Biochem Univ of Tex M D Anderson Hosp & Tumor Inst Houston TX 77025

WARD, DAVID, b Manchester, Eng, July 8, 38; m 64; c 2. HISTORICAL GEOGRAPHY, URBAN GEOGRAPHY. Educ: Univ Leeds, BA, 59, MA, 61; Univ Wis-Madison, MS, 61, PhD(geog), 63. Prof Exp: Lectr geog, Carleton Univ, 63-64; asst prof, Univ BC, 64-66; from asst prof to assoc prof, 66-71, PROF GEOG, UNIV WIS-MADISON, 71-, CHMN DEPT, 74- Concurrent Pos: Guggenheim fel, Eng, 70-71; Am Coun Learned Socs res grant, 72; mem steering comt, NSF-Asn Am Geogr Urban Goals Proj, 72-74; ed, Asn Am Geogr Monogr Series, 72-75; co-ed, J Hist Geog, 75- Mem: Asn Am Geogr; Inst Brit Geogr; Brit Econ Hist Soc; Econ Hist Asn. Res: Historical geography of urbanization during the Industrial Revolution, specifically the changing social geography of cities from 1840 to 1920. Mailing Add: Dept of Geog Univ of Wis Sci Hall Madison WI 53706

WARD, DAVID, b Wakefield, Eng, Aug 5, 40; Can citizen. EXPERIMENTAL NUCLEAR PHYSICS. Educ: Univ Birmingham, BSc, 61; Univ Manchester, PhD(nuclear physics), 65. Prof Exp: Fel nuclear physics, Univ Manchester, 65-66 & Univ Calif, Berkeley, 66-68; PHYSICIST, CHALK RIVER NUCLEAR LABS, ATOMIC ENERGY CAN, 68- Concurrent Pos: Vis scientist nuclear physics, Univ Calif, Berkeley, 74-75. Mem: Am Phys Soc. Res: Reactions and coulomb excitation in heavy ions; lifetimes of short-lived nuclear states; atomic phenomena in nuclear physics; stopping powers for heavy ions. Mailing Add: Nuclear Physics Br Chalk River Nuclear Labs Chalk River ON Can

WARD, DAVID CHRISTIAN, b Sackville, NB, May 22, 41; m 62; c 1. VIROLOGY, BIOCHEMISTRY. Educ: Mem Univ Nfld, BSc, 61; Univ BC, MSc, 63; Rockefeller Univ, PhD(biochem), 69. Prof Exp: ASST PROF MOLECULAR BIOPHYS & BIOCHEM, SCH MED, YALE UNIV, 71- Concurrent Pos: Leukemia Soc Am fel, Imp Cancer Res Fund, Eng, 69-71. Mem: AAAS. Res: Replication and genetic analysis of animal viruses; mammalian genetics, fractionation and properties of metaphase chromosomes; conformational properties of nucleic acid polymerases. Mailing Add: Dept of Molecular Biophys Yale Univ Sch of Med New Haven CT 06510

WARD, DAVID JUSTIN, b Avoca, Nebr, Sept 9, 22; m 47; c 3. PLANT GENETICS. Educ: Univ Nebr, BS, 47, MS, 48; Univ Minn, PhD, 61. Prof Exp: Res agronomist, Crops Res Div, Agr Res Serv, 48-58, agr adminr, 58-61, res coordr, Sci & Educ Staff, 62-71, asst dir, 71-73, asst coordr environ qual activities, Off of Secy, 69-71, ASSOC COORDR ENVIRON QUAL ACTIVITIES & STAFF OFFICER RES PLANNING & COORD, OFF OF SECY, USDA, 73- Res: Small grains; evolutionary aspects of world small grain population; evaluation and utilization of available germplasm; coordination, planning and evaluation of agriculture and forestry research and environmental quality activities involving a broad spectrum of biological and physical sciences. Mailing Add: Off of the Secy USDA Washington DC 20250

WARD, DIANA VALIELA, b Buenos Aires, Arg, Oct 7, 43; m 67. ECOLOGY, INVERTEBRATE ZOOLOGY. Educ: Rutgers Univ, New Brunswick, AB, 65; Duke Univ, AM, 68, PhD(zool), 70. Prof Exp: Asst prof zool, Rutgers Col, Rutgers Univ, New Brunswick, 71-72, asst res prof entom, Col Agr & Environ Sci, 72-74. Concurrent Pos: Vis asst prof, Inst Animal Resource Ecol, Univ BC, 73-74. Mem: Am Soc Limnol & Oceanog; Am Soc Zoologists; Sigma Xi; Ecol Soc Am. Res: Pesticides and salt marsh ecology; invertebrate structure and function; marine ecology; environmental impact analysis. Mailing Add: Inst Animal Resource Ecol Univ BC Vancouver BC Can

WARD, EDMUND WILLIAM BESWICK, b Stockport, Eng, May 30, 30; m 53; c 3. MICROBIOLOGY, PLANT PATHOLOGY. Educ: Univ Wales, BSc, 52; Univ Alta, MSc, 54, PhD, 58. Prof Exp: Res officer, Alta, 57-61, head lab, 61-73, PLANT PATHOLOGIST, LONDON RES INST, PLANT PATH LAB, CAN DEPT AGR, 73- Mem: Bot Soc Am; Am Phytopath Soc; Can Soc Phytopath. Res: Physiology of fungi; diseases resistance in plants. Mailing Add: London Res Inst Can Dept Agr Univ West Ont London ON Can

WARD, EDWARD HILSON, b Milton, Fla, Sept 15, 30; m 54; c 2. INORGANIC CHEMISTRY, ANALYTICAL CHEMISTRY. Educ: Troy State Col, BS, 58; Univ Miss, PhD(chem), 63. Prof Exp: Fel chem, Fla State Univ, 63-65; assoc prof, 65-68, PROF CHEM, TROY STATE UNIV, 68-, CHMN DEPT PHYS SCI, 70- Mem: Am Chem Soc. Res: Inorganic complexes; non-aqueous solvent systems; co-precipitation. Mailing Add: Dept of Phys Sci Troy State Univ Troy AL 36081

WARD, FRANCES ELLEN, b Freedom, Maine, Mar 21, 39. IMMUNOGENETICS. Educ: Clark Univ, AB, 61; Brown Univ, PhD(biol), 65. Prof Exp: Instr, 67-69, asst prof, 69-73, ASSOC PROF IMMUNOL, DUKE UNIV MED CTR, 73- Concurrent Pos: NIH fel statist, Iowa State Univ, 67-69; dir, Transplant Lab, Durham Vet Admin Hosp, 69- Mem: Genetics Soc Am; AAAS; Transplantation Soc; Am Asn Clin Histocompatibility Testing; Am Asn Immunologists. Res: Genetics of the major human histocompatibility complex; immunogenicity of gene products of the major histocompatibility complex as measured by organ and tissue rejection. Mailing Add: Transplant Lab Durham Vet Admin Hosp Fulton St Durham NC 27705

WARD, FRANK KERNAN, b Brockton, Mass, Jan 19, 31; m 56; c 3. ORGANIC CHEMISTRY, POLYMER CHEMISTRY. Educ: Boston Col, BS, 54; Mass Inst Technol, PhD(org chem), 58. Prof Exp: Res asst chem, Mass Inst Technol, 54-57; res chemist, Celanese Corp, 57-59; scientist, Avco Corp, 59-60, sr scientist, 60-61; sr chemist, 61-64, res chemist, 64-67, group leader, 67-71, ENVIRON TECHNOLOGIST, TEXACO INC, BEACON, 71- Mem: Am Chem Soc. Res: Synthetic polymer chemistry; organometallics; ablative materials; fuel and lubricant additives; petrochemicals. Mailing Add: Deerwood Dr RD-3 Hopewell Junction NY 12533

WARD, FREDERICK EDMUND, US citizen. MEDICINAL CHEMISTRY. Educ: Goshen Col, BA, 71; Univ Notre Dame, PhD(org chem), 74. Prof Exp: Asst res chemist, 70-71, RES CHEMIST, MILES LABS, INC, 74- Mem: Am Chem Soc. Res: Antifungal agents; anti-inflammatory agents. Mailing Add: Miles Labs Inc 1127 Myrtle St Elkhart IN 46514

WARD, FREDERICK ROGER, b Cleveland, Miss, Oct 30, 40; m 64; c 2. MATHEMATICS. Educ: Col William & Mary, BS, 62; Univ Colo, MS, 65; Va Polytech Inst & State Univ, PhD(math), 69. Prof Exp: Instr math, Va Polytech Inst & State Univ, 65-69; asst prof, 69-72, ASSOC PROF MATH, BOISE STATE COL, 72- Mailing Add: Dept of Math Boise State Col Boise ID 83707

WARD, FREDERICK WILLIAM, JR, meteorology, see 12th edition

WARD, FREDRICK JAMES, b Alert Bay, BC, Jan 22, 28; m 56; c 3. LIMNOLOGY, FISH BIOLOGY. Educ: Univ BC, BA, 52, MA, 57; Cornell Univ, PhD(conserv), 62. Prof Exp: Biologist, Int Pac Salmon Fisheries Comn, 52-57, proj supvr, 57-64; from asst prof to assoc prof, 64-73, PROF LIMNOL & INVERT ZOOL, UNIV MAN, 73- Mem: Am Fisheries Soc; Am Soc Limnol & Oceanog; Can Soc Wildlife & Fishery Biol; Int Asn Theoret & Appl Limnol. Res: Limnology, particularly secondary production; dynamics of Pacific salmon populations. Mailing Add: Dept of Zool Univ of Man Winnipeg MB Can

WARD, GEORGE A, b Chicago, Ill, Feb 17, 36; m 57; c 4. ANALYTICAL CHEMISTRY. Educ: Univ Ill, BS, 57; Northwestern Univ, PhD(anal chem), 61. Prof Exp: Sr res chemist, 60-75, MGR, ANAL DIV, RES CTR, HERCULES, INC, 75- Mem: Am Chem Soc. Res: Electrochemistry; nuclear magnetic resonance; organic analysis. Mailing Add: Hercules Res Ctr Wilmington DE 19899

WARD, GEORGE HENRY, b Withrow, Wash, Nov 28, 16; m 46; c 2. SYSTEMATIC BOTANY. Educ: State Col Wash, BS, 40, MS, 48; Stanford Univ, PhD(biol), 52. Prof Exp: Teacher, Wash High Sch, 40-42; asst bot & taxon, State Col Wash, 42 & 46-48; asst bot & taxon, Stanford Univ, 48-52, instr biol, 52-54, curatorial asst, Dudley Herbarium, 49; from asst prof to assoc prof, 54-66, PROF BIOL, KNOX COL, ILL, 66- Mem: AAAS. Res: Cyto-taxonomy of Artemisia; arctic flora; stripmine spoilbank ecology. Mailing Add: Dept of Biol Knox Col Galesburg IL 61401

WARD, GEORGE MERRILL, b New Haven, Vt, June 4, 19; m 48; c 3. DAIRYING, NUTRITION. Educ: Univ Vt, BS, 41; Rutgers Univ, MS, 47; Mich State Univ, PhD(dairy nutrit), 51. Prof Exp: Asst, Rutgers Univ, 46; asst, Mich State Univ, 48-50, asst prof dairy, 50-55; assoc prof dairy sci, 55-66, PROF DAIRY SCI, KANS STATE UNIV, 66- Mem: Am Dairy Sci Asn. Res: Dairy cattle nutrition; roughage evaluation;

mineral nutrition. Mailing Add: Dept of Dairy & Poultry Sci Kans State Univ Manhattan KS 66506

WARD, GERALD MADISON, b Thorndike, Maine, Nov 2, 21; m 48; c 2. ANIMAL SCIENCE. Educ: Univ Maine, BS, 47; Univ Wis, MS, 48; Wash State Univ, PhD(animal sci), 51. Prof Exp: Asst prof dairy sci, Univ Maine, 48-49; exten specialist, Kans State Univ, 51-52 & Wash State Univ, 52-53; from asst prof to assoc prof, Colo State Univ, 53-60; scientist, Los Alamos Sci Lab, Univ Calif, 60-61; PROF ANIMAL SCI, COLO STATE UNIV, 61- Concurrent Pos: Head animal prod & health sect, Joint Div Food & Agr Orgn-Int Atomic Energy Agency, UN, Vienna, Austria, 68-70. Mem: AAAS; Am Dairy Sci Asn; Am Inst Nutrit; Health Physics Soc; Am Soc Animal Sci. Res: Applications of radioisotope technique to animal nutrition; environmental problems of animal agriculture; nutrition of ruminant animals; invivo body composition techniques. Mailing Add: Dept of Animal Sci Colo State Univ Ft Collins CO 80521

WARD, GERTRUDE LUCKHARDT, b Mt Vernon, NY, May 27, 23. ENTOMOLOGY. Educ: Mt Holyoke Col, AB, 44; Univ Mich, MS, 48; Purdue Univ, PhD(entom), 70. Prof Exp: Teaching asst zool, Iowa State Univ, 44-46; reporter-ed, Palladium-Item, Ind, 46-47; teaching asst zool, Univ Mich, 48-49; instr biol, 49-61, lectr, 61-67, from asst prof to assoc prof, 67-75, PROF BIOL, EARLHAM COL, 75-, ASST DIR MUS, 52- Mem: Am Micros Soc; Entom Soc Am. Res: Wasp-spider relationships; Chalybion zimmermanni; parasitism of the bagworm; bat populations in Indiana. Mailing Add: Dept of Biol Earlham Col Richmond IN 47374

WARD, GORDON MARSHALL, b Brampton, Ont, Apr 26, 12; m 38; c 4. PLANT PHYSIOLOGY. Educ: Univ Sask, BSc, 32; McMaster Univ, MA, 33; Univ Toronto, PhD(plant physiol), 52. Prof Exp: Sr physiologist, Tobacco Div, Exp Farm Serv, 33-41 & 47-51, chemist, Chem Div, Sci Serv, 51-59, res scientist, Plant Res Inst, 59-60, RES SCIENTIST, RES STA, CAN DEPT AGR, 60- Concurrent Pos: Can mem, Int Comn Protected Crops, Int Soc Hort Sci. Mem: Can Soc Plant Physiol; Chem Inst Can; Agr Inst Can; Prof Inst Pub Serv Can. Res: Physiology and biochemistry of the tobacco plant; mineral metabolism of plants; nutrition of greenhouse crops. Mailing Add: Res Sta Can Dept Agr Harrow ON Can

WARD, GRAY (GANESH), b Des Moines, Iowa, Dec 24, 34; div; c 2. ATMOSPHERIC PHYSICS. Educ: Univ Va, BEE, 56; Univ Calif, Los Angeles, MS, 58; Univ Colo, Boulder, MA, 62; Univ Md, College Park, PhD(physics), 70. Prof Exp: Mem tech staff, Hughes Aircraft Co, Calif, 56-58; physicist, Nat Bur Stand, Colo, 58-62; asst prof physics, Univ Redlands, 62-64, chmn dept, 63-64; asst prog dir, Off Int Sci Activ, NSF, Washington, DC, 64-65, consult, 65-66; asst prof, 70-75, ASSOC PROF PHYSICS & ELEC ENG, UNIV FLA, 75- Concurrent Pos: Consult, Off Int Sci Activ, NSF, 63-64. Mem: Am Phys Soc; Inst Elec & Electronics Engrs; Optical Soc Am; Am Meteorol Soc; Am Asn Physics Teachers. Res: Microwave and plasma physics; scientific applications of laser scattering; remote measurements of atmospheric properties; integrated optics; science education. Mailing Add: Dept of Elec Eng Univ of Fla Gainesville FL 32611

WARD, HAROLD NATHANIEL, b Evanston, Ill, Apr 29, 36; m 59; c 3. MATHEMATICS. Educ: Swarthmore Col, BA, 58; Harvard Univ, MA, 59, PhD(math), 62. Prof Exp: From instr to asst prof math, Brown Univ, 62-67; asst prof, 67-69, ASSOC PROF MATH, UNIV VA, 69- Mem: Am Math Soc; Math Asn Am. Res: Finite groups; representations of groups; coding theory. Mailing Add: Dept of Math Univ of Va Charlottesville VA 22903

WARD, HAROLD ROY, b Mt Vernon, Ill, Nov 3, 35; m 56; c 2. ENVIRONMENTAL CHEMISTRY. Educ: Southern Ill Univ, AB, 57; Mass Inst Technol, PhD(org chem), 61; Harvard Univ, JD, 75. Prof Exp: NSF fel, 61-62; NATO fel, 62-63; from asst prof to assoc prof, 63-71, PROF CHEM, BROWN UNIV, 71- Concurrent Pos: Spec fel, Environ Protection Agency, 72-75. Mem: Am Chem Soc. Res: Public interest chemistry; environmental law. Mailing Add: Dept of Chem Brown Univ Providence RI 02912

WARD, HELEN LAVINA, b Lafayette, Ind, Dec 10, 10. PARASITOLOGY. Educ: Purdue Univ, BS, 33, MS, 36, PhD(zool), 39. Prof Exp: Teacher, High Sch, Ind, 33-34; asst biol, Purdue Univ, 34-40; instr biol sci, Lindenwood Col, 40-43; asst, Sch Med, Ind Univ, 43-44; from instr to asst prof zool, Univ Tenn, 44-62; SCI ANALYST, DIV TECH INFO, AEC, 62- Mem: Am Soc Parasitol; Am Micros Soc. Res: Helminthology; Acanthocephala. Mailing Add: 516 Valparaiso Rd Oak Ridge TN 37830

WARD, HENRY SILAS, JR, b Birmingham, Ala, Dec 10, 14; m 42; c 3. BOTANY, ECOLOGY. Educ: Auburn Univ, BS, 38; Iowa State Univ, MS, 40, PhD(bot), 48. Prof Exp: Collabr plant-soils, USDA & Iowa Agr Exp Sta, 39-41; from asst prof to prof bot, Auburn Univ, 47-60; PROF BIOL, HUNTINGDON COL, 66- Concurrent Pos: Consult, Commodity Credit Corp, USDA, 50-57; AEC fel radiation biol, Univ Kans, 72. Res: Seed physiology during curing and storage; aquatic ecology of streams and lakes of central Alabama. Mailing Add: Dept of Biol Huntingdon Col Montgomery AL 36106

WARD, INGEBORG L, b Rötha, Ger, Aug 14, 40; US citizen; m 63; c 2. PHYSIOLOGICAL PSYCHOLOGY. Educ: Westhampton Col, BS, 60; Tulane Univ, MS, 65, PhD(psychol), 67. Prof Exp: ASSOC PROF PSYCHOL, VILLANOVA UNIV, 66- Concurrent Pos: NSF res grant, 68-69; Nat Inst Child Health & Human Develop res grant, 70-76; mem ment health small grant comt, NIMH, res career develop award, 75-80. Mem: Am Psychol Asn; Soc Study Reproduction; AAAS. Res: Hormonal and environmental determinants of reproductive behavior, neural and pharmacological bases of sexual behavior. Mailing Add: Dept of Psychol Villanova Univ Villanova PA 19085

WARD, JACK A, b Lebanon, Ore, Sept 2, 35. ZOOLOGY, ANIMAL BEHAVIOR. Educ: Willamette Univ, BS, 57; Univ Wash, MS, 60; Univ Ill, Urbana, PhD(animal behav), 65. Prof Exp: Instr zool, Olympic Col, 60-62; from asst prof to assoc prof animal behav, 65-75, PROF ETHOLOGY, ILL STATE UNIV, 75- Concurrent Pos: Res grants, Ill State Univ, Ill Acad Sci, NIH & NSF, 65- Mem: Am Soc Zool; Animal Behav Soc: Am Soc Ichthyologists & Herpetologists; Am Inst Biol Sci. Res: Behavior of fish; development and regulation of behavior in cichlid fish. Mailing Add: Dept of Biol Sci Ill State Univ Normal IL 61761

WARD, JAMES ANDREW, b Pittsburgh, Pa, May 11, 38; m 66; c 5. PHYSICAL CHEMISTRY. Educ: St Vincent Col, BS, 60; Univ Notre Dame, PhD(phys chem), 64. Prof Exp: Res scientist radiation chem, Babcock & Wilcox Co, 64-65; RES SCIENTIST RADIATION CHEM, UNION CARBIDE CORP, 65- Mem: Am Chem Soc; Sigma Xi. Res: Radiation chemistry; effects of radiation on polymers in solution, polymer degradation and stabilization; radiation polymerization, hydrogels. Mailing Add: Union Carbide Corp PO Box 324 Tuxedo NY 10987

WARD, JAMES AUDLEY, b Timmonsville, SC, May 19, 10; m 35; c 3. MATHEMATICS, COMPUTER SCIENCE. Educ: Davidson Col, BA, 31; La State Univ, MS, 34; Univ Wis, PhD(math), 39. Prof Exp: Asst prof math, Davidson Col, 37-38 & Tenn Polytech Inst, 39-40; prof, Delta State Col, 40-42; from assoc prof to prof, Univ Ga, 46-49; prof, Univ Ky, 49-55; mathematician, US Air Force Missile Develop Ctr, 55-57, chief digital comput br, 57-59; spec asst to vpres, Univac Div, Sperry Rand Corp, 59-62; staff specialist electronic comput, Off Dir Defense Res & Eng, Dept of Defense, 62-69; STAFF ASST COMPUT PROGS, NAVAL SEA SYSTS COMMAND, 69- Concurrent Pos: Consult, Proj Scamp, Univ Calif, Los Angeles, 52 & Tube Turns, Ky, 53-55. Mem: Math Asn Am; Am Math Soc; Soc Indust & Appl Math; Asn Comput Mach; sr mem Inst Elec & Electronics Eng. Res: Digital computers; research in hardware and software; mathematical research in linear algebra and in numerical analysis. Mailing Add: 5106 Marlyn Dr Washington DC 20016

WARD, JAMES B, b Clio, Ky, May 5, 31; m 65. POULTRY NUTRITION, BIOCHEMISTRY. Educ: Berea Col, BS, 57; Univ Ky, MS, 58; Mich State Univ, PhD(poultry nutrit), 62. Prof Exp: Proj leader, Exten Poultry, Univ Tenn, 62-66; assoc prof, NC State Univ, 66-70; dir nutrit, Nash Johnson & Sons, 70-72; PROF POULTRY, NC STATE UNIV, 72- Concurrent Pos: Consult nutritionist, All-In-One Feeds, 59-62 & Armour Foods, 74- Mem: Poultry Sci Asn; World Poultry Sci Asn. Mailing Add: Dept of Poultry Sci NC State Univ Box 5307 Raleigh NC 27607

WARD, JAMES EDWARD, III, b Greenville, SC, Sept 20, 39; m 62; c 3. MATHEMATICS. Educ: Vanderbilt Univ, BA, 61; Univ Va, MA, 64, PhD(math), 68. Prof Exp: Asst to pres, George Peabody Col, 61-62; teaching asst math, Univ Va, 62-64, jr instr, 66-68, asst prof, 68-73, ASSOC PROF MATH, BOWDOIN COL, 73-, DIR SR CTR, 71- Mem: Am Math Soc; Math Asn Am. Res: Jordan algebras of characteristic two; structure of two-groups. Mailing Add: Senior Ctr Bowdoin Col Brunswick ME 04011

WARD, JAMES WELLINGTON, b Weathersby, Miss, Oct 27, 04; m 34; c 2. NEUROANATOMY, PARASITOLOGY. Educ: Univ Ala, BA, 29, MA, 32; Miss State Col, MS, 31; Univ Minn, MS, 47; Univ Miss, PhD(neuroanat), 50. Prof Exp: Instr zool & entom, Miss State Univ, 29-32; instr zool, Univ Okla, 32-33; asst prof zool & entom, Miss State Univ, 33-36; instr zool, Univ Minn, 36-38; assoc prof zool & entom, Miss State Univ, 38-47; actg assoc prof anat, Sch Med, Univ Miss, 47-50, from assoc prof to prof, 50-70; prof anat & chmn dept, Col Med, Univ SFla, 70-74; vis prof anat, Col Med, Univ S Ala, 74-75; PROF & CHMN DIV BIOL SCI, WHITWORTH COL, BROOKHAVEN, MISS, 75- Concurrent Pos: Parasitologist, Miss State Univ, 44-47; vis prof, Gulf Coast Res Labs, 47-58 & Belhaven Col, 58-70; actg chmd dept prev med, Sch Med, Univ Miss, 53-55, secy med fac, dir admis & chmn admis & promotions comt, 54-55. Mem: Am Soc Ichthyol & Herpet; Am Soc Parasitol; Am Asn Anat; Pan-Am Asn Anat. Res: Hematology and neuroembryology of fishes; parasites of wild and domestic animals. Mailing Add: 2075 London Ave Jackson MS 39211

WARD, JAMES WILLIAM, b Amarillo, Tex, Jan 16, 09; m 40; c 2. ANATOMY. Educ: Vanderbilt Univ, BA, 30, MS, 31, PhD(anat), 35, MD, 40. Prof Exp: Asst comp anat, 30-31, asst anat, Med Sch, 32-35, from instr to assoc prof, 35-58, actg head dept, 60-63, PROF ANAT, MED SCH, VANDERBILT UNIV, 58-, PROF NEUROL, 69- Concurrent Pos: Mem neurol study sect, NIH, 62-66; mem, Fogarty Int Fel Comt, 70-73. Mem: Asn Res Nerv & Ment Dis; Am EEG Soc. Res: Physiological and anatomical neurology; electroencephalography. Mailing Add: Dept of Anat Vanderbilt Univ Med Sch Nashville TN 37203

WARD, JERROLD MICHAEL, b New York, NY, Oct 29, 42; m 71. VETERINARY PATHOLOGY. Educ: Cornell Univ, DVM, 66; Univ Calif, Davis, PhD(comp path), 70; Am Col Vet Path, dipl. Prof Exp: Res pathologist, Univ Calif, Davis, 66-68; vet pathologist, Environ Protection Agency, 70-72; VET PATHOLOGIST, NAT CANCER INST, 72- Mem: Am Vet Med Asn. Res: Pathology; cancer research; electron microscopy; hematopoietic pathology; pathogenesis of disease; leukemia. Mailing Add: Bldg 37 Rm 5B-22 NIH Nat Cancer Inst Bethesda MD 20014

WARD, JOHN BERNARD, pharmacy, see 12th edition

WARD, JOHN EDWARD, b Chicago, Ill, Feb 7, 23; m 46; c 6. ORGANIC CHEMISTRY. Educ: Wabash Col, AB, 44; Lawrence Col, MS, 48, PhD, 51. Prof Exp: Res chemist, P H Glatfelter Co, Pa, 50-51, from chief chemist paper chem lab to tech mgr for dept, Nopco Chem Co, NJ, 51-59, RESIDENT MGR, DIAMOND SHAMROCK CHIMIE, SA, 59- Mem: Am Chem Soc; Tech Asn Pulp & Paper Indust. Res: Pulp and papter technology, especially the production of pigment coated papers; market research, product development and international marketing of industrial chemical specialties, especially for the particular requirements of various European markets. Mailing Add: 24 Route de la Veveyse CH-1700 Fribourg Switzerland

WARD, JOHN EVERETT, JR, b Mocksville, NC, Feb 23, 41; m 64; c 2. MYCOLOGY, ECOLOGY. Educ: High Point Col, BS, 63; Wake Forest Univ, MA, 65; Univ SC, PhD(biol), 70. Prof Exp: Asst prof biol, Gaston Col, NC, 65-67; asst prof, 70-74, ASSOC PROF BIOL, HIGH POINT COL, 74- Mem: Mycol Soc Am. Res: Taxonomy and ecology of mushrooms; ecology of soil fungi. Mailing Add: Dept of Biol High Point Col High Point NC 27262

WARD, JOHN F, b Northunberland, Eng, Aug 26, 35; m 66; c 2. RADIATION CHEMISTRY, RADIATION BIOCHEMISTRY. Educ: Durham Univ, BSc, 56, PhD(radiation chem), 60. Prof Exp: Demonstr chem, Kings Col, Durham Univ, 60-62; asst res biophysicist, 62-69, ASSOC RES CHEMIST, LAB NUCLEAR MED & RADIATION BIOL, UNIV CALIF, LOS ANGELES, 69-, ASSOC PROF PATH, SCH MED, 74- Concurrent Pos: Mem radiation study sect, NIH, 76-78. Mem: The Chem Soc; Biophys Soc; Radiation Res Soc. Res: Studies of radiation chemical destruction of biologically significant molecules. Mailing Add: Lab Nuclear Med & Radiation Biol 900 Veteran Ave Los Angeles CA 90024

WARD, JOHN FRANK, b London, Eng, May 14, 34; m 60; c 2. PHYSICS. Educ: Oxford Univ, BA, 57, MA & DPhil(physics), 61. Prof Exp: From lectr to asst prof physics, Univ Mich, 61-64; Assoc Elec Industs res fel, Oxford Univ & lectr, Wadham Col, Univ Oxford, 64-67; assoc prof, 67-74, PROF PHYSICS, UNIV MICH, ANN ARBOR, 74- Concurrent Pos: Consult, Lear-Siegler Laser Systs Ctr, 62-64, Royal Radar Estab, 66 & Photon Sources, 67- Mem: Am Phys Soc. Res: Nonlinear optics; lasers. Mailing Add: Dept of Physics Univ of Mich Ann Arbor MI 48104

WARD, JOHN K, b Litchfield, Nebr, July 1, 27; m 53; c 3. ANIMAL NUTRITION. Educ: McPherson Col, BS, 50; Kans State Univ, BS, 54, PhD, 61. Prof Exp: Asst animal husb, Okla State Univ, 54-56; asst prof agr, McPherson Col, 56-66; assoc prof, 67-74, PROF ANIMAL SCI, UNIV NEBR-LINCOLN, 74- Res: Beef cattle management. Mailing Add: Dept of Animal Sci Baker 235 Univ of Nebr Lincoln NE 68508

WARD, JOHN M, b New Brunswick, NJ, July 14, 24; m 53; c 2. PHYSIOLOGY. Educ: Rutgers Univ, BS, 49; Univ Pa, PhD(bot), 54. Prof Exp: Asst instr bot, Univ Pa, 51-52; from asst prof to prof biol, Temple Univ, 54-66, chmn dept, 59-66; dean sch sci & prof bot & chem, Ore State Univ, 66-70; res prof & dir, 70-71, PRES, DESERT RES INST, UNIV NEV SYST, 71- Concurrent Pos: Lalor Found fel, 55; res assoc, Acad Natural Sci Philadelphia, 56-; prog dir metab biol, NSF, 64-65, consult, 63-; consult, USPHS, 66- Mem: AAAS; Am Chem Soc; Am Soc Plant Physiol; Am Soc Microbiol. Res: Intermediary metabolism; biochemical morphogenesis. Mailing Add: Desert Res Inst Reno NV 89507

WARD, JOHN ROBERT, b Salt Lake City, Utah, Nov 23, 23; m 48; c 4. MEDICINE. Educ: Univ Utah, BS, 44, MD, 46; Univ Calif, Berkeley, MPH, 67. Prof Exp: Resident internal med, Salt Lake County Gen Hosp, 49-51; from instr to assoc prof med, 57-70, chmn dept prev med, 66-70, PROF MED & CHIEF ARTHRITIS DIV, COL MED, UNIV UTAH, 70- Concurrent Pos: Res fel physiol, Col Med, Univ Utah, 48-49; res fel med, Salt Lake County Gen Hosp, 53-54; fel med & rheumatic dis, Mass Gen Hosp & res fel rheumatic dis, Harvard Med Sch, 55-57; consult, Vet Admin Hosp, 57-; sr investr, Arthritis & Rheumatism Found, 59; mem arthritis training grants comt, Nat Inst Arthritis & Metab Dis. Mem: AAAS; AMA; Am Rheumatism Asn; Am Col Physicians; NY Acad Sci. Res: Biology of pleuropneumonia-like organisms; experimental arthritis; immunology; epidemiology. Mailing Add: Arthritis Div Univ of Utah Med Ctr Salt Lake City UT 84112

WARD, JOHN WESLEY, b Martin, Tenn, Apr 8, 25; m 47; c 4. PHARMACOLOGY. Educ: George Washington Univ, BS & MS, 55; Georgetown Univ, PhD(pharmacol), 59. Prof Exp: Res assoc pharmacol, Hazleton Labs, Inc, 50-56, head dept pharmacol, 56-58, chief dept pharmacol & biochem, 58, res appln specialist, 59; prin pharmacologist, 59-60, dir pharmacol res, 60-71, dir pharmacol develop, 71-72, DIR TOXICOL, A H ROBINS CO, 73- Concurrent Pos: Lectr, Med Col Va, 60-65. Mem: AAAS; Am Chem Soc; Soc Toxicol; NY Acad Sci; Am Soc Pharmacol & Exp Therapeut. Res: Structure-activity relationships; general pharmacodynamics; autonomics; toxicology. Mailing Add: A H Robins Co 407 Cummings Dr Richmond VA 23220

WARD, JOHN WILLIAM, b Moline, Ill, Oct 16, 29; m 52; c 4. HIGH TEMPERATURE CHEMISTRY. Educ: Augustana Col, Ill, BA, 52; Wash Univ, MA, 55; Univ NMex, PhD(phys chem), 66. Prof Exp: STAFF MEM, LOS ALAMOS SCI LAB, 56- Concurrent Pos: Consult, US Army Nuclear Defense Lab, Edgewood Arsenal, 66-; sr fel, Alexander von Humboldt Found, Inst Transuranium Elements, Ger, 72-73. Mem: Am Chem Soc; Am Vacuum Soc; fel Am Inst Chem. Res: Vapor pressure theory; Monte Carlo computer simulation of experiment; electrochemistry; gas-surface reactions. Mailing Add: Group CMB-5 MS 730 Los Alamos Sci Lab Los Alamos NM 87545

WARD, JOHN WILLIAM, b Wigan, Eng, Aug 4, 37; m 69; c 2. PHYSICAL CHEMISTRY. Educ: Univ Manchester, BSc, 59, MSc, 60; Cambridge Univ, PhD(phys chem), 62. Prof Exp: Res coun Alta fel, 62-63; res scientist, 63-66, sr res scientist, 66-70, RES ASSOC, UNION OIL CO CALIF, 70- Mem: Am Chem Soc; Soc Appl Spectros; The Chem Soc; assoc Royal Inst Chem. Res: Application of spectroscopic techniques to the study of surface chemistry and catalysis; heterogeneous catalysis; hydrocarbon conversions; petroleum processing. Mailing Add: Union Oil Co Calif Res Dept PO Box 76 Brea CA 92621

WARD, JOSEPH RICHARD, b Salt Lake City, Utah, Dec 7, 42. CHEMISTRY. Educ: Univ Del, BS, 64; State Univ NY Stony Brook, PhD(chem), 69. Prof Exp: RES CHEMIST, US ARMY BALLISTIC RES LABS, ABERDEEN PROVING GROUND, 71- Mem: Am Chem Soc. Res: Erosion of gun barrels, combustion of solid propellants; effect of propellant combustion in the near-wake of supersonic projectiles. Mailing Add: Ballistic Res Labs Aberdeen Proving Ground MD 21005

WARD, KEITH BOLEN, JR, b Paducah, Tex, Feb 20, 43. BIOPHYSICS. Educ: Tex A&M Univ, BS, 65; Johns Hopkins Univ, PhD(biophys), 74. Prof Exp: Res assoc physics, Appl Res Lab, Gen Dynamics Corp, 65-66; NAT RES COUN RES ASSOC, LAB STRUCT MATTER, NAVAL RES LAB, 74- Mem: AAAS; Am Crystallog Asn. Res: Study of the relationship between the structure and function of proteins by using x-ray diffraction analysis. Mailing Add: Code 6030 Lab for Struct Matter Naval Res Lab Washington DC 20375

WARD, KYLE, JR, b Beaumont, Tex, Sept 2, 02; m 29. CELLULOSE CHEMISTRY, PULP & PAPER TECHNOLOGY. Educ: Univ Tex, BA & BS, 23; George Washington Univ, MS, 26; Univ Berlin, PhD(chem), 32. Hon Degrees: MS, Lawrence Univ, 68. Prof Exp: Instr, Univ Tex, 23-24; jr chemist, Bur Chem & Soils, USDA, 24-28, collabr, 36-38, sr chemist, Southern Regional Res Lab, Bur Agr & Chem Eng, 38-41, prin chemist, Bur Agr & Indust Chem, 41-51; res chemist, Hercules Powder Co, 28-36; res assoc, 51-59, leader cellulose group, 59-66, chmn dept chem, 59-68, leader carbohydrate group & chmn sect org chem, 66-68, EMER PROF, INST PAPER CHEM, 68- Prof Exp: Res chemist, Chem Found, 36-38; True Mem lectr, 63; consult, Joint Chiefs of Staff & Am Can Co, 73- Honors & Awards: Cert of Appreciation, US Dept Army; Citation, VI Int Symp Carbohydrate Chem, 72. Mem: AAAS; Am Chem Soc; Fiber Soc; fel Tech Asn Pulp & Paper Indust. Res: Cellulose and derivatives; terpenes and related fields; chlorination and oxidation; cotton fiber properties; textiles; high polymers. Mailing Add: 1821 S Carpenter St Appleton WI 54911

WARD, LAIRD GORDON LINDSAY, b Wellington, NZ, Dec 6, 31; US citizen. INORGANIC CHEMISTRY. Educ: Univ NZ, BSc, 56, MSc, 57; Univ Pa, PhD(inorg chem), 61. Prof Exp: Res chemist, Fabrics & Finishes Dept, E I du Pont de Nemours & Co, 61-63; res fel, Mellon Inst Sci, 63-64; res chemist, Int Nickel Co, Inc, NY, 64-71; res assoc chem, Univ Ga, 72; SR CHEMIST, COLONIAL METALS, INC, 72- Mem: Am Chem Soc; fel Am Inst Chemists. Res: Synthesis of inorganic complexes, especially platinum group metals; organometallic chemistry with group five elements; inorganic pigment applications of non-stoichiometric transition metal oxides; nickel based light stabilizers for polyvinyl chloride. Mailing Add: Colonial Metals Inc PO Box 726 Elkton MD 21921

WARD, LAWRENCE MCCUE, b Canton, Ohio, Dec 11, 44; m 67; c 2. PSYCHOPHYSICS, VISION. Educ: Harvard Univ, AB, 66; Duke Univ, PhD(exp psychol), 71. Prof Exp: Asst prof psychol, Rutgers Univ, 70-73; vis asst prof, 73-74, ASST PROF PSYCHOL, UNIV BC, 74- Concurrent Pos: Consult, Dept Hwys, State NJ, 72; assoc, Acoust Eng, Vancouver, BC, 75- Mem: AAAS; Am Psychol Asn; Psychonomic Soc; Psychomet Soc; Can Psychol Asn. Res: Psychophysical scaling; information processing models of psychophysical judgment; sequential dependencies and context effects in psychophysical judgement; psychophysical measurement and psychological effects of noise; feature processing in auditory and visual pattern recognition; visual illusions. Mailing Add: Dept of Psychol Univ of BC Vancouver BC Can

WARD, LEWIS EDES, JR, b Arlington, Mass, July 20, 25; m 49; c 3. MATHEMATICS. Educ: Univ Calif, AB, 49; Tulane Univ, MS, 51, PhD(math), 53. Prof Exp: Instr math, Univ Nev, 53-54; asst prof, Univ Utah, 54-56; mathematician, US Naval Ord Test Sta, Univ Calif, 56-59; assoc prof, 59-65, PROF MATH, UNIV ORE, 65- Mem: Am Math Soc; Math Asn Am. Res: Topology, especially ordered spaces. Mailing Add: Dept of Math Univ of Ore Eugene OR 97403

WARD, LOUIS EMMERSON, b Mt Vernon, Ill, Jan 19, 18; m 42; c 4. MEDICINE. Educ: Univ Ill, AB, 39; Harvard Univ, MD, 43; Univ Minn, MS, 49. Prof Exp: From instr med to prof clin med, Mayo Grad Sch Med, Univ Minn, 51-73, PROF MED, MAYO MED SCH, 73-; CONSULT SECT MED, MAYO CLIN, 50-, CHMN BD GOV, 64- Mem: Nat Inst Med; Am Rheumatism Asn (past pres); AMA; Nat Soc Clin Rheumatol. Res: Rheumatic diseases. Mailing Add: 200 First St SW Rochester MN 55901

WARD, MARTHA COONFIELD, b Columbus, Ga, June 20, 41; m 64; c 1. CULTURAL ANTHROPOLOGY. Educ: Okla State Univ, BA, 62; Vanderbilt Univ, MAT, 63; Tulane Univ, PhD(anthrop), 69. Prof Exp: Asst prof anthrop, Univ New Orleans, 69-70; NIH grant coord field proj Micronesia, Sch Pub Health, Univ NC, 70-72; asst prof, 72-74, ASSOC PROF ANTHROP, UNIV NEW ORLEANS, 74- Mailing Add: Dept of Anthrop Univ of New Orleans New Orleans LA 70122

WARD, MAURICE LESLIE, chemistry, see 12th edition

WARD, MAX, b Mt Zion, WVa, May 14, 14; m; c 2. BOTANY, MORPHOLOGY. Educ: Glenville State Col, AB, 40; Harvard Univ, MA, 47, PhD(bot), 50. Prof Exp: Teacher, pub schs, Calhoun County, WVa, 34-35 & 36-38, high sch, 40-43 & 45-46; prof bot & chmn div sci & math, Glenville State Col, 48-69; vis prof biol, Univ Akron, 69-70; vis prof bot, Ohio Univ, 70-71; MGR STUDENT-ORIENTED PROG, DIV SCI MANPOWER IMPROV, NSF, 71- Concurrent Pos: NSF res grant, Harvard Univ, 58-60. Mem: AAAS; Bot Soc Am; Am Bryol & Lichenological Soc; Int Soc Plant Morphol. Res: Morphogenesis and developmental morphology in the ferns and mosses. Mailing Add: 5406 Connecticut Ave NW Washington DC 20015

WARD, NEWTON EDWIN, physics, see 12th edition

WARD, OSCAR GARDIEN, b Denver, Colo, Feb 16, 32; m 55; c 2. GENETICS. Educ: Univ Ariz, BS, 58, MS, 60; Purdue Univ, PhD(genetics), 66. Prof Exp: Instr biol, Purdue Univ, 60-64; asst prof, 66-74, LECTR BIOL SCI, UNIV ARIZ, 74- Mem: AAAS; Genetics Soc Am; Bot Soc Am. Res: Chemical mutagenesis; chromosomal specificity of chemical mutagens; role of chromosomes during development; human cytogenetics. Mailing Add: Dept of Biol Sci Univ of Ariz Tucson AZ 85721

WARD, PAUL H, b Lawrence, Ind, Apr 24, 28; m 52; c 2. OTOLARYNGOLOGY. Educ: Anderson Col, AB, 53; Johns Hopkins Univ, MD, 57; Am Bd Otolaryngol, dipl, 62. Prof Exp: Intern, Henry Ford Hosp, Detroit, 57-58; resident otolaryngol, Univ Chicago, 58-61, NIH spec res fel, 61-62, asst prof, 62-64; assoc prof surg & chief div otolaryngol, Sch Med, Vanderbilt Univ, 64-68; PROF SURG & CHIEF HEAD & NECK SURG, CTR HEALTH SCI, UNIV CALIF, LOS ANGELES, 68- Concurrent Pos: USPHS res grant, 63-69; Deafness Res Found res grant, 65-67; NIH res grant, 66-70; attend otolaryngologist, Nashville Metrop Gen Hosp, Tenn, 64-68; consult, Thayer Vet Admin Hosp, 64-68 & Surgeon Gen of US Navy, 74-; mem bd dirs, Bill Wilkerson Hearing & Speech Ctr, 64-68. Mem: AAAS; Am Otol Soc; Am Acad Ophthal & Otolaryngol; Am Laryngol Soc; Am Laryngol, Rhinol & Otol Soc. Res: Cochlear, vestibular and laryngeal physiology; temporal bone pathology; velopharyngeal corrective techniques; laryngeal and palatal reconstruction. Mailing Add: Dept of Surg Univ of Calif Sch of Med Los Angeles CA 90024

WARD, PAUL J, b Blairstown, Mo, Oct 30, 21; m 44; c 3. DAIRY BACTERIOLOGY. Educ: Univ Mo, BS, 46, MS, 47. Prof Exp: Res technician cheese, Kraft Foods Res Lab, 48-53, group leader, 53-59; mgr dairy prod res, Res & Develop Div, Nat Dairy Prod Corp, 59-69; prod mgr natural cheese cutting & contract suppliers, 69-71, nat prod mgr contract suppliers, 71-74, NAT PROD MGR CREAM CHEESE & FORZEN FOODS, KRAFTCO CORP, 74- Mem: Am Dairy Sci Asn; Inst Food Technol. Res: Dairy manufacturing; cheese and all other dairy products. Mailing Add: Kraftco Corp 500 Peshtigo Ct Chicago IL 60690

WARD, PETER A, b Winsted, Conn, Nov 1, 34. PATHOLOGY, IMMUNOLOGY. Educ: Univ Mich, BS, 58, MD, 60. Prof Exp: Intern med, Third Div, Bellevue Hosp, New York, 60-61; resident path, Hosp, Univ Mich, Ann Arbor, 61-63; res fel immunopath, Div Exp Path, Scripps Clin & Res Found, La Jolla, Calif, 63-65; chief immunobiol, Armed Forces Inst Path, Washington, DC, 65-70; PROF PATH, SCH MED, UNIV CONN, 70-, CHMN DEPT, 73- Mem: Am Soc Exp Path; Am Asn Immunologists; Am Asn Pathologists & Bacteriologists. Res: Immunopathology; inflammation; biological role of complement; antibody formation. Mailing Add: Dept of Path Univ of Conn Health Ctr Farmington CT 06032

WARD, PETER LANGDON, b Washington, DC, Aug 10, 43; m 65; c 2. SEISMOLOGY, VOLCANOLOGY. Educ: Dartmouth Col, BA, 65; Columbia Univ, MA, 67, PhD(geophys), 70. Prof Exp: Asst scismol, Columbia Univ, 65-70, res scientist, 70-71; geophysicist, 71-75, CHIEF BR SEISMOL, US GEOL SURV, 75- Mem: AAAS; Am Geophys Union; Seismol Soc Am; Geol Soc Am. Res: Analysis of earthquakes related to volcanoes, geothermal areas and tectonic features; earthquake prediction and hazard reduction. Mailing Add: US Geol Surv NCER 345 Middlefield Rd Menlo Park CA 94025

WARD, RAYMOND LELAND, b San Pedro, Calif, Feb 12, 32; m 58; c 4. PHYSICAL BIOCHEMISTRY. Educ: Univ Calif, BSc, 53; Univ Wash, St Louis, PhD(chem), 56. Prof Exp: Chemist, Lawrence Radiation Lab, Univ Calif, 56-64; NSF sr fel, Harvard Univ, 64-65; CHEMIST, LAWRENCE LIVERMORE LAB, UNIV CALIF, 65- Mem: Am Soc Biol Chem; Am Chem Soc. Res: Magnetic resonance studies of molecular interactions. Mailing Add: Dept of Chem Lawrence Livermore Lab Univ Calif Livermore CA 94550

WARD, RICHARD BERNARD, b Felixstowe, Eng, Jan 3, 32; m 63; c 2. ORGANIC CHEMISTRY. Educ: Univ Birmingham, BSc, 53, PhD, 56. Prof Exp: Atomic Energy Auth fel, Univ Birmingham, Eng, 56-58; Corn Indust Res Found fel, Ohio State Univ, 58-59; res org chemist, 59-65, res supvr, 65-68, prod supvr, 68-70, sr res chemist, 70-75, TECH CONSULT, E I DU PONT DE NEMOURS & CO, INC, 75- Mem: Am Chem Soc; The Chem Soc. Res: Permeation through polymers; fluorine; dye manufacture; fluorocarbons; environmental and toxicological consulting. Mailing Add: Freon Prods Lab El du Pont de Nemours & Co Inc Chestnut Run Wilmington DE 19898

WARD, RICHARD FLOYD, b New York, NY, July 5, 27; m 49; c 3. GEOLOGY. Educ: Bradley Univ, BS, 50; NY Univ, MS, 56; Bryn Mawr Col, PhD, 58. Prof Exp: Geologist, Del State Geol Surv, 54-58; ASSOC PROF GEOL, WAYNE STATE UNIV, 59- Mem: Geol Soc Am; Geochem Soc. Res: Metamorphic and igneous petrology; evolution of crystalline terrains. Mailing Add: Dept of Geol Wayne State Univ Detroit MI 48202

WARD, RICHARD JOHN, b Seattle, Wash, Aug 7, 25; m 50; c 3. ANESTHESIOLOGY. Educ: Gonzaga Univ, BSc, 46; St Louis Univ, MD, 49; Seattle Univ, MEd, 72; Am Bd Anesthesiol, dipl. Prof Exp: Chief anesthesiol serv, Air Force Hosp, Weisbaden, Ger, 54-57; asst chief anesthesiol, Lackland AFB, 57-60, chief, 60-61; chief anesthesiol, Ballard Gen Hosp, 62-63; from instr to assoc prof anesthesiol, 63-72, PROF ANESTHESIOL, SCH MED, UNIV WASH, 72- Concurrent Pos: NIH res fel, 64-65; consult, Surgeon Gen, US Air Force in Europe, 54-57; chief surg res lab, Lackland AFB, 58-61; admin officer, Sch Med, Univ Wash, 66; consult, Madigan Gen Hosp, 72- Mem: AAAS; Am Soc Anesthesiol; AMA; Asn Mil Surg US; fel Am Col Anesthesiol. Res: Pharmacology and physiology of anesthetized man. Mailing Add: Dept of Anesthesiol Univ of Wash Sch of Med Seattle WA 98195

WARD, RICHARD S, b Beirut, Lebanon, Oct 9, 20; m 60; c 3. PSYCHIATRY, PEDIATRICS. Educ: Amherst Col, BA, 42; Columbia Univ, MD, 45, cert, 57. Prof Exp: Clin dir, Child Guid Inst, Jewish Bd Guardians, NY, 56-61; assoc prof, 60-63, PROF PSYCHIAT, SCH MED, EMORY UNIV, 63- Concurrent Pos: Rockefeller fel child psychiat, Babies Hosp, Columbia Univ, 48-50. Mem: Am Psychoanal Asn; Am Psychiat Asn; fel Am Orthopsychiat Asn; Am Acad Child Psychiat. Res: Child development; psychoanalysis. Mailing Add: Emory Univ Psychoanal Clin 1711 Aidmore Dr NE Atlanta GA 30307

WARD, RICHARD THEODORE, b Avoca, Nebr, Sept 10, 25; m 53; c 2. PLANT ECOLOGY. Educ: Univ Nebr, BSc, 48; Univ Minn, MSc, 51; Univ Wis, PhD, 54. Prof Exp: Asst plant ecol, Univ Wis, 53-54; from instr to asst prof biol, Beloit Col, 54-57; from asst prof to prof bot & plant path, 57-75, chmn dept, 66-75, PROF PLANT ECOL, COLO STATE UNIV, 75- Mem: Ecol Soc Am; Sigma Xi; Am Inst Biol Sci. Res: Alpine vegetation; ecologic race differentiation; ecology of the American beech. Mailing Add: Dept of Bot & Plant Path Colo State Univ Ft Collins CO 80521

WARD, ROBERT, b Bronxville, NY, Aug 12, 08; m 61; c 1. PEDIATRICS. Educ: Yale Univ, BA, 30, MD, 33. Prof Exp: Intern pediat, New Haven Hosp, 33-34; asst resident, Johns Hopkins Hosp, 34-36, dispensary chief, 36-37, resident pediat, 37-38; asst prof, Sch Med, Temple Univ, 38-40; res assoc, Res Found, Children's Hosp & asst prof, Col Med, Univ Cincinnati, 40-43; asst prof prev med, Sch Med, Yale Univ, 43-45; from assoc prof to prof pediat, Col Med, NY Univ, 45-58; chmn dept, 58-71, PROF PEDIAT, SCH MED, UNIV SOUTHERN CALIF, 58- Concurrent Pos: Mem comn viral infections, Armed Forces Epidemiol Bd, Washington, DC, 43-; physician-in-chief, Children's Hosp, Los Angeles, 58-71, dir res, 71- Mem: Am Pediat Soc; Soc Pediat Res (secy, 47-51, pres, 53); Am Soc Clin Invest; AMA; fel Am Acad Pediat. Res: Viral infections, especially hepatitis. Mailing Add: Los Angeles CA

WARD, ROBERT C, b Mt Clemens, Mich, Mar 9, 32; c 4. OSTEOPATHY. Educ: Kansas City Col Osteop Med, DO, 57. Prof Exp: Intern, Mt Clemens Gen Hosp, 57-58, staff mem family med, 58-71; preceptor, Col Osteop Med & asst prof family med, 70-71, prof family med & chmn dept, 72-74, PROF MED EDUC RES & DEVELOP, MICH STATE UNIV, 74-; STAFF MEM, LANSING GEN HOSP, 72- Concurrent Pos: Off Med Educ fel, Mich State Univ, 72-73. Mem: Am Osteop Asn; Soc Teachers Family Med; Am Acad Osteop. Res: Family medicine curriculum design; osteopathic therapeutics; community based medical education. Mailing Add: Off of Med Educ Res & Develop Mich State Univ East Lansing MI 48823

WARD, ROBERT CLEVELAND, b Sparta, Tenn, Dec 7, 44; m 65; c 2. NUMERICAL ANALYSIS. Educ: Tenn Technol Univ, BS, 66; Col William & Mary, MS, 69; Univ Va, PhD(appl math), 74. Prof Exp: Mathematician, Langley Res Ctr, Nasa, 66-74; RES STAFF MEM MATH, NUCLEAR DIV, UNION CARBIDE CORP, 74- Mem: Soc Indust & Appl Math; Asn Comput Mach. Res: Developing, analyzing and/or improving numerical techniques in the areas of numerical linear algebra, partial differential equations and sensitivity analysis. Mailing Add: Nuclear Div Union Carbide Corp Bldg 9704-1 PO Box Y Oak Ridge TN 37830

WARD, ROBERT PORTER, b Martin, Tenn, Jan 31, 29; m 50; c 4. VERTEBRATE MORPHOLOGY, VERTEBRATE ECOLOGY. Educ: George Peabody Col, BS, 51, MA, 53; Miss State Univ, PhD, 63. Prof Exp: Prof biol, Tenn Wesleyan Col, 53-54; lab asst zool, Mich State Univ, 54-56; from asst prof to assoc prof biol, Millsaps Col, 56-64; ASSOC PROF BIOL, GEORGE PEABODY COL, 64- Concurrent Pos: Zool ed jour, Tenn Acad Sci. Mem: Fel AAAS. Res: Population dynamics. Mailing Add: Dept of Biol George Peabody Col Nashville TN 37203

WARD, ROBERT T, b Jersey City, NJ, Feb 7, 20; m 60. ZOOLOGY, CYTOLOGY. Educ: NJ State Teachers Col, AB, 42; Columbia Univ, MA, 52, PhD(zool), 60. Prof Exp: Res assoc zool, Columbia Univ, 52-57; ASST PROF ANAT, STATE UNIV NY DOWNSTATE MED CTR, 60- Concurrent Pos: NSF res grant, 62-66. Mem: AAAS; Electron Micros Soc Am; Am Soc Cell Biol; Am Asn Anat. Res: Electron microscopy; histochemistry. Mailing Add: Dept of Anat State Univ NY Downstate Med Ctr Brooklyn NY 11203

WARD, RONALD ANTHONY, b New York, NY, Jan 25, 29; m 50; c 2. MEDICAL ENTOMOLOGY. Educ: Cornell Univ, BSc, 50; Univ Chicago, PhD(zool), 55; Univ London, MSc, 67. Prof Exp: Instr biol, Gonzaga Univ, 55-58; MED ENTOMOLOGIST, WALTER REED ARMY INST RES, 58- Concurrent Pos: US Secy of Army fel, London Sch Hyg & Trop Med, 66-67; proj mgr, Med Entom Proj, Smithsonian Inst, Washington, DC, 74- Mem: AAAS; Am Soc Trop Med & Hyg; Am Mosquito Control Asn; Royal Soc Trop Med & Hyg; Entom Soc Am. Res: Genetic and ecologic factors affecting susceptibility and resistance of arthropods to infectious agents; host adaptation of malaria parasites; mosquito biosystematics. Mailing Add: Dept of Entom Walter Reed Army Inst of Res Washington DC 20012

WARD, RONALD WAYNE, b Johnson City, Tenn, Dec 17, 43; m 66. AGRICULTURAL ECONOMICS, ECONOMETRICS. Educ: Univ Tenn, BS, 65; Iowa State Univ, MS, 67, PhD(econ & statist), 70. Prof Exp: Res asst agr econ, Univ Tenn, summer 65; res asst, Iowa State Univ, 65-69; coop agent, USDA, 69-70; RES ECONOMIST, FLA DEPT CITRUS & ASST PROF ECON, DEPT FOOD & RESOURCE ECON, UNIV FLA, 70- Concurrent Pos: USDA mkt struct res grant, Univ Fla, 71-72. Mem: Am Econ Asn; Am Agr Ecòn Asn. Res: Price analysis; marketing; advertising; market structures. Mailing Add: 1104 McCarty Hall Univ of Fla Gainesville FL 32601

WARD, SAMUEL, b Los Angeles, Calif, Sept 29, 44; m 66; c 1. BIOCHEMISTRY. Educ: Princeton Univ, AB, 65; Calif Inst Technol, PhD(biochem), 71. Prof Exp: Tutor biochem sci, Harvard Col, 73-74; ASST PROF BIOL CHEM, HARVARD MED SCH, 72- Concurrent Pos: NSF fel, Med Res Coun Lab Molecular Biol, Cambridge, Eng, 70-71, NIH spec fel, 71-72. Res: Genetic control of the structure and function of simple nervous systems. Mailing Add: Dept of Biol Chem Harvard Med Sch Boston MA 02115

WARD, SAMUEL ABNER, b Binghamton, NY, Apr 27, 23; m 46; c 2. ELECTRONIC PHYSICS. Educ: Cornell Univ, BEE, 44, MEE, 46, PhD(electronics, physics), 53. Prof Exp: Elec radio engr, US Naval Res Lab, 44-45; asst physics, Cornell Univ, 46-50; res engr, Electron Tubes, Nat Union Radio Corp, 50-51, A B Du Mont Labs, Inc, 51-52 & Electron Tubes & Devices, RCA Labs, Inc, 52-56; physicist, Gen Elec Co, 56-57; chief physicist, Machlett Labs, Inc, 57-61; sr eng physicist, Perkin-Elmer Corp, 61-64; sr scientist, CBS Labs, Stamford, Conn, 64-75; SR ENGR, MACHLETT LABS, DIV RAYTHEON, 75- Mem: Am Phys Soc; Inst Elec & Electronics Engrs; Sigma Xi. Res: Cathodoluminescent properties of phosphors; solid state physical electronics; photoelectric emission from semiconductors and metals; photoeffects in solids; ion implantation doping of semiconductors; electron and ion optics. Mailing Add: 25 Ostend Ave Saugatuck Shores Westport CT 06880

WARD, STANLEY HARRY, b Vancouver, BC, Jan 16, 23; m 43; c 1. GEOPHYSICS. Educ: Univ Toronto, BASc, 49, MA, 50, PhD(geophys), 52. Prof Exp: Geophysicist, Int Nickel Co Can, Ltd, 49; chief geophysicist & managing dir, McPhar Geophys, Ltd, 49-53, exec vpres, 53-54; chief geophysicist & managing dir, Nucom Ltd, 54-58; consult geophysicist, Univ Toronto, 58-59; assoc prof mineral explor, Univ Calif, Berkeley, 59-64, prof geophys eng, 64-70; PROF GEOL SCI & CHMN DEPT, UNIV UTAH, 70- Mem: Soc Explor Geophys; Am Geophys Union; Am Inst Mining, Metall & Petrol Engrs; Inst Elec & Electronics Engrs; Can Inst Mining & Metall. Res: Variation magnetic and electric fields; rock physics; mineral exploration. Mailing Add: Dept of Geol Sci Univ of Utah Salt Lake City UT 84112

WARD, THOMAS CARL, physical chemistry, polymer chemistry, see 12th edition

WARD, THOMAS EDMUND, b Los Angeles, Calif, Nov 10, 44; m 67; c 1. NUCLEAR CHEMISTRY, NUCLEAR PHYSICS. Educ: Northeastern State Col, BSEd, 65; Univ Ark, Fayetteville, MS, 69, PhD(nuclear chem), 71. Prof Exp: Res assoc nuclear chem, Brookhaven Nat Lab, 70-72; STAFF CHEMIST, DEPT PHYSICS, IND UNIV, BLOOMINGTON, 72- Mem: Am Chem Soc; Am Phys Soc. Res: Nuclear spectroscopy and radioactive decay; nuclear reactions and fission. Mailing Add: Dept of Physics Ind Univ Bloomington IN 47401

WARD, THOMAS GREYDON, virology, see 12th edition

WARD, THOMAS JOSEPH, physics, see 12th edition

WARD, THOMAS MARSH, b Newark, NJ, Apr 20, 13; m 40; c 3. PHYSICAL CHEMISTRY. Educ: Univ NC, AB, 39; NC State Univ, MS, 62, PhD(crop sci), 64. Prof Exp: Chemist, Chas Pfizer & Co, 39-41; self-employed, 41-56; ASST PROF CHEM, NC STATE UNIV, 56- Mem: AAAS; Am Chem Soc. Res: Surface phenomena; adsorption, desorption and surface behavior of organic compounds at solid-solution interfaces; nuclear magnetic resonance and infrared studies of chemical bonding and charge transfer complexes. Mailing Add: Dept of Chem NC State Univ Raleigh NC 27607

WARD, TRUMAN L, b Ft Worth, Tex, Oct 21, 25; m 45; c 3. PHYSICS, PHYSICAL CHEMISTRY. Educ: Tulane Univ, BS, 48. Prof Exp: Assoc physicist, 48-63, RES PHYSICIST, SOUTHERN REGIONAL LAB, AGR RES SERV, USDA, 63- Honors & Awards: Awards, Sustained Outstanding Performance, Agr Res Serv, 61, Cert Merit & Dept Agr Superior Serv, 64. Mem: Am Chem Soc; Sigma Xi; fel Am Inst Chem; Am Asn Textile Chem & Colorists. Res: Physical properties of vegetable fats and oils; reaction mechanisms and kinetics; synthesis and reactions of thiorane and epoxy compounds; development of new instrumental procedures; low temperature plasmas; polymers; ion exchanges. Mailing Add: Southern Regional Ctr USDA PO Box 19687 New Orleans LA 70179

WARD, WALLACE DIXON, b Pierre, SDak, June 30, 24; m 49; c 4. PSYCHOACOUSTICS. Educ: SDak Sch Mines & Technol, BS, 44; Harvard Univ, PhD(exp psychol), 53. Hon Degrees: ScD, SDak Sch Mines & Technol, 71. Prof Exp: Asst scientist, Rosemount Res Ctr, Univ Minn, 49; asst, Harvard Univ, 49-53; res engr, Baldwin Piano Co, 53-54; res scientist, Cent Inst Deaf, 54-57; res assoc subcomt noise, Comt Conserv Hearing, Am Acad Ophthal & Otolaryngol, 57-62; PROF OTOLARYNGOL & COMMUN DIS, 62- & ENVIRON HEALTH, UNIV MINN, MINNEAPOLIS, 72- Concurrent Pos: Fels, Acoust Soc Am, 61 & Am Speech & Hearing Asn, 66; chmn exec coun, Comt Hearing, Bioacoust & Biomech, Nat Acad Sci-Nat Res Coun, 71-73; consult, Off Noise Abatement & Control, Environ Protection Agency, 72-73, Air Transport Asn Am, 73- & Edgewood Arsenal, US Army. Mem: Am Otol Soc; Psychonomic Soc; Int Soc Audiol; Am Audiol Soc (vpres, 73-75, pres, 76-77); Int Comt Biol Effects Noise (co-chmn, 73-). Res: Auditory fatigue and noise-induced hearing loss; musical perception; musical psychoacoustics. Mailing Add: Hearing Res Lab 2630 Univ Ave SE Minneapolis MN 55414

WARD, WALTER FREDERICK, b Darlington, Wis, June 23, 40; m 59; c 3. PHYSIOLOGY, ENDOCRINOLOGY. Educ: Univ Wis-Platteville, BSc, 64; Marquette Univ, PhD(physiol), 70. Prof Exp: ASST PROF PHYSIOL, COL MED, PA STATE UNIV, 73- Concurrent Pos: USPHS fel, Brown Univ, 71-73. Mem: AAAS; Am Physiol Soc. Res: Regulation of protein metabolism; mechanisms of hormone action. Mailing Add: Dept of Physiol Col of Med Pa State Univ Hershey Med Ctr Hershey PA 17033

WARD, WILBER W, b Chambersburg, Pa, Dec 12, 16; m 39; c 3. FORESTRY. Educ: Pa State Univ, BS, 40, MF, 52; Yale Univ, DFor, 62. Prof Exp: Co-partner, Dothan Lumber Co, 40-41; inspector forest prod, Pa Railroad Co, 41-48; from instr to assoc prof forestry, Pa State Univ, 48-65, resident dir, Mont Alto Br, Sch Forest Resources, 53-59, PROF FORESTRY, PA STATE UNIV, 65-, DIR SCH FOREST RESOURCES, 66- Mem: Soc Am Foresters. Res: Plant science; silviculture; mensuration; forest management. Mailing Add: Sch of Forest Resources 101 Ferguson Bldg Pa State Univ University Park, PA 16802

WARD, WILFRED HAMLIN, biochemistry, see 12th edition

WARD, WILLIAM CRUSE, b Waco, Tex, Apr 26, 33; m 57; c 3. SEDIMENTARY PETROLOGY. Educ: Univ Tex, Austin, BS, 55, MA, 57; Rice Univ, PhD(geol), 70. Prof Exp: Geologist, Humble Oil & Refining Co, 57-66; asst prof, 70-73, ASSOC PROF GEOL, UNIV NEW ORLEANS, 72ccM 73- Mem: Soc Econ Paleontologists & Mineralogists; Am Asn Petrol Geologists. Res: Petrology and diagenesis of Quaternary limestones of eastern Yucatan; petrology of Triassic-Jurassic sandstones of northeastern Mexico. Mailing Add: Dept of Earth Sci Univ of New Orleans New Orleans LA 70122

WARD, WILLIAM FRANCIS, b Erie, Pa, June 19, 28. PARASITOLOGY. Educ: Gannon Col, BS, 50; Univ Notre Dame, MS, 55. Prof Exp: Instr biol, Col St Mary, Utah, 55-57; asst prof, 57-66, from actg chmn dept to chmn dept, 64-72, ASSOC PROF BIOL, ROSEMONT COL, 66- Mem: AAAS; Am Soc Parasitol; Am Inst Biol Sci. Res: Interrelationships between intestinal parasites and the bacterial flora. Mailing Add: Dept of Biol Rosemont Col Rosemont PA 19010

WARD, WILLIAM ROGER, b Kansas City, Kans, Jan 11, 44; m 67; c 2. PLANETARY SCIENCES. Educ: Univ Mo-Kansas City, BS(physics) & BS(math), 68; Calif Inst Technol, PhD(planetary sci), 72. Prof Exp: Res fel planetary sci, Calif Inst Technol, 73; RES ASSOC ASTRON, HARVARD COL OBSERV, 73- Mem: Am Geophys Union; Am Astron Soc. Res: Formation and dynamical evolution of the solar system. Mailing Add: Ctr Astrophys Harvard Col Observ 60 Garden St Cambridge MA 02138

WARDE, CHARLES JOSEPH, b Castlebar, Ireland, July 3, 40; US citizen; m 69; c 3. ELECTROCHEMISTRY. Educ: Univ Col, Dublin, BSc, 63, PhD(electrochem), 69. Prof Exp: Engr, 68-69, SR ENGR, RES LABS, WESTINGHOUSE ELEC CORP, 69- Mem: Electrochem Soc; Am Chem Soc; Fine Particle Soc. Res: Fuel cells, metal-air batteries, hydrogen generation, water splitting and electro-analysis. Mailing Add: 2502 Hollywood Dr Pittsburgh PA 15235

WARDELL, JOE RUSSELL, JR, b Omaha, Nebr, Nov 11, 29; m 52; c 3. PHARMACOLOGY, PHYSIOLOGY. Educ: Creighton Univ, BS, 51; Univ Nebr, MSc, 59, PhD(pharmacol, physiol), 62. Prof Exp: Sr pharmacologist, 62-65, group leader pharmacol, 64-68, asst dir pharmacol, 68-71, assoc dir pharmacol & mission dir cardiopulmonary res area, 71-75, ASSOC DIR BIOL RES & MISSION DIR CARDIOVASC RES AREA, SMITH KLINE & FRENCH LABS, 75- Mem: AAAS; Am Chem Soc; NY Acad Sci; Am Soc Pharmacol & Exp Therapeut; Am Acad Allergy. Res: Cardiovascular and respiratory pharmacology; immunopharmacology; autonomic pharmacology; regulation of biosynthesis and secretion of respiratory mucus. Mailing Add: Dept of Biol Sci PO Box 7929 Smith Kline & French Labs Philadelphia PA 19101

WARDELL, WILLIAM LEWIS, b Glendale, Calif, Nov 19, 40. PLANT PHYSIOLOGY. Educ: Univ Calif, Davis, BS, 62; Univ Wis, Madison, PhD(plant physiol), 68. Prof Exp: NIH fel, Inst Cancer Res, 68-70; NIH fel, Inst Plant Develop, Univ Wis, 70-71, Am Cancer Soc instnl res grant, 71; ASST PROF PLANT PHYSIOL, UNIV MD BALTIMORE COUNTY, 71- Mem: AAAS; Smithsonian Inst; Am Soc Plant Physiol; Soc Develop Biol. Res: Role of nucleic acids in the transition of plants from vegetative to reproductive growth. Mailing Add: Dept of Biol Sci Univ of Md Baltimore County Catonsville MD 21228

WARDELL, WILLIAM MICHAEL, b Christchurch, NZ, Nov 15, 38; m 65; c 2. CLINICAL PHARMACOLOGY. Educ: Oxford Univ, BA, 61, DPhil(pharmacol), 64, BM & BCh, 67, DM, 73. Prof Exp: Intern med, Radcliffe Infirmary, Oxford, 67; intern med & surg, Dunedin Hosp, Univ Otago, NZ, 68; med res officer clin pharmacol & toxicol, NZ Med Res Coun, 69; lectr clin pharmacol, Med Sch, Univ Otago, NZ, 70; instr clin pharmacol, 71-73, ASST PROF PHARMACOL, TOXICOL & MED, MED CTR, UNIV ROCHESTER, 73- Concurrent Pos: Hon clin asst, Otago Hosp Bd, Dunedin, NZ, 69-70. Mem: Am Soc Pharmacol & Exp Therapeut; Am Soc Clin Pharmacol & Therapeut; Am Col Clin Pharmacol; Australasian Soc Clin & Exp Pharmacol. Res: Design, methodology and analysis of drug studies in man; analgesic and hypnotic drugs in man; regulation and drug development; adverse drug reactions. Mailing Add: Dept of Pharmacol & Toxicol Univ of Rochester Med Ctr Rochester NY 14642

WARDEN, HERBERT EDGAR, b Cleveland, Ohio, Aug 30, 20; m 58; c 4. SURGERY, THORACIC SURGERY. Educ: Washington & Jefferson Col, BS, 42; Univ Chicago, MD, 46; Am Bd Surg, dipl, 58; Bd Thoracic Surg, dipl, 63. Prof Exp: Intern, Clin, Univ Chicago, 46-47; asst resident surg, Hosps, Univ Minn, 51-56, res asst, 53-55, res asst physiol, 55-56, clin instr surg, 55-57, chief resident, 56-57, instr surg, 57-60; assoc prof, 60-62, PROF SURG, MED CTR, W VA UNIV, 62- Concurrent Pos: Coordr, USPHS Cardiovasc Surg Training Prog, Univ Minn, 56-60; consult, Anoka State Hosp, Minn, 58-59. Honors & Awards: Cert of Merit, AMA, 55 & 58, Hektoen Gold Medal, 57; Lasker Award, Am Pub Health Asn, 55; Citation of Merit, Am Col Surgeons, 57. Mem: Soc Univ Surgeons; Am Asn Thoracic Surgeons; Am Surg Asn; Soc Thoracic Surgeons; fel Am Col Surgeons. Res: Cardiovascular surgery and physiology. Mailing Add: Dept of Surg WVa Univ Med Ctr Morgantown WV 26506

WARDEN, JOSEPH TALLMAN, b Huntington, WVa, Aug 7, 46. BIOPHYSICAL CHEMISTRY. Educ: Furman Univ, BS, 68; Univ Minn, PhD(phys chem), 72. Prof Exp: Vis scientist biophys, State Univ Leiden, 72-73; chemist, Univ Calif, Berkeley, 73-75; ASST PROF CHEM, RENSSELAER POLYTECH INST, 75- Mem: Am Chem Soc; Am Soc Photobiol; Biophys Soc. Res: Electron spin resonance investigations of electron transfer components and mechanisms in photosynthesis and mitochondrial respiration; structure and function of heme proteins. Mailing Add: Dept of Chem Rensselaer Polytech Inst Troy NY 12181

WARDEN, WILLIAM KENT, b Argyle, Minn, June 18, 18; m 46; c 2. POULTRY NUTRITION. Educ: Univ Ill, BS, 48, MS, 51; Mich State Univ, PhD(poultry nutrit), 59. Prof Exp: Asst swine nutrit, Univ Ill, 50-51; in charge poultry nutrit res, Chas Pfizer & Co, Inc, 51-56; nutritionist, Haynes Milling Co, Inc, 56; asst poultry nutrit, Mich State Univ, 57-59, poultry nutrit specialist, Exten, 59-64; FIELD TECH CONSULT POULTRY & ANIMAL NUTRIT, DAWES LABS INC, 64- Concurrent Pos: Consult, Haynes Milling Co, Inc, Ind, 56-64. Mem: Poultry Sci Asn; Am Feed Mfrs Nutrit Coun; Am Res Coun. Res: Mechanisms of action of antibiotics in growth stimulation of poultry, especially intestinal microorganisms. Mailing Add: Dawes Labs Inc 450 State St Chicago Heights IL 60411

WARDESKA, JEFFREY GWYNN, b Irondale, Ohio, June 13, 41; m 65; c 1. INORGANIC CHEMISTRY. Educ: Mt Union Col, BSc, 63; Ohio Univ, PhD(inorg chem), 67. Prof Exp: Asst prof, 67-73, ASSOC PROF CHEM, E TENN STATE UNIV, 73- Mem: Am Chem Soc. Res: Coordination chemistry; reactions of coordination compounds of transition metals; thiocyanate complexes; aminoalcohol complexes. Mailing Add: Dept of Chem E Tenn State Univ Johnson City TN 37601

WARDI, AHMAD HASSAN, b Baghdad, Iraq, June 8, 29; US citizen; m 59; c 3. BIOCHEMISTRY, ANALYTICAL CHEMISTRY. Educ: Higher Teacher Col, Baghdad, BSC, 51; Univ Mich, Ann Arbor, MSC, 54; Wayne State Univ, PhD(chem), 59. Prof Exp: Scientist, 59-72, DIR BIOCHEM RES DEPT, WARREN STATE HOSP, 72- Concurrent Pos: NIH grant, 65-72. Mem: AAAS; Am Chem Soc. Res: Cerebral and ocular pentose-containing mucoproteins. Mailing Add: Biochem Res Dept Warren State Hosp PO Box 240 Warren PA 16365

WARDLAW, JANET MELVILLE, b Toronto, Ont, June 20, 24. NUTRITION, HOME ECONOMICS. Educ: Univ Toronto, BA, 46; Univ Tenn, MS, 50; Pa State Univ, PhD(nutrit), 63. Prof Exp: Dietition, Can Red Cross Soc, Toronto, 47-49; nutritionist, Mich Dept Health, 50-53 & Toronto Dept Pub Health, 53-56; asst prof nutrit, Fac Household Sci, Univ Toronto, 56-60 & 63-64, assoc prof, Fac Food Sci, 64-66; assoc dean-dean designate, 67-69, PROF NUTRIT, COL FAMILY & CONSUMER STUDIES, UNIV GUELPH, 66-, DEAN, 69- Honors & Awards: Stuart's Branded Foods Ltd Award, Can Dietetic Asn, 71. Mem: AAAS; Am Dietetic Asn; Nutrit Soc Can; Can Dietetic Asn (treas, 65-67). Res: Sodium regulation during pregnancy; body composition and feeding frequency; community nutrition. Mailing Add: Col Family & Consumer Studies Univ of Guelph Guelph ON Can

WARDLAW, NORMAN CLAUDE, b Trinidad, Brit WI, Nov 22, 35. GEOLOGY. Educ: Univ Manchester, BSc, 57; Univ Glasgow, PhD(geol), 60. Prof Exp: Spec lectr sedimentation, Univ Sask, 60-62, from asst prof to assoc prof geol, 62-72; assoc prof, 72-74, PROF GEOL, UNIV CALGARY, 74- Mem: Soc Econ Paleontologists & Mineralogists; Geol Soc Am; Am Asn Petrol Geologists. Res: Sedimentary petrology and diagenesis of limestones; sedimentary rocks; petrology and geochemistry of evaporites. Mailing Add: Dept of Geol Univ of Calgary Calgary AB Can

WARDLAW, WILLIAM PATTERSON, b Los Angeles, Calif, Mar 3, 36; m 63; c 1. MATHEMATICS. Educ: Rice Inst, BA, 58; Univ Calif, Los Angeles, MA, 64, PhD(math), 66. Prof Exp: Asst prof, Univ Ga, 66-72; ASST PROF MATH, US NAVAL ACAD, 72- Mem: Am Math Soc; Math Asn Am. Res: Lie algebras and Chevalley groups; universal algebra. Mailing Add: Dept of Math US Naval Acad Annapolis MD 21401

WARDLE, JOHN FRANCIS CARLETON, b Hemel Hempstead, Eng, May 8, 45; m 71. RADIO ASTRONOMY. Educ: Univ Cambridge, BA, 66; Univ Manchester, MSc, 68, PhD(radio astron), 69. Prof Exp: Res asst radio astron, Nat Radio Astron Observ, 69-71; instr astrophys, 71-73, ASST PROF ASTROPHYS, BRANDEIS UNIV, 73- Mem: Am Astron Soc; Royal Astron Soc. Res: Extragalactic radio astronomy; cosmology. Mailing Add: Dept of Physics Brandeis Univ Waltham MA 02154

WARDNER, CARL ARTHUR, b Fisher, Minn, July 13, 04; m 30; c 2. ORGANIC CHEMISTRY. Educ: Univ NDak, BS, 27, MS, 29; Univ Pittsburgh, PhD, 32. Prof Exp: Res chemist, Daugherty Ref Co, Pa, 31-35 & Flaat Co, NDak, 35-57; from asst prof to prof chem, 58-71, dir sci inst, 67-71, EMER PROF CHEM, UNIV N DAK, 71- Mem: AAAS; Am Chem Soc; Nat Sci Teachers Asn; Am Inst Chem. Res: Fixation of plant nutrients in alkaline soils; humic acids; soil and precipitation nutrients and their limnological significance; science education. Mailing Add: 1919 Chestnut St Grand Forks ND 58201

WARDOWSKI, WILFRED FRANCIS, II, b Pontiac, Mich, May 23, 37; m 74; c 1. HORTICULTURE. Educ: Mich State Univ, BS, 59, MS, 61, PhD(pomol), 66. Prof Exp: Foreman fruit prod, Blossom Orchard, Leslie, Mich, 61-63; midwest rep tech exten agr chem, Agr Div, Upjohn Co, 66-69; ASSOC PROF EXTEN SERV, CITRUS HARVESTING & HANDLING, AGR RES & EDUC CTR, UNIV FLA, 69- Mem: Am Soc Hort Sci. Res: Nutrition and histology of apples; agricultural chemicals research and development; harvesting and handling of fresh market citrus. Mailing Add: Agr Res & Educ Ctr PO Box 1088 Lake Alfred FL 33850

WARDWELL, JAMES FLETCHER, b Rome, NY, Nov 21, 09; m 31; c 2. MATHEMATICS. Educ: Hamilton Col, AB, 31; Johns Hopkins Univ, AM, 34, PhD(math), 35. Prof Exp: Instr math, Johns Hopkins Univ, 31-35; from instr to prof, 35-73, dir div natural sci & math, 58-63 & 66-67, EMER PROF MATH, COLGATE UNIV, 73- Concurrent Pos: Vis lectr, Univ Calif, 54. Mem: AAAS; assoc Am Math Soc; assoc Math Asn Am. Res: Set theoretic topology. Mailing Add: Dept of Math Colgate Univ Hamilton NY 13346

WARE, ALAN ALFRED, b Portsmouth, Eng, Dec 4, 24; US citizen; m 52; c 4. PLASMA PHYSICS. Educ: Imp Col, Univ London, BSc & ARCS, 44, PhD(physics) & DIC, 47. Prof Exp: Res asst plasma physics, Imp Col, Univ London, 47-51; sect leader, Res Lab, Assoc Elec Industs, UK, 51-63; consult, Gen Atomic, San Diego, 60-61; group leader plasma physics, Culham Lab, UK Atomic Energy Authority, 63-65; asst mgr res oper plasma physics, Aerojet Gen Corp, 65-69; RES SCIENTIST PLASMA PHYSICS, UNIV TEX, AUSTIN, 69- Concurrent Pos: Consult, Los Alamos Sci Lab, 71- Mem: Am Phys Soc. Res: Theory of Tokamak plasmas in research aimed at controlled nuclear fusion power. Mailing Add: Fusion Res Ctr Univ of Tex Austin TX 78712

WARE, ARNOLD GRASSEL, b Butler, Ill, June 1, 15; m 41; c 3. BIOCHEMISTRY. Educ: Carthage Col, BA, 37; Univ Colo, MS, 39, PhD(biochem), 42. Prof Exp: Asst biochem, Sch Med, Univ Colo, 38-42; res assoc, Col Med, Wayne State Univ, 46-49; from asst prof to assoc prof biochem, 49-56, PROF PATH, SCH MED, UNIV SOUTHERN CALIF, 56-; PRES, BIO-CONSULTS, INC, 71- Concurrent Pos: Head clin biochemist, Los Angeles County-Univ Southern Calif Med Ctr, 49- Mem: AAAS; Am Soc Biol Chemists; Am Chem Soc; Am Asn Clin Chemists. Res: Blood coagulation; clinical chemistry. Mailing Add: Dept of Path Univ of Southern Calif Sch Med Los Angeles CA 90007

WARE, CAROLYN BOGARDUS, b Baltimore, Md, Oct 15, 30; div; c 1. NEUROANATOMY, PHYSIOLOGICAL PSYCHOLOGY. Educ: Western Reserve Univ, BA, 52; Columbia Univ, cert phys ther, 53; Univ Buffalo, MEd, 63; Duke Univ, PhD(psychol), 71. Prof Exp: Instr phys ther, Univ NC, 63-66; instr anat, State Univ NY Downstate Med Ctr, 70-75, ASST PROF & ASST DEAN, SCH HEALTH RELATED PROFESSIONS, STATE UNIV NY, BUFFALO, 75- Mem: Soc Neurosci; Am Asn Anatomists; Sigma Xi. Res: Comparative neuroanatomy, visual system and cerebellum; central nervous system and behavior, especially visual discrimination. Mailing Add: Sch Health Related Professions State Univ of NY Buffalo NY 14214

WARE, DANIEL MORRIS, animal ecology, see 12th edition

WARE, DONNA MARIE EGGERS, b Springfield, Mo, Oct 1, 42; m 68. PLANT TAXONOMY. Educ: Southwest Mo State Col, BA, 64; Vanderbilt Univ, PhD(biol), 69. Prof Exp: HERBARIUM CUR VASCULAR PLANTS, COL WILLIAM & MARY, 69- Concurrent Pos: NSF grant-in-aid, Highlands Biol Sta, NC, 70. Mem: Am Soc Plant Taxon. Res: Floristics; revisional and biosystematic taxonomy. Mailing Add: Herbarium Dept Biol Col William & Mary Williamsburg VA 23185

WARE, FREDERICK, b Omaha, Nebr, June 16, 28; m 50; c 5. PHYSIOLOGY, INTERNAL MEDICINE. Educ: Univ Nebr, BS, 49, MS, 53, PhD & MD, 56. Prof Exp: Instr physiol, 53-56, instr internal med, 55-56, asst prof physiol & pharmacol, 60-62, assoc prof physiol & asst prof internal med, 62-70, PROF PHYSIOL, BIOPHYS & INTERNAL MED, COL MED, UNIV NEBR, OMAHA, 70- Mem: Am Physiol Soc; Am Soc Nephrology; Int Soc Nephrology; Am Col Physicians; Am Soc Artificial Internal Organs. Res: Membrane electrophysiology of skeletal muscle and heart; principles of electrocardiography; biophysics of renal function. Mailing Add: Dept of Physiol-Biophys Univ of Nebr Col of Med Omaha NE 68105

WARE, GEORGE HENRY, b Avery, Okla, Apr 27, 24; m 55; c 4. PLANT ECOLOGY. Educ: Univ Okla, BS, 45, MS, 48; Univ Wis, PhD(bot), 55. Prof Exp: From asst prof to prof bot, Northwestern State Univ, 48-68; ECOLOGIST & DENDROLOGIST, MORTON ARBORETUM, 68- Mem: AAAS; Ecol Soc Am. Res: Vegetation of the Southeastern United States; ecology of swamp and floodplain forests; ecology of urban trees. Mailing Add: Morton Arboretum Lisle IL 60532

WARE, GEORGE WHITAKER, JR, b Pine Bluff, Ark, Aug 27, 27; m 52; c 3. ENTOMOLOGY. Educ: Univ Ark, BS, 51, MS, 52; Kans State Univ, PhD(entom), 56. Prof Exp: Assoc prof entom, Ohio State Univ, 56-66; PROF ENTOM & HEAD DEPT, UNIV ARIZ, 67- Concurrent Pos: Consult, Environ Protection Agency, Washington & Coun Grad Schs US, 74- Mem: Entom Soc Am; Am Chem Soc; Soc Toxicol. Res: Insect toxicology; insecticide metabolism and residues; radiation effects in insects; pesticide chemistry. Mailing Add: Dept of Entom Univ of Ariz Tucson AZ 85721

WARE, GLEN CHASE, physical chemistry, deceased

WARE, GLENN OREN, b Athens, Ga, Dec 8, 41; m 67; c 1. APPLIED STATISTICS, OPERATIONS RESEARCH. Educ: Univ Ga, BSF, 63, PhD(forest biomet), 68; Yale Univ, MF, 64. Prof Exp: Res forester, Hudson Pulp & Paper Corp, 64-65; ASST PROF STATIST, UNIV GA, 66- Mem: Soc Am Foresters. Res: Application of mathematical and statistical techniques in the physical and biological sciences. Mailing Add: Dept Statist & Comput Sci Univ of Ga Athens GA 30602

WARE, JAMES GARETH, b Baltimore, Md, Aug 19, 29; m 55; c 1. MATHEMATICS. Educ: Duke Univ, BS, 50; George Peabody Col, MA, 51, PhD(math), 62. Prof Exp: Teacher, McCallie Sch, Tenn, 52-54 & 59-60, dept chmn, 60-65; assoc prof, 67-73, PROF MATH, UNIV TENN, CHATTANOOGA, 73-, CHMN DEPT, 68- Mem: Math Asn Am. Res: Mathematical analysis; geometry. Mailing Add: Dept of Math Univ of Tenn Chattanooga TN 37401

WARE, JAMES H, b Detroit, Mich, Oct 27, 41; m 72; c 1. MATHEMATICAL STATISTICS. Educ: Yale Univ, BA, 63; Stanford Univ, MA, 65, PhD(statist), 69. Prof Exp: Instr statist, Calif State Col, Hayward, 69-70; MATH STATISTICIAN, NIH, 71- Concurrent Pos: Assoc ed, J Am Statist Asn, 73-75; adj assoc prof statist, George Washington Univ, 75- Mem: Am Statist Asn; Biomet Soc. Res: Interest in the areas of nonparametric methods and sequential analysis; survival data analysis and methods for data analysis in clinical trials of chronic disease. Mailing Add: 8501 Farrell Dr Chevy Chase MD 20015

WARE, LAWRENCE L, JR, b Montgomery, WVa, Sept 12, 20; m 46; c 3. MEDICAL MICROBIOLOGY. Educ: Roosevelt Univ, BS, 50. Prof Exp: Res asst bact, Univ Chicago, 46-49; bacteriologist, Med Bact Div, US Army Chem Corp, Ft Detrick, Md, 51-53, Process Res Div, 53-57, bio-eng br, Pilot Plant, 59; lab dir, Div Indian Health, USPHS, Ariz, 59-63, area lab dir, 63-65; microbiologist & sci info specialist, 65-72, CHIEF, SCI & TECH INFO DIV, US ARMY MED RES & DEVELOP COMMAND, 72- Concurrent Pos: Ed, newsletter, Nat Registry Microbiol, 65- Mem: Am Soc Microbiol; NY Acad Sci; Asn Mil Surg US; Soc Indust Microbiol; Am Soc Info Sci. Res: Medical microbiology and immunology; microbial fermentations; medical information and documentation storage and retrieval. Mailing Add: Sci & Tech Info Div Info Systs Off US Army Med Res & Develop Command Rm 8G031 Forrestal Bldg Washington DC 20314

WARE, RAY WILSFORD, b Snyder, Tex, Aug 7, 24; m 49; c 2. MEDICAL ADMINISTRATION, BIOENGINEERING. Educ: Univ Tex, BS, 50, MD, 55. Prof Exp: Med intern, William Beaumont Army Hosp, El Paso, Tex, 55-56; instr physiol & flight surgeon, US Air Force Sch Aviation Med, Randolph AFB, Tex, 56-59; res flight surgeon & chief biomed instrumentation sect, US Air Force Sch Aerospace Med, Brooks AFB, 60-64; mgr med instrumentation res, Electronic Systs Res Dept, Southwest Res Inst, 64-70, inst scientist, Dept Bioeng, 70-72; ASSOC PROF SURG, MED CTR, UNIV KY & ASSOC CHIEF STAFF RES, VET ADMIN HOSP, LEXINGTON, 72- Mem: AMA; Aerospace Med Asn; Inst Elec & Electronics Engrs; Sigma Xi; Asn Advan Med Instrumentation. Res: Medical instrumentation research; surgical research. Mailing Add: Vet Admin Hosp Univ of Ky Med Ctr Lexington KY 40507

WARE, ROGER PERRY, b San Francisco, Calif, Apr 2, 42; m 65; c 1. ALGEBRA. Educ: Univ Calif, Berkeley, AB, 65; Univ Calif, Santa Barbara, MA, 68, PhD(math), 70. Prof Exp: Asst prof math, Northwestern Univ, Evanston, 70-72 & Univ Kans, Lawrence, 72-74; ASSOC PROF MATH, PA STATE UNIV, 74- Concurrent Pos: NSF grants, 71-72, 73 & 74. Mem: Am Math Soc; Math Asn Am. Res: Quadratic forms; ring theory; commutative algebra. Mailing Add: Dept of Math Pa State Univ University Park PA 16802

WARE, STANTON JAMES, b Ann Arbor, Mich, Aug 31, 12; m 40; c 3. ENVIRONMENTAL MANAGEMENT. Educ: Univ Mich, Ann Arbor, BS, 34, MA, 37, PhD(geog), 42; US Naval Acad, MS, 44. Prof Exp: Cartog engr, US Army Air Force, 42, photogram engr, 43; meteorologist, US Navy, 44-46; chief water surv, Bur Reclamation, US Dept Interior, 46-51; chief water projs, USPHS, 51-54, pub health engr, Water Supply & Pollution Control Prog, 54-56; prin res scientist, C W Thornthwaite Assocs, 56-58; staff land & water opers, Bur Reclamation, US Dept Interior, 58-62, chief foreign training, 62-64, chief res & tech coord, 65-66, supv phys scientist, Off Water Resources Res, 66-74, chief eng syst div, 74-75; STAFF DIR ENVIRON MANPOWER COMT, NAT RES COUN/NAT ACAD SCI, 75- Honors & Awards: Tech Excellence Award, Off Water Resources Res, US Dept Interior, 72, Distinguished Serv Award, 74. Mem: Conf Fed Environ Engrs (pres, 67-68); Asn Am Geogrs; Am Meteorol Soc; Am Water Works Asn; AAAS. Res: Air and water quality; environmental pollution control and related manpower supply and demand factors; water resources development. Mailing Add: Nat Res Coun/Nat Acad Sci JH-604 2101 Constitution Ave NW Washington DC 20418

WARE, STEWART ALEXANDER, b Stringer, Miss, Aug 20, 42; m 68. PLANT ECOLOGY. Educ: Millsaps Col, BA, 64; Vanderbilt Univ, PhD(biol), 68. Prof Exp: Asst prof, 67-72, ASSOC PROF BIOL, COL WILLIAM & MARY, 72- Concurrent Pos: Ed, Jeffersonia, Va Bot Newsletter, 69-74. Mem: Ecol Soc Am. Res: Vegetation of the southeastern United States; Quercus systematics and ecology; physiological ecology of rock outcrop plants; ecology and distribution of Talinum. Mailing Add: Dept of Biol Col of William & Mary Williamsburg VA 23185

WARE, WALTER ELISHA, b Jacksonville, Fla, June 1, 33; m 55; c 3. PHYSICS. Educ: US Naval Acad, BS, 55; Univ Colo, PhD(physics), 62. Prof Exp: From instr to assoc prof physics, US Air Force Acad, 62-65, tenure assoc prof, 65-66; RES SCIENTIST, NUCLEAR TECHNOL LAB, KAMAN SCI CORP, 66- Concurrent Pos: Proj consult, Kaman Nuclear, 62-66. Mem: Am Asn Physics Teachers; Inst Elec & Electronics Engrs. Res: Nuclear structure theory; effects of nuclear weapons; electromagnetic theory; quantum theory. Mailing Add: Kaman Sci Corp 1500 Garden of the Gods Rd Colorado Springs CO 80907

WARE, WILLIAM ROMAINE, b Portland, Ore, June 13, 31; m 54. PHYSICAL CHEMISTRY. Educ: Reed Col, BA, 53; Univ Rochester, PhD, 58. Prof Exp: Res chemist, Parma Res Ctr, Union Carbide Corp, 57-60; res assoc chem, Univ Minn, Minneapolis, 61; from asst prof to assoc prof, San Diego State Col, 62-66; from assoc prof to prof, Univ Minn, Minneapolis, 67-71; PROF CHEM, UNIV WESTERN ONT, 71- Mem: Am Chem Soc; Am Phys Soc; fel Chem Inst Can. Res: Molecular photochemistry and photophysics. Mailing Add: Dept of Chem Univ of Western Ont London ON Can

WAREHAM, ELLSWORTH EDWIN, b Avinger, Tex, Oct 3, 14; m 50; c 5. THORACIC SURGERY. Educ: Col Med Evangelists, MD, 42; Am Bd Surg, dipl, 54; Bd Thoracic Surg, dipl, 55. Prof Exp: From asst prof to assoc prof, 58-64, PROF SURG, SCH MED, LOMA LINDA UNIV, 64- Concurrent Pos: Attend physician, White Mem, Los Angeles Gen & Olive View Hosps, 55-; mem staff, Glendale Sanitarium & Hosp & Lavina Sanitorium, 55- Mem: Am Thoracic Soc; AMA. Res: Cardiovascular surgery; open heart surgery; use of heart-lung machine. Mailing Add: Dept of Surg Loma Linda Univ Sch of Med Loma Linda CA 92354

WAREHAM, RICHARD THURMAN, botany, see 12th edition

WARES, GORDON WEBB, astrophysics, see 12th edition

WARF, JAMES CURREN, b Nashville, Tenn, Sept 1, 17; m 65; c 3. INORGANIC CHEMISTRY. Educ: Univ Tulsa, BS, 39; Iowa State Univ, PhD(inorg chem), 46. Prof Exp: Jr chemist, Phillips Petrol Co, Okla, 40-41; instr chem, Univ Tulsa, 41-42; group leader, Manhattan Proj, Iowa State Univ, 42-47; Guggenheim fel, Univ Berne, 47-48; asst prof, 48-50, ASSOC PROF CHEM, UNIV SOUTHERN CALIF, 50- Concurrent Pos: Vis prof, Univ Indonesia, 57-59, Airlangga Univ, Indonesia, 62-64, Tech Univ Vienna, 69-70 & Nat Univ Malaysia, Kuala Lumpur, 74-75; consult, Jet Propulsion Lab, Calif Inst Technol. Mem: Am Chem Soc; Fedn Am Sci. Res: Hydrides of heavy metals; chemistry in liquid ammonia; chemistry of europium and ytterbium. Mailing Add: Dept of Chem Univ of Southern Calif Los Angeles CA 90007

WARFEL, DAVID ROSS, b Pana, Ill, Sept 25, 42; m 65; c 1. ORGANIC CHEMISTRY, POLYMER CHEMISTRY. Educ: Carthage Col, BA, 64; Univ Tenn, Knoxville, PhD(chem), 70. Prof Exp: Scientist polymer chem, Koppers Co, Inc, Monroeville, 69-74; SR SCIENTIST POLYMER CHEM, ARCO/POLYMERS, INC, MONROEVILLE, 74- Mem: Am Chem Soc. Res: Influence of catalysis on the reactivity of organometallic compounds. Mailing Add: 40-E Colony Sq Pittsburgh PA 15239

WARFEL, JOHN HIATT, b Marion, Ind, Mar 3, 16; m 42; c 3. ANATOMY. Educ: Capital Univ, BSc, 38; Ohio State Univ, MSc, 41; Western Reserve Univ, PhD, 48. Prof Exp: Asst, Western Reserve Univ, 46-48; instr, 49-54, assoc, 54-56, asst prof, 56-70, ASSOC PROF ANAT, SCH MED, STATE UNIV NY BUFFALO, 70- Mem: Am Asn Anatomists. Res: Gross anatomy. Mailing Add: 153 Walton Dr Buffalo NY 14226

WARFIELD, ALBERT HARRY, b Baltimore, Md, Aug 4, 38; m 63; c 2. ORGANIC CHEMISTRY. Educ: Univ Md, BS, 60, MS, 63, PhD(pharmaceut chem), 65. Prof Exp: Sr chemist, Liggett & Myers Tobacco Co, NC, 65-71, SR CHEMIST, LIGGETT & MYERS INC, 71- Mem: Am Chem Soc; Am Pharmaceut Asn. Res: Alkaloid and terpene synthesis; tobacco chemistry; synthesis and isolation of flavors and other materials of biological interest. Mailing Add: 404 Englewood Ave Durham NC 27701

WARFIELD, GEORGE, b Piombino, Italy, Apr 21, 19; nat US; m 45; c 3. SOLID STATE ELECTRONICS. Educ: Franklin & Marshall Col, BS, 40; Cornell Univ, PhD(physics), 49. Prof Exp: From asst prof to prof elec eng, Princeton Univ, 49-74; PROF ELEC ENG & EXEC DIR, INST ENERGY CONVERSION, UNIV DEL, 74- Mem: Am Phys Soc; Inst Elec & Electronics Engrs; Int Solar Energy Soc. Res: Photovoltaic cells; solid state device physics; insulator electronics; behavior of electrons in insulators. Mailing Add: Inst of Energy Concersion Univ of Del Newark DE 19711

WARFIELD, PETER FOSTER, b Rye, NY, Aug 4, 18; m 42; c 3. ORGANIC CHEMISTRY. Educ: Hamilton Col, BS, 40; Univ Ill, MS, 41, PhD(org chem), 44. Prof Exp: Asst chem, Univ Ill, 42-44; res chemist, Bakelite Corp, NJ, 44-45; from res chemist to sr res chemist, Ansco Div, Gen Aniline & Film Corp, 45-49, develop specialist, 49-52; chemist, 52-60, tech serv rep, 60-63, res chemist, 63-68, SR RES CHEMIST, PHOTO PRODS DEPT, E I DU PONT DE NEMOURS & CO, INC, 69- Mem: Am Chem Soc; Soc Photog Sci & Eng. Res: Hindered Grignard reactions; aliphatic polyamines; phenolic resins; synthetic peptides; restrainers for photographic gelatin; photographic emulsions; cyanine dyes; color photography; color formers; emulsions and processing; photopolymer printing plates. Mailing Add: Photo Prods Dept EI du Pont de Nemours & Co Inc Parlin NJ 08859

WARFIELD, ROBERT BRECKINRIDGE, JR, b Lexington, Ky, Aug 20, 40; m 64; c 2. ALGEBRA. Educ: Haverford Col, BA, 62; Harvard Univ, PhD(math), 67. Prof Exp: Fel, NMex State Univ, 67-68; from asst prof to assoc prof, 68-74, PROF MATH, UNIV WASH, 74- Concurrent Pos: NSF res grant, Univ Wash, 69-; consult, Proj Seed, Seattle Pub Schs, 71. Mem: Math Asn Am; Am Math Soc. Res: Commutative and noncommutative rings; module theory; infinite Abelian groups; nilpotent and solvable groups. Mailing Add: Dept of Math Univ of Wash Seattle WA 98195

WARFIELD, ROBERT WELMORE, b Asbury Park, NJ, Oct 11, 26; m 55; c 2. POLYMER CHEMISTRY. Educ: Univ Va, BS, 50. Prof Exp: Chemist, US Bur Mines, Md, 50-55; chemist, 55-64, SR SCIENTIST, NAVAL SURFACE WEAPONS CTR, WHITE OAK, 64- Honors & Awards: Meritorious Serv Award, US Dept Interior, 54; Honor Citation, Secy Interior, 55; Meritorious Civilian Serv Award, Naval Ord Lab, 61. Mem: Am Chem Soc. Res: Patentee in field; chemistry and physics of the solid state of polymers; compressibility and electrical properties of polymers; transitions of polymers; polymerization kinetics. Mailing Add: 22712 Ward Ave Hereford Hills Germantown MD 20767

WARFIELD, VIRGINIA MCSHANE, b Charlottesville, Va, Sept 30, 42; m 64; c 2. MATHEMATICAL ANALYSIS. Educ: Bryn Mawr Col, AB, 63; Brown Univ, MA, 65, PhD(math), 71. Prof Exp: Seattle dir math, Spec Elem Educ Disadvantaged Proj, 70-73; LECTR MATH, UNIV WASH, 73- Mem: Sigma Xi. Res: Stochastic integrals and stochastic control theory. Mailing Add: Dept of Math Univ of Wash Seattle WA 98195

WARGA, JACK, b Warsaw, Poland, Dec 5, 22; nat US; m 49; c 2. MATHEMATICS. Educ: NY Univ, PhD(math), 50. Prof Exp: Assoc mathematician, Reeves Instrument Corp, 51-52; prin engr, Repub Aviation Corp, 52-53; sr mathematician & head math dept, Electrodata Div, Burroughts Corp, 54-56; fel, Weizmann Inst Sci, Israel, 56-57; sr staff mathematician & mgr math dept, Res & Adv Develop Div, Avco Corp, 57-66; PROF MATH, NORTHEASTERN UNIV, 66- Concurrent Pos: Ed, J Control, Soc Indust & Appl Math. Mem: AAAS; Am Math Soc; Soc Indust & Appl Math. Res: Mathematical control theory. Mailing Add: Dept of Math Northeastern Univ Boston MA 02115

WARGEL, ROBERT JOSEPH, b Evansville, Ind, Dec 24, 40; m 70. BIOCHEMISTRY, MICROBIOLOGY. Educ: Univ Evansville, BA, 66; Northwestern

Univ, PhD(chem), 70. Prof Exp: Chemist, City of Evansville, Ind, 64-66 & Mead Johnson & Co, summer 66; MICROBIOLOGIST, KRAFTCO CORP, 70- Mem: Am Chem Soc; Am Soc Microbiol. Res: Enzyme isolation and purification; preparation and use of solid support enzyme systems; peptide isolation and chemical peptide synthesis. Mailing Add: Kraftco Corp 801 Waukegan Rd Glenview IL 60025

WARHEIT, ISRAEL ALBERT, information science, see 12th edition

WARING, CHARLES EMMETT, b Philadelphia, Pa, Jan 24, 09; m 36. PHYSICAL CHEMISTRY. Educ: Muskingum Col, AB, 31; Ohio State Univ, MS, 34, PhD(chem), 36. Prof Exp: From instr to asst prof phys chem, Polytech Inst Brooklyn, 36-42; tech aide, Nat Defense Res Comt, Washington, DC, 42-43; liaison officer, Off Sci Res & Develop, Eng, 43-44; sci adv, First Allied Airborne Army, US Army, 44-45; tech ed, Summary Reports Group, Off Sci Res & Develop, Div War Res, Columbia Univ, 45-46; head dept chem, 46-66, PROF PHYS CHEM, UNIV CONN, 46- Concurrent Pos: Lalor fel, Oxford Univ, 39-40; asst tech dir res & head, Res Dept, US Naval Ord Test Sta, 61-62, consult, 53-; consult, Qm Corps, US Army, 55-61 & US Naval Propellant Plant; mem, Army Sci Adv Panel, 56-60 & chmn, Chem, Biol & Radiol Subcomt, 58-60; mem bd dirs, Convec Corp, 60-70; vpres res, Raycon Corp, 63-66 & Windsor Nuclear Corp, 66-72. Honors & Awards: Presidential Citation Merit, 46. Mem: AAAS; Am Chem Soc. Res: Kinetics and mechanisms of gas and liquid phase reactions; mechanism of free radical reactions; ultra high pressure kinetics; high energy reactions; chemiluminescence; magnetooptic rotation. Mailing Add: Dept of Chem Univ of Conn Storrs CT 06268

WARING, DEREK MORRIS HOLT, b Northern Ireland, June 16, 25; nat US; m 50; c 2. ORGANIC CHEMISTRY. Educ: Queen's Univ, Belfast, BSCh, 46, MSc, 47, PhD(chem), 49. Prof Exp: Res chemist, Albright & Wilson, Eng, 50-52; from res chemist to res assoc, 52-67, SR SUPVR RES & DEVELOP, E I DU PONT DE NEMOURS & CO, INC, 67- Mem: Am Chem Soc. Res: Stereochemistry; polynuclear aromatic hydrocarbons; polymer intermediates and chemistry. Mailing Add: Plastics Dept Chestnut Run Lab EI du Pont de Nemours & Co Inc Wilmington DE 19898

WARING, GEORGE HOUSTOUN, b Denver, Colo, July 15, 39; m 62; c 3. ANIMAL BEHAVIOR, VERTEBRATE ZOOLOGY. Educ: Colo State Univ, BS, 62, PhD(zool), 66; Univ Colo, MA, 64. Prof Exp: Asst prof, 66-72, ASSOC PROF ANIMAL BEHAV, SOUTHERN ILL UNIV, CARBONDALE, 72- Concurrent Pos: Guest prof, Univ Munich, 72-73; res prog dir, US Marine Mammal Comn, 74-75. Mem: AAAS; Animal Behav Soc; Am Soc Mammal; Ecol Soc Am; Am Ornith Union. Res: Communicative behaviors of vertebrates; development of intraspecies and interspecies relationships; social behaviors; vertebrate natural history; wildlife ecology; applied ethology. Mailing Add: Dept of Zool Southern Ill Univ Carbondale IL 62901

WARING, MARY GRACE, chemistry, see 12th edition

WARING, RICHARD C, b Excelsior Springs, Mo, Mar 25, 36; m 62; c 3. PHYSICS. Educ: William Jewell Col, BA, 58; Univ Ark, MS, 61. Prof Exp: From instr to asst prof, 60-72, ASSOC PROF PHYSICS, UNIV MO-KANSAS CITY, 72- Concurrent Pos: AEC grants, 63 & 66; NSF grant, 67; Off Water Resources Res grant, 72. Mem: Am Asn Physics Teachers. Res: Infrared reflectance spectroscopy. Mailing Add: Dept of Physics Univ of Mo Kansas City MO 64110

WARING, RICHARD H, b Chicago, Ill, May 17, 35; m 57; c 2. PLANT ECOLOGY. Educ: Univ Minn, St Paul, BS, 57, MS, 59; Univ Calif, Berkeley, PhD(bot), 63. Prof Exp: Asst prof, 63-72, ASSOC PROF FOREST ECOL, ORE STATE UNIV, 72- Concurrent Pos: Dep dir, Coniferous Forest Biome, 73- Mem: AAAS; Ecol Soc Am; Am Inst Biol Sci. Res: Ecosystem analysis of watersheds; environmental classification; physiological ecology; plant-water relationships. Mailing Add: Forest Res Lab Ore State Univ Corvallis OR 97330

WARING, ROBERT KERR, JR, b Palmerton, Pa, Aug 18, 28; m 54; c 4. PHYSICS. Educ: Va Mil Inst, BS, 50; Yale Univ, PhD, 55. Prof Exp: RES PHYSICIST, E I DU PONT DE NEMOURS & CO, INC, 55- Mem: Am Phys Soc. Res: Image recording systems; fine particle ferromagnetism. Mailing Add: Cent Res Dept EI du Pont de Nemours & Co Inc Wilmington DE 19898

WARING, THOMAS G, biology, aquatic ecology, see 12th edition

WARING, WILLIAM WINBURN, b Savannah, Ga, July 20, 23; m 52; c 5. PEDIATRICS. Educ: Harvard Univ, MD, 47. Prof Exp: Intern pediat, Children's Hosp, Boston, Mass, 47-48; intern, Johns Hopkins Hosp, Md, 48-49, asst res, 49-50, chief res outpatient dept, 50-51, chief res, Hosp, 51-52; from instr to assoc prof, 57-65, PROF PEDIAT, SCH MED, TULANE UNIV, 65-, LECTR PHYSIOL, 66- Concurrent Pos: Pulmonary dis adv comt, Nat Heart & Lung Inst, 71-73; vchmn gen med & sci adv coun, Nat Cystic Fibrosis Res Found, 72-73. Mem: Am Pediat Soc; Am Acad Pediat; Am Col Chest Physicians; Am Thoracic Soc (vpres, 72-73). Res: Respiratory disease and physiology in infants and children; cystic fibrosis. Mailing Add: Dept of Pediat Tulane Univ Sch of Med New Orleans LA 70112

WARING, WORDEN, b Washington, DC, Jan 8, 15; m 49; c 1. PHYSICAL CHEMISTRY. Educ: Cornell Univ, BChem, 36; Mass Inst Technol, PhD(phys chem), 40. Prof Exp: Instr chem, Tulane Univ, 40-42, asst prof, 42-43; engr, Shell Develop Co, 43-53; engr opers res group, Arthur D Little, Inc, 53-54; chemist, Semiconductor Div, Raytheon Mfg Co, 54-58; head chem sect, Fairchild Semiconductor Corp, 58-64; prin investr human systs design ctr, Rancho Los Amigos Hosp, Downey, Calif, 64-69; assoc prof biomed eng, 69-72, PROF BIOMED ENG, SCHS MED & ENG, UNIV CALIF, DAVIS, 72- Mem: AAAS; Biomed Eng Soc; Am Chem Soc; Electrochem Soc; Inst Elec & Electronics Engrs. Res: Thermodynamics; phase rule; industrial operations research; surface chemistry; diffusion; electrochemistry; biomedical engineering. Mailing Add: 3 Patwin Rd Davis CA 95616

WARITZ, RICHARD STEFEN, b Portland, Ore, Apr 1, 29; m 50; c 4. TOXICOLOGY. Educ: Reed Col, BA, 51; Stanford Univ, PhD(chem), 57. Prof Exp: Actg instr gen chem & biochem, Wash State Univ, 54-55; sr res chemist, E I du Pont de Nemours & Co, Inc, 56-62, sr res scientist, 62-64, sect chief inhalation toxicol, 64-70, res mgr, Biosci Group, 70-75; SR TOXICOLOGIST, HERCULES INC, 75- Mem: AAAS; Am Chem Soc; Am Indust Hyg Asn; Sigma Xi; Soc Toxicol. Res: Pulmonary toxicology and pharmacology; toxicology of green plants; organometallic biochemistry; mechanisms of toxic actions of chemicals. Mailing Add: Med Dept Hercules Inc 910 Market St Wilmington DE 19899

WARK, DAVID QUENTIN, b Spokane, Wash, Mar 25, 18. METEOROLOGY, ASTROPHYSICS. Educ: Univ Calif, AB, 41, PhD(astron), 59. Prof Exp: Aviation forecaster, US Weather Bur, 46-58, res meteorologist, 58-65, SR SCIENTIST, NAT ENVIRON SATELLITE SERV, NAT OCEANIC & ATMOSPHERIC ADMIN, 65- Concurrent Pos: Consult comt atmospheric sci, Nat Acad Sci. Honors & Awards: Gold Medal, Dept Com, 69; NASA Medal for Exceptional Sci Achievement, 69; Second Half Century Award, Am Meteorol Soc, 70; Lloyd V Berkner Space Utilization Award, Am Astron Soc, 70; Robert M Losey Award, Am Inst Aeronaut & Astronaut, 72. Mem: Am Meteorol Soc; Am Astron Soc; Am Inst Aeronaut & Astronaut; Am Geophys Union; Optical Soc Am. Res: Satellite meteorology; radiative transfer; physical meteorology; planetary meteorology; spectroscopy; airglow. Mailing Add: Nat Environ Satellite Serv Nat Oceanic & Atmospheric Admin Washington DC 20233

WARKANY, JOSEF, b Vienna, Austria, Mar 25, 03; US citizen; m 37; c 2. PEDIATRICS. Educ: Univ Vienna, MD, 26. Hon Degrees: DSc, Thomas Jefferson Med Col, 74; DSc, Univ Ill, 75. Prof Exp: Intern, Univ Pediat Clin, Vienna, 26-27; asst, Fed Inst Mothers & Children, 27-31; from asst prof to prof, 32-72, EMER PROF RES PEDIAT, UNIV CINCINNATI, 72-; DIR MENT RETARDATION RES, INST DEVELOP RES, CHILDREN'S HOSP RES FOUND, 66- Concurrent Pos: Scholar, Children's Hosp Res Found, 31-34, asst attend pediatrician, 34-35, attend pediatrician & fel, 35-; prog consult, Nat Inst Child Health & Human Develop. Honors & Awards: Howland Award, Am Pediat Soc, 70; Child Health Award, Charles H Hood Found, 72. Mem: Am Pediat Soc; Soc Pediat Res; hon mem Harvey Soc; cor mem Fr Soc Pediat; hon mem Europ Teratology Soc. Res: Nutritional deficiencies; congenital malformations in children; experimental teratology; mental retardation. Mailing Add: Children's Hosp Res Found Elland Ave Cincinnati OH 45229

WARKENTIN, BENNO PETER, b Man, Can, June 21, 29; m 56; c 3. ENVIRONMENTAL MANAGEMENT. Educ: Univ BC, BSA, 51; Wash State Univ, MS, 53; Cornell Univ, PhD(soils), 56. Prof Exp: Nat Res Coun Can Overseas fel, Oxford Univ, 56-57; asst prof agr physics, 57-62, assoc prof soil sci, 62-70, dir environ studies, 72-75, PROF SOIL SCI, MACDONALD COL, McGILL UNIV, 70- Mem: Am Soc Agron; Can Soc Soil Sci (pres, 65-66); Int Soc Soil Sci. Res: Physical and chemical properties of clay minerals; physical properties of soils; water in clay soils; solid waste disposal. Mailing Add: Macdonald Col McGill Univ Ste Anne de Bellevue PQ Can

WARKENTIN, JOHN, b Grunthal, Man, Aug 18, 31; m 57; c 3. ORGANIC CHEMISTRY. Educ: Univ Man, BS, 54, MS, 55; Iowa State Univ, PhD(chem), 59. Prof Exp: Fel chem, Calif Inst Technol, 59 & Harvard Univ, 59-60; from asst prof to assoc prof, 60-71, PROF CHEM, McMASTER UNIV, 71- Mem: Am Chem Soc; Chem Inst Can. Res: Synthetic and mechanistic investigations in organic chemistry. Mailing Add: Dept of Chem McMaster Univ Hamilton ON Can

WARKENTIN, JOHN HENRY, b Lowe Farm, Man, Mar 3, 28; m 56; c 1. GEOGRAPHY. Educ: Univ Man, BSc, 48; Univ Toronto, MA, 54, PhD, 61. Prof Exp: Instr geog, Univ Md, 56-57; asst prof, United Col, Winnipeg, 57-59, from asst prof to assoc prof, United Col, Man, 59-63; assoc prof, 63-68, PROF GEOG, YORK UNIV, 68- Mem: Asn Am Geog; Can Asn Geog. Res: Historical and settlement geography. Mailing Add: Dept of Geog York Univ Downsview ON Can

WARLICK, CHARLES HENRY, b Hickory, NC, May 08, 30; m 58; c 1. MATHEMATICS, COMPUTER SCIENCE. Educ: Duke Univ, BS, 52; Univ Md, MA, 55; Univ Cincinnati, PhD(math), 64. Prof Exp: Mathematician, US Dept Army, 52-53; programmer, Int Bus Mach Corp, 54; appl mathematician, Gen Elec Co, 55-57, supvr appl math, 57-62, supvr appl math & comput softwar develop, 63-65; LECTR COMPUT SCI & DIR COMPUT CTR, UNIV TEX, AUSTIN, 65- Concurrent Pos: Vpres, VIM Users Orgn Control Data Corp 6000 Series Comput, 68-70, pres, 70-71. Mem: Asn Comput Mach. Res: Numerical solution of partial differential equations; fundamental solutions of finite difference equations; computer executive operating systems; algorithmic languages. Mailing Add: Comput Ctr Univ of Tex Austin TX 78712

WARMACK, ROBERT JOSEPH, b Paris, Tenn, July 18, 47. MOLECULAR PHYSICS. Educ: Univ Tenn, BS, 70, PhD(physics), 75. Prof Exp: FEL MOLECULAR PHYSICS, OAK RIDGE NAT LAB, 75- Res: Crossed beam molecular physics. Mailing Add: Bldg 4500-S Oak Ridge Nat Lab Oak Ridge TN 37830

WARMAN, JAMES CLARK, b Morgantown, WVa, May 27, 27; m 53; c 3. HYDROGEOLOGY. Educ: WVa Univ, BA, 50, MS, 52. Prof Exp: Geologist, US Geol Surv, 52-65; ASSOC PROF CIVIL ENG, AUBURN UNIV, 70-, DIR WATER RESOURCES RES INST, 65- Concurrent Pos: Grants, Auburn Univ, Water Resources Planning, US Water Resources Coun, 68-, Econ of Pollution Abatement, US Dept Interior, 71-73 & Res Mgt, 72-74; mem work group hydrol maps, US Nat Comt for Int Hydrol Decade, 70-73; consult, Study of Nat Water Res Probs & Priorities, Univs Coun Water Resources for US Dept Interior, 71-72 & Harmon Eng, 75-; chmn tech div & vpres, Nat Water Well Asn, 71-72. Honors & Awards: Ross L Oliver Award, Nat Water Well Asn, 74. Mem: Am Water Resources Asn (pres elect, 75, pres, 76); Am Geophys Union; fel Geol Soc Am. Res: Occurrence and availability of ground water; water resources planning; research management; waste heat storage by injection into a confined aquifer. Mailing Add: Water Resources Res Inst Auburn Univ Auburn AL 36830

WARME, JOHN EDWARD, b Los Angeles, Calif, Jan 16, 37; div; c 2. MARINE ECOLOGY. Educ: Augustana Col, Ill, BA, 59; Univ Calif, Los Angeles, PhD(geol), 66. Prof Exp: Fulbright scholar, Scotland, 66-67; ASSOC PROF GEOL, RICE UNIV, 67- Mem: AAAS; Am Asn Petrol Geol; Geol Soc Am; Paleont Soc; Soc Econ Paleont & Mineral. Res: Submarine bio-erosion by invertebrates; burrowing marine invertebrates and trace fossils; population and paleopopulation dynamics; shelled invertebrate community identification and analysis; molluscan ecology; lagoonal and deep marine ecology and sedimentation. Mailing Add: Dept of Geol Rice Univ Houston TX 77001

WARME, PAUL KENNETH, b Westbrook, Minn, Jan 23, 42; m 62; c 1. BIOCHEMISTRY. Educ: Univ Minn, Minneapolis, BCh, 64; Univ Ill, Urbana, PhD(biochem), 69. Prof Exp: Fel protein chem, Cornell Univ, 69-72; ASST PROF BIOCHEM, PA STATE UNIV, UNIVERSITY PARK, 72- Concurrent Pos: NIH fel, Cornell Univ, 69-71. Mem: AAAS; Am Chem Soc. Res: Heme proteins and peptides; conformational energy calculations on proteins; computer applications in biochemistry; chemical synthesis of biologically active polypeptides; structure-function relationships of proteins. Mailing Add: Dept of Biochem Pa State Univ University Park PA 16802

WARMKE, HARRY EARL, b Twin Falls, Idaho, Aug 29, 07; m 33, 46; c 3. PLANT CYTOLOGY, PLANT GENETICS. Educ: Stanford Univ, AB, 31, PhD(plant cytogenetics), 35. Prof Exp: Prof & head dept biol, Seton Hall Col, 35-38; asst genetics, Carnegie Inst, 38-40, cytologist, 41-45; head dept cytogenetics, Inst Trop Agr, Univ PR, 45-47; plant breeder, Fed Exp Sta, 47-53, officer chg, 53-63, PLANT GENETICIST, USDA, PR, 64-; PROF AGRON, UNIV FLA, 64- Mem: Am Soc Nat; Genetics Soc Am; Am Genetic Asn. Res: Experimental polyploidy; cytogenetics of sex determination; tropical agriculture; mechanism of cytoplasmic male sterility in plants; electron microscopy of plant viruses and viral inclusions. Mailing Add: Plant Virus Lab Univ of Fla Gainesville FL 32601

WARMKESSEL, CARL ANDREW, b Allentown, Pa, June 15, 13; m 41; c 2. GEOLOGY. Educ: Lehigh Univ, BA, 35, MA, 41, PhD, 51. Prof Exp: Geologist, Lehigh Portland Cement Co, Pa, 36-71; exec dir, 71-75, SOLID WASTE CONSULT, LEHIGH COUNTY AUTHORITY, 75- Concurrent Pos: Geol consult, 71- Mem: Assoc Paleont Soc; fel Geol Soc Am; Am Inst Mining, Metall & Petrol Eng. Res: Insoluble residues in limestones; stratigraphy and sedimentation; structural geology; nonmetallic mining quarrying. Mailing Add: Box 69 Fogelsville PA 18051

WARN, GEORGE FREDERICK, b Raywood, WVa, Nov 28, 14; m 44; c 4. GEOLOGY, GEOGRAPHY. Educ: Hanover Col, AB, 36; Northwestern Univ, MS, 41. Prof Exp: Geologist, Klein Phoebe Mines, 34-35; lectr pub affairs, Chicago Bd Educ, 38-40; geologist, Ohio Oil Co, 40; standard engr, Solar Aircraft Corp, 41-42; chief analyst, Kaiser Iron & Steel Co, 44-48; from lectr to assoc prof geol & geog, Tex Tech Col, 48-56; aerial photo mapper, 58-62, GEOL ENGR, CALIF DIV HWY, 62- Concurrent Pos: Field foreman, Chicago Planning Comn, 39-40; environ & eng geologist, Calif Dept Transp, 58- Mem: AAAS; Geol Soc Am; Am Meteorol Soc; Am Asn Petrol Geol; Geochem Soc. Res: Structural geology of New Mexico and Texas; sedimentation and stratigraphy of New Mexico and Indiana; eolian sedimentation of the South Plains; aerial photo interpretation; highway materials and slope stability; slope erosion measurement and control. Mailing Add: 3605 Trenton Ave San Diego CA 92117

WARNCKE, DARRYL DEAN, soil science, plant nutrition, see 12th edition

WARNE, RONSON JOSEPH, b East Orange, NJ, June 14, 32; m 50. MATHEMATICS. Educ: Columbia Univ, AB, 53; NY Univ, MS, 55; Univ Tenn, PhD(math), 59. Prof Exp: Asst & instr math, Univ Tenn, 55-59; asst prof, La State Univ, 59-63; assoc prof, Va Polytech Inst, 63-64; prof, WVa Univ, 64-69; PROF MATH, UNIV ALA, BIRMINGHAM, 69- Res: Algebraic theory of semigroups. Mailing Add: Dept of Math Univ of Ala Birmingham AL 35294

WARNE, THOMAS MARTIN, b Chicago, Ill, Sept 27, 39; m 61; c 3. ORGANIC CHEMISTRY, PETROLEUM CHEMISTRY. Educ: Yale Univ, BA, 61; Univ Ill, MS, 63; Northwestern Univ, PhD(org chem), 70. Prof Exp: Asst prof org chem, Alliance Col, 63-66; RES CHEMIST, AMOCO OIL CO, DIV STAND OIL CO IND, 70- Mem: Am Chem Soc; Am Soc Lubrication Engrs. Res: Development of industrial lubricants, particularly hydraulic oils, and additives used therein; development of products based upon waxes and petrolatums. Mailing Add: Amoco Oil Res Ctr PO Box 400 Naperville IL 60540

WARNECKE, MELVIN OSCAR, b Delphos, Ohio, July 26, 39; m 62; c 2. FOOD SCIENCE. Educ: Ohio State Univ, BS, 61, MS, 62; Univ Ga, PhD(food sci), 68; Northwestern Univ, MBA, 74. Prof Exp: Res & develop coordr, Food Div, US Army Natick Labs, Mass, 63-64; scientist, Food Res Div, Armour & Co, Ill, 67-69; group leader, Moffett Tech Ctr, CPC Int, Inc, 69-70, sect leader, 70-74; DIR RES & DEVELOP, E J BRACH & SONS, 74- Mem: AAAS; Inst Food Technologists; Am Meat Sci Asn; Sigma Xi; Am Asn Candy Technol. Res: Research and development in the food science field; fresh and processed meats; carbohydrates in confection, snacks, baking and canning. Mailing Add: E J Brach & Sons Box 802 Chicago IL 60690

WARNELL, JOSEPH LEO, organic chemistry, see 12th edition

WARNER, ALDEN HOWARD, b Central Falls, RI, July 2, 37; m 60; c 4. DEVELOPMENTAL BIOLOGY. Educ: Univ Maine, BA, 59; Univ Southern Ill, MA, 61, PhD(physiol), 64. Prof Exp: USPHS fel biol & biochem, Biol Div, Oak Ridge Nat Lab, 64-65; from asst prof to assoc prof, 65-72, PROF BIOL, UNIV WINDSOR, 72- Concurrent Pos: Consult, Biol Div, Oak Ridge Nat Lab, 74-75. Mem: AAAS; Am Soc Biol Chemists; Soc Develop Biol; Can Soc Cell Biol. Res: Nucleic acid and nucleotide metabolism in relation to control of protein synthesis during embryonic development. Mailing Add: Dept of Biol Univ of Windsor Windsor ON Can

WARNER, ALEXANDER CARL, b Mentor, Ohio, Aug 25, 40; m 60; c 3. PHYSIOLOGY, ECOLOGY. Educ: Cent Mo State Col, BS, 61, MA, 64; Kent State Univ, PhD(biol), 70. Prof Exp: Teacher pub sch, Mo, 61-63; asst prof biol, Col Sch of the Ozarks, 64-66; instr, Kent State Univ, 69-70; asst prof physiol, Southern Ill Univ, Carbondale, 70-74; ASSOC PROF ZOOL, HOWARD UNIV, 75- Mem: AAAS; Am Soc Zoologists. Res: Hormonal control of calcium shifts in arthropods. Mailing Add: Dept of Zool Howard Univ Washington DC 20059

WARNER, ANN MARIE, b Denver, Colo, Mar 31, 44; m 68. CLINICAL CHEMISTRY. Educ: Marymount Col, Kans, BS, 66; Univ Kans, PhD(med chem), 70. Prof Exp: NIH fel, Northeastern Univ, 71-73, res assoc med chem, 71-74; MGR CHEM, LAHEY CLIN FOUND, 74- Concurrent Pos: Lectr, Cardinal Cushing Col, 71-72. Mem: Am Chem Soc; Am Asn Clin Chem. Res: Modification of peptides and radioimmunoassay. Mailing Add: Lahey Clin Found 605 Commonwealth Ave Boston MA 02215

WARNER, CAROL MILLER, b New York, NY, Sept 26, 46; m 66; c 1. BIOCHEMISTRY, IMMUNOBIOLOGY. Educ: Queens Col, NY, BA, 66; Univ Calif, Los Angeles, PhD(biochem), 70. Prof Exp: Fel, Yale Univ, 70-71; ASST PROF BIOCHEM, IOWA STATE UNIV, 71- Mem: Soc Develop Biol; Fedn Am Scientists; Sigma Xi. Res: Preimplantation mouse embryo development; immune response of allophenic mice. Mailing Add: Dept of Biochem Iowa State Univ Ames IA 50011

WARNER, CHARLES, b Haslemere, UK, Aug 1, 38; PHYSICAL METEOROLOGY. Educ: Cambridge Univ, BA, 63, MA, 66; McGill Univ, MSc, 67, PhD(meteorol), 71. Prof Exp: Res asst meteorol, McGill Univ, 67-68; res fel radar meteorol, Univ Birmingham, 72-75; RES ASSOC METEOROL, McGILL UNIV, 75- Mem: Am Meteorol Soc; fel Royal Meteorol Soc; Can Meteorol Soc. Res: Study of structures of Alberta hailstorms using weather radar data; scattering and depolarization of electromagnetic waves by hydrometers. Mailing Add: Dept of Meteorol McGill Univ PO Box 6070 Sta A Montreal PQ Can

WARNER, CHARLES D, b Mt Hope, WVa, Mar 25, 45; m 66; c 1. ORGANIC CHEMISTRY. Educ: Univ Mo-Columbia, BS, 67, PhD(org chem), 71. Prof Exp: Asst prof chem, Mo Valley Col, 71-74; ASST PROF CHEM, HASTINGS COL, 74- Mem: Sigma Xi; Am Chem Soc; Nat Sci Teachers Asn. Res: Synthetic organometallic chemistry. Mailing Add: Dept of Chem Hastings Col Hastings NE 68901

WARNER, CHARLES ROBERT, b Aug 24, 31; Can citizen. MATHEMATICS. Educ: Univ Toronto, BA, 55; Rochester Univ, MS, 57, PhD(math), 62. Prof Exp: Instr math, Univ Conn, 58-59; asst prof, Mich State Univ, 62-64; asst prof, 64-69, ASSOC PROF MATH, UNIV MD, COLLEGE PARK, 69- Concurrent Pos: Vis assoc prof, Univ Calif, Irvine, 70-71. Mem: Am Math Soc; Math Asn Am. Res: Banach algebras and harmonic analysis. Mailing Add: Dept of Math Univ of Md College Park MD 20742

WARNER, CHARLES ROCKWELL, analytical chemistry, organic chemistry, see 12th edition

WARNER, DALE ALFORD, organic chemistry, analytical chemistry, see 12th edition

WARNER, DANIEL DOUGLAS, b Mobile, Ala, Sept 1, 42; m 68; c 2. NUMERICAL ANALYSIS. Educ: Ariz State Univ, BS, 65, MA, 66; Univ Calif, San Diego, PhD(math), 74. Prof Exp: Programmer, Process Comput Sect, Gen Elec Co, 63-65; comput analyst, Airesearch Corp, 66; MEM TECH STAFF, BELL TEL LABS, 74- Mem: Am Math Soc; Soc Indust & Appl Math; Asn Comput Mach. Res: Theory of Hermite interpolation with rational functions; numerical solution of stiff systems of ordinary differential equations. Mailing Add: Comput Sci Res Ctr Bell Labs 600 Mountain Ave Murray Hill NJ 07974

WARNER, DAVID CHARLES, b Granite City, Ill, Apr 27, 42; m 64. MATHEMATICAL PHYSICS, PLASMA PHYSICS. Educ: Univ Mo-Columbia, BS, 65, MS, 67, PhD(physics), 70. Prof Exp: ASST PROF PHYSICS, LINCOLN UNIV, MO, 70- Mem: Am Asn Physics Teachers; Am Phys Soc. Res: Quantum kinetic equations of plasma physics; density matrix formalism applied to damping of plasma waves. Mailing Add: Dept of Physics Lincoln Univ Jefferson City MO 65102

WARNER, DON LEE, b Norfolk, Nebr, Jan 4, 34; m 57; c 2. HYDROGEOLOGY, GEOLOGICAL ENGINEERING. Educ: Colo Sch Mines, Geol Engr, 56, MSc, 61; Univ Calif, Berkeley, PhD(eng sci), 64. Prof Exp: Res geologist, R A Taft Sanit Eng Ctr, Ohio, 64-69; assoc prof, 69-72, PROF GEOL ENG, UNIV MO, ROLLA, 72- Mem: Am Inst Mining, Metall & Petrol Eng; Geol Soc Am; Am Asn Petrol Geol; Am Soc Civil Eng; Am Geophys Union. Res: Mineral exploration and exploitation, areal geology; stratigraphic geology and geophysics; geohydrology; water pollution control; engineering geology. Mailing Add: Dept of Geol Eng Univ of Mo Rolla MO 65401

WARNER, DONALD R, b Winston, Mo, July 6, 18; m 45; c 3. ANIMAL HUSBANDRY, ANIMAL NUTRITION. Educ: Univ Mo, BS, 42, MS, 49, PhD(animal nutrit & ed), 60. Prof Exp: Mem staff, Mo Agr Exten Serv, 45-47; instr animal husb, Univ Mo, 47-49 & Univ Nebr, 49-56; ASSOC PROF ANIMAL HUSB, IOWA STATE UNIV, 60- Mem: Am Soc Animal Sci. Res: Swine feeding investigations; antibiotics; protein supplements for pigs on pasture and in dry lot; methods of feeding and effects on carcass value; bloat studies of sheep; breeding, feeding and management of sheep. Mailing Add: Dept of Animal Sci 119 Kildee Hall Iowa State Univ Ames IA 50010

WARNER, DONALD THEODORE, b Holland, Mich, Apr 7, 18; m 45; c 2. BIOCHEMISTRY. Educ: Hope Col, AB, 39; Univ Ill, PhD(biochem), 43. Prof Exp: Res chemist, Gen Mills, Inc, 43-52; RES CHEMIST, UPJOHN CO, 52- Mem: Am Chem Soc. Res: Amino acids; isolation and synthesis; amino acid diets; isolation and properties of threonine; organic synthesis, 1, 4 addition reactions; polysaccharides; peptide synthesis; protein conformation studies; antiasthmatic drugs. Mailing Add: Res Dept Upjohn Co Kalamazoo MI 49006

WARNER, DWAIN WILLARD, b Cottonwood Co, Minn, Sept 1, 17; m 40, 66; c 5. ZOOLOGY. Educ: Carleton Col, BA, 39; Cornell Univ, PhD(ornith), 47. Prof Exp: Lab asst bot, Carleton Col, 38-39; asst zoologist, Cornell Univ, 41, asst, 42-43 & 46, instr 46-47; from asst prof to assoc prof, 47-67, PROF ZOOL, UNIV MINN, MINNEAPOLIS, 67-, CUR BIRDS, BELL MUS NATURAL HIST, 47- Mem: Assoc Am Ornith Union. Res: Birds of Mexico and New Caledonia; zoogeography; biology and ecology of avifauna. Mailing Add: Bell Mus Natural Hist Univ of Minn Minneapolis MN 55455

WARNER, EDWARD NELSON, b Patchogue, NY, Aug 19, 06; m 30. ICHTHYOLOGY. Educ: Marietta Col, AB, 28; Ohio State Univ, MSc, 30, PhD(zool), 40. Prof Exp: Asst zool, Ohio State Univ, 28-29, instr, 29-43; from asst prof to prof biol, 46-74, head dept, 59-74, EMER PROF BIOL, ST LAWRENCE UNIV, 74- Concurrent Pos: With USDA & Ohio Div Conserv, 29. Mem: AAAS; Am Asn Biol Teachers. Res: Vertebrate embryology; ichthyology; histology. Mailing Add: Dept of Biol St Lawrence Univ Canton NY 13617

WARNER, ELDON DEZELLE, b Whitewater, Wis, Oct 5, 11; m; c 3. ENDOCRINOLOGY. Educ: Wis State Teachers Col, BEd, 32; Univ Wis, PhM, 38, PhD(zool), 41. Prof Exp: Asst prof sci, Adams State Teachers Col, 41-43; from instr to assoc prof, 46-58, chmn dept, 56-58, 59-61 & 66-69, PROF ZOOL, UNIV WIS, MILWAUKEE, 58- Concurrent Pos: NSF fel, 58-59. Mem: AAAS; Am Soc Zool. Res: Lower vertebrate endocrinology. Mailing Add: Dept of Zool Univ of Wis Milwaukee WI 53201

WARNER, EMORY DEAN, b North English, Iowa, July 5, 05; m 30; c 3. PATHOLOGY. Educ: Univ Iowa, BS, 27, MD, 29; Am Bd Path, dipl. Prof Exp: Asst resident path, Univ Rochester, 29-30; asst, Col Med, Univ Iowa, 30-31, from instr to prof, 31-70, head dept, 45-70; prof path, Col Med, Univ Ariz, 70-76. Concurrent Pos: Pathologist, Mercy Hosp, 30-33; asst hosp pathologist, Univ Hosps, Iowa, 33-45, pathologist, 45-70. Mem: AAAS; Am Soc Clin Pathologists; Am Soc Exp Pathologists (pres, 57-58); Soc Exp Biol & Med; Am Asn Pathologists & Bacteriologists. Res: Physiology of blood clotting; hemorrhagic diseases; role of the liver in prothrombin production; methods for measuring prothrombin; vitamin K prothrombin; transfusion nephritis; pathophysiology of hypercoagualability and thrombosis. Mailing Add: 1402 E Court St Iowa City IA 52240

WARNER, FRANCIS JAMES, b Chicago, Ill, Oct 3, 97. ANATOMY. Educ: Loyola Univ Chicago, MD, 19; Univ Iowa, BA, 22; Univ Mich, MA, 25; Am Bd Path, dipl, 52. Prof Exp: Instr neurol, Univ Mich, 32-33; instr med sci, Univ Md, 38-39; mem res staff neuroanat, Med Sch, Univ Calif, 42-44; mem res staff neuropath & neuroanat, Med Sch, Columbia Univ, 44-45; lectr neuroanat, Univ Utah, 46-49; guest investr, Sch Med, Yale Univ, 49-50; fel neurol, Henry Ford Hosp, Detroit, Mich, 51-52; guest investr, Cornell Univ, 53-55 & Univ Col, Univ London, 56-57; instr clin neurol, 57-75, ASST PROF NEUROL, SCH MED, TEMPLE UNIV, 75-; MED RES SCIENTIST NEUROEMBRYOL & NEUROPATH, EASTERN PA PSYCHIAT INST, 64- Mem: Am Psychiat Asn; Am Micros Soc; Am Soc Ichthyologists & Herpetologists; Am Soc Zoologists; Am Asn Anatomists. Res: Comparative neuroembryology; neuropathology of the malformation in the human brain in adult and fetal material; teratological brain malformations in the human form. Mailing Add: PO Box 523 Philadelphia PA 19105

WARNER, FRANK WILSON, III, b Pittsfield, Mass, Mar 2, 38; m 58; c 2. MATHEMATICS. Educ: Pa State Univ, BS, 59; Mass Inst Technol, PhD(math), 63. Prof Exp: Instr math, Mass Inst Technol, 63-64; actg asst prof math, Univ Calif, Berkeley, 64-65, asst prof, 65-68; assoc prof, 68-73, PROF MATH, UNIV PA, 73- Mem: Am Math Soc; Math Asn Am; AAAS. Res: Applications of partial differential equations to differential geometry. Mailing Add: Dept of Math E1 Univ of Pa Philadelphia PA 19174

WARNER, FREDERIC COOPER, b Whitesboro, NY, Jan 28, 15; m 40; c 2. GEOMETRY. Educ: Col Wooster, BA, 37; Univ Buffalo, MA, 50, PhD, 53. Prof

Exp: Teacher high schs, NY, 37-46; instr math, Univ Buffalo, 46-53; from asst prof to assoc prof, 53-60, PROF MATH, ST LAWRENCE UNIV, 60- Mem: Am Math Soc; Math Asn Am. Res: Operations research. Mailing Add: Dept of Math St Lawrence Univ Canton NY 13617

WARNER, FREDERIC WILLIAM, b Dayton, Ohio, May 9, 27. ANTHROPOLOGY, ARCHAEOLOGY. Educ: Yale Univ, BA, 53; Hartford Sem Found, PhD(anthrop), 70. Prof Exp: Asst dir, Childrens' Mus Hartford, 60-64; dir, Wildcliff Youth Mus, 64-65; social studies teacher, Farmington Pub Schs, 65-67; asst chmn dept, 71-75, ASSOC PROF ANTHROP, CENT CONN STATE COL, 67- Mem: Fel AAAS; fel Am Anthrop Asn; fel Royal Anthrop Soc Gt Brit & Ireland; Soc Am Archaeol. Res: Relationships between archaeology and ethnohistory, especially in regard to Connecticut and New England Indians. Mailing Add: Warner Rd Collinsville CT 06022

WARNER, GEORGE S, b Montgomery, Ala, June 18, 20; m 46; c 3. MICROBIOLOGY, IMMUNOLOGY. Educ: Miss State Univ, BS, 42; Univ Md, MS, 46, PhD(microbiol), 51. Prof Exp: Res microbiologist, 50-53, DIR BIOL QUAL CONTROL LAB, HYNSON, WESTCOTT & DUNNING, INC, 53- Mem: Am Soc Microbiol; US Animal Health Asn; Am Asn Lab Animal Sci. Res: Antiseptics; sterility; microbic and biologic test of pharmaceuticals; serologic studies on Neisseria, Brucella, Trichinella and animal diseases. Mailing Add: Hynson Westcott & Dunning Inc BQC Lab Charles & Chase Sts Baltimore MD 21201

WARNER, HARLOW LESTER, b Greenport, NY, Aug 26, 42; m 63; c 3. PLANT PHYSIOLOGY, PLANT BREEDING. Educ: Cornell Univ, BS, 64; Univ Idaho, MS, 66; Purdue Univ, PhD(plant physiol), 70. Prof Exp: SR SCIENTIST PLANT PHYSIOL, ROHM & HAAS CO, 69- Mem: Am Soc Hort Sci; Am Soc Plant Physiologists. Res: Agricultural chemicals, specifically plant growth regulators and herbicides; plant growth regulators for developing hybrid seeds. Mailing Add: Rohm & Haas Co Res Labs Spring House PA 19477

WARNER, HUBER RICHARD, b Glendale, Ohio, May 16, 36; m 61; c 2. BIOCHEMISTRY. Educ: Ohio Wesleyan Univ, BA, 58; Mass Inst Technol, BS, 58; Univ Mich, PhD(biochem), 62. Prof Exp: NSF fel biochem, Mass Inst Technol, 62-64; from asst prof to assoc prof, 64-72, PROF BIOCHEM, UNIV MINN, ST PAUL, 72- Mem: Am Chem Soc; Am Soc Biol Chem; Am Soc Microbiol. Res: Biochemistry of bacteriophage infection and replication; lipid metabolism. Mailing Add: Dept of Biochem Univ of Minn St Paul MN 55108

WARNER, JACOB LARUE, b Ky, Jan 1, 21; m 49; c 2. RESEARCH ADMINISTRATION. Educ: Univ Ky, AB, 48, MS, 50. Prof Exp: Instr physics, Univ Ky, 49-50; jr instr, Johns Hopkins Univ, 50-51; physicist, US Naval Ord Lab, 51-62; PHYSICIST, OFF NAVAL RES, 62- Mem: Am Geophys Union. Res: Nuclear weapons effects and research on detection of nuclear explosions. Mailing Add: Off of Naval Res Arlington VA 22217

WARNER, JAMES HOWARD, b Angola, Ind, Dec 24, 38. BOTANY, PLANT ECOLOGY. Educ: Manchester Col, BS, 61; Univ Wis, Madison, MS, 63; Univ Utah, PhD(biol), 71. Prof Exp: Teaching asst bot, Univ Wis, Madison, 61-63; ASSOC PROF BIOL, UNIV WIS, LA CROSSE, 63- Mem: AAAS; Am Inst Biol Sci; Ecol Soc Am. Res: Development of indices of site quality in forest management and plant association. Mailing Add: Dept of Biol Cowley Hall Univ of Wis 1709 Pine St La Crosse WI 54601

WARNER, JEFFREY, b Queens, NY, May 19, 39. PLANETARY SCIENCES. Educ: City Col New York, BS, 60; Harvard Univ, AM, 62, PhD(geol), 67. Prof Exp: Instr geol, Univ Alaska, 64-65; asst prof, Franklin & Marshall Col, 67-68; geologist, Lunar Receiving Lab, NASA Manned Spacecraft Ctr, 68-74, CHIEF GEOCHEM, JOHNSON SPACE CTR, NASA, 74- Mem: AAAS; Mineral Soc Am; Am Geophys Union. Res: Geology of Maine; metamorphism of calcsilicate rocks, lunar geology; petrology of lunar igneous and metamorphic rocks; petrogenesis of impact produced rocks. Mailing Add: Code TN-7 Johnson Space Ctr Houston TX 77058

WARNER, JOHN CHRISTIAN, b Goshen, Ind, May 28, 97; m 25; c 2. CHEMISTRY. Educ: Univ Ind, AB, 19, MA, 20, PhD(chem), 23. Hon Degrees: Fourteen from US cols & univs. Prof Exp: Res chemist, Barrett Co, Pa, 18 & Cosdon Co, Okla, 20-21; instr chem, Univ Ind, 19 & 21-24; res chemist, Wayne Chem Co, 25-26; from instr to asst prof chem, 26-33, assoc prof theoret chem, 33-36 & metall, 36-38, prof chem & head dept, 38-49, dean grad studies, 45-50, from vpres to pres, 49-65, EMER PRES, CARNEGIE INST TECHNOL, 65- Concurrent Pos: Res assoc, Manhattan Proj, 43-45; consult, Found & Indust Trustee, Carnegie Inst, Carnegie Inst Technol & Mellon Inst; mem gen adv comt, AEC, 52-64; consult, Scaife Family Charities, 65-; chmn, Educ Projs, Inc, 66- Honors & Awards: Pittsburgh Award, Am Chem Soc, 45; Gold Medal, Am Inst Chem, 53. Mem: Nat Acad Sci; AAAS; Am Chem Soc (pres, 56); Electrochem Soc (pres, 52-53); Am Soc Eng Educ. Res: Kinetics of reactions in solutions, salt and medium effects, electrostatic contribution to activation energies; thermodynamic properties of solutions; acid-base properties of mixed solvents; thermodynamics, rates and mechanism of corrosion reactions; educational and research administration; scientific and technological education. Mailing Add: 825 Morewood Ave Apt H-4 Pittsburgh PA 15213

WARNER, JOHN NORTHRUP, b Los Angeles, Calif, May 19, 19; m 42; c 3. PLANT BREEDING, AGRONOMY. Educ: Univ Hawaii, BS, 41; Univ Minn, PhD(plant genetics), 50. Prof Exp: Asst, Univ Minn, 48-50; asst agronomist, Exp Sta, Hawaiian Sugar Planters' Asn, 41-46, from asst geneticist to prin geneticist & head dept genetics & path, 46-66; consult agronomist, 66-68, VPRES & DIR AGR SERV, HAWAIIAN AGRON CO, 68- Mem: Am Soc Agron. Res: Sugarcane breeding; field experiment design; pathology; tropical agriculture and agribusiness management. Mailing Add: Hawaiian Agron Co PO Box 3470 Honolulu HI 96801

WARNER, JOHN R, forestry, see 12th edition

WARNER, JOHN SCOTT, b Woodstown, NJ, Oct 25, 28; m 61; c 4. ORGANIC CHEMISTRY. Educ: Rensselaer Polytech Inst, BS, 49; Cornell Univ, PhD(org chem), 52. Prof Exp: Anal chemist, Socony Vacuum Oil Co, 49; asst, Cornell Univ, 49-52; prin chemist, 52-57, proj leader, 57-62, SR RES CHEMIST, ORG CHEM DIV, BATTELLE COLUMBUS LABS, 62- Mem: AAAS; Am Chem Soc. Res: Natural products; pesticides; pharmaceuticals; organic selenium compounds; gas and liquid chromatography; air and water pollution; petroleum analysis. Mailing Add: Battelle Columbus Labs 505 King Ave Columbus OH 43201

WARNER, JOHN WARD, b Dubuque, Iowa, Oct 17, 44; m 66; c 2. ASTRONOMY. Educ: Kalamazoo Col, BA, 66; Ohio State Univ, PhD(astron), 72. Prof Exp: Res fel terrestrial magnetism, Carnegie Inst Washington, 72-74; ASST PROF ASTRON, SCH PHYSICS & ASTRON, UNIV MINN, MINNEAPOLIS, 74- Mem: Am Astron Soc. Res: Infrared and optical studies of gas and dust and its relation to star formation and death, including planetary nebulae, galactic H II regions and galactic nuclei. Mailing Add: 148 Physics Bldg Sch Physcis & Astron Univ Minn Minneapolis MN 55455

WARNER, JOHN WARD, JR, b Davenport, Iowa, Feb 4, 18; m 42; c 5. MATHEMATICS. Educ: Taylor Univ, AB, 40; Univ Iowa, MS, 58; Ohio State Univ, PhD(math ed), 64. Prof Exp: Assoc athletics dir, Asbury Col, 40-41; teacher math & coach high sch, Tenn, 41-42; from instr to asst prof, Dubuque Univ, 42-45, bus mgr, 45-47; bus mgr, Taylor Univ, 47-48; bus mgr, Dubuque Univ, 48-53; pvt bus, Davenport, Iowa, 53-56; asst math, Univ Iowa, 56-58; from instr to assoc prof, 58-68, PROF MATH, COL WOOSTER, 68- Concurrent Pos: Consult, Ohio Coun Teachers Math, 62-65; Nat Defense Educ Act, Ohio, 62-67. Mem: Math Asn Am. Mailing Add: Dept of Math Col of Wooster Wooster OH 44691

WARNER, JONATHAN ROBERT, b New York, NY, Feb 19, 37; m 58; c 2. MOLECULAR BIOLOGY, CELL BIOLOGY. Educ: Yale Univ, BS, 58; Mass Inst Technol, PhD(biophys), 63. Prof Exp: Res assoc biophys, Mass Inst Technol, 63-64; res assoc biochem, 64-65, from asst prof to assoc prof, 65-74, PROF BIOCHEM & CELL BIOL, ALBERT EINSTEIN COL MED, 74-, DIR SUE GOLDING GRAD SCH, 72- Concurrent Pos: NSF fel, 64-65; career scientist, Health Res Coun, City New York, 65-72; Guggenheim fel, 71-72; Am Cancer Soc fac res award, 72-; mem sci adv bd, Damon Runyon-Walter Winchell Cancer Fund, 73-; mem molecular cytol study sect, NIH, 75- Mem: Am Soc Biol Chemists; Am Soc Cell Biol; Am Soc Microbiol. Res: Structure and function of polyribosomes; synthesis and assembly of eukaryotic ribosomes. Mailing Add: Dept of Biochem Albert Einstein Col of Med Bronx NY 10461

WARNER, JUDITH SAUVE, b San Francisco, Calif. ECOLOGY. Educ: Univ Calif, Davis, BS, 64, MA, 69; Mich State Univ, PhD(zool), 75. Prof Exp: RES ASSOC ECOL, US INT BIOL PROG DESERT BIOME, UTAH STATE UNIV, 76- Mem: AAAS; Ecol Soc; Asn Tropical Biol. Res: The quantitative, rather than qualitative, criteria herbivores use in foraging and choosing habitats in an essentially green world. Mailing Add: Col of Natural Resources Utah State Univ Logan UT 84322

WARNER, KENDALL, b Westfield, Mass, Oct 2, 27; m 70; c 1. FISHERIES. Educ: Univ Maine, BS, 50; Cornell Univ, MS, 52. Prof Exp: Fishery aide, US Fish & Wildlife Serv, 50; asst fishery biol, Cornell Univ, 51-52; regional fishery biologist, 52-68, CHIEF RES BIOLOGIST, FISHERY DIV, MAINE DEPT INLAND FISHERIES & GAME, 68- Mem: Am Fisheries Soc; Am Soc Ichthyol & Herpet; Wildlife Soc; fel Am Inst Fishery Res Biologists. Res: Fresh water fisheries; landlocked salmon; brook trout. Mailing Add: Fishery Off Bldg 34 Idaho Ave Maine Dept Inland Fish & Game Bangor ME 04401

WARNER, LAURANCE BLISS, b Brooklyn, NY, Dec 29, 31; m 52; c 3. NUCLEAR PHYSICS. Educ: Rensselaer Polytech Inst, BS, 53; Johns Hopkins Univ, MS, 58; Fla State Univ, PhD(physics), 62. Prof Exp: US Navy, 53-, instr physics & electronics, US Naval Acad, 56-58, terrier battery officer, USS Long Beach CGN-9, 62-64, exec officer, USS Benjamin Stoddert DDG-22, 64-66, physicist, Lawrence Radiation Lab, 66-69, weapons officer, USS Galveston CLG-3, 69-70, MEM STAFF, CHIEF NAVAL OPERS, WASHINGTON, DC, 70- Concurrent Pos: Instr physics, Fed City Col, Washington, DC, 71- Mem: Am Phys Soc. Res: Nuclear decay and reaction spectroscopy; deformed nuclei; fast electronics instrumentation; computer applications; neutron transport; radiative transfer; missile systems analysis. Mailing Add: 5208 Pommeroy Dr Fairfax VA 22030

WARNER, LAWRENCE ALLEN, b Monroe, Ohio, Apr 20, 14; m 42; c 2. GEOLOGY. Educ: Univ Miami, Ohio, AB, 37; Johns Hopkins Univ, PhD(geol), 42. Prof Exp: Geologist, US Antarctic Exped, Marie Byrd Land, 39-41 & Alaskan Br, US Geol Surv, 42-46; from asst prof to assoc prof, 46-58, PROF GEOL, UNIV COLO, BOULDER, 58- Concurrent Pos: Geologist, US Geol Surv, 46-52; res consult, Univ Boston, 52-54; consult, Denver Water Bd, 55- Honors & Awards: Congressional Medal for Sci & Explor. Mem: AAAS; Geol Soc Am; Am Geophys Union; Asn Eng Geol; Mineral Soc Am. Res: Antarctic and Alaskan geology; structure and tectonics of eastern Rocky Mountains; environmental and engineering geology. Mailing Add: Dept of Geol Sci Univ of Colo Boulder CO 80302

WARNER, LLOYD C, agronomy, see 12th edition

WARNER, LOUISE, b Dixon, Ill, Aug 14, 16. ANATOMY. Educ: Univ Chicago, SB, 38, PhD(anat), 49. Prof Exp: Asst prev med, Col Med, Univ Tenn, 41-42; asst anat, Univ Chicago, 42-45, asst, Metall Lab, 45-46, asst anat, 46-48; physiologist, Naval Med Res Inst, Nat Naval Med Ctr, 48-53; ASST PROF ANAT, SCH MED, GEORGETOWN UNIV, 53- Mem: Am Asn Anatomists; Microcirc Soc. Res: Living capillaries; electron microscopy; cinephotomicrography; electronic image processing; television applied to medical research. Mailing Add: 4644 Reservoir Rd NW Washington DC 20007

WARNER, MARK CLAYSON, b Nephi, Utah, Aug 30, 39; m 62; c 6. AQUATIC BIOLOGY. Educ: Utah State Univ, BS, 66, MS, 69; Okla State Univ, PhD(zool), 72. Prof Exp: Aquatic biologist, Tenn Valley Authority, 70-74; AQUATIC BIOLOGIST, MED BIOENG RES & DEVELOP LAB, ENVIRON PROTECTION RES DIV, US ARMY, FT DETRICK, 74- Mem: Am Fisheries Soc; Wildlife Dis Asn; Am Soc Testing & Mat. Res: Aquatic bioassy toxicity studies; review and evaluate research proposals in areas of aquatic biology. Mailing Add: Environ Protection Res Div USAMBRDL Ft Detrick Frederick MD 21701

WARNER, MARLENE RYAN, b Philadelphia, Pa, Apr 30, 38; m 69; c 2. ENDOCRINOLOGY. Educ: Univ Calif, Berkeley, AB, 61, MA, 66; Univ Calif, Davis, PhD(anat), 72. Prof Exp: Res asst anat, Univ Calif, Davis, 68-69; res assoc cell biol, 72-75, INSTR CELL BIOL, BAYLOR COL MED, 75- Mem: Am Soc Zoologists. Res: Tumor biology; reproductive physiology; differentiation; tumor-endocrinology. Mailing Add: Dept of Cell Biol Baylor Col of Med Houston TX 77025

WARNER, MONT MARCELLUS, b Fillmore, Utah, Oct 9, 19; m 47. GEOLOGY. Educ: Brigham Youn Univ, AB, 47, MA, 49; Univ Iowa, PhD(geol), 63. Prof Exp: Geologist, Shell Oil Co, 49-55; consult petrol geol, La & Utah, 55-59; instr geol, Brigham Young Univ, 59-61; asst prof, Ariz State Univ, 63-67, res grant, 64-66; chmn dept ecol, 67-70, PROF GEOL, BOISE STATE COL, 67- Concurrent Pos: Idaho state rep in geothermal matters, 70-72; res & legis work in geothermal resources, 70-72; mem exec comt & comt info educ, Geothermal Resources Coun, 70-72. Mem: Am Asn Petrol Geol; Nat Asn Geol Teachers; Ecol Soc Am; Nat Inst Prof Geol. Res: Sedimentation; structural and petroleum geology. Mailing Add: Boise State Univ Boise ID 83707

WARNER, NANCY ELIZABETH, b Dixon, Ill, July 8, 23. PATHOLOGY. Educ: Univ Chicago, SB, 44, MD, 49; Am Bd Path, dipl, 54. Prof Exp: From intern to asst resident path, Univ Chicago Clins, 49-50; resident, Cedars of Lebanon Hosp, Los Angeles, Calif, 53-54, asst pathologist, 54-58; from asst prof to assoc prof path, Univ Chicago, 58-65, dir lab surg path, Univ Clins, 59-65; assoc clin prof path, Univ

Southern Calif, 65-66; assoc prof, Sch Med, Univ Wash, 66-67; assoc prof, 67-69, PROF PATH, UNIV SOUTHERN CALIF, 69-, CHMN DEPT, 72- Concurrent Pos: Assoc dir labs, Cedars-Sinai Med Ctr, Los Angeles, 65-66; chief pathologist, Women's Hosp, Los Angeles County-Univ Southern Calif Med Ctr, 68-72, dir labs & path, 72- Mem: Fel Col Am Pathologists; Endocrine Soc; Am Asn Anatomists; Microcirc Soc; fel Am Soc Clin Pathologists. Res: Pathology of endocrine glands; comparative pathology of tumors of the gonads; microcirculation in experimental mammary tumors; effect of toxins on the microcirculation. Mailing Add: Dept of Path Univ of Southern Calif Los Angeles CA 90033

WARNER, PAUL LONGSTREET, JR, b New York, NY, June 20, 40; c 2. MEDICINAL CHEMISTRY. Educ: Ursinus Col, BS, 62; Pa State Univ, MS, 64; State Univ NY Buffalo, PhD(med chem), 69. Prof Exp: DIR PATENTS & PROD LAISION, WESTWOOD PHARMACEUT INC, 68- Mem: AAAS; Am Chem Soc. Res: Antimicrobial agents; sunscreen agents; design and synthesis of antipsoriatics; antimetabolites. Mailing Add: Westwood Pharmaceut Inc 468 Dewitt St Buffalo NY 14213

WARNER, PETER, b Winnipeg, Man, Apr 22, 20; m 52; c 6. MICROBIOLOGY. Educ: Univ London, MB & BS, 44, MD, 48, PhD(path), 51. Prof Exp: Asst pathologist, Bland-Sutton Inst Path, Middlesex Hosp, London, Eng, 43-44 & 46-52; pathologist, Winnipeg Gen Hosp, 53-54; assoc prof path, Fac Med, Univ Man, 54-55; head med res div, Inst Med Sci, Australia, 55-58; ASSOC PROF MED BACT, FAC MED, UNIV MAN, 58- Concurrent Pos: Brit Empire Cancer Campaign traveling scholar, Walter Reed Army Med Ctr, DC, 48-49; asst dep minister health, Man Govt, 67-71; chmn, Man Clean Environ Comn, 71-72; consult, Man Inst Tech; ed, Man Med Rev. Mem: Am Acad Microbiol; fel Am Soc Clin Path; Can Soc Microbiol; Can Asn Path; Path Soc Gt Brit & Ireland. Res: Medical microbiology; experimental pathology; health administration. Mailing Add: 368 Wildwood Park Winnipeg MB Can

WARNER, PHILIP MARK, b New York, NY, Nov 5, 46. ORGANIC CHEMISTRY. Educ: Columbia Univ, BA, 66; Univ Calif, Los Angeles, PhD(chem), 70. Prof Exp: NSF fel, Yale Univ, 70-71; instr, 71-73, ASST PROF ORG CHEM, IOWA STATE UNIV, 73- Mem: Am Chem Soc; The Chem Soc. Res: Strained ring compounds; carbonium ions; synthesis. Mailing Add: Dept of Chem Iowa State Univ Ames IA 50011

WARNER, RAY ALLEN, b Davis, Calif, May 5, 38; m 65. EXPERIMENTAL NUCLEAR PHYSICS. Educ: Univ Calif, Berkeley, BS, 61; Univ Calif, Davis, PhD(physics), 69. Prof Exp: Res assoc, 69-71, ASST PROF CHEM & PHYSICS, CYCLOTRON LAB, MICH STATE UNIV, 72- Mem: Am Phys Soc. Res: Nuclear structure from gamma ray spectra and charged particle reactions; beta decay; nuclear astrophysics; collective motion and residual interactions. Mailing Add: Cyclotron Lab Mich State Univ East Lansing MI 48824

WARNER, RAYMOND M, JR, b Barberton, Ohio, Mar 21, 22; m 48; c 3. PHYSICS. Educ: Carnegie Inst Technol, BS, 47; Case Univ, MS, 50, PhD(physics), 52. Prof Exp: Lab asst, Pittsburgh Plate Glass Co, Ohio, 41 & 42; lab instr physics, Carnegie Inst Technol, 43; jr physicist, Res Lab, Corning Glass Works, NY, 47-48; instr physics, Case Univ, 51-52; mem tech staff semiconductor device develop, Bell Tel Labs, Inc, NJ, 52-59; chief engr diode develop, Semiconductor Prod Div, Motorola, Inc, Ariz, 59-61, mgr mat res, 61-63, dir eng, 63-65; mgr metaloxide-semiconductor prog, Semiconductor Components Div, Tex Instruments Inc, 65-67; dir res, Semiconductor Div, Int Tel & Telegraph Corp, Fla, 67-69; dir technol, Semiconductor Dept, Union Carbide Corp, Calif, 69-70; PROF ELEC ENG & CHMN MICROELECTRONICS COMT, UNIV MINN, MINNEAPOLIS, 70- Mem: AAAS; Am Phys Soc; sr mem Inst Elec & Electronics Eng. Res: Semiconductor device physics and engineering; nuclear physics; solid state electronics. Mailing Add: Dept of Elec Eng Univ of Minn Minneapolis MN 55455

WARNER, RICHARD DUDLEY, petrology, geochemistry, see 12th edition

WARNER, RICHARD G, b Washington, DC, Nov 1, 22; m 49; c 4. ANIMAL NUTRITION. Educ: Ohio State Univ, BSc, 47, MS, 48; Cornell Univ, PhD(animal nutrit), 51. Prof Exp: PROF ANIMAL NUTRIT, CORNELL UNIV, 51- Mem: Am Soc Animal Sci; Am Dairy Sci Asn; Am Inst Nutrit. Res: Calf and ruminant nutrition; laboratory animal nutrition; food intake physiology. Mailing Add: Dept of Animal Sci 324 Morrison Hall Cornell Univ Ithaca NY 14853

WARNER, ROBERT, b Buffalo, NY, Feb 16, 12; m 39; c 2. PEDIATRICS. Educ: Harvard Univ, AB, 35; Univ Chicago, MD, 39; Am Bd Pediat, dipl, 49. Prof Exp: Intern, Buffalo Gen Hosp, 39-40; spec intern pediat, Buffalo Children's Hosp, 40-41, res, 46-47; asst, Cincinnati Children's Hosp & Res Found, 47-48; pvt pract, 48-55; ASSOC PROF PEDIAT, SCH MED, STATE UNIV NY BUFFALO, 56-; MED DIR CHILDREN'S REHAB CTR, BUFFALO CHILDREN'S HOSP, 56- Concurrent Pos: Vis teacher, Buffalo Gen Hosp; attend, Buffalo Children's Hosp; ed, Acta Geneticae Medecae et Genellologiae. Mem: Am Soc Human Genetics; fel Am Pub Health Asn; fel Am Acad Pediat; fel Am Acad Cerebral Palsy; Am Acad Neurol. Res: Etiologics and relationships for prevention and prognostication of congenital anomalies and mental retardation, especially mongolism; endocrinology; human genetics; rehabilitation of children with neuromuscular disease cerebral palsy; juvenile amputee; reading and learning problems; treatment of phenylketonuria. Mailing Add: 936 Delaware Ave Buffalo NY 14209

WARNER, ROBERT COLLETT, b Denver, Colo, Aug 31, 13; m 36, 69; c 3. BIOCHEMISTRY. Educ: Calif Inst Technol, BS, 35; NY Univ, MS, 37, PhD(biochem), 41. Prof Exp: Chemist protein chem, Eastern Regional Lab, USDA, Philadelphia, 41-46; from asst prof to prof biochem, Sch Med, NY Univ, 46-69; PROF MOLECULAR BIOL & BIOCHEM & CHMN DEPT, UNIV CALIF, IRVINE, 69- Concurrent Pos: Mem panel plasma, Nat Res Coun, 52-57; Guggenheim Mem Found fel, Carlsberg Lab, Copenhagen, Denmark, 58; mem study sect biophys & biophys chem, NIH, 60-64; chmn study sect biochem, 72-74; assoc ed, J Biol Chem, 68-72. Mem: Am Soc Biol Chemists; Biophys Soc; Am Chem Soc. Res: Physical biochemistry of nucleic acids and proteins; mechanism of genetic recombination in small DNA containing bacteriophages; properties of circular DNA; bacterial plasmids. Mailing Add: Dept Molecular Biol & Biochem Univ of Calif Irvine CA 92717

WARNER, ROBERT EDSON, b Pomeroy, Ohio, Apr 11, 31; m 54; c 4. PHYSICS. Educ: Antioch Col, BS, 54; Univ Rochester, PhD(physics), 59. Prof Exp: From instr to asst prof physics, Univ Rochester, 59-61; asst prof, Antioch Col, 61-63; from asst prof to assoc prof, Univ Man, 63-65; assoc prof, 65-72, PROF PHYSICS, OBERLIN COL, 72- Concurrent Pos: NSF sci fac fel, Oxford Univ, 71-72. Mem: AAAS; Am Phys Soc; Am Asn Physics Teachers. Res: Nuclear scattering and reactions; nucleon-nucleon scattering; final-state interactions; angular correlations. Mailing Add: Dept of Physics Oberlin Col Oberlin OH 44074

WARNER, ROBERT LEWIS, b Redwood Falls, Minn, June 16, 37; m 64; c 2. AGRONOMY, PLANT PHYSIOLOGY. Educ: Univ Minn, BS, 62, MS, 64; Univ Ill, PhD(plant physiol & agron), 68. Prof Exp: Res asst agron, Univ Minn, 62-64 & Univ Ill, 64-68; asst prof, 68-74, ASSOC PROF AGRON, WASH STATE UNIV, 74- Mem: Am Soc Agron; Am Soc Plant Physiol. Res: Factors involved in cold resistance of alfalfa; inheritance and physiological studies of nitrate reductase. Mailing Add: Dept of Agron & Soils Wash State Univ Pullman WA 99163

WARNER, ROBERT MALCOLM, b Clyde, Ohio, June 16, 08; m 50; c 2. ECOLOGY, PLANT PHYSIOLOGY. Educ: Ohio Wesleyan Univ, BA, 30; Iowa State Univ, MS, 37, PhD(plant ecol), 40. Prof Exp: Instr biol, Am Univ, Beirut, 31-35; instr bot, Iowa State Univ, 37-38 & 40-41, lectr, 44; botanist, Res Dept, Firestone Plantations Co, WAfrica, 41-43; res dir, Calif Fig Inst, 46-61; PROF HORT, COL TROP AGR, UNIV HAWAII, 61- Concurrent Pos: Asst, Ohio Exp Sta, 41; assoc, Exp Sta, Univ Calif, Davis, 47-61; lectr, Fresno State Col, 56-60; consult hort, WAfrica, 68; hort consult, Univ Calif, Lawrence Livermore Lab, 75-76. Mem: Fel AAAS; Am Soc Hort Sci; Int Plant Propagators Soc. Res: Ecology and culture of Hevea, Ficus carica, Macadamia; citrus rootstock-scion interactions; influence of rootstocks and ecological factors on citrus quality in the tropics; banana nutrition; tropical crop ecology. Mailing Add: Dept of Hort Univ of Hawaii 3190 Maile Way Honolulu HI 96822

WARNER, ROBERT RONALD, b Long Beach, Calif, Oct 28, 46. MARINE ECOLOGY. Educ: Univ Calif, Berkeley, AB, 68, Univ Calif, San Diego, PhD(marine ecol), 73. Prof Exp: Fel biol, Smithsonian Trop Res Inst, 73-75; ASST PROF BIOL, UNIV CALIF, SANTA BARBARA, 75- Mem: Am Soc Ichthyologists & Herpetologists. Res: Reproductive strategies of marine organisms. Mailing Add: Dept of Biol Sci Univ of Calif Santa Barbara CA 93106

WARNER, SETH L, b Muskegon, Mich, July 11, 27; m 62; c 3. ALGEBRA. Educ: Yale Univ, BS, 50; Harvard Univ, MA, 51, PhD(math), 55. Prof Exp: From res instr to assoc prof, 55-65, PROF MATH, DUKE UNIV, 65- Concurrent Pos: NSF fel, Inst Advan Study, 59-60; vis res prof, Reed Col, 70-71. Mem: Am Math Soc; Math Asn Am. Res: Topological algebra; abstract analysis. Mailing Add: Dept of Math Duke Univ Durham NC 27706

WARNER, THEODORE BAKER, b Chicago, Ill, Mar 6, 31; m 55; c 3. PHYSICAL CHEMISTRY, CHEMICAL OCEANOGRAPHY. Educ: Williams Col, BA, 52; Univ Ind, Bloomington, PhD(phys chem), 63. Prof Exp: Lectr chem, Univ Ind, 62; Nat Acad Sci resident res assoc, 63-64, res chemist, Electrochem Br, 64-67 & Chem Oceanog Br, 67-69, HEAD ELECTROCHEM SECT, CHEM OCEANOG BR, NAVAL RES LAB, 69- Mem: AAAS; Am Chem Soc; Am Geophys Union; Am Soc Limnol & Oceanog. Res: Geochemistry of fluorine; in situ chemical sensors; specific ion electrodes; mechanisms and kinetics of electrode processes; adsorption; catalysis; fused salts. Mailing Add: Code 8334 Chem Oceanog Br Naval Res Lab Washington DC 20375

WARNER, VICTOR DUANE, b Coulee Dam, Wash, Sept 9, 43; m 68. MEDICINAL CHEMISTRY. Educ: Univ Wash, BS, 66; Univ Kans, PhD(med chem), 70. Prof Exp: Asst prof med chem, 70-74, ASSOC PROF MED CHEM, COL PHARM & ALLIED HEALTH SCI, NORTHEASTERN UNIV, 74-, ACTG CHMN DEPT MED CHEM & PHARMACOL, 75- Mem: Am Chem Soc; Am Pharmaceut Asn. Res: Antibacterial agents for inhibition of dental plaque. Mailing Add: Col Pharm & Allied Health Sci Northeastern Univ 360 Huntington Ave Boston MA 02115

WARNER, WALTER CHARLES, b Barberton, Ohio, June 2, 20; m 47; c 2. POLYMER CHEMISTRY. Educ: Oberlin Col, AB, 41; Case Western Reserve Univ, MS, 43, PhD(org chem), 50. Prof Exp: Chem engr, Firestone Tire & Rubber Co, 43-47; sr res chemist, 49-55, group leader anal & testing, 55-62, sect head, Tech Serv, 62-71, SECT HEAD, PHYS TESTING RES & SERV, GEN TIRE & RUBBER CO, 71- Mem: Am Chem Soc; Soc Rheology; Am Soc Testing & Mat; Int Orgn Stand. Mailing Add: Gen Tire & Rubber Co Akron OH 44329

WARNER, WILLIAM HAMER, b Pittsburgh, Pa, Oct 6, 29; m 57; c 1. APPLIED MATHEMATICS. Educ: Carnegie Inst Technol, BS, 50, MS, 51, PhD(math), 53. Prof Exp: Asst math, Carnegie Inst Technol, 50-53; res assoc appl math, Brown Univ, 53-55; from asst prof to assoc prof mech, 55-68, PROF AEROSPACE ENG & MECH, UNIV MINN, MINNEAPOLIS, 68- Mem: AAAS; Soc Indust & Appl Math; Am Math Soc; Math Asn Am; Soc Natural Philos. Res: Continuum mechanics; dynamic stability; energy methods; nonlinear systems; optimization of structures. Mailing Add: Dept Aerospace Eng & Mech Univ of Minn Minneapolis MN 55455

WARNER, WILLIS L, b Endicott, NY, Jan 28, 30. MEDICINE. Educ: Syracuse Univ, BA, 50; State Univ NY, MD, 60. Prof Exp: Res investr biochem, US Naval Radiol Defense Lab, 52-55; intern, San Francisco Hosp, 60-61; resident obstet, St Mary's Hosp, San Francisco, 61-63; assoc clin res, Baxter Labs, Ill, 63-68, assoc dir clin res, 68-71; dir clin res-biol, Hoechst Pharmaceut, Inc, 71-75; DIR CLIN RES, CUTTER LABS, INC, 75- Mem: Am Soc Pharmacol & Exp Therapeut; Am Soc Hemat; fel Int Col Angiol. Mailing Add: 39 Bret Harte Rd San Rafael CA 94901

WARNES, DENNIS DANIEL, b Stephen, Minn, June 14, 33; m 56; c 3. AGRONOMY, WEED SCIENCE. Educ: NDak State Univ, BSc, 55; Univ Minn, St Paul, MSc, 60; Univ Nebr-Lincoln, PhD(plant breeding), 69. Prof Exp: Technician agron, NDak State Univ, 51-55; teaching asst agron, Univ Minn, Minneapolis, 57-60; instr agron & outstate testing, Univ Nebr-Lincoln, 60-66, supvr, Mead Field Lab, 66-69; AGRONOMIST, W CENT EXP STA, UNIV MINN, MINNEAPOLIS, 69- Mem: Am Soc Agron; Weed Sci Soc Am. Res: Variety testing, weed control, row spacing, plant population, disease and insect control in corn, soybeans, field beans, sunflowers, small grains and forage crops; principles of weed control with specific weeds. Mailing Add: W Cent Exp Sta Morris MN 56267

WARNHOFF, EDGAR WILLIAM, b Knoxville, Tenn, May 5, 29; m 56; c 3. ORGANIC CHEMISTRY. Educ: Univ Wash, St Louis, AB, 49; Univ Wis, PhD(chem), 53. Prof Exp: Nat Res Coun fel, Birkbeck Col, Eng, 53-54; asst scientist, Nat Heart Inst, 54-57; NSF fel, Fac Pharm, Univ Paris, 57-58; res assoc chem, Mass Inst Technol, 58-59; asst prof, Univ Southern Calif, 59-62; from asst prof to assoc prof, 62-66, PROF CHEM, UNIV WESTERN ONT, 66- Concurrent Pos: Mem ed adv bd, J Org Chem, 70-74 & Can J Chem, 75-77. Honors & Awards: Merck, Sharp & Dohme Lect Award, Chem Inst Can, 69. Mem: Chem Inst Can; Am Chem Soc; The Chem Soc. Res: Organic nitrogen chemistry; mechanisms of organic reactions, especially alpha-substituted carbonyl compounds. Mailing Add: Dept of Chem Univ of Western Ont London ON Can

WARNICK, ALVIN CROPPER, b Hinckley, Utah, Nov 15, 20; m 47; c 3. PHYSIOLOGY. Educ: Utah State Agr Col, BS, 42; Univ Wis, MS, 47, PhD(physiol of reprod), 50. Prof Exp: Asst animal husb & genetics, Univ Wis, 46-50; asst prof, Ore State Col, 50-53; from asst physiologist to assoc physiologist, 53-62, PHYSIOLOGIST, UNIV FLA, 62- Mem: Am Soc Animal Sci; Am Genetic Asn. Res: Physiology of reproduction; effect of nutrition on fertility. Mailing Add: Dept of Animal Sci Univ of Fla Gainesville FL 32611

WARNICK, EDWARD GEORGE, b Chicago, Ill, Nov 1, 05; m 30. RADIOLOGY. Educ: Loyola Univ, Ill, MD, 37; Am Bd Radiol, dipl, 42. Prof Exp: Intern, St Agnes Hosp, Fond du Lac, Wis, 36-37; asst phys med, Sch Med, Loyola Univ, Ill, 37-43, from asst clin prof to assoc clin prof, Stritch Sch Med, 46-57, roentgenologist, Univ Hosp, 48-50; ASSOC PROF RADIOL, MED SCH, NORTHWESTERN UNIV, CHICAGO, 57-; CHIEF DIAG RADIOL, VET ADMIN RES HOSP, 57- Concurrent Pos: Assoc, Cook County Hosp, 37-40, fel, 40-43; consult, Induction Ctr, US Army & US Air Force, 43, 46-53; roentgenologist, Cook County Hosp, 46-53; chief dept radiol, West Side Vet Admin Hosp, 53-57. Mem: Radiol Soc NAm; Soc Nuclear Med; AMA; Asn Am Med Cols; fel Am Col Radiol. Res: Clinical diagnostic radiology. Mailing Add: 5316 Wellington Ave Chicago IL 60641

WARNICK, JORDAN EDWARD, b Boston, Mass, Mar 21, 42; m 70. NEUROPHARMACOLOGY. Educ: Mass Col Pharm, BS, 63; Purdue Univ, Lafayette, PhD(pharmacol), 68. Prof Exp: USPHS trainee pharmacol, Sch Med, State Univ NY Buffalo, 68-70 & spec awardee, 70-71, asst prof biochem pharmacol, Sch Pharm, 71-74; ASST PROF PHARMACOL & EXP THERAPEUT, SCH MED, UNIV MD, BALTIMORE CITY, 74- Mem: AAAS; NY Acad Sci; Am Soc Pharmacol & Exp Therapeut; Soc Neurosci. Res: Physiology and pharmacology of muscular dystrophy and allied neuromuscular disorders; trophic influence of nerve on muscle; pharmacology of neurotoxins. Mailing Add: Dept Pharmacol & Exp Therapeut Univ of Md Sch of Med Baltimore MD 21201

WARNKE, DETLEF ANDREAS, b Berlin, Ger, Jan 29, 28; m 64; c 2. EARTH SCIENCES. Educ: Freiburg Univ, dipl, 53; Univ Southern Calif, PhD(geol), 65. Prof Exp: Res asst geol, Aachen Tech Univ, 55-56; jr exploitation engr, Shell Oil Co, 57-58; res asst oceanog, Alan Hancock Found, 59-61; from res assoc to asst prof oceanog & geol, 65-71; asst prof, 71-73, actg chmn dept, 73-74, ASSOC PROF EARTH SCI, CALIF STATE UNIV, HAYWARD, 73- Concurrent Pos: Consult, NASA-Ames Res Ctr, 74- Mem: Am Geophys Union; Soc Econ Paleont & Mineral; Am Soc Limnol & Oceanog; Geol Soc Am; Ger Geol Asn. Res: Antartic oceanography; petrography; beach erosion; geochemistry. Mailing Add: Dept of Earth Sci Calif State Univ Hayward CA 94542

WARNOCK, BARTON HOLLAND, botany, see 12th edition

WARNOCK, JOHN EDWARD, b Freeport, Ill, Aug 20, 32; m 55; c 2. ZOOLOGY. Educ: Univ Ill, BS, 54; Univ Wis, MS, 58, PhD(zool), 63. Prof Exp: Field asst wildlife res, Ill Natural Hist Surv, 54 & Southern Ill Univ & Ill Natural Hist Surv, 54-55; res assoc, Ill Natural Hist Surv & Univ Ill, 62-64; from asst prof to assoc prof, 64-72, PROF BIOL SCI, WESTERN ILL UNIV, 72- Mem: AAAS; Am Soc Mammal; Animal Behav Soc; Ecol Soc Am; Wildlife Soc. Res: Mammalogy; animal behavior; vertebrate ecology; ornithology; ecology of small mammal populations, especially the relationships of behavior, physiological condition, density and physical factors of the environment; biology and management of small game animals. Mailing Add: Dept of Biol Sci Western Ill Univ Macomb IL 61455

WARNOCK, LAKEN GUINN, b Newton Falls, Ohio, Apr 19, 28; m 53; c 2. BIOCHEMISTRY. Educ: Milligan Col, BS, 57; Vanderbilt Univ, PhD(biochem), 62. Prof Exp: Instr biochem, Okla State Univ, 62-64; ASST PROF BIOCHEM, VANDERBILT UNIV, 64-; BIOCHEMIST, VET ADMIN HOSP, 64- Concurrent Pos: Consult, Interdept Comt Nutrit, Nat Defense Nutrit Surv, Lebanon, 61. Mem: Am Inst Nutrit. Res: Carbohydrate metabolism in vitamin deficiencies; vitamin nutriture in hemodialysis. Mailing Add: Res Lab Vet Admin Hosp Nashville TN 37203

WARNOCK, MARTHA L, b Detroit, Mich, July 19, 34; m 59; c 2. PATHOLOGY. Educ: Oberlin Col, AB, 56; Harvard Univ, MD, 60. Prof Exp: From instr to asst prof, 65-72, ASSOC PROF PATH, UNIV CHICAGO, 72- Mem: AAAS; Am Thoracic Soc; Int Acad Path. Mailing Add: Dept of Path Univ of Chicago Chicago IL 60637

WARNOCK, ROBERT G, b Salt Lake City, Utah, Mar 28, 25; m 48; c 3. PARASITOLOGY. Educ: Univ Utah, BS, 49, MS, 51, PhD(parasitol), 62. Prof Exp: Instr biol, Univ Utah, 62-63; from asst prof to assoc prof, 63-71, PROF BIOL, WESTMINSTER COL, UTAH, 71- Mem: Am Soc Parasitol. Mailing Add: Dept of Biol Westminster Col Salt Lake City UT 84105

WARNOCK, ROBERT LEE, b Portland, Ore, Feb 20, 30; m 59; c 2. THEORETICAL HIGH ENERGY PHYSICS. Educ: Reed Col, BA, 52; Harvard Univ, AM, 55, PhD(physics), 59. Prof Exp: Res assoc, Boston Univ, 59-60 & Univ Wash, Seattle, 60-62; from asst prof to assoc prof, 62-69, PROF THEORET PHYSICS, ILL INST TECHNOL, 69- Concurrent Pos: Asst physicist, Argonne Nat Lab, 64-67, assoc physicist, 67-71; vis scientist, Int Ctr Theoret Physics, Italy, 67; vis prof, Imperial Col, Univ London & Univ Bonn, 72. Mem: Am Phys Soc; Soc Indust & Appl Math. Res: Theory of elementary particles; dynamics and symmetries of strong interactions; construction of the S matrix; nonlinear analysis in mathematical physics. Mailing Add: Dept of Physics Ill Inst Technol Chicago IL 60616

WARNOCK, WALTER GEORGE, b Grand Falls, NB, July 8, 04; nat US; m 31; c 1. MATHEMATICS. Educ: Mt Allison Univ, AB, 26; Univ Ill, PhD(math), 31. Prof Exp: Instr math, Univ Ill, 27-30; from instr to asst prof, Ft Hays State Col, 31-38; vis prof, Univ Tenn, 38; from asst prof to assoc prof, Univ Ala, 38-46; from assoc prof to prof, 46-69, EMER PROF MATH, RENSSELAER POLYTECH INST, 69- Concurrent Pos: Supvr, Army Specialized Training Prog & Specialized Training & Reassignment, Univ Ala, 42-45. Res: Triple system of numbers and group structures; geometric interpretations. Mailing Add: RD 1 Box 197 Troy NY 12180

WARNTZ, WILLIAM, b Berwick, Pa, Oct 10, 22; m 47; c 2. GEOGRAPHY. Educ: Univ Pa, BS, 49; Univ Pa, AM, 51, PhD(econ), 55. Prof Exp: From instr to asst prof econ & geog, Univ Pa, 49-56; res assoc astro-phys sci, Princeton Univ, 56-66; prof theoret geog & regional planning, Harvard Univ, 66-71; PROF GEOG & CHMN DEPT, UNIV WESTERN ONT, 71- Concurrent Pos: Mem adv coun, Bur Census, Washington, DC, 56-66; Ford Found fel, Am Geog Soc, 56-61 & Off Naval Res fel, 61-66; vis prof geog, Hunter Col, 56-66 & regional sci, Univ Pa, 58-66; lectr & consult, Brookings Inst, 60-; fac sem assoc anthrop, Columbia Univ, 64-66; Off Naval Res grant, Harvard Univ, 66-67 & NSF grant, 69-71; ed, Harvard Papers in Theoret Geog, Harvard Univ, 66-71, dir, Lab Comput Graphics, 68-71; assoc ed, Geog Anal, Ohio State Univ, 69-; consult, Can Coun, 69-; consult, NSF; Can Coun grant & Dean's grant, Univ Western Ont, 72-76. Mem: AAAS (mem coun, 64-66); Regional Sci Asn (pres, 66-67); Asn Am Geog; Int Geog Union. Res: Theoretical geography as general spatial systems theory; recognition that major spatial concepts reveal identical patterns of structure and process among phenomena judged to differ greatly in their non-spatial aspects. Mailing Add: 85 Argilla Rd Andover MA 01810

WAROBLAK, MICHAEL THEODORE, b Clarksburg, WVa, Dec 2, 40; m 66; c 2. ORGANIC CHEMISTRY. Educ: WVa Wesleyan Col, BS, 62; Univ Del, MS, 64, PhD(org chem), 66. Prof Exp: Res chemist, Pioneering Res Lab, Exp Sta, 66-71, sr res chemist, Spruance Res Lab, 71-72, res supvr, 72-75, PROCESS DEVELOP SUPVR, SPRUANCE PLANT, E I DU PONT DE NEMOURS & CO, INC, 75- Mem: Sigma Xi; Am Chem Soc. Res: Textile fibers areas; polymer preparation; spinning; finished product. Mailing Add: Spruance Plant EI du Pont de Nemours & Co Inc PO Box 1559 Richmond VA 23212

WARR, WILLIAM BRUCE, b Providence, RI, June 24, 33; m 59; c 2. NEUROANATOMY. Educ: Brown Univ, BA, 57, MA, 58, PhD(physiol psychol), 63. Prof Exp: NIH fel neurophysiol & neuroanat, Eaton-Peabody Lab Auditory Physiol, Mass Eye & Ear Infirmary, 63-64, res assoc neurophysiol & neuroanat, 64-67; ASSOC PROF ANAT, SCH MED, BOSTON UNIV, 67- Concurrent Pos: Asst, Harvard Med Sch, 67. Mem: AAAS; Am Asn Anatomists. Res: Neuroanatomy and electrophysiology of sensory pathways and their constituent cell groups, especially the auditory system. Mailing Add: Dept of Anat Boston Univ Sch of Med Boston MA 02118

WARREN, BERT, microbiology, see 12th edition

WARREN, BRUCE ALBERT, b Sydney, Australia, Nov 2, 34; m 64; c 2. PATHOLOGY. Educ: Univ Sydney, BSc, 57, MB, BS, 59; Oxford Univ, PhD(path), 64, MA, 67; FRCPath Australia, 72; MRCPath, UK, 73. Prof Exp: Brit Commonwealth scholar, Sir William Dunn Sch Path, Oxford Univ, 62-64; res fel, Div Oncol, Inst Med Res, Chicago Med Sch, 64-65; lectr cardiovasc res, Nuffield Dept Surg, Oxford Univ, 66-68, tutor path, Oxford Med Sch, 67-68; vis asst prof anat, 68, asst prof path, 68-69, assoc prof, 69-74, PROF PATH, UNIV WESTERN ONT, 74- Concurrent Pos: Consult, Westminster Hosp, London, Ont, 68-; consult, Univ Hosp, London, Ont, 72- Mem: AAAS; Can Asn Pathologists; Microcirc Soc; fel Royal Micros Soc; fel Royal Soc Med. Res: Thrombosis and thrombo-embolism; platelet aggregation; fibrinolysis; function of endothelium; microvasculature of tumors; transplantation and growth of tumors; tumor emboli in bloodstream. Mailing Add: Dept of Path Univ of Western Ont London ON Can

WARREN, BRUCE ALFRED, b Waltham, Mass, May 14, 37. PHYSICAL OCEANOGRAPHY. Educ: Amherst Col, BA, 58; Mass Inst Technol, PhD(phys oceanog), 62. Prof Exp: Res asst phys oceanog, 62-63, asst scientist, 63-67, ASSOC SCIENTIST, WOODS HOLE OCEANOG INST, 67- Mem: Am Geophys Union. Res: Dynamics of ocean currents; water-mass structures; general ocean circulation. Mailing Add: Dept of Phys Oceanog Woods Hole Oceanog Inst Woods Hole MA 02543

WARREN, CHARLES EDWARD, b Portland, Ore, Oct 26, 26; m 48; c 3. FISH BIOLOGY. Educ: Ore State Col, BS, 49, MS, 51; Univ Calif, PhD(zool), 61. Prof Exp: Asst prof fisheries, Ore State Univ & asst biologist, Agr Exp Sta, 53-59, assoc prof & assoc biologist, 59-65, PROF FISHERIES, ORE STATE UNIV, 65-, BIOLOGIST, AGR EXP STA, 65- Res: Fish physiology and ecology. Mailing Add: Dept of Fisheries & Wildlife Ore State Univ Corvallis OR 97331

WARREN, CHARLES O, JR, b Bluefield, WVa, Oct 16, 39; m 60; c 1. MYCOLOGY, PLANT PHYSIOLOGY. Educ: Va Polytech Inst, BS, 61, MS, 64; Univ Fla, PhD(bot), 66. Prof Exp: Asst prof, 66-71, ASSOC PROF BIOL, SOUTHWESTERN AT MEMPHIS, 71- Concurrent Pos: Res fel biochem, St Jude Children's Res Hosp, Memphis, Tenn, 70-71. Mem: AAAS; Mycol Soc Am; Bot Soc Am. Res: Fungal physiology; metabolic bases of morphogenesis in aquatic fungi. Mailing Add: Dept of Biol Southwestern at Memphis Memphis TN 38112

WARREN, CHARLES PRESTON, b Chicago, Ill, Apr 7, 21; m 45; c 2. BIOLOGICAL ANTHROPOLOGY. Educ: Northwestern Univ, BS, 47; Ind Univ, MA, 50; Univ Chicago, MA, 61. Prof Exp: Phys anthropologist, Am Graves Regist Serv Group, US Dept Army, Philippines & Japan, 51-55; instr anthrop, 57-65, ASST PROF ANTHROP, UNIV ILL, CHICAGO CIRCLE, 65-; RES ASSOC, PHILIPPINE STUDIES PROG, UNIV CHICAGO, 56- Concurrent Pos: Phys anthropologist, US Army Cent Identification Lab, Sattahip, Thailand, 73-75. Honors & Awards: Meritorious Civilian Serv Award, Dept Army, US Army Cent Identification Lab, Thailand, 75. Mem: Am Anthrop Asn; Am Asn Phys Anthrop; Am Ethnol Soc; fel Soc Appl Anthrop. Res: Philippine ethnography; forensic and legal anthropology; identification of human remains; cognitive learning in non-western societies, Afro and Latin American urban studies. Mailing Add: Dept of Anthrop Box 4348 Univ of Ill at Chicago Circle Chicago IL 60680

WARREN, CHARLES REYNOLDS, b Kyoto, Japan, Sept 24, 13; US citizen; m 45. GEOLOGY. Educ: Yale Univ, BS, 35, PhD(geol), 39. Prof Exp: Jr geologist, Socony-Vacuum Oil Co, Venezuela, 39-40; jr geologist, US Geol Surv, Calif & Ore, 40-41, asst geologist, NY, 46; instr geol, Yale Univ, 46-47; asst prof, Washington & Lee Univ, 47-52; GEOLOGIST, US GEOL SURV, 52- Mem: Fel AAAS; fel Geol Soc Am. Res: Geomorphology; glacial geology. Mailing Add: 3606 Whitehaven Pkwy NW Washington DC 20007

WARREN, CHRISTOPHER DAVID, b Luton, Eng, Apr 24, 38; m 66; c 2. CARBOHYDRATE CHEMISTRY. Educ: Univ Sheffield, BS, 60, PhD(carbohydrate chem), 63; Royal Inst Chem, ARIC, 64. Prof Exp: Mem sci staff lipid & carbohydrate chem, Med Res Coun, Nat Inst Med Res, London, 63-69; res fel biol chem, 69-70, assoc, 70-73, PRIN ASSOC BIOL CHEM, HARVARD MED SCH, 73-; ASST BIOCHEMIST, MASS GEN HOSP, 72- Concurrent Pos: Res fel biochem, Mass Gen Hosp, 69-72. Mem: Fel The Chem Soc; Am Chem Soc. Res: Synthesis of glycosyl phosphates, polyisoprenyl phosphates and glycosyl polyprenyl phosphate and pyrophosphate diesters, and the investigation of the role of these compounds in the biosynthesis of glycoproteins. Mailing Add: Lab for Carbohydrate Res Harvard Med Sch-Mass Gen Hosp Boston MA 02114

WARREN, CLARENCE GERALD, b Mankato, Kans, Dec 21, 24; m 48; c 3. INORGANIC CHEMISTRY. Educ: Park Col, BA, 49; Univ NMex, PhD, 58. Prof Exp: Mem staff, Los Alamos Sci Lab, 49-55; asst prof chem, Western State Col Colo, 57-61; from asst prof to assoc prof, 61-74, PROF CHEM, COLO STATE UNIV, 74- Concurrent Pos: Mem staff, US Geol Surv, 67- Mem: Am Chem Soc. Res: Chemistry of the rare earths; organophosphorus chemistry; solvent extractions; geochemistry. Mailing Add: Dept of Chem Colo State Univ Ft Collins CO 80521

WARREN, CLAUDE NELSON, b Goldendale, Wash, Mar 18, 32; m 55; c 4. ARCHAEOLOGY, CULTURAL ANTHROPOLOGY. Educ: Univ Wash, BA, 54, MA, 59; Univ Calif, Los Angeles, PhD(anthrop), 64. Prof Exp: Res archaeologist, Univ Calif, Los Angeles, 58-61, lectr anthrop, 60-61; instr anthrop & hwy archaeologist, Idaho State Univ, 62-64, asst prof anthrop, 64-67; asst prof, Univ Calif, Santa Barbara, 67-69; assoc prof, 69-72, PROF ANTHROP, UNIV NEV, LAS VEGAS, 72- Concurrent Pos: NSF res grants, Univ Calif, Santa Barbara & Idaho State Univ, 65-68; Found on Arts & Humanities, Univ Nev, Las Vegas, 71-72; mem bd dir, Archaeol Res, Inc, 68- Mem: Soc Am Archaeol; Am Anthrop Asn. Res: Prehistory of western North America, especially early man; ecology and ethno-history. Mailing Add: Dept of Anthrop Univ of Nev Las Vegas NV 89109

WARREN, CRAIG BISHOP, b Philadelphia, Pa, Oct 21, 39; m 64; c 2. PHYSICAL ORGANIC CHEMISTRY. Educ: Franklin & Marshall Col, AB, 61; Villanova Univ, BS, 63; Cornell Univ, PhD(org chem), 70. Prof Exp: Fel prebiol evolution, Corp Res Dept, Monsanto Co, 68-69, sr res chemist, 69-73, res specialist, 73-75; GROUP LEADER, INT FLAVORS & FRAGRANCES, 75- Mem: Am Chem Soc; Sigma Xi; Am Soc Microbiol. Res: Structure-fragrance relationships of molecules; structure-substantivity relationships of fragrance molecules; biodegradation of small molecules. Mailing Add: Int Flavors & Fragrances 1515 Hwy 36 Union Beach NJ 07735

WARREN, DAVID HENRY, b Ithaca, NY, June 9, 30; m 62; c 4. GEOPHYSICS, SEISMOLOGY. Educ: Rensselaer Polytech Inst, BS, 51; Columbia Univ, AM, 56. Prof Exp: Res asst marine geophys, Lamont Geol Observ, Columbia Univ, 51-53; geophysicist, Stand Oil Soc Calif, 53-61 & Alpine Geophys Assocs, Inc, 61-62; GEOPHYSICIST, US GEOL SURV, 63- Mem: Am Geophys Union; Soc Explor Geophys; Seismol Soc Am. Res: Explosion seismology studies of the earth's crust and upper mantle; seismic data processing and display techniques; geologic interpretation of geophysical data. Mailing Add: 345 Middlefield Rd Menlo Park CA 94025

WARREN, DON CAMERON, b Saratoga, Ind, July 16, 90; m 10; c 1. GENETICS. Educ: Univ Ind, AB, 14, AM, 17; Columbia Univ, PhD(genetics), 23. Hon Degrees: Hon Dr, Univ Ind, 72. Prof Exp: Sci asst, Carnegie Inst, 14-15; field agent entom, Exp Sta, Ala Polytech Inst, 17-19; asst state entomologist, State Bd Entom, Ga, 19-21; from asst prof to prof poultry genetics, Kans State Col, 23-48; nat coord poultry breeding, USDA, 48-56; geneticist, 56-68, EMER GENETICIST, KIMBER FARMS, INC, CALIF, 68- Concurrent Pos: Consult, US Dept of State, India, 55. Honors & Awards: Award, Poultry Sci Asn, 33; Borden Award, 40; Superior Serv Award, USDA, 54; Poultry Hist Soc Hall of Fame, 71; Distinguished Serv Award, Kans State Univ, 72. Mem: Fel AAAS; Am Soc Nat; fel Poultry Sci Asn; Am Genetic Asn. Res: Genetics of Drosophila and fowl; physiology of reproduction in the fowl. Mailing Add: 4446 Grover Dr Fremont CA 94536

WARREN, DONALD W, b Brooklyn, NY, Mar 22, 35; m 56; c 2. DENTISTRY. Educ: Univ NC, BS, 56, DDS, 59; Univ Pa, MS, 61, PhD(physiol), 63. Prof Exp: From asst prof to assoc prof dent, 63-69, prof oral biol, 69-70, dir oral facial & communicative disorders prog, 63-70, PROF DENT ECOL & CHMN DEPT, SCH DENT, UNIV NC, CHAPEL HILL, 70-, PROF, DEPT SURG, SCH MED, 75- Concurrent Pos: Asst secy gen, Int Cong Cleft Palate, 66-69; consult, Joint Comt Dent & Speech Path-Audiol, Am Dent Asn & Am Speech & Hearing Asn, 67-71; pres-elect, Am Cleft Palate Educ Found, 75- Mem: Am Dent Asn; Int Asn Dent Res; fel Am Speech & Hearing Asn; Am Cleft Palate Asn (vpres, 67-68). Res: Physiology of speech; effects of oral-facial disorders on the speech process. Mailing Add: Dept of Dent Ecol Univ of NC Sch of Dent Chapel Hill NC 27514

WARREN, DOUGLAS ROBSON, b Fenelon Falls, Ont, July 16, 16; m 42; c 1. OCCUPATIONAL MEDICINE, ENVIRONMENTAL MEDICINE. Educ: Univ Toronto, MD, 41, dipl pub health, 47. Prof Exp: Indust physician & consult, Ont, 47-67; DIR & PARTNER, INDUST MED CONSULT, LTD, 67- Concurrent Pos: Assoc prof indust health, Sch Hyg, Univ Toronto, 67-75, spec lectr, Fac Med, 68-71, mem staff hearing conserv course, Div Exten, 69-; mem assoc comt sci criteria for environ qual, Nat Res Coun Can, 71-; Can med dir, Occidental Life Calif, 71- Res: Industrial medicine, related occupational and environmental subjects; hearing conservation and noise control; administration studies related to sickness absence control; metals in the environment and health factors. Mailing Add: Apt 3402 85 Thorncliffe Park Dr Toronto ON Can

WARREN, EDWARD PERRIN, animal husbandry, see 12th edition

WARREN, FRANCIS GAFFNEY ROSS, microwave physics, see 12th edition

WARREN, FRANCIS SHIRLEY, b Winnipeg, Man, Oct 26, 20; m 43; c 6. AGRONOMY. Educ: Ont Agr Col, BSA, 46; Univ Minn, MSc, 48, PhD(plant genetics & path), 49. Prof Exp: Asst corn breeding, Univ Minn, 47-49; res officer, Exp Farm Can Dept Agr, Ont, 49-53, head forage & cereal sect, NS, 53-66, RES SCIENTIST, FORAGE CROPS SECT, CENT EXP FARM, CAN DEPT AGR, 66- Mem: Can Soc Agron; Agr Inst Can. Res: Cereal and forage crop production. Mailing Add: Res Sta Can Dept of Agr Ottawa ON Can

WARREN, GEORGE, ceramic engineering, see 12th edition

WARREN, GEORGE FREDERICK, b Ithaca, NY, Sept 23, 13; m 44; c 3. WEED SCIENCE. Educ: Cornell Univ, BS, 35, PhD(veg crops), 45. Prof Exp: Dist county agr agent, Maine, 35-38; asst nutrit veg crops, Cornell Univ, 38-42; asst prof hort, Univ Wis, 45-48; assoc prof, 49-55, PROF HORT, PURDUE UNIV, 55- Concurrent Pos: Exec comt mem, Coun Agr Sci & Technol, 75-76. Honors & Awards: Campbell Award, Am Inst Biol Sci, 66. Mem: Fel AAAS; fel Weed Sci Soc Am (pres, 64-66); fel Am Soc Hort Sci. Res: Basis of selective action of herbicides; fate of herbicides in soil; control of weeds in horticultural crops. Mailing Add: Dept of Hort Purdue Univ West Lafayette IN 47907

WARREN, GEORGE HARRY, b Morrisville, Pa, Sept 7, 16; m 46; c 1. MICROBIOLOGY, CHEMOTHERAPY. Educ: Temple Univ, BA, 39, MA, 40; Princeton Univ, PhD(microbiol), 44. Prof Exp: Lab instr microbiol, Princeton Univ, 41-42; instr bact-immunol, Jefferson Med Col, 44-46; head dept antibiotics-bact, Wyeth Inst Med Res, 47-59, DIR DEPT MICROBIOL, WYETH LABS, 60-; PROF MICROBIOL, JEFFERSON MED COL, 66- Mem: Am Soc Microbiol; Soc Exp Biol & Med; Am Asn Cancer Res; fel Royal Soc Trop Med & Hyg; Infectious Dis Soc Am. Res: Antimicrobial agents and chemotherapy; tumor research; enzymes; mucopolysaccharides; host response mechanisms. Mailing Add: Wyeth Labs Box 8299 Philadelphia PA 19101

WARREN, HALLECK BURKETT, JR, b St Louis, Mo, Sept 3, 22; m 51; c 2. BACTERIOLOGY. Educ: Univ St Louis, BS, 43; Univ Ill, MS, 49, PhD(bact), 51. Prof Exp: Asst dairy bact, Univ Ill, 49-50; res microbiologist, Res Div, Abbott Labs, 51-57; head bact, Res Labs, Pet Inc, 57-68, mgr food sci & eng, 68-70; mgr food sci, Fairmont Foods Co, 70-72; MGR FOOD SCI, INTERSTATE BRANDS CORP, 72- Mem: Am Asn Cereal Chem; Am Soc Microbiol. Res: Microbial physiology of flavor components; bacteriology of foods and dairy products; antibiotic action, production and assay; food preservation. Mailing Add: Interstate Brands Corp PO Box 1627 Kansas City MO 64141

WARREN, HAROLD HUBBARD, b Derry, NH, July 5, 22. ORGANIC CHEMISTRY. Educ: Univ NH, BS, 44, MS, 47; Princeton Univ, MA, 49, PhD(chem), 50. Prof Exp: From instr to prof chem, 50-72, HALFORD R CLARK PROF NATURAL SCI, WILLIAMS COL, 72- Mem: AAAS; Am Chem Soc. Res: Determination of structure of natural products and synthesis and evaluation of structural variants. Mailing Add: Thompson Chem Lab Williams Col Williamstown MA 01267

WARREN, HARRY VERNEY, b Anacortes, Wash, Aug 27, 04; m 34. GEOLOGY, MINERALOGY. Educ: Univ BC, BA, 26, BASc, 27; Oxford Univ, BSc, 28, DPhil, 29. Hon Degrees: DSc, Univ Waterloo, 75. Prof Exp: Commonwealth Fund fel, Calif Inst Technol, 29-32; from lectr to prof, 32-71, PROF MINERAL & PETROL, UNIV BC, 71-; GEOCHEM ADV, PLACER DEVELOP LTD, VANCOUVER, 71- Concurrent Pos: Exec mem, BC & Yukon Chamber Mines, 39-, from vpres to pres, 39-54; exec mem, UN Asn Can, 48-, pres, 55-58. Honors & Awards: Order of Can, CM, 71, OC, 72. Mem: Fel Geol Soc Am; Am Inst Mining, Metall & Petrol Eng; fel Royal Soc Can; fel Geol Asn Can; Can Inst Mining & Metall. Res: Lead and zinc deposits in southwestern Europe; precious and base metal relationships in western and North America; rarer metals; precious and base metal deposits of British Columbia; trace elements in relation to mineral exploration and epidemiology; relationship existing between geology and health. Mailing Add: Dept of Geol Univ of BC Vancouver BC Can

WARREN, HERBERT DALE, b Houston, Tex, Apr 8, 32; m 56; c 1. ANALYTICAL CHEMISTRY, INORGANIC CHEMISTRY. Educ: Rice Univ, BA, 54; Univ Idaho, MS, 59; Ore State Univ, PhD(anal chem), 66. Prof Exp: Tech grad chem, Hanford Atomic Prod Oper, Gen Elec Co, Wash, 56-58, chemist II, 58, tech librn, 59; tech asst chem, Los Alamos Sci Lab, 61; res asst, Union Oil Res Ctr, Calif, 62; from instr to asst prof, 63-74, ASSOC PROF CHEM, WESTERN MICH UNIV, 74- Mem: Am Chem Soc; Hist Sci Soc. Res: Organic reagents for spectrophotometric analysis; equilibrium constants of coordination compounds; extraction chromatography of inorganic systems; history of chemistry. Mailing Add: Dept of Chem Western Mich Univ Kalamazoo MI 49001

WARREN, HERMAN LECIL, b Tyler, Tex, Nov 13, 32; m 63; c 3. PLANT PATHOLOGY. Educ: Prairie View Agr & Mech Col, BS, 53; Mich State Univ, MS, 62; Univ Minn, St Paul, PhD(plant path), 69. Prof Exp: Res scientist plant path, Olin Mathieson Chem Corp, 62-67; PLANT PATHOLOGIST, AGR RES SERV, USDA, 69- Concurrent Pos: Asst prof, Purdue Univ, Lafayette, 71- Mem: Am Phytopath Soc; Mycol Soc Am. Res: Relationship of soilborne diseases to stalk rot of corn; survival mechanism of soilborne pathogens; effects of light and temperature on spore germination, growth and production of fungi; physiology of host parasites and growth and production of fungi. Mailing Add: Dept of Bot & Plant Path Purdue Univ West Lafayette IN 47907

WARREN, HOLLAND DOUGLAS, b Wilkes Co, NC, July 31, 32; m 55; c 3. PHYSICS. Educ: Wake Forest Col, BS, 59; Univ Va, MS, 61, PhD(nuclear physics), 63. Prof Exp: Develop physicist, Celanese Corp Am, 63-64; sr physicist, 64-70, RES SPECIALIST, BABCOCK & WILCOX CORP, 70- Mem: Am Phys Soc; Am Nuclear Soc. Res: Neutron spectroscopy; nuclear physics; nuclear instrumentation; reactor instrumentation. Mailing Add: Lynchburg Res Ctr Babcock & Wilcox Corp Lynchburg VA 24505

WARREN, HUGH EUGENE, mathematics, see 12th edition

WARREN, JAMES C, b Oklahoma City, Okla, May 13, 30; m 51; c 4. ENDOCRINOLOGY, BIOCHEMISTRY. Educ: Univ Wichita, AB, 50; Univ Kans, MD, 54; Univ Nebr, PhD(biochem), 61. Prof Exp: Fel, Nat Inst Child Health & Human Develop, 59-61; from asst prof to prof obstet & gynec, Univ Kans, 61-70, from instr to assoc prof biochem, 61-70; PROF BIOL CHEM, PROF OBSTET & GYNEC & HEAD DEPT, SCH MED, WASHINGTON UNIV, 70- Concurrent Pos: Markle scholar med sci, 61. Res: Biosynthesis; metabolism and mechanism of action of free and conjugated steroids; kinetics of steroid interconverting enzymes; biochemistry of menstruation. Mailing Add: Dept of Obstet & Gynec Washington Univ Sch of Med St Louis MO 63103

WARREN, JAMES DONALD, b Ludlow, Mass, June 10, 48; m 74. MEDICINAL CHEMISTRY, SYNTHETIC ORGANIC CHEMISTRY. Educ: Western New Eng Col, BS, 70; Brown Univ, PhD(chem), 74. Prof Exp: Nat Cancer Inst fel, Temple Univ, 73-74; SR RES CHEMIST, LEDERLE LABS, AM CYANAMID CO, 74- Mem: Am Chem Soc. Res: Organic synthesis and evaluation of antihypertensive drug candidates. Mailing Add: Lederle Labs Am Cyanamid Co Pearl River NY 10965

WARREN, JAMES VAUGHN, b Columbus, Ohio, July 1, 15; m 54. INTERNAL MEDICINE. Educ: Ohio State Univ, BA, 35; Harvard Med Sch, MD, 39; Am Bd Internal Med, dipl. Hon Degrees: DSc, Emory Univ, 74. Prof Exp: Med house officer, Peter Bent Brigham Hosp, Mass, 39-41, asst resident med, 41-42; instr internal med, Sch Med, Emory Univ, 42-46; asst prof med, Sch Med, Yale Univ, 46-47; from assoc prof to prof physiol med, Sch Med, Emory Univ, 47-52; prof med, Sch Med, Duke Univ, 52-58; prof internal med & chmn dept, Univ Tex Med Br, 58-61; PROF MED & CHMN DEPT, COL MED, OHIO STATE UNIV, 61- Mem: Nat Inst Med; Am Soc Clin Invest; Am Physiol Soc; Am Heart Asn; Am Col Physicians. Res: Cardiovascular diseases. Mailing Add: Dept of Med Ohio State Univ Col of Med Columbus OH 43210

WARREN, JOEL, b New York, NY, 1914; m 42; c 2. CANCER RESEARCH. Educ: Yale Univ, AB, 36; Columbia Univ, AM, 38, PhD(bact), 40. Prof Exp: Res assoc pediat, Childrens' Hosp, Cincinnati, Ohio, 29-42; chief virus res sect, US Army Med Sch, DC, 46-51, dept bact, 51-54; sci attache for Scandinavia, US Dept State, Sweden, 54-56; virologist, Div Biol Stand, NIH, 56-58; dir biol res, Chas Pfizer & Co, 58-68; DIR LIFE SCI CTR, NOVA UNIV, 68- Concurrent Pos: Nat Res Coun fel, 42-43; vis prof, Ohio State Univ, 49-50 & Univ Md, 51-58; vis res fel, Univ Uppsala, 51; mem microbiol study sect, Grant-in-Aid Div, NIH, 53-55. Mem: Soc Exp Biol & Med; Am Asn Immunol; Am Acad Microbiol. Res: Viruses and viral diseases; biophysical methodology; toxoplasmosis; instrumentation; immunology of viral infections; cancer chemotherapy and immunology. Mailing Add: Life Sci Ctr Nova Univ Ft Lauderdale FL 33314

WARREN, JOHN BERNARD, b London, Eng, Nov 4, 14; Can citizen; m 51; c 4. PHYSICS. Educ: Univ London, BSc, 34, Imp Col, dipl & PhD(physics), 36. Prof Exp: Jr sci officer, Dept Sci & Indust Res, UK, 37-39; lectr appl physics, Southampton Univ, 39; sr sci officer, Telecommun Res Estab, 40-45; sr sci officer, UK Atomic Energy Proj, 45-46; lectr physics, Glasgow Univ, 46-47; PROF PHYSICS, UNIV BC, 47- Concurrent Pos: Res fel, Australian Nat Univ, 55-56; res fel, Rutherford Lab, Abingdon, Eng, 65-66; dir, TRIUMF Proj, Univ BC, 68-71; vis, Europ Orgn Nuclear Res, 72-73. Mem: Am Phys Soc; Can Asn Physicists; fel Royal Soc Can; Brit Inst Physics. Res: Nuclear physics. Mailing Add: TRIUMF Project Univ of BC Vancouver BC Can

WARREN, JOHN LUCIUS, b Chicago, Ill, Dec 7, 32; m 58; c 2. SOLID STATE PHYSICS, RESEARCH ADMINISTRATION. Educ: Univ Chicago, BA, 53; Univ Md, PhD(physics), 59. Prof Exp: Asst prof physics, De Pauw Univ, 59-61; mem staff, 61-73, asst to assoc dir res, 73-75, ASST DIV LEADER, CTR DIV, LOS ALAMOS SCI LAB, UNIV CALIF, 75- Mem: Am Phys Soc. Res: Lattice dynamics using inelastic neutron scattering; group theory applied to lattice dynamics. Mailing Add: MS 640 PO Box 1663 Los Alamos NM 87545

WARREN, JOHN RUSH, b Columbus, Ohio, July 20, 19. BOTANY. Educ: Marietta Col, AB, 41; Ohio State Univ, MS, 49, PhD, 50. Prof Exp: Asst plant pathologist, Agr Exp Sta, Ohio State Univ, 42-46; asst prof, Duke Univ, 46-52; plant pathologist, Standard Fruit & Steamship Co, Honduras, 52-55, dir trop res, 55-59, dir statist qual control, La, 59-60; assoc prof biol sci & chmn dept, Tenn Technol Univ, 60-61; dean grad sch, 64-72, PROF BIOL SCI, MARSHALL UNIV, 64- Concurrent Pos: Consult, Atomic Energy Comn, 66; Fulbright lectr & consult, Nat Univ Honduras, 68-69. Res: Microbiology; ecology; tropical agriculture. Mailing Add: Dept of Biol Sci Marshall Univ Huntington WV 25701

WARREN, JOHN STANLEY, b Ithaca, NY, Dec 19, 37; m 61; c 1. GEOLOGY. Educ: Cornell Univ, BA, 60; Stanford Univ, PhD(geol), 67. Prof Exp: Asst prof geol, Univ Cincinnati, 65-72; ASSOC PROF GEOL, THOMAS JEFFERSON COL, GRAND VALLEY STATE COL, 72- Mem: Paleont Soc. Res: Invertebrate paleontology; micropaleontology; palynology. Mailing Add: Thomas Jefferson Col Grand Valley State Col Allendale MI 49401

WARREN, KENNETH LYLE, b Battle Creek, Mich, Dec 3, 06; m 28; c 1. PHYSICS. Educ: Battle Creek Col, BS, 28, MS, 30; Mich State Col, PhD(physics), 35. Prof Exp: Instr physics & math, Battle Creek Col, 30-33; instr math, Mich State Col, 35-36; head div phys sci, Grand Rapids Col, 36-42, coordr & instr, Civilian Pilot Training Serv, 39-42; prof math, Millsaps Col, 46-49; assoc prof physics, 49-68, PROF PHYSICS & DIR PLANETARIUM, KENT STATE UNIV, 68- Concurrent Pos: Mem res & develop staff, Goodyear Aircraft Corp, 52-58. Mem: AAAS; Am Phys Soc; Am Math Soc; Am Asn Physics Teachers; Int Soc Planetarium Educators. Res: Molecular physics; heat; mechanics. Mailing Add: Dept of Physics Kent State Univ Kent OH 44242

WARREN, KENNETH S, b New York, NY, June 11, 29; m 59; c 2. TROPICAL MEDICINE. Educ: Harvard Univ, AB, 51, MD, 55; Univ London, dipl, 59. Prof Exp: Intern med, Boston City Hosp, Mass, 55-56; mem staff med res, Lab Parasitic Dis, NIH, 56-63; from asst prof to assoc prof prev med, 63-70, from asst prof to assoc prof med, 63-75, adj prof lib sci, 72-75, ASSOC PROF GEOG MED, SCH MED, CASE WESTERN RESERVE UNIV, 70-, DIR DIV GEOG MED, 73-, PROF MED & PROF LIB SCI, 75- Concurrent Pos: NIH res career develop award, 66-71. Mem: Am Soc Clin Invest; Am Soc Trop Med & Hyg; Am Asn Immunologists; Infectious Dis Soc Am; Asn Am Physicians. Res: Schistosomiasis; pathophysiology; immunology; control. Mailing Add: Wearn Res Bldg Univ Hosp Cleveland OH 44106

WARREN, KENNETH SINCLAIR, organic chemistry, see 12th edition

WARREN, LEONARD, biochemistry, see 12th edition

WARREN, LIONEL GUSTAVE, b New York, NY, May 5, 26; m 52; c 3. PARASITOLOGY, PHYSIOLOGY. Educ: Syracuse Univ, AB, 48, MA, 53; Johns Hopkins Univ, ScD(parasitol), 57. Prof Exp: Res assoc biol, Rice Univ, 57-60; vis Int Atomic Energy Agency prof parasitol, Sci Res Inst, Caracas, Venezuela, 60-63; ASSOC PROF MED PARASITOL, LA STATE UNIV MED CTR, NEW ORLEANS, 63- Concurrent Pos: USPHS grant, 64-; scientist, Charity Hosp La, New Orleans, 67- Mem: Am Soc Parasitologists; Soc Protozoologists; Am Soc Trop Med & Hyg; Am Soc Cell Biol. Res: Carbohydrate and oxidative metabolism of endoparasitic animals; energy metabolism of blood parasites, including plasmodia, hemoflagellates and the metabolism of hookworms; immunology of endoparasites. Mailing Add: Dept of Trop Med La State Univ Med Ctr New Orleans LA 70112

WARREN, LLOYD OLIVER, b Fayetteville, Ark, Dec 27, 15; m 42; c 3. ENTOMOLOGY. Educ: Univ Ark, BS, 47, MS, 48; Kans State Col, PhD, 54. Prof Exp: Instr & jr entomologist, Univ Ark, 47-51; instr entom, Kans State Univ, 53-54; from asst prof & asst entomologist to prof entom & entomologist, 54-73, DIR, ARK AGR EXP STA, UNIV ARK, FAYETTEVILLE, 73- Mem: Entom Soc Am. Res: Forest insects; apiculture. Mailing Add: Agr Exp Sta Univ of Ark Fayetteville AR 72701

WARREN, MASHURI LAIRD, b Findlay, Ohio, Jan 12, 40; m 63; c 2. ENVIRONMENTAL PHYSICS, PLASMA PHYSICS. Educ: Ohio Wesleyan Univ, BA, 61; Univ Calif, Berkely, MA, 63, PhD(plasma physics), 68. Prof Exp: Asst prof physics, Calif State Univ, Hayward, 68-74; SCIENCE WRITING, 74- Mem: Am Phys Soc; Int Solar Energy Soc. Res: Physics of environmental problems; atomic collisions of fusion; transport problems in plasma physics. Mailing Add: 10 Sequoia Rd Fairfax CA 94930

WARREN, MCWILSON, b Wayne Co, NC, Aug 29, 29; m 75. MALARIOLOGY, TROPICAL MEDICINE. Educ: Univ NC, BA, 51, MSPH, 52; Rice Univ, PhD(parasitol), 57. Prof Exp: Asst prof prev med & pub health, Sch Med, Univ Okla, 57-59, assoc prof & vchmn dept, 59-61; scientist, Lab Parasite Chemother, NIH, USPHS, 61-69, scientist, Far East Res Proj, Kuala Lumpur, Malaysia, 61-64, officer-in-chg, 63-64, officer-in-chg, Sect Cytol, Chamblee, Ga, 64-65, head sect chemother, Nat Inst Allergy & Infectious Dis, 66-69, parasitologist, Cent Am Malaria Res Sta, Ctr Dis Control, 69-74, SCIENTIST DIR, CTR DIS CONTROL, 72-, PARASITOLOGIST, VECTOR BIOL & CONTROL DIV, BUR TROP DIS, 74- Concurrent Pos: China Med Bd fel trop med, Cent Am, 57; consult, Univ Hosps, Oklahoma City, 57 & US Naval Med Res Unit III, 66; res assoc, Sch Trop Med & Hyg, Univ London, 67-68. Mem: Am Soc Trop Med & Hyg; Am Soc Parasitol; Soc Protozool; Royal Soc Trop Med & Hyg; Am Mosquito Control Asn. Res: Ecology and immunity of the primate malarias; parasite physiology; pathophysiology of infectious disease agents; global and institutional epidemiology and human ecology; field studies on sero-epidemiology of malaria; genetics of malaria vectors; field problems in chemotherapy of malaria; biology of malaria parasites. Mailing Add: Bur of Trop Dis Ctr for Dis Control Atlanta GA 30333

WARREN, MITCHUM ELLISON, JR, b Paris, Tenn, Nov 10 34; m 61; c 2. ORGANIC CHEMISTRY. Educ: Vanderbilt Univ, BA, 56, PhD(org chem), 63. Prof Exp: NIH fel, 63-66; asst prof, 66-71, ASSOC PROF CHEM, GEORGE PEABODY COL, 71- Mem: AAAS; Am Chem Soc. Res: Stereochemistry; optically active compounds; alkaloids. Mailing Add: Dept of Chem George Peabody Col for Teachers Nashville TN 37203

WARREN, PETER, b New York, NY, Sept 30, 38; m 70. MATHEMATICS, BIOMATHEMATICS. Educ: Univ Calif, Berkeley, BA, 60; Univ Wis-Madison, MA, 65, PhD(math), 70. Prof Exp: Mem tech staff, IBM Nordic Labs, 61-63; invited fel theory of traffic control, Thomas J Watson Res Labs, 64; lectr math, Med Sch, Univ Wis-Madison, 65-66; asst prof, 70-74, ASSOC PROF MATH, UNIV DENVER, 74- Concurrent Pos: Statist consult, 73- Mem: Am Math Soc; Math Asn Am. Res: Probability theory; epidemiology; probability theory in Banuch spaces. Mailing Add: Dept of Math Univ of Denver Denver CO 80210

WARREN, RICHARD, b Wichita, Kans, Apr 30, 99; m 37; c 2. PHYSICS. Educ: Purdue Univ, BS, 24, PhD(ed), 54; Columbia Univ, MA, 34. Prof Exp: Eng & math consult, Statist Bur, Columbia Univ, 28-33; assoc prof physics, Middle Ga Col, 43-44; instr, Univ Ark, 44-45; prof physics & mech eng, WVa Inst Technol, 45-46; instr physics, Purdue Univ, 46-50; asst prof, Mich Col Min & Tech, 50-57; assoc prof, Univ Mo-Rolla, 57-69; PROF MATH, CHOWAN COL, 69- Mem: Am Soc Eng Educ; Am Asn Physics Teachers. Res: Teaching physics for engineering students. Mailing Add: Dept of Math Chowan Col Murfreesboro NC 27855

WARREN, RICHARD HAWKS, b Binghamton, NY, Feb 16, 34; m 61; 61; c 2. MATHEMATICS. Educ: US Naval Acad, BS, 56; Univ Mich, Ann Arbor, MS, 64; Univ Colo, Boulder, PhD(math), 71. maintenance off, 56-62, from instr to assoc prof math, US Air Force Acad, 64-69, Dep dir appl math res lab, Aerospace Res Lab, 72-75, CHIEF, APPL MATH GROUP, AIR FORCE FLIGHT DYNAMICS LAB, US AIR FORCE, 75- Concurrent Pos: US Air Force, 56- Mem: Math Asn Am; Am Math Soc. Res: General topology; proximity spaces; fuzzy topological spaces; applied mathematics. Mailing Add: 5654 Candelight Lane Dayton OH 45431

WARREN, RICHARD JOSEPH, b Lowell, Mass, Dec 25, 31; m 58; c 4. ANALYTICAL CHEMISTRY. Educ: Merrimack Col, BS, 53; Univ Pa, MS, 58. Prof Exp: Anal chemist, 56-61, sr anal chemist, 61-73, SR INVESTR, SMITH, KLINE & FRENCH LABS, 73- Mem: Am Chem Soc; Soc Appl Spectros. Res: Infrared, ultra violet and nuclear magnetic resonance spectroscopy; mass spectroscopy; x-ray diffraction. Mailing Add: Anal Chem Smith Kline & French Labs 1500 Spring Garden St Philadelphia PA 19101

WARREN, RICHARD JOSEPH, b Oklahoma City, Okla, June 30, 33; m 69; c 1. HUMAN GENETICS, CYTOGENETICS. Educ: Okla City Univ, AB, 58; St Louis Univ, PhD(microbiol), 67. Prof Exp: NIH fel, Harvard Med Sch-Mass Gen Hosp, 67; NIH fel, Wash Univ, 67-69, asst prof pediat & med, Med Sch, 67-70; asst prof pediat, Univ Miami, 70-74; DIR, GENETICS ASSOCS, 74- Concurrent Pos: Chmn ad hoc comt, NIH Contract-Cytogenetics Registries, 73-; dir, Palm Beach Genetics Clin, 73-; genetics consult, Bur Health & Rehab, State of Fla, 74- Mem: Am Soc Human Genetics; AAAS; Mammalian Cell Genetics Soc; Am Tissue Cult Asn; Am Soc Cell Biol. Res: Human cytogenetics; molecular biology; cellular regulation. Mailing Add: Genetics Assocs 7805 Coral Way Miami FL 33155

WARREN, RICHARD SCOTT, b Malden, Mass, Oct 21, 42; m 67. PLANT PHYSIOLOGY. Educ: Defiance Col, BA, 65; Univ NH, MS, 68, PhD(plant sci), 70. Prof Exp: Sigma Xi grant-in-aid res, 70-71, ASST PROF BOT, CONN COL, 70- Mem: AAAS; Bot Soc Am; Am Inst Biol Sci. Res: Physiological ecology of Halophytes; physiology of disease resistance. Mailing Add: Conn Col 360 Mahegan Ave New London CT 06320

WARREN, ROBERT HOLMES, b Austin, Tex, Feb 20, 41; div. CELL BIOLOGY. Educ: Rice Univ, BA, 62, MA, 63; Harvard Univ, PhD(cell biol), 69. Prof Exp: NIH fels, Cambridge Univ, 69-70 & Univ Tex, Austin, 70-71; asst prof biol struct, 71-75, ASSOC PROF BIOL STRUCT, MED SCH, UNIV MIAMI, 75- Concurrent Pos: NSF res grant, Med Sch, Univ Miami, 72-73 & 74-76. Mem: Am Soc Cell Biol; Soc Develop Biol. Res: Ultrastructural basis of mechanisms of cell motility and cytomorphogenesis. Mailing Add: 5035 SW 92nd Ave Miami FL 33165

WARREN, ROGER WRIGHT, physics, mathematics, see 12th edition

WARREN, ROSLYN PAUKER, zoology, see 12th edition

WARREN, SHIELDS, b Cambridge, Mass, Feb 26, 98; m 23; c 2. PATHOLOGY. Educ: Boston Univ, AB, 18; Harvard Univ, MD, 23; Am Bd Path, cert clin path, path anat & forensic path. Hon Degrees: ScD, Boston Univ, 49, Western Reserve Univ, 52, Case Inst Technol, 56, Northwestern Univ, 59; LLD, Tulane Univ, 53; Dr, Brazil Univ, 64; DMS, Brown Univ, 69. Prof Exp: Asst path, Boston City Hosp, 23-25; from instr to prof, 25-65, EMER PROF PATH, HARVARD MED SCH, 65- Concurrent Pos: Pathologist, New Eng Deaconess Hosp, 27-63, sci dir cancer res inst, 51-68, consult, 68-; pathologist, Pondville State Hosp, 23-48, New Eng Baptist Hosp, 28-63 & Huntington Mem Hosp, 38-42; consult, House of the Good Samaritan, 27-43, Channing Home, 35-58, US Air Force, 48-62 & Vet Admin, 52-; dir, State Tumor Diag Serv, Mass, 28-55; spec consult, US Dept Defense, 59-62; consult, AEC, 59-; consult space prog adv coun, NASA, 71- Chmn atomic casualty comn & mem exec comt & comt path, Nat Res Coun; mem, Nat Adv Cancer Coun, USPHS, 46-49; dir div biol & med, AEC, 47-52, mem adv comt, 52-58, mem med adv comt, PR Nuclear Ctr, 59-66, consult, 56-; mem sci adv bd, Armed Forces Inst Path, 52-; mem res consult bd, US Navy, 53-; US rep, Sci Comt Effects Atomic Radiation, UN, 55-63; pres, Am Bd Path, 56-58; spec adv & actg US rep, Int Conf Peaceful Uses Atomic Energy, Geneva, 58; chmn corp & bd trustees, Boston Univ, 61; chmn life sci comt, NASA, 71-; mem subcomt somatic effects, Nat Acad Sci, 71- Honors & Awards: Proctor Award, Sci Res Soc Am, 52; Banting Medal, Am Diabetes Asn, 53; Albert Einstein Medal & Award, 62; Citation, AEC, 63, Enrico Fermi Award, 72. Mem: Nat Acad Sci; AAAS (vpres, 48); Am Asn Path & Bact (vpres, 47, pres, 48); Am Asn Cancer Res (vpres, 41, pres, 42-46); AMA. Res: Pathology of diabetes mellitus, tumors and tumor metastases; effects of radiation on normal and neoplastic cells and mammals. Mailing Add: Cancer Res Inst 194 Pilgrim Rd Boston MA 02215

WARREN, WILLIAM A, b Findlay, Ohio, Mar 29, 36; m 59; c 3. BIOCHEMISTRY. Educ: Amherst Col, AB, 58; Western Reserve Univ, MD, 62; Univ Mass, PhD, 68. Prof Exp: Staff assoc biochem, Nat Inst Arthritis & Metab Dis, Md, 66-68; ASSOC RES PHYSICIAN, MARY IMOGENE BASSETT HOSP, COOPERSTOWN, NY, 68- Concurrent Pos: USPHS fel biochem, Dartmouth Med Sch, 64-65; fel, Amherst Col, 65-66. Mem: Am Chem Soc; Am Asn Clin Chemists; Am Soc Biol Chemists. Res: Protein structure and function; chemistry of pyridine nucleotides and glyoxylate; mast cell tumor biochemistry. Mailing Add: Mary Imogene Bassett Hosp Cooperstown NY 13326

WARREN, WILLIAM ERNEST, b Rochester, NY, Aug 11, 30; m 55; c 2. APPLIED MATHEMATICS. Educ: Univ Rochester, BS, 56, MS, 59; Cornell Univ, PhD(eng mech), 62. Prof Exp: From instr to asst prof mech & mat, Cornell Univ, 57-62; STAFF MEM, APPL MATH DIV, SANDIA LAB, 62- Mem: Am Inst Aeronaut & Astronaut; Am Math Soc; Soc Indust & Appl Math. Res: Plane elastic systems; thermal stress concentrations; electric field effects on solid dielectrics, particularly dielectric breakdown; wave propagation; solid-fluid interacting systems. Mailing Add: 7712 La Condesa N E Albuquerque NM 87110

WARREN, WILLIAM MICHAEL, b Bancroft, Mich, July 5, 17; m 45; c 5. ANIMAL HUSBANDRY, ANIMAL BREEDING. Educ: Mich State Col, BS, 40; Agr & Mech Col Tex, MS, 48; Univ Mo, PhD(animal breeding), 52. Prof Exp: From instr to assoc prof animal husb, Tex A&M Univ, 41-55; assoc prof animal husb & assoc animal breeder, 55-57, PROF ANIMAL SCI & HEAD DEPT, AUBURN UNIV, 57- Mem: Am Soc Animal Sci. Res: Improvement of livestock through selection and breeding. Mailing Add: Dept of Animal-Dairy Sci Auburn Univ Auburn AL 36830

WARREN, WILLIAM WILLARD, JR, b Seattle, Wash, Nov 7, 38; m 65; c 2.

NUCLEAR MAGNETIC RESONANCE. Educ: Stanford Univ, BS, 60; Wash Univ, PhD(physics), 65. Prof Exp: Asst res physicist, Univ Calif, Los Angeles, 65-68; MEM TECH STAFF, BELL LABS, INC, 68- Honors & Awards: US Sr Scientist Award, Alexander von Humboldt Found, 74. Mem: AAAS; Am Phys Soc. Res: Application of nuclear magnetic resonance to the study of electronic structure and atomic dynamics of liquids and solids, especially metals and semiconductors; electronic transport properties of liquids. Mailing Add: Bell Labs Inc Dept 1525 Murray Hill NJ 07974

WARRICK, ARTHUR W, b Kellerton, Iowa, Dec 4, 40; m 62; c 2. SOIL PHYSICS, MATHEMATICS. Educ: Iowa State Univ, BS, 62, MS, 64, PhD(soil physics), 67. Prof Exp: Res assoc soil physics, Iowa State Univ, 66-67; asst prof, 67-71, ASSOC PROF SOIL PHYSICS, UNIV ARIZ, 71- Mem: Soil Sci Soc Am; Am Soc Agron; Am Geophys Union. Res: Drainage; soil water flow; porous media flow; potential theory. Mailing Add: Dept of Soils Water & Eng Univ of Ariz Tucson AZ 85721

WARRICK, EARL LEATHEN, b Butler, Pa, Sept 23, 11; m 40; c 2. PHYSICAL CHEMISTRY. Educ: Carnegie Inst Technol, BS, 33, MS, 34, DSc(phys chem), 43. Prof Exp: Asst, Mellon Inst Sci, 35-37, fel organosilicon chem, 37-46, sr fel, 46-56; asst dir res, 57-59, mgr hyper-pure silicon div, 59-62, gen mgr electronic prod div, 62-68, mgr new proj bus, 68-72, SR MGT CONSULT, DOW CORINING CORP, 72- Concurrent Pos: Lectr, Univ Pittsburgh, 47-48. Mem: Am Chem Soc. Res: Glass composition; chemical kinetics; gas phase; organosilicon and radiation chemistry; physical chemistry of polymers. Mailing Add: 508 Crescent Dr Midland MI 48640

WARRICK, PERCY, JR, b South Bend, Ind, Aug 6, 35; m 61; c 2. PHYSICAL ORGANIC CHEMISTRY. Educ: Wabash Col, 57; Univ Rochester, PhD(org chem), 61. Prof Exp: Fel phys org chem, Univ Minn, 60-62; res assoc, Mass Inst Technol, 62-63; asst prof, 63-66, ASSOC PROF CHEM, WESTMINSTER COL, PA, 66- Concurrent Pos: Fel, Univ Utah, 70-71. Mem: Am Chem Soc; The Chem Soc. Res: Solvolysis and rearrangement of alkyl-aryl compounds; mechanisms of reactions between metals and solutions; general-acid catalysis; solvent isotope effects in organic reactions; relaxation kinetics; acid-base reactions in mixed solvents. Mailing Add: Dept of Chem Westminster Col New Wilmington PA 16142

WARRINGTON, PATRICK DOUGLAS, b Winnipeg, Man, Mar 21, 42; m 65; c 2. BOTANY. Educ: Univ BC, BSc, 64, PhD(bot), 70. Prof Exp: Consult bot, 72-73; res officer aerial satellite photog, 73-75, BIOLOGIST, BC GOVT, 75- Res: All aspects of the biology of aquatic plants. Mailing Add: 9210 Cresswell Rd RR 2 Sidney BC Can

WARRINGTON, TERRELL L, b Baltimore, Md, June 5, 40; m 64. PHYSICAL CHEMISTRY, BIOCHEMISTRY. Educ: Yale Univ, BA, 61; Purdue Univ, PhD(phys chem), 66. Prof Exp: ASST PROF CHEM, MICH TECHNOL UNIV, 67- Mem: Am Chem Soc. Res: Physical chemistry of biological macromolecules concentrating mainly on conformational studies. Mailing Add: Dept of Chem Mich Technol Univ Houghton MI 49931

WARSCHAUER, DOUGLAS MARVIN, b Haverstraw, NY, Sept 3, 25; div; c 3. PHYSICS, ENERGY CONVERSION. Educ: Drew Univ, BA, 46; Univ Pa, PhD(physics), 52. Prof Exp: Instr math, Drew Univ, 45; asst physics, NY Univ, 46-47, Univ Pa, 47-51; res physicist, Philco Corp, 52; mem staff physics, Lincoln Lab, Mass Inst Technol, 52-58, res physicist aeronaut res lab, Wright-Patterson Air Force Base, 58-59; prin res scientist, Raytheon Co, 60-65; mgr physics lab, Itek Corp, 65-66; chief electronic components lab, NASA Electronics Res Ctr, Mass, 67-70; RES PHYSICIST, NAVAL WEAPONS CTR, 72- Concurrent Pos: Consult, Meret Inc, 74-; assignee, Energy Res & Develop Admin, 75- Mem: AAAS. Res: High pressure; semiconductors; optical properties of solids; crystal growth; photovoltaic energy conversion. Mailing Add: Res Dept Naval Weapons Ctr China Lake CA 93555

WARSCHAWSKI, STEFAN EMANUEL, b Lida, Russia, Apr 8, 04; nat US; m 47. MATHEMATICS. Educ: Univ Basel, PhD, 30. Prof Exp: Assoc math, Columbia Univ, 34-35; assoc elec eng, Cornell Univ, 35-37; instr math, Univ Rochester, 37-38, Brown Univ, 38-39; from asst prof to assoc prof, Washington Univ, 39-45; prof, Univ Minn, Minneapolis, 45-63, head dept inst technol, 52-63; chmn dept, 63-67, PROF MATH, UNIV CALIF, SAN DIEGO, 63- Concurrent Pos: Sr res mathematician appl math group, Brown Univ, 44-45; vis prof, Univ Calif, Los Angeles, 58-59; vis prof math, San Diego State Univ, 73-74. Mem: Am Math Soc; Math Asn Am. Res: Complex analysis, particularly conformal mapping; potential theory; minimal surfaces—boundary behavior. Mailing Add: Dept of Math Univ Calif San Diego La Jolla CA 92037

WARSH, CATHERINE EVELYN, b San Diego, Calif, Apr 19, 43; m 70. OCEANOGRAPHY. Educ: Old Dominion Col, BA, 65; Fla State Univ, MS, 71. Prof Exp: Teacher math, Kempsville Jr High Sch, Virginia Beach, Va, 65-66; researcher marine biol, Duke Univ, NC, 66-67; teacher marine biol & math, First Colonial High Sch, Virginia Beach, 67-68; researcher limnol, Fla State Univ, Tallahassee, 71-72, res programmer biol oceanog, 72, environ specialist pollution & limnol, Dept Pollution Control, 73; admin oceanogr, Nat Oceanic & Atmospheric Admin-Nat Marine Fisheries Serv, 74-75, RES OCEANOGR, DEPT COMMERCE, NAT OCEANIC & ATMOSPHERIC ADMIN-NAT OCEAN SURV, WASHINGTON, DC, 75- Concurrent Pos: Teacher math & phys sci, Griffin Middle Sch, Tallahassee, Fla, 71; phys oceanogr researcher, Dept Commerce, Nat Oceanic & Atmospheric Admin-Nat Marine Fisheries Serv, Washington, DC, 74; task team oil spill trajectory models, Environ Res Labs-Nat Oceanic & Atmospheric Admin, Boulder, Colo, 75- Res: Study of surface circulation, near-shore dynamics over the continental shelf in the western Gulf of Mexico; study of circulation, water properties and transport in Gulf of California; oil-trajectory modeling. Mailing Add: 5646 Stevens Forest Rd Apt 201 Columbia MD 21045

WARSH, KENNETH LEE, b Chicago, Ill, Oct 17, 36; m 70. PHYSICAL OCEANOGRAPHY. Educ: Univ Notre Dame, BS, 58, MS, 60; Fla State Univ, PhD(physics), 62. Prof Exp: Physicist, Lockheed-Ga Co, 63-64; asst prof physics, Jacksonville Univ, 64-66; asst prof oceanog, Fla State Univ, 66-73; SR STAFF PHYSICIST, APPL PHYSICS LAB, JOHNS HOPKINS UNIV, 73- Concurrent Pos: Lectr, Morehouse Col, 63-64. Mem: Am Meteorol Soc; Am Phys Soc; Am Asn Physics Teachers; Am Geophys Union. Res: Marine meteorology; environmental management. Mailing Add: Appl Physics Lab Johns Hopkins Univ Laurel MD 20810

WARSHAUER, STEVEN MICHAEL, b New York, NY, May 20, 45. INVERTEBRATE PALEONTOLOGY, PALEOECOLOGY. Educ: Queens Col, NY, BA, 67; Univ Cincinnati, MS, 69, PhD(geol), 73. Prof Exp: ASST PROF GEOL, W VA UNIV, 72- Mem: Int Paleont Union; Brit Palaeont Asn; Paleont Soc; Soc Econ Paleontologists & Mineralogists; Sigma Xi. Res: Taxonomy and paleocology of Lower Paleozoic Ostracoda; species diversity and benthic community structure of the Mid-Appalachian Silurian; multivariate statistical methods in paleoecology and biostratigraphy. Mailing Add: Dept of Geol & Geog WVa Univ Morgantown WV 26506

WARSHAVSKY, MORDECHAI S, solid state physics, rheology, see 12th edition

WARSHAW, CHARLOTTE MARSH, b Newark, NJ, Feb 5, 20; m 48. GEOCHEMISTRY, MINERALOGY. Educ: Smith Col, AB, 41; Bryn Mawr Col, MA, 42; Pa State Univ, PhD(geochem), 57. Prof Exp: Chemist, Geophys Lab, Carnegie Inst Technol, 42-46; ed asst, Off Sci Res & Develop, 46; geologist & chemist, US Geol Surv, 46-53; mineralogist, Gulf Res & Develop Co, 53-57; asst, Pa State Univ, 51-53; res assoc, 57-60; sr scientist, Tem-Pres Res, Inc, Pa, 60-61; consult geochemist, 61-71; GEOLOGIST, US GEOL SURV, 71- Mem: Mineral Soc Am. Res: Clay mineralogy; mineral synthesis; phase equilibria in silicate systems; mineralogy of volcanic rocks. Mailing Add: 3703 Stewart Driveway Chevy Chase MD 20015

WARSHAW, MYRON M, biophysical chemistry, clinical chemis try, see 12th edition

WARSHAWSKY, HERSHEY, b Montreal, Que, Feb 6, 38; m 60; c 3. HISTOLOGY. Educ: Sir George Williams Univ, BSc, 59; McGill Univ, MSc, 61, PhD(anat), 66. Prof Exp: Lectr anat, McGill Univ, 63-66; res fel orthop res, Harvard Univ, 66-67; asst prof, 67-70, ASSOC PROF ANAT, McGILL UNIV, 70- Concurrent Pos: Vis prof, Univ Sao Paulo. Mem: Am Asn Anat; Can Asn Anat. Res: Use of the enamel organ in the rat incisor as a model system for structural and radioautographic studies of secretory processes and cell renewal. Mailing Add: Dept of Anat McGill Univ Box 6070 Montreal PQ Can

WARSHEL, ARIEH, b Sde-Nahom, Israel, Nov 20, 40; m 66; c 2. CHEMICAL PHYSICS, MOLECULAR BIOLOGY. Educ: Israel Inst Technol, BSc, 66; Wiezmann Inst Sci, MSc, 67, PhD(chem), 69. Prof Exp: Res assoc chem, Harvard Univ, 70-72; res assoc, Wiezmann Inst, 72-73, sr scientist, 73-74; vis scientist, Med Res Coun Lab Molecular Biol, Cambridge, Eng, 74-76; ASST PROF CHEM, UNIV SOUTHERN CALIF, 76- Res: Theoretical study of the early steps of the visual process; resonance Raman of large molecules; simulation of protein folding; simulation of enzymatic reactions. Mailing Add: Dept of Chem Univ of Southern Calif Los Angeles CA 90007

WARSHOWSKY, BENJAMIN, b New York, NY, Jan 21, 19; m 57; c 3. ANALYTICAL CHEMISTRY. Educ: City Col New York, BS, 40; Univ Minn, MS, 42, PhD(biochem), 45. Prof Exp: Anal res chemist, Publicker Industs, Inc, Pa, 45-46; chief anal chem sect, Biol Labs, US Army Chem Corps, Ft Detrick, 47-54, prog coord officer, 54-56, decontamination br, 56-57, spec asst to dir med res, 57-60, chief phys detection br, 60-64, chief rapid warning officer, 64-72, chief biol defense br, Edgewood Arsenal, 72-75; RETIRED. Mem: Am Chem Soc; Sci Res Soc Am. Res: Microbiological detection techniques; instrument development; administration of chemical and biological research; administration of research and development program on rapid detection of microbiological aerosols. Mailing Add: 315 W College Terr Frederick MD 21701

WARSI, NAZIR AHMED, b Sheopur, Uttar Pradesh, India, June 30, 39; m 66. MATHEMATICAL PHYSICS. Educ: St Andrew's Col, Gorakhpur, India, BSc, 57; Gorakhpur Univ, MSc, 59, PhD(shock wave), 61. Prof Exp: Asst prof math, Gorakhpur Univ, India, 59-63; assoc prof physics & math, Savannah State Col, 63-64; prof physics & math, 64-66, PROF MATH, ATLANTA UNIV, 66-, ACTG CHMN DEPT, 70- Mem: Am Math Soc; Tensor Soc. Res: Shock waves in ideal and magneto-gas-dynamic flows; group theory. Mailing Add: Dept of Math Atlanta Univ Atlanta GA 30314

WARSON, SAMUEL R, b St John, NB, Oct 1, 09; US citizen; c 2. PSYCHIATRY. Educ: McGill Univ, BA, 30, MD, 34. Prof Exp: Instr psychiat, Yale Univ, 37-39; assoc, Univ Louisville, 39-40; asst prof, Wash Univ, 40-50; prof, Ind Univ, 50-54; PROF PSYCHIAT, UNIV FLA, 67- Res: Psychiatric theory and practice. Mailing Add: 3741 Tangler Terr Sarasota FL 33579

WARTELL, ROGER MARTIN, b New York, NY, Feb 24, 45; m 68; c 2. BIOPHYSICAL CHEMISTRY. Educ: Stevens Inst Technol, BSc, 66; Univ Rochester, PhD(physics), 71. Prof Exp: NIH fel & res assoc biochem, Univ Wis-Madison, 71-73; ASST PROF PHYSICS, SCH PHYSICS, GA INST TECHNOL, 74- Mem: Biophys Soc; AAAS; Am Phys Soc. Res: The binding of site specific drugs and proteins to DNA; influence of cooperative interactions along DNA on genetic processes; potential energy calculations of nucleic acid conformations. Mailing Add: Sch Physics Ga Inst Technol Atlanta GA 30332

WARTEN, RALPH MARTIN, b Bielefeld, Ger, Jan 6, 26; US citizen; m 50. MATHEMATICS. Educ: Brooklyn Col, BS, 57; Purdue Univ, MS, 59, PhD(math), 61. Prof Exp: Staff mathematician, IBM Corp, NY, 61-65, adv mathematician, 65-66, staff mem sci ctr, Calif, 66-68; assoc prof, 68-73, PROF MATH, CALIF POLYTECH STATE UNIV, SAN LUIS OBISPO, 73- Mem: AAAS; Am Math Soc; Math Asn Am. Res: Ordinary differential equations and numerical analysis. Mailing Add: Dept of Math Calif Polytech State Univ San Luis Obispo CA 93407

WARTER, JANET KIRCHNER, b Greensburg, Pa, July 27, 33; m 62. PALYNOLOGY, PALEOBOTANY. Educ: Pa State Univ, BS, 55, MEd, 60; La State Univ, PhD(bot), 65. Prof Exp: Lectr bot, Calif State Col, Fullerton, 65-66; LECTR GEOL, CALIF STATE UNIV, LONG BEACH, 66-68, & 70- Concurrent Pos: Res assoc, Los Angeles County Mus Natural Hist. Mem: AAAS; Am Asn Stratig Palynologists. Res: Tertiary palynology; Pleistocene seeds and pollen. Mailing Add: 17841 Still Harbor Lane Huntington Beach CA 92647

WARTER, STUART L, b New York, NY, Apr 9, 34; m 62. ORNITHOLOGY, VERTEBRATE PALEONTOLOGY. Educ: Univ Miami, Fla, BS, 56, MS, 58; La State Univ, PhD(zool), 65. Prof Exp: Instr zool, La State Univ, 64-65; from asst prof to assoc prof, 65-75, PROF BIOL, CALIF STATE UNIV, LONG BEACH, 75- Concurrent Pos: Res assoc vert paleont, Los Angeles County Mus Natural Hist, 66- Mem: Am Ornith Union; Cooper Ornith Soc; Wislon Ornith Soc; Soc Vert Paleont. Res: Avian paleontology, morphology and systematics; osteology and relationships of suboscine passerine birds. Mailing Add: Dept of Biol Calif State Univ 6101 E Seventh St Long Beach CA 90840

WARTERS, MARY, b Rome, Ga, Oct 18, 02. GENETICS. Educ: Shorter Col, AB, 23; Ohio State Univ, MA, 25; Univ Tex, PhD(cytogenetics), 43. Prof Exp: Asst zool, Ohio State Univ, 23-25; instr biol, Winthrop Col, 25-27; from instr to prof, 27-71, head dept, 47-69, EMER PROF ZOOL, CENTENARY COL, 71- Concurrent Pos: Vis scientist, Jackson Mem Lab, 51; res partic biol div, Oak Ridge Nat Lab, 59-61. Mem: Fel AAAS. Res: Bryozoa; cytogenetics of Drosophila; chromosomal aberrations in wild populations of Drosophila; x-autosomal translocations. Mailing Add: 3568 Greenway Place Shreveport LA 71105

WARTERS, WILLIAM DENNIS, b Des Moines, Iowa, Mar 22, 28; m 52; c 2. MICROWAVE ELECTRONICS. Educ: Harvard Univ, AB, 49; Calif Inst Technol, MS, 50, PhD(physics), 53. Prof Exp: Asst, Calif Inst Technol, 50-52; mem tech staff guided wave res, 53-61, head repeater res dept, 61-67, dir transmission systs res ctr, 67-69, exec dir tech staff employment, educ & salary admin div, 69-70, DIR MILLIMETER WAVE SYST LAB, BELL TEL LABS, INC, 70- Mem: Am Phys Soc; sr mem Inst Elec & Electronics Eng. Res: Multi-mode wave guides; millimeter waves; microwaves: transmissions systems; communications satellites. Mailing Add: 514 Sunnyside Rd Lincroft NJ 07738

WARTHEN, JOHN DAVID, JR, b Baltimore, Md, Mar 8, 39; m 69. NATURAL PRODUCT CHEMISTRY. Educ: Univ Md, BS, 60, PhD(pharmaceut chem), 66. Prof Exp: RES CHEMIST, AGR RES SERV, USDA, 65- Mem: AAAS; Am Chem Soc. Res: Investigation and synthesis of biologically active natural products; insect attractants, repellants and insecticides. Mailing Add: USDA South Lab 306 BARC East Rm 316 Beltsville MD 20705

WARTHIN, ALDRED SCOTT, JR, b Ann Arbor, Mich, May 14, 04; m 30, 63; c 1. PALEONTOLOGY. Educ: Univ Mich, BSc, 25; Columbia Univ, PhD(geol), 30. Prof Exp: Asst geol, Columbia Univ, 28-29; from asst prof to prof, 29-69, EMER PROF GEOL, VASSAR COL, 69- Concurrent Pos: Mem staff field sta, Wyoming, 39-47 & Mich, 43-44; assoc geol, US Geol Surv, 44-45; lectr geol, Calif State Col, Long Beach, 69-70. Mem: Am Paleont Soc (pres 55); Soc Vert Paleont; Am Asn Petrol Geol. Res: Paleozoic geology; water conservation. Mailing Add: Dept of Geol Vassar Col Poughkeepsie NY 12601

WARTHIN, THOMAS ANGELL, b Ann Arbor, Mich, Aug 12, 09; m 38; c 3. MEDICINE. Educ: Univ Mich, AB, 30; Harvard Med Sch, MD, 34; Am Bd Internal Med, dipl. Prof Exp: Asst med, Med Sch, Yale Univ, 36-37; instr, Johns Hopkins Univ, 37-39; asst, Harvard Med Sch, 41-42, instr, 49-53, from lectr to clin prof, 53-69, prof med, 70-75. Concurrent Pos: Asst, Mass Gen Hosp, 40-54; asst prof, Sch Med, Tufts Col, 48-50, clin prof, 50-54; sr assoc, Peter Bent Brigham Hosp, 52-; chief med serv, West Roxbury Vet Admin Hosp, 52-74. Mem: AMA; Am Clin & Climat Asn; master Am Col Physicians. Res: Gastroenterology; parasitic and mycotic infections. Mailing Add: 810 Neponset St Norwood MA 02062

WARTIK, THOMAS, b Cincinnati, Ohio, Oct 1, 21; m 52; c 2. INORGANIC CHEMISTRY. Educ: Univ Cincinnati, AB, 43; Univ Chicago, PhD(chem), 49. Prof Exp: From asst prof to prof chem & head dept, 50-71, DEAN COL SCI, PA STATE UNIV, 71- Concurrent Pos: Vis scientist radiation lab, Univ Calif, 57, 59, 61; chmn Fulbright selection comt chem, Nat Acad Sci-Nat Res Coun, 66-72; mem adv bd, Am Chem Soc-Petrol Res Fund, 68-71; consult, Radiation Lab, Callery Chem Co, Koppers Co, Inc & NY Bd Regents, 73-74. Mem: AAAS; Am Chem Soc. Res: Chemistry of boron and aluminum compounds; light metal hydrides; organometallic chemistry. Mailing Add: 211 Whitmore Lab Pa State Univ University Park PA 16802

WARTMAN, WILLIAM BECHMANN, b Philadelphia, Pa, June 26, 07; m 40; c 2. PATHOLOGY. Educ: Univ Pa, BS, 29, MD, 32; Am Bd Path, 41. Prof Exp: Intern, Lankenau Hosp, Philadelphia, Pa, 33-35; demonstr path, Case Western Reserve Univ, 35-37, from instr to asst prof, 37-46; Morrison prof, Northwestern Univ, Chicago, 46-69; PROF PATH, UNIV VA, 69- Concurrent Pos: Pathologist in chg, Univ Hosps, Case Western Reserve Univ, 29-46, Hanna res fel, 38-39; dir labs, Passavant Mem Hosp & Wesley Mem Hosp, 46-; secy-treas, Am Bd Path, 51-55, pres, 59-60; consult, Sci Adv Bd, Armed Forces Inst Path; mem comt path, Nat Res Coun; mem spec adv group, Vet Admin, chmn, 65-66. Mem: Am Soc Exp Pathologists (pres, 58); Am Asn Pathologists & Bacteriologists (pres, 64-65); Am Asn Cancer Res; Am Asn Hist Med; Int Soc Geog Path. Res: Chemotropism of leukocytes; venous blood pressure; occlusion of coronary artery; myocardial infarction, history of tumors; filariasis; carcinogenesis. Mailing Add: Dept of Path Univ of Va Charlottesville VA 22903

WARTMAN, WILLIAM BENJAMIN, JR, b South Hill, Va, July 20, 14; m 48; c 1. ORGANIC CHEMISTRY. Educ: Davidson Col, BS, 36. Prof Exp: Chemist, 39-42, res chemist, 46-56, from res assoc to sr res assoc, 56-65, supvr appl res & develop, 65-67, asst mgr new prod div, 67-71, blend develop mgr, 71-73, ASST MGR NEW PRODS, DEPT RES & DEVELOP, AM TOBACCO CO, 73- Mem: Am Chem Soc; Am Inst Chemists. Res: Composition of tobacco and tobacco smoke; flavor research. Mailing Add: Am Tobacco Co Dept of Res & Develop Box 799 Hopewell VA 23860

WARTMANN, HANS J, organic chemistry, see 12th edition

WARWICK, EVERETT JAMES, b Aledo, Ill, May 2, 17; m 42; c 3. ANIMAL SCIENCE. Educ: Univ Ill, BS, 39; Univ Wis, MS, 42, PhD(genetics, animal husb), 43. Prof Exp: Teacher high sch, Ill, 39-40; asst genetics & physiol, Univ Wis, 40-42; from instr to assoc prof animal husb, Wash State Univ, 43-47; agent, USDA, 47-50, geneticist, 50-55, head cattle res sect, Agr Res Serv, 55-57, chief beef cattle res br, Animal Husb Res Div, 57-68, asst dir div, 68-72, MEM NAT PROG STAFF, AGR RES SERV, USDA, 72- Concurrent Pos: Asst prof, Purdue Univ, 47-50; prof, Univ Tenn, 50-55; livestock adv, Ministry Agr & Natural Resources, Tehran, Iran, 73-75. Mem: AAAS; Am Soc Animal Sci. Res: Endocrinology of reproduction in animals; nutrition and genetics of farm animals; breeding systems. Mailing Add: Nat Prog Staff Agr Res Serv USDA Beltsville MD 20705

WARWICK, JAMES WALTER, b Toledo, Ohio, May 22, 24; m 47, 66; c 6. RADIO ASTRONOMY. Educ: Harvard Univ, AB, 47, AM, 48, PhD(astron), 51. Prof Exp: Asst prof astron, Wellesley Col, 50-52; mem res staff, Sacramento Peak Observ, 52-55; mem sr sci staff, High Altitude Observ, 55-61, PROF ASTRO-GEOPHYS, UNIV COLO, BOULDER, 61- Concurrent Pos: Prin investr, Mariner, Jupiter, Saturn 1977 Proj, NASA, 73-81. Mem: Am Astron Soc; Am Math Soc; Am Geophys Union. Res: Theoretical astrophysics; stellar and planetary magnetism; solar physics; solar-terrestrial physics. Mailing Add: Dept of Astro-Geophys Univ of Colo Boulder CO 80302

WARWICK, WARREN J, b Racine, Wis, Jan 27, 28; m 52; c 2. PEDIATRICS. Educ: St Olaf Col, BA, 50; Univ Minn, MD, 54. Prof Exp: Med fel pediat, 55-57, med fel specialist, 59-60, from instr to asst prof, 60-66, ASSOC PROF PEDIAT, UNIV MINN, MINNEAPOLIS, 66- Concurrent Pos: Alpha Omega Phi fel cardiovasc res, 55-57; Am Heart Asn res fel, 59-60; USPHS res career develop award, 61-66; mem ctr prog comt, Nat Cystic Fibrosis Res Found, 64-66, chmn med care comt, 66-71, coop study comt, 71-72; mem exec bd, Sci-Med Comt, Int Cystic Fibrosis (Mucoviscidosis) Asn, 70; co-chmn, Nat Data Registry Comt, Cystic Fibrosis Found, 72- Res: Pulmonary diseases; experimental pathology; immunology; cystic fibrosis. Mailing Add: Dept of Pediat Box 184 Mayo Mem Univ of Minn Minneapolis MN 55455

WASACZ, JOHN PETER, b Brooklyn, NY, Sept 11, 44; m 70; c 2. ORGANIC CHEMISTRY. Educ: St John's Univ, NY, BS, 65; Univ Pa, PhD(org chem), 69. Prof Exp: Asst prof, 69-74, ASSOC PROF CHEM, MANHATTAN COL, 74- Concurrent Pos: Asst mgr NY sect, Am Chem Soc, 73-; NSF fac fel, Columbia Univ, 75-77. Mem: Am Chem Soc; AAAS; Sigma Xi; Res Soc NAm; Coblentz Soc. Res: Organic synthesis; photochemistry of heterocyclic molecules; synthesis of natural products. Mailing Add: Dept of Chem Manhattan Col Bronx NY 10471

WASAN, MADANLAL T, b Saraisaleh, WPakistan, July 13, 30; m 60; c 4. STATISTICS. Educ: Univ Bombay, BA, 52, MA, 54; Univ Ill, PhD(statist), 60. Prof Exp: From asst prof to assoc prof, 59-68, PROF MATH, QUEEN'S UNIV, ONT, 68- Concurrent Pos: Vis assoc prof, Stanford Univ, 65 & Univ Bombay, 65-66; statist consult, Du Pont of Can, 62-65. Mem: Inst Math Statist. Res: Sequential estimation; stochastic approximation; stochastic processes and applied probability. Mailing Add: Dept of Math Queen's Univ Kingston ON Can

WASBAUER, MARIUS SHERIDAN, b Rockford, Ill, Sept 29, 28; m 69; c 3. INSECT TAXONOMY. Educ: Univ Calif, Berkeley, BS, 50, PhD(entom), 61. Prof Exp: SYST ENTOMOLOGIST, CALIF DEPT FOOD & AGR, 59- Concurrent Pos: Collabr, Animal & Plant Health Inspection Serv, USDA, 59-; res assoc, Univ Calif, Berkeley, 73- Mem: Sigma Xi; Int Orgn Biosystematists. Res: Biosystematics of New World Pompilidae and of North American Tiphiidae; larval systematics and biology of North American Tephritidae. Mailing Add: Lab Serv Entom Calif Dept Food & Agr 1220 N St Sacramento CA 95814

WASCOM, EARL RAY, b Corbin, La, Nov 26, 30; m 51; c 2. ECOLOGY. Educ: Southeastern La State Col, BS, 56; La Polytech Inst, MS, 62; La State Univ, PhD(bot), 67. Prof Exp: From instr to assoc prof, 58-68, PROF BIOL SCI, SOUTHEASTERN LA UNIV, 69-, HEAD DEPT, 68- Mem: Am Inst Biol Sci; Ecol Soc Am; Am Soc Plant Taxon; Bot Soc Am. Res: Taxonomy of angiosperms; plant ecology. Mailing Add: Rte 2 Box 14 Corbin LA 70724

WASE, ARTHUR WILLIAM, b Jersey City, NJ, Nov 15, 19; m 40; c 2. BIOCHEMISTRY. Educ: Columbia Col, AB, 47; Rutgers Univ, PhD, 51. Prof Exp: From instr to assoc prof, 51-68, PROF BIOCHEM, HAHNEMANN MED COL, 68-; SR RES FEL BIOCHEM ENDOCRINOL, MERCK INST THERAPEUT RES, 75- Concurrent Pos: Fulbright prof, Univ Brussels & Inst Jules Bordet, Belgium, 56-57; consult, Colgate Biol Res Labs, 58- & pub sect, Radio Corp Am; res fel biochem endocrinol, Merck Inst Therapeut Res, 63-75. Mem: Am Cancer Soc; Am Heart Asn Coun Thrombosis. Res: Radiobiochemistry as applied to cancer research and neuropsychopharmacology; bone metabolism; biochemistry and endocrinology of atherosclerosis. Mailing Add: 213 Valentine St Highland Park NJ 08904

WASER, JURG, b Zurich, Switz, Dec 23, 16; nat US; m 42; c 3. CHEMISTRY. Educ: Calif Inst Technol, PhD(phys chem), 44. Prof Exp: Instr chem, Calif Inst Technol, 42-46, res fel, 44-45, sr res fel molecular struct, 47-48; from asst prof to prof chem, Rice Inst, 48-58; PROF CHEM, CALIF INST TECHNOL, 58- Concurrent Pos: Mem US Nat Comt Crystallog, 58-, secy-treas, 59-61; Guggenheim fel, 63-64. Mem: Am Chem Soc; Am Crystallog Asn(secy, 55-57, vpres, 59, pres, 60); Swiss Phys Soc. Res: Structure of crystals by x-ray diffraction; structure of organic molecules. Mailing Add: Dept of Chem Calif Inst of Technol Pasadena CA 91109

WASFI, SADIQ HASSAN, b Basrah, Iraq, July 1, 37; m 68; c 2. INORGANIC CHEMISTRY. Educ: Univ Baghdad, BS, 61; Georgetown Univ, MS, 66, PhD(inorg chem), 71. Prof Exp: From lectr to asst prof chem, Col Sci, Basrah Univ, Iraq, 71-75; RES ASSOC CHEM, UNIV HAWAII, MANOA, 75- Mem: Am Chem Soc; Iraqi Chem Soc; Sigma Xi. Res: Transition metal complexes of organic thiols and disulfides; heteropoly tungstate and molybdate anions containing several transition metal ions. Mailing Add: 1615 Wilder Ave Apt 707 Honolulu HI 96822

WASHA, GEORGE WILLIAM, b Milwaukee, Wis, May 6, 09; m 34; c 3. MECHANICS. Educ: Univ Wis, BS, 30, MS, 32, PhD(mech), 38. Prof Exp: Instr, 30-40, PROF MECH, UNIV WIS-MADISON, 40-, CHMN DEPT, 53- . Honors & Awards: Wason Medal, Am Concrete Inst, 41. Mem: Nat Acad Sci; Am Soc Test & Mat; Nat Soc Prof Eng; fel Am Concrete Inst. Res: Durability, permeability and plastic flow of concrete; light weight agregates and concrete; vibrated concrete; masonry cements; properties of ferrous metals; concrete block. Mailing Add: 1114 Shorewood Blvd Madison WI 53705

WASHABAUGH, WILLIAM, b Monongahela, Pa, Jan 14, 45; m 69; c 2. ANTHROPOLOGY. Educ: St Bernard's Col, NY, BA, 66; Univ Conn, MA, 70; Wayne State Univ, PhD(anthrop), 74. Prof Exp: ASST PROF ANTHROP, UNIV WIS-MILWAUKEE, 74- Mem: Linguistic Soc Am. Res: Patterns of variation and change in Creole languages; the history of the development of linguistic and anthropological thought. Mailing Add: Dept of Anthrop Univ Wis Milwaukee WI 53201

WASHAM, CLINTON JAY, b Pryor, Okla, June 23, 41; m 69; c 3. MICROBIOLOGY. Educ: Okla State Univ, BS & MS, 63; Ore State Univ, PhD(microbiol), 68. Prof Exp: Res asst microbiol, Ore State Univ, 63-67, instr, 67-68; vis lectr food technol, Iowa State Univ, 68-70; assoc prof biol, Southwestern Union Col, 70-73; vpres, Tolibia Cheese, Inc, Wis, 73-75; ASST PROF DAIRY SCI, UNIV GA, 75- Concurrent Pos: Res & develop consult, Tolibia Cheese, Inc, 75- Mem: Am Soc Microbiol; Int Asn Milk, Food & Environ Sanitarians; Sigma Xi; Am Dairy Sci Asn; AAAS. Res: Dairy starter cultures; blue cheese; culture media; bacterial metabolism; flavor and aroma compounds in dairy products; artificial cheese; factors affecting cheese ripening. Mailing Add: Dept of Dairy Sci Univ of Ga Athens GA 30602

WASHBURN, ALBERT LINCOLN, b New York, NY, June 15, 11; m 35; c 3. GEOMORPHOLOGY, QUATERNARY GEOLOGY. Educ: Dartmouth Col, AB, 35; Yale Univ, PhD(geol), 42. Prof Exp: Mem, Nat Geog Soc exped, Mt McKinley, 36; asst geologist, Boyd E Greenland exped, 37; geol invests, Can Arctic, 38-41 & 49; exec dir, Arctic Inst NAm, 45-51; dir snow, ice & permafrost res estab, Corps Engrs, US Army, 52-53; prof northern geol, Dartmouth Col, 54-60; prof geol, Yale Univ, 60-70; PROF GEOL, UNIV WASH, 66- Concurrent Pos: Consult, Res & Develop Bd, US Dept Defense, 46-53 & Corps Engrs, US Army, 54-60; hon lectr, McGill Univ, 48-50; mem, US Nat Comt, Int Geophys Year, 53-59, Panel Glaciol, Nat Acad Sci, 59-65 & 67-71 & Polar Res, 61-; mem geol expeds, Greenland, 54-58, 60 & 64; dir, Quaternary Res Ctr, Univ Wash, 67-75. Honors & Awards: Kirk Bryan Award, Geol Soc Am, 71; Medaille Andre H Dumont, Geol Soc Belg, 73. Mem: Fel Am Geog Soc; fel Geol Soc Am; Am Geophys Union; hon mem Arctic Inst NAm. Res: Periglacial studies. Mailing Add: Quaternary Res Ctr Univ of Wash Seattle WA 98195

WASHBURN, HENRY BRADFORD, JR, b Cambridge, Mass, June 7, 10; m 40; c 3. GEOGRAPHY. Educ: Harvard Univ, AB, 33, AM, 60. Hon Degrees: PhD, Univ Alaska, 51; DSc, Tufts Univ, 57, Colby Col, 57 & Northeastern Univ, 58. Prof Exp: Leader, Yukon Exped, Nat Geog Soc, DC, 35; instr geog, Inst Geog Explor, Harvard Univ, 35-42; DIR, BOSTON MUS SCI, 39- Concurrent Pos: Leader, Mt McKinley Flights, Nat Geog Soc, 36 & St Elias Range, 38; chief rep for US Army Air Force, US Army Alaskan Test Exped, Mt McKinley, 42, leader Arctic equip tests, 45; expert consult, Off Qm Gen, 42-45; spec liaison, Off Qm Gen & Comndg Gen, Alaska Defense Command, 42; dir reorgn Army Air Forces Flying Clothing & Personal Equip Prog, Wright Field, 43; spec asst to chief, Personal Equip Lab, Air Tech Serv Command, 44; mem vis comt, Mus Comp Zool, Harvard Univ, 46, overseer, Harvard Col, 55-61; leader, Oper White Tower, 47; sci-dir, Chinese-Am Amnyi Machin Exped, 48; dir, Off Naval Res Surv, Mt McKinley, 49; co-leader, Boston Mus Sci-Univ Alaska-Univ Denver Exped, 51; Nat Geog Soc rep, Int Geog Cong, DC, 52; mem nat sci planning bd, World Sci-Pan Pac Expos, 58; chmn, Mass Comt Selection Rhodes

Scholars, 59-60; dir New Eng Tel & Tel Co, 60. Honors & Awards: Peed Award, Royal Geog Soc, 38; Burr Prize, Nat Geog Soc, 40; Except Civilian Serv Award, US Secy War, 46. Mem: Fel Am Geog Soc; fel Am Acad Arts & Sci; Arctic Inst NAm; fel Royal Can Geog Soc; fel Royal Geog Soc. Res: Aerial photography; mapping; cold-weather clothing and equipment; public science education; climate and climate effects. Mailing Add: Boston Mus Sci Science Pk Boston MA 02114

WASHBURN, KENNETH W, b Martinsville, Va, June 21, 37; m 59; c 2. POULTRY GENETICS. Educ: Va Polytech Inst, BS, 59, MS, 62; Univ Mass, PhD(poultry), 65. Prof Exp: Res asst poultry genetics, Va Polytech Inst, 60-62; instr, Univ Mass, 62-65; ASSOC PROF POULTRY GENETICS, UNIV GA, 65- Honors & Awards: Poultry Sci Jr Res Award, Poultry Sci Asn, 75. Mem: Poultry Sci Asn. Res: Genetic-nutrition interrelationships; compensatory growth; feed efficiency; egg shell strength; egg cholesterol; hemoglobins. Mailing Add: Dept Poultry L P Bldg Univ of Ga Athens GA 30602

WASHBURN, LEE CROSS, b Paducah, Ky, Jan 10, 47; m 69; c 1. ORGANIC CHEMISTRY, NUCLEAR MEDICINE. Educ: Murray State Univ, BA, 68; Vanderbilt Univ, PhD(org chem), 72. Prof Exp: ASSOC SCIENTIST RADIOPHARMACEUT DEVELOP, MED DIV, OAK RIDGE ASSOC UNIVS, 72- Mem: Am Chem Soc. Res: Synthesis and testing of short-lived radiopharmaceutical tumor-scanning and organ-imaging agents; radioprotective drugs. Mailing Add: Oak Ridge Assoc Univs PO Box 117 Oak Ridge TN 37830

WASHBURN, RICHARD HANCORNE, b Cadillac, Mich, Mar 25, 19; m 42; c 4. ECONOMIC ENTOMOLOGY. Educ: Mich State Univ, BS, 41; Cornell Univ, PhD, 48. Prof Exp: Asst entom, Cornell Univ, 45-48; asst prof, Univ Ga, 48-49; entomologist hq, US Army Engrs, Alaska, 49; res entomologist, 50-52, exten entomologist, 52-56, SR ENTOMOLOGIST, USDA INST AGR SCI, UNIV ALASKA, 58-, ASSOC PROF ENTOM, 59- Concurrent Pos: Res entomologist, NC Region Agr Res Serv, USDA, 68- Mem: AAAS; Entom Soc Am; Am Hort Soc; Int Plant Propagators Soc. Res: Insect toxicology, ecology and physiology. Mailing Add: Inst of Agr Sci USDA Univ of Alaska Box AE Palmer AK 99645

WASHBURN, ROBERT HENRY, b Lincoln, Nebr, Nov 27, 36; m 66. STRATIGRAPHY, STRUCTURAL GEOLOGY. Educ: Univ Nebr, BS, 59, MS, 61; Columbia Univ, PhD(geol), 66. Prof Exp: Instr geol, Brooklyn Col, 64-66; asst prof, 66-70, ASSOC PROF GEOL, JUNIATA COL, 70- Mem: AAAS; Geol Soc Am; Am Asn Petrol Geol; Soc Econ Paleont & Mineral. Res: Paleozoic stratigraphy; structural geology of central Nevada and Pennsylvania. Mailing Add: Dept of Geol Juniata Col Huntingdon PA 16652

WASHBURN, SHERWOOD LARNED, b Cambridge, Mass, Nov 26, 11; m 38; c 2. ANTHROPOLOGY. Educ: Harvard Univ, AB, 35, PhD(anthrop), 40. Prof Exp: From instr to asst prof anat, Columbia Univ, 39-47; from assoc prof to prof anthrop, Univ Chicago, 47-58, chmn dept, 53-55; prof, 59-75, CHMN DEPT ANTHROP, UNIV CALIF, BERKELEY, 61-, UNIV PROF, 75- Concurrent Pos: Ed, Am J Phys Anthrop, 55; res assoc, Wenner-Gren Found Anthrop Res. Honors & Awards: Viking Fund Medal, Wenner-Gren Found Anthrop Res, 60. Mem: Am Soc Human Genetics; Am Asn Phys Anthrop (secy-treas, 43-47, pres, 51); Am Asn Anat; fel Am Anthrop Asn (pres, 61-62). Res: Primatology; experimental physical anthropology. Mailing Add: Dept of Anthrop Univ of Calif Berkeley CA 94720

WASHBURN, WILLIAM H, b Milwaukee, Wis, Oct 14, 20; m 42; c 3. ANALYTICAL CHEMISTRY. Educ: Univ Wis, BA, 41. Prof Exp: CHEMIST, ABBOTT LABS, 46- Mem: Soc Appl Spectros; Am Chem Soc; Coblentz Soc; Sigma Xi. Res: Infrared spectroscopy, materials purity and chemical structure analysis. Mailing Add: Dept 482 Abbott Labs North Chicago IL 60064

WASHBURNE, STEPHEN SHEPARD, b Hartford, Conn, Sept 6, 42; m 70. ORGANIC CHEMISTRY. Educ: Trinity Col, Conn, BS, 63; Mass Inst Technol, PhD(org chem), 67. Prof Exp: Asst prof, 67-73, ASSOC PROF CHEM, TEMPLE UNIV, 73-, ASST CHMN DEPT, 74- Concurrent Pos: NSF fel, 64-66; NIH fel, 67; chief consult, Petrarch Systs, Inc, 70- Mem: Am Chem Soc. Res: Organosilicon chemistry; cancer chemotherapy; organometallic chemistry of the elements of group IV; malarial chemotherapy. Mailing Add: Dept of Chem Temple Univ Philadelphia PA 19122

WASCHECK, PAUL HOWARD, b Oklahoma City, Aug 12, 40; m 60; c 2. INDUSTRIAL ORGANIC CHEMISTRY. Educ: Univ Okla, BS, 62, PhD(org chem), 67. Prof Exp: Chemist, 66-70, GROUP LEADER, EXPLOR SECT, PETROCHEM DIV, CONTINENTAL OIL CO, 66- Mem: Am Chem Soc. Res: Organic synthesis in petrochemical field; sesquiterpenoids found in octocorallia; industrial organic chemistry as applied to process and product research and development. Mailing Add: Petrochem Div Continental Oil Co Drawer 1267 Ponca City OK 74601

WASHENBERGER, JAMES K, mathematics, see 12th edition

WASHINGTON, ELMER L, b Houston, Tex, Oct 18, 35; m 60; c 2. PHYSICAL CHEMISTRY. Educ: Tex Southern Univ, BS, 57, MS, 58; Ill Inst Technol, PhD(thermodyn), 66. Prof Exp: Sr proj engr, Pratt & Whitney Aircraft Div, United Aircraft Corp, Conn, 65-67, res assoc advan mat res & develop lab, 67-69; asst prof phys sci, 69-72, dean natural sci & math, 72-74, ASSOC PROF CHEM, CHICAGO STATE UNIV, 72-, DEAN COL ARTS & SCI, 74- Mem: Electrochem Soc; Am Chem Soc. Res: Thermodynamics of non-electrolytes; electrochemistry as related to fuel cell technology. Mailing Add: Col of Arts & Sci Chicago State Univ Chicago IL 60628

WASHINGTON, JOHN A, II, b Istanbul, Turkey, May 29, 36; US citizen; m 59; c 3. CLINICAL MICROBIOLOGY, CLINICAL PATHOLOGY. Educ: Univ Va, BA, 57; Johns Hopkins Univ, MD, 61. Prof Exp: From intern to asst resident surg, Med Ctr, Duke Univ, 61-63; Nat Cancer Inst fel, 63-65; resident clin path, Clin Ctr, NIH, 65-67; assoc consult, 67-68, CONSULT MICROBIOL, MAYO CLIN, 68-, HEAD SECT CLIN MICROBIOL, 71-, ASSOC PROF MICROBIOL & LAB MED, MAYO MED SCH, 72- Concurrent Pos: Asst prof microbiol, Mayo Grad Sch Med, 70-72; ed, J Clin Microbiol, 74-75. Mem: Fel Am Soc Clin Path; Am Soc Microbiol; fel Am Col Physicians; fel Am Acad Microbiol; Infectious Dis Soc Am. Res: Antimicrobial agents; antimicrobial susceptibility tests; methodology in clinical bacteriology. Mailing Add: Sect of Clin Microbiol Mayo Clin 200 SW First St Rochester MN 55901

WASHINGTON, WARREN MORTON, b Portland, Ore, Aug 28, 36. METEOROLOGY. Educ: Ore State Univ, BS, 58, MS, 60; Pa State Univ, PhD(meteorol), 64. Prof Exp: Res asst meteorol, Pa State Univ, 61-63; SCIENTIST, NAT CTR ATMOSPHERIC RES, 63- Concurrent Pos: Adj prof metoerol & oceanog, Univ Mich, 69-71; mem var panels, Nat Acad Sci & NSF, 69-; mem, Gov Sci Adv Comt, State Colo, 75- Mem: AAAS; Am Meteorol Soc; Sigma Xi. Res: Numerical modeling of the atmosphere. Mailing Add: Nat Ctr for Atmospheric Res PO Box 3000 Boulder CO 80303

WASHINGTON, WILLIE JAMES, b Madison, Fla, Dec 26, 42; m 70. PLANT GENETICS, CYTOGENETICS. Educ: Fla A&M Univ, BS, 64; Univ Ariz, MS, 66; Univ Mo-Columbia, PhD(plant genetics, cytogenetics), 70. Prof Exp: Asst prof biol, Tougaloo Col, 70-71 & Cent State Univ, Ohio, 71-72; D F Jones fel agron, NDak State Univ, 72-73; ASST PROF BIOL, CENT STATE UNIV, OHIO, 73- Mem: AAAS. Res: Genetics and cytogenetics of higher plants; application of genetical and cytogentical analysis to the improvement of economic crops; vertebrate tissue culture and immunological assays; environmental mutagenesis. Mailing Add: Dept of Biol Cent State Univ Wilberforce OH 45384

WASHINO, ROBERT K, b Sacramento, Calif, Mar 14, 32; m 56; c 3. ENTOMOLOGY, PUBLIC HEALTH. Educ: Univ Calif, Berkeley, BS, 54; Univ Calif, Davis, MS, 56, PhD(entom), 67. Prof Exp: Assoc sr specialist, Calif State Dept Pub Health, 59-65; assoc specialist, 65-67, LECTR & ASST ENTOMOLOGIST, UNIV CALIF, DAVIS, 67- Concurrent Pos: NIH grant, 71-73. Mem: Entom Soc Am; Am Soc Trop Med & Hyg; Am Mosquito Control Asn. Res: Studies regarding the various aspects of insect biology which affect their role as vectors of human and animal pathogens; sociological as well as entomological studies of mosquito pest problems. Mailing Add: Dept of Entom Univ of Calif Davis CA 95616

WASHKO, FLOYD VICTOR, b New Brunswick, NJ, Oct 17, 22; m 47; c 2. VETERINARY PATHOLOGY. Educ: Mich State Univ, DVM, 44; Purdue Univ, MS, 48, PhD, 50. Prof Exp: Vet pract, NJ, 44-45; assoc prof vet sci, Purdue Univ, 46-53; mem staff, Plum Island Animal Dis Lab, USDA, 53-54; VET PATHOLOGIST, MERCK, SHARP & DOHME RES LABS, 54-, SR INVESTR, 74- Mem: Am Vet Med Asn; NY Acad Sci; Am Asn Avian Pathologists; Am Asn Vet Lab Diagnosticians; US Animal Health Asn. Res: Brucellosis of swine and cattle; virus diseases of the bovine; veterinary therapeutics. Mailing Add: Merck, Sharp & Dohme Res Labs Rahway NJ 07065

WASHKO, JOHN BLASIUS, b Hatfield, Mass, Dec 27, 11; m 41; c 3. AGRONOMY. Educ: Rutgers Univ, BS, 36, MS, 38; Univ Wis, PhD(agron), 41. Prof Exp: Asst agron, Rutgers Univ, 36-38, Univ Wis, 38-41; from asst prof to assoc prof & assoc agronomist exp sta, Univ Tenn, 41-46; assoc prof 46-51, PROF AGRON, PA STATE UNIV, 51- Honors & Awards: Merit Cert Award, Am Forage & Grasslands Coun, 68. Mem: Fel Am Soc Agron; Crop Sci Soc Am; Am Forage & Grassland Coun (pres, 58-61). Res: Forage crops and pasture management. Mailing Add: Dept of Agron 119 Tyson Bldg Pa State Univ University Park PA 16802

WASHKO, WALTER WILLIAM, b New Brunswick, NJ, July 29, 20; m 55; c 4. AGRONOMY. Educ: Rutgers Univ, BS, 41, MS, 47, Univ Wis, PhD(agron, bot), 58. Prof Exp: Agronomist, Tex Res Found, 47-53 & Eastern States Farmers Exchange, 53-64; EXT AGRONOMIST & PROF AGRON, UNIV CONN, 64- Mem: Am Soc Agron; Am Inst Biol Sci. Res: Crop production. Mailing Add: Dept of Plant Sci U-102 Univ of Conn Storrs CT 06268

WASHNOK, RICHARD F, agronomy, plant genetics, see 12th edition

WASHTON, NATHAN SEYMOUR, b New York, NY, Nov 9, 16; m 44; c 3. SCIENCE EDUCATION, ENVIRONMENTAL SCIENCE. Educ: NY Univ, BS, 39, EdD(sci ed), 49; Columbia Univ, MA, 41. Prof Exp: Asst biol, NY Univ, 38; instr sci & math, chmn dept & asst dean, Newark Jr Col, 39-42; chmn dept sci & math, Dwight Sch, NY, 45-46; chmn dept sci, Rutgers Univ, 46-50; PROF SCI EDUC & COORDR, QUEENS COL, NY, 50- Concurrent Pos: Vis prof, Puerto Rico, 48, Upsala Col, 49 & Yeshiva Univ, 59; consult, Libr Sci, 57-58; bd exam, Bd Ed, New York. Mem: Fel AAAS; Nat Asn Res Sci Teaching (vpres, 57-58, pres, 58-59); Nat Sci Teachers Asn; sr mem Inst Environ Sci. Res: Curriculum and evaluation of science programs in schools. Mailing Add: 30 Oaktree Lane Manhasset NY 11030

WASHWELL, EDWARD RICHARD, solid state physics, see 12th edition

WASIELEWSKI, PAUL FRANCIS, b Bay Shore, NY, Oct 21, 41; m 67; c 2. PHYSICS, OPERATIONS RESEARCH. Educ: Georgetown Univ, BS, 63; Yale Univ, PhD(physics), 69. Prof Exp: ASSOC SR RES PHYSICIST, TRAFFIC SCI DEPT, GEN MOTORS RES LABS, 69- Mem: Am Phys Soc; Opers Res Soc Am. Res: Nuclear physics; traffic science. Mailing Add: Traffic Sci Dept Gen Motors Res Labs 12 Mile & Mound Rd Warren MI 48090

WASKO, PETER EDMUND, b Ambridge, Pa, July 1, 16; m 41; c 2. METEOROLOGY, CHEMISTRY. Educ: Crnegie Inst Technol, BS, 38; Columbia Univ, MA, 40; NY Univ, MS, 45. Prof Exp: Vol psychologist, NY State Psychiat Inst & Hosp, 38-39; meteorologist, Pan Am World Airways, Inc, 40-42 & 43-49; instr meteorol, NY Univ, 42-43, Chanute AFB, 49-50; meteorologist & asst chief high level forecasting br, Andrews AFB, 50-52; res asst meteorol, Univ Chicago, 52-54; res assoc, Ohio State Univ, 54-56; assoc scientist, Argonne Nat Lab, 56-61; space scientist, Marshall Space Flight Ctr, 61-62; sr scientist, Douglas Aircraft Co, Inc, 62-70; SAFETY ENGR, OCCUP SAFETY & HEALTH ADMIN, US DEPT LABOR, 71- Mem: AAAS; Am Meteorol Soc; Am Inst Aeronaut & Astronaut; Am Geophys Union; Sigma Xi. Res: Diverse aerospace operational problems for missiles and space systems; micrometeorological turbulent diffusion; northern hemisphere weather; synoptic and dynamic meteorology. Mailing Add: Occup Safety & Health Admin US Dept of Labor 633 W Wisconsin Ave Milwaukee WI 53203

WASLEY, WILLIAM LINGEL, b Chicago, Ill, Aug 22, 13; m 40; c 4. TEXTILE CHEMISTRY, POLYMER CHEMISTRY. Educ: Univ Chicago, BS, 35; La State Univ, MS, 36; Stanford Univ, PhD(org chem), 38. Prof Exp: Instr chem, Armour Inst Technol, 38-40; cancer res chemist, Univ Wis, 40-42; chemist, US Forest Serv, 42-45; biochemist, Univ Tex M D Anderson Hosp & Tumor Inst, 45; asst prof chem, Wash Univ, 45-46; sr res specialist, Ansco Div, Gen Aniline & Film Corp, NY, 46-52; res chemist, Union Oil Co, 52-58; RES CHEMIST, WESTERN REGIONAL RES LAB, USDA, 58- Mem: Am Chem Soc. Res: Textiles. Mailing Add: Western Regional Res Labs USDA 800 Buchanan St Albany NY 94710

WASLIEN, CAROL IRENE, b Mayville, NDak, Sept 24, 40; m 70. NUTRITION, BIOCHEMISTRY. Educ: Univ Calif, Santa Barbara, BA, 61; Cornell Univ, MS, 63; Univ Calif, Berkeley, PhD(nutrit), 68. Prof Exp: NIH res training fel, Vanderbilt Univ, Naval Med Res Unit-3, Egypt, 68-69; res assoc nutrit, Vanderbilt Univ, 69-72; ASSOC PROF & HEAD DEPT NUTRIT & FOODS, AUBURN UNIV, 72- Concurrent Pos: Res consult biochem dept, Vanderbilt Univ, 72- Mem: Soc Nutrit Educ; Am Asn Clin Chem; Am Inst Nutrit; Am Dietetic Asn. Res: Human requirements for protein, vitamins and trace minerals; use of micro-organisms as food sources for man; assessment of nutritional status in health and disease. Mailing Add: Dept of Nutrit & Foods Auburn Univ Auburn AL 36830

WASON, SATISH KUMAR, b Lyallpur, India, Feb 24, 40; m 70; c 3. PHYSICAL INORGANIC CHEMISTRY. Educ: Univ Delhi, BSc, 59, MSc, 61; Cornell Univ, PhD(phys chem), 65. Prof Exp: Scientist, Coun Sci & Indust Res, New Delhi, India, 65-66; res assoc phys chem, Boston Univ, 66-67; res chemist, E I du Pont de

Nemours & Co, Inc, Del, 67-69; res scientist, 69-73, SECT LEADER, J M HUBER CORP, 73- Mem: Am Chem Soc; Soc Cosmetic Chemists; Sigma Xi. Res: High temperature thermodynamic and spectroscopic studies; photochemistry and mercury photosensitized reactions; chemistry of synthetic silicas and silicates. Mailing Add: J M Huber Corp PO Box 310 Havre De Grace MD 21078

WASOW, WOLFGANG RICHARD, b Vevey, Switz, July 25, 09; nat US; m 39, 59; c 3. MATHEMATICS. Educ: NY Univ, PhD(math), 42. Prof Exp: Instr math, Goddard Col, 39-40, Conn Col for Women, 41-42 & NY Univ, 42-46; asst prof, Swarthmore Col, 46-49; mathematician numerical anal res, Univ Calif, Los Angeles, 49-55; math res ctr, US Dept Army, Wis, 56-57; PROF MATH, UNIV WIS-MADISON, 57- Concurrent Pos: Fulbright fel, Rome, Italy, 54-55 & Haifa, Israel, 62. Mem: Am Math Soc; Math Asn Am; Soc Indust & Appl Math. Res: Asymptotic theory and numerical solution of differential equations. Mailing Add: Dept of Math Van Vleck Hall Univ of Wis Madison WI 53706

WASS, MARVIN LEROY, b Worthington, Minn, Apr 24, 22; m; c 3. ZOOLOGY, BOTANY. Educ: Winona State Col, BS, 49; Fla State Univ, MS, 53; Univ Fla, PhD, 59. Prof Exp: Cur, Pinellas County Marine Mus, St Petersburg, Fla, 53-55; asst prof biol, Western Carolina Col, 59-60; assoc marine scientist, 60-71, SR MARINE SCIENTIST, VA INST MARINE SCI, 72- Concurrent Pos: Assoc prof, Sch Marine Sci, Col William & Mary, 60-; partic, US Prog Biol, Int Indian Ocean Exped, 64. Mem: Am Soc Limnol & Oceanog. Res: Marine ecology; marine invertebrates; biogeography; estuarine wetlands, ornithology. Mailing Add: Va Inst of Marine Sci Gloucester Point VA 23062

WASS, WALLACE M, b Lake Park, Iowa, Nov 19, 29; m 53; c 4. VETERINARY MEDICINE, VETERINARY SURGERY. Educ: Univ Minn, BS, 51, DVM, 53, PhD(vet med), 61. Hon Degrees: Vet Med, Nat Univ Colombia, 63. Prof Exp: From instr to asst prof vet med, Univ Minn, 58-63; res assoc lab animal med, Brookhaven Nat Lab, 63-64; PROF VET CLIN SCI & HEAD DEPT, IOWA STATE UNIV, 64- Mem: Am Vet Med Asn. Res: Large animal medicine and surgery; metabolic diseases of domestic animals; bovine porphyria. Mailing Add: Dept of Vet Clin Sci Iowa State Univ Ames IA 50010

WASSER, CLINTON HOWARD, b Phoenix, Ariz, Nov 11, 15; m 39; c 3. RANGE ECOLOGY. Educ: Univ Ariz, BS, 37; Univ Nebr, MS, 47; Colo State Univ, MF, 48. Prof Exp: Res asst southwestern forest & range exp sta, US Forest Serv, 37-38; instr & asst range mgt, 38-43, from asst prof to assoc prof, 43-52, PROF RANGE MGT, COLO STATE UNIV, 52- Concurrent Pos: Asst range conservationist, Colo State Univ, 43-47, actg head dept range mgt, 47-50, head, 50-57, chief range conservationist & chief forestry and range mgt sect, Agr Exp Sta, 47-52, dean col forestry & natural resources, 52-69, pres, Res Found, 57-59; collabr, State Prod & Mkt Admin, Colo, 43-44 & Rocky Mt Forest & Range Exp Sta, US Forest Serv, 54-60; consult, Bowes & Hart, Inc, 47; admin tech rep, McIntire Stennis Coop State Forestry Prog, 63-69. Honors & Awards: Accievement Serv Award, Soc Range Mgt, 69. Mem: Soc Am Foresters; Soc Range Mgt (pres-elect, 64, pres, 65); AAAS; Ecol Soc Am. Res: Range management, ecology and seeding; alpine plant ecology. Mailing Add: Dept of Range Sci Colo State Univ Ft Collins CO 80521

WASSER, RICHARD BARKMAN, b Oshkosh, Wis, Sept 26, 36; m 68; c 2. Educ: Univ Wis, BS, 59; Inst Paper Chem, MS, 61, PhD(paper chem), 64. Prof Exp: From res chemist to sr res chemist, 64-73, PROJ LEADER PAPER CHEM, AM CYANAMID CO, 73- Mem: Tech Asn Pulp & Paper Indust; Am Chem Soc; AAAS. Res: Physical chemistry of paper and its modification with chemical additives. Mailing Add: Stamford Res Labs Am Cyanamid Co 1937 W Main St Stamford CT 06904

WASSERBURG, GERALD JOSEPH, b New Brunswick, NJ, Mar 25, 27; m 51; c 2. GEOLOGY, GEOPHYSICS. Educ: Univ Chicago, BSc, 51, MS, 52, PhD(geol), 54. Prof Exp: Res assoc, Enrico Fermi Inst Nuclear Studies, Chicago, 54-55; from asst prof to assoc prof, 55-62, PROF GEOL & GEOPHYS, CALIF INST TECHNOL, 62- Concurrent Pos: Vis prof, Univ Kiel, 60, Harvard Univ, 62, Univ Berne, 66 & Swiss Fed Inst Technol, 67. Honors & Awards: Arthur L Day Medal, Geol Soc Am, 70; Except Sci Achievement Award, NASA 70 & Distinguished Pub Serv Medal, 72; Leonard Medal, Meteoritical Soc, 75. Mem: Nat Acad Sci; fel Am Geophys Union; Geol Soc Am; fel Am Acad Arts & Sci. Res: Application of methods of chemical physics to geologic problems; measurement of absolute geologic time, solar system time scale; nucleosynthesis and variations in isotopic abundances due to long lived natural radioactivities and cosmic ray interactions in nature. Mailing Add: Arms Lab of Geol Sci Calif Inst of Technol Pasadena CA 91109

WASSERMAN, AARON E, b Philadelphia, Pa, Dec, 28, 21; m 44; c 2. FOOD CHEMISTRY. Educ: Philadelphia Col Pharm, BSc, 42, ScD(bact), 48; Mass Inst Technol, MSc, 47. Prof Exp: Instr bact, Pa State Col Optom, 47; res assoc, Sharp & Dohme, Inc, 48-54; CHEMIST & BIOCHEMIST, EASTERN REGIONAL RES CTR, USDA, 54-, HEAD, MEAT COMPOSITION & QUAL INVEST, MEAT LAB, 63- Mem: Am Meat Sci Asn; Am Chem Soc; Inst Food Technol. Res: Flavor chemistry; organoleptic and sensory evaluation of food products; isolation and identification techniques; food processing; meat technology; bacterial physiology and metabolism; food safety. Mailing Add: 600 E Mermaid Lane Philadelphia PA 19118

WASSERMAN, AARON OSIAS, b New York, NY, Oct 15, 27; m 69; c 1. VERTEBRATE ZOOLOGY. Educ: City Col New York, BS, 51; Univ Tex, PhD(zool), 56. Prof Exp: From asst prof to assoc prof, 63-74, PROF BIOL, CITY COL NEW YORK, 74- Mem: Soc Study Evolution; Am Soc Ichthyol & Herpet; Am Soc Zool; Sigma Xi. Res: Speciation problems in anuran amphibians; cytogenetics of anurans. Mailing Add: Dept of Biol City Col of New York Convent Ave & 138th St New York NY 10031

WASSERMAN, AARON REUBEN, b Philadelphia, Pa, Apr 14, 32; m 63; c 2. BIOCHEMISTRY. Educ: Univ Pa, AB, 53; Univ Wis, MS, 57, PhD(biochem), 60. Prof Exp: Asst Univ Wis, 57-59; fel plant biochem, Johnson Found Med Physics, Univ Pa, 60-61, res assoc, Dept Chem, 61-62; fel, Enzyme Inst, Univ Wis-Madison & Dept Molecular Biol, Vanderbilt Univ, 63-65; asst prof, 65-72, ASSOC PROF BIOCHEM, McGILL UNIV, 72- Mem: Can Biochem Soc. Res: Biomembranes; membrane proteins; cytochromes; bioenergetics; photosynthesis. Mailing Add: Dept of Biochem McGill Univ Montreal PQ Can

WASSERMAN, ALBERT J, b Richmond, Va, Jan 25, 28; m 48; c 3. CLINICAL PHARMACOLOGY, INTERNAL MEDICINE. Educ: Univ Va, BA, 47; Med Col Va, MD, 51. Prof Exp: Instr, 56-57, assoc, 57-60, from asst prof to assoc prof, 60-67, PROF MED, MED COL VA, 67- Mem: Am Col Physicians; Am Col Chest Physicians; Am Soc Clin Pharmacol & Therapeut; Am Fedn Clin Res. Res: Cardiovascular pharmacology; human pharmacology and drug trials. Mailing Add: Med Col of Va Richmond VA 23298

WASSERMAN, ALLEN LOWELL, b New York, NY, Dec 7, 34; m 58; c 2. SOLID STATE PHYSICS. Educ: Carnegie Inst Technol, BS, 56; Iowa State Univ, PhD(physics), 63. Prof Exp: Res assoc, Princeton Univ, 63-65; asst prof, 65-71, ASSOC PROF PHYSICS, ORE STATE UNIV, 71- Concurrent Pos: Consult, Lawrence Radiation Lab, 65- Mem: Am Phys Soc. Res: Optical properties of solids; electron gas models of solid behavior. Mailing Add: Dept of Physics Ore State Univ Corvallis OR 97331

WASSERMAN, ARTHUR GABRIEL, b Bayonne, NJ, Nov 10, 38; m 64. MATHEMATICS. Educ: Mass Inst Technol, BS, 60; Brandeis Univ, PhD(math), 65. Prof Exp: Pierce instr math, Harvard Univ, 65-68; asst prof, 68-70, ASSOC PROF MATH, UNIV MICH, ANN ARBOR, 70- Concurrent Pos: Mem, Inst Advan Study, Princeton Univ, 71-72. Mem: Am Math Soc. Res: Topology; transformation groups. Mailing Add: Dept of Math 3014 Angell Hall Univ of Mich Ann Arbor MI 48104

WASSERMAN, DAVID, b New York, NY, Apr 6, 17; m 46; c 2. POLYMER CHEMISTRY. Educ: Brooklyn Col, AB, 37; Columbia Univ, PhD(org chem), 43. Prof Exp: Irvington Varnish & Insulator Co fel chem, Columbia Univ, 43-46; sr chemist, Irvinton Varnish & Insulator Div, Minn Mining & Mfg Co, 46-58, res assoc, Merck, Sharp & Dohme Res Div, 58-64; sr res scientist, 64-68, PROJ LEADER ETHICON, INC, 68- Honors & Awards: P B Hoffman Award for Res, Johnson & Johnson, Inc, 74. Mem: AAAS; NY Acad Sci; Am Chem Soc. Res: Synthetic polymers; organic synthesis; natural products; pharmaceutical chemistry; developed new synthetic absorbable suture. Mailing Add: Polymer Dept Ethicon Inc Div of Johnson & Johnson Inc Somerville NJ 08876

WASSERMAN, EDEL, b New York, NY, July 29, 32; m 55; c 2. THEORETICAL CHEMISTRY, ORGANIC CHEMISTRY. Educ: Cornell Univ, BA, 53; Harvard Univ, AM, 54, PhD(chem), 59. Prof Exp: MEM TECH STAFF, BELL TEL LABS, INC, 57-; PROF CHEM, RUTGERS UNIV, 67- Concurrent Pos: Vis prof, Cornell Univ, 62-63; adv ed, Chem Phys Letters, 68- & J Am Chem Soc, 71-76; regent's lectr, Univ Calif, Irvine, 73. Mem: Am Chem Soc; Am Phys Soc. Res: Physical organic and theoretical studies of electronic structure of organic molecules; electron spin resonance; monolayers. Mailing Add: 71 Hillcrest Ave Summit NJ 07901

WASSERMAN, EDWARD, b New York, NY, Jan 13, 21. PEDIATRICS. Educ: Johns Hopkins Univ, BA, 41; NY Med Col, MD, 46. Prof Exp: From clin asst to instr, 51-54, asst, 54-55, clin assoc, 55-56, from asst prof to assoc prof, 56-64, PROF PEDIAT, NEW YORK MED COL, FLOWER & FIFTH AVE HOSPS, 64-, CHMN DEPT, 66- Concurrent Pos: Mem pediat adv comt, Dept Health, New York, 64- Mem: Fel Am Fedn Clin Res; Am Acad Pediat. Res: Renal diseases. Mailing Add: Dept of Pediat New York Med Col New York NY 10029

WASSERMAN, HARRY H, b Boston, Mass, Dec 1, 20; m 47; c 3. ORGANIC CHEMISTRY. Educ: Mass Inst Technol, BS, 41; Harvard Univ, MS, 42, PhD(org chem), 49. Prof Exp: Res asst, Off Sci Res & Develop, 45; from instr to assoc prof, 48-62, chmn dept, 62-65, PROF CHEM, YALE UNIV, 62- Concurrent Pos: Guggenheim fel, 59-60; Am ed, Tetrahedron & Tetrahedron Letters, 60- ; mem study sect med chem, NIH, 62-66; postdoctoral rev panel, NSF, 63-65. Mem: Am Acad Arts & Sci; Am Chem Soc. Res: Natural products; reactions of organic systems with oxygen; cyclopropanones. Mailing Add: Sterling Chem Lab Yale Univ New Haven CT 06520

WASSERMAN, KARLMAN, b Brooklyn, NY, Mar 12, 27; m 53; c 4. PHYSIOLOGY, MEDICINE. Educ: Upsala Col, BA, 48; Tulane Univ, PhD(physiol), 51, MD, 58. Prof Exp: Intern, Johns Hopkins Univ, 58-59; sr res fel med, Univ Calif, San Francisco, 59-61; asst prof med, Stanford Univ, 62-67, dir respiratory function lab & chest clin, 62-67; assoc prof, 67-72, PROF MED, MED SCH, UNIV CALIF, LOS ANGELES, 72-; CHIEF DIV RESPIRATORY DIS, LOS ANGELES HARBOR GEN HOSP, TORRANCE, 67- Mem: Am Physiol Soc; Am Fedn Clin Res. Res: Respiratory, circulatory and renal physiology; respiratory disease. Mailing Add: Div Resp Physiol & Med Harbor Gen Hosp 1000 W Carson St Torrance CA 90509

WASSERMAN, LOUIS ROBERT, b New York, NY, July 11, 10; m 39. MEDICINE. Educ: Harvard Univ, AB, 31; Rush Med Col, MD, 35; Am Bd Internal Med, dipl, 46. Prof Exp: Intern, Michael Reese Hosp, 35-37; res fel, Mt Sinai Hosp, 37-39, clin asst, 40-42, adj physician physiol hemat, 47-50, assoc physician, 50-54, hematologist & dir dept hemat, 54-72; prof med, 66-72, chmn dept clin sci, 68-72, DISTINGUISHED SERV PROF MED, MT SINAI SCH MED, 72- Concurrent Pos: Res fel, Donner Lab Med Physics, Univ Calif, 46-49, consult, Radiation Lab, 48-51; asst clin prof, Col Physicians & Surgeons, Columbia Univ, 51-60, assoc prof, 60-66; mem med adv comt, Nat Hemophilia Found, 55-70; res collabr, Brookhaven Nat Lab, 60-72; mem med adv comt, Leukemia Soc Am, 65-69; chmn polycythemia vera study group, Nat Cancer Inst, 67-, mem nat cancer planning comt, 72-73, mem cancer control comt, 72-73; diag rev adv group, 72-, mem cancer treatment adv comt, 72-74 & chmn, 74-76; mem blood & hemopoietic syst res eval comt, Vet Admin, 69-72. Mem: Fel Am Col Physicians; Int Soc Hemat (vpres, 70-74); Am Soc Hemat (vpres, 66, pres, 68-69); fel AMA; Am Asn Cancer Res. Res: Diseases of the blood. Mailing Add: Mt Sinai Sch of Med 19 E 98th St Apt 5-B New York NY 10029

WASSERMAN, MARTIN S, b New York, NY, Jan 19, 38; m 63; c 2. CHILD PSYCHIATRY. Educ: Columbia Col, AB, 59; State Univ NY Downstate Med Ctr, MD, 63. Prof Exp: Intern med, Albany Med Ctr, Union Univ, 64; resident psychiat, Kings County Gen Hosp, 64-67; instr, State Univ NY Downstate Med Ctr, 66-68; fel child psychiat, Univ Mich-Childrens Psychiat Hosp, 69-70; ASST PROF PSYCHIAT, UNIV SOUTHERN CALIF, 70- Concurrent Pos: Asst adj prof soc sci, Long Island Univ, 66-69; staff psychiatrist, Cath Charities, New York, 66-69; consult, Oper Headstart, Off Econ Opportunity, San Diego, Calif, 67-69, Ctr Forensic Psychiat, Mich, 69-70 & Bur Prisons, USPHS, 69-70. Mem: Fel Am Psychiat Asn; Am Col Psychiat; assoc Am Psychoanal Asn; Am Acad Child Psychiat. Res: College health; psychiatric education; socio-cultural factors in human development. Mailing Add: Westwood Ctr 1100 Glenden Ave Los Angeles CA 90024

WASSERMAN, MARVIN, b New York, NY, Feb 2, 29; m 64; c 2. EVOLUTIONARY BIOLOGY. Educ: Cornell Univ, BS, 50; Univ Tex, MA, 52, PhD(biol), 54. Prof Exp: Res assoc biol, Univ Tex, 56-60; sr lectr zool, Univ Melbourne, 60-62; from asst prof to assoc prof, 62-67; PROF BIOL, QUEENS COL, NY, 67- Mem: Am Genetic Asn; Genetics Soc Am; Am Soc Nat; Soc Study Evolution; Soc Syst Zool. Res: Evolution and genetics of Drosophila; chromosomal polymorphism. Mailing Add: Dept of Biol Queen's Col Flushing NY 11367

WASSERMAN, MOE STANLEY, b Hampton, Va, May 31, 27; m 52; c 3. PHYSICAL CHEMISTRY. Educ: Univ Va, BS, 46, MS, 47; Univ Mich, PhD(phys chem), 55. Prof Exp: Develop engr, Gen Elec Co, 47-49; adv res engr, Sylvania Elec Prod, Inc, 55-61; sect head chem electronics, 61-64, GROUP HEAD ADVAN MAT & DEVICES, GTE LABS, INC, 64- Mem: Electrochem Soc; Inst Elec & Electronics Engrs; Am Vacuum Soc; Am Inst Chemists; AAAS. Res: Electron diffraction and microscopy; photoconductivity; electroluminescence; information display; semiconductor and dielectric materials and devices; preparation and electronic properties of thin-film materials. Mailing Add: Electronic Components Lab GTE Labs Inc Waltham MA 02154

WASSERMAN, ROBERT H, b Chicago, Ill, Jan 3, 23; m 48; c 1. APPLIED MATHEMATICS. Educ: Univ Chicago, BS, 43; Univ Mich, MS, 46, PhD(math), 57. Prof Exp: Aeronaut res scientist, Nat Adv Comt Aeronaut, 43-57; from asst prof to assoc prof, 57-70, PROF MATH, MICH STATE UNIV, 70- Mem: Am Math Soc; Math Asn Am. Res: Fluid mechanics. Mailing Add: Dept of Math Mich State Univ East Lansing MI 48823

WASSERMAN, ROBERT HAROLD, b Schenectady, NY, Feb 11, 26; m 50; c 3. PHYSIOLOGY, BIOCHEMISTRY. Educ: Cornell Univ, BS, 49, PhD(nutrit), 53; Mich State Univ, MS, 51. Prof Exp: Asst bact, Mich State Univ, 49-50; asst animal nutrit, Cornell Univ, 51-53; res assoc biochem, Univ Tenn-AEC agr res prog, 53-55; sr scientist, Oak Ridge Inst Nuclear Studies, 55-57; assoc prof, 57-63, PROF PHYS BIOL, NY STATE VET COL, CORNELL UNIV, 63-, ACTG CHMN DEPT, 74- Concurrent Pos: Consult, Oak Ridge Inst Nuclear Studies, 57-59; chmn, Conf Calcium & Strontium Transport, 62; Guggenheim fel, 64 & 72; vis fel, Inst Biol Chem, Denmark, 64-65; Orgn Econ Coop & Develop-NSF fel, 65; mem gen med B study sect, NIH, 68-71; organizing comt, Conf Calcium Transfer Mechanisms, 70. Honors & Awards: Mead-Johnson Award Nutrit, Am Inst Nutrit, 69. Mem: AAAS; Soc Exp Biol & Med; Am Physiol Soc; Biophys Soc; Am Inst Nutrit. Res: Ion transport; mineral metabolism; vitamin D action; intestinal absorption. Mailing Add: Dept of Phys Biol NY State Vet Col Cornell Univ Ithaca NY 14850

WASSERMAN, WILLIAM JACK, b New York, NY, Apr 27, 25; m 59; c 2. ORGANIC POLYMER CHEMISTRY. Educ: Univ Calif, Los Angeles, BS, 47; Univ Southern Calif, MS, 50; Univ Wash, PhD(org chem), 54. Prof Exp: Asst prof chem, Humboldt State Col, 54-57; sr res chemist, Martin-Marietta Corp, Wash, 57-62 & Truesdail Labs, Los Angeles, Calif, 62-63; asst prof chem, San Jose State Col, 63-67; INSTR CHEM, SEATTLE CENT COMMUNITY COL, 67- Concurrent Pos: Dir, NSF Inst Polymer Chem for Col Teachers, San Jose State Col, 67; mem writing team, Am Chem Soc Chemtec Proj, 70-72. 70-72. Mem: AAAS; Am Chem Soc; The Chem Soc. Res: Synthesis and evaluation of condensation polymers; emphasis on epoxies, polyesters, polyamides, polyurethanes; cross-linking agent synthesis and evaluation; aldol condensations with unsaturated ketones; furanoid ring-opening reactions. Mailing Add: Div of Sci & Math Seattle Cent Community Col Seattle WA 98122

WASSERMANN, FELIX EMIL, b Bamberg, Ger, Aug 7, 24; US citizen; m 53; c 3. VIROLOGY, MICROBIAL GENETICS. Educ: Univ Wis, BS, 49, MS, 50; NY Univ, PhD(microbiol), 57. Prof Exp: Res assoc microbiol, Univ Chicago, 58; asst virol, Pub Health Res Inst City New York, Inc, 58-60, from asst to assoc epidemiol, 60-65; asst prof, 65-66, ASSOC PROF VIROL, NY MED COL, 67-, ACTG CHMN DEPT, 70- Mem: Am Soc Microbiol. Res: Virus-host cell interaction; viral genetics; epidemiology. Mailing Add: Dept of Microbiol NY Med Col Valhalla NY 10595

WASSERSUG, RICHARD JOEL, b Boston, Mass, Apr 13, 46. EVOLUTIONARY BIOLOGY. Educ: Tufts Univ, BS, 67; Univ Chicago, PhD(evolutionary biol), 73. Prof Exp: Asst prof syst & ecol & asst cur, Mus Natural Hist, Univ Kans, 73-74; ASST PROF ANAT & COMT EVOLUTIONARY BIOL, UNIV CHICAGO, 74- Mem: Soc Study Evolution; Am Soc Ichthyologists & Herpetologists; AAAS; Ecol Soc Am; Am Soc Zoologists. Res: Feeding adaptations of anuran larvae, or tadpoles, and of fish; studies on the evolution of complex life cycles and on the evolution of diversity. Mailing Add: Dept of Anat Univ of Chicago Chicago IL 60637

WASSHAUSEN, DIETER CARL, b Jena, Ger, Apr 15, 38; US citizen; m 61; c 2. SYSTEMATIC BOTANY. Educ: George Washington Univ, BS, 63, MS, 66, PhD(bot), 72. Prof Exp: Asst cur, 69-72, ASSOC CUR DEPT BOT, MUS NATURAL HIST, SMITHSONIAN INST, 72- Mem: Int Asn Plant Taxon. Res: Taxonomy of Acanthaceae and flowering plants of the New World tropics. Mailing Add: Dept of Bot Mus of Natural Hist Smithsonian Inst Washington DC 20560

WASSMUNDT, FREDERICK WILLIAM, b Oak Park, Ill, Aug 6, 32; m 64; c 1. ORGANIC CHEMISTRY. Educ: DePauw Univ, BA, 53; Univ Ill, PhD(chem), 56. Prof Exp: Instr chem, Univ Calif, 56-58; from instr to asst prof, 58-69, ASSOC PROF CHEM, UNIV CONN, 69- Concurrent Pos: Sabbatical leave, Univ Heidelberg, 66-67; invited lectr, Chemische Gesellschaft zu Heidelberg, 74. Mem: Am Chem Soc; The Chem Soc. Res: Exploratory synthesis; organometallic compounds; diazonium salts; bridged bicyclic compounds. Mailing Add: Dept of Chem Univ of Conn Storrs CT 06268

WASSOM, CLYDE E, b Osceola, Oiwa, Feb, 6, 24; m 45; c 3. AGRONOMY. Educ: Iowa State Col, BS, 49, MS, 51, PhD(crop breeding), 53. Prof Exp: Technician agron, Iowa State Col, 47-51, res assoc, 51-54; asst prof, 54-62, ASSOC PROF KANS STATE UNIV, 62- Concurrent Pos: Temp staff mem, Int Ctr Improv Corn & Wheat, Mexico City, 67. Mem: Am Soc Agron. Res: Forage breeding; research and breeding for improvement of red clover and other legumes; corn breeding and genetics research. Mailing Add: Dept of Agron Kans State Univ Manhattan KS 66506

WASSON, BURTON KENDALL, b Hunters Ferry, NB, June 10, 14; m 42; c 2. MEDICINAL CHEMISTRY. Educ: Mt Allison Univ, BSc, 39, MSc, 40; McGill Univ, PhD 42. Prof Exp: Lectr & res asst, McGill Univ, 42-44; sr res chemist, Charles E Frosst & Co, 44-65, GROUP LEADER MED CHEM, MERCK FROSST LABS, 65- Mem: Am Chem Soc; fel Chem Inst Can. Res: Vitamins and hormones; tetrahydrofurans; pyrimidines; substances with central nervous system activity; pyrolysis of ketones and alcohols; p-cymene; azo and azoxybenzenes; medicinal and synthetic organic chemistry. Mailing Add: Med Chem Merck Frosst Labs PO Box 1005 Pointe Claire-Dorval PQ Can

WASSON, JOHN R, b St Louis, Mo, Aug 22, 41; m 63; c 2. PHYSICAL CHEMISTRY, INORGANIC CHEMISTRY. Educ: Univ Mo-Columbia, BS, 63, MA, 66; Ill Inst Technol, PhD(phys chem), 70. Prof Exp: Instr chem, Southern State Col, 65-66; ASST PROF CHEM, UNIV KY, 69- Mem: Am Chem Soc; The Chem Soc. Res: Electron spin resonance; semi-empirical molecular orbital theory; change-transfer complexation; transition metal complexes; higher oxidation states of silver. Mailing Add: Dept of Chem Univ of Ky Lexington KY 40506

WASSON, JOHN TAYLOR, b Springtown, Ark, July 4, 34; m 60; c 2. COSMOCHEMISTRY, PLANETOLOGY. Educ: Univ Ark, BS, 55; Mass Inst Technol, PhD(nuclear chem), 58. Prof Exp: NSF fel tech physics lab, Munich Technol Univ, 58-59; res chemist geophys res directorate, Air Force Cambridge Res Labs, 59-63; NIH spec fel phys inst, Univ Berne, 63-64; assoc prof, 67-72, PROF, UNIV CALIF, LOS ANGELES, 72- Concurrent Pos: Guggenheim fel, Max Planck Inst Chem, 70-71. Mem: AAAS; Am Phys Soc; Am Geophys Union; Geochem Soc; Meteoritical Soc. Res: Compostition and origin of the meteorites; lunar surface materials; geochemistry; analytical chemistry. Mailing Add: Inst Geophys & Planetary Physics Univ of Calif Los Angeles CA 90024

WASSON, OREN A, b Wooster, Ohio, Mar 27, 35; m 59; c 2. NUCLEAR PHYSICS. Educ: Col Wooster, BA, 57; Yale Univ, MS, 59, PhD(physics), 64. Prof Exp: Res staff physicist, Yale Univ, 63-65; from res assoc physics to assoc physicist, Brookhaven Nat Lab, 65-73; PHYSICIST, NAT BUR STAND, 73- Concurrent Pos: Vis mem, Oak Ridge Nat Lab, 71-72; tech expert, Int Atomic Energy Agency, Greece, 75. Mem: AAAS; Am Phys Soc. Res: Experimental nuclear physics; neutron and photon reaction mechanisms; heavy ion studies; computer programming; neutron cross sections and dosimetry. Mailing Add: Nat Bur Stand Washington DC 20234

WASSON, RICHARD LEE, b Farmington, Ill, May 19, 32; m 55; c 2. ORGANIC CHEMISTRY. Educ: Univ Ill, BS, 53; Mass Inst Technol, PhD, 56. Prof Exp: Asst org chem, Mass Inst Technol, 53-56; sr res chemist, 56-63, res specialist, 63-66, res group leader, 66-71, MGR FLAVOR-FRAGRANCE RES, MONSANTO CO, 71- Mem: Am Chem Soc; Inst Food Technol; Am Inst Chem. Res: Preparation and rearrangements of epoxides; halogenation and carboxylation of aromatic compounds; aromatic nitro compounds; basic condensations; organophosphate compounds; isolation and identification of natural products; food chemistry. Mailing Add: Monsanto Co 800 N Lindbergh St Louis MO 63166

WASSON, ROBERT GORDON, b Great Falls, Mont, Sept 22, 98; m 26; c 2. ETHNOMYCOLOGY. Educ: Columbia Univ, BLitt, 20. Hon Degrees: DSc, Univ Bridgeport, 74. Prof Exp: HON RES FEL ETHNOMYCOL, BOT MUS, HARVARD UNIV, 57- Concurrent Pos: Corresp, Nat Mus Natural Hist, Paris & hon researcher, NY Bot Garden, 57- Honors & Awards: Sarah Gildersleeve Fife Mem Award, NY Bot Garden, 58. Mem: Sigma Xi. Res: Every aspect of ethnomycology. Mailing Add: 42 Long Ridge Rd Danbury CT 06810

WASYLKIWSKYJ, WASYL, b Kiwel, Ufraine, Feb 12, 35; US citizen; m 60; c 2. ELECTROMAGNETICS. Educ: City Col New York, BEE, 57; Polytech Inst Brooklyn, MS, 65, PhD(elec eng), 68. Prof Exp: Microwave component design engr, Missiltron Inc, NY, 59-60; Consult & Designers, NY, 60-62; sr tech specialist, ITT Defense Commun Div, Int Tel & Tel Corp, NJ, 62-69; MEM TECH STAFF, INST DEFENSE ANAL, 69- Concurrent Pos: Consult, ITT Fed Labs, NJ, 63-65; assoc prof lectr, George Washington Univ, 70- Honors & Awards: Spec Commendation Award, Inst Elec & Electronics Engrs Antennas & Propagation Soc, 72. Mem: Inst Elec & Electronics Engrs. Res: Electromagnetic radiation and diffraction, guided wave and cavity theory; antenna theory, particularly antenna arrays; microwave technology, particularly parametric and solid state microwave devices. Mailing Add: Inst for Defense Anal 400 Army Navy Dr Arlington VA 22202

WAT, BO YING, b Honolulu, Hawaii, Feb 15, 25; m 48; c 4. MEDICINE. Educ: Col Med Evangelists, MD, 49; Am Bd Path, dipl. Prof Exp: From asst to assoc prof, 58-62, PROF PATH, SCH MED, LOMA LINDA UNIV, 62- Mem: Am Soc Clin Pathologists; AMA. Mailing Add: Dept of Path Loma Linda Univ Sch of Med Loma Linda CA 92354

WAT, EDWARD KOON WAH, b Honolulu, Hawaii, Aug 27, 40; m 69; c 3. ORGANIC CHEMISTRY. Educ: Univ Hawaii, BA, 62; Stanford Univ, PhD(chem), 66. Prof Exp: NIH fel chem, Harvard Univ, 65-66; RES CHEMIST, CENT RES DEPT, EXP STA, E I DU PONT DE NEMOURS & CO, INC, 66- Mem: Am Chem Soc. Res: Synthetic organic chemistry for agricultural and pharmaceutical applications. Mailing Add: Cent Res Dept E I du Pont de Nemours & Co Inc Wilmington DE 19898

WATABE, NORIMITSU, b Kure City, Japan, Nov 29, 22; m 52; c 2. ELECTRON MICROSCOPY. Educ: Tohoku Univ, Japan, MS, 48, DSc(biocrystallog), 60. Prof Exp: Res investr biocrystallog pearl cult, Fuji Pearl Res Lab, Japan, 48-52; asst fac fisheries, Mie Prefectural Univ, Japan, 52-55, lectr, 55-59, consult, Fisheries Exp Sta, 53-59; res assoc calcification electron micros, Duke Univ, 57-70; assoc prof, 70-72, Nat Inst Dent Res grant, Electron Micros Lab, 71-76, PROF CALCIFICATION ELECTRON MICROS, UNIV SC, 72-, DIR ELECTRON MICROSCOPE LAB, 70- Concurrent Pos: Asst, Tohoku Univ, Japan, 48-49; consult, Ford Found Off Latin Am & Caribbean, 68 & SC State Develop Bd, 75- Mem: AAAS; Electron Micros Soc Am; Am Soc Zool; Royal Micros Soc. Res: Microanatomy; ultrastructural and physiological aspects of mechanism of calcification in invertebrates and algae. Mailing Add: 3510 Greenway Dr Columbia SC 29206

WATADA, ALLEY E, b Platteville, Colo, July 20, 30; m 56; c 2. HORTICULTURE, PLANT PHYSIOLOGY. Educ: Colo State Univ, BS, 52, MS, 53; Univ Calif, Davis, PhD(plant physiol), 65. Prof Exp: Lab technician, Univ Calif, Davis, 56-65; from asst prof to assoc prof hort, WVa Univ, 65-71; invests leader, 71-72, RES LEADER, AGR RES SERV, USDA, 72-, RES FOOD TECHNOLOGIST, 71- Mem: Am Soc Hort Sci; Am Soc Plant Physiol; Inst Food Technol; Am Inst Biol Sci. Res: Postharvest physiology of fruits and vegetables; evaluation of chemical composition and physical and sensory measurement of fruits and vegetables for developing nondestructive method of measuring quality and maturity. Mailing Add: ARMI Bldg 002 Rm 113 Agr Res Serv USDA Beltsville MD 20705

WATANABE, AKIRA, b Vancouver, BC, Sept 17, 35; m 57; c 5. COMMUNICATIONS SCIENCES. Educ: McMaster Univ, BSc, 57; Univ Toronto, MA, 62, PhD(molecular physics), 64. Prof Exp: Sci off, 57-69, RES SCIENTIST, DEPT COMMUN, DEFENCE RES TELECOMMUN ESTAB, 69- Mem: Can Asn Physicists; Optical Soc Am. Res: Optical communications, thin-film waveguides and spectroscopy; laser physics. Mailing Add: Commun Res Ctr Box 11490 Sta H Ottawa ON Can

WATANABE, DANIEL SEISHI, b Honolulu, Hawaii, Oct 30, 40; m 66. COMPUTER SCIENCE. Educ: Harvard Univ, AB, 62, AM, 67, PhD(appl math), 70. Prof Exp: Mathematician, Baird-Atomic, Inc, 63-64; ASST PROF COMPUT SCI, UNIV ILL, URBANA, 70- Mem: Soc Indust & Appl Math. Res: Numerical methods for ordinary and partial differential equations. Mailing Add: Dept of Comput Sci Univ of Ill Urbana IL 61801

WATANABE, KYOICHI A, b Amagasaki, Japan, Feb 28, 35; m 62; c 5. ORGANIC CHEMISTRY, BIOCHEMISTRY. Educ: Hokkaido Univ, BA, 58, MA, 60, PhD(chem), 63. Prof Exp: Lectr chem, Sophia Univ, Japan, 63; res assoc, Sloan-Kettering Inst, 63-66; res fel, Univ Alta, 66-68; assoc, 68-72, ASSOC MEM CHEM, SLOAN-KETTERING INST, 72-; ASSOC PROF, SLOAN-KETTERING DIV, GRAD SCH MED SCI, CORNELL UNIV, 72- Mem: Pharm Soc Japan; Am Chem Soc. Res: Structure, syntheses, reactions and stereochemistry of nitrogen heterocyclics, carbohydrates, nucleosides and antibiotics of potential biological activities. Mailing Add: Walker Lab Sloan-Kettering Inst 145 Boston Post Rd Rye NY 10580

WATANABE, MAMORU, b Vancouver, BC, Mar 15, 33. MEDICINE, BIOCHEMISTRY. Educ: McGill Univ, BSc, 55, MD, CM, 57, PhD, 63; FRCPS(C), 63. Prof Exp: Assoc molecular biol, Albert Einstein Col Med, 65-66, asst prof, 66-67; from assoc prof med & biochem to prof med, Univ Alta, 67-73; PROF MED, UNIV CALGARY, 73-, HEAD DIV MED, 75- Concurrent Pos: Ayerst fel, Endocrine Soc,

63-64; res fel, Am Col Physicians, 64-67; dir dept med, Foothills Hosp, Calgary, Alta. Mem: Am Soc Microbiol; Can Med Asn. Res: Secretion rate of aldosterone in normal and abnormal pregnancy; replication of RNA bacteriophages; transport of steroids across cell membranes, using Pseudomonas testosteroni as a model system. Mailing Add: Dept of Med Univ of Calgary Calgary AB Can

WATANABE, MICHAEL SATOSI, b Tokyo, Japan, May 26, 10; m 37; c 1. THEORETICAL PHYSICS, INFORMATION SCIENCE. Educ: Univ Tokyo, BS, 33, DSc, 40; Univ Paris, DSc, 35. Prof Exp: Mem res staff, Inst Physics, Univ Leipzig, 37-39; assoc prof, Eng Sch, Univ Tokyo, 42-45; adv, Far East Command, Allied Powers, 46-47; prof physics & chmn physics dept, St Paul's Univ, 49-50; assoc prof, Wayne Univ, 50-52; prof, US Naval Postgrad Sch, 52-56; sr physicist, Res Ctr, Int Bus Mach Corp, NY, 56-64; prof eng & philos, Yale Univ, 64-66; PROF PHYSICS & INFO SCI, UNIV HAWAII, 66- Concurrent Pos: Vis prof, Yale Univ, 59-60; adj prof, Columbia Univ, 60-61, univ panel mem, 65- Mem: Fel Am Phys Soc; fel Inst Elec & Electronics Engrs; Int Acad Philos Sci (vpres); Int Soc Study Time (pres, 69-73). Res: Quantum theory; statistical mechanics; information theory; philosophy of science; pattern recognition. Mailing Add: 242 Kaalawi Pl Honolulu HI 96816

WATANABE, RONALD S, BIOCHEMISTRY. Educ: Wash State Univ, BS, 47, MS, 49; Univ Okla, PhD(chem), 62. Prof Exp: Assoc prof, 62-70, PROF CHEM, CALIF STATE UNIV, SAN JOSE, 70- Mem: AAAS; Am Chem Soc; Am Soc Plant Physiol; Phytochem Soc NAm. Res: Enzymes. Mailing Add: Dept of Chem Calif State Univ San Jose CA 95192

WATANABE, SHIZUO, b Japan, Dec 20, 20; m 51; c 3. BIOCHEMISTRY. Educ: Hokkaido Univ, BS, 44, DSc(biochem), 53. Prof Exp: From asst & lectr to asst prof biochem, Hokkaido Univ, 44-58; assoc prof, Dartmouth Med Sch, 58-60; assoc prof in residence, 60-64, PROF BIOCHEM IN RESIDENCE, SCH MED, UNIV CALIF, SAN FRANCISCO, 64- Concurrent Pos: Res assoc physiol, Sch Med, Wash Univ, 53-56; sr res assoc chem, Georgetown Univ, 56-57; estab investr, Am Heart Asn, 60-65. Mem: AAAS; Am Physiol Soc; Am Soc Biol Chemists; Japanese Biochem Soc. Res: Biochemistry of muscular contraction. Mailing Add: Dept of Biochem 960 Med Sci Bldg Univ of Calif Med Ctr San Francisco CA 94122

WATANABE, TOMIYA, b Koriyama, Japan, Aug 3, 27; m 59; c 2. AERONOMY, PLASMA PHYSICS. Educ: Tohoku Univ, Japan, BSc, 53, PhD(geophys), 61. Prof Exp: Res asst physics, Univ Md, 60-61; lectr, 61-63, from asst prof to assoc prof, 63-72, PROF GEOPHYS, UNIV BC, 72- Concurrent Pos: Mem, Can Comn IV, Int Sci Radio Union, 63; mem working group of magnetospheric field variations, comt IV, Int Asn Geomag & Aeronomy, 67-; vis assoc res physicist, Space Sci Lab, Univ Calif, Berkeley, 69; sr res fel, Nat Res Coun Can, 69-70, mem aeronomy subcomt, assoc comt geod & geophys, 70- Mem: Am Geophys Union; fel Royal Astron Soc; Soc Terrestrial Magnetism & Elec Japan; Am Phys Soc; Can Asn Physicists. Res: Geomagnetism; generation and propagation of hydromagnetic waves in the terrestrial magnetosphere; electrostatic waves in the ionosphere. Mailing Add: Dept of Geophys & Astron Univ of BC Vancouver BC Can

WATANABE, WARREN HISAMI, organic chemistry, see 12th edition

WATANAKUNAKORN, CHATRCHAI, b Bangkok, Thailand, Sept 6, 35; m 67; c 2. INTERNAL MEDICINE, INFECTIOUS DISEASES. Educ: Univ Univ Med Sci, Thailand, MD, 61. Prof Exp: Intern med, Chulalongkorn Hosp, 61-62; intern, St Francis Hosp, Pittsburgh, Pa, 62-63; resident, Cook County Hosp, Chicago, Ill, 63-64; NIH fel clin nutrit, Univ Iowa, 64-65, resident med, 65-66; NIH fel infectious dis, 66-68, from instr to asst prof med, 68-72, ASSOC PROF MED, UNIV CINCINNATI, 72- Concurrent Pos: Asst attend physician, Cincinnati Gen Hosp, Ohio, 70-72, attend physician, 73-; attend physician, Holmes Hosp & Vet Admin Hosp, Cincinnati, 71-; consult infectious dis, Drake Mem Hosp & clin microbiol, Vet Admin Hosp, 71- Mem: Am Fedn Clin Res; Infectious Dis Soc Am; fel Am Col Angiol; Am Soc Microbiol. Res: Antibiotics; wall-defective microbial variants. Mailing Add: Col of Med Univ of Cincinnati Cincinnati OH 45267

WATERBURY, ARCHIE MYRON, zoology, see 12th edition

WATERBURY, GLENN RAYMOND, b Canon City, Colo, Oct 28, 20; m 53; c 2. ANALYTICAL CHEMISTRY. Educ: Colo Agr & Mech Col, BS, 44; Iowa State Col, PhD(phys chem), 52. Prof Exp: Mem staff anal chem, 46-48, 52-72, GROUP LEADER, LOS ALAMOS SCI LAB, 72- Mem: Am Chem Soc; fel Am Inst Chemists. Res: Special methods of analysis. Mailing Add: Los Alamos Sci Lab Los Alamos NM 87545

WATERBURY, LOWELL DAVID, b Lansing, Mich, Jan 8, 42. PHARMACOLOGY. Educ: Univ Mich, BS, 64; Univ Vt, PhD(pharmacol), 68. Prof Exp: Technician, Univ Mich, 62-64, asst lab instr bot, 62; from instr to asst prof biochem, Inst Lipid Res, Col Med, Baylor Univ, 67-69; asst prof pharmacol, Bowman Gray Sch Med, Wake Forest Univ, 69-74; STAFF RESEARCHER, DEPT EXP PHARMACOL, SYNTEX RES, 74- Mem: AAAS; Am Soc Pharmacol & Exp Therapeut; Am Chem Soc; Am Found Clin Res. Res: Application of gas phase analytical techniques, especially gas-liquid chromatography and mass spectrometry, to pharmacology; metabolism of biogenic amines and drug metabolism; biochemical pharmacology. Mailing Add: Dept of Exp Pharmacol Syntex Res 3401 Hillview Ave Palo Alto CA 95014

WATERFALL, WALLACE, acoustics, deceased

WATERHOUSE, CHRISTINE, b Kennebunk, Maine, Dec 19, 16; m 52; c 2. MEDICINE. Educ: Mt Holyoke Col, AB, 38; Columbia Univ, MD, 42; Am Bd Internal Med, dipl, 50. Prof Exp: Intern med, Presby Hosp, New York, 42-43; asst, 43-44, from instr to assoc prof, 44-67, PROF MED, SCH MED & DENT, UNIV ROCHESTER, 67- Concurrent Pos: Buswell fel, Sch Med & Dent, Univ Rochester, 45-51, Hochstetter fel, 51-53; res physician, Strong Mem Hosp, 43, asst resident, 44, chief resident, 44-45, asst physician, 45-58, sr assoc physician, 58-; assoc, Atomic Energy Proj, Div Med Serv, Univ Rochester, 48-49. Mem: AAAS; Am Soc Clin Invest; fel Am Col Physicians; Am Fedn Clin Res. Res: Metabolism. Mailing Add: Dept of Med Strong Mem Hosp Rochester NY 14642

WATERHOUSE, HOWARD N, b Bethel, Maine, Apr 19, 32. ANIMAL NUTRITION. Educ: Univ Maine, BS, 54; Univ Ill, Urbana, MS, 58, PhD(animal sci), 60. Prof Exp: Group leader animal nutrit, Gen Mills, Inc, 60-62, proj leader, 62-67; mgr poultry nutrit res & tech serv, Allied Mills Inc, 67-70; dir nutrit & res, Robin Hood Multifoods Ltd, Can, 70-74; SR POULTRY SCIENTIST, CENT SOYA, INC, 74- Mem: Poultry Sci Asn; Am Soc Animal Sci. Res: Amino acid studies with chicks; dog and cat management and nutrition; applied animal husbandry; poultry nutrition; egg quality studies. Mailing Add: RR 4 Oakwood Subdiv Decatur IN 46733

WATERHOUSE, JOHN P, b Kent, Eng, Dec 4, 20; m 44; c 3. PATHOLOGY, HISTOLOGY. Educ: Univ London, MB, BS, 51, BDS, 56, MD, 63. Prof Exp: From jr lectr to sr lectr dent path, London Hosp Med Col, Univ London, 56-66, univ reader oral path, 66-67; actg head dept histol, Col Dent, 68-69, PROF ORAL PATH & HEAD DEPT, COL DENT, UNIV ILL MED CTR, 67-, PROF PATH, COL MED, 68- Mem: Am Acad Oral Pathologists; Brit Soc Periodont (hon ed, 64, hon secy, 66); Int Asn Dent Res (asst secy-treas, 73-). Res: Electron microscopy; cytochemistry; lysosomes; stereology; normal and damaged oral mucosa. Mailing Add: Dept of Oral Path Univ of Ill at the Med Ctr Chicago IL 60612

WATERHOUSE, JOSEPH STALLARD, b Toronto, Ont, Apr 28, 29; m 54; c 3. HUMAN ANATOMY, HUMAN PHYSIOLOGY. Educ: Univ Guelph, BSc, 54; Wash State Univ, MSc, 57, PhD(entom), 62. Prof Exp: Aide entom & zool, Wash State Univ, 54-60; asst prof biol, State Univ NY Col Arts & Sci Plattsburgh, 60-64; Ford of Can res fel zool, Carleton Univ, 64-65; assoc prof biol, 65-72, chmn dept biol sci, 74-75, PROF BIOL, STATE UNIV NY COL PLATTSBURGH, 72- Mem: AAAS; Am Inst Biol Sci; Entom Soc Am; Entom Soc Can. Res: Biology of the garden symphylan, Scutigerella immaculata; chemocarcinogenesis in amphibians; calcium metabolism in leopard frogs; entomology; calcium and phosphate metabolism in amphibians. Mailing Add: Dept of Biol Sci State Univ of NY Col Plattsburgh NY 12901

WATERHOUSE, KEITH R, b Derby, Eng, May 10, 29; US citizen; m 55; c 5. UROLOGY, SURGERY. Educ: Cambridge Univ, BA, 50, MA, 57; Oxford Univ, MB & BChir, 53. Prof Exp: Instr urol, Kings County Hosp Ctr, 59-61; from asst prof to assoc prof, 61-65, PROF UROL SURG, STATE UNIV NY DOWNSTATE MED CTR, 65- Concurrent Pos: Consult, Vet Admin Hosp, Brooklyn, NY, St Mary's Hosp, Passaic, NJ & Paterson Gen Hosp, 65-, Samaritan Hosp, Brooklyn, NY, 66- & St Charles Hosp, Port Jefferson, St Francis Hosp, Poughkeepsie, Arden Hill Hosp, Goshen & Brookhaven Mem Hosp, Patchogue, 67- Mem: Am Urol Asn; Am Col Surgeons; Am Acad Pediat; Am Fertil Soc; Int Soc Urol. Res: Investigation, diagnosis and treatment of congenital anomalies of the urinary tract in children and the subsequent treatment of these patients when they develop chronic renal failure. Mailing Add: Dept of Urol State Univ NY Downstate Med Ctr Brooklyn NY 11203

WATERHOUSE, RICHARD (VALENTINE), b Eng, Feb 28, 24; nat US; m 51; c 4. PHYSICS. Educ: Oxford Univ, MA, 49; Cath Univ, PhD, 59. Prof Exp: Res physicist, Royal Navy Torpedo Exp Estab, Scotland, 44-46, Paint Res Asn, London, 46-49 & Nat Bur Stand, 51-59; PROF PHYSICS, AM UNIV, 61- Concurrent Pos: Vis prof, Univ Calif, Berkeley, 69-70 & Univ Delft, Neth, 75-76; consult, US Navy. Mem: Fel Acoust Soc Am. Res: Waves and vibrations; theoretical physics; acoustics. Mailing Add: Dept of Physics Am Univ Washington DC 20016

WATERHOUSE, WILLIAM CHARLES, b Galveston, Tex, Dec 31, 41. MATHEMATICS. Educ: Harvard Univ, AB, 63, AM, 64, PhD(math), 68. Prof Exp: Res assoc math, Off Naval Res & Cornell Univ, 68-69; asst prof, Cornell Univ, 69-75; ASSOC PROF MATH, PA STATE UNIV, UNIVERSITY PARK, 75- Mem: Am Math Soc; Math Asn Am. Res: Algebraic number theory; algebraic geomtry, especially group schemes; history of mathematics. Mailing Add: Dept of Math Pa State Univ University Park PA 16802

WATERMAN, ALLYN JAY, b Cleveland, Ohio, Apr 19, 02; m 28; c 2. EMBRYOLOGY. Educ: Oberlin Col, BA, 25; Western Reserve Univ, MA, 27; Harvard Univ, PhD(biol), 31. Prof Exp: Asst, Western Reserve Univ, 25-27; instr biol, Williams Col, 27-28; asst, Harvard I)niv, 28-31; CRB Ed Found fel, Brussels, 31-32; instr biol, Brooklyn Col, 32-34; from asst prof to prof, Williams Col, 34-67; staff mem, Nat Inst Child Health & Human Develop, 67-74; RETIRED. Concurrent Pos: Instr, Marine Biol Lab, Woods Hole, 36-41, in charge invert course, 41-42; res assoc, Columbia Univ & Brookhaven Nat Lab, 53-54; mem, Mt Desert Biol Lab, 58-60; vis lectr, Found Advan Educ Sci, NIH, 74-76 & Univ Md, 75; consult, World Pop Soc, 74- Mem: Am Soc Cell Biol; Int Soc Cell Biol; Int Soc Develop Biol; Am Soc Zool; Am Asn Anat. Res: Experimental embryology and morphology. Mailing Add: 5304 Roosevelt St Bethesda MD 20014

WATERMAN, DANIEL, b New York, NY, Oct 24, 27; m 60; c 3. MATHEMATICS. Educ: Brooklyn Col, BA, 47; Johns Hopkins Univ, MA, 48; Univ Chicago, PhD, 54. Prof Exp: Res assoc, Cowles Comn Res in Econ, 51-52; from instr to asst prof math, Purdue Univ, 53-59; asst prof, Univ Univ Wis-Milwaukee, 59-61; prof, Wayne State Univ, 61-69; PROF MATH, SYRACUSE UNIV, 69- Mem: Am Math Soc. Res: Fourier analysis; real variables; functional analysis; orthogonal series. Mailing Add: Dept of Math Syracuse Univ Syracuse NY 13210

WATERMAN, FRANK MELVIN, b Delhi, NY, Nov 3, 38; m 61; c 4. MEDICAL PHYSICS. Educ: Hartwick Col, BA, 60; Clarkson Technol Col, MS, 69, PhD(physics), 73. Prof Exp: Res assoc nuclear physics, Kent State Univ, 72-75; RES ASSOC MED PHYSICS, DEPT RADIOL, UNIV CHICAGO, 75- Mem: Am Phys Soc; Am Asn Physicists Med. Res: Applications of neutron physics to biology and medicine; neutron radiation therapy; neutron dosimetry; neutron radiobiology; nuclear instrumentation. Mailing Add: Dept of Radiol Univ of Chicago Chicago IL 60637

WATERMAN, HAROLD HUGHTON, physics, see 12th edition

WATERMAN, MICHAEL ROBERTS, b Tacoma, Wash, Nov 23, 39; m 66; c 2. PROTEIN CHEMISTRY. Educ: Willamette Univ, BA, 61; Univ Ore, PhD(biochem), 69. Prof Exp: Fel biochem, Johnson Res Found, Sch Med, Univ Pa, 68-70; asst prof, 70-75, ASSOC PROF BIOCHEM, HEALTH SCI CTR, UNIV TEX, 75- Mem: Am Chem Soc; Am Soc Biol Chem; Am Soc Hemat. Res: Hemoglobin structure and function relationship; molecular aspects of sickle cell disease. Mailing Add: Dept of Biochem Univ Tex Health Sci Ctr Dallas TX 75235

WATERMAN, MICHAEL S, b Coquille, Ore, June 28, 42; m 62; c 1. MATHEMATICS, STATISTICS. Educ: Ore State Univ, BS, 64, MS, 66; Mich State Univ, MA, 68, PhD(probability, statist), 69. Prof Exp: Teaching asst math, Ore State Univ, 64-66; from asst prof to assoc prof, Idaho State Univ, 69-75, NSF res grant, 71-73; STAFF MEM, LOS ALAMOS SCI LAB, 75- Mem: Am Statist Asn; AAAS; Math Asn Am; Inst Math Statist. Res: Ergodic theory; probabilistic and computational number theory; mathematical biology. Mailing Add: MS-254 C-5 Los Alamos Sci Lab Los Alamos NM 87545

WATERMAN, PETER CARY, physics, applied mathematics, see 12th edition

WATERMAN, TALBOT HOWE, b East Orange, NJ, July 3, 14. COMPARATIVE PHYSIOLOGY. Educ: Harvard Univ, AB, 36, MA, 38, PhD(zool), 43; Yale Univ, MA Privatim, 58. Prof Exp: Jr fel, Harvard Univ Soc of Fels, 38-40, res assoc, Psychoacoustic Lab, Off Sci Res & Develop, Harvard Univ, 41-43; staff mem, Radiation Lab, Mass Inst Technol, 43-45; sci consult, Off Sci Res & Develop, Off Field Serv, US Navy Dept & US Army Air Force, 45; secy comt res, Sigma Xi, 46; instr biol, 46-47, from asst prof to prof zool, 47-74, PROF ZOOL, YALE UNIV, 74- Concurrent Pos: Exec fel, Trumbull Col, 46-56; instr invert zool, Marine Biol Lab, Woods Hole, 47-52, mem corp, 51-; secy corp, Bermuda Biol Sta, 51-61, trustee, 62-; mem Am Inst Biol Sci Adv Com: Hydrobiol, Off Naval Res, 59-65; Guggenheim fel,

62-63; Sigma Xi nat lectr, 64-65; vis prof, Sch Med, Keio Univ, Japan, 68; assoc ed, J Exp Biol, 71-75; vis prof, Japan Soc Promotion Sci, 74; vis lectr & investr, Woods Hole Oceanog Inst, 76- Mem: AAAS; Soc Gen Physiol; Marine Biol Asn UK; Am Physiol Soc; Am Soc Zool. Res: Visual physiology and spatial orientation; deepsea plankton and vertical migrations; compound eye fine structure and information processing; polarized light behavior in the sea. Mailing Add: Dept of Biol 610 Kline Biol Tower Yale Univ New Haven CT 06520

WATERMEIER, LELAND A, b Carlinville, Ill, Dec 27, 27; m 50; c 4. PHYSICAL CHEMISTRY, PHYSICS. Educ: Blackburn Col, BA, 50; Univ Del, MS, 55; Univ London, DIC, 65. Prof Exp: Res chemist, 51-56, supvry aero-fuels res chemist, 56-60, supvry res chemist, 60-66, asst to tech dir weapons technol, 66-68, CHIEF INTERIOR BALLISTICS LAB, US ARMY BALLISTICS RES LABS, 68- Concurrent Pos: Army mem tech steering comt, Interagency Chem Rocket Propulsion Group, 62-67. Mem: Combustion Inst; Am Inst Aeronaut & Astronaut. Res: Stable and unstable combustion mechanisms in solid and liquid propellants; weapons technology, including large and small caliber weapons; complete interior ballistics and propulsion cycle of all caliber of weapons. Mailing Add: Interior Ballistics Lab US Army Ballistics Res Labs Aberdeen Proving Ground MD 21005

WATERS, AARON CLEMENT, b Waterville, Wash, May 6, 05; m 40; c 2. GEOLOGY. Educ: Univ Wash, Seattle, BSc, 26, MSc, 27; Yale Univ, PhD(geol), 30. Prof Exp: Instr geol, Yale Univ, 28-30; from asst prof to prof, Stanford Univ, 30-52; prof, Johns Hopkins Univ, 52-63 & Univ Calif, Santa Barbara, 63-67; prof, 67-72, EMER PROF GEOL, UNIV CALIF, SANTA CRUZ, 72- Concurrent Pos: Geologist, US Geol Surv, 30, 42-45 & 52-; Guggenheim fel, 37-38; bicentennial lectr, Columbia Univ, 54; S F Emmons lectr, 58; NSF sr fel, 60; Condon lectr, 66. Mem: Nat Acad Sci; Geol Soc Am; Geochem Soc; Am Acad Arts & Sci. Res: Petrology; structural geology; geology of the Columbia River plateau and the Cascade Mountains. Mailing Add: 308 Moore St Santa Cruz CA 95060

WATERS, CHARLES EMORY, b Washington, DC, Nov 17, 08; m 44; c 2. CHEMISTRY. Educ: Johns Hopkins Univ, PhD(chem), 31. Prof Exp: Asst, Nat Bur Stand, 27, jr sci aide, 28; asst, Johns Hopkins Univ, 29-31; asst, Asphalt Shingle & Roofing Inst, 31; res chemist, Johns-Manville Corp, NY, 31-37 & Tide Water Assoc Oil Co, 37-42; chemist, Barrett Div, Allied Chem Corp, 46-52, sr res chemist, Nitrogen Div, 52-57, supvry res chemist, 57-58, asst lab dir, 58-59, lab dir, 59-61, mgr fertilizer res, 61-63, tech serv, 63-65, sr scientist, 65-67, coordr waste control, Agr Div, 67-69; teacher, Colonial Heights High Sch, Va, 70-71; teacher math, chem & physics, Bollingbrook Day Sch, Petersburg, 71- Mem: Am Chem Soc; fel Am Inst Chemists. Res: Fertilizers; bituminous and related products; disposal of plant wastes. Mailing Add: 1747 Berkeley Ave Petersburg VA 23803

WATERS, DONALD HILTON, b Palm Beach, Fla, Dec 31, 43; m 67; c 2. NEUROPHARMACOLOGY. Educ: Philadelphia Col Pharm & Sci, BS, 67; Cornell Univ, PhD(pharmacol), 73. Prof Exp: Instr pharmacol, Mt Sinai Sch Med, 72-73, assoc, 73-74; ASST PROF BIOCHEM PHARMACOL, STATE UNIV NY BUFFALO, 74- Mem: AAAS; NY Acad Sci. Res: Drug abuse; physical dependence; barbiturates; ethanol dependence; neuromuscular pharmacology. Mailing Add: Dept of Biochem Pharmacol State Univ of NY Sch of Pharm Buffalo NY 14214

WATERS, ELMER DALE, b Rigby, Idaho, Dec 15, 30. HEAT TRANSFER, FLUID MECHANICS. Educ: Univ Idaho, BS, 53, MS, 56. Prof Exp: From jr engr to engr, Gen Elec Co, 53-65; res assoc thermal hydraul, Battelle Mem Inst, 65-66; lab scientist energy transfer, Douglas Aircraft Co, 66-67, SR LAB SCIENTIST ENERGY TRANSFER, DOUGLAS LABS, McDONNELL DOUGLAS CORP, 67- Honors & Awards: Aerospace Lit Award, McDonnell Douglas Corp, 70 & 71. Mem: Am Inst Aeronaut & Astronaut; Am Soc Mech Engrs. Res: Heat transport mechanisms, boiling burnout, evaporation-condensation phenomena and heat pipes; thermal power plant systems. Mailing Add: 412 Delafield Richland WA 99352

WATERS, IRVING WADE, b Baldwyn, Miss, June 19, 31; m 54; c 2. PHARMACOLOGY. Educ: Delta State Col, BS, 58; Auburn Univ, MS, 60; Univ Fla, PhD(pharmacol), 63. Prof Exp: From asst prof to assoc prof, 66-74, PROF PHARMACOL, UNIV MISS, 74- Concurrent Pos: Strassenburg Labs grant, Univ Fla, 63-64, Fla Heart Asn grant, 64-65, NIH training grant, 65-66. Mem: Am Physiol Soc. Res: Drug metabolism; toxicology. Mailing Add: Dept of Pharmacol Univ of Miss Sch of Pharm University MS 38677

WATERS, JAMES AUGUSTUS, b Postville, Iowa, June 23, 31; m 57; c 4. ORGANIC CHEMISTRY. Educ: Univ Iowa, BS, 53, PhD(pharmaceut chem), 59; Purdue Univ, MS, 57. Prof Exp: Res assoc org chem, Univ Mich, 59-60; RES CHEMIST, NAT INST ARTHRITIS, METAB & DIGESTIVE DIS, 60- Mem: Am Chem Soc. Res: Structure elucidation of natural products; steroid biosynthesis; photochemistry; electrolytic oxidations; synthesis of nucleosides and polynucleotides as anti-tumor and interferon-inducing agents. Mailing Add: Lab of Chem Nat Inst of Arthritis Metab & Digestive Dis Bethesda MD 20014

WATERS, JAMES FREDERICK, b Oak Park, Ill, Mar 17, 38; m 65. VERTEBRATE ANATOMY. Educ: Stanford Univ, BA, 59; Univ Wash, PhD(zool), 69. Prof Exp: From asst prof to assoc prof, 66-75, PROF ZOOL, HUMBOLDT STATE UNIV, 75- Mem: Am Soc Ichthyol & Herpet; Soc Study Evolution; Soc Vert Paleont; Am Soc Zool. Res: Functional locomotor anatomy of lizards and snakes. Mailing Add: Dept of Biol Humboldt State Univ Arcata CA 95221

WATERS, JAMES WAYNE, physics, see 12th edition

WATERS, JOSEPH HEMENWAY, b Brockton, Mass, Dec 23, 30. VERTEBRATE ZOOLOGY. Educ: Univ Mich, BS, 54, MS, 55; Univ Conn, PhD(zool), 60. Prof Exp: Instr biol, Mass State Col Bridgewater, 59-61, Duke Univ, 62-63, Univ RI, 63-64 & Roanoke Col, 64-65; INSTR BIOL, VILLANOVA UNIV, 65- Mem: Fel AAAS; Am Soc Mammal; Am Soc Ichthyol & Herpet; Sigma Xi. Res: Ecology and systematics of vertebrates in eastern north America. Mailing Add: Dept of Biol Villanova Univ Villanova PA 19085

WATERS, KENNETH HAROLD, b Ipswich, Eng, May 17, 13; nat US; m 39; c 4. EXPLORATION GEOPHYSICS. Educ: Univ London, BS, 35, DIC, 37. Prof Exp: Asst lectr, Univ London, 35-38; party chief field seismic explor, Nat Geophys Co, 39-42, supvr field seismic prospecting, 46-55; dir geophys res, 60-75, RES FEL, CONTINENTAL OIL CO, 75- Mem: Soc Explor Geophys; Seismol Soc Am; Europ Asn Explor Geophys; Am Geophys Union. Res: Research theory and interpretation of seismic prospecting; shear wave transmission in sedimentary layers; digital computer data enhancement techniques; stratigraphic trap delineation by seismic techniques. Mailing Add: Explor Res Div Continental Oil Co Ponca City OK 74601

WATERS, KENNETH LEE, b Monroe, Va, Jan 24, 14; m 39. PHARMACY, MEDICINAL CHEMISTRY. Educ: Lynchburg Col, AB, 35; Univ Ga, MS, 37; Univ Md, PhD(pharmaceut chem), 45. Prof Exp: Instr chem, Transylvania Col, 36-37 & Univ Ga, 37-39; from jr chemist to asst chemist, US Food & Drug Admin, Md, 39-43; fel, Mellon Inst, 43-47; tech dir, Zemmer Co, 47-48; DEAN SCH PHARM, UNIV GA, 48- Concurrent Pos: Lectr, Sch Pharm, Univ Pittsburgh, 45-47. Mem: AAAS; Am Chem Soc; Am Pharmaceut Asn. Res: Drug standardization; preparation of alpha-akloximino acids and their derivatives. Mailing Add: Sch of Pharm Univ of Ga Athens GA 30602

WATERS, LARRY CHARLES, b Glenville, Ga, July 1, 39; m 60; c 1. BIOCHEMISTRY. Educ: Valdosta State Col, BS, 61; Univ Ga, MS, 64, PhD(biochem), 65. Prof Exp: Am Cancer Soc fel biochem res, 65-67, STAFF BIOCHEMIST, BIOL DIV, OAK RIDGE NAT LAB, 67- Concurrent Pos: Lectr, Univ Tenn, Knoxville. Mem: Am Soc Biol Chemists. Res: Chemistry and biochemistry of nucleic acids, especially with regards to their mode of action and involvement in cellular differentiation; t ribonucleic acid as a potential regulator of protein synthesis; viral carcinogenesis. Mailing Add: Biol Div Oak Ridge Nat Lab PO Box Y Oak Ridge TN 37830

WATERS, LEVIN LYTTLETON, b Ruxton, Md, June 7, 10; m 45; c 2. PATHOLOGY. Educ: Princeton Univ, AB, 32; Yale Univ, MD, 37. Prof Exp: From instr to assoc prof, 41-59, PROF PATH, SCH MED, YALE UNIV, 59- Concurrent Pos: Ives fel path, 38-39; Int Cancer Res Found fel, 39-40; tech aide, Comt Med Res, Off Sci Res & Develop & Div Med Sci, Nat Res Coun, 41-45; dir & chmn bd sci adv, Jane Coffin Childs Fund Med Res, 60-63 & 70-; mem coun arteriosclerosis & mem coun high blood pressure res, Am Heart Asn. Honors & Awards: Cert of Merit, US Dept Defense, 45. Mem: AAAS; Am Asn Cancer Res; Am Asn Pathologists & Bacteriologists. Res: Cardiovascular renal disease, particularly atherosclerosis; medical school teaching; cancer research and administration. Mailing Add: Dept of Path Yale Univ Sch of Med New Haven CT 06510

WATERS, M ANN ELIZABETH, physical chemistry, see 12th edition

WATERS, MICHAEL DEE, b Charlotte, NC, Apr 17, 42; m 64; c 2. BIOCHEMISTRY, CELL BIOLOGY. Educ: Davidson Col, BS, 64; Univ NC, Chapel Hill, PhD(biochem), 69. Prof Exp: Lab chief biochem, Biophys Lab, Edgewood Arsenal, Md, 69-71; unit chief cellular physiol, Cellular Biol Sect, 71-72, sect chief cellular physiol, Pathobiol Res Br, 72-75, CHIEF CELLULAR BIOCHEM SECT, BIOMED RES BR, ENVIRON RES CTR, ENVIRON PROTECTION AGENCY, 75- Concurrent Pos: Clin instr, George Washington Univ, 71-72. Mem: AAAS; Am Chem Soc; Tissue Cult Asn; Sigma Xi; Reticuloendothelial Soc. Res: Model systems in tissue culture for studies on enzyme regulation, protein biosynthesis, cytotoxic phenomena. Mailing Add: Biomed Res Br Clin Studies Div Environ Protection Agency Research Triangle Park NC 27711

WATERS, NELSON FENN, b New Haven, Conn, Aug 10, 99; m 34; c 4. GENETICS. Educ: Univ Conn, BSc, 25; Harvard Univ, MSc, 30, DSc(zool), 31. Prof Exp: Res assoc prof animal breeding & path, RI State Col, 25-29; asst animal breeding, Harvard Univ, 29-31; res assoc prof poultry husb, Iowa State Col, 31-39; sr animal geneticist, Regional Poultry Res Lab, Bur Animal Indust, USDA, 39-53, ANIMAL GENETICIST, REGIONAL POULTRY RES LAB, AGR RES SERV, USDA, 53- Honors & Awards: Prize, Poultry Sci Asn, 32. Mem: AAAS; Am Genetic Asn; Poultry Sci Asn. Res: Resistance and susceptibility to disease; growth; embryology; inbreeding domestic fowl; poultry genetics; genetics of avian neoplasms; medical genetics. Mailing Add: PO Box 115 Alton Bay NH 03810

WATERS, NORMAN DALE, b Twin Falls, Idaho, May 1, 22; m 54; c 2. ENTOMOLOGY. Educ: Univ Calif, Berkeley, BS, 49, PhD, 55. Prof Exp: Mem staff for parasite introd, USDA, 49-51; entomologist, Cashewnut Improv Comn, India, 52-53; pest control supvr citrus, Limonera Co, Calif, 55-57; ENTOMOLOGIST, EXP STA, UNIV IDAHO, 57-, ASSOC RES PROF ENTOM, 70- Mem: Entom Soc Am; Int Orgn Biol Control. Res: Biological control; legume forage insects; pollinating insects. Mailing Add: Exp Sta Univ of Idaho Parma ID 83660

WATERS, THOMAS FRANK, b Hastings, Mich, May 17, 26; m 53; c 3. AQUATIC ECOLOGY, FISHERIES. Educ: Mich State Univ, BS, 52, MS, 53, PhD(fishery biol), 56. Prof Exp: Supvr, Pigeon River Trout Res Sta, Mich Dept Conserv, 56-57; from asst prof to assoc prof, 58-68, PROF FISHERY BIOL, UNIV MINN, ST PAUL, 68- Mem: Am Fisheries Soc; Ecol Soc Am; Am Inst Fishery Res Biol; Am Soc Limnol & Oceanog. Res: Limnology; ecology of aquatic invertebrates and stream fish populations. Mailing Add: Dept Entom Fisheries & Wildlife Univ Minn St Paul MN 55108

WATERS, WILLIAM E, b Springfield, Mass, July 2, 22; m 43; c 4. FOREST ENTOMOLOGY. Educ: State Univ NY Col Forestry, Syracuse, BS, 48; Duke Univ, MF, 49; Yale Univ, PhD, 58. Prof Exp: Entomologist, Bur Entom & Plant Quarantine, USDA, 49-53, entomologist, Forest Serv, 53-59, chief div forest insect res, 59-63, div forest insect & dis res, 63-65, asst dir insects & dis, 65-66, chief forest insect & dis lab & prin ecologist, Northeastern Forest Exp Sta, Conn, 66-68, chief forest insect res, Div Forest Insect & Dis Res, 68-73, chief scientist, Pac Southwest Forest & Range Exp Sta, 73-75; DEAN COL NATURAL RESOURCES, UNIV CALIF, BERKELEY & ASSOC DIR CALIF AGR EXP STA, 75- Concurrent Pos: Instr, Quinnipiac Col, 61-64; lectr, Yale Univ, 64-68; mem pop dynamics working party & chmn impact of destructive agents proj group, Int Union Forest Res Orgn; chmn working party on forest insects & dis, Food & Agr Orgn-NAm Forestry Comn; mem fed comn wood protection, Nat Acad Sci-Nat Res Coun. Mem: Entom Soc Am; Soc Am Foresters. Res: Forest insect ecology; population dynamics; biometrics; insect genetics and behavior; forest pest management. Mailing Add: Col of Natural Resources Univ of Calif Berkeley CA 94720

WATERS, WILLIAM EDWARD, b Lexington, Ky, July 16, 23; m 47; c 3. PHYSICS. Educ: Univ Ky, BS, 47, MS, 49; Univ Md, PhD(physics), 57. Prof Exp: Electronics scientist, Diamond Ord Fuse Labs, Washington, DC, 48-56, res supvr microwave tubes, 56-60; sr engr, Microwave Electronics Corp, 60-65, sr engr, Varian Assocs, 65-66; dir res & develop, Phys Electronics Labs, 66-67; PRIN ENGR, AERONUTRONIC-FORD CORP, ENG SERV DIV, PALO ALTO, 67- Concurrent Pos: Secy Army study & res fel, Stanford Univ, 58-59. Res: Microwave electron tubes, particularly oscillators; electron optics, including electron guns, beams and focusing systems; low noise microwave solid-state amplifiers and solid-state oscillators; satellite communications receiving and transmitting equipment. Mailing Add: Autonutronic Ford Corp 3939 Fabian Way Palo Alto CA 94303

WATERS, WILLIAM LINCOLN, b Columbus, Ohio, Nov 7, 38; m 62; c 2. ORGANIC CHEMISTRY, ORGANOMETALLIC CHEMISTRY. Educ: Kenyon Col, BA, 61; Univ Hawaii, PhD(org chem), 66. Prof Exp: Instr & res fel chem, Univ Calif, Irvine, 66-68; asst prof, 68-72, ASSOC PROF CHEM, UNIV MONT, 72- Concurrent Pos: NSF grant, 70-72. Res: Organic mechanisms; nuclear magnetic resonance; electrophilic additions to olefins; organomercurials; stereochemistry; ozone reactions. Mailing Add: Dept of Chem Univ of Mont Missoula MT 59801

WATERS, WILLIE ESTEL, b Smith Town, Ky, Sept 19, 31; m 52; c 3. HORTICULTURE. Educ: Univ Ky, BS, 54, MS, 58; Univ Fla, PhD(veg crops), 60. Prof Exp: Asst county agr agent, Ky, 54; asst soils, Univ Ky, 56-58; asst veg crops, 58-60, from asst ornamental horticulturist to assoc ornamental horticulturist, 60-68, horticulturist & head Ridge Ornamental Hort Lab, 68-70, HORTICULTURIST & DIR AGR RES & EDUC CTR, UNIV FLA, 70- Mem: Am Soc Hort Sci. Res: Soil and plant nutrition; weed control and post harvest physiology of ornamental crops. Mailing Add: Agr Res & Educ Ctr Univ of Fla Bradenton FL 33505

WATERWORTH, HOWARD E, b Randolph, Wis, Sept 3, 36; m 66; c 2. PLANT VIROLOGY. Educ: Univ Wis, BS, 58, PhD(plant path), 62. Prof Exp: RES PLANT PATHOLOGIST, AGR RES SERV, USDA, 64- Concurrent Pos: Ed of News, Am Phytopath Soc. Mem: Brit Asn Appl Biol; Am Phytopath Soc; Indian Phytopath Soc. Res: Testing of new apple, pear, and woody ornamentals for the presence of viruses; basic research on these viruses, including purification, study of chemical and physical properties, preparation of antisera, identification of new viruses and screening for resistance to them. Mailing Add: Plant Introd Sta Agr Res Serv US Dept of Agr Glenn Dale MD 20769

WATHEN, RONALD LARRY, b Evansville, Ind, Feb 20, 36; m 73; c 2. MEDICAL RESEARCH. Educ: Butler Univ, BS, 58; Ind Univ, MA, 61, PhD(physiol), 65; Univ Tex Southwestern Med Sch, Dallas, MD, 65. Prof Exp: Instr physiol, Ind Univ Med Sch, 64-66; from instr to asst prof med, Vanderbilt Univ, 69-72; ASST PROF MED, MED SCH, UNIV MINN, MINNEAPOLIS, 72-, ASST PROF MED, GRAD SCH, 73-, ASST MED DIR NEPHROL SECT, REGIONAL KIDNEY DIS PROG, HENNEPIN COUNTY MED CTR, 73- Concurrent Pos: Grant rev consult, Chronic Uremia Artificial Kidney Prog, Nat Inst Arthritis, Metab & Dig Dis, NIH, 73- Mem: Am Soc Nephrol; Int Soc Nephrol; Am Soc Artificial Internal Organs; Am Fedn Clin Res; Sigma Xi. Res: Clinical studies of dialysis patients to assess adequacy of dialysis and intermediary metabolism; dog research studying intra-renal hormonal mechanisms. Mailing Add: Third Floor 527 Park Ave Minneapolis MN 55415

WATKIN, DONALD MORGAN, b Waterford, NY, June 17, 22; m 46; c 4. INTERNAL MEDICINE, PUBLIC HEALTH. Educ: Hamilton Col, AB, 43; Harvard Univ, MD, 46, MPH, 65; Am Bd Internal Med, dipl, 56; Am Bd Nutrit, dipl. Prof Exp: Intern, Long Island Col Med Div, Kings County Hosp, 46-47; sr investr sect geront, NIH, 51-54, sr investr metab serv, Nat Cancer Inst, 54-60, NIH adv nutrit, Pan-Am WHO, UN, Mex, 60-61; assoc prof nutrit & dir NIH gen med clin res ctr, Mass Inst Technol, 62-64; chief res gastroenterol, nutrit & clin aging, Vet Admin Cent Off, 66-69; staff physician, 69-70, ACTG CHIEF SPINAL CORD INJURY SERV, VET ADMIN HOSP, WEST ROXBURY, 70- Concurrent Pos: Res fel nutrit, Sch Pub Health, Harvard Univ, 47-48; res fel, Columbia Univ, Goldwater Mem Hosp, New York & asst, Col Physicians & Surgeons, Columbia Univ, 48-50; spec res fel, Harvard Med Sch & Thorndike Mem Lab, Boston City Hosp & asst, Hosp, 50-51; asst vis physician, Med Serv, Baltimore City Hosps, 51-54; vis lectr, Sch Hyg & Pub Health, Johns Hopkins Univ, 52-55; clin instr, Med Sch, George Washington Univ, 54-60; assoc, George Washington Med Div, DC Gen Hosp & Hosp, George Washington Univ, 54-60; mem, Surgeon Gen Intravenous Nutriment Prog, US Dept Army, 56-; consult, Cushing Hosp, Mass Dept Ment Health, 63 & USAID, 66-67; Ciba Found lectr, Glasgow Univ, 64; dir, Paraguay Surv, Off Int Res, NIH, 65-67; asst clin prof, Med Sch, Boston Univ & vis physician, Med Serv, Vet Admin Hosp, Boston, 67-69; res assoc, Mallory Inst Path Found, Boston, 69-; vchmn & chmn panel aging, White House Conf Food, Nutrit & Health, 69-70, chmn tech comt nutrit, White House Conf Aging, 70-71, mem, Planning Bd, 70-71 & Post Conf Bd, 72-; mem coun atherosclerosis, Am Heart Asn. Mem: Am Physiol Soc; AMA; Am Col Physicians; Am Inst Nutrit; Am Soc Clin Nutrit; Am Pub Health Asn. Res: Gerontology; nutrition energy metabolism; health promotion in developing regions; lipid metabolism; uric acid; diabetes; work and renal physiology; growth and development; cardiomyopathies. Mailing Add: Vet Admin Hosp 1400 VFW Pkwy West Roxbury MA 02132

WATKIN, JOHN EMRYS, b Wales, Mar 19, 29; m 55; c 2. PLANT BIOCHEMISTRY, ENVIRONMENTAL BIOLOGY. Educ: Cambridge Univ, BA, 52; Univ Col NWales, PhD(biochem), 55. Prof Exp: Fel, Prairie Regional Lab, Nat Res Coun Can, Sask, 55-57, from asst res off to assoc res off, 58-70, RES COUN OFFICER, ENVIRON SECRETARIAT, BIOL SCI DIV, NAT RES COUN CAN, 70- Res: Pollution of environment by metals. Mailing Add: Biol Sci Div Nat Res Coun Ottawa ON Can

WATKINS, ALLEN HARRISON, b Charlottesville, Va, Apr 25, 38; m 62; c 2. SCIENCE ADMINISTRATION. Educ: Va Polytech Inst, BS, 61. Prof Exp: Mem prog mgt staff space syst, NASA Manned Spacecraft Ctr, 62-70, prog mgt earth resources, 70-73; DIR, EARTH RESOURCES OBSERV SYST DATA CTR, DEPT INTERIOR, 73- Mem: Am Soc Photogram; Am Inst Aeronaut & Astronaut. Mailing Add: 2705 W 22nd St Sioux Falls SD 57105

WATKINS, CHARLES, b Shaw, Miss, Jan 9, 16; m 41; c 3. PSYCHIATRY. Educ: Univ Tenn, MD, 38. Prof Exp: Intern, DePaul Hosp, St Louis, Mo, 38-39; resident, St Lawrence State Hosp, Ogdensburg, NY, 41-42; resident, Duke Hosp, Durham, NC, 46-48; staff psychiatrist, Vet Admin Hosp, New Orleans, 48-49; clin asst prof, 49-52, from asst prof to prof, 52-67, head dept, 57-67, dir prog develop, Med Ctr, 65-68, asst to chancellor, 68-73, EMER PROF PSYCHIAT, SCH MED, LA STATE UNIV MED CTR, NEW ORLEANS, 74-; CHIEF PSYCHIAT SERV, VET ADMIN CTR, BILOXI, 74-; PROF PSYCHIAT, COL MED, UNIV S ALA, 75- Mem: Am Col Physicians; Am Asn Hist Med. Res: Communication; schizophrenia; history of medicine. Mailing Add: Psychiat Serv Vet Admin Ctr Biloxi MS 39531

WATKINS, CHARLES LEE, b Fairfield, Ala, Oct 27, 42; m. CHEMISTRY. Educ: Univ Ala, Tuscaloosa, BS, 64; Univ Fla, MS, 66, PhD(chem), 68. Prof Exp: Lab asst gen chem, Univ Ala, Tuscaloosa, 63-64; res assoc, Univ NC, Chapel Hill, 69-70; asst prof, 70-74, ASSOC PROF CHEM, UNIV ALA, BIRMINGHAM, 74- Mem: Am Chem Soc. Mailing Add: Dept of Chem Univ of Ala Birmingham AL 35299

WATKINS, DAVID HYDER, b Denver, Colo, Nov 26, 17; m 41; c 2. THORACIC SURGERY. Educ: Univ Colo, AB, 37, MD, 40; Univ Minn, MS, 47, PhD(surg), 49; Am Bd Surg, dipl; Am Bd Thoracic Surg, dipl. Prof Exp: Asst surgeon, Mayo Clin, 44-49; instr surg, Ohio State Univ, 49-51; from assoc prof to prof, Sch Med, Univ Colo, 51-67; CLIN PROF SURG, COL MED, UNIV IOWA, 67-; DIR SURG, IOWA METHODIST HOSP-BROADLAWNS HOSP, DES MOINES, 68- Concurrent Pos: Attend thoracic surgeon, Colo Gen Hosp, 51-67; chief surg serv, Denver Gen Hosp, 52-67; consult, Fitzsimons Army Hosp, Denver & Vet Admin Ctr, Des Moines. Mem: Am Asn Thoracic Surgeons; fel AMA; fel Am Heart Asn; fel Am Col Surgeons; fel Am Col Chest Physicians. Res: Surgery of thorax, great vessels, heart and esophagus; assisted circulation; shock and hemorrhage; experimental surgery. Mailing Add: 6039 N Waterbury Rd Des Moines IA 50312

WATKINS, DON WAYNE, b Louisville, Ky, June 9, 40; m 66; c 3. PHYSIOLOGY. Educ: Univ Louisville, BChE, 63; Univ Wis-Madison, PhD(physiol), 68. Prof Exp: Lectr physiol, Med Sch, Makerere Univ, Uganda, 68-70; asst prof, 70-75, ASSOC PROF PHYSIOL, MED SCH, GEORGE WASHINGTON UNIV, 75- Mem: Biophys Soc; Am Physiol Soc. Res: Membrane transport, physiology, epithelia; kidney salt and water; frog skin. Mailing Add: Dept of Physiol George Washington Univ Med Sch Washington DC 20006

WATKINS, DUDLEY T, b Youngstown, Ohio, May 2, 38; m 60; c 3. ANATOMY. Educ: Oberlin Col, AB, 60; Western Reserve Univ, MD, 66, PhD(anat), 67. Prof Exp: Fel anat, 66-67, asst prof, 67-72, ASSOC PROF ANAT, HEALTH CTR, UNIV CONN, 72- Mem: AAAS; Am Diabetes Asn; Am Asn Anatomists. Res: Mechanism of action of alloxan in the production of diabetes; mechanism of insulin secretion. Mailing Add: Dept of Anat Univ of Conn Health Ctr Farmington CT 06032

WATKINS, ELTON, JR, b Portland, Ore, Aug 16, 21. SURGERY, CANCER. Educ: Reed Col, BA, 41; Univ Ore, MD, 44. Prof Exp: Asst path, Med Sch, Univ Ore, 41-43, biochem, 43-44, resident thoracic surg, Hosps & Clins, 45-48, instr physiol, Med Sch, 49-50; Ore Heart Asn res fel, Children's Hosp Med Ctr, Boston, Mass, 51-53; from instr to asst prof surg, Harvard Med Sch, 54-57; chmn div res, 64-67, SR SURGEON & DIR SIAS SURG LAB, LAHEY CLIN FOUND, 57- Concurrent Pos: Surgeon, Children's Hosp Med Ctr, Boston, Mass, 56-57; asst surgeon, Peter Bent Brigham Hosp, Boston, Mass, 57; consult, Blood Preservation Lab, US Naval Hosp, Boston, 56-; USPHS fel organ transplantation, Sias Surg Lab, Lahey Clin Found, 59-65, USPHS fel cancer immunother, 72-75; lectr, Royal Col Surgeons Eng, 63. Mem: AAAS; Am Asn Cancer Res; Transplantation Soc; Am Fedn Clin Res; Int Soc Surgeons. Res: Cardiovascular physiology and surgery; pancreatic transplantation; regional cancer chemotherapy; experimental manipulation of host immune response to cancer. Mailing Add: Sias Res Lab Lahey Clin Found 605 Commonwealth Ave Boston MA 02215

WATKINS, GEORGE DANIELS, b Evanston, Ill, Apr 28, 24; m 49; c 3. PHYSICS. Educ: Randolph-Macon Col, BS, 43; Harvard Univ, AM, 47, PhD(physics), 52. Prof Exp: PHYSICIST, RES & DEVELOP CTR, GEN ELEC CO, 52- Concurrent Pos: Adj prof, Rensselaer Polytech Inst, 62-65; NSF fel, Oxford Univ, 66-67; adj prof, State Univ NY Albany, 70-71. Mem: Fel Am Phys Soc. Res: Nuclear and electron spin resonance studies; information theory; solid state physics; radiation effects and point defects in solids. Mailing Add: Gen Elec Res & Dev Ctr PO Box 8 Schenectady NY 12301

WATKINS, GUSTAV MCKEE, b Tehuacana, Tex, Oct 19, 08; m 36; c 1. BOTANY. Educ: Univ Tex, BA, 29, MA, 30; Columbia Univ, PhD(plant cytol), 35. Prof Exp: Asst bot, Univ Tex, 27-30; asst, Columbia Univ, 30-35; plant pathologist, Exp Sta, Agr & Mech Col, Tex, 35-41; assoc prof biol, Sam Houston State Teachers Col, 41-43; pathologist, Emergency Plant Dis Prev Prog, Bur Plant Indust, USDA, 43-44; res assoc prof plant path, Iowa State Col, 44-45; mycologist, US Naval Ord Lab, 46-49; prof plant path & head dept plant physiol & path, 50-58, dean agr, 58-60, prof plant sci, 60-74, EMER PROF PLANT PATH, TEX A&M UNIV, 74- Concurrent Pos: Nat Res Coun fel, 38-39; consult, Ford Found, 62-63; dir agr instr, Tex A&M Univ, 60-65, dir, Dominican Repub prog, 65-69. Honors & Awards: Vaughan Res Award, Am Soc Hort Sci, 60. Mem: AAAS; Bot Soc Am; Am Phytopath Soc; Torrey Bot Club; Am Soc Hort Sci. Res: Host-parasite relations in fungous diseases of plants; breeding and selection for resistance to plant disease. Mailing Add: 1022 Harrington E College Station TX 77840

WATKINS, HARRY MITCHELL SHERMAN, b San Francisco, Calif, July 16, 13; m 42; c 2. BACTERIOLOGY, VIROLOGY. Educ: Univ Southern Calif, AB, 37; Univ Calif, PhD(bact), 61. Prof Exp: Res bacteriologist, Univ Calif, Berkeley, 46-73, asst sci dir naval med res unit 1, 46-57, naval biol lab, 50-57; RES BACTERIOLOGIST, NAVAL BIOL LAB, NAVAL SUPPLY CTR, 73- Concurrent Pos: Consult, US Dept Defense. Mem: AAAS; Am Soc Microbiol; Tissue Cult Asn. Res: Plague; Shigellosis; Colorado tick fever; respiratory and arboviruses; experimental aerobiology; epidemiology of respiratory infections in closed environments; methods for virus isolation and identification in the field; microbiological pathogens of mosquitoes. Mailing Add: Naval Biol Lab Naval Supply Ctr Oakland CA 94625

WATKINS, IVAN WARREN, b Minneapolis, Kans, Jan 14, 34; m 55; c 4. PHYSICS. Educ: Univ Kans, BS, 55, MS, 57; Tex A&M Univ, PhD, 68. Prof Exp: Instr physics, Ft Hays Kans State Col, 58-62; from asst prof to prof physics, 63-74, PROF GEOG & EARTH SCI, ST CLOUD STATE COL, 74- Mem: Am Asn Physics Teachers. Res: Theoretical molecular spectroscopy; student attitudes about sciences and how the attitudes correlate with success in science courses. Mailing Add: St Cloud State Col St Cloud MN 56301

WATKINS, JACKIE LLOYD, b Melvin, Tex, Jan 16, 32; m 55; c 4. GEOLOGY. Educ: Southern Methodist Univ, BS, 52, MS, 54; Univ Mich, PhD(geol), 58. Prof Exp: Assoc prof, 58-72, PROF GEOL, MIDWEST UNIV, 72- Concurrent Pos: Chmn bd, Watkins Mineral Corp. Mem: AAAS; Paleont Soc; Soc Econ Paleont & Mineral; Geol Soc Am; Nat Asn Geol Teachers. Res: Invertebrate paleontology; tetracorals and tabulate corals. Mailing Add: Dept of Geol Sci Midwestern Univ Wichita Falls TX 76308

WATKINS, JOEL SMITH, JR, b Poteau, Okla, May 27, 32; m 56; c 2. GEOPHYSICS, SEISMOLOGY. Educ: Univ NC, AB, 53; Univ Tex, PhD(geol), 61. Prof Exp: Geophysicist, Regional Geophys Br, US Geol Surv, 61-64, Astrogeol Br, 64-66; res assoc geophys, Mass Inst Technol, 66-67; from assoc prof to prof, Univ NC, Chapel Hill, 67-73; PROF GEOPHYS, UNIV TEX MARINE SCI INST, 73- Concurrent Pos: Co-investr, Apollo Active Seismic Exp, 65-72, Apollo Lunar Seismic Profiling Exp, 71-; mem adv comt, Int Phase Ocean Drilling, 74- Honors & Awards: Spec Commendation, Geol Soc Am, 73; Except Sci Achievement Medal, NASA, 73. Mem: Am Geophys Union; Soc Explor Geophysicists; Geol Soc Am. Res: Marine geophysics; explosion seismology. Mailing Add: Univ of Tex Marine Sci Inst Galveston TX 77550

WATKINS, JOHN MAULDIN, JR, physics, see 12th edition

WATKINS, JULIAN F, II, b Marvell, Ark, Mar 18, 36; m 54; c 2. ENTOMOLOGY. Educ: Univ Ark, BSA, 56; Kans State Univ, MS, 62, PhD(entom), 64. Prof Exp: Asst county agent, Ark Agr Exten Serv, 56-60; asst prof, 64-69, ASSOC PROF BIOL, BAYLOR UNIV, 69- Mem: Entom Soc Am. Res: Taxonomy and behavior of army ants. Mailing Add: Dept of Biol Baylor Univ Waco TX 76703

WATKINS, KAY ORVILLE, b Nunn, Colo, Apr 28, 32; m 61; c 2. INORGANIC CHEMISTRY, PHYSICAL CHEMISTRY. Educ: Adams State Col, BA, 55; Vanderbilt Univ, PhD(inorg chem), 61. Prof Exp: From asst prof to assoc prof, 61-70, PROF CHEM, ADAMS STATE COL, 70- Concurrent Pos: NSF acad year exten grant, 63-65 & res grant, 65-68; NSF high sch lectr, 66-67. Mem: Am Chem Soc. Res: Equilibria and kinetic studies of inorganic systems in solution. Mailing Add: Dept of Chem Adams State Col Alamosa CO 81101

WATKINS, KENNETH WALTER, b Philadelphia, Pa, Apr 22, 39; m 62. PHYSICAL CHEMISTRY. Educ: Kans State Univ, BS, 61, PhD(chem), 65. Prof Exp: Fel chem, Univ Wash, 65-66; ASST PROF CHEM, COLO STATE UNIV, 66- Mem: Am Chem Soc; AAAS. Res: Chemical kinetics; chemical education. Mailing Add: Dept of Chem Colo State Univ Ft Collins CO 80523

WATKINS, MARK E, b New York, NY, Apr 13, 37; m 61; c 3. MATHEMATICS. Educ: Amherst Col, AB, 59; Yale Univ, MA, 61, PhD(math), 64. Prof Exp: Instr math, Univ NC, Chapel Hill, 63-64; asst prof, 64-68, ASSOC PROF MATH, SYRACUSE UNIV, 68- Concurrent Pos: Vis assoc prof, Univ Waterloo, 67-68; vis prof, Vienna Tech Univ, 73-74. Mem: Am Math Soc. Res: Problems related to vertex-connectivity in graphs; imbedding of graphs; automorphism groups of graphs. Mailing Add: Dept of Math Syracuse Univ Syracuse NY 13210

WATKINS, MARK HANNA, b Huntsville, Tex, Nov 23, 03; m. ANTHROPOLOGY. Educ: Prairie View Col, SB, 26; Univ Chicago, AM, 30, PhD(anthrop), 33. Prof Exp: Asst prof sociol, Louisville Munic Col, Louisville, 33-34; prof, Fisk Univ, 34-47; prof, 47-74, EMER PROF ANTHROP, HOWARD UNIV, 74- Mem: AAAS; fel Am Anthrop Asn; Am Sociol Asn; Ling Soc Am; Int African Inst. Res: Bantu languages and Yoruba. Mailing Add: Dept of Anthrop Howard Univ Washington DC 20001

WATKINS, MARY LOUISE, b El Paso, Tex, July 5, 35; m 52; c 3. BIOCHEMICAL PHARMACOLOGY. Educ: Carson-Newman Col, BS, 59; Vanderbilt Univ, PhD(pharmacol), 66. Prof Exp: Instr pharmacol, 66-71, instr pediat, 67-71, ASST PROF PEDIAT & PHARMACOL, MEHARRY MED COL, 71-, DIR BIOCHEM SECT, DIAG & TRAINING LAB MENT RETARDATION, 67- Res: Hormonal control of lipid metabolism; endocrinology; biochemical genetics; biochemistry of mental retardation. Mailing Add: Diag & Training Lab Dept of Pediat Meharry Med Col Nashville TN 37208

WATKINS, NORMAN DAVID, b Sunbury-on-Thames, Eng, Feb 15, 34; m 59; c 2. GEOPHYSICS. Educ: Univ London, BSc, 56; Univ Birmingham, MSc, 58; Univ Alta, MSc, 61; Univ London, PhD(geophys), 64. Prof Exp: Geophysicist, Shell Oil Co Can, Ltd, 59-61; res asst geophys, Stanford Univ, 62-63, actg asst prof, 63-64, res assoc, 64; Leverhulme vis fel, Chadwick Lab, Univ Liverpool, 64-65; assoc prof geol, Fla State Univ, 65-70; PROF OCEANOG, UNIV RI, 70- Concurrent Pos: NSF res grants, 63-72. Mem: AAAS; Am Geophys Union; Am Chem Soc; Geol Soc Am; Soc Explor Geophys. Res: Paleomagnetism; iron-titanium oxide mineralogy; vulcanology; marine geology. Mailing Add: Grad Sch of Oceanog Univ of RI Kingston RI 02881

WATKINS, PAUL DONALD, b Carterville, Ill, Oct 25, 40. MICROBIOLOGY, BIOCHEMISTRY. Educ: Southern Ill Univ, Carbondale, BA, 63, PhD(microbiol), 70. Prof Exp: Sr microbiologist, Schering Corp, 69-72; SR MICROBIOLOGIST, HOFFMANN-LA ROCHE INC, 72- Mem: AAAS; Am Soc Microbiol; Genetics Soc Am; Am Genetic Asn. Res: Antibiotics; genetics; regulatory mechanisms. Mailing Add: Hoffmann-La Roche Inc Nutley NJ 07110

WATKINS, ROBERT ARNOLD, b Boston, Mass, Aug 3, 26; m 48; c 2. ELECTROOPTICS, MILITARY SYSTEMS. Educ: Brown Univ, ScB, 47; Ohio State Univ, MS, 48, PhD(physics), 53. Prof Exp: Instr physics, Univ WVa, 48-49; engr, Zenith Radio Corp, 52-54; sr engr, 54-57, sect mgr, 57-60, dept mgr, 60-63, PRIN ENGR, RAYTHEON CO, GOLETA, 63- Mem: Inst Elec & Electronics Engrs; Optical Soc Am. Res: Optical systems design with associated electronic and mechanical configurations; laser effects on materials; combined radio frequency and optical systems. Mailing Add: 4575 Camino Molinero Santa Barbara CA 93110

WATKINS, SALLIE ANN, b Jacksonville, Fla, June 27, 22. PHYSICS. Educ: Notre Dame Col, Ohio, BS, 45; Catholic Univ, MS, 54, PhD(physics), 58. Prof Exp: Instr chem & physics, Notre Dame Acad, Ohio, 45-49, Elyria Dist Cath High Sch, Ohio, 49-50; instr physics, Notre Dame Col, Ohio, 50-53, prof, 57-66; teaching asst, Catholic Univ, 55-56; PROF PHYSICS, UNIV SOUTHERN COLO, 66- Concurrent Pos: NSF res grant biophys, 60-63; Oak Ridge Asn Univs res partic fel, Savannah River Lab, 66-68. Mem: Am Phys Soc; Am Asn Physics Teachers. Res: Nuclear reactor physics; ultrasonics and biophysics. Mailing Add: Dept of Physics Univ of Southern Colo Pueblo CO 81001

WATKINS, SPENCER HUNT, b Mayfield, Ky, Sept 15, 24; m 46; c 3. ORGANIC CHEMISTRY. Educ: Univ Ill, BS, 47; Univ Wis, PhD(chem), 50. Prof Exp: Res chemist, 50-55, res supvr, 55-57, tech rep, 57-60, asst sales mgr, 60-63, mgr tech serv, 63-65, mgr mkt develop, 65-74, DIR DEVELOP, PINE & PAPER CHEM DEPT, HERCULES, INC, 74- Mem: Am Chem Soc; Tech Asn Pulp & Paper Indust; Com Develop Asn; Am Asn Textile Chem & Colorists. Res: Paper chemicals, rosin chemistry; urea-formaldehyde resins. Mailing Add: Pine & Paper Chem Dept Hercules Inc 910 Market St Wilmington DE 19899

WATKINS, STANLEY READ, b University Park, NMex, June 22, 29; m 52; c 2. ANALYTICAL CHEMISTRY, ENVIRONMENTAL CHEMISTRY. Educ: NMex State Univ, BS, 51; Univ Colo, PhD(anal chem), 58. Prof Exp: Chemist, E I du Pont de Nemours & Co, 51-54; PROF CHEM, COE COL, 58- Mem: Am Chem Soc. Res: Human metal binding proteins, nuclear magnetic resonance studies of metal-amino acid complexes, spectroscopic analysis of trace metals. Mailing Add: Dept of Chem Coe Col Cedar Rapids IA 52402

WATKINS, STEVEN F, b Amarillo, Tex, May 14, 40; m 66; c 2. STRUCTURAL CHEMISTRY, CRYSTALLOGRAPHY. Educ: Pomona Col, BA, 62; Univ Wis, PhD(phys chem), 67. Prof Exp: Res fel, Bristol Univ, 67-68; asst prof, 68-73, ASSOC PROF CHEM, LA STATE UNIV, BATON ROUGE, 73- Mem: AAAS; Am Chem Soc; Am Crystallog Asn; The Chem Soc. Res: Molecular structure of organometallic, inorganic and organic biological molecules in the solid state by means of x-ray and neutron crystallography. Mailing Add: Dept of Chem La State Univ Baton Rouge LA 70803

WATKINS, WILLIAM, b Los Angeles, Calif, July 7, 42. MATHEMATICS, STATISTICS. Educ: Univ Calif, Santa Barbara, BA, 64, MA, 68, PhD(math), 69. Prof Exp: ASSOC PROF MATH, CALIF STATE UNIV, NORTHRIDGE, 69- Mem: Am Math Soc; Math Asn Am. Res: Linear and multilinear algebra. Mailing Add: Dept of Math Calif State Univ Northridge CA 91324

WATLING, HAROLD, b Hayle, Eng, July 22, 17; nat US; m 41; c 2. ZOOLOGY. Educ: Univ Idaho, BS, 41, MS, 44; Ore State Univ, PhD, 62. Prof Exp: Instr army specialized training prog, Univ Idaho, 44; from instr to assoc prof comp anat, histol & embryol, 45-65, PROF ZOOL, MONT STATE UNIV, 65- Res: Vertebrate anatomy and embryology. Mailing Add: Dept of Zool & Entom Mont State Univ Bozeman MT 59715

WATLINGTON, CHARLES OSCAR, b Midlothian, Va, Apr 9, 32; m 55; c 2. MEDICINE, PHYSIOLOGY. Educ: Va Polytech Inst, BS, 54; Med Col Va, Va Commonwealth Univ, MD, 58, PhD(physiol), 68. Prof Exp: Intern med, Univ Calif, San Francisco, 58-59, asst resident, 60-62; from instr to asst prof, 62-69, ASSOC PROF MED, MED COL VA, VA COMMONWEALTH UNIV, 69- Concurrent Pos: Fel endocrinol, Univ Calif, San Francisco. Mem: Am Physiol Soc; Am Fedn Clin Res; Endocrine Soc; Am Diabetes Asn; Am Soc Nephrology. Res: Regulation of ion transport; sodium and calcium homeostasis; cellular mechanism of action of hormones. Mailing Add: Endocrine Div Dept of Med Med Col of Va Health Sci Div Richmond VA 23219

WATNE, ALVIN LLOYD, b Shabbona, Ill, Jan 13, 27; m 66; c 3. SURGERY. Educ: Univ Ill, BS, 50, MD, 52, MS, 56. Prof Exp: Intern, Indianapolis Gen Hosp, Ind, 52-53; res asst surg, Univ Ill, 53-54; resident, Res & Educ Hosp, Univ Ill, 54-58; assoc cancer res surgeon, Roswell Park Mem Inst, 58-59, assoc chief cancer res surgeon, 59-62; assoc prof, 62-67, PROF SURG, W VA UNIV, 67-, CANCER COORDR, 62- Mem: Am Asn Cancer Res; Am Col Surgeons; Soc Univ Surgeons; NY Acad Sci. Res: Cancer metastases, including dissemination of tumor cells via the blood and lymph and their lodgement and growth; etiology and prevention of polyposis and coli and colon cancer. Mailing Add: Dept of Surg WVa Univ Med Ctr Morgantown WV 26506

WATNICK, ARTHUR SAUL, b Brooklyn, NY, Feb 4, 30; m 57; c 2. ENDOCRINOLOGY. Educ: Brooklyn Col, BS, 53, MA, 56; NY Univ, PhD(physiol), 63. Prof Exp: Biochemist, Sloan-Kettering Inst, 53-54, Beth-El Hosp, Brooklyn, NY, 54-56; from asst chemist to assoc chemist, 56-62, from scientist to sr scientist, 62-70, ASSOC DIR, BIOL RES, SCHERING CORP, 70- Concurrent Pos: Lectr endocrinol, Fairleigh Dickenson Univ, 66- & Rutgers Univ, 71- Mem: AAAS; Soc Exp Biol & Med; Am Chem Soc; Endocrine Soc; NY Acad Sci; Reticuloendothelial Soc. Res: Reproductive physiology; immunology. Mailing Add: Schering Corp Bloomfield NJ 07003

WATRACH, ADOLF MICHAEL, b Poland, Jan 7, 18; nat US; m 55. VETERINARY PATHOLOGY. Educ: Royal (Dick) Vet Col, Scotland, MRCVS, 48; Glasgow Univ, PhD(path), 58. Prof Exp: Asst vet path, Vet Sch, Glasgow Univ, 49-51; from instr to assoc prof, 51-62, PROF VET MED, COL VET MED, UNIV ILL, URBANA, 62- Mem: AAAS; Electron Micros Soc Am; Am Col Vet Path; Vet Med Asn; Int Acad Path. Res: Ultrastructural pathology; virus-cell relationship; viral oncogenesis; structure and development of viruses; cellular immunity. Mailing Add: Col of Vet Med Univ of Ill Urbana IL 61801

WATREL, WARREN GEORGE, b Brooklyn, NY, Jan 5, 35; m 60; c 3. MICROBIOLOGY, BIOCHEMISTRY. Educ: Syracuse Univ, BS, 57, MS, 58. Prof Exp: Asst to dir of NSF, Syracuse Univ, 57, instr, 57-58; pharmaceut sales & mkt staff, Lederle Labs, Am Cyanamid Co, 60-62; biochem specialist, M&T Chem, Am Can Co, 62-64; sales mgr, Pharmacia Fine Chem, Inc, 64-65, dir mkt & gen mgr, 65-72; vpres, Vineland, Vista Labs, Inc, Ideal & Nickolson Inst, Damon Corp, 72-74; VPRES, PHARMACHEM CORP, 74- Concurrent Pos: Am Cyanamid Co study grants, 57-58. Honors & Awards: Mkt Award, Am Chem Soc. Mem: AAAS; Am Chem Soc; Am Mkt Asn; Am Soc Microbiol; NY Acad Sci. Res: Application research and development of products biologically active for use in commercial products; research and development of products to be used in gel filtration chromatography; development, manufacture and sale of veterinarian, pharmacy and industrial chemicals and medical electronic instruments. Mailing Add: 1174 Sherlin Dr Bridgewater NJ 08807

WATROUS, BLANCHE GREENE, b Cleveland, Tenn, Nov 21, 09; m 32; c 4. ANTHROPOLOGY. Educ: Northwestern Univ, BA, 30, PhD(anthrop), 49. Prof Exp: Exten clin psychol, NShore Health Resort, 49-50; clin psychologist, Herrick House, 51; clin psychol consult, NShore Ment Health Asn, 52-54; psychol consult, Highland Park Family Serv, 53-54; consult clin psychologist proj tests, 54-57; staff psychologist, Lake County Ment Health Clin, Waukegan, 57-64; from assoc prof to prof, 64-74, EMER PROF ANTHROP, E CAROLINA UNIV, 74- Concurrent Pos: Lectr, Northwestern Univ, 61-; NDEA grants, Prog African Studies, 68-70; Nat Endowment for Humanities grant to direct AfriInst for NC Pub Sch Teachers, E Carolina Univ, 72-73; exchange prof, Kansai Univ Foreign Studies, Osaka, Japan, 73-74; coordr, African Area Studies & chmn, African Studies Comt. Mem: Fel Am Anthrop Asn. Res: Personality and culture; psychometric testing; study of child in nonliterate societies; Hawaiian Chinese acculturation research. Mailing Add: Dept of Anthrop E Carolina Univ Greenville NC 27834

WATROUS, GEORGE H, JR, b Elmira, NY, Aug 8, 17. DAIRY HUSBANDRY. Educ: Pa State Univ, BS, 42, MS, 47, PhD(dairy sci), 51. Prof Exp: Lab technician, NJ Dairy Labs, 42-44; from instr to assoc prof dairy mfg, 44-64, PROF FOOD SCI & INDUST, PA STATE UNIV, UNIVERSITY PARK, 64- Mem: Am Dairy Sci Asn. Res: Quality control; factors affecting keeping quality of dairy products; cleaning and sanitizing of food equipment; laboratory procedure relating to dairy products. Mailing Add: 107 Borland Lab Pa State Univ University Park PA 16802

WATROUS, JAMES JOSEPH, b Cleveland, Ohio, July 20, 42; m 70; c 1. PHYSIOLOGY. Educ: Univ Dayton, BSEd, 65, MS, 69; Georgetown Univ, PhD(biol), 72. Prof Exp: Teacher sec sch, Ohio, 64-67; teaching asst biol, Univ Dayton, 67-69 & Georgetown Univ, 69-72; ASST PROF BIOL, ST JOSEPH'S COL, PA, 72- Mem: AAAS; Sigma Xi; Am Soc Zoologists. Res: Transport of materials across biological membranes; nerve muscle physiology. Mailing Add: Dept of Biol St Joseph's Col 54th & City Ave Philadelphia PA 19131

WATROUS, RALPH MELVIN, b Waynesville, Ohio, Sept 27, 16; m 46; c 3. CHEMISTRY, PHYSICS. Educ: Univ NMex, BS, 40. Prof Exp: Tech supvr, Inland Div, Gen Motors Corp, 40-46; from opers chemist to sr res chemist, 46-61, GROUP LEADER, MOUND LAB, MONSANTO RES CORP, 61- Mem: Am Chem Soc. Res: Radiochemistry; hot cell facilities. Mailing Add: 3835 Phillipsburg Rd Union OH 45322

WATSCHKE, THOMAS LEE, b Charles City, Iowa, Apr 12, 44; m 65. AGRONOMY, HORTICULTURE. Educ: Iowa State Univ, BS, 67; Va Polytech Inst & State Univ, MS, 69, PhD(agron), 71. Prof Exp: ASSOC PROF AGRON, PA STATE UNIV, UNIVERSITY PARK, 70- Mem: Am Soc Agron; Crop Sci Soc Am. Res: Turfgrass physiology and weed control. Mailing Add: Dept of Agron 20 Tyson Bldg Pa State Univ University Park PA 16802

WATSON, ALBERT THEODORE, chemistry, see 12th edition

WATSON, ANDREW JOHN, b Mich, Aug 1, 21; m 50. WEED SCIENCE. Educ: Mich State Univ, PhD(soil sci), 49. AGRONOMIST, DOW CHEM CO, 49- Mem: Weed Sci Soc Am. Res: Herbicide development. Mailing Add: Ag-Organics Dept Dow Chem Co PO Box 1706 Midland MI 48640

WATSON, ANDREW SAMUEL, b Highland Park, Mich, May 2, 20; m 67; c 2. PSYCHIATRY. Educ: Univ Mich, BS, 42; Temple Univ, MD, 50, MS, 51. Prof Exp: From instr to asst prof psychiat, Med Sch, Univ Pa, 45-59, assoc prof, Law Sch, 55-

59; from asst prof psychiat & asst prof law to assoc prof psychiat & assoc prof law, 59-66, PROF PSYCHIAT, MED SCH, UNIV MICH, ANN ARBOR, & PROF LAW, LAW SCH, 66- Concurrent Pos: Lectr social work, Sch Social Work, Bryn Mawr Col, 55-59; psychiat consult, Mich Dept Corrections, 59-; comnr, Mich Law Enforcement Criminal Justice Comn, 68-72; mem, Adv Comt Divorce, Nat Conf Comnr Uniform State Laws, 69-70; mem, Surgeon's Gen Sci Adv Comn TV & Social Behav, 69-72. Mem: Group Advan Psychiat. Res: Family treatment; professionalizing process of lawyers; applications of psychiatric concepts to law. Mailing Add: CPH Univ Hosp Ann Arbor MI 48104

WATSON, BARBARA KASCENKO, b Russia, Dec 6, 18; nat US; m 41; c 1. BACTERIOLOGY, MICROBIOLOGY. Educ: Cornell Univ, BA, 41; Yale Univ, PhD, 46. Prof Exp: Asst bact & immunol, 48-49, instr, 49-54, res assoc, 54-70, PRIN RES ASSOC MICROBIOL & MOLECULAR GENETICS, HARVARD MED SCH, 70- Concurrent Pos: Asst bacteriologist, Mass Gen Hosp, 61- Mem: AAAS; Am Asn Immunologists. Res: Virus-host cell interactions. Mailing Add: Microbiol & Molecular Genetics Harvard Med Sch Boston MA 02115

WATSON, BARRY, b Middlesbrough, Eng, Dec 2, 40. BIOPHYSICAL CHEMISTRY. Educ: Univ Bradford, BSc, 65, PhD(phys chem), 68. Prof Exp: Scientist, Unilever Res Ltd, 66-68; Off Saline Water fel, Mellon Inst Sci, 69-72; RES CHEMIST, OWENS-ILL INC, 72- Mem: The Chem Soc; Am Chem Soc. Res: Thermodynamics of aqueous solutions; electrochemistry; biomedical research. Mailing Add: Owens-Ill Inc Tech Ctr 1700 N Westwood Ave Toledo OH 43607

WATSON, BENJAMIN FRANKLIN, b Zeigler, Ill, Apr 17, 30; m 62; c 3. MICROBIOLOGY. Educ: Ind Univ, Bloomington, BS, 57; Univ Minn, Minneapolis, MS, 65, PhD(microbiol), 67. Prof Exp: Fel inst microbiol, Copenhagen Univ, 67-69; ASST PROF MICROBIOL, TEX A&M UNIV, 69- Mem: AAAS; Am Chem Soc; Am Soc Microbiol; NY Acad Sci. Res: Microbial metabolism and enzymology; regulation of nucleic acid syntheses in bacteria; virology. Mailing Add: 1612 Leona St College Station TX 77840

WATSON, BERNARD BENNETT, b Philadelphia, Pa, May 17, 11; m 34; c 2. PHYSICS. Educ: Temple Univ, AB, 32; Calif Inst Technol, PhD(physics), 35. Prof Exp: From instr to assoc prof physics, Ariz State Univ, 35-41; from instr to asst prof, Univ Pa, 41-47; sr res fel, Calif Inst Technol, 47-48; assoc prof, Temple Univ, 48-49; specialist for physics, Div Higher Educ, US Off Educ, 49-52; prof & sci personnel specialist, Defense Manpower Admin, 52-53; opers analyst, Opers Res Off, Johns Hopkins Univ, 53-61 & Res Anal Corp, 61-72; OPERS ANALYST, GEN RES CORP, McLEAN, VA, 72- Concurrent Pos: Ed, Philadelphia Scientist, 46-47; mem comt specialized personnel, Off Defense Mobilization, 52-53; consult, Opers Res Off, Johns Hopkins Univ, 52-53; US-UK exchange analyst & supt weapons & weapons systs div, Army Oper Res Estab, Eng, 62-64. Mem: AAAS; Opers Res Soc Am; Am Phys Soc; Am Asn Physics Teachers. Res: X-ray and gamma ray spectroscopy; scientific manpower; operations research. Mailing Add: 6108 Landon Lane Bethesda MD 20034

WATSON, CECIL JAMES, b Minneapolis, Minn, May 31, 01; m 25. MEDICINE. Educ: Univ Minn, BS, 23, MB, 24, MS, 25, MD, 26, PhD(path), 28; Am Bd Internal Med, dipl, 37. Hon Degrees: MD, Univ Mainz, 63 & Univ Munich, 72. Prof Exp: Instr path, Univ Minn, 26-28; instr path & internal med, Northwest Clin, Minot, NDak, 28-30; res chemist, Univ Munich, 30-32, Nat Res Coun fel, 31-32; fel med & resident physician, Minneapolis Gen Hosp & Univ Minn Hosp, 32-33; from instr to prof med, 33-61, dir div internal med, 36-40, chmn dept med, 43-66, DISTINGUISHED SERV PROF MED, UNIV MINN, MINNEAPOLIS, 61-, REGENTS PROF, 68-, DIR UNIT FOR TEACHING & RES INTERNAL MED, NORTHWEST HOSP MINNEAPOLIS, 66- Concurrent Pos: Pathologist, Minneapolis Gen Hosp, 26-28; vis prof & assoc dir health div, Metall Labs, Univ Chicago, 43-46; mem, Life Ins Med Res Coun, 46-49; dir comn liver dis, Armed Forces Epidemiol Bd, 48-54; mem coun, Nat Inst Arthritis & Metab Dis, 51-54; scholar in residence, Fogarty Int Ctr, NIH, 72; hon prof, Lima & Chile; mem comt med & surg, Nat Res Coun; adv comt metab, Surgeon Gen, US Dept Army; mem, Am Bd Internal Med, 44-49. Honors & Awards: Wilson Medal, Am Clin & Climat Asn, 47; Phillips Medal, Am Col Physicians, 57; Modern Med Award, 59; Order of Merit, Chile, 60; Kober Medal, Asn Am Physicians, 72. Mem: Nat Acad Sci; Am Soc Clin Invest (pres, 46); Am Soc Biol Chemists; Asn Am Physicians (pres, 60-61); master Am Col Physicians. Res: Liver and biliary tract disease; jaundice and anemia; urobilin and porphyrin metabolism. Mailing Add: 3318 Edmund Blvd SE Minneapolis MN 55406

WATSON, CLIFFORD ANDREW, b Barr, Mont, Sept 5, 29; m 54; c 2. CEREAL CHEMISTRY. Educ: Mont State Univ, BS, 56; Kans State Univ, MS, 58, PhD(cereal chem), 64. Prof Exp: From instr to assoc prof cereal chem, Mont State Univ, 56-67; res chemist field crops & animal prod res br, Md, 67-68, invests leader mkt qual res div, Kans, 68-72, actg dir, & location leader & res leader, US Graing Mktg Res Ctr, Manhattan, Kans, 72-74, RES CHEMIST, AGR RES SERV, USDA, N DAK 74- Mem: Sigma Xi. Res: Agricultural biochemistry; biochemistry of wheat and barley. Mailing Add: Dept of Cereal Chem NDak State Univ Fargo ND 58102

WATSON, DAVID GOULDING, b Toronto, Ont, May 7, 29; nat US; m 61; c 5. PEDIATRICS. Educ: Univ Toronto, MD, 52. Prof Exp: From asst prof to assoc prof, 59-75, PROF PEDIAT, MED CTR, UNIV MISS, 75- Concurrent Pos: NIH spec fel, Univ Fla, 72-73. Mem: Am Acad Pediat; Am Col Cardiol; Royal Col Physicians Can. Res: Pediatric cardiology. Mailing Add: Dept of Pediat Univ of Miss Med Ctr Jackson MS 39216

WATSON, DAVID LIVINGSTON, b Burlington, Ont, Oct 22, 26; US citizen; m 51; c 3. ENTOMOLOGY. Educ: Univ Guelph, BSA, 51; Cornell Univ, PhD(entom), 56. Prof Exp: Res specialist, Chevron Chem Co, 55-59, coordr indust res, 59-61, res supvr, 61-66; assoc dir indust res, 66-70, DIR PROD DEVELOP DIV, VELSICOL CHEM CORP, CHICAGO, 70- Mem: AAAS; Entom Soc Am; Am Phytopath Soc; Weed Sci Soc Am. Res: Genetic relationship of phosphate resistance in mites; development of crop protection chemicals in agriculture and related fields. Mailing Add: Velsicol Chem Corp 341 E Ohio Chicago IL 60611

WATSON, DAVID WERNER, b Columbus, Ohio, June 14, 31; m 53; c 3. INTERNAL MEDICINE. Educ: Capital Univ, BS, 54; Ohio State Univ, MD, 59; Univ Mich, Ann Arbor, MS, 66. Prof Exp: Intern, Cincinnati Gen Hosp, Ohio, 59-60; resident med, Med Ctr, Univ Mich, Ann Arbor, 60-63, from instr to assoc prof, 66-70; ASSOC PROF MED, SCH MED, UNIV CALIF, DAVIS, 70- Concurrent Pos: NIH fel, Univ Mich, Ann Arbor, 63-65, NIH grant, 66-70; NIH grant, Univ Calif, Davis, 70-72; consult, Wayne County Gen Hosp, Eloise, Mich, 67-70, Vet Admin Hosp, Ann Arbor, 67-70 & San Joaquin Gen Hosp, Stockton, Calif, 70- Mem: Am Col Physicians; Am Fedn Clin Res; Tissue Cult Asn; Am Gastroenterol Asn. Res: Immunology of gastrointestinal disease. Mailing Add: Dept of Med Univ of Calif Sch of Med Davis CA 95616

WATSON, DENNIS RONALD, b Overton, Tex, Dec 7, 41; m 70; c 1. CHEMISTRY. Educ: Howard Payne Col, BA, 64; Univ Colo, Boulder, MS, 67, PhD(chem), 70. Prof Exp: Asst prof, 70-75, ASSOC PROF CHEM, LA COL, 75-, CHMN DEPT, 70- Res: Air and water pollution topics that can be performed by senior level students dealing with local problems and conditions. Mailing Add: Dept of Chem La Col Pineville LA 71360

WATSON, DENNIS WALLACE, b Morpeth, Ont, Apr 29, 14; nat US; m 41; c 2. MICROBIOLOGY. Educ: Univ Toronto, BSA, 34; Dalhousie Univ, MSc, 37; Univ Wis, PhD(bact), 41. Prof Exp: Asst, Biol Bd Can, NS, 35-37, sci asst, 37-38; asst bact, Univ Wis, 38-41, Alumni Res Found fel, 41-42, res assoc, 42; vis investr, Rockefeller Inst, 42; vis investr, Connaught Labs, Toronto, 42-44; asst prof bact, Univ Wis, 46-49; from assoc prof to prof bact, 49-64, PROF MICROBIOL & HEAD DEPT, MED SCH, UNIV MINN, MINNEAPOLIS, 64- Concurrent Pos: Med consult, Fed Security Agency, Washington, DC, 44; assoc mem comn immunization, Armed Forces Epidemiol Bd, 46-59; vis prof, Med Sch, Univ Wash, 50; mem allergy & immunol study sect, NIH, 54-58, mem bd sci counr, Div Biol Standards, 57-59, mem allergy & immunol training grant comt, Nat Inst Allergy & Infectious Dis, 58-60 & 64-66, chmn, 66, mem nat adv coun, 67-71; USPHS spec res fel, WGer, 60-61; mem ad hoc comt med microbiol, Div Med Sci, Nat Acad Sci. Mem: AAAS; Am Chem Soc; Am Soc Microbiol (vpres, 67-68, pres, 68-69); Soc Exp Biol & Med (pres, 75-76); Am Asn Immunologists. Res: Host-parasite relationships; chemistry and immunology of microbial toxins; pathogenesis of group A streptococci; mechanisms of nonspecific resistance to infection. Mailing Add: Med Sch Dept of Microbiol Univ of Minn Box 196 Mayo Minneapolis MN 55455

WATSON, DONALD GORDON, b Moscow, Idaho, Mar 12, 21; m 50; c 3. RADIATION ECOLOGY. Educ: Univ Wash, BS, 48. Prof Exp: Biologist, Wash State Dept Game, 49 & Gen Elec Co, 49-65; SR SCIENTIST, PAC NORTHWEST LAB, BATTELLE MEM INST, 65- Mem: Am Fisheries Soc; Am Soc Limnol & Oceanog; Am Inst Fishery Res Biol. Res: Effects of ionizing radiation on aquatic organisms; mineral cycling in aquatic systems; effects of thermal power stations on aquatic organisms. Mailing Add: Battelle Mem Inst Pac Northwest Lab PO Box 999 Richland WA 99352

WATSON, DONALD PICKETT, b Port Credit, Ont, May 20, 12; nat US; m 55. HORTICULTURE. Educ: Univ Toronto, BSA, 34; Univ London, MSc, 37; Cornell Univ, PhD(plant anat), 48. Prof Exp: Instr hort, State Univ NY Agr Inst, Long Island, 37-42; from asst prof to prof hort & head ornamental hort, Mich State Univ, 48-63; prof hort, 63-75, head dept, 63-66, urban horticulturist, 66-75, EMER PROF HORT, UNIV HAWAII, 75- Mem: Am Hort Soc. Res: Plant science and anatomy; horticultural television and therapy. Mailing Add: 5443 Drover Dr San Diego CA 92115

WATSON, DOUGLAS F, b Salem, Ore, May 19, 13; m 41; c 3. VETERINARY SCIENCE. Educ: Univ Pa, VMD, 37. Prof Exp: Assoc prof clin vet, 55-57, PROF VET SCI & HEAD DEPT, VA POLYTECH INST & STATE UNIV, 57- Mem: AAAS; Am Soc Animal Sci; Am Vet Med Asn. Res: Veterinary parasitology; clinical pathology. Mailing Add: Dept of Vet Med Va Polytech Inst & State Univ Blacksburg VA 24601

WATSON, DUANE CRAIG, b Enid, Okla, Dec 8, 30; m 51; c 3. ANALYTICAL CHEMISTRY. Educ: Eastern NMex Univ, BA, 51, BS, 56, MS, 57. Prof Exp: Res chemist, El Paso Natural Gas Prod, Tex, 57-63; prof res chemist, Philip Morris, Inc, 63-74, SR SCIENTIST, PHILIP MORRIS, USA, 74- Concurrent Pos: Lectr, Univ Tex, El Paso, 61-63. Res: Smoke chemistry; trace analyses; pollution; process instrumentation; gas chromatography. Mailing Add: Philip Morris USA Box 26583 Richmond VA 23261

WATSON, EDGAR, JR, physical chemistry, see 12th edition

WATSON, EDMOND EVELYN, b Montserrat, BWI, Dec 24, 02; m 31; c 2. PHYSICS. Educ: McGill Univ, BS, 25, MS, 26; Cambridge Univ, PhD(physics), 30. Prof Exp: From lectr to prof, 30-74, chmn undergrad studies, 30-71, EMER PROF PHYSICS, QUEEN'S UNIV, ONT, 74- Concurrent Pos: Hydrographer, Int Passamaquoddy Fisheries Comn, 31-33; res assoc, Oceanog Inst, Woods Hole, 33-60; technician, US Navy, 44-45. Mem: Can Asn Physicists (vpres, 52). Res: Electron scattering in helium; oceanography of Bay of Fundy; deep sea currents; anchoring in deep water; currents in coastal waters; electrical measurements. Mailing Add: 82 Traymoor Ave Kingston ON Can

WATSON, EDNA SUE, b Batesville, Miss, July 8, 45; m 65; c 3. IMMUNOLOGY. Educ: Univ Miss, BA, 67, MS, 70, PhD(biol), 75. Prof Exp: RES ASSOC, RES INST PHARMACEUT SCI, SCH PHARM, UNIV MISS, 74- Mem: AAAS; Can Soc Immunol; Brit Soc Immunol. Res: Poison ivy dermatitis, desensitization and immunoprophylaxis; cellular immunology, manipulation of the immune response by cyclic nucleotides. Mailing Add: Sch of Pharm Univ of Miss University MS 38677

WATSON, EDWARD HANN, geology, deceased

WATSON, FLETCHER GUARD, b Baltimore, Md, Apr 27, 12; m 35; c 4. SCIENCE EDUCATION. Educ: Pomona Col, AB, 33; Harvard Univ, MA, 35, PhD(astron), 38. Prof Exp: Instr & asst astron, Harvard Univ, 33-38, exec secy & res assoc, 38-41; instr, Radcliffe Col, 41; tech aide, Nat Defense Res Corp Radiation Lab, Mass Inst Technol, 42-43; from asst prof to assoc prof sci educ, 46-57, prof educ, 57-66, HENRY LEE SHATTUCK PROF EDUC, HARVARD UNIV, 66- Concurrent Pos: Ford Found fel, Europe, 64-65; co-dir, Harvard Proj Physics, 64- Honors & Awards: Distinguished Serv Citation, Nat Sci Teachers Asn, 72. Mem: AAAS; Nat Sci Teachers Asn; Am Asn Physics Teachers; Nat Asn Res Sci Teaching; Asn Educ Teachers in Sci (pres, 63-64). Res: Development and evaluation of new high school physics course; studies of development of science teachers and influence of various teacher-types on pupils. Mailing Add: 24 Hastings Rd Belmont MA 02178

WATSON, FRANK YANDLE, b Charlotte, NC, May 18, 25; m 49; c 4. PATHOLOGY. Educ: Univ Md, MD, 49; Am Bd Path, dipl, 58. Prof Exp: Res physician path, Charlotte Mem Hosp, NC, 52-56; from instr to asst prof path, Col Med, State Univ NY Downstate Med Ctr, 56-61; assoc pathologist, 61-72, DIR LABS, MOUNTAINSIDE HOSP, NJ, 72-; CLIN ASST PROF PATH, COL MED, STATE UNIV NY DOWNSTATE MED CTR, 61- Concurrent Pos: Surg pathologist, Inst Path, Kings County Med Ctr, NY, 56-61. Mem: Fel Col Am Pathologists; fel Am Soc Clin Pathologists; AMA; Asn Am Med Col. Res: General and surgical pathology. Mailing Add: Mountainside Hosp Montclair NJ 07042

WATSON, GEOFFREY STUART, b Bendigo, Australia, Dec 3, 21; m 53; c 4. MATHEMATICAL STATISTICS. Educ: Univ Melbourne, BA, 42; NC State Col, PhD, 52. Hon Degrees: DSc, Univ Melbourne, 67. Prof Exp: Res officer, Commonwealth Sci & Indust Res Orgn, Australia, 43; tutor math, Trinity Col, Univ Melbourne, 44-47; res officer appl econ, Cambridge Univ, 49-51; sr lectr statist, Univ

Melbourne, 51-54; sr fel, Australian Nat Univ, 55-58; res assoc math, Princeton Univ, 58-59; assoc prof, Univ Toronto, 59-62; prof statist & chmn dept, Johns Hopkins Univ, 62-68; on leave to inst genetics, Univ Pavia, 68-69; PROF STATIST & CHMN DEPT, PRINCETON UNIV, 70- Concurrent Pos: Fulbright travel grant, 58; Carnegie grant study microbiol, Carnegie Labs, NY, 58; consult, Ont Res Found, Can, 59-62 & Res Triangle Inst, NC, 60-62; Off Naval Res contract, 61-; NIH training grant, 64-70, spec fel, 68-69; NSF res grant, 66. Mem: Fel AAAS; fel Inst Math Statist; fel Am Statist Asn; fel Royal Statist Soc; fel Int Statist Inst. Res: Application of mathematics, especially probability theory, stochastic processes and statistics, to science; statistical methods in geophysics. Mailing Add: Dept of Statist Fine Hall Princeton Univ Princeton NJ 08540

WATSON, GEORGE E, III, b New York, NY, Aug 13, 31; m 66; c 2. ZOOLOGY, ORNITHOLOGY. Educ: Yale Univ, BA, 53, MS, 61, PhD(biol), 64. Prof Exp: From asst cur to cur ornith, 62-67, chmn dept vert zool, 67-72, CUR VERT ZOOL, NAT MUS NATURAL HIST, SMITHSONIAN INST, 72-; ASSOC PATHOBIOL, SCH PUB HEALTH & HYG, JOHNS HOPKINS UNIV, 70- Concurrent Pos: Mem, Seabird Comt, Int Ornith Cong, 66- & Comt Res & Explor, Nat Geog Soc, 75- Mem: Fel AAAS; fel Am Ornith Union (secy, 73-); Brit Ornith Union; cor mem Ger Ornith Soc. Res: Marine ornithology; birds of Palearctic and Antarctic. Mailing Add: Dept of Vert Zool Birds Nat Mus of Natural Hist Smithsonian Inst Washington DC 20560

WATSON, GORDON DULMAGE, b Rouleau, Sask, Feb 20, 17; m 43; c 5. ANTHROPOLOGY, ARCHAEOLOGY. Educ: Univ Sask, BA, 40. Prof Exp: Demonstr radar electronics, Univ Western Ont, 40-42; lab asst electronic instrumentation, Artil Proof Estab, Dept Nat Defence, 42-43, dep supt internal ballistics lab, 43-45; sci observer, Joint Army-Air Force Arctic Exercise Musk-Ox, 45-46; supt ballistics wing, Can Armament Res & Develop Estab, Defence Res Bd Can, 46-47, exchange sci officer, Electronics Subdiv, Guided Missile Div, US Air Force, Wright-Patterson AFB, Ohio, 48-49, supt ballistics wing, Can Armament Res & Develop Estab, Defence Res Bd Can, 49-52, guided missile proj officer, 50-54, dir weapons res, Hq, 54-59; sci adv to chief gen staff, Can Army, 59-61; defense res attache, Can Embassy, DC, 61-65; chief personnel, Defence Res Bd Can, 65-69, chief plans, 69-74; ARCHAEOL RESEARCHER, 75- Mem: Fel Can Aeronaut & Space Inst; Asn Sci, Eng & Technol Community Can; Soc Am Archaeol; Can Archaeol Asn. Res: Science policy; excavation and research in Canadian archaeology, particularly pertaining to Ontario Paleo, archaic and woodland periods in the Ottawa, Constance Bay and Rideau Lakes areas. Mailing Add: 2086 Fairbanks Ave Ottawa ON Can

WATSON, HAROLD JOHN, b San Francisco, Calif, July 7, 23; m 50; c 5. ORGANIC CHEMISTRY. Educ: Princeton Univ, AB, 44; Univ Ill, AM, 48, PhD(chem), 50. Prof Exp: Asst chem, Univ Ill, 46-47; chemist, Texaco, Inc, 50-57; res chemist, Dan River Mills Inc, 57-60, group leader, 60-64; asst prof, 64-70, ASSOC PROF CHEM, CALIF POLYTECH STATE UNIV, SAN LUIS OBISPO, 70- Concurrent Pos: Chem consult, 62- Mem: Am Chem Soc; fel Am Inst Chemists; AAAS. Res: Organic synthesis; diffusional processes; lubricant additives; textile chemicals; polymers. Mailing Add: Rte 2 Box 666-P Arroyo Grande CA 93420

WATSON, HUGH ALEXANDER, b Ottawa, Ont, Oct 8, 26; US citizen; m 58; c 2. PHYSICS. Educ: Univ Toronto, BS, 48; McGill Univ, MS, 49; Mass Inst Technol, PhD(physics), 52. Prof Exp: Mem staff div indust coop, Mass Inst Technol, 51-52; mem tech staff, 52-75, DEPT HEAD, MASK LAB & MICROGRAPHICS DEPT, BELL TEL LABS, 73- Mem: Sr mem, Inst Elec & Electronics Engrs; Soc Info Display. Res: Electron lithography and facsimile. Mailing Add: Bell Tel Labs Inc Murray Hill NJ 07974

WATSON, JACK ELLSWORTH, b Robertsdale, Pa, Apr 17, 38; m 62; c 3. GENETICS, HUMAN GENETICS. Educ: Shippensburg State Col, BS, 61; Purdue Univ, MS, 63, PhD(genetics), 67. Prof Exp: Asst prof, 65-72, ASSOC PROF ZOOL & ENTOM, AUBURN UNIV, 72- Concurrent Pos: Lectr, St Bernard Col, 70; consult, Pub Sch Sci Prog, Miss. Mem: Am Inst Biol Sci; Genetics Soc Am; Am Genetic Asn; Genetic Soc Japan. Res: Enzymology of sex limited lethals; quantitative genetics of human finger-prints. Mailing Add: Dept of Zool & Entom Auburn Univ Auburn AL 36830

WATSON, JACK THROCK, b Casey, Iowa, May 2, 39; m 66; c 1. ANALYTICAL CHEMISTRY, PHARMACOLOGY. Educ: Iowa State Univ, BS, 61; Mass Inst Technol, PhD(anal chem), 65. Prof Exp: Res chemist, US Air Force Sch Aerospace Med, 65-69; asst prof, 69-70, ASSOC PROF PHARMACOL, SCH MED, VANDERBILT UNIV, 70- Concurrent Pos: Vis asst prof, Tex A&M Univ, 68; Am Inst Chemists fel, Univ Strasbourg, 68-69. Res: Gas chromatography in separation and mass spectrometry in elucidation of structure of biologically significant molecules; prostaglandins; biogenic amines; selective detection of drugs with gas chromatography-mass spectrometry computer systems. Mailing Add: Dept of Pharmacol Vanderbilt Univ Sch of Med Nashville TN 37232

WATSON, JAMES ARTHUR, JR, b Parkersburg, WVa, May 4, 18; m 46; c 6. CHEMISTRY. Educ: Univ Tex, BA, 40; La State Univ, PhD(chem), 47. Prof Exp: Asst chem, Agr & Mech Col, Tex, 40-41; asst, La State Univ, 41-42, instr, 43-46; instr, Univ Ky, 47-48; asst prof, Tex Tech Col, 48-51; assoc prof, Northeastern La State Col, 51-53, head dept, 52-53; chief chemist, Hancock Plant, Columbian Carbon Co, 53-55; assoc prof, 55-65, PROF CHEM, SOUTHEASTERN LA UNIV, 65- Mem: Am Chem Soc. Res: Deionization of sea waters; determination of gossypol; toxicity of tung meal; syntheses of nitrogenhydroxy amino acids; enzyme kinetics; properties of carbon black; inorganic pigments; radiation chemistry. Mailing Add: Dept of Chem Southeastern La Univ Hammond LA 70401

WATSON, JAMES BENNETT, b Chicago, Ill, Aug 10, 18; m 43; c 2. ANTHROPOLOGY. Educ: Univ Chicago, AB, 41, AM, 45, PhD(anthrop), 48. Prof Exp: Asst prof anthrop, Univ Sao Paulo, 44-45; asst prof, Beloit Col, 45-46; assoc prof, Univ Okla, 46-47; assoc prof, Wash Univ, 47-55; exec officer dept, 55-60, PROF ANTHROP, UNIV WASH, 55- Concurrent Pos: Mem, Intensive Inst Span & Portuguese, Am Coun Learned Socs, 42; Ford Found fel, New Guinea, 53-55; NSF res grant, New Guinea Highlands, 60- Mem: AAAS; fel Am Anthrop Asn; Am Ethnol Soc; Soc Appl Anthrop; fel Royal Anthrop Inst Gt Brit & Ireland. Res: Social and cultural anthropology; ethnology of Oceania, especially Melanesia; Latin American cultures, especially Brazilian; acculturation. Mailing Add: Dept of Anthrop Univ of Wash Seattle WA 98105

WATSON, JAMES C, nuclear physics, deceased

WATSON, JAMES DEWEY, b Chicago, Ill, Apr 6, 28. MOLECULAR BIOLOGY. Educ: Univ Chicago, BS, 47; Ind Univ, PhD, 50. Hon Degrees: DSc, Univ Chicago, 61, Ind Univ, 63 & Brandeis Univ, 63; LLD, Univ Notre Dame, 65. Prof Exp: Nat Res Coun fel, Copenhagen Univ, 50-51 & Cambridge Univ, 51-52; Nat Found Infantile Paralysis fel, 52-53; sr res fel biol, Calif Inst Technol, 53-55; from asst prof to assoc prof, 56-61, PROF BIOL, HARVARD UNIV, 61-; DIR, COLD SPRING HARBOR LAB, 68- Concurrent Pos: Mem, Nat Cancer Bd, 72-74. Honors & Awards: Co-recipient, Nobel Prize in Med, 62; John Collins Warren Prize, Mass Gen Hosp, 59; Eli Lilly Biochem Award, 59; Lasker Prize, Am Pub Health Asn, 60; co-recipient, Res Corp Prize, 62; John J Carty Medal, Nat Acad Sci, 72. Mem: Nat Acad Sci; Am Acad Arts & Sci; Royal Danish Acad. Res: Bacteriophage reproduction; structure of the nucleic acids; protein synthesis; induction of cancer by viruses. Mailing Add: Cold Spring Harbor Lab Cold Spring Harbor NY 11724

WATSON, JAMES RAY, JR, b Anniston, Ala, Dec 6, 35; m 60; c 3. PLANT TAXONOMY, PALEOBOTANY. Educ: Auburn Univ, BS, 57, MS, 60; Iowa State Univ, PhD(bot), 63. Prof Exp: Res asst agron, Auburn Univ, 58-60; asst prof bot, 63-68, ASSOC PROF BOT, MISS STATE UNIV, 68-, HEAD DEPT, 75- Concurrent Pos: NSF fel, 64-65. Mem: Bot Soc Am; Int Orgn Biosyst; Am Soc Plant Taxon; Int Asn Plant Taxon. Res: Tribal and subfamilial characteristics of the Gramineae; vascular flora of Mississippi; carboniferous seeds. Mailing Add: Dept of Bot Miss State Univ PO Drawer BD Mississippi State MS 39762

WATSON, JAMES WREFORD, b Sanyuan, Shensi, China, Feb 8, 15; Can citizen; m 39; c 2. CULTURAL GEOGRAPHY. Educ: Univ Edinburgh, MA, 36; Univ Toronto, PhD(geog), 45. Prof Exp: Prof geog, McMaster Univ, 45-49; chief geogr, Can Govt, 49-54; PROF GEOG, UNIV EDINBURGH, 54- Concurrent Pos: Mem, Brit Nat Comt Geog, 64- & NAm Studies Prog, Univ Edinburgh, 67-; chmn goeg sect, Soc Sci Res Coun UK, 67-; mem, Nature Conservancy, UK, 68- Honors & Awards: Award of Merit, Asn Am Geog, 52; Can Gov-General's Award Can Lit, 54; Murchison Medal, Royal Geog Soc, 56; Res Medal, Royal Scottish Geog Soc, 60. Mem: Asn Am Geog; Am Geog Soc; Royal Scottish Geog Soc (vpres, 64-); fel Royal Soc Can; fel Royal Soc Edinburgh. Res: Cultural geography of the American City; regionalism in America; development of the Canadian North. Mailing Add: Dept of Geog Univ of Edinburgh Edinburgh UK

WATSON, JEFFREY, b Butterknowle, Eng, Oct 4, 40; m 65. RESOURCE MANAGEMENT, INFORMATION SCIENCE. Educ: Univ Durham, BSc, 62, dipl educ, 63, PhD(zool), 66. Prof Exp: Scientist, 66-71 DEP ED, OFF OF THE ED, FISHERIES RES BD CAN, 71- Mem: Am Fisheries Soc; Can Soc Zool; Coun Biol Ed. Mailing Add: Off of the Ed Dept of the Environ Ottawa ON Can

WATSON, JERRY MIKE, b Independence, Kans, June 26, 42; m 62. EXPERIMENTAL HIGH ENERGY PHYSICS. Educ: Univ Chicago, BS, 64, MS, 65, PhD(physics), 71. Prof Exp: ASST PHYSICIST, ARGONNE NAT LAB, 70- Mem: Am Phys Soc; AAAS. Res: The development of streamer chamber technology and the use of large chambers in high energy physics experiments, including the study of hyperon beta decay and meson spectroscopy. Mailing Add: Accelerator Res Facil Div Argonne Nat Lab Argonne IL 60439

WATSON, JOHN ALFRED b Chicago, Ill, May 21, 40; m 60; c 4. BIOCHEMISTRY. Educ: Ill Inst Technol, BS, 64; Univ Ill, Chicago, USPHS fel & PhD(biochem), 67. Prof Exp: USPHS fel biochem, Brandeis Univ, 67-69; ASST PROF BIOCHEM, UNIV CALIF, SAN FRANCISCO, 69-, ASSOC DEAN ADMIS MED, 73- Concurrent Pos: From asst dean, to assoc dean student affairs, 69-73. Mem: Am Oil Chem Soc; Sigma Xi; Tissue Cult Asn; Nat Inst Sci. Res: Regulation of lipid metabolism & cholesterol synthesis in cultured cells. Mailing Add: Dept of Biochem and Biophys Univ Calif San Francisco CA 94143

WATSON, JOHN ERNEST, b Kansas City, Mo, May 31, 25; m 52; c 2. AUDIOLOGY. Educ: Univ Denver, BA, 52, MA, 57, PhD(audiol), 60. Prof Exp: Fel med audiol, Univ Iowa, 60-64; chief audiol serv, Vet Admin Ctr, San Juan, PR, 64-66; dir audiol sect, Speech & Hearing Ctr, Memphis, Tenn, 66-67; CHIEF AUDIOL SERV, VET ADMIN HOSP, 67- Mem: Acoust Soc Am; Am Speech & Hearing Asn; Am Psychol Asn. Res: Auditory physiology; clinical testing of pathological conditions of the vestibular and auditory systems. Mailing Add: Vet Admin Hosp 3801 Miranda Blvd Palo Alto CA 94304

WATSON, JOHN H L, b St Catharines, Ont, May 27, 16; m 42; c 3. PHYSICS, BIOPHYSICS. Educ: McMaster Univ, BA, 39; Univ Toronto, MA, 40, PhD(electron optics & micros), 43. Prof Exp: Nat Res Coun Can fel, Univ Toronto, 43-44, asst, 43-45; res physicist, Shawinigan Chem Ltd, Que, 45-47; CHMN DEPT PHYSICS & BIOPHYS, EDSEL B FORD INST MED RES, HENRY FORD HOSP, 47- Concurrent Pos: Lectr, Univ BC, 43-45. Mem: Electron Micros Soc Am (pres, 57); Am Soc Cell Biol; Am Chem Soc. Res: Electron transmission and scanning; biologic and non-biologic electron microscopy; x-ray and electron diffraction; radiation polymerization; optical problems; ultrastructure of pathological and normal human and animal tissues. Mailing Add: Dept of Physics & Biophys Edsel B Ford Inst Med Res Detroit MI 48202

WATSON, JOHN THOMAS, b Indianapolis, Ind, Jan 9, 40; m 64; c 2. CARDIOVASCULAR PHYSIOLOGY, BIOMEDICAL ENGINEERING. Educ: Univ Cincinnati, BSME, 62; Southern Methodist Univ, MSME, 66; Univ Tex Southwestern Med Sch, PhD(physiol), 72. Prof Exp: Student engr, Indianapolis Power & Light Co, 57-59; design consult, Nat Cash Register Co, 59-62; systs engr, Ling-Temco Vought, Inc, 62-66; teaching asst physiol, 66-69, adj instr, 69-71, instr thoracic & cardiovasc surg & physiol, 71-74, ASST PROF SURG & PHYSIOL, UNIV TEX HEALTH SCI CTR DALLAS, 74-, ASST PROF, GRAD SCH BIOMED SCI, 73- Concurrent Pos: Nat Heart & Lung Inst grant, Univ Tex Southwestern Med Sch, 73-75. Mem: Am Soc Artificial Internal Organs; Am Soc Mech Engrs; Am Fedn Clin Res. Res: Circulatory assistance, ischemic heart disease; respiratory insufficiency; hypotension. Mailing Add: Dept of Surg Univ of Tex Health Sci Ctr Dallas TX 75235

WATSON, JOSEPH ALEXANDER, b Pittsburgh, Pa, July 23, 26; m 50; c 2. RADIOBIOLOGY, MICROBIOLOGY. Educ: Univ Pittsburgh, BS, 50, MS, 52, PhD(microbiol), 62. Prof Exp: Res asst radiation health, Univ Pittsburgh, 51-56; res biochemist, Radioisotope Serv, Vet Admin Hosp, Pittsburgh, Pa, 56-57; res assoc radiation health, 47-59, asst prof, 62-67, ASSOC PROF RADIOBIOL, GRAD SCH PUB HEALTH, UNIV PITTSBURGH, 67-, ASSOC PROF RADIOL, SCH MED, 72- Mem: Radiation Res Soc; Am Soc Microbiol; Am Pub Health Asn. Res: Pulmonary clearance of radioactive dusts; radiation effects on the lungs; metabolic control mechanisms in the cell. Mailing Add: Dept of Radiation Health Univ of Pittsburgh Grad Sch of Pub Health 130 De Soto St Pittsburgh PA 15213

WATSON, KENNETH, b Montreal, Que, July 25, 35; m 59; c 2. GEOPHYSICS. Educ: Univ Toronto, BA, 57; Calif Inst Technol, MS, 59, PhD(geophys), 64. Prof Exp: Geophysicist, 63-76, CHIEF, BR PETROPHYS & REMOTE SENSING, US GEOL SURV, 76- Concurrent Pos: Lectr, Northern Ariz Univ, 66-67. Mem: AAAS; Am Soc Photogram; Am Geophys Union; Optical Soc Am; Sigma Xi. Res: Planetary science; behavior of volatiles on the lunar surface; infrared emission and visible light reflection; remote sensing investigations; thermal modeling. Mailing Add: US Geol Surv MS 964 Denver Fed Ctr PO Box 25046 Denver CO 80225

WATSON, KENNETH DE PENCIER, b Vancouver, BC, July 19, 15; m 41; c 3. GEOLOGY. Educ: Univ BC, BASc, 37; Princeton Univ, PhD(geol), 40. Prof Exp:

Instr econ geol, Princeton Univ, 40-43; assoc mining engr, Dept Mines, BC, 43-46; from assoc prof to prof geol & geog, Univ BC, 46-50; prof geol, Univ Calif, Los Angeles, 50-57; chief geologist, Dome Explor Ltd, Can, 57-58; PROF GEOL, UNIV CALIF, LOS ANGELES, 58- Concurrent Pos: Dir, Sigma Mines Ltd, 66- Mem: Fel Geol Soc Am; fel Mineral Soc Am; Soc Econ Geol; Geochem Soc; Mineral Asn Can. Res: Petrology and mineral deposits. Mailing Add: Dept of Geol Univ of Calif Los Angeles CA 90024

WATSON, KENNETH MARSHALL, b Des Moines, Iowa, Sept 7, 21; m 46; c 2. PHYSICS. Educ: Iowa State Col, BS, 43; Univ Iowa, PhD(physics), 48. Prof Exp: Lab instr, Iowa State Col, 42-43; radio engr, US Naval Res Lab, Washington, DC, 43-46; instr physics, Univ Iowa, 48 & Princeton Univ, 48; AEC fel, Inst Advan Study & Radiation Lab, Univ Calif, 48-50; asst prof physics, Ind Univ, 51-53; assoc prof, Univ Wis, 53-59; PROF PHYSICS, UNIV CALIF, BERKELEY, 59- Concurrent Pos: Staff mem, Lawrence Berkeley Lab; consult, Phys Dynamics, Inc, Stanford Res Inst & Tex Instruments, Inc. Mem: Am Phys Soc; Am Geophys Union; Nat Acad Sci. Res: Scattering theory; statistical mechanics; atomic and plasma physics. Mailing Add: Dept of Physics Univ of Calif Berkeley CA 94720

WATSON, LAWRENCE CRAIG, b La Crosse, Wis, Aug 3, 36; m 68. ETHNOLOGY, CULTURAL ANTHROPOLOGY. Educ: Univ Calif, Los Angeles, BA, 58, PhD(anthrop), 67; Univ Southern Calif, MA, 60. Prof Exp: Res asst anthrop, Latin Am Ctr, Univ Calif, Los Angeles, 64 & 65-67; PROF ANTHROP, SAN DIEGO STATE UNIV, 67- Concurrent Pos: Ford Found grant, Latin Am Ctr, Univ Calif, Los Angeles, 64-65; AID grant, Latin Am Ctr, Univ Calif, Los Angeles & Venezuela, 72; fel, Latin Am Ctr, Univ Calif, Los Angeles, 72-73. Mem: Am Anthrop Asn. Res: Psychology and culture change; urbanization; ethnology, all with special reference to South American Indian cultures. Mailing Add: Dept of Anthrop San Diego State Univ San Diego CA 92115

WATSON, LINVILL, b Philadelphia, Pa, Dec 16, 18; m 49; c 1. CULTURAL ANTHROPOLOGY. Educ: Univ Pa, AB, 39, PhD(anthrop), 53. Prof Exp: Instr sociol, Pa State Col, 43-45; instr anthrop, Univ Minn, 46-47; instr sociol & anthrop, Boston Univ, 48-51; asst prof, Col William & Mary, 53-54; asst prof sociol, Overseas Prog, Univ Md, 54-59; res assoc anthrop, Foreign Area Studies Div, Am Univ, 60-61; asst prof, Carnegie Inst Technol, 61-63; assoc prof, Lafayette Col, 63-66; assoc prof sociol, 66-69, PROF ANTHROP & SOCIOL, UNIV SASK, 69- Mem: Fel Am Anthrop Asn; fel African Studies Asn; fel Soc Appl Anthrop; Can Asn African Studies. Res: Ethnology of Africa and North America; comparative study of board games; social structure; ethic relations; culture change. Mailing Add: Dept of Anthrop & Archaeol Univ of Sask Saskatoon SK Can

WATSON, LLOYD SHERMAN, b Bryan, Tex, Jan 24, 22; m 51; c 2. PHYSIOLOGY. Educ: Alfred Univ, BA, 43; Univ Wis, MS, 51, PhD(physiol), 54. Prof Exp: Res assoc, 54-65, SR RES FEL RENAL PHYSIOL, MERCK INST THERAPEUT RES, WEST POINT, 65- Mem: AAAS; NY Acad Sci. Res: Renal pharmacology; cardiovascular physiology; hypertension; electrolyte homeostatic mechanisms. Mailing Add: 301 Lower Valley Rd North Wales PA 19454

WATSON, MARGARET LIEBE, b Cleveland, Ohio, Mar 25, 12; m 38; c 3. GENETICS. Educ: Col Wooster, AB, 33; Univ Mich, MA, 35, PhD, 37. Prof Exp: PROF BIOL, SIMPSON COL, 37-41 & 55- Concurrent Pos: Comnr, N Cent Asn Cols & Sec Schs, 74-78. Mem: AAAS. Res: Genetics and embryology of the mouse. Mailing Add: Dept of Biol Simpson Col Indianola IA 50125

WATSON, MARSHALL TREDWAY, b Blacksburg, Va, Dec 27; 22; m 52; c 2. PHYSICAL CHEMISTRY, TEXTILE CHEMISTRY. Educ: Va Polytech Inst, BS, 43; Princeton Univ, MA, 48, PhD(phys chem), 49. Prof Exp: Asst, Princeton Univ, 46-48; from res chemist to sr res chemist, 49-63, res assoc, 63-68, HEAD FIVERS RES DIV, TENN EASTMAN CO, 68- Mem: Am Chem Soc; Am Asn Textile Technol. Res: Mechanism of protein denaturation; mechanical and rheological properites of polymers; processing of polymers into fibers; structure and properties of fibers. Mailing Add: Res Labs Tenn Eastman Co Kingsport TN 37662

WATSON, MARTHA F, b Janesville, Wis, Feb 2, 35. MATHEMATICS. Educ: Murray State Col, AB, 56; Univ Ky, MA, 58, PhD(math), 62. Prof Exp: Assoc prof, 62-74, PROF MATH, WESTERN KY UNIV, 74- Mem: Am Math Soc; Math Asn Am. Res: Complex analysis. Mailing Add: Dept of Math Western Ky Univ Bowling Green KY 42101

WATSON, MASON B, physics, see 12th edition

WATSON, MAXINE AMANDA, b New Rochelle, NY, May 8, 47. POPULATION BIOLOGY, MOLECULAR GENETICS. Educ: Cornell Univ, BS, 68; Yale Univ, MPh, 70, PhD(biol), 74. Prof Exp: Res assoc biol, Washington Univ, 74-75; ASST PROF BIOL, UNIV UTAH, 75- Mem: Am Soc Plant Physiologists; Am Bryol Soc; Int Asn Bryologists; Brit Ecol Soc. Res: Adaptive significance of molecular variation in plant populations, particularly mosses; mechanisms of molecular adaptation; genetic structure of natural plant populations and its evolutionary significance. Mailing Add: Dept of Biol Univ of Utah Salt Lake City UT 84112

WATSON, MICHAEL DOUGLAS, b St Thomas, Ont, July 27, 36; m 59; c 2. AERONOMY. Educ: Univ Western Ont, BSc, 57, MS, 59, PhD(physics), 61. Prof Exp: Asst res off, 61-67, ASSOC RES OFFICER, ASTROPHYSICS BR, UPPER ATMOSPHERE RES, NAT RES COUN, CAN, 67- Res: Shock-tube excitation of powdered solids; plasma jet diagnostics; optical studies of aurora; observational studies of infrasonic waves from meteors. Mailing Add: Herzberg Inst of Astrophys Nat Res Coun Ottawa ON Can

WATSON, RALPH A, b Milwaukee, Wis, Aug 16, 21; m 44. EXPLORATION GEOL, MINERAL ECONOMICS. Educ: Univ Wis, BS, 46. Prof Exp: Mining engr, Anaconda Copper Mining Co, 46-48, geologist, 48-50; asst geologist, Great Northern Rwy Co, Minn, 50-52, geologist, 52-65, dir mineral res & develop dept, 65-68; mgr mineral develop dept, FMC Corp, 68-74; CONSULT, 74- Concurrent Pos: Chmn, Gov Raw Mat Adv Comt, Wash, 58-65. Mem: Soc Econ Geol; Am Inst Mining, Metall & Petrol Eng; Geol Soc Am. Res: Mining and economic geology. Mailing Add: Sunset Rd Pocatello ID 83201

WATSON, RAND LEWIS, b Denver, Colo, Aug 29, 40; m 62; c 2. NUCLEAR CHEMISTRY. Educ: Colo Sch Mines, BS, 62; Univ Calif, Berkeley, PhD(nuclear chem), 66. Prof Exp: Res assoc, Lawrence Radiation Lab, Univ Calif, 66-67; asst prof, 67-72, ASSOC PROF CHEM, TEX A&M UNIV, 72- Mem: Am Chem Soc. Res: Ionization phenomena; x-ray and electron spectroscopy. Mailing Add: Dept of Chem Tex A&M Univ College Station TX 77843

WATSON, RICHARD ELVIS, b Carterville, Ill, Apr 9, 12; m 36; c 2. PHYSICS. Educ: Southern Ill Norm Univ, BEd, 32; Univ Ill, AM, 35, PhD(physics), 38. Prof Exp: Instr physics, Eastern Ill Teachers Col, 38-39; sci ed, Coop Test Serv, NY, 39-40; asst prof physics, Southern Ill Teachers Col, 40-42; res technologist, Elec Div, Leeds & Northrup Co, 46-58; PROF PHYSICS, SOUTHERN ILL UNIV, CARBONDALE, 58- Mem: Am Phys Soc; Inst Elec & Electronics Eng. Res: Scattering of neutrons by light nuclei; scattering of fast electrons by Coulomb field; electrometer amplifiers and recorders; automatic electrical controllers; economic loading of power systems. Mailing Add: Dept of Physics Southern Ill Univ Carbondale IL 62903

WATSON, RICHARD LEE, b Philadelphia, Pa, May 3, 43. SEDIMENTOLOGY. Educ: Lehigh Univ, AB, 65; Univ Tex, Austin, MA, 68, PhD(geol), 75. Prof Exp: RES SCI ASSOC GEOL, UNIV TEX MARINE SCI INST, 72- Res: Hydrodynamics of sediment transport; longshore transport systems; inlet stability; coastal engineering geology; natural hazards affecting development of the coastal zone. Mailing Add: Marine Sci Inst Port Aransas TX 78373

WATSON, RICHARD NOBLE, b Moline, Ill, Sept 6, 43; m 65; c 1. ORGANIC CHEMISTRY. Educ: Southern Methodist Univ, BS, 65; Univ Ill, Urbana, PhD(org chem), 70. Prof Exp: RES CHEMIST, FREON PROD DIV, ORG CHEM DEPT, E I DU PONT DE NEMOURS & CO, INC, 70- Mem: Am Chem Soc. Res: Catalyst support fabrication; synthetic dyes; process development of chlorofluorocarbons. Mailing Add: Org Chem Dept Chambers Works E I du Pont de Nemours & Co Inc Deepwater NJ 08023

WATSON, RICHARD WHITE, JR, b Indiana, Pa, Apr 13, 33; m 58; c 3. MICROBIOLOGY. Educ: Cornell Univ, BS, 59; Rutgers Univ, MS, 61, PhD(bact), 64. Prof Exp: Sr lab technician, Rutgers Univ, 61-62; res microbiologist, Anheuser-Busch, Inc, 64-67 & Esso Res & Eng Co, 67-71; LAB DIR, NAT HEALTH LABS, INC, MOUNTAINSIDE, 71- Mem: AAAS; Am Soc Microbiol; Am Chem Soc. Res: Analytical biochemistry. Mailing Add: 16 Walnut St New Providence NJ 07974

WATSON, RICHARD WILLIAM, b Wheeling, WVa, Aug 15, 27. PHYSICS. Educ: Carnegie Inst Technol, BS, 62. Prof Exp: Jr res physicist, Explosives Res Group, Carnegie Inst Technol, 54-57, from asst proj supvr to proj supvr, 57-62; res physicist, Explosives Res Ctr, 62-70, supvry res physicist & proj coord, 70-71, RES SUPVR, PITTSBURGH MINING & SAFETY RES CTR, US BUR MINES, 71- Mem: Am Phys Soc; Combustion Inst; Am Soc Test & Mat. Res: Explosives physics and chemistry; hazardous materials. Mailing Add: 3988 Mimosa Dr Bethel Park PA 15102

WATSON, ROBERT BARDEN, b Champaign, Ill, Apr 14, 14; m 41; c 3. ACOUSTICS, LASERS. Educ: Univ Ill, AB, 34; Univ Calif, Los Angeles, MA, 36; Harvard Univ, PhD(physics), 41. Prof Exp: Asst physics, Univ Calif, Los Angeles, 35-36; asst, Harvard Univ, 36-37, instr, 37-38, physics & commun eng, 38-40, res assoc physics, 40-41, spec res assoc, Underwater Sound Lab, 41-45; res physicist, Defense Res Lab, Univ Tex, 45-52, mil physics res lab, 52-57, from asst prof to assoc prof physics, Univ, 46-60; chief, Physics, Electronics & Mech Br, Phys & Eng Sci Div, Off Chief Res & Develop, 60-72, chief, Phys & Eng Sci Div, 72-75, STAFF OFFICER, DIRECTORATE ARMY RES, OFF DEP CHIEF STAFF RES, DEVELOP & ACQUISITION, DEPT ARMY, 75- Mem: Fel AAAS; fel Acoust Soc Am; Am Asn Physics Teachers; Inst Elec & Electronics Eng; Optical Soc Am. Res: Electronic circuits; architectural and musical acoustics; propagation of high frequency electromagnetic radiation; semiconductors; management of broad research programs in physical and engineering sciences. Mailing Add: 1176 Wimbledon Dr McLean VA 22101

WATSON, ROBERT BRIGGS, b Clemson, SC, Jan 16, 03. MEDICINE. Educ: Univ Tenn, BS, 24, MD, 30; Johns Hopkins Univ, MPH, 39. Prof Exp: Prin epidemiologist, Tenn Valley Authority, 34-45; field staff mem, Int Health Div, Rockefeller Found, 45-48, actg dir, Far East Region, 47, dir, 48-53, dir div med & pub health, SAm, 54-63, assoc dir med & pub health, 55-58, assoc dir med & natural sci, 59-66; prof, 67-74, EMER PROF PARASITOL, UNIV NC, CHAPEL HILL, 74- Concurrent Pos: Mem subcomt trop dis, Nat Res Coun; ed, J Am Soc Trop Med & Hyg, 66-71. Mem: Fel Am Col Physicians; fel Am Col Prev Med; Am Soc Trop Med & Hyg. Res: Malaria; malaria control; anopheline mosquitoes. Mailing Add: 419 Whitehead Circle Chapel Hill NC 27514

WATSON, ROBERT C, b Summer Shade, Ky, Jan 22, 14; m 40; c 2. ORGANIC CHEMISTRY. Educ: Western Ky Teachers Col, BS, 39; Univ Louisville, MS, 49. Prof Exp: Asst chief chemist, Alcohol & Tobacco Tax Lab, Ky, 42-43, chief chemist, 43-49, anal chemist, Alcohol & Tobacco Tax Div, Nat Lab, DC, 49-62, asst lab chief, 62-71, CHIEF CHEM BR, BUR ALCOHOL, TOBACCO & FIREARMS, NAT LAB, INTERNAL REVENUE SERV, DEPT TREAS, 72- Mem: Am Chem Soc; AAAS; fel Am Inst Chem; Asn Off Anal Chem. Res: Narcotic drugs; distilled spirits; denatured alcohols and denaturants. Mailing Add: Chem Br Bur Alcohol Tobacco & Firearms Hq Lab Internal Revenue Bldg Washington DC 20226

WATSON, ROBERT DALE, b Elko, Nev, Dec 19, 34; m 59; c 4. ENVIRONMENTAL PHYSICS. Educ: Ariz State Univ, BS, 56, MS, 61. Prof Exp: Assoc engr, Boeing Airplane Co, Kans, 58-59, res engr, Wash, 61; res asst, Ariz State Univ, 59-61; physicist, Cent Radio Propagation Lab, Nat Bur Stand, Colo, 61-63 & Nat Ctr Atmospheric Res, Colo, 63-67; PHYSICIST, BR EXPLOR RES, US GEOL SURV, 67- Res: Remote sensing of environment; electromagnetic wave propagation in gases and solids; scattering properties of polydispersed suspensions of aerosols; optical properties of terrestial surface; luminescence properties of natural materials. Mailing Add: 601 E Cedar Flagstaff AZ 86001

WATSON, ROBERT FLETCHER, b Charlottesville, Va, Jan 24, 10; m 46; c 2. MEDICINE. Educ: Univ Va, MD, 34. Prof Exp: Intern, House of the Good Samaritan, Boston, Mass, 34; intern, Mass Gen Hosp, 35-36; from asst to instr, Med Col, Cornell Univ, 36-39; from asst mem to assoc mem, Rockefeller Inst, 39-46; assoc prof med, 46-50, assoc prof clin med, 50-60, PROF CLIN MED, MED COL, CORNELL UNIV, 60-; ATTEND PHYSICIAN & CHIEF-OF-SERV, VINCENT ASTOR DIAG SERV, NY HOSP, 51- Concurrent Pos: Asst resident, NY Hosp, 36-38, resident physician, 38, assoc attend physician, 46-50; asst resident, Hosp, Rockefeller Inst, 39-41, resident physician, 41-46. Mem: AAAS; Am Soc Clin Invest; Harvey Soc; Am Heart Asn; Am Rheumatism Asn. Res: Cardiovascular physiology; hemolytic streptococcal infection; rheumatic fever and collagen diseases. Mailing Add: Cornell Univ Med Col 1300 York Ave New York NY 10021

WATSON, ROBERT FRANCIS, b Knoxville, Tenn, Nov 20, 36; m 58; c 3. CHEMISTRY, SCIENCE EDUCATION. Educ: Col Wooster, AB, 58; Univ Tenn, Knoxville, PhD(chem), 63. Prof Exp: From asst prof to assoc prof chem, Memphis State Univ, 63-68; from asst prog dir to assoc prog dir, Undergrad Educ in Sci, 68-73, PROG MGR, OFF EXP PROJ & PROGS, NSF, 73- Concurrent Pos: Nat Acad Sci res assoc, Naval Stores Lab, USDA, Fla, 67-68. Mem: Am Chem Soc. Res: Chemistry of indanes; nonclassical ions; physical properties of sulfoxides. Mailing Add: Undergrad Educ in Sci NSF Washington DC 20550

WATSON, ROBERT JOSEPH, b Detroit, Mich, Apr 5, 31; m 53; c 2. GEOPHYSICS.

Educ: Wayne State Univ, BS, 52; Pa State Univ, PhD(geophys), 58. Prof Exp: Res engr, Boeing Co, 58-59; sr res geophysicist, Mobil Oil Corp, 59-63; asst prof geophys, Pa State Univ, 63-65; sr res geophysicist, Field Res Lab, Mobil Oil Corp, Tex, 65-66, res assoc geophys, 66-67, geophys sect supvr, 67-74, CORP CHIEF GEOPHYSICIST, MOBIL OIL CORP, NY, 74- Mem: Soc Explor Geophys; Sigma Xi; Europ Asn Explor Geophys. Res: Geophysical data analysis, especially applications of information theory. Mailing Add: Mobil Oil Corp 150 E 42nd St New York NY 10017

WATSON, ROBERT LEE, b Scribner, Nebr, Dec 17, 31; m 53; c 4. EPIDEMIOLOGY, PUBLIC HEALTH. Educ: Iowa State Univ, DVM, 55; Univ Minn, MPH, 63, PhD, 73. Prof Exp: Jr asst vet, Ga State Dept Health, USPHS, 55-57, asst vet, Md State Dept Health, 57, vet epidemiologist, 57-60, sr asst vet, Univ Minn, 60-61, vet epidemiologist, Miss State Bd Health, 61-63; trainee epidemiol, Sch Pub Health, Univ Minn, 63-64; instr med, 64-67, asst prof epidemiol, 67-70, asst prof med & prev med, 70-71, ASSOC PROF PREV MED, MED CTR, UNIV MISS, 71- Concurrent Pos: Trainee epidemiol, Sch Pub Health, Univ Minn, 64-67; co-investr, NIH grants, 66-73, contract, 71-76. Mem: Soc Epidemiol Res; Asn Teachers Prev Med. Res: Health and health care statistics; socio-cultural-economic factors related to blood pressure levels; toxemia of pregnancy. Mailing Add: Dept of Prev Med Univ of Miss Med Ctr Jackson MS 39216

WATSON, ROBERT LEE, b Plainview, Ark, Nov 8, 34; m 65; c 2. ENTOMOLOGY. Educ: Univ Ark, Fayetteville, BS, 56, MS, 63; Auburn Univ, PhD(entom), 68. Prof Exp: Instr zool, Auburn Univ, 66-67; asst prof, 67-71, ASSOC PROF BIOL, UNIV ARK, LITTLE ROCK, 71- Mem: Entom Soc Am; Am Inst Biol Sci. Res: Taxonomy and ecology of Tabanidae aquatic ecology. Mailing Add: Dept of Biol Univ of Ark Little Rock AR 72204

WATSON, RONALD ROSS, b Tyler, Tex, Dec 9, 42; m 66; c 3. IMMUNOLOGY, NUTRITION. Educ: Brigham Young Univ, BS, 66; Mich State Univ, PhD(biochem), 71. Prof Exp: Res fel immunol & microbiol, Sch Pub Health, Harvard Univ, 71-73; asst prof microbiol, Med Ctr, Univ Miss, 73-74; ASST PROF MICROBIOL, SCH MED, IND UNIV, INDIANAPOLIS, 74- Concurrent Pos: Collabr & vis scientist, Int Ctr Med Res, Cali, Colombia, 74-; res adv, NIH Postdoctoral Training Prog, Ind Univ, 74- Mem: Am Asn Immunologists; Am Soc Microbiol; Sigma Xi. Res: Malnutrition's effects on secretory immunity; secretory immunity in the urinary tract; synthesis and action of apiose compounds on viruses. Mailing Add: Dept of Microbiol Ind Univ Indianapolis IN 46202

WATSON, ROSCOE DERRICK, b Salt Lake City, Utah, Dec 12, 11; m 36; c 4. PLANT PATHOLOGY. Educ: Utah State Col, BS, 35, MS, 37; Cornell Univ, PhD(plant path), 42. Prof Exp: Asst, Bur Plant Indust, USDA, 35-37, plant pathologist, Emergency Plant Dis Proj, 45; jr range exam, US Forest Serv, 37; asst plant path, State Univ NY Col Agr, Cornell Univ, 37-42; plant pathologist, Substa, Exp Sta, Univ Tex, 42-45; PROF PLANT PATH & PATHOLOGIST, UNIV IDAHO, 45- Concurrent Pos: Plant path adv to Iraq, US Opers Mission. Mem: Am Phytopath Soc. Res: Plant pathology of soil borne diseases; ecology of soil borne organisms. Mailing Add: 415 Residence Moscow ID 83843

WATSON, SPENCER C, chemistry, see 12th edition

WATSON, STANLEY ARTHUR, b Los Angeles, Calif, Aug 30, 15; m 42; c 4. BIOCHEMISTRY. Educ: Pomona Col, AB, 39; Univ Ill, AM, 42, PhD(agron), 49. Prof Exp: Asst bot, Univ Ill, 38-42, agron, 46-48; from jr chemist to asst chemist northern regional res lab, Bur Agr & Indust Chem, USDA, 42-46; sect leader res dept, Corn Prod Co, 48-68, asst dir, 68-69, asst dir explor res, 69-75, RES SCIENTIST AGRON & MILLING DEPT, CPC INT, INC, 75- Mem: Am Chem Soc; Am Soc Agron; Am Asn Cereal Chem; Sigma Xi. Res: Chemistry, agronomy and processing of cereal grains; industrial application of corn wet milled products. Mailing Add: CPC Int Inc Box 345 Argo IL 60502

WATSON, STANLEY W, b Seattle, Wash, Jan 3, 21; m 52. BACTERIOLOGY. Educ: Univ Wash, BS, 49, MS, 51; Univ Wis, PhD(bact), 57. Prof Exp: Fisheries biologist, US Fish & Wildlife Serv, Wash, 52-54; res assoc, 57-71, SR SCIENTIST, WOODS HOLE OCEANOG INST, 71- Mem: Am Soc Microbiol; Brit Soc Gen Microbiol; Am Soc Limnol & Oceanog. Res: Marine microbiology; nitrification; marine slime molds; fish diseases; myxobacteria. Mailing Add: Woods Hole Oceanog Lab Woods Hole MA 02543

WATSON, THEO FRANKLIN, b Plainview, Ark, July 2, 31; m 60; c 4. ENTOMOLOGY, ECOLOGY. Educ: Univ Ark, BS, 53, MS, 58; Univ Calif, Berkeley, PhD(entom), 62. Prof Exp: From asst prof to assoc prof entom, Auburn Univ, 62-66; assoc prof, 66-70, PROF & ENTOMOLOGIST, AGR EXP STA, UNIV ARIZ, 70- Mem: Entom Soc Am. Res: Agricultural entomology and ecology; ecology of cotton insects and the integrated approach to pest control in cotton. Mailing Add: Dept of Entom Univ of Ariz Tucson AZ 85721

WATSON, THOMAS ALASTAIR, b NZ, Mar 3, 14; nat Can; m 44; c 3. MEDICINE. Educ: Univ Otago, NZ, MB & ChB, 37; Univ London, dipl, 40. Prof Exp: CLIN PROF & HEAD DEPT THERAPEUT RADIOL, FAC MED, UNIV WESTERN ONT, 63- Concurrent Pos: Dir, Ont Cancer Found London Clin. Mem: Am Radium Soc; Soc Nuclear Med; Can Med Asn; Can Asn Radiologists; Brit Med Asn. Res: Radiation therapy; nuclear medicine. Mailing Add: Fac of Med Univ of Western Ont London ON Can

WATSON, THOMAS RICHARD, JR, b Portsmouth, NH, Jan 3, 17; m 43; c 2. SURGERY. Educ: Dartmouth Col, AB, 37; Columbia Univ, MD, 41. Prof Exp: Staff surgeon, chief surg & dir pulmonary physiol lab, Vet Admin Hosp, Rutland Heights, Mass, 49-53; instr surg & physiol sci, 53-61, ASST CLIN PROF SURG, DARTMOUTH MED SCH, 61- Concurrent Pos: Asst, Peter Bent Brigham Hosp, Boston, Mass, 53; staff surgeon, Mary Hitchcock Mem Hosp, 53-, dir cardiopulmonary lab, Hitchcock Found, 53-; attend surgeon, Vet Admin Hosp, White River, Vt, 53- Mem: Am Thoracic Soc. Res: Thoracic surgery; cardiopulmonary physiology. Mailing Add: Dept of Surg Dartmouth Med Sch Hanover NH 03766

WATSON, TULLY FRANKLIN, b Boswell, Okla, Oct 11, 02; m 24; c 1. PHYSICS. Educ: Univ Okla, BA, 28, MS, 30; Univ Ill, PhD(physics), 35. Prof Exp: Asst physics, Univ Okla, 28-30 & Univ Ill, 30-35; prof & head dept, Northeastern State Col, 35-43 & Phillips Univ, 43-47; prof physics, Wichita State Univ, 47-75; RETIRED. Concurrent Pos: Instr, Okla Agr & Mech Col, 43-44. Mem: Am Phys Soc. Res: Band spectra; molecular structure. Mailing Add: 1507 Burns Wichita KS 67203

WATSON, VANCE H, b Kennett, Mo, Nov 25, 42; m 64; c 2. AGRONOMY. Educ: Southeast Mo State Univ, BS, 64; Univ Mo-Columbia, MS, 66; Miss State Univ, PhD(agron), 69. Prof Exp: Soil conservationist, Soil Conserv Serv, USDA, 63-64; res asst agron, Univ Mo-Columbia, 64-66; ASSOC PROF AGRON, MISS STATE UNIV, 69- Mem: Am Soc Agron; Crop Sci Soc Am; Am Forage & Grassland Coun. Res: Forage crop ecology and management. Mailing Add: Dept of Agron Miss State Univ Box 5248 Mississippi State MS 39762

WATSON, VELVIN RICHARD, b Streator, Ill, June 2, 32; m 58; c 3. GAS DYNAMICS, NUMERICAL ANALYSIS. Educ: Univ Calif, Berkeley, BS, 59, MS, 61; Stanford Univ, PhD(plasma physics), 69. Prof Exp: RES SCIENTIST GAS DYNAMICS & NUMCERICAL ANAL, AMES RES CTR, NASA, 61- Concurrent Pos: Instr, San Jose State Univ, 74- Mem: Am Inst Aeronaut & Astronaut; Am Phys Soc. Res: Improving our understanding of plasma dynamics and gas dynamics by utilizing numerical analysis and computer simulations. Mailing Add: Mail Stop 245-3 Ames Res Ctr NASA Moffett Field CA 94035

WATSON, WILLIAM CRAWFORD, b Glasgow, Scotland, Dec 20, 27; m 54; c 4. INTERNAL MEDICINE, GASTROENTEROLOGY. Educ: Glasgow Univ, MB, ChB, 50, MD, 60, PhD(med), 64; FRCPS(G), 66. Prof Exp: Resident surg, Ballochmyle Hosp, 50-51; resident med, Glasgow Royal Infirmary, 53, sr house officer, 53-54, sr house officer cardiol, 54-55, sr registr med, 55-60, lectr, 61-66; prof, Univ EAfrica, 66-67; consult, Glasgow Royal Infirmary, 67-69; assoc prof, 69-72, PROF MED, UNIV WESTERN ONT, 72-; DIR GASTROENTEROL, VICTORIA HOSP, 69- Concurrent Pos: EAfrican Med Res Coun fel, Nairobi, Kenyatta Nat Hosp, 66-67. Mem: Brit Soc Gastroenterol; Can Asn Gastroenterol; Can Soc Clin Invest; Am Gastroenterol Asn. Res: Biophysical and biochemical aspects of intestinal structure and function. Mailing Add: Gastroenterol Unit Victoria Hosp London ON Can

WATSON, WILLIAM DOUGLAS, b Memphis, Tenn, Jan 12, 42; m 69. ASTROPHYSICS. Educ: Mass Inst Technol, BS, 64, PhD(physics), 68. Prof Exp: Res assoc, Mass Inst Technol, 68-70; res assoc, Cornell Univ, 70-72; asst prof physics & astron, 72-74, ASSOC PROF PHYSICS & ASTRON, UNIV ILL, URBANA, 74- Concurrent Pos: A P Sloan Res fel, 74. Mem: Am Astron Soc; Am Phys Soc; Int Astron Union. Res: Theoretical astrophysics; theory of the interstellar medium. Mailing Add: Depts of Physics & Astron Univ of Ill Urbana IL 61801

WATSON, WILLIAM HAROLD, JR, b Tex, Sept 2, 31; m 56; c 2. PHYSICAL CHEMISTRY. Educ: Rice Univ, BA, 53, PhD(chem), 58. Prof Exp: From asst prof to assoc prof, 57-64, PROF CHEM, TEX CHRISTIAN UNIV, 64-, DIR FASTBIOS LAB, 72- Mem: AAAS; Am Chem Soc; Am Phys Soc; Am Crystallog Asn; The Chem Soc. Res: Structure of biologically active molecules; magnetic interactions in transition metal complexes. Mailing Add: 6024 Wonder Dr Ft Worth TX 76133

WATSON, WILLIAM HARRISON, b Anniston, Ala, Dec 25, 27; m 50; c 4. POLYMER CHEMISTRY. Educ: Emory Univ, AB, 48, MS, 49, PhD(org chem), 52. Prof Exp: Res chemist, Carothers Res Lab, 52-54, res chemist, Dacron Res Lab, NC, 54-61, sr res chemist, 61-65, res supvr, 65-69, supvr technol, 69-71, sr res chemist, 71-75, RES ASSOC, DACRON RES LAB, E I DU PONT DE NEMOURS & CO, INC, 75- Res: Preparation and evaluation of linear condensation polymers as textile and industrial fibers involving synthetic work in preparation of monomers; product development in synthetic yarns and staples; new polymer preparations and evaluation. Mailing Add: E I du Pont de Nemours & Co Inc PO Box 800 Kinston NC 28501

WATSON, WYNNFIELD YOUNG, b Toronto, Ont, Feb 5, 24; m 50; c 3. ENTOMOLOGY, INVERTEBRATE ZOOLOGY. Educ: Univ Toronto, BA, 50, PhD(entom), 55. Prof Exp: Res officer, Fed Dept Forestry, 50-61; from assoc prof to prof zool, Laurentian Univ, 61-74, chmn dept biol, 67-71, dir grad studies, 69-74; CHMN DEPT BIOL, WILFRID LAURIER UNIV, 74- Concurrent Pos: Nat Res Coun Can grants, 65-67. Mem: Entom Soc Can; Can Soc Zool. Res: Systematics of Coccinellidae; carabidae of eastern North America. Mailing Add: Dept of Biol Wilfrid Laurier Univ Waterloo ON Can

WATT, BERNICE K, b Ames, Iowa, June 10, 10; m 42; c 2. NUTRITION. Educ: Iowa State Univ, BS, 32; Kans State Univ, MS, 33; Columbia Univ, PhD, 40. Prof Exp: Instr food & nutrit, Kans State Univ, 36-38, asst prof, 40-41; res leader, Nutrient Data Res Ctr, Consumer & Food Econ Inst, Agr Res Serv, USDA, 41-75; RETIRED. Honors & Awards: Borden Award, 72; USDA Distinguished Serv Award, 74. Mem: Am Inst Nutrit; Am Dietetic Asn; Inst Food Technology; Am Home Econ Asn. Res: Nutrition and food composition. Mailing Add: 1156 Chain Bridge Rd McLean VA 22101

WATT, BOB E, b Tulsa, Okla, July 20, 27; m 46; c 3. NUCLEAR PHYSICS. Educ: Rice Inst, PhD(physics), 46. Prof Exp: Staff mem, Radiation Lab, Mass Inst Technol, 41-45; staff mem, Geophys Lab, Tex Co, 46-47; STAFF MEM, LOS ALAMOS SCI LAB, 47- Mem: AAAS; Optical Soc Am; Am Phys Soc. Res: Transmutations of the light elements; fission process, particularly the neutron spectrum; neutron reactions; polarized targets and reactions; high power lasers; laser energy measurement; molecular spectroscopy and laser induced transitions. Mailing Add: Group L-2 Los Alamos Sci Lab PO Box 1663 Los Alamos NM 87545

WATT, DEAN DAY, b McCammon, Idaho, Sept 21, 17; m 46; c 5. BIOCHEMISTRY. Educ: Univ Idaho, BS, 42; Iowa State Col, PhD(bact physiol), 49. Prof Exp: Res chemist, Westvaco Chlorine Prod, Inc, 42-44; instr bact, Iowa State Col, 47-49; asst microbiologist, Agr Exp Sta, Purdue Univ, 49-53; assoc prof biochem, Tulane Univ La, 53-60; assoc prof zool, Ariz State Univ, 60-63; mem staff, Midwest Res Inst, Mo, 63-69; PROF BIOCHEM, SCH MED, CREIGHTON UNIV, 69- Concurrent Pos: Sr res biochemist & head dept physiol sci, Southeast La Hosp, Mandeville, La, 53-60. Mem: AAAS; Int Soc Toxinology. Res: Metabolism of bacteria; pigments of fungi; biochemistry of mental disease; chemistry of animal venoms. Mailing Add: Dept of Biochem Creighton Univ Sch of Med Omaha NE 68178

WATT, GEORGE WILLARD, b Bellaire, Ohio, Jan 8, 11; m 35; c 3. CHEMISTRY. Educ: Ohio State Univ, BA, 31, MSc, 33, PhD(inorg chem), 35. Prof Exp: Res chemist, Goodyear Tire & Rubber Co, 35-37; from asst prof to assoc prof chem, Univ Tex, 37-43; sr chemist metall lab, Univ Chicago, 43-44; sr supvr, Hanford Eng Works, Gen Elec Co, Wash, 44-45; PROF CHEM, UNIV TEX, AUSTIN, 46- Concurrent Pos: Consult, Jersey Nuclear Co & E I du Pont de Nemours & Co. Honors & Awards: Southwest Regional Award, Am Chem Soc, 75. Mem: Am Chem Soc. Res: Reactions in liquid ammonia; radiochemistry; inorganic chemistry of nonaqueous solvents; coordination chemistry. Mailing Add: Dept of Chem Univ of Tex Austin TX 78712

WATT, GERALD DEE, physical chemistry, biochemistry, see 12th edition

WATT, JAMES, b Thomasville, Ga, Apr 28, 11; m 39; c 2. EPIDEMIOLOGY. Educ: Davidson Col, AB, 31; Johns Hopkins Univ, MD, 35, DrPH, 36; Am Bd Prev Med, dipl, 49. Prof Exp: Asst epidemiologist, USPHS, 36-37; res, Herman Kiefer Hosp, Detroit, Mich, 37-38; from asst surgeon to surgeon, USPHS, 39-49, med dir, 49-52, dir, Nat Heart Inst, 52-61, dir off int health, USPHS, 61-67, spec asst to surgeon gen of prog rev, 67-68; assoc dir prog develop, Children's Hosp of DC, 68-72; CONSULT EPIDEMIOL, 72- Concurrent Pos: Mem, UN Relief & Rehab Admin Cholera Comn to China, 45; clin prof, Sch Med, La State Univ, 48-53; vis physician, Charity Hosp, New Orleans, La, 48-52; dir enteric dis comn, Armed Forces Epidemiol Br, 48-52; lectr, Sch Pub Health, Johns Hopkins Univ, 50- & Univ Calif, 52-; spec asst to secy aging, US Dept Health, Educ & Welfare, 60-61. Honors & Awards: Ashford Award, Am Soc Trop Med & Hyg, 45. Mem: AAAS; Soc Exp Biol & Med; Am Soc Trop Med & Hyg; Am Epidemiol Soc; fel Am Pub Health Asn. Res: Public health and medical research administration. Mailing Add: Rte 1 Box 1 Ely VT 05044

WATT, JOHN YIN CHIEH, b Tientsin, China, Mar 3, 06; nat US; m 39; c 2. PARASITOLOGY. Educ: Univ Iowa, AB, 31; Univ Ill, MS, 32; Cornell Univ, PhD(med parasitol), 42. Prof Exp: Asst parasitol, Peiping Union Med Col, 32-34; field parasitologist & dir labs, Pub Health Training Inst, Chinese Nat Health Admin, 34-39; asst pub health & prev med, Med Col, Cornell Univ, 40-43, instr, 43, lectr, Med Off, US Navy, 42; med parasitologist, Vet Admin Hosp, Ft Howard, 46-75; RETIRED. Concurrent Pos: Consult parasitologist, Cent Hosp, China, 35-38; prof, Nat Kweiyang Med Col, 38; lectr, Long Island Col Med, 42. Honors & Awards: Vet Admin Distinguished Career Award, 74. Mem: AAAS; Am Soc Parasitol; fel Royal Soc Health. Res: Immunology in parasitology; antibiotics on animal parasites. Mailing Add: 4044 The Alameda Baltimore MD 21218

WATT, KENNETH EDMUND FERGUSON, b Toronto, Ont, July 13, 29; m 55; c 2. ECOLOGY. Educ: Univ Toronto, BA, 51; Univ Chicago, PhD(zool), 54. Hon Degrees: LLD, Simon Fraser Univ, 70. Prof Exp: Biometrician, Ont Dept Lands & Forests, 54-57; sr biometrician, Statist Res & Serv, Res Br, Can Dept Agr, 57-61; head statist res & serv, Can Dept Forestry, 61-63; assoc prof zool, 63-64, PROF ZOOL, UNIV CALIF, DAVIS, 65- Concurrent Pos: Consult, Sci Secretariat, Can Privy Coun, 66; sr fel, East-West Ctr, Honolulu, 75. Honors & Awards: Fish Ecol & Mgt Award, Wildlife Soc, 61; Entom Soc Can Gold Medal, 69. Mem: Ecol Soc Am; Soc Gen Systs Res; Entom Soc Can; Japanese Soc Pop Ecol. Res: Theoretical, experimental and field ecology of fish and insects; biomathematics; applied statistics; computer simulation studies for evaluating resource management strategies; epidemiology; regional and global modelling and simulation. Mailing Add: Dept of Zool Univ of Calif Davis CA 95616

WATT, RICHARD FRANKLIN, plant physiology, forestry, see 12th edition

WATT, ROBERT DOUGLAS, b Santa Paula, Calif, July 7, 19; m 46; c 2. EXPERIMENTAL HIGH ENERGY PHYSICS, PHYSICS ENGINEERING. Educ: Univ Calif, BS, 42. Prof Exp: Physicist, Lawrence Radiation Lab, Univ Calif, 46-67; GROUP LEADER, STANFORD LINEAR ACCELERATOR CTR, STANFORD UNIV, 67- Concurrent Pos: Consult, Argonne Nat Lab & Lawrence Radiation Lab. Mem: AAAS; Am Phys Soc. Res: Particle accelerator development and operation; development and use of liquid hydrogen bubble chambers for detection of high energy particles of nuclear physics; research to prove existence of magnetic monopoles. Mailing Add: 11117 Palos Verde Dr Cupertino CA 95014

WATT, WALTON DELBERT, aquatic ecology, see 12th edition

WATT, WARD BELFIELD, b Washington, DC, Oct 21, 40; m 63; c 2. EVOLUTIONARY BIOLOGY. Educ: Yale Univ, BA, 62, MS, 64, PhD(biol), 67. Prof Exp: Asst prof, 69-75, ASSOC PROF BIOL, STANFORD UNIV, 75- Concurrent Pos: Mem bd trustees, Rocky Mountain Biol Lab, 71-75; mem adv panel syst biol, NSF, 73-75. Mem: Am Soc Naturalists; Genetics Soc Am; Soc Study Evolution; AAAS; Arctic Inst NAm. Res: Study of adaptive mechanisms and microevolutionary processes in natural insect populations from perspectives of physiology, genetics and ecology. Mailing Add: Dept of Biol Sci Stanford Univ Stanford CA 94305

WATT, WILLIAM JOSEPH, b Carbondale, Ill, Dec 15, 25; m 56; c 3. INORGANIC CHEMISTRY. Educ: Univ Ill, BS, 49; Cornell Univ, MS, 51, PhD, 56. Prof Exp: Asst prof chem, Davidson Col, 51-53; from asst prof to assoc prof, 55-65, from asst dean to assoc dean col, 66-71, PROF CHEM, WASHINGTON & LEE UNIV, 65-, DEAN COL, 71- Mem: AAAS; Am Chem Soc; Chem Soc France. Res: Magnesium fluoride gels; inorganic polymers; boron compounds; molten salts. Mailing Add: Washington & Lee Univ Lexington VA 24450

WATT, WILLIAM RUSSELL, b Camden, NJ, June 28, 20; m 46; c 3. ORGANIC CHEMISTRY. Educ: Univ Pa, BA, 49; Univ Del, MS, 52, PhD(chem), 55. Prof Exp: Instr gen chem, Philadelphia Col Textiles & Sci, 49-51; res chemist, Am Viscose Corp, 54-60; sr res chemist, Avisun Corp, 60-64; SR RES ASSOC POLYMER CHEM, AM CAN CO, 64- Mem: Am Chem Soc. Res: Photochemistry; organic coatings; cellulose derivatives; stereospecific polymerization; catalysis. Mailing Add: Am Can Co Princeton Labs PO Box 50 Princeton NJ 08540

WATTENBERG, ALBERT, b New York, NY, Apr 13, 17; m 43; c 3. PHYSICS. Educ: City Col New York, BS, 38; Columbia Univ, MA, 39; Univ Chicago, PhD(physics), 47. Prof Exp: Spectroscopist, Schenley Prod, Inc, NY, 39-41; asst, Off Sci Res & Develop, Columbia Univ, 41-42; group leader, Metall Lab, Univ Chicago, 42-46; sr physicist & group leader, Argonne Nat Lab, 47-50; vis asst prof, Univ Ill, 50-51; res physicist, Nuclear Sci Lab & lectr, Mass Inst Technol, 51-58; res prof, 58-66, PROF PHYSICS, UNIV ILL, URBANA, 66- Concurrent Pos: Actg dir nuclear physics div, Argonne Nat Lab, 49-50; NSF fel, Univ Rome, 62-63. Mem: Am Phys Soc. Res: Spectroscopy; nuclear chain reactors; photoneutron techniques; photonuclear reactions; elementary particle physics. Mailing Add: Dept of Physics Univ of Ill Urbana IL 61803

WATTENBERG, FRANKLIN ARVEY, b New York, NY, May 16, 43; m 64; c 2. MATHEMATICS. Educ: Wayne State Univ, BS, 64; Univ Wis-Madison, MS, 65, PhD(math), 68. Prof Exp: Benjamin Peirce asst prof math, Harvard Univ, 68-71; asst prof, 71-74, ASSOC PROF MATH, UNIV MASS, AMHERST, 74- Mem: Am Math Soc; Asn Symbolic Logic. Res: Differential topology; nonstandard analysis; algebraic topology. Mailing Add: Dept of Math Univ of Mass Amherst MA 01002

WATTENBERG, LEE WOLFF, b New York, NY, Dec 22, 21; m 45; c 6. PATHOLOGY. Educ: City Col New York, BS, 41; Univ Minn, Minneapolis, BM, 49, MD, 50; Am Bd Path, dipl, 56. Prof Exp: From instr to assoc prof, 56-66, Hill prof, 59-66, PROF PATH, MED SCH, UNIV MINN, MINNEAPOLIS, 66- Concurrent Pos: Lederle med fac award, 57-59. Mem: Histochem Soc (vpres, 66); Soc Exp Biol & Med; Am Soc Exp Path; Am Asn Pathologists & Bacteriologists; Am Asn Cancer Res. Res: Histochemistry; cancer research; experimental pathology. Mailing Add: Dept of Path Univ of Minn Sch of Med Minneapolis MN 55455

WATTERS, CHRISTOPHER DEFFNER, b Ironton, Ohio, Dec 7, 39; m 67. CELL BIOLOGY. Educ: Univ Notre Dame, BS, 61; Princeton Univ, MA, 64, PhD(biol), 66. Prof Exp: Instr biol, Princeton Univ, 64-66; res assoc cell biol prog, Univ Minn, St Paul, 66-68; asst prof, 68-73, ASSOC PROF BIOL, MIDDLEBURY COL, 73- Concurrent Pos: Vis scientist, Physiol Lab, Cambridge Univ, 73-74; vis assoc prof, Dartmouth Col, 75. Mem: AAAS; NY Acad Sci; Am Soc Cell Biol; Am Soc Zool; Sigma Xi. Res: Cell motility; structural and functional organization of cell membranes; cellular aspects of development. Mailing Add: Dept of Biol Middlebury Col Middlebury VT 05753

WATTERS, FREDERICK LEWIS, b Lethbridge, Alta, July 19, 17. ENTOMOLOGY. Educ: Univ Man, BSc, 47, MSc, 49, PhD(entom), 64. Prof Exp: Officer in-chg stored prod insect lab, Can Dept Agr, 51-57, entomologist, Res Sta, 57-69; crop storage specialist, Crop Protection Br, Food & Agr Orgn, UN, Italy, 69-71; HEAD CROP PROTECTION SECT, RES STA, CAN DEPT AGR, 72- Concurrent Pos: Hon res prof, Univ Man, 59-69 & 71- Mem: Entom Soc Am; Royal Entom Soc London. Res: Biology and control of insects infesting stored cereals and cereal products. Mailing Add: Can Dept of Agr Res Sta 25 Dafoe Rd Winnipeg MB Can

WATTERS, GORDON VALENTINE, b Winnipeg, Man, Apr 8, 28; m 57; c 3. NEUROLOGY, PEDIATRICS. Educ: Univ Minn, Minneapolis, BA, 51; Univ Man, MD, 57. Prof Exp: Asst prof neurol, Winnipeg Children's Hosp, Univ Man, 63-65; asst prof, Children's Med Ctr, Harvard Univ, 65-69; ASSOC PROF NEUROL & PSYCHIAT, MONTREAL CHILDREN'S HOSP, McGILL UNIV, 69- Mem: Am Acad Neurol; Can Neurol Soc. Res: Degenerative disease of nervous system; cerebrospinal fluid dynamics. Mailing Add: Montreal Children's Hosp Montreal PQ Can

WATTERS, JAMES I, b Broadus, Mont, Apr 4, 08; m 38; c 4. CHEMISTRY. Educ: Univ Minn, BS, 31, PhD(anal chem), 43. Prof Exp: Instr, Cornell Univ, 41-43; res assoc, Metall Lab, Univ Chicago, 43-44, head anal div, 44-45; from asst prof to assoc prof chem, Univ Ky, 45-48; assoc prof, 48-58, PROF CHEM, OHIO STATE UNIV, 58- Mem: AAAS; Am Chem Soc. Res: Equilibria of complex species; mixed ligand complexes; acids and complexes of polyphosphates; theory of electrode reactions; applications of potentiometry; absorption and emission spectroscopy; polarography; theory of titrimetry. Mailing Add: Dept of Chem Ohio State Univ 140 W 18th Ave Columbus OH 43210

WATTERS, KENNETH LYNN, b Iowa City, Iowa, Jan 21, 39; m 64; c 3. INORGANIC CHEMISTRY, BIOINORGANIC CHEMISTRY. Educ: Univ Ill, Urbana, BS, 62; Brown Univ, PhD(chem), 70. Prof Exp: Res assoc chem, State Univ NY Buffalo, 69-70; ASST PROF CHEM, UNIV WIS-MILWAUKEE, 70- Mem: Am Chem Soc; Soc Appl Spectros. Res: Spectroscopic studies of transition metal complexes and metallo-enzymes; Raman, resonance Raman, and infrared spectroscopies; studies of catalytic properties of transition metal complexes and metal cluster compounds. Mailing Add: Dept of Chem Univ of Wis Milwaukee WI 53201

WATTERS, NEIL ARCHIBALD, b Glenbrook, Conn, Sept 14, 21; Can citizen; m 47; c 4. SURGERY. Educ: Univ Toronto, MD, 44; FRCPS(C), 52. Prof Exp: SURGEON-IN-CHIEF, WELLESLEY HOSP, 63-; PROF SURG, UNIV TORONTO, 69- Concurrent Pos: Chmn, Interhosp Comt Gen Surg, Univ Toronto, 70-, dir undergrad training, Dept Surg, 72-; mem exec comt, Ont Cancer Treat & Res Found, 72- Mem: Fel Am Col Surgeons. Res: Clinical research; colon carcinoma; breast carcinoma; pilonidal disease. Mailing Add: 172 Rosedale Heights Dr Toronto ON Can

WATTERS, ROBERT LISLE, b Everett, Wash, June 25, 25; m 48; c 2. RADIOCHEMISTRY, HEALTH PHYSICS. Educ: Univ Wash, BS, 50, PhD(chem), 63; Harvard Univ, MS, 59; Am Bd Health Physics, dipl, 66. Prof Exp: Engr asst, Hanford Atomic Prod Oper, Gen Elec Co, Wash, 50-52, supvr bioassay lab, 52-56, supvr radiation monitoring, 56-58; engr, Boeing Co, 59-60; res specialist radiol hazard eval, Atomic Int Div, NAm Aviation, Inc, Calif, 63-65; assoc prof radiochem, Colo State Univ, 65-72; ENVIRON RADIOACTIVITY SPECIALIST, DIV BIOMED & ENVIRON RES, US ENERGY RES & DEVELOP ADMIN, 72-, GROUP LEADER LAND & FRESHWATER RES, 73- Concurrent Pos: Lectr, Mobile Radioisotope Lab, Oak Ridge Inst Nuclear Studies, 66- Mem: AAAS; Health Physics Soc. Res: Translocation of plutonium and americium in the body; environmental behavior of polonium. Mailing Add: Div of Biomed & Environ Res US Energy R&D Admin Washington DC 20545

WATTERSON, ARTHUR C, JR, b Ellwood City, Pa, Apr 19, 38. ORGANIC CHEMISTRY, POLYMER CHEMISTRY. Educ: Geneva Col, BS, 60; Brown Univ, PhD(org chem), 65. Prof Exp: Res assoc chem, Johns Hopkins Univ, 64-65; PROF CHEM, LOWELL TECHNOL INST, 65- Mem: Am Chem Soc; The Chem Soc. Res: Polymer stereochemistry; nuclear magnetic resonance of macromolecules; synthesis of natural products; decomposition of n-nitroso amides; deamination reactions; nitrogen heterocycles. Mailing Add: Dept of Chem Lowell Technol Inst Lowell MA 01854

WATTERSON, JON CRAIG, b Kalamazoo, Mich, Nov 25, 44; m 67; c 2. PLANT PAHTOLOGY. Educ: Carleton Col, BA, 66; Univ Wis, MS, 71. Prof Exp: Res asst veg path, Univ Wis, 66-71, res assoc cranberry path, 71-72; HEAD PATH DEPT, PETOSEED CO INC, 72- Mem: Am Phytopath Soc. Res: Genetics of disease resistance in vegetable corps; breeding for disease resistance in vegetable crops. Mailing Add: Petoseed Res Ctr Petoseed Co Inc Rte 4 Box 1255 Woodland CA 95695

WATTERSON, KENNETH FRANKLIN, b London, Eng, July 16, 29; m 68; c 2. INORGANIC CHEMISTRY. Educ: Univ London, BSc, 52, PhD(chem), 59. Prof Exp: Lect demonstr chem, Birkbeck Col, Univ London, 52-58, res asst, Imp Col, 58-60; res assoc, Cornell Univ, 60-62; sr res chemist, Pennsalt Chem Corp, 62-66; ENGR, HOMER RES LABS, BETHLEHEM STEEL CORP, 66- Mem: Am Chem Soc; Am Soc Metals; Soc Appl Spectros; Am Soc Mass Spectrometry. Res: Metal-gas reactions; surface analysis; ion microprobe spectrometry; auger spectroscopy; corrosion. Mailing Add: Homer Res Labs Bethlehem Steel Corp Bethlehem PA 18016

WATTERSON, RAY LEIGHTON, b Greene, Iowa, Apr 15, 15; m 41; c 2. EMBRYOLOGY. Educ: Coe Col, AB, 36; Univ Rochester, PhD(embryol), 41. Prof Exp: Asst embryol, Univ Rochester, 37-40; asst, Johns Hopkins Univ, 40-41; instr comp anat & embryol, Dartmouth Col, 41-42; asst prof zool, Univ Calif, 42-46; from asst prof to assoc prof, Univ Chicago, 46-49; from assoc prof to prof biol, Northwestern Univ, 49-61, chmn admin comt, 56-58, chmn dept biol, 58-61; prof exp embryol, 61-72, PROF ZOOL & BASIC MED SCI, UNIV ILL, URBANA, 72- Concurrent Pos: Actg asst prof, Stanford Univ, 45-46; vis prof, Univ Wash, 54 & Univ Iowa, 68; mem fel panel cell biol, NIH, 61-64; Leslie B Arey lectr, Med Sch, Northwestern Univ, 65; guest lectr, Bermuda Biol Sta, 66; secy subcomt embryol, Int Anat Nomenclature Comt, 71-72. Mem: AAAS; Soc Develop Biol; Am Soc Cell Biol; Soc Exp Biol & Med; Am Soc Zoologists. Res: Analysis of development of pigment cells in down feathers; neural tube; intraneural vascularity; glycogen body; vertebral column and ribs; hatching muscle; endocrine glands and hormone dependent structures of avian embryos; drug action liver, heart, kidneys and neural tube; myeloschisis; invertebrate embryology. Mailing Add: Dept of Genetics & Develop Univ of Ill Urbana IL 61801

WATTERSTON, KENNETH GORDON, b Rockville Centre, NY, Apr 9, 34; m 61; c 2. FOREST SOILS, ENVIRONMENTAL MANAGEMENT. Educ: State Univ NY

Col Forestry, Syracuse, BS, 59, MS, 62; Univ Wis, PhD(soils), 66. Prof Exp: Res asst forest soils, State Univ NY Col Forestry, Syracuse, 59-61; res asst soils, Univ Wis, 61-65; from asst prof to assoc prof forest soils, 65-75, PROF FOREST SOILS, STEPHEN F AUSTIN STATE UNIV, 75- Mem: Soil Sci Soc Am; Ecol Soc Am; Soc Am Foresters. Res: Forest soil-site relationships; forest soil classification; soil pollution and reclamation. Mailing Add: Sch of Forestry Box 6109 SFA Sta Nacogdoches TX 75961

WATTHEY, JEFFREY WILLIAM HERBERT, b London, Eng, Dec 6, 37; m 61; c 1. ORGANIC CHEMISTRY, MEDICINAL CHEMISTRY. Educ: Imp Col, Univ London, BSc, 59; St Catherine's Col, Oxford, DPhil(org chem), 62. Prof Exp: SR STAFF SCIENTIST, PHARMACEUT DIV, CIBA-GEIGY CORP, ARDSLEY, 64- Mem: Am Chem Soc; The Chem Soc. Res: Organic synthesis; synthesis of biologically active substances. Mailing Add: 61 Deepwood Dr Chappaqua NY 10514

WATTS, ALVA BURL, b Shreveport, La, Aug 1, 18; m 42; c 3. ANIMAL NUTRITION. Educ: Southwestern La Inst, BS, 39; La State Univ, MS, 41; Okla Agr & Mech Col, PhD(animal nutrit), 50. Prof Exp: Jr biochemist, Southern Regional Res Lab, USDA, 41-43; instr animal indust, 46-49, from asst prof to assoc prof poultry sci, 49-55, PROF POULTRY SCI & HEAD DEPT, LA STATE UNIV, BATON ROUGE, 55- Mem: Fel AAAS; fel Am Inst Chemists; Am Inst Nutrit; Am Chem Soc; Am Soc Animal Sci. Res: Feeding value of cottonseed meal; energy requirements of broilers and laying hens; protein quality investigations. Mailing Add: Dept of Poultry Sci La State Univ Baton Rouge LA 70803

WATTS, CHARLES EDWARD, b Mo, Mar 21, 28; m 57; c 2. MATHEMATICS. Educ: Drury Col, MusB, 50; Univ Calif, MA, 56, PhD, 57. Prof Exp: Instr math, Univ Chicago, 57-69, NSF fel, 60-61; from asst prof to assoc prof, 61-69, PROF MATH, UNIV ROCHESTER, 69- Mem: Am Math Soc. Res: Algebraic topology; homological algebra. Mailing Add: Dept of Math Univ of Rochester Rochester NY 14627

WATTS, DANIEL JAY, b East Cleveland, Ohio, Oct 19, 43. CHEMISTRY, BOTANY. Educ: Ohio State Univ, BSc, 65; Ind Univ, Bloomington, AM, 68, PhD(org chem), 69. Prof Exp: RES INVESTR, ORG CHEM, E R SQUIBB & SONS, INC, 69- Mem: Am Chem Soc; The Chem Soc; Bot Soc Am. Res: Antibiotics; anti-infective agents; organic synthesis; chemotaxonomy; natural products. Mailing Add: E R Squibb & Sons Dept Org Chem Box 4000 Princeton NJ 08540

WATTS, DANIEL THOMAS, b Wadesboro, NC, July 31, 16; m 63; c 2. PHARMACOLOGY. Educ: Elon Col, AB, 37; Duke Univ, PhD(physiol), 42. Prof Exp: Asst zool & physiol, Duke Univ, 39-42; physiologist, Naval Air Exp Sta, Pa, 46-47; from asst prof to assoc prof pharmacol, Med Sch, Univ Va, 47-53; prof & head dept, Med Ctr, WVa Univ, 53-66; PROF PHARMACOL & DEAN SCH BASIC SCI & GRAD STUDIES, MED COL VA, 66- Concurrent Pos: Consult, Walter Reed Army Inst Res, 58-67. Mem: Am Soc Pharmacol & Exp Therapeut; Soc Exp Biol & Med; AAAS. Res: Central nervous system depressants; cardiovascular drugs; experimental shock; aerospace physiology. Mailing Add: Sch of Basic Sci & Grad Studies Med Col of Va Richmond VA 23298

WATTS, DENNIS RANDOLPH, b Riverside, Calif, Dec 7, 43; m 69; c 2. PHYSICAL OCEANOGRAPHY. Educ: Univ Calif, Riverside, BA, 66; Cornell Univ, PhD(physics), 73. Prof Exp: Res assoc phys oceanog, Yale Univ, 72-74; ASST PROF PHYS OCEANOG, SCH OCEANOG, UNIV RI, 74- Mem: AAAS; Am Geophys Union; Am Phys Soc; Sigma Xi. Res: Descriptive and dynamical study of ocean currents, their fluctuation such as eddies, and other processes controlling the oceanic thermocline. Mailing Add: Sch of Oceanog Univ of RI Kingston RI 02881

WATTS, DONALD GEORGE, b Winnipeg, Man, Dec 4, 33; m 58; c 2. APPLIED STATISTICS. Educ: Univ BC, BASc, 56, MASc, 58; Univ London, PhD(elec eng), 62. Prof Exp: Systs analyst, DeHavilland Aircraft Co, Can, 62-64; vis asst prof math, Univ Wis-Madison, 64-65, assoc prof statist, 65-70; PROF STATIST, QUEEN'S UNIV, ONT, 70- Honors & Awards: Heaviside Premium Award, Brit Inst Elec Engrs, 61. Mem: Am Statist Asn. Res: Time series analysis, control and applications of statistics in many disciplines; teaching devices and methods. Mailing Add: Dept of Math Queen's Univ Kingston ON Can

WATTS, EXUM DEVER, b Nashville, Tenn, Mar 19, 26; m 48; c 3. ORGANIC CHEMISTRY. Educ: George Peabody Col, BS & MA, 48; Vanderbilt Univ, PhD, 54. Prof Exp: Instr chem, Florence State Col, 48-49; asst prof, Harding Col, 52-54; from asst prof to assoc prof, 54-60, PROF CHEM, MID TENN STATE UNIV, 60- Mem: AAAS; Am Chem Soc; Nat Sci Teachers Asn. Res: Organic mercurials; ultraviolet spectra; interpretation of organic spectra. Mailing Add: Dept of Chem Mid Tenn State Univ Murfreesboro TN 37130

WATTS, HARRY, b Burnham-on-Sea, Eng, June 23, 28; m 57; c 1. PHYSICAL CHEMISTRY. Educ: Bristol Univ, BSc, 49, PhD(aluminum halide complexes), 53. Prof Exp: Fel, Univ Calif, Berkeley, 52-54; sci officer, Explosives Res & Develop Estab, 54-55, sr sci officer, 55-56; lectr, SAustralian Inst Technol, 56-59, sr lectr, 59-63; researcher, Univ Reading, 63, res chemist, 64-73, sr res specialist, 73-75; GROUP LEADER, DOW CHEM CAN, LTD, 75- Concurrent Pos: Consult, Weapons Res Estab SAustralia, 62. Mem: Fel Chem Inst Can; The Chem Soc; Royal Australian Chem Inst. Res: Calorimetry; thermodynamics of nonelectrolyte solutions; diffusion in liquids and gases; flow and diffusion of gases through porous solids; polymer chemistry. Mailing Add: Dow Chem of Can Ltd R&D Dept PO Box 1012 Sarnia ON Can

WATTS, JOHN CONWAY, organic chemistry, see 12th edition

WATTS, JOHN GORDON, b Bethune, SC, Dec 21, 10; m 36; c 2. ENTOMOLOGY. Educ: Clemson Col, BS, 31; Ohio State Univ, MS, 33, PhD, 56. Prof Exp: Asst entom, Exp Sta, Clemson Col, 31-37, asst entomologist, Edisto Exp Sta, 38-41, from assoc entomologist to entomologist, 46-55; head dept bot & entom, 55-75, PROF ENTOM, N MEX STATE UNIV, 55- Mem: Entom Soc Am. Res: Cotton and range insects; insect biology and ecology. Mailing Add: Dept of Bot & Entom NMex State Univ Las Cruces NM 88003

WATTS, JUDITH ELIZABETH CULBRETH, b Greenville, SC, Apr 28, 43; m 64. CLINICAL CHEMISTRY. Educ: Furman Univ, BS, 65; Univ Md, College Park, PhD(inorg chem), 71. Prof Exp: Teaching asst chem, Univ Md, College Park, 65-68; instr, Northern Va Community Col, 68-69; from jr instr to instr, Univ Md, College Park, 69-72; CLIN CHEMIST, DIAG LABS, 73- Mem: AAAS; Am Chem Soc; Sigma Xi. Res: Computer calculation of group electronegatives; correlation of computed charges to observed physical and chemical properties; development of clinical chemistry assays for drugs; heavy metals. Mailing Add: Box 185 Rte 4 Ft Mill SC 29715

WATTS, LEWIS WILLIAM, JR, organic chemistry, see 12th edition

WATTS, MALCOLM S M, b New York, NY, Apr 30, 15; m 47; c 4. INTERNAL MEDICINE. Educ: Harvard Univ, AB, 37, MD, 41. Prof Exp: From asst clin prof to assoc clin prof med, 53-71, coordr cardiovasc bd, 52-56, actg dir cardiovasc res inst, 56-57, asst dean, 56-66, spec asst to chancellor, 63-71, CLIN PROF MED, SCH MED, UNIV CALIF, SAN FRANCISCO, 71-, ASSOC DEAN, 66-, DIR EXTENDED PROGS MED EDUC, 74- Concurrent Pos: Ed, Calif Med, 68-73; ed, Western J Med, 74- Mem: Nat Inst Med; AAAS; fel Am Col Physicians; Am Soc Internal Med (pres, 65); fel Am Geriat Soc. Res: Private practice of internal medicine; examination of the role of the physician in modern society; medical education. Mailing Add: San Francisco Med Ctr Univ of Calif Sch of Med San Francisco CA 94143

WATTS, PLATO HILTON, JR, physical chemistry, biophysics, see 12th edition

WATTS, RAYMOND ELLSWORTH, b Ft Collins, Colo, Dec 3, 09; m 33; c 2. VETERINARY SURGERY. Educ: Colo State Univ, DVM, 42; Univ Ill, MS, 48. Prof Exp: Instr vet path, Univ Ill, 46-48; asst chief vet, State of Wis, 48-50; pvt pract, 50-52; assoc prof clin & vet med, Wash State Univ, 52-60; assoc pathologist, 60-67, PROF VET SCI & ANIMAL PATHOLOGIST, AGR EXP STA, UNIV ARIZ, 67- Concurrent Pos: Exchange prof, Punjab Univ, Pakistan, 54-56. Mem: Am Vet Med Asn; Am Asn Vet Clinicians (secy-treas, 60). Res: Veterinary obstetrics and reproduction; nutritional diseases; antibiotics in dairy animals. Mailing Add: Dept of Vet Sci Rm 224 Univ of Ariz Agr Sci Bldg Tucson AZ 85721

WATTS, RODERICK KENT, b Houston, Tex, Apr 5, 39; m 62; c 1. SOLID STATE PHYSICS. Educ: Rice Univ, BA, 61, MA, 63, PhD(physics), 65. Prof Exp: MEM TECH STAFF, CENT RES LAB, TEX INSTRUMENTS, INC, 65- Concurrent Pos: Mem staff, Inst Appl Solid State Physics, Freiburg, Ger, 71-72. Mem: Am Phys Soc. Res: Luminescence; infrared-to-visible conversion; spectroscopy; integrated optics; lasers; x-ray lithography. Mailing Add: MS 118 Tex Instruments Inc PO Box 5936 Dallas TX 75222

WATTS, TERENCE LESLIE, b Leicester, Eng, May 5, 35; m 62; c 2. PHYSICS. Educ: Univ London, BSc, 57; Yale Univ, PhD(physics), 63. Prof Exp: Res asst nuclear physics, Yale Univ, 63; res assoc particle physics, Duke Univ, 63-64; mem res staff, Lab Nuclear Sci, Mass Inst Technol, 64-65, asst prof physics, 65-70; ASSOC PROF PHYSICS, RUTGERS UNIV, NEW BRUNSWICK, 70- Mem: Am Phys Soc. Res: Particle physics using bubble chambers and spark chambers; data processing of bubble chamber photographs; interactive use of computers. Mailing Add: Dept of Physics Rutgers Univ New Brunswick NJ 08903

WATTS, WILLIAM WILBUR, b Elmira, NY, Aug 23, 41; m 64; c 2. ACADEMIC ADMINISTRATION, INSTITUTIONAL RESEARCH. Educ: Rensselaer Polytech Inst, BEE, 63, MEE, 64, PhD(systs eng), 69. Prof Exp: Instr physics, Wheaton Col, Ill, 66-69, asst prof, 69-72, ASSOC PROF PHYSICS & CHMN DEPT, KING'S COL, NY, 72-, REGISTRAR & DIR INSTNL RES, 75- Mem: Am Inst Physics; Am Asn Physics Teachers; Am Sci Affil. Res: Systems engineering; history and philosophy of science. Mailing Add: King's Col Briarcliff Manor NY 10510

WATTSON, RICHARD BELL, astrophysics, high energy physics, see 12th edition

WATZKE, ROBERT COIT, b Madison, Wis, Dec 19, 22; m 56; c 2. OPHTHALMOLOGY. Educ: Univ Wis, BS, 50, MD, 52; Am Bd Ophthal, dipl, 57. Prof Exp: Intern, Med Ctr, Ind Univ, 52-53; resident, Univ Wis, 56; res asst ophthal, Harvard Med Sch, 56-57; from asst prof to assoc prof, 66-73, PROF OPHTHAL, COL MED, UNIV IOWA, 73- Mem: Fel Am Acad Ophthal. Res: Clinical ophthalmology; retinal detachment and the vitreous humor of the eye. Mailing Add: Dept of Ophthal Univ of Iowa Col of Med Iowa City IA 52240

WATZMAN, NATHAN, b Powhattan Point, Ohio, Feb 15, 26; m 59; c 3. PSYCHOPHARMACOLOGY, PHYSIOLOGY. Educ: Univ Pittsburgh, BS, 47 & 55; MS, 57, PhD(muscular dystrophy), 61. Prof Exp: Asst prof pharmacol, Sch Pharm, Northeast La State Col, 59-63; res assoc prof, Sch Pharm, Univ Pittsburgh, 63-68; health sci adminr, Div Res Grants, 68-69, health sci adminr, Bur Health Manpower Educ, NIH, 69-74; ASSOC DIR REGIONAL PROGS, DIV ASSOC HEALTH PROFESSIONS, HEALTH RESOURCE ADMIN, 74- Mem: Am Soc Pharmacol & Exp Therapeut; Acad Pharmaceut Sci. Res: Drug interaction with stress and other environmental modifications of behavior. Mailing Add: Health Resource Admin Rm 3C08 Bldg 31 9000 Rockville Pike Bethesda MD 20014

WAUD, DOUGLAS RUSSELL, b London, Ont, Oct 21, 32; m 56; c 3. PHARMACOLOGY. Educ: Univ Western Ont, MD, 56; Oxford Univ, DPhil(pharmacol), 64. Prof Exp: Intern, St Joseph's Hosp, London, Ont, 56-57; instr pharmacol, Harvard Med Sch, 59-60; demonstr, Oxford Univ, 61-63; from assoc to assoc prof, Harvard Med Sch, 63-74; PROF PHARMACOL, MED SCH, UNIV MASS, 74- Concurrent Pos: USPHS career develop award, 66- Mem: Am Soc Pharmacol & Exp Therapeut; AAAS; NY Acad Sci; Brit Pharmacol Soc; Brit Physiol Soc. Res: Mechanisms of drug action at molecular level; autonomic and cardiovascular pharmacology; neuromuscular junction physiology and pharmacology; anesthetic agents; uptake and distribution of drugs. Mailing Add: Dept of Pharmacol Univ of Mass Med Sch Worcester MA 01605

WAUER, ROLAND H, b Idaho Falls, Idaho, Mar 22, 34; m 66. ECOLOGY, ORNITHOLOGY. Educ: San Jose State Col, BS, 57; Sul Ross State Univ, MS, 71. Prof Exp: Park naturalist, Death Valley Nat Monument, Calif, 57-62 & Zion Nat Park, Utah, 62-66; chief naturalist, Big Bend Nat Park, 66-72; REGIONAL SCIENTIST, OFF NATURAL SCI, NAT PARK SERV, 72- Mem: Am Ornith Union; Cooper Ornith Soc; Wilson Ornith Soc. Mailing Add: Off of Natural Sci Nat Park Serv Southwest Region Santa Fe NM 87501

WAUGH, DAVID FLOYD, b Kirkwood, Mo, Jan 6, 15; m 37; c 3. BIOLOGY. Educ: Washington Univ, AB, 35, PhD(physiol), 40. Prof Exp: Asst zool, Washington Univ, 36-38, instr, 39-41; from asst prof to assoc prof phys biol, 41-54, PROF BIOPHYS, MASS INST TECHNOL, 54- Concurrent Pos: Mem comt cardiovasc syst & subcomt thrombosis & hemorrhage, Div Med Sci, Nat Acad Sci-Nat Res Coun; mem biophys & biophys chem study sect, NIH, 58-60, chmn, 60-63. Honors & Awards: Borden Award, Am Chem Soc, 62. Mem: AAAS; Am Chem Soc; Am Physiol Soc; Am Soc Biol Chem; Biophys Soc; Am Acad Arts & Sci. Res: Molecular interactions in the development of fibrin clot structure and the bioconversion of prothrombin into thrombin; human platelet characteristics; processes which lead to thrombus formation on implant surfaces. Mailing Add: Dept of Biol 56-335 Mass Inst Technol Cambridge MA 02139

WAUGH, DONOVAN LLOYD, b Poynette, Wis, July 9, 37; m 57; c 3. SOIL FERTILITY. Educ: Univ Wis-Madison, BS, 59, MS, 61, PhD(soils), 63. Prof Exp: Res assoc soils, Univ Wis-Madison, 63-64; from asst prof to assoc prof, NC State Univ, 64-70, regional dir int soil fertil contract, 64-70, int consult soils, 70-74; SOILS ADV, USAID, PARAGUAY, 75- Mem: Am Soc Agron. Res: Soil fertility evaluation through laboratory, greenhouse and field studies; soil fertility management techniques and systems related to no-till cultural practices in corn, soybeans and wheat. Mailing Add: Rte 1 Poynette WI 53955

WAUGH, DOUGLAS OLIVER WILLIAM, b Hove, Eng, Mar 21, 18; m 46. MEDICAL EDUCATION, PATHOLOGY. Educ: McGill Univ, MD, CM, 42, MSc, 48, PhD(path), 50; Royal Col Physicians & Surgeons Can, cert path, 54; FRCP, 64. Prof Exp: Demonstr & asst surg pathologist, Path Inst, McGill Univ, 46-47, assoc prof path, 51-57, Miranda Fraser assoc prof comp path, 57; assoc prof path, Univ & asst pathologist, Hosp, Univ Alta, 50-51; from assoc prof to prof, Queen's Univ, Can, 58-64; prof & head dept, Dalhousie Univ, 64-70; prof path & dean fac med, Queen's Univ, Ont, 70-75; EXEC DIR, ASN CAN MED COLS, 75- Concurrent Pos: Asst prov pathologist, Alta, 50-51; mem cancer diag clin, Alta Dept Health, 50-51; mem comt consults, Can Tumor Registry, 55-62; med mem adv comt tumor registry, Nat Cancer Inst, 56-62, dir inst, 65, pres, 74-76; dir labs, Hotel Dieu Hosp, Kingston, 58-; chmn, Can Cytol Coun, 64-65. Mem: Am Soc Exp Pathologists; Am Asn Pathologists & Bacteriologists; Can Asn Pathologists; Int Acad Pathologists. Res: Renal diseases; hypertension; lesions of experimental hypersensitivity. Mailing Add: Asn of Can Med Cols 151 Slater St Ottawa ON Can

WAUGH, JOHN BLAKE-STEELE, b Sydney, Australia, Nov 19, 23; m 54; c 3. ELECTRONICS, ENGINEERING PHYSICS. Educ: Univ Sydney, BS, 48; Univ NSW, MS, 56. Prof Exp: Res officer, Atomic Energy Can, Ltd, 56-60; fel electronics, Australian Nat Univ, 60-63; assoc scientist, Brookhaven Nat Lab, 63-64; sr res assoc, Univ Rochester, 64-69; mgr imaging, Xerox Corp, 69-71; CORP STAFF SCIENTIST ELECTRONICS, SINGER CO, FAIRFIELD, 71- Mem: Inst Elec & Electronics Engrs; Am Phys Soc; Optical Soc Am; Brit Inst Physics; Inst Elec Engrs Gt Brit. Res: Computers; signal processing; instrumentation; communications. Mailing Add: 144 Lookout Rd Mountain Lakes NJ 07046

WAUGH, JOHN GEORGE, chemistry, mathematics, see 12th edition

WAUGH, JOHN LODOVICK THOMSON, b Avonhead, Scotland, Nov 13, 22; m 49; c 4. PHYSICAL INORGANIC CHEMISTRY. Educ: Univ Glasgow, BSc, 43, PhD(inorg chem), 49. Prof Exp: Plant supt, Imp Chem Industs, Ltd, 43-46; asst lectr, Univ Glasgow, 46-49; Climax Molybdenum Co res fel, Calif Inst Technol, 49-50, inst res fel, 51-53; asst prof chem, Univ Hawaii, 50-51; res chemist, Pac Coast Borax Co, 53-56; assoc prof chem, 56-72, PROF CHEM, UNIV HAWAII, 72- Concurrent Pos: Prog dir, NSF Undergrad Res Partic Prog, 60-63; assoc res officer, Neutron Physics Br, Atomic Energy Can, Ltd, Ont, 62-63. Mem: NY Acad Sci; fel The Chem Soc; assoc Royal Inst Chem. Res: Isopoly and heteropoly compounds; boron and boron compounds; x-ray crystallography; intermetallic compounds; lattice dynamics. Mailing Add: 3181 Beaumont Woods Pl Honolulu HI 96822

WAUGH, JOHN STEWART, b Willimantic, Conn, Apr 25, 29; m 54; c 2. PHYSICAL CHEMISTRY. Educ: Dartmouth Col, AB, 49; Calif Inst Technol, PhD(chem, physics), 53. Prof Exp: From instr to prof, 53-73, A A NOYES PROF CHEM, MASS INST TECHNOL, 73- Concurrent Pos: Sloan res fel, 58-62; assoc, Retina Found, 61-70; vis scientist, USSR Acad Sci, Moscow, 62 & 75; Guggenheim fel & res assoc physics, Univ Calif, Berkeley, 63-64; consult, Lawrence Radiation Lab, Univ Calif, 64-; ed, Advan Magnetic Resonance, 65-; Robert A Welch Found lectr, Univ Tex, 69; mem chem rev panel, Argonne Nat Lab, 70-74, chmn, 73-74; vis prof & Humboldt fel, Max Planck Inst Med Res, 72-; Falk-Plant lectr, Columbia Univ, 74. Honors & Awards: Irving Langmuir Award, Am Chem Soc, 76. Mem: Nat Acad Sci; Int Soc Mangetic Resonance; fel Am Acad Arts & Sci; Am Phys Soc. Res: Magnetic resonance. Mailing Add: Dept of Chem Mass Inst of Technol Cambridge MA 02139

WAUGH, RICHARD CAMPBELL, b Fayetteville, Ark, Aug 8, 17; m 42; c 4. ORGANIC CHEMISTRY. Educ: Univ Ark, BS, 37; Univ Colo, PhD(org chem), 42. Prof Exp: Chemist, Tenn Eastman Corp, 41 & Stand Oil Co, Ind, 42-46; pres, 46-72, CORP VPRES CHEM MFG, ARAPAHOE CHEM DIV, SYNTEX CORP, 72- Mem: Am Chem Soc. Res: Isomerization of hydrocarbons; general organic synthesis. Mailing Add: Syntex Corp Stanford Indust Park Palo Alto CA 94304

WAUGH, ROBERT KENNETH, animal nutrition, see 12th edition

WAUGH, WILLIAM HOWARD, b New York, NY, Mar 13, 25; m 52; c 3. PHYSIOLOGY, INTERNAL MEDICINE. Educ: Tufts Col, MD, 48. Prof Exp: Intern internal med, Long Island Col Hosp, 48-49, asst resident, 50-51; asst resident med, Univ Md Hosp, 51-52; cardiovasc trainee, Med Col Ga, 54-55, asst res prof physiol, 55-60, USPHS sr res fel, 59-60, assoc med, 57-60; from assoc prof to prof med, Col Med, Univ Ky, 60-71, head renal div, 60-68, Ky Heart Asn chair cardiovasc res, 63-71; PROF MED & PHYSIOL, SCH MED, E CAROLINA UNIV, 71-, DIR CLIN SCI, 71- Concurrent Pos: Estab investr, Ga Heart Asn, Med Col Ga, 58, physician in chg hemodialysis sect, 59-60. Mem: Microcirc Soc; Am Soc Nephrology; Am Physiol Soc; Am Heart Asn; fel Am Col Physicians. Res: Hemodynamics; circulatory and renal physiology; nephrology. Mailing Add: E Carolina Univ Sch of Med Greenville NC 27834

WAVE, HERBERT EDWIN, b Portsmouth, NH, Oct 13, 23; m 47; c 1. ENTOMOLOGY, PLANT PATHOLOGY. Educ: Univ Maine, BS, 52; Rutgers Univ, MS, 60, PhD(entom), 61. Prof Exp: Entomologist, USDA, 52-58; res asst entom, Rutgers Univ, 58-61; entomologist, USDA, 61-62; asst prof entom, Univ Mass, 62-67; ASSOC PROF PLANT & SOIL SCI, UNIV MAINE, 67- Mem: Entom Soc Am; Am Soc Hort Sci; Weed Sci Soc Am. Res: Biology and ecology of potato infesting species of aphids; extension pest control for tree and small fruits; orchard herbicides and growth regulators. Mailing Add: Dept of Plant & Soil Sci Univ of Maine Rangeley Rd Orono ME 04473

WAVRIK, JOHN J, b New York, NY, Mar 17, 41; m 61. GEOMETRY. Educ: Johns Hopkins Univ, AB, 61, MA, 64; Stanford Univ, PhD(math), 66. Prof Exp: Joseph Fells Ritt instr math, Columbia Univ, 66-69; ASSOC PROFMATH, UNIV CALIF, SAN DIEGO, 69- Mem: Am Math Soc. Res: Algebraic geometry; complex analytic spaces and deformations of complex structures. Mailing Add: Dept of Math Univ of Calif San Diego La Jolla CA 92093

WAWNER, FRANKLIN EDWARD, JR, b Petersburg, Va, Dec 12, 33; m 53; c 3. MATERIALS SCIENCE. Educ: Randolph-Macon Col, BS, 59; Univ Va, MS, 68, PhD(mat sci), 71. Prof Exp: Physicist, Army Eng Res & Develop Lab, 59-61; sr physicist, Texaco Exp Inc, 61-66; PRIN SCIENTIST COMPOSITE MAT, DEPT MAT SCI, UNIV VA, 71- Concurrent Pos: Consult, Avco Systs Div, Avco Corp, 69- & Army Foreign Sci & Technol Ctr, 74- Res: Composite materials; chemical vapor deposition; characterization of mechanical properties; fracture; structure; microstructure. Mailing Add: Dept of Mat Sci Thornton Hall Univ of Va Charlottesville VA 22901

WAWSZKIEWICZ, EDWARD JOHN, b North Smithfield, RI, Feb 10, 33. MICROBIOLOGY, BIOCHEMISTRY. Educ: Harvard Univ, AB, 54; Univ Calif, Berkeley, PhD, 61. Prof Exp: Asst microbiol, Hopkins Marine Sta, Stanford Univ, 54-55; res asst microbiol, Univ Calif, Berkeley, 55-61; USPHS fel, Max Planck Inst Cell Chem, Ger, 61-63; resident res assoc, Argonne Nat Lab, 64-66; asst mem, Inst Biomed Res, 66-70; ASSOC PROF MICROBIOL, UNIV ILL MED CTR, 70- Mem: AAAS; Am Soc Microbiol; fel Am Inst Chemists; Am Inst Biol Sci; Am Chem Soc. Res: Metabolism of thiobacilli; erythritol metabolism by propionibacteria; propionate metabolism; erythromycin biosynthesis; mouse salmonellosis, iron metabolism; pacifarins; enology. Mailing Add: Dept of Microbiol Univ of Ill at the Med Ctr Chicago IL 60680

WAWZONEK, STANLEY, b Valley Falls, RI, June 23, 14; m 43; c 3. ORGANIC CHEMISTRY. Educ: Brown Univ, BS, 35; Univ Minn, PhD(org chem), 39. Prof Exp: Chemist, Univ Minn, 39-40; Nat Res Coun fel, Univ Ill, 40-41, instr chem, 41-43; instr, Univ Tenn, 43-44; from asst prof to assoc prof, 44-52, chmn dept, 62-68, PROF CHEM, UNIV IOWA, 52- Mem: AAAS; Am Chem Soc; Electrochem Soc. Res: Organic polarography and synthesis; medicinal chemistry; organic nitrogen heterocycles; aminimides. Mailing Add: Dept of Chem Univ of Iowa Iowa City IA 52240

WAX, HARRY, b Boston, Mass, June 17, 18; m 50; c 2. BIOCHEMISTRY. Educ: Univ Calif, Los Angeles, MA, 41; Iowa State Col, PhD(chem), 49. Prof Exp: Dir fine chems div, Wm T Thompson Co, 40-44; res chemist, Dr Geo Piness Allergy Group, 49-53; res chemist pharmaceuts, Stuart Co, 53-59, mgr prod develop, 59-69, Stuart Pharmaceut Res Dept, Atlas Chem Indust, 69-74; WITH ARCHON PURE PROD CORP, 74- Concurrent Pos: Instr, Univ Calif, Los Angeles, 51-52. Mem: AAAS; Am Chem Soc; Am Pharmaceut Asn. Res: Enzymatic synthesis of peptide bonds; application of ion exchange resins and filter paper electrophoresis to the separation of allergens; organosynthesis; pharmaceutical manufacturing processes. Mailing Add: Archon Pure Prod Corp 345 N Baldwin Park Blvd City of Industry CA 91746

WAX, JOAN, b Detroit, Mich, Sept 14, 21. PHARMACOLOGY. Educ: Wayne State Univ, BA, 43; Univ NC, MS, 47. Prof Exp: From res asst to asst res pharmacologist to assoc res pharmacologist, 47-67, RES PHARMACOLOGIST, PARKE, DAVIS & CO, 67- Mem: AAAS; Am Soc Pharmacol & Exp Therapeut; Biomet Soc; Am Chem Soc. Res: Analgetic and anti-inflammatory agents; narcotic antagonists; drug-induced gastrointestinal ulcerogenesis. Mailing Add: Parke Davis & Co Ann Arbor MI 48106

WAX, RICHARD GERALD, b New York, NY, Oct 24, 34; m 64; c 1. MICROBIAL GENETICS. Educ: Polytech Inst Brooklyn, BChE, 56; Yale Univ, MS, 61; Pa State Univ, PhD(biophys), 63. Prof Exp: Nuclear engr, Combustion Eng Co, 56-57; staff fel, NIH, 64-67; NIH spec fel, Weizmann Inst Sci, 67-69; FEL FERMENTATION RES, MERCK RES LABS, 69- Mem: AAAS; Am Soc Microbiol. Res: Molecular biology of the bacterial spore; protein synthesis; antibiotic production. Mailing Add: Merck Res Labs Rahway NJ 07065

WAX, ROBERT LEROY, b Des Moines, Iowa, July 7, 38. SPACE PHYSICS. Educ: Calif Inst Technol, BS, 60; Univ Ill, Urbana, MS, 61; Univ Calif, Berkeley, PhD(physics), 65. Prof Exp: Mem tech staff, TRW Systs Group, 66-71; physicist, Space Environ Lab, Nat Oceanic & Atmospheric Admin, 71-73; CONSULT PHYSICIST, 73- Mem: Am Geophys Union. Mailing Add: 201 S Poinsettia Manhattan Beach CA 90266

WAX, ROSALIE H, b Des Plaines, Ill, Nov 4, 11; m 49. ANTHROPOLOGY. Educ: Univ Calif, Berkeley, BA, 42; Univ Chicago, PhD(anthrop), 50. Prof Exp: Teaching asst anthrop, Univ Chicago, 46-47, instr soc sci, Col, 47-49, asst prof anthrop, 50-57, Ernest E Quantrell prize, 56-57; vis lectr sociol & anthrop, Univ Miami, 59-62; assoc prof, Univ Kans, 64-69, prof, 70-73; PROF ANTHROP, WASH UNIV, 73- Concurrent Pos: Examr, Soc Sci II, Col, Univ Chicago, 47-55, chmn, 56-57; dir workshop Am Indian affairs, Univ Colo, 59 & 60; Soc Sci Res Coun fac res fel, 61; res assoc, Oglala Sioux Educ Res Proj, Emory Univ, 62-64; Wenner-Gren Found grant, 66; assoc dir Indian educ res proj, Univ Kans, 66-69. Mem: Fel AAAS; fel Am Anthrop Asn; Am Ethnol Soc; fel Am Sociol Asn; fel Soc Appl Anthrop. Mailing Add: Dept of Anthrop Wash Univ St Louis MO 63130

WAXDAL, MYRON JOHN, b San Francisco, Calif, Dec 8, 37; m 62; c 2. BIOCHEMISTRY, IMMUNOLOGY. Educ: Univ Wash, BS, 60, PhD(biochem), 65. Prof Exp: Res assoc, Rockefeller Univ, 65-69, asst prof, 69-72; sr staff fel immunol, 72-74, RES CHEMIST, NAT INST ALLERGY & INFECTIOUS DIS, 74- Mem: AAAS; Am Asn Immunologists; Am Asn Biol Chemists. Res: Structure and function, especially of proteins, cell membranes and organelles; immunology, especially antibody structure and cell activation or stimulation. Mailing Add: Lab of Immunol Nat Inst of Allergy & Infect Dis Bethesda MD 20014

WAXLER, GLENN LEE, b Olney, Ill, Feb 24, 25; m 46; c 2. VETERINARY PATHOLOGY. Educ: Univ Ill, BS, 51, DVM, 53; Mich State Univ, MS, 59, PhD(vet path), 61. Prof Exp: Pvt pract, 53-57; from instr to assoc prof vet path, 57-66, PROF PATH, MICH STATE UNIV, 66- Mem: Am Vet Med Asn; Am Col Vet Path; Asn Gnotobiotics; Conf Res Workers Animal Dis; Am Soc Exp Path. Res: Germ-free research; Enteric disease of swine; histopathology of diseases of domestic animals. Mailing Add: Dept of Path Mich State Univ East Lansing MI 48824

WAXLER, WILLIAM LORNE, chemistry, see 12th edition

WAXMAN, ALAN DAVID, b New York, NY, Mar 9, 38; m 61; c 3. NUCLEAR MEDICINE. Educ: Univ Southern Calif, BA, 58, MD, 63. Prof Exp: Intern med, Los Angeles County Gen Hosp, 63-64; resident med, Los Angeles Vet Admin Hosp, 64-65; res assoc metab, Metab Serv, Cancer Inst, NIH, 65-67; asst prof, 68-70, ASSOC PROF RADIOL, SCH MED, UNIV SOUTHERN CALIF, 70-, STAFF PHYSICIAN RADIOL, LOS ANGELES COUNTY-UNIV SOUTHERN CALIF MED CTR, 68- Mem: Soc Nuclear Med; Am Fedn Clin Res; AMA; fel Am Col Physicians; Am Col Nuclear Physicians. Res: Applications of nuclear medicine technology in the detection and evaluation of disease processes, primarily in oncology and hepatic disease. Mailing Add: LAC/USC Med Ctr Box 1952 1200 N State St Los Angeles CA 90033

WAXMAN, BURTON HARVEY, organic chemistry, photographic science, see 12th edition

WAXMAN, HERBERT SUMNER, b Boston, Mass, Sept 1, 36; m 60; c 3. MEDICINE, HEMATOLOGY. Educ: Mass Inst Technol, BS, 58; Harvard Univ, MD, 62; Am Bd Internal Med, dipl, 69, dipl hemat, 74. Prof Exp: From intern to resident med, Mass Gen Hosp, 62-64; res assoc biochem, Nat Cancer Inst, 64-66; resident, Mass Gen Hosp, Boston, 66-67; from asst prof to assoc prof, 68-75, PROF MED & DEP CHMN DEPT, SCH MED, TEMPLE UNIV, 75- Concurrent Pos: Mead Johnson scholar, Am Col Physicians, 66-67; NIH trainee hemat, Sch Med, Washington Univ, 67-68; advan clin fel, Am Cancer Soc, 68-71. Mem: Am Fedn Clin Res; Am Soc Hemat; fel Am Col Physicians. Res: Control of protein synthesis in blood cells; clinical studies in sickle cell and related diseases. Mailing Add: Dept of Med Temple Univ Sch of Med Philadelphia PA 19140

WAXMAN, MONROE H, physical chemistry, see 12th edition

WAXMAN, SIDNEY, b Providence, RI, Nov 13, 23; m 48; c 3. ORNAMENTAL HORTICULTURE. Educ: Univ RI, BS, 51; Cornell Univ, MS, 54, PhD(ornamental hort), 57. Prof Exp: ASSOC PROF ORNAMENTAL HORT, UNIV CONN, 57- Mem: Am Soc Hort Sci; Am Soc Plant Physiol; Am Asn Bot Gardens & Arboretums. Res: Photoperiodism; plant propagation; seed and bud dormancy. Mailing Add: Dept of Plant Sci Univ of Conn Storrs CT 06268

WAXMAN, STEPHEN GEORGE, b Newark, NJ, Aug 17, 45; m 68; c 2. NEUROLOGY. Educ: Harvard Univ, AB, 67; Albert Einstein Col Med, PhD(med sci), 70, MD, 72. Prof Exp: Clin fel neurol, Boston City Hosp, 72-75; ASST PROF NEUROL, HARVARD MED SCH, 75-; ASST NEUROLOGIST, BETH ISRAEL HOSP, BOSTON, 75- Concurrent Pos: Guest lectr, Upsala Col, 68-70; sci investr, Marine Biol Lab, Woods Hole, 71; NIH fel, Harvard Univ, 72-75; vis res fel biol, Mass Inst Technol, 74-75, vis asst prof, 75-; NIH career develop award, 75. Honors & Awards: Trygve Tuve Annual Mem Award, NIH Found Advan Educ in Sci, 74. Mem: Am Soc Cell Biol; Int Brain Res Orgn; Am Acad Neurol. Res: Basic and clinical neurosciences; electron microscopy; neurophysiology; mathematical modeling of brain function; cybernetics and communication theory. Mailing Add: Harvard Neurol Unit Beth Israel Hosp Boston MA 02215

WAY, E LEONG, b Watsonville, Calif, July 10, 16; m 44; c 2. PHARMACOLOGY, TOXICOLOGY. Educ: Univ Calif, BS, 38, MS, 40, PhD(pharmaceut chem), 42. Prof Exp: Asst pharmacol, Univ Calif, 42; pharmaceut chemist, Merck & Co, Inc, NJ, 42-43; from instr to asst prof pharmacol, Med Sch, George Washington Univ, 43-48; from asst prof to assoc prof, 49-57, vchmn dept, 57-67, PROF PHARMACOL & TOXICOL, MED CTR, UNIV CALIF, SAN FRANCISCO, 57-, CHMN DEPT, 73- Concurrent Pos: USPHS spec res fel, Univ Bern, 55-56; China Med Bd res fel, Univ Hong Kong, 62-63; consult, Attorney Gen, Calif, 59-60 & Dept Corrections, Calif, 64-70; mem, Comt on Probs Drug Safety, Nat Acad Sci-Nat Res Coun, 65-71 & Comt on Probs Drug Dependence, 68-74; Kauffman lectr, Ohio State Univ, 66; mem comt on abuse of depressant & stimulant drugs, Dept Health, Educ & Welfare, 66-68, mem pharmacol study sect, 66-70, chmn, 68-70, mem comt on narcotic addiction & drug abuse rev, 70-74, chmn, 71-74; mem sci adv comt drugs, Bur Narcotics & Dangerous Drugs, 68-74; mem alcohol & drug dependence serv adv group & merit rev bd, Vet Admin, 71-; mem controlled substances adv comt, Food & Drug Admin, 74- Honors & Awards: Am Pharmaceut Found Award, 62; Ebert Prize Cert, Am Pharmaceut Asn, 62. Mem: Fel AAAS; Am Soc Pharmacol & Exp Therapeut; Am Pharmaceut Asn; fel Am Col Neuropsychopharmacol; Int Soc Biochem Pharmacol. Res: Drug metabolism of analgetics; hypothalamic-pituitary adrenal mechanisms; pharmacology of analgetics; drug tolerance and physical dependence mechanisms. Mailing Add: Dept of Pharmacol Univ of Calif Med Ctr San Francisco CA 94143

WAY, FREDERICK, III, b Sewickley, Pa, Jan 4, 25; m 48; c 2. COMPUTER SCIENCE. Educ: Univ Pittsburgh, BS, 50. Prof Exp: Physicist, Babcock & Wilcox Co, 51-54, anal eng, Res Lab, 54-56; assoc prof, 56-70, PROF COMPUT TECHNOL, CASE WESTERN RESERVE UNIV, 70-, ASSOC DIR COMPUT CTR, 56- Concurrent Pos: Consult, Thompson Ramo Wooldridge Co, 59 & Bailey Meter Co, 60- Mem: Soc Indust & Appl Math; Asn Comput Mach; Math Asn Am. Res: Investigation and implementation of automatic programming methods for digital computers. Mailing Add: Dept of Comput & Info Sci Case Western Reserve Univ Cleveland OH 44106

WAY, HAROLD E, b Colchester, Ill, Jan 31, 04; m 26; c 3. PHYSICS. Educ: Knox Col, Ill, BS, 25; Univ Pittsburgh, MS, 28; Univ Iowa, PhD(physics), 36. Hon Degrees: LLD, Univ Pittsburgh, 71 & Knox Col, Ill, 71. Prof Exp: From instr to asst prof physics, Knox Col, Ill, 27-37, prof & chmn dept, 37-48, actg pres, 46, vpres, 47-48; Bailey prof physics & chmn dept, 48-66, dean sci & eng, 64-66, EMER PROF PHYSICS, UNION COL, NY, 66- Concurrent Pos: Consult, NSF, 62-66. Res: Single crystals of zinc. Mailing Add: 5515 E Covina Rd Mesa AZ 85205

WAY, JAMES LEONG, b Watsonville, Calif, Mar 21, 27; m 47; c 3. PHARMACOLOGY, TOXICOLOGY. Educ: Univ Calif, Berkeley, BA, 51; George Washington Univ, PhD(pharmacol), 55. Prof Exp: USPHS fel pharmacol, Univ Wis, 55-57; USPHS sr fel, 57-58; from instr to asst prof pharmacol, Univ Wis, 59-62; assoc prof, Med Col Wis, 62-67; PROF PHARMACOL, WASH STATE UNIV, 67- Concurrent Pos: USPHS career develop award, 58-62, spec res fel, 74-75; mem fed task group toxicol eval pesticide in mammalian species, Environ Protection Agency, 73-75; mem toxicol study sect, NIH, 74-78, mem toxicol data bank rev comt, 74-78; vis scientist, Div Molecular Pharmacol, Nat Inst Med Res, London, 73-75. Mem: AAAS; Am Chem Soc; Am Soc Pharmacol & Exp Therapeut; Soc Toxicol. Res: Cancer; nucleic acid; drug metabolism and anticholinesterase alkylphosphate antagonists; cyanide, nitrite and alkylphosphate poisoning; molecular and marine pharmacology. Mailing Add: Dept of Pharmacol Wash State Univ Pullman WA 99163

WAY, JOHN WILLIAM, organic chemistry, see 12th edition

WAY, KATHARINE, b Sewickley, Pa, Feb 20, 03. PHYSICS. Educ: Columbia Univ, BS, 32; Univ NC, PhD(physics), 38. Prof Exp: Huff res fel, Bryn Mawr Col, 38-39; from instr to asst prof physics, Univ Tenn, 39-42; physicist, US Naval Ord Lab, 42, Manhattan Proj, Oak Ridge Nat Lab, 42-48 & Nat Bur Stand, 49-53; dir nuclear data proj, Nat Res Coun, 53-63 & Oak Ridge Nat Lab, 64-68; ADJ PROF PHYSICS, DUKE UNIV, 68-; ED, ATOMIC DATA & NUCLEAR DATA TABLES, 73- Concurrent Pos: Ed, Nuclear Data Tables, 65-73, ed, Atomic Data, 69-73. Mem: Fel Am Phys Soc. Res: Nuclear fission; radiation shielding; nuclear constants. Mailing Add: Dept of Physics Duke Univ Durham NC 27706

WAY, LAWRENCE WELLESLEY, b St Louis, Mo, Nov 15, 33; m 71; c 1. SURGERY. Educ: Cornell Univ, AB, 55; Univ Buffalo, MD, 59. Prof Exp: Clin instr, 67-69, from asst prof to assoc prof, 69-75, PROF SURG, SCH MED, UNIV CALIF, SAN FRANCISCO, 75-, VCHMN DEPT, 72-; CHIEF SURG SERV, VET ADMIN HOSP, SAN FRANCISCO, 72- Concurrent Pos: Fel gastrointestinal physiol, Univ Calif, San Francisco, 67-68 & Vet Admin Ctr, Univ Calif, Los Angeles, 68-69; dir surg out patient dept, Univ Calif, San Francisco, 69-72. Mem: Am Col Surg; Am Gastroenterol Asn; Asn Acad Surg; Soc Surg Alimentary Tract; Soc Univ Surgeons. Res: Physiology and pathophysiology of gastrointestinal secretion. Mailing Add: Surg Serv Vet Admin Hosp 4150 Clement St San Francisco CA 94121

WAY, ROGER DARLINGTON, b Port Matilda, Pa, Nov 7, 18; m 53; c 4. POMOLOGY. Educ: Pa State Univ, BS, 40, MS, 42; Cornell Univ, PhD(pomol), 53. Prof Exp: Res assoc pomol, Exp Sta, State Univ NY Col Agr, 49-50; asst, Cornell Univ, 50-53; from asst prof to assoc prof, 53-70, PROF POMOL, EXP STA, NY STATE COL AGR & LIFE SCI CORNELL UNIV, 70- Mem: Am Soc Hort Sci. Res: Cherry varieties; winter hardiness; apple varieties and breeding. Mailing Add: NY Agr Exp Sta Geneva NY 14456

WAY, STEWART, mechanics, see 12th edition

WAY, WALTER, b Rochester, NY, June 27, 31; m 55; c 3. ANESTHESIOLOGY, PHARMACOLOGY. Educ: Univ Buffalo, BS, 53; State Univ NY, MD, 57; Univ Calif, MS, 62. Prof Exp: USPHS trainee pharmacol, 60-61; from instr to asst prof anesthesia, 61-63, from asst prof to assoc prof anesthesia & pharmacol, 63-74, PROF ANESTHESIA & PHARMACOL, MED CTR, UNIV CALIF, SAN FRANCISCO, 74- Res: Narcotic and clinical pharmacology. Mailing Add: Dept of Anesthesia & Pharmacol Univ of Calif Med Ctr San Francisco CA 94143

WAYBURN, EDGAR, b Macon, Ga, Sept 17, 06; m 47; c 4. INTERNAL MEDICINE. Educ: Univ Ga, AB, 26; Harvard Univ, MD, 30. Prof Exp: Vol asst path, Univ Berlin, 30-31; from clin instr to assoc clin prof, 34-64, ADJ ASSOC CLIN PROF MED, STANFORD UNIV, 64-; EMER ASSOC CLIN PROF, SCH MED, UNIV CALIF, SAN FRANCISCO, 75- Concurrent Pos: Dir, Chronic Illness Serv Ctr, San Francisco, 49-61, Morrison Rehab Ctr, 66-70 & Garden Hosp-Sullivan Rehab Ctr, 70-; assoc clin prof, Sch Med, Univ Calif, San Francisco, 60-75; chief endocrine & metab clin, Pac Med Ctr, 60-72; pres, Sierra Club Found, 71- Mem: AMA; Am Fedn Clin Res; fel Am Col Physicians; Am Soc Internal Med. Res: Diseases of gastro-intestinal tract, metabolism and chest; environment. Mailing Add: 490 Post St San Francisco CA 94102

WAYGOOD, ERNEST ROY, b Bramhall, Eng, Oct 26, 18; m 50; c 1. PLANT PHYSIOLOGY, BIOCHEMISTRY. Educ: Ont Agr Col, BSc, 41; Univ Toronto, MSc, 47, PhD, 49. Prof Exp: Assoc prof plant physiol, McGill Univ, 49-54; PROF BOT, UNIV MAN, 54- Mem: Can Soc Plant Physiol (pres, 59-60); fel Chem Inst Can; fel Royal Soc Can. Res: Enzyme mechanisms in respiration and photosynthesis; physiology of host parasite relationships. Mailing Add: Dept of Bot Univ of Man Winnipeg MB Can

WAYLAND, BRADFORD B, b Lakewood, Ohio, Dec 14, 39; m 59; c 1. INORGANIC CHEMISTRY. Educ: Western Reserve Univ, AB, 61; Univ Ill, PhD(inorg chem), 64. Prof Exp: From asst prof to assoc prof, 64-75, PROF CHEM, UNIV PA, 75- Mem: Am Chem Soc; The Chem Soc. Res: Transition metal ion complexes; molecular complexes; thermodynamics; magnetic resonance; metalloporphyrin species; metal ions in biological systems; surface complexes; metal site catalysis. Mailing Add: Dept of Chem Univ of Pa Philadelphia PA 19104

WAYLAND, JAMES ROBERT, JR, b Plainview, Tex, May 3, 37; m 61; c 2. ASTROPHYSICS, AGRICULTURAL PHYSICS. Educ: Univ of the South, BS, 59; Univ Ariz, PhD(physics), 67. Prof Exp: Res assoc astrophys, Univ Md, 67-70; asst prof astrophysics & agr physics, Tex A&M Univ, 70-74; MEM TECH STAFF, SANDIA LABS, 74- Mem: Am Phys Soc. Res: Cosmic ray physics; environmental impact of energy generating systems; health physics. Mailing Add: Sandia Labs Albuquerque NM 87115

WAYLAND, ROSSER LEE, JR, b Charlottesville, Va, Dec 30, 30; m 53; c 3. TEXTILE CHEMISTRY. Educ: Univ Va, BS, 49, MS, 50, PhD(chem), 52. Prof Exp: Res chemist, 51-54, group leader, Res Div, 54-60, asst dir res, 60-74, MGR CHEM PROD DEPT, DAN RIVER, INC, 74- Honors & Awards: Olney Medal, Am Asn Textile Chemists & Colorists, 75. Mem: Am Asn Textile Chemists & Colorists; Am Chem Soc. Res: Organic synthesis; thermosetting textile resins. Mailing Add: 319 W Main St Danville VA 24541

WAYLAND, RUSSELL GIBSON, b Treadwell, Alaska, Jan 23, 13; m 43, 65; c 2. MINING GEOLOGY, ENGINEERING. Educ: Univ Wash, BS, 34; Univ Minn, MS, 35, PhD(econ geol), 39; Harvard Univ, MA, 37. Prof Exp: Geologist & engr, Homestake Mining Co, SDak, 30-39; geologist, US Geol Surv, 39-42; minerals specialist, Army-Navy Munitions Bd, 42-45; mining indust control officer, Off Mil Govt for Ger, 45-48; US chmn combined coal control group, Allied High Command, Ger, 48-52; staff engr, Off of Dir, 52-58, regional geologist, Br Mineral Classification, Conserv Div, 58-66, asst chief, Conserv Div, DC, 66-67, CHIEF CONSERV DIV, DC, US GEOL SURV, 67- Concurrent Pos: Asst, Univ Minn, 34-36, instr, 37-39; geologist & engr, Alaska Juneau Gold Mining Co, 37. Honors & Awards: Distinguished Serv Award, Dept Interior, 68. Mem: Fel Mineral Soc Am; Am Inst Mining, Metall & Petrol Engrs; Geol Soc Am; Soc Econ Geologists; Asn Prof Geol Scientists. Res: Industrial minerals; coal and petroleum; mine evaluation; mineral land appraisal. Mailing Add: 4660 N 35th Arlington VA 22207

WAYMAN, COOPER H, b Trenton, NJ, Jan 29, 27; m 51; c 2. ENVIRONMENTAL MANAGEMENT. Educ: Rutgers Univ, BS, 51; Univ Pittsburgh, MS, 54; Mich State Univ, PhD(geochem), 59; Univ Denver, JD, 67. Prof Exp: Mining engr, Am Agr Chem Co Div, Continental Oil Co, 51-52; explor geologist, Lone Star Steel Co, 52-53; supv technologist, Appl Res Lab, US Steel Corp, 53-54, 59-60; res chemist, Denver Lab, US Geol Surv, 60-65; from asst prof to assoc prof chem, Colo Sch Mines, 65-71; regional gen counsel, 72-74, DIR OFF ENERGY, ENVIRON PROTECTION AGENCY, 74- Concurrent Pos: Consult, US Geol Surv, 65-71 & US Fish & Wildlife Serv, 66-72; atty-at-law, Colo, 69-; mem, US Dept Interior Oil Shale Environ Adv Panel, 74- Mem: AAAS; Am Chem Soc; Am Soc Agron; Water Pollution Control Fedn. Res: Waste water chemistry; water pollution research on detergents and pesticides both legal and scientific; production of gases and nutrients in lakes and reservoirs. Mailing Add: Environ Protection Agency 1860 Lincoln St Denver CO 80203

WAYMAN, MORRIS, b Can, Mar 19, 15; m 37; c 2. ORGANIC CHEMISTRY, BIOCHEMICAL ENGINEERING. Educ: Univ Toronto, BA, 36, MA, 37, PhD(chem), 41. Prof Exp: Asst chem, Univ Toronto, 36-40; res chemist, Dye & Chem Co, Can, 41-43 & Can Int Paper Co, 43-48; mem staff, Indust Cellulose Res Ltd, 48-52; tech dir, Columbia Cellulose Co, Ltd, 52-59; res dir, Sandwell & Co, Ltd, 59-63; PROF CHEM ENG & APPL CHEM, UNIV TORONTO, 63-, PROF FORESTRY, 73- Mem: AAAS; Am Chem Soc; Tech Asn Pulp & Paper Indust; Can Soc Microbiologists. Res: Chemistry of wood; lignin; pulp and paper; chemical plant design; planning industrial development; single cell protein; feasibility and economic impact analysis; microbial carbon dioxide and nitrogen fixation. Mailing Add: 17 Noel Ave Toronto ON Can

WAYMAN, OLIVER, b Logan, Utah, Jan 8, 16; m 43; c 5. ANIMAL PHYSIOLOGY, ANIMAL NUTRITION. Educ: Utah State Agr Col, BS, 47; Cornell Univ, PhD, 51. Prof Exp: Asst animal husbandman, 51-52, assoc animal scientist, 52-57, chmn dept animal sci, 54-60, ANIMAL SCIENTIST, UNIV HAWAII, 57- Concurrent Pos: Res assoc psychoenergetic lab, Univ Mo, 60-61; Rockefeller Found grant, Colombian Land & Cattle Inst, 69; USDA Coop State Res Serv grant, Honolulu, 71-74. Mem: Am Soc Animal Sci; Soc Study Reproduction. Res: Influence of tropical environment upon growth, development and reproduction of cattle; improvement of feeding value of tropical forage; use of the whole pineapple plant as ruminant feed. Mailing Add: Dept of Animal Sci Univ of Hawaii Honolulu HI 96822

WAYMENT, STANLEY GLEN, mathematics, see 12th edition

WAYMOUTH, CHARITY, b London, Eng, Apr 29, 15. CELL BIOLOGY,

BIOCHEMISTRY. Educ: Univ London, BSc, 36; Aberdeen Univ, PhD(biochem), 44. Prof Exp: Technician, Crumpsall Hosp, Manchester, Eng, 37-39; biochemist, Manchester Gen Hosps, 39-41; mem sci staff & head tissue cult dept, Chester Beatty Res Inst, Royal Cancer Hosp, 47-53; res assoc, 55-57, staff scientist, 57-63, asst dir training, 69-72, SR STAFF SCIENTIST, JACKSON LAB, 63- Concurrent Pos: Beit mem fel, Aberdeen Univ, 44-47, Nat Inst Med Res, London, 45, Carlsberg Biol Found, Copenhagen, 46 & St Thomas' Hosp Med Sch, Eng, 46-47; Am Cancer Soc-Brit Cancer Campaign exchange fel, Jackson Lab, 52-53, res fel, 53-55; hon lectr, Univ Maine, 64-; ed in chief, In Vitro, Tissue Cult Asn, 69-75; Rose Morgan vis prof, Univ Kans, 71. Mem: AAAS; Tissue Cult Asn (pres, 60-62); NY Acad Sci; Brit Biochem Soc; Royal Soc Med. Res: Nucleic acids and growth; synthetic nutrients for cells; carcinogenesis in vitro; cell metabolism and differentiation. Mailing Add: Jackson Lab Bar Harbor ME 04609

WAYMOUTH, JOHN FRANCIS, b Ingenio Barahona, Dominican Repub, May 24, 26; m 49; c 4. APPLIED PHYSICS. Educ: Univ of the South, BS, 47; Mass Inst Technol, PhD(physics), 50. Prof Exp: Lab asst, Univ of the South, 43-44 & 46-47; asst, Mass Inst Technol, 49-50; sr engr, Sylvania Elec Prod, Inc Div, 50-58, sect head, 58-65, mgr physics lab, 65-69, DIR RES SYLVANIA LIGHTING PROD GROUP, GEN TEL & ELECTRONICS CORP, 69- Concurrent Pos: Consult, Magnet Lab, Mass Inst Technol, 58-65. Honors & Awards: W Elenbaas Award, Eindhoven Tech Univ, Dutch Phys Soc & N V Philips Co, 73 Mem: Am Phys Soc; Illum Eng Soc. Res: Oxide cathodes; electroluminescent phosphors; gaseous electronics. Mailing Add: Sylvania Lighting Prod 100 Endicott St Danvers MA 01923

WAYNE, GEORGE JEROME, b New York, NY, NY, Sept 13, 14; m 41; c 2. PSYCHIATRY, PSYCHOANALYSIS. Educ: Brooklyn Col, BS, 34; Univ Western Ont, MD, 39; Am Bd Psychiat & Neurol, dipl, 50. Prof Exp: Med dir, Los Angeles Neurol Inst, 45-56; CLIN PROF PSYCHIAT, SCH MED, UNIV CALIF, LOS ANGELES, 53-; MED DIR, EDGEMONT HOSP, 56- Concurrent Pos: Consult, Vet Admin Hosp, 50; teaching analyst, Inst Psychoanal Med Southern Calif, 53-; consult, Camarillo & Metrop State Hosp, 57; US Info Agency, 59-; pres, Nat Asn Pvt Psychiat Hosps, 75. Mem: Fel Geront Soc; fel Am Psychiat Asn; Am Psychoanal Asn; Aerospace Med Asn; fel Am Col Physicians. Res: Psychotherapy and somatic therapies in schizophrenia; psychiatric explorations of aged people; cause and treatment of psychoses; research in teaching methods. Mailing Add: 4841 Hollywood Blvd Los Angeles CA 90027

WAYNE, LAWRENCE GERSHON, b Los Angeles, Calif, Mar 11, 26; m 48; c 5. MICROBIOLOGY. Educ: Univ Calif, Los Angeles, BS, 49, MA, 50, PhD(microbiol), 52. Prof Exp: Chief bact res lab, Vet Admin Hosp, San Fernando Calif, 52-71; CHIEF TUBERC RES LAB, VET ADMIN HOSP, LONG BEACH, 71-; ASSOC CLIN PROF MED MICROBIOL, UNIV CALIF, IRVINE-CALIF COL MED, 70- Concurrent Pos: Asst clin prof, Univ Calif, Irvine-Calif Col Med, 52-70; clin instr sch med, Univ Calif, Los Angeles, 58-67, lectr, Sch Pub Health, 68-; mem infectious dis res prog comt, Vet Admin, 61-64, mem pulmonary dis res prog comt, 64; mem lab comt, Vet Admin-Armed Forces Coop Study Chemother Tuberc, 61-; consult, Calif Dept Pub Health, 63-; mem mycobacterium taxon subcomt, Int Asn Microbiol Socs, 66-; mem adv comt actinomycetes, Bergey's Manual Trust, 67-; mem bact & mycol study sect, Nat Inst Allergy & Infectious Dis; coordr, Int Working Group on Mycobacterium Taxon. Mem: AAAS; Am Soc Microbiol; Am Thoracic Soc. Res: Natural history and diagnostic techniques of tuberculosis and fungus diseases; physiology and classification of mycobacteria. Mailing Add: VA Hosp Tuberc Res Lab 5901 E Seventh St Long Beach CA 90801

WAYNE, LOWELL GRANT, b Washington, DC, Nov 27, 18; m 42; c 2. AIR POLLUTION, ENVIRONMENTAL HEALTH. Educ: Univ Calif, Berkeley, BS, 37; Calif Inst Technol, PhD(inorg chem), 49; Am Bd Indust Hyg, dipl. Prof Exp: Fel petrol refining, Mellon Inst, 49-52; sr phys chemist, Stanford Res Inst, 53-54; indust health engr, Univ Calif, Los Angeles, 54-56; res photochemist, Air Pollution Control Dist, Los Angeles, 56-62; res analyst, Allan Hancock Found, Univ Southern Calif, 62-69, res assoc comput sci, 63-65, res biol sci, 65-69, sector head, Air Pollution Control Inst, 65-72, res assoc, Sch Pub Admin, 69-72; VPRES & DIR RES, PAC ENVIRON SERV, INC, 72- Concurrent Pos: Prof consult, Comt Motor Vehicles Emissions, Nat Acad Sci-Nat Res Coun, 71- Mem: Fel AAAS; fel Am Inst Chemists; Am Chem Soc; Air Pollution Control Asn. Res: Kinetics and photochemistry of gas phase reactions, especially chemical reactions in polluted urban atmospheres; oxides of nitrogen; air quality modelling; air quality evaluation; atmospheric chemistry. Mailing Add: 3292 Grand View Blvd Los Angeles CA 90066

WAYNE, RICHARD CONABLE, solid state physics, see 12th edition

WAYNE, WILLIAM JOHN, b Cass Co, Mich, Apr 23, 22; m 46; c 3. GEOLOGY. Educ: Ind Univ, AB, 43, MA, 50, PhD(geol), 52. Prof Exp: Head glacial geologist, State Geol Surv, Ind, 52-68; assoc prof geol, 68-71, PROF GEOL, UNIV NEBR, LINCOLN, 71- Concurrent Pos: Vis prof, Univ Wis, 66-67. Mem: AAAS; Geol Soc Am; Arctic Inst NAm; Ger Quaternary Asn; Am Quaternary Asn. Res: Quaternary stratigraphy and paleontology; geomorphology; environmental geology; age and origin of Nebraska sandhills; Pleistocene stratigraphy in eastern Nebraska. Mailing Add: Dept of Geol 433 Morrill Hall Univ of Nebr Lincoln NE 68588

WAYNE, WINSTON JOE, b Oswego, Ill, Sept 20, 14; m 38; c 2. CHEMISTRY. Educ: Univ Ill, BS, 36; Univ Wis, PhD(org chem), 40. Prof Exp: Asst chem, Univ Wis, 36-40; res chemist, 40-61, SR RES CHEMIST, EXP STA, E I DU PONT DE NEMOURS & CO, INC, 61- Mem: Am Chem Soc. Res: Organic chemistry; agricultural chemicals; pharmaceutical chemistry. Mailing Add: E I du Pont de Nemours & Co Inc Wilmington DE 19898

WAYNER, MATTHEW JOHN, b Clifton, NJ, Sept 7, 27. NEUROSCIENCES, PSYCHOPHARMACOLOGY. Educ: Dartmouth Col, AB, 49; Tufts Univ, MS, 50; Univ Ill, PhD(psychol), 53. Prof Exp: From asst prof to assoc prof, 53-63, PROF PSYCHOL, BRAIN RES LAB, SYRACUSE UNIV, 63- Concurrent Pos: Vis prof, Fla State Univ, 63, Kanazawa Univ, Japan, 67, Ariz State Univ, 69 & Latrobe Univ, Australia, 74; ed-in-chief, Physiol & Behav, Pharmacol, Biochem & Behav & Brain Res Bull. Mem: Am Psychol Asn; Am Physiol Soc; Int Brain Res Orgn; Soc Neurosci; NY Acad Sci. Res: Hypothalamic mechanisms and ingestive behavior; neural mechanisms of ingestive behavior and drug action. Mailing Add: Brain Res Lab 601 University Ave Syracuse Univ Syracuse NY 13210

WAYRYNEN, ROBERT ELLIS, b Lake Norden, SDak, Oct 24, 24; m 46; c 2. PHOTOGRAPHIC CHEMISTRY. Educ: SDak State Col, BS, 48; Univ Utah, PhD(chem), 53. Prof Exp: From chemist to sr chemist, Photo Prod Dept, 52-62, res supvr, 62-65, tech serv group supvr, 65-66, field sales mgr, 66-69, TECH MGR PHOTO PROD DEPT, E I DU PONT DE NEMOURS & CO, INC, 69- Mem: Am Chem Soc; Soc Photog Sci & Eng; Soc Motion Picture & TV Eng. Res: Photosynthesis; photography. Mailing Add: Photo Prod Dept E I du Pont de Nemours & Co Inc Wilmington DE 19898

WAYT, LEWIS KEITH, b Durango, Colo, July 9, 23; m 43; c 2. VETERINARY MEDICINE. Educ: Colo State Univ, DVM, 46, MS, 54; Univ Colo, PhD(med), 61. Prof Exp: Pvt pract, 46-50; from asst prof to prof vet med, Colo State Univ, 50-71, head dept, 59-71; HEAD DEPT VET CLIN SCI, SCH VET MED, LA STATE UNIV, BATON ROUGE, 71- Mem: Am Vet Med Asn; Am Asn Vet Clinicians; Am Col Vet Internal Med; Am Animal Hosp Asn. Res: Infectious and nutritional diseases; structural abnormalities; new drug studies. Mailing Add: Dept of Vet Clin Sci La State Univ Sch of Vet Med Baton Rouge LA 70803

WEAD, WILLIAM BADERTSCHER, b Columbus, Ohio, Mar 11, 40; m 62; c 2. CARDIOVASCULAR PHYSIOLOGY. Educ: Wabash Col, BA, 62; Ohio State Univ, MS, 67, PhD(physiol), 69. Prof Exp: Res & teaching asst physiol, Ohio State Univ, 63-69; ASST PROF PHYSIOL, SCH MED, UNIV LOUISVILLE, 69- Res: Myocardial contractility, stress relaxation. Mailing Add: Dept of Physiol & Biophys Univ of Louisville Sch of Med Louisville KY 40201

WEAKLEY, MARTIN LEROY, b Piedmont, WVa, June 5, 25; m 50; c 4. RESEARCH ADMINISTRATION, AGRICULTURAL CHEMISTRY. Educ: Antioch Col, BS, 51. Prof Exp: Lab asst, Nat Cash Register Co, Ohio, 49-51; chemist, Celanese Corp Am, Tex, 51-54 & John Deere Chem Co, Okla, 54-65; chemist, 65-67, MGR CHEM APPLN, RES & DEVELOP DEPT, NIPAK INC, 67- Mem: Am Chem Soc. Res: Product and process development and use application in agricultural field; food chemistry. Mailing Add: Nipak Inc PO Box 338 Pryor OK 74361

WEAKLIEM, HERBERT ALFRED, JR, b Newark, NJ, Mar 24, 26; m 55; c 4. PHYSICAL CHEMISTRY. Educ: Rutgers Univ, BSc, 53; Cornell Univ, PhD(phys chem), 58. Prof Exp: Assoc chemist, Allied Chem & Dye Corp, 53-54; MEM TECH STAFF, RCA LABS, 58- Concurrent Pos: Vis prof physics, Univ Calif, Los Angeles, 72-73. Mem: Am Chem Soc; Am Phys Soc; Am Crystallog Asn. Res: Structures of solids; solid state spectroscopy. Mailing Add: RCA Labs Princeton NJ 08540

WEAKLY, WARD FREDRICK, b North Platte, Nebr, July 16, 38. ANTHROPOLOGY. Educ: Univ Nebr, BA, 60, MA, 61; Univ Ariz, PhD(anthrop), 68. Prof Exp: Asst prof sociol & anthrop, Tulsa, 67-69; res assoc anthrop, Univ Mo-Columbia, 69-74; ARCHEOLOGIST, BUR RECLAMATION, US DEPT INTERIOR, 74- Mem: AAAS; Am Anthrop Asn; Soc Am Archaeol. Res: Archaeology of the North American Great Plains; tree-ring research in the Great Plains. Mailing Add: Bur of Reclamation PO Box 25007 DFC Denver CO 80225

WEAKS, THOMAS ELTON, b Cumberland City, Tenn, Sept 12, 34; m 59; c 2. PLANT PHYSIOLOGY. Educ: Austin Peay State Univ, BS, 56; George Peabody Col, MA, 60; Univ Tenn, Knoxville, PhD(bot), 71. Prof Exp: High sch instr, Fla, 58-65; instr biol, Brevard Jr Col, 66-67; ASST PROF BOT, MARSHALL UNIV, 71- Concurrent Pos: Sigma Xi res grant, 74; instnl res grant, Marshall Univ, 75. Mem: Am Soc Plant Physiologists; Sigma Xi; Scand Soc Plant Physiologists. Res: Inhibitory action of canavanine in higher plants; phytotoxin effects on fungus diseases of legumes. Mailing Add: Div of Biol Sci Marshall Univ Huntington WV 25701

WEAMER, GEORGE L, b Dunsmuir, Calif, Feb 14, 15; m 40; c 3. PETROLEUM CHEMISTRY. Educ: Univ Calif, BS, 38. Prof Exp: Anal chemist, Standard Oil Co Calif, 38-42, res chemist, Chevron Res Co, 42-56, supv res chemist, 56-62, sr eng assoc, 62-68, staff prod engr, Prod Eng Dept, San Francisco, 68-70, SR ENG ASSOC, CHEVRON RES CO, STANDARD OIL CO CALIF, 70- Mem: Am Chem Soc; Am Inst Chem Eng. Res: Additives for fuels and lubricating oils. Mailing Add: Chevron Res Co 576 Standard Ave Richmond CA 94802

WEAR, JAMES OTTO, b Francis, Okla, Oct 25, 37; m 59; c 2. PHYSICAL CHEMISTRY, BIOPHYSICS. Educ: Univ Ark, BS, 59, MS, 60, PhD(phys chem), 62. Prof Exp: Staff mem, Sandia Corp, 61-65; res chemist, Southern Res Support Ctr, Vet Admin, 65-66, dir opers, 66-68, actg chief, 67-68, CHIEF CENT RES INSTRUMENT PROG, VET ADMIN HOSP, LITTLE ROCK, 66-, DIR ENG TRAINING CTR, 72-; ASSOC PROF BIOMED INSTRUMENTATION TECHNOL & CHMN DEPT, COL HEALTH RELATED PROF, UNIV ARK, LITTLE ROCK, 72- Concurrent Pos: Abstractor, Chem Abstr Servs, 62-65; assoc prof, Philander Smith Col, 66-; asst prof, Grad Inst Technol, Univ Ark, 66-69, asst prof, Med Sch, 68-75, assoc prof, 75-; mem, Vet Admin Comt Res Instrumentation, 67; consult, Ark State Health Planning, 68; mem, Ark Gov Sci & Technol Coun; pres, Ark Sci Assocs, 69-73; chmn, Ark Sci & Technol Coun, 73-75. Mem: AAAS; Am Chem Soc; Asn Advan Med Instrumentation; Soc Advan Med Systs; Instrument Soc Am. Res: Chemical kinetics; electrochemistry; complex ions; solution chemistry; radiochemistry; electron spin resonance; trace metals in biological systems; science education; research support and management; hospital instrumentation maintenance; health and social planning and program evaluation. Mailing Add: Vet Admin Hosp 300 E Roosevelt Rd Little Rock AR 72206

WEAR, JOHN BREWSTER, JR, b Madison, Wis, Oct 27, 29; m 53; c 4. UROLOGY. Educ: Williams Col, BA, 70; Univ Wis-Madison, BS, 51, MD, 54; Am Bd Urol, dipl, 64. Prof Exp: Intern med, Univ Ore Hosp, 54-55; resident surg, Univ Mich Hosp, 55-56, resident urol, 58-61; from instr to assoc prof, 61-71, PROF UROL, SCH MED, UNIV WIS-MADISON, 71-, CHMN SECT UROL, DEPT SURG, 64- Concurrent Pos: Sr consult urol, Vet Admin Hosp, Madison. Mem: AMA; Am Col Surgeons; Am Urol Asn; Asn Am Med Cols; Soc Univ Urol. Res: Neurogenic bladder; tumors of the bladder; testicular carcinoma; urinary tract infection and transurethral surgery. Mailing Add: 463N Univ Hosp 1300 University Ave Madison WI 53706

WEAR, JOHN INGRAM, b Opelika, Ala, Apr 29, 16; m 41; c 2. AGRONOMY. Educ: Ala Polytech Inst, BS, 38, MS, 41; Purdue Univ, PhD(soil chem), 49. Prof Exp: Asst agron, 38-39, from asst soil chemist to assoc soil chemist, 45-59, PROF AGRON & SOIL CHEMIST, EXP STA, AUBURN UNIV, 59- Mem: Soil Sci Soc Am; Am Soc Agron. Res: Clay mineralogy of soils; minor element fertility. Mailing Add: Dept of Agron & Soils Auburn Univ Auburn AL 36830

WEAR, ROBERT LEE, b Princeville, Ill, Feb 28, 24; m 46; c 2. ORGANIC POLYMER CHEMISTRY. Educ: Univ Ill, BS, 46; Univ Nebr, MS, 48, PhD(chem), 50. Prof Exp: Sr res chemist, 49-55, group supvr, 55-61, res specialist, 62-67, SR RES SPECIALIST, MINN MINING & MFG CO, 67- Mem: Am Chem Soc. Res: Condensation polymers. Mailing Add: 93 Kraft Rd West St Paul MN 55118

WEARDEN, STANLEY, b Goliad, Tex, Oct 1, 26; m 51; c 5. STATISTICS. Educ: St Louis Univ, BS, 50; Univ Houston, MS, 51; Cornell Univ, PhD, 57. Prof Exp: Asst biol, Univ Houston, 50-51, instr, 51-53; asst animal husb, Cornell Univ, 53-57; asst prof math, Kans State Univ, 57-59, from assoc prof to prof statist, 59-66; dir div statist, 66-69, PROF STATIST, WVA UNIV, 66-, CHMN DEPT STATIST & COMPUT SCI, 69-, DEAN, GRAD SCH, 72- Concurrent Pos: Hon res fel, Birmingham Univ, 63-64; USPHS spec fel, 63-64. Mem: Am Statist Asn; Am Soc Animal Sci; Biomet Soc. Res: Statistical methods in study of quantitative inheritance; low temperature biology; statistics in agricultural research and genetics. Mailing Add: Off of the Dean Grad Sch WVa Univ Morgantown WV 26506

WEARE, JOHN H, b Boston, Mass, Mar 8, 40; m 63; c 1. CHEMISTRY. Educ: Harvey Mudd Col, BS, 62; Johns Hopkins Univ, PhD(chem), 67. Prof Exp: NSF grant, Air Force Off Sci Res, 68-69; ASST PROF PHYS CHEM, UNIV CALIF, SAN DIEGO, 69- Res: Theoretical chemistry, particularly molecular and atomic structure and interactions and the study of irreversible processes. Mailing Add: Dept of Chem Univ of Calif San Diego La Jolla CA 92037

WEARN, RICHARD BENJAMIN, b Newberry, SC, Aug 1, 16; m 40; c 5. ORGANIC CHEMISTRY. Educ: Clemson Col, BS, 37; Univ Ill, MS, 39, PhD(org chem), 41. Prof Exp: Asst, Nat Starch Prod, Inc, NY, 37-38; asst chem, Univ Ill, 39-40; res chemist, Eastman Kodak Co, NY, 41-42; res chemist, Southern Res Inst, 46-47, head org chem div, 47-48; asst res dir, Res & Develop Dept, Colgate-Palmolive Peet Co, 48-52, assoc dir, Colgate-Palmolive Int, Inc, 52-57, dir res & develop, Household Prod Div, 57-65, dir prod develop, 65-67, tech dir, 67-71, VPRES RES & DEVELOP, COLGATE-PALMOLIVE INT, INC, 71- Mem: Am Chem Soc. Res: Heterogeneous catalytic reactions; terpene derivatives; diene addition products of diaroyl ethylenes; structure of cannabinol and of cannabidiol; isolation and synthesis of active principle in marihuana; utilization of soaps and synthetic detergents. Mailing Add: Colgate-Palmolive Int Inc 300 Park Ave New York NY 10022

WEARN, RICHARD BENJAMIN, JR, b Edgewood, Md, Oct 27, 43; m 67; c 1. OCEANOGRAPHY, FLUID DYNAMICS. Educ: Dartmouth Col, AB, 64; Harvard Univ, AM, 65, PhD(physics), 72. Prof Exp: Res fel phys oceanog, Ctr Earth & Planetary Physics, Harvard Univ, 71-75; MEM FAC, DEPT OCEANOG, UNIV WASH, 75- Res: Pressure measurements in the ocean; ocean circulation; deep ocean instrumentation. Mailing Add: Dept of Oceanog Univ of Wash Seattle WA 98195

WEARRING, DANIEL, organic chemistry, see 12th edition

WEART, RICHARD CLAUDE, b Brandon, Iowa, July 20, 22; m 46; c 3. STRATIGRAPHY, PALEONTOLOGY. Educ: Cornell Col, BA, 43; Syracuse Univ, MS, 48; Univ Ill, PhD(geol), 50. Prof Exp: Asst prof geo, Tex Tech Col, 50-52; from paleontologist & paleogeologist to asst chief geologist, Latin Am Div, 52-60, head staff geologist, 60-62, mgr geol res, 63-71, mgr geol, 71-75, MGR STRATEGIC EXPLOR, SUN OIL CO, 75- Mem: Geol Soc Am; Soc Econ Paleontologists & Mineralogists; Paleont Soc; Am Asn Petrol Geologists. Res: Petroleum geology. Mailing Add: Sun Oil Co PO Box 2880 Dallas TX 75221

WEART, WENDELL D, b Brandon, Iowa, Sept 24, 32; m 54; c 3. GEOPHYSICS. Educ: Cornell Col, BA, 53; Univ Wis, PhD(geophys), 61. Prof Exp: Geophysicist, Ballistics Res Lab, 56-59; geophysicist, Sandia Corp, 59-69, supvr underground physics div, 69-75, DEPT MGR, NUCLEAR WASTE SYSTS, SANDIA LABS, 75- Mem: AAAS; Am Geophys Union; Seismol Soc Am; Soc Explor Geophys. Res: Earth physics relating to underground explosion; nuclear waste disposal in geologic media, particular emphasis on salt. Mailing Add: Sandia Labs Waste Mgt Systs Dept 1140 Albuquerque NM 87115

WEARY, MARLYS E, b Chicago, Ill, Mar 13, 39. ANATOMY, HISTOLOGY. Educ: Valparaiso Univ, BA, 60; Univ Ill, MS, 62. Prof Exp: Pharmacologist, 62-66, SUPVR BIOL CONTROL BAXTER LABS, INC, 66- Mem: AAAS; NY Acad Sci. Res: Amphibian regeneration; anticonvulsant research; biologic control and pyrogen testing. Mailing Add: Baxter Labs Pharmacol Dept 6301 Lincoln Morton Grove IL 60053

WEARY, PEYTON EDWIN, b Evanston, Ill, Jan 10, 30; m 52; c 3. DERMATOLOGY. Educ: Univ Va, MD, 55. Prof Exp: Intern, Univ Hosps Cleveland, 55-56; from asst resident to resident, 58-61, from instr to assoc prof, 61-70, PROF DERMAT, SCH MED, UNIV VA, 70-, VCHMN DEPT, 75- Concurrent Pos: Chmn, Coun Nat Prog Dermat, 72-75. Mem: Am Dermat Asn; Am Acad Dermat; Soc Invest Dermat. Res: Exploration of the keratinolytic abilities of various dermatophyte fungal organisms and investigation of ecology of certain lipophilic yeast organisms on the skin surface. Mailing Add: Dept of Dermat Univ of Va Sch of Med Charlottesville VA 22204

WEAST, CLAIR ALEXANDER, b Modesto, Calif, Oct 13, 13; m 40; c 2. FOOD CHEMISTRY. Educ: Univ Calif, BS, 37, MS, 39, PhD(agr chem), 42. Prof Exp: Asst, Univ Calif, 37-42; chemist, USDA, 42-44; chief chemist, Pacific Can Co, 44-46; RES DIR, TILLIE LEWIS FOODS, INC, 46- Res: Food products; chlorophyllase. Mailing Add: Tillie Lewis Foods Inc Drawer J Stockton CA 95204

WEAST, ROBERT CALVIN, b Alliance, Ohio, Sept 2, 16; m 42; c 2. CHEMISTRY. Educ: Mt Union Col, BS, 38; Syracuse Univ, MS, 40; Univ Ill, PhD(chem), 43. Prof Exp: Anal chemist, State Water Surv, Ill, 40-43; res chemist, Am Gas Asn, Ohio, 43-46; from asst prof to prof chem, Case Inst Technol, 46-62, actg head dept, 56-58; assoc dir res, 62-64, VPRES RES, CONSOL NATURAL GAS CO, 62- Mem: Fel AAAS; Am Chem Soc; Nat Asn Corrosion Eng; Combustion Inst; Am Soc Info Sci. Res: Electrochemistry; corrosion; combustion; hydrogen production; chemistry of solid state; technical book editing. Mailing Add: Consol Natural Gas Co 11001 Cedar Ave Cleveland OH 44106

WEATHERBEE, CARL, b Michigan City, Ind, Nov 21, 16; m 50; c 5. ORGANIC CHEMISTRY. Educ: Hanover Col, AB, 40; Univ Ill, AM, 46; Univ Utah, PhD(chem), 50. Prof Exp: Asst chem, Univ Ill, 46; instr, Reed Col, 50; asst prof, Univ Hawaii, 50-51; PROF CHEM & CHMN DEPT, MILLIKIN UNIV, 52- Concurrent Pos: Researcher, Univ Utah, 59-60. Mem: AAAS; Am Chem Soc; Am Inst Chemists. Res: Mannich bases; nitrogen mustards. Mailing Add: Dept of Chem Millikin Univ Decatur IL 62522

WEATHERBY, GERALD DUNCAN, b Neodesha, Kans, Mar 13, 40; m 68; c 2. BIOCHEMISTRY. Educ: Univ Kans, BS, 62 & 63, PhD(biochem), 69. Prof Exp: Teacher high sch, Kans, 63-65; ASST PROF CHEM, LAKE SUPERIOR STATE COL, 69- Res: Mechanism of riboflavin catalyzed carbon-carbon bond oxidations; extrapolation of these model system studies to elucidate mechanisms of flavoenzyme catalysis. Mailing Add: Dept of Chem Lake Superior State Col Sault Ste Marie MI 49783

WEATHERFORD, THOMAS WALLER, III, b Uriah, Ala, Mar 12, 30; m 57; c 2. PERIODONTOLOGY. Educ: Auburn Univ, DVM, 54; Univ Ala, Birmingham, DMD, 61, MSD, 69. Prof Exp: Intern pedodontics, Sch Dent, 61-62, instr dent, 62-69, asst prof periodont, 69-71, ASSOC PROF PERIODONT, SCH DENT, UNIV ALA, 71-, DIR POSTDOCTORAL EDUC, 70-, ASST PROF COMP MED, SCH MED, 69- Concurrent Pos: NIH trainee, 61-62; staff dentist, Birmingham Vet Admin Hosp, 62-65, resident periodont, 66-68, res assoc, 68-70; resident periodont, Sch Dent, Univ Ala, 66-68; consult, Dept Animal Serv, Univ Ala, 67-69; investr, Inst Dent Res, 70-; mem, Am Asn Dent Schs. Mem: AAAS; Am Dent Asn; Int Asn Dent Res; Am Acad Periodont; Am Asn Lab Animal Sci. Res: Animal models in periodontal research; dental plaque control; histochemistry of the periodontium. Mailing Add: Univ of Ala Sch of Dent 1919 Seventh Ave S Birmingham AL 35233

WEATHERLY, NORMAN F, b Elkton, Ore, June 22, 32; m 52; c 6. MEDICAL PARASITOLOGY. Educ: Ore State Univ, BS, 53, MS, 60; Kans State Univ, PhD(parasitol), 62. Prof Exp: NIH trainee parasitol, 62-63, from asst prof to assoc prof, 63-72, PROF PARASITOL, SCH PUB HEALTH, UNIV NC, CHAPEL HILL, 72- Mem: Am Soc Parasitol; Am Soc Trop Med & Hyg; Am Micros Soc; Am Pub Health Asn. Res: General immunobiology of helminth parasites; cell mediated responses of hosts to helminth parasites. Mailing Add: Dept of Parasitol Sch of Pub Health Univ of NC Chapel Hill NC 27514

WEATHERLY, THOMAS LEVI, b Greenville, Miss, Jan 14, 24; m 52; c 3. PHYSICS. Educ: Ohio State Univ, PhD(physics), 51. Prof Exp: Assoc prof, 52-61, PROF PHYSICS, GA INST TECHNOL, 61- Mem: Fel Am Phys Soc. Res: Microwave and radio-frequency spectroscopy. Mailing Add: Sch of Physics Ga Inst of Technol Atlanta GA 30332

WEATHERRED, JACKIE G, b Pampa, Tex, Mar 14, 34; m 60; c 3. PHYSIOLOGY. Educ: Univ Tex, DDS, 59, PhD(physiol), 65. Prof Exp: Dent consult, Tex Inst Rehab & Res, 60-62; instr physiol & res assoc oral path, Med Col Va, 62-63; from asst prof to assoc prof physiol, Dent Sch, Univ Md, Baltimore, 63-69; PROF & COORDR PHYSIOL, SCH DENT, MED COL GA, 69- Concurrent Pos: Mem test construction comt, Coun Nat Bd Dent Examr; basic sci consult, Coun Dent Educ; Am Col Dent fel, 71. Mem: AAAS; Int Asn Dent Res. Res: Plasma kinins and oral physiology; circulation in of dental pulp; predisposing factors in experimental carcinoma of the hamster cheek pouch; membrane transport in oral epithelium. Mailing Add: Sch of Dent Med Col of Ga Augusta GA 30902

WEATHERS, DWIGHT RONALD, b Milledgeville, Ga, Aug 14, 38; m 75; c 4. ORAL PATHOLOGY. Educ: Emory Univ, DDS, 62, MSD, 66. Prof Exp: Asst prof, 67-70, ASSOC PROF ORAL PATH, SCH DENT, EMORY UNIV, 70- Mem: Am Soc Oral Path; Am Dent Asn; Am Asn Dent Res. Res: Herpes simplex virus; neoplasia of oral cavity. Mailing Add: Emory Univ Sch of Dent 1462 Clifton Rd NE Atlanta GA 30322

WEATHERS, LEWIS GLEN, b Sunset, Utah, July 5, 25; m 46; c 4. PLANT PATHOLOGY. Educ: Utah State Col, BS, 49, MS, 51; Univ Wis, PhD(phytopath), 53. Prof Exp: Asst bot, Utah State Univ, 49-51; asst plant path, Univ Wis, 51-53; vchmn dept plant path, 71-73, PROF PLANT PATH & PLANT PATHOLOGIST, UNIV CALIF, RIVERSIDE, 53-, CHMN DEPT, 73- Concurrent Pos: Guggenheim fel, 61-62; Rockefeller fel, 63; NATO fel, 73. Mem: Am Phytopath Soc; Int Orgn Citrus Virol; Sigma Xi. Res: Citrus and virus diseases; scion and rootstock uncongenialities in citrus; effect of environmental factors on virus diseases; interactions of unrelated viruses in mixed infections. Mailing Add: Dept of Plant Path Univ of Calif Riverside CA 92502

WEATHERS, WESLEY WAYNE, b Homer, Ill, Sept 28, 42. ENVIRONMENTAL PHYSIOLOGY. Educ: San Diego State Col, BS, 64; Univ Calif, Los Angeles, MA, 67, PhD(zool), 69. Prof Exp: USPHS cardiovasc scholar, Sch Med, Univ Calif, Los Angeles, 69-70; from asst prof to assoc prof physiol, Rutgers Univ, New Brunswick, 70-75; ASST PROF PHYSIOL, UNIV CALIF, DAVIS, 75- Mem: Am Physiol Soc; Am Soc Zoologists; Sigma Xi; Cooper Ornith Soc. Res: Comparative physiology of circulation and temperature regulation in vertebrates. Mailing Add: Dept of Avian Sci Univ of Calif Davis CA 95616

WEATHERSBY, AUGUSTUS BURNS, b Pinola, Miss, May 19, 13; m 45; c 2. ENTOMOLOGY. Educ: La State Univ, AB, 38, MS, 40, PhD(entom), 54. Prof Exp: Asst dist entomologist, State Dept Agr, La, 40-42; entomologist-parasitologist, Naval Med Res Inst, 47-62; PROF ENTOM, UNIV GA, 62- Mem: AAAS; Entom Soc Am; Am Soc Parasitol; Am Soc Trop Med & Hyg; Am Mosquito Control Asn. Res: Medical entomology; innate immunity of mosquitoes to malaria; malaria survey, control and parasitology; drug action and immunity in malaria; life cycles of malaria; exoerythrocytic stage tissue culture and time-lapse cinephotomicrography; cryobiology. Mailing Add: Dept of Entom Univ of Ga Athens GA 30601

WEATHERSBY, HAL THOMPSON, b Pineville, La, Feb 17, 14; m 33; c 1. ANATOMY. Educ: La Col, BA, 33; La State Univ, MS, 39; Tulane Univ, PhD(anat), 52. Prof Exp: Instr, La High Schs, 33-39; from instr to assoc prof biol, La Col, 39-50, dean men, 48-50; instr anat, Sch Med, Tulane Univ, 52-53; from asst prof to assoc prof, 53-67, PROF ANAT, UNIV TEX HEALTH SCI CTR DALLAS, 67- Mem: AAAS; Am Asn Anatomists; Int Soc Electromyographic Kinesiology. Res: Gross anatomy; arterial patterns and variations; vascular supply of organs; electromyography. Mailing Add: Dept of Anat Univ of Tex Health Sci Ctr Dallas TX 75235

WEATHERWAX, ROBERT STANTON, b Flora, Ind, Oct 18, 19; m 47; c 3. MICROBIOLOGY. Educ: Ind Univ, AB, 41; Univ Calif, Berkeley, PhD(bact), 52. Prof Exp: Bacteriologist, Radiation Lab, Univ Calif, 49-53; asst prof bact, Univ Ky, 53-56; bacteriologist, Chem Corps Biol Labs, 56-61; assoc prof biol, NMex Highlands Univ, 61-65; MICROBIOLOGIST, ABBOTT LABS, NORTH CHICAGO, 65- Mem: Am Soc Microbiol. Res: Genetics of microorganisms. Mailing Add: 234 Jeannette Pl Mundelein IL 60060

WEAVER, ALBERT BRUCE, b Mont, May 27, 17; m 45; c 3. PHYSICS, ACADEMIC ADMINISTRATION. Educ: Univ Mont, AB, 40; Univ Idaho, MS, 41; Univ Chicago, PhD(physics), 52. Prof Exp: Physicist, US Naval Ord Lab, 42-45; res assoc physics, Univ Chicago, 52-53 & Univ Wash, 53-54; from asst prof to assoc prof, Univ Colo, 54-58, chmn dept, 56-58; prof & head dept, 58-70, assoc dean col lib arts, 61-70, provost acad affairs, 70-72, EXEC VPRES, UNIV ARIZ, 72- Mem: AAAS; fel Am Phys Soc. Res: Cosmic rays. Mailing Add: Admin 512 Univ of Ariz Tucson AZ 85721

WEAVER, ALLEN DALE, b Galesburg, Ill, Nov 15, 11; m 40; c 2. PHYSICS. Educ: Knox Col, BS, 33; Univ Mich, MS, 47; NY Univ, PhD(sci educ), 54. Prof Exp: High sch teacher, Ill, 35-37; jr high sch teacher, 37-40; instr physics, phys sci & math, Md State Teachers Col, Salisbury, 47-55; assoc prof phys sci, 55-60, PROF PHYSICS, NORTHERN ILL UNIV, 60- Mem: Nat Sci Teachers Asn; Am Asn Physics Teachers; Nat Asn Res Sci Teaching. Res: Science education. Mailing Add: Dept of Physics Northern Ill Univ De Kalb IL 60115

WEAVER, ANDREW ALBERT, b Sarasota, Fla, Dec 10, 26; m 51; c 4. ENTOMOLOGY. Educ: Col Wooster, BA, 49; Univ Wis, MS, 51, PhD(zool), 55. Prof Exp: Asst prof biol, 56-66, PROF BIOL, COL WOOSTER, 66- Mem: AAAS; Am Micros Soc; Entom Soc Am; Am Soc Zoologists; Soc Syst Zool. Res: Taxonomy of centipedes and copepods. Mailing Add: Dept of Biol Col of Wooster Wooster OH 44691

WEAVER, CHARLES EDWARD, b Lock Haven, Pa, Jan 27, 25; m 46; c 3. GEOCHEMISTRY, SEDIMENTOLOGY. Educ: Pa State Univ, BS, 48, MS, 50, PhD(mineral), 52. Prof Exp: Res assoc mineral, Pa State Univ, 52; res assoc mineral, Shell Res & Develop Co, 52-55, res scientist, 55-59; res group leader mineral,

Continental Oil Co, 59-63; assoc prof, 63-65, PROF MINERAL, GA INST TECHNOL, 65-, DIR SCH GEOPHYS SCI, 70- Honors & Awards: Mineral Soc Am Award, 58; Sigma Xi Res Award, 72. Mem: Clay Minerals Soc (vpres, 66, pres, 67); Mineral Soc Am; Geochem Soc; Geol Soc Am; Soc Econ Paleontologists & Mineralogists. Res: Clay mineralogy and petrology; geochemistry of sediments; clay-water chemistry; radioactive dating of sediments; potassium cycle. Mailing Add: Sch of Geophys Sci Ga Inst of Technol Atlanta GA 30332

WEAVER, CHARLES R, b Springfield, Mo, June 9, 23; m 54; c 1. BIOLOGY, CHEMISTRY. Educ: Drury Col, BS, 48. Prof Exp: Fishery biologist, Bur Com Fisheries, US Fish & Wildlife Serv, 50-66, prog leader, Fisheries-Eng Res Lab, 66-72, CHIEF FISH FACIL SECT, COLUMBIA FISHERIES PROG OFF, NAT MARINE FISHERIES SERV, 72- Mem: Am Fisheries Soc; Inst Fishery Res Biol. Res: Performance, behavior and swimming abilities of adult salmonids in relation to fish passage problems at dams on the Columbia and tributary rivers. Mailing Add: Fish Facil Sec CFPO NMFS 811 NE Oregon St Box 4332 Portland OR 97208

WEAVER, CLAYTON FRED, physical chemistry, inorganic chemistry, see 12th edition

WEAVER, CLYDE RICHARD, b Akron, Ohio, Oct 13, 20; m 46; c 2. BIOMETRY. Educ: Univ Akron, BS, 42; Ohio State Univ, MSc, 47, PhD, 50. Prof Exp: Asst entomologist, 49-55, STATISTICIAN, OHIO AGR RES & DEVELOP CTR, 56- Concurrent Pos: Vis assoc prof, Iowa State Univ, 58. Mem: Biomet Soc. Res: Design and analysis of experiments. Mailing Add: Dept of Statist Ohio Agr Res & Develop Ctr Wooster OH 44691

WEAVER, DAVID DAWSON, b Twin Falls, Idaho, Feb 12, 39; m 67. MEDICINE, HUMAN GENETICS. Educ: Col Idaho, BS, 61; Univ Ore, MS & MD, 66. Prof Exp: Intern med, Milwaukee County Gen Hosp, Wis, 66-67; biogeneticist, Arctic Health Res Ctr, 67-70, pediat residency, Med Sch, Univ Ore, 70-72; USPHS fel human genetics, Med Sch, Univ Wash, 72-74; MARCH OF DIMES FEL GENETICS & METAB DIS, DEPT PEDIAT, UNIV ORE HEALTH SCI CTR, PORTLAND, 74- Concurrent Pos: Lectr genetics & cell biol, Univ Alaska, 68-70. Mem: Am Soc Human Genetics. Res: Dysmorphology, X-chromosome inactivation and metabolic diseases. Mailing Add: Dept of Pediat Univ of Ore Health Sci Ctr Portland OR 97201

WEAVER, DAVID LEO, b Albany, NY, Apr 18, 37; m 66. THEORETICAL PHYSICS. Educ: Rensselaer Polytech Inst, BS, 58; Iowa State Univ, PhD(physics), 63. Prof Exp: Res assoc physics, Iowa State Univ, 63-64; asst prof, 64-69, ASSOC PROF PHYSICS, TUFTS UNIV, 69- Concurrent Pos: NATO fel physics, Europ Orgn Nuclear Res, Switz, 65-66; Nat Nuclear Energy Comt fel, Frascati Nat Lab, Italy, 68-69. Mem: Am Phys Soc. Res: Theoretical elementary particle physics; mathematical physics; molecular biophysics. Mailing Add: Dept of Physics Tufts Univ Medford MA 02155

WEAVER, EDWIN SNELL, b Hartford, Conn, Jan 30, 33; m 55; c 5. PHYSICAL CHEMISTRY. Educ: Yale Univ, BS, 54; Cornell Univ, PhD(chem), 59. Prof Exp: Asst chem, Cornell Univ, 55-57; from asst prof to assoc prof, 58-71, PROF CHEM, MT HOLYOKE COL, 71-, CHMN DEPT, 72- Concurrent Pos: NSF sci fac fel & vis prof, Univ Calif, San Diego, 64-65; vis prof, Univ Conn, 71-72. Mem: AAAS; Am Chem Soc. Res: Physical chemistry of polymers and proteins; neutron activation analysis. Mailing Add: Dept of Chem Mt Holyoke Col South Hadley MA 01075

WEAVER, ELLEN CLEMINSHAW, b Oberlin, Ohio, Feb 18, 25; m 44; c 3. GENETICS. Educ: Western Reserve Univ, AB, 45; Stanford Univ, MA, 52; Univ Calif, PhD(genetics), 59. Prof Exp: Staff mem, Carnegie Inst Dept Plant Biol, 61-62; res assoc, Biophys Lab, Stanford Univ, 62-67; resident res assoc, Ames Res Ctr, NASA, 67-69; actg dir, 73-74, DIR OFF SPONSORED RES PROJ SERV, SAN JOSE STATE UNIV, 74-, ASSOC PROF BIOL SCI, 69- Mem: AAAS; Am Soc Plant Physiol; Biophys Soc. Res: Mechanisms of photosynthesis, using wild type and mutant strains of algae; light-induced electron transport as monitored by means of electron paramagnetic resonance spectroscopy. Mailing Add: Dept of Biol Sci San Jose State Univ San Jose CA 95192

WEAVER, ERVIN EUGENE, b Centralia, Wash, Mar 12, 23; m 48; c 3. ENVIRONMENTAL CHEMISTRY. Educ: Manchester Col, BA, 45; Univ Ill, MA, 47; Western Reserve Univ, PhD(inorg chem), 51. Prof Exp: Asst, Univ Ill, 45-47; asst prof chem, Baldwin-Wallace Col, 47-51; asst prof, Wabash Col, 51-53, assoc prof, 53-61; res scientist, Chem Dept Sci Lab, 61-66, consult, 60-61, res scientist, Vehicle Emissions, Prod Res, 66-72, SUPVR CATALYSTS DEVELOP & TESTING, ENGINE & FOUNDRY DIV, FORD MOTOR CO, 72-, EMISSION PLANNING ASSOC, AUTOMOTIVE EMISSIONS OFF, 73- Concurrent Pos: Consult, Argonne Nat Lab, 56-59. Mem: Am Chem Soc; Soc Automotive Engrs; Air Pollution Control Asn. Res: Hexafluorides of heavy metals; xenon fluorides; nonaqueous solvents; air pollution; catalytic control of automotive emissions. Mailing Add: 18301 Laurel Ave Livonia MI 48152

WEAVER, GEORGE THOMAS, b Anna, Ill, Mar 11, 39; m 60; c 3. FOREST ECOLOGY, SILVICULTURE. Educ: Southern Ill Univ, Carbondale, BA, 60, MSEd, 62; Univ Tenn, Knoxville, PhD(bot), 72. Prof Exp: Instr bot, 70-71, ASST PROF FORESTRY, SOUTHERN ILL UNIV, CARBONDALE, 71- Mem: Ecol Soc Am; Am Inst Biol Sci. Res: Dry matter production and nutrient cycling in forest ecosystems; soil-site relationships of forests. Mailing Add: Dept of Forestry Southern Ill Univ Carbondale IL 62901

WEAVER, GERALD MACKNIGHT, b Wolfville, NS, Can, June 21, 35; m 56; c 1. PLANT BREEDING, CYTOGENETICS. Educ: Acadia Univ, BSc, 56; Rutgers Univ, PhD, 59. Prof Exp: Asst hort, Rutgers Univ, 56-59; head hort sect, Res Br, 59-75, DIR, FREDERICTON RES STA, CAN DEPT AGR, 75- Concurrent Pos: Plant breeder, Gen Cigar Co, 59-; Nat Res Coun sci exchange award, USSR, 66. Honors & Awards: Paul Howe Shepard Award, Am Pomol Soc, 63. Mem: Am Pomol Soc; Can Inst Food Tech; Prof Inst Pub Serv Can. Res: Fruit breeding and related genetic and radiobiological studies; investigations on winterhardiness of stone fruits and mechanisms governing resistance to insect and disease pests. Mailing Add: Fredericton Res Sta Can Dept Agr PO Box 280 Fredericton NB Can

WEAVER, HAROLD ESHLEMAN, chemistry, see 12th edition

WEAVER, HAROLD FRANCIS, b San Jose, Calif, Sept 25, 17; m 39; c 3. ASTRONOMY. Educ: Univ Calif, AB, 40, PhD(astron), 42. Prof Exp: Nat Res Coun fel, Yerkes Observ, Chicago & McDonald Observ, 42-43; tech aide, Nat Defense Res Comt, DC, 43-44; physicist, Radiation Lab, 44-45, from asst astronr to assoc astronr, Lick Observ, 45-51, assoc prof, 51-56, dir radio astron lab, 58-72, PROF ASTRON, UNIV CALIF, BERKELEY, 56- Concurrent Pos: Mem, US Army Air Force-Nat Geog Soc eclipse exped, Brazil, 47. Mem: Int Astron Union; Am Astron Soc. Res: Spectroscopy of peculiar stars; star clusters; galactic structure; radio astronomy. Mailing Add: Dept of Astron Univ of Calif Berkeley CA 94720

WEAVER, HARRY EDWARD, JR, b Philadelphia, Pa, Feb 1, 23; m 44; c 2. EXPERIMENTAL PHYSICS. Educ: Case Inst Technol, BS, 43, MS, 48; Stanford Univ, PhD(physics), 52. Prof Exp: Physicist, Manhattan Proj, Tenn, 44-46; instr physics, Case Inst Technol, 46-48; asst, Stanford Univ, 48-52; asst, Physics Inst, Zurich, 52-54; physicist, Varian Assocs, Calif, 54-69; PHYSICIST, HEWLETT-PACKARD CO, 69- Concurrent Pos: Vis lectr, Univ Zurich, 59-61. Mem: Am Phys Soc. Res: Nuclear magnetic and electron resonance, especially in solids; cryogenic engineering and application of high field superconductive materials in high field solenoids for high resolution nuclear magnetic resonance; x-ray photoelectron spectroscopy. Mailing Add: Hewlett-Packard Co 1501 Page Mill Rd Palo Alto CA 94304

WEAVER, HARRY TALMADGE, b Brewton, Ala, Dec 15, 38; m 59; c 3. SOLID STATE PHYSICS. Educ: Auburn Univ, BS, 60, MS, 61, PhD(physics), 69. Prof Exp: Mem staff eng, Humble Oil & Refining Co, 62-63; MEM STAFF PHYSICS, SANDIA LABS, 68- Mem: Am Phys Soc. Res: Hyperfine interactions in metals and atomic diffusion in solids studied by nuclear magnetic resonance techniques. Mailing Add: Sandia Labs Albuquerque NM 87115

WEAVER, HENRY D, JR, b Harrisonburg, Va, May 5, 28; m 52; c 4. PHYSICAL CHEMISTRY. Educ: George Washington Univ, BS, 50; Univ Del, MS & PhD(phys chem), 53. Prof Exp: Assoc prof chem, Eastern Mennonite Col, 51-57; assoc prof, 57-71, actg dean, 70-72, PROF CHEM, GOSHEN COL, 71-, PROVOST, 72- Concurrent Pos: Tech adv, Lima, Peru, 64-65; Fulbright lectr, Tribhuvan Univ, Nepal, 69-70. Mem: AAAS; Am Chem Soc; Am Sci Affil (pres, 62). Res: Heterogenous kinetics; metal; acid systems; kinetics of complex ion formation. Mailing Add: Goshen Col Goshen IN 46526

WEAVER, J RITNER, b Allentown, Pa, Jan 17, 15; m 44; c 2. ANALYTICAL CHEMISTRY. Educ: Muhlenberg Col, BS, 37. Prof Exp: Anal chemist, Hoffman-La Roche, Inc, 38-46; dir corp qual control, Warner-Hudnut, Inc, 46-58, dir biochem mfg, Warner-Chilcott Labs, 58-60, dir corp qual control, Warner-Lambert Co, 60-69; DIR CORP QUAL CONTROL, CHESEBROUGH-POND'S, INC, 60- Mem: Inst Food Technologists; Am Chem Soc; Soc Cosmetic Chem. Res: Pharmaceutical chemistry. Mailing Add: Chesebrough-Pond's Qual Control Dept PO Box 298 Merrit Blvd Trumbull CT 06611

WEAVER, JAMES BODE, JR, b Hartwell, Ga, Jan 28, 26; m 49; c 1. GENETICS, AGRONOMY. Educ: Univ Ga, BSA, 50, NC State Univ, MS, 52, PhD(genetics, agron), 55. Prof Exp: Asst county agt, Univ Ga, 52-53, asst prof agron, 55-58; dir cotton res, DeKalb Agr Res Inc, 58-63; dir cotton res, Cotton Hybrid Res, Inc, 63-65; supt grounds & agronomist, 65-67, asst prof agron, 67-72, ASSOC PROF AGRON, UNIV GA, 72- Mem: Am Soc Agron; Crop Sci Soc Am; Entom Soc Am. Res: Basic research on cotton improvement; utilization of hybrid vigor in cotton; insect resistance. Mailing Add: Dept of Agron Univ of Ga Plant Sci Bldg Athens GA 30602

WEAVER, JAMES EDMUND, b Francesville, Ind, Feb 5, 35; m 57; c 7. PHARMACOLOGY. Educ: Purdue Univ, BS, 57, MS, 59, PhD(pharmacol), 64. Prof Exp: Instr pharmacol & toxicol, Univ Cincinnati, 59-62; TOXICOLOGIST, PROCTER & GAMBLE CO, 63- Concurrent Pos: Participant, NIMH grant, 57-59, 62-63. Mem: Am Acad Clin Toxicol; Soc Toxicol; Am Acad Dermat; Int Col Pediat; Am Asn Poison Control Ctrs. Res: Toxicologic evaluation of materials proposed for use in synthetic detergent and soap formulations. Mailing Add: Procter & Gamble Co Sharon Woods Tech Ctr Cincinnati OH 45241

WEAVER, JEREMIAH WILLIAM, b New Orleans, La, Sept 29, 16; m 44; c 4. ORGANIC CHEMISTRY. Educ: Loyola Univ, La, BS, 38; Univ Detroit, MS, 40; Tulane Univ La, PhD(chem), 53. Prof Exp: Instr chem, Univ Detroit, 40-41; asst analyst, State Health Dept, La, 41-42; chemist, Southern Regional Res Lab, USDA, 45-56; res scientist, Koppers Co, Inc, 56-60; asst mgr, Cent Lab, Cone Mills Corp, NC, 60-61, mgr lab, 61-68; PROF TEXTILES, UNIV DEL, 68- Mem: AAAS; Am Chem Soc; Am Asn Textile Chemists & Colorists. Res: Cellulose chemistry; textile flammability. Mailing Add: Alison Hall Col of Home Econ Univ of Del Newark DE 19711

WEAVER, JOHN ARTHUR, b Hemingway, SC, Nov 23, 40; m 70. INORGANIC CHEMISTRY. Educ: Va Union Univ, BS, 64; Howard Univ, MS, 68, PhD(inorg chem), 70. Prof Exp: ASSOC PROF CHEM, NC A&T STATE UNIV, 70- Concurrent Pos: NSF grant, NC A&T State Univ, 72-74. Mem: AAAS; Am Chem Soc; Nat Inst Sci; Am Med Asn. Res: Metal in complexes of biological molecules, especially thiazoles and porphyrins. Mailing Add: Dept of Chem NC A&T State Univ Greensboro NC 27411

WEAVER, JOHN CARL, b Gibson, Ohio, Aug 5, 08; m 40; c 2. CHEMISTRY. Educ: Denison Univ, BS, 30; Univ Cincinnati, AM, 33, PhD(chem), 35. Prof Exp: Asst chem, Univ Cincinnati, 30-34, instr, 35-36; chemist & res coordr, 36-68, dir res, Coatings Group, 68-74, CONSULT, COATINGS GROUP, SHERWIN-WILLIAMS CO, 74- Concurrent Pos: Mem utilization res adv comt, USDA, 63-69; tech ed, J Paint Technol, 63-72; consult, Int Sugar Res Found, 75- Honors & Awards: Am Soc Testing & Mat Merit Award, 68; G B Heckel Award, Fedn Socs Coatings Technol, 69. Mem: Fedn Socs Coatings Technol; Am Chem Soc; Am Soc Testing & Mat. Res: Photographic developers; organic analytical reagents; varnishes; driers; resins. Mailing Add: Sherwin-Williams Co 101 Prospect Ave NW Cleveland OH 44101

WEAVER, JOHN DODSWORTH, b London, Eng, Nov 5, 14; m 41; div; c 3. GEOLOGY. Educ: Univ London, BSc, 49, PhD(geol), 53. Prof Exp: Lectr geol, Columbia Univ, 49-52; asst prof, Mt Holyoke Col, 52-55; PROF GEOL, UNIV PR, MAYAGUEZ, 55- Mem: Geol Soc London; Geol Soc Am; Seismol Soc Am. Res: Caribbean regional geology and tectonics. Mailing Add: Dept of Geol Univ of PR Mayaguez PR 00708

WEAVER, JOHN HERBERT, b Cincinnati, Ohio, Sept 16, 46. SOLID STATE PHYSICS. Educ: Univ Mo, BS, 67, MS, 69; Iowa State Univ, PhD(physics), 73. Prof Exp: Fel physics, Mat Res Ctr, Univ Mo-Rolla, 73; res assoc physics, 74-75, ASST SCIENTIST, SYNCHROTRON RADIATION CTR, PHYS SCI LAB, UNIV WIS, STOUGHTON, 75- Concurrent Pos: Assoc, Ames Lab, US Energy Res & Develop Agency, 75- Mem: Am Phys Soc; Sigma Xi. Res: Optical properties of solids; electronic structure of metals; photoemission and surface physics. Mailing Add: Phys Sci Lab Univ of Wis PO Box 6 Stoughton WI 53589

WEAVER, JOHN MARTIN, b Sallisaw, Okla, Apr 9, 23; m 51, 57; c 2. BIOCHEMISTRY, MICROBIOLOGY. Educ: Univ Okla, BS, 48; Univ Tex, PhD(biochem), 55. Prof Exp: Sr res chemist, Cent Res Dept, Anheuser-Busch Inc, 55-59; res biochemist, Med Res Labs, Chas Pfizer & Co, 59-62, asst dir res, Paul-Lewis Div, 62-63; sr res assoc biol, 63-66, sect chief biochem & microbiol, 66-75, MGR BIOCHEM & MICROBIOL, LEVER BROS CO, 75- Mem: AAAS; Int Asn Dent Res; NY Acad Sci; Am Soc Microbiologists. Res: Biodegradation and bioconsequences of detergent constituents; purine biosynthesis; vitamin B-12 fermentation;

carboxyanhydrides in peptide bond synthesis; commercial enzyme separation; oral biology. Mailing Add: 62 Van Allen Rd Glen Rock NJ 07452

WEAVER, JOHN RICHARD, b Goshen, Ind, Sept 3, 20; m 43; c 7. PHYSICAL CHEMISTRY. Educ: Bluffton Col, AB, 42; Univ Mich, MS, 50, PhD, 54. Prof Exp: From asst prof to assoc prof, 50-60, PROF PHYSICS & CHEM, BLUFFTON COL, 60- Concurrent Pos: Consult, Univ Mich, 56-; vis prof, Univ Pr, 63-64. Mem: Am Chem Soc. Res: Electrode kinetics; streaming mercury electrode; molecular structure; dipole moments. Mailing Add: Dept of Chem & Physics Bluffton Col Bluffton OH 45817

WEAVER, JOHN SCOTT, b Rochester, NY, Jan 13, 40; m 61; c 3. GEOPHYSICS. Educ: Univ Rochester, BS, 63, MS, 70, PhD(geol), 71. Prof Exp: Res asst physics, Univ Rochester, 59-61, instr geol & mineral, 68, asst lectr geol, 68-70; asst prof physics, 71-75, ASSOC PROF PHYSICS, QUEENS COL, NY, 76- Concurrent Pos: Guest technician, Brookhaven Nat Lab, 59-61. Mem: Am Geophys Union; AAAS. Res: Geophysics of the solid earth; theoretical and experimental studies of the equation of state of solids at high pressure. Mailing Add: Dept of Physics Queens Col Flushing NY 11367

WEAVER, JOHN TREVOR, b Birmingham, Eng, Nov 5, 32; m 60; c 2. GEOPHYSICS. Educ: Bristol Univ, BSc, 53; Univ Sask, MSc, 55, PhD(physics), 59. Prof Exp: From instr to asst prof math, Univ Sask, 58-61; leader appl math group, Defence Res Estab Pac, BC, 61-66; from asst prof to assoc prof physics, 66-72, PROF PHYSICS, UNIV VICTORIA, 72- Concurrent Pos: Lectr math, Univ Victoria, 62-64; Nat Res Coun Can res grants, 66-, travel fel, Univ Cambridge, 72-73. Mem: Am Geophys Union; Can Asn Physicists; Brit Inst Physics. Res: Electromagnetic induction in the earth; electromagnetic theory; geomagnetic variations. Mailing Add: Dept of Physics Univ of Victoria Victoria BC Can

WEAVER, KENNETH NEWCOMER, b Lancaster, Pa, Jan 16, 27; m 50; c 2. GEOLOGY. Educ: Franklin & Marshall Col, BS, 50; Johns Hopkins Univ, MA, 52, PhD(geol), 54. Prof Exp: Geologist, Medusa Portland Cement Co, 56-61, mgr geol & quarry dept, 61-63; DIR, MD GEOL SURV, 63- Concurrent Pos: Governor's rep, Interstate Oil Compact Comn, 64- & Interstate Mining Compact, 74-; chmn, Md Land Reclamation Comt, 68-; mem adv bd, Outer Continental Shelf Res, 74- Mem: AAAS; fel Geol Soc Am; Am Inst Mining, Metall & Petrol Eng; Asn Am State Geol (vpres, 71, pres, 73). Res: Geology of industrial minerals; environmental and structural geology; research administration. Mailing Add: Md Geol Surv Johns Hopkins Univ Baltimore MD 21218

WEAVER, LAWRENCE CLAYTON, b Bloomfield, Iowa, Jan 23, 24; m 49; c 4. PHARMACOLOGY. Educ: Drake Univ, BS, 49; Univ Utah, PhD(pharmacol), 53. Prof Exp: Asst pharmacol, Univ Utah, 49-53; pharmacologist, Res Dept, Pitman-Moore Co, 53-58, dir pharmacol labs, 59-60, assoc dir pharmacol res, 60-61, head biomed res, Pitman-Moore Div, Dow Chem Co, 61-64, asst dir res & develop labs, 64-65, asst to gen mgr, 65-66; PROF PHARMACOL & DEAN COL PHARM, UNIV MINN, MINNEAPOLIS, 66- Honors & Awards: Am Pharmaceut Asn Found Res Achievement Award, 63. Mem: Am Soc Pharmacol; Soc Exp Biol & Med; Am Pharmaceut Asn; Biomet Soc. Res: Combinations and assay of anticonvulsant drugs; pharmacology of cardiovascular and central nervous system drugs; health care delivery systems. Mailing Add: 703 Forest Dale Rd New Brighton MN 55112

WEAVER, LEO JAMES, b Springfield, Mo, Apr 18, 24; m 49. ORGANIC CHEMISTRY. Educ: Drury Col, BS, 49; Univ Mich, MS, 50. Prof Exp: Res chemist, Org Chems Div, Monsanto Chem Co, 50-54, res chemist, Inorg Chem Div, 54-56, res group leader detergents & surfactants, 56-60, asst dir res, 60-62, dir prod sales, 62-64, dir commercial develop, 64-65, dir res & develop, 65-69, pres, Monsanto Enviro-Chem Systs, Inc, 69-72; EXEC VPRES, CALGON CORP, 72- Concurrent Pos: Mem, Indust Res Inst. Honors & Awards: Monsanto du Bois Res Award, 54. Mem: AAAS; Am Mgt Asn; Am Chem Soc; NY Acad Sci. Res: Alkylation reactions of aromatics and olefins; sulfonation and sulfation of organics; detergents and surfactants. Mailing Add: Calgon Corp PO Box 1346 Pittsburgh PA 15230

WEAVER, LESLIE O, b St John, NB, Nov 16, 10; nat US; m 39; c 1. PLANT PATHOLOGY. Educ: Ont Agr Col, BSA, 34; Cornell Univ, PhD(plant path), 43. Prof Exp: Instr bot, Univ Toronto, 34-37; asst, NY Exp Sta, Geneva, 37-43; asst exten plant pathologist, Pa State Col, 43-48; EXTEN PLANT PATHOLOGIST, UNIV MD, COLLEGE PARK, 48- Res: Diseases of fruit crops; effect of temperature and relative humidity on occurrence of blossom blight of stone fruits. Mailing Add: Dept of Plant Path Univ of Md College Park MD 20742

WEAVER, MERLE LA ROY, horticulture, see 12th edition

WEAVER, MILO WESLEY, b Lufkin, Tex, Feb 16, 13; m 34; c 5. MATHEMATICS. Educ: Univ Tex, BA, 35, MA, 50, PhD, 56. Prof Exp: Pub sch teacher, Tex, 34-45; from instr to asst prof math, 45-61, ASSOC PROF MATH, UNIV TEX, AUSTIN, 61- Concurrent Pos: NSF res fel, 58-59. Mem: Am Math Soc; Math Asn Am. Res: Theory of semigroups; mappings on a finite set; associative algebras. Mailing Add: Dept of Math Univ of Tex Austin TX 78712

WEAVER, MORRIS EUGENE, b Morrison, Okla, June 10, 29; m 50; c 4. ANATOMY, ZOOLOGY. Educ: York Col, Nebr, BS, 51; Univ Omaha, BS, 53; Ore State Univ, MA, 56, PhD(zool), 59. Prof Exp: From instr to assoc prof, 58-72, PROF ANAT, DENT SCH, UNIV ORE HEALTH SCI CTR, 72- Concurrent Pos: Nat Inst Dent Res fel, Inst Animal Physiol, Eng, 66-67; vis reader, Med Sch, Univ Ibadan, 72-73. Mem: Am Asn Anatomists; Int Asn Dent Res; Sigma Xi. Res: Cell biology and mitosis; dental research using swine as experimental animals; microcirculation and temperature regulation. Mailing Add: Dept of Anat Sch of Dent Univ of Ore Health Sci Ctr Portland OR 97201

WEAVER, NEILL KENDALL, b Tariff, WVa, Oct 3, 19; m 45; c 3. MEDICINE. Educ: Oberlin Col, AB, 41; Harvard Med Sch, MD, 44; Am Bd Internal Med, dipl, 52; Am Bd Prev Med, dipl, 59. Prof Exp: Intern med, Med Sch, Pittsburgh & Allegheny Gen Hosp, 44-45; resident, Med Col, Univ Ala & Jefferson Hillman Hosp, 47-49; NIH fel, Sch Med, Tulane Univ & Charity Hosp, La, 49-51; instr med, Sch Med, Tulane Univ, 51-56, asst prof clin & prev med & pub health, 56-58, assoc prof clin & occup med, 58-66; asst med dir, Baton Rouge Refinery, Humble Oil & Refining Co, 56-66, asst med dir, Tex, 66-69; assoc med dir, Exxon Co, 69-73; MED DIR, AM PETROL INST, 73- Concurrent Pos: Vis physician, Charity Hosp, La, 51-65, sr vis physician, 65-; intern, Esso Standard Oil Co, 51-54, asst med dir, Refinery, 54-56; mem consult staff, Baton Rouge Gen Hosp, 56-66; clin prof med & assoc prof prev med, Sch Med, Tulane Univ, 66- Mem: Am Heart Asn; fel Indust Med Asn; fel Am Col Physicians; Am Fedn Clin Res; Am Acad Allergy. Res: Internal and occupational medicine. Mailing Add: Am Petrol Inst 2101 L St NW Washington DC 20037

WEAVER, NEVIN, b Navasota, Tex, Jan 4, 20; m 62; c 3. INSECT PHYSIOLOGY. Educ: Southwestern Univ, AB, 41; Tex A&M Univ, MS, 43, PhD(entom), 53. Prof Exp: Asst biol, Southwestern Univ, 38-41; asst, Tex A&M Univ, 41-43; beekeeper, 46-48; asst entom, Tex A&M Univ, 48-51, from instr to assoc prof, 41-65; chmn dept biol, 65-68, chmn campus planning, 69-70, PROF BIOL, UNIV MASS, BOSTON, 65- Concurrent Pos: Secy, Am comt, Bee Res Asn, 58; fel biol, Harvard Univ, 59-60. Mem: AAAS; Am Soc Zool; Entom Soc Am; Bee Res Asn; Animal Behav Soc. Res: Honeybee and stingless bee dimorphism, development, physiology, biochemistry, especially lipids, pheromones and behavior. Mailing Add: Dept of Biol Univ of Mass Boston MA 02125

WEAVER, QUENTIN CLIFFORD, b Harrisburg, Pa, Mar 22, 23; m 48; c 4. PAPER CHEMISTRY. Educ: Gettysburg Col, BA, 47; Pa State Univ, MS, 49. Prof Exp: Asst anal & qual org chem, Pa State Col, 47-49; res group leader, 49-53, coordr, Staff Tech Serv, 53-56, prod mgr, 56-57, asst dir res div, 58-72, MGR RES & ENG SERV, SCOTT PAPER CO, 72- Mem: Am Chem Soc; Tech Asn Pulp & Paper Indust. Res: Colloidal properties of cellulose; high polymers; paper coatings; plasticizers; high melting hydrocarbons. Mailing Add: Scott Paper Co Res & Eng Serv Scott Plaza-3 Philadelphia PA 19113

WEAVER, RALPH HOLDER, bacteriology, deceased

WEAVER, RICHARD WAYNE, b Twin Falls, Idaho, June 25, 44; m 64; c 1. SOIL MICROBIOLOGY. Educ: Utah State Univ, BS, 66; Iowa State Univ, PhD(soil microbiol), 70. Prof Exp: ASST PROF SOIL MICROBIOL, TEX A&M UNIV, 70- Mem: Soil Sci Soc Am; Am Soc Agron; Am Soc Microbiol. Res: Soil nitrogen; microbial ecology; nitrogen fixation; waste disposal. Mailing Add: Dept of Soil & Crop Sci Tex A&M Univ College Station TX 77843

WEAVER, ROBERT ELWIN, b Lexington, Ky, Mar 15, 28; m 54; c 2. MEDICAL BACTERIOLOGY. Educ: Univ Ky, BS, 49; Univ Wis, MS, 51, MD, 56, PhD(med microbiol), 57. Prof Exp: Asst, Univ Wis, 49-53; sr asst surgeon, Commun Dis Ctr, Phoenix Field Sta, USPHS, Ariz, 57-59; asst prof microbiol, Ohio State Univ, 60-64, asst prof path, 64-66; RES MED OFFICER, CTR DIS CONTROL, USPHS, 66- Mem: AAAS; Am Soc Microbiol; NY Acad Sci. Res: Diagnostic microbiology. Mailing Add: Spec Bact Sect USPHS Ctr for Dis Control Atlanta GA 30333

WEAVER, ROBERT F, b Topeka, Kans, July 18, 42; m 65; c 2. BIOCHEMISTRY. Educ: Col Wooster, BA, 64; Duke Univ, PhD(biochem), 69. Prof Exp: NIH fel, Univ Calif, San Francisco, 69-71; ASST PROF BIOCHEM, UNIV KANS, 71- Concurrent Pos: NIH grant, 72-75. Res: Structure and function of eucaryotic RNA polymerases; cell differentiation. Mailing Add: Dept of Biochem Univ of Kans Lawrence KS 66044

WEAVER, ROBERT HINCHMAN, b Buckhannon, WVa, Dec 2, 31; m 58; c 1. BIOCHEMISTRY. Educ: WVa Wesleyan Col, BS, 53; Univ Md, MS, 55, PhD(biochem), 57. Prof Exp: Fel, Enzyme Inst, Univ Wis, 57-61; assoc prof, 61-70, PROF CHEM, UNIV WIS-STEVENS POINT, 70- Mem: AAAS; Am Chem Soc. Res: Amine and carbohydrate metabolism; enzyme chemistry. Mailing Add: Dept of Chem Univ of Wis Stevens Point WI 54481

WEAVER, ROBERT JOHN, b Lincoln, Nebr, Sept 23, 17; m 51; c 2. PLANT PHYSIOLOGY. Educ: Univ Nebr, AB, 39, MS, 40; Univ Chicago, PhD(plant physiol), 46. Prof Exp: Res assoc bot, Univ Chicago, 46-48; from asst viticulturist to assoc viticulturist, Exp Sta, 48-58, VITICULTURIST, EXP STA, UNIV CALIF, DAVIS, 58-, PROF VITICULTURE, UNIV, 58- Concurrent Pos: Fulbright sr res scholar, Superior Col Agr, Greece, 56; vis res worker, Res Inst Grapevine Breeding, Ger, 63; Fulbright sr res scholar, Indian Agr Res Inst, New Delhi, 69; res fel, Enol & Viticult Res Inst, SAfrica, 73. Mem: AAAS; Sm Soc Plant Physiol; Bot Soc Am; Am Soc Enol; Am Soc Hort Sci. Res: Plant hormones and regulators; physiology of the grapevine. Mailing Add: Dept of Viticulture & Enol Univ of Calif Davis CA 95616

WEAVER, ROBERT MICHAEL, b Goshen, Ind, June 23, 42. CLAY MINERALOGY, SOIL SCIENCE. Educ: Berea Col, BA, 64; Mich State Univ, MS, 66; Univ Wis-Madison, PhD(soil sci), 70. Prof Exp: ASST PROF SOIL SCI, CORNELL UNIV, 71- Mem: Soil Sci Soc Am; Clay Minerals Soc. Res: Genesis of clay minerals; clay mineralogy of tropical soils. Mailing Add: Dept Agron 707 Bradfield Hall Cornell Univ Ithaca NY 14850

WEAVER, SALLY MAE, b Ft Erie, Ont, Aug 24, 40; m 64. ANTHROPOLOGY. Educ: Univ Toronto, BA, 63, MA, 64, PhD(anthrop), 67. Prof Exp: Asst prof, 66-72, ASSOC PROF ANTHROP, UNIV WATERLOO, 72- Concurrent Pos: Trustee, Nat Mus Can, 73-78. Mem: Can Ethnol Soc (pres, 75); Can Sociol & Anthrop Asn; fel Am Anthrop Asn; fel Soc Appl Anthrop. Res: Political anthropology; medical anthropology; Iroquois; Canadian Indians; status of Canadian Indian women. Mailing Add: Dept of Sociol & Anthrop Univ of Waterloo Waterloo ON Can

WEAVER, SYLVIA SHORT, b Concord, Mass, Oct 22, 27; m 53; c 2. MARINE BIOLOGY. Educ: Smith Col, BA, 49; NY Univ, MS, 70, PhD(marine biol), 75. Prof Exp: Teaching fel invert zool & elem biol, 71-72 & 73-75, RES ASSOC BIOL, NY UNIV, 75-, ADJ ASST PROF, 76-; RES ASSOC, HASKINS LABS, PACE UNIV, 75- Mem: Am Soc Limnol & Oceanog; Int Oceanog Found; Am Littoral Soc; Sigma Xi. Res: Plankton communities in inlets as representative of offshore-estuarine interaction and therefore as monitoring sites for environmental conditions. Mailing Add: 161 W 75th St New York NY 10023

WEAVER, TERRY L, b Johnstown, Pa, Apr 13, 46; m 66; c 2. MICROBIOLOGY. Educ: Indiana Univ Pa, BS, 68; Ohio State Univ, MS, 71, PhD(microbiol), 73. Prof Exp: Res microbiologist, US Army Biol Labs, Ft Detrick, 69-71; res assoc microbiol, Ohio State Univ, 71-73; ASST PROF MICROBIOL, CORNELL UNIV, 73- Mem: Am Soc Microbiol; AAAS. Res: Physiology of microbial growth; microbial one carbon metabolism; methylotrophic microorganisms. Mailing Add: Lab of Microbiol 309 Stocking Hall Cornell Univ Ithaca NY 14853

WEAVER, THOMAS, b Grenville, NMex, May 1, 29; m 50; c 3. SOCIAL ANTHROPOLOGY, APPLIED ANTHROPOLOGY. Educ: Univ NMex, BA, 55, MA, 60; Univ Calif, Berkeley, PhD(anthrop), 65. Prof Exp: Exec secy & ex-officio chmn adv comt, Calif State Adv Comn Indian Affairs, 64; asst prof anthrop, Univ Ky, 64-67; asst prof & Maurice Falk sr fac fel, Univ Pittsburgh, 67-69; assoc prof cult anthrop, 69-75, PROF CULT ANTHROP, UNIV ARIZ, 75-, DIR, BUR ETHNIC RES, 69- Concurrent Pos: Mem bd dirs & gov deleg from Calif, Gov Interstate Indian Coun, 64-65; Calif State Adv Comn on Indian Affairs res grant, 65-66; fac travel grant, Univ Ky, 66 & poverty symp grant, 67; Urban Anthrop Conf grant, Sch Am Res, 68; Polit Orgn & Bus Mgt grant, Gila River Indian Community, Ariz, 70-71; USDA grant lit surv Am Indian & Mex Am youths attend Youth Conserv Corps, 72; US Dept Health, Educ & Welfare grant, Douglas, Ariz, 72-73. Mem: Fel AAAS; Am Anthrop Asn; Am Ethnol Soc; Soc Appl Anthrop (pres, 76). Res: Methodology; techniques; migration; complex societies; role theory; kinship; Latin America; American Indians; ethnic groups; social structure; culture change; political organization; social pathologies; conflict; medical systems. Mailing Add: Bur of Et Ethnic Res Univ of Ariz Dept of Anthrop Tucson AZ 85721

WEAVER, WARREN, b Reedsburg, Wis, July 17, 94; m 19; c 2. MATHEMATICS. Educ: Univ Wis, BS, 16, CE, 17, PhD(math physics), 21. Hon Degrees: LLD, Univ Wis, 48; ScD, Univ Sao Paulo, 49, Drexel Inst, 61, Univ Pittsburgh, 64 & NY Univ, 64; DE, Rensselaer Polytech, 62; LHD, Univ Rochester, 63. Prof Exp: Asst prof math, Throop Col Technol, 17-18 & Calif Inst Technol, 19-20; from asst prof to assoc prof, Univ Wis, 20-28, prof & chmn dept, 28-32; dir natural sci, Rockefeller Found, 32-55, vpres natural & med sci, 55-59; vpres, 59-64, CONSULT SCI AFFAIRS, ALFRED P SLOAN FOUND, 64- Concurrent Pos: Dir natural sci, Gen Educ Bd, 32-37; mem div phys sci, Nat Res Coun, 36-39; chmn sect D-2, Nat Defense Res Comt, Off Sci Res & Develop, 40-42, chief appl math panel, 43-46; chmn, Naval Res Adv Comt, 46-47; mem res adv panel, US War Dept, 46-47; chmn basic res group, Res & Develop Bd, US Dept Defense, 52-53; mem, Nat Sci Bd, 56-60; mem, Nat Adv Cancer Coun, 57-60; vpres, Sloan-Kettering Inst, 58-59; nonresident fel, Salk Inst Biol Studies. Honors & Awards: US Medal for Merit, 48; Officer, Legion of Honor, France, 50; Pub Welfare Medal, Nat Acad Sci, 57; Brit King's Defense Freedom Medal; Kalinga Prize & Arches of Sci Award, 65. Mem: AAAS (pres, 54); Am Soc Naturalists; fel Am Phys Soc; Am Math Soc; Am Philos Soc. Res: Electrodynamics; theory of probability; scientific administration. Mailing Add: 40 Hillis Rd RR 3 New Milford CT 06776

WEAVER, WARREN ELDRED, b Sparrows Point, Md, June 5, 21; m 45; c 4. CHEMISTRY. Educ: Univ Md, BS, 42, PhD(pharmaceut chem), 47. Prof Exp: Asst pharm, Univ Md, 42, asst antigas prep, Off Sci Res & Develop, 42-44; asst insecticides, 44-45; chemist, US Naval Res Lab, 45-50; assoc prof chem & pharmaceut chem, 50-54, from actg chmn to chmn dept, 50-56, PROF CHEM & PHARMACEUT CHEM, MED COL VA, 54-, DEAN SCH PHARM, 56- Concurrent Pos: Mem pharm rev comt, NIH, 68-72; mem bd dirs, Am Found Pharm Educ, 69-73; mem, Secretary's Comn to rev Report of President's Task Force on Prescription Drugs, 70; mem, Am Coun Pharmaceut Educ, 74-80 Mem: AAAS; Am Chem Soc; Am Pharmaceut Asn; Am Inst Hist Pharm; Am Asn Cols Pharm (vpres, 67-68, pres, 68-69). Res: Synthetic peptides; insecticides; correlation of structure with fungicidal activity. Mailing Add: Sch of Pharm Med Col of Va Richmond VA 23298

WEAVER, WILLIAM JUDSON, b Twin Falls, Idaho, May 7, 36; m 65; c 1. CRYOBIOLOGY. Educ: Col Idaho, BS, 58; Univ Ore, MS, 65, PhD(exp path), 71. Prof Exp: RES ASSOC SURG, HEALTH SCI CTR, UNIV ORE, 71- Concurrent Pos: Sci consult, Res Industs Corp, Utah, 75- Mem: AAAS; Soc Cryobiol. Res: Development of a means for long-term storage of whole intact organs for human transplantation. Mailing Add: 24 Del Prado Lake Oswego OR 97034

WEAVER, WILLIAM MICHAEL, b Lima, Ohio, Feb 18, 31; m 66. ORGANIC CHEMISTRY. Educ: Johns Carroll Univ, BS, 53; Purdue Univ, MS, , 56, PhD, 58. Prof Exp: Asst org chem, Purdue Univ, 53-58; from instr to assoc prof, 58-70, PROF ORG CHEM, JOHN CARROLL UNIV, 70- Mem: Am Chem Soc. Res: Aliphatic nitrocompound; kinetics; reactions in aprotic solvents. Mailing Add: Dept of Chem John Carroll Univ Cleveland OH 44118

WEBB, ALAN WENDELL, b Enid, Okla, Sept 20, 39; m 61; c 3. PHYSICAL INORGANIC CHEMISTRY. Educ: Brigham Young Univ, BS, 63, PhD(phys chem), 69. Prof Exp: RES CHEMIST, US NAVAL RES LAB, 68- Concurrent Pos: Nat Acad Sci-Nat Res Coun resident res associateship, 68-70. Mem: Am Chem Soc. Res: High pressure and high temperature synthesis of inorganic compounds; study of solid ionic conductors; layered-structure transition metal dichalcogenides; high pressure and low temperature study of metals. Mailing Add: High Pressure Physics Sect US Naval Res Lab Washington DC 20375

WEBB, ALBERT DINSMOOR, b Victorville, Calif, Oct 10, 17; m 43; c 2. ENOLOGY. Educ: Univ Calif, BS, 39, PhD(chem), 48. Prof Exp: From asst prof to assoc prof enol, Col Agr, 48-60, from asst chemist to assoc chemist, Exp Sta, 48-60, PROF ENOL, COL AGR, UNIV CALIF DAVIS, 60-, CHMN DEPT VITICULTURE & ENOL, 73-, CHEMIST, EXP STA, 60- Concurrent Pos: Fulbright res scholar, Australia, 58, Alko Res Labs, Helsinki, Finland, 69; NATO scholar, Bordeaux, 62; vis res prof, Univ Stellenbosch, SAfrica, 70. Mem: Am Chem Soc; AAAS; Am Soc Enologists (pres, 74-75). Res: Identification of aroma and flavor compounds in grapes and wines. Mailing Add: Dept of Viticult & Enol Univ of Calif Davis CA 95616

WEBB, ALFRED MOHR, b Allentown, Pa, Nov 30, 14. MEDICAL MICROBIOLOGY. Educ: Lehigh Univ, BA, 35, MS, 37; Mass Inst Technol, PhD(microbiol), 47. Prof Exp: Asst pediat, Harvard Med Sch, 38-40; res bacteriologist, Lederle Labs, Am Cyanamid Co, NY, 42-46; res bacteriologist, US Dept Army, Ft Detrick, Md, 47-60; scientist adminr, 60-63, chief reagents br, 63-66, CHIEF PROG PLANNING & PROJECTION STAFF, NAT INST ALLERGY & INFECTIOUS DIS, 66- Concurrent Pos: Mem resources & logistics working group, Nat Cancer Inst, 66-70; deleg, Nat Comn Diabetes Mellitus, 75- Mem: AAAS; Am Soc Microbiol; NY Acad Sci. Res: Infectious diseases and immunology; planning and evaluation of research programs; sociology of science; federal role in the encouragement and support of science. Mailing Add: Off of Dir Nat Inst Allergy & Infectious Dis Bethesda MD 20014

WEBB, ALFREDA JOHNSON, b Mobile, Ala, Feb 21, 23; m 59; c 3. VETERINARY MEDICINE. Educ: Tuskegee Inst, BS, 43, DVM, 49; Mich State Univ, MS, 51. Prof Exp: From instr to assoc prof anat, Tuskegee Inst, 50-59; PROF BIOL, NC A&T STATE UNIV, 59- Mem: Am Vet Med Asn; Am Asn Vet Anat. Res: Histology; cytology; embryology. Mailing Add: Dept of Biol NC A&T State Univ Greensboro NC 27411

WEBB, ALLEN NYSTROM,b Wichita, Kans, Dec 14, 21; m 43; c 3. PHYSICAL CHEMISTRY. Educ: Kans State Univ, BS, 43; Univ Calif, Berkeley, PhD(phys chem), 49. Prof Exp: Jr chemist, Boston Consol Gas Co, 43; res chemist, Stand Oil Co, Ind, 44-46; res asst chem, Univ Calif, 46-49; from res chemist to sr res chemist, 49-66, RES ASSOC, TEXACO, INC, 66- Concurrent Pos: Chmn, Gordon Res Conf Catalysis, 70; mem adv bd, Petrol Res Fund, 70-72. Mem: AAAS; Catalysis Soc; Am Chem Soc. Res: Catalysis; chemisorption; molecular spectra; fuel cells; exchange reactions; kinetics; hydrodesulfurization. Mailing Add: Beacon Res Lab Texaco Inc PO Box 509 Beacon NY 12508

WEBB, ANDREW CLIVE, b Bishop's Stortford, Eng, Feb 17, 47; m 75. DEVELOPMENTAL BIOLOGY. Educ: Southampton Univ, BSc, 69, PhD(biol), 73. Prof Exp: Fel biol, Purdue Univ, West Lafayette. 73-75, ASST PROF BIOL SCI, WELLESLEY COL, 75- Mem: Brit Soc Develop Biol. Res: ultrastructure and biochemistry of amphibian oogenesis. Mailing Add: Dept of Biol Sci Wellesley Col Wellesley MA 02181

WEBB, ARTHUR HARPER, b Washington, DC, Dec 28, 15; m 42; c 2, BACTERIOLOGY. Educ: Univ Ill, AB, 39, MS, 40, PhD(bact), 44. Prof Exp: Asst animal path, Univ Ill, 42-43; from instr to assoc prof bact, Col Med, Harvard Univ, 44-60; prof biol, Md State Col, Princess Anne, 60-61; prof biol & chmn Div Sci, Southern Univ, 61-68; PROF BIOL, FED CITY COL, 68- Concurrent Pos: Bacteriologist & chief, Pub Health Labs, US Opers Mission, Gondar, Ethiopia, 54-56. Mem: AAAS; Am Soc Microbiol. Res: Physiology of yeasts; radioactive amino acids; hospital staphylococci; parasitism; medical education in Ethiopia. Mailing Add: 1005 Lagrande Rd Silver Spring MD 20903

WEBB, BILL D, b Ralls, Tex, June 13, 28; m 54; c 3. PLANT BIOCHEMISTRY. Educ: Tex A&M Univ, BS, 56, MS, 59, PhD(biochem & nutrit), 61. Prof Exp: RES CHEMIST, AGR RES SERV, USDA, 61- Mem: Am Asn Cereal Chem; Inst Food Technologists. Res: Physicochemical properties of the rice grain and their relation to rice milling, cooking, processing and nutritive quality. Mailing Add: USDA Regional Rice Qual Lab Rte 5 Box 784 Beaumont TX 77706

WEBB, BURLEIGH C, b Greensboro, NC, Jan 9, 23; m 49; c 3. AGRONOMY, PLANT PHYSIOLOGY. Educ: Agr & Tech Col, NC, BS, 43; Univ Ill, MS, 47; Mich State Univ, PhD, 52. Prof Exp: Instr agron, Tuskegee Inst, 47-49; asst prof, Ala Agr & Mech Col, 51-52; assoc prof, Tuskegee Inst, 52-59; PROF AGRON, NC A&T STATE UNIV, 59-, DEAN SCH AGR, 63- Mem: AAAS; Am Soc Agron. Res: Plant physiology and ecology of forage crop plants; role of growth regulators in developmental and growth phenomena in crop plants. Mailing Add: Sch of Agr NC A&T State Univ Greensboro NC 27411

WEBB, BYRON HORTON, b New York, NY, May 2, 03; m 32; c 4. DAIRY INDUSTRY. Educ: Univ Calif, BS, 25; George Washington Univ, MS, 28; Cornell Univ, PhD(dairy indust), 31. Prof Exp: Jr dairy mfg specialist, Bur Dairy Indust, USDA, 26-29, from assoc specialist to sr specialist, 30-48, prin dairy technologist, 48-51; sr scientist, Nat Dairy Res Labs, Inc, 51-53, asst dir res, 53-54, dir, 54-60; chief dairy prods lab, Eastern Utilization Res & Develop Div, Agr Res Serv, USDA, DC, 60-72; CONSULT DAIRY & FOOD TECHNOL, 72- Honors & Awards: Borden Award, Am Dairy Sci Asn, 43; Superior Serv Award, USDA, 48. Mem: AAAS; Am Chem Soc; Am Dairy Sci Asn; Inst Food Technologists. Res: Methods of improvement of dairy products; development of new products; by-products from milk. Mailing Add: Bolivar Heights Harpers Ferry WV 25425

WEBB, CHARLES ALAN, b Charlottesville, Va, July 23, 47; m 74. PHYSICAL CHEMISTRY. Educ: Lehigh Univ, BA, 69; Univ Miami, PhD(phys chem), 74. Prof Exp: Chemist, 69-70, RES CHEMIST, TEXTILE FIBERS DEPT, E I DU PONT DE NEMOURS & CO, INC, 75- Mem: Am Chem Soc; Electrochem Soc; Int Oceanog Found. Res: Electrochemical corrosion of copper in chloride and amino acid solutions; spunbonded nonwoven polymer technology; polypropylene stabilization. Mailing Add: R&D Lab Textile Fibers Dept E I du Pont de Nemours & Co Inc Old Hickory TN 37138

WEBB, DAVID KNOWLTON, JR, geology, deceased

WEBB, DONALD WAYNE, b Brandon, Man, July 12, 39; m 61; c 2. ENTOMOLOGY. Educ: Univ Manitoba, BSc, 61, MSc, 63. Prof Exp: Lectr, St Paul's Col, Manitoba, 62-63; asst taxonomist, 66-75, ASSOC TAXONOMIST, ILL NATURAL HIST SURV, 75- Res: Aquatic insect ecology and the taxonomy of the mature and immature stages of the Chironomidae. Mailing Add: Ill Natural Hist Surv Urbana IL 61801

WEBB, EDMUND LESLIE, b Cheltenham, Eng, July 6, 23; US citizen. ANESTHESIOLOGY. Educ: Univ London, MB, BS, 53. Prof Exp: Instr anesthesiol, NY Univ, 60; INSTR ANESTHESIOL, STATE UNIV NY DOWNSTATE MED CTR, 65-; DIR ANESTHESIOL, CUMBERLAND HOSP, BROOKLYN, 71- Concurrent Pos: Asst attend, Methodist Hosp, Brooklyn, 60-; assoc dir med res, Squibb Inst Med Res, 62-70; attend, Brooklyn Cumberland Med Ctr, NY, 70- Mem: Brit Med Asn; Am Soc Anesthesiol; fel Am Col Anesthesiol. Res: Halogenated anesthetic agents; analgesics. Mailing Add: 318 W 78th St New York NY 10024

WEBB, FRANK ERNEST, forestry, see 12th edition

WEBB, FRED, JR, b Hinton, WVa, Feb 23, 35; m 56; c 2. GEOLOGY. Educ: Duke Univ, AB, 57; Va Polytech Inst, MS, 59, PhD(struct geol), 65. Prof Exp: Asst geol, Va Polytech Inst, 59-60, instr, Eng Exp Sta, 61-63, instr, Inst, 63-66, resident dir summer field sta, 65; assoc prof geol, Catawba Col, 66-68; from asst prof to assoc prof.prof, 68-73, PROF GEOL, APPALACHIAN STATE UNIV, 73-, CHMN DEPT, 72- Concurrent Pos: Adj prof, Va Polytech Inst & State Univ, 71- Res: Stratigraphy and tectonics of southern Appalachians; middle Ordovician turbidites; Ordovician paleotopography; modal fold analysis. Mailing Add: Dept of Geol Appalachian State Univ Boone NC 28608

WEBB, GEORGE DAYTON, b Oak Park, Ill, June 22, 34; m 57; c 3. NEUROPHYSIOLOGY. Educ: Oberlin Col, AB, 56; Yale Univ, MAT, 57; Univ Colo, PhD(physiol), 62. Prof Exp: High sch teacher, DC, 57-58; asst prof, 66-69, ASSOC PROF PHYSIOL, UNIV VT, 69- Concurrent Pos: Vis fel biochem, Univ Copenhagen, 62-63; vis fel neurol & biochem, Columbia Univ, 63-66. Mem: Am Physiol Soc; Am Soc Neurochem; Soc Gen Physiol; Int Soc Neurochem; NY Acad Sci. Res: Molecular physiology of synaptic and conducting membranes, especially cholinergic mechanisms and ion transport. Mailing Add: Dept of Physiol & Biophys Given Med Bldg Univ of Vt Burlington VT 05401

WEBB, GEORGE WILLIS, b Indianapolis, Ind, Jan 16, 10. ECONOMIC GEOGRAPHY, CULTURAL GEOGRAPHY. Educ: George Peabody Col, BS, 38, MA, 39; Univ Tenn, PhD(geol), 56. Prof Exp: Assoc prof soc sci, Tenn Technol Univ, 46-57; prof geog & geol & chmn dept, East Tenn State Univ, 58-65; PROF GEOG, IND STATE UNIV, TERRE HAUTE, 65- Mem: Asn Am Geogr; Nat Coun Geog Educ. Res: Settlement geography in the Micronesian Islands. Mailing Add: Dept of Geog & Geol Ind State Univ Terre Haute IN 47809

WEBB, GLENN FRANCIS, b Cleveland, Ohio, Sept 30, 42; m 73. MATHEMATICAL ANALYSIS. Educ: Ga Inst Technol, BS, 65; Emory Univ, MS, 66, PhD(math), 68. Prof Exp: Asst prof, 68-73, ASSOC PROF MATH, VANDERBILT UNIV, 74- Concurrent Pos: Vis assoc prof math, Univ Ky, 73; Ital Nat Coun Res fel, Univ Rome, 74. Mem: Am Math Soc; Math Asn Am. Res: Differential equations and functional analysis. Mailing Add: Dept of Math Vanderbilt Univ Nashville TN 37235

WEBB, GLENN R, b Chicago, Ill, May 22, 18; m 53; c 1. MALACOLOGY. Educ: Univ Ill, BS, 48, MS, 49; Univ Okla, PhD, 60. Prof Exp: Res asst, Univ Ill, 48-49 & Univ Okla, 53-56; instr biol, Henderson State Teachers Col, 56-57; fishery res biologist, US Fish & Wildlife Surv, Ft Worth, Tex, 58-59; instr biol, Coastal Carolina Jr Col, 59-60 & SC, Florence & Conway Branches, 60-62; asst prof, High Point Col, 62-63; PROF BIOL, KUTZTOWN STATE COL, 63- Mem: Am Malacol Union. Res: Pulmonate land snails life histories with special regard to sexology as a clue to phylogeny; snail autoecology; Polygyridae-Helicoidea; inter-species hybridization. Mailing Add: Dept of Biol Kutztown State Col Kutztown PA 19530

WEBB, GREGORY WORTHINGTON, b Englewood, NJ, July 29, 26; m 50; c 4. GEOLOGY. Educ: Columbia Univ, AB, 48, MA, 50, PhD(stratig), 54. Prof Exp: Instr, Rutgers Univ, 51-52; geologist, Standard Oil Co, Calif, 52-56; from instr to asst prof geol, Amherst Col, 56-59; from asst prof to assoc prof, 59-71, PROF GEOL, UNIV MASS, AMHERST, 71- Concurrent Pos: Vis lectr, Univ Mass, Amherst, 58-59. Mem: Geol Soc Am; Am Asn Petrol Geol; Am Geophys Union; Soc Econ Paleont & Mineral. Res: Strike-slip faults and associated structural and stratigraphic geology; petroleum geology; distribution patterns of turbidity current deposits; marine geology of offshore eastern Canada. Mailing Add: Dept of Geol Univ of Mass Amherst MA 01002

WEBB, HELEN MARGUERITE, b Nova Scotia, July 7, 13; nat US. INVERTEBRATE PHYSIOLOGY. Educ: Northwestern Univ, BS, 46, MS, 48, PhD(zool), 50. Prof Exp: Asst prof biol, Boston Col, 50-52; asst prof physiol, 52-59, assoc prof biol sci, 59-66, PROF BIOL SCI, GOUCHER COL, 66- Concurrent Pos: Mem corp, Marine Biol Lab, Woods Hole. Mem: Am Soc Zoologists; Soc Gen Physiol; Ecol Soc Am; Int Soc Chronobiol. Res: Biological rhythms; crustacean endocrinology. Mailing Add: Dept of Biol Sci Goucher Col Towson MD 21204

WEBB, HUBERT JUDSON, agricultural chemistry, see 12th edition

WEBB, IRVING D, b Los Angeles, Calif, Jan 2, 21; m 42; c 4. ORGANIC CHEMISTRY. Educ: Univ Calif, Los Angeles, PhD(chem), 44. Prof Exp: Res chemist, Nat Defense Res Comt, 42-43 & E I du Pont de Nemours & Co, Inc, 46-52; res chemist, 52-56, mgr chem res div, 56-62, mgr Korea Petrochem Complex, 67, MGR WAX SALES, UNION OIL CO CALIF, 62-67 & 68- Mem: Am Chem Soc. Res: Inorganic chemistry; petroleum and petrochemicals; agricultural chemicals. Mailing Add: 1127 F 15th St Santa Monica CA 90403

WEBB, JAMES L A, b Webb, Miss, Nov 17, 17; m 46; c 3. ORGANIC CHEMISTRY. Educ: Washington & Lee Univ, BS, 39; Johns Hopkins Univ, PhD(org chem), 43. Prof Exp: Jr instr chem, Johns Hopkins Univ, 39-43, instr, 43-45; from asst prof to prof, Southwestern at Memphis, 45-59; PROF CHEM, GOUCHER COL, 59-, CHMN DEPT, 65- Concurrent Pos: Res chemist, Chapman Chem Co, 51-59. Mem: Am Chem Soc. Res: Organic heterocyclic compounds; theory and antidotes for heavy metal poisoning; disubstituted pyridines; bipyrryls and pyrrole pigments; asphalt additives. Mailing Add: Dept of Chem Goucher Col Baltimore MD 21204

WEBB, JAMES WOODROW, b Warrenville, SC, Jan 21, 13; m 38; c 5. WILDLIFE MANAGEMENT. Educ: Furman Univ, BS, 36; Ala Polytech Inst, MS, 39. Prof Exp: State biologist, Ala Dept Conserv, 39-44; supt pond, Exp Sta, Ala Polytech Inst, 44-47; proj leader, Farm Game Restoration, SC Game & Fish Dept, 47-49, fed aid coordr, 49-52, asst dir dept, 52-59, dir div game, 59-69, exec dir, SC Wildlife & Marine Resources Dept, 69-74, exec dir emer, 74-75; RETIRED. Res: Fish culture. Mailing Add: 1313 Winyah Dr Columbia SC 29203

WEBB, JERRY GLEN, b Rosefield, La, Feb 17, 38; m 61; c 3. THEORETICAL PHYSICS. Educ: Northeast La State Col, BS, 61; Univ Ark-Fayetteville, MS, 64; Tex A&M Univ, PhD(physics), 69. Prof Exp: Instr physics, Cent State Col, Okla, 62-65; assoc prof, 69-75, PROF PHYSICS, UNIV ARK-MONTICELLO, 75- Concurrent Pos: Res partic, Savannah River Lab, Aiken, SC, 70. Mem: Am Asn Physics Teachers. Res: Three particle scattering problem. Mailing Add: Dept of Physics Univ of Ark Monticello AR 71655

WEBB, JOHN RAYMOND, b Morgantown, WVa, Sept 18, 20. AGRONOMY. Educ: WVa Univ, BS, 40, MS, 42; Purdue Univ, PhD(agron), 53. Prof Exp: Instr & asst soils, WVa Univ, 41-43; soil surveyor, Soil Conserv Serv, USDA, 46-47; instr & asst soils, WVa Univ, 47-48; asst agron, Purdue Univ, 49-52; asst prof, 52-56, ASSOC PROF SOILS, DEPT AGRON & AGR EXP STA, IOWA STATE UNIV, 56- Mem: Am Soc Agron; Soil Sci Soc Am. Res: Soil fertility. Mailing Add: Dept of Agron Iowa State Univ Ames IA 50012

WEBB, JOHN SCHURR, b Philadelphia, Pa, Feb 18, 19; m 43; c 3. ANALYTICAL CHEMISTRY. Educ: Yale Univ, BS, 41, PhD(org chem), 44. Prof Exp: Lab asst, Yale Univ, 41-43; res chemist, 43-44 & 46-48, sr res chemist, 48-53, group leader, Lederle Labs Div, 55-62, dept head struct & anal, 63-75, HEAD, ANAL RES DEPT, LEDERLE LABS DIV, AM CYANAMID CO, 75- Mem: AAAS; Am Chem Soc; Am Microchem Soc; Marine Technol Soc. Res: Synthesis of pharmaceuticals; structure of antibiotics; purification, analysis and structure of organics by instrumental techniques; drugs from the sea. Mailing Add: Dept 977 Lederle Labs Div Am Cyanamid Co Pearl River NY 10965

WEBB, JOHN W, b Staines, Eng, July 29, 26; nat US; m 53; c 1. GEOGRAPHY. Educ: St Andrews Univ, MA, 50; Univ Minn, PhD, 58. Prof Exp: Asst geog, Univ Minn, 52-54; instr, Univ Md, 54-55; from instr to assoc prof, 55-67, chmn dept, 66-69, assoc dean, 71-73, PROF GEOG, UNIV MINN, MINNEAPOLIS, 67- Mem: Fel Am Geog Soc; Asn Am Geog. Res: Western Europe; settlement and population geography; history of geography. Mailing Add: Dept of Geog Univ of Minn Minneapolis MN 55455

WEBB, JULIAN PIERCE, b Rochester, NY, Sept 29, 35; m 64; c 2. EXPERIMENTAL PHYSICS. Educ: Harvard Univ, BA, 57; Stanford Univ, MS, 66, PhD(physics), 68. Prof Exp: Res physicist, Res Labs, Eastman Kodak Co, Rochester, 68-76. Mem: Am Phys Soc. Res: Organic dye lasers; chemical physics; laser applications; photoconductivity research. Mailing Add: 16 Wincanton Dr Fairport NY 14450

WEBB, KEMPTON EVANS, b Melrose, Mass, Dec 28, 31; m 55; c 2. GEOGRAPHY. Educ: Harvard Col, AB, 53; Syracuse Univ, MA, 55, PhD(geog), 58. Prof Exp: Tech consult, Bank of Northeast Brazil, 57; asst prof geog, Ind Univ, 58-61; from asst prof to assoc prof geog, 61-70, actg dir, Inst Latin Am Studies, 62, from assoc dir to dir inst, 65-74, PROF GEOG, COLUMBIA UNIV, 70-, CHMN DEPT, 74- Concurrent Pos: Cor mem humid tropics comn, Int Geog Union, 58-; consult, Fulbright Prog Long Range Planning Group, US Dept State, 66-67. Mem: Asn Am Geog; Latin Am Studies Asn. Res: Cultural and economic geography; Latin America, especially Brazil; historical geography of Brazil. Mailing Add: Dept of Geog Columbia Univ New York NY 10027

WEBB, KENNETH EMERSON, JR, b Hamilton, Ohio, Feb 26, 43; m 66; c 2. ANIMAL NUTRITION. Educ: Ohio Univ, BS, 65; Univ Ky, MS, 67, PhD(animal sci), 69. Prof Exp: ASST PROF RUMINANT NUTRIT, VA POLYTECH INST & STATE UNIV, 69- Mem: Am Soc Animal Sci; Animal Nutrit Res Coun; Agr Res Inst. Res: Ruminant nutrition, protein, mineral and solid waste utilization by ruminants. Mailing Add: Agnew Hall Va Polytech Inst & State Univ Blacksburg VA 24061

WEBB, KENNETH LOUIS, biochemical oceanography, see 12th edition

WEBB, LAWRENCE EDWARD, physical chemistry, see 12th edition

WEBB, LELAND FREDERICK, b Hollywood, Calif, July 27, 41; m 63. MATHEMATICS EDUCATION. Educ: Univ Calif, Santa Barbara, BA, 63; Calif State Polytech Univ, San Luis Obispo, MA, 68; Univ Tex, Austin, PhD(math & educ), 71. Prof Exp: Lectr math & educ, Calif State Polytech Univ, San Luis Obispo, 67-68; curric writer math, Southwest Educ Develop Lab, 70; teaching asst math, Univ Tex, Austin, 71; res assoc math, Res & Develop Ctr Teacher Educ, Austin, 71; asst prof, 71-73, ASSOC PROF MATH, CALIF STATE COL, BAKERSFIELD, 73- Concurrent Pos: Consult, Educ Develop Ctr, Newton, Mass, 71-, Greenfield Sch Dist, Bakersfield, 73- & Maricopa Sch Dist, Calif, 74-75; NSF sec sch math & sci grant, 73-75 & 74-, elem sch sci & math grant, 75- Mem: Sigma Xi; Nat Coun Teachers Math; Am Educ Res Asn; Sch Sci & Math Asn; Math Asn Am. Res: Learning in science and mathematics education. Mailing Add: Dept of Math Calif State Col Bakersfield CA 93309

WEBB, MAURICE BARNETT, b Neenah, Wis, May 14, 26; m 56; c 2. PHYSICS. Educ: Univ Wis, BS, 50, MS, 52, PhD, 56. Prof Exp: With Proj Lincoln, Mass Inst Technol, 52-53; with res lab, Gen Elec Co, 56-61; assoc prof, 61-65, PROF PHYSICS, UNIV WIS-MADISON, 65- Mem: Am Phys Soc. Res: Solid state physics; x-ray scattering; surface physics; low energy electron diffraction. Mailing Add: Dept of Physics Univ of Wis Madison WI 53706

WEBB, MORGAN CHOFIELD, III, b North Bend, Ore, Mar 4, 20; m 44; c 6. ENTOMOLOGY, ECOLOGY. Educ: Park Col, AB, 42; Univ Nebr, PhD(entom), 61. Prof Exp: Surv entomologist, Univ Nebr, 59; asst prof biol, Whitworth Col, Wash, 61-65; asst prof, 65-72, ASSOC PROF BIOL, MORNINGSIDE COL, 72- Mem: Wilson Ornith Soc; Cooper Ornith Soc. Res: Ecological aspects of bumblebee biology, including the distribution of certain species. Mailing Add: Dept of Biol Morningside Col Sioux City IA 51106

WEBB, MYRON QUENTIN, b Hamden, Ohio, Feb 12, 19; m 43; c 2. TEXTILE CHEMISTRY. Educ: Ohio State Univ, BA & BS, 40; Northwestern Univ, PhD(chem), 43. Prof Exp: Nat Defense Res Comt fel, Northwestern Univ, 43-44; res chemist, Textile Fibers Dept, NY, 44-47 & Va, 47-52, res supvr, 52-59, res mgr, Orlon Tech Div, Del, 59-65, prod develop mgr, Pioneering Res Div, 65-70, TECH MGR NEW VENTURES, TEXTILE FIBERS DEPT, E I DU PONT DE NEMOURS & CO, INC, DEL, 70- Mem: Am Chem Soc. Res: Organic synthesis; polymers; fibers; textiles. Mailing Add: Textile Fibers Dept E I du Pont de Nemours & Co Inc Wilmington DE 19898

WEBB, NATHANIEL CONANT, JR, b Glen Ridge, NJ, Aug 29, 27; m 56; c 2. PREVENTIVE MEDICINE, MEDICAL STATISTICS. Educ: Harvard Univ, AB, 50, MD, 54. Prof Exp: Intern & med resident, Colo Gen Hosp, 54-57; instr biophys, Sch Med, Univ Colo, 57-59; from asst prof to assoc prof prev med, NJ Col Med, 59-67; STAFF MEM, HEALTH CARE GROUP, ARTHUR D LITTLE, INC, CAMBRIDGE, 67- Concurrent Pos: Baker lectr, Sch Pub Health, Univ Mich, 65. Res: Cost and benefit evaluation in public health and health care; medical data processing systems; mamagement systems in tuberculosis, including epidemiologic modeling and data systems. Mailing Add: 11 Bennington Rd Lexington MA 02173

WEBB, NEIL BROYLES, b Junta, WVa, May 19, 20; m 53; c 4. BIOCHEMISTRY, MICROBIOLOGY. Educ: WVa Univ, BS, 53; Univ Ill, MS, 57; Univ Mo, PhD(food sci), 59. Prof Exp: Asst prof food sci, Mich State Univ, 59-62; dir food tech, Eckert Packing Co, Ohio, 62-66; ASSOC PROF FOOD SCI, NC STATE UNIV, 66- Concurrent Pos: Consult, Eckert Packing Co, Ohio, 61-63, Curtis Packing Co & Jesse Jones Sausage Co, NC, 66-70; mem sausage comt, Am Meat Inst, 65-66, chmn, Atlantic Fisheries Tech Conf, 69; owner, Foodlab, Inc, 72- Mem: Inst Food Technol; Am Meat Sci Asn; Am Chem Soc. Res: Biochemical relationship to texture and rheology of foods; scientific principles relevant to comminuted food products. Mailing Add: Dept of Food Sci 236 Schaub Hall NC State Univ Raleigh NC 27607

WEBB, NORVAL ELLSWORTH, JR, b Indianapolis, Ind, Oct 17, 27; m 50; c 3. PHARMACY, PHARMACEUTICAL CHEMISTRY. Educ: Purdue Univ, BS, 50, MS, 52, PhD(pharm), 56. Prof Exp: Asst pharm, Purdue Univ, 50-52; from instr to assoc prof, SDak State Col, 52-62; PHARMACEUT CHEMIST, PHARMACEUT RES & DEVELOP DEPT, WM S MERRELL CO, 62- Mem: Am Pharmaceut Asn; Acad Pharmaceut Sci. Res: Interaction of pharmaceutical dose forms with packaging materials; dose form development. Mailing Add: Pharmaceut Res & Develop Dept Wm S Merrell Co 110 E Amity Rd Cincinnati OH 45215

WEBB, PATRICIA ANN, b Cambridge, Eng, Apr 5, 25; US citizen; m 49; c 2. VIROLOGY, PEDIATRICS. Educ: Agnes Scott Col, BA, 45; Tulane Univ, MD, 50. Prof Exp: Rotating intern, St Joseph's Mercy Hosp, Pontiac, Mich, 50-51; from jr resident to sr resident pediat, Kern Gen Hosp, Bakersfield, Calif, 51-53; mem staff pediat, Well-Baby & Specialty Clins, Dept Maternal & Child Welfare, Washington, DC, 53-54; asst instr exp pediat, Univ Md, 54-55; mem staff, US Army Med Res Unit, Inst Med Res, Kuala Lumpur, Malaysia, 55-61; consult virus & rickettsial dis sect, Nat Inst Allergy & Infectious Dis, 61-62, sr surgeon, Lab Infectious Dis, 62-63, sr surgeon, Mid Am Res Unit, 63-65, med dir, Mid Am Res Unit, 65-75, MED DIR, CTR FOR DIS CONTROL, USPHS, 75- Concurrent Pos: Pvt pract pediat, Bakersfield, Calif, 54-55; mem organizing comt, Cent Am Cong Microbiol, Panama, 68. Mem: Am Soc Trop Med & Hyg; Soc Microbiol Panama (vpres, 69); Panamanian Asn Microbiol. Res: Hemorrhagic fevers; arenaviruses; arboviruses. Mailing Add: B112 Lee Rd SW Shellville GA 30278

WEBB, PAUL, b Cleveland, Ohio, Dec 2, 23; m 48; c 2. MEDICAL PHYSIOLOGY, ENVIRONMENTAL MEDICINE. Educ: Univ Va, BA, 43, MD, 46; Univ Wash, MS, 52. Prof Exp: Intern, Mont Gen Hosp, Can, 46-47, asst res med, 47-48; post surgeon, Arctic Training Ctr, US Army, Alaska, 49-50; res assoc, Univ Wash, 50-51; asst prof physiol, Univ Okla, 52-54; res physiologist & med officer, Aero Med Lab, US Air Force, 54-57, chief, Environ Sect, 57-58; PRIN ASSOC, WEBB ASSOCS, INC, 58- Concurrent Pos: Assoc prof, Ohio State Univ. Honors & Awards: AILSA Award, Aerospace Med Asn, 69; Ely Award, Human Factors Soc, 72. Mem: Fel AAAS; Am Physiol Soc; fel Aerospace Med Asn; Brit Ergonomics Res Soc; Undersea Med Soc, (secy, 71 & 72). Res: Environmental physiology; human thermal tolerance; protective clothing; artificial atmospheres; physiology of diving; cold exposure; metabolism; human calorimetry; bioenergetics. Mailing Add: Webb Assocs Box 308 Yellow Springs OH 45387

WEBB, PETER NOEL, b Wellington, NZ, Dec 14, 36; m 64; c 2. GEOLOGY, MICROPALEONTOLOGY. Educ: Victoria Univ Wellington, BSc, 59, MSc, 60; State Univ Utrecht, PhD(geol), 66. Prof Exp: Micropaleontologist, NZ Geol Surv, 60-73; PROF MICROPALEONT & CHMN DEPT, NORTHERN ILL UNIV, 73- Concurrent Pos: Vis micropaleontologist, Hebrew Univ Jerusalem, 67 & Fed Geol Surv, Hanover, Ger, 69; secy, Nat Comt Geol, NZ, 70-73; Late Cenozoic Working Group, Sci Comt Antarctic Res, 72-, convenor, Ice Shelf Drilling Group, 73- Honors & Awards: Hamilton Prize, Royal Soc NZ, 60; Cotton Prize, Victoria Univ

Wellington, 60. Mem: Royal Soc NZ (secy, 62, 68); Soc Econ Paleontologists & Mineralogists; Sigma Xi; Geol Soc New Zealand. Res: Cretaceous, Tertiary and recent Foraminifera from Southern Hemisphere, particularly New Zealand, Antarctica, South America and intervening areas, with special emphasis on biostratigraphy, population compositions and climate relationship. Mailing Add: Dept of Geol Davis Hall 312 Northern Ill Univ De Kalb IL 60115

WEBB, PHILIP GILBERT, b Norwich, NY, Oct 17, 43; m 68; c 2. INDUSTRIAL ORGANIC CHEMISTRY. Educ: Hamilton Col, AB, 65; Univ Rochester, PhD(org chem), 70. Prof Exp: Res chemist org pigments, Pigments Div, Am Cyanamid Co, 69-74; PLANT TECH MGR, PIGMENTS DIV, CHEMETRON CORP, 74- Mem: Am Chem Soc. Res: New and improved organic pigments; dispersability and lightfastness of organic compounds in inks, paints and plastics. Mailing Add: Pigments Div Chemetron Corp 491 Columbia Ave Holland MI 49423

WEBB, PHYLLIS MARIE, b Paw Paw, Mich. IMMUNOLOGY. Educ: Mich State Univ, BS, 57; Univ Minn, MS, 64, PhD(immunol), 73. Prof Exp: Med technologist, Butterworth Hosp, Grand Rapids, Mich, 57-60; res asst immunol, Univ Minn, 60-62; Univ Calif Med Sch, San Francisco vis fac, Univ Airlangga, Indonesia, 64-65; res assoc leucocyte typing, State Univ Leiden, 65-67; ASST PROF IMMUNOL, UNIV NOTRE DAME, 73- Honors & Awards: Nat Warner Chilcott Award, Am Soc Med Technol, 60. Mem: Am Soc Microbiol; AAAS; Asn Gnotobiotics; Am Soc Clin Pathologists; Am Soc Med Technologists. Res: Immunological studies on germfree radiation induced chimeric mice, their mitogen responses and restoration of thymus cell functions; immunological studies on germfree mice fed antigen-free diet; xenogenic transplatnation of bone marrow from donor mice to recipient rats. Mailing Add: Dept of Microbiol Univ of Notre Dame Notre Dame IN 46556

WEBB, RICHARD LANSING, b Mountain Lakes, NJ, July 28, 23; m 49; c 3. POLYMER CHEMISTRY. Educ: Mass Inst Technol, BS, 48; Columbia Univ, MA, 49. Prof Exp: Jr chemist, Cent Labs, Gen Foods Corp, NJ, 48 & 49-50; res chemist, Cent Res Div, 50-60, sr res chemist, 60-69, GROUP LEADER, LEDERLE LABS DIV, AM CYANAMID CO, NY, 69- Mem: Am Chem Soc. Res: Reactions of acrylonitrile and derivatives; hydrogen cyanide polymerization; x-ray induced addition reactions of olefins including polymerization; synthesis of polymers for biological applications. Mailing Add: 5 Stephanie Lane S Darien CT 06820

WEBB, ROBERT BRADLEY, b Guy, Ark, Nov 20, 27; m 50; c 2. MICROBIOLOGY, GENETICS. Educ: Harding Col, BS, 47; Univ Okla, MS, 50, PhD(microbiol), 56. Prof Exp: Instr biol, Harding Col, 49-50 & Univ Tenn, 50-52; Nat Res Coun fel radiation biol, 56-58, ASSOC SCIENTIST, DIV BIOL & MED RES, ARGONNE NAT LAB, 58- Concurrent Pos: Adj assoc prof, Northern Ill Univ. Mem: AAAS; Radiation Res Soc; Biophys Soc; Am Soc Microbiol; Genetics Soc Am. Res: Microbial genetics and photobiology; molecular basis of photodynamic inactivation and mutagenesis with acridine dyes; mechanisms of protection against ultraviolet light with acridine dyes; molecular basis of mutagenesis and lethality by near ultraviolet and visible light. Mailing Add: Div Biol & Med Res Argonne Nat Lab Argonne IL 60439

WEBB, ROBERT CARROLL, b Petersburg, Va, Mar 6, 47; c 3. HIGH ENERGY PHYSICS. Educ: Univ Pa, BA, 68; Princeton Univ, MA, 70, PhD(physics), 72. Prof Exp: Adj asst prof high energy physics, Univ Calif, Los Angeles, 72-74; RES ASSOC HIGH ENERGY PHYSICS, PRINCETON UNIV, 75- Mem: Am Phys Soc. Res: Experiments to test charge symmetry violation at large momentum transfers and to search for new particle states in proton-antiproton annihilations. Mailing Add: Princeton Univ Jadwin Hall PO Box 708 Princeton NJ 08540

WEBB, ROBERT G, b Long Beach, Calif, Feb 18, 27. VERTEBRATE ZOOLOGY. Educ: Univ Okla, BS, 50, MS, 52; Univ Kans, PhD(zool), 60. Prof Exp: Instr biol, WTex State Univ, 57-58; PROF BIOL, UNIV TEX EL PASO, 62- Mem: Am Soc Ichthyol & Herpet; Soc Syst Zool; Soc Study Amphibians & Reptiles. Res: Systematics, zoogeography and evolution of amphibians and reptiles, especially in the southwestern United States and northern Mexico. Mailing Add: Dept of Biol Sci Univ of Tex El Paso TX 79968

WEBB, ROBERT HOWARD, b Burlington, Vt, Oct 17, 34; m 53; c 2. MEDICAL PHYSICS. Educ: Harvard Univ, AB, 55; Rutgers Univ, PhD(physics), 59. Prof Exp: Res assoc physics, Stanford Univ, 59-63; asst prof, Tufts Univ, 63-69; dir biomed res & develop, 69-72, SR STAFF SCIENTIST, BLOCK ENG INC, 72- Mem: Am Phys Soc; Am Asn Physics Teachers. Res: Medical diagnostic instrumentation; optics; chemical and solid state physics; electron and nuclear spin resonance; low temperatures. Mailing Add: 16 Monument Sq Charlestown MA 02129

WEBB, ROBERT JOHNSON, b Ewing, Ill, Feb 4, 16; m 39; c 3. ANIMAL HUSBANDRY. Educ: Univ Ill, BS, 37, MS, 39. Prof Exp: Asst animal sci, 37-39, asst to dean & dir, 39-40, assoc prof agr res & exten, 40-50, dir, Dixon Springs Agr Ctr, 40-72, prof animal sci, 50-72, sr staff mem, Ctr for Zoonoses Res, 63-72, asst dir agr exp sta, 64-72, EMER PROF ANIMAL SCI, UNIV ILL, 72-; AGR CONSULT, 72- Concurrent Pos: Adv, Uttar Pradesh Agr Univ, India, 59-61, 63-64 & 66-67; mem, Am Grassland Coun. Honors & Awards: Hon Master Farmer, State of Ill, 72. Mem: Am Soc Animal Sci. Res: Animal science; agronomy; soil science; forage crops; zoonoses; livestock marketing and meats. Mailing Add: Simpson IL 62985

WEBB, ROBERT LEE, b Topeka, Kans, Nov 19, 26; m 52; c 2. ORGANIC CHEMISTRY. Educ: Washburn Univ, BS, 49. Prof Exp: Anal chemist, I H Milling Co, 49-50; res chemist, Org Chem Dept, Glidden Co, 50-56, mgr res lab, 56-59, prod mgr, 59-61, asst tech dir, 61-62, vpres, Terpene Res Inst, 62-64; asst mgr aromatic chem develop, 64-65, supt mfg tech serv terpenes, 65-68, GEN MGR TERPENE & AROMATIC CHEM, UNION CAMP CORP, 68- Mem: Am Chem Soc. Res: Terpene and resin chemistry; synthesis and manufacture of flavor and perfumery chemicals from terpenes; composition and reconstitution of essential oils. Mailing Add: 2479 Holly Point Rd E Orange Park FL 32073

WEBB, ROBERT MACHARDY, b Hamilton, Ohio, Dec 2, 15; m 62. PHYSICAL GEOGRAPHY. Educ: Memphis State Col, BS, 49; Ohio State Univ, MA, 50; Univ Kans, PhD(geog), 62. Prof Exp: Instr geog, Ohio Wesleyan Univ, 52-53; PROF GEOG, UNIV SOUTHWESTERN LA, 56-, COORDR, 75- Concurrent Pos: Consult, Dept Educ, State of La, 67-68. Mem: Asn Am Geog; Int Geog Union; Am Geog Soc; Nat Coun Geog Educ. Res: Agricultural land use; climatology. Mailing Add: Box 861 Univ of Southwestern La Lafayette LA 70501

WEBB, ROBERT WALLACE, b Los Angeles, Calif, Nov 2, 09; m 33; c 3. GEOLOGY. Educ: Univ Calif, Los Angeles, AB, 31; Calif Inst Technol, MS, 32, PhD(geol), 37. Prof Exp: Asst geol, Univ Calif, Los Angeles, 32-36, assoc, 36-37, from instr to assoc prof, 37-48, coordr, Army Specialized Training Prog, 43-45 & Vet Affairs, 45-47, univ coordr, 47-52; lectr, Exten Div, 36-75, from assoc prof to prof, 48-75, chmn dept phys sci, 53-58, dir Exp Prog, Instrs for Cols, 60-63, EMER PROF GEOL, UNIV CALIF, SANTA BARBARA, 75- Concurrent Pos: Exec secy, Div Geol & Geog, Nat Res Coun, 53; exec dir, Am Geol Inst, Washington, DC, 53. Honors & Awards: Neil Miner Award, Nat Asn Geol Teachers, 73, Robert Wallace Webb Award, 73. Mem: Fel Geol Soc Am; fel Mineral Soc Am; fel Meteoritical Soc (secy, 37-41). Res: Petrology; geomorphology; mineralogy; Southern Sierra Nevada of California. Mailing Add: Dept of Geol Sci Univ of Calif Santa Barbara CA 93106

WEBB, RODNEY A, b Eng, July 23, 46; Can citizen; m 70. INVERTEBRATE PHYSIOLOGY, PARASITOLOGY. Educ: Univ London, BSc, 68; Univ Toronto, PhD(zool), 72. Prof Exp: Demonstr zool, Univ Toronto, 68-72; Nat Res Coun Can fel parasitol, Inst Parasitol, MacDonald Col, McGill Univ, 72-74; asst prof zool, Univ NB, 74-75; RES ASSOC ZOOL, YORK UNIV, 75- Mem: Can Soc Zoologists; Am Soc Parasitologists. Res: Fine structure, physiology and biochemistry of parasitic helminths; neuroendocrinology of invertebrates. Mailing Add: Dept of Biol York Univ Downsview ON Can

WEBB, RYLAND EDWIN, b Dondi, Angola, Africa, Jan 24, 32; US citizen; m 58; c 3. NUTRITIONAL BIOCHEMISTRY, TOXICOLOGY. Educ: Univ Ill, BS, 54, PhD(biochem), 61. Prof Exp: Res scientist, Lederle Labs, Am Cyanamid Co, 61-63; asst prof, 63-68, ASSOC PROF BIOCHEM & NUTRIT, VA POLYTECH INST & STATE UNIV, 68- Mem: Am Inst Nutrit; Soc Toxicol. Res: Nutrition and toxicant interactions; applied nutrition programs in Haiti. Mailing Add: Dept of Biochem & Nutrit Va Polytech Inst & State Univ Blacksburg VA 24061

WEBB, SAWNEY DAVID, b Los Angeles, Calif, Oct 31, 36; m 58; c 2. PALEONTOLOGY, ANATOMY. Educ: Cornell Univ, BA, 58; Univ Calif, Berkeley, MA, 61, PhD(paleont), 64. Prof Exp: Instr paleont, Univ Calif, Berkeley, 63-64; from asst cur & asst prof to assoc cur mus & assoc prof zool, 64-75, CUR FOSSIL VERTEBRATES FLA STATE MUS & PROF GEOL & ZOOL, UNIV FLA, 75- Concurrent Pos: NSF grants; Guggenheim fel; assoc ed, Paleobiology, 75-78. Mem: Soc Vert Paleont; Soc Study Evolution; Am Soc Mammal; Am Soc Zool. Res: Fossil mammals, evolution, paleoecology and functional morphology. Mailing Add: Fla State Museum Univ of Fla Gainesville FL 32601

WEBB, STEPHEN RICHARD, b Franklin, Ind, July 3, 38; div; c 3. ENGINEERING STATISTICS. Educ: Univ Ill, BS, 59; Univ Chicago, MS, 60, PhD(statist), 62. Prof Exp: Specialist statist, Rocketdyne Div, NAm Aviation, Inc, Calif, 62-63, prin scientist, 63-68; br mgr, Math Sci Dept, 68-71, CHIEF SCIENTIST ADVAN MATH, ADVAN INFO SYSTS SUBDIV, MCDONNELL DOUGLAS ASTRONAUT CO, 72- Mem: Am Statist Asn; Inst Math Statist. Res: Statistical design of experiments; signal processing. Mailing Add: Dept 373 Mail Sta 13-2 McDonnell Douglas Astronaut Co Huntington Beach CA 92647

WEBB, SYDNEY JAMES, b London, Eng, July 25, 25; m 48; c 4. BIOPHYSICS, MICROBIOLOGY. Educ: Univ London, BSc, 52, MSc, 54, DIC, 55, PhD(biophys), 59, DSc, 68. Prof Exp: Asst lectr biol, Univ London, 52-55; res sect leader microbiol, Boots Pure Drug Co, Eng, 55-56; sci officer, Defence Res Bd, Can, 56-60; prof microbiol, Univ Sask, 60-74; PROF PHYSICS & BIOPHYSICS, UNIV S FLA, 74- Concurrent Pos: Consult & grantee, Defence Res Bd, Can, 60-; Nat Res Coun Can grantee, 64- Mem: Can Soc Cell Biol; Brit Soc Gen Microbiol; Brit Soc Appl Bact; Biophys Soc; Am Soc Photobiol. Res: Influence of bound water on structure of macromolecules; effects of nutrition on in-vivo energy states and on the action of microwave radiations. Mailing Add: Dept of Physics Univ SFla Tampa FL 33620

WEBB, THEODORE STRATTON, JR, b Oklahoma City, Okla, Mar 4, 30; m 52; c 2. APPLIED PHYSICS, AEROSPACE ENGINEERING. Educ: Univ Okla, BS, 51; Calif Inst Technol, PhD(physics), 55. Prof Exp: Dir aerospace tech dept, 69-74, VPRES RES & ENG, GEN DYNAMICS, 74- Concurrent Pos: Adj prof physics, Tex Christian Univ, 55-71. Mem: Am Phys Soc; Am Inst Aeronaut & Astronaut; AAAS. Res: Management of research and development; space system design and development; propulsion systems; nuclear shielding and reactor theory; low energy particle physics. Mailing Add: Gen Dynamics PO Box 748 Ft Worth TX 76101

WEBB, THOMAS EVAN, b Edmonton, Alta, Mar 4, 32; m 61; c 1. BIOCHEMISTRY. Educ: Univ Alta, BSc, 55, MSc, 57; Univ Toronto, PhD(biochem), 61. Prof Exp: Nat Res Coun Can fels, Nat Res Coun, Ont, 61-63 & Univ Wis-Madison, 63-65; asst prof biochem, Med Sch, Univ Man, 65-66; asst prof, Cancer Res Unit, McGill Univ, 66-70, actg dir, 69-70; assoc prof, 70-74, PROF PHYSIOL CHEM, COL MED, OHIO STATE UNIV, 74- Mem: AAAS; Can Biochem Soc; Am Soc Biol Sci; Fedn Am Socs Exp Biol. Res: Identification of normal cellular controls and the biochemistry of cancer. Mailing Add: Dept of Physiol Chem Ohio State Univ Col of Med Columbus OH 43210

WEBB, THOMAS HOWARD, b Norfolk, Va, July 16, 35; m 57; c 2. ORGANIC CHEMISTRY, LUBRICATION ENGINEERING. Educ: Univ Va, BA, 57; Duke Univ, MA, 60, PhD(org chem), 61. Prof Exp: Res chemist, Sinclair Res, Inc, 61, proj leader lubricant additives, 63-65, group leader, 65-66, asst sect leader indust oils, 66-69; supvr lubricants, asphalts & waxes, BP Oil Corp, 69-70; sr proj leader indust oils res, 70-72, RES ASSOC LUBRICANTS & LUBRICATION, STANDARD OIL CO (OHIO), 72- Concurrent Pos: Sr res scientist, Sinclair Oil Corp, 68. Mem: AAAS; Am Chem Soc; Am Inst Chemists; Am Soc Testing & Mat; Am Soc Lubrication Eng. Res: Additives for lubricants; synthesis and structure-property correlations; lubricant formulation and product design. Mailing Add: Standard Oil Co Cleveland OH 44128

WEBB, THOMPSON, III, b Los Angeles, Calif, Jan 13, 44; m 69; c 2. PALYNOLOGY, CLIMATOLOGY. Educ: Swarthmore Col, BA, 66; Univ Wis-Madison, PhD(meteorol), 71. Prof Exp: Inst Sci & Technol fel, Univ Mich, Ann Arbor, 70-71, asst res paleocologist, Great Lakes Res Div, 71-72; asst res prof, 72-75, ASSOC PROF GEOL SCI, BROWN UNIV, 75- Mem: AAAS; Am Meteorol Soc; Am Quaternary Asn; Am Asn Stratig Palynologists. Res: Use of multivariate statistical techniques for calibrating quaternary pollen data in vegetational and climatic terms; production of paleoclimatic and paleovegetation maps. Mailing Add: Dept of Geol Sci Box 1846 Brown Univ Providence RI 02912

WEBB, WATT WETMORE, b Kansas City, Mo, Aug 27, 27; m 50; c 3. PHYSICS. Educ: Mass Inst Technol, ScB, 47, ScD(metall), 55. Prof Exp: Res engr, Union Carbide Metals Co, 47-52, res metallurgist, 55-59, coordr fundamental res, 59-60, asst dir res, 61; assoc prof eng, 61-65, PROF APPL & ENG PHYSICS, CORNELL UNIV, 65- Concurrent Pos: Consult; mem var adv panels of mat adv bd, NSF, 58- & tech comts, Electrochem Soc, 58-; John Simon Guggenheim Found fel, 74-75; assoc ed, Phys Rev Letters, 75- Mem: Fel Am Phys Soc; Biophys Soc; Metall Soc; Inst Elec & Electronics Engrs. Res: Critical and collective phenomena; fluctuations; superconductivity; membrane and cellular biophysics; biophysical processes; fluid dynamics; chemical kinetics; phase transformations. Mailing Add: Sch Appl & Eng Phys Clark Hall Cornell Univ Ithaca NY 14850

WEBB, WATTS RANKIN, b Columbia, Ky, Sept 8, 22; m 44; c 4. SURGERY. Educ: Univ Miss, BA, 42; Johns Hopkins Univ, MD, 45; Am Bd Surg, dipl, 52; Am Bd Thoracic Surg, dipl, 53. Prof Exp: Chief surgeon, Miss State Sanatorium, 52-63; prof surg & chmn, Div Thoracic & Cardiovasc Surg, Univ Tex Southwest Med Sch Dallas,

64-70; PROF SURG & CHMN DEPT, STATE UNIV NY UPSTATE MED CTR, 70- Concurrent Pos: Prof, Sch Med, Univ Miss, 55-63. Mem: Fel Am Col Surgeons; Am Asn Thoracic Surgeons; fel Am Col Chest Physicians; Am Col Cardiol; Am Soc Artificial Internal Organs. Res: Shock; cardiac-pulmonary transplantation; organ preservation; hypothermia; myocardial physiology. Mailing Add: Dept of Surg State Univ of NY Upstate Med Ctr Syracuse NY 13210

WEBB, WILLIAM ALBERT, b Paterson, NJ, Mar 10, 44; m 66; c 2. MATHEMATICS. Educ: Mich State Univ, BS, 65; Pa State Univ, University Park, PhD(math), 68. Prof Exp: Res instr math, Pa State Univ, 68-69; asst prof, 69-75, ASSOC PROF MATH, WASH STATE UNIV, 75- Mem: Am Math Soc; Math Asn Am. Res: Analytic and elementary number theory; number theoretic questions concerning polynomial rings over finite fields. Mailing Add: Dept of Math Wash State Univ Pullman WA 99163

WEBB, WILLIAM GATEWOOD, b Charleston, SC, July 17, 25; m 52; c 3. MEDICINAL CHEMISTRY. Educ: Univ of South, BS, 50; Univ Rochester, PhD(org chem), 54. Prof Exp: Res chemist, Columbia-Southern Chem Corp, WVa, 54-55; res assoc, Res Div, 55-57 & Patent Div, 57-61, patent agent, 61-65, SR PATENT AGENT, STERLING-WINTHROP RES INST, 65- Mem: Am Chem Soc. Res: Synthetic organic chemistry. Mailing Add: Patent Div Sterling-Winthrop Res Inst Rensselaer NY 12144

WEBB, WILLIAM LEONARD, b Minneapolis, Minn, Oct 5, 13; m 38. VERTEBRATE ECOLOGY. Educ: Univ Minn, BS, 35, MS, 40; Syracuse Univ, PhD(zool), 50. Prof Exp: Jr biologist, US Forest Serv, 35-36; asst, Univ Minn, 36-37; asst prof, 37-42 & 46-50, assoc prof, 50-61, PROF, NY STATE COL ENVIRON SCI & FORESTRY, SYRACUSE UNIV, 61-, DEAN GRAD STUDIES, 65- Concurrent Pos: Jr forest zoologist, Roosevelt Wildlife Forest Exp Sta, 37-42; vis prof, Col Forestry, Univ Philippines, 62-64. Mem: Soc Am Foresters; Wildlife Soc; Am Soc Limnol & Oceanog; Ecol Soc Am. Res: Wildlife management; environmental analysis; animal population. Mailing Add: NY State Col of Environ Sci & Forestry Syracuse NY 13210

WEBB, WILLIAM LOGAN, b Chattanooga, Tenn, Feb 13, 30; c 5. PSYCHIATRY. Educ: Princeton Univ, AB, 51; Johns Hopkins Univ, MD, 55. Hon Degrees: MA, Univ Pa, 71. Prof Exp: From instr to asst prof psychiat, Sch Med, Johns Hopkins Univ, 61-64; from asst prof to prof psychiat, Univ Pa, 64-74; PROF PSYCHIAT & CHMN DEPT, UNIV TENN, MEMPHIS, 74- Mem: Am Psychiat Asn; Am Psychosom Soc. Mailing Add: Tenn Psychiat Hosp & Inst 865 Poplar Ave Memphis TN 38103

WEBB, WILLIAM PAUL, b Bismarck, NDak, Dec 30, 22; m 47; c 4. ORGANIC CHEMISTRY. Educ: Univ Notre Dame, BS, 44 & 47; Univ Minn, PhD(org chem), 51. Prof Exp: SR RES ASSOC, CHEVRON RES CO, STAND OIL CO CALIF, SAN FRANCISCO, 51- Mem: Am Chem Soc. Res: Petrochemicals; patent liaison. Mailing Add: 16 Chestnut Ave San Rafael CA 94901

WEBB, WILLIS KEITH, b McCoy, Va, Apr 21, 28; m 50; c 5. VETERINARY PATHOLOGY. Educ: Tex A&M Univ, DVM, 57. Prof Exp: Vet livestock inspector, Animal Dis Eradication Div, USDA, 57; vet, US Food & Drug Admin, 57-62; res assoc, Mead Johnson & Co, Ind, 62-70; VET MED OFFICER, USDA, 70- Res: Regulatory veterinary medicine; eradication of disease from domestic animals; animal welfare; interstate movement of livestock. Mailing Add: Rte 3 Box 458 Christiansburg VA 24073

WEBB, WILLIS LEE, b Nevada, Tex, July 9, 23; m 42; c 1. METEOROLOGY. Educ: Southern Methodist Univ, BS, 52; Univ Okla, MS, 70; Colo State Univ, PhD, 72. Prof Exp: Meteorol observer, US Weather Bur, 42-43, meteorologist, 46-52, physicist, 52-55; METEOROLOGIST & CHIEF SCIENTIST, ATMOSPHERIC SCI LAB, US ARMY ELECTRONICS COMMAND, 55-; LECTR PHYSICS, UNIV TEX, EL PASO, 63- Concurrent Pos: Mem upper atmosphere rocket res comt, Space Sci Bd, Nat Acad Sci, 59-65; chmn meteorol rocket network comt, Inter-Range Instrumentation Group, 60-; consult, Cath Univ, 60-61; mem air blast subcomt, Inter-Oceanic Canal Studies Group, 64-; Army adv, Nat Comt Clear Air Turbulence, 65-66. Honors & Awards: Meritorious Civilian Serv Award, US Army, 58, Commendation, 61. Mem: Am Meteorol Soc; Am Geophys Union; Am Inst Aeronaut & Astronaut. Res: Synoptic exploration of the 25-100 kilometer region with small rocket vehicles to determine the circulation, thermal and composition structure. Mailing Add: Atmospheric Sci Lab US Army Electronics Command White Sands Missile Range NM 88002

WEBBER, BROOKE BLAND, b Philadelphia, Pa, Oct 28, 29; m 53; c 1. GENETICS. Educ: Lafayette Col, AB, 51; Yale Univ, MS, 57, PhD(bot), 59. Prof Exp: Res assoc, Biol Div, Oak Ridge Nat Lab, 59-60; asst prof zool, Univ Tenn, 60-62; biologist, Oak Ridge Nat Lab, 62-68; ASSOC PROF BIOL, AUGUSTA COL, 68- Mem: AAAS; Genetics Soc Am. Res: Neurospora genetics; mutation; recombination; biochemical genetics. Mailing Add: Dept of Biol Augusta Col Augusta GA 30904

WEBBER, CARROLL A, JR, b Orange, NJ, Nov 13, 26; m 47; c 2. MATHEMATICS. Educ: Calif Inst Technol, BS, 46; Univ Calif, Berkeley, MA, 48 & 59; Yeshiva Univ, PhD, 70. Prof Exp: Teacher high sch, 48-49; beekeeper, family bus, 50-56; teacher math, Col Holy Names, 59-61; ASSOC PROF E CAROLINA UNIV, 61- Mem: Am Math Soc; Asn Symbolic Logic. Res: Algorithms; recursive functions. Mailing Add: Dept of Math E Carolina Univ Greenville NC 27834

WEBBER, DONALD SALYER, b Los Angeles, Calif, Jan 15, 17; m 51; c 2. PHYSICS. Educ: Univ Calif, Los Angeles, BA, 38, MA, 41, PhD(physics), 54; Calif Inst Technol, MS, 42. Prof Exp: Assoc physics, Univ Calif, Los Angeles, 49-54, instr, 54-55; mem tech staff, Ramo-Wooldridge Div, Thompson Ramo Wooldridge, Inc, 55-60, mem sr staff, 60-61, assoc mgr photo equip dept, 61; mgr solar physics prog, Calif Div, Lockheed Aircraft Corp, 61-63, mgr astron sci lab, 64-65; SR STAFF ENGR, TRW SYSTS GROUP, 65- Mem: Am Phys Soc; Optical Soc Am. Res: Infrared spectroscopy of crystals; photo-optical instrumentation; solar physics. Mailing Add: 3551 Knobhill Dr Sherman Oaks CA 91423

WEBBER, EDGAR ERNEST, b Worcester, Mass, Sept 11, 32; m 58; c 2. BOTANY. Educ: Univ Mass, BS, 55, PhD(bot), 67; Cornell Univ, MS, 61. Prof Exp: Instr bot, Wellesley Col, 61-62; asst prof, Pa State Univ, Behrend Campus, 65-67; from asst prof to assoc prof, 67-74, PROF BIOL, KEUKA COL, 74- Concurrent Pos: Am Philos Soc grant, 68. Mem: Phycol Soc Am; Int Phycol Soc; Brit Phycol Soc. Res: Ecology and systematics of benthic salt marsh algae; culture and life history studies; morphology. Mailing Add: Dept of Biol Keuka Col Keuka Park NY 14478

WEBBER, GAYLE MILTON, b Sioux City, Iowa, Aug 30, 31; m 60; c 3. ORGANIC CHEMISTRY. Educ: Morningside Col, BS, 53; Univ Iowa, MS, 58. Prof Exp: Sr res asst, 58-74, RES INVESTR, G D SEARLE & CO, 74- Mem: Am Chem Soc. Res: Steroids. Mailing Add: G D Searle & Co Box 5110 Chicago IL 60680

WEBBER, GEORGE ROGER, b Toronto, Ont, Nov 2, 26; m 54. GEOLOGY. Educ: Queen's Univ, Can, BSc, 49; McMaster Univ, MSc, 52; Mass Inst Technol, PhD(geol), 55. Prof Exp: Res assoc, 55-59, asst prof, 59-66, ASSOC PROF GEOL, McGILL UNIV, 66- Mem: Geol Soc Am; Geochem Soc; Spectros Soc Can; Asn Explor Geochem; Geol Asn Can. Res: Geochemistry; x-ray and optical spectrography. Mailing Add: Dept of Geol Sci McGill Univ Montreal PQ Can

WEBBER, HAROLD HASKELL, b Springfield, Mass, Mar 21, 15; m 39; c 5. AQUATIC BIOLOGY. Educ: Univ Tex, BA, 39, PhD(genetics, biochem), 41. Prof Exp: Instr zool, Univ Vt, 41-42; dir prod res, Nat Cotton Coun Am, 42-46; dir basic res cotton tech, Cotton Res Comt Tex, 46-48; exec dir textile res, Lowell Tech Res Found, 48-52; pres, Groton Assocs, Inc, 52-59; mem sr staff res mgt, Arthur D Little, Inc, 59-64; sr corp scientist, Brunswick Corp, 64-66; PRES, GROTON ASSOCS, INC, 66- Educ: Ed-in-chief, Aquaculture, 72, assoc ed, 73-; exec vpres, Maricultura, SA, San Jose, Costa Rica, 73- Mem: AAAS; Am Soc Testing & Mat; Marine Technol Soc; World Maricult Soc (pres elect, 75-76); Am Fisheries Soc. Res: Fiber technology; marine biology, especially mariculture; aquatic biology, especially aquaculture. Mailing Add: Groton Assocs Inc PO Box 517 Groton MA 01450

WEBBER, HERBERT H, b Vancouver, BC, Oct 15, 41; m 63. BIOLOGY. Educ: Univ BC, BSc, 63, PhD(zool), 66. Prof Exp: NATO fel, Can for Study at Hopkins Marine Sta Stanford, 66-68; asst prof biol, Wake Forest Univ, 68-72; asst prof biol, 72-74, ASSOC PROF MARINE RESOURCES, HUXLEY COL ENVIRON STUDIES, WESTERN WASH STATE COL, 74- Mem: AAAS. Res: Invertebrate physiology; reproductive physiology of Gastropoda Mollusca. Mailing Add: Huxley Col Environ Studies Western Wash State Col Bellingham WA 98225

WEBBER, IRMA ELEANOR, b San Diego, Calif, Aug 16, 04; m 27; c 2. BOTANY. Educ: Univ Calif, AB, 26, MA, 27, PhD(bot), 29. Prof Exp: Asst paleobot, Carnegie Inst, 27-29; agent, Div Blister Rust Control, Bur Plant Indust, USDA, 29-31; asst, Citrus Exp Sta, Univ Calif, 32-33; collabr, Div Western Irrig Agr, Bur Plant Indust, 34, Div Cotton & Other Fiber Crops, 36-43 & Bur Plant Indust, Soils & Agr Eng, 43-50, botanist, 50-51, collabr, 51-53, COLLABR, AGR RES SERV, USDA, 53-; COLLABR, DE LA REVISTA LILLOA, ARG, 37- Res: Paleobotany; ecology; taxonomy; plant pathology and anatomy; popularized botany; economic botany. Mailing Add: 500 Arlington Ave Berkeley CA 94707

WEBBER, JOHN ALAN, b Chicago, Ill, Aug 14, 40. MEDICINAL CHEMISTRY. Educ: Univ Colo, BA, 62; Stanford Univ, PhD(org chem), 67. Prof Exp: RES SCIENTIST, ELI LILLY & CO, 66- Mem: Am Chem Soc; Am Soc Microbiol. Res: Structural modification and synthesis of natural products and complex organic molecules; medicinal chemistry of agents with antibiotic activity. Mailing Add: Eli Lilly & Co PO Box 618 M 705 Indianapolis IN 46206

WEBBER, KARL KEEN, b Wayne City, Ill, July 15, 14; m 46. DENTISTRY. Educ: St Louis Univ, DDS, 37. Prof Exp: Instr pre-clin dent, St Louis Univ, 51-52, from instr to asst prof oper dent, 52-60, from actg dir to dir, 54-60; assoc prof dent hyg, Southern Ill Univ, 60-66; from assoc prof to prof oper dent, Sch Dent, St Louis Univ, 66-70, dir paradent educ & dir dent asst utilization prog, 66-70; ASSOC PROF CLIN DENT & CHMN DEPT DENT AUXILIARY PROGS, FOREST PARK COMMUNITY COL, 70- Concurrent Pos: Asst prof, Sch Dent, Wash Univ. Mem: AAAS; Am Dent Asn. Res: Operative and preventive dentistry; dental psychology; dental restorative materials; dental education methods. Mailing Add: Dept of Dent Auxiliary Progs Forest Park Community Col St Louis MO 63110

WEBBER, MARION GEORGE, b Golden, Colo, Dec 18, 21; m 45; c 9. PHARMACEUTICAL CHEMISTRY. Educ: Univ Colo, BS, 47, MS, 48; Univ Fla, PhD(pharm), 51. Prof Exp: Instr pharm & pharmaceut chem, Univ Fla, 50-51; asst prof pharmaceut chem, 51-52, assoc prof pharm, 53-72, PROF PHARM, UNIV HOUSTON, 72- Mem: Am Col Apothecaries; Am Soc Hosp Pharm; Am Pharmaceut Asn. Res: Tillandsia usneoides; pharmaceutical formulation and stability studies. Mailing Add: Univ of Houston Col of Pharm 3801 Cullen Blvd Houston TX 77004

WEBBER, MILO M, b Los Angeles, Calif, Sept 27, 30; m 55; c 2. RADIOLOGY, NUCLEAR MEDICINE. Educ: Univ Calif, Los Angeles, BA, 52, MD, 55; Am Bd Radiol, dipl, 61; Am Bd Nuclear Med, dipl, 72. Prof Exp: Intern surg serv, 55-56, resident radiol, 56-57 & 59-60, instr, 60-61, lectr, 61-62, from asst prof to assoc prof, 62-74, PROF RADIOL SCI, SCH MED, UNIV CALIF, LOS ANGELES, 74- Concurrent Pos: Consult, Queen of Angels Hosp, Los Angeles, 65-; res radiologist, Lab Nuclear Med & Radiation Biol, Sch Med, Univ Calif, Los Angeles, 65- Mem: Am Col Radiol; AMA; Soc Nuclear Med; Radiol Soc NAm. Res: Development and refinement of organ radioisotope scanning procedures, including development of thrombosis localization techniques; application of telecommunications to the practice of radiology and nuclear medicine. Mailing Add: Nuclear Med AR 144B Ctr Health Sci Univ of Calif Los Angeles CA 90024

WEBBER, MUKTA MALA, b India, Dec 5, 37; m 63; c 1. ONCOLOGY, CELL BIOLOGY. Educ: Agra Univ, BSc, 57, MSc, 59; Queen's Univ, Ont, PhD, 63; Univ Toronto, dipl electron micros, 68. Prof Exp: Cancer res scientist, Roswell Park Mem Inst, NY State Dept Health, 62-63; lectr histol & embryol, Queen's Univ, Ont, 63-65, res assoc urogenital oncol, Queen's Univ, Ont & Kingston Gen Hosp, 65-68; ASST PROF UROGENITAL ONCOL & DIR UROL CANCER RES LAB, SCH MED, UNIV COLO, DENVER, 71- Concurrent Pos: Res assoc, Inst Arctic & Alpine Res, Univ Colo, Boulder, 69- Mem: AAAS; Can Soc Cell Biol; Can Asn Anatomists; Tissue Cult Asn; Can Fedn Biol Socs. Res: Cell biology of urogenital tumors; cell biology and etiology of prostatic neoplasia; cell biology of normal, benign and malignant epithelium in vivo and in vitro; cytogenetics of tumor cell populations; chemical carcinogenesis; cell autonomy and cell surface changes. Mailing Add: Div of Urol Univ of Colo Med Ctr Denver CO 80220

WEBBER, PATRICK JOHN, b Bedfordshire, UK, Feb 24, 38; m 63; c 1. ECOLOGY, BIOLOGY. Educ: Univ Reading, BSc, 60; Queen's Univ, Ont, MSc, 63, PhD, 71. Prof Exp: Asst prof biol, York Univ, 66-69; asst prof, 69-72, ASSOC PROF BIOL, UNIV COLO, BOULDER, 72-, MEM FAC, INST ARCTIC & ALPINE RES, 69- Concurrent Pos: Mem steering comt tundra biome, US Int Biol Prog, 69- & nat comt for phenol, 71- Mem: AAAS; Ecol Soc Am; Arctic Inst NAm; Can Bot Asn. Res: Primary productivity, phenology and phytosociology of arctic and alpine tundras; effects of vehicle damage, snowfences and weather modification on tundra ecosystem; gradient analysis of vegetation. Mailing Add: Inst of Arctic & Alpine Res Univ of Colo Boulder CO 80302

WEBBER, RICHARD HARRY, b Camillus, NY, Jan 2, 24; m 46; c 7. ANATOMY. Educ: St Benedict's Col, BS, 48; Univ Notre Dame, MS, 49; St Louis Univ, PhD(anat), 54. Prof Exp: Instr biol, St Benedict's Col, 48; instr, Niagara Univ, 49-51; asst anat, Sch Med, St Louis Univ, 51-54; asst prof, Sch Med, Creighton Univ, 54-59; assoc prof, Sch Med, Temple Univ, 59-61; assoc prof, 61-70, PROF ANAT, SCH MED, STATE UNIV NY BUFFALO, 70- Concurrent Pos: Lederle med fac scholar, 56-59; from chmn elect to chmn, Sect Anat Sci, Am Asn Dent Schs, 71-73; mem, Am

Benedictine Acad; mem, Human Biol Coun. Mem: Am Asn Anatomists; Am Asn Phys Anthrop. Res: Peripheral autonomic nervous system. Mailing Add: Dept of Anat Sci State Univ of NY Buffalo NY 14214

WEBBER, RICHARD LYLE, b Akron, Ohio, Nov 2, 35; m 56; c 2. PHYSIOLOGICAL OPTICS, DENTISTRY. Educ: Albion Col, AB, 58; Univ Mich, Ann Arbor, DDS, 63; Univ Calif, Berkeley, PhD(physiol optics), 71. Prof Exp: Intern dent, USPHS Hosp, Seattle, Wash, 63-64, clin investr, Mat & Technol Br, Dent Health Ctr, Div Dent Health, USPHS, 64-68, lectr, Univ Calif, Berkeley, 70-71, staff investr, Diag Systs, Oral Med & Surg Br, 71-75, CHIEF CLIN INVEST BR & CHIEF DIAG METHODOLOGY SECT, NAT INST DENT RES, 75- Mem: AAAS; Am Acad Dent Radiol; Int Asn Dent Res; Soc Photo-Optical Instrument Eng. Res: Image processing and the study of factors influencing diagnostic systems. Mailing Add: Clin Invest Br Nat Inst of Dent Res Bethesda MD 20014

WEBBER, ROBERT TRUMBULL, solid state physics, see 12th edition

WEBBER, RUSSELL VINCENT, physical chemistry, see 12th edition

WEBBER, STEPHEN EDWARD, b Springfield, Mo, Sept 19, 40; m 62; c 2. PHYSICAL CHEMISTRY. Educ: Wash Univ, AB, 62; Univ Chicago, PhD(chem), 65. Prof Exp: NSF fel, Univ Col, Univ London, 65-66; asst prof, 66-71, ASSOC PROF CHEM, UNIV TEX, AUSTIN, 71- Res: Molecular excitons; electronic relaxation phenomena. Mailing Add: Dept of Chem Univ of Tex Austin TX 78712

WEBBER, THOMAS GRAY, b Oct 12, 12; US citizen; m 41; c 2. ORGANIC CHEMISTRY. Educ: Brown Univ, ScB, 33; Univ Ill, MS, 34; Harvard Univ, PhD(org chem), 40. Prof Exp: Chemist, Am Cyanamid Co, NJ, 35-36 & Nopco Chem Co, 36-37; asst chem, Harvard Univ, 38-40; chemist, Jackson Lab, E I du Pont de Nemours & Co, Inc, Del, 40-41, supvr, Ill, 41-43, chem engr, 42-43, chemist, Wash, 43-45, chief chem engr, Ind, 45, chemist, 45-52, head, Lake Div, Tech Lab, 52-62, color specialist, Plastics Dept, 62-73; CONSULT & ED, 73- Mem: Am Chem Soc. Res: Polynuclear hydrocarbons; phthalocyanine pigments; dyes and pigments; plastics coloration; color. Mailing Add: 1722 Forest Hill Dr Vienna WV 26105

WEBBER, WILLIAM A, b Nfld, Can, Apr 8, 34; m 58; c 3. HISTOLOGY, PHYSIOLOGY. Educ: Univ BC, MD, 58. Prof Exp: Intern, Vancouver Gen Hosp, 58-59; fel physiol, Med Col, Cornell Univ, 59-61; from asst prof to assoc prof anat, 61-69, PROF ANAT, UNIV BC, 69-, ASSOC DEAN MED, 71- Mem: Am Asn Anatomists; Can Asn Anatomists. Res: Renal physiology; kidney structure. Mailing Add: Dept of Anat Univ of BC Vancouver BC Can

WEBBER, WILLIAM R, b Bedford, Iowa, June 9, 29; m 61; c 1. PHYSICS, ASTROPHYSICS. Educ: Coe Col, BS, 51; Univ Iowa, MS, 54, PhD(physics), 57. Prof Exp: Asst prof physics, Univ Md, 58-59; NSF fel, Imp Col, Univ London, 59-61; from asst prof to assoc prof, Univ Minn, Minneapolis, 61-69, PROF PHYSICS & DIR, SPACE SCI CTR, UNIV NH, 69- Concurrent Pos: Co-ed, J Geophys Res, Am Geophys Union, 61-; mem Fields & Particles Subcomt, NASA, 63-66; consult, Boeing Aircraft Co & NAm Aviation, Inc, 63-; vis prof, Univ Adelaide, 68-69; vis prof, Danish Space Res Ctr, 71. Mem: Am Phys Soc; Am Geophys Union; Am Astron Soc. Res: Charge composition and energy spectrum of cosmic rays; motion of charged particles in the earths magnetic field; solar-terrestrial relationships; particle detectors; x-ray astronomy. Mailing Add: Dept of Physics Univ of NH Durham NH 03824

WEBER, ALBERT VINCENT, b Pittsburgh, Pa, Jan 30, 25; m 48; c 3. BOTANY. Educ: Duquesne Univ, BEd, 50, MSc, 52; Univ Minn, PhD(bot), 57. Prof Exp: Asst bot, Duquesne Univ, 50-52; asst, Univ Minn, 52-56; from instr to assoc prof, 56-63, PROF BIOL, UNIV WIS-LA CROSSE, 64-, ASSOC DEAN, COL LETTERS & SCI, 67-, CO-DIR SUMMER SESSION, 69- Mem: Bot Soc Am. Res: Plant anatomy, morphology and morphogenesis. Mailing Add: Dept of Biol Univ of Wis La Crosse WI 54601

WEBER, ALFONS, b Dortmund, Ger, Oct 8, 27; nat US; m 55; c 3. PHYSICS. Educ: Ill Inst Technol, BS, 51, MS, 53, PhD, 56. Prof Exp: Asst physics, Ill Inst Technol, 51-53, instr, 53-56; fel, Nat Res Coun Can, 56-57; from asst prof to assoc prof, 57-66, chmn dept, 64-69, PROF PHYSICS, FORDHAM UNIV, 66- Mem: Fel Am Phys Soc; Am Asn Physics Teachers; Optical Soc Am. Res: Raman and infrared spectroscopy; molecular mechanics; optics. Mailing Add: Dept of Physics Fordham Univ Bronx NY 10458

WEBER, ALFRED HERMAN, b Philadelphia, Pa, Jan 15, 06; m 32; c 7. NUCLEAR PHYSICS, SPACE PHYSICS. Educ: St Joseph's Col, Pa, AB, 28, MA, 31; Univ Pa, PhD(physics), 36. Hon Degrees: DSc, St Joseph's Col, Pa, 68. Prof Exp: Instr physics & math, St Joseph's Col, Pa, 28-34, from asst prof to prof physics & head dept, 34-39; from asst prof to prof, 39-74, tech dir, US Army Air Force Radio Sch, 42-43, chmn dept physics, 51-74, EMER PROF PHYSICS & CHMN DEPT, ST LOUIS UNIV, 74- Concurrent Pos: Sigma Xi, Res Corp, NSF, Army Res Off, Am Cancer Soc & Dept Defense grants; consult, Argonne Nat Lab, 47-57, US Army Ballistic Missile Agency, 57-60 & Marshall Space Flight Ctr, NASA, 60-70; vchmn, Assoc Midwest Univs, Argonne Nat Lab, 60, chmn, 61; prof physics, Univ Ala, 63-67. Mem: Fel Am Phys Soc; Am Asn Physics Teachers; Sigma Xi. Res: Electron emission and diffraction; properties of thin metallic films; x-ray diffraction; neutron diffraction and scattering; nuclear spectroscopy; plasma and space physics. Mailing Add: 1047 Chevy Chase Dr Sarasota FL 33580

WEBER, ALLEN HOWARD, b Lorenzo, Idaho, May 15, 38; m 59; c 3. MICROMETEOROLOGY. Educ: Brigham Young Univ, BS, 60; Univ Ariz, MS, 62; Univ Utah, PhD(meteorol), 66. Prof Exp: Asst prof meteorol, Univ Okla, 66-68 & meteorol & environ health, 68-69; consult, Nat Severe Storms Lab, 69; asst prof, 69-72, ASSOC PROF METEOROL, NC STATE UNIV, 72- Concurrent Pos: NSF grant, Univ Okla, 68-69; res meteorologist, Environ Protection Agency, Nat Environ Res Ctr, 70-72; Environ Protection Agency grant, NC State Univ & NC Water Resources Res Inst, 72- Mem: Am Meteorol Soc. Res: Atmospheric turbulence; physical meteorology. Mailing Add: Dept of Geosci NC State Univ Raleigh NC 27607

WEBER, ALLEN THOMAS, b Long Prairie, Minn, Sept 7, 43; m 69; c 2. MICROBIOLOGY. Educ: Univ Mich, Ann Arbor, BS, 65; Univ Wis-Madison, MS, 67, PhD(bact), 70. Prof Exp: Asst prof, 70-73, ASSOC PROF BIOL, UNIV NEBR AT OMAHA, 73- Mem: AAAS; Am Soc Microbiol; Brit Soc Gen Microbiol. Res: Morphogenesis and development of microorganisms; cellular slime molds; microbial genetics. Mailing Add: Dept of Biol Univ of Nebr Omaha NE 68101

WEBER, ALVIN FRANCIS, b Hartford, Wis, Mar 13, 18; m 45; c 3. CYTOLOGY. Educ: Iowa State Col, DVM, 44; Univ Wis, BA, 46, MS, 48, PhD(vet sci), 49. Prof Exp: Instr vet sci, Univ Wis, 44-49; from asst prof to assoc prof, 49-55, PROF VET ANAT, UNIV MINN, ST PAUL, 55-, HEAD DEPT, 65- Concurrent Pos: USPHS spec fel, Univ Giessen, 59. Mem: Am Asn Anat. Res: Bovine uterine histology, histological and cytochemical changes of the adrenal gland and structure of secretory components of the udder; electron microscopic studies of the hematopoietic organs of domestic animals. Mailing Add: Dept of Vet Anat Univ of Minn St Paul MN 55101

WEBER, ANNEMARIE, b Rostock, Ger, Sept 11, 23; US citizen. PHYSIOLOGY, BIOCHEMISTRY. Educ: Univ Tübingen, MD, 50. Prof Exp: Res asst physiol, Univ Tübingen, 50-51; res fel biophys, Univ Col, London & physiol, Univ Md, 51; Rockefeller fel phys chem, Harvard Med Sch, 52; res fel physiol, Univ Tübingen, 53-54; res assoc neurol, Columbia Univ, 54-59; asst mem physiol, Inst Muscle Dis, New York, 59-63, assoc mem, 63-65; prof biochem, St Louis Univ, 65-72; PROF BIOCHEM, UNIV PA, 72- Mem: Am Physiol Soc; Soc Gen Physiol; Biophys Soc; Am Soc Biol Chemists. Res: Regulation of muscular activity. Mailing Add: Dept of Biochem Univ of Pa Philadelphia PA 19104

WEBER, ARTHUR GEORGE, b St Joseph, Mo, Jan 14, 03; m 30; c 4. CHEMISTRY. Educ: Univ Kans, BS, 27; Univ Wis, PhD(gen chem), 30. Prof Exp: Asst instr chem, Univ Wis, 27-30; res chemist, Ammonia Dept, Exp Sta, E I du Pont de Nemours & Co, Inc, 30-34, proj leader, 34-39, group leader, 39-44, lab dir, 44-50, mgr, Admin & Serv Sect, Polychem Dept, Res Div, 50-56, mgr personnel sect, Personnel & Employee Rels Div, Plastics Dept, 56-68; SPECIALIST COUNSR, TECH SERV DIV, UNIV DEL, 68- Mem: Am Chem Soc. Res: High pressure synthesis; oxidation and hydrogenation reactions; phosphoric acids; process development. Mailing Add: 1514 Brandywine Blvd Wilmington DE 19809

WEBER, BRUCE HOWARD, b Cleveland, Ohio, June 8, 41. BIOCHEMISTRY, NEUROCHEMISTRY. Educ: San Diego State Univ, BS, 63; Univ Calif, San Diego, PhD(chem), 68. Prof Exp: Am Cancer Soc fel, Molecular Biol Inst, Univ Calif, Los Angeles, 68-70; asst prof, 70-72, ASSOC PROF CHEM, CALIF STATE UNIV, FULLERTON, 72- Concurrent Pos: NIH res fel, 75; res scientist, Div Neurosci, City of Hope Nat Med Ctr, 75- Mem: AAAS; Am Chem Soc; Brit Soc Hist Sci; Am Soc Biol Chemists; Sigma Xi. Res: Structure, function, interactions and evolution of proteins, especially enzymes involved in neurotransmitter metabolism. Mailing Add: Dept of Chem Calif State Univ Fullerton CA 90634

WEBER, CHARLES ROBERT, agriculture, see 12th edition

WEBER, CHARLES WALTER, b Harold, SDak, Nov 30, 31; m 61; c 2. BIOCHEMISTRY, NUTRITION. Educ: Colo State Univ, BS, 56, MS, 58; Univ Ariz, PhD(biochem, nutrit), 66. Prof Exp: Res chemist, Univ Colo, 60-63; from asst prof to assoc prof, 66-73, PROF BIOCHEM NUTRIT, UNIV ARIZ, 73- Mem: AAAS; Poultry Sci Asn; Am Chem Soc; NY Acad Sci; Am Inst Nutrit. Res: Methods for carbohydrate determination; trace mineral utilization; minerals and vitamin relationship; amino acid requirements of mice and Coturnix quail. Mailing Add: Dept Nutrit & Food Sci Rm 332 Agr Sci Bldg Univ of Ariz Tucson AZ 85721

WEBER, CHARLES WALTON, b Phoenixville, Pa, Jan 16, 43; m 66. ANALYTICAL CHEMISTRY, INORGANIC CHEMISTRY. Educ: Philadelphia Col Pharm, BSc, 64; Univ Pa, PhD(kinetics), 69. Prof Exp: ASST PROF CHEM, DEL VALLEY COL, 69- Mem: Am Chem Soc. Res: Chemistry of the silver (II) ion in solution, primarily kinetics and complexes; uses of ion-selective electrodes. Mailing Add: Dept of Chem Rt 202 Del Valley Col Doylestown PA 18901

WEBER, CHARLES WILLIAM, b Streator, Ill, Dec 28, 22; m 48; c 3. ANALYTICAL CHEMISTRY. Educ: Northwestern Univ, BS, 49; Ind Univ, PhD(anal chem), 53. Prof Exp: Anal chemist, 53-58 & Sect Anal Develop, 58-62, DEPT HEAD CHEM ANAL, OAK RIDGE GASEOUS DIFFUSION PLANT, UNION CARBIDE CORP, 62- Mem: AAAS; Sigma Xi; fel Am Inst Chem; Am Chem Soc. Res: Instrument development; gas analysis; automation; laboratory management; process control; personnel development. Mailing Add: 1021 W Outer Dr Oak Ridge TN 37830

WEBER, CLIFFORD EDWARD, chemistry, see 12th edition

WEBER, DARRELL J, b Thornton, Idaho, Nov 16, 33; m 62; c 4. BIOCHEMISTRY, PLANT PATHOLOGY. Educ: Univ Idaho, BS, 58, MS, 59; Univ Calif, PhD(plant path), 63. Prof Exp: Res asst agr chem, Univ Idaho, 57-59; res asst plant path, Univ Calif, 59-63; res assoc biochem, Univ Houston, 63-65, asst prof, 65-69; assoc prof, 69-73, PROF BOT, BRIGHAM YOUNG UNIV, 73- Concurrent Pos: Fels, Univ Wis, 63-65; consult, NASA, 66-67; grants, USDA, 66-68, NASA, 67-68, NIH, 68-71 & NSF, 71- Mem: AAAS; Am Phytopath Soc; Mycol Soc Am; Am Inst Biol Sci; Am Soc Microbiol. Res: Phytochemistry; mode of action of fungicides; metabolism of fungal spores; biochemistry of host-parasite complexes; salt tolerance of plants; toxic compounds in plants. Mailing Add: Dept of Bot 285 Widtsoe Brigham Young Univ Provo UT 84602

WEBER, DAVID ALEXANDER, b Lockport, NY, Mar 6, 39; m 61; c 2. MEDICAL PHYSICS. Educ: St Lawrence Univ, BS, 60; Univ Rochester, PhD(radiation biol), 70. Prof Exp: Teaching asst physics, Univ Buffalo, 60-61; asst attend physicist, Mem Sloan-Kettering Cancer Ctr, 61-68; AEC grad lab fel radiation biol & biophys, 68-70, asst prof radiol, 70-75, ASST PROF RADIATION BIOL & BIOPHYS, SCH MED & DENT, UNIV ROCHESTER, 70-, ASSOC PROF RADIOL, 75- Mem: Soc Nuclear Med; Am Asn Physicists in Med; Health Physics Soc. Res: Metabolic radionuclide studies; medical applications of quantitative neutron activation analysis; nuclear medicine data imaging; systems and instrumentation. Mailing Add: Dept of Radiol Univ of Rochester Sch Med & Dent Rochester NY 14642

WEBER, DAVID FREDERICK, b North Terre Haute, Ind, Nov 18, 39; m 63; c 1. CYTOGENETICS. Educ: Purdue Univ, BS, 61; Ind Univ, MS, 63, PhD(genetics), 67. Prof Exp: Asst prof, 67-72, ASSOC PROF GENETICS, ILL STATE UNIV, 72- Concurrent Pos: AEC contract, 70-74. Mem: AAAS; Genetics Soc Am. Res: Analysis of meiosis in monosomics; cytological behavior of univalents; effects of a monosomic chromosome on recombination; study of the frequency and types of spontaneous chromosome abberations arising in monosomics; determination of effects of omonosomy on lipid content in Zea mays; screening for ultrastructureal differences in monosomics; determination of free amino acid profiles in monosomics. Mailing Add: Dept of Biol Sci Ill State Univ Normal IL 61761

WEBER, DEANE FAY, b Aberdeen, SDak, May 17, 25; m 50; c 3. SOIL MICROBIOLOGY. Educ: Jamestown Col, BS, 50; Kans State Univ, MS, 52, PhD, 59. Prof Exp: Biochemist & bacteriologist, Quain & Ramstad Clin, Bismarck, NDak, 52-55; asst vet bact, pathogenic bact & virol, Kans State Univ, 57-58; soil scientist, Soil & Water Conserv Br, Agr Res Serv, 58-64 & IRI Res Inst, Campinas, Brazil, 64-66, MICROBIOLOGIST SOYBEAN INVESTS, CROPS RES DIV, AGR RES SERV, USDA, 67- Mem: Fel AAAS; Soil Sci Soc Am; Am Soc Microbiol; Am Soc Agron. Res: Legume microbiology; nitrogen fixation. Mailing Add: Cell Cult & Nitrogn Fixation Lab Agr Res Ctr-W USDA Beltsville MD 20705

WEBER, DENNIS JOSEPH, b Kalamazoo, Mich, Mar 30, 34; m 53; c 6. ANALYTICAL CHEMISTRY, PHYSICAL CHEMISTRY. Educ: Western Mich

Univ, BS, 58, MA, 62; Univ Fla, PhD(pharm), 67. Prof Exp: Res asst phys & anal chem, Upjohn Co, 58-62; mgr anal chem, Syntex Labs, Calif, 67-70; RES ASSOC PHYS & ANAL CHEM, UPJOHN CO, 70- Mem: Am Pharmaceut Asn; Acad Pharmaceut Sci; The Chem Soc; Am Chem Soc. Res: Kinetics of hydrolysis of drugs; correlation of spectra and structure of hydrazones; structure and stability constants of metal complexes of thiouracils; partition chromatography of steroids; high pressure liquid chromatography. Mailing Add: Upjohn Co Kalamazoo MI 49001

WEBER, EVELYN JOYCE, b Tower Hill, Ill, Nov 9, 28. LIPID CHEMISTRY. Educ: Univ Ill, BS, 53; Iowa State Univ, PhD(biochem), 61. Prof Exp: Asst biochem, Iowa State Univ, 56-61; res assoc, Univ Ill, 61-65; asst prof, 65-72, ASSOC PROF PLANT BIOCHEM, UNIV ILL, URBANA, 72-, RES CHEMIST, USDA, 65- Mem: AAAS; Am Chem Soc; fel Am Inst Chem; Am Oil Chem Soc; Am Soc Plant Physiol. Res: Identification and characterization of complex lipids; biosynthesis of fatty acids and lipids in corn and other plants. Mailing Add: USDA Agr Res Serv 230 Davenport Hall Univ of Ill Urbana IL 61801

WEBER, FAUSTIN N, b Toledo, Ohio, Nov 5, 11; m 37; c 4. ORTHODONTICS. Educ: Univ Mich, DDS, 34, MS, 36; Am Bd Orthod, dipl. Prof Exp: From asst prof to assoc prof, 36-51, PROF ORTHOD & CHMN DEPT, UNIV TENN CTR HEALTH SCI, MEMPHIS, 51- Concurrent Pos: Dir grad prog orthod & prof dent hyg, Univ Tenn Ctr Health Sci, Memphis; consult, Kennedy Gen Vet Hosp; mem, Am Asn Dent Schs. Mem: Am Soc Dent for Children; Am Dent Asn; Am Asn Orthod; fel Am Col Dent; Int Col Dent. Res: Child growth and development. Mailing Add: Dept of Orthod Univ of Tenn Ctr for Health Sci Memphis TN 38163

WEBER, FLORENCE ROBINSON, b Milwaukee, Wis, Aug 26, 21; m 59. GEOLOGY. Educ: Univ Chicago, BS, 43, MS, 48. Prof Exp: Lab instr & librn, Univ Chicago, 42-43, librn, 47; subsurface geologist, Shell Oil Co, Tex, 43-47 & State Geol Surv, Mo, 47; geologist, Alaska Br, 49-54, DC, 54-59 & Alaska Br, 54-59, GEOLOGIST-IN-CHARGE, COL OFF, ALASKA BR, US GEOL SURV, 59-; DISTINGUISHED LECTR GEOL, UNIV ALASKA, FAIRBANKS, 59- Concurrent Pos: Arctic Inst NAm grant, 56. Mem: AAAS; Geol Soc Am; Am Asn Petrol Geol; Arctic Inst NAm. Res: Stratigraphy; structure; petrology; paleontology; petroleum geology. Mailing Add: Alaskan Geol Br US Geol Surv Box 80586 College AK 99701

WEBER, FRANCIS JOHN, b Philadelphia, Pa, Sept 20, 09; m 38; c 2. MEDICINE. Educ: Univ Pa, AB, 32, MD, 36; Johns Hopkins Univ, MPH, 39, DrPH, 51; Am Bd Prev Med, dipl, 49. Prof Exp: With NIH, Md, 39; asst health officer, Tri County Health Unit, 39-40; gen health consult, Regional Off, Fed Security Agency, USPHS, La, 40; health officer, Muscogee County, Ga, 40-41; dir div venereal dis control, State Dept Health, Mich, 41-44; regional med consult, USPHS, Calif, 44, from asst chief to chief tuberc control div, Bur State Serv, DC, 44-48; resident & psychiat researcher, Sch Hyg & Pub Health, Johns Hopkins Univ & Hosp, 48-51, res assoc, 50-51; regional med dir, Regional Off, USPHS, Ohio, 51-52, regional med dir, Colo, 52-58, chief div radiol health, DC, 58-61; assoc prof prev med & chmn div environ med, Col Med, Univ Utah, 66-70; consult, Radiol Health Div, Environ Control Admin, USPHS, 70-72; CONSULT COMMUNITY & ENVIRON HEALTH FIELDS, 72- Concurrent Pos: Asst clin prof, Sch Med, Georgetown Univ, 46-48; asst clin prof, Sch Med, Univ Colo, 53-58, spec consult radiation health & environ med, WHO, UN & USPHS. Mem: Health Physics Soc; fel Am Pub Health Asn; Am Psychiat Asn; Nat Tuberc & Respiratory Dis Asn; Am Col Chest Physicians. Res: Epidemiologic studies in poliomyelitis; venereal diseases; tuberculosis; mental health, preventive medicine and public health. Mailing Add: 150 McKinley Circle Vacaville CA 95688

WEBER, FRANK E, b Chicago, Ill, Feb 1, 35; m 57; c 3. FOOD SCIENCE. Educ: Ill Inst Technol, BS, 57; Univ Ill, MS, 63, PhD(food sci), 64. Prof Exp: Res scientist-leader chem sect, 64-71, mgr prod develop, 71-73, mgr tech res, 73-76, TECH DIR, R T FRENCH CO, 76- Mem: Am Chem Soc; Inst Food Technologists; Am Asn Cereal Chemists; Sigma Xi. Res: Product development; quality control; natural and artificial flavorings; isolation and analysis of flavor substances; food analysis methodology product development; industrial waste treatment. Mailing Add: 43 Sunset Dr Rochester NY 14618

WEBER, FRED CHARLES, JR, organic chemistry, see 12th edition

WEBER, FREDERICK, JR, b Hilgen, Ger, Feb 14, 23; nat US; m 46; c 3. MICROBIOLOGY, PUBLIC HEALTH. Educ: Univ RI, BS, 48; Pa State Univ, MS, 50; Mich State Univ, PhD, 56. Prof Exp: Instr, Wartburg Col, 50-51 & Wayne State Univ, 51-53; mem res staff, Swift & Co, 56-59; MEM RES STAFF, JOSEPH E SEAGRAM & SONS, INC, 59- Mem: Am Soc Microbiol; Inst Food Technologists. Res: Psychiophiles in dairy products; flavor development by microorganisms in foods; aerobic digestion of wastes; alcoholic fermentations. Mailing Add: 9927 Silverwood Lane Louisville KY 40272

WEBER, GEORGE, b Budapest, Hungary, Mar 29, 22; nat US; m 58; c 3. PHARMACOLOGY, BIOCHEMISTRY. Educ: Queen's Univ, Ont, BA, 50, MD, 52. Prof Exp: Nat Cancer Inst Can fel, Univ BC, 52-53; Cancer Res Soc sr fel, Montreal Cancer Inst, 53-58, head dept path chem, 56-59; assoc prof biochem & microbiol, 59-60, assoc prof pharmacol, 60-61, dir lab exp oncol, 61-74, PROF PHARMACOL, SCH MED, IND UNIV, INDIANAPOLIS, 61-, CANCER COORDR BASIC RES, 62- Concurrent Pos: Ed, Adv in Enzyme Regulation, 62-; Oxford Biochem Soc lectr, 69; rapporteur, Int Cancer Cong, Tex, 70; mem sci adv comt, Pharmacol B Study Sect, USPHS, 70-74, chmn exp therapeut study sect, Nat Cancer Inst, 76-78; mem, Tiberine Acad, 71-; mem sci adv, Damon Runyon Mem Fund, 71-75; mem adv comt instnl grants, Am Cancer Soc, 72-76. Honors & Awards: Alecce Award, Cancer Res, Rome, Italy, 71. Mem: Am Soc Pharmacol & Exp Therapeut; Am Physiol Soc; Am Asn Cancer Res; Am Soc Cell Biol; fel Royal Soc Med. Res: Oncology; biochemical pharmacology; regulation of enzymes and metabolism; neoplasia; chemotherapy; chemical pathology of diseases; liver tumors of different growth rates; hormone action; radiation biology; nutrition; regulation of brain metabolism; cell hybridization. Mailing Add: Lab for Exp Oncol Ind Univ Sch of Med Indianapolis IN 46202

WEBER, GEORGE RUSSELL, b Novinger, Mo, Dec 29, 11; m 47; c 2. BACTERIOLOGY. Educ: Univ Mo, BS, 35; Iowa State Col, PhD(sanit & food bact), 40. Prof Exp: Asst chemist, Exp Sta, Univ Mo, 35-36; instr, Iowa State Col, 40-42; bacteriologist, USPHS, 46, sr asst scientist, 47-49, scientist, 49-53, chief sanitizing agents unit, 51-53; proj leader, 53-73, SR RES MICROBIOLOGIST, RES DIV, US INDUST CHEM CO, NAT DISTILLERS & CHEM CORP, 73- Concurrent Pos: Asst, Iowa State Col, 30-40; lectr, Eve Col, Univ Cincinnati, 69-70. Mem: AAAS; Am Soc Microbiol; fel Am Pub Health Asn; NY Acad Sci; fel Royal Soc Health. Res: Sanitary and food bacteriology; germicidal efficiency of hypochlorites, chloramines and quaternary ammonium compounds; antibiotics; fermentations; yeast hybridization; ruminant and chinchilla nutrition; biological metal corrosion. Mailing Add: 1525 Burney Lane Cincinnati OH 45230

WEBER, GREGORIO, b Buenos Aires, Arg, July 4, 16; m 47; c 3. BIOCHEMISTRY, BIOPHYSICS. Educ: Univ Buenos Aires, MD, 43; Cambridge Univ, PhD(biochem), 47. Prof Exp: Beit mem fel biochem, Cambridge Univ, 48-52; lectr, Univ Sheffield, 53-60, reader biophys, 60-62; PROF BIOCHEM, UNIV ILL, URBANA, 62- Mem: Nat Acad Sci; Am Soc Biol Chem; Am Chem Soc; Brit Biochem Soc; Am Acad Arts & Sci. Res: Physical chemistry of proteins; fluorescence methods; excited states. Mailing Add: 397 Roger Adams Lab Dept of Biochem Univ of Ill Urbana IL 61801

WEBER, HANS JÜRGEN, b Berlin, Ger, May 3, 39; m 66; c 1. THEORETICAL NUCLEAR PHYSICS. Educ: Univ Frankfurt, BS, 60, MS, 61, PhD(theoret physics), 65. Prof Exp: Res assoc theoret nuclear physics, Duke Univ, 66-67; asst prof, 68-71, ASSOC PROF THEORET NUCLEAR PHYSICS, UNIV VA, 71- Concurrent Pos: Sesquicentennial assoc, Max Planck Inst, Mainz & Univ Frankfurt, 72-73. Mem: Am Phys Soc. Res: Medium energy physics; photonuclear physics; group theory. Mailing Add: Dept of Physics Univ of Va Charlottesville VA 22901

WEBER, HEATHER ROSS, b Passaic, NJ, Mar 7, 43; m 63; c 2. BIOCHEMISTRY. Educ: Boston Univ, BA, 65; Univ Southern Calif, PhD(cellular & molecular biol), 74. Prof Exp: Res asst hemat, Peter Bent Brigham Hosp, 65-67; res asst biophys, Harvard Med Sch, 67-68; res assoc biochem, 73-74, NIH cancer training grant, 74-76, NIH FEL BIOCHEM, MED SCH, UNIV SOUTHERN CALIF, 76- Mem: AAAS; Am Chem Soc. Res: Transcriptional regulation of eucaryotic gene expression; nucleic acid metabolism of neoplastic cells; effects of 5-bromodeoxyuridine on rat hepatona tissue culture cells. Mailing Add: Dept of Biochem Univ Southern Calif Med Sch Los Angeles CA 90033

WEBER, HEINZ PAUL, b Winterthur, Switz, Apr 18, 39; m 64; c 1. PHYSICS, OPTICS. Educ: Swiss Fed Inst Technol, dipl, 64; Univ Bern, PhD(quantum electronics), 68. Prof Exp: MEM TECH STAFF, BELL LABS, 69- Mem: Optical Soc Am; Inst Elec & Electronics Engrs. Res: Picosecond pulses and integrated optics. Mailing Add: Bell Labs Holmdel NJ 07733

WEBER, JAMES HAROLD, b Madison, Wis, July 21, 36; m 61; c 3. INORGANIC CHEMISTRY. Educ: Marquette Univ, BS, 59; Ohio State Univ, PhD(chem), 63. Prof Exp: Asst prof, 63-70, ASSOC PROF CHEM, UNIV NH, 70- Mem: Am Chem Soc. Res: Coordination chemistry; organometallic chemistry; chemistry of natural water. Mailing Add: Parsons Hall Univ of NH Durham NH 03824

WEBER, JAMES HERBERT, physical chemistry, see 12th edition

WEBER, JANET CROSBY, b Chicago, Ill, June 1, 23; m 49; c 3. BIOCHEMISTRY, COMPUTER SCIENCE. Educ: Iowa State Univ, BS, 45; Univ Ill, Urbana, PhD(chem), 48. Prof Exp: Asst prof vet res, Mont State Univ, 48-50; res assoc, 60-63, instr, 63-69, ASST PROF OPHTHAL, MED CTR, IND UNIV, INDIANAPOLIS, 69- Mem: Am Chem Soc; Asn Res Vision & Ophthal. Res: Ophthalmic biochemistry; micro-analytical methods; etiology of uveitis; computer methods in clinical and research studies; development of computer programs for statistical analysis. Mailing Add: Dept of Ophthal Ind Univ Med Ctr Indianapolis IN 46202

WEBER, JEAN ROBERT, b Thun, Switz, Apr 28, 25; Can citizen; m 54; c 3. GEOPHYSICS. Educ: Swiss Fed Inst Technol, Prof Eng, 52; Univ Alta, PhD(physics), 60. Prof Exp: Res engr, PTT Res Labs, Berne, Switz, 51-53; lectr, Univ Alta, 57-58; geophysicist-in-chg, Oper Hazen, Int Geophys Year, 58-59; RES SCIENTIST, DOM OBSERV, DEPT ENERGY, MINES & RESOURCES, CAN, 60- Concurrent Pos: Mem Arctic Inst NAm Baffin Island Exped, 53 & Univ Toronto, Salmon Glacier Exped, 56; leader, Dom Observ NPole Exped, 67. Mem: Am Geophys Union; Soc Explor Geophys; fel Arctic Inst NAm; Glaciol Soc. Res: Communications electronics; dosimetry of beta radiation and biological effects on allium cepa roots; regional gravity interpretations; continental margins; upper mantle, Arctic Ocean; application of geophysics to glaciology. Mailing Add: Dom Observ 3 Observ Crescent Ottawa ON Can

WEBER, JEROME BERNARD, b Claremont, Minn, Sept 19, 33; m 56; c 4. SOIL CHEMISTRY, WEED SCIENCE. Educ: Univ Minn, BS, 57, MS, 59, PhD(anal chem), 63. Prof Exp: Assoc prof soil pesticide chem, 62-71, PROF SOIL PESTICIDE CHEM & WEED SCI, NC STATE UNIV, 71- Honors & Awards: Sigma Xi Res Award; Outstanding Paper Award. Mem: AAAS; Am Soc Agron; Soil Sci Soc Am; Weed Sci Soc Am; Clay Minerals Soc. Res: Chemistry of soil, fate and biological availability of applied organic compounds, especially herbicides; effects of pesticides on environmental quality; weed ecology studies; behavior of gases in soil. Mailing Add: Dept of Crop Sci Weed Sci Ctr NC State Univ Raleigh NC 27607

WEBER, JOHN DONALD, b Lagon, La, Nov 1, 34; m 62; c 2. PHARMACEUTICAL CHEMISTRY. Educ: Xavier Univ La, BS, 57; Univ Notre Dame, MS, 61; Georgetown Univ, PhD(org chem), 72. Prof Exp: Instr chem, Southern Univ New Orleans, 62-63; anal chemist, 63-68, RES CHEMIST, FOOD & DRUG ADMIN, 68- Mem: Am Chem Soc; Sigma Xi; Asn Off Anal Chemists; Nat Orgn Black Scientists. Res: Pharmaceutical analyses; chromatograph, polarimetric and fluorimetric techniques; optical purity of drugs; stereochemistry and chemical kinetics; problems of relationships between stereoisomerism and physiological activity. Mailing Add: 7204 Seventh St NW Washington DC 20012

WEBER, JOHN R, b Ft Madison, Iowa, Mar 18, 24; m 48; c 2. PLANT PHYSIOLOGY. Educ: Univ Iowa, AB, 48, MS, 49. Prof Exp: Prin lab technician, Citrus Exp Sta, Univ Calif, 52-56; plant physiologist, 56-67, MEM SPEC PROJS EGG HANDLING SYSTS, CITRUS MACH DIV, FMC CORP, 67- Mem: AAAS; Bot Soc Am; Am Soc Plant Physiol. Res: Post-harvest physiology of fruits and vegetables. Mailing Add: 2102 Oak Crest Dr Riverside CA 92506

WEBER, JON, b Kitchener, Ont, July 8, 35; US citizen; m 59; c 2. GEOCHEMISTRY. Educ: McMaster Univ, BSc, 58, MSc, 59; Univ Toronto, PhD(geochem), 62. Prof Exp: Geologist, Shell Oil Co, 56-58; assoc prof geochem, 61-74, PROF MARINE GEOL, PA STATE UNIV, 74- Mem: AAAS; Geochem Soc; Geol Soc Am; Am Asn Petrol Geol; Soc Econ Paleont & Mineral. Res: Stable isotope and trace element geochemistry; application of digital computers to earth sciences; carbonate sediments and sedimentary rocks; coral reefs. Mailing Add: 224 Mineral Sci Pa State Univ University Park PA 16802

WEBER, JOSEPH, b Paterson, NJ, May 17, 19; m 42, 72; c 4. PHYSICS. Educ: US Naval Acad, BS, 40; Cath Univ Am, PhD(physics), 51. Prof Exp: PROF PHYSICS, UNIV MD, COLLEGE PARK, 48- Concurrent Pos: Consult, Res & Eng Depts, US Naval Ord Lab, 48-; Nat Res Coun & Guggenheim fels, 55-56; fel, Inst Advan Study, 55-56, 62-63 & 69-70 & Lorentz Inst Theoret Physics, State Univ Leiden, 56. Honors & Awards: Sigma Xi Award, 70; Boris Pregel Prize, NY Acad Sci, 73. Mem: Fel Am Phys Soc; fel Inst Elec & Electronics Engrs. Res: General relativity; microwave spectroscopy; irreversibility. Mailing Add: Dept of Physics & Astron Univ of Md College Park MD 20740

WEBER, JOSEPH, b Paterson, NJ, Jan 2, 24; m 59; c 2. ORGANIC CHEMISTRY. Educ: Rutgers Univ, BSc, 44, PhD(org chem), 56; NY Univ, MSc, 52. Prof Exp:

Chemist, Hoffmann-La Roche Inc, 47-54; instr org chem, Rutgers Univ, 54-56; sr chemist, 56-64, group leader chem prod, 64-66, group chief, 67-69, sr group chief, 69-73, assoc dir chem prod, 73-74, dir, 74, VPRES & MEM BD DIRS, TECH SERV, HOFFMANN-LA ROCHE, INC, 74- Mem: AAAS; Am Chem Soc; Sigma Xi; NY Acad Sci. Res: Synthetic organic chemistry in fields of pharmaceutical and medicinal chemistry. Mailing Add: Hoffmann-La Roche Inc 340 Kingsland St Nutley NJ 07110

WEBER, JOSEPH, b Budapest, Hungary, Oct 10, 39; Can citizen; m 64; c 1. VIROLOGY. Educ: Univ BC, BSc, 64, MSc, 66; McMaster Univ, PhD(virol), 69. Prof Exp: Nat Cancer Inst res assoc virol, Ohio State Univ, 69-70; asst prof microbiol, 70-75, ASSOC PROF MICROBIOL, UNIV SHERBROOKE, 75- Concurrent Pos: Lectr, Ohio State Univ, 69-70; Med Res Coun Can scholar, 71-76. Mem: Can Soc Microbiol. Res: Viral oncology; cell transformation; adenovirus genetics. Mailing Add: Dept of Microbiol Univ Ctr Hosp Sherbrooke PQ Can

WEBER, JOSEPH ELLIOTT, b Greenfield, Ind, May 3, 10. CHEMISTRY. Educ: Ind Univ, AB, 32, AM, 33, PhD(chem), 37. Prof Exp: Res chemist, Am Can Co, 37; from asst prof to assoc prof, 37-51, PROF ORG CHEM, BOWLING GREEN STATE UNIV, 51- Mem: Am Chem Soc. Res: Organic preparations; electro-organic chemistry. Mailing Add: 816 E Wooster Bowling Green OH 43402

WEBER, JULIUS, b Brooklyn, NY, Apr 8, 14; m 47; c 4. CYTOLOGY, PHOTOGRAPHY. Hon Degrees: DSc, Jersey City State Col, 74. Prof Exp: Asst, Brooklyn Jewish Hosp, New York, 33-35; chief histol technician, Kingston Ave Hosp Infectious Dis, 35-36 & Israel Zion Hosp, Brooklyn, 36-39; head dept med photog, Columbia-Presby Med Ctr, New York, 39-49; HEAD DEPT PHOTOG RES, BETH ISRAEL HOSP, 49- Concurrent Pos: Mem training div, Inst Inter-Am Affairs, US Dept State, 46; consult, Western Union Tel Co, 46-48, Chem Corps, US Army, 47-50, Ansco Div, Gen Aniline & Film Corp, 47-48, Nat Film Bd Can, 51, US Naval Hosp, St Albans, NY, 51-52, St Francis Hosp, New York, 59, Perkin-Elmer Corp, 60 & Ehrenreich Photo-Optical Industs, Inc, 63; lectr, Sch Med, Univ Calif, Los Angeles, 48; guest lectr, Col Physicians & Surgeons, Columbia Univ, 50, Sch Eng, 60; dir med photog, William Douglas McAdams, Inc, 52-59; head dept med photog, Hosp Joint Dis, 53-73; res assoc, Waldemar Med Res Found, 55-; lectr, Grad Sch Pub Admin, NY Univ, 55-57; head dept photog, Misericordia Hosp, 58-; med photographer, Knickerbocker Hosp, 58-63; assoc, Dept Mineral, Am Mus Natural Hist, 60; ed, Image Dynamics Sci & Med, 69-71; dir, Wildcliff Natural Sci Ctr, 69-74; res assoc, Dept Mineral, Royal Ont Mus, Toronto, 71- Mem: Soc Photog Sci & Eng; assoc Photog Soc Am; fel Biol Photog Asn; fel Royal Micros Soc; fel Royal Photog Soc Gt Brit. Res: Photographic instrumentation for endoscopy, micromineralogy; time lapse photomicrography and cinematography; ultraviolet, infrared and gross photomicrography and photomacrography; neurocytology and neuropathology photoimpregnation techniques.

WEBER, KARL HANSEL, b Towner, NDak, Oct 8, 16; m 46; c 4. ORGANIC CHEMISTRY. Educ: Univ NDak, BA, 37; Univ Wis, PhD(org chem), 41. Prof Exp: Res chemist, Armstrong Cork Co, Pa, 41-44; admin asst to dir res, Kilgore Develop Corp, DC, 44-46; document analyst, Off Tech Serv, US Dept Com, 46-47; consult, Res & Develop Bd, 47-49; chief sci div, off coordr & spec adv, US High Comnr for Ger, 50-54; CONSULT, US GOVT, 54- Mem: Am Chem Soc. Res: Synthetic organic insectifuges and insecticides; synthetic resins from naturally occurring raw materials; synthetic rubber-like polymers. Mailing Add: 10605 Marbury Rd Oakton VA 22124

WEBER, KENNETH C, b Cold Spring, Minn, June 30, 37; m 57; c 4. PULMONARY PHYSIOLOGY. Educ: Univ Minn, BSEE, 63, PhD(physiol), 68. Prof Exp: USPHS trainee physiol, Univ Minn, 62-68; asst prof, 68-72, ASSOC PROF PHYSIOL, WVA UNIV, 72- Concurrent Pos: Chief physiol sect, Appalachian Lab Occup Respiratory Dis, Nat Inst Occup Safety & Health, 68-; chmn, Gordon Res Conf Non-Ventilatory Lung Function, 75. Mem: AAAS; Am Physiol Soc; Am Heart Asn; Am Thoracic Soc; Biophys Soc. Res: Cardiopulmonary hemodynamics; respiration physiology; non-ventilatory lung function; lung metabolism. Mailing Add: Dept Physiol & Biophys WVa Univ Med Ctr Morgantown WV 26506

WEBER, KENNETH EARL, physical chemistry, physics, see 12th edition

WEBER, LAVERN J, b Isabel, SDak, June 7, 33; m 59; c 4. PHARMACOLOGY, ACADEMIC ADMINISTRATION. Educ: Pac Lutheran Univ, BA, 58; Univ Wash, MS, 62, PhD(pharmacol), 64. Prof Exp: From instr to asst prof pharmacol, Sch Med, Univ Wash, 64-69; assoc prof pharmacol & toxicol, 69-75, PROF PHARMACOL & TOXICOL, SCH PHARM, ORE STATE UNIV, 75-, ASST DEAN GRAD SCH, 74- Mem: Am Soc Pharmacol & Exp Therapeut; Soc Toxicol; Soc Exp Biol & Med; Am Asn Lab Animal Sci. Res: Anabolism of 5-hydroxytryptamine; biochemistry of autonomic nervous system; comparative pharmacology and toxicology; comparative pharmacology of the autonomic nervous system. Mailing Add: Dept of Pharmacol & Toxicol Ore State Univ Sch of Pharm Corvallis OR 97331

WEBER, LEON, b Detroit, Mich, Feb 4, 31; m 52; c 3. SURFACE CHEMISTRY, PHYSICAL CHEMISTRY. Educ: Wayne State Univ, BS, 52, MS, 54; Carnegie-Mellon Univ, MS & PhD(phys chem), 67. Prof Exp: Chemist, Wayne County Rd, Comn, 51-54; res chemist, Shell Develop Co, 54-56; scientist nuclear chem, Westinghouse Atomic Power Co, 56-57; sr res chemist, Gulf Res & Develop Co, 57-67; SCIENTIST SURFACE CHEM, GLIDDEN-DURKEE DIV, SCM CORP, 67- Mem: AAAS; Am Chem Soc; Fedn Socs Paint Technol. Res: Colloid and surface chemistry of pigments; gas-solid sorption phenomena; reaction kinetics by thermal analysis. Mailing Add: Glidden-Durke Div SCM Corp 3901 Hawkins Point Rd Baltimore MD 21226

WEBER, LOUIS RUSSELL, b St Joseph, Mo, Oct 15, 01; m 26; c 2. EXPERIMENTAL PHYSICS. Educ: Park Col, AB, 25; Univ Mich, AM, 26, PhD(physics), 31. Prof Exp: Prof physics & head dept, Friends Univ, 26-38; prof & head dept, 38-65, EMER PROF PHYSICS, COLO STATE UNIV, 65- Concurrent Pos: Fulbright lectr physics, Col Sci, Baghdad, Iraq, 52-53, De LaSalle Col, Manila, 65 & Univ of the East, 65-66; adv & prof physics & basic sci, Univ Peshawar, Pakistan, Int Coop Admin, 58-60; consult, Palomar Col, San Marcos, Calif, 68-69; Fulbright vis prof, Univ Antioquia, Colombia, 69-71. Honors & Awards: Distinguished Serv Citation, Am Asn Physics Teachers, 65. Mem: Am Phys Soc; fel Optical Soc Am; Acoust Soc Am; Am Asn Physics Teachers; Sigma Xi. Res: Infrared absorption of water vapor beyond 10 microns; rotational infrared absorption water; piezoelectric-direct study of quartz crystals; spectrographic study of trace metals in plants and soils; atmospheric ozone. Mailing Add: PO Box 448 Ft Collins CO 80522

WEBER, MARVIN JOHN, b Fresno, Calif, Feb 26, 32; m 57; c 2. LASERS, SPECTROSCOPY. Educ: Univ Calif, Berkeley, AB, 54, MA, 56, PhD(physics), 59. Prof Exp: Asst physics, Univ Calif, 54-59; prin res scientist, Res Div, Raytheon Co, 59-73; GROUP LEADER, LAWRENCE LIVERMORE LAB, 73- Concurrent Pos: Vis res assoc, Stanford Univ, 66; consult, Div Mat Res, NSF, 73- Mem: Fel Am Phys Soc. Res: Magnetic resonance; optical spectroscopy of solids; luminescence materials; quantum electronics; lasers. Mailing Add: Lawrence Livermore Lab Univ of Calif PO Box 808 Livermore CA 94550

WEBER, MICHAEL JOSEPH, b New York, NY, Aug 23, 42; m 67. VIROLOGY, CELL BIOLOGY. Educ: Haverford Col, BSc, 63; Univ Calif, San Diego, PhD(biol), 68. Prof Exp: Am Cancer Soc Dernham jr fel, Univ Calif, Berkeley, 68-70; asst prof, 70-75, ASSOC PROF MICROBIOL, UNIV ILL, URBANA, 75- Concurrent Pos: NIH res career develop award, 75. Mem: Am Soc Microbiol; Tissue Cult Asn; Soc Gen Physiologists; Am Soc Biol Chemists. Res: Control of growth of animal cells and malignant transformation; tumor virus-induced cell surface changes. Mailing Add: Dept of Microbiol Univ of Ill Urbana IL 61801

WEBER, MORTON M, b New York, NY, May 26, 22; m 55; c 2. MICROBIOLOGY, BIOCHEMISTRY. Educ: City Col New York, BS, 49; Johns Hopkins Univ, ScD(microbiol), 53. Prof Exp: Instr zool & parasitol, St Francis Col, 49; instr med microbiol, Johns Hopkins Univ, 51-55; instr bact & immunol, Harvard Med Sch, 56-59; from asst prof to assoc prof, 56-63, PROF MICROBIOL, SCH MED, ST LOUIS UNIV, 63-, CHMN DEPT, 64- Concurrent Pos: Am Cancer Soc fel, McCollum-Pratt Inst, Johns Hopkins Univ, 53-56; mem microbial chem study sect, NIH, 69-73; vis prof, Oxford Univ, 70-71. Mem: Fel AAAS; Am Soc Biol Chemists; Am Soc Microbiol; Am Acad Microbiol; NY Acad Sci. Res: Physiology and biochemistry of microorganisms; pathways and mechanisms of electron transport; mode of action of antibiotics and other antimicrobial agents; biochemical regulatory mechanisms. Mailing Add: Dept of Microbiol St Louis Univ Sch of Med St Louis MO 63104

WEBER, NEAL ALBERT, b Towner, NDak, Dec 14, 08; m 40; c 3. ENTOMOLOGY, ECOLOGY. Educ: Univ NDak, AB, 30, MS, 32; Harvard Univ, AM, 33, PhD(zool), 35. Hon Degrees: ScD, Univ NDak, 58. Prof Exp: Assoc prof biol, Univ NDak, 36-43 & anat, Sch Med, 43-47; from assoc prof to prof, 47-74, EMER PROF ZOOL, SWARTHMORE COL, 74- Concurrent Pos: Mem expeds, W Indies, 33-36, Trinidad, BWI, 34-36, Orinoco Delta, 35, Brit Guiana, 35-36, Barro Colo Island, CZ & Columbia, 38; Anglo-Egyptian Sudan, Uganda & Kenya, 39; biologist, Am Mus Natural Hist Exped, Cent Africa, 48, Middle East, 50-52, Trop Am, 54- & Europ Mus, 57; consult, Arctic Res Lab, Alaska, 48-50; mem dept zool, Col Arts & Sci Univ Baghdad, Iraq, 50-52; vis prof, Univ Wis, 55-56; mem panel biol & med sci, Comt Polar Res, Nat Acad Sci, 58-60, panel fels, 64-66; mem & US del spec comt Antarctic res, Int Coun Sci Unions, Australia, 59; sci attache, Am Embassy, US Dept State, Buenos Aires, Arg, 60-62; mem, Latin Am Colloquium, Arg, 65 & Brazil, 68; consult, Venezuelan Univs, 72; adj prof biol sci, Fla State Univ, Tallahassee, 74- Honors & Awards: John F Lewis Prize, Am Philos Soc, 73. Mem: AAAS; fel Entom Soc Am; Ecol Soc Am; Mycol Soc Am; Am Soc Zool. Res: Tropical ecology; fungus-growing ants and their fungi; zoogeography. Mailing Add: 2606 Mission Rd Tallahassee FL 32304

WEBER, NEILL, physical chemistry, see 12th edition

WEBER, PAUL VAN VRANKEN, b Highland Park, Ill, Mar 12, 21; m 48; c 3. PLANT PATHOLOGY, PLANT BREEDING. Educ: Cornell Univ, BS, 43; Univ Wis, PhD(plant path), 49. Prof Exp: Asst plant pathologist, Ohio Agr Exp Sta, 49-52; asst plant pathologist & geneticist, Campbell Soup Co, 52-59; CHIEF, BUR PLANT PATH, STATE DEPT AGR, NJ, 59- Mem: Am Phytopath Soc; Am Inst Biol Sci. Res: Plant disease and insect surveys; administration. Mailing Add: 41 Charles Bossert Dr Bordentown NJ 08505

WEBER, PETER B, b Berlin, Ger, July 31, 34; c 1. BIOCHEMISTRY. Educ: Univ Cologne, DNatSc(biol, chem), 61. Prof Exp: NIH grants, Univ Ill, Chicago, 64-65; NSF grants, State Univ NY Buffalo, 65-68; ASST PROF BIOCHEM, ALBANY MED COL, 68- Res: Biochemistry of membrane glycoproteins. Mailing Add: Dept of Biochem Albany Med Col Albany NY 12208

WEBER, RICHARD GERALD, b Newport News, Va, Dec 20, 39; m 59. ENTOMOLOGY. Educ: Eastern Mennonite Col, BS(biol) & BA(foreign lang), 69; Univ Del, MS, 71; Kans State Univ, PhD(entom), 75. Prof Exp: INSTR INSECT MORPHOL, KANS STATE UNIV, 75- Mem: Entom Soc Am. Res: Shade tree and ornamental plant insects; insect morphology in relation to behavior. Mailing Add: Dept of Entom Kans State Univ Manhattan KS 66506

WEBER, RICHARD RAND, b Columbia, Pa, July 28, 38; m 65; c 3. RADIO ASTRONOMY, INSTRUMENTATION. Educ: Franklin & Marshall Col, AB, 60; Univ Md, MS, 68. Prof Exp: Teacher physics & math, Wilson High Sch, West Lawn, Pa, 60-61; RES SCIENTIST RADIO ASTRON, GODDARD SPACE FLIGHT CTR, NASA, 64- Mem: Am Astron Soc; Int Union Radio Sci. Res: Radio experiments on spacecraft, Radio Astronomy Explorer, lunar orbit; Helios, solar orbit; studying plasma radio emissions resulting from interaction of solar wind with moon, earth, and fast solar electrons. Mailing Add: Goddard Space Flight Ctr Code 693 Greenbelt MD 20771

WEBER, RICHARD ROBERT, b Cincinnati, Ohio, June 6, 26; m 50; c 2. APPLIED MATHEMATICS. Educ: Denison Univ, BA, 47; Brown Univ, ScM, 50; Univ Cincinnati, PhD(physics), 62. Prof Exp: Sr res engr, Phys Res Dept, Cincinnati Milacron, Inc, 54-56, sr res supvr appl mech, 56-64, mgr mgt sci, 64-69, mgr corp procedures, Cincinnati Milacron Ltd, Birmingham, Eng, 69-70, managing dir, Foundry Div, 70-75, DIR ADMIN SERV, CINCINNATI MILACRON LTD, BIRMINGHAM, ENG, 75- Res: Hydraulic control systems analysis; nonlinear mechanics and analysis; analog computation; digital computation; management sciences. Mailing Add: Cincinnati Milacron Inc Cincinnati OH 45209

WEBER, ROBERT EMIL, b Oshkosh, Wis, Dec 17, 30; m 53; c 3. POLYMER CHEMISTRY. Educ: Wis State Col Oshkosh, BS, 53; Univ Iowa, PhD, 59. Prof Exp: Res chemist, Res & Develop Ctr, Wis, 58-66, sr res scientist, 66-68, mgr prod develop lab, Munising Div, Mich, 68-71, group leader, Advan Develop Lab, Res & Eng Ctr, Neenah, 71-75, DIR RES & DEVELOP, MUNISING PAPER DIV, KIMBERLY-CLARK CORP, 75- Mem: Am Chem Soc. Res: Physical properties of polymers solutions; physical-chemical properties of fiber-elastomer composites. Mailing Add: 1807 N McDonald St Appleton WI 54911

WEBER, ROBERT HARRISON, b Wauseon, Ohio, Feb 8, 19; m 41; c 2. GEOLOGY. Educ: Ohio State Univ, BSc, 41; Univ Ariz, PhD(geol), 50. Prof Exp: Geologist, Shell Oil Co, 41-42; econ geologist, 50-66, SR GEOLOGIST, NMEX BUR MINES & MINERAL RESOURCES, 66- Concurrent Pos: Fac assoc, NMex Inst Mining & Technol, 66- Mem: Fel AAAS; Fel Geol Soc Am; Soc Econ Geol; Soc Am Archaeol; Am Quaternary Asn. Res: Mineral deposits; petrography and petrology of volcanic rocks; structural fabric of Southwest; Quaternary stratigraphy and geomorphology of the Southwest; early man in the New World. Mailing Add: Box 2046 Campus Sta Socorro NM 87801

WEBER, ROBERT L, b Dayton, Ohio, Mar 1, 13; m 49; c 4. PHYSICS. Educ: Yale Univ, BA, 34, MA, 36; Pa State Univ, PhD(physics), 38. Prof Exp: Asst, 36-38, from instr to asst prof, 38-47, asst supvr physics exten, US Off Educ, 41-42, ASSOC PROF PHYSICS, PA STATE UNIV, 47- Concurrent Pos: Mem fac, Inst High Sch Teachers, Yale Univ, 60-61. Mem: AAAS; Am Asn Physics Teachers; Brit Inst Physics; Am Asn Univ Prof. Res: Electron microscopy; temperature measurement; spectroscopy; texts and films in college physics. Mailing Add: Dept of Physics Davey Lab PA State Univ University Park PA 16802

WEBER, SHIRLEY MAE, b Portage La Prairie, Man, June 15, 26. NUTRITION. Educ: Univ Man, BSc, 48; Univ Minn, MS, 59; Cornell Univ, PhD(nutrit ed), 65. Prof Exp: Lectr foods & nutrit, Col Home Econ, Univ Sask, 55-56; lectr, 56-59, from asst prof to assoc prof, 59-71, PROF FOODS & NUTRIT, SCH HOME ECON, UNIV MAN, 71-, HEAD DEPT, 66- Concurrent Pos: Man comnr on Prairie Prov Cost Study Royal Comn, 66- Mem: Am Home Econ Asn; Can Home Econ Asn; Can Dietetic Asn; Can Coun Nutrit. Res: Human nutrition; dietary methodology; nutrition education. Mailing Add: Dept of Foods & Nutrit Univ of Man Winnipeg MB Can

WEBER, THOMAS ANDREW, b Tiffin, Ohio, June 8, 44; m 68; c 3. THEORETICAL CHEMISTRY. Educ: Univ Notre Dame, BS, 66; Johns Hopkins Univ, PhD(chem), 70. Prof Exp: MEM TECH STAFF CHEM COMPUT, BELL TEL LABS, INC, 70- Mem: AAAS; Am Phys Soc. Res: Computer simulation and modeling of chemical systems; quantum chemistry; molecular dynamics; kinetics of air pollution in the troposphere; polymer and photoresist modeling; minicomputer and data acquisition systems in chemistry. Mailing Add: Rm 1A365 Bell Labs 600 Mountain Ave Murray Hill NJ 07974

WEBER, THOMAS BYRNES, b Oklahoma City, Okla, Sept 1, 25; m 58; c 4. BIOCHEMISTRY. Educ: Okla State Univ, BS, 48; La State Univ, PhD(biochem), 54. Prof Exp: Res scientist, Animal Dis Res Ctr, Agr Res Ctr, USDA, 54-57; head biochem dept, US Navy Dent Res Inst, 57-59; head atmospheric & gas anal, US Air Force Sch Aerospace Med, 59-62; head adv res med develop, Beckman Instruments, Inc, 62-67; pres, Biosci Planning Inc, 67-69; pres, Med Patents, Inc, 69-74; PRES, WEBER DENT PROD, 74- Concurrent Pos: Life sci consult to indust, 58-62; consult, Life Sci & Instrumentation, US Govt, 63-; mem bd dirs, Metab Dynamic Found, 64- Mem: AAAS; Am Asn Clin Chemists; Am Chem Soc; Am Inst Aeronaut & Astronaut; fel Am Inst Chemists. Res: Life support systems; multiphasic screening methods; biochemical and physiological instrumentation; monitoring in closed ecological systems; body fluid analysis; automation of testing tools; handling of ethical pharmaceuticals. Mailing Add: 7723 Fay Ave La Jolla CA 92037

WEBER, VINSON M, b Swanton, Ohio, Sept 25, 12; m 39. DENTISTRY. Educ: Oberlin Col, AB, 34; Univ Mich, MA, 40; Western Reserve Univ, DDS, 46. Prof Exp: Teacher pub schs, Ohio, 34-40 & NY, 40-43; instr oper dent, Western Reserve Univ, 46-47; asst prof, 47-50, assoc prof dent, 50-67, head postgrad dept, 50-70, PROF DENT, DENT SCH, UNIV ORE, 67-, CHMN CONTINUING EDUC DEPT, 70- Concurrent Pos: Pvt pract, Vancouver, Wash. Mem: Am Soc Prev Dent; Am Acad Gen Pract; Am Soc Dent for Children. Mailing Add: Dept of Dent Univ of Ore Dent Sch Portland OR 97201

WEBER, WALDEMAR CARL, b Chicago, Ill, May 4, 37; m 69. MATHEMATICS. Educ: US Naval Acad, BSc, 59; Univ Ill, Urbana, MSc, 64, PhD, 68. Prof Exp: Asst prof, 68-72, ASSOC PROF MATH, BOWLING GREEN STATE UNIV, 72- Mem: Am Math Soc; Math Asn Am. Res: Geometry; applied mathematics. Mailing Add: Dept of Math Bowling Green State Univ Bowling Green OH 43403

WEBER, WALLACE RUDOLPH, b Murphysboro, Ill, Aug 1, 34; m 60; c 2. SYSTEMATIC BOTANY. Educ: Southern Ill Univ, BA, 56, MS, 59; Ohio State Univ, PhD(bot), 68. Prof Exp: Instr biol, Otterbein Col, 59-62; teaching assoc bot, Ohio State Univ, 63-67; asst prof, 67-70, ASSOC PROF LIFE SCI, SOUTHWEST MO STATE UNIV, 70- Mem: Bot Soc Am; Am Soc Plant Taxon; Int Asn Plant Taxon; Am Inst Biol Sci; Sigma Xi. Res: Floristics of Missouri Ozarks; biosystematic studies. Mailing Add: Dept of Life Sci Southwest Mo State Univ Springfield MO 65802

WEBER, WENDELL W, b Maplewood, Mo, Sept 2, 25; m 52; c 2. PHARMACOLOGY, PEDIATRICS. Educ: Cent Methodist Col, BA, 45; Northwestern Univ, PhD(phys chem), 50; Univ Chicago, MD, 59. Prof Exp: Asst prof chem, Univ Tenn, 49-51; opers res analyst, Off of Chief Chem Officer, Dept Army, Washington, DC, 51-55; from resident to chief resident pediat, Univ Calif, San Francisco, 60-62; NIH spec fel human genetics, Univ Col, Univ London, 62-63; from instr to prof pharmacol, Sch Med, NY Univ, 63-74; PROF PHARMACOL, UNIV MICH, ANN ARBOR, 74- Concurrent Pos: NIH spec fel biochem, Sch Med, NY Univ, 63-65; Health Res Coun NY career scientist award, 65-70 & 70-; mem pharmacol-toxicol comt, Nat Inst Gen Med Sci, 69-73. Mem: Am Chem Soc; Am Soc Pharmacol & Exp Therapeut; Am Soc Human Genetics; NY Acad Sci. Res: Physical chemistry; protein-small ion interactions; human genetics; cytogenetics; biochemical genetics and pharmacogenetics; drug metabolism and toxicity. Mailing Add: Dept of Pharmacol Univ of Mich Ann Arbor MI 48104

WEBER, WILFRIED T, b Rosenheim, Ger, Feb 10, 36; US citizen; m 60; c 2. PATHOLOGY, IMMUNOLOGY. Educ: Cornell Univ, BS, 59, DVM, 61; Univ Pa, PhD(path), 66. Prof Exp: From asst prof to assoc prof, 66-75, PROF PATH, SCH VET MED, UNIV PA, 75- Concurrent Pos: NIH res grants, 67-76. Mem: AAAS; Am Vet Med Asn; Reticuloendothelial Soc; Am Soc Exp Path. Res: Hematology; immunopathology; cancer research; tissue culture of lymphoid cells and macrophages. Mailing Add: Dept of Pathobiol Univ Pa Sch of Vet Med Philadelphia PA 19174

WEBER, WILLES HENRY, b Reno, Nev, Sept 22, 42; m 65. SOLID STATE PHYSICS. Educ: Calif Inst Technol, BS, 64; Univ Wis, MS, 65, PhD(physics), 68. Prof Exp: PRIN RES SCIENTIST ASSOC, ENG & RES STAFF, FORD MOTOR CO, 68- Mem: Am Phys Soc. Res: Low-energy electron diffraction; semiconductor physics, injection phenomena, instabilities, lasers, luminescence; plasma effects in metals and semiconductors; integrated optics, guided wave structures. Mailing Add: Eng & Res Staff Ford Motor Co Dearborn MI 48121

WEBER, WILLIAM ADOLPH, b St Louis, Mo, June 23, 11; m 40; c 3. MICROBIOLOGY, PHYSIOLOGY. Educ: St Louis Univ, PhB, 35, PhD, 54. Prof Exp: Eng designer, Maloney Elec Co, Mo, 42-43; eng designer, White Rodgers Elec Co, 43-44; org chemist, Ditzler Color Col Mich, 44-47; biochemist, Anheuser-Busch, Inc, 47-51, microbologist, 51-55; from asst prof to assoc prof bact, Duquesne Univ, 55-59; assoc prof bact & physiol, Quincy Col, 59-64, head dept biol, 60-64; chmn dept med, microbiol & pub health, Kans City Col Osteopath & Surg, 64-65; head dept biol, 65, dir allied health studies, 65-69, PROF BIOL, MIAMI-DADE JR COL, 69- Mem: Am Soc Microbiol; Am Physiol Soc. Res: Microbial metabolism versus pathogenic mechanisms. Mailing Add: 10742 N Kendall Dr Apt L-8 Miami FL 33156

WEBER, WILLIAM ALFRED, b New York, NY, Nov 16, 18; m 40; c 3. BOTANY. Educ: Iowa State Col, BS, 40; State Col Wash, MS, 42, PhD(bot), 46. Prof Exp: Asst prof, 46-70, PROF NATURAL HIST & CUR HERBARIUM, UNIV COLO, BOULDER, 70- Concurrent Pos: Cur, Lichen Herbarium, Am Bryol & Lichenological Soc, 54-70. Mem: AAAS; Bot Soc Am; Am Bryol & Lichenol Soc; Am Soc Plant Taxon; Int Asn Plant Taxon. Res: Lichen and bryophyte flora of Colorado, Galapagos Islands and Australasia; boreal and arctic elements in the Rocky Mountain flora. Mailing Add: Univ of Colo Mus Boulder CO 80302

WEBER, WILLIAM MARK, b Great Bend, Kans, Nov 24, 41; m 64; c 2. QUATERNARY GEOLOGY. Educ: Colo Col, BS, 63; Mont State Univ, MS, 65; Univ Wash, PhD(geol), 71. Prof Exp: Lab technician, Lincoln DeVore Testing Lab, 59-63, consult geologist, 65-67; instr geol, Colo Col, 67-68; asst prof, Minot State Col, 70-74; MEM STAFF, DEPT GEOL, UNIV MONT, 74- Concurrent Pos: Instr, Univ Colo, Cragmoor Campus, 66-68; co-dir exp col, Minot State Col, 71-72, NSF grant, 71-72. Mem: Geol Soc Am; Am Quaternary Asn. Res: Quaternary geology, especially as it applies to paleoenvironmental interpretation and to the analysis of present engineering and environmental problems. Mailing Add: Dept of Geol Univ of Mont Missoula MT 59801

WEBERG, BERTON CHARLES, b St Paul, Minn, Dec 23, 30; m 58; c 4. ORGANIC CHEMISTRY. Educ: Hamline Univ, BS, 54; Univ Colo, PhD(org chem), 58. Prof Exp: Sr res chemist, Abrasives Lab, Minn Mining & Mfg Co, 58-64; PROF CHEM, MANKATO STATE COL, 64- Mem: Am Chem Soc. Res: Chemistry of hindered ketones and vinyl ethers; polymer chemistry. Mailing Add: Dept of Chem Mankato State Col Mankato MN 56001

WEBERS, GERALD F, b Racine, Wis, Apr 14, 32; m 58; c 2. PALEONTOLOGY. Educ: Lawrence Univ, BS, 54; Univ Minn, MS, 61, PhD(geol), 64. Prof Exp: Res assoc geol, Univ Minn, 64-66; PROF GEOL, MACALESTER COL, 66- Mem: AAAS; Geol Soc Am; Am Geophys Union; Paleont Soc. Res: Evolution, taxonomy and paleoecology of Paleozoic invertebrate fossil faunas, especially trilobites, primitive mollusks and conodonts. Mailing Add: Dept of Geol Macalester Col St Paul MN 55101

WEBERS, VINCENT JOSEPH, b Racine, Wis, Apr 28, 22; m 49; c 6. ORGANIC CHEMISTRY. Educ: Univ Wis, BS, 43; Univ Minn, PhD(org chem), 49. Prof Exp: Lab asst, Univ Minn, 43-45; chemist, Wyeth Inst Appl Biochem, 46; lab asst, Univ Minn, 47; chemist, Cent Res Dept, 49-53, res chemist, Photo Prod Dept, 53-65, res assoc, 65-66, res assoc, Org Chem Dept, 66-67, res supvr, Photo Prod Dept, 67-73, RES ASSOC, PHOTO PROD DEPT, E I DU PONT DE NEMOURS & CO, INC, 73- Honors & Awards: V F Payne Award, Am Chem Soc, 65. Mem: Am Chem Soc; Sigma Xi; Soc Photog Scientist & Engrs. Res: Polymer chemistry; lithography; photochemistry; photopolymerization; sensitometry; photosensitive systems. Mailing Add: Bldg 352 Du Pont Exp Sta Wilmington DE 19898

WEBERT, HENRY S, b New Orleans, La, Jan 21, 29. BOTANY. Educ: Loyola Univ, Ill, BS, 51, MEd, 53; La State Univ, MS, 62, PhD(bot), 65. Prof Exp: Teacher high sch, Ill, 54-59; partic biol, NSF Acad Year Inst, Brown Univ, 59-60; res asst bot, La State Univ, 60-64; from asst prof to assoc prof, 64-70, PROF BIOL, NICHOLLS STATE UNIV, 70- Res: Plant growth and development. Mailing Add: Dept of Biol Sci Nicholls State Univ Thibodaux LA 70301

WEBSTER, BARBARA DONAHUE, b Winthrop, Mass, May 19, 29; m 56; c 1. PLANT MORPHOGENESIS. Educ: Univ Mass, BS, 50; Smith Col, MA, 52; Radcliffe Col, PhD, 57. Prof Exp: Instr plant sci, Vassar Col, 52-54; instr biol, Tufts Univ, 57-58; instr agron, 58-72, RES MORPHOLOGIST & LECTR AGRON & RANGE SCI, UNIV CALIF, DAVIS, 72- Concurrent Pos: Res investr, Brookhaven Nat Lab, 52-53 & 55; NSF fel, Purdue Univ, 58-59. Mem: Bot Soc Am; Am Soc Plant Physiol; Int Soc Plant Morphol. Res: Plant growth regulators and abscission; morphology and physiology of plant growth. Mailing Add: Dept of Agron & Range Sci Univ of Calif Davis CA 95616

WEBSTER, BURNICE HOYLE, b Leeville, Tenn, Mar 3, 10; m 39; c 3. THORACIC DISEASES. Educ: Vanderbilt Univ, BA, 36, MD, 40. Hon Degrees: DSc, Holy Trinity Col, 71. Prof Exp: Prof anat & path, Gupton-Jones Col Mortuary Sci, 36-42; intern & resident, St Thomas Hosp, 40-43; ASSOC INTERNAL MED & CHEST DIS, MED SCH, VANDERBILT UNIV, 43-; PROF MED & MED CONSULT, GUPTON SCH MORTUARY SCI, 44- Concurrent Pos: Vis physician, Chest Clin, Vanderbilt Univ, 43-48; consult, Vet Admin, 44-; dir, Nashville Chest Clin, 45-48; pres med staff, Protestant Hosp, 47; pres, Holy Trinity Col, 69-; lectr, St Thomas & Baptist Hosps. Mem: AAAS; fel Am Thoracic Soc; fel Am Geriat Soc; Am Soc Trop Med & Hyg; fel AMA. Res: Parasitic and fungus disease of the lung; changes in fungi following multiple animal passages; pathogenic mutation; thoracology. Mailing Add: 2315 Valley Brook Rd Nashville TN 37215

WEBSTER, CLYDE LEROY, JR, b Colorado Springs, Colo, Nov 15, 44; m 65; c 2. INORGANIC CHEMISTRY, GEOCHEMISTRY. Educ: Walla Walla Col, BSc, 68; Colo State Univ, PhD(chem), 72. Prof Exp: Consult chem, Accu-Labs Res Inc, 73-74; asst mgr chem, Instrument Anal Div, Com Testing & Eng, 74-75; ASST PROF CHEM, LOMA LINDA UNIV, 75- Concurrent Pos: Consult, Geosci Res Inst, 75- Mem: Am Chem Soc; Soc Appl Spectros. Res: Ore body genesis and processes of fossilization as related to the great deluge theory. Mailing Add: Dept of Chem Loma Linda Univ Riverside CA 92505

WEBSTER, DALE ARROY, b St Clair, Mich, Jan 11, 38; m 59; c 1. BIOCHEMISTRY. Educ: Univ Mich, BS, 60; Univ Calif, Berkeley, PhD(biochem), 65. Prof Exp: Res fel med, Mass Gen Hosp & Harvard Med Sch, 65-68; asst prof, 68-74, ASSOC PROF BIOL, ILL INST TECHNOL, 74- Mem: Soc Develop Biol; Am Soc Plant Physiol; Am Soc Biol Chemists. Res: Respiratory cytochromes in microbes. Mailing Add: Dept of Biol Ill Inst of Technol Chicago IL 60616

WEBSTER, DAVID DYER, b Grand Rapids, Minn, May 27, 18; m 46; c 2. MEDICINE, NEUROLOGY. Educ: Univ Minn, BS, 42, MD, 51; Am Bd Psychiat & Neurol, dipl, 60. Prof Exp: PROF NEUROL, MED SCH, UNIV MINN, MINNEAPOLIS, 57- Concurrent Pos: Staff neurologist & dir neurophysiol lab, Minneapolis Vet Hosp, 55- Mem: AAAS; AMA; Am Acad Neurol. Res: Movement disorders; medical electronics. Mailing Add: Minneapolis Vet Hosp 54th & 48th St Minneapolis MN 55417

WEBSTER, DAVID HENRY, b Berwick, NS, Oct 29, 34; m 60; c 2. HORTICULTURE. Educ: Acadia Univ, BSc, 54, MSc, 55; Univ Calif, Davis, PhD(plant physiol), 65. Prof Exp: Mem res staff, NS Res Found, 58-62; RES SCIENTIST, CAN DEPT AGR, 65- Mem: Am Soc Hort Sci. Res: Fruit tree physiology; nutrition; soil physical properties. Mailing Add: Can Dept of Agr Res Sta Kentville NS Can

WEBSTER, DAVID LEE, b Augusta, Ga, Aug 2, 43; m 69. ANTHROPOLOGY, ARCHAEOLOGY. Educ: Univ Minn, BA, 65, MA, 67, PhD(anthrop), 72. Prof Exp: ASST PROF ANTHROP, PA STATE UNIV, UNIVERSITY PARK, 72- Concurrent Pos: Lectr, Macalester Col, 71, lectr anthrop, Univ Minn, 72. Mem: Soc Am

Archaeol. Res: Mesoamerican archaeology; old world prehistory; origins and effects of food production; warfare and socio-political evolution. Mailing Add: Dept Anthrop 519 Soc Sci Bldg Pa State Univ University Park PA 16802

WEBSTER, DOUGLAS B, b Fond du Lac, Wis, Jan 14, 34; m 55; c 2. OTORHINOLARYNGOLOGY. Educ: Oberlin Col, AB, 56; Cornell Univ, PhD(zool), 60. Prof Exp: Fel psychobiol, Calif Inst Technol, 60-62; from asst prof to prof biol, NY Univ, 62-73, actg chmn dept, 67-68; PROF OTORHINOLARYNGOL & ANAT, LA STATE UNIV MED CTR, NEW ORLEANS, 73-, CLIN PROF AUDIOL & SPEECH PATH, SCH ALLIED HEALTH, 74- Concurrent Pos: NIH grants, NY Univ, 63-65, NY Univ & La State Univ Med Ctr, 65-75 & La State Univ Med Ctr, 73-74; ed, Am Soc Zoologists Newslett, 69- Mem: AAAS; Am Asn Anatomists; Soc Neurosci; Soc Study Evolution; Am Soc Zoologists (secy, 69-). Res: Morphology, behavior and physiology of hearing in vertebrates. Mailing Add: Dept of Otorhinolaryngol La State Univ Med Ctr New Orleans LA 70119

WEBSTER, DWIGHT ALBERT, b Manchester, Conn, Feb 2, 19; m 41; c 3. BIOLOGY. Educ: Cornell Univ, BS, 40, PhD(fishery biol), 43. Prof Exp: Asst aquatic biologist, Conn Bd Fisheries & Game, 37-41; instr limnol, 41-45, asst prof limnol & fisheries, 46-49, assoc prof fishery biol, 49-57, head dept, 67-71, PROF FISHERY BIOL, CORNELL UNIV, 57- Mem: AAAS; Am Fisheries Soc; Am Soc Limnol & Oceanog. Res: Population dynamics of freshwater fish, especially the role of hatchery fish in management; ecology and life history of fishes, especially in the Finger Lakes and Adirondacks of New York State. Mailing Add: Dept of Natural Resources Fernow Hall Cornell Univ Ithaca NY 14850

WEBSTER, EDWARD WILLIAM, b London, Eng, Apr 12, 22; nat US; m 50, 61; c 6. MEDICAL PHYSICS. Educ: Univ London, BSc, 43, PhD(elec eng), 46. Prof Exp: Res engr, Eng Elec Co, 45-49; guest researcher, Mass Inst Technol, 49-50, radiation physicist, 50-51; lectr elec eng & nuclear energy, Queen Mary Col, Univ London, 52-53; PHYSICIST, MASS GEN HOSP, 53-, PROF RADIOL, HARVARD MED SCH, 75- Concurrent Pos: London County Coun Blair traveling fel, Mass Inst Technol, 49-50; from asst to asst prof, Harvard Med Sch, 53-67, assoc clin prof physics in radiol, 67-69, assoc prof radiol, 69-75; USPHS spec fel, 65-66; lectr med radiation physics, Sch Pub Health, Harvard Univ, 71- Consult, Mass Eye & Ear Infirmary, 57- & Int Atomic Energy Agency, 60-61; mem comt radiol, Nat Acad Sci, 62-68; consult, Children's Hosp Med Ctr, Boston, 62-; mem, Nat Coun Radiation Protection & Measurements, 64-; consult, WHO, 65 & 67; consult, USPHS, 61-63; mem radiol health study sect, Dept Health, Educ & Welfare, 69-72, radiol training grant comt, NIH, 69- & adv comt med uses of isotopes, US Atomic Energy Comn, 71- Mem: Am Asn Physicists in Med (pres, 63-64); Radiol Soc NAm; fel Am Col Radiol; Health Physics Soc; Soc Nuclear Med (trustee, 73-). Res: Radiological physics; application of radiation, radioisotopes and electronic methods to medical diagnosis and therapy; radiation protection. Mailing Add: Dept of Radiol Mass Gen Hosp Boston MA 02114

WEBSTER, ELEANOR RUDD, b Cleveland, Ohio, Oct 11, 20. ORGANIC CHEMISTRY, HISTORY OF SCIENCE. Educ: Wellesley Col, AB, 42; Mt Holyoke Col, MA, 44; Radcliffe Col, MA, 50, PhD(chem), 52. Prof Exp: Asst chem, Mt Holyoke Col, 42-44; chemist synthetic org res, Eastman Kodak Co, 44-47; from instr to assoc prof, 52-67, dean of freshman & sophomores, 56-60, chmn dept chem, 64-67, dir inst chem, 64-72, dir continuing educ, 69-70, PROF CHEM, WELLESLEY COL, 67- Concurrent Pos: Fulbright grant, Belg, 51-52; NSF sci fac fel, 60-61. Mem: Am Chem Soc; Hist Sci Soc; Brit Soc Hist Sci. Res: Physical organic chemistry; dissemination of science, the public's understanding since 1850. Mailing Add: Dept of Chem Wellesley Col Wellesley MA 02181

WEBSTER, FERRIS, b St Boniface, Man, Aug 7, 34; m 57. PHYSICAL OCEANOGRAPHY. Educ: Univ Alta, BSc, 56, MSc, 57; Mass Inst Technol, PhD(geophys), 61. Prof Exp: Res asst phys oceanog, 59-62, res assoc, 62-63, asst scientist, 63-65, assoc scientist, 65-70, chmn dept phys oceanog, 71-73, SR SCIENTIST, WOODS HOLE OCEANOG INST, 70-, ASSOC DIR RES, 73- Concurrent Pos: Asst prof, Mass Inst Technol, 66-68. Mem: AAAS; Am Geophys Union. Res: Ocean currents; oceanic variability; oceanographic data processing and time-series analysis. Mailing Add: Woods Hole Oceanog Inst Woods Hole MA 02543

WEBSTER, GARY DEAN, b Hutchinson, Kans, Feb 15, 34; m 64; c 2. GEOLOGY, PALEONTOLOGY. Educ: Univ Okla, BS, 56; Univ Kans, MS, 59; Univ Calif, Los Angeles, PhD(geol), 66. Prof Exp: Geologist, Amerada Petrol Corp, 56-57; geologist, Belco Petrol Corp, 60; geologist, Shell Oil Co, 63; lectr phys geol, Calif Lutheran Col, 63-64; mus scientist, Univ Calif, Los Angeles, 64-65; asst prof geol & paleont, San Diego State Col, 65-68; asst prof, 68-70, ASSOC PROF GEOL, WASH STATE UNIV, 70- Concurrent Pos: Mem, Am Geol Inst, Int Field Inst, Spain, 71. Mem: AAAS; Soc Econ Paleont & Mineral; Paleont Soc; Brit Palaeont Asn. Res: Late Paleozoic paleontology and stratigraphy, especially crinoids and conodonts. Mailing Add: Dept of Geol Wash State Univ Pullman WA 99163

WEBSTER, GEORGE CALVIN, b South Haven, Mich, July 17, 24; m 60; c 2. BIOCHEMISTRY. Educ: Western Mich Univ, BS, 48; Univ Minn, MS, 49, PhD(biol), 52. Prof Exp: Res fel, Calif Inst Technol, 52-55; from assoc prof to prof biochem, Ohio State Univ, 55-61; vis prof enzyme chem, Univ Wis, 61-65; chief chemist, Aerospace Serv Div, Cape Kennedy, Fla, 65-70; PROF BIOL SCI & HEAD DEPT, FLA INST TECHNOL, 70- Concurrent Pos: USPHS spec res fel, 61-63; Am Heart Asn estab investr, 63-65. Mem: AAAS; Am Soc Biol Chem; Am Chem Soc; Am Soc Cell Biol; Brit Biochem Soc. Res: Molecular biology of aging, nitrogen metabolism and protein synthesis. Mailing Add: Dept of Biol Sci Fla Inst of Technol Melbourne FL 32901

WEBSTER, GILBERT THEODORE, b Dalton, Nebr, May 8, 11; m 38; c 4. AGRONOMY. Educ: Univ Nebr, BSc, 32, MSc, 37; Iowa State Col, PhD(plant breeding & genetics), 49. Prof Exp: Exten specialist agron, Univ Nebr, 37-43, asst agronomist, 43-48, assoc prof agron & agronomist, 48-54; head dept, 54-66, PROF AGRON, UNIV KY, 66- Mem: Am Soc Agron. Res: Cytogenetics of grasses. Mailing Add: Dept of Agron Univ of Ky Lexington KY 40506

WEBSTER, GORDON RITCHIE, b Kindersley, Sask, Jan 7, 22; m 50; c 2. SOIL CHEMISTRY. Educ: Univ BC, BSA, 49, MSA, 51; Univ Ore, PhD(soils), 58. Prof Exp: Res officer soil sci, Can Dept Agr, 49-60; assoc prof, 60-70, PROF SOIL SCI, UNIV ALTA, 70- Mem: Can Soc Soil Sci. Res: Equilibrium ion activity ratios in soils; ammonium fixation in soils; poor alfalfa growth second time in rotation; yield as related to several independent variables including weather. Mailing Add: Dept of Soil Sci Univ of Alta Edmonton AB Can

WEBSTER, GRADY LINDER, b Ada, Okla, Apr 14, 27; m 56; c 1. PLANT TAXONOMY. Educ: Univ Tex, BA, 47, MA, 49; Univ Mich, PhD(bot), 54. Prof Exp: Lectr trop bot, Harvard Univ, 53, NSF fel biol, 53-55, instr bot, 55-58; asst prof plant sci, Purdue Univ, 58-62, assoc prof biol sci, 62-66; PROF BOT, UNIV CALIF, DAVIS, 66- Concurrent Pos: Guggenheim fel bot, State Univ Utrecht, 64-65. Mem: Am Soc Plant Taxon; Soc Study Evolution; Bot Soc Am; Asn Trop Biol; Int Asn Plant Taxon. Res: Evolution and systematics of vascular plants, especially Euphorbiaceae; vegetational and floristic plant geography; pollination ecology. Mailing Add: Dept of Bot Univ of Calif Davis CA 95616

WEBSTER, HAROLD FRANK, b Buffalo, NY, June 25, 19; m 51; c 3. PHYSICS. Educ: Univ Buffalo, BA, 41, MA, 44; Cornell Univ, PhD(physics), 53. Prof Exp: Mem staff, Radiation Lab, Mass Inst Technol, 43-45; asst, Cornell Univ, 45-51; PHYSICIST, RES LAB, GEN ELEC CO, 51- Concurrent Pos: US del gen assembly, Int Sci Radio Union, 60. Honors & Awards: Baker Award, Inst Elec & Electronics Eng, 58. Mem: Am Phys Soc; Inst Elec & Electronics Eng. Res: Thermionic emission from single crystal surfaces; metal surface wetting; electron beam dynamics; cesium plasma; energy conversion; ultraviolet sensors; solid state science; microelectronics. Mailing Add: Gen Elec Corp Res & Develop Ctr PO Box 43 Schenectady NY 12301

WEBSTER, HARRIS DUANE, b Lansing, Mich, Jan 14, 20; m 43; c 2. VETERINARY PATHOLOGY. Educ: Mich State Col, DVM, 43, MS, 51. Prof Exp: Vet, US Regional Poultry Res Lab, East Lansing, 44 & 46-47; gen pract, Mich, 47-48; instr animal path, Mich State Col, 48-52; RES SCIENTIST PATH & TOXICOL, UPJOHN CO, 52- Mem: Am Vet Med Asn; Am Col Vet Path; Int Acad Pathologists. Res: Toxicologic, experimental pathologic studies and laboratory animal health consultation and service. Mailing Add: 715 Jenks Blvd Kalamazoo MI 49007

WEBSTER, HENRY DEFOREST, b New York, NY, Apr 22, 27; m 51; c 5. NEUROLOGY, NEUROPATHOLOGY. Educ: Amherst Col, BA, 48; Harvard Med Sch, MD, 52; Am Bd Psychiat & Neurol, dipl & cert neurol, 59. Prof Exp: Intern & asst resident med, Harvard Serv, Boston City Hosp, 52-54; asst resident & resident neurol, Mass Gen Hosp, 54-56, res fel neuropath, 56-58; asst instr & assoc neurol, Harvard Med Sch, 58-66, asst prof neuropath, 66; assoc prof neurol, Sch Med, Univ Miami, 66-69, prof, 69; CHIEF SECT CELLULAR NEUROPATH, NAT INST NEUROL & COMMUN DISORDERS & STROKE, 69- Concurrent Pos: Prin investr, Nat Inst Neurol Dis & Stroke grants, 62-69, partic neuroanat vis scientist prog, 64-65; assoc neurologist & asst neuropathologist, Mass Gen Hosp, 63-66; secy gen, VIII Int Cong Neuropath. Mem: Am Asn Neuropath (vpres, 76-77); Am Soc Cell Biol; Asn Res Nerv & Ment Dis; Am Acad Neurol; Am Neurol Asn. Res: Experimental neuropathology utilizing electron microscopy, especially the formation and breakdown of myelin. Mailing Add: Nat Inst Neurol & Commun Disorders & Stroke Bldg 36 Bethesda MD 20014

WEBSTER, HULON LEX, agronomy science, see 12th edition

WEBSTER, ISABELLA MARGARET, b Chicago, Ill, Jan 22, 11. PUBLIC HEALTH. Educ: Northwestern Univ, BA, 32; Univ Minn, PhD(org chem), 36; Woman's Med Col Pa, MD, 47; Univ Liverpool, dipl, 65; Univ Hawaii, MPH, 69. Prof Exp: Asst chem, Univ Minn, 32-36, instr org chem, 36-37; intern, Mary Immaculate Hosp, NY, 47-48; intern obstet & gynec, Georgetown Univ Hosp, 48; mem med staff, Holy Family Hosp, Mandar, India, 48-54; doctor chg, Archbishop Attipetty Jubilee Mem Hosp, S India, 54-57; chief med servs, Kokofu Leprosarium, Ghana Leprosy Serv, 58-68; consult pub health, Atat Hosp, Ethiopia, Nangina Hosp, Nangina, Kenya, Virika Hosp, Ft Portal, Uganda & Lower Shire Mobile Health Unit, Chiromo, Malawi, 69-71; dist superior, Med Mission Sisters, Ghana, 71-74, MEM CENT MED MISSION SISTERS, ROME, 74- Res: Organic chemistry; bacteriology; obstetrics and gynecology; leprosy. Mailing Add: Suore Medico Missionarie Via de Villa Troili 32 00163 Rome Italy

WEBSTER, JACKSON DAN, b Tacoma, Wash, Feb 26, 19; m 44; c 3. PARASITOLOGY, ORNITHOLOGY. Educ: Whitworth Col, Wash, BSc, 39; Cornell Univ, MSc, 41; Rice Inst, PhD(parasitol), 47. Prof Exp: Field researcher ornith, Univ Alaska, 40 & 46; assoc prof biol, Jamestown Col, 47-49; PROF BIOL, HANOVER COL, 49- Mem: Soc Parasitol; assoc Wilson Ornith Soc; assoc Cooper Ornith Soc; Am Ornith Union. Res: Septematics, distribution and populations of birds; systematics of tapeworms. Mailing Add: Dept of Biol Hanover Col Hanover IN 47243

WEBSTER, JACKSON ROSS, b Brigham City, Utah, May 3, 45; m 68; c 1. ECOLOGY. Educ: Wabash Col, BA, 67; Univ Ga, PhD(ecol, zool), 75. Prof Exp: ASST PROF ZOOL, VA POLYTECH INST & STATE UNIV, 75- Mem: Ecol Soc Am; Am Soc Limnol & Oceanog; Soc Comput Simulation. Res: Nutrient and energy dynamics in ecosystems, especially stream ecosystems; ecosystem modeling; litter decomposition in lakes and streams. Mailing Add: Dept of Biol Va Polytech Inst & State Univ Blacksburg VA 24061

WEBSTER, JAMES ALBERT, b Mineola, NY, June 2, 28; m 51; c 4. ORGANIC CHEMISTRY. Educ: Col Wooster, BA, 51; Univ Pittsburgh, MS, 56. Prof Exp: Res chemist, Res & Eng Div, Monsanto Chem Co, 56-61, res chemist, Monsanto Res Corp, 61-66, sr res chemist, 66-71, RES GROUP LEADER, MONSANTO RES CORP, 71- Mem: Am Chem Soc. Res: Silicone chemistry; polymer synthesis; fluorine chemistry. Mailing Add: 7611 Eagle Creek Dr Dayton OH 45459

WEBSTER, JAMES ALLAN, b Lincoln, Nebr, May 1, 39; m 67; c 2. ENTOMOLOGY. Educ: Univ Ky, BS, 61, MS, 64; Kans State Univ, PhD(entom), 68. Prof Exp: RES ENTOMOLOGIST, AGR RES SERV, USDA, 68- Concurrent Pos: Asst prof, Dept Entom, Mich State Univ, 68-74, assoc prof, 74- Mem: Entom Soc Am; Am Soc Agron. Res: Insect resistance in grain and forage crops. Mailing Add: Dept of Entom Mich State Univ East Lansing MI 48824

WEBSTER, JAMES RANDOLPH, JR, b Chicago, Ill, Aug 25, 31; m 54; c 3. PULMONARY PHYSIOLOGY. Educ: Northwestern Univ, Chicago, BS, 53, MS & MD, 56. Prof Exp: Asst prof, 67-71, ASSOC PROF MED, MED SCH, NORTHWESTERN UNIV, CHICAGO, 71- Concurrent Pos: USPHS fel pulmonary physiol, Med Sch, Northwestern Univ, Chicago, 62-64; assoc dir inhalation ther, Chicago Wesley Mem Hosp, 65-66, dir pulmonary function lab, 66- Mem: Am Thoracic Soc; Am Fedn Clin Res; Am Col Physicians. Res: Diseases of the chest. Mailing Add: Dept of Med Northwestern Univ Med Sch Chicago IL 60611

WEBSTER, JOHN H, b Belleville, Ont, Dec 17, 28; m 53; c 3. RADIOTHERAPY. Educ: Queen's Univ, Ont, MD, 55. Prof Exp: Sr cancer res radiologist, Roswell Park Mem Inst, 59-62, assoc cancer res radiologist, 62-63, assoc chief cancer res radiologist, 63-64, chief cancer res radiologist, 64-74; PROF THERAPEUT RADIOL & CHMN DEPT, McGILL UNIV, 74- Concurrent Pos: USPHS grants; therapeut radiologist-in-chief, Montreal Gen Hosp, Royal Victoria Hosp, Montreal Children's Hosp & Jewish Gen Hosp, 74-; sr consult radiotherapist, Montreal Neurol Hosp, 74- Mem: Am Col Radiol; Am Soc Therapeut Radiol; Can Med Asn; Soc Chmn Acad Radiation Oncol Progs; Can Asn Radiologists. Res: Experimental radiotherapy and allied fields; oncologically related research; cancer patient care systems. Mailing Add: Royal Victoria Hosp Therapeut Radiol Dept 687 Pine Ave W Montreal PQ Can

WEBSTER, JOHN MALCOLM, b Wakefield, Eng, May 5, 36. BIOLOGY, PARASITOLOGY. Educ: Univ London, BSc & ARCS, 58, PhD(zool) & DIC, 62. Prof Exp: Sci officer nematol, Rothamsted Exp Sta, Eng, 61-66; sci officer parasitol,

Can Dept Agr Res Inst, Belleville, 66-67; assoc prof, 67-71, PROF BIOL, SIMON FRASER UNIV, 71-, CHMN DEPT BIOL SCI, 74- Mem: AAAS; Soc Nematol; Soc Invert Path; Entom Soc Can; Brit Soc Exp Biol. Res: Physiology of host-parasite relationships, especially those of nematode parasites of plants and insects; economic applications of plant nematology and invertebrate pathology; ultrastructure of host response. Mailing Add: Dept of Biol Sci Simon Fraser Univ Burnaby Vancouver BC Can

WEBSTER, JOHN ROBERT, b Riverdale, Calif, May 5, 16; m 43; c 3. FOOD SCIENCE. Educ: Calif State Univ, Fresno, AB, 39. Prof Exp: DIR RES & DEVELOP, LINDSAY OLIVE GROWERS, 40- Mem: Am Chem Soc; Inst Food Technologists. Res: Improving nutrition, improving quality, and decreasing process cost of olives, cherries, pimiento and pickled peppers. Mailing Add: 386 Alameda St Lindsay CA 93247

WEBSTER, JOHN THOMAS, b Fond du Lac, Wis, Sept 12, 27; m 55; c 3. STATISTICS. Educ: Ripon Col, BA, 51; Purdue Univ, MS, 55; NC State Col, PhD(statist), 60. Prof Exp: Statistician, Westinghouse Elec Corp, 55-57; asst prof statist, Bucknell Univ, 60-62; ASSOC PROF STATIST, SOUTHERN METHODIST UNIV, 62- Mem: Am Statist Asn; Biomet Soc; Am Soc Qual Control. Res: Design and analysis of experiments. Mailing Add: Dept of Statist Southern Methodist Univ Dallas TX 75275

WEBSTER, LESLIE T, JR, b New York, NY, Mar 31, 26; m 55; c 4. PHARMACOLOGY. Educ: Harvard Med Sch, MD, 48; Amherst Col, BA, 64; Am Bd Internal Med, dipl, 57. Prof Exp: Demonstr med, Sch Med, Case Western Reserve Univ, 55-57, instr, 57-60, from sr instr to asst prof biochem, 58-66, asst prof med, 60-70, from asst prof to assoc prof pharmacol, 66-70; PROF PHARMACOL & CHMN DEPT, MED & DENT SCHS, NORTHWESTERN UNIV, CHICAGO, 70- Concurrent Pos: Nat Vitamin Found Wilder fel, 56-59; sr investr, USPHS, 59-61, res career develop award, 61-69; mem gastroenterol & nutrit training grants comt, Nat Inst Arthritis & Metab Dis, 65-69. Mem: Am Col Physicians; Am Soc Clin Invest; Am Soc Biol Chemists; Am Soc Pharmacol & Exp Therapeut. Res: Enzymology; clinical pharmacology; drug metabolism. Mailing Add: Dept of Pharmacol Northwestern Univ Med Sch Chicago IL 60611

WEBSTER, MARION ELIZABETH, b Ottawa, Ont, Apr 9, 21; nat US; m 51. BIOCHEMISTRY, PHARMACOLOGY. Educ: Fla State Univ, BS, 42; Georgetown Univ, MS, 47, PhD(biochem), 50. Prof Exp: Res chemist prod phosphoric acid, Southern Phosphate Corp, 42-43; assoc chemist, Colgate-Palmolive-Peet Co, 43-44; jr chemist, Bur Entom & Plant Quarantine, USDA, 44-45; chemist, US Indust Chem Co, 46-47; biochemist, Walter Reed Army Inst Res, 47-57; BIOCHEMIST, NAT HEART & LUNG INST, 58- Concurrent Pos: Nat Res Coun Brazil & NIH spec fel, Sao Paulo Sch Med, 68-69. Mem: Am Chem Soc; Soc Exp Biol & Med; Am Soc Pharmacol & Exp Therapeut; Am Physiol Soc; NY Acad Sci. Res: Kallikrines, kinins, SRS-A; isolation; mechanism of activation and metabolism; role in inflammation and lung diseases; relationship to complement, histamine, coagulation factors. Mailing Add: Bethesda MD

WEBSTER, MERRITT SAMUEL, b Cheyney, Pa, June 5, 09; m 36; c 3. MATHEMATICS. Educ: Swarthmore Col, AB, 31; Univ Pa, AM, 33, PhD(math), 35. Prof Exp: Asst instr math, Univ Pa, 31-33; instr, Univ Nebr, 35-38; from instr to prof, 38-75, EMER PROF MATH, PURDUE UNIV, WEST LAFAYETTE, 75- Mem: Am Math Soc; Math Asn Am. Res: Orthogonal polynomials; interpolation. Mailing Add: 225 Connolly St West Lafayette IN 47906

WEBSTER, ORRIN JOHN, b Arkansas City, Kans, June 26, 13; m 36; c 3. AGRONOMY. Educ: Univ Nebr, BSc, 34, MSc, 40; Univ Minn, PhD(plant breeding & genetics), 50. Prof Exp: Agronomist, Soil Conserv Serv, USDA, 35-37, Dryland Agr Div, 36-43 & Cereal Corps Res Br, Agr Res Serv, 43-63, dir-coordr major cereal proj, Orgn For African Unity, 63-71, proj leader corn, sorghum & millet res, Fed Exp Sta, PR, 71-74; ADJ PROF AGRON, UNIV ARIZ, 74- Concurrent Pos: Adv corn & sorghum breeding, Govt Nigeria, Liberia, Sudan & Africa, 51; secy sorghum res comt, USDA, 53-67; mem comt preserv indigenous strains sorghum, Nat Acad Sci, 61- Mem: Am Soc Agron. Res: Plant breeding of sorghum; genetics and cytogenetics of sorghum. Mailing Add: Dept of Agron Univ of Ariz Tucson AZ 85721

WEBSTER, OWEN WRIGHT, b Devils Lake, NDak, Mar 25, 29; m 53; c 5. ORGANIC CHEMISTRY. Educ: Univ NDak, BS, 51; Pa State Univ, PhD(chem), 55. Prof Exp: RES CHEMIST, E I DU PONT DE NEMOURS & CO, INC, 55- Mem: Am Chem Soc; Sigma Xi. Res: Synthetic organic chemistry; adamantanes; cyanocarbons; hydrogen cyanide. Mailing Add: 2106 Navaro Rd Wilmington DE 19803

WEBSTER, PAUL DANIEL, III, b Mt Airy, NC, Apr 26, 30; m 57; c 2. MEDICINE, GASTROENTEROLOGY. Educ: Univ Richmond, BS, 52; Bowman Gray Sch Med, MD, 56. Prof Exp: Fel med, Univ Minn, 60-63; fel gastroenterol, Sch Med, Duke Univ, 63-66, asst prof med, Med Ctr, 67-68; assoc prof, 68-71, PROF MED, MED COL GA, 71- Concurrent Pos: USPHS fel, 65-66; chief med serv, Vet Admin Hosp, Augusta, Ga, 73. Mem: Am Gastroenterol Asn; Am Physiol Soc; Am Inst Nutrit; Am Soc Clin Invest. Res: Hormonal control of pancreatic protein synthesis; pancreatic structure and function; cancer of pancreas. Mailing Add: Dept of Med Med Col of Ga Augusta GA 30902

WEBSTER, PETER JOHN, b Cheshire, Eng, May 30, 42; m 65; c 2. DYNAMIC METEOROLOGY. Educ: Royal Melbourne Inst Technol, fel, 64; Mass Inst Technol, PhD(meteorol), 71. Prof Exp: Meteorologist, Commonwealth Bur Meteorol, Australia, 61-65; res scientist meteorol, Commonwealth Sci & Indust Res Orgn, Australia, 71-72; vis asst prof meteorol, Univ Calif, Los Angeles, 72-73; ASST PROF METEOROL, UNIV WASH, 73- Concurrent Pos: Consult, Jet Propulsion Lab, Calif Inst Technol, 72-73 & Environ Protection Agency, Seattle, Wash, 74; mem, Monsoon Exp Nat Comt, Nat Acad Sci, 74- Mem: Am Meteorol Soc; Am Geophys Union. Res: Large scale atmospheric motions; large scale air and sea interaction; dynamics of monsoon flows. Mailing Add: Dept of Atmospheric Sci Univ of Wash Seattle WA 98195

WEBSTER, PORTER GRIGSBY, b Wheatley, Ky, Nov 25, 29; m 61. MATHEMATICS. Educ: Georgetown Col, BA, 51; Auburn Univ, MS, 56, PhD(math), 61. Prof Exp: PROF MATH, UNIV SOUTHERN MISS, 61- Mem: Am Math Soc; Math Asn Am. Mailing Add: Southern Sta Box 226 Dept Math Univ of Southern Miss Hattiesburg MS 39401

WEBSTER, REX N, b Lansing, Mich, Dec 15, 11. BOTANY. Educ: Butler Univ, AB, 33; Johns Hopkins Univ, PhD(morphol), 38. Prof Exp: From instr to assoc prof biol, Middlebury Col, 38-50, actg chmn dept, 43-45; assoc prof bot, 50-62, PROF BOT, BUTLER UNIV, 62- Mem: AAAS; Sigm Res: Cryptogamic botany; morphology and taxonomy of vascular plants; bacteriology; freshwater algae and fungi. Mailing Add: Dept of Bot Butler Univ Indianapolis IN 46208

WEBSTER, RICHARD CURTIS, b Chicago, Ill, Nov 21, 12; m 41. HISTOLOGY, CYTOLOGY. Educ: Univ Louisville, AB, 38, MS, 40; Univ Kans, PhD(anat), 49. Prof Exp: Instr anat, Univ Louisville, 42-46; instr, Univ Kans, 46-48; asst prof, 49-57, ASSOC PROF ANAT, MED CTR, IND UNIV-PURDUE UNIV, INDIANAPOLIS, 57-, COORDR DENT TEACHING ANAT, 72- Mem: Am Asn Anatomists; hon mem Biol Stain Comn; NY Acad Sci. Res: Histochemistry in cancer research. Mailing Add: Dept of Anat Ind Univ-Purdue Univ Med Ctr Indianapolis IN 46207

WEBSTER, RICHARD MILROY, b Buffalo, NY, Feb 3, 41, m 64; c 2. ANATOMY. Educ: Colgate Univ, BA, 63; State Univ NY Buffalo, PhD(anat), 70. Prof Exp: Instr, 69-72, ASST PROF ANAT, LA STATE UNIV MED CTR, NEW ORLEANS, 72- Mem: AAAS; Pan-Am Asn Anatomists; Am Soc Mammalogists; Soc Cryobiol. Res: Urogenital morphology; mammalian hibernation and hypothermia; Chiroptera; histophysiology; histochemistry. Mailing Add: Dept of Anat La State Univ Med Ctr New Orleans LA 70112

WEBSTER, ROBERT EDWARD, b New Haven, Conn, May 31, 38; m 60; c 3. MOLECULAR BIOLOGY, MOLECULAR GENETICS. Educ: Amherst Col, BA, 59; Duke Univ, PhD(microbiol), 65. Prof Exp: NSF fel genetics, Rockefeller Univ, 65-66, asst prof molecular biol, 66-70; ASSOC PROF BIOCHEM, MED CTR, DUKE UNIV, 70- Mem: AAAS; Am Soc Biol Chem. Res: Protein synthesis; phage genetics and morphogenesis; membrane synthesis. Mailing Add: Dept of Biochem Med Ctr Duke Univ Durham NC 27710

WEBSTER, ROBERT G, b Balclutha, NZ, July 5, 32; Australian citizen; m; c 3. VIROLOGY, IMMUNOLOGY. Educ: Otago Univ, BS, 55, MS, 57; Australian Nat Univ, PhD(microbiol), 62. Prof Exp: Virologist, NZ Dept Agr, 58-59; Fulbright scholar, Dept Epidemiol, Sch Pub Health, Univ Mich, Ann Arbor, 62-63; res fel, Dept of Microbiol, John Curtin Med Sch, Australian Nat Univ, 64-66, fel, 66-67; assoc mem, Lab Immunol, St Jude Children's Res Hosp, 68-69; assoc prof, 68-74, PROF MICROBIOL, UNIV TENN CTR HEALTH SCI MEMPHIS, 74-; MEM LABS VIROL & IMMUNOL, ST JUDE CHILDREN'S RES HOSP, 69- Concurrent Pos: Coordr, US-USSR Joint Comt Health Coop, Ecol of Human Influenza & Animal Influenza Rels to Human Infection, 74- Mem: Am Soc Microbiol; Am Asn Immunologists. Res: Structure and immunology of influenza viruses. Mailing Add: Labs of Virol & Immunol St Jude Children's Res Hosp Memphis TN 38101

WEBSTER, ROBERT K, b Solomonville, Ariz, Jan 15, 38; m 59; c 3. PHYTOPATHOLOGY. Educ: Utah State Univ, BS, 61; Univ Calif, Davis, PhD(plant path), 66. Prof Exp: Res assoc plant path, NC State Univ, 66; from asst prof to assoc prof, 66-75, PROF PLANT PATH, UNIV CALIF, DAVIS, 75- Mem: Mycol Soc Am; Am Phytopath Soc; Am Soc Naturalists. Res: Genetics of plant pathogenic fungi; field crop diseases. Mailing Add: Dept of Plant Path Univ of Calif Davis CA 95616

WEBSTER, STEPHEN RUSSELL, b Centralia, Wash, Oct 27, 42; m 65. FOREST SOILS. Educ: Wash State Univ, BS, 65; Ore State Univ, MS, 67; NC State Univ, PhD(soils), 74. Prof Exp: RES SCIENTIST, FORESTRY RES CTR, WEYERHAEUSER CO, CENTRALIA, 72- Mem: Soil Sci Soc Am; Soc Am Foresters. Res: Investigations of forest fertilization and forest tree nutrition. Mailing Add: 658 State Hwy 60 Chehalis WA 98532

WEBSTER, TERRY R, b Hamilton, Ohio, Feb 10, 38; m 64; c 1. BOTANY, PLANT MORPHOLOGY. Educ: Miami Univ, BA, 60; Univ Sask, MA, 62, PhD(bot), 65. Prof Exp: Asst prof bot, 65-71, ASSOC PROF BIOL, UNIV CONN, 71- Mem: Am Fern Soc (secy, 73-76); Bot Soc Am. Res: Morphology of the genus Selaginella. Mailing Add: Biol Sci Group Univ of Conn Storrs CT 06268

WEBSTER, THOMAS G, b Topeka, Kans, Jan 23, 24; m 48; c 3. PSYCHIATRY, ACADEMIC ADMINISTRATION. Educ: Ft Hays Kans State Col, AB, 46; Wayne State Univ, MD, 49. Prof Exp: Commonwealth fel col ment health, Mass Inst Technol, 54-56; Nat Inst Ment Health career teacher grant, Harvard Med Sch, 56-58, instr psychiat, 59-63; training specialist, Psychiat Training Br, Nat Inst Ment Health, 63-66, chief, Continuing Educ Br, 66-72; PROF PSYCHIAT & BEHAV SCI & CHMN DEPT, GEORGE WASHINGTON UNIV, 72- Concurrent Pos: Dir, Presch Retard Children's Prog Greater Boston, 58-62. Mem: AAAS; Am Psychiat Asn; Am Col Psychiat; Am Acad Child Psychiat; Group Advan Psychiat. Res: Psychiatric education; child development; psychopathology; learning process; program evaluation in fields of manpower and training. Mailing Add: Dept of Psychiat & Behav Sci George Washington Univ Med Ctr Washington DC 20037

WEBSTER, VICTOR STUART, b Manquoketa, Iowa, Nov 4, 08; m 31; c 2. ORGANIC CHEMISTRY. Educ: Univ Iowa, AB, 30, MS, 31, PhD(org chem), 33. Prof Exp: Chemist, State Planning Bd, Iowa, 33-35; instr sci, Minn State Teachers Col, Moorhead, 35-36; from instr to prof & head dept, 36-74, EMER PROF CHEM, SDAK STATE UNIV, 74- Mem: Am Chem Soc. Res: Oxidation of unsaturated acids; condensation of phenols with anhydrides; color test for amines; preparation of alcohols by the Grignard method. Mailing Add: Dept of Chem SDak State Univ Brookings SD 57006

WEBSTER, WILLIAM JOHN, JR, b New York, NY, May 3, 43. PLANETARY SCIENCES. Educ: Univ Rochester, BS, 65; Case Western Reserve Univ, PhD(astron), 69. Prof Exp: Res assoc astron, Nat Radio Astron Observ, Va, 68-70; Nat Acad Sci-Nat Res Coun resident res assoc, Solar Physics Lab, 70-71, staff scientist, Meteorol & Earth Sci Lab, 71-74, STAFF SCIENTIST, GEOPHYS BR, GODDARD SPACE FLIGHT CTR, 74- Concurrent Pos: Data anal scientist, Pac Plat Motion Exp, Goddard Space FLight Ctr, 75- Mem: Am Astron Soc; Am Inst Aeronaut & Astronaut. Res: Radio interferometry; gaseous nebulae; radio spectroscopy; continuum mapping of radio sources; microwave observation of earth; precision geodesy by radio interferometry; earthquake prediction. Mailing Add: Geophys Br Code 922 Goddard Space Flight Ctr Greenbelt MD 20771

WEBSTER, WILLIAM MERLE, b Warsaw, NY, June 13, 25; m 47; c 2. PHYSICS. Educ: Union Univ, NY, BS, 45; Princeton Univ, PhD(elec eng), 54. Prof Exp: Res engr, Labs, 46-54, mgr adv develop, RCA Semiconductor & Mat Div, 54-59, dir electronic res lab, RCA Labs, 59-66, staff vpres mat & devices res, 66-68, VPRES LABS, RCA LABS, RCA CORP, 68- Honors & Awards: Ed Award, Inst Radio Eng, 53. Mem: Fel Inst Elec & Electronics Eng. Res: Solid state and gaseous electronics; electron physics. Mailing Add: RCA Labs Princeton NJ 08540

WEBSTER, WILLIAM PHILLIP, b Mt Airy, NC, Apr 26, 30; m 52; c 4. DENTISTRY. Educ: Univ NC, Chapel Hill, BS, 56, DDS, 59, MS, 68. Prof Exp: Instr prosthodont, 59-61, asst prof periodont & oral path, 61-65, asst prof periodont, oral path & path, 65-67, trainee, 67-68, assoc prof path, 69-71, assoc prof dent ecol, 69-73, PROF PATH & DENT ECOL, SCHS MED & DENT, UNIV NC, CHAPEL HILL, 72-; DIR, HOSP DENT SERV, NC MEM HOSP, 68- Concurrent Pos: Res assoc, Sch Med, Univ NC, Chapel Hill, 59-65; mem med adv bd, Nat Hemophilia Found, 62-72; mem med adv comt, Vet Admin Hosp, Fayetteville, NC & sci adv comt, World Fedn Hemophilia. Mem: Am Acad Forensic Sci; Am Dent Asn; Am Soc

Exp Path; NY Acad Sci; Int Soc Thrombosis & Haemostasis. Res: Transplantation; biology; blood coagulation; pathology. Mailing Add: Dept of Path Univ of NC Sch of Med Chapel Hill NC 27514

WEBSTER, WILLIAM WALLACE, b Lincoln, Nebr, Aug 22, 10; m 35; c 2. ORAL SURGERY. Educ: Univ Nebr, BS & DDS, 34; Am Bd Oral Surg, dipl, 51. Prof Exp: Intern, Boston Univ, 34-35; intern, Col Physicians & Surgeons, Columbia Univ, 43; from instr to assoc prof, 47-57, chmn dept, 50-73, PROF ORAL SURG, COL DENT, UNIV NEBR-LINCOLN, 57-, PROF, COL MED, 53- Concurrent Pos: Attend oral surgeon, Lincoln Gen & St Elizabeth Hosps, 46-, consult, 46-; attend oral surgeon & chmn dept oral surg, Bryan Mem Hosp, 49- Mem: Am Soc Oral Surgeons; Am Dent Asn; fel Am Col Dent; Am Acad Oral Path. Res: Maxillofacial surgery. Mailing Add: Dept of Oral Surg Univ of Nebr Col of Dent Lincoln NE 65803

WECHSLER, HARRY C, b Iasi, Rumania, Dec 24, 19; m 49. POLYMER CHEMISTRY. Educ: Victoria Univ, Eng, BS, 39; Hebrew Univ, Israel, MS, 45; Polytech Inst Brooklyn, PhD(chem), 48. Prof Exp: Res dir, Am Monomer Corp, 53-55; gen mgr polyvinyl chloride dept, Borden Chem Co, 55-58, vpres, 58-61, vpres & gen mgr, Thermoplastics Div, 61-65, group vpres plastics & fabrics, 65-66, exec vpres, 66-68, vpres, Borden Inc & pres chem div, 68-71, EXEC VPRES & CHIEF OPERATING OFF, BEATRICE CHEM DIV, BEATRICE FOODS CO, 72- Mem: Am Chem Soc; Mfg Chem Asn; Soc Plastics Eng. Res: Kinetics and mechanism of polymerization; ionic polymerizations; copolymerizations. Mailing Add: 514 Tumbling Hawk Acton MA 01720

WECHSLER, MARTIN T, b New York, NY, July 27, 21; m 53; c 3. MATHEMATICS. Educ: Queen's Col, NY, BS, 42; Univ Mich, MA, 46, PhD(math), 52. Prof Exp: Physicist, Nat Bur Standards, 42-45; instr math, Wayne State Univ, 51-52 & Princeton Univ, 52-53; from instr to asst prof, Wash State Univ, 53-56; from inst to assoc prof, 56-70, PROF MATH, WAYNE STATE UNIV, 70-, CHMN DEPT, 71-, ASSOC DEAN, COL LIBERAL ARTS, 75- Mem: Am Math Soc. Res: Topology; groups of homeomorphisms. Mailing Add: Off of the Dean Col Liberal Arts Wayne State Univ Detroit MI 48202

WECHTER, MARGARET ANN, b Chicago, Ill, Sept 30, 35. ANALYTICAL CHEMISTRY, RADIOCHEMISTRY. Educ: Mundelein Col, BS, 62; Iowa State Univ, PhD(anal chem), 67. Prof Exp: Fel chem, Purdue Univ, 67-68, asst prof chem, Calumet Campus, 69-73; ASSOC PROF CHEM, SOUTHEASTERN MASS UNIV, 73- Concurrent Pos: Res assoc, Ames Lab, Iowa State Univ, 70- Mem: Am Chem Soc; Sigma Xi. Res: Development, analytical applications and electrochemistry of the tungsten bronze electrodes; activation analysis. Mailing Add: Dept of Chem Southeastern Mass Univ North Dartmouth MA 02747

WECHTER, WILLIAM JULIUS, b Louisville, Ky, Feb 13, 32; m 52; c 3. ORGANIC CHEMISTRY. Educ: Univ Ill, AB, 53, MS, 54; Univ Calif, Los Angeles, PhD, 57. Prof Exp: Asst chem, Univ Ill, 54; asst org chem, Univ Calif, Los Angeles, 55-56; res assoc, Dept Chem, 57-68, RES HEAD HYPERSENSITIVITY DIS, UPJOHN CO, 68- Concurrent Pos: Vis scholar, Depts Chem & Med Microbiol, Stanford Univ, 67-68; vchmn, Gordon Conf Med Chem, 72, chmn, 73; sect ed biol, Annual Reports Med Chem, 72-74; adj prof biochem, Kalamazoo Col, 74- Mem: AAAS; Am Asn Cancer Res; Am Chem Soc; Transplantation Soc. Res: Synthesis, mechanisms and isotopic labeling of nucleic acid components; immuno-supressive and anti-cancer drugs; pharmacological control of cell mediated immunity; gene therapy. Mailing Add: Hypersensitivity Dis Res Upjohn Co Kalamazoo MI 49001

WECK, FRIEDRICH JOSEF, b Puettlingen, Ger, Nov 10, 18; nat US; m 56; c 2. ORGANIC CHEMISTRY, PHYSICAL CHEMISTRY. Educ: Univ Bonn, MS, 50; Univ Saarland, PhD(natural sci), 54. Prof Exp: Assistantship, Univ Saarland, 51-52; res assoc, Nat Ctr Sci Res, France, 52-55; head res & develop lab, Saar Water Asn, Ger, 55-57; from group leader new prod & processes sect, to res specialist, Am Potash & Chem Corp, 57-64; dir chem res, Garett Res & Develop Co, 64; PRES, F J WECK CO, 64- Concurrent Pos: Consult, Union Saarbruecken, 54-57. Honors & Awards: Bronze Medal, Chem Soc France, 57. Mem: Am Chem Soc; Ger Chem Soc; fel Am Inst Chem. Res: Organic-inorganic synthesis; mono-polymers; photosensitive materials; environmental safety; food and water; liquid and solid ion exchangers; saline mineral chemistry; extraction-crystallization of salts; product and process development and testing. Mailing Add: F J Weck Co 14859 E Clark Ave Industry CA 91745

WECKEL, KENNETH GRANVILLE, b Canton, Ohio, Oct 9, 05; m 39. FOOD SCIENCE. Educ: Univ Wis, PhD(dairy indust), 35. Prof Exp: Asst prof dairy indust, 35-40, assoc prof, 41-46, PROF FOOD SCI, UNIV WIS-MADISON, 46- Concurrent Pos: Partic, Nat Conf Interstate Milk Shippers; mem, Res & Develop Assocs, US Army. Honors & Awards: Borden Award, Am Dairy Sci Asn, 38; Educ-Indust Award, Int Asn Milk, Food & Environ Sanit, 75. Mem: Am Dairy Sci Asn; Inst Food Technologists; Int Asn Milk, Food & Environ Sanit; Nat Mastitis Coun; Am Asn Candy Technol. Res: Food products manufacture. Mailing Add: Dept of Food Sci Univ of Wis Madison WI 53706

WECKER, STANLEY C, b New York, NY, Apr 29, 33; m 64. VERTEBRATE ECOLOGY. Educ: City Col New York, BS, 55; Univ Mich, MS, 57, PhD(zool), 62. Prof Exp: Asst prof biol, Hofstra Univ, 62-63; from instr to asst prof, 63-70, ASSOC PROF BIOL, CITY COL NEW YORK, 70- Concurrent Pos: Res grants, Sigma Xi, 63 & NSF, 64-68; US Dept Health, Educ & Welfare Off Educ grant & staff ecologist, Environ Educ Proj, North Westchester-Putnam Coop Educ Serv, NY, 71-72; spec consult, NY State Educ Dept, 72; mem, Environ Defense Fund. Honors & Awards: Award, Am Soc Mammal, 61. Mem: Fel AAAS; Animal Behav Soc; Ecol Soc Am; Am Soc Mammal; Am Soc Zool. Res: Behavioral ecology of deer mice, especially habitat orientation; sensory modalities, geographic variation, genetics and development of habitat selection response. Mailing Add: Dept of Biol City Col of New York New York NY 10031

WECKOWICZ, THADDEUS EUGENE, b Iskorst, USSR, Oct 10, 18; Can citizen; m 66. PSYCHIATRY. Educ: Univ Edinburgh, MB & ChB, 45; Univ Leeds, DPM, 52; Univ Sask, PhD(psychol), 62. Prof Exp: Registr psychiat, Univ Leeds, 53-54, sr registr, Bolton Group Hosps, Eng, 54-55; sr resident, Hosp, Univ Sask, 55-56; res psychiatrist, Weyburn Ment Hosp, Sask, 56-59; res assoc psychiat & sessional lectr psychol, 62, asst prof, 62-67, ASSOC PROF PSYCHIAT & PSYCHOL, UNIV ALTA, 67-, MEM STAFF, CTR ADVAN STUDY THEORET PSYCHOL, 66- Concurrent Pos: Fel, Hosp, Univ Sask, 55-56. Mem: Can Psychiat Asn. Res: Psychological and biological aspects of schizophrenia; psychotropic drug research, particularly hallucinogenic drugs; psychopharmacology; multivariate study of depression. Mailing Add: Dept of Psychol & Psychiat Univ of Alta Edmonton AB Can

WECKWERTH, VERNON ERVIN, b Herman, Minn, Apr 29, 31; m 55; c 5. BIOSTATISTICS, HOSPITAL ADMINISTRATION. Educ: Univ Minn, BS, 54, MS, 56, PhD(biostatist), 63. Prof Exp: Teaching asst biostatist, Univ Minn, 54-56, instr, 56-58; head res & statist, Am Hosp Asn, 58-60; from lectr to assoc prof, 60-69, PROF HOSP & HEALTH CARE ADMIN, COORDR CONTINUING HOSP EDUC & PROF FAMILY PRACT, MED SCH, UNIV MINN, MINNEAPOLIS, 69- Concurrent Pos: Assoc dir, Hosp Res & Educ Trust, 58-60; mem adv comt to Nat Ctr for Health Servs Res & Develop, 69-; chmn bd, Minn Systs Res, Inc. Mem: Am Statist Asn; Biomet Soc; Am Hosp Asn; Am Pub Health Asn. Res: Continuing education of health care workers including professionals; health care delivery systems research; teaching research and statistics in health care. Mailing Add: 1260 Mayo Bldg Univ of Minn Minneapolis MN 55455

WEDBERG, HALE LEVERING, b Redlands, Calif, Feb 10, 33; m 59; c 2. PLANT TAXONOMY, EVOLUTION. Educ: Calif State Col Los Angeles, BA, 56; Univ Calif, Los Angeles, PhD(plant sci), 62. Prof Exp: Instr zool, 59-60, from asst prof to assoc prof bot, 60-67, PROF BOT, CALIF STATE UNIV, SAN DIEGO, 67- Mem: AAAS; Soc Study Evolution. Res: Chromosomal variation in natural populations of Clarkia Williamsonii; cytotaxonomy of Camissonia, section Holostigma. Mailing Add: Dept of Bot Calif State Univ San Diego CA 92115

WEDBERG, STANLEY EDWARD, b Bridgeport, Conn, Aug 28, 13; m 41; c 2. MICROBIOLOGY. Educ: Univ Conn, BS, 37; Yale Univ, PhD(bact), 40. Prof Exp: Instr immunol, Sch Med, Yale Univ, 40-41; from instr to assoc prof bact, 41-59, head dept, 55-66, prof, 59-69, fac coordr educ TV, 67-69, prof biol, 68-69, EMER PROF BIOL, UNIV CONN, 69-; BIOLOGIST, SOUTHESTERN COL, CALIF, 69- Concurrent Pos: Consult bacteriologist, Windham Community Mem Hosp, Conn, 46-57. Mem: AAAS; Am Soc Microbiol; Am Pub Health Asn; Am Acad Microbiol. Res: Microbial thermogenesis; hemagglutination of castor beans; bacterial capsule staining; insect microbiology; effect of antihistamines on antibody production of rabbits; sterilization using germicidal gases; educational television. Mailing Add: 3361 Ullman St San Diego CA 92106

WEDDELL, JAMES BLOUNT, b Evanston, Ill, Apr 29, 27. SPACE PHYSICS. Educ: Drew Univ, AB, 49; Northwestern Univ, MS, 51, PhD(physics), 53. Prof Exp: Asst physics, Northwestern Univ, 49-53; res engr, Westinghouse Elec Corp, 53-57; sr scientist, Martin-Marietta Co, 57-62; MEM TECH STAFF, ROCKWELL INT CORP, 62- Mem: Am Phys Soc; Am Geophys Union; Am Inst Aeronaut & Astronaut; Sci Res Soc Am. Res: Space experiments-spacecraft integration; solar physics; magnetosphere; nuclear reactions; radiation shielding. Mailing Add: 936 S Peregrine Pl Anaheim CA 92806

WEDDING, BRENT (M), b Walnut, Ill, May 3, 36; m 69; c 1. EXPERIMENTAL PHYSICS. Educ: Hamilton Col, AB, 58; Univ Ill, Urbana, MS, 61, PhD(physics), 67. Prof Exp: Lectr physics, Southern Ill Univ, Carbondale, 62-63; SR PHYSICIST, CORNING GLASS WORKS, 67-, RES SUPVR, 75- Mem: Am Phys Soc; Optical Soc Am; Am Asn Physics Teachers. Res: Thermal properties of glass and related materials. Mailing Add: Tech Staffs Div Corning Glass Works Corning NY 14830

WEDDING, RANDOLPH TOWNSEND, b St Petersburg, Fla, Nov 6, 21; m 43; c 2. BIOCHEMISTRY. Educ: Univ Fla, BSA, 43, MS, 47; Cornell Univ, PhD(plant physiol), 50. Prof Exp: Tech asst, Agr Exp Sta, Univ Fla, 41-43, 46-47; teaching asst bot, Cornell Univ, 47-49, instr plant physiol, 49-50; jr plant physiologist, 50-52, asst plant physiologist, 52-58, assoc plant physiologist & lectr, 58-61, assoc prof biochem & assoc biochemist, 61-63, chmn dept, 66-75, PROF BIOCHEM & BIOCHEMIST, UNIV CALIF, RIVERSIDE, 63- Concurrent Pos: Agr Res Coun fel, Oxford Univ, 59-60, Orgn Econ Coop & Develop sr sci fel, 64; sr sci fel, Natural Environ Res Coun, 71-72. Mem: Am Soc Plant Physiol; Am Soc Biol Chem; Brit Soc Exp Biol; Brit Biochem Soc; Scand Soc Plant Physiol. Res: Metabolic control; mechanisms of enzyme action and control. Mailing Add: Dept of Biochem Univ of Calif Riverside CA 92502

WEDDLE, ORVILLE HARVEY, biophysics, see 12th edition

WEDDLETON, RICHARD FRANCIS, b Boston, Mass, Oct 10, 39; m 62; c 3. POLYMER CHEMISTRY. Educ: Mass Inst Technol, BS, 61; Ind Univ, Bloomington, MS, 63, PhD(org chem), 65. Prof Exp: Chemist, Mat & Processes Lab, Gen Elec Co, 65-73; MGR INSULATION ENG, NAT ELEC COIL DIV, McGRAW-EDISON CO, 73- Mem: Am Chem Soc; Inst Elec & Electronics Eng. Res: Development and testing of insulation system and corona control systems for use in electrical rotating machinery. Mailing Add: Nat Elec Coil Co 800 King Ave Columbus OH 43212

WEDEGAERTNER, DONALD K, b Kingsburg, Calif, Sept 9, 36; m 58; c 2. ORGANIC CHEMISTRY. Educ: Univ Calif, Berkeley, BS, 58; Univ Ill, Urbana, PhD(org chem), 62. Prof Exp: Res asst org chem, Iowa State Univ, 62-63; from asst prof to assoc prof, 63-71, PROF ORG CHEM, UNIV OF THE PAC, 71- Mem: Am Chem Soc; The Chem Soc. Res: Stereochemistry; reaction mechanisms; chemistry of allenes. Mailing Add: Dept of Chem Univ of the Pac Stockton CA 95211

WEDEL, ARNOLD MARION, b Lawrence, Kans, Jan 31, 28; m 54; c 3. MATHEMATICS. Educ: Bethel Col, Kans, AB, 47; Univ Kans, MA, 48; Iowa State Col, PhD(math), 51. Prof Exp: PROF MATH, BETHEL COL, KANS, 51- Mem: Math Asn Am; Am Math Soc. Res: Hypergeometric series and volterra transforms. Mailing Add: Dept of Math Bethel Col North Newton KS 67117

WEDEL, RICHARD GLENN, b Oakes, NDak, Apr 22, 41; m 67. ELECTROCHEMISTRY. Educ: NDak State Univ, BS, 63; Pa State Univ, PhD(chem), 68. Prof Exp: SR RES CHEMIST, GEN MOTORS RES LABS, 68- Mem: Am Chem Soc; Electrochem Soc; Am Electroplaters Soc; Soc Automotive Engrs; Sigma Xi. Res: Nature of adhesion between materials; surface chemistry of electrodes. Mailing Add: Electrochem Dept 12 Mile & Mound Gen Motors Tech Ctr Warren MI 48090

WEDEMEYER, ROBERT E, b Audubon Co, Iowa, May 13, 26; m 48; c 4. PHYSICAL CHEMISTRY. Educ: Western Ky State Col, BS, 50; Vanderbilt Univ, AM, 52, PhD(chem), 53; Harvard Univ, MPA, 67. Prof Exp: Res radiochemist, Vanderbilt Univ, 51-53; res chemist, Bjorksten Res Lab, 53-55; TECH ADMINR, OAK RIDGE OPERS OFF, US ENERGY RES & DEVELOP ADMIN, 55- Res: Polarography of organic compounds; vapor-liquid equilibria; radiation stability of ion-exchange resins; high vacuum synthesis of boron hydrides; uranium chemistry and metallurgy; boron isotopes. Mailing Add: 105 Dartmouth Circle Oak Ridge TN 37830

WEDERMEYER, GARY ALVIN, b Fromberg, Mont, Oct 15, 35; m 57; c 2. FISHERIES, BIOCHEMISTRY. Educ: Univ Wash, BS, 57, MS, 63, PhD(fisheries), 65. Prof Exp: Asst prof, 67-73, ASSOC PROF, COL FISHERIES, UNIV WASH, 73-, BIOCHEMIST, FISH PHYSIOL, WESTERN FISH DIS LAB, US DEPT INTERIOR, 65- Concurrent Pos: Instr, Seattle Community Col, Wash, 66- Mem: Am Inst Fishery Res Biol; Am Fisheries Soc; Soc Exp Biol & Med. Res: Biochemistry and physiology of fishes; pollution, disease, aquaculture. Mailing Add: Western Fish Dis Lab Dept Interior Bldg 204 Sand Point Naval Air Sta Seattle WA 98115

WEDGWOOD, RALPH JOSIAH PATRICK, b Eng, May 25, 24; nat US; m 43; c 4.

PEDIATRICS. Educ: Harvard Med Sch, MD, 47; Am Bd Pediat, dipl, 55. Prof Exp: Res fel pediat, Harvard Med Sch, 49-51; sr instr pediat & biochem, Western Reserve Univ, 53-57, asst prof pediat & prev med, Sch Med, 57-62; assoc prof, 62-63, chmn dept, 62-72, PROF PEDIAT, SCH MED, UNIV WASH, 63- Concurrent Pos: Spec consult & mem gen clin res ctr comt, NIH, 62-66, mem nat adv res resources coun, 66-70; mem sci adv comn, Nat Found, 69-; mem sci adv bd, St Jude's Children's Res Hosp, Memphis, 70- Mem: Am Asn Immunologists; Am Rheumatism Asn; Heberden Soc; Am Pediat Soc; Infectious Dis Soc Am. Res: Immunobiology; natural resistance factors; infectious and rheumatic diseases in children; general pediatrics. Mailing Add: Dept of Pediat RD-20 Univ of Wash Sch of Med Seattle WA 98195

WEDIN, WALTER F, b Frederic, Wis, Nov 28, 25; m 55; c 2. AGRONOMY. Educ: Univ Wis, BS, 50, MS, 51, PhD(agron, soils), 53. Prof Exp: Wis Alumni Res Found assistantship agron, Univ Wis, 50-53, proj assoc hort, 53, asst prof, 53-57; res agronomist, Forage & Range Res Br, Crops Res Div, Agr Res Serv, USDA, Minn, 57-61; assoc prof, 61-64, PROF AGRON, IOWA STATE UNIV, 64-, DIR, WORLD FOOD INST, 73- Concurrent Pos: Asst prof, Inst Agr, Univ Minn, 59-61; consult, US Agency Int Develop-Iowa State Univ, Uruguay, 64; consult, Coop States Res Serv, USDA, 64, 70, 72, prin agronomist, 68. Honors & Awards: Merit Cert, Am Forage & Grassland Coun, 72. Mem: AAAS; fel Am Soc Agron; Crop Sci Soc Am; Am Soc Animal Sci. Res: Evaluation of forage crops, especially utilization as pasture, hay or silage; techniques involved in measurement of forage nutritive value; animal intake and digestibility; grassland development, international. Mailing Add: Dept of Agron Iowa State Univ Ames IA 50011

WEDLER, FREDERICK CHARLES OLIVER, b Philadelphia, Pa, June 10, 41; m 67; c 2. BIOCHEMISTRY. Educ: Univ NC, Chapel Hill, BS, 63; Northwestern Univ, Evanston, PhD(biochem), 68. Prof Exp: Molecular Biol Inst fel biochem, Univ Calif, Los Angeles, 68-70; asst prof, 70-73, ASSOC PROF CHEM & BIOCHEM, RENSSELAER POLYTECH INST, 75- Concurrent Pos: Petrol Res Fund grant, Rensselaer Polytech Inst, 71-, NSF res grant, 72-74, 75-; NASA res grant, 74, 75. Mem: AAAS; Am Chem Soc; Am Soc Biol Chemists. Res: Enzymology; kinetics and mechanisms of action of key regulatory enzymes; development of new probe systems for protein structures; thermophilic proteins. Mailing Add: Dept of Chem Rensselaer Polytech Inst Troy NY 12181

WEDLICK, HAROLD LEE, b Detroit, Mich, Feb 26, 36; m 59; c 2. HEALTH PHYSICS, ENVIRONMENTAL HEALTH. Educ: Wayne State Univ, BS, 61, MS, 63; Am Bd Health Physics, cert, 74. Prof Exp: Chemist, City of Detroit Health Dept, Mich, 61-63; res assoc environ health, Univ Mich, 64-66; assoc prof radiol sci, Lowell Technol Inst, 66-75; RES ASSOC, OCCUPATIONAL & ENVIRON SAFETY DEPT, BATTELLE NORTHWEST LABS, 76- Mem: Health Physics Soc; Conf Radiol Health; Am Nuclear Soc. Res: Transfer mechanisms and measurements of radionuclides in the environment. Mailing Add: Battelle Northwest PO Box 999 Richland WA 99352

WEDMAN, ELWOOD EDWARD, b Harper, Kans, Aug 22, 22; m 46; c 3. VETERINARY MICROBIOLOGY. Educ: Kans State Univ, DVM, 46; Univ Minn, MPH, 57, PhD(microbiol), 64. Prof Exp: Pvt pract, 45-47, 49-50; vet in charge Mex-US Campaign Eradication Foot & Mouth Dis, Mex, 47-49; vet in charge Wichita Union Stockyards, Animal Dis Eradication Div, Agr Res Serv, USDA, DC, 52-54, staff vet, Lab Serv, 58-61, res assoc, Univ Minn & USDA, 54-58, chief vet diag serv, Animal Dis Labs, USDA, Iowa, 61-64; assoc dir vet med res inst, Col Vet Med, Iowa State Univ, 64-72; HEAD DEPT VET MED, ORE STATE UNIV, 72- Concurrent Pos: Consult, USDA, 63-64, 65-66 & 67 & Pan-Am Health Orgn, 65, 67; mem, Nat Conf Vet Lab Diagnosticians. Honors & Awards: Cert Merit, USDA, 61, 63 & 66. Mem: Am Vet Med Asn; US Animal Health Asn; Am Col Vet Microbiol; Conf Res Workers Animal Dis. Res: Microbiology and epidemiology of animal diseases. Mailing Add: Dept of Vet Med Ore State Univ Corvallis OR 97331

WEDMID, GEORGE YURI, b Stockach, Ger, Feb 18, 48; US citizen; m 74. BIOLOGICAL CHEMISTRY. Educ: Rutgers Univ, BS, 69, PhD(biochem), 74. Prof Exp: Res asst lipid chem, Rutgers Univ, HORMEL FEL, HORMEL INST, UNIV MINN, AUSTIN, 74- Mem: Am Oil Chemists Soc; Int Soc Magnetic Resonance. Res: Phospholipids and analogs, especially nuclear magnetic resonance and enzyme studies of their structure and function. Mailing Add: Hormel Inst Univ of Minn 801 NE 16th Ave Austin MN 55912

WEE, ELIZABETH LIU, b Manila, Philippines, Nov 7, 41; m 67; c 1. BIOPHYSICAL CHEMISTRY, POLYMER CHEMISTRY. Educ: Mapua Inst Technol, BS, 63; Northeastern Univ, MS, 66; Univ Minn, Minneapolis, PhD(chem), 71. Prof Exp: Res assoc biopolymer & phys chem, Univ Minn, 71; RES ASSOC BIOCHEM CONTRACTILE PROTEINS, CHILDREN'S HOSP RES FOUND, 72- Mem: Am Chem Soc. Res: Immunological and biochemical studies of normal muscle development and hereditary diseases of muscle; mechanism of palate shelf rotation. Mailing Add: Children's Hosp Res Found Elland & Bethesda Aves Cincinnati OH 45229

WEED, HOMER CLYDE, b Sun City, Kans, Mar 30, 20; m 58. PHYSICAL CHEMISTRY. Educ: Univ Ariz, BS, 42; Ohio State Univ, MS, 48, PhD(chem), 57. Prof Exp: Assoc develop chemist, Tenn Eastman Corp, 42-46; asst chem, Ohio State Univ, 46-47, 48, 49, res found, 47-48, 49, 53-57, metall, 50-53; CHEMIST, INORG MAT DIV, LAWRENCE LIVERMORE LAB, UNIV CALIF, 57- Mem: AAAS; Am Chem Soc; Am Inst Chem. Res: High temperature chemistry; electrical properties and high pressure mechanical properties of rock; distribution of radionuclides between earth materials and ground water. Mailing Add: Inorg Mats Div Box 808 Univ Calif Lawrence Livermore Lab Livermore CA 94550

WEED, JOHN CONANT, b Lake Charles, La, July 7, 12; m 39; c 4. GYNECOLOGY, INFERTILITY. Educ: Tulane Univ, BS, 33, MD, 36, MS, 40; Am Bd Obstet & Gynec, dipl, 46. Prof Exp: Instr anat, 39-41, assoc clin prof, 53-60, CLIN PROF OBSTET & GYNEC, SCH MED, TULANE UNIV, 60- Concurrent Pos: Mem staff, Ochsner Clin, 45-; chmn dept obstet & gynec, Ochsner Found Hosp, 63-73; bd trustees, Alton Ochsner Med Found. Mem: AMA; Am Fertil Soc (pres, 73-74); Am Col Obstet & Gynec; Am Col Surgeons. Mailing Add: 1514 Jefferson Hwy New Orleans LA 70121

WEED, MARK BARG, plant physiology, see 12th edition

WEED, ROBERT I, b Bridgeport, Conn, Nov 2, 28; m 53; c 5. HEMATOLOGY. Educ: Yale Univ, BS, 49, MD, 52. Prof Exp: From instr to sr instr med, 58-61, from asst prof to assoc prof, 62-68, from asst prof to assoc prof radiation biol, 62-68, PROF MED, RADIATION BIOL & BIOPHYS, SCH MED, UNIV ROCHESTER, 68-, HEAD HEMAT UNIT, 67- Concurrent Pos: Am Cancer Soc fel, 57-58; USPHS fel, 58-61. Mem: Fel Am Col Physicians; Am Soc Clin Invest; Am Physiol Soc; Am Fedn Clin Res; Int Soc Hemat. Res: Membrane physiology of normal and pathologic blood cells. Mailing Add: Dept of Med Univ of Rochester Med Ctr Rochester NY 14642

WEED, STERLING BARG, b Salt Lake City, Utah, Mar 25, 26; m 49; c 2. SOIL CHEMISTRY. Educ: Brigham Young Univ, AB, 51; NC State Col, MS, 53, PhD(soils), 55. Prof Exp: Asst prof soils, Cornell Univ, 55-56; from asst prof to assoc prof, 56-70, PROF SOILS, N C STATE UNIV, 70- Mem: Soil Sci Soc Am; Clay Minerals Soc; Int Soc Soil Sci; Int Clay Minerals Soc. Res: Clay mineralogy; clay-organic interactions. Mailing Add: Dept of Soil Sci NC State Univ Raleigh NC 27607

WEEDEN, JUDITH S, zoology, see 12th edition

WEEDEN, ROBERT BARTON, b Fall River, Mass, Jan 8, 33; m 59. ZOOLOGY. Educ: Univ Mass, BSc, 53; Univ Maine, MSc, 55; Univ BC, PhD(zool), 59. Prof Exp: Asst, Univ Maine, 53-55; asst, Univ BC, 55-58; instr zool, Wash State Univ, 58-59; RES BIOLOGIST, ALASKA DEPT FISH & GAME, 59-, PROF WILDLIFE MGT, UNIV ALASKA, 68- Mem: Arctic Inst NAm; Cooper Ornith Soc; Wildlife Soc; Am Ornith Union. Res: Avian ecology, particularly of Tetraonidae and alpine-arctic environments. Mailing Add: Box 80425 College AK 99701

WEEDON, GENE CLYDE, b Washington, DC, June 11, 36; m 61; c 3. POLYMER CHEMISTRY. Educ: Va Polytech Inst & State Univ, BS, 60. Prof Exp: Res chemist, Esso Res & Eng Co, 60-66; res chemist, 66-71, RES MGR, FIBERS DIV, ALLIED CHEM CORP, 71- Concurrent Pos: Mem, Carpet Adv Comt, Man-Made Fibers Producers Asn, 75- Mem: Am Chem Soc; Res Soc Am. Res: Development of modified nylon polymers for fiber applications. Mailing Add: Fibers Div Tech Ctr PO Box 31 Petersburg VA 23803

WEEG, GERARD PAUL, b Davenport, Iowa, Oct 29, 27; m 48; c 8. COMPUTER SCIENCE. Educ: St Ambrose Col, BS, 49; Okla State Univ, MS, 50; Iowa State Univ, PhD(math), 55. Prof Exp: Instr math, St Ambrose Col, 50-52; mathematician, Univac Div, Sperry Rand Corp, 54-56; prof math & elec eng, Computer Ctr, Mich State Univ, 56-63; vis prof math, Iowa State Univ, 63-64, chmn dept & dir comput ctr, 64-74, PROF COMPUT SCI, UNIV IOWA, 64-, CCC DIR & HEAD DEPT, 74- Mem: Am Math Soc; Math Asn Am; Asn Comput Mach; Inst Elec & Electronics Eng. Res: Sequential machines; finite automata; turning machines; computer networks. Mailing Add: 2 Crestwood Circle Iowa City IA 52240

WEEGE, RANDALL JAMES, b May 14, 26; US citizen; m 51; c 3. GEOLOGY. Educ: Univ Wis-Madison, BS, 51, MS, 55. Prof Exp: Geologist, Anaconda Co, 55-56; geologist-engr, Uranium Div, Calumet & Hecla, Inc, 56-60, resident geologist, Calumet Div, 60-61, asst chief geologist, 61-65, dir geol, 65-68; dir geol, 68-74, DIR EXPLOR, UNIVERSAL OIL PROD CO, 74- Concurrent Pos: Mineral consult, Parsons-Jurden Corp, 66- Mem: Soc Econ Geologists; Am Inst Mining, Metall & Petrol Eng; Geol Soc Am. Res: Mineral exploration techniques and methods. Mailing Add: Universal Oil Prod Co Car Mine & Red Jacket Rd Calumet MI 49913

WEEKES, TREVOR CECIL, b Dublin, Ireland, May 21, 40; m 64; c 3. ASTROPHYSICS. Educ: Nat Univ Dublin, BSc, 62, PhD(physics), 66. Prof Exp: Lectr III physics, Univ Col, Dublin, 64-66; fel astrophys, 66-67, ASTROPHYSICIST, SMITHSONIAN INST, 67-, RESIDENT DIR, MT HOPKINS OBSERV, 69- Concurrent Pos: Consult, Atomic Energy Res Estab, Harwell, 64-66; vis prof, Dublin Inst Advan Studies, 71 & Univ Ariz, 72 & 74. Mem: Brit Inst Physics; AAAS; Am Astron Soc; Royal Astron Soc. Res: Gamma ray astronomy; cosmic ray physics; meteor detection; atmospheric Cerenkov and fluorescence radiation; transient astronomy. Mailing Add: Mt Hopkins Observ Box 97 Amado AZ 85640

WEEKMAN, GERALD THOMAS, b Jamestown, NY, Apr 2, 31; m 54; c 2. ENTOMOLOGY. Educ: Gustavus Adolphus Col, BS, 53; Iowa State Univ, MS, 56, PhD(entom), 57. Prof Exp: From asst to assoc prof entom, Univ Nebr, 57-66; exten assoc prof, 66-71, EXTEN PROF ENTOM, N C STATE UNIV, 71- Concurrent Pos: Expert, Opers Div, Off Pesticide Progs, US Environ Protection Agency, 74- Mem: Entom Soc Am. Res: Economic entomology. Mailing Add: Entom Ext 2309 Gardner Hall NC State Univ Raleigh NC 27607

WEEKS, ALICE MARY DOWSE, b Sherborn, Mass, Aug 26, 09; m 50. GEOLOGY. Educ: Tufts Col, AB, 30; Radcliffe Col, MA, 34, PhD(geol), 49. Prof Exp: Demonstr geol, Bryn Mawr Col, 35-36; asst, Wellesley Col, 36-37, from instr to assoc prof, 37-51; geologist, Exp Geochem & Mineral Br, US Geol Surv, 49-64; assoc prof, 62-63, chmn dept, 62-74, PROF GEOL, TEMPLE UNIV, 64- Concurrent Pos: Mem, Geol Panel, Bd Civil Serv Exam, 57-63; mem, Nat Comt Women in Geosci, 73-75. Mem: Fel AAAS; Geochem Soc; Nat Asn Geol Teachers; fel Geol Soc Am; fel Mineral Soc Am. Res: Vanadium and uranium mineralogy; geochemistry of uranium deposits in Colorado, Utah, New Mexico and the southern Texas Coastal Plain; diagenesis and petrology of tuffaceous sedimentary rocks; geology of eastern Massachusetts; geology of energy resources and problems of the energy crisis. Mailing Add: Dept of Geol Temple Univ Philadelphia PA 19122

WEEKS, CHARLES MERRITT, b Buffalo, NY, Mar 23, 44. BIOPHYSICS. Educ: Cornell Univ, BS, 66; State Univ NY Buffalo, PhD(biophys), 70. Prof Exp: ASST RES SCIENTIST, X-RAY CRYSTALLOG, MED FOUND BUFFALO, 70- Mem: Am Crystallog Asn. Res: Direct methods of phase determination in x-ray crystallography; crystal structures of steroids and related biological materials. Mailing Add: Med Found of Buffalo 73 High St Buffalo NY 14203

WEEKS, DAVID LEE, b Boone, Iowa, June 24, 30; m 57; c 4. EXPERIMENTAL STATISTICS. Educ: Okla State Univ, BS, 52, MS, 57, PhD(statist), 59. Prof Exp: Asst math, 54-57, from asst to assoc prof statist, 57-66, PROF STATIST, OKLA STATE UNIV, 66- Concurrent Pos: Consult, Air Force Armament Lab, Phillips Petrol Co; NSF fac fel, Cornell Univ, 67-68; statist consult, RCA, Albuquerque, 73. Mem: Am Statist Asn; Biomet Soc; Sigma Xi. Res: Experimental design; linear models; optimization techniques; design and analysis of experiments. Mailing Add: Dept of Statist Okla State Univ Stillwater OK 74074

WEEKS, DONALD PAUL, b Terre Haute, Ind, Feb 15, 41; m 64; c 4. MOLECULAR BIOLOGY, PHYSIOLOGY. Educ: Purdue Univ, BSA, 63; Univ Ill, PhD(agron), 67. Prof Exp: Res assoc, 67-68, NIH fel, 68-70, res assoc, 70-74, ASST MEM MOLECULAR BIOL, INST CANCER RES, 74- Mem: AAAS; Am Soc Plant Physiol. Res: Protein synthesis; gene regulation. Mailing Add: Inst for Cancer Res 7701 Burholme Ave Fox Chase Philadelphia PA 19111

WEEKS, DOROTHY WALCOTT, b Philadelphia, Pa, May 3, 93. PHYSICS. Educ: Wellesley Col, BA, 16; Mass Inst Technol, MS, 23, PhD(math), 30; Simmons Col, SM, 25. Prof Exp: Asst exam, US Patent Off, 17-20; asst physics, Mass Inst Technol, 20-21, res assoc, 21-22, instr, 22-24; employment supvr women, Jordan Marsh Co, Mass, 25-27; instr physics, Wellesley Col, 28-29; prof & head dept, Wilson Col, 30-43, 45-56; physicist, Army Ord Mat Res Off, Watertown Arsenal, 56-62, Army Mat Res Agency, 62-64; SPECTROSCOPIST, HARVARD COL OBSERV, 64- Concurrent Pos: Asst, Nat Bur Standards, 20; instr, Buckingham Sch, Cambridge Univ, 23-24; tech aide, Off Sci Res & Develop, 43-46; Guggenheim fel, 49-50; grants, Am Acad Sci, 38, Am Philos Soc, 40, 48 & Res Corp, 53; consult, NSF, 53-56; lectr, Newton

Col Sacred Heart, 66-71. Mem: Am Phys Soc; Optical Soc Am; Am Asn Physics Teachers. Res: Atomic spectroscopy; Lande g values; vacuum ultraviolet; Zeeman patterns of FeI and FeII; coherency matrices; radiological shielding. Mailing Add: 28 Dover Rd Wellesley MA 02181

WEEKS, GERALD, b Birmingham, Eng, Feb 5, 41. BIOCHEMISTRY. Educ: Birmingham Univ, BSc, 62, PhD(biochem), 66. Prof Exp: Res assoc biochem, Duke Univ Med Ctr, 66-69; res fel, Leicester Univ, 69-71; ASST PROF MICROBIOL, UNIV BC, 71- Mem: Am Soc Microbiol; Can Biochem Soc. Res: Molecular basis of cell-cell interaction during the differentiation of the cellular slime mould, Dictyostelium discoideum. Mailing Add: Dept of Microbiol Univ of BC Vancouver BC Can

WEEKS, GREGORY PAUL, b Seattle, Wash, July 16, 47. TEXTILE CHEMISTRY, INSTRUMENTATION. Educ: Univ Wash, BS, 69; Univ Ill, Urbana, PhD(phys chem), 74. Prof Exp: RES CHEMIST, E I DU PONT DE NEMOURS & CO, INC, 74- Res: Development of instrumentation for on-line automated analysis of polymer solutions used in synthetic textile fiber manufacture; synthetic textile technology generally, with emphasis on infrared engineering technology for instrumentation. Mailing Add: 2137 Forest Dr Waynesboro VA 22980

WEEKS, HARMON PATRICK, JR, b Orangeburg, SC, Oct 4, 44; m 65; c 2. WILDLIFE ECOLOGY. Educ: Univ Ga, BSF, 67, MS, 69; Purdue Univ, PhD(wildlife ecol), 74. Prof Exp: Lectr wildlife ecol, Yale Univ, 73-74; ASST PROF WILDLIFE MGT, PURDUE UNIV, WEST LAFAYETTE, 74- Mem: Wildlife Soc; Am Soc Mammalogists; Am Ornithologists Union. Res: Effects of silvicultural practices on wildlife populations; adaptations of homeothermic vertebrates to sodium deficiencies; avian breeding biology. Mailing Add: Dept Forestry & Nat Resources Purdue Univ West Lafayette IN 47907

WEEKS, IVAN FOREST, b Merton, Wis, June 30, 17; m 44; c 2. BIOPHYSICS, COMPUTER SCIENCE. Educ: Univ Wis, PhB, 41, MPh, 42; Univ Calif, Berkeley, MS, 72, PhD(biophys), 75. Prof Exp: Chemist, Standard Oil Co, 42-43; res engr, Shell Oil Co, 43-45; sr res engr, Calif Inst Technol, 45-50; proj assoc chem, Univ Wis, 50-52; theoret physicist, Atomics Int Div, NAm Aviation, Inc, 52-54; group leader theoret physics, Lawrence Radiation Lab, Univ Calif, 54-62; physicist, Defense Res Corp, 62-63; dept head, Aerospace Corp, 63-64; theoret physicist, Gen Atomic, 64-68; PRES, RADIATION HYDRODYN, 68- Concurrent Pos: Consult, Nat Res Coun, 49-50; lectr, Univ Calif, Berkeley, 57-58. Res: Computer simulation; ecology; biophysics; transport phenomena. Mailing Add: 5839 Heron Dr Oakland CA 94618

WEEKS, JAMES ROBERT, b Des Moines, Iowa, Aug 13, 20; m 43; c 3. PHARMACOLOGY. Educ: Univ Nebr, BSc, 41, MS, 46; Univ Mich, PhD(pharmacol), 52. Prof Exp: From instr pharm to prof pharmacol, 46-57; RES ASSOC, UPJOHN CO, 57- Mem: AAAS; Am Soc Pharmacol & Exp Therapeut; Soc Exp Biol & Med. Res: Cardiovascular pharmacology; hypertension; prostaglandins; experimental addiction. Mailing Add: Exp Biol Div Upjohn Co Kalamazoo MI 49002

WEEKS, JOHN DAVID, b Birmingham, Ala, Oct 11, 43. CHEMICAL PHYSICS. Educ: Harvard Col, BA, 65; Univ Chicago, PhD(chem physics), 69. Prof Exp: Res fel chem, Univ Calif, San Diego, 69-71; res fel physics, Cambridge Univ, 72; MEM TECH STAFF MAT PHYSICS, BELL TEL LABS, 73- Mem: Am Phys Soc. Res: Statistical physics; theory of crystal growth; diffusion in semiconductors. Mailing Add: 1D-148 Bell Labs Murray Hill NJ 07094

WEEKS, JOHN LEONARD, b Bath, Eng, May 10, 26; m 52; c 5. OCCUPATIONAL HEALTH. Educ: Univ London, MB, BS, 53, MD, 75; Royal Col Obstet & Gynaec, dipl, 54; Royal Col Physicians & Surgeons, DIH, 57. Prof Exp: Intern, St Thomas's Hosp, London, Eng, 53-54; sr intern, St Helier Hosp, London, 54-55; indust physician, London Transport Exec, 55-58; med officer, Dept Health, Nfld, 58-60; med officer, Int Nickel Co, 60-62; supt, Health & Safety Br, 62-68, DIR HEALTH & SAFETY DIV, HI DIV, WHITESHELL NUCLEAR ESTAB, ATOMIC ENERGY CAN, LTD, 68- Concurrent Pos: Hon prof, Univ Man, 66. Mem: Permanent Comt & Int Asn Occup Health; Brit Soc Occup Med. Res: Industrial and health toxicology, particularly organic coolants; radiation protection and biology. Mailing Add: Whiteshell Nuclear Res Estab Atomic Energy of Can Ltd Pinawa MB Can

WEEKS, LEO, b Norman Park, Ga, June 18, 25; m 54; c 2. GENETICS. Educ: Ga Southern Col, BS, 48; George Peabody Col, MA, 51; Univ Nebr, PhD(zool), 54. Prof Exp: Teacher & asst prin pub sch, Ga, 48-50; instr biol, George Peabody Col, 50-51; instr & asst, Univ Nebr, 51-54; asst prof biol, Austin Peay State Col, 54-56; prof & head dept, Berry Col, 56-62; prof biol, Ga Southern Col, 62-67; HEAD DEPT BIOL, HIGH POINT COL, 68- Mem: Am Genetic Asn; Am Inst Biol Sci; Sigma Xi. Res: Drosophila melanica and other species. Mailing Add: Dept of Biol High Point Col High Point NC 27262

WEEKS, LESLIE VERNON, b Lazear, Colo, July 24, 18; m 44; c 3. SOIL SCIENCE. Educ: Univ Calif, Berkeley, BS, 52; Univ Calif, Riverside, PhD(soil sci), 66. Prof Exp: Sr lab technician, 51-56, prin lab technician, 56-70, STAFF RES ASSOC SOIL SCI, UNIV CALIF, RIVERSIDE, 70-, LECTR SOIL SCI, 75- Mem: Am Soc Agron; Soil Sci Soc Am. Res: Water movement in liquid and vapor phases due to water potential and thermal gradients in unsaturated soils; use of computers in water movement studies under transient flow conditions. Mailing Add: 5496 Jurupa Ave Riverside CA 92504

WEEKS, MARTIN EDWARD, b Centerville, SDak, Oct 1, 06; m 37; c 3. AGRONOMY. Educ: SDak State Col, BS, 34; Univ Wis, PhD(soils), 37. Prof Exp: Asst soils, Univ Ky, 37-40, asst agronomist, 40-45, prof & agronomist, 45-52, head dept agron, 50-52; asst dir, Div Agr Relations, Tenn Valley Auth, 52-59; prof agron, 59-66, PROF SOIL SCI, UNIV MASS, AMHERST, 66- Mem: Am Soc Agron; Soil Sci Soc Am; Int Soc Soil Sci. Res: Soil fertility, chemistry and liming; crop production; evaluation of effectiveness of manure at high rates on crop production and soil and water quality. Mailing Add: Dept of Plant & Soil Sci Univ of Mass Amherst MA 01002

WEEKS, MAURICE HAROLD, b Germantown, NY, Nov 9, 21; m 48; c 2. TOXICOLOGY. Educ: Union Col, BS, 49, MS, 50. Prof Exp: Collab path, Forest Prod Lab, US Forest Serv, Wis, 49; chemist, Hanford Works, Gen Elec Co, 50-55; pharmacologist, Directorate Med Res, Army Chem Ctr, 55-66; PHARMACOLOGIST, US ARMY ENVIRON HYG AGENCY, EDGEWOOD ARSENAL, 66- Mem: AAAS; Sci Res Soc Am; Am Indust Hyg Asn; Soc Toxicol. Res: Animal metabolism studies of administered chemicals; toxicology and hazard evaluation of inhaled aerosols and vapors. Mailing Add: 27 Idlewild Bel Air MD 21014

WEEKS, OWEN BAYARD, b Boone, Iowa, Mar 22, 16; m 40; c 2. CHEMISTRY. Educ: Iowa State Col, BS, 37; Ore State Col, MS, 40; Ohio State Univ, PhD, 43. Prof Exp: Res assoc limnol, Stone Lab Hydrobiol, Ohio State Univ, 43-46; assoc prof bact, NDak Agr Col, 46-49 & Univ Idaho, 49-64; RES PROF, N MEX STATE UNIV, 64- Concurrent Pos: NSF fac fel, 60-61; fel, Inst Org Chem, Norweg Inst Technol, 65; sabbatical leave, Queen Mary Col, Univ London, 71-72 & Univ Liverpool, 72. Mem: AAAS; Am Soc Microbiol; Brit Soc Gen Microbiol. Res: Bacterial taxonomy, nutrition, growth and pigmentation; aquatic biology; relationship of chemical factors to algal production in fresh water; structure and biosynthesis of carotenoids. Mailing Add: Res Ctr NMex State Univ Las Cruces NM 88001

WEEKS, PAUL MARTIN, b Clinton, NC, June 11, 32; m 57; c 6. SURGERY. Educ: Duke Univ, AB, 54; Univ NC, MD, 58. Prof Exp: From instr to prof surg, Med Ctr, Univ Ky, 64-70; CHIEF DIV PLASTIC SURG, DEPT SURG, SCH MED, WASHINGTON UNIV, 71- Mem: Am Col Surgeons; Soc Univ Surgeons; Am Chem Soc; Plastic Surg Res Coun. Res: Interrelationships of collagen and mucopolysaccharides in determining tissue compliance; effects of environment in cell synthesis, particularly regarding mucopolysaccharide or collagen synthesis; effects of irradiation on collagen and mucopolysaccharide synthesis. Mailing Add: Div of Plastic Surg Washington Univ Sch of Med St Louis MO 63110

WEEKS, ROBERT A, b Birmingham, Ala, Aug 23, 24; m 48; c 4. SOLID STATE PHYSICS. Educ: Birmingham-Southern Col, BS, 48; Univ Tenn, MS, 51; Brown Univ, PhD(physics), 66. Prof Exp: Physicist, Solid State Div, 51-68, PRIN INVESTR LUNAR MAT, OAK RIDGE NAT LAB, NASA, 68- Concurrent Pos: Distinguished vis prof, Am Univ Cairo, 68, 70-71; res fel, Univ Reading, 71; consult, Dept Phys Sci & Mat Eng, Am Univ Cairo, 71-; assoc ed, J Geophys Res, 68-74. Mem: AAAS; Am Geophys Union; Am Phys Soc; Sci Res Soc Am; Am Ceramic Soc. Res: Optical and magnetic properties of intrinsic and extrinsic defects of crystalline and glassy solids; effects of photon and particle irradiation upon the properties of diamagnetic non-conducting solids; magnetic properties of extraterrestrial solids. Mailing Add: Oak Ridge Mat Lab Box X Oak Ridge TN 37830

WEEKS, ROBERT HUGH, physical chemistry, see 12th edition

WEEKS, ROBERT JOE, b Quapaw, Okla, Feb 19, 29; m 48; c 3. MYCOLOGY, MEDICAL MICROBIOLOGY. Educ: Southwest Mo State Col, BS, 61. Prof Exp: Mycologist, Kans City Field Sta, Commun Dis Ctr, 62-64, microbiologist, 64-67, chief soil ecol unit, Mycoses Sect, Kansas City Labs, Ecol Invests Prog, Ctr Dis Control, 67-74, MICROBIOLOGIST, MICROBIOL & SEROL UNIT, PROFICIENCY TESTING SECT, BUR LABS, USPHS, 74- Mem: AAAS; Mycol Soc Am. Res: Relationship of the pathogenic fungi to their soil environment. Mailing Add: 746 Garden View Dr Stone Mountain GA 30083

WEEKS, THOMAS F, b Wheeling, WVa, Apr 12, 35; m 55; c 2. PLANT PHYSIOLOGY. Educ: West Liberty State Col, AB, 63; Purdue Univ, MS, 67, PhD(develop bot), 70. Prof Exp: Mem staff sec educ, Ohio County Bd Educ, WVa, 63-64; guest lectr biol, Purdue Univ, 70-71; ASSOC PROF BIOL, UNIV WIS-LA CROSSE, 71- Mem: AAAS; Bot Soc Am; Am Soc Plant Physiol. Res: Plant growth regulators; control mechanisms in plants. Mailing Add: Dept of Biol Univ of Wis La Crosse WI 54601

WEEKS, THOMAS JOSEPH, JR, b Tarrytown, NY, Aug 31, 41; m 67. PHYSICAL INORGANIC CHEMISTRY. Educ: Colgate Univ, BA, 63; Univ Colo, PhD(chem), 67. Prof Exp: Res chemist, Res & Eng Ctr, Johns-Manville Corp, Manville, NJ, 66-68, sr res chemist, 68-70; sr res chemist, 70-72, proj chemist, 72-73, SR STAFF CHEMIST, UNION CARBIDE CORP, TARRYTOWN, NY, 73- Mem: Am Chem Soc; Catalysis Soc; AAAS. Res: Equilibria and kinetics of transition metal complexes; catalysis, adsorption and mechanisms of organic reactions on zeolite molecular sieves. Mailing Add: 74 Lt Cox Dr Pearl River NY 10989

WEEKS, WILFORD FRANK, b Champaign, Ill, Jan 8, 29; m 52; c 2. GLACIOLOGY, HYDROLOGY. Educ: Univ Ill, BS, 51, MS, 53; Univ Chicago, PhD(geol), 56. Prof Exp: Geologist, Mineral Deposits Br, US Geol Surv, 52-55; glaciologist, US Air Force Cambridge Res Ctr, 55-57; asst prof, Wash Univ, 57-62; GLACIOLOGIST, COLD REGIONS Res & ENG LAB, 62- Concurrent Pos: Adj assoc prof, Dartmouth Col, 62-72, adj prof, 72-; mem, Polar Res Bd & chmn panel on glaciol, Nat Acad Sci; chmn, Div River, Lake & Sea Ice, Int Comn Snow & Ice; Japan soc promotion sci vis prof, Inst Low Temp Sci, Hokkaido Univ, Japan, 73. Mem: Fel Geol Soc Am; fel Arctic Inst NAm; Am Geophys Union; Int Glaciol Soc (vpres, 69-72, pres, 72-75). Res: Geophysics of sea, lake and river ice. Mailing Add: Cold Regions Res & Eng Lab Hanover NH 03755

WEEKS, WILLIAM THOMAS, b Portchester, NY, Mar 28, 32; m 54; c 3. COMPUTER SCIENCES. Educ: Williams Col, BA, 54; Univ Mich, MS, 56, PhD(physics), 60. Prof Exp: Assoc physicist, Data Systs Div, 60-61, staff physicist, Systs Develop Div, 61-64, adv physicist, Components Div, 64-68, SR ENGR COMPONENTS DIV, IBM CORP, 68- Res: Scientific computation and numerical analysis as applied to design of electronic digital computers. Mailing Add: Innsbruck Blvd Hopewell Junction NY 12533

WEEMS, HOWARD VINCENT, JR, b Rome, Ga, Apr 11, 22; m 50; c 5. ENTOMOLOGY. Educ: Emory Univ, BA, 46; Univ Fla, MS, 48; Ohio State Univ, PhD(entom), 53. Prof Exp: Asst entom, Univ Fla, 46-47; instr biol, Univ Miss, 48-49; res asst, Ohio Biol Surv, Ohio State Univ, 49-53; TAXON ENTOMOLOGIST, DIV PLANT INDUST, FLA DEPT AGR & CONSUMER SERV, 53-; PROF ENTOM, UNIV FLA, 73- Concurrent Pos: Cur, Fla State Collection Arthropods, 54-; assoc arthropods, Fla State Mus, 59-; ed, Arthropods of Fla & Neighboring Land Areas, Fla Dept Agr & Consumer Serv, 65-; assoc prof entom, Univ Fla, 66-73. Mem: Fel AAAS; Asn Trop Biol; Entom Soc Am; assoc Ecol Soc Am; Soc Syst Zool. Res: Taxonomy and ecology of syrphid flies; identifications of Florida arthropods. Mailing Add: Fla Dept Agr & Consumer Serv PO Box 1269 Gainesville FL 32602

WEEMS, MALCOLM LEE BRUCE, b Nashville, Kans, Dec 8, 45; m 72; c 1. PHYSICS. Educ: Kans State Teachers Col, BSE, 67, MS, 69; Okla State Univ, PhD(physics), 72. Prof Exp: Lectr physics, Kans State Teachers Col, 68-69; instr, 72-75, DIR SCI LAB PROG FOR BLIND, EAST CENTRAL STATE UNIV, 74-, ASST PROF PHYSICS, 75- Mem: Am Inst Physics; Am Asn Physics Teachers. Res: Stellar atmospheres; educational innovation. Mailing Add: Dept of Phys East Cent State Univ Ada OK 74820

WEERTMAN, JOHANNES, b Fairfield, Ala, May 11, 25; m 50; c 2. PHYSICS. Educ: Carnegie Inst Technol, BS, 48, DSc(physics), 51. Prof Exp: Fulbright fel, Ecole Normal Superieure, Paris, 51-52; solid state physicist, US Naval Res Lab, 52-58. sci liaison officer, US Off Naval Res US Embassy, Eng Lab, 58-59; from assoc prof to prof, 59-68, WALTER P MURPHY PROF MAT SCI, NORTHWESTERN UNIV, EVANSTON, 68-, PROF GEOPHYS, 63- Concurrent Pos: Consult, US Army Cold Regions Res & Eng Lab, 60-75, US Naval Res Lab, 60-67, Bain Lab, US Steel Co, 60-62 & Oak Ridge Nat Lab, 63-68; vis prof, Calif Inst Technol, 64; consult, Los Alamos Sci Lab, 67-; vis prof, Scott Polar Res Inst, Cambridge Univ, 71-72; Guggenheim fel, 71-72. Honors & Awards: Horton Award, Sect Hydrol, Am Geophys Union, 62.

Mem: AAAS; fel Geol Soc Am; fel Am Phys Soc; fel Am Soc Metals; Am Quaternary Asn. Res: Creep of crystals; dislocation theory; internal friction; theory of glacier movement; metal physics; glaciology; geophysics; fatigue. Mailing Add: Dept of Mat Sci Northwestern Univ Evanston IL 60201

WEERTMAN, JULIA RANDALL, b Muskegon, Mich, Feb 10, 26; m 50; c 2. SOLID STATE PHYSICS. Educ: Carnegie Inst Technol, BS, 46, MS, 47, DSc(physics), 51. Prof Exp: Rotary Int fel, Ecole Normale Superieure, Univ Paris, 51-52; physicist, US Naval Res Lab, 52-58; vis asst prof, 72-73, ASST PROF MAT SCI, NORTHWESTERN UNIV, EVANSTON, 73- Mem: Am Phys Soc; Am Inst Phys. Res: Dislocation theory; high temperature fatigue. Mailing Add: Dept of Mat Sci & Eng Northwestern Univ Evanston IL 60201

WEESE, RICHARD HENRY, b Hyer, WVa, Dec 13, 38; m 68. ORGANIC POLYMER CHEMISTRY. Educ: WVa State Col, BS, 64; Bucknell Univ, MS, 67. Prof Exp: Scientist chem, 67-74, SR SCIENTIST CHEM, RES LABS, ROHM AND HAAS CO, BRISTOL, 74- Res: Synthesis and application of organic polymers, including additives or modifiers. Mailing Add: RR 1 Glenwood Dr Washington Crossing PA 18977

WEESNER, WILLIAM ELDRED, b Youngstown, Ohio, Feb 24, 16; m 39; c 1. ORGANIC CHEMISTRY. Educ: Allegheny Col, AB, 37; Purdue Univ, MS, 40, PhD(org chem), 42. Prof Exp: Res chemist, Monsanto Chem Co, 42-50, cent res dept, Res & Eng Div, 50-53, group leader, 53-61; group leader, Monsanto Res Corp, Ohio, 61-69, PATENT SEARCH MGR, MONSANTO CO, 69- Mem: Am Chem Soc. Res: Insecticides; oxidations; chemistry of carbon monoxide; plasticizers; antimalarials; fluorine and silicon chemistry. Mailing Add: Monsanto Co 1911 Jefferson Davis Hwy Arlington VA 22202

WEETALL, HOWARD H, b Chicago, Ill, Nov 17, 36; m 62; c 2. IMMUNOCHEMISTRY, ENZYMOLOGY. Educ: Univ Calif, Los Angeles, BA, 59, MA, 61. Prof Exp: Scientist, Jet Propulsion Lab, Calif Inst Technol, 61-65; sr res biologist, Space Gen Corp, 65-66; immunochemist, Bionetics Res Corp, 66-67; res biochemist, 67-71, res assoc, 71-73, SR RES ASSOC BIOCHEM, CORNING GLASS WORKS, 73- Mem: AAAS; Am Chem Soc; Am Soc Microbiol; NY Acad Sci; Am Asn Immunol. Res: Insolubilized biologically active molecules, including antigens, antibodies and enzymes and the characteristics of such materials. Mailing Add: Corning Glass Works Corning NY 14830

WEETE, JOHN DONALD, b Dallas, Tex, June 14, 42. PLANT PHYSIOLOGY, PLANT BIOCHEMISTRY. Educ: Stephen F Austin State Univ, BS, 65, MS, 68; Univ Houston, PhD(biol), 70. Prof Exp: Fel, Baylor Col Med, 69-70; vis scientist, Lunar Sci Inst, 70-71, staff scientist, 71-72, NASA prin investr lunar sample anal, 72; ASST PROF BOT & MICROBIOL, AUBURN UNIV, 72- Concurrent Pos: Spec lectr, La State Univ, New Orleans, 70-71. Honors & Awards: Res Award, Am Phytopath Soc, 68. Mem: Am Phytopath Soc; Mycol Soc Am; Am Oil Chem Soc; Am Soc Plant Physiol; Inst Soc Study Origin of Life. Res: Fungus physiology and biochemistry; lipid composition and metabolism of fungi and higher plants. Mailing Add: Dept of Bot & Microbiol Auburn Univ Auburn AL 30836

WEETMAN, DAVID G, b Poughkeepsie, NY, Feb 20, 38; m 65; c 1. ORGANIC CHEMISTRY. Educ: Pa State Univ, BS, 59; Univ Minn, PhD(org chem), 68. Prof Exp: RES CHEMIST, TEXACO RES LAB, TEXACO, INC, 68- Mem: Am Chem Soc. Res: Synthesis and reactions of dihalocyclopropanes derived from 2- and 3-methyl substituted 4-ethoxy-2H-1-benzothiopyrans. Mailing Add: Texaco Res Ctr PO Box 509 Beacon NY 12508

WEETMAN, GORDON FREDERICK, b York, Eng, Apr 24, 33; Can citizen. FORESTRY. Educ: Univ Toronto, BScF, 55; Yale Univ, MS, 58, PhD(forestry), 62. Prof Exp: Res forester, Pulp & Paper Res Inst Can, 55-72; ASSOC PROF SILVICULT, FAC FORESTRY, UNIV NB, FREDERICTON, 72- Concurrent Pos: Ed, Forestry Chronicle, Can Inst Forestry, 67-71. Mem: Can Inst Forestry (pres, 73-74). Res: Blockage of the nitrogen cycle by raw humus accumulations in boreal forests; nitrogen fertilization; nutrient losses in logging; silviculture. Mailing Add: Fac of Forestry Univ of NB Fredericton NB Can

WEFEL, JOHN PAUL, b Cleveland, Ohio, Apr 28, 44; m 67; c 1. PHYSICS, ASTROPHYSICS. Educ: Valparaiso Univ, BS, 66; Wash Univ, MA, 68, PhD(physics), 71. Prof Exp: Nat Acad Sci-Nat Res Coun resident res assoc astrophys, Naval Res Lab, 71-73, res physicist, 73-75; ROBERT R McCORMICK FEL, THE ENRICO FERMI INST, UNIV CHICAGO, 75- Mem: AAAS; Am Phys Soc. Res: Cosmic ray astrophysics; utilized both passive and electronic detectors to measure element abundances; studied nuclear fragmentation reactions of importance in cosmic ray propagation calculations. Mailing Add: Enrico Fermi Inst Univ Chicago 5630 Ellis Ave Chicago IL 60637

WEFER, JOHN MICHAEL, organic chemistry, polymer chemistry, see 12th edition

WEFERS, KARL, b Bonn, Ger, Aug 4, 28; m 57; c 2. CRYSTALLOGRAPHY, CHEMISTRY. Educ: Univ Bonn, Dr rer nat, 58. Prof Exp: Group leader crystallog & struct chem, Vereinigte Aluminum Werke, Bonn, Ger, 58-66; SCI ASSOC, ALCOA RES LABS, ALCOA TECH CTR, 67- Res: Physical chemistry and structural chemistry of extractive metallurgy of aluminum; surface chemistry of aluminum. Mailing Add: Alcoa Labs Alcoa Tech Ctr Alcoa Center PA 15069

WEG, RUTH BASS, b New York, NY, Oct 12, 20; c 3. GERONTOLOGY. Educ: Hunter Col, BA, 40; Univ Southern Calif, MS, 54, PhD(zool), 58. Prof Exp: Res assoc, dept biochem, 58-59, biol & physiol, 60-70, biochemist, 62-64, biologist in residence, Air Pollution Control Inst, 66, assoc prof biol, Univ & assoc dir educ & training, 68-74, MEM TEACHING FAC, SUMMER INST STUDY GERONT, ANDRUS GERONT CTR, UNIV SOUTHERN CALIF, 68-, DIR, 69-, ASST DEAN STUDENT AFFAIRS, LEONARD DAVIS SCH GERONT, 75- Concurrent Pos: Consult, Rossmoor-Cortese Inst Study Retirement & Aging, Univ Southern Calif, 67, pre-med adv, Univ, 70- Mem: AAAS; Am Inst Biol Sci; Fedn Am Socs Exp Biol; Geront Soc. Res: Molecular bases for physiological phenomena; blood groups of mammals; iron metabolism in liver; mechanism of action of air pollutants on cellular activities; processes of aging on a molecular/organism level. Mailing Add: Andrus Geront Ctr Univ of Southern Calif Los Angeles CA 90007

WEGE, WILLIAM RICHARD, b Shawano, Wis, Mar 31, 26; m 50; c 3. RADIOLOGY. Educ: Marquette Univ, DDS, 52; Univ Ala, MS, 67. Prof Exp: ASSOC PROF RADIOL, MED COL GA, 67- Concurrent Pos: NIH spec fel, Med Col Ga, 67-68; consult, US Army, Ft Jackson, SC, 67- & Vet Admin, 68- Mem: Fel Am Acad Dent Radiol (treas, 72); Int Asn Dent Res (pres, 71); fel Am Col Dent. Res: Cytogenetics in relation to radiation. Mailing Add: Dept of Radiol Med Col of Ga Augusta GA 30902

WEGENER, HORST ALBRECHT RICHARD, b Brooklyn, NY, May 9, 27; m 50; c 4. SOLID STATE SCIENCE. Educ: Columbia Univ, BS, 51; Polytech Inst Brooklyn, MS, 55, PhD(inorg chem), 65. Prof Exp: Lab asst, Patterson Moos & Co, Inc, 49-50, proj engr, 50-55; scientist, Tung-Sol Elec, Inc, 55-57, supvr res, 57-60, group leader, 60-64; staff mem, 64-65, group leader, 65-66, DEPT HEAD, SPERRY RAND RES CTR, 66- Mem: Sr mem Inst Elec & Electronics Eng. Res: Imperfections in crystals; x-ray diffraction crystal structure synthesis; properties and structure of semiconductor compounds; semiconductor device theory, design and processing; volatile and non-volatile large scale integrated memory circuit design and fabrication. Mailing Add: 105 Wolf Rock Rd Carlisle MA 01741

WEGENER, PETER PAUL, b Berlin, Ger, Aug 29, 17; nat US; m. FLUID PHYSICS, GAS DYNAMICS. Educ: Univ Berlin, Dr rer nat(physics, geophys), 43. Prof Exp: Mem & head basic res group supersonic wind tunnels, Ger, 43-45; mem res group hypersonic wind tunnel design & res, US Naval Ord Lab, 46-53; chief gas dynamics res sect, Jet Propulsion Lab, Calif Inst Technol, 53-59; prof appl sci, 60-72, chmn dept, 66-71, HAROLD HODGKINSON PROF ENG & APPL SCI, YALE UNIV, 72- Honors & Awards: Meritorious Civilian Serv Award, US Navy, 51. Mem: Fel Am Phys Soc; Am Inst Aeronaut & Astronaut. Res: Gas dynamics; fluid dynamics; chemical physics related to flow problems such as chemical kinetics and condensation. Mailing Add: Mason Lab Yale Univ New Haven CT 06520

WEGENER, WARNER SMITH, b Cincinnati, Ohio, June 23, 35; m 58; c 3. MICROBIOLOGY, BIOCHEMISTRY. Educ: Univ Cincinnati, BS, 57, PhD(microbiol), 64. Prof Exp: Asst mem, Res Labs, Albert Einstein Med Ctr, 64-68; asst prof, 68-72, ASSOC PROF MICROBIOL, SCH MED, IND UNIV, INDIANAPOLIS, 72- Concurrent Pos: NIH fel, 65-66. Mem: Am Soc Microbiol. Res: Intermediary metabolism and cellular regulatory processes; biochemical basis of microbial pathogenicity. Mailing Add: Dept of Microbiol Ind Univ Sch of Med Indianapolis IN 46207

WEGMAN, MYRON EZRA, b Brooklyn, NY, July 23, 08; m 36; c 4. PUBLIC HEALTH, PEDIATRICS. Educ: City Col New York, BA, 28; Yale Univ, MD, 32; Johns Hopkins Univ, MPH, 38. Prof Exp: From asst to instr pediat, Sch Med, Yale Univ, 32-36; pediat consult, State Dept Health, Md, 36-41; asst prof child hyg, Sch Trop Med, Univ PR, 41-42; dir training & res, Dept Health, New York, 42-46, dir, Sch Health Serv, 43-46; prof pediat & head dept, Sch Med, La State Univ, 46-52; dir div educ & training, Pan Am Sanit Bur, WHO, 52-56, secy gen, Pan Am Health Orgn, 57-60; dean, Sch Pub Health, 60-74, PROF PUB HEALTH & PROF PEDIAT, MED SCH, UNIV MICH, ANN ARBOR, 60-, EMER DEAN, SCH PUB HEALTH, 74- Concurrent Pos: From intern to resident, New Haven Hosp, Conn, 32-36; lectr, Maternal & Child Health, Johns Hopkins Univ, 39-46; asst prof, Col Physicians & Surgeons, Columbia Univ, 41-44; asst prof, Med Col, Cornell Univ, 42-46; pediatrician-in-chief, Charity Hosp, New Orleans, 46-52, pres, Vis Staff, 50-51; consult, Children's Hosp, Washington, DC, 53-60; spec lectr, Sch Med, George Washington Univ, 54-60; pres, Asn Sch Pub Health, 63-66; WHO vis prof, Univ Malaya, 74; mem, Soc of Scholars, Johns Hopkins Univ, 75; John G Searle prof pub health, Univ Mich, 75-; chmn comt pediat hosp rates, Nat Res Coun, 75- Honors & Awards: Grulee Award, Am Acad Pediat, 58; Townsend Harris Medal, City Col New York, 61; Bronfman Prize, 67; Distinguished Serv Award, Mich Pub Health Asn, 74; Walter P Reuther Award, United Automobile Workers, 74; Sedgwick Medal, Am Pub Health Asn, 74. Mem: Fel AAAS; Soc Pediat Res; Am Pediat Soc; Soc Exp Biol & Med; fel AMA. Res: Prematurity; infant mortality; international health organization. Mailing Add: Sch of Pub Health Univ of Mich Ann Arbor MI 48104

WEGMANN, THOMAS GEORGE, b Milwaukee, Wis, Sept 29, 41; m 65; c 1. GENETICS, IMMUNOLOGY. Educ: Univ Wis, BA, 63, PhD(med genetics), 68. Prof Exp: From asst prof to assoc prof biol, Harvard Biol Labs, 69-74; ASSOC PROF IMMUNOL & PRIN INVESTR, MED RES COUN TRANSPLANTATION UNIT, UNIV ALTA, 74- Concurrent Pos: Consult, Med Sch, Univ Mass, 71- Res: Genetics and the immune response; immunological tolerance and developmental analysis of genetically-marked chimeric systems. Mailing Add: Med Res Coun Transplantation Unit Univ of Alta Edmonton AB Can

WEGNER, GENE H, b Madison, Wis, Aug 30, 30; m 53; c 4. PETROLEUM MICROBIOLOGY. Educ: Univ Wis, BS, 53, MS, 57, PhD(bact, biochem), 62. Prof Exp: Asst bact, Univ Wis, 55-57; microbiologist, Eli Lilly & Co, 57-59; asst bact, Univ Wis, 59-60, dept fel, 60-61; SR RES MICROBIOLOGIST, PHILLIPS PETROL CO, 62- Mem: Am Soc Microbiol. Res: Hydrocarbon microbiology; fatty acid metabolism; microbial lipids; single cell protein; continuous culture; pilot plant development. Mailing Add: 94-H Phillips Res Ctr Phillips Petrol Co Bartlesville OK 74003

WEGNER, HARVEY E, b Tacoma, Wash, Aug 12, 25; m 49; c 3. NUCLEAR PHYSICS. Educ: Univ Puget Sound, BS, 48; Univ Wash, MS, 51, PhD(physics), 53. Prof Exp: Asst physicist, Brookhaven Nat Lab, 53-56; physicist, Los Alamos Sci Lab, 56-62; physicist, 62-66, co-dir, Tandem Van de Graff Facility, 70-75, CONSTRUCTION MGR, TANDEM VAN DE GRAFF FACILITY, BROOKHAVEN NAT LAB, 66-, SR PHYSICIST, 68- Concurrent Pos: Consult, Radiation Dynamics, Inc, 63- & Gen Ionex, 74- Mem: Fel Am Phys Soc; Am Asn Physics Teachers. Res: Accelerator construction; cyclotrons in nuclear physics and machine development; semiconductor detectors; low energy accelerator construction and development; heavy ion reaction and fusion physics. Mailing Add: Dept of Phys Brookhaven Nat Lab Upton NY 11973

WEGNER, KARL HEINRICH, b Pierre, SDak, Jan 5, 30; m 57; c 3. MEDICINE, PATHOLOGY. Educ: Yale Univ, BA, 52; Harvard Med Sch, MD, 59. Prof Exp: PROF PATH & CHMN DEPT, SCH MED, UNIV SDAK, 68-, DEAN SCH MED & VPRES HEALTH AFFAIRS, UNIV, 73-; PATHOLOGIST, LAB CLIN MED & SIOUX VALLEY HOSP, 62- Mem: Col Am Path; Am Soc Clin Path; Int Acad Path. Mailing Add: Sunnymede S Minnesota Rd Sioux Falls SD 57101

WEGNER, KENNETH WARREN, b Beloit, Wis, Nov 19, 07; m 31; c 4. MATHEMATICS. Educ: Univ Wis, AB, 29, AM, 32, PhD(math), 34. Prof Exp: From asst to instr math, Univ Wis, 29-34; head dept math & physics, Whitworth Col, 34-35; instr math, Univ Minn, 35-38; prof, Col St Catherine, 38-43; from assoc prof to prof, 43-73, chmn dept math & astron, 61-64, EMER PROF MATH, CARLETON COL, 73- Concurrent Pos: Fulbright lectr, Taiwan, 59-60 & Liberia, 64-65; vis prof, Va Union Univ, 67-68 & Spelman Col, 69-70. Mem: Math Asn Am. Res: Equivalence of pairs of hermitian matrices; higher algebra. Mailing Add: Dept of Math & Astron Carleton Col Northfield MN 55057

WEGNER, MARCUS IMMANUEL, b South Haven, Mich, Mar 3, 15; m 41; c 4. BIOCHEMISTRY, NUTRITION. Educ: St Norbert Col, BS, 36; Univ Wis, PhD(nutrit biochem), 41. Prof Exp: Asst chemist, Exp Sta, Agr & Mech Col, Univ Tex, 41-43; asst nutritionist, Exp Sta, NDak Col, 43-44; res chemist, Mead Johnson & Co, Ind, 44-48; asst res dir, Res Div, Oscar Mayer & Co, 48-51; nutritionist, Pet Milk Co, 51-59, sect leader new prod develop, 59-60, group mgr, 60-61; res dir, Ward Foods, Inc, 61-64; asst res dir, Best Foods Div, Corn Prod Co, 64-71; br chief & diet appraisal, Consumer & Food Economics Div, 69-71, ASST DIR, EASTERN

REGIONAL RES LAB, AGR RES SERV, USDA, 71- Mem: Am Asn Cereal Chem; Am Chem Soc; Inst Food Technol. Res: Chemistry and nutrition of protein hydrolysates; bakery products; food and diet appraisal; new and wider uses for American farm commodities; meats and dairy products; convenience frozen foods; new product development. Mailing Add: Eastern Regional Res Lab USDA 600 E Mermaid Lane Philadelphia PA 19118

WEGNER, PATRICK ANDREW, b South Bend, Ind, Nov 14, 40. ORGANOMETALLIC CHEMISTRY. Educ: Northwestern Univ, Evanston, BA, 62; Univ Calif, Riverside, PhD(chem), 66. Prof Exp: Res chemist, E I du Pont de Nemours & Co, Inc, 66-68; vis prof chem, Harvey Mudd Col, 68-69; from asst prof to assoc prof, 69-74, PROF CHEM, CALIF STATE UNIV, FULLERTON, 75- Concurrent Pos: Grants, Petrol Res Corp, Calif State Univ, Fullerton, 68-72, Res Corp, 70-72, NASA, 72-74. Mem: Am Chem Soc. Res: Boron hydride chemistry; transition metal organometallic chemistry. Mailing Add: Dept of Chem 800 N State College Calif State Univ Fullerton CA 92634

WEGNER, PETER, b Aug 20, 32; US citizen; m 56; c 4. COMPUTER SCIENCE. Educ: Univ London, BSC, 53, PhD(comput sci), 68; Pa State Univ, MA, 59. Prof Exp: Res assoc, Comput Ctr, Mass Inst Technol, 59-60; asst dir statist lab, Harvard Univ, 60-61; lectr comput sci, London Sch Econ, 61-64; asst prof, Pa State Univ, 64-66; assoc prof, Cornell Univ, 66-69; ASSOC PROF COMPUT SCI, BROWN UNIV, 69- Concurrent Pos: Educ ed, Commun, Asn Comput Mach, 71-72. Mem: Asn Comput Mach; Math Asn Am. Res: Programming language theory and implementation; computer science education; semantics of programming languages; theory of computation. Mailing Add: Dept of Appl Math Brown Univ Providence RI 02912

WEGNER, THOMAS NORMAN, b Cleveland, Ohio, July 8, 32; m 62; c 1. ANIMAL PHYSIOLOGY, BIOCHEMISTRY. Educ: Mich State Univ, BS, 54; Colo State Univ, MS, 56; Univ Calif, Davis, PhD(animal physiol), 64. Prof Exp: Asst animal pathologist, 64-66, ASST PROF DAIRY SCI, UNIV ARIZ, 67- Mem: Am Dairy Sci Asn. Res: Carbohydrate metabolism in ruminant animals; pathogenesis of coccidioidomycosis, biochemistry of the immune responses; physiology of abnormal milk production and heat stress on dairy cows. Mailing Add: Dept of Dairy & Food Sci Univ of Ariz Tucson AZ 85721

WEGRIA, RENE, b Fumal, Belg, June 9, 11; nat US. MEDICINE. Educ: Univ Liege, MD, 36; Columbia Univ, MedScD, 45. Prof Exp: Belg Am Educ Found fel, Vanderbilt Univ, Mayo Found & Western Reserve Univ, 37-39; fel physiol, Western Reserve Univ, 39-40, instr, 40-42; jr asst resident med, Cleveland City Hosp, 42-43; asst resident cardiol, Presby Hosp, New York, 43-46; instr med, Col Physicians & Surgeons, Columbia Univ, 46-47, assoc, 47-49, from asst prof to assoc prof, 49-58; prof med & dir dept internal med, 58-63, PROF PHARMACOL & CHMN DEPT, SCH MED, ST LOUIS UNIV, 63- Concurrent Pos: Vis prof, Lovanium Univ, Leopoldville, 58; Arthur Strauss vis physician, Jewish Hosp St Louis, 59. Mem: Fel AAAS; fel NY Acad Sci; Am Physiol Soc; Am Soc Clin Invest; Soc Exp Biol & Med. Res: Cardiovascular physiology, pharmacology and physiopathology; internal medicine. Mailing Add: Dept of Pharmacol St Louis Univ Sch of Med St Louis MO 63104

WEGRICH, OHMER GODFREY, mycology, see 12th edition

WEGST, WALTER F, JR, b Philadelphia, Pa, Dec 26, 34; m 58; c 2. RADIOLOGICAL HEALTH. Educ: Univ Mich, BSE, 56, MSE, 57, PhD(environ health), 63; Am Bd Health Phys, Dipl, 66; Bd Cert Safety Prof, Cert. Prof Exp: Reactor health physicist, Phoenix Mem Lab, Univ Mich, 57-58, lab health physicist, 58-59, lab supvr, 59-60; inst health physicist, 63-68, safety mgr, 68-71, mgr security, 71-73, MGR SAFETY, CALIF INST TECHNOL, 71- Concurrent Pos: Consult, Alpha & Omega Servs, 74- Mem: Health Phys Soc; Am Indust Hyg Asn; Sigma Xi; AAAS. Res: Radiobiological studies on mammalian cells; secondary electron production by charged particle passage through matter; university health physics problems. Mailing Add: Calif Inst Technol 1201 E California Blvd Pasadena CA 91109

WEGWEISER, ARTHUR E, b New York, NY, Feb 20, 34; m 58; c 1. GEOLOGY. Educ: Brooklyn Col, BA, 55; Hofstra Univ, MS, 58; Wash Univ, St Louis, PhD(geol), 66. Prof Exp: Sci teacher, Island Trees High Sch, 58-61; prof geol, 65-74, PROF EARTH SCI, EDINBORO STATE COL, 74-, CHMN DEPT EARTH SCI, 69-, DIR MARINE SCI CONSORTIUM, 72- Concurrent Pos: Adj lectr, Hofstra Univ, 58-61. Mem: Soc Econ Paleont & Mineral. Res: Micropaleontology-ecology of recent Foraminifera; environments of deposition. Mailing Add: Dept of Earth Sci Edinboro State Col Edinboro PA 16412

WEHAUSEN, JOHN VROOMAN, b Duluth, Minn, Sept 23, 13; m 38; c 4. MECHANICS. Educ: Univ Mich, BS, 34, MS, 35, PhD(math), 38. Prof Exp: Instr math, Brown Univ, 37-38, Columbia Univ, 38-40 & Univ Mo, 40-44; consult opers res group, Off Field Serv, Off Sci Res & Develop, US Dept Navy, 44-46, mathematician, David Taylor Model Basin, 46-49, actg head mech br, Off Naval Res, 49-50; assoc res mathematician, Inst Eng Res, 56-57, res mathematician, Dept Naval Archit, 57-58, assoc prof eng sci, 58-59, PROF ENG SCI, UNIV CALIF, BERKELEY, 59- Concurrent Pos: Lectr, Univ Md, 45-50; exec ed, Math Rev, Am Math Soc, 50-56; vis prof, Univ Hamburg, 60-61. Mem: Am Math Soc; Soc Naval Archit & Marine Eng; Math Asn Am. Res: Fluid mechanics, especially theory of water waves and hydrodynamics of ships. Mailing Add: Dept of Naval Archit Univ of Calif Berkeley CA 94720

WEHINGER, PETER AUGUSTUS, b Goshen, NY, Feb 18, 38; m 67. ASTRONOMY. Educ: Union Col, NY, BS, 60; Ind Univ, MA, 62; Case Western Reserve Univ, PhD(astron), 66. Prof Exp: Res asst astron, Ind Univ, 60-62; NASA fel, Warner & Swasey Observ, Case Western Reserve Univ, 63-65; from instr to assoc prof, Univ Mich, 65-72; assoc prof, Univ Kans, 72; vis assoc prof, Tel-Aviv Univ, 72-75; PRIN RES FEL, ROYAL GREENWICH OBSERV, 75- Concurrent Pos: NSF res grant, Kitt Peak Nat Observ, 66. Mem: Am Astron Soc; Royal Astron Soc; Int Astron Union. Res: Astronomical instrumentation; photoelectric photometry; stellar spectroscopy; Galilean satellites; cometary spectroscopy; electronography. Mailing Add: Royal Greenwich Observ Herstmonceux Castle Hailsham East Sussex England

WEHLAU, AMELIA W, b Berkeley, Calif, Feb 5, 30; m 50; c 4. ASTRONOMY. Educ: Univ Calif, Berkeley, AB, 49, PhD(astron), 53. Prof Exp: Lectr, 65-71, ASST PROF ASTRON, UNIV WESTERN ONT, 71- Concurrent Pos: Mem, Can Comt, Int Astron Union, 67-70. Mem: Am Astron Soc; Royal Astron Soc Can; Can Astron Soc. Res: Variable stars in globular clusters. Mailing Add: Dept of Astron Univ of Western Ont London ON Can

WEHLAU, WILLIAM HENRY, b San Francisco, Calif, Apr 7, 26; m 50; c 4. ASTRONOMY. Educ: Univ Calif, AB, 49, PhD(astron), 53. Prof Exp: Instr astron, Case Inst Technol, 53-55; assoc prof, 55-65, PROF ASTRON, UNIV WESTERN ONT, 65-, HEAD DEPT, 55- Mem: Am Astron Soc; Royal Astron Soc Can; Can Astron Soc. Res: Stellar spectroscopy and photometry; astrophysics. Mailing Add: Dept of Astron Univ of Western Ont London ON Can

WEHMAN, ANTHONY THEODORE, b Jamaica, NY, June 6, 42. ORGANIC CHEMISTRY, ORGANOMETALLIC CHEMISTRY. Educ: Univ Detroit, BChE, 65; Univ Del, PhD(chem), 69. Prof Exp: NIH fel chem, Mass Inst Technol, 69-70; instr, Univ Del, 70-71; ASST PROF CHEM, US COAST GUARD ACAD, 71- Concurrent Pos: Comt mem panel on hazardous mat, Nat Acad Sci, 71- Mem: Am Chem Soc; The Chem Soc. Res: Aromatic metal complexes; organic photochemical reactions; benzyne chemistry. Mailing Add: Dept of Phys Sci US Coast Guard Acad New London CT 06320

WEHMAN, HENRY JOSEPH, b Cincinnati, Ohio, Dec 21, 37; m 69; c 2. COMPARATIVE NEUROLOGY. Educ: Spring Hill Col, BS, 63; Johns Hopkins Univ, PhD(biol), 69. Prof Exp: Lab scientist electron micros, Rosewood State Hosp, 69-73; ASST PROF BIOL, KING'S COL, PA, 73- Concurrent Pos: Res assoc prof pediat, Sch Med, Univ Md, 70-73; mem bd dirs, Tech Info Proj, Washington, DC, 74- Mem: AAAS; Am Soc Cell Biol; Soc Neurosci. Res: Electron microscopy of the developing nervous system in mammalian fetuses; effects of maternally introduced agents on the development of fetal nerves. Mailing Add: Dept of Biol King's Col Wilkes-Barre PA 18702

WEHMANN, ALAN AHLERS, b New York, NY, Dec 28, 40; m 68; c 1. PARTICLE PHYSICS. Educ: Rensselaer Polytech Inst, BS, 62; Harvard Univ, MA, 63, PhD(physics), 68. Prof Exp: Res assoc physics, Univ Rochester, 67-69; PHYSICIST I, MESON LAB SECT, FERMI ACCELERATOR LAB, 69- Mem: Am Phys Soc. Res: Experimental high energy particle physics. Mailing Add: Meson Lab Sect Fermi Accelerator Lab Box 500 Batavia IL 60501

WEHNER, ALFRED PETER, b Wiesbaden, Ger, Oct 23, 26; US citizen; m 55; c 4. MEDICINE. Educ: Gutenberg Univ, Dr Med Dent, 53. Prof Exp: Pvt pract, WGer, 51-53; Dr med dent, Guggenheim Dent Clin, 53-54; Dr med dent, 7100th Hosp, US Air Force, Europe, 54-56; res asst microbiol, Field Res Lab, Mobil Oil Co, Tex, 57-62; sr res scientist biomed res, Biomet Instrument Corp, 62-64; dir & pres, Electro-Aerosol Inst, Inc, 64-67; prof biol & chmn dept sci, Univ Plano, 66-67; RES ASSOC INHALATION TOXICOL, PAC NORTHWEST LABS, BATTELLE MEM INST, 67- Concurrent Pos: Fel clin pedodontia, Guggenheim Dent Clin, 53-54; consult, Vet Admin Hosp, McKinney, Tex, 63-65; US rep, Int Soc Biometeorol, 72- Mem: Int Soc Biometeorol; Am Soc Microbiol; Soc Exp Biol & Med; Sigma Xi; fel Int Soc Med Hydrol & Climat. Res: Inhalation toxicology of air pollutants; biological effects of electro-aerosols and air ions; bioclimatology. Mailing Add: Dept of Biol Pac Northwest Labs Battelle Mem Inst Richland WA 99352

WEHNER, DONALD C, b Middletown, NY, Apr 1, 29; m 50; c 3. WATER POLLUTION, ENVIRONMENTAL SCIENCES. Educ: Univ Bridgeport, BA, 51. Prof Exp: Biologist, Lederle Div, 51-54 & 56-58, res biologist, Indust Chem Div, 58-69, asst to dept head, Water Treating Chem Dept, 69-70. microbiologist-field engr, 70-74, dist sales mgr, 74-75, TECH SPECIALIST, PAPER CHEMS DEPT, AM CYANAMID CO, 75- Concurrent Pos: Mem subcomt biol anal of waters for sub-surface injection, Am Petrol Inst, 59-64. Mem: Soc Indust Microbiol(treas, 66-68, vpres, 69-70, pres, 71-72); Am Inst Biol Sci; Water Pollution Control Fedn; Tech Asn Pulp & Paper Indust. Res: Research and development of industrial algicides, bactericides and fungicides. Mailing Add: Am Cyanamid Co PO Box 868 Mobile AL 36601

WEHNER, GOTTFRIED KARL, b Ger, Sept 23, 10; nat US; m 39; c 3. PHYSICS. Educ: Inst Technol, Munich, Ger, dipl, 36, DrIng, 39. Prof Exp: Physicist, Inst Technol, Munich, 39-45, Wright-Patterson Air Force Base, Ohio, 47-55 & Gen Mills, Inc, 55-63; dir appl sci div, Litton Industs, Inc, 63-68; PROF ELEC ENG, UNIV MINN, MINNEAPOLIS, 68- Honors & Awards: Welch Medal, Am Vacuum Soc, 71. Mem: Fel Am Phys Soc; sr mem Inst Elec & Electronics Engrs; sr mem Am Vacuum Soc; Ger Phys Soc; Europ Phys Soc. Res: Plasma and surface physics; thin films; vacuum physics; sputtering. Mailing Add: Dept of Elec Eng Univ of Minn Minneapolis MN 55455

WEHNER, PHILIP, b Chicago, Ill, July 21, 17; m 43; c 2. INDUSTRIAL ORGANIC CHEMISTRY. Educ: Univ Chicago, PhD(org chem), 43. Prof Exp: Res chemist, Gen Aniline & Film Corp, Pa, 43-46; assoc chemist, Argonne Nat Lab, 46-52; chemist, Ciba States Ltd, 52-55; chemist, Toms River-Cincinnati Chem Corp, 55-60, mgr res & develop, 60-64, vpres, 64-68, pres, 68-72; VPRES PROD & TECH DEVELOP, DYESTUFFS & CHEM DIV, CIBA-GEIGY CORP, 72- Mem: Am Chem Soc; Am Asn Textile Chem & Colorists. Mailing Add: Dyestuffs & Chem Div Ciba-Geigy Corp Box 11422 Greensboro NC 27409

WEHR, HENRY WILLIAM, JR, b Evansville, Ind, July 6, 18; m 42; c 3. ORGANIC CHEMISTRY. Educ: Ind Univ, BS, 40. Prof Exp: Anal chemist, 40-44, res chemist, 46-50, tech serv engr, 50-56, sect head, Plastics, 56-63, SR PROJ SPECIALIST, PLASTICS, DOW CHEM CO, 63- Mem: Am Chem Soc; Soc Plastics Eng. Res: Development of plastics materials for commercial uses. Mailing Add: 36 Lexington Court Midland MI 48640

WEHRENBERG, JOHN P, b Springfield, Ill, Aug 10, 27; m 66. MINERALOGY. Educ: Univ Mo, BS, 50; Univ Ill, MS, 52, PhD, 56. Prof Exp: From asst prof to assoc prof, 55-66, PROF GEOL, UNIV MONT, 66- Mem: Mineral Soc Am; Geochem Soc; Am Crystallog Asn. Res: Solid state processes in geology; crystallography; infrared spectra of minerals. Mailing Add: Dept of Geol Univ of Mont Missoula MT 59801

WEHRENBERG, PAUL JAMES, b St Louis, Mo, Dec 13, 42. ATOMIC PHYSICS, SPACE PHYSICS. Educ: Mass Inst Technol, SB, 63; Univ Wash, MS, 66, PhD(physics), 72. Prof Exp: Engr, ITT Indust Labs, Int Tel & Tel Corp, 62 & Lockheed Aircraft Corp, 63; RES ASSOC ATOMIC & SPACE PHYSICS, UNIV WASH, 72- Mem: Am Inst Physics; Am Geophys Union; Sigma Xi. Res: Experimental studies of atomic and molecular collision processes; photometric studies of auroral phenomena; development of detection systems for single visible photons and low energy ions. Mailing Add: 13706 39th Ave NE Seattle WA 98125

WEHRLE, LOUIS, JR, b Santa Monica, Calif, Apr 2, 35; m 63. FOOD MICROBIOLOGY. Educ: Univ Toledo, BS, 57; Ohio State Univ, MS, 67, PhD(microbiol), 69. Prof Exp: Asst prof microbiol, Ohio State Univ, 69-72; SR RES SCIENTIST, CENT STATES CAN CO, MASSILLON, 72- Mem: Inst Food Technologists. Res: Thermoprocessing of food products. Mailing Add: 1433 Stonington Rd NW North Canton OH 44720

WEHRLE, PAUL F, b Ithaca, NY, Dec 18, 21; m 44; c 4. PEDIATRICS, MICROBIOLOGY. Educ: Univ Ariz, BS, 47; Tulane Univ, MD, 47. Prof Exp: Clin instr pediat, Univ Ill Col Med, 50-51; res assoc epidemiol & microbiol, Grad Sch Pub Health, Univ Pittsburgh, 51-53; res assoc, Poliomyelitis Lab, Johns Hopkins Univ, 53-

55; from asst prof to assoc prof pediat, Col Med, State Univ NY Upstate Med Ctr, 55-61, actg chmn dept microbiol, 59-61; prof, 61-71, HASTINGS PROF PEDIAT, SCH MED, UNIV SOUTHERN CALIF, 71- Concurrent Pos: Asst med supt, Chicago Munic Contagious Dis Hosp, Ill, 50-51; head physician contagious dis serv, Los Angeles County-Univ Southern Calif Med Ctr, 61-63, chief physician, Children's Div, 63- Mem: AAAS; Soc Pediat Res; Am Soc Microbiol; Am Pub Health Asn; Am Asn Immunologists. Res: Viral infections in man, especially enteroviruses; antibiotic action; epidemiology of infectious diseases. Mailing Add: Los Angeles County-Univ of Southern Calif Med Ctr Los Angeles CA 90033

WEHRLI, PIUS ANTON, b Bazenheid, Switz, July 27, 33; m 62; c 1. CHEMISTRY. Educ: Swiss Fed Inst Technol, Dipl Chem Eng, 64, PhD(chem), 67. Prof Exp: Sr chemist, 67-69, res fel, 69-70, res group chief, Chem Res Dept, 70-73, RES SECT CHIEF, CHEM RES DEPT, HOFFMANN-LA ROCHE INC, 73- Mem: Am Chem Soc; Swiss Chem Soc; Soc Ger Chem. Res: Development in synthetic organic chemistry. Mailing Add: Chem Res Dept Kingsland Rd Hoffmann-La Roche Inc Nutley NJ 07110

WEHRLI, ROBERT L, b New York, NY, Feb 2, 22; m 44; c 2. INSTRUMENTATION. Educ: Rensselaer Polytech Inst, BS, 47, MS, 48. Prof Exp: Res physicist, Linde Air Prod Div, Carbide & Carbon Corp, 48-49; res physicist, Reaction Motors, Inc, 49-50, chief physicist, 50-52, chief res, 52-54; dir res, Aeronaut & Instrument Div, Robertshaw-Fulton Controls Co, 54-56, vpres & gen mgr, 56-60, vpres & asst to pres, 60-64, mgt dir, Geneva, Switz, 61-62, exec vpres, 64-67; vpres, Dresser Indust, Inc, Tex, 67-70; pres, Cybercom Corp, 70-71; VPRES & DIV MGR, INSTRUMENT DIV, VARIAN ASSOCS, PALO ALTO, 71- Mem: Am Inst Aeronaut & Astronaut; Instrument Soc Am. Mailing Add: 15 Deep Well Lane Los Altos CA 94022

WEHRMEISTER, HERBERT LOUIS, b Chicago, Ill, Nov 8, 20; m 53; c 3. CHEMISTRY. Educ: Ill Inst Technol, BS, 44; Northwestern Univ, MS, 46, PhD(chem), 48. Prof Exp: Lab asst, Portland Cement Asn, Ill, 38-40; control chemist, W H Barber Co, 40-44; res chemist, Miner Labs, 44-46; RES CHEMIST, COMMERCIAL SOLVENTS CORP, 49- Mem: Am Inst Chem; Am Chem Soc. Res: Organic chemistry; derivatives of sulfenic acids; syntheses in the thiazole series; nitoparaffin and hydroxylamine derivatives; zearalenone and derivatives. Mailing Add: Res Dept Commercial Solvents Corp Terre Haute IN 47808

WEHRY, EARL L, JR, b Reading, Pa, Feb 13, 41. ANALYTICAL CHEMISTRY. Educ: Juniata Col, BS, 62; Purdue Univ, PhD(chem), 65. Prof Exp: Instr chem, Ind Univ, 65-66, asst prof, 66-70; asst prof, 70-72, ASSOC PROF CHEM, UNIV TENN, KNOXVILLE, 72- Mem: AAAS; Soc Appl Spectros; Am Chem Soc. Res: Fluorescence and phosphorescence; photochemistry. Mailing Add: Dept of Chem Univ of Tenn Knoxville TN 37916

WEHUNT, RALPH LEE, b Atlanta, Ga, Dec 2, 22; m 43; c 2. SOIL FERTILITY, PLANT PHYSIOLOGY. Educ: Univ Ga, BSA, 46, MSA, 48; Rutgers Univ, New Brunswick, PhD(soil chem), 53. Prof Exp: Educ mgr agron, Chilean Nitrate Educ Bur, Inc, 53-55; soils & fertilizer specialist, Agr Exten Serv, Univ Ga, 55-59; agronomist, Tenn Valley Auth, 59-62, head educ & commun serv, 62-65; chief agronomist, Cities Serv Co, 65-68, mgr personnel develop, CFS Div, 68-71; ACCT SUPVR, DOANE AGR SERV, 71- Concurrent Pos: Consult with several large corps. Mem: Am Soc Agron; Soil Sci Soc Am; Am Soc Training & Develop. Res: Soil chemistry; educational psychology of farmers; market research. Mailing Add: Doane Agr Servs St Louis MO 63144

WEI, CHIN HSUAN, b Yuanlin, Taiwan, Oct 25, 26; m 54; c 2. X-RAY CRYSTALLOGRAPHY. Educ: Cheng Kung Univ, BS, 50; Purdue Univ, MS, 58; Univ Wis, Madison, PhD(phys chem), 62. Prof Exp: Asst phys chem, Cheng Kung Univ, 50-55, instr gen chem, 55-56; res assoc struct chem, Univ Wis, Madison, 62-65, sr spectroscopist, 65-66; BIOPHYSICIST, OAK RIDGE NAT LAB, 66- Mem: Am Chem Soc; Am Crystallog Asn. Res: Structure determination of organometallic complexes and compounds of biological interest; isolation, purification and characterization of proteins. Mailing Add: Biol Div Oak Ridge Nat Lab Oak Ridge TN 37830

WEI, CHUAN-TSENG, b Kaifeng, China, May 7, 22; m 46; c 2. METALLURGY. Educ: Ord Eng Col, China, BS, 44; Univ Ill, MS, 56, PhD(metall), 59. Prof Exp: Res assoc phys metall, Univ Ill, 59-60; assoc prof, 60-67, PROF METALL, MICH STATE UNIV, 67- Mem: Am Soc Metals; Am Phys Soc; Am Inst Mining, Metall & Petrol Eng. Res: Physical metallurgy and solid state physics, especially in x-ray diffraction, crystal imperfections, plastic deformation, low temperature physics and electronic structure of metals. Mailing Add: Col of Eng Mich State Univ East Lansing MI 48823

WEI, CHUNG-CHEN, b Taiwan, Apr 9, 40; m 65; c 1. ORGANIC CHEMISTRY, PHOTOCHEMISTRY. Educ: Nat Taiwan Univ, BS, 63; Colo State Univ, PhD(org chem), 69. Prof Exp: Fel org chem, Johns Hopkins Univ, 69-71; res assoc bio-org chem, Yale Univ, 71-73; assoc chemist, Midwest Res Inst, 73; SR CHEMIST, HOFFMANN-LA ROCHE INC, 74- Mem: Am Chem Soc. Res: Synthesis of heterocyclic compounds and natural products; organic photochemistry. Mailing Add: Chem Res Dept Hoffmann-La Roche Inc Nutley NJ 07110

WEI, DIANA YUN DEE, b Che-Kiang, China, June 8, 30; m 62; c 1. MATHEMATICS. Educ: Taiwan Norm Univ, BS & BEd, 53; Univ Nebr, MS, 60; McGill Univ, PhD(math), 67. Prof Exp: Lectr math, Taipei Inst Technol, Taiwan, 53-58; teaching asst, Univ Nebr, 58-60 & Univ Wash, 60-62; lectr, McGill Univ, 62-65; asst prof, Marianopolis Col, 65-68 & Sir George Williams Univ, 68-72; prof, Sch Comn Baldwin-Cartier, 72-75; PROF MATH, NORFOLK STATE COL, 75- Concurrent Pos: Nat Res Coun Can grant award, 69, 70, 71. Mem: Am Math Soc; Can Math Cong. Res: Groups, rings and modules; homology; category; linear algebra. Mailing Add: Dept of Math Norfolk State Col Norfolk VA 23502

WEI, EDDIE TAK-FUNG, b Shanghai, China, Dec 6, 44; m 66; c 2. PHARMACOLOGY, TOXICOLOGY. Educ: Univ Calif, Berkeley, AB, 65; Univ Calif, San Francisco, PhD(pharmacol), 69. Prof Exp: Nat Inst Arthritis & Metab Dis fel, Stanford Univ, 69-70; asst prof, 70-75, ASSOC PROF ENVIRON TOXICOL, SCH PUB HEALTH, UNIV CALIF, BERKELEY, 75-; ASSOC PROF TOXICOL, DEPT PHARMACOL, UNIV CALIF, SAN FRANCISCO, 75- Concurrent Pos: Nat Inst Drug Abuse fel, 70-77; Nat Inst Environ Health Sci fel, 74-76. Mem: AAAS; Soc Toxicol; Am Soc Pharmacol & Exp Therapeut. Res: Biological mechanisms of morphine dependence; toxic chemicals and their mechanisms of action. Mailing Add: Sch of Pub Health Univ of Calif Berkeley CA 94720

WEI, GUANG-JONG JASON, b Fu-Chou, China, Mar 14, 46; m 71. PHYSICAL BIOCHEMISTRY. Educ: Cheng Kung Univ, Taiwan, BS, 68; Univ Ill, Urbana, PhD(phys chem), 74. Prof Exp: Teaching officer math, Army Chem Corps Sch, Taiwan, 68-69; res assoc, 74-75, NIH FEL PHYS BIOCHEM, MICH STATE UNIV, 75- Mem: Am Chem Soc. Res: Kinetic and equilibrium studies of interacting macromolecules and their biological functions. Mailing Add: 224 Biochem Mich State Univ East Lansing MI 48824

WEI, LUN-SHIN, b Hou-long, Formosa, Jan 14, 29; m 56; c 4. FOOD SCIENCE. Educ: Taiwan Prov Col Agr, BS, 51; Univ Ill, MS, 55, PhD, 58. Prof Exp: Asst agron, 55-57, sci analyst, 57-59, res assoc food sci, 59-64, ASSOC PROF FOOD SCI, UNIV ILL, URBANA, 64- Honors & Awards: Educ & Res Award, Land of Lincoln Soybean Asn, 73. Mem: AAAS; Am Chem Soc; Inst Food Technologists. Res: Foods and plant materials analysis; food processing and preservation; product development; food utilization of soybeans. Mailing Add: South Wing Hort Field Lab Univ of Ill Urbana IL 61801

WEI, PAX SAMUEL PIN, b Chungking, China, Nov 11, 38; US citizen; m 64; c 3. PHYSICAL CHEMISTRY. Educ: Nat Taiwan Univ, BS, 60; Univ Ill, Urbana, MS, 63; Calif Inst Technol, PhD(chem), 68. Prof Exp: Mem tech staff, Bell Tel Labs, NJ, 67-69; res scientist, Boeing Sci Res Labs, 69-71, RES SCIENTIST, BOEING AEROSPACE CO, 72- Mem: Am Chem Soc; Am Phys Soc. Res: Atomic and molecular spectroscopy; surface science; electron diffraction; laser effects. Mailing Add: Boeing Aerospace Co PO Box 3999 Seattle WA 98124

WEI, PETER ENTIEN, organic chemistry, see 12th edition

WEI, PETER HSING-LIEN, b Shantung, China, Feb 11, 22; m 48; c 4. MEDICINAL CHEMISTRY. Educ: St John's Univ, China, BS, 48; Columbia Univ, MA, 53; Univ Pa, PhD, 69. Prof Exp: Res chemist, Norwich Pharmacal Col, 52-60; RES CHEMIST, WYETH LABS, INC, 60- Mem: Am Chem Soc. Res: Pharmaceuticals; synthesis of heterocyclic compounds of biological interest. Mailing Add: L2-20 Med Chem Dept Wyeth Labs Inc Radnor PA 19087

WEI, STEPHEN HON YIN, b Shanghai, China, Sept 17, 37; US citizen; m 63; c 2. PEDODONTICS, HISTOLOGY. Educ: Univ Adelaide, BDS, 61, Hons, 62, MDS, 65; Univ Ill, Chicago, MS, 67; Univ Iowa, DDS, 71. Prof Exp: Dent surgeon, Royal Adelaide Hosp, Univ Adelaide, 63-64, teaching registr, Queen Elizabeth Hosp, 65; res asst pedodontics, Univ Ill, Chicago, 65-66, instr, 67; from asst prof to assoc prof, 67-74, PROF PEDODONTICS, COL DENT, UNIV IOWA, 74- Concurrent Pos: Dentist, Vet Admin Hosp, 69-, consult, 74; consult, Res Comt, Am Acad Pedodont, 71- Mem: Am Dent Asn; Am Asn Hosp Dent; Int Asn Dent Res; Am Acad Pedodontics; Am Soc Dent for Children. Res: Preventive dentistry; remineralization of teeth; systemic and topical fluoride therapy; electron optical studies of dental and other hard tissues; clinical research in pedodontics and dental caries. Mailing Add: Rm 203 Col of Dent Univ of Iowa Iowa City IA 52242

WEI, WHUA FU, b Laipor, China, Oct 27, 20; US citizen; m 66; c 1. SOLID STATE PHYSICS. Educ: Nat Chekiang Univ, BS, 43; Okla State Univ, MS, 51, PhD(physics), 65. Prof Exp: Instr gen physics, Nat Chekiang Univ, 43-49; physicist, Delco Radio Div, Gen Motors Corp, 55-58; res asst semiconductor, Univ Colo, 58-60; res physicist, Brown Eng, 65-66; ASSOC PROF PHYSICS, ARK STATE UNIV, 66- Res: Magnetic and optical properties of semiconductors. Mailing Add: Div of Math & Physics Ark State Univ State University AR 72467

WEIANT, ELIZABETH ABBOTT, b New Britain, Conn, July 4, 18. PHYSIOLOGY. Educ: Tufts Col, BS & MS, 43; Radcliffe Univ, MA, 52; Boston Univ, EdD, 70. Prof Exp: Asst biol, Tufts Univ, 43-45, res assoc, 45-47, instr endocrinol, 47-56, asst prof biol, 56-61; assoc prof, 61-72, ASSOC PROF BIOL, SIMMONS COL, 72- Concurrent Pos: Res biologist, Arthur D Little, 60-61; mem staff, Prod Validation, Cordis Corp, 70. Mem: Am Soc Zool; NY Acad Sci; Am Inst Biol Sci. Res: Experimental embryology; electrophysiology; effect of high temperature on rats; control of spontaneous activity in certain efferent nerve fibers. Mailing Add: Dept of Biol Simmons Col 300 The Fenway Boston MA 02115

WEIBEL, ARMELLA, b Ewing, Nebr, Feb 7, 20. MATHEMATICS. Educ: Alverno Col, BSE, 46; Univ Wis, MS, 52. Prof Exp: Teacher, St Clara Sch, Ill, 38-48; teacher, Frankenstein High Sch, Mo, 48-50; from instr to asst prof, 52-53, ASSOC PROF MATH, ALVERNO COL, 56- Concurrent Pos: Dir, Alverno Ctr, NSF Minn Math & Sci Teaching Proj, 63-70. Mem: Nat Coun Teachers Math. Res: Math education. Mailing Add: 3401 S 39th St Milwaukee WI 53215

WEIBEL, DALE ELDON, b De Witt, Nebr, Dec 14, 20; m 42; c 4. AGRONOMY. Educ: Univ Nebr, BSc, 42, MSc, 47; Iowa State Univ, PhD(plant breeding, path), 55. Prof Exp: Res agronomist, Div Cereal Crops & Dis, USDA, Kans, 47-53, field crops res br, Tex, 53-58; assoc prof, 58-61, PROF AGRON, OKLA STATE UNIV, 61- Mem: Crop Sci Soc Am; Am Soc Agron. Res: Sorghum breeding and genetics. Mailing Add: Dept of Agron Okla State Univ Stillwater OK 74074

WEIBEL, MICHAEL KENT, b Sioux City, Iowa, May 30, 41; m 65; c 2. BIOCHEMISTRY, ENZYMOLOGY. Educ: Iowa State Univ, BS, 64; Purdue Univ, West Lafayette, PhD(phys chem), 68. Prof Exp: ASST PROF BIOCHEM, SCH MED, UNIV PA, 72- Concurrent Pos: NIH fel, Sch Med, Univ Pa, 68-71. Res: Enzyme technology. Mailing Add: Dept of Biochem Univ of Pa Sch of Med Philadelphia PA 19174

WEIBLEN, PAUL WILLARD, b Miller, SDak, Feb 15, 27; m 67; c 1. GEOLOGY, PETROLOGY. Educ: Wartburg Col, BA, 50; Univ Minn, MA, 52, MS & PhD(geol), 65. Prof Exp: Asst prof, 65-71, ASSOC PROF GEOL, UNIV MINN, MINNEAPOLIS, 71- Mem: Electron Probe Anal Soc Am. Res: Petrology, especially the study of gabbroic rocks and associated mineralization; lunar petrology; application of electron probe analysis to problems in mineralogy, geochemistry and petrology. Mailing Add: Dept of Geol & Geophys Univ of Minn Minneapolis MN 55455

WEIBRECHT, WALTER EUGENE, b New York, NY, June 25, 37. INORGANIC CHEMISTRY. Educ: Franklin & Marshall Col, BS, 59; Cornell Univ, PhD(chem), 64. Prof Exp: Am Oil fel chem, Harvard Univ, 63-64; asst prof, Mich State Univ, 64-66; ASST PROF CHEM, UNIV MASS, BOSTON, 66- Mem: AAAS; Am Chem Soc; The Chem Soc. Res: Studies involving borazine as a Lewis acid; B-hydroxyborazines; synthesis; silicon-nitrogen bond cleavage in symmetrically and unsymmetrically alkoxylated silazanes; silicon, germanium and tin transamination equilibria. Mailing Add: Dept of Chem Univ of Mass Boston MA 02116

WEIBUST, ROBERT SMITH, b Newport, RI, May 6, 42. GENETICS, ZOOLOGY. Educ: Colby Col, AB, 64; Univ Maine, Orono, MS, 66, PhD(zool), 70. Prof Exp: Nat Cancer Inst fel, Jackson Lab, 64-65, Nat Inst Gen Med Sci fel, Jackson Lab & Univ Maine, Orono, 68-70, fel, Jackson Lab, 70; asst prof, 70-75, ASSOC PROF BIOL, MOORHEAD STATE UNIV, 75- Mem: AAAS; Genetics Soc Am. Res: Mammalian genetics. Mailing Add: Dept of Biol Moorhead State Univ Moorhead MN 56560

WEICHEL, HUGO, b Selz, Ukraine, July 23, 37; US citizen; m 61; c 2. PHYSICS. Educ: Portland State Univ, BS, 60; US Air Force Inst Technol, MS, 65; Univ Ariz,

PhD, 72. Prof Exp: US Air Force, 60-, res physicist nuclear rocket propulsion, Edwards Air Force Base, Calif, 61-63, student arc plasmas, Aeronaut Res Lab, Ohio, 64-65, proj officer laser produced plasmas, Air Force Weapons Lab, NMex, 65-69, ASST PROF PHYSICS, US AIR FORCE INST TECHNOL, 72- Res: Lasers; optics; solid state; plasmas. Mailing Add: US Air Force Inst Technol (ENP) Wright-Patterson AFB OH 45433

WEICHENTHAL, BURTON ARTHUR, b Stanton, Nebr, Nov 7, 37; m 60; c 1. ANIMAL NUTRITION. Educ: Univ Nebr, BS, 59; SDak State Univ, MS, 62; Colo State Univ, PhD(nutrit), 67. Prof Exp: BEEF CATTLE EXTEN SPECIALIST, UNIV ILL, URBANA, 67- Mem: Am Soc Animal Sci. Res: Ruminant nutrition and physiology. Mailing Add: Dept of Animal Sci Univ of Ill Urbana IL 61801

WEICHERT, DIETER HORST, b Breslau, Ger, May 2, 32; Can citizen; m 62; c 2. SEISMOLOGY. Educ: Univ BC, BASc, 61, PhD(geophys), 65; McMaster Univ, MSc, 63. Prof Exp: Res asst geophys, Univ BC, 60-61, elec eng, Nat Res Coun Can, 61 & geophys, Univ Toronto, 63; RES SCIENTIST, EARTH PHYSICS BR, DEPT ENERGY & MINES RESOURCES, 65- Concurrent Pos: Vis scientist, Geophys Inst, Univ Karlsruhe, 70; UK Atomic Energy Authority, Geophys Inst, Frankfurt Univ, 71. Mem: Seismol Soc Am; Am Geophys Union; Can Geophys Union. Res: Geophysics. Mailing Add: Div of Seismol Earth Phys Br Dept of Energy Mines & Resources Ottawa ON Can

WEICHLEIN, RUSSELL GEORGE, b Dayton, Ohio, Apr 23, 15; m 48; c 1. MICROBIOLOGY. Educ: St Mary's Univ, Tex, BS, 39; Trinity Univ, Tex, MS, 55. Prof Exp: Tester, Ref Lab, Tex Co, 47; asst biologist, Found Appl Res, 48-52; assoc res bacteriologist, Sanit Sci Dept, Southwest Res Inst, 53-58, dept chem res, 58-65; from instr to asst prof, 65-75, ASSOC PROF BIOL SCI, SAN ANTONIO COL, 75- Res: Microbial genetics; biology; bacteriology. Mailing Add: Dept of Biol Sci San Antonio Col 1300 San Pedro San Antonio TX 78284

WEICHMAN, FRANK LUDWIG, b Liegnitz, Ger, Sept 23, 30; Can citizen; m 58; c 2. EXPERIMENTAL SOLID STATE PHYSICS. Educ: Brooklyn Col, BS, 53; Northwestern Univ, PhD(physics), 59. Prof Exp: From asst prof to assoc prof, 58-70, PROF PHYSICS, UNIV ALTA, 70- Concurrent Pos: Mem fac, Univ Strasbourg, 65-66; vis prof, Technion, Haifa, Israel, 73-74. Mem: Can Asn Physicists; Am Phys Soc. Res: Optical and electrical properties of oxide semiconductors with an emphasis on the role of excitons and defect structure on photoconductivity, luminescence and electroluminescence of O. Mailing Add: Dept of Phys Univ of Alta Edmonton AB Can

WEICHSEL, MORTON E, JR, b Pueblo, Colo, June 17, 33; m 62; c 3. PEDIATRICS, NEUROLOGY. Educ: Univ Colo, Boulder, BA, 55; Univ Buffalo, MD, 62. Prof Exp: Intern pediat, Buffalo Children's Hosp, NY, 63; resident, Med Ctr, Stanford Univ, 63-65, fel pediat neurol, 65-67; fel Med Ctr, Univ Colo, 67-68; clin instr pediat, Med Ctr, Stanford Univ, 68-69, fel develop neurochem, 69-71; asst prof human develop & med, Col Human Med, Mich State Univ, 71-74; ASSOC PROF PEDIAT & NEUROL, HARBOR GEN HOSP & SCH MED, UNIV CALIF, LOS ANGELES, 74- Mem: Soc Neurosci; Am Acad Neurol; Soc Pediat Res; Child Neurol Soc; Am Fedn Clin Res. Res: Developmental neurochemistry. Mailing Add: Harbor Gen Hosp 1000 W Carson St Torrance CA 90509

WEICHSEL, PAUL M, b New York, NY, July 22, 31; m 55; c 3. MATHEMATICS. Educ: City Col New York, BS, 53; NY Univ, MS, 54; Calif Inst Technol, PhD(math), 60. Prof Exp: From instr to asst prof math, Univ Ill, Urbana, 60-65; NATO fel, Math Inst, Oxford Eng, 61-62; res fel, Inst Advan Studies, Australian Nat Univ, 65-66; from asst prof to assoc prof, 66-75, PROF MATH, UNIV ILL, URBANA, 75- Concurrent Pos: Part time consult, Argonne Nat Lab, 63-64; vis prof, Hebrew Univ, Jerusalem, 70-71 & Univ Tel Aviv, Israel, 71-72. Mem: Am Math Soc; Math Asn Am. Res: Theory of finite groups; algebra; theory of graphs. Mailing Add: Dept of Math Univ of Ill Urbana IL 61801

WEICHSELBAUM, THEODORE EDWIN, b Macon, Ga, Mar 22, 08; m 34; c 2. BIOCHEMISTRY. Educ: Emory Univ, BS, 30; Univ Edinburgh, PhD(biochem), 35; Am Bd Clin Chem, dipl. Prof Exp: Assoc biochemist, Jewish Hosp, St Louis, Mo, 36-40; res dir, A S Aloe Co, 46-47; asst prof exp surg, Sch Med, Wash Univ, 48-53, assoc prof, 53-71; CORP SR SCIENTIST, SHERWOOD MED INDUSTS, INC, 66- Concurrent Pos: Consult, St Luke's & St Louis City Hosps, 50- Mem: Am Chem Soc; Am Soc Biol Chem; Soc Exp Biol & Med. Res: Sulfur amino acids; ketone body metabolism; estimation of proteins; instrumentation in flame photometry; electrolyte metabolism; metabolism of fructose in man; metabolism and isolation of adrenocortical steoids. Mailing Add: Sherwood Med Industs Inc 5373 Bermuda Rd St Louis MO 63121

WEICK, CHARLES FREDERICK, b Buffalo, NY, Jan 19, 31; m 59; c 2. INORGANIC CHEMISTRY. Educ: Mt Union Col, BS, 53; Univ Rochester, PhD, 59. Prof Exp: ASSOC PROF CHEM, UNION COL, NY, 58- Concurrent Pos: Fel, Univ Kent, Canterbury, Eng, 71-72. Mem: Am Chem Soc. Res: Coordination chemistry; radiochemistry. Mailing Add: Dept of Chem Union Col Schenectady NY 12308

WEICKMANN, HELMUT K, b Munich, Ger, Mar 10, 15; US citizen; m 45; c 4. ATMOSPHERIC PHYSICS, METEOROLOGY. Educ: Univ Frankfurt, PhD(geophys), 39. Hon Degrees: Dr, Univ Clermont-Ferrand, France, 75. Prof Exp: Flight meteorologist, Ger Flight Inst, 39-45; dir high altitude observ, Ger Weather Serv, 45-49; physicist, Atmospheric Physics Br, US Army Electronics Command, 49-61, chief, 61-65; DIR, ATMOSPHERIC PHYSICS & CHEM LAB, NAT OCEANIC & ATMOSPHERIC ADMIN, 65- Concurrent Pos: Co-ed, J Atmospheric Sci. Honors & Awards: Tech Achievement Award, Signal Corps, US Army, 61, Cert of Achievement, Electronics Command, 62. Mem: Fel Am Meteorol Soc; Am Geophys Union; fel AAAS; hon mem Int Asn Meteorol & Atmospheric Physics; Int Union Geod & Geophys. Res: Meteorological research; cloud and weather modification; geophysics. Mailing Add: Atmospheric Physics & Chem Lab Nat Oceanic & Atmospheric Admin Boulder CO 80303

WEIDANZ, WILLIAM P, b Jackson Heights, NY, Jan 30, 35; m 60; c 6. IMMUNOLOGY, MEDICAL MICROBIOLOGY. Educ: Rutgers Univ, BS, 56; Univ RI, MS, 58; Tulane Univ, PhD(microbiol), 61. Prof Exp: NIH fel, 61-64; asst prof immunol & pathogenic bact, La State Univ, 64-66; asst prof, 66-70, ASSOC PROF MICROBIOL, HAHNEMANN MED COL, 70- Concurrent Pos: NIH res grant, 64-67 & 75-78; fac res coun fel, La State Univ, 65; Eleanor Roosevelt Int Cancer fel, Fibiger-Laboratoriet, Copenhagen, Denmark, 75-76. Mem: Am Asn Immunologists; Am Soc Microbiol. Res: Developmental immunology; immunologic deficiency diseases; immunity to malaria; tumor immunology. Mailing Add: Dept of Microbiol Hahnemann Med Col Philadelphia PA 19102

WEIDE, KENNETH DUANE, b Horton, Kans, May 19, 33; m 56; c 3. PATHOLOGY, VETERINARY MEDICINE. Educ: Kans State Univ, BS, 56, DVM & MS, 58; Mich State Univ, PhD(path), 62. Prof Exp: Pvt pract, Mo, 58-59; res instr, Ohio Agr Exp Sta, 59-62; assoc prof path & dir vet diag lab, Kans State Univ, 62-67; prof vet sci, head dept & dir, SDak Animal Dis Res & Diag Lab, SDak State Univ, 67-71; animal pathologist & exten vet, Univ Ariz, 71-73; DEAN COL VET MED, UNIV MO-COLUMBIA, 73- Concurrent Pos: Gnotobiotic res, Mich State Univ, 61-62; mem, Kans Mastitis Control Comt, 63-66; Kans Hog Cholera Eradication Comt, 63-67; contrib ed, Kans Farmer Mag & secy-coord, Kans SPF Swine Prog, 63-67; consult, Kans State Bd Health, 64-66; NIH grant, Nat Inst Allergy & Infectious Dis, 64-67; expert witness, US Dept Justice, 65-69; mem adv comt, Nat Swine Repop Asn, 66-68; chmn, NCent Conf Vet Lab Diagnosticians, 69-70, Western States Conf Vet Lab Diagnostician, 69-70 & Midwest Interprof Sem Dis Common to Animals & Man, 69-70; sect chmn, Western States Animal Health Conf, 70 & 72; mem comt animal health, Nat Acad Sci-Nat Res Coun, 70-73, task force to develop nat animal dis reporting syst, 71-72; consult, NY State Dept Agr & Markets, 70; consult, Ceres Land Co, Colo, 70-71; dir, Intermountain Vet Med Asn, 74-76. Mem: Am Vet Med Asn; Am Animal Health Asn; Am Asn Vet Lab Diag; Conf Res Workers Animal Dis; NY Acad Sci. Res: Diagnostic veterinary medicine; encephalitides; rabies; oncology; swine diseases; feedlot diseases; veterinary medical education and administration. Mailing Add: Col of Vet Med 104 Conaway Hall Univ Mo-Columbia Columbia MO 65201

WEIDEN, MATHIAS HERMAN JOSEPH, b Narrowsburg, NY, Nov 3, 23; m 54; c 4. INSECT TOXICOLOGY. Educ: Manhattan Col, BS, 46; Cornell Univ, PhD(entom), 54. Prof Exp: Chemist, Lederle Labs, Am Cyanamid Co, 46-47; asst entom, Cornell Univ, 47-48 & 51-52, asst prof insecticide chem, 54-59; entomologist, 59-66, RES SCIENTIST, UNION CARBIDE CORP, 66- Mem: Entom Soc Am; Am Chem Soc; NY Acad Sci. Res: Action of insecticides and detoxication mechanisms in insects; insecticide research and development. Mailing Add: Chems & Plastics Union Carbide Corp Box 8361 South Charleston WV 25303

WEIDENHAMER, HARRY E, b Roaring Spring, Pa, Aug 23, 07; m 31; c 2. GEOLOGY. Educ: Ashland Col, BS, 29; Ohio State Univ, MS, 33. Prof Exp: Chemist & metallurgist, Ohio Brass Co, 29-32, in charge testing & plant res, 34-45; supt processing, Mansfield Plating Co, 45-46; assoc prof chem & chmn dept, Ashland Col, 47-63, assoc prof geol, 63-65, prof & chmn dept, 65-74; RETIRED. Mem: AAAS; Am Chem Soc. Res: Analytical chemistry; Pleistocene geology. Mailing Add: 425 Samaritan Ashland OH 44805

WEIDENSAUL, T CRAIG, b Reedsville, Pa, Apr 4, 39; m 65; c 2. PLANT PATHOLOGY. Educ: Gettysburg Col, BA, 62; Duke Univ, MF, 63; Pa State Univ, PhD(plant path), 69. Prof Exp: Plant pathologist, USDA Forest Serv, 63-66; asst, Pa State Univ, 66-69, scholar, 69-70; asst prof, 70-73, ASSOC PROF PLANT PATH, OHIO STATE UNIV, 73-, DIR LAB ENVIRON STUDIES, OHIO AGR RES & DEVELOP CTR, 70- Concurrent Pos: Consult forester & plant pathologist; mem, Agr Pollution Tech Adv Bd, Ohio & NCent Regional Task Force Environ Qual. Mem: Am Inst Biol Sci; Am Phytopath Soc; Soc Am Foresters; Air Pollution Control Asn. Res: Epidemiology of fusarium canker of sugar maple; insect transmission of Fomes annosus; effects of gaseous and metallic air pollutants on plants and plant disease. Mailing Add: Ohio Agr Res & Develop Ctr Wooster OH 44691

WEIDHAAS, DONALD E, b Northampton Mass, Feb 12, 28; m 53; c 2. MEDICAL ENTOMOLOGY. Educ: Univ Mass, BS, 51; State Univ NY, PhD(entom), 55. Prof Exp: Med entomologist, 56-62, asst chief, 62-67, actg chief, 67, DIR INSECTS AFFECTING MAN RES LAB, ENTOM RES DIV, AGR RES SERV, USDA, 67- Mem: Entom Soc Am; Am Mosquito Control Asn. Res: Chemical and other methods of control, resistance studies, physiology and toxicology of insects affecting man or of medical importance. Mailing Add: Agr Res Serv USDA Box 14565 Gainesville FL 32604

WEIDHAAS, JOHN AUGUST, JR, b Northampton, Mass, Oct 13, 25; m 48; c 5. ENTOMOLOGY. Educ: Univ of Mass, BS, 49, MS, 52, PhD(entom), 59. Prof Exp: Instr entom, Univ Mass, 51 & 53-59; from asst prof to assoc prof, Cornell Univ, 59-67; ASSOC PROF ENTOM & EXTEN SPECIALIST, VA POLYTECH INST & STATE UNIV, 67- Concurrent Pos: Collabr, Agr Res Serv, USDA; consult, Elm Res Inst. Mem: Soc Am Foresters; Entom Soc Am; Entom Soc Can. Res: Forest, ornamental and beneficial insects; Acarina. Mailing Add: Dept of Entom 315 Price Hall Va Polytech Inst & State Univ Blacksburg VA 24061

WEIDIE, ALFRED EDWARD, b New Orleans, La, July 31, 31; m 60; c 3. GEOLOGY. Educ: Vanderbilt Univ, BA, 53; La State Univ, MS, 58, PhD(geol), 61. Prof Exp: From asst prof to assoc prof, 61-71, PROF GEOL, LA STATE UNIV, NEW ORLEANS, 71-, CHMN DEPT EARTH SCI, 67- Res: Structural geology and physical stratigraphy. Mailing Add: Dept of Earth Sci La State Univ New Orleans LA 70122

WEIDIG, CHARLES FERDINAND, b Houston, Tex, Oct 24, 45; m 67; c 2. INORGANIC CHEMISTRY, BIOINORGANIC CHEMISTRY. Educ: Tex Christian Univ, BS, 69, PhD(chem), 74; Sam Houston State Univ, MA, 71. Prof Exp: Anal chemist, Indust Labs, Ft Worth, 66-69; ASSOC BIOCHEM, EDSEL B FORD INST MED RES, 74- Mem: Am Chem Soc; Sigma Xi. Res: Rapid reaction kinetics and mechanisms of enzyme-catalyzed reactions with emphasis on the role of metal; ligand binding studies; subunit interactions; synthesis, kinetics and mechanism of hydrolysis of nitrogen base adducts of substituted boranes. Mailing Add: Edsel B Ford Inst Med Res Dept of Biochem Henry Ford Hosp Detroit MI 48202

WEIDLEIN, EDWARD RAY, JR, b Pittsburgh, Pa, July 12, 18; m 43; c 5. ORGANIC CHEMISTRY. Educ: Princeton Univ, AB, 40; Pa State Col, MS, 43. Prof Exp: Asst, Pa State Col, 41-43; res chemist, US Naval Res Lab, Wash, DC, 44; tech asst to dir polymer res, Reconstruct Finance Corp, Off Rubber Reserve, 45-46; tech asst, Mellon Inst, 46-53, sr fel, 53-59; asst dir res, Chem Div, 59-66, DIR ADMIN, TARRYTOWN TECH CTR, UNION CARBIDE CORP, 66- Concurrent Pos: Mem tech indust intel comt, For Econ Admin, Ger, 45. Mem: Sci Res Soc Am; Am Chem Soc; Soc Chem Indust; Commerical Develop Asn. Res: Carbohydrates; medicinals; rubber; chemical product development; food additives. Mailing Add: Croton Lake Rd Mt Kisco NY 10549

WEIDLER, DONALD JOHN, b Fredericksburg, Iowa, Sept 26, 33; m 58; c 2. INTERNAL MEDICINE. Educ: Wartburg Col, BA, 59; Univ Iowa, MS, 62, MD, 65, PhD(physiol, biophys), 69. Prof Exp: Intern internal med, Boston City Hosp, 65-66; instr physiol & biophys, Univ Iowa, 66-69; from asst prof & mem grad fac to assoc prof physiol & biophys, Univ Nebr Med Ctr, 69-72; resident internal med, 72-74, INSTR PHARMACOL, SCH MED, UNIV MICH, ANN ARBOR, 73-, ASST PROF INTERNAL MED, 74- Concurrent Pos: NIH fel, Univ Iowa, 66-69. Mem: Am Physiol Soc; Am Col Physicians; Am Soc Clin Pharmacol & Therapeut; Soc Neurosci. Res: Cerebral infarction, its cardiac effects and drug therapy. Mailing Add: Upjohn Ctr for Clin Pharmacol Univ Mich Med Ctr Ann Arbor MI 48104

WEIDLICH, JOHN EDWARD, JR, b Akron, Ohio, July 31, 19; m 69; c 1. MATHEMATICS. Educ: Stanford Univ, BS, 48, MS, 50; Univ Calif, Berkeley, PhD(math), 61. Prof Exp: Instr math, Univ Santa Clara, 49-51; from assoc res scientist to res scientist, Lockheed Missile & Space Co, 59-64; from asst prof to assoc prof, 64-70, actg chmn dept, 66-67, res sabbatical, 71-72, PROF MATH, CALIF STATE UNIV, HAYWARD, 70- Concurrent Pos: Consult, Lockheed Missile & Space Co, 64-65. Mem: Am Math Soc; Math Asn Am; Soc Indust & Appl Math. Res: Pure mathematics in analysis, differential equations and asymtotics; mathematical nature of physical phenomena, including work in theoretical orbit mechanics and trajectory studies in the n-body problem. Mailing Add: Dept of Math Calif State Univ Hayward CA 94542

WEIDMAN, DONALD ROBERT, b Washington, DC, Oct 3, 36; m 66; c 1. MATHEMATICS. Educ: Rockhurst Col, BS, 58; Univ Notre Dame, PhD(math), 64. Prof Exp: Asst prof math, Boston Col, 64-66; analyst, US Naval Weapons Lab, 66-71; MEM SR RES STAFF, URBAN INST, WASHINGTON, DC, 71- Concurrent Pos: Fac res grant, Boston Col, 66; mem part-time fac, Univ Va, 71-72. Mem: Am Math Soc; Math Asn Am. Res: Operations analysis; mathematical modeling; finite group theory. Mailing Add: Urban Inst 2100 M St NW Washington DC 20037

WEIDMAN, ROBERT MCMASTER, b Missoula, Mont, Mar 20, 23; m 51; c 4. GEOLOGY. Educ: Calif Inst Technol, BS, 44; Univ Ind, MA, 49; Univ Calif, PhD, 59. Prof Exp: Geologist, Stand Oil Co, Calif, 44-47 & State Geol Surv, Ind, 48; instr, Fresno State Col, 49-50; from instr to assoc prof, S3-68, PROF GEOL, UNIV MONT, 68- Mem: Geol Soc Am; Am Asn Petrol Geol. Res: Structural and petroleum geology. Mailing Add: Dept of Geol Univ of Mont Missoula MT 59801

WEIDMAN, THOMAS A, human anatomy, see 12th edition

WEIDNER, BRUCE VANSCOYOC, b Pottstown, Pa, Oct 29, 08; m 34; c 2. CHEMISTRY. Educ: Pa State Col, BS, 31, MS, 32, PhD(inorg chem), 35. Prof Exp: Asst chem, Pa State Col, 32-35, instr arts & sci, Hazleton Undergrad Ctr, 35-37; instr, Univ Alaska, 37-39, asst prof chem, 39-42; asst prof chem, Middlebury Col, 42-46; assoc prof, Utah State Agr Col, 46-47; assoc prof, 47-62, PROF CHEM, MIAMI UNIV, 62- Mem: Am Chem Soc; Soc Appl Spectros; AAAS. Res: Quantitative and inorganic chemistry. Mailing Add: 308 University Ave Oxford OH 45056

WEIDNER, JERRY R, b Bloomington, Ill, July 19, 38; m 62. GEOCHEMISTRY. Educ: Miami Univ, BA, 60, MS, 63; Pa State Univ, University Park, PhD(geochem), 68. Prof Exp: Res assoc geochem, Stanford Univ, 67-68; ASST PROF GEOL, UNIV MD, COLLEGE PARK, 70- Concurrent Pos: Nat Res Coun fel NASA Goddard Space FLight Ctr, Md, 68-70. Mem: AAAS; Am Geophys Union; Mineral Soc Am; Geochem Soc. Res: Experimental petrology. Mailing Add: Dept of Geol Univ Md College Park MD 20740

WEIDNER, MICHAEL GEORGE, JR, b Birmingham, Ala, July 18, 22; m 60; c 1. SURGERY. Educ: Vanderbilt Univ, BA, 44, MD, 46; Am Bd Surg, dipl, 56. Prof Exp: Instr surg, Sch Med, Vanderbilt Univ, 53-57; from sr instr to asst prof, Sch Med, Western Reserve Univ, 57-62; asst dean student affairs, 70-72, assoc prof, 62-68, PROF SURG, MED UNIV SC, 68-, ASSOC DEAN STUDENT AFFAIRS & PROJ DIR AREA HEALTH EDUC CTR, 72- Concurrent Pos: Res fel, Vanderbilt Univ, 54-57; chief surg serv, Vet Admin Hosp, Charleston, SC, 66-72. Mem: AAAS; Am Col Surgeons; Soc Univ Surgeons; Soc Exp Biol & Med; NY Acad Sci. Res: Hemorrhagic shock; animal and human gastrointestinal physiology, especially ulcer disease. Mailing Add: Dept of Surg Col of Med Med Univ of SC Charleston SC 29401

WEIDNER, RICHARD TILGHMAN, b Allentown, Pa, Mar 31, 21; m 47; c 3. PHYSICS, MAGNETIC RESONANCE. Educ: Muhlenberg Col, B BS, 43; Yale Univ, MS, 43, PhD(physics), 48. Prof Exp: Physicist, US Naval Res Lab, 44-46; lab asst & instr physics, Yale Univ, 43-44, asst, 46-47; from instr to assoc prof, 47-63, asst dean col arts & sci, 66-70, PROF PHYSICS, RUTGERS UNIV, NEW BRUNSWICK, 63-, ASSOC DEAN RUTGERS COL, 70- Mem: Fel Am Phys Soc; Am Asn Physics Teachers; Sigma Xi. Res: Electron spin resonance; texts in elementary physics; general physics. Mailing Add: Dept of Physics Rutgers Univ New Brunswick NJ 08903

WEIDNER, TERRY MOHR, b Allentown, Pa, May 31, 37; m 62; c 3. PLANT PHYSIOLOGY. Educ: West Chester State Col, BSEd, 59; Ohio State Univ, MS, 61, PhD(bot), 64. Prof Exp: Asst prof, 64-72, ASSOC PROF BOT, EASTERN ILL UNIV, 72- Mem: Am Inst Biol Sci; Am Soc Plant Physiol; Japanese Soc Plant Physiol. Res: Carbohydrate translocation in higher plants; active ion uptake by roots of higher plants. Mailing Add: Dept of Bot Eastern Ill Univ Charleston IL 61920

WEIDNER, VICTOR RAY, b Clarion, Pa, Jan 4, 32; m 67. OPTICAL PHYSICS. Educ: Clarion State Col, BS, 60. Prof Exp: PHYSICIST, NAT BUR STANDARDS, 60- Mem: Optical Soc Am. Res: Infrared spectrophotometry; development of research spectrophotometers and spectrophotometric standards. Mailing Add: Nat Bur Stand Washington DC 20234

WEIER, RICHARD MATHIAS, b Streator, Ill, Aug 18, 40; m 69; c 1. ORGANIC CHEMISTRY. Educ: Loras Col, BS, 62; Wayne State Univ, PhD(org chem), 67. Prof Exp: Res chemist, Ash-Stevens, Inc, Mich, 66-67; res chemist, Res & Develop Div, Kraftco Corp, 68-69; RES INVESTR, G D SEARLE & CO, 69- Mem: AAAS; Am Chem Soc. Res: Synthesis of steroids and natural products; aldosterone blockade; diuretics; anti-hypertensives. Mailing Add: Chem Res Dept G D Searle & Co Box 5110 Chicago IL 60680

WEIFFENBACH, GEORGE CHARLES, b Newark, NJ, June 20, 21; m 49; c 5. EXPERIMENTAL PHYSICS. Educ: Harvard Univ, AB, 49; Cath Univ, PhD(physics), 58. Prof Exp: Asst, Cath Univ, 49-51; sr physicist, Appl Physics Lab, Johns Hopkins Univ, 51-69; DIR GEOASTRON PROGS, SMITHSONIAN ASTROPHYS OBSERV, 69- Mem: Am Phys Soc; Am Geophys Union. Res: Microwave spectroscopy; time and frequency standards; space systems engineering; geodesy; geophysics. Mailing Add: 60 Garden St Cambridge MA 02138

WEIGAND, BERNARD LEO, organic chemistry, see 12th edition

WEIGANG, OSCAR EMIL, JR, b Seguin, Tex, Dec 11, 31; m 53; c 3. PHYSICAL CHEMISTRY. Educ: Tex Lutheran Col, BA, 54; Univ Mich, MS, 55; Univ Tex, PhD(phys chem), 57. Prof Exp: Res scientist, Spectros Res, Univ Tex, 56-57; from asst prof to assoc prof chem, Tex Lutheran Col, 57-63; assoc prof, Tulane Univ, 63-71, prof & chmn dept, 71-75; PROF & CHMN DEPT, UNIV MAINE, ORONO, 75- Mem: Am Chem Soc; Am Phys Soc; The Chem Soc. Res: Spectroscopy; optical activity; quantum models of the origins of optical rotatory strength. Mailing Add: Dept of Chem Univ of Maine Orono ME 04473

WEIGEL, CHRISTOPH, b Grimma, Ger, Apr 28, 44. SOLID STATE PHYSICS. Educ: Univ Würzburg, Diplom-Physiker, 70, Dr rer nat(physics), 74. Prof Exp: ASSOC SOLID STATE PHYSICS, STATE UNIV NY ALBANY, 74- Concurrent Pos: Fel, Inst Study Defects in Solids, Dept Physics, State Univ NY Albany, 75-; vis scientist, Phys Lab, Univ Amsterdam, 75. Mem: Ger Phys Soc. Res: Theoretical investigation of the electronic properties of perfect and imperfect crystals. Mailing Add: Dept of Physics State Univ of NY Albany NY 12222

WEIGEL, ROBERT DAVID, b Buffalo, NY, Dec 31, 23. VERTEBRATE ZOOLOGY. Educ: Univ Buffalo, BA, 49, MA, 55; Univ Fla, PhD(zool), 58. Prof Exp: Instr gen biol, Univ Fla, 57; asst prof biol, Howard Col, 58-59; assoc prof, 59-64, PROF BIOL, ILL STATE UNIV, 64- Mem: AAAS; Soc Study Evolution; Soc Vert Paleont; Paleont Soc; Am Soc Mammal. Res: Avian paleontology; ornithology; osteology. Mailing Add: Dept of Biol Ill State Univ Normal IL 61761

WEIGEL, RUSSELL CORNELIUS, JR, plant physiology, see 12th edition

WEIGEND, GUIDO GUSTAV, b Zeltweg, Austria, Jan 2, 20; US citizen; m 47; c 3. GEOGRAPHY. Educ: Univ Chicago, BS, 42, MS, 46, PhD(geog), 49. Prof Exp: Geogr, Mil Intel, 46; instr geog, Univ Ill, Chicago Circle, 46-47; from instr to asst prof, Beloit Col, 47-49; from asst prof to assoc prof, 49-57, chmn dept, 51-67, PROF GEOG, RUTGERS UNIV, NEW BRUNSWICK, 57-, ASSOC DEAN, RUTGERS COL, 72- Concurrent Pos: Rutgers Univ liaison officer to UN, 50-52; mem US Nat Comt, Int Geog Union, 51-58 & 61-65; Off Naval Res fel, France & Ger, 52-54; consult various publ, 53; dir NSF & NDEA Insts, 57-67; Fulbright lectr grant, Barcelona, 60-61; vis prof geog, Columbia Univ, 63-67 & NY Univ, 67; Ford Found fel, Latin Am, 66; Rutgers Res Coun & Am Philos Soc fels & Fulbright travel grant, Netherlands & Belg, 70-71; vis prof geog, Univ Rotterdam, 70-71. Mem: AAAS; Asn Am Geog; Am Geog Soc. Res: Geography of seaports, their developmental stages, traffic patterns, and role in inter- and intraregional economic development; specific problems in the geography of Europe and the Mediterranean. Mailing Add: Dept of Geog Rutgers Col Rutgers Univ New Brunswick NJ 08903

WEIGENSBERG, BERNARD IRVINE, b Montreal, Que, Feb 6, 26; m 50; c 3. EXPERIMENTAL PATHOLOGY. Educ: McGill Univ, BSc, 49, MSc, 51, PhD(biochem), 53. Prof Exp: Res asst biochem, 49-54, res assoc path chem, 54-62, asst prof, 62-74, ASSOC PROF EXP PATH, McGILL UNIV, 74- Concurrent Pos: Res assoc, Can Heart Found, 58-66; mem coun arteriosclerosis & coun thrombosis, Am Heart Asn. Mem: Fel Am Soc Study Arteriosclerosis; Am Soc Exp Path. Res: Surfactants, alkyl phenol formaldehyde, polymers, cyclic adenosine monophosphate derivatives; testing in vivo; white platelet type thrombosis; fibrofatty type atherosclerosis rabbits; cholesterol metabolism; lipid metabolism; connective tissue. Mailing Add: Path Inst McGill Univ 3775 University St Montreal PQ Can

WEIGER, ROBERT W, b Grantwood, NJ, Nov 23, 27; m 57; c 1. MEDICINE. Educ: Northwestern Univ, BS, 51, MD, 55. Prof Exp: Intern med, Passavant Mem Hosp, Northwestern Univ, 55-56; clin assoc & resident, Nat Cancer Inst, 56-58, physician div hosps, USPHS, 58-59, resident internal med, USPHS Hosp, Baltimore, Md, 59-62, med scientist, Cancer Chemother Nat Serv Ctr, Nat Cancer Inst, 62, asst dir inst & asst to dir, NIH, 64-65, PROG COORDR CLIN PHARMACOL, NAT INST GEN MED SCI, 65- Concurrent Pos: Staff fel med, Sch Med, Johns Hopkins Univ & Johns Hopkins Hosp, 62; chief officer pesticides, Bur State Serv, 65-66. Mem: AMA; Am Diabetes Asn. Res: Research and science administration, particularly in medical research. Mailing Add: Pharmacol-Toxicol Prog Nat Inst of Gen Med Sci Bethesda MD 20014

WEIGERT, FRANK JULIAN, chemistry, see 12th edition

WEIGL, JOHN WOLFGANG, b Vienna, Austria, Aug 1, 26; nat US; m 46; c 3. PHYSICAL CHEMISTRY. Educ: Columbia Univ, BS, 46; Univ Calif, PhD(chem), 49. Prof Exp: Res chemist, Lawrence Radiation Lab, Univ Calif, 47-49; fel chem, Univ Minn, 50-51; res assoc physics, Ohio State Univ, 51-55, asst prof, 55-57; sr res chemist & res mgr, Ozalid Div, Gen Aniline & Film Corp, NY, 57-62; res mgr, Imaging Mat Area, 62-72, MEM CORP, RES & DEVELOP STAFF, RES LABS, XEROX CORP, 72- Mem: AAAS; Am Chem Soc; Am Phys Soc; Soc Photog Sci & Eng. Res: Photochemistry; kinetics; isotopic tracers; organic photoconductors and optical sensitization; electrophotography. Mailing Add: 534 Wahlmont Dr Webster NY 14581

WEIGL, PETER DOUGLAS, b New York, NY, Nov 9, 39; m 68. ECOLOGY, ANIMAL BEHAVIOR. Educ: Williams Col, AB, 62; Duke Univ, PhD(vert ecol), 69. Prof Exp: Asst prof, 68-74, ASSOC PROF BIOL, WAKE FOREST UNIV, 74- Mem: AAAS; Am Soc Zool; Am Soc Mammal; Ecol Soc Am; Soc Study Evolution. Res: Vertebrate zoology; evolution; energetics of behavior, species interactions, functional morphology. Mailing Add: Dept of Biol Wake Forest Univ Winston-Salem NC 27109

WEIGLE, JACK LEROY, b Montpelier, Ohio, Sept 5, 25; m 54; c 4. PLANT BREEDING. Educ: Purdue Univ, BS, 50, MS, 54; Mich State Univ, PhD, 56. Prof Exp: Asst prof hort, Colo State Univ, 56-61; from asst prof to assoc prof, 61-73, PROF HORT, IOWA STATE UNIV, 73- Mem: Am Soc Hort Sci; Am Genetic Asn. Res: Genetic, cytogenetic and physiological investigations and development of new cultivars of ornamental plants. Mailing Add: Dept of Hort Iowa State Univ Ames IA 50011

WEIGLE, ROBERT EDWARD, b Shiloh, Pa, Apr 27, 27; m 49; c 1. APPLIED MECHANICS, APPLIED MATHEMATICS. Educ: Rensselaer Polytech Inst, BCE, 51, MS, 57, PhD, 59. Prof Exp: Eng consult, 52-54; construct supt, Long Serv Co, Inc, 54-55; assoc res scientist, Dept Mech, Rensselaer Polytech Inst, 55-59; res dir, Res Lab, 59-62, tech dir, Benet Weapons Lab, 62-69, DIR LAB, WATERVLIET ARSENAL, US ARMY, 69-, CHIEF SCIENTIST, BENET WEAPONS LAB, 70- Concurrent Pos: Consult high pressure technol comt, Nat Res Coun, 60-61 & Army mem gun tube technol comt, Nat Adv Bd, 67-69; mem, Adv Panel Weapons & Mat, NATO, 60-67 & consult, Adv Group Aeronaut Res & Develop-Struct & Mat Panel, 71; mem, Dept Defense Forum Phys & Eng Sci. Honors & Awards: Presidential Citation, Secy Army, 65, Civilian Meritorious Serv Award, Dept Army, 65. Mem: Nat Soc Prof Eng; Soc Exp Stress Anal; Am Soc Mech Eng; Am Soc Testing & Mat; Am Acad Mech. Res: Elastic and plastic analysis of the behavior of deformable bodies under the influence of external loads with particular emphasis on their fatigue and fracture performance. Mailing Add: Watervliet Arsenal Watervliet NY 12189

WEIGLE, WILLIAM O, b Monaca, Pa, Apr, 28, 27; m 48; c 2. IMMUNOLOGY, CELL BIOLOGY. Educ: Univ Pittsburgh, BS, 50, MS, 51, PhD(bact), 56. Prof Exp: Res technician, Dept Path, Sch Med, Univ Pittsburgh, 51-52, res assoc, 55-58, asst res prof, 58-59, asst prof immunochem, 59-61; assoc mem, 61-63, MEM DIV EXP PATH, SCRIPPS CLIN & RES FOUND, 63- Concurrent Pos: USPHS fel, 56-59, sr res fel, 59-61, career res award, 62-; adj prof biol, Univ Calif, San Diego. Honors & Awards: Parke-Davis Award, Am Soc Exp Path, 67. Mem: Am Soc Microbiol; Am Asn Immunologists; Am Soc Exp Path; Am Acad Allergy; NY Acad Sci. Res: Mechanisms involved in immunity and diseases of hypersensitivity; immunochemistry; immunopathology. Mailing Add: Dept of Immunopath Scripps Clin & Res Found La Jolla CA 92037

WEIGMAN, BERNARD J, b Baltimore, Md, Nov 18, 32; m 60; c 4. ENGINEERING PHYSICS. Educ: Loyola Col, Md, BS, 54; Univ Notre Dame, PhD(physics), 59. Prof Exp: From instr to assoc prof physics, 58-67, chmn dept math, 60-64, PROF PHYSICS, LOYOLA COL, MD, 67-, CHMN DEPT PHYSICS & ENG, 64- Concurrent Pos: Consult, Martin Co, 59-61. Mem: Am Asn Physics Teachers. Res: Optical and electrical properties of thin films; behavior of an aerosol system; thermoelectric, photoelectric and field emission of electrons; properties of molecules. Mailing Add: Dept of Physics & Eng Loyola Col 4501 N Charles St Baltimore MD 21210

WEIHAUPT, JOHN GEORGE, b La Crosse, Wis, Mar 5, 30; m 61. ASTROGEOLOGY, OCEANOGRAPHY. Educ: Univ Wis, BS, 52, MS, 53 & 71, PhD(geomorphol), 73. Prof Exp: Teaching asst geol, Univ Wis, 52-53, res asst, 53-54; explor geologist, Anaconda Co, 56-57; seismologist, United Geophys Corp, 57-58; geophysicist, Arctic Inst NAm, 58-63; chmn dept phys & biol sci, US Armed Forces Inst, Dept Defense, 63-73; DEAN ACAD AFFAIRS, SCH SCI, IND UNIV-PURDUE UNIV, 73-, ASST DEAN GRAD SCH, 75- Concurrent Pos: Explor geologist, Am Smelting & Ref Co, 53; codiscoverer, USARP Range, Antarctica, 59-60 & Mt Weihaupt named in his honor, 66; lectr, Univ Wis-Madison, 63-73; hon mem, Exped Polaire Francais; mem, Man/Environ Commun Ctr, Community Coun Pub TV & Int Coun Correspondence Educ; sci consult, McGraw-Hill Book Co, 65, geol consult, 68-; sci consult, Holt, Rinehart & Winston, Inc, 65-; ed & consult, John Wiley & Sons, 68. Mem: AAAS; Am Geophys Union; fel Geol Soc Am; Asn Am Geog; Int Soc Study Time. Res: Antarctic geology and geophysics in regions of Victoria Land and the South Pole in Antarctica; geophysical-biological periodic relationships; marine geology and geophysics; fluvial, glacial and coastal geomorphology; channels and paleoclimate of Mars; meteorite crater and impact phenomena. Mailing Add: Ind Univ-Purdue Univ 1201 E 38th St Indianapolis IN 46205

WEIHE, JOSEPH WILLIAM, b Reno, Nev, Sept 23, 21; m 42; c 2. MATHEMATICS. Educ: Univ Nev, BS, 46, MS, 48; Univ Calif, PhD(math), 54. Prof Exp: Asst math, Univ Nev, 46-47, instr, 47-48; instr, Univ Calif, 48-51, assoc, 51-53; staff mem, Math Res Dept, Sandia Corp, NMex, 54-57, supvr systs anal div, 57-63, mgr math dept, 63-68, mgr systs anal dept, 68-69, MGR ANAL SCI DEPT, SANDIA CORP, CALIF, 69- Concurrent Pos: Vis lectr, Univ NMex, 59-60. Mem: Am Math Soc; Opers Res Soc; Am Soc Indust & Appl Math; Math Asn Am; Asn Comput Mach. Res: Weapons systems analysis; operations research; real variable measure theory and integration; generalized Stieltjes integrals; computer science. Mailing Add: 1166 Dunsmuir Pl Livermore CA 94550

WEIHER, JAMES F, b Waverly, Iowa, Mar 30, 33. SURFACE CHEMISTRY, SURFACE PHYSICS. Educ: Carleton Col, BA, 55; Iowa State Univ, PhD(phys chem), 61. Prof Exp: Res chemist, Study Group Radiochem, Max Planck Inst, Mainz, Germany, 55-57; res asst, Inst Atomic Res, Iowa State Univ, 57-61; fel, 61-62, RES CHEMIST, CENT RES DEPT, E I DU PONT DE NEMOURS & CO, INC, 62- Mem: Fel Am Inst Chem; Am Chem Soc; Am Phys Soc. Res: Nuclear chemistry and radiochemistry; kinetics; molecular structure; solid state; Mössbauer effect; magnetic susceptibility. Mailing Add: Cent Res Dept E I du Pont de Nemours & Co Wilmington DE 19898

WEIHING, JOHN LAWSON, b Rocky Ford, Colo, Feb 26, 21; m 48; c 4. PLANT PATHOLOGY. Educ: Colo Agr & Mech Col, BS, 42; Univ Nebr, MS, 49, PhD(bot), 54. Prof Exp: Assoc plant path, 49-50, exten plant pathologist, 50-64, PROF PLANT PATH, AGR EXTEN, UNIV NEBR, LINCOLN, 60-, DIR PANHANDLE STA, 71- Concurrent Pos: Mem Univ Nebr group, Agency Inst Develop, 64- Mem: Bot Soc Am; Am Phytopath Soc; AAAS; Am Inst Biol Sci. Res: Epidemiology of plant diseases. Mailing Add: Univ of Nebr Panhandle Sta 4502 Ave I Scottsbluff NE 69361

WEIJER, JAN, b Heerenveen, Neth, Jan 28, 24; m 52. GENETICS. Educ: Univ Groningen, BSc, 51, MSc, 52, DSc, 54. Prof Exp: Asst genetics, Univ Groningen, 46-52 & exp embryol, 51-52; geneticist, Firestone Bot Res Inst, 52-54, Can, 55-57; prof hort, 57-59, chmn dept genetics, 67-69, PROF GENETICS, UNIV ALTA, 60- Concurrent Pos: Lectr, Neth, 59; consult, US AEC, 67; mem coun, Int Orgn Pure & Appl Biophys. Mem: AAAS; Radiation Res Soc; Am Genetic Asn; Genetics Soc Am; NY Acad Sci. Res: Microbial genetics and cytology, especially fine structure of the gene; biochemical genetics and mutation; biological assessment of radiation damage in humans. Mailing Add: Dept of Genetics Biol Sci Bldg Univ of Alta Edmonton AB Can

WEIK, KENNETH L, botany, see 12th edition

WEIKEL, JOHN HENRY, JR, b Palmerton, Pa, June 14, 29. TOXICOLOGY. Educ: Trinity Col, Conn, BS, 51; Univ Rochester, PhD(pharmacol), 54. Prof Exp: Res assoc, AEC Proj, Univ Rochester, 51-54; pharmacologist, Div Pharmacol, Food & Drug Admin, 54-56; sr pharmacologist, 56-58, group leader, 58-59, sect leader, 59-61, asst dept dir, Res Ctr, 61-62, dir chem pharmacol & safety eval, 62-68, DIR PATH & TOXICOL, MEAD JOHNSON & CO, 68- Mem: Am Chem Soc; Soc Exp Biol & Med; Am Soc Pharmacol & Exp Therapeut; Soc Toxicol; NY Acad Sci. Res: Inorganic metabolism; drug metabolism; endocrinology. Mailing Add: Mead Johnson Res Ctr Evansville IN 47721

WEIL, ANDRE, b Paris, France, May 6, 06. MATHEMATICS. Educ: Univ Paris, DSc, 28. Prof Exp: Prof, Aligarh Muslim Univ, India, 30-32 & Univ Strasbourg, 33-40; lectr, Haverford Col & Swarthmore Col, 41-42; prof math, Univ Sao Paulo, Brazil, 45-47 & Univ Chicago, 47-58; PROF MATH, SCH MATH, INST ADVAN STUDY, 58- Mem: Math Soc France; Indian Math Soc. Res: Theory of numbers; algebraic geometry; group theory. Mailing Add: Sch of Math Inst for Advan Study Princeton NJ 08540

WEIL, BARBARA HUMMEL, organic chemistry, information science, see 12th edition

WEIL, CARROL SOLOMON, b St Joseph, Mo, Dec 16, 17; m 40; c 2. TOXICOLOGY. Educ: Univ Mo, BA, 39, MA, 40. Prof Exp: Bacteriologist, Anchor Serum Co, Mo, 41-42; unit head, Manhattan Div, Univ Rochester, 43-45; res assoc & toxicologist, 42-43, fel, 45-52, SR FEL, DIV SPONSORED RES, CARNEGIE-MELLON INST RES, CARNEGIE-MELLON UNIV, 52- Mem: Am Chem Soc; Am Indust Hyg Asn; Biomet Soc; Soc Toxicol (secy, 63-67, pres, 68-69). Res: Chemical hygiene and toxicology; biometrics. Mailing Add: Carnegie-Mellon Inst of Res Carnegie-Mellon Univ 4400 Fifth Ave Pittsburgh PA 15213

WEIL, CLIFFORD EDWARD, b East Chicago, Ind, Nov 19, 37; m 60; c 3. MATHEMATICS. Educ: Wabash Col, BA, 59; Purdue Univ, MS, 61, PhD(math), 63. Prof Exp: Instr math, Princeton Univ, 63-64 & Univ Chicago, 64-66; asst prof, 66-69, ASSOC PROF MATH, MICH STATE UNIV, 69- Mem: Am Math Soc; Math Asn Am. Res: Functions of a real variable. Mailing Add: Dept of Math Mich State Univ East Lansing MI 48823

WEIL, EDWARD DAVID, b Philadelphia, Pa, June 13, 28; m 52; c 2. ORGANIC CHEMISTRY. Educ: Univ Pa, BS, 50; Univ Ill, MS, 51, PhD(org chem), 53. Prof Exp: Res & develop chemist, Hooker Electrochem Co, 53-56, supvr agr chem res, Hooker Chem Corp, 56-65; supvr org res, 65-69, SR RES ASSOC, STAUFFER CHEM CO, DOBBS FERRY, 69- Honors & Awards: IR-100 Award, Indust Res Inst, 74. Mem: Am Chem Soc. Res: Vinyl polymers; organic sulfur; chlorine derivatives; pesticides; organic phosphorus; rubber chemicals; flame retardants; textile chemicals. Mailing Add: 6 Amherst Dr Hastings-On-Hudson NY 10706

WEIL, FRANCIS ALPHONSE, b Selestat, France, Nov 5, 38; Can citizen; m 64; c 1. PARTICLE PHYSICS. Educ: Univ Strasbourg, BSc, 58; Univ Paris, DiplEng, 61; Dalhousie Univ, MSc, 62, PhD(physics), 68. Prof Exp: Teaching asst math, Univ Paris at the Sorbonne, France, 60-61; res asst physics, Saclay Nuclear Res Ctr, France, 66-67; PROF PHYSICS, UNIV MONCTON, 69- Concurrent Pos: Fel, Killam Found, 68; Nat Res Coun grant, Univ Moncton, 69-70, Univ Res Coun, 70-72. Mem: Can Asn Physicists; Fr-Can Asn Advan Sci; Am Asn Physics Teachers. Res: Quantum field theory. Mailing Add: Dept of Physics & Math Fac of Sci Univ of Moncton Moncton NB Can

WEIL, JESSE LEO, b Ann Arbor, Mich, Dec 9, 31; m 60; c 2. NUCLEAR PHYSICS. Educ: Calif Inst Technol, BS, 52; Columbia Univ, PhD(physics), 59. Prof Exp: Res asst nuclear physics, Columbia Univ, 54-58; res assoc, Rice Univ, 59-60 & 61-63 & Univ Hamburg, 60-61; from asst to assoc prof, 63-73, PROF NUCLEAR PHYSICS, UNIV KY, 73- Concurrent Pos: Res assoc, Rutherford High Energy Lab, 71-72. Mem: Am Phys Soc. Res: Nuclear reactions; scattering of nucleons and nuclei; nuclear energy level studies; radioactivity and fission studies beta decay and nuclear masses; neutron polarization. Mailing Add: Dept of Physics Univ of Ky Lexington KY 40506

WEIL, JOHN A, b Hamburg, Ger, Mar 15, 29; nat US; m 47; c 2. CHEMICAL PHYSICS. Educ: Univ Chicago, MS, 50, PhD(chem), 55. Prof Exp: Res chemist, Inst Study Metals, Univ Chicago, 49-52, Inst Nuclear Studies, 53-54; Corning fel chem, Princeton Univ, 55-56, instr phys chem, 56-59; assoc scientist, Argonne Nat Lab, Ill, 59-71, sr scientist, 71; PROF CHEM & CHEM ENG, UNIV SASK, 71- Concurrent Pos: Fulbright scholar physics, Univ Canterbury, NZ, 67. Mem: Am Phys Soc; Brit Int Physics & Phys Soc; The Chem Soc. Res: Paramagnetic resonance; quantum chemistry; electronic structure of inorganic complexes; defect structure of solids; organic free radicals. Mailing Add: Dept of Chem & Chem Eng Univ of Sask Saskatoon SK Can

WEIL, JOHN VICTOR, b Detroit, Mich, May 10, 35; m 64; c 1. INTERNAL MEDICINE. Educ: Yale Univ, BS, 57, MD, 61. Prof Exp: Intern path, Sch Med, Yale Univ, 61-62, intern med, 62-63, asst resident, 63-64 & 67-68; asst prof, 68-72, ASSOC PROF MED, MED CTR, UNIV COLO, DENVER, 72- Concurrent Pos: NIH res fel cardiol, Univ Colo, Denver, 66-67. Res: Physiological responses to hypoxia; peripheral circulation, erythropoiesis, ventilation, oxygen transport. Mailing Add: CVP Res Lab Univ of Colo Med Ctr Denver CO 80220

WEIL, JOHN WILLIAM, computer science, nuclear physics, see 12th edition

WEIL, JON DAVID, b Evansville, Ind, Mar 24, 37; m 58; c 2. MOLECULAR GENETICS. Educ: Swarthmore Col, BA, 58; Univ Calif, Davis, PhD(genetics), 63. Prof Exp: NSF fel molecular biol, Univ Ore, 63-65; NIH fel, Harvard Univ, 65-67; asst prof, 67-72, ASSOC PROF MOLECULAR BIOL, VANDERBILT UNIV, 72- Res: Bacteriophage genetics and physiology; mechanism of bacteriophage morphogenesis; bacteriophage-host interactions. Mailing Add: Dept of Molecular Biol Vanderbilt Univ Nashville TN 37235

WEIL, MARVIN LEE, b Gainesville, Fla, Sept 28, 24; m 54; c 3. PEDIATRIC NEUROLOGY, VIROLOGY. Educ: Univ Fla, BS, 43; Johns Hopkins Univ, MD, 46; Am Bd Pediat, dipl; Am Bd Psychiat & Neurol, dipl. Prof Exp: Intern, Duke Univ Hosp, 46-47; asst resident pediat, 47-48; resident, Children's Hosp, Cincinnati, 50-52, res assoc, Res Found, 52-53; clin asst prof pediat, Sch Med, Univ Miami, 58-65; Nat Inst Neurol Dis & Stroke spec fel, Johns Hopkins Univ & Univ Calif, Los Angeles, 65-68; asst prof, 68-72, ASSOC PROF PEDIAT & NEUROL, UNIV CALIF, LOS ANGELES, 72- Concurrent Pos: Attend pediatrician, Cincinnati Gen Hosp, 52-53; pvt pract, 53-65; chief div pediat neurol, Harbor Gen Hosp, 68- Mem: AAAS; Am Asn Immunologists; fel Am Acad Pediat; Am Acad Neurol; Am Asn Ment Deficiency; NY Acad Sci. Res: Growth characteristics, genetics, neurotropic behavior of mammalian viruses; immune responses of central nervous system. Mailing Add: Div of Pediat Neurol Harbor Gen Hosp 1000 W Carson St Torrance CA 90509

WEIL, MAX HARRY, b Baden, Switz, Feb 9, 27; nat US; m 55; c 2. CARDIOVASCULAR DISEASES, BIOMEDICAL ENGINEERING. Educ: Univ Mich, AB, 48; State Univ NY Downstate Med Ctr, MD, 52; Univ Minn, PhD(med), 57; Am Bd Internal Med, dipl, 62. Prof Exp: Intern internal med, Cincinnati Gen Hosp, Ohio, 52-53; resident, Univ Hosps, Univ Minn, 53-55; asst clin prof, 57-59, from asst prof to assoc prof, 59-71, CLIN PROF MED, SCH MED, UNIV SOUTHERN CALIF, 71-, CLIN PROF BIOMED ENG, 72-, DIR, CTR CRITICALLY ILL, 68- Concurrent Pos: Vis prof, Univ Pittsburgh; chief cardiol, City of Hope Med Ctr, Duarte, Calif, 57-59, consult, 59-63; attend physician, Los Angeles County Gen Hosp, 58-, sr attend cardiologist, Children's Div, 68-; consult physician, Cedars-Sinai Med Ctr, 65-; chmn comt emergency med care, Bd Supvrs, County Los Angeles, 68-73; past mem comt shock, Nat Acad Sci-Nat Res Coun; fel coun circulation & coun thrombosis, Am Heart Asn; pres, Int Critical Care Med, Inc, 75- Mem: Fel Am Col Cardiol; fel Am Col Physicians; Am Soc Nephrol; Am Soc Pharmacol & Exp Therapeut; Inst Elec & Electronics Engrs. Res: Clinical cardiology and cardiorespiratory physiology; critical care medicine; studies on circulatory shock, both experimental and clinical; biomedical instrumentation and automation; applications of computer techniques to bedside medicine. Mailing Add: Ctr for the Critically Ill Univ Southern Calif Sch of Med Los Angeles CA 90027

WEIL, NICHOLAS A, b Budapest, Hungary, Apr 15, 26; nat US; m 51; c 2. MECHANICS. Educ: Tech Univ Budapest, BS, 47; Univ Ill, MS, 48, PhD, 52; Univ Santa Clara, MBA, 71. Prof Exp: Asst, Univ Ill, 48-52; develop engr, M W Kellogg Co, NY, 52-56, head mech develop div, 56-57, chief develop engr, 57-59; dir mech res, Res Inst, Ill Inst Technol, 59-63; vpres res, Cummins Engine Co, Ind, 63-68; tech adv to chmn, FMC Corp, 68-71, dir planning & ventures, 71-74; DIR CORP ENG PLANNING, INGERSOLL-RAND CO, 74- Concurrent Pos: From adj asst prof to adj assoc prof, NY Univ, 55-59; lectr, Grad Sch Bus, Univ Santa Clara, 71- Mem: Soc Automotive Eng; Am Soc Mech Eng; Nat Soc Prof Eng; Sigma Xi. Res: Internal combustion engines; drive line components; applied mechanics; rheology of materials; fracture dynamics; fluid flow; combustion; stress analysis; heat transfer; process equipment; materials conveying; fuel injection systems; process kinetics. Mailing Add: Ingersoll-Rand Co Box 301 Princeton NJ 08540

WEIL, PETER H, b Vienna, Austria, Jan 24, 21; US citizen; m 55; c 2. THORACIC SURGERY. Educ: Univ Vienna, MD, 49. Prof Exp: Resident surg, Univ Vienna Hosp, 49-52; Fulbright grant surg res, Jefferson Med Col, 52-53; Fulbright grant surg,

Columbia-Presby Hosp, 53-54; resident surg, Univ Vienna Hosp, 54-56; res fel, Cedars of Lebanon Hosp, Los Angeles, 56-57; resident, Baltimore City Hosp & Johns Hopkins Hosp, 57-59; instr, Johns Hopkins Univ, 58-59; from instr to assoc prof, 59-71, PROF SURG, ALBERT EINSTEIN COL MED, 71-; DIR SURG, BRONX MUNIC HOSP CTR, 72- Concurrent Pos: Fulbright traveling fel, New York, 54; chief resident surg, Baltimore City Hosp, 58-59; attend surgeon, Bronx Vet Admin Hosp, 60-63 & Albert Einstein Col Med Hosp, 66-; dir surg, Lincoln Hosp, Bronx, 63-72. Mem: Am Col Surgeons; Am Soc Artificial Internal Organs; Am Trauma Soc; Int Soc Surg. Res: Experimental pulmonary circulation; clinical trauma. Mailing Add: Dept of Surg Albert Einstein Col of Med Bronx NY 10461

WEIL, RAOUL BLOCH, b La Paz, Bolivia; US citizen. PHYSICS. Educ: Univ La Paz, Bachiller, 45; Univ Ill, Urbana, BSEE, 49; Univ Calif, Riverside, MA, 63, PhD(physics), 66. Prof Exp: Trainee, Allis Chalmers Mfg Co, 49-51; consult engr, Bolivian Com Corp, 51-53, head tech dept, 53-55; prof elec eng, Univ La Paz, 56-60; res asst cosmic ray physics, Univ Chicago, 60-61; solid state physics, Univ Calif, 61-65; res physicist, Monsanto Co, 65-67, proj leader, 67-69; assoc prof elec eng, Washington Univ, 69-71; ASSOC PROF PHYSICS, ISRAEL INST TECHNOL, 71- Concurrent Pos: Res assoc, Cosmic Physics Lab, Chacaltaya, 57-60. Mem: Am Phys Soc; Inst Elec & Electronics Engrs; Israel Laser & Electrooptics Soc (secy, 73-); Israel Phys Soc. Res: Solar cells; optical properties of semiconductors in the infrared; lasers. Mailing Add: Physics Dept Technion Israel Inst Technol Haifa Israel

WEIL, THOMAS ANDRE, b Ft Riley, Kans, June 27, 44; m 70. INORGANIC CHEMISTRY. Educ: State Univ NY Col Oswego, BA, 66; Univ Cincinnati, PhD(chem), 70. Prof Exp: Nat Res Coun fel, US Bur Mines, Pa, 70-71; asst prof chem, Trenton State Col, 71-72; NIH fel, Univ Chicago, 72-74; RES CHEMIST, AMOCO CHEM CORP, 74- Concurrent Pos: NSF exchange fel, US-USSR Sci & Technol, 74. Mem: Am Chem Soc. Res: Homogeneous catalysis; transition metal complexes; metals in organic synthesis; mechanisms of metal catalyzed reactions. Mailing Add: Amoco Chem Corp Box 400 Naperville IL 60540

WEIL, THOMAS P, b Mt Vernon, NY, Oct 2, 32; m 65. PUBLIC HEALTH, HOSPITAL ADMINISTRATION. Educ: Union Col, NY, AB, 54; Yale Univ, MPH, 58; Univ Mich, PhD(med care orgn), 65. Prof Exp: S S Goldwater fel hosp admin, Mt Sinai Hosp, New York, 57-58; assoc consult, John G Steinle & Assocs, Mgt Consults, 58-61; asst prof, Sch Pub Health, Univ Calif, Los Angeles, 62-65; assoc dir & consult, Touro Infirmary, New Orleans, 64-66; prof grad studies health serv mgr & dir, Schs Med, Bus & Pub Admin, Univ Mo-Columbia, 66-71; vpres & prin, E D Rosenfeld Assocs, Inc, New York, 71-75; EXEC VPRES, BEDFORD HEALTH ASSOCS, INC, 75- Concurrent Pos: Consult to numerous hosps, planning agencies & med ctrs; vis prof & W K Kellogg Found grant, Univ New South Wales, 69. Mem: Fel Am Pub Health Asn; Am Hosp Asn; Am Col Hosp Adminr; Am Asn Hosp Consult. Res: Management, organization and financing of health services, especially hospital and medical care administration. Mailing Add: Bedford Health Assocs Inc 223 Katonah Ave Katonah NY 10536

WEIL, WILLIAM B, JR, b Minneapolis, Minn, Dec 3, 24; m 49; c 2. PEDIATRICS. Educ: Univ Minn, BA, 45, BS, 46, MB, 47, MD, 48. Prof Exp: USPHS & univ res fels, Harvard Univ, 51-52; sr instr pediat, Western Reserve Univ, 53-57, asst prof, 57-63; assoc prof, Col Med, Univ Fla, 63-65, E I Dupont prof for handicapped children, 65-68; PROF HUMAN DEVELOP & CHMN DEPT, MICH STATE UNIV, 68- Mem: AAAS; Soc Pediat Res (secy-treas, 62-69, pres, 69-70); Am Pediat Soc; Am Acad Pediat; NY Acad Sci. Res: Renal disease; diabetes; nutrition. Mailing Add: Dept of Human Develop Mich State Univ East Lansing MI 48824

WEILAND, GLENN STATLER, b Marion, Pa, Dec 5, 06; m 29; c 1. BIOCHEMISTRY. Educ: Univ Md, BS, 28, MS, 30, PhD(biochem), 33. Prof Exp: Asst, Exp Sta, Univ Md, 28-31, asst chem, Univ, 31-35, instr, 35-37, asst prof, Sch Med, 42-46, assoc prof biochem, 47-49; asst prof chem, Wittenberg Col, 37-42; assoc prof, 46-47 & 49-60, prof, 60-74, EMER PROF CHEM, GETTYSBURG COL, 74- Mem: Am Chem Soc. Res: Essential terpene oils; insect metabolism; porphyrins, disease and cancer; metalloporphyrins. Mailing Add: Dept of Chem Gettysburg Col Gettysburg PA 17325

WEILAND, HENRY JOSEPH, b Rochester, NY, May 7, 90; m 22; c 2. PHYSICAL CHEMISTRY, ORGANIC CHEMISTRY. Educ: Univ Rochester, BS, 13; Univ Ill, MS, 14, PhD(chem), 17. Prof Exp: Asst phys chem, Univ Ill, 14-17; chief chemist, Intermediate Div, Newport Co, 17-26, asst to dir lab, 26-32; dir chem lab, E I du Pont de Nemours & Co, Wis, 32-35, mem patents & process Div, Org Chem Dept, Jackson Lab, 35-40, patents contract div, 40-55; consult, Scott Paper Co, 55-74; RETIRED. Res: Electrical conductance at infinite dilution; dyestuffs; intermediates; fine chemicals; detergents; aromatics; intelligence; US and foreign patents. Mailing Add: 400 S Chester Rd Swarthmore PA 19081

WEILER, EBERHARDT, b Danzig, Jan 11, 29; m 57; c 2. IMMUNOLOGY, MOLECULAR BIOLOGY. Educ: Univ Tübingen, cand med, 49, Dr rer nat(zool, physics, biochem), 53. Prof Exp: Asst immunol, Max Planck Inst Virus Res, 53-60; asst mem, 60-64, ASSOC MEM IMMUNOL, INST CANCER RES, 64- Concurrent Pos: Brit Coun scholar, Chester Beatty Inst Cancer Res, London, 54-55; res fel, Calif Inst Technol, 56-58. Res: Tissue-specific antigens in normal and malignant cells; immunologic memory in determined cells; genetics of antibody formation. Mailing Add: Inst for Cancer Res 7701 Burholme Ave Philadelphia PA 19111

WEILER, EDWARD JOHN, b Chicago, Ill, Jan 15, 49. ASTRONOMY. Educ: Northwestern Univ, BA, 71, MS, 73, PhD(astron), 76. Prof Exp: Res assoc astron, Avionics Lab, Northwestern Univ, 73-74; RES ASSOC, PRINCETON UNIV-NASA GODDARD SPACE FLIGHT CTR, 76- Mem: AAAS; Am Astron Soc. Res: Chromospheric activity in late-type binary stars and the evolution of galaxies. Mailing Add: Peyton Hall Princeton Univ Observ Princeton NJ 08541

WEILER, ERNEST DIETER, b Neuwied, Ger, June 30, 39; US citizen; m 67. ORGANIC CHEMISTRY. Educ: Univ Minn, BChem, 62; Univ Nebr, PhD(org chem), 66; Temple Univ, MBA, 74. Prof Exp: Fel, Univ Basel, 66-67; chemist, 67-73, chemist animal health res, 73-74, PROJ LEADER ANIMAL HEALTH RES, ROHM AND HAAS CO, 74- Mem: Am Chem Soc. Mailing Add: Rohm and Haas Co Spring House PA 19477

WEILER, HENRY S, physical limnology, see 12th edition

WEILER, IVAN-JEANNE MAYFIELD, b Whitehall, Mont, Sept 19, 34; m 57; c 2. IMMUNOLOGY, PSYCHOBIOLOGY. Educ: Pomona Col, BA, 55; Calif Inst Technol, PhD(psychobiol), 58. Prof Exp: RES ASSOC IMMUNOL, INST CANCER RES, 61- Res: Regeneration of optic nerve; role of small lymphocytes in immunologic memory. Mailing Add: Inst for Cancer Res 7701 Burholme Ave Philadelphia PA 19111

WEILER, JOHN HENRY, JR, b Lincoln, Nebr, July 8, 25; m 58; c 2. PLANT TAXONOMY. Educ: Univ Nebr, BSc, 58; Univ Calif, Berkeley, PhD(taxon), 62. Prof Exp: PROF BOT, CALIF STATE UNIV, FRESNO, 62- Mem: Am Soc Plant Taxon; Bot Soc Am; Int Asn Plant Taxon. Res: Biosystematics; floristics of central California. Mailing Add: Dept of Biol Calif State Univ Fresno CA 93710

WEILER, LAWRENCE STANLEY, b Middleton, NS, July 11, 42; m 64; c 3. ORGANIC CHEMISTRY. Educ: Univ Toronto, BSc, 64; Harvard Univ, PhD(org chem), 68. Prof Exp: ASSOC PROF CHEM, UNIV B C, 68- Mem: Am Chem Soc; Chem Inst Can; The Chem Soc; Swiss Chem Soc. Res: Synthesis and study of novel organic compounds; synthesis of natural products and organic metals. Mailing Add: Dept of Chem Univ of B C Vancouver BC Can

WEILER, ROLAND R, b Estonia, Feb 23, 36; Can citizen; m 65; c 3. GEOCHEMISTRY. Educ: Univ Toronto, BA, 59, MA, 60; Dalhousie Univ, PhD(oceanog), 65. Prof Exp: Res scientist, Bedford Inst Oceanog, 64-67, RES SCIENTIST, CAN CENTRE FOR INLAND WATERS, CAN DEPT OF ENERGY, MINES & RESOURCES, 67- Mem: AAAS; Am Soc Limnol & Oceanog; Int Asn Gt Lakes Res. Res: Chemical limnology and geochemistry of sediments. Mailing Add: Can Centre for Inland Waters PO Box 5050 Burlington ON Can

WEILER, WILLIAM ALEXANDER, b Milwaukee, Wis, Nov 8, 41; m 63; c 3. BACTERIOLOGY, MICROBIAL ECOLOGY. Educ: Dartmouth Col, AB, 63; Purdue Univ, PhD(microbiol), 69. Prof Exp: Instr microbiol, Purdue Univ, 66-69; asst prof, 69-74, ASSOC PROF BOT, EASTERN ILL UNIV, 74- Mem: Am Soc Microbiol. Res: Herbicide effects on soil microflora; petroleum degradation in soil and water ecosystems. Mailing Add: Dept of Bot Eastern Ill Univ Charleston IL 61920

WEILERSTEIN, RALPH WALDO, b San Francisco, Calif, Feb 4, 11; m 35; c 3. THERAPEUTICS, PUBLIC HEALTH. Educ: Univ Calif, AB, 31, MD, 35. Prof Exp: Med officer, Civilian Conserv Corps, Calif, 35-37; actg asst surgeon, USPHS, NY & Calif, 37-38; med officer, US Food & Drug Admin, 38-54, assoc med dir, 54-64, dist med officer, 64-70; pub health med officer, Food & Drugs & exec secy, Cancer Adv Coun, Berkeley, 70-73, CHIEF PREV MED PROG, CALIF DEPT PUB HEALTH, SACRAMENTO, 73-; ASSOC CLIN PROF MED, SCH MED, UNIV CALIF, SAN FRANCISCO, 59- Mem: Am Fedn Clin Res. Res: New drugs; cancer drugs and cancer frauds; medical devices; therapeutic and diagnostic claim evaluation. Mailing Add: Prev Med Prog Calif Dept Pub Health Sacramento CA 95814

WEILL, CAROL EDWIN, b Brooklyn, NY, Dec 12, 18; m 47; c 2. ORGANIC CHEMISTRY. Educ: City Col New York, BS, 39; Columbia Univ, AM, 43, PhD(chem), 44. Prof Exp: Res chemist, Div War Res, Columbia Univ, 44-45 & Takamine Lab, 45-47; from instr to assoc prof, 47-60, PROF CHEM, RUTGERS UNIV, NEWARK, 60- Mem: Am Chem Soc. Res: Enzymes; carbohydrates. Mailing Add: Dept of Chem Rutgers Univ Newark NJ 07102

WEILL, DANIEL FRANCIS, b Paris, France, Nov 29, 31; US citizen; m 57; c 3. PETROLOGY, GEOCHEMISTRY. Educ: Cornell Univ, AB, 56; Univ Ill, MS, 58; Univ Calif, PhD(geol), 62. Prof Exp: Res assoc geochem, Univ Calif, Berkeley, 62-63; asst prof geol, Univ Calif, San Diego, 63-66; assoc prof, 66-70, dir ctr volcanology, 68-70, PROF GEOL, UNIV ORE, 70- Concurrent Pos: Grants, NSF, 65-66, 66-68, NASA, 68-75, 71-75 & Am Chem Soc, 70-72; Fulbright-Hays sr res fel, UK, 72-73. Mem: Mineral Soc Am; The Chem Soc. Res: Physical chemistry of geological systems; silicate liquid density, viscosity; diffusion; lunar sample analysis; redox equilibria in silicate melts; experimental trace element distribution; thermodynamic properties of mineral systems. Mailing Add: Dept of Geol Univ of Ore Eugene OR 97403

WEILL, GEORGES GUSTAVE, b Strasbourg, France, Apr 9, 26. MATHEMATICAL ANALYSIS, APPLIED MATHEMATICS. Educ: Univ Paris, DSc(physics), 55; Univ Calif, Los Angeles, PhD(math), 60. Prof Exp: Res scientist, Gen Radio Co, France, 52-56; res fel, Calif Inst Tech, 56-59; res fel math, Harvard Univ, 60-62; lectr & res fel, Yale Univ, 62-64; asst prof, Yeshiva Univ, 64-65; PROF, POLYTECH INST N Y, 66- Concurrent Pos: Consult, Electro-Optical Systs, Inc, 59-60 & Raytheon Co, 61-62. Mem: Am Math Soc; sr mem Inst Elec & Electronics Eng; Math Soc France. Res: Complex analysis; Riemann surfaces; diffraction theory; antennas; theoretical and applied electromagnetics. Mailing Add: Dept of Math Polytech Inst of N Y 333 Jay St Brooklyn NY 11201

WEILL, HANS, b Berlin, Ger, Aug 31, 33; US citizen; m; c 3. MEDICINE, PHYSIOLOGY. Educ: Tulane Univ, BA, 55, MD, 58; Am Bd Internal Med, dipl & Am Bd Pulmonary Dis, dipl, 65. Prof Exp: From instr to assoc prof, 62-71, PROF MED, SCH MED, TULANE UNIV, 71- Concurrent Pos: Chief pulmonary function lab, Vet Admin Hosp, New Orleans, 63-; consult, USPHS Hosp, 64-; dir, Specialized Ctr Res & consult to task force environ lung dis, Nat Heart & Lung Inst, 72- Mem: Am Thoracic Soc (pres, 76-77); fel Am Col Chest Physicians; fel Am Col Physicians; Am Fedn Clin Res; fel Royal Soc Med. Res: Applied pulmonary physiology; occupational respiratory diseases; environmental health sciences. Mailing Add: Dept of Med Tulane Univ Sch of Med New Orleans LA 70112

WEIL-MALHERBE, HANS, b Stuttgart, Ger, Dec 29, 05; nat US; m 34; c 2. BIOCHEMISTRY. Educ: Univ Heidelberg, MD, 30; Univ Durham, MS, 40, DSc, 45. Prof Exp: Asst biochemist, Cancer Res Inst, Univ Durham, 35-45, lectr biochem, King's Col, 45-47; dir res, Runwell Hosp, Wickford, Eng, 47-58; chief sect neurochem, Div Spec Ment Health Res, NIMH, 58-75; RETIRED. Honors & Awards: Burlingame Prize, Royal Med-Psychol Asn, 55. Mem: AAAS; Am Soc Biol Chemists; Brit Biochem Soc; Int Soc Neurochem. Res: Neuropharmacology; brain metabolism; metabolism and biochemistry of biologically active amines, particularly catecholamines. Mailing Add: W A White Bldg St Elizabeth's Hosp Washington DC 20032

WEILMUENSTER, EARL ADAM, b Lenzburg, Ill, Aug 20, 14; c 2. ORGANIC CHEMISTRY. Educ: Southern Ill Univ, BEd, 37; Univ Ill, MS, 38; St Louis Univ, PhD(org chem), 42. Prof Exp: Asst chem, Univ Ill, 37-38; dir org res, Lubrizol Corp, 42-49; asst prof chem, John Carroll Univ, 49-50; dir org res, Bjorksten Res Lab, 50-51; spec asst to dir chem res, Olin Mathieson Chem Corp, 51-55, mgr chem res, 55-56, assoc dir chem res, 56-57, dir fuels res, 57-59; asst mgr propellant develop br, United Tech Ctr, 59-62, tech coordr res & advan technol & eng dept, 62-74; MEM STAFF, FAST BREEDER REACTOR DEPT, SAFETY CRITERIA AND LICENSING, GEN ELEC CO, SUNNYVALE, CALIF, 74- Concurrent Pos: Res consult, Carl F Prutton & Assocs, 49-50. Res: Oil addents; quaternary ammonium compounds; aliphatic phosphorous esters of thiophosphonic acids; lubricants; radioactive plasticizers; battery additives; hydrazine derivatives; liquid, solid and hybrid fuel systems; light metal; organometallic and organoboron hydrides. Mailing Add: 2041 Louise Lane Los Altos CA 94022

WEIMAR, VIRGINIA LEE, b Condon, Ore, Oct 23, 22. PHYSIOLOGY. Educ: Ore State Col, MS, 47; Univ Pa, PhD(physiol), 51. Prof Exp: Physiologist, US Dept Navy, 51; res assoc biochem, Wills Eye Hosp, Philadelphia, Pa, 52-54; researcher ophthal,

Col Physicians & Surgeons, Columbia Univ, 55-58, res assoc, 58-59, asst prof, 59-61; res assoc, 61-63, asst prof, 63-68, ASSOC PROF OPHTHAL, MED SCH, UNIV ORE, PORTLAND, 68- Concurrent Pos: Nat Cancer Inst fel, Univ Pa, 54-55; NIH travel award; deleg, Int Cong Ophthal, Belg, 58 & Jerusalem Conf Prevention of Blindness, 71. Mem: Harvey Soc; Am Physiol Soc; Asn Res Vision & Ophthal; Soc Gen Physiol; Am Soc Cell Biol. Res: Biochemistry and physiology of trauma; wound healing; inflammation; corneal wound healing; cellular ultramicrochemistry; biomathematics. Mailing Add: Dept of Ophthal Univ of Ore Med Sch Portland OR 97201

WEIMBERG, RALPH, b San Diego, Calif, Dec 22, 24; m 52; c 2. BACTERIOLOGY, BIOCHEMISTRY. Educ: Univ Calif, AB, 49, MA, 51, PhD(bact), 55. Prof Exp: Fel, Biol Div, Oak Ridge Nat Labs, 55-56; instr microbiol, Sch Med, Western Reserve Univ, 56-58; biochemist, Northern Utilization Res & Develop Div, 58-65, BIOCHEMIST US SALINITY LAB, AGR RES SERV, US DEPT AGR, 65- Concurrent Pos: Vis assoc prof, Univ Calif, Davis, 64. Mem: AAAS; Am Soc Microbiol; Am Soc Plant Physiol; Am Inst Chem; NY Acad Sci. Res: Microbial physiology and metabolism; properties and location of enzymes. Mailing Add: US Dept of Agr Salinity Lab PO Box 672 Riverside CA 92502

WEIMER, DAVID, b Marion, Ind, Oct 13, 19; m 44; c 4. GAS DYNAMICS. Educ: Ohio State Univ, BS, 41, MS, 46. Prof Exp: Res asst physics, Princeton Univ, 41-45; prof, Am Col SIndia, 46-47; res assoc, Princeton Univ, 47-52; res engr, Armour Res Found, Ill, 52-55 & Am Mach & Foundry Co, 55-56; sr staff scientist physics, Lockheed Corp, Calif, 56-58, 60-64; asst prof, Ohio Northern Univ, 64-68; staff scientist, Martin-Marietta Corp, Colo, 68-69; ASSOC PROF PHYSICS, OHIO NORTHERN UNIV, 69- Concurrent Pos: Sr staff engr, Martin-Marietta Corp, Colo, 72-73. Mem: Am Phys Soc. Res: Gas physics and shock wave phenomena; dissociation and ionization of gas at high temperatures; radiation of excited species in planetary atmospheres. Mailing Add: Dept of Physics Ohio Northern Univ Ada OH 45810

WEIMER, HENRY EBEN, b Continental, Ohio, May 13, 14; m 41. IMMUNOCHEMISTRY. Educ: Wittenberg Col, AB, 35; Univ Southern Calif, MS, 48, PhD(biochem), 50. Prof Exp: Teacher high sch, Ohio, 35-41; res assoc, 50-51, from instr to asst prof, 51-58, ASSOC PROF MICROBIOL & IMMUNOL, SCH MED, UNIV CALIF, LOS ANGELES, 58- Mem: Am Soc Exp Path; Am Asn Immunol. Res: Acute phase proteins in inflammatory states; embryonic serum proteins in health and disease; tumor and fetal antigens. Mailing Add: Dept of Microbiol & Immunol Sch of Med Univ of Calif Los Angeles CA 90024

WEIMER, KATHERINE E, b Rutherford, NJ, Apr 15, 19; m 42; c 3. PHYSICS. Educ: Purdue Univ, BS, 39; Ohio State Univ, PhD(nuclear physics), 42. Prof Exp: MEM RES STAFF, PLASMA PHYSICS LAB, PRINCETON UNIV, 55-59, 60- Mem: Am Phys Soc. Res: Plasma physics; stability theory and problems related to stellarator and tokamak-type devices in controlled thermonuclear fusion research. Mailing Add: 112 Random Rd Princeton NJ 08540

WEIMER, PAUL KESSLER, b Wabash, Ind, Nov 5, 14; m 42; c 3. PHYSICS. Educ: Manchester Col, AB, 36; Univ Kans, AM, 38; Ohio State Univ, PhD(physics), 42. Hon Degrees: DSc, Manchester Col, 68. Prof Exp: Asst, Univ Kans, 36-37; prof physics, Tabor Col, 37-39; asst, Ohio State Univ, 39-42; res engr, 42-64, FEL TECH STAFF, RCA LABS, 64- Honors & Awards: Award, TV Broadcasters Asn, 46; Zworykin Prize, Inst Elec & Electronics Eng, 59, Morris Liebmann Mem Prize, 66. Mem: Am Phys Soc; fel Inst Elec & Electronics Eng. Res: Nuclear physics; electron optics; photoconductivity; secondary emission; semiconductor devices; television camera tubes and solid state image sensors. Mailing Add: RCA Labs Princeton NJ 08540

WEIMER, ROBERT J, b Glendo, Wyo, Sept 4, 26; m 48; c 3. GEOLOGY. Educ: Univ Wyo, BA, 48, MA, 49; Stanford Univ, PhD, 53. Prof Exp: Dist geologist, Union Oil Co Calif, Mont, 53-54; consult res geologist, 54-57; from asst prof to assoc prof, 57-72, head dept, 64-69, PROF GEOL, COLO SCH MINES, 72- Concurrent Pos: Exchange prof, Univ Colo, 60; Fulbright lectr, Univ Adelaide, 67; vis prof, Univ Calgary, 70; mem res assoc comt, Nat Acad Sci, 70-73. Honors & Awards: Distinguished Serv Award, Am Asn Petrol Geol, 75. Mem: AAAS; Geol Soc Am; Am Asn Petrol Geol; Soc Econ Paleont & Mineral (secy-treas, 66-68, vpres, 71, pres, 72). Res: Stratigraphic research in the application of modern sedimentation studies to the geologic record; regional framework of sedimentation in the Cretaceous, Jurassic and Pennsylvanian rock systems; tectonics and sedimentation; geologic history of the Rocky Mountain region. Mailing Add: Dept of Geol Colo Sch of Mines Golden CO 80401

WEIN, JOHN, organic chemistry, see 12th edition

WEIN, ROSS WALLACE, b Exeter, Ont, Oct 29, 40; m 67. PLANT ECOLOGY. Educ: Univ Guelph, BSA, 65, MSc, 66; Utah State Univ, PhD(ecol), 69. Prof Exp: Can Nat Res Coun fel, Univ Alta, 69-71, vis asst prof plant ecol, 71-72; asst prof, 72-74, ASSOC PROF PLANT ECOL, UNIV NB, FREDERICTON, 74- Mem: Ecol Soc Am; Am Soc Range Mgt; Can Bot Asn. Res: Plant production and nutrient cycling following wildfire; plant community dynamics; boreal and arctic ecology. Mailing Add: Dept of Biol Univ of N B Fredericton NB Can

WEINACHT, RICHARD JAY, b Union City, NJ, Dec 10, 31; m 55; c 6. MATHEMATICS. Educ: Univ Notre Dame, BS, 53; Columbia Univ, MS, 55; Univ Md, PhD(math), 62. Prof Exp: NSF fel math, Courant Inst Math Sci, NY Univ, 62-63; from asst prof to assoc prof, 63-74, PROF MATH, UNIV DEL, 74- Concurrent Pos: Vis assoc prof, Rensselaer Polytech Inst, 69-70. Mem: Am Math Soc; Soc Indust & Appl Math. Res: Partial differential equations. Mailing Add: Dept of Math Univ of Del Newark DE 19711

WEINBACH, EUGENE CLAYTON, b Pine Island, NY, Nov 5, 19; m 38; c 2. BIOCHEMISTRY. Educ: Univ Md, BS, 42, PhD(org med chem), 47. Prof Exp: Res chemist, US Naval Res Lab, Washington, DC, 47; USPHS & Nat Cancer Inst fel, Sch Med, Johns Hopkins Univ, 47-48, instr physiol chem, 48-50; RES CHEMIST, NIH, 50-, HEAD, SECT PHYSIOL & BIOCHEM, LAB PARASITIC DIS, NAT INST ALLERGY & INFECTIOUS DIS, 69- Concurrent Pos: Guest worker, Wenner-Gren Inst, Stockholm, Sweden, 60-62. Mem: Am Chem Soc; Am Soc Biol Chem; Brit Biochem Soc. Res: Biochemical mechanisms of drug action; intermediary metabolism of parasites and their hosts; biological oxidations and phosphorylations; biochemistry of mitochondria. Mailing Add: Lab Parasitic Dis NIH Bethesda MD 20014

WEINBAUM, CARL MATTIN, b Manchester, Eng, Jan 16, 37; US citizen; m 63; c 4. MATHEMATICS. Educ: Queens Col, NY, BS, 58; Harvard Univ, AM, 60; NY Univ, PhD(math), 63. Prof Exp: Asst prof math, Univ Calif, Los Angeles, 63-68; assoc prof, Univ Hawaii, 68-70; ASSOC PROF MATH, HAWAII LOA COL, 70- Mem: Am Math Soc. Res: Infinite groups, particularly word problems and small cancellation and knot groups. Mailing Add: 556 Ululani St Kailua HI 96734

WEINBAUM, GEORGE, b Brooklyn, NY, July 27, 32; m 63; c 4. BIOCHEMISTRY, MICROBIOLOGY. Educ: Univ Pa, AB, 53; Pa State Univ, MS, 55, PhD(biochem), 57. Prof Exp: Chief biochem labs, Geisinger Med Ctr, 57-61; asst mem, Res Labs, 61-66, assoc mem, 66-71, BIOSCIENTIST, DEPT PULMONARY DIS, ALBERT EINSTEIN MED CTR, 71-; RES ASSOC PROF MICROBIOL, HEALTH SCI CTR, TEMPLE UNIV, 73- Concurrent Pos: Fulbright res fel, Tokyo, 59-61; NIH res career develop award, 69. Mem: AAAS; Am Soc Biol Chemists; Am Soc Exp Path; Am Soc Microbiol; Am Chem Soc. Res: Membrane structure, synthesis and function; role of glycoproteins in organization of cell surface; etiology of emphysema; action of proteinases on cell membranes. Mailing Add: Korman Bldg Albert Einstein Med Ctr Philadelphia PA 19141

WEINBERG, ALFRED, physics, see 12th edition

WEINBERG, ALVIN MARTIN, b Chicago, Ill, Apr 20, 15; m 40; c 2. NUCLEAR PHYSICS. Educ: Univ Chicago, AB, 35, MS, 36, PhD(physics), 39. Hon Degrees: LLD, Univ Chattanooga, 63, Alfred Univ & Denison Col, 67; ScD, Univ of the Pac, 66, Worcester Polytech Inst, 71, Univ Rochester, 73 & Wake Forest Univ, 74; EngD, Stevens Inst Technol, 73. Prof Exp: Asst math biophys, Univ Chicago, 39-41, physicist, Metall Lab, 41-45; physicist, Clinton Labs, Tenn, 45-48; dir physics div, Oak Ridge Nat Lab, 48-49, res dir, 49-55, dir, 55-74; dir off energy res & develop, Fed Energy Admin, 74; DIR INST ENERGY ANAL, OAK RIDGE ASSOC UNIVS, 75- Concurrent Pos: Mem sci adv bd, US Air Force, 56-59; mem, President's Sci Adv Comt, 60-62; Regents' lectr, Univ Calif, 66. Honors & Awards: Atoms for Peace Award, 60; Lawrence Mem Award, US AEC, 60; Heinrich Hertz Prize, Univ Karlsruhe, 75. Mem: Nat Acad Sci; Nat Acad Eng; fel Am Phys Soc; fel Am Nuclear Soc (pres, 59-60); Am Acad Arts & Sci. Res: Nuclear energy; mathematical theory of nerve function; science policy. Mailing Add: 111 Moylan Lane Oak Ridge TN 37830

WEINBERG, BERND, organic chemistry, deceased

WEINBERG, BERND, b Chicago, Ill, Jan 30, 40; m 65. SPEECH PATHOLOGY. Educ: State Univ NY, BS, 61; Ind Univ, MA, 63, PhD, 65. Prof Exp: Prof otolaryngol, Sch Med & prof speech path, Sch Dent, Ind Univ, 64-66; res fel speech sci, Nat Inst Dent Res, 66-68; dir speech res lab, Med Ctr, Ind Univ, Indianapolis, 68-74; PROF SPEECH PATH & SPEECH SCI, PURDUE UNIV, WEST LAFAYETTE, 74- Mem: Acoust Soc Am; Am Speech & Hearing Asn; Am Cleft Palate Asn. Res: Speech acoustics; speech physiology. Mailing Add: Dept Speech Path & Speech Sci Purdue Univ West Lafayette IN 47906

WEINBERG, DANIELA, b Prague, Czech, Apr 4, 36; US citizen; m 61. CULTURAL ANTHROPOLOGY. Educ: Columbia Univ, AB, 56, AM, 59; Univ Mich, AM, 65, PhD(anthrop), 70. Prof Exp: Instr anthrop, Hunter Col, City Univ New York, 65-67; asst prof, State Univ NY Binghamton, 70-74; ASSOC PROF ANTHROP, UNIV NEBR-LINCOLN, 74- Concurrent Pos: Commune of Bagnes Switz grant-in-aid ethnog res, Switz, 71-72; State Univ NY Res Found grant-in-aid ethnog res, Switz, 71-73, fel & grant-in-aid, 72- Mem: Am Anthrop Asn; Ethnologia Europa; Swiss Ethnol Soc. Res: Cultural ecology and evolution; general systems theory; peasantry; methodology; computer applications in anthropology; Europe. Mailing Add: Dept of Anthrop Univ of Nebr Lincoln NE 68508

WEINBERG, DAVID SAMUEL, b St Louis, Mo, Feb 15, 38; m 61; c 1. POLYMER CHEMISTRY. Educ: Univ Ariz, BS, 60, PhD(org chem), 64. Prof Exp: Res assoc, Stanford Univ, 64-65; res chemist, Phillips Petrol Co, Okla, 65-70; GROUP LEADER, TECH CTR, OWENS-ILL CO, 70- Mem: Am Chem Soc. Res: Polymer synthesis and characterization; molecular spectroscopy; organic research. Mailing Add: 5713 Candlestick Court Toledo OH 43615

WEINBERG, ELLIOT CARL, b Chicago, Ill, Aug 17, 32; m 62; c 2. MATHEMATICS. Educ: Purdue Univ, BA, 56, MS, 58, PhD(math), 61. Prof Exp: From instr to asst prof, 60-66, ASSOC PROF MATH, UNIV ILL, URBANA, 66- Mem: Am Math Soc. Res: Ordered algebraic structures. Mailing Add: Dept of Math Univ of Ill Urbana IL 61801

WEINBERG, ELLIOT HILLEL, b Duluth, Minn, Dec 28, 24; m 47; c 2. SOLID STATE PHYSICS. Educ: Univ Mich, BS & MS, 47; Univ Iowa, PhD(physics), 53. Prof Exp: Asst, Univ Mich, 42-43, res assoc, 47-48; res physicist, Mass Inst Technol, 48-49; from instr to assoc prof physics, Univ Minn, 49-58; prof & chmn dept, NDak State Univ, 58-62; chief scientist, San Francisco Br, 62-67, dir div phys sci, 67-71, liaison scientist, London, 65-66, DIR RES, OFF NAVAL RES, 71- Concurrent Pos: Mem, Col Physics Comn Rev, 62. Mem: AAAS; Am Phys Soc; Am Soc Eng Educ; Am Asn Physics Teachers. Res: Solid state physics; meteoroology; atmospheric and underwater optics; laser physics and applications. Mailing Add: Off of Naval Res Code 401 Arlington VA 22217

WEINBERG, EUGENE DAVID, b Chicago, Ill, Mar 4, 22; m 49; c 4. MEDICAL MICROBIOLOGY. Educ: Univ Chicago, BS, 42, MS, 48, PhD(microbiol), 50. Prof Exp: Asst instr microbiol, Univ Chicago, 47-50; from instr to assoc prof, 50-61, PROF MICROBIOL, IND UNIV, BLOOMINGTON, 61- Mem: AAAS; Am Soc Microbiol; fel Am Acad Microbiol. Res: Roles of trace metals and of metal-binding agents in microbial physiology and in chemotherapy; nutritional immunity; antimicrobial compounds; environmental control of secondary metabolism. Mailing Add: Dept of Microbiol Ind Univ Bloomington IN 47401

WEINBERG, FRED, b Poland, Apr 3, 25; nat Can; m 56; c 1. PHYSICS. Educ: Univ Toronto, BASc, 47, MA, 48, PhD(metall), 51. Prof Exp: Res scientist, Phys Metall Div, Mines Br, Can Dept Energy, Mines & Resources, 51-67, head metal physics sect, 61-67; PROF METALL, UNIV BC, 67- Concurrent Pos: Vis prof, Cavendish Lab, Cambridge Univ, 62-63. Mem: Am Inst Mining, Metall & Petrol Eng. Res: Metal physics; deformation and solidification of metals. Mailing Add: Dept of Metall Univ of BC Vancouver BC Can

WEINBERG, I JACK, b New York, NY, May 25, 35; m 58; c 2. COMPUTER SCIENCE, APPLIED MATHEMATICS. Educ: Yeshiva Univ, BA, 56; Mass Inst Technol, SM, 58, PhD(math), 61. Prof Exp: Mgr math, Avco Corp, 61-70; assoc prof, 70-73, PROF MATH, LOWELL TECHNOL INST, 73- Concurrent Pos: Instr, Northeastern Grad Sch Arts & Sci, 64-69. Mem: Am Math Soc; Soc Indust & Appl Math; Asn Comput Mach; Math Asn Am. Res: Numerical analysis; differential equations; theory of elasticity. Mailing Add: Dept of Math Lowell Technol Inst Lowell MA 01854

WEINBERG, IRVING, b New York, NY, July 3, 18; m 47; c 1. PHYSICS. Educ: Stanford Univ, BS, 50; Univ Colo, MS, 52, PhD, 58. Prof Exp: Physicist, Midway Labs, Chicago, 52-53; instr physics, Univ Colo, 54-57; sr physicist, Mobil Oil Co, NJ & Tex, 58-59; sr res physicist, Aeronutronic Div, Ford Motor Co, 59-64; res physicist, Jet Propulsion Lab, 64-67; physicist, Div Res, 67-72, PHYSICIST, MAT PROG, NASA, 72- Concurrent Pos: Consult, Magnolia Petrol Co, Tex. Mem: Am Phys Soc; Sigma Xi. Res: Space physics; radio frequency spectroscopy of the solid state;

transport phenomena in solids; energy band theory; interaction of radiation with matter. Mailing Add: NASA Lewis Res Ctr MS 302-1 21000 Brookpark Rd Cleveland OH 44135

WEINBERG, JACOB MORRIS, spectroscopy, thermochemistry, see 12th edition

WEINBERG, JERRY L, b Detroit, Mich, Dec 2, 31; m 61; c 2. ASTROPHYSICS, ATMOSPHERIC PHYSICS. Educ: St Lawrence Univ, BS, 58; Univ Colo, PhD(astrophys, atmospheric physics), 63. Prof Exp: Res asst astrophys, High Altitude Observ, Boulder, Colo, 59-63; astrophysicist, Haleakala Observ, Univ Hawaii, 63-68; astrophysicist, Dudley Observ, 68-73, assoc dir, 70-73, RES PROF ASTRON, STATE UNIV N Y ALBANY, 73-, DIR SPACE ASTRON LAB, 73- Concurrent Pos: Pres, Comn 21, Int Astron Union; mem panel cosmic dust, Comt Space Res, Int Coun Sci Unions; mem working group photom, Int Asn Geomag & Aeronomy. Honors & Awards: Significant Sci Achievement Medal, NASA, 74. Mem: AAAS; Am Astron Soc; Int Astron Union; Int Coun Sci Unions; Int Asn Geomag & Aeronomy. Res: Space astronomy; instrumentation, interplanetary medium; night sky research; atmospheric optics. Mailing Add: Space Astron Lab Executive Park E Albany NY 12203

WEINBERG, MICHAEL C, b New York, NY, June 25, 41; m 62; c 2. STATISTICAL MECHANICS. Educ: Columbia Univ, AB, 62; Yale Univ, MS, 63; Univ Chicago, PhD(chem), 67. Prof Exp: Res assoc, Mass Inst Technol, 67-69; SR RES SCIENTIST, OWENS-ILL, INC, 70- Concurrent Pos: Adj asst prof physics, Univ Toledo, 74- Mem: Am Phys Soc; Sigma Xi. Res: Various aspects of non-equilibrium statistical mechanics; light scattering, relaxation processes, dynamical processes in fluids; amorphous state; dynamics and equilibrium properties. Mailing Add: 2361 Cheltenham Rd Toledo OH 43606

WEINBERG, MYRON SIMON, b New York, NY, July 18, 30; m 54; c 3. TOXICOLOGY, RESEARCH ADMINISTRATION. Educ: NY Univ, BA, 50; Fordham Univ, BS, 54; Univ Md, MS, 56, PhD(med chem, pharmacol), 58. Prof Exp: Res assoc med, Sinai Hosp, Baltimore, 56-58; chemist, Ortho Res Found, NJ, 58-59; chief chemist, Norwalk Hosp, Conn, 59-65; assoc dir biol opers, Food & Drug Res Labs, NY, 65-67; dir biol sci lab, Foster D Snell, Inc, 67-68, exec vpres, 69-70, pres, 70-73; VPRES, CHURCH & DWIGHT CO, INC, 73- Mem: Fel Royal Soc Health; Am Soc Clin Pharmacol & Chemother; Am Chem Soc; AAAS; Am Inst Chem. Res: Human and animal testing of chemicals; pharmaceuticals; cosmetics; governmental liaison and regulatory advice; research management; biology; new uses for inorganic chemicals; new business opportunities. Mailing Add: Church & Dwight Co Inc 2 Pennsylvania Plaza New York NY 10001

WEINBERG, NORMAN L, b Toronto, Ont, May 6, 36; m 59; c 2. ORGANIC CHEMISTRY. Educ: Univ Toronto, BSc, 59, MSc, 60; Univ Ottawa, PhD(chem), 63. Prof Exp: Am Technion Soc fel chem, Israel Inst Technol, 63-64; sr res chemist, Bristol Labs, Que, 64-66; res chemist, Am Cyanamid Co, 66-70; group leader, 70-75, PROG LEADER, HOOKER CHEM & PLASTICS CORP, 75- Concurrent Pos: Prof short courses, Am Chem Soc, 72- Honors & Awards: Inventor of Year Award, Hooker Chem & Plastics Corp, 75. Mem: Am Chem Soc; Electrochem Soc. Res: Electro-organic syntheses; mechanism and stereochemistry of electro-organic reactions; chlor/alkali technology; energy systems. Mailing Add: Hooker Chem & Plastics Corp Grand Island NY 14073

WEINBERG, ROGER, b New York, NY, Jan 1, 31. BIOSTATISTICS. Educ: Univ Tex, PhD(genetics), 54; Univ Mich, Ann Arbor, PhD(comput sci), 70. Prof Exp: USPHS fel microbiol, Calif Inst Technol, 57-58; from instr to asst prof bact, Univ Pittsburgh, 58-65; USPHS spec fel biostatist, Univ Mich, Ann Arbor, 65-68, res assoc logic of comput, 68-70; assoc prof comput sci, Kans State Univ, 70-72; ASSOC PROF BIOMET, MED CTR, LA STATE UNIV, NEW ORLEANS, 72- Mem: Biometric Soc; Asn Comput Mach. Res: Computer applications to medicine. Mailing Add: Dept of Biomet Med Ctr La State Univ New Orleans LA 70112

WEINBERG, SIDNEY B, b Philadelphia, Pa, Sept 13, 23; m 51; c 3. PATHOLOGY. Educ: Univ Buffalo, MD, 50; Am Bd Path, dipl, 59. Prof Exp: Asst med examr forensic path, New York, 54-59; CHIEF MED EXAMR, SUFFOLK COUNTY, 60-; PROF FORENSIC PATH, STATE UNIV NY STONY BROOK, 70- Concurrent Pos: Attend pathologist, Vet Admin Hosp, Brooklyn, 54-55; instr, Col Med & Post-Grad Med Sch, NY Univ, 55-57, asst prof, 57-70; lectr, Columbia Univ, 57-73; res assoc, Med Sch, Cornell Univ, 58-59; consult, Brookhaven Mem & Huntington Hosps, 61-, Southside & Good Samaritan Hosps, 65- & Cent Suffolk Hosp, 66-; dir labs, Brunswick Hosp. Mem: AAAS; fel Am Soc Clin Path; AMA; Col Am Path; fel Am Acad Forensic Sci. Res: Forensic pathology, especially sudden death and coronary artery disease; automotive trauma; narcotic addiction and industrial poisoning. Mailing Add: 34 Winoka Dr Huntington Station NY 11746

WEINBERG, SIDNEY R, b New York, NY, Sept 2, 12; m 38; c 3. UROLOGY. Educ: NY Univ, BS, 33, MD, 37. Hon Degrees: MD, Univ Madrid, 62. Prof Exp: Clin instr urol, Post-Grad Sch, NY Univ, 49-52; from instr to assoc prof, Col Med, State Univ NY Downstate Med Ctr, 54-63, clin prof, 63-74; chief urol serv, Jewish Hosp Brooklyn, 63-74. Concurrent Pos: Asst attend urologist, Gouverneur Hosp, 41-50 & Maimonides Hosp, 45-63; attend, Kings County Hosp, 53-74; asst attend, Long Island Col Hosp, 53-60, attend, 60-74; consult, Vet Admin Hosp, 59 & Samaritan & Peninsula Gen Hosps, 62-74. Mem: Am Urol Asn; AMA; Am Col Surgeons; fel NY Acad Med; Int Soc Urol. Res: Physiology of the ureter; use of the bowel in bladder surgery. Mailing Add: Box 6H Sunny Isle Christiansted St Croix VI 00820

WEINBERG, STEVEN, b New York, NY, May 3, 33; m 54; c 1. THEORETICAL PHYSICS. Educ: Cornell Univ, AB, 54; Princeton Univ, PhD(physics), 57. Prof Exp: Instr physics, Columbia Univ, 57-59; res assoc, Lawrence Radiation Lab, Univ Calif, Berkeley, 59-60, from asst prof to prof physics, 60-69; prof, Mass Inst Technol, 69-73; HIGGINS PROF PHYSICS, HARVARD UNIV, 73- Concurrent Pos: Consult, Inst Defense Anal, 60-73; Sloan Found fel, 61-65; Morris Loeb lectr, Harvard Univ, 66-67; vis prof, Mass Inst Technol, 67-68; consult, US Arms Control & Disarmament Agency, 71-73; counr, Am Phys Soc, 72-75; consult, Stanford Res Inst, 73-; sr scientist, Smithsonian Astrophys Observ, 73-; Richtmeyer lectr, Am Asn Physics Teachers, 74; Scott lectr, Cavendish Lab, Cambridge Univ, 75. Honors & Awards: J R Oppenheimer Prize, Univ Miami, 73. Mem: Nat Acad Sci; Am Acad Arts & Sci; Am Phys Soc; Am Astron Soc. Res: Elementary particles; field theory; cosmology. Mailing Add: Dept of Physics Harvard Univ Cambridge MA 02138

WEINBERGER, DANIEL, physical chemistry, see 12th edition

WEINBERGER, EDWARD BERTRAM, b Pittsburgh, Pa, Mar 21, 21; m 44; c 3. COMPUTER SCIENCE. Educ: Mass Inst Technol, BS, 41; Univ Pittsburgh, MS, 47, PhD(math), 50. Prof Exp: Phys chemist, Gulf Res & Develop Co, 43-51, mathematician, 51-52, head comput anal sect, 53-62, res assoc, 62-66, sr scientist, 66-71; systs analyst comput ctr, 72-73, SR SYSTS ANALYST, INFO SYSTS DEPT, CARNEGIE-MELLON UNIV, 73- Concurrent Pos: Lectr, Univ Pittsburgh, 47-73. Mem: AAAS; Asn Comput Mach. Res: Digital computers; programming languages; numerical analysis; data processing. Mailing Add: 6380 Caton St Pittsburgh PA 15217

WEINBERGER, HANS FELIX, b Vienna, Austria, Sept 27, 28; nat US; m 57; c 3. APPLIED MATHEMATICS. Educ: Carnegie Inst Technol, BS & MS, 48, ScD, 50. Prof Exp: Fel, Inst Fluid Dynamics & Applied Math, Univ Md, 50-51, res assoc, 51-53, from asst res prof to assoc res prof, 53-60; assoc prof, 60-61, head, Sch Math, 67-69, PROF MATH, UNIV MINN, MINNEAPOLIS, 61- Concurrent Pos: Vis mem, Courant Inst Math Sci, NY Univ, 66-67; vis prof, Univ Ariz, 70-71 & Stanford Univ, 72-73. Mem: Am Math Soc; Soc Natural Philos; Soc Indust & Appl Math. Res: Approximation of eigenvalues, quadratic functionals and solutions of partial differential equations. Mailing Add: Dept of Math Univ of Minn Minneapolis MN 55455

WEINBERGER, HAROLD, b New York, NY, Mar 24, 10; m 37; c 4. ORGANIC CHEMISTRY. Educ: Polytech Inst Brooklyn, BSc, 31, MS, 33, PhD(org chem), 36. Prof Exp: Asst chem, Polytech Inst Brooklyn, 31-35; instr, Long Island Univ, 36-42 & City Col New York, 42-44; proj leader, Gen Chem Co, NY, 44-46; chief chemist, Glyco Prods Inc, 46-48 & Heyden Chem Corp, 48-52; dir res, Fine Organics, Inc, 52-59; assoc prof, 59-62, dir career ctr, Col Sci & Eng, 71-73, PROF CHEM, FAIRLEIGH DICKINSON UNIV, 62-, FAC LEADER, TEANECK-HACKENSACK CAMPUS, 75- Concurrent Pos: Asst, Long Island Univ, 31-33; instr, Brooklyn Col, 41-42. Mem: Fel AAAS; Am Chem Soc; fel Am Inst Chem; NY Acad Sci. Res: Resin chemistry; germicides; fatty acid derivatives. Mailing Add: Dept of Chem Fairleigh Dickinson Univ Teaneck NJ 07666

WEINBERGER, JOHN HOWARD, horticulture, see 12th edition

WEINBERGER, LESTER, b Feb 21, 28; US citizen; div; c 1. ORGANIC CHEMISTRY, ELECTROPHOTOGRAPHY. Educ: Rutgers Univ, BS, 49; Stevens Inst Technol, MS, 52; Univ Pa, PhD(org chem), 59. Prof Exp: Fel biopolymers, Columbia Univ, 62-63; sect leader photoconductors, Xerox Corp, 63-69; head chem, SCM Corp, 69-71; consult ionography, Xonics Corp, 71-72; res assoc polymers, Columbia Univ, 72-73; MGR EXPLOR RES CHEM, POLYCHROME CORP, 73- Mem: Am Chem Soc; Soc Photog Scientists & Engrs; AAAS; Sigma Xi. Res: Laser imaging systems; novel organic photoconductors, both monomeric and polymeric; photochemical polymerization; new non-silver halide imaging systems; charge transfer and surface chemistry, x-ray diffraction and quantum chemistry; color reproduction systems; chemical plant design and financial analysis. Mailing Add: 4705 Henry Hudson Pkwy Riverdale NY 10471

WEINBERGER, PEARL, b Derby, Eng, Apr 20, 26; Can citizen; m 48; c 2. PLANT PHYSIOLOGY, ENVIRONMENTAL BIOLOGY. Educ: Univ Manchester, BSc, 48, MSc, 51; Univ Ottawa, PhD(biol), 62. Prof Exp: Demonstr bot, Leeds Univ, 48-49; ed asst, Can J Res, 49-50; res officer, Nat Res Coun, Ottawa, 50-52; from lectr to asst prof, 52-67, ASSOC PROF BOT, UNIV OTTAWA, 67- Mem: Can Soc Plant Physiol; Can Soc Cell Biol; Can Bot Asn; Can Genetic Soc; Fr-Can Asn Advan Sci. Res: Environmental control of plant growth, including temperature, imbibition period, light, herbicides, pesticides, audible and ultrasound, in relation to germination, morphogenesis and biochemical and molecular aspects of change with development. Mailing Add: Dept of Biol Univ of Ottawa Ottawa ON Can

WEINBERGER, PETER JAY, b New York, NY, Aug 6, 42. MATHEMATICS, COMPUTER SCIENCE. Educ: Swarthmore Col, BA, 64; Univ Calif, Berkeley, PhD(math), 69. Prof Exp: Mem tech staff, Bellcomm Inc, 69-70; ASST PROF MATH, UNIV MICH, ANN ARBOR, 70- Concurrent Pos: Consult, Math Rev, 75- Mem: Am Math Soc; Asn Comput Mach; Math Asn Am. Res: Number theory; computation complexity and algorithms; computer-assisted typesetting, especially of mathematics. Mailing Add: Dept of Math Univ of Mich Ann Arbor MI 48104

WEINER, CHARLES, b Brooklyn, NY, Aug 11, 31; m 56; c 1. HISTORY OF PHYSICS, HISTORY OF BIOLOGY. Educ: Case Inst Technol, BS, 60, MA, 63, PhD(hist of sci), 65. Prof Exp: Dir proj hist of recent physics in US, Am Inst Physics, 64-65, dir, Ctr for Hist of Physics, 65-74, PROF HIST OF SCI & TECHNOL, MASS INST TECHNOL, 74-, DIR, ORAL HIST PROG, 75- Concurrent Pos: NSF grants, 65-73 & 75; mem adv comt, Nat Union Catalog of Manuscript Collections, Libr of Cong, 65-71; vis lectr, Polytech Inst Brooklyn, 66; proj dir comt hist of contemp physics, Joint Am Inst Physics-Am Acad Arts & Sci, 66-74; Guggenheim fel, Niels Bohr Inst, Copenhagen, Denmark, 70-71; mem exec comt, Sem Technol & Soc Change, Columbia Univ, 69-75. Honors & Awards: Distinguished Serv Citation, Am Asn Physics Teachers, 74. Mem: Fel AAAS; Hist Sci Soc; Soc Hist Technol; Am Hist Asn; Orgn Am Historians. Res: History of twentieth century science; social impact of science and technology; development of the physical sciences in the United States. Mailing Add: Sch Humanities & Soc Sci Mass Inst Technol Cambridge MA 02139

WEINER, EUGENE ROBERT, b Pittsburgh, Pa, Sept 16, 28; m 52; c 4. PHYSICAL CHEMISTRY. Educ: Ohio Univ, BS, 50; Univ Ill, MS, 57; Johns Hopkins Univ, PhD(chem), 63. Prof Exp: Sect chief appl physics sect, Interior Ballistics Lab, Aberdeen Proving Grounds, Md, 50-55; instr, McCoy Col, 62-63; sr scientist, Johnston Labs, Inc, 63-65; asst prof, 65-67, ASSOC PROF CHEM, UNIV DENVER, 67- Concurrent Pos: Consult, US Geol Surv, 70- Mem: Am Chem Soc; Soc Appl Spectros. Res: Atomic and molecular collisions; radiation chemistry; gas phase ion-molecule reactions, free radical reactions, gas kinetics and energetics; laser Raman spectroscopy; remote sensing of pollutants; solid surface catalysis. Mailing Add: Dept of Chem Univ of Denver Denver CO 80210

WEINER, HENRY, b Cleveland, Ohio, May 18, 37; m 60; c 2. ORGANIC CHEMISTRY, BIOCHEMISTRY. Educ: Case Inst Technol, BS, 59; Purdue Univ, PhD(org chem), 63. Prof Exp: Res assoc biol, Brookhaven Nat Lab, 63-65; NIH res fel biochem, Karolinska Inst, Sweden, 65-66; asst prof, 66-69, ASSOC PROF BIOCHEM, PURDUE UNIV, LAFAYETTE, 69-, ASSOC PROF MED CHEM, 75- Mem: Am Soc Biol Chem; Am Chem Soc; Am Asn Univ Prof. Res: Enzymology; protein chemistry; neurotransmitter metabolism; effects of ethanol on metabolism. Mailing Add: Dept of Biochem Purdue Univ West Lafayette IN 47907

WEINER, HERBERT, b Vienna, Austria, Feb 6, 21; nat US; m 53; c 3. PSYCHIATRY, NEUROLOGY. Educ: Harvard Univ, AB, 43; Columbia Univ, MD, 46. Prof Exp: Exchange fel neurol, 50; USPHS fel psychiat, 52-53; instr psychiat, Sch Med & Dent, Univ Rochester, 53-54; guest lectr, Wash Sch Psychiat, 54-55; from instr to assoc prof, 55-66, PROF PSYCHIAT, ALBERT EINSTEIN COL MED, 66-; ATTEND PSYCHIATRIST & DIR EDUC PROG, DIV PSYCHIAT, MONTEFIORE HOSP & MED CTR, NEW YORK, 65-, CHMN DEPT PSYCHIAT, 69- Concurrent Pos: Asst vis psychiatrist & chief adult psychiat in-patient serv, Bronx Munic Hosp Ctr, 55-56, asst vis physician, 56-61, assoc vis psychiatrist, 56-; consult, Home Care Dept, Montefiore Hosp, 55-59; USPHS fel ment health, 56-58 & res career develop award, 62-65; Stern fel, 59-62; mem bd trustees, Scarborough Country Day Sch, 62-; mem ment health study sect, Div Res Grants, NIH, 62-, chmn, 66-67; fel, Ctr Advan Study Behav Sci, Stanford Univ, Guggenheim

Mem Found fel & Commonwealth Fund grant-in-aid, 72-73; ed-in-chief, J Am Psychosom Soc, 72- Mem: AAAS; Am Acad Neurol; Am Psychosom Soc (pres, 71-72); Asn Res Nerv & Ment Dis; Soc Psychophysiol Study of Sleep. Res: Neurophysiology; psychophysiology. Mailing Add: Dept of Psychiat Montefiore Hosp & Med Ctr New York NY 10467

WEINER, HOWARD JACOB, b Chicago, Ill, Aug 26, 37; m 71. MATHEMATICAL STATISTICS. Educ: Ill Inst Technol, BSEE, 58; Univ Chicago, MS, 60; Stanford Univ, PhD(statist), 64. Prof Exp: Actg asst prof statist, Univ Calif, Berkeley, 64; asst prof, Stanford Univ, 64-65; ASST PROF MATH, UNIV CALIF, DAVIS, 65- Mem: Inst Math Statist. Res: Age dependent branching processes. Mailing Add: Dept of Math Univ of Calif Davis CA 95616

WEINER, IRWIN M, b New York, NY, Nov 5, 30; m 61; c 2. PHARMACOLOGY. Educ: Syracuse Univ, AB, 52; State Univ NY, MD, 56. Prof Exp: Fel pharmacol, Sch Med, Johns Hopkins Univ, 56-58, from instr to asst prof pharmacol & exp therapeut, 58-66; assoc prof, 66-68, PROF PHARMACOL & CHMN DEPT, STATE UNIV NY UPSTATE MED CTR, 68- Concurrent Pos: Fel pharmacol & exp therapeut, Sch Med, Johns Hopkins Univ, 58-60; vis prof molecular biol, Albert Einstein Col Med, 64-65; USPHS res career develop award, 64-66; mem study sect pharmacol & exp therapeut, NIH, 65-69; ed, J Pharmacol & Exp Therapeut, 65-72; consult, Sterling Winthrop Res Inst. Mem: AAAS; Am Soc Pharmacol & Exp Therapeut; NY Acad Sci. Res: Pharmacology of diuretics and uricosuric agents; renal excretion of drugs; intestinal absorption of bile salts; bacterial cell wall biosynthesis. Mailing Add: Dept of Pharmacol State Univ of NY Upstate Med Ctr Syracuse NY 13210

WEINER, JEROME HARRIS, b New York, NY, Apr 5, 23; m 50; c 2. APPLIED MATHEMATICS. Educ: Cooper Union, BME, 43; Columbia Univ, AM, 47, PhD(math), 52. Prof Exp: Asst tech dir heat & mass flow analyzer lab, Columbia Univ, 51-57, asst prof civil eng & eng mech, 52-56, prof mech eng, 56-68; L HERBERT BALLOU UNIV PROF ENG & PHYSICS, BROWN UNIV, 68- Mem: Am Math Soc; Am Phys Soc. Res: Thermal stresses; crystal defects. Mailing Add: Div of Eng Brown Univ Providence RI 02912

WEINER, JOHN, b Malvern, NY, Apr 14, 43. CHEMICAL PHYSICS. Educ: Pa State Univ, BS, 64; Univ Chicago, PhD(chem), 70. Prof Exp: Fel & lectr chem, Yale Univ, 70-73; ASST PROF CHEM, DARTMOUTH COL, 73- Mem: Am Chem Soc; Am Phys Soc; AAAS. Res: Reactive and inelastic collision processes in ion-molecule systems; internal state excitation leading to chemiluminescence phenomena. Mailing Add: Dept of Chem Dartmouth Col Hanover NH 03755

WEINER, LAWRENCE MYRON, b Milwaukee, Wis, May 21, 23; m 44; c 2. MICROBIOLOGY, IMMUNOLOGY. Educ: Univ Wis, BA, 47, MS, 48, PhD(med microbiol), 51. Prof Exp: From instr to assoc prof, 51-65, chmn dept & assoc dean, 70-72, PROF MICROBIOL, SCH MED, WAYNE STATE UNIV, 65-, DEP DEAN, 72- Concurrent Pos: Pres, Mich State Bd Examr in Basic Sci, 62-68. Mem: Am Soc Microbiol; Soc Exp Biol & Med. Res: Immunology of infectious diseases; hypersensitivity to chemical agents; tumor immunology. Mailing Add: Off of Dep Dean Wayne State Univ Sch of Med Detroit MI 48201

WEINER, LOUIS MAX, b Chicago, Ill, Nov 11, 26; m 57; c 3. ALGEBRA. Educ: Univ Chicago, SB, 47, SM, 48, PhD(math), 51. Prof Exp: Asst, Univ Chicago, 48; personnel examr, Chicago Civil Serv Comn, 51-52; asst prof math, DePaul Univ, 52-58; res engr, Mech Res Div, Am Mach & Foundry Co, 58-62, supvr, Anal Sect, Gen Am Res Div, 62-64; assoc prof, 64-68, chmn dept, 68-74, PROF MATH, NORTHEASTERN ILL UNIV, 68- Concurrent Pos: Instr, Amundsen Br, Chicago City Jr Col, 58-64. Mem: Am Math Soc; Math Asn Am. Res: Algebra; linear algebras; operations research; slide rule type computers. Mailing Add: 3144 Greenleaf Wilmette IL 60091

WEINER, MILTON LAWRENCE, b Detroit, Mich, Dec 3, 21; m 50; c 4. POLYMER CHEMISTRY, PLASTICS CHEMISTRY. Educ: Wayne State Univ, BS, 47, PhD, 52. Prof Exp: Res chemist, Ethyl Corp, 52-56 & Gen Elec Co, 56-59; res chemist, 59-69, RES ASSOC, PLASTICS DIV, TECH CTR, MOBIL CHEM CO, 69- Res: Polymers; morphology; rheology. Mailing Add: Plastics Div Tech Ctr Mobil Chem Co Macedon NY 14502

WEINER, MURRAY, b New York, NY, Apr 18, 19; m 51; c 1. MEDICINE. Educ: City Col New York, BS, 39; NY Univ, MS & MD, 43. Prof Exp: From instr to asst prof med, Col Med, NY Univ, 54-72; vpres & dir biol res, Geigy Res Labs, NY, 67-72; V PRES RES CLIN PHARMACOL & LAB SCI, MERRELL-NAT LABS, 72- Concurrent Pos: Asst vis physician & consult, Bellevue Hosp, NY, 49-72, Univ Hosp, 53-72, Long Island Jewish Hosp, 54-57 & North Shore Hosp, NY, 55-72; assoc vis physician, Willard Parker Hosp, NY, 52-57 & Goldwater Mem Hosp, NY, 56-72; vis staff, Cincinnati Gen Hosp & Christian Holmes Hosp, Ohio, 72-; clin prof med, Univ Cincinnati, 72- Mem: Soc Exp Biol & Med; Am Physiol Soc; AMA; Am Heart Asn. Res: Clotting mechanism; anticoagulant and anti-inflammatory drugs; drug disposition. Mailing Add: 8915 Spooky Ridge Lane Indian Hill OH 45242

WEINER, MYRON, b Baltimore, Md, May 27, 43; m 64; c 2. PHARMACOLOGY. Educ: Univ Md, Baltimore City, BS, 66, PhD(pharmacol), 71. Prof Exp: Instr pharmacol, Sch Nursing, Univ Md, Baltimore City, 67-68; instr anat & physiol, Sch Nursing, St Agnes Hosp, 68-70; instr, Catonsville Community Col, Md, 70-71; ASST PROF PHARMACOL, SCH PHARM, UNIV SOUTHERN CALIF, 71- Concurrent Pos: Gen res support grant, Univ Southern Calif, 71-73, Cancer Ctr grant, 74-75; Nat Cancer Inst grant, 73-76. Mem: AAAS; NY Acad Sci; Am Col Clin Pharmacol; Am Pharmaceut Asn. Res: Alteration of drug biotransformation caused by cyclic nucleotides; relationship of cyclic nucleotides and prostaglandins to altered drug metabolism caused by cancer. Mailing Add: Dept of Pharmacol Univ of Southern Calif Sch Pharm Los Angeles CA 90033

WEINER, NORMAN, b Rochester, NY, July 13, 28; m 55; c 5. PHARMACOLOGY. Educ: Univ Mich, BS, 49; Harvard Med Sch, MD, 53. Prof Exp: Intern, Harvard Med Serv, Boston City Hosp, 53-54, instr pharmacol, Harvard Med Sch, 56-58, assoc, 58-61, asst prof, 61-67; PROF PHARMACOL & CHMN DEPT, MED CTR, UNIV COLO, DENVER, 67- Concurrent Pos: Dir neuropharmacol lab, Mass Ment Health Ctr, 64-67. Mem: AAAS; Am Soc Pharmacol & Exp Therapeut; Am Soc Neurochem; Am Col Neuropsychopharmacol; Asn Med Sch Pharmacologists. Res: Metabolism of biologically active amines; regulation of synthesis of neurotransmitters; synthesis, storage and release of tissue amines; effect of drugs on energy metabolism of brain; ionization of drugs, drug distribution and relation to biological activity. Mailing Add: Dept of Pharmacol Univ of Colo Med Ctr Denver CO 80220

WEINER, PAUL HARVEY, physical chemistry, see 12th edition

WEINER, RICHARD, b Brooklyn, NY, Aug 21, 36; m 60; c 2. PHYSIOLOGY. Educ: Long Island Univ, BS, 58; NY Univ, PhD(physiol), 65. Prof Exp: USPHS fel microcirc, NY Univ Med Ctr, 65, asst res scientist, 66; from instr to asst prof, 66-72, ASSOC PROF PHYSIOL, NEW YORK MED COL, 73- Mem: AAAS; Am Physiol Soc; Microcirc Soc. Res: Humoral regulation of blood flow in the microcirculation; vascular permeability and tissue injury; vascular smooth muscle. Mailing Add: Dept of Physiol NY Med Col Basic Sci Bldg Valhalla NY 10595

WEINER, RICHARD IRA, b New York, NY, Nov 6, 40. NEUROENDOCRINOLOGY. Educ: Pa State Univ, BS, 63, MS, 65; Univ Calif, San Francisco, PhD(endocrinol), 69. Prof Exp: Fel, Brain Res Inst, Univ Calif, Los Angeles, 69-71; asst prof physiol, Univ Tenn Med Units, 71-72; asst prof anat, Sch Med, Univ Southern Calif, 72-74; ASST PROF OBSTET & GYNEC & PHYSIOL, SCH MED, UNIV CALIF, LOS ANGELES, 74- Mem: Endocrine Soc; Int Neuroendocrine Soc; Am Physiol Soc. Res: Role of brain catecholamines in the neuroendocrine regulation of the secretion of luteinizing hormone and prolactin. Mailing Add: Dept of Obstet & Gynec Univ of Calif Sch of Med San Francisco CA 94143

WEINER, ROBERT ALLEN, b New York, NY, Apr 3, 40; m 61; c 2. SOLID STATE PHYSICS. Educ: Columbia Univ, AB, 61; Harvard Univ, AM, 62, PhD(physics), 67. Prof Exp: Asst res physicist, Univ Calif, San Diego, 67-69; ASST PROF PHYSICS, CARNEGIE-MELLON UNIV, 69- Mem: Am Phys Soc. Res: Electronic effects in the theory of many-body systems. Mailing Add: Dept Physics Mellon Inst Sci Carnegie-Mellon Univ Pittsburgh PA 15213

WEINER, ROBERT SAMUEL, b New York, NY, June 13, 42; m 67; c 1. PHYSICAL CHEMISTRY. Educ: Polytech Inst Brooklyn, BS, 64; Purdue Univ, MS, 67, PhD(phys chem), 70. Prof Exp: Sr res chemist, Rohm and Haas Co, 69-71, group leader fibers, 71-74; TECH DIR, KEMOS, INC, 74- Mem: Am Chem Soc; Am Asn Textile Chem & Colorists; Am Soc Test & Mat. Res: Development of improved fibers for use in carpet applications; development of new products of fusion bonded carpet. Mailing Add: Kemos Inc 1135 Shallowford Rd Marietta GA 30060

WEINER, RONALD MARTIN, b Brooklyn, NY, May 7, 42; m 64; c 2. MICROBIOLOGY. Educ: Brooklyn Col, BS, 64; LI Univ, MS, 67; Iowa State Univ, PhD(microbiol), 70. Prof Exp: Bacteriologist, Greenpoint, Coney Island Hosps, NY, 64-66; teaching asst, Iowa State Univ, 67-68, instr, 69-70; ASST PROF MICROBIOL, UNIV MD, COLLEGE PARK, 70- Mem: AAAS; Am Soc Microbiol; Am Inst Biol Sci. Res: Morphogenesis of Hyphomicrobium; DNA replication and cell division; spectral studies of nucleic acids; microbial ecology. Mailing Add: Dept of Microbiol Univ of Md College Park MD 20742

WEINER, STEPHEN DOUGLAS, b Philadelphia, Pa, Jan 1, 41; m 61; c 2. APPLIED PHYSICS. Educ: Mass Inst Technol, BS, 61, PhD(physics), 65. Prof Exp: Mem staff, 65-73, ASSOC GROUP LEADER, LINCOLN LAB, MASS INST TECHNOL, 73- Res: Missile defense research; reentry physics; electromagnetic scattering and propagation; operations research. Mailing Add: Lincoln Lab Mass Inst of Technol Lexington MA 02173

WEINER, STEVEN ALLAN, b New York, NY, June 6, 42; m 71; c 1. PHYSICAL ORGANIC CHEMISTRY. Educ: Columbia Univ, AB, 63; Iowa State Univ, PhD(org chem), 67. Prof Exp: Res assoc, Calif Inst Technol, 67-68; staff scientist, 68-74, PROJ MGR, RES STAFF, FORD MOTOR CO, 74- Concurrent Pos: Lectr, Univ Mich-Dearborn, 70-71. Honors & Awards: Leibmann Mem Award, Am Chem Soc, 59. Mem: Am Chem Soc; Am Mgt Asn. Res: Photochemistry; free radical kinetics; combustion products and reactions; sodium-sulfur battery; energy storage and conversion; electric car. Mailing Add: Research Staff Ford Motor Co PO Box 2053 Dearborn MI 48121

WEINFELD, HERBERT, b New York, NY, Feb 7, 21; m 46; c 1. BIOCHEMISTRY. Educ: City Col New York, BS, 42; Univ Mich, MS, 48, PhD(biochem), 52. Prof Exp: Asst biochem, Univ Mich, 47-49, asst, Sloan-Kettering Inst, 53-54; instr, Inst Indust Med, NY Univ-Bellevue Med Ctr, 54-55; assoc cancer res scientist, 55-65, PRIN CANCER RES SCIENTIST, DEPT MED, ROSWELL PARK MEM INST, 65-; RES PROF & CHMN DEPT BIOCHEM, ROSWELL PARK GRAD DIV, STATE UNIV NY BUFFALO, 71- Concurrent Pos: Res prof, State Univ NY Buffalo, 72- Mem: AAAS; Am Chem Soc; Harvey Soc; Am Soc Biol Chemists; Am Asn Cancer Res. Res: Metabolism of nucleosides, nucleotides and nucleic acids; bacterial viruses; cellular agents controlling mitotic events. Mailing Add: Roswell Park Mem Inst 666 Elm St Buffalo NY 14263

WEINGARTEN, FREDERICK W, applied mathematics, computer science, see 12th edition

WEINGARTEN, HAROLD IVAN, organic chemistry, see 12th edition

WEINGARTEN, HARRY, mathematical statistics, see 12th edition

WEINGARTNER, DAVID PETER, b Escanaba, Mich, Mar 13, 39; m 64; c 2. PHYTOPATHOLOGY, PLANT NEMATOLOGY. Educ: Univ Mich, BS, 62; Mich State Univ, PhD(plant path), 69. Prof Exp: Teacher high sch, Inkster, Mich, 62-64; asst prof & asst plant pathologist, 69-75, ASSOC PROF & ASSOC PLANT PATHOLOGIST, AGR RES CTR, INST FOOD & AGR, UNIV FLA, 75- Concurrent Pos: Res award, Fla Fruit & Vegetable Asn, 75. Mem: Soc Nematol; Am Phytopath Soc; Potato Asn Am. Res: Fungus diseases; diseases of vegetables and fruits; nematode control on vegetables; potato corky ringspot. Mailing Add: Agr Res Ctr Univ of Fla Hastings FL 32045

WEINGOLD, ALLAN BYRNE, b New York, NY, Sept 2, 30; m 52; c 4. OBSTETRICS & GYNECOLOGY. Educ: Oberlin Col, BA, 51; New York Med Col, MD, 55; Am Bd Obstet & Gynec, dipl, 64. Prof Exp: From asst prof to prof obstet & gynec, New York Med Col, 70-73, asst chmn dept, 71-73; PROF OBSTET & GYNEC & CHMN DEPT, SCH MED, GEORGE WASHINGTON UNIV, 73- Concurrent Pos: Am Cancer Soc fel gynec malignancy, 60-61; training dir, USPHS grant, 66-68; chief obstet & gynec, Metrop Hosp, New York, 67-70; consult, NIH Cancer Ctr, Walter Reed Army Med Ctr, Columbia Hosp for Women & Fairfax Hosp. Mem: Fel Am Col Obstet & Gynec; fel Am Col Surgeons. Res: Studies on monitoring of the fetal environment by endocrine, biochemical and biophysical indices. Mailing Add: Dept of Obstet & Gynec George Washington Univ Sch Med Washington DC 20037

WEINHEIMER, ALFRED JOHN, organic chemistry, see 12th edition

WEINHEIMER, WILLIAM HENRY, horticulture, cytogenetics, see 12th edition

WEINHOLD, ALBERT RAYMOND, b Evans, Colo, Feb 14, 31; m 52; c 2. PLANT PATHOLOGY. Educ: Colo State Univ, BS, 53, MS, 55; Univ Calif, PhD(plant path), 58. Prof Exp: Asst, Univ Calif, Davis, 55-57, from instr & jr plant pathologist to assoc prof & assoc plant pathologist, 59-72, PROF PLANT PATH & PLANT PATHOLOGIST, EXP STA, UNIV CALIF, BERKELEY, 72- Concurrent Pos: Sr ed,

Phytopath, 70-73, ed-in-chief, 73-75. Mem: AAAS; Am Phytopath Soc. Res: Disease and pathogen physiology; soil-borne pathogens; root diseases; potato diseases. Mailing Add: Dept of Plant Path Univ of Calif Berkeley CA 94720

WEINHOLD, PAUL ALLEN, b Evans, Colo, Sept 23, 35; m 56; c 4. BIOCHEMISTRY. Educ: Colo State Univ, BS, 57; Univ Wis, PhD(biochem), 61. Prof Exp: Asst prof biochem & internal med, 65-72, ASSOC PROF BIOL CHEM, MED SCH, UNIV MICH, ANN ARBOR, 72-; BIOCHEMIST, VET ADMIN HOSP, 65- Concurrent Pos: USPHS res fel biochem, Harvard Med Sch, 63-65. Mem: AAAS; Am Soc Biol Chemists. Res: Biochemistry of development, phospholipid metabolism and control mechanisms in metabolism. Mailing Add: Med Res Vet Admin Hosp 2215 Fuller Rd Ann Arbor MI 48105

WEINHOUS, MARTIN S, b Brooklyn, NY, July 30, 44; m 75. ATOMIC PHYSICS. Educ: Rensselaer Polytech Inst, BS, 66; Univ NH, MS, 70, PhD(physics), 74. Prof Exp: Instr physics, St Francis Col, 73; from instr to asst prof, N Adams State Col, 73-75; ASST PROF PHYSICS, NORWICH UNIV, 75- Mem: Am Asn Physics Teachers; Am Phys Soc. Res: Measurement of polarization of light from atomic transitions, and lifetimes of atomic and molecular excited states. Mailing Add: 88 S Main St Northfield VT 05663

WEINHOUSE, SIDNEY, b Chicago, Ill, May 21, 09; c 3. BIOCHEMISTRY, CANCER. Educ: Univ Chicago, BS, 33, PhD(chem), 36. Hon Degrees: DSc, Med Col Pa, 72. Prof Exp: Coman fel org chem, Univ Chicago, 41-44; head biochem res, Houdry Process Corp, Pa, 44-47; biochem res dir, Res Inst, 47-50, prof chem & biol, chmn div biochem, Inst Cancer Res & head dept metab chem, Lankenau Hosp Res Inst, 50-57, PROF BIOCHEM, SCH MED, TEMPLE UNIV, 57-, DIR FELS RES INST, 63- Concurrent Pos: Mem biochem study sect, NIH, 53-58 & nat adv coun, Nat Cancer Inst, Dept Health, Educ & Welfare, 58-62; assoc dir, Fels Res Inst, 61-63; chmn comt biol chem, Div Chem & Chem Technol, Nat Acad Sci-Nat Res Coun, 62-64; co-ed, Advan Cancer Res, 60-, ed, Cancer Res, 69-; mem sci adv comt, Damon Runyon Mem Fund, 52-60 & Environ Health Sci Adv Comt, 68-72. Honors & Awards: Philadelphia Sect Award, Am Chem Soc, 66; G H A Clowes Award, Am Asn Cancer Res, 72. Mem: Am Chem Soc; Am Soc Biol Chemists; Am Asn Cancer Res; Am Diabetes Asn; Soc Exp Biol & Med. Res: Carbohydrate and fatty acid metabolism in normal and neoplastic cells; dietary and hormonal regulation of enzymes of cardohydrate and fatty acid metabolism in liver tumors; effects of hormones on gluconeogenesis; comparative studies on control of respiration and glycolysis in liver and liver tumors. Mailing Add: Fels Res Inst Temple Univ Sch of Med Philadelphia PA 19140

WEININGER, JOSEPH LEOPOLD, physical chemistry, see 12th edition

WEININGER, STEPHEN JOEL, b New York, NY, Mar 28, 37; m 61; c 3. ORGANIC CHEMISTRY. Educ: Brooklyn Col, BA, 57; Univ Pa, PhD(org chem), 64. Prof Exp: Sr demonstr phys chem, Univ Durham, 64-65; asst prof, 65-70, ASSOC PROF CHEM, WORCESTER POLYTECH INST, 70- Mem: Am Chem Soc; The Chem Soc. Res: Carbene chemistry; chemistry of hot intermediates; photochemistry of sulfur compounds. Mailing Add: Dept of Chem Worcester Polytech Inst Worcester MA 01609

WEINKAM, ROBERT JOSEPH, b Cincinnati, Ohio, Dec 27, 42. MEDICINAL CHEMISTRY. Educ: Xavier Univ, Ohio, BS, 64; Duquesne Univ, PhD(chem), 68. Prof Exp: Fel, Syra Res Inst Calif, 68-69; fel, 69-70, ASST PROF RES, PHARMACEUT CHEM, UNIV CALIF, SAN FRANCISCO, 71- Concurrent Pos: Res career develop award, NIH, 75-80. Mem: Am Chem Soc; Am Soc Mass Spectros. Res: Biomedical application of chemical ionization; mass spectrometry; brain tumor chemotherapeutic agents; bile acids. Mailing Add: Dept of Pharmaceut Chem Univ of Calif San Francisco CA 94122

WEINKE, KARL FREDERICK, biochemistry, botany, see 12th edition

WEINLAND, BERNARD THEODORE, b Scarsdale, NY, July 2, 22; m 47; c 4. APPLIED STATISTICS. Educ: Purdue Univ, BS, 48, MS, 61, PhD(poultry genetics), 66. Prof Exp: Poultry plant supt, Ind Farm Bur Coop, 48-49; flock inspector, State Poultry Asn, Ind, 49-51; asst farm supt, Ind Poultry Breeders, Inc, 51-55; farm supt, Regional Poultry Breeding Lab, Agr Res Serv, 55-66, statistician, Biomet Serv, 66-72, REGIONAL CONSULT STATISTICIAN, NORTHEAST REGION, AGR RES SERV, USDA, 72- Mem: Poultry Sci Asn; Am Dairy Sci Asn; Am Soc Animal Sci; Biomet Soc. Res: Gene frequency stability in a random bred poultry population; genetic correlations between purebred and crossbred progeny of the same sires; statistical problems of animal scientist. Mailing Add: Bldg 173 Agr Res Ctr E Beltsville MD 20705

WEINLESS, MICHAEL HOWARD, b New York, NY, Sept 28, 44; div; c 1. MATHEMATICS. Educ: Mass Inst Technol, BS, 64, PhD(math), 68. Prof Exp: Benjamin Pierce instr math, Harvard Univ, 68-69, asst prof, 69-71; PROF MATH, MAHARISHI INT UNIV, 72- Res: Relationship of mathematics to the science of creative intelligence. Mailing Add: Maharishi Int Univ Fairfield IA 52556

WEINMAN, DAVID, II, b Albuquerque, NMex, Jan 13, 09; m 46. TROPICAL MEDICINE, PARASITOLOGY. Educ: Columbia Univ, AB, 29; Univ Paris, MD, 35; Am Bd Path, dipl, 50. Prof Exp: Asst parasitol, Univ Paris, 31-35; fel comp path, Harvard Med Sch, 36-37, asst, 38, instr comp path & trop med, Sch Med & Pub Health, 39-44; asst prof, Col Physicians & Surgeons, Columbia Univ, 44-46; from asst prof to assoc prof, Sch Med & Grad Sch, 48-57, PROF EPIDEMIOL, YALE UNIV, 57- Concurrent Pos: Fulbright award, Brit E Africa, 56-57; med educr, AID, Vietnam, 63; med officer & div chief, US Army-SEATO Med Res Lab, Bangkok, Thailand, 64-65; vis prof, Univ Med Sci, Bangkok, 65. Consult, Yale-New Haven & Vet Admin Hosps; consult, Coun on Drugs, AMA. Mem expert panel parasitic dis, WHO. Harvard Med Sch exped, Peru, 37 & Liberia, 44. Mem: Fel Soc Exp Biol & Med; fel Am Soc Trop Med & Hyg; fel Royal Soc Trop Med & Hyg. Res: African trypanosomiasis; trypanosomiases of primates, human and sub-human, in South Asia; bartonellosis; toxoplasmosis; malaria; medical diseases caused by protozoa; host-parasite relationships; international medical education in microbiology. Mailing Add: Dept of Epidemiol 318 Brady Yale Univ New Haven CT 06510

WEINMANN, CLARENCE JACOB, b Oakland, Calif, May 27, 25; m 57; c 2. PARASITOLOGY. Educ: Univ Calif, BS, 50, PhD(parasitol), 58. Prof Exp: Instr microbiol, Col Med, Univ Fla, 58-60; res fel biol, Rice Univ, 60-62; from asst prof to assoc prof, 62-75, PROF ENTOM & PARASITOL, UNIV CALIF, BERKELEY, 75- Concurrent Pos: Fel trop med & parasitol, Univ Cent Am, 59. Mem: AAAS; Am Soc Parasitol; Entom Soc Am; Am Soc Trop Med & Hyg; Wildlife Dis Asn. Res: Immunity in helminth infections; arthropod-borne helminthiases. Mailing Add: Div of Entom & Parasitol Univ of Calif Berkeley CA 94720

WEINREB, EVA LURIE, b New York, NY; m 50. BIOLOGICAL STRUCTURE, CELL BIOLOGY. Educ: NY Univ, BA, 48; Univ Wis, MA, 49, PhD(zool), 55. Prof Exp: Asst zool, Univ Wis, 49 & 51-55; res assoc path, Sch Med, Marquette Univ, 55-56; dir basic sci, St Mary's Hosp, Sch Nursing, Milwaukee, 56-57; head animal res, Kolmer Res Ctr, 57-58; asst prof biol, Milwaukee-Downer Col, 59-60; fel cell biol & res assoc anat, Med Col, Cornell Univ, 61-63; asst prof biol, Washington Square Col, NY Univ, 63-69; cell biologist & sr res scientist, Biomed Sect, Geometric Data Corp, 71-72; ASSOC PROF BIOL, COMMUNITY COL PHILADELPHIA, 73- Concurrent Pos: Consult, Volu-Sol Med Industs, Inc, 72-73. Mem: AAAS; Am Asn Anatomists; Am Soc Cell Biol; Am Soc Zool; Am Soc Allied Health Prof. Res: Comparative histology; hematology; pathology; anatomy. Mailing Add: Community Col of Philadelphia 1600 Spring Garden St Philadelphia PA 19130

WEINREB, MICHAEL PHILIP, b Lakewood, NJ, Feb 2, 39; m 66; c 2. ATMOSPHERIC PHYSICS. Educ: Univ Pa, BA, 60; Brandeis Univ, PhD(physics), 66. Prof Exp: Instr physics, Brandeis Univ, 64-65; physicist, Electronics Res Ctr, NASA, Mass, 65-70; PHYSICIST, NAT ENVIRON SATELLITE SERV, NAT OCEANIC & ATMOSPHERIC ADMIN, 70- Mem: AAAS; Optical Soc Am. Res: Radiative transfer theory; atomic and molecular physics; remote sensing of atmospheric temperature and constituent profiles. Mailing Add: FOB-4 Nat Environ Satellite Serv Nat Oceanic & Atmospheric Admin Suitland MD 20233

WEINREB, SANDER, b New York, NY, Dec 9, 36; m 57; c 2. RADIO ASTRONOMY, MICROWAVE ENGINEERING. Educ: Mass Inst Technol, BSEE, 58, PhD(elec eng), 63. Prof Exp: Staff mem, Lincoln Lab, Mass Inst Technol, 63-65; HEAD ELECTRONICS DIV, NAT RADIO ASTRON OBSERV, 65- Mem: Inst Elec & Electronics Engrs. Res: Digital autocorrelation techniques; microwave lines in radio astronomy. Mailing Add: Nat Radio Astron Observ Edgemont Rd Charlottesville VA 22901

WEINREB, STEVEN MARTIN, b Brooklyn, NY, May 10, 41; m 65; c 2. ORGANIC CHEMISTRY. Educ: Cornell Univ, AB, 63; Univ Roehester, PhD(chem), 67. Prof Exp: NIH fel, Columbia Univ, 66-67; NIH fel, Mass Inst Technol, 67-68, res assoc, 68-70; asst prof, 70-75, ASSOC PROF CHEM, FORDHAM UNIV, 75- Concurrent Pos: Res fel, Alfred P Sloan Found, 75-77; res career develop award, NIH, 75-80. Mem: Am Chem Soc; The Chem Soc. Res: Structure determination and synthesis of natural products. Mailing Add: Dept of Chem Fordham Univ Bronx NY 10458

WEINREICH, DANIEL, b Claremont, France, June 6, 42; US citizen; m 65; c 2. NEUROBIOLOGY. Educ: Bethany Col, BS, 64; Univ Utah, PhD(pharmacol), 70. Prof Exp: Res asst neuropharmacol, Sandoz Pharmaceut, Inc, 64-65; NSF fel, City of Hope Nat Med Ctr, 70-72, NIMH fel, 73-74; ASST PROF PHARMACOL, SCH MED, UNIV MD, BALTIMORE, 74- Concurrent Pos: Extramural reviewer, NSF, 73-; NSF res grant, 74. Res: Identification and regulation of neurotransmitter and neuromodulator substances in single nerve cells. Mailing Add: Dept Pharmacol & Exp Therapeut Univ of Md Sch of Med Baltimore MD 21201

WEINREICH, GABRIEL, b Vilna, Poland, Feb 12, 28; nat US; m 51, 71; c 5. PHYSICS. Educ: Columbia Univ, AB, 48, MA, 49, PhD(physics), 54. Prof Exp: Asst physics, Columbia Univ, 49-51; mem tech staff, Bell Tel Labs, 53-60; vis assoc prof, 60, assoc prof, 60-64, PROF PHYSICS, UNIV MICH, ANN ARBOR, 64- Res: Atomic spectra; solid state theory; electron-phonon interactions; nonlinear optics; thermodynamics; electron-atom scattering; atomic beam kinetics. Mailing Add: Randall Lab of Physics Univ of Mich Ann Arbor MI 48104

WEINRICH, MARCEL, b Jendiesow, Poland, July 23, 27; nat US; m 58; c 1. PHYSICS. Educ: Bethany Col, WVa, BS, 46; Univ WVa, MS, 48; Columbia Univ, PhD, 57. Prof Exp: Instr physics, Univ WVa, 47-49; physicist, Res & Develop Ctr, Gen Elec Co, 57-69; PROF PHYSICS & CHMN DEPT, JERSEY CITY STATE COL, 69- Mem: Fel AAAS; Am Inst Physics; Am Asn Physics Teachers; Nat Sci Teachers Asn; Am Phys Soc. Res: Meson and plasma physics; breakdown of parity conservation in meson decays; controlled fusion reactors. Mailing Add: Dept Physics Jersey City State Col 2039 Kennedy Blvd Jersey City NJ 07305

WEINRYB, IRA, b New York, NY, Nov 20, 40; m 67; c 2. BIOCHEMISTRY. Educ: Columbia Univ, BS, 61; Yale Univ, MEng, 62, MS, 65, PhD(molecular biophys), 67. Prof Exp: Nat Acad Sci-Nat Res Coun resident res assoc, Lab Phys Biochem, Naval Med Res Inst, Nat Naval Med Ctr, 67-69; from res investr to sr res investr, Dept Biochem Pharmacol, Squibb Inst Med Res, 69-73, head biochem sect, 73-75, res fel, Dept Pharmacol, 75; HEAD BIOCHEM & DRUG METAB, USV PHARMACEUT CORP, 75- Mem: Am Soc Biol Chem; NY Acad Sci; Am Chem Soc. Res: Enzyme kinetics and mechanism; spectroscopy of biological molecules; biochemical pharmacology. Mailing Add: USV Pharmaceut Corp Tuckahoe NY 10707

WEINSHANK, DONALD JEROME, b Chicago, Ill, Apr 29, 37; m 59; c 2. NATURAL SCIENCE, PROTOZOOLOGY. Educ: Northwestern Univ, BA, 58; Univ Wis-Madison, MS, 61, PhD(biochem), 69. Prof Exp: From instr to asst prof, 67-72, ASSOC PROF NAT SCI, UNIV COL, MICH STATE UNIV, 72- Mem: AAAS; Brit Biochem Soc; Am Chem Soc; Soc Protozool. Res: Science and human values; predation models of protozoa and bacteria. Mailing Add: Dept of Nat Sci Univ Col Mich State Univ East Lansing MI 48823

WEINSHENKER, NED MARTIN, b Brooklyn, NY, Oct 13, 42; m 66; c 2. ORGANIC CHEMISTRY. Educ: Polytech Inst Brooklyn, BSc, 64; Mass Inst Technol, PhD(org chem), 69. Prof Exp: NIH fel, Harvard Univ, 68-69; asst prof chem, Univ Md, College Park, 69-70; dir phys sci & prin scientist, Alza Corp, 70-72; VPRES RES, DYNAPOL, 72- Concurrent Pos: Lectr, Stanford Univ, 71. Mem: Am Chem Soc; The Chem Soc; NY Acad Sci. Res: Synthetic organic chemistry; new synthetic methods and physiologically active compounds; polymeric reagents and membrane structure; creation of new safe non-toxic food additives. Mailing Add: Dynapol 1454 Page Mill Rd Palo Alto CA 94304

WEINSIEDER, ALLAN, b Montevideo, Uruguay, Jan 11, 39; US citizen; m 69; c 1. CELL BIOLOGY. Educ: Bates Col, BS, 61; Univ Vt, MS, 65, PhD(zool), 73. Prof Exp: Asst to dir, Fisheries Res Bd Can Biol Sta, St Andrews, NB, 65-69; Nat Eye Inst fel, Oakland Univ, 72-73; ASST PROF ANAT, SCH MED, WAYNE STATE UNIV, 73- Mem: Asn Res Vision & Ophthal; Am Asn Anatomists; Am Soc Cell Biol. Res: Regulation of growth in the vertebrate lens; wound healing. Mailing Add: Dept of Anat Wayne State Univ Sch of Med Detroit MI 48201

WEINSTEIN, ABBOTT SAMSON, b Troy, NY, Apr 20, 24; m 49. BIOSTATISTICS, PUBLIC HEALTH ADMINISTRATION. Educ: Union Col, BA, 45; Cornell Univ, MA, 47. Prof Exp: Intern pub admin, NY State Dept Audit & Control, 47-48; jr statistician social res, NY State Dept Social Welfare, 49-51; statistician & sr statistician bus res, NY State Dept Com, 50-57; asst dir statist serv, NY State Dept Ment Hyg, 57-61; dir biomet br, Saint Elizabeth's Hosp, Washington, DC, 61-66; DIR STATIST & CLIN INFO SYSTS, NY STATE DEPT MENT HYG, 66- Concurrent Pos: Statist consult oncol & cytol, Albany Med Col, 58-61; statist consult, Judicial Conf DC Circuit, 63-64; grant, Nat Inst Ment Health co-prin investr multi-state info syst psychiat patients, Res Found Ment Hyg, Inc, NY, 67-72; mem epidemiol studies rev comt, Nat Inst Ment Health, 71-75. Honors & Awards: Superior Serv Award, US Dept Health, Educ & Welfare, 64; Distinguished Serv Award, NY State Dept Ment Hyg, 70. Mem: Am Statist Asn; fel Am Pub Health Asn. Res: Statistical research in mental health, mental retardation, and alcoholism; analysis of systems of service for the mentally disabled; epidemiology of mental disability; data processing administration. Mailing Add: 10 Village Dr Delmar NY 12054

WEINSTEIN, ALAN DAVID, b New York, NY, June 17, 43; m 67; c 1. GEOMETRY. Educ: Mass Inst Technol, BS, 64; Univ Calif, Berkeley, MA, 66, PhD(math), 67. Prof Exp: Vis fel, Inst Advan Sci Studies, France, 67; C L E Moore instr math, Mass Inst Technol, 67-68; NATO fel, Math Inst, Univ Bonn, 68-69; asst prof, 69-71, ASSOC PROF MATH, UNIV CALIF, BERKELEY, 71- Concurrent Pos: Sloan fel, 71-73. Mem: Am Math Soc; Math Asn Am. Res: Symplectic manifolds; Hamiltonian dynamical systems; riemannian geometry. Mailing Add: Dept of Math Univ of Calif Berkeley CA 94720

WEINSTEIN, ALAN IRA, b New York, NY, Apr 7, 40; m 66; c 2. METEOROLOGY. Educ: City Col New York, BS, 60; Pa State Univ, MS, 63, PhD(meteorol), 68. Prof Exp: Res meteorologist, Meteorol Res Inc, 63-66; res asst meteorol, Pa State Univ, 66-68; res scientist, Meteorol Res Inc, Cohu Electronics Inc, 69-71; res physicist, 71-74, BR CHIEF STRATIFORM CLOUD PHYSICS, AIR FORCE GEOPHYS LABS, 74- Honors & Awards: Sci Achievement Award, US Air Force, 73. Mem: Am Meteorol Soc; Royal Meteorol Soc; Sigma Xi; AAAS; Weather Modification Asn. Res: Development of operational methods of fog and other stratiform cloud modification. Mailing Add: LYP Air Force Geophys Lab Hanscom AFB MA 01731

WEINSTEIN, ALEXANDER, b Astrakhan, Russia, Nov 22, 93; nat US, m 17; c 1. GENETICS, HISTORY OF SCIENCE. Educ: Columbia Univ, BS, 13, AM, 14, PhD(zool), 17. Prof Exp: Asst zool, Columbia Univ, 16-17, Sigma Xi res fel, 21-22, lectr, 28; res asst genetics, Carnegie Inst, 17-19; mem ed staff, Am Men Sci, 20-21; Johnston scholar, Johns Hopkins Univ, 22-23, Nat Res Coun fel, 23-24, assoc zool & hist sci, 30-36; Int Ed Bd fel, Cambridge Univ, 24-25; assoc zool, Univ Ill, 26-27; prof genetics, Univ Minn, 28-29; researcher, Am Philos Soc grant, Columbia Univ, 37-42; instr physics, City Col New York, 42-49; res fel biol, 51-54, RESEARCHER, COMMONWEALTH FUND GRANT, HARVARD UNIV, 51- Mem: AAAS; Am Soc Nat; Am Soc Zool; Genetics Soc Am; Hist Sci Soc. Res: Crossing over and multiple-strand theory; radiation genetics; human genetics; heredity and development; comparative genetics of Drosophila; history of biology and physics; ancient and modern science; environmental and genetic factors. Mailing Add: Biol Labs Harvard Univ Cambridge MA 02138

WEINSTEIN, ALEXANDER, b Saratoff, Russia, Jan 21, 97; nat US; m 28. MATHEMATICS. Educ: Univ Zurich, PhD(math), 21; Univ Paris, DSc(math), 37. Prof Exp: Privatdocent, Univ Zurich, 27, Univ Hamburg, 28 & Breslau Univ, 28-33; grant, Scientists Comt, Univ Paris, 33-34 & Soc Protection Sci & Learning, 34-36; Maitre de Res, Univ Paris, 36-40; lectr appl math, Univ Toronto, 41-43, asst prof math, 43-45, 45-46; mem appl math group, Harvard Univ, 45; assoc prof, Carnegie Inst Technol, 46-47; prof math, Univ Md, 48-50, res prof, 50-67; prof. 67-72, EMER PROF MATH, GEORGETOWN UNIV, 72- Concurrent Pos: Lectr, Univ Paris, 32-33, 38-39, Univ Geneva, 33, 35, Univs Delft, Leiden, Utrecht & Clermont-Ferrand, 34 & Free Univ Brussels, 36; Ritchie lectr, Univ Edinburgh, 39 & New Sch for Social Res, 42-45; vis prof, Okla Agr & Mech Col, 48; Guggenheim fel, 54, 55; Fulbright fel, 55. Mem: Am Math Soc; Lima Acad Exact, Phys & Natural Sci; Nat Acad Lincei. Res: Hydrodynamics; elasticity; vibrations; analytical mechanics; differential equations; conformal representation. Mailing Add: 9300 Piney Branch Rd Silver Spring MD 20903

WEINSTEIN, ALLAN, b New York, NY, Jan 17, 30; m 51; c 2. PHYSICAL CHEMISTRY. Educ: NY Univ, BA, 51; Brooklyn Col, MA, 54; Pa State Univ, PhD(fuel technol), 61. Prof Exp: Radiol chemist, NY Naval Shipyard, 51-55; res asst fuel technol, Pa State Univ, 57-60; chemist, Prod Res Div, Esso Res & Eng Co, 60-61; fel, Pa State Univ, 61-62, asst prof fuel technol, 62-64; chemist, E I du Pont de Nemours & Co, 64-65; sr chemist, Res Dept, R J Reynolds Tobacco Co, 65-71. Mem: Am Chem Soc. Res: Surface chemistry; aerosol technology; kinetics and mechanisms; gas chromatography. Mailing Add: 2119 Allaire Lane NE Atlanta GA 30345

WEINSTEIN, ARTHUR HOWARD, b Brooklyn, NY, Jan 20, 24; m 46; c 2. ORGANIC POLYMER CHEMISTRY, RUBBER CHEMISTRY. Educ: Queens Col, NY, BS, 44; Ohio State Univ, MS, 48, PhD(org chem), 50. Prof Exp: Asst path chemist, Bellevue Hosp, New York, 46; anal chemist, Dept of Purchase, New York, 50; SR RES CHEMIST, RUBBER DIV, GOODYEAR TIRE & RUBBER CO, 51- Mem: Am Chem Soc. Res: Mixed organic acid anhydrides; aminocellulose derivatives; synthesis of aromatic sulfur compounds; polydienes with terminal functionality; polymerization modifiers; castable elastomers; elastomers self-resistant to oxidation or to combustion; specialty rubbers. Mailing Add: 2400 Cambridge Dr Hudson OH 44236

WEINSTEIN, BENJAMIN, organic chemistry, see 12th edition

WEINSTEIN, BERNARD ALLEN, b Bridgeport, Conn, Nov 15, 46; m 70; c 1. SOLID STATE PHYSICS. Educ: Univ Rochester, BS, 68; Brown Univ, PhD(physics), 74. Prof Exp: Fel, Max Planck Inst Solid State Res, Stuttgart, 71-73; res assoc solid state mat, Nat Bur Standards, 73-75; ASST PROF PHYSICS, PURDUE UNIV, WEST LAFAYETTE, 75- Mem: Am Phys Soc; Fedn Am Scientists; Am Asn Univ Profs. Res: Optical properties of solids, with specific work in Raman scattering, visible and infrared spectroscopy and ultra-high pressure research. Mailing Add: Dept of Physics Purdue Univ West Lafayette IN 47907

WEINSTEIN, BERNARD IRA, b Brooklyn, NY, Sept 27, 40; m 66. BIOCHEMISTRY. Educ: Franklin & Marshall Col, BA, 62; Univ Chicago, PhD(biochem), 69. Prof Exp: NIH fel, Univ Calif, San Diego, 69-71; ASST PROF BIOCHEM, MT SINAI SCH MED, 71- Res: Regulation of gene expression in the differentiation of eukaryotic cells. Mailing Add: Dept of Biochem Mt Sinai Sch of Med New York NY 10029

WEINSTEIN, BORIS, b New Orleans, La, Mar 31, 30; m 53; c 2. ORGANIC CHEMISTRY, BIO-ORGANIC CHEMISTRY. Educ: La State Univ, BS, 51; Purdue Univ, MS, 53; Ohio State Univ, PhD(org chem), 59. Prof Exp: Jr chemist, Motor Fuels Lab, La State Univ, 49-51; chemist, United Gas Corp, 51-52; asst, Res Found, Ohio State Univ, 55-56; fel, Univ Calif, 59-60; chemist, Stanford Res Inst, 60-61; lab dir, Stanford Univ, 61-67; assoc prof, 67-74, PROF CHEM, UNIV WASH, 74- Mem: AAAS; Am Chem Soc; fel NY Acad Sci; The Chem Soc; Am Soc Neurochem. Res: Synthesis of biologically active peptides and proteins; phytochemical and phylogenetic relationships; structure and synthesis of natural products; heterocyclic compounds; applications of nuclear magnetic resonance; marine chemistry; neurochemistry; artificial enzymes. Mailing Add: Dept of Chem BG-10 Univ of Wash Seattle WA 98195

WEINSTEIN, CONSTANCE DE COURCY, b London, Eng, Aug 31, 24; US citizen; m 59; c 3. BIOCHEMISTRY. Educ: Univ London, BSc, 48, PhD(biochem), 53. Prof Exp: Res biochemist, Hosp Sick Children, London, 53-54; res fel biochem, Jefferson Med Col, Philadelphia, 55-56; sr instr biochem, Med Sch, Western Reserve Univ, Cleveland, 56-61; res scientist cancer, City of Hope, Duarte, Calif, 70-74; HEALTH SCI ADMINR, NAT HEART & LUNG INST, NIH, 75- Res: Maintaining an interest in basic science and clinical aspects of research into heart, lung and blood diseases. Mailing Add: Nat Heart & Lung Inst NIH Westwood Bldg Westbard Ave Bethesda MD 20014

WEINSTEIN, DAVID, b New York, NY, June 24, 28; m 54; c 2. CYTOGENETICS. Educ: City Col New York, BS, 50; Brooklyn Col, MA, 54; Purdue Univ, PhD(bact), 59. Prof Exp: Asst bacteriologist, Queens Gen Hosp, New York, 53-54; bacteriologist, US Army Chem Corps, Ft Detrick, Md, 54-55; asst bact, Purdue Univ, 55-59; res assoc microbiol, Sch Med, Duke Univ, 59-62; res assoc, Merck Res Inst, Pa, 62-63; assoc, Wistar Inst, Philadelphia, 63-68; SR CYTOGENETICIST, HOFFMANN-LA ROCHE, INC, 68- Mem: AAAS; Am Soc Microbiol; Tissue Cult Asn; Environ Mutagen Soc; Am Soc Cell Biol. Res: Oncogenic virology; tissue culture; transformation. Mailing Add: 78 Luddington Rd West Orange NJ 07052

WEINSTEIN, EDWIN ALEXANDER, b New York, NY, Feb 18, 09; m 47; c 2. NEUROPSYCHIATRY. Educ: Dartmouth Col, AB, 30; Northwestern Univ, MD, 35. Prof Exp: Consult neuropsychiatrist, Walter Reed Army Inst Res, 54-68; PROF NEUROL, MT SINAI SCH MED, 68-; CHIEF NEUROL, VET ADMIN HOSP, BRONX, 74- Concurrent Pos: Attend neurologist, Mt Sinai Hosp, New York, 47-; psychiatrist, Govt VI, 57-60; consult, Nat Naval Med Ctr, 54- & Caribbean Fedn Ment Health, 65-; fel, William A White Inst Psychiat, 60-; sr consult, Peace Corps, 65-68. Mem: Am Neurol Asn; Am Psychiat Asn; Acad Aphasia. Res: Brain function, language and socio-cultural aspects of behavior. Mailing Add: Dept of Neurol Mt Sinai Sch of Med New York NY 10029

WEINSTEIN, HAREL, b Bucarest, Rumania, June 5, 45; m 67; c 1. BIOPHYSICAL CHEMISTRY, QUANTUM CHEMISTRY. Educ: Israel Inst Technol, BSc, 66, MSc, 68, DSc(quantum chem), 71. Prof Exp: Asst chem, Israel Inst Technol, 66-68, sr res asst, 68-71, lectr, 71-73; res assoc, Johns Hopkins Univ, 73-74; ASST PROF PHARMACOL, MT SINAI SCH MED, 74- Concurrent Pos: Consult res assoc biochem, Tel-Aviv Univ, 73; vis scientist genetics, Med Ctr, Stanford Univ, 74-75. Mem: The Chem Soc; Israel Phys Soc; Europ Phys Soc; Am Chem Soc; NY Acad Sci. Res: Formulation and use of theoretical methods for the study of molecular reactivity and interactions between biological molecules; focus on mechanisms of intermolecular recognition in drug action and molecular energy transfer and storage. Mailing Add: Dept of Pharmacol Mt Sinai Sch of Med New York NY 10029

WEINSTEIN, HOWARD, b New York, NY, Nov 9, 27. NEUROCHEMISTRY, CELL PHYSIOLOGY. Educ: Cornell Univ, BA, 49; State Univ Iowa, PhD(zool), 56. Prof Exp: Instr zool & physiol, Wis State Univ-Stevens Point, 56-57; USPHS fel neuroendocrinol, Case Western Reserve Univ, 58-61; res scientist, City of Hope Med Ctr, Duarte, Calif, 61-74; STAFF SCIENTIST, NAT INST NEUROL DIS & STROKE, 74- Concurrent Pos: USPHS grants, 68- Mem: Soc Neurosci; Am Soc Neurochem; Int Soc Neurochem; Am Soc Zoologists. Res: Membrane transport processes in central nervous system. Mailing Add: Nat Inst of Neurol Dis & Stroke NIH Bethesda MD 20014

WEINSTEIN, HYMAN GABRIEL, b Worcester, Mass, June 15, 20; m. BIOCHEMISTRY, PHYSIOLOGY. Educ: Worcester Polytech Inst, BS, 42; Univ Ill, Urbana, MS, 47; Nat Registry Clin Chem, cert. Prof Exp: Res asst, Univ Ill, Urbana, 48-49; res asst, Rheumatic Fever Res Inst, Med Sch, Northwestern Univ, 49-50, res assoc, 50-53; res biochemist, Neurol Res Lab, Vet Admin Hosp, Hines, Ill, 53-54; supvry res biochemist, Vet Admin West Side Hosp, Chicago, 54-64; supvry res biochemist, Geriat Res Lab, 64-65, CHIEF RES-IN-AGING LAB, VET ADMIN HOSP, DOWNEY, ILL, 65-, LECTR, COUN ON ALCOHOLISM, 68- Concurrent Pos: Lectr, Northeastern Ill Univ, 75- Mem: AAAS; Am Chem Soc; Soc Complex Carbohydrates; NY Acad Sci; Sigma Xi. Res: Proteoglycans and related macromolecules in human development, aging and disease; research administration. Mailing Add: Res-in-Aging Lab Vet Admin Hosp Downey IL 60064

WEINSTEIN, I BERNARD, b Madison, Wis, Sept 9, 30; m 52; c 3. MEDICINE. Educ: Univ Wis, BS, 52, MD, 55. Prof Exp: Nat Cancer Inst spec res fel bact & immunol, Harvard Med Sch, 59-60; ASST ATTEND PHYSICIAN, PRESBY HOSP, NEW YORK, 66- Concurrent Pos: Career scientist, Health Res Coun, City of New York, 61-72; Europ Molecular Biol Orgn travel fel, 70-71; adv lung cancer segment, Carcinogenesis Prog, Nat Cancer Inst, 71-74; mem interdisciplinary commun prog, Smithsonian Inst, 71-74; mem pharmacol B study sect, NIH, 71-75; vis physician, Francis Delafield Hosp, New York. Honors & Awards: Meltzer Medal, 64. Mem: AAAS; Am Soc Microbiol; Am Asn Cancer Res; Am Soc Clin Invest; Am Soc Cell Biol. Res: Oncology; cellular and molecular aspects of carcinogenesis; control of gene expression. Mailing Add: Inst of Cancer Res Columbia Univ New York NY 10032

WEINSTEIN, IRA, b Oak Park, Ill, Jan 30, 28; m 54; c 3. ENDOCRINOLOGY, BIOCHEMICAL PHARMACOLOGY. Educ: Roosevelt Univ, BS, 49; Univ Ill, MS, 52; George Washington Univ, PhD(microbiol), 60. Prof Exp: From instr to asst prof pharmacol, Vanderbilt Univ, 60-69; assoc prof, Sch Med, Univ Fla, 69-75; ASSOC PROF PHARMACOL, SCH MED, UNIV MO-COLUMBIA, 75- Concurrent Pos: Fel pharmacol, Med Sch, Vanderbilt Univ, 60-63; USPHS fels, 60-64; Olson Mem Fund fel, 65-66; vis lectr, Hebrew Univ Israel, 65-66; Am Heart Asn advan res fel, 65-67. Mem: Sigma Xi; Am Soc Microbiol; Am Soc Pharmacol & Exp Therapeut. Res: Bacterial physiology, endocrines and drug effects upon lipid metabolism. Mailing Add: Dept of Pharmacol Univ of Mo Sch of Med Columbia MO 65201

WEINSTEIN, JOSEPH M, b Milwaukee, Wis, May 9, 39; m 71. MATHEMATICS. Educ: Univ Calif, Berkeley, BA, 60; Univ Wis-Madison, MA, 61, PhD(math), 65. Prof Exp: Asst prof math, Univ Calif, Los Angeles, 65-69 & St Mary's Col Md, 72-75; ASST PROF MATH, UNIV WASH, 75- Mem: Am Math Soc; Math Asn Am; Soc Indust & Appl Math; Asn Symbolic Logic. Res: Graph theory; logic; biomathamatics; problem-solving theory; science writing. Mailing Add: Dept of Math Univ of Wash Seattle WA 98195

WEINSTEIN, JULIUS, organic chemistry, spectroscopy, deceased

WEINSTEIN, LEONARD HARLAN, b Springfield, Mass, Apr 11, 26; m 50; c 2. PLANT PHYSIOLOGY, ENVIRONMENTAL BIOLOGY. Educ: Pa State Univ, BS, 49; Univ Mass, MS, 50; Rutgers Univ, PhD(plant physiol), 53. Prof Exp: Fel soils, Rutgers Univ, 53-54; from assoc plant physiologist to plant physiologist, 55-63, prog dir plant chem, 63-69, PROG DIR ENVIRON BIOL, BOYCE THOMPSON INST PLANT RES, 69-, MEM BD DIRS, 73- Mem: AAAS; Am Soc Plant Physiol; Air Pollution Control Asn; Harvey Soc; Am Inst Biol Sci. Res: Air pollution; aromatic metabolism; plant nutrition; plant senescence; environmental biology; effects of atmospheric pollutants on plant growth, development, productivity, and quality.

Mailing Add: Boyce Thompson Inst for Plant Res 1086 N Broadway Yonkers NY 10701

WEINSTEIN, LOUIS, b Bridgeport, Conn, Feb 26, 09; m 34. INTERNAL MEDICINE, INFECTIOUS DISEASES. Educ: Yale Univ, BS, 28, MS, 30, PhD(bact), 31; Boston Univ, MD, 43. Prof Exp: Instr bact, Med Sch, Yale Univ, 37-39; res assoc immunol, Sch Med, Boston Univ, 39-44, asst med, 43-44, from instr to assoc prof, 44-57; lectr pediat, Med Sch, Tufts Univ, 50-57, prof med, 57-75; CHIEF INFECTIOUS DIS SERV, VET ADMIN HOSP, WEST ROXBURY, MASS, 75-; PHYSICIAN, PETER BENT BRIGHAM HOSP, BOSTON, 75- Concurrent Pos: Asst, Harvard Med Sch, 45-46, instr, 46-49, lectr, 49-75, vis prof med, 75-; chief infectious dis serv, Mass Mem Hosp, 47-57 & New Eng Ctr Hosp & Boston Floating Hosp, 57-75; assoc physician, Med Serv, Mass Gen Hosp, 58-75; assoc physician in chief, New Eng Med Ctr Hosps, 62-71. Mem: Am Acad Arts & Sci; AAAS; AMA; Am Soc Microbiol; Soc Invest Dermat. Res: Chemotherapy of infection; host factors in infectious disease. Mailing Add: New Eng Med Ctr Hosp 171 Harrison Ave Boston MA 02111

WEINSTEIN, MARVIN, b Bronx, NY, June 7, 42; m 67; c 1. THEORETICAL HIGH ENERGY PHYSICS. Educ: Columbia Univ, BS, 63, MS, 64, PhD(physics), 67. Prof Exp: Physics mem, Inst Advan Study, 67-69; vis asst prof physics, Yeshiva Univ, 69-70 & NY Univ, 70-72; SR RES ASSOC PHYSICS, STANFORD LINEAR ACCELERATOR CTR, STANFORD UNIV, 72- Res: Current algebra; gauge theories of strong, weak and electromagnetic interactions; non-perturbatic methods in quantum field theory. Mailing Add: Stanford Linear Accelerator Ctr Stanford Univ PO Box 4349 Stanford CA 94040

WEINSTEIN, MARVIN JOSEPH, b New York, NY, Oct 20, 16; m 49; c 2. PHYSIOLOGY, MICROBIOLOGY. Educ: Alfred Univ, BA, 40; NY Univ, MS, 49, PhD(biol), 57. Prof Exp: Chief biol technician, LI Univ, 47-49; asst microbiologist, Squibb Inst Med Res, 49-56; sr microbiologist, 56-60, mgr microbiol dept, 60-67, assoc dir biol res, 67-70, DIR MICROBIOL DIV, SCHERING CORP, 70- Concurrent Pos: Mem antibiotic del, US State Dept-USSR Sci & Cultural Exchange Prog, 59. Mem: AAAS; Am Chem Soc; fel Am Acad Microbiol; Infectious Dis Soc Am; Am Soc Microbiol. Res: Antibiotic screening and development; fermentation; microbial transformation of organic compounds; viral and parasitic chemotherapy; microbiological assays. Mailing Add: Div of Microbiol Schering Corp Bloomfield NJ 07003

WEINSTEIN, MARVIN STANLEY, b New York, NY, May 24, 27; m 52; c 2. ACOUSTICS. Educ: St Louis Univ, BS, 48; Univ Md, MS, 51, PhD, 56. Prof Exp: Asst, Univ Md, 48-49; physicist, US Naval Ord Lab, 49-59; vpres, 59-62, pres, 62-74, CHMN BD, UNDERWATER SYSTS, INC, 75- Mem: AAAS; Am Phys Soc; Acoust Soc Am; Inst Elec & Electronics Eng. Res: Underwater acoustics; propagation at long and short range in deep and shallow water, noise, sinusoidal and explosive signals; ultrasonics; ultrasonic modeling; electronics. Mailing Add: 10807 Lombardy Rd Silver Spring MD 20901

WEINSTEIN, PAUL P, b Brooklyn, NY, Dec 9, 19; m 54; c 2. PARASITOLOGY. Educ: Brooklyn Col, AB, 41; Johns Hopkins Univ, ScD(hyg), 49. Prof Exp: Jr parasitologist, USPHS, St Bd Health, Fla, 42-44, sr asst sanitarian, Washington, DC, Ga & PR, 44-46, from scientist to scientist dir & chief lab parasitic dis, NIH, 49-68; chmn dept, 69-75, PROF BIOL, UNIV NOTRE DAME, 75- Concurrent Pos: Vis scientist, Nat Inst Med Res, Eng, 62-63; mem parasitic dis panel, US-Japan Coop Med Sci Prog, 65-69, chmn, 69-73; mem comt int ctr med res & training, NIH, 70-73; mem adv sci bd, Gorgas Mem Inst Trop Prev Med, 72-; mem nat adv comt, Primate Res Ctr, Univ Calif, Davis, 73- Honors & Awards: Ashford Award, Am Soc Trop Med & Hyg, 57. Mem: Fel AAAS; Am Soc Trop Med & Hyg; Am Soc Parasitol (pres, 72); Japanese Soc Parasitol. Res: Cultivation and physiology of parasitic helminths; host-parasite relationships. Mailing Add: Dept of Biol Univ of Notre Dame Notre Dame IN 46556

WEINSTEIN, RONALD S, b Schenectady, NY, Nov 20, 38; m 64; c 2. EXPERIMENTAL PATHOLOGY, ELECTRON MICROSCOPY. Educ: Union Col, NY, BS, 60; Tufts Univ, MD, 65; Am Bd Path, dipl. Prof Exp: Res asst electron micros, Mass Gen Hosp, 62-63; instr path, Sch Med, Tufts Univ, 67-69, assoc prof, 72-75; PROF PATH & CHMN DEPT, RUSH MED COL, 75- Concurrent Pos: From intern to resident path, Mass Gen Hosp, 65-70, actg head, Mixter Lab Electron Micros, 66-70; teaching fel, Harvard Med Sch, 65-70; investr toxicol, Aerospace Med Res Lab, Wright-Patterson AFB, 70-72. Mem: Soc Develop Biol; NY Acad Sci; Am Asn Anatomists; Soc Toxicol; Am Asn Path & Bact. Res: Development and application of high resolution electron microscopy techniques to the study of biological membrane ultrastructure; comparative and functional studies on normal and neoplastic cell membranes; environmental toxicology. Mailing Add: Dept of Path Rush Med Col Chicago IL 60612

WEINSTEIN, ROY MURRAY, physics, see 12th edition

WEINSTEIN, SAM, b Omaha, Nebr, May 24, 16; m 46; c 2. ORTHODONTICS. Educ: Creighton Univ, DDS, 41; Northwestern Univ, MSD, 48; Am Bd Orthod, dipl. Prof Exp: From asst prof to prof orthod, Col Dent, Univ Nebr, 54-71, chmn dept, 63-71; PROF ORTHOD, UNIV CONN, 71- Concurrent Pos: Mem dent study sect, NIH, 69-73; consult, Coun Dent Educ, Am Dent Asn, 69-; examr, Coun on Educ, Can Dent Asn, 75- Mem: Fel AAAS; Am Dent Asn; Am Asn Orthod; Int Asn Dent Res; Int Soc Cranio-Facial Biol. Res: Theoretical mechanics application to muscle forces and tooth movement; cleft palate embryology; growth. Mailing Add: Univ of Conn Health Ctr Farmington CT 06032

WEINSTEIN, STANLEY EDWIN, b New York, NY, Apr 26, 42; m 64. MATHEMATICS, COMPUTER SCIENCE. Educ: Hunter Col, BA, 62; Mich State Univ, MS, 64, PhD(math), 67. Prof Exp: Teaching asst math, Mich State Univ, 62-67; asst prof, 67-72, ASSOC PROF MATH, UNIV UTAH, 72- Concurrent Pos: US Air Force Off Sci Res grant, Univ Utah, 72-73; vis assoc prof, Dept Math, Ariz State Univ, 72-73. Mem: Soc Indust & Appl Math; Asn Comput Mach; Am Math Soc. Res: Approximation theory; numerical analysis; solution of nonlinear equations. Mailing Add: Dept of Math Univ of Utah Salt Lake City UT 84112

WEINSTEIN, STEPHEN HENRY, b Bronx, NY, Apr 14, 37; m 66; c 1. DRUG METABOLISM. Educ: Queens Col, NY, BS, 58; Adelphi Col, MS, 61; Adelphi Univ, PhD(biochem), 67. Prof Exp: Scientist, Warner-Lambert Res Inst, 67-68; sr biochemist, 68-73, GROUP LEADER, ENDO LABS, INC, 73- Mem: AAAS; NY Acad Sci; Am Soc Pharmacol & Exp Therapeut. Res: Metabolism and function of phosphatides; mechanisms of membrane transport; pharmacokinetics; drug metabolism; biochemical pharmacology. Mailing Add: Endo Labs Inc 1000 Stewart Ave Garden City NY 11530

WEINSTOCK, ALFRED, b Toronto, Ont, May 3, 39; m 65. CELL BIOLOGY, PERIODONTOLOGY. Educ: Univ Toronto, DDS, 62; Harvard Univ, cert periodont, 66; McGill Univ, PhD(anat), 69. Prof Exp: Res fel dent med, Forsyth Dent Ctr, Boston, 62-63; Nat Res Coun Can res fel periodont, Sch Dent Med, Harvard Univ, 63-66; Nat Res Coun Can res fel anat sci, Sch Med, McGill Univ, 66-67, from lectr to asst prof, 67-70; chmn sect periodont, 71-74, ASSOC PROF DENT & ANAT, SCH DENT & SCH MED, CTR HEALTH SCI, UNIV CALIF, LOS ANGELES, 70-, MEM, DENT RES INST, 73- Concurrent Pos: Nat Res Coun Can res scholar, Sch Med, McGill Univ, 67-69, Med Res Coun Can res scholar, 69-70; NIH res grant, Univ Calif, Los Angeles, 71-77; consult, Vet Admin Hosp, Brentwood & Sepulveda, Calif, 72- Honors & Awards: Res Award, Can Dent Asn, 71. Mem: AAAS; Am Acad Periodont; Am Asn Anatomists; Am Soc Cell Biol; Int Asn Dent Res. Res: Structural and functional aspects of secretory cells involved in the elaboration of collagen and other glycoproteins, mainly in mineralizing tissues; histology; experimental pathology; periodontal disease. Mailing Add: Univ of Calif Sch of Dent Ctr for Health Sci Los Angeles CA 90024

WEINSTOCK, BARNET MORDECAI, b Brooklyn, NY, Oct 10, 40; m 66. MATHEMATICS. Educ: Columbia Univ, BA, 62; Mass Inst Technol, PhD(math), 66. Prof Exp: From instr to asst prof math, Brown Univ, 66-75; MEM FAC, DEPT MATH, UNIV KY, 75- Mem: Am Math Soc. Res: Functional analysis; several complex variables. Mailing Add: Dept of Math Univ of Ky Lexington KY 40506

WEINSTOCK, BERNARD, b New York, NY, Dec 7, 17; m 47; c 2. PHYSICAL CHEMISTRY. Educ: Brooklyn Col, BA, 38; Univ Chicago, PhD(chem), 47; Oxford Univ, MA, 58. Prof Exp: Asst, Columbia Univ, 39-41; asst chemist, Manhattan Dist, NY, 41-43, NMex, 43-45; from assoc chemist to sr chemist, Argonne Nat Lab, 47-60; staff scientist, 60-65, sr staff scientist, 65-69, MGR & SR SCIENTIST, FORD MOTOR CO, 69- Concurrent Pos: Consult, Argonne Nat Lab, 60-73; mem tech adv comt, Environ Protection Agency, 74-77; mem adv bd, Petrol Res Fund, Am Chem Soc, 72-75. Mem: Am Chem Soc; fel Am Phys Soc; Sigma Xi; Combustion Inst. Res: Isotope measurement and separation inorganic chemistry; liquid helium; gas kinetics; atmospheric chemistry. Mailing Add: Sci Labs Ford Motor Co PO Box 2053 Dearborn MI 48121

WEINSTOCK, EUGENE VICTOR, physics, see 12th edition

WEINSTOCK, HAROLD, b Philadelphia, Pa, Dec 25, 34; m 61; c 2. SOLID STATE PHYSICS, LOW TEMPERATURE PHYSICS. Educ: Temple Univ, BA, 56; Cornell Univ, PhD(helium three), 62. Prof Exp: From res asst to res assoc physics, Cornell Univ, 62-65; asst prof, Mich State Univ, 65-66; assoc prof, 65-73, PROF PHYSICS, ILL INST TECHNOL, 73- Concurrent Pos: Vis prof, Cath Univ Louvain, 70 & Cath Univ Nijmegen, 72-73. Mem: AAAS; Am Phys Soc; Am Asn Physics Teachers. Res: Thermal and electrical conductivity and specific heat of solids; radiation damage in solids; superconductivity; computer use in science education. Mailing Add: Dept of Physics Ill Inst of Technol Chicago IL 60616

WEINSTOCK, IRWIN MORTON, b New York, NY, July 17, 25; m 56; c 3. BIOCHEMISTRY. Educ: Univ Okla, BS, 47; Univ Ill, MS, 48, PhD(chem), 51. Prof Exp: Res assoc biochem, Dept Psychiat, Med Col, Cornell Univ, 51-57; res assoc, New York Med Col, 57-59; from asst mem to assoc mem, Inst Muscle Dis, 64-74; DIR SPEC NEUROL LAB, NASSAU COUNTY MED CTR, 74- Concurrent Pos: Lectr, Hunter Col, 53-54 & Columbia Univ, 68- Mem: AAAS; Harvey Soc. Res: Intermediary metabolism and enzymology of muscle wasting conditions. Mailing Add: Spec Neurol Lab Nassau County Med Ctr East Meadow NY 11554

WEINSTOCK, JEROME, b Brooklyn, NY, Sept 12, 33; m 55; c 3. PLASMA PHYSICS, FLUID MECHANICS. Educ: Cooper Union, BChE, 55; Cornell Univ, PhD(phys chem), 59. Prof Exp: Sr scientist, Nat Bur Standards, 59-65; SR SCIENTIST, NAT OCEANIC & ATMOSPHERIC ADMIN, 65- Concurrent Pos: Nat Res Coun-Nat Acad Sci res fel, 59-62. Mem: Am Phys Soc; Am Geophys Union. Res: Statistical mechanics; transport theory; fluctuation theory; molecular collision theory; basic research in turbulence theory, atmospheric waves, and plasma physics. Mailing Add: Plasma Physics Nat Oceanic & Atmospheric Admin Boulder CO 80302

WEINSTOCK, JOSEPH, b New York, NY, Jan 30, 28; m 52; c 10. ORGANIC CHEMISTRY. Educ: Rutgers Univ, BS, 49; Univ Rochester, PhD(chem), 52. Prof Exp: Res assoc chem, Northwestern Univ, 52-54, instr, 54-56; sr chemist, 56-62, group leader, 62-67, SR INVESTR, SMITH KLINE & FRENCH LABS, 67- Mem: AAAS; Am Chem Soc; NY Acad Sci. Res: Medicinal and synthetic organic chemistry; organic reaction mechanisms; pteridines; diuretic, anti-inflammatory, antihypertensive agents; drug metabolism and identification of metabolites. Mailing Add: Chem F31 Smith Kline & French Labs 1500 Spring Garden St Philadelphia PA 19101

WEINSTOCK, LEONARD M, b Passaic, NJ, Jan 30, 27; m 60; c 2. ORGANIC CHEMISTRY. Educ: Rutgers Univ, BS, 50; Ind Univ, PhD(chem), 58. Prof Exp: SR RES FEL, MERCK & CO, 58- Mem: Am Chem Soc. Res: Chemistry of heterocyclic compounds and beta-lactam antibiotics. Mailing Add: Merck & Co 126 E Lincoln Ave Rahway NJ 07065

WEINSTOCK, MELVYN, b Toronto, Ont, Feb 23, 41. BIOLOGICAL STRUCTURE. Educ: Univ Toronto, DDS, 64; Harvard Univ, cert, 69; McGill Univ, PhD(anat), 72. Prof Exp: Clin fel dent med, Forsyth Dent Ctr, 64-65; res fel ortho & biol, Harvard Univ, 65-69; ASST PROF ANAT, McGILL UNIV, 72-, LECTR ORTHOD, 73- Concurrent Pos: Nat Res Coun fel, Med Res Coun Can, 65-69, Med Res Coun Can scholar, 73- Mem: Am Asn Anatomists; Am Soc Cell Biol; Int Asn Dent Res; Am Asn Orthodontists; Can Dent Asn. Res: Elaboration of collagen and other matrix components of dentin and bone by odontoblasts and osteoblasts as revealed by electron microscope radioautography and freeze fracture techniques. Mailing Add: Dept of Anat & Fac of Dent McGill Univ PO Box 6070 Montreal PQ Can

WEINSTOCK, ROBERT, b Philadelphia, Pa, Feb 2, 19; m 50; c 2. MATHEMATICAL PHYSICS. Educ: Univ Pa, AB, 40; Stanford Univ, PhD, 43. Prof Exp: Instr physics, Stanford Univ, 43-44, math, 46-50, actg asst prof, 50-54; res assoc radar countermeasures, Radio Res Lab, Harvard Univ, 44-45; from asst prof to assoc prof math, Univ Notre Dame, 54-59; vis assoc prof, 59-60, assoc prof, 60-66, PROF PHYSICS, OBERLIN COL, 66- Concurrent Pos: NSF fel, Oxford Univ, 65-66. Mem: AAAS; Am Phys Soc; Am Asn Physics Teachers. Res: Mathematical physics; statistical mechanics; calculus of variations. Mailing Add: Dept of Physics Oberlin Col Oberlin OH 44074

WEINSWIG, MELVIN H, b Lynn, Mass, Feb 2, 35; m 60; c 3. PHARMACEUTICAL CHEMISTRY. Educ: Mass Col Pharm, BS, 55, MS, 57; Univ Ill, PhD(pharmaceut chem), 61. Prof Exp: From asst prof to assoc prof pharmaceut chem, Butler Univ, 61-69; PROF PHARM & CHMN EXTEN SERV PHARM, SCH PHARM, UNIV WIS-MADISON, 69-, ASSOC DEAN, SCH, 71- Concurrent Pos: Consult, Corn Prod Co, 66- Mem: Am Chem Soc; Am Pharmaceut Asn. Res: Novel analytical approach to combination pharmaceutical products; drug abuse education and research. Mailing Add: Rm 155 Pharm Bldg Sch of Pharm Univ of Wis Madison WI 53706

WEINTRAUB, BRUCE DALE, b Buffalo, NY, Sept 19, 40; m 68; c 1. ENDOCRINOLOGY, INTERNAL MEDICINE. Educ: Princeton Univ, AB, 62; Harvard Med Sch, MD, 66. Prof Exp: Intern med, Peter Bent Brigham Hosp, Boston, Mass, 66-67, resident, 67-68; clin assoc endocrinol, Nat Inst Arthritis & Metab Dis, 68-71; instr med, Harvard Med Sch, 71-72; SR INVESTR ENDOCRINOL, NAT INST ARTHRITIS, METAB & DIGESTIVE DIS, 72- Concurrent Pos: Clin & res fel, Nat Inst Arthritis & Metab Dis, 68-71; USPHS res grant & asst med, Mass Gen Hosp, Boston, 71-72. Mem: Am Fedn Clin Res; Endocrine Soc. Res: Structure and properties of hormones secreted by tumors; applications of affinity chromatography to endocrinology; subunits of glycoprotein hormones. Mailing Add: Bldg 10 8N 316 Nat Inst of Arthritis Metab & Digestive Dis Bethesda MD 20014

WEINTRAUB, HERBERT D, b New York, NY, Feb 17, 30; m 63; c 3. ANESTHESIOLOGY, MEDICAL EDUCATION. Educ: NY Univ, BA, 50; Oxford Univ, MA, 58, BM, BCh, 59. Prof Exp: Intern rotating, Strong Mem Hosp, Univ Rochester, 59-61; resident anesthesiol, Columbia Presby Hosp & Med Ctr, New York, 61-63, NIH res fel, 63-64; asst prof, Med Sch, Univ Pa, 64-69; asst prof, Med Sch, Univ Chicago, 69-71; assoc prof, 71-74, PROF ANESTHESIOL, GEORGE WASHINGTON UNIV, 74-, DIR MED EDUC, DEPT ANESTHESIA, MED CTR, 71- Concurrent Pos: Mem sr staff, Hosp Univ Pa, 64-69; assoc dir, Dept Anesthesiol, Michael Reese Hosp, Chicago, 69-71. Mem: NY Acad Sci; Am Soc Anesthesiol; Int Anesthesia Res Soc; Asn Am Med Cols. Res: Anesthesia for cardio-thoracic surgery; educational methods as applied to anesthesiology. Mailing Add: Dept of Anesthesiol George Washington Univ Med Ctr Washington DC 20037

WEINTRAUB, HOWARD STEVEN, b New York, NY, June 6, 43; m 65; c 2. BIOPHARMACEUTICS. Educ: Columbia Univ, BS, 66; State Univ NY Buffalo, PhD(biopharmaceut), 71. Prof Exp: Sr scientist phys pharm, Ciba-Geigy Corp, 71-73; group leader phys pharm, 73-75, GROUP LEADER BIOPHARMACEUT, ORTHO PHARMACEUT CORP, 75- Concurrent Pos: Adj asst prof biopharmaceut, Col Pharmaceut Sci, Columbia Univ, 74- Mem: AAAS; NY Acad Sci; Am Pharmaceut Asn; Acad Pharmaceut Sci. Res: Metabolism and pharmacodynamics of drug substances in man, with emphasis on the mathematical quantitation of these phenomena. Mailing Add: Ortho Pharmaceut Corp Rte 202 Raritan NJ 08869

WEINTRAUB, JOEL D, b New York, NY, May 2, 42; m 68; c 1. HERPETOLOGY. Educ: City Col New York, BS, 63; Univ Calif, Riverside, PhD(zool), 68. Prof Exp: Dir environ studies, 72-75, ASST PROF BIOL, CALIF STATE UNIV, FULLERTON, 68- Mem: Am Soc Ichthyol & Herpet; Soc Study Amphibians & Reptiles. Res: Ecology and movement patterns of amphibians and reptiles; homing and orientation of vertebrates; ecology of desert vertebrates. Mailing Add: Dept of Biol Calif State Univ Fullerton CA 92634

WEINTRAUB, LEONARD, b New York, NY, Apr 21, 26; m 50; c 2. ORGANIC CHEMISTRY, PHARMACEUTICAL CHEMISTRY. Educ: City Col New York, BS, 48; Polytech Inst Brooklyn, MS, 54, PhD(org chem), 68. Prof Exp: Org chemist, Francis Delafield Hosp, New York, 50-54; org chemist, 54-62, group leader, 62-68, dept head org chem, 68-69, DIR CHEM RES, BRISTOL MYERS CO, 69- Mem: AAAS; Am Chem Soc; The Chem Soc; NY Acad Sci; Am Pharmaceut Asn. Res: Heterocyclic chemistry; molecular complexes in organic chemistry; salicylate chemistry; synthetic methods; pharmaceutical and cosmetic analysis. Mailing Add: Bristol Myers Co 1350 Liberty Ave Hillside NJ 07205

WEINTRAUB, LESTER, b New York, NY, Feb 1, 24; m 50; c 2. ORGANIC POLYMER CHEMISTRY. Educ: City Col New York, BS, 46; Fordham Univ, MS, 49; NY Univ, PhD(org chem), 54. Prof Exp: Sr chemist, Atomic Energy Comn Proj, Columbia Univ, 53-55, group leader, 55-58; group leader, Adv Res Proj Agency Proj, NY Univ, 58-59; sr chemist, Cent Res Labs, Air Reduction Co, Inc, NJ, 59-71; group leader polymer res, 71-73, TECH MGR RESINS, PANTASOTE CO, INC, 73- Mem: Am Chem Soc; Am Inst Chem; Soc Plastics Eng. Res: Organic polymer synthesis and structure determination; polyvinyl chloride technology. Mailing Add: 8 Tarlton Dr Livingston NJ 07039

WEINTRAUB, LEWIS ROBERT, b New York, NY, Aug 15, 34; m 67; c 2. HEMATOLOGY. Educ: Dartmouth Col, AB, 55; Harvard Med Sch, MD, 58. Prof Exp: Intern med, Hosp, Univ Pa, 58-59; asst resident res in med, Hosp, Univ Mich, 59-61; fel hemat, Mt Sinai Hosp, New York, 61-62; res hematologist, Walter Reed Army Inst Res, 62-65; from asst prof to assoc prof med, Sch Med, Tufts Univ, 65-72; ASSOC PROF MED, SCH MED, BOSTON UNIV, 72- Concurrent Pos: Asst chief hemat, Walter Reed Gen Hosp, 62-65; asst hematologist & asst physician, New Eng Med Ctr Hosps, 65-72; assoc vis physician, Boston Univ Hosp, 72- Mem: AAAS; AMA; Am Soc Hemat. Res: Clinical hematology and research in field of iron metabolism. Mailing Add: Dept of Hemat Univ Hosp 750 Harrison Ave Boston MA 02118

WEINTRAUB, MARVIN, b Radom, Poland, Oct 17, 24; nat Can; m 48; c 4. PLANT VIROLOGY. Educ: Univ Toronto, BA, 47, PhD(bot), 50. Prof Exp: Demonstr bot, Univ Toronto, 45-50; prin res scientist & head virus chem & physiol sect, Res Br, Can Dept Agr, 50-71, DIR, RES STA, AGR CAN, 71- Concurrent Pos: Hon prof, Univ BC, 71- Mem: AAAS; Am Phytopath Soc; Can Phytopath Soc; fel NY Acad Sci; Int Soc Plant Morphol. Res: Metabolism and cytology of virus-infected plants; fine structure and electron microscopy; virus inhibitors; movement in plants. Mailing Add: Res Sta Agr Can 6660 NW Marine Dr Vancouver BC Can

WEINTRAUB, PHILIP MARVIN, b Cleveland, Ohio, Feb 22, 39; m 61; c 3. ORGANIC CHEMISTRY. Educ: Ohio State Univ, BSc, 60, MSc, 63, PhD(chem), 64. Prof Exp: Res chemist, Pioneering Res Lab, Textile Fibers Dept, E I du Pont de Nemours & Co, Inc, 64-66; sr chemist, Hess & Clark Div, 66-70, ORG CHEMIST, MERRELL-NAT LABS DIV, RICHARDSON-MERRELL INC, 70- Mem: Am Chem Soc; Marine Technol Soc. Res: Steroids; heterocycles; strained ring polycyclic systems; prostaglandins. Mailing Add: Org Chem Dept Merrell-Nat Labs Div Richardson-Merrell Inc 110 E Amity Cincinnati OH 45215

WEINTRAUB, ROBERT LOUIS, b Washington, DC, May 9, 12; m 38; c 3. CELL PHYSIOLOGY, BIOCHEMISTRY. Educ: George Washington Univ, BS, 31, MA, 33, PhD(plant physiol), 38. Prof Exp: Biochemist, Div Radiation & Organisms, Smithsonian Inst, 37-47; supvry plant physiologist, Army Biol Labs, Ft Detrick, 47-59, phys sci adminr, Army gen staff, 59-63; PROF BOT, GEORGE WASHINGTON UNIV, 63-; PLANT PHYSIOLOGIST, SMITHSONIAN RADIATION BIOL LAB, 66- Mem: AAAS; Am Chem Soc; Am Soc Microbiol; Am Soc Plant Physiol; Sigma Xi. Res: Effects of radiant energy on plants; plant growth regulators; microbial physiology. Mailing Add: Dept of Bot George Washington Univ Washington DC 20052

WEINTRAUB, SOL, b Luxembourg, July 3, 37; US citizen; m 60; c 2. MATHEMATICS. Educ: City Col New York, BS, 58; Temple Univ, MA, 60, PhD(physics), 64. Prof Exp: Sr mathematician, Appl Data Res, Inc, 60-65; asst prof, 65-73, ASSOC PROF MATH, QUEENS COL, NY, 73- Concurrent Pos: Consult, Appl Data Res, Inc, 65-67. Mem: Asn Comput Mach; Am Math Soc. Res: Computational mathematics; number theory. Mailing Add: Dept of Math Queens Col Flushing NY 11367

WEINTRITT, DONALD J, b Columbus, Ohio, Feb 10, 26; m 47; c 4. COLLOID CHEMISTRY, PETROLEUM CHEMISTRY. Educ: Univ Tex, BS, 50; Univ Houston, MS, 57. Prof Exp: Chemist, Baroid Div, Nat Lead Co, 50-57, sect leader res & develop, NJ, 69-70, asst dir, Cent Res Lab, NL Industs, 70-72, mgr chem technol, Corp Res & Develop, 72-75, MGR SPEC CHEM LAB, NL INDUSTS, 75- Mem: Am Ceramic Soc; Am Inst Mining, Metall & Petrol Engrs; Am Soc Oceanog. Res: Petroleum exploration and production; fundamental properties of clay, starches, and other colloidal materials and the effect of their interactions as related primarily to oil well drilling, completion and production. Mailing Add: Spec Chem Lab NL Industs PO Box 700 Hightstown NJ 08520

WEINZWEIG, AVRUM ISRAEL, b Toronto, Ont, Apr 22, 26; m 53, 63; c 5. MATHEMATICS. Educ: Univ Toronto, BASc, 50; Harvard Univ, AM, 53, PhD(math), 57. Prof Exp: Res physicist, Weizmann Inst, 48-49; asst chief geophysicist, Weiss Geophys Corp, 50-52; instr math, Univ Calif, Berkeley, 57-59, actg asst prof, 59-60; asst prof, Northwestern Univ, 60-65; ASSOC PROF MATH, UNIV ILL, CHICAGO CIRCLE, 65- Concurrent Pos: Consult, Solomon Schechter Day Schs, Ill. Mem: Am Math Soc; Math Asn Am; Math Soc France. Res: Algebraic topology, particularly fiber spaces; category theory; learning theory; learning and acquisition of mathematical concepts; mathematics education. Mailing Add: Dept of Math Univ of Ill Chicago Circle Box 4348 Chicago IL 60680

WEIPERT, EUGENE ALLEN, b Monroe, Mich, Nov 17, 31; m 60; c 8. INDUSTRIAL ORGANIC CHEMISTRY. Educ: Univ Detroit, BS, 52, MS, 54; Iowa State Univ, PhD(org chem), 57. Prof Exp: Res supvr, Wyandotte Chem Corp, Mich, 58-72; TECH MGR, SOUTHERN SIZING CO, 72- Res: Amines; organometallics; organic reaction mechanisms; surfactants. Mailing Add: Southern Sizing Co PO Box 90987 East Point GA 30344

WEIR, C EDITH, b Wingham, Ont, Dec 16, 21; nat US. NUTRITION, FOOD TECHNOLOGY. Educ: Univ Toronto, BS, 41; Univ Mass, MS, 43, PhD(food tech), 49. Prof Exp: Nutrit res chemist, H J Heinz Co, Pa, 45-48; head sci dept, Berkshire Hills Col, 48-49; meat technologist, Bur Animal Indust, Agr Res Serv, US Dept Agr, Md, 49-53; chief home econ div, Am Meat Inst Found, Ill, 53-60; asst dir human nutrit res div, 60-72, ASST DIR, BELTSVILLE AGR RES CTR, AGR RES SERV, US DEPT AGR, 72- Concurrent Pos: Lectr, Northwestern Univ, 54-60; prof food & nutrit, Univ Md, 68-69. Mem: AAAS; Am Soc Clin Nutrit; NY Acad Sci; Inst Food Technol; Am Inst Nutrit. Res: Human nutrition. Mailing Add: Agr Res Ctr Agr Res Serv US Dept of Agr Beltsville MD 20705

WEIR, DONALD DOUGLAS, b Sussex, Wis, June 27, 28; m 48; c 3. INTERNAL MEDICINE. Educ: Drake Univ, BA, 48; Univ Iowa, MD, 53. Prof Exp: Intern med, Philadelphia Gen Hosp, 53-54; from asst resident to resident, Johns Hopkins & Baltimore City Hosps, 54-57; fel, Johns Hopkins Hosp, 57-58; from instr to assoc prof med, Sch Med, Univ NC, Chapel Hill, 58-69; MED DIR REHAB CTR, ST LUKE'S METHODIST HOSP, 69-; CLIN ASSOC PROF REHAB MED, COL MED, UNIV IOWA, 69- Concurrent Pos: Attend physician, Hosp Univ Iowa, 69-; consult, Vet Admin Hosp, Iowa City & Iowa Soldiers Home, Marshalltown, Iowa, 69- Mem: AAAS; Am Rheumatism Asn; AMA. Res: Rheumatology; chronic illness and patient care; rehabilitation. Mailing Add: St Luke's Methodist Hosp 1026 A Ave NE Cedar Rapids IA 52402

WEIR, F EUGENE, physical chemistry, see 12th edition

WEIR, JAMES HENRY, III, b East Orange, NJ, Sept 25, 32; m 58; c 3. MEDICAL RESEARCH. Educ: Princeton Univ, AB, 54; Columbia Univ, MD, 58. Prof Exp: From med res assoc to dir med serv, 63-75, DIR MED RES, WARNER-LAMBERT CO, 75- Concurrent Pos: Macy teaching fel obstet-gynec, Col Physicians & Surgeons, Columbia Univ, 62-63. Mem: Am Fertil Soc; Am Fedn Clin Res; Sigma Xi. Res: Clinical investigation of new drugs and new medical instrumentation devices. Mailing Add: Warner-Lambert Co 170 Tabor Rd Morris Plains NJ 07950

WEIR, JOHN ARNOLD, b Saskatoon, Sask, Apr 5, 16; nat US; m 46; c 2. GENETICS. Educ: Univ Sask, BSA, 37; Iowa State Col, MS, 42, PhD(genetics), 48. Prof Exp: Asst animal breeding, Dom Exp Sta, Alta, 37-40; instr, Univ Sask, 42, assoc prof animal husb, 48-50; from asst prof to assoc prof, 50-62, PROF ZOOL, UNIV KANS, 62- Concurrent Pos: Consult, Animal Resources Adv Comt, USPHS, 64-67; USPHS spec fel & hon res assoc hist sci, Harvard Univ, 66-67. Mem: AAAS; Genetics Soc Am; Am Genetic Asn. Res: Mammalian genetics; sex ratio and hematology of mice; history of genetics. Mailing Add: Hall Lab of Mammalian Genetics Div of Biol Univ of Kans Lawrence KS 66045

WEIR, JOHN MARSHALL, b Bonneterre, Mo, Aug 20, 11; m 34; c 3. PUBLIC HEALTH. Educ: Univ Chicago, BSc, 33, MD & PhD(path), 37; Johns Hopkins Univ, MPH, 46. Prof Exp: Asst prof histol & embryol, Med Sch, Univ Miss, 36-37; mem field staff, Int Health Div, Rockefeller Found, 39-51, mem staff, Div Med & Pub Health, 51-53, asst dir div, 54-55, assoc dir med educ & pub health, 55-59, assoc dir med & natural sci, 59-64, dir, 64-73; RETIRED. Mem: Assoc Am Soc Trop Med & Hyg; fel Am Pub Health Asn; NY Acad Sci. Res: Influenza; atypical pneumonia; immunology; yellow fever; tuberculosis; epidemiology. Mailing Add: 6 Hubbard Ct Stamford CT 06902

WEIR, JOHN ROBERT, b Wingham, Ont, Oct 17, 12; m 46; c 2. ENVIRONMENTAL BIOLOGY. Educ: Univ Toronto, BSA, 36; Univ Alta, MSc, 38; Univ Minn, PhD(plant breeding, genetics), 44. Hon Degrees: DSc, Univ Man 66 & Univ Guelph, 74. Prof Exp: From lectr to prof field husb & asst head dept, Ont Agr Col, 40-52; dean fac agr & home econ, Univ Man, 52-65; dep dir sci secretariat, Privy Coun Off, 65-67, dir, 67-69; asst dep minister fisheries, 69-72, CHMN FISHERIES RES BD CAN, 69- Concurrent Pos: Mem assoc comt plant breeding, Nat Res Coun, 49-55; proj officer, Royal Comn Govt Orgn, 61; chmn, Agr Econ Res Coun Can, 62-64; consult, Ford Found Univ Orgn, Brazil, 63 & Rockefeller Found Agr Educ, EAfrica, 66-67; agr educator, Int Bank Reconstruct & Develop, Ireland, 73; consult, Adv Comt Acad Planning, Coun Ont Univs, 73-74. Mem: Fel AAAS; Int Coun Explor Sea (vpres, 71-74); Am Genetic Asn; fel Agr Inst Can (pres, 62); fel Royal Soc Arts. Res: Science policy with particular emphasis on fisheries and marine sciences, including government university and industrial relationships in the national program for research and development in Canada. Mailing Add: Dept of the Environ Ottawa ON Can

WEIR, ROBERT JAMES, JR, b Washington, DC, Nov 26, 24; m 47; c 3. TOXICOLOGY. Educ: Univ Md, College Park, BS, 48, MS, 50, PhD, 55. Prof Exp: Asst physiol, Univ Md, 48-51; res assoc toxicol, Hazleton Labs, Va, 51-56, head agr chem dept, 56-58, res applns specialist, 58-62, dir, Hazleton Labs, SA, Lausanne, Switz, 62-65, vpres, Inst Indust & Biol Res, Cologne, Ger, 63-66, vpres, Mkt,

Hazleton Labs, Va, 67-69; V PRES, LITTON BIONETICS, INC, 69- Mem: AAAS; Am Chem Soc (secy-treas, Agr & Food Div, 67-68); Soc Cosmetic Chemists; Environ Mutagen Soc; Soc Toxicol. Res: Toxicology; biochemistry; pharmacology; safety evaluation of drug, cosmetic, food chemical and pesticide development. Mailing Add: Litton Bionetics Inc 5516 Nicholson Lane Kensington MD 20795

WEIR, THOMAS ROBERT, cartography, population geography, see 12th edition

WEIR, WILLIAM CARL, b Lakeview, Ore, Aug 24, 19; m 46; c 2. ANIMAL NUTRITION. Educ: Ore State Col, BS, 40; Univ Wis, MS, 41, PhD(animal husb, biochem), 48. Prof Exp: Asst, Univ Wis, 45-46, instr, 47; assoc prof animal sci, Ore State Col, 48; from asst prof to prof, 48-73, dean students, 58-65, PROF NUTRIT & CHMN DEPT, UNIV CALIF, DAVIS, 73- Concurrent Pos: Fulbright res grant, Univ Western Australia, 65-66; mem comt sheep nutrit, Nat Acad Sci-Nat Res Coun; vis scientist & Univ Calif rep, Univ Chile-Univ Calif Coop Prog, Santiago, Chile, 70-72. Mem: Am Soc Animal Sci; Soc Range Mgt; Am Inst Nutrit; Soc Nutrit Educ. Res: Purines in food; fiber in diets; new sources of nutrients; sheep nutrition. Mailing Add: Dept of Nutrit Univ of Calif Davis CA 95616

WEIR, WILLIAM DAVID, b Oakland, Calif, Mar 15, 41; c 2. PHYSICAL CHEMISTRY. Educ: Occidental Col, AB, 62; Princeton Univ, AM, 63, PhD(chem), 65. Prof Exp: Instr chem, Harvard Univ, 65-68; asst prof, 68-71, ASSOC PROF CHEM, REED COL, 71- Mem: Am Chem Soc; Electrochem Soc. Res: Chemical kinetics; kinetics and mechanisms of electrochemical reactions and dynamics of membrane function; relaxation methods; instrumentation and computation in chemical research. Mailing Add: Dept of Chem Reed Col Portland OR 97202

WEIRICH, GUNTER FRIEDRICH, b Eisenach, Ger, Feb 17, 34; m 68. BIOLOGICAL CHEMISTRY. Educ: Univ Munich, PhD(zool), 63. Prof Exp: Res assoc endocrinol, Philipps Univ, Marburg, Ger, 64-65; res fel insect physiol, Biol Div, Oak Ridge Nat Lab, 65-66; res assoc endocrinol, Philipps Univ, 66-70; sr scientist biol chem, Zoecon Corp, 70-73; RES ASSOC BIOL CHEM, TEX A&M UNIV, 73- Mem: AAAS; Europ Soc Comp Endocrinol; Ger Soc Biol Chem; Ger Zool Soc. Res: Transport and metabolism of insect hormones. Mailing Add: Inst of Develop Biol Tex A&M Univ College Station TX 77843

WEIRICH, WALTER EDWARD, b Saginaw, Mich, Nov 20, 38; m 60; c 2. VETERINARY SURGERY. Educ: Mich State Univ, BS, 61, DVM, 63; Univ Wis, MS, 70, PhD(vet sci, cardiol), 71. Prof Exp: Officer in chg, Vet Corps, US Army, 63-65; pract vet, Madison Vet Clin, 65-68; NIH fel cardiol, Univ Wis-Madison, 68-71; from asst prof to assoc prof cardiol & surg, 71-75, actg head dept cardiol & surg, 75-76, HEAD DEPT SMALL ANIMAL CLIN, SCH VET MED, PURDUE UNIV, WEST LAFAYETTE, 76- Concurrent Pos: Mem cardiol comt, Am Animal Hosp Assoc, 73-; rev, J Am Vet Med Asn. Mem: Am Vet Med Asn; Am Acad Vet Cardiol; Am Asn Vet Med Cols. Res: Hypothermia for cardiac arrest surgery; myocardial infarctions and vascular surgery in the canine. Mailing Add: Dept of Small Animal Clin Sch of Vet Med Purdue Univ West Lafayette IN 47907

WEIS, DALE STERN, b Cleveland, Ohio, Oct 11, 24; m 70; c 2. MICROBIOLOGY, PROTOZOOLOGY. Educ: Western Reserve Univ, BS, 45, MS, 51; Yale Univ, PhD, 55. Prof Exp: Instr bact, Albertus Magnus Col, 53-54; res assoc plant physiol, Univ Minn, 55-57; res assoc biochem, Univ Chicago, 57-58; from instr to asst prof biol, Univ Col, 58-64; chmn natural sci 2, Shimer Col, 64-66; asst prof, 66-69, ASSOC PROF BIOL, CLEVELAND STATE UNIV, 69- Concurrent Pos: Charles F Kettering fel, Univ Minn, 55-57. Mem: AAAS; Soc Protozool; Am Soc Microbiol; NY Acad Sci. Res: Metabolism of algae; symbiosis; host-symbiote interaction. Mailing Add: Dept of Biol & Health Sci Cleveland State Univ Cleveland OH 44115

WEIS, JERRY SAMUEL, b Salina, Kans, Dec 23, 35; m 61; c 3. BIOLOGY. Educ: Kans Wesleyan Univ, AB, 58; Univ Kans, MA, 60, PhD(bot), 64. Prof Exp: Asst prof biol, Univ Minn, 64-65; NIH fel biol, Yale Univ, 65-66; asst prof, 66-71, asst dir div biol, 69-73, ASSOC PROF BIOL, KANS STATE UNIV, 72-, ASSOC DIR DIV BIOL, 75- Mem: AAAS; Inst Soc Ethics & Life Sci; Am Asn of Higher Educ. Res: Plant tissue culture; chemical control of plant growth and development; bioethics. Mailing Add: Div of Biol Kans State Univ Manhattan KS 66506

WEIS, JOE H, b Coulee Dam, Wash, Sept 7, 42. THEORETICAL HIGH ENERGY PHYSICS. Educ: Calif Inst Technol, BS, 64; Univ Calif, Berkeley, PhD(phsyics), 70. Prof Exp: Res assoc physics, Mass Inst Technol, 70-72 & Univ Wash, 72-73; NATO fel & res assoc physics, Europ Ctr Nuclear Res, 73-74; ASST PROF PHYSICS, UNIV WASH, 74- Concurrent Pos: Sloan Found fel, 74. Res: Strong, electromagnetic and weak interactions of hadrons; Regge theory of strong interactions; multiparticle production at high energy; dual resonance model. Mailing Add: Dept of Physics Univ of Wash Seattle WA 98195

WEIS, JUDITH SHULMAN, b New York, NY, May 29, 41; m 62; c 2. DEVELOPMENTAL BIOLOGY, AQUATIC BIOLOGY. Educ: Cornell Univ, BA, 62; NY Univ, MS, 64, PhD(biol), 67. Prof Exp: Lectr biol, Hunter Col, 64-67; asst prof, 67-71, ASSOC PROF ZOOL, RUTGERS UNIV, NEWARK, 71- Concurrent Pos: Rutgers Res Coun grant, 67- Mem: Am Soc Zoologists; Am Inst Biol Sci; Soc Develop Biol; Am Fisheries Soc. Res: Analysis of development of nervous system; nerve growth factor; limb regeneration; marine biology; effects of pollutants on aquatic animals. Mailing Add: Dept of Zool & Physiol Rutgers Univ Newark NJ 07102

WEIS, LEONARD WALTER, b New York, NY, June 23, 23; m 55; c 2. GEOLOGY. Educ: Harvard Univ, SB, 43; Mass Inst Technol, SM, 47; Univ Wis, PhD(geol), 65. Prof Exp: Res observer, Blue Hill Meteorol Observ, 43; asst meteorol, Mass Inst Technol, 44-47; instr geol & geog, Univ RI, 47-49; asst prof geol & actg chmn dept, Coe Col, 53-54; asst prof, Lawrence Univ, 55-65; ASST PROF GEOL, UNIV WIS CTR-FOX VALLEY, 65- Mem: Am Meteorol Soc; Geol Soc Am; Am Geophys Union; Geochem Soc; NY Acad Sci. Res: Igneous and metamorphic petrology; glacial geology, including petrography of sediments; paleoclimatology; meteorological instruments and observations. Mailing Add: Dept of Geol Univ Wis Ctr-Fox Valley Menasha WI 54952

WEIS, PAUL LESTER, b Chicago, Ill, June 22, 22; m 45, 69; c 2. GEOLOGY. Educ: Univ Wis, BS, 47, PhD(geol), 52. Prof Exp: Geologist, US Geol Surv, 50-51; asst prof geol, Univ Va, 51-53; GEOLOGIST, US GEOL SURV, 53- Mem: Am Inst Prof Geologists; fel Geol Soc Am; Soc Econ Geol. Res: Mineral resources; areal geology of northwestern United States; graphite commodity geology; metallic mineral deposits in sedimentary rocks. Mailing Add: US Geol Surv National Center Reston VA 22092

WEIS, PEDDRICK, b South Paris, Maine, June 4, 38; m 62; c 2. ANATOMY, EMBRYOLOGY. Educ: NY Univ, DDS, 63. Prof Exp: NSF res fel, 63-64; instr anat, Col Dent, NY Univ, 64-67; asst prof, 67-71, ASSOC PROF ANAT, NJ MED SCH, COL MED & DENT NJ, 71- Mem: AAAS; Am Asn Anat; Electron Micros Soc Am; Soc Develop Biol. Res: Ultrastructural and cytochemical aspects of nervous system development and regeneration; effects of pollutants on nervous system development and behavior. Mailing Add: Dept of Anat NJ Med Sch Col of Med & Dent of NJ Newark NJ 07103

WEISBACH, JERRY ARNOLD, b New York, NY, Dec 23, 33; m 58; c 3. ORGANIC CHEMISTRY, MEDICINAL CHEMISTRY. Educ: Brooklyn Col, BS, 55; Harvard Univ, MA, 56, PhD(chem), 59. Prof Exp: Res chemist, Sun Oil Co, 59-60; sr med chemist, 60-65, group leader, 65-67, assoc dir chem, 67-71, assoc dir res, US Pharmaceut Prod, 71-75, DEP DIR RES, US PHARMACEUT PROD, SMITH KLINE & FRENCH LABS, 75- Mem: AAAS; Am Chem Soc; Acad Pharmaceut Sci; NY Acad Sci; Am Soc Microbiol. Res: Structure, isolation and synthesis of natural products, particularly antibiotics, alkaloids and lipids; organic biochemistry and synthetic medicinal chemistry. Mailing Add: Smith Kline & French Labs 1500 Spring Garden St Philadelphia PA 19101

WEISBART, MELVIN, b Toronto, Ont, Dec 28, 38; m 63; c 2. COMPARATIVE ENDOCRINOLOGY, COMPARATIVE PHYSIOLOGY. Educ: Univ Toronto, BSc, 61, MA, 63; Univ BC, PhD(physiol), 67. Prof Exp: Nat Res Coun Can fel, Fisheries Res Bd Can, 68-69; ASST PROF BIOL, WAYNE STATE UNIV, 69- Concurrent Pos: Fac res award, Wayne State Univ, 70. Mem: AAAS; Am Soc Zoologists; Can Soc Zoologists; Sigma Xi. Res: Fish physiology and endocrinology. Mailing Add: Dept of Biol Wayne State Univ Detroit MI 48202

WEISBERG, HERBERT, b New York, NY, June 30, 31. MEDICINE. Educ: City Col New York, BS, 53; Univ Lausanne, MD, 58. Prof Exp: From instr to asst prof med, 64-71, ASSOC PROF ANAT & MED, NY MED COL-FLOWER & FIFTH AVE HOSP, 71- Concurrent Pos: USPHS trainee gastroenterol, NY Med Col-Flower & Fifth Ave Hosp, 62-64, USPHS spec fel electron micros, 66-68; vis prof, Sch Med, NY Univ, 71. Mem: Am Fedn Clin Res; Am Gastroenterol Asn. Res: Intracellular pathway of absorption for nutrients in the intestine, especially vitamin B-12. Mailing Add: Dept of Med NY Med Col Flower & Fifth Ave Hosp New York NY 10029

WEISBERG, HERBERT IRA, b New York, NY, Mar 9, 44. STATISTICS. Educ: Columbia Univ, BA, 65; Harvard Univ, MA, 66, PhD(statist), 70. Prof Exp: Asst prof statist, NY Univ, 69-72; SR RES ASSOC, HURON INST, 72- Concurrent Pos: Vis lectr statist, Univ Kent, 71-72; lectr statist, Boston Univ, 74- Mem: Am Statist Asn; Biomet Soc; Am Educ Res Asn. Res: Social research methodology; design and analysis of social program evaluations; data analysis; statistical inference; analysis of data on child development, child abuse and education. Mailing Add: Huron Inst 119 Mt Auburn St Cambridge MA 02138

WEISBERG, HOWARD LOUIS, b Cleveland, Ohio, Nov 12, 39; m 65; c 1. PHYSICS. Educ: Calif Inst Technol, BS, 60; Brandeis Univ, PhD(physics), 65. Prof Exp: Physicist, Univ Calif, Berkeley, 65-70, NSF fel, 65-67; ASST PROF PHYSICS, UNIV PA, 70- Mem: AAAS; Am Phys Soc. Res: Experimental elementary particle physics. Mailing Add: Dept of Physics Univ of Pa Philadelphia PA 19104

WEISBERG, JERRY, b Brooklyn, NY, Mar 11, 27; m 48; c 2. CLINICAL PHARMACOLOGY. Educ: NY Univ, BA, 48; Brown Univ, ScM, 50; Univ NDak, BS, 56; Univ Buffalo, MD, 58. Prof Exp: Chief chem, Maimonides Hosp Brooklyn, NY, 51-52; asst dir clin invest, Warner Lambert Res Inst, 63-65, assoc dir, 65-69, dir clin pharmacol, 69-70; vpres & med dir, USV Pharmaceut Corp, 70-72; DIR EMERGENCY SERV & CHMN AMBULATORY SERV, MORRISTOWN MEM HOSP, 72- Mem: AMA; Acad Psychosom Med. Res: Clinical medicine; drug evaluation; biopharmaceutics; drug metabolism; biophysics and bioinstrumentation. Mailing Add: 151 West End Ave North Plainfield NJ 07060

WEISBERG, JOSEPH SIMPSON, b Jersey City, NJ, June 7, 37; m 64; c 2. OCEANOGRAPHY, GEOLOGY. Educ: Jersey City State Col, BA, 60; Montclair State Col, MA, 64; Columbia Univ, EdD(earth sci educ), 69. Prof Exp: Teacher pub schs, NJ, 60-64; ASSOC PROF GEOSCI, JERSEY CITY STATE COL, 60-, CHMN DEPT, 73- Concurrent Pos: Sci consult, US Off Educ, 68-69. Mem: AAAS; Geol Soc Am; Am Meteorol Soc; Nat Asn Geol Teachers; Nat Sci Teachers Asn. Res: Use of visual aids in science teaching; inquiry and learning; environmental aspects of the geosciences. Mailing Add: Dept of Geosci Jersey City State Col Jersey City NJ 07305

WEISBERG, ROBERT H, b Brooklyn, NY, May 20, 47. PHYSICAL OCEANOGRAPHY. Educ: Cornell Univ, BS, 69; Univ RI, MS, 72, PhD(phys oceanog), 75. Prof Exp: Res asst, 69-74, RES ASSOC PHYS OCEANOG, GRAD SCH OCEANOG, UNIV RI, 74- Mem: Am Geophys Union. Res: Estuarine and equatorial circulation. Mailing Add: Grad Sch of Oceanog Univ of RI Kingston RI 02881

WEISBERG, SAMUEL MYER, b Boston, Mass, June 2, 05; m 30; c 3. BIOCHEMISTRY. Educ: Tufts Col, BS, 26; Am Univ, MS, 28; Johns Hopkins Univ, PhD(biochem), 31. Prof Exp: Anal chemist, Pac Mills, 26-27; jr chemist, USDA, 27-28; res chemist, Nat Dairy Res Labs, 31-35, chief chemist dairy by-prod, 35-48, dir gen chem div, 49-53, chief feedstuffs nutrit lab, 54-58, mgr indust prod lab, 58-65, dir res & develop, Nat Dairy Prod Corp, Ill, 65-68; exec dir, 68-75, EMER EXEC DIR, LEAGUE FOR INT FOOD EDUC, 75-; ED, NEWSLETTER, 72- Concurrent Pos: Mem res & develop assocs, Mil Food & Packaging Systs; consult, Int Food Technol Problems. Mem: Am Chem Soc; fel Geront Soc; fel Am Inst Chemists; NY Acad Sci; Am Dairy Sci Asn. Res: Dairy by-products utilization; animal nutrition; food technology research administration; food problems in developing countries. Mailing Add: 4000 Massachusetts Ave NW Apt 725 Washington DC 20016

WEISBERG, SANFORD, US citizen. APPLIED STATISTICS. Educ: Univ Calif, AB, 69; Harvard Univ, AM, 70, PhD(statist), 73. Prof Exp: ASST PROF APPL STATIST, UNIV MINN, ST PAUL, 72- Mem: Am Statist Asn; Inst Math Statist; Biomet Soc. Res: Data analysis and methods; order statistics and robust analysis. Mailing Add: Dept of Appl Statist Univ of Minn St Paul MN 55108

WEISBERG, STANLEY HERBERT, b Toronto, Can, June 9, 40. ENVIRONMENTAL BIOLOGY. Educ: Univ Toronto, BSA, 64; Univ Alta, MSc, 66, PhD(genetics), 68. Prof Exp: Nat Res Coun Can fel, Univ Geneva, 68-70; asst prof biol, Brock Univ, 70-73; SR BIOLOGIST, SURVEYOR, MENNINGER & CHENEVERT, INC, 75- Concurrent Pos: Res Coun Can grant, 71-72. Mem: AAAS; Genetics Soc Can; Can Soc Cell Biol; Soc Gen Microbiol; Can Soc Micros. Res: Biological impact studies; air, water and noise pollution. Mailing Add: Surveyor Menninger & Chenevert 1550 de Maisonneuve Blvd W Montreal PQ Can

WEISBERGER, WILLIAM I, b New York, NY, Dec 20, 37; m 61; c 3. THEORETICAL HIGH ENERGY PHYSICS. Educ: Amherst Col, BA, 59; Mass Inst Technol, PhD(physics), 64. Prof Exp: Res assoc physics, Stanford Linear Accelerator Ctr, 64-66; asst prof, Princeton Univ, 66-70; PROF PHYSICS, STATE UNIV NY STONY BROOK, 70- Concurrent Pos: Sloan Found fel, 67-69; vis scientist, Weizmann Inst Sci, 68-69 & 74-75; Guggenheim fel, 74-75. Mem: Fel Am Phys Soc. Mailing Add: Inst for Theoret Physics State Univ of NY at Stony Brook Stony Brook NY 11790

WEISBLAT, DAVID IRWIN, b Coshocton, Ohio, Aug 11, 16; m 42; c 3. ORGANIC CHEMISTRY. Educ: Ohio State Univ, AB, 37, PhD(org chem), 41. Prof Exp: Asst org chem, Ohio State Univ, 37-41, res assoc & Hoffmann-La Roche fel, Res Found, 41-43; res chemist, Nutrit Div, 43-45, chem group leader, 45-47, sect leader, Chem Div, 47-50, head chem dept, 50-52, asst dir res, 52-56, dir chem res, 56-62, dir biochem res, 62-68, VPRES PHARMACEUT RES & DEVELOP, UPJOHN CO, 68-, MEM, BD DIRS, 73- Concurrent Pos: Investr, Tech Indust Intel Comt, Fed Repub Ger. Mem: AAAS; Am Chem Soc. Res: Chemistry of simple carbohydrates, heparin, synthetic vitamins and amino acids; steroids and antibiotics; prostaglandin chemistry and clinical application. Mailing Add: Upjohn Co 7000 Portage Rd Kalamazoo MI 49001

WEISBLUM, BERNARD, biochemistry, microbiology, see 12th edition

WEISBORD, NORMAN EDWARD, b Jersey City, NJ, Oct 1, 01; m 39. GEOLOGY, INVERTEBRATE PALEONTOLOGY. Educ: Cornell Univ, AB, 23, MS, 26. Prof Exp: Paleontologist & geologist, Atlantic Refining Co, 23-32; sr field geologist, Standard Oil Co, Arg, 32-34; sr geologist & asst chief geologist, Standard-Vacuum Oil Co, 34-42; chief geologist, Socony-Vacuum Oil Co, Venezuela, 42-57; res assoc, 57-65, PROF GEOL, FLA STATE UNIV, 65- Concurrent Pos: Instr, Cornell Univ, 26; trustee, Paleont Res Inst, 51-63, vpres, 57-59, pres, 59-61; consult, Mobil Oil Corp. Mem: Fel AAAS; fel Geol Soc Am; fel Am Geog Soc; fel Sigma Xi; Paleont Res Inst. Res: Conchology; invertebrate paleontology; stratigraphy; taxonomy and distribution of the barnacles Cirripedia and corals Scleractinia of Florida. Mailing Add: PO Box 1082 Tallahassee FL 32302

WEISBROD, ALAN RICHARD, b Redmond, Ore, Oct 14, 36; m 60; c 2. VERTEBRATE ZOOLOGY, EVOLUTIONARY BIOLOGY. Educ: Univ Minn, Minneapolis, BA, 59; Cornell Univ, MS, 65, PhD(evolutionary biol), 70. Prof Exp: Asst cur birds, Div Biol Sci, Cornell Univ, 66-69, vertebrates, 69-70; ASST PROF FOREST ZOOL, COL FOREST RESOURCES, UNIV WASH, 70-; RES BIOLOGIST, US NAT PARK SERV, 70- Concurrent Pos: Biologist, Alaska Native Claim Lands Task Force, US Nat Park Serv, 72-74. Mem: AAAS; Am Ornith Union; Am Soc Mammal; Ecol Soc Am; Soc Study Evolution. Res: Vertebrate ecology, behavior and systematics. Mailing Add: US Nat Park Serv Coop Park Stud Col of Forest Resources Univ of Wash AR-10 Seattle WA 98195

WEISBRODT, NORMAN WILLIAM, b Cleves, Ohio, June 30, 42; m 65; c 3. PHYSIOLOGY, PHARMACOLOGY. Educ: Univ Cincinnati, BS, 65; Univ Mich, Ann Arbor, PhD(pharmacol), 70. Prof Exp: USPHS res fel, Univ Iowa, 70-71; from instr to asst prof physiol, 71-75, ASSOC PROF PHYSIOL & PHARMACOL, UNIV TEX MED SCH HOUSTON, 75- Mem: Am Asn Clin Res; Am Physiol Soc; Soc Exp Biol & Med. Res: Smooth muscle physiology and pharmacology; gastrointestinal motility. Mailing Add: Prog in Physiol Univ of Tex Med Sch Houston TX 77025

WEISBROT, DAVID R, b Brooklyn, NY, Dec 29, 31; m 60; c 2. POPULATION GENETICS. Educ: Brooklyn Col, BS, 53, MA, 58; Columbia Univ, PhD(zool), 63. Prof Exp: Substitute instr biol, Brooklyn Col, 56-58; res asst genetics, Univ Conn, 58-59 & Long Island Biol Asn, 59-60; instr biol, City Col New York, 60-62; fel genetics, Univ Calif, Berkeley, 63-64; asst prof biol, Tufts Univ, 64-69; assoc prof, State Univ NY Binghamton, 69-72; ASSOC PROF BIOL, WILLIAM PATERSON COL, NJ, 72- Concurrent Pos: Adj assoc prof, Columbia Univ, 76- Mem: AAAS; Genetics Soc Am; Am Genetic Asn; Am Soc Human Genetics. Res: Genotypic interactions among competing strains of Drosophila; relationship of genetic and morphological differences among sibling species. Mailing Add: 1103 Sussex Rd Teaneck NJ 07666

WEISBROTH, STEVEN H, b New York, NY, Sept 16, 34; m 58; c 3. LABORATORY ANIMAL MEDICINE, COMPARATIVE PATHOLOGY. Educ: Cornell Univ, BS, 58; Wash State Univ, MS, 60, DVM, 64; Am Col Lab Animal Med, dipl. Prof Exp: NIH fel lab animal med, NY Univ Med Ctr, 64-66; asst prof path & dir animal facilities, Rockefeller Univ, 66-69; asst prof path & dir dept lab animal med, 69-70, ASSOC PROF PATH & DIR, DIV LAB ANIMAL RESOURCES, STATE UNIV NY STONY BROOK, 70- Concurrent Pos: NIH fel lab animal med, Rockefeller Univ, 69-; mem coun, Am Asn Accreditation of Lab Animal Care, 75-79. Honors & Awards: Res award, Am Asn Lab Animal Sci, 72. Mem: Am Vet Med Asn; Am Asn Lab Animal Sci; Am Soc Exp Path. Res: Spontaneous diseases in laboratory animals. Mailing Add: Div of Lab Animal Resources State Univ of NY Health Sci Ctr Stony Brook NY 11794

WEISBURGER, ELIZABETH KREISER, b Greenlane, Pa, Apr 9, 24; m 47; c 3. ONCOLOGY, TOXICOLOGY. Educ: Lebanon Valley Col, BS, 44; Univ Cincinnati, PhD(org chem), 47. Prof Exp: Res assoc, Univ Cincinnati, 47-49; res fel, 49-51, res org chemist, Biochem Lab, 51-61, carcinogen screening sect, 61-72, CHIEF, CARCINOGEN METAB & TOXICOL BR, NAT CANCER INST, 73- Concurrent Pos: Asst chief ed, J Nat Cancer Inst, 71- Mem: AAAS; Am Asn Cancer Res; Am Chem Soc; The Chem Soc; Soc Toxicol. Res: Metabolism of chemical carcinogens, carcinogen testing, chemical carcinogenesis and toxicology. Mailing Add: 5309 McKinley St Bethesda MD 20014

WEISBURGER, JOHN HANS, b Stuttgart, Ger, Sept 15, 21; nat US; c 3. BIOCHEMICAL PHARMACOLOGY, ONCOLOGY. Educ: Univ Cincinnati, AB, 47, MS, 48, PhD(org chem), 49. Prof Exp: Fel, Nat Cancer Inst, 49-50, head phys-org chem unit, Lab Biochem, 50-61, head carcinogen screening sect, 61-72, dir bioassay segment, 71-72; VPRES RES, AM HEALTH FOUND, 72- Concurrent Pos: Mem biochem & nutrit rev panel, NIH, 58-59, mem interdept tech panel on carcinogens, 62-71, chmn subcomt, Nat Cancer Progs Strategic Plan, 71-74, chmn carcinogenesis group, Nat Large Bowel Cancer Proj, 72-75; mem expert panel nitrites & nitrosamines, USDA, 73-; res prof path, New York Med Col, 73- Assoc ed, J Nat Cancer Inst, 60-62; assoc ed, Cancer Res, 69-76, ed, Arch Toxicol, 75- Honors & Awards: Nat Defense Serv Medal, USPHS, 64 & Meritorious Serv Medal, 70. Mem: Am Asn Cancer Res; Am Soc Biol Chemists; Am Soc Pharmacol & Exp Therapeut; Soc Exp Biol & Med; Soc Toxicol. Res: Etiology of cancer, mechanisms of carcinogenesis; bioassay and metabolism of carcinogens and drugs; host factors in cancer induction; nutrition, endocrinology, immunology, cancers of the endocrine, digestive and excretory organs; preventive medicine. Mailing Add: Am Health Found Naylor Dana Inst for Dis Prev Valhalla NY 10595

WEISE, CHARLES MARTIN, b Bridgeville, Pa, July 8, 26; m 51; c 5. POPULATION ECOLOGY, ORNITHOLOGY. Educ: Ohio Univ, BS, 50; Univ Ill, MS, 51, PhD(zool), 56. Prof Exp: Asst prof biol, Fisk Univ, 53-56; from asst prof to assoc prof, 56-66, PROF ZOOL, UNIV WIS-MILWAUKEE, 66- Mem: AAAS; Cooper Ornith Soc; Wilson Ornith Soc; Am Soc Zoologists; Am Ornith Union. Res: Annual Physiological and reproductive cycles in birds; field ornithology; animal populations and population ecology. Mailing Add: Dept of Zool Univ of Wis-Milwaukee Milwaukee WI 53201

WEISE, JURGEN KARL, b Nov 7, 37; US citizen; m 64; c 2. POLYMER CHEMISTRY. Educ: Univ Bonn, BS, 60; Polytech Inst Brooklyn, PhD(polymer chem), 66. Prof Exp: Ger Res Asn fel chem, Univ Mainz, 66-67; res chemist, Control Res Dept, 67-71, RES CHEMIST, ELASTOMER CHEM DEPT, EXP STA, E I DU PONT DE NEMOURS & CO, 71- Mem: Am Chem Soc. Res: Reactions of polymers; polymer structures and their effects on properties; ring-opening polymerization; fluoropolymers. Mailing Add: Elastomer Chem Dept Exp Sta E I du Pont de Nemours & Co Wilmington DE 19898

WEISEL, CHARLES ARTHUR, organic chemistry, see 12th edition

WEISEL, GEORGE FERDINAND, JR, b Missoula, Mont, Mar 21, 15; m 50; c 2. ZOOLOGY. Educ: Mont State Univ, BS, 41, MA, 42; Univ Calif, Los Angeles, PhD(zool), 49. Prof Exp: Asst zool, Mont State Univ, 41-42, Univ Mich, 42-43 & Scripps Inst Oceanog, Univ Calif, San Diego, 47-48; from instr to assoc prof comp anat & gen zool, 47-69, PROF COMP ANAT & ICHTHYOL, UNIV MONT, 69- Mem: Am Soc Ichthyol & Herpet; Am Soc Zoologists. Res: Anatomy, histology, sex organs, life histories, osteology and endocrinology of fish. Mailing Add: Dept of Zool Univ of Mont Missoula MT 59801

WEISENBERG, RICHARD CHARLES, b Columbus, Ohio, Apr 2, 41. CELL BIOLOGY. Educ: Univ Calif, Santa Barbara, BA, 63; Univ Chicago, PhD(biophys), 68. Prof Exp: USDA trainee, Brandeis Univ, 68-70; asst prof, 71-73, ASSOC PROF BIOL, TEMPLE UNIV, 73- Concurrent Pos: Investr, Marine Biol Lab, Woods Hole, 71- Mem: Am Soc Cell Biol. Res: Cell division and motility; biochemistry of microtubules. Mailing Add: Dept of Biol Temple Univ Philadelphia PA 19122

WEISENBORN, FRANK L, b Portland, Ore, Feb 26, 25; m 45; c 4. ORGANIC CHEMISTRY. Educ: Reed Col, BA, 45; Univ Wash, PhD(chem), 49. Prof Exp: AEC fel, Harvard Univ, 49-50, USPHS fel, 50-51; sr res chemist, Riker Labs, Inc, 51-53; res assoc, Univ Calif, Los Angeles, 53; sr res chemist, 53-59, from res assoc to sr res assoc, 59-63, DIR ORG CHEM, SQUIBB INST MED RES, E R SQUIBB & SONS, INC, 63- Concurrent Pos: Chmn, Gordon Res Conf Natural Prod, 75-76. Mem: Am Chem Soc; The Chem Soc; Swiss Chem Soc. Res: Structure and synthesis of antibiotics, steroids and alkaloids; natural and synthetic hypotensive agents. Mailing Add: Dept of Org Chem E R Squibb & Sons PO Box 4000 Princeton NJ 08540

WEISER, DANIEL, b St Louis, Mo, July 10, 33; m 54; c 4. MATHEMATICS. Educ: Rice Univ, BA, 54, MA, 56, PhD(math), 58. Prof Exp: Asst math, Rice Univ, 57-58; SR RES MATHEMATICIAN, FIELD RES LAB, MOBIL RES & DEVELOP CO, 58- Mem: Soc Indust & Appl Math. Res: Applied mathematics and political science. Mailing Add: 3851 Rugged Circle Dallas TX 75224

WEISER, DAVID W, b Omaha, Nebr, Sept 13, 21; m 42; c 3. INORGANIC CHEMISTRY. Educ: Drury Col, AB, 42; Univ Chicago, MS, 47, PhD(chem), 56. Prof Exp: Asst prof chem, Drake Univ, 50-51; from instr to asst prof natural sci, Univ Chicago, 51-57; dean fac, Shimer Col, 57-63, chmn dept natural sci, 64-67; vis prof chem, Cornell Univ, 67-68; ASSOC PROF CHEM, STATE UNIV NY STONY BROOK, 69-, DIR CTR CURRICULUM DEVELOP, 68- Concurrent Pos: Consult & examr, NCent Asn Cols & Univs, 58-; NSF res fel, Yale Univ, 63-64. Mem: Am Chem Soc; Am Educ Res Asn. Res: Behavior of the aquocobaltic ion; nature of solutions with ionic solutes; epistemology and history of chemistry; science education. Mailing Add: Dept of Chem State Univ of NY Stony Brook NY 11790

WEISER, HERMAN JOSHUA, JR, b Harrisburg, Pa, Dec 9, 24; m 47; c 2. ANALYTICAL CHEMISTRY. Educ: Lebanon Valley Col, BS, 47; Univ Cincinnati, MS, 49, PhD(chem), 51. Prof Exp: ANAL GROUP LEADER, PROCTER & GAMBLE CO, 51- Mem: Am Chem Soc; Am Oil Chem Soc. Res: Optical brightener analysis; gas chromatography of fats, oils and related materials; analysis of detergent raw materials and products. Mailing Add: Procter & Gamble Co Ivorydale Tech Ctr Cincinnati OH 45217

WEISER, PHILIP CRAIG, b Portland, Ore, Oct 26, 41; m 66; c 1. PHYSIOLOGY. Educ: Univ Wash, BS, 63; Univ Minn, MS, 67, PhD(physiol), 69. Prof Exp: ADJ PROF BIOL, UNIV COLO, BOULDER, 70-; RES FEL, DEPT CLIN PHYSIOL, NAT ASTHMA CTR, DENVER, 75- Concurrent Pos: Res physiologist, Physiol Div, Med Res & Nutrit Lab, Fitzsimons Army Med Ctr, Med Serv Corps, US Army, 68-71, chief mil performance br, 71-73; res fel, Cardiovasc Pulmonary Res Lab, Univ Cols Med Ctr, Denver, 73-75. Mem: AAAS; Am Col Sports Med. Res: Biological adaptation to stress, particularly the psychophysiological factors limiting physical performance in normoxic and hypoxic environments. Mailing Add: Dept of Clin Physiol Nat Asthma Ctr Denver CO 80204

WEISER, RUSSELL SHIVLEY, b Grimes, Iowa, Sept 28, 06; m 31; c 1. BACTERIOLOGY, IMMUNOLOGY. Educ: NDak Col, BS, 30, MS, 31; Univ Wash, PhD(bact), 34. Prof Exp: Assoc bact & path, 34-36, from instr to assoc prof & actg exec officer, 36-45, assoc prof microbiol, 45-49, PROF IMMUNOL, SCHS MED & DENT, UNIV WASH, 49- Concurrent Pos: Mem leprosy res panel, US-Japan Coop Med Sci Prog. Mem: Am Asn Immunologists; Reticuloendothelial Soc; Brit Soc Immunol; Transplantation Soc. Res: Immunology of syphilis; immunology of cancer; immunologic tissue injury; macrophages; cell-mediated immunity. Mailing Add: 5741 60th Ave NE Seattle WA 98105

WEISFELD, LEWIS BERNARD, b Philadelphia, Pa, July 2, 29; m 66. PHYSICAL ORGANIC CHEMISTRY, POLYMER CHEMISTRY. Educ: Univ Pa, BA, 52; Univ Del, MS, 53, PhD(chem), 56. Prof Exp: Res chemist, Gen Elec Co, 56-58; res scientist, US Rubber Co, 58-62, sr res specialist, Conn, 62-63; dir labs, Advan Div, Carlisle Chem Works, Inc, New Brunswick, 63-69, Cincinnati Milacron Chem, Inc, 69-72, sci dir, 72-74, VPRES & SCI DIR, CINCINNATI MILACRON CHEM, INC, 74-, CORP DIR APPL SCI RES & DEVELOP, CINCINNATI MILACRON, INC, 75- Mem: AAAS; Am Chem Soc; Soc Plastics Engrs; fel Am Inst Chemists; Am Soc Testing & Mat. Res: Polymer and plastics science and technology; organometallic and isocyanate chemistry; impact plastics; catalysis; polymerization. Mailing Add: Cincinnati Milacron Inc 4701 Marburg Ave Cincinnati OH 45209

WEISGERBER, CYRUS AARON, b Clearfield, Pa, July 21, 16; m 44; c 2. ORGANIC CHEMISTRY. Educ: Franklin & Marshall Col, BS, 40; Pa State Col, MS, 43, PhD(chem), 44. Prof Exp: Instr chem, Pa State Col, 44-46, res chemist, War Prod Bd & Off Prod Res & Develop, 43-46; RES CHEMIST, RES CTR, HERCULES, INC, 46- Mem: Am Chem Soc. Res: Beckman rearrangements of oximes; catalytic rearrangements of alkyl anilines; synthesis of pyridine compounds; reactions of neophyl chloride with alkaline reagents. Mailing Add: Hercules Inc Research Ctr Wilmington DE 19899

WEISGERBER, GEORGE AUSTIN, b Philadelphia, Pa, Dec 31, 18; m 47; c 2. PETROLEUM CHEMISTRY. Educ: Philadelphia Col Pharm, BSc, 40; Univ Del, MSc, 50, PhD(org chem), 51. Prof Exp: Res chemist, Johnson & Johnson, NJ, 40-48; from res chemist to sr chemist, Esso Res & Eng Co, 51-60, from res assoc to sr res assoc, 60-74, SR RES ASSOC, EXXON RES & ENG CO, 74- Mem: Am Chem Soc; Tech Asn Pulp & Paper Indust; Am Soc Testing & Mat. Res: Petroleum product research; additives, burner fuels, industrial and motor lubricants; wax; asphalt. Mailing Add: Exxon Res & Eng Co PO Box 51 Linden NJ 07036

WEISGRABER, KARL HEINRICH, b Norwich, Conn, July 13, 41; m 64; c 2. ORGANIC CHEMISTRY, BIOCHEMISTRY. Educ: Univ Conn, BA, 64, PhD(org chem), 69. Prof Exp: Staff fel, Nat Inst Arthritis & Metab Dis, 69-70 & Nat Res Coun Res Assoc, USDA, Calif, 70-71; prin scientist, Meloy Labs, Va, 71-72; SR STAFF FEL, NAT HEART & LUNG INST, 72- Mem: Am Chem Soc; The Chem Soc; NY Acad Sci. Res: Experimental atherosclerosis; study of serum lipoproteins; mechanism of transport of serum constituents across aortic endothelium. Mailing Add: Nat Heart & Lung Inst Bldg 10 Bethesda MD 20014

WEISHAUPT, CLARA GERTRUDE, b Lynchburg, Ohio, July 20, 98. BOTANY. Educ: Ohio State Univ, BS, 24, MS, 32, PhD, 35. Prof Exp: Asst bot, Ohio State Univ, 32-35; from asst to assoc prof sci, Ala State Teachers Col, 35-46; instr, 46-51, cur herbarium, 49-67, from asst prof to assoc prof, 51-68, EMER ASSOC PROF BOT, OHIO STATE UNIV, 68- Mem: Am Soc Plant Taxon; Int Asn Plant Taxon. Res: Grasses of Ohio. Mailing Add: 328 N Maple Ave Fairborn OH 45324

WEISIGER, JAMES RICHARD, b Oakwood, Ill, Jan 5, 18; m 44; c 2. BIOCHEMISTRY. Educ: Univ Ill, AB, 38, MA, 39; Johns Hopkins Univ, PhD(physiol chem), 43. Prof Exp: Spec investr, Rockefeller Inst, 45, from asst to assoc, 46-56; prof assoc, Nat Acad Sci-Nat Res Coun, 56-61; sci rep, Richardson-Merrell, Inc, 61-62; METAB PROG DIR, NAT INST ARTHRITIS, METAB & DIGESTIVE DIS, 63- Concurrent Pos: Fel, Harvard Med Sch, 42-45. Mem: Am Soc Biol Chemists. Res: Organic and biological chemistry of nitrogen compounds; isolation, structure and synthesis of natural products; metabolic regulation; polyamine metabolism; pharmaceuticals; biologicals; cystic fibrosis and genetic metabolic diseases. Mailing Add: Extramural Prog Nat Inst of Arthritis Metab & Digestive Dis Bethesda MD 20014

WEISKOPF, RICHARD BRUCE, b Brooklyn, NY, Mar 7, 44; m 65; c 1. RESPIRATORY PHYSIOLOGY, ANESTHESIOLOGY. Educ: Brooklyn Col, BS, 64; Albert Einstein Col Med, MD, 68. Prof Exp: Intern surg, Beth Israel Hosp, Boston, 68-69; resident anesthesia, Univ Calif, San Francisco, 69-72; researcher physiol, US Army Res Inst Environ Med, 72-76; RES ANESTHESIOLOGIST, DIV COMBAT & EXP SURG, LETTERMAN ARMY INST RES, 76- Concurrent Pos: Fel, Cardiovasc Res Inst & Dept Anesthesia, Univ Calif, San Francisco, 70-71; instr anesthesia, Med Sch, Harvard Univ, 72-76; assoc anesthesia, Peter Bent Brigham Hosp, Boston, 72-76. Mem: Am Physiol Soc; Am Soc Anesthesiologists; Am Fedn Clin Res. Res: Respiratory physiology, especially regulation of respiration; altitude physiology. Mailing Add: Div of Combat & Exp Surg Letterman Army Inst of Res Presidio San Francisco CA 94129

WEISLEDER, DAVID, chemistry, see 12th edition

WEISLER, LEONARD, b Rochester, NY, Sept 9, 12; m 42; c 4. ORGANIC CHEMISTRY. Educ: Univ Rochester, BS, 34, MS, 36, PhD(org chem), 39. Prof Exp: Asst, Univ Rochester, 35-38; anal chemist, Distillation Prod, Inc, 42-53; ORG RES & DEVELOP CHEMIST, PAPER SERV DIV, EASTMAN KODAK CO, 53- Concurrent Pos: Sr lectr chem, Univ Rochester, 54- Mem: Am Chem Soc. Res: Alkylation of nitro compounds; chemistry and biochemistry of vitamin E; synthesis of A; fats, oils and hydrocarbons; metallurgy; photographic chemistry. Mailing Add: 6 Eastland Ave Rochester NY 14618

WEISLOW, OWEN STUART, b Cleveland, Ohio, Mar 11, 38. IMMUNOLOGY, MICROBIOLOGY. Educ: Delaware Valley Col Sci & Agr, BS, 65; Thomas Jefferson Univ, MS, 68, PhD(microbiol), 70. Prof Exp: Fel immunol, Albert Einstein Med Ctr, Philadelphia, 70-71; fel immunol & virol, Thomas Jefferson Univ, 71-73, instr, 73-74; assoc found scientist immunol, Southwest Found Res & Educ, 74-76; SCIENTIST II IMMUNOL, FREDERICK CANCER RES CTR, LITTON BIONETICS INC, 76- Mem: Am Soc Microbiol; AAAS. Res: Immunobiology of oncogenic viruses; tumor immunology. Mailing Add: Litton Bionetics Inc Frederick Cancer Res Ctr Frederick MD 21701

WEISMAN, HAROLD, b Brooklyn, NY, Oct 24, 28; m; c 2. ANESTHESIOLOGY. Educ: Univ Calif, Los Angeles, BA, 52, DO, 56, MD, 62. Prof Exp: Intern, Pac Hosp, Long Beach, Calif, 56-57; pvt pract gen med, 57-66; resident anesthesia, 66-68, asst prof in residence surg anesthesia, 69-71, ASST PROF ANESTHESIOL, SCH MED, UNIV CALIF, LOS ANGELES, 71- Concurrent Pos: Parke-Davis grant, 70; McNeil Labs grant, 71; secy-treas, Community Hosp North Hollywood, Calif, 59-61, from asst chief of staff to chief of staff, 61-63; mem consult staff, Bel Air Mem Hosp, 68, Valley Presby Hosp, Van Nuys, 68-69, Vet Admin Wadsworth Gen Hosp, 71 & Mt Sinai Hosp, Los Angeles, 72-; mem fac, Jules Stein Eye Inst. Honors & Awards: Physicians Recognition Award, AMA, 69. Mem: AMA; AAAS; Am Soc Anesthesiologists; Int Anesthesia Res Soc; fel Am Col Anesthesiologists. Res: Effect of various anesthetic agents on the oculocardiac reflex; effect of general anesthetics on intraocular pressure; retinal artery flow during anesthesia and surgery; determination of damage to retina with Doppler microwave energy output in New Zealand albino rabbits exposed chronically. Mailing Add: 4159 Adlon Pl Encino CA 91316

WEISMAN, HARVEY, b Winnipeg, Man, Feb 25, 27; m 57; c 2. HISTOLOGY, NEUROPHARMACOLOGY. Educ: Univ Man, BSc, 53, MSc, 55, PhD(histol), 69. Prof Exp: Lectr zool, 58-65, asst prof, 65-72, ASSOC PROF PHARMACOL, FAC MED, UNIV MAN, 72-, SPEC LECTR PEDIAT, 71- Mem: Can Asn Anatomists. Res: Histological, cytological and biochemical reorganization of the central nervous system in mammals as a result of psychopharmacological drugs. Mailing Add: Dept of Pharmacol & Therapeut Univ of Man Fac of Med Winnipeg MB Can

WEISMAN, IRWIN D, physics, see 12th edition

WEISMAN, ROBERT A, b Kingston, NY, Dec 16, 36; m 63; c 3. BIOCHEMISTRY. Educ: Union Univ, NY, BS, 58; Mass Inst Technol, PhD(biochem), 63. Prof Exp: Staff fel biochem, Sect Cellular Physiol, Lab Biochem, Nat Heart Inst, 63-66; asst prof biochem, Med Col Pa, 66-68; from asst prof to assoc prof, 68-74, PROF BIOCHEM, UNIV TEX HEALTH SCI CTR SAN ANTONIO, 74- Mem: Am Chem Soc; Am Soc Biol Chemists. Res: Endocytosis and differentiation mechanisms; biochemistry of protozoa. Mailing Add: Dept of Biochem Univ of Tex Health Sci Ctr San Antonio TX 78229

WEISMAN, RUSSELL, b Cleveland, Ohio, Jan 20, 22; m 47; c 4. MEDICINE, HEMATOLOGY. Educ: Western Reserve Univ, AB, 44, MD, 46. Prof Exp: Res fel med, 52-54, instr, 54-55, sr instr clin med & path, 55-57, asst prof, 57-63, ASSOC PROF MED, CASE WESTERN RESERVE UNIV, 63- Mem: AAAS; Am Soc Hemat; Am Fedn Clin Res. Res: Hemolytic anemia; immunohematology; relation of the spleen and blood destruction. Mailing Add: Dept of Med Case Western Reserve Univ Cleveland OH 44106

WEISMANN, THEODORE JAMES, b Pittsburgh, Pa, Apr 21, 30. PHYSICAL CHEMISTRY. Educ: Duquesne Univ, BS, 52, MS, 54, PhD(phys chem), 56. Prof Exp: RES DIR, GULF RES & DEVELOP CO, 56- Mem: Am Chem Soc; Geochem Soc; Am Phys Soc; Am Soc Mass Spectrometry; AAAS. Res: Boron chemistry; mass spectrometry; organic and isotopic geochemistry; marine geochemistry; molecular structure of organic radicals; magnetic susceptibilities; geochronometry. Mailing Add: 106 Short St Pittsburgh PA 15237

WEISMILLER, RICHARD A, b Elwood, Ind, Feb 23, 42; m 66; c 2. AGRONOMY. Educ: Purdue Univ, BS, 64, MS, 66; Mich State Univ, PhD(soil chem, clay mineral), 69. Prof Exp: Spec scientist soil stabilization, US Air Force Weapons Lab, 69-73; RES AGRONOMIST, PURDUE UNIV, WEST LAFAYETTE, 73- Mem: Am Soc Agron; Soil Sci Soc Am; Clay Minerals Soc; Soil Conserv Soc Am; Sigma Xi. Res: Relation of spectral reflectance of soils to their physicochemical properties; application of remote sensing to soils mapping, land use inventories and change detection as related to land use. Mailing Add: Lab Applns Remote Sensing Purdue Univ 1220 Potter Dr West Lafayette IN 47906

WEISNER, LOUIS, b New York, NY, May 1, 99; m 23; c 1. MATHEMATICS. Educ: City Col New York, BS, 20; Columbia Univ, AM, 22, PhD(math), 23. Prof Exp: Asst math, Columbia Univ, 21-23; instr, Univ Rochester, 23-26; Nat Res Coun fel, Harvard Univ, 26-27; from instr to assoc prof, Hunter Col, 27-54; prof, 55-69, EMER PROF MATH, UNIV NB, 69- Mem: AAAS; Am Math Soc; Math Asn Am; NY Acad Sci; Can Math Cong. Res: Theory of equations; algebraic variants; special functions. Mailing Add: Dept of Math Univ of NB Fredericton NB Can

WEISS, ADOLPH KURT, b Graz, Austria, Mar 22, 23; nat US; m 48; c 1. PHYSIOLOGY. Educ: Okla Baptist Univ, BS, 48; Univ Tenn, MS, 50; Univ Rochester, PhD(physiol), 53. Prof Exp: Asst zool, Univ Tenn, 48-50; from instr to assoc prof physiol, Univ Miami, 53-61; prof biol & head dept, Oklahoma City Univ, 61-64; assoc prof, 64-66, PROF PHYSIOL & BIOPHYS, COL MED, UNIV OKLA, 67- Concurrent Pos: Investr, Howard Hughes Med Inst, 55-58; vis prof, Univ of the West Indies, 60; NIH res career develop award, 64-69; pres, Int Found Study Rat Genetics & Rodent Pest Control, 74- Mem: Fel AAAS; Am Physiol Soc; Soc Exp Biol & Med; Endocrine Soc; fel Geront Soc (vpres, 70-71). Res: Aging; metabolism; hypertension; stress; thyroid function; cold exposure. Mailing Add: Dept of Physiol & Biophys Univ of Okla Health Sci Ctr Oklahoma City OK 73190

WEISS, ALAN R, analytical chemistry, see 12th edition

WEISS, ANDREW W, b Streator, Ill, Mar 13, 30; m 61; c 2. ATOMIC PHYSICS, MOLECULAR PHYSICS. Educ: Univ Detroit, BS, 52, MS, 54; Univ Chicago, PhD(atomic physics), 61. Prof Exp: PHYSICIST, NAT BUR STANDARDS, 61- Mem: AAAS; Am Phys Soc. Res: Application of electronic computers to the determination of the electronic structure and properties of atoms and simple molecules. Mailing Add: Div 222 07 Nat Bur of Standards Washington DC 20234

WEISS, ARTHUR JACOBS, b Philadelphia, Pa, Apr 11, 25; m 52; c 3. INTERNAL MEDICINE. Educ: Pa State Col, BS, 45; Univ Pa, MD, 50. Prof Exp: Instr, 57-62, ASST PROF MED, JEFFERSON MED COL, 62- Concurrent Pos: Am Heart Asn advan res fel, 57- Mem: Am Asn Cancer Res; Am Soc Hemat; Am Soc Pharmacol & Exp Therapeut; Am Col Physicians; Am Soc Clin Oncol. Res: Long-term storage of various tissues; hematology; malignant diseases; oncology. Mailing Add: Dept of Med Jefferson Med Col 1025 Walnut St Philadelphia PA 19107

WEISS, BENJAMIN, b Newark, NJ, Nov 16, 22; m 47; c 6. BIOCHEMISTRY, ORGANIC CHEMISTRY. Educ: Univ Iowa, BS, 44; Univ Ill, PhD(biochem), 49. Prof Exp: Asst inorg chem, Univ Ill, 47-49; res biochemist, Harper Hosp, 49-53; res assoc, 53-65, ASST PROF BRAIN METAB, COL PHYSICIANS & SURGEONS, COLUMBIA UNIV & MEM STAFF, NY STATE PSYCHIAT INST, 53- Mem: Am Chem Soc; Am Soc Biol Chemists. Res: Chemistry and biochemistry of long chain bases; synthesis of long chain base antimetabolites; lipid metabolism; isolation of natural products. Mailing Add: NY State Psychiat Inst 722 W 168th St New York NY 10032

WEISS, BERNARD, b New York, NY, May 27, 25; div; c 2. PSYCHOPHARMACOLOGY, TOXICOLOGY. Educ: NY Univ, BA, 49; Univ Rochester, PhD(psychol), 53. Prof Exp: Res assoc psychol, Univ Rochester, 53-54; exp & physiol psychologist, Sch Aviation Med, US Air Force, 54-56; instr med, Sch Med, Johns Hopkins Univ, 56-65, from instr to asst prof pharmacol, 59-65; assoc prof radiation biol & biophys & brain res, 65-67, PROF RADIATION BIOL & BIOPHYS, PSYCHOL & BRAIN RES, SCH MED & DENT, UNIV ROCHESTER, 67- Concurrent Pos: Mem behav pharmacol comt, NIMH, 65-67; mem comt biol effects of atmospheric pollutants, Nat Acad Sci-Nat Res Coun, 71-74; partic, US-USSR Environ Health Exchange Prog, 73- Mem: AAAS; Am Soc Pharmacol & Exp Therapeut; Am Psychol Asn; Behav Pharmacol Soc (pres, 61-64); Soc Exp Anal Behav. Res: Chemical influences on behavior. Mailing Add: Dept of Radiation Biol & Biophys Univ Rochester Sch of Med & Dent Rochester NY 14642

WEISS, CHARLES, b Poland, Oct 18, 94; nat US; m 20; c 3. IMMUNOLOGY, MEDICAL ANTHROPOLOGY. Educ: City Col New York, BS, 15; NY Univ, MS, 16; Univ Pa, PhD(immunol), 18, MD, 24. Prof Exp: Assoc biochem & exp path, Res Inst Cutaneous Med, Pa, 19-23; intern, Grad Sch Hosp, Univ Pa, 24-25; Nat Res Coun fel med, Columbia Univ, 25-26; asst prof bact, Columbia Univ & Sch Trop Med, PR, 26-28; assoc prof appl bact & immunol, Sch Med, Washington Univ, 28-32; assoc prof res med, Hooper Found, Univ Calif & lectr pediat, Sch Med, 32-44; dir labs, Jewish Hosp, 44-50; head dept microbiol, Northern Div, 50-59, EMER ASSOC CONSULT, ALBERT EINSTEIN MED CTR, 59- Concurrent Pos: Dir clin & res labs, Presby Hosp, San Juan, PR, 26-28 & Mt Zion Hosp, San Francisco, 32-44; Washington Univ traveling fel, Europe & Tunisia, 39; proj dir, AEC, Philadelphia, 57-60; grants, AMA, Nat Tuberc Asn, Commonwealth Fund, NIH, Columbia Found, AEC, Sigma Xi & Mem Found Jewish Cult. Mem: AAAS; Am Soc Microbiol; fel Am Soc Clin fel Col Am Pathologists; fel NY Acad Sci. Res: Immunology of pneumonia; ocular infections; anaerobic bacteria; tropical sprue; enzymatic studies of inflammation, tuberculosis and effects of radiation. Mailing Add: 412 W Mt Airy Ave Philadelphia PA 19119

WEISS, CHARLES, JR, b San Francisco, Calif, Dec 20, 37; m 69. SCIENCE POLICY, CHEMICAL PHYSICS. Educ: Harvard Univ, AB, 59, PhD(chem physics, biochem), 65. Prof Exp: Teaching fel, Harvard Univ, 62-64; NIH fel biophys, Lab Chem Biodyn, Lawrence Radiation Lab, Univ Calif, 67-69; biophysicist, IBM Watson Lab, Columbia Univ, 69-71; SCI & TECHNOL ADV, INT BANK RECONSTRUCT & DEVELOP, 71- Concurrent Pos: Mem corp bd, Vols Tech Assistance. Mem: AAAS; Am Phys Soc; Am Chem Soc; Soc Int Develop. Res: Science and technology in developing countries; role of multinational organizations in technological research and technology transfer; interpretation of electronic spectra of porphyrins and other large aromatic rings; primary processes of photosynthesis. Mailing Add: Int Bank for Reconstruct & Develop 1818 H St NW Washington DC 20433

WEISS, CHARLES FREDERICK, b Cohoctah, Mich, Apr 2, 21; m 47; c 3. PEDIATRICS, PHARMACOLOGY. Educ: Univ Mich, BA, 42; Vanderbilt Univ, MD, 49; Am Bd Pediat, dipl, 60. Prof Exp: Pvt pract, 54-58; med coordr clin invest in pediat & virus res, Parke, Davis & Co, Mich, 58-69; from instr to assoc prof pharm & pharmacol & assoc prof pediat, Col Med, Univ Fla, 69-73; CHIEF OF STAFF, HOPE HAVEN CHILDREN'S HOSP, 73- Concurrent Pos: Staff physician, Children's Hosp, Mich, 54-69 & Receiving Hosp, Detroit; clin asst prof pediat & commun dis, Univ Hosp, Univ Mich, Ann Arbor; med consult to dir div ment retardation, State of Fla, 69-; consult to Surgeon Gen, US Air Force; med dir & vpres, Fla Spec Olympics. Mem: Aerospace Med Asn; AMA; fel Am Acad Pediat; Am Pub Health Asn; Am Fedn Clin Res. Res: Virus and infectious diseases. Mailing Add: Hope Haven Children's Hosp 5720 Atlantic Blvd Jacksonville FL 32211

WEISS, CHARLES MANUEL, b Scranton, Pa, Dec 7, 18; m 42. ENVIRONMENTAL BIOLOGY. Educ: Rutgers Univ, BS, 39; Johns Hopkins Univ, PhD(biol), 50. Prof Exp: Asst bacteriologist, Woods Hole Oceanog Inst, 40-42, res assoc & biologist in-chg, Miami Beach Sta, 42-46, res assoc marine biol, 46-47; chemist-biologist, Baltimore Harbor Proj, Dept Sanit Eng, Johns Hopkins Univ, 47-50; basin biologist, Div Water Pollution Control, USPHS, 50-52; biologist, Sanit Chem Br, Med Labs, Army Chem Ctr, Edgewood, Md, 52-56; assoc prof sanit sci, 56-62, PROF ENVIRON BIOL, SCH PUB HEALTH, UNIV NC, CHAPEL HILL, 62-, DEP HEAD, DEPT ENVIRON SCI & ENG, 67- Concurrent Pos: Consult, Pan Am Health Orgn. Mem: Fel AAAS; Am Soc Limnol & Oceanog; Ecol Soc Am; Am Soc Microbiol; fel Am Pub Health Asn. Res: Dynamics of response of aquatic biota to environmental stress; water quality criteria and indices; aquatic contaminants; limnology of impoundments; stream pollution; marine and estuarine ecology. Mailing Add: Dept of Environ Sci & Eng Univ of NC Sch of Pub Health Chapel Hill NC 27514

WEISS, CHARLES OWEN, physical inorganic chemistry, chemical engineering, see 12th edition

WEISS, DANIEL LEIGH, b Long Branch, NJ, July 27, 23; m 51; c 3. PATHOLOGY. Educ: Columbia Univ, AB, 43, MD, 46. Prof Exp: Res asst path & exp med, Beth Israel Hosp, 50-51; res asst path, Mt Sinai Hosp, NY, 51-53; dir inst path & lab med, DC Gen Hosp, 53-63; PROF PATH, COL MED, UNIV KY, 63- Concurrent Pos: Clin prof, Georgetown Univ, 53-63, George Washington Univ, 53-63 & Howard Univ, 59-63; consult, NIH, 56- Mem: Am Soc Clin Path; Am Asn Path & Bact; Col Am Path; Am Soc Exp Path; Asn Clin Sci. Res: Experimental virus infection; primary and secondary vasculitis; lymphoma immunobiology; pathology of infection and immunity; skeletal pathology; paleopathology. Mailing Add: Dept of Path Univ of Ky Med Ctr Lexington KY 40506

WEISS, DAVID STEVEN, b Newark, NJ, Mar 3, 44; m 69. ORGANIC CHEMISTRY, PHOTOCHEMISTRY. Educ: Lehigh Univ, BS, 65; Columbia Univ, PhD(chem), 69. Prof Exp: NIH fel, Iowa State Univ, 69-72; ASST PROF CHEM, UNIV MICH, ANN ARBOR, 72- Mem: Am Chem Soc. Mailing Add: Dept of Chem Univ of Mich Ann Arbor MI 48104

WEISS, DAVID WALTER, b Vienna, Austria, July 6, 27; nat US; m 51; c 3. MICROBIOLOGY, IMMUNOLOGY. Educ: Brooklyn Col, BA, 49; Rutgers Univ, PhD(microbiol), 52. Prof Exp: Asst, Rockefeller Inst 52-55; dir med res coun tuberc unit, Oxford Univ, 56-57; from asst prof to prof bact & immunol, Univ Calif, Berkeley, 57-67, res immunologist, Cancer Res Genetics Lab, 62-67; PROF IMMUNOL & CHMN LAUTENBERG CTR GEN RES & TUMOR IMMUNOL, HADASSAH MED SCH, HEBREW UNIV, JERUSALEM, 68- Concurrent Pos: Am Cancer Soc scholar, 62-63; res prof, Miller Inst Basic Res Sci, 66-67; lectr, Univ Calif, 64-68; mem, Midwinter Conf Immunologists. Mem: Am Asn Cancer Res; Transplantation Soc; Am Asn Immunologists; NY Acad Sci; Path Soc Gt Brit & Ireland. Res: Pathogenesis and host-parasite relationships; development of nonliving vaccines; relationship of specific and nonspecific immunogenic activities of microorganisms; tumor immunology; mechanisms of immunological unresponsiveness; oncogenic viruses; antibody formation; endotoxins. Mailing Add: 20A Rehov Radak Rehavia Jerusalem Israel

WEISS, DENNIS, b New York, NY, July 2, 40; m 65; c 2. MICROPALEONTOLOGY, ENVIRONMENTAL GEOLOGY. Educ: City Col New York, BS, 63; NY Univ, MS, 67, PhD(geol), 71. Prof Exp: Lectr geol, 64-71, ASST PROF GEOL, CITY COL NEW YORK, 71- Mem: AAAS; Geol Soc Am; Soc Econ Paleontologists & Mineralogists; Am Asn Stratig Palynologists; Nat Asn Geol Teachers. Res: Quaternary paleo-environments. Mailing Add: Dept Earth & Planetary Sci City Col of New York NY 10031

WEISS, DOUGLAS EUGENE, b Aurora, Ill, July 28, 45. ORGANIC POLYMER CHEMISTRY. Educ: Univ Kans, BS, 68; Univ Nebr, PhD(org chem), 72. Prof Exp: Sr res assoc polymers, Univ E Anglia, 72-73; res assoc org chem, Univ Nebr, 73-74; CHEMIST, ELASTOMER CHEM DEPT, E I DU PONT DE NEMOURS & CO, 74- Mem: Am Chem Soc. Res: Synthesis and development of polyurethane casting systems and carbon-13 nuclear magnetic resonance of polymers. Mailing Add: Elastomer Chem Dept Exp Sta E I du Pont de Nemours & Co Wilmington DE 19809

WEISS, EARLE BURTON, b Waltham, Mass, Nov 23, 32; m 63; c 2. PULMONARY PHYSIOLOGY. Educ: Northeastern Univ, BS, 55; Mass Inst Technol, MS, 57; Albert Einstein Col Med, MD, 61; Am Bd Internal Med, dipl, 69. Prof Exp: Nat Heart Inst fel, Tufts Lung Sta, Boston City Hosp, 64-66; from instr to asst prof, 66-70, ASSOC PROF MED, SCH MED, TUFTS UNIV, 70-; ASSOC PROF MED, MED SCH, UNIV MASS, 71-; DIR DEPT RESPIRATORY DIS, ST VINCENT HOSP, 71- Concurrent Pos: Assoc, Tufts Lung Sta, Boston City Hosp, 66-71, physician chg, Pulmonary Function & Physiol Sect, Hosp, 66-71 & Cent Arterial Blood Gas Lab, 67-71, assisting physician, Tufts Med Serv, 67-70, assoc dir, 69-70, dir respiratory intensive care unit, 70-71; consult physiol, Norfolk County Sanatorium, Mass, 66-69; mem toxicol info prog, Nat Libr Med, 70-; tuberc consult, Mass Dept Pub Health, 72- Mem: AAAS; Am Fedn Clin Res; fel Am Col Physicians; fel Am Col Chest Physicians; Am Soc Internal Med. Mailing Add: Dept of Respiratory Dis St Vincent Hosp 25 Winthrop St Worcester MA 01610

WEISS, EDWARD LEONHARDT, b Washington, Pa, Apr 10, 25; m 48; c 4. PHYSICS, METROLOGY. Educ: Thiel Col, BS, 50. Prof Exp: PRIN SCIENTIST PHYSICS, ADVAN BUS DEVELOP, LEEDS & NORTHRUP CO, 51- Mem: Am Optical Soc Am; Instrument Soc Am; Am Soc Mech Engrs. Res: Use of optical transform measurements of scattered laser light for particle size distribution measurement. Mailing Add: Tech Ctr Leeds & Northrup Co North Wales PA 19454

WEISS, EDWARD SEBASTIAN, b New York, NY, June 11, 16; m 39; c 3. BIOSTATISTICS. Educ: Brooklyn Col, BA, 36; Univ Mich, MS, 38. Prof Exp: Dir health res projs, Works Progress Admin, Mich, 39-42; statistician, US Army Respiratory Dis Comn, 43; biostatistician, Div Tuberc Control, USPHS, 46-50, chief biomet & epidemiol, Arctic Health Ctr, 50-53, biostatistician, Div Occup Health, 53-57, from biostatistician, Div Radiol Health to dep chief populations studies prog, Nat Ctr Radiol Health, 58-68; asst prof community health, Univ Mo-Columbia, 68-71; RETIRED. Concurrent Pos: Lectr chronic dis, Sch Hyg & Pub Health, Johns Hopkins Univ, 61-; govt & acad consult, 71- Mem: Fel AAAS; fel Am Pub Health Asn; Am Statist Asn. Res: Bioassay; epidemiology. Mailing Add: 5510 Phelps Luck Dr Columbia MD 21045

WEISS, EDWIN, b Brooklyn, NY, Aug 23, 27; m 52; c 2. ALGEBRA. Educ: Brooklyn Col, BS, 48; Mass Inst Technol, PhD, 53. Prof Exp: Instr math, Univ Mich, 53-54; NSF fel, Inst Advan Study, 54-55; Benjamin Peirce instr, Harvard Univ, 55-58; asst prof, Univ Calif, Los Angeles, 58-59; mem staff, Lincoln Lab, Mass Inst Technol, 59-65; PROF MATH, BOSTON UNIV, 65- Mem: Am Math Soc. Mailing Add: 16 Warwick Rd Brookline MA 02146

WEISS, EMILIO, b Pakrac, Yugoslavia, Oct 4, 18; nat US; m 43; c 2. MEDICAL MICROBIOLOGY. Educ: Univ Kans, AB, 41; Univ Chicago, MS, 42, PhD(bact), 48. Prof Exp: Asst histol, parasitol & bact, Univ Chicago, 42 & 47-48; res assoc, 48-50; instr bact, Loyola Univ, Ill, 47-48; asst prof, Ind Univ, 50-53; chief virol br, Chem Corps Biol Lab, US Dept Army, Ft Detrick, Md, 53-54; asst head virol div, 54-63, dep dir microbiol dept, 63-72, dir, 72-74, CHMN MICROBIOL DEPT, NAVAL MED RES INST, NAT NAVAL MED CTR, 74- Mem: Am Soc Microbiol; Soc Exp Biol & Med; Am Asn Immunologists; Am Acad Microbiol. Res: Biochemical properties of rickettsiae; microbial physiology. Mailing Add: Dept of Microbiol Naval Med Res Inst Bethesda MD 20014

WEISS, FRED TOBY, b Oakland, Calif, July 24, 16; m 40; c 3. ANALYTICAL CHEMISTRY, ENVIRONMENTAL CHEMISTRY. Educ: Univ Calif, Los Angeles, BS, 38; Harvard Univ, MS, 40, PhD(chem), 41. Prof Exp: Res chemist, Shell Oil Co, Ill, 41-43, group leader fuels res, 43-46, res chemist, Shell Develop Co, Calif, 46-52, res supvr, 52-72, staff res chemist, Bellaire Res Ctr, Tex, 72-75, SR STAFF RES CHEMIST, BELLAIRE RES CTR, SHELL DEVELOP CO, TEX, 75- Concurrent Pos: Mem adv comt, NSF-Res Appl to Nat Needs Study of Petrol Indust in Del Estuary, 74- Mem: AAAS; Am Chem Soc; Am Soc Testing & Mat. Res: Characterization and analysis of organic structures; analytical methods for process control; combined use of chemical and physical methods for analysis of organic compounds; environmental analysis; development of analytical methods for studying fate and effects of petroleum in marine environments. Mailing Add: Shell Dev Co Bellaire Res Ctr PO Box 481 Houston TX 77001

WEISS, GARY BRUCE, b New York, NY, Oct 5, 44; m 73. HEMATOLOGY, BIOCHEMISTRY. Educ: NY Univ, BA, 65, MD, 71, PhD(biochem), 72. Prof Exp: Intern med, Sch Med, Univ Calif, San Francisco, 71-72, resident, 72-73; res assoc hemat, Nat Heart & Lung Inst, Bethesda, Md, 73-76; HEMAT FEL, UNIV WASH, 76- Mem: AAAS; Asn Comput Mach; Math Asn Am; Biomet Soc; Am Statist Asn. Res: Mechanism of action of RNA-directed DNA polymerase. Mailing Add: Dept of Hemat Univ of Wash Seattle WA 98195

WEISS, GEORGE B, b Plainfield, NJ, Apr 29, 35; m 60; c 2. PHARMACOLOGY. Educ: Princeton Univ, AB, 57; Vanderbilt Univ, PhD(pharmacol), 62. Prof Exp: USPHS res fels pharmacol, Vanderbilt Univ, 62 & Univ Pa, 62-64; from asst prof to assoc prof pharmacol, Med Col Va, 64-70; assoc prof, 70-74, PROF PHARMACOL, UNIV TEX HEALTH SCI CTR DALLAS, 74- Mem: Biophys Soc; Soc Neurosci; Am Heart Asn; Am Soc Pharmacol & Exp Therapeut; Am Physiol Soc. Res: Cellular pharmacology; actions of drugs on membrane permeability to ions; excitation-contraction coupling and receptors in smooth and striated muscle; actions of lanthanum; nicotine; drugs and ions in brain area slices. Mailing Add: Dept of Pharmacol Univ of Tex Health Sci Ctr Dallas TX 75235

WEISS, GEORGE HERBERT, b New York, NY, Feb 19, 30; m 61; c 3. APPLIED MATHEMATICS. Educ: Columbia Univ, AB, 51; Univ Md, MA, 53, PhD, 58. Prof Exp: Physicist, US Naval Ord Lab, 51-54; math asst, Ballistic Res Lab, Aberdeen Proving Ground, US Army, 54-56; asst math, Inst Fluid Dynamics & Appl Math, Univ Md, 56-58, res assoc, 59-60, res asst prof, 60-63; Weizmann fel, Weizmann Inst Sci, 58-59; NIH study grant, Rockefeller Inst, 63-64; study grant, 64-67, CHIEF PHYS SCI LAB, DIV COMPUT RES & TECHNOL, NIH, 67- Concurrent Pos: Physicist, US Naval Ord Lab, 56-61; consult, Gen Motors Corp, 60-64; consult, IBM Corp, 64; Fulbright sr fel, Imp Col, Univ London, 68-69; assoc ed, Transp Res, 73- Honors & Awards: Wash Acad Sci Award, 66; Superior Serv Award, NIH, 70. Mem: Opers Res Soc Am. Res: Reliability; traffic theory; statistical mechanics; biometry; biochemical separation techniques, stochastic processes. Mailing Add: 1105 N Belgrade Rd Silver Spring MD 20902

WEISS, GEORGE RAYMOND, colloid chemistry, photochemistry, see 12th edition

WEISS, GERALD, b New York, NY, June 29, 32; m 67. ANTHROPOLOGY. Educ: Columbia Univ, BA, 53; Univ Mich, MA, 57, PhD(anthrop), 69. Prof Exp: Instr anthrop, Wayne State Univ, 60; asst prof, 64-72, ASSOC PROF ANTHROP, FLA ATLANTIC UNIV, 72- Mem: AAAS; fel Am Anthrop Asn; fel Royal Anthrop Inst. Res: Culture theory; ethnology; South America; Peruvian Montana. Mailing Add: Dept of Anthrop Fla Atlantic Univ Boca Raton FL 33432

WEISS, GERALD S, b Boyertown, Pa, July 26, 34; m 58; c 3. INORGANIC CHEMISTRY. Educ: Drexel Inst, BS, 57; Univ Pa, PhD(inorg chem, anal chem), 65. Prof Exp: Res chemist, Rohm & Haas Co, 57-59; instr chem, Drexel Inst, 59-65, asst prof, 65-67; assoc prof, 67-69, PROF CHEM, MILLERSVILLE STATE COL, 69-, CHMN DEPT, 75- Mem: Am Chem Soc. Res: Synthesis of volatile hydrides of group IV elements; infrared spectroscopic analysis of small molecules; heavy metal complex ion synthesis and spectroscopic analysis. Mailing Add: 6 Quarry Dr Millersville PA 17551

WEISS, GERSON, b New York, NY, Aug 1, 39; m 59; c 4. REPRODUCTIVE ENDOCRINOLOGY. Educ: NY Univ, BA, 60, MD, 64; Am Bd Obstet & Gynec, dipl, 71, cert reproductive endocrinol, 74. Prof Exp: Instr, 69-71, ASST PROF OBSTET-GYNEC, SCH MED, NY UNIV, 71- Concurrent Pos: Fel reproductive endocrinol, Sch Med, Univ Pittsburgh, 71-73; John Polachek Found Med Res grant, 75; Nat Inst Child Health & Human Develop res grant, 75. Mem: Endocrine Soc; Am Fertil Soc; Soc Study Reproduction. Res: Control and function of the pregnancy and postpartum corpus luteum by means of measuring prolactin binding and steroid production; control of pituitary gonadotropin secretion. Mailing Add: Dept of Obstet-Gynec NY Univ Med Ctr New York NY 10016

WEISS, GUIDO LEOPOLD, b Trieste, Italy, Dec 29, 28; nat US; m 50. MATHEMATICS. Educ: Univ Chicago, PhB, 49, MS, 51, PhD, 56. Prof Exp: From instr to assoc prof math, DePaul Univ, 55-60; chmn dept, 67-70, PROF MATH, WASH UNIV, 66- Honors & Awards: Chauvenet Prize, Math Asn Am, 67. Mem: Am Math Soc; Math Asn Am. Res: Harmonic analysis; complex and real variables. Mailing Add: Dept of Math Wash Univ St Louis MO 63130

WEISS, HAROLD GILBERT, b Perth Amboy, NJ, Feb 6, 23; m 44; c 2. INORGANIC CHEMISTRY, PHYSICAL CHEMISTRY. Educ: Univ Calif, Los Angeles, BS, 48. Prof Exp: Res chemist, US Naval Ord Test Sta, 48-52; supvr catalysis res, Olin Mathieson Chem Corp, 52-59; lab mgr, Nat Eng Sci Co, 59-61; dir chem, Dynamic Sci Corp, 61-66; PRES, WEST COAST TECH SERV, INC, 66- Mem: Am Chem Soc; Am Inst Chemists; Inst Environ Sci. Res: Boron hydrides; organoboranes; surface chemistry and catalysis; analytical chemistry; mass and infrared spectrometry; high vacuum technology; combustion-fire research; environmental control; analytical instrumentation. Mailing Add: 4016 Montego Dr Huntington Beach CA 92649

WEISS, HAROLD SAMUEL, b New York, NY, Sept 10, 22; m 49; c 4. PHYSIOLOGY. Educ: Rutgers Univ, BSc, 47, MSc, 49, PhD(physiol), 50. Prof Exp: From instr to assoc prof avian physiol, Rutgers Univ, 50-62; assoc prof, 62-67, PROF PHYSIOL, COL MED, OHIO STATE UNIV, 67- Mem: Am Physiol Soc; Poultry Sci Asn; Soc Exp Biol & Med; Aerospace Med Asn; Undersea Med Soc. Res: Cardiopulmonary; blood pressure; atherosclerosis; lung mechanics; environmental physiology; acceleration; temperature control; gaseous environment and respiratory disease. Mailing Add: Dept of Physiol Ohio State Univ Col of Med Columbus OH 43210

WEISS, HARVEY JEROME, b New York, NY, June 30, 29; m 57; c 2. INTERNAL MEDICINE. Educ: Harvard Univ, AB, 51, MD, 55. Prof Exp: Intern, Bellevue Hosp, Columbia Univ, 55-56; resident med, Manhattan Vet Admin Hosp, New York, 56-58; Dazian fel hemat, Mt Sinai Hosp, New York, 58-59; instr, Sch Med, NY Univ, 62-64; asst attend hematologist, Mt Sinai Hosp, 65-69; assoc clin prof, 69-71, assoc prof, 72-74, PROF MED, COL PHYSICIANS & SURGEONS, COLUMBIA UNIV, 75- Concurrent Pos: Consult, Walter Reed Army Med Ctr, 65-70; asst prof med, Mt Sinai Sch Med, 66-69; dir div hemat, Roosevelt Hosp, New York, 69- Mem: Soc Exp Biol & Med; Am Physiol Soc; Am Soc Clin Invest; Am Soc Hemat; Am Fedn Clin Res. Res: Hematology; blood coagulation; disorders of hemostasis; platelet physiology. Mailing Add: Dept of Med Roosevelt Hosp 428 W 59th St New York NY 10019

WEISS, HARVEY RICHARD, b New York, NY, May 13, 43; m 66; c 2. PHYSIOLOGY, PHARMACOLOGY. Educ: City Col New York, BS, 65; Duke Univ, PhD(physiol), 69. Prof Exp: Warner-Lambert joint fel pharmacol, Warner-Lambert Res Inst & Col Physicians & Surgeons, Columbia Univ, 69-71; ASST PROF PHYSIOL, RUTGERS MED SCH, COL MED & DENT NJ, 71- Mem: AAAS; Am Physiol Soc. Res: Physiologic and pharmacologic control of tissue oxygen tension and flow; coronary circulation; control of respiratory-cardiovascular interactions. Mailing Add: Dept of Physiol Rutgers Med Sch Piscataway NJ 08854

WEISS, HERBERT V, b Brooklyn, NY, Nov 16, 21; m 55; c 2. CHEMISTRY. Educ: NY Univ, BA, 42, MS, 49; Univ Cincinnati, PhD(biochem), 52. Prof Exp: Toxicologist, Chief Med Exam Lab, New York, 47-49; fel physiol, Sch Med, Univ Rochester, 52-53; instr indust med, Post-grad Med Sch, NY Univ, 53-56; supvry radiochemist, US Naval Radiol Defense Lab, San Francisco, 56-69; RES CHEMIST, NAVAL UNDERSEA CTR, 69- Concurrent Pos: Adj prof, Calif State Univ, San Diego. Mem: AAAS; Am Chem Soc. Res: Nuclear chemistry; analytical chemistry; industrial hygiene and toxicology; environmental chemistry. Mailing Add: Naval Undersea Ctr San Diego CA 92132

WEISS, IRA PAUL, b New York, NY, Feb 27, 42; m 67; c 2. NEUROSCIENCES. Educ: City Col New York, BS, 65; Syracuse Univ, PhD(physiol, psychol), 69. Prof Exp: Instr psychol, Onondaga Community Col, 68; NIH fel neurophysiol, Callier Hearing & Speech Ctr, Dallas, Tex, 69-70, NIH spec fel neurophysiol & consult res design, 70-72; RES PSYCHOLOGIST NEUROSCI, CHILDRENS HOSP NAT MED CTR, 72-, VCHMN DEPT PSYCHOL, 75- Concurrent Pos: Asst prof lectr, Sch Med, George Washington Univ, 72- Mem: Acoustical Soc Am; assoc Am Physiol Soc; Soc Neurosci; Soc Psychophysiol Res; AAAS. Res: Relationships of cortical auditory and visual evoked potentials to developments, central nervous system function, neurological and sensory disorders and to behavior. Mailing Add: Childrens Hosp Nat Med Ctr EEG Res Lab 2124 13th St NW Washington DC 20009

WEISS, IRMA TUCK, b New York, NY, Aug 5, 13; m 38; c 2. BIOCHEMISTRY, ORGANIC CHEMISTRY. Educ: NY Univ, BS, 33, MS, 36, PhD, 42. Prof Exp: Asst instr chem, Wash Sq Col, 40-42, from instr to asst prof biochem, Col Dent, 43-64, ASSOC PROF BIOCHEM, COL DENT, NY UNIV, 64- Mem: Sigma Xi (secy, Sci Res Soc Am, 59-60, treas, 60-61, vpres, 61-62, pres, 62-63); Am Chem Soc. Res: Organic synthesis; application of biochemistry in clinical dentistry; electrophoresis studies of salivary proteins. Mailing Add: 401 First Ave New York NY 10010

WEISS, IRVING, b New York, Apr 10, 19; m 44; c 3. MATHEMATICAL STATISTICS. Educ: Univ Mich, BS, 41; Columbia Univ, MA, 48; Stanford Univ, PhD(statist), 55. Prof Exp: Res asst statist, Stanford Univ, 51-55; instr math, Lehigh Univ, 55-56; tech staff statistician, Bell Tel Labs, 56-59 & Mitre Corp, 59-62; ASSOC PROF MATH, UNIV COLO, BOULDER, 62- Concurrent Pos: Consult, State Dept Employ, Colo, 63, Beech Aircraft Corp, 63-64, dept biol, Univ Rochester, 64-65, Nat Ctr Atmospheric Res, 64-72 & 75, Behav Res & Eval Corp, 73-74 & US Environ Protection Agency, 75. Mem: Inst Math Statist; Am Statist Asn. Res: Probability; applied probability theory; statistical inference; stochastic processes; control charts; tolerance intervals. Mailing Add: Dept of Math Univ of Colo Boulder CO 80302

WEISS, JACK ALLAN, b Chicago, Ill, June 25, 02; m 46; c 1. SURGERY. Educ: Univ Chicago, BS, 22, MD, 24; Am Bd Ophthal & Otolaryngol, dipl, 31. Prof Exp: Intern, Michael Reese Hosp, 24-26; assoc attend otolaryngologist, Ill Eye & Ear Infirmary, 33-46; assoc clin prof, Chicago Med Sch, 48-58, prof otolaryngol & chmn dept, 58-70. Concurrent Pos: Mem attend staff, Cook County Hosp, 48-70, chmn dept otolaryngol, 60-70; attend otolaryngologist, Michael Reese Hosp, 50- Mem: Am Acad Facial, Plastic & Reconstructive Surg; Am Laryngol, Rhinol & Otol Soc; AMA; Asn Am Med Cols; Am Acad Ophthal & Otolaryngol. Res: Otolaryngology and plastic surgery; medical education. Mailing Add: 3600 N Lake Shore Dr Chicago IL 60613

WEISS, JAMES ALLYN, b West Bend, Wis, Apr 16, 43; m 71. ORGANIC CHEMISTRY. Educ: Univ Wis-Madison, BS, 65; Pa State Univ, University Park, PhD(chem), 71. Prof Exp: ASST PROF CHEM, PA STATE UNIV, WORTHINGTON SCRANTON CAMPUS, 71- Mem: AAAS; Am Chem Soc. Res: Organic synthesis; isolation and structural elucidation of natural products; stereochemistry of organic molecules. Mailing Add: Dept of Chem Pa State Univ Worthington Scranton Campus Dunmore PA 18512

WEISS, JAMES MOSES AARON, b St Paul, Minn, Oct 22, 21; m 46; c 2. PSYCHIATRY. Educ: Univ Minn, AB, 41, BS, 47, MB, 49, MD, 50; Yale Univ, MPH, 51; Am Bd Psychiat & Neurol, dipl, 57. Prof Exp: Asst psychol, Col St Thomas, 41-42; intern med, USPHS Hosp, Seattle, 49-50; from instr to asst prof psychiat, Sch Med, Wash Univ, 54-60; assoc prof & actg chmn dept, 60-61, PROF PSYCHIAT & CHMN DEPT, SCH MED, UNIV MO-COLUMBIA, 61-, PROF COMMUNITY HEALTH & MED PRACT, 71- Concurrent Pos: Clin fel psychiat, Sch Med, Yale Univ, 51-53; fac fel, Inter-Univ Coun Inst Social Geront, Univ Conn, 58; dir training, Malcolm Bliss Ment Health Ctr, City of St Louis, 54-60, dir psychiat clin, 54-59, dir div community psychiat serv, 58-59; vis psychiatrist, St Louis City Hosps, Barnes & Affil Hosps & Wash Univ Clins, 54-60; consult to state, fed & nat agencies & orgn, 60-; vis prof, Inst Criminol, Cambridge Univ, 68-69. Mem: Fel AAAS; fel Am Psychiat Asn; fel Am Pub Health Asn; hon mem Asn Mil Surgeons US; AMA. Res: Social psychiatry and gerontology; psychiatric problems of aging; suicide; homicide; antisocial behavior. Mailing Add: Dept of Psychiat Univ of Mo Columbia MO 65201

WEISS, JAMES OWEN, b Memphis, Tenn, Sept 25, 31; m 55; c 3. ORGANIC CHEMISTRY, POLYMER CHEMISTRY. Educ: Duke Univ, BS, 52; Univ Va, MS, 54, PhD(org chem), 57. Prof Exp: Res chemist, Shell Oil Co, 57-58 & E I du Pont de Nemours & Co, 58-61; res chemist, Chemstrand Res Ctr, 61-66; res mgr fiber develop, Beaunit Fibers, 66-68; group leader fiber develop, Celanese Fibers Co, 68-72; CHEM DEVELOP MGR, HOECHST FIBERS INDUSTS, 73- Mem: Am Chem Soc; Fibers Soc. Res: Polyester fiber research and development; nylon, high temperature resistant fiber, spandex fiber, polypropylene fiber and vinyl polymer and fiber research and development. Mailing Add: Hoechst Fibers Industs PO Box 5887 Spartanburg SC 29304

WEISS, JAMES PAUL, b Marion, La, Jan 10, 15; m 41; c 3. OPTICS. Educ: La State Normal Col, BA, 34; La State Univ, MS, 36; Univ Rochester, PhD(optics), 40. Prof Exp: Physicist & supvr res group, 40-62, RES FEL, PHOTO PROD DEPT, E I DU PONT DE NEMOURS & CO, 62- Mem: Optical Soc Am; Soc Photog Scientists & Engrs. Res: Photographic sensitometry; color photography; image quality; x-ray intensifying screens. Mailing Add: Photo Prod Dept E I du Pont de Nemours & Co Parlin NJ 08859

WEISS, JAY M, b Jersey City, NJ, Mar 20, 41; m 63; c 2. PSYCHOPHYSIOLOGY. Educ: Lafayette Col, BA, 62; Yale Univ, PhD(psychol), 67. Prof Exp: USPHS fel & guest investr, 67-68, asst prof, 69-73, ASSOC PROF PHYSIOL PSYCHOL, ROCKEFELLER UNIV, 73- Res: Psychological factors influencing physiological effects of stress; psychosomatic disorders; motivation. Mailing Add: Rockefeller Univ 66th St & York Ave New York NY 10021

WEISS, JERALD AUBREY, b Cleveland, Ohio, June 9, 22; m 49; c 2. MICROWAVE PHYSICS. Educ: Ohio State Univ, PhD(physics), 53. Prof Exp: Instr math, Univ Wyo, 49-51; mem tech staff magnetic mat, Bell Tel Labs, Inc, 53-61; vpres, Hyletronics Corp, Mass, 61-62; from asst prof to assoc prof, 62-66, PROF PHYSICS, WORCESTER POLYTECH INST, 66- Concurrent Pos: Consult, US Army Natick Develop Ctr, 75 & Lincoln Lab, Mass Inst Technol. Mem: Am Phys Soc; sr mem Inst Elec & Electronics Engrs; AAAS. Res: Theory of atomic and molecular structure; magnetic materials and electromagnetic interactions, theory and applications; magnetic resonance spectroscopy and microwave applications. Mailing Add: Dept of Physics Worcester Polytech Inst Worcester MA 01609

WEISS, JEROME, b Brooklyn, NY, Aug 27, 22; m 43; c 2. PHYSICAL CHEMISTRY. Educ: Cornell Univ, BA, 48; Ind Univ, PhD(phys chem), 51. Prof Exp: Asst chem, Ind Univ, 48-50, instr, Exten, 50; CHEMIST, BROOKHAVEN NAT LAB, 51- Concurrent Pos: Lectr, Hofstra Col, 55; Dewar res fel, Univ Edinburgh, 57; consult, Am Soc Testing & Mat. Mem: AAAS; Radiation Res Soc; The Chem Soc; NY Acad Sci; fel Am Inst Chemists. Res: Organic radiation chemistry; chemical dosimetry; health physics; membrane transport. Mailing Add: 17 Locust Ave Stony Brook NY 11790

WEISS, JONAS, b New York, NY, Feb 17, 34; m 59; c 3. POLYMER CHEMISTRY. Educ: City Col New York, BS, 55; NY Univ, PhD(phys chem), 62. Prof Exp: Chemist, Esso Res & Eng Co, 62-64; supvr org & polymer chem, Am Standard Co, 64-70; group leader polymer chem, Nat Patent Develop Corp, 70-74; GROUP LEADER POLYMER APPLNS, CIBA-GEIGY CORP, 74- Mem: Am Chem Soc; The Chem Soc. Res: Polymers and plastics for coatings, sealants, adhesives, binders; permeability and rate of release of polymers; polymers and plastics for building materials; indicators and controls; semi-permeable membranes; ion-exchange polymers. Mailing Add: Ciba-Geigy Corp Plastics & Additives Res Div Saw Mill River Rd Ardsley NY 10502

WEISS, JONATHAN DAVID, b New York, NY, Mar 3, 46; m 72. EXPERIMENTAL SOLID STATE PHYSICS. Educ: Queens Col, BS, 66; Univ Ill, Urbana, MA, 68, PhD(physics), 73. Prof Exp: Res assoc mat sci, Cornell Univ, 73-75; RES ASSOC PHYSICS, BRIGHAM YOUNG UNIV, 75- Mem: Am Phys Soc. Res: Diffusion of tin in lead to about 25 kilobars for the purpose of comparing the results with a prediction of a theory based upon the diffusion of three types of impurity defects in lead simultaneously. Mailing Add: Dept of Physics & Astron Brigham Young Univ Provo UT 84602

WEISS, JOSEPH FRANKLIN, inorganic chemistry, see 12th edition

WEISS, JOSEPH JACOB, b Detroit, Mich, Mar 22, 34; m 68; c 2. RHEUMATOLOGY. Educ: Univ Mich, BA, 55, MD, 61. Prof Exp: STAFF PHYSICIAN, WAYNE COUNTY GEN HOSP, 70-; STAFF PHYSICIAN, MED SCH, UNIV MICH, 71- Concurrent Pos: Rheumatol consult, Vet Admin Hosp, Ann Arbor, Mich, 75- Mem: Fel Am Col Physicians; Am Soc Clin Res; Am Fedn Clin Res. Res: Investigation of the etiology of frozen shoulder; use of arthrography to delineate the cause and assist the diagnosis of shoulder pain and immobility. Mailing Add: Dept of Med Wayne County Gen Hosp Eloise MI 48132

WEISS, KARL H, b Hamburg, Ger, June 21, 26; nat US; m 48; c 2. PHYSICAL CHEMISTRY. Educ: Columbia Univ, BS, 51; NY Univ, PhD(chem), 57. Prof Exp: Res chemist, Color Res Corp, 47-50, tech adminr, 50-54; instr chem, NY Univ, 56-59, asst prof, 59-61; from asst prof to assoc prof, 61-65, PROF CHEM, NORTHEASTERN UNIV, 65-, CHMN DEPT, 69- Concurrent Pos: NSF sr fel, Quantum Chem Group, Univ Uppsala, 68-69. Mem: Am Chem Soc; NY Acad Sci; The Chem Soc. Res: Photochemistry of complex molecules; laser photochemistry; quantum chemistry; charge transfer interaction. Mailing Add: Dept of Chem Northeastern Univ Boston MA 02115

WEISS, KENNETH MONRAD, b Cleveland, Ohio, Nov 29, 41; m 65; c 1. POPULATION BIOLOGY, BIOLOGICAL ANTHROPOLOGY. Educ: Oberlin Col, BA, 63; Univ Mich, MA, 69, PhD(biol anthrop), 72. Prof Exp: Res assoc human genetics, Med Sch, Univ Mich, 72-73; ASST PROF DEMOG & POP GENETICS, UNIV TEX GRAD SCH BIOMED SCI, HOUSTON, 73- Concurrent Pos: Asst prof, Univ Tex Sch Pub Health, Houston, 74- Honors & Awards: Juan Comas Award, Am Asn Phys Anthropologists, 72. Mem: AAAS; Am Asn Phys Anthropologists; Am Anthrop Asn; Human Biol Coun; Pop Asn Am. Res: Demographic evolution of human populations; genetics of the evolution of social behavior in humans and other animals; demographic genetic epidemiology of degenerative diseases; human evolution and biological anthropology. Mailing Add: Ctr Demog & Pop Genetics Univ Tex Health Sci Ctr Houston TX 77030

WEISS, LAWRENCE R, b Brooklyn, NY, Feb 13, 31; m 59; c 2. PHARMACOLOGY, TOXICOLOGY. Educ: Univ Fla, BSc, 53; Ohio State Univ, MSc, 59, PhD(pharmacol), 62. Prof Exp: Pharmacologist & toxicologist, Pesticide Sect, Div Toxicol Eval, Bur Sci, 63-69, toxicologist, Div Pharmacol & Toxicol, 69-72, PROJ LEADER-GROUP SUPVR DRUG PHARMACOL & TOXICOL, DIV DRUG BIOL, FOOD & DRUG ADMIN, 72- Mem: AAAS; NY Acad Sci; Am Soc Pharmacol & Exp Therapeut; Soc Toxicol; Behav Pharmacol Soc. Res: Neuro- and psychopharmacological effects of drugs and pesticides; behavioral toxicology; potentiation of anticholinesterases, chlorinated hydrocarbons and dithiocarbamates by drugs, especially central nervous systems and cardiovascular effects; drug induced cardiomyopathies; abusive drugs; drug interactions. Mailing Add: Div of Drug Biol Food & Drug Admin Washington DC 20204

WEISS, LEON, b Brooklyn, NY, Oct 4, 25; m 49; c 6. ANATOMY. Educ: Long Island Col Med, MD, 48. Prof Exp: Intern med, Maimonides Hosp, Brooklyn, NY, 48-49, asst resident, 49-50; instr med, Col Med, State Univ NY Downstate Med Ctr, 52-53; lectr, Grad Sch, Univ Md, 54-55; assoc anat, Harvard Med Sch, 55-57, asst prof, 57-61; assoc prof, 61-66, PROF ANAT, SCH MED, JOHNS HOPKINS UNIV, 66- Concurrent Pos: USPHS res fel anat, Harvard Med Sch, 50-52; mem med mission to establish hemat lab, Nat Defense Med Ctr, Formosa, 55. Mem: Electron Micros Soc Am; Histochem Soc; Am Asn Anatomists; Tissue Cult Asn. Res: Microscopic anatomy; electron microscopy; histochemistry; tissue culture; connective tissues; reticuloendothelial system; histophysiology of the lympho-hematopoietic system. Mailing Add: Dept of Anat Johns Hopkins Univ Sch of Med Baltimore MD 21205

WEISS, LEONARD, b London, Eng, June 15, 28; m 51; c 3. PATHOLOGY, CELL BIOLOGY. Educ: Cambridge Univ, BA, 50, MA, MB, BChir, 53, MD, 58, PhD(biol), 63, ScD, 71. Prof Exp: House physician, Westminster Hosp, Univ London, 54-55, resident pathologist, Hosp & res assoc & registr morbid anat, Med Sch, Univ, 55-58; mem sci staff, Med Res Coun, Nat Inst Med Res, London, 58-60 & Strangeways Res Lab, Cambridge Univ, 60-64; DIR CANCER RES, DEPT EXP PATH, ROSWELL PARK MEM INST, 64-; RES PROF DERMAT, MED SCH, STATE UNIV NY BUFFALO, 74- Concurrent Pos: Res prof biophys, Roswell Park Grad Div, State Univ NY Buffalo, 65-74. Mem: Am Soc Exp Path; Path Soc Gt Brit & Ireland; Royal Col Path; fel Brit Inst Biol; Int Soc Cell Biol. Res: Radiation sensitivity of hypothermic mammal; biophysics of cell contact phenomena. Mailing Add: Dept of Exp Path Roswell Park Mem Inst Buffalo NY 14263

WEISS, LIONEL EDWARD, b London, Eng, Dec 11, 27; m 64; c 2. GEOLOGY. Educ: Univ Birmingham, BSc, 49, PhD, 53; Univ Edinburgh, ScD, 56. Prof Exp: Res assoc, 51-53, assoc, 56-57, from asst prof to assoc prof, 57-64, Miller res prof, 65-67, PROF GEOL, UNIV CALIF, BERKELEY, 64- Concurrent Pos: Guggenheim fel, 61 & 69. Mem: Geol Soc Am. Res: Natural and experimental deformation of rocks and minerals. Mailing Add: Dept of Geol & Geophys Univ of Calif Berkeley CA 94720

WEISS, LIONEL IRA, b New York, NY, Sept 5, 23; m 46; c 3. MATHEMATICAL STATISTICS. Educ: Columbia Univ, BA, 43, MA, 45, PhD(math statist), 53. Prof Exp: From asst prof to assoc prof statist, Univ Va, 49-56; assoc prof math, Univ Ore, 56-57; assoc prof, 57-61, PROF OPERS RES, CORNELL UNIV, 61- Concurrent Pos: Mem, Nat Res Coun, 66-69. Mem: Inst Math Statist; NY Acad Sci. Res: Statistical decision theory; asymptotic statistical theory. Mailing Add: Dept of Opers Res Upson Hall Cornell Univ Ithaca NY 14853

WEISS, LOUIS CHARLES, b New Orleans, La, July 10, 25; m 49; c 4. TEXTILE PHYSICS. Educ: Tulane Univ, BS, 47, MS, 53. Prof Exp: Physicist & engr, Waterways Exp Sta, Corps Engrs, US Dept Defense, 47-48; physicist, 48-65, RES PHYSICIST, SOUTHERN REGIONAL RES CTR, AGR RES SERV, USDA, 65- Concurrent Pos: Asst, La State Univ, 54-58. Mem: AAAS; Sigma Xi. Res: Mechanical properties of natural and modified textile fibers, yarns and fabrics; textile processing; beta decay; strain gages; instrumentation physics. Mailing Add: Southern Regional Res Ctr PO Box 19687 New Orleans LA 70179

WEISS, MALCOLM PICKETT, b Washington, DC, June 28, 21; m 43; c 4. GEOLOGY. Educ: Univ Minn, PhD(geol), 53. Prof Exp: From instr to assoc prof geol, Ohio State Univ, 52-67; chmn dept, 67-72, PROF GEOL, NORTHERN ILL UNIV, 67- Mem: Geol Soc Am; Soc Econ Paleontologists & Mineralogists; Am Asn Petrol Geologists. Res: Stratigraphy; sedimentary petrography; carbonate petrology. Mailing Add: Dept of Geol Northern Ill Univ De Kalb IL 60115

WEISS, MARK LAWRENCE, b Brooklyn, NY, Nov 1, 45; m 69; c 1. PRIMATOLOGY, PHYSICAL ANTHROPOLOGY. Educ: State Univ NY Binghamton, BA, 66; Univ Calif, Berkeley, MA, 68, PhD(anthrop), 69. Prof Exp: Asst prof, 69-74, ASSOC PROF ANTHROP, WAYNE STATE UNIV, 74- Concurrent Pos: Adj anat, Sch Med, Wayne State Univ, 69-; fel, William Beaumont Hosp, 74. Mem: AAAS; Am Anthrop Asn; Am Asn Phys Anthropologists; Am Soc Human Genetics; Brit Soc Study Human Biol. Res: Biochemical polymorphisms; primate genetics; primate microevolution. Mailing Add: Dept of Anthrop Wayne State Univ Detroit MI 48202

WEISS, MARTIN, b New York, NY, Jan 21, 19; m 49; c 3. GEOLOGY. Educ: City Col New York, BS, 48; Univ Mich, MS, 51, PhD(geol), 54. Prof Exp: Asst geol, Mus Paleont, Univ Mich, 51-53; geologist, US Geol Surv, 53-63; oceanogr, Nat Oceanog Data Ctr, 63-72, MARINE GEOLOGIST, NAT GEOPHYS & SOLAR TERRESTIAL DATA CTR, NAT OCEANIC & ATMOSPHERIC ADMIN, 68- Mem: Geol Soc Am; Am Geophys Union; Marine Technol Soc. Res: Geological oceanography; military geology; Paleozoic ostracoda. Mailing Add: Marine Geol & Geophys Group Code D62 NOAA/EDS Rockville MD 20852

WEISS, MARTIN GEORGE, b Muscatine, Iowa, Oct 30, 11; m 38; c 2. PLANT BREEDING. Educ: Iowa State Univ, BS, 34, MS, 35, PhD(genetics, plant breeding), 41. Prof Exp: Asst geneticist, Div Forage Crops & Dis, Bur Plant Indust, USDA, 36-41; res prof farm crops, Exp Sta, Iowa State Col, 45-49; soybean proj leader, Div Forage Crops & Dis, USDA, 50-53, chief field crops res br, Agr Res Serv, 53-57, assoc dir, Crops Res Div, 57-63, asst to dep adminr farm res, 64-69, asst to assoc adminr, 70-71, dir int progs div, 71-73; AGR RES ADV, MINISTRY AGR, GOVT IRAN & DEVELOP & RESOURCES CORP, TEHRAN, IRAN, 74- Mem: Fel Am Soc Agron; Genetics Soc Am; Crop Sci Soc Am (pres, 58); hon mem Am Soybean Asn; hon mem Asn Off Seed Certifying Agencies. Res: Early generation testing of soybeans, barley and forage crops; plant breeding methods; improvement of birdsfoot trefoil, orchardgrass, bromegrass and soybeans; agricultural research administration.

WEISS, MARTIN JOSEPH, b New York, NY, May 4, 23; m 51; c 2. ORGANIC CHEMISTRY. Educ: NY Univ, AB, 44; Duke Univ, PhD(chem), 49. Prof Exp: Asst, Duke Univ, 44-47; res fel org chem, Hickrill Chem Res Found, 49-50; res chemist, Pharmaceut Res Dept, Calco Chem Div, 50-54, group leader, Lederle Labs, 54-75, DEPT HEAD, LEDERLE LABS, AM CYANAMID CO, 75- Concurrent Pos: Asst, Comt Med Res, Off Sci Res & Develop, Duke Univ, 44-45. Mem: Am Chem Soc. Res: Synthetic medicinal chemistry in prostaglandins, antibiotics, steroids, nucleosides and carbohydrates, indoles and other heterocyclics; chemotherapy; anti-inflammatory, anti-asthma and hypoglycemic agents. Mailing Add: Lederle Labs Am Cyanamid Co Pearl River NY 10965

WEISS, MARVIN, b New York, NY, Feb 6, 14; m 40; c 3. PHARMACEUTICAL CHEMISTRY, SCIENCE ADMINISTRATION. Educ: Brooklyn Col, BS, 37; Univ Ill, MS, 38. Prof Exp: Dir develop & control labs, Am Pharmaceut Co, 41-57; dir anal lab, Berkeley Chem Corp, 57-66; group leader corp control & anal, Millmaster Onyx Corp, 66, DIR ANAL LAB, BERKELEY CHEM DEPT, MILLMASTER ONYX CORP, 66- Concurrent Pos: Corp consult, Mantrose-Hauser Div, US Printing Ink Div, Onyx Div & Carboquimica SA Div, 65- Mem: Am Chem Soc; NY Acad Sci. Res: Urethanes; nitrogen heterocycles; multiple condensations; acetic acid-ammonium acetate reactions with carbonyl compounds; crossed Cannizzaro syntheses; urea reactions; radiation curable inks and coatings; shellac and shellac derivatives. Mailing Add: 227 Elkwood Ave New Providence NJ 07974

WEISS, MARVIN B, b Chicago, Ill, Oct 21, 15; m 39; c 3. DENTISTRY. Educ: Univ Ill, BS, 37, DDS, 39. Prof Exp: Instr histol, 39-40, instr oral diag, 53-54, res asst oral path, 54-56, res assoc pedodontia, 56-57, from asst prof to assoc prof oper dent, 58-67, PROF OPER DENT, COL DENT, UNIV ILL MED CTR, 67- Mem: Am Acad Oral Med; Am Acad Gen Pract; Int Asn Dent Res. Res: Clinical research in dentistry. Mailing Add: Col of Dent Univ of Ill at the Med Ctr Chicago IL 60612

WEISS, MAX LESLIE, b Salt Lake City, Utah, Aug 12, 33; m; c 4. MATHEMATICS. Educ: Yale Univ, BA, 55; Cornell Univ, MS, 58; Univ Wash, PhD(math), 62. Prof Exp: Instr math, Reed Col, 58-60 & Univ Wash, 62-63; NSF fel, Inst Advan Study, 63-64; from asst prof to assoc prof, 64-72, PROF MATH, UNIV CALIF, SANTA BARBARA, 72-; ASSOC PROVOST, COL CREATIVE STUDIES, 71- Mem: Am Math Soc. Res: Function algebras and complex variables. Mailing Add: Dept of Math Univ of Calif Santa Barbara CA 93106

WEISS, MAX TIBOR, b Hungary, Dec 29, 22; nat US; m 53; c 4. PHYSICS. Educ: Mass Inst Technol, MS, 47, PhD(physics), 51. Prof Exp: Engr, Radio Corp Am, 43-44 & US Naval Ord Lab, 45-46; res assoc, Microwave Spectros Lab, Mass Inst Technol, 46-50; mem tech staff, Bel Tel Labs, Inc, 50-59; assoc dept mgr, Appl Physics Dept, Hughes Aircraft Corp, 59-61; dir electronics lab, Aerospace Corp, 61-63, gen mgr labs div, 63-67, asst mgr eng opers, TRW Systs, 67-68; GEN MGR, ELECTRONICS & OPTICS DIV, AEROSPACE CORP, 68- Concurrent Pos: Lectr, City Col New York, 53. Mem: AAAS; fel Inst Elec & Electronics Engrs; fel Am Phys Soc. Res: Microwaves; magnetics; quantum electronics; communications; microelectronics. Mailing Add: Electronics & Optics Div Aerospace Corp PO Box 92957 Los Angeles CA 90009

WEISS, MELFORD STEPHEN, b Brooklyn, NY, July 25, 37; m 64; c 1. CULTURAL ANTHROPOLOGY. Educ: State Univ Binghamton, BA, 63; Mich State Univ, MA, 69, PhD(anthrop), 71. Prof Exp: Asst prof anthrop, Ball State Univ, 66-67; asst prof, 67-72, ASSOC PROF ANTHROP, CALIF STATE UNIV, SACRAMENTO, 72- Mem: AAAS; fel Am Anthrop Asn; Coun Anthrop & Educ. Res: China culture area; Chinese-American community organization; education and anthropology; contemporary North American society, life styles of swinging singles; military society; social anthropology, especially rituals and symbols. Mailing Add: Dept of Anthrop Calif State Univ Sacramento CA 95819

WEISS, MICHAEL DAVID, b Chicago, Ill, Nov 12, 42. MATHEMATICS. Educ: Brandeis Univ, BA, 64; Brown Univ, PhD(math), 70. Prof Exp: Asst prof math, Wayne State Univ, 69-74; ANALYST, KETRON INC, 74- Mem: Am Math Soc; Soc Indust & Appl Math; Opers Res Soc Am. Res: Probability and statistics; ergodic theory; applications; theory of fuzzy sets; fuzzy systems; operations research. Mailing Add: Ketron Inc 1400 Wilson Blvd Arlington VA 22209

WEISS, MICHAEL KARL, b Hatzfeld, Romania, Nov 11, 28; nat US; m 54; c 3. CHEMISTRY. Educ: Western Reserve Univ, MS, 55. Prof Exp: Res chemist, Harshaw Chem Co, 52-54; supvr anal chem, Repub Steel Corp Res Ctr, 55-62, chief anal sect, 62-67; head anal div, 67-72, MGR QUAL CONTROL DEPT, BUNKER HILL CO, 72- Mem: AAAS; Am Chem Soc; Soc Appl Spectros; Am Soc Testing & Mat. Res: Chemical and instrumental methods of analysis; ion exchange, solvent extraction techniques; trace analysis. Mailing Add: 118 Woodland Dr Post Falls ID 83837

WEISS, MICHAEL STEPHEN, b Queens Co, NY, Mar 20, 43; m 65; c 2. SPEECH & HEARING SCIENCES, SPEECH PATHOLOGY. Educ: Long Island Univ, BA, 64; Purdue Univ, MS, 68, PhD(speech sci), 70. Prof Exp: Coordr res, Cleveland Hearing & Speech Ctr & sr res assoc speech sci, Case Western Reserve Univ, 70-71; asst prof speech, Howard Univ, 71-72; INSTR LARYNGOL & OTOL & SCI DIR, INFO CTR HEARING, SPEECH & DISORDERS HUMAN COMMUN, SCH MED, JOHNS HOPKINS UNIV, 72- Concurrent Pos: Lectr, Gallaudet Col, 74- Mem: AAAS; Am Speech & Hearing Asn; Acoust Soc Am. Res: Speech science; physiological and acoustical events which underlie speech production and perception. Mailing Add: Info Ctr B-2 Wood Basic Sci Bldg Johns Hopkins Med Insts Baltimore MD 21205

WEISS, MITCHELL JOSEPH, b Chicago, Ill, Nov 12, 42. INVERTEBRATE ZOOLOGY. Educ: Brown Univ, ScB, 64; Univ Mich, PhD(zool), 70. Prof Exp: Instr zool, Univ Iowa, 73-74; ASST PROF BIOL, LIVINGSTON COL, RUTGERS UNIV, NEW BRUNSWICK, 74- Concurrent Pos: NIH res fel, Univ Wash, 70-72. Mem: Am Soc Zoologists; Int Asn Meiobenthologists. Res: Functional and comparative anatomy and development of invertebrate central nervous systems, at both microscopic and ultrastructural levels; current emphasis on the insect brain. Mailing Add: Dept of Biol Livingston Col Rutgers Univ New Brunswick NJ 08903

WEISS, MORRIS J, b Manhattan, NY, Dec 3, 36; m 62; c 2. PHYSICAL CHEMISTRY. Educ: Brooklyn Col, BS, 58; Univ Fla, PhD(phys chem), 63. Prof Exp: Res engr, Boeing Aircraft Co, 63-64; res assoc phys chem, Argonne Nat Lab, 64-66; res scientist, Douglas Advan Res Calif, 66-69; chemist, Nat Bur Standards, 69-70; ASST PROF CHEM, HEBREW UNIV, JERUSALEM, 71- Concurrent Pos: Vis scientist, Argonne Nat Lab, 75-76 & 76-77. Mem: Am Chem Soc; Am Phys Soc. Res: Atomic and molecular collisions; mass spectrometry; photoelectron spectroscopy. Mailing Add: Dept of Chem Hebrew Univ of Jerusalem Jerusalem Israel

WEISS, NORMAN JAY, b Brooklyn, NY, May 28, 42; m 65; c 2. MATHEMATICAL ANALYSIS. Educ: Harvard Univ, BA, 63; Princeton Univ, PhD, 66. Prof Exp: Instr math, Princeton Univ, 66-68; asst prof, Columbia Univ, 68-71; asst prof, 71-72, ASSOC PROF MATH, QUEENS COL, NY, 73- Mem: Am Math Soc. Res: Real and Fourier analysis. Mailing Add: Dept of Math Queens Col Flushing NY 11367

WEISS, PAUL, b Sagan, Ger, Apr 9, 11; m 42; c 3. PHYSICS. Educ: Cambridge Univ, PhD(math), 36. Prof Exp: Asst lectr math physics, Queen's Univ, Belfast, 39; lectr appl math, Westfield Col, Univ London, 41-51; math physicist, Electron Lab, Gen Elec Co, 52-57; consult res & develop labs, Avco Corp, 57-58; assoc prof physics & math, 58-62, ASSOC PROF MATH, WAYNE STATE UNIV, 62- Concurrent Pos: Mem, Inst Advan Study, 50-51. Mem: Am Math Soc; Am Phys Soc; London Math Soc. Res: Differential equations of mathematical physics. Mailing Add: Dept of Math Wayne State Univ Detroit MI 48202

WEISS, PAUL ALFRED, b Vienna, Austria, Mar 21, 98; nat US; m 26. BIOLOGY. Educ: Univ Vienna, PhD(biol), 22. Hon Degrees: MD, Univ Frankfurt, 49; ScD, Univ Giessen, 57; Dr Med & Surg, Univ Helsinki, 66; ScD, Univ Notre Dame, 72. Prof Exp: Asst dir biol res inst, Vienna Acad Sci, 22-29; fel, Kaiser Wilhelm Inst, Berlin, 29-31; Sterling fel, Yale Univ, 31-33; from asst prof to prof zool, Univ Chicago, 33-54, chmn div biol, Master's Prog, 47-54; mem prof develop biol & head lab, 54-64, EMER MEM, ROCKEFELLER UNIV, 64-, EMER PROF BIOL, 54- Concurrent Pos: Vienna Acad Sci fel, Zool Sta, Naples, 26; Rockefeller Found fel, Oceanog Inst, Monaco & Kaiser Wilhelm Inst, 27-29, Johnson Found, Univ Pa, 35, Univs, Cambridge, Zurich, Amsterdam & Free Univ Brussels, 37; vis prof, Univs, Frankfurt, 48-49, Washington, 49, Stanford, 50, Nebr, 51 & Mass Inst Technol, 56-57; mem inst basic res, Univ Calif, 56-59; distinguished vis prof, NY Univ, 60; vis prof, Indust Univ Santander, 63; assoc neurosci res prog, Mass Inst Technol, 63-; distinguished vis scholar, Pratt Inst, 64; dean & univ prof, Univ Tex, 64-66; distinguished vis lectr, Univ Ga & Tulane Univ, 66; spec lectr, Col de France, 66; distinguished prof, Tex A&M Univ, 66-67; distinguished vis lectr, State Univ NY Buffalo, 67. Spec consult, US Dept State, 51; chief sci adv, Brussels World Fair, 56-58; mem, President's Sci Adv Comt, 58-60; consult, President's Off Sci & Technol, 59- Ed, Quart Rev Biol, Develop Biol, J Theoret Biol, Perspectives in Biol & Med, Annee Biologique (France) & Biophysik (Ger). Chmn comt neurobiol, Nat Res Coun, 47-53, chmn div biol & agr, 51-55, comt develop biol, Biol Coun, 51-63, comt, Int Union Biol Sci, 53-64, mem-at-large, 55-, chmn comt int biol sta, 63-65. Mem Elliott Medal comt, Nat Acad Sci, 57-60, Hartley Medal comt, 59-62, Agassiz Medal comt, 60-63, adv comt, Off Sci Personnel, 60-, chmn renewable resources, Comt Natural Resources, 61-62, mem Henry Fund Comt, 62-65, adv comt int opers & prog, 62-, mem coun, 64-67. Chmn, Merck fellowship bd; mem bd dir, Am Inst Biol Sci, 48-50; mem, Fulbright Comt for Int Exchange of Persons, 49-52; mem res coun, United Cerebral Palsy Asns, 51-53; adv bd, Mass Gen Hosp, 52-55; mem corp, Marine Biol Lab, Woods Hole & Bermuda Biol Sta; mem, Coun Indust Res & Develop to Gov NY State, 60-64; mem sci adv bd, Pac Sci Ctr, 62-; mem adv bd, Mt Sinai Hosp & Sch Med, New York, 63-66. Chmn US nat comt, Int Union Biol Sci, 52-, chmn US deleg, 53, 55, 58 & 61, chmn policy bd, 53-55, mem adv coun Naples Sta; chmn US deleg, Int Coun Sci Unions, 61. Honors & Awards: Cert of Merit, US Off Sci Res & Develop, 45; Citation of Merit, US Depts Army & Navy, 45; Grand Medal Geoffrey St Hilaire, France, 55; Leitz Award, 57; Pub Serv Cert, US Dept State, 58; Eleven Lect Awards, 58-69; Weinstein Award, United Cerebral Palsy Asns, 59; John F Lewis Prize, Am Philos Soc, 71; Semicentennial Doctor's Dipl, Univ Vienna, 72. Mem: Nat Acad Sci; AAAS (vpres, 53); Soc Develop Biol (secy, Soc Study Develop & Growth, 39-41, pres, 41); Harvey Soc (pres, 62); Int Soc Develop Biol (vpres, 49). Res: Experimental and theoretical analysis of growth and differentiation in animals; tissue culture; nerve development; regeneration and wound healing; electron microscopy; functional adaptation; coordination of nerve centers; biological education and research administration; developmental biology; neuroplasmic flow; systemalogy; humanistic relations in science. Mailing Add: 201 E 66th St New York NY 10021

WEISS, PETER JOSEPH, b Vienna, Austria, Nov 23, 18; nat US; m 48; c 2. PHARMACEUTICAL CHEMISTRY. Educ: George Washington Univ, BS, 51; Georgetown Univ, MS, 53, PhD(biochem), 56. Prof Exp: Chemist, 47-56, chief chem br, Div Antibiotics, 56-70, dep dir, Nat Ctr Antibiotic Anal, 70-71, DIR, NAT CTR ANTIBIOTIC ANAL, FOOD & DRUG ADMIN, 72- Mem: Am Chem Soc; Acad Pharmaceut Sci; Am Pharmaceut Asn. Res: Chemical analysis of antibiotics. Mailing Add: Nat Ctr Antibiotic Anal 200 C St BD-430 Washington DC 20204

WEISS, PETER REIFER, b Portland, Maine, May 13, 15; m; c 1. THEORETICAL PHYSICS. Educ: Bowdoin Col, AB, 35; Harvard Univ, MA, 36, PhD(physics), 40. Prof Exp: Instr, Harvard Univ, 40-42; mem staff, Radiation Lab, Mass Inst Technol, 42-46; from asst prof to assoc prof, 46-57, chmn dept, 64-72, PROF PHYSICS, RUTGERS UNIV, NEW BRUNSWICK, 57- Mem: Fel Am Phys Soc. Res: Theory of the solid state; magnetism; superconductivity; transport theory. Mailing Add: Dept of Physics Rutgers Univ New Brunswick NJ 08903

WEISS, PHILIP, b New York, NY, June 12, 16; m 43; c 2. ORGANIC CHEMISTRY. Educ: NY Univ, BS, 39, MSc, 41, PhD(org 48. Prof Exp: Res chemist, Lederle Labs, Am Cyanamid Co, 41-46; sr res chemist, Wallace & Tiernan Prod, Inc, 46-52; sr proj chemist, Colgate-Palmolive Co, 52-57; sr res scientist, Electrochem Dept, Res Labs, Gen Motors Corp, 57-58, asst head electrochem & polymers dept, 58-59, HEAD POLYMERS DEPT, RES LABS, GEN MOTORS CORP, 59- Concurrent Pos: Instr, Cooper 49-51; mem, Bd Trustees, Paint Res Inst, 61-63 & Plastics Inst Am, 64-69; ed, J Appl Polymer Sci; mem, Comt Critical & Strategic Mat, Nat Mat Adv Bd, Adv Bd, Col Eng, Univ Detroit & Adv Bd, Polymer Prog, Col Eng, Princeton Univ; US rep, Plastics & High Polymers Sect, Int Union Pure & Appl Chem. Honors & Awards: Exner Medal, 69. Mem: Am Chem Soc; Soc Plastics Engrs. Res: Polymer synthesis, research and development in plastics, rubber, adhesives and surface coatings; graft and block copolymers; mechanisms of finish failure; adhesion and cohesion; mechanical behavior of polymers; polymer flammability, aging, waste disposal and processing. Mailing Add: Gen Motors Res Lab Rds Warren MI 48090

WEISS, RAINER, b Berlin, Ger, Sept 29, 32; US citizen; m 59; c 2. PHYSICS. Educ: Mass Inst Technol, BS, 55, PhD(physics), 62. Prof Exp: Asst prof physics, Tufts Univ, 60-62; res assoc, Princeton Univ, 62-64; from asst prof to assoc prof, 64-74, PROF PHYSICS, MASS INST TECHNOL, 74- Res: Experimental relativity, infrared astronomy; atomic physics. Mailing Add: Rm 20F001 Mass Inst of Technol Cambridge MA 02139

WEISS, RICHARD GERALD, b Akron, Ohio, Nov 13, 42. PHOTOCHEMISTRY, PHYSICAL ORGANIC CHEMISTRY. Educ: Brown Univ, ScB, 65; Univ Conn, MS, 67, PhD(chem), 69. Prof Exp: NIH res fel chem, Calif Inst Technol, 69-71; vis prof, Inst Chem, Univ Sao Paulo, 71-74; ASST PROF CHEM, GEORGETOWN UNIV, 74- Concurrent Pos: Nat Acad Sci overseas fel, 71-74. Mem: Am Chem Soc; The Chem Soc; Brazilian Soc Advan Sci. Res: Mechanisms and rates of organic and photochemical reactions; steric effects in electronic energy transfer and in decay of excited states. Mailing Add: 4604 W Virginia Ave Bethesda MD 20014

WEISS, RICHARD JEROME, b New York, NY, Dec 14, 23; m c 2. PHYSICS. Educ: City Col New York, BS, 44; Univ Calif, MA, 47; NY Univ, PhD(physics), 50. Prof Exp: PHYSICIST, US ARMY MAT RES CTR, 50- Concurrent Pos: Vis fel, Cavendish Labs, Cambridge Univ, 56-57; lectr, Mass Inst Technol, 59; Secy Army fel, Imp Col, Univ London, 62-63; chmn, Comn Electron Distributions, Int Union Crystallog, 72-75. Honors & Awards: Rockefeller Pub Serv Award, 56. Mem: Am Phys Soc; Am Crystallog Soc. Res: Solid state, neutron and x-ray physics; electron structure of solids. Mailing Add: US Army Mat Res Ctr Watertown MA 02172

WEISS, RICHARD LAWRENCE, geology, see 12th edition

WEISS, RICHARD LOUIS, b Evanston, Ill, June 24, 44. BIOCHEMISTRY. Educ: Stanford Univ, BS, 66; Univ Wash, PhD(biochem), 71. Prof Exp: USPHS fel, Univ Mich, 71-72, Am Cancer Soc fel, 72-73; ASST PROF CHEM, UNIV CALIF, LOS ANGELES, 74- Mem: Am Chem Soc; Am Soc Microbiol; Genetics Soc Am. Res: Regulation of amino acid metabolism in eucaryotic microorganisms; evolution of diploidy. Mailing Add: Dept of Chem Univ of Calif Los Angeles CA 90024

WEISS, RICHARD RAYMOND, b Takoma Park, Md, Aug 26, 28; m 56; c 2. METEOROLOGY, ELECTRICAL ENGINEERING. Educ: Univ Md, BSE, 52; Univ Mich, Ann Arbor, MSE, 55; Univ Wash, PhD(elec eng), 67. Prof Exp: Engr, Sperry Gyroscope Co, 52-54; cardiovasc trainee, Univ Wash, 59-60; asst prof elec eng, Seattle Univ, 60-64; lectr, 64-67, res asst prof, 67-74, SR RES ASSOC ATMOSPHERIC SCI, UNIV WASH, 74- Mem: Inst Elec & Electronics Engrs; Am Geophys Union. Res: Radar meteorology. Mailing Add: 1501 McGilvra Blvd E Seattle WA 98112

WEISS, ROBERT JEROME, b West New York, NJ, Dec 9, 17; m 45; c 3. PSYCHIATRY. Educ: George Washington Univ, AB, 47; Columbia Univ, MD, 51. Prof Exp: Intern, Columbia Div, Bellevue Hosp, 51, asst resident med, 53; resident psychiat, Columbia Psychoanal Clin, 54-59; chief, Mary Hitchcock Mem Hosp, Hanover, NH, 59-70; vis prof, 70, ASSOC DIR, CTR COMMUNITY HEALTH & MED CARE, HARVARD MED SCH, 70-, ASSOC DEAN HEALTH CARE PROGS, 71- Concurrent Pos: NIMH career teacher trainee, Columbia Univ, 56-58; resident, NY State Psychiat Inst & Presby Hosp, New York, 57; asst attend, Vanderbilt Clin, 57-58; assoc, Col Physicians & Surgeons, Columbia Univ, 57-59; asst attend, Presby Hosp, 58-59; consult, NH Div Ment Health, 59-70, Vet Admin Hosp, White River Junction, Vt, 60-70 & Bur Health Manpower Educ, 72-; prof psychiat & chmn dept, Dartmouth Med Sch, 59-70; chmn adv comn, NH Dept Health & Welfare, 61; mem exec comt, NH Ment Health Planning Proj, 64-66 & Adv Comn Ment Health Construct NH, 65-70; coordr panel, NIMH, 65-67, chmn subcomt psychiat, 67-68; psychiatrist, Beth Israel Hosp, 70- Mem: AAAS; fel Am Psychiat Asn; AMA; Am Acad Psychoanal; NY Acad Sci. Res: Epidemiology; preventive psychiatry; community medicine; health care. Mailing Add: Ctr Community Health & Med Care Harvard Med Sch 25 Shattuck St Boston MA 02115

WEISS, ROBERT JOHN, b Pomona, Calif, Apr 9, 37; m 64; c 2. MATHEMATICS. Educ: La Verne Col, BA, 58; Univ Calif, Los Angeles, MA & PhD(math), 62. Prof Exp: Assoc prof math, Bridgewater Col, 62-68; assoc prof & head dept, 68-74, PROF MATH, MARY BALDWIN COL, 74- Mem: Am Math Soc; Math Asn Am. Res: Algebraic topology. Mailing Add: Dept of Math Mary Baldwin Col Staunton VA 24401

WEISS, ROBERT MARTIN, b New York, NY. UROLOGY, PHARMACOLOGY. Educ: Franklin & Marshall Col, BS, 57; State Univ NY Downstate Med Ctr, MD, 60. Prof Exp: Intern med, Second (Cornell) Med Div, Bellevue Hosp, New York, 60-61; resident gen surg, Beth Israel Hosp, New York, 61-62; resident urol, Columbia-Presby Med Ctr, 63-64; fel, Col Physicians & Surgeons, Columbia Univ, 64-65; resident, Columbia-Presby Med Ctr, 65-67; instr surg-urol, 67-68, asst prof urol, 68-71, ASSOC PROF SURG-UROL, SCH MED, YALE UNIV, 71- Concurrent Pos: Res assoc pharmacol, Col Physicians & Surgeons, Columbia Univ, 67-75, adj assoc prof, 75-; attend, Yale-New Haven Hosp, 67-; consult, West Haven Vet Admin Hosp, 67- & Waterbury Hosp, 76-; mem obstruction & neuromuscular dis comt, Nat Inst Arthritis, Metab & Digestive Dis, 74-75; fel, Timothy Dwight Col, Yale Univ, 74-; asst ed, J Urol, 75-; mem adv panel, US Pharmacopeia & Nat Formulary, 76-81. Mem: Am Physiol Soc; Soc Gen Physiologists; Am Acad Pediat; Am Col Surgeons; Am Urol Asn. Res: Mechanical and electrophysiologic properties of ureteral smooth muscle; role of cyclic nucleotides in smooth muscle function. Mailing Add: Yale-New Haven Hosp 789 Howard Ave New Haven CT 06510

WEISS, ROGER HARVEY, b New York, NY, July 13, 26; m 53; c 2. ANALYTICAL CHEMISTRY. Educ: Col Holy Cross, BNaval Sci, 46; Cornell Univ, AB, 50; Ga Inst Technol, PhD(chem), 68. Prof Exp: Instr chem, Univ Toledo, 54-57; from asst prof to assoc prof, 59-, PROF CHEM, HUMBOLDT STATE UNIV, 74- Concurrent Pos: Fulbright lectr, Univ Sind, Pakistan, 73-74. Mem: Am Chem Soc. Res: Titrimetry with solid titrants; trace analysis; solvent extraction; chelation; absorption spectrophotometry; photometric titrations. Mailing Add: Dept of Chem Humboldt State Univ Arcata CA 95521

WEISS, RONALD, b Chicago, Ill, Jan 29, 37; m 67; c 3. FOOD SCIENCE. Educ: Ariz State Univ, BS, 58, MS, 59; Mich State Univ, PhD(chem), 64, MBA, 67. Prof Exp: Res chemist, 64-67, proj coordr, 67-69, mgr prod develop, 69-73, dir growth & develop, 73-75, DIR PLANNING, MILES LABS, INC, 75- Mem: Am Chem Soc; Inst Food Technologists; Soft Drink Technologists Asn. Mailing Add: Marschall Div Miles Labs Inc Elkhart IN 46514

WEISS, SAMUEL BERNARD, b New York, NY, May 18, 26; m 61; c 2. BIOCHEMISTRY. Educ: City Col New York, BS, 48; Univ Southern Calif, PhD(biochem), 54. Prof Exp: Res assoc biochem, Mass Gen Hosp, Boston, 56-57; asst prof, Rockefeller Inst, 57-58; asst prof, 58-63, PROF BIOCHEM, UNIV CHICAGO, 63- Concurrent Pos: Am Heart Asn fel biochem, Univ Chicago, 54-56; Guggenheim fel, Salk Inst, 70-71; res assoc, Argonne Cancer Res Hosp, 58-, assoc dir, 67- Honors & Awards: Theobold Smith Award, AAAS, 61; Am Chem Soc Award Enzyme Chem, 66. Mem: Am Soc Biol Chemists. Res: Enzymology of reactions in the synthesis of lipids, proteins and nucleic acids. Mailing Add: Franklin McLean Mem Res Inst 950 E 59th St Chicago IL 60637

WEISS, SIDNEY, b New York, NY, Dec 16, 20; m 44; c 3. BIOCHEMISTRY. Educ: Queens Col, NY, BS, 42; Fordham Univ, MS, 46, PhD(biochem), 49; Rutgers Univ, MS, 66. Prof Exp: Res chemist, Food Res Labs, Inc, NY, 42-46; res assoc biochem, Inst Cancer Res, Philadelphia, Pa, 49-58; sr res chemist, 58-59, sect head, 59-63, mgr biol res, 63-74, ASSOC DIR RES, COLGATE-PALMOLIVE CO. Mem: Am Chem Soc; Am Soc Biol Chem. Res: Biological oxidations; enzymatic reactions and microbiological transformations. Mailing Add: Colgate-Palmolive Co 909 River Rd Piscataway NJ 08854

WEISS, SOL, b Austria, Apr 19, 13; US citizen; m 40; c 2. MATHEMATICS. Educ: Brooklyn Col, BS, 34; Columbia Univ, MA, 36. Prof Exp: Teacher, Philadelphia Pub Sch Syst, Pa, chmn dept math, 50-64; asst prof, 64-65, ASSOC PROF MATH, WEST CHESTER STATE COL, 65- Concurrent Pos: Consult, Wilmington Sch Dist, Del, 64, Philadelphia Pub Sch Syst, 64-65 & 67-, Cecil County Pub Schs, Md, 66, Upward Bound Prog, Franklin & Marshall Col, 67 & Lehigh Univ Social Restoration Prog, Pa State Prisons, 73-74. Mem: Math Asn Am. Res: Mathematics education for the low achiever. Mailing Add: Dept of Math West Chester State Col West Chester PA 19380

WEISS, THEODORE JOEL, b Rochester, NY, Aug 16, 19; m 41; c 2. LIPID CHEMISTRY. Educ: Syracuse Univ, AB, 40, PhD(biochem), 53. Prof Exp: Asst chemist feed control, State Inspection & Regulatory Serv, Md, 41-43; res chemist powdered milk, Borden Co, 44-49 & Dairymen's League Coop Asn, 49-52; res chemist & div head, Swift & Co, 52-63; tech dir, Capital City Prod Co, 63-64; sr proj leader, Res & Develop Dept, Hunt-Wesson Foods, Inc, 64-68; res chemist, Southern Utilization Res & Develop Div, Agr Res Serv, USDA, 68-70, Dairy Prod Lab, 70-72; TECH MGR INDUST SALES DEPT, HUNT-WESSON FOODS, INC, 72- Mem: AAAS; Am Chem Soc; Inst Food Technologists; Am Oil Chem Soc. Res: Edible fats and oils; margarine; peanut butter; mayonnaise. Mailing Add: Indust Sales Dept Hunt-Wesson 1645 W Valencia Dr Fullerton CA 92634

WEISS, THOMAS E, b New Orleans, La, June 15, 16; m 50; c 2. MEDICINE. Educ: Tulane Univ, MD, 40. Prof Exp: From instr to assoc prof med, 47-64, PROF CLIN MED, SCH MED, TULANE UNIV, 64- Concurrent Pos: Mem staff, Ochsner Clin; trustee, Alton Ochsner Med Found. Mem: Am Rheumatism Asn (past pres). Res: Rheumatic diseases; gout, clinical observations and correlating clinical finds with test and treatments; studies of radioisotope joint scanning in patients with arthritis. Mailing Add: Ochsner Clin 1514 Jefferson Hwy New Orleans LA 70121

WEISS, ULRICH, b Prague, Czech, Jan 24, 08; nat US; m 37; c 2. ORGANIC CHEMISTRY. Educ: Univ Prague, PhD(chem), 30. Prof Exp: Res chemist, Norgine, Inc, Czech, 30-39; asst to mgr, Norgan, Inc, France, 39-40; res chemist, Endo Prod, Inc, 41-51; chemist, Tuberc Res Lab, USPHS, 51-53; res assoc, Columbia Univ, 53-54; chemist, New York Bot Gardens, 55-57; exec secy, Dent Study Sect, Div Res Grants, NIH, 57-58, CHEMIST, LAB CHEM PHYSICS, NAT INST ARTHRITIS & METAB DIS, Concurrent Pos: Instr, Brooklyn Col, 52-55. Mem: Am Chem Soc; The Chem Soc; NY Acad Sci. Res: Chemistry of natural compounds; medicinal chemistry; biosynthesis; morphine derivatives; chiroptical effects. Mailing Add: Bldg 2 Room B1-22 Nat Insts of Health Bethesda MD 20014

WEISS, VIRGIL WAYNE, b Elgin, Tex, May 22, 40; m 61; c 3. POLYMER CHEMISTRY. Educ: Univ Tex, BS, 62; Univ Ill, MS, 64, PhD(chem), 66. Prof Exp: Res specialist, 66-73, COM DEVELOP MGR, POLYMERS & PETROCHEMS, MONSANTO CO, 73- Res: Spectroscopy, polymers. Mailing Add: Monsanto Polymers & Petrochems 800 Lindbergh Blvd St Louis MO 63166

WEISS, WILLIAM, b New York, NY, June 12, 23; m 56; c 4. BIOSTATISTICS, EPIDEMIOLOGY. Educ: George Washington Univ, BA, 48. Prof Exp: Chief statistician, US Food & Drug Admin, 48-62; asst chief perinatal res br, 62-66, CHIEF, OFF BIOMET, NAT INST NEUROL DIS & STROKE, NIH, 66- Mem: Biomet Soc; Am Statist Asn. Res: Biostatistical applications in neurology; communicative disorders; perinatal research. Mailing Add: 609 Jerry Lane NW Vienna VA 22180

WEISS, WILLIAM, b Philadelphia, Pa, July 29, 19; m 41; c 3. PULMONARY DISEASES. Educ: Univ Pa, BA, 40, MD, 44. Prof Exp: Mem staff, Sch Med & Grad Sch Med, Univ Pa & Med Col Pa, 45-66; assoc prof, 66-70, PROF MED, HAHNEMANN MED COL, 70- Concurrent Pos: Chief tuberc, Harbor Gen Hosp, Torrance, Calif, 49-50; clin dir, Pulmonary Dis Serv, Philadelphia Gen Hosp, 50-74; chest consult, Norristown State Hosp, Pa, 51-60; dir, Philadelphia Pulmonary Neoplasm Res Proj, 57-67; Int Agency Res Cancer travel fel, London Mass Radiography Units, 69; ed, Philadelphia Med, 76- Mem: AMA; Am Thoracic Soc; Am Col Physicians. Res: Pulmonary disease, particularly lung cancer. Mailing Add: 3912 Netherfield Rd Philadelphia PA 19129

WEISS, WILLIAM P, b New York, NY, Nov 5, 29; m 60; c 5. PHARMACOLOGY. Educ: Columbia Univ, MD, 56. Prof Exp: Intern, Boston City Hosp, 56-57, asst med resident, 57-58; surgeon, NIMH, 58-62; from asst res prof to asst prof pharmacol, Sch Med, George Washington Univ, 62-68; CHIEF PROG REV & DEVELOP DIV, DC DEPT PUB HEALTH, 68- Mem: AAAS; NY Acad Sci. Res: Action of drugs and various physiological states on the rate of synthesis and breakdown of proteins. Mailing Add: Human Resources Dept DC Dept Pub Health Washington DC 20004

WEISSBACH, ARTHUR b New York, NY, Aug 27, 27; m 58; c 2. BIOCHEMISTRY. Educ: City Col New York, BS, 47; Columbia Univ, PhD(biochem), 53. Prof Exp: Nat Found Infantile Paralysis fel, NIH, 53-55; asst prof biochem, Albany Med Col, Union Univ, NY, 55-56; biochemist, NIH, 56-68; HEAD DEPT CELL BIOL, ROCHE INST MOLECULAR BIOL, 68- Concurrent Pos: Prof lectr, Georgetown Univ, 57-58; NSF fel, 59-60; prof lectr, George Washington Univ, 61-66; adj prof, Dept Human Genetics & Develop, Columbia Univ, 69- Mem: Am Soc Biol Chemists; Am Soc Microbiol. Res: Biochemistry of animal viruses and cellular nucleic acids. Mailing Add: Dept of Cell Biol Roche Inst Molec Biol Nutley NJ 07110

WEISSBACH, HERBERT, b New York, NY, Mar 16, 32; c 4. BIOCHEMISTRY, MOLECULAR BIOLOGY. Educ: George Washington Univ, MS, 55, PhD, 57. Prof Exp: Biochemist, Nat Heart Inst, 53-58; lectr, George Washington Univ, 62-68; head sect enzymes & metab & actg chief lab clin biochem, NIH, 68-69; ASSOC DIR, ROCHE INST MOLECULAR BIOL & DIR DEPT BIOCHEM, 69- Concurrent Pos: NSF travel grant, Int Cong Biol Chem Socs, Moscow, 61; ed, Arch Biochem Biophys, 67-72, exec ed, 72-; ed, J Pharmacol & Exp Therapeut, 67-72, Int J Neuropharm, 69- & J Biol Chem, 72-; adj prof, Dept Human Genetics, Columbia Univ, 69- Honors & Awards: Superior Serv Award, Dept Health, Educ & Welfare, 68; Am Chem Soc Enzyme Award, Mem: AAAS; Am Soc Biol Chemists; Am Chem Soc; Am Soc Pharmacol & Therapeut; Am Soc Microbiol. Res: Mechanism of enzyme and coenzyme action; protein synthesis. Mailing Add: Dept of Biochem Roche Inst of Molec Biol Nutley NJ 07110

WEISSBART, JOSEPH, b Poland, May 28, 25; nat US; m 55; c 3. ELECTROCHEMISTRY. Educ: Univ Calif, BA, 50; Univ Ore, MA, 54, PhD(phys chem), 56. Prof Exp: Anal chemist, Colgate-Palmolive-Peet Co, 51-52; res chemist, Res Labs, Westinghouse Elec Corp, 55-62; staff scientist & proj leader, Palo Alto Res Labs, Lockheed Missiles & Space Co, 62-68; PRES, APPL ELECTROCHEM INC, SUNNYVALE, CALIF, 68- Mem: Am Ceramic Soc; Electrochem Soc; NY Acad Sci; Int Electrochem Soc. Res: Science administration; product research and development, product marketing; manufacture of solid oxide electrolyte oxygen analyzers; development of automated oxygen uptake and carbon dioxide output meters. Mailing Add: 1401 Holt Ave Los Altos CA 94022

WEISSBERG, ALFRED, b Boston, Mass, Jan 29, 28; m 57; c 3. MATHEMATICS, SCIENCE ADMINISTRATION. Educ: Northeastern Univ, BS, 52; Univ NH, MS, 54. Prof Exp: Prin mathematician, Battelle Mem Inst, 53-60; math statistician, US Food & Drug Admin, 60-64; supvry opers res analyst, Nat Bur Standards, 64-67; head file orgn & statist, Toxicol Info Prog, Nat Libr Med, 67-70; SCI & TECHNOL COMMUN OFF, NAT INST NEUROL DIS & STROKE, NIH, 70- Mem: AAAS; Am Soc Info Sci; Am Statist Asn. Res: Mathematical statistics; information system design and operation; application of computer systems to information retrieval. Mailing Add: 1024 Noyes Dr Silver Spring MD 20910

WEISSBERGER, EDWARD, b Rochester, NY, Jan 1, 41. INORGANIC CHEMISTRY, ORGANOMETALLIC CHEMISTRY. Educ: Brown Univ, ScB, 63; Stanford Univ, PhD(chem), 67. Prof Exp: Res assoc inorg chem, Stanford Univ, 67-69; lectr chem, Univ Calif, Santa Cruz, 69; ASST PROF WESLEYAN UNIV, 69- Mem: Am Chem Soc; The Chem Soc. Res: Reactions of coordinated ligands; homogeneous catalysis; metal catalyzed olefin couplings; autoxidation at metal centers; trapping of unstable, reactive intermediates. Mailing Add: Dept of Chem Wesleyan Univ Middletown CT 06457

WEISSBLUTH, MITCHEL, b Russia, Jan 7, 15; m 40; c 3. PHYSICS. Educ: Brooklyn Col, BA, 36; George Washington Univ, MA, 41; Univ Calif, PhD, 50. Prof Exp: Metallurgist, US Navy Yard, Washington, DC, 37-41; radio engr, Crosley Radio Corp, Ohio, 42-43; sr res engr, Jet Propulsion Lab, Calif Inst Technol, 43-45; asst physics, Univ Calif, 45-49, lectr, 50; res assoc, 50-51, instr, 51-54, ASST PROF RADIOL PHYSICS, STANFORD UNIV, 54-, DIR BIOPHYS LAB, 64-, ASSOC PROF APPL PHYSICS, 67- Concurrent Pos: Physicist, Western Regional Res Lab, Bur Agr & Indust Chem, USDA, 48; Fulbright grant, Weizmann Inst Sci, Israel, 60-61; liaison scientist, Off Naval Res, London, 67-68; consult, Stanford Res Inst. Mem: Am Phys Soc; Int Soc Quantum Biol (pres, 73-75). Res: Electronic states in hemoglobin Jahn-Teller effect; acoustic synchrotron radiation. Mailing Add: Hansen Labs Stanford Univ Stanford CA 94305

WEISSE, ALLEN B, b New York, NY, Dec 6, 29; m 67; c 2. CARDIOLOGY, MEDICINE. Educ: NY Univ, BA, 50; State Univ NY, MD, 58; Am Bd Internal Med, dipl, 65, cert cardiovasc dis, 67. Prof Exp: Res fel cardiol, Sch Med, Univ Utah, 61-63; instr med, Sch Med, Seton Hall Univ, 63-65; from instr to assoc prof, 65-74, PROF MED, NJ MED SCH, COL MED & DENT NJ, 74- Mem: Am Fedn Clin Res; Am Physiol Soc; Am Heart Asn; Am Col Physicians; Am Col Cardiol. Res: Cardiovascular physiology and disease. Mailing Add: NJ Med Sch Col Med & Dent NJ 100 Bergen St Newark NJ 07103

WEISSGERBER, RUDOLPH E, b Ger, June 2, 21; US citizen; m 50; c 1. MEDICINE. Educ: Univ Greifswald, MD, 45; Univ Hamburg, dipl trop dis & med parasitol, 48; Emory Univ, cert bus admin, 65. Prof Exp: Pres, Nordmark Chem Co, NY, 52-55; med sci dir, E R Squibb & Sons, Inc, Europe, 55-63, dir prod planning & develop, E R Squibb Int Co, NY, 63-64, dir res & develop, Squibb Int Co, 65-66; vpres & sci dir, Int Div, Bristol-Myers Co, 67-68; DIR SCI & COM DEVELOP-EUROPE, RICHARDSON-MERRELL INC, 69- Mailing Add: Richardson-Merrell Inc 10 Westport Rd Wilton CT 06897

WEISSKOPF, BERNARD, b Berlin, Ger, Dec 11, 29; US citizen; m 65; c 2. PEDIATRICS, PSYCHIATRY. Educ: Syracuse Univ, BA, 51; State Univ Leiden, MD, 55; Am Bd Pediat, dipl, 65. Prof Exp: Physician, State Univ Leiden, 58; intern, Meadowbrook Hosp, Hempstead, NY, 58-59, resident pediat, 59-60; asst chief pediat, US Air Force Hosp, Maxwell AFB, 60-62; fel pediat & child psychiat, Hosp & Sch Med, Johns Hopkins Univ, 62-64; asst prof pediat, Univ Ill Col Med, 64-66; assoc prof, 66-72, PROF PEDIAT, SCH MED, UNIV LOUISVILLE, 72-, ASSOC PSYCHIAT & DIR CHILD EVAL CTR, 66- Concurrent Pos: Clin coordr, Ill State Pediat Inst, 64-66. Mem: Fel Am Acad Pediat; Am Asn Ment Deficiency. Res: Behavioral aspects of pediatrics, especially learning disorders and mental retardation. Mailing Add: Child Eval Ctr 540 S Preston St Louisville KY 40202

WEISSKOPF, VICTOR FREDERICK, b Vienna, Austria, Sept 19, 08; nat US; m 34; c 2. PHYSICS. Educ: Univ Göttingen, PhD(physics), 31. Hon Degrees: PhD, sixteen from various US & foreign univs, 61-70. Prof Exp: Res assoc, Univ Berlin, 31-32; Rockefeller Found fel, Univs Copenhagen & Cambridge, 32-33; res assoc, Swiss Fed Inst Technol, 33-36 & Univ Copenhagen, 36-37; from instr to asst prof physics, Univ Rochester, 37-43; group leader, Los Alamos Sci Lab, 43-47; prof, 45-74, INST EMER PROF PHYSICS & SR LECTR, MASS INST TECHNOL, 74- Concurrent Pos: Dir gen, Europ Ctr Nuclear Res, 61-65. Honors & Awards: Planck Medal, 56; Prix Mondial Cino Del Duca, 72. Mem: Nat Acad Sci; fel Am Phys Soc (vpres, 59, pres, 60); Fedn Am Sci; cor mem Fr, Austrian, Danish, Scottish & Bavarian Acads Sci. Res: Quantum mechanics; electron theory; theory of nuclear phenomena; nuclear physics. Mailing Add: Dept of Physics Mass Inst of Technol Cambridge MA 02139

WEISSLER, ALFRED, b New York, NY, Mar 13, 17; m 41; c 3. CHEMISTRY. Educ: City Col New York, BS, 36; Univ Wis, MS, 38; Univ Md, PhD(inorg chem), 46. Prof Exp: Chemist, US Customs Lab, NY, 38-39; teacher high sch, NY, 39-42; chemist, Metal Div, US Naval Res Lab, 42-46, phys chemist, Sound Div, 46-51; physicist, Off Chief Ord, US Dept Army, 51-52, chief chem br, Off Ord Res, 52-53, chief, Washington Off, 53-54; chemist, NIH, 54-62; phys chem prog chief, Air Force Off Sci Res, 62-67; asst dir, Bur Sci, 67-70, DIR, DIV COLORS & COSMETIC TECHNOL, FOOD & DRUG ADMIN, DEPT HEALTH, EDUC & WELFARE, 70- Concurrent Pos: Asst prof, Univ Md, 46-52; res assoc physics, Cath Univ Am, 60-63; adj prof, Dept Chem, Am Univ, Washington, DC, 61-67 & 73-; guest worker, Weizmann Inst Sci, 65; chmn, Phys Sci Dept Comt, USDA Grad Sch, 66- Mem: Fel AAAS; Am Phys Soc; fel Acoust Soc Am; Biophys Soc. Res: Ultrasonic-chemical reactions; administration of research; sound velocity in liquids as determined by molecular structure; fluroescence and spectrophotometry in chemical analysis; safety of cosmetics and color additives. Mailing Add: 5510 Uppingham St Chevy Chase MD 20015

WEISSLER, ARNOLD M, b Brooklyn, NY, May 13, 27; c 4. CARDIOLOGY, INTERNAL MEDICINE. Educ: NY Univ, BA, 48; State Univ NY Downstate Med Ctr, MD, 53. Prof Exp: Assoc med, Duke Hosp, 55-59; asst prof, Univ Tex Med Br Galveston, 60-61; from asst prof to prof, Ohio State Univ, 61-71, dir div cardiol, 63-71; PROF MED & CHMN DEPT, WAYNE STATE UNIV, 71- Concurrent Pos: Am Heart Asn res fel, Duke Hosp, Durham, NC, 55-57; mem cardiovasc bd, Am Bd Internal Med, chmn, 75-; fel coun clin cardiol, Am Heart Asn. Mem: Fel Am Col Physicians; Am Clin & Climat Asn; fel Am Col Cardiol; Am Soc Clin Invest; Am Soc Pharmacol & Exp Therapeut. Res: Cardiovascular physiology; congestive heart failure; noninvasive techniques in cardiology; myocardial metabolism. Mailing Add: Dept of Internal Med Wayne State Univ Sch of Med Detroit MI 48201

WEISSLER, GERHARD LUDWIG, b Eilenburg, Ger, Feb 20, 18; nat US; m 53; c 2. PHYSICS. Educ: Univ Calif, MA, 41, PhD(physics), 42; Am Bd Radiol, dipl, 52. Prof Exp: Instr radiol, Med Sch, Univ Calif, 42-44; from asst prof to assoc prof, 44-52, head dept, 51-56, PROF PHYSICS, UNIV SOUTHERN CALIF, 52- Mem: Fel Am Phys Soc; fel Optical Soc Am. Res: Gaseous electronics; vacuum ultraviolet radiation physics; photo-absorption and photo-ionization cross sections; photoelectric effect; optical constants and solid state physics; upper atmosphere and astrophysical problems; nuclear accelerator physics; vacuum ultraviolet spectroscopy of hot gaseous plasmas. Mailing Add: Dept of Physics Univ of Southern Calif Los Angeles CA 90007

WEISSLER, HAROLD EDWARD, b St Louis, Mo, Sept 17, 15; m 38; c 2. FOOD CHEMISTRY. Educ: St Mary's Univ, San Antonio, BS, 46; Trinity Univ, Tex, MS, 55. Prof Exp: Chief chemist, Pearl Brewery, 37-56; asst tech dir, 56-61, asst dir res & develop, 61-69, DIR RES & DEVELOP, FALSTAFF BREWING CORP, 69-

Concurrent Pos: Instr, St Mary's Univ, San Antonio, 46-56; instr, San Antonio Col, 50-56 & US Army Sch, Ft Sam Houston, Tex, 51-55; res adv, Incarnate Word Col, 52-56; lectr, Washington Univ, 57- Honors & Awards: Schwarz Lab Award, Master Brewers Asn Am, 69. Mem: AAAS; Am Soc Brewing Chemists; Am Soc Qual Control; Am Math Soc; Am Soc Microbiol. Res: Colligative properties of liquid solutions; maximum density of aqueous solution and biological fluids; fermentation kinetics. Mailing Add: 61 Tealwood Dr St Louis MO 63141

WEISSMAN, ALBERT, b New York, NY, Aug 1, 33; m 57; c 5. PSYCHOPHARMACOLOGY. Educ: NY Univ, BA, 54; Columbia Univ, MA, 55, PhD, 58. Prof Exp: Instr psychol, Columbia Univ, 57; sr res psychologist, Pfizer Res Labs, 58-61, mgr psychopharmacol, 61-72, ASST DIR PHARMACOL, PFIZER INC, 72- Mem: Am Soc Pharmacol & Exp Therapeut. Res: Catecholamines and indolylalkylamines; learning and memory; addiction. Mailing Add: Med Res Lab Pfizer Inc Groton CT 06340

WEISSMAN, EARL BERNARD, b Detroit, Mich, Feb 21, 42; m 67; c 2. CLINICAL CHEMISTRY. Educ: Wayne State Univ, BS, 65, PhD(biochem), 72. Prof Exp: Fel clin chem, Buffalo Gen Hosp, 72-74; ASST MGR LAB OPERS, LAB PROCEDURES, UPJOHN CO, 74- Concurrent Pos: Consult, Ventura County Gen Hosp, 74-; asst prof, Calif State Univ, 75- Mem: Am Asn Clin Chemists. Res: Porphyrin metabolism; clinical methods development. Mailing Add: Lab Procedures-Upjohn 6330 Variel Woodland Hills CA 91364

WEISSMAN, GERARD SELWYN, b Brooklyn, NY, Oct 18, 21; m 55; c 2. PLANT PHYSIOLOGY. Educ: Brooklyn Col, BA, 42; Columbia Univ, MA, 48, PhD(bot), 50. Prof Exp: Asst plant physiol, Columbia Univ, 46-49, res assoc, 49-50; from instr to assoc prof biol, 50-62, fac fel, 66, PROF BOT, CAMDEN COL ARTS & SCI, RUTGERS UNIV, CAMDEN, 62- Concurrent Pos: Vpres acad affairs on leave, Trenton State Col, 70. Mem: AAAS; Am Soc Plant Physiologists; Bot Soc Am. Res: Enzyme regulation; nitrogen metabolism. Mailing Add: Rutgers Univ Camden Col Arts & Sci Camden NJ 08102

WEISSMAN, HERMAN BENJAMIN, b Chicago, Ill, May 16, 20; m 45; c 1. PHYSICS. Educ: Univ Chicago, SB, 52; Ill Inst Technol, MS, 54, PhD(physics), 59. Prof Exp: Chemist, Qm Corps, US Dept Army, 51-52 & Armour Res Found, Ill Inst Technol, 53-54; instr, Chicago Jr Col, 54; asst prof, Elmhurst Col, 54-56 & Univ Ill, 56-59; sr res physicist, Bell & Howell Co, 59-60; asst prof, 60-63, PROF PHYSICS, UNIV ILL, CHICAGO CIRCLE, 63- Mem: Am Asn Physics Teachers. Res: Atomic and molecular physics. Mailing Add: Dept of Physics Univ of Ill at Chicago Circle Chicago IL 60680

WEISSMAN, NORMAN, b New York, NY, Sept 12, 14; m 37; c 3. CLINICAL CHEMISTRY. Educ: City Col New York, BS, 35; Columbia Univ, PhD(biochem), 41. Prof Exp: Res fel dent med, Harvard Sch Dent Med, 41-43, instr, 43-46; lectr, Johns Hopkins Univ, 46-47, asst prof physiol chem & prev med, Sch Med, 47-51; assoc prof med, Col Med, State Univ NY Downstate Med Ctr, 51-56; assoc prof path & biochem, Col Med, Univ Utah, 56-70; asst dir, 70-74, SR RES SCIENTIST, CHEM DEPT, BIO-SCI LABS, 74- Concurrent Pos: Chemist, Maimonides Hosp, New York, 51-56 & Univ Hosp, Univ Utah, 57-65; consult, Vet Admin Hosp, Salt Lake City, Utah, 60-70. Mem: AAAS; Am Soc Biol Chemists; Am Asn Clin Chemists; Am Acad Forensic Sci; Harvey Soc. Res: Amino acid metabolism; bacterial chemistry; histochemistry; copper and connective tissue; toxicology methods. Mailing Add: Bio-Science Labs 7600 Tyrone Ave Van Nuys CA 91405

WEISSMAN, PAUL MORTON, b New York, NY, Oct 17, 36; m 71; c 2. ORGANOMETALLIC CHEMISTRY, ELECTROCHEMISTRY. Educ: City Col New York, BS, 60; Purdue Univ, Lafayette, PhD(chem), 64. Prof Exp: Res asst air pollution, NIH, Md, 59-60; res chemist, Gen Aniline & Film, Inc, 64-65; Petrol Res Fund grant, Univ Cincinnati, 65-66; asst prof chem, Brock Univ, 66-68; asst prof, 68-72, ASSOC PROF CHEM & CHMN DEPT, FAIRLEIGH DICKINSON UNIV, 72- Concurrent Pos: Nat Res Coun Can grant, Brock Univ, 66-68; NSF fel, Brandeis Univ, 67. Mem: Am Chem Soc; The Chem Soc. Res: Organometallic synthesis and reaction mechanisms; voltammetry in non-aqueous solvents; transition metal complexes containing a metal-metal bond. Mailing Add: Dept of Chem Fairleigh Dickinson Univ Madison NJ 07940

WEISSMAN, SAMUEL ISAAC, b South Bend, Ind, June 25, 12; m 43; c 2. PHYSICAL CHEMISTRY. Educ: Univ Chicago, BS, 33, PhD(phys chem), 38. Prof Exp: Fel, Univ Chicago, 39-41; Nat Res Coun fel, Univ Calif, 41-42; res chemist, Manhattan Proj, Calif, 42-43 & NMex, 43-46; from asst prof to assoc prof, 46-55, PROF CHEM, WASH UNIV, 55- Mem: Am Chem Soc. Res: Chemical spectroscopy; fluorescence; electrical conductivity; paramagnetic resonance. Mailing Add: Dept of Chem Wash Univ St Louis MO 63130

WEISSMAN, SHERMAN MORTON, b Chicago, Ill, Nov 22, 30; m 59; c 4. PHYSIOLOGY, BIOCHEMISTRY. Educ: Northwestern Univ, BS, 50; Univ Chicago, MS, 51; Harvard Univ, MD, 55. Prof Exp: Clin assoc cancer, NIH, 56-58; USPHS fel biochem & nucleic acids, Glasgow Univ, 59-60; sr investr, NIH, 60-67; assoc prof med, 67-71, PROF MED & MOLECULAR BIOPHYS & BIOCHEM, SCH MED, YALE UNIV, 71-, PROF HUMAN GENETICS, 72- Mem: Am Soc Biol Chem; Am Soc Clin Invest; Brit Biochem Soc. Res: Nucleic acid metabolism; molecular genetics. Mailing Add: 610 Hunter Radiation Yale Univ Sch of Med New Haven CT 06510

WEISSMAN, SIGMUND, b Vienna, Austria, July 1, 17; nat US; m 45; c 2. CRYSTALLOGRAPHY. Educ: Polytech Inst Brooklyn, PhD(chem), 52. Prof Exp: Res specialist x-ray diffraction, Col Eng, 49-63, PROF MAT ENG & DIR MAT RES LAB, RUTGERS UNIV, NEW BRUNSWICK, 63- Concurrent Pos: Consult, Lawrence Livermore Lab, Univ Calif & Savannah River Lab. Honors & Awards: Howe Medal, Am Soc Metals, 62. Mem: Am Crystallog Asn; Am Inst Mining, Metall & Petrol Engrs; Am Soc Metals; Am Soc Testing & Mat; NY Acad Sci. Res: X-ray crystallography; crystal imperfections of metals and alloys; irradiation of solids; recrystallization; recovery deformation; creep and fatigue of metals and alloys. Mailing Add: Dept of Mech & Mat Sci Col of Eng Rutgers Univ New Brunswick NJ 08903

WEISSMAN, STANLEY, physical chemistry, see 12th edition

WEISSMAN, WILLIAM, b New York, NY, Oct 12, 18; m 48; c 2. SURFACE CHEMISTRY. Educ: Univ Ala, Tuscaloosa, BA, 39; St John's Univ, NY, dipl chem, 44 & Newark Col Eng, dipl chem, 48. Hon Degrees: PhD, Sci Inst Caracas, 52. Prof Exp: Head insulation dept, Inslx Co, 39-46; lacquer chemist, Maas & Waldstein Co, Inc, 46-53; chief chemist, Nat Foil Co, Inc, 53-58; res chemist, Metro Adhesives Inc, 58-61; teacher chem, Passaic County Tech & Voc High Sch, 61-69; res chemist, US Rubber Reclaiming Co, Inc, 69-73; adhesive chemist, Cataphote Div, Ferro Inc, 73-74; sr scientist, Vicksburg Chem Co, 74-75; CONSULT, COMPLEX CHEMS, INC, 75- Concurrent Pos: Consult, Standard Chem Co, 53-69; Sussex County Bd Educ fel, Sussex County Tech Sch, Newton, NJ, 64-65; Passaic County Bd Educ fel, Montclair State Col, 67-68. Mem: Am Chem Soc; AAAS; Nat Geog Soc. Res: Adhesion and cohesion, organic and inorganic gases, liquids and solids; atomic micro and macro parameters. Mailing Add: 108 Katherine Dr Vicksburg MS 39180

WEISSMAN, WILLIAM KENT, b Jersey City, NJ, Jan 16, 27. NEUROSURGERY, BIOMEDICAL ENGINEERING. Educ: Dartmouth Col, AB, 50; Dartmouth Med Sch, dipl, 51; Jefferson Med Col, MD, 56. Prof Exp: Rotating intern, Mary Hitchcock Mem Hosp, Hanover, NH, 56-57, resident surg, 57-58; resident neurosurg, Vet Admin Hosp, Bronx, NY, 58-62; fel neuropath, Col Physicians & Surgeons, Columbia Univ, 62-64, instr, 63-64; clin instr surg & res assoc anat, 65-67, instr surg, 67-70, ASST PROF SURG, COL MED & DENT NJ, 70-; ADJ RES PROF, NEWARK COL ENG, 71- Concurrent Pos: Attend neurosurgeon, Martland Hosp Univ, Col Med & Dent NJ & US Vet Admin Hosp, East Orange, NJ; consult neurosurgeon, USPHS Hosp, Staten Island, NY. Mem: AAAS; NY Acad Sci; Asn Res Nerv & Ment Dis. Res: Computers of neurologic diganosis; states of consciousness. Mailing Add: Div of Neurosurg Col of Med & Dent of NJ Newark NJ 07103

WEISSMANN, BERNARD, b New York, NY, Dec 2, 17. BIOCHEMISTRY. Educ: City Col New York, BS, 38; Univ Mich, MS, 39, PhD, 51. Prof Exp: Res assoc, Col Physicians & Surgeons, Columbia Univ, 50-53 & Mt Sinai Hosp, New York, 53-57; from asst prof to assoc prof, 58-67, PROF BIOL CHEM, UNIV ILL COL MED, 67- Mem: Am Soc Biol Chem; Am Chem Soc; Brit Biochem Soc. Res: Enzymology of acid mucopolysaccharide catabolism; synthetic carbohydrate chemistry. Mailing Add: Dept of Biol Chem Univ of Ill Col of Med Chicago IL 60612

WEISSMANN, GERALD, b Vienna, Austria, Aug 7, 30; US citizen; m 53; c 2. CELL BIOLOGY, INTERNAL MEDICINE. Educ: Columbia Univ, BA, 50; NY Univ, MD, 54; Am Bd Internal Med, dipl, 63. Prof Exp: Res fel biochem, Arthritis & Rheumatism Found, 58-59; from instr to assoc prof, 59-70, PROF MED, SCH MED, NY UNIV, 70-, DIR DIV RHEUMATOLOGY, 74- Concurrent Pos: USPHS spec res fel, Strangeways Res Lab, Cambridge Univ, 60-61; sr investr, Arthritis & Rheumatism Found, 61-65; career investr, Health Res Coun, New York, 66-70; consult, US Food & Drug Admin & Nat Heart & Lung Inst, NIH; investr & instr physiol, Marine Biol Lab, Woods Hole, Mass, 70-; Guggenheim fel, Ctr Immunol & Physiol, Paris, 73-74; ed-in-chief, Inflammation. Honors & Awards: Alessandro Robecchi Prize for Rheumatology, 72; Marine Biol Lab Prize in Cell Biol, 74. Mem: Am Soc Cell Biol; Am Soc Exp Path; Am Soc Clin Invest; Am Rheumatism Asn; Soc Exp Biol & Med. Res: Study of lysosomes as they relate to cell injury; physiology and pharmacology of lysosomes and of artificial lipid structures. Mailing Add: NY Univ Med Ctr 550 First Ave New York NY 10016

WEIST, KATHERINE MORRETT, b Springfield, Ohio, Jan 6, 37; m 58; c 2. ETHNOLOGY. Educ: Ohio State Univ, BS, 58, MA, 62; Univ Calif, Berkeley, PhD(anthrop), 70. Prof Exp: Instr, 69-71, ASST PROF ANTHROP, UNIV MONT, 71- Concurrent Pos: Fel, Smithsonian Inst, 73-74; co-dir, Northern Plains Ethnohist Proj, Univ Mont, 74-; bk rev ed, Ethnohist, 75- Mem: Fel Am Anthrop Asn; Am Soc Ethnohist; Am Ethnol Soc; Soc Appl Anthrop; Sigma Xi. Res: Northern Plains ethnohistory; Northern Cheyenne culture and history; position of women cross-culturally; culture change. Mailing Add: Dept of Anthrop Univ of Mont Missoula MT 59801

WEIST, WILLIAM GODFREY, JR, b New York, NY, July 6, 31; m 56; c 4. GEOLOGY, HYDROLOGY. Educ: Amherst Col, BA, 53; Univ Colo, MS, 56. Prof Exp: Geologist, Ground Water Br, US Geol Surv, Colo, 56-67, hydrologist, Water Resources Div, NY, 67-70, supvry hydrologist, Ind, 70-73, asst dist chief, Ind Dist, 72-73, CHIEF, HYDROL STUDIES SECT, IND DIST, US GEOL SURV, 73- Mem: Geol Soc Am; Am Asn Petrol Geologists; Am Inst Prof Geologists; Nat Water Well Asn. Res: Ground-water hydrology. Mailing Add: Water Resources Div USGS 1819 N Meridian Indianapolis IN 46202

WEISTROP, DONNA ETTA, b New York, NY, June 10, 44. ASTRONOMY. Educ: Wellesley Col, BA, 65; Calif Inst Technol, PhD(astron), 71. Prof Exp: Vis lectr astron, Tel Aviv Univ, 71-73; univ fel astron, Ohio State Univ, 73-74; ASST ASTRONR, KIT PEAK NAT OBSERV, 74- Mem: Am Astron Soc. Res: Characteristics of disk and halo components of the galaxy as they relate to problems of galactic structure; stellar luminosity functions and density distributions. Mailing Add: Kitt Peak Nat Observ PO Box 26732 Tucson AZ 85726

WEISTROP, JESSIE SYD, b New York, NY, Apr 14, 48. PLANT PHYSIOLOGY, CELL BIOLOGY. Educ: State Univ NY Stony Brook, BS, 69; Univ Mass, Amherst, MA, 71, PhD(bot), 75. Prof Exp: FEL MEMBRANE BIOSYNTHESIS, RADIATION BIOL LAB, SMITHSONIAN INST, 74- Mem: AAAS; Am Soc Plant Physiologists; Bot Soc Am; Am Inst Biol Sci. Res: Biochemical and ultrastructural development of chloroplasts and the biosynthesis of chloroplast membranes; sulfate metabolism in algae. Mailing Add: Radiation Biol Lab 12441 Parklawn Dr Rockville MD 20852

WEISZ, GIDEON, solid state physics, see 12th edition

WEISZ, JUDITH, b Budapest, Hungary, Aug 6, 26; Brit citizen. REPRODUCTIVE PHYSIOLOGY, NEUROENDOCRINOLOGY. Educ: Newnham Col, Eng, BA, 48, MB, BCh, 51. Prof Exp: Clin training, London Hosp, Eng, 48-51; intern, Hadassah Med Sch, Hebrew Univ, Jerusalem, 52-54; second asst internal med, Tel Hashomer Govt & Mil Hosp, Israel, 54-56; registr internal med, London Hosp, Eng, 56-57; first asst internal med, Tel Hashomer Govt & Mil Hosp, Israel, 57-59; res fel endocrinol, Mt Sinai Hosp, New York, 60-62; fel, Training Prog Steroid Biochem, Worcester Found Exp Biol, Mass, 62-63, staff scientist, Training Prog Reproductive Physiol, 63-70, assoc dir, Training Prog Physiol of Reproduction, 68-72, sr scientist, Worcester Found Exp Biol, 70-72; ASSOC PROF, DIV REPRODUCTIVE BIOL, DEPT OBSTET & GYNEC, MILTON S HERSHEY MED CTR, PA STATE UNIV, 72-, HEAD DIV, 75-, SR MEM, GRAD SCH FAC, UNIV, 73- Mem: Brit Med Asn; Endocrine Soc; Int Soc Neuroendocrinol. Res: Neuroendocrine regulation of reproductive function; steroid biochemistry. Mailing Add: Div of Reproductive Biol Milton S Hershey Med Ctr Hershey PA 17033

WEISZ, PAUL B, b Vienna, Austria, Nov 3, 21; nat US; m 45; c 3. EMBRYOLOGY. Educ: McGill Univ, BSc, 43, MSc, 44, PhD(zool), 46. Prof Exp: Demonstr zool, McGill Univ, 43-46, lectr, 46-47, lectr biol, Sir George Williams Col, 43-44, lectr, 44-47; from instr to assoc prof, 47-57, PROF BIOL, BROWN UNIV, 57- Concurrent Pos: Mem, Ed Policies Comn & Comn Undergrad Educ Biol Sci. Mem: Soc Develop Biol; Am Soc Zool; Nat Asn Biol Teachers; Nat Sci Teachers Asn. Res: Embryology and development of amphibians and crustaceans; morphogenesis; cytochemistry and nuclear functions in Protozoa; comparative embryology; science writing. Mailing Add: Div of Biol & Med Sci Brown Univ Providence RI 02912

WEISZ, PAUL BURG, b Pilsen, Czech, July 2, 19; m 43; c 2. CHEMISTRY, CHEMICAL ENGINEERING. Educ: Ala Polytech Univ, BSc, 40; Swiss Fed Inst Technol, ScD, 66. Prof Exp: Asst, Univ Berlin, 38-39 & Bartol Res Found, Pa, 40-46; res assoc, 46-61, sr scientist, 61-67, mgr process res, 67-69, MGR CENT RES DIV,

MOBIL RES & DEVELOP CORP, 69- Concurrent Pos: Instr, Swarthmore Col, 42-43; vis prof, Princeton Univ, 74-; agent, US Patent Off. Honors & Awards: Murphee Award, Am Chem Soc, 72, Pioneer Award, Am Inst Chemists, 74. Mem: Fel Am Phys Soc; Am Chem Soc; The Chem Soc; fel Am Inst Chemists; Am Inst Chem Engrs. Res: Cosmic rays; Geiger counters; electric discharge phenomena; catalysis; diffusion phenomena; petroleum; petroleum processes; basic and interdisciplinary phenomena in the sciences. Mailing Add: Mobil Res & Develop Corp PO Box 1025 Princeton NJ 08540

WEISZ, ROBERT STEPHEN, b New York, NY, May 24, 18; m 56; c 3. INORGANIC CHEMISTRY. Educ: Cornell Univ, AB, 39, PhD(chem), 42. Prof Exp: Asst instr chem, Cornell Univ, 39-41; res fel, Westinghouse Elec & Mfg Co, 42, res engr, 42-46; res engr, Thomas A Edison, Inc, 46-49 & RCA Labs, 49-56; res dir, Telemeter Memories, Inc, 56-60, Ampex Comput Prods Co, 60-61; res dir, Electronics Memories, Inc, 61-69; INDUST CONSULT, 69- Mem: Am Chem Soc; Am Ceramic Soc. Res: Reactions in solids; ceramics; electrical properties of inorganic compounds; magnetic materials. Mailing Add: 15926 Alcima Ave Pacific Palisades CA 90272

WEITKAMP, WILLIAM GEORGE, b Fremont, Nebr, June 22, 34; m 56; c 3. NUCLEAR PHYSICS. Educ: St Olaf Col, BA, 56; Univ Wis, MS, 61, PhD(physics), 65. Prof Exp: Res asst physics, Univ Wis, 59-64; res asst prof, Univ Wash, Seattle, 64-67; asst prof, Univ Pittsburgh, 67-68; sr res assoc, 68-73, RES ASSOC PROF PHYSICS, UNIV WASH, 73-, TECH DIR, NUCLEAR PHYSICS LAB, 68- Mem: Am Phys Soc. Res: Polarization phenomena in reactions involving light nuclei; time reversal invarience in nuclear reactions; isobaric analog states in heavy nuclei; nuclear instrumentation. Mailing Add: Nuclear Physics Lab Univ of Wash Seattle WA 98195

WEITLAUF, HARRY, b Seattle, Wash, July 8, 37; m 64. REPRODUCTIVE PHYSIOLOGY. Educ: Univ Wash, BS, 59, MD, 63. Prof Exp: Fel reproductive physiol, Med Ctr, Univ Kans, 66-68, from asst prof to assoc prof anat, 68-74; ASSOC PROF ANAT, UNIV ORE MED CTR, 74- Mem: Am Asn Anat; Soc Study Reproduction. Res: Blastocyst metabolism during the preimplantation phase of development. Mailing Add: Dept of Anat Univ of Ore Med Ctr Portland OR 97201

WEITSEN, HOWARD ARTHUR, b New York, NY, Sept 23, 44; m 68; c 2. NEUROCHEMISTRY, ANATOMY. Educ: Jacksonville Univ, BA, 66; Med Col Va, PhD(anat), 69. Prof Exp: Instr anat, Med Col Va, 69-70; res assoc neuropharmacol, Lab Neuropharmacol, NIMH, 70-72; ASST PROF ANAT, LA STATE UNIV MED CTR, 72- Mem: Am Asn Anatomists. Res: Localizing adrenergic and cholinergic neurotransmitters in transplanted and pathological organs. Mailing Add: Dept of Anat La State Univ Med Ctr New Orleans LA 70119

WEITZ, ERIC, b New York, NY, Sept 18, 47. CHEMICAL PHYSICS, PHYSICAL CHEMISTRY. Educ: Mass Inst Technol, BS, 68; Columbia Univ, PhD(chem), 72. Prof Exp: Res assoc chem, Univ Calif, Berkeley, 72-74; ASST PROF CHEM, NORTHWESTERN UNIV, EVANSTON, 74- Mem: Am Phys Soc; Am Chem Soc. Res: Vibrational energy transfer in small molecules and the relation of internal energy to chemical reactivity; applications to laser-induced chemistry; isotope separation; studies of matrix isolated molecules. Mailing Add: Dept of Chem Northwestern Univ Evanston IL 60201

WEITZ, JOHN HILLS, b Cleveland, Ohio, Sept 20, 16; m 45; c 3. MINERALOGY, ECONOMIC GEOLOGY. Educ: Wesleyan Univ, BA, 38; Lehigh Univ, MS, 40; Pa State Univ, PhD(geol), 54. Prof Exp: Asst, Johns Hopkins Univ, 40-42; from jr mineral economist to mineral economist, US Bur Mines, Washington, DC, 42-46; coop geologist, State Geol Surv, Pa, 46-52; asst prof geol, Lehigh Univ, 47-52; geologist & secy, 52-61, PRES, INDEPENDENT EXPLOSIVES CO, 61- Mem: AAAS; Geol Soc Am; Mineral Soc Am; Am Inst Mining, Metall & Petrol Engrs. Res: Economic geology; clay minerals. Mailing Add: Independent Explosives Co 20950 Center Ridge Rd Rocky River OH 44116

WEITZ, JOSEPH LEONARD, b Cleveland, Ohio, June 2, 22; m 49; c 3. GEOLOGY. Educ: Wesleyan Univ, BA, 44; Yale Univ, MS, 46, PhD(geol), 54. Prof Exp: Geologist, Fuels Br, US Geol Surv, 48-55 & Independent Explosives Co, Pa, 55-58; asst prof geol, Wesleyan Univ, 58-60 & Colo State Univ, 60-61; assoc prof, Hanover Col, 61-62; assoc prof, 62-67, PROF GEOL, COLO STATE UNIV, 67- Concurrent Pos: Geologist, US Geol Surv, 66-; dir, Earth Sci Curric Proj, Colo, 67-69; mem, Coun Educ Geol Sci, 67-72; ed, J Geol Educ, 71-74. Mem: Geol Soc Am; Am Asn Petrol Geologists. Res: Geology of mineral fuels; Mesozoic stratigraphy; structural geology; geologic compilation. Mailing Add: Dept of Earth Resources Colo State Univ Ft Collins CO 80521

WEITZMAN, ELLIOT D, b Newark, NJ, Feb 4, 29; m; c 2. NEUROLOGY, NEUROPHYSIOLOGY. Educ: Univ Iowa, BA, 50; Univ Chicago, MD, 55. Prof Exp: Instr neuroanat, Univ Chicago, 51-52; instr med physiol, Univ Ill, 52; asst neurol, Columbia Univ, 58-59; assoc, 61-63, from asst prof to assoc prof, 63-69, PROF NEUROL, ALBERT EINSTEIN COL MED, 69-, CHMN DEPT, 71-, PROF NEUROSCI, 74-; CHIEF DIV NEUROL, MONTEFIORE HOSP & MED CTR, 69- Concurrent Pos: Consult, Jacobi Hosp, 61-69, Bronx Vet Admin Hosp, 63-67 & Health Res Coun, New York, 64-69; ed, Sleep Reviews. Mem: Am Neurol Asn; Soc Neurosci; Am Acad Neurol; Asn Res Nerv & Ment Dis; Asn Psychophysiol. Res: Study of sleep. Mailing Add: Dept of Neurol Montefiore Hosp & Med Ctr Bronx NY 10467

WEITZMAN, MARY C, b Norfolk, Va, Oct 13, 14; div; c 2. HISTOLOGY, ENDOCRINOLOGY. Educ: George Washington Univ, BA, 54, MA, 57; Albert Einstein Col Med, PhD(anat), 63. Prof Exp: Instr histol, 63-66, assoc anat, 66-68, asst prof, 68-74, ASSOC PROF ANAT, ALBERT EINSTEIN COL MED, 74- Concurrent Pos: Ed, Bibliographic Neuroendocrinologica. Mem: NY Acad Sci; Am Asn Anatomists; Europ Soc Comp Endocrinol; Am Soc Zoologists; Int Soc Neuroendocrinol. Res: Neuroendocrine mechanism in crustaceans. Mailing Add: Dept of Anat Albert Einstein Col of Med Bronx NY 10461

WEITZMAN, STANLEY HOWARD, b Mill Valley, Calif, Mar 16, 27; m 48; c 2. ICHTHYOLOGY. Educ: Univ Calif, AB, 51, AM, 53; Stanford Univ, PhD(biol), 60. Prof Exp: Sr lab technician, Univ Calif, 50-56; instr anat, Sch Med, Stanford Univ, 57-62; assoc cur, 63-67, CUR, DIV FISHES, US NAT MUS, SMITHSONIAN INST, 67- Mem: Am Soc Ichthyol & Herpet; Am Fisheries Soc; Soc Syst Zool; Soc Study Evolution. Res: Evolution, taxonomy and morphology of fishes. Mailing Add: Div of Fishes US Nat Mus Smithsonian Inst Washington DC 20560

WEITZNER, HAROLD, b Boston, Mass, May 19, 33; m 62; c 2. APPLIED MATHEMATICS, PLASMA PHYSICS. Educ: Univ Calif, AB, 54; Harvard Univ, AM, 55, PhD(physics), 58. Prof Exp: NSF fel, 58-59; assoc res scientist, 59-60, res scientist, 60-62, from asst prof to assoc prof, 62-69, PROF MATH, COURANT INST MATH SCI, NY UNIV, 69- Concurrent Pos: Consult, Los Alamos Sci Lab, Univ Calif, 62- & Lawrence Radiation Lab, Univ Calif, Berkeley, 67- Mem: Am Phys Soc. Res: Wave propagation problems in magnetohydrodynamics; kinetic theory and plasma oscillation problems; magnetohydrodynamic equilibrium; stability theory. Mailing Add: NYU Courant Inst of Math Sci 251 Mercer St New York NY 10012

WEITZNER, STANLEY, b New York, NY, Mar 11, 31; m 61; c 3. PATHOLOGY. Educ: NY Univ, BA, 51; Univ Geneva, MD, 56. Prof Exp: Intern, Queens Gen Hosp, Jamaica, NY, 57-58; resident path, Vet Admin Hosp, New York, 58-62; assoc pathologist, Southside Hosp, Bay Shore, 63-64; dir lab, Monticello Hosp, 65-66; from instr to assoc prof path, Sch Med, Univ NMex, 67-74; assoc prof, Univ Tex Health Sci Ctr, San Antonio, 74-75; ASSOC PROF PATH & DIR SURG PATH, UNIV MISS MED CTR, JACKSON, 75 Concurrent Pos: Instr, Sch Med, NY Univ, 60-61; dep med examr, Suffolk County, NY, 64; assoc, Col Physicians & Surgeons, Columbia Univ, 65-67; asst chief lab, Vet Admin Hosp, Albuquerque, NMex, 66-74; chief anat path, Vet Admin Hosp, San Antonio, 74-75. Mem: Int Acad Path; Am Soc Clin Path; Am Acad Oral Path; NY Acad Sci. Res: Laboratory medicine with prime interest in surgical and anatomical pathology. Mailing Add: Dept of Path Univ of Miss Med Ctr Jackson MI 39216

WEITZNER, STANLEY WALLACE, b Brooklyn, NY, Feb 12, 29; m 55; c 4. ANESTHESIOLOGY. Educ: NY Univ, BA, 49, MD, 53. Prof Exp: From asst prof to assoc prof, 59-70, PROF ANESTHESIOL, COL MED, STATE UNIV NY DOWNSTATE MED CTR, 70- Concurrent Pos: Vis anesthesiologist, Kings County Hosp, 56- Mem: Am Soc Anesthesiol; AMA; NY Acad Sci. Res: Artificial mechanical ventilation; acid-base balance; analog computation. Mailing Add: Dept of Anesthesiol State Univ NY Downstate Med Ctr Brooklyn NY 11203

WEKSLER, MARC EDWARD, b New York, NY, Apr 16, 37. MEDICINE, IMMUNOLOGY. Educ: Swarthmore Col, BA, 58; Columbia Univ, MD, 62. Prof Exp: Asst prof, 70-75, ASSOC PROF MED, MED COL, CORNELL UNIV, 70- Concurrent Pos: USPHS fel, Transplantation Unit, St Mary's Hosp, London, Eng, 67-68 & spec fel, New York Hosp, 68-70; asst attend physician, New York Hosp, 70-; dir, Hemodialysis Unit, Mem Hosp, New York. Mem: Am Soc Clin Invest; fel Am Col Physicians; Am Asn Immunologists. Res: Cellular immunology; renal transplantation. Mailing Add: Dept of Med Cornell Univ Med Col New York NY 10021

WEKSTEIN, DAVID ROBERT, b Boston, Mass, Feb 26, 37; m 58; c 4. PHYSIOLOGY. Educ: Boston Univ, AB, 57, MA, 58; Univ Rochester, PhD(physiol), 62. Prof Exp: From instr to asst prof, 62-68, ASSOC PROF PHYSIOL, COL MED, UNIV KY, 68- Mem: AAAS; Soc Exp Biol & Med; Geront Soc; Am Soc Zoologists; Am Physiol Soc. Res: Developmental physiology; physiology of temperature regulation; biology of aging; circadian rhythms. Mailing Add: Dept of Physiol & Biophys Univ of Ky Col of Med Lexington KY 40506

WELANDER, ARTHUR DONOVAN, b Chicago, Ill, July 4, 08; div; c 2. FISHERIES, RADIOBIOLOGY. Educ: Univ Wash, BS, 34, MS, 40, PhD(fisheries), 46. Prof Exp: Sci asst, Int Fishery Comn, Wash, 35-37; assoc instr ichthyol, 37-43, instr zool, 43-44, res assoc appl fisheries, 44-46, from instr to assoc prof, 46-58, PROF ICHTHYOL, UNIV WASH, 58- Concurrent Pos: Res assoc, AEC, 47-54, radiobiologist, Oper Crossroads, Bikini, 46, Bikini Sci Resurv, 47-58, Bikini-Eniwetok Resurv, 49 & 64, Opers Ivy, 52, Castle, 54, Redwing, 56, Hardtack, 58, Dominic, 62. Mem: Am Soc Icthyol & Herpet. Res: Biological effects of radiation; anatomy, taxonomy and histology of fishes; life history of Pacific salmon; effects of roentgen rays on growth and development of embryos and larvae of fishes; fishes of Central Pacific. Mailing Add: 108 Fisheries Ctr Univ of Wash Col of Fisheries Seattle WA 98195

WELBAND, WILBUR A, b Regina, Sask, July 11, 31; m 54; c 4. ANATOMY, NEUROLOGY. Educ: George Williams Col, BS, 54, MS, 56; Loyola Univ, Ill, PhD(anat), 61. Prof Exp: Instr anat, Stritch Sch Med, Loyola Univ, Ill, 61-63; asst prof anat, 63-69, instr neurol, 66-68, ASST PROF NEUROL, MED COL GA, 68-, ASSOC PROF ANAT, 69- Concurrent Pos: Grants, USPHS, 63 & Damon Runyon Mem, 65-; consult physiologist, Vet Admin Hosp, Augusta, Ga, 68- Mem: AAAS; Am Asn Anatomists; assoc Am Acad Neurol. Res: Clinical and experimental studies in electromyography; myo-neural junction anatomy and pathology. Mailing Add: Dept of Anat Med Col of Ga Augusta GA 30902

WELBECK, PAA-BEKOE HENRY, b Koforidua, Ghana, Apr 13, 40; m 74; c 1. COMMUNICATIONS SCIENCE, PUBLIC HEALTH EDUCATION. Educ: Livingstone Col, BA, 65; Boston Univ, MS, 68; Mich State Univ, PhD(educ technol), 71. Prof Exp: Asst prof psychiat & dir dept, Mich State Univ, 71-72; DIR LEARNING RESOURCES, COL DENT, HOWARD UNIV, 73- Mem: Am Asn Suicidology; Asn Biomed Commun Dirs; Health Educ Media Asn; Asn Educ Commun & Technol. Res: Role of biomedical communications in the prevention of, treatment of, teaching of, and evaluation of disease; instructional development in the health professions. Mailing Add: Div of Learning Resources Howard Univ Col of Dent Washington DC 20059

WELBER, BENJAMIN, b Czech, Nov 17, 20; m 49; c 3. HIGH PRESSURE PHYSICS. Educ: Yeshiva Univ, BA, 42; Columbia Univ, MA, 47, PhD(physics), 56. Prof Exp: Mem staff, NASA, Cleveland, Ohio, 49-60; guest scientist, Clarendon Lab, Oxford Univ, 58-59; UNESCO expert, Paris & UNESCO fel, Hebrew Univ, Israel, 60-61; SR STAFF MEM, THOMAS J WATSON RES CTR, IBM CORP, 61- Mem: AAAS; fel Am Phys Soc; Inst Elec & Electronics Engrs; Sigma Xi. Res: Solid state; electronic properties of solids at high pressures; low temperature and optical physics; paramagnetic resonance. Mailing Add: T J Watson Res Ctr IBM Corp PO Box 218 Yorktown Heights NY 10598

WELBORN, JOSEPH F, b Horton, Kans, July 23; m 44; c 5. ORAL SURGERY. Educ: Univ Nebr, BS, 47, DDS, 48; Northwestern Univ, MSD, 54; Am Bd Oral Surg, dipl, 62. Prof Exp: Chief oral surg, US Air Force Hosps, 50-66; ASSOC PROF ORAL SURG & DIR GRAD TRAINING, DENT SCH, NORTHWESTERN UNIV, CHICAGO, 66- Concurrent Pos: Attend oral surg, Cook County Hosp, Chicago, Ill, 67- & Vet Admin Res Hosp, Chicago, 67-; consult, Vet Admin Hosp, Downey, Ill. Mem: Fel Am Col Dent; Am Soc Oral Surg; Am Dent Asn; assoc Irish Dent Asn. Res: Oral cancer; synthetic bone paste in connection with autogenour bone grafting. Mailing Add: Dept of Oral Surg Northwestern Univ Dent Sch Chicago IL 60611

WELBOURNE, FRANK FITZHUGH, b Columbia, SC, Aug 22, 34. BOTANY, ZOOLOGY. Educ: Univ SC, BS, 56, MS, 58; Univ Okla, PhD(bot), 62. Prof Exp: Asst prof bot, Univ SDak, 61-63; asst prof biol, Memphis State Univ, 63-65 & Ga Southwestern Col, 65-69; assoc prof biol, Florence Regional Campus, Univ SC, 69-70; ASSOC PROF BIOL, FRANCIS MARION COL, 70- Mem: AAAS; Ecol Soc Am. Res: Plant ecology. Mailing Add: Dept of Biol Francis Marion Col Florence SC 29501

WELBY, CHARLES WILLIAM, b Bakersfield, Calif, Oct 9, 26; m 48; c 2. HYDROGEOLOGY, STRATIGRAPHY. Educ: Univ Calif, BS, 48, MS, 49; Mass Inst Technol, PhD(geol), 52. Prof Exp: Geologist, Calif Co, 52-54; asst prof geol, Middlebury Col, 54-58, Trinity Col, Conn, 58-61 & Rensselaer Polytech Inst, 61-62; assoc prof geol & chmn dept, Southern Miss Univ, 62-65; ASSOC PROF GEOL, NC

STATE UNIV, 65- Mem: AAAS; fel Geol Soc Am; Soc Econ Paleont & Mineral; Asn Eng Geologists; Int Asn Hydrologists. Res: Occurrence and management of ground water; structure and stratigraphy; occurrence and exploration for petroleum; geochemistry of sediments; water resource planning and management; remote sensing. Mailing Add: Dept of Geosci NC State Univ Raleigh NC 27607

WELCH, AARON WADDINGTON, b Georgetown, Md, July 25, 16; m 41; c 1. BOTANY. Educ: Univ Md, BS, 37; Iowa State Col, PhD(plant path), 42. Prof Exp: Mgr, Southeastern Exp Sta, Iowa, 40-42; assoc pathologist, Div Forage Crops, USDA, 45-47; plant pathologist, Exp Sta, 47-60, mgr res & develop farm, Clayton, 60-73, DIR RES & DEVELOP FARM, CLAYTON, E I DU PONT DE NEMOURS & CO, INC, 73- Mem: Am Phytopath Soc. Res: Plant pathology; helminthology; herbicides. Mailing Add: Res & Develop Farm 420 N Baylan E I du Pont de Nemours & Co Inc Raleigh NC 27603

WELCH, ARNOLD DEMERRITT, b Nottingham, NH, Nov 7, 08; m 33, 66; c 3. PHARMACOLOGY, CHEMOTHERAPY. Educ: Univ Fla, BS, 30, MS, 31; Univ Toronto, PhD(pharmacol), 34; Washington Univ, MD, 39. Hon Degrees: DSc, Univ Fla, 73. Prof Exp: Asst physiol, Exp Sta, Univ Fla, 29-31; asst pharmacol, Sch Med, Washington Univ, 35-36, instr, 36-40; dir pharmacol res, Sharpe & Dohme, Inc, 40-44, asst dir res, 42-43, dir, 43-44; prof pharmacol & dir dept, Sch Med, Western Reserve Univ, 44-53; prof & chmn dept, Sch Med & Grad Sch, Yale Univ, 53-67, Eugene Higgins prof, 57-67; dir, Squibb Inst Med Res, 67-72, pres, 72-74, vpres res & develop, E R Squibb & Sons, Inc, 67-74; CHMN DIV BIOCHEM & CLIN PHARMACOL, ST JUDE CHILDREN'S RES HOSP, 75-; PROF PHARMACOL, UNIV TENN SCH MED, 75- Concurrent Pos: Fulbright scholar, Oxford Univ, 52; Commonwealth scholar, Univ Frankfurt, 64-65; mem, Sci Adv Comt, Leonard Wood Mem Found, 48-53; mem comts, Nat Res Coun, 48-56, chmn comt growth, 52-54; Herter lectr, NY Univ, 52; consult to Surgeon Gen, USPHS, 52-64; Carter lectr, Columbia Univ, 53; mem, Div Comt Biol & Med Res, NSF, 53-55; Rockwood lectr, Univ Iowa, 55; mem, Pharmcol & Exp Therapeut Study Sect, NIH, 59-63, chmn, 60-63, chmn chemother study sect, 63-65; mem, Sci Adv Bd, St Jude Hosp, 69-72; mem, Sci Adv Bd Biol Sci, Princeton Univ, 69-75; emer chmn bd dirs, Squibb Inst Med Res, 74- Honors & Awards: Sollmann Award, Am Soc Pharmacol & Exp Therapeut, 66. Mem: AAAS; Am Soc Pharmacol & Exp Therapeut; Am Soc Biol Chemists; Am Chem Soc. Res: Cellular localization of pressor amines; sulfonamides; filariasis; structure and action of chlorine-like compounds; biosynthesis of labile methyl group; biosynthesis and antagonism of utilization of nucleic acid precursors; metabolic approaches to cancer and virus chemotherapy; inhibition of enzyme induction. Mailing Add: St Jude Children's Res Hosp 332 N Lauderdale Memphis TN 38101

WELCH, BILLY EDWARD, b West, Tex, Sept 16, 29; m 56; c 4. PHYSIOLOGY, BIOCHEMISTRY. Educ: Abilene Christian Col, BS, 50; Agr & Mech Col Tex, PhD, 54. Prof Exp: Physiologist, Human Eng Br, Northrop Aircraft, Inc, 57-59; physiologist, Dept Physiol, US Air Force Sch Aviation Med, 59, chief environ systs br, 59-68, chief environ systs div, US Air Force Sch Aerospace Med, 68-71; SPEC ASST ENVIRON QUAL TO ASST SECY AIR FORCE INSTALLATIONS & LOGISTICS, PENTAGON, 72- Mem: Fel Aerospace Med Asn; Am Physiol Soc. Res: Environmental physiology and quality. Mailing Add: SAFILE The Pentagon Washington DC 20330

WELCH, BRUCE L, b Atlanta, Ga, Nov 18, 31; m 59. ECOLOGY, NEUROPHARMACOLOGY. Educ: Auburn Univ, BA, 51; Duke Univ, PhD(biol), 62. Prof Exp: Asst prof biol, Col William & Mary, 62-66; sr scientist, Mem Res Ctr, Univ Tenn, Knoxville, 66-69; sr investr environ neurobiol, 69-70, DIR ENVIRON NEUROBIOL, FRIENDS PSYCHIAT RES, INC, BALTIMORE, 70-; ASSOC PROF PSYCHIAT & BEHAV SCI, SCH MED, JOHNS HOPKINS UNIV, 70- Concurrent Pos: Grants, Found Fund Res Psychiat, Air Force Off Sci Res, US Army Med Res & Develop Command, NIH & NASA, 62-; mem ad hoc adv comt physiol indicators stress, Dept Army, 63- Mem: Am Soc Naturalists; Ecol Soc Am; Soc Exp Biol & Med; Am Soc Pharmacol & Exp Therapeut; Soc Toxicol. Res: Functional synecology; population dynamics; neuroendocrine feedback mechanisms; neurochemistry and neuropharmacology of aggression; physiological effects of noise; ecology of homicide and suicide; environmental neurobiology. Mailing Add: Dept of Psychiat & Behav Sci Johns Hopkins Univ Sch of Med Baltimore MD 21205

WELCH, CHARLES DARREL, b Albright, WVa, Sept 2, 19; m 48; c 3. AGRONOMY. Educ: Univ WVa, BS, 42; NC State Univ, MS, 48, PhD(soils), 60. Prof Exp: Instr soil fertil, NC State Univ, 48-49; agronomist, NC Dept Agr, 50-63; EXTEN SOIL CHEMIST, TEX A&M UNIV, 63- Mem: AAAS; Am Soc Agron; Soil Sci Soc Am. Res: Plant nutrient response in field and greenhouse soils; soil tests for nitrogen. Mailing Add: Soil Testing Lab Tex A&M Univ College Station TX 77843

WELCH, CHARLES STUART, b Newburyport, Mass, Oct 4, 09; m 41; c 3. SURGERY. Educ: Tufts Col, MD, 32; Univ Minn, MS & PhD(surg), 37. Prof Exp: Asst prof surg, Albany Med Col, 37-47; prof, Med Sch, Tufts Univ, 45-50; assoc prof, 50-52, PROF SURG, ALBANY MED COL, 52- Concurrent Pos: Attend surgeon, Albany Hosp. Mem: Fel Am Surg Asn; fel Am Col Surg. Res: General surgery. Mailing Add: 11 Northern Blvd Albany NY 12210

WELCH, CLARK MOORE, b Mountain Grove, Mo, Mar 3, 25; m 66; c 2. ORGANIC CHEMISTRY. Educ: Univ Tenn, BS, 46, MS, 47, PhD(chem), 52. Prof Exp: Res chemist, Monsanto Chem Co, 47-48; asst prof chem, La State Univ, 52-56; RES LEADER, EXPLOR RES, COTTON CHEM REACTIONS LAB, USDA, 56- Mem: AAAS; Am Chem Soc; Am Asn Textile Chemists & Colorists. Res: Polycondensation reactions; germicidal fabrics; flameproofing; ionic and coordination catalysis; sulfonamides; inorganic polymers; cellulose chemistry; synthesis of organophosphorus compounds; textile finishing; free radical reactions of vinyl and silicone polymers. Mailing Add: 1036 Homestead Ave Metairie LA 70005

WELCH, CLAUDE ALTON, b Flint, Mich, Oct 24, 21; m 49; c 2. ZOOLOGY. Educ: Mich State Univ, BS, 48, PhD(zool), S7. Prof Exp: From instr to prof natural sci, Mich State Univ, 53-69; prof, 69-73, O T WALTER PROF BIOL, MACALESTER COL, 73-, CHMN DEPT, 69- Concurrent Pos: Supvr blue version writing team, Biol Sci Curriculum Study, 61, consult, Adaptation Comt, Japan, 63-65 & Turkey, 65. Mem: AAAS; Nat Asn Biol Teachers (pres, 72); Nat Sci Teachers Asn. Res: Cell physiology. Mailing Add: Dept of Biol Macalester Col St Paul MN 55101

WELCH, CLETUS NORMAN, b Convoy, Ohio, Feb 2, 37; m 60; c 2. PHYSICAL INORGANIC CHEMISTRY. Educ: Bowling Green State Univ, BS, 61; Ohio State Univ, MS, 64, PhD(chem), 66. Prof Exp: SR RES CHEMIST, PPG INDUSTS, INC, 66- Mem: Am Chem Soc. Res: Halogen and halogen-oxygen chemistry; synthesis of electrocatalysts; electrode processes; electrochemical cell design and materials of construction; electroless, electrolytic and vapor deposition of metallic coatings. Mailing Add: 1217 Meadowlark Dr Clinton OH 44216

WELCH, DEAN EARL, b Aledo, Ill, Aug 5, 37; m 58; c 6. ORGANIC CHEMISTRY. Educ: Monmouth Col, BA, 59; Mass Inst Technol, PhD(org chem), 63. Prof Exp: Chemist, Escambia Chem Corp, 63-64; assoc scientist, 64-66, scientist, 66-68, qual assurance dir, 68-69, chem res dir, 69-71, res dir, 71-72, VPRES RES, SALSBURY LABS, 72- Mem: Am Chem Soc; The Chem Soc. Res: Organic synthesis; research and development administration. Mailing Add: Salsbury Labs Rockford Rd Charles City IA 50616

WELCH, DONALD RAY, b Buena Vista, Ohio, June 6, 34; m 64. PLANT PATHOLOGY, BIOCHEMISTRY. Educ: Univ Miami, AB, 59; Pa State Univ, MS, 61, PhD(plant path), 63. Prof Exp: From instr to asst prof, 63-71, ASSOC PROF BIOL, BELOIT COL, 71- Concurrent Pos: Resident res assoc & Assoc Cols Midwest prof, Argonne Nat Lab, 65-66. Mem: AAAS; Am Phytopath Soc; Am Soc Microbiol; Am Soc Plant Physiol. Res: Structure of plant viruses; methods of preserving purified viruses. Mailing Add: Dept of Biol Beloit Col Beloit WI 53511

WELCH, ELDRED, organic chemistry, see 12th edition

WELCH, FELIX PERRY, mathematics, see 12th edition

WELCH, FRANK JOSEPH, b Fresno, Calif, Aug 5, 29; m 54; c 5. ORGANIC CHEMISTRY. Educ: Stanford Univ, PhD(org chem), 54. Prof Exp: Res chemist polymers, Union Carbide Chem Co, 54-62, group leader latex polymerization, Res & Develop Dept, Union Carbide Corp, 62-71; DIR RES & ENG, AVERY LABEL CO, AVERY PROD CORP, 71- Mem: Tech Asn Pulp & Paper Indust; Am Chem Soc. Res: Ionic and free radical polymerization kinetics; organo-metallic and organophosphorus chemistry; latex polymerization; coatings and printing inks. Mailing Add: Avery Label Co 777 E Foothill Blvd Azusa CA 91702

WELCH, GARTH LARRY, b Brigham City, Utah, Feb 14, 37; m 60; c 6. ACADEMIC ADMINISTRATION, INORGANIC CHEMISTRY. Educ: Univ Utah, BS, 59, PhD(inorg chem), 63. Prof Exp: Fel, Univ Calif, Los Angeles, 62-64; from asst prof to assoc prof, 64-72, PROF CHEM, WEBER STATE COL, 72-, DEAN, SCH NATURAL SCI, 74- Mem: Am Chem Soc. Res: Kinetics of complex ions; reversion rates of polymers. Mailing Add: Sch of Natural Sci Weber State Col Ogden UT 84408

WELCH, GEORGE IRA, physics, see 12th edition

WELCH, GORDON E, b Sabinal, Tex, Aug 20, 33; m 57; c 3. MICROBIOLOGY. Educ: Southwest Tex State Univ, BS, 60, MA, 62; Tex A&M Univ, PhD(microbiol), 66. Prof Exp: Chmn dept biol & chem, Southwest Tex Jr Col, 60-62; assoc prof biol, San Antonio Col, 66-67; assoc prof, 67-73, PROF BIOL, ANGELO STATE UNIV, 73- Concurrent Pos: Consult, Bexar Co Hosp Dist, San Antonio, Tex, 66. Mem: AAAS; Am Soc Microbiol; Am Inst Biol Sci. Res: Virology, particularly myxoviruses, their nature, properties and pathogenicity. Mailing Add: Dept of Biol Angelo State Univ San Angelo TX 76901

WELCH, GRAEME P, b Los Angeles, Calif, Nov 25, 17; m 45; c 4. BIOPHYSICS. Educ: Univ Calif, Los Angeles, AB, 40; Univ Calif, Berkeley, MA, 50, PhD(biophys), 57. Prof Exp: Assoc physicist, Div War Res, Univ Calif, 41-45, physicist, Donner Lab, 48-59; engr biophys, Saclay Nuclear Res Ctr, France, 59-61; BIOPHYSICIST, DONNER LAB, LAWRENCE RADIATION LAB, UNIV CALIF, BERKELEY, 61- Mem: Radiation Res Soc. Res: Biological effects of radiation; physics and dosimetry of heavy-particle radiations. Mailing Add: Lawrence Radiation Lab Univ of Calif Berkeley CA 94720

WELCH, HAROLD FRANCIS, b Ilion, NY, Mar 25, 27; m 49; c 5. SURGERY. Educ: Union Col, NY, BA, 48; Tufts Univ, MD, 53. Prof Exp: Clin investr surg, Vet Admin Hosp, 59-62, chief, 62-67; PROF SURG, ALBANY MED COL, 67-, ATTEND SURGEON, MED CTR, 67- Mem: Fel Am Col Surg. Mailing Add: Dept of Surg Albany Med Col Albany NY 11208

WELCH, HUGH GORDON, physiology, see 12th edition

WELCH, J PHILIP, b Macclesfield, Eng, June 18, 33; m 58; c 4. HUMAN GENETICS. Educ: Univ Edinburgh, MB & ChB, 58; Johns Hopkins Univ, PhD, 69. Prof Exp: Fel med, Johns Hopkins Hosp, 63-67; asst prof, 67-72, ASSOC PROF PEDIAT, DALHOUSIE UNIV, 72- Concurrent Pos: Consult, Children's Hosp, Victoria Gen Hosp, Grace Maternity Hosp & Halifax Infirmary, 67- Mem: Am Soc Human Genetics; Am Fedn Clin Res; Am Asn Advan Aging Res; Can Genetics Soc; Soc Study Social Biol. Res: Biochemical, behavioral and cytogenetic aspect of mental abilities and aberrations. Mailing Add: Dept of Pediat Dalhousie Univ Halifax NS Can

WELCH, JAMES ALEXANDER, b Versailles, Ky, May 25, 24; m 51; c 2. ANIMAL HUSBANDRY, ANIMAL PHYSIOLOGY. Educ: Univ Ky, BS, 47; Univ Ill, PhD(animal sci), 52. Prof Exp: From asst prof to assoc prof animal husb, 52-60, actg chmn dept animal sci, 62, PROF ANIMAL SCI, UNIV WVA & ANIMAL SCIENTIST, EXP STA, 60- Concurrent Pos: From asst animal husbandman to assoc animal husbandman, 52-60; chief party, WVa Univ-AID contract team, Uganda & lectr, Vet Training Inst, Entebbe, 63-66. Mem: Fel AAAS; Am Soc Animal Sci. Mailing Add: Div of Animal & Vet Sci WVa Univ Morgantown WV 26506

WELCH, JAMES EDWARD, b San Rafael, Calif, July 19, 11; m 37; c 3. GENETICS. Educ: Univ Calif, BS, 34, MS, 35; Cornell Univ, PhD(genetics), 42. Prof Exp: Trainee agron, Soil Conserv Serv, USDA, 35, jr agr aide, 35-36, agr aide, 36; jr olericulturist, Exp Sta, Univ Hawaii, 36-38, asst olericulturist, 38-39; asst, Maize Genetics Coop, Cornell Univ, 40-42; asst horticulturist, Exp Sta, La State Univ, 42-43; assoc horticulturist, Regional Veg Breeding Lab, USDA, SC, 43-47; asst olericulturist, 47-52, ASSOC OLERICULTURIST, EXP STA, UNIV CALIF, DAVIS, 52- Concurrent Pos: Lectr veg crops, Col Agr, Univ Calif, Davis, 49-59; collabr, Plant Sci Res Div, USDA, 58-74. Honors & Awards: Western Res Man of the Year Award, Pac Seedmen's Asn, 65. Mem: Fel AAAS; Am Inst Biol Sci; Am Soc Hort Sci; Am Genetic Asn. Res: Plant morphological characters associated with earworm resistance in maize; genetic linkage in autotetraploid maize; genetics of male sterility in the carrot; breeding improved varieties of vegetable crops, especially sweet corn, lima beans, carrots, celery, tomatoes and lettuce. Mailing Add: Dept of Veg Crops Univ of Calif Davis CA 95616

WELCH, JAMES GRAHAM, b Ithaca, NY, Aug 16, 32; m 56; c 4. ANIMAL NUTRITION. Educ: Cornell Univ, BS, 55; Univ Wis, MS, 57, PhD(biochem, animal husb), 59. Prof Exp: Asst, Univ Wis, 55-59; asst prof animal husb, Rutgers Univ, 59-63, assoc prof nutrit, 63-68; assoc prof, 68-70, PROF ANIMAL SCI, UNIV VT, 70- Mem: AAAS; Am Soc Animal Sci; Am Dairy Sci Asn; Am Inst Nutrit. Res: Ruminant nutrition. Mailing Add: Dept of Animal Sci Univ of Vt Burlington VT 05401

WELCH, JASPER ARTHUR, JR, physics, see 12th edition

WELCH, LESTER MARSHALL, b Shepherdsville, Ky, Dec 28, 17; m 42; c 2. PETROLEUM CHEMISTRY. Educ: Taylor Univ, AB, 39; Purdue Univ, MS, 41,

PhD(chem), 42. Prof Exp: Res chemist, E I du Pont de Nemours & Co, Va, 42-44; res chemist, Stand Oil Develop Co, 44-47, res group head, 47-50; tech dir, Carter Bell Mfg Co, 50-55; dir res, 55-60, VPRES & DIR RES & DEVELOP, PETRO-TEX CHEM CORP, 60- Mem: Am Inst Chem Eng; Am Chem Soc; Am Inst Chem. Res: Distillation; polymers; rubber; resins; effect of azeotopism on rectification; petrochemicals. Mailing Add: Petro-Tex Chem Corp Box 2548 Houston TX 77001

WELCH, LIN, b Tahoka, Tex, Dec 9, 27; m 50; c 3. SPEECH PATHOLOGY. Educ: WTex State Univ, BS, 48; Baylor Univ, MA, 49; Univ Mo, PhD(speech path), 60. Prof Exp: Instr speech, Blue Mountain Col, 52-53; instr, Univ Mo, 53-56; PROF SPEECH PATH, CENT MO STATE UNIV, 56-, HEAD DEPT SPEECH PATH & AUDIOL, 70- Concurrent Pos: Proj dir, Neurol & Sensory Dis Prof grants, 64-66; dir, Cleft Palate Camp, Mo Crippled Children's Serv, 66, 67, 68 & 69; chmn, Continuing Educ Comt, Am Speech & Hearing Asn, 72-73. Mem: Am Speech & Hearing Asn. Res: Cleft palate; stuttering. Mailing Add: Dept of Speech Path & Audiol Cent Mo State Univ Warrensburg MO 64093

WELCH, LLOYD RICHARD, b Detroit, Mich, Sept 28, 27; m 53; c 3. MATHEMATICS. Educ: Univ Ill, BS, 51; Calif Inst Technol, PhD(math), 58. Prof Exp: Sr res engr, Jet Propulsion Lab, Calif Inst Technol, 57-59; staff mathematician, Inst Defense Anal, 59-65; assoc prof elec eng, 65-68, PROF ELEC ENG, UNIV SOUTHERN CALIF, 68- Mem: Am Math Soc; Math Asn Am; Soc Indust & Appl Math. Res: Probability theory; combinatorics; communication theory. Mailing Add: Dept of Elec Eng Univ of Southern Calif Los Angeles CA 90007

WELCH, LOUIS FREDERICK, soil fertility, see 12th edition

WELCH, MELVIN BRUCE, b Hood River, Ore, Feb 24, 45; m 67; c 2. INORGANIC CHEMISTRY, POLYMER CHEMISTRY. Educ: Linfield Col, BA, 67; Univ Utah, PhD(inorg chem), 74. Prof Exp: RES CHEMIST, PHILLIPS PETROL CO, 74- Mem: Am Chem Soc; Soc Plastics Engrs. Res: Olefin polymerization and organometallic chemistry. Mailing Add: 83-E PRC Phillips Petrol Co Bartlesville OK 74004

WELCH, MICHAEL JOHN, b Stoke-on-Trent, Eng, June 28, 39; m 67. RADIOCHEMISTRY, NUCLEAR MEDICINE. Educ: Cambridge Univ, BA, 61, MA, 64; Univ London, PhD(radiochem), 65. Prof Exp: Res assoc chem, Brookhaven Nat Lab, 65-67; from asst prof to assoc prof, 67-74, PROF RADIATION CHEM, SCH MED, WASHINGTON UNIV, 74- Mem: The Chem Soc; Am Chem Soc; Soc Nuclear Med; Radiation Res Soc. Res: Hot atom chemistry; isotopes in medicine. Mailing Add: Dept of Radiol Washington Univ Sch of Med St Louis MO 63110

WELCH, PETER D, b Detroit, Mich, May 19, 28; m 53; c 2. MATHEMATICAL STATISTICS. Educ: Univ Wis, MS, 51; NMex State Univ, MS, 56; Columbia Univ, PhD(math statist), 63. Prof Exp: Assoc mathematician, Phys Sci Lab, NMex State Univ, 51-56; RES STAFF MEM PROBABILITY & STATIST, IBM RES CTR, 56- Concurrent Pos: Adj prof, Columbia Univ, 64-65, 74-75. Mem: Inst Elec & Electronics Engrs. Res: Queueing theory; time series analysis; signal processing. Mailing Add: 85 Croton Ave Mount Kisco NY 15049

WELCH, QUINTIN B, b Sterling Co, Tex, June 28, 35; m 56; c 2. GENETICS, APPLIED STATISTICS. Educ: Univ Ark, BS, 57; Okla State Univ, MS, 60; Univ Calif, Berkeley, PhD(genetics), 69. Prof Exp: Instr res, Okla State Univ, 58-60; biometrician, Animal Breeding Consult, Calif, 60-62; sr statistician, Inst Personality Assessment & Res, Univ Calif, Berkeley, 62-66 & Family Planning Proj, Sch Pub Health, 66-67; lectr biomet, Univ Calif, Santa Barbara, 69-70; res assoc, Univ Md Int Ctr Med Res & Training, Lahore, WPakistan & Kuala Lumpur, Malaysia, 70-72; asst res geneticist, Univ Calif Int Ctr Med Res & Training, Kuala Lumpur, Malaysia, 72-73; ASST PROF GENETICS, UNIV MO-KANSAS CITY, 73- Mem: AAAS; Biomet Soc; Am Genetics Asn; World Poultry Sci Asn. Res: Human population genetics. Mailing Add: Dept of Biol Univ Mo-Kansas City Kansas City MO 64110

WELCH, RICHARD MARTIN, b Brooklyn, NY, Nov 4, 33; m 54; c 3. BIOCHEMICAL PHARMACOLOGY. Educ: St John's Univ, NY, BS, 57; Jefferson Med Col, MS, 60, PhD(pharmacol), 62. Prof Exp: Asst prof pharmacol, Jefferson Med Col, 62-63; GROUP LEADER MEDICINAL BIOCHEM, BURROUGHS WELLCOME RES LABS, 65- Concurrent Pos: Fel, Albert Einstein Col Med, 63-65. Mem: Am Soc Pharmacol & Exp Therapeut. Res: Pharmacodynamics; absorption, distribution and metabolism of drugs. Mailing Add: Med Biochem Burroughs Wellcome Res Labs Research Triangle Park NC 27609

WELCH, ROBERT MCCLAM, b Sumter Co, SC, Sept 20, 12; m 59; c 2. GENETICS, CELL BIOLOGY. Educ: Col Charleston, AB, 32; Univ Tex, PhD(zool), 56. Prof Exp: Res assoc, Genetics Found, Univ Tex, 56-63; NIH spec fel, Inst Med Cell Res & Genetics, Karolinska Inst, Sweden, 63-66; assoc prof biol, 67-72, PROF BIOL, W GA COL, 72- Mem: Am Soc Cell Biol; Genetics Soc Am; Sigma Xi. Res: Histochemistry; cytochemistry. Mailing Add: Dept of Biol WGa Col Carrollton GA 30117

WELCH, RODNEY CHANNING, b Ayrshire, Iowa, Dec 16, 04; m 37; c 3. FOOD SCIENCE. Educ: State Col Wash, BS, 28, MS, 31; Pa State Col, PhD(dairy husb), 35. Prof Exp: Asst county agent, Exten Serv, State Col Wash, 28-30; asst, Pa State Col, 31-34; dir lab, Sylvan Seal Milk, Inc, 34-43; supt, Wilbur Chocolate Co, 43-57, vpres, 55-71, dir, 59-71, tech consult, 71-73; CONSULT, 73- Concurrent Pos: Volunteer exec, Int Exec Serv Corps, 75- Mem: Am Asn Cereal Chem; NY Acad Sci. Res: Milk and chocolate products. Mailing Add: 415 S Cedar St Lititz PA 17543

WELCH, RONALD MAURICE, b Chicago, Ill, Dec 30, 43; m 67. ATMOSPHERIC PHYSICS, METEOROLOGY. Educ: Calif State Univ, Long Beach, BS, 65, MA, 67; Univ Utah, PhD(physics), 71, PhD(meteorol), 76. Prof Exp: Engr-scientist, Missile & Space Systs Div, Douglas Aircraft Co, Calif, 66; res assoc geophys, Univ Utah, 71-72, res assoc physics, 72, teaching assoc, 72-73, assoc instr, Meteorol Dept, 73-74, res assoc, 74-76; RES ASSOC, DEPT ATMOSPHERIC SCI, COLO STATE UNIV, 76- Mem: AAAS; Am Phys Soc; Am Geophys Union; Sigma Xi; Am Meteorol Soc. Res: Radiative transfer in planetary atmospheres; climatology; electronic properties of materials. Mailing Add: Dept Atmospheric Sci Colo State Univ Fort Collins CO 80521

WELCH, ROSS MAYNARD, b Lancaster, Calif, May 8, 43; m 65; c 2. PLANT NUTRITION, PLANT PHYSIOLOGY. Educ: Calif State Polytech Col, San Luis Obispo, 66; Univ Calif, Davis, MS, 69, PhD(soil sci), 71. Prof Exp: Res assoc plant mineral nutrit, Dept Agron & US Plant, Soil & Nutrit Lab, Cornell Univ, 71-72, PLANT PHYSIOLOGIST, AGR RES SERV, US PLANT, SOIL & NUTRIT LAB, USDA, 72- Concurrent Pos: Asst prof agron, Cornell Univ, 74- Mem: AAAS; Soil Sci Soc Am; Am Soc Plant Physiol; Am Soc Agron; Soc Environ Geochem & Health. Res: Plant mineral nutrition; ion transport by plant tissues and cells; trace element physiology; physiological form and bioavailability of mineral elements in plants to animals and humans. Mailing Add: US Plant Soil & Nutrit Lab Cornell Univ Tower Rd Ithaca NY 14853

WELCH, ROY ALLEN, b Waukesha, Wis, Nov 14, 39; m 67. GEOGRAPHY, PHOTOGRAMMETRY. Educ: Carroll Col, BS, 61; Univ Okla, Norman, MA, 65; Univ Glasgow, PhD(geog, photogram), 68. Prof Exp: Photo-analyst geog, US Govt, 62-64; mgr earth sci dept, Itek Corp, 68-69; Nat Acad Sci-Nat Res Counc res assoc, US Geol Surv, 69-71; ASSOC PROF GEOG, UNIV GA, 71- Concurrent Pos: Consult, AID, 72; mem working group optical transfer function-modulation transfer function, Comn I, Int Soc Photogram, 72-76; mem remote sensing comt, Asn Am Geogrs, 73-75; remote sensing specialist, US Geol Surv, 73-; dep dir Photog Div, Am Soc Photogram, 74-, dir Soc, 74-77. Honors & Awards: III Talbert Abrams, Am Soc Photogram, 71, Presidential Citation, 75. Mem: Am Soc Photogram; Brit Photogram Soc; Asn Am Geogrs; Brit Cartog Soc; Sigma Xi. Res: Analyses of the quality and applications of aircraft and satellite imagery for geographic tasks, with particular reference to land use mapping and cartography. Mailing Add: Dept of Geog Univ of Ga Athens GA 30602

WELCH, STANLEY L, b Rockport, Utah, Sept 7, 28; m 49; c 8. PLANT TAXONOMY. Educ: Brigham Young Univ, BS, 51, MS, 57; Iowa State Univ, PhD(plant taxon), 60. Prof Exp: From asst prof to assoc prof, 60-67, PROF BOT, BRIGHAM YOUNG UNIV, 67- Concurrent Pos: Consult, Wetherill Mesa Proj, Mesa Verde Nat Park, Colo, 61-63 & Alaska Flora Proj, Iowa State Univ, 65- Mem: Am Soc Plant Taxon. Res: Flora of Alaska and Yukon; Astragalus and Oxytropis in Alaska; legumes of Utah; viscid Oxytropis of northern hemisphere. Mailing Add: Rm 113 B49 Herbarium Brigham Young Univ Provo UT 84601

WELCH, STEVEN CHARLES, b Inglewood, Calif, Feb 18, 40; m 62; c 2. CHEMISTRY. Educ: Univ Calif, Los Angeles, BS, 64; Univ Southern Calif, PhD(chem), 68. Prof Exp: NIH fel, Calif Inst Technol, 68-70; ASST PROF CHEM & PHARM, UNIV HOUSTON, 70- Concurrent Pos: Nat Inst Gen Med Sci grant, Univ Houston, 71-77, Welch Found grant, 72-78. Mem: Am Chem Soc; The Chem Soc. Res: Synthetic organic chemistry; synthesis of molecules of theoretical and biological interest. Mailing Add: Dept of Chem Univ of Houston Houston TX 77004

WELCH, THOMAS HARRIS, b Westfield, NY, Dec 4, 15; m 38; c 3. ORGANIC CHEMISTRY. Educ: Cornell Univ, AB, 39. Prof Exp: Group leader, Linde Dev, 40-50, develop supvr, 50-56, mgr prod develop, Silicones Div, 56-65, tech dir, 65-66, DIR RES & DEVELOP, CHEM & PLASTICS DIV, UNION CARBIDE CORP, 67- Mem: Am Chem Soc. Res: Organo-silicon compounds. Mailing Add: Tarrytown Res & Develop Lab Union Carbide Corp Tarrytown NY 10591

WELCH, WALTER RAYNES, b Rumford, Maine, Oct 25, 20; m 44; c 3. MARINE BIOLOGY. Educ: Univ Maine, BS, 47, MS, 50. Prof Exp: Fisheries res biologist, Clam Prog, US Nat Marine Fisheries Serv, 49-58, prog leader, 58-64, proj leader lobster prog, 64-67, prog leader, 67-71, asst lab dir, 71-73, consult div water resources mgt & marine adv serv, 71-73; MARINE RESOURCE SCIENTIST & PROJ LEADER, MAINE DEPT MARINE RESOURCES, 73- Honors & Awards: Superior Performance Award, US Bur Com Fisheries, 69. Mem: Nat Shellfisheries Asn. Res: Molluscan and crustacean ecology; environmental measurement and interpretation. Mailing Add: Fisheries Res Sta Maine Dept of Marine Resources West Boothbay Harbor ME 04575

WELCH, WILLARD MCKOWAN, JR, b Frankfort, Ky, Mar 1, 44; m 66; c 1. ORGANIC CHEMISTRY. Educ: Mass Inst Technol, BS, 65; Rice Univ, PhD(org chem), 69. Prof Exp: RES CHEMIST, MED RES LABS, PFIZER, INC, 70- Res: Synthetic organic chemistry; central nervous system drugs; heterocyclic chemistry. Mailing Add: Med Res Labs Pfizer Inc Groton CT 06340

WELCH, WILLIAM HENRY, JR, b Los Angeles, Calif, Dec 13, 40; m 65; c 2. BIOCHEMISTRY. Educ: Univ Calif, Berkeley, BA, 63; Univ Kans, PhD(biochem), 69. Prof Exp: NIH fel, Brandeis Univ, 69-70; ASST PROF BIOCHEM, UNIV NEV, RENO, 70-, ASST BIOCHEMIST, 74- Mem: AAAS; Am Chem Soc. Res: Enzymology; molecular basis of adaptive phenomena; mechanism of small ion effectors of enzymes. Mailing Add: Dept of Biochem Univ of Nev Reno NV 89507

WELCH, WILLIAM JOHN, b Chester, Pa, Jan 17, 34; m 55; c 3. RADIO ASTRONOMY. Educ: Stanford Univ, BS, 55; Univ Calif, Berkeley, PhD(eng sci), 60. Prof Exp: From asst prof to prof elec eng, 60-72, PROF ELEC ENG & ASTRON & DIR RADIO ASTRON LAB, UNIV CALIF, BERKELEY, 72- Concurrent Pos: Mem, NSF Atron Adv Panel, 73-76; trustee at large, Assoc Univ Inc, 74-78. Mem: Am Astron Soc; Inst Elec & Electronics Eng; Int Sci Radio Union; AAAS; Int Astron Union. Res: Radio astronomical studies of the planets, interstellar medium and extragalactic radio sources; instrumenation for radio astronomy at millimeter wave lengths. Mailing Add: Radio Astron Lab Univ of Calif Berkeley CA 94720

WELCH, WINONA HAZEL, b Goodland, Ind, May 5, 96. BOTANY. Educ: DePauw Univ, AB, 23; Univ Ill, AM, 25; Ind Univ, PhD(bot), 28. Prof Exp: Asst bot, Univ Ill, 23-26; head dept biol, Cent Norm Col, Ind, 26-27; instr bot, Ind Univ, 28-30; from asst prof to prof, 30-61, actg head dept bot & bact, 39-40, 46-47, 53, head dept, 56-61, EMER PROF BOT, DePAUW UNIV, 61-, CUR HERBARIUM, 64- Concurrent Pos: Res grants, Ind Acad, 37, 41, 51, 54-55, 57-59, 64-, Am Philos Soc grant, 37, 53, grad coun grant, DePauw Univ, 40-, Sigma Xi grant, 63-; lectr-consult, Pigeon Lake Field Sta in Cryptogamic Botany, Wis Field Biol Prog, 64 & Plymouth State Col, 71. Mem: AAAS; Bot Soc Am; Am Soc Plant Taxon; Am Bryol & Lichenological Soc (vpres, 36-38, secy-treas, 41-54, pres, 54-56); Torrey Bot Club. Res: Plant ecology and taxonomy; bryophytes of Indiana; Fontinalaceae monographs for North America and the World; Hookeriaceae monographs for North America, West Indies and Central America. Mailing Add: Dept of Bot & Bact 8 Harrison Hall DePauw Univ Greencastle IN 46135

WELCH, ZARA D, b North Manchester, Ind, June 28, 15. ORGANIC CHEMISTRY. Educ: Manchester Col, AB, 37; Purdue Univ, MS, 39, PhD(org chem), 42. Prof Exp: Asst, Purdue Univ, 37-41; res chemist, Va Smelting Co, 42; res fel chem, 42-49, from instr to asst prof, 49-63, admin asst to head, 49-68, ASSOC PROF CHEM, PURDUE UNIV, 63-, ASST TO HEAD DEPT, 68- Concurrent Pos: Off Sci Res & Develop fel, 42-43. Mem: Am Chem Soc. Res: Dehydrohalogenation; halogenation; fluorine chemistry; recovery of uranium. Mailing Add: Dept of Chem Purdue Univ West Lafayette IN 47907

WELCHER, FRANK JOHNSON, analytical chemistry, see 12th edition

WELCHER, RICHARD PARKE, b Hartford, Conn, July 21, 19; m 50; c 5. CHEMISTRY. Educ: Trinity Col, Conn, AB, 41; Mass Inst Technol, BS, 43, PhD(org chem), 47. Prof Exp: Control supvr anal dept, Tenn Eastman Corp, Tenn, 44-45; SR RES CHEMIST, AM CYANAMID CO, STAMFORD, 47- Mem: Am Chem Soc; Sigma Xi. Res: Heterocyclic compounds; organophosphorus, nitrogen and sulfur chemistry; nitriles, amides, polyamines, cyanogen, s-triazines, dicyandiamide; biocides, ion and electron exchange resins; surfactants; mining chemicals; chemistry of cationic polymers. Mailing Add: 16 Watch Tower Lane Old Greenwich CT 06870

WELD, CHARLES BEECHER, b Vancouver, BC, Feb 3, 99; m 30; c 3. HUMAN PHYSIOLOGY. Educ: Univ BC, BA, 22, MA, 24; Univ Toronto, MD, 29; Dalhousie Univ, LLD, 70. Prof Exp: Asst, Connaught Labs, Univ Toronto, 24, res assoc & lectr physiol, 30-36; prof, 36-69, EMER PROF PHYSIOL, DALHOUSIE UNIV, 69- Concurrent Pos: Physiologist, Hosp Sick Children, Toronto, 31-36; dir, NS Mus, 50- Honors & Awards: Starr Gold Medal, 33. Mem: Am Physiol Soc; Can Physiol Soc (pres, 46); fel Royal Soc Can. Res: Diphtheria toxoid; freezing of fish; parathyroid; acute intestinal obstruction; pneumothorax; aqueous humor; lipemia; intestinal secretion; interstitial fluid pressures. Mailing Add: 6550 Waegwoltic Ave Halifax NS Can

WELDEN, ARTHUR LUNA, b Birmingham, Ala, Jan 27, 27; m 50; c 2. MYCOLOGY. Educ: Birmingham-Southern Col, AB, 50; Univ Tenn, MS, 51; Univ Iowa, PhD(bot), 54. Prof Exp: Asst prof biol, Milliken Univ, 54-55; from instr to assoc prof bot, 55-68, PROF BOT, TULANE UNIV, 68- Concurrent Pos: Am Philos Soc grant, 57; NSF grantee, 60- Mem: Mycol Soc Am; Am Soc Plant Taxon; Int Asn Plant Taxon; Asn Trop Biol; Bot Soc Mex. Res: Myxomycetes; tropical fungi; Thelephoraceae. Mailing Add: Dept of Biol Tulane Univ New Orleans LA 70118

WELDER, FRANK A, b Victoria, Tex, Jan 12, 23; m 55; c 3. GEOMORPHOLOGY, STRATIGRAPHY. Educ: Univ Tex, BS, 49; La State Univ, PhD(geol), 55. Prof Exp: Res asst coastal geomorphol, La State Univ, 51-54; groundwater geologist, US Geol Surv, 54-58; asst prof geol, Northeast La State Col, 58-64; HYDROGEOLOGIST, US GEOL SURV, 64- Concurrent Pos: NSF grant, 63; consult. Mem: Geol Soc Am. Res: Groundwater geology. Mailing Add: US Geol Surv Water Resources Div Fed Ctr Denver CO 80225

WELDON, JOHN WILLIAM, b Milwaukee, Wis, May 26, 26; m 56; c 3. CHEMISTRY. Educ: Marquette Univ, BS, 46; Univ Calif, Berkeley, MS, 48, PhD(org chem), 51. Prof Exp: Anal chemist, Mare Island Naval Shipyard, 48-49; res chemist, Tee Pak, 52-54; from instr to asst prof, Chicago Undergrad Div, Univ Ill, 54-65; assoc prof, 65-71, PROF CHEM, GRAND VALLEY STATE COL, 71-, CHMN DEPT, 66- Mem: Am Chem Soc. Res: Natural products; kinetics. Mailing Add: Dept of Chem Grand Valley State Col Allendale MI 49401

WELDON, VIRGINIA V, b Toronto, Ont, Sept 8, 35; US citizen; m 63; c 2. PEDIATRIC ENDOCRINOLOGY, MEDICAL ADMINISTRATION. Educ: Smith Col, AB, 57; State Univ NY Buffalo, MD, 62. Prof Exp: Intern, resident & fel, Sch Med, Johns Hopkins Univ & Hosp, 62-67, instr pediat, Univ, 67-68; from instr to asst prof, 68-73, ASSOC PROF PEDIAT, SCH MED, WASHINGTON UNIV, 73-, ASST TO VCHANCELLOR MED AFFAIRS, 75-; CO-DIR DIV METAB & ENDOCRINOL, ST LOUIS CHILDREN'S HOSP, 73- Concurrent Pos: Consult, Adv Comt Endocrinol & Metab, Food & Drug Admin, 73-; mem, State of Mo Health Manpower Planning Task Force, 76- Mem: AAAS; Sigma Xi; Endocrine Soc; Soc Pediat Res. Res: Aldosterone secretion in children; disorders of growth and growth hormone secretion in children. Mailing Add: St Louis Children's Hosp 500 S Kingshighway St Louis MO 63110

WELFORD, NORMAN TRAVISS, b London, Eng, Feb 5, 21; nat US; m 44; c 3. BIOMEDICAL ENGINEERING. Educ: Cambridge Univ, BA, 41, MA & MB, BCh, 45. Prof Exp: Intern & resident, Addenbrooks Hosp, Cambridge, Eng, 49-55; res assoc biophys, Univ Western Ont, 55-56; assoc prof psychophysiol, Antioch Col & res assoc psychophysiol-neurophysiol, Fels Res Inst, 56-66; DIR BIOMED ENG & FAC ASSOC PSYCHIAT, UNIV TEX MED BR, GALVESTON, 66- Concurrent Pos: USPHS res fel, Univ Western Ont, 55-56. Mem: Soc Psychophysiol Res; Inst Elec & Electronics Engrs; Brit Ergonomics Res Soc; sr mem Instrument Soc Am; Asn Advan Med Instrumentation. Res: Automation of data gathering and handling and stimulus presentation; psychophysiology of human sensory-motor performance and fetal heart rate behavior. Mailing Add: Med Electronics Lab Univ of Tex Med Br Galveston TX 77550

WELGE, HENRY JOHN, b St Louis, Mo, Aug 15, 07; m 40; c 2. FLUID DYNAMICS. Educ: Univ Ill, BSc, 29; Calif Inst Technol, MSc, 32, PhD(phys chem), 36. Prof Exp: Anal chemist, Richfield Oil Co, Calif, 29-31; res chemist, Tex Co, 37; from instr to asst prof chem, Agr & Mech Col, Tex, 37-44; sr res chemist, Exxon Prod Res Co, Houston, Tex, 44-72; CONSULT PETROL PROD, 72- Mem: Am Inst Mining, Metall & Petrol Eng. Res: Photochemistry; reaction kinetics and equilibria and diffusion; capillarity in petroleum production; single phase and multiphase fluid flow in porous media. Mailing Add: 11637 Green Oaks Houston TX 77024

WELIKY, IRVING, b Mt Vernon, NY, Aug 29, 24; m 51; c 4. CLINICAL PHARMACOLOGY. Educ: Ill Wesleyan Univ, BS, 48; Columbia Univ, PhD(biochem), 58. Prof Exp: Asst biochem, Sloan-Kettering Inst Cancer Res, 49-52 & Mass Gen Hosp, 61-62; asst prof, Univ Pittsburgh, 62-68; res group leader, 68-72, ASST DIR CLIN PHARMACOL, SQUIBB INST MED RES, 72- Concurrent Pos: Res fel biochem, Harvard Med Sch, 57-60. Mem: Am Soc Clin Pharmacol & Therapeut; Am Col Clin Pharmacol; AAAS; NY Acad Sci. Res: Intermediary metabolism of purines and pyrimidines; biosynthesis of porphyrins; biosynthesis and metabolism of steroid hormones in tumor and normal tissues; pharmacology; pharmacokinetics. Mailing Add: Squibb Inst for Med Res PO Box 4000 Princeton NJ 08540

WELIKY, NORMAN, b New York, NY, Nov 1, 19; m 55; c 2. ORGANIC CHEMISTRY, IMMUNOCHEMISTRY. Educ: City Col New York, BChE, 39; Polytech Inst Brooklyn, PhD(chem), 57. Prof Exp: Chemist, Mineral Pigments Corp, 46-47 & Reichhold Chem, Inc, 50-52; res fel hemoglobins, Dept Chem & Chem Eng, Calif Inst Technol, 57-59, group supvr molecular struct & synthesis, Jet Propulsion Lab, 59-65; mem tech staff, Biosci Dept, TRW Systs, 66-67, asst mgr biosci & electrochem dept, 68-71, sr scientist, 71-75. Mem: AAAS; Am Chem Soc; The Chem Soc; NY Acad Sci. Res: Physical chemistry; enzyme chemistry; specific insoluble adsorbents; biological specificity; cancer research; environmental quality criteria. Mailing Add: 1072 Ridge Crest St Monterey Park CA 91754

WELKER, CAROL, b New York, NY, Jan 28, 37; m 65; c 1. NEUROPHYSIOLOGY. Educ: Brooklyn Col, BA, 56; Univ Chicago, MA, 60, PhD(anthrop), 62. Prof Exp: NIH fel neurophysiol, Univ Wis-Madison, 62-63; instr physiol, New York Med Col, 64-65; RES SCIENTIST NEUROPHYSIOL, CENT WIS COLONY & TRAINING SCH, WIS DEPT HEALTH & SOCIAL SERV, 65- Concurrent Pos: Nat Inst Neurol Dis & Stroke grant neurophysiol, Cent Wis Colony & Training Sch, 67-73. Mem: AAAS; Am Asn Anatomists; Soc Neurosci; Am Soc Zool; Animal Behav Soc. Res: Neurophysiology and neuroanatomy of the somatic sensory systems in mammals; development of the mammalian brain and behavior. Mailing Add: Cent Wis Colony & Training Sch Res Dept 317 Knutson Dr Madison WI 53704

WELKER, EVERETT LINUS, b Greenview, Ill, Mar 30, 11; m 33; c 2. MATHMATICAL STATISTICS. Educ: Univ Ill, AB, 30, AM, 31, PhD(math statist), 38. Prof Exp: Asst math, Univ Ill, 35-38, instr, 38-42, assoc, 42-44, from asst prof to assoc prof, 44-47, assoc math, Bur Med Econ Res, AMA, 47-52; statistician, Dept Defense, 52-57; mgr adv studies dept, Arinc Res Corp, 57-63; mgr syst effectiveness anal prog, Tempo, Gen Elec Ctr Advan Studies, Calif, 63-71; STAFF SCIENTIST, TRW SYSTS, 71- Mem: Am Statist Asn; Am Math Soc; Economet Soc; Inst Math Statist; Am Inst Aeronaut & Astronaut. Res: Correlation theory; vital and engineering statistics; biometrics; reliability. Mailing Add: 365 San Roque Dr Escondido CA 92025

WELKER, GEORGE W, b Cumberland City, Tenn, July 4, 23; m 50; c 3. PARASITOLOGY, BACTERIOLOGY. Educ: Mid Tenn State Univ, BS, 44; George Peabody Col, MA, 50; Ohio State Univ, PhD, 62. Prof Exp: Teacher high sch, Ohio, 46-49; asst biol, George Peabody Col, 49-50; from asst prof to assoc prof, 50-65, PROF BIOL, BALL STATE UNIV, 65- Concurrent Pos: Danforth teacher study grant, 57-58. Mem: AAAS; Am Sci Affil. Res: Helminth parasites and microbial physiology. Mailing Add: Dept of Biol Ball State Univ Muncie IN 47306

WELKER, NEIL ERNEST, b Batavia, NY, Apr 21, 32; m 54; c 3. BIOCHEMISTRY. Educ: Univ Buffalo, BA, 58; Western Reserve Univ, PhD(microbiol), 63. Prof Exp: Fel, Univ Ill, Urbana, 63-64; from asst prof to assoc prof bio sci, 64-74, PROF BIOL SCI & ASSOC DEAN COL ARTS & SCI, NORTHWESTERN UNIV, 74- Mem: AAAS; Am Soc Microbiol; Brit Soc Gen Microbiol; Am Soc Biol Chemists. Mailing Add: Dept of Biochem & Molecular Biol Northwestern Univ Evanston IL 60201

WELKER, WALLACE I, b Batavia, NY, Dec 17, 26. NEUROPHYSIOLOGY, NEUROANATOMY. Educ: Univ Chicago, PhD(psychol), 54. Prof Exp: Asst, Yerkes Labs Primate Biol, 52-54; PROF NEUROPHYSIOL, MED SCH, UNIV WIS-MADISON, 67- Concurrent Pos: NIH fel neurophysiol, Univ Wis-Madison, 54-56, Sister Kenny Found scholar, 57-62, NIH career develop fel, 62-67. Mem: Am Physiol Soc; Psychonomic Soc; Am Asn Anatomists; Am Soc Zoologists. Res: Psychology; neural correlates of behavior; comparative physiology. Mailing Add: Dept of Neurophysiol Med Sch Univ Wis-Madison Madison WI 53706

WELKER, WILLIAM V, JR, b Milwaukee, Wis, Nov 12, 28; m 51; c 4. WEED SCIENCE, HORTICULTURE. Educ: Univ Wis, BS, 52, PhD(hort, plant physiol), 62. Prof Exp: RES HORTICULTURIST, AGR RES SERV, USDA, 59-; RES PROF SOILS & CROPS, RUTGERS UNIV, NEW BRUNSWICK, 59- Mem: Weed Sci Soc Am; Am Soc Hort Sci. Res: Chemical control of weeds in horticultural crops; influence of long term use of herbicides upon crops; fate of herbicides in soil.

WELKIE, GEORGE WILLIAM, b Hazleton, Pa, Apr 11, 32; m 57; c 2. VIROLOGY, PLANT PHYSIOLOGY. Educ: Pa State Univ, BS, 52, MS, 54; Univ Wis, PhD(plant path), 57. Prof Exp: Asst prof, 57-62, ASSOC PROF BOT & PLANT PATH, UTAH STATE UNIV, 62- Concurrent Pos: NSF fel virol, Rothamsted Exp Sta, Eng. Mem: Am Phytopath Soc. Res: Virus infection and synthesis; effect of virus infection on host metabolism; mineral nutrition of plants with relation to metabolism. Mailing Add: Dept of Bot Utah State Univ Logan UT 84322

WELLAND, GRANT VINCENT, b Toronto, Ont, June 25, 40; m 66. MATHEMATICS. Educ: Purdue Univ, BS, 63, MS, 65, PhD(math), 66. Prof Exp: Teaching asst math, Purdue Univ, 63-65; from asst prof to assoc prof, 66-70; ASSOC PROF MATH, UNIV MO-ST LOUIS, 70- Concurrent Pos: Dir, NSF undergrad res prog, 67-68; NSF res grant, 68-70; vis prof, Univ Madrid, 70. Mem: AAAS; Am Math Soc. Res: Harmonic analysis and differentiation. Mailing Add: Dept of Math Univ of Mo St Louis MO 63121

WELLAND, ROBERT ROY, b Toronto, Ont, Jan 31, 33; m 57; c 3. MATHEMATICS. Educ: Univ Okla, BS, 56, MA, 57; Purdue Univ, PhD, 60. Prof Exp: Instr math, Univ Okla, 56-57; NSF asst, Purdue Univ, 57-58, from instr to asst prof, 58-60; asst prof, Ohio State Univ, 60-63; asst prof, 63-74, ASSOC PROF MATH, NORTHWESTERN UNIV, EVANSTON, 74- Concurrent Pos: Vis instr, Univ Chicago, 61. Mem: Am Math Soc. Res: Functional analysis; special spaces; nonlinear analysis. Mailing Add: Dept of Math Northwestern Univ Evanston IL 60201

WELLAR, BARRY SHELDON, b Latchford, Ont, Feb 3, 40; m 64; c 2. URBAN GEOGRAPHY, COMMUNICATIONS. Educ: Queen's Univ, BA(econ, com), 64 & BA(geog), 65; Northwestern Univ, MSc, 67, PhD(geog), 69. Prof Exp: Asst prof geog & res assoc info systs, Univ Kans, 69-72; SR RES OFFICER, MINISTRY OF STATE-URBAN AFFAIRS, GOVT CAN, 72- Concurrent Pos: Mem, Hwy Res Bd, Nat Acad Sci-Nat Res Coun, 67-; consult, Am Pub Health Asn, 69-72 & City Kansas City, Kans, 72; mem bd dirs, Urban & Regional Info Systs Asn, 72- Mem: AAAS; Can Asn Geog; Urban & Regional Info Systs Asn; Am Inst Planners. Res: Urban and regional information systems; mathematical modeling; measurement of housing quality; transportation system research and development. Mailing Add: Ministry of State for Urban Affairs 355 River Rd Ottawa ON Can

WELLDON, PAUL BURKE, b Nashua, NH, May 10, 16; m 41; c 2. ORGANIC CHEMISTRY. Educ: Dartmouth Col, AB, 37, AM, 39; Univ Ill, PhD(org chem), 42. Prof Exp: Instr chem, Dartmouth Col, 37-39; res chemist, 42-51, mgr cellulose prod res div, 52-60, tech asst to dir res, 60-62, mgr appln res div, 62-65, MGR PERSONNEL DIV, EXP STA, HERCULES INC, 67- Mem: AAAS; Am Chem Soc; Soc Platics Eng. Res: Chlorination of organic compounds; chemistry of cellulose derivatives; polyolefins; polymer applications. Mailing Add: 118 Meriden Dr Canterbury Hills R D 2 Hockessin DE 19707

WELLENREITER, RODGER HENRY, b Bloomington, Ill, Oct 23, 42; m 63; c 2. POULTRY NUTRITION. Educ: Ill State Univ, BS, 64; Mich State Univ, MS, 67, PhD(animal husb), 70. Prof Exp: SR SCIENTIST POULTRY NUTRIT, LILLY RES LABS, ELI LILLY & CO, 70- Concurrent Pos: Sigma Xi res award, 70. Mem: Poultry Sci Asn; World Poultry Sci Asn; Sigma Xi. Res: Means of improving the efficiency of conversion of animal feedstuffs into products for human comsumption. Mailing Add: Lilly Res Labs Box 708 Greenfield IN 46140

WELLER, CHARLES STAGG, JR, b Nashville, Tenn, Dec 28, 40; m 66; c 2. PHYSICS. Educ: Mass Inst Technol, BS, 62; Univ Pittsburgh, PhD(physics), 67. Prof Exp: RES PHYSICIST, NAVAL RES LAB, 67-, SECT HEAD, 74- Mem: Am Phys Soc; Am Geophys Union; Am Astron Soc. Res: Space science; upper atmospheric studies; interplanetary medium; ultraviolet optics. Mailing Add: Naval Res Lab Washington DC 20375

WELLER, DAVID LLOYD, b Munfordville, Ky, Sept 28, 38; m 68; c 1. BIOCHEMISTRY, MOLECULAR BIOLOGY. Educ: Rochester Inst Technol, BS, 62; Iowa State Univ, PhD(bio-chem), 66. Prof Exp: Res fel molecular biol, Children's Cancer Res Found, Boston, Mass, 66-67; asst prof agr biochem, 67-71, chmn biol sci prog, 71-75, chmn cell biol, 72-75, ASSOC PROF BIOCHEM & MICROBIOL, UNIV VT, 71-, ASST DEAN, COL AGR & ASSOC DIR, AGR EXP STA, 75- Mem: Am Chem Soc; Biophys Soc; NY Acad Sci; Am Inst Chem; Soc Protozoologists. Res: Ribosomes and RNAases of Entamoeba; isoelectric focusing of proteins. Mailing Add: Morrill Hall Univ of Vt Burlington VT 05401

WELLER, EDWIN MATTHEW, b Winnipeg, Man, Mar 1, 25; US citizen; m 57; c 2. CHEMICAL EMBRYOLOGY, DEVELOPMENTAL PHYSIOLOGY. Educ: Univ Calif, Los Angeles, BA, 47, MA, 49, PhD(zool), 57. Prof Exp: Asst prof biol, Randolph-Macon Woman's Col, 57-59; res instr anat, Col Med, Univ Utah, 59-62; asst prof anat, Col Med, Baylor Univ, 62-66; ASSOC PROF ANAT, UNIV ALA, BIRMINGHAM, 66-, ASST PROF ENG BIOPHYS, 68- Mem: Am Asn Anatomists; Soc Study Reprod; Teratology Soc; Sigma Xi; AAAS. Res: Comparative aspects of the role of phase-specific proteins in differentiation, embryogenesis, and oncogenesis; hormonal and environmental effects on fetal function, especially of cardiovascular and central nervous system organ systems. Mailing Add: Dept of Anat Univ of Ala Univ Sta Box 317 Birmingham AL 35294

WELLER, GLENN PETER, b New Orleans, La, Dec 7, 43; m 69; c 1. MATHEMATICS. Educ: Tulane Univ, BA, 64; Univ Chicago, SM, 65, PhD(math), 68. Prof Exp: Asst prof math, Roosevelt Univ, 68-69 & Univ Ill, Chicago Circle, 69-75; ASST PROF, KENNEDY-KING COL, 75- Mem: Am Math Soc. Res: Geometric topology. Mailing Add: 5326 S Hyde Park Blvd Chicago IL 60615

WELLER, GUNTER ERNST, b Haifa, Palestine, June 14, 34; m 63; c 2. METEOROLOGY. Educ: Univ Melbourne, BSc, 62, MSc, 65, PhD(meteorol), 67. Prof Exp: Meteorologist, Commonwealth Bur Meteorol, 60-61; glaciologist, Australian Nat Antarctic Res Exped, 62-66; NSF res fel, Univ Melbourne, 67; from asst prof to assoc prof geophys, 68-73, prog mgr polar meteorol, Off Polar Progs, NSF, 72-74, PROJ MGR, NAT OCEANIC & ATMOSPHERIC AGENCY, OUTER CONTINENTAL SHELF, ARCTIC PROJ OFF, UNIV ALASKA, 75-, PROF GEOPHYS, GEOPHYS INST, 73- Concurrent Pos: US rep, Working Group Meteorol, Sci Comt Antarctic Res, 74-; mem, Polar Res Bd, Nat Acad Sci, 75- & Int Comn Polar Meteorol, Int Union Geod & Geophys, 75- Honors & Awards: Polar Medal, Commonwealth of Australia; Antarctic Serv Medal, US Govt, 74. Mem: Am Meteorol Soc; Am Geophys Union; Arctic Inst NAm; Glaciol Soc. Res: Polar meteorology and climatology; micrometeorology; glacio-meteorology; biometeorology. Mailing Add: Geophys Inst Univ of Alaska Fairbanks AK 99701

WELLER, HARRY, b Philadelphia, Pa, Feb 7, 16; m 42; c 3. ZOOLOGY. Educ: Univ Pa, BA, 37, PhD(zool), 55; Univ Del, MA, 51. Prof Exp: Res assoc physiol, Sch Med, Univ Va, 54-57; ASSOC PROF ZOOL, MIAMI UNIV, 57- Mem: AAAS; Am Soc Zool; NY Acad Sci; Sigma Xi. Res: Cell phsiology; comparative endocrinology. Mailing Add: Dept of Zool Miami Univ Oxford OH 45056

WELLER, HENRY RICHARD, b East Rutherford, NJ, Mar 15, 41; m 64; c 2. NUCLEAR PHYSICS. Educ: Fairleigh Dickinson Univ, BS, 62; Duke Univ, PhD(nuclear physics), 68. Prof Exp: Fel physics, 67-68, asst prof, 68-73, ASSOC PROF PHYSICS, UNIV FLA, 73- Mem: Am Phys Soc. Res: Experimental nuclear structure studies, especially reaction mechanism of helium-3 induced reactions on light nuclei. Mailing Add: Dept of Physics Univ of Fla Gainesville FL 32601

WELLER, JAMES MARVIN, b Chicago, Ill, Aug 1, 99; m 23; c 1. GEOLOGY. Educ: Univ Chicago, BS, 23, PhD(geol), 27. Prof Exp: Geologist, Chanute Spelter Co, 23 & State Geol Surv, Ky, 24-25; geologist & head div stratig & paleont, Ill State Geol Surv, 25-45; prof invert paleont, 45-65, EMER PROF INVERT PALEONT, UNIV CHICAGO, 65-; RES ASSOC, FIELD MUS NATURAL HIST, 64- Concurrent Pos: Asst prof, Univ Ill, 36-37; geologist, Stand Vacuum Oil Co, China, 37-38; ed, J Paleont, Soc Econ Paleont & Mineral, 42-51; mem comt stratig, Nat Res Coun, 44-62; geologist, Mutual Security Agency & Foreign Opers Admin, US Geol Surv, Philippines, 52-54; vis prof, Univ Tex, 56. Mem: Fel Geol Soc Am; fel Paleont Soc; Soc Econ Paleont & Mineral (pres, 64-65); Am Asn Petrol Geol. Res: Geology of Illinois, Kentucky and neighboring states, Pakistan, China and Philippines; Carboniferous stratigraphy and paleontology; cycles of sedimentation; Carboniferous and Permian trilobites; fluorspar deposits; coal geology; compaction of sediments. Mailing Add: 65 Eliseo Dr Greenbrae CA 94904

WELLER, JOHN MARTIN, b Ann Arbor, Mich, Mar 4, 19; m 71; c 2. INTERNAL MEDICINE, NEPHROLOGY. Educ: Univ Mich, AB, 40; Harvard Med Sch, MD, 43; Am Bd Internal Med, dipl, 53. Prof Exp: Asst & instr med, Harvard Med Sch, 50-53; from asst prof to assoc prof, 53-63, PROF INTERNAL MED, UNIV MICH, ANN ARBOR, 63- Concurrent Pos: Am Col Physicians Stengel res fel, 48-49; res & teaching fel biochem, Harvard Med Sch, 48-50; Nat Res Coun fel med sci, 49-50; asst & jr assoc med, Peter Brent Brigham Hosp, Boston, 50-53. Mem: Soc Exp Biol & Med; Am Physiol Soc; Am Heart Asn; Am Fedn Clin Res; fel Am Col Physicians. Res: Electrolyte metabolism; renal physiology and diseases; hypertension. Mailing Add: B2902 Univ Hosp Ann Arbor MI 48104

WELLER, LOWELL ERNEST, b Continental, Ohio, Apr 17, 23; m 44; c 2. ORGANIC CHEMISTRY, BIOCHEMISTRY. Educ: Bowling Green State Univ, BS, 48; Mich State Univ, MS, 51, PhD(chem), 56. Prof Exp: Asst chem, Bowling Green State Univ, 46-48; from instr to asst prof biochem, Mich State Univ, 48-57; assoc prof chem, 57-58, PROF CHEM, UNIV EVANSVILLE, 58-, HEAD DEPT, 57- Mem: AAAS; Am Chem Soc; The Chem Soc; Sigma Xi. Res: Organic synthesis, reaction mechanisms and structure determination especially by spectroscopic methods. Mailing Add: Dept of Chem Univ of Evansville Evansville IN 47701

WELLER, MILTON WEBSTER, b St Louis, Mo, May 23, 29; m 47; c 1. ANIMAL ECOLOGY. Educ: Univ Mo, AB, 51, MA, 54, PhD(zool), 56. Prof Exp: Instr zool, Univ Mo, 56-57; from asst prof to prof, 57-74, chmn fisheries & wildlife sect, 67-74; PROF ZOOL & HEAD DEPT ENTOM, FISHERIES & WILDLIFE, UNIV MINN, ST PAUL, 74- Concurrent Pos: NSF res grants, 64-65 & 70-72. Mem: Fel AAAS; Ecol Soc Am; Cooper Ornith Soc; Wildlife Soc; fel Am Ornith Union. Res: Avian ecology, especially reproductive behavior, productivity, habitat and human relationships of water fowl with special emphasis on prairie and polar habitats. Mailing Add: Dept of Entom Fisheries & Wildlife Univ of Minn St Paul MN 55108

WELLER, PAUL FRANKLIN, b Kankakee, Ill, Aug 30, 35; m 58; c 2. INORGANIC CHEMISTRY. Educ: Univ Ill, BS, 57; Cornell Univ, PhD(chem), 62. Prof Exp: Res scientist, T J Watson Res Ctr, Int Bus Mach Corp, 61-65; from asst prof to prof chem, State Univ NY Col Fredonia, 65-75, chmn dept, 74-75; PROF CHEM & DEAN ARTS & SCI, WESTERN ILL UNIV, 75- Mem: AAAS; Am Chem Soc; NY Acad Sci; Mat Res Soc. Res: Wide band gap semiconductors; defect oxides; solid state materials characterization; crystal growth. Mailing Add: Dean of Arts & Sci Western Ill Univ Macomb IL 61455

WELLER, RICHARD IRWIN, b Newark, NJ, Mar 3, 21; m 53; c 2. PHYSICS, ACADEMIC ADMINSTRATION. Educ: City Col New York, BEE, 44; Union Col, BS, 48; Fordham Univ, MS, 50, PhD(physics), 53. Prof Exp: Elec engr, NAm Phillips Co, 44, Guy F Atkinson Co, 44-45, NAm Phillips Co, 45-46, Crow, Lewis & Wick, 47, Edward E Ashley, 48, Allied Processes Co, 50, V L Falotico & Assocs, 50-51 & Singmaster & Breyer, 51; instr physics, Brooklyn Col, 52-53; asst prof, State Univ, NY Maritime Col, 53-54; med physicist, Brookhaven Nat Lab, 54-57; prof physics, Franklin & Marshall Col, 57-70, chmn dept, 57-61; PROF PHYSICS & DEAN SCH SCI & MATH, EDINBORO STATE COL, 70- Concurrent Pos: Instr, Manhattan Col, 49-50, Fordham Univ, 50-52 & Newark Col Eng, 52-53; consult, Brookhaven Nat Lab, Fairchild Camera & Instrument Corp & Nuclear Sci & Eng Corp; mem Nat Res Coun subcomt, Nat Acad Sci. Mem: AAAS; Am Asn Physics Teachers; Am Phys Soc; Am Soc Eng Educ; Health Physics Soc. Res: Atmospheric electricity; radioactivity; radioisotopes dosimetry and instrumentation; biophysics. Mailing Add: Sch of Sci & Math Edinboro State Col Edinboro PA 16444

WELLER, THOMAS HUCKLE, b Ann Arbor, Mich, June 15, 15; m 45; c 4. TROPICAL MEDICINE, INFECTIOUS DISEASES. Educ: Univ Mich, AB, 36, MS, 37; Harvard Med Sch, MD, 40. Hon Degrees: LLD, Univ Mich, 56. Prof Exp: Res fel comp path & trop med, Harvard Med Sch, 40, teaching fel bact, 40-41, Milton fel pediat, 47-48; USPHS fel, 48; instr comp path & trop med, 48-49, from asst prof to assoc prof trop pub health, 49-54, STRONG PROF TROP PUB HEALTH, SCH PUB HEALTH, HARVARD UNIV & HEAD DEPT, 54-, DIR CTR PREV INFECTIOUS DIS, 66- Concurrent Pos: Intern, Children's Hosp, 41-42, asst res, 46, asst dir res, Div Infectious Dis, Children's Med Ctr, 49-56, assoc physician, 49-55; area consult, US Vet Admin, 49-64; consult & mem trop med & parasitol study sect, USPHS, 53-55; dir, Comn Parasitic Dis, Armed Forces Epidemiol Br, 53-59, mem, 59-72. Honors & Awards: Nobel Prize Physiol & Med, 54; Mead Johnson Award, Am Acad Pediat, 53; co-winner, Kimble Methodol Award, 54; Ledlie Prize, 63. Mem: Nat Acad Sci; Am Soc Trop Med & Hyg (pres, 64); Am Asn Immunologists; Asn Am Physicians. Res: In vitro cultivation of viruses, especially poliomyelitis, mumps and Coxsackie viruses; etiology of epidemic pleurodynia, varicella and Herpes zoster; cytomegalic inclusion and Rubella; helminth infections, especially schistosomiasis and enterobiasis. Mailing Add: Dept of Trop Pub Health Harvard Sch of Pub Health Boston MA 02115

WELLERSON, RALPH, JR, b New York, NY, Dec 12, 24; m 51; c 2. MICROBIOLOGY. Educ: Hobart Col, BS, 49; Rutgers Univ, MS, 51; Purdue Univ, PhD(bact), 54. Prof Exp: Sr scientist, 54-62, RES FEL MICROBIOL, DIV MICROBIOL, ORTHO RES FOUND, ORTHO PHARMACEUT CORP, 62- Mem: Am Soc Microbiol. Res: Metabolism and nutrition of microorganisms; microbial fermentations; chemotherapy of Trichomonas vaginalis; immunological control of fertility. Mailing Add: Div of Microbiol Ortho Pharmaceut Corp Raritan NJ 08869

WELLES, HARRY LESLIE, b Rockville, Conn, Jan 2, 45; m 66; c 1. PHARMACEUTICS. Educ: Cent Conn State Col, BS, 67; Univ Conn, MS, 69, PhD(pharmaceut), 74. Prof Exp: Res assoc pharmaceut, Sch Pharm, Univ Wis-Madison, 72-74; ASST PROF PHARMACEUT, COL PHARM, DALHOUSIE UNIV, 74- Mem: Am Pharmaceut Asn; Sigma Xi. Res: Solution thermodynamics and interfacial phenomena as they effect pharmaceutical systems. Mailing Add: Col of Pharm Dalhousie Univ Halifax NS Can

WELLES, SAMUEL PAUL, b Gloucester, Mass, Nov 9, 07; m 31; c 3. VERTEBRATE PALEONTOLOGY. Educ: Univ Calif, AB, 30, PhD(vert paleont), 40. Prof Exp: Field & lab asst, 31-39, asst cur reptiles & amphibians, 39-42, sr mus cur, 42-46, prin mus cur, 46-47, prin mus paleontologist & lectr, 46-74, actg dir, 47-48, RES ASSOC, MUS PALEONT, UNIV CALIF, BERKELEY, 74- Concurrent Pos: Mem, US Nat Mus exped, 30 & Univ Calif expeds, 31-42, 45; Fulbright scholar, Univ NZ, 69-70. Mem: Fel Geol Soc Am; Paleont Soc; Soc Vert Paleont. Res: Triassic labyrinthodonts; Cretaceous Plesiosaurs; dinosaurs. Mailing Add: Mus of Paleont Univ of Calif Berkeley CA 94720

WELLHAUSEN, EDWIN JOHN, b Fairfax, Okla, Sept 10, 07; m 37; c 1. PLANT BREEDING. Educ: Univ Idaho, BS, 32; Iowa State Col, PhD(plant genetics), 36. Hon Degrees: DSc, San Carlos Univ, Guatemala, 60. Prof Exp: Gen Ed Bd fel genetics, Rockefeller Inst & Univ Calif, 36-37; assoc agronomist, Exp Sta, Mont State Col, 37-38; assoc prof agron, Univ WVa & assoc geneticist, Exp Sta, 39-43; geneticist, 43-51, local dir, Mex Agr Prog, 52-58, dir Inter-Am Corn Improv Prog, 59-66, dir int maize & wheat improv ctr, 66-71, assoc dir agr sci, 59-73, SPEC STAFF MEM, ROCKEFELLER FOUND, MEX, 73- Honors & Awards: Aztec Eagle Award, Govt of Mex, 69; Distinguished Achievement Citation, Iowa State Univ, 69; Int Agron Award, Am Soc Agron, 69. Mem: AAAS; Genetics Soc Am; Am Phytopath Soc; Bot Soc Am; fel Am Soc Agron. Res: Genetics of disease resistance; genetic diversification and origin of corn types in Mexico and Central America; corn breeding methods for Latin America; research adminstration. Mailing Add: Rockefeller Found Calle Londres 40 Despacho 101 Mexico DF Mexico

WELLING, DANIEL J, b Kansas City, Mo, May 1, 37. THEORETICAL PHYSICS, BIOPHYSICS. Educ: Rockhurst Col, BS, 58; St Louis Univ, PhD(physics), 63. Prof Exp: Asst prof physics, Southern Ill Univ, 64; AEC fel theoret physics, Argonne Nat Lab, 64-66; asst prof & assoc scientist, Univ Wis-Milwaukee, 66-72; ASSOC PROF THEORET PHYSICS, UNIV KANS MED CTR, 72- Mem: Am Phys Soc; Am Soc Nephrology. Res: Medical physics; mathematical modeling of transport in the nephron. Mailing Add: Dept of Physiol Univ of Kans Med Ctr Kansas City KS 66103

WELLINGS, IAN, b Glasgow, Scotland, Oct 25, 33; m 60; c 2. ORGANIC CHEMISTRY, MEDICINAL CHEMISTRY. Educ: Glasgow Univ, BSc, 56, PhD(org chem), 59. Prof Exp: Res asst org chem, Royal Inst Technol, Sweden, 59-62; res assoc, Johns Hopkins Univ, 62-63; res chemist, Indust & Biochem Dept, Exp Sta, 63-69 & Stine Lab, 69-73, SR RES CHEMIST, E I DU PONT DE NEMOURS & CO, INC, 73- Mem: Fel The Chem Soc. Res: Medicinal and heterocyclic chemistry; chemistry and biological activity of synthetic analogues of nicotine. Mailing Add: Biochem Dept Exp Sta E I du Pont de Nemours & Co Inc Wilmington DE 19898

WELLINGTON, GEORGE HARVEY, b Springport, Mich, Sept 19, 15; m 39; c 3. ANIMAL SCIENCE. Educ: Mich State Univ, BS, 37, PhD, 54; Kans State Univ, MS, 40. Prof Exp: Asst, Kans State Univ, 38-40; agr exten agent, Univ Mich, 45-46; from asst prof to assoc prof animal sci, 47-57, PROF ANIMAL SCI, CORNELL UNIV, 57- Concurrent Pos: Ford Found consult livestock in Syria, 62; vis prof, Univ Aleppo, Syria, 65-66. Honors & Awards: Signal Serv Award, Am Meat Sci Asn, 66. Mem: Am Soc Animal Sci; Am Meat Sci Asn; Inst Food Technol. Res: Meat animal development and composition; meat quality factors and processing. Mailing Add: Morrison Hall Cornell Univ Ithaca NY 14850

WELLINGTON, JOHN SESSIONS, b Glendale, Calif, Sept 28, 21; m 44; c 2. PATHOLOGY. Educ: Univ Calif, Berkeley, AB, 42; Univ Calif, San Francisco, MD, 45. Prof Exp: Instr path, Sch Med, Univ Calif, 53-55, lectr, 55-56; vis asst prof, Univ Indonesia, 56-58; assoc prof, 58-69, PROF PATH, SCH MED, UNIV CALIF, SAN FRANCISCO, 69-, ASSOC DEAN, 65- Mem: Fel Col Am Path; Am Soc Exp Path; Int Acad Path. Res: Leukemia; injury and repair in hematopoietic tissue. Mailing Add: Dept of Path Univ of Calif Sch of Med San Francisco CA 94122

WELLINGTON, WILLIAM GEORGE, b Vancouver, BC, Aug 16, 20; m 59; c 2. INSECT ECOLOGY, BIOMETEOROLOGY. Educ: Univ BC, BA, 41; Univ Toronto, MA, 45, PhD(zool, entom), 47. Prof Exp: Meteorol officer, Meteorol Serv, Can Dept Transport, 42-45; forest biologist, Forest Biol Div, Can Dept Agr, 46-60; head bioclimat sect, Can Dept Forestry, 60-64, prin scientist insect ecol, 64-68; prof ecol, Univ Toronto, 68-70; PROF PLANT SCI & RESOURCE ECOL, UNIV BC,70-, DIR INST ANIMAL RESOURCE ECOL, 73- Honors & Awards: Gold Medal for Outstanding Achievement in Can Entom, Entom Soc Can, 68; Outstanding Achievement Award Bioclimat, Am Meteorol Soc, 69. Mem: AAAS; Am Meteorol Soc; Can Soc Zoologists; Entom Soc Am; Entom Soc Can. Res: Micro- and synoptic climatology; interdisciplinary associations of meteorology and the life sciences; population biology; physiological, behavioral and social variations affecting population dynamics; insect behavior; human ecology. Mailing Add: Inst of Animal Resource Ecol Univ of BC Vancouver BC Can

WELLISCH, ERIC, chemistry, see 12th edition

WELLMAN, ANGELA MYRA, b Sudbury, Suffolk, Eng, June 13, 35; m 59. MICROBIAL PHYSIOLOGY. Educ: Bristol Univ, BSc, 56, PhD(mycol), 61. Prof Exp: Demonstr bot, Bristol Univ, 56-59; Nat Res Coun Can fel, 59-61, instr, 61-62, lectr, 62-64, asst prof, 64-68, asst dean sci, 74-75, ASSOC DEAN SCI, UNIV WESTERN ONT, 75-, ASSOC PROF BOT, 68- Mem: Bot Soc Am; Can Bot Asn; Brit Mycol Soc; Soc Cryobiol. Res: Cryobiology; survival of fungal spores after exposure to low temperature; purine metabolism in Neurospora; hydrocarbon oxidation by micro-organisms. Mailing Add: Off Dean of Sci Natural Sci Ctr Univ of Western Ont London ON Can

WELLMAN, RICHARD HARRISON, b Mountain View, Alta, May 25, 14; US citizen; m 36; c 2. PLANT PATHOLOGY. Educ: Wash State Col, BS, 35, PhD(plant path), 39. Prof Exp: Lab asst, Wash State Col, 38; investr & fel plant path, Crop Protection Inst, Boyce Thompson Inst, 39-41, res assoc, 41-44; head biol res div, Union Carbide Chem Co, 45-54, mgr agr chem div, 55-66, vpres & gen mgr agr chem, Union Carbide Corp, 66-74; MANAGING DIR, BOYCE THOMPSON INST, 74- Mem: AAAS; Am Phytopath Soc; Am Chem Soc; NY Acad Sci. Res: Development of fungicides; insecticides and herbicides. Mailing Add: Boyce Thompson Inst 1086 N Broadway Yonkers NY 10701

WELLMAN, ROBERT PERSEY, chemical physics, geophysics, see 12th edition

WELLMAN, RUSSEL ELMER, b New Berlin, NY, July 16, 22; m 48; c 2. PHYSICAL CHEMISTRY. Educ: Howard Col, BA, 50; Univ Rochester, PhD(phys chem), 55. Prof Exp: Res chemist, Callery Chem Co, 55, group leader appl chem, 56-60; sr chemist, Inorg & Phys Chem Sect, Southern Res Inst, 60-66; unit mgr polymer technol, 66-67. scientist, 67-73, SR CHEMIST, XEROX CORP, 73- Mem: AAAS; Am Chem Soc; NY Acad Sci; Am Inst Chem. Res: Viscometry; physical chemistry of polymers; adhesion of polymers; microencapsulation. Mailing Add: Xerox Corp Rochester NY 14644

WELLMAN, WILLIAM EDWARD, b Ft Wayne, Ind, Oct 22, 32; m 66; c 2. ORGANIC CHEMISTRY. Educ: Col Wooster, BA, 54; Ohio State Univ, PhD(org chem), 60. Prof Exp: Chemist, Esso Res & Eng Co, 60-63, sr chemist, 63-64, proj leader, 64-69, res assoc, 68-73, sect head, 69-73, MGR OXYGEN SOLVENTS & LOWER ALCOHOLS, SOLVENTS TECHNOL DIV, EXXON CHEM CO, 73- Mem: Am Chem Soc. Res: Petrochemicals, solvents and fiber intermediates; process development. Mailing Add: Exxon Chem Co PO Box 536 Linden NJ 07036

WELLMAN, WILLIAM WENDELL, physical chemistry, see 12th edition

WELLMANN, KLAUS FREDERICH, b Gerbstedt, Ger, Feb 18, 29; US citizen; m 56. ANATOMIC PATHOLOGY, CLINICAL PATHOLOGY. Educ: Heidelberg Col, MD, 54. Prof Exp: Intern & resident path, surg & pediat, var Ger hosps, 55-57; intern med & surg, Wheeling Hosp, WVa, 57; resident path, Mem Hosp, Charlotte, NC, 58, County Hosp, Milwaukee, Wis, 58-59, Path Inst, Univ Cologne, 60 & Civic Hosp, Ottawa, Ont, 60-62; resident, Jewish Chronic Dis Hosp, Brooklyn, NY, 62-64, asst pathologist, 63-68; asst dir lab, 68-72, ASSOC DIR LAB, KINGSBROOK JEWISH MED CTR, 72-, RES ASSOC, ISAAC ALBERT RES INST, 64- Concurrent Pos: From clin asst prof to clin assoc prof path, State Univ NY Downstate Med Ctr, 65-73, clin prof, 73- Honors & Awards: Gold Award, Am Soc Clin Pathologists-Col Am Pathologists, 68. Mem: Am Soc Clin Pathologists; AMA; Am Soc Exp Path; Am Asn Pathologists & Bacteriologists; Fedn Am Socs Exp Biol. Res: Ultrastructure of pancreatic pathology; radiation damage; experimental diabetes; fluorescence microscopy; autopsy pathology. Mailing Add: Isaac Albert Res Inst Kingsbrook Jewish Med Ctr Brooklyn NY 11203

WELLNER, DANIEL, b Antwerp, Belg, June 9, 34; US citizen; m 62; c 1. BIOCHEMISTRY. Educ: Harvard Col, AB, 56; Tufts Univ, PhD(biochem), 61. Prof Exp: Instr biochem, Sch Med, Tufts Univ, 62-63, sr instr, 63-65, asst prof, 65-67; asst prof, 67-69, ASSOC PROF BIOCHEM, MED COL, CORNELL UNIV, 69- Concurrent Pos: NATO fel biochem, Weizmann Inst Sci, 61-62; Lederle med fac award, 64-67. Mem: AAAS; Am Chem Soc; NY Acad Sci; Harvey Soc; Am Soc Biol Chemists. Res: Ribonuclease; amino acid oxidases; flavoproteins; mechanism of enzyme action. Mailing Add: Dept of Biochem Cornell Univ Med Col New York NY 10021

WELLNER, MARCEL, b Antwerp, Belg, Feb 8, 30; US citizen; m 61; c 2. PHYSICS. Educ: Mass Inst Technol, BS, 52; Princeton Univ, PhD(physics), 58. Prof Exp: Instr physics, Princeton Univ, 56-58 & Brandeis Univ, 58-59; mem, Inst Advan Study, 59-60; res assoc physics, Univ Ind, 60-63; NSF fel, Atomic Energy Res Estab, Eng, 63-64; from asst prof to assoc prof, 64-71, PROF PHYSICS, SYRACUSE UNIV, 71- Concurrent Pos: Physicist, Cavendish Lab, Cambridge Univ, Eng, 68-69. Mem: Am Phys Soc. Res: Quantum field theory and mathematical methods. Mailing Add: Dept of Physics Syracuse Univ Syracuse NY 13210

WELLNER, VAIRA PAMILJANS, b Aluksne, Latvia, Jan 28, 36; US citizen; m 62; c 1. BIOCHEMISTRY. Educ: Boston Univ, AB, 58; Tufts Univ, PhD(biochem), 64. Prof Exp: Res assoc biochem, Sch Med, Tufts Univ, 66-67; RES ASSOC BIOCHEM, COL MED, CORNELL UNIV, 72- Concurrent Pos: Res fel biochem, Sch Med, Tufts Univ, 63-66 & Col Med, Cornell Univ, 67-72. Mem: Am Chem Soc. Res: Mechanisms of enzyme action. Mailing Add: Dept of Biochem Cornell Univ Med Col New York NY 10021

WELLOCK, LOIS MARGARET, b Harbor Beach, Mich. PHYSICAL MEDICINE. Educ: Eastern Mich Univ, BS, 39; Northwestern Univ, cert phys ther, 43; Univ Mich, MS, 48, PhD(educ), 65. Prof Exp: Phys therapist, Hosps & Pub Schs, 43-47; asst prof health & phys educ, Bowling Green State Univ, 48-54; asst prof phys ther, Med Sch, Northwestern Univ, 54-65; ASST PROF PHYS THER, UNIV MICH, ANN ARBOR, 65- Mem: Am Phys Ther Asn. Res: Therapeutic exercise; educational preparation. Mailing Add: Curric in Phys Ther Univ of Mich Ann Arbor MI 48104

WELLONS, JESSE DAVIS, III, b Roanoke, Va, Apr 4, 38; m 58; c 4. WOOD TECHNOLOGY, POLYMER SCIENCE. Educ: Duke Univ, BS, 60, MF, 63, PhD(wood technol, polymer sci), 66. Prof Exp: Res chemist, Res Triangle Inst, NC, 62-65, res assoc wood chem, Iowa State Univ, 65-66, from asst prof to assoc prof, 66-70; ASSOC PROF FOREST PROD CHEM, ORE STATE UNIV, 70- Mem: Forest Prod Res Soc; Soc Wood Sci & Technol. Res: Effect of radiation on cellulose; radiation grafting to cellulose; wood adhesives; use of plastics to modify properties of wood; sorption and diffusion of monomers in wood. Mailing Add: Sch of Forestry Ore State Univ Corvallis OR 97331

WELLS, ADONIRAM JUDSON, b Chicago, Ill, Apr 1, 17; m 37; c 6. PHYSICAL CHEMISTRY. Educ: Harvard Univ, SB, 38, AM, 40, PhD(phys chem), 41. Prof Exp: Res chemist, Ammonia Dept, 41-46, res supvr, 46-48, mgt asst, 48-50, mgr film develop, 50-53, dir, Yerkes Res Lab, NY, 53-55, asst res dir film dept, Del, 55-59, res dir electrochem dept, 59-69, DIR INDUST PRODS DIV, FABRICS & FINISHES DEPT, E I DU PONT DE NEMOURS & CO, DEL, 69- Mem: Am Chem Soc. Res: Infrared and Raman spectroscopy; thermodynamics; process and market development; polymer chemistry; high temperatures. Mailing Add: Fabrics & Finishes Dept E I du Pont de Nemours & Co Wilmington DE 19898

WELLS, BENJAMIN B, JR, b Rochester, Minn, May 31, 41; m 67. MATHEMATICAL ANALYSIS. Educ: Univ Mich, BS, 61, MS, 62; Univ Calif, Berkeley, PhD(math), 67. Prof Exp: From instr to asst prof math, Univ Ore, 67-70; Fulbright lectr, Univ Santiago, Chile, 70-71; ASSOC PROF MATH, UNIV HAWAII, 71- Mem: Am Math Soc. Res: Measure theory; harmonic analysis. Mailing Add: Dept of Math Univ of Hawaii Honolulu HI 96822

WELLS, BOBBY R, b Wickliffe, Ky, July 30, 34; m 60; c 1. SOIL CHEMISTRY, SOIL FERTILITY. Educ: Murray State Univ, BS, 59; Univ Ark, MS, 61; Univ Mo, PhD(soil chem), 64. Prof Exp: Asst soils, Univ Ark, 59-60; asst county agent, Univ Mo, 60-61, asst soils, 61-64; asst prof agr, Murray State Univ, 64-66; asst prof, Rice Br Exp Sta, 66-74, ASSOC PROF AGRON, UNIV ARK, FAYETTEVILLE, 74- Mem: Am Soc Agron; Soil Sci Soc Am. Res: Investigations into soil-plant relationships for rice. Mailing Add: Dept of Agron Univ of Ark Fayetteville AR 72701

WELLS, CHARLES EDMON, b Dothan, Ala, May 19, 29; m 62; c 3. PSYCHIATRY, NEUROLOGY. Educ: Emory Univ, AB, 48, MD, 53. Prof Exp: From intern to asst resident med, New York Hosp, 53-54; clin assoc neurol, NIH, 54-56; resident, New York Hosp, 57-58; assoc prof neurol, 61-75, psychiat, 68-72, PROF NEUROL, SCH MED, VANDERBILT UNIV, 75-, PROF PSYCHIAT & VCHMN DEPT, 72- Concurrent Pos: NIH fel neurol, New York Hosp-Cornell Med Ctr, 58-59. Mem: Am Psychiat Asn; Am Col Psychiat; Am Neurol Asn; Asn Res Nerv & Ment Dis. Res: Dementia; use of literature in teaching of psychiatry. Mailing Add: Dept of Psychiat Vanderbilt Univ Sch of Med Nashville TN 37232

WELLS, CHARLES FREDERICK, b Atlanta, Ga, May 4, 37; m 62; c 2. ALGEBRA. Educ: Oberlin Col, AB, 62; Duke Univ, PhD(math), 65. Prof Exp: Asst prof math, 65-72, ASSOC PROF MATH, CASE WESTERN RESERVE UNIV, 72- Concurrent Pos: Prin investr, NSF grants, 65-67; guest, Math Res Inst, Swiss Fed Inst Technol, Zurich, 75-76. Mem: Math Asn Am; Am Math Soc. Res: Algebraic theory of small categories; generalized wreath products; permutations of finite fields. Mailing Add: Dept of Math Case Western Reserve Univ Cleveland OH 44106

WELLS, CHARLES HENRY, b Chicago, Ill, Dec 6, 31; m 56; c 2. PHYSIOLOGY. Educ: Randolph-Macon Col, BA, 54; Mich State Univ, MS, 60, PhD(physiol), 63. Prof Exp: Instr physiol, Univ Tex Med Br, 62-64; asst prof, Med Col SC, 64-67; asst prof, 67-73, ASSOC PROF PHYSIOL, UNIV TEX MED BR, GALVESTON, 73- Concurrent Pos: Consult, US Air Force, 69-; chief, Physiol Div, Shriners Burns Inst. Mem: Aerospace Med Asn; Undersea Med Soc; Am Physiol Soc; Am Burn Asn; Int Soc Burn Injury. Res: Blood rheology, microcirculation in stress; thermal injury; decompression sickness; computer aided instructional systems development. Mailing Add: Dept of Physiol Univ of Tex Med Br Galveston TX 77550

WELLS, CHARLES PRENTISS, b Tabor, Iowa, Jan 23, 09; m 41; c 2. MATHEMATICS. Educ: Simpson Col, BA, 30; Iowa State Col, MS, 32, PhD(math), 35. Prof Exp: Asst, Iowa State Col, 30-34; instr math, NDak Col, 34-38; from instr to asst prof, Mich State Univ, 38-44; analyst, US Dept Army Air Force, 44-45; assoc prof, 45-51, PROF MATH, MICH STATE UNIV, 51-, CHMN DEPT, 60- Concurrent Pos: Lectr, Brown Univ, 48-49. Mem: Am Math Soc; Math Asn Am. Res: Quantum mechanics; electromagnetic waves; heat transmission; elasticity. Mailing Add: Dept of Math Mich State Univ East Lansing MI 48823

WELLS, CHARLES ROBERT EDWIN, b New Brighton, Pa, Aug 6, 13; m 41; c 4. PEDIATRIC CARDIOLOGY. Educ: Grove City Col, BS, 35; Temple Univ, MD, 40, MS, 49. Prof Exp: Pvt pract, 49-62; PROF PEDIAT, SCH MED, TEMPLE UNIV, 67- Concurrent Pos: Attend pediatrician & attend cardiologist, St Christopher's Hosp for Children, 49-, chief dept cardiol, 62-72, sr consult, Dept Cardiol, 72-; consult, Episcopal & Wilmington Gen Hosps. Mailing Add: St Christopher's Hosp for Children 2600 N Lawrence St Philadelphia PA 19133

WELLS, CHARLES VAN, b Summerton, SC, July 1, 37; m 66; c 2. BOTANY. Educ: Presby Col, SC, BS, 59; Appalachian State Univ, MA, 63; Univ Ariz, PhD(bot), 69. Prof Exp: Teacher & coach, High Sch, SC, 59-62; teaching asst biol, Appalachian State Univ, 62-63; asst prof biol, Newberry Col, 63-64; teaching asst biol, Univ Ariz, 64-69; ASSOC PROF BIOL, LENOIR RHYNE COL, 64- Res: Effects of radiation on vegetative morphology of sirogonium. Mailing Add: Dept of Biol Lenoir Rhyne Col Box 473 Hickory NC 28601

WELLS, DANA, b Wick, WVa, Sept 16, 07. GEOLOGY, PALEONTOLOGY. Educ: Univ WVa, AB, 28; Univ Kans, AM, 30; Columbia Univ, PhD(geol), 49. Prof Exp: Asst instr phys geol, Univ Kans, 28-29; cur mus, Univ Cincinnati, 29-30; from instr to prof, 30-74, chmn dept, 61-74, EMER PROF GEOL, W VA UNIV, 74- Concurrent Pos: Coop paleontologist, WVa Geol & Econ Surv, 34- Mem: Geol Soc Am; Paleont Soc; Am Asn Petrol Geol. Res: Invertebrate paleontology and stratigraphy of the Greenbrier series of the Mississippian system in West Virginia. Mailing Add: Dept of Geol WVa Univ Morgantown WV 26506

WELLS, DANIEL R, b New York, NY, May 2, 21; m 43; c 2. PHYSICS. Educ: Cornell Univ, BME, 42; NY Univ, MS, 55; Stevens Inst Technol, PhD(physics), 63. Prof Exp: Res assoc plasma physics, Princton Univ, 55-64; assoc prof physics, 64-67, PROF PHYSICS, UNIV MIAMI, 67- Concurrent Pos: Res assoc, Stevens Inst Technol, 61-64; assoc prof, Seton Hall, 62-64; res grants, AEC & Air Force Off Sci Res, 64-74. Mem: Am Phys Soc. Res: Controlled thermonuclear research. Mailing Add: Dept of Physics Univ of Miami Coral Gables FL 33124

WELLS, DARRELL GIBSON, b Pierre, SDak, Feb 21, 17; m 46; c 3. PLANT BREEDING. Educ: SDak State Col, BS, 41; State Col Wash, MS, 43; Univ Wis, PhD, 49. Prof Exp: Asst agronomist, State Col Wash, 43-45; asst, Univ Wis, 45-49; from assoc prof to prof agron, Miss State Univ, 49-62; PROF WHEAT BREEDING, S DAK STATE UNIV, 62- Concurrent Pos: Mem tech asst mission, Western Region, Nigeria, Int Develop Serv, 58-60. Mem: Am Soc Agron. Res: Small grain genetics and breeding; cowpea and lima bean breeding. Mailing Add: Dept of Plant Sci SDak State Univ Brookings SD 57006

WELLS, DARTHON VERNON, b Saline Co, Ark, Oct 11, 29; m 50; c 3. ORGANIC CHEMISTRY. Educ: Florence State Col, BS, 54; Univ Miss, MS, 57; Univ SC, PhD(chem), 60. Prof Exp: Res assoc chem, Brown Univ, 59-61; from asst prof to assoc prof, 61-72, PROF CHEM, LA STATE UNIV, ALEXANDRIA, 72- Mem: Am Chem Soc; Sigma Xi. Res: Chemistry of organic peroxide decomposition; nucleophilic displacement of phosphorus; carbanion rearrangement reactions. Mailing Add: Dept of Chem La State Univ Alexandria LA 71301

WELLS, DONALD O, b McKeesport, Pa, Apr 3, 33; m 55; c 6. NUCLEAR PHYSICS. Educ: Stanford Univ, BS, 55, MS, 56, PhD(physics), 63. Prof Exp: Scientist nuclear physics, Lockheed Missile & Space Co, 56-57; res asst, Stanford Univ, 57-61; asst prof physics, Univ Ore, 61-67; assoc prof, 67-75, VPRES ADMIN, UNIV MAN, 75- Concurrent Pos: Consult, Lockheed Missile & Space Co, 57-61. Mem: Am Phys Soc; Can Asn Physicists. Res: Experimental studies of nuclear reaction mechanisms, nuclear structure and radioactive decay schemes; solid state detectors; online data analysis with computers; nuclear model calculations. Mailing Add: Univ of Man Winnipeg MB Can

WELLS, EDWARD HENRY, b Chefoo, China, Mar 25, 01; US citizen; m 33; c 3. MATHEMATICS. Educ: Col Wooster, BS, 22; Lafayette Col, AM, 26. Prof Exp: Instr math, Lafayette Col, 23-26 & Princeton Univ, 26-29; asst prof, Univ NH, 29-31; home off rep, Prudential Life Ins Co, NJ, 31-35, asst mathematician, 35-39, mathematician, 39-41; asst actuary, Mutual Life Ins Co, NY, 41-48, actuary, 48-52, vpres, 52-66; lectr math, comput technol & econ statist, Bloomfield Col, 66-71; actuarial consult, 70-73; RETIRED. Concurrent Pos: Int Exec Serv Corps assignment, Beneficial Life Ins Co, Manila, Philippines, 66, Sao Paulo Campanhia Nacional de Seguros, Brazil, 68 & 72. Honors & Awards: Prize, Soc Actuaries, 40. Mem: Fel Soc Actuaries. Res: Actuarial mathematics. Mailing Add: 20 Berkeley Rd Millburn NJ 07041

WELLS, EDWARD JOSEPH, b Sydney, Australia, Oct 10, 36; m 62; c 2. PHYSICAL CHEMISTRY. Educ: Univ Sydney, BSc, 58, MSc, 60; Oxford Univ, DPhil(magnetic resonance), 62. Prof Exp: Gowrie travelling scholar, 59-61; fel magnetic resonance, Univ BC, 61-63; instr phys chem, 63-64; res assoc magnetic resonance, Univ Ill, Urbana, 64-65; asst prof chem, 65-67, ASSOC PROF CHEM, SIMON FRASER UNIV, 67-, ACTG CHMN DEPT, 75-, ASSOC MEM PHYSICS DEPT, 69- Concurrent Pos: With Australian Nat Serv, 55-58. Mem: Chem Inst Can; fel The Chem Soc; Am Inst Physics; fel Royal Soc Arts; Fedn Am Scientists. Res: Chemical and biological applications of nuclear magnetic resonance. Mailing Add: Dept of Chem Simon Fraser Univ Burnaby BC Can

WELLS, EUGENE ERNEST, JR, b Greenville, SC, June 9, 41; m 63; c 2. ORGANIC CHEMISTRY, ELECTROCHEMISTRY. Educ: Davidson Col, BS, 63; Univ Fla, PhD(org chem), 67. Prof Exp: Chemist, Power Sources Div, US Army Electronics Command, Ft Monmouth, NJ, 69-71; RES CHEMIST, ELECTROCHEM BR, NAVAL RES LAB, 71- Mem: Am Chem Soc; Am Inst Chem; Electrochem Soc. Res: Chemistry at electrode surfaces; application of computers to control of laboratory experiments. Mailing Add: Code 6160 Electrochem Br Naval Res Lab Washington DC 20390

WELLS, EUGENE HADLEY, b Maryville, Mo, May 20, 09; m 38, 48, 55; c 3. CHEMISTRY. Educ: Northwestern Mo State Teachers Col, BA, 31; Univ Iowa, MS, 33, PhD(org chem), 35. Prof Exp: Prof chem, Buena Vista Col, 35-36; assoc chemist, Food & Drug Admin, Fed Security Agency, 36-45; dir control, Frederick Stearns Co Div, Sterling Drug Co, 45-47; asst dir control, Winthrop-Stearns, Inc, 47-58, dir control, Winthrop Labs Div, Sterling Drug Inc, 58-74; RETIRED. Mem: Emer mem Am Chem Soc. Res: Organic research; methods of pharmaceutical analysis and quality control in the pharmaceutical industry. Mailing Add: Hill Crest Acres Box 67 Rt 1 Womelsdorf PA 19567

WELLS, FRANK EDWARD, b Granby, Mo, Mar 29, 25; m 47; c 2. MICROBIOLOGY, FOOD SCIENCE. Educ: Southwest Mo State Col, BS, 51; Purdue Univ, MS, 58, PhD(microbiol), 61. Prof Exp: Med technologist, Vet Admin Hosp, Springfield, Mo, 51-52 & US War Dept, Camp Crowder, Mo, 52-53; bacteriologist, Henningsen Foods Inc, 53-56; instr poultry prod tech, Agr Exp Sta, Purdue Univ, 56-61; tech dir food prod, Monark Egg Corp, Mo, 61; PRIN MICROBIOLOGIST, MIDWEST RES INST, 61- Concurrent Pos: Lectr, Univ Mo, Kansas City, 66- Mem: AAAS; Soc Invert Path; Wildlife Dis Asn. Res: Food microbiology; viral and bacterial diseases of insects; industrial fermentations. Mailing Add: Midwest Res Inst 425 Volker Blvd Kansas City MO 64110

WELLS, FRANKLIN BURNHAM, b Chicago, Ill, June 9, 07; m 27; c 2. ORGANIC CHEMISTRY. Educ: Univ Ill, BS, 30; McGill Univ, MSc, 31, PhD(org chem), 33. Prof Exp: Asst eng exp sta, Univ Ill, 29-30; demonstr, McGill Univ, 30-31, 32-33; analyst textile oils div, Richards Chem Works, NJ, 34-37, asst to dir lubrication div, 37-39; dir develop, Swan-Finch Oil Corp, NJ, 39-40; dir lubrication div, Richards Chem Works, NJ, 40-42; res chemist, Ellis Foster Co, 42-46; dir sci labs, Bloomfield Col, 46-51; asst mgr prod develop dept, Trojan Powder Co, Pa, 51-63; res chemist, Picatinny Arsenal, 63-72; CONSULT, DEPT DEFENSE, 72- Concurrent Pos: Abstractor, Chem Abstr & Apicult Abstr, Gt Brit. Mem: Fel AAAS; fel Am Inst Chem; fel The Chem Soc; hon mam Bee Res Asn; Am Chem Soc. Res: Lubricants; rust inhibitors; wetting agents; modified oil varnish bases; jojoba oil derivatives; polyol preparation and utilization; silage preservatives; commercial, military and primary explosives. Mailing Add: Friendship Village 1400 N Drake Rd Apt 191 Kalamazoo MI 49007

WELLS, FREDERICK JOSEPH, b Dayton, Ohio, Nov 13, 44. SEISMOLOGY. Educ: Univ Dayton, BS, 66; Brown Univ, ScM, 68, PhD(seismol), 72. Prof Exp: Res geophysicist, 72-75, ADVAN RES GEOPHYSICIST, DENVER RES CTR, MARATHON OIL CO, 75- Mem: Soc Explor Geophysicists; Seismol Soc Am. Res: Empirical and theoretical research of the seismic response of sedimentary rocks as a function of rock type, pore fluid and seismic frequency from the megahertz to the hertz range. Mailing Add: Denver Res Ctr Marathon Oil Co Littleton CO 80120

WELLS, GARLAND RAY, b El Dorado, Ark, Oct 10, 36; m 56; c 1. FORESTRY, ECONOMICS. Educ: La Polytech Inst, BS, 58; NC State Univ, MF, 61; Duke Univ, DF, 68. Prof Exp: Instr forestry, La Polytech Inst, 62-63, asst prof, 63-65; asst prof, 65-75, ASSOC PROF FORESTRY, UNIV TENN, KNOXVILLE, 75- Mem: Soc Am Foresters; Am Econ Asn. Res: Land ownership research especially the practice of forestry by private, non-industrial forest owners. Mailing Add: Dept of Forestry Univ of Tenn Knoxville TN 37901

WELLS, GEORGE SHERMAN, b Romeo, Mich, July 4, 17; m 52; c 3. BIOCHEMISTRY. Educ: Olivet Col, AB, 39; Univ Mich, Ann Arbor, MS, 41, PhD(biochem), 51. Prof Exp: Res asst, Arthritis Unit, Univ Hosp, Univ Mich, 41-42; instr biochem, Med Sch, Tufts Univ, 51-52; res asst, Dermat Unit, Univ Hosp, Univ Mich, 52-53; biochemist, Path Dept, Vet Admin Hosp, Dearborn, Mich, 54-62; BIOCHEMIST, PATH DEPT, GEISINGER MED CTR, 62- Concurrent Pos: Pres, Cent Pa Coun Res Ment Retardation, 68-69. Mem: Am Asn Clin Chemists; Am Chem Soc. Mailing Add: Path Dept Geisinger Med Ctr Danville PA 17821

WELLS, HENRY BRADLEY, b Ridgeland, SC; m 47; c 4. BIOSTATISTICS. Educ: Emory Univ, BA, 50; Univ NC, MSPH, 53, PhD(pub health), 59. Prof Exp: Chief statistician, Ga State Dept Pub Health, 50-55; statistician, NC State Bd Health, 56-58, consult, 58-64; from instr to assoc prof biostatist, 58-69, PROF BIOSTATIST, SCH PUB HEALTH, UNIV NC, CHAPEL HILL, 69- Concurrent Pos: Nat Inst Child Health & Human Develop spec fel award, 70-71. Mem: Am Statist Asn; Biomet Soc; Am Pub Health Asn; Pop Asn Am; Int Union Sci Study Pop. Res: Measurement of population change; survey methods in demographic research; evaluation of family planning programs. Mailing Add: Dept of Biostatist Univ of NC Sch of Pub Health Chapel Hill NC 27514

WELLS, HERBERT, b New Haven, Conn, July 27, 30; m 59; c 3. PHARMACOLOGY, ORTHODONTICS. Educ: Yale Univ, BA, 52; Harvard Univ, DMD, 56. Prof Exp: Res assoc orthod, Sch Dent Med, Harvard Univ, 59-60, assoc pharmacol, 60-63, asst prof dent & dir, Dent Clin, 63-68; PROF PHARMACOL, SCH GRAD DENT, BOSTON UNIV, 68- Concurrent Pos: Res fel orthod & pharmacol, Sch Dent Med, Harvard Univ, 56-59; mem, Panel Drugs Dent, Nat Res Coun, 66- & Dent Study Sect, NIH, 74- Honors & Awards: Lord-Chaim Res Award, 60; Oral Sci Prize, Int Asn Dent Res, 64. Mem: Am Soc Pharmacol & Exp Therapeut; Int Asn Dent Res. Res: Growth and secretion of exocrine and endocrine glands; salivary glands; parathyroid glands; experimental teratology; cleft palate formation. Mailing Add: Oral Pharmacol Lab Boston Univ Sch of Grad Dent Boston MA 02118

WELLS, HOMER DOUGLAS, b Blaine, Ky, Nov 11, 23; m 42; c 1. PLANT PATHOLOGY. Educ: Univ Ky, BS, 48, MS, 49; NC State Col, PhD(plant path), 54. Prof Exp: Tech asst agron agr exp sta, Univ Ky, 48-49, asst, 49-50; asst plant path, NC State Col, 50-52; asst agronomist, 52-53, PATHOLOGIST FORAGE CROPS, GA COASTAL PLAIN EXP STA, USDA, 53- Mem: Am Phytopath Soc. Res: Turf and forage crop disease problems. Mailing Add: Ga Coastal Plain Exp Sta Tifton GA 31794

WELLS, IBERT CLIFTON, b Fayette, Mo, Apr 12, 21; m 48; c 6. BIOCHEMISTRY. Educ: Cent Methodist Col, AB, 42; St Louis Univ, PhD(biochem), 48. Prof Exp: From instr to assoc prof biochem, Col Med, State Univ NY Upstate Med Ctr, 50-61; PROF BIOCHEM & CHMN DEPT, SCH MED, CREIGHTON UNIV, 61- Concurrent Pos: Nat Res Coun fel, Calif Inst Technol, 48-50. Honors & Awards: Com Solvents Corp Award, 52. Mem: Am Soc Biol Chemists; Am Chem Soc; Soc Exp Biol & Med. Res: Cholesterol metabolism; choline and one-carbon metabolism. Mailing Add: Dept of Biochem Creighton Univ Sch of Med Omaha NE 68131

WELLS, J GORDON, b Salt Lake City, Utah, Sept 18, 18; m 41; c 1. PHYSIOLOGY. Educ: Pepperdine Col, BS, 46; Univ Southern Calif, PhD, 51. Prof Exp: Mem fac phys educ, Pepperdine Col, 46-48; asst aviation physiol, Univ Southern Calif, 48-49; aviation physiologist, US Air Force Sch Aviation Med, 49-56; supvr human eng, Norair Div, Northrop Corp, 56-61; asst dir life sci, Appl Sci, Space & Info Systs Div, NAm Aviation, Inc, 61-62, mgr Apollo Crew Systs, Apollo Eng, 62-66, asst to vpres & gen mgr life sci opers, 66-67, dir space progs, Life Sci Opers, 67, mgr life sci & systs, Sci & Technol, Res Eng & Test, Space Div, 67-68, mgr life sci & systs, 68-71, SUPVR LIFE SCI, SYSTS ENG & TECHNOL, RES & ENG & TEST, SPACE DIV, N AM ROCKWELL CORP, 71- Mem: AAAS; Human Factors Soc; Am Astronaut Soc; assoc fel Am Inst Aeronaut & Astronaut; fel Aerospace Med Asn (vpres, 59-60). Res: Physiological aspects of aerospace medicine; human factors and bioastronautics. Mailing Add: 30443 La Vista Verde Dr Rancho Palos Verdes CA 90274

WELLS, JACK E, b Kansas City, Kans, Jan 14, 22; m 42; c 2. PEDODONTICS. Educ: Univ Mo-Kansas City, DDS, 51, MSD, 57. Prof Exp: From assoc prof to prof dent & chmn dept pedodontics, Sch Med, Univ Mo-Kansas City, 57-73; PROF PEDODONTICS & DEAN, COL DENT, UNIV TENN CTR HEALTH SCI, MEMPHIS, 73- Concurrent Pos: Dir undergrad pedodontics, Univ Mo-Kansas City, 57-58, coordr chairside dent asst prog, 57-73, asst dean, Sch Dent, 60-63, proj dir chronically ill, Aged & Handicapped Dent Proj & assoc dean, 63-73; chief dent serv, Children's Mercy Hosp, 59-73. Mem: Am Col Dent; Am Acad Pedodontics; Int Asn Dent Res. Res: Dentistry. Mailing Add: Col of Dent Univ Tenn Ctr Health Sci Memphis TN 38163

WELLS, JACK NULK, b McLouth, Kans, May 17, 37; m 60; c 2. MEDICINAL CHEMISTRY, PHARMACOLOGY. Educ: Park Col, BA, 59; Univ Mich, MS, 62, PhD(med chem), 63. Prof Exp: From asst prof to assoc prof med chem, Purdue Univ, 63-67; vis scholar, 72-73, asst prof physiol, 73-75, ASST PROF PHARMACOL, SCH MED, VANDERBILT UNIV, 75- Concurrent Pos: Fel, Ohio State Univ, 63. Mem: AAAS; Am Chem Soc. Res: Phosphodiesterase; smooth muscle physiology. Mailing Add: Dept of Pharmacol Vanderbilt Univ Sch of Med Nashville TN 37232

WELLS, JACQUELINE GAYE, b Pittsburgh, Pa, May 17, 31; m 51; c 2. ALGEBRA. Educ: Univ Pittsburgh, BS, 52, MS, 64, PhD(math), 72. Prof Exp: ASST PROF MATH, PA STATE UNIV, McKEESPORT, 64- Mailing Add: Dept of Math Pa State Univ McKeesport PA 15132

WELLS, JAMES EDWARD, b Akron, Ohio, Dec 12, 43; m 65. PHYSICAL INORGANIC CHEMISTRY. Educ: Univ Cincinnati, BS, 65; Purdue Univ, PhD(chem), 72. Prof Exp: Res chemist, Goodyear Atomic Corp, 65-67; RES CHEMIST, HOUDRY LABS, AIR PROD & CHEM INC, 73- Res: Heterogeneous catalysis, including oxidation catalysts, specialty catalysts, and various petroleum refining catalysts; preparation, properties and applications of such catalysts. Mailing Add: Houdry Labs Air Prod & Chem Inc PO Box 427 Marcus Hook PA 19061

WELLS, JAMES HOWARD, b Howe, Tex, June 20, 32; m 53; c 2. MATHEMATICS. Educ: Tex Tech Col, BS, 52, MS, 54; Univ Tex, PhD(math), 58. Prof Exp: Instr math, Univ Tex, 57-58; from instr to asst prof, Univ NC, 58-59; vis asst prof, Univ Calif, Berkeley, 60-61; assoc prof, 62-69, PROF MATH, UNIV KY, 69- Mem: Am Math Soc; Math Asn Am. Res: Analysis. Mailing Add: Dept of Math Univ of Ky Lexington KY 40506

WELLS, JAMES RAY, b Delaware, Ohio, May 28, 32; m 58; c 2. PLANT TAXONOMY. Educ: Univ Tenn, BS, 54, MS, 56; Ohio State Univ, PhD(bot), 63. Prof Exp: Asst prof biol, ECarolina Col, 63-64 & Old Dominion Col, 64-66; BOTANIST & ASST TO DIR, CRANBROOK INST SCI, 66- Concurrent Pos: Vis prof, Stephen F Austin State Col, 64; NSF grant, 67-69; adj assoc prof biol, Oakland Univ, 69- & Wayne State Univ, 73-; chmn Mich Natural Areas Coun, 70-72. Mem: Am Soc Plant Taxon; Bot Soc Am; Am Soc Nat; Int Soc Plant Taxon. Res:

Biosystematics of the genus Polymnia including chemotaxonomy. Mailing Add: Cranbrook Inst of Sci 500 Lone Pine Rd Bloomfield Hills MI 48013

WELLS, JAMES ROBERT, b Moundsville, WVa, Apr 5, 40; m 62; c 2. CHEMISTRY. Educ: Wheeling Col, BS, 62; Univ Pittsburgh, PhD(chem), 67. Prof Exp: Chemist plastics dept, Exp Sta, 67-71, sr res chemist, Sabine River Lab, 71-72, res supvr, Polymer Intermediates Dept, 72-74, LAB DIR, POLYMER INTERMEDIATES DEPT, SABINE RIVER LAB, E I DUPONT DE NEMOURS & CO, INC, 74- Mem: Am Chem Soc. Mailing Add: Polymer Interm Dept Sabine River Lab E I du Pont de Nemours & Co Inc Orange TX 77630

WELLS, JANE FRANCES, b Davenport, Iowa, Feb 24, 44. MATHEMATICS. Educ: Marycrest Col, BA, 66; Univ Iowa, MS, 67, PhD(math), 70. Prof Exp: Asst prof math, Purdue Univ, Ft Wayne, 70-74; UNIV PROF, GOVERNORS STATE UNIV, 74- Mem: Am Math Soc; Math Asn Am. Res: Noether lattices. Mailing Add: Col Bus & Pub Serv Governors State Univ Park Forest IL 60466

WELLS, JAY BYRON, b Chicago, Ill, July 3, 26; m 48; c 5. PHYSIOLOGY. Educ: Univ Denver, AB, 51, MS, 52; Univ Rochester, PhD(physiol), 64. Prof Exp: Res physiologist, Med Neurol Br, 54-59, 63-75, RES PHYSIOLOGIST, LAB BIOPHYS, NAT INST NEUROL & COMMUN DIS & STROKE, 76- Res: Nerve-muscle physiology and pharmacology; muscle mechanics; respiratory physiology. Mailing Add: Lab Biophys NINCDS Marine Biol Lab Woods Hole MA 02543

WELLS, JERRY SCOTT, microbiology, biology, see 12th edition

WELLS, JOHN ARTHUR, b Clearwater, Fla, Oct 2, 35. PLANT CHEMISTRY. Educ: Univ Fla, BSCh, 58; Auburn Univ, PhD(chem), 70. Prof Exp: Chemist, Naval Stores Lab, 58-61, res chemist, 70-73, RES CHEMIST, US VEGETABLE, USDA, 73- Mem: Am Chem Soc. Res: Chemistry of plant material developed in a vegetable breeding program; biochemical nature of plant host resistance to diseases and insects; nutrients and toxicants in new breeding lines. Mailing Add: US Vegetable Lab PO Box 3348 Charleston SC 29407

WELLS, JOHN CALHOUN, JR, b Tampa, Fla, May 12, 41; m 63; c 2. EXPERIMENTAL NUCLEAR PHYSICS. Educ: Fla State Univ, BS, 61; Johns Hopkins Univ, PhD(physics), 68. Prof Exp: Nat Acad Sci-Nat Res Coun assoc & res physicist, US Naval Ord Lab, Md, 68-70; asst prof physics, 70-75, ASSOC PROF PHYSICS, TENN TECHNOL UNIV, 75- Concurrent Pos: Consult physics div, Oak Ridge Nat Lab, 71- Mem: Am Phys Soc; Am Asn Physics Teachers. Res: Nuclear charged-particle reactions; heavy-ion reactions; gamma-ray spectroscopy; neutron reactions. Mailing Add: Dept of Physics Tenn Technol Univ Cookeville TN 38501

WELLS, JOHN CLARENCE, b Philadelphia, Pa, Jan 12, 14; m 44; c 4. PHYSICS. Educ: Colgate Univ, AB, 37; Columbia Univ, MA, 40, EdD(sci ed), 52. Prof Exp: Teacher, Pleasant Hill Acad, Tenn, 37-39; teacher high sch, NY, 39-42; ballistic supvr, Radford Ord Works, 42-45; teacher high sch, NY, 45-46; asst, Dept Natural Sci, Teachers Col, Columbia Univ, 46-47; assoc prof physics, 47-53, PROF PHYSICS, MADISON COL, VA, 47- Mem: AAAS; Nat Sci Teachers Asn; Nat Asn Res Sci. Res: Teaching. Mailing Add: Dept of Physics Madison Col Harrisonburg VA 22801

WELLS, JOHN MILTON, plant pathology, plant physiology, see 12th edition

WELLS, JOHN MORGAN, JR, b Hopewell, Va, Apr 12, 40. MARINE BIOLOGY. Educ: Randolph-Macon Col, BS, 62; Univ Calif, San Diego, PhD(marine biol), 69. Prof Exp: Res physiologist, Wrightsville Marine Bio-Med Lab, NC, 69-72; asst prof physiol, Sch Med, Univ NC, 70-72; SCI COORDR MARINE BIOL, MANNED UNDERSEA SCI & TECHNOL OFF, NAT OCEANIC & ATMOSPHERIC ADMIN, 72- Res: Blood function at high hydrostatic and inert gas pressures; physiological symbiosis between algae and invertebrates; community metabolism of marine benthic communities. Mailing Add: Manned Undersea Sci & Technol Off NOAA 11400 Rockville Park Rockville MD 20852

WELLS, JOHN WEST, b Philadelphia, Pa, July 15, 07; m 32; c 1. PALEONTOLOGY, MARINE ZOOLOGY. Educ: Univ Pittsburgh, BS, 28; Cornell Univ, MA, 30, PhD(paleont), 33. Prof Exp: Instr geol, Univ Tex, 29-31; Nat Res Coun fel, Brit Mus, Paris & Berlin, 33-34; asst sci, NY State Norm Sch, 37-38; from instr to prof geol, Ohio State Univ, 38-48; PROF GEOL, CORNELL UNIV, 48- Concurrent Pos: Geologist, US Geol Surv, 46-, Off Strategic Serv, 44-45, Bikini Sci Resurv, 47 & Pac Sci Bd, Arno Atoll Exped, 50; Fulbright lectr, Univ Queensland, 54; pres, Paleont Res Inst, 61-64. Honors & Awards: Paleont Medal, Paleont Soc, 74. Mem: Nat Acad Sci; fel Geol Soc Am; Paleont Soc (pres, 61-62); Soc Vert Paleont; Soc Syst Zool. Res: Invertebrate paleontology and paleoecology; vertebrate paleontology; stratigraphy; invertebrate zoology. Mailing Add: Dept of Geol Sci Cornell Univ Ithaca NY 14850

WELLS, JOSEPH, b Boston, Mass, Oct 6, 34; m 56; c 5. NEUROANATOMY, NEUROBIOLOGY. Educ: Univ RI, BS, 56; Duke Univ, PhD(anat), 59. Prof Exp: Instr anat, Duke Univ, 59-61 & Yale Univ, 61-63; asst prof, Sch Med, Univ Md, Baltimore, 63-68; ASSOC PROF ANAT, COL MED, UNIV VT, 68- Concurrent Pos: NIH fel, 59-61; Nat Inst Neurol Dis & Blindness res grant, 64-67; Lederle Med Fac Award, Univ Md, 68. Mem: Am Asn Anatomists; Soc Neurosci. Res: Neuroanatomy using silver stains; invertebrate neurobiology using electron microscopy; neuroanal plasticity using autoradiography. Mailing Add: Dept of Anat Univ of Vt Col of Med Burlington VT 05401

WELLS, JOSEPH ALBERT, b Wellsville, Mo, Apr 6, 16; m 70; c 4. PHARMACOLOGY, PHYSIOLOGY. Educ: Univ Denver, BS, 37, MS, 38; Northwestern Univ, PhD(physiol, pharmacol), 42, BM, 45, MD, 47. Prof Exp: Instr physiol, Univ Denver, 41-44, assoc, 44-46, from asst prof to assoc prof, 46-50; prof pharmacol, Northwestern Univ, 50-70, assoc dean, 64-70, chmn dept pharmacol, 60-66; DEAN, STRITCH SCH MED, LOYOLA UNIV, ILL, 70- Mem: AAAS; Soc Exp Biol & Med; Am Soc Pharmacol & Exp Therapeut. Res: Autonomic drugs; histamines; anaphylaxis; pyrogens. Mailing Add: Off of the Dean Stritch Sch of Med Loyola Univ Maywood IL 60153

WELLS, JOSEPH S, b Meade, Kans, Mar 19, 30; m 56; c 4. PHYSICS. Educ: Kans State Univ, BS, 56, MS, 58; Univ Colo, PhD(physics), 64. Prof Exp: Proj leader microwave noise, Electronics Calibration Ctr, 59-62, physicist, Radio & Microwave Mat Sect, 62-64, RES PHYSICIST, QUANTUM ELECTRONICS SECT, NAT BUR STANDS, 64- Concurrent Pos: Res assoc, Univ Colo, 64-66, lectr, 66- Mem: Am Phys Soc; Sci Res Soc Am. Res: Paramagnetic and antiferromagnetic resonance; microwave measurements; infrared frequency synthesis; laser stabilization and frequency measurements; infrared spectroscopy. Mailing Add: Quantum Electronics Sect Nat Bur Stands Denver Fed Bldg Denver CO 80202

WELLS, KENNETH, b Portsmouth, Ohio, July 24, 27; m 54; c 2. MYCOLOGY. Educ: Univ Ky, BS, 50; Univ Iowa, MS & PhD(bot), 57. Prof Exp: Asst bot, Univ Iowa, 54-57; instr & jr botanist, 57-58, from asst prof & asst botanist to assoc prof & assoc botanist, 59-72, PROF BOT & BOTANIST, UNIV CALIF, DAVIS, 72- Concurrent Pos: Fulbright res fel, Univ Brazil, 63-64; Nat Acad Sci exchange scientist, Romania, 71-72. Mem: Mycol Soc Am; Brit Mycol Soc; Mycol Soc Japan. Res: Taxonomy of the fungi, especially the Tremellales; ultrastructure of the fungi, especially of the basidium; general mycology. Mailing Add: Dept of Bot Univ of Calif Davis CA 95616

WELLS, KENNETH LINCOLN, b Lone Mountain, Tenn, May 28, 35; m 60. SOIL SCIENCE, AGRONOMY. Educ: Univ Tenn, BS, 57, MS, 59; Iowa State Univ, PhD(soil sci), 63. Prof Exp: Res assoc soils, Iowa State Univ, 59-63; agriculturist, Tenn Valley Authority, 63-65, agronomist, 65-69; asst prof, 69-72, ASSOC PROF AGRON, UNIV KY, 72- Mem: Am Soc Agron; Soil Sci Soc Am. Res: Soil fertility, crop production, genesis and classification. Mailing Add: Dept of Agron Univ of Ky Lexington KY 40506

WELLS, MARION ROBERT, b Jackson, Miss, Feb 9, 37; m 59; c 3. CELL PHYSIOLOGY. Educ: Memphis State Univ, BS, 60, MA, 63; Miss State Univ, PhD(zool), 71. Prof Exp: Instr biol, Troy State Col, 63-64; ASSOC PROF BIOL, MID TENN STATE UNIV, 64- Mem: AAAS; Sigma Xi. Res: Binding of insecticides to cell particulate. Mailing Add: Dept of Biol Mid Tenn State Univ Murfreesboro TN 37130

WELLS, MARTHA CAROL, b Albuquerque, NMex, Mar 15, 47. MICROBIAL PHYSIOLOGY. Educ: Univ Tex, Austin, BA, 69, PhD(microbiol), 72. Prof Exp: Instr microbiol, Univ Tex, Austin, 72-74; MICROBIOLOGIST, CENT RES & DEVELOP DEPT, EXP STA, E I DU PONT DE NEMOURS & CO, INC, 75- Concurrent Pos: Res assoc, Univ Tex, Austin, 72-75. Mem: AAAS; Am Soc Microbiol. Res: Industrial fermentations, enzymology and microbial hydroxylation reactions. Mailing Add: Cent Res & Develop Dept Exp Sta E I du Pont de Nemours & Co Inc Wilmington DE 19898

WELLS, MICHAEL ARTHUR, b Los Angeles, Calif, Nov 8, 38; m 58; c 3. BIOCHEMISTRY. Educ: Univ Southern Calif, BA, 61; Univ Ky, PhD(biochem), 65. Prof Exp: Am Cancer Soc fel biochem, Univ Wash, 65-67; asst prof, 67-73, ASSOC PROF BIOCHEM, UNIV ARIZ, 72- Concurrent Pos: Macy fac scholar, Josiah Macy Found, 75-76. Mem: AAAS; Am Chem Soc. Res: Complex lipids, their synthesis, chemical and physical properties, metabolism and interaction with proteins. Mailing Add: Dept of Biochem Univ of Ariz Tucson AZ 85721

WELLS, MICHAEL BYRON, b Kansas City, Mo, July 3, 22; m 48; c 4. OPTICAL PHYSICS, RADIATION PHYSICS. Educ: Univ Mo-Kansas City, BA, 48, MA, 50. Prof Exp: Instr, Univ Mo-Kansas City, 50-56; proj nuclear engr, Gen Dynamics, Ft Worth, Tex, 56-63; VPRES, RADIATION RES ASSOCS, INC, 63- Concurrent Pos: Instr, Tex Christian Univ, 58-63. Mem: Am Nuclear Soc; Optical Soc Am. Res: Nuclear radiation transport calculations; radiation shielding; atmospheric optics; ultraviolet, visible and infrared radiation transport in planetary atmospheres. Mailing Add: Radiations Res Assoc Inc 3550 Hulen St Ft Worth TX 76107

WELLS, MILTON ERNEST, b Calera, Okla, Nov 28, 32; m 55; c 2. ANIMAL BREEDING. Educ: Okla State Univ, BS, 55, MS, 59, PhD(animal breeding), 62. Prof Exp: Asst prof animal sci, Imp Ethiopian Col Agr, 61-65; asst prof dairy sci, 65-74, ASSOC PROF ANIMAL SCI & INDUST, OKLA STATE UNIV, 74- Mem: Am Soc Animal Sci; Am Dairy Sci Asn. Res: Reproductive physiology; vibriosis; international animal agriculture; acrosome of sperm cells. Mailing Add: Dept of Animal Sci Okla State Univ Stillwater OK 74074

WELLS, OSBORN O, forest genetics, see 12th edition

WELLS, OTHO SYLVESTER, b Burgaw, NC, Sept 15, 38; m 68; c 1. HORTICULTURE. Educ: NC State Univ, BS, 61; Mich State Univ, MS, 63; Rutgers Univ, PhD(hort), 66. Prof Exp: Asst prof, 66-71, ASSOC PROF PLANT SCI, UNIV NH, 71-, EXTEN HORTICULTURIST, VEGETABLES, 74- Mem: Am Soc Hort Sci. Res: Crop production under environmentally controlled conditions. Mailing Add: Dept of Plant Sci Univ of NH Durham NH 03824

WELLS, OUIDA CAROLYN, b Atlanta, Ga, July 23, 33. ZOOLOGY. Educ: Agnes Scott Col, BA, 55; Emory Univ, MS, 56, PhD(genetics), 58. Prof Exp: Asst, Emory Univ, 55-57; res assoc, Oak Ridge Nat Lab, 58-60; asst prof biol, 60-65, assoc prof natural sci, 65-68, asst dean, 69-71, PROF BIOL, LONGWOOD COL, 68-, ASSOC DEAN, 71- Mem: Genetics Soc Am; Soc Protozool; Am Soc Cell Biol; Am Soc Zool. Res: Genetics, cytology, physiology and biochemistry of microorganisms; radiation biology of ciliate Tetrahymena pyriformis, especially survival and death of irradiated cells using high levels of ionizing radiation. Mailing Add: Longwood Col Farmville VA 23901

WELLS, PATRICK HARRINGTON, b Palo Alto, Calif, June 19, 26; m 51; c 3. BIOLOGY. Educ: Univ Calif, AB, 48; Stanford Univ, PhD(biol), 51. Prof Exp: Asst prof zool, Univ Mo, 51-57; from asst prof to assoc prof biol, 57-71, PROF BIOL, OCCIDENTAL COL, 71- Concurrent Pos: Res assoc, Univ Calif, 67-72; ed, Southern Calif Acad Sci Bull, 72- Mem: AAAS; Bee Res Asn; Am Soc Zool; Lepidop Soc; Am Inst Biol Sci. Res: Physiology, experimental biology and animal behavior; physiology of learning in invertebrate animals; foraging behavior and recruitment in honey bees; natural history; population biology, ecology and physiology of butterflies and other arthropods. Mailing Add: Dept of Biol Occidental Col Los Angeles CA 90041

WELLS, PATRICK ROLAND, b Liberty, Tex, Apr 1, 31. PHARMACOLOGY. Educ: Tex Southern Univ, BS, 57; Univ Nebr, MS, 59, PhD(pharmaceut sci), 61. Prof Exp: Asst prof pharmacol, Fordham Univ, 61-63; from asst prof to assoc prof & actg chmn dept, Univ Nebr, 63-70; PROF PHARMACOL & DEAN SCH PHARM, TEX SOUTHERN UNIV, 70- Mem: AAAS; Am Pharmaceut Asn. Res: Cardiovascular screening of plant tissue culture. Mailing Add: Sch of Pharm Tex Southern Univ Houston TX 77004

WELLS, PAULA PARKER, b Sandusky, Ohio, Jan 1, 43; m 65. ORGANIC CHEMISTRY. Educ: Ohio Univ, BS, 65; Purdue Univ, PhD(org chem), 73. Prof Exp: Anal chemist, Mead Paper Co, 65-67; anal chemist, Dept Agron, Purdue Univ, 67-68; RES CHEMIST, ROHM AND HAAS CO, 73- Mem: Am Chem Soc. Res: Process development for agriculture, animal and human health products. Mailing Add: Rohm and Haas Co Norristown & McKean Rds Spring House PA 19477

WELLS, PHILIP VINCENT, b Brooklyn, NY, Apr 24, 28; m 59; c 3. BOTANY. Educ: Brooklyn Col, BA, 51; Univ Wis, MS, 56; Duke Univ, PhD(bot), 59. Prof Exp: Asst, Univ Wis, 54-55 & Duke Univ, 55-58; instr bot, Univ Calif, Santa Barbara, 58-59 & Calif Polytech Col, 59-60; res assoc biol, NMex Highlands Univ, 60-62; from asst prof to assoc prof bot, 62-71, PROF BOT & SYST ECOL, UNIV KANS, 71- Concurrent Pos: Actg dir, Bot Garden & vis assoc prof, Univ Calif, Berkeley, 66-67. Mem: AAAS; Soc Study Evolution; Bot Soc Am; Ecol Soc Am. Res: Pleistocene paleobotany; systematics, ecology and evolution in Arctostaphylos; vegetation of

North America; physiological ecology. Mailing Add: Dept of Bot Univ of Kans Lawrence KS 66044

WELLS, PHILLIP RICHARD, b Northampton, Mass, May 23, 31; m 62; c 3. FOOD SCIENCE, DAIRY SCIENCE. Educ: Univ Mass, BS, 57; Pa State Univ, MS, 59; Rutgers Univ, PhD(dairy sci), 62. Prof Exp: Proj leader food prod develop, Colgate Palmolive Co, 62-66; sect leader new prod develop, Corn Prod Co, 66-72, asst to dir res, 72-75, ASST DIR TECH SERV, BEST FOODS DIV, CPC INT, 75- Mem: Am Dairy Sci Asn; Inst Food Technol. Res: Dried and concentrated dairy products; snack products; stabilizers and emulsifiers; protein based foods; dried eggs; research management. Mailing Add: Best Foods Res Ctr CPC Int 1120 Commerce Ave Union NJ 07083

WELLS, RALPH GORDON, b Newark, Ohio, Sept 24, 15; m 42. MATERIALS SCIENCE. Educ: Muskingum Col, BA, 39; Ohio State Univ, MSc, 47; Univ Mich, PhD(mineral), 51. Prof Exp: Res metallurgist, US Steel Corp, 47-51, supvry technologist, 51-55; assoc res engr, Univ Mich, 56-59; SR RES SCIENTIST, MAT RES CTR, COLT INDUSTS CRUCIBLE INC, 59- Mem: Am Phys Soc; sr mem Inst Elec & Electronics Engrs; Am Inst Mining, Metall & Petrol Eng. Res: Application of Tools and techniques of mineralogy and crystallography to metallurgy and ceramics. Mailing Add: Mat Res Ctr Colt Industs Crucible Inc Box 88 Pittsburgh PA 15230

WELLS, RAYMOND O'NEIL, JR, b Dallas, Tex, June 12, 40; m 63; c 1. MATHEMATICS. Educ: Rice Univ, BS, 62; NY Univ, MS, 64, PhD(math), 65. Prof Exp: From asst prof to assoc prof, 65-74, PROF MATH, RICE UNIV, 74- Concurrent Pos: Vis asst prof, Brandeis Univ, 67-68; mem, Inst Advan Study, 70-71; mem Regional Conf Bd, Conf Bd Math Sci, 74-77; Guggenheim fel, John Simon Guggenheim Mem Found, 74-75; US sr scientist award, Alexander von Humboldt Found, 74-75; vis prof math, Univ Göttingen, 74-75. Mem: Am Math Soc; Math Asn Am. Res: Analytic continuation and approximation theory in several complex variables; theory of complex manifolds and spaces; automorphic functions; algebraic geometry. Mailing Add: Dept of Math Rice Univ Houston TX 77001

WELLS, ROBERT DALE, b Uniontown, Pa, Oct 2, 38; m 60; c 2. BIO-CHEMISTRY, MOLECULAR BIOLOGY. Educ: Ohio Wesleyan Univ, BS, 60; Univ Pittsburgh, PhD(biochem), 64. Prof Exp: NIH fel, Univ Pittsburgh, 64; NIH fel, Enzyme Inst, 64-66, from asst prof to assoc prof biochem, 66-73, PROF BIOCHEM, UNIV WIS-MADISON, 73- Concurrent Pos: Mem regional coun, Am Inst Biol Sci, 66-68. Mem: AAAS; Am Chem Soc; Am Soc Biol Chem. Res: Polynucleotides synthesis; DNA physical chemistry, synthesis and replication; polypeptide synthesis; tumor virology. Mailing Add: Dept of Biochem Univ of Wis-Madison Madison WI 53706

WELLS, ROE, b Washington, DC, Apr 17, 22; m 46; c 3. INTERNAL MEDICINE. Educ: Cornell Univ, AB, 44, MD, 46. Prof Exp: Intern med, New York Hosp, 46-47; asst resident med, Hopkins Hosp, 51-52; sr asst resident, Peter Brent Brigham Hosp, Boston, 52-53; instr, 53-56, clin assoc, 56-59, assoc, 59-65, asst prof, 65-70, ASSOC PROF MED, HARVARD MED SCH, 70-; CHIEF STAFF, VET ADMIN HOSP, BOSTON, 74- Concurrent Pos: Nat Heart Inst trainee, Johns Hopkins Hosp, Md, 49-50; jr assoc med, Peter Brent Brigham Hosp, 53-55, assoc, 55-; investr, Howard Hughes Med Inst, 55-67; exec med dir, Mass Rehab Hosp, 72-74. Mem: Am Thoracic Soc; Soc Rheol; Am Heart Asn; fel Am Col Physicians; fel Am Col Cardiol. Res: Physiology and cardiology. Mailing Add: Vet Admin Hosp Boston MA 02132

WELLS, RUSSELL FREDERICK, b Brooklyn, NY, Oct 24, 37; m 75; c 4. BIOLOGY. Educ: Lafayette Col, BA, 59; Springfield Col, MS, 62; Univ NC, Chapel Hill, MA, 65; Purdue Univ, PhD(biol educ), 70. Prof Exp: Teacher & coach, pvt sch, NC, 62-65; NSF acad year fel zool, Univ NC, Chapel Hill, 65-66; asst prof biol, Montclair State Col, 66-68; asst prof biol sci, Purdue Univ, 70-71; asst prof, 71-74, ASSOC PROF BIOL, ST LAWRENCE UNIV, 74- Mem: Am Inst Biol Sci; Nat Asn Biol Teachers. Res: Development of biological teaching materials for individualized instruction, primarily audio-tutorial; development of time-lapse cine studies of prolonged biological phenomena. Mailing Add: Dept of Biol St Lawrence Univ Canton NY 13617

WELLS, STEWART ALDERSON, b Saskatoon, Sask, May 16, 20; wid; c 3. GENETICS. Educ: Univ Sask, BSA, 42, MSc, 45; Univ Alta, PhD(genetics), 55. Prof Exp: Asst, Univ Sask, 42-45; res officer, Exp Sta, Sask, 45-48 & Alta, 48-66, RES SCIENTIST, BARLEY BREEDING, LETHBRIDGE RES STA, CAN DEPT AGR, ALTA, 66- Mem: Can Soc Agron; Genetics Soc Can; Agr Inst Can. Res: Cereal breeding. Mailing Add: Plant Sci Sect Res Sta Can Dept of Agr Lethbridge AB Can

WELLS, WARREN F, b Des Moines, Iowa, May 16, 26; m 50; c 3. BIOCHEMISTRY. Educ: Univ Northern Iowa, BA, 50, MA, 55; Univ Ill, PhD(biochem), 59. Prof Exp: Instr sci, Clear Lake High Sch, Iowa, 50-52; instr chem, Undergrad Div, Univ Ill, 53-54, res asst biochem, Col Med, 54-59; from instr to asst prof, 59-70, ASSOC PROF BIOCHEM, MED & DENT SCHS, NORTHWESTERN UNIV, CHICAGO, 70- Concurrent Pos: Consult, Col Am Path, 62-64; biochemist, Vet Admin Hosp, 63-65; consult, Vet Admin Hosps, 64-74; lectr, Dept Ment Health, State of Ill, 67-69. Mem: AAAS; Am Chem Soc; NY Acad Sci; fel Am Inst Chemists; Am Acad Neurol. Res: Nucleic acid and protein metabolism of normal and pathologic nervous tissue; biochemical changes accompanying periodontal disease. Mailing Add: Dept of Biochem Northwestern Univ McGaw Med Ctr Chicago IL 60611

WELLS, WILLARD H, b Austin, Tex, Feb 23, 31; m 56; c 2. OPTICAL PHYSICS. Educ: Univ Tex, BS, 52; Calif Inst Technol, PhD(physics), 59. Prof Exp: Scientist, Jet Propulsion Lab, Calif Inst Technol, 59-62, res group supvr, 62-67; CHIEF SCIENTIST, TETRA TECH, INC, 67- Mem: Am Phys Soc. Res: Quantum electronics and mechanics; applied mathematics; spacecraft mechanics and space communications; underwater optical systems; optical communications including fiber optics. Mailing Add: Tetra Tech Inc 630 N Rosemead Pasadena CA 91107

WELLS, WILLIAM T, b Wytheville, Va, Aug 9, 33; m 55; c 3. APPLIED STATISTICS, APPLIED MATHEMATICS. Educ: Col William & Mary, BS, 54; NC State Univ, MS, 56, PhD(exp statist), 59. Prof Exp: Res asst appl math, NC State Col, 58-59; statistician, Res Triangle Inst, 59-62; engr guided missiles range div, Pan Am World Airways, Inc, 62-65; sr consult appl statist, 65-67, mgr appl sci dept, 67-68, VPRES & DIR APPL SCI DIV, WOLF RES & DEVELOP CORP, 68- Mem: Inst Math Statist; Am Statist Asn. Res: Statistical methods in trajectory determination; geodesy; instrumentation evaluation; biomedical data analysis. Mailing Add: Appl Sci Div 6801 Kenilworth Ave Wolf Res & Develop Corp Riverdale MD 20840

WELLS, WILLIAM WOOD, b Traverse City, Mich, June 8, 27; m 50; c 4. BIOCHEMISTRY. Educ: Univ Mich, BS, 49, MS, 51; Univ Wis, PhD, 55. Prof Exp: Res assoc, Upjohn Co, 51-52; asst biochem, Univ Wis, 52-55; from instr to assoc prof, Univ Pittsburgh, 55-66; PROF BIOCHEM, MICH STATE UNIV, 66- Concurrent Pos: Mem, Metab Study Sect, NIH, 66-70. Mem: Am Chem Soc; Am Soc Biol Chem; Am Soc Neurochem; Int Soc Neurochem; Brit Biochem Soc. Res: Sterol structure and metabolism; relationship of sterol metabolism to experimental atherosclerosis; galactose metabolism and mental retardation; brain energy metabolism. Mailing Add: Dept of Biochem Mich State Univ East Lansing MI 48823

WELLSO, STANLEY GORDON, b Oshkosh, Wis, Feb 13, 35; m 57; c 3. ENTOMOLOGY. Educ: Univ Wis, BS, 57; Tex A&M Univ, MS, 62, PhD(entom), 66. Prof Exp: Asst prof entom, Colo State Univ, 65-67; entomologist, 67-73, RES LEADER ENTOM & SMALL GRAINS, AGR RES SERV, USDA, MICH STATE UNIV, 73- Mem: AAAS; Entom Soc Am. Res: Diapause induction and termination in insects; influence of photoperiod on insect's growth and development; rearing insects on artificial diets; ecological studies with grasshoppers; taxonomy and ecology of buprestids; cereal leaf beetle project; insect behavior and feeding during oviposition. Mailing Add: USDA Agr Res Serv Dept of Entom Mich State Univ East Lansing MI 48824

WELLUM, GLYN RICHARD, b Darington, Eng, July 18, 44; m 71. CHEMISTRY. Educ: King's Col, Univ London, BSc, 66; Royal Holloway Col, PhD(chem), 70. Prof Exp: Fel, Northeastern Univ, 70; res fel biochem, 70-73, ASST CHEM NEUROSURGERY, MASS GEN HOSP, 73- Mem: Am Chem Soc. Res: Cancer therapy by boron neutron capture; polyhedral boranes; carboranes; immunochemistry. Mailing Add: Warren 465 Mass Gen Hosp Boston MA 02114

WELLWOOD, ARNOLD AUGUSTUS, b Lockhartville, NS, Jan 29, 14; m 42; c 2. CYTOGENETICS, PLANT MORPHOLOGY. Educ: Acadia Univ, BSc, 39, BA, 40; Cornell Univ, PhD(cytogenetics, plant morphol), 56. Prof Exp: Lectr bot, McGill Univ, 42-47; asst prof genetics, Ont Agr Col, 47-58; res geneticist, US Int Coop Admin, Nigeria, 58-60; asst prof cytogenetics, Univ Conn, 60-63; assoc prof genetics, 63-70, PROF GENETICS & FLORISTICS & CUR HERBARIUM, WILFRID LAURIER UNIV, 70- Mem: Can Bot Asn; Bot Soc Am; Am Genetic Asn; Genetics Soc Can. Res: Inheritance of cytological characters in plants; floristics; genetics of disease resistance in flax; chemical taxonomy; vascular flora of Perth County, Ontario. Mailing Add: Dept of Biol Wilfrid Laurier Univ Waterloo ON Can

WELLWOOD, ROBERT WILLIAM, b Vancouver, BC, Mar 11, 12; m 41; c 4. WOOD SCIENCE & TECHNOLOGY. Educ: Univ BC, BASc, 35; Duke Univ, MF, 39, PhD(wood sci), 43. Prof Exp: Wood technologist, Commonwealth Plywood Co Ltd, Que, 43-46; assoc prof, 46-51, actg dean, 61-62 & 64-65, PROF FORESTRY, UNIV BC, 51- Concurrent Pos: Vis prof, Yale Univ, 58. Mem: Forest Prod Res Soc; Can Inst Forestry; Commonwealth Forestry Asn; Soc Wood Sci & Technol. Res: Wood utilization and product development; wood anatomy; properties of coniferous woods in relation to conditions of tree growth; feasibility studies in the developing countries. Mailing Add: Fac of Forestry Univ of BC Vancouver BC Can

WELMERS, EVERETT THOMAS, b Orange City, Iowa, Oct 27, 12; m 38; c 2. MATHEMATICS. Educ: Hope Col, AB, 32; Univ Mich, PhD(math), 37. Hon Degrees: ScD, Hope Col, 66. Prof Exp: From instr to asst prof math, Mich State Col, 37-44; flight res engr, Bell Aircraft Corp, 44, flutter engr, 44-46, group leader dynamic anal, 46-49, chief dynamics, 49-57, dir, L D Bell Res Ctr, 57-60, asst to pres, Corp, 58-60; group dir satellite systs, 60-63, asst for tech oper, Manned Systs Div, 64-67, asst to gen mgr, El Segundo Tech Opers, 67-68, ASST TO PRES, AEROSPACE CORP, 68- Concurrent Pos: With Nat Adv Comt Aeronaut, 44-59; prof, Millard Fillmore Col, 45-59; on leave to Inst Defense Anal & Adv Res Proj Agency, 59-60; comnr, Community Redevelop Agency, City of Los Angeles, 68- Mem: Am Math Soc; Math Asn Am; Am Inst Aeronaut & Astronaut; Inst Elec & Electronics Engrs. Res: Integration theory; jet propulsion; flutter; applied mathematics; aircraft and helicopter dynamics; operations analysis; computers; system analysis; satellites. Mailing Add: 1626 Old Oak Rd Los Angeles CA 90049

WELNA, CECILIA, b New Britain, Conn. MATHEMATICS. Educ: St Josephs Col, Conn, BS, 49; Univ Conn, MA, 52, PhD(ed), 60. Prof Exp: Instr, Mt St Joseph Acad, 49-50; asst instr math, Univ Conn, 50-55; instr, Univ Mass, 55-56; from asst prof to assoc prof, 56-68, PROF MATH & CHMN DEPT, UNIV HARTFORD, 68- Mem: Math Asn Am; Am Math Soc; Sigma Xi. Mailing Add: 176 Smith St New Britain CT 06053

WELSCH, CLIFFORD WILLIAM, JR, b St Louis, Mo, Sept 10, 35; m 58; c 3. PHYSIOLOGICAL CHEMISTRY, ONCOLOGY. Educ: Univ Mo, BS, 57, MS, 62, PhD(physiol chem), 65. Prof Exp: Instr physiol chem, Univ Mo, 63-65; asst prof natural sci, 65-66, asst prof anat, 68-71, ASSOC PROF ANAT, COL HUMAN MED, MICH STATE UNIV, 71- Concurrent Pos: Nat Cancer Inst res fel oncol, Mich State Univ, 66-68; NIH career development award, 71-76. Mem: AAAS; Am Asn Cancer Res; Soc Exp Biol & Med; Int Soc Neuroendocrinol; Am Physiol Soc. Res: Investigations in mammary gland carcinogenesis. Mailing Add: 2173 Belding Ct Okemos MI 48864

WELSCH, FEDERICO, b Sevilla, Spain, Dec 26, 33; US citizen; m 59; c 4. MOLECULAR BIOLOGY, MEDICINE. Educ: Univ Barcelona, BA, 50; Univ Valencia, MD, 55, DMedSci, 57; Ctr Res & Advan Study, Mex, MS, 65; Dartmouth Col, PhD(molecular biol), 68. Prof Exp: Intern internal med, Univ Hamburg, 54-55; from instr to asst prof physiol chem, Univ Valencia, 55-57; prof physiol, Univ Guadalajara, 58-60; prof biochem, Univ Chihuahua, 60-63; instr, Ctr Advan Res & Study, Mex, 63-65; asst prof, Dartmouth Med Sch, 68-70; assoc dir, 70-74, EXEC DIR, WORCESTER FOUND EXP BIOL, 74-; ASSOC RES PROF BIOCHEM, MED SCH, UNIV MASS, 70- Concurrent Pos: USPHS & Am Heart Asn grants, Dartmouth Med Sch & Worcester Found, 68-; hon collabr, Span Res Coun, 58. Mem: AAAS; Am Chem Soc; Asn Am Med Cols. Res: Molecular biology and medicine. Mailing Add: Worcester Found for Exp Biol 222 Maple Ave Shrewsbury MA 01545

WELSCH, ROY ELMER, b Kansas City, Mo, July 31, 43. STATISTICS. Educ: Princeton Univ, AB, 65; Stanford Univ, MS, 66, PhD(math), 69. Prof Exp: ASST PROF OPERS RES, MASS INST TECHNOL, 69- Concurrent Pos: Res assoc, Nat Bur Econ Res, 73-; assoc ed, Annals Statist, Inst Math Statist. Mem: Am Statist Asn; Inst Math Statist. Res: Data analysis; robust statistics; multiple comparisons; graphics. Mailing Add: Sloan Sch of Mgt Mass Inst of Technol Cambridge MA 02139

WELSER, JOHN RALPH, veterinary medicine, anatomy, see 12th edition

WELSH, CHARLES EDWARD, chemistry, see 12th edition

WELSH, DAVID ALBERT, b Pittsburgh, Pa, Oct 25, 42; m 64; c 3. ORGANIC CHEMISTRY, POLYMER CHEMISTRY. Educ: Carnegie Mellon Univ, BS, 64, MS, 68, PhD(org chem), 69. Prof Exp: Res chemist, Kippers Co Res Labs, 68-69, Edgewood Arsenal, 69-71 & Koppers Co Res Labs, 71-74; SR RES SCIENTIST, RES DEPT, ARCO/POLYMERS INC, 74- Mem: Am Chem Soc. Res: Free radical polymerization and copolymerization; polymer properties; organic synthesis. Mailing Add: 1042 Harvard Rd Monroeville PA 15146

WELSH, FRANCIS EUGENE, spectroscopy, physical chemistry, see 12th edition

WELSH, GEORGE W, III, b New York, NY, Aug 7, 20; m 50; c 2. INTERNAL MEDICINE, ENDOCRINOLOGY. Educ: Yale Univ, BA, 42; Univ Rochester, MD, 50; Am Bd Internal Med, dipl, 57. Prof Exp: Asst med, Med Col, Cornell Univ, 51-52, Dartmouth Med Sch, 52-54 & Sch Med, Univ Wash, 54-55; from instr to asst prof med, 56-64, dir continuing med educ, 65-68, dir off continuing educ health sci, 68-74, ASSOC PROF MED, COL MED, UNIV VT, 64- Concurrent Pos: Fel med, Mary Hitchcock Hosp, Hanover, NH, 52-53; Nat Inst Arthritis & Metab Dis trainee, Sch Med, Univ Wash, 54-55, USPHS res fel, 55; USPHS grants, 62-66; chmn interdept coun aging, State of Vt, 63-69; mem adv comt, Northeast Regional Med Libr Serv, 70-74, chmn, 72; adv coun, Off Health Care Educ, Northeast Ctr Continuing Educ, 71- Mem: Fel Am Col Physicians; Am Diabetes Asn; Endocrine Soc; Am Fedn Clin Res; Am Soc Internal Med. Mailing Add: Rowell Bldg Div of Health Sci Univ of Vt Burlington VT 05401

WELSH, HARRY L, b Aurora, Ont, Mar 23, 10; m 42. ATOMIC PHYSICS, MOLECULAR PHYSICS. Educ: Univ Toronto, BA, 30, MA, 31, PhD(physics), 36. Hon Degrees: DSc, Univ Windsor, 64 & Mem Univ Nfld, 68. Prof Exp: From asst prof to assoc prof, 42-54, chmn dept, 62-68, PROF PHYSICS, UNIV TORONTO, 54- Honors & Awards: Medal of Serv, Order of Can, 72. Mem: Fel Am Phys Soc; fel Royal Soc Can; Royal Astron Soc; Can Asn Physicists (vpres, 72-73); fel Royal Soc. Res: Infrared and Raman spectra, low temperature and high pressure spectroscopy. Mailing Add: Dept of Physics Univ of Toronto Toronto ON Can

WELSH, JAMES FRANCIS, b Pittsburgh, Pa, June 21, 30; m 55; c 5. ZOOLOGY. Educ: State Univ NY, BA, 53; Univ Calif, Los Angeles, PhD, 59. Prof Exp: Instr biol, St Mary's Col, 57-58; instr zool & human anat, Univ Calif, Los Angeles, 58-59; asst prof anat & physiol, 59-67, assoc prof physiol, 67-70, PROF ZOOL, HUMBOLDT STATE UNIV, 70- Mem: AAAS; Am Soc Parasitol; Wildlife Dis Asn. Res: Immunological research on Hymenolepis nana; enzyme systems of trematodes. Mailing Add: Dept of Zool Humboldt State Univ Arcata CA 95521

WELSH, JAMES RALPH, b Langdon, NDak, Sept 4, 33; m 52; c 4. PLANT BREEDING, PLANT GENETICS. Educ: NDak State Univ, BS, 56; Mont State Univ, PhD(plant genetics), 63. Prof Exp: Exten agt agron, NDak Exten Serv, 56-60; from asst prof to assoc prof, Mont State Univ, 63-68; assoc prof, 68-72, PROF AGRON, COLO STATE UNIV, 72- Mem: Am Soc Agron; Crop Sci Soc Am; Sigma Xi. Res: Wheat breeding and genetics; drought tolerance in winter wheat; high yielding cultivars. Mailing Add: Dept of Agron Colo State Univ Ft Collins CO 80523

WELSH, JOHN ELLIOTT, SR, b Berea, Ky, Nov 4, 27; m 65; c 3. GEOLOGY. Educ: Berea Col, AB, 50; Univ Wyo, MA, 51; Univ Utah, PhD(geol), 59. Prof Exp: Jr geologist, Magnolia Petrol Corp, 51; geologist, Shell Oil Co, 53-56; asst prof geol, Western State Col Colo, 56-61 & Colo State Univ, 61-62; RES GEOLOGIST, KENNECOTT EXPLOR INC, SALT LAKE CITY, 70- Concurrent Pos: Independent consult geologist, 56-70. Mem: Am Asn Petrol Geol; Soc Econ Geol. Res: Structural and stratigraphic analysis of mining districts. Mailing Add: 4780 Bonair St Holladay UT 84117

WELSH, JOHN HENRY, b Boothbay, Maine, Aug 25, 01; m 31; c 3. ZOOLOGY. Educ: Berea Col, AB, 22; Harvard Univ, MA, 28, PhD(biol), 29. Prof Exp: Prin high sch, Maine, 22-23; sci master, Berkshire Sch, Mass, 24-27; tutor biol, 28-47, instr zool, 29-32, fac instr, 32-40, from assoc prof to prof, 40-68, chmn dept biol, 47-50, dir biol labs, 56-59, EMER PROF ZOOL, HARVARD UNIV, 68- Concurrent Pos: Instr, Lesley Sch, 28-31; secy corp, Bermuda Biol Sta Res, Inc, 37-47, trustee, 38-68; assoc, Kirkland House, 48-68; USPHS res grant, 47-68; Guggenheim fel, Eng, 52; Fulbright res award, Univ Sao Paulo, 60; Rockefeller Found travel grant, 60. Mem: AAAS; Am Physiol Soc; Am Soc Zool; Soc Gen Physiol; fel Am Acad Arts & Sci. Res: Comparative neurophysiology; neurohormones and neurosecretion; comparative neuropharmacology; venoms and their modes of action. Mailing Add: Dover Rd Boothbay ME 04537

WELSH, LAWRENCE B, b Santa Barbara, Calif, Oct 21, 39; m 61; c 1. SOLID STATE PHYSICS. Educ: Pomona Col, BA, 61; Univ Calif, Berkeley, PhD(physics), 66. Prof Exp: Fel physics, Univ Pa, 66-68; asst prof, Northwestern Univ, Evanston, 68-74; GROUP LEADER MAT SCI, UOP, INC, 74- Concurrent Pos: Vis asst prof physics, Northwestern Univ, 74-75. Mem: Am Phys Soc. Res: Ceramics; thin film depositions; fuel cell technology; nuclear magnetic resonance in metals. Mailing Add: Corp Res Ctr UOP Inc Des Plaines IL 60016

WELSH, MAURICE FITZWILLIAM, b Cardonald, Scotland, Nov 2, 16; m 52; c 5. HORTICULTURE. Educ: Univ BC, BSA, 38; Univ Toronto, PhD(plant path), 42. Prof Exp: Lab asst bot, Univ Toronto, 38-39; plant pathologist, Plant Path Lab, Can Dept Agr, 45-56, officer-in-charge, 56-59, head plant path sect, Res Sta, 59-75; DIST HORTICULTURIST, BC DEPT AGR, SUMMERLAND, BC, 75- Concurrent Pos: Mem staff, E Malling Res Sta, Eng, 61-62; vis prof, Cornell Univ, NY State Exp Sta, 73-74. Mem: Am Phytopath Soc; Can Phytopath Soc (vpres, 65-66, pres, 66-67); Agr Inst Can. Res: Horticulture and plant pathology of tree fruits, grapes and vegetables. Mailing Add: BC Dept Agr PO Box 198 Summerland BC Can

WELSH, MICHAEL S, geography, applied statistics, see 12th edition

WELSH, RICHARD STANLEY, b Philadelphia, Pa, Nov 15, 21; m 56; c 2. BIOCHEMISTRY. Educ: Harvard Col, SB, 43; Univ Pa, MS, 46; Stanford Univ, PhD(biochem, phys chem), 52. Prof Exp: Asst instr inorg anal, Univ Pa, 44-45; technician virus res, Rockefeller Inst, 46-47; res asst, Stanford Univ, 57-59; Am Heart Asn advan res fel, Univ Redlands, 59-61, Am Heart Asn estab investr, 61-64; Am Heart Asn estab investr, Univ Calif, Riverside, 64-65, Lab Molecular Biol, NIH, 65-66 & Brookhaven Nat Lab, 66-67; assoc scientist, Div Microbiol, Brookhaven Nat Lab, 67-68; RES ASSOC, INST MED, ATOMIC RES INST, W GER, 68- Concurrent Pos: Am Heart Asn res fel virus res, Univ Redlands, 57-59. Mem: Biophys Soc; NY Acad Sci. Res: Molecular characterization of DNA subunit fractions isolated nondegradatively from calf thymus, liver and other sources; polymerization reactions of the subunits with specific phosphopeptides, enzymes, adenosinetriphosphate and their biological significance; characterization of the phosphopeptides split out of DNA during its cleavage into subunits by reaction with chelating reagents and mechanism of the reaction. Mailing Add: Inst for Med Atomic Res Inst 517 Jülich West Germany

WELSH, ROBERT EDWARD, b Pittsburgh, Pa, Oct 1, 32; m 56; c 3. PHYSICS. Educ: Georgetown Univ, BS, 54; Pa State Univ, PhD(physics), 60. Prof Exp: Res physicist, Carnegie Inst Technol, 60-63, asst dir nuclear res ctr, 62-63; assoc prof physics, 63-68, asst dir, Space Radiation Effects Lab, 67-72, PROF PHYSICS, COL WILLIAM & MARY, 68- Concurrent Pos: Consult, Langley Res Ctr, NASA, 64-66; guest scientist, Argonne Nat Lab & Brookhaven Nat Lab, 71-; res assoc, Rutherford Lab, Eng, 72-73. Mem: Fel Am Phys Soc; Am Asn Univ Prof. Res: Experimental nuclear physics; muon and pion physics; decay and capture of muon and pions; mesic x-rays; antiprotonic atoms. Mailing Add: 213 Kingswood Dr Williamsburg VA 23185

WELSH, RONALD, b Houston, Tex, Oct 13, 26; m 50; c 3. PATHOLOGY. Educ: Univ Tex, BA, 47, MD, 50; Am Bd Path, dipl, 55. Prof Exp: Asst prof path, Univ Tex Med Br Galveston, 55-57; from asst prof to assoc prof, 57-63, PROF PATH, SCH MED, LA STATE UNIV NEW ORLEANS, 63- Concurrent Pos: Pathologist, Univ Tex Med Br Hosps, Galveston, 55-57; vis pathologist, Charity Hosp, New Orleans, 57-, sect dir surg path; pathologist, Coroner's Off, Orleans Parish; path consult, Vet Admin Hosp, New Orleans. Mem: Am Soc Clin Path; Am Soc Exp Path; Am Asn Pathologists & Bacteriologists; Col Am Path. Res: Thyroid disease; electron microscopy, particularly of basic activities of inflammatory cells; oncology. Mailing Add: Dept of Path La State Univ Med Ctr New Orleans LA 70112

WELSH, THOMAS LAURENCE, pharmaceutical chemistry, pharmacy, see 12th edition

WELSHIMER, HERBERT JEFFERSON, b West Mansfield, Ohio, Feb 23, 20; m 46; c 3. BACTERIOLOGY. Educ: Ohio State Univ, BSc, 43, PhD(bact), 47. Prof Exp: Asst med bact, Ohio State Univ, 44-46; instr bact, Ind Univ, 47-49; from asst prof to assoc prof, 49-68, PROF BACT, MED COL VA, VA COMMONWEALTH UNIV, 68- Concurrent Pos: USPHS fel, Ohio State Univ, 47; attend bacteriologist, Johnston-Willis Hosp, 54- Mem: AAAS; Am Soc Microbiol; fel Am Acad Microbiol; NY Acad Sci. Res: Clinical bacteriology; immunology; lysozyme; bacteriophage; bacterial cytology; listeriosis. Mailing Add: 7400 Biscayne Rd Richmond VA 23229

WELSHONS, WILLIAM JOHN, b Pitcairn, Pa, July 18, 22; m 49; c 4. GENETICS. Educ: Univ Calif, PhD(zool), 54. Prof Exp: Res assoc, Biol Div, Oak Ridge Nat Lab, 54-55, mem staff, 55-65; PROF GENETICS & HEAD DEPT, IOWA STATE UNIV, 65- Concurrent Pos: NSF sr fel, Santiago, Chile, 63-64. Mem: Genetics Soc Am. Res: Pseudoallelism and crossing over in Drosophila; sex determination and cytogenetics in the mouse. Mailing Add: Dept of Genetics Iowa State Univ Ames IA 50010

WELSTEAD, WILLIAM JOHN, JR, b Newport News, Va, July 17, 35; m 57; c 2. ORGANIC CHEMISTRY. Educ: Univ Richmond, BS, 57; Univ Va, PhD(chem), 62. Prof Exp: NSF grant, Iowa State Univ, 63-64; res chemist, 64-72, assoc dir chem res, 72-73, DIR CHEM RES, A H ROBINS CO, 73- Concurrent Pos: Org chemist, Army Chem Ctr, MD, 62-63. Mem: AAAS; Am Chem Soc. Res: Synthetic organic and medicinal chemistry; heterocyclics. Mailing Add: 1211 Sherwood Ave Richmond VA 23220

WELSTED, JOHN EDWARD, b Norwich, Eng, Dec 6, 35; m 62; c 2. PHYSICAL GEOGRAPHY. Educ: Bristol Univ, BSc, 58, cert educ, 61, PhD(geog), 71; McGill Univ, MSc, 60. Prof Exp: Asst master geog, Maidenhead Grammar Sch, Eng, 61-62; teacher, Oromocto High Sch, NB, 62-64; demonstr geog, Bristol Univ, 64-65; asst prof, 65-71, ASSOC PROF GEOG, BRANDON UNIV, 71- Mem: Can Asn Geog; Asn Am Geog; Brit Geog Asn. Res: Changes of sea level in Nova Scotia and New Brunswick, Canada; use of air photographs for interpreting coastal morphology. Mailing Add: Dept of Geog Brandon Univ Brandon MB Can

WELT, ISAAC DAVIDSON, b Montreal, Que, May 13, 22; nat US; m 45; c 3. INFORMATION SCIENCE, DOCUMENTATION. Educ: McGill Univ, BSc, 44, MSc, 45; Yale Univ, PhD(physiol chem), 49. Prof Exp: Instr chem, Sir George Williams Col, 46; lab asst anat, Yale Univ, 46-47, asst, Nutrit Lab, 47-48; instr chem, New Haven YMCA Jr Col, Conn, 48-49; asst, Div Physiol & Nutrit, Pub Health Res Inst New York, Inc, 49-51; asst prof biochem, Col Med, Baylor Univ, 51-53; res assoc pharmacol, Chem-Biol Coord Ctr, Nat Res Coun, Washington, DC, 53-55, dir cardiovasc lit proj, Div Med Sci, Nat Acad Sci-Nat Res Coun, 55-61; assoc dir & chief, Wash Br, Inst Advan Med Commun, 61-64; PROG DIR SCI & TECH INFO SYSTS, CTR TECHNOL & ADMIN, AM UNIV, 64-, PROF INFO SCI, 64- Concurrent Pos: Asst dir radioisotope unit, Vet Admin Hosp, Tex, 51-53; prof lectr chem, 56-61. Mem: AAAS; Am Chem Soc; Am Soc Info Sci; Am Asn Univ Profs. Res: Endocrine influences and isotopes in intermediary; metabolism; nutrition; medical and biological literature research; chemical-biological correlations; research administration; education in information science and documentation; information storage and retrieval systems. Mailing Add: Ctr Technol & Admin Am Univ Washington DC 20016

WELT, LOUIS GORDON, medicine, deceased

WELT, MARTIN A, b Brooklyn, NY, Oct 7, 32; m 62; c 1. NUCLEONICS. Educ: Clarkson Col, BChE, 54; Iowa State Univ, MS, 55; Mass Inst Technol, SM, 57; NC State Univ, PhD(physics), 64. Prof Exp: Reactor physicist, US AEC, Washington, DC, 57-58; supvr energy conversion sect, Chance Vought Corp, 59-61; pres, Int Sci Corp, NC, 61-67; PRES, RADIATION TECHNOL, INC, 68- Concurrent Pos: Lectr, George Washington Univ, 58; adj prof, Southern Methodist Univ, 60-62; asst prof, NC State Univ, 64-67; mem, Ames Lab, US AEC, 54-55; aeronaut res scientist, Lewis Lab, Nat Adv Comt Aeronaut, 56; aeronaut res scientist, Union Carbide Co, Tenn, 56-66; dir, Adv Res Assocs, 62-63; mem, Working Comt, Proj Starfire, Southern Interstates Nuclear Bd, 64-65; dir, Nuclear Reactor Proj, NC State Univ, 64-66. Mem: AAAS; Am Nuclear Soc; Am Chem Soc; Am Phys Soc; Nat Soc Prof Eng. Res: Saline water conversion; radioisotope and radiation physics; plasma oscillations; radiation processing; design and analysis of nuclear facilities; hazards evaluation; thermoelectric energy conversion; nuclear research administration. Mailing Add: Radiation Technol Inc Lake Denmark Rd Rockaway NJ 07866

WELTER, ALPHONSE NICHOLAS, b Dudelange, Luxembourg, Apr 8, 25; US citizen; m 54; c 5. PHYSIOLOGY, ANATOMY. Educ: Loras Col, AB, 52; Univ Ill, MS, 57, PhD(physiol), 59. Prof Exp: Instr physiol, Sch Med, Marquette Univ, 59-62; res physiologist, Lederle Div, Am Cyanamid Co, 62-67; RES SPECIALIST, 3M CO, 67- Concurrent Pos: USPHS fel, 59-61; Nat Heart Inst res fel, 61-62; grants, Wis Heart Asn, 60-61 & Am Heart Asn, 60-62. Mem: AAAS; assoc mem Am Physiol Soc. Res: Cardiovascular physiology and pharmacology, specifically pulmonary circulation; respiratory and renal physiology. Mailing Add: Bldg 218-2 3M Co 2501 Hudson Rd St Paul MN 55101

WELTER, C JOSEPH, b Tiffin, Ohio, June 11, 32; m 55; c 3. MICROBIOLOGY, IMMUNOLOGY. Educ: King's Col, BS, 54; Univ Notre Dame, MS, 56; Mich State Univ, PhD, 59. Prof Exp: Asst zool, Univ Notre Dame, 54-56; asst parasitol, Mich State Univ, 56-59; dir res, Diamond Labs, Inc, 59-69, vpres res, 69-74; PRES, AMBICO, INC, 74- Mem: AAAS; Am Soc Parasitol; US Animal Health Asn; Am Soc Trop Med & Hyg; Am Soc Microbiol. Res: Protozoology; parasitology; virology. Mailing Add: Ambico Inc PO Box M Dallas Center IA 50063

WELTER, DAVE ALLEN, b Lorain, Ohio, Aug 7, 36; m 64; c 2. ANATOMY, CYTOGENETICS. Educ: Univ Ga, BS, 61; Med Col Ga, MS, 62, PhD(anat), 70. Prof Exp: Dir cytogenetics lab, Gracewood Hosp, 62-69; instr, 70-72, ASST PROF ANAT, MED COL GA, 72- Concurrent Pos: Cytogenetic consult, Ft Gordon Hosp, 65 & Gracewood Hosp, Ga, 69- Res: Birth defects; neuroanatomy; gross anatomy; embryology. Mailing Add: Dept of Anat Med Col of Ga Augusta GA 30902

WELTMAN, A STANLEY, b Brooklyn, NY, Apr 20, 19. PHYSIOLOGY,

ENDOCRINOLOGY. Educ: Brooklyn Col, BA, 41; Columbia Univ, MA, 49; Univ Mo, PhD(zool), 56. Prof Exp: Res biologist endocrinol, USDA, Beltsville Res Ctr, 42-43; ASSOC PROF PHARMACOL & RES, LABS THERAPEUT RES, BROOKLYN COL PHARM, 56- Concurrent Pos: Assoc prof, Biol Dept Grad Fac, Long Island Univ, 70-; res grants, Coun Tobacco Res, USA Inc, 72 & 73. Mem: Am Physiol Soc; Am Soc Pharmacol & Exp Therapeut; Endocrinol Soc; Am Soc Zoologists; Am Asn Lab Animal Sci. Res: Investigations of behavioral and endocrine effects of environmental stresses, noise, vibration, isolation, simulated-subway stress, hallucinogenic agents, nicotine, vasectomy in rats and/or mice and studies with spontaneously hypertensive rats. Mailing Add: Labs for Therapeut Res Brooklyn Col of Pharm 600 Lafayette Ave Brooklyn NY 11216

WELTMAN, CLARENCE A, b New York, NY, Mar 17, 19; m 43; c 2. PHYSICAL CHEMISTRY, ORGANIC CHEMISTRY. Educ: NY Univ, BA, 40. Prof Exp: Assoc chemist, Explosives Res Lab, Nat Defense Res Comt, 41-45; res chemist, 45-49, chief chemist, 49-54, exec vpres & tech dir, 54-60, PRES, ALOX CORP, 60- Mem: AAAS; Am Chem Soc; Am Soc Testing & Mat; Am Soc Lubrication Eng; Nat Asn Corrosion Eng. Res: Development of organic surface active agents and their application to problems of lubrication and corrosion prevention. Mailing Add: Alox Corp Buffalo Ave & Iroquois St Niagara Falls NY 14302

WELTMAN, JOEL KENNETH, b New York, NY, May 22, 33; m 56; c 2. IMMUNOLOGY, BIOCHEMISTRY. Educ: State Univ NY, MD, 58; Univ Colo, PhD(microbiol), 63; Brown Univ, MA, 72. Prof Exp: Intern, Ind Univ, 58-59; instr microbiol, Univ Colo, 63; asst prof, 66-70, ASSOC PROF MED, BROWN UNIV, 70- Mem: Am Chem Soc; Biophys Soc; Am Soc Biol Chemists; Am Soc Microbiol. Res: Immunology and protein chemistry. Mailing Add: Div of Biomed Sci Brown Univ Providence RI 02912

WELTNER, WILLIAM, JR, b Baltimore, Md, Dec 8, 22; m 47; c 3. PHYSICAL CHEMISTRY. Educ: Johns Hopkins Univ, BE, 43; Univ Calif, PhD(chem), 50. Prof Exp: Res asst, Hercules Powder Co, Del, 43-44 & Manhattan Proj, Columbia Univ, 44-46; fel, Univ Minn, 50; from instr to asst prof chem, Johns Hopkins Univ, 50-54; fel, Harvard Univ, 54-56; res chemist, Union Carbide Res Inst, Tarrytown, NY, 56-66; PROF CHEM, UNIV FLA, 66- Concurrent Pos: Mem opers res group, US Army Chem Ctr, 51-52; consult, Nat Bur Stand, 54. Mem: Am Chem Soc; Am Phys Soc. Res: Adsorption; thermodynamics; quantum and hith temperature chemistry; spectroscopy. Mailing Add: Dept of Chem Univ of Fla Gainesville FL 32601

WELTON, DONALD ELDON, organic chemistry, see 12th edition

WELTON, RICHARD FREDERICK, b Sidney, Nebr, Apr 25, 33; m 57; c 3. AGRICULTURAL EDUCATION. Educ: Colo State Univ, BS, 59, MEd, 66; Ohio State Univ, PhD(agr educ), 71. Prof Exp: Teacher voc agr, Lyman High Sch, Nebr, 59-62, Kearney High Sch, 62-63 & Eaton High Sch, Colo, 63-69; res assoc agr educ, Ohio State Univ, 69-70, acad adv, 70-71; ASST PROF AGR EDUC, DEPT AGR INDUST, SOUTHERN ILL UNIV, 71- Concurrent Pos: Consult, Dept Agr Educ, Univ Nebr, 67 & Nat Future Farmers Am, 70; agr educ specialist, UN Develop Prog, Santa Maria, Brazil, 71-73; chmn, Int Progs Comt, Future Farmers Am, 75-77. Mem: Am Asn Teacher Educ Agr; Am Voc Asn. Res: Teacher education; instructional materials for organizing, conducting and evaluating young farmer programs. Mailing Add: Rt 6 Carbondale IL 62901

WELTON, THEODORE ALLEN, b Saratoga Springs, NY, July 4, 18; m 43; c 4. THEORETICAL PHYSICS. Educ: Mass Inst Technol, BS, 39; Univ Ill, PhD(physics), 43. Prof Exp: Instr physics, Univ Ill, 43-44; jr scientist theoret physics, Los Alamos Sci Lab, 44-45; res assoc, Mass Inst Technol, 46-48; asst prof, Univ Pa, 48-50; prin physicist, 50-59, SR PHYSICIST, OAK RIDGE NAT LAB, 59-; FORD FOUND PROF PHYSICS, UNIV TENN, 63- Mem: Fel AAAS; fel Am Phys Soc; Electron Micros Soc Am. Res: Quantum theory of fields; theoretical nuclear physics; quantum theory of irreversible processes; theory of nuclear reactors and shielding; theory of particle accelerators; theory of lasers; theory of electron microscopy. Mailing Add: Physics Div PO Box X Oak Ridge Nat Lab Oak Ridge TN 37830

WELTON, WILLIAM ARCH, b Fairmont, WVa, June 21, 28; m 56; c 2. DERMATOLOGY. Educ: Harvard Univ, AB, 50; Univ Md, MD, 54. Prof Exp: CHMN DIV DERMAT, SCH MED, W VA UNIV, 61- Concurrent Pos: Osborne fel dermal path, 59-60. Mem: Am Acad Dermat. Res: Skin pathology. Mailing Add: Dept of Med WVa Univ Sch of Med Morgantown WV 26506

WELTY, JOSEPH D, b Marion, Ind, Nov 22, 31; m 56; c 4. PHARMACOLOGY, PHYSIOLOGY. Educ: Purdue Univ, BS, 58; Univ SDak, MA, 62, PhD(pharmacol), 63. Prof Exp: Asst scientist, Dr Salisbury Labs, 58-61; instr pharmacol, 63-64, from asst prof to assoc prof physiol, 64-72, PROF PHYSIOL, SCH MED, UNIV S DAK, VERMILLION, 72-; DIR HYPERTENSION EDUC PROG & CORONARY CARE TRAINING PROG, S DAK REGIONAL MED PROGRSPROGS, 74- Concurrent Pos: Consult staff, Sacred Heart Hosp, Yankton, SDak, 67-; mem, Int Study Group Res Cardiac Metab; assoc dir, Coronary Care Training Prog, SDak Regional Med Progs, 70-74. Mem: Soc Exp Biol & Med; Am Chem Soc. Res: Cardiovascular physiology, contractile proteins in congestive heart failure and antiarrhythmias. Mailing Add: Dept of Physiol Univ of SDak Sch of Med Vermillion SD 57069

WELTY, RONALD EARLE, b Winona, Minn, Dec 7, 34; m 62; c 2. PLANT PATHOLOGY. Educ: Winona State Col, BS, 56; Univ Minn, MS, 61, PhD(plant path), 65. Prof Exp: Teacher high sch, Minn, 56-57 & 58-59; res asst plant path, Univ Minn, 59-61 & 62-65; instr bot & plant path, La State Univ, 62; res assoc plant path, NC State Univ, 65-66; PLANT PATHOLOGIST, AGR RES SERV, USDA, 66-, ASSOC PROF PLANT PATH, NC STATE UNIV, 73- Mem: AAAS; Am Soc Microbiol; Mycol Soc Am; Am Phytopath Soc. Res: Diseases of forage crops; soil-borne fungus diseases; general phytopathology. Mailing Add: USDA Dept of Plant Path NC State Univ Raleigh NC 27607

WEMPE, LAWRENCE KYRAN, b Hutchinson, Kans, Oct 3, 41; m 65; c 1. ORGANIC CHEMISTRY. Educ: Rockhurst Col, BA, 63; Univ Kans, PhD(org chem), 68. Prof Exp: Lab asst water treatment chem, Deady Chem Co, Kans, 62-63; SR CHEMIST, ROHM & HAAS CO, 68- Concurrent Pos: Instr, Montgomery County Community Col, Blue Bell, Pa, 73- Mem: Am Chem Soc. Res: Organic synthesis; reaction mechanisms; natural products; biogenetic cyclizations; free radical reactions; polymer chemistry; coatings; textile and paper chemicals. Mailing Add: Rohm & Haas Co Spring House PA 19477

WEMPNER, GERALD A, mechanics, applied mathematics, see 12th edition

WEMYSS, COURTNEY TITUS, JR, b Arlington, NJ, Dec 30, 22; m 51; c 3. ZOOLOGY. Educ: Swarthmore Col, AB, 47; Rutgers Univ, PhD, 51. Prof Exp: Asst zool, Rutgers Univ, 47-51; res fel bact & immunol, Harvard Med Sch, 51-52; asst prof biol, Loyola Univ, 53-54; instr physiol, NY Med Col, 54-60; assoc prof, 60-70, PROF BIOL, HOFSTRA UNIV, 70- Concurrent Pos: Guest investr, Rockefeller Inst, 60-

Mem: NY Acad Sci. Res: Invertebrate immunity; comparative serology; tissue specificity. Mailing Add: Dept of Biol Hofstra Univ Hempstead NY 11550

WEN, CHI-PANG, b Taipei, Taiwan, Oct 23, 40; m 67; c 1. PREVENTIVE MEDICINE, FAMILY MEDICINE. Educ: Nat Taiwan Univ, MD, 66; Harvard Univ, MPH, 69, DrPH, 72. Prof Exp: ASST PROF HEALTH SERV EDUC & RES & HUMAN DEVELOP, COL HUMAN MED, MICH STATE UNIV, 72- Mem: Fel Am Col Prev Med; Soc Nutrit Educ. Res: Epidemiology; international health; medicine in China; coronary heart disease; lactose intolerance; glutamate metabolism. Mailing Add: Off of Health Serv Educ & Res Mich State Univ East Lansing MI 48824

WEN, KWAN-SUN, geophysics, fluid mechanics, see 12th edition

WEN, RICHARD YUTZE, b Shanghai, China, Mar 17, 30; m 62; c 2. ORGANIC POLYMER CHEMISTRY. Educ: Wesleyan Univ, BA, 51; Univ Mich, MS, 53; Ind Univ, PhD(org chem), 62. Prof Exp: Chemist, Nalco Chem Co, 53-56; res chemist, Dow Chem Co, Mich, 62-69; sr chemist, 69-74, RES SPECIALIST, CENT RES LABS, 3M CO, 74- Mem: AAAS; Am Chem Soc. Res: Application of polymers to life sciences and coatings. Mailing Add: Cent Res Labs Box 33221 3M Co St Paul MN 55133

WEN, SUNG-FENG, b Hsinchu, Taiwan, Mar 3, 33; US citizen; m 66; c 2. MEDICINE. Educ: Nat Taiwan Univ, MB, 58. Prof Exp: Intern med, Univ Louisville, 62-63; resident, Chicago Med Sch, 63-64; res fel nephrol, Univ Wis-Madison, 64-66, instr, 66-67; res fel renal physiol, McGill Univ, 67-70; asst prof, 70-74, ASSOC PROF MED, UNIV WIS-MADISON, 74-, RENNEBOHM PROF, 75- Concurrent Pos: Mem coun kidney cardiovasc dis, Am Heart Asn. Mem: Am Fedn Clin Res; Am Soc Nephrol; Int Soc Nephrol; Nat Kidney Found. Res: Renal physiology and pathophysiology, especially related to renal transport of sodium, potassium, phosphate and glucose under normal and abnormal conditions using micropuncture techniques. Mailing Add: Dept of Med Univ Hosps 1300 University Ave Madison WI 53706

WEN, WEN-YANG, b Hsin-tsu, Taiwan, Mar 7, 31; m 59; c 2. PHYSICAL CHEMISTRY. Educ: Nat Taiwan Univ, BS, 53; Univ Pittsburgh, PhD(chem), 57. Prof Exp: Res assoc, Univ Pittsburgh, 57-58; res fel, Northwestern Univ, 58-60; asst prof chem, DePaul Univ, 60-62; from asst prof to assoc prof, 62-73, PROF CHEM, CLARK UNIV, 73- Concurrent Pos: Humboldt scholar, Univ Karlsruhe, 70-71. Mem: Am Chem Soc; Am Asn Univ Prof; AAAS; Sigma Xi. Res: Structure of water; themodynamic properties of large ions in solutions; tetraalkylammonium salts and hydrophobic bonds; nuclear magnetic resonance; Azoniaspiroalkane ions. Mailing Add: Dept of Chem Clark Univ Worcester MA 01610

WENAAS, PAUL EMIL, b Butte, Mont, Oct 4, 10; m 38; c 3. CHEMISTRY. Educ: Mont State Col, BS, 31; Univ Chicago, PhD(chem), 34. Prof Exp: Chemist, Simoniz Co, 34-35, chief chemist, 45-51, mgr tech res, 51-57, vpres, 57-60; instr, Univ Ill, 60-62; asst prof chem, Carnegie Inst Technol, 62-66; ASSOC PROF CHEM, UNIV WIS-WHITEWATER, 66- Mem: Am Chem Soc. Res: Colloid chemistry. Mailing Add: Dept of Chem Univ of Wis Whitewater WI 53190

WEND, DAVID VAN VRANKEN, b Poughkeepsie, NY, Oct 18, 23; m 53; c 3. MATHEMATICS. Educ: Univ Mich, BS, 45, MA, 46, PhD(math), 55. Prof Exp: Instr math, Reed Col, 49-51 & Iowa State Univ, 52-55; from asst prof to assoc prof, Univ Utah, 55-66; assoc prof, 66-67, PROF MATH, MONT STATE UNIV, 67- Mem: Am Math Soc; Math Asn Am; Soc Indust & Appl Math. Res: Functions of a complex variable; differential equations. Mailing Add: Dept of Math Mont State Univ Bozeman MT 59715

WENDE, CHARLES DAVID, b Wilmington, Del, Dec 4, 41; m 65; c 1. SPACE PHYSICS. Educ: Mass Inst Technol, BS, 63; Univ Iowa, MS, 66, PhD(physics), 68. Prof Exp: Res assoc space physics, Univ Iowa, 68-69; ASTROPHYSICIST, GODDARD SPACE FLIGHT CTR, NASA, 69- Mem: AAAS; Am Geophys Union; Int Union Radio Sci. Res: Solar x-ray and radio astronomy; x-ray astronomy; application of interactive computing to modeling experiment hardware and to data reduction and analysis. Mailing Add: Code 601 NASA Goddard Space Flight Ctr Greenbelt MD 20771

WENDEL, CARLTON TYRUS, b Fredericksburg, Tex, Oct 6, 39; m 63; c 2. ANALYTICAL CHEMISTRY. Educ: Tex Lutheran Col, BS, 62; Tex Tech Col, MS, 65, PhD(chem), 67. Prof Exp: Instr, 67-69, ASST PROF CHEM, TEX WOMAN'S UNIV, 69- Mem: Am Chem Soc; Sigma Xi. Res: Chemical aspects of water pollution. Mailing Add: Dept of Chem Tex Woman's Univ Denton TX 76204

WENDEL, HERBERT A, b Ludwigshafen, Ger, Feb 17, 14; nat US; m 44; c 3. PHARMACOLOGY, INTERNAL MEDICINE. Educ: Univ Berlin, DrMed, 39. Prof Exp: Res asst pharmacol, Univ Berlin, 46-49; lectr, Univ Mainz, 50-51; res assoc, Univ Pa, 52-56; asst sect head, Smith Kline & French Labs, 56-60; consult, Farbenfabriken Bayer, Ger, 60-61; mgr clin eval, E I du Pont de Nemours & Co, 61-64; dir res, Warren-Teed Pharmaceut Inc, Ohio, 64-68; ASSOC PROF PHARMACOL, MED SCH, UNIV ORE, 68- Mem: AAAS; Am Soc Pharmacol & Exp Therapeut; Am Soc Clin Pharmacol & Therapeut; Ger Pharmacol Soc. Res: Drug research; clinical drug investigation. Mailing Add: Dept of Pharm Univ of Ore Med Sch Portland OR 97201

WENDEL, JAMES GUTWILLIG, b Portland, Ore, Apr, 18, 22; m 44; c 6. MATHEMATICS. Educ: Reed Col, BA, 43; Calif Inst Technol, PhD(math), 48. Prof Exp: Asst Nat Defense Res Comt, Calif Inst Technol, 42-45, instr, 45-48; instr math, Yale Univ, 48-51; assoc mathematician, Rand Corp, 51-52; from asst prof to assoc prof math, La State Univ, 52-55; from asst prof to assoc prof, 55-61, assoc chmn dept, 68-70, PROF MATH, UNIV MICH, ANN ARBOR, 61-, ASSOC CHMN DEPT, 73- Concurrent Pos: Vis prof, Aarhus Univ, 62-64 & Univ London, 70-71. Mem: Am Math Soc; Math Asn Am; Inst Math Statist. Res: Probability theory. Mailing Add: Dept of Math Univ of Mich Ann Arbor MI 48104

WENDEL, OTTO WILLIAM, biophysical chemistry, see 12th edition

WENDEL, SAMUEL REECE, b Charleston, Ill, Sept 1, 44; m 67; c 2. BIOINORGANIC CHEMISTRY, ORGANOMETALLIC CHEMISTRY. Educ: Univ Ill, Urbana, BS, 66; Univ Mont, PhD(org chem), 73. Prof Exp: Chemist, 66-69, PROJ SPECIALIST ORGANOSILICON CHEM, DOW CORNING CORP, 73- Mem: AAAS; Am Chem Soc. Res: Design and synthesis of bioactive organosilicon compounds; organosilicon heterocyclic compounds. Mailing Add: Dow Corning Corp Midland MI 48640

WENDEL, WILLIAM BEAN, b Manchester, Tenn, Apr 10, 01; m 38; c 2. BIOCHEMISTRY. Educ: Emory Univ, BS, 23; Wash Univ, PhD(biochem), 32. Prof Exp: Instr chem, Col Med, Dent & Pharm, Univ Tenn, 23-28, from asst prof to assoc prof, Cols & Schs Biol Sci & Nursing, 33-45; asst biochem, Sch Med, Wash Univ, 28-

32, instr, 32-33, asst prof biochem & chemist, 35-37; prof biochem & chmn dept, 45-71, EMER PROF BIOCHEM, TULANE UNIV, 71- Concurrent Pos: Mem biochem test comt, Nat Bd Med Exam, 59-63. Honors & Awards: Distinguished Serv Award, Inst Int Ed & Readers Digest Found, 66. Mem: Asn Am Med Cols; Am Chem Soc; Am Soc Biol Chem; fel Am Asn Clin Chem. Res: Metabolism of normal erthyrocytes and malarial parasites; formation and reduction of methemoglobin; antimalarials; synthesis of pyridine nucleotides; basic sciences in health-related educational institutes in developing countries. Mailing Add: 2510 Adams St New Orleans LA 70125

WENDEN, HENRY EDWARD, b New York, NY, Nov 24, 16; m 43; c 4. MINERALOGY. Educ: Yale Univ, BS, 38; Harvard Univ, MA, 50, PhD, 58. Prof Exp: Instr physics & geol, Boston Univ, 49-53; asst prof mineral & geol, Tufts Univ, 53-57; from asst prof to assoc prof mineral, 57-63, PROF MINERAL, OHIO STATE UNIV, 63- Concurrent Pos: Res assoc, Harvard Univ, 58. Honors & Awards: Neil Miner Award, Nat Asn Geol Teachers, 61. Mem: Fel Mineral Soc Am; Nat Asn Geol Teachers; Mineral Asn Can. Res: Crystal chemistry; history of mineralogy and crystallography; x-ray crystallography; crystal morphology; physical and electrical properties of crystals. Mailing Add: Dept of Geol & Mineral Ohio State Univ 104 W 19th Ave Columbus OH 43210

WENDER, IRVING, b New York, NY, June 19, 15; m 42; c 3. FUEL SCIENCE, ORGANOMETALLIC CHEMISTRY. Educ: City Col New York, BS, 36; Columbia Univ, MA, 45; Univ Pittsburgh, PhD(chem), 50. Prof Exp: Chemist & res assoc, Manhattan Proj, Univ Chicago, 44-46; org chemist, Pittsburgh Coal Res Ctr, US Bur Mines, 46-53, chief chem sect, 53-71, res dir, Pittsburgh Energy Res Ctr, 71-75; DIR PITTSBURGH ENERGY RES CTR, ENERGY RES & DEVELOP ADMIN, 75- Concurrent Pos: Lectr, Univ Pittsburgh, 63-69, adj prof, 69- Honors & Awards: Bituminous Coal Res Award, 56, 60; H H Storch Award, 64; Gold Medal Distinguished Serv Award, US Dept Interior, 66; Pittsburgh Award, Am Chem Soc, 68; K K Kelley Award, 69. Mem: Am Chem Soc; The Chem Soc. Res: Chemistry of carbon monoxide, metal carbonyls, coal conversion; catalysis; reactions at high pressures; synthetic fuels from coal; carbon monoxide chemistry. Mailing Add: Energy Res & Develop Admin 4800 Forbes Ave Pittsburgh PA 15213

WENDER, PAUL H, b New York, NY, May 12, 34; m 70; c 3. PSYCHIATRY, CHILD PSYCHIATRY. Educ: Harvard Univ, AB, 55; Columbia Univ, MD, 59. Prof Exp: Intern, Barnes Hosp, New York, 59-60; resident adult psychiat, Mass Ment Health Ctr, 60-62; resident, St Elizabeth's Hosp, 62-63; resident child psychiat, Johns Hopkins Univ, 64-67, asst prof pediat & psychiat, 67-73; PROF PSYCHIAT, COL MED, UNIV UTAH, 73- Concurrent Pos: NIMH fel, NIH, Bethesda, Md, 64-66, res psychiatrist, 67-73. Honors & Awards: Hofheimer Award, Am Psychiat Asn, 74. Mem: Am Psychiat Asn; Am Acad Child Psychiat; Psychiat Res Soc. Res: Genetics and schizophrenia; minimal brain dysfunction in children. Mailing Add: Dept of Psychiat Univ of Utah Col of Med Salt Lake City UT 84132

WENDER, SIMON HAROLD, b Dalton, Ga, Sept 4, 13; m 42; c 3. BIOCHEMISTRY. Educ: Emory Univ, AB, 34, MS, 35; Univ Minn, PhD(agr biochem), 38. Prof Exp: Res assoc, Med Sch, Emory Univ, 38-39; assoc chemist, Exp Sta, Agr & Mech Col, Tex, 39-41; from instr to asst prof chem, Univ Ky, 41-46; assoc prof biochem, 46-49, RES PROF BIOCHEM, UNIV OKLA, 53-, CONSULT PROF, HEALTH SCI CTR, 71- Concurrent Pos: Former mem bd dirs, Oak Ridge Assoc Univs; past chmn coun, Oak Ridge Inst Nuclear Studies, Okla, rep to coun, 52-64 & 71-; vis res assoc, Argonne Nat Lab, 54-64; vis prof, Univ Wis, 66. Mem: Fel AAAS; Am Chem Soc; Am Soc Biol Chem; Soc Exp Biol & Med; Am Soc Plant Physiol. Res: Chromatography; plant phenolics and plant and animal oxidoreductases. Mailing Add: Dept of Chem 620 Parrington Oval Univ of Okla Norman OK 73069

WENDHAL, RONALD, b Seattle, Wash, Dec 29, 25; m; c 2. AUDIOLOGY, SPEECH PATHOLOGY. Educ: Univ Wash, BS, 50, MS, 55; Univ Iowa, PhD(speech path), 57. Prof Exp: Jr res psychologist, Univ Wash, 50-51; instr speech path, Ind Univ, 56-58; dir speech lab, Wilkerson Hearing & Speech Ctr, 58-60; res dir, Houston Speech & Hearing Ctr, 60-63; assoc prof speech & dir commun sci lab, Univ Minn, Minneapolis, 63-67; prof speech & dir commun sci lab, Univ Houston, 67-75; MEM FAC, DEPT AUDIOL, BAYLOR COL MED, 75- Concurrent Pos: Prin investr, NIH grants, 63-65 & 66-68. Mem: Am Speech & Hearing Asn; Acoust Soc Am. Res: Perceptual correlates of voice and speech. Mailing Add: Dept of Audiol Baylor Col of Med Houston TX 77025

WENDLAND, RAY THEODORE, b Minneapolis, Minn, July 11, 11; m 46; c 1. ORGANIC CHEMISTRY. Educ: Carleton Col, BA, 33; Iowa State Univ, PhD(chem), 37. Prof Exp: Res chemist, Universal Oil Prod Co, 38-39; instr org chem & biochem, Coe Col, 39-42; asst prof, Middlebury Col, 42-43; res chemist synthetic rubber, War Prod Bd, Univ Minn, 43-44; asst prof org & biol chem, Lehigh Univ, 44-47; prof org chem, NDak State Univ, 47-55; res fel petrol chem, Mellon Inst, 55-58; prof chem & chmn div sci & math, Winona State Col, 58-63; PROF CHEM, CARROLL COL, WIS, 63- Mem: AAAS; Am Chem Soc. Res: Organic synthesis; polymer and petroleum chemistry. Mailing Add: 101 Morningside Dr Waukesha WI 53186

WENDLAND, WAYNE MARCEL, b Beaver Dam, Wis, Aug 9, 34; m 56; c 4. METEOROLOGY, CLIMATOLOGY. Educ: Lawrence Col, BA, 56; Univ Wis-Madison, MS, 65, PhD(meteorol), 72. Prof Exp: Proj supvr meteorol, 66-70, instr climat, 70, ASST PROF CLIMAT, UNIV WIS-MADISON, 70- Mem: AAAS; Am Meteorol Soc; Am Quaternary Asn; Asn Am Geog; Tree Ring Soc (vpres, 74-). Res: Past climatic patterns; climatic episodes of the Holocene; radiocarbon-calender anomolies; climatic reconstructions from tree rings. Mailing Add: Dept of Geog Univ of Wis-Madison Madison WI 53706

WENDLANDT, WESLEY W, b Galesville, Wis, Nov 20, 27. INORGANIC CHEMISTRY. Educ: Wis State Col, River Falls, BS, 50; Univ Iowa, MS, 52, PhD(chem), 54. Prof Exp: From asst prof to prof chem, Tex Tech Col, 54-66; chmn dept, 66-72, PROF CHEM, UNIV HOUSTON, 66- Concurrent Pos: Vis prof, NMex Highlands Univ, 61. Honors & Awards: Mettler Award, 70. Mem: AAAS; Am Chem Soc; The Chem Soc; NAm Thermal Anal Soc; Int Confedn Thermal Anal. Res: Rare-earth chemistry; coordination compounds; metal chelates; thermogravimetry; differential thermal analysis; high pressure chemistry; solid state chemistry; reflectance spectroscopy. Mailing Add: Dept of Chem Univ of Houston Houston TX 77004

WENDLER, GERD DIERK, b Hamburg, WGer, June 16, 39; m 69; c 2. METEOROLOGY. Educ: Innsbruck Univ, PhD(meteorol), 64. Prof Exp: Data process asst meteorol, Inst Meteorol, Innsbruck Univ, 60-64, res asst, 65-66; asst geophysicist, 66-67, asst prof meteorol, 67-70, ASSOC PROF METEOROL, GEOPHYS INST, UNIV ALASKA, FAIRBANKS, 70- Concurrent Pos: NSF grant, McCall Glacier, Geophys Inst, Univ Alaska, Fairbanks, 69-, Sea grant Arctic Ocean, 71- & NASA satellite grant cent Alaska, 72- Mem: Am Meteorol Soc; Am Geophys Union; Glaciol Soc; Arctic Inst NAm; Ger Soc Polar Res. Res: Meteorology in the arctic, especially of Alaska. Mailing Add: Geophys Inst Univ of Alaska Fairbanks AK 99701

WENDLER, HENRY O, b Red Bluff, Calif, Apr 30, 24; m 47; c 4. FISH BIOLOGY, FISHERIES MANAGEMENT. Educ: Univ Wash, BS, 51. Prof Exp: Aquatic biologist, 51-64, sr biologist, 64-75, MGR OFF INTERGOVT OPERS, WASH DEPT FISHERIES, 75- Concurrent Pos: Lectr, Univ Wash, 68- Mem: Fel Am Inst Fishery Res Biol. Mailing Add: Wash Dept of Fisheries Gen Admin Bldg Olympia WA 98504

WENDLER, NORMAN LORD, b Stockton, Calif, Dec 18, 15; m 47; c 2. ORGANIC CHEMISTRY. Educ: Middlebury Col, AB, 37; Rutgers Univ, MSc, 39; Univ Mich, PhD(org chem), 44. Prof Exp: Res chemist, Merck & Co, Inc, 39-41; Lilly fel, Harvard Univ, 45-46; Swiss-Am exchange fel, Univ Basel, 46-47; res chemist, 47-69, sr investr, Res Labs, 69-74, SR SCIENTIST, MERCK SHARP & DOHME RES LABS, MERCK & CO, INC, 74- Res: Synthesis of steroids, vitamins, griseofulvin, zearalenone and prostaglandins; rearrangement phenomena in steroids. Mailing Add: Merck Sharp & Dohme Res Labs Rahway NJ 07065

WENDORF, FRED, b Terrell, Tex, July 31, 24; m 45; c 4. ANTHROPOLOGY, ARCHAEOLOGY. Educ: Univ Ariz, AB, 47; Harvard Univ, AM, 50, PhD, 53. Prof Exp: Res assoc anthrop, Lab Anthrop, Mus NMex, 51-58, dir anthrop, 58-64; chmn dept, 68-74, prof anthrop, 64-74, HENDERSON-MORRISON PROF ANTHROP, SOUTHERN METHODIST UNIV, 74- Mem: AAAS; Am Anthrop Asn; Soc Am Archaeol (treas, 74-78). Res: American archaeology; early man in the New World; Pleistocene geology; prehistory of North Africa, especially Sudan and Egypt. Mailing Add: Dept of Anthrop Southern Methodist Univ Dallas TX 75222

WENDRICKS, ROLAND N, b Casco, Wis, July 26, 30; m 52; c 3. PHYSICAL CHEMISTRY. Educ: St Norbert Col, BS, 52; Northwestern Univ, MS, 61. Prof Exp: Group supvr blow molding process, 52-67, supvr blow molding plastics res & develop, 67-70, MGR PLASTICS EQUIP ENG, AM CAN CO, 70- Mem: Soc Plastics Eng. Res: Processing of thermoplastic polymers. Mailing Add: Am Can Co American Lane Greenwich CT 06830

WENDROFF, BURTON, b New York, NY, Mar 10, 30. MATHEMATICS. Educ: NY Univ, BA, 51, PhD(math), 58; Mass Inst Technol, SM, 52. Prof Exp: Staff mem, Los Alamos Sci Lab, 52-66; from assoc prof to prof math, Univ Denver, 66-74; GROUP LEADER & STAFF MEM, LOS ALAMOS SCI LAB, 73- Mem: Am Math Soc; Soc Indust & Appl Math. Res: Applied mathematics; numerical analysis. Mailing Add: Los Alamos Sci Lab PO Box 1663 Los Alamos NM 87545

WENDT, ARNOLD, b Red Bud, Ill, Jan 14, 22; m 43; c 1. MATHEMATICS. Educ: Univ Wis, PhD(math), 52. Prof Exp: PROF MATH, WESTERN ILL UNIV, 52- Mem: AAAS; Am Math Soc; Math Asn Am. Res: Analysis and applied mathematics. Mailing Add: Dept of Math Western Ill Univ Macomb IL 61455

WENDT, CHARLES WILLIAM, b Plainview, Tex, July 12, 31; m 55; c 5. SOIL PHYSICS. Educ: Tex A&M Univ, BS, 51, PhD(soil physics), 66; Tex Tech Col, MS, 57. Prof Exp: Res asst agron, Tex Tech Col, 53-55, from instr to asst prof, 57-63; res asst soil physics, 63-65, res assoc, 65-66, from asst prof to assoc prof, 66-74, PROF SOIL PHYSICS, TEX A&M UNIV, 74- Concurrent Pos: Off Water Resources Res US Dept Interior res grant, Agr Res & Exten Ctr, Tex A&M Univ, 67-70 & Environ Protection Agency res grant, Veg Res Sta, 70- Mem: AAAS; Am Soc Agron; Soil Sci Soc Am; Am Soc Plant Physiol. Res: Water quality of irrigation return flows as affected by irrigation and fertilization practices; plant modification for more efficient water use. Mailing Add: Tex Agr Exp Sta Tex A&M Univ Agr Res & Exten Ctr Rt 3 Lubbock TX 79401

WENDT, GEORGE FRANCIS, marine biology, microbiology, see 12th edition

WENDT, GERHARD RUDOLF, organic chemistry, biochemistry, see 12th edition

WENDT, RICHARD P, b St Louis, Mo, Oct 6, 32; m 70; c 1. PHYSICAL CHEMISTRY. Educ: Washington Univ, AB, 54; Univ Wis, PhD(phys chem), 61. Prof Exp: Asst prof chem, La State Univ, 62-66; assoc prof, 66-73, PROF CHEM, LOYOLA UNIV, NEW ORLEANS, 73- Concurrent Pos: Consult res div, Vet Admin Hosp, New Orleans, 66-, NIH res fel, 71; consult, Gulf South Res Inst, 73- Mem: AAAS. Res: Diffusion in liquids; nonequilibrium thermodynamics; mass transfer across synthetic membranes. Mailing Add: Dept of Chem Loyola Univ New Orleans LA 70118

WENDT, ROBERT CHARLES, b Aurora, Ill, July 5, 29; m 53; c 4. SURFACE CHEMISTRY. Educ: NCent Col, Ill, BA, 51; Univ Ill, PhD(phys chem), 55. Prof Exp: Res chemist, Yerkes Lab, Film Dept, 55-64, staff scientist, 64-69, STAFF SCIENTIST, EXP STA LAB, FILM DEPT, E I DU PONT DE NEMOURS & CO, INC, 69- Mem: Am Chem Soc. Res: Diffusion phenomena; polymer physical properties; adhesion; polymer surface chemistry and physics. Mailing Add: 3316 Coachman Rd Wilmington DE 19803

WENDT, THEODORE MIL, b Ft Collins, Colo, Sept 14, 40; m 63; c 2. MICROBIOLOGY, CHEMISTRY. Educ: Colo State Univ, BS, 64, MS, 66, PhD(microbiol), 68. Prof Exp: RES MICROBIOLOGIST, US ARMY NATICK LABS, 68- Mem: Am Soc Microbiol; Sigma Xi; AAAS. Res: Microbiological deterioration of materials, especially polymers; water pollution abatement through biological activity; relationship of chemical structure to biological susceptibility; aquatic microbial ecology. Mailing Add: 3 Grandview Dr Franklin MA 02038

WENE, GEORGE PETER, entomology, see 12th edition

WENESER, JOSEPH, b New York, NY, Nov 23, 22; m 56. THEORETICAL PHYSICS. Educ: City Col, BS, 42; Columbia Univ, MA, 48, PhD(physics), 52. Prof Exp: Asst physics, Manhattan Proj, Columbia Univ, 42-46; assoc physicist, Brookhaven Nat Lab, 52-55; asst prof, Univ Ill, 55-57; chmn dept physics, 70-75, SR PHYSICIST, BROOKHAVEN NAT LAB, 57- Mem: Fel Am Phys Soc. Res: Theoretical nuclear physics. Mailing Add: Dept of Physics Bldg 510A Brookhaven Nat Lab Upton NY 11973

WENGEL, RAYMOND WILLIAM, b Cambridge, Wis, Dec 27, 28; m 51; c 2. SOIL PHYSICS. Educ: Univ Wis, BS, 54, MS, 56, PhD(soils), 57. Prof Exp: From asst prof to assoc prof agron, 57-71, PROF AGRON, UNIV CONN, 71- Mem: Am Soc Agron. Res: Soil structure; soil oxygen; soil moisture; animal waste disposal; solid waste disposal; groundwater pollution. Mailing Add: Dept of Plant Sci Univ of Conn Storrs CT 06268

WENGER, BYRON SYLVESTER, b Russell, Kans, Oct 13, 19; m 47; c 4. DEVELOPMENTAL BIOLOGY. Educ: Univ Wyo, BS, 40, MS, 41; Washington Univ, PhD(zool), 49. Prof Exp: NIH fel pharmacol, Washington Univ, 49-51; from asst prof to assoc prof anat, Univ Kans, 51-62, from assoc prof comp biochem & physiol to prof biochem, 62-69; PROF ANAT, UNIV SASK, 69- Concurrent Pos: Vis assoc prof, Washington Univ, 64-66. Mem: Am Asn Anatomists; Am Soc Zoologists; Soc Develop Biol; Am Soc Neurochemists; Am Soc Cell Biol. Res: Experimental and biochemical studies of differentiation in normal chick and mouse embryos and during genetic and drug induced teratogenesis. Mailing Add: Dept of Anat Univ of Sask Saskatoon SK Can

WENGER, DAVID ARTHUR, b Philadelphia, Pa, June 3, 42; m 66; c 1. BIOCHEMICAL GENETICS, PEDIATRICS. Educ: Temple Univ, BS, 64, PhD(biochem), 68. Prof Exp: ASST PROF PEDIAT & NEUROL, B F STOLINSKY RES LABS, MED CTR, UNIV COLO, DENVER, 71- Concurrent Pos: Fel, Weizmann Inst Sci, Rehovot, Israel, 68-69; Multiple Sclerosis fel, Sch Med, Univ Calif, San Diego, 69-71. Mem: AAAS; Am Soc Neurochem; Soc Carbohydrate Res; Int Soc Neurochem. Res: Lipid biochemistry; sphingolipidoses; genetic disease of children; lysosomal enzymes. Mailing Add: Dept of Pediat Univ of Colo Med Ctr Denver CO 80220

WENGER, ELEANOR LERNER, b New York, NY, July 18, 21; m 47; c 4. DEVELOPMENTAL BIOLOGY. Educ: Brooklyn Col, BA, 40; Oberlin Col, MA, 43; Washington Univ, PhD(zool), 48. Prof Exp: Res assoc path, Washington Univ, 47-48, res assoc & lectr zool, 49-51; res assoc anat, Univ Kans, 51-52, instr, 52-53, res assoc, 53-62, res assoc comp biochem & physiol, 62-64; res assoc biol, Washington Univ, 64-66; res assoc comp biochem & physiol, Univ Kans, 66-69; assoc prof anat, Univ Sask, 69-71, prof res assoc, 71-73, asst prof biol, 73-75. Mem: AAAS; Am Soc Zoologists; Soc Develop Biol; Am Inst Biol Scientists. Res: Experimental and biochemical analysis of sites and modes of gene action. Mailing Add: Dept of Biol Univ of Sask Saskatoon SK Can

WENGER, FRANZ, b Bern, Switz, Nov 28, 25; nat US; m 55. PHYSICAL CHEMISTRY. Educ: Univ Bern, Lic phil nat, 53, PhD(chem), 54. Prof Exp: Chemist, Lonza, Inc, Switz, 55; fel photochem, Nat Res Coun Can, 55-57; fel polymer sci, Mellon Inst, 58-63; sr staff assoc, Cent Res Lab, Celanese Corp, NJ, 63-66; mgr spec prod res, Polaroid Corp, Mass, 66-69; GROUP VPRES-IN-CHG RES & DEVELOP & ENG, ENGELHARD INDUSTS DIV, ENGELHARD MINERALS & CHEM CORP, NEWARK, 69- Mem: Am Chem Soc. Res: Reaction kinetics; structure-properties relationship of polymeric materials; photographic technology; process research and development. Mailing Add: 363 Cherry Hill Rd Mountainside NJ 07092

WENGER, HERBERT CHARLES, b Silverdale, Pa, May 17, 26; m 48; c 4. PHARMACOLOGY, PHYSIOLOGY. Educ: Goshen Col, BA, 52. Prof Exp: Res asst pharmacol, Sharp & Dohme Med Res Labs, 54-55; res assoc, 56-65, RES PHARMACOLOGIST, MERCK INST THERAPEUT RES, 66- Res: Pharmacodynamics of cardiovascular drugs including antiarrhythmic and antifibrillatory drugs and compounds affecting the renin-angiotensin system; serotonin antagonists; antihistaminics. Mailing Add: Dept of Pharmacol Merck Inst for Therapeut Res West Point PA 19486

WENGER, JULIUS, internal medicine, see 12th edition

WENGER, KARL FREDERICK, forestry, plant ecology, see 12th edition

WENGER, NANETTE KASS, b New York, NY, Sept 3, 30. MEDICINE, CARDIOLOGY. Educ: Hunter Col, BA, 51; Harvard Med Sch, 54. Prof Exp: Intern, Mt Sinai Hosp, New York, 54-55, resident med, 55-56, chief resident cardiol, 56-57; sr asst resident med, Grady Mem Hosp, Atlanta, 58; instr med, 59-62, assoc, 62-64, from asst prof to assoc prof, 64-71, PROF MED, SCH MED, EMORY UNIV, 71-, DIR, CARDIAC CLINS, GRADY MEM HOSP, ATLANTA, 60- Concurrent Pos: Fel cardiol, Sch Med, Grady Mem Hosp, Emory Univ, 58-59; fel coun clin cardiol, Am Heart Asn, 70; dir, Proj Cardiac Eval & Med & Voc Rehab, 66-; mem, Rehab Comt, Inter-Soc Comn Heart Disease Resources, 69-75; consult, Streptokinase Urokinase Myocardial Infarction Trial Planning Group, Myocardial Infarction Br, Nat Heart & Lung Inst, 70-71, mem, Nat Thrombosis Adv Comt, 71-74 & Heart Panel, Heart, Lung & Blood Vessel Dis Act, 72; consult, Int Div Social & Rehab Serv, Dept Health, Educ & Welfare, 70- & J Chest, 70-; mem, Cent Comt, Am Heart Asn, 71, Exec Comt, 71-, Coun Clin Cardiol & Coun Thrombosis, 71-, chmn, Prog Comt, 75- Mem: AMA; fel Am Col Cardiol; Am Heart Asn; Am Fedn Clin Res; Am Thoracic Asn (vpres, 75-). Res: Urokinase streptokinase pulmonary embolism trial; sudden death in myocardial infarction; ischemic heart disease in young adults; clinical evaluation of myocardial infarction patients; evaluation of patient education programs; evaluation of cholesterol binding resin in hypercholesterolemia. Mailing Add: Dept of Med Emory Univ Sch of Med Atlanta GA 30322

WENGER, RONALD HAROLD, b Dayton, Ohio, Nov 30, 37; m 63. MATHEMATICS. Educ: Miami Univ, Ohio, AB, 59; Mich State Univ, MS, 61, PhD(math), 65. Prof Exp: ASST PROF MATH, UNIV DEL, 65-, ASST TO PROVOST FOR ACAD PLANNING, 69-, ASSOC DEAN, COL ARTS & SCI, 72- Concurrent Pos: Asst dean, Col Arts & Sci, 68-69; Am Coun Educ fel acad admin, 70-71. Mem: Am Math Soc; Math Asn Am. Res: Semigroup rings. Mailing Add: Dean's Off Col of Arts & Sci Univ of Del Newark DE 19711

WENGER, SHERMAN ALEXANDER, b Millersburg, Ohio, Feb 17, 15; m 40; c 4. GEOLOGY. Educ: Col Wooster, AB, 36; Harvard Univ, AM, 38, PhD(geol), 47. Prof Exp: Mem staff seismol, Shell Oil Co, Okla, 37; mining geologist, Ramshorn Mining Co, Idaho, 38; petrol geologist, Shell Oil Co, 40-42, res geologist, 45-47; from asst prof to prof geol, Univ NMex, 47-76; RETIRED. Concurrent Pos: Consult res geologist, 47-76; mem adv bd, Energy Equities Inc, 70-74; dir, Pub Lands Explor Inc, 71-76. Ed, Bull, Am Asn Petrol Geologists, 57-59; nat ed, Am Inst Prof Geologists, 65-66. Honors & Awards: Award, Am Asn Petrol Geologists, 48. Mem: Fel Geol Soc Am; Am Soc Photogram; Soc Econ Paleont & Mineral; Nat Asn Geol Teachers; Mex Asn Petrol Geologists. Res: Stratigraphic analysis; petroleum exploration and geology; hydrogeology; sedimentology; geomorphology; photogeology; reefing limestones of Majuro; Pennsylvanian stratigraphy of Four Corners region. Mailing Add: 1040 Stanford Dr NE Albuquerque NM 87106

WENIG, JEFFREY, b New York, NY, Jan 16, 37; m 64; c 2. PHARMACOLOGY, PHYSIOLOGY. Educ: Syracuse Univ, BS, 57; NY Univ, MS, 61, PhD(physiol), 65. Prof Exp: Sr res assoc, Med Col, Cornell Univ, 58-66; sr investr & group leader pharmacol, 66-69, ASST DIR TOXICOL, ENDO RES LABS, 69- Concurrent Pos: Instr, Eve Div, Hunter Col, 63-67; asst prof, Eve Grad Div, Hofstra Univ, 69-72, assoc prof, 72-; mem adv bd biol, State Univ NY Agr & Tech Col Farmingdale. Res: Pharmacological and toxicological evaluation of potential therapeutic agents in the anti-inflammatory, cardiovascular, analgesic and central nervous system areas. Mailing Add: Endo Res Labs 1000 Stewart St Garden City NY 11530

WENJEN, CHIEN, b China, Feb 16, 11; m 60. MATHEMATICS. Educ: Nat Cent Univ, China, BA, 31; Univ Calif, Los Angeles, PhD, 53. Prof Exp: Asst prof math, Tex Tech Col, 56-59; from asst prof to assoc prof, 59-68, PROF MATH, CALIF STATE UNIV LONG BEACH, 68- Mem: AAAS; Am Math Soc; Math Asn Am. Res: Functional analysis; Banach algebras; topology. Mailing Add: Dept of Math Calif State Univ 6101 E Seventh St Long Beach CA 90840

WENK, EUGENE J, b New York, NY, Oct 21, 27; m 54; c 3. ANATOMY. Educ: Columbia Univ, AB, 50, AM, 51; New York Med Col, PhD(anat), 72. Prof Exp: Instr, 72-73, ASST PROF ANAT, NEW YORK MED COL, 73- Mem: AAAS; Am Asn Anatomists. Res: Ultrastructure and function of lymphoid system. Mailing Add: Dept of Anat New York Med Col Valhalla NY 10595

WENK, HANS-RUDOLF, b Zurich, Switz, Oct 25, 41; m 70. CRYSTALLOGRAPHY, STRUCTURAL GEOLOGY. Educ: Univ Basel, BA, 63; Univ Zurich, PhD(crystallog), 65. Prof Exp: NSF fel, Inst Geophys, Univ Calif, Los Angeles, 66-67; from asst prof to assoc prof geol, 67-73, PROF GEOL & GEOPHYS, UNIV CALIF, BERKELEY, 73- Concurrent Pos: Co-worker, Swiss Geol Comn, 66-; Miller prof, Miller Inst Basic Res, 71-72. Mem: Mineral Soc Am; Swiss Mineral & Petrog Soc; Swiss Soc Natural Hist; Asn Study Deep Zones Earth's Crust. Res: Structural geology of metamorphic belts; experimental rock deformation; preferred orientation; crystal chemistry of silicates, plagioclase, crystal structures and refinements; lunar minerals. Mailing Add: Dept of Geol & Geophys Univ of Calif 497 Earth Sci Bldg Berkeley CA 94720

WENKERT, ERNEST, b Vienna, Austria, Oct 16, 25; nat US; m 48; c 4. NATURAL PRODUCTS CHEMISTRY. Educ: Univ Wash, BS, 45, MS, 47; Harvard Univ, PhD(chem), 51. Prof Exp: Instr chem, Lower Columbia Jr Col, 47-48; from asst prof to prof org chem, Iowa State Univ, 51-61; prof, Ind Univ, Bloomington, 61-69, Herman T Briscoe prof, 69-73; E D BUTCHER PROF CHEM, RICE UNIV, 73- Concurrent Pos: Lectr var US & foreign orgn & acad insts, 51-; actg head dept org chem, Weizmann Inst, 64-65; Guggenheim fel, 65-66; mem NIH med chem B study sect, 71-72, chmn, 72-75. Honors & Awards: Ernest Guenther Award, Am Chem Soc. Mem: Am Chem Soc; The Chem Soc; Swiss Chem Soc. Mailing Add: Dept of Chem Rice Univ Houston TX 77001

WENNEMER, JAY, b South Weymouth, Mass, Apr 1, 47; m 71; c 2. MARINE ECOLOGY. Educ: Northeastern Univ, BA, 70. Prof Exp: Res scientist ichthyol, 70-75, SUPVR ADMIN & ICHTHYOPLANKTON, BATTELLE MEM INST, WATERFORD, 75- Mem: Am Fisheries Soc; Estuarine Res Fedn. Res: Effect of power generation on finfish populations; ichthyoplankton, benthic and intertidal communities; local marine ecosystems. Mailing Add: 82 Atlantic Dr Old Saybrook CT 06475

WENNER, ADRIAN MANLEY, b Roseau, Minn, May 24, 28; m 57; c 2. ZOOLOGY. Educ: Gustavus Adolphus Col, BS, 51; Chico State Col, MA, 55; Univ Mich, MS, 58, PhD(zool), 61. Prof Exp: Prin elem sch, Ore, 54-55; teacher high sch, Calif, 55-56; fel zool, Univ Mich, 56-60; from asst prof biol sci to assoc prof biol, 60-73, PROF NATURAL HIST, UNIV CALIF, SANTA BARBARA, 73- Concurrent Pos: Consult, Teledyne, Inc, 62-64 & Autonetics Div, NAm Aviation, Inc, 64-65. Mem: AAAS; Am Soc Zoologists; Sigma Xi; Am Soc Naturalists. Res: Problems of growth in marine crustaceans as it occurs in nature; natural history of marine crustacea; animal communication. Mailing Add: Marine Sci Inst Univ of Calif Santa Barbara CA 93106

WENNER, BRUCE RICHARD, b Lancaster, Pa, Apr 25, 38; m 65; c 3. TOPOLOGY. Educ: Col Wooster, BA, 60; Duke Univ, PhD(math), 64. Prof Exp: Asst prof math, Univ Vt, 64-68; asst prof, 68-70, ASSOC PROF MATH, UNIV MO-KANSAS CITY, 70- Concurrent Pos: NASA res grant, 65-66. Mem: Am Math Soc; Math Asn Am. Res: Dimension theory with regard to topological dimension functions, especially metrizable spaces. Mailing Add: Dept of Math Univ of Mo Kansas City MO 64110

WENNER, CHARLES EARL, b Lattimer, Pa, May 2, 24; m 48; c 2. BIOCHEMISTRY. Educ: Temple Univ, BA, 48, PhD(chem), 53. Prof Exp: Res fel, Lankenau Hosp Res Inst & Inst Cancer Res, 50-54, res assoc, 54-55; sr scientist, Dept Exp Biol, 56-61, assoc scientist, 61-65, PRIN SCIENTIST, DEPT EXP BIOL, ROSWELL PARK MEM INST, 65-; PROF BIOCHEM, GRAD SCH, STATE UNIV NY BUFFALO, 70- Concurrent Pos: Runyon fel, Inst Cancer Res, 52-54; from asst prof to assoc prof biochem, Grad Sch, State Univ NY Buffalo, 58-70; Johnson Res Found vis res prof, Univ Pa, 65-66. Mem: Am Chem Soc; Am Asn Cancer Res; Biophys Soc; Fedn Am Soc Exp Biol. Res: Energy control mechanisms; membrane biochemistry; peptide-induced ion transport; membrane transport bioenergetics. Mailing Add: Dept of Exp Biol 666 Elm St Roswell Park Mem Inst Buffalo NY 14203

WENNER, DAVID BRUCE, b Flint, Mich, May 28, 41; m 68. GEOCHEMISTRY, GEOLOGY. Educ: Univ Cincinnati, BS, 63; Calif Inst Technol, MS, 66, PhD(geochem), 71. Prof Exp: Vis asst prof, 71, ASST PROF GEOL, UNIV GA, 71- Mem: AAAS; Geol Soc Am; Am Geophys Union. Res: Stable isotope geochemistry with applications to igneous and metamorphic rocks. Mailing Add: Dept of Geol Univ of Ga Athens GA 30601

WENNER, HERBERT ALLAN, b Drums, Pa, Nov 14, 12; m 42; c 4. MEDICINE. Educ: Bucknell Univ, BSc, 33; Univ Rochester, MD, 39; Am Bd Microbiol, dipl, 62. Prof Exp: Instr prev med, Sch Med, Yale Univ, 44-46; from asst prof to assoc prof pediat & bact, Univ Kans, 46-51, res prof pediat, 51-69; DISTINGUISHED PROF PEDIAT, CHILDREN'S MERCY HOSP, UNIV MO-KANSAS CITY, 69- Concurrent Pos: Babbott fel, Sch Med, Yale Univ, 41-42; Nat Res Coun fel, Yale Univ & Johns Hopkins Univ, 43-44; NIH res career award, 62-69; assoc physician, Dept Internal Med, New Haven Hosp, Conn, 44-46; consult, Mo State Bd Health & Nat Commun Dis Ctr; mem, Echovirus Subcomt, Picornavirus Study Group, NIH; clin prof pediat, Univ Kans Med Ctr, Kansas City, 73- Honors & Awards: Presidential & Distinguished Serv Awards, Nat Found Infantile Paralysis; Cert Serv Award, Panel Picornaviruses, NIH. Mem: Fel AAAS; fel Am Acad Pediat; fel Am Pub Health Asn; Soc Pediat Res; Am Pediat Soc. Res: Etiology, pathogenesis and epidemiology of infectious diseases. Mailing Add: Children's Mercy Hosp Univ of Mo-Kansas City Kansas City MO 64108

WENNERSTEN, DWIGHT L, b O'Neill, Nebr, May 16, 12; m 48. PHYSICS. Educ: Morningside Col, AB, 33; Univ Nebr, MS, 37. Prof Exp: Instr sci & math, Wayland Acad & Jr Col, 38-40; instr physics, Grinnell Col, 40-42; physicist ord lab, US Dept Navy, 42-47, bur ord, 47-56; chief, Gen Physics Div, 56-74, CONSULT SCI TO TECHNOL TRANSFER, TECH PLANNING DIV, OFF SCI RES, US DEPT AIR FORCE, 74- Mem: AAAS; Soc Physics Students. Res: Atomic and molecular physics, astrophysics, microwave physics and applied physics. Mailing Add: 1901 Wyoming Ave NW Washington DC 20009

WENRICH-VERBEEK, KAREN JANE, b Lebanon, Pa, Apr 9, 47; m 69. GEOLOGY. Educ: Pa State Univ, BS, 69, MS, 71, PhD(geol), 75. Prof Exp: Geologist, Molybdenum Corp Am, 69; instr geol, Bucknell Univ, 73-74; GEOLOGIST, US GEOL SURV, 74- Concurrent Pos: US Geol Surv adv, Energy Resource & Develop Admin Nat Uranium Resource Eval Prog, 75- Mem: Geol Soc Am; Sigma Xi; Am Geophys Union. Res: Uranium exploration, specifically the use of uranium in water, stream sediments, soils and modern decaying plant materials as a tool for exploration; trace element geochemistry in volcanic rocks. Mailing Add: Mail Stop 916 Fed Ctr US Geol Surv Denver CO 80225

WENSKA, TOM MARION, b Honolulu, Hawaii, May 2, 45. MATHEMATICS. Educ: Univ Hawaii, BA, 66; Univ Southern Calif, PhD(math), 70. Prof Exp: Asst prof, 70-74, ASSOC PROF MATH, UNIV HAWAII, 74- Mem: Am Math Soc. Res: Numerical algorithms for solutions of Fredholm integral equations. Mailing Add: Dept of Math Univ of Hawaii Honolulu HI 96822

WENSLEY, RALPH NELSON, plant pathology, see 12th edition

WENT, FRITS WARMOLT, b Utrecht, Neth, May 18, 03; nat US; m 27; c 2. BOTANY. Educ: Univ Utrecht, AB, 22, MA, 25, PhD(bot), 27. Hon Degrees: PhD, Univ Paris, 56, Methodist Cent Col, 63 & Univ Upsala, 68; DSc, McGill Univ, 59; LLD, Univ Edmonton, 71. Prof Exp: Asst, Univ Utrecht, 22-27; plant physiologist, Bot Gardens, Java, 28-30, head foreigners lab, 30-32; instr bot, Med Col, Batavia, Java, 30-31; from asst prof to prof plant physiol, Calif Inst Technol, 33-58; dir, Mo Bot Garden, 58-63; prof bot, Washington Univ, 63-65; distinguished prof, 65-75, EMER RES PROF BOT, DESERT RES INST, UNIV NEV, RENO, 75- Concurrent Pos: With USDA; trustee, Calif Arboretum Found, 47-58; pres, Am Inst Biol Sci, 62. Honors & Awards: Hodgkins Award, Smithsonian Inst, 67. Mem: Nat Acad Sci; AAAS; Am Soc Plant Physiol (pres, 47); Bot Soc Am (pres, 58); Ecol Soc Am. Res: Plant physiology; growth, climatic responses, translocation and water relationships; hormones; relation between climate and vegetation; physiology of crop plants; volatile substances of plant origin in atmosphere. Mailing Add: Desert Res Inst Univ of Nev Reno NV 89507

WENT, HANS ADRIAAN, b Bogor, Indonesia, Dec 3, 29; nat US; m 51; c 2. PHYSIOLOGY. Educ: Univ Calif, AB, 51, MA, 53, PhD(zool), 58. Prof Exp: From instr to asst prof, 59-69, ASSOC PROF ZOOL, WASH STATE UNIV, 69- Mem: Am Soc Zool. Res: Cell division, especially molecular origin of mitotic apparatus; cell physiology. Mailing Add: Dept of Zool & Zoophysiology Wash State Univ Pullman WA 99164

WENTE, HENRY CHRISTIAN, b New York, NY, Aug 18, 36. MATHEMATICS. Educ: Harvard Univ, BA, 58, MA, 59, PhD(math), 66. Prof Exp: From instr to asst prof, Tufts Univ, 63-70, lectr, 70-71; asst prof, 71-74, ASSOC PROF MATH, UNIV TOLEDO, 74- Concurrent Pos: Univ grant & res assoc, Math Inst, Univ Bonn, 72-73. Mem: AAAS; Am Math Soc. Res: Existence theorems in the calculus of variations, especially those arising from two-dimensional parametric surfaces immersed in Euclidean space; surfaces minimizing area subject to a volume constraint. Mailing Add: Dept of Math Univ of Toledo Toledo OH 43606

WENTINK, TUNIS, JR, b Paterson, NJ, Feb 3, 20; m 68. PHYSICAL CHEMISTRY. Educ: Rutgers Univ, BS, 41; Cornell Univ, PhD(chem), 54. Prof Exp: Res chemist, Photoprods Div, E I du Pont de Nemours & Co, 41-43; from res assoc to mem staff, Div Indust Co-op, Mass Inst Technol, 47-48; from res assoc to specialist microwave spectros, Brookhaven Nat Lab, 48-50; asst, Cornell Univ, 50-54; physicist, Gen Elec Co, 53-55; prin res scientist & supvr, Chem Lab, Avco-Everett Res Lab, Avco Corp, 55-59, from prin scientist to sr consult scientist, Adv Res & Develop Div, 59-67; prin scientist, GCA Corp, 67-68; head exp physics dept, Panametrics, Inc, 68-70; assoc dir, 70-72, DIR INST ARCTIC ENVIRON ENG, 72-, PROF PHYSICS, UNIV ALASKA, 70- Concurrent Pos: NSF vis prof, Geophys Inst, Univ Alaska, 68; consult. Mem: Am Phys Soc; Sigma Xi. Res: Spectroscopy; radiation chemistry of high temperature gases and solids; ultraviolet to microwave regions and instrumentation; molecular absolute intensities and radiative lifetimes; re-entry physics and signatures; ablation; arctic environmental technology. Mailing Add: Dept of Physics Univ of Alaska Fairbanks AK 99701

WENTLAND, MARK PHILIP, b New Britain, Conn, Jan 22, 45; m 70. ORGANIC CHEMISTRY. Educ: Cent Conn State Col, BS, 66; Rice Univ, PhD(chem), 70. Prof Exp: ASSOC RES CHEMIST, STERLING-WINTHROP RES INST, 70- Concurrent Pos: Adj assoc prof, Rensselaer Polytech Inst, 71- Mem: Am Chem Soc. Res: The design and synthesis of potentially useful medicinal agents. Mailing Add: Sterling-Winthrop Res Inst Rensselaer NY 12144

WENTLAND, STEPHEN HENRY, b New Britain, Conn, May 1, 40; m 65; c 2. BIOCHEMISTRY, ORGANIC CHEMISTRY. Educ: Rensselaer Polytech Inst, BS, 62; Yale Univ, MS, 64, PhD(chem), 68. Prof Exp: NIH fel chem, Ind Univ, Bloomington, 68-70; sr org chemist, Smith Kline & French, Inc, 70-72; res assoc biochem, 72-74, INSTR MED, UNIV COLO MED CTR, DENVER, 74- Mem: Am Chem Soc. Res: Isolation and structure determination of macromolecules; synthesis of organic compounds. Mailing Add: Dept of Lab Med Univ of Colo Med Ctr Denver CO 80220

WENTORF, ROBERT H, JR, b Wis, May 28, 26; m 49; c 3. PHYSICAL CHEMISTRY. Educ: Univ Wis, BSChE, 48, PhD(chem), 52. Prof Exp: Asst chem, Univ Wis, 46; RES ASSOC CHEM, CORP RES & DEVELOP CTR, GEN ELEC CO, 52- Concurrent Pos: Brittingham vis prof, Univ Wis, 67-68. Honors & Awards: Ipatieff Prize, 65. Res: High pressure chemistry and physics, diamond synthesis, energy systems, solar energy utilization. Mailing Add: 383 Vly Rd Schenectady NY 12309

WENTWORTH, BERNARD C, b Freedom, Maine, Feb 16, 35; m 60; c 4. AVIAN PHYSIOLOGY. Educ: Univ Maine, Orono, BS, 57; Univ Mass, MS, 60, PhD(avian physiol), 63. Prof Exp: Physiologist, US Dept Interior, 63-69; assoc prof, 69-73, PROF POULTRY SCI, UNIV WIS-MADISON, 73- Mem: AAAS; Poultry Sci Asn; Soc Study Reproduction; NY Acad Sci. Res: Basic physiology of birds and comparative endocrinology of animals as related to applied benefits to biomedicine and agriculture. Mailing Add: Dept of Poultry Sci Animal Sci Bldg Univ of Wis Madison WI 53706

WENTWORTH, BERTTINA BROWN, b Rockfort, Ill, Aug 9, 19. MICROBIOLOGY. Educ: Univ Ky, BS, 41; Ohio State Univ, MS, 58; Univ Calif, Los Angeles, PhD(microbiol), 64. Prof Exp: Microbiologist, Ohio Dept Health, Columbus, 49-59; NIH fel, Australian Nat Univ, 64-66; assoc prof pathobiol, Sch Pub Health, Univ Wash, 67-73; COORD MICROBIOLOGIST, BUR DIS CONTROL & LAB SERV, MICH DEPT PUB HEALTH, 73- Concurrent Pos: Adj assoc prof, Mich State Univ, 73- Mem: Am Soc Microbiol; Am Pub Health Asn; Am Asn Immunol. Res: Virology; immunology. Mailing Add: Bur Lab Mich Dept Pub Health 3500 N Logan Lansing MI 48914

WENTWORTH, CARL M, JR, b New York, NY, Feb 8, 36; m 68. ENVIRONMENTAL GEOLOGY, SEDIMENTOLOGY. Educ: Dartmouth Col, AB, 58; Stanford Univ, MS, 60, PhD(geol), 67. Prof Exp: GEOLOGIST, US GEOL SURV, 63- Mem: Geol Soc Am; Soc Econ Paleontologists & Mineralogists; Seismol Soc Am. Res: Major geologic hazards of United States; active faults and movement histories; landslides and slope stability; engineering character of geologic materials; geology of California Coast Ranges. Mailing Add: US Geol Surv 345 Middlefield Rd Menlo Park CA 94025

WENTWORTH, GARY, b Orange, Mass, Aug 3, 39; m 61; c 1. ORGANIC CHEMISTRY, POLYMER CHEMISTRY. Educ: Rensselaer Polytech Inst, BS, 61; Ga Inst Technol, PhD(org chem), 66. Prof Exp: Res chemist, Union Carbide Corp, 66-68; sr res chemist, 68-73, SR RES SPECIALIST, MONSANTO DEVELOP CTR, 73- Mem: Am Chem Soc; The Chem Soc. Res: Fiber chemistry; free radical polymerization; alternating copolymerization. Mailing Add: Monsanto Develop Ctr PO Box 12274 Research Triangle Park NC 27709

WENTWORTH, RUPERT A D, b Hattiesburg, Miss, Nov 5, 34; m 56, 72; c 3. INORGANIC CHEMISTRY. Educ: Fordham Univ, BS, 55; Mich State Univ, PhD(inorg chem), 63. Prof Exp: Fel with Prof T S Piper, Univ Ill, 63-65; from asst prof to assoc prof, 65-74, PROF CHEM, COL ARTS & SCI, GRAD SCH, IND UNIV, BLOOMINGTON, 74- Mem: Am Chem Soc. Res: Chemistry of the lower oxidation states of transition metal ions; chemistry of polynuclear complexes. Mailing Add: Dept of Chem Ind Univ Grad Sch Col of Arts & Sci Bloomington IN 47401

WENTWORTH, STANLEY EARL, b Natick, Mass, July 13, 40; m 67; c 2. ORGANIC POLYMER CHEMISTRY. Educ: Northeastern Univ, BS, 63, PhD(org chem), 67. Prof Exp: Res chemist, Army Natick Labs, 67-68, RES CHEMIST, ORG MAT LAB, ARMY MAT & MECH RES CTR, 70- Mem: Am Chem Soc. Res: Organic synthesis in the areas of organofluorine compounds; diazoalkanes and monomers for high temperature resins; synthesis of polyphenylquinoxalines and polyurethanes. Mailing Add: Org Mat Lab Army Mat & Mech Res Ctr Watertown MA 02172

WENTWORTH, WAYNE, b Rochester, Minn, May 29, 30; m 54; c 4. ANALYTICAL CHEMISTRY, PHYSICAL CHEMISTRY. Educ: St Olaf Col, BA, 52; Fla State Univ, PhD(chem), 57. Prof Exp: Res mathematician, Radio Corp Am, 56-59; from asst prof to assoc prof, 59-68, PROF CHEM, UNIV HOUSTON, 68- Mem: Am Chem Soc. Res: Electron attachment to molecules; molecular complexes; molecular spectroscopy. Mailing Add: Dept of Chem Univ of Houston 3801 Cullen Blvd Houston TX 77004

WENTZ, WILLIAM BUDD, b Philadelphia, Pa, Aug 9, 24; m 45; c 3. OBSTETRICS & GYNECOLOGY, ONCOLOGY. Educ: Univ Pa, BA, 51, MA, 53; Western Reserve Univ, MD, 58; Am Bd Obstet & Gynec, dipl, 66. Prof Exp: Prin investr physiol, US Naval Aviation Med Acceleration Lab, 53-54; intern med, Lankenau Hosp, Philadelphia, Pa, 48-49, prin investr gynec & oncol, 59-63; asst prof obstet & gynec, Hahnemann Med Col, 63-66; assoc prof obstet & gynec, 66-71, PROF REPRODUCTIVE BIOL, SCH MED, CASE WESTERN RESERVE UNIV, 71- Concurrent Pos: Am Cancer Soc grants, Lankenau Hosp, Philadelphia, 59-63; Nat Cancer Inst gramts, 64- Mem: Am Col Obstet & Gynec; Soc Gynec Oncol; Am Soc Cytol. Res: Experimental gynecological pathology; carcinogenesis; cancer research treatment of malignant and premalignant disease. Mailing Add: Dept of Reproductive Biol Case Western Reserve Univ Cleveland OH 44106

WENTZEL, DONAT GOTTHARD, b Zürich, Switz, June 25, 34; US citizen; m 59; c 1. ASTROPHYSICS. Educ: Univ Chicago, BA, 54, BS, 55, MS, 56, PhD(physics), 60. Prof Exp: From instr to assoc prof astron, Univ Mich, 60-66; assoc prof, 67-74, PROF ASTRON, UNIV MD, COLLEGE PARK, 74- Concurrent Pos: Alfred P Sloan res fel, 62-66; vis lectr, Princeton Univ, 64; vis prof, Tata Inst Fundamental Res, Bombay, 73. Mem: Fel AAAS; Am Astron Soc; Int Astron Union. Res: Effects of magnetic fields on fluid dynamics and charged particles on the sun and in interplanetary and interstellar space; astronomy education. Mailing Add: Astron Prog Univ of Md College Park MD 20742

WENTZEL, GREGOR, b Dusseldorf, Ger, Feb 17, 98; nat US; m 29; c 1. THEORETICAL PHYSICS. Educ: Univ Munich, PhD(physics), 21. Hon Degrees: DSc, Swiss Fed Inst Technol, 66. Prof Exp: Privatdocent theoret physics, Univ Munich, 22-26; prof, Univ Leipzig, 26-28 & Univ Zurich, 28-48; prof physics, Univ Chicago, 48-69, EMER PROF PHYSICS, UNIV CHICAGO, 69- Concurrent Pos: Vis prof, Univ Wis, 30, Purdue Univ, 47, Stanford Univ, 49, Inst Fundamental Res, India, 51 & 56, Univ Calif, 54 & Europ Orgn Nuclear Res, Geneva, 58; Solvay Prof, Free Univ Brussels, 70. Honors & Awards: Max Planck Medal, Ger Phys Soc, 75. Mem: Nat Acad Sci; fel Am Phys Soc; Swiss Phys Soc; Swiss Nature Soc. Res: Atomic and nuclear physics; elementary particles. Mailing Add: 77 via Collina 6612 Ascona Switzerland

WENZ, DONALD A, inorganic chemistry, see 12th edition

WENZEL, ALAN RICHARD, b Port Chester, NY, Feb 8, 38. APPLIED MATHEMATICS. Educ: NY Univ, BAE, 60, MS, 62, PhD(math), 70. Prof Exp: Mem res staff, Wyle Labs, 65-68; res assoc, Univ Miami, 70-73; Nat Res Coun assoc, NASA Ames Res Ctr, 73-75, VIS SCIENTIST, INST COMPUT APPLN SCI & ENG, NASA LANGLEY RES CTR, 75- Mem: Acoust Soc Am. Res: Theoretical research in fluid dynamics and wave propagation. Mailing Add: ICASE M/S 132C NASA Langley Res Ctr Hampton VA 23665

WENZEL, BERNICE MARTHA (MRS WENDELL E JEFFREY), b Bridgeport, Conn, June 22, 21; m 52. NEUROPSYCHOLOGY, MEDICAL EDUCATION. Educ: Beaver Col, AB, 42; Columbia Univ, AM, 43, PhD(psychol), 48. Prof Exp: Instr psychol, Newcomb Col, Tulane Univ, 45-46; from instr to asst prof, Barnard Col, Columbia Univ, 46-55; from asst prof to assoc prof physiol, 59-69, vchmn dept, 71-73, PROF PHYSIOL, SCH MED, UNIV CALIF, LOS ANGELES, 69-, PROF PSYCHIAT, 71-, ASST DEAN, 74- Concurrent Pos: Fel, Ment Health Training Prog, Sch Med, Univ Calif, Los Angeles, 57-59. Mem: Fel AAAS; fel Am Psychol Asn; Am Physiol Soc; Int Brain Res Orgn; Soc Neurosci. Res: Behavioral physiology; olfaction; autonomic nervous system. Mailing Add: Dept of Physiol Univ of Calif Sch of Med Los Angeles CA 90024

WENZEL, DUANE GREVE, b Wausau, Wis, Sept 18, 20; m 43; c 4. PHARMACOLOGY. Educ: Univ Wis, BS, 45, PhD, 48. Prof Exp: From asst prof to assoc prof, 48-56, PROF PHARMACOL, SCH PHARM, UNIV KANS, 56- Mem: Tissue Cult Asn; Am Pharmaceut Asn; Soc Toxicol. Res: Cause, treatment and prophylaxis of cardiovascular disease; pharmacology and toxicology of environmental agents in cultured cells. Mailing Add: Dept of Pharmacol & Toxicol Univ of Kans Sch of Pharm Lawrence KS 66044

WENZEL, FREDERICK J, b Marshfield, Wis, Aug 5, 30; m 52; c 6. BIOCHEMISTRY, BIOLOGY. Educ: Univ Wis-Stevens Point, BS, 56. Prof Exp: Res asst, St Joseph's Hosp, Marshfield, 50-53; dir labs, Marshfield Clin, 53-65, secy found, 58-64, EXEC DIR, MARSHFIELD MED FOUND, 65- Mem: Am Chem Soc; Am Inst Chem; Am Fedn Clin Res. Res: Hypersensitivity phenomenon in the lung, such as farmer's lung and maple bark disease; natural history of pulmonary thromboembolism, especially diagnosis and treatment and studies of the fibrinolytic process in this disease. Mailing Add: Marshfield Med Found 510 N St Joseph Ave Marshfield WI 54449

WENZEL, RICHARD LOUIS, b Marietta, Ohio, Sept 4, 21; m 47; c 3. PUBLIC HEALTH ADMINISTRATION, PREVENTIVE MEDICINE. Educ: Marietta Col,

AB, 43; Ohio State Univ, MD, 46; Univ Mich, MPH, 47; Am Bd Prev Med, cert pub health, 63. Prof Exp: Intern, Jersey City Med Ctr, 46-47; resident obstet & gynec, St Ann's Maternity Hosp, Columbus, Ohio, 47; chief commun dis & dep health off, Columbus Dept Health, 53-58; health officer, Marietta & Washington County, 58-60; assoc prof pub health admin, Sch Pub Health, Univ Mich, Ann Arbor, 60-70; HEALTH COMNR, TOLEDO & LUCAS COUNTY HEALTH DEPTS, 70- Concurrent Pos: Asst prof, Col Med, Ohio State Univ, 54-57; consult, Div Health Mobilization, USPHS, 61-72, Div Indian Health, 64-67 & Bur Med Serv, 66-67; mem, Emergency Health Preparedness Adv Comt, 67-72; consult, Nat Comn Community Health Serv, 64-66; assoc clin prof, Dept Social Med, Med Col Ohio; adj prof pub health admin, Sch Pub Health, Univ Mich, Ann Arbor, 70- Mem: Fel Am Pub Health Asn; fel Am Col Prev Med; US Conf City Health Offs. Res: Survey, assessment, and evaluation of community health services. Mailing Add: Toledo Health Dept 635 N Erie St Toledo OH 43624

WENZEL, RUPERT LEON, b Owen, Wis, Oct 16, 15; m 40; c 3. ZOOLOGY. Educ: Cent YMCA Col, AB, 38, Univ Chicago, PhD(zool), 62. Prof Exp: Asst zool, Cent YMCA Col, 36-38; res asst, Univ Chicago, 37-40; asst cur insects, 40-50, cur insects, 51-70, CHMN DEPT ZOOL, FIELD MUS NATURAL HIST, 70- Concurrent Pos: Lectr, Roosevelt Col, 46-47 & Univ Chicago, 62-; res asso biol, Northwestern Univ, 59-, vis prof, 63. Honors & Awards: Order of Vasco Nunez de Balboa, Panama, 67. Mem: AAAS; Entom Soc Am; Soc Study Evolution; Soc Syst Zool; Asn Trop Biol. Res: Taxonomy of streblid and nycteribiid batflies and histerid beetles; zoogeography; evolution of ectoparasites of terrestrial vertebrates. Mailing Add: Dept of Zool Field Mus of Natural Hist Chicago IL 60605

WENZEL, WILLIAM ALFRED, b Cincinnati, Ohio, Apr 30, 24; m 55; c 2. ELEMENTARY PARTICLE PHYSICS. Educ: Williams Col, AB, 44; Calif Inst Technol, MS, 48, PhD(physics), 52. Prof Exp: Res fel physics, Calif Inst Technol, 52-53; assoc dir physics, 70-73, RES PHYSICIST, LAWRENCE BERKELEY LAB, UNIV CALIF, 53- Mem: Fel Am Phys Soc; AAAS. Res: Experimentation in high energy physics; study of weak and electromagnetic interactions and of hadron phenomena at high transverse momenta; development of electronic instrumentation and accelerator facilities. Mailing Add: Physics Div Lawrence Berkeley Lab Univ Calif Berkeley CA 94720

WENZINGER, GEORGE ROBERT, b Newport News, Va, July 24, 33. CHEMISTRY. Educ: Washington Univ, AB, 55; Univ Rochester, PhD(chem), 60. Prof Exp: NSF fel, Yale Univ, 62-63; ASST PROF ORG CHEM, UNIV S FLA, 63- Mem: AAAS; Am Chem Soc. Res: Conjugate elimination reactions; cyclopropyl chemistry; synthesis. Mailing Add: Dept of Chem Univ of SFla Tampa FL 33620

WEPPELMAN, ROGER MICHAEL, b Pittsburgh, Pa, Nov 4, 44; m 71. BIOCHEMISTRY, PARASITOLOGY. Educ: Univ Pittsburgh, BS, 65, PhD(microbiol), 70. Prof Exp: Instr microbiol, Univ Pittsburgh, 70-71; Am Cancer Soc fel biochem, Univ Calif, Berkeley, 71-73; SR RES BIOCHEMIST, DEPT BASIC ANIMAL SCI RES, MERCK & CO, 73- Mem: Am Soc Microbiol. Res: Protozoan metabolic pathways; genetics of drug resistance of parasitic protozoa. Mailing Add: 1232 Sunnyfield Lane Scotch Plains NJ 07076

WERBEL, BURTON, physical chemistry, see 12th edition

WERBEL, LESLIE MORTON, b New York, NY, Mar 31, 31; m 58; c 3. ORGANIC CHEMISTRY. Educ: Queens Col, NY, BS, 51; Columbia Univ, AM, 52; Univ Ill, PhD(chem), 57. Prof Exp: Res chemist, 57-67, sr res chemist, 67-75, SR RES SCIENTIST, PARKE, DAVIS & CO, 75- Concurrent Pos: Lectr col pharm, Univ Mich, 67, adj prof, 71. Mem: Am Chem Soc; The Chem Soc; Int Soc Heterocyclic Chem. Res: Medicinal chemistry; chemotherapy of parasitic infections; relation of intermediary metabolism of host and invading organism to drug action; novel heterocyclic ring systems. Mailing Add: Parke Davis & Co 2800 Plymouth Rd Ann Arbor MI 48106

WERBER, ERNA ALTURE, b Vienna, Austria, Dec 9, 09; nat US. VIROLOGY. Educ: Univ Vienna, PhD, 29. Prof Exp: Mem staff, Rudolf & Child Hosp, Vienna, Austria, 30-36 & Beth Israel Hosp, Boston, 37-39; asst microbiologist, Squibb Inst, NJ, 40-42; microbiologist, Wallace Labs, 43-44; chief microbiologist, Loewe Res Labs, 44-58; virologist & dir, Virus Lab, Delaware, Inc, 59-61; res assoc infectious dis, Med Sch, Cornell Univ & Bellevue Hosp, 61-68; res assoc pediat microbiol, Med Sch, NY Univ & Bellevue Hosp, 68-71; MYCOLOGIST, MONTEFIORE HOSP MED CTR, 71- Mem: Soc Exp Biol & Med; Am Soc Microbiol; Am Asn Immunol; Mycol Soc Am; Am Thoracic Soc. Res: Chemotherapy; antibiotics; tissue culture; bacteriology; serology; granulomous diseases; mutation of Cryptococci neoformans. Mailing Add: Apt 3D 170 W 73rd St New York NY 10023

WERBER, FRANK XAVIER, b Vienna, Austria, Apr 8, 24; nat US; m 50; c 2. ORGANIC CHEMISTRY, POLYMER CHEMISTRY. Educ: Queens Col, NY, BS, 44; Univ Ill, MS, 47, PhD(chem), 49. Prof Exp: Asst, Queens Col, NY, 46; res chemist, Org Res Dept, Res Ctr, B F Goodrich Co, 49-55, sr res chemist, 55-56; dir polymer res, Res Div, W R Grace & Co, 57-63, vpres org & polymer res, 63-67; VPRES RES & DEVELOP, J P STEVENS & CO, INC, 67- Concurrent Pos: Chmn exec comt, Textile Res Inst, 72- Mem: AAAS; Am Chem Soc; NY Acad Sci; Am Asn Textile Technol; Am Asn Textile Chemists & Colorists. Res: Industrial organic chemistry; modification of polymers; adhesives; condensation polymers; textile chemistry, technology, machinery and engineering.

WERBIN, HAROLD, b New York, NY, Oct 2, 22; m 50. BIOLOGICAL CHEMISTRY. Educ: Brooklyn Col, BS, 44; Polytech Inst Brooklyn, MS, 47, PhD(chem), 50. Prof Exp: Asst, Brooklyn Jewish Hosp, 44-46; res assoc, Polytech Inst Brooklyn, 47-48; dir labs, Hillside Hosp, 50-53; res assoc biochem, Argonne Cancer Res Hosp, Chicago, 53-56; res biochemist dept physiol & soils & plant nutrit, Univ Calif, Berkeley, 57-66; ASSOC PROF BIOL, UNIV TEX, DALLAS, 66- Concurrent Pos: AEC contract, 67-, Energy Res & Develop Admin contract, 75-76, NSF grant, 67-70; ed, Southwest Retort, 68-; Robert A Welch Found grant, 71- Mem: Am Soc Biol Chem; Am Chem Soc; Am Soc Photobiol. Res: Synthesis and biochemistry of steroid hormones; photobiology of RNA phages; photobiochemical studies and purification of DNA-photolyase; photochemistry of electron transport quinines to probe membrane architecture. Mailing Add: Prog in Biol Univ of Tex Dallas PO Box 688 Richardson TX 75080

WERBLIN, FRANK SIMON, b New York, NY, Jan 24, 37; c 1. BIOENGINEERING. Educ: Mass Inst Technol, BS, 58, MS, 62; Johns Hopkins Univ, PhD(bioeng), 68. Prof Exp: From asst prof to assoc prof, 70-75, PROF ELEC ENG, UNIV CALIF, BERKELEY, 75- Concurrent Pos: Guggenheim fel, 74; Miller Professorship, Miller Found, 76. Res: Neurophysiological and biochemical basis for function of the vertebrate retina; mechanisms of light transduction, contrast detection, gain control studied in terms of molecular events in photoreceptors, ionic events at synapses, membrane events in neurons. Mailing Add: Dept of Elec Eng Univ of Calif Berkeley CA 94720

WERBOFF, JACK, b Brooklyn, NY, Apr 24, 28; m 51; c 5. PSYCHOBIOLOGY. Educ: Brooklyn Col, BA, 49; Columbia Univ, MA, 50; Washington Univ, PhD(psychol), 57. Prof Exp: Asst hosp spec surg, New York, 48-49; asst, Univ Minn, 50-51; instr, Fla State Univ, 52-54; asst, Washington Univ, 54-57; head animal behav lab, Lafayette Clin, Detroit, 57-63; assoc prof, Wayne State Univ, 57-63; staff scientist, Jackson Lab, 63-68; PROF PSYCHOL, UNIV CONN, 68- Concurrent Pos: Lectr, Univ Maine, 64-67, prof, 67-68; consult, Mt Desert Island Child Guid Clin, Bar Harbor, 65-68. Mem: AAAS; Am Psychol Asn; Teratology Soc; Soc Res Child Develop; Soc Psychophysiol Res. Res: Developmental psychobiology; physiological bases of behavior; prenatal, neurophysiological, endocrinological and pharmacological effects on behavior and development. Mailing Add: Dept Behav Sci/Community Health Univ Conn Health Ctr Box U-155 Storrs CT 06268

WERBROUCK, ALBERT EUGENE, nuclear physics, see 12th edition

WERDEGAR, DAVID, b New York, NY, Sept 16, 30; m 61; c 2. COMMUNITY HEALTH. Educ: Cornell Univ, AB, 51, MA, 52; New York Med Col, MD, 56; Univ Calif, Berkeley, MPH, 70. Prof Exp: Consult, Calif State Dept Pub Health, 64-65; asst prof family & community med, 65-69, assoc prof, 69-75, PROF MED, DIV AMBULATORY COMMUNITY MED, SCH MED, UNIV CALIF, SAN FRANCISCO, 75-, CHMN SOCIAL MED UNDERGRAD CURRIC PATHWAY, 70-, ACTG ASSOC DEAN & DIR, FRESNO MED EDUC PROG, 75- Concurrent Pos: NIMH spec fel, Univ Calif, San Francisco, 63-64, Dept Health, Educ & Welfare Div Chronic Dis fel, 64-67; co-dir, Regional Med Prog Cardiovasc Dis Prev Northwest Calif, 67-70; planning officer, Regional Med Prog, Calif; med dir, Univ Calif Home Care Serv, 65- Mem: AAAS; Am Pub Health Asn; Asn Teachers Prev Med; Am Fedn Clin Res; Am Col Physicians. Res: Family medicine; health policy; social aspects of health care; organization of health care services; evaluation of quality of health care. Mailing Add: Dept of Community Med Univ of Calif Sch of Med San Francisco CA 94112

WERDEL, JUDITH ANN, b Lackawanna, NY, June 22, 37. INFORMATION SCIENCE, SCIENCE POLICY. Educ: State Univ NY Buffalo, BA, 58; Mt Holyoke Col, MA, 69. Prof Exp: Sci info specialist, Shell Develop Co, Calif, 59-63; prof asst info sci, Off Doc, 63-70, prof assoc, Bd Int Orgn & Progs & Bd Sci & Technol for Int Develop, 70-74, STAFF OFFICER, COMT INT SCI & TECH INFO PROGS, COMN INT RELS, NAT ACAD SCI, 74- Concurrent Pos: Secy bd dirs, Doc Abstr, Inc, 66-68. Mem: Am Soc Info Sci. Res: National and international policies and programs for the development of scientific and technical information systems; technical assistance in scientific and technical information systems; promotion and development of the field of information science. Mailing Add: Comn on Int Rels Nat Acad of Sci Washington DC 20418

WERDER, ALVAR ARVID, b Sweden, Mar 12, 17; nat US; m 44; c 2. MICROBIOLOGY. Educ: Univ Minn, BA, 45, MS, 47, PhD, 49. Prof Exp: From instr to asst prof bact, Univ Minn, 49-52; assoc prof, 52-55, PROF MICROBIOL, SCH MED, UNIV KANS, 55-, CHMN DEPT, 61- Mem: Am Soc Microbiol; Am Soc Exp Path. Res: Oncogenic viruses; studies on germfree animals. Mailing Add: Dept of Microbiol Univ of Kans Sch of Med Kansas City KS 66103

WERESUB, LUELLA KAYLA, b Russia, Mar 29, 18; nat Can. MYCOLOGY. Educ: Queen's Univ, Ont, BA, 50; Univ Toronto, MA, 52, PhD(mycol), 57. Prof Exp: Copy writer & announcer, Radio Sta CJIC, Can, 39-41 & Radio Sta CHML, 41-47; asst prof bot, Univ Manitoba, 52-55; MYCOLOGIST, RES BR, CAN DEPT AGR, 57- Mem: Mycol Soc Am; Can Bot Asn; Brit Mycol Soc; Int Asn Plant Taxon. Res: Taxonomy, with special reference to the Aphyllophorales; botanical nomenclature. Mailing Add: Biosystematics Res Inst Agr Can Cent Exp Farm Ottawa ON Can

WERGEDAL, JON E, b Eau Claire, Wis, Feb 19, 36. BIOCHEMISTRY. Educ: St Olaf Col, BA, 58; Univ Wis, MS, 60, PhD(biochem), 63. Prof Exp: Res instr, 63-71, RES ASST PROF MED, SCH MED, UNIV WASH, 71-; BIOCHEMIST, VET ADMIN HOSP, SEATTLE, 62- Mem: AAAS. Res: Bone metabolism. Mailing Add: Dept of Med Univ of Wash Sch of Med Seattle WA 98105

WERGIN, WILLIAM PETER, b Manitowoc, Wis, Apr 20, 42; m 62; c 2. CYTOLOGY. Educ: Univ Wis-Madison, BS, 64, PhD(bot), 70. Prof Exp: Res cytologist plant path, 70-72, res cytologist weed sci, 72-74, RES CYTOLOGIST NEMATOL & PROJ LEADER ANIMAL REPRODUCTION, AGR RES SERV, USDA, 74- Honors & Awards: Diamond Award, Bot Soc Am, 75. Mem: Am Soc Cell Biol; Bot Soc Am; Electron Micros Soc Am; Am Soc Plant Physiologists; Soc Nematologists. Res: Current transmission and scanning electron microscopic examinations include host-parasite interactions between higher plants and nematodes and problems of sperm transport and fetal mortality relating to animal reproduction. Mailing Add: Beltsville Agr Res Ctr-E Agr Res Serv USDA Beltsville MD 20705

WERKEMA, GEORGE JAN, b Vancouver, Wash, Nov 24, 36; m 69. PHYSICAL CHEMISTRY, HEALTH PHYSICS. Educ: Calvin Col, BS, 58; Univ Colo, PhD(chem), 65. Prof Exp: From res chemist to sr res chemist, Dow Chem Co, 63-72, res mgr, Rocky Flats Div, 72-75; HEALTH PHYSICIST, US ENERGY RES & DEVELOP ADMIN, 75- Mem: Am Chem Soc; Sigma Xi; Health Physics Soc. Res: X-ray crystallography; computer programming; instrument development; health physics; environmental control and monitoring; industrial hygiene. Mailing Add: 3504 Embudito Dr NE Albuquerque NM 87111

WERKEMA, MARILYN S, b Holland, Mich, July 28, 37; div. PHYSICAL CHEMISTRY. Educ: Calvin Col, AB, 54; Univ Colo, PhD(phys chem), 65. Prof Exp: Sr qual control engr, Martin Marietta Corp, 65-68; res chemist, Dow Chem USA, 68-75; RES & DEVELOP GROUP LEADER, ROCKWELL INT, 75- Mem: Am Soc Metals; Soc Women Engrs. Res: Metal fabrication and characterization, Pu, Be, U. Mailing Add: Box 888 Golden CO 80401

WERKHEISER, ARTHUR H, JR, b Easton, Pa, May 2, 35; m 62; c 2. LASERS. Educ: Lafayette Col, BS, 57; Univ Tenn, MS, 59, PhD(physics), 65. Prof Exp: Instr physics, Univ Tenn, 64; physicist, Res & Develop Directorate, 65-75, BR CHIEF, MODELING & ANAL DIV, ARMY HIGH ENERGY LASERS RES & ENG DIRECTORATE, US ARMY MISSILE COMMAND, 75- Mem: Am Phys Soc. Res: Isomer shifts of solid solutions; conduction electron polarization; Mössbauer effect physics; mathematical modeling of high energy laser systems. Mailing Add: Modeling & Anal Div Army High Energy Lasers Res & Eng Direct US Army Missile Command Redstone Arsenal AL 35809

WERKING, ROBERT JUNIOR, b Richmond, Ind, Jan 21, 30; m 51; c 2. SCIENCE EDUCATION. Educ: Hillsdale Col, BS, 61; Syracuse Univ, MS, 65; Ind Univ, Bloomington, EdD(sci educ), 71. Prof Exp: Asst prof physics, 65-69, asst prof sci educ, 71-73, CHMN DIV NATURAL SCI & MATH, MARION COL, 73- Mem: Am Asn Physics Teachers; Nat Sci Teachers Asn; Am Sci Affil. Res: Science instruction; instructional objectives; teaching effectiveness; instructional methods. Mailing Add: Div of Natural Sci & Math Marion Col Marion IN 46952

WERKMAN, SIDNEY LEE, b Washington, DC, May 3, 27; c 1. MEDICINE. Educ: Williams Col, AB, 48; Cornell Univ, MD, 52. Prof Exp: Assoc prof psychiat, Med Sch, George Washington Univ, 61-69; assoc prof, 69-72, PROF PSYCHIAT, UNIV COLO MED CTR, DENVER, 72- Concurrent Pos: Commonwealth Fund fel, Florence, Italy, 63-64; res consult, USPHS, 60-68; assoc dir, Joint Comn Ment Health Children, 67-68; consult, NIMH, 74- Mem: AAAS; Am Psychiat Asn; Am Acad Child Psychiat; Am Orthopsychiat Asn; Group Advan Psychiat. Res: Psychological factors in nutrition and development; brain dysfunction in children; attitude studies; psychological adjustment of Americans overseas. Mailing Add: Dept of Psychiat Univ of Colo Med Ctr Denver CO 80220

WERMAN, ROBERT, b Brooklyn, NY, May 2, 29; m 54; c 4. NEUROPHYSIOLOGY. Educ: NY Univ, AB, 48, MD, 52. Prof Exp: Intern, Montefiore Hosp, New York, 52-53; resident neurol, Mt Sinai Hosp, 53-54 & 56-58; asst prof, Col Physicians & Surgeons, Columbia Univ, 60-61; prof psychiat, Sch Med, Ind Univ, Indianapolis, 61-69, prof anat & physiol, Ind Univ Bloomington, 64-69; PROF NEUROPHYSIOL, HEBREW UNIV, JERUSALEM, 69- Concurrent Pos: Nat Inst Neurol Dis & Blindness trainee neurophysiol, Columbia Univ, 58-60; vis scientist, Cambride Univ, 60-61. Mem: AAAS. Res: Electrophysiology; neuromuscular junction; membrane properties and ionic movements; synaptic physiology. Mailing Add: Inst of Life Sci Hebrew Univ of Jerusalem Jerusalem Israel

WERMUND, EDMUND GERALD, JR, b Arlington, NJ, Apr 15, 26; m 51; c 2. GEOLOGY. Educ: Franklin & Marshall Col, BS, 48; La State Univ, PhD(geol), 61. Prof Exp: Instr geol, La State Univ, 52-57; sr res technologist, Field Res Lab, Mobil Oil Corp, 57-68, res assoc, Mobil Res & Develop Corp, 68-70; tech mgr, Remote Sensing, Inc, 70-71; res scientist, Bur Econ Geol, 71-73, ASSOC DIR, BUR ECON GEOL, UNIV TEX, AUSTIN, 73- Mem: AAAS; Am Asn Petrol Geologists; fel Geol Soc Am; Soc Econ Paleontologists & Mineralogists; Sigma Xi. Res: Petroleum geology, environmental geology; regional geology. Mailing Add: Box X Bur of Econ Geol Univ of Tex Austin TX 78712

WERMUS, GERALD R, b St Paul, Minn, May 8, 38; m 65; c 2. BIOCHEMISTRY, CLINICAL CHEMISTRY. Educ: Col St Thomas, BS, 60; Ariz State Univ, MS, 63; Iowa State Univ, PhD(biochem), 66. Prof Exp: Formulation chemist, Econ Lab, Inc, 60-61; res chemist, Photo Prod Dept, 61-72, res supvr, 72-74, RES MGR, PHOTO PROD DEPT, INSTRUMENT PROD DIV, E I DU PONT DE NEMOURS & CO, INC, 74- Concurrent Pos: Mem, Nat Comt Clin Lab Standards. Mem: Am Asn Clin Chem; Am Chem Soc; Asn Advan Med Instrumentation. Res: Development of clinical laboratory methodology; bacterial metabolism; lipid peroxidation; laboratory administration. Mailing Add: 114 Venus Dr Newark DE 19711

WERMUTH, JEROME FRANCIS, b Madison, Wis, Oct 19, 36; m 64; c 5. DEVELOPMENTAL BIOLOGY. Educ: Univ Wis-Madison, BS, 57, MS, 60; Ind Univ, PhD(zool), 68. Prof Exp: Instr biol, Rockhurst Col, 61-64; asst prof, St Joseph's Col, Ind, 65-66; res assoc, Univ Notre Dame, 68-69; ASST PROF BIOL, PURDUE UNIV, CALUMET CAMPUS, 76- Concurrent Pos: Vis asst prof & Nat Cancer Inst spec res fel life sci, Ind State Univ, Terre Haute, 72-73; vis asst prof zool, Ind Univ, Bloomington, 75-76. Mem: NY Acad Sci. Res: Effects of x-irradiation on the developmental physiology of colonial marine cnidarians. Mailing Add: Dept of Biol Purdue Univ Calumet Campus Hammond IN 46323

WERNEKE, MICHAEL FRANCIS, b Spartanburg, SC, Oct 7, 43; m 66; c 2. INORGANIC CHEMISTRY. Educ: St Bonaventure Univ, BS, 65; Clarkson Col Technol, PhD(inorg chem), 71. Prof Exp: Res chemist, 70-74, TECH SUPVR, AM CYANAMID CO, 74- Honors & Awards: Boris Pregel Award, NY Acad Sci, 71. Mem: Am Chem Soc; Am Inst Mining, Metall & Petrol Eng. Res: Chemistry and physics of mineral beneficiation and ore processing; synthetic, water-soluble polyelectrolytes for water purification; catalytic reactions of metals and metal complexes. Mailing Add: Am Cyanamid Co 1937 W Main St Stamford CT 06904

WERNER, ALLEN CARL, organic chemistry, see 12th edition

WERNER, CHRISTIAN, b Hamburg, WGer, July 31, 35; m 69. GEOGRAPHY. Educ: Free Univ Berlin, Fed Repub Ger, MA(math) & MA(geog), 60, MA(philos), 61, PhD(geog), 65. Prof Exp: Wissenschaftlicher asst geog, Free Univ Berlin, 65-66; from asst prof to assoc prof, Northwestern Univ, 66-69, fac assoc, Transp Ctr, 67-69; assoc prof, 69-72, PROF GEOG, UNIV CALIF, IRVINE, 72-, DEAN, SCH SOC SCI, 74- Concurrent Pos: NSF res grant, Univ Calif, Irvine, 70-72; reviewer var founds & proj jcurs; consult var agencies & corps; Wallace J Eckert vis scientist hydrol, T J Watson Res Ctr, IBM Corp, NY, 72-73. Mem: Asn Am Geog; Regional Sci Asn. Res: Applications of mathematics in geography; network theory; transportation theory. Mailing Add: Sch of Soc Sci Univ of Calif Irvine CA 92664

WERNER, ERVIN ROBERT, JR, b Philadelphia, Pa, May 2, 32; m 53; c 5. ORGANIC CHEMISTRY. Educ: Haverford Col, BS, 54; Univ Md, MS, 58; Univ Pa, PhD(org chem), 60. Prof Exp: Chemist, Rohm & Haas Co, 57-60; res chemist, 60-69, staff chemist, 69-72, RES ASSOC, E I DU PONT DE NEMOURS & CO, INC, 72- Mem: Am Chem Soc. Res: Organic synthesis; flame retardant polyesters; organophosphorus chemistry; water solution polymers; aqueous wire enamels; emulsion systems and house paints; interior emulsion paints. Mailing Add: E I du Pont de Nemours & Co Inc 3500 Grays Ferry Ave Philadelphia PA 19146

WERNER, FLOYD GERALD, b Ottawa, Ill, June 1, 21; m 52; c 3. ENTOMOLOGY. Educ: Harvard Univ, SB, 43, PhD(biol), 50. Prof Exp: From asst prof to assoc prof zool, Univ Vt, 50-54; in-chg econ insect surv, 54, asst prof entom, 54-58, assoc prof, 58-60, PROF ENTOM, UNIV ARIZ, 60-, ENTOMOLOGIST, EXP STA, 60- Mem: AAAS; Entom Soc Am; Soc Syst Zool. Res: Taxonomy of Coleoptera, Meloidae and Anthicidae; monitoring insect parasitoids in crop plants. Mailing Add: Dept of Entom Univ of Ariz Tucson AZ 85721

WERNER, GERHARD, b Vienna, Austria, Sept 28, 21; m 58. PHARMACOLOGY. Educ: Univ Vienna, MD, 45. Prof Exp: From instr to asst prof pharmacol, Univ Vienna, 43-52; prof, Univ Calcutta, 52-54 & Med Sch, Univ Sao Paulo, 55-57; assoc prof, Med Col, Cornell Univ, 57-61 & Med Sch, Johns Hopkins Univ, 61-65; chmn dept, 65-75, PROF PHARMACOL, SCH MED, UNIV PITTSBURGH, 65-, DEAN, SCH MED, 75- Concurrent Pos: Mem adv bd, Indian Coun Med Res, 52-54; mem pharmacol study sect, NIH, 65-69, mem chem-biol info handling prog, 68-71; NSF consult, 69-71. Mem: Am Soc Pharmacol & Exp Therapeut; Soc Exp Biol & Med; Am Physiol Soc; Int Brain Res Orgn; Asn Res Nerve & Ment Dis. Res: Neuropharmacology; neurophysiology; computer applications. Mailing Add: Dept of Pharmacol Univ of Pittsburgh Sch of Med Pittsburgh PA 15261

WERNER, HARRY JAY, b Brooklyn, NY, June 7, 21; m 43; c 3. GEOLOGY. Educ: Syracuse Univ, AB, 47, PhD(geol), 56; Washington Univ, AM, 49. Prof Exp: Field geologist, State Geol Surv, Va, 48-51; asst prof geol, St Lawrence Univ, 51-55; res geologist, Pan Am Petrol Corp, 55-60, res sect supvr, 60-63; ASSOC PROF GEOL, UNIV PITTSBURGH, 63- Mem: Geol Soc Am; Am Asn Petrol Geologists. Res: Stratigraphy; petrology; structural geology; carbonate sedimentation; underwater drilling techniques. Mailing Add: Dept of Geol Univ of Pittsburgh Pittsburgh PA 15213

WERNER, HENRY JAMES, b Hartford, Wis, Oct 9, 14; m 47. HISTOLOGY, CYTOLOGY. Educ: Marquette Univ, BS, 39, MS, 41; Univ Md, PhD(histol, cytol), 48. Prof Exp: Lab asst, Marquette Univ, 39-41; head dept sci, Shenandoah Col, 41-45; instr histol & cytol, Univ Md, 46-48; assoc prof, Otterbein Col, 48-49; assoc prof zool, 49-67, PROF ZOOL, LA STATE UNIV, BATON ROUGE, 67- Mem: Am Soc Cell Biol; Electron Micros Soc Am. Res: Electron microscopy; human fungi. Mailing Add: Dept of Zool La State Univ Baton Rouge LA 70803

WERNER, JOAN KATHLEEN, b Adams, Wis, Aug 13, 32. NEUROANATOMY, GROSS ANATOMY. Educ: Univ Wis, BS, 54, PhD(anat), 72. Prof Exp: Staff phys therapist, St Luke's Hosp, Milwaukee, Wis, 54-55; staff phys therapist, Vet Admin Hosp, Iron Mountain, 55-58; teaching asst anat, Med Sch, Univ Wis, 58-60; asst chief phys ther, Vet Admin Hosp, Madison, 60-65; chief phys ther, Vis Nurse Serv, Madison, 65-66; instr phys ther, 66-68, ASST PROF ANAT, MED SCH, UNIV WIS-MADISON, 72- Concurrent Pos: NIH fel, Univ Wis, 74-76; examr, State Wis Phys Ther Exam Coun, 74-77. Mem: AAAS; Am Asn Anatomists; Soc Neurosci; Am Phys Ther Asn. Res: Dependency of the somatosensory receptors upon normal innervation during development. Mailing Add: Dept of Anat 366 Bardeen Labs Univ of Wis Madison WI 53706

WERNER, JOHN KIRWIN, b Conrad, Mont, Sept 14, 41; m 67; c 2. HERPETOLOGY, ECOLOGY. Educ: Carroll Col, Mont, BA, 63; Univ Notre Dame, PhD(biol), 68. Prof Exp: Chief dept med zool, 406th Med Lab, US Med Command, Japan, 68-71; asst prof biol, 71-75, ASSOC PROF BIOL & HEAD DEPT, NORTHERN MICH UNIV, 75- Concurrent Pos: Assoc investr, US Army Res & Develop Protocol, 68-71; US Forest Serv grants amphibian ecol, 71-75. Mem: AAAS; Am Inst Biol Sci; Ecol Soc Am; Am Soc Ichthyologists & Herpetologists. Res: Ecology and reproduction of amphibians and reptiles. Mailing Add: 1303 Kimber Ave Marquette MI 49855

WERNER, LINCOLN HARVEY, b New York, NY, Feb 19, 18; m 44; c 2. ORGANIC CHEMISTRY. Educ: Swiss Fed Inst Technol, dipl, 41, DrTechSci, 44. Prof Exp: Asst, Swiss Fed Inst Technol, 44-46; res chemist, Ciba Basle, Ltd, 46-47, res chemist, Ciba Pharmaceut Prods, Inc, NJ, 47-68, MGR CHEM RES ADMIN, CIBA PHARMACEUT CO, 68- Mem: Am Chem Soc. Res: Pharmaceuticals; cardiovascular drugs; diuretics. Mailing Add: Pharmaceut Div Ciba-Geigy Corp 556 Morris Ave Summit NJ 07901

WERNER, MARIO, b Zurich, Switz, Aug 21, 31; US citizen; m 68. LABORATORY MEDICINE. Educ: Univ Zurich, MD, 56, DrMed, 60. Prof Exp: Asst physician, Kantonsspital St Gallen, Switz, 57-58 & Univ Basel, 58-61; NIH res fel metab, Swiss Acad Med Sci, 61-62; NIH res fel physiol, Rockefeller Univ, 62-64; head physician, Med Sch Essen, WGer, 64-66; asst prof lab med, Univ Calif, San Francisco, 67-70; assoc prof med path, Wash Univ, 70-72; PROF LAB MED, GEORGE WASHINGTON UNIV, 72- Concurrent Pos: Consult, Univ Tex M D Anderson Hosp & Tumor Inst, 70- & Nat Heart & Lung Inst, 73-; chmn health care delivery comt, Bi-State Regional Med Prog, 71-72; trustee, Found Interdisciplinary Biocharacterizations Pop, 71-75. Mem: Am Soc Clin Path; Col Am Path; Acad Clin Lab Physicians & Sci; Am Asn Clin Chem; Am Fedn Clin Res. Res: Laboratory data processing and diagnostic discrimination by computer; blood protein and lipoprotein metabolism. Mailing Add: Div of Lab Med George Washington Univ Med Ctr Washington DC 20037

WERNER, MICHAEL WOLOCK, b Chicago, Ill, Oct 9, 42; m 67; c 1. ASTROPHYSICS. Educ: Haverford Col, BA, 63; Cornell Univ, PhD(astron), 68. Prof Exp: Res physicist, US Naval Res Lab, 63-64; vis fel, Inst Theoret Astron, Cambridge, Eng, 68-69; lectr physics, Univ Calif, Berkeley, 69-72; ASST PROF PHYSICS, CALIF INST TECHNOL, 72- Mem: AAAS; Am Astron Soc. Res: Observational infrared astronomy; astrophysics of the interstellar medium. Mailing Add: Dept of Physics Calif Inst of Technol Pasadena CA 91109

WERNER, NEWTON DAVIS, organic chemistry, see 12th edition

WERNER, OSWALD, b Rimavska Sobota, Czech, Feb 26, 28; US citizen; m 57; c 3. ANTHROPOLOGY, LINGUISTICS. Educ: Stuttgart Tech Univ, BS, 50; Syracuse Univ, MA, 61; Univ Ind, PhD(anthrop, ling), 63. Prof Exp: From asst prof to assoc prof, 69-71, PROF ANTHROP & LING, NORTHWESTERN UNIV, EVANSTON, 71- Concurrent Pos: NIMH study grant, 63-; NSF fel, 64; mem cult anthrop fel comt, NIMH, 68-72. Mem: Am Anthrop Asn; Ling Soc Am. Res: Semantics; ethnoscience; Navaho Indians; artificial intelligence; medical translation lexicography. Mailing Add: Dept of Anthrop Col Arts & Sci Northwestern Univ Evanston IL 60201

WERNER, RAYMOND EDMUND, b Cincinnati, Ohio, Apr 18, 19; m 41; c 2. ORGANIC CHEMISTRY. Educ: Univ Cincinnati, ChE, 41, MS, 43, ScD(org chem), 45; Xavier Univ, Ohio, MBA, 69. Prof Exp: Res chemist, Interchem Corp, Ohio, 45-46; develop chemist, 46-55, dir develop, 55-75, VPRES RES & DEVELOP, TECH CTR, HILTON DAVIS DIV, STERLING DRUG, INC, 75- Mem: Am Chem Soc. Res: Administration of research and development in areas of fine organics, dyes, pharmaceuticals, pigments and graphic arts materials. Mailing Add: Tech Ctr Hilton Davis Div Sterling Drug Inc 2235 Langdon Farm Rd Cincinnati OH 45237

WERNER, RICHARD ALLEN, b Reading, Pa, Feb 20, 36. ENTOMOLOGY. Educ: Pa State Univ, BS, 58 & 60; Univ Md, MS, 66; NC State Univ, PhD(entom), 71. Prof Exp: Forester, 57-59, RES ENTOMOLOGIST, FORESTRY SCI LAB, US FOREST SERV, USDA, 60- Concurrent Pos: NIH res grant, Univ Md, 64-65. Mem: Soc Am Foresters; Entom Soc Am. Res: Aggregation behavior of bark beetles to host- and insect-produced chemicals; absorption, translocation and metabolism of systemic insecticides in loblolly pine; insecticide-forest ecosystem relationships. Mailing Add: Forestry Sci Lab USDA US Forest Serv Box 12254 Research Triangle Park NC 27709

WERNER, ROBERT GEORGE, b Plymouth, Ind, Mar 6, 36; m 58; c 2. ZOOLOGY. Educ: Purdue Univ, BS, 58; Univ Calif, Los Angeles, MA, 63; Ind Univ, PhD(zool), 66. Prof Exp: Asst prof zool, State Univ NY Col Forestry, Syracuse, 66-69; asst prof fisheries, Cornell Univ, 69-70; ASSOC PROF ZOOL, STATE UNIV NY COL ENVIRON SCI & FORESTRY, 70- Mem: Am Fisheries Soc; Am Soc Limnol & Oceanog; Ecol Soc Am. Res: Ecology of larval freshwater fish. Mailing Add: Dept Forest Zool State Univ NY Col Environ Sci & Forestry Syracuse NY 13210

WERNER, RUDOLF, b Königsberg, Ger, Dec 17, 34; m 61; c 2. CHEMISTRY. Educ: Univ Freiburg, Dipl(chem), 60, Dr rer nat(chem), 63. Prof Exp: Res asst, Inst Macromolecular Chem, Univ Freiburg, 61-65; res assoc, Carnegie Inst Genetics Res Unit, 65-67; sr staff investr, Cold Spring Harbor Lab Quant Biol, 68-70; ASSOC PROF BIOCHEM, SCH MED, UNIV MIAMI, 70- Concurrent Pos: Estab investr,

Am Heart Asn, 69-74. Mem: AAAS; Am Soc Biol Chemists. Res: Molecular biology; DNA replication. Mailing Add: Dept of Biochem Univ of Miami Sch of Med Miami FL 33152

WERNER, SAMUEL ALFRED, b Elgin, Ill, Jan 5, 37; m 61; c 1. SOLID STATE PHYSICS, NUCLEAR ENGINEERING. Educ: Dartmouth Col, AB, 59, MS, 61; Univ Mich, Ann Arbor, PhD(nuclear eng), 65. Prof Exp: Instr solid state physics & mat, Thayer Sch Eng, Dartmouth Col, 60-61; staff scientist, Physics Dept, Sci Lab, Ford Motor Co, 65-70, sr scientist, 70-75; PROF PHYSICS, UNIV MO-COLUMBIA, 75- Concurrent Pos: Consult, Argonne Nat Lab, 68-; adj assoc prof, Univ Mich, 69-; guest scientist, Aktiebolaget Atomenergi, Sweden, 70. Mem: AAAS; fel Am Phys Soc. Res: Neutron scattering; magnetism; physics of metals and alloys. Mailing Add: Dept of Physics Univ of Mo Columbia MO 65201

WERNER, SANFORD BENSON, b Newark, NJ, Jan 5, 39; m 68; c 2. PREVENTIVE MEDICINE. Educ: Rutgers Univ, BA, 60; Wash Univ, MD, 64; Univ Calif, MPH, 70. Prof Exp: Intern internal med, Vanderbilt Univ Hosp, Nashville, Tenn, 64-65; med epidemiologist, Epidemic Intel Serv, Ctr Dis Control, Ga, 65-67; resident internal med, Univ Wash Hosps, Seattle, 67-69; internist, Alaska Clin, Anchorage, 69; MED EPIDEMIOLOGIST INFECTIOUS DIS CONTROL, CALIF STATE DEPT HEALTH, 70- Concurrent Pos: Resident prev med, Sch Pub Health, Univ Calif, Berkeley, 70-71, lectr infectious dis epidemiol, 70- Mem: Fel Am Col Prev Med. Res: Epidemiology of botulism, typhoid, salmonellosis and other enteric diseases. Mailing Add: 2151 Berkeley Way Berkeley CA 94704

WERNER, SIDNEY CHARLES, b New York, NY, June 29, 09; m 47; c 4. MEDICINE. Educ: Columbia Univ, AB, 29, MD, 32, ScD(med, internal med), 37. Prof Exp: ASST MED, COL PHYSICIANS & SURGEONS, COLUMBIA UNIV, 34-PROF CLIN MED, 63- Concurrent Pos: From intern to attend physician, Presby Hosp, New York, 32-74, consult med, 74-; spec consult, Endocrine Study Sect, NIH, 52-55, endocrine panel, Nat Cancer Inst, 55-58, Cancer Chemother Serv Ctr, 58-62; Jacobaeus lectr, Finland, 68; consult, Var Hosps, NY & Conn. Honors & Awards: Wilson Award, 60; Stevens Triennial Award, 66; Distinguished Serv Award, Am Thyroid Asn, 69; Distinguished Thyroid Scientist, Int Thyroid Cong, 75. Mem: Fel AAAS; Am Thyroid Asn (pres, 72-73); Am Soc Clin Invest; Endocrine Soc; Asn Am Physicians. Res: Thyroid physiology and disease. Mailing Add: Presby Hosp 620 W 168th St New York NY 10032

WERNER, THOMAS CLYDE, b York, Pa, June 19, 42; m 65; c 2. ANALYTICAL CHEMISTRY. Educ: Juniata Col, BS, 64; Mass Inst Technol, PhD(anal chem), 69. Prof Exp: Fel chem, Harvard Med Sch & Mass Gen Hosp, Boston, 69-70; fel med sch, Tufts Univ, 70-71; ASST PROF CHEM, UNION COL, NY, 71- Mem: Am Chem Soc. Res: Application of absorption and luminescence spectroscopy to the study of molecular structure. Mailing Add: Dept of Chem Union Col Schenectady NY 12308

WERNER, WILLIAM ERNEST, JR, b Mt Marion, NY, June 30, 25; m 47; c 2. ECOLOGY. Educ: State Univ NY, BA, 50, MA, 51; Cornell Univ, PhD(mammal), 54. Prof Exp: Instr biol, State Univ NY Col Teachers, Albany, 51-52; PROF BIOL, BLACKBURN COL, 54- Concurrent Pos: Nat Heart Inst fel, Marine Inst, Miami, 63-64. Mem: AAAS; Am Soc Mammalogists; Am Soc Ichthyologists & Herpetologists; Ecol Soc Am; Sigma Xi. Res: Ecology of mammals, reptiles, amphibians and marine invertebrates. Mailing Add: Dept of Biol Blackburn Col Carlinville IL 62626

WERNICK, JOEL, b Newark, NJ, July 12, 39. AUDIOLOGY, SPEECH PATHOLOGY. Educ: Adelphi Univ, AB, 61; Univ Okla, MS, 64; Stanford Univ, PhD, 67. Prof Exp: Psychologist audition lab, Civil Aeromed Res Inst, 62-63; res assoc dept speech, Ind Univ, Bloomington, 63-64; NIH fel speech path & audiol, Med Ctr, Stanford Univ, 67; assoc neurobiol sci, Univ Fla, 67-68; ASST PROF SPEECH PATH & AUDIOL, UNIV IOWA, 68- Concurrent Pos: Res assoc, Commun Sci Lab, 67-68. Mem: Acoust Soc Am. Mailing Add: 127A/SHC Dept of Speech Path Univ of Iowa Iowa City IA 52240

WERNICK, ROBERT J, b New York, NY, June 19, 28; m 54; c 3. MATHEMATICS. Educ: Univ Mich, Ann Arbor, BSE, 49; Stevens Inst Technol, MS, 52; Rensselaer Polytech Inst, PhD(math), 69. Prof Exp: Res engr, Stevens Inst Technol, 51-53; tech asst, Panel Hydrodyn Submerged Bodies, Comt Undersea Warfare, Nat Res Coun, Washington, DC, 54; mathematician, Gen Eng Labs, Gen Elec Co, NY, 59-62 & Mech Technol, Inc, 62-64; asst prof math, State Univ NY Albany, 64-68; ASSOC PROF MATH, STATE UNIV NY OSWEGO, 69- Concurrent Pos: Consult, Advan Technol Labs, Gen Elec Co, NY, 63-66 & Mgch Technol, Inc, 65; referee, J Appl Mech, 69- Mem: Am Math Soc; Soc Indust & Appl Math; Math Asn Am. Res: Numerical analysis; game theory applied to determining the spectra of linear operators. Mailing Add: Dept of Math State Univ of NY Oswego NY 13126

WERNICK, WILLIAM, b New York, NY, Dec 18, 10; m 33; c 2. MATHEMATICS. Educ: NY Univ, BS, 33, MS, 34, PhD(math), 41. Prof Exp: Teacher, New York Bd Educ, 37-46, chmn dept math, 46-65; ASSOC PROF MATH, CITY COL NEW YORK, 65- Concurrent Pos: Lectr, Hunter Col, 46-50; assoc, Columbia Univ, 50-65. Mem: Am Math Soc; Math Asn Am; Asn Symbolic Logic. Res: Symbolic logic; general topology; calculus of operators; complete sets of logical functions; functional dependence in the calculus of propositions; distributive properties of set operators. Mailing Add: Dept of Math City Col of New York Convent Ave at 138th St New York NY 10031

WERNIMONT, GRANT (THEODORE), b Geneva, Nebr, Nov 30, 04; m 34; c 1. CHEMISTRY. Educ: Nebr Wesleyan Univ, AB, 27; Univ Utah, AM, 29; Purdue Univ, PhD(anal & inorg chem), 34. Prof Exp: Asst chem, Univ Utah, 27-29; asst, Purdue Univ, 29-34, instr, 34-35; anal & res chemist, Eastman Kodak Co, 35-69; vis prof chem, Purdue Univ, Lafayette, 69-71; CONSULT TO MANAGING DIR, AM SOC TESTING & MAT, 74- Concurrent Pos: Mem adv comt, Bur Med Devises & Diag Prod, Food & Drug Admin, 73-76. Honors & Awards: Am Soc Testing & Mat Award, 66. Mem: Am Chem Soc; Am Soc Qual Control. Res: Physical properties of liquid nitrous oxide; determination of the zirconium-hafnium ratio; chemical determination of water; determination of selenium in urine; electrometric titrations by the deadstop end point method; statistical quality control; electronic data processing. Mailing Add: 1736 Williamsburg Apts West Lafayette IN 47906

WERNSMAN, EARL ALLEN, b Vernon, Ill, Nov 4, 35; m 59; c 1. CROP BREEDING. Educ: Univ Ill, Urbana, BS, 58, MS, 60; Purdue Univ, PhD(genetics), 63. Prof Exp: Res assoc genetics, Iowa State Univ, 63-64; from asst prof to assoc prof, 64-72, PROF GENETICS, NC STATE UNIV, 72- Mem: Am Soc Agron; Am Genetic Asn; Genetics Soc Can. Res: Cytogenetics of Lotus; genetics of Nicotiana and alkaloid production; physiology of cytoplasmic male-sterility in Nicotiana; interspecific hybridization in Nicotiana. Mailing Add: Dept of Crop Sci NC State Univ Raleigh NC 27607

WERNSTEDT, FREDERICK LAGE, b Portland, Ore, Feb 8, 21; m 48; c 2. GEOGRAPHY OF SOUTHEAST ASIA. Educ: Univ Calif, Los Angeles, BA, 48, PhD(geog), 53; Syracuse Univ, MA, 50. Prof Exp: From asst prof to assoc prof, 52-67, PROF GEOG, PA STATE UNIV, 67- Concurrent Pos: NSF grant, 64-67; Fulbright Found grant, Univ Malaya, 68-69; assoc prof, Ariz State Univ, 61-62. Res: Climatology; population and migrations in Southeast Asia. Mailing Add: Dept of Geog Pa State Univ University Park PA 16802

WERNTZ, CARL H, b Washington, DC, Aug 7, 31; m 58; c 2. THEORETICAL NUCLEAR PHYSICS, ASTROPHYSICS. Educ: George Washington Univ, BS, 53; Univ Minn, MS, 55, PhD(physics), 60. Prof Exp: Res assoc physics, Univ Wis, 60-62; assoc prof, 62-70, PROF PHYSICS, CATH UNIV AM, 70-, CHMN DEPT, 75- Mem: Am Phys Soc. Res: Theory of interaction of pions with nuclei; study of solar nuclear reactions. Mailing Add: Dept of Physics Cath Univ of Am Washington DC 20017

WERNTZ, HENRY OSCAR, b Atlantic City, NJ, Jan 19, 30; m 57; c 3. ZOOLOGY. Educ: Rutgers Univ, BS, 52; Yale Univ, PhD(zool), 58. Prof Exp: Instr biol, Harvard Univ, 57-60; asst prof biol, 60-64, actg chmn dept, 66-67, ASSOC PROF BIOL, NORTHEASTERN UNIV, 64-, CHMN DEPT, 75- Mem: AAAS; Am Soc Zoologists. Res: Osmotic and ionic regulation in invertebrates; environmental physiology. Mailing Add: Dept of Biol Northeastern Univ Boston MA 02115

WERNTZ, JAMES HERBERT, JR, b Wilmington, Del, Sept 3, 28; m 55; c 4. PHYSICS. Educ: Oberlin Col, BA, 50; Univ Wis, MS, 52, PhD(physics), 57. Prof Exp: From asst prof to assoc prof physics, 56-68, PROF PHYSICS, UNIV MINN, MINNEAPOLIS, 68-, DIR CTR EDUC DEVELOP, 67- Mem: Am Phys Soc; Am Asn Physics Teachers. Res: Physics of very low temperatures, especially liquid helium phenomena; science education. Mailing Add: Sch of Physics & Astron Univ of Minn Minneapolis MN 55455

WERNY, FRANK, b Ger, June 6, 36; US citizen; wid; c 3. NATURAL PRODUCTS CHEMISTRY, TEXTILE TECHNOLOGY. Educ: Univ Puget Sound, BS, 58; Univ Hawaii, PhD(org chem), 63. Prof Exp: Ford Found fel, Free Univ Berlin, 62-63; res chemist, Carothers Res Lab, 63-67, sr res chemist, Textile Res Lab, 67-72, SR RES CHEMIST, CHRISTINA LABS, E I DU PONT DE NEMOURS & CO, INC, 72- Mem: Am Chem Soc. Res: Polyamide polymers and fibers; flammability of polymers and textiles; irradiation grafting and melt blending of polymers; polymer rheology; carpet yarns and carpets; nonwoven fabric research and development. Mailing Add: 3305 Hermitage Rd Devonshire Wilmington DE 19810

WERSHAW, ROBERT LAWRENCE, b Norwalk, Conn, Sept 17, 35; m 63; c 2. HYDROLOGY, GEOCHEMISTRY. Educ: Tex Western Col, BS, 57; Calif Inst Technol, MS, 59; Univ Tex, PhD(geol), 63. Prof Exp: HYDROLOGIST, US GEOL SURV, 63- Mem: Geochem Soc; Am Chem Soc. Res: Geochemistry of naturally occurring polyelectrolytes and their interaction with water pollutants. Mailing Add: US Geol Surv Rm 34 Bldg 56 Fed Ctr Denver CO 80225

WERT, CHARLES ALLEN, b Battle Creek, Iowa, Dec 31, 19; m 42; c 2. PHYSICS. Educ: Morningside Col, BA, 41; Univ Iowa, MS, 43, PhD(physics), 48. Prof Exp: Mem staff radiation lab, Mass Inst Technol, 43-45; asst, Univ Iowa, 46-48; instr inst study metals, Univ Chicago, 48-50; from assoc prof to prof mining & metall eng, 50-72, PROF METALL & MINING ENG, UNIV ILL, URBANA, 72-, HEAD DEPT, 66- Mem: Am Phys Soc; Am Inst Mining, Metall & Petrol Engrs. Res: Internal friction of metals; diffusion in solids; alloying behavior of metals with carbon, nitrogen, oxygen and carbon. Mailing Add: Dept of Metall & Mining Eng Univ of Ill Urbana IL 61801

WERT, JONATHAN MAXWELL, JR, b Port Royal, Pa, Nov 8, 39; m 62; c 3. ENVIRONMENTAL SCIENCES. Educ: Austin Peay State Univ, BS, 66, MS, 68; Univ Ala, PhD(educ planning), 74. Prof Exp: Chief naturalist, Pa Dept Environ Resources, 68-69; chief naturalist, Bays Mountain Park, 69-71; supvr environ educ sect, Tenn Valley Authority, 71-75; CONSULT ENERGY CONSERV, ENVIRON CTR, UNIV TENN, KNOXVILLE, 75- Concurrent Pos: Mem adv bd, Environ Educ Report, Inc, 74-; mem nat adv coun environ educ, Dept Health, Educ & Welfare, 75-78. Mem: Conserv Educ Asn; Int Soc Educ Planners; Am Soc Ecol Educ; Nat Audubon Soc; Nat Educ Asn. Res: Development of methodologies for carrying out comprehensive environmental planning, including technological assessment, cost-benefit analysis and environmental assessment; energy conservation; systems planning. Mailing Add: Environ Ctr S Stadium Hall Univ of Tenn Knoxville TN 37916

WERT, RICHARD THOMAS, b Berwyn, Ill, Nov 30, 38; m 63; c 1. PHYSICAL OCEANOGRAPHY. Educ: Univ Calif, Berkeley, BS, 61; Tex A&M Univ, MS, 68, PhD(oceanog), 70. Prof Exp: Engr aeronutronic div, Ford Motor Co, 61-64 & NAm Aviation, 64-65; engr aeronutronic div, Ford Motor Co, 65-66; sci officer oceanog, Off Naval Res, 70-72; ACAD ADMINR, SCRIPPS INST OCEANOG, 72- Res: Academic administration; data management. Mailing Add: 13654 Calais Dr Del Mar CA 92014

WERTH, GLENN CONRAD, b Denver, Colo, July 21, 26; m 50; c 3. PHYSICS. Educ: Univ Colo, BS, 49; Univ Calif, Los Angeles, MS, 50, PhD(physics, acoustics), 53. Prof Exp: Physicist, Arthur D Little, Inc, 53-54 & Calif Res Corp, 54-59; physicist, 59-66, ASSOC DIR, LAWRENCE LIVERMORE LAB, UNIV CALIF, 66- Mem: Acoust Soc Am; Soc Explor Geophys; Am Geophys Union. Res: Seismology; geophysical exploration for oil; use of nuclear explosives in industry and commerce; development of alternate energy supplies. Mailing Add: 2062 Bldg 121 Lawrence Livermore Lab Univ of Calif Livermore CA 94551

WERTH, JEAN MARIE, b Rochester, NY, Jan 21, 43. MICROBIAL PHYSIOLOGY. Educ: Nazareth Col Rochester, BS, 64; Syracuse Univ, MS, 69, PhD(microbiol), 73. Prof Exp: ASST PROF BIOL, WILLIAM PATERSON COL, 72- Mem: AAAS; Am Soc Microbiol; Sigma Xi; Am Inst Biol Sci. Res: Effects of antibiotics and hormones on growth and extracellular lysin production in strains of Streptococcus zymogenes and various soil bacteria; biochemical identification of the hemolysis inhibitor released by antibiotic treated cells. Mailing Add: Dept of Biol William Paterson Col Wayne NJ 07470

WERTH, JOHN ST CLAIR, JR, b Bluefield, WVa, Nov 22, 40; m 61; c 2. MATHEMATICS. Educ: Emory Univ, BS, 62, MS, 63; Univ Wash, PhD(math), 68. Prof Exp: Lectr math, Emory Univ, 62-63; res instr, NMex State Univ, 68-69, asst prof, 69-74; ASST PROF MATH, UNIV NEV, 74- Concurrent Pos: Reviewer, Math Rev, 71-72. Mem: Am Math Soc. Res: Abelian groups; ring theory. Mailing Add: Dept of Math Univ of Nev Las Vegas NV 89109

WERTH, RICHARD GEORGE, b Markesan, Wis, Feb 5, 20; m 43; c 1. ORGANIC CHEMISTRY. Educ: Wartburg Col, BA, 42; Univ Wis, MS, 48, PhD, 50. Prof Exp: Jr chemist, Electrochem Dept, E I du Pont de Nemours & Co, 42-44 & 46; Alumni Res Found asst, Univ Wis, 48-50; assoc prof chem, 50-53, head dept, 61-69, PROF CHEM, CONCORDIA COL, MOORHEAD, MINN, 53-, CHMN DEPT, 74-

Concurrent Pos: Res partic, Oak Ridge Nat Lab, 62-63; vis fel, Cornell Univ, 70-71. Mem: AAAS; Am Chem Soc; Am Inst Chemists; Soc Appl Spectros. Res: Methods of organic synthesis of polycyclic compounds; formaldehyde products; reaction mechanism for dehydration of bicycloheptanediols; mass spectra of vinylogous imides. Mailing Add: 1207 S Seventh St Moorhead MN 56560

WERTH, ROBERT JOSEPH, b Hays, Kans, Apr 4, 40; m 67; c 1. ZOOLOGY. Educ: St Benedict's Col, Kans, BS, 61; Univ Mo-Kansas City, MS, 65; Univ Colo, Boulder, PhD(zool), 69. Prof Exp: ASST PROF BIOL, PURDUE UNIV, CALUMET CAMPUS, 69- Mem: AAAS; Am Soc Ichthyologists & Herpetologists. Res: Reptilian ecology and physiology; correlation of environmental requirement with physiological adaptation. Mailing Add: Dept of Biol Purdue Univ Calumet Campus Hammond IN 46323

WERTHAMER, NATHAN RICHARD, physics, see 12th edition

WERTHAMER, SEYMOUR, b New York, NY, July 12, 24; m 44; c 3. PATHOLOGY, COMPUTER SCIENCE. Educ: NY Univ, BA, 45; Chicago Med Sch, MB & MD, 49. Prof Exp: Cancer res fel, Univ Berne, 52-53; pathologist, Mt Sinai Hosp, Hartford, Conn, 54; asst pathologist, Beth Israel Hosp, New York, 55; dir labs, Paul Kimball Hosp, Lakewood, NJ, 57-63; dir clin path, Brooklyn-Cumberland Med Ctr, 63-66; DIR PATH & CLIN LABS, METHODIST HOSP BROOKLYN, 66- Concurrent Pos: Instr, Seton Hall Col, 57-63; county physician, Ocean County, 58-61; consult pathologist, Deborah Hosp, NJ, 59-63; from asst clin prof to assoc clin prof, State Univ NY Downstate Med Ctr, 63-73, clin prof path, 73- Mem: Fel Col Am Path; fel Am Soc Clin Path; fel Am Col Physicians; Am Soc Exp Path; Am Asn Path & Bact. Res: Computer system design and application to medical communication and information; mechanisms of protein synthesis in leucocytes in vitro. Mailing Add: Dept of Path Methodist Hosp of Brooklyn Brooklyn NY 11215

WERTHEIM, ARTHUR ROBERT, b Newark, NJ, July 5, 15; m 47. MEDICINE. Educ: Dartmouth Col, AB, 35; Jefferson Med Col, MD, 39; Am Bd Internal Med, dipl, 50. Prof Exp: Intern, Philadelphia Gen Hosp, 39-41, resident, 41-43; resident, Goldwater Mem Hosp, 46-47; asst, Long Island Col Med, 48; instr, Col Med, State Univ NY Downstate Med Ctr, 49-51; from asst prof to assoc prof, 51-69, PROF MED, COL PHYSICIANS & SURGEONS, COLUMBIA UNIV, 69- Concurrent Pos: Res fel, Goldwater Mem Hosp, 47; asst med, Col Physicians & Surgeons, Columbia Univ, 47; from clin assoc to dir med serv, Maimonides Hosp, Brooklyn, 48-51; res assoc, Goldwater Mem Hosp, 51-54, vis physician, 54-68; actg chief, Med Serv, Francis Delafield Hosp, New York, 71-74, chief, 74-75; attending physician, Presby Hosp, NY. Mem: Harvey Soc; AMA; fel Am Col Physicians. Res: Clinical and laboratory research in chronic diseases; metabolic aspects of hypertension and atherosclerosis. Mailing Add: Col of Physicians & Surgeons Columbia Univ 630 W 168th St New York NY 10032

WERTHEIM, GUNTHER KLAUS, b Berlin, Ger, Feb 26, 27; nat US; m 56; c 3. SOLID STATE PHYSICS. Educ: Stevens Inst Technol, ME, 51; Harvard Univ, AM, 52, PhD(physics), 55. Prof Exp: Assoc phys oceanog, Woods Hole Oceanog Inst, 54; mem tech staff, Bell Tel Labs, Inc, 55-62; HEAD CRYSTAL PHYSICS RES DEPT, BELL LABS, 62- Concurrent Pos: Adj prof, Stevens Inst Technol, 66-68; mem vis comts physics, Harvard Univ, 69-75 & Bartol Res Found, 72- Mem: Am Phys Soc; Sigma Xi. Res: X-ray photo-electron spectroscopy; Mössbauer effect; angular correlations; magnetism; semiconductors. Mailing Add: Bell Labs Murray Hill NJ 07974

WERTHEIM, MICHAEL STEPHEN, physics, see 12th edition

WERTHEIM, ROBERT HALLEY, b Carlsbad, NMex, Nov 9, 22; m 46; c 2. NUCLEAR PHYSICS. Educ: US Naval Acad, BS, 45; Mass Inst Technol, SM, 54. Prof Exp: US Navy, 45-, staff officer, Armed Forces Spec Weapons Proj, 48-49, spec asst nuclear applns, Spec Projs Off, 56, head re-entry body sect, 56-61, weapons develop dept, Naval Ord Test Sta, 61-62, mem staff, Dir of Defense Res & Eng, 62-65, head missile br, Spec Projs Off, 65-67, dep tech dir, 67-68, TECH DIR, SPEC PROJS OFF, US NAVY, 68- Honors & Awards: William S Parsons Award, 71; Legion of Merit, 72; Gold Medal, Am Soc Naval Engrs, 73. Mem: Am Soc Naval Engrs; assoc fel Am Inst Aeronaut & Astronaut. Res: Weapons systems research, development and testing, especially ballistic missiles and nuclear applications. Mailing Add: 4105 Parkedge Ln Annandale VA 22003

WERTHEIMER, ALAN LEE, b Cleveland, Ohio, Dec 22, 46; m 69; c 1. OPTICS. Educ: Univ Rochester, BS, 68, PhD(optics), 74. Prof Exp: Optical designer lens design, Itek Corp, 68-69; SR SCIENTIST OPTICS RES, LEEDS & NORTHRUP CO, 74- Mem: Optical Soc Am; Soc Photo-Optical Instrumentation Eng. Res: Use of light scattering from small particles to analyze particle distributions in the one to one hundred micron size range; light scattering to detect bacterial contamination in medical fluids. Mailing Add: Leeds & Northrup Co Tech Ctr North Wales PA 19454

WERTHEIMER, ALBERT I, b Buffalo, NY, Sept 14, 42; m 65; c 2. PUBLIC HEALTH ADMINISTRATION. Educ: Univ Buffalo, BS, 65; State Univ NY Buffalo, MBA, 67; Purdue Univ, PhD(med sociol, pharm), 69. Prof Exp: Asst prof pharm, Sch Pharm, State Univ NY Buffalo, 69-71, asst prof mkt, Sch Mgt, 70-71; researcher med care, Social Security Admin, Dept Health, Educ & Welfare, 72 & USPHS 72-73; ASSOC PROF PHARM ADMIN & DIR GRAD PROG, UNIV MINN, MINNEAPOLIS, 73- Concurrent Pos: WHO fel, 75. Mem: Am Pub Health Asn; Am Pharmaceut Asn; Am Sociol Asn; AMA. Res: Drug use process; drug ecology studies. Mailing Add: Col of Pharm Univ of Minn Minneapolis MN 55455

WERTHEIMER, FREDERICK WILLIAM, b Saginaw, Mich, May 21, 25; m 56; c 3. PERIODONTOLOGY, ORAL PATHOLOGY. Educ: Univ Mich, DDS, 49; Northwestern Univ, MSD, 54; Georgetown Univ, MS, 60; Am Bd Periodont, dipl. Prof Exp: Pvt practr, 52-53 & 54-58; investr dent, Nat Inst Dent Res, 60-61; sr assoc periodont, Henry Ford Hosp, Detroit, 61-68; assoc prof path, 65-68, PROF PATH & CHMN DEPT, DENT SCH, UNIV DETROIT, 68- Concurrent Pos: NIH res & training grants, 65-; adj assoc prof, Med Sch, Wayne State Univ; consult, Vet Admin Hosp & Detroit Gen Hosp, Detroit. Mem: AAAS; fel Am Acad Oral Path; fel Am Col Dent; fel Int Col Dent; Am Acad Periodont. Res: Histochemistry of periodontal disease and oral neoplasia. Mailing Add: Dept of Path Univ of Detroit Dent Sch Detroit MI 48207

WERTHESSEN, NICHOLAS THEODORE, b New York, Nov 12, 11; m; c 5. ENDOCRINOLOGY. Educ: Harvard Univ, BS, 33, PhD(endocrinol), 37. Prof Exp: Res assoc, Cambridge & Clark Univs, 37-39; dir, Endocrine Lab, New Eng Med Ctr, 39-45; mem staff, Worcester Found, 45-50; chmn dept physiol & biol, Southwest Found Res & Educ, 50-62; assoc res prof obstet & gynec & assoc prof physiol, Sch Med, Univ Okla, 62-65; SR BIOSCIENTIST, OFF NAVAL RES, 65- Concurrent Pos: Res assoc, Civil Aeronaut Admin, Nat Res Coun & Works Progress Admin, Clark Univ, 42; dir fertil clin lab, Beth Israel Hosp, 44-45; mem coun arteriosclerosis, Am Heart Asn. Mem: AAAS; fel Am Physiol Soc; Am Asn Cancer Res; Biomet Soc;

Am Fertil Soc. Res: Steroid hormone metabolism in normal and malignant tissues; lipid metabolism in arterial wall. Mailing Add: Off of Naval Res 495 Summer St Boston MA 02210

WERTWIJN, GEORGE, b Amsterdam, Neth, June 2, 30; nat US; m 53; c 2. PHYSICS, CHEMISTRY. Educ: Univ Amsterdam, BS, 50, Nat Phil Drs, 52; Univ Chicago, MBA, 65. Prof Exp: With N V Phillips Lamp Works, Neth, 53-57, Zenith Radio Corp, Ill, 57-60 & Fansteel Metall Corp, 61-63; vpres & gen mgr, Scully-Int, Inc, Bendix Corp, 63-70; EXEC VPRES, CHEMAPERM MAGNETICS, INC, 72- Mem: Fel Am Inst Chemists; sr mem Inst Elec & Electronics Eng; Royal Neth Acad Sci. Res: High pressure physics and thermodynamics; semiconductor, solid state and antifriction devices. Mailing Add: RR 1 Box 206B Elburn IL 60119

WERTZ, DAVID LEE, b Hammond, Ind, Feb 3, 40; m 61; c 2. INORGANIC CHEMISTRY, PHYSICAL CHEMISTRY. Educ: Ark State Univ, BS, 62; Univ Ark, PhD(chem), 66. Prof Exp: From asst prof to assoc prof, 66-74, PROF CHEM, UNIV SOUTHERN MISS, 74- Mem: Am Chem Soc; Am Phys Soc. Res: X-ray diffraction and spectral studies of solute-solvent interactions in concentrated solutions; x-ray diffraction studies of liquid structure. Mailing Add: Dept of Chem Univ of Southern Miss SS Box 464 Hattiesburg MS 39401

WERTZ, DENNIS WILLIAM, b Reading, Pa, Mar 21, 42; m 62; c 2. PHYSICAL CHEMISTRY. Educ: Univ Md, BS, 64; Univ SC, PhD(chem), 68. Prof Exp: Res assoc spectros, Mass Inst Technol, 68-69; ASST PROF CHEM, NC STATE UNIV, 69- Mem: Am Chem Soc; Am Phys Soc. Res: Spectroscopic investigations of small ring molecules; infrared and Raman investigations of organometallic compounds. Mailing Add: Dept of Chem NC State Univ Raleigh NC 27607

WERTZ, GAIL T WILLIAMS, b Washington, DC, Oct 31, 43; m 66. VIROLOGY. Educ: Col William & Mary, BS, 66; Univ Pittsburgh, PhD(microbiol), 70. Prof Exp: NIH fel, Med Sch, Univ Mich, Ann Arbor, 70-71; sr res assoc virol, 71-73; ASST PROF BACT & IMMUNOL, MED SCH, UNIV NC, CHAPEL HILL, 73- Concurrent Pos: NSF & NIH res grants. Mem: AAAS; Am Soc Microbiol; Am Tissue Cult Asb; Sigma Xi. Res: Virus-host interactions; viral replication; mechanism of viral nucleic acid synthesis and replication; effect of virus on host macromolecular synthesis and host response to infection-interference phenomenon. Mailing Add: 608 Laurel Hill Rd Chapel Hill NC 27514

WERTZ, JAMES RICHARD, b Kingman, Ariz, Feb 20, 44; m 67; c 1. THEORETICAL ASTROPHYSICS, ASTRONAUTICS. Educ: Mass Inst Technol, SB, 66; Univ Tex, Austin, PhD(physics), 70. Prof Exp: Asst prof physics & astron, Moorhead State Col, 70-73, NSF inst grant, 71-73; SR ANALYST, COMPUTER SCI CORP, 73- Concurrent Pos: Mem task group educ astron, Am Astron Soc, 74- Mem: AAAS; Brit Interplanetary Soc; Am Phys Soc; Am Astron Soc. Res: Spacecraft attitude determination; interstellar travel and navigation; hierarchical cosmology. Mailing Add: Computer Sci Corp 8728 Colesville Rd Silver Spring MD 20910

WERTZ, JOHN EDWARD, b Denver, Colo, Dec 4, 16; m 43; c 3. CHEMISTRY. Educ: Univ Denver, BS, 37, MS, 38; Univ Chicago, PhD(chem), 48. Prof Exp: Asst prof chem, Augustana Col, 41-44; instr physics, Gustavus Adolphus Col, 44-45; instr mech eng, 45-47, from asst prof to prof phys chem, 48-74, PROF CHEM, UNIV MINN, MINNEAPOLIS, 74- Concurrent Pos: Fulbright res scholar & Guggenheim fel, Clarendon Lab, Oxford Univ, 57-58. Mem: Fel Am Phys Soc; Brit Inst Physics; Am Chem Soc; The Chem Soc. Res: Fluorescence of crystals; surface energy of solids; adsorption of vapors on solids; magnetic susceptibility of adsorbed layers; nuclear and paramagnetic resonance; structure and electronic properties of defects in solids; infrared detectors. Mailing Add: Dept of Chem Univ of Minn Minneapolis MN 55455

WESCHE, ROLF JUERGEN, b Hamburg, Ger, Feb 8, 41; m 66. GEOGRAPHY. Educ: Univ NC, MA, 64; Univ Fla, PhD(geog), 67. Prof Exp: Asst prof geog, 67-72, secy dept, 69-71, interim chmn dept, 71-72, ASSOC PROF GEOG, UNIV OTTAWA, 72- Concurrent Pos: Can Coun res grant, Eastern Peru, 69 & Eastern Columbia, 70-71; Can Coun leave fel, Colombia, Peru & Brazil, 73-74. Mem: Asn Am Geog; Can Asn Geog; Conf Latin Am Geog. Res: Economic development of Andean countries, specifically human migration, agricultural colonization and regional development in Upper Amazonia; internal migration; migrant adaptation and organizational forms of frontier colonization in the Amazon rainforest. Mailing Add: Dept of Geog Univ of Ottawa Ottawa ON Can

WESCHLER, CHARLES JOHN, b Youngstown, Ohio, Jan 29, 48; m 71. PHYSICAL INORGANIC CHEMISTRY. Educ: Boston Col, BS, 69; Univ Chicago, MS, 72, PhD(chem), 74. Prof Exp: Fel phys inorg chem, Northwestern Univ, 74-75; MEM TECH STAFF CHEM, BELL LABS, 75- Concurrent Pos: Lab grad partic, Argonne Nat Lab, 72-73. Mem: Am Chem Soc. Res: Dust studies, with particular emphasis on the role dust plays in corrosion processes; redox chemistry; synthetic oxygen carriers. Mailing Add: Bell Labs Crawfords Corner Rd Holmdel NJ 07733

WESCHLER, JOSEPH ROBERT, b Erie, Pa, Oct 3, 25; m 71; c 5. ORGANIC CHEMISTRY. Educ: Gannon Col, BS, 49; Case Western Reserve Univ, MS, 53, PhD(chem), 55. Prof Exp: Chemist, Permacel Div, Johnson & Johnson, 55-59; dir mkt, Ciba Prod Corp, 59-71; mgr tech mkt, 71-73, MGR AUTOMOTIVE FINISHES, SHERWIN WILLIAMS CO, 73- Mem: Am Chem Soc; Am Inst Chemists. Res: Coatings; plastics; adhesives. Mailing Add: 1005 Balfour Rd Grosse Pointe MI 48230

WESCHLER, MARY CHARLES, b Erie, Pa, Feb 22, 20. ENVIRONMENTAL CHEMISTRY. Educ: Mercyhurst Col, BS, 40; Univ Notre Dame, MS, 51; Carnegie Inst Technol, PhD(phys chem), 56. Prof Exp: Teacher, Erie Diocesan High Schs, 40-46; instr physics & chem, 46-63, chmn div natural sci & math, 63-75, DIR CHEM DEPT, MERCYHURST COL, 75- Concurrent Pos: Petrol Res Fund grant, 58-62; dir, NSF res partic grant, 65-66; NSF fac fel, Argonne Nat Lab, 70-71. Mem: Am Chem Soc. Res: Kinetics of hydrolysis of benzene polycarboxylic esters. Mailing Add: Div of Natural Sci & Math Mercyhurst Col Erie PA 16501

WESCOE, W CLARKE, b Allentown, Pa, May 3, 20; m 44; c 3. PHARMACOLOGY, EXPERIMENTAL MEDICINE. Educ: Muhlenberg Col, BS, 41; Cornell Univ, MD, 44. Prof Exp: Intern med, New York Hosp, 44-45, asst resident physician, 45-46; from instr to asst prof pharmacol, Med Col, Cornell Univ, 48-51; prof, Sch Med, Univ Kans, 51-69, dean sch med, 52-60, dir med ctr, 53-60, chancellor, 60-69; vpres med affairs, 69, vpres corp, 69-71; exec vpres res & med affairs, 71-74, CHMN, BD DIRS, STERLING DRUG, INC, 74- Concurrent Pos: Markle scholar med sci, 49-54; mem, Coun Med Educ, AMA, 57-67, chmn, 63-67. Mem: Fel AAAS; fel Am Col Physicians; Am Soc Pharmacol & Exp Therapeut; AMA. Res: Neuromuscular transmission and drugs affecting it. Mailing Add: Sterling Drugs Inc 90 Park Ave New York NY 10016

WESCOTT, ELBERT WAYNE, botany, see 12th edition

WESCOTT, LYLE DUMOND, JR, b Hackensack, NJ, Jan 27, 37; m 59; c 2. ORGANIC CHEMISTRY. Educ: Ga Inst Technol, BS, 59; Pa State Univ, PhD(chem), 63. Prof Exp: Res fel, Pa State Univ, 63-64; res chemist, Baytown Labs, Esso Res & Eng Co, Tex, 64-68; asst prof chem, 68-71, ASSOC PROF CHEM, CHRISTIAN BROS COL, 71-, HEAD DEPT, 74- Mem: Am Chem Soc; The Chem Soc. Res: Low temperature reactions of vapor species of refractory materials; stabilization of polymeric materials. Mailing Add: Dept of Chem Christian Bros Col 650 E Pkwy S Memphis TN 38104

WESCOTT, MASON EATON, mathematics, see 12th edition

WESCOTT, RICHARD BRESLICH, b Chicago, Ill, July 8, 32; m 54; c 3. VETERINARY PARASITOLOGY. Educ: Univ Wis, BS, 54, MS, 64, PhD(vet med), 65; Univ Minn, DVM, 58; Am Col Lab Animal Med, dipl, 67. Prof Exp: USPHS fel parasitol, Univ Wis, 62-65; assoc prof vet microbiol, Sch Vet Med, Univ Mo-Columbia, 65-71; PROF VET PATH, WASH STATE UNIV, 71-, ACTG CHMN DEPT, 75- Mem: Am Vet Med Asn. Res: Laboratory animal medicine; nematode-virus interactions in gnotobiotic host animals. Mailing Add: Dept of Vet Path Wash State Univ Pullman WA 99163

WESCOTT, WILLIAM B, b Pendleton, Ore, Nov 10, 22; m 69; c 2. ORAL PATHOLOGY. Educ: Univ Ore, DDS, 51, MS, 62. Prof Exp: Dir oral tumor registry, Dent Sch, Univ Ore, 53-54; pvt practr gen dent, Ore, 53-59; asst gen path, Dent Sch, Univ Ore, 59-60; from asst prof to assoc prof path, 62-69, clin assoc oral & dent med, 66-67, sr clin instr, 567-69, prof path & assoc dean admin affairs, 69-72; co-dir res & educ training prog, Oral Dis Res Lab, Vet Admin Hosp, Houston, 72-75; CHIEF DENT SERV, VET ADMIN HOSP, DURHAM, NC, 75- Concurrent Pos: Am Cancer Soc fel, 59-61; USPHS fel, 61-63; prof path, Univ Tex Dent Br, Houston, 72-75. Mem: Am Dent Asn; Int Asn Dent Res; fel Am Acad Oral Path; Int Asn Microbiol; fel Am Col Dent. Res: Correlation of clinical and histopathologic findings; fluorescent antibody technic; foreign body reaction; bacteriologic and fungal aspects; bacteriologic and fungal changes under bizarre atmospheres and increased pressures; salivary glands. Mailing Add: Vet Admin Hosp 508 Fulton St Durham NC 27705

WESELI, DONALD FENTON, b Cleveland, Ohio, May 25, 31. IMMUNOLOGY. Educ: Ohio State Univ, BSc, 53, MSc, 54, PhD, 58. Prof Exp: Asst prof dairy sci, Ohio State Univ, 60-64; ASSOC PROF ANIMAL SCI, TEX A&M UNIV, 64- Mem: AAAS; Am Soc Animal Sci; Am Dairy Sci Asn. Res: Immunogenetics; biochemical genetics. Mailing Add: Dept of Animal Sci Tex A&M Univ College Station TX 77843

WESELOH, RONALD MACK, b Los Angeles, Calif, June 30, 44. ENTOMOLOGY. Educ: Brigham Young Univ, BS, 66; Univ Calif, Riverside, PhD(entom), 70. Prof Exp: ASST AGR SCIENTIST, CONN AGR EXP STA, 70- Mem: Entom Soc Am; Entom Soc Can; Int Orgn Biol Control. Res: Control of insect pests by means of parasites; insect behavior and ecology. Mailing Add: Dept of Entom Conn Agr Exp Sta New Haven CT 06504

WESEMEYER, HARALD, b Hamburg, Ger, Aug 29, 24; nat Can; m 53; c 4. EXPERIMENTAL PHYSICS. Educ: Univ Hamburg, dipl, 53; Univ BC, PhD(physics), 58. Prof Exp: Sci officer, Can Dept Mines & Tech Surv, 58-61; RES PHYSICIST, TELEFONADTIEBOLAGET L M ERICSSON, 61- Mem: Swed Asn Tech Physicists. Res: Magnetic resonances; low temperature and solid state physics; semiconductors; dielectrics; thin films; microcircuits. Mailing Add: Telefonaktiebolaget L M Ericsson Stockholm-Tyresö 1 Sweden

WESENBERG, DARRELL, b Madison, Wis, Dec 6, 39; m 68; c 3. AGRONOMY, GENETICS. Educ: Univ Wis-Madison, BS, 62, MS, 65, PhD(agron), 68. Prof Exp: RES AGRONOMIST, BR EXP STA, AGR RES SERV, USDA, 68- Mem: Crops Sci Soc Am; Am Soc Agron; Am Genetic Asn. Res: Plant breeding and plant genetics in cereal crops. Mailing Add: Br Exp Sta Agr Res Serv USDA Aberdeen ID 83210

WESER, ELLIOT, b New York, NY, Jan 12, 32; m 55; c 1. MEDICINE, GASTROENTEROLOGY. Educ: Columbia Univ, AB, 53, MD, 57; Am Bd Internal Med, dipl, 64; Am Bd Nutrit, cert, 71. Prof Exp: Med intern & asst resident, Sch Med & King County Hosp, Univ Wash, 57-59; sr resident med, Bronx Munic Hosp Ctr, Albert Einstein Col of Med, 59-60; clin assoc gastroenterol, Nat Inst Arthritis & Metab Dis, 61-63; from instr to asst prof med, Med Col, Cornell Univ, 63-67; PROF PHYSIOL & MED, UNIV TEX MED SCH, SAN ANTONIO, 67-, DEP CHMN DEPT, 69-, HEAD SECT GASTROENTEROL, 67- Concurrent Pos: Res fel med, New York Hosp-Cornell Med Ctr, 60-61; New York Res Coun career scientist award, 63-67; asst attend physician, New York Hosp, 64-67; asst vis physician, Bellevue Hosp, 64-67; attend physician, Bexar County Hosps, 67- Mem: AAAS; fel Am Col Physicians; Am Fedn Clin Res; Am Gastroenterol Asn; AMA. Mailing Add: Dept of Gastroenterol Univ of Tex Med Sch San Antonio TX 78229

WESLER, OSCAR, b Brooklyn, NY, July 12, 21. MATHEMATICS, STATISTICS. Educ: City Col New York, BS, 42; NY Univ, MS, 43; Stanford Univ, PhD(math statist), 55. Prof Exp: Asst math, NY Univ, 42-43; asst & instr, Princeton Univ, 43-46; res assoc math statist, Stanford Univ, 52-55, actg asst prof statist, 55-56; from asst prof to assoc prof math, Univ Mich, 56-64; PROF STATIST & MATH, NC STATE UNIV, 64- Concurrent Pos: Consult inst sci & technnol, Univ Mich, 56-64; vis assoc prof, Stanford Univ, 60, vis prof, 62-63, 73 & 74; NSF nat lectr, 63-; prof on-site studies prog, Int Bus Mach Corp, 66-; vis scholar, Univ Calif, Berkeley, 72-73. Mem: Am Math Soc; Inst Math Statist. Res: Probability and statistics; stochastic processes; statistical decision theory; functional analysis. Mailing Add: Dept of Statist NC State Univ Raleigh NC 27607

WESLEY, DEAN E, b Flint, Mich, Feb 26, 37; m 59; c 4. SOIL FERTILITY. Educ: Mich State Univ, BS, 61; SDak State Univ, PhD(soil fertil), 65. Prof Exp: Exten soil specialist & asst prof soil res, Univ Nebr, 65-66; from asst prof to assoc prof agron, 66-74, ASSOC PROF AGR, WESTERN ILL UNIV, 74- Mem: Am Soc Agron; Soil Sci Soc Am; Soil Conserv Soc Am. Res: Plant nutrition; nutrients such as phosphorus, zinc, iron and sulfur. Mailing Add: Dept of Agr Western Ill Univ Macomb IL 61455

WESLEY, JAMES PAUL, theoretical physics, see 12th edition

WESLEY, MARVIN LARRY, b Cedar Bluffs, Nebr, May 5, 44; m 68; c 2. AGRICULTURAL METEOROLOGY. Educ: Univ Nebr, Lincoln, BS, 65; Univ Wis-Madison, MS, 68, PhD(soil sci), 70. Prof Exp: Physicist, US Army Ballistic Res Labs, 70-73; asst meteorologist, 73-76, METEOROLOGIST, ARGONNE NAT LAB, 76- Mem: Am Meteorol Soc; Am Soc Agron; Am Geophys Union. Res: Studies of turbulent transfer of heat, momentum and pollutants in the lower atmosphere; remote sensing of turbulence using line-of-sight optical techniques; nonspectral measurements of solar radiation components. Mailing Add: Argonne Nat Lab Bldg 181 Argonne IL 60439

WESLEY, NEWTON K, b Westport, Ore, Oct 1, 17; m 40; c 2. OPTOMETRY. Educ: NPac Col Optom, OD, 39; Monroe Col Optom, DOS, 46. Prof Exp: Pres, Am Optom Ctr, 46-55; CO-OWNER, WESLEY-JESSEN INC, & SUBSIDIARIES, 48-; PRES & DIR, NAT EYE RES FOUND, 59- Mem: AAAS; Am Optom Asn. Res: Contact lenses. Mailing Add: Wesley-Jessen Inc 18 S Michigan Ave Chicago IL 60603

WESLEY, ROBERT COOK, b Jamestown, Ky, Aug 19, 26; m 53; c 2. DENTISTRY. Educ: Berea Col, AB, 50; Univ Louisville, DMD, 54. Prof Exp: Pvt practr, 54-67; asst prof prosthodont, 67-72, ASSOC PROF PROSTHODONT, COL DENT, UNIV KY, 72-, ASSOC PROF FAMILY PRACT & DIR DIV ORAL HEALTH, DEPT FAMILY PRACT, COL MED, 75- Concurrent Pos: Consult, Vet Admin Hosp, Lexington, Ky, 72. Mem: Int Asn Dent Res; Am Dent Asn. Res: Complete dentures especially related to geriatric patients. Mailing Add: Col of Dent Univ of Ky Med Ctr Lexington KY 40506

WESLEY, ROY LEWIS, b Liberty, Ky, May 1, 29; m 54; c 5. FOOD SCIENCE. Educ: Berea Col, BS, 55; Purdue Univ, MS, 58, PhD, 66. Prof Exp: From asst prof to assoc prof poultry sci, 58-74, PROF POULTRY SCI, VA POLYTECH INST & STATE UNIV, 74- Mem: Poultry Sci Asn; Inst Food Technol. Res: Physiology. Mailing Add: Dept of Food Sci Va Polytech Inst & State Univ Blacksburg VA 24061

WESLEY, WALTER GLEN, b Ft Worth, Tex, Oct 12, 38; m 68; c 5. THEORETICAL PHYSICS. Educ: Tex Christian Univ, BA, 61; Univ NC, PhD(physics), 70. Prof Exp: From asst prof to assoc prof physics, 66-75, PROF PHYSICS, MOORHEAD STATE UNIV, 75- Mem: AAAS; Hist Sci Soc; Philos of Sci Asn. Res: Gravitational theory and quantum gravitation; interaction of science and society. Mailing Add: Dept of Physics Moorhead State Univ Moorhead MN 56560

WESLEY, WILLIAM KEITH, b Griffin, Ga, Oct 24, 46; m 66; c 2. AGRONOMY. Educ: Univ Ga, BSA, 69, MS, 70; Univ Tenn, PhD(plant & soil sci), 73. Prof Exp: EXTEN AGRONOMIST, COOP EXTEN SERV, UNIV GA, 73- Mem: Am Soc Agron; Crop Sci Soc Am; Am Forage & Grassland Coun. Res: Education with respect to all aspects of forage and feed grain production; the use of applied research plots as teaching tools; current emphasis on fertilizer use for both commodities and forage management. Mailing Add: Coop Exten Serv Univ of Ga PO Box 1209 Tipton GA 31794

WESNER, GORDON EUGENE, JR, b Kansas City, Mo, Oct 11, 36; m 57; c 3. PHYSIOLOGY. Educ: DePauw Univ, BA, 58; Purdue Univ, MS, 60, cert, 62; Univ Nebr, PhD(physiol), 66. Prof Exp: Teacher pub sch, Ind, 59-60; instr zool & physiol, Kansas City, Kans Jr Col, 60-62; from asst prof to assoc prof, Nebr Wesleyan Univ, 62-68; assoc prof biol & chmn dept, 68-71, assoc dean, dir grad studies & dir instnl res, 70-71, VPRES & DEAN, ROCKFORD COL, 71- Concurrent Pos: Consult sch radiation technol, St Elizabeth Hosp, 66-68. Mem: AAAS. Res: Human environmental physiology; cardiovascular studies on peripheral blood flow and other related phenomena; cellular environment. Mailing Add: Rockford Col 5050 E State St Rockford IL 61101

WESOLOSKI, GEORGE D, b Berwyn, Ill, Mar 12, 40; m 71; c 2. ANIMAL NUTRITION. Educ: Millikin Univ, BS, 62; Univ Ill, MS, 70, PhD(animal nutrit), 73. Prof Exp: Res scholar animal nutrit, Ctr Int Trop Agr, 71-73; INT NUTRITIONIST, CENT SOYA CO, 73- Mem: Am Soc Animal Sci; Latin Am Soc Animal Sci. Res: Use of tropical ingredients, such as mandioca, bananas, coffee and sugar by-products, in animal feeds. Mailing Add: Cent Soya Co 1200 N Second St Decatur IN 46733

WESOLOWSKI, S ADAM, b Saugus, Mass, Feb 6, 23; m 45; c 5. SURGERY. Educ: Tufts Col, MD, 48, MS, 51. Prof Exp: Intern surg, Johns Hopkins Hosp, 48-49; asst resident, New Eng Ctr Hosp, 49-52; resident & asst instr, Col Med, State Univ NY Downstate Med Ctr, 54-56; sr registr, Guy's Hosp, London, Eng, 56-57; from instr to assoc prof, 57-64, CLIN PROF SURG, COL MED, STATE UNIV NY DOWNSTATE MED CTR, 64-; DIR CARDIOVASC RES LAB, MERCY HOSP, 66-, CHIEF THORACIC-CARDIOVASC SERV, 74- Concurrent Pos: Teaching fel, Tufts Col, 51-52; Am Col Surgeons Third Res Scholar, 56-59; chmn, Dept Surg, Meadowbrook Hosp, 64-66; mem, US Nat Comt Eng In Med & Biol; chmn, US Tech Adv Group, Int Stand Orgn, Tech Comt 150 & mem, Subcomt 2 Cardiovasc Implants. Mem: Nat Acad Eng; Am Soc Artificial Internal Organs (pres, 66-67); Am Soc Testing & Mat; Soc Vascular Surg; Soc Thoracic Surg. Res: Cardiovascular surgery, physiology and dynamics; screening and development of vascular prosthetic materials. Mailing Add: Cardiovasc Res Lab Mercy Hosp Rockville Centre NY 11570

WESSEL, CARL JOHN, b Pittsburgh, Pa, Oct 5, 11; m 42; c 5. BIOCHEMISTRY. Educ: Canisius Col, BS, 34; Univ Detroit, MS, 38; Cath Univ, PhD(biochem), 41. Prof Exp: Chemist, Pratt & Lambert Co, NY, 34-36; instr chem, Univ Detroit, 36-38 & 41; assoc dir control, Gelatin Prod Corp, Mich, 41-46; res assoc, Prev Deterioration Ctr, Nat Acad Sci-Nat Res Coun, 46-48, from asst dir to dir, 48-65; sci info coordr, Food & Drug Admin, Dept Health, Educ & Welfare, 65-66; sr vpres sci info & libr systs, John I Thompson & Co, 66-68; chief scientist, 68-72, CHIEF SCIENTIST & VPRES, TRACOR JITCO, INC, 72-, DIR CARCINOGENESIS BIOASSAY OPERS, 74- Concurrent Pos: Instr, Wayne Univ, 44-46; mem bd dirs, Nat Fedn Abstracting & Indexing Serv, 62-64, secy, 63-65; mem corrosion res coun, Eng Found, 62-65. Mem: Fel AAAS; Am Chem Soc; fel Am Inst Chemists; Soc Indust Microbiol; fel Inst Environ Sci. Res: Fungicides; deterioration of materials; effects of environments on materials and equipment; information and data storage and retrieval systems; carcinogenesis bioassays. Mailing Add: Tracor Jitco Inc 1776 E Jefferson St Rockville MD 20852

WESSEL, GUNTER KURT, b Berlin, Ger, Mar 29, 20; US citizen; m 53; c 3. PHYSICS. Educ: Tech Hochsch, Berlin, BS, 40; Univ Göttingen, dipl, 47, PhD(physics), 48. Prof Exp: Physicist, Lorenz Radio AG, Berlin, 43-45; fel, Nat Res Coun Can, 51-53; consult physics, Gen Elec Co, NY, 53-61; PROF PHYSICS, SYRACUSE UNIV, 61- Res: Atomic physics; spectroscopy; masers; lasers; magnetic resonance. Mailing Add: Dept of Physics Syracuse Univ Syracuse NY 13210

WESSEL, HANS U, b Duisburg, Ger, Apr 18, 27; US citizen; m 55; c 4. CARDIOLOGY, PHYSIOLOGY. Educ: Univ Freib urg, MD, 53. Prof Exp: Instr med, 60-63, assoc, 63-65, asst prof, 65-67, asst prof pediat, 67-71, ASSOC PROF PEDIAT, MED SCH, NORTHWESTERN UNIV, CHICAGO, 71-, ASSOC PROF ENG SCI, UNIV, 73- Concurrent Pos: NIH cardiovasc res trainee, Med Sch, Northwestern Univ, Chicago, 61-63; Am Heart Asn estab investr, 65-70, fel coun circulation, 65- Mem: AAAS; assoc fel Am Col Cardiol; assoc mem Inst Elec & Electronics Engrs; sr mem Instrument Soc Am; Biomed Eng Soc (secy-treas). Res: Bioengineering; effect of pulmonary vascular disease on pulmonary gas exchange; instrumentation; indicator dilution techniques; thermal velocity probes; patient monitoring system; exercise physiology in children. Mailing Add: 63 Mulberry Rd Deerfield IL 60015

WESSEL, JOHN EMMIT, b Los Angeles, Calif, Mar 8, 42. CHEMICAL PHYSICS, FORENSIC SCIENCE. Educ: Univ Calif, Los Angeles, BS, 65; Univ Chicago, PhD(chem), 70. Prof Exp: Fel chem, Univ Pa, 69-72, instr, 72-74; MEM TECH STAFF, CHEM & PHYSICS LAB, AEROSPACE CORP, 74- Mem: Am Phys Soc;

Sigma Xi. Res: Laser spectroscopy applied to energy relaxation processes and excited state structure in molecular systems; applications to selective photochemistry, detection and forensic sciences. Mailing Add: Chem & Physics Lab Aerospace Corp PO Box 92957 Los Angeles CA 90009

WESSEL, RICHARD DEATON, b Greenville, Ohio, July 3, 24; m 52; c 2. ENTOMOLOGY. Educ: Wittenberg Univ, AB, 47; Ohio State Univ, MSc, 49, PhD, 51. Prof Exp: Asst, Ohio Agr Exp Sta, 49-51; res entomologist, Calif Spray-Chem Corp, 52-57, field res supvr, Ortho Div, Calif Chem Co, 57-62, res scientist-int, Ortho Div, Chevron Chem Co, Richmond, 63-65, mgr field res & tech serv int, 66-68, mgr res & develop, 68-73, PRES DIR GEN, CHEVRON CHEM SAF, PARIS, 73- Mem: Entom Soc Am. Res: Agricultural chemical research and development. Mailing Add: 9 Rue Avice 92310 Sevres France

WESSELLS, NORMAN KEITH, b Jersey City, NJ, May 11, 32; m 55; c 3. DEVELOPMENTAL BIOLOGY. Educ: Yale Univ, BS, 54, PhD(zool), 60. Prof Exp: Am Cancer Soc fel, 60-62; from asst prof to assoc prof biol sci, 62-70, PROF BIOL SCI, STANFORD UNIV, 71-, CHMN DEPT, 72-, ACTG DIR, HOPKINS MARINE STA, 72- Concurrent Pos: Am Cancer Soc scholar cancer res, Dept Biochem, Univ Wash, 68-69; Guggenheim Found fel, 75-76. Honors & Awards: Herman Beerman Award, Soc Investigative Dermatol, 71. Mem: Am Soc Zoologists; Soc Develop Biol. Res: Embryonic induction; cytodifferentiation; chemistry, ultrastructure of skin, pancreas development; development of nerve cells, axons. Mailing Add: Dept of Biol Sci Stanford Univ Stanford CA 94305

WESSELS, KENNETH EDWIN, b Creston, Iowa, June 29, 21; m; c 3. DENTISTRY. Educ: Univ Iowa, DDS, 46, MS, 48. Prof Exp: Instr pedodont, Univ Iowa, 46-48, asst prof & actg head dept, 48-49, assoc prof, 49-52, prof & head dept, 52-60, clin prof pediat, Col Med, 52-60, dir, Children's Dent Clins, 48-60; asst secy coun dent educ, Am Dent Asn, 60-62, secy, 62-66; prof pedodont & chmn dept, Sch Dent, St Louis Univ, 66-69; chief party, Am Dent Asn Educ Proj, Univ Saigon, 69-72; PROF CHILD DEVELOP & CHIEF DEPT DENT, CHILD DEVELOP CTR, UNIV TENN, MEMPHIS, 72-, CHIEF CRANIOFACIAL ANOMALIES CLIN, 73- Concurrent Pos: Supvr dent sect, Iowa Hosp Sch Severely Handicapped Children, 52-60; Fulbright lectr, Australia, 57; consult, WHO; mem, Clin Adv Comt, United Cerebral Palsy. Mem: Am Dent Asn; fel Am Col Dent; fel Int Col Dent; Int Asn Dent Res. Res: Epidemiology and nutrition in dental caries. Mailing Add: Child Develop Ctr 711 Jefferson Ave Memphis TN 38105

WESSELY, HARRY WILLIAM, physics, see 12th edition

WESSENBERG, HARRY SANDERS, b Billings, Mont, Aug 26, 13; m 42; c 3. PROTOZOOLOGY, PARASITOLOGY. Educ: Univ Calif, AB, 36, MA, 51, PhD(zool), 58. Prof Exp: PROF BIOL, SAN FRANCISCO STATE UNIV, 51- Concurrent Pos: NIH fel trop med & parasitol, Cent Am, 60. Mem: Soc Protozool; Am Soc Parasitol. Res: Opalinid protozoa; ultrastructure of protozoa. Mailing Add: Dept Biol San Francisco State Univ 160 Holloway Ave San Francisco CA 94132

WESSLER, STANFORD, b New York, NY, Apr 20, 17; m 42; c 3. MEDICINE. Educ: Harvard Univ, BA, 38; NY Univ, MD, 42; Am Bd Internal Med, dipl. Prof Exp: Asst med, Harvard Med Sch, 49-51, instr, 51-54, assoc, 54-57, asst prof & tutor, 57-64; prof med, Sch Med, Washington Univ, 64-74, John E & Adeline Simon prof, 66-74; PROF MED, SCH MED & ASSOC DEAN POST-GRAD SCH MED, NY UNIV, 74- Concurrent Pos: Res fel med, Harvard Med Sch, 46-49; Nat Heart Inst trainee, 49-51; Am Heart Asn estab investr, 54-59; James F Mitchell Award heart & vascular res, 72-; asst, Beth Israel Hosp, Boston, 46-49, assoc, 49-56, physician, Vasc Clin, 54-64, assoc vis physician, 57-58, vis physician, 59-64, head, Anticoagulation Clin, 60-64, dir, Clin Res Ctr Thrombosis & Atherosclerosis, 61-64, assoc dir, 64-; mem, Comt Thrombosis & Hemorrhage, Nat Res Coun, 60-64; physician-in-chief, Jewish Hosp, St Louis & assoc physician, Barnes Hosp, 64-75; mem, Med Adv Bd, Coun Circulation, Am Heart Asn, 64- & Coun Stroke, 67-, vchmn, Coun Thrombosis, 71-74, chmn, 74-, chmn, Publ Comt, 72-, vpres coun, 74-; vchmn heart training comt, Nat Heart Inst, 65- & mem, Thrombosis Adv Comt, 67-71; chmn, Subgroup Thromboembolism, Inter-Soc Comn Heart Dis Resources, 69-; dir, Nat Heart & Lung Inst Thrombosis Ctr, Washington Univ Sch Med, 71-74; mem, Comn Stroke, Nat Inst Neurol Dis & Stroke, 72-74; attend physician, NY Univ Med Ctr, Univ Hosp, New York, 74- & Bellevue Hosp Ctr, 74- Mem: Am Fedn Clin Res: Am Physiol Soc; Am Soc Clin Invest; fel Am Col Physicians; Asn Am Physicians. Res: Blood coagulation; peripheral vascular disease. Mailing Add: NY Univ Post-grad Med Sch 550 First Ave New York NY 10016

WESSLING, RITCHIE A, b Iowa, Sept 15, 32; m 61; c 5. PHYSICAL CHEMISTRY, POLYMER SCIENCE. Educ: Mich State Univ, BS, 57, MS, 59; Univ Pa, PhD(phys chem), 62. Prof Exp: Chemist, Wyandotte Chem Co, 59; chemist polymer res lab, 62, ASSOC SCIENTIST, POLYMER SCI GROUP, PHYS RES LAB, DOW CHEM CO, 68- Mem: Am Chem Soc. Res: Relationship between physical properties of polymeric materials and their chemical structure. Mailing Add: Phys Res Lab Dow Chem Co 1702 Bldg Midland MI 48640

WESSLING, WOLFGANG HEINRICH, b Hakeborn, Ger, Nov 27, 28; US citizen; m 58. GENETICS, PLANT PATHOLOGY. Educ: Hannover Tech Univ, dipl Gartner, 53; Cornell Univ, MS, 55; NC State Univ, PhD(genetics), 57. Prof Exp: Foreign aid adv plant breeding, Ger Govt, 57-58; dir exp sta, Anderson Clayton Co, Brazil, 59-63; plant breeder, Farmers Forage Res, Ind, 63-65 & Delta & Pine Land Co, Courtaulds, Ltd, Miss, 65-71; mgr, 71-73, ASSOC DIR, COTTON, INC, 73- Concurrent Pos: Adj prof crop sci dept, NC State Univ; mem res comt, Plains Cotton Growers, Inc. Mem: Am Soc Agron; Am Genetics Asn. Res: Insect resistance; cotton genetics and pathology. Mailing Add: 4505 Creedmoor Rd Raleigh NC 27612

WESSMAN, GARNER ELMER, b Cokato, Minn, Sept 14, 20; m 43, 69; c 5. BACTERIOLOGY. Educ: Buena Vista Col, BS, 42; Iowa State Univ, PhD(physiol bact), 52. Prof Exp: Instr bact, Iowa State Univ, 49-52; med bacteriologist, Safety Div, Ft Detrick, 52-53, med bact div, 54-61; bacteriologist, Inst Paper Chem, 53-54; RES BACTERIOLOGIST, NAT ANIMAL DIS CTR, USDA, 61- Mem: Am Acad Microbiol; Am Soc Microbiol; Brit Soc Gen Microbiol; Conf Res Workers Animal Dis. Res: Nutrition and physiology of Pasteurella; immunochemistry of streptococci. Mailing Add: Nat Animal Dis Ctr Ames IA 50010

WESSON, JAMES ROBERT, b Jackson Gap, Ala, Nov 1, 21; m 43; c 4. MATHEMATICS. Educ: Birmingham-Southern Col, BS, 49; Vanderbilt Univ, MA, 49, PhD(math), 53. Prof Exp: Instr math, Univ Tenn, 49-50; from asst prof to assoc prof, Birmingham-Southern Col, 51-57; from asst prof to assoc prof, 57-66, PROF MATH, VANDERBILT UNIV, 66-, DIR UNDERGRAD STUDIES, 70- Mem: Am Math Soc; Math Asn Am. Res: Projective planes; abstract algebra; numerical solutions of differential equations. Mailing Add: Dept of Math Vanderbilt Univ Box 1595 Sta B Nashville TN 37235

WESSON, LAURENCE GODDARD, JR, b Midland, Mich, Oct 18, 17; m 48; c 4. PHYSIOLOGY, INTERNAL MEDICINE. Educ: Haverford Col, AB, 38; Harvard Univ, MD, 42. Prof Exp: From instr to asst prof physiol, Col Med, NY Univ, 46-50, from instr to assoc prof med, Postgrad Med Sch, 50-62; PROF MED, JEFFERSON MED COL, 62- Mem: Am Physiol Soc; Am Fedn Clin Res: Am Soc Nephrology. Res: Renal and cellular physiology; renal diseases. Mailing Add: Dept of Med Jefferson Med Col Philadelphia PA 19107

WEST, ARTHUR JAMES, II, b Boston, Mass, Dec 14, 27; div; c 3. PARASITOLOGY, MARINE BIOLOGY. Educ: Suffolk Univ, BS, 51, MA, 56; Univ NH, MS, 62, PhD(zool), 64. Prof Exp: From instr to assoc prof biol, Suffolk Univ, 52-65, prof & co-chmn dept, 65-68; dean div natural sci, New Eng Col, 68-70; chmn dept biol, 70-73, prog dir, Pre-Col Educ in Sci, NSF, 72-73, PROF BIOL, SUFFOLK UNIV, 70-, DIR, ROBERT S FRIEDMAN COBSCOOK BAY LAB, 75- Concurrent Pos: Instr sci & chmn dept, Emerson Col, 56-59; asst prof biol & chmn dept, Mass Col Optom, 57-60; mem, Gov Comn Ocean Mgt, Mass; NSF dir, Marine Sci Interinstnl Prog, NH Col & Univ Coun, 68-70. Mem: AAAS; Am Soc Parasitol; Am Soc Zoologists; Am Inst Biol Scientists; Am Soc Limnol & Oceanog. Res: Embryology and histochemistry of larval forms of Acanthocephala; morphology of priapulids; life cycle studies of marine parasites. Mailing Add: Suffolk Univ 41 Temple St Boston MA 02114

WEST, BILLY, microbiology, see 12th edition

WEST, BOB, b Ellenville, NY, Mar 7, 31; m 57; c 3. PHARMACOLOGY, TOXICOLOGY. Educ: Union Univ, NY, BS, 52; Purdue Univ, MSc, 54, PhD(pharmacol), 56. Prof Exp: Asst pharmacol, Purdue Univ, 52-54; res assoc, Rohm and Haas Co, 56-58; res pharmacologist, Am Cyanamid Co, 58-60; asst dir, Rosner-Hixon Labs, 60-65, vpres pharmacol & toxicol, 65-68; dir sci info, Vick Chem Co, Richardson-Merrell Inc, 68-72, dir sci & regulatory affairs, 72-75; PRES, BOB WEST ASSOCS, 75- Concurrent Pos: Res assoc, Jefferson Med Col, 56-58. Mem: Soc Toxicol; Am Soc Pharmacol & Exp Therapeut; Drug Info Asn; Acad Pharmaceut Sci; Am Pharmaceut Asn. Res: Pharmacology of antispasmodics, anticholinesterases and analgesics; chelating agents; heavy metal poisoning; depigmentation agents; topical anti-infectives; drug and cosmetic safety evaluation; pharmaceuticals. Mailing Add: Bob West Assocs PO Box 2001 Stamford CT 06906

WEST, BRUCE DAVID, b Madison, Wis, July 10, 35; m 57; c 2. BIOCHEMISTRY. Educ: Univ Wis, BS, 57, MS, 61, PhD(biochem), 62. Prof Exp: Fel biochem, Univ Wis, 62-63; asst prof chem, Univ NMex, 63-69; asst prof, 69-74, ASSOC PROF CHEM, EASTERN MICH UNIV, 74- Mem: Am Chem Soc; The Chem Soc. Res: Coumarin anticoagulants; structure-activity relationship, synthesis, biodegradation. Mailing Add: Dept of Chem Eastern Mich Univ Ypsilanti MI 48197

WEST, CHARLES ALLEN, b Greencastle, Ind, Nov 4, 27; m 52; c 2. BIOCHEMISTRY. Educ: DePauw Univ, AB, 49; Univ Ill, PhD(chem), 52. Prof Exp: Instr chem, 52 & 55, from asst prof to assoc prof, 56-67, vchmn dept, 70-75, PROF CHEM, UNIV CALIF, LOS ANGELES, 67- Concurrent Pos: Guggenheim fel, 61-62. Mem: Am Chem Soc; Am Soc Biol Chemists; Am Soc Plant Physiol. Res: Chemistry and biosynthesis of natural products of physiological importance, including gibberellins and other plant growth regulators; metabolic regulation. Mailing Add: Dept of Chem Univ of Calif Los Angeles CA 90024

WEST, CHARLES DAVID, b Riverside, Calif, July 25, 37; m 63; c 1. ANALYTICAL CHEMISTRY. Educ: Pomona Col, BA, 59; Mass Inst Technol, PhD(anal chem), 64. Prof Exp: From res chemist to sr res chemist, Beckman Instruments, Inc, 64-67; asst prof, 67-74, ASSOC PROF CHEM, OCCIDENTAL COL, 74- Mem: AAAS; Am Chem Soc; Optical Soc Am; Soc Appl Spectros. Res: Emission spectroscopy; flame photometry; atomic absorption instrument design; atomic and molecular fluorescence instrumentation. Mailing Add: Dept of Chem Occidental Col 1600 Campus Rd Los Angeles CA 90041

WEST, CHARLES DONALD, b Ogden, Utah, Oct 25, 20; m 46; c 4. BIOCHEMISTRY. Educ: Univ Utah, BA, 41, MD, 44, PhD, 50. Prof Exp: Instr med, Med Col, Cornell Univ, 50-54, res assoc, Sloan-Kettering Div, 52-53, asst prof, 53-57; from asst res prof to assoc res prof biochem, 57-71, assoc prof med, 65-69, PROF MED, MED CTR, UNIV UTAH, 69-, PROF BIOCHEM, 71-, CO-DIR, CLIN RES CTR, 66- Concurrent Pos: From asst to assoc, Sloan-Kettering Inst Cancer Res, 50-57; asst dir prof serv res & assoc chief staff, Vet Admin Hosp, Salt Lake City, 57-65. Mem: AAAS; Endocrine Soc; Asn Cancer Res; Fedn Clin Res: Harvey Soc. Res: Endocrinology. Mailing Add: Univ of Utah Med Ctr 50 N Medical Dr Salt Lake City UT 84132

WEST, CHARLES P, b New York, NY, Aug 28, 16; m 42; c 4. ORGANIC CHEMISTRY. Educ: City Col New York, BS, 36; Univ Chicago, PhD(org chem), 39. Prof Exp: Dir res, Neo Metal & Chem Co, 39-41; vpres, Nat Chem & Color Co, 46-54; PRES & DIR, RESIN RES LABS, INC, NEWARK, NJ, 54- Concurrent Pos: Guest lectr, Newark Col Eng, 59-60. Mem: AAAS; Am Chem Soc; Soc Plastics Eng; Nat Asn Corrosion Eng; fel Am Inst Chem. Res: Resins and polymers, particularly emulsion and dispersion polymers. Mailing Add: 60 Sharon Ct Metuchen NJ 08840

WEST, CLARK DARWIN, b Jamestown, NY, July 4, 18; m 44; c 3. IMMUNOLOGY, NEPHROLOGY. Educ: Col Wooster, AB, 40; Univ Mich, MD, 43. Prof Exp: Intern surg, Univ Mich Hosp, 43-44, resident pediat, 44-46; from asst prof to assoc prof, 51-62, PROF PEDIAT, COL MED, UNIV CINCINNATI, 62- Concurrent Pos: Children's Hosp Res Found scholar, 48-49, fel, 53-; Nat Res Coun sr fel pediat, Children's Hosp Res Found, 49-50 & Cardiopulmonary Lab, Bellevue Hosp, New York, 50-51; res assoc, Children's Hosp Res Found, 51-53, assoc dir, 63-, supv biochemist, Hosp, 51-65, attend pediatrician, 51-; attend pediatrician, Cincinnati Gen Hosp, 53-; mem, Gen Clin Res Ctr Comt, Div Res Facilities & resources, NIH, 65-69; mem, Urol & Renal Dis Training Comt, Nat Inst Arthritis, Metab & Digestive Dis, 72-73. Mem: Am Physiol Soc; Soc Pediat Res (secy-treas, 58-62, pres, 63-64); Am Soc Nephrology; Am Asn Immunologists; Am Pediat Soc. Res: Complement; glomerulonephritis. Mailing Add: Children's Hosp Res Found Elland Ave & Bethesda Cincinnati OH 45229

WEST, DAVID ARMSTRONG, b Beirut, Lebanon, Apr 9, 33; US citizen; m 58; c 3. GENETICS. Educ: Cornell Univ, BA, 55, PhD(vert zool), 59. Prof Exp: Asst prof zool, Cornell Univ, 59-60; NATO fel, 60-61; USPHS fel, 61-62; asst prof zool, 62-68, ASSOC PROF ZOOL, VA POLYTECH INST & STATE UNIV, 68- Concurrent Pos: Sci Res Coun sr vis fel, 66; ed, Va J Sci, 74- Mem: Lepidopterist Soc; AAAS; Am Ornith Union; Genetic Soc Am; Am Genetic Asn. Res: Evolutionary genetics; species evolution in birds; genetics of natural populations of isopods and butterflies; polymorphisms. Mailing Add: Dept of Biol Va Polytech Inst & State Univ Blacksburg VA 24061

WEST, DONALD COREY, b Harrington Harbour, Que, Jan 19, 18; m 44; c 2. COMPUTER SCIENCE. Educ: Acadia Univ, BSc, 39, BA, 40; Univ Toronto, MA, 41, PhD(physics), 52. Prof Exp: Instr physics, Acadia Univ, 45-47; physicist, NS Res

Found, 48-53 & Cent Res Lab, Can Industs, Ltd, Que, 53-70; ASSOC PROF COMPUT SCI & DIR COMPUT CTR, LOYOLA CAMPUS, CONCORDIA UNIV, 70-, CHMN DEPT COMPUT SCI, 74- Mem: Asn Comput Mach; Can Info Processing Soc; Am Phys Soc. Res: Naval radar; photoelectricity; high polymer plastics, especially molecular weight determinations and rheology; electrical digital and analog computers; operations research. Mailing Add: Dept Comput Sci Loyola Campus Concordia Univ 7141 Sherbrooke St W Montreal PQ Can

WEST, DONALD K, b Providence, RI, May 14, 29; m 57; c 3. ASTROPHYSICS. Educ: Univ RI, BS, 57; Rutgers Univ, MS, 60; Univ Wis, PhD(astron), 64. Prof Exp: ASTROPHYSICIST, GODDARD SPACE FLIGHT CTR, NASA, 64- Mem: Am Astron Soc. Res: Physics of emission line stars; astronomical observations from space telescopes. Mailing Add: Code 672 Goddard Space Flight Ctr Greenbelt MD 20771

WEST, DONALD MARKHAM, b Pasadena, Calif, Apr 22, 25; m 48. ANALYTICAL CHEMISTRY. Educ: Stanford Univ, BS, 49, PhD(chem), 58. Prof Exp: Actg instr chem, Stanford Univ, 54-55; from asst prof to assoc prof, 56-65, PROF CHEM, SAN JOSE STATE UNIV, 65- Mem: Am Chem Soc. Res: Analysis of organic compounds. Mailing Add: Dept of Chem San Jose State Univ 125 S Seventh St San Jose CA 95192

WEST, DOUGLAS XAVIER, b Tacoma, Wash, June 11, 37; m 64; c 2. INORGANIC CHEMISTRY. Educ: Whitman Col, AB, 59; Wash State Univ, PhD(chem), 64. Prof Exp: Instr chem, Upsala Col, 64-65; from asst prof to prof, Cent Mich Univ, 65-75, dir univ honors progs, 70-72; PROF INORG CHEM & CHMN DEPT, ILL STATE UNIV, 75- Mem: Am Chem Soc; Sigma Xi. Res: Chemistry of pentacyanometallates, kinetics and mechanisms; transition metal complexes of n-oxides and sulfoxides; electron spin resonance. Mailing Add: Dept of Chem Ill State Univ Normal IL 61761

WEST, EDMUND CARY, b Santa Ana, Calif, Mar 18, 36. EXPERIMENTAL HIGH ENERGY PHYSICS. Educ: Stanford Univ, BS, 58; Univ Wis, MS, 60, PhD(physics), 66. Prof Exp: Lectr physics & res asst, 65-66, asst prof, 67-72, SYST MGR, POLLY PROJ ANAL SCI PHOTOGRAPHS, UNIV TORONTO, 72- Mem: Phys Soc. Res: Bubble chamber physics; use of computers in measurement of various types of scientific photographs. Mailing Add: Dept of Physics Univ of Toronto Toronto ON Can

WEST, EDWARD STAUNTON, b Stuart, Va, Sept 9, 96; m 20. BIOCHEMISTRY. Educ: Randolph-Macon Col, AB, 17; Kans State Col, MS, 20; Univ Chicago, PhD(org chem), 23. Hon Degrees: LLD, Randolph-Macon Col, 65; DSc, Univ Portland, 69. Prof Exp: From asst to instr chem, Kans State Col, 17-22; asst org chem, Univ Chicago, 22-23; from instr to assoc prof biochem, Sch Med, Wash Univ, 23-34; prof biochem & head dept, Sch Med, 34-66, EMER PROF BIOCHEM, SCH MED, UNIV ORE, 66-; ASST TO DIR, ORE REGIONAL PRIMATE RES CTR, 66- Concurrent Pos: Shaffer lectr med sch, Washington Univ, 56; mem adv coun, Life Ins Med Res Fund, 56-59, chmn, 59-60; mem biochem test comt, Nat Bd Med Examr, 57-58, chmn, 58-61, mem bd, 58-61; sr scientist, Ore Regional Primate Res Ctr, 60-66. Mem: AAAS; Am Chem Soc; Am Soc Biol Chemists; Soc Exp Biol & Med. Res: Carbohydrate chemistry and metabolism; chemistry and metabolism of ascorbic acid; lipid metabolism; ketosis; tissue phosphates. Mailing Add: Ore Regional Primate Res Ctr 505 NW 185th Ave Beaverton OR 97005

WEST, ERIC NEIL, b Montreal, Que, Mar 28, 41; m 65; c 2. STATISTICS, COMPUTER SCIENCE. Educ: Royal Mil Col Can, BSc, 63; Iowa State Univ, MS, 67, PhD(statist), 70. Prof Exp: Res assoc statist & comput, Iowa State Univ, 67-70; asst prof comput sci, Univ Alta, 70-72; assoc prof statist & comput sci, 72-73, ASSOC PROF QUANT METHODS, SIR GEORGE WILLIAMS CAMPUS, CONCORDIA UNIV, 73- Concurrent Pos: Consult, Pro Data Serv, 70-; pvt consult, 70- Mem: Am Statist Asn; Inst Math Statist; Sigma Xi. Res: Statistical inference; computation systems analysis; business decision making; forecasting. Mailing Add: Fac of Quant Methods Sir George Williams Campus Concordia Univ 1455 de Maisonneuve W Montreal PQ Can

WEST, ESTAL DALE, physics, physical chemistry, see 12th edition

WEST, FELICIA EMMINGER, b Chicora, Pa, Sept 14, 26; m 48; c 2. SCIENCE EDUCATION. Educ: J B Stetson Univ, BS, 48; Univ Fla, MEd, 65, EdD(sci educ , geol), 71. Prof Exp: Teacher high sch, Fla, 60-64; instr physics & elem sci, Miami Dade Jr Col, 65-66; instr elem sci & phys sci, St Johns River Jr Col, 66-67; teacher & student gen sci, Lab Sch, Univ Fla, 67-69, teacher gen sci & elem sci, 69-72; staff assoc, AAAS, 72-75; CHMN DIV NATURAL SCI, MATH & PHYS EDUC, FLA JR COL, S CAMPUS, 75- Concurrent Pos: Coord ed, Fla Asn Sci Teachers J, 75- Res: Development of unified science curriculum materials for use at community college level; development of field guides to specific sites in Florida for use by secondary and community college instructors. Mailing Add: 424 Oceanwood Dr Neptune Beach FL 32233

WEST, FRED RALPH, JR, b Pittsburgh, Pa, June 7, 25. PHARMACOLOGY. Educ: Hampton Inst, BS, 47; Tuskegee Inst, MS, 48; Univ Chicago, PhD(pharmacol), 56; Howard Univ, MD, 63. Prof Exp: Instr chem, St Augustine's Col, 48-50; asst, George W Carver Found, 50-53; instr pharmacol, 56-65, ASST PROF PHARMACOL, SCH MED, HOWARD UNIV, 65- Mem: AAAS. Res: Cultivation of animal and plant cells in vitro. Mailing Add: Dept of Pharmacol Howard Univ Sch of Med Washington DC 20001

WEST, FREDERICK HADLEIGH, archaeology, biological anthropology, see 12th edition

WEST, FREDERICK RICHARD, astronomy, see 12th edition

WEST, GEORGE CURTISS, b Newton, Mass, May 13, 31; m; c 4. PHYSIOLOGICAL ECOLOGY, ORNITHOLOGY. Educ: Middlebury Col, BA, 53; Univ Ill, Urbana, MS, 56, PhD(physiol ecol), 58. Prof Exp: Fel div biosci, Nat Res Coun Can, 59-60; asst prof zool, Univ RI, 60-63; from asst prof to prof zoophysiol, Inst Arctic Biol, 63-74, actg dean, Col Biol Sci & Renewable Resources & actg dir, Div Life Sci, 74-75, ACTG DIR, INST ARCTIC BIOL, UNIV ALASKA, 74- Concurrent Pos: Alexander von Humboldt Found fel, Aschoff Div, Max Planck Inst Physiol of Behav, 71-72; mem US-USSR bilateral exchange working group protection northern ecosysts, 75- Mem: Am Ornith Union; Am Physiol Soc; Ecol Soc Am; Wildlife Soc; Wilson Ornith Soc. Res: Bioenergetics and temperature regulation of birds; migration, fat deposition, food habits of birds; fatty acid analysis of plant and animal lipids. Mailing Add: Inst of Arctic Biol Univ of Alaska Fairbanks AK 99701

WEST, GORDON FOX, b Toronto, Ont, Apr 21, 33. GEOPHYSICS. Educ: Univ Toronto, BASc, 55, MA, 57, PhD(geophys), 60. Prof Exp: Geophysicist, Dom Gulf Co, 55-56; lectr geophys, 58-66, assoc prof, 66-72, PROF PHYSICS, UNIV TORONTO, 72- Mem: Soc Explor Geophysicists; Am Geophys Union; Geol Asn Can; Can Asn Physicists. Res: Electromagnetic geophysical methods; geophysical studies of precambrian shields. Mailing Add: Geophys Lab Dept of Physics Univ of Toronto Toronto ON Can

WEST, HAROLD D, b Flemington, NJ, July 16, 04; m 27; c 2. BIOCHEMISTRY. Educ: Univ Ill, AB, 25, MS, 30, PhD(biochem), 37. Hon Degrees: LLD, Morris Brown Col, 55; DSc, Meharry Med Col, 69. Prof Exp: Head dept sci, Morris Brown Col, 25-27; from assoc prof to prof physiol chem, 27-52, chmn dept, 38-67, pres & trustee, 52-67, PROF BIOCHEM, MEHARRY MED COL, 52- Mem: Am Soc Biol Chem; Soc Exp Biol & Med. Res: Metabolism of pantothenic acid; effect of low protein diets on growth and development of oral structure; iron uptakes in pathological conditions using radio-iron as a tracer; radioactive silver; metabolism of diphenyl; mercapturic acid series. Mailing Add: 3519 Geneva Circle Nashville TN 37209

WEST, HARRY IRWIN, JR, b Foley, Ala, Dec 3, 25; m 56; c 4. SPACE PHYSICS, NUCLEAR PHYSICS. Educ: Auburn Univ, BS, 46, MS, 47; Stanford Univ, PhD(physics), 55. Prof Exp: PHYSICIST, LAWRENCE LIVERMORE LAB, UNIV CALIF, 55- Mem: Fel Am Phys Soc; Am Geophys Union. Res: Nuclear spectroscopy and measurements of charged particles in the earth's radiation belts. Mailing Add: L-232 Lawrence Livermore Lab Univ of Calif Livermore CA 94550

WEST, JAMES EDWARD, b Grinnell, Iowa, May 1, 44; m 65; c 1. MATHEMATICS. Educ: La State Univ, BS, 64, PhD(math), 67. Prof Exp: Mem, Inst Advan Study, 67-68; asst prof math, Univ Ky, 68-69; asst prof, 69-72, NSF res grant, 70-73, ASSOC PROF MATH, CORNELL UNIV, 72- Concurrent Pos: Asst prof, Univ Ky, 69-71. Mem: Am Math Soc. Res: Topology of infinite-dimensional spaces and manifolds; geometric and point-set topology. Mailing Add: Dept of Math Cornell Univ Ithaca NY 14853

WEST, JERRY LEE, b North Wilkesboro, NC, Nov 1, 40; m 65; c 2. FISH BIOLOGY, ZOOLOGY. Educ: Appalachian State Univ, BS, 62; NC State Univ, MS, 65, PhD(zool), 68. Prof Exp: Asst prof biol, 67-73, ASSOC PROF BIOL, WESTERN CAROLINA UNIV, 73-, ACTG HEAD DEPT, 74- Mem: Am Fisheries Soc; Am Inst Fishery Res Biologists. Res: Effects of sediment on trout streams. Mailing Add: Dept of Biol Western Carolina Univ Cullowhee NC 28723

WEST, JOHN B, b Adelaide, SAustralia, Dec 27, 28. PHYSIOLOGY, MEDICINE. Educ: Univ Adelaide, MB, BS, 52, MD, 58; Univ London, PhD(appl physiol), 60. Prof Exp: Res assoc respiratory physiol, Royal Postgrad Med Sch, London, 54-60; physiologist, Himalayan Sci & Mountaineering Exped, 60-61; asst prof physiol, Univ Buffalo, 61-62; lectr med, Royal Postgrad Sch, London, 63-68; PROF MED, UNIV CALIF, SAN DIEGO, 69- Concurrent Pos: Mem, Cardiovasc Study Sect, NIH, 71-75. Mem: Am Physiol Soc; Am Soc Clin Invest; Am Thoracic Soc; Brit Physiol Soc. Res: Respiratory function in health and disease. Mailing Add: Dept of Med Univ of Calif at San Diego La Jolla CA 92307

WEST, JOHN LESLIE, b Lockwood, Mo, Oct 17, 11; m 30; c 1. PATHOLOGY, BACTERIOLOGY. Educ: Kans State Univ, DVM, 36; Univ Wis, MS, 48, PhD, 52. Prof Exp: Asst prof path & parasitol, Ala Polytech Univ, 37-40; vet, Div Vet Sci, Agr & Mech Col, Tex, 40-42; asst path & virol, Univ Wis, 46-50; assoc prof path, Univ Tenn, 50-53; asst prof poultry dis, Pa State Univ, 53; prof path, parasitol & pub health, Col Vet Med, Kans State Univ, 53-68; SCIENTIST & VET PATHOLOGIST, AGR RES LAB, UNIV TENN-AEC, 68- Mem: Am Vet Med Asn; Am Asn Avian Path; Conf Res Workers Animal Dis. Res: Poultry and fur animal diseases; distemper; vitamin A deficiency lesions in ruminants; fluorine intoxication in ruminants; avian encephalomyelitis. Mailing Add: Agr Res Lab Univ of Tenn-AEC 1299 Bethel Valley Rd Oak Ridge TN 37830

WEST, JOHN WYATT, b Decaturville, Tenn, Oct 18, 23; m 47; c 3. POULTRY NUTRITION. Educ: Univ Tenn, BSA, 47, MS, 48; Purdue Univ, PhD(poultry nutrit), 51. Prof Exp: Asst dir feeds res, Security Mills, Inc, Tenn, 51; from assoc prof to prof poultry husb, Miss State Univ, 52-56; prof poultry sci & head dept, Okla State Univ, 56-68; ASSOC DEAN, SCH AGR & NATURAL RESOURCES, CALIF POLYTECH STATE UNIV, SAN LUIS OBISPO, 68- Mem: Poultry Sci Asn. Res: Arsenic compounds and vitamin-amino acid interrelationships in poultry nutrition; nutritional value of cottonseed meal in broiler and turkey rations; antibiotic-protein interrelationships in broiler rations. Mailing Add: Sch of Agr & Natural Resources Calif Polytech State Univ San Luis Obispo CA 93407

WEST, JOSEPH ANTHONY, organic chemistry, see 12th edition

WEST, KEITH P, b Simla, Colo, Aug 20, 20; m 46; c 3. ZOOLOGY, RADIATION BIOLOGY. Educ: Chico State Col, AB, 42; Stanford Univ, MA, 48. Prof Exp: Instr biol & chem, Vallejo Col, 47-48; from instr to PROF BIOL SCI, DREXEL UNIV, 48- Concurrent Pos: Asst dean col eng & sci, Drexel Univ, 67-68, actg head dept biol sci, 71-72. Mem: AAAS; Health Physics Soc. Res: Biological effects of radiation; sanitary quality control of food products. Mailing Add: Dept of Biol Sci Drexel Univ Philadelphia PA 19104

WEST, KELLY M, b Oklahoma City, Okla, May 31, 25. MEDICINE. Educ: Univ Okla, MD, 48; Am Bd Internal Med, dipl, 54. Prof Exp: Clin asst, 52, from instr to assoc prof, 53-68, PROF MED, SCH MED, UNIV OKLA, 68- Concurrent Pos: Consult, Nat Libr Med, NIH & WHO. Mem: Fel Am Col Physicians. Res: Diabetes; migration of scientists and physicians; epidemiology of diabetes. Mailing Add: Univ of Okla Health Sci Ctr 800 NE 13th St Oklahoma City OK 73104

WEST, KENNETH CALVIN, b Broken Bow, Nebr, Apr 1, 35; m 60; c 3. ANALYTICAL CHEMISTRY. Educ: Wheaton Col, Ill, BS, 56; Ind Univ, PhD(anal chem), 67. Prof Exp: Asst prof chem, 67-75, ASSOC PROF CHEM, ST LAWRENCE UNIV, 75- Mem: Am Chem Soc. Res: Instrumentation. Mailing Add: Dept of Chem St Lawrence Univ Canton NY 13617

WEST, LOUIS JOLYON, b New York, NY, Oct 6, 24; m 44; c 3. PSYCHIATRY, NEUROLOGY. Educ: Univ Minn, BS, 46, MB, 48, MD, 49; Am Bd Psychiat & Neurol, dipl, 54. Prof Exp: Intern med, Univ Hosps, Univ Minn, 48-49; asst psychiat, Med Col, Cornell Univ, 49-52; prof psychiat & head dept psychiat, neurol & behav sci, Sch Med, Univ Okla, 54-69; PROF PSYCHIAT & CHMN DEPT, SCH MED, UNIV CALIF, LOS ANGELES, 69-, MED DIR, NEUROPSYCHIAT INST, 69-, PSYCHIATRIST IN CHIEF, UNIV CALIF HOSPS & CLINS, 69- Concurrent Pos: Resident, Payne Whitney Clin, New York Hosp, 49-52; res coordr, Okla Alcoholism Asn & chief behav sci, Okla Med Res Found, 56-69; consult, Oklahoma City Vet Admin Hosp, 56-69, US Air Force Hosp, Tinker AFB, Okla, 56-66, US Air Force Aero-Space Med Ctr, 61-66 & Peace Corps, 62-63; nat consult, Surgeon Gen, US Air Force, 57-62, mem adv coun, Behav Sci Div, Air Force Off Sci Res, 56-58; mem prof adv coun, Nat Asn Ment Health, 59-64; consult, US Info Agency, 60-61; mem exec coun adv comt behav res, Nat Acad Sci-Nat Res Coun, 61-63; mem nat adv comt psychiat, neurol & psychol, Spec Med Adv Group, US Vet Admin; mem, Nat Adv Ment Health Coun, NIMH, 65-69, White House Conf Civil Rights, 66 & Nat Adv

Comt Alcoholism, Dept Health, Educ & Welfare; consult ed, Med Aspects Human Sexuality, 67-; mem bd dir, Kittay Found, 72-; mem residency rev comt psychiat & neurol, AMA, 73-; mem adv panel res & develop, US Army, 74- Mem: Fel Am Col Neuropsychopharmacol; fel Am Col Psychiat; fel Am Psychiat Asn; Pavlovian Soc NAm (pres, 74-75); Soc Biol Psychiat. Res: Experimental psychopathology, especially relating to disturbances of perception and altered states of consciousness; psychophysiological correlates in clinical practice; alcohol and drug abuse; life-threatening behavior; interaction of biological, psychological and sociocultural factors in personality development and function. Mailing Add: 760 Westwood Plaza Los Angeles CA 90024

WEST, MICHAEL STEVEN, anthropology, see 12th edition

WEST, NEIL ELLIOTT, b Portland, Ore, Dec 17, 37; m 63; c 1. ECOLOGY. Educ: Ore State Univ, BS, 60, PhD(plant ecol), 64. Prof Exp: From asst prof to assoc prof plant ecol, Dept Range Sci & Ecol Ctr, 64-75, PROF RANGE SCI, UTAH STATE UNIV, 75- Concurrent Pos: Forest ecologist, Ore Forest Res Lab, Ore State Univ, 63; NSF fel & vis prof, Inst Ecol, Univ Ga, 70-71; consult, Nat Park Serv, Occidental Petrol, Argonne Nat Lab & Environ Protection Agency. Mem: Int Asn Ecol; Ecol Soc Am; Brit Ecol Soc; Int Soc Plant Geog & Ecol. Res: Plant ecology theory and its application to wildland resource management, particularly synecology and soil-vegetation relationships; systems ecology; community structure succession, productivity, nutrient cycling in arid, semi-arid, riparian and woodland ecosystems. Mailing Add: Dept of Range Sci Utah State Univ Logan UT 84322

WEST, PHILIP WILLIAM, b Crookston, Minn, Apr 12, 13; m 35; c 2. CHEMISTRY. Educ: Univ NDak, BS & MS, 35; Univ Iowa, PhD(chem), 39. Hon Degrees: DSc, Univ NDak, 58. Prof Exp: Asst chemist, State Geol Surv, NDak, 35-36; asst sanit chem, Univ Iowa, 36-37; asst chemist, State Dept Health, Iowa, 37-40; res chemist & microchemist, Econ Lab, Inc, Minn, 40; from instr to prof, 40-53, BOYD PROF CHEM, LA STATE UNIV, BATON ROUGE, 53-, DIR ENVIRON SCI INST, 66- Concurrent Pos: Smith lectr, 55; consult, Ethyl Corp, A D Little Co & USPHS; consult, Kem-Tech Labs, Inc & chmn bd, 66-; ed, Analytica Chemica Acta; co-ed, Sci Total Environ; mem working party 1, Sci Comt Probs Environ; adj prof, Environ Protection Agency; mem chem panel, NSF; pres comt new reactions, Int Union Pure & Appl Chem & pres anal chem div, 66-, mem sect toxicol & indust hyg; mem study sect, USPHS, 60-65. Honors & Awards: Southwest Award, Am Chem Soc, 54 & Coates Award, 67. Mem: Am Chem Soc; hon mem Brit Soc Anal Chem; hon mem Austrian Asn Microchem & Anal Chem; hon mem Japanese Soc Anal Chem. Res: Water treatment and analysis; polarized light microscopy; spot tests; organic reagents; complex ions; analysis of petroleum; polarography; chromatography; high frequency titrations; inorganic extractions; catalyzed and induced reactions; air pollution. Mailing Add: Dept of Chem La State Univ Baton Rouge LA 70803

WEST, RICHARD FUSSELL, b Rockland, Maine, Aug 27, 17; m 42; c 2. FORESTRY, WOOD TECHNOLOGY. Educ: Rutgers Univ, BS, 40; Yale Univ, MF, 42. Prof Exp: Wood technologist, Chance Vought Aircraft Corp, 43-44; chemist, Crocker-Burbank & Co, 44-46; asst prof wood technol, Sch Forestry, La State Univ, 46-53; assoc prof & head dept, Col Agr, 53-60, PROF WOOD TECHNOL & HEAD FORESTRY SECT, COOK COL, RUTGERS UNIV, NEW BRUNSWICK, 60- Concurrent Pos: Consult wood technologist, 53- Mem: Soc Wood Sci & Technol; Soc Am Foresters; Forest Prod Res Soc. Res: Wood properties, treatment and use; forest policy and forest land management and use. Mailing Add: Forestry Sect Cook Col Rutgers Univ New Brunswick NJ 08903

WEST, RICHARD LOWELL, b Quincy, Fla, Mar 20, 34; m 56; c 4. ORGANIC CHEMISTRY. Educ: Univ of the South, BS, 55; Univ Rochester, PhD(chem), 61; Univ Del, MBA, 73. Prof Exp: Sr chemist, Atlas Chem Ind, Inc, 59-66, res chemist, 66-70, res supvr chem, ICI Am Inc, 70-74, MGR CHEM & POLYMER RES, ICI UNITED STATES INC, 75- Mem: Am Chem Soc; Am Asn Textile Chemists & Colorists; Am Oil Chemists Soc; NY Acad Sci. Res: Organic polymer synthesis; organic synthesis; surfactants; textile adjuncts; industrial chemical development. Mailing Add: ICI United States Inc Wilmington DE 19897

WEST, ROBERT, b Glen Ridge, NJ, Mar 18, 28; m 50; c 2. ORGANOMETALLIC CHEMISTRY. Educ: Cornell Univ, BA, 50; Harvard Univ, AM, 52, PhD(chem), 54. Prof Exp: Asst prof inorg chem, Lehigh Univ, 54-56; from instr to assoc prof chem, 56-63, PROF CHEM, UNIV WIS-MADISON, 63- Concurrent Pos: Abbott lectr, Univ NDak, 64; Fulbright vis prof, Kyoto & Osaka Univs, 64-65; vis prof & NIH sr fel, Univ Würzburg, 68-69; Seydel-Wooley lectr, Ga Inst Technol, 70; Sun Oil lectr, Ohio Univ, 71; Edgar C Britton lectr, Midland, Mich, 71; vis prof, Haile Sellassie Univ, 72; consult, Allied Chem Co & NIH. Honors & Awards: F S Kipping Award, Am Chem Soc, 70. Mem: AAAS; Am Chem Soc; The Chem Soc; Chem Soc Japan. Res: Organometallic compounds, especially of silicon and lithium; new aromatic species; polyquinonoid compounds; educational innovation. Mailing Add: 5702 Old Sauk Rd Madison WI 53705

WEST, ROBERT COOPER, b Enid, Okla, June 30, 13; m 68. GEOGRAPHY OF LATIN AMERICA, HISTORICAL GEOGRAPHY. Educ: Univ Calif, Los Angeles, AB, 35, MA, 38; Univ Calif, Berkeley, PhD(geog), 46. Prof Exp: Cartographer, Off Strategic Serv, 41-45; cult geographer, Smithsonian Inst, 46-47; from asst prof to prof, 48-70, BOYD PROF GEOG, LA STATE UNIV, BATON ROUGE, 70- Concurrent Pos: Guggenheim fel, Seville, Spain, 55-56; NSF grant, Mexico, 69-70; vis lectr, Univ Wis-Madison, 66. Honors & Awards: Outstanding Achievement in Geog Award, Asn Am Geog, 73. Mem: AAAS; Am Geog Soc; Asn Am Geog; Latin Am Studies Asn. Res: Regional geography of Mexico and Central America; historical geography of Latin America; cultural geography and ethnography of Indian and Negro groups in Latin America; biogeography of the humid tropics. Mailing Add: Dept of Geog & Anthrop La State Univ Baton Rouge LA 70803

WEST, ROBERT ELMER, b Blackfoot, Idaho, Apr 2, 38; div; c 3. EXPLORATION GEOPHYSICS. Educ: Univ Idaho, BS, 61; Univ Ariz, MS, 70, PhD(geosci), 72. Prof Exp: Physicist, Phillips Petrol Co, 61-62, 65; res assoc geophys, Univ Ariz, 68-69, 71-72; geophysicist, Humble Oil & Refining Co, 72-74; GEOPHYSICIST, MINING GEOPHYS SURV, 74- Mem: Soc Explor Geophys. Res: Application of gravity, magnetics and electrical methods to exploration for mineral and ore deposits. Mailing Add: Mining Geophys Surv 2400 E Grant Rd Tucson AZ 85719

WEST, ROBERT MACLELLAN, b Appleton, Wis, Sept 1, 42; m 65; c 1. VERTEBRATE PALEONTOLOGY. Educ: Lawrence Col, BA, 63; Univ Chicago, SM, 64, PhD(evolutionary biol), 68. Prof Exp: Res assoc geol & geophys sci, Princeton Univ, 68-69; asst prof biol, Adelphi Univ, 69-74; CUR GEOL, MILWAUKEE PUB MUS, 74- Concurrent Pos: Adj assoc prof, Dept Geol Sci, Univ Wis-Milwaukee, 74- Mem: Geol Soc Am; Soc Study Evolution; Am Soc Mammal. Res: Asiatic mammalian evolution; paleontologic aspects of plate tectonics; evolution of early Tertiary mammals and mammalian communities; biostratigraphy of Tertiary deposits of intermontane basins in North America. Mailing Add: Dept of Geol Milwaukee Pub Mus 800 W Wells St Milwaukee WI 53233

WEST, RONALD ROBERT, b Centralia, Ill, Nov 14, 35; m 58; c 2. PALEOBIOLOGY, PALEOECOLOGY. Educ: Univ Mo-Rolla, BS, 58; Univ Kans, MS, 62; Univ Okla, PhD(paleoecol geol), 70. Prof Exp: Stratigrapher, Shell Oil Co, Okla, 56, micropaleontologist, La, 58-59; invertebrate paleontologist, Kans Geol Surv, 60, geologist, 61; paleobiologist & paleoecologist, Humble Oil & Refining Co, Tex & Okla, 61-67; instr geol, Univ Okla, 67-68; asst prof paleobiol, 69-74, ASSOC PROF PALAEOBIOL, DEPT GEOL, KANS STATE UNIV, 74-, ANCILLARY PROF BIOL DIV, 74- Concurrent Pos: Am Chem Soc-Petrol Res Fund grant paleobiol; mem adv coun, Friends of Woodrow Wilson Nat Fel Found; consult res lab, Amoco Prod Co, Okla, 74- Honors & Awards: Geol Soc Am Award, 72. Mem: AAAS; Paleont Soc; Soc Econ Mineralogists & Paleontologists; Soc Syst Zool; Geol Soc Am. Res: Paleoecology and paleozoology of upper paleozoic invertebrates; structure and dynamics of benthic fossil communities; population studies and ecology of modern and fossil brachiopods and bivalves; carbonate sedimentation; recent marine invertebrate ecology; biometrics. Mailing Add: Palaeobiol Lab Dept of Geol Kans State Univ Manhattan KS 66506

WEST, ROSE GAYLE, b Pascagoula, Miss, Oct 31, 43; m 62; c 1. PHYSICAL CHEMISTRY. Educ: Univ Southern Miss, BA, 65, PhD(phys chem), 69. Prof Exp: Asst prof chem, 68-73, actg chmn dept, 72-74, ASSOC PROF CHEM, WILLIAM CAREY COL, 73-, CHMN DEPT, 74- Mem: Am Chem Soc. Res: Thermo chemistry, heats of combustion and resonance energies of aromatic hydrocarbons and 5- and 6-membered aromatic nitrogen heterocyclic compounds; special projects for undergraduate physical chemistry laboratories. Mailing Add: PO Box 196 William Carey Col Hattiesburg MS 39401

WEST, SEYMOUR S, b New York, NY, Apr 19, 20; m 47; c 3. BIOPHYSICS, ANATOMY. Educ: City Col New York, BEE, 50; Western Reserve Univ, PhD(biophys, anat), 63. Prof Exp: Sr engr, A B Dumont Labs, Inc, 54-57; asst prof biomed eng, Western Reserve Univ, 62-64; sr scientist & head dept phys biol, Melpar, Inc, 64-66; assoc prof biomed eng & actg head dept, 66-67, assoc prof eng biophys & actg chmn dept, 67-69, PROF ENG BIOPHYS & CHMN DEPT, MED CTR, UNIV ALA, 69- Concurrent Pos: Mem, Automation in Med Lab Sci Rev Comt, NIH. Mem: Fel Royal Micros Soc; Biophys Soc; Inst Elec & Electronics Engrs; Biomed Eng Soc; Int Asn Dent Res. Res: Relation of physical optical properties of living cells and their morphological components to cytochemistry and physical chemistry; television fluorescence spectrophotometry and microspectropolarimetry. Mailing Add: Dept of Eng Biophys Univ of Ala Med Ctr Univ Sta Birmingham AL 35294

WEST, SHERLIE HILL, b Forbus, Tenn, Feb 18, 27; m 49; c 2. PLANT PHYSIOLOGY, AGRONOMY. Educ: Tenn Polytech Univ, BS, 49; Univ Ky, MS, 54; Univ Ill, PhD(agron, bot), 58. Prof Exp: Asst agronomist, Univ Ky, 54-55; res agronomist, USDA, 58-60; from asst agronomist to assoc agronomist, 58-70, AGRONOMIST, UNIV FLA, 70-, PROF AGRON, 74-, ASST DEAN RES, INST FOOD & AGR SCI, 72-; PLANT PHYSIOLOGIST, USDA, 60- Concurrent Pos: Consult, O M Scott & Sons, Ohio, 59-64. Mem: Am Chem Soc; Am Soc Plant Physiol; Am Soc Agron. Res: Nucleic acid metabolism and growth due to environmental factors; mechanism of hormone action; drought and cold tolerance; development of genetic criteria of selection of superior plants; cool temperature effects on carbohydrate metabolism; mechanisms of cool temperature dormancy in tropical grasses. Mailing Add: Fla Agr Exp Sta 1022 McCarty Hall Univ of Fla Gainesville FL 32611

WEST, STANLEY A, b Bath, NY, May 21, 43. APPLIED ANTHROPOLOGY. Educ: Syracuse Univ, BS, 65, PhD(anthrop), 73. Prof Exp: Asst prof anthrop, Western Mich Univ, 70-74; ASST PROF CIVIL ENG, MASS INST TECHNOL, 74- Concurrent Pos: Fel, Rockefeller Found, 74-77; consult, AAAS, 75. Mem: Fel Am Anthrop Asn; fel Soc Appl Anthrop; AAAS; Int Soc Technol Assessment. Res: Development and testing of technological aids to human judgment; analysis and design of sociotechnical systems. Mailing Add: 26 Townsend Rd Belmont MA 02178

WEST, THEODORE CLINTON, b Central, SC, May 17, 19; m 42; c 3. PHARMACOLOGY. Educ: Univ Wash, BS, 48, MS, 49, PhD(pharmacol), 52. Prof Exp: From instr to prof pharmacol, Univ Wash, 49-68, asst chmn dept, 63-68, asst dean planning, 66-68; PROF MED EDUC & PHARMACOL & DIR OFF MED EDUC, SCH MED, UNIV CALIF, DAVIS, 68- Mem: Am Soc Pharmacol & Exp Therapeut. Res: Pharmacology of cardiac and smooth muscle; medical education. Mailing Add: Off of Med Educ Univ of Calif Sch of Med Davis CA 95616

WEST, WALTER SCOTT, b Fayette, Wis, Mar 12, 12; m 40; c 3. ECONOMIC GEOLOGY, GEOCHEMISTRY. Educ: Cornell Col, AB, 34; Univ Wis-Platteville, BE, 35; Univ Tenn, MS, 37. Prof Exp: Prin high sch, Mo, 37-38; instr geol, Univ NC, 40-42; eng aide & cartographer, Alaskan Geol Br, 42-46, geologist, 46-54, secy geol names comt, Geol Div, 54-67, CHIEF, WIS LEAD-ZINC PROJ, EASTERN MINERAL RESOURCES BR, GEOL DIV, US GEOL SURV, 66- Mem: Am Inst Mining, Metall & Petrol Eng; Arctic Inst NAm; Soc Econ Geologists. Res: Radioactive deposits, cement raw materials, pumice and riprap in Alaska; stream sediment, trace element and lead isotope studies; mineralogy and genesis of lead and zinc deposits in Wisconsin and Tennessee; geologic and geochemical mapping and topical studies in Upper Mississippi Valley zinc-lead district, Wisconsin, Illinois and Iowa. Mailing Add: US Geol Surv Rountree Hall Univ of Wis Platteville WI 53818

WEST, WARWICK REED, JR, b Evington, Va, Feb 9, 22; m 46; c 3. INVERTEBRATE ZOOLOGY. Educ: Lynchburg Col, BS, 43; Univ Va, PhD(biol), 52. Prof Exp: Instr biol, Lynchburg Col, 46-49 & 52; from asst prof to assoc prof, 52-66, PROF BIOL, UNIV RICHMOND, 66-, CHMN DEPT, 65- Mem: AAAS. Res: Milipede anatomy. Mailing Add: Dept of Biol Univ of Richmond Richmond VA 23173

WEST, WILLIAM ALVIN, b Essex Junction, Vt, May 3, 22; m 46; c 4. ORGANIC CHEMISTRY. Educ: Univ Vt, BS, 43; Rensselaer Polytech Inst, MS, 47, PhD(chem), 50. Prof Exp: Instr chem, Rensselaer Polytech Inst, 46-50; org chemist pharmaceut, Carter Prod, Inc, 50-54; RES ASSOC, E I DU PONT DE NEMOURS & CO, INC, 54- Mem: Am Chem Soc. Res: Organic and inorganic pigment research and development. Mailing Add: 7 Casho Mill Rd Newark DE 19711

WEST, WILLIAM IRVIN, b Dallas, Tex, Apr 26, 16; m 42. FORESTRY. Educ: Univ Wash, Seattle, BSF, 39, MF, 41. Prof Exp: Res assoc, Col Forestry, Univ Wash, Seattle, 41-42; assoc prof, 46-56, head dept, 46-67, PROF FOREST PROD, ORE STATE UNIV, 56- Mem: Soc Am Foresters; Forest Prod Res Soc; Am Soc Wood Sci & Technol. Res: Recovery of cork from Douglas fir bark; structure of pit membranes in bordered pit-pairs of coniferous species; giant tranverse resin canals which appear in Douglas fir; forest products industries. Mailing Add: Dept of Forest Prod Ore State Univ Corvallis OR 97331

WEST, WILLIAM LIONEL, b Charlotte, NC, Nov 30, 23; m 72. ZOOLOGY, PHARMACOLOGY. Educ: J C Smith Univ, BS, 47; Univ Iowa, PhD, 55. Prof Exp: Asst zool, Univ Iowa, 49-55, res assoc radiation, Col Med, 55-56; from instr to assoc prof pharmacol, 56-73, PROF PHARMACOL & RADIOL & CHMN DEPT PHARMACOL, COL MED, HOWARD UNIV, 73- Mem: Am Soc Zool; Am Soc Pharmacol & Exp Therapeut; Am Physiol Soc; Am Nuclear Soc; NY Acad Sci. Res: Biochemical and endocrine pharmacology; cellular physiology. Mailing Add: Dept of Pharmacol Howard Univ Col of Med Washington DC 20001

WEST, WILLIAM T, b Holyoke, Mass, June 26, 25; m 50; c 2. HISTOLOGY, ANATOMY. Educ: Am Inst Col, BA, 49; Univ Rochester, PhD(anat), 56. Prof Exp: Fel anat, Univ Rochester, 56-57, instr, 57-58; assoc staff scientist, Jackson Mem Lab, 58-62; asst prof, 62-69, ASSOC PROF ANAT, STATE UNIV NY DOWNSTATE MED CTR, 69- Res: Histopathology; human anatomy; pathology; endocrinology. Mailing Add: Dept of Anat State Univ NY Downstate Med Ctr Brooklyn NY 11203

WEST, WILLIAM WARREN, organic chemistry, petroleum chemistry, see 12th edition

WESTALL, FREDERICK CHARLES, b Pasadena, Calif, Nov 6, 43; m 68. BIOCHEMISTRY. Educ: Univ Calif, Los Angeles, BS, 64; San Diego State Col, MS, 66; Univ Calif, San Diego, PhD(chem), 70. Prof Exp: Multiple Sclerosis Soc res fel biochem, 70-72, res assoc, 72-73, ASST RES PROF BIOCHEM, SALK INST, 73-; SR FEL, LINUS PAULING INST SCI & MED, MENLO PARK, CALIF, 75- Mem: Am Chem Soc; Tissue Cult Soc; Soc Neurosci. Res: Biochemistry of neurological diseases; solid phase peptide synthesis; chromatographic techniques of disease identification; aging; immunological effects of adjuvants. Mailing Add: 10010 N Torrey Pine Rd La Jolla CA 92037

WESTAWAY, KENNETH C, b Hamilton, Ont, Aug 14, 38; m 62; c 3. PHYSICAL ORGANIC CHEMISTRY. Educ: McMaster Univ, BSc, 62, PhD(phys org chem), 68. Prof Exp: Asst prof chem, 68-75, ASSOC PROF CHEM, LAURENTIAN UNIV, 75-, CHMN DEPT, 74- Mem: Chem Inst Can; Am Chem Soc. Res: Mechanisms of organic reactions, with particular emphasis on nucleophilic substitution reactions and elimination reactions; effects of isotopic mass on the rates of organic reactions; identification and analysis of organic compounds. Mailing Add: Dept of Chem Laurentian Univ Sudbury ON Can

WESTBERG, KARL ROGERS, b Norwalk, Conn, Dec 17, 39; m 71. CHEMISTRY, PHYSICS. Educ: Bowdoin Col, BA, 61; Brown Univ, PhD(chem), 69. Prof Exp: Engr, Perkin-Elmer Corp, 61; MEM TECH STAFF, AEROSPACE CORP, 68- Mem: Am Chem Soc. Res: Chemical kinetics; air pollution chemistry; atmospheric chemistry; aerospace sciences. Mailing Add: Aerospace Corp PO Box 92957 Los Angeles CA 90009

WESTBROOK, DAVID REX, b London, Eng, May 12, 37; m 60; c 2. APPLIED MATHEMATICS. Educ: Univ London, BSc, 58, PhD, 61. Prof Exp: Lectr math, Univ Singapore, 61-64; sr lectr, Univ Melbourne, 64; lectr, Univ Nottingham, 64-65; vis mem, NY Univ, 65-66; asst prof, 66-71, ASSOC PROF MATH, UNIV CALGARY, 71- Concurrent Pos: Nat Res Coun Can res grant, 67-68. Mem: Am Math Soc; Soc Indust & Appl Math. Res: Theory of elastic plates and shells. Mailing Add: Dept Math Statist & Comput Sci Univ of Calgary Calgary AB Can

WESTBROOK, FRED EMERSON, b Arlington, Tenn, Dec 19, 16; m 44; c 2. AGRONOMY, PLANT SCIENCE. Educ: Tenn State Univ, BS, 46, MS, 47; Mich State Univ, PhD(crop sci), 54. Prof Exp: Instr high sch, Tenn, 47-48; prof agron, Tenn State Univ, 48-57, prof plant sci & head dept, 57-72; AGRONOMIST, FED EXTEN SERV, USDA, 72- Mem: Am Soc Agron; Soil Conserv Soc Am; Crop Sci Soc Am; Am Forage & Grassland Coun; Weed Sci Soc Am. Res: Soil fertility; micronutrient in crop production; soil and water pollution studies. Mailing Add: Fed Exten Serv USDA 5925 S Agr Bldg Washington DC 20250

WESTBROOK, GEORGE FRANKLIN, b Jacksonville, Fla, Nov 25, 15; m 38; c 2. FOOD CHEMISTRY. Educ: Univ Fla, BSA, 37, PhD(chem), 57. Prof Exp: Asst chemist, Exp Sta, Univ Fla, 34-37; CHIEF CHEMIST, FRUIT & VEG INSPECTION DIV, STATE DEPT AGR, FLA, 37- Mem: Am Chem Soc; Inst Food Technologists; Asn Food & Drug Officials. Res: Citrus chemistry and technology, especially fresh fruit and processed citrus products. Mailing Add: Fruit & Veg Inspection Div State Dept of Agr Winter Haven FL 33880

WESTBROOK, RUSSELL DAVID, b Cleveland, Ohio, Mar 21, 21; m 49; c 3. SOLID STATE PHYSICS. Educ: Col Wooster, BA, 42; Ohio State Univ, MS, 48. Prof Exp: Mem staff, US Naval Res Lab, 42-43; technician, Brush Develop Co, 46; group leader res labs, Union Carbide Corp, 48-66; PHYSICIST, SOLID STATE DIV, OAK RIDGE NAT LAB, 66- Mem: Inst Elec & Electronics Engrs. Res: Physics of carbon and graphite; intermetallic compounds; silicon; germanium; growth and analysis of semiconductor single crystals. Mailing Add: Solid State Div Oak Ridge Nat Lab PO Box X Oak Ridge TN 37830

WESTBY, CARL A, b Los Angeles, Calif, Feb 8, 36; m 58; c 4. MICROBIAL PHYSIOLOGY. Educ: Univ Calif, Riverside, AB, 58; Univ Calif, Davis, PhD(bact), 64. Prof Exp: Fel bact physiol, Sch Med, Univ Pa, 64-67; asst prof bact, Utah State Univ, 67-73; ASST PROF MICROBIOL, S DAK STATE UNIV, 73- Concurrent Pos: Med prod consult, Med Prod Div, 3M Co, 74- Mem: Am Soc Microbiol. Res: Bacterial physiology and genetics. Mailing Add: Dept of Microbiol SDak State Univ Brookings SD 57006

WESTCOTT, DONALD ELVIN, b Springfield, Mass, Apr 15, 29; m 49; c 3. FOOD TECHNOLOGY. Educ: Univ Mass, BS, 50, MS, 51, PhD(food technol), 54. Prof Exp: Instr food technol, Univ Mass, 52-54; chemist plastics develop, Pro-Phy-Lac-Tic Brush Co, 54-55; prod eval mgr, Cryovac Div, W R Grace & Co, 55-62; lab dir, United Fruit & Food Corp, 62-64; chief, Cereal & Gen Prod, 64-69, CHIEF, PLANT PROD DIV, US ARMY NATICK LABS, 69- Mem: Inst Food Technologists. Res: Formulation, processing, preservation and stability of plant products for military rations; flexible packaging materials; freeze-drying; food specifications; food research management; food flavor and texture. Mailing Add: 75 Charter Rd Acton MA 01720

WESTCOTT, PETER WALTER, b Mt Vernon, NY, Nov 5, 38; m 64. ECOLOGY, ZOOLOGY. Educ: Amherst Col, BA, 60; Univ Ariz, MS, 62; Univ Fla, PhD(zool), 70. Prof Exp: Asst prof zool, Univ Nebr, Lincoln, 70-71; PROF ECOL, BIOL SCI CTR, LONDRINA STATE UNIV, BRAZIL, 72- Mem: Am Ornith Union; Cooper Ornith Soc. Res: Field studies of avian social and breeding behavior and ecology; field studies of tropical avian social structures; experimental studies of animal mimicry. Mailing Add: Biol Sci Ctr Londrina State Univ Univ Campus PO Box 1530 Londrina 86-100 Parana Brazil

WESTCOTT, WAYNE LESLIE, b Blue Hill, Nebr, Sept 22, 18; m 41; c 2. BIOCHEMISTRY. Educ: Univ Nebr, BSc, 47; Univ Utah, MS, 49, PhD(biochem), 53. Prof Exp: Tech asst to supt biol & biochem mfg, E R Squibb & Sons, 53-55, head dept biol & biochem mfg, 55-59, sr res scientist, Squibb Inst Med Res Div, Olin Mathieson Chem Corp, 59-64, res supvr, Biol Dept, 64-68; head natural prod develop, Armour Pharmaceut Co, Ill, 68-72, dir mfg processes, 72-75; MGR PROD & PROCESS IMPROV, SCI PROD DIV, ABBOTT LABS, 75- Mem: Am Chem Soc; Soc Cryobiol; NY Acad Sci. Res: Isolation of natural products from tissues of animal origin; protein isolation; blood fractionation. Mailing Add: Sci Prod Div Abbott Labs 5555 Valley Blvd Los Angeles CA 90023

WEST EBERHARD, MARY JANE, b Pontiac, Mich, Aug 20, 41; m. ANIMAL BEHAVIOR, EVOLUTION. Educ: Univ Mich, AB, 63, MS, 65, PhD(zool), 67. Prof Exp: Res fel biol, Harvard Univ, 67-69; RES FEL BIOL, UNIV VALLE, COLOMBIA, 69- Concurrent Pos: Assoc, Smithsonian Trop Res Inst, 73- Mem: AAAS; Animal Behav Soc. Res: Evolution of social behavior; social biology of polistine wasps. Mailing Add: Dept of Biol Univ of Valle Cali Colombia

WESTENBARGER, GENE ARLAN, b Lancaster, Ohio, July 25, 35; m 58; c 4. PHYSICAL CHEMISTRY. Educ: Ohio Univ, BS, 57; Univ Calif, Berkeley, PhD(phys chem), 63. Prof Exp: Chemist, Battelle Mem Inst, 57; res & develop coordr, Qm Food & Container Inst, 58-59; asst prof, 63-67, ASSOC PROF PHYS CHEM, OHIO UNIV, 67- Mem: AAAS; Am Phys Soc; Am Chem Soc. Res: Thermodynamic and magnetic studies at low temperature. Mailing Add: Dept of Chem Ohio Univ Athens OH 45701

WESTENBERG, ARTHUR AYER, b Menomonie, Wis, Mar 1, 22; m 45; c 1. CHEMICAL PHYSICS. Educ: Carleton Col, AB, 43; Harvard Univ, AM, 48, PhD(chem), 50. Prof Exp: Chemist, Mayo Clin, 50; asst prof chem, Lafayette Col, 50-52; sr staff mem, 52-58, PRIN STAFF MEM, APPL PHYSICS LAB, JOHNS HOPKINS UNIV, 58-, SUPVR CHEM PHYSICS RES, 63- Concurrent Pos: Consult, Proj Squid, Off Naval Res, 60-65; mem adv comt, Army Res Off, 66-71; mem comt assess environ effects of supersonic transport, US Dept Com, 71. Honors & Awards: Hillebrand Award, Am Chem Soc, 66; Silver Medal, Combustion Inst, 66. Mem: Am Chem Soc; Am Phys Soc; Combustion Inst. Res: Chemical kinetics; high temperature gas properties; combustion; air pollution chemistry; electron spin resonance spectroscopy. Mailing Add: Appl Physics Lab Johns Hopkins Univ Laurel MD 20810

WESTER, ELBERT TRUMAN, b Burneyville, Okla, Jan 10, 19; m 37; c 3. MATHEMATICS. Educ: Southeastern State Col, BS, 50; Univ Okla, MA, 48, EdD, 59. Prof Exp: Supvr math, Southeastern State Col, 42-44; instr & asst math, Univ Okla, 47-48; asst prof, Cent State Col, Okla, 47-57, assoc prof math & registr, 57-60; educ specialist, Fed Aviation Agency Acad, Okla, 60-65; dir, Consult Ctr, Univ Okla, 65-67; vpres acad affairs, 67-72, PRES, GRAYSON COUNTY COL, 72- Mem: Math Asn Am. Mailing Add: Grayson County Col 6101 Hwy 691 Denison TX 75020

WESTER, JOHN WALTER, JR, b Youngstown, Ohio, Sept 30, 20; m 43; c 7. APPLIED MATHEMATICS. Educ: Dartmouth Col, AB, 42. Prof Exp: Res asst elec eng, Mass Inst Technol, 46-47; mathematician & assoc dir, Lab Appl Sci, Univ Chicago, 47-62; exec dir, Acad Intersci Methodology, 62-69, res dir, 69-73. Concurrent Pos: Chmn bd & pres, Wester Fuel & Supply Co, Indianola Realty Inc, Burnkey Inc & Wholesale Distrib Co, Youngstown, Ohio. Mem: Opers Res Soc Am; Soc Indust & Appl Math. Res: Planning and analysis of governmental and business operations management. Mailing Add: 18456 Dundee Ave Homewood IL 60430

WESTER, LYNDON LEONARD, b Adelaide, Australia, June 30, 45. GEOGRAPHY. Educ: Univ Adelaide, BA(hons), 67; Univ Calif, Los Angeles, MA, 69, PhD(geog), 75. Prof Exp: ASST PROF GEOG, UNIV HAWAII, 71- Mem: AAAS; Asn Am Geogrs; Inst Australian Geogrs; Soc Econ Bot. Res: Human modification of vegetation communities; establishment and spread of exotic species. Mailing Add: Dept of Geog Univ of Hawaii Honolulu HI 96822

WESTER, ROBERT EMERSON, b Miami, Fla, Dec 31, 05; m 31; c 3. HORTICULTURE. Educ: George Washington Univ, AB, 30, MA, 37. Prof Exp: Asst sci aide, USDA, 26-37; jr toxicologist, US War Dept, 37-38; jr plant physiologist, USDA, 38-42, asst horticulturist, 42-53, assoc horticulturist, Veg Sect, Hort Crops Res Br, Agr Res Serv, 53-58, horticulturist, 58-60, sr horticulturist, 60-64, prin horticulturist, 64-72, collabr, Appl Plant Path Lab, Plant Protection Inst, Agr Res Ctr, 72-75; RETIRED. Honors & Awards: Meritorious Serv Award, Bean Improv Coop Conv, 71; Superior Serv Award, USDA, 69; Citation of Merit at Farmers Home Week, Univ Del, 68. Mem: AAAS; Am Soc Hort Sci; Am Inst Biol Sci. Res: Breeding lima beans for resistance to heat, nematodes, downy mildew and anthracnose; effects of plastics on weed control, plant growth and greenhouse; suburban, home and farm vegetable gardening bulletins; minigardens. Mailing Add: 4314 Howard Rd Beltsville MD 20705

WESTERBERG, MARTHA ROSALIE, b Rockford, Ill, Feb 20, 14; m 44. MEDICINE, NEUROLOGY. Educ: Rockford Col, AB, 35; Univ Chicago, MD, 41; Am Bd Pyschiat & Neurol, dipl, 47. Prof Exp: From instr to assoc prof, 45-62, PROF NEUROL, MED SCH, UNIV MICH, ANN ARBOR, 62- Concurrent Pos: Consult, Vet Admin Hosp, Ann Arbor, 54-55 & Wayne County Gen Hosp; mem, Med Adv Bd, Myasthenia Gravis Found. Mem: AAAS; AMA; Am Neurol Asn; fel Am Acad Neurol. Res: Myasthenia gravis; migraine. Mailing Add: Dept of Neurol Univ of Mich Med Sch Ann Arbor MI 48104

WESTERDAHL, CAROLYN ANN LOVEJOY, b Oklahoma City, Okla, Apr 16, 35; m 61. SURFACE CHEMISTRY. Educ: Univ Chicago, BA, 55, BS, 57; Univ Calif, Berkeley, PhD(chem), 61. Prof Exp: CHEMIST, PICATINNY ARSENAL, 67- Mem: Sigma Xi. Res: Adhesion. Mailing Add: Mat Eng Div Picatinny Arsenal Dover NJ 07801

WESTERDAHL, RAYMOND P, b Chicago, Ill, Mar 22, 29; m 61. PHYSICAL CHEMISTRY. Educ: Univ Ill, BS, 51; Univ Chicago, MS, 59, PhD(phys chem), 62. Prof Exp: Res chemist, Esso Res & Eng Co, Standard Oil Co, NJ, 62-67; RES CHEMIST, FELTMAN RES LAB, PICATINNY ARSENAL, US ARMY, 67- Mem: AAAS; fel Am Inst Chemists; Am Chem Soc; Soc Appl Spectros. Res: Mechanisms of pyrotechnic reactions; high temperature Raman spectroscopy; structure and spectra of flames; chemiluminescent reactions; reactions in fused salts; analysis of air pollutants; theory of ignition. Mailing Add: 7 Comanche Trail Denville NJ 07834

WESTERFELD, WILFRED WIEDEY, b St Charles, Mo, Dec 13, 13; m 38; c 5. BIOCHEMISTRY. Educ: Univ Mo, BS, 34; St Louis Univ, PhD(biochem), 38. Prof Exp: Tutor, Harvard Univ, 40-41, instr biochem, Med Sch, 41-42, assoc, 42-44, asst prof, 44-45; PROF BIOCHEM, COL MED, STATE UNIV NY UPSTATE MED CTR, 45- Concurrent Pos: Nat Res Coun res fel med sci, Oxford Univ, 38-39 & Columbia Univ, 39-40; teaching fel, Harvard Univ, 40-41; actg dean, Col Med, State Univ NY Upstate Med Ctr, 56-57, assoc dean, 60-64, dean grad studies basic sci, 64-68, actg pres, Med Ctr, 67-68. Mem: AAAS; Am Chem Soc; Am Soc Biol Chemists; Soc Exp Biol & Med. Res: Estrogens; xanthine oxidase; alcohol metabolism; thyroid. Mailing Add: Dept of Biochem State Univ of NY Upstate Med Ctr Syracuse NY 13210

WESTERFIELD, CLIFFORD, b Ohio Co, Ky, Apr 7, 08; m 36; c 4. VETERINARY ANATOMY. Educ: Western Ky State Teachers Col, BS, 30; Univ Ky, MS, 32; Mich

State Col, DVM, 38. Prof Exp: Instr sci, Western Ky State Teachers Col, 30-31; high sch teacher, 33-35; jr vet, Bur Animal Indust, USDA, Washington, DC, 38-39; asst vet, Exp Sta, Univ Ky, 39-40, assoc animal pathologist, 45-46; asst prof vet anat, Ohio State Univ, 40-44; prof vet anat & histol & head dept, Univ Ga, 46-53; asst prof, Iowa State Col, 53-54; prof vet anat & histol & head dept, 54-62, prof vet anat, 62-75, EMER PROF VET ANAT, COL VET MED, UNIV GA, 75- Mem: AAAS; Am Asn Vet Anat (secy, 58, pres, 60); Am Vet Med Asn; World Asn Vet Anat. Res: Equine diseases; histopathology; veterinary histology. Mailing Add: 116 Whitehead Rd Athens GA 30601

WESTERFIELD, EVERETT COMMODORE, b Beda, Ky, Apr 3, 01; m 30; c 2. PHYSICS. Educ: Univ Colo, BA, 28, MA, 30, PhD(physics), 40. Prof Exp: Apparatus asst, Univ Colo, 35-39, asst, 40; lab technician & tech adv, Douglas Aircraft Co, 41-43, phys test engr, 43-45; assoc physicist war res, Univ Calif, 45-46; physicist, US Navy Electronic Lab, 46-70, res physicist, Naval Underwater Ctr, 70-75; RETIRED. Mem: AAAS; assoc Am Phys Soc; assoc Am Math Soc; Soc Indust & Appl Math; Inst Elec & Electronics Engrs. Res: Underwater sound; electronics; communication and information theory. Mailing Add: Box 6206 Point Loma Sta San Diego CA 92106

WESTERHOUT, GART, b The Hague, Neth, June 15, 27; m 56; c 5. ASTRONOMY. Educ: State Univ Leiden, Drs, 54, PhD(astron), 58. Prof Exp: Res asst astron, Univ Observ, State Univ Leiden, 52-54, sci officer, 54-59, chief sci officer, 59-62; dir astron prog, 62-73, chmn div math, phys & eng sci, 72-73, PROF ASTRON, UNIV MD, COLLEGE PARK, 62- Concurrent Pos: NATO fel, 59; mem user's comt, Nat Radio Astron Observ, 65-; mem US nat comt, Int Astron Union, 71-74; trustee-at-large, Assoc Univs Inc, 71-74; mem comt radio frequencies, Nat Res Coun, 72-; mem US nat comt, Int Sci Radio Union, 72-78, chmn comn radio-astron, 75-78; Humboldt Found sr fel, Ger, 73; mem, Inter-Union Comn for Allocation of Frequencies, 75- Mem: Am Astron Soc; Royal Astron Soc; Int Astron Union. Res: Radio astronomy; 21-centimeter line research; galactic structure. Mailing Add: Astron Prog Univ of Md College Park MD 20742

WESTERLUND, LAWRENCE HECKER, physics, see 12th edition

WESTERMAN, LOWELL, inorganic chemistry, see 12th edition

WESTERMANN, D T, b July 4, 41; US citizen. SOIL SCIENCE, SOIL CHEMISTRY. Educ: Colo State Univ, BS, 63; Ore State Univ, MS, 65, PhD, 68. Prof Exp: SOIL SCIENTIST, SNAKE RIVER CONSERV RES CTR, AGR RES SERV, USDA, 68- Concurrent Pos: Affil prof, Univ Idaho, 71- Res: Plant nutrition. Mailing Add: Snake River Conserv Res Ctr USDA Rte 1 Box 186 Kimberly ID 83341

WESTERMANN, GERD ERNST GEROLD, b Berlin, Ger, May 11, 27; m 56; c 3. GEOLOGY. Educ: Brunswick Tech Univ, BSc, 50; Univ Tübingen, MSc & PhD(geol, paleont), 53. Prof Exp: Paleontologist, Geol Surv Ger, 53-57; from lectr to assoc prof, 57-69, PROF GEOL, McMASTER UNIV, 69- Concurrent Pos: Consult, 58-62; mem, Chilean Jurassic Comt. Mem: Paleont Res Inst; Am Paleont Soc; Soc Econ Paleont & Mineral; Can Palaeont Asn; Int Paleont Union (secy-gen). Res: Mesozoic Mollusca, especially Jurassic Ammonoidea and Triassic Pectinacea; functional morphology; taxonomy; intercontinental biochronology. Mailing Add: Dept of Geol McMaster Univ Hamilton ON Can

WESTERN, DONALD WARD, b Poland, NY, May 7, 15; m 43; c 5. MATHEMATICS. Educ: Denison Univ, BA, 37; Mich State Col, MA, 39; Brown Univ, PhD(math), 46. Prof Exp: Instr math, Mich State Col, 37-39; from instr to asst prof, Brown Univ, 39-48; from assoc prof to prof math, 48-74, chmn dept math & astron, 52-72, CHARLES A DANA PROF MATH, FRANKLIN & MARSHALL COL, 74- Concurrent Pos: NSF fac fel, 60. Mem: Am Math Soc; Math Asn Am. Res: Inequalities in the complex plane. Mailing Add: Dept of Math & Astron Franklin & Marshall Col Lancaster PA 17604

WESTERVELT, CLINTON ALBERT, JR, b Portland, Ore, June 15, 36; m 65. INVERTEBRATE ZOOLOGY, PARASITOLOGY. Educ: Lewis & Clark Col, BA, 58; Univ Ariz, MS, 61, PhD(zool), 66. Prof Exp: Asst prof, 65-69, ASSOC PROF BIOL, CHAPMAN COL, 69- Mem: Am Soc Parasitologists. Res: Biology and systematics of decapod crustaceans, primarily Anomurans. Mailing Add: Dept of Biol Chapman Col Orange CA 92666

WESTERVELT, DONALD RAMSEY, b New York, NY, May 6, 26; m 49; c 5. PHYSICS. Educ: Calif Inst Technol, BS, 49. Prof Exp: Res engr, NAm Aviation, Inc, 49-55; mem staff, 55-57, assoc group leader, Test Div, 57-59, alternate group leader, 60-66, GROUP LEADER, TEST DIV, LOS ALAMOS SCI LAB, 66- Concurrent Pos: Consult, Adv Res Projs Agency, 65; assoc sci dep air drop prog, Joint Task Force Eight, Defense Atomic Support Agency & chmn aircraft use comt, Nev Opers Off, AEC, 68-; mem US deleg, Conf Comt on Disarmament, 74-75. Mem: Am Phys Soc; Int Inst Strategic Studies. Res: Radiation effects in ionic crystals; spectroscopic and diagnostic experimentation in nuclear weapons test program; detection of nuclear explosions; physics of upper atmosphere and high altitude nuclear explosions; molecular physics. Mailing Add: Los Alamos Sci Lab Box 1663 Los Alamos NM 87544

WESTERVELT, PETER JOCELYN, b Albany, NY, Dec 16, 19. THEORETICAL PHYSICS. Educ: Mass Inst Technol, BS, 47, MS, 49, PhD(physics), 51. Prof Exp: Mem staff, Radiation Lab, Mass Inst Technol, 40-41 & Underwater Sound Lab, 41-45, asst physics, 46-47, res assoc, 48-50; from asst prof to assoc prof, 51-70, PROF PHYSICS, BROWN UNIV, 70- Concurrent Pos: Consult to asst attache for res, US Navy, Am Embassy, London, 51-52, Bolt, Beranek & Newman, Inc & Appl Res Labs, Univ Tex, Austin, 71- Mem subcomt aircraft noise, NASA, 54-59; mem comt hearing & bio-acoust, Armed Forces-Nat Res Coun, 57-, mem exec coun, 60-61, chmn, 67-68; mem sonic boom comt, Nat Acad Sci, 68-71. Mem: Fel Acoust Soc Am; fel Am Phys Soc; Am Astron Soc. Res: Physical effects of high amplitude sound waves; air acoustics; underwater sound; general relativity. Mailing Add: Dept of Physics Brown Univ Providence RI 02912

WESTFAHL, JEROME CLARENCE, b Milwaukee, Wis, Jan 10, 20; m 42; c 3. ORGANIC CHEMISTRY. Educ: Univ Wis, BS, 42; Cornell Univ, PhD(chem), 50. Prof Exp: Chemist, B F Goodrich Co, 42-47; asst, Baker Lab, Cornell Univ, 47-50; res chemist, 50-60, res assoc, 60-68, SR RES ASSOC, RES & DEVELOP CTR, B F GOODRICH CO, 68- Mem: Am Chem Soc. Res: Synthesis of substituted 1,3-butadienes; chemistry of vinylidene cyanide; nuclear magnetic resonance and electron paramagnetic resonance spectroscopy; computer applications in chemistry. Mailing Add: B F Goodrich Res & Develop Ctr Brecksville OH 44141

WESTFALL, ARTHUR OSCAR, b Santa Monica, Calif, Feb 17, 18; m 45; c 4. HYDROLOGY, AGRICULTURAL ENGINEERING. Educ: Calif State Polytech Col, BS, 52. Prof Exp: Hydrol engr, Water Resources Div, US Geol Surv, 53-64, US Agency Int Develop tech adv, Off Int Activ, Surv, Afghanistan, 64-69, hydraul engr, 69-72, ASST DIST CHIEF HYDROLOGIST, WATER RESOURCES DIV, US GEOL SURV, 72- Res: River basin hydrology; planning and development of nationwide river gaging station network in Afghanistan. Mailing Add: US Geol Surv 975 W Third Ave Columbus OH 43212

WESTFALL, BERTIS ALFRED, b Halfway, Mo, June 4, 07; m 33. PHARMACOLOGY, PHYSIOLOGY. Educ: Univ Mo, AB, 33, MA, 34, PhD(physiol, pharmacol), 38. Prof Exp: From instr to assoc prof physiol & pharmacol, 36-48, prof pharmacol, 48-72, chmn dept physiol & pharmacol, 53-65, chmn dept pharmacol, 65-72, EMER PROF PHARMACOL, SCH MED, UNIV MO-COLUMBIA, 72- Concurrent Pos: From assoc aquatic physiologist to aquatic physiologist, US Fish & Wildlife Serv, 34-46. Mem: Am Physiol Soc; Soc Exp Biol & Med; Am Soc Pharmacol & Exp Therapeut. Res: Arsenic storage; water balance; pyruvic acid; barbiturates; glycogen; plant extractives; comparative pharmacology. Mailing Add: Dept of Pharmacol Univ of Mo Sch of Med Columbia MO 65201

WESTFALL, DAVID PATRICK, b Harrisburg, WVa, June 9, 42; m 65; c 2. PHARMACOLOGY. Educ: Brown Univ, BA, 64; WVa Univ, MS, 66, PhD(pharmacol), 68. Prof Exp: Demonstr pharmacol, Oxford Univ, 68-70; asst prof, 70-73, ASSOC PROF PHARMACOL, MED SCH, WVA UNIV, 73- Concurrent Pos: J H Burn fel pharmacol, Oxford Univ, 68-70. Mem: AAAS; Am Soc Pharmacol & Exp Therapeut. Res: Pharmacology and physiology of smooth and cardiac muscle; factors governing the sensitivity of muscle to drugs. Mailing Add: Dept of Pharmacol WVa Univ Med Sch Morgantown WV 26506

WESTFALL, DWAYNE GENE, b Aberdeen, Idaho, Nov 21, 38; m 61; c 2. SOIL CHEMISTRY, AGRONOMY. Educ: Univ Idaho, BS, 61; Wash State Univ, PhD(soils), 68. Prof Exp: Res asst soils, Wash State Univ, 66-67; from asst prof to assoc prof soil chem, 67-74, ASSOC PROF SOIL & CROP SCI, AGR RES & EXTEN CTR, TEX A&M UNIV, 74- Mem: Am Chem Soc; Am Soc Agron; Soil Sci Soc Am. Res: Soil chemistry of submerged soils; fertility of rice and pastures with speciality in micronutrients and nitrogen efficiency and utilization; pollution and quality of irrigation waters. Mailing Add: Agr Res & Exten Ctr Tex A&M Univ Rte 5 Box 366 Beaumont TX 77706

WESTFALL, JANE ANNE, b Berkeley, Calif, June 21, 28. NEUROSCIENCES. Educ: Col Pac, AB, 50; Mills Col, MA, 52; Univ Calif, Berkeley, PhD(zool), 65. Prof Exp: Res asst zool, Univ NC, 52-53 & Univ Calif, Berkeley, 55-56; lab technician cancer res, Univ Calif, Berkeley, 57-58, lab technician zool, 58-65, asst res zoologist, 65-67; asst prof anat, 67-70, ASSOC PROF PHYSIOL SCI & DIR ULTRASTRUCT RES LAB, KANS STATE UNIV, 70- Concurrent Pos: Vis prof molecular, cellular & develop biol, Univ Colo, Boulder, 74-75. Mem: Electron Micros Soc Am; Am Soc Cell Biol; Am Soc Zoologists; Soc Neurosci; Am Asn Anat. Res: Electron microscopy, scanning, conventional and high voltage, of synapses and neuromuscular junctions in simple systems in muscular dystrophy. Mailing Add: Ultrastruct Res Lab Kans State Univ Manhattan KS 66506

WESTFALL, JOHN EDWARD, b San Francisco, Calif, Aug 16, 38; m 65; c 2. HISTORICAL GEOGRAPHY. Educ: Univ Calif, Berkeley, BA, 60; George Washington Univ, MA, 64, PhD(geog), 69. Prof Exp: Cartogr, US Coast & Geod Surv, 60-64; asst prof, 68-71, chmn dept, 70-71, ASSOC PROF GEOG, SAN FRANCISCO STATE UNIV, 71- Concurrent Pos: Dept Health, Educ & Welfare grant, San Francisco State Univ, 69. NSF fel, 69-70, Carnegie Corp fel, 72-73; planning consult, Appalachian Study Proj, Litton Industs, 65 & Marcou-O'Leary Assocs, 66; res asst, Urban Land Inst, 65-66. Mem: AAAS; fel Am Geog Soc; Asn Am Geog. Res: Quantitative applications in historical and cultural geography, especially historical demography and urban history; computer applications in geography. Mailing Add: Dept Geog San Francisco State Univ 1600 Holloway Ave San Francisco CA 94132

WESTFALL, JONATHAN JACKSON, b Buckhannon, WVa, July 10, 08; m 38; c 2. BOTANY. Educ: WVa Wesleyan Col, BS, 32; Univ Chicago, MS, 38, PhD(bot), 39. Prof Exp: High sch teacher, WVa, 32-36; asst bot, Univ Chicago, 38-39; instr biol, Moorehead State Teachers Col, 39-47, actg head div, 46-47; head dept bot, 47-62, actg chmn div biol sci, 59-60, dir, NSF Acad Year Inst, 59-71, actg chmn div biol sci, 63-64, PROF BOT, UNIV GA, 47- Concurrent Pos: Consult, Am Inst Biol Sci, 61. Mem: AAAS; Bot Soc Am; Ecol Soc Am; Am Genetic Asn; Nat Asn Biol Teachers. Res: Plant morphology; cytology; cytogenetics. Mailing Add: 160 Terrell Dr Athens GA 30601

WESTFALL, MINTER JACKSON, JR, b Orlando, Fla, Jan 28, 16; m 45; c 3. BIOLOGY. Educ: Rollins Col, BS, 41; Cornell Univ, PhD(nature study), 47. Prof Exp: Wildlife technician, Ala Coop Wildlife Res Unit, 37; asst dir mus, Rollins Col, 37-40; from asst to sr asst biol, Cornell Univ, 42-47; from asst prof to assoc prof, 47-69, PROF ZOOL & ENTOM, UNIV FLA, 69- Concurrent Pos: US dep game warden, 36-70; Howell fel, Highlands Biol Sta, NC, 53. Mem: Entom Soc Am. Res: Wildlife management of mourning dove; bird migration and movement; taxonomy, ecology, zoogeography, life histories of the Odonata. Mailing Add: Dept of Zool Univ of Fla Gainesville FL 32601

WESTFALL, ROBERT JUDSON, b Pictou, Colo, Aug 24, 13; m 41; c 2. BIOCHEMISTRY. Educ: Utah State Col, BS, 35; Mich State Col, MS, 37; Purdue Univ, PhD(biochem), 47. Prof Exp: Biochemist, Merck Sharp & Dohme Res Labs, 42-57; sr biochemist, Process Develop Dept, 57-71, SR BIOCHEMIST, REHEIS RES & DEVELOP DEPT, ARMOUR PHARMACEUT CO, 71- Mem: AAAS; Am Chem Soc. Res: Protein chemistry; bovine plasma fractions. Mailing Add: Armour Pharmaceut Co PO Box 511 Kankakee IL 60901

WESTFALL, THOMAS CREED, b Latrobe, Pa, Oct 31, 37; m 61; c 1. PHARMACOLOGY. Educ: WVa Univ, AB, 59, MS, 61, PhD(pharmacol), 62. Prof Exp: From instr to asst prof pharmacol, WVa Univ, 62-65; from asst prof to assoc prof, 65-74, PROF PHARMACOL, SCH MED, UNIV VA, 69- Concurrent Pos: Nat Heart Inst fel physiol, Karolinska Inst, Sweden, 63-64; dir grad studies, Dept Pharmacol, Sch Med, Univ Va, 68-; IUPHAR int fel, 74; vis fac scholar, Group Biochem Neuropharmacol, Lab Molecular Biol, Col France, Paris, 74-75. Mem: AAAS; Am Soc Pharmacol & Exp Therapeut; Soc Exp Biol & Med; Soc Neurosci; Am Heart Asn. Res: Autonomic, cardiovascular, biochemical and neuropharmacology; influence of drugs on the syntheses, uptake, storage, metabolism and receptor interaction of biogenic amines, particularly catecholamines; cardiovascular and autonomic actions of nicotine; neurotransmitter physiology and pharmacology. Mailing Add: Dept of Pharmacol Univ of Va Sch of Med Charlottesville VA 22903

WESTGARD, JAMES BLAKE, b Billings, Mont, Feb 12, 35; m 66; c 2. ELEMENTARY PARTICLE PHYSICS. Educ: Reed Col, BA, 57; Syracuse Univ, PhD(physics), 63. Prof Exp: Res fel physics, Carnegie Inst Technol, 63-66; asst prof, 66-73, ASSOC PROF PHYSICS, IND STATE UNIV, TERRE HAUTE, 66- Mem: Am Phys Soc. Res: Particle physics. Mailing Add: Dept of Physics Ind State Univ Terre Haute IN 47809

WESTHAUS, PAUL ANTHONY, b St Louis, Mo, Dec 10, 38. ATOMIC PHYSICS, MOLECULAR PHYSICS. Educ: St Louis Univ, BS, 61; Washington Univ, PhD(physics), 66. Prof Exp: Proj assoc, Theoret Chem Inst, Univ Wis, 66-67; ASSOC PROF PHYSICS, OKLA STATE UNIV, 68- Concurrent Pos: Res staff scientist, Yale Univ, 69; NIH career develop award, 72-77. Mem: Am Phys Soc. Res: Atomic structure and radiative transitions; electronic structure of molecules and intermolecular forces; quantum biology; many-electron problem. Mailing Add: Dept of Physics Okla State Univ Stillwater OK 74074

WESTHEAD, EDWARD WILLIAM, JR, b Philadelphia, Pa, June 19, 30; m; c 2. BIOCHEMISTRY. Educ: Haverford Col, BA, 51, MA, 52; Polytech Inst Brooklyn, PhD(polymer & phys chem), 55. Prof Exp: Am Scand Found res fel, Biochem Inst, Univ Uppsala, 55-56, NSF res fel, 56-57; res assoc physiol chem, Univ Minn, 58-60; asst prof biochem, Dartmouth Med Sch, 61-66; assoc prof & actg head dept, 66-71, PROF BIOCHEM, UNIV MASS, AMHERST, 71- Concurrent Pos: NIH career develop award, 61-66; NIH spec fel, Oxford Univ, 72-73. Mem: Am Chem Soc; Am Soc Biol Chemists. Res: Enzyme mechanisms and control; biochemistry of catecholamine secretion. Mailing Add: Dept of Biochem Univ of Mass Amherst MA 01002

WESTHEIMER, FRANK HENRY, b Baltimore, Md, Jan 15, 12; m 37; c 2. ORGANIC CHEMISTRY, ENZYMOLOGY. Educ: Dartmouth Col, AB, 32; Harvard Univ, MA, 33, PhD(chem), 35. Hon Degrees: DSc, Dartmouth Col, 61 & Univ Chicago, 73. Prof Exp: Nat Res Coun fel chem, Columbia Univ, 35-36; res assoc org chem, Univ Chicago, 36-37, from instr to asst prof chem, 37-44; res supvr, Explosives Res Lab, Nat Defense Res Comt, Pa, 44-45; from assoc prof to prof chem, Univ Chicago, 46-54; prof chem, 54-60, chmn dept, 59-62, LOEB PROF CHEM, HARVARD UNIV, 60- Concurrent Pos: Vis prof chem, Harvard Univ, 53-54; chmn comt surv chem, Nat Acad Sci, 64-65, mem coun, 72-75; mem, President's Sci Adv Comt, 67-70. Honors & Awards: Willard Gibbs Medal, 70; James Flack Norris Award, Am Chem Soc, 70. Mem: Am Chem Soc. Res: Electrostatic effects in organic chemistry; mechanism of nitration; chemical and biochemical oxidation; theory of steric effects; enzymic and chemical decarboxylation; phosphate esters; photoaffinity labeling. Mailing Add: Conant Lab Harvard Univ Cambridge MA 02138

WESTHEIMER, GERALD, b Berlin, Ger, May 13, 24. PHYSIOLOGICAL OPTICS. Educ: Univ Sydney, BSc, 47; Ohio State Univ, PhD(physics), 53. Prof Exp: Optometrist, Australia, 45-51; assoc optom, Ohio State Univ, 51-52; from asst prof to assoc prof physiol optics, 54-60; prof, Univ Houston, 53-54; assoc prof physiol optics & optom, 60-63, prof physiol optics, 63-68, chmn physiol optics group, 64-67, PROF PHYSIOL, UNIV CALIF, BERKELEY, 68- Concurrent Pos: Vis researcher, Physiol Lab, Cambridge Univ, 58-59; mem, Nat Acad Sci-Nat Res Coun Comt Vision, 57-72, mem exec coun, 69-72; mem visual sci study sect, NIH, 66-70, mem vision res training comt, Nat Eye Inst, 70-74. Mem: Fel AAAS; Optical Soc Am; Soc Neurosci; assoc Brit Physiol Soc. Res: Biophysics and physiology of visual system; neurophysiology. Mailing Add: Dept of Physiol-Anat Univ of Calif Berkeley CA 94720

WESTHOFF, DENNIS CHARLES, b Jersey City, NJ, Nov 20, 42; m 63; c 1. FOOD MICROBIOLOGY. Educ: Univ Ga, BSA, 66; NC State Univ, MS, 68, PhD(food sci), 71. Prof Exp: ASST PROF DAIRY SCI, UNIV MD, COLLEGE PARK, 70- Mem: Am Soc Microbiol; Am Dairy Sci Asn; NY Acad Sci. Res: Bioprocessing of foods; waste utilization. Mailing Add: Dept of Dairy Sci Univ of Md College Park MD 20742

WESTING, ARTHUR HERBERT, b New York, NY, July 18, 28; m 56; c 2. BOTANY. Educ: Columbia Univ, AB, 50, Yale Univ, MF, 54, PhD, 59. Prof Exp: Res forester, US Forest Serv, 54-55; asst prof forestry, Purdue Univ, 59-64; assoc prof tree physiol, Univ Mass, 64-65; assoc prof biol, Middlebury Col, 65-66; assoc prof bot, 66-71, chmn dept biol, 66-74, PROF BOT, WINDHAM COL, 71- Concurrent Pos: Fel forest biol, NC State Col, 60; Bullard fel, Harvard Univ, 63-64, res fel, 70; fel bot, Univ Mass, 66; trustee, Vt Wild Land Found, 66- & Vt Acad Arts & Sci, 67-71; fel nuclear sci, St Augustine's Col, 68. Mem: Fel AAAS; Bot Soc Am; Soc Am Foresters; Fedn Am Sci. Res: Tree physiology; forest botany; plant growth and development; ecology of war. Mailing Add: Dept of Biol Windham Col Putney VT 05346

WESTKAEMPER, REMBERTA, b Spring Hill, Minn, Jan 2, 90. BOTANY. Educ: Col St Benedict, Minn, BA, 19; Univ Minn, MS, 22, PhD(bot), 29. Prof Exp: Asst prof bot, 17-22, prof biol, 22-57, head dept, 29-57, pres, 57-61, prof biol, 61-73, EMER PROF BIOL, COL ST BENEDICT, MINN, 73- Mem: AAAS. Res: Vitamin content of marine algae; flora of Stearns County. Mailing Add: Col of St Benedict St Joseph MN 56374

WESTLAKE, DONALD WILLIAM SPECK, b Woodstock, Ont, Feb 27, 31; m 54. MICROBIOLOGY. Educ: Univ BC, BS, 53, MS, 55; Univ Wis, PhD, 58. Prof Exp: From asst res officer to assoc res officer, Nat Res Coun Can, 58-66; assoc prof microbiol, 66-69, PROF MICROBIOL & CHMN DEPT, UNIV ALTA, 69- Mem: Can Soc Microbiol; Chem Inst Can; Can Soc Biochem; Brit Soc Gen Microbiol. Res: Microbial biochemistry; interconversion of aromatic compounds; metabolic pathways; fermentations. Mailing Add: Dept of Microbiol Fac of Sci Univ of Alta Edmonton AB Can

WESTLAKE, HARRY EDWARD, JR, b Jersey City, NJ, Oct 13, 15; m 43; c 2. ORGANIC CHEMISTRY. Educ: Princeton Univ, AB, 37, AM, 39, PhD(chem), 40; NY Univ, JD, 58. Prof Exp: Fel, Mellon Inst Indust Res, 40-44, sr fel, 44-46; chemist, Am Cyanamid Co, NJ, 46-52, tech patent supvr, 52-54, patent agt, Conn, 54-58, patent attorney, 58-60; patent sect chief, 60-66, asst patent counsel, 66-69, patent counsel, 69-72, ASST DIR PATENTS, MERCK & CO, INC, 72- Mem: Am Chem Soc; The Chem Soc. Res: Organic sulfur compounds; sulfur containing resins; vat dyestuffs; sulfur dyes; patent law. Mailing Add: Patent Dept Merck & Co Rahway NJ 07065

WESTLAKE, PHILIP RADCLIFFE, information science, neuroscience, see 12th edition

WESTLAKE, ROBERT ELMER, b Jersey City, NJ, Oct 2, 18; m 44; c 3. INTERNAL MEDICINE, CARDIOLOGY. Educ: Princeton Univ, AB, 40; Columbia Univ, MD, 43. Prof Exp: From instr to asst prof med, Col Med, State Univ NY Upstate Med Ctr, 49-52, from clin asst prof to clin prof, 52-67; DIR PROF SERV, COMMUNITY-GEN HOSP, 67- Concurrent Pos: Mem, Adv Comt Heart, Cancer & Stroke, Surgeon Gen, 66-67; mem, Adv Coun, US Dept Defense, 69-74; electrocardiographer, Syracuse Med Ctr; pres, PSRO of Cent NY; chmn, Bd Dirs, Blue Shield Cent NY. Mem: AAAS; Am Col Physicians; Am Fedn Clin Res: Am Soc Internal Med (pres, 65-66); fel Soc Advan Med Systs. Res: Electrocardiography. Mailing Add: 600 E Genesee Syracuse NY 13202

WESTLAKE, WILFRED JAMES, b Weston-super-Mare, Eng, Mar 21, 29; US citizen; m 69. MATHEMATICS, STATISTICS. Educ: Univ London, BSc, 50, PhD(math), 53. Prof Exp: Sci officer, Govt Commun Hq, UK, 52-56, sr sci officer, 56-60; sr statistician, E I du Pont de Nemours & Co, 60-63; staff specialist, Control Data Corp, 63-66; prin scientist, Booz-Allen Appl Res, Inc, Calif, 66, res dir, 66-68; asst dir biostatist, 68-70, assoc dir, 70-73, DIR BIOSTATIST, SMITH KLINE & FRENCH LABS, 73- Concurrent Pos: Adj assoc prof biomet, Sch Med, Temple Univ, 69-; assoc ed, Biometrics, 75- Mem: Fel Am Statist Asn; Inst Math Statist; Biomet Soc. Res: Design of experiments; analysis of variance; regression analysis. Mailing Add: Dept of Biostatist Smith Kline & French Labs Philadelphia PA 19101

WESTLAKE, WILLIAM ELLIS, b Melrose, Iowa, Dec 3, 07; m 35; c 2. PESTICIDE CHEMISTRY. Educ: Mont State Col, BSc, 33; Univ Minn, PhD(biochem), 42. Prof Exp: Entomologist, Bur Entom & Plant Quarantine, USDA, 38-41; chemist, Fed Cartridge Corp, 42-44; chemist, Entom Res Div, USDA, 44-65; res chemist, Dept Entom, Univ Calif, Riverside, 65-75; ED, ARCH ENVIRON CONTAMINATION & TOXICOL, 74- Mem: AAAS; Am Chem Soc; Entom Soc Am. Res: Pesticide residues and methods for their determination in foods and the environment. Mailing Add: PO Box 1225 Twain Harte CA 95383

WESTLAND, ALAN DUANE, b Toledo, Ohio, Dec 29, 29; m 56. INORGANIC CHEMISTRY. Educ: Univ Toronto, BA, 53, MA, 54, PhD(chem), 56. Prof Exp: Fel, Univ Münster, 56-58; from asst prof to assoc prof, 58-69, PROF INORG CHEM, UNIV OTTAWA, 69- Mem: Am Chem Soc; Chem Inst Can; The Chem Soc. Res: Preparative and physical inorganic chemistry of the transition elements. Mailing Add: Dept of Chem Univ of Ottawa Ottawa ON Can

WESTLAND, ROGER DEAN, organic chemistry, see 12th edition

WESTLEY, JOHN LEONARD, b Wilsonville, Nebr, Aug 29, 27; m 56; c 3. ENZYMOLOGY, PHYSICAL BIOCHEMISTRY. Educ: Univ Chicago, PhB, 48, PhD(biochem), 54. Prof Exp: NSF fel biochem, Calif Inst Technol, 54-55, res fel, 55-56; from instr to assoc prof, 56-64, PROF BIOCHEM, UNIV CHICAGO, 64- Concurrent Pos: USPHS res career develop award, 62-72. Mem: Am Soc Biol Chemists. Res: Mechanisms of enzyme action; prebiotic catalysis; sulfur metabolism; kinetic analysis; enzyme regulation. Mailing Add: Dept of Biochem Univ of Chicago Chicago IL 60637

WESTLEY, JOHN WILLIAM, b Cambridge, Eng, Feb 5, 36; m 59; c 2. ORGANIC CHEMISTRY, BIOCHEMISTRY. Educ: Univ Nottingham, BSc, 58, PhD(org chem), 61. Prof Exp: Res assoc org chem, Stanford Univ, 61-62 & Med Ctr, Univ Calif, San Francisco, 62-63; res assoc org chem & biochem, Sch Med, Stanford Univ, 63-68; sr chemist, 68-71, RES FEL, DEPT MICROBIOL, HOFFMANN-LA ROCHE INC, 71- Concurrent Pos: Squibb Inst Med Res & NIH fels, 61-62; NIH fel, 62-63; NASA grant, 63-68. Mem: Am Chem Soc; The Chem Soc; NY Acad Sci. Res: Structure determination, chemistry and biosynthesis of antibiotics; optical resolution of asymmetric compounds by gas chromatography; structure and biological activity. Mailing Add: Dept of Microbiol Hoffmann-La Roche Inc Nutley NJ 07110

WESTLEY, RONALD E, b Tacoma, Wash, Apr 10, 28; m 52; c 2. MARINE BIOLOGY, FISHERIES. Educ: Univ Wash, BS, 51. Prof Exp: From biologist I to biologist III, 51-66, BIOLOGIST IV & LAB DIR, WASH STATE DEPT FISHERIES, 66- Mem: Inst Fishery Res Biol; Nat Shellfisheries Asn. Res: Oyster research; marine ecology; aquatic primary productivity; reservoir limnology; water pollution; resource management and development. Mailing Add: State Shellfish Lab Star Rte 2 Box 600 Brinnon WA 98320

WESTLUND, ROY ALEX, JR, physical chemistry, see 12th edition

WESTMAN, ALBERT ERNEST ROBERTS, b Ottawa, Ont, May 31, 00; m 25; c 3. PHYSICAL INORGANIC CHEMISTRY. Educ: Univ Toronto, BA, 21, MA, 22, PhD(electrochem), 24. Prof Exp: Res assoc ceramics, Eng Exp Sta, Univ Ill, 24-28; assoc prof, Rutgers Univ, 28-29; dir chem dept, Ont Res Found, 29-58, dir res, 58-65; appl res consult, 65-76; RETIRED. Concurrent Pos: Researcher, Carborundum Co & Dept Mines, Ont; asst, Univ Toronto. Honors & Awards: Forrest Award, Am Ceramic Soc, 54. Mem: Electrochem Soc; fel Am Ceramic Soc; fel Am Soc Qual Control; fel Brit Soc Glass Technol; fel Am Statist Asn. Res: Physical and industrial chemistry; refractories; electric furnace arcs; statistics; plating; corrosion; phosphates; glass. Mailing Add: 35 Glenayr Rd Toronto ON Can

WESTMAN, JACK CONRAD, b Cadillac, Mich, Oct 28, 27; m 53; c 3. CHILD PSYCHIATRY. Educ: Univ Mich, BS, 49, MD, 52, MS, 59. Prof Exp: Intern, Duke Univ Hosp, 52-53; resident psychiat, Univ Mich, 56-57; from instr to assoc prof, Med Sch, Univ Mich, 58-65; coordr diag & treat unit, Ctr Ment Retardation & Human Develop, 66-74, PROF PSYCHIAT, MED SCH, UNIV WIS-MADISON, 65-, DIR CHILD PSYCHIAT DIV, UNIV HOSPS, 65- Concurrent Pos: Fel child psychiat, Univ Mich, 57-59; lectr, Sch Social Work, Univ Mich, 59-64; mem, Sr Staff, Children's Psychiat Hosp, Ann Arbor, Mich, 59-65, dir outpatient serv, 61-62, dir outpatient & day care serv, 62-65. Mem: Fel Am Orthopsychiat Asn; Asn Am Med Cols; fel Am Acad Child Psychiat; Am Asn Ment Deficiency; Soc Prof Child Psychiatrists. Res: Efficacy of medication of hyperkinetic system; psychiatric aspects of learning disabilities and mental retardation; the role of child psychiatry in divorce; predicting later adjustment from nursery school behavior; individual differences in children. Mailing Add: 427 Lorch St Madison WI 53706

WESTMAN, JAMES ROSS, b Pasadena, Calif, Apr 5, 10; m 39; c 2. BIOLOGY. Educ: Brown Univ, PhB, 32; Cornell Univ, PhD(limnol, fisheries), 41. Prof Exp: Asst biol sci, Univ Rochester, 33-36; sci investr, State Biol Surv, NY, 41; biologist & spec rep, S B Penick & Co, NY, 42-43; aquatic biologist, US Fish & Wildlife Serv, 43-45; sr aquatic biologist, State Conserv Dept, NY, 45-50; assoc prof wildlife conserv & assoc res specialist, 50-52, prof wildlife conserv, chmn dept & res specialist, 52-64, RES PROF ENVIRON RESOURCES, RUTGERS UNIV, NEW BRUNSWICK, 64- Mem: AAAS; Am Fisheries Soc; fel NY Acad Sci. Res: Aquatic resources; fishery. Mailing Add: Dept of Environ Resources Rutgers Univ New Brunswick NJ 08903

WESTMAN, RAGNAR THEOPHILE, b Kramfors, Sweden, Oct 2, 05; US citizen; m 32. HOSPITAL ADMINISTRATION, PUBLIC HEALTH ADMINISTRATION. Educ: Hamline Univ, BS, 27; Univ Minn, MB, 30, MD, 31; Johns Hopkins Univ, MPH, 36, DrPH, 39; Am Bd Prev Med, dipl. Prof Exp: Intern, Minneapolis Gen Hosp, 31; epidemiologist & asst to comnr health, Minneapolis Div Pub Health, 32-37; county dir health, Bay County, Mich, 37-38; asst pub health adminr, Johns Hopkins Univ, 38-39; prof hyg & prev med & head dept, Sch Med, Univ Kans, 39-41; actg dir health, USPHS, Jackson County, Miss, 42, actg asst dir health, epidemiologist & dir tuberc control bur, New Orleans Bd Health, 42-43, actg comnr health, Seattle, Wash, 43-45, dist off, Kansas City, Mo, 45-46, cent off, Washington, DC, 46-47; asst area med dir & area dir prof serv, US Vet Admin, Mo, 46-47; dir prof in quiries, Cent Off, US Vet Admin, DC, 65-73; RETIRED. Mem: AAAS; fel AMA; fel Am Pub Health Asn; Asn Mil Surgeons US; Col Prev Med. Res: Epidemiology of undulant fever and typhoid fever. Mailing Add: 1401 Blair Mill Rd Silver Spring MD 20910

WESTMAN, THOMAS LOUIS, b Herndon, Kans, Oct 10, 34; m 62. ORGANIC CHEMISTRY. Educ: Univ Wash, BS, 57; Univ Calif, Berkeley, PhD(org chem), 61. Prof Exp: Asst prof chem, Univ Fla, 61-64; res specialist, Cent Res Dept, Monsanto Co, Mo, 64-69; mgr chem res, Sherwood Med Industs, 69-71; DIR BIOCHEM RES, McGAW LABS DIV, AM HOSP SUPPLY CORP, 71- Concurrent Pos: Esso Co fel, Univ Ill, 60-61; Am Chem Soc-Petrol Res Fund & NSF res grants, Univ Fla, 62-64; assoc prof, Eve Div, Univ Mo-St Louis, 65-70. Mem: AAAS; Am Chem Soc; The Chem Soc. Res: Biochemical, nutritional, hematological and physiological research and development; medical products, including new drug and device systems. Mailing Add: McGaw Labs Div PO Box 11887 Santa Ana CA 92711

WESTMEYER, PAUL, b Dillsboro, Ind, Dec 9, 25; m 47; c 5. SCIENCE EDUCATION. Educ: Ball State Univ, BS, 49, MS, 53; Univ Ill, EdD, 60. Prof Exp: Pub sch teacher, Ind, 49-54; instr chem & sci, Univ Ill High Sch, 54-61; from asst prof to assoc prof sci educ, Univ Ill, 60-63; assoc prof, Univ Tex, 63-66; prof, Fla State Univ, 66-73; PROF SCI EDUC, UNIV TEX, SAN ANTONIO, 73- Concurrent Pos: Instr, Exten, Purdue Univ, 51-54; consult, Chicago Sch Bd, Ill, 54; proj assoc, Earlham Col, 59-62; partic, Int Seminar Chem, Dublin, Ireland, 60; consult, CBA Insts, 61-63; NSF in-serv grant, 64-65, coop col-sch sci grants, 64-66, summer inst grants & in-serv grants, 67-73. Mem: AAAS; Am Chem Soc; Nat Asn Res Sci Teaching; Nat Sci Teachers Asn; Asn Educ Teachers Sci (pres, 71-72). Res: Course development and evaluation; computer-assisted-instruction in chemistry; development of laboratory materials, tests and instructional procedures in chemistry; better utilization of staff in science teaching. Mailing Add: Dept of Sci Educ Univ of Tex San Antonio TX 78285

WESTMORE, JOHN BRIAN, b Welling, Eng, Apr 23, 37; m 61; c 2. PHYSICAL CHEMISTRY, INORGANIC CHEMISTRY. Educ: Univ London, BSc, 58, PhD(phys chem), 61. Prof Exp: Fel chem, Nat Res Coun Can, 61-63; asst prof, 63-69, ASSOC PROF CHEM, UNIV MAN, 69- Mem: Chem Inst Can; The Chem Soc; Am Soc Mass Spectrometry. Res: High temperature chemistry by mass spectrometry and other techniques; ionization and dissociation of molecules; mass spectrometry of metal chelates and nucleosides; thermodynamics of metal chelation. Mailing Add: Dept of Chem Univ of Man Winnipeg MB Can

WESTMORELAND, JAMES EDWARD, III, physics, see 12th edition

WESTMORELAND, WINFRED WILLIAM, b Santa Maria, Calif, Feb 7, 19; m 42; c 3. DENTISTRY. Educ: Col Physicians & Surgeons San Francisco, DDS, 42; Univ Calif, MPH, 57; Am Bd Dent Pub Health, dipl. Prof Exp: Extern oral surg, San Francisco Hosp, 46-47; clin instr oper dent, Col Physicians & Surgeons, Univ of Pac, 49-53, asst clin prof prosthetic dent & lectr dent mat, 53-58, lectr pub health dent, 58-69, asst clin prof dent pub health, 66-69; SR DENT CONSULT, CALIF DEPT HEALTH, OAKLAND, 67- Concurrent Pos: Pub health dent officer, Calif State Dept Health, 55-67; consult, Resident Training Prog, US Army, Ft Ord, Calif, 67-72; pvt practr. Mem: Am Dent Asn; Am Pub Health Asn; Int Asn Dent Res. Res: Epidemiology of dental caries; periodontal disease; cleft lip and palate; radiation exposure; dental fluorosis; dental health administration and education. Mailing Add: 644 S N St Livermore CA 94550

WESTNEAT, DAVID FRENCH, b Oradell, NJ, June 18, 29; m 58; c 3. ANALYTICAL CHEMISTRY. Educ: Allegheny Col, BS, 50; Univ Pittsburgh, PhD(chem), 56. Prof Exp: Res chemist, E I du Pont de Nemours & Co, 56-60; asst prof chem, Akron Univ, 60-65; chmn dept, 68-72, ASSOC PROF CHEM, WITTENBERG UNIV, 65- Concurrent Pos: Vis prof, Sci & Soc Prog, Cornell Univ, 72-73. Mem: AAAS; Am Chem Soc. Res: Instrumental analysis, especially absorption spectroscopy and gas chromatography. Mailing Add: Dept of Chem Wittenberg Univ Springfield OH 45501

WESTON, ALAN JAY, b New York, NY, Feb 7, 40; m 62; c 3. AUDIOLOGY, SPEECH PATHOLOGY. Educ: Univ Ala, BA, 63; MA, 65; Univ Kans, PhD(speech path & audiol), 69. Prof Exp: Instr speech, Univ Ala, 65-67; asst prof, Univ Kans, 69-70; chmn dept & dir, Memphis Speech & Hearing Ctr, 70-76, PROF AUDIOL & SPEECH PATH, MEMPHIS STATE UNIV, 72- Concurrent Pos: Off Educ fel, 69-70; mem, Cleft Palate Clin & Oto-audiol Clin Teams, State Crippled Children's Serv & Univ Ala, 65-67; mem, Fels Adv Bd, US Off Educ, 69-70, consult, 71-; spec asst, Bur Educ Handicapped, Off Educ, Dept Health, Educ & Welfare, 69-70; adj prof otolaryngol & maxillofacial surg & coordr audiol & speech path, Univ Tenn Med Units-Memphis City Hosps, 71- Mem: Am Speech & Hearing Asn; Coun Except Children; Am Psychol Asn. Res: Use of paired stimuli in the modification of articulatory behavior. Mailing Add: 807 Jefferson Ave Memphis TN 38105

WESTON, ARTHUR WALTER, b Smiths Falls, Ont, Feb 13, 14; nat US; m 40; c 4. ORGANIC CHEMISTRY. Educ: Queen's Univ, Ont, BA, 34, MA, 35; Northwestern Univ, PhD(org chem), 38. Prof Exp: Asst chem, Northwestern Univ, 35-37, res assoc, 38-40; from res chemist to asst head org res, 40-54, asst to dir develop, 54-55, asst dir, 55-57, dir res, 57-59, dir res & develop, 59-61, dir co, 59-69, vpres res & develop, 61-68, VPRES SCI AFFAIRS, ABBOTT LABS, 68- Concurrent Pos: Mem exec comt, Div Chem & Chem Technol, Nat Res Coun, 61-65; mem ad hoc comt chem agts, Dept of Defense, 61-65; mem, Indust Res Inst & Dirs Indust Res; mem indust panel sci & technol, NSF, 74- Mem: Am Chem Soc; Pharmaceut Mfrs Asn; Chem Inst Can. Res: Organic medicinals; antibiotics; plant processes. Mailing Add: Abbott Labs 1400 Sheridan Rd North Chicago IL 60064

WESTON, CHARLES RICHARD, b South Gate, Calif, Apr 24, 33; m 53; c 4. DEVELOPMENTAL BIOLOGY, MICROBIAL ECOLOGY. Educ: Univ Calif, Santa Barbara, BA, 57; Princeton Univ, PhD(biol), 65. Prof Exp: Res scientist, E R Squibb Inst Med Res, 60-61; res assoc & asst prof, Univ Rochester, 62-66; Nat Acad Sci-Nat Res Coun resident res assoc, Jet Propulsion Lab, Calif Inst Technol, 66-68; lectr, 68-69, ASSOC PROF BIOL, CALIF STATE UNIV, NORTHRIDGE, 70- Mem: AAAS; Bot Soc Am; Am Soc Microbiol. Res: Morphogenesis of fungal mycelium; development of instrumentation for remote life detection; interactions and distribution of soil microorganisms. Mailing Add: Dept of Biol Calif State Univ Northridge CA 91324

WESTON, EDWIN BENJAMIN, astronomy, astrophysics, see 12th edition

WESTON, HENRY GRIGGS, JR, b Hemet, Calif, Apr 7, 22; m 47; c 3. ECOLOGY, VERTEBRATE ZOOLOGY. Educ: San Diego State Col, BA, 43; Univ Calif, Berkeley, MA, 47; Iowa State Col, PhD(zool), 50. Prof Exp: Asst prof biol, Grinnell Col, 50-55; PROF BIOL, SAN JOSE STATE UNIV, 55- Mem: Wildlife Soc; Cooper Ornith Soc; Am Ornith Union; Am Soc Mammal. Res: Birds of California; bird-banding; field ecology. Mailing Add: Dept of Biol San Jose St Univ 125 S Seventh St San Jose CA 95192

WESTON, HENRY MORGAN, b Fairfield, Ala, Aug 13, 26; m 49; c 2. CHEMISTRY. Educ: Fla Southern Col, BS, 50; Southwestern Baptist Theol Sem, Tex, MDiv, 72. Prof Exp: Jr metallurgist, Tenn Coal, Iron & RR Co, 48-49, metallurgist, 50-51; chemist, Ala Power Co, 51-52; res chemist, Reynolds Metals Co, Ala, 52-63, sr test scientist, 63-69; PHOTOCHEMIST, MARCEL'S, INC, 72- Mem: Am Chem Soc. Res: Aluminum spectroscopy; photomicrography; instrumental analysis; technical documentation. Mailing Add: 2002 Jacocks Lane Ft Worth TX 76115

WESTON, JAMES A, b Washington, DC, June 20, 36; m 58; c 2. DEVELOPMENTAL BIOLOGY, CELL BIOLOGY. Educ: Cornell Univ, AB, 58; Yale Univ, PhD(biol), 62. Prof Exp: USPHS fel zool, Univ Col, Univ London, 62-64; from asst prof to assoc prof biol, Case Western Reserve Univ, 64-70; assoc prof, 70-74, PROF BIOL, UNIV ORE, 74- Mem: AAAS; Am Soc Zoologists; Soc Develop Biol (secy, 73-76); Int Soc Develop Biol; Am Soc Cell Biol. Res: Cellular control of morphogenetic movements in vertebrate development; regulation of cellular phenotypic expression of neural crest cells in vivo and in vitro; properties of cell surfaces in normal and transformed states. Mailing Add: Dept of Biol Univ of Ore Eugene OR 97403

WESTON, JAMES T, b Newark, NJ, May 16, 24; m 52; c 2. MEDICINE, PATHOLOGY. Educ: Cornell Univ, MD, 48; Am Bd Path, dipl, cert anat path, 67, forensic path, 68. Prof Exp: Intern path, Children's Hosp, Boston, 48-49; pathologist, Off Coroner, San Diego, 52-53, chief med dept, 54-61; asst med examr, Off Med Examr, Pa, 61-67; from asst prof to assoc prof path, Col Med, Univ Utah, 70-73; PROF PATH, SCH MED, UNIV N MEX, 73-; CHIEF MED INVESTR, STATE N MEX, 73- Concurrent Pos: Fel path, Mem Hosp Cancer, New York, 53-54; assoc pathologist, San Diego County Hosp & Community Hosp, Chula Vista, Calif, 56-61; vis lectr, Jefferson Med Col, 62-67; chief med examr, Utah State Dept Health, 67-73; consult & mem, Sci Adv Bd Consult, Armed Forces Inst Path, 71- Mem: Fel Am Acad Forensic Sci (secy-treas, 71-75, pres elect, 75); Am Soc Clin Path; Col Am Pathologists; Am Asn Pathologists & Bacteriologists. Res: Child abuse patterns; sociologic and pathologic correlations; environmental pathology; forensic pathology. Mailing Add: Dept of Path Univ of NMex Albuquerque NM 87131

WESTON, JEAN KENDRICK, b Spokane, Wash, Nov 30, 06; m 33. EXPERIMENTAL PATHOLOGY. Educ: Northern State Teachers Col, Mich, AB, 30; Univ Mich, MS, 31, PhD(neuroanat), 33; Temple Univ, MD, 41. Prof Exp: Asst anat, Univ Mich, 30-31, instr, 35-37; asst, Univ Groningen, 37-38; asst, Sch Med, Temple Univ, 38-43, from assoc prof to prof, 43-51; lab dir path res, Parke, Davis & Co, 51-57, dir clin invest, 57-62; med dir, Burroughs-Wellcome & Co, 62-64; dir dept drugs, AMA, 64-66; vpres, Nat Pharmaceut Coun, Inc, 66-75; ADJ PROF CLIN ENG, MED SCH, GEORGE WASHINGTON UNIV, 75-; PRES, WESTON RES LABS INC, TOXICOL CONSULT, 69- Mem: AAAS; Am Asn Anatomists; NY Acad Sci. Res: Comparative microscopic anatomy of the vertebrate nervous system; histochemistry; neuropathology; toxicology; drug development; clinical drug investigations; industrial medicine. Mailing Add: Dept of Clin Eng George Washington Univ Med Sch Washington DC 20006

WESTON, JOHN COLBY, b Dover-Foxcroft, Maine, Dec 31, 26; m 51; c 2. HISTOLOGY, EMBRYOLOGY. Educ: Bowdoin Col, AB, 51; Syracuse Univ, MS, 54, PhD(zool), 56. Prof Exp: From instr to assoc prof anat, Col Med, Ohio State Univ, 55-67; assoc prof, 67-69, PROF BIOL, MUHLENBURG COL, 69- Mem: Am Asn Anat; Electron Micros Soc Am; Soc Develop Biol. Res: Biochemical, histochemical and electron microscopic analyses of development. Mailing Add: Dept of Biol Muhlenburg Col Allentown PA 18104

WESTON, KENNETH W, b Milwaukee, Wis, Feb 1, 29; m 66; c 2. ALGEBRA. Educ: Univ Wis, BS, 53, MS, 55, PhD(math), 63. Prof Exp: Instr math, Univ Wis-Milwaukee, 61-63; asst prof, Univ Notre Dame, 63-69; assoc prof, Marquette Univ, 69-71; ASSOC PROF MATH, UNIV WIS-PARKSIDE, 71- Mem: Am Math Soc; Math Asn Am. Res: Groups satisfying Engel condition and connections between ring and group theory; model theory. Mailing Add: Div of Sci Univ of Wis-Parkside Kenosha WI 53140

WESTON, NORMAN ERNEST, physical chemistry, polymer chemistry, see 12th edition

WESTON, RALPH E, JR, b San Francisco, Calif, Nov 9, 23; m 51; c 3. CHEMICAL KINETICS. Educ: Univ Calif, BS, 46; Stanford Univ, PhD(chem), 50. Prof Exp: Asst, Stanford Univ, 47-49; from assoc chemist to chemist, 51-65, SR CHEMIST, BROOKHAVEN NAT LAB, 65- Concurrent Pos: Vis scientist, Saclay Nuclear Res Ctr, France, 60-61; vis lectr, Univ Calif, Berkeley, 68-69; lectr, Columbia Univ, 71-72; mem comt chem kinetics, Nat Res Coun, 75- Mem: AAAS; Am Chem Soc; Sigma Xi. Res: Kinetics of gas phase reactions; isotope effects in reaction kinetics and equilibria; photochemistry; application of lasers to chemical problems. Mailing Add: Chem Dept Brookhaven Nat Lab Upton NY 11973

WESTON, VAUGHAN HATHERLEY, b Parry Sound, Ont, May 1, 31; m 54; c 3. APPLIED MATHEMATICS, THEORETICAL PHYSICS. Educ: Univ Toronto, BA, 53, MA, 54, PhD, 56. Prof Exp: Lectr math, Univ Toronto, 57-58; res assoc, Radiation Lab, Univ Mich, 58-59, from assoc res mathematician to res mathematician, 59-69; PROF MATH, PURDUE UNIV, WEST LAFAYETTE, 69- Mem: Am Math Soc; Soc Indust & Appl Math. Res: Electromagnetic theory; diffraction; plasmas; inverse scattering. Mailing Add: Div of Math Sci Purdue Univ West Lafayette IN 47907

WESTOVER, JAMES DONALD, b Clarkdale, Ariz, Sept 22, 34; m 59; c 4. ORGANIC CHEMISTRY. Educ: Ariz State Univ, BS, 50, MS, 62; Brigham Young Univ, PhD(chem), 66. Prof Exp: Res chemist, Dacron Res Lab, E I du Pont de Nemours & Co, Inc, NC, 65-70; ASST 65-70; ASST PROF CHEM, CALIF POLYTECH STATE UNIV, SAN LUIS OBISPO, 70- Res: Synthetic organic chemistry in area of nitrogen heterocyclic compounds; polyester fibers. Mailing Add: Dept of Chem Calif Polytech State Univ San Luis Obispo CA 93407

WESTPHAL, JAMES ADOLPH, b Dubuque, Iowa, June 13, 30. PLANETARY SCIENCES. Educ: Univ Tulsa, BS, 53. Prof Exp: Sr res fel, 66-71, ASSOC PROF PLANETARY SCI, CALIF INST TECHNOL, 71- Mem: Optical Soc Am; Am Astron Soc. Res: Infrared astronomy; infrared atmospheric properties; planetary astronomy; astronomical instrumentation. Mailing Add: 176 Mudd Bldg Calif Inst of Technol Pasadena CA 91125

WESTPHAL, MILTON C, JR, b Philadelphia, Pa, June 2, 26; m 51; c 4. MEDICINE, PEDIATRICS. Educ: Yale Univ, BS, 47; Univ Pa, MD, 51; Am Bd Pediat, dipl, 57. Prof Exp: Intern, Univ Pa Hosp, 51-52; resident, Children's Hosp Philadelphia, 54-56; asst instr pediat, Sch Med, Univ Pa, 56-57, instr, 57-61, assoc prof, 65-67; PROF PEDIAT & CHMN DEPT, COL MED, MED UNIV SC, 67- Concurrent Pos: Chief resident, Children's Hosp Philadelphia, 56-57, from asst physician to assoc physician, 57-61, mem, Res Dept, 57-61; asst pediatrician to outpatients, Pa Hosp, 57-58, hosp, 58-59, assoc pediatrician, 59-61; asst attend, Children's Hosp Buffalo, 61-64, assoc attend, 64-; dir, Buffalo Poison Control Ctr, 61-67; proj dir & chmn, Comt Prin Investrs, Collab Study Cerebral Palsy, Ment Retardation & Other Neurol & Sensory

Dis Infancy & Childhood, 63-64. Mem: Am Pediat Soc. Res: Neonatology; perinatal area. Mailing Add: Dept of Pediat Med Univ of SC Col of Med Charleston SC 29401

WESTPHAL, ULRICH FRIEDRICH, b Göttingen, Ger, May 3, 10; nat US; m 40; c 3. BIOCHEMISTRY. Educ: Univ Göttingen, PhD(chem), 33; Univ Berlin, Drhabil (org chem), 41. Prof Exp: Asst org chem, Danzig Tech Univ, 33-36; res assoc & asst dir, Kaiser Wilhelm Inst Biochem, 36-45; chief, Chem Dept, Med Clin, Univ Tübingen, 45-49; chief protein & steroid sect, US Army Med Res Lab, 49-59, dir biochem div, 59-61; PROF BIOCHEM, SCH MED, UNIV LOUISVILLE, 61- Concurrent Pos: USPHS res career award, 62-; docent, Univ Tübingen, 46, prof extraordinary, 48; adj assoc prof, Univ Louisville, 53-59, adj prof, 59-; mem, Study Sect Biochem, USPHS, 61-65. Mem: AAAS; Am Chem Soc; Am Soc Biol Chemists; Soc Exp Biol & Med; Endocrine Soc. Res: Chemistry and metabolism of steroid hormones; serum proteins; steroid-protein interactions. Mailing Add: Dept of Biochem Univ of Louisville Sch of Med Louisville KY 40201

WESTPHAL, WARREN HENRY, b Easton, Pa, Feb 19, 25; m 46; c 3. EXPLORATION GEOPHYSICS. Educ: Columbia Univ, AB, 47. Prof Exp: Jr mining engr, NJ Zinc Co, NJ, 47-48, geologist, 48-49, res geologist, Pa, 49-50, geophysicist, Colo, 50-55; sr geologist, Tidewater Assoc Oil Co, NMex, 55; chief geophysicist, Utah Construct & Mining Co, 56-59; sr geophysicist, Stanford Res Inst, 59-66, chmn earth sci dept, 66-69; VPRES, INTERCONTINENTAL ENERGY CORP, 69- Mem: AAAS; Soc Econ Geologists; Am Asn Petrol Geologists; Am Inst Mining, Metall & Petrol Engrs; Soc Mining Engrs. Res: Earthquake seismology. Mailing Add: Intercontinental Energy Corp 600 S Cherry St Denver CO 80222

WESTRUM, EDGAR FRANCIS, JR, b Albert Lea, Minn, Mar 16, 19; m 43; c 4. PHYSICAL CHEMISTRY, THERMODYNAMICS. Educ: Univ Minn, BChem, 40; Univ Calif, PhD(phys chem), 44. Prof Exp: Res chemist, Metall Lab, Univ Chicago, 44-46 & Radiation Lab, Univ Calif, 46; from asst prof to assoc prof, 46-57, PROF PHYS CHEM, UNIV MICH, ANN ARBOR, 58- Concurrent Pos: Chmn comt data for sci & technol, Nat Bur Standards; chmn comn I.2 & phys chem div, Int Union Pure & Appl Chem; ed, J Chem Thermodyn & Bull Thermodyn & Thermochem; assoc ed, Comt on Data for Sci & Technol Bull, 74- Mem: Fel AAAS; fel Am Inst Chemists; Am Chem Soc; The Chem Soc; fel Am Phys Soc. Res: Thermodynamics, actinide, lanthanide, and transition compounds; thermochemistry; cryogenic calorimetry; molecular dynamics plastic crystals; thermophysics of phase ordering, Schottky, transitions. Mailing Add: Dept of Chem Univ of Mich Ann Arbor MI 48104

WESTRUM, LESNICK EDWARD, b Tacoma, Wash, Oct 19, 34. NEUROSURGERY, BIOLOGICAL STRUCTURE. Educ: Wash State Univ, BS, 58; Univ Wash, MD, 63; Univ London, PhD(anat), 66. Prof Exp: NIH fel, Univ London, 63-66, hon res asst anat, Univ Col, 64-66; res asst prof, 66-67, ASST PROF SURG & BIOL STRUCT, SCH MED, UNIV WASH, 67- Mem: AAAS; Am Asn Anatomists; Soc Neurosci. Res: Studies of synapses in normal and experimental conditions; emphasis on trigeminal and limbic systems. Mailing Add: Dept of Neurol Surg Univ of Wash Sch of Med Seattle WA 98195

WESTURA, EDWIN EUGENE, b Morristown, NJ, Aug 29, 30; m 53; c 5. INTERNAL MEDICINE, CARDIOVASCULAR DISEASES. Educ: St Peter's Col, NJ, BA, 52; Creighton Univ, MD, 57; Am Bd Internal Med, dipl, 66. Prof Exp: Chemist, Picatinny Arsenal, 52-53; intern, DC Gen Hosp, Washington, DC, 57-58, resident internal med, 58-59; instr phys diag, Med Br, Univ Tex, 61-62; NIH fel, Hosp & Sch Med, Georgetown Univ, 62-63, chief resident internal med, 63-64, from instr to asst prof internal med & phys diag, 66-68; ASSOC PROF INTERNAL MED, SCH MED, ST LOUIS UNIV, 68-, ASSOC PROF PHYSIOL & HEAD SECT CARDIOVASC DIS, 71- Concurrent Pos: NIH fel cardiol, Med Br, Univ Tex, 61-62; instr jr student prog & ward attend, Georgetown Univ Hosp & DC Gen Hosp, 64-; chief, Cardiovasc Lab, Georgetown Clin Res Inst, Off Aviation Med, Fed Aviation Agency, 64-66, flight surgeon, Med Res Planning Br, Res & Educ Div, 65, chief, Med Res Planning Br, Med Applns Div, 65-66, actg chief, Georgetown Clin Res Inst, 66, consult, Fed Air Surg; mem attend staff, Morris Cafritz Mem Hosp, Washington, DC, 65 & Columbia Hosp for Women, 66-; chief, Appl Physiol Lab, Heart Dis Control Prog, Div Chronic Dis Control, Nat Ctr Chronic Dis Control, USPHS, 66-, actg chief, Peripheral Vascular Dis Sect, 66- Mem: Am Heart Asn. Res: Clinical cardiovascular research; hemodynamics; exercise physiology; bio-engineering. Mailing Add: Dept of Internal Med St Louis Univ Sch of Med St Louis MO 63103

WESTVELD, RUTHFORD HENRY, b Grand Rapids, Mich, Mar 18, 00; m 35; c 1. FORESTRY. Educ: Mich State Col, BS, 22, PhD(soils), 46; Yale Univ, MF, 25. Prof Exp: Jr forester, US Forest Serv, NMex, 22-24, asst silviculturist, Univ Ore, 25-28; from asst prof to assoc prof forestry, Mich State Col, 28-36; asst prof, Univ Mo, 36-38; prof silvicult, Univ Fla, 38-46; prof forestry, Ala Polytech Inst, 46-47; prof, 47-70, EMER DIR SCH FORESTRY, UNIV MO-COLUMBIA, 65-, EMER PROF FORESTRY, 70- Mem: Soc Am Foresters. Res: Soils; forest relationships; nutrition of slash pine. Mailing Add: 1312 SW 22nd St Boynton Beach FL 33435

WESTWICK, ROY, b Vancouver, BC, May 23, 33; m 59; c 2. MATHEMATICS. Educ: Univ BC, BA, 56, MA, 57, PhD(math), 60. Prof Exp: Nat Res coun Can overseas fel, math, Univ Col, London, 60-62; from asst prof to assoc prof, 62-70, PROF MATH, UNIV BC, 70- Concurrent Pos: Can Coun sr fel, 67-68. Mem: Am Math Soc; Can Math Cong; London Math Soc. Res: Linear and multilinear algebra; measure theory. Mailing Add: Dept of Math Univ of BC Vancouver BC Can

WESTWOOD, MELVIN (NEIL), b Hiawatha, Utah, Mar 25, 23; m 46; c 4. POMOLOGY, PLANT PHYSIOLOGY. Educ: Utah State Univ, BS, 53; Wash State Univ, PhD(pomol, plant physiol), 56. Prof Exp: Asst field botanist, Utah State Univ, 51-52, supt, Hort Res Field Sta, 52-53; asst, Wash State Univ, 53-55; from asst res horticulturist to res horticulturist, USDA, 55-60; assoc prof, 60-67, PROF HORT, ORE STATE UNIV, 67- Honors & Awards: Gourley Award, Am Soc Hort Sci, 58 & Stark Award, 69; Paul Howe Shepard Award, Am Pomol Soc, 68. Mem: AAAS; fel Am Soc Hort Sci (pres, 74-75); Am Soc Plant Physiologists; Scand Soc Plant Physiol. Res: Deciduous fruit tree and rootstock physiology; growth dynamics; plant hormones; chemical thinning; high density orchard systems. Mailing Add: Dept of Hort Ore State Univ Corvallis OR 97331

WESTWOOD, WILLIAM DICKSON, b Kirkcaldy, Scotland, Jan 4, 37; m 61; c 1. PHYSICS. Educ: Univ Aberdeen, BSc, 59, PhD(physics), 62. Hon Degrees: PhD(physics), Univ Adelaide, 66. Prof Exp: Scientist, Northern Elec Co Ltd, 62-65; lectr physics, Flinders Univ, Australia, 66-69; scientist, Northern Elec Co Ltd, 69-70; MGR THIN FILM PHYSICS, BELL-NORTHERN RES LTD, 71- Mem: Fel Brit Inst Physics; Can Asn Physicists; Am Vacuum Soc. Res: Thin film physics; sputtering; spectroscopy of gas discharges; integrated optics. Mailing Add: Dept 5C30 Bell-Northern Res Ltd Box 3511 Sta C Ottawa ON Can

WESWIG, PAUL HENRY, b St Paul, Minn, July 13, 13; m 40; c 2. ANIMAL NUTRITION. Educ: St Olaf Col, BA, 35; Univ Minn, MS, 39, PhD(biochem), 41. Prof Exp: Instr chem, St Olaf Col, 36-37; asst biochem, Univ Minn, 38-41; asst prof biochem, Univ & asst chemist, Exp Sta, 41-42, assoc prof biochem, 46-47, PROF BIOCHEM, ORE STATE UNIV, 57- Mem: Am Chem Soc; Am Inst Nutrit; Am Dairy Sci Asn; Am Soc Animal Sci. Res: Trace mineral metabolism in ruminants and laboratory animals including requirement, interrelationship, toxicity, tissue enzyme activity and element concentration including selenium, copper and heavy metals. Mailing Add: Dept of Agr Chem Ore State Univ Corvallis OR 97331

WETENKAMP, HARRY R, b Forest Park, Ill, Jan 25, 19; m 42; c 2. MECHANICS. Educ: Univ Ill, BS, 42, MS, 50. Prof Exp: Ord inspector, Rochester Ord Dist, 42-43; res asst, 46-47, res assoc, 47-50, from asst prof to assoc prof, 50-59, PROF THEORET & APPL MECH, UNIV ILL, URBANA, 59- Mem: Am Soc Eng Educ; Am Soc Metals. Res: Thermal damage of railway car wheel material. Mailing Add: Dept of Theoret Appl Mech Univ of Ill Urbana IL 61801

WETHERBEE, DAVID KENNETH, vertebrate zoology, see 12th edition

WETHERELL, DONALD FRANCIS, b Manchester, Conn, Nov 25, 27; m 49; c 4. PLANT PHYSIOLOGY. Educ: Univ Conn, BA, 51; Univ Md, MS, 53, PhD(plant physiol), 56. Prof Exp: Res assoc & lectr plant physiol, Univ Md, 56-58; from asst prof to assoc prof plant physiol, 58-67, actg chmn dept bot, 61-64, chmn regulatory biol sect, 67-68, PROF PLANT PHYSIOL, UNIV CONN, 67- Concurrent Pos: Guggenheim fel, 65-66. Mem: Am Soc Plant Physiologists; Bot Soc Am; Int Asn Plant Tissue Cult. Res: Regulatory mechanisms of growth and development in plants. Mailing Add: Biol Sci Group Univ of Conn Storrs CT 06268

WETHERELL, HERBERT RANSON, JR, b Chicago, Ill, Jan 25, 27; m 62; c 1. PHARMACOLOGY, TOXICOLOGY. Educ: Yale Univ, BS, 49, PhD(pharmacol), 54. Prof Exp: Asst prof physiol & pharmacol, Med Col, Univ Nebr, 53-61; toxicologist, Wayne County Med Examr Off, 61-72; TOXICOLOGIST, CRIME LAB, MICH DEPT PUB HEALTH, 72- Mem: Am Chem Soc; Am Acad Forensic Sci; The Chem Soc. Res: Relationship between chemical structure and pharmacologic activity; toxicology and drug metabolism; chlorophyll chemistry; infrared spectrophotometry. Mailing Add: Crime Lab Mich Dept Pub Health 3500 N Logan St Lansing MI 48914

WETHERILL, GEORGE WEST, b Philadelphia, Pa, Aug 12, 25; m 50. GEOPHYSICS. Educ: Univ Chicago, PhB, 48, SB, 49, MS, 51, PhD(physics), 53. Prof Exp: Mem staff, Carnegie Inst, Washington Dept Terrestrial Magnetism, 53-60; prof geophys & geol, Univ Calif, Los Angeles, 60-75, chmn dept planetary & space sci, 68-72; DIR DEPT TERRESTRIAL MAGNETISM, CARNEGIE INST WASHINGTON, 75- Concurrent Pos: Vis prof, Calif Inst Technol, 59. Mem: Nat Acad Sci; fel Am Acad Arts & Sci; fel Am Geophys Union; Meteoritical Soc (vpres, 72-74); Geochem Soc (vpres, 73-74, pres, 74-75). Res: Planetology; geochronology; meteorites; origin and evolution of solar system; lunar history; Precambrian geology; kinetics of human lead metabolism. Mailing Add: Dept of Terrestrial Magnetism Carnegie Inst Washington DC 20015

WETHERINGTON, RONALD K, b St Petersburg, Fla, Nov 27, 35; m 58; c 2. BIOLOGICAL ANTHROPOLOGY. Educ: Tex Tech Univ, BA, 58; Univ Mich, MA, 60, PhD(anthrop), 64. Prof Exp: PROF ANTHROP, SOUTHERN METHODIST UNIV, 64- Concurrent Pos: Dir, Fort Burgwin Res Ctr, NMex, 76- Mem: AAAS; Am Asn Phys Anthrop. Res: Skeletal growth and development in man; human ecology and evolutionary theory; adaptive aspects of human demography. Mailing Add: Dept of Anthrop Southern Methodist Univ Dallas TX 75222

WETLAUFER, DONALD BURTON, b New Berlin, Apr 4, 25; m 50; c 2. BIOCHEMISTRY. Educ: Univ Wis, BS, 46, PhD(biochem), 54. Prof Exp: Jr chemist, Argonne Nat Lab, 44 & 46-47; res chemist, Bjorksten Res Labs, Inc, 48-50; asst anal chem, Univ Wis, 44-46, asst biochem, 50-52, res assoc enzymol, Inst Enzyme Res, 54; res assoc, Children's Cancer Res Found & Dept Biol Chem, Harvard Med Sch, 58-61; asst prof biochem, Sch Med, Ind Univ, Indianapolis, 61-62; from assoc prof to prof biochem, Med Sch, Univ Minn, Minneapolis, 62-75; DUPONT PROF CHEM & CHMN DEPT, UNIV DEL, 75- Concurrent Pos: Nat Found Infantile Paralysis fel protein chem, Carlsberg Lab, Denmark, 55-56; Am Heart Asn fel, Biol Labs, Harvard Univ, 56-58; tutor, Harvard Univ, 58-61. Mem: Am Chem Soc; Am Soc Biol Chemists; Biophys Soc; fel Am Inst Chemists. Res: Chemical and physical basis of structure, stability and reactivity of proteins; acquisition of three-dimensional structure of macromolecules; mechanisms of enzyme action. Mailing Add: Dept of Chem Univ of Del Newark DE 19711

WETMORE, ALEXANDER, b North Freedom, Wis, June 18, 86; m 12, 53; c 1. ORNITHOLOGY. Educ: Univ Kans, AB, 12; George Washington Univ, MS, 16, PhD(zool), 20. Hon Degrees: ScD, George Washington Univ, 32, Univ Wis, 46, Centre Col, 47 & Ripon Col, 59. Prof Exp: Agent, Bur Biol Surv, USDA, 10-13, from asst biologist to biologist, 13-24; supt, Nat Zool Park, 24-25; from asst secy to secy, 25-52, RES ASSOC, SMITHSONIAN INST, 53- Concurrent Pos: Trustee, Textile Mus Washington, 28-52, Am Schs Orient Res, 44-49, Nat Art Gallery, 45-52, George Washington Univ, 45-64, Pac War Mem, 46 & Sci Serv, 46-53; dir, CZ Biol Area, 40-46, Res Corp, 47-53 & Gorgas Mem Inst Trop & Prev Med, 49- Mem, Am Comt Int Wildlife Protection; Comt Protection Cult Treasures in War Areas, 42-46; vchmn, Nat Adv Comt Aeronaut, 45-52; nat adv comn, State Dept Educ, Sci & Cult Coop, 46-52; comn, Inst Nat Parks, Belgium Congo, 46-52; adv bd, Arctic Res Lab, 48-52. Deleg, Int Ornith Cong, Amsterdam, 30, Oxford, 34, chmn, US deleg, Rouen, 38, pres, 50; secy-gen, Eighth Am Sci Cong; US rep, Inter-Am Comn Experts Nat Protection Wildlife Preserves, 40. Honors & Awards: Isidore Geoffroy St Hilaire Medal, Soc Nat d'Acclimitation France, 27; Herman Medal, Hungarian Ornith Asn, 31; Order of Merit, Carlos Manuel de Cespedes, Cuba, 48; Brewster Medal, Am Ornith Union, 59, Coues Award, 72. Mem: Nat Acad Sci (home secy, 51-55); AAAS; Am Philos Soc; Am Soc Mammal; Am Ornith Union (vpres, 23-25, pres, 26-29). Res: Taxonomy and classification of higher groups, ecology and geographic distribution of birds; fossil birds. Mailing Add: Div of Birds Smithsonian Inst Washington DC 20560

WETMORE, CLIFFORD MAJOR, b Akron, Ohio, June 18, 34; m 59; c 2. LICHENOLOGY. Educ: Mich State Univ, BS, 56, MS, 59, PhD(bot), 65. Prof Exp: From instr to assoc prof biol, Wartburg Col, 64-70; asst prof, 70-72, ASSOC PROF BOT, UNIV MINN, MINNEAPOLIS, 72- Concurrent Pos: NSF res grants, 66-68 & 71-73, fel, 70. Mem: AAAS; Am Bryol & Lichenological Soc; Int Asn Plant Taxonomists; Am Inst Biol Sci. Res: Lichens of the Black Hills, South Dakota, and Minnesota; desert lichens; ultrastructure of lichens; lichen genera; distributions of lichens; herbarium computer techniques. Mailing Add: Dept of Bot Univ of Minn St Paul MN 55108

WETMORE, DAVID EUGENE, b Stella, Nebr, Dec 18, 35; m 59; c 2. ORGANIC CHEMISTRY, PETROLEUM CHEMISTRY. Educ: Park Col, BA, 58; Univ Kans, MA, 62; Tex A&M Univ, PhD(org chem), 65. Prof Exp: Res chemist, Sun Oil Co, 65-67; asst prof, 67-70, ASSOC PROF CHEM, ST ANDREWS PRESBY COL, 70-, CHMN CHEM PROG, 68- Mem: Am Chem Soc. Res: Asphaltic hydrocarbons;

nitrogen compounds in shale oil; beta-lactone syntheses and reactions. Mailing Add: Dept of Chem St Andrews Presby Col Laurinburg NC 28352

WETMORE, ORVILLE CHASE, chemistry, see 12th edition

WETMORE, RALPH HARTLEY, b Yarmouth, NS, Apr 27, 92; nat US; m 23, 40; c 2. BOTANY, MORPHOLOGY. Educ: Acadia Univ, BSc, 21; Harvard Univ, AM, 22, PhD(bot), 24. Hon Degrees: DSc, Acadia Univ, 48. Prof Exp: Nat Res Coun fel bot, Harvard Univ, 24-25; asst prof, Acadia Univ, 25-26; from asst prof to prof bot, 26-62, chmn dept, 32-34, dir bot labs, 33-34, chmn dept biol, 46-47, dir biol labs, 52-56. Concurrent Pos: Grants, Milton Fund, Harvard Univ, Am Cancer Soc, 48-58 & NSF, 52-66; vis res prof, Dartmouth Col, 66-67 & Simon Fraser Univ, 72. Honors & Awards: Jeanette S Pelton Award in Exp Plant Morphol, 69. Mem: Nat Acad Sci; fel Am Acad Arts & Sci; fel AAAS; Bot Soc Am (pres, 53); Am Soc Plant Physiologists. Res: Plant antomy; morphogenesis; development of vascular plants. Mailing Add: 12 Francis Ave Cambridge MA 02138

WETMORE, STANLEY IRWIN, JR, b Queens, NY, June 1, 39; m 61; c 2. ORGANIC CHEMISTRY. Educ: Rensselaer Polytech Inst, BS, 60, MS, 62; State Univ NY Buffalo, PhD(org chem), 73. Prof Exp: Chemist, Mobil Oil Corp, 62-63; from instr to asst prof, 64-74, ASSOC PROF CHEM, VA MIL INST, 75- Mem: Am Chem Soc; Sigma Xi. Res: Synthesis and properties of unique carbenes; photochemistry of small ring heterocyclic compounds. Mailing Add: Dept of Chem Va Mil Inst Lexington VA 24450

WETMUR, JAMES GERARD, b New Castle, Pa, July 1, 41; m 65; c 2. BIOPHYSICAL CHEMISTRY. Educ: Yale Univ, BS, 63; Calif Inst Technol, PhD(chem), 67. Prof Exp: Asst prof chem & biochem, Univ Ill, Urbana, 69-75; MEM FAC, DEPT BIOCHEM, MT SINAI SCH MED, CITY UNIV NEW YORK, 75- Mem: Am Chem Soc; Biophys Soc. Res: Kinetics of renaturation of DNA; electron microscopic studies of topological changes in DNA. Mailing Add: Dept Biochem Mt Sinai Sch of Med City Univ of New York New York NY 10029

WETS, ROGER J B, b Uccle, Belgium, Feb 20, 37; m 61; c 2. MATHEMATICS. Educ: Free Univ Brussels, BA, 59; Univ Calif, Berkeley, PhD(appl math), 64. Prof Exp: Staff mem, Boeing Sci Res Lab, Wash, 64-70; prof math, Univ Chicago, 70-71; PROF MATH, UNIV KY, 71- Concurrent Pos: Vis lectr, Univ Wash, 66, Univ Calif, Berkeley, 67 & Inst Info & Automation Res, Paris, 69; vis res, Math Ctr, Montreal, 70. Mem: Am Math Soc; Soc Indust & Appl Math. Res: Mathematical programming; stochastic optimization. Mailing Add: Dept of Math Univ of Ky Lexington KY 40506

WETSEL, FERD R, chemistry, deceased

WETSMAN, ALLEN WILLIAM, b Greenbelt, Md, July 11, 40; m 61; c 4. MATHEMATICS. Educ: Syracuse Univ, BS, 62, MS, 64, PhD(math), 68. Prof Exp: From asst prof to assoc prof, 68-74, PROF MATH, PURDUE UNIV, LAFAYETTE, 74- Concurrent Pos: NSF grants, 69-75; Sloan Found grant, 72. Res: Classical function theory. Mailing Add: Dept of Math Purdue Univ Lafayette IN 47907

WETSTONE, DAVID MAJOR, physics, chemistry, see 12th edition

WETSTONE, HOWARD J, b Hartford, Conn, Apr 27, 26; m 47; c 4. INTERNAL MEDICINE, BIOCHEMISTRY. Educ: Wesleyan Univ, BA, 47; Tufts Univ, MD, 51. Prof Exp: Intern med, New Eng Ctr Hosp, 51-52, jr asst resident, 52-53; asst resident, 53-54, resident, 54-55, dir liver enzyme lab, 58-60, dir biochem res lab, 60-63, dir med res, 63-66, asst dir dept med, 66-70, dir outpatient dept, 70-72, DIR AMBULATORY SERV, HARTFORD HOSP, 72-, SR PHYSICIAN INTERNAL MED, 70- Concurrent Pos: Assoc prof med & community med & health care, Med Sch, Univ Conn, 71- Mem: Sigma Xi; Am Col Physicians; Am Col Emergency Physicians. Res: Pharmacogenetics; hypertension; enzymology; community health systems; primary care. Mailing Add: Hartford Hosp Hartford CT 06105

WETTACK, F SHELDON, b Coffeyville, Kans, Dec 5, 38; m 56; c 4. PHYSICAL CHEMISTRY. Educ: San Jose State Col, AB, 60, MA, 62; Univ Tex, Austin, PhD(chem), 68. Prof Exp: High sch teacher, Calif, 61-64; from asst prof to assoc prof, 67-72, PROF CHEM, HOPE COL, 72-, DEAN NATURAL SCI, 74- Concurrent Pos: Camille & Henry Dreyfus Found Teacher-Scholar Award, 70-75. Mem: Am Chem Soc. Res: Photochemistry; energy transfer; fluorescence spectroscopy. Mailing Add: Dept of Chem Hope Col Holland MI 49423

WETTE, REIMUT, b Mannheim, Ger, May 12, 27; US citizen; m 50; c 5. BIOSTATISTICS, BIOMATHEMATICS. Educ: Univ Heidelberg, MS, 52, DSc(natural sci), 55. Prof Exp: Sci asst biomet, Zool Inst, Univ Heidelberg, 52-61; asst biometrician, Univ Tex, 61-64, mem grad fac, 65-66, assoc prof biomath, Univ Tex M D Anderson Hosp & Tumor Inst, 64-66; PROF BIOSTATIST & DIR DIV BIOSTATIST, SCH MED & PROF APPL MATH, SCH ENG & APPL SCI, WASHINGTON UNIV, 66- Mem: Am Statist Asn; Biomet Soc; Inst Math Statist; NY Acad Sci. Res: Mathematical approaches to basic science and medical aspects of neoplasia; application and problem-oriented development of mathematical-statistical methodology in biomedical research. Mailing Add: Div of Biostatist Dept of Prev Med Washington Univ Sch of Med St Louis MO 63110

WETTEMANN, ROBERT PAUL, b New Haven, Conn, Nov 12, 44; m 68; c 2. REPRODUCTIVE PHYSIOLOGY. Educ: Univ Conn, BS, 66; Mich State Univ, MS, 68, PhD(dairy), 72. Prof Exp: Asst prof, 72-75, ASSOC PROF ANIMAL SCI, OKLA STATE UNIV, 75- Honors & Awards: Richard Hoyt Award, Am Dairy Sci Asn, 71. Mem: Am Soc Animal Sci; Am Dairy Sci Asn; Soc Study Fertil; Soc Study Reproduction. Res: Influence of the environment on endocrine and reproductive function in animals. Mailing Add: Dept of Animal Sci & Indust Okla State Univ Stillwater OK 74074

WETTER, LESLIE ROBERT, b Millet, Alta, Apr 17, 17; m 42; c 1. BIOCHEMISTRY. Educ: Univ Alta, BSc, 44, MSc, 46; Univ Wis, PhD(biochem), 50. Prof Exp: Asst plant biochem, Univ Alta, 46; from asst res officer to sr res officer, 50-70, PRIN RES OFFICER, PRAIRIE REGIONAL LAB, NAT RES COUN CAN, 70-, ASSOC MEM, COUN, 72- Mem: Am Soc Plant Physiologists; Can Soc Plant Physiologists; Chem Inst Can. Res: Biosynthesis of thioglucosides in higher plants; enzymes in higher plants, particularly those related to thioglucosides; utilization of isotopes in plant metabolism; toxic properties in rape meal; plant cell cultures. Mailing Add: Prairie Regional Lab Nat Res Coun Saskatoon SK Can

WETTERAU, FRANK P, b Hicksville, NY, May 16, 19; m 43; c 3. ANALYTICAL CHEMISTRY. Educ: Polytech Inst Brooklyn, BS, 41. Prof Exp: Anal chemist, 41-70, SUPVR, GEN ANAL GROUP, ANAL SECT, CHEM RES & DEVELOP LABS, ICI US INC, 70- Mem: Am Chem Soc; Am Oil Chemists Soc; fel Am Inst Chemists. Res: Separation science, including chromatography, ion exchange and extraction, particularly determination of additives in food products. Mailing Add: Chem Res & Develop Labs ICI US Inc Wilmington DE 19897

WETTERSTROM, EDWIN, mechanical engineering, see 12th edition

WETTSTEIN, FELIX O, b Uerikon, Switz, Jan 1, 32; m 60; c 3. MOLECULAR BIOLOGY. Educ: Swiss Fed Inst Technol, BS, 56, PhD(agr chem), 60. Prof Exp: Fel, Univ Pittsburgh, 60-64; asst res biochemist, Univ Calif, Berkeley, 64-67; assoc prof, 67-74, PROF MOLECULAR BIOL, MED MICROBIOL & IMMUNOL, UNIV CALIF, LOS ANGELES, 74- Mem: AAAS; Am Soc Microbiol. Res: Regulation of RNA and protein biosynthesis in differentiating and transformed animal cells. Mailing Add: Dept of Med Microbiol & Immunol Univ of Calif Los Angeles CA 90024

WETZEL, ALLAN BROOKE, b Dayton, Ohio, May 29, 33; m 59; c 3. NEUROPSYCHOLOGY. Educ: Univ Ky, BS, 54; Ohio State Univ, MA, 63, PhD(psychol), 65. Prof Exp: Trainee biosci, Stanford Univ, 65-67; trainee neurophysiol, Univ Wis-Madison, 67-69, Nat Inst Neurol Dis & Stroke spec fel, 68-69; instr surg, 70-74, ASST PROF SURG & OTO-MAXILLOFACIAL SURG, MED SCH, NORTHWESTERN UNIV, CHICAGO, 74- Concurrent Pos: Res investr neuropsychol, Neurosurg Res Lab, Northwestern Mem Hosp, 70-74. Mem: AAAS; Am Psychol Asn; Soc Neurosci. Res: Brain function; neurophysiology; neuroendocrinology. Mailing Add: Dept of Otolaryngol Northwestern Univ Med Sch Chicago IL 60611

WETZEL, FRANKLIN HUFF, b Kewanee, Ill, Dec 2, 25; m 51; c 2. CHEMISTRY. Educ: Univ Ill, BS, 47; Univ Pa, MS, 49, PhD(chem), 51. Prof Exp: Asst instr chem, Univ Pa, 47-51; res chemist, 51-59, res supvr, 59-62, asst to dir res, 62-64, asst mgr corp planning, 64-68, mgr planning, 68-70, gen mgr, Haskon, Inc, 70-72, dir new bus develop, Polymers Dept, 72-73, dir res & develop, Coatings & Spec Prod Dept, 73-75, DIR, NEW ENTERPRISE DEPT, HERCULES, INC, 75- Res: Water soluble polymers; product development; strategic planning. Mailing Add: New Enterprise Dept Hercules Inc Wilmington DE 19899

WETZEL, JOHN EDWIN, b Hammond, Ind, Mar 6, 32; m 62. MATHEMATICS. Educ: Purdue Univ, Lafayette, BS, 54; Stanford Univ, PhD(math), 64. Prof Exp: From instr to asst prof, 61-68, ASSOC PROF MATH, UNIV ILL, URBANA-CHAMPAIGN, 68- Mem: Am Math Soc; Math Asn Am. Res: Geometric inequalities; combinatorial geometry; mathematics education. Mailing Add: Dept of Math Univ of Ill at Urbana-Champaign Urbana IL 61801

WETZEL, JOHN WILLIAM, organic chemistry, see 12th edition

WETZEL, KARL JOSEPH, b Waynesboro, Va, May 29, 37; m 68. EXPERIMENTAL NUCLEAR PHYSICS. Educ: Georgetown Univ, BS, 59; Yale Univ, MS, 60, PhD(physics), 65. Prof Exp: NSF fel, Inst Tech Nuclear Physics, Darmstadt, Ger, 65-66, Ger Govt guest res fel, 66-67; appointee, Argonne Nat Lab, 67-69; asst prof, 69-72, ASSOC PROF PHYSICS, UNIV PORTLAND, 72- Mem: Am Phys Soc. Res: Neutron capture gamma rays; electron and photon scattering; photonuclear processes. Mailing Add: Dept of Physics Univ of Portland Portland OR 97203

WETZEL, LEWIS BERNARD, b Milwaukee, Wis, Mar 19, 25; m 49; c 1. APPLIED PHYSICS. Educ: Northwestern Univ, BS, 49; Harvard Univ, MA, 50, PhD(appl physics), 57. Prof Exp: Mem tech staff electron device develop, Bell Tel Labs, NJ, 56-59; asst prof, Div Eng, Brown Univ, 59-62, assoc prof, 62-65; mem res staff ionospheric physics, Inst Defense Anal, 65-68; supt commun sci div, 68-72, SR SCIENTIST ELECTRONICS AREA, NAVAL RES LAB, 72- Concurrent Pos: Mem adv group aerospace res & develop, Electromagnetic Propagation Panel, NATO. Mem: AAAS; Am Geophys Union; Int Union Radio Sci. Res: Wave scattering by obstacles and rough surfaces; atmospheric and ionospheric propagation. Mailing Add: Naval Res Lab Washington DC 20375

WETZEL, NICHOLAS, b Jacksonville, Fla, July 17, 20; m 45; c 6. NEUROSURGERY. Educ: Princeton Univ, AB, 42; Northwestern Univ, MD, 46, MS, 50, PhD, 58. Prof Exp: Clin asst, 52-54, instr, 54-55, assoc, 55-57, asst prof, 57-63, ASSOC PROF SURG, MED SCH, NORTHWESTERN UNIV, CHICAGO, 63- Mem: AAAS; Am Asn Neurol Surg; AMA; Am Col Surgeons. Res: Human stereotaxic surgery for movement disorders; intractable pain; human olfactory system. Mailing Add: Dept of Surg Northwestern Univ Med Sch Chicago IL 60611

WETZEL, RALPH MARTIN, b Macomb, Ill, June 30, 17; m 47; c 3. MAMMALOGY. Educ: Western Ill Univ, BE, 38; Univ Ill, MS, 39, PhD, 49. Prof Exp: Spec asst conserv educ, Univ Ill, 39-40, asst zool, 46-49; from instr to assoc prof, 49-66, PROF ZOOL, UNIV CONN, 66- Concurrent Pos: Sr clerk, Ord Dept, US War Dept, 41; scientist in residence, US Nat Mus, Smithsonian Inst, 68-71, res assoc, Dept Vert Zool, 73-76. Mem: Am Soc Mammalogists; Soc Syst Zool; Ecol Soc Am; Sigma Xi. Res: Neotropical mammals; speciation and zoogeography; evolution of South American Tayassuidae. Mailing Add: Dept of Syst & Evolutionary Biol Univ of Conn Biol Sci Group Storrs CT 06268

WETZEL, ROBERT GEORGE, b Ann Arbor, Mich, Aug 16, 36; m 59; c 4. LIMNOLOGY. Educ: Univ Mich, BSc, 58, MSc, 59; Univ Calif, Davis, PhD(limnol), 62. Prof Exp: Tech asst, Univ Mich, 58-59; res technician, US Fish & Wildlife Serv, Mich, 59; res & tech asst, Univ Calif, 59-62; res assoc, Ind Univ, 62-65; from asst prof to assoc prof, 65-71, PROF BOT, MICH STATE UNIV, 71- Concurrent Pos: Res fel, Aquatic Res Unit, Ind Dept Natural Resources, 63-64; NSF res fel, 63-65 & 67-69, travel award, Int Asn Limnol Cong, 65 & 68; partic, AEC Contract, 65-; co-ed, Communications, 65-; Off Water Resources Res res fel, 69-71; int consult, Int Biol Prog, Int Coun Sci Unions, 69-; NSF grant, 72-73. Mem: Fel AAAS; Am Inst Biol Sci; Am Soc Limnol & Oceanog; Ecol Soc Am; Int Asn Theoret & Appl Limnol (gen secy & treas, 68-). Res: Biological productivity of California, Indiana and Michigan lakes; physiological ecology of algae and aquatic macrophytes. Mailing Add: W K Kellogg Biol Sta Mich State Univ Hickory Corners MI 49060

WETZIG, CALVIN ULYSSES, b Clint, Tex, June 17, 09; m 33; c 2. PURE MATHEMATICS. Educ: Sul Ross State Col, BS, 31; Univ Tex, MA, 37. Prof Exp: Teacher, Clint High Sch, 31-34, Shamrock Jr High Sch, 34-35 & Seymour High Sch, 35-38; prof & head dept, 38-75, EMER PROF MATH, SOUTHERN STATE COL, ARK, 75- Mem: Math Asn Am. Res: Cauchy fields; differential equations. Mailing Add: Dept of Math Southern State Col Magnolia AR 71753

WETZLER, THEODORE FRANCIS, public health, see 12th edition

WEWERKA, EUGENE MICHAEL, b St Paul, Minn, Nov 7, 38; m 59; c 4. PHYSICAL ORGANIC CHEMISTRY. Educ: Univ Minn, BA, 62, PhD(org chem), 65. Prof Exp: STAFF MEM, LOS ALAMOS SCI LAB, 65- Concurrent Pos: Adj prof, Univ NMex, 72- Mem: Fel Am Inst Chemists; Am Chem Soc. Res: Chemical kinetics; mechanism studies; polymer characterization; analytical methods; physical chemistry of polymers; chemistry of coal and oil shale conversion processes. Mailing Add: Los Alamos Sci Lab Univ of Calif Los Alamos NM 87544

WEXELL, DALE RICHARD, b Corning, NY, Apr 10, 43. INORGANIC

CHEMISTRY. Educ: Fordham Univ, BS, 64; Georgetown Univ, MS, 69, PhD(inorg chem), 71. Prof Exp: Instr chem, Georgetown Univ, 64-70; pres, Premium Enterprises, Inc, 65-70; fel, 71-72, RES SCIENTIST GLASS CHEM, CORNING GLASS WORKS, 72- Mem: AAAS; Am Chem Soc; The Chem Soc; Am Ceramic Soc; Sigma Xi. Res: Aqueous silicates; heteropoly electrolytes; high-temperature materials; surface chemistry of glass; inorganic coatings technology; opal glasses; microwave processing; photochromism; electrical, magnetic and optical behavior in glass and ceramics. Mailing Add: Sullivan Park Corning Glass Works Corning NY 14830

WEXLER, ARTHUR SAMUEL, b New York, NY, Oct 27, 18; m 47; c 3. ANALYTICAL CHEMISTRY. Educ: Kans State Col, BS, 40; Polytech Inst Brooklyn, PhD(chem), 51. Prof Exp: Chemist, C F Kirk & Co, 47-49 & Nopco Chem Co, 49-51; sr chemist, Pepsi-Cola Co, 51-60; MGR ANAL LAB, DEWEY & ALMY CHEM DIV, W R GRACE & CO, CAMBRIDGE, 60- Mem: Am Chem Soc; fel Am Inst Chemists. Res: Analytical methods for monomers and polymers; physical organic chemistry; biochemistry; infrared spectroscopy. Mailing Add: 4 Marshall Rd Lexington MA 02173

WEXLER, BERNARD CARL, b Boston, Mass, May 1, 23; m 46; c 3. EXPERIMENTAL MEDICINE. Educ: Univ Ore, BS, 47; Univ Calif, MA, 48; Stanford Univ, PhD(anat, biochem), 52. Prof Exp: Asst anat, Sch Med, Stanford Univ, 49-52; mem res staff, Baxter Labs, 52-55; res assoc, May Inst Med Res, 55-61, asst dir, 61-64; from asst prof to assoc prof exp path, 55-71, exp med, 71-75, PROF EXP MED & PATH, COL MED, UNIV CINCINNATI, 75-; DIR, MAY INST MED RES, 64- Concurrent Pos: Am Heart Asn advan res fel, 60-62; Nat Heart Inst res career develop award, 62-72; lectr, Dominican Col, 50-52; res assoc, Stanford Res Inst, 51-52; mem, Coun Arteriosclerosis & Coun Basic Sci, Am Heart Asn, 62-, Coun Stroke & Coun High Blood Pressure Res; mem, Coun Arteriosclerosis & Ischemic Heart Dis, Int Soc Cardiol; mem med staff, Jewish Hosp, Cincinnati. Mem: Am Soc Physiologists; AAAS; Am Diabetes Asn; Asn Am Med Cols; Endocrine Soc. Res: Pituitary-adrenal physiology; experimental pathology. Mailing Add: May Inst for Med Res 421 Ridgeway Ave Cincinnati OH 45229

WEXLER, CHARLES, b Fall River, Mass, July 5, 06; m 27; c 1. MATHEMATICS. Educ: Harvard Univ, AB, 27, AM, 29, PhD(math), 30. Prof Exp: Instr math, Harvard Univ, 27-30; chmn dept, 30-58, PROF MATH, ARIZ STATE UNIV, 30- Concurrent Pos: Instr, Navy Pre-Flight, Ga, 42; lectr, Univ Ga, 42 & George Washington Univ, 46; scientist, Los Alamos Sci Labs, 47; res assoc, Univ Calif, Los Angeles, 52; regional dir sec sch prog, NSF-Math Asn Am, 60-61. Res: Number theory; analysis. Mailing Add: Dept of Math Ariz State Univ Tempe AZ 85281

WEXLER, RAYMOND, meteorology, see 12th edition

WEXLER, SOLOMON, b Milwaukee, Wis, June 16, 19; m 51; c 2. CHEMICAL PHYSICS. Educ: Univ Chicago, BS, 41, PhD(chem), 48. Prof Exp: Jr chemist, Metall Lab, Univ Chicago, 42-44; jr scientist, Los Alamos Sci Lab, 44-46; assoc chemist, Oak Ridge Nat Lab, 46; SR CHEMIST, ARGONNE NAT LAB, 48- Concurrent Pos: Mem panel molecular beam exp for space shuttle, NASA, 74-75. Mem: AAAS; Am Chem Soc; Am Phys Soc. Res: Chemical effects of nuclear transformation; high-pressure mass spectrometry; interactions of ions with molecules; molecular beam-magnetic resonance spectroscopy; analytical chemistry of uranium and plutonium; chemical accelerators; ion cyclotron resonance spectroscopy; chemi-ionization reactions in accelerated crossed-molecular beams. Mailing Add: Argonne Nat Lab 9700 S Cass Ave Argonne IL 60439

WEYBREW, JOSEPH ARTHUR, b Wamego, Kans, July 13, 15; m 42; c 2. PLANT CHEMISTRY. Educ: Kans State Col, BS, 38, MS, 39; Univ Wis, PhD(plant physiol), 42. Prof Exp: Res dir, W J Small Co, 42; asst nutritionist, Exp Sta, Kans State Col, 42-43; chief chemist, Indust Hyg Div, State Bd Health, Kans, 43; assoc res prof animal nutrit, 46-49, assoc res prof agron, 49-51, res prof, 51-56, res prof chem, 56-60, actg head chem res, 57-60, WILLIAM NEALS REYNOLDS DISTINGUISHED PROF AGR, NC STATE UNIV, 57-, RES PROF FIELD CROPS, 60- Mem: AAAS; Am Chem Soc. Res: Tobacco biochemistry, especially fluecuring and quality evaluation; tobacco biogenetics. Mailing Add: 112 Pineland Circle Raleigh NC 27606

WEYENBERG, DONALD RICHARD, organic chemistry, inorganic chemistry, see 12th edition

WEYH, JOHN ARTHUR, b Havre, Mont, Sept 9, 42; m 62; c 4. ANALYTICAL CHEMISTRY, INORGANIC CHEMISTRY. Educ: Col Great Falls, BA, 64; Wash State Univ, MS, 66, PhD(chem), 68. Prof Exp: Instr chem, Wash State Univ, 66-67; asst prof, 68-71, ASSOC PROF CHEM, WESTERN WASH STATE COL, 71- Mem: Am Chem Soc. Res: Synthesis, characterization and kinetic studies on coordination compounds. Mailing Add: Dept of Chem Western Wash State Col Bellingham WA 98225

WEYHMANN, WALTER VICTOR, b Roanoke, Va, Nov 27, 35; m 57; c 1. SOLID STATE PHYSICS, LOW TEMPERATURE PHYSICS. Educ: Duke Univ, BS, 57; Harvard Univ, AM, 58, PhD(physics), 63. Prof Exp: Res fel physics, Harvard Univ, 63-64; from asst prof to assoc prof 64-75, PROF PHYSICS & HEAD SCH PHYSICS & ASTRON, UNIV MINN, MINNEAPOLIS, 75- Mem: Am Phys Soc. Res: Nuclear magnetic resonance measurement of sublattice magnetizations; nuclear orientation studies of dilute impurities in metals; production of very low temperatures. Mailing Add: Sch of Physics & Astron Univ of Minn Minneapolis MN 55455

WEYL, F JOACHIM, b Zurich, Switz, Feb 19, 15; nat US; m 40; c 2. MATHEMATICS. Educ: Swarthmore Col, BA, 35; Princeton Univ, MA, 37, PhD(math), 39. Prof Exp: Asst math, Univ Ill, 37-39; instr, Univ Md, 39-40 & Ind Univ, 40-44; phys sci mathematician, Bur Ord, Washington, DC, 44-47; mathematician, US Off Naval Res, 47-49, head math br, 49-51; sci liaison officer, US Embassy, London, 51-52; head math br, US Off Naval Res, 52-53; dir math sci div, US Embassy, Washington, DC, 53-57, dir naval anal group, 58, res dir, 58-61, chief scientist & dep chief naval res, 61-66; spec consult to pres, Nat Acad Sci, Washington, DC, 66-68; DEAN SCI & MATH, HUNTER COL, NY, 68- Honors & Awards: Meritorious Civilian Serv Award, US Navy; Nat Civil Serv Award, 64; Distinguished Serv Award, Dept Defense, 65. Mem: Am Math Soc; Math Asn Am; Soc Indust & Appl Math (pres, 61); Am Phys Soc. Res: Applied mathematics. Mailing Add: Hunter Col 695 Park Ave New York NY 10021

WEYL, PETER K, b Ger, May 6, 24; nat US; m 47; c 3. ENVIRONMENTAL MANAGEMENT. Educ: Univ Chicago, ScM, 51, PhD(physics), 53. Prof Exp: Lectr physics, Roosevelt Col, 51-53; asst prof, Brazilian Ctr Phys Res, 53-54; physicist, Explor & Prod Res Labs, Shell Develop Co, 54-59, sr physicist, 59-63; prof oceanog, Ore State Univ, 63-66; PROF OCEANOG, STATE UNIV NY STONY BROOK, 66- Concurrent Pos: Lectr, Univ Houston, 55-62; vis prof, Hebrew Univ Jerusalem, 72-73. Mem: AAAS; Am Geophys Union; Am Soc Limnol & Oceanog. Res: Chemical and physical oceanography; ocean-climate interaction; environmental stability; coastal zone management. Mailing Add: 90 Christian Ave Stony Brook NY 11790

WEYLAND, JACK ARNOLD, b Butte, Mont, June 12, 40; m 65; c 3. PHYSICS. Educ: Mont State Univ, BS, 62, PhD(solid state physics), 69. Prof Exp: ASST PROF PHYSICS, SDAK SCH MINES & TECHNOL, 68-, RES CORP RES GRANT, 70- Mem: AAAS; Am Phys Soc; Am Asn Physics Teachers. Res: High pressure diffusion and magnetic susceptibility studies. Mailing Add: Dept of Physics SDak Sch of Mines & Technol Rapid City SD 57701

WEYMANN, HELMUT DIETRICH, b Duisburg, Ger, Aug 2, 27; nat US; m 54; c 3. RHEOLOGY. Educ: Aachen Tech Univ, DSc(physics), 54. Prof Exp: Res assoc fluid dynamics, Univ Md, 54-55; privat docent appl mech, Aachen Tech Univ, 56-57; asst res prof fluid dynamics, Univ Md, 57-60; assoc prof mech eng, 60-65, PROF MECH & AEROSPACE SCI, UNIV ROCHESTER, 65- Concurrent Pos: Consult, Nat Bur Standards, Washington, DC, 59-61; vis prof, Univ Md, 67-68 & Aachen Tech Univ, 75-76. Mem: Am Phys Soc; Soc Rheol; Ger Phys Soc; Ger Soc Appl Math & Mech. Res: Hemorheology. Mailing Add: Dept of Mech & Aerospace Sci Univ of Rochester Rochester NY 14627

WEYMANN, RAY J, b Los Angeles, Calif, Dec 2, 34; m 56; c 3. ASTRONOMY. Educ: Calif Inst Technol, BS, 56; Princeton Univ, PhD(astron), 59. Prof Exp: Res fel astron, Calif Inst Technol, 59-61; from asst prof to assoc prof, 61-67, PROF ASTRON, UNIV ARIZ, 67-, HEAD DEPT ASTRON & DIR & ASTRONR, ASTRONOMER STEWARD OBSERV, 70- Mem: Am Astron Soc; Royal Astron Soc. Res: Theoretical astrophysics; stellar spectroscopy. Mailing Add: Steward Observ Univ of Ariz Tucson AZ 85721

WEYMOUTH, JOHN WALTER, b Palo Alto, Calif, Jan 14, 22; m 66; c 3. SOLID STATE PHYSICS, ARCHAEOMETRY. Educ: Univ Calif, AB, 43, MA, 50, PhD(physics), 51. Prof Exp: Instr physics, Vassar Col, 52-54; from asst prof to assoc prof, Clarkson Col Technol, 54-58; from asst prof to assoc prof, 58-64, PROF PHYSICS, UNIV NEBR-LINCOLN, 64- Mem: Am Phys Soc; Soc Am Archaeol; Am Crystallog Asn. Res: Solid state and crystal physics; x-ray diffraction; lattice dynamics; physical methods in archaeology. Mailing Add: Dept of Physics Univ of Nebr Lincoln NE 68588

WEYMOUTH, PATRICIA PERKINS, b Birmingham, Mich, Dec 31, 18; div; c 3. NATURAL SCIENCE, HISTORY OF SCIENCE. Educ: Russell Sage Col, AB, 40; Univ Cincinnati, PhD(biochem), 44. Prof Exp: Asst biochem, Armored Med Res Lab, Ft Knox, 44-46; res assoc med physics, Univ Calif, 46-49; res assoc radio, Stanford-Lane Hosp, San Francisco, 49-52, Vassar Col, 52-54 & Clarkson Col Technol, 54-58; res assoc biochem & nutrit, Univ Nebr, 58-67; res assoc, 67-69, assoc prof, 69-75, PROF NATURAL SCI, MICH STATE UNIV, 75- Mem: Fel AAAS; NY Acad Sci; Sigma Xi. Res: Bacteriological and cancer biochemistry; nucleic acid and enzyme studies; interpenetrance of science and other disciplines. Mailing Add: Dept of Natural Sci Mich State Univ East Lansing MI 48823

WEYMOUTH, RICHARD J, b Brewer, Maine, July 19, 28; m; c 1. ANATOMY, ENDOCRINOLOGY. Educ: Univ Maine, BS, 50; Univ Mich, MS & PhD, 55; Marquette Univ, MD, 63. Prof Exp: Instr anat, Miami Univ, 55-59; asst prof, Sch Med, Marquette Univ, 59-61; intern, Univ Mich Hosps, 63-64; from assoc prof to prof anat, Med Col Va, 64-75; PROF ANAT & CHMN DEPT, SCH MED, UNIV SC, 75- Concurrent Pos: Lectr med, Med Col Va, 64-75. Res: Electron microscopy, endocrine glands and kidney. Mailing Add: Dept of Anat Sch of Med Univ of SC Columbia SC 29208

WEYNA, PHILIP LEO, b Chicago, Ill, May 11, 32; m 54; c 4. ORGANIC CHEMISTRY, POLYMER CHEMISTRY. Educ: Loyola Univ, Ill, BS, 54; Univ Wis, PhD(org chem), 58. Prof Exp: Res chemist, 58-64, supvr polymer res, 64-66, dir, 66-73, MGR CHEM SPECIALIZATION, MORTON CHEM CO, 73- Mem: Am Chem Soc; Tech Asn Pulp & Paper Indust. Res: Polymeric coatings and adhesives. Mailing Add: Morton Chem Co 1275 Lake Ave Woodstock IL 60098

WEYTER, FREDERICK WILLIAM, b Philadelphia, Pa, Oct 7, 34; m 65; c 2. BIOCHEMICAL GENETICS. Educ: Univ Pa, AB, 56; Amherst Col, MA, 58; Univ Ill, PhD(biochem), 62. Prof Exp: From instr to asst prof, 62-73, ASSOC PROF BIOL, COLGATE UNIV, 73- Concurrent Pos: Fulbright lectr, Afghanistan, 65-66. Mem: Am Chem Soc. Res: Drug metabolism in microorganisms; regulation of arginine biosynthesis in E coli. Mailing Add: Dept of Biol Colgate Univ Hamilton NY 13346

WHALEN, DAVID MAYO, organic chemistry, see 12th edition

WHALEN, JAMES WILLIAM, b Enid, Okla, Mar 16, 23; m 46; c 2. SURFACE CHEMISTRY. Educ: Univ Okla, BS, 46, MS, 47, PhD(chem), 51. Prof Exp: Res assoc, Mobil Oil Corp, 50-68; chmn dept chem, 68-72, dean col sci, 72-75, PROF CHEM, UNIV TEX, EL PASO, 68- Res: Surface phenomena; adsorption; calorimetry. Mailing Add: Dept of Chem Univ of Tex El Paso TX 79968

WHALEN, JOSEPH WILSON, b Battle Creek, Mich, May 27, 23; m 54; c 2. MICROBIOLOGY. Educ: Mich State Univ, BS, 49, MS, 51, PhD(microbiol), 55. Prof Exp: Bacteriologist, Arthur S Kimball Sanatorium, Battle Creek, 48-50, lab dir, 54-55; bacteriologist, Calhoun County Health Dept, Mich, 50-52; bacteriologist, Biol Labs, Pitman-Moore Co, 55-56, head bact dept, 56-64, mgr bact & immunochem depts, 64-71; asst to dir, Biol Labs, 72-73, res specialist, 73-76, SR RES SPECIALIST, DOW CHEM CO, 76- Mem: Sigma Xi; Am Soc Microbiol. Res: Bacteriological and immunochemical investigations in tuberculosis; bacterins and vaccines; antimicrobial agents; chemotherapy. Mailing Add: 6014 Sturgeon Creek Pkwy Midland MI 48640

WHALEN, WILLIAM JAMES, b Ft Dodge, Iowa, July 9, 15; m 46; c 3. PHYSIOLOGY. Educ: Stanford Univ, BA, 48, MA, 49, PhD(physiol), 51. Prof Exp: Asst physiol, Stanford Univ, 48-49, res assoc, 49-51, instr, 50-51; from instr to asst prof, Univ Calif, Los Angeles, 51-60; assoc prof, Col Med, Univ Iowa, 60-67; DIR RES, ST VINCENT CHARITY HOSP, 67- Concurrent Pos: Fulbright scholar, 58-59; prof, Case Western Reserve Univ, 67-68, adj prof, 68- Mem: AAAS; Microcirc Soc; Am Physiol Soc; Cardiac Muscle Soc (pres, 66-68); NY Acad Sci. Res: Cardiovascular research; cardiac function in isolated preparations; respiratory control mechanisms; autonomic pharmacology; tissue oxygen tension and cell metabolism. Mailing Add: Res Dept St Vincent Charity Hosp Cleveland OH 44115

WHALEY, HOWARD ARNOLD, b Iroquois, Ill, Sept 1, 34; m 54; c 4. ORGANIC CHEMISTRY. Educ: Univ Ill, Urbana, BS, 56; Univ Wis-Madison, PhD(org chem), 61. Prof Exp: Res chemist, Lederle Labs, Am Cyanamid Co, 61-66; RES CHEMIST, UPJOHN CO, 66- Mem: Am Chem Soc. Res: Finding, isolating and studying new antibiotics and modifying known antibiotics. Mailing Add: Infectious Dis Dept Upjohn Co Kalamazoo MI 49002

WHALEY, JULIAN WENDELL, b Parkersburg, WVa, Aug 12, 37; m 61; c 3. PLANT PATHOLOGY, PLANT SCIENCE. Educ: West Liberty State Col, BS, 59; WVa Univ, MS, 61; Univ Ariz, PhD(plant path), 64. Prof Exp: Sr plant pathologist, Eli Lilly & Co, 64-70; ASSOC PROF PLANT SCI, CALIF STATE UNIV, FRESNO,

70- Mem: Am Phytopath Soc; Mycol Soc Am. Res: Plant protection; pesticides effect on soil fungi; mycorrhizal fungi. Mailing Add: Dept of Plant Sci Calif State Univ Fresno CA 93710

WHALEY, PETER WALTER, b Baltimore, Md, June 27, 37; m 60; c 3. GEOLOGY. Educ: Ohio Wesleyan Univ, BA, 59; Univ Ky, MS, 64; La State Univ, PhD(geol), 69. Prof Exp: Asst prof, 68-72, ASSOC PROF GEOL, MURRAY STATE UNIV, 72- Mem: Soc Econ Paleontologists & Mineralogists; Geol Soc Am; Sigma Xi. Res: Modern depositional environments; carboniferous system of the Eastern United States. Mailing Add: Dept of Chem & Geol Murray State Univ Murray KY 42071

WHALEY, RANDALL MCVAY, b Hastings, Nebr, Aug 21, 15; m 39; c 3. PHYSICS. Educ: Univ Colo, BA, 38, MA, 40; Purdue Univ, PhD(physics), 47. Hon Degrees: DSc, Philadelphia Col Pharm & Sci, 72. Prof Exp: Asst Univ Colo, 38-40; tech asst to sales mgr, G-M Labs, Inc, Ill, 40-41; from instr to prof physics, Purdue Univ, 45-60, exec asst head dept, 55-57, assoc dean sch sci, educ & humanities & actg dir res found, 59-60; vpres grad studies & res, Wayne State Univ, 60-65; chancellor, Univ Mo-Kansas City, 65-67; consult, Am Coun Educ, Washington, DC, 67-68; prin, Cresap, McCormick & Paget Inc, NY, 68-70; PRES, UNIVERSITY CITY SCI CTR, 70- Concurrent Pos: Exec dir adv bd educ, Nat Acad Sci, 57-59; vchmn, Uni-Coll Corp, 74- Mem: Fel AAAS; Am Soc Eng Educ; Am Asn Physics Teachers; Am Sci Film Asn (pres, 63-); Int Sci Film Asn (pres, 64-). Res: Cosmic rays; semiconductors; electron synchrotron; educational administration. Mailing Add: University City Sci Ctr 3508 Science Center Philadelphia PA 19104

WHALEY, ROSS SAMUEL, b Detroit, Mich, Nov 7, 37; c 3. FOREST ECONOMICS. Educ: Univ Mich, BS, 59; Colo State Univ, MS, 61; Univ Mich, PhD(natural resource econ), 69. Prof Exp: Instr forestry, Colo State Univ, 60-61; res forester, Southern Forest Exp Sta, US Forest Serv, 61-63; asst prof natural resource econ, Utah State Univ, 65-67, prof forest sci & head dept, 67-69; assoc dean col forestry & natural resources, Colo State Univ, 70-75; HEAD DEPT LANDSCAPE ARCHIT & REGIONAL PLANNING, UNIV MASS, 75- Concurrent Pos: Consult, Intermountain Forest & Range Exp Sta, USD Forest Serv, 66-67, Rocky Mountain Forest & Range Exp Sta, 67, Pub Land Law Rev Comn, 70, Wallace, McHarg, Roberts & Todd, Joe Meheen Eng & Geddes, Brecher, Qualls, Cunningham, Architects. Mem: Soc Am Foresters; Am Econ Asn. Res: Application of economic theory to problems of natural resources policy and regional planning. Mailing Add: Dept of Landscape Archit Univ of Mass Amherst MA 01072

WHALEY, THOMAS PATRICK, inorganic chemistry, physical chemistry, see 12th edition

WHALEY, THOMAS WILLIAMS, b Albuquerque, NMex, June 13, 42. ORGANIC CHEMISTRY. Educ: Univ NMex, BS, 67, MS, 69, PhD(chem), 71. Prof Exp: MEM STAFF CHEM, LOS ALAMOS SCI LAB, UNIV CALIF, 71- Concurrent Pos: Adj asst prof chem, Univ NMex, Los Alamos Grad Ctr, 73-; ed, J Labelled Compounds & Radiopharmaceut, 74- Mem: Am Chem Soc; Sigma Xi; AAAS. Res: Organic synthesis with stable isotopes. Mailing Add: Los Alamos Sci Lab PO Box 1663 MS 890 Los Alamos NM 87545

WHALEY, WILLIAM GORDON, b New York, NY, Jan 16, 14; m 38; c 1. BIOLOGY. Educ: Univ Mass, BS, 36; Columbia Univ, PhD(biol sci), 39. Prof Exp: Chemist, Arbuckle Bros, NY, 36-38; lectr bot, Barnard Col, Columbia, 39-40, instr, 40-43; sr geneticist, USDA, 43-46; assoc prof, 46-48, chmn dept, 49-62, assoc dean grad sch, 54-57 & dean, 57-72, dir, Univ Res Inst, 57-72, PROF BOT, UNIV TEX, AUSTIN, 48-, DIR CELL RES INST, 47-, ASHBEL SMITH PROF CELL BIOL & BOT & MEM CTR FOR HIGHER EDUC, 72- Concurrent Pos: Mem adv comt educ, NSF, 64-69, chmn, 67-69; vis prof, Rockefeller Univ, 64-72. Mem: Am Soc Cell Biol; Bot Soc Am; Genetics Soc Am; Soc Develop Biol; Am Soc Plant Physiol; Torrey Bot Club (treas, 41-43). Res: Growth and development; Golgi apparatus function; developmental ultrastructure. Mailing Add: Biol Labs 220 Univ of Tex Austin TX 78712

WHALEY, WILSON MONROE, b Baltimore, Md, July 21, 20; m 56; c 1. TEXTILE CHEMISTRY. Educ: Univ Md, BS, 42, MS, 44, PhD(chem), 47. Prof Exp: Org chemist, Naval Res Lab, 44-47; fel chem, Univ Ill, 47-49; asst prof, Univ Tenn, 49-53; asst dir res labs, Pabst Brewing Co, 53-55; sect head chem, Res Ctr, Gen Foods Corp, 55-59; asst tech dir, Midwest Div, Arthur D Little, Inc, 59-62; mgr indust develop, IIT Res Inst, 62-65, mgr org chem, 63-65; dir res & planning, Burlington Industs, Inc, 65-71; pres, Whaley Assocs, NY, 71-75; PROF TEXTILE CHEM & HEAD DEPT, NC STATE UNIV, 75- Concurrent Pos: Consult, Oak Ridge Nat Lab, 51-55; adj assoc prof, Cornell Univ, 74-75. Mem: Am Asn Textile Chemists & Colorists; Am Chem Soc; Am Inst Chemists. Res: Heterocyclic and organophosphorus compounds; proteins; carbohydrates; polymer syntheses; textile fibers, finishes and processes; plastics, resins and composite structures; textile chemicals, polymers and processes. Mailing Add: Sch of Textiles NC State Univ Raleigh NC 27607

WHALIN, EDWIN ANSIL, JR, b Barlow, Ky, Mar 6, 24; m 48; c 4. PHYSICS. Educ: Univ Ill, BS, 45, MS, 47, PhD(physics), 54. Prof Exp: From asst prof to prof physics, Univ NDak, 54-66; assoc prof, 66-70, PROF PHYSICS, EASTERN ILL UNIV, 70- Mem: Am Phys Soc. Res: Nuclear physics. Mailing Add: Dept of Physics Eastern Ill Univ Charleston IL 61920

WHALIN, ROBERT WARREN, oceanography, coastal engineering, see 12th edition

WHALING, WARD, b Dallas, Tex, Sept 29, 23. NUCLEAR PHYSICS. Educ: Rice Inst, BA, 44, MA, 47, PhD(physics), 49. Prof Exp: Fel, 49-52, from asst prof to assoc prof, 52-62, PROF PHYSICS, CALIF INST TECHNOL, 62- Mem: Am Phys Soc. Res: Penetration of charged particles through matter; atomic spectroscopy. Mailing Add: Dept of Physics 106-38 Calif Inst of Technol Pasadena CA 91109

WHALLEY, EDWARD, b Darwen, Eng, June 20, 25; m 56; c 3. PHYSICAL CHEMISTRY. Educ: Univ London, BSc & ARCS, 45, PhD(phys chem), 49, DSc, 63; Imp Col, dipl, 49. Prof Exp: Lectr chem, Royal Tech Col, Salford, Eng, 48-50; fel, 50-52, asst res officer, Pure Chem Div, 52-53, from asst res officer to sr res officer, Appl Chem Div, 53-61, PRIN RES OFFICER, APPL CHEM DIV, NAT RES COUN CAN, 62-, HEAD HIGH PRESSURE SECT, 62- Concurrent Pos: Vis prof, Univ Western Ont, 67 & Kyoto Univ, Japan, 75-76; assoc mem comn thermodyn & thermochem, Int Union Pure & Appl Chem, 73-; chmn, Can Nat Comt for Int Asn Properties of Steam, 74-; treas, Int Asn Advan High Pressure Sci & Technol, 75- Mem: Am Phys Soc; Chem Inst Can; fel Royal Soc Can (assoc hon treas, 69-71, assoc hon secy, 71-74, hon secy, 76-); The Chem Soc. Res: High pressure physical chemistry; far infrared and Raman spectroscopy. Mailing Add: Div of Chem Nat Res Coun Ottawa ON Can

WHALLON, ROBERT, JR, b Boston, Mass, Apr 23, 40; m 62; c 1. ANTHROPOLOGY, ARCHAEOLOGY. Educ: Harvard Univ, BA, 61; Univ Chicago, MA, 63, PhD(anthrop), 66. Prof Exp: Asst prof, 66-67, ASSOC PROF ANTHROP, UNIV MICH, ANN ARBOR, 67-, CUR MEDITER PREHIST, UNIV MUS ANTHROP, 66-, ASSOC DIR, 75- Mem: Am Anthrop Asn; Soc Am Archaeol; Sigma Xi. Res: Prehistory of the eastern United States, Europe and the Near East; archaeological methodology. Mailing Add: Dept of Anthrop Univ of Mich Ann Arbor MI 48104

WHAM, GEORGE SIMS, b Laurens, SC, Jan 27, 20; m 47; c 3. TEXTILE CHEMISTRY. Educ: Clemson Univ, BS, 41; Univ Tenn, Knoxville, MS, 47; Pa State Univ, PhD(textile chem), 51. Prof Exp: Textile technologist, USDA, 47-49; res asst textile finishes, Pa State Univ, 49-51; assoc prof textiles & prof textile chem, Tex Woman's Univ, 51-54; dir textiles, Good Housekeeping Inst & Mag, 54-60; dir res & develop textile fabric & finish res, Phillips-Van Heusen Corp, 60-62; VPRES & TECH DIR, GOOD HOUSEKEEPING MAG & INST, 62- Concurrent Pos: Mem US deleg, Pan-Am Standards Conf Textiles, Peru, 62; indust adv comt textile info, Underwriters' Labs, 67-; leader US deleg, Int Orgn Standarization Colorfastness of Textiles, Ger, 68 & Mass, 71; panel mem, Nat Acad Sci-Nat Acad Eng, 70- Mem: Am Asn Textile Chemists & Colorists (vpres, 69-72, pres, 75); Am Soc Textiles & Mat; Am Nat Standards Inst; Am Asn Textile Technol. Res: Application research in field of textile fibers, fabrics and finishes. Mailing Add: 959 Eighth Ave New York NY 10019

WHAN, RUTH ELAINE, b Brownsville, Pa, Oct 28, 31; m 55; c 3. PHYSICAL CHEMISTRY. Educ: Allegheny Col, BS, 52; Carnegie Inst Technol, MS, 54; Univ NMex, PhD(phys chem), 61. Prof Exp: Res assoc phys chem, Callery Chem Co Div, Mine Safety Appliances Co, 54-57; fel, NSF grant chem to Dr Glenn Crosby, Univ NMex, 61-62; DIV SUPVR, MAT ANAL DIV II, SANDIA LABS DIV, WESTERN ELEC CORP, 62- Res: Molecular spectroscopy; luminescence of rare earth chelates; optical studies of radiation damage in semiconductors; effect of ion implantation in semiconductors; scanning electron microscopy; carbon composite analysis. Mailing Add: Div 5522 Sandia Labs Albuquerque NM 87115

WHANG, ROBERT, b Honolulu, Hawaii, Mar 7, 28; m 56; c 4. INTERNAL MEDICINE, NEPHROLOGY. Educ: St Louis Univ, BS, 52, MD, 56; Am Bd Internal Med, dipl, 65, cert nephrol, 72 & 74. Prof Exp: Intern med, Johns Hopkins Univ Hosp, 56-57; asst resident, Baltimore City Hosps, 57-59, resident, 59-60; from instr to assoc prof med, Sch Med, Univ NMex, 63-71; prof, Sch Med, Univ Conn, 71-73, assoc dean, Vet Admin Hosp Affairs, 72-73; PROF MED, SCH MED, IND UNIV, INDIANAPOLIS, 73-; ASST DEAN VET ADMIN HOSP AFFAIRS, 73- Concurrent Pos: Life Ins Med res fel, Univ NC, 60-62, USPHS trainee renal dis, 62-63; chief metab, Vet Admin Hosp, Albuquerque, 66-71; chief staff, Vet Admin Hosp, Newington, Conn, 71-73 & Indianapolis, Ind, 73- Mem: Fel Am Col Physicians; Am Fedn Clin Res; Int Soc Nephrology; Am Soc Nephrology. Res: Magnesium deficiency, interrelationship of magnesium and potassium, electrolyte changes in uremia. Mailing Add: Vet Admin Hosp 1481 W Tenth St Indianapolis IN 46202

WHANG, SUKOO JACK, b Seoul, Korea, Feb 3, 34; US citizen; m 63; c 3. MEDICAL MICROBIOLOGY, IMMUNOLOGY. Educ: Univ Calif, Los Angeles, MS, 60, PhD(med microbiol, immunol), 63, MD, 72; Am Bd Med Microbiol, dipl, 75. Prof Exp: Asst prof microbiol, Calif State Polytech Univ, 63-64; chief microbiol & serol dept, Providence Hosp, Southfield, Mich, 64-65; chief microbiol & immunol dept, Ref Lab, Div Abbott Labs, Calif, 65-69; CHIEF MICROBIOL & IMMUNOL DIV, CLIN LAB & CHMN INFECTION CONTROL COMT, WHITE MEM MED CTR, LOS ANGELES, 72- Concurrent Pos: Chief microbiol & Immunol Div, Clin Lab, White Mem Med Ctr, 69-70. Mem: NY Acad Sci; Am Soc Microbiol; Sigma Xi; Am Col Physicians; Am Soc Clin Pathologists. Res: Clinical microbiology and serology; syphilis serology; diagnostic tests for the detection of inborn errors of metabolism; fluorescent antibody testing. Mailing Add: 1325 Via Del Ray South Pasadena CA 91030

WHANGER, PHILIP DANIEL, b Lewisburg, WVa, Aug 30, 36; m 64; c 2. NUTRITIONAL BIOCHEMISTRY. Educ: Berry Col, BS, 59; WVa Univ, MS, 61; NC State Univ, PhD(nutrit), 65. Prof Exp: Res assoc biochem, Mich State Univ, 65-66; asst prof, 66-70, ASSOC PROF NUTRIT & BIOCHEM, ORE STATE UNIV, 70- Concurrent Pos: NIH res fel, Mich State Univ, 66-67, res grants selenium & myopathies, Ore State Univ, 68- & environ cadmium, 71-; NIH spec fel, 72; assoc staff, Harvard Med Sch, 72-73. Mem: Am Inst Nutrit; Am Soc Animal Sci; Soc Environ Geochem & Health. Res: Altered metabolic pathways under sulfur deficiency; relationships of vitamin E and selenium in myopathies, biochemical properties of selenium and cadmium metallo-proteins, metabolic pathways in rumen microbes. Mailing Add: Dept of Agr Chem Ore State Univ Corvallis OR 97331

WHAPLES, GEORGE WILLIAM, b Neponset, Ill, Nov 27, 14; m 49; c 3. MATHEMATICS. Educ: Knox Col, AB, 35; Univ Wis, AM, 37, PhD(math), 39. Prof Exp: Asst math, Univ Wis, 37-39, asst prof, 46-47; fel, Ind Univ, 39-41; asst, Inst Advan Study, 41-42; instr, Johns Hopkins Univ, 42-43; instr, Ind Univ, 43; asst prof, Univ Pa, 43-46; from assoc prof to prof, Ind Univ, 47-66; PROF MATH, UNIV MASS, AMHERST, 66- Mem: Am Math Soc. Res: Algebraic number theory; class field theory; ring theory. Mailing Add: Dept of Math Univ of Mass Amherst MA 01002

WHARTON, CHARLES BENJAMIN, b Gold Hill, Ore, Mar 29, 26; m 53; c 3. PLASMA PHYSICS. Educ: Univ Calif, Berkeley, BSEE, 50, MS, 52. Prof Exp: Proj engr, Lawrence Radiation Lab, Univ Calif, 50-62; staff mem exp physics, Gen Atomic Div, Gen Dynamics Corp, Calif, 62-67; dir lab plasma studies, 72-73, PROF PLASMA PHYSICS, CORNELL UNIV, 67- Concurrent Pos: Tech advisor, UN Conf on Peaceful Uses of Atomic Energy, Geneva, Switz, 58; sci engr, Max Planck Inst Physics & Astrophys, Ger, 59-60; consult, Aerojet-Gen Nucleonics Div, Gen Tire & Rubber Co, 60-62 & US Naval Res Lab, Washington, DC, 70-; controlled fusion res mem eval panel on quantum electronics & plasma physics, Nat Res Coun, 70-73; vis scientist, Max Planck Inst Plasma Physics, Munich, Ger, 73-74; von Humboldt Found sr scientist award, 73; consult, Lawrence Livermore Lab, 75- Mem: Fel Am Phys Soc; Inst Elec & Electronics Engrs; Nuclear & Plasma Sci Soc (vpres, 75). Res: Plasma diagnostics; waves in plasmas; plasma instabilities; microwave technology; electronic circuitry; nonlinear waves; relativistic electron beams; plasma heating; controlled fusion research. Mailing Add: Phillips Hall Cornell Univ Ithaca NY 14853

WHARTON, CHARLES HEIZER, b Minneapolis, Minn, July 28, 23. VERTEBRATE ZOOLOGY, ECOLOGY. Educ: Emory Univ, AB, 51; Cornell Univ, MS, 54; Univ Fla, PhD, 58. Prof Exp: Pvt zoologist, Philippines, 46-47; spec consult, Int Div, USPHS, 50; mem scrub-typhus team, US Dept Army, Borneo, 51; dir, Forest Cattle Surv Exped Southeast Asia, Coolidge Found, 51-52; asst prof, 58-66, PROF BIOL, GA STATE UNIV, 66- Concurrent Pos: Prin investr, Joint Kouprey Venture, Nat Acad Sci, 63-64; chmn, Ga Coun for Preservation of Natural Areas, 66-68. Mem: Am Soc Mammal; Am Soc Ichthyol & Herpet. Res: Mammalogy; herpetology; international and natural area conservation; vertebrate epidemiology; biology of vertebrates; Southern River swamp ecosystems. Mailing Add: Ga State Univ 33 Gilmer St Atlanta GA 30303

WHARTON, DAVID CARRIE, b Avoca, Pa, Nov 3, 30; m 61; c 2. BIOCHEMISTRY.

Educ: Pa State Univ, BS, 52, MS, 54, PhD(bot), 56. Prof Exp: Asst plant path, Pa State Univ, 52-56; fel enzyme chem, Enzyme Inst, Univ Wis, 59-61; res scientist, E I du Pont de Nemours & Co, 61-62; asst prof biochem, Univ Wis, 62-64 & Sch Med, Univ Va, 64-66; from asst prof to assoc prof, Cornell Univ, 66-73; PROF BIOCHEM, UNIV TEX HEALTH SCI CTR, SAN ANTONIO, 73- Mem: Am Soc Biol Chemists; Am Chem Soc; Brit Soc Gen Microbiol. Res: Electron transport; metalloenzymes. Mailing Add: Dept of Biochem Univ of Tex Health Sci Ctr San Antonio TX 78284

WHARTON, FERDINAND DECATUR, JR, b Henderson, NC, June 27, 19; m 44; c 4. ANIMAL NUTRITION, POULTRY NUTRITION. Educ: Agr & Tech Col NC, BS, 39; Univ Conn, MS, 48. Prof Exp: Asst animal husb, Agr & Tech Col NC, 39-40; high sch instr, NC, 40-42; prof poultry & head dept, Princess Anne Col, 42-43; asst dir race rels, Triumph Explosives, Inc, 43-44; head poultry dept, Olds Farms Convalescent Hosp, Avon, Conn, 44-47; asst poultry husb, Univ Conn, 47-48; dir nutrit, Res Lab, Dawe's Labs, Inc, 48-62; USAID adv, Ghana Ministry Agr, 60-64; sr proj mgr, 64-68, develop mgr, New Enterprise Div, 68-72, mgr environ health & regulatory practices, Fabricated Prod Div, 72-74, MGR ENVIRON AFFAIRS, CONTAINER BUS GROUP, MONSANTO COMMERCIAL PROD CO, 75- Mem: AAAS; Sigma Xi; Poultry Sci Asn; Animal Nutrit Res Coun; Am Inst Nutrit. Res: Nutrient requirements of poultry and livestock; evaluation of protein food stuffs for humans and animals; ecological aspects of polymeric food and beverage containers. Mailing Add: Container Bus Group Monsanto Commercial Prod Co St Louis MO 63166

WHARTON, GEORGE WILLARD, JR, b Belleville, NJ, Jan 25, 14; m 38, 54; c 3. ZOOLOGY. Educ: Duke Univ, BS, 35, PhD(zool), 39. Prof Exp: Res aide, US Bur Fisheries, 35-36; from instr to assoc prof zool, Duke Univ, 39-53; prof & head dept, Univ Md, 53-61; prof zool & entom & chmn dept, 61-68, DIR, ACAROLOGY LAB, OHIO STATE UNIV, 69- Concurrent Pos: Biologist, Norfolk Naval Shipyard, 41-43; collab, USDA, 46-; Guggenheim fel, 50-51. Mem: AAAS; Am Soc Parasitologists; Soc Syst Zool (secy-treas, 48, pres, 56); Am Soc Zoologists. Res: Acarology; water-balance. Mailing Add: Acarology Lab Ohio State Univ Columbus OH 43210

WHARTON, HARRY WHITNEY, b Watertown, NY, May 4, 31; m 55; c 2. ANALYTICAL CHEMISTRY. Educ: Iowa State Univ, BS, 53, MS, 58, PhD(anal chem), 60. Prof Exp: Res chemist, Rath Packing Co, 56; asst anal chem, Iowa State Univ, 56-60; res chemist, 60-61, group leader anal chem, 61-64, group leader, Soap Prod Div, 65, sect head anal chem, Food Prod Div, 65-70, sect head new prod res, 70-75, SECT HEAD FOODS ANAL, FOOD PROD DIV, PROCTER & GAMBLE CO, 75- Mem: AAAS; Am Chem Soc; Am Oil Chem Soc. Res: Micro methods of analysis involving spectrophotometry, microdiffusion, spectrophotometric and nonaqueous titrations, polarography and inorganic oxidation-reduction reactions; development of analytical methods for food products; development of new food products. Mailing Add: Procter & Gamble Co Food Prod Div 6000 Ctr Hill Rd Cincinnati OH 45224

WHARTON, JAMES DUMONT, b Kentland, Ind, May 27, 14; m 37; c 4. PUBLIC HEALTH, PREVENTIVE MEDICINE. Educ: DePauw Univ, AB, 37; Univ Chicago, MD, 40; Johns Hopkins Univ, MPH, 47; Am Bd Prev Med, dipl, 55. Prof Exp: Officer in chg epidemiol unit, US Navy, Pa, 44-45; prev med officer, 14th Naval Dist, 45-46, asst exec officer, Naval Med Res Inst, 48-50, exec officer, Naval Med Field Res Lab, 50-51, officer in chg, Fleet Epidemic Dis Control Unit, 51-53, chief prev med div, US Naval Med Sch, 53-55; chief health & sanit div, Int Coop Admin, Brazil, USPHS, 55-60, mem staff, Bur State Serv, 60-65, dir div community health serv, Bur Health Serv, 65-67; comnr health, City Cincinnati, 67-70; assoc clin prof, Col Med, Univ Cincinnati, 68-70; assoc dir, Div Med Care Serv & Eval, 70-71, ASST COMNR MED CARE SERV & EVAL, NY STATE DEPT HEALTH, 72- Mem: Am Pub Health Asn; Am Col Prev Med. Res: Epidemiology; public health administration. Mailing Add: NY State Dept of Health Empire State Plaza Tower 612 Albany NY 12237

WHARTON, JAMES HENRY, b Mangum, Okla, July 23, 37; m 56; c 2. PHYSICAL CHEMISTRY. Educ: Northeast La Univ, BS, 59; La State Univ, PhD(phys chem), 62. Prof Exp: Asst prof, 62-63 & 65-69, assoc dean, Col Chem & Physics, 69-71, ASSOC PROF CHEM, LA STATE UNIV, BATON ROUGE, 69-, DEAN GEN COL, 71- Concurrent Pos: Consult, Univ Tex, San Antonio, 70-71. Mem: Am Chem Soc. Res: Molecular spectroscopy; electron spin resonance. Mailing Add: Box 12 Dept of Chem La State Univ Baton Rouge LA 70803

WHARTON, LENNARD, b Boston, Mass, Dec 10, 33; m 57; c 3. PHYSICAL CHEMISTRY. Educ: Mass Inst Technol, BS, 55; Univ Cambridge, MA, 60; Harvard Univ, PhD(chem), 63. Prof Exp: Asst prof chem, 63-68, ASSOC PROF CHEM, UNIV CHICAGO, 68- Concurrent Pos: Sloan fel, 64-66, consult, ITE-Imp Corp, Pa, vpres eng, 72-73. Mem: Am Phys Soc; sr mem Inst Elec & Electronics Engrs. Res: Molecular beams and structure; spectroscopy; chemical kinetics; experimental physical chemistry; scattering phenomena; surface sciences; solar energy conversion; electrical power transmission and distribution. Mailing Add: James Franck Inst Univ of Chicago Chicago IL 60637

WHARTON, MARION AGNES, b Cayuga, Ont, Nov 17, 10; nat US. NUTRITION. Educ: Univ Toronto, BA, 33; Univ Western Ont, MS, 34; Mich State Univ, PhD(nutrit), 47; Am Bd Nutrit, dipl. Prof Exp: Instr, Univ Toronto, 36-37; dietician, Prov Dept Health, Ont, 37-39; asst, Mich State Univ, 40-46; res nutritionist, Univ WVa, 46-48; asst prof foods & nutrit, Ohio State Univ, 48-52; from asst prof to assoc prof, NDak Agr Col, 52-55; prof, Univ Southern Ill, 55-61 & Univ RI, 61-63; assoc prof, Iowa State Univ, 63-65; assoc prof, 65-70, PROF FOODS & NUTRIT, CALIF STATE UNIV, LONG BEACH, 70- Mem: AAAS; Am Chem Soc; Am Dietetic Asn; Am Home Econ Asn. Mailing Add: Dept of Home Econ Calif State Univ Long Beach CA 90840

WHARTON, PETER STANLEY, b Oxford, Eng, May 9, 31; m 55; c 4. ORGANIC CHEMISTRY. Educ: Cambridge Univ, BA, 52, MA, 57; Yale Univ, MS, 57, PhD, 59. Prof Exp: Fel, Columbia Univ, 58-60; from instr to prof org chem, Univ Wis-Madison, 60-68; PROF CHEM, WESLEYAN UNIV, 68- Honors & Awards: Frederick Gardner Cottrell Award, 61. Mem: Am Chem Soc; The Chem Soc. Res: Synthetic and mechanistic alicyclic chemistry. Mailing Add: Dept of Chem Wesleyan Univ Middletown CT 06457

WHARTON, WALTER WASHINGTON, b Boone, Ky, Mar 25, 26; m 47; c 5. PHYSICAL CHEMISTRY. Educ: Georgetown Col, AB, 50; Univ Ky, MS, 52, PhD, 55. Prof Exp: Res chemist, 54-59, SUPVR RES CHEM & CHIEF LIQUID PROPULSION TECH BR, PROPULSION LAB, US ARMY MISSIBLE COMMAND, REDSTONE ARSENAL, 59- Concurrent Pos: Instr, Exten Ctr, Univ Ala, 56- Mem: Am Chem Soc. Res: Propulsion technology; propellant chemistry; combustion kinetics; laser technology. Mailing Add: 2811 Barcody Rd SE Huntsville AL 35801

WHARTON, WILLIAM RAYMOND, b Knoxville, Tenn, Mar 30, 43; m 67; c 3. EXPERIMENTAL NUCLEAR PHYSICS. Educ: Stanford Univ, BS, 65; Univ Wash, PhD(nuclear physics), 72. Prof Exp: Res assoc exp nuclear physics, Argonne Nat Lab, 72-74 & Rutgers Univ, 74-75; ASST PROF PHYSICS, CARNEGIE-MELLON UNIV, 75- Mem: Am Phys Soc. Res: Experimental medium energy nuclear physics involving the use and study of nuclei interacting with pions, kaons and muons. Mailing Add: Dept of Physics Carnegie-Mellon Univ Pittsburgh PA 15213

WHATLEY, ALFRED T, b Denver, Colo, Apr 20, 22; div; c 4. PHYSICAL CHEMISTRY. Educ: Princeton Univ, AB, 48, AM, 50, PhD(chem), 52. Prof Exp: Chemist, Hanford Works, Gen Elec Co, 52-55, eng consult, Aircraft Nuclear Propulsion, 55-57, physicist, Vallecitos Atomic Lab, 57-61; staff engr, Martin Co, 61-62; sr sci specialist, EG&G, 62-70; EXEC DIR, WESTERN INTERSTATE NUCLEAR BD, LAKEWOOD, CO, 70- Mem: Am Nuclear Soc. Res: Nuclear science. Mailing Add: PO Box 540 Breckenridge CO 80424

WHATLEY, BOOKER TILLMAN, b Alexandria, Ala, Nov 5, 15; m 43. HORTICULTURE, PLANT PHYSIOLOGY. Educ: Ala Agr & Mech Col, BS, 41; Rutgers Univ, PhD, 57. Prof Exp: Agr exten agent, Butler County, Ala, 46-47; prin high sch, Ala, 47-50; tech oper officer, Chofu Hydroponic Farm, Japan, 50-54; assoc prof & head dept hort, Southern Univ, 57-60; adv hort, US Opers Mission, Ministry Agr, Ghana, 60-62; prof hort, Southern Univ, 62-68; PROF PLANT & SOIL SCI, TUSKEEGEE INST, 68- Mem: AAAS; Am Soc Hort Sci; Am Soc Plant Physiol. Res: The effect of budding methods, wrapping materials and hormones on Myristica fragrans and its vegetative propagation. Mailing Add: Dept of Plant & Soil Sci Tuskeegee Institute AL 36088

WHATLEY, JAMES ARNOLD, b Calvert, Tex, Feb 26, 16; m 39; c 2. ANIMAL BREEDING. Educ: Agr & Mech Col, Tex, BS, 36; Iowa State Col, MS, 37, PhD(animal breeding), 39. Prof Exp: From asst prof to prof animal husb, 39-64, from assoc dir to dir, Agr Exp Sta, 64-74, dir agr res, 66-68, ASSOC DIR, AGR EXP STA, OKLA STATE UNIV, 74-, DEAN AGR, 68- Concurrent Pos: With bur animal indust, USDA, 44. Mem: AAAS; Genetics Soc Am; Am Soc Animal Sci. Res: Swine breeding. Mailing Add: 2221 W Eighth Stillwater OK 74074

WHATLEY, MALCOLM CLIFFORD, particle physics, social sciences, see 12th edition

WHATLEY, THOMAS ALVAH, b Midland, Ark, Aug 23, 32; m 54; c 4. PHYSICAL CHEMISTRY, INSTRUMENTATION. Educ: Fresno State Col, BS, 53; Univ Ore, PhD(phys chem), 61. Prof Exp: Sr scientist, Lockheed Aircraft Corp, 58-61; inorg res group head, United Tech Div, United Aircraft Corp, 61-65; sr res chemist, F&M Div, Hewlett-Packard Co, Pa, 65-68; eng specialist, 68-74, MGR APPLN, APPL RES LABS, 74- Mem: Am Chem Soc; Am Soc Mass Spectrometry; Microbeam Anal Soc; Am Vacuum Soc. Res: Ion probe mass spectrometry; analytical instrumentation development; material science; solid state device structure analysis; ion-solid interactions. Mailing Add: Appl Res Labs PO Box 4729 Van Nuys CA 91412

WHAYNE, TOM FRENCH, b Columbus, Ky, Dec 26, 05; m 34; c 2. MEDICINE. Educ: Univ Ky, AB, 27; Washington Univ, MD, 31; Harvard Univ, MPH, 49, DrPH, 50; Am Bd Prev Med, dipl, 49. Prof Exp: Intern, Mo Baptist Hosp, St Louis, 31-32; house physician, Mo Pac Hosp, 32-33; surgeon, Civilian Conserv Corps, 33-34; physician, Fitzsimons Gen Hosp, Denver, 34; physician, CZ, Med Corps, US Army, 38-41, physician, Off Surgeon Gen, 41-43, asst mil attache med, US Embassy, London, 43-44, chief prev med sect, 12th Army Group, 44-45; chief prev med sect, Off Chief Surgeon, Europe, 45-46, dep chief prev med div, Off Surgeon Gen, Washington, DC, 46-47, chief, 47-48, chief dept training doctrines, Army Med Serv Grad Sch, Walter Reed Army Med Ctr, 50-51, chief prev med div, Off Surgeon Gen, DC, 51-55; prof prev med & pub health, Sch Med, Univ Pa, 55-63, assoc dean, 58-63; asst vpres med ctr, 63-67, actg dean col med, 66-67, prof community med & assoc dean col med, 63-72, EMER PROF COMMUNITY MED, UNIV KY, 72- Concurrent Pos: Consult, Surgeon Gen, US Army; adv, US Deleg, World Health Assembly, Geneva, 48 & 53; mem, Bd Dirs, Gorgas Mem Inst Trop & Prev Med. Mem: Fel Am Soc Trop Med & Hyg; Am Epidemiol Soc; fel Am Pub Health Asn; Asn Am Med Cols; fel Am Col Prev Med. Res: Epidemiology. Mailing Add: 623 Tateswood Dr Lexington KY 40502

WHEALTON, JOHN HOBSON, b Brooklyn, NY, Apr 27, 43; m 72. PLASMA PHYSICS. Educ: Univ Lowell, BS, 66; Univ Del, MS, 68, PhD(physics), 71. Prof Exp: Res assoc Div Eng, Brown Univ, 71-72, res assoc, Dept Chem, 72-73; res assoc, Joint Inst Lab Astrophys, Univ Colo-Nat Bur Stand, 73-75, MEM STAFF, THERMONUCLEAR DIV, OAK RIDGE NAT LABS, 75- Mem: Am Phys Soc. Res: Analysis of drift tube swarm experiments; kinetic theory of collision-dominated weakly ionized gases in presence of strong fields; space charge ion optics relevent to neutral beam plasma heating. Mailing Add: Thermonuclear Div Oak Ridge Nat Labs Oak Ridge TN 37830

WHEALY, ROGER DALE, b Colman, SDak, Aug 21, 08; m 33; c 1. PHYSICAL CHEMISTRY, ANALYTICAL CHEMISTRY. Educ: East Nor Sch, BS, 30; Univ Colo, MS, 37; Univ Ore, MS, 48; Univ Colo, PhD(chem), 53. Prof Exp: Teacher high sch, Tex, 37-43; instr physics, Carnegie Inst Technol, 43-44; from asst prof to prof chem, WTex State Col, 47-58; prof, 58-75, EMER PROF ANAL CHEM, TEX A&M UNIV, 75- Concurrent Pos: Vis prof, Univ Alaska, 66-67. Mem: AAAS; Am Chem Soc. Res: Complex ions in solution. Mailing Add: Dept of Chem Tex A&M Univ College Station TX 77843

WHEAT, JOE BEN, b Van Horn, Tex, Apr 21, 16; m 47. ANTHROPOLOGY, ARCHAEOLOGY. Educ: Univ Calif, BA, 37; Univ Ariz, MA, 49, PhD(anthrop), 53. Prof Exp: Field dir archaeol, Works Prog Admin, Tex Tech Col, 39-41; archaeologist, River Basin Surv, Smithsonian Inst, 47; instr anthrop, Univ Ariz, 48-50; field foreman, Archaeol Field Sch, 49-50; ranger-archaeologist, US Nat Park Serv, 52-53; from asst prof to assoc prof, 53-62, PROF NATURAL HIST, UNIV COLO, BOULDER, 62-, CUR ANTHROP, MUS, 53- Concurrent Pos: Mem proj rev bd, NSF, 60-, grant, Univ Colo Mus, 61-65 & 68-69; NSF & Smithsonian Inst res grant, Univ Colo & Repub of Sudan, 62-66; sr archaeologist Nubian Exped, 62-66, prin investr, Nubian Exped, 66; Smithsonian Inst res grant, Univ Colo & Tunisia, 66-67. Mem: Fel AAAS; Soc Am Archaeol (secy, 60-64, pres-elect, 65-66, pres, 66-67); fel Am Anthrop Asn; Am Ethnol Soc; Sigma Xi. Res: Early man in North America; southwestern archaeology; Colorado archaeology; material culture; southwest native textiles. Mailing Add: 1515 Baseline Rd Boulder CO 80302

WHEAT, JOHN DAVID, b Ranger, Tex, July 12, 21; m 50; c 3. ANIMAL GENETICS. Educ: Agr & Mech Col, Tex, BS, 42, MS, 51; Iowa State Col, PhD(animal breeding, genetics), 54. Prof Exp: From asst prof to assoc prof, 54-69, PROF ANIMAL SCI & INDUST, KANS STATE UNIV, 69- Concurrent Pos: Beef cattle breeding adv, Ministry of Animal & Forest Resources, US Agency Int Develop-Kans State Univ Contract Team, Northern Nigeria, 66-68; mem fac, Ahmadu Bello Univ, Nigeria, 66-68, livestock breeding consult, Taiwan, 72. Mem: Am Soc Animal Sci; Am Genetic Asn. Res: Population genetics; muscling selection research in swine. Mailing Add: Dept of Animal Sci & Indust Kans State Univ Manhattan KS 66506

WHEAT, JOSEPH ALLEN, b Charlottesville, Va, Mar 31, 13; m 42; c 1. CHEMISTRY. Educ: Univ Va, BS, 34; Cornell Univ, MS, 36, PhD(inorg chem), 39. Prof Exp: Instr chem, Trinity Col, Conn, 39-40; microchemist, Biochem Res Found, Del, 40-41; chemist, Celanese Corp Am, NJ, 41-49; spectroscopist, Air Reduction Co, 49-53; CHEMIST, SAVANNAH RIVER LAB, ATOMIC ENERGY DIV, E I DU PONT DE NEMOURS & CO, INC, 53- Concurrent Pos: Prof, Voorhees Col, 67- Mem: Am Chem Soc; Soc Appl Spectros; Am Inst Chemists. Res: Analytical chemistry; instrumental analysis; infrared, emission and atomic absorption spectroscopy; application of computers and programmable calculators to reduction of spectroscopic data. Mailing Add: 1478 Canterbury Ct SE Aiken SC 29801

WHEAT, MYRON WILLIAM, JR, b Sapulpa, Okla, Mar 24, 24; m 49, 70; c 5. THORACIC SURGERY, CARDIOVASCULAR SURGERY. Educ: Washington Univ, AB, 49, MD, 51; Am Bd Surg, dipl, 58; Bd Thoracic Surg, dipl, 58. Prof Exp: Intern surg, Barnes Hosp, St Louis, Mo, 51-52, asst resident, 52-55, resident, 55-56; instr surg, Washington Univ, 56-58; from asst prof to prof, Col Med, Univ Fla, 58-72, chief div thoracic & cardiovasc surg, 58-72; CHIEF DEPT SURG, DIAG CLIN, CHIEF THORACIC & CARDIOVASC SURG, DIAG CLIN, LARGO, FLA, 72- Concurrent Pos: Clin fel chest surg, Barnes Hosp, St Louis, Mo, 56-58; consult, Vet Admin Hosp, Lake City, Fla, 58-; mem, Bd Thoracic Surg, 69-75; chief div thoracic & cardiovasc surg, Sch Med, Univ Louisville, 72-75. Mem: Fel Am Col Cardiol; fel Am Col Surgeons; Am Asn Thoracic Surg (secy); Am Surg Asn; Soc Thoracic Surg. Res: Ultrastructure of the heart; localization of radioactive digitalis in heart muscle; relationship between lysosomes and heart disease in humans; treatment of aneurysms of aortic root and of dissecting aneurysms; surgical treatment of acquired heart disease. Mailing Add: Dept of Surg Diag Clin 1551 W Bay Dr Largo FL 33540

WHEAT, PERCY WAYNE, b Bogalusa, La, Apr 18, 40; m 63; c 2. INDUSTRIAL ORGANIC CHEMISTRY. Educ: Miss Col, BS, 62; Univ Ala, Tuscaloosa, PhD(org chem), 70. Prof Exp: Chemist, Liggett & Myers Tobacco Co, NC, 63-65; instr org chem, Univ Ala, Tuscaloosa, 68-69; SR DEVELOP CHEMIST, CIBA-GEIGY CORP, 70- Mem: Am Chem Soc. Res: Reaction mechanisms; organic synthesis and process development for textile auxiliaries; specialty and agricultural chemicals. Mailing Add: Develop Dept Ciba-Geigy Corp Saraland AL 36553

WHEAT, ROBERT WAYNE, b Springfield, Mo, Nov 10, 26; m 48; c 3. MICROBIOLOGY, BIOCHEMISTRY. Educ: Wash Univ, PhD(microbiol), 55. Prof Exp: USPHS fel biochem, NIH, Md, 55-56; instr biochem, 56-58, assoc, 58-60, assoc prof microbiol, 66-74, ASST PROF BIOCHEM, SCH MED, DUKE UNIV, 60-, PROF MICROBIOL, 74- Concurrent Pos: NIH consult, 69-72. Mem: Am Chem Soc; Am Soc Biol Chem; Am Soc Microbiol. Res: Biochemistry of microorganisms, amino sugars and polysaccharides; cell surface antigens. Mailing Add: Dept of Microbiol & Biochem Duke Univ Med Ctr Durham NC 27710

WHEATCROFT, MERRILL GORDON, b Utica, Kans, Oct 20, 14; m 39; c 4. DENTAL PATHOLOGY. Educ: Univ Kansas City, DDS, 39. Prof Exp: Dent res officer, Naval Med Res Inst, Md, 49-52, head dent res, Naval Med Res Unit, Egypt, 52-55, head res & sci, Naval Dent Sch, 55-60; assoc prof dent res, 60-67, PROF PATH, UNIV TEX DENT BR, HOUSTON, 67- Mem: AAAS; Am Dent Asn; fel Am Col Dent; Int Asn Dent Res. Res: Changes in the contents of the gingival crevice and associated tissued in periodontal disease. Mailing Add: PO Box 20068 Houston TX 77025

WHEATLAND, DAVID ALAN, b Boston, Mass, Aug 27, 40; m 65; c 1. INORGANIC CHEMISTRY. Educ: Brown Univ, ScB, 63; Univ Md, PhD(inorg chem), 67. Prof Exp: Asst prof chem, Bowdoin Col, 67-73; RES CHEMIST, S D WARREN RES LAB, 73- Concurrent Pos: Petrol Res Fund grant, 68-70. Mem: Am Chem Soc. Res: Reprographic research and development. Mailing Add: S D Warren Res Labs Westbrook ME 04092

WHEATLEY, GEORGE MILHOLLAND, b Baltimore, Md, Mar 21, 09; m 33; c 4. PREVENTIVE MEDICINE. Educ: Cath Univ Am, BS, 29; Harvard Univ, MD, 33; Columbia Univ, MPH, 42. Prof Exp: Intern, Hartford Hosp, 33-35; pediat house officer, Harriet Lane Home, Johns Hopkins Univ, 35-36; asst dir, Astoria Sch Health Study, New York Dept Health, 37-39, prin pediatrician, 39-40; asst med dir, Metrop Life Ins Co, 41-45, asst vpres med, 45-49, third vpres, 49-61, third vpres & med dir, 61-65, sr med dir, 65-69, vpres & chief med dir, 69-74; MED DIR, DEPT SOCIAL SERV, SUFFOLK COUNTY, NY, 74- Concurrent Pos: Mead-Johnson fel, NY Postgrad Hosp, 36-37; consult childhood accident prev prog, Europ Nations, WHO, Switz, 56 & Belg, 58. Honors & Awards: Grulee Award, Am Acad Pediat, 64. Mem: Fel Am Pediat Soc; fel AMA; fel Am Pub Health Asn; fel Am Acad Pediat (pres, 60-61); fel NY Acad Med. Res: Medicine. Mailing Add: Dept Social Serv Med Asst Admin 10 Oval Dr Hauppauge NY 11787

WHEATLEY, JOHN CHARLES, b Tucson, Ariz, Feb 17, 27; m 49; c 2. PHYSICS. Educ: Univ Pittsburgh, PhD(physics), 52. Hon Degrees: DSc, Leiden Univ, Neth, 75. Prof Exp: From instr to prof, Univ Ill, 52-67; PROF PHYSICS, UNIV CALIF, SAN DIEGO, 67- Concurrent Pos: Guggenheim fel, Univ Leiden, 54-55. Honors & Awards: Simon Prize, 66; Ninth Fritz London Award, 75. Mem: Nat Acad Sci. Res: Cryogenics; low temperature physics; experimental research at millikelvin temperatures with emphasis on properties of helium. Mailing Add: Dept of Physics Univ of Calif, San Diego La Jolla CA 92037

WHEATLEY, QUENTIN DE LATTICE, physical chemistry, see 12th edition

WHEATLEY, VICTOR RICHARD, b London, Eng, Nov 4, 18; m 43; c 2. BIOCHEMISTRY. Educ: Univ London, BSc, 47, PhD(org chem), 50, DSc(chem, biochem), 68. Prof Exp: Biochemist, St Bartholomew's Hosp Med Col, London, 48-57; res assoc dermat, Univ Chicago, 57-59; sr res assoc, Stanford Univ, 59-62; ASSOC PROF DERMAT, MED CTR, NY UNIV, 62- Concurrent Pos: NIH fels, Med Ctr, NY Univ, 62-68. Honors & Awards: Bronze Medal, Am Acad Dermat, 58; Spec Award, Soc Cosmetic Chem, 62. Mem: Am Soc Biol Chemists. Res: Biochemistry of skin, especially the lipid metabolism of skin. Mailing Add: Dept of Dermat NY Univ Med Ctr New York NY 10016

WHEATLEY, WILLIAM BACON, b Penn Yan, NY, Feb 24, 20; m 42; c 5. ORGANIC CHEMISTRY. Educ: Colgate Univ, AB, 42; Ohio State Univ, PhD(org chem), 47. Prof Exp: Tech asst, Am Can Co, NY, 41; asst org chem, Ohio State Univ, 42-44, res assoc, Res Found, 44-46, asst org chem, Ohio State Univ, 46-47; sr res chemist, 47-50, proj leader, 50-57, asst to sci dir, 57-65, dir library & info serv, 65-73, DIR CHEM DOC, BRISTOL LABS, INC, 73- Mem: Am Chem Soc. Res: Synthetic organic chemistry; aromatic hydrocarbons; therapeutically active compounds; information storage and retrieval. Mailing Add: Bristol Labs Inc PO Box 657 Syracuse NY 13201

WHEATON, BURDETTE CARL, b Mankato, Minn, July 3, 38; m 68; c 3. ALGEBRA. Educ: Mankato State Col, BS, 59; Univ Iowa, MS, 61, PhD(math), 65. Prof Exp: Instr math, Univ Iowa, 59-63; asst prof, Western Ill Univ, 63-65; asst prof, 65-72, PROF MATH, MANKATO STATE UNIV, 72- Mem: Am Math Soc; Math Asn Am; Sigma Xi. Res: Abstract algebra, particularly group theory and group representations. Mailing Add: Dept of Math Mankato State Univ Mankato MN 56001

WHEATON, JONATHAN EDWARD, b Fullerton, Calif, Jan 22, 47; m 67; c 2. NEUROENDOCRINOLOGY. Educ: Univ Calif, Davis, BS, 69; Ore State Univ, MS, 70, PhD(animal physiol), 73. Prof Exp: Fel neuroendocrinol, Southwestern Med Sch, 73-75; ASST PROF PHYSIOL, UNIV MINN, ST PAUL, 75- Mem: Am Soc Animal Sci; Sigma Xi. Res: Neurohormones, their control and effects on anterior pituitary gland hormones. Mailing Add: Dept of Animal Sci Univ of Minn St Paul MN 55108

WHEATON, ROBERT MILLER, b Danbury, Ohio, Oct 11, 19; m 43, 67; c 5. INDUSTRIAL CHEMISTRY. Educ: Oberlin Col, AB, 41. Prof Exp: Chemist, Celotex Corp, 41; chemist & process specialist, Trojan Powder Co, 42-43, head process specialists, 44-45; chemist, 46-49, group leader res, 50-55, div leader, 56-71, ASSOC SCIENTIST, WESTERN DIV RES, DOW CHEM CO, 72- Concurrent Pos: Chmn, Gordon Res Conf Ion Exchange, 63. Mem: Am Chem Soc; Sigma Xi. Res: Synthesis, applications and properties of ion exchange resins. Mailing Add: 156 Warwick Dr Walnut Creek CA 94598

WHEATON, THOMAS ADAIR, b Orlando, Fla, Apr 5, 36; m 61; c 2. PLANT PHYSIOLOGY, HORTICULTURE. Educ: Univ Fla, BS, 58, MS, 60; Univ Calif, Davis, PhD(plant physiol), 63. Prof Exp: Asst horticulturist, 63-70, ASSOC PROF HORT & ASSOC HORTICULTURIST, CITRUS EXP STA, INST FOOD & AGR SCI, AGR RES & EDUC CTR, UNIV FLA, 70- Mem: Am Soc Plant Physiol; Am Soc Hort Sci. Res: Chilling injury in plants; nitrogen metabolism and growth regulation in citrus. Mailing Add: Citrus Exp Sta Inst Food & Agr Sci Agr Res & Educ Ctr Univ of Fla Lake Alfred FL 33850

WHEBELL, CHARLES FREDERICK JOHN, b Nassau, Bahamas, Mar 20, 30; Can citizen; m 54; c 3. GEOGRAPHY. Educ: Univ Western Ont, BA, 52, MSc, 55; Univ London, PhD, 61. Prof Exp: Instr, 57-59, lectr 59-61, asst prof, 61-65, ASSOC PROF GEOG, UNIV WESTERN ONT, 65- Concurrent Pos: Vis lectr, New Eng, Australia, 67. Mem: Asn Am Geog. Res: General cultural geography, especially settlement and political aspects; administrative areas and local government. Mailing Add: Dept of Geog Univ of Western Ont London ON Can

WHEBY, MUNSEY S, b Roanoke, Va, Nov 19, 30; m 55; c 3. INTERNAL MEDICINE. Educ: Roanoke Col, BS, 51; Univ Va, MD, 55. Prof Exp: Asst chief hemat, Walter Reed Army Inst Res & Walter Reed Gen Hosp, 59-61, chief gastroenterol, Walter Reed Army Inst Res, 61-62, chief med div, US Army Trop Res Med Lab, 62-65; assoc prof med, Rutgers Med Sch, 65-66; assoc prof, 66-72, PROF MED, SCH MED, UNIV VA, 72- Mem: Am Fedn Clin Res; AMA; Am Soc Hemat. Res: Gastrointestinal absorption of iron; folic acid and B-12 metabolism. Mailing Add: Dept of Internal Med Univ of Va Sch of Med Charlottesville VA 22901

WHEDON, GEORGE DONALD, b Geneva, NY, July 4, 15; m 42; c 2. MEDICAL RESEARCH ADMINISTRATION, AEROSPACE MEDICINE. Educ: Hobart Col, AB, 36; Univ Rochester, MD, 41; Am Bd Internal Med, dipl, 50; Am Bd Nutrit, cert, 68. Hon Degrees: ScD, Hobart Col, 67. Prof Exp: Intern med, Mary Imogene Bassett Hosp, Cooperstown, NY, 41-42; asst, Sch Med & asst resident, Strong Mem Hosp, Univ Rochester, 42-44; from instr to asst prof, Cornell Univ, 44-52; chief metab dis br, 52-65, asst dir inst, 56-62, DIR, NAT INST ARTHRITIS, METAB & DIGESTIVE DIS, 62- Concurrent Pos: USPHS fel, NY Hosp-Cornell Med Ctr, 51; from asst physician to physician, Outpatient Dept, NY Hosp-Cornell Med Ctr, 44-52; mem, Subcomt Calcium Comt Dietary Allowances, Food & Nutrit Bd, Nat Res Coun, 59-64; consult space med div, NASA, 63-; mem, Med Alumni Coun, Sch Med, Univ Rochester, 71-, Trustees Coun, 71-, vchmn, 73, chmn, 74-75; chmn, Adv Panel Med Prog to NASA, Am Inst Biol Sci, 71-; chmn, Life Sci Comt, NASA, 74-, mem, Space Prog Adv Coun, 74- Honors & Awards: Super award, USPHS, 67; Cit Alumni Award, Univ Rochester, 71; Ayerst Award, Endocrine Soc, 74; Except Sci Achievement Award, NASA, 74. Mem: AAAS; Am Asn Physicians; Endocrine Soc; Am Diabetes Asn; Am Fedn Clin Res. Res: Metabolic and physiological aspects of convalescence and immobilization; metabolic and kinetic studies of disorders of bone; space medicine, particularly metabolism. Mailing Add: Nat Inst of Arthritis Metab & Digestive Dis Bethesda MD 20014

WHEEDEN, RICHARD LEE, b Baltimore, Md, Nov 29, 40; m 62; c 2. MATHEMATICS. Educ: Johns Hopkins Univ, AB, 61; Univ Chicago, MS, 62, PhD(math), 65. Prof Exp: Instr math, Univ Chicago, 65-66; mem, Inst Adv Study, 66-67; from asst prof to assoc prof, 67-74, PROF MATH, RUTGERS UNIV, NEW BRUNSWICK, 74- Concurrent Pos: NSF fel, 66-67. Res: Harmonic analysis. Mailing Add: Dept of Math Rutgers Univ New Brunswick NJ 08903

WHEELER, ALAN CLEMENT, b Concord, NH, Sept 24, 40; m 62. STATISTICS, OPERATIONS RESEARCH. Educ: Harvard Univ, BA, 62; Stanford Univ, MS, 64, PhD(statist), 68. Prof Exp: Asst prof appl math, Wash Univ, 66-70; asst prof comput sci & opers res, 70-72, ASSOC PROF COMPUT SCI & OPERS RES, SOUTHERN METHODIST UNIV, 72- Concurrent Pos: Statist consult, Nat Ctr Drug Anal, 69-70; mem transportation systs planning group, Hwy Res Bd, 71- Mem: Am Inst Indust Eng; Am Statist Asn; Opers Res Soc Am; Inst Math Statist; Inst Mgt Sci. Res: Mathematical inventory theory; stochastic processes in operations research; stochastic models in health care; mathematical analysis of transportation and water resource systems. Mailing Add: Comput Sci/Opers Res Ctr Southern Methodist Univ Dallas TX 75222

WHEELER, ALBERT HAROLD, b St Louis, Mo, Dec 11, 15; m 38; c 3. BACTERIOLOGY. Educ: Lincoln Univ, AB, 36; Iowa State Col, MS, 37; Univ Mich, MSPH, 38, DrPH, 44. Prof Exp: Clin technician, Col Med, Howard Univ, 38-40; asst, 41-44, res assoc, Univ Hosp, 44-52, asst prof bact, 52-58, ASSOC PROF BACT, MED SCH, UNIV MICH, ANN ARBOR, 59-, ASSOC PROF MICROBIOL, 74- Concurrent Pos: Consult, Serol Lab, Univ Hosp, Univ Mich. Mem: Soc Exp Biol & Med; Am Asn Immunol. Res: Active and passive immunity in experimental syphilis; serodiagnosis of syphilis; serology of biologic false positive reactions in syphilis; treponemicidal activity of various animal sera and complements. Mailing Add: Dept of Bact & Dermatol Univ of Mich Med Ctr Ann Arbor MI 48103

WHEELER, ALFRED GEORGE, JR, b Nebraska City, Nebr, Apr 11, 44. ENTOMOLOGY. Educ: Grinnell Col, BA, 66; Cornell Univ, PhD(insect ecol), 71. Prof Exp: ENTOMOLOGIST, BUR PLANT INDUST, PA DEPT AGR, 71- Concurrent Pos: Consult, Dames & Moore, 74- Mem: Entom Soc Am; Entom Soc Can. Res: Life history studies of Hemiptera-Heteroptera, especially Miridae; biology of insects affecting ornamental plants; study of insect-plant associations. Mailing Add: Bur of Plant Indust Pa Dept of Agr 2301 N Cameron St Harrisburg PA 17120

WHEELER, ALLAN GORDON, b Gary, Ind, July 12, 23; m 49; c 4. PHARMACOLOGY, PHYSIOLOGY. Educ: Valparaiso Univ, BA, 48; Univ Wis, MA, 50. Prof Exp: Asst pharmacol & anesthesiol, Univ Wis, 50-54; assoc pharmacologist, Res Ctr, Mead Johnson & Co, Ind, 54-58, sr pharmacologist, 54-59, group leader toxicol, 59-68; SUPVR INDUST TOXICOL, BIO-MED RES LAB, ICI US, INC, 68- Mem: Am Indust Hyg Asn; Drug Info Asn; Environ Mutagen Soc; Sigma Xi; Am Asn Lab Animal Sci. Res: Anesthesiology; toxicology. Mailing Add: Bio-Med Res Lab ICI US Inc Concord Pike & New Murphy Rd Wilmington DE 19897

WHEELER, BARBARA CANTY, physics, see 12th edition

WHEELER, BERNICE MARION, b Winsted, Conn, June 30, 15. ZOOLOGY. Educ: Conn Col, AB, 37; Smith Col, MA, 39; Yale Univ, PhD(zool), 48. Prof Exp: Asst zool, Smith Col, 37-39; instr, Westbrook Jr Col, 39-42; asst, Yale Univ, 42-47; from instr to assoc prof, 47-66, PROF ZOOL, CONN COL, 66- Concurrent Pos: Ford Found fel, 54-55. Mem: Am Genetic Asn. Res: Genetics; ecology; evolution. Mailing Add: Dept of Zool Conn Col PO Box 1553 New London CT 06320

WHEELER, CHARLES HORATIO, III, b Wheeling, WVa, Oct 30, 04; m 40; c 3. MATHEMATICS. Educ: Washington & Jefferson Col, BS, 26; Johns Hopkins Univ, PhD(math), 33. Hon Degrees: DSc, Washington & Jefferson Col, 44. Prof Exp: Asst, Johns Hopkins Univ, 26-27, jr instr, 27-28; from asst prof to assoc prof, 28-41, PROF MATH, UNIV RICHMOND, 41-, SECY & TREAS, 42- Mem: AAAS; Am Math Soc; Math Asn Am. Res: Topology; pure mathematics; statistics. Mailing Add: University of Richmond Richmond VA 23173

WHEELER, CHARLES MERVYN, JR, b Moundsville, WVa, Oct 29, 21; m 43; c 6. PHYSICAL CHEMISTRY. Educ: WVa Univ, BS, 47, MS, 49, PhD(chem), 51. Prof Exp: Asst prof, 50-55, dean students, 60-61, ASSOC PROF CHEM, UNIV NH, 55- Concurrent Pos: Consult, Lima, 57, 59 & 60; chmn postgrad chem dept, Am Col, Madurai, India, 68-69. Mem: Am Chem Soc. Res: Metal-ion complexes. Mailing Add: Parsons Hall Univ of NH Durham NH 03824

WHEELER, CLAYTON EUGENE, JR, b Viroqua, Wis, June 30, 17; m 52; c 3. DERMATOLOGY. Educ: Univ Wis, BA, 38, MD, 41; Am Bd Dermat, dipl, 51. Prof Exp: Resident & instr internal med, Med Sch, Univ Mich, 42-44, resident & instr dermat, 49-51; from asst prof to prof, Sch Med, Univ Va, 51-62; chief div, 62-72, PROF DERMAT, SCH MED, UNIV NC, CHAPEL HILL, 62-, CHMN DEPT, 72- Concurrent Pos: Res fel endocrinol & metab, Univ Mich, 47-48; mem, Comn Cutaneous Dis, Armed Forces Epidemiol Bd, Subcomt Dermat, Surgeon Gen Adv Comt Gen Med & Dermat Training Grant Comt, NIH; mem & dir, Am Bd Dermat; chmn, Residency Rev Comt Dermat; rep, Am Bd Med Specialties; mem, Coun Nat Prog Dermat. Mem: Soc Invest Dermat; Asn Profs Dermat (pres); AMA; Am Acad Dermat. Res: Viral diseases of skin, especially Herpes simplex; tissue culture of skin and contact sensitivity. Mailing Add: Dept of Dermat NC Mem Hosp Chapel Hill NC 27514

WHEELER, DARRELL DEANE, b West Liberty, Ky, Feb 24, 39; m 63; c 2. PHYSIOLOGY. Educ: Transylvania Col, AB, 62; Univ Ky, PhD(physiol), 67. Prof Exp: Asst prof, 68-75, ASSOC PROF PHYSIOL, MED UNIV SC, 75- Concurrent Pos: NIH fel physiol & biophys, Univ Ky, 67-68; NIH res grants, 69-72 & 75- Res: Cell physiology; aging and membrane transport in the nervous system. Mailing Add: Dept of Physiol Med Univ of SC Charleston SC 29401

WHEELER, DAVID LAURIE, b Saginaw, Mich, July 30, 34; m 58; c 2. GEOGRAPHY OF THE MEDITERRANEAN, HISTORICAL GEOGRAPHY. Educ: Univ Mich, BA, 56, MA, 58, PhD(geog), 62. Prof Exp: From asst to assoc prof geog, Ill State Univ, 61-72, asst dean student serv, 67-68, assoc dean, 68-69, assoc dean grad sch, 69-72; PROF GEOG & DEAN GRAD SCH & RES, W TEX STATE UNIV, 72- Concurrent Pos: Fac res grant, Ill State Univ, 64-65 & 70-71; consult, Fiedler Publ Co, Mich, 62, McGraw-Hill Publ Co, NY, 69 & 70 & Van Nostrand-Reinhold Co, 72. Mem: Asn Am Geog; Nat Coun Geog Educ. Res: Agrarian reform and economic development in the Mediterranean Basin, with an emphasis on Italy; historical geography of the United States; beef cattle industry in the United States. Mailing Add: Grad Sch WTex State Univ Canyon TX 79016

WHEELER, DESMOND MICHAEL SHERLOCK, b Northwich, Cheshire, Eng, Apr 18, 29; m 53. ORGANIC CHEMISTRY. Educ: Nat Univ Ireland, BSc, 50, PhD(chem), 55; Univ Dublin, MA, 54. Prof Exp: Dep lectr phys chem, Trinity Col, Dublin, 52-53, asst lectr org chem, 53-55; res fel chem, Harvard Univ, 55-58; asst prof chem, Univ Nebr, 58-59 & Univ SC, 59-61; from asst prof to assoc prof, 61-66, PROF CHEM, UNIV NEBR, LINCOLN, 66- Concurrent Pos: Res fel, Univ Sussex, 67-68; NATO sr fel, Sch Pharm, Univ London, 70. Mem: Am Chem Soc; The Chem Soc; Inst Chem Ireland. Res: Chemistry of natural products; synthesis of diterpenoid acids; stereochemistry of reductions; structures of plant extractives; naturally occurring quinones; photochemistry. Mailing Add: Dept of Chem Univ of Nebr Lincoln NE 68588

WHEELER, DONALD ALSOP, b Philadlephia, Pa, Aug 16, 31; m 53; c 4. GENETICS. Educ: Mich State Univ, BS, 53, MS, 56; Cornell Univ, PhD(plant breeding), 61. Prof Exp: Instr biol, Delta Col, 61-65, head dept, 63-65; assoc prof, 65-73, asst head dept, 71-73, PROF BIOL, EDINBORO STATE COL, 73- Mem: AAAS. Mailing Add: Sherrod Hill Rd Edinboro PA 16412

WHEELER, DONALD BINGHAM, JR, b Cleveland, Ohio, May 24, 17. PHYSICS. Educ: Lehigh Univ, BS, 38; Calif Inst Technol, PhD(physics), 47. Prof Exp: Instr physics, Occidental Col, 41-42; asst prof, 47-57, ASSOC PROF PHYSICS, LEHIGH UNIV, 57- Mem: Am Phys Soc. Res: Electric dipole moment determinations; microwave propagation; dispersion and absorption of electromagnetic waves in fatty acids. Mailing Add: Dept of Physics Lehigh Univ Bethlehem PA 18015

WHEELER, DONALD DEAN, organic chemistry, see 12th edition

WHEELER, DONALD JEFFERSON, b Oklahoma City, Okla, Sept 30, 44; m 66; c 1. APPLIED STATISTICS, MATHEMATICAL STATISTICS. Educ: Univ Tex, BA, 66; Southern Methodist Univ, MS, 68, PhD(statist), 70. Prof Exp: ASSOC PROF STATIST, UNIV TENN, KNOXVILLE, 70- Mem: Am Statist Asn. Res: Applications of statistics in marketing and behavioral science. Mailing Add: Dept of Statist Univ of Tenn Knoxville TN 37916

WHEELER, DONALD RUSSELL, solid state physics, see 12th edition

WHEELER, EDWARD LOCKWOOD, organic chemistry, see 12th edition

WHEELER, EDWARD NORWOOD, b Yancey, Tex, Oct 11, 27; m 50; c 5. ORGANIC CHEMISTRY. Educ: Tex Col Arts & Indust, BS, 47, BSCE, 49; Univ Tex, MA, 51, PhD(org chem), 53. Prof Exp: Res chemist, 53-55, group leader, 55-62, sect head, 62-67, dir chem res, 67-69, dir res, Tech Ctr, 69-72, dir develop, 72-74, DIR DEVELOP & RES & PLANNING, CELANESE CHEM CO, 74- Mem: AAAS; Am Chem Soc. Res: Acrylic acid; vinyl monomers; propiolactone reactions; palladium-olefin reactions; liquid phase oxidation of carbonyl compounds and olefins; process development and synthesis of petrochemicals. Mailing Add: Celanese Chem Co 1121 Ave of the Americas New York NY 10036

WHEELER, EDWARD STUBBS, b Philadelphia, Pa, June 3, 27; m 52; c 4. RESEARCH ADMINISTRATION. Educ: Haverford Col, AB, 48; Cornell Univ, PhD(chem), 52. Prof Exp: Asst chem, Cornell Univ, 48-51; assoc chemist org chem res, Atlantic Ref Co, 52-53, supv chemist, 53-59; mgr adhesives div, Amchem Prods, Inc, 59-62; mgr thermosetting polymer develop, Insulating Mat Dept, Gen Elec Co, 63-66, mgr-engr, Insulator Dept, 66-71, consult, Corp Exec Staff, 71-75; VPRES TECHNOL, LAPP DIV, INTERPACE CORP, 75- Concurrent Pos: Mem bd dirs, Am Nat Metric Coun, 74-75. Mem: AAAS; Am Chem Soc; Soc Plastics Eng; Inst Elec & Electronics Engrs; Am Soc Testing & Mat. Res: Synthetic organic chemistry; polymers; electrical insulation; insulators; engineering standards; metric conversion. Mailing Add: Lapp Div Interpace Corp Gilbert St LeRoy NY 14482

WHEELER, ELLSWORTH HAINES, JR, b Geneva, NY, Sept 19, 35; m 56; c 3. BIOLOGICAL OCEANOGRAPHY. Educ: Dartmouth Col, AB, 57; Univ RI, PhD(oceanog), 68. Prof Exp: Instr oceanog, Univ RI, 67-68; actg asst prof biol, Hopkins Marine Sta, Stanford Univ, 68-70; ASST PROF ZOOL, UNIV NH, 70- Mem: AAAS; Am Soc Limnol & Oceanog; Arctic Inst NAm. Res: Deep-sea biological oceanography; deep sea as an environment for pelagic organisms including the physiology and distribution of zooplankton and fishes below the euphotic zone. Mailing Add: Dept of Zool Univ of NH Durham NH 03824

WHEELER, ELMER PERLEY, b Bow, NH, Feb 23, 16; m 38; c 3. INDUSTRIAL HYGIENE. Educ: Univ NH, BS, 36. Prof Exp: Health inspector, City Health Dept, Concord, NH, 36-37; indust hygienist & trainee, State Health Dept, NY, 38-41; indust hygienist, 47-50, ASST DIR MED DEPT, MONSANTO CO, 50- Mem: Am Chem Soc; Sigma Xi; Am Pub Health Asn; Am Indust Hyg Asn (pres, 59-60); Air Pollution Control Asn (vpres, 55-56). Res: Physiological, biological and engineering sciences. Mailing Add: Monsanto Co 800 N Lindbergh Blvd St Louis MO 63166

WHEELER, EVERETT PEPPERRELL, II, petrology, deceased

WHEELER, FRANK CARLISLE, b Millinocket, Maine, Jan 26, 17; m 46; c 2. PHARMACEUTICAL CHEMISTRY. Educ: Mass Col Pharm, BS, 40, MS, 42; Purdue Univ, PhD(pharmaceut chem), 49. Prof Exp: Anal chemist, Burroughs Wellcome & Co, 42-43; pharmaceut chemist, 49-58, chief ampoule pilot plant, 58-65, head, 65-66, dir, Parenteral Opers Div, 66-75, DIR QUAL CONTROL & TECH SERV, ELI LILLY & CO, 75- Mem: AAAS; Am Chem Soc; Am Pharmaceut Asn; Pharmaceut Mfrs Asn; NY Acad Sci. Res: Pharmaceutical development; manufacture and control of chiefly parenteral products. Mailing Add: Eli Lilly & Co 307 E McCarty St Indianapolis IN 46206

WHEELER, GEORGE CARLOS, b Bonham, Tex, Apr 10, 97; m 21, 41; c 2. ENTOMOLOGY. Educ: Rice Inst, AB, 18; Harvard Univ, MS, 20, ScD(entom), 21; Univ NDak, LLD, 70. Prof Exp: From instr to asst prof zool, Syracuse Univ, 21-26; prof biol, 26-65, head dept, 26-63, univ prof, 65-67, UNIV EMER PROF BIOL, UNIV NDAK, 67-; RES ASSOC, DESERT RES INST, UNIV NEV SYST, 67- Mem: AAAS; fel Entom Soc Am; Ecol Soc Am; Soc Syst Zool. Res: Morphology and taxonomy of ant larvae; ants of North Dakota; desert ants. Mailing Add: Desert Res Inst Univ of Nev Syst Reno NV 89507

WHEELER, GEORGE EDWARD, b Pittsburgh, Pa, Dec 14, 14; m 59. PLANT ANATOMY. Educ: Univ Pittsburgh, BS, 36, MS, 41; Columbia Univ, PhD(bot), 53. Prof Exp: Assoc prof biol, 47-71, PROF BIOL, BROOKLYN COL, 71- Mem: AAAS; Bot Soc Am; Torrey Bot Club. Mailing Add: Dept of Biol Brooklyn Col Brooklyn NY 11213

WHEELER, GEORGE WILLIAM, b Dedham, Mass, Dec 23, 24; m 57; c 2. EXPERIMENTAL PHYSICS. Educ: Union Col, BS, 49; Yale Univ, MS & PhD(physics), 53. Prof Exp: Asst nuclear physics, Yale Univ, 51-53; res assoc marine acoustics, Woods Hole Oceanog Inst, 53-54; asst prof physics, Yale Univ, 54-58, sr res assoc, 59-64; physicist & head alternating gradient synchroton conversion div, Accelerator Dept, Brookhaven Nat Lab, 64-72; physicist high energy physics prog, Div Phys Res, US Energy Res & Develop Admin, Washington, DC, 72-74; PROF PHYSICS & DEAN SCI, LEHMAN COL, CITY UNIV NEW YORK, 74- Concurrent Pos: Consult, Los Alamos Sci Lab, 57-72 & US Energy Res & Develop Admin, 74- Mem: AAAS; Am Phys Soc; Inst Elec & Electronics Engrs. Res: Particle accelerators; experimental nuclear physics; marine acoustics; high power electronics. Mailing Add: PO Box 443 Port Jefferson NY 11777

WHEELER, GEORGE WILLIS, b Mantorville, Minn, Sept 16, 22; m 44; c 2. PHYSICS. Educ: Macalester Col, BA, 44; Harvard Univ, SM, 48, PhD, 50. Hon Degrees: ScD, Macalester Col, 65. Prof Exp: Asst appl physics, Harvard Univ, 49-50; mem tech staff, 50-53, spec systs engr, 53-55, dir transmission systs develop, 55-59, dir mil commun systs eng, 59-64, dir spec defense studies, Madison, 64-74, MEM STAFF, BELL TEL LABS, ILL, 74- Mem: Sr mem Inst Elec & Electronics Eng. Res: Microwaves and antenna; communications and radar systems. Mailing Add: Bell Tel Labs Naperville Rd Naperville IL 60540

WHEELER, GILBERT VERNON, b Sour Lake, Tex, July 5, 22; m 42; c 2. CHEMISTRY. Educ: Millikin Univ, BS, 44; Univ Ill, MS, 47. Prof Exp: Instr physics, Millikin Univ, 43-44; physicist, Res & Develop Dept, Phillips Petrol Co, 48-51, physicist, Atomic Energy Div, 52-53, reactor engr, 51-52, supvr spectrochem lab, 53-63, supvr spectros sect, 63-65; SUPVR SPECTROS SECT, ALLIED CHEM CORP, 65- Concurrent Pos: Int del, Soc Appl Spectros, 75. Mem: Soc Appl Spectros (pres, 73); Am Chem Soc; Spectros Soc Can. Res: Graphite furnaces; inductive coupled plasmas and sputter sources for spectroscopy; isotopic and isotope dilution mass spectrometry. Mailing Add: Allied Chem Corp 550 Second St Idaho Falls ID 83401

WHEELER, GLYNN PEARCE, b Milan, Tenn, Oct 13, 19; m 43; c 2. BIOCHEMISTRY. Educ: Vanderbilt Univ, AB, 41; Univ Akron, MS, 47; Vanderbilt Univ, PhD(org chem), 50. Prof Exp: Anal chemist, Tenn Coal, Iron & RR Co, Ala, 41; shift supvr, Ala Ord Works, 42; res chemist, B F Goodrich Co, Ohio, 42-46; chemist, 46-48, biochemist, 50-56, head intermediary metab sect, 56-66, HEAD CANCER BIOCHEM DIV, SOUTHERN RES INST, 66- Mem: AAAS; Am Chem Soc; Am Asn Cancer Res; Am Soc Biol Chem. Res: Cancer biochemistry; chemotherapy of canerr; nucleic acids. Mailing Add: Southern Res Inst Biochem Dept 2000 Ninth Ave S Birmingham AL 35205

WHEELER, HARRY ERNEST, b WCharleston, Vt, Jan 25, 19; m 44. PHYTOPATHOLOGY. Educ: Univ Vt, BS, 41; La State Univ, MS, 47, PhD(bot), 49. Prof Exp: Lab asst bot & bact, Univ Vt, 39-41; asst bot & plant path, La State Univ, 46-49, from asst prof to prof bot & plant path, 49-67; PROF PLANT PATH, UNIV KY, 67- Concurrent Pos: Vis investr & res partic, Biol Div, Oak Ridge Nat Lab, 49-50; Guggenheim fel, Biol Labs, Harvard Univ, 58. Mem: AAAS; Bot Soc Am; Am

Phytopath Soc; Mycol Soc Am. Res: Genetics and cytology of fungi; host relations of plant path- ogens; electron microscopy. Mailing Add: Dept of Plant Path Univ of Ky Lexington KY 40506

WHEELER, HARRY EUGENE, b Pipestone, Minn, Feb 1, 07; m 38; c 3. STRATIGRAPHY. Educ: Univ Ore, BS, 30; Stanford Univ, AM, 32, PhD(geol), 35. Prof Exp: Asst, Univ Ore, 28-30; field asst, US Geol Surv, 30, field asst topog br, 35; from instr to assoc prof geol, Univ Nev, 35-48; assoc prof, 48-51, PROF GEOL, UNIV WASH, 51- Concurrent Pos: Geologist, State Anal Lab, Nev, 36-42; stratigrapher, State Bur Mines, 36-47; consult, Gulf Oil Corp, 50-58; mem Permian subcomt, Comt Stratig, Nat Res Coun; mem French Nat Cent Sci Res Symposium, 57 & Am Stratig Comn, 58-60; vis prof, Ind Univ, 56-57; Gulf Oil Corp res fel, 59; vis lectr, Univ Tex, 61; Nat Acad Sci exchange scientist to USSR, 63; vis prof, Southern Methodist Univ, 66; consult, Northwest Europe & North Sea, Mobil Oil Corp, 73. Mem: Fel Geol Soc Am; fel Paleont Soc; Am Asn Petrol Geol; Soc Econ Paleontologists & Mineralogists. Res: Basic stratigraphic concepts, principles and procedures, and their application to interregional geological analysis and historical synthesis. Mailing Add: Dept of Geol Sci Univ of Wash Seattle WA 98195

WHEELER, HARRY OGDEN, b Little Rock, Ark, Oct 11, 28; m 49; c 2. MICROBIOLOGY, IMMUNOLOGY. Educ: Univ Houston, BS, 54, MS, 56; Agr & Mech Col, Tex, PhD(nutrit), 59. Prof Exp: Res instr, Agr & Mech Col, Tex, 58; asst nutritionist, Clemson Col, 59-63, assoc nutritionist, 63-66; aerospace technologist, 66-68, ASST CHIEF, BIOMED SPECIALTIES BR, NASA-JOHNSON SPACE CTR, 68- Mem: AAAS. Res: Unidentified growth factors; protein and amino acid supplementation; microbial aspects of crew and spacecraft during manned space flights, immunological responses due to spaceflight hygiene. Mailing Add: Code DB-3 NASA Johnson Space Ctr Houston TX 77058

WHEELER, HENRY ORSON, b Los Angeles, Calif, Apr 7, 24; m 47; c 2. MEDICINE. Educ: Harvard Med Sch, MD, 51. Prof Exp: Assoc prof med, Col Physicians & Surgeons, Columbia Univ, 62-68; PROF MED, SCH MED, UNIV CALIF, SAN DIEGO, 68- Mem: AAAS; Am Soc Clin Invest; Am Physiol Soc. Res: Hepatic physiology; bile formation; gallbladder; ion transport. Mailing Add: Dept of Med Sch of Med Univ of Calif at San Diego La Jolla CA 92037

WHEELER, JAMES DONLAN, b St Louis, Mo, July 19, 23. PHARMACEUTICAL CHEMISTRY, BIOCHEMISTRY. Educ: St Louis Univ, AB, 47, PhL, 48, MS, 52, STL, 56; Univ Mo-Kansas City, PhD(pharmaceut chem), 65. Prof Exp: Instr chem high sch, St Louis Univ, 50-51; from instr to assoc prof, 56-74, head dept, 67-74, PROF CHEM, ROCKHURST COL, 74- Mem: AAAS; Am Chem Soc; NY Acad Sci. Res: Biochemistry and physiology of the effects of training and exercise; lactose hydrolysis. Mailing Add: Rockhurst Col 5225 Troost Kansas City MO 64110

WHEELER, JAMES ENGLISH, b Durham, NC, May 5, 38; m 66; c 3. PATHOLOGY. Educ: Harvard Univ, AB, 58; Johns Hopkins Univ, MD, 62. Prof Exp: Intern med, Johns Hopkins Univ, 62-63, resident path, 63-66; resident path, State Univ NY Upstate Med Ctr, 69-70; assoc, 70-72, ASST PROF PATH, SCH MED, UNIV PA, 72- Concurrent Pos: USPHS cancer control sr clin trainee, Mem Hosp Cancer & Allied Dis, New York, 66-67. Mem: Fel Int Acad Path. Res: Surgical and gynecological pathology. Mailing Add: Dept of Path Hosp of the Univ of Pa Philadelphia PA 19104

WHEELER, JAMES ORTON, b Muncie, Ind, Mar 7, 38; m 60; c 4. TRANSPORTATION GEOGRAPHY, ECONOMIC GEOGRAPHY. Educ: Ball State Univ, BS, 59; Ind Univ, MA, 63, PhD(geog), 66. Prof Exp: Instr geog, Ohio State Univ, 64-65; instr, Western Mich Univ, 65-67; from asst prof to assoc prof, Mich State Univ, 67-71; assoc prof, 71-73, PROF GEOG, UNIV GA, 73-, HEAD DEPT, 75- Concurrent Pos: Mem, Hwy Res Bd, Nat Acad Sci-Nat Res Coun. Mem: Asn Am Geog; Am Geog Soc; Regional Sci Asn; Nat Coun Geog Educ. Res: Transportation geography, specifically person movement in urban areas; industrial location; urban geography. Mailing Add: Dept of Geog Univ of Ga Athens GA 30602

WHEELER, JAMES WILLIAM, JR, b Clarksburg, WVa, Oct 2, 34; m 57; c 1. ORGANIC CHEMISTRY. Educ: Antioch Col, BS, 57; Stanford Univ, MS, 59, PhD(chem), 62. Prof Exp: NSF fel chem, Cornell Univ, 63-64, NIH trainee, 64; from asst prof to assoc prof, 64-71, NIH spec fel, 71-72, PROF CHEM, HOWARD UNIV, 71- Mem: AAAS; Am Chem Soc; Am Soc Mass Spectros. Res: Chemistry of arthropod pheromones, small ring compounds and monoterpenes. Mailing Add: Dept of Chem Howard Univ Washington DC 20001

WHEELER, JEANETTE NORRIS, b Newton, Iowa, May 21, 18; m 41; c 1. ENTOMOLOGY. Educ: Univ NDak, BA, 39, MS, 56, PhD, 62. Prof Exp: Res assoc biol, Univ NDak, 65-67; RES ASSOC, DESERT RES INST, UNIV NEV SYST, RENO, 67- Mem: Entom Soc Am. Res: Taxonomy and morphology of the ant larvae; desert ants. Mailing Add: Desert Res Inst Univ of Nev Syst Reno NV 89507

WHEELER, JESSE HARRISON, JR, b Scottsboro, Ala, Nov 24, 18; m 42; c 3. GEOGRAPHY, HISTORICAL GEOGRAPHY. Educ: Auburn Univ, BS, 39, MS, 41; Univ Chicago, PhD(geog), 50. Prof Exp: Teacher eng, Boys High Sch, Atlanta, Ga, 40-41; from instr to assoc prof, 49-64, PROF GEOG, UNIV MO-COLUMBIA, 64- Concurrent Pos: Ford Found fac fel, Univ Mo, 51-52. Mem: Asn Am Geog; Am Geog Soc; Soc Hist Discoveries; Agr Hist Soc. Res: Historical geography of North America, especially geographical change in agriculture; geography of US South, Appalachia; world regional geography. Mailing Add: Dept of Geog Univ of Mo Columbia MO 65201

WHEELER, JOHN ARCHIBALD, b Jacksonville, Fla, July 9, 11; m 35; c 3. THEORETICAL PHYSICS. Educ: Johns Hopkins Univ, PhD(physics), 33. Hon Degrees: ScD, Univ NC, 59, Yale Univ, 74; PhD, Univ Uppsala, 75. Prof Exp: Nat Res Coun fel, NY Univ & Copenhagen Univ, 33-35; from asst prof to assoc prof physics, Univ NC, 35-38; asst prof, Princeton Univ, 38-42, physicist atomic energy proj, 39-42; physicist, Metall Lab, Univ Chicago, 42-43, E I du Pont de Nemours & Co, Del, 43-44, Hanford Eng Works, Wash, 44-45 & Los Alamos Sci Lab, 50-53; from assoc prof to prof, 45-66, JOSEPH HENRY PROF PHYSICS, PRINCETON UNIV, 66- Concurrent Pos: US rep cosmic ray comn, Int Union Pure & Appl Physics, Poland, 47, vpres, Union, 51-54; Guggenheim fel, Univ Paris & Copenhagen Univ, 49-50; dir proj Matterhorn, Princeton Univ, 51-53; Lorentz prof, Univ Leiden, 56; mem adv comt, Oak Ridge Nat Lab, 57-67; sci adv, US Senate Del, Conf NATO Parliamentarians, France, 57; mem, Joint Congressional Comt Atomic Energy, 58; proj chmn, Dept Defense Advan, Res Proj Agency, 58; consult, AEC; trustee, Battelle Mem Inst, 60-; Fulbright prof, Kyoto Univ, 62; chmn joint comt on hist theoret physics in 20th century, Am Phys Soc-Am Philos Soc, 62-; vis fel, Clare Col, Cambridge Univ, 64; chmn, US Gen Adv Comt Arms Control & Disarmament, 69-; Battelle Mem prof, Univ Wash, 75. Honors & Awards: Morrison Prize, NY Acad Sci, 46; Einstein Prize, Strauss Found, 66; Enrico Fermi Award, AEC, 68; Franklin Medal, Franklin Inst, 69; Nat Medal Sci, 71. Mem: Nat Acad Sci; fel Am Phys Soc (pres, 66); Am Philos Soc (vpres, 71-73); Am Math Soc; Am Acad Arts & Sci. Res: Atomic and nuclear physics; scattering theory; fission; nuclear chain reactors; direct electromagnetic interaction between particles; mathematics of semiclassical analysis of physical processes; mu-meson; relativity; space-time and geometrodynamics. Mailing Add: Joseph Henry Labs Princeton Univ Princeton NJ 08540

WHEELER, JOHN C, b Urbana, Ill, Mar 26, 41; m 63; c 1. THEORETICAL CHEMISTRY, CHEMICAL PHYSICS. Educ: Oberlin Col, BA, 63; Cornell Univ, PhD(theoret chem), 68. Prof Exp: NSF fel chem, Harvard Univ, 67-69; ASST PROF CHEM, UNIV CALIF, SAN DIEGO, 69- Concurrent Pos: Alfred P Sloan Found fel, 72- Mem: Am Phys Soc. Res: Statistical mechanics and thermodynamics of single and multi component systems, phase transitions and critical phenomena; rigorous bounds in statistical mechanics and thermodynamics. Mailing Add: Dept of Chem Univ of Calif Box 109 La Jolla CA 92037

WHEELER, JOHN CRAIG, b Glendale, Calif, Apr 5, 43; m 66; c 2. THEORETICAL ASTROPHYSICS. Educ: Mass Inst Technol, BS, 65; Univ Colo, PhD(physics), 69. Prof Exp: Res fel, Calif Inst Technol, 69-71; asst prof astron, Harvard Univ, 71-74; ASSOC PROF ASTRON, UNIV TEX, AUSTIN, 74- Mem: Am Astron Soc; Sigma Xi; Int Astron Union. Res: High energy and relativistic astrophysics; supernova hydrodynamics; black hole physics; active nuclei of galaxies; compact objects in binary systems. Mailing Add: Dept of Astron Univ of Tex Austin TX 78712

WHEELER, JOHN OLIVER, b Mussoorie, India, Dec 19, 24; m 52; c 2. GEOLOGY. Educ: Univ BC, BASc, 47; Columbia Univ, PhD(geol), 56. Prof Exp: Asst geol, Columbia Univ, 49-51; tech officer, 52-55, geologist, 56-68, head cordilleran & Pac margins sect, 68-70, chief regional & econ geol div, 70-74, DEP DIR, GEOL SURV CAN, 74- Concurrent Pos: Vis prof, Univ Toronto, 72. Mem: Fel Royal Soc Can; fel Geol Soc Am; Geol Asn Can (pres, 70-71); Can Inst Mining & Metall. Res: Geological mapping in Central and Southern Yukon and Southeastern British Columbia; glacial geology in Southern Yukon; tectonics and structure of southern part of Western Canadian cordillera; recent glacier fluctuations of Selkirk Mountains; tectonics of Canadian cordillera and Canada. Mailing Add: Geol Surv of Can 601 Booth St Ottawa ON Can

WHEELER, JOHN RUSSELL, b Miami, Fla, Nov 11, 44. ECOLOGY, ENVIRONMENTAL CHEMISTRY. Educ: Union Col, AB, 66; Dalhousie Univ, PhD(oceanog), 72. Prof Exp: Res scientist, Fisheries Res Bd Can, 72-73; RES ASSOC ECOL, UNIV GA, 74- Concurrent Pos: Mem, Inst Ecol, Univ Ga, 76- Mem: AAAS; Sigma Xi; Am Chem Soc. Res: Ecology and chemistry of wetlands and ocean waters; chemistry and microbiology of the atmosphere; responses of organisms to environmental chemical parameters; remote sensing; scientific data in litigation and regulatory agencies. Mailing Add: Univ of Ga Marine Inst Sapelo Island GA 31327

WHEELER, KEITH WILSON, b Iowa City, Iowa, Jan 9, 18; m 40; c 2. INFORMATION SCIENCE. Educ: Knox Col, AB, 38; Purdue Univ, MS, 40, PhD(org chem), 44. Prof Exp: Asst, Purdue Univ, 39-43; res chemist, William S Merrell Co, Ohio, 43-56, head records off, 57-64, head sci info dept, 64-71; SR PRIN INVESTR, TECH INFO SERV DEPT, MEAD JOHNSON RES CTR, 71- Mem: Fel AAAS; fel Am Inst Chemists; Sigma Xi; Am Chem Soc; Drug Info Asn. Res: Documentation. Mailing Add: Tech Info Serv Dept Mead Johnson Res Ctr Evansville IN 47721

WHEELER, KENNETH THEODORE, JR, b Dover, NH, Sept 11, 40. BIOPHYSICS, RADIATION BIOLOGY. Educ: Harvard Univ, BA, 62; Wesleyan Univ, MAT, 63; Univ Kans, PhD(radiation biophys), 70. Prof Exp: Asst radiation biologist, Colo State Univ, 70-72, asst prof radiation biol, 72; ASST PROF NEUROL SURG & RADIOL, MED SCH, UNIV CALIF, SAN FRANCISCO, 72- Mem: Radiation Res Soc; Biophys Soc; Health Physics Soc. Res: In vivo DNA damage and repair in normal and nondividing tissue and tumor tissue. Mailing Add: M784 Dept of Neurol Surg Univ of Calif Med Ctr San Francisco CA 94122

WHEELER, LARRY MEADE, b Bethany, Mo, June 4, 18; m 45; c 3. PHARMACEUTICAL CHEMISTRY. Educ: Univ Iowa, BS, 40, MS, 41, PhD(pharmaceut chem), 43. Prof Exp: Control chemist, Great Lakes Pipe Line Co, Iowa, 38-40; asst pharmacist hosps, Iowa, 40-41; sr researcher, 43-48, from asst dir to dir prod develop, 48-72, DIR PHARMACEUT RES & DEVELOP, PARKE, DAVIS & CO, 72- Mem: AAAS; Am Chem Soc; Am Pharmaceut Asn. Res: Chemotherapeutic agents for treatment of venereal and tropical infections; chemical inhibitors of hyaluronidase, especially in rheumatic diseases; pharmaceutical formulations; synthetic and biochemical development; pharmaceuticals and development of products. Mailing Add: 71 Radnor Circle Grosse Point Farms MI 48236

WHEELER, LAWRENCE, b Indianapolis, Ind, Apr 4, 23; m 46; c 2. PSYCHOPHYSICS, PSYCHOLOGY. Educ: Ind Univ, AB, 48, MA, 50, PhD(psychol), 62. Prof Exp: From res asst to teaching asst psychol, Ind Univ, 45-50; indust engr electronics, Sarkes Tarzian Inc, 53-58; from res assoc to instr psychol, Ind Univ, 58-63; from asst prof to assoc prof psychol, Calif State Col Hayward, 64-69; PROF PSYCHOL & OPTICAL SCI, UNIV ARIZ, 69-, CHMN DEPT PSYCHOL, 75- Concurrent Pos: Dir behav res, Archonics Corp, 58-; fel life sci, Carnegie Inst Wash, Nat Phys Lab, Teddington, Eng, 63-64. Mem: Am Psychol Asn; Colour Group Gt Brit; Sigma Xi; AAAS. Res: Visual psychophysics for evaluation of image quality and information retrieval; applied environmental psychology; assessment of user responses to constructed environments. Mailing Add: Dept of Psychol Col of Lib Arts Univ Ariz Tucson AZ 85721

WHEELER, LOUIS CUTTER, b Claremont, Calif, Aug 2, 10; m 35; c 2. SYSTEMATIC BOTANY. Educ: Univ Calif, Los Angeles, AB, 33; Claremont Col, MA, 34; Harvard Univ, PhD(bot), 39. Prof Exp: Asst to technician, US Forest Serv, 34-35; asst, Gray Herbarium, Harvard Univ, 36-39; instr bot, Univ Mo, 39-40; instr bot & biol, American Univ, 40-41; instr bot, Univ Pa, 41-42; rubber technologist, Bur Agr & Indust Chem, USDA, 43-44; res asst, Off Sci Res & Develop, Calif Inst Technol, 45; head dept bot, 45-53, from asst prof to assoc prof, 45-56, assoc prof biol, 56-61, prof bot, 61-74, EMER PROF BOT, UNIV SOUTHERN CALIF, 74- Mem: AAAS; Am Soc Plant Taxon; Bot Soc Am; Int Asn Plant Taxon. Res: Taxonomy of the Euphorbiaceae of North America and Ceylon; botanical nomenclature; phytogeochemistry; geobotanical prospecting; forensic botany; botanical bibliography; Californian flora. Mailing Add: Dept of Biol Sci Univ of Southern Calif Los Angeles CA 90007

WHEELER, MARGARET CAMERON, b Toronto, Ont, Sept 3, 23; nat US; m 57; c 2. ANTHROPOLOGY. Educ: Univ Toronto, BPHE, 43, BA, 46, MA, 48; Yale Univ, PhD(anthrop), 57. Prof Exp: Instr anthrop, Univ Toronto, 50-53, lectr, 53-56, asst prof, 56-57; res assoc poverty prog, Neighborhood Improvement Proj, 63-65; asst prof, 65-71, asst dean col arts & sci, 66-69, ASSOC PROF ANTHROP, STATE UNIV NY STONY BROOK, 72- Concurrent Pos: Res fel, Yale Univ, 57-63; NIMH grant, 63-65; forum mem, White House Conf Children, 70. Mem: AAAS; fel Am Anthrop Asn; Am Ethnol Soc; Soc Appl Anthrop; Am Asn Phys Anthrop. Res: Urban anthropology;

community studies; Jewish studies; ethnicity; consumer research; marketing problems; poverty. Mailing Add: Dept of Anthrop State Univ NY Stony Brook NY 11790

WHEELER, MARSHALL RALPH, b Carlinville, Ill, Apr 7, 17; m 44, 66; c 3. ZOOLOGY, GENETICS. Educ: Baylor Univ, BA, 39; Univ Tex, PhD(genetics), 47. Prof Exp: From instr to assoc prof, 47-60, PROF ZOOL, UNIV TEX, AUSTIN, 61- Concurrent Pos: Gosney fel, Calif Inst Technol, 49-50. Mem: Genetics Soc Am; Am Soc Nat; Soc Study Evolution; Soc Syst Zool; Entom Soc Am. Res: Speciation and taxonomy in Drosophila; biology of acalyptrate Diptera; insect cytogenetics. Mailing Add: Dept of Zool Univ of Tex Austin TX 78712

WHEELER, MARY FANETT, b Cuero, Tex, Dec 21, 38; m 63; c 1. NUMERICAL ANALYSIS. Educ: Univ Tex, BA, 60, MA, 63, PhD(math), 71. Prof Exp: Programmer math, Univ Tex Comput Ctr, 61-65; from programmer to instr, 65-73, ASST PROF MATH SCI, RICE UNIV, 73- Mem: Am Math Soc; Soc Indust & Appl Math. Res: Numerical solution of partial and ordinary differential equations. Mailing Add: Dept of Math Sci Rice Univ Houston TX 77001

WHEELER, NICHOLAS ALLAN, b The Dalles, Ore, May 24, 33; m 62; c 2. MATHEMATICAL PHYSICS. Educ: Reed Col, BA, 55; Brandeis Univ, PhD(physics), 60. Prof Exp: NSF fel, State Univ Utrecht & Europ Orgn Nuclear Res, 60-62; res assoc physics, Brandeis Univ, 62-63; asst prof, 63-65, ASSOC PROF PHYSICS, REED COL, 65- Res: Structure and interconnections among physical theories, especially classical and quantum dynamics, classical field theories, statistical mechanics and thermodynamics. Mailing Add: Dept of Physics Reed Col Portland OR 97202

WHEELER, NORMAN C, anesthesiology, see 12th edition

WHEELER, NORRIS GENE, organic chemistry, see 12th edition

WHEELER, ORA LEON, b Custer, Okla, Sept 7, 13; m 41; c 3. CHEMISTRY. Educ: Col Puget Sound, BS, 37; Univ Wash, PhD(chem), 42. Prof Exp: Chemist, Shawinigan Resins Corp, Mass, 45-52; tech dir, Colton Chem Co, Ohio, 52-58 & Air Reduction Co, Inc, 58-65; gen mgr, Cellate, Inc, 65-67; vpres, Colloids, Inc, 67-70; VPRES, CHEM INSTALLERS CO, 71- Concurrent Pos: Vpres, Propak Corp, 72-75; asst prof chem, County Col Morris, 75- Mem: Am Chem Soc; fel Am Inst Chemists. Res: Colloidal electrolytes; synthetic resins. Mailing Add: Chem Installers Co PO Box 59 Liberty Corner NJ 07938

WHEELER, PHILIP RIDGLY, b Grand Rapids, Mich, July 21, 07; m 34. FORESTRY. Educ: Univ Mich, BSF, 29, MF, 30. Prof Exp: Aide to organizer, Brazilian Forest Serv, 30-31; jr forester, Southern Forest Exp Sta, US Forest Serv, 31-34, asst forest economist, 34-35, assoc forest economist & comput chief, Southern Forest Surv, 35-39, forest economist in chg mensurational anal, 39-42, prog leader, Ozark Br Sta, 45-47, chief div forest mgt res, 47-51, chief div forest econ, 51-62; consult, 62-66; forest resource economist & consult, Food & Agr Orgn UN, 66; resource analyst, Southern Forest Resource Anal Comt, 67-69; CONSULT, 69- Honors & Awards: Superior Service Award, Secy Agr, 62; Sir William Schlich Mem Award, Soc Am Foresters, 70. Res: Regional forest resources and growth, forest survey mensurational methods, procedures and instruments and wood, using plant location studies. Mailing Add: 2616 Jefferson Ave New Orleans LA 70115

WHEELER, RALPH JOHN, b Devine, Tex, Sept 14, 29; m 57; c 1. ANALYTICAL CHEMISTRY. Educ: Trinity Univ, San Antonio, Tex, BS, 63. Prof Exp: Res chemist, Southwest Res Inst, 63-68; anal chemist, 68-73, mgr anal chem, 73-74, ASSOC DIR, LIFE SCI DIV, GULF SOUTH RES INST, 74- Mem: Am Chem Soc; Am Asn Lab Animal Sci; Soc Res Adminrs. Res: Carcinogenesis bioassay of pesticides and other environmental chemicals; development of new chromatographic instrumentation and methodology for the analysis of airborne polynuclear arenes. Mailing Add: Gulf South Res Inst PO Box 1177 New Iberia LA 70560

WHEELER, RICHARD HUNTING, b Brooklyn, NY, Jan 30, 31; m 54; c 3. FOREST HYDROLOGY. Educ: Univ Maine, Orono, BS, 53; Colo State Univ, MF, 69. Prof Exp: Forester, Savannah River Proj, US Forest Serv-AEC, SC, 57-59; forester, Ouachita Nat Forest, US Forest Serv, 59-62, forester & staff consult water resources, Roosevelt Nat Forest, 62-64 & Arapaho Nat Forest, 64-66, hydrologist & staff consult, Northern Region, Div Soil, Air & Water Mgt, 66-74; FOREST HYDROLOGIST & CONSULT, FOOD & AGR ORGN UN, MAE SA WATERSHED PROJ, CHIANG MAI, THAILAND, 74- Mem: Soc Am Foresters; Am Forestry Asn; Am Geophys Union. Res: Wildlife water quality and water resource management; general forest management. Mailing Add: UN Develop Prog PO Box 618 Bangkok Thailand

WHEELER, ROBERT LEE, b Minneapolis, Minn, Jan 17, 44; m 67; c 1. MATHEMATICS. Educ: Univ Minn, BS, 66; Univ Wis-Madison, MA, 69, PhD(math), 71. Prof Exp: ASST PROF MATH, UNIV MO-COLUMBIA, 71- Concurrent Pos: Vis asst prof math, Iowa State Univ, 74-75. Mem: Am Math Soc; Soc Indust & Appl Math; Math Asn Am. Res: Volterra integral equations; tauberian theory; functions of one complex variable. Mailing Add: Dept of Math Univ of Mo Columbia MO 65201

WHEELER, ROBERT REID, b Youngstown, Ohio, Apr 16, 17; m 40; c 2. PETROLEUM GEOLOGY. Educ: Johns Hopkins Univ, AB, 38; Harvard Univ, PhD(stratig), 42. Prof Exp: Subsurface geologist, Shell Oil Co, Tex, 42; field geologist, Superior Oil Co, NMex, 42-44, res geologist, Okla, 45; chief geologist, Eason Oil Co, 45-47 & C P Burton Oil Co, 47-54; mgr, Pyramid Oil & Gas Co, 54-59; lectr, Southern Methodist Univ, 63-65; asst prof, East Tex State Univ, 65-67; ASSOC PROF GEOL, LAMAR UNIV, 67- Concurrent Pos: Planning chmn, Int Oil & Gas Ed Ctr, Southern Methodist Univ; consult, geologist, 47- Mem: Fel AAAS; Geol Soc Am; Am Asn Petrol Geol. Res: Paleozoic stratigraphy, especially of Champlain Valley and midcontinent; geology and oil of Anadarko Basin; midcontinental tectonics; structural maps of Oklahoma and midcontinent; oil prospects of Red Sea, Ethiopia; oil from prospect to pipeline. Mailing Add: Dept of Geol Lamar Univ Beaumont TX 77710

WHEELER, ROBERT STEVENSON, b Woodstock, Ill, Apr 1, 15; m 42; c 2. AVIAN PHYSIOLOGY. Educ: Univ Chicago, BS, 39, PhD(zool), 42. Prof Exp: Asst prof zool, 45-46, from asst prof to assoc prof poultry husb, 46-50, head dept & chmn poultry div, 50-54, assoc dir resident instr, Col Agr, 50-54, PROF POULTRY SCI, UNIV GA, 50-, DIR RESIDENT INSTR, COL AGR, 54- Mem: AAAS; Endocrine Soc; Am Soc Zoologists; Poultry Sci Asn. Res: Avian endocrine physiology; physiological genetics and reproductive physiology of the fowl. Mailing Add: Rm 102 Conner Hall Col Agr Univ Ga Athens GA 30602

WHEELER, RURIC E, b Grayson Co, Ky, Nov 30, 23; m 46; c 2. MATHEMATICAL STATISTICS. Educ: Western Ky State Col, AB, 47; Univ Ky, MS, 48, PhD(math, statist), 52. Prof Exp: Instr math & statist, Univ Ky, 48-52; asst prof, Fla State Univ, 52-53; assoc prof, 53-55, head dept math & eng, 56-65, chmn div natural sci, 65-70, asst to acad dean, 67-68, dean, Howard Col Arts & Sci, 68-70, PROF MATH, SAMFORD UNIV, 56-, VPRES ACAD AFFAIRS, 70- Concurrent Pos: Consult, Dynamics Dept, Hayes Int Corp, 56-67; trustee, Mid-South Technol Inst, 58-67; dir, NSF vis sci prog, 63-67 & coop prog, 65-67; trustee, Gorgas Found, 68-; mem, Am Conf Acad Deans. Mem: Am Math Soc; Math Asn Am; Am Statist Asn; Inst Math Statist. Res: Statistical distributions; stochastic processes. Mailing Add: Dept of Math Samford Univ Birmingham AL 35206

WHEELER, SAMUEL CRANE, JR, b Montclair, NJ, June 3, 13; m 42; c 6. PHYSICS. Educ: Miami Univ, AB, 42; Univ Ill, MS, 43; Ohio State Univ, PhD(physics), 60. Prof Exp: Asst, Miami Univ, 39-42, instr physics, 43-44; asst, Univ Ill, 42-43; asst physicist, Div War Res, Univ Calif, San Diego, 44-46; physicist, US Navy Electronics Lab, 46-48; from instr to assoc prof, 48-64, chmn dept physics & astron, 60-70, PROF PHYSICS, DENISON UNIV, 64- Concurrent Pos: NSF fac fel, 59-60 & prog dir, Div Instnl Progs, 63-64, consult, Div Undergrad Educ Sci, 66-; mem exam bd, NCent Asn Cols & Sec Schs, Comn Higher Educ, 70- Mem: Am Phys Soc; Am Asn Physics Teachers; Am Astron Soc. Res: Underwater sound calibration techniques; sonic properties of materials; theory of infrared spectra of polyatomic molecules; theoretical molecular physics. Mailing Add: Dept of Physics & Astron Denison Univ Granville OH 43023

WHEELER, TAMARA STECH, b New Castle, Ind, July 4, 46; m 67. ANTHROPOLOGY. Educ: Bryn Mawr Col, AB, 67, PhD(archaeol), 73. Prof Exp: Vis scholar, 74-76, RES ASST PROF METALL, UNIV PA, 76- Mem: Archaeol Inst Am; Am Schs Oriental Res; Brit Inst Archaeol Ankara; Brit Sch Archaeol Iraq. Res: Ancient metallurgy, scientific analysis of artifacts to reconstruct history of copper alloying and development of iron and steel technology in eastern Mediterranean and Near East. Mailing Add: Metall & Mat Sci LRSM Univ of Pa Philadelphia PA 19174

WHEELER, THOMAS NEIL, b Ocala, Fla, Feb 6, 43; m 67. ORGANIC CHEMISTRY. Educ: Univ Fla, BS, 64; Cornell Univ, PhD(org chem), 69. Prof Exp: From asst prof to assoc prof chem, Fla Technol Univ, 69-75; RES CHEMIST, UNION CARBIDE CORP, 75- Concurrent Pos: Petrol res fund type B grant, Fla Technol Univ, 70-72. Mem: Am Chem Soc. Res: Exploratory synthesis of pesticides. Mailing Add: Union Carbide Corp Tech Ctr S Charleston WV 25303

WHEELER, WALTER HALL, b Syracuse, NY, Dec 21, 23; m 45; c 2. VERTEBRATE PALEONTOLOGY. Educ: Univ Mich, BS, 45, MS, 48; Yale Univ, PhD(geol), 51. Prof Exp: From instr to assoc prof, 51-68, PROF GEOL, UNIV NC, CHAPEL HILL, 68- Concurrent Pos: Vis scholar, Univ Calif, Berkeley, 66-67. Mem: Geol Soc Am; Am Inst Prof Geol; Soc Vert Paleont; Am Asn Petrol Geol; Paleont Soc. Res: Large early Tertiary mammals; stratigraphy and paleontology of the North Carolina coastal plain; the Triassic of North Carolina. Mailing Add: Dept of Geol Univ of NC Chapel Hill NC 27514

WHEELER, WARREN E, b Youngstown, Ohio, May 10, 09; m 39; c 3. MEDICINE. Educ: Mt Union Col, BS, 29; Harvard Med Sch, MD, 33. Prof Exp: Resident pediat, Boston Children's Hosp, Mass, 33-37; asst med dir, Children's Hosp, Detroit, Mich, 41-45; prof pediat, Col Med, Ohio State Univ, 45-63 & bact, 49-63; prof & chmn dept, 63-74, EMER PROF PEDIAT, MED CTR, UNIV KY, 74- Concurrent Pos: Chief ed, Am J Dis Children, 59-63. Mem: Soc Pediat Res; Am Pediat Soc; Infectious Dis Soc Am. Res: Blood group immunology; infectious diseases in children; pediatric education. Mailing Add: 12 Lansdowne Estates Lexington KY 40502

WHEELER, WILLIAM CARL, microbiology, deceased

WHEELER, WILLIAM HOLLIS, b Akron, Ohio, Feb 10, 46; m 68. MATHEMATICAL LOGIC. Educ: Vanderbilt Univ, BA, 68; Yale Univ, PhD(math), 72. Prof Exp: ASST PROF MATH, IND UNIV, BLOOMINGTON, 72- Concurrent Pos: Vis lectr math, Bedford Col, Univ London, 73-74. Mem: Am Math Soc; Asn Symbolic Logic; London Math Soc. Res: Model theory; metamathematics of algebra; applications of logic to algebra. Mailing Add: Dept of Math Ind Univ Bloomington IN 47401

WHEELER, WILLIAM JOE, b Flora, Ind, Oct 14, 40; m 62; c 2. MEDICINAL CHEMISTRY. Educ: Purdue Univ, BS, 62; Butler Univ, MS, 66; Purdue Univ, PhD(med chem), 70. Prof Exp: Teacher math & sci, Northwestern Sch Corp, Henry County, Ind, 62-63; anal chemist, Allison Div, Gen Motors Corp, Ind, 63-65; org chemist, Eli Lilly & Co, 65-67; asst, Purdue Univ, Lafayette, 67-68, fel, 68-70; SR PHARMACEUT CHEMIST, ELI LILLY & CO, 70- Mem: Am Chem Soc; Am Soc Microbiol. Res: Beta-lactam antibiotics, anti-inflammatory compounds. Mailing Add: 1555 N Huber St Indianapolis IN 46219

WHEELER, WILLIAM RALEIGH, b Indianapolis, Ind, Jan 10, 16; m 44; c 2. ORGANIC CHEMISTRY, CHEMICAL ENGINEERING. Educ: Univ Ill, BS, 36; Pa State Col, MS, 37, PhD, 41. Prof Exp: Res & develop chemist, Reilly Tar & Chem Corp, 39-42, 45-47; res chemist, Cincinnati Milling Mach Co, 47-49; res & develop chemist, 49-74, DIR, REILLY LABS, REILLY TAR & CHEM CORP, 74- Mem: AAAS; Am Inst Chem Eng; Am Chem Soc; The Chem Soc. Res: Heterogeneous catalytic reactions; process development and plant design. Mailing Add: 502 W 77th St Indianapolis IN 46260

WHEELER, WILLIS BOLY, b Oakland, Calif, June 13, 38; m 64; c 2. BIOCHEMISTRY. Educ: George Washington Univ, BS, 61, MS, 63; Pa State Univ, PhD(biochem), 66. Prof Exp: Res asst cancer, George Washington Univ, 61-63; instr chem pesticides, Pa State Univ, 64-66; asst prof 66-72, ASSOC PROF PESTICIDES, UNIV FLA, 72- Mem: AAAS; Am Chem Soc. Res: Disappearance of chemical from plant and plant environment; behavior of pesticides in soil ecosystems; movement through soils; effects on soil microbial populations; metabolism; residue detection methodology. Mailing Add: Pesticide Res Lab Dept Food Sci Univ of Fla Gainesville FL 32611

WHEELESS, LEON LUM, JR, b Jackson, Miss, Nov 6, 35; m 57; c 3. ELECTRICAL ENGINEERING, BIOMEDICAL ENGINEERING. Educ: Mass Inst Technol, SB, 58; Univ Rochester, MS, 62, PhD(elec eng), 65. Prof Exp: Sect head, Electronics Dept, Bausch & Lomb, Inc, 58-61, tech specialist, Biophys Dept, 61-65, res scientist, Cent Res Lab, 65-68, sr res scientist, Biomed Res Dept, Anal Systs Div, 68-69, dir biomed res, 69-71; ASSOC PROF PATH & ELEC ENG, MED CTR, UNIV ROCHESTER, 71-, DIR, CYTOPATHOLOGY AUTOMATION DIV, DEPT PATH, 75- Mem: Inst Elec & Electronics Engrs; Am Soc Cytol. Res: Biomedical instrumentation; computerized systems for automatic recognition of abnormal cells; automated cytopathology instrumentation; pattern recognition. Mailing Add: Dept of Path Univ of Rochester Med Ctr Rochester NY 14642

WHEELIS, MARK LEWIS, b Chelsea, Mass, Jan 8, 44; m 65; c 2. MICROBIOLOGY, GENETICS. Educ: Univ Calif, Berkeley, AB, 65, MA, 67, PhD(bacteriol), 69. Prof Exp: Lab technician bacteriol, Univ Calif, Berkeley, 65-66; res assoc, Univ Ill, Urbana-

Champaign, 69-70; ASST PROF BACTERIOL, COL LETTERS & SCI, UNIV CALIF, DAVIS, 70- Concurrent Pos: USPHS res grant, Univ Calif, Davis, 72-73. Mem: Am Soc Microbiol; Genetics Soc Am; Brit Soc Gen Microbiol. Res: Genetics of non-enteric prokaryotes; dissimilation of aromatic acids. Mailing Add: Dept of Bacteriol Univ of Calif Col of Letters & Sci Davis CA 95616

WHEELOCK, EARLE FREDERICK, b New York, NY, Feb 19, 27; m 55; c 3. VIROLOGY. Educ: Mass Inst Technol, BS, 50; Columbia Univ, MD, 55; Rockefeller Inst, PhD(biol), 61. Prof Exp: Intern med, Clins, Univ Chicago, 55-56; resident, Strong Mem Hosp, Rochester, NY, 56-57; fel biol & virol, Rockefeller Inst, 57-61; from asst prof to assoc prof prev med, Sch Med, Western Reserve Univ, 61-71; PROF MICROBIOL, JEFFERSON MED COL, 71- Concurrent Pos: USPHS res career develop award, 66-71. Mem: Am Soc Clin Invest; Am Asn Immunol; Soc Exp Biol & Med; Am Soc Microbiol; Am Asn Cancer Res. Res: Animal virology; mechanism of host resistance to viral infections; role of leucocytes and interferon in human viral infections; effect of nontumor viruses on virus-induced leukemia in mice; suppression of leukemia viral infections. Mailing Add: Dept of Microbiol Jefferson Med Col Philadelphia PA 19107

WHEELOCK, KENNETH STEVEN, b Kansas City, Mo, Sept 18, 43; m 72. PETROLEUM CHEMISTRY, INORGANIC CHEMISTRY. Educ: Univ Mo-Kansas City, BS, 65; Tulane Univ, PhD(chem), 70. Prof Exp: Chemist, 69-72, RES CHEMIST, EXXON RES & DEVELOP LABS, 72- Concurrent Pos: Consult, Dept Chem, Tulane Univ, 70- Honors & Awards: Award, Am Chem Soc, 65. Mem: Am Chem Soc. Res: Theoretical aspects of transition metal chemistry and catalysis; low valent complexes of transition metals with unsaturated ligands; quantum chemistry of catalysis; theory of finely divided metals. Mailing Add: Exxon Res & Develop Labs PO Box 2226 Baton Rouge LA 70821

WHEELOCK, MARK CARROLL, b Sioux City, Iowa, June 14, 05; m 31; c 1. PATHOLOGY. Educ: Univ Iowa, MD, 30; Am Bd Path, dipl. Prof Exp: Intern, Western Pa Hosp, Pittsburgh, 30-31; pvt pract, Sioux City, Iowa, 31-37; asst demonstr & instr path, Western Reserve Univ, 37-40; sr resident, Pondville Hosp for Cancer, Walpole, Mass, 40-41; demonstr, Harvard Med Sch, 41-42; asst prof path & bact, Univ Ala, 42-43; asst path, St Louis Univ, 46-47; assoc prof, Med Sch & lectr, Dent Sch, Northwestern Univ, 47-65; ASSOC PROF PATH, SCH MED, UNIV MIAMI, 65- Concurrent Pos: Asst path, Univ Hosp, Western Reserve Univ & resident, City Hosp, Cleveland, 37-40; asst pathologist, New Eng Deaconess, Baptist & Palmer Mem Hosps, Boston, 41-42; pathologist, Passavant Mem Hosp, Chicago, 47-65; pathologist, Tumor Clin, Northwestern Univ, 47-65; attend pathologist, Jackson Mem Hosp, Miami, Fla, 66- Mem: Am Soc Clin Path; Am Asn Path & Bact; Asn Mil Surgeons US; Asn Clin Scientists; Col Am Path. Res: Tumor pathology. Mailing Add: 210 Seaview Dr 511 Miami FL 33149

WHEELON, ALBERT DEWELL, b Moline, Ill, Jan 18, 29; m 53; c 2. THEORETICAL PHYSICS. Educ: Stanford Univ, BSc, 49; Mass Inst Technol, PhD(physics), 52. Prof Exp: Asst, Res Lab Electronics, Mass Inst Technol, 51-52; sr mem tech staff, Ramo-Wooldridge Corp, 53-62; with US Govt, 62-66; VPRES & GROUP EXEC SPACE & COMMUN GROUP, HUGHES AIRCRAFT CO, 66- Concurrent Pos: Mem, Defense Sci Bd. Mem: Nat Acad Eng; Int Union Radio Sci; Am Phys Soc; fel Inst Elec & Electronics Engrs. Res: Meson theory; general relativity; turbulence theory; analysis of ballistic missile and space systems; electromagnetic propagation and radio signal statistics. Mailing Add: 320 S Canyon View Dr Los Angeles CA 90049

WHEELWRIGHT, EARL J, b Rexburg, Idaho, Mar 26, 28; m 47; c 6. NUCLEAR CHEMISTRY. Educ: Brigham Young Univ, BS, 50; Iowa State Univ, PhD, 55. Prof Exp: Chemist, Hanford Atomic Prod Oper, Gen Elec Co, 55-60, sr scientist, 60-64; RES ASSOC, PAC NORTHWEST LABS, BATTELLE MEM INST, 65- Mem: Am Chem Soc. Res: Separation chemistry, ion exchange and chelation chemistry as applied to lanthanides, actinides and fission products. Mailing Add: Chem Tech Dept Pac Northwest Labs Battelle Mem Inst Richland WA 99352

WHELAN, BARBARA JEAN KING, b Reno, Nev, Nov 12, 39; m 70. ORGANIC CHEMISTRY. Educ: Univ Calif, Davis, BS, 61; Mass Inst Technol, PhD(org chem), 65. Prof Exp: NIH fel chem, Brandeis Univ, 65-67; from asst prof to assoc prof, Fairleigh Dickinson Univ, 67-75; RES ASSOC, WOODS HOLE OCEANOG INST, 75- Mem: AAAS; Am Chem Soc. Res: Organic geochemistry; petroleum genesis; reactions of organic compounds with clay minerals. Mailing Add: Dept of Chem Woods Hole Oceanog Inst Woods Hole MA 02543

WHELAN, JAMES ARTHUR, b Steele Co, Minn, Sept 25, 28; m 50; c 3. ECONOMIC GEOLOGY. Educ: Univ Minn, BMinE, 49, MS, 56, PhD, 59. Prof Exp: Instr mining eng, Univ Minn, 57-59; from asst prof to prof mineral, Univ Utah, 59-68, prof mining & geol eng, 68-69; dep off in chg construct, US Navy, Marianas, 69-71; PROF GEOL & GEOPHYS SCI, UNIV UTAH, 71- Mem: Geol Soc Am; Am Inst Mining, Metall & Petrol Eng. Res: Geochemistry; mineralogy. Mailing Add: Utah Geol & Mineral Surv Univ of Utah Salt Lake City UT 84112

WHELAN, JOHN MICHAEL, b Lyndhurst, NJ, Sept 12, 21; m 43; c 3. POLYMER CHEMISTRY. Educ: Stevens Inst Technol, ME, 41, MS, 43; Polytech Inst Brooklyn, PhD(org chem), 59. Prof Exp: Res chemist, 41-53, group leader, 53-55, sect head, 55-63, asst dir res & develop, 63-72, RES ASSOC, UNION CARBIDE CORP, 72- Mem: Am Chem Soc. Res: Synthetic polymers. Mailing Add: Union Carbide Res & Develop Labs 1 River Rd Bound Brook NJ 08805

WHELAN, THOMAS, III, b Houston, Tex, Dec 21, 44; m 68; c 1. MARINE GEOCHEMISTRY. Educ: Austin Col, BA, 66; Univ Tex, Austin, MA, 68; Tex A&M Univ, PhD(chem oceanog), 71. Prof Exp: Asst prof marine sci, 71-75, ASSOC PROF MARINE SCI, COASTAL STUDIES INST, LA STATE UNIV, BATON ROUGE, 75- Mem: Geochem Soc; AAAS; Am Asn Plant Physiologists. Res: Organic geochemistry of marine environments; geochemistry of natural gases; effects of oil in coastal environment. Mailing Add: Dept of Marine Sci La State Univ Baton Rouge LA 70803

WHELAN, WILLIAM JOSEPH, b Salford, UK, Nov 14, 24; m 51. BIOCHEMISTRY. Educ: Univ Birmingham, BSc, 45, PhD, 48, DSc(org chem), 55. Prof Exp: Sr lectr, Univ Col NWales, 48-55; sr mem, Lister Inst Prev Med, London, Eng, 56-64; prof biochem, Royal Free Hosp Sch Med, Univ London, 64-67; PROF BIOCHEM & CHMN DEPT, SCH MED, UNIV MIAMI, 67-, ASSOC DEAN UNIV AFFAIRS, 72- Concurrent Pos: Secy-gen, Fedn Europ Biochem Socs, 65-67 & Pan-Am Asn Biochem Socs, 69-72; mem physiol chem study sect, NIH, 71-75, chmn, 73-75; gen-secy, Int Union Biochem, 73- Honors & Awards: Carl Lucas Alsberg Lectr, 67; Ciba Medal & Lectr, 69; Diplome d'Honneur, Fedn Europ Biochem Socs, 74; Award of Merit, Japanese Soc Starch Sci, 75. Mem: Brit Biochem Soc; Am Soc Biol Chem; Am Chem Soc. Res: Glycogen and starch, structure and metabolism; extracellular microbial enzymes. Mailing Add: Dept of Biochem Sch of Med Univ Miami PO Box 520875 Miami FL 33152

WHELAN, WILLIAM PAUL, JR, b Brooklyn, NY, Sept 22, 23; m 55; c 3. ORGANIC POLYMER CHEMISTRY. Educ: Holy Cross Col, AB, 43, MS, 47; Columbia Univ, PhD(chem), 52. Prof Exp: Res scientist, 52-66, SR RES SCIENTIST, OXFORD MGT & RES CTR, CORP RES CTR, UNIROYAL, INC, MIDDLEBURY, 66- Mem: Am Chem Soc; Sigma Xi. Res: Polymer flammability; rocket motor insulators; organic reaction mechanisms; synthesis and solvolysis of bridgehead-substituted bicyclics; synthesis of polyurethane elastomers; vinyl polymerization; organic synthesis; crystallization kinetics; mechanism of ablative insulation. Mailing Add: Orchard Lane Woodbury CT 06798

WHELTON, BARTLETT DAVID, b San Francisco, Calif, Dec 2, 41. MEDICINAL CHEMISTRY. Educ: Univ San Francisco, BS, 63; Univ Wash, PhD(med chem), 69. Prof Exp: Fel med chem, Univ Alta, 69-71; asst prof, Univ of the Pac, 71-74; ASST PROF MED CHEM, EASTERN WASH STATE COL, 74- Mem: Am Chem Soc; Sigma Xi. Res: Synthesis of antineoplastic agents; heavy metal toxicology. Mailing Add: Dept of Chem Eastern Wash State Col Cheney WA 99004

WHEREAT, ARTHUR FINCH, b New York, NY, June 30, 27; m 53; c 3. CARDIOVASCULAR DISEASES. Educ: Williams Col, BA, 47; Univ Pa, MD, 51. Prof Exp: Intern & med resident internal med, Hosp Univ Pa, 51-54, fel cardiol, 54-56; assoc in biochem, 56-60, asst prof med, 60-68, ASSOC PROF MED, SCH MED, UNIV PA, 68-; CHIEF OF STAFF, VET ADMIN HOSP, PHILADELPHIA, 76- Concurrent Pos: Mem coun arteriosclerosis, Am Heart Asn, 64 & coun clin cardiol, 69. Mem: Fel Am Col Physicians; fel Am Col Cardiol; Am Soc Biol Chemists; Am Physiol Soc; Am Heart Asn. Res: Biochemical changes in arterial wall associated with atherogenesis; biochemical changes in heart muscle during hypoxia and ischemia. Mailing Add: Vet Admin Hosp 38th & University Ave Philadelphia PA 19104

WHETSEL, KERMIT BAZIL, b Tenn, Dec 9, 23; m 49; c 4. ANALYTICAL CHEMISTRY. Educ: East Tenn State Univ, BS, 43; Univ Tenn, MS, 47, PhD(chem), 50. Prof Exp: From chemist to sr chemist, 50-66, develop assoc, 66-73, SR DEVELOP ASSOC, TENN EASTMAN CO DIV, EASTMAN KODAK CO, 73- Concurrent Pos: Nat Acad Sci-Nat Res Coun resident res assoc, US Naval Res Lab, 60-61. Mem: AAAS; Am Chem Soc; Soc Appl Spectros; Coblentz Soc. Res: Instrumental analysis of organic compounds; correlation of absorption spectra with structure; solvent effects on infrared spectra; spectroscopic study of hydrogen bonded complexes. Mailing Add: Tenn Eastman Co Kingsport TN 37662

WHETSTONE, RICHARD ROY, b Merced Co, Calif, June 28, 16; m 47; c 2. PESTICIDE CHEMISTRY. Educ: George Washington Univ, BS, 40; Harvard Univ, MA, 41, PhD(org chem), 42. Prof Exp: Jr chemist, USDA, Washington, DC, 36-40; chemist, Shell Develop Co, Calif, 42-52, supvr res, 52-54, mgr, Org Chem Dept, 54-66, asst dir phys sci, 66-68, mgr res & develop, Agr Div, Shell Chem Co, 68-72, SPEC ASSIGNMENT, AGR DIV, SHELL CHEM CO, 72- Mem: Am Chem Soc. Res: Research and development of new pesticides, particularly organophosphorus insecticides. Mailing Add: Agr Div Shell Chem Co 2401 Crow Canyon Rd San Ramon CA 94583

WHETSTONE, STANLEY L, JR, b Newark, NJ, Aug 30, 25; m 52; c 4. EXPERIMENTAL NUCLEAR PHYSICS. Educ: Williams Col, BA, 49; Univ Calif, Berkeley, PhD(physics), 55. Prof Exp: Staff mem & physicist, Los Alamos Sci Lab, 55-70 & 75-76; physics sect head, Int Atomic Energy Agency, Vienna, Austria, 70-75; NUCLEAR PHYSICIST, US ENERGY RES & DEVELOP ADMIN, 76- Concurrent Pos: Vis lectr, Univ Wash, 67-68. Mem: Am Phys Soc. Mailing Add: Div of Phys Res US Energy Res & Develop Admin Washington DC 20545

WHETTEN, JOHN T, b Willimantic, Conn, Mar 16, 35; m 60; c 3. GEOLOGY, OCEANOGRAPHY. Educ: Princeton Univ, AB, 57; Univ Calif, Berkeley, MA, 59; Princeton Univ, PhD(geol), 62. Prof Exp: Fulbright fel, Australia & NZ, 62-63; res instr oceanog, 63-64, res asst prof, 64-65, from asst prof to assoc prof, 65-72, assoc dean grad sch, 68-69, chmn dept, 69-74, PROF GEOL SCI, UNIV WASH, 72- Mem: AAAS; Geol Soc Am; Soc Econ Paleontologists & Mineralogists. Res: Sedimentology; sedimentary petrology; marine geology. Mailing Add: Dept of Geol Sci Univ of Wash Seattle WA 98195

WHETTEN, NATHAN REY, b Provo, Utah, Aug 11, 28; m 53; c 3. PHYSICS. Educ: Yale Univ, BS, 49, MS, 50, PhD(physics), 53. Prof Exp: PHYSICIST, RES LAB, GEN ELEC CO, 53- Concurrent Pos: Vis lectr, Union Col, 64-65, adj prof, 67- Mem: Am Phys Soc; Am Vacuum Soc (pres-elect, 75, pres, 76). Res: Cosmic rays; secondary electron emission; surface physics; high vacuum; mass spectrometry; electron physics; medical physics. Mailing Add: Res & Develop Ctr Gen Elec Co PO Box 8 Schenectady NY 12305

WHICKER, FLOYD WARD, b Cedar City, Utah, July 24, 37; m 57; c 3. RADIATION BIOLOGY, ECOLOGY. Educ: Colo State Univ, BS, 62, PhD(radiation biol), 65. Prof Exp: Asst prof, 65-72, ASSOC PROF RADIATION BIOL, COLO STATE UNIV, 72-, PROF RADIOL, 74- Concurrent Pos: Consult, US AEC, El Paso Nat Gas Co, CER Geonuclear Corp & Utah Int Co; lectr, Colo State Univ Speakers Bur, 67- Mem: AAAS; Ecol Soc Am; Wildlife Soc; Health Physics Soc. Res: Radiation ecology; radionuclide accumulation in fish and wildlife and radiation effects on plant and animal populations. Mailing Add: Dept of Radiol & Radiation Biol Colo State Univ Ft Collins CO 80521

WHICKER, LAWRENCE R, b Bristol, Va, Oct 3, 34; m 58; c 1. MICROWAVE PHYSICS, ELECTROMAGNETISM. Educ: Univ Tenn, BS, 57, MS, 58; Purdue Univ, PhD(elec eng), 64. Prof Exp: Teaching asst, Univ Tenn, 58; sr engr, Microwave Electronics Div, Sperry Rand Corp, 58-61; fel engr, Surface Div, Westinghouse Elec Corp, 64-65, assoc dir appl physics, 65-66, mgr microwave physics group, Aerospace Div, 66-68, adv engr microwave & antenna group, 68-69; HEAD MICROWAVE TECHNIQUES BR, ELECTRONICS DIV, NAVAL RES LAB, 69- Concurrent Pos: Lectr, Univ Md, 64-; vpres res & develop, I-Tel. Inc, 67-68. Mem: Sr mem Inst Elec & Electronics Engrs. Res: Millimeter-coupled mode techniques; microwave filters; electromagnetic propagation studies; microwave solid state techniques, including microwave latching phasers, acoustics and integrated circuits. Mailing Add: 1218 Balfour Dr Arnold MD 21012

WHIDBY, JERRY FRANK, b Baltimore, Md, Oct 29, 43; m 67; c 2. ANALYTICAL CHEMISTRY, PHYSICAL CHEMISTRY. Educ: NGa Col, BS, 65; Univ Ga, PhD(anal chem), 70. Prof Exp: Res chemist, Gen Elec Co, Mo, 71-72; res assoc anal chem, 72-75, RES CHEMIST RES & DEVELOP, PHILIP MORRIS INC, 75- Res: Proton exchange kinetics; environmental research-sensors; kinetics of filter action. Mailing Add: 714 N Pinetta Dr Richmond VA 23235

WHIDDEN, HELEN LOUISE, b East Brewster, Mass, Dec 6, 07. PHYSICAL CHEMISTRY. Educ: Wellesley Col, AB, 29; Smith Col, MA, 37; Univ Mass, PhD(chem), 53. Prof Exp: Instr chem, Hood Col, 29-35; asst, Med Sch, Harvard Univ, 37-39; chemist, Newton Hosp, 39-41; from instr to assoc prof, 41-56, PROF CHEM, RANDOLPH-MACON WOMAN'S COL, 56-, CHMN DEPT, 63-

Concurrent Pos: Res chemist, US Naval Res Proj, Randolph-Macon Woman's Col, 47-49; instr chem, Seven Hills Sch, 63- Mem: Am Chem Soc; Nat Sci Teachers Asn; Am Asn Higher Educ; fel Am Inst Chem; NY Acad Sci. Res: Extraction and colorimetric analysis of ketosteroids; synthesis of protocatchuic aldehyde glucosides; synthesis and kinetics of emulsion hydrolysis of nitrogen-containing phenolic glucosides; copper chelates of 1, 3-dicarbonyl compounds; acylation of methyl cyclopropyl ketone; electrochemical studies in nonaqueous solvents; hydrogen bonding in dicarbonyl compounds. Mailing Add: Dept of Chem Randolph-Macon Woman's Col Lynchburg VA 24504

WHIDDEN, PHILLIPS, applied mathematics, see 12th edition

WHIFFEN, JAMES DOUGLASS, b New York, NY, Jan 16, 31; m 60; c 1. SURGERY, BIOENGINEERING. Educ: Univ Wis, BS, 52, MD, 55; Am Bd Surg, dipl, 63. Prof Exp: Intern, Ohio State Univ, 55-56; resident, 56-57 & 59-62, from instr to assoc prof, 62-71, actg chmn dept, 72-74, PROF SURG, MED SCH, UNIV WIS-MADISON, 71-, ASST DEAN MED SCH, 75- Concurrent Pos: Nat Heart Inst res fel, 62-64; res career develop award, 65-75; Markle Scholar, 66- Mem: Am Col Surg; Am Soc Artificial Internal Organs; Am Soc Test & Mat. Res: Cardiovascular prostheses; cardiopulmonary support devices; biomaterials. Mailing Add: Univ Hosps 1300 University Ave Madison WI 53706

WHIGHAM, DAVID KEITH, b Blanchard, Iowa, Aug 15, 38; m 64; c 3. AGRONOMY. Educ: Iowa State Univ, BS, 66, MS, 69, PhD(crop prod), 71. Prof Exp: Instr agron, Iowa State Univ, 68-71; agronomist, USDA, 71-73; ASST PROF AGRON, UNIV ILL, URBANA, 73- Mem: Am Soc Agron; Crop Sci Soc Am; Asn Advan Agr Sci Africa. Res: Crop production of economic crops; international soybean research. Mailing Add: Dept of Agron Univ of Ill Urbana IL 61801

WHIKEHART, DAVID RALPH, b Pittsburgh, Pa, Aug 21, 39; m 69; c 1. BIOCHEMISTRY. Educ: WVa Univ, PhD(biochem), 69. Prof Exp: Res assoc, Harvard Med Sch, 71-72; asst biochemist, McLean Hosp, Belmont, Mass, 71-72; spec fel, 72-74, SR STAFF FEL OPHTHALMIC BIOCHEM, NAT EYE INST, 74- Concurrent Pos: Res fel neurochem, Harvard Med Sch, 69-71, fel neurochem, McLean Hosp, Belmont, Mass, 69-71. Mem: Asn Res Vision & Ophthal. Res: Metabolism and transport in the cornea. Mailing Add: Lab of Vision Res Bldg 6 Rm 207 Nat Eye Inst Bethesda MD 20014

WHIPKEY, KENNETH LEE, b Cortland, Ohio, June 5, 32; m 62. MATHEMATICS, STATISTICS. Educ: Kent State Univ, AB, 53, MA, 58; Case Western Reserve Univ, PhD(educ statist), 69. Prof Exp: Instr high sch, 54-57; asst math, Kent State Univ, 57-58; instr high sch, 58-59; asst prof, Youngstown Univ, 59-67; ASSOC PROF MATH, WESTMINSTER COL, PA, 68- Concurrent Pos: Instr, In-Serv Insts Teachers, NSF, 64-67; lectr & vis scientist, Ohio Acad Sci, 64-66; lectr, Holt, Rinehart & Winston, 65. Mem: Math Asn Am. Res: Factor analysis; calculus. Mailing Add: 456 Bradley Lane Youngstown OH 44504

WHIPP, ARTHUR ANDREW, b Opelousas, La, May 9, 25; m 45; c 2. ENTOMOLOGY. Educ: La State Univ, BS, 47; Univ Wis, MS, 50, PhD(econ entom), 51. Prof Exp: Res entomologist, 51-64, SUPVR FIELD RES, ORTHO DIV, CHEVRON CHEM CO, 64- Res: Field testing of new insecticides, fungicides and herbicides; new controls for major pests on citrus and vegetables. Mailing Add: Chevron Chem Co 940 Hensley St Richmond CA 94804

WHIPP, SHANNON CARL, b Jacksonville, Fla, May 3, 31; m 56; c 5. VETERINARY PHYSIOLOGY, VETERINARY SURGERY. Educ: Univ Minn, BS, 57, DVM, 59, PhD(physiol), 65. Prof Exp: Field vet, Minn Livestock Bd, 59-60; instr physiol, Univ Minn, 60-65; RES VET, NAT ANIMAL DIS CTR, AGR RES SERV, USDA, 65- Concurrent Pos: NIH fel, 62-65. Mem: Am Soc Vet Physiol & Pharmacol; Am Vet Med Asn; Comp Gastroenterol Soc; Conf Res Workers Animal Dis; NY Acad Sci. Res: Mechanisms of diarrhea; adrenal function in domestic animals; secretory diarrhea; enteric colibacillosis. Mailing Add: Nat Animal Dis Ctr PO Box 70 Ames IA 50010

WHIPPEY, PATRICK WILLIAM, b Reading, UK, Feb 18, 40. SOLID STATE PHYSICS. Educ: Univ Reading, BSc, 62, PhD(physics), 66. Prof Exp: Asst lectr physics, Univ Reading, 65-66; lectr, 66-67, asst prof, 67-72, ASSOC PROF PHYSICS, UNIV WESTERN ONT, 72- Mem: Can Asn Physicists. Res: Electron spin resonance in diamond; optical properties of rare earth doped calcium fluoride. Mailing Add: Dept of Physics Univ of Western Ont London ON Can

WHIPPLE, EARL BENNETT, b Thomson, Ga, Apr 9, 30; m 51; c 4. PHYSICAL CHEMISTRY. Educ: Emory Univ, BS, 51, PhD(phys chem), 59. Prof Exp: Chemist, Va-Carolina Chem Corp, Va, 52-53; res scientist chem, Union Carbide Res Inst, 59-66, group leader chem, 66-70, mgr res, Cent Sci Lab, Tarrytown Tech Ctr, Union Carbide Corp, 70-74; RES ADV, CENT RES, PFIZER, INC, 74- Concurrent Pos: Adj prof, Rockefeller Univ, 69-74. Mem: NY Acad Sci; Am Chem Soc. Res: Chemical applications of nuclear and electron spin resonance; structure and electronic properties of molecules. Mailing Add: 4 Dwayne Rd Old Saybrook CT 06475

WHIPPLE, ELDEN C, JR, physics, space science, see 12th edition

WHIPPLE, FRANCIS OLIVER, b Antwerp, Belg, Dec 26, 25; m 49; c 1. FOREST PRODUCTS, RESOURCE MANAGEMENT. Educ: Reed Col, BA, 48; Ore State Col, MS, 51, PhD(chem, 53. Prof Exp: Res chemist, Crown Zellerbach Corp, 53-59, from mgr tech serv to dir res & develop, Crown Zellerbach Can, Ltd, 59-72, VPRES & GEN MGR, VEN DEV ENTERPRISES, LTD, 72- Mailing Add: Crown Zellerbach Can Ltd 1030 W Georgia St Vancouver BC Can

WHIPPLE, FRED LAWRENCE, b Red Oak, Iowa, Nov 5, 06; m 46; c 3. ASTRONOMY, SPACE PHYSICS. Educ: Univ Calif, Los Angeles, AB, 27; Univ Calif, Berkeley, PhD(astron), 31. Hon Degrees: MA, Harvard Univ, 45; DSc, Am Int Col, 48 & Temple Univ, 61; DLitt, Northeastern Univ, 61. Prof Exp: Instr astron, Univ, 32-38, lectr, 38-45, from assoc prof to prof, 45-70, chmn dept, 49-56, PHILLIPS PROF ASTRON, OBSERV, HARVARD UNIV, 70-, MEM STAFF, 31- Concurrent Pos: Res assoc radio res lab, Off Sci Res & Develop, 42-45; leader, Harvard proj upper atmospheric & meteor res, Bur Ord, US Navy, 46-51, Air Res & Develop Command, US Air Force, 48-62, Off Naval Res, 51-57, Off Ord Res, US Army, 53-57, dir, Harvard radio meteor proj, Nat Bur Stand, 57-61, NSF, 60-63 & NASA, 63-; Lowell lectr, Lowell Technol Inst, 47; dir astrophys observ, Smithsonian Inst, 55-73, sr scientist astrophys observ, 73-; ed, Smithsonian Contrib Astrophys, 56-; regional ed, Planetary & Space Sci, 58- Mem, US Rocket & Satellite Res Panel, 46-; subcomt, Nat Adv Comt Aeronaut, 46-52; panel upper atmosphere, US Res & Develop Bd, 47-52; adv panel astron, NSF, 52-55, chmn, 54-55, mem div comt math & phys sci, 64-; sci adv bd, US Air Force, 53-62, assoc adv, 63-67, mem geophys & space tech panels; comts meteorol & atmospheric sci, Nat Acad Sci-Nat Res Coun, 58-, space sci bd, 58- & subcomt potential contamination & interference from space exp, 63-; dir optical satellite tracking proj & proj dir orbiting astron observ, NASA, 58-72, mem space sci working group orbiting astron observ, 59-69, consult aeronomy subcomt, 61-63, comt planetary atmospheres, 62-63, dir meteorite photog & recovery prog, 62-73, mem working group geod satellite prog, 63- & mem comet & asteroid sci adv comt, 71-72, chmn, 73-74; mem joint bio-astronaut comt, Armed Forces-Nat Res Coun, 59-61; spec consult comt sci & astronaut, US House Rep, 60-73; mem working groups geod satellites & tracking, telemetry & dynamics, Comt Space Res, 60-, chmn sci coun geod uses artificial satellites, 65-; chmn, Gordon Res Conf Chem & Physics Space, 63; trustee-at-large, Univ Corp Atmospheric Res, Colo, 64-68 & mem, Comt Nat Ctr Atmospheric Res Staff-Univ rels, 65-68. Mem, vpres & pres comns, Int Astron Union, 32-, voting rep, 52 & 55; deleg, Inter-Am Astrophys Cong, Mex, 42; mem comn 3, U S nat comt, Int Sci Radio Union, 49-61; Int Astron Fedn, 55-; mem working group satellite tracking & comput & chief investr proj optical tracking artificial earth satellites, Int Geophys Yr, 55-58, mem tech panel earth satellite prog & tech panel rocketry, 55-59. Honors & Awards: Donohue Medals; President's Cert Merit; Smith Medal, Nat Acad Sci, 49; Except Serv Award, US Air Force, 60; Medal, Univ Liege, 60; Award, Am Astronaut Soc, 61; Comdr, Order of Merit Res & Invention, France, 62; President's Distinguished Fed Civilian Serv Award, 63; Space Pioneers Medallion, 68; Pub Serv Award, NASA, 69; Leonard Medal, Meteoritical Soc, 70; Kepler Medal, AAAS, 71; Career Serv Award, Nat Civil Serv League, 72; Henry Medal, Bd Regents, Smithsonian Inst, 73. Mem: Nat Acad Sci; fel Am Astron Soc (vpres, 48-50); fel Am Astronaut Soc (vpres, 62-64); fel Am Geophys Union; fel Am Inst Aeronaut & Astronaut. Res: Photometry; computation of comet discoveries; colors of external galaxies; novae; meteor orbits; earth's upper atmosphere; stellar and planetary evolution. Mailing Add: 35 Elizabeth Rd Belmont MA 02178

WHIPPLE, GEORGE HOYT, pathology, deceased

WHIPPLE, GEORGE HOYT, JR, b San Francisco, Calif, May 4, 17; m 43, 68; c 5. RADIOLOGICAL HEALTH. Educ: Wesleyan Univ, BS, 39; Univ Rochester, PhD(biophys), 53. Prof Exp: Mem staff, Div Indust Coop, Mass Inst Technol, 42-47; group leader health physics, Hanford Works, Gen Elec Co, Wash, 47-50; from instr to asst prof, Atomic Energy Proj, Univ Rochester, 50-57; assoc prof, 57-72, PROF RADIOL HEALTH, UNIV MICH, ANN ARBOR, 72- Concurrent Pos: Consult various govt & indust agencies. Mem: Health Physics Soc; Am Indust Hyg Asn. Res: Effects of radiations on simple biological systems, particularly luminous bacteria; practical radiation hazard control and protection; environmental aspects of nuclear power generation. Mailing Add: Univ of Mich Sch of Pub Health Ann Arbor MI 48104

WHIPPLE, GERALD HOWARD, b Calif, Feb 6, 23; m 47; c 5. MEDICINE. Educ: Harvard Univ, SB, 43; Univ Calif, MD, 46; Am Bd Internal Med, dipl, 60. Prof Exp: Instr, Sch Med, Boston Univ, 56-57, assoc, 57-60, from asst prof to assoc prof, 60-68; assoc prof, 68-71, PROF MED, COL MED, UNIV CALIF, IRVINE, 71-, CHIEF CARDIOL DIV, 74- Concurrent Pos: Res fel med, Harvard Univ, 53-56; vol asst med, Congenital Heart Clin, Children's Med Ctr, Boston, 54-56; physician in chg, EKG Lab, Univ Hosp, Boston Univ, 56-58, physician, Cardiac Care Univ, 65-68, assoc vis physician, 56-68, assoc mem, Evans Mem Res Found, 56-68, sect head clin cardiol res; consult cardiol, Congenital Heart Clin, Boston City Hosp, 58-59; res consult, Providence Vet Admin Hosp, RI, 60-68; res assoc, Mass Inst Technol, 63-66; fel coun clin cardiol, Am Heart Asn; heart coordr area VIII, Regional Med Prog Cancer, Heart & Stroke, 68-; physician & dir intensive cardiac care unit, Orange County Med Ctr, 69-70; chief med serv, Vet Admin Hosp, Long Beach, 70-74. Mem: Fel Am Col Physicians; fel Am Col Cardiol. Res: Academic cardiology; cardiology; cardiac arrhythmias; epidemiologic evaluation of acute ischemic heart disease and sudden death. Mailing Add: 101 City Dr S Orange CA 92668

WHIPPLE, ROYSON NEWTON, b Buffalo, NY, May 28, 12; m 35; c 2. FOOD TECHNOLOGY. Educ: Univ Mich, BS, 35; Cornell Univ, MS, 39. Prof Exp: Instr high sch, NY, 35-45; prof & head div food technol, 45-57, PRES, STATE UNIV NY AGR & TECH COL, MORRISVILLE, 57- Mem: Inst Food Technologists. Res: Background needs for food technologists engaged in fruit and vegetable processing. Mailing Add: Box 206 Morrisville NY 13408

WHISLER, FRANK DUANE, b Burton, WVa, Nov 20, 34; m 56; c 4. SOIL PHYSICS. Educ: Univ WVa, BS, 57, MS, 58; Univ Ill, PhD(soil physics), 64. Prof Exp: Soil scientist, Agr Res Serv, USDA, Ill, 58-69, soil scientist, Water Conserv Lab, 69-73; PROF AGRON & AGRONOMIST, MISS STATE UNIV, 73- Honors & Awards: Award, Soil Sci Soc Am, 63. Mem: Am Soc Agron; Soil Sci Soc Am. Res: Water movement into and through soils or other porous material from both a theoretical and experimental point of view. Mailing Add: Dept of Agron Miss State Univ Box 5248 Mississippi State MS 39759

WHISLER, HOWARD CLINTON, b Oakland, Calif, Feb 4, 31; m 53; c 2. BOTANY, MICROBIOLOGY. Educ: Univ Calif, Berkeley, BSc, 54, PhD(bot), 61. Prof Exp: NATO fel, Univ Montpellier, 60-61; asst prof bot, McGill Univ, 61-63; from asst prof to assoc prof, 63-73, PROF BOT, UNIV WASH, 73- Concurrent Pos: NSF fel, Univ Geneva, 68-69. Mem: Mycol Soc Am; Phycol Soc Am; Soc Protozool; Bot Soc Am. Res: Development of the aquatic phycomycetes; insect microbiology. Mailing Add: Dept of Bot Univ of Wash Seattle WA 98105

WHISLER, KENNETH EUGENE, b Davenport, Iowa, Feb 5, 37; m 62; c 3. PHYSIOLOGY, CLINICAL CHEMISTRY. Educ: Augustana Col, AB, 58; Univ Wis-Madison, MS, 61, PhD(physiol), 64. Prof Exp: Res asst oncol, Sch Med, Univ Wis-Madison, 58-62, asst physiol, 62-64, instr, 64-68; CHIEF BIOCHEM, MILWAUKEE CHILDREN'S HOSP, 68- Concurrent Pos: Asst prof physiol, Med Col, Univ Wis-Milwaukee, 68-71; consult path, Northwest Gen Hosp, 70-72. Mem: Am Chem Soc; Am Asn Clin Chem. Res: Biochemistry of metabolic disorders in children; methods for the detection of acute lead intoxication. Mailing Add: Dept of Chem Milwaukee Children's Hosp Milwaukee WI 53233

WHISLER, WALTER WILLIAM, b Davenport, Iowa, Feb 9, 34; m 59; c 2. BIOCHEMISTRY, NEUROSURGERY. Educ: Augustana Col, AB, 55; Univ Ill, MD, 59, PhD(biochem), 69; Am Bd Neurol Surg, dipl, 67. Prof Exp: USPHS fel, 64-65; PROF BIOCHEM & NEUROSURG & CHMN DEPT NEUROSURG, RUSH MED COL, 70- Concurrent Pos: Attend neurosurgeon & chmn dept neurosurg, Presby-St Luke's Hosp, 70-; clin assoc prof, Univ Ill Col Med, 70- Mem: AMA; Am Asn Neurol Surg; Cong Neurol Surg; Int Soc Res Stereoencephalotomy. Res: Mechanisms of catecholomine oxidation; metabolism of the psychotomimetic amines; biochemistry of brain tumors. Mailing Add: Dept of Neurosurg Presby-St Luke's Hosp Chicago IL 60612

WHISNANT, JACK PAGE, b Little Rock, Ark, Oct 26, 24; m 44; c 3. MEDICINE, NEUROLOGY. Educ: Univ Ark, BS, 48, MD, 51. Prof Exp: From instr to assoc prof, 56-69, PROF NEUROL, MAYO MED SCH, UNIV MINN, 69-, CHMN DEPT, 71- Concurrent Pos: Consult, Mayo Clin, 55-, head sect neurol, 63-71. Mem: Am Acad Neurol; Am Neurol Asn. Res: Clinical neurology, especially vascular diseases of the nervous system. Mailing Add: Dept Neurol Mayo Med Sch Univ of Minn Rochester MN 55901

WHISONANT, ROBERT CLYDE, b Columbia, SC, Apr 20, 41; m 63; c 2. STRATIGRAPHY, SEDIMENTOLOGY. Educ: Clemson Univ, BS, 63; Fla State Univ, MS, 65, PhD(geol), 67. Prof Exp: Petrol geologist, Humble Oil & Ref Co, 67-71; asst prof, 71-72, ASSOC PROF GEOL, RADFORD COL, 72- Mem: Geol Soc Am; Am Asn Petrol Geol; Soc Econ Paleontologists & Mineralogists; Sigma Xi. Res: Paleozoic rocks of the southern Appalachians, specifically, sedimentary petrography and paleocurrent features; paleoenvironmental determinations; stratigraphic analysis. Mailing Add: Dept of Geol Radford Col Radford VA 24142

WHISSELL-BUECHY, DOROTHY Y E, b St Louis, Mo, Apr 12, 26; m 56; c 3. HUMAN GENETICS. Educ: Wellesley Col, AB, 48; Univ Tex Southwestern Med Sch Dallas, MD, 56; Univ Calif, Berkeley, PhD(genetics), 68. Prof Exp: INSTR GENETICS IN PUB HEALTH, UNIV CALIF, BERKELEY, 64-, ASST RES GENETICIST, INST HUMAN DEVELOP, 69- Concurrent Pos: Instr, Univ Exten, 71- Res: Genetics of quantitative traits; assessment of health at various stages of development, especially middle age; genetics of olfaction. Mailing Add: 1203 Tolman Hall Univ Calif Inst Human Develop Berkeley CA 94720

WHISTLER, DAVID PAUL, b Summit, NJ, June 15, 47; 40; m 64; c 2. VERTEBRATE PALEONTOLOGY. Educ: Univ Calif, Riverside, BA, 63, MA, 65; Univ Calif, Berkeley, PhD(paleont), 69. Prof Exp: Mus scientist paleont, Univ Calif Mus Paleont, 67-69; asst prof geol, Tex Tech Univ, 69-70; SR CUR VERT PALEONT, NATURAL HIST MUS, LOS ANGELES COUNTY, 69- Mem: Soc Vert Paleont; Paleont Soc; Geol Soc Am; Am Soc Mammalogists. Res: Evolution, taxonomy and biostratigraphy of smaller vertebrates, amphibians, reptiles rodents and insectivores in later Tertiary. Mailing Add: 900 Expos Blvd Los Angeles CA 90007

WHISTLER, ROY LESTER, b Morgantown, WVa, Mar 21, 12; m 35; c 2. CHEMISTRY. Educ: Heidelberg Col, BS, 34; Ohio State Univ, MS, 35; Iowa State Univ, PhD, 38. Prof Exp: Instr chem, Iowa State Col, 35-38; fel, Nat Bur Stand, 38-40; head starch struct sect, USDA, 40-45 & Northern Regional Res Lab, Bur Agr Chem & Eng, 45-46; asst head dept, 48-60, PROF BIOCHEM, PURDUE UNIV, LAFAYETTE, 46-, CHMN INST AGR UTILIZATION RES, 61- Concurrent Pos: Vis lectr, Univ Witwatersrand, 61, Cape Town, 65, NZ & Australia, 67 & 74, Czech Acad Sci & Hungarian Acad Sci, 68, Taiwan, 70 & France & Poland, 75. Honors & Awards: Hudson Award, Am Chem Soc, 60, Payen Award, 67; Annual Res Award, Japanese Tech Soc Starch, 67; Alsberg Schoch Award, Am Asn Cereal Chem, 70, Osborn Medal, 74; German Saare Medal, 74. Mem: AAAS; Am Chem Soc; Am Soc Biol Chemists; Am Asn Cereal Chem (pres, 72-73); Int Cartog Asn (pres, 72-74). Res: Chemistry and biochemistry of carbohydrates, particularly polysaccharides; polysaccharide chemistry. Mailing Add: Dept of Biochem Purdue Univ West Lafayette IN 47907

WHITACRE, DAVID MARTIN, b Mariemont, Ohio, Dec 4, 43; m 66; c 1. ENVIRONMENTAL SCIENCES. Educ: Wilmington Col, Ohio, AB, 65; Ohio State Univ, MSc, 66; Univ Ariz, PhD(entom, zool), 69. Prof Exp: Asst prof biol, Univ PR, Mayagüez, 69-71; State of Ariz res grant entom, Univ Ariz, 71-72; proj mgr entom toxicol, 72-73, DIR ENVIRON RES, VELSICOL CHEM CORP, 73- Mem: AAAS; Entom Soc Am; Am Chem Soc. Res: Metabolism of pesticides in plants and animals. Mailing Add: Velsicol Chem Corp 341 E Ohio St Chicago IL 60611

WHITACRE, FRANCIS MARION, b Salamonia, Ind, Apr 10, 05; m 29. EXPERIMENTAL BIOLOGY, CHEMISTRY. Educ: Ind Univ, AB, 27, AM, 28, PhD(org chem), 34. Prof Exp: Instr gen chem, Case Western Reserve Univ, 29-31, chem eng, 31-34 & org chem, 34-35, from asst prof to assoc prof org chem, 35-45, prof & secy grad sch, 45; sci dir, Schieffelin & Co, NY, 45-52; assoc dir chem develop, Olin Mathieson Chem Corp, 52-54, dep dir energy div, 54-58, dir prod develop, 58-59; CHMN BD, APPLN RES CORP, 59-; PRES, WHITACRE PEDAG INC, 72- Concurrent Pos: Dir, GAP Instrument Co, 61-72; consult gen sci, Portledge Sch, Locust Valley, NY, 70- Mem: Fel AAAS; Am Chem Soc; Ny Acad Sci. Res: Natural and synthetic drugs; agricultural chemicals; science education, chemistry and biology. Mailing Add: 8 Brookdale Rd Glen Cove NY 11542

WHITAKER, ARTHUR CHARLES, organic chemistry, see 12th edition

WHITAKER, CLAY WESTERFIELD, b Greenville, Ky, Apr 17, 24; m; c 4. OTOLARYNGOLOGY. Educ: Berea Col, BA, 48; Western Reserve Univ, MD, 52; Am Bd Otolaryngol, dipl, 58. Prof Exp: Asst otolaryngologist, Highland View Hosp, Cleveland, Ohio, 52; asst to dir ear, nose & throat dept, St Luke's Hosp, Cleveland, 56; asst otolaryngologist, Univ Hosp & demonstr & clin instr otolaryngol, Med Sch, Western Reserve Univ, 56-61; from asst prof to assoc prof, 64-70, PROF SURG, SCH MED, UNIV SOUTHERN CALIF, 70-; PROF & DIR EAR, NOSE & THROAT RES TRAINING, LOS ANGELES COUNTY-UNIV SOUTHERN CALIF MED CTR, 64-, CHIEF PHYSICIAN & DIR DEPT OTOLARYNGOL, 65- Concurrent Pos: Consult, Highland View Hosp, Cleveland, 56-61, Dept Pub Social Serv-Med Sci Div Hearing Aid Prog, Los Angeles County, 65-, Porterville State Hosp, Calif, 66-, tech adv comt on hearing aids, Dept Health Care Serv, State of Calif, 67-, Children's Hosp, Los Angeles, 67- & Hollywood Presby Hosp, Los Angeles, 67- Mem: AMA; Am Acad Ophthal & Otolaryngol. Res: Otorhinolaryngology; experimental surgery for device as practical prosthetic larynx; etiologic factors influencing Bell's Palsy. Mailing Add: Dept of Otolaryngol LAC-USC Med Ctr Box 795 Los Angeles CA 90033

WHITAKER, DONALD ROBERT, b Winnipeg, Man, July 12, 19; m 47; c 2. BIOCHEMISTRY. Educ: Univ Man, BSc, 41; Univ London, PhD, 48, DSc, 63. Prof Exp: From assoc to prin res off, Div Biosci, Nat Res Coun Can, 48-71; PROF BIOCHEM, UNIV OTTAWA, 71- Concurrent Pos: Chmn, Pan-Am Asn Biochem Socs. Mem: Am Chem Soc; Am Soc Biol Chemists; Can Biochem Soc (pres, 70-71); Brit Biochem Soc. Res: Structure, function and biosynthesis of microbial extracellular enzymes, particularly proteases; bacteriolytic enzymes and cellulases. Mailing Add: 97 Belmont Ave Ottawa ON Can

WHITAKER, ELLIS HOBART, b Salem, Mass, Dec 2, 08; m 35, 66; c 2. PLANT PHYSIOLOGY. Educ: Worcester Polytech Inst, BS, 30; Cornell Univ, MS, 36, PhD(plant physiol), 49. Prof Exp: Engr, Gilbert & Barker Mfg, Mass, 30-31; teacher, Monson Acad, 33-35 & Westover Sch, Conn, 36-39; asst gen bot, Cornell Univ, 40-41; instr phys sci & gen biol, State Univ NY Col Oneonta, 41-48, asst prof chem, 48-54, assoc prof biol, 54-64; assoc prof, 64-71, PROF BIOL, SOUTHEASTERN MASS UNIV, 71- Mem: Fel AAAS. Res: General biology; enzymes in insectivorous plants. Mailing Add: 32 Prospect St South Dartmouth MA 02748

WHITAKER, EWEN A, b London, Eng, June 22, 22; m 46; c 3. PLANETARY SCIENCES. Educ: Brit Inst Mech Eng, cert, 42. Prof Exp: Lab asst chem anal & phys testing, Siemens Bros & Co Ltd, 40-41, lab asst spectrochem anal, 41-49; sci asst, Royal Greenwich Observ, 49-53, from asst exp officer to exp officer, 53-58; res assoc, Yerkes Observ, Univ Chicago, 58-60; RES ASSOC, LUNAR & PLANETARY LAB, UNIV ARIZ, 60- Concurrent Pos: Co-experimenter, Ranger moonshots, 62-66; mem TV exp team, Surveyor Spacecraft, 64-68; mem site selection team, Orbiter 5, 67-68; mem, Apollo Orbital Sci Photo Team, 69-73. Mem: Int Astron Union; Am Astron Soc; fel Royal Astron Soc. Res: Study of moon, particularly of surface features and properties by earthbased telescopic observations and research and spacecraft; standardization of lunar nomenclature. Mailing Add: Lunar & Planetary Lab Univ of Ariz Tucson AZ 85721

WHITAKER, HARRY ALLEN, b Rochester, Minn, Jan 26, 39; m 71. NEUROPSYCHOLOGY. Educ: Univ Calif, Los Angeles, PhD(ling), 69. Prof Exp: Asst prof ling & psychol, 69-73, asst prof neurol, 71-74, ASSOC PROF PSYCHOL & NEUROL, UNIV ROCHESTER, 74- Concurrent Pos: NIH fel, Mayo Clin, 71-72; mem adj staff phys med, Rochester Gen Hosp, 70-; ed-in-chief, Brain & Lang, 73- Mem: Acad Aphasia; Soc Neurosci; Ling Soc Am; Psychonomic Soc. Res: Language disorders; behavior and brain correlations; animal communication systems; neurolinguistics. Mailing Add: Depts of Psychol & Neurol Univ of Rochester Rochester NY 14627

WHITAKER, JOE RUSSELL, b Cynthiana, Ky, Jan 30, 00; m 22; c 3. GEOGRAPHY. Educ: Univ Chicago, BS, 21; Univ Wis, MS, 23; Univ Chicago, PhD(geog), 30. Prof Exp: Head dept geog, Mich State Teachers Col, Marquette Univ, 24-30; from asst prof to assoc prof, Univ Wis, 30-40; prof & head dept, 40-68, EMER PROF GEOG, PEABODY COL, 68-; PROF GEOG, UNIV TENN, NASHVILLE, 68- Concurrent Pos: Nat Res Coun Can grant, 35; consult, US War Dept, 42-47, US State Dept, 42-44 & Field Enterprises Ed Corp, 56-67; mem, Nat Res Coun, 39-54. Honors & Awards: Nat Coun Geog Ed Award, 55. Mem: Asn Am Geog (pres, 53); Nat Coun Geog Ed (pres, 38); hon fel Am Geog Soc. Res: Historical and cultural geography of eastern United States; educational geography; Anglo-America. Mailing Add: Dept of Geog Univ of Tenn Nashville TN 37203

WHITAKER, JOHN O, JR, b Oneonta, NY, Apr 22, 35; m 57; c 3. VERTEBRATE ECOLOGY, MAMMALOGY. Educ: Cornell Univ, BS, 57, PhD(vert zool), 62. Prof Exp: From asst prof to assoc prof, 62-71, PROF LIFE SCI, IND STATE UNIV, TERRE HAUTE, 71- Mem: Fel AAAS; Am Soc Mammal; Ecol Soc Am; Soc Study Amphibians & Reptiles; Am Inst Biol Sci. Res: Food habits, parasites, habitats and interrelations of species. Mailing Add: Dept of Life Sci Ind State Univ Terre Haute IN 47809

WHITAKER, JOHN ROBERT, b Lubbock, Tex, Sept 13, 29; m 52; c 4. BIOCHEMISTRY. Educ: Berea Col, AB, 51; Ohio State Univ, PhD(agr biochem), 54. Prof Exp: Lab asst chem, Berea Col, 50-51; asst agr biochem, Ohio State Univ, 51-52, asst instr, 53-54, fel, 54; from instr to assoc prof, 56-67, PROF FOOD SCI & TECHNOL, UNIV CALIF, DAVIS, 67-, ASSOC DEAN, COL AGR & ENVIRON SCI, 75- Concurrent Pos: NIH spec fel, Northwestern Univ, 63-64; vis prof, Vet Col, Norway, 72 & Univ Bristol, Eng, 72-73. Honors & Awards: Centennial Achievement Award, Ohio State Univ, 70; William V Cruess Award, Inst Food Technologists, 73. Mem: Am Chem Soc; Inst Food Technologists; Am Soc Biol Chemists. Res: Water in relation to the structure of proteins; relationship between structure and function in enzymes; chemical improvement of proteins for food use. Mailing Add: Dept of Food Sci & Technol Univ of Calif Davis CA 95616

WHITAKER, JOSEPH SAMUEL, organic chemistry, see 12th edition

WHITAKER, LESLIE A, b Denver, Colo, Sept 25, 23; m 51; c 3. ANALYTICAL CHEMISTRY. Educ: Univ Denver, BS, 49, MA, 56; Univ of the Pac, PhD(anal chem), 68. Prof Exp: Teacher chem & physics, Littleton High Sch, 52-54; instr chem, Scottsbluff Col, 56-57; asst prof & dean of fac, Cottey Col, 57-61; instr, Col of Desert, 62-64; assoc prof, Calif Polytech State Univ, San Luis Obispo, 67-72; teacher & bus mgr, Desert Sun Sch, 72-75; DIR LABS CHEM, UNIV SOUTHERN CALIF, 75- Mem: Am Chem Soc. Mailing Add: Dept of Chem Univ of Southern Calif Los Angeles CA 90007

WHITAKER, MACK LEE, b Forest City, NC, Dec 2, 31; m 56; c 2. MATHEMATICS. Educ: Appalachian State Teachers Col, BS, 53, MA, 56; Fla State Univ, EdD, 61. Prof Exp: Teacher, Piedmont High Sch, 56-58; from assoc prof to prof math, Radford Col, 60-68, chmn dept, 62-66; assoc prof math educ, Auburn Univ, 68-69; PROF MATH, RADFORD COL, 69- Mem: Math Asn Am. Res: Mathematics education; abstract algebra; foundations of mathematics. Mailing Add: Dept of Math Radford Col Radford VA 24141

WHITAKER, ROBERT DALLAS, b Tampa, Fla, Mar 5, 33; m 60; c 3. INROGANIC CHEMISTRY. Educ: Univ Washington & Lee, BS, 55; Univ Fla, PhD(inorg chem), 59. Prof Exp: Asst prof chem, Washington & Lee Univ, 59-62; from asst prof to assoc prof, 62-74, PROF CHEM, UNIV S FLA, TAMPA, 74- Mem: Am Chem Soc. Res: Molecular addition compounds. Mailing Add: Dept of Chem Univ of SFla Tampa FL 33620

WHITAKER, SIDNEY HOPKINS, b Spring Valley, Ill, Apr 7, 40. GEOLOGY. Educ: Oberlin Col, BA, 62; Univ Ill, PhD(geol), 65. Prof Exp: Fel, 65-67, asst res officer, 67-71, ASSOC RES OFFICER GEOL, SASK RES COUN, 71- Mem: AAAS; Am Soc Photogram; Glaciol Soc; Geol Asn Can; Geol Soc Am. Res: Groundwater geology; techniques of groundwater exploration; development and pollution; field geology; mapping; subsurface exploration and stratigraphy; lignite exploration. Mailing Add: Geol Div Sask Res Coun Saskatoon SK Can

WHITAKER, THOMAS WALLACE, b Monrovia, Calif, Aug 13, 05; m 31; c 2. GENETICS. Educ: Univ Calif, BS, 27; Univ Va, MS, 29, PhD(genetics, cytol), 31. Prof Exp: Asst bot, Bussey Inst & Arnold Arboretum fel, Harvard Univ, 31-34; assoc prof bot, Agnes Scott Col, 34-36; from assoc geneticist to geneticist, Bur Plant Indust, 36-52, sr geneticist, Hort Crops Res Br, 52-56, prin geneticist & invest leader, Crops Res Div, 56-61, res geneticist & invest leader, 61-73, COLLABR, CROPS RES DIV, AGR RES SERV, USDA, 73-; RES ASSOC MARINE BIOL, SCRIPPS INST OCEANOG, UNIV CALIF, SAN DIEGO, 64- Concurrent Pos: Guggenheim Mem Found fel, Wash Univ, 46-47 & Univ Calif, Davis, 59; consult, Peabody Archaeol Proj, Tehuacan, Mex, 62; lectr, Tulane Univ, 67 & Univ Ariz, 71; ed, Hort Sci, Am Soc Hort Sci, 69, Jour, 69-72; mem comt genetic vulnerability of major crops, Nat Res Coun, 71; adj prof biol, Univ Calif, San Diego, 73- Mem: Fel Am Soc Hort Sci (pres, 74); hon life mem Can Soc Hort Sci; fel Torrey Bot Club; Soc Econ Botanists (pres, 69); Am Plant Life Soc (exec secy, 75-). Res: Plant breeding; vegetable crops; genetics and cytology of Lactuca and the Cucurbitaceae; origin and domestication of cucurbits as related to cultural history; development of lettuce cultivars resistant to cabbage looper. Mailing Add: PO Box 150 La Jolla CA 92038

WHITAKER, WILLIAM ARMSTRONG, b Little Rock, Ark, Jan 10, 36; m 57; c 2. COMPUTER SCIENCE, NATURAL SCIENCE. Educ: Tulane Univ, BS, 55, MS, 56; Univ Chicago, PhD(physics), 63. Prof Exp: Asst physics, Tulane Univ, 54-56; US Air Force, 56-, prof off res & develop, 57-68, chief high altitude group, Weapons Lab, 68-70, chief scientist, 70-72, mil asst res, Off Dir Defense Res & Eng, 73-75, SPEC ASST TO DIR DEFENSE ADVAN RES PROJ AGENCY, US AIR FORCE, 75- Mem: Am Phys Soc; Am Astron Soc; Am Geophys Union; Am Meteorol Soc; Inst Elec & Electronics Engrs. Res: Direction of advanced research, defense software management, common high order computer programming languages. Mailing Add: 3813 Col Ellis Ave Alexandria VA 22304

WHITBY, OWEN, b Luton, Eng, Feb 24, 42; Can & UK citizen; m 70. STATISTICS. Educ: McMaster Univ, BSc Hons, 64; Stanford Univ, MS, 66, PhD(statist), 72. Prof Exp: Asst prof, 71-75, ASSOC PROF STATIST, TEACHERS COL, COLUMBIA UNIV, 75- Mem: AAAS; Am Statist Asn; Asn Comput Mach; Biomet Soc; Inst Math Statist. Res: Biostatistics; mathematical statistics; teaching. Mailing Add: Prog in Statist Teachers Col Columbia Univ New York NY 10027

WHITCHER, WENDELL JENNISON, b South Newbury, Vt, July 7, 09; m 39, 59; c 4. CHEMISTRY. Educ: Dartmouth Col, AB, 31, AM, 33; Harvard Univ, PhD(chem), 40. Prof Exp: Instr chem, Dartmouth Col, 31-34; chemist, E I du Pont de Nemours & Co, 40-49; assoc prof chem, Grove City Col, 49-52; asst prof, 52-56, ASSOC PROF CHEM, UNIV VT, 56- Mem: Am Chem Soc. Res: Process development in anthraquinone and azo dyestuffs; fluorine chemicals. Mailing Add: Dept of Chem Univ of Vt Burlington VT 05401

WHITCOMB, CARL ERWIN, b Independence, Kans, Oct 26, 39; m 63; c 2. HORTICULTURE, PLANT ECOLOGY. Educ: Kans State Univ, BSA, 64; Iowa State Univ, MS, 66, PhD(hort, plant ecol), 69. Prof Exp: Asst prof ornamental hort, Univ Fla, 67-72; ASSOC PROF HORT, OKLA STATE UNIV, 72- Mem: Am Soc Hort Sci; Ecol Soc Am; Am Soc Agron; Soc Am Foresters. Res: Plant interactions in man-made or man-managed landscapes; production, establishment and maintenance of landscape plants. Mailing Add: Dept of Hort Okla State Univ Stillwater OK 74074

WHITCOMB, DONALD LEROY, b Ilion, NY, Feb 1, 25; m 48; c 4. ANALYTICAL CHEMISTRY. Educ: Univ Rochester, BS, 52, PhD(phys chem), 56. Prof Exp: Res chemist, Res Labs, Eastman Kodak Co, NY, 46-53; mem tech staff, Bell Tel Labs, Pa, 56-63; mem tech staff, Microwave Tube Div, Hughes Aircraft Co, 63-64 & Space Systs Div, 64-66; chemist, Motorola Semiconductor Prod Inc, 66-70; CHIEF CHEMIST, ARIZ STATE DEPT HEALTH, 70- Mem: Am Chem Soc; Am Phys Soc. Mailing Add: 4414 N Dromedary Rd Phoenix AZ 85018

WHITCOMB, GORDON PUTNAM, b Spencer, Mass, Feb 26, 13; m 44. ORGANIC CHEMISTRY. Educ: Worcester Polytech Inst, BS, 34, MS, 36; Yale Univ, PhD(org chem), 39. Prof Exp: Res chemist, Am Cyanamid Co, 39-42, priorities specialist, 42-45, tech personnel, 45-49, col rels rep, 49-51, asst personnel dir, 51-56, admin mgr, Eng Construct Div, 56-60, coordr col rels, 60-63, mgr col rels, 63-75; RETIRED. Mem: Am Chem Soc. Res: Pharmaceuticals; dyes; textile chemicals. Mailing Add: Wooded Rd Watchung NJ 07060

WHITCOMB, STUART ESTES, b Brooklyn, NY, Jan 3, 11; m 40; c 1. PHYSICS. Educ: Antioch Col, BS, 34; Syracuse Univ, MS, 36; Ohio State Univ, PhD(physics), 39. Prof Exp: Asst physics, Syracuse Univ, 34-36 & Ohio State Univ, 36-39; instr, Ga Sch Technol, 39-42; from asst prof to prof, Kans State Col, 42-57; physicist, Sandia Corp, 57-64; PROF PHYSICS, EARLHAM COL, 64- Concurrent Pos: Ballistician, Aberdeen Proving Ground, 44-45; head dept, Kans State Col, 53-57; mem physics comt, Col Entrace Exam Bd, 65-71; consult, Phys Sci for Non-Sci Students Proj, 66- Mem: AAAS; Am Phys Soc; Am Asn Physics Teachers. Res: Infrared spectroscopy; dynamics of impacts. Mailing Add: Dept of Physics Earlham Col Richmond IN 47374

WHITCOMB, WALTER HENRY, b Enid, Okla, Jan 26, 28; m 46; c 3. INTERNAL MEDICINE, NUCLEAR MEDICINE. Educ: Univ Okla, BA, 50, MD, 53; Am Bd Nuclear Med, dipl, 72. Prof Exp: Clin asst, Sch Med & chief res med, Med Ctr, Univ Okla, 56-57, chief exp med group, Radiobiol Lab, Univ Tex-US Air Force, 58-60; investr & instr radiobiol, Bionucleonics Dept, US Air Force Sch Aerospace Med, 60-62; from asst prof to assoc prof, 62-70, PROF MED, SCH MED, UNIV OKLA, 70-, ASSOC PROF RADIOL, 69-, ASST DEAN VET AFFAIRS, 70-; CHIEF STAFF, VET ADMIN HOSP, 70- Concurrent Pos: Res fel hemat, Univ Okla, 57-58; clin investr, Southwest Cancer Chemother Study Group, Vet Admin Hosp, 62-66, chief radioisotopes serv & hemat sect, 62-70, assoc chief staff, 63-67; asst prof radiol, Sch Med, Univ Okla, 64-69; mem eval comt res hemat, Vet Admin, 65- Mem: Am Col Physicians; Am Fedn Clin Res; Soc Nuclear Med; Am Soc Hemat; Am Physiol Soc. Res: Biological effects of radiation; control of erythropoiesis; physiology of erythropoietin and erythropoietin inhibitor factors. Mailing Add: Off of the Chief of Staff Vet Admin Hosp Oklahoma City OK 73104

WHITCOMB, WILLARD HALL, b Manchester, NH, July 2, 15; m 43. ENTOMOLOGY. Educ: Bates Col, BS, 38; Agr & Mech Col, Tex, MS, 42; Cornell Univ, PhD(entom), 47. Prof Exp: Entomologist, Ministry Agr, Venezuela, 47-52 & Shell Co, Venezuela, 52-56; prof entom, Univ Ark, 56-57; prof entom, Big Bend Hort Lab, Univ Fla, 67-69, PROF ENTOM, UNIV FLA, 69- Mem: Entom Soc Am; Entom Soc Can; Int Orgn Biol Control. Res: Biological control of arthropods; ecology and population dynamics; pest management; tropical entomology. Mailing Add: Dept of Entom Univ of Fla Gainesville FL 32601

WHITE, ABRAHAM, b Cleveland, Ohio, Mar 8, 08; m 37. BIOCHEMISTRY. Educ: Univ Denver, AB, 27, MA, 28; Univ Mich, PhD(physiol chem), 31. Hon Degrees: LHD, Yeshiva Univ, 59; DSc, Univ Denver, 75. Prof Exp: Asst chem, Univ Denver, 27-28; asst physiol chem, Univ Mich, 28-30; Sterling fel, Yale Univ, 31-32, Am Physiol Soc Porter fel, 32-33, instr sch med, 33-37, from asst prof to assoc prof, 37-48, exec officer dept, 47-48; prof & chmn dept, Sch Med, Univ Calif, Los Angeles, 48-51; vpres & dir res, Chem Specialties Co, Inc, NY, 51-53; prof biochem, chmn dept & assoc dean, Albert Einstein Col Med, Yeshiva Univ, 53-72; sr scientist, 72-73, DISTINGUISHED SCIENTIST, SYNTEX RES, 73- Concurrent Pos: Am Physiol Soc fel, Int Physiol Cong, USSR, 35; consult, US Vet Admin, Calif, 49-51; vis prof biochem, Med Sch, Stanford Univ, 72-73, consult prof biochem, 73-; vis prof biochem, Albert Einstein Col Med, Yeshiva Univ, 72- Honors & Awards: Lilly Prize, Am Chem Soc, 38. Mem: Nat Acad Sci; Am Chem Soc; Soc Exp Biol & Med; fel NY Acad Sci; Am Soc Cell Biol. Res: Chemistry and metabolism of amino acids, functions of lymphoid tissue; biochemistry of hormones. Mailing Add: 580 Arastradero Rd Apt 507 Palo Alto CA 94306

WHITE, ALAN GEORGE CASTLE, b Boston, Eng, Aug 12, 16; nat US; m 45; c 2. BACTERIOLOGY. Educ: RI State Col, BS, 38; Pa State Col, MS, 40; Iowa State Col, PhD(bact), 47. Prof Exp: Technician, State Dept Health, RI, 38; res bacteriologist, J E Seagram & Sons, Inc, Ky, 39-40; tech asst, Rockefeller Inst, 40-43; instr bact, Iowa State Univ, 46-47; from asst prof to prof biochem, Sch Med, Tulane Univ, 47-67; PROF BIOL & HEAD DEPT, VA MIL INST, 67- Mem: Am Soc Biol Chemists; Am Soc Microbiol; Soc Exp Biol & Med; Sigma Xi. Res: Microbial fermentations; tracer studies in metabolism; microbial metabolism of arginine. Mailing Add: Dept of Biol Va Mil Inst Lexington VA 24450

WHITE, ALAN JONATHON, b South Norwalk, Conn, Apr 21, 48; m 74. INORGANIC CHEMISTRY, ORGANOMETALLIC CHEMISTRY. Educ: Univ Vt, BSc, 70; Mass Inst Technol, PhD(chem), 74. Prof Exp: Lectr chem, Tex A&M Univ, 74-75; ASST PROF CHEM, AGNES SCOTT COL, 75- Mem: Am Chem Soc; The Chem Soc. Res: Synthesis and characterization of organometallic compounds of the transition elements; mechanisms of organometallic reactions. Mailing Add: Dept of Chem Agnes Scott Col Decatur GA 30030

WHITE, ALBERT CORNELIUS, b Clearwater, Fla, July 17, 27; m 49; c 3. ENTOMOLOGY. Educ: Clemson Col, BS, 51; Univ Wis, MS, 53. Prof Exp: Res entomologist, Ortho Div, 53-69, INT RES SPECIALIST, CHEVRON CHEM CO, 69- Mem: Entom Soc Am; Am Mosquito Control Asn; Weed Sci Soc Am. Res: Administration of area research programs for all pesticides. Mailing Add: Chevron Chem Int Inc 13-3 Roppongi 3 Chome Minato-ku Tokyo Japan

WHITE, ALBERT GEORGE, JR, b Centralia, Ill, July 16, 40; m 67; c 4. MATHEMATICS. Educ: Southern Ill Univ, Edwardsville, BA, 62; Univ Mo-Columbia, MA, 64; St Louis Univ, PhD(math), 68. Prof Exp: Asst prof math, Ill State Univ, 67-69; ASSOC PROF MATH, ST BONAVENTURE UNIV, 69-, CHMN DEPT, 70- Mem: Am Math Soc; Math Asn Am. Mailing Add: Dept of Math St Bonaventure Univ St Bonaventure NY 14778

WHITE, ALBERT M, b Derby, Conn, June 12, 26; m 55; c 6. CLINICAL PHARMACY. Educ: Univ Conn, BS, 48, MS, 52. Prof Exp: Asst chem & pharm, Univ Conn, 50-52; from instr to assoc prof, 56-72, PROF PHARM, ALBANY COL PHARM, 72-, ASST DEAN, 74-, DIR PHARM SERV TERESIAN HOUSE, 75- Concurrent Pos: Clin instr admin med, State Univ NY Upstate Med Ctr; assoc clin prof, Albany Vet Admin Hosp; consult, Whitney M Young Health Ctr & Villa Mary Immaculate Nursing Home; state dir, Am Bd Dipl in Pharm. Mem: Am Pharmaceut Asn; Am Soc Hosp Pharmacists; Am Soc Consult Pharmacists; Am Asn Cols Pharm. Res: Delivery of clinical pharmaceutical services to institutionalized and health center patients; practice of clinical pharmacy by community pharmacists; evaluation of practice experience programs. Mailing Add: Albany Col of Pharm Albany NY 12208

WHITE, ALLEN INGOLF, b Silverton, Ore, July 10, 14; m 38; c 3. PHARMACEUTICAL CHEMISTRY. Educ: Univ Minn, BS,37, MS, 38, PhD(pharmaceut chem), 40. Prof Exp: With bur plant indust, USDA, 38-40; from instr to assoc prof, 40-48, PROF PHARM, WASH STATE UNIV, 48-, DEAN COL PHARM, 60- Concurrent Pos: Dir, Am Found Pharmaceut Educ, 75- Mem: AAAS; Am Chem Soc; Am Pharmaceut Asn; Am Asn Cols Pharm (pres, 74-75); Sigma Xi. Res: Organic synthesis of medicinal products. Mailing Add: Col of Pharm Wash State Univ Pullman WA 99163

WHITE, ALVIN MURRAY, b New York, NY, June 21, 25; m 46; c 2. MATHEMATICS, SCIENCE EDUCATION. Educ: Columbia Univ, AB, 49; Univ Calif, Los Angeles, MA, 51; Stanford Univ, PhD, 61. Prof Exp: Asst prof math, Univ Santa Clara, 54-61; mem math res ctr, US Army, Wis, 61-62; ASSOC PROF MATH, HARVEY MUDD COL, 62- Concurrent Pos: Fac fel, Danforth Found, 75-76; vis scientist, Div Study & Res in Educ, Mass Inst Technol, 76. Mem: Am Math Soc; Math Asn Am; Fedn Am Scientists; AAAS; Am Asn Univ Profs. Res: Function theoretical aspects of partial differential equations; quasiconformal mapping; nature of scientific creativity; nurture of scientific creativity. Mailing Add: Dept of Math Harvey Mudd Col Claremont CA 91711

WHITE, ANDREW WILSON, b Thomaston, Ga, Aug 1, 27; m 50; c 2. SOIL CONSERVATION, SOIL FERTILITY. Educ: Univ Ga, BS, 49, MS, 58, PhD, 69. Prof Exp: Soil scientist, Soil & Water Conserv Res Div, 51-61, res soil scientist, Southern Piedmont Conserv Res Ctr, 61-75, SOIL SCIENTIST, SOUTHEASTERN FRUIT & NUT TREE RES STA, AGR RES SERV, USDA, 75- Mem: Am Soc Agron; Soil Sci Soc Am; Soil Conserv Soc Am. Res: Soil management problems related to pesticides and soil and water conservation; soil fertility and management problems related to pecan production. Mailing Add: Southeast Fruit & Nut Tree Res Sta Agr Res Serv USDA Byron GA 31008

WHITE, ANTA M, b Strasbourg, France, Feb 2, 33; m 57; c 2. ANTHROPOLOGY. Educ: Univ Paris, Lic es Litt, 54, dipl ES, 55; Univ Bordeaux, Dr es Litt(pre-hist), 65. Prof Exp: Res asst prehist, Institut de Paleontologie Humaine, 55-57; lectr anthrop & res assoc, Mus Anthrop, Univ Mich, 62-65; lectr, 66-67, asst prof, 67-71, ASSOC PROF ANTHROP, UNIV KANS, 71- Concurrent Pos: NSF res grant, 67-69. Res: Old World prehistory, especially Europe-Mediterranean; primitive technology. Mailing Add: Dept of Anthrop Univ of Kans Lawrence KS 66044

WHITE, ARNOLD ALLEN, b New York, NY, Oct 13, 23; m 53; c 5. BIOCHEMISTRY. Educ: Univ Iowa, AB, 47, MS, 49; Georgetown Univ, PhD(biochem), 54. Prof Exp: From instr to asst prof biochem, Georgetown Univ, 52-56; asst prof, 56-60, ASSOC PROF BIOCHEM, UNIV MO-COLUMBIA, 60-, INVESTR, DALTON SCI RES CTR, 66- Mem: AAAS. Res: Mechanism of hormone action; cyclic nucleotide research. Mailing Add: Dalton Res Ctr Univ of Mo Columbia MO 65201

WHITE, ARTHUR C, b Williamsburg, Ky, Aug 1, 25; m 49; c 3. INTERNAL MEDICINE, INFECTIOUS DISEASES. Educ: Univ Ky, BS; Harvard Univ, MD, 52. Prof Exp: Instr med, Vanderbilt Univ, 53-58; from instr to asst prof, Univ Louisville, 58-63; assoc prof, Med Col Ga, 63-67; PROF MED, SCH MED, IND UNIV, INDIANAPOLIS, 67- Concurrent Pos: Consult, Vet Admin Hosps, Louisville, Ky, 59-63 & Augusta, Ga, 63-67; drug efficacy study, Nat Acad Sci, 66. Mem: Am Soc Microbiol; Am Fedn Clin Res; Am Col Physicians; Infectious Dis Soc Am. Res: Staphylococcal epidemiology and immunology; immunoglobulins and their activity; immunology of gram negative infections; histamine release. Mailing Add: Infectious Dis Div Ind Univ Sch of Med Indianapolis IN 46207

WHITE, ARTHUR THOMAS II, b Orange, NJ, Oct 7, 39; m 61. MATHEMATICS. Educ: Oberlin Col, AB, 61; Mich State Univ, MS, 66, PhD(math), 69. Prof Exp: Actuarial trainee, Home Life Ins Co, 61-62; asst prof, 69-73, ASSOC PROF MATH, WESTERN MICH UNIV, 73- Concurrent Pos: Asst, Mich State Univ, 65-69; NSF grant, 73-74. Mem: Am Math Soc; Math Asn Am; Sigma Xi. Res: Topological graph theory. Mailing Add: 2502 Law Ave Kalamazoo MI 49008

WHITE, AUGUSTUS AARON, III, b Memphis, Tenn, June 4, 36; m 74; c 1. ORTHOPEDIC SURGERY, BIOMEDICAL ENGINEERING. Educ: Brown Univ, BA, 57; Stanford Univ, MD, 61; Karolinska Inst, Sweden, DrMedSc, 69. Prof Exp: Intern, Univ Hosp, Ann Arbor, Mich, 61-62; from instr to asst prof, 65-72, ASSOC PROF ORTHOP SURG, SCH MED, YALE UNIV, 72-, DIR BIOMECH RES, SECT ORTHOP SURG, 73- Concurrent Pos: Chief resident orthop surg, Vet Admin Hosp, West Haven, Conn, 66; attend orthop surgeon, Yale-New Haven Hosp, 69-; consult, Vet Admin Hosp, West Haven & Hill Health Ctr, New Haven, 69-; mem bioeng res comt, Int Coun Sports & Phys Educ, 74; vpres health affairs, Soulville Found, 76. Honors & Awards: Outstanding Orthop Res Award, Am Acad Orthop Surgeons, 76. Mem: Orthop Res Soc; Cervical Spine Res Soc; Int Soc Study Lumbar Spine; Am Acad Orthop Surgeons; Nat Med Asn. Res: Mechanical studies on the entire human spine designed to provide knowledge and technology applicable to clinical problems; development of an engineering system which will accelerate fracture healing. Mailing Add: Yale Univ Sch of Med SLOG 217 333 Cedar St New Haven CT 06510

WHITE, BEN ELWOOD, chemistry, see 12th edition

WHITE, BERNARD J, b Portland, Ore, Jan 8, 37; m 63; c 5. BIOCHEMISTRY. Educ: Univ Portland, BS, 58; Univ Ore, MA, 61, PhD(biochem), 63. Prof Exp: Asst prof chem, Loras Col, 63-68; asst prof, 68-74, ASSOC PROF BIOCHEM, IOWA STATE UNIV, 74- Mem: AAAS; Am Chem Soc. Res: Protein structure; biochemical evolution; enzymology. Mailing Add: Dept of Biochem & Biophys Iowa State Univ Ames IA 50010

WHITE, BLANCHE BABETTE, b Cumberland, Md, July 25, 05. CHEMISTRY. Educ: Goucher Col, AB, 25; Univ Chicago, MS, 27. Prof Exp: Asst chem, Goucher Col, 25-26; chemist, Celanese Corp Am, 27-45, sect head cellulose derivatives, 46-56; head tech info, Charlotte Develop Labs, NC, 56-59; asst librn, Res Div, W R Grace & Co, 59-63, lit scientist, 63-70; CHEM LIT CONSULT, 70- Mem: Am Chem Soc. Res: Scientific information retrieval; cellulose chemistry. Mailing Add: 8201 16th St Silver Spring MD 20910

WHITE, BOOKER TALIAFERRO, b Tryon, NC, Sept 2, 07; m 43. BIOCHEMISTRY. Educ: WVa State Col, BS, 29; Ohio State Univ, MS, 37, PhD(chem), 45. Prof Exp: Instr chem, Kiltrell Col, 30-32; prin high sch, 32-36; instr jr col, SC, 37-38, Halifax County Training Sch, 38-40, Morristown Jr Col, 40-41 & Ala Agr & Mech Col, 45-47; prof chem & chmn dept, Agr & Tech Col, NC, 47-67, dir res, 53-67; chmn dept natural sci, Ala A&M Univ, 67-70, prof chem, 67-75, actg chmn dept chem, 70-75. Mem: Am Chem Soc; Nat Inst Sci. Res: Nutrition of albino rats; bitter principle of boneset, Eupatorium perfoliatum; colorimetric estimation of lactotenulin; contaminated milk. Mailing Add: 710 Mobile St Greensboro NC 27406

WHITE, BRIAN, b Brigg, Eng, Feb 19, 36; m 62; c 2. GEOLOGY. Educ: Univ Wales, BSc, 63, PhD(geol), 66. Prof Exp: Fel, Dalhousie Univ, 66-68; from instr to asst prof, 68-73, ASSOC PROF GEOL & CHMN DEPT, SMITH COL, 73- Mem: Norweg Geol Soc; Soc Econ Paleontologists & Mineralogists; Geol Soc Am. Res: Stratigraphy, sedimentary petrology and micropaleobotany of precambrian sedimentary rocks; impact of urban development on the hydrogeology of small river basins. Mailing Add: Dept of Geol Smith Col Northampton MA 01060

WHITE, BRIGGS JOHNSTON, b Bristol, Colo, Jan 5, 11; m 41; c 2. CHEMISTRY. Educ: Sterling Col, BA, 32; Univ Colo, MA, 37, PhD(chem), 40. Hon Degrees: DCL, Sterling Col, 65. Prof Exp: Asst chem, Univ Colo, 36-40; jr chemist, Lab, Fed Bur Invest, 40-41, asst chemist, 41-42, anal chemist, 42-43, prin anal chemist, 43-44, res chemist, 44-53, adv chemist, 53-57, chief physics & chem sect, 57-60, asst chief lab, 61-73, asst dir in charge lab div, 73-75; RETIRED. Concurrent Pos: Chmn gov bd, Am Soc Crime Lab Dirs, 74-75. Mem: Am Chem Soc; Am Acad Forensic Sci (pres, 74-75); Am Soc Crime Lab Dirs. Res: Serology of dried blood factors; toxicology x-ray spectrometry; forensic science administration and management. Mailing Add: 2601 Buckney Lane Washington DC 20031

WHITE, BRUCE LANGTON, b Wellington, NZ, Mar 2, 31; m 54; c 2. PHYSICS. Educ: New Zealand, BSc, 52; Univ London, DIC & PhD, 56. Prof Exp: Res fel, 56-59, res assoc, 59-60, from asst prof to assoc prof, 60-70, PROF PHYSICS, UNIV BC, 70- Mem: Am Phys Soc. Res: Experimental low energy nuclear physics; experimental cosmology and gravitation; Mössbauer effect. Mailing Add: Dept of Physics Univ of BC Vancouver BC Can

WHITE, CARLOS A, entomology, see 12th edition

WHITE, CHARLES A, JR, b San Diego, Calif, Aug 1, 22; m 60; c 3. OBSTETRICS & GYNECOLOGY. Educ: Colo Agr & Mech Col, DVM, 45; Univ Utah, MD, 55; Am Bd Obstet & Gynec, dipl, 64. Prof Exp: Pvt pract vet med, 45-51; intern, Salt Lake County Gen Hosp, 55-56; resident obstet & gynec, Dee Mem Hosp, Ogden, Utah, 56-57; resident, Univ Hosp, Univ Iowa, 59-61, assoc, Col Med, 61-62, from asst prof to prof, 62-74; PROF OBSTET & GYNEC & CHMN DEPT, WVA UNIV, 74- Concurrent Pos: Examr, Am Bd Obstet & Gynec, 71- Mem: AMA; Am Col Obstet & Gynec; NY Acad Sci. Mailing Add: Dept of Obstet & Gynec WVa Univ Morgantown WV 26506

WHITE, CHARLES HENRY, b Birmingham, Ala, Mar 15, 43; m 65; c 1. DAIRY MICROBIOLOGY. Educ: Miss State Univ, BS, 65, MS, 69; Univ Mo, PhD(dairy microbiol), 71. Prof Exp: Sr food scientist, Archer Daniels Midland Co, 71-72; ASST PROF DAIRY MICROBIOL, UNIV GA, 72- Concurrent Pos: Consult, Long Life Dairy Prod, 73-; partic, Comt to Revise Stand Methods for Exam Dairy Prod, 74- Mem: Am Dairy Sci Asn; Inst Food Technol; Int Asn Milk & Food Sanitarians; Nat Environ Health Asn; Cult Dairy Prod Inst. Res: Psychrotrophic bacteria and relationship with shelf-life of dairy products, including measurement of proteolytic activity of raw milk as well as determination of heat-stable protease from the psychrotrophs; diacetyl reductases. Mailing Add: Dairy Sci Bldg Univ of Ga Athens GA 30602

WHITE, CHARLEY MONROE, b Rose Hill, Ill, Dec 9, 32; m 60; c 3. ECOLOGY, WILDLIFE BIOLOGY. Educ: Eastern Ill Univ, BS, 60; Purdue Univ, MS, 62, PhD(ecol), 68. Prof Exp: Asst prof, 66-71, ASSOC PROF BIOL, UNIV WIS-STEVENS POINT, 71- Mem: Wildlife Soc; Ecol Soc Am; Am Soc Mammal. Res: Productivity of White-tailed deer; population dynamics. Mailing Add: Dept of Biol Univ of Wis Stevens Point WI 54481

WHITE, CHRISTOPHER CLARKE, b Medford, Mass, June 24, 37. MATHEMATICS. Educ: Bowdoin Col, AB, 59; Miami Univ, MA, 63; Univ Ore, PhD(math), 67. Prof Exp: Asst prof math, Univ NH, 67-70; ASSOC PROF MATH, CASTLETON STATE COL, 70-, CHMN DEPT, 74- Res: Banach algebras; harmonic analysis; calculus of variations. Mailing Add: Dept of Math Castleton State Col Castleton VT 05735

WHITE, CLARK WOODY, physics, see 12th edition

WHITE, CLAYTON M, b Afton, Wyo, Apr 19, 36; m 59; c 5. VERTEBRATE ZOOLOGY, ECOLOGY. Educ: Univ Utah, AB, 61, PhD(zool), 68. Prof Exp: Instr zool & cur birds, Univ Kans, 65-66; instr zool & res fel, Cornell Univ, 68-70; asst prof, 70-74, ASSOC PROF ZOOL, BRIGHAM YOUNG UNIV, 74- Concurrent Pos: Consult, Columbus Labs, Battelle Mem Inst, 72- Honors & Awards: Francis B Roberts Award, 68. Mem: AAAS; Am Ornith Union; Soc Syst Zool; Cooper Ornith Union; Wilson Ornith Soc. Res: Avian evolution and systematics; ecology of raptorial birds; impact of environmental pollution in avian populations. Mailing Add: Dept of Zool Brigham Young Univ Provo UT 84601

WHITE, CLAYTON SAMUEL, b Ft Collins, Colo, Oct 11, 12; m 41; c 3. MEDICINE, PHYSIOLOGY. Educ: Univ Colo, AB, 34, MD, 42; Oxford Univ, BA, 38; Am Bd Prev Med, dipl, 53. Hon Degrees: DSc, Univ NMex, 74. Prof Exp: Head sect aviation med & internal med, Lovelace Clin, 47-50, dir res, Lovelace Found Med Educ & Res, 50-74, pres-dir, 65-74; PRES, OKLA MED RES FOUND, 74- Concurrent Pos: Mem aeromed & biosci panel, Sci Adv Bd, US Air Force, 55-60, chmn, 58-60, consult, Air Force Weapons Lab, 58 & Defense Nuclear Agency, 65-74; mem ad hoc subcomt initial effects & struct to protect against them, Adv Bd Civil Defense, Nat Acad Sci, 66-, mem comt space biol & med, Space Sci Bd, 69- Honors & Awards: US Air Force Except Serv Award, 60; Tuttle Award, Aerospace Med Asn, 62; Spec Aerospace Med Honor Citation, AMA, 62; NMex Distinguished Pub Serv Award, 73. Mem: Fel AAAS; fel Aerospace Med Asn; AMA; fel Am Col Prev Med. Res: Aerospace medicine and physiology; blast biology; biodynamics; environmental medicine; industrial health and safety. Mailing Add: Okla Med Res Found 825 NE 13th St Oklahoma City OK 73104

WHITE, COLIN, b Australia, Aug 25, 13; m 43; c 2. BIOMETRY. Educ: Univ Sydney, BSc, 35, MSc, 36, MB, BS, 40. Prof Exp: Intern, Sydney Hosp, Australia, 41; lectr physiol, Univ Sydney, 42; med officer, Australian Inst Anat, 43-46; lectr physiol, Univ Birmingham, 46-48; asst prof, Univ Pa, 48-50; lectr physiol, Univ Birmingham, 50-53; from asst prof to assoc prof, 53-62, PROF BIOMET, SCH MED, YALE UNIV, 62- Mem: Am Statist Asn; Biomet Soc; Royal Statist Soc; Int Epidemiol Soc. Res: Eipidemiology of chronic diseases; vital statistics. Mailing Add: 107 Thornton St Hamden CT 06517

WHITE, DAVID, b Russia, Jan 14, 25; nat US; m 45; c 3. PHYSICAL CHEMISTRY. Educ: McGill Univ, BSc, 44; Univ Toronto, PhD(chem), 47. Prof Exp: Asst, Univ Toronto, 44-47; fel, Ohio State Univ, 48-50, asst dir cryogenic lab, 50-53; asst prof chem, Syracuse Univ, 53-54; from asst prof to prof, Ohio State Univ, 54-66; PROF CHEM & CHMN DEPT, UNIV PA, 66- Concurrent Pos: Vis prof, Technion & Weizmann Insts, Israel, 63-64; Fulbright fel & vis prof, Univ Kyoto & Univ Tokyo, Japan, 65; Nat Ctr Sci Res fel & vis prof, Inst Appl Quantum Mech, France, 74-75. Mem: Am Chem Soc; Am Phys Soc. Res: Low temperature thermodynamics and solid state nuclear magnetic resonance; molecular structure of inorganic vapor species; infrared and Raman spectroscopy. Mailing Add: Dept of Chem Univ of Pa Philadelphia PA 19104

WHITE, DAVID, b Boston, Mass, Apr 26, 36; m 59; c 3. MICROBIOLOGY. Educ: Brandeis Univ, AB, 58, PhD(biol), 65. Prof Exp: Res scientist, Exobiol Div, Ames Res Ctr, NASA, 63-65; res assoc microbial physiol, Med Sch, Univ Minn, 65-67; asst prof, 67-74, ASSOC PROF MICROBIOL, IND UNIV, BLOOMINGTON, 74- Concurrent Pos: Res grants, Am Cancer Soc, 68-70, NSF, 68-70, 70-72. Mem: Am Soc Microbiol. Res: Microbial physiology; microbial development; myxobacteria. Mailing Add: Dept of Microbiol Ind Univ Bloomington IN 47401

WHITE, DAVID ARCHER, b Philadelphia, Pa, Jan 22, 27; m 52; c 3. GEOLOGY. Educ: Dartmouth Col, BA, 50; Univ Minn, MS, 51, PhD(geol), 54. Prof Exp: RES ADV, EXXON PROD RES CO, 54- Mem: Geol Soc Am; Am Asn Petrol Geol. Res: Geology of the Mesabi range, Minnesota; geochemistry; stratigraphy; hydrocarbon assessment. Mailing Add: Box 2189 Houston TX 77001

WHITE, DAVID ARNOLD, b Salt Lake City, Utah, Jan 20, 36; m 60; c 2. ZOOLOGY, AQUATIC ECOLOGY. Educ: Brigham Young Univ, BS, 61, MS, 64; Univ Wis, PhD(zool), 67. Prof Exp: Wildlife aid migratory waterfowl res, Bear River Refuge, US Fish & Wildlife Serv, 61-62; asst prof biol, Wis State Univ, Stevens Point, 65-66; asst prof zool, Brigham Young Univ, 67-75; MEM FAC, DEPT BIOL SCI, WESTERN ILL UNIV, 75- Mem: Am Inst Fish Res Biologists; Int Soc Limnol; Freshwater Biol Soc UK; Am Fisheries Soc; Sigma Xi. Res: Biology; aquatic invertebrates of Mississippi River; energy flow in aquatic systems. Mailing Add: Dept of Biol Sci Western Ill Univ Macomb IL 61445

WHITE, DAVID C, b Davenport, NY, Sept 6, 24; m 46; c 4. PATHOLOGY, RADIOBIOLOGY. Educ: Duke Univ, MD, 47. Prof Exp: MedC, US Army, 48-67, chief radiation path, 60-67, REGISTR, REGISTRY RADIATION PATH, ARMED FORCES INST PATH, 67- Res: Effects of ionizing irradiation upon biological systems with special reference to sequelae of therapeutic radiation. Mailing Add: Registry of Radiation Path Armed Forces Inst of Path Washington DC 20305

WHITE, DAVID CLEAVELAND, b Moline, Ill, May 18, 29; m 56; c 3. BIOCHEMISTRY. Educ: Dartmouth Col, AB, 51; Tufts Univ, MD, 55; Rockefeller Univ, PhD(biochem), 62. Prof Exp: Intern, Univ Hosp, Univ Pa, 55-56, instr physiol, 56-58, res assoc med, 58; from asst prof to prof biochem, Univ Ky, 62-72; PROF BIOL & ASSOC DIR, PROG MED SCI, FLA STATE UNIV, 72- Mem: AAAS; Am Chem Soc; Am Soc Limnol & Oceanog; Am Diabetic Asn; Gulf Estuarine Res Soc. Res: Microbial ecology of estuaries. Mailing Add: Prog in Med Sci Dept Biol Sci Fla State Univ Tallahassee FL 32306

WHITE, DAVID EVANS, b Syracuse, NY, Dec 13, 32; m 52; c 4. FOREST ECONOMICS. Educ: State Univ NY Col Forestry, Syracuse Univ, BS, 59, MS, 60, PhD(econ), 65. Prof Exp: Forester, Crown-Zellerbach Corp, 60-61; instr forest econ, State Univ NY Col Forestry, Syracuse Univ, 61-64; from asst prof to assoc prof, 64-71, PROF FOREST ECON, W VA UNIV, 71-, DIR DIV FORESTRY, 66- Mem: Soc Am Foresters. Res: Forest resources policy and administration; multi-disciplinary studies in environmental decision-making; natural resources economics; land use planning. Mailing Add: Div of Forestry WVa Univ Morgantown WV 26506

WHITE, DAVID GOVER, b Woodbury, NJ, Sept 21, 27; m 59. INORGANIC CHEMISTRY. Educ: Cornell Univ, BChE, 50; Harvard Univ, PhD(chem), 54. Prof Exp: From asst prof to assoc prof, 53-62, PROF CHEM, GEORGE WASHINGTON UNIV, 62- Concurrent Pos: NSF fel, 60. Mem: AAAS; Am Chem Soc; The Chem Soc. Res: Organometallic chemistry; boron-nitrogen compounds; metal complexes. Mailing Add: Dept of Chem George Washington Univ Washington DC 20006

WHITE, DAVID HALBERT, b Midland, Mich, Jan 25, 45. PHYSICAL ORGANIC CHEMISTRY, EXOBIOLOGY. Educ: Mich State Univ, BS, 67; Calif Inst Technol, PhD(chem), 72. Prof Exp: Asst prof chem, Lawrence Univ, 73-75; ASST PROF CHEM, UNIV SANTA CLARA, 75- Mem: AAAS; Am Chem Soc; Fedn Am Scientists. Res: Homoaromaticity; carbene and diradical intermediates; origin of the genetic code. Mailing Add: Dept of Chem Univ of Santa Clara Santa Clara CA 95053

WHITE, DAVID HYWEL, b Cardiff, Wales, June 4, 31; m 54; c 2. EXPERIMENTAL HIGH ENERGY PHYSICS. Educ: Univ Wales, BSc, 53; Univ Birmingham, PhD(physics), 56. Prof Exp: Res fel physics, Univ Birmingham, 56-58, asst lectr, 58-59; res assoc, Univ Pa, 59-62, asst prof, 62-65; assoc prof, 65-69, PROF PHYSICS, CORNELL UNIV, 69- Mem: Am Phys Soc; AAAS. Res: Experimental high energy physics; weak interactions; strong interactions; electromagnetic interactions. Mailing Add: Dept of Physics Cornell Univ Ithaca NY 14850

WHITE, DAVID LEON, organic chemistry, see 12th edition

WHITE, DAVID RAYMOND, b Oak Park, Ill, Sept 20, 40; m 64; c 1. ORGANIC CHEMISTRY. Educ: St John's Univ, BA, 62; Univ Wis, PhD(org chem), 66. Prof Exp: Grant, NIH, 66-68; MEM STAFF, UPJOHN CO, 68- Mem: Am Chem Soc; The Chem Soc. Res: Biogenetic type synthesis; steroid reactions; new synthetic methods; prostaglandin synthesis. Mailing Add: 1815 Greenlawn Kalamazoo MI 49007

WHITE, DEAN KINCAID, b Tulsa, Okla, Aug 27, 44; m 67; c 1. DENTAL PATHOLOGY. Educ: Univ Okla, BS, 66; Univ Mo-Kansas City, DDS, 70; Ind Univ, MSD, 72; Am Bd Oral Path, dipl, 75. Prof Exp: ASST PROF PATH, SCH DENT, TEMPLE UNIV, 72- Mem: Am Dent Asn; Am Acad Oral Path. Res: Clinical research in oral neoplasia. Mailing Add: Dept of Path Temple Univ Sch of Dent Philadelphia PA 19140

WHITE, DENIS NALDRETT, b Bristol, Eng, June 10, 16; m 38; c 4. MEDICINE. Educ: Cambridge Univ, BA, 37, MA, MB & BCh, 40, MD, 56; FRCP, FACP. Prof Exp: First asst med, London Hosp, 43-46; sr registr, Univ London, 46-48; from asst prof to assoc prof, 48-60, PROF MED, QUEEN'S UNIV, ONT, 60- Concurrent Pos: Neurologist, Kingston Gen Hosp; head EEG Dept, Kingston Gen Hosp & Ont Hosp, Smiths Falls. Mem: Fel Am Col Physicians; Am Electroencephalog Soc; Am Acad Neurol; Can Soc Electroencephalog; Am Inst Ultrasonics in Med (past pres). Res: Neurology; medical ultrasonics; echoencephalography. Mailing Add: Dept of Med Etherington Hall Queen's Univ Kingston ON Can

WHITE, DONALD BENJAMIN, b Framingham, Mass, Feb 15, 30; m 53; c 6. ORNAMENTAL HORTICULTURE, GENETICS. Educ: Univ Mass, BS, 56; Iowa State Univ, PhD(hort genetics, breeding), 61. Prof Exp: Res assoc hort, Iowa State Univ, 56-59, res asst, 59-61; from asst prof to assoc prof, 61-69, PROF HORT, UNIV MINN, ST PAUL, 69-, PROF LANDSCAPE ARCHIT, 74- Mem: Am Soc Hort Sci; Am Soc Agron; Soil Sci Soc Am. Res: Physiology of cold acclimation and dwarfing of woody plants; breeding and genetics of grasses; physiology of chemical growth regulation of monocots. Mailing Add: Dept of Hort Sci Univ of Minn St Paul MN 55101

WHITE, DONALD EDWARD, b Dinuba, Calif, May 7, 14; m 41; c 3. GEOLOGY. Educ: Stanford Univ, AB, 36; Princeton Univ, PhD(econ geol, petrol), 39. Prof Exp: Assoc geologist, Geol Surv Nfld, Can, 37-38; geologist, 39-63, RES GEOLOGIST, US GEOL SURV, 63- Concurrent Pos: Asst chief mineral deposits br, US Geol Surv, DC, 58-60. Honors & Awards: Distinguished Serv Award, US Dept Interior, 71. Mem: Nat Acad Sci; fel Geol Soc Am; Fel Soc Econ Geologists; fel Mineral Soc Am; Geochem Soc. Res: Origin and geochemistry of thermal and mineral springs and their relations to volcanism and ore deposits; geothermal energy; origin and nature of ore-forming fluids; origin and characteristics of geysers; isotope geology of waters; rock alteration; abnormal geothermal gradients. Mailing Add: US Geol Surv 345 Middlefield Rd Menlo Park CA 94025

WHITE, DONALD GLENN, b Charleston, WVa, Mar 16, 46; m 68; c 1. PLANT PATHOLOGY. Educ: Marshall Univ, BA, 68, MS, 70; Ohio State Univ, PhD(plant path), 73. Prof Exp: Lectr plant path, Ohio State Univ, 73-74; ASST PROF PLANT PATH, UNIV ILL, URBANA, 74- Mem: Am Phytopath Soc; Crop Sci Soc Am; Am Soc Agron. Res: Fungal diseases of field crops; stalk rot, ear rot, storage molds of corn; mycotoxins. Mailing Add: 246 Davenport Hall Univ of Ill Urbana IL 61801

WHITE, DONALD HARVEY, b Berkeley, Calif, Apr 30, 31; m 53; c 5. NUCLEAR PHYSICS, PARTICLE PHYSICS. Educ: Univ Calif, Berkeley, BA, 53; Cornell Univ, PhD(physics), 60. Prof Exp: Asst Cornell Univ, 53-57, part-time instr, 57-58, asst, 59-60; res physicist, Lawrence Livermore Lab, Univ Calif, 60-71; PROF PHYSICS, ORE COL EDUC, 71- Concurrent Pos: Lectr, Univ Calif, Berkeley, 70; consult, Lawrence Livermore Lab, 71- Mem: Am Phys Soc; Am Asn Physics Teachers. Res: Experimental nuclear physics; bubble chamber; thermal-neutron capture processes; nuclear spectroscopy. Mailing Add: Dept Natural Sci & Math Ore Col Educ Monmouth OR 97361

WHITE, DONALD L, solid state physics, see 12th edition

WHITE, DONALD PERRY, b New York, NY, May 19, 16; m 46; c 3. FORESTRY, SOIL SCIENCE. Educ: State Univ NY, BS, 37; Univ Wis, MS, 40, PhD(forest soils), 51. Prof Exp: Asst soils, Univ Wis, 38-40; forester, US Indian Serv, 40-42 & 46-47; asst soils, Univ Wis, 47-48; from instr to asst prof silvicult, State Univ NY Col Forestry, Syracuse Univ, 48-56; assoc prof forestry, 57-64, mem adv bd forest sci, 59-65, PROF FORESTRY, MICH STATE UNIV, 65- Concurrent Pos: Collabr, US Forest Serv, 59-64; comnr, NAm Foreign Soils Conf, 68- Mem: Soc Am Foresters; Soil Sci Soc Am. Res: Forest soils and fertilization; forest hydrology; watershed management; herbicides; reforestation techniques; Christmas trees. Mailing Add: Dept of Forestry Mich State Univ East Lansing MI 48824

WHITE, DONALD ROBERTSON, b Schenectady, NY, Sept 27, 24; m 47; c 4. OPTICAL PHYSICS. Educ: Union Col, BS, 48; Princeton Univ, MA, 50, PhD(physics), 51. Prof Exp: Res asst physics, Princeton Univ, 51-52; physicist, 52-68, MGR OPTICAL PHYSICS BR, CORP RES & DEVELOP CTR, GEN ELEC CO, 68- Concurrent Pos: Adj prof, Rensselaer Polytech Inst, 60-65. Mem: Fel Am Phys Soc; Combustion Inst. Res: Shock tubes and shock wave phenomena; gaseous detonation; optically pumped lasers; light scattering sensors. Mailing Add: Garnsey Rd Rexford NY 12148

WHITE, DOUGLAS RICHIE, b Minneapolis, Minn, Mar 13, 42; m 71. ANTHROPOLOGY. Educ: Univ Minn, BA, 64; Univ Minn, MA, 67, PhD(anthrop), 69. Prof Exp: Dir cross-cult res, Societal Res Arch Syst, 65-67; asst prof, 67-71, ASSOC PROF ANTHROP, UNIV PITTSBURGH, 72-, ASSOCPROF SOCIOL, 73- Concurrent Pos: Ed consult, J Ethnol, 67; NSF fel, Cross-cult Cumulative Coding Ctr, 68-72; mem comt infor, Div Soc Sci, Nat Acad Sci-Nat Res Coun, 68-; res assoc anthrop, Proj Impress, Dartmouth Col, 70; mem adv panel anthrop, Behav Sci Div, NSF, 70-72; Ling Inst Am fel, 71; methodologist, Lang Attitudes Res Proj, Ireland, 71-73; ed consult, Human Orgn, 72- Mem: Soc Gen Systs Res; Am Anthrop Asn; Soc Appl Anthrop. Res: Mathematical, cross-cultural and urban anthropology; the organization of cultural diversity, including political economic studies based on cognition, social identity and decision-making analysis. Mailing Add: Dept of Anthrop Univ of Pittsburgh 4200 Fifth Ave Pittsburgh PA 15213

WHITE, DWAIN MONTGOMERY, b Minneapolis, Minn, Feb 16, 31; m 56; c 4. ORGANIC CHEMISTRY, POLYMER CHEMISTRY. Educ: Univ Wis, BS, 53, PhD(chem), 56. Prof Exp: RES CHEMIST, RES & DEVELOP CTR, GEN ELEC CO, 56- Mem: Am Chem Soc; The Chem Soc. Res: Organic synthesis and structure determination; oxidative coupling reactions; heterocyclics; synthesis and reactions of polyphenylene oxides. Mailing Add: Gen Elec Res & Develop Ctr PO Box 8 Schenectady NY 12301

WHITE, EDGAR C, b Fielden, Ky, Dec 14, 12; m 40; c 2. SURGERY. Educ: Univ Louisville, MD, 37. Prof Exp: Asst instr clin surg, Sch Med, Univ Louisville, 39-41, instr, 46-48; asst chief surg serv, Vet Admin Hosp, Louisville, 48-49; HEAD DEPT SURG & CHIEF SECT GEN SURG, UNIV TEX M D ANDERSON HOSP & TUMOR INST, HOUSTON, 49-, PROF SURG, UNIV TEX MED SCH, HOUSTON, 63- Concurrent Pos: Clin asst prof, Col Med, Baylor Univ, 50-66, clin assoc prof, 66-; from assoc prof to prof, Postgrad Sch Med, Univ Tex, 50-63; consult, Tex Children's & St Luke's Hosps, Houston, 55- Mem: Am Cancer Soc; Am Col Surg; AMA. Res: Clinical oncology; surgical physiology and pathology. Mailing Add: Univ of Tex M D Anderson Hosp & Tumor Inst Houston TX 77025

WHITE, EDWARD, b Florence, SC, Nov 23, 33; m 55; c 5. IMMUNOLOGY, ENDODONTICS. Educ: Emory Univ, DDS, 58; Med Univ SC, MS, 66; Univ Calif, Los Angeles, PhD(microbiol, immunol), 69. Prof Exp: Asst prof microbiol & immunol, Sch Dent, Univ Southern Calif, 69-72; PROF ENDODONTICS & CHMN DEPT, COL DENT MED, MED UNIV SC, 72- Mem: Transplantation Soc; Am Soc Microbiol. Res: Immunology of transplantation; etiology of dental pulpal disease. Mailing Add: Dept of Endodontics Med Univ of SC Col of Dent Med Charleston SC 29401

WHITE, EDWARD AUSTIN, b Brooklyn, NY, Nov 28, 15; m 48; c 2. BIOCHEMISTRY, NUTRITION. Educ: Fordham Univ, BS, 37, MS, 40, PhD(biochem), 46. Prof Exp: Instr anal chem, Fordham Univ, 37-40; anal res chemist, Calco Chem Co, NJ, 40-42 & Winthrop Chem Co, NY & DC, 42-43; prof biochem, Col Mt St Vincent, 43-46; actg head dept chem, Inst Appl Arts & Sci, NY, 46-47; chief chem gen lab, Japan, 47-50, ADV MED SCI, US DEPT ARMY, WASHINGTON, DC, 50- Mem: Fel AAAS; Am Chem Soc; fel Am Inst Chem. Res: Nutrition in animals; analytical methods; pharmaceuticals; scientific translation. Mailing Add: 5307 Sangamore Rd Washington DC 20016

WHITE, EDWARD LEWIS, b Boston, Mass, Jan 8, 47; m 70. NEUROANATOMY. Educ: Clark Univ, AB, 68; Georgetown Univ, PhD(anat), 72. Prof Exp: Premier asst anat, Inst Anat Normale, Switz, 73-75; ASST PROF ANAT, SCH MED, BOSTON UNIV, 75- Mem: Am Asn Anatomists. Res: Ultrastructure and synaptic organization in mammalian central nervous systems. Mailing Add: Boston Univ Sch of Med 80 E Concord St Boston MA 02118

WHITE, EDWARD RODERICK, b Richwood, WVa, Nov 1, 31; m 54; c 3. PHYSICAL CHEMISTRY. Educ: Marietta Col, BS, 53; Ohio State Univ, PhD(phys chem), 57. Prof Exp: Res chemist, Union Carbide Olefins Corp, WVa, 58-62; res physicist, IIT Res Inst, 62-64; sr scientist, Nuclear Sci & Eng Corp, Pa, 64-65; res assoc radiation chem & biol, Grad Sch Pub Health, Univ Pittsburgh, 65-67, asst res prof, 67-69; sr anal chemist, 69-72, SR RES INVESTR, ANAL & PHYS CHEM DEPT, SMITH KLINE & FRENCH LABS, 72- Mem: AAAS; Am Chem Soc. Res: Gas, high pressure liquid and thin layer chromatography; pharmaceutical analysis; radiation chemistry; electron impact phenomena; kinetics; tracer and vacuum techniques; spectroscopy. Mailing Add: Smith Kline & French Labs 1500 Spring Garden St Philadelphia PA 19101

WHITE, EDWIN HENRY, b Gouverneur, NY, Dec 22, 37; m 61; c 3. FORESTRY, SOIL SCIENCE. Educ: State Univ NY Col Forestry, Syracuse Univ, BS, 62, MS, 64; Auburn Univ, PhD(soils), 69. Prof Exp: Technician forest soils, State Univ NY Col Forestry, Syracuse Univ, 61-62; instr forestry, Auburn Univ, 64-65, res asst, 65-68; fel forestry & soils, Univ Fla, 68-69; res soil scientist, US Forest Serv, Miss, 69-70; asst prof forestry, Univ Ky, 70-74; ASSOC PROF FOREST RESOURCES, UNIV MINN, 74- Mem: Soil Sci Soc Am; Soil Conserv Soc Am; Soc Am Foresters. Res: Forest soils and silviculture; soil-site-species relationships; tree planting research. Mailing Add: Cloquet Forestry Ctr Cloquet MN 55720

WHITE, ELIZABETH LLOYD, b Norfolk, Va, Sept 28, 16. EXPERIMENTAL EMBRYOLOGY, MOLECULAR BIOLOGY. Educ: Goucher Col, AB, 37; Bryn Mawr Col, MA, 38, PhD(embryol), 47. Prof Exp: Researcher, Wistar Inst, Univ Pa, 40-42; chemist, Res Dept, Gen Refractories Co, 42-46; instr zool, Wash Univ, 47-49; from asst prof to assoc prof zool, 49-61, PROF BIOL, WHEATON COL, MASS, 61- Concurrent Pos: NSF fel hist of sci, Johns Hopkins Univ & Cambridge Univ, 58-59; NSF fel & vis prof biol, Johns Hopkins Univ, 66-67, sr res assoc, 73-74. Mem: Fel AAAS; Soc Develop Biol; Am Asn Anat. Res: Experimental embryology of teleosts; history of science; specialized nuclear RNA in chick limb-bud differentiation. Mailing Add: Dept of Biol Wheaton Col Norton MA 02766

WHITE, ELWOOD VERNON, organic chemistry, see 12th edition

WHITE, EMIL HENRY, b Akron, Ohio, Aug 17, 26. ORGANIC CHEMISTRY. Educ: Univ Akron, BS, 47; Purdue Univ, MS, 48, PhD(chem), 50. Prof Exp: Fel, Univ Chicago, 50-51 & Harvard Univ, 51-52; instr org chem, Yale Univ, 52-56; from asst prof to assoc prof, 57-64, PROF ORG CHEM, JOHNS HOPKINS UNIV, 64- Concurrent Pos: Guggenheim fel, 58-59; NIH sr fel, 65-66 & 72-73. Mem: AAAS; Am Chem Soc. Res: Mechanism of reaction in organic chemistry; synthesis; chemiluminescence and bioluminescence; deamination reactions. Mailing Add: Dept of Chem Dunning Hall Johns Hopkins Univ Baltimore MD 21218

WHITE, EUGENE WILBERT, b Indiana, Pa, Jan 23, 33; m 52; c 3. INSTRUMENTATION, MINERALOGY. Educ: Pa State Univ, BS, 55, MS, 58, PhD(solid state tech), 65. Prof Exp: Res asst mineral, Pa State Univ, 55-59; head, X-ray Diffraction Applns Lab, Picker X-ray Co, Ohio, 59-61; design engr, Tem-Pres Res, Inc, Pa, 61; res asst electron microprobe, 62-65, from asst prof to assoc prof solid state sci, 65-74, PROF SOLID STATE SCI, MAT RES LAB, PA STATE UNIV, 74- Mem: Am Crystallog Asn. Res: Electron microprobe research; x-ray spectroscopy. Mailing Add: Mat Res Lab Res Bldg No 1 Pa State Univ University Park PA 16802

WHITE, FLORENCE ROY, b Newcastle, Pa, Mar 6, 09; m 32; c 4. BIOCHEMISTRY. Educ: Univ Ill, AB, 30, AM, 31; Univ Mich, PhD(chem), 35. Prof Exp: Asst physiol, Sch Med, Univ Mich, 35-36; res asst, Sch Med, Yale Univ, 38-39; asst chemist, Bur Home Econ, USDA, 42; chemist, NIH, 42-45; res asst, Nat Acad Sci, 56-58; chemist, Cancer Chemo Nat Serv Ctr, 58-66, HEAD BIOCHEM SECT, DRUG EVAL BR, NAT CANCER INST, 66- Concurrent Pos: Res fel, Univ Mich, 36-38. Mem: Am Soc Biol Chem; Am Asn Cancer Res. Res: Biochemistry of cancer. Mailing Add: Drug Eval Br Drug Res & Develop Nat Cancer Inst Bethesda MD 20014

WHITE, FRANCIS MICHAEL, b Indianapolis, Ind, Aug 6, 18; m 41; c 5. ZOOLOGY, PARASITOLOGY. Educ: Earlham Col, AB, 39; Purdue Univ, MS, 41. Prof Exp: Asst biol, Purdue Univ, 39-41; instr biol, 41-44 & 46-47, from asst prof to assoc prof zool, 47-64, PROF BIOL, PHILADELPHIA COL PHARM & SCI, 64- Concurrent Pos: Lectr, Wagner Free Inst Sci, 67- Mem: AAAS; Am Soc Parasitol; Am Inst Biol Sci. Res: Nematode parasites of fish; interrelationships of parasites as determined by chromatographic tests; drug testing on rats. Mailing Add: Dept of Biol Sci Philadelphia Col of Pharm & Sci Philadelphia PA 19104

WHITE, FRANKLIN ESTABROOK, b Denver, Colo, Mar 26, 22; m 44; c 2.

PHYSICS. Educ: Univ Denver, BS, 48, MA, 51; Univ Mich, MS, 55. Prof Exp: Res assoc atmospheric infrared studies, Univ Denver, 49-51; infrared instrumentation, Univ Mich, 51-55; teacher, 59-62, RES PHYSICIST, UNIV DENVER, 55- Concurrent Pos: Consult, Air Force Opers Anal Off, 56-71. Mem: Am Phys Soc. Res: Atmospheric infrared absorption; balloon and rocket instrumentation for energetic particle measurements; radio propagation and telemetry; operations analysis; atmospheric electricity. Mailing Add: Denver Res Inst Univ of Denver Denver CO 80210

WHITE, FRANKLIN HENRY, b Alton, Ill, Feb 11, 19; m 47; c 2. VETERINARY MICROBIOLOGY. Educ: Shurtleff Col, BS, 42; Univ Ill, MS, 48, PhD(bact), 55; Am Bd Med Microbiol, cert pub health & med lab microbiol, 74. Prof Exp: Bacteriologist, State Dept Pub Health, Ill, 46-49; instr bact, Col Vet Med, Univ Ill, 49-55; from asst bacteriologist to assoc bacteriologist, 55-67, assoc prof bact, 61-67, PROF BACT & BACTERIOLOGIST, UNIV FLA, 67- Mem: Am Soc Microbiol; Conf Res Workers Animal Dis; Wildlife Dis Asn. Res: Pathogenic microbiology and immunology; leptospirosis; vibriosis; wildlife diseases; epizootiology. Mailing Add: Col of Vet Med Univ of Fla Gainesville FL 32611

WHITE, FRED DONALD, meteorology, see 12th edition

WHITE, FRED G, b Spanish Fork, Utah, Jan 19, 28; m 54; c 7. PLANT BIOCHEMISTRY. Educ: Brigham Young Univ, BS, 52, MS, 56; Univ Calif, PhD(biochem), 61. Prof Exp: From asst prof to assoc prof, 61-72, PROF CHEM, BRIGHAM YOUNG UNIV, 72- Concurrent Pos: Estab investr, Am Heart Asn, 65-70. Mem: AAAS; Am Chem Soc; Am Inst Biol Sci; NY Acad Sci; Sigma Xi. Res: Mechanisms of enzymes and co-enzymes; biological nitrogen fixation; plant growth regulators; plant biochemistry; plant genetic engineering. Mailing Add: Grad Sect Biochem Brigham Young Univ Provo UT 84602

WHITE, FRED NEWTON, b Yelgar, La, June 17, 27; m 51. CARDIOVASCULAR PHYSIOLOGY, COMPARATIVE PHYSIOLOGY. Educ: Univ Houston, BS, 50, MS, 51; Univ Ill, PhD(zool), 55. Prof Exp: Asst prof biol, Univ Houston, 55-57; asst prof exp med, Southwestern Med Sch, Univ Tex, 58-59 & 62-63; assoc prof physiol, Am Univ Beirut, 59-62; PROF PHYSIOL, SCH MED, UNIV CALIF, LOS ANGELES, 63- Concurrent Pos: Am Physiol Soc cardiovasc training fels, 57 & 58. Mem: Am Physiol Soc; Soc Exp Biol & Med. Res: Control of renin secretion; peripheral circulation; comparative aspects of vertebrate circulation; environmental physiology. Mailing Add: Dept of Physiol Univ of Calif Sch of Med Los Angeles CA 90024

WHITE, FREDERICK ANDREW, b Detroit, Mich, Mar 11, 18; m 42; c 4. PHYSICS. Educ: Wayne Univ, BS, 40; Univ Mich, MS, 41; Univ Wis, PhD, 59. Prof Exp: Sr inspector, Rochester Army Ord Plant, NY, 41-43; asst physicist, Manhattan Proj, Univ Rochester, 43-45, grad instr, Univ, 45-46; asst physics, Knolls Atomic Power Lab, Gen Elec Co, 47-48, res assoc, 48-57, consult physicist, 59-62; PROF NUCLEAR ENG & INDUST LIAISON SCIENTIST, RENSSELAER POLYTECH INST, 62- Concurrent Pos: Consult, Bell Tel Labs, 69. Mem: AAAS; Am Inst Aeronaut & Astronaut; Am Nuclear Soc; Am Phys Soc; Am Soc Eng Educ. Res: Mass spectrometry and isotopic abundance measurements; acoustics; diffusion in metals; semiconductor devices; interaction of ions with matter; plasma diagnostics. Mailing Add: Dept of Nuclear Eng Rensselaer Polytech Inst Troy NY 12181

WHITE, FREDERICK ELMER, b Peabody, Mass, Jan 21, 09; m 43. PHYSICS. Educ: Boston Col, AB, 30; Brown Univ, ScM, 32, PhD(physics), 34. Prof Exp: Instr physics, Williams Col, Mass, 35-36; asst prof, Boston Col, 36-42; physicist, Duke Univ, 42-45; prof physics, Boston Col, 45-74; RETIRED. Mem: AAAS; Am Phys Soc; Acoust Soc Am; Am Asn Physics Teachers. Res: Acoustic filters; interdeterminacy principle; atmospheric acoustics; ultrasonics; acoustics; theoretical physics; mechanics. Mailing Add: 12 Columbia Rd Beverly MA 01915

WHITE, FREDERICK HOWARD, JR, b Washington, DC, Jan 19, 26. PROTEIN CHEMISTRY. Educ: Univ Va, BS, 49; Univ Md, MS, 52; Univ Wis, PhD(biochem), 57. Prof Exp: Asst chem, Univ Md, 51-52; chemist, Nat Heart Inst, 52-53; asst biochem, Univ Wis, 53-56; chemist, Lab Biochem, 56-75, CHEMIST, LAB CELL BIOL, NAT HEART & LUNG INST, 75- Mem: AAAS; Am Chem Soc; Am Soc Biol Chem; Radiation Res Soc. Res: Chemistry of sulfur in proteins; protein conformation; radiolysis of proteins. Mailing Add: Lab of Cell Biol Nat Heart & Lung Inst Bethesda MD 20014

WHITE, FREDRIC PAUL, b New York, NY, July 24, 42; m 64; c 1. BIOCHEMISTRY, NEUROSCIENCES. Educ: Purdue Univ, BChE, 64; Ind Univ, PhD(biol chem), 71. Prof Exp: Chem engr, E I du Pont de Nemours & Co, Inc, 64-66; biochem trainee, NIH, 67-71; res scientist neurochem, Med Sch, Univ Colo, 72-73; vis asst prof neurophys, Ind State Univ, 73-74; ASST PROF BIOCHEM, MEM UNIV MFLD, 74- Mem: Soc Neurosci. Res: Studies aimed at elucidating which proteins are being transported by fast axoplasmic flow in neurons and how these proteins become incorporated into preexisting neural membrane elements. Mailing Add: Fac of Med Mem Univ of Nfld St John's NF Can

WHITE, GEORGE ALBERT, b Bentley, Ill, Dec 25, 30; m 56; c 2. AGRONOMY, BOTANY. Educ: Univ Ill, BS, 57, MS, 58; Univ Minn, St Paul, PhD(agron, agr bot), 61. Prof Exp: Res agronomist, 62-64, invests leader, 64-72, CHIEF GERMPLASM RESOURCES, AGR RES CTR, USDA, 72- Honors & Awards: Crambe Award, USDA, 66. Mem: Am Soc Agron; Crop Sci Soc Am; Soc Econ Bot. Res: Agronomic development of new crops such as sources of paper pulp, seed oils and gums; coordination of plant germplasm on national basis. Mailing Add: Germplasm Resources Lab USDA Agr Res Ctr Beltsville MD 20705

WHITE, GEORGE CHARLES, JR, b Coatesville, Pa, Sept 14, 18; m 43; c 2. PHYSICS. Educ: Villanova Univ, BS, 39. Prof Exp: Physicist, Physics & Math Br, Pitman-Dunn Res Labs, 46-52, chief, 53-57, asst dir, Physics Res Lab, 57-58, asst spec missions off, 59-60, tech asst, Inst Res, 60-64, tech asst, Off of Dir, Pitman-Dunn Res Labs, 65-71, DEP DIR, PITMAN-DUNN RES LABS, FRANKFORD ARSENAL, US DEPT ARMY, 72- Mem: Am Nuclear Soc; Sigma Xi. Res: Nuclear, radiation and solid state physics; materials research. Mailing Add: 705 Avondale Rd Erdenheim PA 19118

WHITE, GEORGE MATTHEWS, b Salt Lake City, Utah, Dec 7, 41. COMMUNICATION SCIENCE. Educ: Mich State Univ, BS, 64; Univ Ore, PhD(chem phys), 68. Prof Exp: NIH res fel, Dept Comput Sci, Stanford Univ, 68-70; RES SCIENTIST COMPUT SCI, XEROX PALO ALTO RES CTR, 70- Mem: Inst Elec & Electronic Engrs; Asn Comput Mach; Acoust Soc Am; Pattern Recognition Soc. Res: Pattern recognition, machine perception, artificial intelligence, automatic speech recognition, signal processing and perceptual psychology. Mailing Add: Xerox Palo Alto Res Ctr 3333 Coyote Hill Rd Palo Alto CA 94304

WHITE, GEORGE NICHOLS, JR, b Concord, Mass, July 1, 19; m 48; c 3. APPLIED MATHEMATICS. Educ: Harvard Univ, BS, 41; Brown Univ, MS, 48, PhD(appl math), 50. Prof Exp: Trainee, Phys Test Lab, T Mason Co, 41-42; technician radar, US Civil Serv, 42-45; technician electronics, Oceanog Inst, Woods Hole, 45-46; res assoc appl math, Brown Univ, 48-50; MEM STAFF APPL MATH, LOS ALAMOS SCI LAB, 50- Concurrent Pos: Prof, Univ NMex, 57-60. Res: Mathematics theory of plasticity; hydrodynamics; elasticity. Mailing Add: 119 Tunyo Los Alamos NM 87544

WHITE, GEORGE THOMAS, organic chemistry, see 12th edition

WHITE, GEORGE V S, b Opelousas, La, Dec 15, 23; m 48; c 3. VERTEBRATE ZOOLOGY. Educ: Univ Southwestern La, BS, 44; La State Univ, MS, 47, PhD, 54. Prof Exp: Instr zool, Westminster Col, Mo, 50-53; asst prof, Memphis State Univ, 54-55; PROF BIOL, McNEESE STATE UNIV, 55-, HEAD DEPT, 70- Mem: Am Soc Zool. Res: Embryology; comparative anatomy; mammalian histology. Mailing Add: Dept of Biol McNeese State Univ Lake Charles LA 70601

WHITE, GEORGE WILLARD, b North Lawrence, Ohio, July 8, 03; m 28. GEOLOGY. Educ: Otterbein Col, AB, 21; Ohio State Univ, MA, 25, PhD(geol), 33. Hon Degrees: ScD, Otterbein Col, 49, Univ NH, 55, Bowling Green State Univ, 63. Prof Exp: Instr geol, Univ Tenn, 25-26; from instr to prof, Univ NH, 26-41, actg dean grad sch, 40; prof geol, Ohio State Univ, 41-47; prof & head dept, 47-54, res prof, 65-71, EMER RES PROF GEOL, UNIV ILL, URBANA, 71-; CONSULT, GEOL SURV OHIO, 73- Concurrent Pos: Geologist, Water Resources Br, US Geol Surv, 42-46 & 49-69; state geologist, Ohio, 46-47; consult, State Geol Surv, Ill, 48-; Gurley lectr, Cornell Univ, 55; vpres NAm, Int Comn Hist Geol, 65-; chmn, US Nat Comt Hist Geol. Honors & Awards: Orton Award, Ohio State Univ, 61. Mem: AAAS (vpres, 51); fel Geol Soc Am; Am Asn Petrol Geol; Am Inst Mining, Metall & Petrol Eng; Hist Sci Soc. Res: Ground water and glacial geology; history of American geology. Mailing Add: Dept of Geol Univ of Ill Urbana IL 61801

WHITE, GIFFORD, b San Saba, Tex, Feb 17, 12; m 35; c 2. PHYSICS. Educ: Univ Tex, BA & MA, 39. Prof Exp: Geophysicist, Humble Oil & Refining Co, 34-38; res engr, Sperry Gyroscope Co, 41-47; vpres, Statham Instruments, Inc, 47-52; PRES, WHITE INSTRUMENTS, INC, 53- Honors & Awards: Cert of Appreciation, US War Dept, 45. Mem: Am Phys Soc; Soc Explor Geophys; fel Inst Elec & Electronics Eng. Res: Instrumentation for physical measurements; circuit theory. Mailing Add: White Instruments Inc Box 698 Austin TX 78767

WHITE, GILBERT FOWLER, b Chicago, Ill, Nov 26, 11; m 45; c 3. RESOURCE GEOGRAPHY. Educ: Univ Chicago, SB, 32, SM, 34, PhD, 42. Hon Degrees: L, Hamilton Col, 54, Swarthmore Col, 56, Earlham Col, 68 & Mich State Univ, 70; ScD, Haverford Col, 56. Prof Exp: Geogr, Miss Valley Comt, Nat Resources Comt & Nat Resources Planning Bd, 34-40; geogr, Bur Budget, Exec Off President, 40-42; geogr, Am Friends Serv Comt, 42-46; pres, Haverford Col, 46-55; prof geog, Univ Chicago, 56-69; PROF GEOG & DIR, INST BEHAV SCI, UNIV COLO, BOULDER, 70- Concurrent Pos: Mem, Hoover Comn Task Force Natural Resource, 48; vchmn, President's Water Resources Policy Comn, 50; mem adv comt arid zone res, UNESCO, 53-56; chmn, UN Panel Integrated River Develop, 57-58; consult, Lower Mekong Coord Comt, Cambodia, Laos, Thailand & Vietnam, 61-62 & 70; vis prof geog, Oxford Univ, 62-63; chmn, Am Friends Serv Comt, 63-69; chmn comt water, Nat Acad Sci-Nat Res Coun, 64-68; mem, Spec NSF Comn Weather Modification, 64-69; chmn steering comt, High Sch Geog Proj, 64-70; chmn, Bur Budget Task Force Fed Flood Policy, 65-66; sci adv to adminr, UN Develop Prog, 66-70; mem adv comt natural resources res, UNESCO, 67-71; mem adv comt environ sci, NSF, 68-71; chmn comn man & environ, Int Geog Union, 69-; mem sci comt probs of environ, Int Coun Sci Unions, 70-; chmn adv bd, Energy Policy Proj, 72-74; chmn int environ progs comt, Nat Acad Sci-Nat Res Coun, 72-; trustee, Pop Coun; trustee & chmn, Resources for Future; mem exec bd, Int Union Conserv Nature & Natural Resource, 72-74; mem technol assessment adv coun, US Cong, 74-75. Honors & Awards: Distinguished Serv Award, Asn Am Geogr, 55 & 73; Daly Medal, Am Geog Soc, 71; Eben Award, Am Water Resources Asn, 72. Mem: Nat Acad Sci; Asn Am Geogr (pres, 61-62). Res: Human aspect of resources management; flood plain use; domestic water use; river basin management. Mailing Add: Inst Behav Sci Bldg 1 Univ of Colo Boulder CO 80302

WHITE, GORDON ALLAN, b Vancouver, BC, Nov 8, 32; m 59. PLANT PHYSIOLOGY, BIOCHEMISTRY. Educ: Univ BC, BA, 54, MA, 55; Iowa State Univ, PhD, 59. Prof Exp: Fel plant biochem, Prairie Regional Lab, Nat Res Coun Can, 59-60; asst prof chem, Ore State Univ, 60-62; PLANT BIOCHEMIST, LONDON RES INST, CAN DEPT AGR, 62- Res: Oxidative enzymes and their role in metabolism; biochemistry of fungi, including obligate parasites; carbohydrate catabolism in microorganisms. Mailing Add: London Res Inst Can Dept of Agr University Sub Post Off London ON Can

WHITE, GORDON JUSTICE, b Elgin, Ill, May 3, 31; m 53; c 4. IMMUNOCHEMISTRY, BIOCHEMISTRY. Educ: Univ Ill, BS, 53; Northwestern Univ, MS, 58; Brandeis Univ, PhD(biol), 62. Prof Exp: Res assoc virol, 62-68, RES ASSOC IMMUNOL, UPJOHN CO, 68- Res: Pathogenesis of upper respiratory virus disease; enzyme modification of biological activity of gamma globulin; studies of inhibitors of serum complement and of production and detection of reagin-like antibodies in the rat; modulation of immediate hypersensitivity reactions by control of intracellular cyclic nucleotide levels. Mailing Add: Dept 7244 The Upjohn Co Kalamazoo MI 49006

WHITE, GORDON M, analytical chemistry, see 12th edition

WHITE, HALBERT CONSTANTINE, b Dinuba, Calif, Feb 27, 16; m 43; c 4. ORGANIC CHEMISTRY. Educ: Stanford Univ, AB, 37, PhD(org chem), 40. Prof Exp: Res chemist, 40-45, group leader, Britton Res Lab, 45-68, head dept chem, Dow Human Health Res & Develop, 68-72, head dept proj monitors, 72-74, MGR PLANNING & COORD, DOW HUMAN HEALTH RES & DEVELOP, DOW CHEM CO, 74- Mem: AAAS; Am Chem Soc; Sigma Xi. Res: Synthesis and reaction of amino acids; pharmaceutical chemistry. Mailing Add: 6406 Knyghton Rd Indianapolis IN 46220

WHITE, HARLAN E, agronomy, see 12th edition

WHITE, HAROLD BIRTS, JR, b Little Rock, Ark, Mar 13, 29; m 58; c 2. BIOCHEMISTRY. Educ: Columbia Univ, AB, 51, MA, 53, PhD(physiol), 57. Prof Exp: Fel, Purdue Univ, 57-59, asst prof biochem, 59-61; from asst prof to assoc prof, 61-68, PROF BIOCHEM, SCH MED, UNIV MISS, 68- Concurrent Pos: Vis scientist, Univ Milan, 68-69. Mem: Am Chem Soc; Am Soc Biol Chem. Res: Brain lipid modification; poxvirus influence on lipid metabometabolism. Mailing Add: Dept of Biochem Univ of Miss Sch of Med Jackson MS 39216

WHITE, HAROLD J, b New York, NY, Jan 4, 20. PATHOLOGY Educ: Harvard Univ, BS, 41; Univ Geneva, MD, 52. Prof Exp: Instr path, Sch Med, Yale Univ, 57-58; asst prof, 58-61, PROF PATH, SCH MED, UNIV ARK, LITTLE ROCK, 66-

Concurrent Pos: Mem staff, Vet Admin Hosp, 60- Mem: NY Acad Med; Int Acad Path. Res: Role of mucopolysaccharides in pathologic conditions. Mailing Add: Dept of Path Univ of Ark Little Rock AR 72206

WHITE, HAROLD KEITH, b Straughn, Ind, July 11, 23; m 47; c 3. ORGANIC CHEMISTRY, BIOCHEMISTRY. Educ: Butler Univ, BS, 47; Purdue Univ, MS, 50; Ind Univ, PhD(org chem), 54. Prof Exp: Asst chemist, State Chem Off Ind, 47-49; res chemist, Mead Johnson & Co, 53-55; assoc prof, 55-57, PROF CHEM, HANOVER COL, 57- Mem: Am Chem Soc. Res: Stereochemistry; synthesis of polycyclics; medicinal chemistry. Mailing Add: Dept of Chem Hanover Col Hanover IN 47243

WHITE, HAROLD MCCOY, b Camden, SC, Feb 1, 32; m 55; c 3. ORGANIC CHEMISTRY. Educ: Clemson Univ, BS, 54, PhD(org chem), 62. Prof Exp: Chemist, SC Agr Res Sta, 54-55; fel ozone chem, Univ Tex, 62-64; from asst prof to assoc prof chem, 64-71, PROF CHEM, SOUTHWESTERN STATE COL, 71- Concurrent Pos: Welch Found fel, 63-64. Mem: Am Chem Soc; Sigma Xi. Res: Reactions of ozone with organic compounds; mechanism of the ozonation of hydrocarbons and reactions of ozone in basic media. Mailing Add: Rte 2 Davis Rd Weatherford OK 73096

WHITE, HARRY HOUSTON, b Batesville, Ark, Jan 21, 34; m 57; c 3. NEUROLOGY. Educ: Univ Kans, AB, 55, MD, 58. Prof Exp: From instr to assoc prof neurol, Sch Med, Univ Kans, 64-72; PROF NEUROL & CHMN DEPT, SCH MED, UNIV MO-COLUMBIA, 72- Concurrent Pos: Markle scholar acad med, 65- Mem: AAAS; Am Acad Neurol; Asn Res Nerv & Ment Dis; Am Neurol Asn. Res: Genetically-determined diseases of the nervous system; inborn errors of metabolism. Mailing Add: Sch of Med Univ of Mo Columbia MO 65201

WHITE, HARRY JOSEPH, b Philadelphia, Pa, Feb 19, 31; m 56; c 4. ORGANIC CHEMISTRY. Educ: LaSalle Col, BA, 54; Univ Notre Dame, PhD(org chem), 58. Prof Exp: Res chemist, 58-67, coordr PhD recruiting, 67-68, asst mgr manpower & employment, 68-72, MGR RECRUITING & PLACEMENT, ROHM AND HAAS CO, 72- Mem: Am Chem Soc. Res: Technical recruiting, placement and manpower planning; petroleum additives; polymer chemistry. Mailing Add: Rohm and Haas Co Independence Mall W Philadelphia PA 19105

WHITE, HARVEY ELLIOTT, b Parkersburg, WVa, Jan 28, 02; m 28; c 3. PHYSICS. Educ: Occidental Col, AB, 25; Cornell Univ, PhD(physics), 29. Hon Degrees: ScD, Occidental Col, 61. Prof Exp: From asst to instr physics, Cornell Univ, 25-29; Int Res fel, Phys & Tech Inst, Ger, 29-30; from asst prof to prof physics, Univ Calif, Berkeley, 30-60; dir, Lawrence Hall Sci, 60-74; RETIRED. Concurrent Pos: Guggenheim fel, 48. Honors & Awards: Edison, Parents Mag, Peabody TV & Sylvania Awards; Han Christian Oersted Medal, 67. Mem: Fel Am Phys Soc; fel Optical Soc Am; Am Asn Physics Teachers. Res: Ultraviolet and infrared optics. Mailing Add: 543 Spruce St Berkeley CA 94707

WHITE, HELEN LYNG, b Oceanside, NY, Oct 25, 30; m 55; c 2. BIOCHEMISTRY. Educ: Russell Sage Col, BA, 52; Univ Del, MS, 63; Univ NC, Chapel Hill, PhD(biochem), 67. Prof Exp: Chemist, E I du Pont de Nemours & Co, Inc, 52-56; res assoc med chem, Univ NC, 67-70; SR RES PHARMACOLOGIST, WELLCOME RES LABS, 70- Mem: Am Chem Soc. Res: Enzyme mechanisms and inhibitors; neurochemistry. Mailing Add: Dept of Pharmacol Wellcome Res Labs Research Triangle Park NC 27709

WHITE, HENRY W, b Blytheville, Ark, Dec 20, 41; m 62; c 2. PHYSICS. Educ: Pepperdine Col, BA, 63; Univ Calif, Riverside, MS, 65, PhD(physics), 69. Prof Exp: ASSOC PROF PHYSICS, UNIV MO-COLUMBIA, 69- Mem: Am Phys Soc; Am Asn Physics Teachers. Res: Low temperature thermal properties of magnetic semiconductors. Mailing Add: Dept of Physics Univ of Mo Columbia MO 65201

WHITE, HORACE FREDERICK, b Fresno, Calif, Apr 25, 25; m 52; c 3. PHYSICAL CHEMISTRY. Educ: Fresno State Col, AB, 47; Ore State Col, MS, 50; Brown Univ, PhD(phys chem), 53. Prof Exp: Fel, Univ Minn, 52-54; spectroscopist, Res Dept, M W Kellogg Co Div, Pullman, Inc, 54-56, instrumental methods supvr, 56-57; spectroscopist, Res Dept, Union Carbide Chem Co, 57-65; from asst prof to assoc prof chem, 65-75, actg chmn dept, 66-68, PROF CHEM, PORTLAND STATE UNIV, 75- Mem: Am Chem Soc. Res: Molecular structure using infrared spectroscopy and nuclear magnetic resonance spectrometry techniques; mass spectrometry; x-ray crystallography for structural determinations. Mailing Add: Portland State Univ Dept Chem PO Box 751 Portland OR 97207

WHITE, HOWARD JULIAN, JR, b Batavia, NY, Nov 20, 20; m 49; c 2. SCIENCE ADMINISTRATION. Educ: Princeton Univ, AB, 42, PhD(chem), 47; Univ Wis, MS, 44. Prof Exp: Asst chem, Univ Wis, 42-44; from res chemist to assoc dir res, Textile Res Inst, NJ, 47-57, dir, 57-60; sr phys chemist, Stanford Res Inst, 60-61; spec asst res to Asst Secy Navy, Res & Develop, 61-64; asst chief phys chem div, 64-66, PROG MGR, OFF STANDARD REFERENCE DATA, NAT BUR STANDARDS, 66- Mem: AAAS; Am Chem Soc; Fiber Soc; Calorimetry Conf. Res: Surface chemistry; solutions, swelling, adsorption and dyeing of fibers; reference data on thermodynamics and transport properties and colloid and surface properties. Mailing Add: 8028 Park Overlook Dr Bethesda MD 20034

WHITE, HOWARD SORREL, b Philadelphia, Pa, Oct 11, 34; m 59; c 2. ORGANIC CHEMISTRY. Educ: Philadelphia Col Pharm & Sci, BSc, 56; Temple Univ, MA, 58, PhD(org chem), 63. Prof Exp: Res chemist, Archer-Daniels-Midland Co, 63-67; res chemist, Gillette Co, 67-70; RES CHEMIST, ARMSTRONG CORK CO, 70- Mem: AAAS; Am Chem Soc. Res: Structure-property relationships in polymers; water soluble polymers; polyurethane chemistry and applications; cellular and foam structures. Mailing Add: Armstrong Cork Co 2500 Columbia Ave Lancaster PA 17604

WHITE, JACK EDWARD, b Stuart, Fla, July 24, 21; m 45; c 5. MEDICINE. Educ: Fla Agr & Mech Univ, BA, 41; Howard Univ, MD, 44; Am Bd Surg, dipl, 51. Prof Exp: Intern med & surg, Freedmen's Hosp, 45, asst resident surgeon surg, 45-46, chief resident surgeon chest surg, 47-48, chief resident gen surg, 48-49; asst resident surgeon, Mem Ctr, New York, 49-50, resident surgeon, 50-51; from asst prof to assoc prof, 51-63, PROF SURG, MED SCH, HOWARD UNIV, 63-, DIR CANCER TEACHING PROJ, 51-, DIR CANCER RES CTR, 72- Concurrent Pos: Consult, Nat Cancer Inst, 53-56; surgeon, Freedmen's Hosp, 51-, chmn dept oncol, 73-; attend surgeon, Washington Hosp Ctr, 63-; mem rev comt, Support Serv, Div Cancer Control, Nat Cancer Inst, 74- Mem: Nat Med Asn; fel Am Col Surg; Am Radium Soc; Soc Surg Alimentary Tract; Soc Head & Neck Surg. Res: Cancer of tongue, breast, mesentery, soft parts and lymphatic organs; penetrating trauma of abdomen; cancer epidemiology; retroperitonitis; cancer chemotherapy; peptic ulcer; pheochromocytoma. Mailing Add: Dept of Surg Howard Univ Med Sch Washington DC 20001

WHITE, JACK LEE, b Los Angeles, Calif, Oct 29, 25; m 50; c 1. MATERIALS SCIENCE. Educ: Calif Inst Technol, BS, 49; Carnegie Inst Technol, BS, 50; Imp Col Univ London, dipl, 55; Univ Calif, PhD(metall), 55. Prof Exp: Res engr, Univ Calif, 55; Nat Acad Sci res assoc, US Naval Res Lab, 55-57, consult, Phys Metall Br, 57-58; mem res staff, Gen Atomic Div, Gen Dynamics Corp, 58-67; res off, Petten Ctr, Europ Atomic Energy Comn, 67-69; vis scientist, Gulf Gen Atomic, 69-70; assoc prof mat sci, Univ Calif, Davis, 71-72; vis scientist, 72-73, STAFF SCIENTIST, AEROSPACE CORP, 73- Concurrent Pos: Consult, Gulf Gen Atomic, 70-72. Mem: AAAS; Am Chem Soc; Am Ceramic Soc; Am Inst Mining, Metall & Petrol Eng; Brit Inst Metals. Res: Carbonaceous and graphitic materials; high-temperature materials; high-temperature physical chemistry; metallurgical thermodynamics. Mailing Add: 690 Rimini Rd Del Mar CA 92014

WHITE, JAMES ANTHONY, geology, mineral economics, see 12th edition

WHITE, JAMES CARL, b Ft Wayne, Ind, Mar 1, 22; m 46; c 4. ANALYTICAL CHEMISTRY. Educ: Ind Univ, BS, 43; Ohio State Univ, MS, 48, PhD(chem), 50. Prof Exp: Chemist, Joslyn Mfg Co, 46; asst, Ohio State Univ, 46-50; asst div dir, 50-67, assoc dir, Anal Chem Div, 67-72, DIR, ANAL CHEM DIV, OAK RIDGE NAT LAB, 72- Mem: AAAS; Am Chem Soc; Int Union Pure & Appl Chem; Fedn Anal Chem & Spectros. Res: Research administration; molten salts; separations; reference materials. Mailing Add: 5425 Shenandoah Trail Knoxville TN 37919

WHITE, JAMES CARRICK, b Scobey, Mont, Oct 29, 16; m 41; c 3. FOOD MICROBIOLOGY. Educ: Cornell Univ, PhD(bact), 44. Prof Exp: Dir res, Borden Cheese Co, 44-46; assoc prof, 46-51, PROF DAIRY INDUST, CORNELL UNIV, 51-, PROF HOTEL ADMIN, 72- Mem: Inst Food Technol; Int Asn Milk, Food & Environ Sanitarians. Res: Food poisoning; waste technology; food sanitation. Mailing Add: Statler Hall Cornell Univ Ithaca NY 14850

WHITE, JAMES CLARENCE, b Hodge, La, July 7, 36; m 57; c 2. PLANT PATHOLOGY. Educ: La Polytech Inst, BS, 59; La State Univ, MS, 61, PhD(plant path), 63. Prof Exp: Asst prof bot, Southeastern La Col, 63-65; asst prof, 65-70, PROF BOT, LA TECH UNIV, 70- Mem: Am Phytopath Soc. Res: Pathological histology and studies of Tabasco pepper plants infected with tobacco etch virus. Mailing Add: Dept of Bot & Bacteriol La Tech Univ Ruston LA 71270

WHITE, JAMES DAVID, b Bristol, Eng, June 14, 35; m 60; c 2. ORGANIC CHEMISTRY. Educ: Cambridge Univ, BA, 59; Univ BC, MSc, 61; Mass Inst Technol, PhD(org chem), 65. Prof Exp: From instr to asst prof chem, Harvard Univ, 65-71; ASSOC PROF CHEM, ORE STATE UNIV, 71- Mem: Am Chem Soc; The Chem Soc. Res: Organic synthesis and photochemistry; chemistry of natural products; heterocyclic compounds. Mailing Add: Dept of Chem Ore State Univ Corvallis OR 97331

WHITE, JAMES EDWARD, b Cherokee, Tex, May 10, 18; m 41; c 4. GEOPHYSICS. Educ: Univ Tex, BS, 40, MA, 46; Mass Inst Technol, PhD(phsycis), 49. Prof Exp: Physicist, Underwater Sound Lab, Mass Inst Technol, 41-45, Defense Res Lab, Univ Tex, 45-46, Mobil Oil Co, 49-55, Marathon Oil Co, 55-69 & Globe Universal Sci, Inc, 69-72; mem faculty, Colo Sch Mines, 72-73; L A NELSON PROF GEOL SCI, UNIV TEX, EL PASO, 73- Concurrent Pos: Nat Acad Sci exchange scientist, USSR & Yugoslavia, 73-74; mem space appl bd, Nat Acad Eng, 73- Mem: Fel Acoust Soc Am; Am Phys Soc; Seismol Soc Am; Soc Explor Geophys (past pres); Am Geophys Union. Res: Seismic prospecting; waves in solids; engineering geophysics; earthquake dynamics. Mailing Add: Dept of Geol Sci Univ of Tex El Paso TX 79968

WHITE, JAMES EDWIN, b Pittsburgh, Pa, June 4, 35; m 60; c 2. BIOLOGY, ECOLOGY. Educ: Dartmouth Col, AB, 57; Rutgers Univ, PhD(zool), 61. Prof Exp: Asst prof biol, Parsons Col, 61-62; from instr to assoc prof, 62-74, PROF BIOL, KEUKA COL, 74-, CHMN DEPT, 69- Concurrent Pos: Consult, NY State Scholar Exam, 64-65. Mem: AAAS; Am Soc Mammal; Ecol Soc Am. Res: Mammal population ecology; animal behavior; small mammal parasites. Mailing Add: Dept of Biol Keuka Col Keuka Park NY 14478

WHITE, JAMES MINOR, chemistry, mathematics, see 12th edition

WHITE, JAMES PATRICK, b Indianapolis, Ind, Sept 13, 39; m 62; c 4. MICROBIAL PHYSIOLOGY. Educ: Marian Col, Ind, BS, 62; Univ Ark, MS, 65, PhD(microbiol), 67. Prof Exp: Asst prof, 67-70, ASSOC PROF MICROBIOL, ST BONAVENTURE UNIV, 70-, ASSOC PROF BIOL, 74- Mem: AAAS; Am Soc Microbiol; Mycol Soc Am. Res: Physiology and nutrition of pigment formation in Helminthosporium species; effects of trace elements in nitrogen metabolism of microorganisms. Mailing Add: Dept of Biol St Bonaventure Univ St Bonaventure NY 14778

WHITE, JAMES RUSHTON, b Ft Benning, Ga, July 28, 23; m 55; c 2. BIOCHEMISTRY. Educ: Stanford Univ, BS, 48, PhD(chem), 53. Prof Exp: Res chemist, Pioneering Res Lab, E I du Pont de Nemours & Co, Inc, 53-59; res assoc biochem, Univ Pa, 59-62; from asst prof to assoc prof, 62-71, PROF BIOCHEM, SCH MED, UNIV NC, CHAPEL HILL, 71- Concurrent Pos: NSF fel, 60-62. Mem: AAAS; Am Soc Biol Chem; Am Chem Soc; Biophys Soc. Res: Protein biosynthesis and macromolecular metabolism in bacteria; antibacterial action of antibiotics and other inhibitors; complexes of nucleic acids with low molecular weight ligands. Mailing Add: Dept of Biochem Univ of NC Med Sch Chapel Hill NC 27514

WHITE, JAMES RUSSELL, b Elgin, Ill, July 13, 19; m 44; c 1. PHYSICAL CHEMISTRY. Educ: Ind Univ, BS, 42; Yale Univ, PhD(phys chem), 44. Prof Exp: Phys chemist, Tenn Eastman Corp, 44-47; res assoc physics, Socony-Vacuum Oil Co, 47-59, supvr nuclear res group, Mobil Res & Develop Corp, 59-69, MGR CHEM & LUBRICANTS RES SECT, MOBIL RES & DEVELOP CORP, 69- Mem: AAAS; Am Chem Soc; Am Nuclear Soc. Res: Mass spectrometry; radio chemical tracers; phycial chemistry of solutions; ionization potentials; lubricants and lubrication; thermal diffusion; radiation chemistry. Mailing Add: Mobil Res & Develop Corp Box 1025 Princeton NJ 08540

WHITE, JAMES WILSON, b Salisbury, NC, May 29, 14; m 42; c 3. PHYSICS. Educ: Davidson Col, BS, 34; Univ NC, MS, 36, PhD(physics), 38. Hon Degrees: DSc, King Col, 65. Prof Exp: Instr physics, Emory Jr Col, 38-39; prof, King Col, 39-42; instr, Univ Tenn, 42-44; res physicist, Fulton Sylphon Co, 44-45; from asst prof to assoc prof, 45-58, PROF PHYSICS, UNIV TENN, KNOXVILLE, 58- Mem: Am Phys Soc; Am Asn Physics Teachers. Res: Instrumentation. Mailing Add: Dept of Physics-Astron Univ of Tenn Knoxville TN 37916

WHITE, JANE VICKNAIR, b Houma, La, Feb 10, 47; m 68; c 1. NUTRITION. Educ: St Mary's Diminican Col, BS, 68; Univ Tenn, Knoxville, PhD(nutrit), 75. Prof Exp: ASST PROF NUTRIT, DEPT FAMILY PRACT MED, UNIV TENN, KNOXVILLE, 75- Concurrent Pos: Lectr, Dept Food Sci, Nutrit, Food Syst Admin, Univ Tenn, 75- Mem: Am Dietetic Asn; Am Home Econ Asn; Nat Educ Asn. Res: Effect of sulfur nutrition on the enzymes of carbohydrate and fat metabolism; nutrition education in family practice residency programs. Mailing Add: 249 Newport Rd Knoxville TN 37922

WHITE, JERRY EUGENE, b Mt Vernon, Ill, Oct 6, 46; m 68. ORGANIC CHEMISTRY. Educ: Southern Ill Univ, Carbondale, BA, 68; Vanderbilt Univ, PhD(chem), 72. Prof Exp: Res chemist polymer synthesis, Army Mat & Mech Res Ctr, 73-76; SR RES CHEMIST SYNTHESIS, FOAM PROD RES LAB, EASTERN DIV, DOW CHEM USA, 76- Concurrent Pos: Co-chmn, Army Mat & Mech Res Ctr Technol Assessment & Planning Conf, 75. Mem: Am Chem Soc; Sigma Xi. Res: Synthetic, structural and mechanistic studies in the chemistry of novel organic sulfur and organic phosphorus compounds; preparation and characterization of high molecular weight poly organo phosphazenes; nuclear magnetic resonance spectroscopy of macromolecules. Mailing Add: Dow Chem USA Eastern Div Res Ctr Granville OH 43023

WHITE, JESSE EDMUND, b Indianapolis, Ind, June 9, 27; m 50. PHYSICAL INORGANIC CHEMISTRY. Educ: Va Mil Inst, BS, 49; Ind Univ, PhD(chem), 58. Prof Exp: Asst prof chem, Lafayette Col, 55-59; from asst prof to assoc prof, 59-71, PROF CHEM, SOUTHERN ILL UNIV, EDWARDSVILLE, 71- Concurrent Pos: Petrol Res Fund fac award advan sci study, Mass Inst Technol, 63-64. Mem: Am Chem Soc; Nat Sci Teachers Asn. Res: Transition metal complexes. Mailing Add: Dept of Chem Southern Ill Univ Edwardsville IL 62026

WHITE, JESSE STEVEN, b Cleveland, Miss, May 9, 17; m. PARASITOLOGY. Educ: Delta State Col, BS, 40; Miss State Col, MS, 49; Univ Ala, PhD, 59. Prof Exp: Asst prof, 46-59, head div sci, 59-70, PROF BIOL, DELTA STATE UNIV, 59- Concurrent Pos: NSF fel, 58-59. Res: Medical entomology. Mailing Add: Dept of Biol Delta State Univ Cleveland MI 38732

WHITE, JOAN FULTON, zoology, developmental biology, deceased

WHITE, JOE LLOYD, b Pierce, Okla, Nov 8, 21; m 45; c 5. SOIL MINERALOGY, SOIL CHEMISTRY. Educ: Okla State Univ, BS, 44, MS, 45; Univ Wis, PhD(soil chem), 47. Prof Exp: From asst prof to assoc prof, 47-60, PROF AGRON, PURDUE UNIV, WEST LAFAYETTE, 60- Concurrent Pos: Rockefeller fel natural sci, Nat Res Coun, 53-54; NSF sr fel, Louvain, 65-66; Soil Sci Soc Am Rep, Earth Sci Div, Nat Res Coun, 70-73; Fulbright res scholar, Athens, 72-73; Guggenheim fel, Versailles, 72-73. Honors & Awards: Soil Sci Award, Am Soc Agron, 69. Mem: AAAS; Am Chem Soc; fel Am Soc Agron; fel Mineral Soc Am; Clay Minerals Soc (treas, 69-72). Res: Weathering of micaceous minerals; pesticide-soil colloid interactions; application of infrared spectroscopy to study of aluminosilicates; structure and properties of aluminum hydroxide gels. Mailing Add: Dept of Agron Purdue Univ West Lafayette IN 47906

WHITE, JOE WADE, b Dill City, Okla, Aug 22, 40; m 62; c 2. PHYSICAL ORGANIC CHEMISTRY. Educ: Okla State Univ, BS, 63; Univ Ariz, PhD(chem), 67. Prof Exp: Sr chemist, 67-71, res specialist chem, Indust Tape Lab, 71-72, res supvr, Indust Specialties Lab, 72-73, RES MGR, INDUST SPECIALTIES LAB, 3M CO, 73- Mem: Am Chem Soc; Soc Plastics Eng. Res: Kinetics and rheology of gelling polymerizations. Mailing Add: Indust Specialties Lab 3M Co St Paul MN 55101

WHITE, JOHN ANDERSON, b Bahia, Brazil, Oct 18, 19; m 48; c 1. VERTEBRATE ZOOLOGY, PALEONTOLOGY. Educ: William Jewell Col, AB, 42; Univ Kans, PhD(zool), 53. Prof Exp: Instr biol, William Jewell Col, 46-47 & Univ Ill, 53-55; prof, Calif State Col Long Beach, 55-66; CUR VERT PALEONT, MUS, IDAHO STATE UNIV, 66-, PROF BIOL, 70- Mem: Fel AAAS; Am Soc Mammal; Soc Syst Zool; Soc Vert Paleont; Paleont Soc. Res: Systematics, evolution and ecology of late Tertiary and Quaternary rodents. Mailing Add: Dept of Biol Idaho State Univ Pocatello ID 83201

WHITE, JOHN ARNOLD, b Chicago, Ill, Jan 30, 33; m 64; c 3. ATOMIC PHYSICS. Educ: Oberlin Col, BA, 54; Yale Univ, MS, 55, PhD(physics), 59. Prof Exp: Instr physics, Yale Univ, 58-59; instr, Harvard Univ, 59-62; res assoc, Yale Univ, 62-63; physicist, Nat Bur Standards, 63-64; res assoc physics, Univ Md, 65-66; assoc prof, 66-68, PROF PHYSICS, AM UNIV, 68- Concurrent Pos: Consult, Nat Bur Standards, 66-; NSF grants, Am Univ, 66, 67, 69 & 71; vis scientist, Mass Inst Technol, 72; res contracts, Off Naval Res, 73 & 74. Mem: AAAS; Am Phys Soc. Res: Atomic beams; magnetism of rare earth ions in solids; lasers; spontaneous emission in external fields; speed of light; unified time-length standardization; laser light scattering; critical point phenomena in fluids. Mailing Add: Dept of Physics Am Univ Washington DC 20016

WHITE, JOHN DAVID, b Newark, NJ, Feb 14, 28; m 51; c 1. BACTERIOLOGY. Educ: Univ Buffalo, BA, 48, MA, 50; Vanderbilt Univ, PhD(bact), 53. Prof Exp: Asst bact, Univ Buffalo, 48-50; asst bot, Vanderbilt Univ, 50-53; res bacteriologist, US Army Hosp, Camp Kilmer, 54-55; bacteriologist, Armed Forces Inst Path, 55-56; bacteriologist, Path Div, US Dept Army, Ft Detrick, 56-59, chief clin path br, 59-68, actg chief path div, 68-71; MICROBIOLOGIST, PATH DIV, US ARMY MED RES INST INFECTIOUS DIS, 71- Mem: Am Soc Microbiol; Sigma Xi; Am Soc Exp Path; NY Acad Sci; Electron Micros Soc Am. Res: Pathogenesis of infectious disease; immunology; fluorescent antibody methods; electron microscopy. Mailing Add: US Army Med Res Inst Infect Dis Path Div Ft Detrick Frederick MD 21701

WHITE, JOHN FRANCIS, b New Orleans, La, Feb 9, 21; m 50; c 4. GEOLOGY. Educ: Univ Calif, BS, 47, PhD, 55. Prof Exp: Geologist, Mining Co Guatemala, 47-48 & Consol Coppermines Corp, 48-51; consult, Hydrothermal Res Proj, 55-59, from asst prof to assoc prof, 55-71, PROF GEOL, ANTIOCH COL, 71- Mem: Geol Soc Am; Geochem Soc. Res: Economic geology; geomorphology; petrology. Mailing Add: Dept of Geol Antioch Col Yellow Springs OH 45387

WHITE, JOHN FRANCIS, b Indianapolis, Ind, July 21, 44; m; c 1. PHYSIOLOGY. Educ: Marian Col, BS, 66; Ind Univ, PhD(physiol), 70. Prof Exp: ASST PROF PHYSIOL, EMORY UNIV, 73- Concurrent Pos: NSF training fel, Univ Rochester, 71-72; NIH fel, Univ BC, 72-73. Mem: Biophys Soc. Res: Mechanisms of intestinal cotransport of sugars and amino acids with ions by absorptive cells; intracellular ionic activities and compartmentalization of ions. Mailing Add: Dept Physiol Div Basic Health Sci Emory Univ Atlanta GA 30322

WHITE, JOHN GREVILLE, b Saltcoats, Scotland, Mar 27, 22; nat US; m 53; c 3. CHEMISTRY. Educ: Glasgow Univ, BSc, 44, PhD(chem), 47. Prof Exp: Asst chem, Glasgow Univ, 45-47; from instr to asst prof, Princeton Univ, 47-55; mem tech staff, Radio Corp Am, 56-66; PROF PHYS CHEM, FORDHAM UNIV, 66- Mem: Am Chem Soc; Am Crystallog Asn; NY Acad Sci. Res: X-ray crystal structure analysis; complex organic structures; accurate small organic structures; inorganic structures; neutron diffraction. Mailing Add: Dept of Chem Fordham Univ Rose Hill Campus Bronx NY 10458

WHITE, JOHN IRVING, b Jerseyville, Ill, May 17, 18; m 46; c 3. PHYSIOLOGY. Educ: Univ Ill, AB, 39; Rutgers Univ, PhD(physiol, biochem), 50. Prof Exp: Res assoc, Med Sch, 50-52, from asst prof to assoc prof, 53-60, PROF PHYSIOL, DENT SCH, UNIV MD, BALTIMORE CITY, 60- Concurrent Pos: Guggenheim Mem Found fel, 58-59. Mem: AAAS; Am Chem Soc; Biophys Soc; Am Physiol Soc. Res: Physiology and chemistry of muscle contraction; protein metabolism and nutrition; salivary gland and cardiovascular physiology. Mailing Add: Dept of Physiol Univ of Md Sch of Dent Baltimore MD 21201

WHITE, JOHN JOSEPH, III, b Arlington, Mass, Apr 24, 39; m 68; c 2. SOLID STATE PHYSICS, APPLIED PHYSICS. Educ: Col William & Mary, BS, 60; Univ NC, PhD(physics), 65. Prof Exp: Res assoc physics, Univ NC, 65; asst prof, Univ Ga, 67-73; sr engr, BDM Corp, 73-74; RES SCIENTIST, COLUMBUS LABS, BATTELLE MEM INST, 74- Concurrent Pos: Consult, Oak Ridge Nat Lab, 72-73; dir, Ga State Sci Fair, 73. Mem: Am Phys Soc; Optical Soc Am; Am Asn Physics Teachers; Am Defense Preparedness Asn. Res: Applied mechanics; plasticity of metals; explosion containment; systems analysis; optical properties of silver halides, high resolution specific heat measurements; solid state of polymers, methods of data analysis. Mailing Add: Battelle Columbus Labs 505 King Ave Columbus OH 43220

WHITE, JOHN MARVIN, b Martin, Tenn, June 9, 37; m 56; c 2. GENETICS, ANIMAL BREEDING. Educ: Univ Tenn, BS, 59; Pa State Univ, MS, 64; NC State Univ, PhD(animal breeding), 67. Prof Exp: From asst prof to assoc prof, 67-74, PROF DAIRY SCI, VA POLYTECH INST & STATE UNIV, 74- Mem: Biomet Soc; Am Dairy Sci Asn; Am Soc Animal Sci. Res: Quantitative genetics; measurement of response to single and multiple trait selection and correlated responses in mice and dairy cattle. Mailing Add: Dept of Dairy Sci Va Polytech Inst & State Univ Blacksburg VA 24061

WHITE, JOHN MICHAEL, b Danville, Ill, Nov 26, 38; m 60; c 1. CHEMICAL PHYSICS. Educ: Harding Col, BS, 60; Univ Ill, MS, 62, PhD(chem), 66. Prof Exp: Asst prof, 66-70, ASSOC PROF CHEM, UNIV TEX, AUSTIN, 70- Concurrent Pos. Vis staff mem, Los Alamos Sci Lab. Mem: Am Chem Soc; Am Phys Soc. Res: Gas phase kinetics; energy transfer processes; photochemistry of simple molecules; surface chemistry. Mailing Add: Dept of Chem Univ of Tex Austin TX 78712

WHITE, JOHN ROBERT, b Kansas City, Mo, Nov 27, 45; m 67. COMPUTER SCIENCES, INFORMATION SCIENCES. Educ: Univ Calif, Santa Barbara, BA, 67, MS, 68, PhD(comput sci), 73. Prof Exp: From syst programmer res to res asst comput, Comput Syst Lab, Univ Calif, Santa Barbara, 67-71; Nat Defense Educ Act fel, 71-73, ASST PROF COMPUT SCI, ELEC ENG & COMPUT SCI, UNIV CONN, 73- Mem: Asn Comput Mach; Inst Elec & Electronics Engrs; Comput Soc. Res: Programming languages for enhancing software reliability and decreasing software development costs. Mailing Add: Comput Sci Group Univ of Conn Storrs CT 06268

WHITE, JOHN THOMAS, b El Paso, Tex, Aug 23, 31; m 58; c 3. MATHEMATICS. Educ: Univ Tex, BA, 52, MA, 53, PhD(math), 62. Prof Exp: Spec instr math, Univ Tex, 59-62; asst prof, Univ Kans, 62-65; ASSOC PROF MATH, TEX TECH UNIV, 65- Concurrent Pos: Fel, Tex Ctr Res, 66-; assoc dir, Comt Undergrad prog math, Univ Calif, Berkeley, 70-71. Mem: Math Asn Am; Am Math Soc; Soc Indust & Appl Math. Res: Distribution theory and transform analysis; integral transform theory. Mailing Add: Dept of Math Tex Tech Univ Lubbock TX 79409

WHITE, JOHN UNDERHILL, b Philadelphia, Pa, Dec 7, 11; m 48; c 5. PHYSICS. Educ: Harvard Univ, AB, 34; Univ Calif, PhD(physics), 40. Prof Exp: Physicist, Standard Oil Develop Co, NJ, 39-44 & Perkin-Elmer Corp, 44-49; PHYSICIST, WHITE DEVELOP CORP, 49- Mem: Am Phys Soc; fel Optical Soc Am; Soc Appl Spectros; Am Soc Testing & Mat. Res: Ultraviolet and infrared spectroscopy; flame and fluorescence photometry; design and development of optical systems and apparatus. Mailing Add: White Develop Corp 80 Lincoln Ave Stamford CT 06902

WHITE, JOHN W, b Ardmore, Okla, Aug 9, 33; m 54; c 3. HORTICULTURE. Educ: Okla State Univ, BS, 55; Colo State Univ, MS, 57; Pa State Univ, PhD(hort), 64. Prof Exp: From asst prof to assoc prof 64-75, PROF FLORICULT, PA STATE UNIV, UNIVERSITY PARK, 75-; rev ed, Jour Am Soc Hort Sci, 71- Honors & Awards: Garland Award, Am Carnation Soc. Mem: Am Soc Hort Sci. Res: Physical and chemical properties of soil; experimental designs for greenhouse structures; effects of the environment on plant growth. Mailing Add: Dept of Hort 101 Tyson Bldg Pa State Univ University Park PA 16802

WHITE, JONATHAN WINBORNE, JR, b State College, Pa, Sept 29, 16; m 43; c 2. AGRICULTURAL CHEMISTRY. Educ: Pa State Univ, BS, 37; Purdue Univ, MS, 39, PhD(agr chem), 42. Prof Exp: Asst agr chem, Purdue Univ, 37-42; from asst chemist to assoc chemist, Eastern Regional Res Lab, Bur Agr & Indust Chem, USDA, 42-44; assoc chemist, Off Censorship, 44-45; assoc chemist, Eastern Regional Res Lab, Bur Agr & Indust Chem, 45-47, chemist, 47-50, sr chemist & head honey invests, 50-55, prin chemist, Eastern Mkt & Nutrit Res Div, Agr Res Serv, 55-63, chief plant prod lab, 65-74, CHIEF CHEMIST, EASTERN MKT & NUTRIT RES DIV, AGR RES SERV, USDA, 63-, LEADER HONEY RES, 75- Mem: AAAS; Am Chem Soc; Inst Food Technol. Res: Carotenoid pigments of plants; rubber recovery from domestic plants; composition of apple essence; chemistry and utilization of honey, fruits and vegetables. Mailing Add: Eastern Regional Res Lab USDA Agr Res Serv Philadelphia PA 19118

WHITE, JOSEPH MALLIE, b Dallas, Tex, Dec 4, 21; m 50; c 2. ANESTHESIOLOGY. Educ: Univ Tex, MD, 47; Univ Iowa, MS, 50. Prof Exp: Instr anesthesiol, Univ Iowa Hosps, 50-51; asst prof, Sch Med, Univ Wash, 54-56; prof, Sch Med, Univ Okla, 56-68, assoc dean res & admin. 60-68, assoc dir med ctr, 65-68, from assoc dean to dean med fac, 65-68; vpres acad affairs & dean med, Univ Tex Med Br Galveston, 68-73; PROVOST HEALTH AFFAIRS, UNIV MO-COLUMBIA, 73- Concurrent Pos: Consult, Vet Admin Hosp, Oklahoma City, 56-68. Mem: Am Soc Anesthesiol; NY Acad Sci. Res: Pharmacology related to anesthesiology; pulmonary physiology. Mailing Add: Off Provost Health Affairs 111 Jesse Hall Univ of Mo Columbia MO 65201

WHITE, JULIUS, b Pittsburgh, Pa, Mar 23, 04; m 32; c 4. ORGANIC CHEMISTRY. Educ: Univ Denver, BA, 25, MA, 26; Univ Ill, PhD(org chem), 31. Prof Exp: Instr chem, Univ Ill, 26-31, fel, 31-32; Nat Res Coun fel med, Med Sch, Univ Mich, 32-34, res assoc biol chem, 34-38; fel physiol chem, Sch Med, Yale Univ, 38-39; fel, Nat Cancer Inst, 39-42, sr biochemist, 42, prin biochemist, 45-46, head chemist, 47-54, chief lab physiol, 55-70; PROF CHEM, MONTGOMERY COL, 71- Mem: AAAS; Am Chem Soc; Soc Exp Biol & Med; Am Asn Cancer Res; Am Inst Nutrit. Res: Isotopes as tracers in biological reactions; biological effects of ionizing radiation; tumor-host relationships. Mailing Add: Dept of Chem Montgomery Col Rockville MD 20852

WHITE, JUNE BROUSSARD, b Elizabeth, La, Aug 27, 24; div; c 2. INORGANIC CHEMISTRY, ANALYTICAL CHEMISTRY. Educ: La Polytech Univ, BS, 44; Univ Southwestern La, BS, 59, MS, 61; La State Univ, Baton Rouge, PhD(inorg chem), 70. Prof Exp: Anal chemist, Cities Serv Refining Corp, La, 44-45; chemist, Esso Lab, Standard Oil Co, NJ, La, 45-48; teacher, Parish Sch Bd, La, 53-59; asst prof chem, Univ Southwestern La, 61-66; NSF res partic, La State Univ, Baton Rouge, 64 & 66, instr, 66-68; chmn div natural sci, 69-71, PROF CHEM & CHMN DEPT, UNION UNIV, TENN, 68- Mem: Am Chem Soc. Res: Transition metal complexes which have d-2 electronic system; electron spin resonance; electronic transitions; magnetic properties. Mailing Add: Dept of Chem Union Univ Jackson TN 38301

WHITE, KENNETH L, SR, b Oct 15, 40; m 62; c 3. GEOGRAPHY. Educ: Calif State Univ, Fullerton, BA, 69, MA, 70; Univ Calif, Riverside, ABD, 74. Prof Exp: Lectr geog, San Bernardino Col, 72-73; lectr, 73-75, ASST PROF GEOG, CALIF STATE COL, SAN BERNARDINO, 75- Mem: Asn Am Geogr; Soil Sci Soc Am; Asn Pac Coast Geogr; Int Soc Soil Sci. Res: Relationships between clay genesis and geomorphology. Mailing Add: Dept of Geog Calif State Col San Bernardino CA 92507

WHITE, KERR LACHLAN, b Winnipeg, Man, Jan 23, 17; nat US; m; c 2. EPIDEMIOLOGY. Educ: McGill Univ, BA, 40, MD & CM, 49; Am Bd Internal Med, dipl, 57. Prof Exp: Personnel asst, RCA Victor Co, Can, 41-42; intern & resident med, Mary Hitchcock Mem Hosp, Hanover, NH, 49-52; from asst prof to assoc prof med & prev med, Sch Med, Univ NC, 53-62; prof epidemiol & community med & chmn dept, Col Med, Univ Vt, 62-64; chmn dept, 64-72, PROF HEALTH CARE ORGN, SCH HYG & PUB HEALTH, JOHNS HOPKINS UNIV, 64- Concurrent Pos: Hosmer fel med & psychiat, Royal Victoria Hosp, McGill Univ, 52-53; Commonwealth Fund advan fel, Med Res Coun Gt Brit & Sch Hyg & Trop Med, Univ London, 59-60; consult, Nat Ctr Health Statist & Health Resources Admin, Dept Health, Educ & Welfare, 66-; chmn, US Nat Comn Vital & Health Statist, 75-; mem expert panel orgn med care, WHO, 67; mem bd dirs, Found Child Develop, NY, 69-; consult, NSF, 74-; trustee, Case Western Reserve Univ, 74- Mem: Inst of Med of Nat Acad Sci; AAAS; AMA; fel Am Pub Health Asn; fel Am Col Physicians. Res: Medical care and education. Mailing Add: Dept Hlth Care Orgn Johns Hopkins Univ Sch Hyg & Pub Health Baltimore MD 21205

WHITE, KEVIN JOSEPH, b Queens, NY, Aug 28, 36; m 66. MOLECULAR PHYSICS. Educ: Georgetown Univ, BS, 58; Duke Univ, PhD(physics), 65. Prof Exp: Physicist, Naval Ord Lab, 58; res assoc physics, Duke Univ, 65; PHYSICIST, BALLISTIC RES LABS, 65- Mem: Am Phys Soc. Res: Millimeter wave microwave spectroscopy; Stark effect in rotational spectra; electron spin resonance; radiation damage in oxidizers; radical formation by atom addition and abstraction reactions; supersonic molecular beams for studying high pressure chemical reactions. Mailing Add: 406 Fowler Ct Joppa MD 21085

WHITE, LARRY DALE, b Sayre, Okla, Nov 24, 40; m 64; c 2. RANGE MANAGEMENT. Educ: Northern Ariz Univ, BS, 63; Univ Ariz, MS, 65, PhD(range ecol), 68. Prof Exp: Asst forester, US Forest Serv, 63; range adv, Near East Found, Kenya Govt, 67-69; ASST PROF RANGE ECOSYST MGT, UNIV FLA, 70- Mem: Soc Range Mgt; Ecol Soc Am. Res: Management and use of range ecosystems in Florida; livestock-wildlife habitat relationships; effects of forestry practices on understory vegetation; manipulation of range ecosystems. Mailing Add: Sch Forest Resources & Conserv Univ of Fla Gainesville FL 32611

WHITE, LARRY MELVIN, b Duchesne, Utah, May 6, 39; m 59; c 5. RANGE SCIENCE. Educ: Utah State Univ, BS, 61; Mont State Univ, MS, 71, PhD(crop & soil sci), 72. Prof Exp: Range conservationist, 61-66, RANGE SCIENTIST, AGR RES SERV, USDA, 66- Mem: Soc Range Mgt; Am Soc Agron; Crop Sci Soc Am; Weed Sci Soc Am. Res: Study of the phenology, physiology, morphology of growth and development of major range plant species in the northern Great Plains in relationship to range management practices. Mailing Add: North Plains Soil & Water Res Ctr PO Box 1109 Sidney MT 59270

WHITE, LAWRENCE S, b Chelsea, Mass, Mar 9, 23; m 46; c 2. PHYSICS. Educ: Mass Inst Technol, BS, 47. Prof Exp: Physicist, Res Lab, Titanium Div, Nat Lead Co, South Amboy, 48-62, sr technologist, 62-70; ASSOC PHYSICIST, HOFFMANN-LA ROCHE INC, NUTLEY, 70- Mem: Am Chem Soc. Res: Physical properties and colorimetry of titania pigments; electron microscopy and diffraction; light scattering; surface properties of pharmaceutical solids; scanning electron microscopy. Mailing Add: Hoffmann-La Roche Inc 340 Kingsland St Nutley NJ 07110

WHITE, LELAND DARRELL, b Porterville, Utah, Jan 31, 31; m 53; c 3. ENTOMOLOGY, BOTANY. Educ: Brigham Young Univ, BS, 57, MS, 59; Utah State Univ, PhD(agr entom), 64. Prof Exp: Asst prof entom, SDak State Univ, 63-65; RES ENTOMOLOGIST, AGR RES SERV, USDA, 65- Mem: Entom Soc Am. Res: Ecology, distribution and taxonomy of phytophagous mites; insect autocidal studies; release of Cobalt-60 sterilized codling moths in apple orchards of Pacific Northwest. Mailing Add: Yakima Agr Res Lab USDA 3706 Nob Hill Blvd Yakima WA 98902

WHITE, LELAND MARION, physical chemistry, see 12th edition

WHITE, LENDELL AARON, b Sabetha, Kans, Nov 10, 26; m 48; c 2. BACTERIOLOGY. Educ: Univ Kans, BA, 51, MA, 55; Nat Registry Microbiol, regist. Prof Exp: Bacteriologist, State Pub Health Lab, Kans, 51-53; asst, Virol Lab, Univ Kans, 53-54; bacteriologist, Spec Res Unit, Lab Br, 55-57, virus diag unit, 57-59, encephalitis sect, Tech Br, 59-60, venereal dis res lab, 60-63 & viral reagents unit, 63-71, SUPVRY RES MICROBIOLOGIST, CTR DIS CONTROL, USPHS, 71- Mem: AAAS; Am Soc Microbiol; Sigma Xi. Res: Virology; isolation, identification, typing and determination of antigenic relationships of viruses with established strains; serologic and antigenic relationships among the arthropod-borne encephalitides; infectivity and fluorescent antibody studies with Neisseria gonorrhoeae; production of viral reagents; immunization procedures for reference antisera. Mailing Add: 2534 Wilson Woods Dr Decatur GA 30033

WHITE, LEROY ALBERT, b New York, NY, June 24, 29; m 58; c 2. PHYSICAL CHEMISTRY. Educ: Mass Inst Technol, BS, 50; Columbia Univ, MS, 51. Prof Exp: Res chemist, Monsanto Chem Co, 51-55; PROJ LEADER, DeBELL & RICHARDSON, CONSULTS, HAZARDVILLE, 55- Mem: Am Chem Soc. Res: Organic synthesis; vinyl polymerization; nylon and epoxy reactions; general polymer development; coatings; membrane technology; photodegradable plastics; adhesives. Mailing Add: 78 Root Rd Somers CT 06071

WHITE, LEWIS L, b DeRidder, La, Oct 25, 24; m 48; c 3. EMBRYOLOGY. Educ: Southern Univ, BS, 48; Duquesne Univ, MS, 50; Univ Iowa, PhD(zool), 57. Prof Exp: Assoc prof embryol, 50-60, head dept biol, 60-70, PROF BIOL, SOUTHERN UNIV, BATON ROUGE, 60-, DEAN COL SCI, 70- Mem: Am Soc Zool. Res: Development and differentiation of sex. Mailing Add: Off of the Dean Col of Sci Southern Univ Baton Rouge LA 70813

WHITE, LEWIS THEODORE, forest pathology, natural science, see 12th edition

WHITE, LOCKE, JR, b South Boston, Va, Mar 5, 19. CHEMICAL PHYSICS. Educ: Davidson Col, BS, 39; Univ NC, PhD(phys chem), 43. Prof Exp: Asst anal chem, Davidson Col, 38-39; lab asst, Univ NC, 39-42; chemist, Naval Res Lab, DC, 42-45; chemist & head phys div, Southern Res Inst, 45-48, asst dir, 48-61; PROF PHYSICS, DAVIDSON COL, 61- Concurrent Pos: Lectr, Howard Col, 48; vis prof, Birmingham-Southern Col, 53-54 & 59; vis scientist, Univ Reading, 70-71. Mem: AAAS; Am Phys Soc; Am Asn Physics Teachers. Res: Properties of aerosols; electronic instrumentation; adsorption; solid state physics. Mailing Add: Dept of Physics Davidson Col Davidson NC 28036

WHITE, LOWELL DEANE, b New York, NY, Apr 18, 25; m 48; c 3. PHYSICS. Educ: Princeton Univ, AB, 49, MA, 51, PhD(phys chem), 56. Prof Exp: Instr physics, Princeton Univ, 53-55; MEM TECH STAFF, BELL TEL LABS, INC, 55- Mem: Am Phys Soc; Inst Elec & Electronics Eng. Res: Microwaves; impedance measurements. Mailing Add: Bell Labs 2525 Shadeland Ave Indianapolis IN 46206

WHITE, LOWELL ELMOND, JR, b Tacoma, Wash, Jan 16, 28; m 47; c 3. NEUROSURGERY, MEDICAL EDUCATION. Educ: Univ Wash, BS, 51, MD, 53. Prof Exp: Asst neurosurg, Sch Med, Univ Wash, 54-57, from instr to assoc prof, 57-70, assoc dean, 65-68; prof neurol surg & chief div, Univ Fla, 70-72; PROF NEUROSCI & CHMN DIV, UNIV S ALA, 72- Concurrent Pos: Guggenheim Found fel, Univ Oslo, 57-58; consult, Div Res Facil & Resources & chmn nat adv comt animal resources, NIH, 66-67; consult, Div Hosp & Med Facil, USPHS & grants admin adv comt, Dept of Health Educ & Welfare, 66-70. Mem: Am Asn Neuropath; Am Asn Anat; Am Asn Neurol Surg; AMA; Asn Am Med Cols. Res: Neuroanatomy. Mailing Add: Div of Neurosci Univ of S Ala Mobile AL 36608

WHITE, MACK, b Petersburg, Va, June 30, 21; m 47. MICROBIOLOGY. Educ: Va State Col, BS, 52. Prof Exp: Bacteriologist, US Army Biol Warfare Ctr, Ft Detrick, Md, 52-63; microbiologist, 63-70, MEM STAFF, ANTIBIOTICS & INSULIN CERT, US FOOD & DRUG ADMIN, 70- Mem: Am Soc Microbiol. Res: Biological agents; sterility testing of antibiotics; environmental control studies. Mailing Add: Antibiotics & Insulin Cert US Food & Drug Admin FB8 Rm 2858 Washington DC 20025

WHITE, MALCOLM LUNT, b Schenectady, NY, Aug 16, 27; m 51; c 3. PHYSICAL CHEMISTRY. Educ: Colgate Univ, BA, 49; Northwestern Univ, PhD(chem), 53. Prof Exp: Investr geochem, NJ Zinc Co, 53-59; res chemist, Am Cyanamid Co, 59-61; MEM TECH STAFF, BELL TEL LABS, 61- Mem: Am Chem Soc. Res: Nucleation; geochemistry; physical chemistry of colloid systems; surface chemistry; materials for electron device technology; coating and encapsulation of solid state devices. Mailing Add: Bell Tel Labs 555 Union Blvd Allentown PA 18103

WHITE, MARIAN E, anthropology, see 12th edition

WHITE, MARSHALL, wildlife biology, wildlife management, see 12th edition

WHITE, MAURICE LEOPOLD, b New York, NY, Sept 30, 28; m 51; c 2. MICROBIOLOGY. Educ: Univ Calif, Los Angeles, BA, 51, PhD, 57; Am Bd Microbiol, dipl, 66. Prof Exp: Chief dept bact, Cedars of Lebanon Hosp, Los Angeles, Calif, 57-63; CHIEF MICROBIOL UNIT, LAB SERV, WADSWORTH HOSP, VET ADMIN CTR, LOS ANGELES, 63-; INSTR MED MICROBIOL & IMMUNOL, SCH MED, UNIV CALIF, LOS ANGELES, 63- Concurrent Pos: Consult, Vet Admin Hosp, San Fernando, Calif, 60- Mem: AAAS; Am Soc Microbiol; Am Pub Health Asn; Brit Soc Gen Microbiol. Res: Staphylococcal phosphatase; nutrition and bacteriophage studies of Bordetella pertussis; taxonomy of Brucellaceae and Enterbacteriaceae; clinical microbiology. Mailing Add: Lab Serv Wadsworth Hosp Vet Admin Ctr Los Angeles CA 90073

WHITE, MAX GREGG, economic geology, see 12th edition

WHITE, MILTON GRANDISON, b Claremont, Calif, Jan 12, 10; m 35; c 3. EXPERIMENTAL HIGH ENERGY PHYSICS. Educ: Univ Calif, AB, 31, PhD(physics), 35. Prof Exp: Asst, Univ Calif, 31-33; Nat Res Coun fel physics, Princeton Univ, 35-37, from instr to asst prof, 37-40; consult, Nat Defense Res Comt, Radiation Lab, Mass Inst Technol, 40-42, div head transmitters, 42-44, div head airborne radar, 44-46; prof, 46-49, EUGENE HIGGINS PROF PHYSICS, PRINCETON UNIV, 49- Concurrent Pos: Consult, Brookhaven Nat Lab, 46- & NSF, 54; dir, Princeton Univ-Univ Pa Accelerator, 57-; trustee, Assoc Univs, Inc, 67-; vis, Inst Advan Studies, 71-; trustee, Univs Res Asn, 73-, chmn bd, 75. Mem: AAAS; Am Phys Soc. Res: High energy accelerators; nuclear and elementary particle physics; radiobiological effects of high energy heavy ions; microwave radar. Mailing Add: Joseph Henry Labs Jadwin Hall Princeton Univ PO Box 708 Princeton NJ 08540

WHITE, MYRON EDWARD, b Boston, Mass, May 1, 20; m 48; c 4. MATHEMATICS, COMPUTER SCIENCE. Educ: Wesleyan Univ, AB, 41; Columbia Univ, AM, 50, PhD(math), 62. Prof Exp: From instr to assoc prof math, 53-73, dir sec sci training progs, 63-73, PROF MATH, STEVENS INST TECHNOL, 73-, DIR MOVE AHEAD PROG, 74- Concurrent Pos: Teacher, NSF Math Insts. Mem: Am Math Soc; Math Asn Am; Math Soc France; Indian Math Soc. Res: Computer programming languages. Mailing Add: Dept of Math Stevens Inst Technol Hoboken NJ 07030

WHITE, NORMAN EDWARD, b Springfield, Ohio, Jan 20, 17; m 54; c 1. PHYSICAL CHEMISTRY. Educ: Wittenberg Univ, AB, 38; Univ Pa, MS, 41, PhD(phys chem), 54. Prof Exp: From instr to prof chem, Drexel Univ, 47-65; chmn dept, 65-71, PROF CHEM, BLOOMSBURG STATE COL, 65- Mem: AAAS; Am Chem Soc. Res: Hydrogen bond association; molecular weights in solution. Mailing Add: 6 Kent Rd RD 4 Bloomsburg PA 17815

WHITE, PAUL A, b Hollywood, Calif, Aug 21, 15; m 39; c 4. MATHEMATICS. Educ: Univ Calif, Los Angeles, AB, 37, MA, 39; Univ Va, PhD(math), 42. Prof Exp: Asst math, Univ Calif, Los Angeles, 37-39; instr, Univ Va, 39-42; asst prof, La State Univ, 42-46; from asst prof to assoc prof, 46-53, PROF MATH, UNIV SOUTHERN CALIF, 53- Concurrent Pos: Asst prof, Tulane Univ, 44; mathematician, Ballistics Res Lab, Aberdeen Proving Ground, 45; vis prof, Univ Innsbruck, 60-61; writer & lectr, African Ed Proj, 66-68; NSF lectr, India, 67; writer, UNESCO Arab Math Proj, 69-70; writer & adv bd mem, Sec Sch Math Curriculum Improv Study, 70-72; mem adv bd, Sch Math Study Group, 70-72. Mem: Am Math Soc. Res: Topology; R-regular convergence spaces. Mailing Add: Dept of Math Univ of Southern Calif Los Angeles CA 90008

WHITE, PAUL CHAPIN, b Boston, Mass, Oct 2, 41; m 64; c 2. PHYSICS. Educ: Harpur Col, BA, 63; State Univ NY Binghamton, MA, 66; Univ Tex, Austin, PhD(physics), 70. Prof Exp: Asst prof physics, St Edward's Univ, 69-74, chmn div phys & biol sci, 72-74; MEM STAFF, LOS ALAMOS SCI LAB, 74- Mem: AAAS; Am Asn Physics Teachers. Res: General relativistic astrophysics; inhomogeneous cosmologies; relativistic transport theory; radiation biophysics. Mailing Add: Los Alamos Sci Lab PO Box 1663 Los Alamos NM 87544

WHITE, PETER, b Philadelphia, Pa, June 12, 30; m 53; c 4. HEMATOLOGY. Educ: Yale Univ, BA, 51; Univ Pa, MD, 55. Prof Exp: From assoc to asst prof med, Sch

Med, Univ Pa, 63-69, assoc dir clin res ctr, 67-69; assoc prof med, 69-72, PROF MED & CHIEF DIV HEMAT, MED COL OHIO, 72-, DEP CHMN DEPT MED, 69- Concurrent Pos: USPHS res fel hemat, Sch Med, Univ Pa, 63-65; mem res in nursing in patient care rev comt, Bur Health Prof Educ & Manpower Training, NIH, 70-74. Mem: Am Col Physicians; AAAS; Am Soc Hemat; Am Fedn Clin Res. Res: Hemoglobin metabolism. Mailing Add: Dept of Med Med Col of Ohio PO Box 6190 Toledo OH 43614

WHITE, PHILIP CLEAVER, b Chicago, Ill, May 10, 13; m 39; c 3. CHEMISTRY. Educ: Univ Chicago, BS, 35, PhD(org chem), 38. Prof Exp: Res chemist, Stand Oil Co, Ind, 38-45, group leader, 45, chief chemist, 46-50, div dir, 50-51, mgr res, 56-58, gen mgr res & develop, 58-60; mgr res & develop, Pan Am Refining Corp, 51-56, gen mgr res & develop, Am Oil Co, 61-65, vpres res & develop, 66-69; GEN MGR RES, AMOCO RES CTR, STAND OIL CO, IND, 69- Concurrent Pos: Pres, Indust Res Inst, 71-72, vpres coord res coun, 71-72. Mem: Am Chem Soc; Soc Automotive Eng; Am Inst Chem Eng. Res: Petroleum products, processes and analysis; research administration. Mailing Add: Amoco Res Ctr Box 400 Stand Oil Co Warrinville Rd Naperville IL 60540

WHITE, PHILIP TAYLOR, b Anderson, Ind, Apr 15, 21; m 44; c 2. NEUROLOGY. Educ: George Washington Univ, MD, 46; Am Bd Psychiat & Neurol, dipl, 54. Prof Exp: Resident psychiat, Marion County Gen Hosp, Ind, 49-50, resident neurol, 50-51; fel neurol & EEG, Mayo Clin, Minn, 51-53; from asst prof to prof neurol, Med Ctr, Ind Univ, 53-64; chmn div allied sci, Barrow Neurol Inst, Phoenix, Ariz, 64-66; chmn dept neurol, 66-72, PROF NEUROL, MED COL WIS, 66-, ASSOC DEAN, 72- Concurrent Pos: Consult, Wood Vet Admin Hosp, Milwaukee; mem adv comt neurol & sensory dis, State Bd Health, Wis; mem, Nat Review Comt Regional Med Prog Servs, Dept Health, Educ & Welfare; asst examr, Am Bd Psychiat & Neurol, 59- Mem: Am Epilepsy Soc; Am Electroencephalog Soc; Am Neurol Asn; Asn Res Nerv & Ment Dis; Am Acad Neurol. Res: Clinical and electrographic aspects of epilepsy; educational methods in medicine; isotope methods; learning and behavior disorders. Mailing Add: Off of the Dean Med Col of Wis Milwaukee WI 53233

WHITE, R MILFORD, b Stewartsville, Mo, Jan 14, 32. PHYSICAL CHEMISTRY. Educ: Baker Univ, BA, 55; Univ Kans, PhD(phys chem), 60. Prof Exp: Asst prof chem, Jacksonville Univ, 60-62; from asst prof to assoc prof, 62-65, PROF CHEM, BAKER UNIV, 65- Concurrent Pos: Guest appointee, Oak Ridge Nat Lab, 71-72. Mem: Am Chem Soc. Res: Hot-atom chemistry; radiochemistry; mass spectrometry; nuclear chemistry; photoelectron spectroscopy. Mailing Add: Dept of Chem Baker Univ Baldwin KS 66006

WHITE, RALPH LAWRENCE, JR, b Troy, NC, June 19, 41; m 68. SYNTHETIC ORGANIC CHEMISTRY. Educ: Univ NC, Chapel Hill, BS, 63; Ind Univ, Bloomington, PhD(org chem), 67. Prof Exp: Fel, 67-68; instr med chem, Sch Pharm, Univ NC, Chapel Hill, 68-69; SR RES CHEMIST, NORWICH PHARMACAL CO, 69- Mem: Am Chem Soc. Res: Chemistry and synthesis of thiophene compounds; synthesis of potential biologically active compounds. Mailing Add: 2 Hillview Dr Norwich NY 13815

WHITE, RAY HENRY, b Lakewood, Ohio, Apr 28, 36; m 62; c 3. SOLID STATE PHYSICS. Educ: Calif Inst Technol, BS, 57; Univ Calif, Berkeley, PhD(physics), 64. Prof Exp: Res asst physics, Univ Calif, 58-63; lectr, Univ Singapore, 63-67; asst prof, Calif State Polytech Col, 67-68; asst prof physics, 68-70, chmn dept, 70-72, chmn dept sci & math, 72-75, ASSOC PROF PHYSICS, UNIV SAN DIEGO, 70- Concurrent Pos: Vis fel, Univ Singapore, 75-76. Mem: Am Phys Soc. Res: Superconductivity. Mailing Add: Dept of Physics Univ of San Diego San Diego CA 92110

WHITE, RAYMOND CYRUS, chemistry, see 12th edition

WHITE, RAYMOND E, b Freeport, Ill, May 6, 33; m 56; c 3. ASTRONOMY. Educ: Univ Ill, Urbana, BS, 55, PhD(astron), 67. Prof Exp: From instr to asst prof astron, 64-72, asst dir, Steward Observ, 72-74, RES ASSOC & LECTR ASTRON, STEWARD OBSERV, UNIV ARIZ, 74- Mem: AAAS; Am Astron Soc; Royal Astron Soc; Int Astron Union; Sigma Xi. Res: Observational astronomy; structure of the Milky Way Galaxy, particularly with respect to the identification and distribution of Population II stellar component. Mailing Add: Steward Observ Univ of Ariz Tucson AZ 85721

WHITE, RAYMOND GENE, b Elana, WVa, Oct 10, 30; m 52; c 2. VETERINARY SCIENCE. Educ: Okla State Univ, BS, 58, DVM, 60; Univ Nebr, MS, 71. Prof Exp: Res vet, Chemagro Corp, 64-69; field res vet, 69-72, EXTEN & RES VET, NORTH PLATTE STA, UNIV NEBR, 72- Mem: Am Vet Med Asn; Am Asn Exten Vet; Am Asn Bovine Practitioners; Am Asn Swine Practitioners; Am Soc Animal Sci. Res: Initiating and supervising field research activities involving animal health products, pesticides and anthelmintics. Mailing Add: Univ of Nebr North Platte Sta North Platte NE 69101

WHITE, RAYMOND PETRIE, JR, b New York, NY, Feb 13, 37; m 61; c 2. ANATOMY, ORAL SURGERY. Educ: Med Col Va, DDS, 62, PhD(anat), 67; Am Bd Oral Surg, dipl, 74. Prof Exp: Intern oral surg, Med Col Va, 64-65, from asst resident to resident, 65-67; from asst prof to assoc prof, Col Dent, Univ Ky, 67-71, chmn dept, 69-71; asst dean admin affairs & prof oral surg, Va Commonwealth Univ, 71-74; DEAN SCH DENT, UNIV NC, CHAPEL HILL, 74- Concurrent Pos: Consult, NIMH, 67-71, Vet Admin Hosp, Lexington, 67-71 & Huntington, 68-71 & McGuire Vet Admin Hosp, Richmond, Va, 71-74; mem adv comt, Am Bd Oral Surg, 74- Mem: AAAS; Am Acad Oral Path; Int Asn Dent Res; Am Soc Oral Surg; Sigma Xi. Res: Premalignant and malignant oral mucosa; correction facial deformity with surgery-orthodontic therapy. Mailing Add: Off of the Dean Sch of Dent Univ NC Chapel Hill NC 27514

WHITE, RICHARD ALAN, b Philadelphia, Pa, Oct 25, 35; m 65. DEVELOPMENTAL ANATOMY, MORPHOLOGY. Educ: Temple Univ, BS & MEd, 57; Univ Mich, MA, 59, PhD(bot), 62. Prof Exp: NSF fel, Univ Manchester, 62-63; from asst prof to assoc prof, 63-73, PROF PLANT ANAT, DUKE UNIV, 73- Mem: AAAS; Bot Soc Am; Torrey Bot Club; Am Fern Soc; Int Soc Plant Morphol. Res: Comparative morphology of tracheary cells of ferns; developmental studies of fern stelar patterns; comparative and developmental studies of lower vascular plants. Mailing Add: Dept of Bot Duke Univ Durham NC 27706

WHITE, RICHARD ARNOLD, physics, see 12th edition

WHITE, RICHARD EARL, b Akron, Ohio, Aug 23, 33. SYSTEMATIC ENTOMOLOGY. Educ: Univ Akron, BS, 57; Ohio State Univ, MSc, 59, PhD(entom), 63. Prof Exp: Asst prof zool, Union Col, Ky, 64-65; RES SCIENTIST, USDA, 65- Mem: AAAS; Entom Soc Am. Res: Taxonomy of Coleoptera, especially the families Anobiidae and Chrysomelidae. Mailing Add: Syst Entom Lab c/o US Nat Mus Natural Hist Washington DC 20560

WHITE, RICHARD PAUL, b Gary, Ind, May 27, 25; m 47; c 2. PHARMACOLOGY. Educ: Ind State Univ, BS, 47; Univ Kans, MS, 49, PhD(physiol), 54. Prof Exp: Instr physiol, Univ Kans, 49-54; psychophysiologist, Galesburg State Res Hosp, Ill, 54-56; from instr to assoc prof, 56-70, PROF PHARMACOL, MED UNITS, UNIV TENN, MEMPHIS, 70- Concurrent Pos: NIH career develop award, 59- Mem: Soc Biol Psychiat; Int Col Neuropsychopharmacol; Soc Neurosci; Int Soc Biochem Pharmacol. Res: Neuropharmacology. Mailing Add: Dept of Pharmacol Univ of Tenn Med Units Memphis TN 38163

WHITE, RICHARD T, physics, see 12th edition

WHITE, RICHARD WALLACE, b Buffalo, NY, June 6, 30; m 60; c 1. UNDERWATER ACOUSTICS. Educ: Univ Idaho, BS, 53; Wash State Univ, PhD(physics), 70. Prof Exp: Physicist, Atomic Energy Div, Phillips Petrol Corp, Idaho, 60-61 & Stanford Res Inst, 61-63; PHYSICIST, NAVAL UNDERSEA CTR, 68- Res: Wave propagation and shock phenomena; acoustic ray tracing in inhomogeneous moving media.

WHITE, RICHARD WILLIAM, b Rudyard, Mich, Mar 6, 36; m 67. GEOLOGY. Educ: Mich Col Mining & Technol, BS, 61; Univ Calif, Berkeley, PhD(geol), 65. Prof Exp: Asst prof geol, Lake Superior State Col, Mich Technol Univ, 66; GEOLOGIST, US GEOL SURV, 66- Res: Petrology of ultramafic and basaltic rocks; geology of Liberia, West Africa; geochemistry of weathering of igneous rocks; environmental geochemistry of the Snake River Plain; geochemistry and health. Mailing Add: Br Regional Geochem US Geol Surv Denver CO 80225

WHITE, ROBERT B, b Ennis, Tex, Jan 5, 21; m 42; c 3. PSYCHIATRY. Educ: Tex A&M Univ, BS, 41; Univ Tex, MD, 44; Western New Eng Inst Psychoanal, cert, 59. Prof Exp: From asst psychiatrist to sr psychiatrist, Austen Riggs Ctr, Inc, 51-62; assoc prof psychiat, 62-67, PROF PSYCHIAT, UNIV TEX MED BR GALVESTON, 67- Concurrent Pos: Teaching analyst, New Orleans Psychoanal Inst, 62-66, training analyst psychoanal, 66-; consult, Alcoholism Res Proj, Col Med, Baylor Univ, 63-65 & Hedgecroft Hosp, Houston, 63-65; mem gen planning comt comprehensive statewide ment health prog planning, Tex State Dept Health, Austin, 63-65; mem, Residency Review Comt for Psychiatry & Neurology, 69-; training analyst, Houston-Galveston Psychoanalytic Sch, 74- Honors & Awards: David Rapaport Prize, Western New Eng Inst Psychoanal, 59. Mem: AAAS; Am Psychoanalytic Asn; fel Am Psychiatric Asn; fel Am Col Psychiatrists; fel Am Col Psychoanal. Res: Psychoanalysis. Mailing Add: Dept of Neurol & Psychiat Univ of Tex Med Br Galveston TX 77550

WHITE, ROBERT F, statistics, see 12th edition

WHITE, ROBERT J, b Duluth, Minn, Jan 21, 26; m 50; c 10. SURGERY, NEUROPHYSIOLOGY. Educ: Univ Minn, BS, 51, PhD(neurosurg physiol), 62; Harvard Univ, MD, 53. Prof Exp: Intern surg, Peter Bent Brigham Hosp, 53-54; resident, Boston Children's Hosp & Peter Bent Brigham Hosp, 54-55; asst to staff, Mayo Clin, 58-59, res assoc neurophysiol, 59-61; from asst prof to assoc prof, 61-66, PROF NEUROSURG, SCH MED, CASE WESTERN RESERVE UNIV, 66-, CO-CHMN DEPT, 72-; DIR NEUROSURG & BRAIN RES LAB, CLEVELAND METROP GEN HOSP, 61- Concurrent Pos: Fel neurosurg, Mayo Clin, 55-58; assoc neurosurgeon, Univ Hosps & sr attend neurosurgeon, Vet Admin Hosp, 61-; gen ed, Int Soc Angiol J, 66; lectr, USSR, 66, 68, 70 & 71; ed for western hemisphere, Resuscitation; mem med adv bd, Nat Paraplegia Found; assoc ed, Surg Neurol. Honors & Awards: Res Award, Mayo Clin, 58. Mem: Am Asn Neurol Surg; Cong Neurol Surg; Am Acad Neurol; Soc Univ Neurosurg; AMA. Res: Special neurosurgical techniques for vascular disease of brain utilizing hypothermia and extracorporeal perfusion systems; development of isolated primate brain; technique of brain transplantation and neurochemical and brain circulation studies; utilizing hypothermia. Mailing Add: Dept of Neurosurg Cleveland Metrop Gen Hosp Cleveland OH 44109

WHITE, ROBERT KELLER, b Greeneville, Tenn, Mar 3, 30; m 52; c 2. PSYCHOPHYSIOLOGY, RADIOBIOLOGY. Educ: Milligan Col, BA, 52; Univ Tex, PhD(psychol), 62. Prof Exp: Res scientist, Radiobiol Lab, Balcones Res Ctr, Univ Tex, 56-61; asst prof psychol, Tex Tech Col, 61-65; proj dir, Armed Forces Radiobiol Res Lab, Defense Atomic Support Agency, 65-67; mem tech staff space res, Bellcomm, Inc, 67-70; chmn dept, 70-71, PROF PSYCHOL, WILLIAM PATERSON COL NJ, 70- Concurrent Pos: Lectr, USDA Grad Sch, 65- & Col Gen Studies, George Washington Univ, 66-67. Mem: Am Psychol Asn; Simulation Coun; Am Soc Cybernet. Res: Effects of whole body irradiation upon the physiology and behavior of various species; in uterine irradiation of rats; gamma neutron pulse irradiation upon the psychophysiology of monkeys. Mailing Add: Dept of Psychol William Paterson Col of NJ Wayne NJ 07470

WHITE, ROBERT LEE, b Plainfield, NJ, Feb 14, 27; m 52; c 4. PHYSICS. Educ: Columbia Univ, BA, 49, MA, 51, PhD(physics), 54. Prof Exp: Asst geophys, Columbia Univ, 49, asst physics, 49-51, lectr physics & asst chem, 51-52, asst physics, 52-54; res physicist, Res Labs, Hughes Aircraft Co, 54-61; head magnetics dept, Labs, Gen Tel & Electronics Corp, 61-63; PROF ELEC ENG & MAT SCI, STANFORD UNIV, 63-, DIR, INST ELECTRONICS IN MED, 73- Concurrent Pos: Indust consult, 63-; Guggenheim fel, Oxford Univ, 69-70. Mem: Fel Am Phys Soc; sr mem Inst Elec & Electronics Eng. Res: Microwave spectroscopy; solid state physics, especially magnetics; neurophysiology; neural prostheses, especially auditory. Mailing Add: 23930 Jabil Ln Los Altos Hills CA 94022

WHITE, ROBERT M, b Boston, Mass, Feb 13, 23; m 48; c 2. METEOROLOGY. Educ: Harvard Univ, BA, 44; Mass Inst Technol, MS, 49, ScD, 50. Prof Exp: Asst meteorol, Mass Inst Technol, 48-50; chief large scale processes sect, Air Force Cambridge Res Ctr, 52-58 & meteorol develop lab, 58-59; res assoc, Mass Inst Technol, 59; assoc dir res, Travelers Ins Co, 59-60, pres, Travelers Res Ctr, Inc, 60-63; chief weather bur, 63-65 & Environ Sci Serv Admin, 65-70, ADMINISTR, NAT OCEANIC & ATMOSPHERIC ADMIN, US DEPT COM, 71- Concurrent Pos: US Dept Com rep, Comt Water Resources Res, Fed Coun Sci & Technol; mem exec comt, US rep & chmn ad hoc comt, World Meteorol Orgn; mem, Vis Comt, Div Eng & Appl Physics, Harvard Univ; mem, Vis Comt, Dept Meteorol & Earth & Planetary Sci, Mass Inst Technol; mem, Vis Comt, Col Eng, Univ Okla; mem bd govs, US Power Squadron. Honors & Awards: Rosenberger Medal, Univ Chicago; Cleveland Abbe Award, Am Meteorol Soc. Mem: Am Geophys Union; Am Meteorol Soc; Am Soc Oceanog; Marine Technol Soc; Royal Meteorol Soc. Res: Atmospheric and ocean sciences; environmental science. Mailing Add: Nat Oceanic & Atmospheric Admin Washington Sci Ctr US Dept Com Washington DC 20852

WHITE, ROBERT MANSON, b Mokanshan, China, July 14, 16; US citizen; m 41; c 3. ANTHROPOMETRICS, PHYSICAL ANTHROPOLOGY. Educ: Haverford Col, BS, 39. Prof Exp: Anthropologist, US Army Quartermaster Climatic Res Lab, Mass, 48-52, anthropologist, Environ Protection Div, US Army Quartermaster Res & Develop Ctr, 52-61, res anthropologist, Pioneering Res Div, US Army Natick Labs, 61-70, RES ANTHROPOLOGIST, CLOTHING, EQUIPMENT & MAT ENG LAB,

US ARMY NATICK DEVELOP CTR, 70- Concurrent Pos: Consult anthropologist, NATO Anthropometric Surv Turkey, Greece & Italy, US Air Force, 60-61; anthropologist in charge, Anthropometric Surv Thailand & Vietnam, Advan Res Proj Agency, 62-63, consult anthropologist, Anthropometric Surv Latin Am, 65-70 & Iran, 68-69. Mem: Fel Human Factors Soc; Am Asn Phys Anthropologists; fel Am Anthrop Asn; Sigma Xi. Res: Anthropometry and physical anthropology of United States military personnel and civilian populations; anthropometry of foreign military and civilian populations; applications of anthropometric data and human engineering in clothing, vehicles, aircraft, and man-equipment systems. Mailing Add: 343 Holden Wood Rd Concord MA 01742

WHITE, ROBERT MARSHALL, b Reading, Pa, Oct 2, 38. SOLID STATE PHYSICS. Educ: Mass Inst Technol, BS, 60; Stanford Univ, PhD(physics), 64. Prof Exp: Res physicist, Lincoln Lab, Mass Inst Technol, 60; res assoc, Microwave Lab, Stanford Univ, 63-64; NSF fel, Univ Calif, Berkeley, 65-66; asst prof physics, Stanford Univ, 66-71; NSF sr rel, Cambridge Univ, 71-72; PRIN SCIENTIST, XEROX CORP, 72- Mem: Inst Elec & Electronics Engrs. Res: Theory of nonlinear phenomena in ferrites; spinwave theory; magneto-optical phenomena; metal-ion-metal transition; magnetic properties of amorphous materials. Mailing Add: Xerox Palo Alto Res Ctr 3333 Coyote Hill Palo Alto CA 94304

WHITE, ROBERT ROSS, microbiology, see 12th edition

WHITE, ROBERT STEPHEN, b Elsworth, Kans, Dec 28, 20; m 42; c 4. ASTROPHYSICS, SPACE PHYSICS. Educ: Southwestern Col, Kans, AB, 42; Univ Ill, MS, 43; Univ Calif, PhD(physics), 51. Hon Degrees: DSc, Southwestern Col, Kans, 71. Prof Exp: Asst physics, Univ Ill, 42-44; asst, Univ Calif, 46-48; physicist, Lawrence Radiation Lab, Univ Calif, 48-61; physicist & head particles & fields dept, Space Physics Lab, Aerospace Corp, 61-67; chmn dept physics, 70-73, PROF PHYSICS, UNIV CALIF, RIVERSIDE, 67-, DIR, INST GEOPHYS & PLANETARY PHYSICS, 67- Concurrent Pos: Lectr, Univ Calif, 53-54 & 57-59; NSF sr fel, 61-62. Mem: AAAS; Am Phys Soc; Am Geophys Union; Am Astron Soc. Res: Space physics and astrophysics; radiation belts of Earth and Jupiter; albedo neutrons from Earth; solar neutrons and gamma-rays; gamma-rays from astrophysical sources. Mailing Add: Dept of Physics Univ of Calif Riverside CA 92502

WHITE, ROBERT WINSLOW, b Somerville, Mass, Mar 28, 34; m 59; c 2. ORGANIC CHEMISTRY, POLYMER CHEMISTRY. Educ: Mass Inst Technol, SB, 55; Univ Ill, PhD(chem), 59. Prof Exp: Chemist, Rohm and Haas Co, Pa, 58-69; dept head, Nat Lead Co, NY, 69-72; tech dir, Baker Castor Oil Co, 72-74; BUS MGR, INDUST CHEM DIV, NL INDUST, 74- Mem: Am Chem Soc; The Chem Soc; Swiss Chem Soc. Res: Organic synthesis; process and product development; coatings technology, urethanes, waxes, surface active agents. Mailing Add: NL Indust PO Box 700 Hightstown NJ 08520

WHITE, ROBERTA JEAN, b Loup City, Nebr, Dec 8, 26. VIROLOGY. Educ: Univ Nebr, AB, 48, MS, 54; Univ Calif, PhD(bact), 61. Prof Exp: Technician, Mayo Clin, 48-51; technician, Col Med, Univ Nebr, 51-52, instr microbiol, 54-56; asst bact, Univ Calif, 56-60; asst prof, 61-68, ASSOC PROF VIROL, COL MED, UNIV NEBR, OMAHA, 68- Mem: AAAS; Am Soc Microbiol. Res: Pathogenesis of virus diseases and multiplication. Mailing Add: Dept of Med Microbiol Univ of Nebr Med Ctr Omaha NE 68105

WHITE, RONALD JEROME, b Wibaux, Mont, Oct 31, 36; m 61; c 2. ZOOLOGY, PHYSIOLOGY. Educ: Calif State Polytech Col, BS, 59; Ore State Univ, MS, 61, PhD(physiol), 68. Prof Exp: Asst prof, 65-69, ASSOC PROF BIOL, EASTERN WASH STATE COL, 69- Mem: AAAS; Am Soc Zool. Res: Physiology of reproduction in the pigtail macaque. Mailing Add: Dept of Biol Eastern Wash State Col Cheney WA 99004

WHITE, RONALD JOSEPH, b Opelousas, La, Dec 4, 40; m 63; c 3. THEORETICAL CHEMISTRY, APPLIED MATHEMATICS. Educ: Univ Southwestern La, BS, 63; Univ Wis, PhD(phys chem), 68. Prof Exp: NSF fel, Oxford Univ, 67-68; assoc, Bell Tel Labs, 68-70; asst prof math, Univ Southwestern La, 70-73; res assoc physiol & biophys, Med Ctr, Univ Miss, 73-75; ASSOC PROF MATH & DIR UNIV HONORS PROG, UNIV SOUTHWESTERN LA, 75- Mem: Am Phys Soc; Sigma Xi. Res: Perturbation theory; quantum mechanics of small atoms and molecules; mathematical models of physical systems; physiological models. Mailing Add: Univ Honors Prog Univ Southwestern La Lafayette LA 70501

WHITE, RONALD KEITH, b Benton, Ill, Mar 1, 36; m 70. ENERGY CONVERSION, ENVIRONMENTAL PHYSICS. Educ: Purdue Univ, BSEE, 57, MS, 64, PhD(physics), 69; Cornell Univ, MBA, 59. Prof Exp: Res physicist, Lawrence Radiation Lab, Univ Calif, 69-71; RES PHYSICIST, STANFORD RES INST, 71- Mem: Am Phys Soc. Res: Interaction of radiation and matter; energy and environmental technology. Mailing Add: Stanford Res Inst 333 Ravenswood Ave Menlo Park CA 94025

WHITE, RONALD PAUL, SR, b Coral Gables, Fla, Mar 1, 35; m 60; c 4. SOIL CHEMISTRY, SOIL FERTILITY. Educ: Mass State Col Bridgewater, BS, 61; Univ Mass, MS, 63; Mich State Univ, PhD(soil sci), 68. Prof Exp: Instr soil sci, Mich State Univ, 65-66; RES SCIENTIST, RES STA, CAN DEPT AGR, 68- Mem: Am Soc Agron; Soil Sci Soc Am; Agr Inst Can. Res: Soil fertility requirements and crop management practices of corn and potatoes; soil manganese levels and plant manganese toxicity; soil phosphorus; plant root cation exchange capacity. Mailing Add: Forage Sect Res Sta Can Dept of Agr PO Box 1210 Charlottetown PE Can

WHITE, ROSCOE BERYL, b Freeport, Ill, Dec 20, 37; m 66. HIGH ENERGY PHYSICS, PLASMA PHYSICS. Educ: Univ Minn, BS, 59; Princeton Univ, PhD(physics), 63. Prof Exp: Res asst, Princeton Univ, 62, instr, 62-63; res assoc, Univ Minn, 63; US Acad Sci exchange scientist, Lebedev Inst, Moscow, 63-64; vis scientist, Int Ctr Theoret Physics, Italy, 64-66; asst prof physics, Univ Calif, Los Angeles, 66-72; mem, Inst Advan Study, 72-74; RES PHYSICIST, PRINCETON UNIV, 74- Res: Theoretical plasma physics. Mailing Add: Dept of Physics Princeton Univ Princeton NJ 08540

WHITE, ROSEANN SPICOLA, b Tampa, Fla, Aug 4, 43; m 65. BIOCHEMISTRY, MICROBIOLOGY. Educ: Univ Fla, BS, 65; Univ Tex Southwestern Med Sch, Dallas, PhD(biochem), 70. Prof Exp: Asst prof, 69-72, ASSOC PROF MICROBIOL, FLA TECHNOL UNIV, 73- Concurrent Pos: Clin chemist, Orange Mem Hosp, 69-72; Sigma Xi grant. Mem: AAAS; Am Soc Microbiol; Sigma Xi. Res: Control vitamin B-6 biosynthesis; vitamin B-6 regulation of apoenzyme levels; trace elemental analysis of plants; chemical characterization of H capsulatum antigens. Mailing Add: Dept of Biol Sci Fla Technol Univ Orlando FL 32816

WHITE, RUDOLPH CARTER, inorganic chemistry, see 12th edition

WHITE, RUSSELL ALAN, b New York, NY, Dec 9, 36; m 58; c 2. GEOGRAPHY. Educ: Hunter Col, BA, 58; Columbia Univ, MA, 62, PhD(geog), 68. Prof Exp: Instr geog, Hofstra Univ, 62-68; asst prof, 68-72, ASSOC PROF GEOG & CHMN DEPT GEOL & GEOG, HUNTER COL, 73- Mem: Asn Am Geogr; Am Geog Soc. Res: Political historical geography and water resources of the North Mexican states. Mailing Add: Dept Geol & Geog Hunter Col Box 1709 New York NY 10021

WHITE, SIDNEY EDWARD, b Manchester, NH, Mar 14, 16; m 46; c 1. GEOLOGY. Educ: Tufts Col, BS, 39; Harvard Univ, MA, 42; Syracuse Univ, PhD(geol), 51. Prof Exp: Lab asst, Tufts Col, 37-40, lab instr, 41-42, instr geol, 47-48; lab instr, Harvard Univ, 42; asst instr geol, Syracuse Univ, 48-51; from asst prof to assoc prof, 51-67, PROF GEOL, OHIO STATE UNIV, 67- Concurrent Pos: Recorder, US Geol Surv, 41-42, geologist, Mass, 46-48; ed, Geol Soc Am Jour, 62-71; geologist, NSF Projs, Univ Colo, 64-66, assoc prof, 65-66; Univ Colo Men & Women Scholastic Honoraries award, 65-66. Mem: Geol Soc Am; Am Quaternary Asn; Mex Geol Soc. Res: Geomorphology; glacial and Pleistocene geology; glaciology; volcanology; alpine and periglacial mass movement studies. Mailing Add: Dept of Geol & Mineral Ohio State Univ Columbus OH 43210

WHITE, STANLEY C, b Lebanon, Ohio, Jan 13, 26; m 48; c 5. PREVENTIVE MEDICINE. Educ: Miami Univ, AB, 45; Univ Cincinnati, MD, 49; Johns Hopkins Univ, MPH, 53. Prof Exp: Asst flight surgeon, Eglin Air Force Base, Fla, 50-51; Med Corps, US Air Force, 51-, resident aviation med, Sch Aviation Med, Tex, 51-52, Sch Pub Health & Hyg, Johns Hopkins Univ, 52-53 & Hq Tactical Air Command, Langley Air Force Base, 53-54, chief respiration sect, Physiol Br, Aeromed Lab, Wright Air Develop Ctr, Ohio, 54-58, chief life systs div space task group & Manned Spacecraft Ctr, NASA, Langley Air Force Base & Houston, Tex, 54-63, dir bioastronaut & systs support, Aerospace Med Div, Brooks Air Force Base, 63-64, asst dep res & develop, 64-66, asst bioastronaut to prog dir, Manned Orbiting Lab Prog, DC, 66-69, dir biomed res, Hq Air Force Systs Command, Andrews Air Force Base, Md, 69-70, Skylab med opers officer, Hq NASA, DC, 70-71, asst dir life sci for sci activities, Hq, NASA, 71-74, MIL ASST FOR MED & LIFE SCI & ASST DIR ENVIRON & LIFE SCI, OFF DIR DEFENSE RES & ENG, PENTAGON, 74- Concurrent Pos: Mem bioastronaut comt, Nat Res Coun, 59-60; space surgeon, US Dept Defense, 60. Honors & Awards: Boyington Award, Am Astronaut Soc, 60; Louis G Bauer Award, Aerospace Med Asn, 62, Theodore C Lyster Award, 74 & Hubertus Strughold Award, Space Med Br, 69; Honor Citation, AMA, 62 & Air Force Asn, 63; Laureate Award, Int Acad Aviation & Space Med, 62. Mem: Fel Am Col Prev Med; fel Aerospace Med Asn; Int Acad Astronaut; Int Acad Aviation & Space Med. Res: Aviation medicine leading to manned aerospace flight; vehicle design and operational flight support. Mailing Add: 2215 Calhoun St Oxon Hill MD 20022

WHITE, STANTON M, b Gardner, Mass, Mar 10, 35; m 58; c 2. SEDIMENTARY PETROLOGY, MARINE GEOLOGY. Educ: Univ Mass, Amherst, BS, 56; Univ Rochester, MS, 60; Univ Wash, PhD(marine geol), 67. Prof Exp: Instr geol, Columbia Basin Col, 60-64; instr oceanog, Univ Wash, 65-67; from asst prof to assoc prof geol, 67-74, PROF GEOL, CALIF STATE UNIV, FRESNO, 74- Concurrent Pos: Res scientist, Deep Sea Drilling Proj, 72- Mem: AAAS; Geol Soc Am; Soc Econ Paleont & Mineral. Res: Mineralogy, petrology and geochemistry of continental shelf sediments; marine sediments. Mailing Add: Dept of Geol Calif State Univ Fresno CA 93710

WHITE, STEPHEN HALLEY, b Wewoka, Okla, May 14, 40; m 61; c 6. BIOPHYSICS, PHYSIOLOGY. Educ: Univ Colo, BA, 63; Univ Wash, MS, 65, PhD(physiol, biophys), 69. Prof Exp: Asst prof, 72-75, vchmn dept, 74-75, ASSOC PROF PHYSIOL, UNIV CALIF, IRVINE, 75- Concurrent Pos: USPHS grant biochem, Univ Va, 71-72; res grants, NSF & NIH. Mem: AAAS; Biophys Soc; NY Acad Sci; Am Physiol Soc. Res: Structure of biological membranes and the physical properties of lipid bilayer membranes. Mailing Add: Dept of Physiol Col of Med Univ of Calif Irvine CA 92717

WHITE, SUSAN RUTH, b Ft Wayne, Ind, Oct 13, 42; m 64; c 1. NEUROSCIENCES. Educ: Purdue Univ, BS, 64; Ind Univ, PhD(psycho-pharmacol), 71. Prof Exp: Fel neurophysiol, Ind State Univ, 72-74; res scientist, 74-75, ASST PROF NEUROSCI, MEM UNIV NFLD, 75- Mem: Soc Neurosci; Sigma Xi. Res: Investigating the neurophysiological and behavioral changes which accompany the development of experimental allergic encephalomyelitis. Mailing Add: Fac of Med Mem Univ of Nfld St John's NF Can

WHITE, THEODORE ELMER, vertebrate paleontology, see 12th edition

WHITE, THOMAS DAVID, b Sarnia, Ont, Apr 8, 43; m 65; c 3. NEUROCHEMISTRY, NEUROPHARMACOLOGY. Educ: Univ Western Ont, BSc, 65, MSc, 67; Bristol Univ, PhD(pharmacol), 70. Prof Exp: ASST PROF PHARMACOL, FAC MED, DALHOUSIE UNIV, 71- Concurrent Pos: Med Res Coun Can fel, Univ Alta, 70-71. Mem: Can Pharmacol Soc. Res: Physiology and pharmacology of brain synapses. Mailing Add: Dept of Pharmacol Dalhousie Univ Fac of Med Halifax NS Can

WHITE, THOMAS GAILAND, b Artesia, NMex, Oct 12, 32; m 58; c 2. PLANT BREEDING. Educ: NMex State Univ, BS, 54; Tex A&M Univ, MS, 58, PhD(plant breeding), 62. Prof Exp: From instr to asst prof cotton genetics, Tex A&M Univ, 60-65; res geneticist, USDA, 65-67; dir biol res, Occidental Petrol Corp, 67-71; crops develop mgr, 71-74, PLANNING COORDR, GILROY FOODS, INC, 74- Concurrent Pos: Leader monosomic res proj, Nat Cotton Coun Grant, Found Cotton Res & Educ, 63-66; mem, Agr Res Inst. Mem: Am Soc Hort Sci; Am Mgt Asn. Res: Basic genetic and cytogenetic research in cotton; applied research in controlled atmosphere use in transport of perishable commodities; genetics and improvement of onions; management, research and development. Mailing Add: Gilroy Foods Inc PO Box 1088 Gilroy CA 95020

WHITE, THOMAS GEORGE, b Sulphur, Okla, Sept 10, 22; m 46; c 1. IMMUNOPATHOLOGY. Educ: Univ Okla, BS, 48; Harvard Univ, MS, 52; Univ Minn, PhD(bact, immunol), 57. Prof Exp: Fel brucellosis, Univ Minn, 57-58; sr res microbiologist, Nat Animal Dis Lab, Agr Res Serv, USDA, 58-63; IMMUNOLOGIST, US VET ADMIN, 63- Mem: Am Soc Microbiol. Res: Non-specific serum agglutinins; kinetics of heat inactivation of serum agglutinins; isolation and immunological characterization of bacterial antigens; serological typing and relationship to immunogenicity; pathogenesis of rheumatoid arthritis. Mailing Add: Vet Admin Hosp 4801 Linwood Blvd Kansas City MO 64128

WHITE, THOMAS TAYLOR, b New York, NY, 20. SURGERY, PHYSIOLOGY. Educ: Harvard Univ, BS, 42; NY Univ, MD, 45; Am Bd Surg, dipl, 53. Prof Exp: Instr surg, NY Univ, 50-52; from instr to assoc prof, 53-67, PROF SURG, SCH MED, UNIV WASH, 67- Concurrent Pos: Columbia Univ Lambert traveling fel, Europe, 52-53; Am Cancer Soc res grant, 59-60; USPHS grant, 59-; from Guggenheim Found fel, Univ Lyon, 64-65. Honors & Awards: Mott Medal, NY Univ, 45. Mem: AAAS; Soc Univ Surg; Am Col Surg; Soc Exp Biol & Med; Am Gastroenterol Asn. Res: Gastrointestinal physiology; breast cancer; surgical infections; burns; surgery and physiology of the stomach, bile ducts, liver and pancreas; hernia. Mailing Add: 1115 Columbia St Seattle WA 98104

WHITE, THOMAS WAYNE, b Caldwell Co, Ky, Dec 9, 34; m 60; c 4. RUMINANT NUTRITION. Educ: Univ Ky, BS, 56, MS, 57; Univ Mo-Columbia, PhD(animal nutrit), 60. Prof Exp: Asst dist salesman, Nat Oats Co, 61-62; assoc prof, 62-72, PROF ANIMAL SCI, LA STATE UNIV RICE EXP STA, 72- Concurrent Pos: On leave, Purdue Univ, 71. Mem: Am Soc Animal Sci; Am Dairy Sci Asn. Res: Level and source of roughage in ruminant rations as measured by digestibility and feedlot performance; whole shelled corn with various protein supplement for finishing cattle; influence of type and variety of sorghum grain on ration digestibility. Mailing Add: La State Univ Rice Exp Sta PO Box 1429 Crowley LA 70526

WHITE, WALLACE FLETCHER, b Louisville, Ky, May 11, 08; m 39; c 4. PHARMACOLOGY. Educ: Butler Univ, BS, 30; Univ Iowa, MS, 32; Yale Univ, PhD(pharmacol), 49. Prof Exp: Teacher high sch, Ky, 30-31; asst zool, Univ Iowa, 31-32; asst, Evansville Col, 35 & Ind Univ, 35-37; instr zool & pharmacol, Conn Col Pharm, 37-41; from instr to assoc prof, Col Pharm, Univ Conn, 41-49; assoc prof, 49-55, PROF PHARMACOL, COL PHARM, UNIV MINN, MINNEAPOLIS, 55- Mem: AAAS; fel Am Pub Health Asn. Res: Cardiotonic drugs; iron absorption; quantitative pharmacology; programmed learning techniques in pharmacology. Mailing Add: 2217 Folwell St St Paul MN 55108

WHITE, WALTER FINCH, b Norfolk, Va, June 24, 11; m 36; c 1. HYDROLOGY. Educ: NC State Col, BSChE, 34. Prof Exp: Jr chemist, US Geol Surv, DC, 34-37; asst chemist, State Engrs Off, NMex, 37-42; chemist, US Geol Surv, DC, 42-57, staff hydrologist, Va, 57-66, DIST CHIEF WATER RESOURCES DIV, US GEOL SURV, 66- Mem: AAAS; Am Chem Soc; Am Water Works Asn. Res: Chemical characteristics of natural waters; industrial utility of surface water of Pennsylvania and Ohio; quality of waters in Pecos River Basin; water resources of Maryland, Delaware and District of Columbia. Mailing Add: Water Resources Div US Geol Surv 8809 Satyr Hill Rd Parkville MD 21234

WHITE, WALTER STANLEY, b Cambridge, Mass, Apr 13, 15; m 41; c 2. ECONOMIC GEOLOGY. Educ: Harvard Univ, AB, 36, PhD(struct geol), 46; Calif Inst Technol, MS, 37. Prof Exp: Asst, Harvard Univ, 37-40; asst chief mineral deposits br, 54-56, asst chief geologist, 60-63, GEOLOGIST, US GEOL SURV, 39- Concurrent Pos: Mem, Nat Res Coun, 58-68, mem exec comt earth sci div, 59-61 & 65-68. Mem: Fel Geol Soc Am; Soc Econ Geol (vpres, 76). Res: Structural geology; structural geology of ore deposits. Mailing Add: 3514 Hamlet Pl Chevy Chase MD 20015

WHITE, WALTER THOMAS, electrical engineering, see 12th edition

WHITE, WARREN HUMPHREYS, b New York, NY, July 14, 41. MATHEMATICS, AIR POLLUTION. Educ: Calif Inst Technol, BS, 63; Univ Wis-Madison, MS, 64, PhD(math), 67. Prof Exp: Asst prof math, Ariz State Univ, 67-70; vis prof, Inst Pure & Appl Math, Brazil, 70-72; fel environ eng, Calif Inst Technol, 72-74; SCIENTIST AIR QUAL, METEOROL RES INC, 74- Concurrent Pos: Consult, Meteorol Res Inc, 72-74 & TRW Inc, 74. Mem: Air Pollution Control Asn; Am Math Soc; Am Meteorol Soc; Math Asn Am. Res: Radiative transfer in polluted atmospheres; aerosols and atmospheric chemistry; dynamical systems. Mailing Add: Meteorol Res Inc 464 W Woodbury Rd Altadena CA 91001

WHITE, WILLARD WORSTER, III, b Perth Amboy, NJ, July 6, 44; m 72, MAGNETOHYDRODYNAMICS, SOLID STATE PHYSICS. Educ: Univ Del, BS, 66; Rensselaer Polytech Inst, PhD(physics), 70. Prof Exp: RES PHYSICIST, MISSION RES CORP, 72 Concurrent Pos: Assoc, Rensselaer Polytech Inst, 70-72. Mem: Am Phys Soc. Res: Electromagnetic phenomena in the ionosphere; surface physics of materials; radiation damage in solids. Mailing Add: 926 Mission Ridge Rd Santa Barbara CA 93103

WHITE, WILLIAM, b Millbrook, Ont, Apr 1, 28; m 60; c 3. MEDICAL PHYSICS, RESEARCH ADMINISTRATION. Educ: Queen's Univ, Ont, BSc, 50; McGill Univ, PhD(physics), 61. Prof Exp: Res physicist, Bldg Prod Ltd, Montreal, Can, 51-56 & Am Radiator & Standard Sanit Res Lab, NJ, 61-63; DIR RES, SEARLE MED INSTRUMENTATION GROUP, SEARLE ANAL, INC DIV, G D SEARLE & CO, 63- Mem: Am Phys Soc; Soc Nuclear Med. Res: Nuclear spectroscopy and instrumentation in the physical and biological sciences; penetration of charged particles; blocking and channeling of atomic particles in crystals; automated clinical chemistry apparatus; nuclear medical gamma-ray cameras; methods for the organization of research. Mailing Add: Searle Anal Inc 2000 Nuclear Dr Des Plaines IL 60018

WHITE, WILLIAM ALEXANDER, b Paterson, NJ, June 15, 06; m; c 2. GEOLOGY. Educ: Duke Univ, AB, 30; Univ NC, MA, 31, PhD(geol), 38; Mont Sch Mines, MS, 34. Prof Exp: Petrogr, Lago Petrol Corp, Venezuela, 38-40; assoc geologist, Soil Conserv Serv, USDA, 40-42; asst state geologist, Dept Conserv & Develop, NC, 42-44; assoc prof geol, 44-50, PROF GEOL, UNIV NC, CHAPEL HILL, 50- Mem: Fel Am Geol Soc. Res: Geomorphology; glacial geology. Mailing Add: Dept of Geol Univ of NC Chapel Hill NC 27514

WHITE, WILLIAM ARTHUR, b Sumner, Ill, Dec 9, 16; m 41. MINERALOGY, PETROLOGY. Educ: Univ Ill, BS, 40, MS, 47, PhD(geol), 55. Prof Exp: Spec asst chem, 43-44, asst geol, 44-47, from asst geologist to assoc geologist, 47-54, head clay resources & clay mineral tech sect, 58-73, GEOLOGIST, ILL STATE GEOL SURV, 54- Mem: AAAS; Mineral Soc Am; Am Chem Soc; fel Am Geol Soc; Geochem Soc. Res: Physical properties of clays as related to soil mechanics; ceramic properties of clays and sediments; clay mineralogy of sediments and the environments in which they were accumulated; the role of clay minerals in environmental geology. Mailing Add: Ill State Geol Surv Urbana IL 61801

WHITE, WILLIAM BLAINE, b Huntingdon, Pa, Jan 5, 34; m 59; c 2. GEOCHEMISTRY, MATERIALS SCIENCE. Educ: Juniata Col, BS, 54; Pa State Univ, PhD(geochem), 62. Prof Exp: Res assoc chem physics, Mellon Inst, 54-58; res assoc geochem, 62-63, from asst prof to assoc prof, 63-72, PROF GEOCHEM, PA STATE UNIV, 72- Honors & Awards: Outstanding Serv Award, Nat Speleol Soc, 75. Mem: AAAS; Am Geophys Union; Am Ceramic Soc; Am Mineral Soc; Nat Speleol Soc (exec vpres, 65-67). Res: High temperature chemistry; infrared and optical spectroscopy of solids; mineralogy; ground water hydrogeology; environmental geomorphology; solid state chemistry; glass science; infrared, optical and luminescence spectroscopy; geomorphology. Mailing Add: 210 Mat Res Lab Pa State Univ University Park PA 16802

WHITE, WILLIAM CALVIN, b Burkeville, Va, Apr 22, 28; m 54; c 2. SOIL FERTILITY, PLANT PHYSIOLOGY. Educ: Va Polytech Inst, BS, 49; Iowa State Univ, MS, 53, PhD(soil fertil), 57. Prof Exp: Asst prof agron, Va Polytech Inst, 49-51; assoc prof soil sci, NC State Univ, 57-63; asst dir div sci serv, Nat Plant Food Inst, 63-70, VPRES MEM SERV, FERTILIZER INST, 70- Concurrent Pos: Mem, Am Forage & Grassland Coun, 64- Mem: AAAS; Soil Sci Soc Am; Am Soc Agron. Res: Nitrogen nutrition of plants; nature of residual fertilizer nitrogen in soils; fertilizer technology and fertilizer industry structure. Mailing Add: Fertilizer Inst 1015 18th St NW Washington DC 20036

WHITE, WILLIAM CHARLES, b Jacksonville, Fla, May 12, 22; m 52. PHYSICS, ASTRONOMY. Educ: Ohio Wesleyan Univ, BA, 48; Ohio State Univ, MS, 50. Prof Exp: Physicist, 50-73, OPERS RES ANALYST, NAVAL WEAPONS CTR, 73- Concurrent Pos: Consult, Astrophys Observ, Smithsonian Inst, 59-60 & Dearborn Observ, Northwestern Univ, 60-62; sci observer, Stargazer Balloon Flight, 61; partic, Aerial Photog Eclipse of Quiet Sun, 62, NASA Mobile Launch Exped, 65, Sandia Eclipse Exped, 65 & Oceanog & Geophys Exped, SAm, 67. Mem: Am Astron Soc; NY Acad Sci; Sigma Xi. Res: Astrophysical research with infrared detectors and balloon-bourne observatories; atmospheric physics research with high altitude balloons: ozone as a function of latitude. Mailing Add: Naval Weapons Ctr Code 127 China Lake CA 93555

WHITE, WILLIAM HAROLD, b Windsor, Ont, May 12, 15; m 37; c 2. PHYSICAL CHEMISTRY, ORGANIC CHEMISTRY. Educ: Univ Western Ont, BA, 37, MA, 40; McGill Univ, PhD(phys chem), 43. Prof Exp: Res chemist, Nat Res Coun Can, 37-43; dir res, Frank W Horner Co, 43-45; res coordr, 45-57, asst mgr prod res, 57-59, tech adv, 59-64, RES ADMINR, RES DEPT, IMP OIL ENTERPRISES LTD, 64- Mem: Fel Chem Inst Can. Res: Petroleum research administration; petroleum refining and production technology. Mailing Add: Res Dept Imp Oil Enterprises Ltd Sarnia ON Can

WHITE, WILLIAM NORTH, b Walton, NY, Sept 16, 25; m 51; c 2. PHYSICAL ORGANIC CHEMISTRY. Educ: Cornell Univ, AB, 50; Harvard Univ, MA, 51; PhD(org chem), 53. Prof Exp: Nat Res Coun fel, Crellin Labs, Calif Inst Technol, 53-54; from asst prof to assoc prof chem, Ohio State Univ, 54-63; chmn dept, 63-71 & 75-76, PROF CHEM, UNIV VT, 63- Concurrent Pos: NSF sr fel biol, Brookhaven Nat Labs, 63-64; NSF sr fel chem, Harvard Univ, 65; vis scholar biochem, Brandeis Univ, 74-75. Mem: AAAS; Am Chem Soc; The Chem Soc. Res: Reaction mechanisms; rearrangements; structure reactivity correlations; molecular complexes; bio-organic chemistry; carbanion and carbonyl group chemistry. Mailing Add: Dept of Chem Univ of Vt Burlington VT 05401

WHITE, WILLIAM P, b Mobile, Ala, Aug 30, 43; m 68; c 1. CHEMICAL PHYSICS. Educ: Auburn Univ, BS, 67; Ohio State Univ, PhD(phys chem), 71. Prof Exp: NSF grant, State Univ NY Binghamton, 71-74; MEM STAFF, LOCKHEED ELECTRONICS CO, INC, HOUSTON, TEX, 74- Mem: Optical Soc Am; Am Chem Soc. Res: High resolution moleculalr spectroscopy of transient species; laser spectroscopy; astronomical spectroscopy. Mailing Add: Lockheed Electronics Co Inc C17 16811 El Camino Real Houston TX 77058

WHITE, WILLIAM WALLACE, b Cleveland, Ohio, Sept 7, 39; m 64; c 2. OPERATIONS RESEARCH. Educ: Princeton Univ, AB, 61; Univ Calif, Berkeley, MS, 63, PhD(opers res), 66. Prof Exp: Staff mem mgt sci, Philadelphia Sci Ctr, 66-74, advan appl adv, Advan Syst Develop Div, 74-75, PROD ANAL ADV, SYST PROD DIV, IBM CORP, 75- Concurrent Pos: Adj assoc prof math, Columbia Univ, 74- Mem: Opers Res Soc Am; Asn Comput Mach; Math Prog Soc. Res: Theory and practice of mathematical programming, with application to computer science. Mailing Add: IBM Corp PO Box 390 Poughkeepsie NY 12602

WHITE, WINIFRED SHARLENE, b Pittsburgh, Pa, July 17, 13. BIOLOGY. Educ: Millikin Univ, AB, 34; Ind Univ, AM, 35; Univ Mich, PhD(zool), 42. Prof Exp: Asst zool, Univ Mich, 37-40; instr biol, Grand Rapids Jr Col, 41-42; instr zool, Milwaukee-Downer Col, 42; jr geneticist, Heredity Clin, Univ Mich, 43-44; instr biol, Sullins Col, 44-45; assoc prof, Okla Col Women, 45-47; teacher, Santiago Col, Chile, 47-48; prof, Okla Col Women, 53-65, chmn div sci, 58-65, head dept biol, 60-65; PROF BIOL, SCH OF THE OZARKS, 65-, HEAD DEPT, 67- Concurrent Pos: Fulbright exchange teacher, Univ Wales, 54-55. Mem: Sigma Xi. Res: Genetics of the chrysanthemum aphid; human genetics. Mailing Add: Dept of Biol Sch of the Ozarks Point Lookout MO 65726

WHITE, ZEBULON WATERS, b Baltimore, Md, Oct 3, 15; m 39; c 4. FORESTRY. Educ: Dartmouth Col, AB, 36; Yale Univ, MF, 38. Prof Exp: Consult forester, Pomeroy & McGowin, Ark, 40-58; prof indust forestry, Yale Univ, 58-72, assoc dean sch forestry, 65-71; PRES, ZEBULON WHITE & CO, 72- Mem: Soc Am Foresters. Res: Industrial forest management. Mailing Add: Zebulon White & Co 207 S Holly St Hammond LA 70401

WHITECOTTON, JOSEPH WALTER, anthropology, see 12th edition

WHITED, DEAN ALLEN, b Nebraska City, Nebr, Mar 28, 40; m 64; c 2. GENETICS, AGRONOMY. Educ: Univ Nebr, Lincoln, BS, 62, MS, 64; NDak State Univ, PhD(agron), 67. Prof Exp: Agency Int Develop grant & res assoc wheat qual, Univ Nebr, Lincoln, 67-68; asst prof agron, 68-73, ASSOC PROF AGRON, NDAK STATE UNIV, 73-, CHMN GENETICS INST, 71- Mem: Am Soc Agron; Am Genetic Asn. Res: Genetics, especially soybean genetics and production. Mailing Add: Dept of Agron NDak State Univ Fargo ND 58102

WHITEFORD, ANDREW HUNTER, b Winnipeg, Man, Sept 1, 13; US citizen; m 39; c 4. ETHNOLOGY. Educ: Beloit Col, BA, 37; Univ Chicago, MA, 43, PhD(anthrop), 50. Prof Exp: Res archaeologist, Univ Tenn, 38-42; cur, Logan Mus Anthrop, 43-46, actg dir, 46-51, dir, 51-74, from instr to asst prof anthrop, 43-55, George L Collie prof, 55-74, EMER PROF ANTHROP, BELOIT COL, 74- Concurrent Pos: Wenner-Gren Found Anthrop Res fel, 49-50 & 74; Soc Sci Res Coun-Am Coun Learned Socs Joint Comt Latin Am Studies fel, Colombia, 61-62; NSF fac res fel, Spain & SAm, 61-62; Cullister Found grant, Upper Amazon, 65; NSF grant, Oaxaca, Mex, 66; NIMH grant, Popayan, Colombia, 70 & Queretaro, Mex, 76; Penrose Fund res grant, Am Philos Soc, 75. Mem: Fel Am Anthrop Asn; Am Ethnol Soc; Asn Sci Mus Dir. Res: Social structure and culture of contemporary Latin America with focus upon secondary urban areas; primitive art and material culture, concentrating upon traditional products of American Indian cultures. Mailing Add: 2550 Hawthorne Dr Beloit WI 53511

WHITEFORD, ROBERT DANIEL, b Atlanta, Ga, Sept 23, 22; m 56; c 1. VETERINARY MEDICINE. Educ: Univ Ga, DVM, 51; Iowa State Univ, MS, 56, PhD, 64. Prof Exp: Clinician, Ont Vet Col, 51-52; vet, Brunswick Animal Hosp, Ga, 52-54; instr vet anat, Iowa State Univ, 54-57; fel comp ophthal, Col Med, Univ Iowa, 57-59; assoc prof vet anat, Sch Vet Med, Auburn Univ, 59-64, prof anat, 64-69; assoc prof biomed sci, 69-72, PROF BIOMED SCI, UNIV GUELPH, 72- Mem: Asn Res Vision & Ophthal; Am Asn Vet Anat. Res: Veterinary histology; comparative neuroanatomy; neurophysiology; neuro-ophthalmology; clinical neurology. Mailing Add: Dept of Biomed Sci Ont Vet Col Univ Guelph Guelph ON Can

WHITEHAIR, CHARLES KENNETH, b Abilene, Kans, Mar 3, 16; m 58; c 2. NUTRITION, PATHOLOGY. Educ: Kans State Col, DVM, 40; Univ Wis, MS, 43, PhD(nutrit), 47. Prof Exp: Instr animal dis, Univ Wis, 40-47; from asst prof to prof animal nutrit, Okla State Univ, 47-53, head vet res & prof physiol & nutrit, 54-56; PROF PATH, MICH STATE UNIV, 56- Concurrent Pos: Assoc prof, Univ Ill, 52-53. Mem: Am Soc Exp Path; Soc Exp Biol & Med; Am Inst Nutrit; Conf Res Workers Animal Dis (vpres, 64, pres, 65); Am Vet Med Asn. Res: Nutritional pathology; swine nutrition and diseases; metabolic diseases of livestock; relation of nutrition to diseases. Mailing Add: Dept of Path Mich State Univ East Lansing MI 48824

WHITEHAIR, LEO A, b Abilene, Kans, June 13, 29; m 58; c 3. VETERINARY MEDICINE, FOOD SCIENCE. Educ: Kans State Univ, BS & DVM, 53; Univ Wis, MS, 54, PhD(food sci), 62; Am Bd Vet Pub Health, dipl. Prof Exp: Vet off nutrit br, Aeromed Lab, Wright-Patterson AFB, Ohio, 54-58; lab vet nutrit br, Food & Container Inst Armed Forces, Ill, 61-62; vet food technologist, Biol Br, Div Biol & Med, US AEC, 62-67; health scientist adminr, 68-75, DIR PRIMATE RES CTR PROG, ANIMAL RESOURCES BR, DIV RES RESOURCES, NIH, 75- Concurrent Pos: Consult advisor, Food & Agr Orgn-UN-WHO-Int Atomic Energy Agency joint expert comt meeting, Rome, 64; tech adv int prog irradiation fruit & fruit juices, Inst Biol & Agr, Seibersdorf Reactor Ctr, Austria, 65; vet off dir, Am Vet Med Asn. Mem: Am Vet Med Asn; Am Bd Vet Pub Health (secy-treas, 75-); Am Soc Animal Sci; Inst Food Technologists; Am Asn Vet Nutritionists. Res: Animal nutrition; wholesomeness and public health safety aspects of irradiated foods; laboratory animal resources; animal models for biomedical research. Mailing Add: Animal Resources Br Div of Res Resources Bldg 31 5B-30 NIH Bethesda MD 20014

WHITEHEAD, ANDREW BRUCE, b Quebec, Que, Oct 18, 32; m 62; c 3. PLANETOLOGY, PHYSICS. Educ: Univ NB, BSc, 53; McGill Univ, MSc, 55, PhD(physics), 57. Prof Exp: Res fel nuclear physics, Atomic Energy Res Estab, Harwell, Eng, 57-60; res fel nuclear physics, 61-62, res specialist, 63-65, group supvr physics, 65-67, sect mgr physics, 67-69, mgr lunar & planetary sci sect, 69-71, asst proj scientist, Mariner 9, 71-73, actg asst mgr, Space Sci Div, 73-75, STAFF SCIENTIST, SPACE SCI DIV, JET PROPULSION LAB, 75- Mem: Am Phys Soc. Res: Nuclear structure; particle detection; secondary electron emission; atomic stopping. Mailing Add: 565 Paulette Pl La Canada CA 91011

WHITEHEAD, ARMAND T, b Reno, Nev, May 19, 36; m 54; c 4. ENTOMOLOGY. Educ: Brigham Young Univ, BS, 65; Univ Calif, Berkeley, PhD(entom), 69. Prof Exp: ASST PROF ZOOL, BRIGHAM YOUNG UNIV, 69- Mem: AAAS; Am Soc Zool. Res: Neurophysiology and morphology of insect salivary glands; neurophysiological effects of pesticides. Mailing Add: Dept of Zool Brigham Young Univ Provo UT 84601

WHITEHEAD, DONALD REED, b Quincy, Mass, Sept 14, 32; m 55; c 2 BOTANY PALEOECOLOGY. Educ: Harvard Univ, AB, 54, AM, 55, PhD(biol), 58. Prof Exp: From instr to assoc prof biol, Williams Col, 59-67; assoc prof biol, 67-74, PROF ZOOL, IND UNIV, BLOOMINGTON, 74- Concurrent Pos: Fulbright fel, Geol Surv Denmark, 58-59, res prof, 63; consult, Jersey Prod Res Co, 60; res prof bot mus, Univ Bergen, 63; NSF grants, 63. Mem: AAAS; Ecol Soc Am; Geol Soc Am; Am Ornith Union. Res: Pleistocene environmental changes in unglaciated regions; pollen morphology. Mailing Add: Div of Biol Sci Ind Univ Bloomington IN 47401

WHITEHEAD, EUGENE IRVING, b Canton, SDak, Mar 4, 18; m 55; c 2. PLANT BIOCHEMISTRY. Educ: SDak State Col, BS, 39, MS, 41. Prof Exp: Lab asst, 40-42, sta analyst, 42-43, asst agr chemist, 43-46, assoc chemist, 46-60, assoc prof sta biochem, Grad Fac, 60-67, PROF STA BIOCHEM, GRAD FAC, SDAK STATE UNIV, 67- Mem: Am Soc Plant Physiol; Am Chem Soc. Res: Nitrogen metabolism of cereal crops; winter hardiness of cereal plants. Mailing Add: Dept Chem Sect Exp Sta Biochem SDak State Univ Brookings SD 57006

WHITEHEAD, FLOY EUGENIA, b Athens, Ga, Feb 10, 13. NUTRITION. Educ: Univ Ga, BS, 36, MS, 42; Harvard Univ, DSc, 51; Am Bd Nutrit, dipl, 52. Prof Exp: Teacher, High Schs, Ga, 36-40; asst prof home econ, WGa Col, 40-42; assoc dir health educ, State Dept Pub Health, Ga, 42-43; assoc prof home econ, La State Univ, 44-48 & Miss State Col, 48-49; fel, Sch Pub Health, Harvard Univ, 49-52; dir nutrit, Wheat Flour Inst, Ill, 52-53; dir nutrit educ, Nat Dairy Coun, 53-55; chmn dept home econ, 55-71, PROF HOME ECON, UNIV IOWA, 55- Concurrent Pos: Vis lectr, Harvard Univ, 52-54; co-dir nutrit educ res, Pub Schs, Mo, 52-55; pres elect, Nat Coun Adminr Home Econ, 66-67, pres, 67-68; res grant off nutrit, AID, US Dept State. . Honors & Awards: Roberts Award, Am Dietetic Asn, 56. Mem: AAAS; Am Pub Health Asn; Am Home Econ Asn; Am Dietetic Asn; Soc Nutrit Educ. Res: Dietary surveys; nutrition education; analysis of nutrition education research, 1900-1970. Mailing Add: 306 Ferson Ave Iowa City IA 52240

WHITEHEAD, FRED, b Walland, Tenn, Aug 24, 05; m 35; c 2. CHEMISTRY. Educ: Ky Wesleyan Col, AB, 31; Univ Tenn, MS, 33; Univ Mich, PhD(chem), 45. Prof Exp: Prof chem, Ky Wesleyan Col, 32-50, dean & registr, 44-50; PROF CHEM, HUNTINGDON COL, 50- Mem: Am Chem Soc. Res: Rate of dissociation of pentaarylethanes; physics; mathematics. Mailing Add: 3496 Cloverdale Rd Montgomery AL 36111

WHITEHEAD, GEORGE WILLIAM, b Bloomington, Ill, Aug 2, 18; m 47. MATHEMATICS. Educ: Univ Chicago, SB, 37, SM, 38, PhD(math), 41. Prof Exp: Instr math, Univ Tenn, 39, Purdue Univ, 41-45 & Princeton Univ, 45-47; from asst prof to assoc prof, Brown Univ, 47-49; from asst prof to assoc prof, 49-57, PROF MATH, MASS INST TECHNOL, 57- Concurrent Pos: Guggenheim fel & Fulbright res scholar, 55-56; vis prof, Princeton Univ, 58-59; NSF sr fel, 65-66. Mem: Nat Acad Sci; fel Am Acad Arts & Sci; Am Math Soc; Math Asn Am. Res: Algebraic topology, especially homotopy theory. Mailing Add: Dept of Math Rm 2-284 Mass Inst of Technol Cambridge MA 02139

WHITEHEAD, HOWARD ALLAN, b Toronto, Ont, Apr 17, 27; m 52; c 4. BACTERIOLOGY. Educ: Mt Allison Univ, BSc, 49; McGill Univ, MSc & PhD(bact), 53. Prof Exp: RES MGR, KIMBERLY-CLARK CORP, 53- Mem: Tech Asn Pulp & Paper Indust; Soc Cosmetic Chem; Can Soc Microbiol; Can Pub Health Asn. Res: Lethal and mutagenic effects of ultraviolet irradiation of bacterial microorganisms; menstruation, physiology and feminine hygiene. Mailing Add: Kimberly-Clark Corp 2100 N Winchester Rd Neenah WI 54956

WHITEHEAD, JAMES RENNIE, b Clitheroe, Eng, Aug 4, 17; m 44; c 2. PHYSICS. Educ: Univ Manchester, BSc, 39; Cambridge Univ, PhD(physics), 49. Prof Exp: Scientist, Telecommun Res Estab, Eng, 39-51; assoc prof physics, McGill Univ, 51-55; dir res, RCA Victor Co, Ltd, 55-65; dep dir sci secretariat, Govt Can, 65-67, prin sci adv, 67-71, asst secy int affairs, Ministry of State for Sci & Technol, 71-73, spec adv, 73-75; SR ADV, INT DEVELOP RES CTR, GOVT CAN, 75- Concurrent Pos: Sr sci officer, Brit Air Comn, Washington, DC, 44-45 & Cambridge Univ, 46-49; consult, Defence Res Bd Can, 52-54; consult sci policy in Venezuela, UNESCO, 70; Can deleg to sci & technol policy comt, Orgn Econ Coop & Develop, vchmn comt sci policy, 73-75; mem sci comt, NATO; consult, Philip A Lapp Ltd, 75- Mem: Fel Royal Soc Can; sr mem Inst Elec & Electronics Eng; fel Brit Inst Elec Eng; fel Brit Inst Physics; fel Can Aeronaut & Space Inst. Res: Physical electronics; circuits; systems, propagation; electron microscopy; friction; science policy. Mailing Add: 1368 Chattaway Ave Ottawa ON Can

WHITEHEAD, KATHLEEN BUTCHER, mathematics, see 12th edition

WHITEHEAD, MARIAN NEDRA, b Calif, Sept 5, 22. NUCLEAR PHYSICS, PARTICLE PHYSICS. Educ: Reed Col, AB, 44; Columbia Univ, MS, 45; Univ Calif, PhD(physics), 52. Prof Exp: Physicist, US Naval Ord Test Sta, 45-46 & Radiation Lab, Univ Calif, 49-60; Fulbright sr res fel, Inst Physics, Bologna, Italy, 61-62; physicist, Stanford Linear Accelerator Ctr, Stanford Univ, 62-64; assoc prof, 64-67, chmn dept, 69-75, PROF PHYSICS, CALIF STATE UNIV, HAYWARD, 67- Mem: Fel Am Phys Soc; Am Asn Physics Teachers (treas, 74-). Res: Meson and cosmic physics. Mailing Add: Dept of Physics Calif State Univ Hayward CA 94542

WHITEHEAD, MARVIN DELBERT, b Paoli, Okla, Dec 18, 17; m 40; c 1. PHYTOPATHOLOGY, MYCOLOGY. Educ: Okla State Univ, BS, 39, MS, 46; Univ Wis, PhD(plant path, mycol), 49. Prof Exp: Asst agr aide, Soil Conserv Serv, USDA, Okla, 36-38; asst agron, Okla State Univ, 39-40; sr seed analyst, Fed State Seed Lab, Ala, 40-42; asst plant path, Univ Wis, 46-48; asst prof, Tex A&M Univ, 49-55; assoc prof, Univ Mo, 55-60; prof bot, Edinboro State Col, 60-63; prof plant path, Ga Southern Col, 63-68; PROF BOT & PLANT PATH, GA STATE UNIV, 68- Concurrent Pos: Consulting plant pathologist, US Army, Ft McPherson, Ga, 74- Mem: AAAS; Am Phytopath Soc; Mycol Soc Am; Bot Soc Am; Am Inst Biol Sci. Res: Field crop disease pathology; soil borne and seed borne diseases; phytopathological histology and techniques; fungus and smut taxonomy; yield loss from plant disease; disease resistance; antibiotics in control of Dutch elm disease. Mailing Add: Dept of Bot Ga State Univ Atlanta GA 30303

WHITEHEAD, MICHAEL ANTHONY, b London, Eng, June 30, 35; m 68. THEORETICAL CHEMISTRY, QUANTUM CHEMISTRY. Educ: Univ London, BSc, 56, PhD(phys chem), 60, DSc(theoret & phys chem, 74. Prof Exp: Asst lectr chem, Queen Mary Col, Univ London, 58-60; Fulbright scholar, Univ Cincinnati, 60-62, fel, 60-61, asst prof, 61-62; from asst prof to assoc prof, 62-75, PROF CHEM, McGILL UNIV, 75- Concurrent Pos: Nat Res Coun Can travel fel & vis prof, Cambridge Univ, 71-72; vis prof theoret chem, Oxford Univ, 72-74. Mem: AAAS; Am Phys Soc; Can Inst Chem; The Chem Soc. Res: Nuclear quadrupole and electron sping resonance; electronegativity theory; Gaussian orbitals; inorganic polymers; molecular orbital calculations involving complete neglect of differential overlap; Pade type approximants; beyond Hartree-Fock calculations. Mailing Add: Dept of Chem McGill Univ PO Box 6070 Montreal PQ Can

WHITEHEAD, THOMAS HILLYER, b Maysville, Ga, Sept 5, 04; m 31; c 2. ANALYTICAL CHEMISTRY. Educ: Univ Ga, BS, 25; Columbia Univ, AM, 28, PhD(colloid chem), 30. Prof Exp: Teacher, High Sch, Ga, 25-27; asst chem, Columbia Univ, 27-30; adj prof, 30-35, prof, 39-72, asst head dept, 46-60, coord instr insts, 60-68, actg dean grad sch, 68-69, dean, 69-72, EMER PROF CHEM, UNIV GA, 72- Concurrent Pos: Consult & mem adv coun, Chem Corps, US Dept Army, 48-65, AEC, 52-, US Off Educ, Washington, DC, 65- & F W Dodge Div, McGraw-Hill Inc, 65- Mem: AAAS; Am Chem Soc; Sigma Xi. Res: Colloid chemistry; oxidation-reduction indicators; inorganic colloidal systems; complex inorganic compounds; indicators in analytical chemistry; iodometric method for copper. Mailing Add: 236 Henderson Ave Athens GA 30601

WHITEHEAD, WALTER DEXTER, JR, b San Diego, Calif, Nov 30, 22; m 49; c 2. NUCLEAR PHYSICS, ATOMIC PHYSICS. Educ: Univ Va, BS, 44, MS, 46, PhD(physics), 49. Prof Exp: Asst, Univ Va, 43-45; physicist, Bartol Res Found, 49-53; from asst prof to assoc prof physics, NC State Col, 53-56; from asst prof to assoc prof, 56-61, chmn dept, 68-69, dean fac arts & sci, 71-72, PROF PHYSICS, UNIV VA, 61-, DIR CTR ADVAN STUDIES, 65-, DEAN GRAD SCH ARTS & SCI, 69- Concurrent Pos: Vis scientist, Inst Nuclear Physics, 59-60; mem bd admin, Va Inst Marine Sci, 71-; mem Nat Res Coun eval panel, Ctr Radiation Res, Inst Basic Standards. Mem: AAAS; fel Am Phys Soc; Am Asn Physics Teachers. Res: Nuclear spectroscopy; neutron scattering; photonuclear reactions; x-ray interactions. Mailing Add: 444 Cabell Hall Univ of Va Charlottesville VA 22901

WHITEHEAD, WAYNE LEE, plant genetics, plant breeding, see 12th edition

WHITEHEAD, WILLIAM EARL, b Martin, Tenn, May 24, 45; m 68; c 1. PSYCHOPHYSIOLOGY. Educ: Ariz State Univ, BA, 67; Univ Chicago, PhD(psychol), 73. Prof Exp: ASST PROF PSYCHIAT, COL MED, UNIV CINCINNATI, 73- Concurrent Pos: Consult, Vet Admin Hosp, 75-; adj asst prof psychol, Univ Cincinnati, 75- Mem: AAAS; Soc Psychophysiol Res; assoc Am Psychosom Soc. Res: Biofeedback treatment of psychophysiologic disorders; etiology of psychosomatic disorders; psychotropic drug evaluation. Mailing Add: Univ Cincinnati Col of Med 231 Bethesda Ave Cincinnati OH 45267

WHITEHILL, ALVIN RICHARD, b Groton, Vt, Mar 18, 15; m 43; c 4. BACTERIOLOGY. Educ: Dartmouth Col, AB, 37; Cornell Univ, PhD(bact), 42. Prof Exp: Asst bact, Cornell Univ, 38-41; instr biol, Ill Inst Technol, 41-44; chemist, Lederle Labs, Am Cyanamid Co, 44-60; PROF BACT, UNIV MAINE, ORONO, 61-, PROF MICROBIOL, 74- Res: Variability of the group Bacillus Pasteurii; intestinal synthesis of B-12. Mailing Add: 256 Hitchner Hall Univ of Maine Orono ME 04473

WHITEHILL, JULES LEONARD, b New York, NY, Mar 7, 12; m 43; c 3. SURGERY, MEDICAL EDUCATION. Educ: City Col New York, BSc, 32; NY Univ, MD, 35; Am Bd Surg, dipl, 43. Prof Exp: Resident surg, Mt Sinai Hosp, New York, 35-39; chief surg, Cent Manhattan Med Group, 46-48; pvt pract, Ariz, 48-63; from assoc prof to prof surg, 63-73, asst dean clin progs, 63-67, assoc dean, 67-69, chmn dept surg, 68-70, EMER PROF SURG, CHICAGO MED SCH, 73- Concurrent Pos: Fel exp surg, Mt Sinai Hosp, New York, 40-42; consult, Vet Admin Hosp, Tucson, Ariz, 48-63, West Side Vet Admin, 68-70, Elgin State Hosp, Ill, 65- & Strategic Air Command, US Air Force, 66-; med dir, dir med educ & attend surgeon, Mt Sinai Hosp Med Ctr, Chicago, 63-68; mem fac, Medcom; ed consult, Excerpta Medica; former co-chmn bd regents & int bd gov, Int Col Surg; vis prof surg, Univ Zagreb; mem sci adv comt & mem bd trustees, Chapman Col, Orange, Calif. Honors & Awards: Achievement Award, US Air Force, 67; Jules Leonard Whitehill Chair-in-Surg estab, Sch Med, NY Univ, 73. Mem: Fel Am Col Surg; AMA; fel Royal Soc Med; Int Col Surg; Sigma Xi. Res: Electromagnetic measurement of blood flow; prophylaxis of surgical keloids with steroids; automatic device for intestinal anastomosis. Mailing Add: 1716 El Camino del Teatro La Jolla CA 92037

WHITEHORN, DAVID, b Ann Arbor, Mich, Nov 17, 41; m 64. NEUROPHYSIOLOGY. Educ: Univ Mich, BA, 63; Univ Wash, PhD(physiol), 68. Prof Exp: USPHS fel, Univ Utah, 68-70; ASST PROF PHYSIOL, COL MED, UNIV VT, 70- Mem: Am Physiol Soc; Soc Neurosci. Res: Central nervous system;

organization and information processing. Mailing Add: Dept of Physiol Given Med Bldg Univ of Vt Burlington VT 05401

WHITEHORN, WILLIAM VICTOR, b Detroit, Mich, Oct 3, 15; m 38; c 2. PHYSIOLOGY, MEDICINE. Educ: Univ Mich, AB, 36, MD, 39. Prof Exp: Asst physiol, Univ Mich, 40-42; res assoc, Ohio State Univ, 42-44, from instr to asst prof physiol & med, 44-47; asst prof physiol, Col Med, Univ Ill, Chicago, 47-50, prof, 54-70; dir div health sci & spec asst to pres for med affairs, Univ Del, 70-74; ASST COMNR PROF & CONSUMER PROG, FOOD & DRUG ADMIN, 74- Mem: AAAS; Am Physiol Soc; Soc Exp Biol & Med; Am Heart Asn. Res: Cardiac metabolism and function; applied physiology of respiration and circulation. Mailing Add: FDA Prof & Consumer Prog 5600 Fishers Lane Rockville MD 20852

WHITEHOUSE, BRUCE ALAN, b Henderson, Ky, Sept 6, 39; m 61; c 2. POLYMER CHEMISTRY, TEXTILE CHEMISTRY. Educ: Col Charleston, BS, 63; Ga Inst Technol, PhD(phys chem), 67. Prof Exp: Res chemist, Plastics Dept, Polyolefins Div, Orange Tex, 67-72, res chemist, Textile Fibers Dept, Nylon Tech Div, Chattanooga, Tenn, 72-73, sr res chemist, 73, RES SUPVR, DACRON RES LAB, TEXTILE FIBERS DEPT, E I DU PONT DE NEMOURS & CO, INC, 73- Mem: Am Chem Soc. Res: Solid state structure and properties; solid state nuclear magnetic and quadrupole resonance; structure and properties of polymers; chromatography; polymer synthesis. Mailing Add: Dacron Res Lab E I du Pont de Nemours & Co Kinston NC 28501

WHITEHOUSE, FRANK, JR, b Ann Arbor, Mich, Nov 20, 24; m 51; c 4. MICROBIOLOGY, IMMUNOLOGY. Educ: Univ Mich, BA & MD, 53. Prof Exp: Intern, Blodgett Mem Hosp, Grand Rapids, Mich, 53-54; from instr to asst prof, 54-67, ASSOC PROF MICROBIOL, UNIV MICH, ANN ARBOR, 67- Concurrent Pos: Lectr, Ohio State Univ, 59; preprof counsr, Univ Mich, 60-; exec dir, Nat Asn Adv Health Professions. Mem: NY Acad Sci; Soc Exp Biol & Med; Am Inst Biol Sci; Am Soc Microbiol; Asn Am Med Cols. Res: Enzymatic degradation of antibodies; tumor immunology; chemotherapy. Mailing Add: Dept Microbiol Med Sch Bldg II Univ of Mich Ann Arbor MI 48109

WHITEHOUSE, HARRY S, organic chemistry, see 12th edition

WHITEHOUSE, RONALD LESLIE S, b Birmingham, Eng, Aug 9, 37; Can citizen; m 59; c 2. ELECTRON MICROSCOPY, MICROBIOLOGY. Educ: Univ Nottingham, BSc, 59; Univ Alta, MSc, 61, PhD(plant physiol, biochem), 65. Prof Exp: Nat Res Coun Can overseas fel, Bot Lab, Univ Bergen, 66; prof assoc food microbiol, 67-68, asst prof med bact & electron micros, 68-74, ASSOC PROF BACT, UNIV ALTA, 74- Mem: Am Soc Microbiol. Res: Electron microscopic methods for biological materials; in vitro cultivation of Mycobacterium lepraemurium; bacterial ultrastructure. Mailing Add: Dept of Med Bact Univ of Alta Edmonton AB Can

WHITEHOUSE, ULYSSES GRANT, b Henderson, Ky, July 27, 17; m 45, 54; c 1. PHYSICAL CHEMISTRY. Educ: Univ Ky, BS, 40, MS, 41; Tex A&M Univ, PhD, 53. Prof Exp: Asst, Photog Lab, Lafayette Studios, Eastman Kodak Co, Ky, 40-41; develop & control supvr, US Rubber Co, Pa, 42-43; res scientist, Manhattan Proj, Columbia, 43-45; assoc res dir, Manhattan Proj, Carbide & Carbon Chem Corp, 45-46; instr gen & phys chem, Univ Iowa, 47-48; fel, Carnegie Inst Technol, 48-50; Dow fel chem oceanog, 50-51, Am Petrol Inst chem sedimentologist, 51-53, CHEM OCEANOGR, RES FOUND & DIR ELECTRON MICROS LABS, TEX A&M UNIV, 53-, PROF ELECTRON MICROS & COLLOID SCI, 56-, PROF BIOCHEM & BIOPHYS, 59-, COOP PROF CHEM & PHYSICS, 60-, ASSOC PROF BIOL, 74- Concurrent Pos: Consult, Walnut Hill Dairy, 45-50; res consult, NIH, 59- Honors & Awards: Award, Div War Res, Columbia Univ, 46. Mem: Am Chem Soc; Am Soc Limnol & Oceanog; Electron Micros Soc; Am Fedn Am Sci; Geochem Soc. Res: Colloid science; electron optics; medical biochemistry; crystallography; Raman spectroscopy; molecular spectra; electrochemistry; clay mineralogy; marine sediments; photographic developers; quantum theory of valences; vacuum systems; isotope separation; flocculation processes; x-ray diffraction; trinitrotolvene; capillary flow; laboratory design. Mailing Add: Dept of Biol Tex A&M Univ PO Box 5748 S Sta College Sta TX 77843

WHITEHOUSE, WALTER MACINTIRE, b Millersburg, Ohio, Jan 28, 16; m 45; c 3. RADIOLOGY. Educ: Eastern Mich Univ, AB, 36; Univ Mich, MS, 37, MD, 41. Prof Exp: From instr to assoc prof, 51-63, PROF RADIOL, MED SCH, UNIV MICH, ANN ARBOR, 63-, CHMN DEPT, 65- Mem: Roentgen Ray Soc; Am Thoracic Soc; Am Col Radiol; Radiol Soc NAm. Res: Diagnostic radiology with emphasis on thoracic, gastrointestinal and obstetric areas; development and evaluation of radiologic and pararadiologic diagnostic equipment; evaluation of teaching methods in radiology. Mailing Add: Dept of Radiol Univ Hosp Ann Arbor MI 48104

WHITEHURST, DARRELL DUAYNE, b Vernon, Ill, July 8, 38; m 67; c 2. ORGANIC CHEMISTRY. Educ: Bradley Univ, AB, 60; Univ Iowa, MS, 63, PhD(org chem), 64. Prof Exp: Res chemist, 64-67, SR RES CHEMIST, MOBIL RES & DEVELOP CORP, 67-, PROG CHMN, 71- Concurrent Pos: Mem comt task force motor fuel & photochem smog, Am Petrol Inst; res assoc, BPRI, 73, prin investr fundamental coal chem study, 75. Mem: AAAS; Am Chem Soc. Res: Organic syntheses; acetylene oxidations and coordination compounds of platinum; catalysis by ion exchange resins; catalysis by transition metals and compounds thereof; catalysis by zeolites; homogeneous-heterogeneous catalysts interconversion; metal plating; petrochemicals. Mailing Add: Mobil Res & Develop Corp Princeton NJ 08540

WHITEHURST, HARRY BERNARD, b Dallas, Tex, Sept 13, 22; m 48; c 2. PHYSICAL CHEMISTRY. Educ: Rice Inst, BA, 44, MA, 48, PhD(chem), 50. Prof Exp: Fel, Univ Minn, 50-51; res chemist, Owens-Corning Fiberglas Corp, 51-59; assoc prof, 59-71, PROF CHEM, ARIZ STATE UNIV, 71- Mem: Fel AAAS; Am Chem Soc; fel Am Inst Chemists. Res: Radiochemistry; adsorption; surface chemistry of glass; electrical properties of oxides. Mailing Add: Dept of Chem Ariz State Univ Tempe AZ 85281

WHITEHURST, ROBERT NEAL, b Edwin, Ala, Oct 24, 22; m 50; c 4. PHYSICS. Educ: Univ Ala, BS, 43, MS, 48; Stanford Univ, PhD(physics), 58. Prof Exp: Instr physics, Univ Ala, 48-49; asst microwave, Stanford Univ, 52-54; from asst prof to assoc prof, 54-61, actg chmn dept, 58-59, PROF PHYSICS, UNIV ALA, 61- Concurrent Pos: Consult, Redstone Arsenal, US Army Rocket & Guided Missile Agency, 54-64. Mem: Am Phys Soc; Am Asn Physics Teachers; Am Astron Soc. Res: Electromagnetic theory, especially radiation from high-speed electrons; radio astronomy. Mailing Add: Dept of Physics Univ of Ala University AL 35486

WHITEHURST, SANFORD HUEY, b Dallas, Tex, July 22, 14; m 42; c 1. SOIL CONSERVATION. Educ: Agr & Mech Col, Tex, BS, 38. Prof Exp: Jr soil surveyor, USDA, Tex, 38-39, jr agronomist, Soil Conserv Serv, 39-40, jr soil conservationist, 40; asst soil conservationist, 41, technician in charge, Soil Conserv Proj, 41-42, conservationist, Ellis-Prairie Soil Conserv Dist, 42, work unit conservationist, Dalworth Soil Conserv Dist, 46-52; agronomist, Agr Rels Div & mgr demonstration farm, 52-56, admin asst to dir, 56-57, asst dir & prin agronomist, 58-61, chief agronomist, asst dir & head, Hoblitzelle Agr Lab, 61-63, ADMINSTR, TEX RES FOUND, 63-, DIR EXTEN, 66-; SUPT & COORDR AGR PROGS, RES & EXTEN CTR AT DALLAS, TEX A&M UNIV, 72- Mem: Am Soc Agron; Soil Conserv Soc Am. Res: Water conservation; farm, pasture and rangeland management; application of conservation measures; developing of farming systems. Mailing Add: Tex A&M Univ Res & Exten Ctr Box 43 Renner TX 75079

WHITEKER, MCELWYN D, b Harrison Co, Ky, Aug 4, 29; m 50; c 3. ANIMAL SCIENCE. Educ: Univ Ky, BS, 51, MS, 57, PhD(nutrit, biochem), 61. Prof Exp: From asst prof to assoc prof animal sci, Iowa State Univ, 61-67; LIVESTOCK EXTEN SPECIALIST, UNIV KY, 67-, PROF ANIMAL SCI, 69- Mem: Am Soc Animal Sci. Res: Nutrition; animal breeding. Mailing Add: Dept of Animal Sci Univ of Ky Col of Agr Lexington KY 40506

WHITEKER, ROY ARCHIE, b Long Beach, Calif, Aug 22, 27; m 60; c 1. ANALYTICAL CHEMISTRY. Educ: Univ Calif, Los Angeles, BS, 50, MS, 52; Calif Inst Technol, PhD(chem), 56. Prof Exp: Instr chem, Mass Inst Technol, 55-57; from asst prof to prof, Harvey Mudd Col, 57-74, actg chmn dept, 69-71; dep exec secy, 71-72, actg exec secy, 72, EXEC SECY, COUN INT EXCHANGE OF SCHOLARS, 72- Concurrent Pos: NSF sci fac fel, Royal Inst Technol, Sweden, 63-64; vis assoc prof, Univ Calif, Riverside, 67; assoc dir fel off, Nat Res Coun, 67-68; dir summer session, Claremont Grad Sch, 69-70. Mem: Am Chem Soc. Res: Electroanalytical chemistry; complex ions. Mailing Add: 1615 Evers Dr McLean VA 22101

WHITELAW, DONALD MACKAY, b Vancouver, BC, Oct 20, 13; m 40; c 3. INTERNAL MEDICINE. Educ: Univ BC, BA, 34; McGill Univ, MD & CM, 39; FRCP(C), 47. Prof Exp: Clin instr med, Univ BC, 52-53, from assoc prof to prof, 53-61; assoc prof, Univ Toronto, 61-64; PROF MED, UNIV BC, 64- Concurrent Pos: Physician, Cancer Control Agency BC. Mem: Am Soc Hemat; Can Med Asn; Int Soc Hemat. Res: Hematology and oncology; cell proliferation; cancer chemotherapy. Mailing Add: BC Cancer Inst 2656 Heather St Vancouver BC Can

WHITELAW, NEILL GORDON, physics, deceased

WHITELEY, ARTHUR HENRY, b Dowagiac, Mich, Dec 17, 16; m 44. ZOOLOGY. Educ: Kalamazoo Col, BA, 38; Univ Wis, MA, 39; Princeton Univ, PhD(biol), 45. Prof Exp: Asst zool, Univ Calif, 39-42; asst biol, Off Sci Res & Develop, Princeton Univ, 42-45, res assoc, Nat Defense Res Comt, 45-46; res assoc zool, Sch Med, Univ Tex, 46; Nat Res Coun fel, Calif Inst Technol, 46-47; from asst prof to assoc prof, 47-57, PROF ZOOL, UNIV WASH, 57- Concurrent Pos: Guggenheim fel, Europe, 55-56. Mem: Am Soc Zool; Soc Gen Physiol; Soc Develop Biol; Int Soc Develop Biol; Int Soc Cell Biol. Res: Physiology of fertilization of marine invertebrates; active transport in eggs; genetic expression in developing invertebrate embryos and regenerating stentors; physiology of interspecies hybrids of sea urchins. Mailing Add: Dept of Zool Univ of Wash Seattle WA 98125

WHITELEY, ELI LAMAR, b Florence, Tex, Dec 10, 13; m 49; c 5. SOIL PHYSICS. Educ: Tex A&M Univ, BS, 41, PhD(soil physics), 59; NC State Univ, MS, 49. Prof Exp: Instr, 46-59, ASSOC PROF AGRON, TEX A&M UNIV, 59- Concurrent Pos: Mem, Nat Coord Comt New Crops, 65- Mem: Am Soc Agron; Soil Sci Soc Am; Crop Sci Soc Am; Int Soc Soil Sci. Res: Soil and crop management; new crops, development of new crops for the production of oils, gums and paper pulp. Mailing Add: Dept of Soil & Crop Sci Tex A&M Univ College Station TX 77834

WHITELEY, HELEN RIABOFF, b Harbin, China, June 8, 22; nat US; m 44. MICROBIAL PHYSIOLOGY. Educ: Univ Calif, Berkeley, BA, 42; Univ Tex, MA, 46; Univ Wash, PhD, 51. Prof Exp: AEC fel, 51-53, res assoc microbiol, 53-57, from res asst prof to res assoc prof, 57-64, PROF MICROBIOL, UNIV WASH, 64- Concurrent Pos: Mem panel, NIH, 66- Mem: Am Soc Microbiol; Brit Soc Gen Microbiol; Am Soc Biol Chem. Res: Allosteric enzymes; synthesis of RNA; echinoderm development; mixed phage infections. Mailing Add: Dept of Microbiol Univ of Wash Seattle WA 98105

WHITELEY, NORMAN MCKEE, b Portland, Ore, Dec 20, 46. CLINICAL BIOCHEMISTRY. Educ: Calif Inst Technol, BS, 68; Harvard Univ, PhD(biol chem), 74. Prof Exp: BIOCHEMIST, PHOTO PROD DEPT, E I DU PONT DE NEMOURS & CO, INC, 74- Mem: AAAS. Res: Development of biochemical tests for use on the automatic clinical analyzer. Mailing Add: ACA Div Photo Prod E I du Pont de Nemours & Co Inc Wilmington DE 19898

WHITELEY, THOMAS EDWARD, b Dunsmuir, Calif, Oct 14, 32; m 53; c 3. ORGANIC CHEMISTRY. Educ: Univ Colo, BA, 54; Univ SC, MS, 56; Ohio State Univ, PhD(carbohydrate chem), 60. Prof Exp: Res chemist, Eastman Kodak Co, 56-57; res asst carbohydrate chem, Ohio State Univ, 57-59; sr res chemist, 60-63, res assoc photog chem, 63-69, ASST DIV HEAD, EMULSION RES DIV, EASTMAN KODAK CO, 69- Mem: Soc Photog Sci & Eng; Sigma Xi. Res: Polymer science and technology; photgraphic chemistry related to the silver halide emulsion. Mailing Add: Kodak Park Res Labs Eastman Kodak Co Rochester NY 14650

WHITEMAN, ALBERT LEON, b Philadelphia, Pa, Feb 15, 15; m 45. MATHEMATICS. Educ: Univ Pa, AB, 36, AM, 37, PhD(math), 40. Prof Exp: Asst instr math, Univ Pa, 38-40; Benjamin Pierce instr, Harvard Univ, 40-42; instr, Purdue Univ, 46; res mathematician, US Dept Navy, 46-48; from asst prof to assoc prof, 48-56, PROF MATH, UNIV SOUTHERN CALIF, 56- Concurrent Pos: Mem, Inst Advan Study, 52-54, 59-60 & 67-68; chmn res conf theory numbers, NSF, 55 & 63; ed, Pac J Math, 57-62; res mathematician, Inst Defense Anal, 60-61. Mem: Am Math Soc; Math Asn Am. Res: Theory of numbers; combinatorial analysis. Mailing Add: Dept of Math Univ of Southern Calif Los Angeles CA 90007

WHITEMAN, CHARLES E, b Eldred, Ill, Sept 28, 18; m 43; c 3. VETERINARY PATHOLOGY. Educ: Kans State Col, DVM, 43; Iowa State Univ, PhD(vet path), 60; Am Col Vet Pathologists, dipl. Prof Exp: Assoc prof vet path, Mich State Univ, 60-61; assoc prof, 61-71, PROF VET PATH, COLO STATE UNIV, 71- Mem: Am Vet Med Asn; Am Asn Avian Pathologists. Res: Placental pathology; respiratory diseases; poultry diseases. Mailing Add: Dept of Path Colo State Univ Ft Collins CO 80521

WHITEMAN, ELDON EUGENE, b Tarentum, Pa, May 5, 13; m 39; c 3. ZOOLOGY. Educ: Greenville Col, BS, 36; Mich State Univ, MS, 41, PhD(zool), 65. Prof Exp: Asst dir, Kellogg Bird Sanctuary, 39-41; chmn natural sci div, 63-71, PROF BIOL, SPRING ARBOR COL, 46-, DIR ENVIRON STUDIES, 72- Mem: Nat Audubon Soc; Nat Wildlife Soc. Res: Development of a summer travel course for the college student in the area of environmental studies. Mailing Add: Spring Arbor Col Spring Arbor MI 49283

WHITEMAN, JOE V, b Walkerville, Ill, July 13, 19; m 45; c 1. ANIMAL BREEDING. Educ: NMex State Univ, BS, 43; Okla State Univ, MS, 51, PhD, 52. Prof Exp: From asst prof to assoc prof, 52-63, PROF ANIMAL SCI, OKLA STATE

UNIV, 63- Mem: AAAS; Am Soc Animal Sci; Biomet Soc. Res: Genetic and environmental factors governing the growth and development of meat animals. Mailing Add: Dept of Animal Sci & Indust Okla State Univ Stillwater OK 74074

WHITEMAN, JOHN DAVID, b Darby, Pa, May 24, 43; m 69; c 2. PHYSICAL CHEMISTRY. Educ: LaSalle Col, BA, 65; Univ Pa, PhD(phys chem), 71. Prof Exp: Sr res chemist anal chem, 71-75, SR RES CHEMIST PHARMACEUT, ROHM AND HAAS CHEM CO, 75- Concurrent Pos: Fel, Dept Chem, Univ Pa, 71-72. Mem: Am Phys Soc; Am Chem Soc. Res: Energy band structure of molecular crystals; polymer physics; structure activity relationships in agricultural and pharmaceutical chemicals; analytical chemistry. Mailing Add: Rohm and Haas Res Labs Norristown & McKean Rds Spring House PA 19477

WHITEMORE, HOWARD LLOYD, b Dallas, Wis, Dec 3, 35; m 62; c 2. VETERINARY MEDICINE. Educ: Okla State Univ, BS, 58, DVM, 60; Univ Wis, PhD(vet sci), 73; Am Col Theriogenologists, dipl. Prof Exp: Vet pvt pract, 60-69; ASSOC PROF COL VET MED, UNIV MINN, ST PAUL, 74- Honors & Awards: Burr Beach Award, Dept Vet Sci, Univ Wis, 73. Mem: Am Vet Med Asn. Res: Fertility; abortion and pregnancy diagnosis in dairy cattle. Mailing Add: Col of Vet Med Univ of Minn St Paul MN 55108

WHITENBERG, DAVID CALVIN, b Duffau, Tex, Feb 6, 31; m 51; c 1. PLANT PHYSIOLOGY, BIOCHEMISTRY. Educ: Tex A&M Univ, BS, 57, MS, 59, PhD(plant physiol, biochem), 62. Prof Exp: Res plant physiologist, USDA, 61-65; asst prof, 65-67, ASSOC PROF BIOL, SOUTHWEST TEX STATE UNIV, 67- Mem: Am Soc Plant Physiologists; fel Am Inst Chemists. Res: Seed physiology and biochemistry. Mailing Add: Dept of Biol Southwest Tex State Univ San Marcos TX 78666

WHITER, PAUL FRANCIS, b London, Eng, Aug 28, 34; m 58; c 4. RESEARCH ADMINISTRATION. Educ: Univ London, BSc, 60, PhD(org chem), 65. Prof Exp: From res scientist to prin scientist, Wilkinson Sword Res Ltd, 63-68, chem proj mgr, Wilkinson Sword, Inc, 68-70, res mgr NAm, 70-73, SPEC PROD MGR, WILKINSON SWORD, INC, 73- Concurrent Pos: Asst lectr, Northwest Kent Col Technol, 62-63. Mem: AAAS; Am Chem Soc; The Chem Soc; Soc Cosmetic Chem. Res: Commercialization and management. Mailing Add: Wilkinson Sword Inc 100 Industrial Rd Berkeley Heights NJ 07922

WHITESCARVER, JACK EDWARD, b Palestine, Tex, May 16, 37. MEDICAL MICROBIOLOGY. Educ: Sam Houston State Univ, Huntsville, BS, 59; Col Med & Dent NJ, PhD(microbiol), 74. Prof Exp: Teaching fel biol, Sam Houston Univ, 62-64; res assoc oncol, Univ Tex, M D Anderson Hosp & Tumor Inst, 64-66; res assoc oncol, Southern Calif Cancer Ctr, Los Angeles, 66-71; teaching fel microbiol, Col Med & Dent NJ, 71-74; RES FEL MICROBIOL, HARVARD UNIV SCH PUB HEALTH, 74- Concurrent Pos: Fel, Albert Soiland Cancer Found, 67-70. Mem: Am Soc Microbiol; Tissue Cult Asn. Res: Morphology and ultrastructural immunology of Rickettsiae; cell culture studies and host-parasite relationship. Mailing Add: Harvard Sch Pub Health-Microbiol 665 Huntington Ave Boston MA 02115

WHITESELL, JAMES JUDD, b Philadelphia, Pa, Oct 14, 39; m 65; c 1. Educ: Dickinson Col, BS, 62; Univ Fla, MEd, 67, MS, 69, PhD(entom), 74. Prof Exp: Teacher sci, James S Rickards Jr High Sch, 63-67; res assoc lovebug res, Dept Entom, Univ Fla, 73-74; TEACHER BIOL & ZOOL, SNEAD STATE JR COL, 74-, CHMN SCI & MATH DIV, 75- Mem: Sigma Xi; Entom Soc Am. Res: Systematics and ecology of sound producing insects with emphasis on copiphorine katydids. Mailing Add: Div of Sci & Math Snead State Jr Col Boaz AL 35957

WHITESELL, JAMES KELLER, b Philadelphia, Pa, Nov 2, 44; m 66; c 2. SYNTHETIC ORGANIC CHEMISTRY. Educ: Pa State Univ, BS, 66; Harvard Univ, PhD(chem), 71. Prof Exp: Fel chem, Woodward Res Inst, 70-73; ASST PROF CHEM, UNIV TEX, AUSTIN, 73- Mem: Am Chem Soc; The Chem Soc. Res: Total synthesis of naturally occuring and theoretically interesting molecules. Mailing Add: Dept of Chem Univ of Tex Austin TX 78712

WHITESELL, WILLIAM JAMES, b Newnan, Ga, Dec 23, 27; m 60; c 5. THEORETICAL PHYSICS. Educ: Univ SC, BS, 48; Purdue Univ, MS, 51, PhD(physics), 59. Prof Exp: From instr to asst prof physics, Brooklyn Col, 58-63; asst prof, 63-69, ASSOC PROF PHYSICS, ANTIOCH COL, 69- Concurrent Pos: Sr lectr, Victoria Univ, Wellington, 70-72. Mem: AAAS; Am Asn Physics Teachers. Mailing Add: Dept of Physics Antioch Col Yellow Springs OH 45387

WHITESIDE, BOBBY GENE, b Keota, Okla, June 16, 40; m 64; c 2. FISHERIES. Educ: Okla State Univ, BS, 62, MS, 64, PhD(fisheries), 67. Prof Exp: Tex Water Develop Bd grant, 72-73, ASSOC PROF BIOL, SOUTHWESTERN TEX STATE UNIV, 67- Mem: Am Fisheries Soc. Res: Fisheries management; population dynamics; ecology. Mailing Add: Dept of Biol Southwest Tex State Univ San Marcos TX 78666

WHITESIDE, CHARLES HUGH, b Grapevine, Tex, June 25, 32; m 56; c 2. ANALYTICAL CHEMISTRY, ENVIRONMENTAL SCIENCES. Educ: Tex A&M Univ, BS, 53, MS, 58, PhD(biochem), 60. Prof Exp: Robert A Welch Found res fel, Dept Biochem & Nutrit, Tex A&M Univ, 60-61; sr scientist, Mead Johnson & Co, 61-64; teacher chem, Kilgore Col, 64-67, chmn dept, 67-71; PRES, ANA-LAB CORP, 67- Mem: Am Chem Soc; Am Oil Chemists' Soc. Res: Water quality and waste water technology; animal nutrition. Mailing Add: 2600 Dudley Rd Kilgore TX 75662

WHITESIDE, EUGENE PERRY, b Champaign, Ill, Oct 18, 12; m 36; c 3. SOIL SCIENCE. Educ: Univ Ill, BS, 34; Univ Mo, PhD(soils), 44. Prof Exp: Asst soil physics & soil surv, Exp Sta, Univ Ill, 34-38; asst soils, Univ Mo, 38-39; assoc soil surv, Exp Sta, Univ Ill, 39-43; assoc soils, Univ Tenn, 43-44; asst chief soil surv & asst prof soil physics, Exp Sta, Univ Ill, 44-49; assoc prof, 49-53, PROF SOIL SCI, MICH STATE UNIV, 53- Concurrent Pos: Assoc soil surv, Emergency Rubber Proj, US Forest Serv, Calif, 43; consult, Natural Resources Sect, Agr Div, Gen Hqs, Supreme Comdr Allied Powers, Japan, 46-47, Rockefeller Found, Mex, 61 & Agency Int Develop, Taiwan, 62-63 & Arg, 72. Mem: Soil Sci Soc Am. Res: Soil mineralogy, chemistry, geography, classification and genesis. Mailing Add: Dept of Crop & Soil Sci Col of Agr & Natural Resources Mich State Univ East Lansing MI 48824

WHITESIDE, HAVEN, b Boston, Mass, Sept 3, 31; m 58; c 4. PHYSICS. Educ: Middlebury Col, AB, 52; Harvard Univ, SM, 53, PhD(appl physics), 63. Prof Exp: Asst prof physics, Amherst Col & Oberlin Col, 62-68; assoc prof, 68-70, PROF PHYSICS, FED CITY COL, 70- Concurrent Pos: NSF res grants, 64-69; NASA res grants, 69-; Cong fel, Am Phys Soc, 74-75; prof staff mem, US Senate Comt Pub Works, 75- Mem: Am Phys Soc; Am Asn Physics Teachers; Sigma Xi. Res: High energy astrophysics; environmental physics and public policy. Mailing Add: 5511 30th St NW Washington DC 20015

WHITESIDE, JACK OLIVER, b Barnstaple, Eng, June 5, 28; m 51; c 2. PLANT PATHOLOGY. Educ: Univ London, BSc, 48, PhD(plant physiol), 53. Prof Exp: Plant physiologist, Ministry Agr, Rhodesia, Africa, 48-53, from plant pathologist to chief plant pathologist, 53-67; assoc plant pathologist, Citrus Exp Sta, 68-73, PROF PLANT PATH, UNIV FLA, 73- Mem: Am Phytopath Soc. Res: Identification and control of plant diseases present in Rhodesia; behavior and control of fungus diseases of citrus. Mailing Add: Agr Res & Educ Ctr Box 1088 Univ of Fla Lake Alfred FL 33850

WHITESIDE, MELBOURNE C, b Washington, DC, Dec 16, 37; m 61; c 2. AQUATIC ECOLOGY. Educ: Willamette Univ, BA, 62; Ariz State Univ, MS, 64; Ind Univ, Bloomington, PhD(zool), 68. Prof Exp: Res fel limnol, Univ Minn, 68-69; asst prof ecol & limnol, Calif State Univ, Fullerton, 69-72; ASST PROF ZOOL, UNIV TENN, KNOXVILLE, 72- Mem: Am Soc Limnol & Oceanog; Ecol Soc Am. Res: Paleolimnology; community ecology and population dynamics of aquatic organisms; sampling problems in aquatic environments. Mailing Add: Dept of Zool Univ of Tenn Knoxville TN 37916

WHITESIDE, ROBERTA EMERSON, b Birmingham, Ala, Sept 1, 27. IMMUNOLOGY. Educ: Univ Ala, BS, 47; Boston Univ, MA, 55, PhD(microbiol), 57. Prof Exp: Asst prof biol, Simmons Col, 57-62; ASST RES PROF MICROBIOL, SCH MED, BOSTON UNIV, 57- Mem: Am Soc Microbiol; Am Asn Immunologists. Mailing Add: Dept of Microbiol Boston Univ Sch of Med Boston MA 02215

WHITESIDE, THERESA L, b Katowice, Poland, Mar 10, 39; US citizen; m 61; c 1. IMMUNOLOGY, IMMUNOPATHOLOGY. Educ: Columbia Univ, BS, 62, MA, 64, PhD(microbiol), 67. Prof Exp: NIH fel, Sch Med, NY Univ, 67-69, lectr microbiol & assoc res scientist, 69-70; res assoc ophthal, Col Physicians & Surgeons, Columbia Univ, 70-73; ASST PROF PATH & ASSOC DIR CLIN IMMUNOPATH, MED SCH, UNIV PITTSBURGH, 73- Concurrent Pos: NIH spec fel ophthal, Col Physicians & Surgeons, Columbia Univ, 72-73. Mem: AAAS; Am Soc Microbiol. Res: Immunology of surface-associated antigens; structure and biochemistry of biological membranes; clinical immunopathology; lymphocyte membrane receptors. Mailing Add: Med Sch Dept of Path Univ of Pittsburgh Pittsburgh PA 15213

WHITESIDE, WESLEY C, b Milan, Ill, Aug 22, 27. BOTANY. Educ: Augustana Col, BA, 51; Univ Ill, MS, 56; Fla State Univ, PhD, 59. Prof Exp: Instr bot & gen biol, Montgomery Jr Col, 59-60; from asst prof to assoc prof, 60-70, PROF BOT, EASTERN ILL UNIV, 70- Concurrent Pos: Res grant, Highlands Biol Sta, 59. Mem: Mycol Soc Am. Res: Morphology; cytology and taxonomy of the ascomycete fungi, especially the Pyrenomycetes; taxonomy of lichens. Mailing Add: Dept of Bot Eastern Ill Univ Charleston IL 61920

WHITESIDES, GEORGE MCCLELLAND, b Louisville, Ky, Aug 3, 39. ORGANIC CHEMISTRY. Educ: Harvard Univ, AB, 60; Calif Inst Technol, PhD(chem), 64. Prof Exp: From asst prof to assoc prof, 63-74, PROF CHEM, MASS INST TECHNOL, 74- Mem: Am Chem Soc; Am Phys Soc. Res: Inorganic chemistry; mechanisms and structure. Mailing Add: Rm 18-298 Dept of Chem Mass Inst of Technol Cambridge MA 02139

WHITESITT, JOHN ELDON, b Stevensville, Mont, Jan 15, 22; m 44; c 3. MATHEMATICS. Educ: Mont State Univ, AB, 43; Univ Ill, AM, 49, PhD(math), 54. Prof Exp: From instr to assoc prof, 46-61, head dept, 61-66, PROF MATH, MONT STATE UNIV, 61- Mem: Am Math Soc; Math Asn Am. Res: Ring theory; linear algebra; Boolean algebra. Mailing Add: Dept of Math Mont State Univ Bozeman MT 59715

WHITE-STEVENS, ROBERT HENRY, b London, Eng, June 3, 12; nat US; m 37; c 4. ENVIRONMENTAL BIOLOGY. Educ: McGill Univ, BSA, 33, MSc, 36; Cornell Univ, PhD(physiol, biochem, genetics), 42. Prof Exp: Asst horticulturist, McGill Univ, 33-35; asst veg crops, Cornell Univ, 35-40, res instr, 40-42, asst res prof, 42-46; dir foods & nutrit, Hosp Bur Stand, New York, 46-48; res dir, Ky Chem Industs, Inc, Ohio, 48-52; dept head nutrit & physiol sect, Res Div, Am Cyanamid Co, 52-57, dir plant indust sect, 57-59, dir nutrit, physiol & biochem sect, Agr Div, 59-61, asst to dir res & develop, 61-69; PROF BIOL, CHMN BUR CONSERV & ENVIRON SCI & ASST DIR EXP STA, COOK COL, RUTGERS UNIV, NEW BRUNSWICK, 69- Mem: Fel AAAS; Am Chem Soc; Soil Conserv Soc Am; Am Soc Plant Physiologists; Poultry Sci Asn. Res: Physiology and toxicity of chemicals in environment; experimental design and biometrics. Mailing Add: PO Box 231 Cook Col Rutgers Univ New Brunswick NJ 08903

WHITEWAY, STIRLING GIDDINGS, b Stellarton, NS, May 17, 27; m 52; c 2. PHYSICAL CHEMISTRY, INORGANIC CHEMISTRY. Educ: Dalhousie Univ, BSc, 47, dipl, 48, MSc, 49; McGill Univ, PhD(phys chem), 53. Prof Exp: Fel photochem, Pure Chem Div, 52-53, from asst res officer to assoc res officer, 53-68, SR RES OFFICER, ATLANTIC REGIONAL LAB, NAT RES COUN CAN, 68- Concurrent Pos: Spec lectr, Dalhousie Univ, 54-55. Mem: Chem Inst Can; Can Inst Mining & Metall. Res: Chemistry of high temperature reactions. Mailing Add: Nat Res Coun of Can Atlantic Regional Lab 1411 Oxford St Halifax NS Can

WHITFIELD, CAROL FAYE, b Altoona, Pa, May 14, 39. PHYSIOLOGY, BIOCHEMISTRY. Educ: Juniata Col, BS, 61; Syracuse Univ, MS, 64; George Washington Univ, PhD(physiol), 68. Prof Exp: Teaching asst zool, Syracuse Univ, 61-63; res assoc, 68-70, ASST PROF PHYSIOL, MILTON S HERSHEY MED CTR, COL MED, PA STATE UNIV, HERSHEY, 70- Mem: AAAS; Biophys Soc; Am Physiol Soc. Res: Metabolic and hormonal regulation of carrier-mediated sugar transport in erythrocytes and muscle; regulation of membrane permeability; membrane structure; cardiac muscle metabolism; mechanisms of hormone action. Mailing Add: Dept of Physiol Milton S Hershey Med Ctr Pa State Univ Hershey PA 17033

WHITFIELD, CAROLYN DICKSON, b Indianapolis, Ind, Aug 21, 41; m 65. BIOCHEMISTRY. Educ: Wellesley Col, AB, 63; Univ Chicago, MS, 65; George Washington Univ, PhD(biochem), 69. Prof Exp: Wellcome Found fel, Univ Edinburgh, 69-70; Am Cancer Soc fel, 70-72, asst res biol chemist, 70-74, SCHOLAR HUMAN GENETICS, MED SCH, UNIV MICH, ANN ARBOR, 74- Res: Mechanism of action of flavoproteins; galactose mutants in animal cells; methionine biosynthesis. Mailing Add: Dept of Human Genetics Univ of Mich Med Sch Ann Arbor MI 48104

WHITFIELD, GEORGE BUCKMASTER, JR, b Newark, NJ, Dec 4, 23; m 44; c 3. BIOCHEMISTRY. Educ: Cornell Col, BA, 46; Univ Ill, MS, 51, PhD(chem), 53. Prof Exp: Jr res scientist, 47-51, sr res scientist, 53-59, sect head microbiol, 59-66, mgr, 66-68, INFECTIOUS DIS RES MGR, UPJOHN CO, 68- Mem: Am Chem Soc; Am Soc Microbiol; NY Acad Sci. Res: Isolation and characterization of new antibiotics and antitumor agents; paper chromatography; microbiological assay; tissue culture; in vitro, in vivo and clinical evaluation of new antibiotics, antifungal and antiparasitic agents. Mailing Add: Unit 7254-126-2 Upjohn Co Kalamazoo MI 49001

WHITFIELD, GEORGE DANLEY, b New York, NY, Dec 12, 30; m 54; c 2. PHYSICS. Educ: City Col New York, BS, 54; Columbia Univ, MA, 59, PhD(physics), 61. Prof Exp: Asst physics, Watson Labs, Int Bus Mach Corp, 54-59; res assoc, Univ

Ill, Urbana, 60-61, res asst prof, 61-62; asst prof, Princeton Univ, 62-65; ASSOC PROF PHYSICS, PA STATE UNIV, UNIVERSITY PARK, 65- Mem: Am Phys Soc. Res: Theoretical solid state physics; electron-phonon interaction. Mailing Add: Dept of Physics Pa State Univ University Park PA 16802

WHITFIELD, GRAHAM FRANK, b Cheam, Eng, Feb 8, 42; m 71. SURGERY. Educ: Univ London, BSc, 63, ARIC, 65, PhD(org chem), 69; New York Med Col, MD, 76. Prof Exp: Res chemist, Unilever Res Lab, Hertfordshire, Eng, 63-66; fel chem, Temple Univ, 69-71, from instr to asst prof, 71-73; SURG RESIDENT, NEW YORK MED COL-METROP HOSP CTR, 76- Mem: Am Chem Soc; AMA. Res: Surgical oncology; early detection of breast cancer. Mailing Add: Metrop Hosp Ctr 1901 First Ave New York NY 10029

WHITFIELD, HARVEY JAMES, JR, b Chicago, Ill, Apr 10, 40; m 65. MOLECULAR BIOLOGY. Educ: Univ Ill, Urbana, BS, 61; Univ Ill Col Med, MD, 64. Prof Exp: Intern, Res & Educ Hosp, Chicago, 64-65; staff asst molecular biol, NIH, 65-69; USPHS spec fel, Med Res Coun Microbial Genetics Unit, Univ Edinburgh, 69-70; asst prof, 70-75, ASSOC PROF BIOCHEM, MED SCH, UNIV MICH, 75- Concurrent Pos: USPHS grant, Univ Mich, Ann Arbor, 71- Mem: AAAS; Am Soc Biol Chemists; Am Soc Microbiol; Genetics Soc Am. Res: Replication and segregation of episomal DNA; microbial genetics. Mailing Add: Dept of Biol Chem Univ of Mich Ann Arbor MI 48104

WHITFIELD, JAMES F, b Sarnia, Ont, July 1, 31; m 51; c 4. CELL PHYSIOLOGY, CANCER. Educ: McGill Univ, BSc, 51; Univ Western Ont, MSc, 52, PhD(bact, immunol), 55. Prof Exp: Res officer bact & viral genetics, Atomic Energy Can Ltd, 55-58, res officer cellular radiobiol, 58-62; sect chief, Europ Joint Res Ctr, Europ Atomic Energy Comn, Italy, 62-65; head cell physiol sect, Radiation Biol Div, 65-72, HEAD ANIMAL & CELL PHYSIOL SECT, BIOL SCI DIV, NAT RES COUN CAN, 72- Mem: Tissue Cult Asn; Am Soc Cell Biol. Res: Control of cell proliferation; in vivo and in vitro effects of calcium, hormones and cyclic nucleotides. Mailing Add: Biol Sci Div Nat Res Coun of Can Ottawa ON Can

WHITFIELD, JOHN HOWARD MERVYN, b Thessalon, Ont, Sept 11, 39; m 60; c 4. MATHEMATICS. Educ: Abilene Christian Col, BA, 61; Tex Christian Univ, MA, 62; Case Inst Technol, PhD(math), 66. Prof Exp: Asst prof, 65-70, chmn dept, 72-75, ASSOC PROF MATH, LAKEHEAD UNIV, 70- Concurrent Pos: Vis scholar, Univ Wash, 71-72. Mem: Am Math Soc; Math Asn Am; Can Math Cong. Res: Functional analysis; differentiable functions and norms; geometry of Banach spaces. Mailing Add: Dept of Math Sci Lakehead Univ Thunder Bay ON Can

WHITFIELD, ROBERT EDWARD, b Waverly, Tenn, Aug 11, 21; m 43; c 4. CHEMISTRY. Educ: Univ Tenn, BS, 43; Harvard Univ, AM, 48, PhD(org chem), 49. Prof Exp: Res chemist, Shell Develop Co, 43-46, Am Cyanamid Co, 49-51 & Dow Chem Co, 51-58; SR RES CHEMIST, WESTERN REGIONAL RES LAB, USDA, ALBANY, 58- Mem: AAAS; Am Chem Soc. Res: Organic reaction mechanisms; spectra and structure of molecules; chelates; petrochemicals; physical organic, polymer, fiber and protein chemistry. Mailing Add: 1841 Pleasant Hill Rd Pleasant Hill CA 94523

WHITFIELD, WILLIS JAMES, physics, mathematics, see 12th edition

WHITFILL, DONALD LEE, b Madill, Okla, Mar 13, 39; m 60; c 2. PHYSICAL INORGANIC CHEMISTRY. Educ: Southeastern State Col, BS, 61; Univ Okla, PhD(chem), 66. Prof Exp: Instr chem, Univ Okla, 66-67; res scientist, Plant Foods Res Div, 67-70, RES SCIENTIST, PROD RES DIV, CONTINENTAL OIL CO, 70- Mem: Am Chem Soc; Soc Petrol Engrs. Res: Transition metal chemistry; electrochemistry; drilling fluid and cement technology. Mailing Add: 1700 Cedar Lane Ponca City OK 74601

WHITFORD, ALBERT EDWARD, b Milton, Wis, Oct 22, 05; m 37; c 3. ASTROPHYSICS. Educ: Milton Col, BA, 26; Univ Wis, MA, 28, PhD(physics), 32. Prof Exp: Asst, Washburn Observ, Univ Wis, 32-33; Nat Res Coun fel, Mt Wilson Observ & Calif Inst Technol, 33-35; res assoc astron, Washburn Observ, Univ Wis, 35-38, asst prof astrophys, 38-46, assoc prof astron, 46-48, prof & dir observ, 48-58; astronomer & dir, 58-68, astronomer & prof, 68-73, EMER PROF ASTRON, LICK OBSERV, UNIV CALIF, SANTA CRUZ, 73- Concurrent Pos: Mem staff, Radiation Lab, Mass Inst Technol, 41-46. Mem: Nat Acad Sci; Am Astron Soc (vpres, 65-67, pres, 67-70); Am Acad Arts & Sci. Res: Photoelectric instrumentation; interstellar absorption; spectrophotometry of stars and galaxies; stellar population of galaxies. Mailing Add: Lick Observ Univ of Calif Santa Cruz CA 95064

WHITFORD, GARY M, b Gouveneur, NY, Mar 9, 37; m 65; c 2. TOXICOLOGY, PHYSIOLOGY. Educ: Univ Rochester, BS, 65, MS, 69, PhD(toxicol), 72; Med Col Ga, DMD, 75. Prof Exp: Instr oral biol, NJ Dent Sch, 71-72; ASST PROF ORAL BIOL, MED COL GA, 72- Mem: Int Asn Dent Res. Res: Metabolism, biological effects and toxicology of fluoride. Mailing Add: Dept of Oral Biol Med Col of Ga Augusta GA 30902

WHITFORD, HOWARD WAYNE, b Benavides, Tex, Apr 6, 40; m 65; c 2. VETERINARY MICROBIOLOGY. Educ: Tex A&M Univ, BS, 63, DVM, 64, PhD(vet microbiol), 76; Am Col Vet Microbiologists, dipl, 73. Prof Exp: Vet lab officer res, US Army Med Unit, Frederick, Md, 65-68; vet officer, Rocky Mountain Lab, Nat Inst Allergy & Infectious Dis, USPHS, 68-70; NIH fel vet microbiol, Sch Vet Med, Tex A&M Univ, 70-71, from grad asst to instr, 71-74; BACTERIOLOGIST, DIAG SERV, TEX VET MED DIAG LAB, 74- Mem: Am Col Vet Microbiologists; Am Asn Vet Lab Diagnosticians; Am Vet Med Asn; US Animal Health Asn. Res: Bacteriologic diagnostic techniques; infectious diseases of sheep and goats; pathogenic bacteriology, immunology of infectious diseases. Mailing Add: Drawer 3040 College Station TX 77840

WHITFORD, LARRY ALSTON, b Ernul, NC, Apr 11, 02; m 28; c 4. BOTANY. Educ: NC State Col, BS, 25, MS, 29; Ohio State Univ, PhD(bot), 41. Prof Exp: High sch teacher, 25-26; from instr to prof, 26-68, EMER PROF BOT, NC STATE UNIV, 68- Concurrent Pos: Consult, Duke Power Co Environ Labs, 73- & Aquatic Control, 75- Mem: AAAS; Phycol Soc Am (vpres, 56, pres, 57); Int Phycol Soc. Res: Floristics of fresh-water algae, especially in southeastern United States. Mailing Add: Dept of Bot NC State Univ Raleigh NC 27607

WHITFORD, PHILIP BURTON, b Argyle, Minn, Jan 9, 20; m 46; c 2. PLANT ECOLOGY. Educ: Northern Ill State Teachers Col, BEd, 41; Univ Ill, MS, 42; Univ Wis, PhD(bot), 48. Prof Exp: Asst bot, Univ Wis, 46-48; ed asst conserv, State Bd Natural Resources, Md, 48-49; from asst prof to assoc prof, 49-61, chmn dept, 66-69, PROF BOT, UNIV WIS-MILWAUKEE, 61- Mem: Fel AAAS; Ecol Soc Am; Am Inst Biol Sci. Res: Population and distribution of plants of the prairie-forest border; successions of native plants; resource management; applied ecology and conservation. Mailing Add: Dept of Bot Univ of Wis Milwaukee WI 53201

WHITFORD, WALTER GEORGE, b Providence, RI, June 12, 36; m 59, 69; c 3. ECOLOGY. Educ: Univ RI, BA, 61, PhD(zool), 64. Prof Exp: From asst prof to assoc prof, 64-72, PROF BIOL, N MEX STATE UNIV, 72- Concurrent Pos: Coordr Joranada Site & mem desert biome sect, Int Biol Prog, Nat Res Coun; ecol consult, Pub Serv Co NMex, 71- & Union Oil Co Calif, 74-; ed, Ecol Soc Am, 75- Mem: AAAS; Am Soc Ichthyologists & Herpetologists; Am Soc Naturalists; Ecol Soc Am; Entom Soc Am. Res: Desert ecology; ecology of social insects; arthropod physiological ecology. Mailing Add: Dept of Biol NMex State Univ Las Cruces NM 88001

WHITHAM, GERALD BERESFORD, b Halifax, Eng, Dec 13, 27; m 51; c 3. APPLIED MATHEMATICS. Educ: Univ Manchester, BSc, 48, MSc, 49, PhD, 53. Prof Exp: Res assoc, Inst Math Sci, NY Univ, 51-53; lectr math, Univ Manchester, 53-56; assoc prof, Inst Math Sci, NY Univ, 56-59; prof, Mass Inst Technol, 59-62; PROF MATH, CALIF INST TECHNOL, 62- Mem: Fel Am Acad Arts & Sci; fel Royal Soc. Res: Fluid dynamics. Mailing Add: Dept of Appl Math Calif Inst of Technol Pasadena CA 91125

WHITHAM, KENNETH, b Chesterfield, Eng, Nov 6, 27; m 53; c 3. GEOPHYSICS. Educ: Cambridge Univ, BA, 48, MA, 52; Univ Toronto, MA, 49, PhD(geophys), 51. Prof Exp: Geophysicist, Dom Observ, Ottawa, 51-64; chief div seismol, 64-73, DIR GEN, EARTH PHYSICS BR, CAN DEPT ENERGY, MINES & RESOURCES, 73- Mem: Fel Royal Soc Can. Res: Seismology; physics of the earth's interior, including geothermal studies; geomagnetism. Mailing Add: Earth Physics Br 1 Observatory Crescent Ottawa ON Can

WHITING, ALFRED FRANK, b Burlington, Vt, May 22, 12; c 2. ETHNOBOTANY. Educ: Univ Vt, BS, 33; Univ Mich, MA, 34. Prof Exp: Asst prof anthrop, Univ Ore, 43-47; anthropologist, US Trust Territory of the Pac, Ponape Dist, 51-54; cur anthrop, 55-74, EMER CUR ANTHROP, DARTMOUTH COL MUS, 74-; RES ETHNOBOTANIST, MUS NORTHERN ARIZ, 74- Mem: Am Anthrop Asn. Res: Ethnobotany of Hopi, Havasupai and Western Apache; Hopi material culture. Mailing Add: Mus of Northern Ariz Box 1389 Flagstaff AZ 86001

WHITING, ANNA RACHEL, b Saugerties, NY, Feb 19, 92; m 18. GENETICS. Educ: Smith Col, BA, 16; Univ Iowa, PhD(zool), 24. Hon Degrees: ScD, Smith Col, 64. Prof Exp: Teacher, Conn High Sch, 16-18; assoc prof biol, Catawba Col, 27-28; prof & head dept, Pa Col Women, 28-36; guest investr zool, Univ Pa, 36-43; assoc prof, Swarthmore Col, 43-44; instr, Univ Pa, 44-46; assoc prof, Swarthmore Col, 48; res assoc, Univ Pa, 48-62; consult, Oak Ridge Nat Lab, 53-72; CONSULT, MARINE BIOL LAB, WOODS HOLE, 72- Mem: AAAS; Am Soc Naturalists; Am Soc Zoologists; Genetics Soc Am. Res: Sex determination and heredity in Hymenoptera; nature of x-ray induced injury in the cell; x-rays and nutrition. Mailing Add: 326 Hawthorn Rd Baltimore MD 21210

WHITING, ANNE MARGARET, b Morrisville, Vt, May 17, 41. VERTEBRATE ANATOMY, EMBRYOLOGY. Educ: Eastern Nazarene Col, AB, 63; Univ Ill, Urbana, MS, 65; Pa State Univ, PhD(zool), 69. Prof Exp: From asst prof to assoc prof, 68-73, PROF BIOL, HOUGHTON COL, 73- Mem: AAAS; Am Soc Zoologists; Am Inst Biol Sci. Res: Morphology, histology and histochemistry of squamate cloacal glands. Mailing Add: Dept of Biol Houghton Col Houghton NY 14744

WHITING, FRANK M, b Tucson, Ariz, Dec 5, 32; m 58; c 2. NUTRITION, BIOCHEMISTRY. Educ: Univ Ariz, BS, 56, MS, 68, PhD(agr biochem, nutrit), 71. Prof Exp: Field man qual control, United Dairymen Ariz, 56-60, res technician pesticide residues, 60-65; res asst, 65-71, ASST PROF ANIMAL SCI & ASST ANIMAL SCIENTIST, UNIV ARIZ, 71- Mem: Am Dairy Sci Asn. Res: Pesticide chemistry; ruminant nutrition; lipid metabolism; pesticide residues in feeds and animal products. Mailing Add: Dept of Animal Sci Univ of Ariz Col of Agr Tucson AZ 85721

WHITLA, WILLIAM ALEXANDER, b Galt, Ont, Oct 16, 38; m 64; c 3. STRUCTURAL CHEMISTRY. Educ: McMaster Univ, BSc, 60, PhD(inorg chem), 65. Prof Exp: Nat Res Coun overseas fel x-ray crystallog, Oxford Univ, 65-66; teaching fel, Univ BC, 66-67; asst prof, 67-75, ASSOC PROF CHEM, MT ALLISON UNIV, 75- Mem: Am Crystallog Asn; Chem Inst Can; The Chem Soc. Res: Bonding in inorganic adducts using x-ray crystallography. Mailing Add: Dept of Chem Mt Allison Univ Sackville NB Can

WHITLEY, JAMES R, b Jamesport, Mo, Apr 21, 21; m 42. BIOCHEMISTRY, NUTRITION. Educ: Univ Mo, AB, 42, MS, 47, PhD(agr chem), 52. Prof Exp: Pvt herbicide bus, 52-62; SUPVR WATER QUAL INVEST AQUATIC BIOL, MO DEPT CONSERV, 62- Mem: Am Fisheries Soc; Weed Sci Soc Am; Water Pollution Control Fedn. Res: Water quality investigations related to ecology of fish and other aquatic organisms. Mailing Add: Mo Dept of Conserv 1110 College Ave Columbia MO 65201

WHITLEY, JOSEPH EFIRD, b Albemarle, NC, Mar 22, 31; m 58; c 2. RADIOLOGY. Educ: Wake Forest Col, BS, 51; Bowman Gray Sch Med, MD, 55; Am Bd Radiol, dipl, 60, cert nuclear med, 72. Prof Exp: Intern, Pa Hosp, Philadelphia, 55-56; resident radiol, NC Baptist Hosp, Winston-Salem, 56-69; from asst prof to assoc prof, 62-69, PROF RADIOL, BOWMAN GRAY SCH MED, 69- Concurrent Pos: James Picker Found scholar radiol res, Bowman Gray Sch Med, 59-61, advan fel acad radiol, Karolinska Inst, Sweden & Mass Inst Technol, 61-62; consult, Epilepsy Prog, Nat Inst Neurol Dis & Blindness, 67- Mem: Am Fedn Clin Res; Asn Univ Radiol; Radiol Soc NAm; Soc Nuclear Med; Am Roentgen Ray Soc. Res: Description and evaluation of cardiovascular phenomena by angiographic and radioisotopic techniques, particularly reno-vascular hypertension and pulmonary embolism; evaluation of modes of medical education. Mailing Add: Dept of Radiol Bowman Gray Sch of Med Winston-Salem NC 27103

WHITLEY, LARRY STEPHEN, b Mattoon, Ill, Jan 30, 37; m 58; c 3. ZOOLOGY. Educ: Eastern Ill Univ, BS, 58; Purdue Univ, MS, 60, PhD(ecol), 63. Prof Exp: Instr environ biol, Purdue Univ, 63; from asst prof to assoc prof, 63-71, PROF ZOOL, EASTERN ILL UNIV, 71- Concurrent Pos: NIH res grants, 65-67; Fed Water Qual Admin grant, Dept Interior, 69-71. Mem: AAAS; Ecol Soc Am; Soc Syst Zool. Res: Physiology and systematics of tubificid worms; biology of polluted aquatic ecosystems and the tolerance mechanisms of aquatic organisms. Mailing Add: Dept of Zool Eastern Ill Univ Charleston IL 61920

WHITLEY, NANCY O'NEIL, b Winston-Salem, NC, Feb 21, 32; m 58; c 2. RADIOLOGY. Educ: Bowman Gray Sch Med, MD, 57. Prof Exp: Intern, Jefferson Davis Hosp, Houston, Tex, 57-58; cardiovasc trainee, Bowman Gray Sch Med, 59-61; physician, Med Dept, Western Elec Co, 63-66; resident, 66-69, from instr to asst prof, 69-74, ASSOC PROF RADIOL, BOWMAN GRAY SCH MED, 74- Concurrent Pos: Fel cardiol, Bowman Gray Sch Med, 58-59. Mem: Am Col Radiol; AMA. Res: Techniques and procedures of angiography. Mailing Add: Dept of Radiol Bowman Gray Sch of Med Winston-Salem NC 27103

WHITLEY, WILLIAM THURMON, b Deland, Fla, Oct 24, 41; m 68. MATHEMATICS. Educ: Stetson Univ, BS, 63; Univ NC, Chapel Hill, MA, 66; Va Polytech Inst & State Univ, PhD(math), 69. Prof Exp: Instr math, Va Polytech Inst & State Univ, 69-70; ASSOC PROF MATH, MARSHALL UNIV, 70- Mem: Am Math Soc; Math Asn Am. Res: Deleted products of topological spaces; rings of continuous real-valued functions. Mailing Add: Dept of Math Marshall Univ Huntington WV 25701

WHITLOCK, DAVID GRAHAM, b Portland, Ore, Aug 26, 24; m 48; c 3. NEUROANATOMY, NEUROPHYSIOLOGY. Educ: Ore State Col, BS, 46; Univ Ore, MD, 49, PhD, 51. Prof Exp: Instr anat, Med Sch, Univ Ore, 50-51; asst prof, Univ Calif, Los Angeles, 51-54; from asst prof to prof, State Univ NY Upstate Med Ctr, 55-67, chmn dept, 66-67; PROF ANAT & CHMN DEPT, UNIV COLO MED CTR, DENVER, 67- Concurrent Pos: Fulbright res scholar, Inst Physiol, Pisa, Italy, 51-52; consult, US Sci Exhibit, 61 & Neurol Study Sect, USPHS, 60-64; chmn, Neurol B Study Sect, Nat Inst Neurol Dis & Blindness, 66-67. Mem: Am Asn Anatomists; Int Brain Res Orgn. Res: Anatomy and physiology of peripheral and central nervous system pathways. Mailing Add: Dept of Anat Univ of Colo Med Ctr Denver CO 80220

WHITLOCK, GAYLORD PURCELL, b Mt Vernon, Ill, July 7, 17; m 41; c 2. AGRICULTURE, BIOCHEMISTRY. Educ: Southern Ill Univ, BEd, 39; Pa State Col, MS, 41, PhD(agr, biochem), 42. Prof Exp: Res asst, Iowa State Col, 43-46, asst prof, 46-47; specialist nutrit serv, Merck & Co, Inc, 47-56; dir health ed, Nat Dairy Coun, 56-61; PROG LEADER FAMILY & CONSUMER SCI, AGR EXTEN SERV, UNIV CALIF, BERKELEY, 61-, AGRICULTURIST COOP EXTEN & VPRES AGR SCI, 73- Mem: Am Chem Soc; Inst Food Technol; Am Home Econ Asn; Am Pub Health Asn. Res: Vitamins; human and animal nutrition; foods. Mailing Add: 3980 Rockville Rd Suisun CA 94585

WHITLOCK, JOHN HENDRICK, b Medicine Hat, Alta, Sept 10, 13; US citizen; m 35; c 2. VETERINARY PARASITOLOGY, PARASITOLOGY. Educ: Iowa State Univ, DVM, 34; Kans State Univ, MS, 35. Prof Exp: Asst zool, Kans State Univ, 34-35, from instr to asst prof path, 35-44; from asst prof to assoc prof, 44-51, PROF PARASITOL, NY STATE VET COL & DIV BIOL SCI, CORNELL UNIV, 51- Concurrent Pos: Mem bd trustees, Cornell Univ. Mem: Fel AAAS; Am Soc Parasitologists; Am Vet Med Asn; Biomet Soc. Res: Experimental epidemiology; environmental biology of parasitisms; nematode taxonomy, physiology and pathogenesis; diseases of sheep. Mailing Add: Dept of Path NY State Vet Col Cornell Univ Ithaca NY 14850

WHITLOCK, L RONALD, b Troy, Pa, July 6, 44; m 68; c 1. ANALYTICAL CHEMISTRY. Educ: Pa State Univ, BS, 66; Univ Mass, PhD(anal chem), 71. Prof Exp: Fel polymer sci, Univ Mass, 70-72; SR RES CHEMIST, EASTMAN KODAK CO RES LABS, 72- Mem: Am Chem Soc; Sigma Xi. Res: Development of new methods for chemical analysis of polymers and chemicals using modern analytical instruments. Mailing Add: Eastman Kodak Co Res Labs 1669 Lake Ave Rochester NY 14650

WHITLOCK, LAPSLEY CRAIG, b Lebanon, Ky, Aug 31, 42; m 62; c 3. EXPERIMENTAL NUCLEAR PHYSICS, SPECTROSCOPY. Educ: Georgetown Col, BS, 64; Vanderbilt Univ, PhD, 69. Prof Exp: Asst prof, 69-75, ASSOC PROF PHYSICS, MISS COL, 75-, HEAD DEPT, 70- Mem: Am Phys Soc; Am Asn Physics Teachers. Res: Gamma ray spectroscopy in decay of radioactive nuclides. Mailing Add: Dept of Physics Miss Col Clinton MS 39058

WHITLOCK, LEIGH STUART, b Mayaguez, PR, May 14, 26; m 49; c 4. INFORMATION SCIENCE, COMPUTER SYSTEMS. Educ: Univ Colo, BA, 55; La State Univ, MS, 57, PhD, 63. Prof Exp: Agriculturist golden nematode proj, Agr Res Serv, USDA, 53-54; biol aide nematol sect, 55-57, nematologist, 57-63; res assoc dept biostatist, Sch Pub Health, Univ Mich, Ann Arbor & sr systs analyst data processing, Univ Hosp, 65-68, asst prof biostatist, Sch Pub Health, 68-72; DIR SCI & TECH INFO, HOECHST PHARMACEUT, INC, 72- Mem: AAAS; Pharmaceut Mfrs Asn; Am Soc Info Sci; Sigma Xi; Drug Info Asn. Res: Information systems; pharmaceutical computing; computer information retrieval and clinical research data processing; statistical analysis; data base design. Mailing Add: 69 Claire Dr Bridgewater NJ 08807

WHITLOCK, RICHARD T, b Columbus, Ohio, July 8, 31; m 60; c 1. THEORETICAL PHYSICS. Educ: Capital Univ, BS, 58; Western Reserve Univ, MS, 61, PhD(physics), 63. Prof Exp: Asst physics, Western Reserve Univ, 58-60, instr, 62-63; from asst prof to assoc prof, Thiel Col, 63-67; ASSOC PROF PHYSICS, UNIV NC, GREENSBORO, 67- Mem: Am Asn Physics Teachers. Res: Many-body boson problem with applications to the theory of liquid helium; two-fluid hydrodynamics with applications to the theory of liquid helium; light and sound interactions. Mailing Add: Dept of Physics Univ of NC Greensboro NC 27412

WHITLOCK, ROBERT HENRY, b Canton, Pa, July 28, 41; m 63; c 3. VETERINARY MEDICINE, PATHOLOGY. Educ: Cornell Univ, DVM, 65, PhD(nutrit path), 70. Prof Exp: Intern vet med, 65-67, NIH spec fel, 69-70, ASST PROF VET MED, NY STATE VET COL, CORNELL UNIV, 70- Mem: Comp Gastroenterol Soc; Am Soc Vet Clin Path; Am Vet Med Asn. Res: Pathogenesis of metabolic diseases in domestic animals. Mailing Add: NY State Col of Vet Med Cornell Univ Ithaca NY 14850

WHITMAN, ANDREW PETER, b Detroit, Mich, Feb 28, 26. MATHEMATICS. Educ: Tulane Univ, BS, 45; Cath Univ, MS, 58, PhD(math), 61; Woodstock Col, Md, STL, 64. Prof Exp: Instr civil eng, Tulane Univ, 46-51; asst prof math, Loyola Univ, La, 65-66, actg chmn dept, 66-67; asst prof, 67-71, ASSOC PROF MATH, UNIV HOUSTON, 71- Concurrent Pos: NSF res grant, 66-68. Mem: Am Math Soc; Math Asn Am. Res: Differential topology and geometry. Mailing Add: Dept of Math Univ of Houston Houston TX 77004

WHITMAN, CHARLES INKLEY, b New York, NY, Mar 17, 25; m 50; c 5. PHYSICAL CHEMISTRY. Educ: Yale Univ, BS, 44, PhD(phys chem), 49. Prof Exp: From instr to asst prof chem, NY Univ, 49-53; sr engr, Atomic Energy Div, Sylvania Elec Prod, Inc, 52-54, adv res engr, 54-55, sect head, 55-57, mgr mkt, Sylvania-Corning Nuclear Corp, 57-60, eng specialist, Gen Tel & Electronics Labs, 60-61; asst tech dir, Int Copper Res Asn, Inc, NY, 61-64, tech dir chem, 64-70; dir res & develop, Phelps Dodge Industs, Inc, 70-75; DIR RES & DEVELOP, METAL GROUP, GLIDDEN-DURKEE DIV, SCM CORP, 75- Mem: Am Chem Soc; Am Soc Metals; Am Inst Mining, Metall & Petrol Engrs. Res: Metal fabrication; powder metallurgy; electrical conductors. Mailing Add: Glidden-Durkee Div of SCM Corp 900 Union Commerce Bldg Cleveland OH 44114

WHITMAN, DONALD RAY, b Ft Wayne, Ind, Nov 7, 31; m 55; c 2. THEORETICAL CHEMISTRY, PHYSICAL CHEMISTRY. Educ: Case Western Reserve Univ, BS, 53; Yale Univ, PhD(phys chem), 57. Prof Exp: From asst prof to assoc prof phys chem, 57-70, assoc vpres, 72-73, V PRES, CASE WESTERN RESERVE UNIV, 73-, PROF PHYS CHEM, 70- Mem: Am Phys Soc. Res: Molecular quantum mechanics. Mailing Add: Adelbert Hall Case Western Reserve Univ Cleveland OH 44106

WHITMAN, ERWIN N, b Jersey City, NJ, May 21, 27; m 47; c 3. RESEARCH ADMINISTRATION, CLINICAL PHARMACOLOGY. Educ: Rutgers Univ, New Brunswick, BS, 48; Univ Chicago, MS, 51, MD, 54. Prof Exp: Res assoc anat, Sch Med, Univ Chicago, 51-52; intern med-surg, Presby Hosp, New York, NY, 54-55; pvt pract, NJ, 55-56; staff physician, Burroughs Wellcome & Co, NY, 56-62; staff assoc, Dept Clin Pharmacol, 62-64, assoc dir dept, 64-68, dir dept med pharmacol, 68-72, DIR DEPT CLIN PHARMACOL, HOFFMANN-LA ROCHE, INC, 72- Concurrent Pos: Fel psychiat, Essex County Overbrook Hosp, Cedar Grove, NJ, 66-67; asst med, Newark City Hosp, NJ, 62-69; clin asst clin pharmacol, Newark Beth Israel Med Ctr, 62-, asst attend neuropsychiat, 65- Mem: AAAS; Am Soc Clin Pharmacol & Therapeut; NY Acad Sci; Acad Psychosom Med; Soc Invest Dermat. Res: Clinical pharmacokinetics; objective measurement of drug effects in man, especially in psychiatry and dermatology; clinical investigation methodology; medical electronics and biophysics. Mailing Add: Dept of Clin Pharmacol Hoffmann-La Roche Inc Nutley NJ 07110

WHITMAN, GERALD MESSNER, b Galesburg, Ill, Apr 17, 15; m 39; c 2. ORGANIC CHEMISTRY. Educ: Univ Ill, BS, 36; Univ Wis, PhD(org chem), 40. Prof Exp: Res chemist, Exp Sta, 40-49, res supvr, 49-53, lab dir, 53-56, asst res dir org chem dept, 57-60, dir pioneering res, 60-62, asst gen dir res, 62-73, DIR RES & DEVELOP, ORG CHEM DEPT, E I DU PONT DE NEMOURS & CO, INC, 73- Mem: AAAS; Am Chem Soc. Res: Vapor phase catalysis; high pressure synthesis; catalytic reactions of hydrocarbons; gas phase oxidation. Mailing Add: 8494 Nemours Bldg Org Chem Dept E I du Pont de Nemours & Co Inc Wilmington DE 19898

WHITMAN, NELSON, chemistry, see 12th edition

WHITMAN, PHILIP MARTIN, b Pittsburgh, Pa, Dec 23, 16. MATHEMATICS. Educ: Haverford Col, BS, 37; Harvard Univ, AM, 38, PhD(math), 41. Prof Exp: Instr math, Harvard Univ, 38-41 & Univ Pa, 41-44; scientist, Los Alamos Sci Lab, Univ Calif, 44-46; asst prof math, Tufts Col, 46-48; mathematician, Appl Physics Lab, Johns Hopkins Univ, 48-61; chmn dept, 61-67, PROF MATH, RI COL, 61- Concurrent Pos: Consult, Weapons Systs Eval Group, Off Secy Defense, 51-55; Parsons fel, Johns Hopkins Univ, 58-59; consult, Opers Eval Group, Off Chief Naval Opers, 60-61. Mem: AAAS; Am Math Soc; Opers Res Soc Am; Math Asn Am. Res: Lattice theory; operations research. Mailing Add: Dept of Math RI Col Providence RI 02908

WHITMAN, ROY MILTON, b New York, NY, June 16, 25; m 68; c 4. PSYCHIATRY. Educ: Ind Univ, BS, 44, MD, 46. Prof Exp: Intern, Kings County Hosp, 46-47; resident psychiat, Duke Hosp, 47-48; from instr to asst prof, Univ Chicago, 52-54; from asst prof to assoc prof neurol & psychiat, Med Sch, Northwestern Univ, 54-57, assoc prof, 57; assoc prof, 57-67, PROF PSYCHIAT, COL MED, UNIV CINCINNATI, 67- Concurrent Pos: USPHS fel, Clins, Univ Chicago, 50-52; chief neurol & psychiat, Vet Admin Res Hosp, Chicago, Ill, 54-57; consult, Vet Admin Hosp, Cincinnati, Ohio, 57-, Ill State Psychiat Inst, 63-68, Cent Clin, Cincinnati, Ohio & Vet Admin Res Hosp, Chicago; clinician, Cincinnati Gen Hosp. Mem: Soc Personality Assessment; Am Psychiat Asn; Am Psychoanal Asn. Res: Psychophysiology of dreaming; sex research; psychoanalytic methods; values in psychoanalysis; psychosomatic medicine. Mailing Add: Dept of Psychiat Cincinnati Gen Hosp Cincinnati OH 45267

WHITMAN, VERNON ELEAZER, physics, see 12th edition

WHITMAN, WALTER WILLIAM, b Pittsfield, Mass, Apr 11, 37; m 60; c 2. MATHEMATICS, ENGINEERING PHYSICS. Educ: Cornell Univ, BA, 59, PhD(math), 64. Prof Exp: Asst prof statist, Univ Calif, Berkeley, 64-66; assoc prof math, 66-69, PROF MATH, COLO SCH MINES, 69- Res: Probability theory, specifically Random Walks; statistics, applications including time series analysis. Mailing Add: Dept of Math Colo Sch of Mines Golden CO 80401

WHITMAN, WARREN CHARLES, b Fargo, NDak, Jan 15, 11; m 41; c 2. RANGE SCIENCE. Educ: NDak Col, BS, 35, MS, 36; Univ Wis, PhD(plant physiol), 39. Prof Exp: Asst botanist, Exp Sta, NDak Col, 38-42; range ecologist, US Forest Serv, 46-47; assoc botanist, Exp Sta, 47-51, BOTANIST, N DAK STATE UNIV, 51-, PROF BOT, 58- Mem: AAAS; Ecol Soc Am; Soc Range Mgt. Res: Botanical composition and production of range and pastures; autecology of range plants; grassland microclimate. Mailing Add: Dept of Bot NDak State Univ Fargo ND 58102

WHITMER, JOHN CHARLES, b Kingfisher, Okla, Jan 28, 39. PHYSICAL CHEMISTRY. Educ: Univ Rochester, BS, 60; Univ Mich, MS, 62, PhD(chem), 65. Prof Exp: Asst prof chem, Western Wash State Col, 65-66; lectr, Univ EAfrica, 67-69; asst prof, 69-72, ASSOC PROF CHEM, WESTERN WASH STATE COL, 72- Mem: AAAS; Am Chem Soc. Res: Molecular spectroscopy. Mailing Add: Dept of Chem Western Wash State Col Bellingham WA 98225

WHITMER, ROBERT MOREHOUSE, b Battle Creek, Mich, June 14, 08; m 36, 51; c 2. PHYSICS, ELECTRICAL ENGINEERING. Educ: Univ Mich, BA, 28, MA, 34, PhD(physics), 39. Prof Exp: Mem staff, Bell Tel Labs, Inc, 28-32; mem fac, Amherst Col, 32-33; engr, Philco Radio & TV Corp, Pa, 35-36; engr, Hercules Powder Co, 36; instr physics, Purdue Univ, 37-41; mem staff, Radiation Lab, Mass Inst Technol, 41-46; prof physics, Rensselaer Polytech Inst, 46-56; sr staff physicist, TRW Systs, 56-73; CONSULT, 73- Concurrent Pos: Consult, US Air Force, 52-53; mem, Security Resources Panel, Off Defense Mobilization, 57; mem, Reentry Body Identification Group, Off Secy Defense, 58. Mem: AAAS; Am Phys Soc; sr mem Inst Elec & Electronics Engrs. Res: Electromagnetics; wave propagation; electromagnetic shielding; operations analyses. Mailing Add: 724 Tenth St Manhattan Beach CA 90266

WHITMER, ROMAYNE FLEMMING, b Chicago, Ill, Sept 5, 25; m 48; c 3. OPTICS. Educ: Univ NMex, BS, 46, Univ Wash, MS, 48; Polytech Inst Brooklyn, PhD(elec eng), 55. Prof Exp: Res engr, Sylvania Elec Prod, NY, 48-51; res scientist, Hazeltine Electronics Corp, 51; staff mem, Los Alamos Sci Lab, 52-56; head plasma physics dept, Microwave Physics Lab, Gen Tel & Electronics Labs, Calif, 56-60, lab mgr & res dir, 60-63; asst mgr, Electronic Sci Lab, 63-65, mgr & res dir, 65-72, dir commun & info sci, 72-75, PROG MGR, ARMY LASER PROGS, LOCKHEED MISSILES & SPACE CO, 75- Concurrent Pos: Mem US Nat Comt, Comn VI, Int Sci Radio Union, 66- Mem: Fel Am Phys Soc; Inst Elec & Electronics Engrs; fel Am Phys Soc. Res: Electromagnetics; network theory; electron optics; traveling wave tubes; plasma physics; interaction of electromagnetic waves with ionized gases; laser and optical communications. Mailing Add: Lockheed Missiles & Space Co 3251 Hanover St Palo Alto CA 94034

WHITMIRE, CARRIE ELLA, b Electra, Tex, Oct 17, 26. BACTERIOLOGY. Educ:

Univ Tex, BA, 46; Univ Kans, MA, 53, PhD(bact), 55. Prof Exp: Bacteriologist, Parkland Hosp, Dallas, Tex, 46-47, Vet Admin Hosp, 47-48 & US Army Chem Ctr, Ft Detrick, Md, 48-52; asst virologist, Univ Kans, 52-55; assoc scientist, Ortho Res Found, NJ, 55-61; proj supvr, Merck Sharp & Dohme Biol Div, Pa, 61-64; tech asst, Winthrop Labs, Biol, NY, 64-67; proj dir & virologist, 67-75, DIR DEPT EXP ONCOL, MICROBIOL ASSOCS, INC, 75- Mem: Am Soc Microbiol; Tissue Cult Asn; Soc Exp Biol & Med. Res: Medical virology and bacteriology; tissue culture; cancer research. Mailing Add: Dept of Exp Oncol Microbiol Assocs Inc Washington DC 20014

WHITMORE, DONALD HERBERT, JR, b Buffalo, NY, May 6, 44; m 70. COMPARATIVE PHYSIOLOGY. Educ: Ind Univ, BA, 66; Northwestern Univ, PhD(biol sci), 71. Prof Exp: NIH fel insect physiol, Northwestern Univ, 71-73; ASST PROF BIOL, UNIV TEX, ARLINGTON, 73- Mem: Sigma Xi. Res: The role of environmental influences on the physiology and biochemistry of animals, particularly how animals adapt to environmental stress. Mailing Add: Dept of Biol Univ of Tex Arlington TX 76019

WHITMORE, EDWARD HUGH, b Ottawa, Ill, Feb 26, 26; m 49; c 2. GEOMETRY. Educ: Ill State Univ, BS, 48, MS, 51; Ohio State Univ, PhD(math ed, math), 56. Prof Exp: Instr high sch, Ill, 48-51; asst prof math, Northern Ill Univ, 55-56; from asst prof to assoc prof, San Francisco State Col, 56-6S; chmn dept, 65-74, PROF MATH, CENT MICH UNIV, 65- Mem: Math Asn Am. Res: Mathematics education; sequences in elementary geometry and their history. Mailing Add: Dept of Math Cent Mich Univ Mt Pleasant MI 48858

WHITMORE, FRANK CHARLES, physics, see 12th edition

WHITMORE, FRANK CLIFFORD, JR, b Cambridge, Mass, Nov 17, 15; m 39; c 4. VERTEBRATE PALEONTOLOGY. Educ: Amherst Col, AB, 38; Pa State Col, MS, 39; Harvard Univ, AM, 41, PhD(geol), 42. Prof Exp: Asst paleont, Harvard Univ, 40; instr geol, RI State Col, 42-44; geologist, 44-46, chief mil geol br, 46-59, RES GEOLOGIST, US GEOL SURV 59- Concurrent Pos: Res assoc, Smithsonian Inst, 67-; mem comt res & explor, Nat Geog Soc, 70- Honors & Awards: Medal of Freedom. Mem: AAAS; fel Geol Soc Am; Soc Vert Paleont; Paleont Soc. Res: Mammalian paleontology; Fossil Cetacea. Mailing Add: Paleont & Stratig Br US Geol Sur Rm E-501 Nat Mus of Natural Hist Washington DC 20244

WHITMORE, FRANK WILLIAM, b Ponca City, Okla, May 15, 32; m 55; c 3. FOREST PHYSIOLOGY. Educ: Okla State Univ, BS, 54; Univ Mich, MF, 56, PhD(forestry), 64. Prof Exp: Res forester, US Forest Serv, 57-61, plant physiologist, 64-65; res assoc forestry, Univ Mich, 65-67; asst prof, 67-69, ASSOC PROF FORESTRY, OHIO AGR RES & DEVELOP CTR, 69- Mem: AAAS; Soc Am Foresters; Am Soc Plant Physiologists. Res: Silviculture; water relations in trees; physiology of wood formation. Mailing Add: Dept of Forestry Ohio Agr Res & Develop Ctr Wooster OH 44691

WHITMORE, GEORGE E, b Zionsville, Ind, Jan 25, 16; m 42; c 6. VETERINARY MEDICINE, TOXICOLOGY. Educ: Mich State Univ, DVM, 41. Prof Exp: Asst animal pathologist, Univ Ill, 41-44; vet meat inspector, USDA, Indianapolis, Ind, 44-52, vet, Agr Res Ctr, Beltsville, Md, 52-63; vet, Pub Health Serv, US Food & Drug Admin, 63-71; VET MED OFFICER, ENVIRONMENTAL PROTECTION AGENCY, 71- Res: Safety of pesticide and food additive residues; veterinary drug pharmacology; regulatory veterinary medicine; bovine infertility; domestic animal infectious and parasitic diseases. Mailing Add: 10425 43rd Ave Beltsville MD 20705

WHITMORE, GORDON FRANCIS, b Saskatoon, Sask, June 29, 31; m 54; c 2. RADIOBIOLOGY, CANCER. Educ: Univ Sask, BA, 53, MA, 54; Yale Univ, PhD(biophys), 57. Prof Exp: From asst prof to assoc prof, 60-65, PROF BIOPHYS, UNIV TORONTO, 65-, HEAD DEPT, 71-, ASSOC DEAN FAC MED, 74-; PHYSICIST, ONT CANCER INST, 56-, ASSOC DIR PHYS DIV, 57- Concurrent Pos: Vis prof, Pa State Univ, 63; mem, Nat Cancer Inst Grants Panel, 63- & Nat Res Coun Assoc Comt Radiobiol, 64-; mem radiation study sect, NIH, 65- Honors & Awards: Ernest Berry-Anderson Prize, Royal Soc Edinburgh, 66. Mem: Biophys Soc; Radiation Res Soc; Can Soc Cell Biol; Can Asn Physicists; Royal Soc Can. Res: Radiation physics; radiobiology of mammalian cells in vitro; action of chemotherapeutic agents; mammalian cell genetics. Mailing Add: Ont Cancer Inst 500 Sherbourne St Toronto ON Can

WHITMORE, JAMES HOBSON, nutrition, see 12th edition

WHITMORE, LESTER MCCLELLAN, JR, biochemical engineering, see 12th edition

WHITMORE, MARY (ELIZABETH) ROWE, b Oakland, Calif, Oct 26, 36; m 61; c 3. ANATOMY, ZOOLOGY. Educ: Univ Calif, Berkeley, BA, 59; Smith Col, MA, 61; Univ Minn, Minneapolis, PhD(anat), 69. Prof Exp: Teaching asst zool, Smith Col, 59-61; teaching asst anat, Univ Minn, Minneapolis, 61-66; vis asst prof, 70-71, ASST PROF ZOOL, UNIV OKLA, 72- Mem: AAAS; Am Soc Zool. Res: Comparative morphology and histology of endocrine glands in the lower vertebrates; biology of cyclostomes. Mailing Add: Dept of Zool Univ of Okla Norman OK 73069

WHITMORE, ROBERT ARTHUR, plant physiology, see 12th edition

WHITMORE, ROY ALVIN, JR, b Baltimore, Md, Aug 14, 28; m 53; c 3. FORESTRY. Educ: Univ Mich, BSF, 52, MF, 54. Prof Exp: Forest economist, Cent States Forest Exp Sta, USDA, 52-58; PROF FORESTRY, UNIV VT, 58- Mem: Soc Am Foresters; Forest Prod Res Soc. Res: Forest products utilization and marketing. Mailing Add: Dept of Forestry Univ of Vt Burlington VT 05401

WHITMORE, STEPHEN CARR, b Holyoke, Mass, Oct 17, 31; m 61; c 3. PHYSICS. Educ: Amherst Col, BA, 54; Univ Minn, PhD(liquid helium), 66. Prof Exp: Res assoc physics, Univ Mish, 66-69; ASST PROF PHYSICS, UNIV OKLA, 69- Mem: Am Phys Soc. Res: Low temperature physics; liquid helium. Mailing Add: Dept of Physics Univ of Okla Norman OK 73069

WHITMORE, WILLIAM FRANCIS, b Boston, Mass, Jan 6, 17; m 46; c 4. MATHEMATICS. Educ: Mass Inst Technol, SB, 38; Univ Calif, PhD(math), 41. Prof Exp: With US Naval Ord Lab, Washington, DC, 41-42; instr physics, Mass Inst Technol, 42-46; opers analyst, Opers Eval Group, US Dept Navy, 46-57, chief scientist, Spec Projs Off, Bur Ord, 57-59; consult scientist, Chief Scientist's Staff, Lockheed Aircraft Corp, 59-62, dep chief scientist, 62-64, asst to pres, 64-69, SR CONSULT SCIENTIST, LOCKHEED MISSILES & SPACE CO, 69- Concurrent Pos: Spec consult, Air Forces Eval Bd, 45; sci analyst, Commanding Gen, 1st Marine Air Wing, Korea, 53; consult, Defense Sci Bd, 66-68 & Marine Bd, Nat Acad Eng, 72-; chmn, Navy Lab Adv Bd, Ord, 67-75. Honors & Awards: Meritorious Pub Serv Citation, US Dept Navy, 61. Mem: Am Math Soc; Optical Soc Am; Opers Res Soc Am; Math Asn Am; assoc fel Am Inst Aeronaut & Astronaut. Res: Classical boundary value problems in physics; weapon systems analysis. Mailing Add: 14120 Miranda Ave Los Altos Hills CA 94022

WHITNEY, AMBROSE GRUNHAGEN, b St Paul, Minn, Dec 25, 14; m 40; c 2. ORGANIC CHEMISTRY. Educ: Col St Thomas, BS, 36; Univ Minn, PhD, 40. Prof Exp: Mem staff res dept, Com Solvents Corp, 40-44, mem staff tech develop div, 44-49, mem staff res dept, 49-51; mem staff, Armour & Co, 51; mem staff develop planning, Davison Chem Div, 51-57, mem staff bus develop div, 57-61, mem staff com develop dept, 61-71, DEVELOP ASSOC, W R GRACE & CO, 71- Mem: Am Chem Soc; Chem Mkt Res Asn; Soc Chem Indust. Res: Rosenmund-von-Braum nitrile synthesis; preparation of coumarone by cyclization; reactions of nitroparaffins; research project evaluation; market studies. Mailing Add: Res Div W R Grace & Co Columbia MD 21044

WHITNEY, ARTHUR SHELDON, b Oberlin, Ohio, Oct 31, 33; m 64; c 2. AGRONOMY. Educ: Ohio State Univ, BS, 55; Cornell Univ, MS, 58; Univ Hawaii, PhD(soil sci), 66. Prof Exp: Res instr agron, Univ Philippines, 59-60; asst agronomist, 65-71, ASSOC AGRONOMIST, HAWAII AGR EXP STA, 71- Mem: Crop Sci Soc Am; Am Soc Agron; Am Forage & Grassland Coun. Res: Pasture management; legume agronomy; plant nutrition. Mailing Add: Maui Res Ctr Hawaii Agr Exp Sta PO Box 187 Kula Maui HI 96790

WHITNEY, CHARLES ALLEN, b Milwaukee, Wis, Jan 31, 29; m 51; c 5. ASTROPHYSICS. Educ: Mass Inst Technol, BS, 51; Harvard Univ, AM, 53, PhD(astron), 55. Prof Exp: Assoc prof, 63-68, PROF ASTRON, HARVARD UNIV, 68-; PHYSICIST, SMITHSONIAN ASTROPHYS OBSERV, 56- Concurrent Pos: Guggenheim Found fel, 71. Mem: Int Astron Union; Am Astron Soc; Am Acad Arts & Sci. Res: Stellar atmospheres; theory of variable stars and associated problems of gas dynamics. Mailing Add: Smithsonian Astrophys Observ Cambridge MA 02138

WHITNEY, CHARLES CANDEE, JR, b Newfane, Vt, Oct 12, 39; m 62; c 2. DRUG METABOLISM, BIOPHARMACEUTICS. Educ: Northeastern Univ, AB, 62; Middlebury Col, MS, 64; Univ Calif, Davis, PhD(org chem), 68. Prof Exp: Res chemist, E I du Pont de Nemours & Co, Inc, 68; asst chief clin chem, 3rd US Army Med Lab, 68-70; SR RES BIOCHEMIST, STINE LAB, E I DU PONT DE NEMOURS & CO, INC, 70- Mem: Am Chem Soc; Sigma Xi. Res: Pharmacokinetics, bioavailability and metabolic fate of drugs in the body; analytical methods for the determination of drugs and their metabolites. Mailing Add: 130 Timberline Dr Newark DE 19711

WHITNEY, COLIN GORDON, b London, Eng, July 2, 44; US citizen; m 66; c 2. ELECTROOPTICS. Educ: Mass Inst Technol, BS, 66, PhD(physics), 70. Prof Exp: Asst prof elec eng, Mass Inst Technol, 70-71; group leader acoustooptics, Isomet Corp, 71-72; MEM TECH STAFF, SPERRY RES CTR, SPERRY-RAND CORP, 72- Mem: Sigma Xi; Optical Soc Am. Res: Optical pattern recognition, particularly fingerprint identification; optical character recognition and printed circuit board inspection; ellipsometry and magneto-optics. Mailing Add: Sperry Res Ctr 100 North Rd Sudbury MA 01776

WHITNEY, CYNTHIA KOLB, b Cumberland, Md, July 11, 41; m 63; c 2. ATMOSPHERIC PHYSICS, MATHEMATICAL PHYSICS. Educ: Mass Inst Technol, SB, 63, SM, 65, PhD(physics), 68. Prof Exp: STAFF PHYSICIST, CHARLES STARK DRAPER LAB, INC, 67- Concurrent Pos: Consult, Advan Appln Flight Exp Prog, NASA, 75. Mem: AAAS; Am Meteorol Soc; Optical Soc Am; Am Inst Physics; Sigma Xi. Res: Mathematical and computer modeling of light transport in an absorbing and multiply scattering medium; statistical description of complex physical systems; mathematical formalisms in fundamental physics. Mailing Add: Charles Stark Draper Lab Inc Cambridge MA 02139

WHITNEY, DANIEL DEWAYNE, b Alma, Mich, July 25, 37; m 59; c 2. ANTHROPOLOGY. Educ: Mich State Univ, BA, 62, MA, 63, PhD(anthrop), 68; Western State Univ, JD, 76. Prof Exp: Instr soc sci, Lansing Community Col, 65-66; from asst prof to assoc prof anthrop, San Diego State Col, 66-72; assoc dean col arts & letters, 69-71, PROF ANTHROP, SAN DIEGO STATE UNIV, 72- Concurrent Pos: Community develop coordr, Jamaica Peace Corps Training Proj, 67; prog chmn annual meeting, Am Anthrop Asn, 70; ed anthrop newsletter, Am Anthrop Asn, 74- Mem: Am Anthrop Asn; Soc Appl Anthrop; Asn Asian Studies. Res: Communication of information within the boundaries of a society, especially innovation; the influence of social structure on values and patterns of behavior; Japan; Okinawa, southeast Asia; law and society. Mailing Add: Dept of Anthrop San Diego State Univ San Diego CA 92182

WHITNEY, DAVID EARLE, b Springfield, Vt, June 18, 40; m 66; c 2. MARINE ECOLOGY. Educ: Univ Vt, BA, 63, MS, 65; Univ Del, PhD(biol), 73. Prof Exp: Res assoc biol, Brookhaven Nat Lab, 72-75; RES ASSOC BIOL, MARINE INST, UNIV GA, 75- Mem: AAAS; Am Soc Limnol & Oceanog; Sigma Xi; Ecol Soc Am. Res: Algal primary productivity and nutrient cycling in estuarine and coastal ecosystems. Mailing Add: Univ of Ga Marine Inst Sapelo Island GA 31327

WHITNEY, DONALD RANSOM, b Cleveland Heights, Ohio, Nov 27, 15; m 39; c 4. STATISTICS, MATHEMATICAL STATISTICS. Educ: Oberlin Col, BA, 36; Princeton Univ, MA, 39; Ohio State Univ, PhD(math), 48. Prof Exp: Instr math, Mary Washington Col, 39-42; prof math, 48-70, PROF STATIST & CHMN DEPT, OHIO STATE UNIV, 70- Concurrent Pos: Consult, Burgess & Niple Ltd, Ohio Bell Tel Co, Pub Utilities Comn, Cincinnati Bell Tel Co, NAm Aviation & Goodyear Atomic Corp. Mem: Fel AAAS; fel Am Statist Asn; Inst Math Statist; Biomet Soc; Am Math Soc. Res: Non-parametric statistics; general statistical methodology. Mailing Add: Dept of Statist Ohio State Univ Columbus OH 43210

WHITNEY, DOROTHY MCCARTNEY (MRS LUDWIK ANIGSTEIN), b Salina, Kans, May 16, 02; m 58. MICROBIOLOGY. Educ: Univ Kans, AB, 27, MA, 34. Prof Exp: Asst bact, Univ Kans, 30-34; bacteriologist, C W Wilson Mem Hosp, NY, 34-41; bacteriologist, Res Lab, Pillsbury Mills, Inc, Minn, 41-45; res assoc, Univ Tex Med Br, Galveston, 45-50, asst prof prev med & pub health, 50-58, asst prof microbiol, 58-67, res scientist, Dept Prev Med & Pub Health, 67-70; RETIRED. Concurrent Pos: McLaughlin traveling fel, Italy, 54. Mem: Am Soc Microbiol; Soc Exp Biol & Med. Res: Bacteriology of flour; chemotherapy and immunology of Rickettsia; antitissue antibodies; immunity and experimental tumors. Mailing Add: 28 Manor Way Galveston TX 77550

WHITNEY, ELEANOR NOSS, b Plainfield, NJ, Oct 5, 38; div; c 3. NUTRITION. Educ: Harvard Univ, BA, 60; Washington Univ, PhD(biol), 70. Prof Exp: Instr biol, Fla A&M Univ, 70-72; instr, 72-73, ASST PROF NUTRIT, FLA STATE UNIV, 73- Concurrent Pos: Res assoc chem, Fla State Univ, 70-72. Mem: NY Acad Sci; Am Dietetic Asn. Res: Incidence of malnutrition in alcoholics and its effect on their ability to respond to rehabilitation therapy. Mailing Add: Dept of Food & Nutrit Fla State Univ Sch of Home Econ Tallahassee FL 32306

WHITNEY, ELLSWORTH DOW, b Buffalo, NY, Sept 17, 28; m 54; c 2. PHYSICAL CHEMISTRY. Educ: Univ Buffalo, BA, 50; NY Univ, PhD(phys chem), 54. Prof Exp: Chemist, Charles C Kawin Co, 46-50; res chemist, Olin Mathieson Chem Corp, 54-57, chem res proj specialist, 57-59; sr res chemist, Carborundum Co, 59-62, sr res assoc, 62-70; assoc prof, 70-75, PROF MAT SCI & ENG, CERAMICS DIV, UNIV FLA, 75-, DIR CTR RES IN MINING & MINERAL RESOURCES, 72- Concurrent Pos: Asst prof, Erie County Technol Inst, 63-68; lectr, Eve Sch, State Univ NY Buffalo, 66-69; partner, Whitney & Onoda, Consult, 74- Mem: Am Ceramic Soc; Am Chem Soc; Am Soc Mining, Metall & Petrol Engrs; Soc Mfg Engrs. Res: Crystal growth; kinetics of surface exchange; heterogeneous catalysis; boron and metal hydrides; borohydrides; fluorine oxidizers; high energy propellants; ultrahigh pressure solid state phenomena; phase transformations in solids; ceramic cutting tools; solid state reaction kinetics, refractory and wear-resistant materials; mining and mineral research. Mailing Add: Ceramic Div Dept Mat Sci & Eng Univ of Fla Gainesville FL 32601

WHITNEY, ELVIN DALE, b West Bountiful, Utah, Mar 23, 28; m 58; c 3. PLANT PATHOLOGY, PLANT BREEDING. Educ: Utah State Univ, BS, 50; Cornell Univ, PhD(plant path), 65. Prof Exp: PLANT PATHOLOGIST, AGR RES SERV, USDA, 65- Mem: Am Phytopath Soc; Int Soc Plant Pathologists; Am Soc Sugar Beet Technologists. Res: Fungal, bacterial and nematode diseases of sugar beet; breeding for disease resistance. Mailing Add: Agr Res Serv US Dept of Agr PO Box 5098 Salinas CA 93901

WHITNEY, GEORGE CROSIER, organic chemistry, see 12th edition

WHITNEY, GEORGE STEPHEN, b Wheatland, Wyo, Feb 5, 34; m 59; c 3. ORGANIC CHEMISTRY. Educ: Univ Colo, BA, 55; Northwestern Univ, PhD(org chem), 62. Prof Exp: Asst prof org chem, Wabash Col, 61-62; from asst prof to assoc prof, 62-68, PROF ORG CHEM, WASHINGTON & LEE UNIV, 73- Concurrent Pos: Swiss-Am Found fel, Univ Basel, 64-65; Sloan-Washington & Lee fel, Univ Bristol, 70-71. Res: Physical organic chemistry and mechanisms; free-radical additions of organic sulfur compounds; cycloalkenes; bicyclic compounds; stereochemistry; organic synthesis. Mailing Add: 823 Thorn Hill Rd Lexington VA 24450

WHITNEY, HARVEY STUART, b Langdon, Alta, Oct 14, 35; m 62; c 2. PHYTOPATHOLOGY, MYCOLOGY. Educ: Univ Sask, BSA, 56, MSc, 58; Univ Calif, Berkeley, PhD(plant path), 63. Prof Exp: Res officer seedling dis, Forest Biol Div, Can Dept Agr, 58-61; RES SCIENTIST, CAN FOREST SERV, CAN DEPT ENVIRON, 61- Concurrent Pos: Can Forest Serv fel, Univ Calif, Berkeley, 70-71. Mem: AAAS; Am Phytopath Soc; Mycol Soc Am; Can Phytopath Soc; Am Acad Arts & Sci. Res: Taxonomy and heterokaryosis in Rhizoctonia; insect-fungus-tree relationships in conifers attacked by bark beetles; tree resistance and predisposition; symbiology. Mailing Add: 1925 Casa Marcia Victoria BC Can

WHITNEY, HASSLER, b New York, NY, Mar 23, 07; m 30, 55; c 2. MATHEMATICS. Educ: Yale Univ, PhB, 28, MusB, 29; Harvard Univ, PhD, 32. Hon Degrees: ScD, Yale Univ, 47. Prof Exp: Instr math, Harvard Univ, 30-31, Nat Res Coun fel & lectr, 32-33, from instr to prof, 33-52; PROF MATH, INST ADVAN STUDY, 52- Concurrent Pos: Mem math panel, Nat Defense Res Comt, 43-45. Mem: Nat Acad Sci; Am Math Soc (vpres, 48-50); Am Philos Soc. Res: Topology; manifolds; integration theory; analytic varieties. Mailing Add: Sch of Math Inst for Advan Study Princeton NJ 08540

WHITNEY, JAMES EARL, inorganic chemistry, physical chemistry, see 12th edition

WHITNEY, JOEL GAYTON, b Cambridge, Mass, Oct 13, 37; m 71. ORGANIC CHEMISTRY. Educ: Harvard Univ, AB, 59; Mass Inst Technol, PhD(org chem), 63. Prof Exp: SR RES CHEMIST, BIOCHEM DEPT, E I DU PONT DE NEMOURS & CO, INC, 63- Mem: Am Chem Soc. Res: Amino acid syntheses; medicinal chemistry, especially heterocyclic chemistry; synthesis of central nervous system agents.

WHITNEY, JOHN BARRY, JR, b Augusta, Ga, June 25, 16; m 41; c 3. PLANT PHYSIOLOGY. Educ: Univ Ga, BS, 35; NC State Col, MS, 38; Ohio State Univ, PhD(plant physiol), 41. Prof Exp: Tutor, Univ Ga, 35-36; asst, Ohio State Univ, 38-41; plant physiologist, Cent Fibre Corp, 41-43 & 46; from asst prof to assoc prof, , PROF BOT, CLEMSON UNIV, 55- Concurrent Pos: Oak Ridge Inst Nuclear Studies fel, Univ Tenn AEC Agr Res Prog, 52-53; area consult, Biol Sci Curric Study SC, 69-72. Mem: Am Soc Plant Physiol. Res: Water relations of plants; structure and microchemistry of plant cell walls; plant microchemistry; nutrition of microorganisms; radioisotope tracer applications. Mailing Add: Dept of Bot Clemson Univ Clemson SC 29631

WHITNEY, JOHN EDWARD, b Casper, Wyo, July 6, 26; m 49; c 5. PHYSIOLOGY. Educ: Univ Calif, Berkeley, AB, 47, MA, 48, PhD(physiol), 51; Cambridge Univ, PhD(biochem), 56. Prof Exp: Res assoc physiol, Cedars of Lebanon Hosp, 51-52; res assoc biochem, Univ Calif, Los Angeles, 52-54; from asst prof to assoc prof, 56-62, actg head dept, 59-62, PROF PHYSIOL & HEAD DEPT, SCH MED, UNIV ARK, LITTLE ROCK, 62- Concurrent Pos: USPHS fel, Univ Calif, Los Angeles, 52-53; Am Cancer Soc fel, Cambridge Univ, 54-56; fel biochem, Univ Calif, Los Angeles, 52-54. Mem: AAAS; Am Physiol Soc; Endocrine Soc; Soc Exp Biol & Med; Am Diabetes Asn. Res: Endocrinology and metabolism, especially pancreatic-pituitary hormone interrelationships. Mailing Add: Dept of Physiol Univ of Ark Sch of Med Little Rock AR 72201

WHITNEY, JOHN FRANKLIN, physical chemistry, crystallography, see 12th edition

WHITNEY, JOHN GLEN, b Ponca City, Okla, June 4, 39; m 58; c 4. MICROBIOLOGY. Educ: Okla State Univ, BS, 61, PhD(microbiol), 67. Prof Exp: Sr microbiologist, 67-71, head fermentation prod res & microbiol res depts, 72-73, DIR MICROBIOL & FERMENTATION PROD RES, ELI LILLY & CO, 73- Mem: Soc Indust Microbiol; Am Soc Microbiol. Res: Discovery, isolation and evaluation of fermentation products; control and regulation of secondary metabolism. Mailing Add: Fermentation Prod Res Eli Lilly & Co Indianapolis IN 46206

WHITNEY, MARION ISABELLE, b Austin, Tex, Apr 23, 11. GEOLOGY. Educ: Univ Tex, BA, 30, MA, 31, PhD(geol, paleont), 37. Prof Exp: Teacher pub sch, 33-36; asst prof geol, Kans State Teachers Col, 37-42; teacher geol & biol, Kilgare Col, 42-46; asst prof geol, Tex Christian Univ, 46-51 & Sul Ross State Col, 51-52; prof geol & biol, Ark Polytech Col, 52-54; assoc prof geol, Tulane Univ, 54-55; assoc prof, La Tech Inst, 55-60; teacher biol, Texarkana Col, 60-61; assoc prof, 61-69, PROF BIOL, CENT MICH UNIV, 69- Mem: Am Asn Petrol Geol; Soc Econ Paleont & Mineral; Geol Soc Am. Res: Description of the fauna of the Glen Rose formation of Texas; development of new data concerning the method of aerodynamic erosion of rock, dunes and snow. Mailing Add: Dept of Biol Cent Mich Univ Mt Pleasant MI 48858

WHITNEY, NORMAN JOHN, b Langdon, Alta, July 24, 25; m 51; c 4. MYCOLOGY, PLANT PATHOLOGY. Educ: Univ Alta, BSc, 47; Univ Western Ont, MSc, 49; Univ Toronto, PhD(mycol, plant path), 53; McGill Univ, BD, 64. Prof Exp: Lectr bot, Univ Toronto, 50-52; plant pathologist, Res Sta, Can Dept Agr, 52-61; lectr bot, McGill Univ, 61-64; lectr biol, 65-73, student counsr, 66-73, ASSOC PROF BIOL & COUNSR STUDENT SERV, UNIV NB, 73- Mem: AAAS; Can Phytopath Soc. Res: Soil-borne diseases of plants; fungal degradation of water pollutants; science and religion; spore germination in the phyllosphere. Mailing Add: Dept of Biol Univ of NB Fredericton NB Can

WHITNEY, PHILIP LAWRENCE, b Oberlin, Ohio, Nov 5, 37; m 60; c 4. BIOCHEMISTRY. Educ: Ohio State Univ, BS, 60; Duke Univ, PhD(biochem), 65. Prof Exp: ASST PROF BIOCHEM, SCH MED, UNIV MIAMI, 69- Concurrent Pos: USPHS fel biochem, Gothenberg Univ, 65-66 & Harvard Univ, 66-68; res fel, Ind Univ, Bloomington, 68-69. Res: Characterization of glycoproteins of respiratory mucus and chemical modifications of specific groups of carbonic anhydrase. Mailing Add: Dept Biochem Sch Med Univ Miami PO Box 520875 Miami FL 33152

WHITNEY, PHILIP ROY, b Providence, RI, Nov 10, 35; m 59; c 4. GEOCHEMISTRY, PETROLOGY. Educ: Mass Inst Technol, BS, 56, PhD(geol), 62. Prof Exp: Asst prof geochem, State Univ NY Col Ceramics, Alfred Univ, 62-67; asst prof geol, Rensselaer Polytech Inst, 67-70; ASSOC SCIENTIST GEOCHEM, NY STATE MUS & SCI SERV, 70- Concurrent Pos: Adj assoc prof, Rensselaer Polytech Inst, 75- Mem: Geol Soc Am; Geochem Soc. Res: Geochemistry and petrology of anorthosite; geology of the Adirondack area; geochemistry of freshwater manganese oxides. Mailing Add: Geol Surv NY State Mus & Sci Serv Educ Bldg Albany NY 12234

WHITNEY, RAE, b Rochester, NY, June 4, 14. BIOLOGY. Educ: Alfred Univ, BA, 36; Brown Univ, MS, 40; Univ Wis, PhD(zool), 44. Prof Exp: Asst biol, Alfred Univ, 34-38, from asst prof to assoc prof, 43-49; chmn dept, Jamestown Col, 49-51; res assoc, Boston Univ, 51-57, asst res prof, 57-59; dir animal prod, Bio-Res Consults, 59-61; assoc prof, 61-67, PROF BIOL, ELMIRA COL, 67- Concurrent Pos: Asst, Brown Univ, 38-40 & Univ Wis, 41-42; actg chmn dept biol, Alfred Univ, 44-45; mem comt standards, Inst Lab Animal Resources, 60. Mem: AAAS; Am Physiol Soc; NY Acad Sci. Res: Physiology of mammalian reproduction in females; regeneration and function of transplanted adrenal glands; genetics of Syrian hamsters; plankton studies of the Finger Lakes. Mailing Add: Dept of Biol Elmira Col Elmira NY 14901

WHITNEY, RICHARD RALPH, b Salt Lake City, Utah, June 29, 27; m 50; c 4. FISHERY BIOLOGY. Educ: Univ Utah, MS, 51; Iowa State Col, PhD(fisheries mgt), 55. Prof Exp: Res biologist, Salton Sea Invest, Univ Calif, 54-57; proj leader, Susquehanna Fishery Study, State Dept Res & Educ, Md, 58-60; chief tuna behav invests, Tuna Resources Lab, US Bur Com Fisheries, 61-67; UNIT LEADER, WASH COOP FISHERY UNIT, 67- Concurrent Pos: Tech adv & medium of communication, George H Boldt, Sr Judge US Dist Court, Tacoma, 74-; consult, Conn Yankee Atomic Power Co, 65- Mem: AAAS; Inst Fishery Res Biol; Am Fisheries Soc. Res: Aquatic ecology; fisheries. Mailing Add: Col of Fisheries Univ of Wash Seattle WA 98195

WHITNEY, ROBERT ARTHUR, JR, b Oklahoma City, Okla, July 27, 35; m 58; c 5. LABORATORY ANIMAL MEDICINE, COMPARATIVE MEDICINE. Educ: Okla State Univ, BS, 57, DVM, 59; Ohio State Univ, MS, 65. Prof Exp: US Army fel & resident lab animal med, Ohio State Univ, 63-65, chief animal resources br, US Army Edgewood Arsenal, 65-70, dir lab animal training prog, US Army Vet Corps, 68-70, commanding officer, 4th Med Detachment, Viet Nam, 70-71; proj officer animal resources br, 71-72, CHIEF VET RESOURCES BR, DIV RES SERV, NIH, 72- Concurrent Pos: Consult lab animal med, US Army Surgeon Gen Off, 67-70. Mem: Am Col Lab Animal Med; Am Vet Med Asn; Am Asn Lab Animal Sci; Am Asn Lab Animal Practitioners; Sigma Xi. Res: Diseases of laboratory animals; primatology. Mailing Add: Vet Resources Br Nat Insts of Health Bethesda MD 20014

WHITNEY, ROBERT BYRON, b Minneapolis, Minn, July 28, 05; m 33, 48; c 4. ORGANIC CHEMISTRY. Educ: Univ Minn, BA, 24, PhD(org chem), 27. Prof Exp: Instr chem, Harvard Univ, 28-30; from instr to prof, actg dean fac, 66, EMER PROF CHEM, AMHERST COL, 71- Concurrent Pos: Pratt fel, Univs Heidelberg & Halle, 36-37; consult, Radiation Lab, Mass Inst Technol, 45-46; staff mem, Radiation Lab, Univ Calif, 54, Petrol Res Fund fel, Oxford Univ, 60-61; coordr, Amherst, Smith & Mt Holyoke Cols & Univ Mass, 63-66. Mem: AAAS; Am Chem Soc. Res: Kinetics of disproportionation of free radicals; reactions of alkali metals with organic substances; thermodynamics of electrocapillarity; acid-base catalysis. Mailing Add: Dept of Chem Amherst Col Amherst MA 01002

WHITNEY, ROBERT C, b Seattle, Wash, July 20, 19; m 42; c 2. SCIENCE EDUCATION, PHYSICS. Educ: Univ Wash, BS, 47; Cornell Univ, MS, 58, PhD(sci educ, physics), 63. Prof Exp: Teacher, Wash High Sch, 47-55, 56-57 & 58-59; assoc dir shell merit fels, Shell Found, Cornell Univ, 59-61, assoc dir acad year inst, NSF, 61-63; assoc prof, 63-66, PROF PHYS SCI, CALIF STATE UNIV, HAYWARD, 66- Concurrent Pos: Consult, Murray, Fremont & Palo Alto Sch Dist, Calif, 65-66 & Livermore Sch Dist, 67; NSF fel, Univ Wash, 71-72. Mem: AAAS; Am Asn Physics Teachers; Nat Sci Teachers Asn. Res: Improvement of high school physics facilities; improvement in the teaching of high school physics and elementary science. Mailing Add: Dept of Earth Sci Calif State Univ Hayward CA 94542

WHITNEY, ROBERT MCLAUGHLIN, b St Paul, Minn, Sept 28, 11; m 34, 67; c 2. FOOD CHEMISTRY. Educ: Augustana Col, SDak, AB, 36; Univ Ill, PhD(phys chem), 44. Prof Exp: Instr math, Augustana Col, SDak, 36-37; high sch teacher, NDak, 37-38 & Ill, 38-40; asst chemist, Ill State Water Surv, 40-42; instr chem, Univ Ill, 42-44; res chemist, Dean Milk Co, 44-46; assoc prof dairy mfg res, 46-50, assoc prof dairy technol, 50-59, PROF DAIRY TECHNOL, UNIV ILL, URBANA, 59-, PROF FOOD CHEM, 73- Honors & Awards: Borden Co Found Res Award, Am Dairy Sci Asn, 61. Mem: Am Chem Soc; Am Dairy Sci Asn; fel Am Inst Chemists; Inst Food Technologists. Res: Chemical analysis of dairy products; investigation of flavors in dairy products; ultrasonic bactericidal effects; physical-chemical state of milk proteins; investigation of the proteins in the milk fat-globule membrane. Mailing Add: Dept of Food Sci Univ of Ill Urbana IL 61801

WHITNEY, ROBERT SHANNON, soil science, see 12th edition

WHITNEY, ROY DAVIDSON, b Langdon, Alta, Dec 30, 27; m 53; c 4. FOREST PATHOLOGY. Educ: Univ BC, BSF, 51; Yale Univ, MF, 54; Queen's Univ, Ont, PhD(forest path), 60. Prof Exp: RES SCIENTIST, CAN FORESTRY SERV, 51- Mem: Am Phytopath Soc; Can Phytopath Soc; Can Inst Forestry. Res: Investigations of root rots of conifers, including identification of causal fungi, symptomatology, infection courts and spore germination; determination of pathogenic potentials by inoculations. Mailing Add: Can Forestry Serv Box 490 Sault Ste Marie ON Can

WHITNEY, THOMAS ALLEN, b Toledo, Ohio, June 22, 40; m 62; c 3. ORGANIC CHEMISTRY. Educ: Northwestern Univ, Evanston, BA, 62; Univ Calif, Los Angeles, PhD(chem), 67. Prof Exp: SR RES CHEMIST, CORP RES LAB, EXXON RES &

ENG CO, 67- Mem: Am Chem Soc. Res: Homogeneous catalysis; asymmetric synthesis; organic reactions. Mailing Add: Exxon Res & Eng Co Corp Res Lab PO Box 45 Linden NJ 07036

WHITNEY, WENDELL KEITH, b Miltonvale, Kans, Nov 27, 27; m 45; c 2. ENTOMOLOGY, AGRICULTURE. Educ: Kans State Univ, BS, 56, MS, 58, PhD(entom, zool), 62. Prof Exp: Biol aide, Stored Prod Insect Br, USDA, Kans, 51-56, entomologist, 56-58; instr entom, Kans State Univ, 58-62; entomologist, Bioprod Dept, Dow Chem Co, 62-68; Ford Found entomologist, Int Inst Trop Agr, Nigeria, 68-73; CHIEF ENTOMOLOGIST, PLANT PROD RES & DEVELOP, CYANAMID INT, 74- Concurrent Pos: Res grants, 58-62; consult, Industs & USDA, 59-62. Mem: Entom Soc Am; Asn Advan Agr Sci Africa; Entom Soc Nigeria; Nigerian Soc Plant Protection; Am Mosquito Control Asn. Res: Plant pest and disease control; control of stored products pests. Mailing Add: Plant Prod Res & Develop Cyanamid Int PO Box 400 Princeton NJ 08540

WHITNEY, WILLIAM BERNARD, b Kansas City, Kans, Oct 16, 05; m 37; c 3. ORGANIC CHEMISTRY. Educ: Univ Tex, AB, 33, AM, 34, PhD(org chem), 37. Prof Exp: From assoc prof to prof chem, Univ Tex, Arlington, 37-43; chief chemist, NAm Aviation, Inc, 43-45; sr res chemist, Res Dept, Phillips Petrol Co, 45-69; ASSOC PROF CHEM, TEX WOMAN'S UNIV, 69- Mem: AAAS; Electrochem Soc; Am Chem Soc; fel Am Inst Chemists. Res: Hydantoins; ketones; laminated plastics and adhesives; electroplating; lubricants; diolefins; polymerization; fuel cells; electro-organic chemistry. Mailing Add: Dept of Chem Tex Woman's Univ Denton TX 76204

WHITNEY, WILLIAM MERRILL, b Coeur d'Alene, Idaho, Dec 5, 29; m 50; c 2. PHYSICS. Educ: Calif Inst Technol, BS, 51; Mass Inst Technol, PhD(physics), 56. Prof Exp: From instr to asst prof physics, Mass Int Technol, 56-63; mem tech staff, 63-67, mgr guid & control res sect, 67-70, MGR ASTRIONICS RES SECT, JET PROPULSION LAB, 70-, TECH LEADER, ROBOT RES PROG, 71- Mem: AAAS; Am Phys Soc. Res: Low temperature and semiconductor physics; computer science. Mailing Add: Astrionics Res Sect 198-229 Jet Propulsion Lab Pasadena CA 91103

WHITRIDGE, JOHN, JR, b Baltimore, Md, Nov 25, 08; m 36; c 2. OBSTETRICS & GYNECOLOGY. Educ: Yale Univ, AB, 30; Johns Hopkins Univ, MD, 34. Prof Exp: Intern, Union Mem Hosp, Baltimore, Md, 34-35; from intern to asst resident obstet, Johns Hopkins Hosp, 35-38, resident, 38-39; from instr to assoc prof obstet & gynec, 38-73, lectr pub health admin, 47-66, EMER ASSOC PROF GYNEC & OBSTET, SCH MED, JOHNS HOPKINS UNIV, 73-, LECTR POP & FAMILY HEALTH, SCH HYG & PUB HEALTH, 66- Concurrent Pos: Consult, State Dept Health, Md, 45-53, chief div maternal & child health, 53-56, chief bur prev med, 56-66; prog consult, Planned Parenthood Asn Md, Inc. Mailing Add: 1003 Poplar Hill Rd Baltimore MD 21210

WHITSEL, BARRY L, b Mt Union, Pa, Aug 26, 37; m 60; c 2. NEUROPHYSIOLOGY, NEUROPHARMACOLOGY. Educ: Gettysburg Col, AB, 59; Univ Pa, MS, 63; Univ Ill, PhD(pharmacol), 66. Prof Exp: Res asst psychopharmacol, Wyeth Inst, 59-61; res assoc pharmacol, Sch Med, Univ Pittsburgh, 65-66, from instr to asst prof, 66-74; ASSOC PROF DENT RES & PHYSIOL, SCH MED, UNIV NC, CHAPEL HILL, 74- Concurrent Pos: Res scientist develop award, NIMH, 68-73. Res: Pharmacology. Mailing Add: Dept of Physiol Univ of NC Sch of Med Chapel Hill NC 27514

WHITSELL, JOHN CRAWFORD, II, b St Joseph, Mo, Dec 21, 29; m 65. SURGERY. Educ: Grinnell Col, AB, 50; Wash Univ, MD, 54; Am Bd Surg, dipl, 62; Am Bd Thoracic Surg, dipl, 64. Prof Exp: Instr surg, Med Col, Cornell Univ, 63-66, asst attend surgeon, New York Hosp, 63-68, from asst prof to assoc prof, 66-70, PROF SURG, MED COL, CORNELL UNIV, 70-, SURG DIR RENAL TRANSPLANT UNIT, NEW YORK HOSP-CORNELL MED CTR, 68- Concurrent Pos: Assoc attend surgeon, New York Hosp, 68-70, attend surgeon, 70- Mem: AMA; Am Col Surg; Transplantation Soc; NY Acad Sci; Harvey Soc Res: Renal transplantation. Mailing Add: 517 E 71st St New York NY 10021

WHITSETT, THOMAS L, b Tulsa, Okla, July 14, 36; m 59; c 2. INTERNAL MEDICINE, CLINICAL PHARMACOLOGY. Educ: Pasadena Col, BA, 58; Univ Okla, MD, 62. Prof Exp: Clin asst, Med Ctr, Univ Okla, 67-68; vis asst prof med, Sch Med, Emory Univ, 69-70; asst prof, 70-72, ASSOC PROF MED, MED CTR, UNIV OKLA, 72-, ASST PROF PHARMACOL, 70- Concurrent Pos: Found fac develop award, Pharmaceut Mfr Asn, 71; trainee clin pharmacol, Med Ctr, Univ Okla, 67-68 & Sch Med, Emory Univ, 68-70. Mem: Am Heart Asn; Am Fedn Clin Res; Am Soc Pharmacol & Exp Therapeut. Res: Early phases of new drug investigation, especially cardiovascular and respiratory agents. Mailing Add: Univ of Okla Health Sci Ctr Oklahoma City OK 73104

WHITSON, GARY LEE, zoology, protozoology, see 12th edition

WHITSON, MONA MCCLURG, plant physiology, microbiology, see 12th edition

WHITSON, ROBERT EDD, b Spearman, Tex, Apr 30, 42; m 63; c 2. AGRICULTURAL ECONOMICS, RANGE MANAGEMENT. Educ: Tex Tech Univ, BS, 65, MS, 67, PhD(agr econ), 74. Prof Exp: Area economist, 69-71, ASST PROF RANGE ECON, TEX AGR EXP STA, TEX A&M UNIV, 74- Mem: Am Soc Agr Econ; Soc Range Mgt. Res: Examination of risk management alternatives for ranchers and an evaluation of changing feed price relationships on efficient ranch organizations. Mailing Add: Dept of Range Sci Tex A&M Univ College Station TX 77843

WHITT, CARLTON DENNIS, b Elkmont, Ala, July 4, 19; m 42; c 5. INDUSTRIAL CHEMISTRY, ORGANIC POLYMER CHEMISTRY. Educ: Univ Ala, AB, 41, MS, 42, PhD(inorg chem), 71. Prof Exp: High sch teacher, 40-41; instr org chem, Exten, Univ Ala, 43-45; from asst res chemist to asst assoc res chemist, Tenn Valley Auth, 45-52; chem engr, Chemstrand Corp, 52-64; CHEM ENGR, MONSANTO CO, 64- Mem: Am Chem Soc. Res: Chemical warfare agents; vapor pressures and fundamental data on phosphoric acids; fixation of fertilizers on soils and clays; crystal structure determination by x-ray methods. Mailing Add: Rte 9 Box 552 Athens AL 35611

WHITT, DARNELL MOSES, b Greensboro, NC, Apr 30, 13; m 36; c 1. SOIL PHYSICS, FIELD CROPS. Educ: NC State Univ, BS, 34; Univ Mo, AM, 35, PhD(crops), 52. Prof Exp: Soil surveyor, Soil Conserv Serv, USDA, 35-36, res agronomist, 36-42 & 46-52, res soil conservationist, Agr Res Serv, 52-55, regional liaison officer, Agr Res Serv & Soil Conserv Serv, 55-56, nat liaison officer, 56-59, dir conserv planning, Soil Conserv Serv, 59-63, dir plant sci div, 63-72, dep adminr, 72-75; COORDR RIVER BASIN STUDIES, INT JOINT COMN, 75- Concurrent Pos: Mem, Nat Comt Res Needs in Soil & Water Conserv, 58-59; consult, Repub of Nauru & SPac Comn, New Caledonia. Mem: Am Soc Agron; Soil Sci Soc Am; Soc Range Mgt; Soil Conserv Soc Am; Int Soc Soil Sci. Res: Water pollution. Mailing Add: Int Joint Comn Suite 800 100 Ouellette Ave Windsor ON Can

WHITT, DIXIE DAILEY, b Longmont, Colo, Mar 9, 39; m 63. MICROBIAL ECOLOGY, MICROBIAL PHYSIOLOGY. Educ: Colo State Univ, BS, 61, PhD(zool), 65. Prof Exp: USPHS fel biochem genetics, Yale Univ, 65-69; RES ASSOC MICROBIOL, UNIV ILL, URBANA, 69- Concurrent Pos: Mem comt status women microbiologists, Am Soc Microbiol, 74-78. Mem: AAAS; Genetics Soc Am; Am Soc Microbiol. Res: Structure and function of pyridoxal phosphate dependent enzymes; biochemical genetics of microorganisms; host-parasite relationships as an expression of the host's environmental conditions; host-intestinal microflora interactions. Mailing Add: Dept of Microbiol Univ of Ill Urbana IL 61801

WHITT, GREGORY SIDNEY, b Detroit, Mich, June 13, 38; m 63. DEVELOPMENTAL GENETICS, BIOCHEMICAL GENETICS. Educ: Colo State Univ, BS, 62, MS, 65; Yale Univ, PhD(biol), 70. Prof Exp: Asst prof, 69-72, ASSOC PROF ZOOL, UNIV ILL, URBANA, 72- Mem: AAAS; Genetics Soc Am; Am Inst Biol Sci; Soc Develop Biol; Am Soc Naturalists. Res: Genetic and epigenetic regulation of isozyme synthesis during cytodifferentiation; lactate dehydrogenase polypeptide assembly and evolution; genetic and molecular bases of isozymes; gene duplication and evolution. Mailing Add: Dept of Genetics & Develop Univ of Ill Urbana IL 61801

WHITT, WARD, b Buffalo, NY, Jan 29, 42; m 68. OPERATIONS RESEARCH. Educ: Dartmouth Col, AB, 64; Cornell Univ, PhD(opers res), 69. Prof Exp: Vis asst prof opers res, Stanford Univ, 68-69; asst prof admin sci, 69-74, ASSOC PROF STATIST, SCH OF ORGN & MGT, YALE UNIV, 74- Concurrent Pos: NSF res initiation grant admin sci, Yale Univ, 71-73, jr fac fel, 72-73. Res: Probability theory and its applications; mathematical models in the social sciences. Mailing Add: Sch of Orgn & Mgt Yale Univ 2 Hillhouse Ave New Haven CT 06520

WHITTAKER, ARTHUR GREENVILLE, physical chemistry, see 12th edition

WHITTAKER, FREDERICK HORACE, b Columbus, Ohio, Mar 9, 28; m 52; c 2. PARASITOLOGY. Educ: Otterbein Col, BA, 51; Univ Ga, MSc, 56; Univ Ill, PhD(zool, parasitol), 63. Prof Exp: Instr biol & chem, Spartanburg Jr Col, 57-58; res biologist, Abbott Labs, Ill, 63-64; from asst prof to assoc prof, 64-72, PROF ZOOL, UNIV LOUISVILLE, 72- Concurrent Pos: Consult, Abbott Labs, 64-65. Mem: AAAS; Am Soc Parasitologists. Res: Effects of fermentation liquors on invertebrates; taxonomy and life cycles of trematodes and cestodes; electron microscopy and scanning electron microscopy of cestodes and trematodes; systematics and ecology of helminths of cavefishes. Mailing Add: 401 Deerfield Lane Louisville KY 40207

WHITTAKER, JAMES CURTISS, b Danville, Ill, Nov 9, 36; m 61; c 1. FORESTRY. Educ: Purdue Univ, West Lafayette, BS, 58, MS, 60; Ohio State Univ, PhD(agr econ), 65. Prof Exp: Res forester, US Forest Serv, 61-68; PROF FORESTRY, UNIV MAINE, ORONO, 68- Mem: Soc Am Foresters. Res: Recreation and land use planning. Mailing Add: 247 Nutting Univ of Maine Orono ME 04473

WHITTAKER, JAMES VICTOR, b Los Angeles, Calif, Aug 1, 31. MATHEMATICS. Educ: Univ Calif, Los Angeles, BA, 53, MA, 54, PhD, 58. Prof Exp: Assoc math, Univ Calif, Los Angeles, 57-58; from instr to assoc prof, 58-69, PROF MATH, UNIV BC, 69- Mem: Am Math Soc; Math Asn Am; Can Math Cong. Res: Group theory; point set topology; topological groups. Mailing Add: Dept of Math Univ of BC Vancouver BC Can

WHITTAKER, JOHN RICHARD, b Cornwall, Ont, Aug 19, 34; m 62. EMBRYOLOGY. Educ: Queen's Univ, Ont, BA, 58, MSc, 59; Yale Univ, PhD(develop biol), 62. Prof Exp: Asst prof zool, Univ Calif, Los Angeles, 62-67; ASSOC MEM, WISTAR INST ANAT & BIOL, 67-; ASSOC PROF ANAT, SCH MED, UNIV PA, 71- Honors & Awards: MBL Award, Marine Biol Lab, Woods Hole, 71. Mem: Am Soc Zool; Soc Develop Biol; Int Soc Develop Biol; Tissue Cult Asn. Res: Pigment cell differentiation; melanin biochemistry; morphogenetic substances in mosaic embryos; ascidian embryology; tunicate phylogeny. Mailing Add: Wistar Inst of Anat & Biol 36th St at Spruce Philadelphia PA 19104

WHITTAKER, MACK PAGE, b Richfield, Utah, Aug 13, 40; m 60; c 3. INORGANIC CHEMISTRY. Educ: Brigham Young Univ, BS, 62; Univ Utah, PhD(inorg chem), 66. Prof Exp: Res chemist, Great Lakes Res Corp, Tenn, 66-67, sect head, 67-72, asst tech dir, Great Lakes Carbon Corp, 72-73, TECH DIR, GREAT LAKES CARBON CORP, 73- Mem: Am Chem Soc; Am Inst Mining, Metall & Petrol Engrs. Res: Fast reaction kinetics; kinetics of inorganic polymerization systems; crystal structure evaluation; x-ray diffraction; high temperature chemistry; carbon and graphite technology. Mailing Add: 57 Ocean Dr W Stamford CT 06902

WHITTAKER, ROBERT HARDING, b Wichita, Kans, Dec 27, 20; m 54; c 3. ECOLOGY. Educ: Washburn Univ, AB, 42; Univ Ill, PhD(zool), 48. Prof Exp: From instr to asst prof zool, Wash State Univ, 48-51; sr scientist, Biol Sect, Hanford Lab, Gen Elec Co, Wash, 51-54; from instr to assoc prof biol, Brooklyn Col, 54-64; vis scientist, Brookhaven Nat Lab, 64-66; prof biol, Univ Calif, Irvine, 66-68; PROF BIOL, CORNELL UNIV, 68- Mem: Ecol Soc Am; Am Soc Limnol & Oceanog; Am Soc Zoologists; Torrey Bot Club; Brit Ecol Soc. Res: Theory of natural communities; vegetation analysis and comparison of animal communities; production analysis of forests and nutrient circulation in ecosystems. Mailing Add: Ecol & Systs Langmuir Lab Cornell Univ Ithaca NY 14853

WHITTED, HAROLD HORATIO, b Washington, DC, Aug 25, 09; m 35; c 2. MEDICINE, PUBLIC HEALTH. Educ: Howard Univ, BS, 31, MD, 36; Harvard Univ, MPH, 41; Am Bd Prev Med, dipl, 50. Prof Exp: From instr to assoc prof, 48-72, PROF PUB HEALTH & PREV MED, COL MED, HOWARD UNIV, 72-, ACTG CHMN DEPT COMMUNITY HEALTH PRACT, 70- Concurrent Pos: Am Cancer Soc grant, 59-61. Mem: AAAS; fel Am Pub Health Asn; fel Royal Soc Health; dipl mem Pan-Am Med Asn; AMA. Res: Community health practice; preventive medicine. Mailing Add: Howard Univ Col of Med 520 W St NW Washington DC 20001

WHITTEMBURY, GUILLERMO, b Trujillo, Peru, Nov 17, 29; m 61; c 3. BIOPHYSICS. Educ: San Marcos Univ, Lima, BM, 55; Univ Cayetano Heredia, Peru, MD, 65. Prof Exp: Instr anat, San Marcos Univ, Lima, 49-50, asst prof med, 55-57, asst prof biophys, 60-62; sr scientist, 62-67, head dept gen physiol, 67-70, MEM STAFF, VENEZUELAN INST SCI RES, 70- Concurrent Pos: Res fel, Biophys Lab, Harvard Med Sch, 57-60; Rockefeller Found fel, 57-59; Helen Hay Whitney Found fel, 59-60; Daniel Carrion Price fel, Peru, 65; vis prof, Yale Univ, 70; mem, Int Union Pure & Appl Physics, 63; dir, Latin Am Ctr Biol, 73; fel, Churchill Col, Cambridge, 76- Mem: Am Soc Nephrology; Biophys Soc; Peruvian Nephrology Soc. Res: Transport processes across membranes; kidney physiology. Mailing Add: Venezuelan Inst Sci Res PO Box 1827 Caracas Venezuela

WHITTEMORE, ALICE S, b New York, NY, July 5, 36; m 58; c 2. BIOMATHEMATICS, BIOSTATISTICS. Educ: Marymount Manhattan Col, BS, 58; Hunter Col, MA, 64; City Univ New York, PhD(math), 67. Prof Exp: From asst prof to assoc prof math, Hunter Col, 67-74; adj assoc prof environ med, Med Ctr, NY Univ, 74-76; FAC MEM, DEPT STATIST, STANFORD UNIV, 76- Concurrent Pos: City Univ New York res grants, 69 & 70; Sloan Found res grant, Soc Indust & Appl Math Inst Math & Soc, 74-76; Rockefeller Found res grant, 76-77. Mem: AAAS; Soc Indust & Appl Math; Sigma Xi; Am Math Soc; Math Asn Am. Res: Environmental carcinogenesis. Mailing Add: Dept of Statist Stanford Univ Stanford CA 94305

WHITTEMORE, CHARLES ALAN, b Grand Junction, Colo, Dec 14, 35; m 63; c 2. ORGANIC CHEMISTRY. Educ: Stanford Univ, BSc, 57; Univ Colo, PhD(org chem), 63. Prof Exp: Sr chemist, Cent Res Labs, Minn Mining & Mfg Co, 63-69; asst prof, 69-74, ASSOC PROF CHEM, COLO WOMEN'S COL, 74- Mem: Am Chem Soc. Res: Elimination reactions; organic reaction mechanisms; Friedel-Crafts reactions; organic synthesis. Mailing Add: Dept of Chem Colo Women's Col Denver CO 80220

WHITTEMORE, DONALD OSGOOD, b Pittsburgh, Pa, May 4, 44; m 71. GEOCHEMISTRY. Educ: Univ NH, BS, 66; Pa State Univ, University Park, PhD(geochem), 73. Prof Exp: ASST PROF GEOL, KANS STATE UNIV, 72- Mem: AAAS; Geol Soc Am; Geochem Soc; Sigma Xi. Res: Low temperature, pressure aqueous geochemistry, especially water resource and pollution geochemistry; chemistry and mineralogy of ferric oxyhydroxides. Mailing Add: Dept of Geol Thompson Hall Kans State Univ Manhattan KS 66506

WHITTEMORE, FREDERICK WINSOR, b Boston, Mass, Apr 8, 16; m 41; c 3. ENTOMOLOGY. Educ: Mass State Col, BS, 37, MS, 38, PhD(entom), 41; Johns Hopkins Univ, MPH, 48. Prof Exp: Res entomologist, E L Bruce Co, 39-41; entomologist, US Army, 41-62; sr scientist, Pan Am Health Orgn, 62-64; pesticides specialist, Food & Agr Orgn, 64-69, chief, Crop Protection Br, 69-71, sr officer, Plant Protection Serv, 71-73; DEP DIR OPERS DIV, OFF PESTICIDE PROGS, US ENVIRON PROTECTION AGENCY, 73- Mem: Entom Soc Am; Am Mosquito Control Asn. Res: Promotion of international agreement on pesticide tolerances and specifications; establishment of laboratory and field test facilities for pesticides in developing countries. Mailing Add: Opers Div Off Pesticide Progs US Environ Protection Agency Washington DC 20460

WHITTEMORE, IRVILLE MERRILL, b Berkeley, Calif, June 12, 28; m 51; c 2. PETROLEUM CHEMISTRY. Educ: Univ Calif, Berkeley, BS, 52; Syracuse Univ, PhD(chem), 64. Prof Exp: Chemist, Arthur D Little, Inc, 52-55 & Lawrence Radiation Lab, 55-61; res assoc, Syracuse Univ, 61-63; res chemist, 63-69, SR RES CHEMIST, CHEVRON RES CO, 69- Mem: Am Chem Soc. Res: Radiochemistry; kinetics; gas chromatography; photochemistry. Mailing Add: Chevron Res Co 576 Standard Ave Richmond CA 94802

WHITTEMORE, RUTH, b Cambridge, Mass, June 11, 17. PEDIATRICS, CARDIOLOGY. Educ: Mt Holyoke Col, BA, 38; Johns Hopkins Univ, MD, 42; Am Bd Pediat, dipl, 53, cert pediat cardiol, 61. Prof Exp: Intern & resident pediat, New Haven Hosp, 42-44; resident, Johns Hopkins Hosp, 44-45, asst physician, Harriet Lane Cardiac Clin, 45-47; physician, Div Crippled Children, 47-59; dir, New Haven Rheumatic Fever & Cardiac prog, 47-60, sr pediatrician, New Haven Pediat Cardiac Res Prog, 59-66, PEDIAT CARDIOLOGIST & DIR NEW HAVEN PEDIAT CARDIAC RES PROG, STATE DEPT HEALTH, CONN, 66- Concurrent Pos: From asst clin prof to assoc clin prof, Sch Med, Yale Univ, 47-66, clin prof pediat, 66-; mem, Coun Rheumatic Fever & Congenital Heart Dis, Am Heart Asn, 55-, chmn, Comt Congenital Heart Dis, 56-60; chmn task force heart dis & youth, Conn Heart Asn, 75- Mem: Fel Am Acad Pediat; fel Am Col Cardiol. Res: Rheumatic fever; etiology and prevention of congenital heart defects; diagnostic services and care of the pediatric cardiac patient; pregnancy in the congenital cardiac, growth and development of offspring. Mailing Add: Dept of Pediat Yale Univ Sch of Med New Haven CT 06510

WHITTEMORE, WILLIAM LESLIE, b Skowhegan, Maine, Sept 25, 24; m 50. PHYSICS. Educ: Colby Col, AB, 45; Harvard Univ, MA, 47, PhD(physics), 49. Prof Exp: Assoc scientist, Brookhaven Nat Lab, 48-50, physicist, 50-56; physicist, Gen Atomic Div, Gen Dynamics Corp, 57-67, STAFF PHYSICIST, TRIGA REACTORS FACIL, GEN ATOMIC CO, 67- Concurrent Pos: Vis lectr, Harvard Univ, 50-51; sci consult, Korean Atomic Energy Res Inst, 60- Mem: Am Phys Soc; Am Nuclear Soc. Res: Utilization of reactors; neutron research; neutron radiography. Mailing Add: Triga Reactors Facil Gen Atomic Co PO Box 81608 San Diego CA 92138

WHITTEN, BERTWELL KNEELAND, b Boston, Mass, Apr 1, 41; m 62; c 3. ENVIRONMENTAL PHYSIOLOGY, COMPARATIVE PHYSIOLOGY. Educ: Middlebury Col, AB, 62; Purdue Univ, MS, 64, PhD(environ physiol), 66. Prof Exp: Res physiologist, US Army Med Res & Nutrit Lab, Fitzsimons Gen Hosp, 66-68, res physiologist, Res Inst Environ Med, Army Natick Labs, 68-72; assoc prof, 72-74, PROF BIOL SCI, MICH TECHNOL UNIV, 74- Mem: AAAS; Am Soc Zoologists; Am Physiol Soc. Res: Cardiovascular adaptations to hypoxia; protein and lipid metabolism in hibernators; effect of hypoxia on intermediary metabolism in animals and man. Mailing Add: Dept of Biol Sci Mich Technol Univ Houghton MI 49931

WHITTEN, CHARLES A, JR, b Harrisburg, Pa, Jan 20, 40; m 65. NUCLEAR PHYSICS, INTERMEDIATE ENERGY PHYSICS. Educ: Yale Univ, BS, 61; Princeton Univ, MA, 63, PhD(physics), 66. Prof Exp: Res physicist, A W Wright Nuclear Struct Lab, Yale Univ, 65-68; asst prof nuclear physics, 68-74, ASSOC PROF PHYSICS, UNIV CALIF, LOS ANGELES, 74- Mem: Am Phys Soc. Res: Direct reaction spectroscopy; isobaric analogue studies; intermediate energy nuclear physics. Mailing Add: Dept of Physics Univ of Calif Los Angeles CA 90024

WHITTEN, DAVID G, b Washington, DC, Jan 25, 38; m 60; c 2. PHYSICAL ORGANIC CHEMISTRY, BIOPHYSICAL CHEMISTRY. Educ: Johns Hopkins Univ, BA, 59, MA, 61, PhD(org chem), 63. Prof Exp: Sr scientist, Jet Propulsion Lab, Calif Inst Technol, 63-65, NIH fel chem, Inst, 65-66; from asst prof to assoc prof, 66-73, PROF CHEM, UNIV NC, CHAPEL HILL, 73- Concurrent Pos: Consult, Sci Data Systs, Inc, 66 & Tenn Eastman Co, 66-; Alfred P Sloan Found fel, 70-; Alexander von Humboldt fel, Max Planck Inst Biophys Chem, 72-73. Honors & Awards: Sr US Scientist Award, Alexander von Humboldt Found, 74-75. Mem: Am Chem Soc; The Chem Soc; Am Soc Photobiol. Res: Photobiology; photochemistry in organized monolayer assemblies; solid state and surface chemistry; chemistry of N-heterocyclic compounds; porphyrins and organometallic compounds. Mailing Add: Dept of Chem Univ of NC Chapel Hill NC 27514

WHITTEN, DAVID NELSON, physiology, see 12th edition

WHITTEN, ELMER HAMMOND, b Stoughton, Mass, Feb 18, 27; m 50; c 2. MEDICAL PHYSIOLOGY. Educ: Northeastern Univ, BS, 52; Mass State Col, Bridgewater, MEd, 67; Colo State Univ, PhD(physiol), 70. Prof Exp: Med serv rep drug sales, Pitman-Moore Co, Dow Chem Co, 54-56; admin asst sales, Metals & Controls, Inc, 56-58; head customer serv, Tex Instruments Inc, 58-66; instr human physiol, Colo State Univ, 70; from asst prof med physiol to assoc prof physiol & pharmacol, 70-72, ASSOC DEAN ACAD AFFAIRS, KANSAS CITY COL OSTEOP MED, 72-, CHMN DEPT PHYSIOL, 71- Mem: NY Acad Sci. Res: Neonatal enteritis; transport phenomena across the intestinal wall during stages in the progress of enteritis as it effects electrolytes and water. Mailing Add: Dept of Physiol Kans City Col Osteop Med Kansas City MO 64124

WHITTEN, ERIC HAROLD TIMOTHY, b Ilford, Eng, July 26, 27; m 53; c 1. GEOLOGY. Educ: Univ London, BSc, 48, PhD(geol), 52, DSc, 68. Prof Exp: Managerial chief clerk, Rex Thomas, Ltd, 43-45; lectr geol, Queen Mary Col, Univ London, 48-58; assoc prof, 58-62, PROF GEOL, NORTHWESTERN UNIV, EVANSTON, 62- Concurrent Pos: Vis assoc prof, Univ Calif, Berkeley, 57 & 60, Univ Calif, Santa Barbara, 59 & Univ Colo, 61 & 63. Mem: AAAS; Geol Soc Am; Geol Soc London; Brit Geol Asn; Int Asn Math Geol. Res: Structural geology and petrology of granitic and deformed rocks; application of statistical analysis to quantitative geology problems. Mailing Add: Dept of Geol Sci Northwestern Univ Evanston IL 60201

WHITTEN, HARRELL DAVID, b Columbus, Miss, July 27, 45; m 74; c 1. IMMUNOLOGY. Educ: Miss Col, BS, 67; Univ Miss, MS, 69, PhD(immunol), 74. Prof Exp: Asst zool, Univ Miss, 67, res fel immunol, 70-74; fel microbiol, Univ Ala, Birmingham, 74-75; ASST PROF BIOL, OLD DOMINION UNIV, 75-; ASST PROF PATH, EASTERN VA MED SCH, 75- Mem: AAAS; NY Acad Sci; Sigma Xi. Res: Tumor research, particularly the nature of the immune cells augmentable toward tumor rejection. Mailing Add: Sch of Sci Old Dominion Univ Box 6173 Norfolk VA 23508

WHITTEN, JERRY LYNN, b Bartow, Fla, Aug 13, 37; m 62; c 1. THEORETICAL CHEMISTRY. Educ: Ga Inst Technol, BS, 60, PhD(chem), 64. Prof Exp: Res assoc chem, Princeton Univ, 63-65, instr, 65; asst prof, Mich State Univ, 65-67; from asst prof to assoc prof, 67-73, PROF CHEM, STATE UNIV NY STONY BROOK, 73- Concurrent Pos: Res grants, Petrol Res Fund, 66-67 & 74-76 & NSF, 67-72; Alfred P Sloan fel, 69-71. Mem: Am Phys Soc; Am Chem Soc. Res: Theoretical studies of molecular structure and bonding; ab initio construction of electronic wave functions; studies of excited electronic states and magnetic interactions. Mailing Add: Dept of Chem State Univ of NY Stony Brook NY 11790

WHITTEN, JOAN MARGARET, developmental biology, deceased

WHITTEN, KENNETH WAYNE, b Collinsville, Ala, Feb 4, 32. INORGANIC CHEMISTRY. Educ: Berry Col, AB, 53; Univ Miss, MS, 58; Univ Ill, PhD(inorg chem), 65. Prof Exp: Instr chem, Univ Miss, 55-56; asst prof, Berry Col, 56-58; instr, Univ Southwestern La, 58-59; asst prof, Miss State Col Women, 59-60 & Univ Ala, 63-66; asst prof chem & coord gen chem, 67-70, ASSOC PROF CHEM, UNIV GA, 70- Mem: Am Chem Soc. Res: Synthesis in fused salt media; chemical education; theories of testing. Mailing Add: Dept of Chem Univ of Ga Athens GA 30601

WHITTEN, MAURICE MASON, b Providence, RI, Oct 1, 23. ANALYTICAL CHEMISTRY, HISTORY OF SCIENCE. Educ: Colby Col, AB, 45; Columbia Univ, MA, 49; Ohio State Univ, PhD, 71. Prof Exp: Sci teacher, Wilton Acad, 45-48 & Maine High Sch, 48-55; instr phys sci, Gorham State Teachers Col, 55-59; TV sci teacher, State Dept Educ, Maine, 59-60; asst prof, 61-63, assoc prof phys sci & chem, 64-71, PROF PHYS SCI & CHEM, UNIV MAINE, PORTLAND-GORHAM, 71- Concurrent Pos: Lectr, Cent Maine Gen Hosp, Lewiston, 52-53. Honors & Awards: Elizabeth Thompson Award, Am Acad Arts & Sci, 54. Mem: AAAS; Am Chem Soc; Nat Sci Teachers Asn; fel Am Inst Chemists; Hist Sci Soc. Res: Science education at the college level, especially critical thinking and scientific literacy; air pollution; history of gun powder mills of Maine. Mailing Add: Dept of Phys Sci & Eng Univ of Maine at Portland-Gorham Gorham ME 04038

WHITTEN, NORMAN EARL, JR, b Orange, NJ, May 23, 37; m 62. ANTHROPOLOGY. Educ: Colgate Univ, BA, 59; Univ NC, MA, 61, PhD(anthrop), 64. Prof Exp: Res assoc anthrop, Tulane Univ, 64-65; from asst prof to assoc prof, Wash Univ, 65-70; assoc prof, 70-73, PROF ANTHROP, UNIV ILL, URBANA, 73-, DIR RES, CTR LATIN AM & CARIBBEAN RES, 74- Concurrent Pos: Nat Inst Ment Health res fel, Tulane Univ & Univ Valle, Colombia, 64-65; lectr, Peace Corps, 66; vis assoc prof anthrop, Univ Calif, Los Angeles, 69-70; NSF res grants, Eastern Ecuador, 70-72 & 72-75. Mem: AAAS; Am Anthrop Asn; Am Sociol Asn; Inter-Am Indian Inst; Royal Anthrop Inst. Res: Social organization; ritual; cultural change; Afro-American adaptation; tropical forest Indian strategies in South America; ethnic stratification. Mailing Add: Dept of Anthrop Univ of Ill Urbana IL 61801

WHITTEN, ROBERT CRAIG, JR, b Bristol, Va, Dec 6, 26; m 53; c 2. AERONOMY. Educ: US Merchant Marine Acad, BS, 47; Univ Buffalo, BA, 55; Duke Univ, MA, 58, PhD(theoret physics), 59; San Jose State Univ, MS, 71. Prof Exp: Asst, Duke Univ, 55-57, instr, 57-58, asst, 58-59; from physicist to sr physicist, Stanford Res Inst, 59-67; RES SCIENTIST, NASA-AMES RES CTR, 67- Concurrent Pos: Lectr, Stanford Univ, 61-62 & 64-66, Univ Santa Clara, 64 & 69 & San Jose State Univ, 72. Mem: Am Geophys Union. Res: Structure, chemistry and dynamics of planetary atmospheres and ionospheres; chemistry and meteorology of the stratosphere; the quantum mechanical three body problem; atomic theory. Mailing Add: Mail Stop 245-3 Space Sci Div NASA-Ames Res Ctr Moffett Field CA 94035

WHITTENBERGER, JAMES LAVERRE, b Dahinda, Ill, Feb 12, 14; m 43; c 3. PHYSIOLOGY. Educ: Univ Chicago, SB, 37, MD, 38. Hon Degrees: AM, Harvard Univ, 51. Prof Exp: Intern, Cincinnati Gen Hosp, 38-39; Smith fel surg, Univ Chicago, 39-40; asst resident, Thorndike Mem Lab, Boston City Hosp, 40-42, house physician, 4th Med Serv, 42-43; assoc physiol, 46-47, from asst prof to assoc prof, 47-50, PROF PHYSIOL, SCH PUB HEALTH, HARVARD UNIV, 51-, HEAD DEPT, 48-, JAMES STEVENS SIMMONS PROF PUB HEALTH, 58-, ASSOC DEAN, 66- Concurrent Pos: Res fel med, Harvard Med Sch, 40-42; Commonwealth Fund fel med & physiol, Sch Med, NY Univ, 43; asst, Peter Bent Brigham Hosp, 46-; consult, Children's Hosp, 48- Mem: AAAS; Am Physiol Soc; Am Soc Clin Invest; Am Indust Hyg Asn. Res: Respiratory physiology; occupational medicine; environmental health. Mailing Add: 52 Gun Club Lane Weston MA 02193

WHITTEN-WOLFE, BARBARA L, b Minneapolis, Minn, Sept 26, 46. MATHEMATICAL PHYSICS. Educ: Carleton Col, BA, 68; Univ Rochester, MA, 71, PhD(physics), 76. Prof Exp: INSTR INTERDISCIPLINARY STUDIES, WESTERN COL, MIAMI UNIV, 74- Mem: Am Asn Physics Teachers; Am Math Asn. Res: Algebraic statistical mechanics to the measurement problem in quantum mechanics. Mailing Add: Western Col Miami Univ Oxford OH 45056

WHITTIER, ANGUS CHARLES, b Ottawa, Ont, Oct 17, 21; m 48; c 4. PHYSICS. Educ: Queen's Univ, Ont, BSc, 48; McGill Univ, MSc & PhD(physics), 52. Prof Exp: Asst res officer, Atomic Energy Can, Ltd, 52-55; supv physicist, Atomic Power Dept, Can Gen Elec, 55-67, mgr reactor anal, 67-72; SUPT SHIELDING & COMPUT BR POWER PROJS, ATOMIC ENERGY CAN LTD, SHERIDAN PARK, 72- Mem: Am Nuclear Soc. Res: Nuclear physics, particularly reactor physics. Mailing Add: 2493 Vineland Rd Mississauga ON Can

WHITTIER, DEAN PAGE, b Worcester, Mass, July 2, 35; m 58; c 2. PLANT MORPHOLOGY. Educ: Univ Mass, BS, 57; Harvard Univ, AM, 59, PhD(biol), 61. Prof Exp: Asst prof bot, Va Polytech Inst, 61-64; NIH fel biol, Harvard Univ, 64-65; asst prof, 65-68, ASSOC PROF BIOL, VANDERBILT UNIV, 68-, CHMN DEPT GEN BIOL, 75- Mem: Am Inst Biol Sci; Bot Soc Am; Am Fern Soc (treas, 74-75); Soc Develop Biol; Int Soc Plant Morphologists. Res: Morphogenesis; apomixis in lower vascular plants. Mailing Add: Dept of Gen Biol Vanderbilt Univ Nashville TN 37235

WHITTIER, HENRY O, b Schenectady, NY, Sept 1, 37; m 59; c 1. BOTANY, BRYOLOGY. Educ: Miami Univ, BS, 59, MA, 61; Columbia Univ, PhD(biol), 68. Prof Exp: Res asst bot, Miami Univ Schooner Col Rebel Exped to SPac, 60; instr, Univ Hawaii, 62-64; res asst bryol, NY Bot Garden, 64-68; asst prof, 68-72, ASSOC PROF BIOL SCI, FLA TECHNOL UNIV, 72- Mem: Am Bryol & Lichenological Soc; Am Inst Biol Sci; Sigma Xi; Am Bot Soc; Int Soc Plant Taxonomists. Res: Tropical botany, taxonomy, ecology and geography, especially Pacific islands Bryophyta. Mailing Add: Dept of Biol Sci Fla Technol Univ Orlando FL 32816

WHITTIER, HERBERT LINCOLN, b Corry, Pa, Mar 26, 41; m 65; c 1. ANTHROPOLOGY. Educ: Univ SFla, BA, 63; Fla State Univ, MS, 65; Mich State Univ, MA, 69, PhD(anthrop), 73. Prof Exp: Researcher anthrop, 70-75, INSTR ANTHROP, MICH STATE UNIV, 75- Mem: Fel Borneo Res Coun, 68- Res: Study of social organization of hill tribes in Borneo; culture change and modernization in Indonesia and Malaysia; the effects of modernization on hill tribes, rural-urban migrations; integration in pluralistic societies. Mailing Add: Dept of Anthrop Mich State Univ East Lansing MI 48824

WHITTIER, JOHN RENSSELAER, b Washington, DC, Aug 7, 19; m 50; c 2. NEUROLOGY, PSYCHIATRY. Educ: Harvard Univ, BA, 39; Columbia Univ, MD, 43; Am Bd Psychiat & Neurol, dipl. Prof Exp: Intern, Gorgas Hosp, CZ, 43-44; asst neurol, Col Physicians & Surgeons, Columbia Univ, 46-48, asst resident, Neurol Inst, 48-49; resident psychiatrist, Vet Admin Hosp, Bronx, 49-51; DIR PSYCHIAT RES, CREEDMOOR INST PSYCHOBIOL STUDIES, CREEDMOOR PSYCHIAT CTR, 54- Concurrent Pos: Asst clin prof neurol, Col Physicians & Surgeons, Columbia Univ, 54-55, asst clin prof psychiat, 55-; guest lectr, NJ State Hosps; mem & dir, NY State Res Found Ment Hyg. Mem: AAAS; Am Psychiat Asn; Am Asn Anat; Am Acad Neurol. Res: Aging; degenerative diseases; neural systems; heredity; research administration. Mailing Add: Creedmoor Inst Station 60 Queens Village NY 11427

WHITTIG, LYNN D, b Meridian, Idaho, Jan 16, 22; m 45; c 3. SOIL CHEMISTRY, SOIL MINERALOGY. Educ: Univ Wis, BS, 49, MS, 50, PhD(soil sci), 54. Prof Exp: Soil scientist, Soil Conserv Serv, USDA, 54-56; from asst prof to assoc prof, 57-70, res assoc, 63-64, PROF SOIL SCI, UNIV CALIF, DAVIS, 70-, V CHMN DEPT LAND, AIR & WATER RESOURCES, 75- Mem: Fel Am Soc Agron; Soil Sci Soc Am; Clay Minerals Soc. Res: Clay mineralogy and mineral weathering processes; chemistry, morphology and genesis of salt-affected soils; x-ray diffraction and fluorescence methods of analysis. Mailing Add: Dept Land, Air & Water Resources Univ of Calif Davis CA 95616

WHITTINGHAM, ALVA DAY, systematic botany, plant cytology, see 12th edition

WHITTINGHAM, DAVID JAMES, b Halifax, NS, Oct 25, 23; m 47; c 1. ORGANIC CHEMISTRY. Educ: Univ NB, BS, 44, MSc, 47; Univ Col, Univ London, PhD(org chem), 50. Prof Exp: Nat Res Coun Can fels, 50-52; lectr org chem, Univ NB, 52-53; Nat Res Coun Can fels, 53-55; res chemist, Monsanto Can Ltd, 55-59; asst ed org indexing, 59-61, assoc ed, 62, asst dept head, 63-64, actg dept head, 64, asst managing ed subj indexes, 65-69, mgr indexing, 69-70, mgr gen subj indexing, 70-72, asst managing ed publ div, 72-75, SR ED PUBL DIV, CHEM ABSTR SERV, 75- Mem: AAAS; Am Chem Soc; The Chem Soc. Res: Organic reaction mechanisms; synthetic organics; infrared spectroscopy of steroids; natural product structure elucidation; indexing of chemical literature; administration. Mailing Add: Chem Abstr Serv Ohio State Univ Columbus OH 43210

WHITTINGHAM, MICHAEL STANLEY, b Nottingham, Eng, Dec 22, 41; m 69; c 2. SOLID STATE CHEMISTRY. Educ: Oxford Univ, BA, 64, MA, 67, DPhil(chem), 68. Prof Exp: Res assoc mat sci, Stanford Univ, 68-72; mem sci staff, 72-75, HEAD CHEM PHYSICS GROUP, CORP RES LABS, EXXON RES & ENG CO, 75- Concurrent Pos: Demonstr, Dept Inorg Chem, Oxford Univ, 65-67. Honors & Awards: Young Author Award, Electrochem Soc, 71. Mem: Am Chem Soc; Electrochem Soc; The Chem Soc. Res: Chemical properties of highly non-stoichiometric materials; fast ion transport in solids; electrochemical control of the properties of materials; solid state electrochemistry; high energy-density batteries. Mailing Add: 32 Arlene Ct Fanwood NJ 07023

WHITTINGHAM, WILLIAM FRANCIS, b Beaver Dam, Wis, Feb 23, 26; m 49; c 2. MYCOLOGY. Educ: Univ Wis, BS, 50, MS, 52, PhD(bot), 54. Prof Exp: Res assoc bact, 54-56, from instr to assoc prof, 56-69, PROF BOT, UNIV WIS-MADISON, 69- Mem: Bot Soc Am; Brit Soc Gen Microbiol; Brit Mycol Soc. Res: Physiology of fungi; ecology of soil fungi; fungal parasite-host relationships. Mailing Add: Dept of Bot Univ of Wis Madison WI 53706

WHITTINGHILL, MAURICE, b St Joseph, Mo, May 15, 09; m 32, 55; c 2. ZOOLOGY. Educ: Dartmouth Col, AB, 31; Univ Mich, PhD(zool), 37. Prof Exp: Instr, Dartmouth Col, 31-33; asst, Univ Mich, 35; Nat Res Coun fel biol sci, Calif Inst Technol, 36-37; fel biol, Bennington Col, 37-42; assoc prof zool, 42-52, prof, 52-64, vis prof, 74, EMER PROF ZOOL, UNIV NC, CHAPEL HILL, 74- Prof Exp: Prof, Univ Mich, 46; sr biologist, Oak Ridge Nat Lab, 49; Wachtmeister vis prof biol, Va Mil Inst, 76. Mem: Genetics Soc Am; Am Soc Zool; Am Soc Nat; Am Soc Human Genetics; Am Genetic Asn (vpres, 72, pres, 73). Res: Genetics of Drosophila; irradiation and temperature effects; mutation and crossing over; spondylitis. Mailing Add: 1905 S Lake Shore Dr Chapel Hill NC 27514

WHITTINGTON, MELVIN OTHAL, JR, b Shawnee, Okla, Aug 19, 44; m 72; c 1. MICROBIOLOGY. Educ: Okla Baptist Univ, BS, 67; NMex State Univ, MS, 69; Univ Minn, PhD(microbiol), 76. Prof Exp: ASST PROF BIOL, OKLA BAPTIST UNIV, 75- Concurrent Pos: Consult & dir bact anal, James Sausage Co, 70- Mem: Am Soc Microbiol; Sigma Xi. Res: Study of DNA-protein complexes in determining conformation and biological activity of Bacillus subtilis bacteriophage phi 29. Mailing Add: Dept of Biol Okla Baptist Univ Shawnee OK 74801

WHITTINGTON, STUART GORDON, b Chesterfield, Eng, Apr 16, 42; Can & UK citizen; m 64; c 1. PHYSICAL CHEMISTRY. Educ: Cambridge Univ, BA, 63, PhD(chem), 72. Prof Exp: Scientist chem, Unilever Res Lab, UK, 63-66; res fel, Univ Calif, San Diego, 66-67; res fel, Univ Toronto, 67-68; scientist, Unilever Res Lab, UK, 68-70; asst prof, 70-75, ASSOC PROF CHEM, UNIV TORONTO, 75- Res: Statistical mechanics; Monte Carlo methods; excluded volume effect in polymers; conformational statistics of polysaccharides. Mailing Add: Dept of Chem Univ of Toronto Toronto ON Can

WHITTLE, BETTY ANN, b Dothan, Ala, Dec 14, 43. NUTRITION. Educ: Ala Col, BS, 65; Univ Tenn, Knoxville, PhD(nutrit), 70. Prof Exp: Asst nutrit, Univ Tenn, Knoxville, 69-70; asst prof, Auburn Univ, 70-74; ASST PROF NUTRIT DEPT HUMAN NUTRIT & FOOD MGT & CHIEF NUTRIT NISONGER CTR MENTAL RETARDATION & DEVELOP DISABILITIES, OHIO STATE UNIV, 74- Mem: Sigma Xi. Res: Utilization of inorganic sulfate and its relationship to the metabolism of the sulfur-containing amino acids; nutritional needs and ways of meeting them of persons with developmental disabilities. Mailing Add: 1580 Cannon Dr Nisonger Ctr Ohio State Univ Columbus OH 43210

WHITTLE, CHARLES EDWARD, JR, b Brownsville, Ky, Mar 8, 31; m 52; c 10. PHYSICS, APPLIED MATHEMATICS. Educ: Centre Col, AB, 49; Washington Univ, PhD(nuclear physics), 53. Prof Exp: Fulbright & Res Corp grants, State Univ Leiden, 53-54; res scientist, Union Carbide Corp, 54-56; asst & assoc prof physics, Western Ky Univ, 56-60, prof & chmn dept, 60-62; coordr res, 62-64, from assoc dean to dean, 64-72, PROF PHYSICS, CENTRE COL KY, 62-, MATTON CHAIR APPL MATH, 72- Mem: Am Asn Physics Teachers; Am Phys Soc. Res: Nuclear and optical spectroscopy; applied mathematics in biological and social sciences. Mailing Add: Dept of Physics-Appl Math Centre Col of Ky Danville KY 40422

WHITTLE, GEORGE PATTERSON, b Eufaula, Ala, July 1, 25; m 63; c 1. ANALYTICAL CHEMISTRY. Educ: Ga Inst Technol, BChE, 46, BIE, 47; Univ Fla, MS, 64, PhD(chem), 66. Prof Exp: Chem engr, Hercules Powder Co, 47-49; self-employed, Whittle Lumber Co, 50-53; chemist, Swift & Co, 53-55; chief chemist, Allied Chem Co, 55-57; res engr, Tenn Corp, 57-63; PROF ENVIRON ENG, UNIV ALA, 67- Mem: Am Chem Soc; Am Water Works Asn; Am Soc Civil Eng. Res: Bioenvironmental engineering; water treatment and chemistry, pollution control; colloids; analytical chemistry of water and wastewater; reaction kinetics of halogen residuals in water. Mailing Add: Dept of Civil & Mineral Eng Univ of Ala PO Box 1468 University AL 35486

WHITTLE, JOHN ANTONY, b Settle, Yorks, Eng, Mar 13, 42; m 68. ORGANIC CHEMISTRY, BIOCHEMISTRY. Educ: Univ Glasgow, BSc, 64; Imp Col, dipl, & Univ London, PhD(org chem), 67. Prof Exp: Fel, Rutgers Univ, NB, 67-69; ASST PROF CHEM, LAMAR UNIV, 69- Mem: Am Chem Soc; The Chem Soc. Res: Biosynthesis of sesquiterpenoids and other natural products; synthesis of sesquiterpenoid ring systems. Mailing Add: Dept of Chem Lamar Univ PO Box 10022 Lamar Univ Sta Beaumont TX 77710

WHITTLE, PHILIP RODGER, b Russell Springs, Ky, July 11, 43; m 67; c 2. ORGANIC CHEMISTRY, FORENSIC CHEMISTRY. Educ: Univ Ky, BS, 65; Iowa State Univ, PhD(org chem), 69. Prof Exp: NIH fel, Univ Colo, Boulder, 69-70; ASSOC PROF CHEM, MO SOUTHERN STATE COL, 70-, DIR, REGIONAL CRIMINALISTICS LAB, 72- Mem: Am Chem Soc (secy-treas, 66-); Am Soc Crime Lab Dirs. Res: Electrocyclic cyclopropane ring openings; toxicology; modern drug analysis; trace evidence; forensic applications. Mailing Add: Dept of Chem Mo Southern State Col Joplin MO 64801

WHITTLESEY, EMMET FINLAY, b Winchester, Mass, Oct 9, 23; m 66; c 3. MATHEMATICS. Educ: Princeton Univ, AB, 48, MA, 55, PhD(math), 56. Prof Exp: Instr math, Pa State Univ, 50-51 & Bates Col, 51-54; from instr to assoc prof, 54-65, PROF MATH, TRINITY COL, CONN, 65- Concurrent Pos: NSF fel, 62-63. Mem: Am Math Soc; Math Asn Am. Res: Combinatorial topology; functional analysis. Mailing Add: Dept of Math Trinity Col Hartford CT 06106

WHITTLESEY, JOHN RB, b Los Angeles, Calif, July 21, 27; m 66. EXPLORATION GEOPHYSICS, MATHEMATICAL STATISTICS. Educ: Calif Inst Technol, BS, 48, MS, 50. Prof Exp: Instr physics & astron, Univ Nev, 50; Ford Found behav sci grant & res asst, Univ NC, 52-54; res mathematician res clin neuropsychiat inst, Med Ctr, Univ Calif, Los Angeles, 57-59, data processing analyst, 59-62, analyst brain res inst, 62-64; mem res staff seismic explor data processing, Ampex Corp, Calif, 64-72; mem res & tech staffs seismic data processing res, Ray Geophys Div, Mandrel Industs, 64-72, MEM RES STAFF SEISMIC DATA PROCESSING RES, PETTY RAY GEOPHYS, INC, SUBSID GESOURCE, INC, 72- Concurrent Pos: Statist consult numerous behav scientists, Calif, 58-71; NIMH spec res fel brain res inst, Univ Calif, Los Angeles, 63-64. Honors & Awards: Award, Soc Explor Geophys, 65. Mem: Fel AAAS; Soc Explor Geophys. Res: Mathematics and digital computers applied to psychiatry, brain research, bombing-range instrumentation and exploration geophysics; data processing; time-series analysis; seismic modeling; laser fusion. Mailing Add: 5439 Del Monte Dr Houston TX 77027

WHITTON, LESLIE, b New Bedford, Mass, Sept 1, 23; m 47; c 4. PLANT CYTOLOGY, PLANT GENETICS. Educ: Utah State Univ, BS, 49; Univ Calif, MS, 53; Cornell Univ, PhD, 64. Prof Exp: Asst prof hort, Univ Maine, Orono, 56-62; asst prof, 62-64, asst prof bot, 64-68, ASSOC PROF BOT, BRIGHAM YOUNG UNIV, 68- Mem: Bot Soc Am. Res: Cytology, genetics and breeding of small fruit species and native shrub species of the Rocky Mountain region. Mailing Add: Dept of Bot & Range Sci Brigham Young Univ Provo UT 84601

WHITTOW, GEORGE CAUSEY, b Milford Haven, UK, Feb 28, 30; m 55; c 1. PHYSIOLOGY. Educ: Univ London, BSc, 52; Univ Malaya, PhD(physiol), 57. Prof Exp: Asst lectr physiol, Univ Malaya, 52-54, lectr, 54-59; sr sci officer, Hannah Res Inst, Ayr, Scotland, 59-65; assoc prof physiol, Rutgers Univ, New Brunswick, 65-68; PROF PHYSIOL, SCH MED, UNIV HAWAII, 68- Concurrent Pos: USPHS fel avian physiol, Rutgers Univ, 62. Mem: Am Physiol Soc; Am Soc Zool; Ecol Soc Am; Brit Inst Biol. Res: Physiology of thermoregulation; physiological ecology. Mailing Add: Sch of Med Univ of Hawaii Honolulu HI 96822

WHITTY, ELMO BENJAMIN, b Lee, Fla, Mar 6, 37. AGRONOMY. Educ: Univ Fla, BSA, 59, MSA, 61; NC State Univ, PhD(soil sci), 65. Prof Exp: Asst prof, 66-71, ASSOC PROF AGRON, UNIV FLA, 71- Mem: Am Soc Agron; Soil Sci Soc Am. Res: Crop production; cultural practices; variety testing; plant growth regulators; nutrition of plants. Mailing Add: Dept of Agron Univ of Fla Gainesville FL 32611

WHITWORTH, CLYDE W, b Paulding Co, Ga, Oct 9, 26; m 51; c 3. PHARMACY. Educ: Univ Ga, BS, 50, MS, 56; Univ Fla, PhD(pharm), 63. Prof Exp: From instr to asst prof pharm, Univ Ga, 54-60; asst prof, Northeast La State Col, 63-66; ASSOC PROF PHARM, UNIV GA, 66- Concurrent Pos: Mead Johnson res award, 65-66; William A Webster Co grant prod stability, 72- Mem: Am Pharmaceut Asn. Res: Factors influencing drug absorption and drug release from external preparations. Mailing Add: Univ of Ga Sch of Pharm Athens GA 30601

WHITWORTH, WALTER RICHARD, b La Crosse, Wis, Feb 22, 34; m 57; c 4. AQUATIC BIOLOGY. Educ: Wis State Col, Stevens Point, BS, 58; Okla State Univ, MS, 61, PhD(zool), 63. Prof Exp: Fish biologist, Southeastern Fish Control Lab, 63-64; asst prof fisherfisheries, 64-69, ASSOC PROF FISHERIES, UNIV CONN, 69- Mem: Am Fisheries Soc; Am Soc Ichthyol & Herpet; Ecol Soc Am; Am Soc Limnol & Oceanog. Res: Effects of the environment on fish; primary productivity; fish taxonomy and toxicology. Mailing Add: U-87 Col of Agr & Natural Res Univ of Conn Storrs CT 06268

WHORTON, RAYBURN HARLEN, b London, Ark, Apr 28, 31; m 56; c 3. PAPER CHEMISTRY. Educ: Ark Polytech Col, BS, 53; Univ Ark, MS, 56. Prof Exp: Instr chem, Ark Polytech Col, 55-56; res chemist, Crossett Co, 57-62; proj supvr, Ga-Pac Corp, 62-63; sr proj chemist, 64-69, SECT LEADER PAPER DEVELOP, ERLING RIIS RES, INT PAPER CO, 69- Mem: AAAS; Tech Asn Pulp & Paper Indust. Res: Papermaking; surface and internal sizing; printability coatings; paper-plastic combinations; converting processes. Mailing Add: Res Dept Int Paper Co Mobile AL 36601

WHYBROW, PETER CHARLES, b Hertfordshire, Eng, June 13, 39; nat US; m 63; c 2. PSYCHIATRY. Educ: Univ London, MB & BS, 62; Royal Col Physicians, dipl, 62; Conjoint Bd Physicians & Surgeons Eng, dipl psychol med, 68. Hon Degrees: MA, Dartmouth Col, 74. Prof Exp: House physician, Med Res Coun-Univ Col Hosp, London, 62; house surgeon, St Helier Hosp, Surrey, Eng, 63; sr house physician, Univ Col Hosp, London, 63-64; house physician, Prince of Wales Hosp, London, 64; resident psychiat, Univ NC, 65-67, instr, 67-68; sci officer, Med Res Coun, Eng, 68-69; from asst prof to assoc prof, 69-71, PROF PSYCHIAT & CHMN DEPT, DARTMOUTH MED SCH, 71- Concurrent Pos: NIMH res fel, Univ NC, 67-68; lectr, Univ Col Hosp Med Sch, London, 68-69; dir res training, Dartmouth Hitchcock Affil Hosps, 69-71, dir psychiat, 70-; consult, Vet Admin Hosp, 70- Mem: Brit Med Asn; Royal Col Psychiat; Brit Soc Psychosom Res; fel Am Psychiat Asn. Res: Psychobiology of affective disorders, particularly pharmacologic and endocrinologic aspects. Mailing Add: Dept of Psychiat Dartmouth Med Sch Hanover NH 03755

WHYBURN, KENNETH GORDON, mathematical analysis, see 12th edition

WHYBURN, LUCILLE ENID, b Lewisville, Tex, July 31, 05; m 25; c 1. MATHEMATICS. Educ: Univ Tex, BA, 27, MA, 36. Prof Exp: Res assoc math, Johns Hopkins Univ, 31-34; lectr, Univ Va, 44-45, asst prof, 46-47; assoc prof, Sweet Briar Col, 60-62; asst prof, 62-67, ASSOC PROF MATH, UNIV VA, 67- Concurrent Pos: Lectr, Univ Tex, 75-76. Mem: Am Math Soc; Math Asn Am. Res: Rotation groups about a set of fixed points; biographical research into the R L Moore Collection at the University of Texas. Mailing Add: 133 Bollingwood Rd Charlottesville VA 22903

WHYMARK, ROY R, b Brighton, Eng, Apr 16, 23; m 49; c 1. ACOUSTICS. Educ: London Univ, BSc, 43. Prof Exp: Res scientist, Mullard Res Labs, Eng, 48-53, group leader ultrasonics, 53-56; mgr acoustics, Vickers Group Res Estab, 56-58; group leader acoustics res, IIT Res Inst, 58-60, mgr, 60-62, asst dir physics res, 62-63, res dir fluid dynamics, 63-65; dir appl physics, Tracor Inc, 65-69; pres, Interand Corp, 69-73; PRES, INTERSONICS INC, 73- Concurrent Pos: Res analyst, US Air Force, 66-67; int ed, Ultrasonics J, 70- Mem: Acoust Soc Am. Res: Ultrasonics; transducers; magnetostrictor theory and analysis; materials evaluation and testing; sonic processing; NASA space processing of materials. Mailing Add: 194 E Grand Ave Fox Lake IL 60020

WHYTE, DONALD EDWARD, b Regina, Sask, Jan 22;m 18; nat US; m 43; c 4. ORGANIC CHEMISTRY. Educ: Univ Sask, BA, 39, MA, 41; Columbia Univ, PhD(chem), 43. Prof Exp: Chemist naval stores, Hercules Powder Co, 43-46; head org res sect, 46-52, res serv mgr, 52-57, tech serv mgr, 57-59, dir serv prod develop, 59-61, appl res dir, 61-64, res mgr, 65-71, prod res mgr int opers, 71-74, DIR RES & DEVELOP INT OPERS, S C JOHNSON & SON, INC, 74- Mem: Am Chem Soc; Am Oil Chem Soc. Am Soc Qual Control. Res: Analytical and market research; new product evaluation; new product development of polishes, coating and porelon; polymers; insecticides; insect repellants; synergists and microbiology. Mailing Add: S C Johnson & Son Inc Racine WI 53403

WHYTE, JAMES HOWDEN, plant physiology, deceased

WHYTE, MICHAEL PETER, b New York, NY, Dec 19, 46; m 74. INTERNAL MEDICINE. Educ: Washington Square Col, NY Univ, BA, 68; State Univ NY, MD, 72. Prof Exp: From intern to resident internal med, Dept Med, Bellevue Hosp, New York, 72-74; clin assoc metab neurol, Nat Inst Neurol & Commun Dis & Stroke, NIH, 74-76; FEL ENDOCRINOL, SCH MED, WASHINGTON UNIV, 76- Res: Calcium metabolism and metabolic bone disease; metabolic neurology; anticonvulsant drug metabolism. Mailing Add: Dept of Med Jewish Hosp 216 Kings Hwy St Louis MO 63178

WHYTE, THADDEUS EVERETT, JR, b Washington, DC, Dec 8, 37; m 59; c 2. PHYSICAL CHEMISTRY. Educ: Georgetown Univ, BS, 60; Howard Univ, MS, 62, PhD(phys chem), 65. Prof Exp: Phys chemist, Nat Bur Standards, 62-63; res chemist, Howard Univ, 63-64; lectr chem, Sacramento State Col, 65-67; sr res chemist, Mobil Res & Develop Corp, 67-73, group leader, 69-72, proj leader, Reforming & Spec Process Group, 73-75 & Res Planning & Econ, Res Planning Group, 75-76; SECT MGR AMINES PROD, CHEM GROUP, AIR PROD & CHEM INC, 76- Concurrent Pos: Lectr, Dept Sci & Math, Gloucester County Col, NJ, 69-70. Mem: Am Chem Soc; Am Inst Chemists. Res: Heterogeneous catalysis; structure determination of heterogeneous catalysts; small angle x-ray scattering; petroleum catalysis. Mailing Add: Chem Group Air Prod & Chem Inc Allentown PA 18105

WIANT, HARRY VERNON, JR, b Burnsville, WVa, Nov 4, 32; m 54; c 2. FORESTRY. Educ: Univ WVa, BSF, 54; Univ Ga, MF, 59; Yale Univ, PhD(forestry), 63. Prof Exp: Jr forester, US Forest Serv, 57; asst prof forestry, Humboldt State Col, 61-65; prof & asst to dean, Stephen F Austin State Col, 65-72; PROF FORESTRY, WVA UNIV, 72- Mem: Soc Am Foresters. Res: Concentration of carbon dioxide near forest floor; ecology and silviculture of redwood; chemical and mechanical control of undesirable hardwoods; prediction of site quality; volume determinations; silviculture of southern forest trees; dendrological techniques. Mailing Add: Div of Forestry WVa Univ Morgantown WV 26506

WIBERG, GEORGE STUART, b Winnipeg, Man, Apr 1, 20; m 55; c 2. TOXICOLOGY. Educ: Univ Man, BSc, 50, MSc, 51; Univ Alta, PhD(biochem), 55. Prof Exp: Chemist physiol & hormone sect food & drug labs, 55-62, res scientist path & toxicol sect, Food & Drug Dir, 62-70, sci adv & actg head hazardous prod sect, Div Toxicol, Food Adv Bur, Health Protection Br, 70-73, HEAD TOXICOL EVAL ENVIRON CONTAMINANTS DIV, ENVIRON HEALTH DIRECTORATE, HEALTH PROTECTION BR, CAN DEPT NAT HEALTH & WELFARE, 73- Concurrent Pos: Consult comt experts transport dangerous goods, UN, 74. Mem: Can Biochem Soc; Soc Toxicol; Can Soc Res Toxicol. Res: Toxicology of household products including dermal, cutaneous and ocular toxicity and inhalation toxicology of household aerosols and solvents. Mailing Add: Health Protection Br Can Dept of Nat Health & Welfare Ottawa ON Can

WIBERG, JOHN SAMUEL, b Plaistow, NH, Dec 4, 30; m 52; c 3. BIOCHEMISTRY, GENETICS. Educ: Trinity Col, Conn, BS, 52; Univ Rochester, PhD(pharmacol), 56. Prof Exp: Clin lab officer, Wright-Patterson Air Force Base, Ohio, 57-58; res assoc biochem, Mass Inst Technol, 59-63; asst prof, 63-70, ASSOC PROF RADIATION BIOL & BIOPHYS, SCH MED & DENT, UNIV ROCHESTER, 70- Concurrent Pos: NIH res fel, Mass Inst Technol, 59-60. Mem: AAAS; Am Soc Biol Chem; Genetics Soc Am; Am Soc Microbiol; Am Chem Soc. Res: Metal ion interactions with nucleic acids and enzymes; biochemical genetics of bacterial viruses; nucleic acid function and metabolism; regulation of enzyme synthesis. Mailing Add: Dept Radiation Biol & Biophys Univ Rochester Sch Med & Dent Rochester NY 14642

WIBERG, KENNETH BERLE, b Brooklyn, NY, Sept 22, 27; m 51; c 3. ORGANIC CHEMISTRY. Educ: Mass Inst Technol, BS, 48; Columbia Univ, PhD(chem), 50. Prof Exp: Instr chem, Univ Wash, Seattle, 50-52, from asst prof to assoc prof, 52-57; vis prof, Harvard Univ, 57-58; prof, Univ Wash, Seattle, 58-60; prof, 60-68, chmn dept, 68-71, WHITEHEAD PROF CHEM, YALE UNIV, 68- Concurrent Pos: Sloan fel, 58-62; Boomer Mem lectr, Univ Alta, 59; Guggenheim fel, 61-62. Honors & Awards: Award, Am Chem Soc, 62, J F Norris Award Phys Org Chem, 73. Mem: Nat Acad Sci; AAAS; Am Chem Soc; The Chem Soc. Res: Stereochemistry and kinetics of organic reactions, particularly oxidation reactions and molecular rearrangements; synthesis and reactions of highly strained compounds. Mailing Add: Dept of Chem Yale Univ 225 Prospect St New Haven CT 06520

WIBERLEY, STEPHEN EDWARD, b Troy, NY, May 31, 19; m 42; c 2. ANALYTICAL CHEMISTRY. Educ: Williams Col, AB, 41; Rensselaer Polytech Inst, MS, 48, PhD(chem), 50. Prof Exp: Sr chemist, Congoleum Nairn, Inc, 41-44; anal chemist, Gen Elec Corp, 46-48; instr chem, 46-48, res assoc, US AEC contract, 48-50, from asst prof to assoc prof anal chem, 50-57, assoc dean, Grad Sch, 64-65, PROF ANAL CHEM, RENSSELAER POLYTECH INST, 57-, DEAN, GRAD SCH, 65-, VPROVOST GRAD PROGS & RES, 69- Concurrent Pos: Vis physicist, Brookhaven Nat Labs, 50; consult, Imp Color Chem & Paper Corp, Socony-Mobil Oil Co, Inc, Huyck Felt Co, Schnectady Chem, Inc & Nat Gypsum Co. Mem: AAAS; Am Chem Soc. Res: Instrumental analysis; infrared and Raman spectroscopy; analysis of radioactive elements. Mailing Add: Grad Sch Rensselaer Polytech Inst Troy NY 12180

WIBKING, ROBERT KENTON, b Moweaqua, Ill, Mar 17, 26; m 50; c 5. GEOGRAPHY. Educ: Eastern Ill State Col, BS, 49; Univ Ky, MA, 51; Univ Nebr, PhD(geog), 63. Prof Exp: Instr geog, Univ Mo, 51-54; cartographer, US Army Map Serv, 54-55; instr geog, Univ NDak, 57-58; asst prof, Kans State Teachers Col, 58-62; PROF GEOG, AUSTIN PEAY STATE UNIV, 62- Mem: Asn Am Geog. Res: Economic and physical geography; cartography; Europe. Mailing Add: Dept of Geog Austin Peay State Univ Clarksville IL 37040

WICH, GROSVENOR SEARLES, b Brooklyn, NY, July 12, 24. ORGANIC CHEMISTRY. Educ: Univ Rochester, BS, 49; Carnegie Inst Technol, MSc, 52, PhD(chem), 54. Prof Exp: Asst, Carnegie Inst Technol, 49-52; res chemist, Jackson Lab, 53-56, patent chemist, 56-60, PATENT CHEMIST EXP STA, E I DU PONT DE NEMOURS & CO, INC, 60- Mem: AAAS; Am Chem Soc; Am Inst Chem; The Chem Soc; Sigma Xi. Res: Macrocyclic polyethers; macromolecules; coordination catalysis; alpha olefin copolymers. Mailing Add: Elastchem Exp Sta E I du Pont de Nemours & Co Inc Wilmington DE 19898

WICHELHAUSEN, RUTH HECHLER, b Stettin, Ger, June 7, 08; nat US; m 37. LABORATORY MEDICINE, CHEMOTHERAPY. Educ: Univ Göttingen, MD, 33. Prof Exp: Intern univ hosp, Univ Göttingen, 33-34, res, 34-35; intern, Franklin Square Hosp, Baltimore, Md, 36; res physician, Montgomery County Gen Hosp, Md, 36-39; asst bact sch med, Johns Hopkins Univ, 39-42, instr, 42-46; dir arthritis res, Vet Admin Hosp, Washington, DC, 46-50, chief arthritis res unit, 50-58; assoc chief tuberc coop study control lab, Vet Admin Hosp, Atlanta, Ga, 59-61, chief, 61-75. RETIRED. Concurrent Pos: Assoc sch med, George Washington Univ, 48-54, asst clin prof, 54-58. Mem: Am Soc Microbiol; Am Thoracic Soc; Am Rheumatism Asn; Am Acad Microbiol; NY Acad Sci. Res: Cultivation and serology of oral spirochetes; clinical and bacteriological arthritis; pleuropneumonia-like organisms; tuberculosis. Mailing Add: 4835 Northway Dr NE Atlanta GA 30342

WICHER, ENOS R, b Dixon, Ill, Feb 16, 11; m; c 1. PHYSICS. Educ: St Ambrose Col, BS, 33; Univ Iowa, MS, 34. Prof Exp: Asst physics, Univ Wis, 36-38; instr physics & math, Olivet Col, 40-41; physicist, Columbia Univ, 44-45; physicist, Spec, Inc, 46-49; physicist, Univ Ga, 51-53; physicist, Univ of Am, 54-61; actg chmn dept, 68-70, dir freshman div, 70-72, PROF PHYSICS, HARVEY MUDD COL, 61- Mem: Am Phys Soc. Res: Mechanics. Mailing Add: Dept of Physics Harvey Mudd Col Claremont CA 91711

WICHMANN, EYVIND HUGO, b Stockholm, Sweden, May 30, 28; nat US; m 51; c 2. THEORETICAL PHYSICS. Educ: Inst Tech, Finland, AB, 50; Columbia Univ, AM, 53, PhD, 56. Prof Exp: Mem staff physics, Inst Advan Study, 55-57; from asst prof to assoc prof, 57-67, PROF PHYSICS, UNIV CALIF, BERKELEY, 67- Mem: Am Phys Soc. Res: Quantum field theory and quantum electrodynamics. Mailing Add: Dept of Physics Univ of Calif Berkeley CA 94720

WICHMANN, ROBERT W, b Seattle, Wash, Jan 20, 26; m 48; c 4. VETERINARY MICROBIOLOGY. Educ: Univ Calif, BS, 51, DVM, 55, PhD, 60. Prof Exp: Asst specialist exp sta, Univ Calif, 55-60; DIR & PRES, POULTRY HEALTH LABS, 60- Mem: AAAS; Am Vet Med Asn; Am Asn Avian Path; US Animal Health Asn; Poultry Sci Asn. Res: Immunology. Mailing Add: Poultry Health Labs PO Box 204 Davis CA 95616

WICHOLAS, MARK L, b Lawrence, Mass, June 11, 40; m 65; c 2. INORGANIC CHEMISTRY. Educ: Boston Univ, AB, 61; Mich State Univ, MS, 64; Univ Ill, PhD(chem), 67. Prof Exp: Asst prof chem, 67-72, ASSOC PROF CHEM, WESTERN WASH STATE COL, 72- Mem: Am Chem Soc. Res: Physical inorganic chemistry; coordination chemistry of transition metals; bio-inorganic chemistry. Mailing Add: Dept of Chem Western Wash State Col Bellingham WA 98225

WICHSER, FRANK WELLINGTON, chemistry, see 12th edition

WICHT, MARION CAMMACK, b Eastabutchie, Miss, Mar 5, 14; m 40; c 3. MATHEMATICS. Educ: Miss Southern Col, BS, 35; Vanderbilt Univ, MA, 36; Auburn Univ, PhD, 57. Prof Exp: Master math, Montgomery Bell Acad, 36-38; actuary, Liberty Mutual Ins Co, 38-40; master, Montgomery Bell Acad, 40-42; instr airplane mechs, US Army Air Forces, 42-43; instr math, La State Univ, 46-50; PROF MATH & HEAD DEPT, NGA COL, 50- Concurrent Pos: Exchange scholar, Univ Berlin, 39-40; instr, Vanderbilt Univ, 40-42; vis prof, US Naval Postgrad Sch, 63-64; vis prof, Univ Ariz, 71. Mem: Math Asn Am; Am Math Soc; Asn Comput Mach; Soc Indust & Appl Math. Res: Poles and polars; foundations of Riemannian geometry; polyharmonics of polynomials. Mailing Add: 415 College Ave Dahlonega GA 30533

WICHT, MARION CAMMACK, JR, entomology, biology, see 12th edition

WICHTERMAN, RALPH, b Philadelphia, Pa, Sept 8, 07; m 33; c 2. ZOOLOGY. Educ: Temple Univ, BS, 30; Univ Pa, MA, 32, PhD(zool), 36. Prof Exp: Asst instr, 29-32, instr biol, 32-39, from asst prof to assoc prof, 39-49, prof, 50-74, EMER PROF BIOL, TEMPLE UNIV, 75- Concurrent Pos: Guest investr, Dry Tortugas Lab, 39; vis prof, Univ Pa, 51; AEC protozoologist, Acad Natural Sci, Philadephia, 52; lectr, Nenski Inst Exp Biol, Warsaw, Poland, 62; Am Philos Soc-NSF award, Zool Sta, Naples, Italy, 62 & 69; mem corp, Marine Biol Lab, Woods Hole. Mem: Fel AAAS; Am Soc Zool; Am Soc Parasitol; Micros Soc Am; Soc Protozool. Res: Protozoology; parasitology; cytology; histology; parasitic protozoa; sexual processes in ciliates; biology of Paramecium; gamma and x-radiation of protozoa. Mailing Add: 31 Buzzards Bay Ave Woods Hole MA 02543

WICK, ARNE N, biochemistry, see 12th edition

WICK, EMILY LIPPINCOTT, b Youngstown, Ohio, Dec 9, 21. ORGANIC CHEMISTRY, ACADEMIC ADMINISTRATION. Educ: Mt Holyoke Col, AB, 43, MA, 45; Mass Inst Technol, PhD(org chem), 51. Hon Degrees: ScD, Mt Holyoke Col, 72. Prof Exp: Instr chem, Mt Holyoke Col, 45-46; res assoc org chem, Mass Inst Technol, 51-53; org chemist flavor lab, Arthur D Little, Inc, 53-57; res assoc food sci, Mass Inst Technol, 57-59, asst prof food chem, 59-63, from assoc prof to prof, 63-73, assoc dean student affairs, 65-72; PROF CHEM & DEAN FAC, MT HOLYOKE COL, 73- Mem: AAAS; Am Chem Soc; Inst Food Technol. Res: Chemistry of food and natural products. Mailing Add: Mary Lyon Hall Mt Holyoke Col South Hadley MA 01075

WICK, GIAN CARLO, b Torino, Italy, Oct 15, 09; nat US; m 43; c 2. PHYSICS. Educ: Univ Torino, PhD(physics), 30. Prof Exp: Asst prof theoret physics, Univ Palermo, 37-38; assoc prof, Univ Padova, 38-40; prof, Univ Rome, 40-45; prof, Univ Notre Dame, 46-48; prof, Univ Calif, 48-50; prof, Carnegie Inst Technol, 51-57; sr physicist, Brookhaven Nat Lab, 57-65; PROF PHYSICS, COLUMBIA UNIV, 65- Mem: Nat Acad Sci; fel Am Phys Soc; Am Acad Arts & Sci. Res: Nuclear physics; elementary physics. Mailing Add: 109 Low Mem Libr Dept of Physics Columbia Univ New York NY 10027

WICK, JAMES ROY, b Henry Co, Iowa, Dec 17, 12; m 42; c 2. ENTOMOLOGY. Educ: Iowa Wesleyan Col, BS, 48; Kans State Col, MS, 50; Iowa State Col, PhD(entom), 54. Prof Exp: Instr biol, Iowa State Col, 52-54, asst prof, 54-59; assoc prof, 59-64, PROF ZOOL & CHMN DEPT BIOL SCI, NORTHERN ARIZ UNIV, 64- Mem: AAAS; Am Inst Biol Sci; Am Entom Soc. Res: Insect histology and developmental anatomy. Mailing Add: Dept of Biol Sci Box 5640 Northern Ariz Univ Flagstaff AZ 86001

WICK, LAWRENCE BERNARD, b Abbey, Sask, Can, May 2, 17; US citizen; m 46; c 2. ORGANIC CHEMISTRY. Educ: Univ Mich, BS, 40, MS, 43, PhD, 48. Prof Exp: Res chemist, Gen Motors Corp, 41-42 & Manhattan Proj, Tenn, 44-45; asst prof chem, Northern Ill State Col, 47-49, Coe Col, 49-51 & Kans State Col, 51-52; res chemist, Glidden Co, 52-58; assoc prof chem, Drury Col, 58-60; assoc prof, 60-72, PROF CHEM, OHIO WESLEYAN UNIV, 72- Concurrent Pos: Res assoc, Kings Col, Univ London, 71. Mem: Am Chem Soc; The Chem Soc; Swiss Chem Soc. Res: Synthesis and reactions of cyclic ketones; synthesis of steroids; organo-phosphorus chemistry. Mailing Add: Dept of Chem Ohio Wesleyan Univ Delaware OH 43015

WICKE, CHARLES ROBINSON, b Roanoke, Va, Apr 13, 28; m 65; c 2. ANTHROPOLOGY, ARCHAEOLOGY. Educ: Univ Va, BA, 48; Mex City Col, MA, 54; Univ Ariz, PhD(anthrop), 65. Prof Exp: Instr anthrop, Mex City Col, 58-61; asst prof, Univ Am, 64-66; Fulbright lectr, Nat Univ Paraguay, 66-67; assoc prof, Northern Ill Univ, 68-69; ASSOC PROF ANTHROP, HEALTH SCI CTR, UNIV OKLA, 69- Concurrent Pos: Mem bd, Museo Frissell, Mitla, Oaxaca, Mex, 64- Mem: Fel AAAS; fel Am Anthrop Asn; fel Royal Anthrop Inst Gt Brit & Ireland; Mex Soc Anthrop; Inst Study Universal Hist Through Arts & Artificats. Res: Human ecology context of precolumbian art; differential access to power in social stratification. Mailing Add: Dept Human Ecol Health Sci Ctr Univ of Okla Box 26901 Oklahoma City OK 73190

WICKE, HOWARD HENRY, b Chicago, Ill, Aug 29, 24; m 45; c 4. TOPOLOGY. Educ: Uniiv Iowa, PhD(math), 52. Prof Exp: Instr math, Lehigh Univ, 52-54; mem staff, Sandia Corp, 54-61, supvr, 61-70; PROF MATH, OHIO UNIV, 70- Mem: Am Math Soc; Math Asn Am. Res: General topology; point-set topology; set theory; applied mathematics. Mailing Add: Dept of Math Ohio Univ Athens OH 45701

WICKELGREN, WARREN OTIS, b Munster, Ind, Oct 15, 41; m 65; c 2. PHYSIOLOGY, PSYCHOLOGY. Educ: Univ Mich, Ann Arbor, AB, 63; Yale Univ, PhD(psychol), 67. Prof Exp: NIH trainee, Yale Univ, 67-69, asst prof physiol, 69-70; ASST PROF PHYSIOL, UNIV COLO MED CTR, DENVER, 70- Concurrent Pos: NIH res career develop award, Univ Colo Med Ctr, Denver, 71. Mem: Soc Neurosci; Am Physiol Soc. Res: Organization of simple nervous systems; neurophysiology of learning. Mailing Add: Dept of Physiol Univ of Colo Med Ctr Denver CO 80220

WICKER, BERTHOLD ROBERT, b Grand Island, Nebr, Nov 12, 06; m 35; c 2. MATHEMATICS. Educ: Grand Island Col, AB, 28; Univ Nebr, MA, 30; Univ Iowa, PhD(math), 48. Prof Exp: Instr math & drawing, St Mary's Col, Minn, 30-33; instr drawing, St Thomas Col, 33-34; instr math & drawing, Rockhurst Col, 34-38; asst math, Univ Iowa, 38-39; instr, 39-43, head dept, 43-70, PROF MATH, LOYOLA UNIV LOS ANGELES, 43- Res: Second order differential equation of the first degree containing two parameters. Mailing Add: Dept of Math Loyola Univ Los Angeles CA 90045

WICKER, ED FRANKLIN, b Upper Tygart, Ky, Aug 21, 30; m 53; c 2. PLANT PATHOLOGY, FORESTRY. Educ: Wash State Univ, BS, 59, PhD(plant path), 65. Prof Exp: Res forester, 59-63, PLANT PATHOLOGIST, INTERMOUNTAIN FOREST & RANGE EXP STA, US FOREST SERV, 63- Concurrent Pos: Mem, Western Int Forest Dis Work Conf. Honors & Awards: Japanese Govt Res Award Foreign Spec, Govt Japan, 74. Mem: Am Phytopath Soc; Mycol Soc Am; Soc Am Foresters. Res: Biology and control of dwarf mistletoes; biological control of forest tree diseases. Mailing Add: Forestry Sci Lab 1221 S Main Moscow ID 83843

WICKER, EVERETT E, b Lockport, NY, Apr 6, 19; m 43; c 2. NUCLEAR SCIENCE. Educ: Univ Pittsburgh, BS, 41; Carnegie Inst Technol, MS, 57. Prof Exp: Physicist, Kennametal, Inc, 41-42 & 46-47; from asst technologist to technologist, 47-54, from supvry technologist to res technologist, 54-64, ASSOC RES CONSULT, RES LAB, US STEEL CORP, 64- Mem: Am Phys Soc; Am Nuclear Soc. Res: Neutron and charged particle activation analysis; nuclear reactor materials; nuclear and reactor physics; industrial and research uses of radioisotopes and radiation. Mailing Add: US Steel Corp Res Lab MS 73 125 Jamison Lane Monroeville PA 15146

WICKER, ROBERT KIRK, b Altoona, Pa, Mar 4, 38; m 61; c 3. PHYSICAL INORGANIC CHEMISTRY. Educ: Juniata Col, BS, 60; Univ Del, MS, 63, PhD(phys chem), 66. Prof Exp: Asst prof chem, Davis & Elkins Col, 65-67; asst prof, 67-70, ASSOC PROF CHEM, WASHINGTON & JEFFERSON COL, 70- Mem: AAAS; Am Chem Soc. Res: Thermodynamic properties of nonaqueous electrolyte solutions; preparation and structure determinations of copper complexes. Mailing Add: Dept of Chem Washington & Jefferson Col Washington PA 15301

WICKER, THOMAS HAMILTON, JR, b Orlando, Fla, Nov 19, 23; m 49; c 3. ORGANIC CHEMISTRY, POLYMER CHEMISTRY. Educ: Univ Fla, BS, 44, MS, 48, PhD(org chem), 51. Prof Exp: From assoc res chemist to res chemist, 51-58, SR RES CHEMIST, TENN EASTMAN CO, 58- Mem: Am Chem Soc. Res: 2-cyanoacrylate adhesives; condensation polymers. Mailing Add: 4619 Mitchell Rd Kingsport TN 37664

WICKERHAUSER, MILAN, b Zemun, Yugoslavia, Aug 28, 22; m 56; c 2. BIOCHEMISTRY. Educ: Chem engr, Univ Zagreb, 46, PhD, 61. Prof Exp: Develop chemist, Inst Immunol, Yugoslavia, 46-53, head dept serum & toxoid purification, 53-57, head dept human plasma fractionation, 57-62; immunochemist, Immunol, Inc, Ill, 63; res assoc fractionation plasma protein & blood coagulation studies, Hyland Labs, Calif, 64-66; SR RES SCIENTIST, AM NAT RED CROSS, 66-, HEAD PLASMA FRACTIONATION SECT, 68-, DIR, AM RED CROSS NAT FRACTIONATION CTR, BLOOD RES LAB, 70- Concurrent Pos: WHO fel, Wellcome Physiol Res Lab, Eng, Lister Inst Prev Med, London & State Serum Inst, Copenhagen, 50. Mem: AAAS; Am Chem Soc; Int Soc Thrombosis and Haemostasis. Res: Isolation and characterization of plasma proteins with special emphasis on the large scale methodology; blood coagulation; immunoglobulins; development of large-scale plasma fractionation methods. Mailing Add: Red Cross Nat Fractionation Ctr 9312 Old Georgetown Rd Bethesda MD 20014

WICKERSHAM, ARTHUR FRANK, JR, b San Francisco, Calif, June 20, 22; m 48; c 3. PHYSICS. Educ: Univ Calif, AB, 43, PhD(physics), 54. Prof Exp: Physicist radiation lab, Univ Calif, 49-54; asst prof physics, San Jose State Col, 55; eng specialist, Sylvania Electronic Defense Lab Div, Gen Tel & Electronics Corp, 56-63; sr physicist, Stanford Res Inst, 63-74. Res: Radio antennas and propagation; radio surface waves; acoustic-gravity waves; ionospheric physics; short-pulse and subterranean radar. Mailing Add: 336 Torino Dr Apt 3 San Carlos CA 94070

WICKERSHAM, EDWARD WALKER, b Kelton, Pa, Apr 26, 32; m 59; c 3. REPRODUCTIVE PHYSIOLOGY, HUMAN SEXUALITY. Educ: Pa State Univ, BS, 57, MS, 59; Univ Wis, PhD(dairy physiol), 62. Prof Exp: NIH trainee endocrinol, Univ Wis, 62-63; asst prof biol, WVa Univ, 63-64; asst prof zool, 64-68, ASSOC PROF BIOL, PA STATE UNIV, 68- Mem: Am Soc Animal Sci; Brit Soc Study Fertil; Soc Study Reproduction; Am Asn Sex Educators & Counselors. Res: Mammalian reproductive physiology and endocrinology; factors influencing corpus luteum function; biological aspects of human sexuality. Mailing Add: Dept of Biol 417 Life Sci Bldg Pa State Univ University Park PA 16802

WICKERSHEIM, KENNETH ALAN, b Fullerton, Calif, Mar 4, 28; m 52, 67; c 2. SOLID STATE PHYSICS, SPECTROSCOPY. Educ: Univ Calif, Los Angeles, AB, 50, MA, 56, PhD(physics), 59. Prof Exp: Mem staff, Los Alamos Sci Lab, 53-55; asst physics, Univ Calif, Los Angeles, 55-58; mem staff res labs, Hughes Aircraft Co, 58-61; res physicist, Palo Alto Labs, Gen Tel & Electronics Lab, Inc, 61-63; assoc prof mat sci, Stanford Univ, 63-64; staff scientist solid state physics, Lockheed Palo Alto Res Labs, 64-65, sr mem, 65-66, head adv electronics, 66-70; PRES, SPECTROTHERM CORP, 70- Concurrent Pos: Res assoc sch earth sci, Stanford Univ, 64-; consult, Appl Physics Corp, 63-64. Mem: AAAS; fel Am Phys Soc. Res: Optical properties of materials; spectra of rare earth ions in solids; infrared spectra of crystals; optical and infrared instrumentation; medical and industrial thermography. Mailing Add: 3895 Middlefield Rd Palo Alto CA 94303

WICKES, GLENN FRENCH, b Cleveland, Ohio, May 8, 18; m 42; c 3. ZOOLOGY, CHEMISTRY. Educ: Baldwin-Wallace Col, BS, 41. Prof Exp: Lab asst, Gen Elec Co, 36-37; shift foreman, 41-42, supvr natural estrone, 45-47, assoc pharmaceut develop, 47-50, mgr sterile prod, 50-55, dir sterile prod mgr, 55-60, prod mgr, 60-64, VPRES PROD & DEVELOP, BEN VENUE LABS, INC, 64- Mem: Soc Cryobiol. Res: Development of lyophilized dosage forms of special drugs to be used in cancer chemotherapy. Mailing Add: Ben Venue Labs Inc 270 Northfield Rd Bedford OH 44146

WICKES, HARRY E, bPortland, Ore, June 24, 25; m 49; c 3. MATHEMATICS. Educ: Brigham Young Univ, BS, 50, MEd, 54; Harvard Univ, MEd, 62; Colo State Col, EdD, 67. Prof Exp: Teacher high sch, Mont, 50-51; teacher, Idaho, 51-54; prin elem & high sch, Idaho, 54-57; instr math, 57-63, from asst prof to assoc prof, 64-75, PROF MATH, BRIGHAM YOUNG UNIV, 75- Mem: Math Asn Am; Nat Coun Teachers Math. Res: Mathematics education. Mailing Add: Dept of Math Brigham Young Univ Provo UT 84602

WICKES, WILLIAM CASTLES, b Lynwood, Calif, Nov 25, 46; m 71; c 1. COSMOLOGY. Educ: Univ Calif, Los Angeles, BS, 67; Princeton Univ, MA, 69, PhD(physics), 72. Prof Exp: From res assoc to instr physics, 72-75, ASST PROF PHYSICS, PRINCETON UNIV, 75- Mem: Am Astron Soc; Sigma Xi. Res: Double star interferometry; experimental and theoretical cosmology. Mailing Add: Dept of Physics Princeton Univ Jadwin Hall Princeton NJ 08540

WICKHAM, DONALD G, b Beaverton, Ore, Feb 24, 22; m 54; c 2. INORGANIC CHEMISTRY. Educ: Univ Denver, BS, 47, MS, 50; Mass Inst Technol, PhD(inorg chem), 54. Prof Exp: Chemist, Lincoln Lab, Mass Inst Technol, 54-57, sect leader, 57-60; mem staff, Res Labs, Hughes Aircraft Co, 60-61; mgr mat res & develop, Components Div, 61-65, MGR FERRITE MEMORY-CORE DEVELOP, AMPEX COMPUT PROD CO, 65- Mem: Fel Am Inst Chem; Am Crystallog Asn; Am Phys Soc. Res: Inorganic solid state chemistry; magnetic materials; inorganic syntheses. Mailing Add: Ampex Comput Prods Co 2245 Pontius Ave Los Angeles CA 90064

WICKHAM, JAMES EDGAR, JR, b Glen Allen, Va, Apr 7, 33; m 53; c 2. ANALYTICAL CHEMISTRY, INORGANIC CHEMISTRY. Educ: Randolph-Macon Col, BS, 57. Prof Exp: Asst chemist, Philip Morris, Inc, 57-59, group leader, 59-62, supvr, 62-69, head cigarette testing, 69-71, sr scientist, 71-74, MGR CIGARETTE TESTING SERV, PHILIP MORRIS USA, 74- Mem: AAAS; Am Chem Soc. Res: Wet and instrumental methods development; gas-liquid chromatography tobacco and smoke chemistry; quality control operations. Mailing Add: Philip Morris USA Develop Div PO Box 26583 Richmond VA 23261

WICKHAM, WILLIAM TERRY, JR, b Cleveland, Ohio, May 28, 29; m 52; c 3. POLYMER CHEMISTRY, POLYMER PHYSICS. Educ: Heidelberg Col, AB, 51; Case Inst Technol, MS, 54, PhD(org chem), 56. Prof Exp: Instr, Case Inst Technol, 55-56; res chemist, Owens-Ill Glass Co, 56-58; group leader, Dow Chem Co, 58-62; tech mgr, Celanese Plastics Co, 62-67; dir res, Southern Div, 67-72, VPRES RES & DEVELOP, DAYCO CORP, 72- Mem: Am Chem Soc; Am Phys Soc; Am Inst Chemists. Res: Polymer morphology; synthesis; stability; structure; manufacture. Mailing Add: Dayco Corp Waynesville NC 28786

WICKLIFF, JAMES LEROY, b Knoxville, Iowa, Nov 14, 31; m 56; c 3. PLANT PHYSIOLOGY, PHOTOBIOLOGY. Educ: Iowa State Univ, BS, 55, PhD(plant physiol), 62. Prof Exp: Assoc bot & plant path, Iowa State Univ, 62-65; asst prof bot & bact, 65-69, ASSOC PROF BOT & BACT, UNIV ARK, FAYETTEVILLE, 69- Concurrent Pos: Consult mat identification & develop proj comn undergrad educ biol sci, NSF, 65-66. Mem: AAAS; Am Soc Plant Physiol; Am Inst Biol Sci; Am Soc Photobiol. Res: Chlorophyll biochemistry; photosynthesis; photophysiology of higher plants. Mailing Add: Dept of Bot & Bact Univ of Ark Fayetteville AR 72701

WICKLUND, ARTHUR BARRY, b Dec 8, 42; US citizen; m 74. EXPERIMENTAL HIGH ENERGY PHYSICS. Educ: Harvard Univ, BA, 64; Univ Calif, Berkeley, PhD(physics), 70. Prof Exp: Res asst high energy physics, Univ Calif, Berkeley, 65-70; fel, 70-73, ASST PHYSICIST, ARGONNE NAT LAB, 73- Mem: Am Phys Soc. Res: Strong interaction phenomenology; production mechanisms in few body reactions. Mailing Add: Argonne Nat Lab Bldg 362 Argonne IL 60639

WICKMAN, FRANS ERIK, b Stockholm, Sweden, Mar 21, 15. GEOCHEMISTRY. Educ: Univ Stockholm, MS, 38, PhD(mineral, geol), 43. Hon Degrees: DSc, Univ Stockholm, 52. Prof Exp: Physicist cent res lab, Boliden Mining Co, 44-46; lectr mineral, Univ Stockholm, 46-47; res prof, Swedish Mus Natural Hist, 47-67; prof geochem, Pa State Univ, University Park, 67-74; PROF GEOCHEM, UNIV STOCKHOLM, SWEDEN, 74- Concurrent Pos: Res assoc, Univ Chicago, 47-48; vis prof, Harvard Univ, 65-66. Mem: Royal Swedish Acad Sci; Royal Norweg Acad Sci; Royal Danish Acad Sci. Res: Mathematical models in geology and geochemistry. Mailing Add: Box 6801 11386 Stockholm Sweden

WICKMAN, HERBERT HOLLIS, b Omaha, Nebr, Sept 30, 36; m 57; c 2. PHYSICAL CHEMISTRY, BIOPHYSICS. Educ: Univ Omaha, AB, 59; Univ Calif, Berkeley, PhD(chem), 64. Prof Exp: Mem tech staff chem physics res lab, Bell Tel Labs, NJ, 64-70; vis scientist, Democritos Nuclear Res Ctr, Greece, 69-70; ASSOC PROF CHEM, ORE STATE UNIV, 70- Mem: Am Chem Soc; Am Phys Soc; Sigma Xi. Res: Cooperative phemonea in condensed phases; solid state spectroscopy. Mailing Add: Dept of Chem Ore State Univ Corvallis OR 97331

WICKREMA SINHA, ASOKA J, b Colombo, Ceylon, Sept 8, 37; US citizen. BIOCHEMISTRY, ORGANIC CHEMISTRY. Educ: Univ London, BSc, 61; Univ Birmingham, MSc, 64, PhD(org chem), 66. Prof Exp: Fel biochem, Univ Birmingham, 66-67; staff scientist, Worcester Found Exp Biol, 67-69; res scientist chem & biochem, 69-75, SR RES SCIENTIST, RES DIV, UPJOHN CO, 75- Mem: Am Chem Soc; The Chem Soc; Brit Biochem Soc. Res: Chemical synthesis; carbonium ion chemistry; biosynthesis and metabolism of steroids; in vivo and in vitro metabolism; analytical methods and assay development; metabolism absorption, distribution and excretion of drugs; isolation and structure elucidation; radiotracer techniques. Mailing Add: Res Div Upjohn Co Kalamazoo MI 49001

WICKS, FREDERICK JOHN, b Winnipeg, Man, Nov 22, 37; m 67. MINERALOGY. Educ: Univ Man, BSc, 60, MSc, 65; Oxford Univ, DPhil(mineral), 69. Prof Exp: Geologist, Giant Yellowknife Mines Ltd, 60 & 61; consult mineralogist, Govt & Indust, 62; mineralogist, Man Hwys Br, 63-65 & Geol Surv Can, 67; asst cur, 70-75, ASSOC CUR MINERAL, ROYAL ONT MUS, 75- Mem: Mineral Asn Can (secy, 73-75); Geol Asn Can; Mineral Soc Am; Clay Minerals Soc. Res: Structure, chemistry and paragenesis of the serpentine minerals; asbestos deposites; nickel sulfide ore deposits associated with serpentinites; minerals associated with serpentine miners. Mailing Add: Dept of Mineral & Geol Royal Ont Mus 100 Queen's Park Toronto ON Can

WICKS, WESLEY DOANE, b Providence, RI, Feb 13, 36; m 59; c 3. BIOCHEMISTRY, PHARMACOLOGY. Educ: Bates Col, BS, 57; Harvard Univ, MA, 59, PhD(med sci), 64. Prof Exp: Staff mem biochem, Biol Div, Oak Ridge Nat Lab, Tenn, 65-69; staff mem, Div Res, Nat Jewish Hosp, Denver, 69-72; ASSOC PROF PHARMACOL, MED CTR, UNIV COLO, DENVER, 72- Concurrent Pos: Am Cancer Soc fel, Biol Div, Oak Ridge Nat Lab, Tenn, 63-65; hon fac mem, Dept Biosci, Fed Univ Pernambuco, Recife, Brazil, 68- Mem: AAAS; Am Soc Biol Chem; Am Chem Soc; Endocrine Soc. Res: Regulation of specific protein synthesis by hormones and cyclic adenosine phosphate; use of cultured cells for studies of biochemical regulatory mechanisms. Mailing Add: Dept of Pharmacol Univ of Colo Med Ctr Denver CO 80220

WICKS, ZENO W, JR, b Port Jervis, NY, July 24, 20; m 41; c 6. POLYMER CHEMISTRY. Educ: Oberlin Col, AB, 41; Univ Ill, PhD(org chem), 44. Prof Exp: Res chemist, Interchem Corp, 44-47, dist tech dir finishes div, 48-49, res dir textile colors div, 49-51, assoc dir cent res labs, 51-54, dir, 54-59, mgr com develop, 59-61, vpres planning, 62-63, vpres & dir corp, 64-69; mem staff, Inmont Corp, NY, 69-72; PROF POLYMERS & COATINGS & CHMN DEPT, NDAK STATE UNIV, 72- Mem: Am Chem Soc; Soc Coatings Technol; Oil & Colour Chemists Asn. Res: Organic surface coatings and related polymer research. Mailing Add: Dept of Polymers & Coatings NDak State Univ Fargo ND 58102

WICKSON, EDWARD JAMES, b New York, NY, Jan 25, 20; m 52; c 5. PLASTICS CHEMISTRY. Educ: Univ Calif, Berkeley, BS, 42. Prof Exp: Anal chemist, Gen Chem Co, Calif, 42-43; chemist, Celanese Corp Am, NJ, 46-50; admin asst to lab dir, Vitro Corp Am, 51-54; sr chemist, Chicopee Mfg Co, 54-55; sr chemist, Enjay Labs, Esso Res & Eng Co, 55-56, group leader, 56-60, res assoc, 60-61, head chem sect, 61-69, vinyl indust assoc, Enjay Chem Co, 69-71, sr res assoc, Enjay Chem Lab, 71-75, CHIEF SCIENTIST PLASTICIZERS, ESSOCHEM EUROPE, 75- Mem: Am Chem Soc; sr mem Soc Plastics Eng. Res: Plasticizers; chemical specialties; oxo alcohols; trialylacetic acids; propylene polymers. Mailing Add: Ave des Quatre Saisons 21 1410 Waterloo Belgium

WICKSTROM, ERIC, b Chicago, Ill, Dec 21, 46; m 67; c 2. BIOPHYSICAL CHEMISTRY. Educ: Calif Inst Technol, BS, 68; Univ Calif, Berkeley, PhD(chem), 72. Prof Exp: Res asst chem, Univ Calif, Berkeley, 68-72; res assoc molecular, cellular & develop biol, Univ Colo, Boulder, 73-74; ASST PROF CHEM, UNIV DENVER, 74- Mem: Am Chem Soc. Res: Tertiary structure of transfer RNA; protein synthesis initiation. Mailing Add: Dept of Chem Univ of Denver Denver CO 80210

WICKSTROM, JACK, b Omaha, Nebr, Aug 7, 13; m 40; c 3. MEDICINE. Educ: Univ Nebr, AB, 35, MD, 39. Prof Exp: Assoc prof, 46-56, PROF ORTHOP & CHMN DEPT, SCH MED, TULANE UNIV, 56- Concurrent Pos: Fel path, Univ Nebr, 44; fel orthop surg, Sch Med, Tulane Univ, 44-46. Honors & Awards: Gold Medal, Am Acad Orthop Surg, 51; Scudder Orator, Am Col Surg, 74; Surgeons Award, Nat Safety Coun, 75. Mem: AAAS. Res: Bone metabolism and growth; biomechanics; surgical implants and injury control. Mailing Add: 1430 Tulane Ave New Orleans LA 70112

WIDDEN, PAUL RODNEY, b London, Eng, Sept 23, 43; m 67; c 2. SOIL MICROBIOLOGY. Educ: Univ Liverpool, BSc Hons, 65; Univ Calgary, PhD(mycol), 71. Prof Exp: ASST PROF MICROBIAL ECOL, CONCORDIA UNIV, LOYOLA CAMPUS, 73- Concurrent Pos: Nat Res Coun Can operating grant, 75-78. Mem: Can Soc Microbiologists. Res: The distribution of fungi in tundra and temperate forest soils; effects of environment on the distribution of soil fungi; effects of temperature on fungal growth and respiration. Mailing Add: Dept of Biol Loyola Campus Concordia Univ 7141 Sherbrooke W Montreal PQ Can

WIDDER, DAVID VERNON, b Harrisburg, Pa, Mar 25, 98; m 39; c 2. MATHEMATICS. Educ: Harvard Univ, AB, 20, AM, 23, PhD(math), 24. Prof Exp: Assoc math, Bryn Mawr Col, 24-26, assoc prof & head dept, 27-30, prof, 30-32; Nat Res Coun fel, Univ Chicago & Rice Inst, 26-27; asst prof, 31-32, from assoc prof to prof, 32-68, chmn dept, 42-48, dir acad year inst, 63-66, EMER PROF MATH, HARVARD UNIV, 68- Concurrent Pos: Guggenheim Found fel, Cambridge Univ, 35-36; Fulbright fel, Italy, 55-56 & Australia, 62-63; consult, Rand Corp, 49-69. Mem: Am Math Soc; Math Asn Am; fel Am Acad Arts & Sci. Res: Series expansions; Laplace transform; integral transforms; heat conduction. Mailing Add: 30 Gould Rd Arlington MA 02174

WIDDER, JAMES STONE, b Cleveland, Ohio, Feb 28, 35; m 60; c 1. IMMUNOLOGY, BIOCHEMISTRY. Educ: Ohio State Univ, BS, 57, MS, 59, PhD(immunol), 62. Prof Exp: Res microbiologist, Miami Valley Lab, 62-64, prod researcher, Winton Hill Tech Ctr, 64-74, SECT HEADER BASIC SKIN RES & PROJ, MIAMI VALLEY LAB, PROCTER & GAMBLE CO, 74- Concurrent Pos: Lectr, Univ Cincinnati, 62- Mem: AAAS; Am Soc Microbiol; Int Asn Dent Res (secy, 66-67). Res: Product development of biologically oriented products and associated governmental problems. Mailing Add: Miami Valley Lab Procter & Gamble Co Cincinnati OH 45224

WIDDOWSON, DAVID CARL, b Pittsburgh, Pa, July 19, 11; m 42; c 5. BIOLOGY. Educ: Univ Ala, BS, 39, MA, 47; George Peabody Col, MEd, 51, EdD, 53. Prof Exp: Head dept biol, Berry Col, 47-54; from asst prof to assoc prof sci, 54-66, PROF SCI & CHMN BIO-SOCIAL CORE STAFF, TROY STATE UNIV, 66- Concurrent Pos: Lectr & consult, Sch Med Tech, St Margaret's Hosp, Montgomery, Ala, 59- Res: Medical technology; sanitary science. Mailing Add: Dept of Sci Troy State Univ Troy AL 36081

WIDEBURG, NORMAN EARL, b Chicago, Ill, Mar 8, 33; m 58; c 4. INDUSTRIAL MICROBIOLOGY. Educ: Ill Inst Technol, BS, 54; Univ Wis, MS, 56. Prof Exp: BIOCHEMIST, ABBOTT LABS, 58- Mem: AAAS; Am Chem Soc. Res: Microbial transformations; fermentation; antibiotics. Mailing Add: Abbott Labs North Chicago IL 60064

WIDEMAN, CHARLES JAMES, b Walkermine, Calif, Feb 7, 36; m 63; c 2. GEOPHYSICS. Educ: Colo Sch Mines, BSc, 58, MSc, 67, PhD(geophys), 75. Prof Exp: Sr geophysicist, Westinghouse Elec Corp, 67-68; asst prof, 68-73, ASSOC PROF GEOPHYS & CHMN DEPT, MONT COL MINERAL SCI & TECHNOL, 73- Mem: Seismol Soc Am. Res: Local seismicity; seismic risk analysis and earth strain studies; gravity investigations over and near the Boulder Batholith of Southwestern Montana. Mailing Add: Mont Col Mineral Sci & Technol Butte MT 59701

WIDEMAN, LAWSON GIBSON, b Morrelton, Mo, July 17, 43; m 65; c 3. ORGANIC CHEMISTRY. Educ: Univ Mo-Rolla, BS, 66, MS, 67; Univ Akron, PhD(chem), 71. Prof Exp: Staff res chemist, 67-71, SR RES CHEMIST, RES DIV, GOODYEAR TIRE & RUBBER CO, 71- Mem: Am Chem Soc; Sigma Xi. Res: Organometallic homogeneous catalysis; chemistry of carbanions in solution; selective reactions of organoboranes; homo- and heterogeneous hydrogenation reactions. Mailing Add: Goodyear Tire & Rubber Co 142 Goodyear Blvd Akron OH 44316

WIDESS, MOSES B, b Sverdlovsk, Russia, Sept 21, 11; nat US; m 35; c 2. GEOPHYSICS. Educ: Calif Inst Technol, BS, 33, MS, 34, PhD(elec eng), 36. Prof Exp: Party chief, Western Geophys Co, Calif, 36-42; consult geophysicist, Amoco Prod Co, 42-73; CONSULT, 73- Mem: Soc Explor Geophys; Seismol Soc Am; Am Asn Petrol Geol; Am Geophys Union; Europ Asn Explor Geophys. Res: Geophysical interpretation and methods. Mailing Add: 11617 Monica Lane Houston TX 77024

WIDGER, WILLIAM KNOWLTON, JR, b Lynn, Mass, Jan 23, 21; m 43; c 2. METEOROLOGY. Educ: Univ NH, BS, 42; Mass Inst Technol, ScD(meteorol), 49. Prof Exp: Asst meteorol, Mass Inst Technol, 46-49; asst prof, State Univ NY Col Agr, Cornell, 49-51; chief oper tech sect atmospheric anal lab, Geophys Res Dir, Air Force Cambridge Res Labs, 52-58, chief satellite meteorol br atmospheric circulations lab, 58-60; asst chief meteorol progs off space flight progs hq, NASA, 60-61, chief oper meteorol satellite progs off appln, 61-63; dir satellite meteorol res, Allied Res Assocs, Inc, 63-67; prof atmospheric physics, Drexel Inst Technol, 67-69; pres, Belknap Col, 69-72; DIR GEOPHYS SERV, BIOSPHERIC CONSULT INT, INC, MEREDITH, NH, 73- Concurrent Pos: Mem comt meteorol aspects satellites space sci bd, Nat Acad Sci-Nat Res Coun, 58-62, mem panel meteorol serv comt int progs in atmospheric sci & hydrol, 62-63; mem, Joint Meteorol Satellite Adv Comt, 59-63, secy, 60-63; lectr meteorol, Lyndon State Col, Vt, 74-; vis prof, New Eng Col, NH, 73. Mem: Fel AAAS; Am Meteorol Soc; Am Geophys Union; Sigma Xi. Res: Satellite, research, applied meterology; climatology; physical limnology. Mailing Add: RD 2 Hillrise Lane Meredith NH 03253

WIDGOFF, MILDRED, b Buffalo, NY, Aug 24, 24; m 45; c 2. ELEMENTARY PARTICLE PHYSICS. Educ: Univ Buffalo, BA, 44; Cornell Univ, PhD(physics), 52. Prof Exp: Asst physics, Manhattan Proj, 44-45; assoc physics, Brookhaven Nat Lab, 52-54; res fel, Harvard Univ, 55-58; res asst prof physics, 58-66, res assoc prof, 66-74, PROF PHYSICS, BROWN UNIV, 74-, EXEC OFFICER DEP, 68- Mem: Am Phys Soc. Res: Cosmic rays; medium and high energy particle physics; studies of interactions of elementary experimental particles at medium and high energies. Mailing Add: Dept of Physics Brown Univ Providence RI 02912

WIDHOLM, JACK MILTON, b Watseka, Ill, Mar 11, 39; m 64; c 2. PLANT PHYSIOLOGY, GENETICS. Educ: Univ Ill, BS, 61; Calif Inst Technol, PhD(biochem), 66. Prof Exp: Res chemist, Int Minerals & Chem Corp, 65-68; asst prof physiol dept agron, 68-73, ASSOC PROF PLANT PHYSIOL DEPT AGRON, UNIV ILL, URBANA, 73- Mem: AAAS; Am Soc Plant Physiol; Tissue Cult Asn; Scand Soc Plant Physiol; Am Soc Agron. Res: Plant biochemistry and genetics, especially genetic manipulation, control of amino acid biosynthesis and photorespiration. Mailing Add: Dept of Agron Univ of Ill Urbana IL 61801

WIDIGER, ALEXANDER HERMAN, chemistry, see 12th edition

WIDING, KENNETH GORDON, b St Paul, Minn, Apr 15, 27. SOLAR PHYSICS, ATOMIC SPECTROSCOPY. Educ: Wesleyan Univ, AB, 50; Ind Univ, MA, 52; Univ Calif, Berkeley, PhD(astron), 59. Prof Exp: MEM STAFF, US NAVAL RES LAB, 59- Mem: Am Astron Soc; Int Astron Union. Res: Ultraviolet solar spectrum; analysis of solar extreme ultra-violet data obtained from Apollo Telescope Mount/Skylab. Mailing Add: Code 7144 US Naval Res Lab Washington DC 20375

WIDMANN, FRANCES KING, b Boston, Mass, July 23, 35; m 58; c 2. PATHOLOGY. Educ: Swarthmore Col, BA, 56; Western Reserve Univ, MD, 60; Am Bd Path, dipl & cert anat & clin path, 65, cert immunohemat, 73. Prof Exp: Intern, Cleveland Metrop Gen Hosp, Ohio, 60-61; resident anat & clin path, Sch Med, Univ NC, Chapel Hill, 61-64; resident clin path, Norfolk Gen Hosp, Va, 64-65, staff pathologist, 65-66; from instr to asst prof path, Sch Med, Univ NC, Chapel Hill, 66-70; asst prof, 71-73, assoc dir, Sch Med Technol, 71-73, ASSOC PROF PATH, SCH MED, DUKE UNIV, 73-, DIR, SCH MED TECHNOL, 73-; ASST CHIEF LAB SERV, VET ADMIN HOSP, 72- Res: Blood banking and medical education, especially in clinical pathology and medical technology training. Mailing Add: Vet Admin Hosp 508 Fulton St Durham NC 27705

WIDMARK, RUDOLPH M, b Horn, Austria, July 18, 25; m 55; c 1. IMMUNOLOGY, MEDICAL MICROBIOLOGY. Educ: Univ Timisoara, Romania, MD, 52; Univ Bucharest, PhD(immunol, microbiol), 56. Prof Exp: Instr path, Sch Med, Univ Timisoara, 49-52, asst prof virol, 52; asst prof microbiol, Sch Med, Univ Bucharest, 53-55; assoc prof, Cantacuzino Inst, Bucharest, 56-62; dep dir clin lab, Hanusch Hosp, Vienna, 62-63; trainee med & arthritis, State Univ NY Downstate Med Ctr, 64, asst prof microbiol & immunol, 64-67; dir immunol, Denver Chem Mfg Co, 67-68; staff consult life sci, Technicon Corp, 68-71; dir biol res, BioQuest, 71; ASST DIR CLIN RES, AYERST LABS, 71- Res: Antigen-antibody reactions; passive hemagglutination; heterophil antibodies; preservation of erythrocytes for antigen-antibody reactions. Mailing Add: Ayerst Labs 685 Third Ave New York NY 10017

WIDMAYER, DOROTHEA JANE, b Washington, DC, Oct 10, 30. ZOOLOGY, MICROBIAL GENETICS. Educ: Wellesley Col, BA, 53, MA, 55; Ind Univ, PhD(zool), 62. Prof Exp: Instr biol, Simmons Col, 55-57; instr zool, 61-63, from asst prof to assoc prof biol, 63-74, prof, 74-75, KENAN PROF BIOL, WELLESLEY COL, 75-, CHMN DEPT BIOL SCI, 72- Concurrent Pos: NSF sci fac fel inst animal genetics, Univ Edinburgh, 67-68; grant, Ascent of Man, Res Corp, 74. Mem: AAAS; Genetics Soc Am; Am Soc Zool; Soc Protozool. Res: Gene action and cytoplasmic inheritance in Paramecium aurelia. Mailing Add: Dept of Biol Sci Wellesley Col Wellesley MA 02181

WIDMER, CARL, b Kingston, Pa, Mar 3, 24; m 59; c 2. BIOCHEMISTRY. Educ: Pa State Univ, BS, 48; Agr & Mech Col, Tex, MS, 49; Univ Rochester, PhD(biochem), 52. Prof Exp: USPHS fel, Ore State Col, 54-56; fel inst enzyme res, Univ Wis, 56-57; asst prof chem, Humboldt State Col, 57-61; assoc prof, Agrarian Univ, Peru, 62-64; asst prof biochem & nutrit, Tex A&M Univ, 65-66; ASSOC PROF CHEM, ELBERT COVELL COL, UNIV OF PAC, 66- Mem: Am Soc Limnol & Oceanog. Res: Human nutrition; limnology. Mailing Add: Elbert Covell Col Univ of the Pac Stockton CA 95204

WIDMER, ELMER ANDREAS, b Dodge, NDak, Apr 27, 25; m; c 2. HELMINTHOLOGY, MEDICAL PARASITOLOGY. Educ: Union Col, Nebr, BA, 51; Univ Colo, MA, 56; Colo State Univ, PhD(zool), 65; Univ NC, MPH, 74. Prof Exp: Teacher, High Sch, 52-53; instr biol, La Sierra Col, 53-58, from asst prof to assoc prof, 58-67; assoc prof environ & trop health, 67-71, PROF ENVIRON & TROP HEALTH, LOMA LINDA UNIV, 71-, CHMN DEPT, 67- Concurrent Pos: Fel sch med, La State Univ, 66; WHO fel, Africa, 71. Mem: Am Pub Health Asn; Royal Soc Trop Med & Hyg; Am Soc Parasitol; Am Soc Trop Med & Hyg; Wildlife Dis Asn. Res: Reptilian parasitology; tropical helminthology; host-parasite interactions between cestodes. Mailing Add: Dept of Environ & Trop Health Sch of Health Loma Linda Univ Loma Linda CA 92354

WIDMER, HANS, b Bauma, Switz, Dec 2, 36; m 64; c 2. POLYMER CHEMISTRY. Educ: Swiss Fed Inst Tech, MS, 60, PhD(phys chem), 63. Prof Exp: Sr physicist, Goodyear Tire & Rubber Co, Ohio, 63-66, Europ res rep, 66-69, MGR EUROP CHEM TECH CTR, GOODYEAR INT CO, 69- Mem: Am Chem Soc; Am Inst Chem Eng; Ger Chem Soc; Int Inst Synthetic Rubber Producers. Res: Gas chromatography; electron paramagnetic resonance and mass spectrometry; decomposition of hydrocarbons under high energy radiation; vulcanization; graft, solution and emulsion polymerization; research liaison; technical service; research and development management. Mailing Add: Cie Francaise Goodyear Quartier de Courtaboeuf B P 31 Orsay 91402 France

WIDMER, KEMBLE, b New Rochelle, NY, Feb 26, 13; m 39; c 2. GEOLOGY. Educ: Lehigh Univ, AB, 37; Princeton Univ, MA, 47, PhD(geol), 50. Prof Exp: From instr to asst prof geol, Rutgers Univ, 48-50; assoc prof & chmn dept, Champlain Col, NY, 50-53; prin geologist, 53-58, STATE GEOLOGIST, DIV RESOURCE DEVELOP, BUR GEOL & TOPOG, NJ DEPT CONSERV & ECON DEVELOP, 58-, NUCLEAR INDUST COORDR, 63- Concurrent Pos: Tech consult, US Mil Acad, 63-; seminar assoc, Columbia Univ. Mem: AAAS; Geol Soc Am; Am Inst Mining, Metall & Petrol Eng. Res: Areal, economic, Pleistocene and engineering geology. Mailing Add: NJ State Dept of Environ 1474 Prospect St Box 2809 Trenton NJ 08625

WIDMER, RICHARD ERNEST, b West New York, NJ, June 19, 22; m 49; c 3. HORTICULTURE. Educ: Rutgers Univ, BS, 43, MS, 49; Univ Minn, PhD(hort), 55. Prof Exp: Instr hort, 49-55, from asst prof to assoc prof, 55-64, PROF HORT, UNIV MINN, ST PAUL, 64- Concurrent Pos: Fulbright study grant, Agr Inst, Ireland, 68-69; AID consult, Hassan II Inst Agron & Vet Med, Rabat, Morocco, 73. Mem: Am Soc Hort Sci; Int Hort Soc. Res: Physiological studies of commercial greenhouse crops; breeding of garden chrysanthemums; ornamental horticulture. Mailing Add: Dept of Hort Sci Univ of Minn St Paul MN 55101

WIDMOYER, FRED BIXLER, b Grandfield, Okla, Nov 25, 20; m 61; c 3. HORTICULTURE. Educ: Tex Tech Col, BA, 42, MS, 50; Mich State Univ, PhD(ornamental hort), 54. Prof Exp: Instr bot, Tex Tech Col, 46-50; asst prof & exten specialist hort, Mich State Univ, 54-60; assoc prof landscape design & nursery mgt, Univ Conn, 60-63; PROF HORT & HEAD DEPT, NMEX STATE UNIV, 63- Concurrent Pos: Mem adv comt, Nat Arboretum. Honors & Awards: Esther Longyear Murphy Medal, 61. Mem: Am Asn Bot Gardens & Arboretums (secy-treas, 61); Am Soc Hort Sci. Res: Ornamental horticulture; plant anatomy; growth regulators and plant propagation; developmental morphology. Mailing Add: Dept of Hort NMex State Univ Box 3530 Las Cruces NM 88003

WIDNELL, CHRISTOPHER COURTENAY, b London, Eng, May 19, 40; m 65; c 2. CELL BIOLOGY, BIOCHEMISTRY. Educ: Cambridge Univ, BA, 62; Univ London, PhD(biochem), 65. Prof Exp: Res assoc cell biol, Rockefeller Univ, 66-68; staff mem biochem, Nat Inst Med Res, London, Eng, 68-69; ASSOC PROF ANAT & CELL BIOL, SCH MED, UNIV PITTSBURGH, 69- Concurrent Pos: Jane Coffin Childs fel, Univ Chicago, 65-66; ed, Arch Biochem & Biophys, 72- Mem: Am Soc Biol Chem; Am Soc Cell Biol; Brit Biochem Soc. Res: Membrane structure and function; cytochemical localization of membrane proteins; synthesis and assembly of membrane components; cellular aging. Mailing Add: Dept of Anat & Cell Biol Univ of Pittsburgh Sch of Med Pittsburgh PA 15261

WIDNER, JIMMY NEWTON, b Clovis, NMex, Feb 10, 42; m 64; c 2. CROP BREEDING. Educ: NMex State Univ, BS, 64; NDak State Univ, PhD(agron), 68. Prof Exp: Plant breeder, Great Western Sugar Co, Colo, 68-72; res mgr, Northern Ohio Sugar Co, 72-75; SR PLANT BREEDER, GREAT WESTERN SUGAR CO, 75- Mem: Am Soc Agron; Crop Sci Soc Am; Soc Sugar Beet Technol. Res: Development of improved varieties and hybrids of sugar beets. Mailing Add: Agr Res Ctr Great Western Sugar Co Longmont CO 80501

WIDNER, WILLIAM RICHARD, b Baxter Co, Ark, Apr 24, 20; m 43; c 1. PHYSIOLOGY, BACTERIOLOGY. Educ: Eastern NMex Univ, AB, 42; Univ NMex, MS, 48, PhD, 52. Prof Exp: Lab asst biol, Eastern NMex Univ, 42; asst, Univ NMex, 46-48; biomed researcher, Los Alamos Sci Lab, 48-50; asst, Univ NMex, 50-52; indust hygienist, Sandia Corp, 52-55; teacher, Albuquerque Indian Sch, NMex, 55-56; prof biol & head dept, Howard Payne Col, 56-59; asst prof biol & bact, 59-64, PROF BIOL, BAYLOR UNIV, 64- Mem: AAAS; Am Soc Microbiol; Sigma Xi (treas, 73-74). Res: Cell mitoses and growth of normal and malignant tissues; effects of ionizing radiations on living cells; radiation produced cataracts; bacterial metabolism. Mailing Add: 111 Turtle Creek Dr Waco TX 76710

WIDOM, BENJAMIN, b Newark, NJ, Oct 13, 27; m 53; c 3. PHYSICAL CHEMISTRY. Educ: Columbia Univ, AB, 49; Cornell Univ, PhD(chem), 53. Prof Exp: Res assoc chem, Univ NC, 52-54; instr, 54-55, from asst prof to assoc prof, 55-63, PROF CHEM, CORNELL UNIV, 63- Concurrent Pos: Guggenheim & Fulbright fels, 61-62; NSF sr fel, 65; Guggenheim fel, 69; van der Waals prof, Univ Amsterdam, 72. Mem: Nat Acad Sci; Am Phys Soc; Am Chem Soc. Res: Chemical kinetics; phase transitions; statistical mechanics. Mailing Add: Dept of Chem Cornell Univ Ithaca NY 14853

WIDOM, HAROLD, b Newark, NJ, Sept 23, 32; m 55; c 3. MATHEMATICS. Educ: Univ Chicago, SM, 52, PhD(math), 55. Prof Exp: From instr to prof math, Cornell Univ, 55-68; PROF MATH, UNIV CALIF, SANTA CRUZ, 68- Concurrent Pos: Res fels, NSF, 59-60 & Sloan Found, 61-63; Guggenheim res fel, 67-68 & 72-73. Mem: Am Math Soc. Res: Integral equations; orthogonal polynomials. Mailing Add: Dept of Math Univ of Calif Santa Cruz CA 95060

WIDRA, ABE, b Philadelphia, Pa, Jan 17, 24; m 52; c 4. MICROBIOLOGY. Educ: Brooklyn Col, BA, 48; Univ Fla, MS, 52; Univ Pa, PhD(med microbiol), 54; Am Bd Med Microbiol, dipl, 69. Prof Exp: Tech asst bact & serol, Philadelphia Gen Hosp, 49-50; res assoc cytol & cytogenetics, Univ Pa, 54-55; instr bact & immunol, Univ NC, 55-59, asst prof, 59-64; ASSOC PROF MICROBIOL, MED CTR, UNIV ILL, CHICAGO, 64- Concurrent Pos: Consult, Presby-St Luke's Hosp, Chicago, 66- Mem: Am Soc Microbiol. Res: Medical mycology. Mailing Add: 307 Keystone Ave River Forest IL 60305

WIDSTROM, NEIL WAYNE, b Hecla, SDak, Nov 11, 33; m 60; c 2. GENETICS, PLANT BREEDING. Educ: SDak State Univ, BS, 59, PhD(plant breeding), 62. Prof Exp: Fel genetics, NC State Univ, 63-64; RES GENETICIST PLANTS, AGR RES SERV, USDA, 64- Mem: AAAS; Am Soc Agron; Crop Sci Soc Am; Genetics Soc Am; Am Genetic Asn. Res: Plant genetics; genetics of resistance to insects by plants. Mailing Add: Southern Grain Insects Res Lab Coastal Plain Exp Sta Tifton GA 31794

WIEBE, HAROLD T, b Chicago, Ill, Aug 22, 13; m 40; c 1. ZOOLOGY, BIOLOGY. Educ: Greenville Col, BA, 39; Univ Ill, MA, 40, PhD(zool), 50. Prof Exp: Teacher, Tabor Col & Acad, 40-41; prin, Baxter Sch, Colo, 41-42; assoc prof biol, Greenville Col, 42-46; from asst prof to prof zool, Taylor Univ, 49-52, actg dean, 51-52; assoc prof, 52-53, dir grad studies, 52-59, dean grad sch, 59-71, PROF ZOOL, SEATTLE PAC COL, 53- Mem: AAAS; Am Sci Affil; Am Inst Biol Sci. Res: Development of Daphnia pulex; studies on the parmacology and toxicology of the chlorinated hydrocarbon, chloradane; theory of organic evolution versus divine creation. Mailing Add: Dept of Zool Seattle Pac Col Seattle WA 98119

WIEBE, HERMAN HENRY, b Newton, Kans, May 30, 21; m 51; c 2. BOTANY, PLANT PHYSIOLOGY. Educ: Goshen Col, BA, 47; Univ Iowa, MS, 49; Duke Univ, PhD(bot), 53. Prof Exp: Instr bot, NC State Col, 49-50 & 52-53; res partic, Oak Ridge Inst Nuclear Studies, 53-54; assoc prof bot, 54-62, PROF BOT, UTAH STATE UNIV, 62- Concurrent Pos: NSF fel, Bot Inst, Stuttgart-Hohenheim, Ger, 64-65; Fulbright Hays vis prof, Trinity Col, Dublin, 73-74. Mem: AAAS; Am Soc Plant Physiol; Scand Soc Plant Physiol. Res: Plant water relations; air pollution inuury; mineral nutrition. Mailing Add: Dept of Bot Utah State Univ Logan UT 84322

WIEBE, JOHN, b Sask, June 3, 26; m 47; c 6. HORTICULTURE. Educ: Ont Agr Col, BSA, 51; Cornell Univ, MS, 53, PhD, 55. Prof Exp: RES SCIENTIST, HORT RES INST, ONT DEPT AGR, 55- Res: Viticulture and physiology. Mailing Add: Hort Res Inst Ont Dept Agr Vineland Station ON Can

WIEBE, JOHN PETER, b Neu-Schönsee, Ukraine, Aug 28, 38; Can citizen; m 64; c 2. PHYSIOLOGY, ENDOCRINOLOGY. Educ: Univ BC, BSc, 63, PhD(physiol), 67. Prof Exp: Nat Res Coun Can res fel zool, Univ Leeds, 68-69; asst prof endocrinol, Tex A&M Univ, 70-72; ASST PROF PHYSIOL, UNIV WESTERN ONT, 72-, ZOOL, 74- Mem: AAAS; Am Soc Zool; Can Soc Zool. Res: Mechanisms of hormone action; environmental endocrinology; reproductive physiology; photic effects on cellular components. Mailing Add: Dept of Zool Univ of Western Ont London ON Can

WIEBE, LEONARD IRVING, b Swift Current, Sask, Oct 14, 41; c 3. BIONUCLEONICS. Educ: Univ Sask, BSP, 63, MSc, 66; Univ Sydney, PhD(drug metab), 69. Prof Exp: Asst prof pharmaceut chem bionucleonics, 70-73, chmn bionucleonics div, 74-75, ASSOC PROF BIONUCLEONICS, UNIV ALTA, 73-, CHMN RES REACTOR COMT, 74- Concurrent Pos: Sessional lectr, Univ Sask, 65-66; fel, Univ Alta, 69-70; sessional lectr, Univ Sydney, 73. Res: Production of short-lived radionuclides for incorporation into radiopharmaceuticals for use in diagnostic nuclear medicine. Mailing Add: Div of Bionucleonics & Radiopharm Univ of Alta Edmonton AB Can

WIEBE, RICHARD PENNER, b Pittsburgh, Pa, Jan 5, 28; m 56; c 2. MATHEMATICAL LOGIC. Educ: Univ Ill, Urbana, BS, 49, MS, 51; Univ Calif, Berkeley, PhD(philos), 64. Prof Exp: Instr philos, Johns Hopkins Univ, 60-62; ASST PROF MATH, ST MARY'S COL, CALIF, 63- Mem: Asn Symbolic Logic. Res: Foundations of mathematics; semantics; philosophy of science and mathematics. Mailing Add: Dept of Math St Mary's Col of Calif Moraga CA 94575

WIEBE, ROBERT A, b San Meteo, Calif, Nov 2, 39; m 65. PETROLOGY. Educ: Stanford Univ, BS, 61; Univ Wash, Seattle, MS, 63; Stanford Univ, PhD, 66. Prof Exp: Asst prof geol, 66-73, ASSOC PROF GEOL, FRANKLIN & MARSHALL COL, 73- Concurrent Pos: NATO fel, Univ Edinburgh, 72-73. Mem: Geol Soc Am; Mineral Soc Am. Res: Igneous and metamorphic petrology; mineralogy; plutonic igneious rocks of the Northern Appalachians; the Nain Anorthosite-Adamellite complex. Mailing Add: Dept of Geol Franklin & Marshall Col Lancaster PA 17604

WIEBELHAUS, VIRGIL D, b Fairfax, SDak, Feb 19, 16; m 51; c 1. BIOCHEMISTRY, PHARMACOLOGY. Educ: SDak State Col, BS, 39; Purdue Univ, MS, 41; Univ Wis, PhD(biochem), 48. Prof Exp: Res assoc, Sharp & Dohme, 49-56; sr scientist, 56-61, group leader, 61-67, asst dir, 67-71, ASSOC DIR, SMITH KLINE & FRENCH LABS, 71- Mem: Am Soc Nephrology; Am Soc Pharmacol & Exp Therapeut; Biophys Soc; Brit Biochem Soc; Int Soc Biochem Pharmacol. Res: Oxidative phosphorylation; carbohydrate metabolism; thyroxine toxicity; renal transport mechanisms; diuretic pharmacology; catecholamine metabolism; anti-inflammatory physiology; gastric acid biochemistry. Mailing Add: Smith Kline & French Labs 1500 Spring Garden St Philadelphia PA 19101

WIEBERS, JOYCE ADAMS, b Lancaster, Ohio, Nov 16, 26; m 57; c 2. ANALYTICAL BIOCHEMISTRY. Educ: Col Mt St Joseph, BS, 47; Univ Cincinnati, MS, 50; Purdue Univ, PhD(bact), 58. Prof Exp: Asst bact, 54, res assoc, 58-65, asst prof org chem, 66-70, res biochemist, Dept Biol Sci, 70-74, ASSOC PROF BIOL CHEM, PURDUE UNIV, LAFAYETTE, 74- Mem: Am Chem Soc; Am Soc Biol Chem; Am Soc Mass Spectrometry. Res: Mass spectrometry of nucleic acids; synthesis of defined oligoribonucleotides; enzymology; DNA sequence by mass spectrometry. Mailing Add: PO Box 2477 West Lafayette IN 47906

WIEBUSCH, F B, b Brenham, Tex, Aug 26, 23. PERIODONTOLOGY. Educ: Univ Tex, BA, 43, DDS, 47; Am Bd Periodont, dipl. Prof Exp: Pub health dent consult, State Health Dept, Tex, 47-51; prof oral diag & therapeut & chmn dept, 54-71, PROF PERIODONT & ASST DEAN CONTINUING EDUC, MED COL VA, 71- Concurrent Pos: Consult, Vet Admin Hosps, Richmond & Salem, Va. Mem: Am Dent Asn; Am Acad Periodont; Am Col Dent; Am Acad Dent Med. Res: Periodontics. Mailing Add: 520 N 12th St Richmond VA 23298

WIEBUSH, JOSEPH ROY, b Lancaster, Pa, Oct 18, 20; m 43; c 1. ANALYTICAL CHEMISTRY, MARINE CHEMISTRY. Educ: Franklin & Marshall Col, BS, 41; Univ Md, MS, 51, PhD(chem), 55. Prof Exp: Supvr explosives dept, Hercules Powder Co, 41-43, chemist, 46-48; asst chem, Univ Md, 48-51; res chemist, Mead Corp, Ohio, 55-56; dir res, Nat Inst Drycleaning, 56-60; assoc prof chem, 60-66, PROF CHEM & CHMN DEPT, US NAVAL ACAD, 66- Mem: Am Chem Soc; Sigma Xi. Res: Fluorescence analysis; toxicity of organic solvents; analytical techniques for trace elements; marine corrosion and fouling; oceanographic applications; environmental pollution. Mailing Add: Dept of Chem US Naval Acad Annapolis MD 21402

WIECH, NORBERT LEONARD, b Chicago, Ill, Mar 13, 39; m 61; c 3. BIOCHEMISTRY, NUTRITION. Educ: Univ Notre Dame, BS, 60, MS, 63; Tulane Univ, PhD(biochem), 66. Prof Exp: Res assoc nutrit, Sch Pub Health, Harvard Univ, 66-67; sect head biochem pharm, Merrell Int, Strasbourg, France, 72-74, SECT HEAD DEPT METAB DIS, MERRELL NAT LABS DIV, RICHARDSON-MERRELL, INC, 67- Mem: AAAS; Am Oil Chemists Soc; Am Chem Soc; NY Acad Sci; Am Diabetes Asn. Res: Insulin-glucagon secretory regulation; membrane receptors; carbohydrate-lipid metabolism in health and disease. Mailing Add: Merrell Nat Labs Div Richardson-Merrell Inc Cincinnati OH 45215

WIECHELMAN, KAREN JANICE, b Central City, Nebr, Apr 30, 47; m 70. BIOPHYSICAL CHEMISTRY. Educ: Univ Nebr, BS, 69, PhD(biochem), 73. Prof Exp: Res assoc biophys, Univ Pittsburgh, 73-76; ASST PROF CHEM, UNIV SOUTHWESTERN LA, 76- Res: Use of biophysical techniques to investigate various biological systems. Mailing Add: Dept of Chem Univ of Southwestern La Lafayette LA 70504

WIECKERT, DAVID ALVIN, dairy science, animal behavior, see 12th edition

WIECZOROWSKI, ELSIE, b Chicago, Ill, July 4, 06. PEDIATRICS. Educ: Northwestern Univ, BS, 29, MS, 31, PhD(path), 37, MD, 45. Prof Exp: Asst, Med Sch, Northwestern Univ, 38-42; fel, Mayo Clin, 44-46; instr & Abt fel, 46-49, assoc, 49-59, ASST PROF PEDIAT, MED SCH, NORTHWESTERN UNIV, CHICAGO, 59- Mem: Fel AMA; fel Am Acad Pediat. Mailing Add: 932 Wolfram St Chicago IL 60657

WIED, GEORGE LUDWIG, b Carlsbad, Czech, Feb 7, 21; m 49. OBSTETRICS & GYNECOLOGY. Educ: Charles Univ, Prague, MD, 44. Prof Exp: Intern, County Hosp, Carlsbad, Czech, 45; resident obstet & gynec, Univ Munich, 46-48; asst, Univ Berlin, 48-52, co-chmn dept, 53; from asst prof to assoc prof, 54-65, actg chmn dept, 74-75, PROF OBSTET & GYNEC, SCH MED, UNIV CHICAGO, 65-, PROF PATH, 67-, DIR SCHS CYTOTECHNOL & CYTOCYBERNET, 59- Concurrent Pos: Ed-in-chief, Acta Cytologica, 57-; ed, Monogr Clin Cytol, 64 & J Reproductive Med, 67- Honors & Awards: Surgeon Gen Cert Merit, 52; Goldblatt Cytol Award, 61; George N Papanicolaou Cytol Award, 70. Mem: Am Soc Cytol (pres, 65-66); Am Soc Cell Biol; Int Acad Cytol (pres elect, 74-77); Ger Soc Obstet & Gynec; Ger Soc Cytol. Res: Cytopathology; exfoliative cytology; biological image processing. Mailing Add: Dept of Obstet & Gynec Univ of Chicago Sch of Med Chicago IL 60637

WIEDEMAN, MARY PURCELL, b Lexington, Ky, June 21, 19. PHYSIOLOGY. Educ: Univ Ky, MS, 43; Ind Univ, PhD(physiol), 53. Prof Exp: Assoc physiol, Woman's Med Col Pa, 53-56; from asst prof to assoc prof, 56-67, PROF PHYSIOL, SCH MED, TEMPLE UNIV, 67- Honors & Awards: Honor Achievement Award, Angiol Res Found, 65. Mem: Am Physiol Soc; Microcirc Soc; Am Asn Anat; NY Acad Sci. Res: Microcirculatory physiology. Mailing Add: Dept of Physiol Temple Univ Sch of Med Philadelphia PA 19140

WIEDEMAN, VARLEY EARL, b Oklahoma City, Okla, Mar 14, 33; m 63; c 2. BOTANY, ECOLOGY. Educ: Univ Okla, BS, 57, MS, 60; Univ Tex, PhD(bot), 64. Prof Exp: Chemist-biologist, USPHS, 59-61; from asst prof to assoc prof, 64-74, PROF PLANT ECOL, UNIV LOUISVILLE, 74- Mem: Bot Soc Am; Am Phycol Soc; Am Soc Limnol & Oceanog; Int Phycol Soc. Res: Ecology of algae with relation to water pollution, water treatment and sewage treatment. Mailing Add: Dept of Biol Univ of Louisville Louisville KY 40208

WIEDEMEIER, HERIBERT, b Steinheim, WGer, Aug 4, 28; nat US. INORGANIC CHEMISTRY. Educ: Univ u:nster, BS, 54, MSc, 57, DSc, 60. Prof Exp: Asst inorg & phys chem, Univ u:nster, 56-58, instr, 58-60; res assoc chem, Univ Kans, 60-62; res assoc, Univ u:nster, 62-63; res assoc, Univ Kans, 63-64; asst prof, 64-67, assoc prof phys chem, 67-72, PROF CHEM, RENSSELAER POLYTECH INST, 72- Mem: AAAS; Am Chem Soc; Ger Chem Soc. Res: Growth of single crystals of metal chalcogenides; crystal growth mechanism and morphology; thermodynamic and kinetic studies of condensation and vaporization processes of inorganic materials at elevated temperatures; crystal growth in zero-gravity. Mailing Add: Dept of Chem Rensselaer Polytech Inst Troy NY 12181

WIEDENBECK, MARCELLUS LEE, b Lancaster, NY, Oct 11, 19; m 46; c 6. NUCLEAR PHYSICS. Educ: Canisius Col, BS, 41; Univ Notre Dame, MS, 42, PhD(physics), 45. Prof Exp: Instr physics, Univ Notre Dame, 44-46; from asst prof to assoc prof, 46-55, PROF PHYSICS, UNIV MICH, ANN ARBOR, 55- Mem: Am Phys Soc. Res: Nuclear spectroscopy; beta ray and alpha ray spectra; coincidence methods; spectroscopy of some heavy nuclei. Mailing Add: Dept of Physics Univ of Mich Ann Arbor MI 48104

WIEDENHEFT, CHARLES JOHN, b Sandusky, Ohio, Oct 23, 41. CHEMISTRY. Educ: Capital Univ, BS, 63; Case Western Reserve Univ, MS, 65, PhD(chem), 67. Prof Exp: RES SPECIALIST, MONSANTO RES CORP, 67- Mem: AAAS; Am Chem Soc. Res: Coordination compounds of the actinide ions. Mailing Add: Monsanto Res Corp Mound Lab Miamisburg OH 45342

WIEDENMANN, LYNN G, b Moline, Ill, Apr 21, 28; m 56; c 4. POLYMER CHEMISTRY, ORGANIC CHEMISTRY. Educ: Ill Wesleyan Univ, BS, 50; Univ Iowa, MS, 52, PhD(org chem), 55. Prof Exp: Asst chemist, Rocky Mountain Arsenal, 55-57; res chemist, Tex-US Chem Corp, 57-60; res chemist, Rock Island Arsenal, 60-69; PROF CHEM & HEAD DEPT, BLACK HAWK COL, 69- Mem: AAAS; Am Chem Soc. Res: Polymer synthesis; high temperature resistant elastomers; boron and stereoregular butadiene polymers; butadiene derivatives; antioxidants; organic phosphorus compounds; dibenzopyrylium compounds. Mailing Add: Dept of Chem Black Hawk Col Moline IL 61265

WIEDER, GRACE MARILYN, b New York, NY, May 10, 28. PHYSICAL CHEMISTRY. Educ: Univ Vt, BA, 49; Mt Holyoke Col, AM, 51; Polytech Inst Brooklyn, PhD(phys chem), 61. Prof Exp: Res assoc chem, Univ Southern Calif, 60-62; instr, 62-65, ASST PROF CHEM, BROOKLYN COL, 66- Concurrent Pos: Vis scientist, Univ Wash, 70-71. Mem: Am Chem Soc; Am Phys Soc. Res: Infrared and Raman spectroscopy. Mailing Add: Dept of Chem Brooklyn Col Brooklyn NY 11210

WIEDER, HAROLD, b Cleveland, Ohio, July 18, 27; m 63; c 3. OPTICAL PHYSICS. Educ: Univ Rochester, BS, 50, MA, 57, MS, 58; Case Inst Technol, PhD(physics), 64. Prof Exp: Engr, Sarnoff Res Ctr, RCA Labs, 50-54; physicist, Parma Res Ctr, Union Carbide Corp, 57-61; physicist, Watson Res Ctr, 63-68, PHYSICIST, SAN JOSE RES LAB, IBM CORP, 68- Mem: Am Phys Soc. Res: Optical, magneto-optic, photoconductive, and pyroelectric properties of ordered and disordered films; transient thermal and thermomagnetic techniques; mode coupling and intra-cavity laser effects; level crossing and anticrossing spectroscopy. Mailing Add: 20175 Knollwood Dr Saratoga CA 95070

WIEDER, IRWIN, b Cleveland, Ohio, Sept 26, 25; m 53; c 4. PHYSICS, BIOPHYSICS. Educ: Case Inst Technol, BS, 50; Stanford Univ, PhD(physics), 56. Prof Exp: Asst, Stanford Univ, 51-56; res physicist, Res Lab, Westinghouse Elec Corp, 56-60 & Varian Assocs, 60-61; dir res, Interphase Corp, 61-66; prin scientist, Carver Corp, 66-69; NIH spec fel, Dept Biol Sci, Stanford Univ, 70-71; VIS PROF, DEPT ELECTRONICS, WEIZMANN INST SCI, 71- Mem: Am Phys Soc; Inst Elec & Electronics Engrs. Res: Magnetic resonance; microwave-optical effects; optical pumping in gases, liquids and solids; masers and lasers; energy transfer in biological systems; cellular communication; cell membranes and surfaces; fluorescent antibody spectroscopy. Mailing Add: Dept of Electronics Weizmann Inst of Sci PO Box 26 Rehovot Israel

WIEDER, SOL, b Bronx, NY, Jan 6, 40; m 63; c 3. PHYSICS. Educ: City Col New York, BS, 61; NY Univ, MS, 62, PhD(physics), 66. Prof Exp: Lectr physics, City Col New York, 62-64; instr, NY Univ, 64-65; instr, Bronx Community Col, 65-66; mem tech staff, Bell Tel Labs, Inc, 66-67; asst prof, NY Univ, 67; assoc prof, 67-75, PROF PHYSICS, FAIRLEIGH DICKINSON UNIV, 75- Mem: Am Phys Soc; Am Geophys Union; Am Asn Physics Teachers. Res: Many-particle physics; geophysics; solar energy. Mailing Add: Dept of Physics Fairleigh Dickinson Univ Teaneck NJ 07666

WIEDERHIELM, CURT ARNE, b Motala, Sweden, Dec 11, 23; US citizen; div; c 2. CARDIOVASCULAR PHYSIOLOGY. Educ: Univ Wash, PhD(physiol), 61. Prof Exp: Res asst, 48-50, chief lab asst surg, 51-53, chief lab asst cardiol, 53-57, asst dir cardiovasc training prog, 56-66, from instr to assoc prof, 61-70, PROF PHYSIOL & BIOPHYS, SCH MED, UNIV WASH, 70- Concurrent Pos: USPHS career res develop award, 64-74. Mem: Microcirc Soc; Am Physiol Soc; Int Soc Biorheol; Int Soc Hemorheol; Simulation Coun. Res: Physiology and biophysics of the microcirculation; viscoelastic wall properties of microscopic blood vessels; transcapillary and interstitial transport phenomena; peripheral control of blood flow and exchange processes; blood rheology. Mailing Add: Dept of Physiol & Biophys SJ-40 Univ of Wash Seattle WA 98195

WIEDERHOLD, PIETER RIJK, b Malang, Indonesia, Jan 24, 28; US citizen; m 56; c 3. PHYSICS, ELECTRICAL ENGINEERING. Educ: Delft Univ Technol, Ir, 53. Prof Exp: Sr proj engr, Sylvania Elec Prod, Inc, 53-61; mgr energy conversion, Ion Physics Corp, Inc, 61-64; mgr cryogenics, Magnion, Inc, 64-66; div mgr space physics, Comstock & Wescott, Inc, 66-74; PRES, GEN EASTERN CORP, 74- Mem: Am Phys Soc; Inst Elec & Electronics Engrs; Am Inst Aeronaut & Astronaut; Am Meteorol Soc; Instrument Soc Am. Res: Superconductivity; magnetics; energy conversion; humidity instruments. Mailing Add: Gen Eastern Corp 36 Maple St Watertown MA 02172

WIEDERHOLT, WIGBERT C, b Warmbrunn, Ger, Apr 22, 31; US citizen; m 60; c 3. NEUROLOGY, NEUROPHYSIOLOGY. Educ: Univ Freiburg, MD, 56. Prof Exp: Asst to staff neurol, Mayo Clin, 65; from asst prof to assoc prof med, Ohio State Univ, 66-72, chief clin neurophysiol, 69-72; PROF NEUROSCI, UNIV CALIF, SAN DIEGO, 72-, NEUROLOGIST-IN-CHIEF, DEPT NEUROSCI, 73- Honors & Awards: S Weir Mitchel Award, Am Acad Neurol, 65. Mem: AAAS; fel Am Acad Neurol; Am Neurol Asn; Am Electroencephalog Soc; Am Asn Electromyog & Electrodiag (secy-treas, 71-). Res: Effect of anoxia on neurophysiologic parameters of brain function. Mailing Add: Univ Hosp 225 W Dickinson St San Diego CA 92103

WIEDMAN, HAROLD W, b Palermo, Calif, Jan 11, 30. PHYTOPATHOLOGY. Educ: Chico State Col, AB, 52; Ore State Col, PhD(bot), 56. Prof Exp: Asst, Ore State Col, 53-56; asst plant pathologist, NMex State Univ, 56-58 & State Dept Agr, Calif, 58-59; asst prof bot, Humboldt State Col, 59-61; from asst prof to assoc prof, 61-69, PROF BIOL SCI, CALIF STATE UNIV, SACRAMENTO, 69- Mem: Am Phytopath Soc; Bot Soc Am. Res: Soil-borne diseases; biological control; diseases of vegetables and cotton. Mailing Add: Dept of Biol Sci Calif State Univ Sacramento CA 95819

WIEDMANN, JEROME LEE, analytical chemistry, see 12th edition

WIEDMEIER, VERNON THOMAS, b Harvey, NDak, Jan 10, 35; m 57; c 4. PHYSIOLOGY. Educ: NDak State Teachers Col, Valley City, BS, 59; NDak State Univ, MS, 61; Marquette Univ, PhD(physiol), 68. Prof Exp: Instr biol, NPark Col, 60-61; asst prof, St Ambrose Col, 61-64; asst prof, 71-75, ASSOC PROF PHYSIOL, MED COL GA, 75- Concurrent Pos: NIH trainee, Univ Va, 69-70 & fel, 70-71.

Mem: Am Physiol Soc. Res: Myocardial metabolism and the regulation of coronary blood flow. Mailing Add: Dept of Physiol Med Col of Ga Augusta GA 30902

WIEGAND, CLYDE EDWARD, b Long Beach, Wash, May 23, 15. EXPERIMENTAL PHYSICS. Educ: Willamette Univ, BA, 40; Univ Calif, Berkeley, PhD(physics), 50. Prof Exp: STAFF SCIENTIST PHYSICS, LAWRENCE RADIATION LAB, UNIV CALIF, BERKELEY, 41- Mem: Am Phys Soc. Res: Study of kaonic atoms. Mailing Add: Lawrence Berkeley Lab Berkeley CA 94720

WIEGAND, CRAIG LOREN, b Santa Rosa, Tex, Jan 11, 33; m 62; c 2. SOIL PHYSICS, REMOTE SENSING. Educ: Agr & Mech Col, Tex, BS, 55, MS, 56; Utah State Univ, PhD(soil physics), 60. Prof Exp: Res soil scientist, 60-73, TECH ADV, SOIL & WATER CONSERV RES CTR, AGR RES SERV, USDA, 73- Concurrent Pos: Mid-career fel, Woodrow Wilson Sch Pub & Int Affairs, 74-75. Honors & Awards: Superior Performance Award, AAAS, 70. Mem: Fel AAAS; Soil Sci Soc Am; Soil Conserv Soc Am; Am Soc Agron; Int Soil Sci Soc. Res: Meteorology; hydrology; plant physiology; physical chemistry; micrometeorology; plant-water relations; principal investigator on NASA contracts dealing with earth observation satellite data applications to agriculture. Mailing Add: Soil & Water Conserv Res Ctr Agr Res Serv Box 267 Weslaco TX 78596

WIEGAND, DONALD ARTHUR, b Rochester, NY, July 21, 27; m 59; c 2. SOLID STATE PHYSICS. Educ: Cornell Univ, BEE, 52, MEE, 53, PhD(eng physics), 56. Prof Exp: Asst & assoc physics, Cornell Univ, 55-56; res physicist, Carnegie-Mellon Univ, 56-59, from asst prof to assoc prof physics, 59-68; RES PHYSICIST, FELDMAN RES LAB, PICATINNY ARSENAL, 68- Concurrent Pos: Fulbright grant, Darmstadt Tech Univ, WGer, 60-61. Mem: Am Phys Soc. Res: Imperfections in solids; luminescence; photo-conductive processes; optical absorption; x-ray diffraction; x-ray photoelectron spectroscopy; metastable solids. Mailing Add: Feldman Res Lab Picatinny Arsenal Dover NJ 07801

WIEGAND, GAYL, b Estherville, Iowa, July 18, 39; m 63. ORGANIC CHEMISTRY. Educ: Univ Iowa, BS, 61; Univ Mass, PhD(org chem), 65. Prof Exp: Asst prof, 65-72, ASSOC PROF CHEM, IDAHO STATE UNIV, 72- Mem: Am Chem Soc. Res: Mechanisms of organic like reactions occurring at elements other than carbon; organosulfur chemistry. Mailing Add: Dept of Chem Idaho State Univ Pocatello ID 83201

WIEGAND, OSCAR FERNANDO, b Mex, Nov 3, 21; US citizen; m 49; c 5. PLANT PHYSIOLOGY. Educ: Univ Tex, BA, 50, MA, 52, PhD(cell physiol, chem), 56. Hon Degrees: Dr, Univ Guadalajara, 65. Prof Exp: Asst prof biol, ETex State Col, 56-57; asst prof pharmacol, Univ Tex Southwest Med Sch, 57-60, vis lectr, Univ Tex, Austin, 60-62, ASST PROF ZOOL, UNIV TEX, AUSTIN, 62- Concurrent Pos: Smith-Mundt fel & vis prof, Univ Guadalajara, 61-62, Distinguished prof, 62, gen coord model univ develop prog, 63-67; mem study group for reform & improv educ, Latin Am Univ, 65-; head consult, Univ Reform Model, Cath Univ Rio de Janeiro, 66-; consult univ reform prog, Agency Int Develop-Govt Brazil, 66- Mem: Am Soc Pharmacol & Exp Therapeut; Soc Gen Physiol. Res: Photomorphogenesis, growth physiology and tropisms; water and electrolyte equilibria in tissues; histamine reactions; tracer technique; respiromtery; growth methods for plant tissues. Mailing Add: Dept of Zool Univ of Tex Austin TX 78712

WIEGAND, RONALD GAY, b Chicago, Ill, Dec 28, 29; m 56; c 4. PHARMACOLOGY, BIOCHEMISTRY. Educ: Mass Inst Technol, BS, 52, MS, 53; Emory Univ, PhD(pharmacol), 56; Univ Chicago, MBA, 71. Prof Exp: Assoc histochemist, Armed Forces Inst Path, Walter Reed Army Med Ctr, 57-58; asst pathologist, Med Res & Nutrit Labs, Fitzsimons Army Hosp, Denver, 58; res pharmacologist, 58-61, group leader, 61-63, head chem pharmacol sect, 63-70, mgr chem pharmacol, 70-73, dir prod planning, 73-75, DIR RES & DEVELOP, AGR & VET PROD DIV, ABBOTT LABS, 75- Mem: Am Chem Soc; Am Soc Pharmacol. Res: Drug absorption, distribution, excretion and metabolism; pharmacokinetics; biopharmaceutics; mechanism of drug action. Mailing Add: Abbott Labs Agr & Vet Prod Div North Chicago IL 60064

WIEGAND, ROY VERNON, b Dodson, Mont, May 13, 14; m 42; c 4. ATOMIC SPECTROSCOPY. Educ: Mont State Univ, BS, 37, MS, 40; Pa State Univ, PhD(physics), 46. Prof Exp: Asst physics, Mont State Univ, 37-40; asst, Pa State Univ, 40-43, asst petrol res, 43-46, instr, 46-47; from asst prof to assoc prof physics, 47-57, PROF PHYSICS, MONT STATE UNIV, 57- Mem: Optical Soc Am; Am Asn Physics Teachers. Res: Raman spectroscopy; photoelectric Raman spectrography for analysis of hydrocarbon mixtures. Mailing Add: Dept of Physics Mont State Univ Bozeman MT 59715

WIEGAND, SYLVIA MARGARET, b Cape Town, SAfrica, Mar 8, 45; US citizen; m 66; c 1. ALGEBRA. Educ: Bryn Mawr Col, AB, 66; Univ Wash, MA, 67; Univ Wis, PhD(algebra), 72. Prof Exp: Comput programmer, Univ Wis, 65 & Bryn Mawr Col, 65-66; teaching asst math, Univ Wis, 67-72; from instr to asst prof, 72-76, ASSOC PROF MATH, UNIV NEBR, LINCOLN, 76- Mem: Am Math Soc; Math Asn Am; Asn Women Math. Res: Commutative algebra. Mailing Add: Dept of Math Univ of Nebr Lincoln NE 68588

WIEGEL, WILLIAM EDWARD, b Shullsburg, Wis, July 3, 33; m 61; c 2. GEOLOGY, COMPUTER SCIENCE. Educ: Univ Ark, BS, 55, MS, 59. Prof Exp: From asst res geologist to assoc res geologist, Cities Serv Res & Develop Co, 58-66; res geologist, 66-69, SYSTS ANALYST, CITIES SERV OIL CO, 69- Mem: Am Asn Petrol Geol. Res: Lithofacies mapping; stratigraphy; computer applications in geology and petroleum exploration; statistical studies in geology; heavy minerals of Atoka formation in Arkansas. Mailing Add: Cities Serv Oil Co Box 300 Tulsa OK 74102

WIEGERT, PHILIP E, b Antigo, Wis, Apr 7, 27; m 59; c 6. ORGANIC CHEMISTRY, PHYSICAL CHEMISTRY. Educ: Univ Wis, BS, 50; Univ Ill, MS, 51, PhD, 54. Prof Exp: Chemist, Mallinckrodt Chem Works, 54-61, group leader, 61-66, asst dir pharmaceut chem, 66-74, PLANT MGR, NAT CATHETER CORP, DIV MALLINCKRODT CHEM WORKS, 74- Mem: Am Chem Soc; The Chem Soc. Res: X-ray contrast media; opium alkaloids; pharmaceutical chemicals; medical devices. Mailing Add: Hook Rd Nat Catheter Corp Argyle NY 12801

WIEGERT, RICHARD G, b Toledo, Ohio, Sept 9, 32; m 55; c 2. ECOLOGY. Educ: Adrian Col, BS, 54; Mich State Univ, MS, 58; Univ Mich, PhD(zool), 62. Prof Exp: Instr zool, Univ Mich, 61-62; from asst prof to assoc prof, 62-71, PROF ZOOL, UNIV GA, 71- Prof Exp: NSF grants, Yellowstone Nat Park, 68-76. Mem: AAAS; Am Soc Mammal; Ecol Soc Am; Brit Ecol Soc; Am Soc Naturalists. Res: Plant and animal ecolog, particularly problems of population and community energy utilization; population density regulation; interspecies competition; systems ecology and the dynamics of thermal spring and estuarine communities. Mailing Add: Dept of Zool Univ of Ga Athens GA 30602

WIEGMANN, NORMAN ARTHUR, b Los Angeles, Calif, Apr 13, 20. ALGEBRA. Educ: Univ Southern Calif, AB, 41; Univ Wis, MA, 43, PhD(math), 47. Prof Exp: Asst math, Univ Wis, 41-47; instr, Univ Mich, 47-51; mem staff, Nat Bur Stand, 51-53; from assoc prof to prof, Cath Univ Am, 53-60; prof, George Washington Univ, 60-66; PROF MATH, CALIF STATE COL DOMINGUEZ HILLS, 66- Mem: Am Math Soc. Res: Abstract algebra; matrix theory. Mailing Add: Dept of Math Calif State Col 1000 E Victoria St Dominguez Hills CA 90747

WIELAND, BRUCE WENDELL, b Carroll, Iowa, Apr 15, 37; m 66. BIOMEDICAL ENGINEERING. Educ: Iowa State Univ, BS, 60; Ohio State Univ, PhD(nuclear eng), 73. Prof Exp: Engr, Oak Ridge Nat Lab, 60-66; res engr, Battelle Mem Inst, 67-68; NIH spec res fel, Ohio State Univ, 69-73; SCIENTIST MED RADIOISOTOPES, OAK RIDGE ASSOC UNIVS, 74- Concurrent Pos: Res assoc, Sch Med, Washington Univ, 71-73; consult, Oak Ridge Assoc Univs, 72-73. Mem: Soc Nuclear Med. Res: Applications of accelerator-produced radioisotopes in nuclear medicine. Mailing Add: Med & Health Sci Div Oak Ridge Assoc Univs PO Box 117 Oak Ridge TN 37830

WIEN, RICHARD W, JR, b Bay County, Fla, May 17, 45; m 68; c 2. PHYSICAL CHEMISTRY. Educ: Purdue Univ, BS, 67; Stanford Univ, PhD(phys chem), 71. Prof Exp: NSF fel phys chem, Stanford Univ, 67-71; SR RES CHEMIST PHOTOG SCI, EASTMAN KODAK RES LAB, 71- Mem: Am Chem Soc; AAAS. Res: Improvement of photographic speed of color reversal films; use of electron spectroscopy to study silver halide systems; development of new color photographic systems. Mailing Add: Eastman Kodak Co Bldg 59 Kodak Park Rochester NY 14650

WIENER, ALEXANDER S, b Brooklyn, NY, Mar 16, 07; m 32; c 2. IMMUNOHEMATOLOGY, MEDICINE. Educ: Cornell Univ, AB, 26; State Univ NY, MD, 30; Am Bd Path, dipl, 37; FRCP. Hon Degrees: Dr, Univ Toulouse, 69. Prof Exp: Head div genetics & biomet, Jewish Hosp, Brooklyn, 33-35; DIR, WIENER LABS, 35- Concurrent Pos: NIH res grants, Lab Exp Med & Surg in Primates, Sch Med, NY Univ, 61-; pvt pract, 32-; head blood transfusion div, Jewish Hosp, 32-52, attend immunohematologist, 52-; mem comt human heredity, Nat Res Coun, 38-41; mem fac, Postgrad Med Sch, NY Univ, 38-40, lectr, 40-49, asst prof forensic med, 49-55, adj assoc prof, 56-66, assoc prof, 66-68, prof, 68-; bacteriologist, Off Chief Med Exam, New York, 38-49, head serol div, 48-; head blood transfusion div, Adelphi Hosp, 49-52, attend immunohematologist, 53-, pres med staff, 63- Honors & Awards: Silver Medal, Am Soc Clin Path, 41, Burdick Medal, 46; Alvarenge Prize, Col Physicians Philadelphia, 45; Lasker Award, Am Pub Health Asn, 46, Passano Award, 51; Mod Med Award, 53; Landsteiner Award, 56; Carlos J Finlay Medal, 58; Joseph P Kennedy, Jr Int Award, 66; Allan Award, Am Soc Human Genetics, 75. Mem: Fel AAAS; Int Soc Blood Transfusion; Ger Soc Forensic & Social Med; Soc Study Blood (pres, 45); fel Royal Col Physicians. Res: Human blood groups, including discovery of Rh factor with K Landsteiner; discovery of I-i blood types; Rh-Hr blood types, heredity, serology, nomenclature; applications of blood grouping in clinical medicine, forensic medicine and anthropology; blood groups of apes and monkeys, including discovery of V-A-B and C-E-F blood group systems of chimpanzees; forensic serology. Mailing Add: 64 Rutland Rd Brooklyn NY 11225

WIENER, BENJAMIN, b New York, NY, Oct 5, 20; m 42; c 2. BIOCHEMISTRY. Educ: Queens Col, NY, BS, 41. Prof Exp: Tech dir, Triumph Explosives Co, Md, 42-43; chem engr, Nat Bur Stand, 43-44; sr res scientist, Columbia Univ, 44-45; sr res chemist, Tenn Eastman Corp, 45-47; instr zool, Univ Tenn, 47; tech adv & copywriter, Lamex Corp, NY, 47-48; tech rep, Wyeth, Inc, Pa, 48-50; pres, Guardian Pharmacal, Inc, 50-56; PRES & CHMN BD, ZENITH LABS, INC, 56- Concurrent Pos: Mem fac, New Sch Social Res, 50-54 & Univ Berne, 54-55; psychotherapist, Theodore Reik Clin & Inst Psychother, 56-58; pres, Aberdeen Pharmacal Corp; dir, Paramount Surg Supply Co, Inc, Continental Vitamin Corp, Pralex Corp, Island Traders Corp, Pace Bond Drug Co & Aetna Pharmacal Corp; mem, Nat UN Comt. Mem: AAAS; Am Chem Soc; NY Acad Sci; Am Inst Mgt; Nat Psychol Asn for Psychoanal. Res: Genetics of Drosophila; analytical and uranium chemistry; instrumentation; physical methods; polarography; spectrophotometry; high vacuum technique; rheology; colloid and ball milling; pharmaceutical chemistry and manufacture; chemistry of psychologic processes; psychoanalysis. Mailing Add: Zenith Labs Inc 140 La Grand Ave Northvale NJ 07647

WIENER, CHARLES, biochemistry, see 12th edition

WIENER, HOWARD LAWRENCE, b Portland, Ore, Mar 16, 37; m 62; c 1. STATISTICS, OPERATIONS RESEARCH. Educ: Univ Ore, BS, 59; Northwestern Univ, Evanston, MS, 61; Cath Univ Am, PhD(math), 71. Prof Exp: Sci analyst opers eval group, Ctr Naval Anal, Va, 61-63, opers res analyst-mathematician, Naval Ord Lab, MD, 63-71, opers res analyst, 71-75, HEAD OPERS ANAL & PLANNING SECT, SYSTS RES BR, NAVAL RES LAB, WASHINGTON, DC, 75- Mem: Am Statist Asn; Opers Res Soc Am. Res: Applications of statistics and probability to operational problems; data analysis; time series analysis. Mailing Add: 602 West View Terr Alexandria VA 22301

WIENER, JOSEPH, b Toronto, Ont, Sept 21, 27; m 54; c 2. PATHOLOGY, CELL BIOLOGY. Educ: Univ Toronto, MD, 53. Prof Exp: Assoc path, Col Physicians & Surgeons, Columbia Univ, 60-63, asst prof, 63-68; PROF PATH & ATTEND PATHOLOGIST, NY MED COL, 68- Mem: Am Soc Cell Biol; Am Soc Exp Path; Am Asn Path & Bact. Res: Experimental pathology. Mailing Add: Dept of Path NY Med Col Basic Sci Bldg Valhalla NY 10595

WIENER, LESLIE, b Brooklyn, NY, July 8, 34; m 57; c 3. CARDIOLOGY. Educ: Univ Ill, BS, 55; Chicago Med Sch, MD, 59. Prof Exp: Intern, DC Gen Hosp, 59-60, asst resident med, 60-61, chief resident med, 62-63; head cardiopulmonary lab, US Naval Hosp, 63-66, head cardiopulmonary br, Dept Med, 66-68; asst prof med & dir clin cardiovasc res, Hahnemann Med Col & Hosp, 68-69; assoc prof, 71-73, PROF MED, JEFFERSON MED COL, 73-, DIR CARDIAC CARE UNIT, THOMAS JEFFERSON UNIV & HOSP, 69- Concurrent Pos: Fel med, DC Gen Hosp, 61-62; Wash Heart Asn fel cardiol, 61-62. Mem: AMA; Am Col Physicians; Am Heart Asn; Am Col Cardiol; NY Acad Sci. Res: Pulmonary physiology. Mailing Add: Cardiac Care Unit Thomas Jefferson Univ Hosp Philadelphia PA 19107

WIENER, ROBERT NEWMAN, b New York, NY, Aug 27, 30; m 54; c 3. CHEMISTRY. Educ: Harvard Univ, AB, 51; Univ Pa, MS, 53, PhD, 56. Prof Exp: Asst instr chem, Univ Pa, 51-54; instr, Rutgers Univ, 55-58; asst prof, 58-62, ASSOC PROF CHEM, NORTHEASTERN UNIV, 62- Mem: Am Phys Soc. Res: Physical chemistry; molecular spectroscopy. Mailing Add: Dept of Chem Northeastern Univ Boston MA 02115

WIENER, SIDNEY, b New York, NY, Nov 17, 22; m 44; c 2. MATERIALS SCIENCE. Educ: Univ Calif, Los Angeles, BS, 47; Univ Calif, Berkeley, PhD(biochem), 52. Prof Exp: Exploitation engr, Prod Lab, Shell Oil Co, 52-56; mem tech staff, Airborne Systs Labs, 57-60, group head org mat, Res & Develop Div, 60-62, sect head, 62-65, asst dept mgr mat tech, 65-70, mgr space & commun group support activity, 70-73, MGR PROJ CONTROL, COMPONENTS & MAT LABS, HUGHES AIRCRAFT CO, 73- Mem: AAAS; Am Chem Soc; Sigma Xi. Res:

Physical chemistry and elucidation of structure of nucleic acid using enzymatic reactions and acid-base relations; technical administration in materials. Mailing Add: 5609 Edgemere Dr Torrance CA 90503

WIENER, STANLEY L, b New York, NY, Nov 5, 30; m 53; c 3. INTERNAL MEDICINE, EXPERIMENTAL PATHOLOGY. Educ: Univ Rochester, AB, 52; Sch Med, Univ Rochester, MD, 56. Prof Exp: From asst prof to assoc prof med, State Univ NY Stony Brook, 71-72; assoc dir res & educ, 72-73, ASSOC DIR MED, LONG ISLAND JEWISH-HILLSIDE MED CTR, 73- Concurrent Pos: Chmn res comt, Am Heart Asn, NY State Affil, 7375. Mem: Am Soc Exp Path; Am Soc Exp Biol & Med; Am Fedn Clin Res; Am Rheumatism Asn. Res: In vivo studies of fibroblast activation and studies of neutrophil chemotaxis and enzyme release into inflammatory liquid. Mailing Add: Long Island Jewish-Hillside Med Ctr Dept of Med New Hyde Park NY 11040

WIENKER, CURTIS WAKEFIELD, b Seattle, Wash, Feb 3, 45; m 66; c 1. PHYSICAL ANTHROPOLOGY. Educ: Univ Wash, BA, 67; Univ Ariz, MA, 70, PhD(anthrop), 75. Prof Exp: Instr anthrop, Shoreline Community Col, summer, 69; lectr, 72-75, ASST PROF ANTHROP, UNIV S FLA, 75- Mem: Am Asn Phys Anthrop Sigma Xi; Human Biol Coun. Res: Human population genetics; human evolution; Black population biology; cultural influences on human biology. Mailing Add: Dept of Anthrop Univ of SFla Tampa FL 33620

WIENS, DELBERT, b Munich, NDak, July 9, 32; m 55; c 3. SYSTEMATIC BIOLOGY, EVOLUTIONARY BIOLOGY. Educ: Pomona Col, BA, 5 55; Univ Utah, MS, 57; Claremont Grad Sch, PhD(bot), 61. Prof Exp: Instr biol, Univ Colo, 60-62, asst prof, 62-64; asst prof bot, 64-66, assoc prof biol, 67-74, PROF BIOL, UNIV UTAH, 74- Concurrent Pos: Fulbright lectr & hon prof, Univ Guayaquil, 64-65; mem, Flora of Ceylon Proj, 68; vis lectr, Flinders Univ SAustralia, 72. Mem: AAAS; Am Soc Plant Taxon; Bot Soc Am; Int Asn Plant Taxon; Soc Study Evolution. Res: Systematics, biogeography, chromosome systems and pollination ecology of flowering plants, particularly the mistletoe family. Mailing Add: Dept of Biol Univ of Utah Salt Lake City UT 84112

WIENS, JACOB HENRY, b Bradshaw, Nebr, Dec 3, 10; m 34, 70; c 2. PHYSICS, ELECTRONICS. Educ: Univ Calif, AB, 37, MS, 43, PhD(physics), 44. Prof Exp: In exp radio, US Park Serv, 32; head, US Army Signal Corps Sch, Calif, 41-42, Manhattan Proj, 42-44 & US Navy Res Prog, 44-45; dir, 63-74, EMER DIR, COL OF THE AIR, KCSM-FM, KCSM-TV, COL SAN MATEO, 74-, DIR TECH EDUC, 66-, DIR SYSTS EDUC, 66-; OWNER, WIENS ELECTRONICS LABS, 45- Concurrent Pos: In-chg radio & electronics, San Mateo Jr Col, 39-55; consult elec engr, 48-; chmn physics dept, Col San Mateo, 55-59, coordr technician prog develop, 59-65; chief engr, KCSM-FM Educ Broadcasting Sta, 52-64; mem. State Steering Comt Voc Educ Comt Electronics Educ, 60-63; dir, Bay Area Coun Electronics Educ, 60-64 & Eng Technician Surv, 60-62; secy, NCoast Indust Educ Admin Coun, 61-62; mem, Calif Pub Schs Instr TV Comt; Bay Region instr, TV for Educ; mem comt educ TV, Calif Jr Col Asn. Mem: Nat Asn Educ Asn Broadcasters. Res: Production of the isotope of Mercury of Mass 198 for a primary standard of wavelength; mass production of man-made element by use of neutrons from the under charge cyclotron; nuclear physics; spectroscopy; television; color television; technician education; college administration. Mailing Add: Col of the Air Col of San Mateo 1700 W Hillsdale Blvd San Mateo CA 94402

WIENS, JOHN ANTHONY, b Moscow, Idaho, Sept 29, 39; m 61; c 2. ANIMAL ECOLOGY, ANIMAL BEHAVIOR. Educ: Univ Okla, BS, 61; Univ Wis, MS, 63, PhD(zool), 66. Prof Exp: From asst prof to assoc prof zool, 66-75, PROF ZOOL, ORE STATE UNIV, 75- Concurrent Pos: NSF res grant, 67-69 & 74-; Am Philos Soc res grant, 72-75; vis prof, Colo State Univ, 73-74. Mem: Am Soc Naturalists; fel Am Ornith Union (treas, 74-); Ecol Soc Am; Cooper Ornith Soc; Wilson Ornith Soc. Res: Avian community structure and function; population modeling and analysis; vertebrate behavioral ecology; methods of habitat description and analysis. Mailing Add: Dept of Zool Ore State Univ Corvallis OR 97331

WIER, CHARLES EUGENE, bJasonville, Ind, May 15, 21; m 49; c 3. ECONOMIC GEOLOGY. Educ: Ind Univ, AB, 43, AM, 50, PhD(econ geol), 55. Prof Exp: Geologist & head coal sect, Ind Geol Surv, 49-75; assoc prof geol, Ind Univ, Bloomington, 65-75; MGR COAL EXPLOR, AMAX INT GROUP, INC, 75- Concurrent Pos: Consult, AMAX Int Coal, Inc, 74-75. Mem: Geol Soc Am; Soc Econ Geol; Am Asn Petrol Geol; Am Inst Mining, Metall & Petrol Engrs. Res: Pennsylvanian stratigraphy; coal resources and coal petrology; environmental geology. Mailing Add: AMAX Int Group AMAX Ctr Greenwich CT 06830

WIER, DONALD RAYMOND, soil science, physical chemistry, see 12th edition

WIER, JACK KNIGHT, b Cairo, Nebr, Aug 31, 23; m 47. PHARMACOGNOSY. Educ: Univ Wis, PhB, 45; Univ Nebr, BS, 56; Univ Wash, Seattle, MS, 59, PhD(pharmacog), 61. Prof Exp: Asst prof pharmacog, 61-67, ASSOC PROF PHARMACOG, UNIV NC, CHAPEL HILL, 67- Concurrent Pos: Consult, F W Dodge Co Div, McGraw-Hill, Inc, 65-69. Mem: Am Soc Pharmacog (secy, 70-); Am Pharmaceut Asn; Acad Pharmaceut Sci. Res: Metabolic products of macrofungi; biosynthesis of indole alkaloids in higher plants. Mailing Add: Beard Hall Univ of NC Chapel Hill NC 27514

WIER, JAMES ARISTA, b Newberry, Ind, Aug 27, 16; m 43; c 2. MEDICINE. Educ: Univ Louisville, MD, 38; Am Bd Internal Med, dipl, 53. Prof Exp: Med Corps, US Army, 41-, asst chief med consult div, Off Surgeon Gen, 51-53, chief pulmonary dis serv, Fitzsimons Army Hosp, 54-60 & chief prof serv, 62-64, chief prof serv, Letterman Gen Hosp, 64-65, surgeon, Vietnam, 66-67, cmndg officer, William Beaumont Gen Hosp, 67-68, mem staff, Off Dep Asst Secy Defense for Health, 68-69, CMNDG GEN, FITZSIMONS GEN HOSP, US ARMY, 69- Concurrent Pos: Res fel cardio-pulmonary dis, Buffalo Gen Hosp, NY, 47-49; asst clin prof, Sch Med, Univ Colo, 55-60; consult, Surgeon Gen, US Army, 58- Mem: Am Thoracic Soc; AMA; Am Col Physicians. Res: Chemotherapy of tuberculosis and other pulmonary diseases; etiology and medical management of pulmonary diseases; military medical history. Mailing Add: Fitzsimons Gen Hosp Denver CO 80240

WIER, KAREN, b Eau Claire, Wis, Dec 31, 37; m 66; div 74. GEOLOGY. Educ: Univ Wash, Seattle, BS, 59; Bryn Mawr Col, PhD(petrol), 63. Prof Exp: Teaching asst, Bryn Mawr Col, 59-61; GEOLOGIST, US GEOL SURV, 65- Res: Petrology; petrography; structure of metamorphic rocks. Mailing Add: US Geol Surv Nat Ctr Stop 928 Reston VA 22092

WIERBICKI, EUGEN, b Krasnae, Byelorussia, Jan 4, 22; nat US; m 49; c 2. BIOCHEMISTRY, AGRICULTURAL CHEMISTRY. Educ: Munich Tech Univ, DrAgr Sci, 49; Ohio State Univ, PhD(biochem), 53. Prof Exp: Asst, Ohio State Univ, 50-51, res assoc meat biochem, Res Found, 54-56; res scientist & proj leader, Rath Packing Co, 56-60, mgr meat res, 61-62; chief irradiated food prod div, Food Lab, US Army Natick Labs, 62-74, HEAD IRRADIATED FOOD PROD GROUP, RADIATION PRESERV FOODS DIV, FOOD ENG LAB, US ARMY NATICK DEVELOP CTR, 74-; Concurrent Pos: Abstr, Chem Abstracts, 54-64; mem USDA indust food res team, USSR, 60; tour dir, USSR Food Processing Del to US, 64. Mem: Fel AAAS; Am Chem Soc; Inst Food Technol; NY Acad Sci; Am Meat Sci Asn. Res: Radiation preservation of foods; product technology, radiation processing, process specifications and quality control of irradiated foods; meat science and technology. Mailing Add: Food Eng Lab US Army Natick Develop Ctr Natick MA 01760

WIERCINSKI, FLOYD JOSEPH, b Cicero, Ill, Dec 8, 12; m 40; c 3. CELL PHYSIOLOGY. Educ: Univ Chicago, BS, 36, MS, 38; Univ Pa, PhD(zool), 43. Prof Exp: Asst physiol, Univ Pa, 40-42, instr, 42-43; instr biol, Cath Univ, 43-44; assoc pharmacol, Stritch Sch Med, Loyola, Ill, 44-46; chmn dept biol, Lewis Col, 45-47; instr physiol, Ill Inst Technol, 47-48; asst prof, Hahnemann Med Col, 48-57; assoc prof biol sci & in charge radiation biol, Drexel Inst, 57-63; PROF BIOL, NORTHEASTERN ILL UNIV, 64- Concurrent Pos: Partic radioisotope methodology, Hahnemann Med Col, 56; res fel, Lankenau Hosp, Philadelphia, 58-59; partic radiation biol, Univ Rochester, 61; partic AEC prog phys chem, Univ Minn, 63-64; assoc prof biol sci eve div, Northwestern Univ, 65-67, prof, 67-; mem corp, Marine Biol Lab, Woods Hole, 56. Honors & Awards: Recognition Award, Am Inst Ultrasonics Med, 64. Mem: Am Physiol Soc; Soc Gen Physiol. Res: Supersonic effects in living organisms; contraction modes in muscle fibers; effect of trypsin on muscle protoplasm; pH of animal cells; adenosine triphosphate and cations in muscle fiber; intracellular pH; contraction sites in muscle fiber; mechanisms in muscle contraction. Mailing Add: Dept of Biol Northeastern Ill Univ Chicago IL 60625

WIERENGA, PETER J, b Uithuizen, Neth, June 27, 34; m 63; c 3. SOIL PHYSICS, AGRONOMY. Educ: State Agr Univ, Wageningen, BS, 61, MS, 63; Univ Calif, Davis, PhD(soil sci), 68. Prof Exp: Res water scientist, Univ Calif, Davis, 65-68; asst prof agron, 68-72, ASSOC PROF AGRON, NMEX STATE UNIV, 72- Concurrent Pos: Sabattical leave, Nat Ctr Sci Res, Inst Mech, Grenoble, France, 75. Mem: Am Soc Agron; Soil Sci Soc Am; Am Geophys Union; Neth Royal Soc Agr Sci. Res: Simulation of transfer processes in soils, such as movement of water, heat and salts; quality of irrigation return flow. Mailing Add: Dept of Agron NMex State Univ PO Box 3Q Las Cruces NM 88003

WIERENGA, WENDELL, b Hudsonville, Mich, Feb 5, 48; m 68; c 1. ORGANIC CHEMISTRY. Educ: Hope Col, BA, 70; Stanford Univ, PhD(org chem), 73. Prof Exp: RES SCIENTIST ORG CHEM, EXP CHEM RES, UPJOHN CO, 74- Concurrent Pos: Fel, Am Cancer Soc, Dept Chem, Stanford Univ, 73-74. Mem: Am Chem Soc; The Chem Soc; Royal Inst Chem. Res: Design and synthesis of biologically and medicinally important organic compounds with particular emphasis on agents that interact with membranes and membrane transport. Mailing Add: Exp Chem Res Upjohn Co Kalamazoo MI 49001

WIERENGO, CYRIL JOHN, JR, b Picayune, Miss, Mar 7, 40; m 62; c 2. ORGANIC CHEMISTRY. Educ: Univ Southern Miss, BA, 62, MS, 64; Miss State Univ, PhD(chem), 74. Prof Exp: Res chemist, Dow Chem Co, 64-67; ASSOC PROF CHEM, MISS UNIV FOR WOMEN, 67- Mem: Am Chem Soc. Res: The synthesis and chemistry of bis-heterocyclic compounds. Mailing Add: Miss Univ of Women Box W119 Columbus MS 39701

WIERSEMA, RICHARD JOSEPH, b St Louis, Mo, Aug 9, 41. INORGANIC CHEMISTRY. Educ: St Louis Univ, BS, 65; Univ Kans, PhD(inorg chem), 69. Prof Exp: Fel inorg chem, Univ Calif, Los Angeles, 69-74; SR CHEMIST, RES DIV, ROHM AND HAAS CO, 74- Mem: Am Chem Soc. Res: Basic development and evaluation of materials to be used in the context of industrial coatings. Mailing Add: Res Div Rohm and Haas Co Norristown & McKean Rds Springhouse PA 19477

WIERSMA, CORNELIS ADRIANUS GERRIT, b Naaldwyk, Holland, Oct 10, 05; m 32. PHYSIOLOGY. Educ: Leiden Univ, BS, 26; Univ Utrecht, MS, 29, PhD(comp physiol), 33. Prof Exp: Asst comp physiol, Univ Utrecht, 29-30 & 31; asst med physiol, 32-34; from asst prof to assoc prof biol, 34-47, PROF BIOL, CALIF INST TECHNOL, 47- Concurrent Pos: Guggenheim Mem fel, 57; vis prof, Cambridge Univ, 57-58; mem physiol training comt, NIH, 60-65. Mem: Am Physiol Soc; Soc Exp Biol & Med; Soc Gen Physiol; cor mem Royal Neth Acad Sci; Neth Royal Zool Soc. Res: Neuromuscular physiology; vertebrates and invertebrates; neurophysiology; crustaceans; applied neurophysiology. Mailing Add: Div of Biol Calif Inst of Technol Pasadena CA 91125

WIERSMA, DANIEL, b Volga, SDak, Nov 4, 16; m 43; c 2. SOIL SCIENCE. Educ: SDak State Univ, BS, 42; Univ Wyo, MS, 52; Univ Calif, PhD, 56. Prof Exp: County agr exten agent, Butte County, SDak, 46-52; asst, Univ Calif, 52-55; from asst prof to assoc prof agron, 56-64, PROF AGRON, PURDUE UNIV, 64-, DIR WATER RESOURCES RES CTR, 65- Concurrent Pos: Consult Rockefeller agr prog, Colombia, SAm, 63 & 65. Mem: Am Soc Agron; Soil Sci Soc Am; Soil Conserv Soc Am. Res: Water resources development; plant, soil and water relations; response of plants to water conditions; use of water by plants as affected by soil, water and climate. Mailing Add: Dept of Agron Agr Exp Sta Purdue Univ West Lafayette IN 47906

WIERSMA, JAMES H, b Beaver Dam, Wis, Jan 4, 40; m 61; c 1. ANALYTICAL CHEMISTRY. Educ: Wis State Univ, Oshkosh, BS, 61; Univ Mo, Kansas City, MS, 65, PhD(chem), 68. Prof Exp: Clin chemist, Mercy Hosp, Oshkosh, 61-62; USPHS traineeship water chem, 67-68; asst prof chem, 68-72, ASSOC PROF CHEM, UNIV WIS- GREEN BAY, 72- Mem: Am Chem Soc. Res: Environmental sciences especially related chemistry and development of analytical methods. Mailing Add: Sci & Environ Change Univ of Wis 120 SUniversity Circle Dr Green Bay WI 54301

WIESBOECK, ROBERT A, b Frankfurt, Ger, Jan 19, 30; m 50; c 2. ORGANOMETALLIC CHEMISTRY. Educ: Munich Tech Univ, BS, 55, PhD(chem), 57. Prof Exp: NSF fel phys org chem, Ga Inst Technol, 58-59; res chemist, Redstone Res Div, Rohm and Haas Co, 59-63; staff scientist, 63-68, res scientist, 68-70, mgr chem, 70-74, MGR ATLANTA RES CTR, US STEEL CORP, 74- Mem: AAAS; Am Chem Soc. Res: Organic and inorganic fluorine chemistry of nitrogen, phosphorous and sulphur. Mailing Add: US Steel Corp Atlanta Res Ctr 685 Indust Way Decatur GA 30033

WIESE, ALLEN F, b Eyota, Minn, Dec 16, 25; m 48; c 3. WEED SCIENCE. Educ: Univ Minn, BS, 49, MS, 51, PhD(agron), 53. Prof Exp: PROF AGRON, AGR EXP STA, TEX A&M UNIV, 53- Mem: AAAS; Am Soc Agron; Weed Sci Soc Am. Res: Weed control methods in crop production. Mailing Add: Tex Agr Exp Sta Bushland TX 79012

WIESE, ALVIN CARL, b Milwaukee, Wis, Aug 13, 13; m 44; c 2. BIOCHEMISTRY. Educ: Univ Wis, BS, 35, MS, 37, PhD(biochem), 40. Prof Exp: Asst biochem, Univ Wis, 35-40; instr chem, Okla Agr & Mech Col, 40-42; spec res assoc, Univ Ill, 42-45, spec asst animal nutrit, 45-46; prof agr biochem & head dept, 46-72, PROF

BIOCHEM, UNIV IDAHO, 72- Mem: AAAS; Am Chem Soc; Soc Exp Biol & Med; Poultry Sci Asn; Am Inst Nutrit. Res: Nutritional biochemistry; enzymology; effect of fluorides on enzymes; air pollution; trace minerals. Mailing Add: Dept of Bact & Biochem Univ of Idaho Moscow ID 83843

WIESE, HELEN JEAN COLEMAN, b San Antonio, Tex, Dec 10, 41; m 75. MEDICAL ANTHROPOLOGY. Educ: Univ Wis-Milwaukee, BA, 63; Stanford Univ, MA, 64; Univ NC, Chapel Hill, PhD(anthrop), 72. Prof Exp: ASST PROF BEHAV SCI, COL MED, UNIV KY, 72- Mem: Soc Appl Anthrop; Soc Med Anthrop; Am Anthrop Asn; Asn Behav Sci Med Educ; Inst Soc Ethics Life Sci. Res: Cross-cultural variation in acceptable body image; effects of pharmaceutical counseling on patient compliance with chemotherapy for congestive heart failure; attitudes toward various contraceptive devices; rates of gonorrhea in two Kentucky counties. Mailing Add: Dept of Behav Sci Univ of Ky Col of Med Lexington KY 40506

WIESE, JOHN HERBERT, b Los Angeles, Calif, Jan 15, 17; m 42; c 2. GEOLOGY. Educ: Univ Calif, Los Angeles, AB, 40, MA, 41, PhD(struct geol), 47. Prof Exp: Geologist, US Geol Surv, 41-48; geologist, Richfield Oil Co, 48-59, supvr explor res, 59-66, sr geologist, Atlantic Richfield Co, 67-73; CONSULT GEOLOGIST, 73- Concurrent Pos: Vis indust prof, Southern Methodist Univ, 70-73. Res: Geology of Nevada; petroleum exploration; sedimentology; landslides; continental shelf resources; geology of central California coast. Mailing Add: 1595 Los Osos Valley Rd 16-C Los Osos CA 93402

WIESE, MAURICE VICTOR, b Columbus, Nebr, Sept 22, 40; m 63; c 3. PLANT PATHOLOGY, PLANT PHYSIOLOGY. Educ: Univ Nebr, BS, 63, MS, 65; Univ Calif, PhD(plant path), 69. Prof Exp: Asst prof plant path, 69-74, ASSOC PROF PLANT PATH, MICH STATE UNIV, 74- Mem: Am Phytopath Soc; Am Soc Plant Physiol. Res: Pathogenesis, etiology and control of wheat diseases. Mailing Add: Dept of Bot & Plant Path Mich State Univ East Lansing MI 48823

WIESE, RICHARD ANTON, b Howells, Nebr, Apr 3, 28; m 54; c 7. SOIL FERTILITY. Educ: Univ Nebr, BS, 54, MS, 56; NC State Univ, PhD, 61. Prof Exp: From asst prof to assoc prof soil fertil, Univ Wis, 61-67; assoc prof agron, 67-74, PROF AGRON, UNIV NEBR, LINCOLN, 74- Mem: Am Soc Agron; Soil Sci Soc Am; Am Chem Soc; Soil Conserv Soc Am. Res: Plant nutrition as effected by soil release of nutrients. Mailing Add: Dept of Agron Univ of Nebr Lincoln NE 68503

WIESE, ROBERT GEORGE, JR, b Boston, Mass, Sept 14, 33; m 58; c 4. GEOLOGY. Educ: Yale Univ, BS, 55; Harvard Univ, AM, 57, PhD(geol), 61. Prof Exp: Explor geologist, New Park Mining Co, 60-63; explor geologist, US Smelting Refining & Mining Co, 63-64; mem staff geol dept, 64-70, PROF GEOL & CHMN DEPT, MT UNION COL, 70- Concurrent Pos: Consult geologist. Mem: AAAS; Geol Soc Am; Am Inst Mining, Metall & Petrol Eng; Am Geophys Union; Soc Econ Geol. Res: Petrology and geochemistry of White Pine copper deposit, Michigan; petrology of wallrock alteration; genesis of mineral deposits; coal geology, exploration, development. Mailing Add: 135 Overlook Dr Alliance OH 44601

WIESE, WOLFGANG LOTHAR, b Tilsit, Ger, Apr 21, 31; US citizen; m 57; c 2. ATOMIC PHYSICS, PLASMA PHYSICS. Educ: Kiel Univ, BS, 54, PhD(physics), 57. Prof Exp: Res assoc physics, Univ Md, 58-59; physicist, 60-62, SECT CHIEF PHYSICS, NAT BUR STANDARDS, 63- Concurrent Pos: Guggenheim fel, 66-67. Mem: Fel Optical Soc Am; Am Phys Soc. Res: Experimental plasma spectroscopy; stabilized arcs and pulsed plasma sources; atomic transition probabilities; spectral line broadening. Mailing Add: Plasma Spectros Sect Nat Bur of Standards Washington DC 20234

WIESEL, BENJAMIN, b New York, NY, Apr 26, 11; c 2. PSYCHIATRY, NEUROLOGY. Educ: NY Univ, BA, 31, MD, 36. Prof Exp: Dir dept psychiat, Hartford Hosp, 50-72; PROF PSYCHIAT & CHMN DEPT, HEALTH CTR, UNIV CONN, HARTFORD, 73- Mem: Am Psychiat Asn. Res: Clinical psychiatry with emphasis on depressions. Mailing Add: McCook Hosp Univ of Conn Sch of Med Hartford CT 06112

WIESEL, TORSTEN NILS, b Upsala, Sweden, June 3, 24; div. PHYSIOLOGY. Educ: Karolinska Inst, Sweden, MD, 54. Prof Exp: Instr physiol, Royal Caroline Medico-Surg Inst & asst, Dept Child Psychiat, Hosp, 54-55; asst prof ophthal physiol, Johns Hopkins Univ, 58-59; assoc neurophysiol & neuropharmacol, 59-60, asst prof, 60-64, asst prof neurophysiol, Dept Psychiat, 64-67, prof physiol, 67-68, prof neurobiol, 68-74, ROBERT WINTHROP PROF NEUROBIOL, HARVARD MED SCH, 74-, CHMN DEPT, 73- Concurrent Pos: Fel ophthal, Med Sch, Johns Hopkins Univ, 55-58. Honors & Awards: Jules Stein Award, Trustees Prev of Blindness, 71; Rosenstiel Award, 72. Mem: AAAS; Am Physiol Soc; Am Acad Arts & Sci; Swed Physiol Soc. Res: Neurophysiology, especially the visual system. Mailing Add: Dept of Neurobiol Harvard Med Sch Boston MA 02115

WIESENDANGER, HANS ULRICH DAVID, b Zurich, Switz, Jan 13, 28; nat US; m 54; c 4. PHYSICAL CHEMISTRY. Educ: Swiss Fed Inst Technol, dipl, 51, DrScTech, 54. Prof Exp: Tech adv inst phys therapy, Zurich Univ, 53-55; fel, Univ Calif, Los Angeles, 55-56; sr res chemist, Kaiser Aluminum & Chem Corp, 57-59; phys chemist, Stanford Res Inst, 59-66; mkt mgr sci instrument dept, Electronics Assocs, Inc, 66-70; dir mkt, Uthe Technol Int, 70-72; mgr int div, Barnes-Hind Pharmaceut, Inc, 72-74; CONSULT, 74- Concurrent Pos: Consult, 54-55 & 71-74. Mem: Am Chem Soc; Am Vacuum Soc. Res: Surface chemistry; ultra high vacuum; radiochemistry; isotopes; tracer methods; catalysis; instrumentation; mass spectrometry; technoeconomics; international marketing; long range planning; technology assessment and transfer; new ventures; acquisitions and business opportunites analysis. Mailing Add: 1151 Buckingham Dr Los Altos CA 94022

WIESENFELD, JOHN RICHARD, b New York, NY, July 26, 44; m 66. CHEMICAL KINETICS, PHOTOCHEMISTRY. Educ: City Col New York, BS, 65; Case Inst Technol, PhD(chem), 69; Cambridge Univ, MA, 70. Prof Exp: US Air Force fel phys chem, Cambridge Univ, 69-70, NSF fel, 70-71, Stokes res fel, Pembroke Col, 70-72; ASST PROF CHEM, CORNELL UNIV, 72- Concurrent Pos: US Hon Ramsay fel, Ramsay Mem Trust, UK, 71. Mem: AAAS; Am Chem Soc. Res: Gas phase kinetics of electronically excited atoms and molecules in defined quantum states; energy storage and transfer in atoms and simple molecules. Mailing Add: Dept of Chem Cornell Univ Ithaca NY 14853

WIESER, HELMUT, b Austria, July 4, 35; Can citizen; m 67; c 2. CHEMISTRY. Educ: Univ BC, BSc, 62; Univ Alta, PhD(chem), 66. Prof Exp: Session instr, 66-68, asst prof, 68-74, ASSOC PROF CHEM, UNIV CALGARY, 74- Mem: AAAS; Chem Inst Can. Res: Molecular spectroscopy and structure; infrared and Raman spectroscopy. Mailing Add: Dept of Chem Univ of Calgary Calgary AB Can

WIESLER, DONALD PAUL, b Milwaukee, Wis, Dec 24, 29; m 56; c 6. ORGANIC CHEMISTRY. Educ: Univ Wis, BS, 51; Purdue Univ, MS, 53, PhD(chem), 57. Prof Exp: Asst chem, Purdue Univ, 51-55; instr, Wabash Col, 55-56; asst prof, Carroll Col, 56-59; prof, Lincoln Mem Univ, 59-64; assoc prof, Pfeiffer Col, 64-65; prof, Polk Jr Col, 66-68; vis asst prof, Univ Fla, 68-70; asst prof, Mercer Univ, 70-74; INSTR DEPT SCI & MATH, CALDWELL COMMUNITY COL, 74- Res: Peroxy acid oxidation of ketones. Mailing Add: Dept of Sci & Math Caldwell Community Col Lenoir NC 28645

WIESMEYER, HERBERT, b Chicago, Ill, Jan 12, 32; m 54; c 2. MICROBIOLOGY. Educ: Univ Ill, BS, 54; Wash Univ, St Louis, PhD, 59. Prof Exp: NSF fel, Johns Hopkins Univ, 59-61; fel, McCollum-Pratt Inst, 61; NATO fel, 61-62; asst prof, 62-67, ASSOC PROF MOLECULAR BIOL, VANDERBILT UNIV, 67- Mem: Am Soc Microbiol; Genetics Soc Am. Res: Bacterial physiology. Mailing Add: Dept of Molecular Biol Vanderbilt Univ Nashville TN 37203

WIESNER, HAROLD JEROME, physical chemistry, see 12th edition

WIESNER, KAREL, b Prague, Czech, Nov 25, 19; Can citizen; m 42; c 1. CHEMISTRY. Educ: Charles Univ, Prague, RNDr, 45. Hon Degrees: DSc, Univ NB, Fredericton, 70 & Univ Western Ont, 72. Prof Exp: Asst phys chem, Charles Univ, Prague, 45-46; fel, Swiss Fed Inst Technol, 46-47, Rockefeller fel, 47-48; from asst prof chem to prof org chem, Univ NB, 48-62; dir chem res, Ayerst Labs, Montreal, 62-64; RES PROF CHEM, UNIV N B, FREDERICTON, 64- Honors & Awards: Medal, Chem Inst Can, 63. Mem: Am Chem Soc; Chem Inst Can; fel Royal Soc Can; Brit Chem Soc; Swiss Chem Soc. Res: Synthesis and structure determination of natural products; study of fast reactions by polarography. Mailing Add: Dept of Chem Univ of NB Fredericton NB Can

WIESNER, LOREN ELWOOD, b Estilline, SDak, Nov 13, 38; m 59; c 3. SEED PHYSIOLOGY. Educ: SDak State Univ, BS, 60, MS, 63; Ore State Univ, PhD(agron), 71. Prof Exp: Asst agron, SDak State Univ, 63, asst county agt, 63-64; res asst agron, Ore State Univ, 68-70; asst prof seed technol, 64-68, ASSOC PROF SEED PHYSIOL, MONT STATE UNIV, 70- Mem: Am Soc Agron; Crop Sci Soc Am; Asn Off Seed Analysts; Asn Off Seed Cert Agencies. Res: Seed research related to production, technology, physiology and ecology. Mailing Add: Dept of Plant & Soil Sci Mont State Univ Bozeman MT 59715

WIESNER, RAKOMA, b New York City, NY, May 21, 20; m 58; c 1. BIOCHEMISTRY. Educ: Brooklyn Col, BA, 40, MA, 50; Columbia Univ, PhD(biochem), 62. Prof Exp: Technician clin lab, Brooklyn Jewish Hosp, 42-47; res asst zool, Columbia Univ, 47-50, res worker biochem, 50-62; res assoc enzymol, Inst Muscle Dis, 62-67; res assoc biochem, Mt Sinai Sch Med, 69-74; RES ASSOC BIOCHEM, COLUMBIA UNIV, 74- Mem: Am Chem Soc. Res: Nucleic acid biochemistry; control mechanisms of enzymes.

WIESNET, DONALD RICHARD, b Buffalo, NY, Feb 7, 27; m 52; c 4. GEOLOGY, HYDROLOGY. Educ: Univ Buffalo, BA, 50, MA, 51. Prof Exp: Asst, Univ Buffalo, 50-51; geologist, US Geol Surv, 52-54, chief manuscript rev sect, 54-55, chief geophys br, 56-57, asst to geol map ed, 57-59, asst chief br tech illustrations, 59-61, geohydrol map ed, 61-64, proj geologist, 65-67; oceanogr, Naval Oceanog Off, 67-68, res hydrologist, 68-71, SR RES HYDROLOGIST, NAT ENVIRON SATELLITE SERV, 71- Concurrent Pos: Mem comt hydrol, US Water Resources Coun, 68-71; rapporteur remote sensing of hydrol elements, Comn Hydrol, World Meteorol Orgn, 72-; mem work group remote sensing in hydrol, US Nat Com, Int Hydrol Decade, 72-; mem remote sensing comt, Int Field Year on Great Lakes, 72-75. Mem: Fel Geol Soc Am; Am Soc Photogram; Int Glaciol Soc; Am Water Resources Asn; Am Geophys Union. Res: Satellite hydrology; remote sensing of hydrologic parameters such as snow, ice, soil moisture, floods, coastal hydrology; hydrologic maps. Mailing Add: Nat Oceanic & Atmospheric Admin Nat Environ Satellite Serv s-33 Washington DC 20233

WIEST, EMIL GABRIEL, b Winona, Minn, Sept 10, 15; m 43; c 3. CHEMISTRY. Educ: Univ Minn, BChem, 37; Ohio State Univ, PhD(org chem), 40. Prof Exp: Res chemist, 40-47, group leader, 47-51, supvr develop, 51-52, chief supvr chem control, 52-57, supt process dept, 57-71, ANAL COORDR ENVIRON DIV, E I DU PONT DE NEMOURS & CO, INC, 71- Mem: Fel Am Inst Chem; Air Pollution Control Asn; Am Chem Soc; Am Soc Test & Mat. Res: Azo and vat dyes; fluorine chemicals; textile, paper and leather auxiliaries; analytical control and development; structure of diolefines. Mailing Add: 3210 Coachman Rd Surrey Park Wilmington DE 19803

WIEST, WALTER GIBSON, b Price, Utah, Feb 16, 22; m 48; c 7. BIOCHEMISTRY. Educ: Brigham Young Univ, AB, 48; Univ Wis, MS, 51, PhD(biochem), 52. Prof Exp: Asst biochem, Univ Wis, 48-52; from instr to assoc prof, Univ Utah, 52-64; assoc prof, 64-68, PROF BIOCHEM IN OBSTET & GYNEC, SCH MED, WASH UNIV, 68- Concurrent Pos: USPHS spec fel, Univ Cologne, 59-60. Mem: AAAS; Am Soc Biol Chem; Endocrine Soc; Soc Gynec Invest; Int Soc Res Reproduction. Res: Biosynthesis, metabolism and mode of action of steroid hormones, especially progesterone; application of radioisotopic techniques to steroid biochemistry. Mailing Add: 4911 Barnes Hosp Plaza St Louis MO 63110

WIETING, TERENCE JAMES, b Chicago, Ill, Sept 4, 35; m 70; c 2. PHYSICS. Educ: Mass Inst Technol, SB, 57; Harvard Univ, BD, 62; Cambridge Univ, PhD(physics), 69. Prof Exp: Res staff mem physics, Naval Supersonic Lab, Mass Inst Technol, 58-60; res scientist, Mithras, Inc, Mass, 62-63; Nat Acad Sci-Nat Res Coun res assoc, 69-71, RES PHYSICIST, NAVAL RES LAB, 71- Mem: Am Phys Soc; Sigma Xi. Res: Optical properties of metals; physics of layered compounds; Raman scattering; lattice dynamics. Mailing Add: Metal Physics Br Naval Res Lab Washington DC 20390

WIEWIOROWSKI, TADEUSZ KAROL, b Sopot, Poland, Nov 3, 35; US citizen; m 62; c 2. CHEMISTRY. Educ: Loyola Univ, La, BS, 59; Tulane Univ La, PhD(chem), 65. Prof Exp: SUPT CHEM RES & DEVELOP, FREEPORT MINERALS CO, 59- Mem: Am Chem Soc. Res: Inorganic and physical chemistry; process development; hydrometallurgy; management of chemical research and development; solvent extraction. Mailing Add: Res & Develop Lab Freeport Minerals Co Belle Chasse LA 70037

WIGAND, JEFFREY STEPHEN, b Manhattan, NY, Dec 17, 42; m 70. ENDOCRINOLOGY, BIOCHEMISTRY. Educ: State Univ NY Buffalo, BA, 68, MA, 71, PhD(endocrinol), 72; NY Univ, MBA, 76. Prof Exp: DIR LICENSING & CORP DEVELOP HEALTH CARE, PFIZER INC, 75- Concurrent Pos: Mem comt criteria & purity biol compounds, Nat Acad Sci-Nat Res Coun, 73- Mem: AAAS; Am Chem Soc; NY Acad Sci; Fedn Am Socs Exp Biol; Endocrine Soc. Res: Protein chemistry related to hormones; role of nucleotides and cyclic adenosinemonophosphate in nuerogenic diseases. Mailing Add: Pfizer Inc 235 E 42nd St New York NY 10017

WIGEN, PHILIP E, b LaCrosse, Wash, May 11, 33; m 54; c 3. SOLID STATE PHYSICS, MAGNETISM. Educ: Pac Lutheran Col, BA, 55; Mich State Univ, PhD(physics), 60. Prof Exp: Assoc res scientist, Lockheed Res Labs, Calif, 60-63, res scientist, 63-65; assoc prof physics, 65-71, PROF PHYSICS, OHIO STATE UNIV,

71- Concurrent Pos: Consult, Res Labs, Battelle Mem Inst, 67- & Drackett Co, 71- Mem: Am Phys Soc; Am Asn Physics Teachers; Inst Elec & Electronics Eng. Res: Magnetism in metals and insulators; paramagnetic resonance. Mailing Add: Dept of Physics Ohio State Univ Columbus OH 43210

WIGFIELD, DONALD COMPSTON, b Godalming, Eng, June 13, 43; m 66. ORGANIC CHEMISTRY. Educ: Univ Birmingham, BSc, 64; Univ Toronto, PhD(org chem), 67. Prof Exp: Fel, Univ BC, 67-68, teaching fel, 68-69; asst prof chem, 69-72, ASSOC PROF CHEM, CARLETON UNIV, 72- Concurrent Pos: Vis assoc prof, Univ Victoria, 75-76. Mem: Am Chem Soc; fel The Chem Soc; Chem Inst Can. Res: Determination of the nature of organic transition states; organic reaction mechanisms; biosynthesis of indole alkaloids; liquid scintillation counting. Mailing Add: Dept of Chem Carleton Univ Ottawa ON Can

WIGGANS, DONALD SHERMAN, b Lincoln, Nebr, July 14, 25; m 51; c 4. BIOCHEMISTRY. Educ: Univ Nebr, BSc, 49; Univ Ill, PhD(chem), 52. Prof Exp: Instr biochem, Yale Univ, 52-54; from asst prof to assoc prof, 54-61, PROF BIOCHEM, UNIV TEX HEALTH SCI CTR DALLAS, 61- Concurrent Pos: Vis prof, Southern Methodist Univ, 64-66 & Univ Tex, Arlington, 67-74. Mem: Am Chem Soc; Am Soc Biol Chemists; Soc Exp Biol & Med; Am Inst Nutrit. Res: Intermediary metabolism of amino acids and peptides; mechanism of protein synthesis. Mailing Add: Dept of Biochem Univ of Tex Health Sci Ctr Dallas TX 75235

WIGGANS, SAMUEL CLAUDE, b Lincoln, Nebr, Sept 2, 22; m 57; c 2. PLANT PHYSIOLOGY. Educ: Univ Nebr, BS, 47; Univ Wis, MS, 48, PhD(plant physiol), 51. Prof Exp: Asst agron, Univ Wis, 47-49, asst bot, 49-51; asst prof agron & bot, Iowa State Univ, 51-58; assoc prof hort, Okla State Univ, 58-62; chmn dept hort sci, 62-65, PROF HORT, UNIV VT, 62-, CHMN DEPT PLANT & SOIL SCI, 65- Mem: Am Soc Hort Sci; Am Soc Agron; Am Soc Plant Physiol; Scandinavian Soc Plant Physiol. Res: Mineral nutrition; growth regulation; fertilizer placement and techniques; photoperiod and temperature; radioisotopes. Mailing Add: Dept of Plant & Soil Sci Univ of Vt Burlington VT 05401

WIGGER, H JOACHIM, b Hagen, Ger, May 29, 28; m 57; c 2. MEDICINE, PATHOLOGY. Educ: Johanneum Col, Lueneburg, Ger, BA, 49; Univ Hamburg, DMSc, 54. Prof Exp: Assoc dir labs, Children's Hosp, Washington, DC, 62-64; assoc, 64-67, asst prof, 67-69, ASSOC PROF PATH, COLUMBIA UNIV, 69- Concurrent Pos: Consult, USPHS Hosp, Staten Island, 66-; asst attend pathologist, Presby Hosp, New York, 67-69, assoc attend pathologist, 69- Mem: NY Acad Sci. Res: Pediatric and developmental pathology. Mailing Add: Dept of Path Columbia Univ New York NY 10032

WIGGERS, HAROLD CARL, b Ann Arbor, Mich, Sept 1, 10; m 35; c 2. PHYSIOLOGY. Educ: Wesleyan Univ, BA, 32; Western Reserve Univ, PhD(physiol), 36. Hon Degrees: ScD, Union Col, NY, 59. Prof Exp: Am Physiol Soc Porter fel, Harvard Med Sch, 36-37; instr, Col Physicians & Surgeons, Columbia Univ, 37-42; asst prof, Sch Med, Western Reserve Univ, 42-43; assoc prof, Univ Ill, 43-47; prof physiol & pharmacol & chmn dept, 47-53, prof physiol & dean, 53-74, exec vpres, 66-74, EMER DEAN, ALBANY MED COL, 74-; CONSULT, SCH MED, E CAROLINA UNIV, 74- Concurrent Pos: Mem bd trustees, Rensselaer Polytech; mem panel for reviewing specialized ctrs of res applns in hypertension, Nat Heart & Lung Inst; chmn deans comt, Albany Vet Admin Hosp; mem bd dirs, Blue Cross of Northeastern NY, bd trustees, Regional Hosp Review & Planning Coun, Albany, NY & bd dirs, Bankers Trust Co, NY. Mem: AAAS; Nat Soc Med Res; Am Physiol Soc; Am Heart Asn; Asn Am Med Cols. Res: Electrophysiology; audiology; ovulation; cardiovascular lesions; medical education. Mailing Add: Albany Med Col 47 New Scotland Ave Albany NY 12208

WIGGERS, KENNETH DALE, b Davenport, Iowa, Oct 17, 42; m 70. NUTRITION. Educ: Iowa State Univ, BS, 65, PhD(nutrit physiol), 71. Prof Exp: Fel path, 71-72, training fel path, Univ BC, 72-73; assoc nutrit, 73-75, ASST PROF NUTRIT, DEPT ANIMAL SCI, IOWA STATE UNIV, 75- Mem: Am Dairy Sci Asn; Am Inst Biol Sci; Sigma Xi. Res: Experimental atherosclerosis and lipid metabolism in young ruminants. Mailing Add: 313 Kildee Hall Iowa State Univ Ames IA 50011

WIGGIN, EDWIN ALBERT, b Exeter, NH, Aug 11, 21; m 47; c 2. CHEMISTRY. Educ: Univ NH, BS, 43. Prof Exp: Asst, SAM Labs, Columbia Univ, 43-45; res chemist, Carbide & Carbon Chem Co, 45-48; chief tech develop br, Isotopes Div, AEC, 48-54; EXEC VPRES, ATOMIC INDUST FORUM, 54- Concurrent Pos: Mem adv comt indust info, AEC, 53-54, mem adv comt isotope & radiation develop, 58-60 & 70-72. Mem: AAAS; Am Chem Soc; Am Nuclear Soc; Inst Nuclear Mat Mgt. Res: Application of atomic energy and radioactive by-products. Mailing Add: 17 Meadowcroft Ct Gaithersburg MD 20760

WIGGIN, HENRY CARVEL, b Canton, SDak, Sept 20, 23; m 49; c 3. GENETICS, PLANT BREEDING. Educ: Univ SDak, BA, 47; Purdue Univ, MS, 50; Kans State Col, PhD, 54. Prof Exp: Asst bot & plant path, Purdue Univ, 50; sci aide wheat breeding, Rockefeller Found, Mex, 50-51; asst coop corn trials, Kans State Col, 53; cereal crops specialist, Int Coop Admin, Ecuador, 54-56, wheat geneticist & corn breeder, 56-58, dir, La Tamborada Exp Sta, US Opers Mission, Bolivia, 59-60, res agronomist, Suakoko Exp Sta, Liberia, 60-62 & Dar-es-Salaam, 62-64, soils adv, Ibadan, Nigeria, 64-65, agron adv, 65-70; vis lectr, Iowa State Univ, 70-71; agron adv res, Kabul, Afghanistan, 71-73, RES PROJ MGR, US AID, DAR-ES-SALAAM, TANZANIA, 73- Mem: AAAS; Am Genetic Asn; Am Soc Agron; Crop Sci Soc Am. Res: Genetics of disease resistance in wheat; development of corn synthetics; high lysine corn. Mailing Add: US AID PMB 9130 Dar-es-Salaam Tanzania

WIGGIN, NORMAN JACK BRIDGMAN, b Syracuse, NY, Oct 17, 20; Can citizen; m 44; c 5. IMMUNOLOGY. Educ: Queen's Univ, Ont, MD & CM, 44, MSc, 53; Cambridge Univ, PhD(immunol), 55. Prof Exp: Head bact sect, Kingston Lab, Defence Res Bd Can, 55-58, dep chief supt, Kingston Lab & Chem Labs, 57-58, chief supt, Med Labs, Ont, 58-67, chief of plans, 67-69; mem bilingual & bicult develop prog, Laval Univ, 69-70; dep chmn & chmn univ grants rev comt & adv comt defence indust res, Defence Res Bd Can, 70-72; DIR GEN RES, HEALTH DEPT, DEPT NAT HEALTH & WELFARE, CAN, 72- Concurrent Pos: Mem defence med res adv comt, Dept Nat Defence, Can, 58-70; mem coun, Int Biomed Electronics, Univ Toronto, 63-67; mem, Defence Res Coun Can, 67-72; mem interdept comt innovation, adv comt sci & technol in the North, adv comt prog advan indust technol & adv comt indust res assistance prog, Can Govt, 70-72; mem, Nat Health Grant Rev Comt, 72-; mem, Can Inst Int Affairs & Med Coun Can. Mem: NY Acad Sci; Can Soc Microbiol. Res: Medicine; anticomplementary action; conglutination; biological research administration; public health and health care delivery research. Mailing Add: 82 Rothwell Dr Ottawa ON Can

WIGGINS, ALVIN DENNIE, b Harrisburg, Ill, May 5, 22; m 50; c 6. MATHEMATICAL STATISTICS. Educ: Univ Calif, Berkeley, AB, 51, MA, 53, PhD(statist), 58. Prof Exp: Res asst statist & assoc biostatist, Sch Pub Health, Univ Calif, Berkeley, 54-57; instr math, Ctr Grad Studies, Univ Wash, 58-63; assoc res biostatistician & lectr biostatist, Sch Pub Health, Univ Calif, Berkeley, 63-69; asst prof biostatist, 69-73, ASSOC PROF BIOSTATIST, UNIV CALIF, DAVIS, 73- Concurrent Pos: Sr statistician, Hanford Labs Oper, Gen Elec Co, 57-63. Mem: Inst Math Statist; Biomet Soc; Am Statist Asn; fel Royal Statist Soc. Res: Mathematical models of biological phenomena; statistical theory of estimation; application of stochastic processes to problems of health, medicine and biology; design and analysis of experiments; stochastic differential equations in biology. Mailing Add: Dept of Epidem & Prev Med Univ of Calif Sch of Vet Med Davis CA 95616

WIGGINS, EARL LOWELL, b Ringwood, Okla, July 11, 21; m 45; c 4. PHYSIOLOGY. Educ: Okla State Univ, BS, 47, MS, 48; Univ Wis, PhD(physiol of reprod), 51. Prof Exp: Asst, Univ Wis, 48-50; animal geneticist, Sheep Exp Sta, USDA, Idaho, 50-56; assoc prof, 56-73, PROF ANIMAL SCI, AUBURN UNIV, 73- Mem: AAAS; Am Soc Animal Sci; Am Genetic Asn; Soc Study Reproduction. Res: Puberty and related phenomena in sheep and swine; artificial insemination in swine; causes of reproductive failure in ewes; factors affecting the breeding season in ewes; reproduction, breeding and genetics in farm animals. Mailing Add: Dept of Animal & Dairy Sci Auburn Univ Auburn AL 36830

WIGGINS, ERNEST JAMES, b Trenton, Ont, Nov 25, 17; m 45; c 2. PHYSICAL CHEMISTRY, CHEMICAL ENGINEERING. Educ: Queen's Univ, Ont, BSc, 38; McGill Univ, PhD(phys chem), 46. Prof Exp: Supt eng develop sect, Atomic Energy Proj, Nat Res Coun, 46-48; head munitions & eng sect, Suffield Exp Sta, Defence Res Bd, 48-52; sr chemist, Stanford Res Inst, 52-58; head chem div, Sask Res Coun, 58-61; asst dir, Chem Div, Ont Res Found, 61-62; DIR RES, RES COUN ALTA, 62- Mem: Am Chem Soc; Am Inst Aeronaut & Astronaut; Arctic Inst NAm; Chem Inst Can; Brit Soc Chem Indust. Res: Chemical process development; environmental studies; transportation systems. Mailing Add: Res Coun of Alta 11315 87th Ave Edmonton AB Can

WIGGINS, GLENN BLAKELY, b Toronto, Ont, Jan 29, 27; m 49; c 3. ENTOMOLOGY, FRESH WATER BIOLOGY. Educ: Univ Toronto, BA, 49, MA, 50, PhD, 58. Prof Exp: Asst biologist, Nfld Fisheries Res Sta, Fisheries Res Bd Can, 50-51; asst cur dept entom & invertebrate zool, 52-61, from asst cur in chg to assoc cur in chg, 61-64, CUR DEPT ENTOM & INVERTEBRATE ZOOL, ROYAL ONT MUS, 64-; PROF ZOOL, UNIV TORONTO, 68- Mem: Soc Syst Zool; Entom Soc Am; Entom Soc Can; Can Soc Zool. Res: Systematic entomology; systematics and ecology of freshwater invertebrates, especially Trichoptera. Mailing Add: Royal Ont Mus 100 Queen's Park Toronto ON Can

WIGGINS, JAMES WENDELL, b Fayette, Ala, May 9, 42; m 64; c 2. BIOPHYSICS. Educ: Univ Ala, BS, 63; Johns Hopkins Univ, PhD(physics), 68. Prof Exp: Res assoc physics, 68-69, biophysics, 69-73, NIH spec fel, 73-74, ASST PROF BIOPHYSICS, JOHNS HOPKINS UNIV, 74- Mem: Am Phys Soc; AAAS; Electron Micros Soc Am. Res: Unconventional electron microscopy directed at structures below twenty angstroms, especially DNA sequence and organization of proteins in membranes. Mailing Add: 2708 Whitney Ave Baltimore MD 21215

WIGGINS, JAMES WILLIAM, b Paris, Ark, Mar 5, 40. INORGANIC CHEMISTRY. Educ: Univ Ark, Fayetteville, BS, 62; Univ Fla, PhD(chem), 66. Prof Exp: Res grant, Univ Calif, Riverside, 66-68; interim asst prof chem, Univ Fla, 68-69; ASST PROF CHEM & INORG CHEM, UNIV ARK, LITTLE ROCK, 69- Mem: Am Chem Soc. Res: Boron-nitrogen-carbon chemistry, synthesis of compounds; mechanisms of reactions leading to unusual structures; water quality in Arkansas and the effect of changing the stream beds on the water quality. Mailing Add: Dept of Chem Univ of Ark Little Rock AR 72204

WIGGINS, JAY ROSS, b Baltimore, Md, Apr 12, 47; m 72. PHARMACOLOGY. Educ: Bucknell Univ, BS, 69; Columbia Univ, PhD(pharmacol), 75. Prof Exp: Res assoc, Rockefeller Univ, 73-75; ASST PROF PHARMACOL, UNIV S FLA, 76- Mem: AAAS; Biophys Soc. Res: Electrophysiology and pharmacology of cardiac arrhythmias; mechanisms of excitation-contraction coupling in cardiac muscle. Mailing Add: Dept of Pharmacol Univ of S Fla Col of Med Tampa FL 33612

WIGGINS, JOHN SHEARON, b Chicago, Ill, Feb 8, 15. SPACE PHYSICS. Educ: Earlham Col, AB, 36; Calif Inst Technol, MS, 38; Univ Southern Calif, PhD(physics), 56. Prof Exp: Lectr physics, Univ Southern Calif, 41-43; from instr to asst prof, Univ Redlands, 44-46; asst prof, Univ Okla, 46-50; lectr physics, Univ Southern Calif, 50-56, asst prof, 57-58; mem tech staff, Semiconductor Div, Hughes Aircraft Co, 58-63; MEM TECH STAFF, SPACE SCI DEPT, TRW SYSTS, 65- Concurrent Pos: UNESCO vis prof, Concepcion Univ, Chile, 64. Mem: Am Phys Soc; Am Asn Univ Prof; Sigma Xi. Res: Photoelectricity; electron microscopy; optical, beta-ray and gamma-ray spectroscopy; linear accelerator; semiconductor devices; space physics; space science instrumentation; payload design, test and integration; radiation damage and measurement; spacecraft charging. Mailing Add: Space Sci Dept R5/1280 TRW Systs Redondo Beach CA 90278

WIGGINS, RALPH AMBROSE, b Broadwater, Nebr, Apr 4, 40. GEOPHYSICS. Educ: Colo Sch Mines, GpE, 61; Mass Inst Technol, PhD(geophys), 65. Prof Exp: Proj dir geophys, Geoscience, Inc, 65-66; res assoc, Mass Inst Technol, 66-70; asst prof, Univ Toronto, 70-73; assoc prof, Univ BC, 73-75; SR RES GEOPHYSICIST, WESTERN GEOPHYS CO, 75- Mem: Am Geophys Union; Soc Explor Geophys; Seismol Soc Am; Geol Soc Am; fel Royal Astron Soc. Res: Seismology; computer applications for interpretation and inversion of geophysical observations. Mailing Add: Western Geophys Co PO Box 2469 Houston TX 77001

WIGGINS, THOMAS ARTHUR, b Indiana, Pa, Feb 24, 21; m 53; c 2. OPTICS. Educ: Pa State Univ, BS, 42, PhD(physics), 53; George Washington Univ, MS, 49. Prof Exp: Instr physics, George Washington Univ, 48-50; from asst prof to assoc prof, 53-63, PROF PHYSICS, PA STATE UNIV, UNIVERSITY PARK, 63- Mem: Fel Am Phys Soc; fel Optical Soc Am. Res: High resolution and high precision molecular spectra of simple molecules, especially secondary wavelength standards; theory of rotation-vibration of molecules in the infrared; spontaneous and stimulated light scattering. Mailing Add: Osmond Lab Dept of Physics Pa State Univ University Park PA 16802

WIGGINS, VIRGIL DALE, b Tulsa, Okla, June 25, 31; m 52; c 3. PALYNOLOGY. Educ: Univ Okla, BS, 57, MS, 62. Prof Exp: Sr palynological technician, Sun Oil Co Prod Res, 58-59; explor palynologist, 59-69, SR EXPLOR PALYNOLOGIST, PAC NORTHWEST DIV, STANDARD OIL CO CALIF, WESTERN OPERS INC, 69- Concurrent Pos: Alaskan mem, Int Palynological Comn Working Group P3, 74- Mem: Am Asn Stratig Palynologists. Res: Application of palynology to arctic petroleum exploration. Mailing Add: Standard Oil Co of Calif Western Opers Inc PO Box 3862 San Francisco CA 94119

WIGGS, ALFRED JAMES, b Duncan, BC, Feb 11, 35; m 56; c 3. ZOOLOGY, PHYSIOLOGY. Educ: Univ BC, BS, 59, MSc, 62; Univ Alta, PhD(zool), 67. Prof Exp: Asst prof, 67-74, ASSOC PROF BIOL, UNIV NB, FREDERICTON, 74- Mem: AAAS; Can Soc Zool; Am Soc Zool. Res: Effects of temperature on metabolism;

acclimatization to temperature change; fish endocrines and their role in acclimitization; iodine metabolism and thyroid function in fish. Mailing Add: Dept of Biol Univ of NB Fredericton NB Can

WIGH, RUSSELL, b Weehawken, NJ, Nov 17, 14; m 39; c 3. MEDICINE, RADIOLOGY. Educ: Rutgers Univ, BS, 35; Harvard Med Sch, MD, 39. Prof Exp: Asst demonstr radiol, Jefferson Med Col, 46-49, instr, 49-50, assoc, 50, asst prof, 50-52; asst prof, Col Physicians & Surgeons, Columbia Univ, 52-54, assoc prof, 54-56; prof & chmn dept, Med Col Ga, 56-63; dir dept radiol, Bartholomew County Hosp, 63-71; ASSOC PROF RADIOL, SCH MED, IND UNIV, INDIANAPOLIS, 72- Concurrent Pos: Consult, Vet Admin Hosps, New York, 55-56 & Augusta, Ga, 56-63 & Battey State Hosp, Rome, 56-62; clin prof, Sch Med, Univ Louisville, 67-72. Honors & Awards: Cert of Merit, Am Roentgen Ray Soc, 51. Mem: Radiol Soc NAm; AMA; fel Am Col Radiol. Res: Photofluorographic detection of silent gastric neoplasms; clinical radiological investigations of various body systems. Mailing Add: 2767 Lafayette Ave Columbus IN 47201

WIGHT, HEWITT GLENN, b Murray, Utah, Feb 8, 21; m 43; c 4. SYNTHETIC ORGANIC CHEMISTRY. Educ: Univ Utah, BS, 43; Univ Calif, PhD(chem), 55. Prof Exp: PROF CHEM, CALIF POLYTECH STATE UNIV, SAN LUIS OBISPO, 52- Mem: Am Chem Soc. Res: Chemical education; organic syntheses. Mailing Add: Dept of Chem Calif Polytech State Univ San Luis Obispo CA 93407

WIGHT, HOWARD MORGAN, b Corvallis, Ore, Dec 4, 23; m 45; c 3. WILDLIFE ECOLOGY. Educ: Ore State Univ, BS, 48; Pa State Univ, MS, 50. Prof Exp: Res biologist, Mo Conserv Comn, 50-61; migratory bird pop sta, US Bur Sport Fisheries & Wildlife, 61-64; assoc prof, 64-70, PROF FISHERIES & WILDLIFE, ORE STATE UNIV, 70-, LEADER, ORE COOP WILDLIFE RES UNIT, 71- Mem: AAAS; Wildlife Soc. Res: Avian populations. Mailing Add: Dept of Fisheries & Wildlife Ore State Univ Corvallis OR 97331

WIGHT, JERALD ROSS, b Brigham City, Utah, Oct 5, 31; m 54; c 7. RANGE SCIENCE. Educ: Utah State Univ, BS, 53, MS, 59; Univ Wyo, PhD(range sci), 66. Prof Exp: Lab technician olericult, Univ Calif, 58-63; RANGE SCIENTIST, AGR RES SERV, USDA, 65- Mem: Soil Sci Soc Am; Soil Conserv Soc Am; Am Soc Agron; Soc Range Mgt. Res: Plant, soil, climate and animal relationships in range ecosystems, with emphasis on improving productivity. Mailing Add: North Plains Soil & Water Res Ctr USDA Agr Res Serv Box 1109 Sidney MT 59270

WIGHTMAN, ARTHUR STRONG, b Rochester, NY, Mar 30, 22; m 45; c 1. MATHEMATICAL PHYSICS. Educ: Yale Univ, BA, 42; Princeton Univ, PhD(physics), 49. Hon Degrees: DSc, Swiss Fed Inst Technol, 69. Prof Exp: Instr physics, Yale Univ, 43-44; from instr to assoc prof, 49-60, prof math physics, 60-71, THOMAS D JONES PROF MATH PHYSICS, PRINCETON UNIV, 71- Concurrent Pos: Nat Res Coun fel, Inst Theoret Physics, Copenhagen, 51-52; NSF fel, Naples, 56-57; vis prof, Paris, 57; vis prof, Inst Advan Study Sci, Bures-sur-Yvette, 63-64 & 68-69. Mem: Nat Acad Sci; AAAS; Am Math Soc; Am Phys Soc; Fedn Am Sci (treas, 54-56). Res: Elementary particle physics; quantum field theory; mathematical physics; functional analysis. Mailing Add: Dept of Physics Princeton Univ Box 708 Princeton NJ 08540

WIGHTMAN, FRANK, b Padiham, Eng, Jan 22, 28; Can citizen; m 56; c 5. PLANT PHYSIOLOGY. Educ: Univ Leeds, BSc, 48, PhD(plant physiol), 54. Prof Exp: Sr sci officer plant physiol, Agr Res Coun Unit, Wye Col, Univ London, 52-58; Nat Res Coun Can fel, Nat Res Coun Lab, Univ Sask, Can, 58-59; assoc prof, 60-66, chmn dept, 68-71, PROF BIOL, CARLETON UNIV, 66- Concurrent Pos: Nuffield Found vis fel, Univ Col, Univ London, 65-66; vis prof, Univ Lausanne, 66. Mem: Can Soc Plant Physiol; Am Soc Plant Physiol; Brit Soc Exp Biol. Res: Biosynthesis and physiological activity of indole and phenyl plant growth hormones; characterization of enzymes catalyzing amino acid metabolism; biochemistry and physiology of bud dormancy and bud burst. Mailing Add: Dept of Biol Carleton Univ Ottawa ON Can

WIGHTMAN, JAMES PINCKNEY, b Ashland, Va, May 14, 35; m 56; c 4. PHYSICAL CHEMISTRY. Educ: Randolph-Macon Col, BS, 55; Lehigh Univ, MS, 58, PhD(chem), 60. Prof Exp: Res assoc fuel sci, Pa State Univ, 60-62; PROF CHEM, VA POLYTECH INST & STATE UNIV, 62- Concurrent Pos: Vis res prof, Univ Bristol, 75-76. Mem: Am Chem Soc; Am Asn Univ Prof; Am Vacuum Soc; Int Microwave Power Inst. Res: Adsorption of gases and liquids on solids; reactions of atoms with solids; thermodynamics of adhesion; electron spectroscopic chemical analysis of solids surfaces. Mailing Add: 1300 Westover Dr Blacksburg VA 24060

WIGHTMAN, KEITH JOHN ROY, b Sandwich, Ont, May 12, 14; m 41; c 3. INTERNAL MEDICINE. Educ: Univ Toronto, MD, 37; FRCP(C), 42; FRCP(London). Hon Degrees: DSc, Mem Univ Nfld. Prof Exp: From demonstr to prof med & head dept, Fac Med, Univ Toronto, 46-69, dir & assoc dean, Div Postgrad Med Educ, 69-74; MED DIR, ONT CANCER TREATMENT & RES FOUND, 74- Concurrent Pos: Mem drug adv comt, Food & Drugs Div, Can Dept Nat Health & Welfare, 50-59; chmn panel antibiotics, Defence Res Bd, 52-61; mem coun, Royal Col Physicians Can, 64; pres, Gairdner Found; bd mem, Muskoka Hosp Res Fund, Hepatic Found Can & Ont Cancer Treatment & Res Found; pres, Royal Col Physicians & Surgeons Can, 74-76. Mem: Am Soc Hemat; Am Clin & Climat Asn; fel Am Col Physicians; Can Soc Clin Invest; Can Med Asn. Res: Intestinal absorption; antimicrobial agents; chemotherapy of malignant disease; hematology. Mailing Add: 181 Glencairn Ave Toronto ON Can

WIGHTMAN, ROBERT HARLAN, b Ottawa, Ont, Jan 24, 37; m 61; c 3. ORGANIC CHEMISTRY. Educ: Univ NB, BSc, 58, PhD(org chem), 62. Prof Exp: Nat Res Coun Can overseas fel org chem, Imp Col, Univ London, 62-63; res assoc, Stanford Univ, 63-65; asst prof, 65-69, ASSOC PROF ORG CHEM, CARLETON UNIV, 69- Mem: Am Chem Soc; Chem Inst Can. Res: New synthetic organic methods; syntheses of organic compounds of biological and theoretical interest. Mailing Add: Dept of Chem Carleton Univ Ottawa ON Can

WIGINTON, CARROLL LAMAR, b Burnsville, Miss, June 5, 35; m 57; c 4. MATHEMATICS. Educ: Univ Tenn, BS, 59, MS, 61, PhD(math), 64. Prof Exp: Consult math, Oak Ridge Nat Lab, 61-65; asst prof, Space Inst, Univ Tenn, 64-65; asst prof, 65-69, ASSOC PROF MATH, UNIV HOUSTON, 69- Concurrent Pos: Consult, Math Res Inc, Houston, Tex, 65- Mem: Am Math Soc; Math Asn Am. Res: Applied mathematics; engineering; functional analysis. Mailing Add: Dept of Math Univ of Houston Houston TX 77004

WIGLE, ERNEST DOUGLAS, b Windsor, Ont, Oct 30, 28; m 58; c 5. MEDICINE, CARDIOLOGY. Educ: Univ Toronto, MD, 53; FRCPS(C), 58. Prof Exp: McLaughlin Found fel cardiol, Univ Toronto, 59-60; sr res assoc, Ont Heart Found, 63-66; from asst prof to assoc prof, 66-72, PROF MED, UNIV TORONTO, 72- Concurrent Pos: Dir cardiovasc unit, Toronto Gen Hosp, 64-72, dir div cardiol, 72-; fel coun clin cardiol, Am Heart Asn, 65- Mem: Fel Am Col Physicians; Am Soc Clin Invests; Am Fedn Clin Res; Can Soc Clin Invest. Res: Muscular subaortic stenosis, hemodynamics, pharmacology and electrocardiography; hemodynamics of acute valvular insufficiency; cardiomyopathy; ventricular aneurysm; heart catheterization; automated assessment of left ventricular function. Mailing Add: Cardiovasc Unit Toronto Gen Hosp Toronto ON Can

WIGLER, PAUL WILLIAM, b New York, NY, Aug 26, 28; m 52; c 4. BIOCHEMISTRY. Educ: Queens Col, NY, BS, 50; Brooklyn Col, MA, 52; Univ Calif, Berkeley, PhD(biochem), 58. Prof Exp: Jr res biochemist, Virus Lab, Univ Calif, Berkeley, 58; NIH fel biochem, Univ Wis, 58-60; asst prof, Sch Med, Univ Okla, 60-63, assoc prof, 63-66; assoc prof res, 66-68, RES PROF BIOCHEM, MEM RES CTR, UNIV TENN, KNOXVILLE, 68- Concurrent Pos: Biochemist, Okla Med Res Found, 60-63, asst head cancer sect, 63-65, assoc head, 65-66; Nat Cancer Inst res career develop award, 66. Mem: Am Soc Biol Chemists; Am Chem Soc; Am Asn Cancer Res. Res: Mechanism of cell membrane transport of nucleosides and relationship to cell growth; carcinogenesis. Mailing Add: Mem Res Ctr Univ of Tenn Knoxville TN 37920

WIGLEY, NEIL MARCHAND, b Mt Vernon, Wash, Feb 16, 36; m 68; c 4. MATHEMATICAL ANALYSIS. Educ: Univ Calif, Berkeley, BA, 59, PhD(math), 63. Prof Exp: Staff mem math, Los Alamos Sci Lab, 63-65; asst prof, Univ Ariz, 65-67; assoc prof, Univ NC, 67-68; fel, Alex V Humboldt Found, Univ Bonn, 68-70; assoc prof, 70-74, PROF MATH, UNIV WINDSOR, 74- Mem: Am Math Soc. Res: Partial differential equations. Mailing Add: Dept of Math Univ of Windsor Windsor ON Can

WIGLEY, PERRY BRASWELL, geology, paleontology, see 12th edition

WIGLEY, ROLAND L, b Blawenburg, NJ, Oct 4, 23; m 56; c 2. MARINE ECOLOGY. Educ: Univ Maine, BS, 49; Cornell Univ, PhD(ichthyol), 53. Prof Exp: Fishery biologist, Dept Conserv & Econ Develop, NJ, 52-53; FISHERY RES BIOLOGIST, NAT MARINE FISHERIES SERV, 53- Concurrent Pos: Consult, Aquatic Sci Info Retrieval Ctr, Taft Lab, RI, 61-64 & John Wiley & Sons, NY, 66-; US deleg, Protein Resources Panel, US-Japan Coop Prog Natural Resources; shellfisheries consult, Food & Agr Orgn, UN; mem adv panel 5, Int Comn, Northwest Atlantic Fisheries Comn; US rep shellfish & benthos comt, Int Coun Explor Sea, 73-, US mem introd nonindigenous marine organisms panel, 74-; assoc ed, Nat Shellfisheries Asn Proc; assoc ed, Fishery Bull. Mem: Am Soc Ichthyologists & Herpetologists; Soc Syst Zool; Am Fisheries Soc; Am Soc Limnol & Oceanog. Res: Ecological aspects of offshore marine benthonic animal communities; taxonomy and geographic distribution of marine fishes and invertebrate organisms. Mailing Add: Nat Marine Fisheries Serv Woods Hole MA 02543

WIGNER, EUGENE PAUL, b Budapest, Hungary, Nov 17, 02; nat US; m 36, 41; c 2. MATHEMATICAL PHYSICS. Educ: Tech Hochsch, Berlin, DrIng, 25. Hon Degrees: Nineteen from US & foreign cols & univs, 49-73. Prof Exp: Asst, Tech Hochsch, Berlin, 26-27, privatdocent, 28-30, N B Ausserord prof theoret physics, 30-33; asst, Univ Göttingen, 27-28; lectr math physics, Princeton Univ, 30, prof, 30-36; prof physics, Univ Wis, 37-38; Thomas D Jones prof math physics, Palmer Phys Lab, 38-71, EMER THOMAS D JONES PROF MATH PHYSICS, PRINCETON UNIV, 71- Concurrent Pos: Sci guest, Kaiser Wilhelm Inst Berlin, 31 & Metall Lab, Chicago, 42-45; dir res & develop, Clinton Labs, Tenn, 46-47; Lorentz lectr, Inst Lorentz, Leiden, 57; dir harbor proj civil defense, Nat Acad Sci, 63; dir course 29, Int Sch Physics Enrico Fermi, 63; dir, Civil Defense Res Proj, Oak Ridge, Tenn, 64-65; Kramers prof, State Univ Utrecht, 75. Consult, Off Sci Res & Develop, 41-42, Oak Ridge Nat Lab & Exxon Nuclear Co; mem vis comt, Nat Bur Stand, 47-51; gen adv comt, AEC, 52-57, 59-64. Honors & Awards: Nobel Prize in Physics, 63; Medal for Merit, 46; Franklin Medal, Franklin Inst, 50; Fermi Award, 58; Atoms for Peace Award, 60; Max Planck Medal, Ger Phys Soc, 61; George Washington Award, Am Hungarian Studies Found, 64; Semmelweiss Medal, Am Hungarian Med Asn, 65; Nat Medal Sci, 69; Albert Einstein Award, 72. Mem: Nat Acad Sci; Am Math Soc; fel Am Phys Soc (vpres, 55, pres, 66); Am Acad Arts & Sci; Am Philos Soc. Res: Application of group theory of quantum mechanics; rate of chemical reactions; theory of metallic cohesion; nuclear structure and reactions; philosophical implications of quantum mechanics. Mailing Add: Dept of Math Physics Jadwin Hall Princeton Univ PO Box 708 Princeton NJ 08540

WIGTON, ROBERT SPENCER, b Omaha, Nebr, Nov 1, 11; m 37; c 2. MEDICINE. Educ: Univ Nebr, BSc, 32, MA & MD, 35. Prof Exp: Instr, Sch Med, Univ Pa, 38-42; PROF NEUROL & PSYCHIAT, COL MED, UNIV NEBR, OMAHA, 46- Concurrent Pos: Fel neurol, Hosp Univ Pa, 37-40; resident, Pa Hosp, 40-42; consult, Union Pac RR, 53- Mem: AAAS; AMA; Am Psychiat Asn. Res: Clinical neuropsychiatry; neurophysiology in relation to behavior. Mailing Add: Dept of Neurol & Psychiat Univ of Nebr Col of Med Omaha NE 68105

WIITANEN, WAYNE ALFRED, b Detroit, Mich, May 6, 35; m 73; c 1. BIONICS, COMPUTER SCIENCES. Educ: Harvard Univ, AB, 68, MA, 69, PhD(biol), 72. Prof Exp: Consult comput sci, 67-68; vpres, Mgt Eng Inc, 69-71; ASST PROF BIOL, UNIV ORE, 71-, ASST PROF COMPUT SCI, 73- Mem: AAAS; Inst Elec & Electronics Engrs; Soc Comput Simulation. Res: Applications of computers and mathematics to biological problems with special emphasis on the mammalian nervous system; dynamical biological systems simulation. Mailing Add: Dept of Biol Univ of Ore Eugene OR 97403

WIJANGCO, ANTONIO ROBLES, b Manila, Philippines, Apr 6, 44; m 67. HIGH ENERGY PHYSICS. Educ: Ateneo de Manila Univ, AB, 65; Columbia Univ, MS, 71, PhD(physics), 76. Prof Exp: RES ASSOC PHYSICS, NEVIS LABS, COLUMBIA UNIV, 75- Res: Photon beams of energy, 100-400 Giga-electron volts as a probe of nuclear and sub-nuclear matter. Mailing Add: Nevis Labs PO Box 137 Irvington NY 10533

WIJGA, PIETER JOHANNES, chemical engineering, organic chemistry, see 12th edition

WIJNBERG, LOUIS, theoretical atomic physics, applied mathematics, see 12th edition

WIJNEN, JOSEPH M H, b Wittem, Netherlands, Sept 22, 20; US citizen; m 67; c 2. PHYSICAL CHEMISTRY, PHOTOCHEMISTRY. Educ: Cath Univ Louvain, Lic Sci, 46, Dr Sci(chem), 48. Prof Exp: Lectr chem, Cath Univ Louvain, 48-49; Nat Res Coun Can fel photochem, 49-51; res assoc, NY Univ, 51-53; Nat Res Coun Can res officer, 53-55; res assoc chem, Celanese Corp Am, 55-58; sr fel photochem, Mellon Inst, 58-63; PROF CHEM, HUNTER COL, 63- Concurrent Pos: Consult, US Bur Mines, Pittsburgh, Pa, 61-63; NSF res grants, 67-69 & 70-72; vis prof, Univ Bonn, 69-70. Mem: Am Chem Soc; Nat Combustion Inst; fel Am Inst Chemists. Res: Primary processes in photochemical reactions; free radical reactions; kinetics of free radical induced polymerization reactions. Mailing Add: Dept of Chem Hunter Col New York NY 10021

WIJSMAN, ROBERT ARTHUR, b Hague, Netherlands, Aug 20, 20; m 53; c 3. MATHEMATICAL STATISTICS. Educ: Delft Inst Technol, Netherlands, Ir, 45;

Univ Calif, PhD(physics), 52. Prof Exp: Lectr med physics, Univ Calif, 52-53, instr math, 53-54, res statistician, 54-55, lectr statist & pub health, 55-56, actg asst prof statist, 56-57; from asst prof to assoc prof, 57-65, PROF STATIST, UNIV ILL, URBANA, 65- Concurrent Pos: Vis prof, Columbia Univ, 67-68. Mem: Inst Math Statist; Am Math Soc. Res: Sequential and multivariate analysis. Mailing Add: 310 Altgeld Hall Univ of Ill Urbana IL 61801

WIKJORD, ALFRED GEORGE, b Flin Flon, Man, July 15, 43; m 68. NUCLEAR CHEMISTRY. Educ: Univ Manitoba, BSc, 64, MSc, 65; McGill Univ, PhD(chem), 69. Prof Exp: NATO sci fel, Strasbourg Macromolecular Res Ctr, France, 69-70; RES OFFICER, ANAL SCI BR, WHITESHELL NUCLEAR RES ESTAB, ATOMIC ENERGY CAN LTD, 70- Mem: Chem Inst Can. Res: Chemistry of nuclear reactors; heavy water production. Mailing Add: Anal Sci Br Whiteshell Nuclear Res Estab Pinawa MB Can

WIKLER, ABRAHAM, b New York, NY, Oct 12, 10; m 35; c 4. PSYCHIATRY. Educ: Columbia Univ, AB, 30; Long Island Col Med, MD, 35; Am Bd Psychiat & Neurol, dipl, 44. Prof Exp: Intern med & surg, Brooklyn Jewish, Montefiore, Kingston Ave & US Marine Hosps, NY & resident psychiat, Hosp, Ky, 40-42; from exp neuropsychiatrist to chief neuropsychiat sect, Addiction Res Ctr, NIMH, 43-63; PROF PSYCHIAT & PHARMACOL, COL MED, UNIV KY, 63- Concurrent Pos: USPHS res fel neurophysiol, Ill Neuropsychiat Inst, Lab of Physiol, Yale Univ & Rockefeller Inst, 42-43; fel, Ctr Advan Study Behav Sci, 70-71; asst prof, Col Med, Univ Cincinnati, 48-62; consult, Rockefeller Univ, 63-69, Addiction Res Ctr, NIMH, Ky, 63- & Div Res Grants, 57-59 & 67-69; mem expert adv panel, Comt Dependence-Producing Drugs, WHO, 63- Mem: Asn Res Nerv & Ment Dis (pres, 66); Am Acad Neurol. Res: Drug addiction; neurophysiological and psychological pharmacodynamics; electroencephalography; clinical neurology and psychiatry. Mailing Add: Dept of Psychiat Univ of Ky Med Ctr Lexington KY 40506

WIKMAN-COFFELT, JOAN, b Chicago, Ill, June 9, 33; m 70. BIOCHEMISTRY. Educ: Alverno Col, BS, 59; St Mary's Col, Minn, MS, 64; St Louis Univ, PhD(biochem), 68. Prof Exp: Asst prof biochem, Med Sch, Univ Okla, 70-72; ASST PROF INTERNAL MED & BIOL CHEM, MED SCH, UNIV CALIF, DAVIS, 72- Concurrent Pos: NIH fel, Baylor Col Med, 68-70. Res: Regulation and biological significance of myosin synthesis, phosphorylation, and enzymatic alterations. Mailing Add: Dept of Internal Med Univ of Calif Davis CA 95616

WIKNER, EMIL GORDON, physics, see 12th edition

WIKOFF, HELEN LANDMAN, b Columbus, Ohio, July 9, 00. BIOCHEMISTRY. Educ: Ohio State Univ, BA, 21, MS, 22, PhD(org chem), 24. Prof Exp: Asst chem, 21-24, from instr to prof physiol chem, 25-69, EMER PROF PHYSIOL CHEM, COL MED, OHIO STATE UNIV, 69- Mem: AAAS; Am Chem Soc. Res: Blood analysis and analytical methods applied to other body tissues and fluids; composition of foods; biochemistry computer programs. Mailing Add: 357 Arden Rd Columbus OH 43214

WIKSWO, MURIEL ANASTASIA, b Teaneck, NJ. CELL BIOLOGY. Educ: Sweet Briar Col, AB, 66; Northwestern Univ, Evanston, PhD(biol), 70. Prof Exp: NIH res fel anat & physiol, Harvard Univ, 70-72; res fel dermat, Sch Med, Yale Univ, 72-74; MEM STAFF, DEPT ORAL HISTOPATH, SCH DENT MED, HARVARD UNIV, 74- Mem: AAAS; Am Soc Cell Biol; Am Soc Zool. Res: Pigment cell biology; biochemistry; physiology; electron microscopy. Mailing Add: Dept of Oral Histopath Harvard Univ Sch of Dent Med Boston MA 02115

WIKTOR, TADEUSZ JAN, b Stryj, Poland, Sept 9, 20; US citizen; m 48; c 3. VIROLOGY. Educ: Sch Vet Med, Alfort & Univ Paris, DrVetMed, 46; Inst Trop Med, Alfort, France, cert trop vet med, 46. Prof Exp: Asst path, Sch Vet Med, Alfort, France, 46-47; dir lab microbiol, Vet Lab, Kisenyi, Belgian Congo, 47-49, Vet Lab, Stanleyville, 49-55, Vet Lab, Astrida, 55-58 & Vet Lab, Elizabethville, 58-60; Food & Agr Orgn expert virol, Animal Husb Res Inst, WPakistan, 60-61; assoc mem virol rabies, 61-75, MEM, WISTAR INST ANAT & BIOL, 75- Concurrent Pos: Mem, expert comt rabies, WHO, 63; assoc prof, Sch Vet Med, Univ Pa, 67. Res: Animal pathology, viral and bacterial diseases, epidemiology, vaccine development and production; research on rabies virus. Mailing Add: 9 Downs Circle Wynnewood PA 19096

WIKUM, DOUGLAS ARNOLD, biology, ecology, see 12th edition

WILANSKY, ALBERT, b St John's, Nfld, Sept 13, 21; nat US; m 69; c 5. MATHEMATICS. Educ: Dalhousie Univ, MA, 44; Brown Univ, PhD(math), 47. Prof Exp: Demonstr physics, Dalhousie Univ, 42-44; instr math, Brown Univ, 46-48; from asst prof to assoc prof, 48-57, PROF MATH, LEHIGH UNIV, 57- Concurrent Pos: Consult, Frankford Arsenal, 57-58; Fulbright vis prof, Reading Univ, 72-73. Honors & Awards: Ford Prize, Math Asn Am, 69. Mem: Am Math Soc; Math Asn Am. Res: Pure mathematics; analysis; summability; linear topological space; Banach algebra; functional analysis. Mailing Add: Dept of Math Lehigh Univ Bethlehem PA 18015

WILBAND, JOHN TRUAX, b Montreal, Que, May 26, 38; m 60; c 3. PETROLOGY, GEOCHEMISTRY. Educ: Univ NB, BSc, 60, MSc, 62; Univ Ill, PhD(geochem), 65. Prof Exp: Asst prof, 65-69, ASSOC PROF GEOL, UNIV TOLEDO, 69- Concurrent Pos: Consult & res scientist, Owens-Ill Co, 66- Mem: Geochem Soc; Mineral Soc Am; Mineral Asn Can. Res: Geochemistry of ore bodies; distribution of trace elements during metamorphism; x-ray absorption edge spectroscopy. Mailing Add: Dept of Geol Univ of Toledo Toledo OH 43606

WILBANKS, JOHN RANDALL, b Foreman, Ark, June 10, 38; m 62; c 1. PETROLOGY, STRUCTURAL GEOLOGY. Educ: NMex Inst Mining & Technol, BS, 60; Tex Tech Univ, MS, 66, PhD(geol), 69. Prof Exp: Geologist, NMex State Hwy Dept, 61-63; teaching asst geol, Tex Tech Univ, 63-65, res assoc geol, Marie Byrd Land, 67-69, co-investr & NSF Antarctic grant, 69-70, vis asst prof geol, 70-71; ASSOC PROF & CHMN DEPT GEOSCI, UNIV NEV, LAS VEGAS, 71- Honors & Awards: US Antarctic Serv Medal, Off Polar Progs, NSF, 70. Mem: Geol Soc Am. Res: Petrology and structure of metamorphic complexes; Marie Byrd Land, Antarctica; morphology of metamorphic zircons; tectonics of the southern great basin. Mailing Add: Dept of Geosci Univ of Nev Las Vegas NV 89154

WILBANKS, THOMAS JOHN, b Texarkana, Ark, Aug 4, 38; m 59; c 3. GEOGRAPHY. Educ: Trinity Univ, BA, 60; Syracuse Univ, MA, 67, PhD(geog), 69. Prof Exp: Asst prof geog, Syracuse Univ, 69-73, exec dir, Urban Transp Inst, 71-72, res dir environ policy proj, 72-73; ASSOC PROF GEOG & CHMN DEPT, UNIV OKLA, 73- Mem: Asn Am Geog; Regional Sci Asn. Res: Regional development; applications of pattern analysis; location and public policy; areal subdivision and public administration. Mailing Add: Dept of Geog Univ of Okla Norman OK 73069

WILBARGER, EDWARD STANLEY, JR, b Billings, Mont, Feb 21, 31; m 59; c 2. PHYSICS. Educ: Va Mil Inst, BS, 52; US Naval Postgrad Sch, MS, 56. Prof Exp: Chief indust hyg sect, Off Surgeon Gen, US Army, 56-58; engr physicist, Proj Res Aviation Med, US Naval Med Res Inst, 58-59; engr physicist, Bioastronaut Res Unit, Ord Missile Command, US Army, Redstone Arsenal, 59, chief inspections, Health & Safety Br, Off Chief Engrs, 59-60; head bioinstrumentation group, Corp, 60-62, sr res physicist, Aerospace Opers Dept, AC Electronics Defense Res Labs, 62-71, SR RES PHYSICIST, DELCO ELECTRONICS DIV, GEN MOTORS CORP, 71- Mem: Am Acad Mech. Res: Control engineering; measurement methods; physiological response to stress; design and fabrication of control systems; mobility systems analysis for off-road vehicles; analysis and design of auto safety systems; hypervelocity interior ballistics and fluid dynamics. Mailing Add: Delco Electronics Div 6767 Hollister Ave Goleta CA 93017

WILBER, CHARLES GRADY, b Waukesha, Wis, June 18, 16; m 52; c 6. PHYSIOLOGY. Educ: Marquette Univ, BSc, 38; Johns Hopkins Univ, MA, 41, PhD(gen physiol), 42. Prof Exp: Lab asst zool, Marquette Univ, 38-39; asst, Johns Hopkins Univ, 40-42; instr, St Louis Univ, 42, assoc prof physiol & dir biol labs, 49-52; asst prof, Fordham Univ, 46-49; chief animal ecol br, Chem Corps Med Labs, US Army Chem Ctr, Md, 52-56, comp physiol br, Chem Res & Develop Labs, 56-60; prof biol & dean grad sch, Kent State Univ, 61-64; dir marine labs & prof, Univ Del, 64-67; PROF ZOOL, COLO STATE UNIV, 67-, DIR FORENSIC SCI LAB, 74- Concurrent Pos: Leader, Fordham Arctic Exped, 48; assoc, Univ Pa, 53-60; prof lectr, Loyola Col, 57-60; mem corp, Marine Biol Lab, Woods Hole; mem panel environ physiol, US Dept Army; mem life sci comt, Nat Acad Sci-Air Res & Develop Command; consult, USPHS; dep coroner, Larimer County, Colo; dir, Ecol Consults, Inc, 72-74; toxicologist, Thorne Ecol Inst, 72- Mem: Fel Am Acad Forensic Sci; Am Physiol Soc; fel NY Acad Sci. Res: Biochemistry of body fluids; chemistry of metabolism; comparative aspects of environmental physiology; climatic adaption; forensic biology; wound ballistics; environmental quality; environmental pathology; comparative toxicology; oceanography. Mailing Add: 900 Edwards Ft Collins CO 80521

WILBER, JOE CASLEY, JR, b Jonesboro, Ark, Feb 28, 29; m 51. CHEMISTRY, SCIENCE EDUCATION. Educ: Memphis State Col, BS, 50, MA, 53; Univ Ga, EdD, 61. Prof Exp: Teacher pub schs, Tenn, 50-51 & 53-56; teacher & chmn sci dept, pub sch, Ga, 56-59; from asst prof to assoc prof chem, Ga Southern Col, 60-70; teacher chem & chmn dept sci, Wingate Col, 70-73; ASSOC PROF CHEM, PAUL D CAMP COMMUNITY COL, 73- Mem: Am Chem Soc; AAAS. Res: General chemistry; qualitative analysis, a non-sulfide scheme. Mailing Add: Dept of Chem Paul D Camp Community Col PO Box 737 Franklin VA 23851

WILBER, LAURA ANN, b Memphis, Tenn, May 26, 34. AUDIOLOGY, SPEECH PATHOLOGY. Educ: Univ Southern Miss, BS, 55; Gallaudet Col, MS, 58; Northwestern Univ, PhD(audiol), 64. Prof Exp: Teacher hard of hearing & deaf, McKinley Elem Sch, Bakersfield, Calif, 55-57; speech therapist & coordr spec educ, US Army Dependent Sch Syst, Heidelberg, Ger, 57-61; res asst audiol, Northwestern Univ, 61-64; asst res audiologist, Univ Calif, Los Angeles, 64-70; asst prof, 70-75, ASSOC PROF REHAB MED & DIR HEARING & SPEECH SERV, ALBERT EINSTEIN COL MED, 75- Concurrent Pos: Spec instr, Calif State Col, Los Angeles & Univ Southern Calif, 65-70; dir audiol clin, Hosp, Univ Calif, Los Angeles, 68-69; chmn, Clin Sch-Coun New York, 72-73; mem, Dir Hosp Speech & Hearing, Prog Asn & Soc Ear, Nose & Throat Advan in Children; US rep, Int Stand Orgn; mem, Am Nat Stand Inst; adj assoc prof, City Univ New York, 74- Mem: Am Speech & Hearing Asn; Acoust Soc Am. Mailing Add: Rm 3C 37 VE Albert Einstein Col of Med Bronx NY 10461

WILBERT, JOHANNES, b Cologne, Ger, June 23, 27; m 55; c 2. ANTHROPOLOGY. Educ: Univ Cologne, BA, 51, PhD(anthrop), 55. Prof Exp: Lectr lang, Univ Birmingham, Eng, 51-52; dir anthrop, La Salle Found Natural Sci, dir latin Am Ctr, 63-74, PROF ANTHROP, UNIV CALIF, LOS ANGELES, 63- Concurrent Pos: Ed, Antropologica. Res: Cultural anthropology of Venezuelan and other South American Indians; blood groups of the Amerindian and Mongoloids. Mailing Add: Dept of Anthrop Univ of Calif Los Angeles CA 90024

WILBORN, WALTER HARRISON, b Arbyrd, Mo, May 20, 35; m 57; c 3. HUMAN ANATOMY. Educ: Harding Col, BA, 57; St Louis Univ, MS, 62; Univ Tenn, PhD(anat), 67. Prof Exp: From asst prof to assoc prof anat, Med Ctr, Univ Ala, Birmingham, 67-73; ASSOC PROF ANAT, COL MED, UNIV S ALA, 73- Concurrent Pos: NIH res grant, 67-72. Mem: Int Asn Dent Res; Am Asn Anat; Am Soc Cell Biol; Histochem Soc. Res: Cytology and cytochemistry of mammalian salivary glands; secretory mechanisms; reproductive biology. Mailing Add: Dept of Anat Univ SAla Col of Med Mobile AL 36688

WILBUR, DAVID WESLEY, b Hinsdale, Ill, Dec 15, 37; m 68; c 1. BIOPHYSICS. Educ: Pac Union Col, BA, 61; Univ Calif, Berkeley, PhD(biophys), 65; Loma Linda Univ, MD, 71; Am Bd Internal Med, cert, 75. Prof Exp: Biophysicist, Lawrence Radiation Lab, Univ Calif, 65-67; asst prof physiol & biophys, Sch Med, Loma Linda Univ, 67-69, med intern, 71-72, med resident, 72-74, res assoc biomath, 69-74; Am Cancer Soc med oncol clin fel, Roswell Park Mem Inst, 74-76; MEM, MED ONCOL, DEPT INTERNAL MED, LOMA LINDA UNIV, 76- Mem: AAAS. Res: Cancer chemotherapy; computer simulation of cell population kinetics; role of immunology in cancer. Mailing Add: Dept of Internal Med Loma Linda Univ Loma Linda CA 92354

WILBUR, DONALD ALDRICH, b Mt Pleasant, Mich, Feb 18, 04; m 38; c 3. PHYSICS. Educ: Cent Mich Col Educ, AB, 25; Univ Mich, MS, 28, PhD(physics), 32. Prof Exp: Instr physics, Grinnell Col, 32-33; assoc prof physics & elec eng, Rensselaer Polytech Inst, 33-43; res assoc, 43-60, mgr res super power microwave tube lab, 60-66, CONSULT, MICROWAVE TUBE BUS SECT, GEN ELEC CO, 66- Mem: Sr mem Inst Elec & Electronics Engrs. Res: Wavebeam interactions in microwave tubes. Mailing Add: 369 Sixth Ave N Tierra Verde FL 33715

WILBUR, DONALD LEE, b Chicago, Ill, Apr 28, 66; c 2. NEUROENDOCRINOLOGY. Educ: Ind State Univ, BS, 68, MA, 70; Med Univ SC, PhD(anat), 74. Prof Exp: ASST PROF ANAT, SCH MED, TEX TECH UNIV, 74- Concurrent Pos: Consult, Med Univ SC, 74-75. Mem: Am Asn Anatomists. Res: Electron microscopic, immunocytochemical studies of the pituitary gland and circumventricular organs, correlated with radioimmunoassayable levels of circulating hormones; mechanisms of hormone synthesis and release. Mailing Add: Dept of Anat Tex Tech Univ Sch of Med Lubbock TX 79409

WILBUR, FRANK HALL, botany, developmental physiology, see 12th edition

WILBUR, HENRY MILES, b Bridgeport, Conn, Jan 25, 44; m 67; c 1. ZOOLOGY, ECOLOGY. Educ: Duke Univ, BS, 66; Univ Mich, Ann Arbor, PhD(zool), 71. Prof Exp: Univ Mich Soc Fels jr fel, Div Reptiles & Amphibians, Mus Zool, Univ Mich, Ann Arbor, 71-73; ASST PROF ZOOL, DUKE UNIV, 73- Concurrent Pos: Edwin S George scholar, Edwin S George Reserve, Mich, 68-69. Honors & Awards: Stove Award, Am Soc Ichthyol & Herpet, 70. Mem: AAAS; Ecol Soc Am; Soc Study Evolution; Brit Ecol Soc; Am Soc Ichthyol & Herpet. Res: Evolutionary ecology;

evolution of species interactions and life histories. Mailing Add: Dept of Zool Duke Univ Durham NC 27706

WILBUR, JAMES MYERS, JR, b Philadelphia, Pa, Oct 31, 29; m 60; c 3. ORGANIC CHEMISTRY. Educ: Muhlenberg Col, BS, 51; Univ Pa, PhD, 59. Prof Exp: Res chemist, J T Baker Chem Co, NJ, 51-53; NIH fel cancer chemother, Univ Minn, 58-60; res chemist, E I du Pont de Nemours & Co, 60-62; fel, Univ Ariz, 62-63; assoc prof chem, 63-66, PROF CHEM, SOUTHWEST MO STATE COL, 66- Mem: Am Chem Soc. Res: Medicinal chemistry; cancer chemotherapy; organic mechanisms; polymers. Mailing Add: Dept of Chem Southwest Mo State Col Springfield MO 65802

WILBUR, KARL MILTON, b Binghamton, NY, Jay 7, 12; m 46; c 2. PHYSIOLOGY. Educ: Ohio State Univ, BA, 35, MA, 36; Univ Pa, PhD(zool), 40. Prof Exp: Asst zool, Ohio State Univ, 35-36; instr, Univ Pa, 39-40; Rockefeller fel, NY Univ, 40-41; instr zool, Ohio State Univ, 41-42, asst prof physiol, Med Sch, Dalhousie Univ, 42-44; assoc prof zool, 46-50, PROF ZOOL, DUKE UNIV, 50- Concurrent Pos: Physiologist, AEC, 52-53. Mem: Am Physiol Soc; Am Soc Naturalists; Soc Gen Physiol; Am Soc Zoologists. Res: Cellular physiology; calcification in marine organisms; cell division. Mailing Add: Dept of Zool Duke Univ Durham NC 27706

WILBUR, ROBERT DANIEL, b Glendale, Calif, May 7, 31; m 52; c 3. ANIMAL NUTRITION, ANIMAL PHYSIOLOGY. Educ: Calif State Polytech Col, BS, 54; Iowa State Univ, PhD(animal nutrit, bact), 59. Prof Exp: Asst nutrit & bact, Iowa State Univ, 54-59, fel, 59; RES NUTRITIONIST, AM CYANAMID CO, 59- Mem: AAAS; Am Soc Animal Sci. Res: Nutrition and physiology of domesticated animals. Mailing Add: Agr Div Am Cyanamid Co Princeton NJ 08540

WILBUR, ROBERT LYNCH, b Annapolis, Md, July 4, 25; m 55; c 6. PLANT TAXONOMY. Educ: Duke Univ, BS, 46, AM, 48; Univ Mich, PhD, 52. Prof Exp: Asst bot, Duke Univ, 46-47, Univ Hawaii, 47-48 & Univ Mich, 48-52; asst prof, Univ Ga, 52-53; asst prof & cur herbarium, NC State Col, 53-57; from asst prof to assoc prof bot, 57-70, PROF BOT, DUKE UNIV, 70-, CHMN DEPT, 71-, CUR, HERBARIUM, 57- Mem: Am Soc Plant Taxon; Int Asn Plant Taxon. Res: Taxonomy and phytogeography of vascular plants; flora of the southeastern United States and Central America. Mailing Add: 265 Biol Sci Bldg Duke Univ Dept of Bot Durham NC 27706

WILBUR, WILLY JOHN, mathematics, see 12th edition

WILBURN, ADOLPH YARBROUGH, science manpower development, academic administration, see 12th edition

WILBURN, RICHARD LEE, b Heyworth, Ill, Oct 18, 26; m 47; c 2. PHYSICAL CHEMISTRY. Educ: Univ Wash, BS, 48, MS, 51, PhD(phys chem), 54. Prof Exp: Group leader polymers, Morse Labs, Calif, 55-56; phys chemist, 56-57, asst chief phys chem sect, 57-58, CHIEF PHYS CHEM SECT, DUGWAY PROVING GROUNDS, US DEPT ARMY, 58- Honors & Awards: Dept Army Res & Develop Achievement Award, 65. Mem: Am Chem Soc. Res: Polymer development; wood preservation; semiconducting organic compounds; electrochemical double-layer phenomena; kinetics and mechanism of chemiluminescence; physical properties of chemical warfare agents; ultraviolet, visible and infrared spectroscopy; instrumental analysis; liquid state. Mailing Add: 1930 Laurelhurst Dr Salt Lake City UT 84108

WILCE, ROBERT THAYER, b Carbondale, Pa, Dec 9, 24; m 56. BOTANY. Educ: Univ Scranton, BS, 50; Univ Vt, MS, 52; Univ Mich, PhD(bot), 57. Prof Exp: Instr bot, Univ Mich, 57-58, fel, Horace Rackham Grad Sch, 58-59; from instr to asst prof, 59-68, ASSOC PROF BOT, UNIV MASS, AMHERST, 68- Mem: Phycol Soc Am. Res: Systematic morphology, distribution and ecology of the attached algae of arctic and subarctic areas, especially the Canadian eastern arctic and northwest Greenland. Mailing Add: Dept of Bot Univ of Mass Amherst MA 01003

WILCHINSKY, ZIGMOND WALTER, b New York, NY, Aug 26, 15; m 40; c 1. POLYMER PHYSICS. Educ: Rutgers Univ, BS, 37, MS, 39; Mass Inst Technol, PhD(physics), 42. Prof Exp: Asst physics, Rutgers Univ, 37-39; mem staff radiation lab, Mass Inst Technol, 42-45; sect head, US Naval Res Lab, Washington, DC, 43-46; SR RES ASSOC, EXXON CHEM CO, 46- Mem: AAAS; Am Phys Soc; Am Chem Soc; Am Crystallog Asn; NY Acad Sci. Res: Structure of plastics; rubber technology; x-ray diffraction; physical chemistry of catalysts; adsorption; development of microwave generators; vacuum tube development. Mailing Add: Exxon Chem Co PO Box 45 Linden NJ 07036

WILCOMB, MAXWELL JEFFERS, JR, zoology, see 12th edition

WILCOX, ARCHER CARL, b Milwaukee, Wis, Apr 22, 20; m 48; c 3. CHEMISTRY, BIOCHEMISTRY. Educ: Univ Kans, BS, 46, MS, 48; Kans State Univ, PhD(biochem eng), 53. Prof Exp: Instr chem eng, Univ Kans, 48-50; res engr, Universal Food Prod, Inc, 53-55, Charles A. Krause Milling Co, 55-57 & Dairyland Food Labs, 57-59; lab dir food anal, Wil-Kil Food Anal & Res Labs, 59-61; assoc prof biol & chem, Milwaukee Downer Col, 61-64; assoc prof chem, 64-69, PROF CHEM, UNIV WIS-EAU CLAIRE, 69- Mem: Am Chem Soc; Am Soc Microbiol. Res: Microbial biochemistry. Mailing Add: Dept of Biochem Univ of Wis Eau Claire WI 54701

WILCOX, BENSON REID, b Charlotte, NC, May 26, 32; m 59; c 4. CARDIOVASCULAR SURGERY, THORACIC SURGERY. Educ: Univ NC, Chapel Hill, BA, 53, MD, 57. Prof Exp: From instr to assoc prof, 63-71, PROF SURG, UNIV NC, CHAPEL HILL, 71-, CHIEF DIV CARDIOVASC & THORACIC SURG, 69- Concurrent Pos: NIH fel, Univ NC, Chapel Hill, 63-64 & grant, 68-74; Markle scholar, 67-72; consult, NC Sanatorium Syst, 64- Mem: Am Asn Thoracic Surg; Am Col Surg; Am Surg Asn; Soc Thoracic Surg; Soc Univ Surg. Res: Application of biomathematical and engineering principles to the study of the circulation; pulmonary circulation in heart disease. Mailing Add: Div Cardiothoracic Surg Univ NC Sch Med Box 7 Div Health Affairs Chapel Hill NC 27514

WILCOX, CALVIN HAYDEN, b Cicero, NY, Jan 29, 24; m 47; c 3. MATHEMATICS. Educ: Harvard Univ, AB, 51, AM, 52, PhD(math), 55. Prof Exp: Mathematician, Air Force Cambridge Res Ctr, 53-55; from instr to assoc prof math, Calif Inst Technol, 55-61; prof math & mem US Army Math Res Ctr, Univ Wis, 61-66; prof math, Univ Ariz, 66-69 & Univ Denver, 69-71; PROF MATH, UNIV UTAH, 71- Concurrent Pos: Vis prof, Inst Theoret Physics, Univ Geneva, 70-71; ed, Rocky Mountain J Math, 75- Mem: AAAS; Am Math Soc; Soc Indust & Appl Math. Res: Applied mathematics and mathematical physics, especially theories of wave propagation and scattering in classical and quantum physics; boundary value problems for partial differential equations. Mailing Add: Dept of Math Univ of Utah Salt Lake City UT 84112

WILCOX, CHARLES FREDERICK, JR, b Providence, RI, July 20, 30; m 57; c 3. PHYSICAL CHEMISTY, ORGANIC CHEMISTRY. Educ: Mass Inst Technol, BS, 52; Univ Calif, Los Angeles, PhD(org chem), 57. Prof Exp: NSF fel, Harvard Univ, 57; from instr to assoc prof, 57-74, PROF CHEM, CORNELL UNIV, 74- Concurrent Pos: Guggenheim fel, 66-67; vis prof, Calif Inst Technol, 67; asst ed, J Am Chem Soc, 65-66. Mem: Am Chem Soc; The Chem Soc. Res: Physical aspects of organic chemistry. Mailing Add: Dept of Chem Cornell Univ Ithaca NY 14850

WILCOX, CHARLES HAMILTON, b Rochester, NY, May 21, 29. THEORETICAL PHYSICS, ENGINEERING MANAGEMENT. Educ: Duke Univ, BS, 50; Univ Ill, MS, 52; Univ Southern Calif, 70. Prof Exp: Res physicist & lectr physics, Eng Exp Sta, Ga Inst Technol, 52-53; sr mem tech staff & assoc mgr theoret studies dept, Res Labs, 53-67, mgr tech planning, Aerospace Group, 67-70, dir corp independent res & develop, 70-74, DIR ENG, AEROSPACE GROUPS, HUGHES AIRCRAFT CO, 74- Concurrent Pos: Lectr, Univ Southern Calif, 55-59 & Univ Calif, Los Angeles, 61-63; consult, Stanford Res Inst, 73- Mem: AAAS; Am Phys Soc; Inst Elec & Electronics Eng; Sigma Xi; Inst Mgt Sci. Res: Scattering and diffraction of electromagnetic waves; radiowave propagation and geophysics; technology planning and development. Mailing Add: 10520 Draper Ave Los Angeles CA 90064

WILCOX, CHARLES JULIAN, b Harrisburg, Pa, Mar 28, 30; m 55; c 2. DAIRY SCIENCE. Educ: Univ Vt, BS, 50; Rutgers Univ, MS, 55, PhD(animal breeding), 59. Prof Exp: Res asst dairy sci, Rutgers Univ, 50 & 53-55; owner & mgr dairy farm, 55-56; res asst dairy sci, Rutgers Univ, 56-59; asst prof dairy sci & assoc geneticist, 65-71, PROF DAIRY SCI & GENETICIST, UNIV FLA, 71- Mem: Am Dairy Sci Asn; Am Soc Animal Sci; Am Inst Biol Sci; Latin Am Asn Animal Prod. Res: Quantitative genetics of productive traits of farm animals, including milk yield and composition, reproductive performance, birth weights, gestation lengths, heat tolerance, type conformation, disease resistance, life span and livability. Mailing Add: Dept of Dairy Sci Univ of Fla Gainesville FL 32611

WILCOX, CLIFFORD LAVAR, b Archer, Idaho, Apr 15, 25; m 45; c 5. DAIRY HUSBANDRY. Educ: Utah State Univ, BS, 51; Univ Minn, MS, 57, PhD(dairy husb), 59. Prof Exp: Asst dairying, 56-59, instr dairy husb, 60, exten dairy specialist, 60-65, asst dir agr exp sta, 68-72, SUPT, AGR EXP STA, UNIV MINN, 65- Mem: Am Dairy Sci Asn. Res: Dairy cattle breeding. Mailing Add: Inst of Agr Univ of Minn Agr Exp Sta Rosemount MN 55068

WILCOX, ETHELWYN BERNICE, b Wyoming, Iowa, Mar 19, 06. NUTRITION. Educ: Iowa State Univ, BS, 31, MS, 37, PhD(nutrit), 42. Prof Exp: Teacher high sch, 32-36; asst & supvr animal nutrit lab, Iowa State Univ, 37-42; asst home economist, Exp Sta, Wash State Univ, 42-43; from asst prof to prof nutrit, 43-71, fac hon res lectr, 59, head dept food & nutrit, 65-71, EMER PROF NUTRIT, UTAH STATE UNIV, 71- Honors & Awards: Spec Fac Award, Utah State Univ, 64. Mem: Am Home Econ Asn; Am Dietetic Asn; Inst Food Technologists; Am Inst Nutrit. Res: Nutritional status of population groups; lipid metabolism; chemical components of venison flavor. Mailing Add: 788 Hillcrest Ave Logan UT 84321

WILCOX, FLOYD LEWIS, b Lodi, NY, Dec 17, 37; m 57; c 2. SCIENCE EDUCATION. Educ: Houghton Col, BS, 65; Univ Miami, Coral Gables, MS, 67; State Univ NY Binghamton, PhD(chem), 71. Prof Exp: Assoc prof, 70-75, PROF SCI CHEM, CENT WESLEYAN COL, 75- Mem: Am Chem Soc; Am Sci Affil; Nat Sci Teachers Asn. Res: Water pollution analysis aimed at monitoring local streams and lakes; new and novel teaching methods and aids; new methods for the removal of foreign substances from natural or industrial waters. Mailing Add: Cent Wesleyan Col Central SC 29630

WILCOX, FRANK H, b Norwich, Conn, June 15, 27; m 60; c 2. GENETICS. Educ: Univ Conn, BS, 51; Cornell Univ, MS, 53, PhD(animal genetics), 55. Prof Exp: Assoc prof poultry physiol, Univ Md, 55-67; ASSOC PROF LIFE SCI, IND STATE UNIV, TERRE HAUTE, 67- Concurrent Pos: Poultry Sci travel award to World Poultry Cong, Australia, 62; USPHS res fel, 75; vis investr, Jackson Lab, 75-76. Mem: Genetics Soc Am; World Poultry Sci Asn. Res: Biochemical genetics, especially electrophoretic variants in vertebrates. Mailing Add: Dept of Life Sci Ind State Univ Terre Haute IN 47809

WILCOX, GARY LYNN, b Ventura, Calif, Jan 7, 47; m 70. BIOCHEMICAL GENETICS. Educ: Univ Calif, Santa Barbara, BA, 69, MA, 72, PhD(molecular biol), 72. Prof Exp: Res assoc biol, Univ Calif, Santa Barbara, 72-74; ASST PROF BACT, UNIV CALIF, LOS ANGELES, 74- Mem: Am Soc Microbiol; Genetics Soc Am; Am Soc Biol Chemists. Res: Molecular basis of positive and negative control of L-arabinose utilization in E coli; protein nucleic acid interactions. Mailing Add: Dept of Bact Univ of Calif Los Angeles CA 90024

WILCOX, GERALD EUGENE, b Wautoma, Wis, July 17, 25; m 49; c 3. SOIL FERTILITY. Educ: Univ Wis, BS, 49, MS, 51, PhD, 53. Prof Exp: Asst agronomist, Northern La Hill Farm Exp Sta, La State Univ, 53-57; assoc prof, 57-71, PROF HORT & AGRON, PURDUE UNIV, LAFAYETTE, 71- Concurrent Pos: Plant nutritionist, US Agency Int Develop, Brazil, 75-77. Mem: Soil Sci Soc Am; Am Soc Hort Sci; Int Soil Sci Soc. Res: Mineral nutrition of vegetable crops, especially fertilization of tomatoes and potatoes; soil fertility; culture and mechanization of tomato production. Mailing Add: Dept of Hort Purdue Univ West Lafayette IN 47906

WILCOX, HAROLD EDWIN, b Sidney, Ohio, June 3, 13; m 41; c 2. CHEMISTRY. Educ: Ohio Wesleyan Univ, BA, 34; Ohio State Univ, MSc, 37, PhD(chem), 39. Prof Exp: High sch teacher, Ohio, 34-35; chemist, Gummed Prod Co, 35-36; from asst prof to assoc prof chem, Howard Col, 39-46; prof, Birmingham-Southern Col, 46-62, head dept, 47-62; actg chmn dept, 65-66, PROF CHEM, OHIO WESLEYAN UNIV, 62- Concurrent Pos: Instr eng sci & mgt war training, Ala Polytech Inst, 42-45; vis prof, Howard Col, 46-47; consult, Norwood Clin, 52-53, 54-59 & 60-61; Ford Found grant, Univ Calif, 53-54; NSF grant, Calif Inst Technol, 59-60, vis assoc, Div Chem & Chem Eng, 66-67; Am Chem Soc-Div Chem Educ vis scientist, cols, 60-; consult, Med Col Ala, 61-62; adv chem, Grad Rec Exam Comt, 61-72, chmn, 68-72. Mem: AAAS; Am Chem Soc; Am Asn Clin Chem. Res: Chemical tests of nutritional status; antifoam action; use of oscillography in polarograph analysis; clinical chemistry. Mailing Add: Dept of Chem Ohio Wesleyan Univ Delaware OH 43015

WILCOX, HARRY HAMMOND, b Canton, Ohio, May 31, 18; m 41; c 3. ANATOMY. Educ: Univ Mich, BS, 39, MS, 40, PhD(zool), 48. Prof Exp: Assoc prof biol, Morningside Col, 47-48; assoc anat, Sch Med, Univ Pa, 48-52; from asst prof to prof, 52-67, GOODMAN PROF ANAT, UNIV TENN, MEMPHIS, 67- Mem: Am Soc Zool; Am Asn Anat; Am Acad Neurol. Res: Effects of aging on the nervous system; internal ear; central nervous system pathways. Mailing Add: Dept of Anat Univ of Tenn Ctr for Health Sci Memphis TN 38163

WILCOX, HENRY G, b Hornell, NY, Jan 26, 33; m 66; c 2. BIOCHEMISTRY, PHARMACOLOGY. Educ: Univ Fla, PhD(biochem), 64. Prof Exp: Asst prof, 68-74, ASSOC PROF PHARMACOL, VANDERBILT UNIV, 74- Concurrent Pos: NIH fel pharmacol, Vanderbilt Univ, 64-67. Res: Plasma lipoprotein metabolism, structure and function; methodology for lipoprotein isolation and analysis; hormonal control of lipid metabolism and transport. Mailing Add: Dept of Pharmacol Vanderbilt Univ Nashville TN 37203

WILCOX, HOWARD ALBERT, b Minneapolis, Minn, Nov 9, 20; m 43; c 3. PHYSICS, ENVIRONMENTAL MANAGEMENT. Educ: Univ Minn, BA, 43; Univ Chicago, MA & PhD(physics), 48. Prof Exp: Instr physics, Harvard Univ & Radcliffe Col, 43-44; jr scientist, Los Alamos Sci Lab, NMex, 44-46; asst, Inst Nuclear Studies, Univ Chicago, 46-48; from instr to asst prof & mem staff radiation lab, Univ Calif, 48-50; res physicist & head guided missile develop div, US Naval Ord Test Sta, 50-55, head weapons develop dept, 55-58, asst tech dir res & head res dept, 58-59; dep dir defense res & eng, Off Secy Defense, Washington, DC, 59-60; dir res & eng defense res labs, Gen Motors Corp, 60-66, tech dir adv power systs, Res Labs, 66-67; physicist, US Naval Weapons Ctr, China Lake, 71-74; MGR OCEAN FOOD & ENERGY FARM PROJ, US NAVAL UNDERSEA CTR, 74-; TECH & MGT CONSULT, 67- Concurrent Pos: Vpres, Minicars, Inc, Goleta, 68-74. Mem: AAAS; fel Am Phys Soc. Res: Production of mesons in nuclear collisions; guided missile system engineering; oceanography; hypervelocity flight physics; lunar and terrestrial vehicles; advanced power systems; technical management; earth's energy balance. Mailing Add: 882 Golden Park Ave San Diego CA 92106

WILCOX, HOWARD JOSEPH, b Plattsburgh, NY, Oct 20, 39. MATHEMATICS. Educ: Hamilton Col, AB, 61; Univ Rochester, PhD(math), 66. Prof Exp: Asst prof math, Univ Conn, 65-67; asst prof, Amherst Col, 67-70; asst prof, 70-72, ASSOC PROF MATH, WELLESLEY COL, 72- Mem: Am Math Soc; Math Asn Am; Sigma Xi. Res: Topological groups; general topology. Mailing Add: Dept of Math Wellesley Col Wellesley MA 02181

WILCOX, HUGH EDWARD, b Manchester, Calif, Sept 2, 16; m 38; c 5. PLANT PHYSIOLOGY. Educ: Univ Calif, BS, 38, PhD, 50; Syracuse Univ, MS, 40. Prof Exp: Asst, State Univ NY Col Forestry, Syracuse, 38-40; technician, Dept Forestry, Univ Calif, 41-42; physicist, Radiation Lab, 42-45; physicist & ord engr, US Naval Ord Test Sta, 45-46; assoc prof forest prod & wood technologist, Ore State Col, 46-50; res assoc & proj leader, Res Found, State Univ NY, 50-54; assoc prof forestry, 54-59, PROF FORESTRY, STATE UNIV NY COL ENVIRON SCI & FORESTRY, 59- Mem: AAAS; Soc Am Foresters; Bot Soc Am; Am Soc Plant Physiol. Res: Growth periodicity; dormancy; physiology of cambial activity; wound healing and regeneration; growth and differentiation of roots; mycorrhiza. Mailing Add: Dept of Forest Bot & Path State Univ of NY Col of Environ Sci & Forestry Syracuse NY 13210

WILCOX, JAMES RAYMOND, b Minneapolis, Minn, Jan 20, 31; m 55; c 2. PLANT BREEDING, GENETICS. Educ: Univ Minn, BA, 53, MS, 59; Iowa State Univ, PhD(plant genetics), 61. Prof Exp: Res geneticist, Inst Forest Genetics, Forest Serv, USDA, 61-66, RES GENETICIST, USDA AGR RES SERV, PURDUE UNIV, WEST LAFAYETTE, 66- Mem: AAAS; Am Soc Agron; Am Genetic Asn; Crop Sci Soc Am. Res: Soybean breeding and genetics. Mailing Add: Room 2-318 Lilly Hall Purdue Univ Dept of Agron West Lafayette IN 47906

WILCOX, JOHN BROWN, physics, see 12th edition

WILCOX, JOHN MARSH, b Iowa City, Iowa, Jan 31, 25; m 55; c 2. SOLAR PHYSICS. Educ: Iowa State Col, BS, 48; Univ Calif, Berkeley, PhD(physics), 54. Prof Exp: Physicist, Lawrence Radiation Lab, Univ Calif, Berkeley, 51-64, res physicist, Space Sci Lab, 64-71; RES PHYSICIST & ADJ PROF, INST PLASMA RES, STANFORD UNIV, 71- Concurrent Pos: Vis physicist, Royal Inst Technol, Stockholm, Sweden, 61-62; assoc ed, J Geophys Res. Mem: AAAS; Am Phys Soc; Am Geophys Union; Am Astron Soc; Royal Astron Soc. Res: Solar and interplanetary magnetic fields; solar wind; photospheric magnetic fields; photospheric supergranulation and chromospheric network; solar wind interactions with geomagnetic field. Mailing Add: Stanford Univ Inst Plasma Res Via Crespi Stanford CA 94305

WILCOX, JOSEPH CLIFFORD, b McLean, Ill, June 18, 30; m 52; c 2. FOOD SCIENCE, FOOD MICROBIOLOGY. Educ: Univ Ill, BS, 52, MS, 54; Inst Mgt, Ill Benedictine Col, grad, 70. Prof Exp: Bacteriologist, 56-58, SECT HEAD SAUSAGE DEVELOP, RES & DEVELOP DEPT, FOOD RES DIV, ARMOUR & CO, 58- Mem: AAAS; Inst Food Technologists. Res: Bacteriological, chemical and radiological warfare decontamination; application of bacteriological principles in development and study of food products and associated problems; research in fresh, semidry and dry sausage items; utilization of nonmeat proteins. Mailing Add: Armour & Co Food Res Div 801 W 22nd St Oak Brook IL 60521

WILCOX, LEE R, physics, see 12th edition

WILCOX, LEE ROY, b Chicago, Ill, June 8, 12; m 40; c 2. ALGEBRA, SCIENCE EDUCATION. Educ: Univ Chicago, SB, 32, SM, 33, PhD(math), 35. Prof Exp: Mem sch math, Inst Advan Study, 35-36, asst, 36-38; instr math, Univ Wis, 38-40; from asst prof to assoc prof, 40-58, PROF MATH, ILL INST TECHNOL, 58-, DIR CTR EDUC DEVELOP, 69- Mem: Am Math Soc; Am Soc Eng Educ; Math Asn Am. Res: Theory of lattices; foundations of mathematics; abstract algebra; mathematics education. Mailing Add: 1404 Forest Ave Wilmette IL 60091

WILCOX, LOUIS VAN INWEGEN, JR, b Orange, NJ, Aug 24, 31; m 56; c 3. ECOLOGY. Educ: Colgate Univ, AB, 53; Cornell Univ, MS, 58, PhD(plant path), 61. Prof Exp: Asst plant physiol, Cornell Univ, 55-57; asst plant path, 57-61; asst prof biol, Lycoming Col, 61-65 & Earlham Col, 65-71; dir, Fahkahatchee Environ Studies Ctr, 71-73; DIR ENVIRON QUAL PROG, HAMPSHIRE COL, 73- Mem: Ecol Soc Am. Res: Mangrove ecology. Mailing Add: Sch of Natural Sci & Math Hampshire Col Amherst MA 01002

WILCOX, MARION ALLEN, b Dillon, SC, Oct 12, 17; m 42; c 3. ZOOLOGY. Prof Exp: Mus technician, Med Mus, Armed Forces Inst Path, 50-60; med plastic exhibit specialist, NIH, 60-69, HEAD MED MODELS & MOULAGE LAB, DIV RES SERVS, MED ARTS SECT, NIH, 69- Res: Applications of plastics to medical technology; museum technology. Mailing Add: Med Arts Sect Bldg 10 NIH Div of Res Serv Bethesda MD 20014

WILCOX, MERRILL, b Milwaukee, Wis, Oct 10, 29; m 62; c 2. PLANT PHYSIOLOGY. Educ: Univ Md, BS, 52, MS, 54; NC State Univ, PhD(agron, plant physiol), 61. Prof Exp: Biol aide marine biol, Chesapeake Biol Lab, Univ Md, 56-57; from asst prof to assoc prof agron, 60-72, PROF AGRON, UNIV FLA, 72- Concurrent Pos: Grants, Am Cancer Soc, 63-66, NSF, 63-65 & Geigy Chem Corp, 70-74. Mem: AAAS; Am Soc Plant Physiol; Weed Sci Soc Am; Scand Soc Plant Physiol; Am Chem Soc. Res: Structure-activity relationships and metabolism of herbicides and plant growth regulators; abscission by Glyoxime. Mailing Add: Herbicide Lab Dept of Agron Univ of Fla Gainesville FL 32611

WILCOX, NETTIE ELANE, medicine, see 12th edition

WILCOX, PAUL WILLIAM, pharmacy, see 12th edition

WILCOX, RAY EVERETT, b Janesville, Wis, Mar 31, 12; m 42; c 4. GEOLOGY. Educ: Univ Wis, PhB, 33, PhM, 37, PhD(geol), 41. Prof Exp: Geologist, State Geol Surv, Wis, 35-39 & Jones & Laughlin Steel Corp, 41-42; GEOLOGIST, US GEOL SURV, 46- Mem: AAAS; fel Geol Soc Am; fel Mineral Soc Am; Brit Mineral Soc; Geochem Soc. Res: Igneous petrology; volcanology; volcanic ash chronology; petrographic methods; optical crystallography. Mailing Add: US Geol Surv PO Box 25046 Denver CO 80225

WILCOX, ROBERTA ARLENE, b Hopkinton, RI, Nov 12, 32. MEDICAL STATISTICS. Educ: Univ RI, BS, 54; Johns Hopkins Univ, ScM, 58. Prof Exp: Biostatistician, State Dept Health, NY, 57-59; res statistician, Lederle Labs, Am Cyanamid Co, 59-66; sr biostatistician, Med Div, Ciba-Geigy Corp, 66-72; BIOSTATISTICIAN, PFIZER CENT RES, 72- Mem: Am Statist Asn; Biomet Soc; Inst Math Statist; NY Acad Sci. Res: Experimental design and analysis applicable to medical and drug research. Mailing Add: Pfizer Cent Res Groton CT 06340

WILCOX, RONALD BRUCE, b Seattle, Wash, Sept 23, 34; m 58; c 2. BIOCHEMISTRY, ENDOCRINOLOGY. Educ: Pac Union Col, BS, 57; Univ Utah, PhD(biochem), 62. Prof Exp: Res fel med, Mass Gen Hosp & Harvard Med Sch, 62-65; from asst prof to assoc prof, 65-73, PROF BIOCHEM, SCH MED, LOMA LINDA UNIV, 73- Mem: AAAS; Endocrine Soc. Res: Biochemistry and metabolism of hormones and drugs. Mailing Add: Dept of Biochem Loma Linda Univ Sch of Med Loma Linda CA 92354

WILCOX, RONALD ERWIN, b Ft Wayne, Ind, Jan 6, 29; m 59; c 3. GEOLOGY. Educ: Iowa State Univ, BS, 50, MS, 52; Columbia Univ, PhD(petrol), 58. Prof Exp: Asst geol, Iowa State Univ, 50-52 & Columbia Univ, 52-54; res geologist, Humble Oil & Refining Co, 56-64; sr res geologist, Esso Prod Res Co, 64-72; CONSULT GEOLOGIST, 72- Concurrent Pos: Lectr, Univ Houston, 72-75, adj prof, 75- Honors & Awards: President's Award, Am Asn Petrol Geologists, 75. Mem: Fel AAAS; Geochem Soc; fel Geol Soc Am; Am Asn Petrol Geologists; Am Geophys Union. Res: Structural geology; petrology; structure of continental margins; salt tectonics; orogenic belts; metamorphism. Mailing Add: PO Box 1230 Bellaire TX 77401

WILCOX, WEBSTER WAYNE, b Berkeley, Calif, Oct 28, 38; m 60; c 2. FOREST PATHOLOGY, FOREST PRODUCTS. Educ: Univ Calif, Berkeley, BS, 60; Univ Wis-Madison, MS, 62, PhD(plant path), 65. Prof Exp: Plant pathologist, US Forest Prod Lab, Wis, 60-64; asst forest prod pathologist, 64-72, ASSOC FOREST PROD PATHOLOGIST, UNIV CALIF, BERKELEY, 72-, LECTR, 64- Concurrent Pos: Fulbright-Hays sr fel, Ger, 73-74. Honors & Awards: Forest Prod Res Soc Award, 65. Mem: Forest Prod Res Soc; Am Phytopath Soc; Soc Wood Sci & Technol; Am Inst Biol Sci. Res: Wood deterioration; microscopy of wood decay. Mailing Add: Forest Prod Lab Univ of Calif Richmond CA 94804

WILCOX, WESLEY C, b St Anthony, Idaho, July 19, 25; m 48; c 4. MICROBIOLOGY. Educ: Univ Utah, BA, 51, MS, 55; Univ Wash, PhD(microbiol), 58. Prof Exp: Donner fel med res, Western Reserve Univ, 58-59, USPHS fel prev med, 59-60; assoc microbiol, Univ Pa, 60-62, asst prof, 62-63; asst prof, Univ Vt, 63-65; PROF MICROBIOL, UNIV PA, 65- Concurrent Pos: Res career develop award, Univ Pa, 60-63. Mem: Am Asn Immunol; Am Soc Microbiol. Res: Virology; immunology; biochemistry. Mailing Add: Dept of Microbiol Univ of Pa Sch of Vet Med Philadelphia PA 19104

WILCOX, WESLEY CRAIN, b Bloomington, Ill, Apr 8, 26; m 48; c 5. AGRONOMY, BOTANY. Educ: Univ Ill, BS, 50, MS, 51. Prof Exp: Field supvr, Found Dept, 51-55, plant breeder, Res Dept, 55-66, mgr spec proj res, 66-71, MGR QUAL CONTROL, QUAL CONTROL DEPT, FUNK SEEDS INT, 71- Mem: AAAS; Am Inst Biol Sci; Am Soc Agron; Genetics Soc Am; Crop Sci Soc Am. Res: High quality seed of hybrid corn and sorghum. Mailing Add: Funk Seeds Int Qual Control Dept 1300 W Washington St Bloomington IL 61701

WILCOX, WILLIAM JENKINS, JR, b Harrisburg, Pa, Jan 26, 23; m 46; c 3. ENGINEERING PHYSICS. Educ: Washington & Lee Univ, BA, 43; Univ Tenn, MS, 58. Prof Exp: Chemist, Tenn Eastman Corp, 43-48; chemist, 48-49, tech asst to lab dir, 49-55, head dept physics, 55-67, prog mgr, 67-69, PROD PLANTS TECH DIR, NUCLEAR DIV, UNION CARBIDE CORP, 69- Mem: AAAS; Am Chem Soc; Sigma Xi; fel Am Inst Chemists; NY Acad Sci. Res: Isotope separation processes, research and development; structure of porous materials; materials development. Mailing Add: 412 New York Ave Oak Ridge TN 37830

WILCOXSON, ROY DELL, b Columbia, Utah, Jan 12, 26; m 49; c 4. PLANT PATHOLOGY. Educ: Utah State Univ, BS, 53; Univ Minn, MS, 55, PhD(plant path), 57. Prof Exp: Asst prof, 57-66, PROF PLANT PATH, UNIV MINN, ST PAUL, 66- Concurrent Pos: Spec staff mem, Rockefeller Found; vis prof, Indian Agr Res Inst, New Delhi. Mem: AAAS; Am Phytopath Soc; Am Soc Agron; Indian Phytopath Soc. Res: Diseases of forage crops and cereal crops; cereal rust diseases; physiology of fungi; virology. Mailing Add: Dept of Plant Path Univ of Minn St Paul MN 55101

WILD, BRADFORD WILLISTON, b Fall River, Mass, Dec 5, 27. OPTOMETRY. Educ: Brown Univ, AB, 49; Columbia Univ, BS, 51, MS, 52; Ohio State Univ, PhD(physiol optics), 59. Prof Exp: From instr to asst prof optom & physiol optics, Ohio State Univ, 59-63, assoc prof optom & physiol optics, 63-69; dean, Col Optom, Pac Univ, 69-74; ASSOC DEAN, SCH OPTOM, MED CTR, UNIV ALA, BIRMINGHAM, 74- Concurrent Pos: Res psychologist, Gen Vision Sect, US Naval Med Res Lab, Conn. Mem: Optical Soc Am; Am Optom Asn. Res: Physiological optics, especially retinal interaction, border phenomena and problems of visibility. Mailing Add: Sch of Optom Univ of Ala Med Ctr Birmingham AL 35294

WILD, GAYNOR (CLARKE), b Winner, SDak, Nov 10, 34; m 73; c 2. BIOCHEMISTRY. Educ: SDak Sch Mines & Technol, BS, 55; Tulane Univ, PhD(biochem), 62. Prof Exp: Fel biochem, Clayton Found Biochem Inst, Univ Tex, 62-63; res assoc, Rockefeller Univ, 63-65, asst prof, 65-67; ASST PROF BIOCHEM, SCH MED, UNIV NMEX, 67- Mem: AAAS. Res: Neurochemistry; mechanism of action of serotonin; enzymology; lipid biochemistry. Mailing Add: Dept of Biochem Univ of NMex Sch of Med Albuquerque NM 87131

WILD, GENE MURIEL, b Fremont, Nebr, Oct 15, 26; m 48; c 4. BIOCHEMISTRY. Educ: Iowa State Univ, BS, 48, MS, 50, PhD(biochem), 53. Prof Exp: Sr biochemist, 53-73, RES SCIENTIST, ELI LILLY & CO, 73- Res: Purification process research in antibiotics; chemical analysis and paper chromatography of antibiotics and related materials. Mailing Add: Antibiotics Develop Dept Eli Lilly & Co Indianapolis IN 46206

WILD, JACK WILLIAM, b Woonsocket, SDak, June 8, 23; m 45; c 2. PHYSICS. Educ: SDak State Col, BS, 48; Univ Kans, MS & PhD(physics), 52. Prof Exp: Asst, Univ Kans, 48-52; physicist, Nat Bur Stand, 52-53; proj leader, Diamond Ord Fuze

Labs, 53-54; mgr br off, Frederick Res Corp, 54-56; sr engr, Cook Res Labs, 56-57, staff engr, 57-61; adv engr, Westinghouse Elec Corp, 61-63, dir space systs, 63-67, mgr space sci, 67; prin engr, Hq, NASA, 67-68, dir mission planning & opers, 68-70, DIR ADVAN PROG STUDIES, HQ, NASA, 70- Mem: Am Inst Aeronaut & Astronaut; Am Astronaut Soc. Res: Effect of external treatments on electret charges; contact electrification of dusts; aircraft cockpit and space systems instrumentation; space systems integration. Mailing Add: NASA Hq Code MTE Washington DC 20546

WILD, JAMES ROBERT, b Sedalia, Mo, Nov 24, 45. MOLECULAR BIOLOGY. Educ: Univ Calif, Davis, BA, 67; Univ Calif, Riverside, PhD(biol), 71. Prof Exp: Res biochemist, Univ Calif, Riverside, 72; microbiologist, Naval Med Res Inst, Nat Naval Med Ctr, 72-75; ASST PROF GENETICS, TEX A&M UNIV, 75- Mem: Am Soc Microbiol; Genetics Soc Am; Undersea Med Soc; Sigma Xi; Fedn Am Scientists. Res: Effects of stress on regulatory aspects of microbial physiology; gene expression, mutation effects and metabolite transport. Mailing Add: Genetics Div Tex A&M Univ Plant Sci Dept College Station TX 77843

WILD, JOHN FREDERICK, b Erie, Pa, June 20, 42; m 66; c 1. NUCLEAR CHEMISTRY. Educ: Pa State Univ, BS, 64; Mass Inst Technol, PhD(nuclear chem), 68. Prof Exp: Res chemist, Knolls Atomic Power Lab, 68-69; RES CHEMIST, LAWRENCE LIVERMORE LAB, 69- Mem: Am Chem Soc. Res: Nuclear chemistry with emphasis on decay and chemical properties of isotopes of elements above Z 96. Mailing Add: Lawrence Livermore Lab Livermore CA 94550

WILD, JOHN FREDERICK, b Wallingford, Conn, Nov 30, 26; m 65; c 2. PHYSICS. Educ: Yale Univ, BS, 50, MS, 51, PhD(physics), 58. Prof Exp: From instr to asst prof physics, Trinity Col, Conn, 57-62; asst prof, 62-67, ASSOC PROF PHYSICS, WORCESTER POLYTECH INST, 67- Mem: Am Phys Soc; Am Asn Physics Teachers. Res: Wave functions for valence electron of neutral caesium for Fermi-Thomas central field; quantum mechanics; color vision. Mailing Add: Physics Dept Worcester Polytech Inst Worcester MA 01609

WILD, ROBERT LEE, b Sedalia, Mo, Oct 9, 21; m 43; c 3. SOLID STATE PHYSICS. Educ: Cent Mo State Col, BS, 43; Univ Mo, MA, 48, PhD(physics), 50. Prof Exp: Asst instr physics, Univ Mo, 49; asst prof, Univ NDak, 50-53; from asst prof to assoc prof, 53-63, chmn dept, 63-68, PROF PHYSICS, UNIV CALIF, RIVERSIDE, 63- Concurrent Pos: NSF fel, Univ Ill, 59-60; vis prof, Tech Univ Denmark, 67-68 & Univ Münster, 75. Mem: Am Phys Soc; Am Asn Physics Teachers; Sigma Xi. Res: Small angle x-ray scattering by liquids and solids; optical and transport properties of solids. Mailing Add: Dept of Physics Univ of Calif Riverside CA 92502

WILD, WAYNE GRANT, b Waterville, Kans, Aug 9, 17; m 39; c 4. PHYSICS, MATHEMATICS. Educ: SDak State Univ, BS, 40; Univ Wis, MS, 48; Univ Ill, MA, 67. Prof Exp: Prof physics & head dept, Buena Vista Col, 48-67, chmn natural sci div, 53-67; assoc prof physics, 67-69, ASSOC PROF MATH, UNIV WIS-STEVENS POINT, 69- Mem: Am Phys Soc; Fedn Am Scientists. Res: Thermionic emission. Mailing Add: Dept of Math Univ of Wis Stevens Point WI 54481

WILDASIN, HARRY LEWIS, b York Co, Pa, Oct 10, 23; m 45, 70; c 2. BIOCHEMISTRY. Educ: Pa State Univ, PhD(dairy), 50. Prof Exp: Asst prof dairying, Univ Conn, 49-52; dir qual control, Whiting Milk Co, Boston, 52-57; DIR QUAL CONTROL & GOVT RELS, H P HOOD, INC, BOSTON, 57- Mem: AAAS; Am Dairy Sci Asn; NY Acad Sci; Nat Environ Health Asn; Am Pub Health Asn. Res: Frozen milk; lactose; milk proteins; surface active agents; antibiotics; salmonella; radioactive elements in milk.

WILDBERGER, WILLIAM CAMPBELL, b Boston, Mass, Dec 8, 14; m 42; c 2. MEDICAL ADMINISTRATION, PSYCHIATRY. Educ: Boston Univ, BS, 37, MD, 40. Prof Exp: Mem pract group med, Perry Clin, 46-50; clin dir pediat & psychiat, Woodward State Hosp-Sch, 58-61, actg supt admin & psychiat, 61-63, supt psychiat admin, 63-69; CHIEF ADMITTING, VET ADMIN HOSP, DES MOINES, 69- Concurrent Pos: Mem, White House Conf on Ment Retardation, 62. Mem: Fel Am Asn Ment Deficiency; AMA; Am Hosp Asn; Am Psychiat Asn. Res: Child psychiatry, pediatrics and administration in mental retardation.

WILDE, ANTHONY FLORY, b New York, NY, May 16, 30; m 72. PHYSICAL CHEMISTRY. Educ: Yale Univ, BS, 52; Ind Univ, PhD(phys chem), 59. Prof Exp: Res chemist, Monsanto Res Corp, 59-63; res chemist, US Army Natick Labs, 63-68, RES CHEMIST, ARMY MAT & MECH RES CTR, 68- Mem: Am Chem Soc; Soc Exp Stress Anal; NY Acad Sci. Res: Polymer rheology, especially dynamic mechanical and optical properties of organic polymers; stress wave propagation and fracture in materials; dielectric and piezoelectric properties of organic polymers. Mailing Add: Org Mat Lab Army Mat & Mech Res Ctr Watertown MA 02172

WILDE, CARROLL ORVILLE, b Elmhurst, Ill, June 5, 32; m 52, 71; c 4. MATHEMATICS. Educ: Ill State Univ, BS, 58; Univ Ill, Urbana, PhD(math), 64. Prof Exp: Instr math, SDak Sch Mines & Technol, 58-59 & Col Wooster, 59-61; asst prof, Univ Minn, Minneapolis, 64-68; assoc prof, 68-75, PROF MATH, NAVAL POSTGRAD SCH, 75- Mem: Am Math Soc; Math Asn Am; Sigma Xi. Res: Functional analysis, especially invariant means for semigroups; topology; probability theory; digital image processing. Mailing Add: Dept of Math Naval Postgrad Sch Monterey CA 93940

WILDE, CHARLES EDWARD, JR, b Boston, Mass, Nov 5, 18; m 44; c 3. BIOLOGY. Educ: Dartmouth Col, AB, 40; Princeton Univ, MA, 47, PhD(biol), 49. Hon Degrees: MA, Univ Pa, 72. Prof Exp: Instr zool, Dartmouth Col, 40-41 & Princeton Univ, 46-49; from asst prof to prof zool, Sch Dent Med, Univ Pa, 49-75, prof embryol, Dept Biol & Dept Path, 57-75; PROF ZOOL, UNIV RI, 75-, CHMN DEPT, 75- Concurrent Pos: Guggenheim Mem Found fel, 57-58; guest investr, Strangeways Res Lab, Cambridge Univ, 57-58; consult, Dept Animal Genetics, Univ Edinburgh, 57; trustee, Mt Desert Island Biol Lab, currently, dir, 67-70. Mem: AAAS; Am Soc Zoologists; Soc Develop Biol; NY Acad Sci; Int Inst Embryol. Res: Differentiation of organs in vitro; tissue culture; metabolite control of cell differentiation; embryology of the head; cytochimeras of muscle; temporal relations of energy flow; RNA and protein synthesis in morphogenesis and differentiation; molecular and genomic control of symmetry in embryogenesis. Mailing Add: Dept of Zool Univ of RI Kingston RI 02881

WILDE, EDWIN FREDERICK, JR, b Lombard, Ill, Jan 14, 31; m 52; c 3. MATHEMATICS. Educ: Ill State Univ, BS, 52, MS, 53; Univ Ill, MA, 55, PhD(math), 59. Prof Exp: Vpres planning, 71-74, PROF MATH, BELOIT COL, 55- Concurrent Pos: NSF sci fac fel, 64; consult, AID, India, 65 & Inst Int Educ, EPakistan, 69; consult & evaluator, NCent Asn Cols & Schs, 74- Mem: Math Asn Am. Mailing Add: RR 1 Box 200 Beloit WI 53511

WILDE, GARNER LEE, b Spring Creek, Tex, Sept 29, 26; m 51; c 2. GEOLOGY. Educ: Tex Christian Univ, BA, 50, MA, 52. Prof Exp: Jr geologist, Humble Oil & Refining Co, 52-53, from assoc paleontologist to paleontologist, 53-63, sr res geologist, 63-67, prof geologist, 67-71, PROF GEOLOGIST, HQ STAFF, EXXON CO, USA, 71- Concurrent Pos: Lectr, Case Western Reserve Univ, 63; vis lectr, Tex Tech Univ, 67 & Univ Mo, 69; mem bd dirs, Cushman Found Foraminiferal Res, 70-75, pres, 74-75; vis lectr, Kans State Univ, 71 & Rensselaer Polytech Inst, 71. Mem: Fel Geol Soc Am; Soc Econ Paleontologists & Mineralogists; Am Asn Petrol Geologists. Res: Stratigraphic and paleontological studies on late Paleozoic Fusulinid Foraminifera, Calcareous algae and Mesozoic Calcareous Nannofossils. Mailing Add: Exxon Co USA PO Box 2189 Houston TX 77001

WILDE, GERALD ELDON, b Ballinger, Tex, Dec 7, 39. ENTOMOLOGY. Educ: Tex Tech Col, BS, 62; Cornell Univ, PhD(entom), 66. Prof Exp: Res asst entom, Cornell Univ, 62-66; ASSOC PROF ENTOM & RES ENTOMOLOGIST AGR RES STA, KANS STATE UNIV, 66- Mem: Entom Soc Am. Res: Economic entomology; field crops insects. Mailing Add: Dept of Entom Kans State Univ Manhattan KS 66502

WILDE, KENNETH ALFRED, b Cedar City, Utah, Mar 4, 29; m 61; c 2. PHYSICAL CHEMISTRY. Educ: Univ Utah, BS, 50, PhD(chem), 53. Prof Exp: Res chemist, Redstone Res Labs, Rohm and Haas Co, 53-70, res chemist, Res Div, 70-75; SR SCIENTIST, RADIAN CORP, 75- Mem: Am Chem Soc. Res: Chemical kinetics in electric discharges; combustion wave theory; high temperature thermodynamics and kinetics; aerothermodynamics; mass transfer process simulation; solution thermodynamics and statistical mechanics. Mailing Add: Radian Corp 8500 Shoal Creek Blvd Austin TX 78766

WILDE, PAT, b Chicago, Ill, Sept 25, 35. OCEANOGRAPHY. Educ: Yale Univ, BS, 57; Harvard Univ, AM, 61, PhD(geol), 65. Prof Exp: Geologist, Shell Oil Co, 57-59; res geologist, Scripps Inst Oceanog, 60-62; lectr ocean eng, Univ Calif, Berkeley, 64-68, asst prof, 68-75, res oceanogr, 66-75, res engr, 64-66; RES SCIENTIST & OCEANOGR, LAWRENCE BERKELEY LAB, UNIV CALIF, BERKELEY, 75- Concurrent Pos: Consult, Coastal Res Panel Earthquake Eng, Nat Acad Eng, 65-67; adv tech adv bd, Dept Eng, City & County of San Francisco, 70- Mem: AAAS; Geol Soc Am; Am Geophys Union; Marine Technol Soc; Geochem Soc. Res: Marine electrochemistry; sediment transport in marine environments. Mailing Add: Lawrence Berkeley Lab Univ of Calif Berkeley CA 94720

WILDE, RICHARD EDWARD, JR, b Los Angeles, Calif, Jan 7, 31; m 60; c 3. SOLID STATE CHEMISTRY. Educ: Univ Calif, Los Angeles, BS, 56; Univ Wash, PhD(chem), 61. Prof Exp: Res assoc, Johns Hopkins Univ, 61-63; asst prof chem, 63-67, ASSOC PROF CHEM, TEX TECH UNIV, 67- Mem: AAAS; Am Chem Soc; Am Phys Soc; The Chem Soc. Res: Infrared and Raman spectroscopy of nonmetal hydrides; phase transitions in solids; dynamics of defect solids. Mailing Add: Dept of Chem Tex Tech Univ Lubbock TX 79409

WILDE, WALTER HERBERT ATKINSON, entomology, see 12th edition

WILDE, WALTER SAMUEL, b Toronto, Ont, Feb 12, 09; US citizen; m 36; c 3. MEDICAL PHYSIOLOGY. Educ: Miami Univ, AB, 31; Univ Minn, MA, 33, PhD(zool), 37. Prof Exp: Asst zool & physiol, Univ Minn, 31-33; instr zool, Miami Univ, 33-34; asst zool & physiol, Univ Minn, 34-37; teaching fel physiol, Univ Rochester, 37-38; instr zool, Univ Wyo, 38-39; from instr to asst prof physiol, Sch Med, La State Univ, 39-45; res assoc, Carnegie Inst, 45-47; sr physiologist, NIH, 47; from assoc prof to prof physiol, Tulane Univ, 47-56; prof, 56-75, EMER PROF PHYSIOL, UNIV MICH, ANN ARBOR, 75- Concurrent Pos: Guest lectr, Mt Desert Island Biol Lab, 56. Mem: Fel AAAS; fel Soc Exp Biol & Med; fel Am Physiol Soc. Res: Interstitial and capillary albumin; ion transport and kidney; renal stop flow method; blood-brain barrier and aqueous humor. Mailing Add: 151 Bayview Ave Naples FL 33940

WILDEMAN, THOMAS RAYMOND, b Madison, Wis, May 11, 40; m 65; c 2. ANALYTICAL CHEMISTRY, GEOCHEMISTRY. Educ: Col St Thomas, BS, 62; Univ Wis, PhD(phys chem), 67. Prof Exp: Lectr chem, Univ Wis, 66-67; asst prof, 67-73, ASSOC PROF CHEM, COLO SCH MINES, 73- Concurrent Pos: Consult, US Geol Surv, 70- Mem: AAAS; Am Geophys Union; Geochem Soc. Res: Properties of trace elements in inorganic solids; isotopic, radiochemical and atomic absorption analysis; geochemistry of trace elements in rocks and waters. Mailing Add: Dept of Chem Colo Sch of Mines Golden CO 80401

WILDENTHAL, BRYAN HOBSON, b San Marcos, Tex, Nov 4, 37; m 57; c 3. PHYSICS. Educ: Sul Ross State Col, BA, 58; Univ Kans, PhD(physics), 64. Prof Exp: Res assoc physics, Rice Univ, 64-66; US Atomic Energy Comn fel, Oak Ridge Nat Lab, 66-68; asst prof, Tex A&M Univ, 68-69; assoc prof, 69-72, PROF PHYSICS, MICH STATE UNIV, 72- Mem: Am Phys Soc; Am Asn Physics Teachers. Res: Study of the low lying quantum states of atomic nuclei via direct reaction experiments and shell model theory. Mailing Add: Cyclotron Lab Mich State Univ East Lansing MI 48823

WILDENTHAL, KERN, b San Marcos, Tex, July 1, 41; m 64; c 2. PHYSIOLOGY, INTERNAL MEDICINE. Educ: Sul Ross Col, BA, 60; Univ Tex Southwestern Med Sch Dallas, MD, 64; Cambridge Univ, PhD(cell physiol), 70. Prof Exp: Intern, Bellevue Hosp-NY Univ, 64-65; resident, Parkland Hosp-Univ Tex Southwestern Med Sch Dallas, 65-66; vis scientist, Strangeways Res Lab, Cambridge Univ, 68-70; from asst prof to assoc prof, 70-75, PROF PHYSIOL & INTERNAL MED, UNIV TEX HEALTH SCI CTR DALLAS, 75- Concurrent Pos: Guggenheim Found fel, Univ Cambridge, Eng, 75-76. Mem: Am Soc Clin Invest; Cardiac Muscle Soc; Am Physiol Soc; Am Fedn Clin Res; Am Heart Asn. Res: Cardiovascular physiology, fetal physiology; myocardial organ culture and metabolism. Mailing Add: Cardiopulmonary Div Univ of Tex Health Sci Ctr Dallas TX 75235

WILDER, CLEO DUKE, JR, b Macon, Ga, Sept 24, 25; m 50; c 2. VERTEBRATE ZOOLOGY. Educ: Univ NC, AB, 48; Univ Tenn, MS, 51; Univ Fla, PhD, 62. Prof Exp: Instr biol, Presby Col, SC, 51-53; asst, Univ Fla, 55-57 & 58-59; asst prof, Memphis State Univ, 59-62; asst zool, Va Polytech Inst, 62-69; ASSOC PROF BIOL, MURRAY STATE UNIV, 69- Mem: Am Soc Ichthyologists & Herpetologists; Am Soc Mammal; Soc Study Amphibians & Reptiles; Animal Behav Soc; Herpetologists' League. Res: Taxonomy, ecology, distribution, behavior and evolution of amphibians, reptiles and mammals; aggression in vertebrates; ecology of stream drainage systems in western Kentucky. Mailing Add: Dept of Biol Murray State Univ Murray KY 42071

WILDER, DONALD GEORGE, b Kingston, Ont, May 23, 16; m 49; c 4. FISHERIES MANAGEMENT. Educ: Queen's Univ, Ont, BA, 38; Univ Toronto, MA, 40, PhD(morphol), 44. Prof Exp: Asst biol, Queen's Univ, Ont, 36-38; asst biol, Univ Toronto, 39-42; sci asst, Fisheries Res Bd Can, 42-45, biologist, 46-47, sr biologist & asst dir biol sta, 48-50, prin scientist, 57-72, RES SCIENTIST, FISHERIES & MARINE SERV, CAN DEPT ENVIRON, 72- Res: Population dynamics and management of marine invertebrate fisheries, particularly lobsters. Mailing Add: Biol Sta Fisheries & Marine Serv St Andrews NB Can

WILDER, DONALD RICHARD, b Kodaikanal, SIndia, Apr 20, 32; US citizen; m 56; c 3. MATHEMATICS, OPTICS. Educ: Oberlin Col, BA, 53; Univ Rochester, MS, 64. Prof Exp: With apparatus & optical div, 53-54, 57-60, res physicist, 60-67, SR RES PHYSICIST, RES LABS, EASTMAN KODAK CO, 67- Mem: AAAS; Optical Soc Am; Math Asn Am; Soc Photog Sci & Eng. Res: Geometrical optics; theory of Hamilton's characteristic functions as applied to lens design. Mailing Add: Eastman Kodak Co Res Labs 1669 Lake Ave Rochester NY 14650

WILDER, GENE RAY, organic chemistry, see 12th edition

WILDER, MARTIN STUART, b Brooklyn, NY, May 20, 37; m 64; c 1. MICROBIOLOGY. Educ: Brooklyn Col, BS, 60; Univ Kans, MA, 63, PhD(microbiol), 66. Prof Exp: Res aide microbiol, Sloan-Kettering Inst Cancer Res, 60-61; Nat Acad Sci-Nat Res Coun res assoc, Microbiol Div, US Dept Army, 66-68; asst prof microbiol, 68-74, ASSOC PROF MICROBIOL, UNIV MASS, AMHERST, 74- Concurrent Pos: Nat Acad Sci grant, 74. Mem: Am Soc Microbiol; Reticuloendothelial Soc; NY Acad Sci. Res: Pathogenesis and pathology of infectious diseases; interactions of platelets, bacteria and leukocytes. Mailing Add: Dept of Microbiol Univ of Mass Amherst MA 01002

WILDER, PELHAM, JR, b Americus, Ga, July 20, 20; m 45; c 3. ORGANIC CHEMISTRY. Educ: Emory Univ, AB, 42, MA, 43; Harvard Univ, MA, 47, PhD(org chem), 50. Prof Exp: From instr to prof chem, 49-68, PROF CHEM & PHARMACOL, DUKE UNIV, 68- Concurrent Pos: Consult, NSF, 60-68, E I du Pont de Nemours & Co, Inc, 66-69 & Res Triangle Inst, 68-; mem advan placement chem comt, Col Entrance Exam Bd, 68-, chmn, 70-, mem advan placement standing comt, 69-72. Mem: Am Chem Soc; The Chem Soc. Res: Stereochemical studies; kinetic, thermodynamic control and mechanism of organic reactions; pharmacology. Mailing Add: Dept of Chem Duke Univ Durham NC 27706

WILDER, RAYMOND LOUIS, b Palmer, Mass, Nov 3, 96; m 21; c 4. MATHEMATICS. Educ: Brown Univ, PhB, 20, MS, 21; Univ Tex, PhD(math), 23. Hon Degrees: DSc, Bucknell Univ, 55 & Brown Univ, 58. Prof Exp: Asst math, Brown Univ, 20-21; instr, Univ Tex, 21-24; asst prof, Ohio State Univ, 24-26; from asst prof to prof, 26-47, res prof, 47-67, EMER PROF MATH, UNIV MICH, ANN ARBOR, 67-; VIS PROF, UNIV CALIF, SANTA BARBARA, 69- Concurrent Pos: Researcher, Inst Advan Study, 33-34; Guggenheim fel, Univ Tex, 40-41; vis prof, Univ Southern Calif, 47; res assoc, Calif Inst Technol, 49-50; Taft Mem lectr, Univ Cincinnati, 58; vis res prof, Fla State Univ, 61-62; mem comt sci & pub policy, Nat Acad Sci, 65-67. Honors & Awards: Lester R Ford Award & Distinguished Serv Award, Math Asn Am, 73. Mem: Nat Acad Sci; AAAS (vpres math sect, 48); Am Math Soc (vpres, 50, pres, 55-56); Asn Symbolic Logic; Am Anthrop Asn. Res: Cultural anthropology; culturological history of mathematics. Mailing Add: Dept of Math Univ of Calif Santa Barbara CA 93106

WILDER, RUSSELL MORSE, b Chicago, Ill, Oct 14, 12; m 58; c 4. MEDICINE, PSYCHIATRY. Educ: Princeton Univ, BS, 34; Harvard Med Sch, MD, 38; Univ Minn, PhD, 43; Am Bd Internal Med, dipl, 48. Prof Exp: Intern med, Univ Chicago Clins, 38-39; fel med, Mayo Found, 39-42, first asst med, Mayo Clin, 42-44; clin assoc, Univ Minn, 46-54; chief physician, Med Serv, Topeka Vet Admin Hosp, Kans, 55-58; coordr psychiat training prog for med practitioners, Menninger Found, Topeka, 60-71; PROF PSYCHIAT & MED, SCH MED, UNIV MINN, MINNEAPOLIS, 71-; CHIEF SECT PSYCHOL MED, VET ADMIN HOSP, MINNEAPOLIS, 71- Concurrent Pos: Chief psychosom serv, Vet Admin Hosp, Topeka, 58-63; internist, Menninger Clin, Topeka, 62-71; consult prof, Student Counseling Serv, Kans State Univ, 67-71; coordr psychiat training prog for med students, Menninger Found, 67-70; physician-in-residence, Vet Admin Hosps, 67-72. Mem: Fel Am Col Physicians. Mailing Add: Med Serv Vet Admin Hosp Minneapolis MN 55417

WILDER, VIOLET MYRTLE, b Granville, Iowa, Apr 8, 08. BIOCHEMISTRY. Educ: Univ Nebr, BA, 28, MA, 34, PhD(biochem), 38. Prof Exp: Res chemist, Dr G A Young, Omaha, 38-40; instr biochem, Univ Ark, 40; dir, Lab Maternal & Child Health, Univ Nebr, 40-43 & 46-47; instr physiol chem, Woman's Med Col Pa, 43-46; from asst prof to assoc prof, 47-73, EMER ASSOC PROF BIOCHEM, COL MED, UNIV NEBR MED CTR, OMAHA, 73- Mem: AAAS; Am Chem Soc; fel Am Asn Clin Chem. Res: Enzyme-hormone relationships; clinical chemistry. Mailing Add: 2045 S 18th St Lincoln NE 68502

WILDER, WILLIAM BAYLOR, b Walden, Ky, Mar 21, 15; m 41; c 1. BOTANY, GENETICS. Educ: Western Ky State Col, BS, 47; Purdue Univ, MS, 50, PhD(genetics), 57. Prof Exp: Agronomist, USDA, Univ Alaska, 51-54; instr pub schs, Alaska, 54-55; instr biol, 55-60, PROF BIOL & HEAD DEPT, CUMBERLAND COL, 60- Mem: Am Genetic Asn. Res: Development of forage crops adapted to long, cold winters of Alaska. Mailing Add: Dept of Biol Cumberland Col Box 269 Williamsburg KY 40769

WILDEY, ROBERT LEROY, b Los Angeles, Calif, Aug 22, 34; m 59; c 2. ASTRONOMY, ASTROPHYSICS. Educ: Calif Inst Technol, 57, MS, 58, PhD(astron), 62. Prof Exp: Res engr, Jet Propulsion Lab, Calif Inst Technol, 59-60, res fel astron & geol, Mt Wilson & Palomar Observ & div geol sci, Calif Inst Technol, 62-65, lectr geol, 64-65; astronr & astrophysicist, Ctr Astrogeol, Us Geol Serv, 65-72; ASSOC PROF ASTROPHYS & ASTRON, NORTHERN ARIZ UNIV, 72- Concurrent Pos: Consult, United Electrodynamics Corp, 62-63, Aeronutronics Div, Ford Motor Co, 63-64, World Book Encycl Sci Serv, 63-64 & US Geol Surv, 72-; vis prof, Univ Calif, Berkeley, 66; mem planetary astron panel, Space Sci Bd, Nat Acad Sci, 67- Mem: Am Astron Soc; Am Geophys Union; fel Geol Soc Am; Int Astron Union; Am Asn Physics Teachers. Res: Infrared photometry for very cold celestial bodies; discovery of Jupiter's hot shadows, far-infrared stellar radiation and lunar night time hotspots; laser selenodesy; lunar photometry; chemistry and ages of star clusters; line blanketing; Apollo and Mariner-Mars experimenter; gravitation and cosmology; photoclinometry; automated digital photogrammetry and photoclinometry; radiative transfer theory. Mailing Add: Observ for Astron & Astrophys Northern Ariz Univ Flagstaff AZ 86001

WILDFEUER, MARVIN EMANUEL, b Bronx, NY, Apr 16, 36; m 67; c 2. BIOCHEMISTRY. Educ: Queen's Col, NY, BS, 57; Iowa State Univ, MS, 59; Univ Del, PhD(chem), 63. Prof Exp: NIH fel molecular biol, Univ Calif, San Diego, 63-65; res scientist, Sansum Clin & Res Found, 65-67; SR CHEMIST, ELI LILLY & CO, 68- Mem: Am Chem Soc. Res: Purification of fermentation products; antibiotics. Mailing Add: Dept TL969 Eli Lilly & Co Lafayette IN 47902

WILDI, BERNARD SYLVESTER, b Columbus, Ohio, May 23, 20. ORGANIC CHEMISTRY. Educ: Ohio State Univ, BSc, 43, PhD(org chem), 48. Prof Exp: Res chemist, Nat Defense Res Comt, Ohio State Univ, 43-44 & Manhattan Proj, Los Alamos Sci Lab, NMex, 44-47; Nat Res Coun fel, Harvard Univ, 48-49; mem fac, Fla State Univ, 49-50; mem staff, 50-53, group leader, 53-65, mgr life sci, 65-69, DISTINGUISHED SCI FEL, MONSANTO CO, 69- Mem: AAAS; Am Chem Soc. Res: Structure of natural products; chemical spectroscopy; organic synthesis. Mailing Add: 1234 Folger Kirkwood MO 63122

WILDING, LAWRENCE PAUL, b Winner, SDak, Oct 1, 34; m 56; c 4. SOIL SCIENCE, AGRONOMY. Educ: SDak State Univ, BSc, 56, MSc, 59; Univ Ill, PhD(soils), 62. Prof Exp: Asst agron, SDak State Univ, 56-59; Campbell Soup Co fel plant sci, 59-62; from asst prof to assoc prof agron, 62-69, PROF AGRON, OHIO STATE UNIV, 69- Concurrent Pos: Fel, Univ Guelph, 72. Mem: Fel Am Soc Agron; Soil Sci Soc Am; Int Soc Soil Sci; Clay Minerals Soc; Soil Conserv Soc Am. Res: Soil classification and genesis among different climatic, chronologic and topographic sequences; origin, depth distributions, properties and radiocarbon age of soil opal phytoliths; statistical variability in soil physical and chemical parameters; clay mineralogy; sediment mineralogy and soil erosion. Mailing Add: Dept of Agron Ohio State Univ Columbus OH 43210

WILDISH, DAVID JOHN, b Kent, UK, June 26, 39; m 64; c 3. ZOOLOGY, ECOLOGY. Educ: London Univ, BSc, 65, PhD(estuarine ecol), 69. Prof Exp: Nat Res Coun Can fel, Biol Sta, Fisheries Res Bd Can, 69-70, RES SCIENTIST, BIOL STA, CAN DEPT ENVIRON, 70- Mem: Marine Biol Asn UK; Can Soc Zool; Estuarine & Brackish-Water Sci Asn. Res: Marine biology; effects of environmental contaminants on enzymes, individuals and ecosystems; effects of pollution in estuaries and brackish water. Mailing Add: Dept of Environ Biol Sta St Andrews NB Can

WILDMAN, GARY CECIL, b Middlefield, Ohio, Nov 25, 42; m 65; c 2. POLYMER CHEMISTRY. Educ: Thiel Col, AB, 64; Duke Univ, PhD(phys chem), 70. Prof Exp: Res chemist, Hercules Inc, 68-71; ASSOC PROF POLYMER SCI & CHMN DEPT, UNIV SOUTHERN MISS, 71- Mem: Am Chem Soc; Am Crystallog Asn; Sigma Xi; Fedn Socs Paint Technol; Soc Plastics Engrs. Res: Structure-property relationships of synthetic polymers; x-ray diffraction studies of single crystals and polymeric materials; surface coatings. Mailing Add: Univ South Miss Polymer Sci Dept Box 276 Southern Sta Hattiesburg MS 39401

WILDMAN, RUTH BOWMAN, b Wilkes-Barre, Pa, July 16, 24; m 47; c 2. AQUATIC BIOLOGY. Educ: Mt Holyoke Col, AB, 46; Univ Ill, Urbana, MS, 47; Iowa State Univ, PhD(cell biol, bot), 69. Prof Exp: Asst chemist, Ill Geol Surv, 47-49; res asst biochem, Univ Wis-Madison, 49-50; asst prof bot & biol, 69-72, ASSOC PROF BOT & BIOL, IOWA STATE UNIV, 72-, ASST DEAN COL SCI & HUMANITIES, 75- Concurrent Pos: Univ res grants, Iowa State Univ, 70-71 & 74-75; Iowa State Water Resources Res Inst grant, 72-75. Mem: Am Soc Cell Biol; Phycol Soc Am; Bot Soc Am; Am Inst Biol Sci. Res: Role of blue-green algae in aquatic ecology; algal cytology and physiology; bioconcentration of chemical and energy-associated pollutants; effects of strip-mining on aquatic systems. Mailing Add: Dept of Bot & Plant Path Iowa State Univ Ames IA 50010

WILDMAN, WILLIAM COOPER, b Oak Park, Ill, Mar 20, 23; m 47; c 2. ORGANIC CHEMISTRY. Educ: DePauw Univ, AB, 43; Univ Ill, MS, 47, PhD(chem), 49. Prof Exp: Asst, DePauw Univ, 43-44; proj assoc org res, Univ Wis, 49-50; asst prof org chem, Princeton Univ, 50-53; chemist, Nat Heart Inst, 53-62; PROF CHEM, IOWA STATE UNIV, 62- Concurrent Pos: Lectr, Georgetown Univ, 60; Guggenheim fel, 60-61. Mem: Am Chem Soc; Am Soc Biol Chemists. Res: Natural products; instrumental techniques; biogenesis. Mailing Add: Dept of Chem Iowa State Univ Ames IA 50011

WILDNAUER, RICHARD HARRY, b New Kensington, Pa, Feb 14, 40; m 66. DERMATOLOGY, POLYMER CHEMISTRY. Educ: St Vincent Col, BS, 62; WVa Univ, PhD(biochem), 66; Rider Col, MBA, 74. Prof Exp: Fel, Univ Kans, 66-67; sr scientist, 67-73, SR GROUP LEADER, JOHNSON & JOHNSON RES LABS, 73- Mem: NY Acad Sci; Soc Invest Dermat; Am Chem Soc. Res: Enzyme kinetics and mechanisms; skin physiology and biochemistry; membrane transport properties; wound healing; physical polymer characterizations. Mailing Add: Skin Biol Dept Johnson & Johnson Res Labs New Brunswick NJ 08903

WILDS, ALFRED LAWRENCE, b Kansas City, Mo, Mar 1, 15; m 37. ORGANIC CHEMISTRY. Educ: Univ Mich, BS, 36, MS, 37, PhD(org chem), 39. Prof Exp: Asst chem, Univ Mich, 37-39, DuPont fel, 39-40; from instr to assoc prof chem, 40-46, PROF CHEM, UNIV WIS-MADISON, 46- Concurrent Pos: Co-off investr, Nat Defense Res Comt, Univ Wis-Madison, 42-45; Guggenheim fel, 57. Mem: AAAS; Am Chem Soc; The Chem Soc; Swiss Chem Soc. Res: Modern organic synthesis; new methods, integrated syntheses, stereochemistry of catalytic hydrogenations, metal reductions; synthesis of natural products, steroids, hormone analogs; reactions of diazoketones; nuclear magnetic resonance studies. Mailing Add: Dept of Chem Univ of Wis Madison WI 53706

WILDS, PRESTON LEA, b Aiken, SC, Dec 18, 26; m 50, 63; c 4. OBSTETRICS & GYNECOLOGY. Educ: Yale Univ, BA, 49; Univ Pa, MD, 53; Am Bd Obstet & Gynec, dipl, 62. Prof Exp: Asst obstet & gynec, Sch Med, La State Univ, 54-57; pvt pract, SC, 57-59; clin instr, 59, from instr to assoc prof, 59-67, PROF OBSTET & GYNEC, MED COL GA, 67- Mem: Fel Am Col Obstet & Gynec; AMA; Asn Am Med Cols. Res: Programmed instruction; gynecologic oncology. Mailing Add: Dept of Obstet & Gynec Med Col of Ga Augusta GA 30902

WILDT, DAVID EDWIN, b Jacksonville, Ill, Mar 12, 50; m 70. REPRODUCTIVE PHYSIOLOGY. Educ: Ill State Univ, BS, 72; Mich State Univ, MS, 73, PhD(animal husb, physiol), 75. Prof Exp: Res asst reproductive physiol, Endocrine Res Univ, Mich State Univ, 72-75; FEL REPRODUCTIVE PHYSIOL, INST COMP MED, BAYLOR COL MED, 75- Mem: Sigma Xi; Am Soc Animal Sci; Soc Study Reproduction; Int Primatological Soc. Res: Establishment of ovarian morphological reproductive patterns in canine, feline and nonhuman primates using laparoscopy; development and use of frozen semen techniques for the preservation of captive wild mammal species. Mailing Add: Inst of Comp Med Baylor Col of Med Tex Med Ctr Houston TX 77025

WILDT, RUPERT, astronomy, deceased

WILDUNG, RAYMOND EARL, b Van Nuys, Calif, Feb 24, 41; m 61; c 2. SOIL SCIENCE, ENVIRONMENTAL CHEMISTRY. Educ: Calif State Polytech Col, San Luis Obispo, BS, 62; Univ Wis-Madison, MS, 64, PhD(soil sci), 66. Prof Exp: NIH fel, Univ Wis-Madison, 66-67; sr res scientist, Soil-Sediment Sci, 67-71, PROG LEADER ENVIRON CHEM, BATTELLE PAC NORTHWEST LABS, 71- Concurrent Pos: Grants, USDA, Battelle Pac Northwest Labs, 68-70, Environ Protection Agency, 68-71, US Energy Res & Develop Admin, 68-, Nat Inst Environ Health Sci, 71- Mem: AAAS; Am Soc Agron; Int Soc Soil Sci; Soil Sci Soc Am. Res: Soil-sediment science; fate and behavior of pollutants, including pesticides, petroleum residuals, metals and metal complexes in soils, plants, sediments and waters. Mailing Add: Battelle Pac Northwest Labs PO Box 999 Richland WA 99352

WILE, HOWARD P, b New York, NY, Jan 4, 11; m 35; c 2. RESEARCH

ADMINISTRATION. Educ: Dartmouth Col, AB, 32. Prof Exp: Mem staff admin, Mass Inst Technol, 44-46; adminr res, Polytech Inst Brooklyn, 46-65; EXEC DIR COMT GOVT RELS, NAT ASN COL & UNIV BUS OFFICERS, 65- Concurrent Pos: Consult, Am Coun Educ; chmn grant admin adv comt, Dept Health, Educ & Welfare, 67-73; mem res & develop study group, Comn Govt Procurement, 71-72. Res: Governmental relations. Mailing Add: Nat Asn of Col & Univ Bus Officers One Dupont Circle NW Washington DC 20036

WILEMSKI, GERALD, b Dunkirk, NY, Oct 15, 46. PHYSICAL CHEMISTRY. Educ: Canisius Col, BS, 68; Yale Univ, PhD(chem), 72. Prof Exp: Res assoc, Dept Eng & Appl Sci, Yale Univ, 72-74; RES ASSOC CHEM, DARTMOUTH COL, 74- Res: Statistical mechanics; polymer solutions; Brownian motion; nucleation. Mailing Add: Dept of Chem Dartmouth Col Hanover NH 03755

WILEN, SAMUEL HENRY, b Brussels, Belgium, Mar 6, 31; nat US; m 60; c 2. ORGANIC CHEMISTRY. Educ: City Col New York, BS, 51; Univ Kans, PhD(chem), 56. Prof Exp: Asst instr chem, Univ Kans, 51-52, asst, 53-55; res assoc, Univ Notre Dame, 55-57; from instr to assoc prof, 57-71, PROF CHEM, CITY COL NEW YORK, 71- Concurrent Pos: Sci assoc, State Univ Groningen, 68-69. Mem: AAAS; Am Chem Soc; The Chem Soc. Res: Chemistry of heterocyclic compounds; resolving agents and optical resolutions; reaction of aluminum halides with esters; organic synthesis. Mailing Add: Dept of Chem City Col of New York New York NY 10031

WILES, ALFRED BARKSDALE, b Flora, Miss, Oct 8, 18; m 49; c 1. PLANT PATHOLOGY. Educ: Miss State Col, BS, 40; Univ Ark, MS, 42; Univ Wis, PhD(plant path), 51. Prof Exp: Asst plant path, Univ Ark, 40-42 & Univ Wis, 48-50; assoc plant pathologist, Agr Exp Sta, 50-64, ADJ PROF PLANT PATH, MISS STATE UNIV, 64-; RES PLANT PATHOLOGIST, USDA, 64- Concurrent Pos: Assoc plant pathologist, USDA, 50-64. Mem: AAAS; Am Phytopath Soc. Res: Diseases of cotton; disease resistance in cotton; seed treatment chemicals; pathology of seed. Mailing Add: Dept of Plant Path Miss State Univ 204 C Dorman Forestry Plant Sci Bldg Mississippi State MS 39762

WILES, DAVID M, b Springhill, NS, Dec 28, 32; m 57; c 2. POLYMER CHEMISTRY. Educ: McMaster Univ, BSc, 54, MSc, 55; McGill Univ, PhD(phys chem), 57. Prof Exp: Nat Res Coun Can & Can Ramsay Mem fels, Univ Leeds, 57-59; asst res officer, High Polymer Sect, 59-61, assoc res officer, 61-67, sr res officer, 67-74, HEAD TEXTILE CHEM SECT, NAT RES COUN CAN, 66-, DIR DIV CHEM, 75- Mem: Am Chem Soc; Fiber Soc; Chem Inst Can; Can Inst Textile Sci; The Chem Soc. Res: Polymerization kinetics and mechanisms; synthesis of stereoregular polymers; polymer structure; photodegradation of fiber forming macromolecules; polymer stabilization; fiber morphology; modification of fibers; polymer surface studies; microbiological deterioration. Mailing Add: Div of Chem Nat Res Coun of Can Ottawa ON Can

WILES, DONALD ROY, b Truro, NS, Aug 30, 25; m 52; c 3. INORGANIC CHEMISTRY. Educ: Mt Allison Univ, BSc, 46, BEd, 47; McMaster Univ, MSc, 50; Mass Inst Technol, PhD(chem), 53. Prof Exp: Chemist, Eldorado Mining & Refining Ltd, Can, 47-48; res assoc radiochem, Chem Inst, Oslo, Norway, 53-55; res assoc metall chem, Univ BC, 55-59; from asst prof to assoc prof nuclear inorg chem, 59-69, PROF NUCLEAR INORG CHEM, CARLETON UNIV, 69- Concurrent Pos: Vis scientist, Inst Hot Atom Chem, Nuclear Res Ctr, Karlsruhe, Ger, 69-70. Mem: Am Chem Soc; Chem Inst Can; The Chem Soc; Norweg Chem Soc. Res: Dissolution kinetics of metals and oxides; hot atom chemistry in organic solids; nuclear fission; radiochemistry; Mössbauer spectroscopy; ion implantation. Mailing Add: Dept of Chem Carleton Univ Ottawa ON Can

WILES, JOSEPH ST CLAIR, b Brooklyn, NY, July 27, 14; m 45; c 2. PHARMACOLOGY, TOXICOLOGY. Educ: Morris Brown Col, AB, 41; Columbia Univ, 50. Prof Exp: Res technician, Water Lab, Corps Engrs, US Army, Ft Benning, Ga, 42-43, bacteriologist, Bio-Anal Br, Aerosol Sect, Army Chem Ctr, Chem Corps, Md, 46-50, biologist, Field Toxicol Br, Directorate Med Res, 51-56, pharmacologist, Chem Warfare Labs, Army Chem Ctr, 56-58, PHARMACOLOGIST, PHYSIOTOXICOL BR, TOXICOL DIV, EDGEWOOD ARSENAL BIOMED LAB, ABERDEEN PROVING GROUND, US ARMY, 58- Honors & Awards: Dept of Army Awards, 59, 62, 66 & 67. Mem: AAAS; Sigma Xi; Soc Toxicol; Am Ord Asn; NY Acad Sci. Res: Evaluation of the influence of environmental variables and levels of protection on the toxicity of chemical agents in various animals; health hazard safety evaluations of military, household or environmental chemicals in animals. Mailing Add: 1423 N Ellamont St Baltimore MD 21216

WILES, MICHAEL, b Sheffield, Eng, May 8, 40; m 63; c 2. PARASITOLOGY, MARINE BIOLOGY. Educ: Univ Leeds, BSc, 62, PhD(zool), 65. Prof Exp: Res scientist, Fisheries Res Bd, Can, 65-67; asst prof freshwater ecol, 67-72, ASSOC PROF FRESHWATER ECOL, ST MARY'S UNIV, NS, 72- Mem: Wildlife Dis Asn; Soc Invert Path; Can Soc Zool; Brit Soc Parasitol. Res: Parasitic larvae of freshwater molluscs; parasites of seal stomachs and freshwater fishes of Eastern Canada. Mailing Add: Dept of Biol St Mary's Univ Halifax NS Can

WILES, ROBERT ALLAN, b Quincy, Mass, Apr 6, 29; m 51; c 5. INDUSTRIAL ORGANIC CHEMISTRY. Educ: Univ NH, BS, 51, MS, 55; Mass Inst Technol, PhD(chem), 58. Prof Exp: Res chemist, Sun Oil Co, 57-59; res chemist, Solvay Process Div, 59-64, res supvr, 64-70, res supvr indust chem div, 66-68, MGR PROCESS RES, ALLIED CHEM CORP, BUFFALO, 71-, MGR SPECIALTY CHEM DIV, 68- Mem: Am Chem Soc. Res: Urethane chemicals and fluorochemicals. Mailing Add: 169 Euclid Ave Hamburg NY 14075

WILES, WILLIAM WALTER, b Glen Ridge, NJ, Sept 26, 27; m 54; c 1. GEOLOGY, MICROPALEONTOLOGY. Educ: Dartmouth Col, AB, 50; Columbia Univ, PhD(geol), 60. Prof Exp: From instr to asst prof, 57-66, ASSOC PROF GEOL, NEWARK COL ARTS & SCI, RUTGERS UNIV, 66- Mem: Geol Soc Am; Paleont Soc. Res: Planktonic Foraminifera, especially as they occur in deep sea sediments, and can be used as indices to past climatic conditions. Mailing Add: Dept of Geol Newark Col of Arts & Sci Rutgers Univ Newark NJ 07102

WILEY, BILL BEAUFORD, b St Joseph, Mo, Nov 12, 23; m 45; c 2. MICROBIOLOGY, IMMUNOLOGY. Educ: Univ Kans, BA, 49, MA, 50; Univ Rochester, PhD(microbiol), 56. Prof Exp: Asst dir Rochester Health Bur Labs, Med Ctr, Univ Rochester, 50-56; asst prof microbiol, Univ Sask, 56-62; asst prof, 62-74, ASSOC PROF MICROBIOL, MED CTR, UNIV UTAH, 74- Concurrent Pos: Med Res Coun Can fel, Univ Sask, 57-62; Nat Inst Allergy & Infectious Dis fel, Univ Utah, 63- Mem: AAAS; Can Soc Microbiol; Am Soc Microbiol; NY Acad Sci. Res: Encapsulation and virulence of staphylococcus aureus staphylococcal scalded skin syndrome. Mailing Add: Dept of Microbiol Univ Utah Med Ctr Salt Lake City UT 84132

WILEY, DON CRAIG, b Akron, Ohio, Oct 21, 44. BIOPHYSICS. Educ: Tufts Univ, SB, 66; Harvard Univ, PhD(biophys), 71. Prof Exp: Asst prof biochem & molecular biol, 71-75, ASSOC PROF BIOCHEM, HARVARD UNIV, 75- Concurrent Pos: Jane Sloan Coffin Fund grant, Harvard Univ, 72-73. Mem: AAAS. Res: X-ray diffraction; structure of macromolecules and assemblies of macromolecules; metabolic and genetic regulation; viral membrane glycoproteins. Mailing Add: Dept of Biochem Harvard Univ Cambridge MA 02135

WILEY, DOUGLAS WALKER, b Shanghai, China, Apr 8, 29; m 50; c 4. ORGANIC CHEMISTRY. Educ: Univ Richmond, BS, 49; Columbia Univ, MA, 52; Yale Univ, PhD(chem), 55. Prof Exp: RES CHEMIST, CENT RES DEPT, E I DU PONT DE NEMOURS & CO, INC, 55- Mem: Am Chem Soc. Res: Synthesis, structure and mechanism in area of fluorocarbons, cyanocarbons and energy-transforming organic solids. Mailing Add: Cent Res Dept Exp Sta E I du Pont de Nemours & Co Inc Wilmington DE 19898

WILEY, GROVE GEORGE, microbiology, see 12th edition

WILEY, JOHN DUNCAN, b Nashville, Tenn, Mar 23, 42. SOLID STATE PHYSICS. Educ: Ind Univ, BS, 64; Univ Wis, MS, 65, PhD(physics), 68. Prof Exp: Mem tech staff, Optical & Magnetic Mat Dept, Bell Tel Labs, 68-74; RES ASSOC, MAX PLANCK INST SOLID BODIES RES, 74- Mem: Am Phys Soc; Am Asn Physics Teachers. Res: Transport properties of semiconductors; infrared absorption in semiconductors; growth and characterization of semiconductor crystals; growth of magnetic garnets. Mailing Add: Max Planck Inst Solid Bodies Res PO Box 1099 Stuttgart West Germany

WILEY, JOHN HERBERT, b Mendota, Ill, Jan 14, 21; m 48; c 3. SPEECH PATHOLOGY, AUDIOLOGY. Educ: Northern Ill State Col, BE, 41; Univ Iowa, MA, 42; Univ Southern Calif, PhD(speech, psychol), 50. Prof Exp: Asst, Speech Clin, Univ Iowa, 41-42; res voice commun, Nat Defense Res Coun Proj, Univ Tex, 44-45; acoustic technician, Borden Gen Hosp, 46; lectr speech path, Univ Southern Calif, 47-48; dir speech & hearing labs & asst to assoc prof, Univ Nebr, 48-60, assoc prof neurol & psychiat, Col Med & Nebraska Psychiat Inst, 60-65; prof speech & coordr children's servs, Speech Clin, 65-70, PROF PHYS MED & REHAB, UNIV MICH, ANN ARBOR, 70- Concurrent Pos: Consult, Plymouth State Home & Training Sch, 66- Mem: Fel Am Speech & Hearing Asn; Am Psychol Asn; Acoust Soc Am; Am Acad Ment Retardation; Am Asn Ment Deficiency. Res: Speech and language problems of the retarded; diagnosis of speech and language problems; communicative disorders in the multiple handicapped. Mailing Add: Speech Clin Univ of Mich 1111 E Catherine St Ann Arbor MI 48104

WILEY, JOHN ROBERT, b San Angelo, Tex, Oct 10, 46; m 68; c 1. NUCLEAR CHEMISTRY. Educ: Univ Houston, BS, 69; Purdue Univ, PhD(nuclear chem), 74. Prof Exp: RES CHEM SEPARATIONS, SAVANNAH RIVER PLANT, E I DU PONT DE NEMOURS & CO, INC, 74- Mem: Am Chem Soc; AAAS. Res: Management of nuclear plant wastes. Mailing Add: Savannah River Lab E I du Pont de Nemours & Co Inc Aiken SC 29801

WILEY, LORRAINE, b Sacramento, Calif. PLANT PHYSIOLOGY. Educ: Sacramento State Col, AB, 64; Univ Calif, Davis, MS, 66, PhD(plant physiol), 71. Prof Exp: Asst prof bot, Howard Univ, 71-72; ASST PROF BIOL, CALIF STATE UNIV, FRESNO, 72- Concurrent Pos: Res botanist, Univ Calif, Davis, 71. Mem: Am Soc Plant Physiol. Res: Plant protein metabolism; seed physiology; plant hormone physiology. Mailing Add: Dept of Biol Calif State Univ Fresno CA 93740

WILEY, LYNN M, b Tucson, Ariz, Feb 24, 47; m 69. DEVELOPMENTAL BIOLOGY. Educ: Univ Calif, Irvine, BS, 68, MS, 71; Univ Calif, San Francisco, PhD(anat), 75. Prof Exp: Lab technician electron micros, Univ Calif, Irvine, Calif Col Med, 69-71; CELL BIOLOGIST MAMMALIAN DEVELOP, SAN FRANCISCO MED CTR, UNIV CALIF, 75- Mem: AAAS; Fedn Am Scientists; Soc Study Reproduction; Am Asn Anatomists. Res: Cell surface in early mammalian development; origin of primary germ layers; regulation of cell determination. Mailing Add: Dept of Radiobiol San Francisco Med Ctr San Francisco CA 94143

WILEY, MICHAEL DAVID, b Long Beach, Calif, Nov 28, 39; m 61; c 2. ORGANIC CHEMISTRY. Educ: Univ Southern Calif, BS, 61; Univ Wash, PhD(org chem), 69. Prof Exp: Asst prof, 68-74, ASSOC PROF CHEM, CALIF LUTHERAN COL, 74- Mem: AAAS; Am Chem Soc; The Chem Soc. Res: Reaction mechanisms; carbonium ions. Mailing Add: Dept of Chem Calif Lutheran Col Thousand Oaks CA 91360

WILEY, PAUL FEARS, b Sullivan, Ill, June 21, 16; m 38; c 3. ORGANIC CHEMISTRY. Educ: Univ Ill, BS, 38, MS, 39; Univ Minn, PhD(org chem), 44. Prof Exp: Jr chemist, Merck & Co, Inc, NJ, 39-41; sr res chemist, Allied Chem Corp, 44-46; res chemist, Eli Lilly & Co, 46-58, res assoc, 58-60; RES ASSOC, RES LABS, UPJOHN CO, 60- Mem: Am Chem Soc. Res: Chemistry of antibiotics and natural products. Mailing Add: Res Labs Upjohn Co Kalamazoo MI 49001

WILEY, RICHARD HAVEN, b Mattoon, Ill, May 10, 13; m 40; c 2. CHEMISTRY. Educ: Univ Ill, AB, 34, MS, 35; Univ Wis, PhD(chem), 37; Temple Univ, LLB, 43. Prof Exp: Res chemist, E I du Pont de Nemours & Co, Inc, 37-45; assoc prof chem, Univ NC, 45-49; prof & chmn dept, Univ Louisville, 49-65; exec officer doctoral prog, 65-68, PROF CHEM, HUNTER COL, 65- Concurrent Pos: NSF sr fel, Imp Col, Univ London, 57-58; vis prof, City Col New York, 63-64; consult; res proj dir with NSF, AEC, NIH, Off Naval Res, Off Ord Res, Bur Naval Ord & NASA. Honors & Awards: Award, Am Chem Soc, 65. Mem: AAAS; Am Chem Soc. Res: Polymer chemistry; organic synthesis; mass spectrometry. Mailing Add: Dept of Chem Hunter Col 695 Park Ave New York NY 10021

WILEY, RICHARD HAVEN, JR, b Wilmington, Del, June 14, 43; m 71. ETHOLOGY, ECOLOGY. Educ: Harvard Univ, BA, 65; Rockefeller Univ, PhD(animal behav), 70. Prof Exp: Fel animal behav, Rockefeller Univ, 70-71; ASST PROF ZOOL, UNIV NC, CHAPEL HILL, 71- Mem: AAAS; Animal Behav Soc; Ecol Soc Am; Soc Study Evolution; Am Soc Naturalists. Res: Comparative studies of vertebrate social organization; behavioral mechanisms in aggression and affiliation; ecology of social organization; vocal communication. Mailing Add: Dept of Zool Univ of NC Chapel Hill NC 27514

WILEY, ROBERT A, b Ann Arbor, Mich, Sept 5, 34; m 55; c 3. MEDICINAL CHEMISTRY. Educ: Univ Mich, BS, 55; Univ Calif, San Francisco, PhD(pharmaceut chem), 62. Prof Exp: From asst prof to assoc prof, 62-71, PROF MED CHEM, UNIV KANS, 71- Mem: Am Chem Soc; Am Pharmaceut Asn; Soc Toxicol. Res: Relationship between biological activity and chemical properties among drugs; chemical aspects of drug metabolism. Mailing Add: Dept of Med Chem Univ of Kans Lawrence KS 66045

WILEY, ROBERT CRAIG, b Washington, DC, Nov 14, 24; m 51; c 3. FOOD SCIENCE, HORTICULTURE. Educ: Univ Md, BS, 49, MS, 50; Ore State Univ, PhD(food tech), 53. Prof Exp: Asst, Ore State Univ, 51-52; food specialist, US Dept

Navy, 53; from asst prof to assoc prof, 53-69, PROF HORT, UNIV MD, COLLEGE PARK, 69- Honors & Awards: Woodbury Res Award co-recipient, 61, 62. Mem: Am Soc Hort Sci; Inst Food Technol. Res: Measurement of structural and textural characteristics of foods; measurement of polysaccharides, fatty acids and enzymes in fruits and vegetables; aroma analyses by gas liquid chromatography; thermal processing of foods. Mailing Add: Dept of Hort Univ of Md Col of Agr College Park MD 20742

WILEY, RONALD LEE, b Dayton, Ohio, Oct 4, 37; m 61; c 1. RESPIRATORY PHYSIOLOGY, PULMONARY PHYSIOLOGY. Educ: Univ Miami, Ohio, BS, 59; Univ Ky, PhD(physiol, biophys), 66. Prof Exp: Teacher, Talawanda High Sch, Ohio, 59-62; instr physiol & NIH fel, Marquette Univ, 66-67; asst prof, 67-71, ASSOC PROF ZOOL & PHYSIOL, MIAMI UNIV, 71-; ASST PROF, WRIGHT STATE UNIV MED SCH, DAYTON, 74- Concurrent Pos: Mem med staff, McCullough-Hyde Mem Hosp, Oxford, Ohio, 73- Mem: Am Physiol Soc; Sigma Xi. Res: Control of respiration; respiratory response to added resistance; exercise physiology. Mailing Add: Dept of Zool Miami Univ Oxford OH 45056

WILEY, SAMUEL L, b Springfield, Ohio, Dec 24, 37; div. PHYSICS. Educ: Capital Univ, BS, 59; Ohio State Univ, PhD(physics), 68. Prof Exp: Instr physics, Capital Univ, 64-67; asst prof, 68-71, chmn dept physics & info sci, 73-75, ASSOC PROF PHYSICS, CALIF STATE COL, DOMINGUEZ HILLS, 71- Concurrent Pos: Lectr, Otis Art Inst, 72-73. Mem: AAAS; Am Asn Physics Teachers; Asn Comput Mach. Res: Theoretical physics; phase transitions and cooperative phenomena; Ising and Heisenberg models of magnetic systems; computer applications in physics teaching. Mailing Add: Dept of Physics & Info Sci Calif State Col Dominguez Hills CA 90747

WILEY, WILLIAM HENRY, b Cherokee Co, Tex, Feb 19, 13; m 36; c 1. GENETICS, AGRICULTURE. Educ: Agr & Mech Col, Tex, BS, 36, MS, 37, PhD(genetics), 49. Prof Exp: Asst poultry husb, Agr & Mech Col, Tex, 36-37; from instr to asst prof animal indust, Univ r, 37-42 & 45-47; from assoc prof to prof, poultry husb, RI Univ, 47-62, assoc dir exp sta, 51-59, dean col agr & dir exp sta, 59-62; dean, 62-72, PROF POULTRY SCI, COL AGR & BIOL SCI, CLEMSON UNIV, 72- Mem: Sigma Xi; Poultry Sci Asn. Res: Genetics; development of annotated, retrievable bibliography on turkeys in National Agriculture Library. Mailing Add: 104 Lakeside Ct Clemson SC 29631

WILEY, WILLIAM LEE, b Durham, NC, June 3, 28; m 57; c 2. ORGANIC CHEMISTRY. Educ: Princeton Univ, AB, 49; Univ NC, PhD(org chem), 53. Prof Exp: DIV HEAD, RES & DEVELOP ORG CHEM DEPT, E I DU PONT DE NEMOURS & CO, INC, 53-55 & 57- Mem: Am Chem Soc. Res: Plastics. Mailing Add: Fairfield Rd Chadds Ford PA 19317

WILEY, WILLIAM RODNEY, b Oxford, Miss, Sept 5, 32; m 52; c 1. MICROBIOLOGY, BIOCHEMISTRY. Educ: Tougaloo Col, BS, 54; Univ Ill, MS, 60; Wash State Univ, PhD(bact), 65. Prof Exp: Coord Life Sci Prog, Pac Northwest Labs, 65-71 & Battelle Mem Inst, 71-74, MGR, BIOL DEPT, BATTELLE-NORTHWEST, 74- Concurrent Pos: Adj prof, 68-75. Mem: AAAS; Fedn Am Scientists; Am Soc Microbiol; Am Soc Biol Chem. Res: Microbial metabolism, particularly the factors which control intracellular pool formation and membrane transport of amino acids and sugars in microorganisms and mammalian cells. Mailing Add: Biol Dept Pac Northwest Lab Battelle Mem Inst PO Box 999 Richland WA 99352

WILF, HERBERT, b Phila, Pa, June 13, 31; m 52; c 3. MATHEMATICS. Educ: Mass Inst Technol, BS, 52; Columbia Univ, MS, 54, PhD, 58. Prof Exp: Mathematician, Nuclear Develop Corp Am, NY, 54-59; asst prof math, Univ Ill, 59-62; assoc prof, 62-65, PROF MATH, UNIV PA, 65- Concurrent Pos: Consult, Argonne Nat Lab. Mem: Am Math Soc; Soc Indust & Appl Math. Res: Numerical and combinatorial analysis. Mailing Add: Dept of Math Univ of Pa Philadelphia PA 19104

WILFERTH, ROBERT, physical chemistry, see 12th edition

WILFONG, ROBERT EDWARD, b Wayne Co, Ill, Jan 3, 20; m 38; c 5. CHEMISTRY. Educ: Univ Wis, BS, 41, MS, 42, PhD(phys chem), 44. Prof Exp: Res chemist, Va, 44-48, res supvr, 48-51, res mgr, 51-53, tech supt, 53-59, lab dir, 59-64, tech mgr, 64-71, TECH DIR, E I DU PONT DE NEMOURS & CO, INC, WILMINGTON, DEL, 71- Honors & Awards: Award of Merit, Nat Defense Res Comt. Mem: AAAS; Am Chem Soc; NY Acad Sci. Res: Photosynthesis; submarine detection; kinetics of ro rocket propellant decomposition; modified rocket propellants; infrared spectroscopy; new textile fibers, Orlon, Nylon and Dacron; aromatic polyamides. Mailing Add: RD 2 Kennett Square PA 19348

WILFORD, BILL HOWARD, forest entomology, see 12th edition

WILFRET, GARY JOE, b Sacramento, Calif, Oct 13, 43; m 68; c 2. PLANT BREEDING, GENETICS. Educ: Univ Hawaii, BS, 65, PhD(hort), 68. Prof Exp: Asst prof biol, Ga South Col, 68-69; asst geneticist, 69-74, ASSOC GENETICIST, AGR RES & EDUC CTR, UNIV FLA, BRADENTON, 74- Mem: Am Soc Hort Sci; Am Asn Trop Biol; Bot Soc Am; Tissue Cult Asn. Res: Breeding of ornamental plants for disease resistance and adaptation to subtropical conditions; hybridizing Dendrobium species to determine sexual compatability, to examine meiotic behavior and to clarify genome relationships. Mailing Add: Agr Res & Educ Ctr Univ of Fla 5007 60th St E Bradenton FL 33505

WILGRAM, GEORGE FRIEDERICH, b Vienna, Austria, Apr 12, 24; nat US; m 56; c 3. PHYSIOLOGY, DERMATOLOGY. Educ: Univ Vienna, MD, 51; Univ Toronto, MA, 53, PhD, 57. Prof Exp: Lectr physiol, Univ Toronto, 57-58; asst prof exp path, Univ Chicago, 58-59; res assoc dermat, Harvard Med Sch, 59-60, asst prof, 61-67; PROF DERMAT, TUFTS UNIV, 67- Mem: Am Soc Exp Path; Soc Exp Biol & Med; Am Heart Asn. Res: Genetics of keratinization and pigmentation. Mailing Add: New Eng Ctr Hosp 260 Tremont St Boston MA 02111

WILGUS, DONOVAN RAY, b La Plata, Mo, Aug 11, 21; m 52; c 3. ORGANIC CHEMISTRY. Educ: Northeast Mo State Teachers Col, AB, 42; Univ Colo, PhD, 51. Prof Exp: From assoc res chemist to supv res chemist, 51-66, SR RES ASSOC, CHEVRON RES CO, 66- Mem: Am Chem Soc. Res: Diels-Alder reaction; lubricating oil additives; synthetic oils. Mailing Add: 2912 Cindy Ct Richmond CA 94803

WILGUS, HERBERT SEDGWICK, b Brooklyn, NY, Dec 2, 05; m 26; c 6. POULTRY SCIENCE. Educ: Cornell Univ, BS, 26, PhD(poultry nutrit), 30. Prof Exp: From asst poultry nutrit to investr, Cornell Univ, 26-36; prof poultry husb & head dept, Colo Agr & Mech Col & chief poultry sect, Exp Sta, 36-50, assoc dir Exp Sta, 45-47; dir res & nutrit, Peter Hand Found, 50-58; res, Ray Ewing Co, 58-61, vpres tech sales serv & res, 61-62; mgr tech agr serv, Hoffmann-La Roche, Inc, 62-70; CONSULT, 71- Mem: AAAS; Am Soc Animal Sci; Soc Exp Biol & Med; Poultry Sci Asn; World Poultry Sci Asn. Res: Poultry nutrition; calcium; phosphorus; manganese; iodine; riboflavin; protein quality; egg quality; incubation; vitamins. Mailing Add: 377 Highland Ave San Rafael CA 94901

WILHELM, ALAN ROY, b Buffalo, NY, Oct 30, 36. MICROBIOLOGY, VIROLOGY. Educ: Stanford Univ, AB, 58; Univ Wis, Madison, MS, 65, PhD(bact), 67. Prof Exp: Microbiologist, US Army, 67-69; asst prof, 69-72, ASSOC PROF BIOL SCI, CALIF STATE UNIV, CHICO, 72- Mem: AAAS; Am Soc Microbiol. Res: Arboviruses; viral infection of poikilothermic cells. Mailing Add: Dept of Biol Sci Calif State Univ First & Normal St Chico CA 95926

WILHELM, DALE LEROY, b Greenview, Ill, June 20, 26; m 51; c 4. INORGANIC CHEMISTRY. Educ: Univ Ill, BS, 51; Univ Tenn, MS, 52, PhD(chem), 54. Prof Exp: From asst prof to assoc prof chem, Univ WVa, 54-63; assoc prof, Cornell Col, 63-66; PROF CHEM, OHIO NORTHERN UNIV, 66-, ASST DEAN, LIBERAL ARTS COL, 73-, CHMN CHEM DEPT, 74- Mem: AAAS; Am Chem Soc; The Chem Soc. Res: Heteropolyanions; electrophoresis in stabilized media. Mailing Add: Dept of Chem Ohio Northern Univ Ada OH 45810

WILHELM, EUGENE J, JR, b St Louis, Mo, July 25, 33. BIOGEOGRAPHY, CULTURAL GEOGRAPHY. Educ: St Louis Univ, BS, 59; La State Univ, Baton Rouge, MA, 61; Tex A&M Univ, PhD(geog), 71. Prof Exp: Instr geog, DePaul Univ, 62-63; asst prof, St Louis Univ, 63-65; asst prof, McGill Univ, 65-68; vis lectr, Univ Va, 68-69 & 71-72; ASSOC PROF GEOG, SLIPPERY ROCK STATE COL, 72- Mem: Asn Am Geog. Res: Cultural and natural history of Appalachia; ecological problems in national park areas; ethnobiology; folk geography. Mailing Add: Dept of Geog Slippery Rock State Col Slippery Rock PA 16057

WILHELM, HARLEY ALMEY, b Ellston, Iowa, Aug 5, 00; m 23; c 4. PHYSICAL CHEMISTRY. Educ: Drake Univ, AB, 23; Iowa State Univ, PhD(phys chem), 31. Hon Degrees: LLD, Drake Univ, 61. Prof Exp: Teacher high sch, Iowa, 23-24, 25-26; instr, Intermountain Union Col, 26-27; asst, 27-28, from instr to prof, 28-71, EMER PROF CHEM, IOWA STATE UNIV, 71-; PRIN SCIENTIST, INST ATOMIC RES & AMES LAB, US ATOMIC ENERGY COMN, 66- Concurrent Pos: Assoc dir, Ames Lab, US Atomic Energy Comn, 46-66; mat consult, Off Saline Water, US Dept Interior, 66-67; consult, Div Reactor Licensing, US Atomic Energy Comn, DC, 68 & Bendix Corp, 71-72; vis prof, Univ Waterloo, 69; chmn panel depleted uranium, Nat Res Coun, DC, 70-71; del, mission to Europe & SAm, US State Dept. Honors & Awards: Gold Medal Award, Am Chem Soc, 54; Eisenman Award, Am Soc Metals, 62. Mem: Am Chem Soc; Am Soc Metals; Am Inst Mining, Metall & Petrol Eng; Am Nuclear Soc. Res: Metallurgy of uranium, thorium and other metals and alloys for nuclear energy; applied physical chemistry in areas pertaining to metallurgy and engineering. Mailing Add: 513 Hayward Ave Ames IA 50010

WILHELM, JAMES MAURICE, b Redfield, SDak, May 20, 40; m 69. MICROBIOLOGY, MOLECULAR BIOLOGY. Educ: SDak Sch Mines & Technol, BS, 62; Case Western Reserve Univ, PhD(biochem), 68. Prof Exp: Am Cancer Soc fel biophys, Univ Chicago, 68-70; from assoc prof to asst prof, microbiol, Sch Med, Univ Pa, 70-73; ASST PROF MICROBIOL, SCH MED, UNIV ROCHESTER, 73- Mem: Am Soc Cell Biol; Am Soc Microbiol. Res: Protein synthesis; control of viral replication; cell biology. Mailing Add: Dept of Microbiol Univ of Rochester Sch of Med Rochester NY 14627

WILHELM, JOHN ARTHUR, biochemistry, see 12th edition

WILHELM, ROBERT CARL, molecular biology, biochemistry, see 12th edition

WILHELM, RUDOLF ERNST, b Hanover, Ger, Dec 26, 26; US citizen; m 52; c 3. ALLERGY, IMMUNOLOGY. Educ: Univ Ill, Chicago, MS, 51; Am Bd Internal Med, dipl, 61. Prof Exp: Asst resident internal med, Detroit Receiving Hosp, Mich, 52-53; resident allergy, Roosevelt Hosp Inst, NY, 57; resident internal med, Henry Ford Hosp, Detroit, 58-59; instr med, Sch Med, La State Univ, New Orleans, 59-60; asst prof med, 60-64, asst prof dermat, 64-66, ASSOC PROF DERMAT, SCH MED, WAYNE STATE UNIV, 66-; CHIEF, ALLERGY SECT, VET ADMIN HOSP, ALLEN PARK, 60- Concurrent Pos: Consult, USPHS Hosp, Detroit, 62-69; secy, Dept Med, Oakwood Hosp, Dearborn, 65-67. Mem: Fel Am Col Physicians; fel Am Acad Allergy. Res: Delayed-type allergic skin reactions such as atopic eczema and contact dermatitis; methods and mechanics of allergy hyposensitization injections; anti-allergic drug treatment; methods of medical education in allergy and internal medicine. Mailing Add: Wilhelm Allergy Clin 751 S Military Rd Dearborn MI 48124

WILHELM, STEPHEN, b Imperial Co, Calif, Apr 19, 19; m 44; c 2. PLANT PATHOLOGY. Educ: Univ Calif, AB, 42, PhD(plant path), 48. Prof Exp: Instr plant path & jr plant pathologist, Exp Sta, 48-50, asst prof & asst plant pathologist, 50-56, assoc prof & assoc plant pathologist, 56-60, PROF PLANT PATH & PLANT PATHOLOGIST, UNIV CALIF, BERKELEY, 60- Concurrent Pos: Guggenheim fel, 58-59. Mem: AAAS; fel Am Phytopath Soc. Res: Verticilium wilt; diseases of small fruit; root infecting fungi; soil fumigation. Mailing Add: 12770 Skyline Blvd Oakland CA 94619

WILHELM, WALTER EUGENE, b St Louis, Mo, May 16, 31; m 61; c 1. PARASITOLOGY, PROTOZOOLOGY. Educ: Harris Teachers Col, BA, 55; Univ Ill, MS, 59; Univ Southern Ill, PhD(zool), 65. Prof Exp: Lectr embryol, Univ Southern Ill, 62-63; asst prof, 64-68, ASSOC PROF BIOL, MEMPHIS STATE UNIV, 68- Concurrent Pos: NSF fel parasitol, Univ Ill, 67. Mem: Soc Protozool; Am Soc Parasitol. Res: Parasitic helminths and protozoa, particularly those which may be used as agents of biological control; ecology of free-living protozoa; biology of limax amoebae. Mailing Add: Dept of Biol Memphis State Univ Memphis TN 38152

WILHELMI, ALFRED ELLIS, b Lakewood, Ohio, Sept 28, 10; m 40, 73; c 2. BIOCHEMISTRY, ENDOCRINOLOGY. Educ: Oxford Univ, BA, 35, DPhil(biochem), 37; Western Reserve Univ, AB, 33. Prof Exp: From instr to assoc prof physiol chem, Sch Med, Yale Univ, 39-50; prof, 50-60, CHARLES HOWARD CHANDLER PROF BIOCHEM, EMORY UNIV, 60-, CHMN DEPT, 50- Concurrent Pos: Alexander Browne Coxe fel physiol chem, Yale Univ, 38-39; Upjohn scholar, Endocrine Soc, 60; mem panel regulatory biol, NSF, 50-56; mem & chmn endocrinol study sect, NIH, 55-60; mem, Nat Adv Arthritis & Metab Dis Coun, 62-65; mem expert panel biol standardization, WHO, 65-; mem adv coun, Nat Endowment for Humanities, 67-69. Mem: Endocrine Soc (pres, 68-69); Am Soc Biol Chem; Am Zool Soc. Res: Metabolism; chemistry and physiology of anterior pituitary growth hormone. Mailing Add: Dept of Biochem Emory Univ Atlanta GA 30322

WILHELMSEN, PAUL CHADWICK, b Salt Lake City, Utah, Dec 15, 27; m 51; c 3. PHYSICAL CHEMISTRY. Educ: Univ Utah, BS, 50, PhD(phys chem), 54. Prof Exp: Res chemist, Shell Oil Co, 54-56 & Shell Develop Co, Calif, 56-72; dir res, Berkeley Appl Res Corp, 72-74; PRES, SONTEL INC, 74- Mem: Am Chem Soc; Am Inst Chem. Res: Nucleation; ellipsometry; reaction kinetics; process stream analyzers; analytical instrumentation; market research; product development. Mailing Add: 281 Livorna Heights Alamo CA 94507

WILHELMY, JERRY BARNARD, b Sewickley, Pa, July 31, 42; m 64. NUCLEAR CHEMISTRY, NUCLEAR PHYSICS. Educ: Univ Ariz, BSChE, 64; Univ Calif,

Berkeley, PhD(nuclear chem), 69. Prof Exp: Fel nuclear chem, Lawrence Berkeley Lab, 69-72; MEM STAFF NUCLEAR CHEM, LOS ALAMOS SCI LAB, 72- Mem: Am Chem Soc; Am Phys Soc. Res: Properties of nuclear fission; fission barriers; fission product spectroscopy, fission produced neutrons. Mailing Add: CNC-11 MS-514 Los Alamos Sci Lab Los Alamos NM 87545

WILHITE, DOUGLAS LEE, quantum chemistry, see 12th edition

WILHM, JERRY L, b Kansas City, Kans, Apr 27, 30; m 55; c 2. LIMNOLOGY, ECOLOGY. Educ: Kans State Teachers Col, BS, 52, MS, 56; Okla State Univ, PhD(zool), 65. Prof Exp: Teacher high sch, Kans, 56-62; US AEC fel, Oak Ridge, Tenn, 65-66; PROF ZOOL, SCH BIOL SCI, OKLA STATE UNIV, 66- Concurrent Pos: Biol consult, Am Inst Biol Sci Film Series, 61-62. Mem: Am Soc Limnol & Oceanog; Ecol Soc Am. Res: Biological effects of oil refinery effluents. Mailing Add: Sch of Biol Sci Okla State Univ Stillwater OK 74074

WILHOFT, DANIEL C, b Newark, NJ, Nov 16, 30; m 51; c 2. ZOOLOGY. Educ: Rutgers Univ, AB, 56; Univ Calif, Berkeley, MA, 58, PhD(zool), 63. Prof Exp: From instr to assoc prof, 62-69, PROF ZOOL, RUTGERS UNIV, NEWARK, 69-, CHMN DEPT ZOOL & PHYSIOL, 69- Concurrent Pos: NIH res grant, 64-67. Mem: AAAS; Am Soc Zool; fel Zool Soc London. Res: Reproductive cycles in lizards; reptilian endocrinology; endocrine control of thermoregulation in reptiles. Mailing Add: Dept of Zool & Physiol Rutgers Univ Newark NJ 07102

WILHOIT, EUGENE DENNIS, b Frankfort, Ky, Jan 28, 31; m 58; c 1. PHYSICAL CHEMISTRY. Educ: Univ Ky, BS, 53, PhD(phys chem), 56. Prof Exp: From res chemist to sr res chemist, Polychem Dept, 56-67, admin asst technol dept, 67, asst div supt, 67-69, div supt res, 69-75, GEN TECH SUPT, E I DU PONT DE NEMOURS & CO, INC, 75- Mem: AAAS; Am Chem Soc. Res: Electrochemistry and electrolytic conductance; reactions and synthesis of polymer intermediates; nonaqueous solutions; oxidation mechanisms. Mailing Add: 226 Greenwich Lane Wilmington NC 28401

WILHOIT, RANDOLPH CARROLL, b San Antonio, Tex, Oct 16, 25; m 48; c 3. THERMODYNAMICS, PHYSICAL BIOCHEMISTRY. Educ: Trinity Univ, Tex, AB, 47; Univ Kans, MA, 49; Northwestern Univ, PhD, 52. Prof Exp: Fel phys chem, Univ Ind, 52-53; asst prof, Tex Tech Col, 53-57; assoc prof, NMex Highlands Univ, 57-60, prof, 60-64; ASSOC PROF CHEM & ASSOC DIR THERMODYN RES CTR, TEX A&M UNIV, 64- Concurrent Pos: Assoc ed, J Chem & Eng Data of Am Chem Soc, 71- Mem: AAAS; Am Chem Soc; Calorimetry Conf. Res: Thermochemistry; molecular structure; energetics of biochemical reactions. Mailing Add: Dept of Chem Tex A&M Univ College Station TX 77843

WILIMOVSKY, NORMAN JOSEPH, b Chicago, Ill, Sept 9, 25; m 47; c 4. ICHTHYOLOGY, FISHERIES. Educ: Univ Mich, BS, 48, MA, 49; Stanford Univ, PhD, 56. Prof Exp: Head fish & game off, Mil Govt, Bavaria, Ger, 46; assoc ichthyologist, Fisheries Surv Brazil, 50-51; prin investr, Arctic Invests, Stanford Univ, 51-54, res assoc, 55-56; chief marine fisheries invests, US Fish & Wildlife Serv, Alaska, 56-60; assoc prof, Fisheries & Zool, 60-64, dir, Inst Fisheries, 63-66, PROF FISHERIES, UNIV BC, 64- Concurrent Pos: Mem comt proj Chariot, AEC; staff specialist, US Coun Marine Resources & Eng Develop, 67-68; mem, Environ Protection Bd, 70- Mem: Fel AAAS; Am Soc Ichthyol & Herpet; Am Fisheries Soc; Am Soc Limnol & Oceanog; fel Arctic Inst NAm. Res: Systematics of fishes; fisheries; ecology of ice and arctic hydrobiology; underwater instrumentation; history of exploration in the Arctic; science policy formulation. Mailing Add: Inst Resource Ecol Univ of BC Vancouver BC Can

WILIP, ELMAR KONSTANTIN, b St Petersburg, Russia, Apr 12, 10; nat US; m 45. ORGANIC CHEMISTRY. Educ: Tartu State Univ, Estonia, dipl, 38; Breslau Univ, Ger, Dr rer nat, 41. Prof Exp: Res asst org chem, Breslau Univ, 42-44; group leader, Ger Nat Res Coun, 44-45; chief chemist, Chemische Fabrik Marienfelde GmbH, 46-49; res chemist, Monsanto Ltd, Can, 50-55 & Dewey & Almy Chem Co div, WR Grace & Co, Mass, 55-59; sr res chemist, Nalco Chem Co, Ill, 59-62; res chemist, Armour Res Found, 62-63; tech ed, Argonne Nat Lab, 63-69; lit chemist, Res Ctr, Phillips Petrol Co, 69-75; RETIRED. Mem: Am Chem Soc; Ger Chem Soc. Res: Organic synthesis; natural products; polymer chemistry; nuclear science. Mailing Add: 312 S Santa Fe Apt Bartlesville OK 74003

WILK, MARTIN BRADBURY, statistics, see 12th edition

WILK, SHERWIN, b New York, NY, Aug 25, 38; m 63; c 2. BIOCHEMISTRY, PHARMACOLOGY. Educ: Syracuse Univ, BS, 60; Purdue Univ, MS, 62; Fordham Univ, PhD(biochem), 67. Prof Exp: Res asst biochem, Mt Sinai Hosp, 62-67; ASSOC PROF PHARMACOL, MT SINAI SCH MED, 69- Concurrent Pos: NIH fel, Sch Med, Cornell Univ, 67-69; NIH res career develop award, Mt Sinai Sch Med, 69- Mem: AAAS; Am Soc Pharmacol & Exp Therapeut; Am Soc Neurochem; Am Chem Soc. Res: Biochemistry of catecholamines; relationship to affective disorders; metabolism of catecholamines in central nervous system; amino acid transport and the gamma glutamyl cycle. Mailing Add: Dept of Pharmacol Mt Sinai Sch of Med New York NY 10029

WILK, WILLIAM DAVID, b Pittsburgh, Pa, Mar 6, 42; m 70; c 3. INORGANIC CHEMISTRY. Educ: Thiel Col, BA, 64; Northwestern Univ, PhD(chem), 68. Prof Exp: Asst prof anal & inorg chem, 68, ASSOC PROF CHEM, CALIF STATE COL, DOMINGUEZ HILLS, 68- Mem: The Chem Soc; Nat Sci Teachers Asn. Res: Ligand substitution effects on cobalt III complexes. Mailing Add: 730 Muskingum Ave Pacific Palisades CA 90272

WILKE, FREDERICK WALTER, b Pana, Ill, Sept 13, 33; m 56; c 4. MATHEMATICS. Educ: Drury Col, AB, 54; Wash Univ, MA, 59; Univ Mo, Columbia, MA, 60, PhD(math), 66. Prof Exp: Instr math, Southwest Mo State Col, 63-65; asst prof, 65-73, ASSOC PROF MATH, UNIV MO, ST LOUIS, 73- Mem: Am Math Soc; Math Asn Am. Res: Finite projective planes, especially translation planes. Mailing Add: Dept of Math Univ of Mo 8001 Natural Bridge Rd St Louis MO 63121

WILKEN, DAVID RICHARD, b Amarillo, Tex, Feb 20, 34; m 55; c 3. BIOCHEMISTRY. Educ: Blackburn Col, BA, 55; Univ Ill, MS, 58; Mich State Univ, PhD(biochem), 60. Prof Exp: From asst prof to assoc prof biochem, Okla State Univ, 62-66; asst prof, 66-72, ASSOC PROF PHYSIOL CHEM, UNIV WIS, MADISON, 72-; RES CHEMIST, LAB EXP PATH, VET ADMIN HOSP, 66- Concurrent Pos: Nat Found fel, Inst Enzyme Res, Univ Wis, 60-62. Mem: Am Soc Biol Chem; Am Chem Soc. Res: Diabetes; glycoprotein biosynthesis; pantothenic acid metabolism; enzymology. Mailing Add: Vet Admin Hosp 2500 Overlook Terr Madison WI 53705

WILKEN, DIETER H, b Los Angeles, Calif, Apr 12, 44. SYSTEMATIC BOTANY. Educ: Calif State Univ, Los Angeles, BA, 67; Univ Calif, Santa Barbara, PhD(biol), 71. Prof Exp: Res asst, Los Angeles State & County Arboretum, 66-67; asst prof biol, Occidental Col, 71-73; ASST PROF BIOL, COLO STATE UNIV, 73- Concurrent Pos: Cur, Colo State Univ Herbarium, 73- Mem: Bot Soc Am; Am Soc Plant Taxon; Int Soc Plant Taxon. Res: Systematics of higher plants within field of cytology, biochemistry, anatomy and breeding behavior; ecology and evolutionary significance of pollination. Mailing Add: Dept of Bot & Plant Path Colo State Univ Ft Collins CO 80521

WILKEN, DONALD RAYL, b New Orleans, La, Apr 25, 38; m 58; c 1. PURE MATHEMATICS. Educ: Tulane Univ, BS, 58, PhD(math), 65; Univ Calif, Los Angeles, MA, 62. Prof Exp: NSF fel, Brandeis Univ, 65-66; instr math, Mass Inst Technol, 66-68; asst prof, 68-70, ASSOC PROF MATH, STATE UNIV N Y ALBANY, 70- Res: Functional and complex analysis. Mailing Add: Dept of Math State Univ of N Y Albany NY 12222

WILKEN, GENE C, b Los Angeles, Calif, July 4, 28; m 66; c 3. GEOGRAPHY. Educ: Univ Calif, Berkeley, BS, 51, MA, 63, PhD(geog), 67. Prof Exp: Assoc prof, 67-73, PROF GEOG, COLO STATE UNIV, 73- Concurrent Pos: Colo State Univ res fel, 67 & 70; Wenner-Gren Found Anthrop Res fel, 69-70; Soc Sci Res Coun-Am Coun Learned Soc fel, 73-74; NSF fel, 73-75. Mem: AAAS; Am Geog Soc; Asn Am Geog. Res: Cultural geography; farming systems; development. Mailing Add: Dept of Econ Colo State Univ Ft Collins CO 80523

WILKEN, LEON OTTO, JR, b Waterbury, Conn, Oct 21, 24; m 46. PHARMACEUTICS, BIOPHARMACEUTICS. Educ: Loyola Univ, La, BS, 51; Univ Tex, MS, 53, PhD(pharm), 63. Prof Exp: Spec instr pharm, Univ Tex, 53-63; assoc prof, 63-72, PROF PHARM, AUBURN UNIV, 72-, HEAD, PHARMACEUT DIV, 73- Concurrent Pos: Pharm consult, Vet Admin Hosp, Tuskegee, Ala, 74- Mem: Am Asn Cols Pharm; Am Pharmaceut Asn; Am Chem Soc. Res: Enteric coated and sustained release dosage forms; investigations of plants known in folk-medicine for actual therapeutic value; time-dose distribution of selected mycotoxins in the rat; bioavailability and stability related to dosage forms. Mailing Add: 1321 Azalea Dr Auburn AL 36830

WILKEN, PHILIP HENRY, organic chemistry, see 12th edition

WILKENFELD, BYRON ELI, pharmacology, see 12th edition

WILKENFELD, JASON MICHAEL, b Brooklyn, NY, May 28, 39; m 62; c 2. SOLID STATE PHYSICS. Educ: Columbia Col, AB, 60; NY Univ, MS, 65, PhD(physics), 72. Prof Exp: Programmer analyst, Syst Develop Corp, 63; res asst physics, Radiation & Solid State Lab, NY Univ, 65-70; fel physics, New Eng Inst, 70-72; STAFF PHYSICIST, IRT CORP, 73- Concurrent Pos: Lectr physics, Hunter Col, City Univ New York, 66-69 & NY Univ, 69. Mem: Am Phys Soc. Res: Radiation effects in materials; electrical properties of dielectrics, especially polymeric insulators; positron annihilation as a morphological probe. Mailing Add: IRT Corp PO Box 80817 San Diego CA 92138

WILKENING, GEORGE MARTIN, b New York, NY, Dec 31, 23; m 50; c 5. ENVIRONMENTAL HEALTH. Educ: Queen's Col, NY, BS, 49; Columbia Univ, MS, 50. Prof Exp: Indust hygienist, State Health Dept, Va, 50-51; indust hygienist, Esso Res & Eng Co, 52-56, sr indust hygienist, Esso Standard Oil Co, 56-61; asst chief indust hygienist, Humble Oil & Ref Co, 61-63; head indust hyg & safety admin, 63-64, HEAD ENVIRON HEALTH & SAFETY DEPT, BELL LABS, INC, 64- Concurrent Pos: Lectr, Columbia Univ, 59-; mem, Safety Tech Adv Bd, Am Nat Stand Inst, 65-; chmn, Laser Hazards Stand Comt, 68-; mem, Environ Radiation Adv Comt, Environ Protection Agency, 67-72; tech electronic prod radiation stand comt mem, Dept Health, Educ & Welfare, 68-; mem, Nat Coun Radiation Protection & Measurements, 73-79; chmn, Tech Comt Lasers, Int Electrotech Comn, 73-; chmn, Sci Comt Microwaves, Nat Coun Radiation Protection & Measurements, 75- Mem: AAAS; Sigma Xi; Acoust Soc Am; Am Indust Hyg Asn; fel NY Acad Sci. Res: Dosimetry of exposure to electromagnetic radiations, including acoustical noise; industrial toxicology; biological effects of chemical, physical and biological agents in the environment; development of standards for permissible levels of exposure to environmental agents. Mailing Add: Environ Health & Safety Dept Bell Labs Murray Hill NJ 07974

WILKENING, LAUREL LYNN, b Richland, Wash, Nov 23, 44. METEORITICS, PLANETARY SCIENCES. Educ: Reed Col, BA, 66; Univ Calif, San Diego, PhD(chem), 70. Prof Exp: Am Asn Univ Women fel, Tata Inst Fundamental Res, India & Max Planck Inst Chem, 71; res assoc chem, Enrico Fermi Inst, Univ Chicago, 72-73; ASST PROF CHEM, UNIV ARIZ, 73- Honors & Awards: Nininger Meteorite Award, Ariz State Univ, 70. Mem: AAAS; Meteoritical Soc; Am Geophys Union; Am Astron Soc. Res: Chemistry and mineralogy of meteorites; chronology of the solar system; fission track dating. Mailing Add: Lunar & Planetary Lab Univ of Ariz Tucson AZ 85721

WILKENING, MARVIN C, b Malone, Tex, July 1, 20; m 43; c 4. ANIMAL NUTRITION, BIOCHEMISTRY. Educ: Univ Tex, AB, 41; Agr & Mech Col, Tex, MS, 47. Hon Degrees: ScD, Athens Col, 56. Prof Exp: Res chemist, Dow Chem Co, Tex, 41-44; asst to dir res, Security Mills, Tenn, 47-50; dir res, Ala Flour Mills, 50-66; DIR TECH SERV, CONAGRA, INC, 67- Mem: Fel AAAS; Am Chem Soc; Am Soc Animal Sci; Poultry Sci Asn; Am Dairy Sci Asn. Res: Application of basic animal nutrition and management research. Mailing Add: ConAgra Inc 3801 Harney St Omaha NE 68131

WILKENING, MARVIN HUBERT, b Oak Ridge, Mo, Mar 13, 18; m 42; c 2. ATMOSPHERIC PHYSICS. Educ: Southeastern Mo Univ, BS, 39; Ill Inst Technol, MS, 43, PhD(nuclear physics), 49. Prof Exp: Teacher high sch, Mo, 39-41; physicist, Manhattan Proj, 42-45; instr physics, Ill Inst Technol, 46-48; assoc prof physics, 48-52, prof physics & geophys & head dept, 52-68, PROF PHYSICS & DEAN GRAD STUDIES, N MEX INST MINING & TECHNOL, 67- Concurrent Pos: Consult to subcomn ions, Aerosols & Radioactivity, Int Comn Atmospheric Elec. Honors & Awards: Cert Serv, Off Sci Res Develop, 46. Mem: Fel AAAS; Am Phys Soc; Am Asn Phys Teachers; Am Geophys Union; Am Meteorol Soc. Res: Natural atmospheric radioactivity; atmospheric physics-ions, aerosols, tracing studies of cloud dynamics. Mailing Add: 1218 South Dr Socorro NM 87801

WILKENS, HANS J, b Eschershausen, Ger, Apr 2, 15; US citizen; div; c 1. BIOCHEMICAL PHARMACOLOGY. Educ: Vet Col, Hanover, Ger, DVM, 40, DrMedVet, 49. Prof Exp: Res assoc animal husb, Biol Sta, Hamelin, Ger, 48-49; pvt pract vet med, Thiede, Brunswick, Ger, 49-52; cancer res scientist, Roswell Park Mem Inst, Buffalo, NY, 56-64; ASSOC PROF BIOCHEM PHARMACOL, SCH PHARM, STATE UNIV NY BUFFALO, 64- Mem: AAAS; Am Soc Hemat; Am Soc Exp Pharmacol & Therapeut; Am Pharmaceut Asn. Res: Physiologic and pathophysiologic significance of the hypotensive polypeptide system; role of the fibrinolytic system in the pathogenesis of obstructive vascular disease. Mailing Add: 17 Old Glenwood Rd West Falls NY 14170

WILKENS, JEWEL L, b Lorraine, Kans, Aug 5, 37; m 68. INSECT PHYSIOLOGY.

Educ: Univ Ottawa, Kans, BA, 59; Tulane Univ, MSc, 61; Univ Calif, Los Angeles, PhD(zool), 67. Prof Exp: NIMH fel, Brain Res Inst, Univ Calif, Los Angeles, 67-68; asst prof, 69-72, ASSOC PROF BIOL, UNIV CALGARY, 72- Mem: AAAS; Biophys Soc. Res: Neuroendocrinology of invertebrates; neuronal activity associated with hormone release; neurophysiology of crustacean motor systems. Mailing Add: Dept of Biol Univ of Calgary Calgary AB Can

WILKENS, LON ALLAN, b Ellsworth, Kans, Sept 7, 42; m 65; c 3. NEUROBIOLOGY. Educ: Univ Kans, Lawrence, BA, 65; Fla State Univ, PhD(physiol), 70. Prof Exp: Fel neurobiol, Univ Tex, Austin, 70-73; asst prof biol, Bryn Mawr Col, 73-75; ASST PROF BIOL, UNIV MO-ST LOUIS, 75- Mem: Am Soc Zoologists; Soc Neurosci; AAAS. Res: Neural basis of behavior, including circulatory, somatosensory and respiratory physiology in decapod crustaceans and in bivalve and cephalopod molluscs, specifically related to the function of central interneurons and mechanisms of integration. Mailing Add: Dept of Biol Univ of Mo St Louis MO 63121

WILKENS, WALTER FREDERICK, biochemistry, see 12th edition

WILKERSON, CLARENCE WENDELL, JR, b Laredo, Tex, Aug 12, 44; m 65; c 1. TOPOLOGY. Educ: Rice Univ, BA, 66, PhD(math), 70. Prof Exp: Asst prof math, Univ Hawaii, Monoa, 70-72; res assoc, Swiss Fed Inst Technol, 72-73; res assoc, Carleton Univ, 73-74; instr, 74-75, ASST PROF MATH, UNIV PA, 75- Mem: Am Math Soc. Res: Algebraic topology and homotopy theory of Lie groups, H-spaces and associated spaces. Mailing Add: Dept of Math E-1 Univ of Pa Philadelphia PA 19174

WILKERSON, JOHN CHRISTOPHER, b Washington, DC, Mar 15, 26; m 58; c 2. PHYSICAL OCEANOGRAPHY. Educ: Univ Md, BS, 51. Prof Exp: Phys oceanogr, 60-62, from proj scientist to sr proj scientist, 62-67, PROJ MGR, US NAVAL OCEANOG OFF, 67- Concurrent Pos: Proj mgr data mgt, US Naval Oceanog Off, 74- Res: Remote sensing; aircraft platforms; data management. Mailing Add: 4834 Butterworth Pl NW Washington DC 20016

WILKERSON, ROBERT C, b Orange, Tex, June 25, 18; m 42; c 1. PHYSICS, MATHEMATICS. Educ: Univ Okla, BS, 41. Prof Exp: Seismic comput, Geophys Party, Stanolind Oil & Gas Co, Okla, 41-43; spectroscopist, Sinclair Rubber Inc, Tex, 43-47; res & develop div, 47-49, group leader phys instruments, 49-55, anal res, 55, head anal sect, 55-65, ADMIN MGR, TECH CTR, CELANESE CHEM CO, NY, 65- Mem: Am Chem Soc; Coblentz Soc. Res: Application of physical instruments in analytical support of organic chemistry research and development; development of application of computer systems for management information data and for technical information storage and retrieval systems. Mailing Add: Celanese Chem Co Tech Ctr PO Box 9077 Corpus Christi TX 78408

WILKERSON, ROBERT DOUGLAS, b Wilson, NC, Aug 5, 44; m 65; c 2. PHARMACOLOGY. Educ: Univ NC, Chapel Hill, BS, 67; Med Univ SC, MS, 69, PhD(pharmacol), 72. Prof Exp: MEM FAC PHARMACOL, COL MED, UNIV S ALA, 73- Concurrent Pos: USPHS fel, Sch Med, Tulane Univ, 71-73. Res: Cardiovascular pharmacology; pulmonary pharmacology. Mailing Add: Dept of Pharmacol Univ of S Ala Col of Med Mobile AL 36688

WILKERSON, THOMAS DELANEY, b Detroit, Mich, Feb 18, 32; div; c 3. PLASMA PHYSICS, SPACE PHYSICS. Educ: Univ Mich, BS, 53, MS, 54, PhD(physics), 62. Prof Exp: Consult, Proj Matterhorn, Princeton Univ, 59-60, mem proj res staff plasma, 60-61; from asst prof to assoc prof, 61-68, actg dir dept, 68-69, PROF PLASMA & SPACE PHYSICS, INST FLUID DYNAMICS & APPL MATH, UNIV MD, COLLEGE PARK, 68- Concurrent Pos: Consult, Radiation Div, US Naval Res Lab, 62-64; physics consult, Atlantic Res, 68-69; lectr, von Karman Inst, Brussels, 69; vpres, Versar, Inc, 69-70, consult, 70-; vis prof, Stanford Univ, 70; consult, Stanford Res Inst, 71; vis, Desert Res Inst, 71. Mem: Am Phys Soc; Am Astron Soc. Res: Laser-guided electrical discharges in gases, high-resolution absorption spectra of atmospheric gases; laser sounding of the atmosphere, infrared schlieren and interferometry; collision-induced spectroscopy in dense gases. Mailing Add: Inst Fluid Dynamics Univ of Md College Park MD 20742

WILKES, ALFRED, b Birmingham, Eng, Jan 7, 09; m 37; c 1. BIOLOGY. Educ: Univ Toronto, BSA, 35, MSA, 38, PhD(zool, entom), 40. Prof Exp: Asst, Univ Toronto, 35-40; lectr entom, Queen's Univ, Ont, 43-45; officer-in-chg, Entom Lab, 48-55, AGR SCIENTIST, ENTOM RES INST, CAN DEPT AGR, 55- Mem: AAAS; Entom Soc Am; Am Genetic Asn; Entom Soc Can; Genetics Soc Can. Res: Insect genetics and biological control. Mailing Add: Can Dept of Agr Entom Res Inst Exp Taxon Sect Cent Exp Farm Ottawa ON Can

WILKES, ANN BROOM, b Jackson, Miss, Aug 12, 43; m 68; c 1. PHARMACOLOGY. Educ: Webster Col, BS, 64; Univ Miss, PhD(pharmacol), 69. Prof Exp: Asst prof pharmacol, Med Sch, Univ Miss, 69-72; asst prof, 72-75, ASSOC PROF PHARMACOL, SCH MED, LA STATE UNIV, SHREVEPORT, 75- Concurrent Pos: Miss Heart Asn res fel, Med Ctr, Univ Miss, 69-71 & grant-in-aid, 71-72; La Heart Asn, La State Univ Found & Stiles Found grants. Res: Cardiac muscle contraction and ultrastructure. Mailing Add: Dept of Pharmacol La State Univ Sch of Med Shreveport LA 71130

WILKES, BENJAMIN GARRISON, JR, organic chemistry, see 12th edition

WILKES, CHARLES EUGENE, b Worcester, Mass, Oct 9, 39; m 61; c 4. CHEMISTRY. Educ: Worcester Polytech Inst, BS, 61; Princeton Univ, PhD(phys chem), 64. Prof Exp: From res chemist to sr res chemist, 64-69, sect leader, 69-73, sr res assoc, 73-74, SECT MGR, CORP RES NEW TECHNOL, RES CTR, B F GOODRICH CO, 74- Mem: Am Chem Soc. Res: Chemical physics; polymer characterization; x-ray diffraction; molecular spectroscopy; lab automation. Mailing Add: B F Goodrich Co Res Ctr Brecksville OH 44141

WILKES, GLENN RICHARD, b Houtzdale, Pa, Mar 25, 37; m 61; c 2. INORGANIC CHEMISTRY. Educ: Pa State Univ, BS, 60; Univ Wis, PhD(inorg chem), 65. Prof Exp: Scholar, Univ Calif, Los Angeles, 65-66; RES SCIENTIST, EASTMAN KODAK CO, 66- Mem: Am Chem Soc; Am Crystallog Asn; The Chem Soc. Res: Crystal and molecular structure determination of organometallic compounds by single crystal x-ray diffraction studies; synthesis of metal carbonyls and their derivatives. Mailing Add: 1102 Shoemaker Rd Webster NY 14580

WILKES, HILBERT GARRISON, JR, b Los Angeles, Calif, Oct 2, 37; m 65. PLANT GENETICS. Educ: Pomona Col, BA, 59; Harvard Univ, PhD(biol), 66. Prof Exp: Asst prof biol, Tulane Univ, 66-70; ASST PROF BIOL, UNIV MASS, BOSTON, 70- Concurrent Pos: Mem, World Maize Germplasm Comn, Rockefeller Found & Food & Agr Orgn; exec comt, Assembly Life Sci, Nat Res Coun; mem ed bd, J Econ Bot. Mem: Soc Study Econ Bot (secy, 73-75 & 75-77); Soc Study Evolution; Bot Soc Am; Soc Econ Bot; Indian Bot Soc. Res: Evolution under domestication in cultivated plants, especially maize and its wild relatives, teosinte and Tripsacum. Mailing Add: Dept of Biol Col II Univ of Mass Boston MA 02125

WILKES, JAMES C, b Mar 13, 21; US citizen; m 45; c 6. BIOLOGY. Educ: Troy State Col, BS, 48; Univ Tenn, MS, 50; Univ Ala, PhD(bot), 54. Prof Exp: Prof biol, Tenn Wesleyan Col, 50-51; prof & head dept, Jacksonville State Univ, 52-56; ed adv med sci sch aviation med, Air Univ, 56-57; prof & head dept biol, Huntingdon Col, 57-60 & Miss State Col Women, 60-66; PROF BIOL, TROY STATE UNIV, 66-, HEAD DEPT, 68- Mem: Am Bryol & Lichenological Soc. Res: Bryophytes; radiation biology. Mailing Add: Dept of Biol Troy State Univ Troy AL 36081

WILKES, JOHN BARKER, b Berkeley, Calif, Jan 17, 16; m 38; c 2. PETROLEUM CHEMISTRY. Educ: Univ Calif, BS, 37; Stanford Univ, PhD(chem), 48. Prof Exp: Chemist, Poultry Prod of Cent Calif, 37-40; from res chemist to sr res chemist, 48-62, res assoc, 62-65, SR RES ASSOC, CHEVRON RES CO, 65- Mem: AAAS; Am Chem Soc. Res: Kinetics; thermodynamics; catalysis; chemicals from petroleum; solubility theory. Mailing Add: 2935 Oxford Ave Richmond CA 94806

WILKES, JOHN STUART, b Panama, CZ, Mar 6, 47; m 71. BIOCHEMISTRY. Educ: State Univ NY Buffalo, BA, 69; Northwestern Univ, Evanston, MS, 71, PhD(chem), 73. Prof Exp: Res chemist, Frank J Seiler Res Lab, US Air Force Acad, 73-75; ASST PROF CHEM, UNIV COLO, DENVER, 76- Mem: AAAS; Am Chem Soc. Res: Synthesis, characteristics and utilization of analog polynucleotides; enzymology of nucleotide analogs; synthesis of stable isotope labeled propellants. Mailing Add: Dept of Chem Univ of Colo 1100 14th St Denver CO 80202

WILKES, RICHARD JEFFREY, b Chicago, Ill, Oct 18, 45; m 70. EXPERIMENTAL HIGH ENERGY PHYSICS, COSMIC RAY PHYSICS. Educ: Univ Mich, BSE, 67; Univ Wis, MS, 69, PhD(physics), 74. Prof Exp: RES ASSOC & INSTR PHYSICS, COSMIC RAY LAB, UNIV WASH, 74- Res: High energy interactions, especially hadronic multiparticle production using counter/spark chamber and nuclear emulsion techniques. Mailing Add: Dept of Physics FM-15 Univ of Wash Seattle WA 98195

WILKES, STANLEY NORTHRUP, b Corvallis, Ore, Jan 3, 27; m 58; c 3. PARASITOLOGY, MARINE ZOOLOGY. Educ: Ore State Univ, BS, 50, MS, 57, PhD(zool), 66. Prof Exp: Aquatic biologist, Res Div, Ore Fish Comn, 50-51 & 56-60; asst prof biol, E Carolina Col, 65-66; asst prof, 66-71, ASSOC PROF ZOOL, NORTHERN ARIZ UNIV, 71- Mem: Am Soc Parasitol; Am Micros Soc; Marine Biol Asn UK. Res: Taxonomy, life history studies and distribution of parasitic copepods of fishes, elasmobranchs and invertebrates; taxonomy of monogenetic Trematoda; taxonomy and ecology of marine invertebrates and fishes. Mailing Add: Dept of Biol Sci Box 5640 Northern Ariz Univ Flagstaff AZ 86001

WILKES, WILLIAM ROY, b Harvey, Ill, Feb 15, 39; m 59; c 3. CRYOGENICS. Educ: DePauw Univ, AB, 59; Univ Ill, MS, 61, PhD(physics), 66. Prof Exp: Asst prof physics, Wake Forest Univ, 65-67; sr res physicist, 67-74, RES SPECIALIST, MOUND LAB, MONSANTO RES CORP, 74- Mem: AAAS; Am Phys Soc. Res: Cryogenic isotope separation; Tritian technology; liquid helium; superconductivity. Mailing Add: Mound Lab Monsanto Res Corp Miamisburg OH 45342

WILKIE, CHARLES ARTHUR, b Detroit, Mich, Nov 21, 41; m 64; c 3. INORGANIC CHEMISTRY. Educ: Univ Detroit, BS, 63; Wayne State Univ, PhD(inorg chem), 67. Prof Exp: Asst prof, 67-74, ASSOC PROF CHEM, MARQUETTE UNIV, 74- Mem: Am Chem Soc; The Chem Soc. Res: Organometallic chemistry, particularly of groups I, II and III. Mailing Add: Dept of Chem Marquette Univ Milwaukee WI 53233

WILKIE, DONALD W, b Vancouver, BC, June 20, 31; m 56; c 3. MARINE BIOLOGY. Educ: Univ BC, BA, 60, MSc, 66. Prof Exp: Asst res biologist, BC Fish & Game Dept, 60-61; asst cur, Vancouver Pub Aquarium, 61-63; cur, Philadelphia Aquarium, Inc, 63-65; AQUARIUM DIR, SCRIPPS INST OCEANOG, 65- Mem: Am Soc Ichthyol & Herpet. Res: Pigmentation and coloration of fishes; environmental physiology as it relates to maintenance of marine fishes and cetaceans; aquariology and methods of public education in aquaria and museums. Mailing Add: Aquarium-Mus A-007 Scripps Inst Oceanog La Jolla CA 92093

WILKIE, JOHN BROWN, physical chemistry, biochemistry, see 12th edition

WILKIE, RICHARD W, b Idaho Falls, Idaho, Apr 14, 38; m 69; c 1. GEOGRAPHY. Educ: Univ Wash, BA, 60, MA, 63, PhD(geog), 68. Prof Exp: From instr to asst prof, 68-72, ASSOC PROF GEOG, UNIV MASS, AMHERST, 73- Concurrent Pos: Fulbright fel, Arg, 65-67; chmn, Five Col Latin Am Studies Coun, 73. Mem: Asn Am Geog; Pop Asn Am; Latin Am Studies Asn. Res: micro-spatial behavior of peasants; the migration process; environmental perception. Mailing Add: Dept of Geol & Geog Univ of Mass Amherst MA 01002

WILKIN, LOUIS ALDEN, b Bath, Maine, Mar 5, 39; m 60; c 3. ORGANIC CHEMISTRY. Educ: The Citadel, BS, 60; Clemson Univ, MS, 62, PhD(org chem), 65. Prof Exp: From chemist to sr chemist, 65-75, DEVELOP ASSOC, TENN EASTMAN CO, 75- Mem: AAAS. Res: Novel rearrangements of acetylenic alcohols and esters upon treatment with basic alumina. Mailing Add: 1122 Watauga St Kingsport TN 37660

WILKINS, BRUCE TABOR, b Greenport, NY, June 21, 31; m 56; c 3. RESOURCE MANAGEMENT. Educ: Cornell Univ, BS, 52, PhD(resource planning), 67; Mont State Univ, MS, 56. Prof Exp: Res biologist, Mont Fish & Game Dept, 56-59; county agr agent, Broome County Exten Serv, NY, 59-63; exten specialist, 63-67, asst prof, 67-73, ASSOC PROF CONSERV, CORNELL UNIV, 73- Concurrent Pos: Prog leader, NY Sea Grant Adv Serv, 72-75, assoc dir, 75-; vis prof, Univ BC, 74; vis Sea Grant prof, Ore State Univ, 75. Mem: Wildlife Soc; Ecol Soc Am; Am Fisheries Soc; Am Inst Biol Sci; Nat Recreation & Park Asn. Res: Relationships of human demands and natural resources, especially fish and wildlife resources; resource policy, particularly recreational policy issues. Mailing Add: Dept of Natural Resources Fernow Hall Cornell Univ Ithaca NY 14850

WILKINS, CHARLES LEE, b Los Angeles, Calif, Aug 14, 38; m 66. ANALYTICAL CHEMISTRY. Educ: Chapman Col, BS, 61; Univ Ore, PhD(chem), 66. Prof Exp: Asst prof, 67-72, ASSOC PROF CHEM, UNIV NEBR, LINCOLN, 72- Concurrent Pos: Vis assoc prof, Univ NC, Chapel Hill, 74-75. Mem: Am Chem Soc; Am Soc Test & Mat; The Chem Soc; Soc Appl Spectros. Res: Pattern recognition application in chemistry; mass spectrometry; nuclear magnetic resonance spectroscopy and computer applications to chemical problems and laboratory automation. Mailing Add: Dept of Chem Univ of Nebr Lincoln NE 68508

WILKINS, CORNELIUS K, JR, organic chemistry, see 12th edition

WILKINS, CURTIS C, b La Crosse, Wis, Oct 28, 35; m 54; c 3. PHYSICAL CHEMISTRY. Educ: Wis State Univ, BS, 57; Mich State Univ, PhD(chem), 64. Prof

Exp: Asst prof chem, WVa Wesleyan Col, 62-65; assoc prof, 65-70, PROF CHEM, WESTERN KY UNIV, 70- Concurrent Pos: NSF res participation fel, Univ Tenn, 65. Mem: Am Chem Soc. Res: Dilute solution properties of stereo-regular polymers; flame photometry studies of trace elements. Mailing Add: Dept of Chem Western Ky Univ Bowling Green KY 42101

WILKINS, EBTISAM A M SEOUDI, b Monofia, Egypt, Mar 10, 45; US citizen; m 74; c 1. BIOENGINEERING, CHEMICAL ENGINEERING. Educ: Cairo Univ, BSc, 65, MSc, 68; Univ Va, MSc, 73, PhD(chem eng), 76. Prof Exp: Res eng metall, Nat Res Ctr, Cairo, Egypt, 65-68; res specialist biomed eng, Div Biomed Eng, Univ Va, 69-73, res asst chem eng, Dept Chem Eng, 73-76; FEL BIOENERGETICS, DEPT KINESIOLOGY, SIMON FRASER UNIV, 76- Mem: Am Inst Chem Engrs. Res: Drug effects at cellular level, using cell culture, artificial lipid membrane techniques, especially carotenoids and skeletal muscle cells; effect of electric magnetic fields on bone fracture repair; modelling of blood flow in bones. Mailing Add: Dept of Kinesiology Simon Fraser Univ Burnaby BC Can

WILKINS, EUGENE MORRILL, b Hugo, Okla, Oct 5, 22; m 49; c 2. METEOROLOGY. Educ: Univ Chicago, MS, 49; Univ Okla, PhD(meteorol), 63. Prof Exp: Res scientist, US Weather Bur, 50-59; sr scientist, Ling-Temco-Vought, Inc, Tex, 59-71, SR SCIENTIST, ADVAN TECHNOL CTR, LTV AEROSPACE CORP, 71- Concurrent Pos: NSF res grants, 61-63, 65-67, 71-73; adj prof, Univ Okla, 65-; air pollution consult, Tex Pub Health Serv. Mem: Am Meteorol Soc; Am Geophys Union; Am Inst Aeronaut & Astronaut. Res: Atmospheric turbulence and diffusion; physics of the high atmosphere; severe storm research; satellite meteorology; weather modification. Mailing Add: Advan Technol Ctr Inc PO Box 6144 Dallas TX 75222

WILKINS, HAROLD, b Cobden, Ill, Nov 3, 33. HORTICULTURE, PLANT PHYSIOLOGY. Educ: Univ Ill, BS, 56, MS, 57, PhD(hort, plant physiol), 65. Prof Exp: Res horticulturist, Gulf Coast Exp Sta, Univ Fla, 65-66; from asst prof to assoc prof hort, 66-74, PROF HORT SCI & LANDSCAPE ARCHIT, UNIV MINN, ST PAUL, 74- Concurrent Pos: Consult, Gas Chromatography Sch, Fisk Univ, 65. Mem: Am Soc Hort Sci. Res: Post harvest physiology; respiration and ethylene emanation in aging floral tissue; physiology of lilies in response to photoperiod and cold treatments. Mailing Add: Dept of Hort Col of Agr Univ of Minn St Paul MN 55101

WILKINS, HOMER CLIFTON, b Caneyville, Ky, May 28, 23; m 46; c 2. PHYSICS. Educ: Harvard Univ, SB, 44, AM, 48; Wash Univ, PhD(physics), 52. Prof Exp: Instr physics, Robert Col, Istanbul, 44-46 & Grinnell Col, 48-50; asst, Wash Univ, 50-51; from assoc prof to prof, Alfred Univ, 52-58, chmn dept, 52-58; assoc prof, Mt Holyoke Col, 58-65, chmn dept, 61-65; planning specialist educ divs, NSF, 65-66, assoc prog dir, Div Undergrad Sci, 66-68; prof physics, Fed City Col, 68-69; adj prof & coordr 20-col proj, Univ NC, Chapel Hill, 69-72; dean, Dag Hammarskjold Col, 72-75; MEM FAC, DEPT PHYSICS, UNIV NC, 75- Mem: AAAS; Am Phys Soc; Am Asn Physics Teachers. Res: High energy physics; nuclear instrumentation; science education; educational administration. Mailing Add: Dept of Physics Univ of NC Chapel Hill NC 27514

WILKINS, J ERNEST, JR, b Chicago, Ill, Nov 27, 23; m 47; c 2. MATHEMATICS. Educ: Univ Chicago, SB, 40, SM, 41, PhD(math), 42; NY Univ, BME, 57, MME, 60. Prof Exp: Instr math, Tuskegee Inst, 43-44; from assoc physicist to physicist, Manhattan Proj, Metall Lab, Univ Chicago, 44-46; mathematician, Am Optical Co, 46-50; sr mathematician, Nuclear Develop Corp Am, 50-55, mgr physics & math dept, 55-57, asst mgr res & develop, 58-59 & mgr, 59-60; asst chmn, Theoret Physics Dept, Gen Atomic Div, Gen Dynamics Corp, 60-65, asst dir lab, 65-70; DISTINGUISHED PROF APPL MATH PHYSICS, HOWARD UNIV, 70- Mem: AAAS; Am Math Soc; Optical Soc Am; Am Nuclear Soc; Soc Indust & Appl Math. Res: Differential and integral equations; Bessel functions; nuclear reactors; calculus of variation. Mailing Add: Dept of Physics Howard Univ Washington DC 20059

WILKINS, JUDD RICE, b Chicago, Ill, Dec 12, 20; m 50; c 2. BACTERIOLOGY. Educ: Univ Ill, BS, 46, MS, 47, PhD(bact), 50. Prof Exp: Asst prof bact, Med Sch, Univ SDak, 50-51; res investr, Upjohn Co, 51-57; res scientist, Booz Allen Appl Res, Inc, 57-64; head dept microbiol, Eye Res Found, Md, 64-66; STAFF BIOLOGIST, NASA, LANGLEY RES CTR, 66- Mem: Am Soc Microbiol. Res: Medical bacteriology; antibiotics; experimental pathology; aerospace medicine; operations research. Mailing Add: 313 Beechmount Dr Hampton VA 23369

WILKINS, LAWRENCE EUGENE, physical chemistry, see 12th edition

WILKINS, MICHAEL GRAY, b Northampton, Eng, Sept 9, 38; m 74; c 1. BIOMEDICAL ENGINEERING. Educ: Univ Manchester, BSc, 61; Univ Ill, Urbana, PhD(biophys), 70. Prof Exp: Consult comput sci, 63-65; asst prof biomed eng & physiol, Univ Va, 69-75; INDEPENDENT CONSULT, 75-; PRES, MIDDLE EAST TECHNOL INC, 75-; ASSOC DIR, OMNITECHNOL CTR, INC, 75- Concurrent Pos: Nat Acad Sci travel grant, Moscow Biophys Cong, Kiev, 72. Mem: Math Asn Am; Asn Comput Mach; Inst Elec & Electronics Eng. Res: Structure and analysis of large biological systems; computer applications in life sciences; biostatistics; systems analysis in life science; technology transfer. Mailing Add: PO Box 5144 Barracks Rd Charlottesville VA 22903

WILKINS, ORIN PERRY, b Williston, NDak, Feb 4, 15; m 48; c 2. ZOOLOGY, BOTANY. Educ: Univ Tex, BA, 46, MA, 48, PhD(zool, bot), 55. Prof Exp: Vis asst prof zool, comp anat, histol & embryol, Tex Western Col, 51-53; res scientist, Army Res & Develop Bd grant for spec proj, Ft Sam Houston Med Lab, 53-55; PROF BIOL SCI, CENTENARY COL, LA, 55- Res: Medical entomology and field ecology; ecology of mosquitoes and of freshwater lakes and ponds. Mailing Add: Centenary Col of La Shreveport LA 71104

WILKINS, PETER OSBORNE, b Toronto, Ont, June 28, 21; US citizen; m 47. MICROBIOLOGY. Educ: Rochester Univ, BM, 49; Univ Hawaii, MS, 58; Univ Pa, PhD(microbiol), 62. Prof Exp: From instr to asst prof microbiol, NJ Col Med, 62-68; asst prof, 68-75, ASSOC PROF MICROBIOL, UNIV WESTERN ONT, 75- Concurrent Pos: NSF res fel, 62-64; NIH res grant, 64-; Med Res Coun Can res grant, 70- Mem: AAAS; Am Soc Microbiol. Res: Microbial physiology; membrane transport; psychrotrophic bacteria. Mailing Add: Dept of Bact & Immunol Univ of Western Ont London ON Can

WILKINS, RALPH G, b Southampton, Eng, Jan 7, 27; m 51; c 2. INORGANIC CHEMISTRY. Educ: Univ Southampton, BSc, 47, PhD(chem), 50; Univ London, DSc(chem), 61. Prof Exp: Res chemist, Imp Chem Indust, Eng, 49-52; res assoc inorg chem, Univ Southern Calif, 52-53; from lectr to sr lectr, Sheffield Univ, 53-62; guest prof, Max Planck Inst Phys Chem, 62-63; prof, State Univ NY Buffalo, 63-73; HEAD DEPT CHEM, N MEX STATE UNIV, 73- Mem: Am Chem Soc; The Chem Soc. Res: Mechanisms of transition metal complexes and metalloenzyme reactions. Mailing Add: Dept of Chem NMex State Univ Las Cruces NM 88003

WILKINS, RAYMOND LESLIE, b Boston, Mass, Jan 13, 25; m 50; c 2. ORGANIC CHEMISTRY. Educ: Univ Chicago, AB, 51, MS, 54, PhD(chem), 57. Prof Exp: Sr scientist, 56-68, head instrument technol lab, 68-73, MGR CHEM PROCESS RES DEPT, ROHM AND HAAS CO, 74- Concurrent Pos: Mem, Pa Governor's Sci Adv Comt, 69-, chmn health care delivery panel, 69-71. Mem: Electron Micros Soc Am; NY Acad Sci; fel Royal Micros Soc; Am Chem Soc. Res: Correlation of the microstructure of heterogeneous organic plastics, polymers and emulsions with their gross properties; mechanisms of polymer formation. Mailing Add: Rohm & Haas Co Lab 48 5000 Richmond St Philadelphia PA 19137

WILKINS, ROGER LAWRENCE, b Newport News, Va, Dec 14, 28; m 55; c 2. CHEMICAL PHYSICS. Educ: Hampton Inst, BS, 51; Howard Univ, MS, 52; Univ Southern Calif, PhD(chem physics), 67. Prof Exp: Aeronaut scientist, NASA, Ohio, 52-55; sr tech specialist, Rocketdyne Div, NAm Aviation, Inc, 55-60; STAFF SCIENTIST, AEROPHYS DEPT, AERODYN & PROPULSION RES LAB, AEROSPACE CORP, 60- Mem: Combustion Inst. Res: Chemical lasers; application of quantum mechanics and statistical mechanics to treatment of energy transfer processes in chemical reactions; application of computers to calculate properties of molecules from first principles. Mailing Add: Aerophys Dept Aerodynam & Propul Res Lab Aerospace Corp 2400 El Segundo Blvd El Segundo CA 90245

WILKINS, SAMUEL AUSTELL, JR, b Dallas, NC, Aug 21, 13; m 47; c 5. SURGERY. Educ: Univ NC, AB, 34; Cornell Univ, MD, 38. Prof Exp: Resident surg, Mem Hosp, New York, 46-50; from instr to assoc, 50-60, from asst prof to assoc prof, 60-70, assoc dir, Robert Winship Mem Clin, 60-66, PROF SURG, SCH MED, EMORY UNIV, 70-, DIR ROBERT WINSHIP MEM CLIN, 66- Concurrent Pos: Fel surg, Mem Hosp, New York, 46-50; trustee & mem staff, Wesley Woods Convalescent Ctr; consult, Third Army, US Army, 56-60. Mem: AMA; Am Col Surg; James Ewing Soc (treas, 62-67, pres, 68-69); Soc Head & Neck Surg. Res: Problems of urinary diversion; management of dysphagia; surgical techniques, particularly as applicable to surgery of the pelvis and the head and neck; chemotherapy of solid tumors; immunotherapy of neoplasms. Mailing Add: Emory Univ Clin Atlanta GA 30322

WILKINS, TRACY DALE, b Sparkman, Ark, July 25, 43; m 65. MICROBIOLOGY. Educ: Univ Ark, BS, 65; Univ Tex, PhD(microbiol), 69. Prof Exp: Fel germ-free res, Med Ctr, Univ Ky, 69-71; asst prof, 72-75, ASSOC PROF MICROBIOL, ANAEROBE LAB, VA POLYTECH INST & STATE UNIV, 75- Res: Anaerobic bacteriology; clinical methodology; antimicrobial susceptibility testing; intestinal microbiology; antibiotics; colon cancer. Mailing Add: Anaerobe Lab Va Polytech Inst & State Univ PO Box 49 Blacksburg VA 24061

WILKINSON, ARTHUR, b Saginaw, Mich, Aug, 14, 25; m 62. MATHEMATICS, COMPUTER SCIENCE. Educ: Temple Univ, BS, 50, EdM, 54; Lehigh Univ, EdD, 72. Prof Exp: Teacher high sch, NJ, 51-52 & Pa, 52-53, head dept, 53-65; assoc prof, 65-75, PROF MATH & COMPUT SCI, E STROUDSBURG STATE COL, 75- Concurrent Pos: Dir res, Stroudsburg Res Assocs, Inc, 74- Mem: Math Asn Am. Res: Applied mathematics; statistics; system analysis; computer sciences; military systems; logical design and scientific programming of electronic computer systems; military applications of electronic computer technology. Mailing Add: Dept of Math E Stroudsburg State Col East Stroudsburg PA 18301

WILKINSON, BRUCE H, b Lancaster, Pa, June 2, 42. SEDIMENTOLOGY. Educ: Univ Wyo, BS, 65, MS, 67; Univ Tex, PhD(geol), 74. Prof Exp: Geologist asst oil shale, US Geol Surv, 65; geologist petrol, Gulf Oil Co, 67-69; ASST PROF GEOL, UNIV MICH, ANN ARBOR, 73- Mem: Geol Soc Am; Sigma Xi. Res: Source and distribution of Holocene sediments of the Texas Gulf Coast and of Michigan. Mailing Add: Dept of Geol & Mineral Univ of Mich Ann Arbor MI 48104

WILKINSON, CHARLES BROCK, b Richmond, Va, Jan 16, 22; m 45; c 1. MEDICINE, PSYCHIATRY. Educ: Va Union Univ, BS, 41; Howard Univ, MD, 44; Univ Colo, MS, 50. Prof Exp: From intern to resident internal med, Freedmen's Hosp, Washington, DC, 45-47; resident psychiat, Univ Colo, 47-50; from instr to asst prof neuropsychiat, Col Med, Howard Univ, 50-55; staff physician, Rollman Receiving Ctr, Cincinnati, Ohio, 58-59; dir adult outpatient serv, Greater Kansas City Ment Health Found, 59-60, dir training, 60-69, assoc dir found, 63-68; clin assoc prof, 59-65, chmn dept, 67-69, PROF PSYCHIAT, SCH MED, UNIV MO-KANSAS CITY, 65-, ASST DEAN, 71-; EXEC DIR, GREATER KANSAS CITY MENT HEALTH FOUND, 68- Concurrent Pos: Lectr, Sch Social Work, Howard Univ, 53-55; asst prof, Med Ctr, Univ Kans, 59-65; consult, Family & Children's Serv, Kans, 62-65; mem, Nat Adv Coun, NIMH, 66-70 & US Nat Comt Vital & Health Statist, 72-; consult to ed staff, Psychiat Annals, 70-; consult, Region IV, Fed Aviation Admin, 70-; mem bd trustees, Am Psychiat Asn, 71-74; mem bd govs, Group Advan Psychiat, 73-75; consult, Psychiat Educ Br, NIMH, 74- Mem: Fel Am Psychiat Asn; AMA; Nat Med Asn; Am Group Psychother Asn; Pan-Am Med Asn. Res: Family dynamics and family therapy; psychiatry and the community; evaluation of psychiatric efforts in community mental health; racism and its effects. Mailing Add: 600 E 22nd St Kansas City MO 64108

WILKINSON, CHRISTOPHER FOSTER, b Yorkshire, Eng, Feb 9, 38; m 60; c 2. ENTOMOLOGY, ORGANIC CHEMISTRY. Educ: Univ Reading, BSc, 61; Univ Calif, Riverside, PhD(entom), 65. Prof Exp: UK Civil Serv Comn sr res fel insecticide chem, Pest Infestation Lab, Agr Res Coun, Eng, 65-66; asst prof, 66-71, ASSOC PROF INSECT TOXICOL, CORNELL UNIV, 71- Mem: Entom Soc Am; Brit Biochem Soc. Res: Structure-activity relationships and mode of action of insecticide synergists; microsomal drug metabolism. Mailing Add: Dept of Entom Cornell Univ Ithaca NY 14850

WILKINSON, DANIEL R, b Glasgow, Ky, May 30, 38; m 61; c 2. PLANT PATHOLOGY, PLANT BREEDING. Educ: Western Ky Univ, BS, 61; Clemson Univ, MS, 63; Univ Ill, PhD(plant path), 67. Prof Exp: PLANT PATHOLOGIST, PIONEER HI-BRED INT, INC, 67- Mem: Am Phytopath Soc. Res: Breeding for disease and insect resistance; genetics. Mailing Add: Dept of Corn Breeding Pioneer Hi-Bred Int Inc Johnston IA 50131

WILKINSON, DAVID IAN, b Cookstown, Northern Ireland, Dec 17, 32; m 57; c 2. BIOCHEMISTRY. Educ: Queens Univ, Belfast, BS, 54, PhD(chem), 57. Prof Exp: USPHS res fel chem, Wayne State Univ, 57-58 & Univ Calif, Los Angeles, 58-59; res chemist, Brit Drug Houses, Eng, 59-61; NIH res grant & res fel chem, res assoc biochem, 63-73, SR SCIENTIST DERMAT, SCH MED, STANFORD UNIV, 73- Concurrent Pos: Fulbright fel, 57-59; NIH res grant dermat, Sch Med, Stanford Univ, 71-73. Mem: The Chem Soc; Am Chem Soc; Soc Invest Dermat. Res: Skin lipids; prostaglandins; metabolism of fatty acids in skin; polyunsaturated fatty acids. Mailing Add: Dept of Dermat Stanford Univ Sch of Med Stanford CA 96305

WILKINSON, DAVID TODD, b Hillsdale, Mich, May 13, 35; m 58; c 2. PHYSICS, COSMOLOGY. Educ: Univ Mich, BSE, 57, MSE, 59, PhD(physics), 62. Prof Exp: Lectr physics, Univ Mich, 62-63; from instr to assoc prof, 63-71, PROF PHYSICS, PRINCETON UNIV, 72- Concurrent Pos: Alfred P Sloan Found fel, 66-68. Mem:

Fel Am Phys Soc; Am Asn Physics Teachers. Res: Atomic physics, properties of electrons and positrons; gravitation and relativity; cosmic microwave radiation. Mailing Add: Jadwin Hall Princeton Univ Princeton NJ 08540

WILKINSON, GRANT ROBERT, b Derby, Eng, Aug 27, 41; m 64; c 2. PHARMACOLOGY. Educ: Univ Manchester, BSc, 63; Univ London, PhD(pharmaceut chem), 66. Prof Exp: Asst prof pharm, Col Pharm, Univ Ky, 68-71; ASSOC PROF PHARMACOL, SCH MED, VANDERBILT UNIV, 71- Concurrent Pos: USPHS fel, Univ Calif, San Francisco, 66-68. Mem: AAAS; Am Soc Pharmacol & Exp Therapeut; NY Acad Sci; Am Pharmaceut Asn. Res: Clinical pharmacology; application of analytical methodology, drug metabolism and pharmacokinetics. Mailing Add: Dept of Pharmacol Vanderbilt Univ Sch of Med Nashville TN 37232

WILKINSON, GUY ERNEST, soil physics, agronomy, see 12th edition

WILKINSON, JACK DALE, b Ottumwa, Iowa, Jan 27, 31; m 53; c 5. MATHEMATICS, EDUCATION. Educ: Univ Northern Iowa, BA, 52, MA, 58; Iowa State Univ, PhD(math, educ), 70. Prof Exp: PROF MATH, UNIV NORTHERN IOWA, 62- Concurrent Pos: Consult, 64-; mem comt affiliated groups, Nat Coun Teachers Math, 74-77. Mem: Am Educ Res Asn. Res: Activity learning; attitudes toward mathematical learning; learning styles in mathematics education; problem solving and applications of mathematics in the elementary and junior high schools. Mailing Add: Dept of Math Univ of Northern Iowa Cedar Falls IA 50613

WILKINSON, JAMES FREEMAN, b Providence, RI, June 9, 47; m 68; c 3. AGRONOMY. Educ: Univ RI, BS, 69, MS, 71; Mich State Univ, PhD(agron), 73. Prof Exp: Exten assoc crop sci, Mich State Univ, 72-73; ASST PROF AGRON, OHIO STATE UNIV, 74- Mem: Am Soc Agron; Crop Sci Soc Am. Res: Management, physiology and nutrition of turfgrass species; specific areas include shade adaptation, irrigation management and applied nutrition research. Mailing Add: Dept of Agron Ohio State Univ Columbus OH 43210

WILKINSON, JOHN WESLEY, b Bexley, Ont, Nov 1, 28; m 53; c 1. MATHEMATICAL STATISTICS. Educ: Queen's Univ, Ont, BA, 50, MA, 52; Univ NC, PhD(math statist), 56. Prof Exp: Statistician, Can Industs, Ltd, 52-53; asst prof math, Queen's Univ, Ont, 56-58; res mathematician, Res Labs, Westinghouse Elec Corp, 58-64, fel mathematician, 64-65; prof statist, 65-70; PROF OPERS RES & STATIST, RENSSELAER POLYTECH INST, 70- & CHMN MGT, 70- Concurrent Pos: Consult, Can Industs, Ltd, 56-58, Watervliet Arsenal, Bendix Corp, Kamyr, Inc, 72-, Shaker Res Corp, 72-, NY State Depts Transp, Health, Budget, 72- & NY State Legis Comn Expenditure Rev; assoc ed, Technometrics, 70- Mem: Inst Math Statist; Inst Mgt Sci; Am Inst Decision Sci; Am Soc Qual Control; Am Statist Asn. Res: Statistical design of experiments; statistical inference; mathematical modeling; statistical applications to problems of environment and energy. Mailing Add: Dept of Mgt Rensselaer Polytech Inst Troy NY 12181

WILKINSON, JOSEPH RIDLEY, b Palatka, Fla, Sept 24, 17; m 41; c 2. ANALYTICAL CHEMISTRY. Educ: The Citadel, BS, 41; Univ Ga, MS, 49; Fla State Univ, PhD, 55. Prof Exp: Instr gen chem, 46-49, asst prof gen & anal chem, 49-55, assoc prof anal chem, 55-60, PROF ANAL CHEM, THE CITADEL, 60-, HEAD DEPT CHEM, 68- Mem: Am Chem Soc. Res: Nuclear inorganic chemistry; gamma ray spectroscopy; colorimetric methods of analysis. Mailing Add: Dept of Chem The Citadel Charleston SC 29409

WILKINSON, MICHAEL KENNERLY, b Palatka, Fla, Feb 9, 21; m 44; c 3. PHYSICS, SOLID STATE PHYSICS. Educ: The Citadel, BS, 42; Mass Inst Technol, PhD(physics), 50. Prof Exp: Res assoc, Res Lab Electronics, Mass Inst Technol, 48-50; res physicist, 50-64, assoc dir, 64-72, DIR SOLID STATE DIV, OAK RIDGE NAT LAB, 72- Concurrent Pos: Neely vis prof, Ga Inst Technol, 61-62 & prof, 62- Mem: AAAS; fel Am Phys Soc; Am Crystallog Asn; Sigma Xi. Res: Neutron diffraction and spectrometry; magnetic properties of solids; dynamical properties of crystal lattices; x-ray diffraction; physical electronics. Mailing Add: Solid State Div Oak Ridge Nat Lab PO Box X Oak Ridge TN 37830

WILKINSON, PAUL KENNETH, b Oneonta, NY, Oct 19, 45; m 67; c 2. BIOPHARMACEUTICS, PHARMACODYNAMICS. Educ: Univ Conn, BS Pharm, 69; Univ Mich, Ann Arbor, MS & PhD(pharm), 75. Prof Exp: Dep chief pharmacist, USPHS, 69-71; asst, Sch Pharm, Univ Mich, Ann Arbor, 71-75; ASST PROF PHARMACEUT, SCH PHARM, AUBURN UNIV, 75- Mem: Am Pharmaceut Asn; Acad Pharmaceut Sci; Am Asn Cols Pharm; Sigma Xi. Res: Study of ethyl alcohol concentrations in various segments of the vascular system and evaluation of the kinetics of the oral absorption of ethanol in fasting subjects. Mailing Add: Sch of Pharm Auburn Univ Auburn AL 36830

WILKINSON, PAUL R, b Calcutta, India, Apr 16, 19; m 47; c 5. ENTOMOLOGY. Educ: Cambridge Univ, BA, 41, MA, 45, PhD, 68. Prof Exp: Entomologist, Colonial Insecticide Res Unit, Uganda, 46-49; res officer, Commonwealth Sci & Indust Res Orgn, Australia, 50-62; RES SCIENTIST, CAN DEPT AGR, 62- Mem: Ecol Soc Am; Entom Soc Can. Res: Acarology, ecology, physiology and control of ticks and biting flies. Mailing Add: Can Dept of Agr Res Sta Lethbridge AB Can

WILKINSON, RALEIGH JAMES, b East St Louis, Ill, Mar 24, 34; m 56; c 2. FOOD SCIENCE, FOOD CHEMISTRY. Educ: Southern Ill Univ, BS, 60, MS, 61; Mich State Univ, PhD(food sci), 65; Univ Chicago, MBA, 74. Prof Exp: Proj leader, 65-66, group leader, 66-70, sect mgr, 70, mgr prod eval, 70-74, ASST DIR TECH SERV, QUAKER OATS CO, BARRINGTON, 74- Mem: Inst Food Technologists. Res: Relationships between cooked temperature and tenderness or juiciness of boneless turkey rolls as measured by physical sensory and chemical methods. Mailing Add: 1306 W Sigwalt Arlington Heights IL 60005

WILKINSON, RALPH RUSSELL, b Portland, Ore, Feb 20, 30; m 56. ANALYTICAL CHEMISTRY, ENVIRONMENTAL MANAGEMENT. Educ: Reed Col, BA, 53; Univ Ore, Eugene, PhD(phys chem), 62; Univ Mo-Kansas City, MBA, 74. Prof Exp: Sr res chemist, Sprague Elec Co, Tektronix Inc, MacDermid Inc & Chemagro Corp, Div Baychem Corp, 61-72; res chemist, US Vet Hosp, 73-75; ASSOC CHEMIST TECHNOL ASSESSMENT, MIDWEST RES INST, 75- Mem: Am Chem Soc; Sigma Xi. Res: Technology assessment; pesticides. Mailing Add: Midwest Res Inst 425 Volker Blvd Kansas City MO 64110

WILKINSON, RAYMOND GEORGE, b Duluth, Minn, June 2, 22; m 48; c 3. ORGANIC CHEMISTRY. Educ: Harvard Univ, BS, 43; Univ Mich, MS, 48, PhD(org chem), 52. Prof Exp: From res chemist to sr res chemist, Lederle Div, Am Cyanamid Co, 51-62, group leader process improv, 62-66; asst to managing ed, Subject Index Div, Chem Abstracts Serv, Ohio State Univ, 66-68; SR RES CHEMIST, LEDERLE DIV, AM CYANAMID CO, 68- Mem: Am Chem Soc. Res: Synthesis of steroids, tetracyclines and their degradation products; antituberculosis agents. Mailing Add: Lederle Labs Am Cyanamid Co Middletown Rd Pearl River NY 10965

WILKINSON, ROBERT CLEVELAND, JR, b Grand Rapids, Mich, Oct 2, 23; m 48; c 3. ENTOMOLOGY, ECOLOGY. Educ: Mich State Univ, BS, 49, MS, 50; Univ Wis, PhD(entom), 61. Prof Exp: Entomologist, State Dept Agr, Mich, 51-53; supvr, Off State Entomologist, Wis, 53-57; res asst entom, Univ Wis, 57-60; from asst entomologist to assoc entomologist, 60-70, PROF ENTOM, UNIV FLA, 71- Concurrent Pos: Res grants, Buckeye Cellulose Corp, Ford Found, Southern Forest Dis & Insect Res Coun, US Forest Serv. Mem: AAAS; Entom Soc Am; Soc Am Foresters; Entom Soc Can. Res: Forest entomology; bionomics of pine sawflies and bark beetles. Mailing Add: Dept of Entom & Nemat 204 Newell Hall Univ of Fla Gainesville FL 32611

WILKINSON, ROBERT ELZWORTH, b Mt Ayr, Iowa, July 27, 16; m 43; c 3. PLANT PATHOLOGY. Educ: State Col Iowa, BA, 38; Iowa State Univ, MS, 40; Cornell Univ, PhD(plant path), 48. Prof Exp: Asst prof, 48-52, ASSOC PROF PLANT PATH, N Y STATE COL AGR & LIFE SCI, CORNELL UNIV, 52- Concurrent Pos: Plant virologist, Int Coop Admin, Israel, 55-56; veg pathologist, Food & Agr Orgn UN, Egypt, 63-64, consult, 64-70; vis prof, Fed Univ Vicosa, 71-72. Mem: Am Phytopath Soc. Res: Diseases of vegetables; virus diseases; disease resistance in vegetables. Mailing Add: Dept of Plant Path NY State Col Agr Cornell Univ Ithaca NY 14850

WILKINSON, ROBERT EUGENE, b Oilton, Okla, Oct 24, 26; m 51; c 2. PLANT PHYSIOLOGY, WEED SCIENCE. Educ: Univ Ill, BS, 50; Univ Okla, MS, 52; Univ Calif, Davis, PhD(plant physiol), 56. Prof Exp: Plant physiologist, Crops Res Div, Agr Res Serv, USDA, Ark, 57-62 & NMex, 62-65, assoc agronomist, 65-73, AGRONOMIST, EXP STA, USDA, GA, 73- Concurrent Pos: Sr Fulbright-Hayes Lectr appl ecol, Univ Turku, Finland, 74-75. Mem: AAAS; Weed Sci Soc Am; Am Soc Plant Physiol. Res: Absorption; translocation; herbicide response; environmental response of plants and effect of herbicides on metabolism. Mailing Add: Dept of Agron Ga Exp Sta Experiment GA 30212

WILKINSON, ROBERT FOSTER, b Halifax, Eng, Sept 23, 20; Can citizen; m 50. PHYSICAL CHEMISTRY. Educ: Univ London, BPharm, 40, BSc, 41. Prof Exp: Sci officer, UK Armament Res Estab, 41-47, group leader pyrotech, 47-49, prin sci officer, 50-53; UK Armament Res Estab exchange officer, US Army Ord Corp, Frankford Arsenal, Pa, 49-50; from supt propulsion wing to dep chief supt, Can Armament Res & Develop Estab, 53-64; dir develop policy & planning, Can Forces Hq, 64-65, sci asst to chief tech serv, 65-69; dir gen, Defence Res Bd, Defence Res Estab, Ottawa, 69-75; DEP CHIEF RES & DEVELOP, DEPT NAT DEFENCE, CAN, 75- Concurrent Pos: Can rep, Group Experts Explosives & Propellants, NATO, 54-59. Mem: Fel Chem Inst Can; assoc fel Can Aeronaut & Space Inst. Res: Physics, chemistry and technology related to weapons systems and military equipment; explosives; propulsion; research and development management. Mailing Add: 416 Briar Ave Ottawa ON Can

WILKINSON, RONALD CRAIG, b Augusta, Ga, Oct 1, 43; m 67. FOREST GENETICS, PHYSIOLOGY. Educ: Univ Wash, BS, 65; Yale Univ, MF, 66; Mich State Univ, PhD(forest genetics), 70. Prof Exp: RES PLANT GENETICIST, NORTHEASTERN FOREST EXP STA, US FOREST SERV, 70- Mem: AAAS; Soc Am Foresters; Phytochem Soc NAm. Res: Controlled and natural hybridization; comparative physiology of species, ecological races and hybrids; natural variation and adaptation; genetic and physiological resistance to insects and diseases; biochemical systematics. Mailing Add: US Forest Serv Northeastern Forest Exp Sta Box 640 Durham NH 03824

WILKINSON, STANLEY R, b West Amboy, NY, Mar 28, 31; m 57; c 3. AGRONOMY, SOIL SCIENCE. Educ: Cornell Univ, BS, 54; Purdue Univ, MS, 56, PhD(soil fertil, plant nutrit), 61. Prof Exp: Instr soil fertil & plant nutrit, Purdue Univ, 57-60; res soil scientist, Pasture Res Lab, 60-65, RES SOIL SCIENTIST, SOUTHERN PIEDMONT CONSERV RES CTR, SOIL & WATER CONSERV RES DIV, AGR RES SERV, USDA, 65- Mem: Am Soc Agron; Soil Sci Soc Am; Crop Sci Soc Am; Int Soc Soil Sci. Res: Mineral nutrient requirements of corn, soybeans, forage grasses and legumes; biuret toxicity to corn; root growth response to fertilizer; competitive phenomena between forage species; land application of wastes and environmental quality. Mailing Add: Southern Piedmont Conserv Res Ctr Box 555 Watkinsville GA 30677

WILKINSON, THOMAS PRESTON, b Gisburn, Eng, Mar 14, 41; m 66; c 2. PHYSICAL GEOGRAPHY. Educ: Univ Durham, BSc, 63; Univ Newcastle, PhD(geomorphol), 72. Prof Exp: From lectr to asst prof, 67-73, ASSOC PROF GEOG, CARLETON UNIV, 73- Concurrent Pos: Vis lectr, Univ Liverpool, 72-73. Res: Fluvial geomorphology; geography curricula. Mailing Add: Dept of Geog Carleton Univ Ottawa ON Can

WILKINSON, THOMAS ROSS, b Baltimore, Md, Aug 20, 37; m 65; c 3. MICROBIOLOGY. Educ: Univ Notre Dame, BS, 59; Univ Md, College Park, MS, 62; Wash State Univ, PhD(microbiol), 70. Prof Exp: Technician aerobiol, Naval Biol Lab, Univ Calif, Berkeley, 65-66; asst prof, 70-73, agr exp sta grant, 71-74, ASSOC PROF PATH & IMMUNOL, S DAK STATE UNIV, 75-, HEAD MICROBIOL DEPT, 75- Mem: AAAS; Sigma Xi; Am Soc Microbiol; NY Acad Sci. Res: Rapid isolation technique for Listeria monocytogenes; survival of pathogens on metal surfaces; miniature cell systems for virus isolation and epidemiology of enclosed environments; epidemiology of Listeria and pathogenesis of Listeria L-forms. Mailing Add: Dept of Microbiol SDak State Univ Brookings SD 57006

WILKINSON, WALTER SHEPARD, biochemistry, nutrition, see 12th edition

WILKINSON, WILLIAM KENNETH, b Newcastle, Ind, Jan 17, 18; m 42; c 5. ORGANIC CHEMISTRY, POLYMER CHEMISTRY. Educ: DePauw Univ, AB, 40; Northwestern Univ, MS, 47, PhD(org chem), 48. Prof Exp: Training supvr, Trojan Powder Co, 41-43; res chemist, Firestone Tire & Rubber Co, 43-45; from res chemist to res assoc, 48-65, RES FEL, BENGER LAB, TEXTILE FIBERS DEPT, E I DU PONT DE NEMOURS & CO, 65- Mem: Am Chem Soc. Res: Rubber and textile fibers; explosives; acetylene chemistry. Mailing Add: 792 Northgate Ave Waynesboro VA 22980

WILKINSON, WILLIAM LYLE, b Sikeston, Mo, Jan 18, 21; m 45, 69; c 6. OPERATIONS RESEARCH. Educ: US Naval Postgrad Sch, BS & MS, 55. Prof Exp: SR STAFF SCIENTIST, GEORGE WASHINGTON UNIV, 65- Concurrent Pos: Lectr, Univ Calif, Los Angeles, 65-66. Mem: Opers Res Soc Am. Res: Transportation networks in logistics research; computer-based management information systems; quantitative evaluation of weapon systems and tactics; man-computer systems; military sciences. Mailing Add: 1309 Alps Dr McLean VA 22101

WILKNISS, PETER EBERHARD, b Berlin, Ger, Sept 28, 34; US citizen; m 63; c 2. OCEANOGRAPHY, RADIOCHEMISTRY. Educ: Munich Tech Univ, MS, 59, PhD(radiochem), 61. Prof Exp: Res chemist, US Naval Ord Sta, 61-66, RES CHEMIST, US NAVAL RES LAB, 66- Mem: AAAS; Sigma Xi; Am Geophys

Union. Res: Physical chemistry of the air/sea interface; radiochemistry applied to oceanography. Mailing Add: US Naval Res Lab Code 8330 4555 Overlook Ave Washington DC 20390

WILKOFF, LEE JOSEPH, b Youngstown, Ohio, Oct 17, 24; m 53; c 1. MICROBIOLOGY, BIOCHEMISTRY. Educ: Roosevelt Univ, BS, 48; Univ Chicago, PhD(microbiol), 63. Prof Exp: Chemist, H Kramer & Co, 48-49; res asst biochem, Ben May Lab Cancer Res, Univ Chicago, 49-52 & Dept Med, 54-60; biochemist, Vet Admin Hosp, Hines, Ill, 52-54; dir microbiol lab, Woodard Res Corp, 63-64; sr microbiologist, 64-70, HEAD CELL BIOL DIV, SOUTHERN RES INST, 70- Mem: AAAS; Am Asn Cancer Res; Soc Exp Biol & Med; Am Soc Microbiol; Tissue Cult Asn. Res: Cell biology and chemotherapy of tumor cells; effect of anticancer drugs on the kinetic behavior of tumor cells; cellular sites of action of anticancer agents. Mailing Add: Cell Biol Div Southern Res Inst 2000 Ninth Ave S Birmingham AL 35205

WILKOWSKE, HOWARD HUGO, b Zachow, Wis, Sept 10, 17; m 48; c 3. DAIRY BACTERIOLOGY. Educ: Tex Tech Col, BS, 40, MS, 42; Iowa State Col, PhD, 49. Prof Exp: Assoc prof dairy mfg & assoc dairy technologist, Agr Exp Sta, 50-57, asst dir, Agr Exp Stas, 57-68, ASST DEAN RES, INST FOOD & AGR SCI, UNIV FLA, 68- Concurrent Pos: Mem adv coun, Ctr Trop Agr, US AID dairy specialist, Costa Rica, 58, Ghana, 69 & Venezuela, 71. Mem: AAAS; Am Soc Microbiol; Am Dairy Sci Asn. Res: Antibiotics in dairy products; continuous and automatic manufacture of fermented dairy products; bacteriophage of dairy microorganisms; agricultural research administration. Mailing Add: Inst Food & Agr Sci Univ of Fla 1022 McCarty Gainesville FL 32611

WILKS, CHARLES EDWARD, b Taylor, Tex, Feb 28, 39; m 63; c 1. MATHEMATICS. Educ: Univ Tex, Austin, BA, 62, MA, 64, PhD(math), 69. Prof Exp: Asst prof math, Wilkes Col, 69-72; asst prof, 72-73, ASSOC PROF MATH, SEAVER COL, PEPPERDINE UNIV, 73- Mem: Math Asn Am; Sigma Xi. Res: Real analysis; measure and integration; products of borel measures; regularity of product measures using regular conditional measures. Mailing Add: Dept of Natural Sci Pepperdine Univ Malibu CA 90265

WILKS, JOHN WILLIAM, b Kenosha, Wis, July 5, 44. REPRODUCTIVE ENDOCRINOLOGY. Educ: Univ Wis-Madison, BS, 66; Cornell Univ, PhD(physiol), 71. Prof Exp: RES SCIENTIST REPRODUCTIVE ENDOCRINOL, FERTIL RES, UPJOHN CO, 70- Mem: AAAS; Soc Study Reproduction; Am Soc Animal Sci. Res: Physiologic and pharmacologic control of ovarian function, especially the corpus luteum; endocrinology of the menstrual cycle. Mailing Add: Fertil Res Upjohn Co Kalamazoo MI 49001

WILKS, LOUIS PHILLIP, b Dayton, Ohio, June 28, 13; m 38; c 3. CHEMISTRY. Educ: Univ Dayton, BS, 35. Prof Exp: Instr chem, Univ Dayton, 35; res chemist, Thomas & Hochwalt Lab, 35-38; res chemist, 38-40, dir labs, 40-46, tech dir, 46-56, vpres, 56-66, dir new prod develop, 56-63, dir long range tech planning, 63-70, DIR CORP LONG-RANGE PLANNING, VELSICOL CHEM CORP, 70- Mem: AAAS; Am Chem Soc; Commercial Develop Asn; Planning Exec Inst. Res: Products from petrochemical by-products; agricultural chemicals; research management and corporate planning. Mailing Add: Velsicol Chem Corp 341 E Ohio St Chicago IL 60611

WILKS, PHILLIP HOWARD, b Dayton, Ohio, Apr 13, 34; m 55; c 3. HIGH TEMPERATURE CHEMISTRY. Educ: Univ Cincinnati, ChE, 56; Northwestern Univ, Evanston, MS, 57; Univ Pittsburgh, PhD(inorg chem), 66. Prof Exp: Tech engr nuclear reactor mat, Aircraft Nuclear Propulsion, Gen Elec Co, 57-62; adv engr, Astronuclear Lab, Westinghouse Corp, 62-69; tech dir indust plasma chem, Humphreys Corp, 69-75; PRES, PLASMA MAT INC, 75- Concurrent Pos: Secy sect comt N-6 on nuclear safety stand, Am Nat Stand Inst, 59-64; mem plasma chem indust liaison comt, 73-; chmn plasma chem Gordon conf, 76. Mem: Am Chem Soc. Res: Industrial plasma chemistry; high temperature behavior of halides; high temperature materials. Mailing Add: 130 North Bend Dr Manchester NH 03104

WILKS, WILLIAM TAYLOR, b Berea, Ky, Jan 14, 11; m 35; c 1. PHYSICS. Educ: Ala Polytech Inst, BS, 30, MS, 35; Columbia Univ, EdD, 49. Prof Exp: Teacher high sch, Ala, 30-42; instr phys sci & math, Goddard Col, 42-43; asst prof physics, Ala Polytech Inst, 43-44; head physics dept, Marion Inst, 44-46; assoc prof sci, 47-55, prof & head dept, 55-68, VPRES ACAD AFFAIRS, TROY STATE UNIV, 68-, VPRES UNIV, 72- Mem: AAAS; Nat Sci Teachers Asn. Res: Evaluation of science teaching programs; education of science teachers. Mailing Add: Troy State Univ Troy AL 36081

WILL, CLIFFORD MARTIN, b Hamilton, Ont, Can, Nov 13, 46; m 70; c 2. THEORETICAL ASTROPHYSICS. Educ: McMaster Univ, BSc, 68; Calif Inst Technol, PhD(physics), 71. Prof Exp: Instr physics, Calif Inst Technol, 71-72; fel, Enrico Fermi Inst, Univ Chicago, 72-74; ASST PROF PHYSICS, STANFORD UNIV, 74- Concurrent Pos: Res fel, Calif Inst Technol, 71-72; Sloan Found res fel, 75. Mem: Am Phys Soc; Am Aston Soc; Sigma Xi. Res: General relativity theory and its applications to astrophysics. Mailing Add: Dept of Physics Stanford Univ Stanford CA 94305

WILL, FRITZ, III, b Richmond, Va, Oct 24, 26; m 54; c 2. ANALYTICAL CHEMISTRY. Educ: Univ Va, BS, 49, MS, 51, PhD(chem), 53. Prof Exp: Asst chem, Univ Va, 47-51; res chemist, Res Labs, Aluminium Co Am, 53-65; res chemist, 65-69, MGR ANAL CHEM DIV, PHILIP MORRIS RES CTR, 69- Honors & Awards: Medal, Am Inst Chemists, 49. Mem: Am Chem Soc; Sigma Xi. Res: Analytical methods; spectrophotometry; ultraviolet and visual absorption spectroscopy; nuclear magnetic resonance spectroscopy. Mailing Add: Philip Morris Res Ctr PO Box 26583 Richmond VA 23261

WILL, FRITZ GUSTAV, b Breslau, Ger, Jan 12, 31; m 58; c 3. ELECTROCHEMISTRY. Educ: Munich Tech Univ, BS, 53, MS, 57, PhD(phys chem), 59. Prof Exp: Res scientist, Eng Res & Develop Labs, US Dept Army, Va, 59-60; electrochemist, 60-69, mgr electrochem mat & reactions unit, 69-72, MEM RES STAFF, RES & DEVELOP CTR, GEN ELEC CO, 72- Concurrent Pos: Div ed, J Electrochem Soc, 74- Honors & Awards: Battery Res Award, Electrochem Soc, 64. Mem: Electrochem Soc. Res: Electrochemical instrumentation; electrocatalysis; electrode kinetics; chemisorption; porous electrodes; fuel cells; batteries; nickel-cadmium; solid electrolyte; zinc-halogen; utility battery systems. Mailing Add: Gen Elec Res & Develop Ctr PO Box 8 Schenectady NY 12301

WILL, JAMES ARTHUR, b Wauwatosa, Wis, Nov 2, 30; m 53; c 3. PHYSIOLOGY. Educ: Univ Wis, BS, 52, MS, 53, PhD(vet sci), 67; Kans State Univ, DVM, 60. Prof Exp: Vet, Columbus Vet Hosp, Wis, 60-67; from asst prof to assoc prof, 67-74, PROF VET SCI & CHMN DEPT, MED SCH, UNIV WIS-MADISON, 74-, MEM STAFF, CARDIOVASC RES LAB, 74- Concurrent Pos: NIH spec fel, New Med Sch, Univ Liverpool, 72-73; mem coun basic sci & circulation, Am Heart Asn; mem comt primary pulmonary hypertension, WHO. Mem: AAAS; Am Vet Med Asn; sr mem Am Fedn Clin Res; Am Physiol Soc. Res: Cardiopulmonary physiopathology, particularly relationship between function and disease under natural and altered environmental conditions or with impairment of function by a disease process. Mailing Add: Dept of Vet Sci Univ of Wis Madison WI 53706

WILL, JOHN JUNIOR, b Cincinnati, Ohio, Aug 13, 24; m 46; c 3. HEMATOLOGY. Educ: Univ Cincinnati, MD, 47. Prof Exp: Intern med, Presby Hosp, New York, 47-48; resident internal med, Cincinnati Gen Hosp, 48-50; asst chief med serv, Travis AFB, 50-52; from instr to assoc prof, 52-65, PROF MED, COL MED, UNIV CINCINNATI, 65- Concurrent Pos: Fel, Nutrit & Hemat Lab, Cincinnati Gen Hosp, 52-55, co-dir, Hemat Lab, 57-, clinician, Outpatient Dept & chief clinician, Hemat Clin. Mem: Am Fedn Clin Res. Res: Nutrition; relationship of vitamin B-12 folic acid and ascorbic acid to the absorption and utilization of versene in iron deficiency anemia; nuclease inhibitors in human white blood cells in leukemia and other disease states; effect of antimetabolites on leukemia in rats and humans; cancer chemotherapy. Mailing Add: Dept of Med Univ of Cincinnati Cincinnati OH 45221

WILL, OTTO ALLEN, JR, b Caldwell, Kans, Apr 26, 10; m 53; c 2. PSYCHIATRY. Educ: Stanford Univ, AB, 33, MD, 40. Prof Exp: Intern med, Stanford Univ, 39-40, asst resident pediat, 40-41 & med, 41-42; psychiatrist, St Elizabeth's Hosp, DC, 43-47; psychiatrist, Chestnut Lodge, Rockville, 47-54, dir psychother, 54-67; MED DIR, AUSTEN RIGGS CTR, 67- Concurrent Pos: Vis prof, Dept Psychiat, Chicago, 63-64; clin prof psychiat, Sch Med, Univ Md, 56-57 & Cornell Univ, 67-; vis prof, Univ Cincinnati, 72-; lectr, Johns Hopkins Univ, Honors & Awards: Reichmann Award, 62. Mem: Am Psychoanal Asn; fel Am Psychiat Asn; fel Am Acad Psychoanal. Res: Psychoanalysis; psychotherapy; psychotherapeutic intervention in schizophrenia. Mailing Add: Austen Riggs Ctr Stockbridge MA 01262

WILL, THEODORE A, b Orange, NJ, Aug 9, 37. SOLID STATE PHYSICS. Educ: Johns Hopkins Univ, AB, 59; Univ Chicago, SM, 61; Case Western Reserve Univ, PhD(physics), 68. Prof Exp: Asst prof physics, Heidelberg Col, 62-64 & Grinnell Col, 68-72; vis prof, Mat Res Ctr, Nat Univ Mex, 72-74; MEM FAC, CIUDAD UNIV, 74- Mem: Am Phys Soc; Am Asn Physics Teachers. Res: Electron phonon interaction in PbTl and PbBi from superconductor tunneling measurements; electron tunneling. Mailing Add: Ciudad Univ AP 70-337 Mexico 20 DF Mexico

WILLARD, DANIEL, b Baltimore, Md, Aug 22, 26; m 58; c 2. PHYSICS. Educ: Yale Univ, BS, 49, MS, 50; Mass Inst Technol, PhD(physics), 54. Prof Exp: Res assoc physics, Brookhaven Nat Lab, 54-55; instr, Swarthmore Col, 55-58; assoc prof, Va Polytech Inst, 58-61; opers res analyst, Opers Res Off, Johns Hopkins Univ, 61 & Res Anal Corp, Va, 61-63; OPERS RES ANALYST, OFF UNDER SECY US ARMY, 63- Concurrent Pos: Consult, Langley Res Ctr, NASA, 59-61. Mem: Am Phys Soc; Royal Astron Soc. Res: Cosmic rays; heavy unstable particles; radio astronomy; mathematical models of combat; operations research; systems analysis. Mailing Add: Off Under Secy Army Dept of the Army Washington DC 20310

WILLARD, HAROLD JAMES, JR, b Washington, DC, Apr 24, 38; m; c 4. SOLID STATE PHYSICS, NUCLEAR ENGINEERING. Educ: Col William & Mary, BS, 60; Rensselaer Polytech Inst, BME, 60, MS, 62, PhD(physics), 68. Prof Exp: Sr scientist, Bettis Atomic Power Lab, 67-74, SUPVR TRAINING IMPROVEMENT GROUP, NAVAL REACTOR FACIL, BETTIS ATOMIC POWER LAB, WESTINGHOUSE ELEC CORP, 74- Mem: Am Phys Soc; Am Nuclear Soc; Sigma Xi; AAAS. Res: Improved methods of training operations personnel; radiation effects on mechanical behavior of metals; ultrasonics. Mailing Add: 1281 Herring St Idaho Falls ID 83401

WILLARD, HARVEY BRADFORD, b Worcester, Mass, Aug 9, 25; m 50; c 2. NUCLEAR PHYSICS. Educ: Mass Inst Technol, SB, 48, PhD(physics), 50. Prof Exp: Physicist, Oak Ridge Nat Lab, 50-57, co-dir, High Voltage Lab, 57-63, assoc dir physics div, 63-67; chmn dept physics, 67-71, PROF PHYSICS, CASE WESTERN RESERVE UNIV, 67-, DEAN SCI, 70-, VPROVOST, CASE INST TECHNOL, 70- Mem: Fel Am Phys Soc. Res: Nuclear scattering; reaction and polarization phenomena; energy levels of nuclei; the few nucleon problem; Van de Graaff accelerators; medium energy studies of proton-proton scattering and meson production with polarized beams and targets. Mailing Add: Crawford Hall Case Western Reserve Univ Cleveland OH 44106

WILLARD, JAMES MATTHEW, b St Johnsbury, Vt, Nov 18, 39; m 61; c 3. BIOCHEMISTRY. Educ: St Michael's Col, Vt, AB, 61; Cornell Univ, PhD(biochem), 67. Prof Exp: Res assoc biochem, Case Western Reserve Univ, 66-69; asst prof biochem, Col Med, Univ Vt, 69-75; ASST PROF BIOL, CLEVELAND STATE UNIV, 75- Res: Preparation and effect of certain analogues of phosphoribosyl pyrophosphate on de novo purine synthesis; phosphoenolpyruvate carboxytransphosphorylase from propionate bacteria. Mailing Add: Dept of Biol & Health Sci Cleveland State Univ Cleveland OH 44115

WILLARD, JOE RAYMOND, b Moline, Ill, Apr 26, 25; m 46; c 5. ORGANIC CHEMISTRY. Educ: Univ Ill, BS, 48; Univ Nebr, MS, 51, PhD(org chem), 53. Prof Exp: Proj leader, Westvaco Chem Div, 52-54 & Niagara Chem Div, 54-56, group leader org synthesis, 56-58, supvr, 58-62, lab mgr, 62-66, MGR TECH INFO & PATENTS, NIAGARA CHEM DIV, FMC CORP, 66- Mem: Am Chem Soc; Entom Soc Am; Am Soc Info Sci. Res: Chemistry of pesticides; information retrieval; nomenclature of organic compounds. Mailing Add: Res & Develop Dept FMC Corp Niagara Chem Div 100 Niagara St Middleport NY 14105

WILLARD, JOHN ELA, b Oak Park, Ill, Oct 31, 08; m 37; c 4. PHYSICAL CHEMISTRY. Educ: Harvard Univ, SB, 30; Univ Wis, PhD(phys chem), 35. Prof Exp: Instr chem, Avon Sch, 30-32 & Haverford Col, 35-37; from instr to prof, 37-63, dean grad sch, 58-63, chmn dept chem, 70-72, VILAS RES PROF CHEM, UNIV WIS-MADISON, 63- Concurrent Pos: Assoc sect chief, Plutonium Chem Sect, Metall Lab, Univ Chicago, 42-44, dir pile chem div, 45-46; area supvr, Hanford Eng Works, E I du Pont de Nemours & Co, Inc, 44-45; consult, Oak Ridge Nat Lab, 46-49; mem, Phys Chem Panel, Off Naval Res, 48-50, Surv Comt, AEC, 49 & Isotope Distrib Adv Comt, AEC, 53-57; secy, Nuclear Chem Sect, Int Cong Pure & Appl Chem, 51; mem partic inst exec bd, Argonne Nat Lab, 50-53, chmn, 52-53, chem div vis comt, 58-64; mem adv bd, Gordon Res Confs, 55-60, chmn conf radiation chem, 68; mem bd vis chem div, Brookhaven Nat Lab, 56-59; mem panel on basic res & grad educ, President's Sci Adv Comt, 59-60 & on basic res & nat goals, Nat Acad Sci, 64-65. Honors & Awards: Award, Nuclear Appl in Chem, Am Chem Soc, 59. Mem: Am Chem Soc (chmn div phys & inorg chem, 57); Am Phys Soc; Radiation Res Soc; AAAS. Res: Radiation chemistry; photochemistry; chemical effects of nuclear transformations; nature and reactions of trapped intermediates. Mailing Add: Dept of Chem Univ of Wis Madison WI 53706

WILLARD, JOHN JAY, b Woodstock, Maine, May 22, 34; m 67. ORGANIC CHEMISTRY. Educ: Clarkson Col Technol, BS, 56; Princeton Univ, MA, 58, PhD(chem), 60. Prof Exp: Fel chem, State Univ NY Col Forestry, Syracuse, 59-61; NIH fel, Univ Birmingham, 61-63; sr scientist, Textile Res Inst, 63-66; res assoc chem, JP Stevens & Co, Inc, 66-73; CONSULT, 73- Mem: Am Chem Soc; Fiber Soc. Res: Structural studies of carbohydrate derivatives, mucoproteins and cellulose derivatives, especially cotton. Mailing Add: 400 Main St Bethel ME 04217

WILLARD, JOHN ROYAL, b Dunchurch, Ont, Aug 22, 38; m 65. ENTOMOLOGY, ECOLOGY. Educ: Univ Toronto, BScAg, 62; Ore State Univ, MA, 64; Univ Adelaide, PhD(entom, animal ecol), 69. Prof Exp: Fel subsoil invert, Matador Proj, Int Biol Prog, Univ Sask, 69-74; ASST RES OFFICER, SASK RES COUN, 74- Mem: Entom Soc Australia; Australian Entom Soc; Entom Soc Can. Res: Environmental impact assessment; methods of dispersal of a wingless insect; dispersal and population dynamics of insects; behavior of insects; insecticide control of crop insects. Mailing Add: Chem Div Sask Res Coun 30 Campus Dr Saskatoon SK Can

WILLARD, JOHN WESLEY, b Davenport, Iowa, June 23, 07; m 28; c 2. CHEMISTRY. Educ: Purdue Univ, BS, 29, MS, 40, PhD(chem), 43. Prof Exp: Res chemist, du Pont Viscoloid Co, NJ, 28-30; chemist, Anderson-Pritchard Oil Corp, Okla, 30-32; vpres, Cleaner Corp Am, Ill, 32-35; asst, Purdue Univ, 34-40; from instr to asst prof chem, Va Mil Inst, 40-46; assoc prof, 46-50, PROF CHEM, S DAK SCH MINES & TECHNOL, 50-; DIR RES, PATENT DEVELOP CORP, 72- Honors & Awards: Western Elec Award, 67. Mem; Am Chem Soc; fel Am Inst Chem. Res: Soils; removal of paraffin-type sludges from oil wells and equipment removal of carbonaceous materials from metal and minerals. Mailing Add: Dept of Chem SDak Sch of Mines & Technol Rapid City SD 57701

WILLARD, PAUL EDWIN, b Ogdensburg, NY, Sept 11, 19; m 43; c 2. ORGANIC CHEMISTRY. Educ: Univ Chicago, BS, 41. Prof Exp: Res chemist, Gen Labs, US Rubber Co, NJ, 41-44 & Celanese Corp Am, 44-47; res chemist, Ohio-Apex, Inc, 47-50, from asst res dir to res dir, Ohio-Apex Div, 50-57, tech mgr, 57-58, mgr applns, Tech Serv Lab, Org Chem Div, 58-62, asst dir govt liaison, NJ, 62-65, mgr ceramic fibers res & develop, 65-69, sr res chemist, 69-74, RES ASSOC, FMC CORP, 74- Mem: Am Chem Soc; Am Soc Testing & Mat; Am Asn Textile Chemists & Colorists. Res: Hydrogen peroxide; thermosetting plastics, especially rheology; plastics testing and applications; composite materials; ceramic fibers; thermal analysis. Mailing Add: FMC Corp PO Box 8 Princeton NJ 08540

WILLARD, PAUL W, b Marshalltown, Iowa, Mar 21, 33; m 52; c 4. PHYSIOLOGY, PHARMACOLOGY. Educ: Iowa State Univ, BS, 55; Univ Iowa, PhD(physiol), 59; Ind Univ, MBA, 68. Prof Exp: Nat Heart Inst fel physiol, Lankenau Hosp, Philadelphia, Pa, 59-61; sr scientist, Div Pharmacol Res, Eli Lilly & Co, 61-69; sr clin res coordr, 69, SPECIALIST, TECH FORECASTING & PLANNING, MED PRODS DIV, 3M CO, 71- Mem: Am Physiol Soc; Am Soc Pharmacol & Exp Therapeut; Soc Toxicol; Am Col Cardiol. Res: Cardiovascular physiology and pharmacology; market and technical planning. Mailing Add: Med Prods Div 3M Co 3M Ctr 2501 Hudson Rd St Paul MN 55101

WILLARD, ROBERT JACKSON, b Brockton, Mass, Mar 21, 29; m 54; c 2. GEOLOGY. Educ: Boston Univ, AB, 51, AM, 53, PhD(geol), 58. Prof Exp: Instr geol, Wellesley Col, 56-57; from instr to asst prof, Univ Ark, 57-63; geologist, 63-67, HEAD FABRIC ANAL LAB, TWIN CITIES MINING RES CTR, US BUR MINES, 68- Concurrent Pos: Nat Park Serv study grant, 60-61. Mem: Geol Soc Am; Am Soc Mining, Metall & Petrol Eng. Res: Structural geology and stratigraphy of northwest Arkansas and northwest Maine; petrofabrics; electron fractography; relation of rock fabric to various deformation and fragmentation tests. Mailing Add: US Bur of Mines PO Box 1660 St Paul MN 55111

WILLARD, STEPHEN, b Syracuse, NY, Nov 1, 41; m 63; c 2. MATHEMATICS. Educ: Univ Rochester, AB, 62, MA, 64, PhD(math), 65. Prof Exp: Asst prof math, Lehigh Univ, 65-66 & Case Western Reserve Univ, 66-69; ASSOC PROF MATH, UNIV ALTA, 69- Mem: Am Math Soc; Math Asn Am. Res: Convergence structures; mapping properties of topological spaces; generalizations of compactness; absolute Borel and analytic sets in topological spaces; set theory. Mailing Add: Dept of Math Univ of Alta Edmonton AB Can

WILLARD, THOMAS MAXWELL, b Beaumont, Tex, Aug 30, 37; m 59; c 2. INORGANIC CHEMISTRY, ANALYTICAL CHEMISTRY. Educ: Lamar State Col, BS, 59; Tulane Univ, PhD(inorg chem), 64. Prof Exp: Fac mem, 64-65, CHMN DEPT CHEM, FLA SOUTHERN UNIV, 65- Concurrent Pos: Consult, Wellman-Power Gas, Inc, Fla, 66- & Standard Spray & Chem Co, 68-; Rotary Int group study exchange fel, Japan, 73. Mem: Am Chem Soc. Res: Interactions between very weak acids and very weak bases in non-polar media. Mailing Add: Dept of Chem Fla Southern Univ Lakeland FL 33802

WILLARD, WILLIAM KENNETH, b Hagerstown, Md, Nov 5, 29; m 61; c 2. ECOLOGY. Educ: Univ Ga, BSF, 57, MS, 60; Univ Tenn, PhD(zool), 65. Prof Exp: From asst prof to assoc prof zool, Clemson Univ, 65-75; PROF ZOOL & CHMN DEPT BIOL, TENN TECHNOL UNIV, 75- Mem: AAAS; Ecol Soc Am. Res: Effects of radiations on populations; fate of radioactive materials in the environment; population dynamics; ecosystem analysis; radiation ecology; bioenergetics of food chain relationships; ecological strategies in mammalian population dynamics. Mailing Add: Dept of Biol Tenn Technol Univ Cookeville TN 38501

WILLARD, WILLIAM ROBERT, b Seattle, Wash, Nov 20, 08; m 36; c 3. PUBLIC HEALTH. Educ: Yale Univ, BS, 31, MD, 34, DrPh, 37; Am Bd Prev Med, dipl, 48. Hon Degrees: DSc, Transylvania Univ, Lexington, Ky, 59. Prof Exp: Intern pediat, Hopkins Hosp, 34-35; asst res, Strong Mem Hosp, Rochester, NY, 35-36; assoc med economist, Social Security Bd, 47; dep state health officer, State Dept Health, Md, 37-44; from asst prof to prof pub health, Yale Univ, 46-51, asst dean in charge, Post-Grad Med Ed, 48-51; prof pub health & dean, Col Med, State Univ NY Upstate Med Ctr, 51-56; vpres, Med Ctr, Univ Ky, 56-70, prof community med, Col Med, 60-72, spec asst to pres health affairs, 70-72, dean col med, 56-66; DEAN COL COMMUNITY HEALTH SCI, UNIV ALA, TUSCALOOSA, 72- Concurrent Pos: Mem, Hosp adv coun, Ky State Health Dept, 57- & Blue Shield Adv Comn, Ky State Med Asn, 59-62; chmn, Health Resources Adv Comt, Off Emergency Planning, Exec Off President, 62-; consult, USPHS, 64-; mem, Panel Fed Use of Health Manpower, Nat Adv Comn Health Manpower, 66- & Health Adv Comt, Appalachian Regional Comn, 66- Honors & Awards: Abraham Flexner Award, 72; Distinguished Serv Award, AMA, 75. Mem: AMA; fel Am Pub Health Asn; Asn Am Med Cols; fel Am Col Physicians; fel Am Col Prev Med. Mailing Add: Col Community Health Sci Univ of Ala PO Box 6291 University AL 35486

WILLARDSON, ROBERT KENT, b Gunnison, Utah, July 11, 23; m 47; c 3. SOLID STATE PHYSICS. Educ: Brigham Young Univ, BS, 49; Iowa State Col, MS, 51. Prof Exp: Instr physics, Brigham Young Univ, 47-48 & Iowa State Col, 48-49; res physicist, Ames Lab, AEC, 49-51; prin physicist, Battelle Mem Inst, 51-56, asst chief, Phys Chem Div, 56-60; chief scientist, Res Ctr, Bell & Howell Co, 60-64, dir solid state res, 64-67, dir mat res, 67-69, gen mgr electronic mat div, 69-73, pres, Electronic Mat Corp, 73; ASST TO GEN MGR, DIV ELECTRONIC MAT, COMINCO AM, INC, 73- Mem: Am Phys Soc; Electrochem Soc; Am Chem Soc; Inst Elec & Electronics Eng. Res: Preparation, electrical and optical properties of high purity metals, alloys and semiconductors; analysis, control and effects of impurities and lattice defects in these materials; electronic transport phenomena in semiconductors. Mailing Add: Div Electronic Mat Cominco Am Inc 101 Spokane Ind Park Spokane WA 99216

WILLAUER, WHITING R, astronomy, see 12th edition

WILLCOTT, MARK ROBERT, III, b Muskogee, Okla, July 23, 33; m 55; c 4. ORGANIC CHEMISTRY. Educ: Rice Univ, BA, 55; Yale Univ, MS, 59, PhD(chem), 63. Prof Exp: Asst prof chem, Emory Univ, 62-64; from asst prof to assoc prof, 65-73, PROF CHEM, UNIV HOUSTON, 73- Concurrent Pos: NIH fel, Univ Wis, 64-65; Guggenheim fel, 72-73; consult, Upjohn Co, 65- & Aldrich Chem Co, 72- Mem: Am Chem Soc; The Chem Soc. Res: Thermal rearrangements of organic compounds; nuclear magnetic resonance spectroscopy. Mailing Add: Dept of Chem Univ of Houston Houston TX 77004

WILLCOX, ALFRED BURTON, b Sioux Rapids, Iowa, Sept 18, 25; m 48; c 4. MATHEMATICS. Educ: Yale Univ, MA, 49, PhD(math), 53. Prof Exp: From instr to prof math, Amherst Col, 53-68, exec dir comt undergrad prog math, 63-64; EXEC DIR, MATH ASN AM, 68- Concurrent Pos: Vis asst prof, Univ Chicago, 57-58; vis lectr, Uppsala Univ, Sweden, 67-68. Mem: AAAS; Am Math Soc; Math Asn Am (vpres, 64-66). Res: Banach algebras. Mailing Add: Math Asn Am 1225 Connecticut Ave NW Washington DC 20036

WILLDEN, CHARLES RONALD, b Neola, Utah, Sept 30, 29; m 50; c 3. GEOLOGY. Educ: Univ Utah, BS, 51, MS, 52; Stanford Univ, PhD(geol), 60. Prof Exp: Geologist, US Geol Surv, Colo, 52-68; sr geologist, Vanguard Explor Co, 68-72; GEOLOGIST, SILVER RESOURCES CORP, 72- Mem: Geol Soc Am; Soc Econ Geol; Am Asn Petrol Geol. Res: Structural and economic geology. Mailing Add: 8750 Kings Hill Dr Salt Lake City UT 84121

WILLE, JOHN JACOB, JR, b New York, NY, June 24, 37; m 61; c 4. CELL BIOLOGY. Educ: Cornell Univ, BA, 60; Univ Ind, PhD(genetics), 65. Prof Exp: Resident res assoc cell biol, Biol Div, Argonne Nat Lab, 65-66, fel, 66-68; asst prof biol sci, Univ Cincinnati, 68-72; NIH spec fel, Univ Chicago, 72-73, res assoc, Dept Biophys, 73-75; ASST PROF ZOOL & PHYSIOL, LA STATE UNIV, 75- Res: Developmental genetics of unicellular organisms; molecular biology of biological rhythms. Mailing Add: Dept of Zool & Physiol La State Univ Baton Rouge LA 70803

WILLEBOORDSE, FRISO, b Bandung, Indonesia, July 31, 33; m 58; c 4. ANALYTICAL CHEMISTRY. Educ: Univ Indonesia, BS, 54; Univ Amsterdam, Drs, 57, PhD(anal chem), 69. Prof Exp: Lectr inorg chem, Univ Natal, 60-61, sr lectr anal chem, 61-62; res chemist, WVa, 62-69, res scientist, Bound Brook, 69-72, GROUP LEADER ANAL RES CHEM & PLASTICS, UNION CARBIDE CORP, BOUND BROOK, NJ, 72- Concurrent Pos: Vis prof, Rutgers Univ, 73-74. Mem: Am Chem Soc; Royal Netherlands Chem Soc. Res: Polymer characterization, polarography; differential chemical kinetics. Mailing Add: 6 Heritage Dr Warren NJ 07060

WILLEFORD, BENNETT RUFUS, JR, b Greenville, SC, Oct 28, 21. PHYSICAL CHEMISTRY. Educ: Emory Univ, BA, 43; Univ Wis, MS, 49, PhD(phys chem), 50. Prof Exp: Jr chemist, Shell Develop Co, 43-46; asst, Univ Wis, 46-50; from asst prof to assoc prof, 50-61, PROF CHEM, BUCKNELL UNIV, 61- Concurrent Pos: Res fel, Univ Minn, 56-57; consult, US Fish & Wildlife Serv, 60-64; NSF sci fac fel, Univ Munich, 62-63; res assoc, Univ NC, Chapel Hill, 69-70. Mem: Am Chem Soc. Res: Structure of metal coordination compounds; organometallic chemistry. Mailing Add: Dept of Chem Bucknell Univ Lewisburg PA 17837

WILLEMOT, CLAUDE, b Ghent, Belg, Dec 26, 33; Can citizen; m 66; c 2. PLANT PHYSIOLOGY. Educ: McGill Univ, MSc, 63, PhD(plant physiol), 64. Prof Exp: Nat Res Coun Can fel, Nat Inst Agr Res, Versailles, France, 64-65 & Univ Calif, Davis, 65-67; RES SCIENTIST PLANT PHYSIOL, RES STA, CAN DEPT AGR, 67- Concurrent Pos: Lectr, Fac Agr, Laval Univ, 68-; assoc ed, Can J Biochem, 75- Mem: Can Soc Plant Physiol (secy-treas, 71-72, secy, 72-73); Am Soc Plant Physiol; Soc Cryobiol. Res: Mechanism of plant frost hardiness; plant lipid metabolism. Mailing Add: Res Sta Can Dept Agr 2560 chemin Gomin Ste-Foy PQ Can

WILLEMS, EMILIO, b Cologne, Ger, Aug 18, 05; nat US; m 32; c 3. ANTHROPOLOGY. Educ: Univ Berlin, MA, 28, PhD, 30. Prof Exp: Asst prof sociol, Univ Sao Paulo, 37-41, prof anthrop, Univ & Sch Sociol, 41-49; PROF ANTHROP, VANDERBILT UNIV, 49- Concurrent Pos: Guggenheim fel, 51; vis prof, Mich State Univ, 52 & Univ Mich, 52-53; Soc Sci Res Coun grant, 54; vis prof, Nat Univ Colombia, 62-63; consult, UNESCO, 51-53; ed, Sociologia, 39-49. Mem: Am Anthrop Asn; Am Studies Asn. Res: Culture change in Latin America; physical anthropology of Brazilian Indians. Mailing Add: Dept of Anthrop Vanderbilt Univ Nashville TN 37203

WILLEMSEN, ROGER WAYNE, b Oskaloosa, Iowa, Jan 14, 44; m 66; c 2. PHYSIOLOGICAL ECOLOGY, BOTANY. Educ: Cent Col, Iowa, BA, 66; Kans State Col, Pittsburg, Kans, MS, 68; Univ Okla, PhD(bot), 71. Prof Exp: Grants, 72-73, ASST PROF BOT, RUTGERS UNIV, NEW BRUNSWICK, 71- Mem: Ecol Soc Am; Am Inst Biol Sci; Torrey Bot Club; Bot Soc Am; Weed Sci Soc Am. Res: The physiology and ecology of weed seed germination; allelopathy; old-field succession. Mailing Add: Dept of Bot Rutgers Univ New Brunswick NJ 08903

WILLENBROCK, FREDERICK KARL, b New York, NY, July 19, 20; m 44. PHYSICS. Educ: Brown Univ, BS, 42; Harvard Univ, MA, 47, PhD(appl physics), 50. Prof Exp: Res fel & lectr, Harvard Univ, 50-55, from asst dir labs, Div Eng & Appl Physics to dir labs & assoc dean div, 54-67; prof & provost fac eng & appl sci, State Univ NY, Buffalo, 67-70; DIR INST APPL TECHNOL, NAT BUR STAND, 70- Mem: Nat Acad Eng; Am Phys Soc; Am Soc Eng Educ; Inst Elec & Electronics Eng. Res: Solid state electronics and electromagnetic phenomena. Mailing Add: Inst Appl Technol Nat Bur Stand Washington DC 20234

WILLENKIN, ROBERT L, b Oceanside, NY, Feb 12, 31; m 64; c 4. ANESTHESIOLOGY, PHYSIOLOGY. Educ: Hofstra Univ, BA, 51; State Univ NY, MD, 55. Prof Exp: Fel anesthesiol, Yale Univ, 60-61; fel physiol, Univ Wash, 62-63; asst prof anesthesiol, Sch Med, Yale Univ, 63-68; assoc prof, Albany Med Col, 68-74; ASSOC PROF ANESTHESIOL, SCH MED, YALE UNIV, 74- Mem: AAAS; Am Soc Anesthesiol; NY Acad Sci. Res: Cardiovascular physiology. Mailing Add: Dept of Anesthesiol Yale Univ Sch of Med New Haven CT 06520

WILLERDING, MARGARET FRANCES, b St Louis, Mo, Apr 26, 19. MATHEMATICS. Educ: Harris Teachers Col, BS, 40; St Louis Univ, MA, 43, PhD(math), 47. Prof Exp: Instr math, Wash Univ, 47-48; asst prof, Harris Teachers Col, 48-56; assoc prof, 56-66, PROF MATH, CALIF STATE UNIV, SAN DIEGO, 66- Mem: Am Math Soc; Math Asn Am. Res: Number theory; mathematics education. Mailing Add: 10241 Vivera Dr La Mesa CA 92041

WILLETT, DOUGLAS W, b Adams County, NDak, May 25, 37; m 59; c 5. MATHEMATICS. Educ: SDak Sch Mines & Technol, BS, 59; Calif Inst Technol, PhD(math), 63. Prof Exp: From asst prof to assoc prof math, Univ Alta, 62-66; assoc prof, 66-72, PROF MATH, UNIV UTAH, 72- Concurrent Pos: Vis prof, Univ Alta, 71-72. Mem: Soc Indust & Appl Math; Can Math Cong. Res: Ordinary differential equations and mathematical analysis. Mailing Add: Dept of Math Univ of Utah Salt Lake City UT 84112

WILLETT, HILDA POPE, b Decatur, Ga, July 15, 23; m 56; c 2. MICROBIOLOGY. Educ: Woman'sCol Ga, AB, 44; Duke Univ, MA, 46, PhD(microbiol), 49. Prof Exp: Instr microbiol, 48-50, assoc, 50-52, from asst prof to assoc prof, 52-64, PROF BACT, SCH MED, DUKE UNIV, 64- Mem: AAAS; Am Soc Microbiol; Soc Exp Biol & Med; Am Acad Microbiol. Res: Vitamin and amino acid metabolism of virulent and avirulent Mycobacterium tuberculosis; mode of action of isoniazid. Mailing Add: Dept of Microbiol Duke Univ Med Ctr Durham NC 27710

WILLETT, HURD CURTIS, b Providence, RI, Jan 1, 03; m 35, 49; c 5. METEOROLOGY. Educ: Princeton Univ, BS, 24; George Washington Univ, PhD(meteorol), 29. Prof Exp: Observer, US Weather Bur, Washington, DC, 24-27, asst meteorologist, 28-29, meteorologist, 41; from instr to prof, 29-68, EMER PROF METEOROL, MASS INST TECHNOL, 68- Concurrent Pos: Lectr, Harvard Univ, 39-45; expert consult, Weather Div, US Air Force, 42-45 & Res & Develop Bd, 47-52. Mem: AAAS; Am Meteorol Soc; Am Asn Geog; Am Geophys Union; Am Acad Arts & Sci. Res: Fog and haze; synoptic meteorology, particularly long range weather forecasting; climatic changes. Mailing Add: Dept Meteorol Rm 54-1414 Mass Inst Technol 77 Massachusetts Ave Cambridge MA 02139

WILLETT, JOSEPH ERWIN, b Albany, Mo, June 9, 29; m 55; c 3. PLASMA PHYSICS. Educ: Univ Mo, BA, 51, MA, 53, PhD(physics), 56. Prof Exp: From asst to instr physics, Univ Mo, 53-55, Stewart fel, 55; physicist & aeronaut res engr, US Naval Ord Lab, Md, 56-58; res scientist, McDonnell Aircraft Corp, 58-61, instr, McDonnell Eve Sch, 60-61; staff scientist, Gen Dynamics/Ft Worth, Tex, 61-65; ASSOC PROF PHYSICS, UNIV MO-COLUMBIA, 65- Concurrent Pos: Instr & adj prof, Tex Christian Univ, 62-64. Honors & Awards: Superior Accomplishment Award, US Naval Ord Lab, 58. Mem: Am Phys Soc. Mailing Add: Dept of Physics Univ of Mo Columbia MO 65201

WILLETT, LYNN BRUNSON, b Colorado Springs, Colo, Aug 2, 44; m 66; c 1. ANIMAL PHYSIOLOGY, DAIRY SCIENCE. Educ: Colo State Univ, BS, 66; Purdue Univ, Lafayette, MS, 68, PhD(animal physiol), 71. Prof Exp: ASST PROF DAIRY SCI, OHIO AGR RES & DEVELOP CTR, OHIO STATE UNIV, 71- Mem: Am Dairy Sci Asn; Am Soc Animal Sci. Res: Elimination and metabolism of polychlorinated biphenyl residues by cattle; steroid hormone relationships in cattle. Mailing Add: Dept Dairy Sci Ohio Agr Res & Develop Ctr Ohio State Univ Wooster OH 44691

WILLETT, NORMAN P, b Paterson, NJ, Apr 13, 28; m 56; c 3. MICROBIOLOGY. Educ: Rutgers Univ, BS, 49; Syracuse Univ, MS, 52; Mich State Univ, PhD(microbiol), 55. Prof Exp: Asst dent med, Harvard Univ, 55-57; sr res microbiologist, Squibb Inst Med Res, 57-60; res microbiologist, Bzura, Inc, 60-62; sr res assoc, Lever Brothers, 62-63; chief microbiologist, Food & Drug Res Inc, 63; res assoc, Sch Vet Med, Univ Pa, 63-66; assoc prof, Sch Pharm, 66-73, PROF MICROBIOL, SCH DENT, TEMPLE UNIV, 73-, HEAD DEPT, 67- Concurrent Pos: Mem, Am Asn Dent Schs. Mem: AAAS; Am Soc Microbiol; Am Chem Soc; Soc Indust Microbiol; NY Acad Sci. Res: Biochemical basis of pathogenicity; physiology of streptococci; antibiotic biosynthesis; biochemistry and microbiology of saliva; organic acid and antibiotic fermentation; chemotherapy. Mailing Add: 604 Pine Tree Rd Jenkintown PA 19046

WILLETT, RICHARD MICHAEL, b Louisville, Ky, May 2, 45; m. MATHEMATICS. Educ: US Air Force Acad, BS, 67; NC State Univ, MA, 69, PhD(math), 71. Prof Exp: Teaching asst math, NC State Univ, 67-71, instr, 71-72; ASST PROF MATH, UNIV NC, GREENSBORO, 72- Res: Finite field theory; error-correcting codes. Mailing Add: Dept of Math Univ of NC Greensboro NC 27412

WILLETT, ROBERT A, plasma physics, see 12th edition

WILLETT, ROGER, b Northfield, Minn, July 13, 36; m 57; c 6. PHYSICAL CHEMISTRY, CHEMICAL PHYSICS. Educ: St Olaf Col, BA, 58; Iowa State Univ, PhD(chem, physics), 62. Prof Exp: From instr to assoc prof, 62-72, PROF CHEM, WASH STATE UNIV, 72-, CHMN DEPT, 74- Mem: Am Chem Soc; Am Crystallog Asn; Sigma Xi. Res: X-ray diffraction and crystallography; magnetic susceptibility and interactions; chemical bonding; molecular and electronic structure; electronic spectra and electron spin resonance studies of transition metal ions. Mailing Add: Dept of Chem Wash State Univ Pullman WA 99163

WILLETTE, GORDON LOUIS, b Dighton, Mass, Dec 19, 33. PETROLEUM CHEMISTRY. Educ: Brown Univ, ScB, 55; Univ Minn, PhD(org chem), 59. Prof Exp: CHEMIST, RES LABS, ROHM AND HAAS CO, 59- Mem: Am Chem Soc. Res: Polymer synthesis. Mailing Add: Res Labs Rohm & Haas Co Norristown & McKean Rds Spring House PA 19477

WILLETTE, ROBERT EDMOND, b Grand Rapids, Mich, Aug 15, 33; m 59; c 3. MEDICINAL CHEMISTRY, ORGANIC CHEMISTRY. Educ: Ferris State Col, BS, 55; Univ Minn, PhD(pharmaceut chem), 60. Prof Exp: Spec instr med chem, Ferris State Col, 59-61; NIH res fel, Australian Nat Univ, 61-63; res officer, Div Org Chem, Commonwealth Sci & Indust Res Orgn, 63-64; res assoc med chem, Univ Mich, 65-66; from asst prof to assoc prof, Sch Pharm, Univ Conn, 66-72; CHEMIST, DIV RES, NAT INST DRUG ABUSE, 72- Mem: AAAS; Am Pharmaceut Asn; Am Chem Soc; The Chem Soc. Res: Isolation and structure determination of natural products; synthesis of heterocyclic compounds of medicinal interest, particularly analgetics and narcotic antagonists and alkylating and acylating agents related to cancer. Mailing Add: Rockwall Bldg Rm 666 11400 Rockville Pike Rockville MD 20852

WILLEY, ANN MORRIS, b Rome, NY, Nov 21, 49; m 70; c 1. HUMAN GENETICS. Educ: Cornell Univ, BS, 71; Univ Minn, PhD(genetics), 74. Prof Exp: Res assoc med genetics, Univ Minn Med Sch, 74-75; ASST PROF BIOL, DEPT BIOL SCI, DOUGLASS COL, RUTGERS UNIV, NEW BRUNSWICK, 75- Concurrent Pos: NIH fel, Med Genetics Inst Training Grant, 74-75. Res: The molecular organization of eukaroytic genome, particularly the organization of the different classes of DNA sequences in the human chromosome. Mailing Add: Dept of Biol Sci Douglass Col Rutgers Univ New Brunswick NJ 08903

WILLEY, CLIFF RUFUS, b Hornell, NY, Nov 20, 35; m 57; c 5. SOIL PHYSICS. Educ: Cornell Univ, BS, 57, MS, 59; Univ Wis-Madison, PhD(soil physics), 62. Prof Exp: Soil scientist, Agr Res Serv, USDA, 62-73; chief solid waste serv, 73-74, CHIEF TECH SERV, MD ENVIRON SERV, DEPT NATURAL RESOURCES, 74- Concurrent Pos: Assoc prof agr eng, NC State Univ, 67- Mem: Soil Sci Soc Am; Am Soc Agron; AAAS; Sigma Xi. Mailing Add: Tawes State Off Bldg Annapolis MD 21401

WILLEY, ROBERT BRUCE, b Long Branch, NJ, Sept 15, 30; m 56. ANIMAL BEHAVIOR. Educ: NJ State Teachers Col, BA, 52; Harvard Univ, PhD(biol), 59. Prof Exp: Teacher high sch, NJ, 52-54; from asst prof to assoc prof biol, Ripon Col, 59-65; ASSOC PROF BIOL, UNIV ILL, CHICAGO CIRCLE, 65- Concurrent Pos: Res grants, Sigma Xi, 62 & 67, NSF, 64-66 & 72-74; mem bd trustees, Rocky Mountain Biol Lab, 63-, vpres, 72-76. Mem: Am Soc Zool; Soc Study Evolution; Ecol Soc Am; Acridological Asn; Animal Behav Soc. Res: Invertebrate behavior; interspecific behavior and evolution of sympatric insect populations; animal communication systems. Mailing Add: Dept of Biol Sci Box 4348 Univ of Ill at Chicago Circle Chicago IL 60680

WILLEY, RUTH LIPPITT, b Wickford, RI, May 11, 28; m 56. ENTOMOLOGY. Educ: Wellesley Col, BA, 50; Radcliffe Col, PhD(biol), 56. Prof Exp: Docent mus educ, Peabody Mus Natural Hist, Yale Univ, 50-52; instr zool, Wellesley Col, 56-57; res fel opthal, Mass Eye & Ear Infirmary, 57-58; comm histologist, Triarch Prod, 59-60; asst prof, 65-71, ASSOC PROF BIOL SCI, UNIV ILL, CHICAGO, 71- Concurrent Pos: Sigma Xi res grant, 53 & 69; NSF grant, 64-67 & 72-75; grad res bd grant, Univ Ill, 67-70 & 72; mem, Rocky Mountain Biol Lab, Colo, 58-, secy, 66-68, mem bd trustees, 75-. Mem: AAAS; Entom Soc Am; Am Soc Prof Micros. Res: Evolution, ecology and zoogeography of insects, especially the Odonata; ultrastructure and ecology of symbiotic algae and protozoa. Mailing Add: Dept of Biol Sci Univ of Ill Box 4348 Chicago IL 60680

WILLHAM, RICHARD LEWIS, b Hutchinson, Kans, May 4, 32; m 54; c 2. ANIMAL BREEDING. Educ: Univ Okla, BS, 54; Iowa State Univ, MS, 55, PhD(animal breeding), 60. Prof Exp: Asst prof animal breeding, Iowa State Univ, 59-63; assoc prof, Okla State Univ, 63-66; PROF ANIMAL BREEDING, IOWA STATE UNIV, 66- Concurrent Pos: AEC grant, Iowa State Univ, 59-63. Mem: Biomet Soc; Am Soc Animal Sci. Res: Beef cattle breeding; evaluation of the results of selection and crossbreeding; beef improvement federation work with national sire evaluation program development in beef industry. Mailing Add: Dept of Animal Sci Iowa State Univ Ames IA 50010

WILLHOIT, DONALD GILLMOR, b Kansas City, Mo, Feb 5, 34; m 56; c 4. RADIATION HEALTH. Educ: William Jewell Col, AB, 56; Univ Wash, 58; Univ Pittsburgh, ScD(radiation health), 64. Prof Exp: Assoc scientist & health physicist, Westinghouse Testing Reactor, 58-60; radiation safety off, Univ Pittsburgh, 60-61, teaching fel, 61-64; asst prof radiol health, Sch Pub Health, 64-68, ASSOC PROF RADIATION BIOPHYS, UNIV NC, CHAPEL HILL, 68-, DIR HEALTH & SAFETY, 74- Concurrent Pos: Consult, Health Physics Div, Oak Ridge Nat Lab. Mem: Health Physics Soc; Radiation Res Soc; Am Indust Hyg Asn. Res: Tumor and normal cell kinetics; radiation dose-rate and fractionation effects, late effects. Mailing Add: Health & Safety Off Univ of NC Chapel Hill NC 27514

WILLIAMS, AARON, JR, b Newark, NJ, Jan 29, 42. PHYSICAL GEOGRAPHY. Educ: Fla State Univ, BS, 65; Univ Mo, MA, 67; Univ Okla, PhD(geog), 71. Prof Exp: Meteorologist, US Weather Bur, 66; instr geog, 67-69, ASST PROF GEOG, UNIV S ALA, 71- Mem: Am Asn Geog; Am Meteorol Soc. Res: Climatology of coastal and tropical environments; radar climatology. Mailing Add: Dept of Geol & Geog Univ of SAla Mobile AL 36608

WILLIAMS, ALAN DAVID, chemistry, see 12th edition

WILLIAMS, ALBERT DORAN, b Hornell, NY, Mar 28, 20; m 43; c 2. NUCLEAR MEDICINE. Educ: Alfred Univ, AB, 48; Univ Southern Calif, MS, 49, PhD(biochem), 55. Prof Exp: Res assoc, Sch Med, Univ Ill, 53-56; res assoc, Univ Calif, Los Angeles, 56-58; RADIOISOTOPE BIOCHEMIST, NUCLEAR MED SERV, VET ADMIN HOSP, LONG BEACH, 58- Concurrent Pos: Asst prof, Calif State Univ, Long Beach, 57-59; asst clin prof, Univ Calif, Irvine-Calif Col Med, 65- Mem: AAAS; Am Asn Clin Chemists; Health Physics Soc; Int Radiation Protection Asn. Res: Radioimmunoassay; thyroid hormone biosynthesis; clinical chemistry methodology. Mailing Add: 1112 Stevely Ave Long Beach CA 90815

WILLIAMS, ALBERT JAMES, III, b Philadelphia, Pa, Oct 17, 40; m 63. OCEANOGRAPHY, OCEAN ENGINEERING. Educ: Swarthmore Col, BA, 62; Johns Hopkins Univ, PhD(physics), 69. Prof Exp: Investr, 69-71, asst scientist, 71-75, ASSOC SCIENTIST, OCEAN ENG, WOODS HOLE OCEANOG INST, 75- Mem: AAAS; Am Geophys Union; Marine Technol Soc. Res: Ocean microstructure, mixing and thermohaline convection; benthic boundary layer processes; oceanographic, optical and electronic instrumentation. Mailing Add: Woods Hole Oceanog Inst Woods Hole MA 02543

WILLIAMS, ALBERT LLOYD, b Vineland, NJ, Apr 14, 20; m 43; c 2. ORGANIC CHEMISTRY, PETROLEUM CHEMISTRY. Educ: Univ Pa, AB, 42, MS, 47, PhD(chem), 50. Prof Exp: Chemist, E I du Pont de Nemours & Co, 42-43; chemist, Philco Corp, 43-45; instr chem, Univ Rochester, 49-51; sr chemist, 51-65, RES ASSOC, MOBIL OIL CORP, 65- Mem: AAAS; Am Chem Soc. Res: Composition of petroleum; oxidation and antioxidation of petroleum and hydrocarbons; synthesis and stabilization of lubricants. Mailing Add: Mobil Oil Corp Mobil Tech Ctr Princeton NJ 08540

WILLIAMS, ALBERT SIMPSON, b York Co, SC, Jan 23, 24; m 46. PLANT PATHOLOGY. Educ: Emory Univ, AB, 48; Univ Tenn, MS, 49; NC State Univ, PhD(plant path), 54. Prof Exp: Asst prof biol, Athens Col, 50; plant pathologist, State Plant Bd, Miss, 50-51; asst, NC State Univ, 51-54; assoc prof plant path, Va Polytech Inst, 54-68; exten prof plant path, 68-75, PROF HORT & CHMN DEPT, UNIV KY, 75- Mem: Am Phytopath Soc; Am Soc Nematol; Soc Europ Nematol. Mailing Add: Dept of Hort Univ of Ky Lexington KY 40506

WILLIAMS, ALLAN, RAWSON, b Jericho, Vt, Dec 1, 18; div; c 1. ORGANIC CHEMISTRY. Educ: Univ Vt, BS, 40, MS, 41; NY Univ, PhD(chem), 52. Prof Exp: Res chemist, Stamford Res Labs, Am Cyanamid Co, 47; res chemist, US Rubber Co, 52-59, sr res scientist, 59-60, mat res, Tire Div, 60-64, develop compounding, 64-75, mgr compounding & elastomers, Develop Dept, Uniroyal Tire Co, 71-73. Mem: Am Chem Soc. Res: Monomers; rubber chemicals; heterocyclics. Mailing Add: Develop Dept 6600 E Jefferson Uniroyal Tire Co Detroit MI 48232

WILLIAMS, ANNA MARIA, b Tampa, Fla, June 29, 27. MICROBIOLOGY. Educ: Univ Ala, BS, 48; Univ Wis, MS, 51, PhD(bact), 54. Prof Exp: Antibiotics lab supvr, Merck & Co, Inc, 48-50; res asst bact, Univ Wis, 50-54; Fulbright res grant, biophys res group, Univ Utrecht, 54-55; proj assoc, McArdle Mem Inst Cancer Res, Univ Wis-Madison, 55-57, res assoc med, Sch Med, 57-64, asst prof, 64-69, ASSOC PROF LIFE SCI, UNIV WIS-PARKSIDE, 69- Mem: AAAS; Am Chem Soc; Am Soc Microbiol; Am Inst Biol Sci. Res: Nucleic acid metabolism; leukemia; cancer chemotherapy; bacterial enzymes. Mailing Add: Dept of Life Sci Univ of Wis-Parkside Kenosha WI 53140

WILLIAMS, ANTHONY VEARNCOMBE, b Blaenrhondda, Wales, Apr 3, 38; US citizen; m 64; c 1. GEOGRAPHY, APPLIED STATISTICS. Educ: Wayne State Univ, AB, 58; Ohio State Univ, MA, 59; Mich State Univ, PhD(geog), 68. Prof Exp: Prog mil logistics, US Army, 60-63; asst prof, 66-72, ASSOC PROF GEOG, PA STATE UNIV, UNIVERSITY PARK, 72- Concurrent Pos: Nat Inst Child Health & Human Develop Pop Issues Res Off, Pa State Univ, 72-74; asst prof, Dept Comput Sci, Pa State Univ, 66-70; bk ed, Data Base Spec Interest Group for Bus Data Processing, Asn Comput Machinery, 71- Mem: AAAS; Asn Am Geog; Asn Comput Mach; World Future Soc; Int Peace Res Soc. Res: Application of computers to geographic research and instruction; human population movements; simulation modeling. Mailing Add: Dept of Geog Pa State Univ 409 Deike University Park PA 16802

WILLIAMS, ARDIS MAE, b Boston, Mass, June 8, 19; m 49; c 1. PHYSICAL CHEMISTRY. Educ: Mt Holyoke Col, AB, 41; Vassar Col, AM, 46. Prof Exp: Chemist, Harvard Med Sch, 41-42; teacher, Low Heywood Sch, 42-44; lectr chem, Vassar Col, 44-46; lectr, Barnard Col, 46-48; res chemist, Merck & Co, 48-49; res chemist, Med Col Va, 50-54; teacher, Tatnall Sch, 63-67; ASSOC PROF CHEM, WEST CHESTER STATE COL, 67- Mem: AAAS; Am Chem Soc. Res: Surface area and adsorption by activated carbon; relationship of structure and adsorption of organic molecules. Mailing Add: Dept of Chem West Chester State Col West Chester PA 19380

WILLIAMS, ARTHUR H, physics, see 12th edition

WILLIAMS, ARTHUR OLNEY, JR, b Providence, RI, Apr 7, 13; m 38; c 1. PHYSICS. Educ: Mass Inst Technol, BS, 34; Brown Univ, ScM, 36, PhD(physics), 37. Prof Exp: From instr to asst prof physics, Univ Maine, 37-42; from asst prof to assoc prof, 42-51, chmn dept, 56-60, 62-63, PROF PHYSICS, BROWN UNIV, 51- Concurrent Pos: Mem, RI AEC, 55-68. Mem: Fel Am Phys Soc; Am Asn Physics Teachers; fel Acoust Soc Am. Res: Physical acoustics. Mailing Add: Dept of Physics Brown Univ Providence RI 02912

WILLIAMS, ARTHUR ROBERT, b Feb 20, 41; US citizen; m 62; c 3. SOLID STATE PHYSICS. Educ: Dartmouth Col, AB, 62; Harvard Univ, PhD(solid state physics), 69. Prof Exp: Appl mathematician, Info Res Inc, Mass, 67-68; fel, 68-69, STAFF PHYSICIST, WATSON RES CTR, IBM CORP, 69- Mem: Am Phys Soc. Res: Effective one-electron theory of Fermi surface and optical data for solids. Mailing Add: IBM Watson Res Ctr PO Box 218 Yorktown Heights NY 10598

WILLIAMS, AUBREY WILLIS, JR, b Madison, Wis, July 31, 24; c 2. ANTHROPOLOGY, ETHNOGRAPHY. Educ: Univ NC, BA, 55, MA, 57; Univ Ariz, PhD(anthrop), 64. Prof Exp: Res asst, Human Rels Area Files, Univ NC, 57; work camp dir, Am Friends Serv Comt, Mex, 57-58; res asst Mohave Indians, Bur Ethnic Res, Univ Ariz, 59-60; anthropologist, Navajo Tribe, Window Rock, Ariz, 61; asst prof anthrop, 62-67, dir, Anthrop Div, 67-71, ASSOC PROF ANTHROP, UNIV MD, COLLEGE PARK, 67- Mem: Fel AAAS; fel Am Anthrop Asn; Soc Am Archaeol; Am Ethnol Soc. Res: Ethnology of indigenous and mestizo people of the new world, especially those inhabiting arid regions; effect of water control on present social, economics and political systems. Mailing Add: Div of Anthrop Univ of Md College Park MD 20742

WILLIAMS, AUGUSTUS KENNETH, biochemistry, bacteriology, see 12th edition

WILLIAMS, AUSTIN BEATTY, b Plattsburg, Mo, Oct 17, 19; m 46; c 1. SYSTEMATIC ZOOLOGY. Educ: McPherson Col, AB, 43; Univ Kans, PhD(zool), 51. Prof Exp: Asst genetics, Univ Wis, 43-44; teacher pub schs, Kans, 44-46; shrimp investr, Inst Fisheries Res, Univ NC, 51-52, asst prof, 52-55; asst prof, Univ Ill, 55-56; assoc prof, Inst Fisheries Res, Univ NC, 56-63, prof, Inst Marine Sci, 63-71; SYST ZOOLOGIST, NAT SYSTS LAB, NAT MARINE FISHERIES SERV, 71- Concurrent Pos: Adj prof, Univ NC, Chapel Hill, 71- Mem: AAAS; Am Soc Zool; Ecol Soc Am; Am Soc Limnol & Oceanog; Soc Study Evolution. Res: Taxonomy; ecology; life histories of decapod crustacea; special reference to western Atlantic region; estuarine ecology. Mailing Add: Nat Systs Lab US Nat Mus Nat Marine Fisheries Serv Washington DC 20560

WILLIAMS, BENJAMIN HAYDEN, b Davenport, Iowa, Dec 18, 21; m 46; c 5. ANATOMY, ORTHODONTICS. Educ: Ohio State Univ, DDS, 46, MSc, 49; Univ Ill, MS, 51; Am Bd Orthod, dipl. Prof Exp: Fel, Univ Ill, 50-51; from instr to assoc prof, 51-64, PROF ORTHOD & CHMN DEPT, COL DENT, OHIO STATE UNIV, 64- Concurrent Pos: Staff mem, Children's Hosp, Columbus, Ohio, 61-; mem bd med, State Crippled Children's Servs, Ohio, 66-; mem orthod sect, Am Asn Dent Schs. Mem: Int Asn Dent Res; Am Asn Orthod. Res: Cranio-facial growth and development; growth predictions; dental development and eruption; normal and abnormal growth. Mailing Add: Dept of Dent Ohio State Univ Col of Dent Columbus OH 43210

WILLIAMS, BENNIE B, b Scranton, Tex, Jan 16, 22; m 48; c 4. MATHEMATICS. Educ: Howard Payne Col, BA, 48; Univ Tex, MA, 53, PhD(math), 66. Prof Exp: From instr to asst prof math, Howard Payne Col, 48-61; asst prof, 66-71, ASSOC PROF MATH, UNIV TEX, ARLINGTON, 71- Mem: Am Math Soc; Math Asn Am. Res: Foundations of mathematics; ordinary differential equations. Mailing Add: Dept of Math Univ of Tex Arlington TX 76019

WILLIAMS, BERNARD LEO, physical chemistry, polymer chemistry, see 12th edition

WILLIAMS, BETTY JEAN, b Homerville, Ga, Feb 13, 37. PHARMACOLOGY. Educ: Ga State Col Women, BA, 58; Emory Univ, MS, 63, PhD(pharmacol), 67. Prof Exp: Fel, Univ Wash, 66-67, res instr, 67-68; ASST PROF PHARMACOL & TOXICOL, UNIV TEX MED BR GALVESTON, 68- Mem: AAAS. Res: Molecular mechanisms of hormone action with emphasis on catecholamines. Mailing Add: Dept of Pharmacol & Toxicol Univ of Tex Med Br Galveston TX 77550

WILLIAMS, BOBBY JOE, b Idabel, Okla, Nov 3, 30; m 57; c 4. PHYSICAL ANTHROPOLOGY, POPULATION GENETICS. Educ: Univ Okla, BA, 53, MA, 57; Univ Mich, PhD(anthrop, human genetics), 65. Prof Exp: Asst prof anthrop, Univ Wis, Milwaukee, 63-65; asst prof, 65-72, ASSOC PROF ANTHROP, UNIV CALIF, LOS ANGELES, 72- Mem: AAAS; Am Anthrop Asn; Am Asn Phys Anthrop. Res: Human population genetics; human evolution; population processes in simple societies. Mailing Add: Dept of Anthrop Univ of Calif Los Angeles CA 90024

WILLIAMS, BROWN F, b Evanston, Ill, Dec 22, 40; m 69. SOLID STATE PHYSICS. Educ: Univ Calif, Riverside, BA, 62, MA, 64, PhD(physics), 66. Prof Exp: Mem tech staff, 66-68, leader electron emission, 68-70, mgr electro-optics lab, 70-73, HEAD QUANTUM ELECTRONICS RES, RCA LABS, 73- Honors & Awards: Outstanding Achievement Award, RCA Corp, 68, David-Sarnoff Award, 70. Mem: AAAS; Am Phys Soc. Res: Electro-optic devices; optical information recording; solar energy. Mailing Add: Head Quantum Electronics RCA Labs Princeton NJ 08540

WILLIAMS, BYRON BENNETT, JR, b Donaldsonville, Ga, Sept 2, 22; m 49; c 2. PHARMACOLOGY. Educ: Univ Fla, BS, 45, MS, 47, PhD, 51. Prof Exp: PROF PHARMACOL, AUBURN UNIV, 51- Mem: Am Pharmaceut Asn. Res: Antihistamines; autonomics; psychopharmacology. Mailing Add: Box 70 RR 3 Cox Rd Auburn AL 36830

WILLIAMS, BYRON LEE, JR, b Guantanamo Bay, Cuba, Aug 29, 20; m 42; c 3. ORGANIC CHEMISTRY. Educ: E Tex State Teachers Col, BS, 40, MS, 42; Univ Okla, PhD(org chem), 53. Prof Exp: Teacher pub sch, Tex, 40-42; chem engr, Chem Warfare Serv, US Dept Army, 42-44, chem engr, Chem Corps, 46-47; assoc prof chem, E Tex State Teachers Col, 47-41; asst, Res Found, Univ Okla, 51-53; assoc prof chem, E Tex State Teachers Col, 53-54; res chemist, 54-59, from asst dir res to assoc dir res, Plastics Div, 59-64, dir process tech & eng dept, Hydrocarbons Div, 64-65, dir res, Hydrocarbons & Polymers Div, 65-67, DIR CORP RES DEPT, MONSANTO CO, 67- Mem: Am Chem Soc. Res: Isolation, chemical characterization and identification of flavonoid type chemical compounds from selected natural products using ion exchange, adsorption and paper chromatography; synthesis of pigments and demethylation studies; hydrolytic enzyme studies of natural glycosides. Mailing Add: 609 Mosley Rd Creve Coeur MO 63141

WILLIAMS, CALVIT HERNDON, JR, b Houston, Tex, Dec 28, 36; m 59; c 4. PHYSICAL CHEMISTRY, ANALYTICAL CHEMISTRY. Educ: Univ St Thomas, Tex, BA, 58; Brown Univ, PhD(chem), 64. Prof Exp: Fel chem, Rice Univ, 64-66; tech staff mem, Sandia Labs, 66-70; ASST PROF CHEM, STATE UNIV CAMPINAS, BRAZIL, 71- Mem: AAAS; Am Chem Soc; Am Phys Soc; Am Inst Chem; Am Soc Mass Spectrometry. Res: High energy crossed-molecular-beam reaction kinetics; thermodynamics of high temperature processes; mass spectrometry and gas chromatography applied to environmental and biomedical analysis. Mailing Add: Inst Chem State Univ Campinas 13.100 Campinas Sao Paulo Brazil

WILLIAMS, CARL JAMES, JR, b Parkersburg, WVa, Feb 5, 30; m 56; c 3. PHOTOGRAPHIC CHEMISTRY. Educ: Marietta Col, BS, 51; Ohio State Univ, MS, 55; Univ Notre Dame, PhD(org chem), 58. Prof Exp: Chemist, 51-52, from res chemist to sr res chemist, 58-68, res assoc, 68-70, TECH ASSOC, EASTMAN KODAK CO, 70- Mem: Am Chem Soc. Res: Development of new and improved color photographic products, embodying all phases of photographic chemistry. Mailing Add: Eastman Kodak Co FEPMD Bldg 30 Rochester NY 14650

WILLIAMS, CAROL ANN, b Stratford, NJ, Oct 3, 40. ASTRONOMY, CELESTIAL MECHANICS. Educ: Conn Col, BA, 62; Yale Univ, PhD(astron), 67. Prof Exp: Assoc res engr & consult, Jet Propulsion Lab, 64 & 65; part-time instr physics, Conn Col, 66-67; res staff astronr, Yale Univ, 67-68; asst prof, 68-73, ASSOC PROF ASTRON, UNIV S FLA, TAMPA, 73- Mem: Am Astron Soc; Sigma Xi. Res: Resonance in celestial mechanics; astrometry. Mailing Add: Dept of Astron Univ of SFla Tampa FL 33620

WILLIAMS, CARROLL BURNS, JR, b St Louis, Mo, Sept 24, 29; m 58; c 3. FOREST ENTOMOLOGY. Educ: Univ Mich, BS, 55, MS, 57, PhD(forest entom), 63. Prof Exp: Entomologist, Pac Northwest Forest & Range Exp Sta, US Forest Serv, Ore, 57-58, res forester, 58-60, forestry sci lab, 61-65, res entomologist, Pac Southwest Forest & Range Exp Sta, Calif, 65-68, leader insect impact proj, Forest Insect & Dis Lab, Northeastern Exp Sta, Conn, 68-72, res entomologist, Pac Southwest Forest & Range Exp Sta, Calif, 72-75, PROJ LEADER, PIONEERING RES UNIT, INTEGRATED MGT SYSTS FOREST INSECT & DIS, PAC SOUTHWEST FOREST & RANGE EXP STA, CALIF, US FOREST SERV, 75- Concurrent Pos: Lectr, Sch Forestry, Yale Univ, 69-71; consult, NSF, 71-74. Mem: Soc Am Foresters; Ecol Soc Am; Entom Soc Am. Res: Evaluate and predict impact of forest insect and disease on forest resources; modelling pest management systems; computer simulation experiments of insect and disease control techniques and strategies; forest management decision models. Mailing Add: 1960 Addison St PO Box 245 Berkeley CA 94701

WILLIAMS, CARROLL MILTON, b Richmond, Va, Dec 2, 16; m 41; c 3. BIOLOGY. Educ: Univ Richmond, BS, 37; Harvard Univ, AM, 38, PhD(biol), 41, MD, 46. Hon Degrees: DSc, Univ Richmond, 60. Prof Exp: Fel, Soc Fels, 41-46, from asst prof to prof zool, 46-66, chmn dept, 59-62, chmn subdept cellular & develop biol, 72-73, BENJAMIN BUSSEY PROF BIOL, HARVARD UNIV, 66- Concurrent Pos: Lowell lectr, 48; Harvey lectr, 52; Dakin lectr, 64; Thomas lectr, 67; Nieuwland lectr, 67; Murlin lectr, 67; Delafield lectr, 70; David Rivett Mem lectr, 73; consult, NSF, 54-59; Guggenheim fel, 55-56; trustee, Radcliffe Col, 61-63. Honors & Awards: Res Prize, AAAS, 50; Boylston Medal, Harvard Med Sch, 61, George Ledlie Prize, Harvard Univ, 67; Howard Taylor Ricketts Award, Univ Chicago, 69. Mem: Nat Acad Sci (chmn, sect zool, 70-73, coun, 73-76); Inst Med Nat Acad Sci; fel AAAS; fel Am Acad Arts & Sci (mem coun, 52-55 & 74-78); fel Entom Soc Am. Res: Physiology of insects with special reference to endocrinology of growth and metamorphosis. Mailing Add: Biol Lab Harvard Univ Cambridge MA 02138

WILLIAMS, CECIL R, b Hydro, Okla, Sept 4, 12; m 34; c 2. BIOLOGY, CHEMISTRY. Educ: Phillips Univ, BA, 34; Univ Wis, MA, 36; Univ Okla, PhD(chem, ecol), 54. Prof Exp: Instr physics & chem, Woodward High Sch & Jr Col, 37-41; chemist, Sinclair Oil Co, 42-45; prof biol & chmn dept, 45-71, PROF ZOOL, PHILLIPS UNIV, 71- Res: Developmental anatomy, particularly genetics and embryology. Mailing Add: Dept of Zool Phillips Univ Enid OK 73701

WILLIAMS, CHARLES EDWIN, b Oil City, Pa, Aug 14, 29; m 57; c 2. GEOGRAPHY. Educ: Pa State Univ, BS, 55, MS, 57; Northwestern Univ, PhD(geog), 62. Prof Exp: Instr Geog, Univ Ark, 57-58; instr, Southeast Mo State Col, 58-59; instr, Univ Wis, 62-63; assoc prof, 63-67, PRCF GEOG, SOUTHEAST MO STATE UNIV, 67- Res: Physical geography; zoogeography. Mailing Add: Dept of Earth Sci Southeast Mo State Univ Cape Girardeau MO 63701

WILLIAMS, CHARLES HADDON, JR, b Washington, DC, June 29, 32; m 62; c 2. BIOCHEMISTRY. Educ: Univ Md, BS, 56; Duke Univ, PhD(biochem), 61. Prof Exp: Am Cancer Soc fel, Dept Biochem, Sheffield Sci Sch, 61-63; from instr to asst prof, 63-70, ASSOC PROF BIOL CHEM, UNIV MICH, ANN ARBOR, 70-; RES BIOCHEMIST, GEN MED RES, VET ADMIN HOSP, ANN ARBOR, 63- Mem: AAAS; Am Chem Soc; Am Soc Biol Chemists. Res: Mechanism of action and structure of flavoproteins; roles of various amino acid residues in catalysis by flavoproteins and in their structures. Mailing Add: Gen Med Res Vet Admin Hosp 2215 Fuller Rd Ann Arbor MI 48105

WILLIAMS, CHARLES HERBERT, b Aurora, Mo, Jan 21, 35; m 56; c 3. BIOCHEMISTRY. Educ: Univ Mo-Columbia, BS, 57, MS, 67, PhD(agr chem), 68. Prof Exp: Asst prof biochem, Dept Anesthesiol, Univ Wis-Madison, 70-73 & Inst

Enzyme Res, 70-72; ASSOC PROF BIOCHEM & MED, UNIV MO-COLUMBIA, 73- Concurrent Pos: Trainee, Inst Enzyme Res, Univ Wis-Madison, 68-70; Fulbright lectr, Peru, 71. Res: Malignant hyperpyrexia and surfactant system of the lung; biochemistry of malignant hyperthermia; role of glycoproteins in the lung. Mailing Add: 412 Manor Columbia MO 65201

WILLIAMS, CHARLES MELVILLE, b Regina, Sask, Mar 18, 25; m 53; c 3. PHYSIOLOGY, GENETICS. Educ: Univ BC, BSA, 49, MSA, 52, Ore State Col, PhD(genetics), 55. Prof Exp: Assoc prof animal physiol, 55-67, HEAD DEPT ANIMAL SCI, UNIV SASK, 67- Mem: Am Soc Animal Sci; Can Soc Animal Prod; fel Agr Inst Can. Res: Effect of low environmental temperatures on farm animals. Mailing Add: Dept of Animal Sci Univ of Sask Saskatoon SK Can

WILLIAMS, CHARLES WILEY, b Augusta, Ky, Mar 24, 12; m 45; c 5. MATHEMATICS. Educ: Harvard Univ, AB, 33; Univ Md, MA, 36; Univ Va, PhD(math), 49. Prof Exp: Instr math, Washington & Lee Univ, 37-39; instr, Armstrong Jr Col, 39-42; instr US Navy Pre-Flight Sch, 42-43; lectr, Univ NC, 43-45; instr, NC State Col, 45-46; from asst prof to assoc prof, 48-64, PROF MATH, WASHINGTON & LEE UNIV, 64- Mem: Math Asn Am; Am Math Soc. Res: Computer programming. Mailing Add: Dept of Math Washington & Lee Univ Lexington VA 24450

WILLIAMS, CHRISTOPHER NOEL, b York, Eng, Dec 25, 35; Can citizen; m 60; c 4. GASTROENTEROLOGY. Educ: Royal Col Physicians & Surgeons London, MRCS & LRCP, 60; Royal Col Physicians & Surgeons Can, FRCP(C), 68. Prof Exp: Intern med, Univ Pa, 69-71; lectr, 69-72, ASST PROF MED, DALHOUSIE UNIV, 72-; ASST PHYSICIAN MED, VICTORIA GEN HOSP, 69- Concurrent Pos: MacLaughlin fel, Univ Pa, 69-70, Med Res Coun fel, 70-71; dir, Gastrointestinal Res Lab, Dalhousie Univ, 71-; consult gastroenterol, Camp Hill Hosp, Halifax, 73-; grants in aid, Med Res Coun Can, 71 & Nat Health & Welfare Can, 73. Mem: Can Med Asn; Can Asn Gastroenterol; Can Soc Clin Res; Am Gastroenterol Asn; Am Asn Study Liver Dis. Res: Detailed kinetic studies of bile acid metabolism in health and disease, particularly liver and inflammatory bowel disease; dietary and biochemical factors and prevalence of gallstones in Caucasians and Canadian Indians. Mailing Add: 5849 University Ave Rm C-D 1 Clin Res Ctr Halifax NS Can

WILLIAMS, CHRISTOPHER P S, b Medford, Ore, Oct 12, 31; m 57; c 3. PEDIATRICS, MEDICINE. Educ: Univ Ore, BA, 53, MD, 58. Prof Exp: Asst prof pediat, Sch Med, Univ Wash, 62-68; ASSOC PROF PEDIAT, CRIPPLED CHILDREN'S DIV, MED SCH, UNIV ORE, 68- Res: Medical education. Mailing Add: Pediat-Crippled Children's Div Univ of Ore Med Sch Portland OR 97201

WILLIAMS, CLARA HINTON, b Baltimore, Md, Dec 28, 14; m 38; c 3. BIOCHEMISTRY. Educ: Goucher Col, AB, 35; Univ Rochester, MS, 38, PhD(biochem), 40. Prof Exp: Lab asst chem & physiol, Goucher Col, 35-37, instr inorg chem, 44; bacteriologist, Md State Dept Health, 40-42; instr inorg chem, Adult Educ Ctr, Baltimore, 46-49; res biochemist, Vet Admin Hosp, Perry Point, Md, 51-64; res chemist, Div Toxicol & Pharmacol, US Food & Drug Admin, 64-68, chief residue toxicol br, Div Pesticides, 68-71; chief toxicol br, Pesticides Tolerances Div, Environ Protection Agency, 71-74; RETIRED. Mem: Soc Toxicol. Res: Biochemical manifestations of pesticide toxicology. Mailing Add: Box 294N Rte number 3 Harpers Ferry WV 25425

WILLIAMS, CLAYTON DREWS, b St Louis, Mo, Oct 22, 35; m 59; c 4. THEORETICAL PHYSICS, SOLID STATE PHYSICS. Educ: Rice Inst, BA, 57; Wash Univ, PhD(physics), 61. Prof Exp: Asst prof, 61-64, ASSOC PROF PHYSICS, VA POLYTECH INST & STATE UNIV, 64- Mem: Am Phys Soc; Am Asn Physics Teachers. Res: Many-body problem; solid-state theory. Mailing Add: Dept of Physics Va Polytech Inst & State Univ Blacksburg VA 24061

WILLIAMS, CLYDE MICHAEL, b Marlow, Okla, Oct 8, 28; m 53; c 4. PHYSIOLOGY. Educ: Rice Inst, BA, 48; Baylor Univ, MD, 52; Oxford Univ, DPhil(physiol), 54. Prof Exp: Tech dir radioisotope lab, Vet Admin Hosp, Pittsburgh, Pa, 58-60; res radiol, 60-63, from asst prof to assoc prof, 63-65, PROF RADIOL & CHMN DEPT, COL MED, UNIV FLA, 65- Mem: Am Physiol Soc; Soc Nuclear Med; Am Roentgen Ray Soc; Radiol Soc NAm. Res: Metabolism of aromatic amines; gas chromatography of aromatic acids; use of computer for medical diagnosis. Mailing Add: Dept of Radiol Univ of Fla Col of Med Gainesville FL 32610

WILLIAMS, COLIN JAMES, b London, Eng, May 12, 38. PHYSICAL CHEMISTRY, INORGANIC CHEMISTRY. Educ: Univ London, BSc, 60, PhD(phys chem, inorg chem), 60 & Imp Col, dipl, 63. Prof Exp: Res chemist, Cent Res Div, Mobil Oil Corp, NJ, 63-66; RES CHEMIST, CAHN INSTRUMENT CO, 67- Mem: Am Chem Soc; assoc mem The Chem Soc. Res: Physical and inorganic chemical properties of zeolites and their spectroscopic and catalytic properties; thermoanalysis; diffuse reflectance spectroscopy; physical adsorption. Mailing Add: Cahn Instrument Co 16207 S Carmenita Rd Cerritos CA 90701

WILLIAMS, CURTIS, b Tupelo, Okla, Jan 17, 37; m 59; c 3. AGRONOMY, PLANT BREEDING. Educ: Okla State Univ, BS, 59, MS, 60; Univ Ark, Fayetteville, PhD(agron), 70. Prof Exp: Asst prof, 69-73, ASSOC PROF AGRON, LA STATE UNIV, BATON ROUGE, 73- Mem: Am Soc Agron; Crop Sci Soc Am; Am Genetic Asn; Am Inst Biol Sci. Res: Soybean breeding for higher yield, quality seed, nematode resistance; inheritance studies of soybeans. Mailing Add: 220 Parker Agr Ctr La State Univ Baton Rouge LA 70803

WILLIAMS, CURTIS ALVIN, JR, b Moorestown, NJ, June 26, 27; m 60; c 3. BIOCHEMISTRY. Educ: Pa State Univ, BS, 50; PhD(zool), 54. Prof Exp: USPHS fel, Pasteur Inst, Paris, 53-54; Carlsberg Lab, Copenhagen, 54-55; res assoc microbiol, Rockefeller Inst, 55-57; Nat Inst Allergy & Infectious Dis, 57-60; from asst prof to assoc prof biochem genetics, Rockefeller Univ, 60-69; PROF BIOL & DEAN NATURAL SCI, STATE UNIV NY COL PURCHASE, 69- Concurrent Pos: Adj prof, Rockefeller Univ, 70- Mem: Am Soc Microbiol; Soc Neurosci; Am Asn Immunol. Res: Immunology; immunochemistry; protein chemistry; genetics; microbiology; neuroscience; biology. Mailing Add: Off of the Dean of Natural Sci State Univ of NY Col Purchase NY 10577

WILLIAMS, DALE GORDON, b Chicago, Ill, Aug 9, 29; m 52. PHYSICAL CHEMISTRY. Educ: Beloit Col, BS, 51; Univ Minn, MS, 54; Univ Iowa, PhD, 57. Prof Exp: Engr, Linde Co Div, Union Carbide Corp, 56-58; res aide, Inst Paper Chem, 59-62, res assoc & assoc prof chem, 62-75; RES SCIENTIST, UNION CAMP CORP, 75- Mem: Am Chem Soc; Tech Asn Pulp & Paper Indust; Sigma Xi. Res: The application of physical chemistry to paper science and technology. Mailing Add: Union Camp Corp PO Box 412 Princeton NJ 08540

WILLIAMS, DANIEL CHARLES, b Compton, Calif, Dec 15, 44; m 64; c 2. CELL BIOLOGY, ELECTRON MICROSCOPY. Educ: Calif State Univ, Long Beach, BS, 67; Iowa State Univ, PhD(cell biol), 72. Prof Exp: Asst microbiologist, Purex Corp, 66-67; res engr microbiol, NAm Aviation, Inc, 67-72; instr cell & develop biol, Kans State Univ, 7274; ASST PROF DEVELOP BIOL, UNIV NOTRE DAME, 74- Concurrent Pos: NIH fel, 67-71 & 72-74; consult, BioInfo Assocs, 72-74. Mem: Sigma Xi; Am Soc Cell Biol; Electron Micros Soc Am. Res: Structure-function relationships and control mechanisms associated with cellular and developmental processes especially processes associated with biological mineralization. Mailing Add: Dept of Biol Univ of Notre Dame Notre Dame IN 46556

WILLIAMS, DANIEL FRANK, b Redmond, Ore, Nov 20, 42; c 2. MAMMALOGY. Educ: Cent Wash State Col, BA, 66; Univ NMex, MS, 68, PhD(zool), 71. Prof Exp: Asst prof, 71-75, ASSOC PROF BIOL SCI, CALIF STATE COL, STANISLAUS, 75- Mem: Am Soc Mammal; Soc Study Evolution; Soc Syst Zool. Res: Systematics and evolution of mammals; ecology of mammals; evolution of chromosome morphology in mammals. Mailing Add: Dept of Biol Sci 800 Monte Vista Calif State Col Stanislaus Turlock CA 95380

WILLIAMS, DARRELL DEAN, b Kansas City, Mo, Mar 23, 35; m 64; c 2. PHYSIOLOGY. Educ: Univ Mo, BA, 60, MA, 62, PhD(physiol), 65. Prof Exp: Res physiologist, US Air Force Arctic Aeromed Space Lab, 65-67; lectr physiol, 68-71, asst prof med sci, 71-74; Dept Health Educ & Welfare res physiologist, Arctic Health Res Ctr, 67-74, ASSOC PROF MED SCI, UNIV ALASKA, FAIRBANKS, 74- Concurrent Pos: Vis scientist, Dept Physiol & Biophys & Regional Primate Res Ctr, Univ Wash, Seattle. Mem: Am Soc Zool; Can Physiol Soc; Int Soc Biometeorol. Res: Environmental physiology, primarily cold adaptation, hibernation, lipid metabolism and temperature regulation. Mailing Add: Dept of Med Sci Univ of Alaska Fairbanks AK 99701

WILLIAMS, DAVID ALLEN, b Wakefield, Mass, Dec 22, 38; m 62; c 4. MEDICINAL CHEMISTRY. Educ: Mass Col Pharm, BS, 60, MS, 62; Univ Minn, Minneapolis, PhD(med chem), 68. Prof Exp: Sr scientist, Med Chem Div, Mallinckrodt Chem Works, 67-69; ASST PROF BIOCHEM, MASS COL PHARM, 69- Mem: AAAS; Am Chem Soc; Am Pharmaceut Asn. Res: Stereochemistry of drug action; application of structure activity relationships to derivatives of biogenic amines; analgesics. Mailing Add: 179 Longwood Ave Boston MA 02115

WILLIAMS, DAVID C, b Portland, Ore, Nov 15, 22; m; c 2. PHYSICS. Educ: Reed Col, BA, 45. Prof Exp: Physicist, US Naval Ord Test Sta, 45-46; physicist, Sandia Corp, 47-56, supvr appl sci div, 56-61; staff physicist, Kaman Aircraft, Colo, 61-69, sr res scientist, Kaman Nuclear shock & Structure Lab, 69-74; MEM STAFF, SANDIA LAB, KIRKLAND AFB, 74- Mem: AAAS; Am Phys Soc. Res: New weapons systems; properties of explosives; stress wave propagation in solids. Mailing Add: Sandia Lab Kirkland AFB Box 5800 Code 5222 Albuquerque NM 87115

WILLIAMS, DAVID CARY, b Santa Monica, Calif, June 22, 35; m 62. NUCLEAR CHEMISTRY, NUCLEAR PHYSICS. Educ: Harvard univ, AB, 57; Mass Inst Technol, PhD(nuclear chem), 62. Prof Exp: Fel nuclear chem, Princeton Univ, 62-64; mem staff, Los Alamos Sci Lab, 64-66; MEM TECH STAFF, SANDIA LABS, 66- Mem: Am Nuclear Soc; AAAS; Am Phys Soc; Am Chem Soc. Res: Nuclear reactions, nuclear decay schemes and nuclear reaction spectroscopy; fast reactor safety research; statistical models of nuclear reactions; atmospheric tracer studies. Mailing Add: Div 5315 Sandia Labs Albuquerque NM 87115

WILLIAMS, DAVID DOUGLAS F, b Bushey, Eng, May 13, 30; US citizen; m 52; c 6. HORTICULTURE. Educ: Univ Reading, BSc & dipl hort, 52; Univ Wis, Madison, PhD(bot & hort), 62. Prof Exp: Technician hort, Cent Exp Farm, Ottawa, Can, 52-53; res off fruit breeding, Exp Sta, Morden, Man, 53-56; proj asst hort, Univ Wis, Madison, 56-61; asst horticulturalist & supt hort, Maui Br Sta, Univ Hawaii, 62-67; plant breeder, Pineapple Res Inst Hawaii, 68-71, dir, 71-73; RES DIR & QUAL CONTROL MGR, MAUI LAND & PINEAPPLE CO, 73- Concurrent Pos: Lectr, Maunaolu Col Maui, 67. Mem: Am Soc Hort Sci. Res: Effect of cultural, handling and cannery practices on the yield and quality of pineapple products. Mailing Add: RR1 Box 517 Makawao HI 96768

WILLIAMS, DAVID EDWARD, b Ottawa, Can, Feb 24, 25; m 47; c 3. FORESTRY. Educ: Univ Toronto, BScF, 51. Prof Exp: DIR, FOREST FIRE RES INST, ENVIRON CAN, 76- Concurrent Pos: Chmn, Forestry Comn, Food & Agr Org, UN, 74-75. Mem: Can Inst Forestry. Res: Forest fire problems in Canada. Mailing Add: Environ Can Ottawa ON Can

WILLIAMS, DAVID EMERTON, b Victor, Idaho, Apr 12, 15; m 42; c 4. SOIL CHEMISTRY. Educ: Univ Utah, BS, 39, MS, 41; Univ Minn, PhD(soil chem), 46. Prof Exp: Asst chem, Univ Utah, 39-41; asst soil chem, Univ Minn, 41-44; from asst soil chemist to assoc soil chemist, 46-63, SOIL CHEMIST, UNIV CALIF, BERKELEY, 63- Mem: Soil Sci Soc Am; Sigma Xi. Res: Soil as a treatment system for municipal organic wastes; study of soil organic matter; fixed potassium in soils and method by which plants extract same; manganese, boron and silicon interactions in plant nutrition. Mailing Add: Dept of Soils & Plant Nutrit Univ of Calif 108 Hilgard Hall Berkeley CA 94720

WILLIAMS, DAVID FRANCIS, b New Orleans, Lam Sept 4, 38; m 64; c 2. MEDICAL ENTOMOLOGY. Educ: Univ Southwest La, BS, 64, MS, 67; Univ Fla, PhD(entom), 69. Prof Exp: Asst prof biol, Greensboro Col, 69-71; res entomologist, WFla Arthropod Res Lab, State Fla, 71-74; LOCATION & RES LEADER, FED EXP STA, AGR RES SERV, USDA, ST CROIX, 74- Mem: Entom Soc Am; Am Mosquito Control Asn; Am Registry Prof Entomologists. Res: Bioloy, population dynamics and control of biting flies affecting man and animals; use of the sterile male techniques to control Stomoxys calcitrans on an island involving mass rearing, irradiating, releasing and monitoring results. Mailing Add: USDA Agr Res Serv Fed Sta PO Box H Kingshill St Croix VI 00850

WILLIAMS, DAVID JAMES, b Syracuse, NY, Feb 20, 43; m 65; c 3. PHYSICAL CHEMISTRY. Educ: Le Moyne Col, NY, BS, 64; Univ Rochester, PhD(phys chem), 68. Prof Exp: Scientist, 68-75, MGR PHYS CHEM, CORP RES LABS, XEROX CORP, 75- Mem: Am Chem Soc; Am Phys Soc. Res: Mechanistics of photogeneration and transport of electronic charge in organic and polymeric materials; pulsed nuclear magnetic resonance; electron spin resonance; electrical measurements; optical spectroscopy. Mailing Add: Corp Res Labs Xerox Corp Rochester NY 14603

WILLIAMS, DAVID JOHN, III b Cordele, Ga, Sept 22, 27; m 48; c 3. THERIOGENOLOGY. Educ: Univ Ga, DVM, 53, BSA, 61; Auburn Univ, MS, 63; Royal Vet Col, Sweden, FRVC, 65; Am Col Theriogenologists, dipl, 71. Prof Exp: Pvt pract vet med, 53-60; instr vet med & surg, Sch Vet Med, Univ Ga, 61; from instr to assoc prof, Auburn Univ, 61-66; assoc prof, 66-73, PROF VET MED & SURG, COL VET MED, UNIV GA, 73- Concurrent Pos: Am Vet Med Asn fel, 64-65; Auburn Univ grant, 65-66; Animal Dis res grants, 67-70, 71-72. Mem: Am Vet Med Asn; Vet Soc Study Breeding Soundness; Am Col Theriogenologists. Res: Problems of behavior; bovine and equine reproduction; fat necrosis of bovine as influenced by ecological system. Mailing Add: Dept of Med & Surg Univ of Ga Col of Vet Med Athens GA 30602

WILLIAMS, DAVID LEE, b Oakland, Calif, Oct 11, 39; div; c 1. GEOPHYSICS, OCEANOGRAPHY. Educ: Univ Tex, BA, 62; Mass Inst Technol & Woods Hole Oceanog Inst, PhD(marine geophys), 74. Prof Exp: Fel, Woods Hole Oceanog Inst, 74; GEOPHYSICIST, US GEOL SURV, 74- Mem: Sigma Xi; Am Geophys Union. Res: Terrestial heat flow; thermal evolution of the Earth; geothermal energy. Mailing Add: Br of Regional Geophys Stop 964 Box 25046 Fed Ctr Denver CO 80225

WILLIAMS, DAVID LLEWELYN, b Hawarden, Wales, Apr 25, 37; m 62; c 1. METAL PHYSICS. Educ: Univ Col NWales, BSc, 57; Cambridge Univ, PhD(superconductivity), 60. Prof Exp: Nat Res Coun Can fel, 60-62, from instr to assoc prof, 62-71, PROF PHYSICS, UNIV B C, 71- Concurrent Pos: Nat Res Coun sr fel, Copenhagen Univ & Bristol Univ, 69-70. Mem: Am Phys Soc; Can Asn Physicists. Res: Nuclear magnetic resonance; positron annihilation in metal single crystals. Mailing Add: Dept of Physics Univ of B C Vancouver BC Can

WILLIAMS, DAVID LLOYD, b Springfield, Mass, Aug 15, 35; m 64. DENTAL RESEARCH. Educ: Trinity Col, BS, 57; Northwestern Univ, PhD(anal chem), 62. Prof Exp: Sr res chemist, Monsanto Res Corp, Monsanto Co, 61-67, New Enterprise Div, 67-69; sr res chemist, Am Hosp Supply Corp, 69-70; res assoc dent, Tufts Univ, 70; PROG MGR BIOMED, ABCOR INC, 70- Concurrent Pos: Consult, Joslin Diabetes Found, 74- Mem: Electrochem Soc; Am Chem Soc; Int Asn Dent Res; AAAS. Res: Dental caries (enzymes and fluoride delivery), and caries measurement instrumentation; delivery of progesterone for contraception and carcinogens for research purposes. Mailing Add: ABCOR Inc 850 S Main St Wilmington MA 01887

WILLIAMS, DAVID NOEL, b Lewisburg, Tenn, Oct 10, 34; m 56; c 2. MATHEMATICAL PHYSICS. Educ: Maryville Col, BA, 56; Univ Calif, Berkeley, PhD(theoret physics), 64. Prof Exp: Engr, Lockheed Missile Systs Div, Calif, 56-58; fel, Swiss Fed Inst Technol, 63-65, Nuclear Res Ctr, Saclay, France, 65-66 & Inst Adv Study, Princeton Univ, 66-67; asst prof theoret physics, 67-74, ASSOC PROF PHYSICS, UNIV MICH, ANN ARBOR, 74- Mem: AAAS; Am Phys Soc. Res: Holomorphic, Lorentz covariant functions; analytic S matrix theory; analytic parametrization of higher spin scattering amplitudes; applications of functional analysis in the scattering theory and quantum field theory of elementary particles. Mailing Add: Dept of Physics & Astron Univ of Mich Ann Arbor MI 48104

WILLIAMS, DAVID TREVOR, b Slough, Eng, Oct 4, 40; m. ENVIRONMENTAL CHEMISTRY. Educ: Univ Bristol, BSc, 61; Queen's Univ, MSc, 63, PhD(chem), 66. Prof Exp: Res scientist food chem, Foods Directorate, 69-75, RES SCIENTIST ENVIRON CHEM, ENVIRON HEALTH DIRECTORATE, HEALTH & WELFARE, CAN, 75- Mem: The Chem Soc; Chem Inst Can. Res: Identification of organic contaminants in air and drinking water; effects of water treatment procedures on the organic contaminants of drinking water. Mailing Add: Environ Health Directorate Tunneys Pasture Ottawa ON Can

WILLIAMS, DAVID TYNDALE, b Manila, Philippines, Feb 24, 07; m; c 3. APPLIED PHYSICS. Educ: Columbia Univ, AB, 30, AM, 33; NY Univ, PhD(physics), 38. Prof Exp: Instr physics, Carleton Univ, 39; physicist, Nat Adv Comt Aeronaut, 39-45; res engr, Battelle Mem Inst, 45, supvr & dept consult, 47-57; assoc prof aeronaut eng, Univ Mich, 46-47; PROF AEROSPACE ENG & PHYSICS, UNIV FLA, 57- Concurrent Pos: Consult, Smithsonian Astrophys Observ, 57-58, Battelle Mem Inst, 57-, Rand Corp, 59 & Eglin AFB,71- Mem: Am Inst Aeronaut & Astronaut; Optical Soc Am. Res: Aeronautical engineering; aerothermodynamics; nuclear engineering; medical physics; dielectrics; photoconductivity; electromagnetic wave propagation; fluid physics; meteoritics; combustion; spectroscopy; acoustics; microwave radiometry. Mailing Add: Dept of Eng Sci Univ of Fla Gainesville FL 32601

WILLIAMS, DEAN E, b Iowa City, Iowa, Feb 18, 24; m 44; c 3. SPEECH PATHOLOGY, AUDIOLOGY. Educ: Univ Iowa, BA, 47, MA, 49, PhD(speech path), 52. Prof Exp: From asst prof to assoc prof speech path, Fla State Univ, 49-53; res assoc, Univ Iowa, 51-52; asst prof, Ind Univ, 53-58; PROF SPEECH PATH, UNIV IOWA, 58- Concurrent Pos: Mem adv panel speech path & audiol, US Voc Rehab Admin, 60-62; consult rev panel speech & hearing, Neurol & Sensory Dis Serv Proj, 64-67 & perinatal res, Nat Inst Neurol Dis & Blindness; mem bd dirs, Am Bd Exam Speech Path & Audiol. Mem: Fel Am Speech & Hearing Asn; Int Soc Gen Semantics. Res: Stuttering, especially onset and development of the problem and conditions under which stuttering frequency increases or decreases. Mailing Add: Johnson Speech & Hearing Ctr Woolf Ave Iowa City IA 52240

WILLIAMS, DENIS R, b London, Eng, Feb 4, 41; m 64; c 2. THEORETICAL CHEMISTRY, ACADEMIC ADMINISTRATION. Educ: Sheffield Sci Sch, BSc, 62, PhD(chem), 65. Prof Exp: Res assoc chem, Vanderbilt Univ, 65-67; asst prof, 67-72, ASSOC PROF CHEM, ASSOC DEAN GRAD SCH & COORDR RES & SERV, UNIV COLO, DENVER, 72- Concurrent Pos: Univ Colo educ experimentation grant, 67-70; NSF res grant, 69-71 & 71-73. Honors & Awards: Pub Understanding of Sci-Mus Exhibit, NSF, 74. Res: Substituent effects in aromatic hydrocarbons; high resolution ultraviolet spectroscopy of aromatic hydrocarbons; lattice energy of crystalline hydrocarbons; D-orbital participation in cyclic organic sulfides. Mailing Add: Grad Sch Univ of Colo 1100 14th St Denver CO 80202

WILLIAMS, DIGBY FREDERICK, b London, Eng, Oct 20, 35; m 63. PHYSICAI CHEMISTRY, SOLID STATE CHEMISTRY. Educ: Univ Nottingham, BSc, 57, Univ Man, MSc, 61, PhD(phys chem), 63. Prof Exp: Asst tech off, Imp Chem Industs Ltd, Eng, 57-60; res asst high temperature liquids NATO fel, Imp Col, Univ London, 63-64; fel, 64-66, ASST RES OFF SOLID STATE CHEM, CHEM DIV, NAT RES COUN CAN, 66- Res: Optical and electrical properties of solid organic state. Mailing Add: 1133c Chem Div Nat Res Coun of Can Sussex Dr Ottawa ON Can

WILLIAMS, DONALD BENJAMIN, b New York, NY, Aug 8, 33; m 56; c 2. BIOLOGY. Educ: Maryville Col, BA, 55; Emory Univ, MS, 57, PhD(biol), 59. Prof Exp: From asst prof to assoc prof biol, Maryville Col, 58-61; asst prof zool, 61-67, ASSOC PROF BIOL, VASSAR COL, 67- Mem: Soc Protozool; NY Acad Sci. Res: Ecology, physiology, cytology and genetics of ciliated protozoa; effects of radiation and antimetabolites on various protozoan life stages. Mailing Add: Dept of Biol Vassar Col Poughkeepsie NY 12601

WILLIAMS, DONALD ELMER, b Kansas City, Mo, Mar 7, 30. X-RAY CRYSTALLOGRAPHY, PHYSICAL CHEMISTRY. Educ: William Jewell Col, BA, 50; Iowa State Univ, PhD(chem), 64. Prof Exp: Res asst chem, Iowa State Univ, 57-62, from asst chemist to assoc chemist, 62-67; assoc prof, 67-71, PROF CHEM, UNIV LOUISVILLE, 71- Mem: AAAS; Am Chem Soc; Am Crystallog Asn; Am Phys Soc. Res: Molecular packing analysis; intermolecular forces; stable free radicals. Mailing Add: Dept of Chem Univ of Louisville Louisville KY 40208

WILLIAMS, DONALD ERROL, b Detroit, Mich, May 12, 28; m 53; c 3. PHYSICAL CHEMISTRY. Educ: Colgate Univ, BA, 51; Princeton Univ, PhD(phys chem), 57. Prof Exp: Asst chem, Princeton Univ, 51-54; sr chemist, Merck Sharp & Dohme Res Labs, 54-65; head phys & anal chem sect, Strasenburgh Labs Div, Wallace & Tiernan, Inc, 65-68; mgr chem develop, Xerox MDO, 68-70; asst dir qual control, Ayerst Labs Div, Am Home Prod Corp, 70-71, dir, 71-74; DIR QUAL CONTROL, BRISTOL LABS DIV, BRISTOL-MYERS CO, 74- Mem: Am Chem Soc. Res: Physical and analytical chemistry of pharmaceuticals; laboratory automation. Mailing Add: Bristol Labs Div Bristol-Myers Co Syracuse NY 13201

WILLIAMS, DONALD HOWARD, b Ellwood City, Pa, Mar 9, 38; m 60; c 2. INORGANIC CHEMISTRY. Educ: Muskingum Col, BS, 60; Ohio State Univ, PhD(inorg chem), 64. Prof Exp: Asst prof chem, Univ Ky, 64-69; PROF CHEM, HOPE COL, 69- DIR INST ENVIRON QUAL, 70-, DIR SUMMER SCH, 72- Concurrent Pos: Grants, Res Corp, NY, 65-; water resources inst, US Dept Interior, 66-68; Petrol Res Found & W K Kellogg Found. Mem: Am Chem Soc; Inst Environ Sci. Res: Stereochemistry of transition metal complexes; environmental chemistry. Mailing Add: Dept of Chem Hope Col Holland MI 49423

WILLIAMS, DONALD J, b Fitchburg, Mass, Dec 25, 33; m 53; c 3. SPACE PHYSICS, NUCLEAR PHYSICS. Educ: Yale Univ, BS, 55, MS, 58, PhD(nuclear physics), 62. Prof Exp: Sr staff physicist, Appl Physics Lab, Johns Hopkins Univ, 61-65; sect head auroral & trapped radiation, Goddard Space Flight Ctr, NASA, 65-68, head particle physics sect, 68-69, head particle physics br, 69-70; DIR, SPACE ENVIRON LAB, ENVIRON RES LABS, NAT OCEANIC & ATMOSPHERIC ADMIN, 70- Concurrent Pos: Lectr, Univ Colo, Boulder. Honors & Awards: Leigh Page Mem Prize, Yale Univ, 58. Mem: Am Geophys Union; Am Phys Soc. Res: Nuclear scattering and excitations; earth's trapped particle population and magnetic field configuration; solar flares and cosmic rays; interplanetary physics; interaction of interplanetary medium with Earth's environment; space plasma instabilities. Mailing Add: Space Environ Lab NOAA Environ Res Labs R43 Boulder CO 80302

WILLIAMS, DUDLEY, b Covington, Ga, Apr 12, 12; m 37; c 2. MOLECULAR SPECTROSCOPY, PLANETARY ATMOSPHERES. Educ: Univ NC, AB, 33, MA, 34, PhD(physics), 36. Prof Exp: Instr physics, Univ Fla, 36-38, asst prof phys sci, 38-41; staff mem, Radiation Lab, Mass Inst Technol, 41-43; asst prof physics, Univ Okla, 43-44; staff mem, Los Alamos Sci Lab, Calif, 44-46; from assoc prof to prof physics, Ohio State Univ, 46-63, actg chmn dept, 52-53 & 58-59; prof & head dept, NC State Univ, 63-64; REGENT'S DISTINGUISHED PROF PHYSICS, KANS STATE UNIV, 64- Concurrent Pos: Guggenheim fel, Univ Amsterdam & Oxford Univ, 56; NSF sr fel, Univ Liege, 61-62. Mem: Fel Am Phys Soc; fel Optical Soc Am; Am Asn Physics Teachers. Res: Infrared spectroscopy; microwave transmission; mass spectroscopy; nuclear and atmospheric physics; planetary atmospheres; determination of nuclear magnet moments. Mailing Add: Dept of Physics Kans State Univ Manhattan KS 66506

WILLIAMS, EARL R, b Alix, Ark, Apr 28, 15; m 37; c 2. PHYSICS, MATHEMATICS. Educ: Col of Ozarks, BS, 38; Univ Ark, MS, 50, EdD, 71. Prof Exp: Instr math & physics & chmn dept, Col of Ozarks, 45-57; instr, Ark Polytech Col, 57-60; from instr to assoc prof, 60-71, PROF MATH & PHYSICS, JOHN BROWN UNIV, 71-, CHMN DEPT, 60- Mem: Am Asn Physics Teachers. Res: Gamma ray spectroscopy; higher education in science. Mailing Add: Dept of Physics & Math John Brown Univ Siloam Springs AR 72761

WILLIAMS, EBENEZER DAVID, JR, b Nanticoke, Pa, June 30, 27; m 54; c 1. TEXTILE CHEMISTRY. Educ: Swarthmore Col, AB, 47; Univ Pa, MA, 49, PhD(org chem), 52. Prof Exp: Asst instr chem, Univ Pa, 47-52; from res chemist to sr res chemist, 52-63, tech supvr, 63-66, res assoc, 66-74, DEVELOP FEL, TEXTILE RES LAB, TEXTILE FIBERS DEPT, E I DU PONT DE NEMOURS & CO, INC, 74- Mem: Am Chem Soc; Sigma Xi. Res: Polymer chemistry; synthetic textiles; mechanism of dyeing; dyeing technology of synthetic fibers. Mailing Add: Textile Res Lab E I du Pont de Nemours & Co Inc Wilmington DE 19898

WILLIAMS, EDDIE ROBERT, b Chicago, Ill, Jan 6, 45; m; c 2. MATHEMATICS. Educ: Ottawa Univ, BA, 66; Columbia Univ, PhD(math), 71. Prof Exp: Instr math, Intensive Summer Studies Prog, Columbia Univ, 70; ASST PROF MATH, NORTHERN ILL UNIV, 70- Mem: Am Math Soc. Res: Pure mathematics; several complex variable theory; mathematics education; mathematics for the disadvantaged student. Mailing Add: Dept of Math Northern Ill Univ DeKalb IL 60115

WILLIAMS, EDWARD FOSTER, JR, b Carroll Co, Ga, Jan 27, 06; m 30; c 2. BIOCHEMISTRY. Educ: Auburn Univ, BS, 25, MS, 29; Univ Tenn, PhD(biochem), 32. Prof Exp: Asst res chemist, Exp Sta, Auburn Univ, 25-27; instr chem, Univ Tenn, 27-29; fel biochem, NY Post-Grad Med Sch, Columbia Univ, 32-33; res biochemist, Killian Res Labs, 33-35; from instr to assoc prof chem, 35-62, prof biochem, 62-72, EMER PROF BIOCHEM, UNIV TENN, MEMPHIS, 72- Concurrent Pos: Fulbright lectr, Univ Shiraz, 60-61. Mem: Am Chem Soc; fel Am Asn Clin Chem. Res: Acid-base balance; ammonia metabolism; hemoglobin and its derivatives; gastric digestion; radioactive isotope tracer studies of metabolism; calcium and lead metabolism. Mailing Add: Dept of Biochem Univ of Tenn Memphis TN 38163

WILLIAMS, EDWARD JAMES, physical biochemistry, see 12th edition

WILLIAMS, EDWIN BRUCE, b Ladoga, Ind, Nov 3, 18; m 40; c 1. PHYTOPATHOLOGY. Educ: Wabash Col, AB, 50; Purdue Univ, MS, 52, PhD(plant path), 54. Prof Exp: From asst prof to assoc prof, 54-69, PROF PLANT PATH, PURDUE UNIV, LAFAYETTE, 69-, PLANT PATHOLOGIST, 54- Mem: Am Phytopath Soc; Am Pomol Soc. Res: Genetics of Venturia inaequalis; breeding apples for disease resistance. Mailing Add: Dept of Bot & Plant Path Purdue Univ West Lafayette IN 47906

WILLIAMS, ELEANOR RUTH, b Ropesville, Tex, Apr 23, 24. NUTRITION, FOODS. Educ: Tex Woman's Univ, BS, 45; Iowa State Univ, MS, 47; Cornell Univ, PhD(nutrit), 63. Prof Exp: Res assoc nutrit, Tex Agr Exp Sta, 48-49; instr food & nutrit, Southern Methodist Univ, 49-51; from instr to asst prof, Cornell Univ, 51-59; assoc prof, Univ Nebr, 63-65; assoc prof nutrit, Teachers Col, Columbia Univ, 65-72; assoc prof, State Univ NY Col Buffalo, 72-74; ASSOC PROF NUTRIT, UNIV MD, 74- Mem: AAAS; Am Pub Health Asn; Am Dietetic Asn; Am Home Econ Asn. Res: B vitamins and reproduction in the rat; non-specific nitrogen intake; adequacy of cereal protein for man; nutrition education. Mailing Add: Dept of Food Nutrit & Inst Admin Univ of Md College Park MD 20742

WILLIAMS, ELIOT CHURCHILL, b Chicago, Ill, Nov 9, 13; m 45; c 4. INVERTEBRATE ZOOLOGY, ANIMAL ECOLOGY. Educ: Cent YMCA Col, BA, 35; Northwestern Univ, PhD(zool), 40. Prof Exp: Instr zool, Cent YMCA Col, 35-36; asst dir, Chicago Acad Sci, 40-47; asst prof biol, Roosevelt Univ, 47-48; assoc prof, 48-57, PROF ZOOL, WABASH COL, 57-, CHMN DIV SCI, 76- Concurrent Pos: Lectr, Roosevelt Univ, 46-47; Ford Found fel, Johns Hopkins Marine Sta, Stanford Univ, 54-55; consult, Fed Chem Co, 54- Mem: AAAS; Am Soc Zool; Ecol Soc Am;

Am Inst Biol Sci. Res: Animal populations; cave animals; pigmentation in cave planarians; taxonomy and ecology of Symphyla; radioisotope cycling; energy relationships in ecosystems. Mailing Add: Dept of Biol Wabash Col Crawfordsville IN 47933

WILLIAMS, ELMER LEE, b Ironton, Ohio, Apr 14, 29; m 57; c 3. PHYSICAL CHEMISTRY. Educ: Ohio Univ, BS, 51, MS, 55; Ind Univ, PhD(phys chem), 59. Prof Exp: Asst chem, Ind Univ, 55-58; develop engr, Sylvania Elec Prods, Inc Div, Gen Tel & Electronics Corp, 58-60; PHYS CHEMIST, OWENS-ILL, INC, 60- Mem: Am Chem Soc; The Chem Soc; Am Ceramic Soc; Electrochem Soc. Res: Diffusion of ions and atoms in glass and molten silicates; oxygen of mass 18 work and tracer work in the solid state; semiconductors; gas lasers; gas discharge displays. Mailing Add: Tech Ctr Owens-Ill Inc 1700 N Westwood Toledo OH 43666

WILLIAMS, EMMETT LEWIS, JR, b Lynchburg, Va, June 6, 33; m 57; c 3. SOLID STATE PHYSICS. Educ: Va Polytech Inst, BS, 56, MS, 62; Clemson Univ, PhD(mat eng), 66. Prof Exp: Assoc aircraft engr, Lockheed-Ga Co, 56-57; mat engr, Atomic Energy Div, Babcock & Wilcox Co, 57-59; asst prof metall eng, Va Polytech Inst, 59-64; res asst ceramic eng, Clemson Univ, 64-65; res scientist, Union Carbide Nuclear Corp, 65-66; PROF PHYSICS, BOB JONES UNIV, 66-, CHMN DEPT, 73- Concurrent Pos: Consult, Inland Motors, 61-64, Leaders Am Sci, 66, Continental Tel Labs, 70-72, Polysci Corp & Electrotech Corp. Mem: Am Asn Physics Teachers; hon res mem, Sigma Xi. Res: Solid state, surface physics; thermodynamics; formation of limestone stalactites in laboratory; thermodynamics of living organisms. Mailing Add: Box 34606 Dept of Physics Bob Jones Univ Greenville SC 29614

WILLIAMS, ERNEST EDWARD, b Easton, Pa, Jan 7, 14. HERPETOLOGY, ECOLOGY. Educ: Lafayette Col, BS, 33; Columbia Univ, PhD(zool), 49. Prof Exp: Asst zool, Columbia Univ, 40-42 & 46-48; from instr to assoc prof, 49-70, PROF BIOL, HARVARD UNIV, 70-, ALEXANDER AGASSIZ PROF ZOOL, 72-, CUR REPTILES & AMPHIBIANS, MUS COMP ZOOL, 57- Concurrent Pos: Guggenheim fel, 52-53. Mem: AAAS; Am Soc Ichthyol & Herpet; Soc Syst Zool; Soc Study Evolution; fel Am Acad Arts & Sci. Res: Taxonomy, paleontology and morphology of reptiles; West Indian paleontology and zoogeography; evolution. Mailing Add: Mus of Comp Zool Harvard Univ Cambridge MA 02138

WILLIAMS, ERNEST YOUNG, b Nevis, BWI, Feb 24, 00. NEUROLOGY, PSYCHIATRY. Educ: Howard Univ, BS, 27, MD, 30. Prof Exp: Gen Educ Bd fel, Columbia Univ, 31-32 & Bellevue Hosp, New York, 32-33; from asst prof to prof, 33-70, head dept, 50-70, EMER PROF NEUROL & PSYCHIAT, COL MED, HOWARD UNIV, 70- Concurrent Pos: Chief neuropsychiat serv, Freedman's Hosp. Mem: Am Psychiat Asn; Asn Res Nerve & Ment Dis; Nat Med Asn; Soc Biol Psychiat. Res: Treatment of multiple sclerosis; management of chronic alcoholism; incidence of mental disease in the Negro; histopathology of the granular layer of the cerebellum; treatment of heroin addiction. Mailing Add: Dept of Neurol & Psychiat Howard Univ Col of Med Washington DC 20011

WILLIAMS, ERVIN RAY, organic chemistry, see 12th edition

WILLIAMS, EUGENE G, b New Haven, Conn, June 9, 25; m 52; c 2. GEOLOGY, MINERALOGY. Educ: Lehigh Univ, BA, 50; Univ Ill, MS, 52; Pa State Univ, PhD, 57. Prof Exp: Instr geol, Kent State Univ, 52-53; instr, 54-55, res assoc, 56-57, from asst prof to assoc prof, 57-71, PROF GEOL, PA STATE UNIV, 71- Mem: Geol Soc Am; Am Asn Petrol Geol. Res: Stratigraphy and petrography of upper Paleozoic rocks of eastern United States. Mailing Add: Dept of Geol 339 Deike Bldg Pa State Univ University Park PA 16802

WILLIAMS, EVAN THOMAS, b New York, NY, May 17, 36; m 59; c 2. NUCLEAR CHEMISTRY. Educ: Williams Col, BA, 58; Mass Inst Technol, PhD(chem), 63. Prof Exp: Civil engr, Res Estab Ris , Roskilde, Denmark, 63-65; asst prof, 65-71, ASSOC PROF CHEM, BROOKLYN COL, 71- Concurrent Pos: Consult, Geosci Instruments Corp, 66-72; mem exec bd, New York City Coun on the Environ 73- Mem: AAAS; Am Chem Soc; Am Phys Soc; Geochem Soc; NY Acad Sci. Res: Special methods of activation analysis; environmental applications; neutron cross sections of radioactive nuclides; fast-neutron reactions; radiocarbon dating. Mailing Add: Dept of Chem Brooklyn Col Brooklyn NY 11210

WILLIAMS, FERD ELTON, b Erie, Pa, June 9, 20; m 48; c 5. PHYSICS. Educ: Univ Pittsburgh, BS, 42; Princeton Univ, MA, 45, PhD(phys chem), 46. Prof Exp: Res chemist, Res Labs, Radio Corp Am, 42-46; asst prof chem, Univ NC, 46-48; res assoc, Res Lab, Gen Elec Co, 48-49, mgr, Light Prod Sect, 49-59, theoret physicist, Gen Physics Dept, 59-61; prof, 61-62, H FLETCHER BROWN PROF PHYSICS, 62-, CHMN DEPT, 61- Concurrent Pos: Prin investr, Army Res Off & Army Engrs res grants, 62-; mem adv comt solid state physics, Oak Ridge Nat Labs, 63-67; ed-in-chief, J Luminescence, 69-; indust & govt consult, Picatinny Arsenal. Mem: Fel AAAS; fel Optical Soc Am; fel Am Phys Soc; Am Chem Soc; Am Asn Physics Teachers. Res: Solid state physics; semiconductivity; luminescence; chemical physics. Mailing Add: 1008 Dixon Dr Christine Manor Newark DE 19711

WILLIAMS, FLOYD JAMES, b Lowell, Ohio, Oct 26, 30; m 53; c 3. PHYTOPATHOLOGY. Educ: Ohio State Univ, BSc, 55, MSc, 58, PhD(phytopath), 61. Prof Exp: Asst prof bot & phytopath, Univ Md, 61-65; res scientist, USDA, India, 65-70; RES ADV, US AID, PAKISTAN, 70- Concurrent Pos: NSF fel, 61. Honors & Awards: Meritorious Honor Award, US AID, 74. Res: Research administration and research development in developing countries. Mailing Add: Dept of State Washington DC 20520

WILLIAMS, FLOYD JAMES, b Electra, Tex, Jan 9, 20; m 55; c 2. GEOLOGY. Educ: Univ Calif, BS, 43; Colo Sch Mines, MS, 51; Columbia Univ, PhD(geol), 58. Prof Exp: Mining engr, Bradley Mining Co, Idaho, 43; mining engr, Idaho-Md Mines Corp, Calif, 46-47; explosives engr, Hercules Powder Co, 47-48; chief reconnaissance sect, Salt Lake eExplor Br, Div Raw Mat, USAEC, 52-54; geologist, Standard Oil Co, Calif, 56-58; assoc prof geol, Univ Redlands, 58-66; supvr spectrog, Kaiser Steel Corp, Calif, 66-72; HEAD DEPT GEOL, SAN BERNADINO VALLEY COL, 72- Concurrent Pos: NSF fel & res assoc geol & geophys, Univ Calif, Berkeley, 63-64; consult geol, City of San Bernardino, Calif, 74- Mem: Geol Soc Am; AAAS. Res: Criteria for active faults, earthquake prediction. Mailing Add: 130 Sunridge Way Redlands CA 92373

WILLIAMS, FRANCIS, b Whitley Bay, Eng, May 14, 27; m 57. FISHERIES, BIOLOGICAL OCEANOGRAPHY. Educ: Univ Durham, BSc, 51, MSc, 55; Univ Newcastle, Eng, DSc(marine fisheries), 64. Prof Exp: Prin sci officer, Pelagic Fish & Fisheries, E African Marine Fisheries Res Orgn, Zanzibar, 51-62; dir Guinean Trawling Surv, Orgn African Unity, Lagos, Nigeria, 62-66; mgr pelagic fisheries surv W Africa, dept fisheries, Food & Agr Orgn UN, Rome, Italy, 66-68; assoc res biologist, Scripps Inst Oceanog, Univ Calif, San Diego, 68-73; chmn div fisheries & appl estuarine ecol, 73-74, PROF MARINE SCI, ROSENSTIEL SCH MARINE & ATMOSPHERIC SCI, UNIV MIAMI, 73-, CHMN DIV BIOL & LIVING RESOURCES, 74- Mem: Sci fel Zool Soc London; Marine Biol Asn UK; Challenger Soc; Am Fisheries Soc; fel Am Inst Fishery Res Biol. Res: Exploratory fishing surveys and fish taxonomy; correlation of fish and fisheries with environment. Mailing Add: Rosentiel Sch Marine & Atmos Sci Univ of Miami Miami FL 33149

WILLIAMS, FRANCIS MERRILL, mathematics, see 12th edition

WILLIAMS, FRANCIS TRUEMAN, b Youngstown, Ohio, Sept 24, 31; m 53. ORGANIC CHEMISTRY. Educ: Denison Univ, BS, 53; Ohio State Univ, PhD(chem), 58. Prof Exp: Asst chem, Ohio State Univ, 53-56; asst prof, 58-65, chmn dept, 66-68, ASSOC PROF CHEM, ANTIOCH COL, 65- Concurrent Pos: Consult, Dow Chem Co, Mich, 65-66; sabbatical, Univ Tübingen, 68-69. Mem: Am Chem Soc. Res: Physical organic chemistry; aliphatic nitro compounds; history and chemical technology; enzyme kinetics. Mailing Add: Dept of Chem Antioch Col Yellow Springs OH 45387

WILLIAMS, FRED DEVOE, b New York, NY, Dec 16, 36; m 57; c 3. BACTERIOLOGY, BIOCHEMISTRY. Educ: Rutgers Univ, BA, 60, MS, 62, PhD(bact), 64. Prof Exp: Instr bact, Rutgers Univ, 63-64; from asst prof to assoc prof, 64-75, PROF BACT, IOWA STATE UNIV, 75- Mem: AAAS; Am Soc Microbiol. Res: Biosynthesis of folic acid-like compounds by microorganisms and the metabolic regulation of these biosynthetic pathways; physiology and biochemistry of microbial behavior. Mailing Add: Dept of Bact Iowa State Univ Ames IA 50011

WILLIAMS, FRED EUGENE, b Wichita Falls, Tex, Oct 23, 41; m 64; c 2. PHYSIOLOGY. Educ: Arlington State Col, BS, 66; Baylor Univ, PhD(physiol), 72. Prof Exp: ASST PROF PHYSIOL, BAYLOR COL DENT, 72- Mem: Sigma Xi. Res: Effects of micropotentials upon bone formation and healing rates after fracture. Mailing Add: Dept of Physiol Baylor Col of Dent Dallas TX 75226

WILLIAMS, FREDERICK MCGEE, b Washington, DC, Jan 10, 34; m 55; c 4. BIOLOGY. Educ: Stanford Univ, AB, 55; Yale Univ, PhD(biol), 65. Prof Exp: Asst prof biol, Lehigh Univ, 63-64; from asst prof to assoc prof zool, Univ Minn, Minneapolis, 64-70; ASSOC PROF BIOL, PA STATE UNIV, UNIVERSITY PARK, 70- Concurrent Pos: NASA-Am Inst Biol Sci fel, 65-66; chmn grad prog ecol, Pa State Univ, 74- Mem: AAAS; Ecol Soc Am; Am Soc Zool; Am Inst Biol Sci. Res: Theoretical and experimental population dynamics; theory of ecosystem structure and stability; environmentally induced shape changes in algal cells; mathematical biology; mechanisms of competition and predation. Mailing Add: Dept of Biol Pa State Univ University Park PA 16802

WILLIAMS, FREDERICK WALLACE, b Cumberland, Md, Sept 24, 39; m 64; c 2. ANALYTICAL CHEMISTRY. Educ: Univ Ala, BS, 61, MSc, 63, PhD(chem), 65. Prof Exp: Nat Acad Sci-Nat Res Coun fel, 65-66, res chemist, 66-73, SUPV RES CHEMIST, US NAVAL RES LAB, DC, 73- Mem: AAAS; Am Chem Soc; Combustion Inst. Res: Fundamental mechanisms of combustion. Mailing Add: 13408 Colwyn Rd Oxon Hill MD 20022

WILLIAMS, FREDRICK DAVID, b Winnipeg, Man, Sept 1, 37; m 64; c 2. POLYMER CHEMISTRY, PHYSICAL CHEMISTRY. Educ: Univ Man, BSc, 59, MSc, 61, PhD(chem), 62. Prof Exp: Ital Govt res scholarship, Inst Indust Chem, Milan Polytech Inst, 62-63; res chemist, Allis-Chalmers Mfg Corp, 63-65; asst prof, 65-69, ASSOC PROF CHEM, MICH TECHNOL UNIV, 69- Mem: Am Chem Soc. Res: Polymer chemistry. Mailing Add: Dept of Chem Mich Technol Univ Houghton MI 49931

WILLIAMS, GARETH, b Rhos, Wales, Apr 28, 37; US citizen; m 65; c 2. APPLIED MATHEMATICS. Educ: Univ Wales, BSc, 59, PhD(math), 62. Prof Exp: Asst prof math, Univ Fla, 62-65; from asst prof to assoc prof, Univ Denver, 65-73; ASSOC PROF MATH, STETSON UNIV, 73- Mem: Am Math Soc; Tensor Soc Gt Brit. Res: Relativity; differential geometry; mathematical models; computer science; linear algebra. Mailing Add: Dept of Math Stetson Univ Deland FL 32720

WILLIAMS, GARY MURRAY, b Regina, Sask, May 7, 40; US citizen; m 66; c 2. PATHOLOGY. Educ: Washington & Jefferson Col, BA, 63; Univ Pittsburgh, MD, 67. Prof Exp: Instr path, Med Sch, Harvard Univ, 67-69; staff assoc carcinogenesis, Nat Cancer Inst, 69-71; asst prof path, Fels Res Inst & Med Sch, Temple Univ, 71-75; RES ASSOC PROF PATH, NEW YORK MED COL, 75-; CHIEF DIV EXP PATH, NAYLOR-DANA INST DIS PREV, AM HEALTH FOUND, 75- Concurrent Pos: Int Agency Res on Cancer res training fel, Wenner-Gren Inst, Stockholm, Sweden, 71-72; from intern to resident path, Mass Gen Hosp, 67-69. Honors & Awards: Sheard-Sanford Award, Am Soc Clin Path, 67. Mem: Am Asn Cancer Res; Am Soc Exp Path. Res: Chemical carcinogenesis; cell culture. Mailing Add: Naylor-Dana Inst for Dis Prev Div of Exp Path Am Health Found Valhalla NY 10595

WILLIAMS, GENE R, b Yuba City, Calif, Nov 10, 32; m 54, 69; c 2. PLANT PHYSIOLOGY, PLANT BIOCHEMISTRY. Educ: Univ Calif, BS, 57, MS, 59, PhD(plant physiol), 63. Prof Exp: Lectr bot, Univ Calif, 61-62; Am Cancer Soc fel biochem, Biol Div, Oak Ridge Nat Lab, 63-65; asst prof, 65-68, ASSOC PROF BOT, IND UNIV, BLOOMINGTON, 68- Mem: AAAS; Am Soc Plant Physiol. Res: Plant metabolism, protein and nucleic acid synthesis, amino acid activation and chloroplast development; effects of light on plant development. Mailing Add: Dept of Bot Ind Univ Bloomington IN 47401

WILLIAMS, GENEVA HYLAND, b Champaign, Ill, Dec 25, 34. DEVELOPMENTAL BIOLOGY. Educ: Univ Mo, AB, 55; Vanderbilt Univ & Peabody Col, MAT, 56; Wash Univ, PhD(embryol), 67. Prof Exp: Teacher high sch, Mo, 56-59; asst prof zool, Univ Toronto, 64-70; res assoc pharmacol, 70-74, INSTR OBSTET & GYNEC, PA STATE UNIV, 74- Concurrent Pos: Nat Res Coun Can oper res grant, 66-69. Mem: AAAS; NY Acad Sci; Am Inst Biol Sci; Tissue Cult Asn. Res: Regulation of enzyme activity in tissues of developing vertebrates; tissue interactions in post-embryonic organs. Mailing Add: Dept of Obstet & Gynec Pa State Univ Hershey Med Ctr Hershey PA 17033

WILLIAMS, GEORGE, JR, b Benton, La, June 15, 31; m 57. BOTANY. Educ: Southern Univ, BS, 57; Univ NH, MS, 59, PhD(bot), 63. Prof Exp: Asst bot, Univ NH, 57-63; PROF BIOL, SOUTHERN UNIV, BATON ROUGE, 63- Mem: Am Soc Plant Physiol. Res: Plant growth as modified by light quality and chemical factors, particularly growth hormones. Mailing Add: Dept of Biol Southern Univ Baton Rouge LA 70813

WILLIAMS, GEORGE ABIAH, b Brooklyn, NY, Apr 1, 31; div; c 3. SOLID STATE PHYSICS. Educ: Colgate Univ, BA, 52; Univ Ill, PhD, 56. Prof Exp: Res assoc physics, Stanford Univ, 56-59; mem tech staff, Bell Tel Labs, NJ, 59-63; vis asst prof physics, Cornell Univ, 63-64; assoc prof, 64-70, PROF PHYSICS, UNIV UTAH, 70- Concurrent Pos: NSF res grant, 72-74; Air Force Off Sci Res grant, 65-70. Mem: Am Phys Soc. Res: Wave propagation in solid state plasmas; plasma effects in solids;

superconductivity; solid helium-3. Mailing Add: Dept of Physics Univ of Utah Salt Lake City UT 84112

WILLIAMS, GEORGE ARTHUR, b Wilcox, Ariz, Jan 14, 18; m 43; c 3. GEOLOGY, ENGINEERING. Educ: Tex Western Col, BS, 43; Univ Ariz, PhD(geol), 51. Prof Exp: Engr, Asarco Mining Co, Inc, Mex, 46; geologist, Peru Mining Co, 46-48, supt, Kearney Mine, 48; geologist, US Geol Surv, 51-57; assoc prof, 57-65, PROF GEOL ENG & HEAD DEPT GEOL, UNIV IDAHO, 65- Concurrent Pos: Gov appointed, Control Bd, Idaho Bur Mines & Geol. Mem: Fel AAAS; fel Geol Soc Am; Soc Econ Paleont & Mineral; Soc Econ Geol; Am Asn Petrol Geol. Res: Economic geology; sedimentation; stratigraphy. Mailing Add: Dept of Geol Univ of Idaho Moscow ID 83843

WILLIAMS, GEORGE CHRISTOPHER, b Charlotte, NC, May 12, 26; m 50; c 4. ZOOLOGY. Educ: Univ Calif, Berkeley, AB, 49, MA, Los Angeles, 52, PhD, 55. Prof Exp: From instr to asst prof natural sci, Mich State Univ, 55-60; assoc prof, 60-66, PROF BIOL SCI, STATE UNIV N Y STONY BROOK, 66- Mem: AAAS; Soc Study Evolution; Am Soc Ichthyol & Herpet; Am Soc Limnol & Oceanog; Am Fisheries Soc. Res: Evolution; marine ecology; ichthyology; animal behavior; population genetics. Mailing Add: Dept of Biol Sci State Univ of NY Stony Brook NY 11790

WILLIAMS, GEORGE JACKSON, III, b Corpus Christi, Tex, July 14, 38; m 61; c 2. Educ: Tex A&I Univ, BS, 61; George Peabody Col, MA, 66; Univ Tex, Austin, PhD(bot), 69. Prof Exp: Asst prof gen biol, Univ Denver, 69-72; asst prof, 72-75, ASSOC PROF GEN BIOL & BOT, WASH STATE UNIV, 75- Concurrent Pos: NSF grant grassland biome, Int Biol Prog, Pawnee Nat Grasslands, Colo State Univ, 71-72 & 74-75, grant physiol ecol C3 and C4 grasses, 74-76. Mem: AAAS; Bot Soc Am; Am Sco Plant Physiol; Ecol Soc Am. Res: Physiological and biochemical adaptations of plant populations; colonizing plant species and ecotypic differentiation. Mailing Add: Prog in Gen Biol & Dept of Bot Wash State Univ Pullman WA 99163

WILLIAMS, GEORGE KENNETH, b Detroit, Mich, July 8, 32; m 54; c 4. MATHEMATICS. Educ: Univ Ky, BAE, 55, MA, 58; Univ Va, PhD(math), 64. Prof Exp: Teacher high sch, Mich, 55-56; asst prof math, Madison Col, 58-60 & Univ Notre Dame, 64-68; assoc prof, 68-72, PROF MATH, SOUTHWESTERN AT MEMPHIS, 72- Mem: Am Math Soc; Math Asn Am. Res: Complex analysis; topology. Mailing Add: Dept of Math Southwestern at Memphis Memphis TN 38112

WILLIAMS, GEORGE PATTESON, JR, physics, see 12th edition

WILLIAMS, GEORGE RAINEY, b Atlanta, Ga, Oct 25, 26; m 50; c 4. MEDICINE. Educ: Northwestern Univ, BS, 47, BMed, 50, MD, 51. Prof Exp: Instr surg, Johns Hopkins Hosp, 57-58; from asst prof to assoc prof, 58-63, PROF SURG, UNIV OKLA, 63-, CHMN DEPT SURG, 74- Concurrent Pos: Markle scholar, 60. Mem: Soc Univ Surg; Am Surg Asn; Soc Vascular Surg; Am Asn Thoracic Surg. Res: Cardiovascular surgery. Mailing Add: Dept of Surg Univ of Okla Health Sci Ctr Oklahoma City OK 73190

WILLIAMS, GEORGE ROBERTSON, b Comanche, Tex, Dec 20, 19; m 41; c 3. HORTICULTURE. Educ: Agr & Mech Col, Tex, BS, 47; Va Polytech Inst, MS, 51. Prof Exp: Tech asst, Agr & Mech Col, Tex, 48-49, exten horticulturist, 54-56; assoc exten horticulturist, 51-54, assoc horticulturist, Winchester Fruit Lab, 57-71, PROF HORT, VA POLYTECH INST & STATE UNIV, 71- Mem: Am Soc Hort Sci. Res: Apple rootstocks, particularly dwarf and semi-dwarf types. Mailing Add: Dept of Hort Va Polytech Inst & State Univ Blacksburg VA 24061

WILLIAMS, GEORGE RONALD, b Liverpool, Eng, Jan 4, 28; m 52; c 3. BIOCHEMISTRY. Educ: Univ Liverpool, PhD(biochem), 51, DSc, 69. Prof Exp: Worshipful Co Goldsmith's traveling fel, Banting & Best Dept Med Res, Univ Toronto, 52-53; res assoc, Johnson Found Med Biophys, Univ Pa, 53-55; Med Res Coun res assoc path, Oxford Univ, 55-56; asst prof biochem, Banting & Best Dept Med Res, 56-61, assoc prof, 61-66, PROF BIOCHEM, UNIV TORONTO, 66-, CHMN DEPT, 70- Mem: Am Soc Biol Chem; Can Biochem Soc (pres, 71-72); Can Soc Cell Biol; Brit Biochem Soc. Res: Cellular bio- energetics; mode of action of oxidizing enzymes; mitochondrial organization; metabolic control systems; geobiochemistry. Mailing Add: Dept of Biochem Univ of Toronto Med Sci Toronto ON Can

WILLIAMS, GEORGE W, b Nashville, Tenn, Oct 31, 46; m 68. BIOSTATISTICS. Educ: Bucknell Univ, BS, 68; George Washington Univ, MA, 70; Univ NC, PhD(biostatist), 72. Prof Exp: ASST PROF BIOSTATIST, UNIV MICH, ANN ARBOR, 72- Mem: Inst Math Statist; Am Statist Asn; Biomet Soc. Res: Nonparametric methods; multivariate analysis. Mailing Add: Dept of Biostatist Univ of Mich Ann Arbor MI 48104

WILLIAMS, GEORGE ZUR, b Chemulpo, Korea, Apr 7, 07. CLINICAL PATHOLOGY. Educ: Univ Colo, MD, 31. Hon Degrees: DSc, Univ Colo. Prof Exp: Mem staff, Inst Path, Univ Colo, 32-36; from assoc to prof, Med Col Va, 36-54, head dept clin path & hosp labs & dir dept oncol, 46-53; chief dept clin path, Clin Ctr, NIH, 53-69; DIR INST HEALTH RES, INST MED SCI, 69- Concurrent Pos: Dir labs, Nat Jewish Hosp, 34-36; consult, Surgeon Gen, US Air Force. Honors & Awards: Distinguished Serv Award, Dept Health, Educ & Welfare; Ward Burdic Award, Am Soc Clin Path. Mem: AAAS; Am Asn Cancer Res; Col Am Path. Res: Silicosis; effects of x-ray on malignant tissue; cytochemistry; normal human chemistry and physiology; laboratory automation and computerization. Mailing Add: Williams Inst of Health Res Inst of Med Sci 2340 Clay St San Francisco CA 94115

WILLIAMS, GERALD ALBERT, b Plankinton, SDak, Apr 1, 21; m 50; c 3. MEDICAL RESEARCH, ENDOCRINOLOGY. Educ: SDak State Col, BS, 45; George Washington Univ, MD, 49. Prof Exp: Instr med, Sch Med, Univ Va, 57-59; from asst prof to assoc prof, 59-69, PROF MED, UNIV ILL COL MED, 69- Concurrent Pos: Chief nuclear med serv & endocrinol sect, Vet Admin West Side Hosp, Chicago, 59-; attend physician, Univ Ill Hosp, 59-, chief endocrinol, 67- Mem: Am Fedn Clin Res; fel Am Col Physicians; Endocrine Soc; Soc Exp Biol & Med. Res: Parathyroid physiology; calcium metabolism; thyroid disorders. Mailing Add: Vet Admin West Side Hosp 820 S Damen Ave PO Box 8195 Chicago IL 60680

WILLIAMS, GERALD GORDON, b Victoria, Tex, Dec 10, 17; m 41; c 7. PLANT PHYSIOLOGY, SOIL SCIENCE. Educ: Pepperdine Col, BA, 43; Univ Calif, BS, 49, MS, 50; Purdue Univ, PhD(plant physiol, soils), 54. Prof Exp: Asst supt prod control, Plomb Tool Co, 41-44; instr math, Univ Calif, 46-48; instr high sch, Calif, 48-50; asst prof agron, Purdue Univ, 50-55; mgr irrigation res & develop, Olin Mathieson Chem Corp, 55-57; soil scientist & in-chg Southern Piedmont exp sta, Eastern Soil & Water Mgt Res Br, Agr Res Serv, USDA, 57-61; dir div agr rels, 61-62, DIR DIV AGR DEVELOP, TENN VALLEY AUTHORITY, 62- Concurrent Pos: Comt chmn coun agr sci & technol, Agr Res Inst, 74-75. Honors & Awards: Distinguished Serv Award, Nat Limestone Inst, 72; Founder's Award, Pepperdine Univ, 75; Meritorious Serv Award, Ohio State Univ Alumni Asn, 75. Mem: Am Soc Agron; Soil Sci Soc Am; Soil Conserv Soc Am; Int Soil Sci Soc. Res: Soils and fertilizers. Mailing Add: Div of Agr Develop Tenn Valley Authority T-218 NFDC Muscle Shoals AL 35660

WILLIAMS, GRAHEME JOHN BRAMALD, b Auckland, NZ, Jan 8, 42; m 69; c 1. STRUCTURAL CHEMISTRY, CRYSTALLOGRAPHY. Educ: Univ Auckland, BSc, 66, MSc Hons, 67; Univ Alta, PhD(bio; PhD(biochem), 72. Prof Exp: Fel chem, Univ Montreal, 72-73; RES ASSOC CHEM, BROOKHAVEN NAT LAB, 73- Mem: Am Crystallog Asn; NY Acad Sci. Res: Application of crystallography to structural problems in chemistry and biochemistry, enzymes, drugs and metabolites; crystallographic technique, improved computational methods and data base management. Mailing Add: Dept of Chem Brookhaven Nat Lab Upton NY 11973

WILLIAMS, HAROLD, b St John's, Nfld, Mar 14, 34; m 58; c 3. GEOLOGY. Educ: Mem Univ Nfld, BSc, 56, MSc, 58; Univ Toronto, PhD(geol), 61. Prof Exp: Res scientist, Geol Surv Can, Ont, 61-68; assoc prof, 68-71, PROF GEOL MEM UNIV NFLD, 71- Mem: Royal Soc Can; Geol Soc Am; fel Geol Asn Can. Res: Regional geology; ophiolite suites; continental margins; Appalachian geology. Mailing Add: Dept of Geol Mem Univ of Nfld St John's NF Can

WILLIAMS, HAROLD EDWARD, b Stites, Idaho, Dec 11, 13; m 38; c 2. PLANT PATHOLOGY. Educ: Wash State Univ, BS, 40; Ore State Col, PhD(bot), 56. Prof Exp: Field aide, USDA, 41-42 & 46-47; jr plant pathologist, Wash State Univ & State Dept Agr, 47-52; asst, Ore State Col, 52-56; PLANT PATHOLOGIST, STATE DEPT AGR, CALIF, 56- Mem: Am Phytopath Soc; Plant Propagators Soc. Res: Host range and seed transmission of stone fruit viruses; Rubus, ornamental, pome fruit and rose virus diseases; virus free nursery stock; plant virus disease diagnosis. Mailing Add: 44 Casa Mobile Circle West Sacramento CA 95691

WILLIAMS, HAROLD HAMILTON, b Roanoke, Va, Dec 23, 11. ORNAMENTAL HORTICULTURE. Educ: Hampton Inst, BS, 33; Cornell Univ, MS, 40, PhD(ornamental hort), 44. Prof Exp: Instr agr, Agr & Tech Col, NC, 35-39; head dept ornamental hort, Hampton Inst, 41-45; landscape architect, Liberian Centennial Expos, Washington, DC, 45-46; supv groundsman, Los Angeles Housing Authority, 48-55, landscape consult, 55-63; plant physiologist & res specialist, Los Angeles State & County Arboretum, 63-71; TURF CONSULT, CHACON HORT RES LABS, 72- Concurrent Pos: Instr, Los Angeles Trade-Tech Col, 64-; instr, Pasadena City Col, 68-; mem bd dirs, Southern Calif Turfgrass Coun; exec secy-treas, Calif Federated Turfgrass Coun; partic, Int Turfgrass Res Cong, 69. Mem: Am Soc Agron; Am Soc Hort Sci. Res: Turfgrass management practices; community beautification in Southern California. Mailing Add: 1128 E 80th St Los Angeles CA 90001

WILLIAMS, HAROLD HENDERSON, b Blanchard, Pa, Aug 29, 07; m 35; c 3. BIOCHEMISTRY. Educ: Pa State Univ, BS, 29; Cornell Univ, PhD(nutrit), 33. Prof Exp: Asst, Cornell Univ, 29-33; Sterling fel, Yale Univ, 33-35; res assoc, Children's Fund Mich, 35-39, from asst dir to assoc dir res lab, 39-45; prof, 45-73, head dept, 55-64, EMER PROF BIOCHEM, CORNELL UNIV, 73- Concurrent Pos: Mem nutrit res adv comt, USDA, 51-61; comt amino acids, Food & Nutrit Bd, Nat Acad Sci-Nat Res Coun, 52-72; nutrit study sect, NIH, 58-62, study sect comt nutrit & med res, Brazil, 62; spec organizing comt, conf fish & nutrit, Food & Agr Orgn, UN, Italy, 61, expert panel milk qual, 63-73; grad educ grants panel, US Bur Com Fisheries, 65-67 & exec comt, Off Biochem Nomenclature, 65-; vis comt biol & phys sci, Western Reserve Univ, 65-67; overseas corresp, Nutrit Abstr & Rev, 57-71; Am Soc Biol Chem rep, div biol & agr, Nat Res Coun, 65-67. Honors & Awards: Borden Award, Am Inst Nutrit, 53. Mem: AAAS; Am Soc Biol Chem; Am Chem Soc; Am Inst Nutrit. Res: Amino acid and protein metabolism; selenium metabolism in microorganisms. Mailing Add: 1060 Highland Rd Ithaca NY 14850

WILLIAMS, HARRY DOUGLAS, b Madisonville, Ky, Apr 4, 20. ORGANIC CHEMISTRY. Educ: Murray State Col, BS, 41; Ind Univ, MA, 47; Univ Del, PhD(chem), 56. Prof Exp: Physicist, Ind Ord Works, E I du Pont de Nemours & Co, Inc, 41-43, supvr, 43-44, res chemist, Burnside Lab, 47-55, sect head, Eastern Lab, NJ, 55-65, head patent & intel sect, 65-72, patent chemist, exp sta, 72-73, RES STAFF CHEMIST, EXP STA, E I DU PONT DE NEMOURS & CO, INC, WILMINGTON, DEL, 73- Mem: AAAS; Am Chem Soc; Sigma Xi. Res: Cellulose and nitric acid chemistry; polymer intermediates; catalysis; nylon intermediates; explosives. Mailing Add: 473 Hollywood Penns Grove NJ 08069

WILLIAMS, HARRY THOMAS, b Hampton, Va, July 22, 41; m 75; c 5. THEORETICAL PHYSICS. Educ: Univ Va, BS, 63, PhD(physics), 67. Prof Exp: Res assoc nuclear physics, Nat Bur Stand, 67-69; guest prof, Univ Erlangen-Nürenberg, 70; staff scientist, Kaman Sci div, Kaman Sci Corp, 71-73; ASST PROF PHYSICS, WASHINGTON & LEE UNIV, 74- Mem: Sigma Xi. Res: Effect of baryon resonance admixtures in nuclear wave function upon nuclear properties and reactions; response of circular loop antennae to electromagnetic fields. Mailing Add: Dept of Physics Washington & Lee Univ Lexington VA 24450

WILLIAMS, HENRY LANE, JR, otolaryngology, deceased

WILLIAMS, HENRY WARRINGTON, b Dallas, Tex, July 10, 34; m 58; c 4. ZOOLOGY, ANIMAL BEHAVIOR. Educ: Southern Methodist Univ, BS, 55; Utah State Univ, MS, 61, PhD(behavior), 66. Prof Exp: Assoc prof, 64-70, PROF BIOL, WESTMINSTER COL, MO, 70-, CHMN DIV NATURAL SCI & MATH, 73- Mem: AAAS; Animal Behav Soc; Am Ornith Union; Cooper Ornith Soc; Am Asn Univ Prof (pres, 67-68 & 74-75). Res: Investigations in the field of animal behavior with particular concern for sound communication in avian species. Mailing Add: Div of Natural Sci & Math Westminster Col Fulton MO 65251

WILLIAMS, HERBERT H, b Fentonville, NY, May 19, 21; m 48; c 3. ANTHROPOLOGY. Educ: Western Reserve Univ, BA, 48; Univ Pittsburgh, MS, 49, PhD(anthrop), 58. Prof Exp: Psychologist, Mid E Res Proj, Columbia Univ, 49-51; res analyst, Inst Int Educ, 51-53; PROF ANTHROP, CALIF STATE UNIV SAN FRANCISCO, 55- Res: Culture and personality; social structure; Middle East. Mailing Add: Dept of Anthrop Calif State Univ San Francisco CA 94132

WILLIAMS, HIBBARD E, b Utica, NY, Sept 28, 32; m; c 2. MEDICAL GENETICS. Educ: Cornell Univ, AB, 54, MD, 58. Prof Exp: Intern & asst resident med, Mass Gen Hosp, 58-60; clin assoc arthritis & metab dis & sr asst surgeon, NIH, 60-62; resident med, Mass Gen Hosp, 62-63, chief resident & teaching asst, Sch Med, Harvard Univ, 63-64, instr, 64-65; from asst prof to assoc prof, 65-72, PROF MED, SCH MED, UNIV CALIF, SAN FRANCISCO, 72- Concurrent Pos: Markle scholar, 68-73; chief med serv, San Francisco Gen Hosp. Mem: Am Soc Clin Invest (secy-treas); Asn Am Physicians; Am Fedn Clin Res. Res: Inborn errors of metabolism. Mailing Add: San Francisco Gen Hosp B 40 B4401 San Francisco CA 94110

WILLIAMS, HOWARD PERSON, analytical chemistry, see 12th edition

WILLIAMS, HOWEL, b Liverpool, Eng, Oct 12, 98; nat US; m 30. GEOLOGY. Educ:

Liverpool Univ, MA, 24, DSc, 28; Univ London, DIC, 24. Prof Exp: Demonstr geol, Royal Sch Mines, London, 29-30; assoc prof geol, 30-37, prof & chmn dept, 45-70, EMER PROF GEOL, UNIV CALIF, BERKELEY, 70- Concurrent Pos: Condon lectr, Univ Ore, 45; expert consult, Res & Develop Bd; with US Geol Surv, 44-46. Mem: Nat Acad Sci; fel Geol Soc Am; Seismol Soc Am. Res: Igneous petrography; volcanology; regional geology. Mailing Add: Dept of Geol 493 Earth Sci Univ of Calif Berkeley CA 94720

WILLIAMS, HUGH COWIE, b London, Ont, July 23, 43; m 67; c 1. MATHEMATICS, COMPUTER SCIENCE. Educ: Univ Waterloo, BSc, 66, Math, 67, PhD(math), 69. Prof Exp: Nat Res Coun Can fel, York Univ, 69-70; ASSOC PROF COMPUT SCI, UNIV MAN, 70- Res: Application of the computer to problems arising in the theory of numbers. Mailing Add: Dept of Comput Sci Univ of Man Winnipeg MB Can

WILLIAMS, HUGH HARRISON, b Boston, Mass, Dec 4, 44; m 70; c 1. EXPERIMENTAL HIGH ENERGY PHYSICS. Educ: Haverford Col, BS, 66; Stanford Univ, PhD(physics), 71. Prof Exp: Res assoc physics, Brookhaven Nat Lab, 71-73, assoc physicist, 73-74; ASST PROF PHYSICS, UNIV PA, 74- Mem: Am Phys Soc. Res: Experimental study of elementary particles, their nature and interactions, with particular emphasis on the study of weak interactions. Mailing Add: Dept of Physics Univ of Pa Philadelphia PA 19174

WILLIAMS, HULEN BROWN, b Lauratown, Ark, Oct 8, 20; m 42; c 2. PHYSICAL CHEMISTRY. Educ: Hendrix Col, AB, 41; La State Univ, MS, 43, PhD(chem), 48. Prof Exp: From instr to assoc prof, 43-57, dir lab stores, 48-56, admin asst to dean, 52-56, PROF CHEMISTRY, LA STATE UNIV, BATON ROUGE, 57- HEAD DEPT, 56-, DEAN COL CHEM & PHYSICS, 68- Honors & Awards: Coates Award, Am Chem Soc, 63. Mem: Am Chem Soc; Electron Micros Soc Am; Am Inst Chemists. Res: Light scattering of latices; proteins; protein metal complexes; organic reaction mechanisms. Mailing Add: 470 Castle Kirk Ave Baton Rouge LA 70808

WILLIAMS, JACK A, b Wichita, Kans, June 29, 26; m 49; c 4. ORGANIC GEOCHEMISTRY. Educ: Univ Kans, AB, 50, PhD(org chem), 54. Prof Exp: Chemist, Standard Oil Co, Ind, 53-57; CHEMIST, RES CTR, AMOCO PROD CO, 57- Res: Organic geochemistry of petroleum and associated sedimentary substances. Mailing Add: 7317 E 59th St Tulsa OK 74145

WILLIAMS, JACK L R, b Edmonton, Alta, Oct 25, 23; nat US; m 50; c 5. ORGANIC CHEMISTRY. Educ: Univ Alta, BSc, 46; Univ Ill, PhD(org chem), 48. Prof Exp: Spec asst, Off Rubber Reserve, Univ Ill, 46-48; Du Pont fel, Univ Wis, 48-49; res chemist, 49-55, from res assoc to sr res assoc, 55-68, SR LAB HEAD, EASTMAN KODAK CO, 68- Mem: Am Chem Soc. Res: Organic synthesis; rubber chemistry; high pressure reactions; oxo synthesis; catalytic hydrogenation; high polymer chemistry; organic photochemistry; photochemistry of boron. Mailing Add: Div of Chem Eastman Kodak Co Res Labs Rochester NY 14650

WILLIAMS, JACK MARVIN, b Delta, Colo, Sept 26, 38; m 58; c 3. STRUCTURAL CHEMISTRY, INORGANIC CHEMISTRY. Educ: Lewis & Clark Col, BS, 60; Wash State Univ, MS, 63, PhD(phys chem), 66. Prof Exp: Resident res assoc neutron & x-ray diffraction anal, Argonne Nat Lab, 66-68, from asst chemist to assoc chemist, Chem Div, 68-72, CHEMIST, CHEM DIV, ARGONNE NAT LAB, 72- Mem: Am Crystallog Asn; Am Chem Soc; Am Phys Soc. Res: Neutron and x-ray diffraction as applied to the elucidation of the nature of chemical bonding; neutron diffraction as applied to the study of biologically related materials; chemical bonding. Mailing Add: Chem Div Argonne Nat Lab 9700 S Cass Ave Argonne IL 60439

WILLIAMS, JAMES ALFRED, veterinary medicine, veterinary pathology, see 12th edition

WILLIAMS, JAMES CARL, b Covington, La, 35; m 63; c 4. ANIMAL PARASITOLOGY. Educ: Southeastern La Col, BS, 57; La State Univ, Baton Rouge, MS, 62; La State Univ, New Orleans, PhD(med parasitol), 69. Prof Exp: From instr to asst prof, 57-73, ASSOC PROF VET PARASITOL, LA STATE UNIV, BATON ROUGE, 73- Mem: Am Soc Parasitol. Res: Immunology of helminth infections; immunologic aspects of host-parasite relationships; epidemiology of parasitism in ruminants. Mailing Add: Dept of Vet Sci La State Univ Baton Rouge LA 70803

WILLIAMS, JAMES EARL, JR, b Freesport, Pa, June 1, 38; m 58; c 3. PHYSICAL CHEMISTRY. Educ: Univ Pittsburgh, BS, 65, MS, 72. Prof Exp: Scientist chem, 65-73, sr scientist, 73-75, GROUP LEADER CHEM, ALUMINUM CO AM, 73- Mem: Am Chem Soc; Fedn Socs Coating Technol; Aluminum Asn; Steel Struct Painting Coun. Res: Development of aluminum-pigmented coatings, development of coatings and coating processes for automotive and cooking utensil applications, and development of corrosion resistant maintenance and engineering coatings. Mailing Add: Alcoa Labs Alcoa Tech Ctr Alcoa Center PA 15069

WILLIAMS, JAMES GARNER, b Atascedero, Calif, Nov 4, 44; m 66. MATHEMATICS. Educ: Carleton Col, BA, 66; Univ Calif, Berkeley, PhD(math), 73. Prof Exp: Asst prof, 72-75, ASSOC PROF MATH, BOWLING GREEN STATE UNIV, 75- Mem: Am Math Soc; Math Asn Am. Res: Topology and uniform spaces; elementary category theory; logic. Mailing Add: Dept of Math Bowling Green State Univ Bowling Green OH 43403

WILLIAMS, JAMES HENRY, JR, b Los Angeles, Calif, July 14, 18; m 39; c 2. AGRONOMY. Educ: Ore State Col, BS, 49; Iowa State Col, MS, 50, PhD(agron), 52. Prof Exp: Res assoc, Iowa State Col, 50-52; from asst prof to assoc prof agron, Univ Nebr, Lincoln, 52-72, PROF AGRON, UNIV NEBR, LINCOLN, EAST CAMPUS, 72- Mem: Am Soc Agron; Crop Sci Soc Am; Soc Econ Bot. Res: Soybean breeding. Mailing Add: 342 Keim Hall Univ of Nebr East Campus Lincoln NE 68583

WILLIAMS, JAMES HORACE, b Manchester, Eng, Jan 1, 08; nat US; m 32; c 3. ORGANIC CHEMISTRY. Educ: Worcester Polytech Inst, BSc, 29; NY Univ, PhD(org chem), 33. Prof Exp: Mem, Patent Dept, Allied Chem & Dye Corp, NY, 33-35; chemist, Swann Chem Co, Ala, 35-36; res chemist, Stamford Labs, Am Cyanamid Co, 37-40, tech adv, Patent Dept, 40-45, admin dir res, Lederle Labs, 45-48, dir res, 48-53, chem & biol, 53-55, asst to dir labs, 56-63, asst to gen mgr, 63-73; RETIRED. Honors & Awards: Citation, US Dept State, 58. Mem: Fel AAAS; Am Chem Soc; fel Am Inst Chemists; fel NY Acad Sci. Res: Pharmaceutical products; patents; licenses; natural and synthetic chemotherapy products; sulfa drugs; preparation of thiocholines. Mailing Add: 146 Melrose Pl Ridgewood NJ 07450

WILLIAMS, JAMES HUTCHISON, b Westerville, Ohio, Feb 20, 22; m 43; c 4. OBSTETRICS & GYNECOLOGY. Educ: Otterbein Col, AB, 44; Ohio State Univ, MD, 46, MMSc, 52; Am Bd Obstet & Gynec, dipl. Prof Exp: From instr to assoc prof, 55-70, assoc dir, Inst Perinatal Studies, 60-64 & Ctr Perinatal Studies, 65-70, PROF OBSTET & GYNEC, OHIO STATE UNIV, 70-, ASSOC DEAN COL MED, 61- Mem: Fel Am Col Surg; fel Am Col Obstet & Gynec. Res: Perinatal morbidity and mortality; selection of medical students; medical student evaluation in education. Mailing Add: Col of Med Ohio State Univ Columbus OH 43210

WILLIAMS, JAMES LOVON, JR, b Salem, Ind, May 16, 29; m 52; c 3. WEED SCIENCE, PLANT PHYSIOLOGY. Educ: Purdue Univ, Lafayette, BS, 57, MS, 59, PhD(plant path), 61. Prof Exp: From instr to assoc prof, 60-71, PROF WEED SCI, PURDUE UNIV, LAFAYETTE, 71- Mem: Weed Sci Soc Am; Am Soc Agron; Crop Sci Soc Am. Res: Weed control systems to minimize pollution potential from weeds and their control; effects of soil properties and climatic factors on control systems; fate of herbicides in soil and water. Mailing Add: Dept of Bot & Plant Path Purdue Univ Lilly Hall Life Sci West Lafayette IN 47907

WILLIAMS, JAMES RAYMOND, b Centralia, Mo, Oct 30, 36. ANTHROPOLOGY, ARCHAEOLOGY. Educ: Northeast Mo State Univ, BS, 58; Univ Mo-Columbia, MA, 64, PhD(anthrop), 71. Prof Exp: Teacher hist, Vandalia Pub Schs, Mo, 58-60; teacher, Mex Pub Schs, Mo, 60-62; asst archaeol, Univ Mo-Columbia, 62-64, asst instr anthrop, 64-66, res assoc archaeol, 66-69; asst prof, 69-75, ASSOC PROF ANTHROP, UNIV S FLA, 75- Mem: Soc Am Archaeol; fel Am Anthrop Asn; fel AAAS. Res: Archaeology of the southeastern United States, Mississippi Valley, Southeast Missouri and Florida; prehistoric environmental utilization patterns; middle eastern archaeology. Mailing Add: Dept of Anthrop Univ of S Fla Tampa FL 33620

WILLIAMS, JAMES STANLEY, b Coronado, Calif, Oct 12, 34; m 56; c 3. STATISTICS. Educ: Wash State Univ, BS, 55; Agr & Mech Col Tex, MS, 57; NC State Col, PhD(exp & math statist), 61. Prof Exp: Instr genetics, Agr & Mech Col, Tex, 56-57; statistician, Res Triangle Inst, NC, 60-62; PROF STATIST, COLO STATE UNIV, 62- Concurrent Pos: Assoc ed, Biometrics, 72- Mem: Fel Am Statist Asn; Biomet Soc; Inst Math Statist. Res: Theory of statistical inference; multivariate analysis; indexing theory; population genetics. Mailing Add: Dept of Statist Colo State Univ Ft Collins CO 80521

WILLIAMS, JEAN PAUL, b New York, NY, Dec 29, 18; m 45. ANALYTICAL CHEMISTRY. Educ: Kent State Univ, BS, 40; Univ NC, PhD(chem), 50. Prof Exp: Chemist, Nat Bur Standards, 42-45; instr chem, Univ NC, 45-50; res chemist, 50-57, mgr tech serv res dept, 57-64, mgr instrumental anal res dept, 64-73, MGR ANAL SERV RES, CORNING GLASS WORKS, 73- Mem: AAAS; Am Chem Soc; Am Ceramic Soc; Soc Appl Spectros; Microbeam Anal Soc. Res: Inorganic chemical, instrumental and x-ray analysis; classical wet methods; polarography; flame spectrophotometry; glass property measurements; microscopy; mass spectrometry; electron microprobe analysis; atomic absorption and emission. Mailing Add: Corning Glass Works Sullivan Park Corning NY 14830

WILLIAMS, JEFFREY F, b Bristol, Eng, Aug 28, 42; m 64. PARASITOLOGY, IMMUNOLOGY. Educ: Univ Bristol, BVSc, 64; Univ Pa, PhD(parasitol), 68. Prof Exp: Parasitologist, Pan-Am Health Orgn, Buenos Aires, Arg, 68-71; asst prof, 71-73, ASSOC PROF MICROBIOL, MICH STATE UNIV, 74- Mem: Brit Soc Immunol; Am Soc Parasitol; Royal Soc Trop Med; Am Soc Trop Med. Res: Mechanisms of resistance to helminth infections in domestic animals and man. Mailing Add: Dept of Microbiol & Pub Health Mich State Univ East Lansing MI 48824

WILLIAMS, JEROME, b Toronto, Ont, July 15, 26; m 53; c 2. PHYSICAL OCEANOGRAPHY. Educ: Univ Md, BS, 50; Johns Hopkins Univ, MA, 52. Prof Exp: Res staff asst, Chesapeake Bay Inst, Johns Hopkins Univ, 52-56; physicist, Vitro Labs, 56-57; asst prof physics, 57-64, assoc prof oceanog, 64-72, res prof environ protection, 72-74, PROF OCEANOG, US NAVAL ACAD, 74- Concurrent Pos: Res assoc, Chesapeake Bay Inst, Johns Hopkins Univ, 57-71; tech ed, Naval Inst, 66- Mem: AAAS; Marine Technol Soc; Am Geophys Union; Instrument Soc Am; Estuarine Res Fedn (vpres, 71-73, secy, 73-75). Res: Underwater transparency; oceanographic instrumentation; environmental protection. Mailing Add: Dept of Environ Sci US Naval Acad Annapolis MD 21402

WILLIAMS, JESSE BASCOM, b Lone Oak, Tex, Oct 24, 17; m 44; c 3. ANIMAL SCIENCE. Educ: Okla State Univ, BS, 47; Pa State Univ, MS, 48, PhD(dairy husb), 50. Prof Exp: Res asst dairy husb, Pa State Univ, 48-50; asst prof, NDak State Univ, asst dairy husbandman, Agr Exp Sta, 50-55; from asst prof to assoc prof, 55-61, PROF ANIMAL SCI, UNIV MINN, ST PAUL, 61- Mem: Fel AAAS; Am Soc Animal Sci; Am Dairy Sci Asn. Res: Infant ruminant nutrition; synthetic diets; mechanical feeding devices; heat treatment of dried skim milk powders; immunoglobulin absorption patterns. Mailing Add: Dept of Animal Sci Univ of Minn St Paul MN 55101

WILLIAMS, JESSE NOAH, JR, biochemistry, see 12th edition

WILLIAMS, JIMMY CALVIN, b Palestine, Tex, Oct 26, 43; m 67; c 3. MICROBIOLOGY, BIOCHEMISTRY. Educ: Tex A&M Univ, BS, 69, MS, 71, PhD(biochem), 73. Prof Exp: Res asst cancer, Lab Exp Oncol, Riley Cancer Wing, Ind Sch Med, 73-74; MICROBIOLOGIST, NAVAL MED RES INST, NAT NAVAL MED CTR, 74- Concurrent Pos: Vis instr, Lab Exp Oncol, Ind Sch Med, 74, vis asst prof, 75-76. Mem: Sigma Xi; Nat Am Soc Microbiol. Res: Basic biology of rickettsiae entailing detailed analysis of the biosynthesis of the purines and pyrimidines; basic biology of rat hepatomas with emphasis on the biosynthesis of pyrimidine enzymes and a combined chemotherapy approach to the problems of rapid cellular proliferation. Mailing Add: Dept of Microbiol Naval Med Res Inst Bethesda MD 20014

WILLIAMS, JOEL LAWSON, b Sarecta, NC, Nov 10, 41; m 62; c 2. POLYMER CHEMISTRY. Educ: NC State Univ, BS, 65, MS, 67, PhD(polymer sci), 70. Prof Exp: Sr chemist polymer res, Camille Dreyfus Lab, Research Triangle Inst, 62-74; HEAD MAT SCI DEPT, BECTON DICKINSON RES CTR, RESEARCH TRIANGLE PARK, NC, 74- Concurrent Pos: Adj prof chem eng, NC State Univ, 72- Mem: Am Chem Soc. Res: Permeability and diffusion in membranes, polymer synthesis and characterization with special emphasis on the utilization of radiation chemistry as a tool for graft modification of polymeric substrates, ionic polymerization, high-energy irradiation applications, irradiation grafting and blood compatibility of polymers. Mailing Add: Becton Dickinson Res Ctr Box 12016 Research Triangle Park NC 27709

WILLIAMS, JOEL MANN, JR, b Suffolk, Va, Apr 6, 40; m 62; c 2. FUEL SCIENCE, PHYSICAL ORGANIC CHEMISTRY. Educ: Col William & Mary, BS, 62; Northwestern Univ, Evanston, PhD(org chem), 66. Prof Exp: NSF fel, Univ Minn, Minneapolis, 66-67, asst prof chem, 67-68; res chemist, Benger Lab, E I du Pont de Nemours & Co, Inc, Va, 68-72; MEM STAFF, LOS ALAMOS SCI LAB, UNIV CALIF, 72- Mem: Am Chem Soc; Am Carbon Soc. Res: Research on coal and coal wastes, especially trace elements of environmental concern; oil-shale chemistry and morphology; carbon fiber-aluminum composites; chemical vapor deposition; synthetic graphites; elastomers; textile fibers; proton transfer and intramolecular displacement reactions. Mailing Add: CMB-8 MS 734 Los Alamos Sci Lab Univ of Calif Los Alamos NM 87545

WILLIAMS, JOEL QUITMAN, b Lake Charles, La, Mar 6, 22; m 47; c 2. PHYSICS. Educ: Centenary Col, BS, 43; Ga Inst Technol, MS, 48; Duke Univ, PhD(physics), 52. Prof Exp: From asst prof to assoc prof, 46-49, 51-70, PROF PHYSICS, GA INST TECHNOL, 70- Mem: Am Phys Soc; Am Asn Physics Teachers. Res: Microwave spectroscopy. Mailing Add: Sch of Physics Ga Inst of Technol Atlanta GA 30332

WILLIAMS, JOHN ALBERT, b Springfield, Ill, Mar 28, 37; m 59; c 2. ASTRONOMY. Educ: Univ Mich, AB, 49; Univ Calif, Berkeley, PhD(astron), 63. Prof Exp: NSF fel, Univ Calif, Berkeley & Princeton Univ, 63-64; from instr to asst prof astron, Univ Mich, Ann Arbor, 64-70; ASSOC PROF PHYSICS, ALBION COL, 70- Mem: AAAS; Am Astron Soc. Res: Photometry of astronomical objects; quantitative spectral classification; interstellar matter. Mailing Add: Dept of Physics Albion Col Albion MI 49224

WILLIAMS, JOHN ANDREW, b Des Moines, Iowa, Aug 3, 41; m 65; c 2. PHYSIOLOGY. Educ: Cent Wash State Col, BA, 63; Univ Wash, MD & PhD(physiol, biophys), 68. Prof Exp: Staff assoc, Clin Endocrinol Br, Nat Inst Arthritis & Metab Dis, 69-71; asst prof, 72-74, ASSOC PROF PHYSIOL, UNIV CALIF, SAN FRANCISCO, 74- Concurrent Pos: NIH fel, Dept Pharmacol, Univ Utah, 68-69; Helen Hay Whitney Found fel, Univ Cambridge, 71-72; USPHS grant, Univ Calif, San Francisco, 73-77, Nat Cystic Fibrosis Found grant, 74-77. Mem: Endocrine Soc; Am Physiol Soc. Res: Cellular physiology; endocrinology. Mailing Add: Dept of Physiol Univ of Calif Sch of Med San Francisco CA 94143

WILLIAMS, JOHN C, b Hazard, Ky, June 18, 25; m 57; c 1. ZOOLOGY. Educ: Mich State Univ, BS, 53; Univ Ky, MS, 57; Univ Louisville, PhD(biol), 63. Prof Exp: Instr biol, Mary Washington Col, Univ Va, 56-57; asst prof, Transylvania Col, 57-59; from asst prof to prof, Murray State Univ, 62-69; assoc prof, 69-72, PROF BIOL, EASTERN KY UNIV, 72- Concurrent Pos: Mussel Fishery Invests grant, Tenn, Ohio & Green Rivers, 66-69; Commercial Fisheries Invests of Ky River grant, 72-74. Mem: Am Fisheries Soc; Am Soc Ichthyol & Herpet. Res: Fisheries biology; mammalogy; herpetology; natural history of freshwater mussels. Mailing Add: Dept of Biol Sci Eastern Ky Univ Richmond KY 40475

WILLIAMS, JOHN CASWELL, b Rocky Mount, NC, Apr 9, 27; m 49; c 2. PLANT GENETICS, STATISTICS. Educ: NC State Univ, BS, 49, MS, 56; Iowa State Univ, PhD(crop breeding, statist), 60. Prof Exp: Res instr plant breeding, NC State Univ, 49-57, asst prof, 60-61; asst agronomist & lectr plant breeding, Univ Calif, Davis, 61-66; asst prof genetics, Univ Ky, 66-70; ASSOC PROF STATIST, AUBURN UNIV, 70- Mem: Am Soc Agron; Crop Sci Soc Am; Biomet Soc. Mailing Add: Res Data Anal Auburn Univ Auburn AL 36830

WILLIAMS, JOHN CHARLES, ceramic engineering, see 12th edition

WILLIAMS, JOHN COVINGTON, b Oxford, Ohio, Feb 11, 11; m 32; c 3. CHEMISTRY. Educ: Oberlin Col, AB, 31; Miami Univ, MA, 33; Iowa State Univ, PhD(colloid chem), 37. Prof Exp: Instr chem, Iowa State Univ, 35-38; res chemist, Champion Paper & Fibre Co, 38-42; fel, Mellon Inst, 42-45; res dir, Hawley Prod Co, 45-62; res dir, Cuno Eng Corp, 62-69, vpres res, 69-72; RES OFFICER, PRESERV OFF, LIBR CONG, 72- Mem: Am Chem Soc; Tech Asn Pulp & Paper Indust. Res: Pulp molding processes applied to filter production; resin sizing of pulp in beater; conservation and permanence of library materials. Mailing Add: Res & Testing Off Preserv Off Libr of Cong Washington DC 20540

WILLIAMS, JOHN EDWARD, fishery biology, see 12th edition

WILLIAMS, JOHN F, JR, b Louisville, Ky, Oct 25, 31; m 66. INTERNAL MEDICINE, CARDIOLOGY. Educ: Ind Univ, Indianapolis, MD, 56. Prof Exp: Intern med, Univ Minn Hosps, 56-57; resident internal med, Med Ctr, Ind Univ, Indianapolis, 57-59, from asst prof to assoc prof internal med, Sch Med, 65-70; PROF MED & DIR DIV CARDIOL, UNIV TEX MED BR GALVESTON, 70- Concurrent Pos: Am Heart Asn res fel cardiol, Med Ctr, Ind Univ, Indianapolis, 59-61; USPHS res fel, 61-63; fel, Cardiol Br, Nat Heart Inst, 63-65; chief cardiovasc res lab, Vet Admin Hosp, Indianapolis, 65-70. Mem: AAAS; Am Physiol Soc; Am Fedn Clin Res; fel Am Col Cardiol; fel Am Col Physicians. Res: Cardiovascular physiology and pharmacology. Mailing Add: Div of Cardiol Ziegler 103 Univ of Tex Med Br Galveston TX 77550

WILLIAMS, JOHN FREDERICK, b York, SC, May 14, 23; m 45; c 2. CHEMISTRY. Educ: Univ SC, BS, 44; Clemson Univ, MS, 51; Univ Va, PhD(chem), 54. Prof Exp: Instr chem, Clemson Univ, 49-51; sr chemist, anal chem, res dept, Liggett & Myers Tobacco Co, 54-60, RES SUPVR ANAL CHEM, RES DEPT, LIGGETT & MYERS INC, 60- Mem: AAAS; Am Chem Soc; Coblentz Soc; Soc Appl Spectros. Res: Development and application of chromatographic, spectrophotometric, automatic and classical methods of analysis in the study of natural products. Mailing Add: Res Dept Liggett & Myers Inc Durham NC 27702

WILLIAMS, JOHN PETER, b London, Eng, Apr 17, 39; m 63; c 2. PLANT BIOCHEMISTRY, CYTOLOGY. Educ: Univ Leicester, BSc, 60; Univ London, PhD(plant physiol & cytol) & dipl, Imp Col, 63. Prof Exp: Asst prof, 63-68, ASSOC PROF BOT, UNIV TORONTO, 68- Mem: Can Soc Plant Physiol. Res: Structure and function of chloroplasts; quinone, nucleic acid and lipid content and development; aspects of photosynthesis. Mailing Add: Dept of Bot Univ of Toronto Toronto ON Can

WILLIAMS, JOHN RODERICK, b Birmingham, Eng, July 5, 40; m 74. SYNTHETIC ORGANIC CHEMISTRY. Educ: Univ Western Australia, BSc, 62, PhD(org chem), 66. Prof Exp: Vis fel org photochem, NIH, 66-67; NIH fel & res assoc, Columbia Univ, 67-68; asst prof, 68-74, ASSOC PROF ORG CHEM, TEMPLE UNIV, 74- Mem: Am Chem Soc; assoc mem Royal Australian Chem Inst; fel The Chem Soc. Res: Photosensitivity and photobiology; structure determination of natural products; marine and steroid chemistry. Mailing Add: Dept of Chem Temple Univ Philadelphia PA 19122

WILLIAMS, JOHN SIMEON, b Demopolis, Ala, Oct 29, 14; m; c 4. AGRICULTURE, PLANT ECOLOGY. Educ: Tex A&M Univ, BS, 40; Univ Mo, MA, 41; Univ Nebr, PhD(plant ecol), 47. Prof Exp: Soil surveyor, US Chem & Soils, 41; head agr, Tarleton State Col, 41-42; agronomist, Exp Sta, Tex A&M, 42-45; range conservationist mgr, US Forest Serv, 47-49; dir res & prof agr res work, Univ Houston, 50-58; chief soils, crops & livestock, US Dept Mission to Govt Iran, 58-60; agr attache abominable snowman exped, Southwest Res Found, 60-61; ASSOC PROF BIOL SCI, UNIV TEX, EL PASO, 61- Concurrent Pos: Organized Res Fund grant, Univ Tex, El Paso, 63-67. Honors & Awards: Award, Univ Houston, 57. Mem: Am Soc Agron; Soc Range Mgt. Res: Mineral and protein nutrition of range livestock; taxonomy and plant ecology of Chihuahuan Desert. Mailing Add: Apdo 6-945 Guadalajara Jalisco Mexico

WILLIAMS, JOHN T, b Bristol Twp, Ohio, Aug 2, 23; m 49; c 2. PHYSICAL CHEMISTRY. Educ: Hamline Univ, BS, 44; Univ Minn, MS, 49; Iowa State Col, PhD(chem), 54. Prof Exp: From asst prof to assoc prof, 54-63, PROF CHEM, COLO SCH MINES, 63- Mem: Am Chem Soc; AAAS. Res: Thermodynamics; phase equilibrium. Mailing Add: Dept of Chem & Geochem Colo Sch of Mines Golden CO 80401

WILLIAMS, JOHN WARREN, b Woburn, Mass, Feb 10, 98; m 25; c 1. PHYSICAL BIOCHEMISTRY. Educ: Worcester Polytech Inst, BS, 21; Univ Wis, MS, 22, PhD(chem), 25. Prof Exp: Asst chem, 21-25, from instr to prof, 25-68, EMER PROF CHEM, UNIV WIS-MADISON, 68- Concurrent Pos: Nat Res Coun fel, Copenhagen Univ & Univ Leipzig, 27-28; Int Educ Bd fel, Univ Uppsala, 34-35, Nobel guest prof, 68; vis prof, Calif Inst Technol, 46-47 & 53-54; Guggenheim fel, Copenhagen Univ, Oxford & Calif Inst Technol, 56-57; mem comt colloid sci, Nat Res Coun. Honors & Awards: Kendall Award, Am Chem Soc, 55. Mem: Nat Acad Sci; Am Chem Soc; Am Soc Biol Chem. Res: Physical chemistry of the proteins and high polymers. Mailing Add: Dept of Chem Univ of Wis Madison WI 53706

WILLIAMS, JOHN WATKINS, III, b Alexandria, La, Mar 11, 42; m 70. CYTOGENETICS. Educ: Univ Southwestern La, BS, 65; La State Univ, MS, 68, PhD(genetics zool), 71. Prof Exp: Res assoc genetics, La State Univ, 68-70; asst prof, 71-74, ASSOC PROF GENETICS & EMBRYOL, TUSKEGEE INST, 74- Mem: AAAS; Genetics Soc Am; Sigma Xi. Res: Nature of DNA sequences in lateral loop axes of T viridescens lampbrush chromosomes and the chromosomal incorporation of exogenous DNA in Drosophila polytene chromosomes. Mailing Add: Dept of Biol Tuskegee Inst Tuskegee Institute AL 36088

WILLIAMS, JOHN WHARTON, b Wichita, Kans, May 3, 45; m 68. GEOLOGY. Educ: Col William & Mary, BS, 67; Stanford Univ, MS, 68, PhD(geol), 70. Prof Exp: GEOLOGIST, CALIF DIV MINES & GEOL, 71- Mem: AAAS; Geol Soc Am; Asn Eng Geologists. Res: Detection, analysis and delineation of geologic hazards so as to provide for the proper location and construction of man's civil engineering works. Mailing Add: 1021 Crestview Dr San Carlos CA 94070

WILLIAMS, JOSEPH BLAIR, physical organic chemistry, see 12th edition

WILLIAMS, JOSEPH JOHN, b Toronto, Ont, Oct 7, 43; m 68; c 1. MATHEMATICS. Educ: Univ Toronto, BS, 66, MS, 67, PhD(math), 70. Prof Exp: ASST PROF APPL MATH, FAC SCI, UNIV MAN, 70- Concurrent Pos: Nat Res Coun Can operating grant, Univ Man, 71- Mem: Can Math Cong; Am Math Soc; Math Asn Am. Res: Functional analysis; Von Neumann algebras; C algebras. Mailing Add: Dept of Math Fac of Sci Univ of Man Winnipeg MB Can

WILLIAMS, JOSEPH LEE, b New Bern, NC, Nov 2, 36; m 62; c 4. FOOD CHEMISTRY, LIPID CHEMISTRY. Educ: Morehouse Col, BS, 60; Tuskegee Inst Technol, MS, 62; Univ Ill, Urbana, PhD(food sci), 70. Prof Exp: George Washington Carver fel, Carver Found, Tuskegee Inst Technol, 60-62; chemist, Monsanto Co, 63-66; USPHS fel, Burnsides Res Lab, Univ Ill, 68-70; dir multidisciplinary labs, Sch Vet Med, Tuskegee Inst Technol, 70-72; SR SCIENTIST, RES & DEVELOP DIV, KRAFTCO CORP, 72- Concurrent Pos: Consult & mgr audiovisual & multimedia learning resource ctr for sci & med stud & individualized study progs, Tuskegee Inst Technol, 70-72; Ninth Annual George Washington Carver lectr, 71. Mem: Fel AAAS; Am Chem Soc; Am Oil Chem Soc; fel Am Inst Chemists; Inst Food Technologists. Res: Food science and lipid chemistry as it relates to the feeding of the public; biochemical utilization by man and the nutritional impact upon man; designing and managing multidisciplinary science laboratories and multimedia learning programs. Mailing Add: Edible Oil Prod Lab R&D Div Kraftco Corp 801 Waukegan Rd Glenview IL 60025

WILLIAMS, JOSEPHINE LOUISE, b Bowling Green, Ky, May 23, 26. PHYSICAL CHEMISTRY, SURFACE CHEMISTRY. Educ: Western Ky Univ, BS, 47; Northwestern Univ, MS, 50; Univ Cincinnati, BS, 58. Prof Exp: Res assoc heterocyclics, Dept Chem, Western Ky Univ, 47-48; sr chemist, Cimcool Div, 50-55, sr res chemist, Cent Res Div, 55-57, sr res supvr abrasives, Metal Working Fluids, 57-71, sr res supvr, Com Develop Dept, 71-75, SR RES ASSOC, CINCINNATI MILACRON, INC, 75- Mem: Am Chem Soc; Am Ceramic Soc. Res: Chemistry of friction, lubrication and wear; mechanisms of wear of abrasives and bonded abrasives; metal working fluids and processes; corrosion, surfactants, emulsions, electrode processes. Mailing Add: Dept 41D Cincinnati Milacron Inc Cincinnati OH 45209

WILLIAMS, JOY ELIZABETH P, b Blackshear, Ga, June 12, 29; m 59; c 2. BACTERIOLOGY, BIOCHEMISTRY. Educ: Univ Ga, BS, 50, MS, 58, PhD(bact), 61. Prof Exp: Teacher high sch, Ga, 52-57; USPHS fel biol div, Oak Ridge Nat Lab, 61-63; res assoc bact, 63-66, asst prof, 66-68, mgr biol automated info, Comput Ctr, 68-70, ASST PROF SCI EDUC & ASST DIR HONS PROG, UNIV GA, 70- Concurrent Pos: Mem, Nat Collegiate Hons Coun. Mem: AAAS; Am Soc Microbiol; NY Acad Sci. Res: Bacterial metabolism of alginic and mannuronic acids; formic hydrogenlyase systems in bacteria; bacterial degradation of organic synthetic sulfur compounds; sulfur metabolism in bacteria. Mailing Add: Hons Prog Univ of Ga Athens GA 30601

WILLIAMS, JULIAN, b Oswestry, Eng, Mar 1, 38; m 66. GEOENVIRONMENTAL SCIENCE. Educ: Cambridge Univ, BA, 60; Univ Canterbury, PhD(soil chem), 66. Prof Exp: Sci officer, Rothamsted Exp Sta, Eng, 66-67; res scientist, Res Br, Can Dept Agr, Man, 67-68; vis asst prof soil sci, Univ Wis-Madison, 68-70; RES SCIENTIST, CAN CENTRE FOR INLAND WATERS, 70- Mem: Soil Sci Soc Am. Res: Forms of inorganic phosphate; phosphorus in lake waters and sediments; forms of iron, manganese, sulphur and minor elements in lake sediments. Mailing Add: Can Centre for Inland Waters Burlington ON Can

WILLIAMS, KENNETH BOCK, b Petersburg, Tex, Jan 18, 30; m 52; c 2. PLANT TAXONOMY, ZOOLOGY. Educ: Abilene Christian Col, BS, 50; Univ Tex, MA, 59; Univ Ariz, PhD(bot), 67. Prof Exp: Asst prof, 67-73, ASSOC PROF BIOL, ABILENE CHRISTIAN COL, 73- Res: Biosystematic studies in the Gramineae. Mailing Add: Dept of Biol Abilene Christian Col Sta ACC Abilene TX 79601

WILLIAMS, KENNETH C, inorganic chemistry, organometallic chemistry, see 12th edition

WILLIAMS, KENNETH IRVINE HARVEY, organic chemistry, see 12th edition

WILLIAMS, KENNETH L, b Saybrook, Ill, Sept 4, 34; m 54; c 3. ZOOLOGY. Educ: Univ Ill, Urbana, BS, 60, MS, 61; La State Univ, Baton Rouge, PhD(zool), 70. Prof Exp: Instr comp anat & biol, Millikin Univ, 62-64; ASSOC PROF ZOOL & BIOL, NORTHWESTERN UNIV, 66- Concurrent Pos: Sigma Xi grant, La State Univ, 66; NSF fel, Northwestern State Univ, 68, Sigma Xi grant, 71. Mem: Am Soc Ichthyol & Herpet; Soc Study Amphibians & Reptiles; Soc Syst Zool. Res: Systematics and anatomy. Mailing Add: Dept of Biol Sci Northwestern State Univ Natchitoches LA 71457

WILLIAMS, KENNETH ROGER, b New York, NY, Aug 23, 22; m 49; c 1. TEXTILE PHYSICS. Educ: Columbia Univ, AB, 48, MA, 49. Prof Exp: Res physicist, 49-63, RES ASSOC, E I DU PONT DE NEMOURS & CO, INC, 63- Mem: Opers Res Soc Am; Am Phys Soc. Res: Mechanical properties of fibers; high polymer physics; economic analysis of new ventures; design of experiments. Mailing Add: Textile Res Lab E I du Pont de Nemours & Co Inc Wilmington DE 19898

WILLIAMS, KENNETH STUART, b Croydon, Eng, Aug 20, 40; m 62; c 3. MATHEMATICS. Educ: Univ Birmingham, BSc, 62; Univ Toronto, MA, 63, PhD(math), 65. Prof Exp: Lectr math, Univ Manchester, 65-66; from asst prof to assoc prof, 66-75, PROF MATH, CARLETON UNIV, 75- Mem: Math Asn Am; Can Math Cong; Am Math Soc. Res: Theory of numbers. Mailing Add: Dept of Math Carleton Univ Ottawa ON Can

WILLIAMS, LANSING EARL, b Spencer, WVa, Aug 8, 21; m 46; c 2. PLANT PATHOLOGY. Educ: Morris Harvey Col, BSc, 50; Ohio State Univ, MSc, 52, PhD(bot, plant path), 54. Prof Exp: Lab asst, Morris Harvey Col, 49-50; lab asst, Ohio State Univ, 50-52, from instr to assoc prof, 54-65, PROF BOT & PLANT PATH, OHIO STATE UNIV & OHIO AGR RES & DEVELOP CTR, 65-, ASSOC CHMN DEPT, 68- Mem: AAAS; Am Phytopath Soc; Sigma Xi; Nat Res Soc. Res: Corn viruses and stalk rot; mycotoxins; relation of soil fungal flora to soil-borne plant pathogens. Mailing Add: Dept of Plant Path Ohio Agr Res & Develop Ctr Wooster OH 44691

WILLIAMS, LARRY GALE, b Lincoln, Nebr, Sept 28, 39; m 62; c 2. GENETICS. Educ: Univ Nebr, Lincoln, BS, 61, MS, 63; Calif Inst Technol, PhD(biochem), 68. Prof Exp: NIH fel bot, Univ Mich, Ann Arbor, 67-71; ASST PROF BIOL, KANS STATE UNIV, 71- Mem: AAAS; Am Soc Microbiol. Res: Pyrimidine metabolism and enzymatic studies of cytoplasmic mutants of Neurospura crassa. Mailing Add: Div of Biol Kans State Univ Manhattan KS 66502

WILLIAMS, LAURENCE L, organic chemistry, see 12th edition

WILLIAMS, LAWRENCE ERNEST, b Youngstown, Ohio, Nov 29, 37; m 66; c 2. NUCLEAR MEDICINE, BIOPHYSICS. Educ: Carnegie-Mellon Univ, BS, 59; Univ Minn, Minneapolis, MS, 62, PhD(physics), 65. Prof Exp: Sr sci officer, Rutherford High Energy Lab, Eng, 65-68; asst prof physics, Western Ill Univ, 68-70; ASST PROF RADIOL, UNIV MINN, MINNEAPOLIS, 73- Concurrent Pos: NIH spec fel nuclear med, Nuclear Med Clin, Univ Minn, Minneapolis, 71-73; NIH grant, 74. Mem: Am Phys Soc; Am Asn Physicists in Med. Res: Computers; nuclear giant resonance; emphysema and physical models of the human lung; transit time distributions of physiological systems; image enhancement. Mailing Add: Box 382 Mayo Bldg Univ of Minn Hosps Minneapolis MN 55455

WILLIAMS, LEAH ANN, b Clarksburg, WVa, July 20, 32. DEVELOPMENTAL BIOLOGY. Educ: WVa Univ, AB, 54, MS, 58, PhD(biol), 70. Prof Exp: Instr anat & physiol, Exten, Pa State Univ, 58-59; instr gen zool, anat & physiol, W Liberty State Col, summer 59; from instr to asst prof, 59-73, ASSOC PROF BIOL, W VA UNIV, 73-, ASSOC CHMN DEPT, 75- Mem: AAAS; Am Soc Zool; Soc Develop Biol. Res: Regeneration; control mechanisms in the regenerative processes in the eyes of newts. Mailing Add: Dept of Biol WVa Univ Morgantown WV 26506

WILLIAMS, LEAMON DALE, b Flippin, Ark, Sept 28, 35; m 61; c 2. FOOD SCIENCE, CHEMISTRY. Educ: Univ Ark, BS, 58, MS, 61; Mich State Univ, PhD(food sci, biochem), 63. Prof Exp: Res chemist, Foods Div, Anderson, Clayton & Co, Tex, 63-69; DIR RES, FOOD & CHEM RES DEPT, CENT SOYA CO, 69- Honors & Awards: MacGee Award, Am Oil Chem Soc, 63. Mem: Poultry Sci Asn; Inst Food Technologists; Am Oil Chem Soc. Res: Organic chemistry of lipids; esterifiability of hydroxyls; interesterification and development of fat based derivatives; protein chemistry. Mailing Add: Food & Chem Res Dept Cent Soya Co 1825 N Laramie Chicago IL 60639

WILLIAMS, LELAND HENDRY, b Columbia, SC, Feb 24, 30; m 52; c 2. MATHEMATICS, COMPUTER SCIENCE. Educ: Univ SC, BS, 50; Univ GA, MS, 51; Duke Univ, PhD(math), 61. Prof Exp: Mathematician, Redstone Arsenal, 51-53; res assoc math & vis asst prof, Duke Univ, 60-62; math consult comput, Fla State Univ, 62-64, asst dir comput ctr, 64-66, asst prof math, Univ, 62-66; dir comput ctr & assoc prof math, Auburn Univ, 66-70; PRES & DIR, TRIANGLE UNIVS COMPUT CTR, 70- Concurrent Pos: Lectr, NSF math inst, Stetson Univ, 61- consult, US Army Res Off-Univ Durham, 63; assoc dir & lectr, NSF comput inst, Fla State Univ, 66; adj assoc prof, Duke Univ, Univ NC, Chapel Hill & NC State Univ, 70- Mem: Math Asn Am; Am Sci Affil; Am Math Soc; Asn Comput Mach. Res: Numerical analysis; nonnumeric mathematical computation; computation center management. Mailing Add: Triangle Univs Comput Ctr Box 12076 Research Triangle Park NC 27709

WILLIAMS, LESLEY LATTIN, b New Bedford, Mass, Aug 10, 39. PHYSICAL CHEMISTRY. Educ: Hollins Col, AB, 61, Univ Wis-Madison, PhD(chem), 68. Prof Exp: Asst prof, 68-74, ASSOC PROF CHEM, CHICAGO STATE UNIV, 74- Concurrent Pos: Lectr, Univ Md, Munich Campus, 71-72. Mem: AAAS; Am Phys Soc. Res: Nuclear magnetic resonance relaxation mechanisms in inorganic fluorides, including solvent effects; hexafluorides. Mailing Add: Dept of Phys Sci Chicago State Univ Chicago IL 60628

WILLIAMS, LEWIS DAVID, b Hopkinsville, Ky, Apr 2, 44; m 71. CHEMISTRY. Educ: Univ Chicago, BS, 66; Harvard Univ, PhD(org chem), 71. Prof Exp: Atholl McBean fel chem, Stanford Res Inst, 70-71; Presidential Intern, Western Regional Res Lab, Agr Res Serv, USDA, Albany, 62-73; BIOCHEMIST, DIAG DATA, INC, MT VIEW, 73- Mem: Am Chem Soc. Res: Physical organic chemistry; structure-reativity relationships; biochemistry. Mailing Add: 11770 Par Ave Los Altos CA 94022

WILLIAMS, LLEWELYN, economic botany, see 12th edition

WILLIAMS, LLOYD BAYARD, b Corvallis, Ore, Sept 28, 13; m 41; c 2. MATHEMATICAL ANALYSIS. Educ: Reed Col, BA, 35; Univ Chicago, SM, 39. Prof Exp: Instr math, Ga Inst Technol, 40-42; from instr to asst prof, Hamilton Col, 42-47; from asst prof to assoc prof, Reed Col, 47-57; assoc prof, Wesleyan Univ, 57-58; PROF MATH, REED COL, 58- Mem: Math Asn Am. Res: Preparation of undergraduate mathematics courses, primarily in analysis; convolution quotients in differential equations. Mailing Add: Dept of Math Reed Col Portland OR 97202

WILLIAMS, LLOYD K, mathematics, see 12th edition

WILLIAMS, LORING RIDER, b Buckhannon, WVa, Jan 6, 07; m 41; c 2. INORGANIC CHEMISTRY. Educ: WVa Wesleyan Col, BS, 27; WVa Univ, MS, 32; Univ Ill, PhD(inorg chem), 39. Prof Exp: Teacher high sch, WVa, 27-31; instr chem, Alderson-Broaddus Col, 32-34; teacher high sch, WVa, 34-38; teacher chem univ high sch, Univ Ill, 38-39; from instr to prof chem, 39-72, chmn dept, 57-61, EMER PROF CHEM, UNIV NEV, RENO, 72- Mem: AAAS; Am Chem Sco. Res: Distribution of selenium in plants and soils. Mailing Add: 4975 Malapi Way Sparks NV 89431

WILLIAMS, LOUIS FRANCIS, JR, b Washington, DC, Dec 17, 32; m 65. COMPUTER SCIENCES. Educ: The Citadel, BS, 54; Univ Mich, Ann Arbor, MSE, 60; Univ Fla, PhD(math), 70. Prof Exp: Instr math, Univ Fla, 70-71; ASST PROF MATH, UNIV S ALA, 71- Mem: Am Math Soc; Soc Indust & Appl Math. Res: Applying computers to mathematics to aid in the learning of mathematical concepts; efficient numerical and statistical computing. Mailing Add: Dept of Math Univ of SAla Mobile AL 36688

WILLIAMS, LOUIS GRESSETT, b Owensboro, Ky, Oct 28, 13; m 42; c 2. FRESH WATER ECOLOGY, ALGOLOGY. Educ: Marshall Univ, AB, 37; Duke Univ, MA, 40, PhD(biol), 48. Prof Exp: Asst, Marshall Col, 37-38; teacher high sch, Fla, 39-40 & NC, 40-41; asst, Duke Univ, 46-47; instr bot, Univ NC, 48; assoc prof biol, Furman Univ, 48-58; in charge, USPHS Plankton Prog, Nat Water Qual Network, Ohio, 58-65; in charge plankton prog, Nat Water Qual Lab, Minn, 65-67; PROF BIOL, UNIV ALA, TUSCALOOSA, 67- Concurrent Pos: Carnegie grant, 49; Ford Found fel, Univ Calif, 51-52. Honors & Awards: Jefferson Award, 51. Mem: Fel AAAS; Bot Soc Am; Am Soc Limnol & Oceanog; Ecol Soc Am; Phycol Soc Am. Res: Plankton ecology of rivers and Great Lakes; ecology of freshwater diatoms; marine algae of tropical Atlantic Ocean; science educational television; cycling of heavy metals between ooze and plankton; movement of radionuclides between aquatic food chains; cycling of heavy metals and radionuclides in food webs. Mailing Add: 1246 Northwood Lake Northport AL 35476

WILLIAMS, LOUIS OTHO, b Jackson, Wyo, Dec 16, 08; m 34. SYSTEMATIC BOTANY. Educ: Univ Wyo, AB, 32, MS, 33; Wash Univ, PhD(bot), 36. Prof Exp: Asst, Bot Mus, Harvard Univ, 36-37, res assoc, 37-46; botanist, United Fruit Co, 46-57; econ botanist, Agr Res Serv, USDA, Md, 57-60; from curator to chief curator, Cent Am Bot, 60-70, chmn dept bot, 71-74, EMER CURATOR, FIELD MUS NATURAL HIST, 75- Concurrent Pos: Res assoc, Mass Inst Technol, 47-51; Consul of Guatemala, Chicago, 67-74. Res: Flora of Guatemala; native and economic flora of Central America. Mailing Add: Field Mus of Natural Hist E Roosevelt Rd & S Lakeshore Chicago IL 60605

WILLIAMS, LUTHER STEWARD, b Sawyerville, Ala, Aug 19, 40; m 63. MOLECULAR BIOLOGY. Educ: Miles Col, BA, 61; Atlanta Univ, MS, 63; Purdue Univ, PhD(molecular biol), 68. Prof Exp: Lab instr biol, Spelman Col, 61-62; lab instr, Atlanta Univ, 62-63, instr, 63-64; teaching asst, Purdue Univ, 64-66; Am Cancer Soc fel, State Univ NY Stony Brook, 68-69; asst prof, Atlanta Univ, 69-70; asst prof biol sci, 70-73, ASSOC PROF BIOL, PURDUE UNIV, WEST LAFAYETTE, 73- Concurrent Pos: NSF teaching asst, 62-63; NIH career develop award, Purdue Univ, 71-75; assoc prof biol, Mass Inst Technol, 73-74; chmn MARC Prog, Nat Inst Gen Med Sci, 75-76. Mem: AAAS; Am Soc Microbiol; Am Chem Soc; Am Soc Biol Chemists. Res: Physiological role of aminoacyl-transfer RNA synthetases in bacterial metabolism. Mailing Add: Dept of Biol Sci Purdue Univ West Lafayette IN 47907

WILLIAMS, LYMAN O, b State College, Pa, Apr 1, 34; m 63; c 2. STRUCTURAL GEOLOGY. Educ: Univ Ga, BS, 56, Univ Iowa, MS, 59, PhD(geol), 62. Prof Exp: Explor geologist, Calif Co, 61-63; asst prof geol, Monmouth Col, 63-64; assoc prof, Eastern Tenn State Univ, 64-69; assoc prof, 69-73, PROF GEOL, MONMOUTH COL, 73- Mem: Geol Soc Am. Res: Petrology and structure of crystalline rock terranes; remote sensing of environment. Mailing Add: Dept of Geol Monmouth Col Monmouth IL 61462

WILLIAMS, LYNN DOLORES, b Seattle, Wash, Mar 20, 44. MATHEMATICAL ANALYSIS. Educ: Lewis & Clark Col, BS, 66; Univ Ore, MS, 70, PhD(math), 72. Prof Exp: Vis asst prof math, Univ Ore, 72-73; ASST PROF MATH, LA STATE UNIV, BATON ROUGE, 73- Mem: Am Math Soc; Sigma Xi; Math Asn Am; Asn Women in Math. Res: Functional analysis, especially Banach algebras of operators, Fredholm operator theory and approximate identities in Banach algebras. Mailing Add: Dept of Math La State Univ Baton Rouge LA 70808

WILLIAMS, LYNN ROY, b Detroit, Mich, Apr 23, 45; m 67; c 1. MATHEMATICS. Educ: King Col, BA, 67; Univ Ky, MA, 68, PhD(math), 71. Prof Exp: Asst prof math, La State Univ, Baton Rouge, 71-75; ASST PROF MATH, IND UNIV, SOUTH BEND, 75- Mem: Am Math Soc. Res: Functional analysis; Hp theory; harmonic analysis. Mailing Add: Dept of Math Ind Univ South Bend IN 46615

WILLIAMS, MARION ERVIN, b Huntsville, Ala, Apr 24, 27; m 55; c 1. MYCOLOGY. Educ: Ind Univ, BA, 47, MA, 48; Univ Iowa, PhD, 54. Prof Exp: Instr biol, Dillard Univ, 48-50; asst bot, Univ Iowa, 51-54; assoc prof biol, Prairie View Agr & Mech Col, 54-55; assoc prof sci, Ala State Col, 55-56; from asst prof to assoc prof, 56-71, PROF BIOL, FISK UNIV, 71- Mem: Mycol Soc Am; Bot Soc Am. Res: Tremellaceous fungi. Mailing Add: Dept of Biol Fisk Univ Nashville TN 37203

WILLIAMS, MARION JACK, b Hochheim, Tex, Oct 18, 28; m 46; c 2. THORACIC SURGERY, SURGERY. Educ: Univ Houston, BA, 49; Univ Tex, MD, Dallas, 53. Prof Exp: Resident, Univ Tex Southwestern Med Sch Dallas, 56-60; resident US Air Force, 60-, asst chief surg, Carswell AFB Hosp, Tex, 60-62, chief, Itazuke Air Base Hosp, Japan, 62-65, resident thoracic-cardiovasc surg, Wilford Hall Med Ctr, 65-67, chief thoracic surg, Med Ctr, Keesler AFB, 67-74, chmn dept surg & dir surg residency, 68-74; PROF SURG, MED SCH, TULANE UNIV, 74- ACTG CHIEF SURG, 75- Concurrent Pos: Res fel surg, Univ Tex Southwestern Med Sch Dallas, 54; assoc clin prof surg, Med Sch, Tulane Univ, 67-; sr consult, Surgeon Gen, US Air Force, 68-; consult surg, Charity Hosp, New Orleans, Vet Admin Hosp, Alexandria, La, Huey P Long Charity Hosp & Lallie Kemp Charity Hosp. Mem: Soc Thoracic Surg; fel Am Col Surg; AMA. Res: Fluid and electrolyte balance; clinical surgery. Mailing Add: Dept of Surg Tulane Univ Med Sch New Orleans LA 70112

WILLIAMS, MARION PORTER, b Salem, Ind, Jan 24, 46; m 68. FOOD SCIENCE. Educ: Purdue Univ, BS, 68, PhD(food sci), 73. Prof Exp: From sr food scientist to sr res scientist, 73-75, MGR INT RES & PROD DEVELOP, RES LABS, CARNATION CO, 75- Mem: Inst Food Technologists. Res: Coordination and planning of international research and development. Mailing Add: Carnation Res Labs 8015 Van Nuys Blvd Van Nuys CA 91412

WILLIAMS, MARSHALL HENRY, JR, b New Haven, Conn, July 15, 25; m 48; c 4. PHYSIOLOGY, INTERNAL MEDICINE. Educ: Yale Univ, BS, 45, MD, 47. Prof Exp: Intern, Presby Hosp, New York, 47-48, asst resident med, 48-49; asst resident, New Haven Hosp, Conn, 49-50, asst, 50; chief respiratory sect, Dept Cardiorespiratory Dis, Army Med Serv Grad Sch, Walter Reed Army Hosp, 52-55; dir cardiorespiratory lab, Grasslands Hosp, Valhalla, NY, 55-59; vis asst prof physiol, 55-59, assoc prof med & physiol, 59-66, PROF MED, ALBERT EINSTEIN COL MED, 66- Concurrent Pos: NIH trainee, New Haven Hosp, Conn, 50; dir chest serv, Bronx Munic Hosp Ctr, New York, 59- Mem: AAAS; Am Physiol Soc; Am Thoracic Soc; Am Soc Clin Invest; Am Heart Asn. Res: Respiratory and clinical cardiopulmonary physiology. Mailing Add: Albert Einstein Col of Med Yeshiva Univ New York NY 10461

WILLIAMS, MARTIN BARBOUR, b Centreville, Ala, Nov 19, 17. CHEMISTRY. Educ: Univ Ala, BS, 39, MS, 40. Prof Exp: Chemist, Int Paper Co, Ark, 40-41; instr chem, Univ Ala, 41-42 & 44; chemist, Tenn Copper Co, 42; prof, Whitworth Col, Spokane, Wash, 42-43; commodity specialist, For Econ Admin, Washington, DC, 44-46; chem commodities specialist, For Trade Div, Econ & Sci Sect, Gen Hq, Supreme Comdr Allied Powers, US Army, Japan, 46-47, chem engr, Indust Div, 47-50, chief chemist, Criminal Invest Lab, Gen Hq, Far East Command, 50-53; dir res, Thomas Ala Kaolin Co, 54; mat engr, Naval Air Sta, Fla, 55-56; chemist, Adv Propellant Res & Develop, Propulsion Lab, Res & Develop Directorate, US Army Missile Command, Redstone Arsenal, 56-62; chief, Huntsville Tech Opers Div, Armed Serv Tech Info Agency, Ala, 62; develop engr, Westing-Arc Div, Westinghouse Elec Corp, 63-64; indust rep, Ala State Planning & Indust Develop Bd, 65; INSTR CHEM, MARION INST, 65- Concurrent Pos: Instr, Tokyo Armed Forces Educ Ctr, 50-51 & Univ Ala, Huntsville Ctr, 57-59; coordr, Joint Army-Navy-Air Force-Advan Res Projs Agency-NASA solid propellant group meeting, Colo, 61; dir, Am Inst Chem, 61-67. Mem: Am Chem Soc; The Chem Soc; Am Inst Chem. Res: Chemical economics; analytical chemistry; high-energy rocket propellants; forensic science; industrial development; fine chemicals; kraft paper; welding rod coatings. Mailing Add: 549 Walnut St Centreville AL 35042

WILLIAMS, MARTIN WESLEY, b Columbus, Ohio, May 11, 21; m 62; c 5. TOXICOLOGY. Educ: Anderson Col, BA, 48; Ohio State Univ, PhD, 53. Prof Exp: Instr physiol, Univ Vt, 53-56; toxicologist, US Food & Drug Admin, 56-58; assoc prof pharmacol, Univ Vt, 58-61; toxicologist, Vet Admin Hosp, Tucson, Ariz, 62-70; PROF BIOL & CHMN DEPT, PIKEVILLE COL, 70- Mem: AAAS; Am Soc Pharmacol & Exp Therapeut; Soc Exp Biol & Med; Soc Toxicol; Am Soc Clin Pharmacol & Therapeut. Res: Nutritional factors on tumor growth; toxicity of ant venom; ischemic pain and analgesia; remote telemetry of physiological parameters. Mailing Add: Box 45 Pikeville Col Pikeville KY 41501

WILLIAMS, MARY ANN, b Albany, NY, May 18, 25. NUTRITION, BIOCHEMISTRY. Educ: Iowa State Col, BS, 46; Cornell Univ, MS, 50; Univ Calif, PhD, 54. Prof Exp: Asst pathologist, Univ Ky, 49-51; asst nutrit, Univ Calif, 51-54; res assoc, McCollum-Pratt Inst, John's Hopkins Univ, 54-55; from instr to asst prof, 55-63, assoc prof, 63-75, PROF NUTRIT, UNIV CALIF, BERKELEY, 75- Concurrent Pos: Guggenheim fel, 63-64. Mem: AAAS; Am Chem Soc; Am Inst Nutrit; Soc Exp Biol & Med. Res: Essential fatty acid metabolism and functions; metabolic effects of feeding patterns. Mailing Add: Dept of Nutrit Sci 119 Morgan Hall Univ of Calif Berkeley CA 94720

WILLIAMS, MARY BEARDEN, b Lexington, Ky, Aug 29, 36. EVOLUTIONARY BIOLOGY, PHILOSOPHY OF SCIENCE. Educ: Reed Col, BA, 58; Univ Pa, MA, 61; Univ London, PhD(math biol) & DIC, 67. Prof Exp: Res assoc biomath, Univ Tex M D Anderson Hosp & Tumor Inst, 63-64; asst prof, NC State Univ, 67-73; ASST PROF PHILOS OHIO STATE UNIV, 74- Concurrent Pos: Vis asst prof hist & philos of sci, Ind Univ, 73-74. Mem: Am Soc Nat; Soc Study Evolution; Math Asn Am; Philos Sci Asn; Soc Syst Zool. Res: Axiomatization of evolutionary theory; logical status of evolutionary predictions; evolution of population self-regulation; philosophy of biology; bioethics. Mailing Add: Dept of Philos Ohio State Univ Columbus OH 43210

WILLIAMS, MARY CARR, b Port Arthur, Tex, Dec 25, 26; m 51; c 1. STEROID CHEMISTRY. Educ: Tex Woman's Univ, BA & BS, 49; St Mary's Univ, MS, 75. Prof Exp: Res chemist lipid metab, Dept Biochem & Biophys, Tex A&M Univ, 49-63; RES SCIENTIST STEROID METAB, DEPT ENDOCRINOL, SOUTHWEST FOUND RES & EDUC, 63- Mem: Am Chem Soc. Res: Metabolism of natural and synthetic steroid hormones. Mailing Add: Southwest Found Res & Educ PO Box 28147 San Antonio TX 78284

WILLIAMS, MARY ELIZABETH, b Augusta, Ky, Nov 30, 09. MATHEMATICS. Educ: Radcliffe Col, AB, 33; Univ Ky, MA, 36. Instr high sch & jr col, Ky, 36-41; asst prof, Marshall Col, 42-46; assoc prof, 46-57, PROF MATH, SKIDMORE COL, 57- Mem: Am Math Soc; Math Asn Am. Res: American women in mathematics in the nineteenth century. Mailing Add: Dept of Math Skidmore Col Saratoga Springs NY 12866

WILLIAMS, MARY LOUISE MONICA FRITTS, b Detroit, Mich, Apr 16, 40; m 68; c 1. ANATOMY, EMBRYOLOGY. Educ: Univ Detroit, BS, 62, MS, 64; Wayne State Univ, PhD(embryol, anat), 71. Prof Exp: Instr microbiol & embryol, Mercy Col, Mich, 64-67, asst prof, 67-68; asst prof microbiol, Univ Detroit, 68-69; ASST PROF ANAT, MED SCH, WAYNE STATE UNIV, 71- Mem: AAAS; Genetics Soc Am; Am Soc Microbiol; Soc Develop Biol. Res: Primordial germ cells; gonadal development; tissue culture. Mailing Add: Dept of Anat Wayne State Univ Detroit MI 48201

WILLIAMS, MARYON JOHNSTON, JR, b Griffin, Ga, Jan 14, 45; m 68; c 2. BIOMEDICAL ENGINEERING, ELECTRICAL ENGINEERING. Educ: Ga Inst Technol, BEE, 68; Rutgers Univ, MS, 70, PhD(biomed eng), 72. Prof Exp: Electronic engr radar, US Army Electronics Command, Ft Monmouth, NJ, 68-69; instr biomed eng, 72-74, LECTR PHYSIOL, MED COL GA, 73-, ASST PROF HEALTH SYSTS & INFO SCI, BIOMED ENG SECT, 74- Concurrent Pos: Ga Heart Asn res grant, Med Col Ga, 73-75. Mem: Inst Elec & Electronics Eng; Asn Advan Med Instrumentation; Biomed Eng Soc. Res: Heart assist devices; computer monitoring; blood glucose dynamics. Mailing Add: Sect of Biomed Eng Med Col of Ga Augusta GA 30902

WILLIAMS, MAX BULLOCK, b Provo, Utah, Mar 30, 15; m 41; c 5. ANALYTICAL CHEMISTRY. Educ: Univ Utah, BS, 36, MS, 38; Cornell Univ, PhD(phys chem), 41. Prof Exp: Instr, 41-44, from asst prof to assoc prof, 44-57, PROF CHEM, ORE STATE UNIV, 57- Mem: Am Chem Soc. Res: X-ray crystallography; azeotropic distillation; spectrophotometry; chemical microscopy; chelating ion exchange resins. Mailing Add: Dept of Chem Ore State Univ Corvallis OR 97331

WILLIAMS, MAX W, b Cardston, Alta, Aug 24, 30; US citizen; m 54; c 4. PLANT PHYSIOLOGY, HORTICULTURE. Educ: Utah State Univ, BSc, 54, MSc, 57; Wash State Univ, PhD(hort), 61. Prof Exp: Res asst, Utah State Univ, 54-55; actg supt, Utah Tree Fruit Exp Sta, 55-58; res asst, Wash State Univ, 58-61; RES LEADER, TECH ADV & PLANT PHYSIOLOGIST, AGR RES SERV, USDA, 61- Mem: Am Soc Hort Sci; Int Soc Hort Sci. Res: Chemical thinning of apples; growth retardants; cytokinins; auxins. Mailing Add: Box 99 PO Annex 106 Wenatchee WA 98801

WILLIAMS, MELVIN DONALD, b Pittsburgh, Pa, Feb 3, 33; m 58; c 3. ANTHROPOLOGY. Educ: Univ Pittsburgh, BA, 55, MA, 69, PhD(anthrop), 73. Prof Exp: Instr sociol & anthrop, Carlow Col, 69-72, asst prof, 72-76, chmn dept sociol & anthrop, 73-76; ASSOC PROF ANTHROP, UNIV PITTSBURGH, 76- Concurrent Pos: Assoc ed, J of Ethnol, 76- Mem: AAAS; fel Am Anthrop Asn; Am Sociol Asn; African Studies Asn. Res: Ethnicity; urban studies; Africa; Afro-Americans; psychological anthropology; social psychiatry, North West Coast. Mailing Add: Dept of Anthrop Univ of Pittsburgh Pittsburgh PA 15260

WILLIAMS, MERLIN CHARLES, b Howard, SDak, July 20, 31; m 59; c 4. METEOROLOGY, ENGINEERING. Educ: SDak State Univ, BS, 53; Univ Chicago, cert, 54; Univ Wyo, MS, 62. Prof Exp: Instr civil eng, SDak State Univ, 57-58; instr & res asst weather eng, 58-59; engr, US Bur Reclamation, 59-61; asst civil eng, Univ Wyo, 61-62, proj dir weather modification res, 62-66; dir weather modification res, Fresno State Col Found, 66-71; dir, SDak State Weather Control Comn, 71-74; DIR, OFF WEATHER MODIFICATION PROGS, ENVIRON RES LABS, NAT OCEANIC & ATMOSPHERIC ADMIN, 74- Concurrent Pos: Consult adv bd weather modification, NSF, 75- Mem: Am Soc Civil Eng; Am Meteorol Soc; Am Geophys Union; Weather Modification Asn (pres, 69). Res: Water resources research to investigate increasing water supplies, including weather modification research, fluid mechanics and hydrology; basic hydrometeorological studies; mountain meteorology; snow physics; hurricane modification (abatement), boundary layer dynamics. Mailing Add: ERL-NOAA Off Weather Modification Progs Boulder CO 80302

WILLIAMS, MEURIG W, organic chemistry, see 12th edition

WILLIAMS, MICHAEL EUGENE, b Ina, Ill, Aug 4, 40; m 68. VERTEBRATE PALEONTOLOGY. Educ: Mo Sch Mines & Metall, Rolla, BS, 63; Univ Kans, MS, 72. Prof Exp: CUR VERT PALEONT, CLEVELAND MUS NATURAL HIST, 76- Mem: Soc Vert Paleont; Paleont Soc; Int Paleont Union. Res: Paleozoic fishes with special emphasis on chondrichthyans; sedimentation and environments of deposition of various black shale units. Mailing Add: Cleveland Mus of Natural Hist Wade Oval Univ Circle Cleveland OH 44106

WILLIAMS, MICHAEL JOHN, b July 2, 42; Brit citizen; m 73; c 1. TEXTILE CHEMISTRY. Educ: Univ Leeds, BSc, 64, PhD(textile chem), 67. Prof Exp: Assoc res scientist, Dept Textiles, 67-69, res scientist, Dept Textiles Clothing & Footwear, 69-72, ASST DIR, DEPT TEXTILES, CLOTHING & FOOTWEAR, ONT RES FOUND, 72- Mem: Soc Res Adminrs; Inst Textile Sci. Res: Textile flammability; specification development; development of fabrics and finishes for specialized end uses; effects of pollution on textiles; control of prevention of static electricity. Mailing Add: Ont Res Found Sheridan Park Mississauga ON Can

WILLIAMS, MICHAEL LEDELL, b Paragould, Ark, Sept 11, 43; m 63; c 2. ENTOMOLOGY, SYSTEMATICS. Educ: Ark State Univ, BS, 67; Va Polytech Inst & State Univ, MS, 69, PhD(entom), 72. Prof Exp: Asst entomologist, Md Dept Agr, 71-73; ASST PROF, DEPT ZOOL-ENTOM, AUBURN UNIV, 73- Concurrent Pos: Sigma Xi res award, 69. Mem: Entom Soc Am; Sigma Xi. Res: Insular speciation of scale insects of the Galapagos Islands; systematics and morphology of New World Coccidae of the suborder homoptera; natural host plant resistance to scale insects; scale insects of Alabama. Mailing Add: Dept of Zool-Entom Auburn Univ Auburn AL 36830

WILLIAMS, MILES COBURN, b Osage City, Kans, Jan 28, 29; m 53; c 2. PLANT PHYSIOLOGY. Educ: Kansas State Univ, BS & MS, 51; Univ Ill, PhD(agron), 56. Prof Exp: PLANT PHYSIOLOGIST, AGR RES SERV, USDA, 56- Mem: Weed Sci Soc Am; Soc Range Mgt. Res: Biochemical and physiological research on poisonous range weeds, especially methods of chemical and biological control; Astragalus, Delphinium. Mailing Add: Agr Res Serv USDA Dept of Biol Utah State Univ Logan UT 84322

WILLIAMS, MYRA NICOL, b Dallas, Tex, June 8, 41; m 68; c 2. MOLECULAR BIOPHYSICS. Educ: Southern Methodist Univ, BS, 64; Yale Univ, MS, 65, PhD(molecular biophys), 68. Prof Exp: RES FEL BIOPHYS, MERCK INST THERAPEUT RES, MERCK & CO, INC, 69- Mem: Biophys Soc. Res: Structure and function of proteins; biophysical and biochemical studies on dihydrofolate reductase, including nuclear magnetic resonance, fluorescence and ultraviolet difference spectroscopic studies, chemical modifications and enzyme kinetics. Mailing Add: Merck Inst Therapeut Res Rahway NJ 07065

WILLIAMS, NEAL THOMAS, b East Orange, NJ, Mar 16, 21; m 48; c 2. ENGINEERING PHYSICS. Educ: Cornell Univ, AB, 48. Prof Exp: Supvr magnetron eng, Westinghouse Elec Co, 42-44, develop engr, 48-51; res assoc, Radiation Lab, Columbia Univ, 44-48, assoc res physicist, 52; mem tech staff, Bell Tel Labs, Inc, 51-52; chief engr, L L Constantin & Co, 52-53; res engr, T A Edison, Inc, 53-60; chief engr, Seal-A-Metic, Inc, 60-65; div mgr, Platronics, Inc, 65-66; PRES, PLATRONICS-SEALS, INC, CLIFTON, 66- Mem: Am Phys Soc; sr mem Inst Elec & Electronics Eng; NY Acad Sci. Res: Microwave magnetrons and electronics; traveling wave tubes and backward oscillators; radar duplexers; gas discharges; metal-ceramic seals; low voltage x-rays. Mailing Add: 36 Hawthorne Ave Bloomfield NJ 07003

WILLIAMS, NELSON NOEL, b Dayton, Ohio, Mar 10, 30; m 52; c 5. HUMAN ECOLOGY. Educ: Ohio State Univ, BSc, 55, MSc, 57, PhD(pollen anal), 62. Prof Exp: Botanist, US Geol Surv, 58-59; from instr to asst prof biol, Nev Southern Univ, 61-68; CONSULT HUMAN-ENVIRON PROBS LAB, 68-; DIR WIXOM ALLERGY CLIN, 70- Mem: AAAS; Ecol Soc Am. Res: Pollen analysis and airborne pollen and spores; radio-ecology. Mailing Add: 10363 La Cienega Las Vegas NV 89119

WILLIAMS, NORMAN, b Mirfield, Eng, June 15, 18; m 42; c 3. OCCUPATIONAL MEDICINE. Educ: Univ London, MB & BS, 47; London Sch Hyg & Trop Med, dipl pub health, 48, dipl indust health, 49; Univ Sask, MD, 62; Am Bd Prev Med, dipl & cert occup med, 59. Prof Exp: House surgeon, St Mary's Hosp, London, 42, supernumerary med registr, 46-47; med officer, London Transport Exec, 49-54; sr med officer, East Midlands Gas Bd, 54-57; dir occup health, Sask Dept Pub Health, Can, 57-67; PROF OCCUP MED, JEFFERSON MED COL, 67- Concurrent Pos: Lectr, Univ Sask, 61-67; mem adv comt x-ray safety stand, Can Dept Health & Welfare, 64-67; consult, Philadelphia Dept Pub Health, 67- Mem: Fel Indust Med Asn. Res: Segmental vibration exposure effects; toxicity of methyl normal butyl ketone; polymer-fume fever. Mailing Add: Dept of Commun Health & Prev Med Jefferson Med Col Philadelphia PA 19107

WILLIAMS, NORMAN DALE, b Nebr, Nov 4, 24; m 47; c 2. PLANT GENETICS. Educ: Univ Nebr, BS, 51, MS, 54, PhD(agron), 56. Prof Exp: Assoc genetics, Agronne Nat Lab, 54-56, res assoc, 56; GENETICIST, AGR RES SERV, USDA, 56- Concurrent Pos: Adj Prof, NDak State Univ, 61- Mem: AAAS; Am Soc Agron; Am Genetic Asn; Genetics Soc Am. Res: Genetic studies of host-parasite relationships, especially wheat and wheat stem rust; mutation induction. Mailing Add: USDA State Univ Sta Fargo ND 58102

WILLIAMS, NORMAN EUGENE, b Grove City, Pa, July 29, 28; m 53; c 2.

ZOOLOGY. Educ: Youngstown State Univ, AB, 52; Brown Univ, ScM, 54; Univ Calif, Los Angeles, PhD(zool), 58. Prof Exp: Instr, 57-59, from asst prof to assoc prof, 59-67, PROF ZOOL, UNIV IOWA, 67- Concurrent Pos: NIH ser fel, Carlsberg Found, Denmark, 63-64 & Dept Biol Struct, Univ Wash, 66-67. Mem: Soc Protozool; Am Soc Cell Biol. Res: Cellular development; synthesis and assembly of organellar proteins; electron microscopy. Mailing Add: Dept of Zool Univ of Iowa Iowa City IA 52240

WILLIAMS, NORRIS HAGAN, botany, see 12th edition

WILLIAMS, OLWEN, b Union, Conn, Jan 19, 17. VERTEBRATE ECOLOGY. Educ: Alfred Univ, BFA, 41; Univ Colo, MA, 51, PhD(zool), 52. Prof Exp: Instr sci, Putney Sch, Vt, 42-48; from instr to assoc prof biol, 51-67, PROF BIOL, UNIV COLO, BOULDER, 67- Concurrent Pos: Fulbright lectr, Pierce Univ Col, Athens, Greece, 65-66. Mem: AAAS; Am Ornith Union; Am Soc Mammal; Animal Behav Soc; Ecol Soc Am. Res: Avain and mammalian population and behavioral ecology; ecoenergetics of rodents and shrews; micotine biology. Mailing Add: Dept of EPO Biol Univ of Colo Boulder CO 80302

WILLIAMS, OREN FRANCIS, b Oakland City, Ind, Mar 5, 20; m 50; c 2. INORGANIC CHEMISTRY. Educ: Univ Toledo, BE, 43; Univ Ill, MS, 47, PhD(chem), 51. Prof Exp: Chemist, State Geol Surv, Ill, 44-47; res chemist, Food Mach & Chem Corp, 51-59; chemist, NIH, 59-61; asst prog dir chem, 61-64, PROG DIR INORG CHEM, NSF, DC, 64- Mem: AAAS; Am Chem Soc. Res: Organic chemistry; agricultural chemicals; organic fluorine and coordination compounds; research administration. Mailing Add: Prog Dir Inorg Chem NSF 1800 G St NW Washington DC 20006

WILLIAMS, OWEN WINGATE, b Trouville, France, Aug 24, 24; US citizen; m 46; c 5. GEODESY, GEOPHYSICS. Educ: Kalamazoo Col, AB, 48. Prof Exp: Geod engr, Army Map Serv, Washington, DC, 48-55; phys scientist, Aeronaut Chart & Info Center, Dept Air Force, 55-57, phys scientist & chief, Terrestrial Sci Lab, Air Force Cambridge Res Labs, Hanscom AFB, 57-72; ASST DEP DIR PLANS, REQUIREMENTS & TECHNOL, DEFENSE MAPPING AGENCY, WASHINGTON, DC, 72- Concurrent Pos: Guest lectr, Acad Sci USSR, 66. Honors & Awards: Meritorious Civilian Serv Award, US Air Force, 73. Mem: Fel AAAS; Am Geophys Union; Am Inst Aeronaut & Astronaut; Instrument Soc Am; Armed Forces Mgt Asn. Res: Theory, instrumentation and test of optical and laser celestial geodetic techniques; gravimetry; cartography; crustal physics. Mailing Add: 4703 Ponderosa Dr Annandale VA 22003

WILLIAMS, PATRICIA BELL, b Detroit, Mich. PHARMACOLOGY. Educ: Col Pharm, Univ Mich, BS, 68; Med Col Va, Va Commonwealth Univ, PhD(pharmacol), 72. Prof Exp: Lab asst pharmacol, Health Sci, Med Col Va, Va Commonwealth Univ, 68-70, teaching asst, 70-71; ASST PROF PHARMACOL, EASTERN VA MED SCH, 72- Concurrent Pos: Consult, United Drug Abuse Coun & Health Adv Coord Comt, Model Cities Comprehensive Health Sci Proj, 72; lectr, Sch Continuing Educ, Univ Va, 72; asst prof nursing & dent hyg, Old Dom Univ, 72-74; Tidewater Heart Asn res grant, 75. Mem: AAAS; Asn Am Med Cols; Asn Women in Sci; Am Fedn Clin Res. Res: Cardiovascular pharmacology and physiology of vascular smooth muscle with particular interest in the etiology and treatment of hypertension and peripheral vascular disease. Mailing Add: Dept of Pharmacol Eastern Va Med Sch PO Box 1980 Norfolk VA 23501

WILLIAMS, PATRICK KELLY, b San Angelo, Tex, July 31, 43; m 68. ECOLOGY. Educ: Univ Tex, Austin, BA, 66; Univ Minn, Minneapolis, MS, 69; Ind Univ, Bloomington, PhD(zool), 73. Prof Exp: ASST PROF BIOL, UNIV DAYTON, 73- Mem: Ecol Soc Am; Am Soc Limnol & Oceanog; Am Inst Biol Sci; Am Soc Ichthyologists & Herpetologists; AAAS. Res: Experimental population ecology on rodents with emphasis on natural regulation and management. Mailing Add: Dept of Ecol Univ of Dayton Dayton OH 45469

WILLIAMS, PAUL HUGH, b Vancouver, BC, May 6, 38; m 63. PLANT PATHOLOGY, PLANT PHYSIOLOGY. Educ: Univ BC, BSA, 59; Univ Wis, PhD(plant path), 62. Prof Exp: From asst prof to assoc prof, 62-71, PROF PLANT PATH, UNIV WIS-MADISON, 71- Mem: Am Phytopath Soc. Res: Physiology of host-parasite relations and resistance breeding for disease resistance in vegetables; light and electron microscope cytology of host-parasite relations. Mailing Add: Dept of Plant Path Univ of Wis Madison WI 53706

WILLIAMS, PAUL LINCOLN, geology, see 12th edition

WILLIAMS, PETER M b New York, NY, July 17, 27; m 58; c 3. CHEMICAL OCEANOGRAPHY. Educ: Washington & Lee Univ, BS, 49; Univ Calif, Los Angeles, MS, 58, PhD(oceanog), 60. Prof Exp: Asst res chemist, Smith, Kline & French Labs, 49-51; lab technician, Citrus Exp Sta, Univ Calif, Riverside, 53-54; asst prof marine chem, Inst Oceanog, Univ BC, 60-63; asst res chemist, 63-69, ASSOC RES CHEMIST, INST MARINE RESOURCES, UNIV CALIF, SAN DIEGO, 69- Mim: AAAS; Am Chem Soc; Am Soc Limnol & Oceanog. Res: Organic chemistry of sea water with respect to dissolved and patriculate organic matter derived from marine organisms which is present as dissolved and patriculate constituents in sea water. Mailing Add: Inst of Marine Resources A-018 Univ of Calif San Diego La Jolla CA 92093

WILLIAMS, PHILIP, b Providence, RI, June 17, 08; m 64; c 4. PROSTHODONTICS. Educ: Tufts Univ, DMD, 32. Prof Exp: Fel prosthodont, 32-33, instr, 33-43, asst prof oral prosthetics, 46-50, from assoc prof to prof prosthodont, 51-69, chmn dept, 57-69, PROF RESTORATIVE DENT, GRAD DIV, TUFTS UNIV, 69- Concurrent Pos: Clin researcher, Stewart Growth Study, Harvard Univ, 33-43, lectr, Sch Dent Med, 71-72; consult, Vet Admin, 54-72; lectr, World Cong-Int Dent Fedn, Mexico City, 72. Mem: AAAS; hon fel Am Col Dent; Am Prosthodont Soc; Am Acad Crown & Bridge Prosthodont (pres, 72). Res: Oral and dental occlusion, electronic transmission of dental materials. Mailing Add: 793 Lynnfield St Lynn MA 01904

WILLIAMS, PHILIP, JR, b San Francisco, Calif, Nov 2, 15; m 43; c 5. METEOROLOGY. Educ: Univ Calif, Berkeley, AB, 38; Univ Calif, Los Angeles, MA, 43. Prof Exp: Observer, 38-42, forecaster, 42-56, res forecaster, 56-64, asst chief sci serv dir, Regional Hq, 64-73, CHIEF METEOROL SERV DIV, NAT WEATHER SERV, 73- Honors & Awards: Nat Weather Serv Awards, 58-59, 67; Bronze Medal, US Dept Com, 67. Mem: Am Meteorol Soc. Res: Synoptic meteorology and short range forecasting. Mailing Add: Nat Weather Serv Box 11188 Salt Lake City UT 84106

WILLIAMS, PHILIP SIDNEY, b Abingtion, Va, June 15, 08; m 34; c 4. PHYSICS, GEOPHYSICS. Educ: Col of William & Mary, BS, 30; Wash Univ, MS, 32, PhD(physics), 34. Prof Exp: Asst physics, Wash Univ, 30-32, res assoc, Sch Med, 34-35; res geophysicist, Carter Oil Co, 35-37, chief geophys res, 37-45, chief res, 45-58; sr res adv, Jersey Prod Res Co, 58-62; ASSOC PROF PHYSICS, UNIV TULSA, 62- Mem: Am Phys Soc; Soc Explor Geophys. Res: Diffuse scattering of x-rays from crystals; spectrographic analysis of biological materials; geophysical prospecting for oil; oil finding and production research. Mailing Add: Dept of Physics Univ of Tulsa Tulsa OK 74104

WILLIAMS, PHLETUS P, b Junior, WVa, Aug 3, 33; m 60; c 3. MICROBIOLOGY, BIOCHEMISTRY. Educ: Davis & Elkins Col, BS, 55; Univ Md, MS, 59; NDak State Univ, PhD(animal nutrit), 68. Prof Exp: Microbiologist beef cattle res br, Animal Husb Res Div, Md, 59-60, dairy cattle res br, 60-61, beef cattle res br, 61-64, prof bact, 72-73, MICROBIOLOGIST, METAB & RADIATION RES LAB, N DAK STATE UNIV, AGR RES SERV, USDA, 64- ADJ PROF BACT, 73- Mem: AAAS; Am Soc Animal Sci; Am Soc Microbiol; Brit Soc Gen Microbiol. Res: Development of rumen protozoal controlled bovines; chemical, physiological, cultural and metabolical study of rumen bacteria and protozoa; microbial metabolic fate studies with lipoidal and pesticidal compounds. Mailing Add: Metab & Radiation Res Lab USDA State Univ Sta Fargo ND 58102

WILLIAMS, RALPH BENJAMIN, b Salt Lake City, Utah, Aug 19, 10; m 30; c 2. PUBLIC HEALTH, MICROBIOLOGY. Educ: Univ Utah, BS, 35; Univ NC, MPH, 62, DPH, 64; Am Bd Med Microbiol, dipl. Prof Exp: Res bacteriologist, Dept Bact & Path, Univ Utah, 35-37; bacteriologist, State Dept Health, Utah, 37-39, chief bacteriologist, 39-40; dir lab, State Dept Health, Wyo, 40-44; div labs, Territory Dept Health, State of Alaska, 44-59, chief br labs, Dept Health & Welfare, 59-70, dir div pub health & chief br community health, 68-70; PUB HEALTH CONSULT IN MICROBIOL & PARASITOL, 70- Concurrent Pos: Ornithologist & field naturalist, Southeastern Alaska. Mem: Am Soc Microbiol; Cooper Ornith Soc; Am Ornith Union; Arctic Inst NAm. Res: Enteric bacteriology; antibiotics; rabies, tubercle bacilli and ornithology; parasitology; distribution and migration of birds in Western North America; microbiological studies and ecto- and endo-parasitological investigations of birds and mammals. Mailing Add: PO Box 2354 Juneau AK 99803

WILLIAMS, RALPH C, JR, b Washington, DC, Feb 17, 28; m 51; c 4. INTERNAL MEDICINE, IMMUNOLOGY. Educ: Cornell Univ, AB, 50, MD, 54. Prof Exp: Guest investr immunol, Rockefeller Inst, 61-63; from asst prof to assoc prof med, Med Sch, Univ Minn, Minneapolis, 63-69; PROF MED & CHMN DEPT, SCH MED, UNIV N MEX, 69- Concurrent Pos: Consult, Bur Hearings & Appeals, Soc Security Admin, 65- Mem: Am Rheumatism Asn; Am Fedn Clin Res; Am Soc Clin Invest; Am Asn Immunol; Soc Exp Biol & Med. Res: Rheumatic diseases; immunopathology; immunoglobulin abnormalities and their relation to disease. Mailing Add: Dept of Med Seventh Floor Bernalillo County Med Ctr Albuquerque NM 87106

WILLIAMS, RAYMOND CRAWFORD, b Kansas City, Mo, Sept 22, 24; m 59; c 1. VETERINARY ANATOMY. Educ: Kans State Col, DVM, 46; Cornell Univ, MS, 55, PhD, 61. Prof Exp: Instr, 46-54, from asst prof to assoc prof, 54-64, PROF ANAT & HISTOL, SCH VET MED, TUSKEGEE INST, 64- Mem: AAAS; Am Vet Med Asn; Am Asn Vet Anat; World Asn Vet Anat. Res: Descriptive vertebrate anatomy; fetal size and age relationships; dentition development; anatomical museum methods. Mailing Add: Dept of Anat Sch of Vet Med Tuskegee Inst AL 36088

WILLIAMS, RAYMOND FRANCIS X, inorganic chemistry, see 12th edition

WILLIAMS, REED CHESTER, b Chicago, Ill, June 10, 41. ANALYTICAL CHEMISTRY. Educ: Lawrence Univ, BA, 63; Univ Wash, PhD(chem), 68. Prof Exp: RES CHEMIST, E I DU PONT DE NEMOURS & CO, INC, 68- Mem: Am Chem Soc; Sigma Xi. Res: Analytical chemistry; application of high speed liquid column chromatography to the separation and quantitation of complex mixtures. Mailing Add: Exp Sta Bldg 173 E I du Pont de Nemours & Co Inc Wilmington DE 19898

WILLIAMS, RICHARD, b Chicago, Ill, Aug 5, 27; m 61; c 3. PHYSICAL CHEMISTRY. Educ: Miami Univ, AB, 50; Harvard Univ, PhD(phys chem), 54. Prof Exp: Instr chem, Harvard Univ, 55-58; MEM TECH STAFF, RCA LABS, 58- Concurrent Pos: Fulbright lectr, Sao Carlos Sch Eng, 69. Mem: Fel Am Phys Soc; Brazilian Acad Sci. Res: Electrical properties of insulators; liquid crystals; luminescence of organic molecules; physical chemistry of surfaces. Mailing Add: RCA Labs Princeton NJ 08540

WILLIAMS, RICHARD ANDERSON, b Akron, Ohio, July 21, 31; m 51; c 5. TEXTILE CHEMISTRY. Educ: Wabash Col, AB, 53; Univ Rochester, PhD(chem), 57. Prof Exp: Res chemist, Patent Div, 56-60, sr res chemist, Nylon Tech Div, 60-68, supvr res, Qiana Tech Div, 68-71, supvr res & develop, Orlon-Lycra Tech Div, 71-72, PATENT SUPVR, PATENT LIAISON DIV, TEXTILE FIBERS DEPT, E I DU PONT DE NEMOURS & CO, 72- Mem: Am Asn Textile Chem & Color; Am Chem Soc. Res: Olefin-forming elimination reactions; polymer chemistry; synthetic fibers and applications. Mailing Add: RD 2 Box 36 Hockessin DE 19707

WILLIAMS, RICHARD BIRGE, b Monticello, NY, Feb 5, 29; m 51, 66; c 2. MARINE ECOLOGY. Educ: Univ Buffalo, BA, 51; Univ Wis, MS, 52; Harvard Univ, PhD(biol), 62. Prof Exp: Oceanogr, US Naval Hydrographic Off, 52-54; biologist, Biophys Div, Army Chem Ctr, 54-57; fishery biologist, Radiobiol Lab, US Bur Commercial Fisheries, 62-67; health physicist, Oak Ridge Nat Lab, 67-68; fishery biologist, Atlantic Estuarine Fisheries Ctr, Nat Marine Fisheries Serv, 68-72; PROG DIR BIOL OCEANOG, NSF, 72- Concurrent Pos: Adj asst prof, NC State Univ, 66-70, adj assoc prof, 70- Mem: AAAS; Am Soc Limnol & Oceanog; Ecol Soc Am. Res: Measurement of rate of production of estuarine plants; analysis of estuarine food webs. Mailing Add: Biog Oceanog NSF Washington DC 20550

WILLIAMS, RICHARD JOHN, b Hazleton, Pa, May 24, 44; m 66. GEOCHEMISTRY. Educ: Lehigh Univ, BA, 66; Johns Hopkins Univ, MA, 68, PhD(geochem), 70. Prof Exp: Teaching asst thermodyn, Johns Hopkins Univ, 69-70; space scientist lunar studies, 70-73, SR SPACE SCIENTIST LUNAR & PLANETARY STUDIES, JOHNSON SPACE CTR, NASA, 73- Mem: Am Geophys Union; Geochem Soc; Mineral Soc Am; Meteoritical Soc. Res: Application of the techniques of theoretical and experimental petrology to the solution of geological and geochemical problems. Mailing Add: TN7 Johnson Space Ctr NASA Houston TX 77058

WILLIAMS, RICHARD KELSO, b Chattanooga, Tenn, Oct 20, 38; m 66. MATHEMATICS. Educ: Vanderbilt Univ, BA, 60, MA, 62, PhD(math), 65. Prof Exp: Asst prof, 65-69, ASSOC PROF MATH, SOUTHERN METHODIST UNIV, 69- Mem: Math Asn Am; Am Math Soc. Res: Complex function theory; topology. Mailing Add: Dept of Math Southern Methodist Univ Dallas TX 75275

WILLIAMS, RICHARD SUGDEN, JR, b New York, NY, Dec 6, 38; m 60; c 2. QUATERNARY GEOLOGY. Educ: Univ Mich, Ann Arbor, BS, 61, MS, 62; Pa State Univ, PhD(geol), 65. Prof Exp: Proj scientist geol, Air Force Cambridge Res Labs, 65-68, res geologist, 68-69, br chief, 69-71; GEOLOGIST, US GEOL SURV, 71- Concurrent Pos: Geologist, UNESCO, 71- Mem: AAAS; fel Geol Soc Am; Am Geophys Union; Am Soc Photogram; Int Asn Volcanology & Chem Earth's Interior.

Res: Remote sensing of dynamic environmental phenomena; chiefly photogeologic and thermal infrared studies of volcanoes, geothermal areas and glaciers with aircraft and satellite (LANDSAT, NOAA) sensors with particular emphasis on Iceland. Mailing Add: US Geol Surv Reston VA 20242

WILLIAMS, RICHARD TAYLOR, b Tarboro, NC, May 27, 46. SOLID STATE PHYSICS. Educ: Wake Forest Univ, BS, 68; Princeton Univ, MA, 71, PhD(physics), 74. Prof Exp: PHYSICIST, NAVAL RES LAB, 69- Mem: Am Phys Soc; AAAS. Res: Effects of ionizing radiation in insulating solids, particularly time-resolved studies of exciton self-trapping and defect formation in halide crystals. Mailing Add: Insulator Physics Br Mat Sci Div Naval Res Lab Washington DC 20375

WILLIAMS, RICKEY JAY, b Muskogee, Okla, May 13, 42. PHYSICAL INORGANIC CHEMISTRY. Educ: Tex Christian Univ, BA, 64, MD, PhD(phys chem), 68. Prof Exp: Fel Los Alamos Sci Lab, 68-70 & Baylor Univ, 70-71; asst prof, 71-74, ASSOC PROF CHEM, MIDWESTERN STATE UNIV, 74- Mem: Am Crystallog Asn; Am Chem Soc. Res: Crystal structure studies of inorganic compounds. Mailing Add: Dept of Chem Midwestern State Univ Wichita Falls TX 76308

WILLIAMS, ROBERT ALLEN, b Cleveland, Ohio, Apr 25, 45. NUCLEAR CHEMISTRY. Educ: Oberlin Col, BA, 66; Carnegie-Mellon Univ, MS, 69, PhD(nuclear chem), 72. Prof Exp: Res assoc nuclear chem, 72-73, presidential intern, 73-74, STAFF MEM NUCLEAR CHEM, LOS ALAMOS SCI LAB, UNIV CALIF, 74- Mem: Am Chem Soc; Sigma Xi. Res: Pionic and muonic nuclear reactions; neutron activation analysis; nuclear spectroscopy. Mailing Add: Los Alamos Sci Lab Mail Stop 824 Los Alamos NM 87545

WILLIAMS, ROBERT BRUCE, b Washington, DC, Apr 30, 38; m 61; c 3. PHYSICAL OCEANOGRAPHY. Educ: San Diego State Col, BS, 64; Univ Calif, San Diego, MS, 68, PhD(eng sci), 74. Prof Exp: Lab technician, Univ Calif, San Diego, 60-66, assoc engr, 66-74; OCEANOGRAPHER, NATO, ITALY, 74- Res: Oceanographic underwater acoustics, turbulence and thermal microstructure; small computer systems for data processing; ultra high frequency atmospheric propagation. Mailing Add: NATO Saclant Res Ctr APO New York NY 09019

WILLIAMS, ROBERT CALVIN, b Key West, Fla, May 1, 44; m 69. ANALYTICAL CHEMISTRY, PHYSICAL CHEMISTRY. Educ: Univ Kans, BS, 66; Univ Wis-Madison, PhD(phys chem), 72. Prof Exp: Instr chem, Univ Nebr-Lincoln, 72-74; SR CHEMIST, CENT RES LABS, 3M CO, 74- Concurrent Pos: Res assoc, Univ Nebr-Lincoln, 72-74. Mem: Am Chem Soc; Am Phys Soc; Soc Appl Spectros; Coblentz Soc; AAAS. Res: Computer coupled instrumentation and instrumental methods of analysis; molecular spectroscopy, particularly infrared and mass spectroscopy; infrared normal coordinate analysis. Mailing Add: Cent Res Labs 3M Co St Paul MN 55133

WILLIAMS, ROBERT CASE, chemistry, see 12th edition

WILLIAMS, ROBERT DEE, veterinary microbiology, immunology, see 12th edition

WILLIAMS, ROBERT E, b Rexburg, Idaho, Aug 21, 19; m 41; c 4. RANGE MANAGEMENT, ECOLOGY. Educ: Univ Idaho, BS, 41; Univ Ga, MS, 62. Prof Exp: Range conservationist, Tex, 42-44 & 46-48, work unit conservationist, 48-49, range conservationist, La, 49-51 & 53-56, area conservationist, 51-53, regional range conservationist, Ga, 56-61, soil conservationist, Washington, DC, 61-65, head range conservationist, 65-72, ASST TO THE ADMINR FOR ENVIRON DEVELOP, SOIL CONSERV SERV, USDA, 72- Concurrent Pos: Mem, Am Del Int Grassland Cong, Brazil, 64. Mem: Soil Conserv Soc Am; Soc Range Mgt; Ecol Soc Am. Res: Development of policy and guidelines on environmental development; training programs for personnel on the environment. Mailing Add: Soil Conserv Serv USDA Washington DC 20250

WILLIAMS, ROBERT EARL, b Cleveland, Ohio, Apr 6, 30; m 58. PHYSICAL CHEMISTRY. Educ: Loyola Univ, La, BS, 52; Okla State Univ, PhD(chem), 59. Prof Exp: Asst chemist, Los Alamos Sci Lab, 55; res engr, Humble Oil & Refining Co, 57-64, sr res chemist, Esso Prod Res Co, 64-69, new prod develop mgr, Humble Oil & Refining Co, 69-71, sr environ chemist, Esso Prod Res Co, 71-72, RES ENGR, EXXON PROD RES CO, 72- Mem: Am Chem Soc; Am Inst Mining, Metall & Petrol Engrs. Res: Surface and colloid chemistry; interfacial phenomena; oil recovery from porous media; air pollution ambient air analysis. Mailing Add: Exxon Prod Res Co PO Box 2189 Houston TX 77001

WILLIAMS, ROBERT EDWARD, b Greencastle, Ind, June 28, 25; m 57; c 2. APPLIED CHEMISTRY. Educ: DePauw Univ, BA, 48; Univ NMex, PhD(chem), 52. Prof Exp: Res assoc, Univ NMex, 52-53; res chemist, Olin Mathieson Chem Corp, 53-59; sr staff res supvr, Nat Eng Sci Co, 59-62; dir organometallic chem res, Space-Gen Corp, 62-69; PRES, CHEM SYSTS INC, 69- Concurrent Pos: Consult, Off Naval Res Contract, 62-64. Mem: Am Chem Soc. Res: Carborane and boron hydride chemistry; boron nuclear magnetic resonance spectroscopy; mass spectroscopy; vacuum techniques; electron deficient structure, rearrangements, polymerization and semiconducting polymers; refractory filaments; vapor deposition. Mailing Add: Chem Systs Inc 1852 McGaw Ave Irvine CA 92714

WILLIAMS, ROBERT ELLIS, b Grenada, Miss, Apr 9, 30; m 67; c 3. BIOLOGY. Educ: Memphis State Univ, BA, 56; Ohio State Univ, MA, 57, PhD, 61. Prof Exp: Res assoc, Ohio State Res Found, 60; entomologist, Army Biol Labs, 61-63, chief insect biol res br, 63-65; asst head microbiol br, Off Naval Res, 65-66, prof officer biol countermeasures, 66-67, head virol dept, Naval Med Res Unit 3, Cairo, Egypt, 68-73; LIBRARIAN, ANTIOCH COL, 73- Mem: Am Soc Trop Med & Hyg; Entom Soc Am; Am Mosquito Control Asn. Res: Entomology; arbovirology; ecology. Mailing Add: Librarian Antioch Col Yellow Springs OH 45387

WILLIAMS, ROBERT ELVIN, b Bloomfield, Mo, Dec 28, 35; m 58; c 3. MATHEMATICS. Educ: Univ Mo, BS, 59, MA, 61, PhD(math), 65. Prof Exp: Asst prof math, Westminster Col, Mo, 63-65; ASST PROF MATH, KANS STATE UNIV, 65- Mem: Am Math Soc; Math Asn Am. Res: Near-rings; prime rings and quotient rings; divisibility in Cohn rings. Mailing Add: Dept of Math Kans State Univ Manhattan KS 66506

WILLIAMS, ROBERT FONES, b Bessemer, Ala, July 27, 28; m 52; c 1. TOPOLOGY. Educ: Univ Tex, BA, 48; Univ Va, PhD(math), 54. Prof Exp: Asst prof math, Fla State Univ, 54-55; vis lectr, Univ Wis, 55-56; asst prof, Purdue Univ, 56-59; NSF fel & mem Inst Adv Study, 59-61; asst prof, Univ Chicago, 61-63; assoc prof, 63-67, PROF MATH, NORTHWESTERN UNIV, EVANSTON, 67- Concurrent Pos: NSF grant, Univ Geneva, 68-69 & Inst Advan Sci Study, Bures-sur-Yvette, France, 70, 72-73. Mem: Am Math Soc. Res: Transformation groups; toplogical dynamics; global analysis; differentiable dynamical systems. Mailing Add: Dept of Math Northwestern Univ Evanston IL 60201

WILLIAMS, ROBERT FRANK, b Cleveland, Ohio, July 13, 18; m 44; c 4. MEDICINE. Educ: Dartmouth Col, AB, 40; Case Western Reserve Univ, MD, 43. Prof Exp: Intern med, Univ & Univ Hosps, 43-44, asst resident, 44-45 & 47-48, researcher med & prev med, 48-49, instr med, Sch Med, 49-50, sr instr, 50-53, asst prof, 53-61, asst dean, Sch Med, 57-61, ASSOC PROF MED, SCH MED, CASE WESTERN RESERVE UNIV, 61-, DIR CLINS & EMERGENCY, UNIV HOSPS, 69- Concurrent Pos: Physician in chg outpatient dept, Univ Hosps, Case Western Reserve Univ, 49-69, assoc vis physician, 70- Mem: AMA; Asn Am Med Cols. Res: Internal medicine. Mailing Add: Sch of Med Case Western Reserve Univ Cleveland OH 44106

WILLIAMS, ROBERT GLENN, b Teaneck, NJ, Oct 14, 37; m 65; c 3. OCEANOGRAPHY. Educ: NY Univ, BA, 60, MS, 65, PhD(oceanog), 71. Prof Exp: Asst res scientist oceanog, NY Univ, 60-61 & 63-65; oceanogr, US Naval Underwater Systs Ctr, New London, Conn, 65-72; prog coordr oceanog, Nat Oceanic & Atmospheric Admin, Rockville, Md, 72-74, dep chief scientist, Ship Oceanogr, 74, OCEANOGR, CTR EXP DESIGN & DATA ANAL, NAT OCEANIC & ATMOSPHERIC ADMIN, 74- Mem: Am Geophys Union; Am Soc Limnol & Oceanog. Res: Air-sea interaction and dynamic studies of the tropical oceans, and continental shelves, with emphasis on exchange of heat and moisture with the atmosphere on time scales of several weeks. Mailing Add: Ctr for Exp Design & Data Anal NOAA/EDS Page Bldg 2 3300 Whitehaven St NW Washington DC 20235

WILLIAMS, ROBERT H, biochemistry, see 12th edition

WILLIAMS, ROBERT HACKNEY, b Providence, RI, Jan 3, 15; m 42; c 2. ORGANIC CHEMISTRY. Educ: Univ NC, AB, 35, MA, 37; Temple Univ, PhD(chem), 53. Prof Exp: Res chemist, Res & Develop Lab, 38-42 & 46-59, sr res chemist, Cent Res Div Lab, 59-72, ASST TO ADMIN MGR, CENT RES DIV, MOBIL RES & DEVELOP CORP, 72-, SR RES CHEMIST, 73- Mem: Am Chem Soc; Am Inst Chem. Res: Petroleum additives; hydrocracking; lube oil manufacture and composition; radiation chemistry of hydrocarbons; application of nuclear radiation to petroleum processing; radiation and photochemical induced organic chemical reactions; oxidation of hydrocarbons. Mailing Add: Cent Res Div Mobil Res & Develop Corp Box 1025 Princeton NJ 08540

WILLIAMS, ROBERT HARDIN, b Savannah, Tenn, Sept 27, 09; m 42; c 3. MEDICINE. Educ: Washington & Lee Univ, BA, 29; Johns Hopkins Univ, MD, 34. Hon Degrees: DSc, Washington & Lee Univ. Prof Exp: Intern path, Mallory Inst Path, Boston City Hosp, 34-35; from intern to chief resident, Vanderbilt Univ Hosp, 35-39; instr, Harvard Med Sch, 40-44, assoc, 44-46, asst prof, 46-48; chmn dept, 48-63, prof med, 48-75, FIRST ROBERT H WILLIAMS PROF MED, UNIV WASH, 75- Concurrent Pos: Res fel, Mass Gen Hosp, 39-40; mem coun, Nat Inst Arthritis & Metab Dis; dir, Diabetes Res Inst, Univ of Wash, 72- Honors & Awards: Silver Medal, AMA, 44; Stengel Award, Am Col Physicians; Outstanding Chmn of Med Award, Asn Prof Med; Distinguished Leadership Award, Endocrine Soc; Banting Award; Outstanding Achievement Award, Modern Med. Mem: AAAS; Am Soc Clin Invest (pres); Asn Am Physicians; Am Prof Med (pres); Am Diabetes Asn. Res: Clinical medicine; endocrinology; biochemistry. Mailing Add: Dept of Med Univ of Wash Sch of Med Seattle WA 98105

WILLIAMS, ROBERT HAWORTH, b Seattle, Wash, Mar 31, 14; m 39, 54; c 3. PLANT PHYSIOLOGY. Educ: Univ Wash, Seattle, BS, 35, MS, 40; Cornell Univ, PhD(plant physiol), 41. Prof Exp: Asst bot, Cornell Univ, 36-38, instr, 38-41; from asst prof, Univ Miami, Fla, 41-54, prof marine biol, 50-54, asst dir marine lab, 42-51; prof marine sci, Fla State Univ, 54-55; PROF BIOL, UNIV MIAMI, 55- Concurrent Pos: Asst, US Frozen Pack Lab, Wash, 35. Mem: Am Soc Limnol & Oceanog; Phycol Soc Am. Res: Carbon dioxide absorption of roots; carbohydrates of the large brown algae; physiology of fern spore germination; ecology of marine algae of Florida. Mailing Add: Dept of Biol Univ of Miami Coral Gables FL 33124

WILLIAMS, ROBERT JACKSON, b Iowa City, Iowa, June 13, 31; div, m 74; c 3. PHYSIOLOGY, ECOLOGY. Educ: Univ Wis-Madison, BA, 53, MA, 61; Univ Md, College Park, PhD(zool), 67. Prof Exp: Cryobiologist, Am Found Biol Res, Wis, 60-62; biologist, Naval Med Res Inst, Md, 63-67; RES SCIENTIST CRYOBIOL, BLOOD RES LAB, AM NAT RED CROSS, 67- Mem: AAAS; Am Physiol Soc; Soc Cryobiol; Am Soc Plant Physiol; Biophys Soc. Res: Mechanisms of cell freezing injury; effects of low temperature on geographical distribution; instrumentation. Mailing Add: Blood Res Lab Am Nat Red Cross 9312 Old Georgetown Rd Bethesda MD 20014

WILLIAMS, ROBERT K, b Ft Worth, Tex, Jan 6, 28; m 52; c 3. ENTOMOLOGY, CELL BIOLOGY. Educ: Agr & Mech Col Tex, BS, 48, MS, 56, PhD(entom), 59. Prof Exp: Asst county agent in training, Agr Exten Serv, Agr & Mech Col, Tex, 48, asst county agent, Agr Exten Serv & Wood County, 48, Agr exten Serv & Bowie County, 50 & Agr Exten Serv & Eastland County, 52-54, res asst entom, Col, 54-58; PROF BIOL, E TEX STATE UNIV, 58- Concurrent Pos: NSF Col Sci Improv Prog grant, Dept Physiol Chem, Univ Wis, 69-70. Res: Physiology; cell physiology; molecular biology. Mailing Add: Dept of Biol East Tex State Univ Commerce TX 75428

WILLIAMS, ROBERT L, b Buffalo, NY, July 22, 22; m 49; c 2. PSYCHIATRY, NEUROLOGY. Educ: Alfred Univ, BA, 44; Albany Med Col, Union Univ, NY, MD, 46. Prof Exp: Chief, Air Force Neurol Ctr, Lackland AFB Hosp, 52-55, chief neuropsychiat serv, 53-55, chief consult, Off Surgeon Gen, US Air Force, 55-58; from assoc prof to prof psychiat & neurol, Col Med, Univ Fla, 58-72, chmn dept psychiat, 64-72; PROF PSYCHIAT & CHMN DEPT, BAYLOR COL MED, 72- Concurrent Pos: Mem, Nat Adv Ment Health Coun & Nat Adv Neurol Dis & Blindness Coun, 55-58; Fla rep, Comn Ment Illness, Southern Regional Educ Bd, 64-72, chmn, 71-72; psychiat consult, Indust Security Prog, Dept Defense & consult psychiat & neurol, Surgeon Gen, US Air Force, 66- Mem: Fel AAAS; AMA; fel Am Psychiat Asn; fel Am Acad Neurol; Am Electroencephalog Soc. Res: Psychophysiology of sleep; medical education. Mailing Add: Baylor Col of Med Tex Med Ctr Houston TX 77025

WILLIAMS, ROBERT LEROY, b St Thomas, Ont, July 9, 28; m 51; c 3. PHYSICS. Educ: Univ Western Ont, BSc, 51; Univ BC, MA, 52, PhD(physics), 56. Prof Exp: Sci officer, Defense Res Bd Can, 55-59; group leader physics, RCA Victor Co, 59-63; chief physicist, Simtec Ltd, 63-65, gen mgr, 65-66; mem sci staff physics, 66-69, BR MGR DETECTOR DEVELOP, TEX INSTRUMENTS INC, 69- Mem: Am Phys Soc; sr mem Inst Elec & Electronics Eng. Res: Physics of infrared and nuclear particle detectors; development of detectors; diode, transistor and bulk structures. Mailing Add: 830 Northlake Dr Richardson TX 75080

WILLIAMS, ROBERT PIERCE, b Chicago, Ill, Oct 27, 20; m 44; c 2. MICROBIOLOGY. Educ: Dartmouth Col, AB, 42; Univ Chicago, SM, 46, PhD(bact, parasitol), 49; Am Bd Med Microbiol, cert. Prof Exp: Asst bact, Univ Chicago, 46-47, cur, 47-49; instr, Univ Southern Calif, 49-51; from asst prof to assoc prof, 51-63, actg chmn dept, 61-66, PROF MICROBIOL, BAYLOR COL MED, 63- Concurrent Pos: Lectr, Univ Houston, 64; vis mem grad fac, Col Vet Med, Tex A&M Univ, 65-;

consult, Vet Admin Hosp & M D Anderson Hosp, 57- Mem: AAAS; Am Soc Microbiol; Am Chem Soc; fel Am Acad Microbiol; Brit Soc Gen Microbiol. Res: Bacterial pigments, particularly prodigiosin and porphyrins; virulence versus avirulence of microorganisms; microbial nutrition; host-parasite relationships in bacterial infections. Mailing Add: Dept of Microbiol Baylor Col of Med Houston TX 77025

WILLIAMS, ROBERT WALTER, b Palo Alto, Calif, June 3, 20; m 46, 58, 69; c 3. PHYSICS. Educ: Stanford Univ, AB, 41; Princeton Univ, MA, 43; Mass Inst Technol, PhD(physics), 48. Prof Exp: Lab asst, Princeton Univ, 41-42; from jr physicist to assoc physicist, Manhattan Proj, Princeton Univ & Los Alamos Sci Lab, 42-46; res assoc, Mass Inst Technol, 46-48, from asst prof to assoc prof, 48-59; PROF PHYSICS, UNIV WASH, 59- Mem: Fel Am Phys Soc; fel Am Acad Arts & Sci. Res: Nuclear high energy physics; cosmic rays; elementary particles. Mailing Add: Physics BJ-10 Univ of Wash Seattle WA 98195

WILLIAMS, ROBLEY COOK, b Santa Rosa, Calif, Oct 13, 08; m 31; c 2. BIOPHYSICS, MOLECULAR BIOLOGY. Educ: Cornell Univ, AB, 31, PhD(physics), 35. Prof Exp: Asst physics, Cornell Univ, 29-35; from instr to asst prof astron, Observ, Univ Mich, 35-45, from assoc prof to prof physics, 45-50; prof biophys, 50-59 & virol, 59-64, chmn dept molecular biol, 64-69, PROF MOLECULAR BIOL, UNIV CALIF, BERKELEY, 64- ASSOC DIR VIRUS LAB & RES BIOPHYSICIST, 50- Concurrent Pos: Vpres & consult, Evaporated Metal Films Corp, 35-56; res assoc, Off Sci Res & Develop, 41-42; coun mem, Int Union Pure & Appl Biophys, 61-69, pres, Comn Molecular Biophys, 61-69; Nat Acad Sci rep, UN Educ Sci & Cultural Orgn, 63-69; mem bd trustees, Deep Springs Col, 68-, chmn, 71- Honors & Awards: Longstreth Medal, Franklin Inst, 39, Scott Award, 54. Mem: Nat Acad Sci; Biophys Soc(pres, 58 & 59); Electron Micros Soc Am(pres, 51). Res: Electron microscopy of biological objects; development of techniques for electron microscopy of virus particles. Mailing Add: Dept of Molecular Biol Univ of Calif Berkeley CA 94720

WILLIAMS, ROBLEY COOK, JR, b Ann Arbor, Mich, Oct 15, 40; m 68; c 1. PHYSICAL BIOCHEMISTRY. Educ: Cornell Univ, BA, 62; Rockefeller Univ, PhD(phys biochem), 68. Prof Exp: Nat Inst Arthritis & Metab Diseases fel, State Univ NY Buffalo, 67-68; asst prof biol, 69-74, ASSOC PROF BIOL, YALE UNIV, 74- Mem: Am Chem Soc; Biophys Soc. Res: Protein-protein association and structure-function relationships in proteins. Mailing Add: Dept of Biol Yale Univ New Haven CT 06520

WILLIAMS, ROGER JOHN, b Ootacumund, India, Aug 14, 93; US citizen; m 16, 53; c 4. BIOCHEMISTRY, NUTRITION. Educ: Univ Redlands, BS, 14; Univ Chicago, MS, 18, PhD(biochem), 19. Hon Degrees: DSc, Univ Redlands, 34, Columbia Univ, 42, Ore State Univ, 56. Prof Exp: Res chemist, Fleischmann Co, Ill, 19-20; from asst prof to prof chem, Univ Ore, 20-32; prof, Ore State Univ, 32-39; prof chem, 39-71, dir, Clayton Found Biochem Inst, 40-63, EMER PROF CHEM, UNIV TEX, AUSTIN, 71-, RES SCIENTIST, CLAYTON FOUND BIOCHEM INST, 71- Concurrent Pos: Mem, Food & Nutrit Bd, 49-53; comt probs alcohol, Nat Res Coun, 49-53; med adv bd, Muscular Dystrophy Asn Am, Inc, 52-; res & med comt, Nat Multiple Sclerosis Soc, 60- Honors & Awards: Mead Johnson Award, Am Inst Nutrit, 41; Chandler Medal, Columbia Univ, 42. Mem: Nat Acad Sci; AAAS; Am Soc Biol Chem; Am Chem Soc(pres, 57); Soc Exp Biol & Med. Res: Pantothenic acid yeast nutrilites; folic acid; avidin; microbiological assay methods for B vitamins; B vitamins in normal and malignant tissues; nutrition of fungi; etiology of alcoholism; humanics; biochemical individuality. Mailing Add: Clayton Found Biochem Inst Univ of Tex Austin TX 78712

WILLIAMS, ROGER TERRY, b Covina, Calif, June 15, 36; m 64; c 3. DYNAMIC METEOROLOGY. Educ: Univ Calif, Los Angeles, AB, 59, MS, 61, PhD(meterol), 63. Prof Exp: Ford Found fel, Univ Calif, Los Angeles, 63-64; res assoc meteorol, Mass Inst Technol, 64-66; asst prof, Univ Utah, 66-68; assoc prof, 68-74, PROF METEOROL, NAVAL POSTGRAD SCH, 74- Concurrent Pos: Mem, FGGE Adv Panel, Nat Res Coun, 74- Mem: AAAS; Am Geophys Union; Am Meteorol Soc; Royal Meteorol Soc; Meteorol Soc Japan. Res: Numerical weather prediction; dynamics of the atmosphere and other geophysical systems; application of numerical methods; dynamics of atmospheric waves and fronts. Mailing Add: Code 51 Wu Naval Postgrad Sch Monterey CA 93940

WILLIAMS, ROGER WRIGHT, b Great Falls, Mont, Jan 24, 18; m 43; c 2. MEDICAL ENTOMOLOGY, PARASITOLOGY. Educ: Univ Ill, BS, 39, MS, 41; Columbia Univ, PhD(med entom), 47; London Sch Hyg & Trop Med, cert appl parasitol & entom, 57. Prof Exp: Jr sci asst, Div Cereal Crops, Bur Entom & Plant Quarantine, USDA, 41; asst biol, Cornell Univ, 42; from res asst to res assoc parasitol, 44-48, from asst prof to assoc prof med entom, 48-66, actg head div trop med, 70, PROF PUB HEALTH, MED ENTOM, SCH PUB HEALTH, COLUMBIA UNIV, 66- Concurrent Pos: NSF sr fel, 56; fel trop med, La State Univ, 57; spec consult, USPHS, Alaska, 49, consult, Ga, 52; corp mem, Bermuda Biol Sta Res, 58-; consult, US Nat Park Serv, Virgin Island Govt & Jackson Hole Preserve, Inc, 59 & 61; consult, Rockefeller Found, WI, 63, mem field staff, Nigeria, 64-65; consult, WHO, Burma, 66. Mem: Fel AAAS; Entom Soc Am; Am Soc Parasitol; Am Soc Trop Med & Hyg. Res: Biology, control and taxonomy of arthropods of medical importance; biology and chemotherapy of helminths. Mailing Add: Div of Trop Med Sch of Pub Health Columbia Univ Col of Phys & Surg New York NY 10032

WILLIAMS, RONALD ALVIN, b Somerset, Pa, Aug 4, 42; m 64; c 2. BIOMEDICAL ENGINEERING, ELECTRICAL ENGINEERING. Educ: Pa State Univ, BS, 64, MS, 66, PhD(elec eng), 70. Prof Exp: Res bioengr, Columbus Labs, Battelle Mem Inst, 70-72, assoc chief bioeng sci & biochem div, 72-74; MGR PROD EXPLORATION, FENWAL LABS, 74- Mem: Inst Elec & Electronics Eng; Asn Advan Med Instrumentation. Res: Respiratory physiology; mathematical modeling and computer simulation of biological systems; medical instrumentation development. Mailing Add: Fenwal Labs Div of Baxter Labs Rte 120 & Wilson Rd Round Lake IL 60073

WILLIAMS, RONALD LEE, b Koleen, Ind, June 26, 36; m 57; c 3. PHARMACOLOGY, PHYSIOLOGY. Educ: Butler Univ, BS, 59, MS, 61; Tulane Univ, PhD(pharmacol), 64. Prof Exp: Asst prof, 66-70, ASSOC PROF PHARMACOL, LA STATE UNIV MED CTR, NEW ORLEANS, 70- Concurrent Pos: La Heart Asn grants-in-aid, 65-66 & 67-68; USPHS res contract, 68. Mem: AAAS; Am Pharmaceut Asn; Am Soc Pharmacol & Exp Therapeut; NY Acad Sci. Res: Effect of autonomic drugs and neurotransmitters upon renal function and their relationships between hemodynamics and tubular changes. Mailing Add: Dept of Pharmacol La State Univ Med Ctr New Orleans LA 70112

WILLIAMS, RONALD LLOYDE, b Northfield, Minn, May 7, 44; m 64; c 2. PHYSICAL CHEMISTRY, ENVIRONMENTAL CHEMISTRY. Educ: St Olaf Col, BA, 66; Iowa State Univ, PhD(phys chem), 70. Prof Exp: Fel, Univ Calif, Irvine, 70-72; RES CHEMIST, GEN MOTORS RES LAB, 72- Mem: Am Chem Soc; Soc Automotive Engrs. Res: Reaction kinetics; hot atom reactions; non-exhaust automobile emissions. Mailing Add: Environ Sci Dept Gen Motors Res Labs Warren MI 48090

WILLIAMS, RONALD WENDELL, b Atlanta, Ga, Nov 9, 39; m 63; c 2. SOLID STATE PHYSICS. Educ: Christian Bros Col, BSc, 62; Iowa State Univ, PhD(physics), 66. Prof Exp: Instr physics, Iowa State Univ, 66-67; staff scientist, Oak Ridge Nat Lab, 67-70; ASSOC PROF ELEC ENG, UNIV VT, 70- Res: Band structure and transport properties of metals. Mailing Add: Dept of Elec Eng Univ of Vt Burlington VT 05401

WILLIAMS, ROSS EDWARD, b Carlinville, Ill, June 28, 22; m 58; c 3. PHYSICS. Educ: Bowdoin Col, BS, 43; Columbia Univ, MA, 47, PhD(physics), 55. Prof Exp: Sr res engr, Sperry Prod Inc, 47-49; consult physicist, Paul Rosenberg Assocs, 53-60; sr res assoc, Oceanog Acoustics & Signal Processing, 60-65, asst dir, 65-66, ASSOC DIR, OCEANOG ACOUSTICS & SIGNAL PROCESSING, HUDSON LABS, COLUMBIA UNIV, 66-, PROF OCEAN ENG, 68-, LECTR ELEC ENG, 66- Concurrent Pos: Mem, Comt Undersea Warfare, Nat Res Coun-Nat Acad Sci; consult, Naval Res Lab & Nat Acad Sci; chmn bd dirs, Ocean & Atmospheric Sci, Inc, 68- Mem: Acoust Soc Am; Am Phys Soc; Am Soc Photogram; sr mem Inst Elec & Electronics Eng. Res: Oceanography; acoustic propagation; surveillance system design; signal processing techniques; aerial reconnaissance; automatic mapping; optical data processing; electronic design. Mailing Add: 23 Alta Pl Yonkers NY 10710

WILLIAMS, ROSS ERNEST, bio-organic chemistry, see 12th edition

WILLIAMS, ROY EDWARD, b Cookeville, Tenn, Feb 12, 38; m 59. HYDROGEOLOGY. Educ: Ind Univ, Bloomington, BSc, 61, MA, 62; Univ Ill, Urbana, PhD(hydrogeol), 66. Prof Exp: Teaching asst phys geol, Ind Univ, Bloomington, 63-64; teaching asst eng geol, Univ Ill, Urbana, 64-66; asst prof, 66-70, PROF HYDROGEOL & HYDROGEOLOGIST, UNIV IDAHO, 70- Concurrent Pos: Res asst, Ill State Geol Surv, 64-66; grants, Idaho Water Resources Res Inst & Univ Idaho Res Comt, 66- & Idaho Short Term Appl Res Fund, 68- Mem: Am Geophys Union. Res: Studies of pollution of ground and surface water and the relation between ground water flow systems and certain engineering problems. Mailing Add: Col of Mines Univ of Idaho Moscow ID 83843

WILLIAMS, ROY LEE, b Portsmouth, Va, Feb 20, 37; m 57; c 1. ORGANIC CHEMISTRY. Educ: Col William & Mary, BS, 60; Univ Del, PhD(org chem), 65. Prof Exp: Res chemist, Am Cyanamid Co, NJ, 64-65; from asst prof to assoc prof, 65-73, PROF CHEM, OLD DOM UNIV, 73- Concurrent Pos: Res grant, Army Med Res Inst, Walter Reed Hosp, Washington, DC, 66-68; consult, Chem & Physics Br, Langley Res Ctr, NASA, Va, 65- Mem: Am Chem Soc; fel The Chem Soc; Int Soc Heterocyclic Chem. Res: Heterocyclic, organic synthesis, including heterocyclic polymers; medicinals; synthetics. Mailing Add: Alfriend Lab Dept of Chem Old Dom Univ Norfolk VA 23508

WILLIAMS, RUSSELL RAYMOND, b Lost Creek, WVa, Apr 11, 26; m 53; c 3. PARASITOLOGY, INVERTEBRATE ZOOLOGY. Educ: Ohio State Univ, BSc, 55, MSc, 57, PhD(zool), 63. Prof Exp: Lab unit operator, B F Goodrich Chem Co, 45-50; from asst instr to instr zool, Ohio State Univ, 57-63; from asst prof to assoc prof, 63-67, PROF BIOL, WAYNESBURG COL, 67-, CHMN DEPT, 70-, PREMED ADV, 73- Concurrent Pos: Res Corp grant, 64-66; vis prof, Univ Northern Colo, 68-; partic, Res Corp Conf for New Sci Chairmen, 71. Mem: AAAS; Am Soc Parasitol; Wildlife Dis Asn; Am Micros Soc; Am Inst Biol Sci. Res: Life history and taxonomic studies on trematodes. Mailing Add: Dept of Biol Waynesburg Col Waynesburg PA 15370

WILLIAMS, SCOTT LANSING, b Marlow, Okla, Feb 21, 18; m 45; c 2. METEOROLOGY. Educ: Okla State Univ, BS, 40; NY Univ, MS, 51. Prof Exp: Weather officer, US Air Force, 48-67, weather forecaster & aerial observer, 48-50, 51-52, class supvr, Weather Forecast Course, Chanute AFB, Ill, 52-55, analyst, Joint Numerical Weather Prediction Unit, 57, weather officer, Directorate Sci Serv, Air Weather Serv, 57-60, br chief, Directorate Aerospace Sci, 60-67; meteorologist, Off Fed Coord Meteorol Serv & assoc res, Dept Com, 67-68; data processing mgr, 68-72, TECH ASST, CTR EXP DESIGN AND DATA ANAL, NAT OCEANIC & ATOMSPHERIC ADMIN, US DEPT OF COM, 72- Mem: Am Meteorol Soc. Res: Numerical weather prediction. Mailing Add: Rx8 Ctr Exp Design & Data Anal Nat Oceanic & Atmospheric Admin Rockville MD 20852

WILLIAMS, SCOTT WARNER, b Staten Island, NY, Apr 22, 43. MATHEMATICS. Educ: Morgan State Col, BS, 64; Lehigh Univ, MS, 67, PhD(math), 69. Prof Exp: Instr, Pa State Univ, 68-69; res assoc, Pa State Univ, 69-71; ASST PROF MATH, STATE UNIV NY BUFFALO, 71- Mem: Am Math Soc. Res: General topology, completeness, paracompactness, linearly ordered spaces and Baire spaces; algebra, groups and categories. Mailing Add: Dept of Math State Univ NY Buffalo Amherst NY 14226

WILLIAMS, SIDNEY, b Peabody, Mass, Apr 4, 15. ANALYTICAL CHEMISTRY. Educ: Univ Mass, BS, 37, MS, 39. Prof Exp: Chemist, 41-57 & Food & Drug Off, 57-59, supvry chemist, 59-63, asst to chief, Pesticides Br, 63-65, asst chief, 65-71, chief, Indust Chem Practices Br, 71-72, DEP DIR, DIV CHEM TECHNOL, US FOOD & DRUG ADMIN, 72- Mem: AAAS; Am Chem Soc; Asn Off Anal Chem. Res: Industrial chemical residues in foods; food and drug analysis. Mailing Add: Div of Chem Technol Food & Drug Admin Washington DC 20204

WILLIAMS, SIDNEY ARTHUR, b Ann Arbor, Mich, Dec 26, 33; m 57; c 1. MINERALOGY. Educ: Mich Technol Univ, BS, MS, 57; Univ Ariz, PhD(mineral), 62. Prof Exp: Instr mineral, Mich Technol Univ, 60-61, asst prof, 61-63; mineralogist, Silver King Mines, Inc, 63-65; DIR RES EXPLOR GEOL, PHELPS DODGE CORP, 65- Concurrent Pos: Mineralogist, Brit Mus Natural Hist, 71. Mem: Fel Mineral Soc Am; Mineral Asn Can; Brit Mineral Soc; Soc Econ Geol. Res: Descriptive mineralogy and crystallography; petrology of altered rocks related to ore deposits. Mailing Add: Phelps Dodge Corp PO Drawer 1217 Douglas AZ 85607

WILLIAMS, STANLEY A, b Lawrence, Kans, May 14, 32; m 58; c 2. THEORETICAL PHYSICS. Educ: Nebr Wesleyan Univ, BA, 54; Rensselaer Polytech Inst, PhD(physics), 62. Prof Exp: NSF fel, Univ Birmingham, 62-63; asst prof, 63-67, ASSOC PROF PHYSICS, IOWA STATE UNIV, 67- Concurrent Pos: Assoc scientist, Ames Lab, 63-67, scientist, 67- Mem: Am Phys Soc; Am Asn Physics Teachers. Res: Mathematical physics, principally the application of group theoretic techniques to nuclear and elementary particle physics; structure of fission fragment nuclei. Mailing Add: Dept of Physics Iowa State Univ Ames IA 50010

WILLIAMS, STANLEY CLARK, b Long Beach, Calif, Aug 24, 39; m 65; c 2. ECOLOGY. Educ: San Diego State Col, AB, 61, MA, 63; Ariz State Univ, PhD(zool, 68. Prof Exp: Asst prof, 67-70, ASSOC PROF BIOL, SAN FRANCISCO STATE UNIV, 70- Concurrent Pos: Res assoc, Calif Acad Sci, 67-; NSF grants, Mexico, 68-72; lectr, Moss Landing Marine Sta, 72-; mem Int Ctr Arachnological Documentation; ecol consult, Mill Valley, Calif, Mem: AAAS; Ecol Soc Am; Am Entom Soc. Res:

Invertebrate ecology; scorpion systematics. Mailing Add: Dept of Biol San Francisco State Univ San Francisco CA 94132

WILLIAMS, STEPHEN, b Minneapolis, Minn, Aug 28, 26; m 62; c 2. ANTHROPOLOGY, ARCHAEOLOGY. Educ: Yale Univ, BA, 49; Univ Mich, MA, 50; Yale Univ, PhD(anthrop), 54. Hon Degrees: MA, Harvard Univ, 62. Prof Exp: Res fel anthrop, Peabody Mus, Harvard Univ, 55-58; asst prof anthrop, Harvard Univ, 58-62, assoc prof, 62-67, exec dir res & planning, Peabody Mus, 67, chmn dept anthrop & actg dir, Peabody Mus, 67-69, PROF ANTHROP, HARVARD UNIV, 67-, CURATOR N AM ARCHAEOL, PEABODY MUS, 62-, DIR, 69- Concurrent Pos: Researcher hist & archaeol res, Caldo Indian Case, Dept Justice, 54-55; mem, Adv Comt Pre-Columbian Art, 69- & Conf Hist Site Archaeol. Mem: Soc Am Archaeol; Am Asn Museums; Am Anthrop Asn; Archaeol Inst Am; Soc Hist Archaeol. Res: Archaeology of the Eastern United States with special emphasis on the Lower Mississippi Valley; historic archaeology; museum planning and organization for research use. Mailing Add: Peabody Mus Harvard Univ 11 Divinity Ave Cambridge MA 02138

WILLIAMS, STEPHEN EDWARD, b St Louis, Mo, Oct 9, 42; m 68. PLANT PHYSIOLOGY. Educ: Cent Col, Mo, BA, 64; Univ Tenn, Knoxville, MS, 66; Wash Univ, PhD(biol), 71. Prof Exp: Lectr plant physiol, Cornell Univ, 70-73; ASST PROF BIOL, LEBANON VALLEY COL, 73- Mem: AAAS; Am Soc Plant Physiol. Res: Electrophysiology, plant sensory physiology, excitable plant cells; carnivorous plants, especially Droseraceae, electron microscopy; canavanine synthesis and metabolism in legumes. Mailing Add: Dept of Biol Lebanon Valley Col Annville PA 17003

WILLIAMS, STEVEN FRANK, b Tacoma, Wash, May 8, 44; m 66. FISHERIES, ICHTHYOLOGY. Educ: Univ Wash, BS, 66; Univ Calif, Los Angeles, MA, 68; Ore State Univ, PhD(fisheries), 74. Prof Exp: Res biologist fisheries, US Peace Corps, Chile, 68-70; ASST PROF BIOL, ST CLOUD STATE UNIV, 74- Concurrent Pos: Res asst fisheries, Ore State Univ, 70- Mem: AAAS; Am Fisheries Soc. Res: Natural distribution and abundance of fishes and factors which affect them; fish culture, especially optimum culture conditions. Mailing Add: Dept of Biol St Cloud State Univ St Cloud MN 56301

WILLIAMS, SUSAN CATHERINE FRARY, b Los Angeles, Calif, May 9, 45; m 70; c 1. BIOCHEMISTRY. Educ: Bennington Col, BA, 66; Univ Vt, PhD(biochem), 72. Prof Exp: Res assoc, Scripps Inst Oceanog, Univ Calif, 72-73; res scientist, Lamont-Doherty Geol Observ, Columbia Univ, 73-76; CLIN RES ANALYST, LEDERLE LABS, AM CYANAMID CORP, 76- Mem: AAAS; Am Chem Soc. Mailing Add: Lederle Labs Pearl River NY 10965

WILLIAMS, T GLYNE, b Franklin Co, Tenn, May 16, 16; m 41; c 3. PSYCHIATRY. Educ: Univ of the South, BS, 39; Vanderbilt Univ, MD, 43. Prof Exp: Intern, Baroness Erlanger Hosp, Chattanooga, Tenn, 43-44; psychotherapist, Vet Admin Ment Hyg Clin, Md, 46-50; chief chronic serv, Spring Grove State Hosp, Md, 50-53, clin dir, 53-56; asst prof psychiat & pub health, Sch Med, Yale Univ, 56-60; assoc prof, Sch Med, Univ Okla, 60-63; asst comnr, Dept Ment Hyg, Md, 63-69; supt, Rosewood State Hosp, Mills, Md, 64-72; assoc dir, Md State Psychiat Res Ctr, Catonsville, 72-75; SPEC ASST, SPRING GROVE HOSP CTR, CATONSVILLE, MD, 75- Concurrent Pos: Lectr, Cath Univ, 47-48; instr, Sch Med, Univ Md, 53-56; dir, State Dept Ment Health, Okla, 60-63; consult, Dept Ment Health, Conn, 60-63 & Friends of Psychiat Res, Inc, Md, 63- Mem: AAAS; Am Asn Ment Deficiency; fel Am Orthopsychiat Asn; fel Am Psychiat Asn. Res: Administrative psychiatry; mental retardation; schizophrenia; public health. Mailing Add: 3414 Pierce Dr Ellicott City MD 21043

WILLIAMS, TERENCE HEATON, b Oldham, Eng, Jan 5, 29; m 56; c 3. NEUROANATOMY, ELECTRON MICROSCOPY. Educ: Univ Manchester, MB, ChB, 53; Univ Wales, PhD(anat), 60. Prof Exp: House surgeon, Manchester Univ & Royal Infirmary, 53-54; jr registr surg, London Hosp, 55-56; asst lectr anat, Univ Col, Dublin, 57-58; lectr, Univ Wales, 58-61; lectr & sr lectr exp neurol, Univ Manchester, 61-68; vis lectr electron microscopy of nerv syst, Harvard Med Sch, 65-66; prof neuroanat, Sch Med, Tulane Univ, 68-73; PROF ANAT & HEAD DEPT, COL MED, UNIV IOWA, 73- Concurrent Pos: Brit Med Res Coun traveling fel, Harvard Med Sch, 64-65; Peck Sci Res award; NIH res awards, 69- Mem: Soc Neurosci; Am Asn Anat; Anat Soc Gt Brit & Ireland. Res: Ultrastructure of nervous tissues; plasticity of nervous system; promoting CNS regeneration; catecholamine neurons, central and ganglionic. Mailing Add: Dept of Anat Univ of Iowa Col of Med Iowa City IA 52242

WILLIAMS, THEODORE P, b Marianna, Pa, May 24, 33; m 56; c 5. BIOPHYSICS, PSYCHOBIOLOGY. Educ: Muskingum Col, BS, 55; Princeton Univ, MA, 57, PhD(phys chem), 59. Prof Exp: Res assoc chem, Brown Univ, 59-61, fel psychol, 61-63, asst prof biol & med sci, 63-66; assoc prof biol sci, 66-73, actg chmn dept, 70-71, PROF BIOL SCI, FLA STATE UNIV, 73-, CO-DIR PSYCHOBIOL PROG, 71- Concurrent Pos: NIH grant, 63-74; Nat Acad Sci exchange fel, USA-USSR, 73; NSF grant, 74-77. Mem: Biophys Soc; Asn Res Vision Ophthalmol. Res: Visual processes; sensory mechanisms; fast chemical reactions. Mailing Add: Rm 511 Inst of Molecular Biophys Fla State Univ Tallahassee FL 32306

WILLIAMS, THEODORE ROOSEVELT, b Washington, DC, Oct 23, 30; m 54; c 4. ANALYTICAL CHEMISTRY. Educ: Howard Univ, BS, 52; Pa State Univ, MS, 54; Univ Conn, PhD, 60. Prof Exp: Asst instr chem, Univ Conn, 56-59; from instr to assoc prof, 59-66, PROF CHEM, COL WOOSTER, 66- Concurrent Pos: Res assoc, Harvard Univ, 67-68, Sloan vis prof chem, 69-70; vis prof, Univ Conn, 72-73. Mem: Am Chem Soc. Res: Electroanalytical chemistry. Mailing Add: Dept of Chem Col of Wooster Wooster OH 44691

WILLIAMS, THEODORE SHIELDS, b Kansas City, Kans, June 2, 11; m 36; c 2. VETERINARY MEDICINE. Educ: Kans State Univ, DVM, 35; Iowa State Univ, MS, 46. Prof Exp: Col vet, Prairie View State Col, 36; vet inspector, Meat Inspection Div, USDA, 36-45; head dept, 45-55, dean sch, 47-72; PROF PATH, SCH VET MED, TUSKEGEE INST, 45- Concurrent Pos: Consult, Vet Admin Hosp, Tuskegee, Ala. Honors & Awards: Distinguished Serv award, Kans State Univ, 59. Mem: Am Vet Med Asn. Res: Pathological lesions associated with tissue invading migratory parasites in animals. Mailing Add: Dept of Path Sch of Vet Med Tuskegee Inst Tuskegee Inst AL 36088

WILLIAMS, THOMAS ELLIS, b Plymouth, Pa, Feb 11, 29; m 52; c 2. GEOLOGY. Educ: Rochester Univ, BA, 51; Southern Methodist Univ, MS, 57; Yale Univ, PhD(geol), 62. Prof Exp: Instr, 59-62, from asst prof to assoc prof, 62-70, PROF GEOL, SOUTHERN METHODIST UNIV, 70-, CHMN, 74- Mem: Paleont Soc; Geol Soc Am. Res: Paleontology of fusuline foraminifers. Mailing Add: Dept of Geol Sci Southern Methodist Univ Dallas TX 75275

WILLIAMS, THOMAS FFRANCON, b Denbighshire, Wales, Jan 30, 28; m 59; c 2. PHYSICAL CHEMISTRY. Educ: Univ London, BSc, 49, PhD(chem), 60. Prof Exp: From sci officer to prin sci officer, Chem Div, Atomic Energy Res Estab, Eng, 49-61; from asst prof to prof, 61-74, DISTINGUISHED SERV PROF CHEM, UNIV TENN, KNOXVILLE, 74- Concurrent Pos: Assoc, Northwestern Univ, 57-59; NSF vis scientist, Kyoto Univ, 65-66; Guggenheim fel, Royal Inst Technol, Sweden, 72-73. Mem: Am Chem Soc; The Chem Soc. Res: Electron spin resonance studies of trapped radicals; free radical reactions at low temperature; electronic structure of inorganic radicals; radiation chemistry; radiation-induced ionic polymerization. Mailing Add: Dept of Chem Univ of Tenn Knoxville TN 37916

WILLIAMS, THOMAS FRANKLIN, b Belmont, NC, Nov 26, 21; m 51; c 2. INTERNAL MEDICINE. Educ: Univ NC, BS, 42; Columbia Univ, MA, 43; Harvard Univ, MD, 50. Prof Exp: Asst chem, Columbia Univ, 42-43; asst med, Johns Hopkins Univ, 51-53 & Boston Univ, 53-54; from instr to prof, Sch Med, Univ NC, 56-68; PROF MED, UNIV ROCHESTER, 68-; MED DIR, MONROE COMMUNITY HOSP, 68- Concurrent Pos: Res fel, Sch Med, Univ NC, 54-56, Markle scholar, 57-61; fel physiol, Vanderbilt Univ, 66-67. Mem: AAAS; Soc Exp Biol & Med; Am Fedn Clin Res; Am Pub Health Asn; Am Col Physicians. Res: Diseases of metabolism, especially for chronic illness; metabolic and renal physiology. Mailing Add: Monroe Community Hosp 435 E Henrietta Rd Rochester NY 14603

WILLIAMS, THOMAS HENRY, b Jamaica, WI, Apr 21, 34; m 67; c 2. STRUCTURAL CHEMISTRY. Educ: Univ WI, BSc, 56; Yale Univ, MS, 60, PhD(chem), 61. Prof Exp: Fel Univ Notre Dame, 61-62; RES CHEMIST, HOFFMANN-LA ROCHE INC, NUTLEY, 63- Mem: Am Chem Soc. Res: Structural elucidation of natural products; applications of nuclear magnetic resonance spectroscopy in organic chemistry. Mailing Add: 25 Lynwood Rd Cedar Grove NJ 07009

WILLIAMS, THOMAS RHYS, b Martins Ferry, Ohio, June 13, 28; m 52; c 3. CULTURAL ANTHROPOLOGY, ETHNOGRAPHY. Educ: Miami Univ, BA, 51; Univ Ariz, MA, 56; Syracuse Univ, PhD(anthrop), 56. Prof Exp: From asst prof to assoc prof anthrop, Sacramento State Col, 56-65, sr fac mem dept sociol and anthrop, 65-67, chmn dept anthrop, 67-71, PROF ANTHROP, OHIO STATE UNIV, 65- Concurrent Pos: NSF fel Borneo ethnographis res, 59-60 & 62-63; Am Coun Learned Soc-Soc Sci Res Coun Asian studies fel, 58-59 & 63-64. Mem: AAAS; fel Am Anthrop Asn. Res: Socialization and enculturation process; cultural learning; educational anthropology; peoples and cultures of Island Asia, southeast Asia; nature of human nature; cognatic social organization; cognitive processes; microcultural communication. Mailing Add: Dept of Anthrop Page Hall Ohio State Univ 65 S Oval Dr Columbus OH 43210

WILLIAMS, THOMAS WALLEY, JR, b Ventnor, NJ, Nov 28, 09; m 35; c 2. HUMAN ANATOMY. Educ: Univ Pittsburgh, BS, 34, MS, 36, PhD(anat), 39. Prof Exp: Instr anat, Sch Med, Univ Pittsburgh, 39-44; asst prof anat & histol, Sch Med, Univ WVa, 44-48, assoc prof anat, 48-56, prof microanat & chmn dept, 56-65, prof, 66-72, EMER PROF ANAT, UNIV WVA, 72- Mem: Am Asn Anat; Microcirc Soc. Res: Microcirculation, angioarchitecture and human microanatomy. Mailing Add: Dept of Anat WVa Univ Med Ctr Morgantown WV 26506

WILLIAMS, TIMOTHY CHENEY, b New York, NY, May 7, 42; m 64; c 2. ANIMAL BEHAVIOR. Educ: Swarthmore Col, BA, 64; Harvard Univ, AM, 66; Rockefeller Univ, PhD(biol), 68. Prof Exp: Fel, Woods Hole Oceanog Inst, 68-69; ASST PROF BIOL, STATE UNIV NY BUFFALO, 69- Concurrent Pos: Vis investr, Woods Hole Oceanographic Inst, 75- Mem: Am Ornith Union; Am Soc Mammalogists; Animal Behav Soc. Res: Homing and migratory behavior of birds and bats; sensory basis of oriented behavior. Mailing Add: Dept of Biol State Univ of NY Buffalo NY 14214

WILLIAMS, TODD ROBERTSON, b Washington, DC, Nov 3, 45; m 67; c 1. MEDICINAL CHEMISTRY. Educ: Cornell Univ, AB, 67; Univ Calif, Los Angeles, PhD(org chem), 71. Prof Exp: Fel, Syntex Res Co, 71-72; SR CHEMIST MED CHEM, 3M CO, 72- Mem: Am Chem Soc. Res: Synthesis of prostaglandin analogs. Mailing Add: Cent Res Labs 3M Co PO Box 33221 St Paul MN 55101

WILLIAMS, TOM VARE, b Philadelphia, Pa, Dec 27, 38; m 63; c 3. PLANT BREEDING, PLANT PATHOLOGY. Educ: Univ Conn, BS, 60; Rutgers Univ, MS, 63, PhD(plant breeding), 66. Prof Exp: Agronomist, Soil Conserv Serv, USDA, 66-67; res horticulturist, Birds Eye Div, Gen Foods Corp, 67-70; RES DIR SEED DEPT, AGR CHEM DIV, FMC CORP, 70- Mem: Am Soc Agron; Am Soc Hort Sci; Am Genetic Asn. Res: Vegetable variety development. Mailing Add: Agr Chem Div FMC Corp Box 2508 El Macero CA 95618

WILLIAMS, VERNON, b Augusta, Ga, Nov 10, 26; m 69; c 1. MATHEMATICS, EDUCATION. Educ: Paine Col, BA, 49; Univ Mich, Ann Arbor, MA, 54; Okla State Univ, EdD, 69. Prof Exp: Instr, Paine Col, 49-54 & Fla A&M Univ, 54-56; from asst prof to assoc prof, 56-69, PROF MATH, SOUTHERN UNIV, BATON ROUGE, 69- Concurrent Pos: Adv, Math Sect, La Educ Asn, 71- Mem: Am Math Asn. Res: Number theory; educational technology as devoted to higher education as well as secondary education. Mailing Add: Dept of Math Southern Univ Baton Rouge LA 70813

WILLIAMS, VICK FRANKLIN, b Pittsburg, Tex, Apr 30, 36; m 62; c 2. ANATOMY. Educ: Austin Col, BA, 58; Univ Tex, MD & PhD(anat), 64. Prof Exp: Intern path, Charity Hosp La, New Orleans, 64-65; from instr to asst prof anat, Univ Tex Southwestern Med Sch Dallas, 65-70; ASSOC PROF ANAT, DENT SCH, UNIV TEX HEALTH SCI CTR SAN ANTONIO, 70- & UNIV TEX MED SCH SAN ANTONIO, 70- Honors & Awards: Borden Award, 64. Mem: AAAS; Am Asn Anat; Am Soc Cell Biol; Electron Micros Soc Am; Soc Neurosci. Res: Ultrastructure of the central nervous system of mammals. Mailing Add: Dept of Anat Univ of Tex Health Sci Ctr San Antonio TX 78284

WILLIAMS, WALLACE REID, biochemistry, physiology, see 12th edition

WILLIAMS, WALTER FORD, b Yazoo City, Miss, Oct 7, 27; m 50; c 4. PHYSIOLOGY. Educ: Univ Mo, BS, 50, MS, 51, PhD(dairy physiol), 55. Prof Exp: Asst, Univ Mo, 53-55, res assoc, 55-57; from asst prof to assoc prof, 57-69, PROF DAIRY PHYSIOL, UNIV MD, COLLEGE PARK, 69- Concurrent Pos: USPHS fel, Nat Cancer Inst, 55-57. Mem: AAAS; Am Dairy Sci Asn; Soc Study Reproduction; Am Soc Animal Sci. Res: Reproductive processes; endocrine regulation of metabolism in relation to reproduction, growth and lactation in domestic and wild mammals. Mailing Add: Dept of Dairy Sci Animal Sci Ctr Univ of Md College Park MD 20742

WILLIAMS, WAYNE THOMAS, plant pathology, botany, see 12th edition

WILLIAMS, WELLS ELDON, b Cadillac, Mich, July 8, 19; m 44. FISHERIES. Educ: Mich State Univ, BS, 53, MS, 54, PhD(fisheries, wildlife), 58. Prof Exp: Asst fisheries & wildlife, 54-56; from instr to asst prof, 56-70, ASSOC PROF NATURAL SCI, MICH STATE UNIV, 70- Res: Freshwater ecology; habitat improvement; food conversion and growth rates of fishes; pond culture; physiology of freshwater fishes. Mailing Add: Dept of Natural Sci Mich State Univ East Lansing MI 48823

WILLIAMS, WENDELL STERLING, b Lake Forest, Ill, Oct 27, 28; m 52; c 2. SOLID STATE PHYSICS. Educ: Swarthmore Col, BA, 51; Cornell Univ, PhD(physics), 56. Prof Exp: Physicist, Leeds & Northrup Co, 51; asst, Cornell Univ, 52-55; physicist, Union Carbide Corp, 56-65, 66-67; sr res visitor, Dept Metall, Cambridge Univ, 65-66; assoc prof, 67-69, co-chmn, Bioeng Comt, 71, PROF PHYSICS & CERAMICS, UNIV ILL, URBANA, 69-, PRIN INVESTR, MAT RES LAB, 67- Concurrent Pos: Nat Sci Foun lectr, Univ PR, 61; mem, Mat Sci Comt, Argonne Ctr Educ Affairs, Argonne Nat Lab, 71-; mem adv comt, Metals & Ceramics Div, Oak Ridge Nat Lab, 72-; task coordr energy res, Div Materials Res, Nat Sci Found, 74-75; consult, Kennametal Inc, 75- Mem: AAAS; Am Phys Soc; Am Ceramic Soc; Am Inst Mining, Metall & Petrol Eng; NY Acad Sci. Res: Electrical, thermal and mechanical properties of refractory hard metals; high-strength fibers; defects in solids; low and high temperature thermal conductivity; field ion microscopy; electrical properties of bone; implant materials. Mailing Add: Materials Res Lab Univ of Ill Urbana IL 61801

WILLIAMS, WILLIAM ARNOLD, b Johnson City, NY, Aug 2, 22; m 43; c 3. AGRONOMY. Educ: Cornell Univ, BS, 47, MS, 48, PhD(agron), 51. Prof Exp: Instr, 51-53, from asst prof to assoc prof, 54-64, PROF AGRON, UNIV CALIF, DAVIS, 65- Concurrent Pos: Fulbright scholar, Australia, 60; Rockefeller fel, Cent & SAm, 66; mem staff, Res Vessel Alpha Helix, Amazon Exped, 67. Mem: Am Soc Agron; Crop Sci Soc Am; Soil Sci Soc Am; Am Soc Plant Physiol; Am Ecol Soc. Res: Physiological ecology of Leguminosae used for soil and pasture improvement; computer modeling. Mailing Add: Dept of Agron & Range Sci Univ of Calif Davis CA 95616

WILLIAMS, WILLIAM DONALD, b Macon, Ga, Apr 22, 28; m 52; c 4. PHYSICAL CHEMISTRY. Educ: Harding Col, BS, 50; Univ Ky, MS, 52, PhD(chem), 54. Prof Exp: Assoc prof chem, 54-63, PROF CHEM & CHMN DEPT PHYS SCI, HARDING COL, 63- Mem: AAAS; Am Chem Soc. Res: Chemistry and kinetics of flames. Mailing Add: Dept of Chem Harding Col Searcy AR 72143

WILLIAMS, WILLIAM HOWARD, b Hamilton, Ont, Can, June 11, 31; m 58; c 3. MATHEMATICS, STATISTICS. Educ: McMaster Univ, BA, 54; Iowa State Univ, MS, 56, PhD(statist), 58. Prof Exp: Asst prof statist, Iowa State Univ, 57-58; asst prof math, McMaster Univ, 58-60; MEM STAFF, MATH RES DEPT, BELL TEL LABS, INC, 60- Concurrent Pos: Consult, Off Sci & Tech, White House, 63-65 & US Bur of Census; prof, Univ Mich, 69-70; vis prof statist, Univ of Calif, Berkeley, 74-75. Mem: Math Asn Am; Inst Math Statist; fel Am Statist Asn; Biomet Soc; Economet Soc. Res: Theory and applications of statistics; econometrics. Mailing Add: Bell Tel Labs Murray Hill NJ 07974

WILLIAMS, WILLIAM JOSEPH, b Bridgeton, NJ, Dec 8, 26; m 50; c 3. MEDICINE, BIOCHEMISTRY. Educ: Univ Pa, MD, 49. Prof Exp: Intern, Hosp Univ Pa, 49-50; sr instr microbiol, Sch Med, Western Reserve Univ, 52; resident med, Hosp Univ Pa, 54-55; assoc med, Sch Med, Univ Pa, 55-56, from asst prof to prof, 56-70; PROF MED & CHMN DEPT, STATE UNIV NY UPSTATE MED CTR, 70- Concurrent Pos: Am Cancer Soc fel physiol chem, Sch Med, Univ Pa, 50-52, Am Philos Soc Daland fel res clin med, 55-57; Markle scholar, 57-62; USPHS res career develop award, 63-68; asst prof, Sch Med, Wash Univ, 59-60; mem hemat training comt, Nat Inst Arthritis & Metab Dis, 64-68, res career prog comt, 68-72 & thrombosis adv comt, 69-73, chmn, 71-73; mem adv coun, Nat Arthritis Metab & Digestive Dis, NIH, 75- Mem: Am Soc Hemat; Am Soc Clin Invest; Am Fedn Clin Res; Am Soc Biol Chem; Asn Am Physicians. Res: Internal medicine; hematology; blood coagulation; blood cell metabolism. Mailing Add: Dept of Med State Univ of NY Upstate Med Ctr Syracuse NY 13210

WILLIAMS, WILLIAM LANE, b Rock Hill, SC, Dec 23, 14. ANATOMY. Educ: Wofford Col, BS, 35; Duke Univ, MA, 39; Yale Univ, PhD(anat), 41. Prof Exp: Asst anat, Sch Med, Yale Univ, 39-40; instr, Sch Med & Dent, Univ Rochester, 41-42; instr, Sch Med, Yale Univ, 42-43; asst prof, Sch Med, La State Univ, 43-45; from asst prof to assoc prof, Univ Minn, Minneapolis, 45-58; PROF ANAT & CHMN DEPT, MED CTR, UNIV MISS, 58- Concurrent Pos: Donner Found fel anat, Sch Med, Yale Univ, 42-43; asst vchancellor, Med Ctr, Univ Miss, 75- Mem: Am Soc Exp Path; Am Physiol Soc; Soc Exp Biol & Med; Am Asn Anat; Am Inst Nutrit. Res: Endocrinology, experimental pathology; nutrition; cardiovascular disease; hepatic liposis. Mailing Add: Dept of Anat Univ of Miss Med Ctr Jackson MS 39216

WILLIAMS, WILLIAM LAWRENCE, b St Cloud, Minn, June 14, 19; m 46; c 2. BIOCHEMISTRY. Educ: Univ Minn, BS, 42; Univ Wis, MS, 47, PhD(biochem), 49. Prof Exp: Asst prof biochem, NC State Univ, 49-50; res biochemist, Lederle Lab, 50-59; res assoc prof, 59-60, RES PROF BIOCHEM, UNIV GA, 60- Concurrent Pos: NIH career develop award, 62-, res grant, 64-67, training grant, 65-70; indust consult; dir, Reproduction Res Labs. Mem: Am Soc Biol Chem; Brit Soc Study Fertil; Soc Study Reproduction; Am Fertil Soc; Am Physiol Soc. Res: Animal reproduction. Mailing Add: Dept of Biochem Boyd Grad Studies Ctr Univ of Ga Athens GA 30601

WILLIAMS, WILLIAM LEE, b Chickasha, Okla, May 28, 37; m 59; c 2. ATOMIC PHYSICS. Educ: Rice Inst, BA, 59; Dartmouth Col, MA, 61; Yale Univ, PhD(physics), 65. Prof Exp: From instr to asst prof, 65-69, ASSOC PROF PHYSICS, UNIV MICH, ANN ARBOR, 69- Mem: Am Phys Soc. Res: High precision atomic physics of hydrogenic atoms; atomic lifetime measurements. Mailing Add: Dept of Physics Univ of Mich Ann Arbor MI 48104

WILLIAMS, WILLIAM MICHAEL, b Columbus, Ga, Sept 15, 44. ORGANIC CHEMISTRY. Educ: Auburn Univ, BS, 66; Univ Fla, PhD(chem), 71. Prof Exp: Fel chem, Ga Inst Technol, 71-73; RES CHEMIST, E I DU PONT DE NEMOURS & CO, INC, 73- Mem: Am Chem Soc. Res: New dyes for synthetic fibers. Mailing Add: Jackson Lab E I du Pont Co Deepwater NJ 08023

WILLIAMS, WILLIAM NORMAN, b Detroit, Mich, June 18, 39. SPEECH PATHOLOGY. Educ: Fla State Univ, BA, 63, MS, 64; Univ Fla, PhD(speech path audiol), 69. Prof Exp: Res assoc speech physiol, Dept of Commun Dis, 65-66, ASST PROF SPEECH PHYSIOL, DEPT BASIC DENT SCI, COL DENT, UNIV FLA, 71- Concurrent Pos: Res clin consult, Audiol-speech path, Vet Admin Hosp, Gainesville, Fla, 71-; res consult, Lancaster Cleft Palate Clin, Pa, 75-, WHO fel, Sweden & WGer, 75. Mem: Am Speech & Hearing Asn; Am Cleft Palate Asn; Am Asn Phonetic Sci. Res: Radiographic assessment of oral motor function of normal subjects and individuals with oral facial-communicative disorders; develop instrumentation and measurement strategies for assessing oral sensory-perceptual integrity. Mailing Add: Dept of Basic Dent Sci Univ of Fla Col of Dent Gainesville FL 32610

WILLIAMS, WILLIAM ORVILLE, b Carlsbad, NMex, Oct 19, 40; m 60. MATHEMATICS, MECHANICS. Educ: Rice Univ, BA, 62, MS, 63; Brown Univ, PhD(appl math), 67. Prof Exp: Assoc res engr, Houston Res Lab, Humble Oil & Ref Co, 64; asst prof, 66-70, ASSOC PROF MATH, CARNEGIE-MELLON UNIV, 70- Mem: Soc Natural Philos; Am Math Soc; Math Asn Am. Res: Foundations of continuum mechanics; thermodynamics. Mailing Add: Dept of Math Carnegie-Mellon Univ Pittsburgh PA 15213

WILLIAMS, WILLIAM THOMAS, b San Marcos, Tex, Dec 22, 24; m 51; c 1. BIOCHEMISTRY, CELL PHYSIOLOGY. Educ: Southwest Tex State Univ, BS, 47; Tex A&M Univ, MS, 51, PhD(biochem), 65. Prof Exp: Instr chem, Tex A&M, 47-48; res asst biochem, Tex Agr Exp Sta, 58-61, res chemist, Marine Lab, Tex A&M Univ, 61-62; res biologist, US Air Force Sch Aerospace Med, Brooks AFB, 63-64, res chemist, 64-72, ASSOC FOUND SCI, SOUTHWEST FOUND RES EDUC, 73- Mem: Fel Am Inst Chemists; Am Inst Biol Sci; fel AAAS; Am Chem Soc; Am Soc Microbiol. Res: Regulation of cell metabolism and growth; nitrogen nutrition and metabolism; analytical biochemistry. Mailing Add: 116 Elm Spring Lane San Antonio TX 78231

WILLIAMS, WILLIAM WILSON, b Washington, DC, Aug 26, 15; m 44. ORGANIC CHEMISTRY. Educ: Univ Md, BS, 36; Univ Ill, PhD(org chem), 39. Prof Exp: Technician, Univ Md, 33-36; asst chem, Univ Ill, 36-37; res chemist, Colgate-Palmolive-Peet Co, NJ, 38; chemist, Jackson Lab, E I du Pont de Nemours & Co, NJ, 39-43; sr chemist, Gen Aniline & Film Corp, NY, 43-47, sect leader, Cent Res Lab, Pa, 47-52, res mgr, 52-55, tech asst to dyestuff sales mgr, 55-56, for tech rep, 56-59, mgr for liaison, 59-62, mgr tech develop, 62-63; attache & sci officer, Int Sci & Tech Affairs, Ger, 63-70, ATTACHE & SCI OFFICER, INT SCI & TECH AFFAIRS, US DEPT STATE, INDIA, 71-, PHYS SCI OFFICER, ADMIN SECT, AM EMBASSY, 74- Concurrent Pos: Instr, Exten Serv, Rutgers Univ, 41-42. Mem: AAAS; Am Chem Soc; Soc Ger Chem; Swiss Chem Soc. Res: Dyestuffs; intermediates; dyeing auxiliaries; vulcanizing agents; polymers; detergents; action of sulfur dioxide on acetylenes; diazotype processes; optical bleaching agents; international scientific and technological affairs. Mailing Add: Admin Sect US Dept State Am Embassy Shanti Path Chanakyapuri 21 New Delhi India

WILLIAMS, WILLIE ELBERT, b Jacksonville, Tex, June 6, 27; m 51; c 2. MATHEMATICS, STATISTICS. Educ: Huston-Tillotson Col, BS, 52; Tex Southern Univ, MS, 53; Mich State Univ, PhD(math educ), 72. Prof Exp: Teacher math, Lufkin Independent Schs, 53-59 & Case Western Reserve Univ, 64-73; dept chmn, Cleveland Bd Educ, 60-73; ASSOC PROF MATH, FLA INT UNIV, 73- Mem: Math Asn Am; Nat Coun Teachers Math. Res: Teacher effectiveness in mathematics and how children learn mathematics. Mailing Add: Fla Int Univ Tamiami Trail Miami FL 33199

WILLIAMS-ASHMAN, HOWARD GUY, b London, Eng, Sept 3, 25; nat US; m 59; c 4. BIOCHEMISTRY. Educ: Cambridge Univ, BA, 46; Univ London, PhD, 59. Prof Exp: Biochemist, Chester Beatty Res Inst, Eng, 49-50; from asst prof to prof biochem, Univ Chicago, 53-64; prof pharmacol & exp therapeut & prof reprod biol, Sch Med, Johns Hopkins Univ, 64-69; PROF BIOCHEM & PHYSIOL, PRITZKER SCH MED, UNIV CHICAGO, 69- Concurrent Pos: Am Cancer Soc scholar, 53-57; USPHS res career award, 62-64. Mem: Am Soc Biol Chem; Am Soc Pharmacol & Exp Therapeut; fel Am Acad Arts & Sci; Brit Biochem Soc; Soc Gen Physiol. Res: Mechanism of hormone action; reproductive physiology; chemical pathology. Mailing Add: Ben May Lab for Cancer Res Pritzker Sch of Med Univ Chicago Chicago IL 60637

WILLIAMSON, ADDISON HEATON, organic chemistry, see 12th edition

WILLIAMSON, ARTHUR ELRIDGE, JR, b Montgomery, Ala, July 6, 26; m 51; c 2. ELECTRO OPTICS. Educ: Auburn Univ, BEP, 50, MS, 51. Prof Exp: Res engr, NAm Aviation, Inc, 51-52; instr physics, Univ Richmond, 52-53; res physicist, Southern Res Inst, 53-55; asst prof physics & res proj dir, Ga Inst Technol, 55-59; chief, Electrooptics Lab, Martin Marietta Corp, 59-73; HEAD, ELECTRO OPTICS SECT, SOUTHERN RES INST, 73- Mem: Am Phys Soc; Optical Soc Am. Res: Optics. Mailing Add: Southern Res Inst 2000 Ninth Ave S Birmingham AL 35205

WILLIAMSON, ARTHUR TANDY, b Kingston, Ont, Sept 18, 06; m 33; c 2. PHYSICAL CHEMISTRY. Educ: Queen's Univ, Ont, BA, 27, MA, 28; Princeton Univ, AM, 29, PhD(chem), 31. Prof Exp: Res assoc chem, Princeton Univ, 31-32; Nat Res Coun fel, Harvard Univ, 32-33; Int Res fel, Oxford Univ, 33-34; res chemist, Imp Chem Industs, 34-41, asst sect mgr, Alkali Group, Res Dept, 41-47, sect mgr, 47; res chemist & group leader, Cent Res Lab, Can Industs, Ltd, Que, 47-68 & Tech Info Off, Atomic Energy Can Ltd, Man, 68-71; tech consult, 71-73; RETIRED. Mem: Fel The Chem Soc; fel Chem Inst Can. Res: Catalysis and reaction kinetics; physical chemistry of solutions; thermodynamics. Mailing Add: Box 407 Pinawa MB Can

WILLIAMSON, CHARLES EDWARD, b Newport, Ind, May 29, 15; m 42; c 3. PLANT PATHOLOGY. Educ: Wabash Col, AB, 37; Cornell Univ, PhD(plant path), 49. Prof Exp: Asst prof, 48-72, ASSOC PROF PLANT PATH, NY STATE COL AGR & LIFE SCI, CORNELL UNIV, 72- Mem: Am Phytopath Soc. Res: Diseases of ornamentals, including those caused by nematodes, soil fumigants and soil treatments. Mailing Add: Cornell Ornamentals Res Lab State Univ Campus Farmingdale NY 11735

WILLIAMSON, CHARLES ELVIN, b Portsmouth, Va, Dec 5, 26; m 52; c 6. BIO-ORGANIC CHEMISTRY, ONCOLOGY. Educ: Col William & Mary, BS, 50; Johns Hopkins Univ, PhD(bio-org chem), 70. Prof Exp: RES CHEMIST, RES LABS, EDGEWOOD ARSENAL, 52- Concurrent Pos: Res assoc, Siani Hosp Baltimroe, 55-; res assoc, Johns Hopkins Univ Sch Med, 72- Res: Microenvironmental forces at biologic binding sites; reactions at cell surfaces; hydrophobic and electrostatic catalyses; neoplastic changes and cancer chemotherapy. Mailing Add: 210 E Ring Factory Rd Bel Air MD 21014

WILLIAMSON, CHARLES WESLEY, b Blue Earth, Minn, Sept 2, 24; m 44; c 5. PLASTICS CHEMISTRY. Educ: Kenyon Col, AB, 49; Rensselaer Polytech Inst, PhD(phys chem), 53. Prof Exp: Chemist, E I du Pont de Nemours & Co, Inc, 53-56, plastic technologist, 56-59; sr chemist, 59-63, res assoc polymer processing, 63-67, sr res assoc plastics fabrication, 68-74, SR RES ASSOC, PLASTICS TECHNOL DIV, EXXON CHEM CO, 74- Mem: Am Chem Soc; Soc Plastics Engrs. Res: Polymerization of polyolefins; viscoelastic properties of polymer melts; morphology of themoplastics; polyolefin film fabrication process development; film fabrication and solution casting of high temperture plastics. Mailing Add: Exxon Chem Co Plastics Technol Div PO Box 4255 Baytown TX 77520

WILLIAMSON, CLARENCE KELLY, b McKeesport, Pa, Jan 19, 24; m 51; c 2. MICROBIOLOGY. Educ: Univ Pittsburgh, BS, 49, MS, 51, PhD, 55. Prof Exp: Instr bact, Sch Pharm, Univ Pittsburgh, 51-55; from asst prof to assoc prof, 55-63, chmn dept microbiol, 62-72, PROF MICROBIOL, MIAMI UNIV, 63-, DEAN, COL ARTS & SCI, 71- Concurrent Pos: Consult, Warren-Teed Prod Co, 54-64; mem, Com Arts & Sci, Nat Asn State Univ & Land-Grant Col, 75- Mem: AAAS; Am Soc Microbiol; fel Am Acad Microbiol; Reticuloendothelial Soc; Brit Soc Gen Microbiol. Res: Microbic dissociation, Pseudomonas aeruginosa; classification and polysaccharides of viridans streptococci; post-streptococcal nephritis. Mailing Add: Col of Arts & Sci Miami Univ Oxford OH 45056

WILLIAMSON, CLAUDE F, b Henderson, Tex, Mar 29, 33; m 59. NUCLEAR PHYSICS. Educ: Univ Tex, BS, 55, MA, 56, PhD(physics), 59. Prof Exp: Physicist, Saclay Nuclear Res Ctr, France, 60-62; res asst prof nuclear physics, Nuclear Physics Lab, Univ Wash, 62-66; res physicist, Lab Nuclear Sci, 66-75, SR RES SCI, DEPT PHYSICS, MASS INST TECHNOL, 75- Mem: AAAS; Am Phys Soc. Res: Fast neutron physics; nuclear reaction gamma rays; nuclear structure by electron scattering. Mailing Add: Rm 26-431 Mass Inst Technol 77 Massachusetts Ave Cambridge MA 02139

WILLIAMSON, DARRELL LEROY, entomology, forest management, see 12th edition

WILLIAMSON, DAVID GADSBY, b Honolulu, Hawaii, June 12, 41; m 63; c 2. CHEMICAL KINETICS. Educ: Univ Colo, Boulder, BA, 63; Univ Calif, Los Angeles, PhD(phys chem), 66. Prof Exp: Chemist, Nat Bur Standards, Colo, 63; teaching & res asst chem, Univ Calif, Los Angeles, 63-66; fel, Nat Res Coun Can, 67-68; asst prof, 68-72, ASSOC PROF CHEM, CALIF POLYTECH STATE UNIV, SAN LUIS OBISPO, 72- Concurrent Pos: Res grant, Environ Protection Agency, 72. Mem: Am Chem Soc. Res: Ozone chemistry and the chemistry of free radicals of importance to atmospheric chemistry; development of energy sources alternate to petroleum products. Mailing Add: Dept of Chem Calif Polytech State Univ San Luis Obispo CA 93401

WILLIAMSON, DAVID LEE, b Humboldt, Nebr, July 17, 30; m 68; c 2. GENETICS. Educ: Nebr State Teachers Col, Peru, AB, 52; Univ Nebr, MS, 55, PhD(zool), 59. Prof Exp: Instr biol, Dana Col, 55-56; Fulbright scholar, Lab Genetics, Gif-sur-Yvette, France, 59-60; asst prof genetics, Univ Utah, 60-61; NIH fel, Yale Univ, 61-64; res fel, Med Col Pa, 64-66, asst prof, 66-71; ASSOC PROF GENETICS, STATE UNIV NY STONY BROOK, 71- Mem: AAAS; Am Soc Nat; Genetics Soc Am; Soc Study Evolution; Int Soc Mycoplasmologists. Res: Maternally inherited traits in Drosophila; biology of spiroplasmas. Mailing Add: Dept of Anat Sci State Univ NY Stony Brook NY 11790

WILLIAMSON, DENIS GEORGE, b Trail, BC, June 9, 41; m 62; c 3. BIOCHEMISTRY. Educ: Univ BC, BSc, 63, PhD(biochem), 68. Prof Exp: Med Res Coun fel, 68-71, lectr, 71-72, ASST PROF BIOCHEM, UNIV OTTAWA, 72- Mem: Can Biochem Soc. Res: Metabolism of steroid hormones; steroid conjugates; purification and characterization of steroid dehydrogenases. Mailing Add: Dept of Biochem Univ of Ottawa Ottawa ON Can

WILLIAMSON, DONALD ELWIN, b Lansing, Mich, Oct 24, 13; m 40; c 3. INSTRUMENTATION, BIOMEDICAL ENGINEERING. Educ: Carleton Col, AB, 35; Univ Mich, MS, 36. Prof Exp: Res physicist, Dept Eng Res, Univ Mich, 36, res engr, 44-45; mgr, Profilometer Div, Physicists Res Co, 36-44; res engr, Lincoln Park Industs, 45-47; chief engr & assoc dir res, Baird Assocs, Inc, Mass, 47-52; pres & treas, Williamson Develop Co, 53-60; SCI ADV TO PRES, CORDIS CORP, 60- Concurrent Pos: Chmn, Gordon Res Conf Instrumentation, 56. Mem: Am Soc Mech Eng; Optical Soc Am. Res: Roughness measurement; optics; infrared instruments; physiological and cardiovascular instrumentation. Mailing Add: 13001 Old Cutler Rd Miami FL 33156

WILLIAMSON, DOUGLAS BLEECKER, b Brooklyn, NY, Nov 16, 18; m 41; c 3. PHYSICS. Educ: Columbia Univ, BA, 39, MA, 41; EdD, 56. Prof Exp: Asst prof math & physics, Patterson Sch, 39-40 & Gordon Mil Col, 41-43; asst prof math, Fla Mil Acad, 40-41; asst prof, 46-60, ASSOC PROF PHYSICS, UNDERGRAD & GRAD SCHS, W VA UNIV, 60- Mem: Am Meteorol Soc; Am Asn Physics Teachers. Res: Atmospheric pollution. Mailing Add: Dept of Physics WVa Univ Morgantown WV 26506

WILLIAMSON, DOUGLAS HARRIS, b Croydon, Eng, Mar 3, 24; m 52; c 2. GEOLOGY. Educ: Aberdeen Univ, BSc, 50, PhD, 52. Prof Exp: Lectr geol, Aberdeen Univ, 50-53; asst prof, Mt Allison Univ, 53-54, Sr James Dunn prof & head dept, 54-66; head dept, 66-74, PROF GEOL, LAURENTIAN UNIV, 66-, ASSOC DEAN SCI, 69-, Concurrent Pos: Nat Res Coun Can major equip grant, 65, operating grants, 65 & 66; Geol Surv Can grant, Univ NB, 62, NB Res & Productivity Coun grant, 62-66; mem subcomt, Nat Adv Comt Res Geol Sci in Can, 56-57. Mem: AAAS; Nat Asn Geol Teachers; Can Inst Mining & Metall; fel Brit Geol Soc. X-ray crystallography; petrogenesis and synthesis of metamorphic mineral assemblages; mineralogy of fluorspar; exploration of mineral resources of Ontario; geology of Pre-Carboniferous rocks of Southern New Brunswick. Mailing Add: Dept of Geol Rm 319 Sci II Laurentian Univ Sudbury ON Can

WILLIAMSON, FRANCIS SIDNEY LAINER, b Little Rock, Ark, Feb 6, 27; m 48, 68; c 3. ZOOLOGY, PUBLIC HEALTH ADMINISTRATION. Educ: San Diego State Col, BS, 50; Univ Calif, MA, 55; Johns Hopkins Univ, ScD, 68. Prof Exp: Mus technician vert zool, Univ Calif, 53-55; med biologist, Arctic Health Res Ctr, USPHS, 55-64; dir, Chesapeake Bay Ctr Environ Studies, Smithsonian Inst, 68-75; NIH fel & res assoc, 64-68, ASSOC PATHOBIOL, SCH HYG & PUB HEALTH, JOHNS HOPKINS UNIV, 68-; COMMISSIONER, DEPT HEALTH AND SOCIAL SERV, ALASKA, 75- Concurrent Pos: Consult, Avain ecol, AEC, 59-61; instr, Anchorage Community Col, Univ Alaska, 61-64; consult ecol, Battelle Mem Inst, 67-; adv, Chesapeake Bay Study, US Army Corps Engrs, 69-75; ed, Biosci, Am Inst Biol Sci, 70 & Condor, Cooper Ornith Soc, 71-; consult, Md Dept Nat Resources, 72-73; adv environ qual comn, Gov Sci Adv Coun, Md, 72-73. Honors & Awards: Exceptional Serv Award, Smithsonian Inst, 75. Mem: Fel AAAS; Am Inst Biol Sci; Cooper Ornith Soc; Fel Am Ornith Union; Wilson Ornith Soc. Res: Taxonomy, behavior, distribution, parasites and viral diseases of birds; taxonomy, distribution, and life histories of helminths; medical helminthology; estuarine ecology; environmental planning. Mailing Add: Pouch H01 Dept of Health and Social Serv Juneau AK 99811

WILLIAMSON, FRANK SHAVER, JR, b Springfield, Mass, Dec 12, 26; m 61; c 3. INORGANIC CHEMISTRY. Educ: Middlebury Col, BA, 48; Univ Wis, PhD(chem), 54. Prof Exp: Instr chem, Dartmouth Col, 54-56, asst prof, 56-62; univ rep, Gen Elec Res Lab, 62-66; ADMIN OFF, DEPT CHEM, WASH UNIV, 66- Mem: AAAS; Am Chem Soc. Res: Kinetics and mechanisms of reactions in solutions. Mailing Add:

WILLIAMSON, FREDERICK DALE, b Poplar, Mont, Apr 15, 38; m 59; c 3. POLYMER CHEMISTRY. Educ: NDak State Univ, BS, 60, MS, 62, PhD(polymers & coatings chem), 66. Prof Exp: Chemist, Glidden Co, 62; res chemist, Missiles & Space Systs Div, McDonnell Douglas Corp, 66-68; tech dir paints & coatings, Midland Div, Dexter Corp, 68-71; TECH DIR PAINTS & COATINGS, O'BRIEN CORP, 71- Mem: Am Chem Soc; Fedn Soc Paint Technol; Nat Paint & Coatings Asn. Res: Polymer synthesis; high temperature-resistant coatings; coatings rheology; high energy-curable coatings systems; water-borne resins and coating systems. Mailing Add: O'Brien Corp 2001 W Wash PO Box 4037 South Bend IN 46634

WILLIAMSON, HAROLD EMANUEL, b Racine, Wis, Aug 8, 30; m 57; c 3. PHARMACOLOGY. Educ: Univ Wis, BS, 53, PhD(pharmacol, toxicol), 59. Prof Exp: Asst pharmacol, Univ Wis, 55-59, proj assoc, 59-60; from instr to assoc prof, 60-70, PROF PHARMACOL, COL MED, UNIV IOWA, 70- Mem: AAAS; Am Soc Pharmacol & Exp Therapeut; Am Fedn Clin Res; Soc Exp Biol & Med; Am Soc Nephrology. Res: Renal pharmacology and physiology, especially the effect of diuretics and hormones on electrolyte and water transport. Mailing Add: Dept of Pharmacol Univ of Iowa Col of Med Iowa City IA 52242

WILLIAMSON, HUGH A, b Kemp, Tex, Aug 11, 32; m 56; c 4. PHYSICS. Educ: North Tex State Univ, BA, 54; Univ Tex, PhD(physics, math), 62. Prof Exp: Res scientist, Molecular Physics Res Lab & Mil Physics Res Lab, Univ Tex, 60-63, res fel, 62-63; res scientist, Res Lab, United Aircraft Corp, 63-65, sr res scientist, 65-67; from asst prof to assoc prof, 67-74, PROF PHYSICS, CALIF STATE UNIV, FRESNO, 74- Res: Atomic and molecular structure; gaseous electronics; electron scattering processes off neutral atoms including elastic, inelastic and free-free scattering processes. Mailing Add: Dept of Physics Calif State Univ Fresno CA 93726

WILLIAMSON, HUGH A, b Williamsport, Pa, Apr 20, 27; m 60; c 2. ORGANIC CHEMISTRY, SCIENCE EDUCATION. Educ: Bucknell Univ, AB, 60, MA, 61; Cornell Univ, EdD(sci ed, chem), 66. Prof Exp: Teacher high sch, Pa, 50-55; PROF CHEM, LOCK HAVEN STATE COL, 55-, ASSOC DEAN ARTS & SCI, 73- Mem: Nat Asn Res Sci Teaching. Res: Synthesis of cyclooctatetraene compounds; reductions of epoxides. Mailing Add: Dept of Chem Lock Haven State Col Lock Haven PA 17745

WILLIAMSON, JAMES LAWRENCE, b Rebecca, Ga, Feb 28, 29; m 49; c 3. ANIMAL SCIENCE, ANIMAL NUTRITION. Educ: Univ Ga, BSA, 51; Univ Ill, MS, 52, PhD, 57. Prof Exp: Mgr, Beef Cattle & Sheep Res Div, 58-59, mgr, Livestock Res Dept, 59-60, dir res, 64-67, VPRES & DIR RES, CHOW DIV, RALSTON PURINA CO, 67- Concurrent Pos: Mem nutrit coun, Am Feed Mfrs Asn. Mem: Am Soc Animal Sci; Am Dairy Sci Asn; Poultry Sci Asn. Res: Animal nutrition and management. Mailing Add: Chow Div Ralston Purina Co Checkerboard Sq St Louis MO 63188

WILLIAMSON, JERRY ROBERT, b Danville, Ill, Feb 14, 38; m 65; c 1. ORGANIC CHEMISTRY, POLYMER CHEMISTRY. Educ: Univ Ill, Urbana, BA, 60; Univ Iowa, MS, 63, PhD(org polymer chem), 64. Prof Exp: Petrol Res Fund res fel polymer res, Univ Iowa, 61-62, teaching asst gen & org chem, 62-64; asst prof chem & actg chmn div sci, Jarvis Christian Col, 64-66; res fel, Tex Christian Univ, 66-67; asst prof, 67-70, ASSOC PROF CHEM, EASTERN MICH UNIV, 70- Concurrent Pos: Partic, State Tech Serv Prog, Mich, 67- Mem: AAAS; Am Chem Soc. Res: Organic polymer chemistry, thermally stable materials; polymer analysis via gel permeation chromatography. Mailing Add: Dept of Chem Eastern Mich Univ Ypsilanti MI 48197

WILLIAMSON, JOHN HYBERT, b Clarkton, NC, Apr 28, 38; m 63; c 2. GENETICS. Educ: NC State Col, BS, 60; Cornell Univ, MS, 63; Univ Ga, PhD(zool), 66. Prof Exp: Fel biol, Oak Ridge Nat Lab, 66-67; mem fac life sci, Univ Calif, Riverside, 67-69; asst prof, 69-72, ASSOC PROF ZOOL, UNIV CALGARY, 72-, ACAD ADMIN OFFICER BIOL DEPT, 74- Mem: Genetics Soc Am. Res: Chromosome mechanics; radiation biology. Mailing Add: Dept of Biol Univ of Calgary Calgary AB Can

WILLIAMSON, JOHN RICHARD, b Coventry, Eng, Sept 18, 33; m 61; c 3. BIOCHEMISTRY, BIOPHYSICS. Educ: Oxford Univ, BA, 56, MA, 59, PhD(biochem), 60. Prof Exp: Dept demonstr biochem, Oxford Univ, 60-61; independent investr, Baker Clin Res Lab, Harvard Univ, 61-63; assoc phys biochem, 63-65, from asst prof to assoc prof, 65-69, PROF PHYS BIOCHEM, JOHNSON RES FOUND, 69- Concurrent Pos: USPHS fel, 61-63; Am Heart Asn grant, 66-72; NIH res grants & contract, 71-; estab investr, Am Heart Asn, 67-72, mem coun basic sci. Mem: Am Soc Biol Chem; Am Diabetes Asn; Brit Biochem Soc; NY Acad Sci; Am Physiol Soc. Res: Mode of action of hormones and drugs; control of metabolic pathways; effect of hormones on cells and hormone interactions in animal cells; role of anion transport across mitochondrial membranes; myocardial ischemia. Mailing Add: Dept of Biochem & Biophys Richards Bldg G4 Univ of Pa Philadelphia PA 19174

WILLIAMSON, KENNETH DALE, b Drumright, Okla, Sept 4, 20; m 46; c 3. PHYSICAL CHEMISTRY. Educ: Univ Okla, BS, 47, MS, 48; Univ Tex, PhD(chem), 54. Prof Exp: Chemist, Petrol Exp Sta, US Bur Mines, Okla, 48-50; spec instr chem, Univ Tex, 53; res chemist, 53-66, group leader res, 66-71, RES SCIENTIST, UNION CARBIDE CORP, 71- Concurrent Pos: Asst prof, Morris Harvey Col, 56-59 & WVa State Col, 62-64. Mem: AAAS; Am Chem Soc; Sigma Xi. Res: Physical properties of gas hydrates; physical properties of pure compounds and mixtures; thermodynamics; calorimetry; kinetics of pyrolysis of hydrocarbons; catalysis. Mailing Add: 1022 Sand Hill Dr St Albans WV 24177

WILLIAMSON, KENNETH LEE, b Tarentum, Pa, Apr 13, 34; m 56; c 3. ORGANIC CHEMISTRY, SPECTROSCOPY. Educ: Harvard Univ, BA, 56; Univ Wis, PhD(org chem), 60. Prof Exp: NIH fel, Stanford Univ, 60-61; from asst prof to assoc prof, 61-69, PROF CHEM, MT HOLYOKE COL, 69- Concurrent Pos: Mem grad faculty, Univ Mass, 62-; vis prof, Cornell Univ, 66; NSF sci faculty fel & fel Univ Liverpool, 68-69; vis assoc, Calif Inst Technol & secy, Exp Nuclear Magnetic Resonance Spectroscopy Confs, 73-; fel, John Simon Guggenheim Found, 75-76. Mem: Am Chem Soc; Sigma Xi; AAAS. Res: Conformational analysis by means of nuclear magnetic resonance spectroscopy. Mailing Add: Dept of Chem Mt Holyoke Col South Hadley MA 01075

WILLIAMSON, LUTHER HOWARD, b Osyka, Miss, Oct 9, 36; m 57; c 3. PHYSICAL CHEMISTRY. Educ: La State Univ, BS, 59, MS, 62, PhD(phys chem), 65. Prof Exp: Res chemist corrosion, 65-70, SR RES CHEMIST CORROSION, WATER CHEM, MOBIL RES & DEVELOP CORP, 70- Mem: Am Chem Soc; Nat Asn Corrosion Engrs. Res: Surface chemistry; corrosion, corrosion inhibition, electrochemistry of corrosion; hydrogen embrittlement; sulfide stress corrosion cracking; water chemistry, oilfield chemistry, chemistry of scale formation; water pollution, air pollution; environmental science. Mailing Add: Mobil Res & Develop Corp 3600 Duncanville Rd Dallas TX 75211

WILLIAMSON, MARTIN JOHN, b Ryde, Eng, Apr 19, 40. PHYSICAL ORGANIC CHEMISTRY, ANALYTICAL CHEMISTRY. Educ: Univ Exeter, BS, 61, PhD(chem), 65. Prof Exp: Nat Res Coun Can fel, McGill Univ, 64-66; sci officer, Govt Analyst Dept, Guyana, SAm, 67; SR CHEMIST, ROHM AND HAAS CO, 68- Mem: The Chem Soc. Res: Kinetics of nitration in sulphuric acid; properties of organic compounds in acidic solutions; acidity functions; development of analytical methods for pesticides using gas and liquid chromatography; ultra violet, infrared, nuclear magnetic resonance and mass spectroscopy. Mailing Add: Rohm and Haas Res Labs Spring House PA 19477

WILLIAMSON, NATHAN KIBLER, physics, deceased

WILLIAMSON, PENELOPE ROSE, b New York, NY, Feb 6, 43; m 68; c 1. PSYCHOSOMATIC MEDICINE, HUMAN ECOLOGY. Educ: Antioch Col, BA, 65; Johns Hopkins Univ, ScD(ecol), 69. Prof Exp: ASST PROF MENT HYG, SCH HYG & PUB HEALTH, JOHNS HOPKINS UNIV, 69-, ASST PROF MED PSYCHOL, DEPT PSYCHIAT & BEHAV SCI, 73-, ASST PROF, SCH HEALTH SERV, 74- Concurrent Pos: Smithsonian Inst fel, Chesapeake Bay Ctr Environ Studies, Edgewater, Md, 69-70, Nat Audubon Soc grant, 70-72; asst prof, Eve Col, Johns Hopkins Univ, 73- Mem: Ecol Soc Am; Environ Design Res Asn. Res: Application of ecological methods and theory to the study of human behavior in urban areas; evaluation of behavioral science clinical teaching on health practitioner competencies and patient outcome. Mailing Add: Sch of Health Serv Johns Hopkins Univ 624 N Broadway Baltimore MD 21205

WILLIAMSON, PIERCE MACDONALD, b Iredell, Tex, July 12, 05; m 34. PHYSICAL CHEMISTRY. Educ: Rice Inst Technol, BA, 29; Univ Tex, MA, 41, PhD(chem), 43. Prof Exp: Res chemist, Humble Oil & Ref Co, Tex, 30-37; instr chem, Univ Tex, 39-43 & La State Univ, 43-45; develop chemist, Ethyl Corp, La, 45-46; assoc prof chem, Sampson Col, 46-49 & Wagner Lutheran Col, 49-50; res chemist, Chem Res Br, Aeronaut Res Lab, 51-52, asst chief, 52-61, res chemist, 61-64, res chemist, Fibrous Mat Br, Air Force Mat Lab, Wright-Patterson AFB, US Air Force, 64-70; RETIRED. Mem: Am Chem Soc. Res: Properties of continuous surfaces and thin films; volume and surface characteristics and properties of the fiber form. Mailing Add: PO Box 8333 Park Cities Br Dallas TX 75205

WILLIAMSON, RALPH EDWARD, b Wilson, NC, Dec 28, 23; m 46; c 2. PLANT PHYSIOLOGY. Educ: NC State Univ, BS, 48; Univ Wis, MS, 50, PhD(bot), 58. Prof Exp: Botanist, Chem Corps, Dept Army, 48-49, plant physiologist, 50-51, 53-57 & Soil & Water Conserv Res Div, 57-74, PLANT PHYSIOLOGIST, TOBACCO RES LAB, USDA, OXFORD, NC, 74- Mem: Am Soc Plant Physiol; Am Soc Agron; Soil Sci Soc Am. Res: Determine major organic constituents of leaf tobacco that, if eliminated or reduced through breeding, cultural, processing, or curing techniques, would result in tobacco varieties less harmful to the consumer. Mailing Add: 716 Currituck Dr Raleigh NC 27609

WILLIAMSON, RALPH ELMORE, b Tulsa, Okla, July 8, 17; m 53; c 1. ASTRONOMY. Educ: Phillips Univ, AB, 38; Drake Univ, MA, 39; Univ Chicago, PhD(theoret astrophys), 43. Prof Exp: Instr astron, Yerkes Observ, Univ Chicago, 42-43, instr physics, US Air Force, 43-44; instr astron, US Navy, Cornell Univ, 44-46; lectr, David Dunlap Observ, Univ Toronto, 46-47, from asst prof to assoc prof, 47-53; MEM STAFF, LOS ALAMOS SCI LAB, 53- Concurrent Pos: Consult, Radio Astron Proj, Cornell Univ, 47-53. Mem: Am Astron Soc. Res: Theory of stellar atmospheres; stellar dynamics and interiors; radio astronomy; electronic computers; weapons physics and design. Mailing Add: Los Alamos Sci Lab Los Alamos NM 87544

WILLIAMSON, RICHARD EDMUND, b Chicago, Ill, May 23, 27; m 50. MATHEMATICS. Educ: Dartmouth Col, AB, 50; Univ Pa, AM, 51, PhD, 55. Prof Exp: Res asst, Univ Pa, 55-56; instr, 56-58, from asst prof to assoc prof, 58-66, PROF MATH, DARTMOUTH COL, 66- Concurrent Pos: Res fel, Harvard Univ, 60-61. Mem: Am Math Soc. Res: Analysis. Mailing Add: Dept of Math Dartmouth Col Hanover NH 03755

WILLIAMSON, ROBERT EMMETT, b Ashland, Kans, June 9, 37; m 58; c 3. GEOMETRY, TOPOLOGY. Educ: Univ Ariz, BS, 59; Univ Calif, Berkeley, PhD(math), 63. Prof Exp: Mem, Inst Adv Study, 63-65; vis prof, Univ Warwick, 65-66; asst prof, Yale Univ, 66-69; ASSOC PROF MATH, CLAREMONT GRAD SCH, 69- Concurrent Pos: Nat Acad Sci-Air Force Off Sci Res fel, 63-64. Mem: Am Math Soc. Res: Algebraic model of surgery; singularities of smooth maps. Mailing Add: Dept Math Claremont Grad Sch Claremont CA 91711

WILLIAMSON, ROBERT LESLIE, entomology, biochemistry, see 12th edition

WILLIAMSON, ROBERT MARSHALL, b Madison, Wis, Feb 2, 23; m 50. PHYSICS. Educ: Univ Fla, BS, 43, PhD(physics), 51. Prof Exp: Res assoc, Duke Univ, 51-53, from asst prof to assoc prof physics, 53-62; PROF PHYSICS, OAKLAND UNIV, 62- Concurrent Pos: Fulbright lectr, Univ Catania, 59-60. Mem: Am Phys Soc. Res: Nuclear spectroscopy. Mailing Add: Dept of Physics Oakland Univ Rochester MI 48063

WILLIAMSON, ROBERT SAMUEL, b Cincinnati, Ohio, June 18, 22. PHYSICS. Educ: Queens Col, BS, 45; NY Univ, MS, 48; Polytech Inst Brooklyn, PhD(physics), 57. Prof Exp: Tutor, 52-56, instr, 56-60, from asst prof to assoc prof, 60-68, from asst dean to assoc dean admin, 68-72, PROF PHYSICS, QUEENS COL, NY, 68- Mem: Am Phys Soc; Am Asn Physics Teachers; Am Crystallog Asn; Inst Elec & Electronics Eng. Res: X-ray crystallography; electronics. Mailing Add: Dept of Physics Queens Col Flushing NY 11367

WILLIAMSON, ROGER CLINTON, organic chemistry, physical chemistry, see 12th edition

WILLIAMSON, SAMUEL JOHNS, b West Reading, Pa, Nov 6, 39; m 66. LOW TEMPERATURE PHYSICS. Educ: Mass Inst Technol, SB, 61, ScD(physics), 65. Prof Exp: Res staff mem solid state physics, Nat Magnet Lab, Mass Inst Technol, 65-66 & Nat Acad Sci-Nat Res Coun fel, Dept Physics of Solids, Fac Sci, Orsay, France, 66-67; res staff mem, Sci Ctr Div, NAm Aviation, 67-70; lectr, Univ Calif, Santa Barbara, 70-71; ASSOC PROF PHYSICS, NY UNIV, 71- Mem: Am Phys Soc; AAAS; NY Acad Sci. Res: Polarimetry; anti-ferromagnetic resonance; de Haas-Van Alphen effect and Fermi surface studies; superconductivity; ultra low temperature physics; fundamentals of air pollution. Mailing Add: Dept of Physics NY Univ 4 Washington Pl New York NY 10003

WILLIAMSON, STANLEY ELLSWORTH, b Sabetha, Kans, Apr 30, 10; m 35; c 3. SCIENCE EDUCATION. Educ: Nebr Wesleyan Univ, BA, 31; Columbia Univ, MA, 36; Univ Ore, EdD, 56. Prof Exp: Chmn dept sci high sch, Nebr, 31-36; sci supvr, Univ High Sch, Eugene, Ore, 36-44, prin, 44-46; dean sch educ, Ore State Univ, 73-75; PROF SCI EDUC & CHMN DEPT, ORE STATE UNIV, 46- Concurrent Pos: Dir Nat Sci Found Prof Sec Teachers, 57-74; mem steering comt, Biol Sci Curriculum Proj, 64-70; consult, Minister of Educ, SAustralia, 71. Honors & Awards: Robert Carleton Award, Nat Sci Teachers Asn, 73. Mem: Fel AAAS; Nat Sci Teachers Asn(pres, 63-64); Nat Asn Biol Teachers (vpres, 69). Res: Biology; international studies in science education. Mailing Add: 3335 NW McKivley Corvallis OR 97330

WILLIAMSON, STANLEY GILL, b Manhattan, Kans, Aug 28, 38; m 65. MATHEMATICS. Educ: Calif Inst Technol, BS, 60; Stanford Univ, MS, 62; Univ Calif, Santa Barbara,PhD(math), 65. Prof Exp: From asst prof to assoc prof, 65-75, PROF MATH, UNIV CALIF, SAN DIEGO, 75- Mem: Soc Indust Appl Math. Res: Combinatorial analysis; computation. Mailing Add: Dept of Math Univ of Calif San Diego La Jolla CA 92037

WILLIAMSON, STANLEY MORRIS, b Chattanooga, Ten, Mar 18, 36; m 66. CHEMISTRY. Educ: Univ NC, BS, 58; Univ Wash, PhD(chem), 61. Prof Exp: Asst prof chem, Univ Calif, Berkeley, 61-65, from asst prof to assoc prof, 65-74, PROF CHEM, UNIV CALIF, SANTA CRUZ, 74-, DEAN GRAD DIV, 72- Mem: AAAS; Am Chem Soc; The Chem Soc. Res: Fluorine chemistry of compounds of sulfur, nitrogen, oxygen and xenon including preparations and properties. Mailing Add: Div of Natural Sci Univ of Calif Santa Cruz CA 95060

WILLIAMSON, SUSAN, b Boston, Mass, Dec 29, 36. MATHEMATICS. Educ: Radcliffe Col, AB, 58; Brandeis Univ, AM, 61, PhD(math), 63. Prof Exp: Instr, Cardinal Cushing Col, 62-63; asst prof math, Boston Col, 63-64; scholar hist sci, Harvard Univ, 64-65; from asst prof to assoc prof, 65-71, dean col, Regis Col, 73-75, PROF MATH, REGIS COL, MASS, 71- Mem: Am Math Soc; Math Asn Am; Asn Women Math. Res: Associative algebras; commutative rings. Mailing Add: Dept of Math Regis Col Weston MA 02193

WILLIAMSON, THURMOND A, b Angleton, Tex, Dec 11, 22; m 48; c 3. ORGANIC CHEMISTRY. Educ: Univ Tex, El Paso, BA, 43; Columbia Univ, MA, 45, PhD(org chem), 46. Prof Exp: Res chemist, E I du Pont de Nemours & Co, 46-49; partner & chemist, Agr Specialties Co, 49-59; VPRES, THURON INDUST, INC, 59- Mem: Am Chem Soc. Res: Organic synthesis and fomulation of pharmaceuticals and insecticides and related fields, including legal, patent and contract work; business management. Mailing Add: Thuron Industs Inc 12200 Denton Dr Dallas TX 75234

WILLIAMSON, WAYNE TIAMON, microbiology, see 12th edition

WILLIAMSON, WEAVER M, veterinary medicine, see 12th edition

WILLIAMSON, WILLIAM, JR, b Newport, RI, Jan 20, 34; m 57; c 2. ATOMIC PHYSICS, ASTROPHYSICS. Educ: San Francisco State Col, BA, 55; Univ Calif, Berkeley, MA, 58; Univ Colo, PhD(physics), 63. Prof Exp: Physicist, US Naval Radiol Defense Lab, Calif, 56-58; Fulbright fel, Frascati Labs, Italy, 61-62; asst physics, Univ Colo, 62-63; fel, Inst Sci & Tech, Univ Mich, 63-64, instr, 64-65; asst prof physics & astron, 65-69, assoc prof, 69-75, PROF PHYSICS, UNIV TOLEDO, 75- Concurrent Pos: Vis prof, Univ Adelaide, 71-72. Mem: Am Astron Soc; Am Phys Soc; Am Asn Physics Teachers; Italian Phys Soc. Mailing Add: Ritter Astrophys Res Ctr Univ of Toledo Toledo OH 43606

WILLIAMS-THORNTON, MYRA, zoology, see 12th edition

WILLIFORD, WILLIAM OLIN, b San Pedro, Calif, July 18, 33; m 71. MATHEMATICAL STATISTICS. Educ: Pepperdine Col, BA, 57; Fla State Univ, MS, 59; Va Polytech Inst & State Univ, PhD(statist), 67. Prof Exp: Instr math, Fla State Univ, 59-62; asst prof, Roanoke Col, 62-63; NIH fel, 64-67; ASST PROF STATIST, UNIV GA, 67-, COMPUT SCI, 74- Mem: Am Statist Asn; Inst Math Statist; Biomet Soc; Math Asn Am. Res: Bayesian inference; accident statistics. Mailing Add: Dept of Statist Univ of Ga Athens GA 30601

WILLIG, FRANK JEWELL, physics, see 12th edition

WILLIGER, ERVIN JOHN, b Szeged, Hungary, June 18, 27; US citizen; m 51; c 4. PLASTICS CHEMISTRY, RUBBER CHEMISTRY. Educ: Budapest Tech Univ, dipl chem eng, 50. Prof Exp: Res assoc polymers, Inst Plastics Res, Budapest Tech Univ, 50-56; res & develop chemist, Naugatuck Chem, US Rubber Co, 57-63; polymer chemist, Lucidol Div, Wallace & Tiernan, Inc, 63-65; sr res chemist, Res Ctr, Gen Tire & Rubber Co, 65-69; mgr tech serv & mkt, Union Process Co, Akron, 69-73; SR RES CHEMIST, RES & DEVELOP CTR, B F GOODRICH CO, 73- Mem: Nat Soc Prof Eng; Am Chem Soc; Soc Plastics Eng. Res: Preparative polymer chemistry of thermoplastics and thermosets; structure behavior study of reinforced plastics; preparation and application of unsaturated polyesters; study of peroxides and other free radical sources; toughening and fatigue of reinforced composites; technology of liquid elastomers. Mailing Add: 665 Fairwood Dr Tallmadge OH 44278

WILLIGES, GEORGE GOUDIE, b Sioux City, Iowa, May 18, 24; m 47; c 1. PLANT PATHOLOGY, PLANT TAXONOMY. Educ: Univ Corpus Christi, BA, 55; Tex A&I Univ, MA, 59; Tex A&M Univ, PhD(plant path), 69. Prof Exp: Teacher biol, Sinton Independent Sch Dist, 55-60; ASSOC PROF BIOL, TEX A&I UNIV, 61-, CUR HERBARIUM, 71- Mem: AAAS; Am Phytopath Soc. Res: Pathogenic variability, physiological and environmental effects on growth and reproduction of Sclerotium rolfsii; Fusarium diseases of cacti. Mailing Add: Dept of Biol Tex A&I Univ Kingsville TX 78363

WILLIHNGANZ, EUGENE, organic chemistry, physical chemistry, see 12th edition

WILLINGHAM, ALLAN KING, b Washington, DC, July 11, 41; m 67. BIOCHEMISTRY. Educ: George Washington Univ, BS, 63; St Louis Univ, PhD(biochem), 70. Prof Exp: Fel biochem, Res Inst Hosp Joint Dis, New York, 70-71; instr, 71-75, ASST PROF BIOCHEM, COL MED, UNIV NEBR, OMAHA, 75- Concurrent Pos: USPHS res grant, Nat Heart & Lung Inst, 74-77. Res: Interconversion of phylloquinone and its 2, 3- epoxide and its relationship to the vitamin K-dependent carboxylation of glutamic acid residues to form active clotting proteins. Mailing Add: Dept of Biochem Univ of Nebr Med Ctr Omaha NE 68105

WILLINGHAM, CHARLES ALLEN, b Longview, Tex, Apr 17, 34; m 55; c 2. MARINE ECOLOGY, MARINE BIOLOGY. Educ: NTex State Univ, BA, 58, MA, 59; Univ Miami, PhD(marine sci), 65. Prof Exp: Instr biol, ECent State Col, Okla, 59-60; res scientist microbiol petrol prod, Jersey Prod Res Co, Esso, 60; MARINE BIOLOGIST MARINE ECOL, WILLIAM F CLAPP LABS, BATTELLE MEM INST, 65- Mem: Am Soc Microbiol; AAAS; Ecol Soc Am; Am Fisheries Soc; Am Inst Biol Sci. Res: Fundamental and applied research on the microbiology of natural marine products, biofouling ecology, materials deterioration and estuarine ecology. Mailing Add: William F Clapp Labs Inc Battelle Mem Inst Drawer AH Duxbury MA 02332

WILLINGHAM, CHARLES BAYNARD, b Macon, Ga, May 13, 09; m 38; c 2. PHYSICAL CHEMISTRY. Educ: Mercer Univ, AB, 34, BS, 35; Univ Md, MS, 39. Prof Exp: Res assoc, Am Petrol Inst Proj, Nat Bur Stand, 36-45, chemist, 45-49, sr chemist, 49-50; head dept phys measurement, Mellon Inst, 50-54, dept phys chem, 54-57, head prof rels, 57-71, asst dir, Ctr Spec Studies & asst dir div, Sponsored Res, Carnegie-Mellon Inst, 71-75; RETIRED. Mem: AAAS; Am Chem Soc; Am Inst Chemists. Res: Methods for separation, purification, and analysis of complex mixtures by distillation, extraction and chromatography; research administration. Mailing Add: 777 Lebanon Ave Pittsburgh PA 15228

WILLINGTON, ROBERT PETER, b Vancouver, BC, Apr 25, 43; m 66. FOREST

HYDROLOGY. Educ: Univ BC, BSF, 67, MSc, 69, PhD(forest hydrol), 71. Prof Exp: Lectr, 70-71, ASST PROF FOREST HYDROL, UNIV BC, 71- Concurrent Pos: Consult, Fisheries Serv, Can Dept Environ, 71-73, researcher, Forestry Serv, 71-74. Mem: Can Inst Forestry; Can Soc Wildlife & Fisheries Biol. Res: Effects of forest land management on water quantity and quality with specific reference to fish habitat. Mailing Add: Fac of Forestry Univ of BC Vancouver BC Can

WILLIS, CARL BERTRAM, b Charlottetown, PEI, Nov 27, 37; m 62; c 2. PLANT PATHOLOGY. Educ: McGill Univ, BSc, 59; Univ Wis-Madison, PhD(plant path), 62. Prof Exp: RES SCIENTIST PLANT PATH, CAN DEPT AGR, 62- Mem: Can Phytopath Soc; Am Phytopath Soc; Agr Inst Can; Soc Nematol. Res: Forage crops diseases; factors affecting root rots of forage legumes. Mailing Add: Can Dept of Agr Res Sta PO Box 1210 Charlottetown PE Can

WILLIS, CARL RAEBURN, JR, b Madison, Wis, Apr 5, 39; m 60; c 4. PHARMACY,JPHARMACEUTICAL CHEMISTRY. Educ: Purdue Univ, BS, 61, MS, 64, PhD(indust pharm), 66. Prof Exp: Teaching asst bionucleonics & mfg pharm, Purdue Univ, 61-62, teaching assoc pharmaceut chem, 62-63; sr pharmaceut chemist, Warren-Teed Pharmaceut, Inc, Rohm and Haas Co, 66-69; supvr prod develop,Pharm Res & Develop Div, Ciba Pharmaceut Co, 69, mgr process & mat technol, 70-71; mgr prod develop, Pharmaceut Div, Ciba-Geigy Corp, 71-72; ASSOC DIR, DRUG REGULATORY AFFAIRS, STERLING DRUG INC, 72- Mem: AAAS; Am Pharmaceut Asn; Acad Pharmaceut Sci; Am Chem Soc. Res: Industrial pharmacy; pharmacokinetics; biopharmaceutics; pharmaceutical dosage form research and development; package materials research and development; drug regulatory affairs. Mailing Add: Sterling Drug Inc 90 Park Ave New York NY 10016

WILLIS, CHARLES RICHARD, b Westertown, NY, July 7, 28; m 54; c 2. PHYSICS. Educ: Syracuse Univ, BA, 51, PhD, 58. Prof Exp: From asst prof to assoc prof, 57-68, PROF PHYSICS, BOSTON UNIV, 68- Mem: Am Phys Soc. Res: Statistical mechanics; laser physics; classical many-body problems; quantum optics. Mailing Add: Dept of Physics Boston Univ Boston MA 02215

WILLIS, CHRISTOPHER JOHN, b Sutton, Eng, June 6, 34; m 60; c 3. INORGANIC CHEMISTRY. Educ: Cambridge Univ, BA, 55, PhD(chem), 58, MA, 59. Prof Exp: Fel chem, Univ BC, 58-60, lectr, 60-61; from lectr to asst prof, 61-66, ASSOC PROF, UNIV WESTERN ONT, 66- Concurrent Pos: Nat Res Coun Can res grant. Mem: Chem Inst Can; The Chem Soc. Res: Synthesis and study of fluorinated alchols, alkoxides and related fluorinated ligands. Mailing Add: Dept of Chem Univ of Western Ont London ON Can

WILLIS, CLIFFORD LEON, b Chanute, Kans, Feb 20, 13; m 47; c 1. GEOLOGY. Educ: Univ Kans, BS, 39; Univ Wash, PhD(geol), 50. Prof Exp: Geophysicist, Carter Oil Co, 39-42, geologist, 46-47; instr geol, Univ Wash, 50-52, asst prof, 53-54; chief geologist, 54-68, VPRES, HARZA ENG CO, CHICAGO, 68- Concurrent Pos: Consult geologist, US, Turkey, Greece & Iraq, 51- Honors & Awards: Haworth Distinguished Alumnus Award, Univ Kans, 63. Mem: Geol Soc Am: Am Asn Petrol Geol; Am Inst Mining, Metall & Petrol Eng; Brit Geol Asn; Int Soc Rock Mech. Res: Petrology; structural and engineering geology. Mailing Add: 16 Briar Rd Golf IL 60029

WILLIS, DAVID EDWIN, b Cleveland, Ohio, Mar 13, 26; m 48; c 4. GEOPHYSICS, GEOLOGY. Educ: Western Reserve Univ, BS, 50; Univ Mich, Ann Arbor, MS, 57, PhD, 68. Prof Exp: Computer seismic explor, Keystone Explor Co, 50-52, party chief, 52-54, asst supvr, 54-55; res assoc, Univ Mich, Ann Arbor, 55-60, assoc res geophysicist, 60-63, res geophysicist, 63-68, actg head geophys lab, 65-67, assoc prof geol & head geophys lab, 68-70; assoc prof, 70-73, PROF GEOL, UNIV WIS-MILWAUKEE, 73-, CHMN DEPT GEOL SCI, 72- Concurrent Pos: Lectr, Univ Mich, Ann Arbor, 63-64. Mem: Soc Explor Geophys; Am Geophys Union; Seismol Soc Am; Geol Soc Am. Res: Seismic and acoustic wave propagation; earthquake seismology; ground vibration studies. Mailing Add: Dept of Geol Sci Univ of Wis Milwaukee WI 53201

WILLIS, DAVID LEE, b Pasadena, Calif, Mar 15, 27; m 50; c 3. BIOLOGY, RADIATION BIOLOGY. Educ: Biola Col, BTh, 49, BA, 51; Wheaton Col, Ill, BS, 52; Calif State Col Long Beach, MA, 54; Ore State Univ, PhD(radiation biol), 63. Prof Exp: Teacher high sch, Calif, 52-57; instr biol, Fullerton Jr Col, 57-61; from asst prof to assoc prof, 62-71, PROF BIOL, ORE STATE UNIV, 71-, CHMN DEPT, 69- Concurrent Pos: Consult, Comn Undergrad Educ Biol Sci, NSF, 66-70; vis investr, Oak Ridge Nat Lab, 68-69; consult, Portland Gen Elec Co, 70-72. Mem: Radiation Res Soc; Health Physics Soc; AAAS; fel Am Sci Affiliation (pres, 75). Res: Freshwater radioecology; radionuclide cycling in amphibians; radiation effects on reptiles and amphibians; general applications of radiotracer techniques to biology; development of modern laboratory work for general biology courses. Mailing Add: Dept of Gen Sci Ore State Univ Corvallis OR 97331

WILLIS, DAWN BUTLER, b Memphis, Tenn, Sept 16, 37; c 2. MICROBIOLOGY, BIOCHEMISTRY. Educ: Memphis State Univ, BS, 57; Univ Tenn, MS, 65, PhD(microbiol), 68. Prof Exp: Chemist, Nat Heart Inst, Bethesda, Md, 57-58; technician, Baptist Hosp, Memphis, Tenn, 58-59; microbiologist, Buckman Labs, 60; res assoc physiol, Univ Tenn, 61-62, instr med, 70-72; ASST MEM VIROL, ST JUDE HOSP, 72- Concurrent Pos: Fel microbiol, Univ Tenn, 63-65, NIH fel, Lab Molecular Biol, 68-72; res trainee, St Jude Hosp, 65-68. Mem: Am Soc Microbiol. Res: Protein, RNA biosynthesis; biochemistry of viral infection. Mailing Add: Virol Lab St Jude Hosp Memphis TN 38103

WILLIS, DONALD EDWARD, analytical chemistry, see 12th edition

WILLIS, EDWIN O'NEILL, b Russellville, Ala, Jan 18, 35; m 70; c 1. ORNITHOLOGY, ETHOLOGY. Educ: Va Polytech Inst, BS, 56; La State Univ, Baton Rouge, MS, 58; Univ Calif, Berkeley, PhD(zool), 64. Prof Exp: Frank M Chapman fel, Am Mus Natural Hist, 64-66; asst prof zool, San Diego State Col, 67-68; asst prof, Univ Wash, 68-69; asst prof biol-psychol, Oberlin Col, 69-71; res assoc biol, Princeton Univ, 71-75; RES ASSOC & ASST PROF BIOL, GRAD SCH, UNICAMP, CAMPINAS, BRAZIL, 75- Concurrent Pos: Res assoc, Am Mus Natural Hist, 66-69; NSF grant, Oberlin Col & Princeton Univ, 70-71; NSF grant, Brazil, Princeton Univ, 72-74. Mem: Am Ornith Union; Wilson Ornith Soc; Cooper Ornith Soc; Ecol Soc Am. Res: Field studies of behavior and ecology of neotropical birds, especially birds that follow army ants. Mailing Add: Biologia UNICAMP Caixa Postal 1170 13100 Campinas Est Sao Paulo Brazil

WILLIS, EDWIN ROY, b New York, NY, Feb 23, 11; m 43; c 1. ZOOLOGY, ENTOMOLOGY. Educ: La State Univ, BS, 38, MS, 39; Ohio State Univ, PhD(entom), 47. Prof Exp: Asst zool, Ohio State Univ, 39-41; entomologist, Pioneering Res Div, Qm Res & Develop Ctr, US Dept Army, 47-59, Cent Res Lab, United Fruit Co, 59-62 & Div Trop Res, Tela R R Co, Honduras, 62; PROF ENTOM, ILL STATE UNIV, 62- Concurrent Pos: Mem staff, USDA. Mem: Fel AAAS; Entom Soc Am; Am Soc Zool; Animal Behav Soc; Asn Trop Biol. Res: Insect behavior and biology; behavior of cockroaches. Mailing Add: Dept of Biol Ill State Univ Normal IL 61761

WILLIS, ERIC HERBERT, b Bebington, Eng, Nov 20, 27; US citizen; m 53; c 2. GEOPHYSICS, GEOCHEMISTRY. Educ: Univ London, BSc, 47; Cambridge Univ, MA, PhD(geophys), 58. Prof Exp: Asst dir res radiocarbon dating, Cambridge Univ, 57-64; vpres & dir res, Teledyne Isotopes, 64-70; dir nuclear monitoring res, Advan Res Projs Agency, 70-75, DIR, GEOTHERMAL ENERGY DIV, ENERGY RES ADMIN & DEVELOP, DEPT DEFENSE, 75- Concurrent Pos: Pres comn absolute dating, Int Union Quaternary Res, 64- Honors & Awards: Meritorious civilian service medal, Secy Defense, 74. Mem: Geochem Soc; Am Geophys Union. Res: Radiocarbon dating; atmospheric chemistry; seismic verification problems associated with underground nuclear test detection; exploitation of geothermal energy from hydrothermal, geopressured and hot dry rock resources. Mailing Add: 3413 Wessynton Way Alexandria VA 22309

WILLIS, FRENCH HOKE, b Carlisle, Ind, Apr 9, 04; m 64. PHYSICS. Educ: Purdue Univ, BSEE, 25; Harvard Univ, SME, 28; NY Univ, PhD(physics), 43. Prof Exp: Engr, Cia Minera de Penoles, Mex, 25-26 & Am Tel & Tel Co, 26-34; mem tech staff, Bell Tel Labs, NJ, 34-69; PHYSICIST, OFF TELECOMMUN, DEPT COMMERCE, BOULDER, COLO, 69- Concurrent Pos: Mem, Int Radio Consultative Comt, 48-, US Exec Comt, 57-61 & US Del Plenary Assembly, Warsaw, 56, Los Angeles, 59. Mem: Inst Elec & Electronics Eng; Am Phys Soc; Int Union Radio Sci. Res: Ultrasonics; radio propagation; antennas; communications. Mailing Add: 1 New St Mendham NJ 07945

WILLIS, GEORGE MIRRON, b Alto, Tex, Oct 6, 31; m 64; c 2. PLANT PATHOLOGY. Educ: Prairie View Agr & Mech Col, BS, 52; Ohio State Univ, MS, 59, PhD(bot, plant path), 62. Prof Exp: Asst bot, Ohio State Univ, 55-58 & plant path, 58-59, res asst, Ohio Agr Res & Develop Ctr, 59-62, State of Ohio fel, 62; res plant pathologist, US Army Biol Labs, Ft Detrick, Md, 62-68; prof bot, Cent State Univ, Ohio, 68-70; grants assoc, 70-71, PROG DIR, GEN RES SUPPORT BR, NIH, 71- Concurrent Pos: Vis scientist, US Army Biol Labs, Ft Detrick, Md, 63; lectr, Frederick Community Col, 64-67. Mem: AAAS; Am Phytopath Soc; Sigma Xi. Res: Study of etiology and epidemiology of plant diseases. Mailing Add: Gen Res Support Br Div Res Resources NIH Bethesda MD 20014

WILLIS, GROVER C, JR, b Kansas City, Mo, May 25, 21; m 41; c 3. PHYSICAL CHEMISTRY, ELECTROCHEMISTRY. Educ: Whittier Col, BA, 52; Univ Ore, MA, 55, PhD(chem), 57. Prof Exp: Petrol inspector, Gen Petrol Corp, Calif, 41-43; chemist, W C Hardesty Co, 46-51; assoc, Univ Ore, 53-55, res assoc, 55-57; from asst prof to assoc prof, 57-67, PROF PHYS & ANAL CHEM, CALIF STATE UNIV, CHICO, 67- Mem: Am Chem Soc. Res: Anodic oxide formation kinetics; mechanism; reaction rates, thermodynamics, adsorption phenomena and diffusion processes at dropping mercury electrodes; electrochemical instrumentation. Mailing Add: Dept of Chem Calif State Univ Chico CA 95926

WILLIS, GUYE HENRY, b Los Angeles, Calif, July 1, 37; m 60; c 2. SOIL CHEMISTRY. Educ: Okla State Univ, BS, 61; Auburn Univ, MS, 63, PhD(soil chem), 65. Prof Exp: RES SOIL SCIENTIST, SOIL & WATER CONSERV RES DIV, AGR RES SERV, USDA, 65- Mem: Am Soc Agron; Soil Sci Soc Am. Res: Fate of agricultural chemicals, including pesticides and fertilizers, in the environment; soil chemistry-plant nutrition relationships. Mailing Add: Soil & Water Conserv Res Div Agr Res Serv USDA PO Drawer U Univ Sta Baton Rouge LA 70803

WILLIS, HAROLD LESTER, b McPherson, Kans, Oct 20, 40. ENTOMOLOGY. Educ: Kans State Teachers Col, BA, 62; Univ Kans, PhD(entom), 66. Prof Exp: Asst prof, 66-67, ASSOC PROF ZOOL, UNIV WIS-PLATTEVILLE, 67- Mem: AAAS. Res: Bionomics, taxonomy and zoogeography of Nearctic tiger beetles. Mailing Add: Dept of Biol Univ of Wis Platteville WI 53818

WILLIS, ISAAC, b Albany, Ga, July 13, 40; m 65; c 2. DERMATOLOGY. Educ: Morehouse Col, BS, 61; Howard Univ, MD, 65. Prof Exp: Assoc dermat, Sch Med, Univ Pa, 69-70; head internal med res team & dermatologist, Letterman Army Inst, US Army Med Corps, 70-72; res assoc & clin instr dermat, Sch Med, Univ Calif, 70-72; asst prof med dermat, Sch Med, Johns Hopkins Univ, 72-73; ASST PROF MED DERMAT, SCH MED, EMORY UNIV & CHIEF DERMAT, VET ADMIN HOSP, ATLANTA, 73- Concurrent Pos: Asst attend physician, Philadelphia Gen Hosp, 69-70; Dermat Found res award, Univ Pa, 70; attend physician, Univ Calif Med Ctr, 70-72; attend physician, Johns Hopkins Hosp, Baltimore City Hosps & Good Samaritan Hosp, 72-73; consult & lectr dermat, Bur Med & Surg, US Dept Navy, 72-75; consult asst to Prof Dermat, Howard Univ Col Med, 72-; mem Formulary Task Force, Nat Prog Dermat, Am Fedn Clin Res. Honors & Awards: Dohme Chem Co Award, Dermat Sect, Nat Med Asn, 67. Mem: Am Soc Photobiol; Soc Invest Dermat; Am Acad Dermat; Am Fedn Clin Res; Dermat Found. Res: Phototherapy, photochemotherapy; acute and chronic effects of ultraviolet light, including carcinogenesis; effects of light on bacteria and fungi, and effects of heat and humidity on skin. Mailing Add: 4690 Millbrook Dr NW Atlanta GA 30327

WILLIS, JAMES E, biochemistry, see 12th edition

WILLIS, JAMES STEWART, JR, b West Point, NY, Feb 9, 35; m 59; c 2. PHYSICS. Educ: US Mil Acad, BS, 58; Rensselaer Polytech Inst, MS, 64, PhD(physics), 66. Prof Exp: US Army, 58-, from instr to asst prof, 66-69, ASSOC PROF PHYSICS, US MIL ACAD, 70- Mem: AAAS; Am Asn Physics Teachers. Res: Type II superconductivity. Mailing Add: Dept of Physics US Mil Acad West Point NY 10996

WILLIS, JEANNE ELEANOR, b Ironton, Ohio, Feb 20, 30. PALEOBOTANY. Educ: Ohio Univ, BSEd, 49, MS, 50; Univ Ill, PhD, 54. Prof Exp: Asst bot, Ohio Univ, 49-50 & Univ Ill, 50-54; asst prof biol & geol, Wesleyan Col, 54-55; assoc prof, 55-71, PROF BIOL & GEOL & CHMN DEPT LIFE & EARTH SCI, OTTERBEIN COL, 71- Mem: AAAS; Bot Soc Am; Torrey Bot Club. Res: Structure and evolutionary relationships of fossil ferns and gymnosperms. Mailing Add: Dept of Life & Earth Sci Otterbein Col Westerville OH 43081

WILLIS, JOHN STEELE, b Long Beach, Calif, Jan 19, 35; m 58. PHYSIOLOGY, ZOOLOGY. Educ: Univ Calif, Berkeley, AB, 56; Harvard Univ, AM, 58, PhD(biol), 61. Prof Exp: Nat Heart Inst fel biochem, Oxford Univ, 61-62; from asst prof to assoc prof, 62-72, PROF PHYSIOL, UNIV ILL, URBANA, 72- Concurrent Pos: Nat Inst Gen Med Sci res grants, 63-72; mem physiol study sect, NIH, 69-73; consult, Basic Sci Rev Bd, Vet Admin, 75-78. Mem: Brit Soc Exp Biol; Soc Gen Physiol; Am Soc Zool; Am Physiol Soc. Res: Cold resistance of tissues of hibernating mammals; active cation transport. Mailing Add: Dept of Physiol Univ of Ill Urbana IL 61801

WILLIS, JUDITH HORWITZ, b Detroit, Mich, Jan 2, 35; m 58. PHYSIOLOGY, DEVELOPMENTAL BIOLOGY. Educ: Cornell Univ, AB, 56; Radcliffe Col, AM, 57, PhD(biol), 61. Prof Exp: USPHS res fel, Harvard Univ, 60-61 & Oxford Univ, 61-62; from instr to asst prof, 63-68, ASSOC PROF ENTOM, UNIV ILL, URBANA,

68- Mem: AAAS; Am Soc Cell Biol; Soc Develop Biol; Am Soc Zool. Res: Endocrine action in insect metamorphosis. Mailing Add: Dept of Entom 320 Morrill Hall Univ of Ill Urbana IL 61801

WILLIS, JUDITH IONE, b Tacoma, Wash, June 15, 48. IMMUNOBIOLOGY, HUMAN ANATOMY. Educ: Pac Lutheran Univ, BS, 70; Univ Wash, PhD(anat, immunol), 74. Prof Exp: ASST PROF ANAT, COL MED, OHIO STATE UNIV, 74- Mem: AAAS; Reticuloendothelial Soc; An Asn Anatomists; Sigma Xi. Res: Thymus-dependent lymphocyte differentiation and tumor immunobiology. Mailing Add: Dept of Anat Graves Hall Ohio State Univ Columbus OH 43210

WILLIS, PARK WEED, III, b Seattle, Wash, Nov 18, 25; m 48; c 6. INTERNAL MEDICINE, CARDIOVASCULAR DISEASES. Educ: Univ Pa, MD, 48. Prof Exp: Intern, Pa Hosp, 48-50; resident internal med, Univ Hosp & Med Sch, 52-53, jr clin instr, Med Sch, 53-54, from instr to assoc prof, 54-65, asst prof postgrad med, 57-59, PROF INTERNAL MED, MED SCH, UNIV MICH, ANN ARBOR, 65-, DIR DIV CARDIOL, 69- Concurrent Pos: Attend physician, Vet Admin Hosp, Ann Arbor, 54-59, consult, 59-; consult health serv, Univ Mich, 56- Mem: AAAS; Cent Soc Clin Res; fel Am Col Cardiol; AMA; fel Am Col Physicians. Res: Clinical cardiology. Mailing Add: 1016 Martin Pl Ann Arbor MI 48104

WILLIS, PHYLLIDA MAVE, b Wallington, Eng, Mar ll, 18; nat US. PHYSICAL CHEMISTRY. Educ: Mt Holyoke Col, AB, 38; Smith Col, AM, 40; Columbia Univ, PhD(phys chem), 46. Prof Exp: Teacher sci, Knox Sch, NY, 40-42; asst chem, Columbia Univ, 42-44; from instr to asst prof, Wellesley Col, 46-54; assoc prof & head dept, Newcomb Col, Tulane Univ, 54-60; WHITAKER PROF CHEM & CHMN DEPT CHEM, PHYSICS & ASTRON, HOOD COL, 60- Concurrent Pos: Am Asn Univ Women fel, Oxford Univ, 51-52; NSF fac fel, Univ Minn, 58. Mem: AAAS; Am Chem Soc; Soc Appl Spectros; Am Asn Physics Teachers. Res: Molecular spectroscopy. Mailing Add: Dept of Chem Physics & Astron Hood Col Frederick MD 21701

WILLIS, ROLAND GEORGE, b Glasgow, Scotland, Dec 22, 37; m 63; c 4. PHYSICAL ORGANIC CHEMISTRY. Educ: Glasgow Univ, BSc, 59, PhD(org chem), 62. Prof Exp: NIH fel phys chem, Cornell Univ, 62-64; RES ASSOC, EASTMAN KODAK CO, 64- Res: Physical organic chemistry as related to photography; redox processes at metal-solution interfaces. Mailing Add: Res Labs Eastman Kodak Co 343 State St Rochester NY 14650

WILLIS, RONALD PORTER, b Cowley, Wyo, Sept 20, 26; m 53; c 5. GEOLOGY. Educ: Univ Wyo, BS, 52, MA, 53; Univ Ill, PhD(geol), 58. Prof Exp: Geologist, Richfield Oil Corp, 53-55; asst, Univ Ill, 55-57; geologist, Richmond Explor Co, 58-61; chief geologist, Bahrain Petrol Co, 61-65; regional geologist, Amoseas, Tripoli, Libya, 65; geologist, Chevron Oil Co, Okla, 65-67; ASSOC PROF GEOL, UNIV WIS-EAU CLAIRE, 67- Mem: Geol Soc Am; Soc Econ Paleont & Mineral; Am Asn Petrol Geol. Res: Stratigraphy; sedimentation; paleontology; petroleum geology. Mailing Add: Dept of Geol Univ of Wis Eau Claire WI 54701

WILLIS, VICTOR MAX, b Weston, Mo, Aug 3, 27; m 52; c 2. ORGANIC CHEMISTRY. Educ: Univ Ill, BS, 50. Prof Exp: MGR DEVELOP SECT, SHERWIN-WILLIAMS CO, 49- Mem: Am Chem Soc; AAAS; Fedn Socs Paint Technol. Res: Water dispersed coatings systems; pigment dispersion systems; surfactant systems; organic physical chemistry related to coatings pigments and polymers and properties. Mailing Add: 604 Forsythe Ave Calumet City IL 60409

WILLIS, WAYNE OWEN, b Paonia, Colo, Jan 21, 28; m 51; c 4. SOIL PHYSICS. Educ: Colo State Univ, BS, 52; Iowa State Univ, MS, 53, PhD(soil physics), 56. Prof Exp: Asst soils res, Iowa State Univ, 52-53; agent soils res, Iowa, 53-56, res soil scientist, Wash, 56-57 & Salinity Lab, Calif, 57-58, res soil scientist, Northern Great Plains Res Ctr, 58-68, SUPVRY SOIL SCIENTIST, NORTHERN GREAT PLAINS RES CTR, AGR RES SERV, USDA, 68- Mem: Am Soc Agron; Soil Sci Soc Am; Can Soc Soil Sci; Int Soil Sci Soc; Soil Conserv Soc Am. Res: Interrelationships of soil water, soil temperature, plant growth and frozen soils. Mailing Add: Northern Great Plains Res Ctr PO Box 459 Mandan ND 58554

WILLIS, WILLIAM DARRELL, JR, b Dallas, Tex, July 19, 34; m 60; c 1. NEUROPHYSIOLOGY, NEUROANATOMY. Educ: Tex A&M Univ, BS & BA, 56; Univ Tex, MD, 60; Australian Nat Univ, PhD(physiol), 63. Prof Exp: Student instr neurophysiol, Univ Tex Southwestern Med Sch Dallas, 58-, asst prof anat, 63-64, prof & chmn dept, 64-70; PROF ANAT & PHYSIOL, UNIV TEX MED BR GALVESTON, 70-, CHIEF COMP MARINE NEUROBIOL, MARINE BIOMED INST, 70- Concurrent Pos: NIH res fel, Australian Nat Univ, 60-62 & Univ Pisa, 62-63; Nat Inst Neurol Dis & Blindness res grant, 63; mem neurol B study sect, NIH, 68-72, chmn, 70-72, mem neurol disorders prog, proj rev comt, 72- Mem: AAAS; Am Asn Anat; Am Physiol Soc; Soc Exp Biol & Med; Soc Neurosci. Res: Electrophysiology of the mammalian spinal cord; somatic sensory pathways. Mailing Add: Neurobiol Lab Mar Biomed Inst 200 University Blvd Galveston TX 77550

WILLIS, WILLIAM HILLMAN, b Trenton, Tenn, Oct 28, 08; m 34; c 2. SOILS, BACTERIOLOGY. Educ: Union Univ, Tenn, AB, 30; Iowa State Col, MS, 31, PhD(soil bact), 33. Prof Exp: Asst soil technologist, Soil Conserv Serv, USDA, 34-38, assoc soil scientist flood control surv, 38-42; from assoc agronomist to agronomist, 42-66, assoc prof, 42-48, PROF AGRON, LA STATE UNIV, BATON ROUGE, 48-, HEAD DEPT, 66- Concurrent Pos: Chmn, Southern Regional Soil Res Comt, 73-75. Mem: Am Chem Soc; Soil Sci Soc Am; Am Soc Agron. Res: Soil microbiology; nitrogen fertilization and nutrition of rice; biological nitrogen fixation; microbial nitrate reduction. Mailing Add: Dept of Agron La State Univ Baton Rouge LA 70803

WILLIS, WILLIAM J, b Ft Smith, Ark, Sept 15, 32; m 58; c 3. PHYSICS. Educ: Yale Univ, BS, 54, PhD, 58. Prof Exp: Physicist, Brookhaven Nat Lab, 58-65; PROF PHYSICS, YALE UNIV, 65- Concurrent Pos: Physicist, Europ Orgn Nuclear Res, 61-62. Mem: Am Phys Soc. Res: Elementary particle physics; weak interactions of strange particles, resonances and high energy collisions. Mailing Add: Sloane Physics Lab Yale Univ New Haven CT 06520

WILLIS, WILLIAM RUSSELL, b Moundsville, WVa, Feb 14, 26; m 46; c 4. PHYSICS. Educ: WVa Wesleyan Col, BS, 48; Okla State Univ, MS, 50, PhD(chem physics), 54. Prof Exp: Chemist, Oak Ridge Nat Lab, 52-55; assoc prof physics, W Liberty State Col, 55-56; prof, WVa Wesleyan Col, 56-65; asst prog dir, NSF, 65-66; sci fac fel, Univ Colo, 66-67; PROF PHYSICS & CHMN DEPT, NORTHERN ARIZ UNIV, 67- Concurrent Pos: Consult, Oak Ridge Nat Lab, 55-56, 58-65 & NSF, 67- Mem: AAAS; Am Asn Physics Teachers. Res: Diffusion in solids; molecular physics; ellipsometry. Mailing Add: Dept of Physics Northern Ariz Univ Flagstaff AZ 86001

WILLIS, WILLIAM SHEDRICK, JR, b Waco, Tex, July 11, 21; m 49. ANTHROPOLOGY. Educ: Howard Univ, AB, 42; Columbia Univ, PhD(anthrop), 55. Prof Exp: Asst prof anthrop, Bishop Col; from asst prof to assoc prof, Southern Methodist Univ, 65-75; MEM FAC, DEPT ANTHROP, COLUMBIA UNIV, 75- Mem: Fel AAAS; fel Am Anthrop Asn; Am Soc Ethnohist. Res: Cultural anthropology; ethnohistory; history of anthropology; New World Indians and Negroes. Mailing Add: Dept of Anthrop Columbia Univ New York NY 10027

WILLIS, WILLIAM VAN, b Morganton, NC, Oct 15, 37. INORGANIC CHEMISTRY, ANALYTICAL CHEMISTRY. Educ: Ga Inst Technol, BS, 60; Univ Tenn, MS, 63, PhD(chem), 66. Prof Exp: Res assoc radiation & radiochem, Eng Exp Sta, Ga Inst Technol, 59-61; sci writer nuclear decontamination, Univ Tenn, 63-64; USAEC res fel, 66-67; ASSOC PROF CHEM, CALIF STATE UNIV, FULLERTON, 67- Concurrent Pos: Chmn, State Regional Water Qual Control Bd. Mem: Am Chem Soc. Res: Neutron activation analysis; radiochemical tracer analysis; transition metal transport in biological systems. Mailing Add: Dept of Chem Calif State Univ Fullerton CA 92631

WILLIS, WILLIAM W, b Fredericksburg, Va, July 9, 37; m 54; c 3. NUCLEAR PHYSICS, SOLID STATE ELECTRONICS. Educ: George Washington Univ, BA, 58; Univ Va, MS, 61, PhD(physics), 63. Prof Exp: Physicist, Knolls Atomic Power Lab, Gen Elec Co, 63-67; MEM NUCLEAR & ENVIRON STAFF, AEROSPACE CORP, 67- Res: Neutron polarization; scintillation detectors; reactor design; vulnerability and hardening analyses and tests of missile and space systems components, especially electronics. Mailing Add: PO Box 92957 Los Angeles CA 90009

WILLISTON, JOHN STODDARD, b Ft Madison, Iowa, July 23, 34; m 61. NEUROPHYSIOLOGY. Educ: Univ Wis-Madison, BS, 61; Calif State Univ, San Francisco, MA, 65; Univ Southern Calif, PhD(physiol, psychol), 68. Prof Exp: NIMH fel, Univ Calif, San Francisco, 68-70; ASST PROF PHYSIOL & BEHAV BIOL, SAN FRANCISCO STATE UNIV, 70- Concurrent Pos: Res physiologist, Univ Calif, San Francisco, 75- Mem: AAAS; Am Psychol Asn; Soc Neurosci; Am Soc Zool; Animal Behav Soc. Res: Neurological substrates of behavioral plasticity; neuroelectrical activity; neuroethology; psychotrophic drugs. Mailing Add: Dept of Physiol & Behav Biol San Francisco State Univ San Francisco CA 94132

WILLITS, CHARLES HAINES, b Camden, NJ, June 25, 23; m 45; c 2. ORGANIC CHEMISTRY. Educ: Wheaton Col, Ill, BS, 44; Ohio State Univ, MS, 48; Ore State Univ, PhD(org chem), 55. Prof Exp: Res engr, Battelle Mem Inst, 48-51; from asst prof to assoc prof, 55-70, actg chmn dept, 60-69, PROF CHEM, RUTGERS UNIV, CAMDEN, 70- Mem: Am Chem Soc; Am Sci Affil. Res: Synthesis of purine derivatives; mechanism of Hofmann degradation of amides; Fries rearrangement of higher esters. Mailing Add: Dept of Chem Col of Arts & Sci Rutgers Univ 406 Penn St Camden NJ 08102

WILLITS, LYLE WILMOT, b Kansas City, Mo, Mar 20, 16; m 44. PHARMACY. Educ: Univ Kans, AB, 40; Univ Kansas City, BS, 49, MS, 56. Prof Exp: From instr to assoc prof, 50-66, asst dean, 66-69, PROF PHARM, UNIV MO-KANSAS CITY, 66-, ASSOC DEAN SCH PHARM, 69- Mem: Am Pharmaceut Asn; Am Col Apothecaries. Res: Effects of certain surfactants on suspending agents. Mailing Add: Sch of Pharm Univ of Mo Kansas City MO 64110

WILLITS, RICHARD ELLIS, b Indianapolis, Ind, Dec 5, 37; m 60; c 2. DAIRY MICROBIOLOGY, BIOCHEMISTRY. Educ: Purdue Univ, BS, 61, MS, 62, PhD(dairy microbiol), 64. Prof Exp: Res assoc flavor chem, Ore State Univ, 64-65; prod develop scientist, Pillsbury Co, Ind, 65-67, scientist, 67-69, proj leader prod develop, 69-75, MGR, EUROP REFRIGERATED FOODS RES & DEVELOP-PILLSBURY CO, 75- Mem: Inst Food Technol. Res: Development of methods for detection of abnormal milk; analytical procedures for food flavor research; microbiological investigation of pathogens in dough products; commercialization of refrigerated dough products. Mailing Add: Refrig Foods Res & Develop Pillsbury Co 311 Second St SE Minneapolis MN 55414

WILLKE, HERBERT LOUIS, JR, b Toledo, Ohio, Jan 17, 37. APPLIED MATHEMATICS. Educ: Mass Inst Technol, BS, 59, PhD(physics), 63. Prof Exp: Res assoc phys acoustics, Res Lab Electronics, Mass Inst Technol, 63-65, instr math, 65-67, res assoc, 67-68; asst prof, Boston Univ, 68-73; resident visitor, Bell Labs, 73-74; CONSULT, SONOTECH INC, 74- Res: Stability of dynamical systems, especially hydrodynamic and magnetohydrodynamic flow; sound generation in mechanical-fluid dynamical systems. Mailing Add: Apt 18 55 Magazine St Cambridge MA 02139

WILLKE, THOMAS ALOYS, b Rome City, Ind, Apr 22, 32; m 54; c 6. MATHEMATICAL STATISTICS. Educ: Xavier Univ, Ohio, AB, 54; Ohio State Univ, MS, 56, PhD(math), 60. Prof Exp: Res mathematician, Nat Bur Stand, 61-63; asst prof math, Univ Md, 63-66; assoc prof, 66-72, PROF MATH, OHIO STATE UNIV, 72-, DIR STATIST LAB, 71- Concurrent Pos: Lectr, Univ Md, 61-63; res mathematician, Nat Bur Stand, 62-66. Mem: Math Asn Am; Inst Math Statist; Am Statist Asn; Int Asn Statist Phys Sci. Res: Design and analysis of experiments. Mailing Add: Col of Arts & Sci Ohio State Univ Columbus OH 43210

WILLMAN, HAROLD BOWEN, b Newcastle, Ind, July 30, 01; m 31; c 2. GEOLOGY. Educ: Univ Ill, AB, 26, MA, 28, PhD(geol), 31. Prof Exp: Field asst, Ky State Geol Surv, 26; asst instr geol, Univ Ill, 26-28; from asst geologist to assoc geologist, 27-43, head div stratig & areal geol, 45-69, GEOLOGIST, ILL STATE GEOL SURV, 43-, EMER HEAD DIV STRATIG & AREAL GEOL, 69- Mem: AAAS; fel Geol Soc Am. Res: Ordovician, Silurian, Pennsylvanian and Pleistocene stratigraphys; geologic mapping. Mailing Add: State Geol Surv Natural Resources Bldg Urbana IL 61801

WILLMAN, VALLEE L, b Greenville, Ill, May 4, 25; m 52; c 9. SURGERY. Educ: St Louis Univ, MD, 51; Am Bd Surg, dipl, 57; Bd Thoracic Surg, dipl, 61. Prof Exp: Sr instr, 57-58, from asst prof to assoc prof, 58-64, PROF SURG, SCH MED, ST LOUIS UNIV, 64-, CHMN DEPT, 69- Concurrent Pos: McBride fel cancer, Sch Med, St Louis Univ, 56-57; attend physician, Vet Admin Hosp, 57- & St Louis City Hosp, 60- Mem: Soc Univ Surg; Am Surg Asn; Am Physiol Soc; Int Cardiovasc Soc. Res: Cardiovascular surgery and extracorporeal circulation. Mailing Add: 1325 S Grand Blvd St Louis MO 63104

WILLMAN, WARREN WALTON, b Chicago, Ill, May 24, 43. APPLIED MATHEMATICS. Educ: Univ Mich, BA, 65, BSE, 65; Harvard Univ, PhD(appl math), 69. Prof Exp: Mathematician, Shell Develop Co, Shell Oil Co, 69-70; OPERS RES ANALYST, US NAVAL RES LAB, 71- Res: Stochastic optimal control theory, estimation theory. Mailing Add: US Naval Res Lab Code 7931 Washington DC 20375

WILLMANN, ROBERT B, b Seguin, Tex, May 7, 31; m 62. PARTICLE PHYSICS. Educ: Tex A&M Univ, BS, 54; Univ Wis, PhD(physics), 60. Prof Exp: Res assoc physics, Univ Wis, 60-61; from asst prof to assoc prof, 61-71, PROF PHYSICS, PURDUE UNIV, LAFAYETTE, 71- Mem: Am Phys Soc. Res: Weak and strong interactions in particle physics. Mailing Add: Dept of Physics Purdue Univ Lafayette IN 47907

WILLMES, HENRY, b Bocholt, Ger, Aug 30, 39; US citizen; m 66; c 2. NUCLEAR PHYSICS. Educ: Univ Calif, Los Angeles, BS, 61, MA, 62, PhD(physics), 66. Prof Exp: Res physicist, Aerospace Res Labs, Wright-Patterson AFB, 65-68; asst prof, 69-73, ASSOC PROF PHYSICS, UNIV IDAHO, 73-, CHMN DEPT, 75- Mem: Am Phys Soc. Res: Few nucleon systems; nuclear structure; applications of nuclear technology. Mailing Add: Dept of Physics Univ of Idaho Moscow ID 83843

WILLMON, THOMAS L, b Cleburne, Tex, Sept 10, 06; m 33; c 1. MEDICINE, PHYSIOLOGY. Educ: Abilene Christian Col, BA, 27; Univ Chicago, MD, 31; Am Bd Prev Med, dipl, 56. Prof Exp: Mem staff physiol res & res admin, Submarine Sch, Med Corps, US Navy, 33, Deep Sea Diving Sch & Exp Diving Unit, 37-41 & Res Div, Bur Med, 46-47, officer in charge, Med Res Lab, 47-51, exec officer, Med Res Inst, Nat Naval Med Ctr, 51-56; med & res dir, Nat Multiple Sclerosis Soc, NY, 56-64; assoc dir, Reed Neurol Res Ctr, Ctr Health Sci, Univ Calif, Los Angeles, 65-74; RETIRED. Mem: AAAS; Am Physiol Soc; Am Acad Neurol; NY Acad Sci. Res: Research administration; multiple sclerosis and the demyelinating diseases; respiratory physiology; submarine medicine; deep-sea diving and underwater swimming. Mailing Add: 347 E Palace Ave Santa Fe NM 87501

WILLMORTH, JOHN H, b Gooding, Idaho, Dec 15, 35; m 59; c 4. SOLID STATE ELECTRONICS. Educ: Princeton Univ, BSE, 57; Univ Wash, MSEE, 59, PhD, 72. Prof Exp: Asst prof, 58-63, ASSOC PROF PHYSICS & HEAD DEPT, COL OF IDAHO, 63- Mem: Am Asn Physics Teachers. Res: Nuclear quadrupole resonance studies of deuterated hydrogen-bonded ferroelectric materials; ferroelectric mechanism in crystals. Mailing Add: Dept of Physics Col of Idaho Caldwell ID 83605

WILLMS, CHARLES RONALD, b Rupert, Idaho, June 26, 33; m 55; c 3. BIOCHEMISTRY. Educ: Univ Tex, BA, 55; Southwest Tex State Col, MA, 56; Tex A&M Univ, PhD(biochem), 59. Prof Exp: Asst prof chem, Southwest Tex State Col, 59-62; res scientist assoc, Clayton Found Biochem Inst, Univ Tex, 62-64; assoc prof, 64-68, chmn dept, 68-75, PROF CHEM, SOUTHWEST TEX STATE UNIV, 68- Mem: AAAS; Am Chem Soc; NY Acad Sci. Res: Enzyme and protein chemistry; carbohydrate metabilism; characterization and isolation of proteolytic enzymes. Mailing Add: Dept of Chem Southwest Tex State Univ San Marcos TX 78666

WILLNER, DAVID, b Vienna, Austria, July 2, 30; m 54; c 2. ORGANIC CHEMISTRY. Educ: Hebrew Univ, Israel, MSc, 56, PhD(org chem), 59. Prof Exp: From res asst to res assoc org chem, Weizmann Inst, 59-64; scientist, New Eng Inst Med Res, Conn, 64-66; SR RES SCIENTIST, BRISTOL LABS, 66- Concurrent Pos: Asst org chem, Bar-Ilan Univ, Israel, 57-59, lectr org chem & reaction mechanism, 62-63; fel, Dept Chem, Univ Southern Calif, 59-61 & Calif Inst Technol, 61-62. Mem: Am Chem Soc. Res: Medicinal chemistry; structure elucidation and synthesis of natural products and physiological active compounds, antibiotics and lipids; reaction mechanisms; photochemistry. Mailing Add: Bristol Labs Syracuse NY 13201

WILLNER, DOROTHY, b New York, NY, Aug 26, 27. CULTURAL ANTHROPOLOGY. Educ: Univ Chicago, PhB, 47, MA, 53, PhD(anthrop), 61. Prof Exp: Anthropologist, Land Settlement Dept, Jewish Agency, Israel, 55-58; expert community develop, UN, 58; asst prof anthrop, Univ Iowa, 59-60; Res Ctr Econ Develop & Cult Change fel, Univ Chicago, 61-62; asst prof, Univ NC, 62-63; asst prof, Hunter Col, 64-65; assoc prof, 65-70, PROF ANTHROP, UNIV KANS, 70- Concurrent Pos: Consult, President's Comt Juv Delinq & Youth Crime, Washington Action for Youth, United Planning Asn, 64; NIMH spec fel, Univ Rochester, 68-69. Mem: Fel Am Anthrop Asn; fel Soc Appl Anthrop; Am Ethnol Soc. Res: Change and stability in complex societies; political anthropology; social structure; organizations and decision-making; myth and ritual. Mailing Add: Dept of Anthrop Univ of Kans Lawrence KS 66044

WILLOUGHBY, DONALD S, b Napaneee, Ont, Sept 20, 24; m 49; c 3. MEDICAL BACTERIOLOGY, IMMUNOLOGY. Educ: Queen's Univ, Ont, BA, 47, Hons, 48, MA, 49; Univ Minn, Minneapolis, PhD(bact), 60. Prof Exp: Asst bact, Fisheries Res Bd Can, 49-51; control chemist, Anglo-Can Drug Co, Ont, 52-54; scientist, Defence Res Bd Can, 54-56; supvr, Prov Pub Health Lab, Sask, 59-61; res scientist, Defence Res Bd Can, 61-71; SUPVR REGIONAL PUB HEALTH LABS, PROV ONT, 71-; DIR LAB SERV BR, ONT MINISTRY HEALTH, 74- Mem: Am Soc Microbiol; Can Soc Microbiol; Can Pub Health Asn. Res: Medical microbiology. Mailing Add: Labs Br Ministry of Health Box 9000 Terminal A Toronto ON Can

WILLOUGHBY, RALPH ARTHUR, b Santa Rosa, Calif, Aug 15, 23; m 47; c 2. MATHEMATICS. Educ: Univ Calif, AB, 47, PhD(math), 51. Prof Exp: From asst prof to assoc prof math, Ga Inst Technol, 51-55; mem staff, Atomic Energy Div, Babcock & Wilcox Co, 55-57; MEM STAFF, MATH SCI DEPT, THOMAS J WATSON RES CTR, IBM CORP, 57- Concurrent Pos: Res partic, Math Panel, Oak Ridge Nat Lab, 54, consult, 54-55. Mem: Soc Indust & Appl Math. Res: Numerical analysis; numerical mathematics of noisy physical data. Mailing Add: 14 Garey Dr Chappaqua NY 10514.

WILLOUGHBY, RUSSELL A, b Tilston, Man, July 7, 33; m 54; c 3. VETERINARY MEDICINE. Educ: Univ Toronto, DVM, 57; Cornell Univ, PhD(vet path), 65. Prof Exp: Pvt pract vet med, Grenfell, Sask, 57-61; asst prof clin vet med, Ont Vet Col, Toronto, 61-62; res asst vet path, Cornell Univ, 62-65; assoc prof, 65-67, PROF CLIN VET MED, ONT VET COL, UNIV GUELPH, 67- Mem: Am Vet Med Asn; Am Asn Vet Clinicians; Can Vet Med Asn; Am Col Vet Internal Med (secy, 72-). Res: Environmental effects on animals, including heavy metal toxicity, the effects of intensification and the interaction between pollutants and infectious agents. Mailing Add: Dept of Clin Studies Ont Vet Col Univ of Guelph Guelph ON Can

WILLOUGHBY, STEPHEN SCHUYLER, b Madison, Wis, Sept 27, 32; m 54; c 2. MATHEMATICS EDUCATION. Educ: Harvard Univ, AB, 53, AMT, 55; Columbia Univ, EdD(math educ), 61. Prof Exp: Teacher math & sci, Newton Pub Schs, Mass, 54-57; teacher math, Greenwich Pub Schs, Conn, 5759; instr educ & math, Univ Wis-Madison, 60-61, asst prof, 61-65; PROF EDUC & MATH, NY UNIV, 65-, CHMN DEPT MATH EDUC, 67- Concurrent Pos: Consult, NSF, 73- Mem: Math Asn Am; Conf Bd Math Sci; Nat Coun Teachers Math. Res: Learning and teaching mathematics. Mailing Add: 23 Press Bldg NY Univ New York NY 10003

WILLOUGHBY, WILLIAM FRANKLIN, b Washington, DC, Feb 4, 36. PATHOLOGY, IMMUNOLOGY. Educ: Johns Hopkins Univ, AB, 57, MD & PhD(microbiol), 65. Prof Exp: Arthritis Found fel, Scripps Clin & Res Found, 67-69; asst prof path & microbiol, Case Western Reserve Univ, 69-72; DIR, VIRGINIA MASON RES CTR, 72- Concurrent Pos: NIH gen res support grant. Res: Pulmonary immunopathology. Mailing Add: Virginia Mason Res Ctr 1000 Seneca St Seattle WA 98101

WILLOWS, ARTHUR OWEN DENNIS, b Winnipeg, Man, Mar 26, 41; m 63; c 3. NEUROPHYSIOLOGY. Educ: Yale Univ, BS, 63; Univ Ore, PhD(biol), 67. Prof Exp: Asst prof, Univ Ore, 67-68, res assoc neurophysiol, 68-69; from asst prof to assoc prof, 69-75, PROF ZOOL, UNIV WASH, 75-, DIR FRIDAY HARBOR LABS, 73- Mem: Soc Gen Physiol; Soc Neurosci; Am Physiol Soc. Res: Neuroethology; neurophysiological basis of behavior. Mailing Add: Friday Harbor Labs Friday Harbor WA 98250

WILLS, BERNT LLOYD, b Drake, NDak, July 24, 09; m 37; c 2. GEOGRAPHY. Educ: Valley City State Col, BA, 34; Univ Mont, MA, 37; Northwestern Univ, Evanston, PhD(geog), 53. Prof Exp: Teacher social studies pub schs, Mont & NDak, 28-42; from asst prof to assoc prof geog, 43-54, PROF GEOG, UNIV N DAK, 54- Concurrent Pos: Vis prof geog, Univ Sydney, 68. Mem: Nat Coun Geog Educ; Asn Am Geog. Res: Geography of Puerto Rico and North Dakota, the northern prairie state. Mailing Add: Dept of Geog Univ of NDak Grand Forks ND 58201

WILLS, CHRISTOPHER J, b London, Eng, Mar 23, 38; m 65. GENETICS, BIOLOGY. Educ: Univ BC, BA, 60, MSc, 62; Univ Calif, Berkeley, PhD(genetics), 65. Prof Exp: NIH fel genetics, Univ Calif, Berkeley, 65-66; asst prof biol, Wesleyan Univ, 66-72; ASSOC PROF BIOL, UNIV CALIF, SAN DIEGO, 72- Concurrent Pos: NIH res grant, 67- Mem: AAAS; Genetics Soc Am; Am Soc Naturalists. Res: Maintenance of genetic variability in natural populations; production, through selection in the laboratory, and characterization of isoenzymes in yeast. Mailing Add: Dept of Biol Univ of Calif at San Diego La Jolla CA 92037

WILLS, DONALD L, b Peoria, Ill, May 12, 24; m 46; c 2. GEOLOGY. Educ: Univ Ill, BS, 49, MS, 51; Univ Iowa, PhD(geol), 71. PROF GEOL & CHMN DEPT, MONMOUTH COL, ILL, 51- Concurrent Pos: Environ consult, 71- Mem: Am Inst Prof Geol; Sigma Xi; Nat Asn Geol Teachers; Geol Soc Am. Res: Biostratigraphic studies of Mississippian Chesterian series. Mailing Add: Dept of Geol Monmouth Col Monmouth IL 61462

WILLS, FRANKLIN KNIGHT, b Bristol, Pa, July 11, 25; m 51; c 3. VETERINARY BACTERIOLOGY, PATHOLOGY. Educ: Univ Pa, VMD, 50; Agr & Mech Col Tex, MS, 55; Univ Conn, PhD(vet path), 66. Prof Exp: Instr vet bact, Agr & Mech Col Tex, 52-55; asst prof poultry dis res, Univ Md, 55-57; instr vet path, Univ Conn, 57-62; assoc prof poultry dis res, Univ Md, College Park, 62-69; ASSOC DIR LAB, STERWIN LABS, INC, 69- Mem: Am Vet Med Asn. Res: Diseases of poultry; reproductive diseases of cattle; bacteriology, virology and pathology. Mailing Add: Sterwin Labs Inc Millsboro DE 19966

WILLS, IRVIN ANDREWS, b Mendota, Ill, Dec 21, 04; m 34; c 3. ZOOLOGY. Educ: Wheaton Col, Ill, BS, 27; Univ Iowa, MS, 32, PhD, 35. Prof Exp: Instr zool, Valparaiso Univ, 27-29; from asst instr to asst, Univ Iowa, 30-34, res assoc, 34-35; dean, 34-55, chmn div natural sci, 55-72, prof biol, 35-74, EMER PROF BIOL, JOHN BROWN UNIV, 74- Mem: AAAS; Am Soc Zoologists; Nat Asn Biol Teachers. Res: Biology of sex; metabolism in relation to sex determination; endocrinology; non-specificity of pituitary hormone in amphibians; embryology of snakes. Mailing Add: Dept of Biol John Brown Univ Siloam Springs AR 72761

WILLS, JAMES E, JR, b Tucumcari, NMex, Mar 20, 16. PHYSICS. Educ: Miss Col, BA, 36; Univ Va, MA, 38; Univ Tex, PhD(physics), 56. Prof Exp: Instr physics, Ga Sch Technol, 38-39; asst prof, Baylor Univ, 46-51; from assoc prof to prof, Stetson Univ, 56-64; chmn dept, 65-72, PROF PHYSICS, UNIV NC, ASHEVILLE, 64- Mem: Am Phys Soc. Res: Fast neutron spectroscopy. Mailing Add: Dept of Physics Univ of NC Ashevelle NC 28803

WILLS, JAMES HENRY, b Richmond, Va, Aug 7, 12; m 48. PHARMACOLOGY. Educ: Va Polytech Inst, BS, 34; Med Col Va, MS, 36; Univ Rochester, PhD(physiol), 41. Prof Exp: Assoc pharmacol, Sch Med & Dent, Univ Rochester, 41-46; asst prof, Univ Tenn, 46-47; pharmacologist, Med Labs, Chem Corps, US Army, 47-48, actg chief pharmacol br, 48-49, chief, 49-57, asst chief physiol div, Directorate Med Res, 56-61, chief, 61-65; assoc dir, Inst Exp Path & Toxicol & prof pharmacol, 65-75, prof toxicol, 70-75, ADJ PROF PHARMACOL & TOXICOL, ALBANY MED COL, 75-; SR CRITERIA REVIEWER, NAT INST OCCUP SAFETY & HEALTH, 75- Concurrent Pos: USSecy Army fel, 58-59; mem toxicol subcomt, Food Protection Comt, Nat Acad Sci-Nat Res Coun, 52-70; exec secy, NY State Pesticide Control Bd, 65-70; mem toxicol study sect, Div Res Grants, NIH, 68-72; mem AIB5 med adv comt to NASA, Off of Life Sci, 72-; mem hazardous mat adv comt, Environ Protection Agency, 75- Honors & Awards: Dept Army Decoration for Exceptional Civilian Serv, 65. Mem: Am Chem Soc; Am Soc Pharmacol & Exp Therapeut; Soc Exp Biol & Med; Soc Toxicol. Res: Factors in the secretion of saliva and affecting urine formation; effects of organic nitrates on blood pressure; mechanisms of actions of chemicals and antagonists of toxic effects. Mailing Add: Nat Inst Occup Safety & Health 5600 Fishers Lane Rockville MD 20852

WILLS, JOHN G, b Greeley, Colo, Feb 4, 31; m 54; c 5. THEORETICAL PHYSICS, NUCLEAR PHYSICS. Educ: San Diego State Col, AB, 53; Univ Wash, MS, 56, PhD(physics), 63. Prof Exp: Staff mem, Los Alamos Sci Lab, 56-60; res asst, 63-64, asst prof, 64-67, ASSOC PROF PHYSICS, IND UNIV, BLOOMINGTON, 67- Res: Nuclear theory. Mailing Add: Dept of Physics Ind Univ Bloomington IN 47401

WILLS, KLAUS DIETER, organic chemistry, see 12th edition

WILLS, WIRT HENRY, b Petersburg, Va, Feb 12, 24; m 54; c 4. PLANT PATHOLOGY. Educ: Univ Richmond, BA, 50; Duke Univ, MA, 52, PhD(bot), 54. Prof Exp: From asst prof to assoc prof, 54-68, PROF PLANT PATH, VA POLYTECH INST & STATE UNIV, 69- Mem: AAAS; Am Phytopath Soc. Res: Ecology of root diseases. Mailing Add: Dept of Plant Path & Physiol Va Polytech Inst & State Univ Blacksburg VA 24061

WILLSON, ALLAN E, horticulture, see 12th edition

WILLSON, CLYDE D, b Omaha, Nebr, May 7, 35; m 54; c 4. ORGANIC CHEMISTRY, MOLECULAR BIOLOGY. Educ: Univ Calif, Berkeley, BA, 56, PhD(chem), 60. Prof Exp: NIH fel bact genetics & protein synthesis, Pasteur Inst, Paris, 60-62; asst prof biochem, Univ Calif, Berkeley, 62-68, res fel entom, Miller Inst, 68-69; INSTR LIFE SCI, LANEY COL, 69- Concurrent Pos: Vis prof biol, Brandeis Univ, 74-75. Res: Heterocyclic organic chemistry; bacterial enzyme regulation and genetic control; characterization of messenger RNA; biochemistry of communication substances in insects. Mailing Add: Dept of Life Sci Laney Col 1001 Third Ave Oakland CA 94606

WILLSON, DAN LEROY, b Parson, Kans, Aug 25, 23. BIOLOGY. Educ: Kans State Teachers Col, BS, 48, MS, 49; Univ Okla, PhD(bot), 58. Prof Exp: Instr biol, Okla Mil Acad, 48-55; assoc prof, Cent State Col, 58-61 & Cent Wash State Col, 61-68; PROF BIOL & DEAN ARTS & SCI, WINONA STATE COL, 68- Mem: AAAS; Ecol Soc Am; Am Physcol Soc Am. Res: Cytology, morphology and taxonomy of algae, especially ecology of soil algae. Mailing Add: Winona State Col Winona MN 55987

WILLSON, DONALD BRUCE, b Bloomington, Ind, Oct 25, 41; m 65; c 4.

PHYSICAL INORGANIC CHEMISTRY, EXTRACTIVE METALLURGY. Educ: Geneva Col, BA, 63; Tufts Univ, PhD(chem), 69. Prof Exp: Lab instr chem & res asst, Tufts Univ, 64-69; res assoc Air Force Off Sci Res, Geneva Col, 69; chemist, Kawecki-Berylco Industs, Inc, Pa, 70-73, proj leader-group leader, 74; TECH MGR, M&R REFRACTORY METALS, INC, WINSLOW, 75- Mem: AAAS; Am Chem Soc; fel The Chem Soc. Res: Extractive metallurgy and physical, inorganic and analytical chemistry of the refractory, transition, rare earth and noble metals and their compounds. Mailing Add: 210 Springfield Ave Merchantville NJ 08109

WILLSON, JOHN ELLIS, b Scranton, Pa, May 4, 29; m 55; c 3. VETERINARY MEDICINE, TOXICOLOGY. Educ: Pa State Univ, BS, 50; NY State Vet Col, DVM, 54. Prof Exp: Intern, Angell Mem Animal Hosp, Boston, Mass, 57-58, mem staff, 58-61; head dept pharmacol, John L Smith Mem Cancer Res, Chas Pfizer & Co, Inc, NJ, 61-63; sr pathologist, 63-66, ASST DIR, JOHNSON & JOHNSON RES FOUND, NEW BRUNSWICK, 66- Concurrent Pos: Mem coun accreditation, Am Asn Accreditation of Lab Animal Care, 72-; mem adv coun, Inst Lab Animal Resources, Div Biol Sci, Assembly of Life Sci, Nat Res Coun-Nat Acad Sci, 74-77. Honors & Awards: Philip B Hoffman Res Scientist Award, Johnson & Johnson, 72. Mem: Am Vet Med Asn; Soc Toxicol; Am Asn Lab Animal Sci. Res: Toxicology of foods, drugs and cosmetics; ethylene oxide sterilant residues; laboratory animal husbandry and medicine. Mailing Add: 42 Addison Dr Basking Ridge NJ 07920

WILLSON, JOHN TUCKER, b Bismarck, NDak, Aug 26, 24; m 47; c 3. ANATOMY. Educ: George Washington Univ, BS, 48, MS, 49; Univ Colo, PhD(anat), 53. Prof Exp: Asst, 51-53, from instr to assoc prof, 53-75, PROF ANAT, SCH MED, UNIV COLO, DENVER, 75- Mem: AAAS; Am Asn Anat; Am Soc Cell Biol; Microcirc Soc. Res: Microcirculation, intravascular erythrocyte agglutination, reproduction, fertility and sterility and fine structure. Mailing Add: Dept of Anat Univ of Colo Med Ctr Denver CO 80220

WILLSON, KARL STUART, b Cleveland, Ohio, June 11, 10; m 33; c 3. ELECTROCHEMISTRY, CORROSION. Educ: Western Reserve Univ, BA, 31, MA, 32, PhD(chem), 35. Prof Exp: Asst quant anal, Western Reserve Univ, 31-35; res chemist, Ansul Chem Co, 35-44; res chemist, Manhattan Proj, Harshaw Chem Co, 44-46, res chemist & supvr electroplating res, 46-52; mgr res & develop div, Gen Dry Batteries, Inc, 52-59; group leader phys chem, Lubrizol Corp, 59-62; assoc dir electroplating res & develop, Harshaw Chem Co, 62-69, sr res assoc, 70-75; CHEM CONSULT, 75- Mem: Am Chem Soc; Electrochem Soc; Am Electroplaters Soc; Am Inst Chem. Res: Dry battery technology; refrigerant and other gases; uranium compounds; electroplating and corrosion; fire extinguisher chemistry; chemistry of gases; lubricating materials; phosphate coatings. Mailing Add: c/o Dept of Chem Case Western Reserve Univ Cleveland OH 44106

WILLSON, LEE ANNE MORDY, b Honolulu, Hawaii, Mar 14, 47; m 69; c 1. ASTRONOMY. Educ: Harvard Univ, AB, 68; Univ Mich, MS, 70, PhD(astron), 73. Prof Exp: Instr, 73-75, ASST PROF ASTROPHYS, DEPT PHYSICS, IOWA STATE UNIV, 75- Mem: Am Astron Soc. Res: Problems of stellar atmospheres, particularly theories of mass loss, extended atmospheres, and variable stars. Mailing Add: Dept of Physics Iowa State Univ Ames IA 50011

WILLSON, MARY FRANCES, b Madison, Wis, July 28, 38; m 72. ECOLOGY, EVOLUTION. Educ: Grinnell Col, BA, 60; Univ Wash, PhD(zool), 64. Prof Exp: Res assoc zool, Duke Univ, 63; instr biol, Simmons Col, 64-65; asst prof, 65-71, ASSOC PROF ZOOL, UNIV ILL, URBANA-CHAMPAIGN, 71- Concurrent Pos: Res grants, Chapman Fund, Am Mus Natural Hist, 64 & Univ Ill Res Bd, 66-69. Mem: Ecol Soc Am; Am Ornith Union; Brit Ecol Soc; Cooper Ornith Soc; Am Soc Naturalists. Res: Evolutionary ecology; ecological niches and community structure. Mailing Add: Dept Zool Vivarium Bldg Univ of Ill Champaign IL 61820

WILLSON, PHILIP JAMES, b Detroit, Mich, Apr 23, 26; m 48; c 3. HIGH TEMPERATURE CHEMISTRY, CERAMICS. Educ: Wayne State Univ, BA, 51. Prof Exp: Technician electronics, US Navy, 45-46 & Gen Motors Corp, 48-50; chemist, 51-55, RES SUPVR CHEM, CHRYSLER RES, 56- Mem: Am Ceramic Soc; Soc Automotive Engrs. Res: High temperature structural ceramic materials for turbine engine applications; fracture mechanics characterization of material. Mailing Add: Chrysler Corp Res C I M S 418-19-30 PO Box 1118 Detroit MI 48231

WILLSON, RICHARD ATWOOD, b Minneapolis, Minn. GASTROENTEROLOGY. Educ: Univ Minn, BA, 58, BS, 59, MD, 62, MS, 69. Prof Exp: Intern, Mary Fletcher Hosp, 62-63; resident internal med, Univ Vt, 63-64; resident, Mayo Clin, 66-68, NIH res fel gastroenterol, 68-71; res fel, Liver Unit, Dept Med, King's Col Hosp Med Sch, London, 72-73; ASST PROF MED, UNIV WASH, 73-; HEAD DIV GASTROENTEROL, HARBORVIEW MED CTR, SEATTLE, 73- Mem: Am Gastroenterol Asn; Am Asn Study Liver Dis; Am Fedn Clin Res. Res: Treatment of acute fulminant hepatic failure and the study of hepatic injury secondary to drugs and drug metabolism. Mailing Add: Dept of Med Univ of Wash Seattle WA 98195

WILLWERTH, LAWRENCE JAMES, b Melrose, Mass, Oct 3, 32; m 56; c 3. PLASTICS CHEMISTRY. Educ: Lowell Technol Inst, BS, 72, MS, 75. Prof Exp: Jr chemist plastics, Nat Polychem Inc, Mass, 55-60; chemist, Avco Corp, Mass, 60-66; TECH MGR CHEM-PLASTICS, K J QUINN & CO INC, MALDEN, MASS, 66- Concurrent Pos: Instr polymer characterization, Eve Div, Lowell Technol Inst, 72- Mem: Am Chem Soc; Am Inst Chemists; Soc Plastics Engrs. Res: Research and development of polyurethane plastics; attainment of specific properties through rearrangement and addition of various species to the polymer backbone. Mailing Add: 19 Post Rd North Hampton NH 03862

WILLY, WILLIAM EDWARD, chemistry, see 12th edition

WILM, HAROLD GRIDLEY, b Topeka, Kans, May 6, 07; m 29, 70; c 2. FORESTRY. Educ: Colo Col, BS, 29; Cornell Univ, MF, 30, PhD(silvics), 32. Prof Exp: Asst & instr forestry, Cornell Univ, 29-32; res technician, Lake States Forest Exp Sta, US Forest Serv, 32, Calif Forest & Range Exp Sta, 32-38, Rocky Mountain Forest & Range Exp Sta, 38-48, Southern Forest Exp Sta, 48-49, Pac Northwestern Exp Sta, 49-51, chief div forest influences res, Washington, DC, 51-53; assoc dean, 53-59, EMER PROF FORESTRY & EMER ASSOC DEAN, STATE UNIV NY COL FORESTRY, SYRACUSE UNIV, 66-; CONSULT, 70- Concurrent Pos: Comnr, State Conserv Dept, NY, 59-66; asst dir, Fed Water Resources Coun, 66-70; prof forestry & dir water resources res ctr, Univ Vt, 71-75. Mem: Int Asn Sci Hydrol (pres, 57-60); fel Soc Am Foresters; Am Geophys Union. Res: Statistical methods in watershed research; watershed management and forest influences; open channel hydraulics; forestry education; public administration. Mailing Add: Riverside Dr Milton VT 05468

WILMARTH, VERL RICHARD, b De Smet, SDak, Mar 13, 21; m 49; c 4. ECONOMIC GEOLOGY. Educ: SDak Sch Mines & Technol, Geol Eng, 48. Prof Exp: Geologist, US Geol Surv, 48-65; chief planetology, NASA, 65-69, CHIEF LUNAR SCIENTIST, MANNED SPACECRAFT CTR, NASA, 69- Mem: Geol Soc Am; Soc Econ Geol; Asn Eng Geol; Geochem Soc. Res: Planetology. Mailing Add: Code TA NASA Manned Spacecraft Ctr Houston TX 77058

WILMER, MICHAEL EMORY, b Washington, DC, Oct 11, 41. INFORMATION SCIENCE. Educ: Cath Univ Am, BSEE, 63, MSEE, 67, PhD(elec eng), 68. Prof Exp: PRIN SCIENTIST IMAGE PROCESSING, PALO ALTO RES CTR, XEROX CORP, 67- Mem: Sigma Xi; Inst Elec & Electronics Engrs. Res: Digital processing of images and speech for enhancement, compression and recognition. Mailing Add: Xerox Palo Alto Res Ctr 3333 Coyote Hill Rd Palo Alto CA 94304

WILMOT, GEORGE BARWICK, b Waterbury, Conn, Oct 27, 28; m 53; c 7. PHYSICAL CHEMISTRY. Educ: Rensselaer Polytech Inst, BS, 51; Mass Inst Technol, PhD(phys chem), 54. Prof Exp: Res chemist, Naval Ord Sta, 54-73, RES CHEMIST, NAVAL SURFACE WEAPONS CTR, 73- Mem: AAAS; Am Chem Soc; Am Phys Soc. Res: Infrared and Raman spectroscopy; propellant combustion; thermodynamics; lasers. Mailing Add: Propellant Sci Div Naval Surface Weapons Ctr Indian Head MD 20640

WILMOT, WILLIAM HEYWOOD, II, physical chemistry, see 12th edition

WILMOTH, JAMES HERDMAN, b Burchard, Nebr, May 31, 10; m 37. BIOLOGY. Educ: Monmouth Col, BS, 32; Kans State Univ, MS, 34; NY Univ, PhD(parasitol), 42. Prof Exp: Asst parasitol, Kans State Univ, 33-34; instr biol & actg head dept, Monmouth Col, 34-35; asst, NY Univ, 37-39; from tutor to asst prof, Brooklyn Col, 39-47; from asst prof to assoc prof, Triple Cities Col, Syracuse, 47-50, chmn dept, 47-56, chmn div sci & math, 52-57; assoc prof, 50-54, PROF BIOL, STATE UNIV NY BINGHAMTON, 54- Mem: Fel AAAS; Am Micros Soc; Soc Parasitol; Am Soc Zoologists; fel NY Acad Sci. Res: Invertebrate structure and physiology; physiology of parasites; in vitro culture of platyhelminths; investigations on Tardigrada. Mailing Add: Dept of Biol State Univ of NY Binghamton NY 13901

WILMSEN, EDWIN NORMAN, b Galveston, Tex, Mar 13, 32; m 52; c 5. ANTHROPOLOGY. Educ: Tex A&M, BArch, 58; Mass Inst Technol, MArch, 59; Univ Ariz, MA, 66, PhD(anthrop), 67. Prof Exp: Instr archit, Univ Tex, Austin, 59-60; asst prof, Univ Ariz, 60-64; res assoc anthrop, Ariz State Mus & Smithsonian Inst, 67-68; from asst prof to assoc prof anthrop, 68-74, cur, Great Lakes Div, Mus Anthrop, 68-74, ASSOC RES SCIENTIST, MUS ANTHROP, UNIV MICH, ANN ARBOR, 74- Concurrent Pos: NSF res grant, 67-70. Mem: AAAS; fel Am Anthrop Asn; Soc Am Archaeol. Res: The anthropology of Pleistocene hunting societies in North America and northern Eurasia; comparison of these with modern hunting-gathering societies. Mailing Add: Mus of Anthrop Univ Mich Ann Arbor MI 48104

WILNER, BURTON IRWIN, virology, see 12th edition

WILNER, GEORGE DUBAR, b New York, NY, Dec 7, 40. HEMATOLOGY, PATHOLOGY. Educ: Northwestern Univ, BS, 62, MD, 65. Prof Exp: From instr to asst prof, 69-75, ASSOC PROF PATH, COL PHYSICIANS & SURGEONS, COLUMBIA UNIV, 75- Mem: Am Heart Asn. Res: Hemostasis and thrombosis. Mailing Add: Col of Physicians & Surgeons Columbia Univ New York NY 10032

WILNER, JACOB, b USSR, May 29, 09; nat Can; m 39. PLANT PHYSIOLOGY. Educ: Univ Sask, BSA, 36; Univ Minn, MSc, 47, PhD, 52. Prof Exp: Asst hort, Univ Sask, 35-37; tree planting supvr, Forest Nursery Sta, Sask, 37-47, tree physiologist, 47-56, PLANT PHYSIOLOGIST, PLANT RES INST, CAN DEPT AGR, CENT EXP FARM, ONT, 56- Concurrent Pos: Resident, Cornell Univ, 53; hon fel, Grad Sch, Univ Minn, 54- Mem: AAAS; Soc Cryobiol; Can Bot Asn; Am Soc Hort Sci; Can Soc Hort Sci. Res: Fundamentals of frost hardiness in plants; methods of testing ability of overwintering plants to resist frost injury. Mailing Add: Plant Res Inst Cent Exp Farm 1193 Cline Crescent Ottawa ON Can

WILNER, JEROME, b Brooklyn, NY, May 29, 23; m 49; c 2. DAIRY INDUSTRY. Educ: Cornell Univ, BS, 47. Prof Exp: Supvr plate assay labs, Div Antibiotics & Insulin Cert, Bur Drugs, 47-67, asst br chief, Microbiol Assay Br, 67-69, asst chief, Microbiol Assay Br, Nat Ctr Antibiotics & Insulin Anal, Div Pharmaceut Sci, 69-72, SPEC ASST TO DIR, NAT CTR ANTIBIOTICS ANAL, OFF PHARMACEUT RES & TESTING, BUR DRUGS, FOOD & DRUG ADMIN, 72- Honors & Awards: Superior Serv Award, Dept Health, Educ & Welfare, 62; Award of Merit, Food & Drug Admin, 65. Res: Antibiotics; dairy manufacturing. Mailing Add: 1330 Massachusetts Ave NW Washington DC 20005

WILPIZESKI, CHESTER ROBERT, b Forty Fort, Pa, Aug 20, 30; m 56; c 3. OTOLOGY, SPEECH SCIENCE. Educ: Univ Pa, AB, 57, PhD(psychol), 65. Prof Exp: Res assoc otolaryngol, Jefferson Med Col, 61-65; res psychologist, Wilmington, Del, 66-69; ASSOC PROF OTOLARYNGOL, THOMAS JEFFERSON UNIV, 69- Concurrent Pos: NIH res grants, Jefferson Med Col, 69-70 & 72- Mem: Am Neurotology Soc; Int Neuropsychol Soc; Pavlovian Soc NAm, Res: Animal models and techniques for the study of experimental deafness. Mailing Add: Dept of Otolaryngol Thomas Jefferson Univ Philadelphia PA 19107

WILSDORF, DORIS KUHLMANN, b Bremen, Ger, Feb 15, 22; US citizen; m 50; c 2. MATERIALS SCIENCE. Educ: Univ Göttingen, Dipl, 46, Dr rer nat, 47; Univ Witwatersrand, DSc(phys metall), 55. Prof Exp: Res assoc, Dept Metall, Univ Göttingen, 48 & H H Wills Phys Lab, Bristol Univ, 49-50; lectr physics, Univ Witwatersrand, 50-56; from assoc prof to prof metall eng, Univ Pa, 57-63; prof eng physics, 63-66, UNIV PROF APPL SCI, UNIV VA, 66- Mem: Fel Am Phys Soc; Am Soc Eng Educ; Ger Metall Soc; Am Inst Mining & Metall Engrs. Res: Theory of crystal defects, crystal plasticity, radiation damage; theory of liquids. Mailing Add: 109 Dept of Physics Univ of Va Charlottesville VA 22901

WILSEY, NEAL DAVID, b Tunkhannock, Pa, July 27, 37; m 61; c 4. PHYSICS. Educ: Hartwick Col, BA, 61; Colo State Univ, MS, 64, PhD(physics), 67. Prof Exp: Res physicist solid state physics, 67-74, HEAD ELECTRONIC MAT SECT, RADIATION EFFECTS BR, US NAVAL RES LAB, 74- Concurrent Pos: Instr, Univ Md, 71-72. Mem: Am Phys Soc. Res: Radiation effects in solids; magnetic materials. Mailing Add: US Naval Res Lab Code 6627 Washington DC 20375

WILSHER, RUDOLPH STANLEY, organic chemistry, polymer chemistry, see 12th edition

WILSHIRE, HOWARD GORDON, b Shawnee, Okla, Aug 19, 26; m 54; c 3. PETROLOGY. Educ: Univ Okla, BA, 52; Univ Calif, Berkeley, PhD(geol), 56. Prof Exp: Lectr geol, Univ Sydney, 56-60; res fel, Australian Nat Univ, 61; GEOLOGIST, US GEOL SURV, 61- Mem: Geol Soc Am; Am Geophys Union. Res: Structure and petrology of igneous rocks; petrology and stratigraphy of lunar rocks; mechanical erosion of terrestrial deposits. Mailing Add: US Geol Surv 345 Middlefield Rd Menlo Park CA 94025

WILSKA, ALVAR P, b Parikkala, Finland, Mar 14, 11; m 35; c 5. PHYSICS. Educ:

Univ Helsinki, DrMed, 38. Prof Exp: Lectr physiol, Univ Helsinki, 40-44, prof, 44-58; vis prof cell res, La State Univ, 59-60; PROF PHYSICS, UNIV ARIZ, 60- Concurrent Pos: Fel, Rockefeller Inst Med Res, 40-41; head, Wihuri Res Inst, Finland, 44-47; consult, Philips Electronic Instruments, Inc, 60- Mem: Fel AAAS. Res: Experimental physiology; electrophysiology; microelectrodes; medical physics; optics; electron optics. Mailing Add: Dept of Physics Univ of Ariz Tucson AZ 85721

WILSON, ALBERT GEORGE, astronomy, philosophy of science, see 12th edition

WILSON, ALLAN CHARLES, b Ngaruawahia, NZ, Oct 18, 34; m 58; c 2. BIOCHEMISTRY, ZOOLOGY. Educ: Univ Otago, NZ, BSc, 55; Wash State Univ, MS, 57; Univ Calif, Berkeley, PhD(biochem), 61. Prof Exp: Fel biochem, Brandeis Univ, 61-64; from asst prof to assoc prof, 64-72, PROF BIOCHEM, UNIV CALIF, BERKELEY, 72- Concurrent Pos: NSF grants, Univ Calif, Berkeley, 65-75, NIH grant, 71-75; mem, Alpha Helix Exped, New Guinea, 69; Guggenheim Mem Found award, Weizmann Inst Sci & Univ Nairobi, 72-73. Mem: Am Soc Biol Chem; Am Soc Zool; Soc Study Evolution; Am Ornith Union; Genetics Soc Am. Res: Isoenzymes; immunochemistry of proteins; human evolution; phylogenetic relationships of higher taxonomic categories of animals and microbes; molecular and genetic basis of organismal evolution. Mailing Add: 401 Biochem Bldg Univ of Calif Berkeley CA 94720

WILSON, ALMA MCDONALD, b Colonia Pacheco, Chihuahua, Mex, Aug 29, 29; US citizen; m 52; c 7. PLANT PHYSIOLOGY. Educ: Brigham Young Univ, BA, 54; Univ Calif, Davis, PhD, 62. Prof Exp: Plant physiologist, Agr Res Serv, USDA, Wash State Univ, 62-72, PLANT PHYSIOLOGIST, AGR RES SERV, USDA, COLO STATE UNIV, 72- Mem: Am Soc Plant Physiologists; Soc Range Mgt; Am Soc Agron. Res: Physiology of seed germination; metabolic effects of temperature and moisture stress on higher plants; forage species in relation to seedling establishment on semiarid lands. Mailing Add: Agr Res Serv USDA Colo State Univ Ft Collins CO 80523

WILSON, ALPHA EDMOND, physics, see 12th edition

WILSON, ANDREW HASTIE, b Penicuik, Scotland, May 30, 28; Can citizen; m 59; c 4. SCIENCE POLICY, MECHANICAL ENGINEERING. Educ: Univ Glasgow, BSc, 49, MA, 54. Prof Exp: Apprentice mech eng, Harland & Wolff Ltd, 46-49; design engr, MacTaggart, Scott Ltd, 49-50; sales engr, SKF Ball Bearing Co Ltd, 54-57; sr design engr, Atomic Energy Can Ltd, 58-60, res adminr, 60-64; chief res officer sci policy, Econ Coun Can, 64-68; SCI ADV SCI POLICY, SCI COUN CAN, 68- Concurrent Pos: Mem, Can Eng Manpower Coun, 75- Mem: Eng Inst Can; Am Soc Mech Engrs; Inst Mech Engrs. Res: Science policy, invention and innovation in the industrial context and at the industry-government interface. Mailing Add: Sci Coun of Can 150 Kent St Ottawa ON Can

WILSON, ANDREW ROBERT, b Dublin, Ireland, Sept 13, 41; US citizen; m 68. PLASMA PHYSICS. Educ: Trinity Col, Univ Dublin, BA, 62, MA, 65; Oxford Univ, DPhil(physics), 68. Prof Exp: Fr Govt boursier nuclear physics, Inst Fourier, Grenoble, 62-63; fel solid state physics, Lincoln Lab, Mass Inst Technol, 68-70; systs engr, Elec Supply Bd, Dublin, Ireland, 70-71; PROG MGR PLASMA PHYSICS, SYSTS, SCI & SOFTWARE, 71- Mem: Am Phys Soc; Inst Elec & Electronics Engrs. Res: Spacecraft charging analysis; system generated electromagnetic theory. Mailing Add: Systs Sci & Software PO Box 1620 La Jolla CA 92038

WILSON, ANDREW WILKINS, b Santa Ana, Calif, Aug 13, 13; m 46; c 2. GEOGRAPHY. Educ: Pomona Col, BA, 35; Stanford Univ, MBA, 37; Syracuse Univ, PhD(soc sci), 55. Prof Exp: Instr com, Calif State Univ, Fresno, 46-48; from instr to assoc prof bus admin, Col Bus & Pub Admin, 50-61, PROF GEOG, REGIONAL DEVELOP & URBAN PLANNING, COL BUS & PUB ADMIN, UNIV ARIZ, 61- Concurrent Pos: Planning consult, City of Flagstaff, 58-59 & City of Sierra Vista, 60-61; mem, Comn Geog of Arid Lands, Int Geog Union, 68-72 & vchmn, Working Group Desertification, 72-76; mem bd regents, Am Indust Develop Coun, 70-76. Mem: AAAS; Asn Am Geogr; Nat Coun Geog Educ; Am Asn Univ Prof; Am Soc Planning Officials. Res: Urban and economic geography; arid lands. Mailing Add: 4435 Timrod St Tucson AZ 85711

WILSON, ANGUS, b Mexico, Maine, Aug 13, 20. RUBBER CHEMISTRY. Educ: Georgetown Univ, BS, 41. Prof Exp: Control & anal chemist, E I du Pont de Nemours & Co, Inc, 41-43; prod supvr, Joseph E Seagrams & Sons, Inc, 47-48; control lab supvr, Govt Lab, Univ Akron, 49-52; rubber chemist, 52-68, head rubber technol group, 68-74, HEAD RUBBER & PLASTICS GROUP, US ARMY NATICK DEVELOP CTR, 74- Mem: Am Chem Soc; Sigma Xi. Res: Development of methods of testing rubber and elastomeric materials; compounding of fluorocarbon elastomers; evaluation of experimental elastomers for possible end item applications. Mailing Add: MA Div CE & ME Lab US Army Natick Develop Ctr Natick MA 01760

WILSON, ARCHIE SPENCER, b Tekoa, Wash, Jan 19, 21; m 44; c 3. CHEMISTRY. Educ: Iowa State Univ, BS, 46; Univ Chicago, MS, 50, PhD, 51. Prof Exp: Asst chem, Iowa State Univ, 43-46; res assoc, Gallium Proj, US Dept Navy, 48-49; asst, Univ Chicago, 49-50; instr chem, Univ Nebr, 50-51; sr scientist, Gen Elec Co, 51-65; sr res scientist, Pac Northwest Labs, Battelle Mem Inst, 64-71; PROF CHEM & ASSOC CHMN DEPT, UNIV MINN, MINNEAPOLIS, 71- Concurrent Pos: US sci adv, Int Conf Peaceful Uses Atomic Energy, Geneva, 58; mem, US-UK Ruthenium Conf, 58. Mem: Am Chem Soc. Res: Crystal structure of compounds of uranium; ruthenium chemistry; solvent extraction of the actinide elements. Mailing Add: Dept of Chem Univ of Minn Minneapolis MN 55455

WILSON, ARMIN GUSCHEL, b Sapulpa, Okla, Dec 13, 16; m 43; c 2. ORGANIC CHEMISTRY, MEDICINAL CHEMISTRY. Educ: Rice Inst, BA, 39, MA, 41; Harvard Univ, PhD(org chem), 45. 45. Prof Exp: Org chemist, Off Sci Res & Develop, Harvard Univ, 45-47 & Merck & Co, Inc, NJ, 47-52; dept head res div, Bristol-Myers Co, 52-68; chmn math & sci, Mercer County Community Col, 68-69; teacher & counr urban univ prog, Grad Sch Educ, 69-72, PROF ACAD FOUND, LIVINGSTON COL, RUTGERS UNIV, NEW BRUNSWICK, 72- Concurrent Pos: Instr, Union Jr Col; chmn, Gordon Conf Med Chem, 67. Mem: Am Chem Soc; fel NY Acad Sci. Res: Synthetic organic chemistry; structure-action relationships of drugs; reaction mechanisms; photochemistry; nature of science; relationship of science and poetry. Mailing Add: 249 Harrison Ave Highland Park NJ 08904

WILSON, ARTHUR MOBERG, analytical chemistry, see 12th edition

WILSON, BARRY WILLIAM, b Brooklyn, NY, Aug 20, 31; m 56. CELL BIOLOGY, NEUROBIOLOGY. Educ: Univ Chicago, BA, 50; Ill Inst Technol, BS & MS, 57; Univ Calif, Los Angeles, PhD(zool), 62. Prof Exp: Asst zool, Ill Inst Technol, 56-57; asst, Univ Calif, Los Angeles, 57-58, USPHS cardiovasc trainee, 58, fel, 59-61, jr res zoologist, 62; asst prof poultry husb & animal physiol & asst biologist, Exp Sta, 62-68, assoc prof avian sci & animal physiol & assoc biologist, 68-72, PROF AVIAN SCI & ANIMAL PHYSIOL, UNIV CALIF, DAVIS, 72- Mem: Am Soc Cell Biol; Soc Neurosci; Soc Develop Biol; Am Soc Naturalists; Tissue Cult Asn. Res: Cell growth and development; emphasis on muscle and nerve using cell culture and intact animals; regulation of acetylcholinesterase and other molecules of nerve, muscle; muscular dystrophy and pesticide action. Mailing Add: Dept of Avian Sci Univ of Calif Davis CA 95616

WILSON, BENJAMIN JAMES, b Pennsboro, WVa, Jan 7, 23; m 44; c 4. MICROBIOLOGY. Educ: Univ WVa, AB, 43, MS, 47; George Washington Univ, PhD(microbiol), 55. Prof Exp: Med bacteriologist, Biol Warfare Labs, Ft Detrick, 49-51, chief microbiol br, 51-59; assoc prof biol, David Lipscomb Col, 59-65; asst prof, 63-67, ASSOC PROF BIOCHEM, SCH MED, VANDERBILT UNIV, 67- Concurrent Pos: Consult nutrit sect, Off Int Res, NIH, 65-67 & Food & Drug Admin, 72-; mem subcomt toxicants occurring naturally in foods, Nat Acad Sci, 68-74; contrib ed, Nutrit Reviews, 73- Mem: Am Soc Microbiol; Am Chem Soc; Soc Toxicol; NY Acad Sci. Res: Mycotoxins; natural toxicants; microbial toxins. Mailing Add: Dept of Biochem Vanderbilt Univ Sch of Med Nashville TN 37232

WILSON, BILLY RAY, b SC, Apr 1, 22; m 47; c 1. ENTOMOLOGY. Educ: Clemson Col, BS, 42; Rutgers Univ, PhD(entom), 59. Prof Exp: Entomologist, Bur Entom & Plant Quarantine, USDA, 48-51 & Qm Res & Develop Labs, US Dept Army, 51-54; asst, 54-59, assoc res specialist, 59-66, chmn bur conserv & environ sci, 63-69, PROF ENTOM, RUTGERS UNIV, NEW BRUNSWICK, 63-, CHMN DEPT ENTOM & ECON ZOOL, 71- Mem: Entom Soc Am; Am Inst Biol Sci. Res: Insect physiology; insect resistance to insecticides; biology; biochemistry; environmental science. Mailing Add: 709 N Second Ave Highland Park NJ 08904

WILSON, BOBBY EUGENE, b Denton, Tex, Apr 16, 32; m 56; c 2. MICROBIOLOGY, ECOLOGY. Educ: North Tex State Univ, BS, 52, MS, 56; Univ Tex, PhD(microbiol), 63. Prof Exp: Instr biol, Col Arts & Indust, 56-59; from asst prof to assoc prof, 63-70, PROF BIOL, E TEX STATE UNIV, 70- Mem: Sigma Xi; Am Inst Biol Sci. Res: Cytotoxicity of staphylococcal toxins towards mammalian cells in vitro; changes in serum proteins following x-irradiation; effects of electromagnetic fields on plants and animals. Mailing Add: Dept of Biol ETex State Univ ETex Sta Commerce TX 75428

WILSON, BRAYTON F, b Cambridge, Mass, May 27, 34; m 60; c 2. BOTANY, FORESTRY. Educ: Harvard Univ, AB, 55, MF, 57; Australian Forestry Sch, dipl forestry, 59; Univ Calif, Berkeley, PhD(bot), 61. Prof Exp: Forest botanist, Harvard Univ, 61-67; asst prof, 67-72, ASSOC PROF FORESTRY, UNIV MASS, AMHERST, 72- Mem: AAAS; Bot Soc Am; Am Soc Plant Physiol. Res: Tree growth. Mailing Add: Dept of Forestry Univ of Mass Amherst MA 01002

WILSON, BRIAN GRAHAM, b Belfast, Ireland, Apr 9, 30; m 59; c 3. X-RAY ASTRONOMY. Educ: Queen's Univ, Belfast, BSc, 52; Nat Univ Ireland, PhD(cosmic radiation), 56. Prof Exp: Fel, Nat Res Coun Can, 55-57, officer in charge, Sulphur Mt Lab, Banff, 57-60, assoc res officer, 59-60; assoc prof physics, Univ Calgary, 60-65, prof, 65-70, dean arts & sci, 67-70; PROF ASTRON & ACAD VPRES, SIMON FRASER UNIV, 70- Mem: Can Asn Physicists. Res: Space physics; cosmic radiation; x-ray astronomy. Mailing Add: Off of Acad VPres Simon Fraser Univ Burnaby BC Can

WILSON, BRUCE NOLAN, polymer chemistry, see 12th edition

WILSON, BURTON DAVID, b Los Angeles, Calif, Oct 20, 32; m 58; c 3. ORGANIC CHEMISTRY. Educ: Univ Calif, Los Angeles, BS, 54; Univ Ill, PhD, 58. Prof Exp: CHEMIST, EASTMAN KODAK CO, 57- Mem: Am Chem Soc. Res: Development and production problem solving on chemicals for photographic end uses. Mailing Add: Eastman Kodak Co 343 State St Rochester NY 14650

WILSON, BYRON J, b Jackson, Wyo, Feb 2, 31; m 58; c 7. INORGANIC CHEMISTRY. Educ; Idaho State Univ, BS, 56; Southern Ill Univ, MA, 58; Univ Wash, PhD(chem), 61. Prof Exp: Asst prof chem, Vanderbilt Univ, 61-65; from asst prof to assoc prof, 65-72, PROF CHEM, BRIGHAM YOUNG UNIV, 72- Mem: AAAS; Am Chem Soc. Res: Inorganic free-radical research. Mailing Add: 545 E 3050 North Provo UT 84601

WILSON, CARL C, b Halfway, Ore, Mar 17, 15; m 42; c 2. FOREST MANAGEMENT. Educ: Univ Idaho, BS, 39; Univ Calif, Berkeley, MS, 41. Prof Exp: Fire control asst, Lassen Nat Forest, US Forest Serv, Calif, 46-49 & Plumas Nat Forest, 49, forester, Angeles Nat Forest, 49-55, res forester & proj leader forest fire res, Calif Forest Exp Sta, 56-57, chief forest fire res div, 57-62, asst dir forest fire & eng res & chief forest fire lab, Pac Southwest Forest & Range Exp Sta, 62-73, ASST DIR & NAT FIRE SPECIALIST, COOP FIRE CONTROL, STATE & PVT FORESTRY, US FOREST SERV, 73- Concurrent Pos: Assoc, Exp Sta, Univ Calif, Riverside, 64-; consult fire mgt, Food & Agr Orgn-UN Environ Prog, UN, Rome, Italy, 75. Mem: Soc Am Foresters; Am Soc Range Mgt. Res: Forest fire science; detection and control of forest fires for the protection of the human environment. Mailing Add: US Forest Serv PO Box 245 Berkeley CA 94701

WILSON, CARROLL KLEPPER, b Denton, Tex, Aug 24, 17; m 41; c 2. MATHEMATICS. Educ: NTex State Col, BA, 37; Univ Tex, MA, 40. Prof Exp: Teacher pub schs, Tex, 37-41; instr radar, US Civil Serv, 41-43; actg chmn div math & natural sci, 56-60, ASSOC PROF MATH, EASTERN NMEX UNIV, 46-, HEAD DEPT, 60- Mem: Am Math Soc; Math Asn Am; Asn Comput Mach. Res: Analysis. Mailing Add: Dept of Math Eastern NMex Univ Portales NM 88130

WILSON, CHARLES B, b Neosho, Mo, Aug 31, 29; m 56; c 3. MEDICINE. Educ: Tulane Univ, BS, 51, MD, 54. Prof Exp: Resident path, Tulane Univ, 55-56, instr neurosurg, 60-61; resident, Ochsner Clin, 56-60; instr, La State Univ, 61-63; from asst prof to prof, Univ Ky, 63-68; PROF NEUROSURG, UNIV CALIF, SAN FRANCISCO, 68- Mem: Am Asn Neurol Surg; Am Asn Neuropath; Soc Neurol Surg. Res: Brain tumor chemotherapy; tissue culture. Mailing Add: Dept of Neurosurg Univ of Calif Med Ctr San Francisco CA 94143

WILSON, CHARLES L, b Bristol, Va, Apr 9, 32; m 58; c 4. PLANT PATHOLOGY. Educ: Univ Va, AB, 53; WVa Univ, MS, 56, PhD, 58. Prof Exp: Asst, WVa Univ, 55-58; from asst prof to prof plant path, Univ Ark, 58-69; invests leader shade tree & farm windbreak invests, Agr Res Serv, USDA, 69-72, res leader, Shade Tree & Ornamental Plants Lab, 72-74, RES PLANT PATHOLOGIST & ADJ PROF, AGR RES SERV, USDA & OHIO AGR RES & DEVELOP CTR, 74- Mem: Am Phytopath Soc; Int Soc Arboricult. Res: Tree wilt pathogens; insect-fungus relationships; tree nursery diseases; wood-staining fungi; pathological histology of wood; fungal cytology; lysosomal system and mechanisms for resistance in plant cells. Mailing Add: Dept of Plant Path Ohio Agr Res & Develop Ctr Wooster OH 44691

WILSON, CHARLES MARSHALL, b East St Louis, Ill, July 22, 15; m 42; c 1. FOOD BACTERIOLOGY. Educ: Univ Ill, AB, 38, MS, 40, PhD(bact), 46. Prof Exp: Asst prof bact, Univ Ill, 46-51; dir qual control, Spec Div, 51-68, DIR QUAL

ASSURANCE LABS, BORDEN CO, 68- Mem: Am Soc Microbiol; Inst Sanit Mgt; Inst Food Technol. Res: Food spoilage and poisoning. Mailing Add: 990 Kingsmill Pkwy Columbus OH 43229

WILSON, CHARLES MAYE, b Mt Olive, NC, Oct 16, 16. MYCOLOGY. Educ: Univ Va, BSc, 41, MA, 42; Harvard Univ, PhD(biol), 50. Prof Exp: Nat Res Coun fel, Univ Calif, 50-51; instr biol, Harvard Univ, 51-53; from asst prof to assoc prof bot, 53-62, chmn dept biol, 62-70, PROF BOT, McGILL UNIV, 62- Concurrent Pos: Vis prof, Univ Calif, 57. Mem: Bot Soc Am; Mycol Soc Am. Res: Cytology and life cycles of the lower fungi. Mailing Add: Dept of Biol McGill Univ Montreal PQ Can

WILSON, CHARLES OREN, b Salt Lake City, Utah, May 9, 26; div; c 4. CLINICAL CHEMISTRY. Educ: Stanford Univ, BS, 49. Prof Exp: Prod chemist, Transandino Co, Calif, 49-51; res chemist, Nat Bur Stand, 51-52; sr res chemist, Olin Mathieson Chem Corp, NY, 52-54 & Calif, 54-59; sr res chemist, Nat Eng Sci Corp, 59-61 & Am Potash & Chem Corp, 61-68; chemist, Electro-Optical Systs, Inc, 68-69 & Int Chem & Nuclear Corp, 69-70; MGR QUAL CONTROL, REAGENTS QUAL CONTROL, ANAL SYSTS, ABBOTT DIAG DIV, 70- Mem: Am Chem Soc. Res: Organoboron and organophosphorus chemistry; infrared spectroscopy; mass spectrometry; rare earth research; solvent cal extraction; quality control of clinical diagnostic reagents. Mailing Add: Abbott Diag Div Anal Systs Reagents Qual Control 820 Mission St South Pasadena CA 91030

WILSON, CHARLES OWENS, b Bothell, Wash, Jan 25, 11; m 36; c 4. PHARMACEUTICAL CHEMISTRY. Educ: Univ Wash, BS, 34, MS, 35, PhD(org chem), 38. Prof Exp: Asst prof pharmaceut chem, George Washington Univ, 38-40; from asst prof to prof, Univ Minn, 40-48; prof, Univ Tex, 48-59; PROF PHARMACEUT CHEM & DEAN SCH PHARM, ORE STATE UNIV, 59- Mem: AAAS; Am Chem Soc; Am Pharmaceut Asn. Res: Synthesis of organic medicinal compounds; hypnotics and local anesthetics. Mailing Add: 4010 SW Fairhaven Dr Corvallis OR 97330

WILSON, CHARLES R, b Baltimore, Md, Jan 25, 29; m 59; c 3. GEOPHYSICS. Educ: Case Inst Technol, BS, 51; Univ NMex, MS, 56; Univ Alaska, PhD(geophys), 63. Prof Exp: PROF PHYSICS, GEOPHYS INST, UNIV ALASKA, 59- Concurrent Pos: Fulbright grant, Paris, 63-64; vis scientist, Nat Ctr Atmospheric Res, 68-69. Mem: Am Geophys Union. Res: Magnetic storms; geomagnetic micropulsations; auroral infrasonics. Mailing Add: Geophys Inst Univ of Alaska Fairbanks AK 99701

WILSON, CHARLES WOODSON, III, b Columbus, Ohio, Nov 20, 24; m 48; c 4. PHYSICS. Educ: Univ Mich, BSE, 47, MS, 48; Wash Univ, PhD(physics), 52. Prof Exp: Res assoc, Stanford Univ, 52; res physicist, Prod Dept, Res Div, Texaco Inc, 52-56; physicist, Res & Develop Dept, Chem Div, Union Carbide Corp, 56-62, res scientist, 62-64, group leader, 64-65; PROF PHYSICS & POLYMER SCI, HEAD PHYSICS DEPT & RES ASSOC, INST POLYMER SCI, UNIV AKRON, 65- Mem: AAAS; Am Phys Soc; Am Asn Physics Teachers. Res: Nuclear and electron spin resonance; molecular and high polymer physics. Mailing Add: Dept of Physics Univ of Akron Akron OH 44235

WILSON, CHRISTINE SHEARER, b Orleans, Mass. NUTRITION. Educ: Brown Univ, BA, 50; Univ Calif, Berkeley, PhD(nutrit, anthrop), 70. Prof Exp: Asst ed, Nutrit Rev, Sch Pub Health, Harvard Univ, 51-56; nutrit analyst, USDA, 57-58; asst res nutritionist, 70-71, res assoc nutrit anthrop, 71-74, ASST RES NUTRITIONIST, DEPT INT HEALTH, UNIV CALIF, SAN FRANCISCO, 74 & 75-, LECTR, 74- Concurrent Pos: Consult, Soc Nutrit Educ, 71-72; USPHS spec res fel, 72-73; lectr, Dept Anthrop, Univ Calif, Riverside, 73; contrib ed, J Nutrit Educ, 73-74; consult, Children's TV Workshop, 73-74. Mem: Am Anthrop Asn; Am Inst Nutrit; Soc Med Anthrop; Soc Nutrit Educ; Am Pub Health Asn. Res: Food in the culture; social influences on nutritional status; ethnographic field research on diet and nutritional health; dietary factors in breast cancer. Mailing Add: Dept of Int Health Univ of Calif San Francisco CA 94143

WILSON, CHRISTOPHER LUMLEY, b Leeds, Eng, Aug 31, 09; nat US; m 37; c 2. CHEMISTRY. Educ: Univ Leeds, BSc, 30; Univ London, PhD(chem), 32, DSc, 39. Prof Exp: Demonstr chem, Univ London, 32-33, from asst lectr to lectr, 33-46; prof, Univ Notre Dame, 46-48, Ohio State Univ, 48-54 & Univ Notre Dame, 54-61; prof chem & physics, High Point Col, 61-74; PRES, UNIPOINT INDUSTS, INC, 66- Concurrent Pos: Acad rels officer, Imp Chem Industs, Eng, 39-42; dir res, Revertex, Ltd, 42-46; dir, Nease Chem Co, 51-61; vpres & dir, Phillips-Foscue Corp, 58-61; dir, Unipoint Industs, Inc, Consolidated Pappas Enterprises Inc & Eng for Indust Inc. Mem: Am Chem Soc; fel The Chem Soc; Electrochem Soc; Soc Plastics Indust; Asn Consult Chemists & Chem Engrs. Res: Mechanism of tautomeric change; use of deuterium and radio-isotopes in organic reactions; long-wave spectroscopy of organic molecules; electrolytic reduction; gas phase catalytic chemistry; cellular synthetic resin products for use as insulation and sponges; polyurethanes; chemistry of furan derivatives. Mailing Add: Unipoint Industs Inc PO Box 2082 High Point NC 27261

WILSON, CHRISTOPHER PAUL, b Austin, Tex, July 29, 47. ASTRONOMY. Educ: Harvey Mudd Col, BS, 68; Univ Calif, Berkeley, PhD(astron), 73. Prof Exp: Fel astron, Hale Observs, 73-75; LECTR ASTRON, YALE UNIV, 75- Mem: Am Astron Soc. Res: Stellar dynamics; structure of galaxies. Mailing Add: Yale Univ Dept of Astron Box 2023 Yale Sta New Haven CT 06520

WILSON, CLAUDE E, b Starkey, NY, Jan 29, 39; m 60; c 1. ANALYTICAL CHEMISTRY, ELECTROCHEMISTRY. Educ: Harpur Col, BA, 60; Columbia Univ, MA, 61, PhD(chem), 66. Prof Exp: Instr chem, Drew Univ, 64-65; asst prof, Univ Pittsburgh, 65-71; ASST PROF CHEM, IND UNIV-PURDUE UNIV, INDIANAPOLIS, 71- Mem: AAAS; Am Chem Soc. Res: Electrochemical analysis of metals in natural water systems; applications of liquid chromatography to water analysis. Mailing Add: Dept of Chem Ind Univ-Purdue Univ Indianapolis IN 46205

WILSON, CLIFTON ARLIE, b Preston, Miss, June 12, 16; m 41; c 2. ENTOMOLOGY. Educ: Miss State Univ, BS, 40; Iowa State Univ, MS, 42. Prof Exp: Asst entom, Iowa State Univ, 40-42; med entomologist, USPHS, 42; res assoc entom, Exp Sta, Rutgers Univ, 46-48; from asst prof to assoc prof, 48-66, PROF ENTOM, MISS STATE UNIV, 66- Concurrent Pos: Asst entomologist, State Dept Agr, Iowa, 41. Mem: Entom Soc Am. Res: Insecticidal control of cotton insects; taxonomy of Lygaeidae; biology and control of cotton insects; taxonomy of aquatic Hemiptera; bee culture; turfgrass insect research. Mailing Add: Dept of Entom Miss State Univ Mississippi State MS 39762

WILSON, CLYDE LIVINGSTON, b Ohio, July 29, 22; m 51; c 2. SOIL PHYSICS. Educ: Ohio State Univ, BS, 47, BAE, 48, PhD(agron), 52. Prof Exp: Asst soils, Ohio State Univ, 48-52; from res agronomist to res specialist, 52-72, SR RES SPECIALIST, MONSANTO CO, 72- Mem: Am Soc Agron; Am Soc Agr Engrs; Soil Sci Soc Am; Weed Sci Soc Am. Res: Saturated water flow in tiled lands; chemical coil conditioners; herbicide investigations. Mailing Add: 1530 Lynkirk Lane Kirkwood MO 63122

WILSON, COLON HAYES, JR, b Marshallberg, NC, Apr 10, 32; m 56; c 3. RHEUMATOLOGY, INTERNAL MEDICINE. Educ: Duke Univ, BA, 52, MD, 56. Prof Exp: From intern to asst resident med, Univ Va Hosp, 56-58; resident, Edward J Meyer Mem Hosp, Buffalo, NY, 61-63; res instr, State Univ NY Buffalo, 64-66; from asst prof to assoc prof rheumatology, 66-73, asst prof phys med, 66-74, DIR DIV RHEUMATOLOGY, MED SCH, EMORY UNIV, 66-, PROF MED, 74- Concurrent Pos: Fel, Buffalo Gen Hosp, State Univ NY Buffalo, 63-66; Nat Inst Arthritis & Metab Dis prog grant, Edward J Meyer Mem Hosp, Buffalo. Mem: Am Fedn Clin Res; Am Rheumatism Asn; Reticuloendothelial Soc; fel Am Col Physicians. Res: Significance of various patterns of antinuclear antibody fluorescence with respect to specific diagnosis and prognosis on various collagen vascular diseases; the efficacy of early synovectomy in rheumatoid arthritis in prevention of late deformity and preservation of function. Mailing Add: Dept of Med Emory Univ Sch of Med Atlanta GA 30303

WILSON, COYT TAYLOR, b Fulton, Miss, July 27, 13; m 36; c 2. PLANT PATHOLOGY. Educ: Ala Polytech Inst, BS, 38, MS, 41; Univ Minn, PhD(plant path), 46. Prof Exp: Instr bot, Ala Polytech Inst, 40-41; instr plant path, Univ Minn, 41-43; asst plant pathologist, Exp Sta, Auburn Univ, 44-47, prof plant path & plant pathologist, 47-51, asst dean, Col Agr & assoc dir, Agr Exp Sta, 51-64; from assoc dir to dir, Va Agr Exp Sta, 64-66; assoc dean, 66-71, EXEC ASSOC DEAN, RES DIV, VA POLYTECH INST & STATE UNIV, 71-, DIR AGR & LIFE SCI RES, 66- Concurrent Pos: AID short term res consult, Ministry Agr, Iran, 60, Turkey, 66 & EPakistan. Mailing Add: Res Div Va Polytech Inst & State Univ Blacksburg VA 24061

WILSON, CURTIS MARSHALL, b Stillwater, Minn, Sept 11, 26; m 52; c 4. PLANT BIOCHEMISTRY. Educ: Univ Minn, BS, 48, MS, 51; Univ Wis, PhD(bot), 54. Prof Exp: Asst, Univ Minn, 48-51; from asst prof to assoc prof plant physiol, Rutgers Univ, 54-59; assoc prof, 66-72, RES CHEMIST, AGR RES SERV, USDA, 59-; PROF PLANT PHYSIOL, UNIV ILL, URBANA, 72- Mem: AAAS; Am Soc Plant Physiol; Scand Soc Plant Physiol. Res: Plant biochemistry; protein synthesis and nucleic acid metabolism; seed development. Mailing Add: Dept of Agron Univ Ill Urbana IL 61801

WILSON, CYNTHIA, b Gillingham, Eng, Aug 31, 26; Can citizen. CLIMATOLOGY. Educ: Univ London, BA, 47, teachers dipl, 48; McGill Univ, MSc, 58; Laval Univ, PhD, 72. Prof Exp: Teacher high sch, Eng, 48-54; res asst meteorol & climat, Meteorol Res Group, McGill Univ, 54-62; asst prof climat, Inst Geog, 64-67, ASSOC PROF CLIMAT & RESEARCHER, CTR NORDIC STUDIES, LAVAL UNIV, 67- Concurrent Pos: Expert, Cold Regions Res & Eng Lab, US Army, 62-66; under contract to Meteorol Serv Can, 66-68. Honors & Awards: Darton Prize, Royal Meteorol Soc, 63. Mem: Am Meteorol Soc; Can Meteorol Soc; Can Asn Geog; Royal Meteorol Soc. Res: Meteorology and climatology of cold regions. Mailing Add: Ctr Nordic Studies Univ Laval Quebec PQ Can

WILSON, DANA E, b Chicago, Ill, Oct 5, 37; m 60; c 3. MEDICINE, METABOLISM. Educ: Oberlin Col, AB, 57; Western Reserve Univ, MD, 62. Prof Exp: Intern & asst resident internal med, Boston City Hosp, Mass, 62-64, sr resident, 66-67; clin assoc allergy & infectious dis, NIH, 64-66; asst prof clin nutrit & human metab, Mass Inst Technol, 69-71; asst prof, 71-74, ASSOC PROF MED, COL MED, UNIV UTAH, 74- Concurrent Pos: Fel diabetes & metab, Thorndike Mem Lab, Harvard Med Sch, 67-69. Mem: AAAS; Am Fedn Clin Res; Am Diabetes Asn. Res: Diabetes and metabolism; lipoproteins and lipid transport. Mailing Add: Dept of Med Univ of Utah Col of Med Salt Lake City UT 84112

WILSON, DARCY BENOIT, b Rhinebeck, NY, May 14, 36; m 57; c 3. IMMUNOLOGY, IMMUNOBIOLOGY. Educ: Harvard Univ, AB, 58; Univ Pa, PhD(zool), 62. Prof Exp: Assoc, 63-65, from asst prof to assoc prof path & med genet, 66-74, PROF PATH & HUMAN GENET, SCH MED, UNIV PA, 74- Concurrent Pos: Res fel transplantation immunol, Wistar Inst, Univ Pa, 62-63, res fel med genet, Sch Med, 65-66; Helen Hay Whitney Found fel, 64-67; USPHS career develop award, 67-72. Mem: Am Asn Immunol. Res: Immunology of tissue transplantation, particularly immunologic behavior of lymphoid cells in vitro and in vivo. Mailing Add: Dept of Path Univ of Pa Sch of Med Philadelphia PA 19174

WILSON, DAVID BUCKINGHAM, b Cambridge, Mass, Jan 15, 40; m 63; c 3. BIOCHEMISTRY. Educ: Harvard Univ, BA, 61; Stanford Univ, PhD(biochem), 65. Prof Exp: Jane Coffin Childs fel biochem, Sch Med, Johns Hopkins Univ, 65-67; asst prof, 67-74, ASSOC PROF BIOCHEM, CORNELL UNIV, 74- Concurrent Pos: NIH grant, 67-73. Mem: Am Soc Biol Chemists; Am Soc Microbiol. Res: Control of protein synthesis; purification and characterization of the galactose enzymes in Escherichia coli; mechanism and regulation of sugar transport in bacteria; the regulation of the galactose enzymes in yeast. Mailing Add: Dept of Biochem Wing Hall Cornell Univ Ithaca NY 14850

WILSON, DAVID E, b Meade, Kans, Aug 5, 29; m 56; c 2. MATHEMATICS. Educ: Kans State Col, Pittsburg, AB, 51, MS, 54; Univ Kans, PhD(math), 67. Prof Exp: Asst prof math, Univ Hawaii, 61-65; asst prof, 66-70, ASSOC PROF MATH, WABASH COL, 70- Mem: Am Math Soc; Math Asn Am. Res: Quasicon-formal mappings in n-space. Mailing Add: Dept of Math Wabash Col Crawfordsville IN 47933

WILSON, DAVID EVERETT, b Greenwich, Conn, May 27, 29; m 61; c 2. CELL PHYSIOLOGY, PROTOZOOLOGY. Educ: Univ Fla, BS, 61; Univ Calif, Los Angeles, PhD(zool), 66. Prof Exp: Technician, Fla State Plant Bd, 58-59; res asst protozool, Phelps Sanit Eng Lab, Univ Fla, 60-61 & Univ Calif, Los Angeles, 61-62; ASSOC PROF BIOL, CENT COL, IOWA, 66- Concurrent Pos: NIH fel protozool, 66; NSF grant comput curric develop in biol sci, 70-72; chmn Conduit Biol Comt, 72-74. Mem: AAAS; Soc Protozool; Am Soc Zoologists. Res: Physiology and taxonomy of the Heliozoa; electrophysiology; physiology of meditation. Mailing Add: Dept of Biol Central Col Pella IA 50219

WILSON, DAVID F, b Wray, Colo, Mar 28, 38; m 62. BIOCHEMISTRY. Educ: Colo State Univ, BS, 59; Ore State Univ, PhD(biochem), 64. Prof Exp: USPHS fel phys biochem, Johnson Res Found, 64-67, Pa Plan scholar, Sch Med, 67-69, res assoc, 67-68, asst prof, 68-72, ASSOC PROF BIOCHEM & BIOPHYS, MED SCH, UNIV PA, 72- Honors & Awards: Eli Lilly Award, Am Chem Soc, 71. Mem: Am Soc Biol Chemists; Biophys Soc. Res: Mitochondrial electron transport and energy conservation; low temperature spectrophotometry. Mailing Add: Dept Biochem & Biophys Univ Pa Med Sch Philadelphia PA 19174

WILSON, DAVID FRANKLIN, b Queens Village, NY, Feb 23, 41; m 65; c 2. NEUROPHYSIOLOGY. Educ: Hofstra Univ, BA, 63; Univ Del, MA, 66, PhD(biol sci), 68. Prof Exp: USPHS fel, Northwestern Univ, 68-69; asst prof, 69-73, ASSOC PROF ZOOL, MIAMI UNIV, 73- Concurrent Pos: USPHS grants, 70-72 & 73-75. Mem: Am Physiol Soc; Sigma Xi. Res: Examination of neuromuscular transmission using intracellular recording techniques. Mailing Add: Dept of Zool Miami Univ Oxford OH 45056

WILSON, DAVID GEORGE, b Spokane, Wash, Dec 15, 19; m 42; c 3. RANGE CONSERVATION, RESEARCH ADMINISTRATION. Educ: Univ Idaho, BS, 47; Agr & Mech Col, Tex, MS, 50, PhD(range mgt), 61. Prof Exp: Dep fire warden, Clearwater Timber Protective Asn, 47; instr range mgt, Agr & Mech Col, Tex, 47-48, asst prof, 49-50; from instr to assoc prof range mgt & plants, Univ Ariz, 53-64; range consult, 64-65; RES COORDR, US BUR LAND MGT, 65- Mem: Soc Range Mgt. Res: Natural resources. Mailing Add: US Bur Land Mgt Denver Fed Ctr Bldg 50 Denver CO 80225

WILSON, DAVID GEORGE, b St Catharines, Ont, Dec 3, 21; m 44; c 2. PLANT BIOCHEMISTRY. Educ: Univ Toronto, BA, 44; Queen's Univ, Ont, BA, 49, MA, 50; Univ Wis, PhD(biochem), 53. Prof Exp: Agr scientist, Conn Agr Exp Sta, 53-56; PROF BOT, UNIV WESTERN ONT, 56- Mem: AAAS; Am Soc Plant Physiol; Can Soc Plant Physiol. Res: Organic acids and amino acid metabolism in plants; biochemistry of cellulose and lignin degradation. Mailing Add: Dept of Plant Sci Univ of Western Ont London ON Can

WILSON, DAVID J, b Ames, Iowa, June 25, 30; m 52; c 5. PHYSICAL CHEMISTRY. Educ: Stanford Univ, BS, 52; Calif Inst Technol, PhD(chem), 58. Prof Exp: From instr to prof chem, Univ Rochester, 57-69; PROF CHEM, VANDERBILT UNIV, 69- Concurrent Pos: Alfred P Sloan fel, 64-66. Mem: Am Chem Soc; Am Phys Soc. Res: Energy transfer in gases; homogeneous gas reactions; pesticide and heavy metal residues; surface chemistry. Mailing Add: Dept of Chem Vanderbilt Univ Nashville TN 37235

WILSON, DAVID LOUIS, b Washington, DC, Jan 11, 43; m 67. NEUROSCIENCES, MOLECULAR BIOLOGY. Educ: Univ Md, College Park, BS, 64; Univ Chicago, PhD(biophys), 69. Prof Exp: ASST PROF PHYSIOL & BIOPHYS, MED SCH, UNIV MIAMI, 72- Concurrent Pos: Helen Hay Whitney fel, Calif Inst Technol, 69-72; NIH grant, 72-74 & 75-78. Mem: AAAS; Am Physiol Soc; Biophys Soc; Soc Neurosci; Fedn Am Sci. Res: Protein synthesis in neurons; theoretical neuroscience. Mailing Add: Univ of Miami Sch of Med PO Box 520875 Miami FL 33152

WILSON, DAVID ORIN, plant nutrition, see 12th edition

WILSON, DEAN GEORGE, b Payson, Utah, Apr 8, 29; m 51; c 6. PHYSICAL CHEMISTRY, INORGANIC CHEMISTRY. Educ: Brigham Young Univ, BS, 51, MS, 52; Univ Utah, PhD(fuels eng), 58. Prof Exp: Res chemist, Kennecott Copper Corp, 52-54; sr metallurgist, Geneva Works, 57-59, chief chemist, 59-61, gen supvr coal & chem res, Raw Mat Res Lab, 61-64, supt nitrogen plant, 64-66, div supt coke & coal chem, Utah, 66-69, asst gen supt, Clairton ks, 69-73, GEN SUPT, CLAIRTON WORKS, US STEEL CORP, 73- Mem: Am Chem Soc; Asn Iron & Steel Eng; Am Iron & Steel Inst. Res: Combustion and explosives; coal carbonization; coal chemicals and related products; fertilizers. Mailing Add: Clairton Works US Steel Corp 400 State St Clairton PA 15025

WILSON, DONALD ALAN, b San Francisco, Calif, Sept 16, 30; m 57; c 3. BIOCHEMISTRY, MICROBIAL PHYSIOLOGY. Educ: San Jose State Col, BA, 59; Western Reserve Univ, PhD(microbiol), 65. Prof Exp: Fel microbiol, Pioneering Res Div, US Army Natick Labs, 65-66; assoc, 66-67, asst prof, 67-70, ASSOC PROF MICROBIOL, CHICAGO MED SCH, 70- Res: Microbial enzymology. Mailing Add: Dept of Microbiol Chicago Med Sch Chicago IL 60612

WILSON, DONALD ALVIN, physical chemistry, see 12th edition

WILSON, DONALD DEAN, organic chemistry, physical chemistry, see 12th edition

WILSON, DONALD ELSWORTH, food technology, see 12th edition

WILSON, DONALD LAURENCE, b Hamilton, Ont, Oct 2, 21; m 45; c 5. MEDICINE. Educ: Queen's Univ, Ont, MD, CM, 44; Univ Toronto, MA, 48; FRCP(C), 51. Prof Exp: From asst prof to assoc prof, 52-67, PROF MED, QUEEN'S UNIV, ONT, 67- Concurrent Pos: Attend physician, Kingston Gen Hosp, 52-; consult, Can Forces Hosp, Kingston; pres, Coun Col Physicians & Surgeons, Ont, 65-66; pres, Ont Med Asn, 73-74. Mem: Endocrine Soc; Am Diabetes Asn; NY Acad Sci; Can Soc Clin Invest; Can Physiol Soc. Res: Endocrinology and diseases of metabolism. Mailing Add: Dept of Med Queen's Univ Kingston ON Can

WILSON, DONALD RICHARD, b Plaistow, NH, Feb 8, 36; m 56; c 3. ORGANIC CHEMISTRY, POLYMER CHEMISTRY. Educ: Univ Wash, BS, 58; Univ Calif, Los Angeles, PhD(org chem), 62. Prof Exp: Mgr control lab, Am Marietta Co, Wash, 57-58; from res chemist to sr res chemist, E I du Pont de Nemours & Co, Inc, Del, 61-67; sr scientist, Xerox Corp, Webster, 67, res mgr org & polymer chem, 68, develop mgr org & polymer mat, 68-70, technol prog mgr advan xerography, 70-71, prin scientist xerographic mat, 71-72, mgr explor graphic sci, 72-75; RES DIR CHEM & CATALYSIS, CELANESE RES CO, 75- Mem: Am Chem Soc. Res: Organic and polymer synthesis-reaction mechanisms; stereochemistry; organometallics; textile fiber chemistry; high temperature fibers; xerography; xerographic imaging materials; triboelectricity. Mailing Add: Celanese Res Co PO Box 1000 Summit NJ 07901

WILSON, DONALD ROBERT, b Edmonton, Alta, Jan 20, 13; m 42; c 4. INTERNAL MEDICINE. Educ: Oxford Univ, BA, 37; McGill Univ, MD & CM, 39; FRCPS(C). Prof Exp: Mem fac med, 47-49, from asst prof to assoc prof med, 49-54, head dept, 54-68, PROF MED, FAC MED, UNIV ALTA, 54-, DIR, R S McLAUGHLIN RES & EXAM CTR, 68- Concurrent Pos: Markle scholar, Univ Alta, 49-54; consult, Royal Can Air Force, 50-; consult internal med, Armed Forces Med Coun Can; mem adv med comt, Nat Res Coun Can, 52-55; mem, Rhodes Scholarships Selection Comt, Alta; mem coun, Royal Col Physicians & Surgeons Can, vpres div med, 66-68; chief serv, Dept Vet Affairs, Alta, 61. Mem: Endocrine Soc; Am Fedn Clin Res; fel Am Col Physicians; Can Soc Clin Invest; Can Diabetic Asn. Res: Endocrinology. Mailing Add: Dept of Med 8113 Clin Sci Bldg Univ of Alta Edmonton AB Can

WILSON, DORIS BURDA, b Cleveland, Ohio, July 1, 37; m 68. ANATOMY, EMBRYOLOGY. Educ: Ohio Wesleyan Univ, BA, 59; Radcliffe Col, MA, 60; Harvard Univ, PhD(biol), 63. Prof Exp: Teaching fel biol, Harvard Univ, 60-62; asst prof zool, San Diego State Col, 63-65; asst prof anat, Sch Med, Stanford Univ, 65-69; res anatomist & lectr, Sch Med, Univ Calif, San Diego, 69-73; assoc prof anat, Sch Med, Univ Calif, Davis, 73-75; ASSOC PROF SURG/ANAT, SCH MED, UNIV CALIF, SAN DIEGO, 75- Concurrent Pos: Fulbright vis prof, Taiwan, 70-71. Mem: AAAS; Histochem Soc; Teratology Soc; Am Soc Zool; Am Asn Anat. Res: Developmental biology; neuroembryology; teratology. Mailing Add: Dept of Surg/Anat Univ of Calif San Diego La Jolla CA 92093

WILSON, DWIGHT ELLIOTT, JR, b Greensburg, Pa, June 7, 32; m 53; c 4. VIROLOGY. Educ: Yale Univ, BS, 53, PhD(biophys), 56. Prof Exp: From asst prof to assoc prof, 56-68, PROF BIOL, RENSSELAER POLYTECH INST, 68- Concurrent Pos: NIH fel, 66-67. Mem: Biophys Soc; Am Soc Microbiol. Res: Mechanisms of animal virus replication. Mailing Add: Dept of Biol Rensselaer Polytech Inst Troy NY 12181

WILSON, EDGAR BRIGHT, b Gallatin, Tenn, Dec 18, 08; m 35 & 55; c 6. CHEMICAL PHYSICS. Educ: Princeton Univ, BS, 30, AM, 31; Calif Inst Technol, PhD(phys chem), 23. Hon Degrees: Dr, Free Univ Brussels, 75. Prof Exp: Fel, Calif Inst Technol, 33-34; jr fel, 34-36, from asst prof to prof, 36-48, RICHARDS PROF CHEM, HARVARD UNIV, 48- Concurrent Pos: Res dir, Underwater Explosives Res Lab, Woods Hole, 42-44; Fulbright grant & Guggenheim fel, Oxford Univ, 49-50; res dir, Weapons Systs Eval Group, 52-53. Honors & Awards: Prize, Am Chem Soc, 37; Medal for Merit, 48; Cert of Appreciation, US Dept Defense, 53; Debye Award, Am Chem Soc, 62, Norris Award, 66, G N Lewis Award, Calif Sect, Pauling Award, 72; Rumford Medal, Am Acad Arts & Sci, 73. Mem: Nat Acad Sci; Am Chem Soc; fel Am Phys Soc; Am Acad Arts & Sci; Int Acad Quantum Molecular Sci. Res: Quantum mechanics in chemistry; molecular dynamics; microwave spectroscopy. Mailing Add: Harvard Chem Labs 12 Oxford St Cambridge MA 02138

WILSON, EDMOND WOODROW, JR, b Selma, Ala, Jan 18, 40; m 65; c 1. PHYSICAL CHEMISTRY. Educ: Auburn Univ, BS, 62; Univ Ala, Tuscaloosa, MS, 65, PhD(phys chem), 68. Prof Exp: Temporary instr gen chem, Univ Ala, Tuscaloosa, 66-68; res assoc biophys chem, Univ Va, 68-70; asst prof, 70-73, ASSOC PROF PHYS CHEM, HARDING COL, 73- Mem: Am Chem Soc. Res: Calorimetry and spectroscopy of biological molecules containing transition metal ions. Mailing Add: Dept of Phys Sci Harding Col Searcy AR 72134

WILSON, EDWARD C, b Dallas, Tex, May 18, 33; m 59; c 2. INTERNAL MEDICINE, GASTROENTEROLOGY. Educ: Randolph-Macon Col, BS, 54; Med Col Va, MD, 58. Prof Exp: Fel clin pharmacol, 62-63; res fel gastroenterol, 63-65; ASST PROF MED, SCH MED, UNIV VA, 65- Res: Fat metabolism of the liver as it related to human hepatic disease; mechanisms of small intestinal absorption of amino acids and carbohydrates. Mailing Add: Univ of Va Sch of Med Charlottesville VA 22903

WILSON, EDWARD CARL, b Daytona Beach, Fla, Jan 25, 29. INVERTEBRATE PALEONTOLOGY. Educ: Univ Calif, Berkeley, BA, 58, MA, 60, PhD(paleont), 67. Prof Exp: Cur paleont & chmn div invert, Natural Hist Mus San Diego, 64-67; CUR INVERT PALEONT, LOS ANGELES COUNTY MUS NATURAL HIST, 67- Mem: Paleont Soc; Brit Paleont Asn. Res: Late Paleozoic corals. Mailing Add: Los Angeles Count Mus Natural Hist 900 Exposition Blvd Los Angeles CA 90007

WILSON, EDWARD MATTHEW, b Content, Jamaica, Dec 19, 37; m 62; c 2. DAIRY SCIENCE. Educ: McGill Univ, BScAgr, 64, MSc, 66; Ohio State Univ, PhD(dairy sci), 69. Prof Exp: Res asst animal genetics, McGill Univ, 64-66; res assoc dairy sci, Ohio State Univ, 66-69; asst prof animal sci, Tuskegee Inst, 69-73; prin physiologist, Coop State Res Serv, USDA, 73-74; DEAN COOP RES & EXTEN & HEAD DEPT AGR & NATURAL RESOURCES, LINCOLN UNIV, 74- Concurrent Pos: Mem, Task Force to Repub SAfrica, 74. Mem: Am Dairy Sci Asn. Res: Genetic polymorphisms of bovine blood and milk proteins; immunological properties of seminal proteins and their implications in the reproductive process. Mailing Add: Dept of Agr & Natural Resources Lincoln Univ Jefferson City MO 65101

WILSON, EDWARD OSBORNE, b Birmingham, Ala, June 10, 29; m 55; c 1. BEHAVIORAL BIOLOGY, EVOLUTIONARY BIOLOGY. Educ: Univ Ala, BS, 49, MS, 50; Harvard Univ, PhD(biol), 55. Prof Exp: Biologist, State Dept Conserv, Ala, 49; Soc Fels jr fel, 53-56, from asst prof to assoc prof, 56-64, PROF ZOOL, HARVARD UNIV, 64- Concurrent Pos: Mem exped, WIndies & Mex, 53, New Caledonia, 54, Australia & New Guinea, 55, Ceylon, 55 & Surinam, 61; Charles and Martha Hitchcock Prof, Univ Calif, Berkeley, 72. Honors & Awards: Cleveland Prize, AAAS, 67; Mercer Award, Ecol Soc Am, 71; Founders' Mem Award, Entom Soc Am, 72. Mem: Nat Acad Sci; fel Am Acad Arts & Sci; Soc Study Evolution (pres, 73). Res: Classification, ecology and behavior of ants; speciation; chemical communication in animals; biogeography. Mailing Add: Mus of Comp Zool Harvard Univ Cambridge MA 02138

WILSON, EDWIN E, b Peabody, Mass, June 16, 20; m 44. PHARMACY. Educ: Mass Col Pharm, BS, 41, MS, 43. Prof Exp: Pharmacist, Panama Canal Dept Health Div, Gorgas Hosp, Ancon, CZ, 43-44 & Hutchinson Drug Co, Inc, Mass, 46-54; from instr to prof pharm, Mass Col Pharm, 54-74; RETIRED. Mem: Am Pharmaceut Asn; Am Col Apothecaries. Res: Lipolytic activity of pancreatin; methylcellulose as a satisfactory dispersing agent; dermatologic and ophthalmic preparations; sterile solutions and techniques; prescription writing and interpretation; pharmaceutical incompatibilities and formulation; improved pharmaceutical techniques. Mailing Add: 107 Lafayette St Marblehead MA 01945

WILSON, ELIZABETH, b Missoula, Mont, Feb 17, 15. MEDICAL MICROBIOLOGY. Educ: Earlham Col, AB, 37; Purdue Univ, MS, 51, PhD(bact), 54. Prof Exp: Bacteriologist, State Dept Health, Mich, 37-42; res asst oral bact & dent caries, Dent Sch, Univ Mich, 42-43; bacteriologist, Sanit Eng Ctr, USPHS, Ohio, 54-60, Commun Dis Ctr, Ga, 60-65 & Streptococcus Dis Sect, Colo, 65-71; CHIEF, LAB CONSULTATION & DEVELOP, STATE HYG LAB, UNIV IOWA, 71- Mem: Am Soc Microbiol; Am Pub Health Asn; NY Acad Sci. Res: Clinical microbiol education; bacterial pathogens in foods; fluorescent antibody technique; streptococcus identification and epidemiology. Mailing Add: State Hyg Lab Univ of Iowa Iowa City IA 52242

WILSON, ELWOOD JUSTIN, JR, b New York, NY, Nov 28, 17; m 41; c 4. ORGANIC CHEMISTRY. Educ: Princeton Univ, AB, 38, MA, 40, PhD(chem), 41. Prof Exp: Corn Industs Res Found fel, NIH, 41-42; proj engr, Sperry Gyroscope Co, NY, 42-44; sr chemist, Exp, Inc, 46-47, secy, 47-49, vpres, 49-54; res dir, Detroit Controls Co, 54-59; pres, Adv Tech Labs Div, Am Radiator & Stand Sanit Corp, 59-64; PRES, E J WILSON ASSOCS, INC, 64-; PRES, EPOXON PRODS, INC, 71- Concurrent Pos: Vpres, Flight Res, Inc, 53-54; chmn & dir, Data Cartridge, Inc, 65-67. Mem: AAAS; Am Chem Soc. Res: Proteins; carbohydrates; fuels; combustion; synthetic organic chemistry; interior ballistics; rockets; petrochemicals; aerospace instruments; nuclear reactors; general and technical management; building materials. Mailing Add: 1125 Westridge Dr Portola Valley CA 94025

WILSON, ERIC LEROY, b Sharon, Pa, Mar 17, 35; m 59; c 2. MATHEMATICS. Educ: Westminster Col, Pa, BS, 57; Vanderbilt Univ, PhD(math), 66. Prof Exp: From instr to asst prof, 62-70, ASSOC PROF MATH, WITTENBERG UNIV, 70-, CHMN DEPT, 73- Concurrent Pos: Asst prof, Univ of the South, 67-68. Mem: Math Asn Am; Nat Speleol Soc. Res: Loop isotopy; graph theory. Mailing Add: Dept of Math Wittenberg Univ Springfield OH 45501

WILSON, EUGENE M, b Buckhannon, WVa, May 4, 28; m 51; c 2. PLANT PATHOLOGY. Educ: Univ WVa, BS, 51, MS, 54; Univ Calif, PhD(plant path), 58. Prof Exp: Asst plant path, Univ WVa, 51 & Univ Calif, 54-58; plant pathologist, Cent Res Lab, United Fruit Co, Mass, 58-60; technologist, Shell Chem Co, 60-72; plant pathologist pesticide regulation, 72-73, chief plant path sect, Off Pesticide Progs, 73-74, PROD MGR, US ENVIRON PROTECTION AGENCY, 74- Mem: Am Phytopath Soc. Res: Physiology of fungi; plant disease control; host parasite relationship; biological control of plant pests. Mailing Add: 5 De Forest Ave New City NY 10956

WILSON, EUGENE MURPHEY, b Augusta, Ga, Aug 27, 31; m 55; c 3. GEOGRAPHY. Educ: Univ Ala, BS, 60; La State Univ, MS, 62, PhD, 69. Prof Exp: Instr, Univ Ala, 63-69; asst prof, 69-74, ASSOC PROF GEOG, UNIV S ALA, 74- Concurrent Pos: Univ SAla res grants, Yucatan, Mex, 69, 71 & 72. Mem: Asn Am Geog. Res: Settlement geography; karst morphology; Middle America. Mailing Add: Dept of Geol & Geog Univ of SAla Mobile AL 36608

WILSON, EVA DONELSON, b Ogden, Iowa, Aug 8, 05; m 47. NUTRITION. Educ: Iowa State Col, BS, 27; Univ Chicago, PhD(nutrit), 34. Prof Exp: Mem staff, Merrill Palmer Sch, 27-31 & Children's Fund of Mich, Detroit, 31-32; from instr to assoc prof home econ, Univ Minn, 34-46; prof, Ohio State Univ, 46-48 & Pa State Univ, 48-60; prof, 60-72, EMER PROF HOME ECON, OHIO STATE UNIV, 72-; VIS PROF, UNIV SAO PAULO, 72- Concurrent Pos: Consult, Univ Sao Paulo, 75-76. Mem: Am Home Econ Asn; Am Dietetic Asn; Am Inst Nutrit. Res: Metabolic balance studies; food analysis; nutritional status of children and adults; dietary studies; vitamin B and G content of human milk as influenced by the consumption of brewers yeast. Mailing Add: SQs 103 Bloco S Apt 21 Brasilia DF Brasil

WILSON, EVELYN H, b Philadelphia, Pa, Oct 8, 21; m 43; c 2. ORGANIC CHEMISTRY. Educ: Bryn Mawr Col, AB, 42; Radcliffe Col, AM, 44, PhD(org chem), 46. Prof Exp: Res chemist, Merck & Co, Inc, 46-53; sr scientist, Johnson & Johnson, 53-59; lectr chem, Westfield Sr High Sch, NJ, 49-65; sci supvr, New Brunswick Pub Schs, 65-67; assoc prof sci educ, 67-72, PROF SCI EDUC & DIR BUDGET & PLANNING, RUTGERS UNIV, NEW BRUNSWICK, 72- Res: Educational planning and administration; philosophy of science; science education. Mailing Add: Off of Budget & Planning Rutgers Univ New Brunswick NJ 08903

WILSON, EVERETT D, b Covington, Ind, July 13, 28; m 52; c 4. PHYSIOLOGY, ENDOCRINOLOGY. Educ: Ind State Teachers Col, BS, 50, MS, 51; Purdue Univ, PhD(physiol, endocrinol), 60. Prof Exp: Teacher pub sch, Ind, 48-50 & 55-56, supvr, 56-57; asst biol, Purdue Univ, 57-58, res asst, 58-60; asst prof zool, Southern Ill Univ, 60-61; assoc prof, 62-64, PROF, SAM HOUSTON STATE UNIV, 64- DEAN COL SCI, 65- Concurrent Pos: Lalor res fel, 61; NATO fel, Nat Med Res Inst, 62; chief, Grants Br Pop & Reproduction Ctr, Nat Inst Child Health, NIH, 71-72; adv, Oak Ridge Pop Res Inst, 72-74. Mem: Am Soc Zoologists; Endocrine Soc; Am Soc Animal Sci; Brit Soc Study Fertil. Res: Factors affecting mammalian reproduction. Mailing Add: Col of Sci Sam Houston State Univ Huntsville TX 77340

WILSON, F WESLEY JR, b Washington, DC, Apr 22, 39. MATHEMATICS. Educ: Univ Md, College Park, BS, 61, PhD(math), 64. Prof Exp: Res assoc math, Div Appl Math, Brown Univ, 64-66; asst prof, Univ Mich, Ann Arbor, 66-67; asst prof, 67-69, ASSOC PROF MATH, UNIV COLO, BOULDER, 69- Concurrent Pos: Vis assoc prof, Univ Md, College Park, 70-71. Mem: Am Math Soc. Res: Applications of differential topology to problems of nonlinear ordinary differential equations; analytical problems in differential topology. Mailing Add: Dept of Math Univ of Colo Boulder CO 80302

WILSON, FLOYD DEE, b Fresno, Calif, Dec 28, 39. EXPERIMENTAL PATHOLOGY, HEMATOLOGY. Educ: Univ Calif, Davis, BS, 68, DVM, 70. Prof Exp: ASSOC RES VET & EXP PATHOLOGIST & HEMATOLOGIST, RADIOBIOL LAB, UNIV CALIF, DAVIS, 70- Res: Mechanisms of radiation-induced leukemogenesis using in vitro clonogenic techniques; development of in vitro techniques for investigating the hematopoietic inductive microenvironment; effects of particulates on bone marrow progenitors of macrophages and on pulmonary alveolar macrophages; use of lymphoid clonogenic techniques for detection of radiation-induced injury. Mailing Add: Radiobiol Lab Univ of Calif Davis CA 95616

WILSON, FRANCES G, b Chariton, Iowa, Feb 14, 15. ZOOLOGY, BIOLOGY. Educ: Drake Univ, BS, 36, MS, 37. Prof Exp: Prof biol sci, N Park Col, 38-40; instr, Oak Park & River Forest Twp High Sch, 40-49; PROF ZOOL & BIOL, E LOS ANGELES COL, 49- Concurrent Pos: Instr, Am Dependents Schs, Far East Command, Misawa Air Base, Japan, 54-55; vis teacher, USSR, 56; write & give shows, Gardens & Zoos 'Round the World, Scope, Educ TV, Channel 7, 69-; study of Stone Age Tribes of New Guinea, Los Angeles Mus Natural Hist, 71; vis teacher, Sikkim & Bhutan, 75. Honors & Awards: Nat Freedoms Found Award, Future Teachers Am, 52. Mem: Nat Audubon Soc; Wilderness Soc; Am Soc Zoologists; Soc Develop Biol; Soc Woman Geogr. Res: Embryonic development of Peyer's Patches in selected mammals; embryology. Mailing Add: Dept of Life Sci East Los Angeles Col Los Angeles CA 90022

WILSON, FRANK B, b Detroit, Mich, Jan 8, 29; m 50; c 7. SPEECH, AUDIOLOGY. Educ: Bowling Green Univ, BS, 50; Northwestern Univ, PhD, 56. Prof Exp: Speech & hearing clinician, Cerebral Palsy Ctr, Ohio, 51-53, actg dir, 52-53; res asst, Lang Inst, Northwestern Univ, 55-57; asst prof speech, St Louis Univ, 57-59; coordr speech & hearing, Spec Dist for Educ & Training Handicapped Children, St Louis County, Mo, 59-65; dir div speech path, Dept Otolaryngol, Jewish Hosp St Louis, Mo, 66-72; DIR RES, SPEC SCH DIST ST LOUIS COUNTY, 73- Concurrent Pos: Consult, US Off Res, 63-; assoc prof, Wash Univ, 68. Mem: Fel Am Speech & Hearing Asn; Am Cleft Palate Asn. Res: Articulatory behavior of the retarded child and the efficacy of speech therapy; hearing deviation among orthopedically handicapped children; basis of nonorganic articulation disorders in children; voice disorders in school-age children. Mailing Add: Spec Sch Dist St Louis County 12110 Clayton Rd St Louis MO 63131

WILSON, FRANK CRANE, b Rome, Ga, Dec 29, 29; m 51; c 3. MEDICINE, ORTHOPEDIC SURGERY. Educ: Vanderbilt Univ, AB, 50; Med Col Ga, MD, 54; Am Bd Orthop Surg, dipl, 67. Prof Exp: Instr orthop surg, Columbia Univ, 63; from instr to assoc prof, 64-71, PROF ORTHOP SURG, SCH MED, UNIV NC, CHAPEL HILL, 71- Concurrent Pos: Markle scholar, 66-; consult, Watts Hosp, Durham, NC, 65-; chief div orthop surg, NC Mem Hosp, 67- Mem: AAAS; Am Col Surg; Am Rheumatism Asn. Res: Trauma; infections of bones and joints; rheumatoid arthritis. Mailing Add: Div Orthop Surg NC Mem Hosp Chapel Hill NC 27514

WILSON, FRANK DOUGLAS, b Salt Lake City, Utah, Dec 17, 28; m 50; c 8. PLANT GENETICS. Educ: Univ Utah, BS, 50, MS, 53; Wash State Univ, PhD(bot), 57. Prof Exp: Asst biol & genetics, Univ Utah, 51-53; asst bot, Wash State Univ, 53-56, asst agron, 56, jr animal scientist, 56-57; PLANT GENETICIST, WESTERN COTTON RES LAB, AGR RES SERV, USDA, 57- Concurrent Pos: Consult, Agron Exp Sta, Int Coop Admin, Cuba, 59- Mem: AAAS; Crop Sci Soc Am; Asn Taxon Study Trop African Flora. Res: Insect resistance in cotton; taxonomy of Hibiscus section Furcaria. Mailing Add: Western Cotton Res Lab 4135 E Broadway Phoenix AZ 85040

WILSON, FRANK JOSEPH, b Pittsburgh, Pa; c 2. ANATOMY, CELL BIOLOGY. Educ: St Vincent Col, BA, 64; Univ Pittsburgh, PhD(anat, cell biol), 69. Prof Exp: Muscular Dystrophy Asn Am, Inc fel, Univ Birmingham, 70-72; ASST PROF ANAT, RUTGERS MED SCH, COL MED & DENT NJ, 72- Res: Immunochemistry and biochemistry of the contractile proteins of muscle. Mailing Add: Dept of Anat Rutgers Med Sch Piscataway NJ 08854

WILSON, FRANK MACDONNELL, b Los Angeles, Calif, Apr 15, 25; m 45; c 4. MEDICAL ADMINISTRATION. Educ: Ore State Col, BS, 50, MS, 51; Univ Ore, DMD, 55. Prof Exp: Asst prof diag-surg, Dent Sch, Univ Ore, 66-68, assoc prof surg, 68-69, assoc prof dent & dir hosp internships, Med Sch, 69-71; PROF & DEAN-DIR, DENT & ALLIED HEALTH TECHNOLOGICS, ORE TECH INST, 71- Concurrent Pos: Adj prof, Dent Sch, Univ Ore, 71-, clinician, Postgrad Dept, 67-; mem, Am Asn Dent Schs, 67- Mem: AAAS; Am Dent Asn; Am Asn Hosp Dentists. Res: Allied health education programs; career ladder systems; college-hospital articulations. Mailing Add: Div of Allied Health Ore Tech Inst Klamath Falls OR 97601

WILSON, FREDDIE ELTON, b Lenexa, Kans, Dec 23, 37; m 61; c 1. ENDOCRINOLOGY. Educ: Univ Kans, BA, 58, MA, 60; Wash State Univ, PhD(zoophysiol), 65. Prof Exp: Instr biol, Lewis & Clark Col, 60-61; asst prof zool, 65-71, PHYSIOLOGIST, AGR EXP STA, KANS STATE UNIV, 65-, ASSOC PROF BIOL, UNIV, 71- Mem: AAAS; Am Soc Zoologists; Int Soc Neuroendocrinol. Res: Avian reproductive physiology; neuroendocrine control of annual reproductive cycles; photoperiodism. Mailing Add: Div of Biol Kans State Univ Manhattan KS 66506

WILSON, FREDERICK SUTPHEN, b Trenton, NJ, Feb 12, 27; m 50; c 3. FAMILY MEDICINE. Educ: Dickinson Col, ScB, 48; Thomas Jefferson Univ, MD, 53. Prof Exp: Dir clin invest, McNeil Labs, 69-71, dir med serv, 71-74; ASST PROF COMMUNITY MED, MED CTR, TEMPLE UNIV, 74- Concurrent Pos: Physician-in-chief & dir family pract ctr/family pract residency prog, Abington Mem Hosp, 74- Honors & Awards: AMA Physicians Recognition Award, 72. Mailing Add: Abington Mem Hosp 1200 Old York Rd Abington PA 19001

WILSON, GEOFFREY LEONARD, b London, Eng, Oct 26, 24; nat US; m 55; c 2. ACOUSTICS. Educ: Oxford Univ, BA, 45, BSc, & MA, 49. Prof Exp: Sci off, Royal Naval Sci Serv, Brit Admiralty, Clarendon Lab, Oxford Univ, 44-48, H M Underwater Detection Estab, 48-51 & Torpedo Exp Estab, 52-53; design engr, Can Westinghouse Co, Ont, 53-59; asst prof, 59-65, ASSOC PROF ENG RES, PA STATE UNIV, 65- Concurrent Pos: Vis res fel, Loughborough Univ Technol, Eng, 68-69. Mem: Acoust Soc Am; Audio Eng Soc; Inst Elec & Electronics Engrs; Brit Inst Elec Eng; Brit Inst Acoust. Res: Acoustics, especially underwater acoustics and transducer design. Mailing Add: Appl Res Lab Pa State Univ Box 30 State College PA 16801

WILSON, GEORGE DONALD, b Chatham, Ont, Jan 17, 25; nat US; m 50; c 3. BIOCHEMISTRY. Educ: Mich State Univ, BS, 49; Univ Wis, MS, 50, PhD(biochem), 53. Prof Exp: Instr, Univ Wis, 52-53; chief div food technol, Am Meat Inst Found, 53-61; tech dir, Klarer of Ky, 61-66, vpres, 66-70; DIR MFG SERV, HY-GRADE FOOD PROD CORP, 70- Concurrent Pos: Mem, Am Meat Inst Comn. Honors & Awards: Inst Food Technol Award, 60. Mem: Am Meat Sci Asn; Inst Food Technol. Res: Production, processing and distribution of fresh and processed meats. Mailing Add: PO Box 4771 Detroit MI 48219

WILSON, GEORGE PORTER, III, b Flint, Mich, Nov 10, 27; m 56; c 4. VETERINARY SURGERY, IMMUNOLOGY. Educ: Univ Ill, BS, 51; Univ Pa, VMD, 55; Ohio State Univ, MSc, 59. Prof Exp: Intern vet med, Angell Mem Animal Hosp, 55-56; from instr to assoc prof, 56-69, PROF VET SURG, COL VET MED, OHIO STATE UNIV, 69-; ASSOC PROF MICROBIOL, COL BIOL SCI, 67- Concurrent Pos: Ohio State Univ Develop Fund grant, 67-; Mark Morris Found grant, 68; Am Cancer Soc instnl grant, 68-69. Mem: AAAS; Am Vet Med Asn; Am Col Vet Surg. Res: Veterinary surgery, especially cancer research as related to immunology and chemotherapy. Mailing Add: 1935 Coffey Rd Columbus OH 43210

WILSON, GEORGE RODGER, b Commercial Point, Ohio, July 10, 23; m 46; c 4. AGRICULTURE, ANIMAL SCIENCE. Educ: Ohio State Univ, BS, 48, MS, 56, PhD(animal sci), 63. Prof Exp: County agent, Butler County, Ohio, 48-54; from asst prof to assoc prof, 54-72, PROF ANIMAL SCI, OHIO STATE UNIV, 72- Mem: Am Soc Animal Sci. Res: Animal breeding, physiology and production. Mailing Add: Dept of Animal Sci Ohio State Univ Columbus OH 43210

WILSON, GEORGE SPENCER, b Bronxville, NY, May 23, 39; m 64. ANALYTICAL CHEMISTRY. Educ: Princeton Univ, BA, 61; Univ Ill, MS, 63, PhD, 65. Prof Exp: NIH fel, Univ Ill, 65-66, instr, 66-67; asst prof, 67-72, ASSOC PROF CHEM, UNIV ARIZ, 72- Res: Analytical applications of biochemical reactions; electrochemical synthesis of unusual or unstable products; oxidation-reduction and acid-base reactions in nonaqueous solvents; development of computer-controlled instrumentation. Mailing Add: Dept of Chem Univ of Ariz Tucson AZ 85721

WILSON, GERALD EARL, plasma physics, nuclear engineering, see 12th edition

WILSON, GLENN RHODES, b Altamont, Ill, Nov 10, 21. ORGANIC CHEMISTRY, BIOCHEMISTRY. Educ: Univ Ill, BS, 43; Univ Iowa, PhD(org chem), 51. Prof Exp: Res chemist, Res Labs, Ethyl Corp, Mich, 51-54 & Nat Cylinder Gas Co, Ill, 54-55; sr res engr, Sci Lab, Ford Motor Co, Mich, 54-59; scientist, Monsanto Res Corp, Mass, 59-65 & Dayton Lab, Ohio, 65-72, SCIENTIST, MONSANTO TEXTILE CO, 72- Mem: AAAS; Am Chem Soc; NY Acad Sci. Res: Fatty alcohols; lubricants; organosilicon and polymer chemistry. Mailing Add: 4311 Bayou Blvd Pensacola FL 32503

WILSON, GORDON, JR, b Bowling Green, Ky, Sept 13, 25; m 53; c 2. ORGANIC CHEMISTRY. Educ: Western Ky State Col, BS, 47; Univ Ky, MS, 49; Purdue Univ, PhD(org chem), 58. Prof Exp: Instr chem, Univ Minn, Duluth, 49-54; res chemist, Dow Chem Co, 57-61; assoc prof, 61-65, PROF CHEM & HEAD DEPT, WESTERN KY UNIV, 65- Mem: AAAS; Am Chem Soc. Res: Organic fluorine compounds; polymer chemistry, especially polyelectrolytes. Mailing Add: Dept of Chem Western Ky Univ Bowling Green KY 42101

WILSON, GRAEME STEWART, b Edinburgh, Scotland. VISUAL PHYSIOLOGY. Educ: Manchester Univ, MSc, 65; Univ Calif, Berkeley, PhD(physiol optics), 71. Prof Exp: Biologist, Brit Antarctic Surv, 65-67; ASST PROF PHYSIOL OPTICS, MED CTR, UNIV ALA, BIRMINGHAM, 72- Res: Corneal physiology; tear drainage; vision of aquatic mammals; pulfrichology. Mailing Add: Sch of Optom Univ of Ala Med Ctr Birmingham AL 35294

WILSON, GRANT IVINS, b Salt Lake City, Utah, Oct 29, 21; m 57; c 4. ANIMAL PARASITOLOGY. Educ: Univ Utah, BS, 48, MS, 55; Univ Md, PhD(parasitol), 64. Prof Exp: Fish culturist, US Fish & Wildlife Serv, 51-53; res parasitologist, Agr Res Serv, USDA, 53-63; asst prof zool, Univ Mont, 63-65; sr res parasitologist, 65-73, ZOOLOGIST, AGR RES SERV, USDA, 73- Res: Life history, pathogenicity and immunological studies on the helminth parasites of ruminants; ectoparasites of livestock. Mailing Add: Agr Res Serv USDA Box 705 Albuquerque NM 87103

WILSON, GUSTAVUS EDWIN, JR, b Philadelphia, Pa, Oct 6, 39; m 61; c 4. BIO-ORGANIC CHEMISTRY. Educ: Mass Inst Technol, SB, 61; Univ Ill, Urbana, PhD(chem), 64. Prof Exp: From instr to assoc prof, 64-75, PROF CHEM, POLYTECH INST NEW YORK, 75- Concurrent Pos: Petrol Res Fund res grants, 64-68; adj assoc prof, Rockefeller Univ, 72-; vis assoc prof, Univ Pa, 74-75; res scholar, Am Cancer Soc, 74-75. Mem: AAAS; Am Chem Soc; The Chem Soc; fel NY Acad Sci. Res: Mechanistic and synthetic studies in organic sulfur chemistry; application of organic chemistry to biological problems; magnetic resonance in chemistry and biochemistry. Mailing Add: Dept of Chem Polytech Inst of New York Brooklyn NY 11201

WILSON, HAROLD ALBERT, b Tilton, Ill, Oct 10, 05; m 37; c 2. MICROBIOLOGY. Educ: La State Univ, BS, 32, MS, 33; Iowa State Col, PhD(soil microbiol), 37. Prof Exp: Instr agr, Panhandle Agr & Mech Col, 35-36; instr & asst, Iowa State Col, 37-38; prof agron, Southwestern La Inst, 38-44; soil conservationist, Soil Conserv Serv, USDA, 44-47; assoc prof bact, WVa Univ & assoc bacteriologist, Exp Sta, 47-57, prof & bacteriologist, 57-72, EMER PROF BACT, WVA UNIV, 72- Mem: Am Soc Agron; Am Soc Microbiol; Soil Sci Soc Am. Res: Microbiology of sanitary landfills and sewage decomposition in acid mine water. Mailing Add: 1297 Fairlawns Morgantown WV 26505

WILSON, HAROLD FREDERICK, b Columbiana, Ohio, Aug 15, 22; m 49; c 4. ORGANIC CHEMISTRY. Educ: Oberlin Col, AB, 47; Univ Rochester, PhD(chem), 50. Prof Exp: Res chemist, 50-57, lab head, 57-63, res supvr, 63-68, from asst dir res to dir res, 68-74, V PRES, ROHM AND HAAS CO, 74- Res: Insecticides; herbicides; growth regulators. Mailing Add: 924 Rydal Rd Jenkintown PA 19046

WILSON, HARRY DAVID BRUCE, b Winnipeg, Man, Nov 10, 16; m 41; c 3. ECONOMIC GEOLOGY. Educ: Univ Man, BSc, 36; Calif Inst Technol, MS, 39, PhD, 42. Prof Exp: Res & explor geologist, Int Nickel Co, Can, 41-47; asst prof geol, Univ Man, 47-49; geologist, Africa & Europe, 49-51; assoc prof, 51-57, head dept, 65-72, PROF GEOL, UNIV MAN, 57- Concurrent Pos: Mem, Nat Res Coun Can, 69-72; consult, Falconbridge Nickel Mines Ltd & Selco Mining Co Ltd. Mem: Soc Econ Geologists (pres elect, 75); Geol Soc Am; Geol Asn Can (pres, 65-66); Can Inst Mining & Metall; Royal Soc Can. Res: Geology and geochemistry of ore deposits; structure and origin of continental crust. Mailing Add: Dept of Earth Sci Fac Sci Univ of Man Winnipeg Man Can

WILSON, HARWELL, b Lincoln, Ala, May 23, 08; m 41; c 3. SURGERY. Educ: Vanderbilt Univ, AB, 28, MD, 32. Prof Exp: Intern, Univ Chicago, 32-33, asst resident surg, 34-37, resident surg path, 37-38, chief resident & instr surg, 38-39; from instr to asst prof, 39-48, PROF SURG, UNIV TENN, MEMPHIS, 48- Concurrent Pos: Res fel, Univ Chicago, 33-34; Elkin lectr, Emory Univ, 61; Phemister lectr, Univ Chicago, 62; Pilcher lectr, Vanderbilt Univ, 62; vis prof & Oestricher lectr, NY Univ, 74; vis prof & Crawford Long lectr, Atlanta Univ, 74; vis prof & Edwards mem lectr, Univ Md, 75; vis prof & guest lectr, Univ Ala, 75. Honors & Awards: Distinguished Serv Award, Tenn State Med Asn, 75; Distinguished Serv Award, Univ Chicago, Med Alumni Asn, 52. Mem: AMA; Am Surg Asn; Soc Surg Alimentary Tract (past pres); Int Soc Surg; Am Col Surg (treas, 62-). Res: Research surgical trauma; vascular surgery; clinical research regarding problems of gastrointestinal tract and neoplasms; production of surgical teaching motion pictures. Mailing Add: 951 Court Memphis TN 38103

WILSON, HENRY E, b Marshalltown, Iowa, Aug 24, 10; m 46; c 3. INTERNAL MEDICINE, HEMATOLOGY. Educ: Univ Ill, AB, 32; McGill Univ, MD & CM, 37. Prof Exp: Assoc prof med, Ohio State Univ, 44-46 & Northwestern Univ, 46-54; PROF MED, OHIO STATE UNIV, 54- Mem: Am Col Physicians; Am Fedn Clin Res; Am Soc Hemat; Int Soc Hemat. Res: Oncology; immunology and immunotherapy of cancer; chemotherapy of leukemia, lymphoma and solid tumors. Mailing Add: Dept of Med Ohio State Univ Columbus OH 43210

WILSON, HENRY HAMILTON, nuclear physics, astronomy, see 12th edition

WILSON, HENRY R, b Webbville, Ky, Mar 6, 36; m 59; c 2. REPRODUCTIVE PHYSIOLOGY, ENVIRONMENTAL PHYSIOLOGY. Educ: Univ Ky, BS, 57, MS, 59; Univ Md, PhD(poultry physiol), 62. Prof Exp: Asst prof poultry, 62-67, asst poultry physiologist, 62-67, assoc prof & assoc poultry physiologist, 67-74, PROF & POULTRY PHYSIOLOGIST, UNIV FLA, 74- Mem: Soc Exp Biol & Med; Soc Study Reproduction; Wildlife Soc; Poultry Sci Asn; World Poultry Sci Asn. Res: Reproduction in male chickens; heat tolerance in chickens; delaying sexual maturity in chickens; management techniques, fertility and hatchability; game birds. Mailing Add: Dept of Poultry Sci Univ of Fla Gainesville FL 32611

WILSON, HOWARD LE ROY, b Salem, Ore, Dec 8, 32; m 60; c 3. MATHEMATICS. Educ: Willamette Univ, BA, 54; Univ Ill, MS, 60, PhD(educ), 66. Prof Exp: Teacher high sch, Ore, 55-56; from instr to asst prof math, Eastern Ore Col, 56-64; asst prof, 64-68, ASSOC PROF MATH & SCI EDUC, ORE STATE UNIV, 68- Concurrent Pos: Expert in math, Field Sta, Papua, New Guinea, UN Educ Sci & Cult Orgn, 71-73. Res: Mathematics education and teacher training. Mailing Add: Dept of Sci Educ Ore State Univ Corvallis OR 97331

WILSON, HOWELL KENNETH, b Savannah, Ga, Aug 28, 37; m 66. MATHEMATICS. Educ: Ga Inst Technol, BS, 60; Univ Minn, PhD(math), 64. Prof Exp: From asst prof to assoc prof math, Ga Inst Technol, 64-69; assoc prof, 69-73, PROF MATH, SOUTHERN ILL UNIV, EDWARDSVILLE, 73- Mem: Am Math Soc; Math Asn Am. Res: Ordinary differential equations. Mailing Add: Dept of Math Studies Southern Ill Univ Edwardsville IL 62025

WILSON, IRWIN B, b Yonkers, NY, May 8, 21; m 52; c 2. BIOCHEMISTRY. Educ: City Col New York, BS, 41; Columbia Univ, AM, 47, PhD(phys chem), 48. Prof Exp: Jr chemist, Picatinny Arsenal, NJ, 42; res assoc, Columbia Univ, 42-45; chemist, Union Carbide & Carbon Corp, 45; instr chem, City Col New York, 46-49; assoc, Col Physicians & Surgeons, Columbia Univ, 48-49, from asst prof to prof biochem, 49-66; PROF CHEM, UNIV COLO, BOULDER, 66- Honors & Awards: Asn Res Nerv & Ment Dis Award, 58. Mem: Am Chem Soc; Am Soc Biol Chem; Am Acad Neurol. Res: Enzymology; protein chemistry; nerve function; peptide hormones. Mailing Add: Dept of Chem Univ of Colo Boulder CO 80302

WILSON, JACK BELMONT, b Morgantown, WVa, Dec 1, 21; m 43; c 6. PLANT PATHOLOGY. Educ: WVa Univ, BS, 53, MS, 54, PhD(plant path), 57. Prof Exp: Instr plant path, Univ Md, 56-57, asst prof, 57-62; res plant pathologist, Potato Handling Res Ctr, USDA, Maine, 62-67; assoc prof plant path & exten plant pathologist & entomologist, WVa Univ, 67-69; asst to br chief, Hort Crops Res Br, Mkt Qual Res Div, 69-73, ASST AREA RES DIR, US PLANT & SOIL NUTRITION LAB, CORNELL UNIV, AGR RES SERV, USDA, 73- Concurrent Pos: Assoc prof, Univ Maine, 66-67. Mem: Am Phytopath Soc; Potato Asn Am; Europ Asn Potato Res. Res: Diseases of potatoes and ornamental plants; plant disease and insect diagnosis. Mailing Add: US Plant & Soil Nutrit Lab Cornell Univ ARS USDA Tower Rd Ithaca NY 14850

WILSON, JACK CHARLES, b Waterloo, Iowa, Dec 17, 28; m 48; c 4. ALGEBRA. Educ: Iowa State Teachers Col, BA, 51; Univ Iowa, MS, 54; Case Western Reserve Univ, PhD(math), 60. Prof Exp: Instr math, Cent Col, Iowa, 53-56; asst prof, Fenn Col, 56-59; assoc prof, Cent Col, Iowa, 59-65 & Earlham Col, 65-70; PROF MATH, UNIV NC, ASHEVILLE, 70- Mem: Am Math Soc; Math Asn Am. Res: Pure mathemtics; functions on algebras. Mailing Add: Dept of Math Univ of NC Asheville NC 28801

WILSON, JACK HAROLD, b Toronto, Ont, May 2, 43; m 65; c 1. BIOCHEMISTRY, MICROBIOLOGY. Educ: McGill Univ, BSc, 64; McMaster Univ, PhD(molecular biol), 68. Prof Exp: Res scientist, 68-73, RES BIOCHEMIST, RES LABS, UNIROYAL LTD, 73- Concurrent Pos: Consult, Int Bank Reconstruct & Develop, 72. Mem: Soc Econ Bot; Inst Food Technol; Am Phytopath Soc; NY Acad Sci. Res: Antimicrobial agents; plant pathology; mode of agricultural chemical action; plant protein; novel protein sources; plant disease losses. Mailing Add: Res Labs Uniroyal Ltd 120 Huron St Guelph ON Can

WILSON, JACK LOWERY, b Looxahoma, Miss, June 24, 43; m 66; c 3. ANATOMY. Educ: Univ Southern Miss, BS, 64; Univ Miss, MS, 67, PhD(anat), 68. Prof Exp: From instr to asst prof, 68-74, ASSOC PROF ANAT, MED UNITS, UNIV TENN, MEMPHIS, 74- Mem: Am Asn Anat. Res: Relations of age and sex hormones to the dietary induction, high fat and low protein components of cardiac and hepatic lesions in mice, also study of the fine structure of these lesions; cerebrovascular spasm. Mailing Add: Dept of Anat Univ of Tenn Med Units Memphis TN 38103

WILSON, JACK MARTIN, b Camp Atterbury, Ind, June 29, 45; m 69; c 2. CHEMICAL PHYSICS, BIOPHYSICS. Educ: Thiel Col, AB, 67; Kent State Univ, MA, 70, PhD(physics), 72. Prof Exp: Res asst physics liquid crystals, Liquid Crystal Inst, Kent State Univ, 69, instr physics, Columbiana Co Br, 69-70, res asst physics liquid crystals, Liquid Crystal Inst, 70-72; ASST PROF PHYSICS, SAM HOUSTON STATE UNIV, 72- Honors & Awards: Sigma Xi Res Award, 74. Mem: Am Inst Physics; Sigma Xi; AAAS. Res: Studying the structure of Sickle Cell hemoglobin; digital filtering of experimental data and Mössbauer spectroscopy of liquid crystals. Mailing Add: Dept of Physics Sam Houston State Univ Hunstville TX 77340

WILSON, JAMES, JR, b Glasgow, Scotland, Sept 7, 18; US citizen; m 41; c 3. ORGANIC CHEMISTRY. Educ: NY Univ, BA, 43; Univ Rochester, EdM(guid), 52. Prof Exp: Prod chemist, Merck & Co, 36-43; anal res chemist, Standard Oil Co, 43-46; PROF ORG CHEM, ROCHESTER INST TECHNOL, 46- Mem: Am Chem Soc. Res: Qualitative organic analysis reagents. Mailing Add: Dept of Chem Rochester Inst of Technol One Lomb Memorial Dr Rochester NY 14623

WILSON, JAMES ALBERT, b Boston, Mass, Jan 28, 29. PHYSIOLOGY, BIOCHEMISTRY. Educ: Northeastern Univ, BS, 53; Univ Mich, MS, 55, PhD(zool), 59. Prof Exp: Res assoc zool, Univ Mich, 58-59; asst prof physiol, 59-72, ASSOC PROF PHYSIOL, OHIO UNIV, 72- Mem: AAAS; Am Inst Biol Sci; Am Soc Zoologists. Res: Pressure-temperature pH effects on the activity of adenosine-triphosphatases from rabbit tissues; contractility of glycerol-extracted muscle fibers and related enzyme activity; factors affecting muscle relaxation. Mailing Add: Dept of Zool & Microbiol Ohio Univ Athens OH 45701

WILSON, JAMES ALEXANDER, b Walters, Okla, Nov 19, 30; m 54; c 4. PLANT BREEDING. Educ: Okla State Univ, BS, 52, MS, 54; Tex A&M Univ, PhD(plant breeding), 58. Prof Exp: Instr agron, Tex A&M Univ, 54-57; asst prof, Ft Hays Exp Sta, Kans State Univ, 57-61; RES AGRONOMIST, DEKALB AGRESEARCH INC, 61- Mem: Am Asn Cereal Chem; Am Soc Agron. Res: Wheat breeding; genetics; cytology and management. Mailing Add: DeKalb AgResearch Inc 1831 Woodrow Ave Wichita KS 67203

WILSON, JAMES BLAKE, b Albion, Mich, Feb 9, 24; m 49; c 4. APPLIED MATHEMATICS. Educ: Univ Fla, BS, 48, PhD(appl math), 57; Cornell Univ, MS, 51. Prof Exp: Instr mech, Cornell Univ, 50; instr, Dept Ord, US Mil Acad, 51-54; asst math, Univ Fla, 54-56; asst prof, 57-63, ASSOC PROF MATH, NC STATE UNIV, 63- Mem: Am Math Soc; Math Asn Am. Res: Elasticity; mechanics; numerical analysis. Mailing Add: 1311 Greenwood Circle Cary NC 27511

WILSON, JAMES BRUCE, b Washington, DC, Sept 22, 29. COMPUTER SCIENCE, BIOMEDICAL ENGINEERING. Educ: George Washington Univ, BEE, 58. Prof Exp: From res asst to res assoc, George Washington Univ, 58-60; SR RES SCIENTIST, NAT BIOMED RES FOUND, 60- Mem: Pattern Recognition Soc (secy, 66-). Res: Systems and circuit design of biomedical scanner; computer programming for pattern recognition; switching circuit theory and logical design; Boolean algebra; information retrieval; research-planning methodology. Mailing Add: Nat Biomed Res Found Georgetown Univ Med Ctr Washington DC 20007

WILSON, JAMES DENNIS, b The Dalles, Ore, Feb 28, 40; m 62; c 2. ORGANIC CHEMISTRY. Educ: Harvard Univ, AB, 62; Univ Wash, PhD(chem), 66. Prof Exp: SR RES CHEMIST, CENT RES DEPT, MONSANTO CO, 66- Mem: Am Chem Soc. Res: Correlation of molecular structure with physical properties, especially solid-state electrical, magnetic and olfactory properties; synthesis of unsaturated nitrogen- and sulfur-compounds. Mailing Add: Cent Res Dept Monsanto Co 800 N Lindbergh Blvd St Louis MO 63166

WILSON, JAMES FRANKLIN, b Christopher, Ill, Oct 27, 20; m 43; c 2. MICROBIOLOGY, GENETICS. Educ: Southern Ill Univ, BS, 44; Iowa State Col, MS, 46; Stanford Univ, PhD, 59. Prof Exp: Instr biol, Hartnell Col, 46-64; PROF BIOL, UNIV NC, GREENSBORO, 64- Mem: AAAS; Genetics Soc Am. Res: Application of microsurgical techniques to the study of heterocaryosis and cytoplasmic heredity in Neurospora crassa. Mailing Add: Dept of Biol Univ of NC Greensboro NC 27412

WILSON, JAMES GRAVES, b Clarksdale, Miss, Apr 2, 15; m 41; c 3. EMBRYOLOGY, TERATOLOGY. Educ: Miss Col, BA, 36; Univ Richmond, MA, 38; Yale Univ, PhD(anat), 42. Hon Degrees: DSc, Med Col Wis, 75. Prof Exp: Asst anat & primate biol, Sch Med, Yale Univ, 39-42; from instr to asst prof anat, Sch Med & Dent, Univ Rochester, 42-50; from assoc prof to prof, Col Med, Univ Cincinnati, 50-55; prof & head dept, Col Med, Univ Fla, 55-66; PROF RES PEDIAT & ANAT, COL MED, UNIV CINCINNATI & DIV HEAD PATH EMBRYOL, INST DEVELOP RES, CHILDREN'S HOSP RES FOUND, 66- Concurrent Pos: Mem comt postdoctoral fels med sci, Nat Res Coun, 57-61; mem develop biol training comt, NIH, 59-62, human embryol & develop study sect, 62-66 & anat sci training comt, 66-; mem fac, Teratol Workshop, Univ Fla, 64, Univ Calif, 65, Univ Colo, 66 & int workshops, Copenhagen Univ, 66 & Kyoto Univ, 69; mem & chmn sci group testing drugs teratogenicity, WHO, 66; mem adv comt protocols safety eval, Food & Drug Admin, 67-; mem toxicol adv comt, Food & Drug Admin, 75-; mem sci adv bd, Nat Ctr Toxicol Res, 73-; mem Basil O'Conner Starter Grant Selection Comt, NFS. Mem: AAAS; Teratology Soc; Environ Mutagen Soc; Europ Teratology Soc; Am Asn Anat. Res: Experimental and human teratology, especially mechanisms of drug teratogenicity; use of non-human primates in teratological studies; embryology of malformations; experimental production of malformations with vitamin A deficiency; mechanisms of teratogenesis; x-rays; anemia. Mailing Add: Children's Hosp Res Found Elland Ave & Bethesda Cincinnati OH 45229

WILSON, JAMES LARRY, b Jackson, Tenn, Feb 9, 42; m 73. FISHERIES MANAGEMENT. Educ: Union Univ, BS, 64; Univ Fla, MS, 67; Univ Tenn, PhD(zool), 70. Prof Exp: Asst prof, 70-75, ASSOC PROF FISHERIES, DEPT AGR BIOL, UNIV TENN INST AGR, 75- Mem: Am Fisheries Soc. Res: Culture and management techniques associated with the improvement of sport fishing in farm ponds and lakes; maximizing production of commercially raised fishes. Mailing Add: Dept of Agr Biol PO Box 1071 Univ Tenn Inst Agr Knoxville TN 37901

WILSON, JAMES LEE, b Waxahachie, Tex, Dec 1, 20; m 44; c 3. GEOLOGY. Educ: Univ Tex, BA, 42, MA, 44; Yale Univ, PhD(geol), 49. Prof Exp: Jr geologist, Carter Oil Co, 43-44; assoc prof geol, Univ Tex, 49-52; res geologist, Shell Develop Co & Shell Int Res, 52-66; WIESS PROF GEOL, RICE UNIV, 66- Mem: Fel Geol Soc Am; Paleont Soc; Soc Econ Paleont & Mineral (pres, 75-76); Am Asn Petrol Geol. Res: Cambrian paleontology; Paleozoic biostratigraphy; carbonate petrography and petrology; sedimentology of carbonate strata. Mailing Add: 5005 Holt Bellaire TX 77401

WILSON, JAMES LESTER, b Nashville, Tenn, July 18, 25; m 48; c 2. ZOOLOGY. Educ: George Peabody Col, BS, 51, MA, 52; Vanderbilt Univ, PhD, 59. Prof Exp: Instr biol & phys sci, Ark State Col, 52-54; asst, Vanderbilt Univ, 55-56; from assoc prof to prof biol & chmn div sci & math, Belmont Col, 56-67; assoc prof, 67-70, PROF ZOOL, UNIV TENN, NASHVILLE, 70- Honors & Awards: Sullivan Award, George Peabody Col, 51. Mem: Fel AAAS; Acarological Soc Am. Res: Taxonomy; life history and ecology of water mites; hydracarology. Mailing Add: Dept of Zool Univ of Tenn Nashville TN 37203

WILSON, JAMES NEWTON, b Erie, Pa, Aug 9, 15; m 35; c 3. GEOGRAPHY. Educ: Edinboro State Col, BS, 38; Columbia Univ, MA, 42, EdD(geog), 49. Prof Exp: Instr geog, Teachers Col, Columbia Univ, 46-49, Hunter Col, 49 & Towson State Col, 49-50; from asst prof to assoc prof, 50-67, chmn dept, 62-64, dir Nat Defense Educ Act Inst Geog, 66, chmn dept geog, 71-72, PROF GEOG, CALIF STATE UNIV, LONG BEACH, 67- Concurrent Pos: State coordr, Calif Coun Geogr, 65- Mem: Nat Coun Geog Ed; Asn Am Geog. Res: Political geography; geography of western Europe and California; geography in public schools. Mailing Add: Dept of Geog Calif State Univ Long Beach CA 90840

WILSON, JAMES R, b Berkeley, Calif, Oct 21, 22; m 49; c 5. THEORETICAL ASTROPHYSICS. Educ: Univ Calif, BS, 43, PhD(physics), 52. Prof Exp: Physicist, Sandia Corp, 52-53; PHYSICIST, LAWRENCE LIVERMORE LAB, UNIV CALIF, 53- Mem: Fel Am Phys Soc; AAAS. Res: Astrophysics, structure of rotating stars, relativity; design of nuclear explosives. Mailing Add: 737 S M St Livermore CA 94550

WILSON, JAMES RUSSELL, b Pittsburgh, Pa, Jan 10, 33. BEHAVIORAL BIOLOGY. Educ: Univ Calif, Berkeley, AB, 59, PhD(psychol), 68. Prof Exp: Res psychologist, Univ Calif, Berkeley, 63-66; res assoc psychol, Univ Colo, Boulder, 66-68; instr psychol, Univ Calif, Berkeley, 69; asst prof, 69-74, ASSOC PROF PSYCHOL & BIOL, UNIV COLO, BOULDER, 74- Concurrent Pos: Assoc researcher, Univ Hawaii, 74- Mem: Genetics Soc Am; Behav Genetics Asn. Res: Genetic analysis of behavioral phenotypes, including reading disability, aggression, alcohol use and cognitive abilities. Mailing Add: Inst for Behav Genetics Univ of Colo Boulder CO 80302

WILSON, JAMES STEWART, b Mich, June 3, 32; m 54; c 3. BOTANY. Educ: Kalamazoo Col, BA, 54; Univ Mich, MS, 57, PhD(bot), 61. Prof Exp: From asst prof to assoc prof, 59-70, PROF BIOL, EMPORIA KANS STATE COL, 70- Mem: AAAS; Am Soc Plant Taxon; Bot Soc Am; Int Asn Plant Taxon. Res: Biosystematics of flowering plants with emphasis on population ecology. Mailing Add: Dept of Biol Emporia Kans State Col Emporia KS 66801

WILSON, JAMES TINLEY, b Claremont, Calif, Nov 13, 14; m 43; c 1. GEOPHYSICS. Educ: Univ Calif, AB, 35, PhD(seismol), 39. Prof Exp: Teaching asst geol, Univ Calif, 35-39, Harvard Univ, 39-40 & Univ Mich, 40-43; from instr to assoc prof, 45-55, chmn dept, 56-61, assoc dir, 60-62, PROF GEOL & MINERAL, UNIV MICH, ANN ARBOR, 55-, DIR INST SCI & TECHNOL, 62- Concurrent Pos: Chmn comt on seismol, Nat Acad Sci-Nat Res Coun, 60-65, chmn comt on remote sensing of environ, 64-68; consult, Atomic Energy Comn, 67-75 & Off Sci & Technol, 70; trustee, Univs Space Res Asn, 70-75; mem panel on strong-motion seismol, Nat Acad Sci-Nat Res Coun, 72-; consult, Nuclear Regulatory Comn, 75- & Energy Res & Develop Admin, 75-; mem panel on gas reserve estimation, Nat Acad Sci-BMR, 75- Mem: Fel Am Geophys Union (pres seismol sect, 74-76); Earthquake Eng Res Inst; Geol Soc Am; Royal Astron Soc; Am Mining, Metall & Petrol Eng. Res: Structure of the earth; local earthquakes; elastic waves; seismicity. Mailing Add: Inst of Sci & Technol Univ of Mich Ann Arbor MI 48105

WILSON, JAMES WALTER, b Caneyville, Ky, Aug 1, 35; m 63; c 2. PATHOLOGY. Educ: Murray State Univ, AB, 57; Western Ky Univ, BS, 58; Univ Ky, MS, 59, PhD(anat), 65; Duke Univ, MD, 67. Prof Exp: Teaching asst anat, Univ Ky, 63-65; instr, 65-68, assoc path, 68-70, from asst prof to assoc prof path, Med Ctr, Duke Univ, 70-75; PROF PATH, HEALTH SCI CTR, UNIV TENN, 75- Concurrent Pos: NIH res training fel, Med Ctr, Duke Univ, 68-71; vis prof, Rutgers Med Sch, 66, Med Univ SC, 68 & Col Med, Univ Fla, 72. Honors & Awards: Trent Prize, Duke Univ, 67; Ruth Gray Mem Lectr, Northwestern Univ, Evanston & Ruth Gray Award, Evanston Hosp, 72. Mem: Am Asn Anat; Am Soc Clin Path; Am Soc Med Technol; Soc Automotive Eng; Biol Photog Asn. Res: Pulmonary vascular microcirculation; lung shock; steroids in shock; mechanical ventilators; pharmacologic respiratory stimulants; congenital heart disease; cellular morphology and physiology; medical history; pulmonary and cardiovascular pathology. Mailing Add: Univ of Tenn Ctr for Health Sci 858 Madison Ave Box 51 Memphis TN 38163

WILSON, JAMES WILLIAM, b Rice, Va, Oct 21, 19; m 50; c 4. MEDICINAL CHEMISTRY. Educ: Hampden-Sydney Col, BS, 41; Univ Va, MS, 44, PhD(org chem), 46. Prof Exp: Asst chem, Univ Va, 41-43; res chemist, 46-53, asst sect head org chem, 53-54, head med chem sect, 54-66, staff dir, 66-67, ASSOC DIR CHEM, SMITH KLINE & FRENCH LABS, 67- Mem: AAAS; Am Chem Soc; NY Acad Sci; Soc Chem Indust. Res: Medicinal chemistry; analgesics; cardiovascular drugs; psychopharmacological, diuretic and anti-inflammatory agents. Mailing Add: 15 Kinterra Rd Wayne PA 19087

WILSON, JAMES WOODROW, b Yakima, Wash, Dec 5, 17; m 42; c 1. CHEMISTRY. Educ: Wash State Univ, BS, 39; Univ Rochester, PhD(org chem), 42. Prof Exp: Res chemist, B F Goodrich Co, 42-44; res chemist, 44-49, group leader, 49-55, res supvr, 55-65, mgr ref res, 65-69, SR PLANNER, CORP PLANNING, UNION OIL CO, CALIF, 69- Mem: Am Chem Soc; Soc Petrol Eng. Res: Organic rearrangements; synthetic rubber; lubricating greases; petroleum reservoir behavior; catalysts and processes for petroleum refining. Mailing Add: Union Oil Co of Calif Box 7600 Los Angeles CA 90051

WILSON, JANE AUSTELL, b Gaffney, SC, June 10, 40; m 60; c 1. ZOOLOGY, PHYSIOLOGY. Educ: Limestone Col, BS, 62; MS, Clemson Univ, 66, PhD(zool), 68. Prof Exp: Teacher high sch, SC, 62-64; res asst zool, Clemson Univ, 64-66; ASST PROF ZOOL, AUBURN UNIV, 68- Mem: Am Inst Biol Sci. Res: Histology and physiology of reproductive structure of housefly. Mailing Add: Dept of Zool & Entom Auburn Univ Auburn AL 36830

WILSON, JEAN DONALD, b Wellington, Tex, Aug 26, 32. INTERNAL MEDICINE. Educ: Univ Tex, BA, 51, MD, 55; Am Bd Internal Med, dipl, 64. Prof Exp: From intern to asst resident internal med, Parkland Mem Hosp, Dallas, 55-58; clin assoc clin biochem, Nat Heart Inst, 58-60; from instr to assoc prof internal med, 60-68, PROF INTERNAL MED, UNIV TEX HEALTH SCI CTR DALLAS, 68- Concurrent Pos: Estab investr, Am Heart Asn, 60-65; mem metab study sect, NIH, 64-68. Mem: Am Soc Clin Invest; Soc Exp Biol & Med; Am Soc Biol Chem; Endocrine Soc; Asn Am Physicians. Res: Intermediary metabolism; regulation of cholesterol metabolism in the intact animal; mechanism of action of steroid hormones. Mailing Add: Dept of Internal Med Univ of Tex Health Sci Ctr Dallas TX 75235

WILSON, JERRY DICK, b Coshocton, Ohio, May 6, 37; m 63; c 2. SCIENCE WRITING, PHYSICS. Educ: Ohio Univ, BS, 62, PhD(physics), 70; Union Col, MS, 65. Prof Exp: Mat behav physicist dielectrics, Gen Elec Co, 63-66; asst prof physics, Ohio Univ, 70-75; ASST PROF PHYSICS, LANDER COL, 75- Concurrent Pos: Fel, Ohio Acad Sci, 72. Mem: Am Asn Physics Teachers; Nat Sci Teachers Asn; Am Med Technologists. Res: Science writing and science teaching, particularly for non-science students. Mailing Add: Dept of Sci Lander Col Greenwood SC 29646

WILSON, JERRY LEE, b Heavener, Okla, Jan 30, 38; m 61; c 2. BIOCHEMISTRY. Educ: Okla State Univ, BS, 61; Univ Okla, PhD(chem), 67. Prof Exp: USPHS res fel, Univ Calif, Davis, 67-69; asst prof, 69-74, ASSOC PROF CHEM, CALIF STATE UNIV, SACRAMENTO, 74- Mem: Am Chem Soc. Res: Plant biochemistry; enzymology; protein chemistry. Mailing Add: Dept of Chem Calif State Univ Sacramento CA 95819

WILSON, JIM DEGNAN, wood chemistry, see 12th edition

WILSON, JOE BRANSFORD, b Dallas, Tex, June 29, 14; m 44; c 2. BACTERIOLOGY. Educ: Univ Tex, BA, 39; Univ Wis, MS, 41, PhD(bact), 47; Am Bd Microbiol, dipl. Prof Exp: Instr bact, Univ Tex, 39; asst, 39-42, from instr to assoc prof, 46-55, assoc dean grad sch, 65-69, chmn dept, 68-73, PROF BACT, UNIV WIS-MADISON, 55- Concurrent Pos: Mem tech adv panel, Off Asst Secy Defense, 52-63. Mem: Fel AAAS; Am Soc Microbiol; Soc Exp Biol & Med; Am Asn Immunol; fel Am Acad Microbiol. Res: Metabolism and pathogenesis of Brucella, Cocci, Vibrio and Leptospira. Mailing Add: Dept of Bacteriol Univ of Wis Madison WI 53706

WILSON, JOHN ANDREW, b Lawrence, Mass, Nov 3, 14; m 38; c 3. GEOLOGY. Educ: Univ Mich, BA, 37, PhD(geol), 41. Prof Exp: Instr geol, Univ Idaho, 40-46; from asst prof to assoc prof, 46-60, PROF GEOL, UNIV TEX, AUSTIN, 60- Concurrent Pos: Am Philos Soc grant, 58. Honors & Awards: Award, Soc Econ Paleont & Mineral, 59. Mem: AAAS; fel Geol Soc Am; Soc Vert Paleont (secy-treas, 49-51, pres, 52); Soc Study Evolution; Paleont Soc. Res: Vertebrate paleontology; stratigraphy. Mailing Add: Dept of Geol Univ of Tex Austin TX 78712

WILSON, JOHN ANTHONY, b Halifax, Eng, July 16, 38; m 63; c 2. SOLID STATE SCIENCE. Educ: Cambridge Univ, BA, 61, MA, 64, PhD(chem physics), 68. Prof Exp: Acad solid state res, Cambridge Univ, 68-72; MEM STAFF, BELL TEL LABS, 72- Concurrent Pos: Turner & Newall res fel, Cambridge Univ, 69-72. Mem: Am Phys Soc; Am Chem Soc. Res: Solid state chemico-physics of transition metal compounds, particularly layer compounds; charge density wave, excitonic insulator and Mott transition materials pioneered. Mailing Add: Rm 1E-304 Bell Tel Labs Murray Hill NJ 07974

WILSON, JOHN CHARLES, b Yonkers, NY, Sept 3, 43; m 64; c 2. ORGANIC POLYMER CHEMISTRY. Educ: Union Col, BS, 65; Rochester Inst Technol, MS, 69. Prof Exp: SR RES CHEMIST POLYMER CHEM, EASTMAN KODAK CO, 75- Mem: Am Chem Soc. Res: Organic synthesis directed toward polymer structure-property relationships. Mailing Add: Res Labs Eastman Kodak Co 1669 Lake Ave Rochester NY 14650

WILSON, JOHN CLELAND, b Galt, Ont, June 8, 35; m 64; c 3. COMPUTER SCIENCE. Educ: Univ Toronto, BASc, 58; Univ Waterloo, MSc, 62, PhD(math), 66. Prof Exp: Analyst, KCS Data Control Ltd, Ont, 58-62; asst prof math, Univ Waterloo, 66-70, assoc dir comput ctr, 68-70; ASSOC PROF COMPUT SCI & DIR COMPUT CTR, UNIV TORONTO, 70- Concurrent Pos: Nat Res Coun fels, Univ Waterloo, 67 & Univ Toronto, 72; Dept Univ Affairs fel, Univ Waterloo, 68. Mem: Asn Comput Mach; Can Info Processing Soc. Res: Computer hardware monitoring. Mailing Add: Rm 143 Sandford Fleming Bldg Comput Ctr Univ of Toronto Toronto ON Can

WILSON, JOHN COE, b Manhattan, Kans, July 16, 31; m 55; c 2. GEOLOGY. Educ: Calif Inst Technol, BS, 53, PhD(geol), 61; Univ Kans, MS, 55. Prof Exp: Explor geologist, Bear Creek Mining Co, 61-64, res geologist, 64-68, chief, Geol Res Div, 68-71, DIR EXPLOR SERVS DEPT, KENNECOTT EXPLOR INC, KENNECOTT COPPER CORP, 71- Mem: Geol Soc Am; Soc Econ Geol; Can Inst Mining & Metall; Am Inst Mining, Metall & Petrol Eng. Res: Ore deposit geology; exploration geology. Mailing Add: Explor Serv Dept 2300 W 1700 South St Salt Lake City UT 84104

WILSON, JOHN DRENNAN, b Peoria, Ill, Mar 29, 38; m 67; c 1. PHYSIOLOGY, RADIOBIOLOGY. Educ: Carleton Col, BA, 60; Univ Ill, Urbana, MS, 63, PhD(physiol), 66. Prof Exp: Res assoc radiobiol, Univ Tex, Austin, 66-72; ASST PROF RADIOBIOL, MED COL VA, 72- Mem: AAAS; Radiation Res Soc. Res: Lethal and mutagenic effects of radiation on microorganisms; effects of accelerated particles on mammalian systems. Mailing Add: Dept of Radiol Med Col of Va Richmond VA 23219

WILSON, JOHN EDWARD, b Ft Wayne, Ind, Apr 27, 39; m 64; c 2. NEUROCHEMISTRY, ENZYMOLOGY. Educ: Univ Notre Dame, BS, 61; Univ Ill, MS, 62, PhD(biochem), 64. Prof Exp: From asst prof to assoc prof, 67-75, PROF BIOCHEM, MICH STATE UNIV, 75- Concurrent Pos: NSF fel, Univ Ill, Urbana, 64-65. Mem: Am Soc Biol Chem; Am Soc Neurochem; Int Soc Neurochem; Am

Chem Soc. Res: Brain hexokinase; brain mitochondria; regulation of energy metabolism in brain. Mailing Add: Dept of Biochem Mich State Univ East Lansing MI 48824

WILSON, JOHN ERIC, b Champaign, Ill, Dec 13, 19; m 47; c 3. BIOCHEMISTRY. Educ: Univ Chicago, SB, 41; Univ Ill, MS, 44; Cornell Univ, PhD(biochem), 48. Prof Exp: Asst chem, Univ Ill, 41-44; asst biochem, Med Col, Cornell Univ, 44-48, res assoc, 48-50; from asst prof to assoc prof biochem, 50-65, PROF BIOCHEM, SCH MED, UNIV NC, CHAPEL HILL, 65- Mem: Fel AAAS; Am Chem Soc; Am Soc Biol Chem; Am Soc Neurochem; Harvey Soc. Res: Effects of experience and behavior on brain metabolism; neurochemistry. Mailing Add: Dept of Biochem Univ of NC Sch of Med Chapel Hill NC 27514

WILSON, JOHN F, b Niagara Falls, NY, Dec 23, 22; m 50; c 8. MEDICINE, PATHOLOGY. Educ: Univ Cincinnati, MD, 52. Prof Exp: Intern pediat, Univ Ark Hosp, 52-53; resident, Children's Hosp, Cincinnati, Ohio, 55-57; instr, Univ Cincinnati, 57-58; from instr to asst prof pediat, 58-68, instr path, 67-68, ASST PROF PATH, UNIV UTAH, 68-, ASSOC PROF PEDIAT, 73-; PATHOLOGIST & DIR LABS, PRIMARY CHILDREN'S HOSP, 69- Concurrent Pos: Smith Kline & French fel hemat, 57-59; from co-prin investr to prin investr gastrointestinal tract in iron deficiency anemia NIH grants, 61-66, prin investr, copper metab in acute leukemia, 63-64; mem comn child nutrit, Food & Nutrit Bd, Nat Res Coun, 64-66; resident path, Univ Utah, 66-69; assoc prog dir, Children's Cancer Study Group A, NIH, Univ Utah, 70- Mem: AMA; Am Soc Hemat; Am Fedn Clin Res; Am Soc Clin Path; Col Am Path. Res: Iron and copper metabolism in iron deficiency; childhood malignanceis. Mailing Add: Primary Children's Hosp 320 12th Ave Salt Lake City UT 84103

WILSON, JOHN HOWARD, b July 27, 44; US citizen; m 65; c 1. BIOCHEMISTRY, GENETICS. Educ: Wabash Col, AB, 66; Calif Inst Technol, PhD(biochem, genet), 72. Prof Exp: ASST PROF BIOCHEM, BAYLOR COL MED, 73- Concurrent Pos: Damon Runyon fel biochem, Med Ctr, Stanford Univ, 71-73. Res: Molecular biology of tumourogenic viruses. Mailing Add: Dept of Biochem Baylor Col of Med Houston TX 77025

WILSON, JOHN HUMAN, b Wills Point, Tex, Feb 24, 00; m 24; c 2. EXPLORATION GEOPHYSICS. Educ: Colo Sch Mines, EM, 23. Prof Exp: Geologist, Midwest Ref Co, Colo, 23-26; geologist & geophysicist, Huasteca Petrol Co, Mex, 26-27; asst prof geophys, Colo Sch Mines, 28-29; consult geologist & geophysicist, 29-34; pres, Colo Geophys Corp, 34-37; vpres, Independent Explor Co, Piper Petrol Co, Woodson Oil Co & Wilson Explor Co, 38-61; PRES, PIPER PETROL CO & WILSON EXPLOR CO, 61- Concurrent Pos: Explor consult, 50- Mem: Soc Explor Geophys (secy-treas, 36); Am Asn Petrol Geol; assoc Am Inst Mining, Metall & Petrol Eng. Res: Design of geophysical equipment; incipient metamorphism of sediments; seismic velocities; exploration techniques. Mailing Add: 1212 W El Paso St Ft Worth TX 76102

WILSON, JOHN M, b Apr 4, 40; US citizen; m 63; c 2. FOOD SCIENCE. Educ: West Chester State Col, BS, 63; Univ Md, College Park, MS, 68, PhD(food sci, biochem), 71. Prof Exp: Fac res asst food sci, Univ Md, College Park, 68-71; ASST PROF FOOD SCI, UNIV MAINE, ORONO, 71- Mem: Inst Food Technol. Res: Food toxicology and nutrition; liquid waste disposal; infrared analysis; food processing technology. Mailing Add: Dept of Food Sci Univ of Maine Orono ME 04473

WILSON, JOHN NEVILLE, b Portland, Maine, June 13, 18; m 44; c 3. PHYSICS, RESOURCE MANAGEMENT. Educ: Rice Inst, BA, 40; Harvard Univ, AM, 41. Prof Exp: Res chemist, Va, 41, res physicist, 42-43, tech specialist, Manhattan Dist, Del, 43-44, sr supvr, Wash, 44-45, res physicist, Va, 45-50, tech specialist, Del, 50-51, SC, 51-53, res supvr, Appl Physics Div, 53, res mgr, 54-70, SUPT, PLANNING AND ANAL DEPT, SAVANNAH RIVER PLANT, E I DU PONT DE NEMOURS & CO, 70- Concurrent Pos: Asst physicist, Nat Defense Res Comt, Radio & Sound Lab, Univ Calif, 41-42; assoc physicist, Clinton Labs, Tenn, 44. Mem: Soc Rheol. Res: Rayon spinning, yarn structure and physical testing; health physics; oceanography; radiation and chemical process instrumentation; electronics; non-destructive testing; computer models for finance and control of large industrial plant and laboratory. Mailing Add: 1208 Abbeville Ave NW Aiken SC 29801

WILSON, JOHN PHILIP, b Knoxville, Ill, Dec 24, 35. ANTHROPOLOGY. Educ: Univ Ill, BS, 58, MA, 61; Harvard Univ, PhD(anthrop), 69. Prof Exp: Cur archeol, Mus NMex, 63-66, cur hist archaeol, 66-73; staff archaeologist, Nat Hist Sites Serv Can, 73-75; RES ARCHAEOLOGIST, N MEX STATE UNIV, 75- Mem: Soc Am Archaeol; fel Am Anthrop Asn; Am Soc Ethnohist; Soc Hist Archaeol; Western Hist Asn. Res: Southwestern US archaeology and history; military history; historical archaeology; analysis of cultural resources; environmental impact assessments; Spanish-colonial studies. Mailing Add: Dept of Sociol-Anthrop Box 3BV N Mex State Univ Las Cruces NM 88003

WILSON, JOHN PHILLIPS, b Stamford, Conn, June 5, 16; m 40; c 1. MATHEMATICS. Educ: Univ Southern Miss, BA, 57; Johns Hopkins Univ, MEd, 60, MS, 70. Prof Exp: Res staff asst, Ballistic Anal Lab, Inst Coop Res, Johns Hopkins Univ, 57-62, res assoc, 62-67, res scientist, 67-69; SR RES ANALYST, THOR DIV, FALCON RES & DEVELOP CO, BALTIMORE, 69- Mem: Am Ord Asn. Res: Military operations analysis, particularly target vulnerability and weapons systems evaluation; geo and celestial navigation for surface vessels; merchant marine industry-shipboard operations. Mailing Add: 722 Shelley Rd Towson MD 21204

WILSON, JOHN RANDALL, b Miami, Fla, June 12, 34; m 59; c 3. PHYSICAL CHEMISTRY. Educ: Univ Fla, BS, 56; Univ Wis, MS, 59, PhD(chem), 65. Prof Exp: Asst prof chem, Franklin Col, 59-62, Miami Univ, 64-67 & Asheville-Biltmore Col, 67-68; assoc prof, 68-72, PROF CHEM, SHIPPENSBURG STATE COL, 72- Mem: AAAS; Am Chem Soc. Res: Radiation chemistry; mechanisms of exchange reactions; photochemistry. Mailing Add: Dept of Chem Shippensburg State Col Shippensburg PA 17257

WILSON, JOHN SHIRLEY, b Mattoon, Ill, May 22, 20; m 46; c 2. INORGANIC CHEMISTRY. Educ: Eastern Ill State Col, BEd, 42; Univ Ill, MS, 49; Tulane Univ, PhD(chem), 52. Prof Exp: Control chemist, E I du Pont de Nemours & Co, 42-43; instr chem, La Col, 46-48; assoc prof, 51-52, dean men, 46-48, 51-52; assoc prof, McNeese State Col, 52-54; group leader, Cent Lab, 54-59, Electrochem Res Lab, 59-62, dir Ga Explosives Res Lab, 62-65, res specialist, Polyol Latex Lab, 66-68, RES SPECIALIST, CONTRACT RES DEPT, DOW CHEM CO, FREEPORT, 68- Mem: Am Chem Soc; Nat Asn Corrosion Eng; Fel Am Inst Chem. Res: Complex ions as related to chelates formed by dye molecules with metal ions; chemistry and flashing characteristics of geothermal water; testing of chelate-type high pressure boiler cleaning agents; corrosion of metals in sea water and brines at elevated temperatures. Mailing Add: 131 Daisy Lake Jackson TX 77566

WILSON, JOHN T, b Gainesville, Tex, Apr 27, 38; m 62; c 3. PEDIATRICS, PHARMACOLOGY. Educ: Tulane Univ La, BS, 60, MS & MD, 63. Prof Exp: From intern to resident clin pediat, Palo Alto-Stanford Med Ctr, Palo Alto, Calif, 63-65; res assoc biochem pharmacol, Univ Iowa, 65-66; res assoc biochem pharmacol & endocrinol, Nat Inst Child Health & Human Develop, Bethesda, Md, 66-68; attend pediatrician & dir lab perinatal med pharmacol, Children's Hosp, San Francisco, 69-70; ASSOC PROF PEDIAT & PHARMACOL, MED SCH, VANDERBILT UNIV, 70- Concurrent Pos: Fel neonatal med & dir lab develop pharmacol, Children's Hosp, San Francisco, 68-69; NIH res career develop award, 69 & 72; lectr, Med Ctr, Univ Calif, San Francisco, 69-70; res assoc, J F Kennedy Ctr, 70- Mem: AAAS; Soc Pediat Res; Am Soc Pharmacol & Exp Therapeut; Am Soc Clin Pharmacol & Therapeut; Am Acad Pediat. Res: Pediatric clinical pharmacology, drug metabolism. Mailing Add: Dept of Pediat Vanderbilt Univ Sch of Med Nashville TN 37232

WILSON, JOHN THOMAS, JR, b Birmingham, Ala, June 2, 24; m 68. ENVIRONMENTAL MEDICINE, PREVENTIVE MEDICINE. Educ: Howard Univ, BS, 46; Columbia Univ, MD, 50; Univ Cincinnati, ScD(indust med), 56. Prof Exp: Physician, Div Indust Hyg, NY State Dept Labor, 55-56 & Sidney Hillman Health Ctr, New York, 56-57; chief bur occup health, Santa Clara County Health Dept, Calif, 57-61; life sci adv, Lockheed Aircraft Corp, 61-67, head biol sci res labs, Lockheed Missiles & Space Co, 67-69; asst prof community & prev med, Sch Med, Stanford Univ, 69-71; prof community health pract & chmn dept, Col Med, Howard Univ, 71-74; PROF ENVIRON HEALTH & CHMN DEPT, SCH PUB HEALTH & COMMUNITY MED, UNIV WASH, 74- Concurrent Pos: Nat Med Fel fel, 53-55; fel indust med, Univ Cincinnati, 53-56; lectr, Sch Pub Health, Univ Calif, Berkeley, 59-61. Mem: Fel Indust Med Asn; fel Am Col Physicians; Am Acad Occup Med; Am Indust Hyg Asn. Res: Occupational and environmental medicine; toxicology; industrial hygiene. Mailing Add: Sch Pub Health & Community Med Univ Wash Dept Environ Health Seattle WA 98195

WILSON, JOHN TUZO, b Ottawa, Ont, Oct 24, 08; m 38; c 2. GEOPHYSICS, TECTONICS. Educ: Univ Toronto, BA, 30; Cambridge Univ, MA, 32, ScD, 58; Princeton Univ, PhD(geol), 36; Carleton Univ, LLD, 58. Hon Degrees: DSc, Univ Western Ont, 58, Acadia Univ Mem Univ Newf, 68; ScD, Franklin & Marshall Col, 69; DSc, McGill Univ, 74; DUniv, Univ Calgary, 74. Prof Exp: Asst geologist, Geol Surv Can, 36-38; prof geophys, Univ Toronto, 46-74; prin, Erindale Col, 67-74; DIR-GEN, ONT SCI CENTRE, 74- Concurrent Pos: Vis prof, Australian Nat Univ, 50-65; pres, Int Union Geol & Geophys, 57-60; mem, Nat Res Coun Can, 58-64; mem, Defence Res Bd Can, 60-66; distinguished lectr, Univ Toronto, 74- Honors & Awards: Miller Medal, Royal Soc Can, 56; Blaylock Medal, Can Inst Mining & Metal, 59; Bucher Medal, Am Geophys Union, 68; Companion Award, Order Can, 74; J J Carty Medal, Nat Acad Sci, 75. Mem: Foreign assoc Nat Acad Sci; fel Geol Soc Am; fel Royal Soc Can (pres, 72-73); fel Royal Soc; foreign mem Am Philos Soc. Res: Physics of the earth; continental structure. Mailing Add: Ont Sci Centre 770 Don Mills Road Don Mills ON Can

WILSON, JOHN WILLIAM, b Arkansas City, Kans, Aug 6, 40; m 62; c 1. THEORETICAL HIGH ENERGY PHYSICS, HEALTH PHYSICS. Educ: Kans State Univ, BS, 62; Col William & Mary, MS, 69, PhD(physics), 75. Prof Exp: Aerospace technologist simulation, 63-70, SPACE SCIENTIST SPACE PHYSICS, LANGLEY RES CTR, NASA, 70- Honors & Awards: NASA spec achievement, 75. Res: High-energy heavy ion reaction theory; high-energy transport theory; health physics aspects of high-altitude aircraft and space operations and dosimetry. Mailing Add: NASA Langley Res Ctr Mail Stop 160 Hampton VA 23365

WILSON, JOHN WILLIAM, JR, b Albany, NY, Apr 1, 16; m 40; c 2. PETROLEUM CHEMISTRY, MICROSCOPY. Educ: Amherst Col, AB, 38. Prof Exp: From chemist to sr chemist, Res Dept, 40-60, Paulsboro Lab, 60-68, supv chemist, 68-69, mgr indust lubricants div, Res & Tech Serv Dept, Mobil Oil Co Ltd, Eng, 69-70, SUPV CHEMIST RES DEPT, PAULSBORO LAB, MOBIL RES & DEVELOP CORP, 70- Mem: Am Chem Soc; Electron Micros Soc Am; NY Acad Sci; fel Am Inst Chem; fel Brit Inst Petrol. Res: Color of petroleum products; colorimetric methods of analysis for petroleum products and related materials; development of specialized lubricating greases; microscopic studies of lubricating grease structure; fundamental studies mechanism of lubrication. Mailing Add: Mobil Res & Develop Corp Res Dept Paulsboro NJ 08066

WILSON, JOHN WILLIAM, III, b New York, NY, May 10, 43; m 66. ZOOGEOGRAPHY, PALEONTOLOGY. Educ: Amherst Col, BA, 66; Univ Chicago, PhD(evolutionary biol), 72. Prof Exp: ASST PROF BIOL, GEORGE MASON UNIV, 72- Mem: AAAS; Soc Study Evolution; Soc Vert Paleont; Am Soc Mammal; Ecol Soc Am. Res: Zoogeography of mammals; latitudinal gradients; paleoecology of mammals; changes in resource utilization of mammals during late Cretaceous and Cenozoic; Pleistocene extinctions. Mailing Add: Dept of Biol George Mason Univ Fairfax VA 22030

WILSON, JOSEPH EDWARD, b Hannibal, Mo, Jan 8, 20; m 45; c 4. PHYSICAL CHEMISTRY. Educ: Univ Chicago, BS, 39; Univ Rochester, PhD(phys chem), 42. Prof Exp: Res chemist, Goodyear Aircraft Corp, 42-46, Argonne Nat Lab, 46-47 & Firestone Tire & Rubber Co, 47-50; sr chemist, Bakelite Co Div, Union Carbide & Carbon Corp, 50-57 & J T Baker Chem Co, 57; develop supvr, Atlas Powder Co, 58-61; proj mgr, Kordite Co, 61-64; res dir, Pollock Paper Div, St Regis Paper Co, 64-67; assoc prof, 67-71, PROF PHYS CHEM, BISHOP COL, 71- Concurrent Pos: Plastics consult, 67- Mem: Am Chem Soc. Res: Photochemistry; polymerization; stability of polymers; radiation chemistry of plastics. Mailing Add: Dept of Chem Bishop Col Dallas TX 75241

WILSON, JOSEPH WILLIAM, b Massena, NY, Apr 11, 34. ORGANIC CHEMISTRY. Educ: Mass Inst Technol, BS, 56; Ind Univ, PhD(chem), 61. Prof Exp: Res assoc, Univ Wis, 61-63; asst prof, 63-70, ASSOC PROF CHEM, UNIV KY, 70- Mem: AAAS; Am Chem Soc. Res: organic photochemistry. Mailing Add: Dept of Chem Univ of KY Lexington KY 40506

WILSON, KARL A, b Buffalo, NY, Jan 19, 47; m 74. BIOCHEMISTRY. Educ: State Univ NY Buffalo, BA, 69, PhD(biochem), 73. Prof Exp: Res assoc biochem, Roswell Park Mem Inst, 73-74; RES ASSOC BIOCHEM, PURDUE UNIV, WEST LAFAYETTE, 74- Concurrent Pos: NSF grad fel, 69-72. Mem: AAAS; Sigma Xi. Res: Mechanism and physiology of proteases and their protein inhibitors; molecular evolution of proteins and protein sequencing. Mailing Add: Dept of Chem Purdue Univ West Lafayette IN 47907

WILSON, KATHERINE SCHMITKONS, b Lorain, Ohio, Jan 22, 13; m 36. GENETICS. Educ: Oberlin Col, AB, 33; Northwestern Univ, MS, 35; Yale Univ, PhD(bot), 44. Prof Exp: Asst, Northwestern Univ, 33-35; instr biol, Muskingum Col, 35-41 & Day Sch, Conn, 41-42; Seessel res fel, Yale Univ, 48-49, Organon fel bot, 49-50, instr, 53-56; biologist, 56-62, EXEC SECY GENETICS STUDY SECT, DIV RES GRANTS, NAT INSTS HEALTH, 56-, HEALTH SCIENTIST ADMINR, 63- Mem: Fel AAAS; Bot Soc Am; Soc Develop Biol; Genetics Soc Am; Am Soc Human

Genetics. Res: Plant morphogenesis and physiology; microbiology; cell biology; genetics. Mailing Add: Div Res Grants Nat Insts of Health Bethesda MD 20014

WILSON, KATHERINE WOODS, b Los Angeles, Calif, Feb 8, 23. AIR POLLUTION. Educ: Univ Calif, BS, 44, MS, 45; Univ Calif, Los Angeles, PhD(chem), 48. Prof Exp: Asst prof chem, WVa Univ, 48-53 & Pepperdine Col, 53-54; mem staff, Los Angeles County Air Pollution Control Dist, 54-55; lectr chem & assoc res chemist, Univ Calif, Los Angeles, 55-62; phys chemist, Stanford Res Inst, 62-71; staff officer, Univ Calif, Riverside, 71-72; supv air pollution chemist, Air Pollution Control Serv, San Diego County, 73-74; DIR AIR QUAL STUDIES, COPLEY INT CORP, 74- Concurrent Pos: Mem, Air Pollution Chem & Physics Adv Comt, Environ Protection Agency, 73- Mem: Am Chem Soc; Air Pollution Control Asn. Res: Air pollution; cotton chemistry; chemical analysis of agricultural products. Mailing Add: Copley Int Corp PO Box 1530 La Jolla CA 92038

WILSON, KENNETH ALLEN, b Rio de Janeiro, Brazil, Apr 15, 28; US citizen. PLANT MORPHOLOGY, SYSTEMATIC BOTANY. Educ: Univ Miami, BA, 51; Univ Hawaii, MS, 53; Univ Mich, PhD(bot), 58. Prof Exp: Botanist, Gray Herbarium & Arnold Arboretum, Harvard Univ, 57-60; from asst prof to assoc prof, 60-67, assoc dean sch letters & sci, 66-73, PROF BOT, CALIF STATE UNIV, NORTHRIDGE, 67- Mem: Bot Soc Am; Am Soc Plant Taxon; Am Fern Soc; Int Asn Plant Taxon. Res: Taxonomy; pteridophytes. Mailing Add: Dept of Bot Calif State Univ Northridge CA 91324

WILSON, KENNETH GEDDES, b Waltham, Mass, June 8, 36. PHYSICS. Educ: Harvard Univ, AB, 56; Calif Inst Technol, PhD(physics), 61. Prof Exp: Jr fel, Harvard Univ, 59-62; Ford Found fel, Europ Orgn Nuclear Res, Geneva, Switz, 62-63; from asst prof to assoc prof, 63-71, PROF PHYSICS, CORNELL UNIV, 71- Concurrent Pos: Mem staff, Stanford Linear Accelerator Ctr, 69-70. Mem: Am Phys Soc. Res: Elementary particle theory. Mailing Add: Lab of Nuclear Studies Cornell Univ Ithaca NY 14850

WILSON, KENNETH MACKENZIE, b Liverpool, Eng, Oct 16, 12; nat US; m 40; c 3. PHYSIOLOGY, CHEMISTRY. Educ: Liverpool Col Technol, ARIC, 39. Prof Exp: Exp officer, Chem Defense Exp Estab, Eng, 40-47, sr sci officer, 47-52, prin sci officer, 52-57; physiologist, Chem Res & Develop Labs, Army Chem Ctr, 57-69, PHYSIOLOGIST, MED RES LAB, EDGEWOOD ARSENAL, DEPT ARMY, 69- Concurrent Pos: Instr, Sch Med, Johns Hopkins Univ, 60-73, asst prof, 73- Mem: Am Chem Soc; Am Physiol Soc; Am Inst Chem; NY Acad Sci; AAAS. Res: Physiological and pharmacological actions of chemical agents; factors governing penetration of biological membranes by chemicals; relationship of chemical structure to biological activity; medical apparatus and instruments; mechanism of action of anesthetics. Mailing Add: Med Res Lab Appl Physiol Sect Bldg 3100 APG Edgewood Arsenal MD 21010

WILSON, KENNETH SHERIDAN, b Waterloo, Iowa, May 1, 24; m 48, 62; c 1. MYCOLOGY, PLANT PATHOLOGY. Educ: Colo Col, BS, 49; Univ Wyo, MS, 50; Purdue Univ, PhD(mycol), 54. Prof Exp: Asst, Colo Col, 46-49 & Univ Wyo, 49-50; asst, 52-53, from instr to assoc prof, 54-71, PROF BIOL SCI, PURDUE UNIV, CALUMET CAMPUS, 71- Mem: AAAS; Mycol Soc Am; Bot Soc Am; Soc Indust Microbiol; Am Soc Microbiol. Res: Mycological taxonomy; fungus physiology; plant morphology; microbiological ecology. Mailing Add: Dept of Biol Sci Purdue Univ Calumet Campus Hammond IN 46323

WILSON, KENT RAYMOND, b Philadelphia, Pa, Jan 14, 37. CHEMICAL PHYSICS, ENVIRONMENTAL SCIENCES. Educ: Harvard Col, AB, 58; Univ Strasbourg, dipl, 59; Univ Calif, Berkeley, PhD(chem), 64. Prof Exp: Res fel chem, Harvard Univ, 64-65; res chemist, Nat Bur Stand, 65; asst prof phys chem, 65-71, ASSOC PROF PHYS CHEM, UNIV CALIF, SAN DIEGO, 71- Concurrent Pos: Sloan res fel, 70-72; mem comt comput in chem, Div Chem & Chem Technol, Nat Res Coun, 70-72; mem comput & biomath sci study sect, NIH, 71-74; mem panel photochem oxidants, ozone & hydrocarbons, Nat Res Coun, 73-74. Mem: Am Phys Soc; Am Chem Soc; Air Pollution Control Asn. Res: Molecular dynamics of chemical reactions; specialized computer systems for solution of scientific problems; computer animation; archaeological chemistry; air pollution. Mailing Add: Dept of Chem Univ of Calif San Diego La Jolla CA 92093

WILSON, L KENNETH, b Sacramento, Calif, Sept 22, 10; m 35; c 1. EXPLORATION GEOLOGY. Educ: Stanford Univ, AB, 32. Prof Exp: Geologist gold mining, Mother Lode Mining Dist, Calif, 32-35; geologist, Calumet & Hecla Copper Co, 35-38; mgr, Auburn Chicago Mine, Calif, 38-39; geologist, Cord Mining Interests, 39-43 & Am Smelting, 43-60; CONSULT GEOLOGIST, 60- Concurrent Pos: Gov app, Western Gov Mining Adv Coun, 67- Mem: Fel Geol Soc Am; Am Inst Prof Geologists; Am Inst Mining, Metall & Petrol Engrs; Am Asn Petrol Geologists; Soc Econ Geologists. Res: Exploration geology, domestic mineral resources; assistance to legal counsel in mining law, litigation, utilization and acquisition of mineral property. Mailing Add: PO Box 7123 Menlo Park CA 94025

WILSON, LARRY DAVID, zoology, see 12th edition

WILSON, LARRY EUGENE, b Wapakoneta, Ohio, Nov 17, 35; m 58; c 2. ANALYTICAL CHEMISTRY. Educ: Ohio State Univ, BSc, 57, PhD(anal chem), 62. Prof Exp: Anal chemist, Dow Chem Co, Mich, 63-64; asst prof chem, Mich State Univ, 64-65; anal chemist, Dow Chem Co, Mich, 65-66, supvr control lab, 66-67, coordr lab technician training, 67-69; asst prof, 69-72, ASSOC PROF CHEM, LANCASTER BR, OHIO UNIV, 72- Mem: Am Chem Soc. Res: Acid-base equilibria; methods of teaching. Mailing Add: Dept of Chem Lancaster Br Ohio Univ Lancaster OH 43130

WILSON, LAUREN R, b Yates Center, Kans, May 4, 36; m 59; c 2. INORGANIC CHEMISTRY, ENVIRONMENTAL CHEMISTRY. Educ: Baker Univ, BS, 58; Univ Kans, PhD(inorg chem), 63. Prof Exp: Asst prof, 63-70, PROF CHEM & CHMN DEPT, OHIO WESLEYAN UNIV, 70- Concurrent Pos: Mem staff, Oak Ridge Nat Lab, 72. Mem: Am Chem Soc; The Chem Soc. Res: Synthesis of transition metal compounds, thermochemistry of complex compounds; metal ions in natural and biological systems. Mailing Add: Dept of Chem Ohio Wesleyan Univ Delaware OH 43015

WILSON, LAURENCE EDWARD, b Aberdeen SDak, June 29, 30; m 57; c 3. INORGANIC CHEMISTRY. Educ: Western Wash Col Educ, BA, 52; Univ Wash, PhD(chem), 57. Prof Exp: Instr chem, Amherst Col, 56-59; from asst prof to assoc prof, San Jose State Col, 59-63; ASSOC PROF CHEM, KALAMAZOO COL, 63- Mem: AAAS; Am Chem Soc. Mailing Add: Dept of Chem Kalamazoo Col Kalamazoo MI 49001

WILSON, LELAND LESLIE, b Williamsburg, Ky, June 13, 14; m 38; c 3. CHEMISTRY. Educ: Eastern Ky State Col, BS, 34; Univ Ky, MS, 41; Peabody Col, PhD(sci educ), 51. Prof Exp: Instr high sch, Ky, 35-36 & Fla, 36-42; radio theory, US Air Force Tech Sch, Univ Chicago, 42-43; asst prof sci, Eastern Ky State Col, 46-49, 51-52; prof physics, Ga Teachers Col, 52-55; assoc prof, 55-59, head dept chem, 68-75, PROF CHEM & PHYS SCI, UNIV NORTHERN IOWA, 59- Mem: AAAS; Am Chem Soc; Nat Sci Teachers Asn. Res: Chemical instrumentation; metal chelate stability; molecular models and demonstrations for chemistry teaching. Mailing Add: Dept of Chem Univ of Northern Iowa Cedar Falls IA 50613

WILSON, LEON WILLIAM, JR, b Pittsburgh, Pa, Sept 18, 35; m 59; c 2. POLYMER CHEMISTRY. Educ: Carnegie-Mellon Univ, BS, 57; Univ Nebr, MS, 61, PhD(org chem), 63. Prof Exp: Chemist, Goodyear Tire & Rubber Co, Ohio, 57-58; res chemist, Marshall Lab, E I du Pont de Nemours & Co, Inc, Pa, 63-65; SR RES CHEMIST, RES CTR, US STEEL CORP, 65- Mem: Am Chem Soc. Res: Free radical and condensation polymerization; gas-liquid chromatography; phase contrast microscopy. Mailing Add: US Steel Corp Res Ctr MS 79 Monroeville PA 15146

WILSON, LEONARD GILCHRIST, b Orillia, Ont, June 11, 28. HISTORY OF MEDICINE. Educ: Univ Toronto, BA, 49; Univ London, MSc, 55; Univ Wis, PhD(hist sci), 58. Prof Exp: Lectr biol, Mt Allison Univ, 50-53; vis instr hist sci, Univ Calif, 58-59; vis asst prof, Cornell Univ, 59-60; from asst prof to assoc prof hist med, Sch Med, Yale Univ, 60-67; PROF HISTORY OF MED, UNIV MINN, MINNEAPOLIS, 67- Mem: AAAS; Am Asn Hist Med; Am Hist Asn; Hist Sci Soc; Soc Hist Technol. Res: History of biology; history of physiology in the seventeenth century. Mailing Add: 510 Diehl Hall Biomed Libr Univ of Minn Minneapolis MN 55455

WILSON, LEONARD RICHARD, b Superior, Wis, July 23, 06; m 30; c 2. GEOLOGY, PALYNOLOGY. Educ: Univ Wis, PhB, 30, PhM, 32, PhD(bot), 36. Prof Exp: Asst, Wis Geol & Natural Hist Surv, 31-35; from instr to prof geol & bot, Coe Col, 35-46; prof geol & head dept geol & mineral, Univ Mass, 46-56; prof geol, NY Univ, 56-57; prof, 57-62, res prof, 62-67, GEORGE LYNN GROSS RES PROF GEOL & GEOPHYS, UNIV OKLA, 68-, CUR MICROPALEONT & PALEOBOT, STOVALL MUS, 71-; GEOLOGIST, OKLA GEOL SURV, 57- Concurrent Pos: Melhaup scholar, Ohio State Univ, 39-40; consult, Carter Oil Co, 46-56; leader, Am Geog Soc Greenland Ice Cap Exped, 53; res assoc, Am Mus Natural Hist, 57-; adj prof, Univ Tulsa; consult, Jersey Prod Res Corp, 56-62, Humble Oil Co, 63-64 & Sinclair Oil Co, 63-69. Honors & Awards: VI Gunnar Erdtman Int Medal, Palynology Soc India, 73. Mem: Fel Geol Soc Am; Am Bot Soc; Am Asn Petrol Geol; Soc Econ Paleont & Mineral; hon mem Am Asn Stratig Palynologists. Res: Stratigraphic and paleoecologic palynology. Mailing Add: Dept of Geol Univ of Okla Norman OK 73069

WILSON, LESLIE, b Boston, Mass, June 29, 41; m 64; c 1. PHARMACOLOGY, BIOLOGY. Educ: Mass Col Pharm, BS, 63; Tufts Univ, PhD(pharmacol), 67. Prof Exp: Asst prof pharmacol, Sch Med, Stanford Univ, 69-75; ASSOC PROF BIOL, UNIV CALIF, SANTA BARBARA, 75- Concurrent Pos: USPHS fel, Univ Calif, Berkeley, 67-69; Nat Inst Neurol Dis & Stroke res grant, Sch Med, Stanford Univ, 70-; Am Cancer Soc grant. Mem: AAAS; Am Soc Pharmacol & Exp Therapeut; Am Soc Cell Biol. Res: Biochemical and pharmacological properties of microtubule proteins; mechanism of action of antimitotic chemical agents. Mailing Add: Dept of Biol Sci Univ of Calif Santa Barbara CA 93106

WILSON, LESTER A, JR, b Charleston, SC, Apr 30, 17; m 45; c 4. MEDICINE. Educ: Col William & Mary, BS, 38; Med Col SC, MD, 42. Prof Exp: From asst prof to assoc prof, 51-65, PROF OBSTET & GYNEC, SCH MED, UNIV VA, 65- Mailing Add: Dept of Obstet & gynec Univ Va Sch of Med Charlottesville VA 22903

WILSON, LINDA S WHATLEY, b Washington, DC, Nov 10, 36; m 57, 70; c 1. CHEMISTRY, RESEARCH ADMINISTRATION. Educ: Tulane Univ, BA, 57; Univ Wis, PhD(inorg chem), 62. Prof Exp: Res assoc inorg chem, Univ Wis, 62; Nat Inst Dent Res trainee phys chem & res assoc molecular spectros, Univ Md, 62-64, res asst prof, Molecular Spectros, 64-67; vis res fel, Univ Southampton, 67; vis asst prof, Univ Mo-St Louis, 67-68; asst to vchancellor res, Wash Univ, 68-69, asst vchancellor res, 69-74, assoc vchancellor res, 74-75; ASSOC VCHANCELLOR RES, UNIV ILL, URBANA, 75- Concurrent Pos: Mem, Gen Res Support Adv Comt, NIH, 71-75, chmn, 74-75; mem, Comt Govt Relations, Nat Asn Cols & Univ Bus Officers, 71-, chmn, 73-75. Mem: AAAS; Am Chem Soc. Res: Molecular spectroscopy; spectroscopic studies of molecular interactions; charge transfer complexes; coordination compounds and hydrogen bonded species; optical studies at high pressures. Mailing Add: Rm 338 Admin Bldg Univ of Ill Urbana IL 61801

WILSON, LORENZO GEORGE, b Appleton, NY, July 25, 38; m 62; c 3. HORTICULTURE, VEGETABLE CROPS. Educ: Cornell Univ, BS, 61; Wash State Univ, MS, 64; Mich State Univ, PhD(hort), 69. Prof Exp: Res assoc postharvest physiol, United Fruit Co, 63-66; plant physiologist, United Brands Co, 69-75; EXTEN HORT SPECIALIST VEG CROPS, HORT DEPT, NC STATE UNIV, 75- Concurrent Pos: Consult postharvest produce handling, United Brands Co, Cent & SAm. Mem: Am Soc Hort Sci; Int Soc Hort Sci; Sigma Xi. Res: Investigations to determine optimum cultural practices for Irish potato and sweet potato production in North Carolina, including the use of fertilizers, pesticides and harvesting, handling and storage techniques for enhanced quality maintenance. Mailing Add: Dept of Hort Sci NC State Univ Raleigh NC 27607

WILSON, LORNE GRAHAM, b Saskatoon, Sask, Oct 23, 29; US citizen; m 57; c 2. SOIL PHYSICS, HYDROLOGY. Educ: Univ BC, BS, 51; Univ Calif, MS, 57, PhD(soil sci), 62. Prof Exp: Asst specialist irrig drainage, Univ Calif, 56-58; from asst hydrologist, to assoc hydrologist, 62-67, HYDROLOGIST, WATER RESOURCES RES CTR, UNIV ARIZ, 67- Res: Survey of drainage problems in San Joaquin Valley, California; simultaneous flow of air and water during infiltration in soils; subsurface flow characteristics during natural and artificial recharge in stratified sediments. Mailing Add: Water Resources Res Ctr Univ of Ariz Tucson AZ 85721

WILSON, LOUIS FREDERICK, b Milwaukee, Wis, Nov 22, 32; m 56; c 4. ENTOMOLOGY. Educ: Marquette Univ, BS, 55, MS, 57; Univ Minn, PhD(entom), 62. Prof Exp: Instr cytol & parasitol, Marquette Univ, 54-57; state entomologist, Minn, 58; PRIN INSECT ECOLOGIST, N CENT FOREST EXP STA, US FOREST SERV, 58-; asst prof forestry, 67-73, asst prof entom, 69-73, ASSOC PROF FORESTRY, MICH STATE UNIV, 73-, ASSOC PROF ENTOM, 73- Honors & Awards: Outstanding Performance Award, N Cent Forest Exp Sta, 65. Mem: Entom Soc Am; Soc Am Foresters; Entom Soc Can. Res: Insect ecology and behavior; population dynamics; insect impact. Mailing Add: 900 Longfellow Dr East Lansing MI 48823

WILSON, LOWELL D, b Pampa, Tex, May 11, 33; m 60; c 2. ENDOCRINOLOGY, BIOLOGICAL CHEMISTRY. Educ: Univ Calif, Berkeley, AB, 55; Univ Chicago, MD, 60; Univ Southern Calif, PhD(biochem), 68. Prof Exp: From instr to asst prof med, Sch Med, Univ Southern Calif, 66-68; asst prof med & biol chem, 68-72, ASSOC PROF MED & BIOL CHEM, SCH MED, UNIV CALIF, DAVIS, 72- Mem: Endocrine Soc; Am Fedn Clin Res; Am Chem Soc; Am Soc Biol Chem. Res:

Biochemistry; metabolic control processes; hormone action. Mailing Add: Dept of Internal Med Univ of Calif Sch of Med Davis CA 95616

WILSON, LOWELL L, b Egan, Ill, Jan 3, 36; m 55; c 3. ANIMAL SCIENCES. Educ: Wis State Univ, BS, 60; SDak State Univ, MS, 62, PhD(animal sci), 64. Prof Exp: Asst geneticist, Jacques Seed Corn Co, 58-60; res asst animal genetics, SDak State Univ, 60-64; livestock specialist, Purdue Univ, 64-66; assoc prof animal prod, 66-71, PROF ANIMAL SCI, PA STATE UNIV, UNIVERSITY PARK, 71- Honors & Awards: Meat Animal Mgt Award, Am Soc Animal Sci, 73. Mem: AAAS; Am Genetic Asn; Am Soc Animal Sci; Am Meat Sci Asn. Res: Selection indices for beef cattle and estimations of genetic parameters from use of selected sires; ranch x sire and sex x sire interactions; beef cattle and sheep behavior. Mailing Add: 238 Webster Dr State College PA 16802

WILSON, LYNN HAROLD, b Clinton, Mich, Feb 8, 35; m 57; c 3. POLYMER CHEMISTRY. Educ: Univ Colo, AB, 57, PhD(org chem), 61. Prof Exp: Asst chem, Univ Colo, 56-59, res assoc org chem, 59-61; sr chemist, Minn Mining & Mfg Co, 61-66, sr scientist, 66-68; staff chemist, IBM Corp, 68-69, mgr res & develop magnetic media, 69-75, ADV SCIENTIST, IBM CORP, 75- Mem: Am Chem Soc; Sigma Xi. Res: Organic polymer science; surface chemistry as related to adsorption phenomena and adhesion; organic fluorine chemistry; friction and wear phenomena at high energy interfaces. Mailing Add: 7363 Cortez Lane Boulder CO 80303

WILSON, LYNN OLSON, b Wilmington, Del, July 9, 44. APPLIED MATHEMATICS. Educ: Oberlin Col, AB, 65; Univ Wis, PhD(appl math), 70. Prof Exp: MEM TECH STAFF MATH RES, BELL LABS, 70- Mem: Am Phys Soc; Soc Indust & Appl Math. Res: Partial differential equations; asymptotics; electromagnetic theory. Mailing Add: Bell Labs Ctr 121 600 Mountain Ave Murray Hill NJ 07974

WILSON, MABEL FLOREY, b Omaha, Nebr, Mar 17, 06; m 29; c 2. ANALYTICAL CHEMISTRY. Educ: Buena Vista Col, BS, 27; Mich State Col, MS, 30, PhD(phys chem), 37. Prof Exp: Teacher high schs, Iowa, 28-29; lab asst, Mich State Col, 35-36; spec analyst, Burgess Labs, 37-38; spectroscopist, Diamond Alkali Co, 38-44, sr res chemist, 44-52; sr spectroscopist, Res Lab, Air Reduction Co, Inc, 52-56, head instrumental anal sect, 56-61; group leader plastics div res, Allied Chem Corp, 61-67, res assoc, 67-70; FORENSIC SPECTROSCOPIST, LAB OF STATE MED EXAMR, NEWARK, NJ, 70- Honors & Awards: Award of Merit, Am Soc Test & Mat, 71. Mem: Am Chem Soc; Soc Appl Spectros; fel Am Inst Chem; fel Am Soc Test & Mat. Res: Trace metals in nonmetallic materials, particularly body tissues and fluids; plastics by means of instrumental analysis including optical and x-ray spectroscopy, x-ray diffraction, atomic, infrared and ultraviolet absorption.

WILSON, MARION EVANS, b Irwin, Pa, Nov 6, 16. MEDICAL MICROBIOLOGY. Educ: Temple Univ, BSMT, 39; Smith Col, MA, 47; Boston Univ, PhD(med sci), 51. Prof Exp: Instr microbiol & immunol, State Univ NY Downstate Med Ctr, 51-54; asst dir microbiol & serol, St Luke's Hosp, New York, 54-65; dir microbiol & serol, Mt Sinai Hosp, Miami Beach, 65-66; chief microbiologist, 66-71, ASST DIR BUR LABS, NEW YORK CITY DEPT HEALTH, 71- Concurrent Pos: Consult microbiol, Social Security Admin, Bur Health Ins, US Dept Health, Educ & Welfare, 72; mem adv comt, Prof Exam Serv, Nat Clin Lab Technologist Exam, 74-75. Mem: Fel & dipl Am Acad Microbiol; Am Soc Microbiol; Am Pub Health Asn; fel NY Acad Sci. Res: Areas of public health interest, particularly gonorrhea detection; development and conduct of laboratory improvement methods, especially proficiency testing techniques. Mailing Add: New York City Dept of Health Bur of Labs 455 First Ave New York NY 10016

WILSON, MARK CURTIS, b Ware, Mass, Sept 19, 21. ENTOMOLOGY. Educ: Univ Mass, BS, 44; Ohio State Univ, MS, 46. Prof Exp: Field Aide, Div Truck Crops Invests, Bur Entom & Plant Quarantine, US Dept Agr, 45; asst zool, Ohio State Univ, 44-47; from asst prof to assoc prof, 47-69, PROF ENTOM, PURDUE UNIV, W LAFAYETTE, 69- Concurrent Pos: Consult, Adv Comt, Alfalfa Seed Coun, Food & Agr Orgn, Rumania, 71. Mem: Entom Soc Am. Res: Insect pest management; economic insect thresholds; host plant resistance. Mailing Add: Dept of Entom Purdue Univ West Lafayette IN 47906

WILSON, MARK FERLIN, b Akron, Ohio, Aug 24, 39; m 62; c 2. MASS SPECTROMETRY. Educ: Union Col, NY, BS, 61; Ohio State Univ, MSc, 62, PhD(low temperature physics), 67. Prof Exp: Res asst physics, Low Temperature Lab, Ohio State Univ, 61-67; res physicist, Plastics Dept, Exp Sta, Del, 67-70, & Polymer Intermediates Dept, Sabine River Lab, Tex, 70-73, TECH REP, DUPONT INSTRUMENTS, E I DU PONT DE NEMOURS & CO, INC, BEAUMONT, 73- Mem: Am Phys Soc; Sigma Xi. Res: Superconductivity; osmotic pressure of dilute solutions of liquid helium-3 and helium-4.

WILSON, MARK VINCENT HARDMAN, b Toronto, Ont, Feb 11, 46; m 70; c 2. VERTEBRATE PALEONTOLOGY. Educ: Univ Toronto, BSc, 68, MSc, 70, PhD(geol), 74. Prof Exp: Asst prof biol, Queen's Univ, Kingston, Ont, 74-75; ASST PROF ZOOL, UNIV ALTA, 75- Concurrent Pos: Res assoc, Dept Vert Palaeont, Royal Ont Mus, 74- Mem: Soc Vert Paleont; Can Soc Zoologists; Am Soc Ichthyologists & Herpetologists; Soc Syst Zool; Soc Study Evolution. Res: Fossil fishes, especially Tertiary freshwater teleosts; zoological systematics, especially phenetics; paleoecology of freshwater lacustrine sediments; Tertiary insects; anatomy and evolution of recent fishes. Mailing Add: Dept of Zool Univ of Alta Edmonton AB Can

WILSON, MARTIN, b Berlin, Ger, June 12, 13; US citizen; m 47; c 2. CHEMISTRY. Educ: Univ Geneva, DSc(constitution of starch), 39. Prof Exp: Res chemist, Palestine Potash Co, 43-48, Bonneville Ltd, Utah, 48-50, Kennecott Copper Corp, Utah, 50-55 & Nat Potash Co, NMex, 56; from res chemist to sr res chemist, 56-62, SR SCIENTIST, US BORAX RES CORP, 62- Mem: Am Chem Soc. Res: Potash refining; phase equilibrium. Mailing Add: US Borax Res Corp 412 Crescent Way Anaheim CA 92801

WILSON, MARVIN CRACRAFT, b Wheeling, WVa, Aug 7, 43; m 66; c 2. PHARMACY, PHARMACOLOGY. Educ: WVa Univ, BS, 66; Univ Mich, Ann Arbor, PhD(pharmacol), 70. Prof Exp: Res assoc, Dept Psychiat, Univ Chicago, 70; asst prof, 70-73, ASSOC PROF PHARMACOL, SCH PHARM, UNIV MISS, 73- Mem: AAAS; Am Asn Cols Pharm; Acad Pharmaceut Sci; Am Soc Clin Pharmacol & Therapeut; Behav Pharmacol Soc. Res: Neurochemical, neurophysiological and neuropharmacological factors which mediate psychomotor stimulant self-administration behavior; pharmacokinetics of stimulant self-administration; effects of cocaine on positively and negatively reinforced behavior and group behavior of non-human primates. Mailing Add: Dept of Pharmacol Univ of Miss Sch of Pharm University MS 38677

WILSON, MARY HELEN, biochemistry, see 12th edition

WILSON, MATHEW KENT, b Salt Lake City, Utah, Dec 22, 20; m 44; c 1. PHYSICAL CHEMISTRY. Educ: Univ Utah, BS, 43; Calif Inst Technol, PhD(phys chem), 48. Prof Exp: Asst, Off Sci Res & Develop, Calif Inst Technol, 43-46; instr chem, Harvard Univ, 48-51, asst prof, 51-56; prof & chmn dept, Tufts Univ, 56-66; head chem sect, 66-74, head, Off Energy-Related Gen Res, 74; DEP ASST DIR PLANNING & EVAL, MATH & PHYS SCI & ENG DIRECTORATE, NAT SCI FOUND, 75- Concurrent Pos: Guggenheim fel & Fulbright scholar, King's Col, London, 54-55; mem, Adv Coun Col Chem, 63-66; ed, Spectrochimia Acta, 64- Mem: AAAS; Am Chem Soc; Optical Soc Am; fel Am Acad Arts & Sci. Res: Molecular spectroscopy. Mailing Add: Nat Sci Found 1800 G St NW Washington DC 20550

WILSON, MATTHEW WOODROW, b Schenectady, NY, Mar 27, 13; m 40; c 2. PHYSICAL CHEMISTRY. Educ: Union Col, BS, 38; Univ Wis, MS, 41; Western Reserve Univ, PhD(phys chem), 46. Prof Exp: Asst phys & anal chem, Univ Wis, 38-40; res chemist, 40-58, SR DEVELOP ENGR, B F GOODRICH CO, 58- Mem: Am Chem Soc; Fiber Soc; Sigma Xi. Res: Electrochemistry; colloids; textiles; adhesives; tire cord. Mailing Add: 3520 Edgewood Dr Stowe OH 44224

WILSON, MCCLURE, b Ogden, Utah, July 30, 24; m 50; c 2. RADIOLOGY. Educ: Univ Ark, BS, 47, MD, 48. Prof Exp: From asst prof to assoc prof radiol, Univ Tex Med Br, 55-63; radiologist, Scott & White Clin, Temple, Tex, 63-64; assoc prof radiol, 64-69, PROF RADIOL, UNIV TEX MED BR GALVESTON, 69- Mem: AMA; Radiol Soc NAm; Am Roentgen Ray Soc. Mailing Add: Dept of Radiol Univ of Tex Med Br Galveston TX 77550

WILSON, MELVIN SMITH, physics, see 12th edition

WILSON, MICHAEL FRIEND, b Morgantown, WVa, Jan 13, 27; m 54; c 5. PHYSIOLOGY, BIOPHYSICS. Educ: WVa Univ, AB, 49; Univ Pa, MD, 53. Prof Exp: From intern to resident med, Presby Hosp, Philadelphia, 53-55; resident physician internal med, Med Ctr, Temple Univ, 55-57; from asst prof to assoc prof physiol & biophys, Col Med, Univ Ky, 60-65; PROF PHYSIOL & BIOPHYS & CHMN DEPT, MED CTR, WVA UNIV, 65-, CLIN PROF MED, 73- Concurrent Pos: Fel physiol & cardiol, Med Ctr, Temple Univ, 57-58; res fel physiol & biophys, Sch Med, Univ Wash, 58-60; vis prof, Sch Med, Univ Nottingham, 72-73. Mem: Fel Am Col Cardiol; Am Physiol Soc; Biophys Soc; Pavlovian Soc NAm; Inst Elec & Electronics Eng. Res: Neurocirculatory control, cardiovascular dynamics and behavior correlates; renal blood flow and function; coronary artery disease. Mailing Add: Dept of Physiol & Biophys WVa Univ Med Ctr Morgantown WV 26506

WILSON, MICHAEL JOHN, b Iowa City, Iowa, June 3, 42; m 69. ENDOCRINOLOGY, CELL BIOLOGY. Educ: St Ambrose Col, BA, 64; Univ Iowa, MS, 67, PhD(zool), 71. Prof Exp: NIH fel biochem, Harvard Univ, 71-73; res assoc, 73-75, ASST PROF LAB MED & PATH, UNIV MINN, MINNEAPOLIS, 75- Mem: Am Soc Zoologists; Am Soc Cell Biol; AAAS. Res: Site and mode of action of hormones; in particular, the subcellular sites of hormone interaction in induction of specific effects and the cellular organelle response and role in mediating these effects. Mailing Add: Toxicol Res Lab Vet Admin Hosp Minneapolis MN 55417

WILSON, MICHAEL ROBERT, b Oakland, Calif, July 26, 40; m 66; c 2. GEOGRAPHY. Educ: Leland Stanford Jr Univ, BA, 62; Univ Ore, PhD(geog), 72. Prof Exp: Lectr, 70-73, ASST PROF GEOG, UNIV SASK, 73- Mem: Am Geog Soc; Asn Am Geogrs. Res: Culture-environment relationships in non-industrial cultures, known as cultural ecology. Mailing Add: Dept of Geog Univ of Sask Saskatoon SK Can

WILSON, MIRIAM GEISENDORFER, b Yakima, Wash, Dec 3, 22; m 47; c 5. MEDICINE, PEDIATRICS. Educ: Univ Wash, BS, 44, MS, 45; Univ Calif, MD, 50; Am Bd Pediat, dipl. Prof Exp: From intern to resident pediat, Los Angeles County Hosp, 50-54; pvt pract, 54-56; sr physician, Bur Maternal & Child Health, Los Angeles County Health Dept, 57-58; asst prof pediat, Sch Med, Univ Calif, Los Angeles, 58-65, asst prof maternal & child health, Sch Pub Health, 60-65; assoc prof pediat, 65-69, PROF PEDIAT, SCH MED, UNIV SOUTHERN CALIF, 69-; CHIEF GENETICS DIV & DIR CYTOGENETICS LAB, PEDIAT PAVILION, LOS ANGELES COUNTY-UNIV SOUTHERN CALIF MED CTR, 65- Concurrent Pos: Mem, President's Comt Ment Retardation, 72-75. Mem: Am Acad Pediat; Am Pediat Soc; Am Pub Health Asn; Am Soc Human Genetics. Res: Medical genetics and cytogenetics; medical problems of the newborn and premature infant; growth and development of the infant and child; maternal and child health. Mailing Add: Dept of Pediat Univ of Southern Calif Med Ctr Los Angeles CA 90033

WILSON, MONTE DALE, b Pomeroy, Wash, Nov 16, 38; m 62; c 2. GEOLOGY. Educ: Brigham Young Univ, BS, 62; Univ Idaho, MS, 68, PhD(geol), 70. Prof Exp: Geophysicist, Can Magnetic Reduction, Ltd, Alta, 62-63 & US Army Ballistics Res Labs, 63-65; teacher high schs, Idaho, 65-67; instr geol, Univ Idaho, 68-69; PROF GEOL, BOISE STATE UNIV, 69- Concurrent Pos: Mem, Idaho Bd Registration Prof Geologists, 75-80. Mem: Am Asn Petrol Geol; Am Quaternary Asn; Nat Asn Geol Teachers; Geol Soc Am. Res: Glacial and periglacial geomorphology; field geology and mapping in Northern Rocky Mountains. Mailing Add: Dept of Geol Boise State Univ Boise ID 83725

WILSON, NIXON ALBERT, b Litchfield, Ill, May 20, 30; m 63; c 2. ACAROLOGY. Educ: Earlham Col, BA, 52; Univ Mich, MWM, 54; Purdue Univ, PhD(entom), 61. Prof Exp: Animal ecologist, Plague Res Labs, Hawaii State Dept Health, 61-62; acarologist, B P Bishop Mus, 62-69; from asst prof to assoc prof, 69-75, PROF BIOL, UNIV NORTHERN IOWA, 75- Concurrent Pos: USPHS res grants, B P Bishop Mus, 67-69; grant, Univ Northern Iowa, 69-72; res assoc, Fla State Collection of Arthropods, Fla Dept Agr & Consumer Servs, 70- Mem: Am Soc Mammal; Am Soc Parasitol; Wildlife Dis Asn; Acarological Soc Am. Res: Ectoparasites of vertebrates, especially mites; ecology of bats. Mailing Add: Dept of Biol Univ of Northern Iowa Cedar Falls IA 50613

WILSON, OLIN CHADDOCK, b San Francisco, Calif, Jan 13, 09; m 43; c 2. ASTRONOMY, ASTROPHYSICS. Educ: Univ Calif, AB, 29; Calif Inst Technol, PhD(astrophys), 34. Prof Exp: Computer & asst, 31-36, asst astronomer, 36-51, ASTRONOMER, MT WILSON OBSERV, 51- Mem: Nat Acad Sci; Am Astron Soc. Res: Radial velocity programs; stellar atmospheres; interstellar matter; Wolf-Rayet stars; planetary nebulae; stellar chromospheres. Mailing Add: Mt Wilson & Palomar Observs 813 Santa Barbara St Pasadena CA 91101

WILSON, ONSLOW HARUS, b Basseterre, St Kitts, July 12, 34; Can citizen; m 58; c 5. CELL BIOLOGY. Educ: Mt Allison Univ, BA, 57; McGill Univ, MSc, 63, PhD(biochem), 66. Prof Exp: Can Med Res Coun grant, City of Hope Med Ctr, Calif, 66-68; ASST PROF MEMBRANE BIOL, CLIN RES INST MONTREAL, 68- Mem: AAAS: Am Soc Microbiol; Can Soc Immunol. Res: Role of lipids in expression of membrane phenotypes; membrane and the immune response with emphasis on the thymus; role of the thymus in hypertensive disease. Mailing Add: Clin Res Inst of Montreal 110 Pine Ave W Lab 233 Montreal PQ Can

WILSON, OSCAR BRYAN, JR, b Tex, Aug 15, 22; m 45; c 3. PHYSICS. Educ: Univ Tex, BS, 44; Univ Calif, Los Angeles, MA, 48, PhD, 51. Prof Exp: Mem tech staff, Hughes Aircraft Co, 51-52; physicist, Soundrive Engine Co, 52-57; assoc prof, 57-62, PROF PHYSICS, NAVAL POSTGRAD SCH, 62- Mem: AAAS; Am Phys Soc; Acoust Soc Am; Inst Elec & Electronics Engrs. Res: Physical acoustics; underwater acoustics. Mailing Add: Dept of Physics Naval Postgrad Sch Monterey CA 93940

WILSON, P DAVID, b Roswell, NMex, Oct 4, 33; m 65; c 1. STATISTICS, BIOMATHEMATICS. Educ: Univ Colo, Boulder, BA, 56; Univ Minn, Minneapolis, MS, 63; Johns Hopkins Univ, PhD(biostatist), 70. Prof Exp: Res assoc biostatist, Med Sch, Univ Md, Baltimore, 64-66; consult, Nat Coun Stream Improvement, 67; consult, Biostatist & Biomath, Ctr Study Trauma, Med Sch, Univ Md, Baltimore, 67-71; asst prof biomet, Med Col Va, 70-71; math statistician, Bur Drugs, Food & Drug Admin, Dept Health, Educ & Welfare, 71-72; asst prof biostatist & biomath, Ctr Study Trauma, 72-74, ASST PROF, DEPT SOC & PREV MED, MED SCH, UNIV MD, BALTIMORE, 74- Concurrent Pos: Consult, Huntingdon Res Ctr, Baltimore, 73-; lectr, Dept Biostatist, Sch Hyg & Pub Health, Johns Hopkins Univ, 73- Mem: Biomet Soc; Am Statist Asn. Res: Estimation and prediction in physiological systems. Mailing Add: 206 Upnor Rd Baltimore MD 21212

WILSON, PATRICIA A, physical chemistry, physics, see 12th edition

WILSON, PAUL ROBERT, b Chicago, Ill, Oct 25, 39. MATHEMATICS. Educ: Univ Cincinnati, BA, 61, MA, 62; Univ Ill, PhD(math), 67. Prof Exp: Asst prof math, Univ Nebr, Lincoln, 67-71; ASST PROF MATH, ALMA COL, 71- Mem: Am Math Soc. Res: Algebraic number theory. Mailing Add: Dept of Math Alma Col Alma MI 48801

WILSON, PEGGY MAYFIELD DUNLAP, b Austin, Tex, Mar 24, 27; m 75. SURFACE CHEMISTRY. Educ: Univ Tex, BS, 48, PhD(chem), 52. Prof Exp: Spec instr chem, Univ Tex, 52-53; sr res technologist, 53-67, RES ASSOC, MOBIL RES & DEVELOP CORP, 67- Concurrent Pos: State Republican Committeewoman, 71- Mem: Am Chem Soc. Res: Interfacial tension and contact angles; tertiary oil recovery. Mailing Add: Mobil Res & Develop Corp PO Box 900 Dallas TX 75221

WILSON, PERRY BAKER, b Norman, Okla, Feb 24, 27; m 51; c 3. HIGH ENERGY PHYSICS. Educ: Wash State Univ, BS, 50, MS, 52; Stanford Univ, PhD(physics), 58. Prof Exp: Staff physicist, Linfield Res Inst, Ore, 58-59; res assoc accelerator physicist, High Energy Physics Lab, Stanford Univ, 59-64, assoc dir opers, 64-68; vis scientist, Europ Orgn Nuclear Res, Geneva, Switzerland, 68-69; SR RES ASSOC, STANFORD LINEAR ACCELERATOR CTR, STANFORD UNIV, 69- Concurrent Pos: Consult, Gen Atomic Div, Gen Dynamics Corp, Calif, 63-64, Phys Electronics Labs, 64 & US Naval Postgrad Sch, 64-65; adj prof, Stanford Linear Accelerator Ctr, Stanford Univ, 74-; consult, Varian Assocs, 74- Mem: AAAS; Am Phys Soc; Inst Elec & Electronics Eng. Res: Theory and design of linear electron accelerators and storage rings for high energy particle physics; design of medical accelerators; theory and design of superconducting accelerators. Mailing Add: Stanford Linear Accelerator Ctr PO Box 4349 Stanford CA 94305

WILSON, PERRY WILLIAM, b Bonanza, Ark, Nov 25, 02; m 29; c 2. BACTERIOLOGY, BIOCHEMISTRY. Educ: Univ Wis, BS, 28, MS, 29, PhD(bact, biochem), 32. Prof Exp: Res assoc, Univ Wis-Madison, 32-34, from asst prof to prof agr bact, 34-74; RETIRED. Concurrent Pos: Guggenheim Mem Found fel, Cambridge Univ & Biochem Inst, Helsinki, 36-37; consult, Fed Security Admin; ed, Bact Rev, Am Soc Microbiol, 52-58. Mem: Nat Acad Sci; Am Soc Microbiol (pres, 57); Am Chem Soc; Am Soc Biol Chemists; Soc Exp Biol & Med. Res: Use of heavy nitrogen isotope nitrogen-15 in study of mechanism of biological nitrogen fixation; enzyme systems in biological nitrogen fixation. Mailing Add: Dept of Bact Univ of Wis Madison WI 53706

WILSON, PHILO CALHOUN, b Westfield, Mass, Jan 29, 24; m 47; c 3. STRATIGRAPHY, MARINE GEOLOGY. Educ: Williams Col, AB, 48; Cornell Univ, MS, 50; Wash State Univ, PhD(stratig), 54. Prof Exp: From geologist to staff geologist, Sohio Petrol Co, 54-60; area geologist, Champlin Oil Co, 60-63; assoc prof, 63-71, chmn dept, 67-73, PROF EARTH SCI, STATE UNIV NY COL ONEONTA, 71- Concurrent Pos: Mem selection panels rev of NSF Proposals, 65-66 & 70. Mem: Geol Soc Am; Am Asn Petrol Geol. Res: Sedimentation; regional stratigraphic analysis; Pennsylvanian system of Wyoming. Mailing Add: Dept of Earth Sci State Univ of NY Col Oneonta NY 13820

WILSON, R MARSHALL, organic chemistry, see 12th edition

WILSON, RAPHAEL, b Trenton, NJ, Apr 25, 25. MEDICAL BACTERIOLOGY. Educ: Univ Notre Dame, BS, 48; Univ Tex, MA, 51, PhD(bact), 54. Prof Exp: From instr to assoc prof biol, St Edward's Univ, 48-59, dean col, 51-58, dir testing & guid, 58-59; from asst prof to assoc prof biol, Univ Notre Dame, 59-71; PROF PEDIAT, BAYLOR COL MED, 71- Concurrent Pos: Vis prof, Univ Ulm, Ger, 69-70 & Baylor Col Med, 70-71. Mem: Transplantation Soc; Radiation Res Soc; Am Soc Microbiol; Soc Exp Biol & Med; Soc Exp Hemat. Res: Germfree life; protection against radiation damage; role of the thymus; clinical gnotobiology and immunology; gastrointestinal microflora. Mailing Add: Dept of Pediat Baylor Col Med Houston TX 77025

WILSON, RAY FLOYD, b Lee Co, Tex, Feb 20, 26; m 57; c 2. ANALYTICAL CHEMISTRY, PHYSICAL CHEMISTRY. Educ: Houston-Tillotson Col, BS, 50; Tex Southern Univ, MS, 61; Univ Tex, PhD(chem), 53. Prof Exp: Asst, Univ Tex, 51-53; assoc prof, 53-57, PROF CHEM, TEX SOUTHERN UNIV, 57- Concurrent Pos: Grants, Res Corp, 53-55, NSF, 54, 56 & Welch Found, 57, 59. Mem: Am Chem Soc. Res: Interaction of platinum elements with certain organic reagents. Mailing Add: Dept of Chem Tex Southern Univ Houston TX 77004

WILSON, RAYMOND B, analytical chemistry, see 12th edition

WILSON, RAYMOND EDGAR, physics, see 12th edition

WILSON, RAYMOND HIRAM, JR, b Gap, Pa, Feb 14, 11; m 40; c 1. ASTRONOMY, APPLIED MATHEMATICS. Educ: Swarthmore Col, AB, 31; Univ Pa, AM, 33, PhD(astron), 35. Prof Exp: Res asst astron, Sproul Observ, Swarthmore Col, 29-30 & Flower Observ, Univ Pa, 31-35; instr math, Southern Methodist Univ, 35-36 & Gettysburg Col, 37-38; res assoc astron, Univ Minn, 38-39, Sproul Observ, Swarthmore Col & Flower Observ, Univ Pa, 39-40; jr astronr, US Naval Observ, 40-46; asst prof math & astron, Temple Univ, 46-51; asst prof math, Univ Louisville, 51-54; physicist, Proj Vanguard, Naval Res Lab, 54-58; space scientist & chief appl math, NASA, 58-71; prof astron & appl math, Univ of the Aegean, Turkey, 71-74; ASTRONR, ARMAGH OBSERV, 74- Concurrent Pos: Res assoc astron, Flower Observ, 37; officer instr math, US Naval Acad, 42-46; vis asst prof astron, Haverford Col, 48-49; consult phys chem, Res Inst, Temple Univ, 50-54; prof lectr astron, Georgetown Univ, 62-68; liaison mem for NASA, Div Math, Nat Acad Sci-Nat Res Coun, 63-68. Honors & Awards: Patent Appln Award, NASA, 63. Mem: Fel AAAS; Am Astron Soc; Math Asn Am; Soc Indust & Appl Math; Sigma Xi. Res: Orbits and measurement of binary stars; interferential methods for astrometry; numerical methods for solution of orbits. Mailing Add: Armagh Observ Armagh BT61 9DG Northern Ireland

WILSON, RICHARD, b London, Eng, Apr 29, 26; m 52; c 6. PHYSICS. Educ: Oxford Univ, BA, 46, MA & DPhil(physics), 49; Harvard Univ, MA, 56. Prof Exp: Res lectr physics, Christ Church Col, Oxford Univ, 48-53, res off, Clarendon Lab, 53-55; from asst prof to assoc prof, 55-61, PROF PHYSICS, HARVARD UNIV, 61- Concurrent Pos: Res assoc, Univ Rochester, 50-51 & Stanford Univ, 51-52; Guggenheim fel, 61 & 69; Fulbright fel, 61 & 69; trustee, Univs Res Asn, 68-74; consult, Energy Res Develop Agency, 75-, Nuclear Regulatory Comn, Nat Res Coun, 75- & Electric Power Res Inst, 75- Mem: Am Phys Soc; Am Acad Arts & Sci. Res: Elementary particle physics; environmental physics. Mailing Add: Lyman 231 Harvard Univ Cambridge MA 02138

WILSON, RICHARD BARR, b Lincoln, Nebr, Apr 21, 21; m 48; c 2. PATHOLOGY. Educ: Univ Nebr, AB, 43, MD, 45; Am Bd Path, dipl, 57. Prof Exp: Staff physician, Univ Nebr, Lincoln, 47-52; resident path, Univ Hosp, 53-57, assoc, Col Med, 57-62, from asst prof to assoc prof, 62-70, PROF PATH, COL MED & HEAD ELECTRON MICROS SECT, EPPLEY INST CANCER RES, UNIV NEBR, OMAHA, 70- Concurrent Pos: Attend pathologist, Vet Admin Hosp, Omaha, 60- Mem: Col Am Path; Am Soc Clin Path; Electron Micros Soc Am; Int Acad Path; Am Soc Nephrology. Res: Renal disease and biopsies; electron microscopy of human biopsies; morphological studies; animal carcinogenesis; ultrastructure of human neoplasms. Mailing Add: Eppley Inst for Cancer Res Univ of Nebr Med Ctr Omaha NE 68105

WILSON, RICHARD ENOS, physical chemistry, see 12th edition

WILSON, RICHARD FAIRFIELD, b Pittsburgh, Pa, Nov 26, 30; m 52; c 6. GEOLOGY. Educ: Yale Univ, BS, 52; Stanford Univ, MS, 54, PhD(geol), 59. Prof Exp: Geologist, US Geol Surv, 55-62; asst prof geol, 62-66, ASSOC PROF GEOSCI, UNIV ARIZ, 66- Mem: AAAS; Paleont Soc; Geol Soc Am. Res: Stratigraphy; sedimentation; study of sedimentary rocks. Mailing Add: Dept of Geosci Univ of Ariz Tucson AZ 85721

WILSON, RICHARD FERRIN, b Dundurn, Sask, Jan 8, 20; nat US; m 43; c 4. ANIMAL SCIENCE. Educ: Iowa State Col, BS, 43; Univ Ill, MS, 47, PhD(animal sci), 49. Prof Exp: Asst animal sci, Univ Ill, 46-49; assoc prof animal husb, SDak State Col, 49-52, actg chmn dept, 51-52; assoc prof, 52-56, PROF ANIMAL SCI, OHIO STATE UNIV, 57-, IN CHARGE SWINE, 52- Mem: Am Soc Animal Sci. Res: Swine management, nutrition, physiology and breeding; swine production with emphasis on nutrition and physiology involving toxic feeds resulting from molds. Mailing Add: 2029 Fyffe Rd Columbus OH 43210

WILSON, RICHARD GARTH, b Montreal, Que, July 30, 45; m 67; c 1. CLIMATOLOGY. Educ: McGill Univ, BSc, 66, MSc, 68; McMaster Univ, PhD(climat), 71. Prof Exp: Lectr climat, Dept Geog, Queen's Univ, 67-68; lectr, Dept Geog, McGill Univ, 70-71, asst prof, 71-75; SUPVR CLIMATE INVENTORY, CLIMATE & DATA SERV, BC ENVIRON & LAND USE COMT SECRETARIAT, 75- Mem: Am Meteorol Soc; Can Meteorol Soc. Res: Solar and terrestrial radiation, evaporation, snow melt, snow mapping, topoclimatology; climate network design. Mailing Add: BC Environ & Land Use Comt Parliament Bldg Victoria BC Can

WILSON, RICHARD HANSEL, b Madison, Wis, Aug 18, 39; m 64; c 2. PLANT PHYSIOLOGY, BIOCHEMISTRY. Educ: Carleton Col, BA, 61; NC State Univ, MS, 64; Ore State Univ, PhD(plant physiol), 67. Prof Exp: USPHS fel plant physiol, Univ Ill, Urbana, 67-68; asst prof, Univ Tex, Austin, 68-74; RES SPECIALIST, MONSANTO CHEM CO, 74- Mem: Am Soc Plant Physiol; Japanese Soc Plant Physiol; Scandinavian Soc Plant Physiol. Res: Plant growth regulation. Mailing Add: Monsanto Chem Co 800 N Lindbergh Blvd St Louis MO 63166

WILSON, RICHARD HEILBRON, b Decatur, Ohio, May 15, 19; m 42; c 2. AGRICULTURAL EDUCATION, AGRICULTURAL ECONOMICS. Educ: Ohio State Univ, BS, 43, MA, 51, PhD, 55. Prof Exp: Instr high sch, 47-50; res fel agr educ, 51-53, instr, 53-57, from asst prof to assoc prof, 57-69, PROF AGR EDUC, OHIO STATE UNIV, 69- Res: Educational evaluation and research; adult education; farm management; agronomy. Mailing Add: Dept of Agr Educ Ohio State Univ Columbus OH 43210

WILSON, RICHARD HOWARD, b Spearville, Kans, June 24, 42; m 62; c 1. ZOOLOGY, ANIMAL BEHAVIOR. Educ: Kans State Univ, BS, 64, MS, 65; Utah State Univ, PhD(zool), 71. Prof Exp: Instr, Prince Makonnen Sec Sch, 65-66; instr, 66-70, ASST PROF BIOL, UNIV WIS-STOUT, 70- Mem: AAAS; Animal Behav Soc; Am Ornith Union; Cooper Ornith Soc; Wilson Ornith Soc. Res: Animal communication; display postures and signaling mechanisms in birds. Mailing Add: Dept of Biol Univ of Wis-Stout Menomonie WI 54751

WILSON, RICHARD J, b Indianapolis, Ind, May 27, 21; m 46; c 2. PHYSICS, OPTICS. Educ: Univ Rochester, BS, 42. Prof Exp: Res physicist, Univ Rochester, 42-43; optical engr, Argus Cameras Div Sylvania Elec Prod, Inc, Gen Tel & Electronics Corp, 43-50, chief tech engr, 50-58, chief engr, 58-60; asst dir res, 60-61, mgr fiber optics dept, 61-63, gen mgr, Fecker Div, 63-67, instrument div, 67-68, vpres, 68-74, PRES, AM OPTICAL CORP, 74- Mem: Optical Soc Am. Mailing Add: Am Optical Corp Eggert & Sugar Rds Buffalo NY 14215

WILSON, RICHARD LEE, b Marshalltown, Iowa, Sept 18, 39; m 60; c 3. ENTOMOLOGY. Educ: Univ Northern Iowa, BA, 61; Tex A&M Univ, MS, 65; Iowa State Univ, PhD(entom), 71. Prof Exp: Teacher, Independent Sch Dist, Iowa, 65-68; RES ENTOMOLOGIST, AGR RES SERV, US DEPT AGR, 71- Mem: AAAS; Entom Soc Am. Res: Host plant resistance; development of resistant cotton varieties to several cotton pests. Mailing Add: Western Cotton Res Lab 4135 E Broadway Rd Phoenix AZ 85040

WILSON, RICHARD MAC, b Canton, Ohio, Oct 30, 30; m 55; c 3. ELECTROCHEMISTRY, PHYSICAL CHEMISTRY. Educ: Mt Union Col, BS, 52; Ohio State Univ, MS, 56, PhD(chem), 59. Prof Exp: Asst prof chem, Denison Univ, 59-60; sr chemist, Delco-Remy Div, Gen Motors Corp, 60-63; sr tech res, 63-68, proj mgr, 68-70, GROUP LEADER, CONSUMER PRODS DIV, UNION CARBIDE CORP, 70- Mem: Electrochem Soc. Res: Fuel cells and secondary batteries such as silver-zinc, nickel-cadmium and lead acid. Mailing Add: Consumer Prods Div Union Carbide Corp PO Box 6116 Cleveland OH 44101

WILSON, RICHARD MICHAEL, b Gary, Ind, Nov 23, 45; m 66. MATHEMATICS. Educ: Ind Univ, Bloomington, AB, 66; Ohio State Univ, MS, 68, PhD(math), 69. Prof Exp: From asst prof to assoc prof, 69-74, PROF MATH, OHIO STATE UNIV, 74- Concurrent Pos: Res fel, A P Sloan Found, 75-77. Mem: Math Asn Am; Am Math Soc; Soc Indust & Appl Math. Res: Combinatorial mathematics, with emphasis on combinatorial designs and related structures. Mailing Add: Dept of Math Ohio State Univ Columbus OH 43210

WILSON, ROBERT BURTON, b Salt Lake City, Utah, June 29, 36; m 62; c 1. EXPERIMENTAL PATHOLOGY. Educ: Utah State Univ, BS, 58; Wash State Univ, DVM, 61; Univ Toronto, PhD(physiol), 67. Prof Exp: Intern vet med, Angell Mem Animal Hosp, 62-63; asst prof animal sci, Brigham Young Univ, 63-64; res investr, Hosp for Sick Children, 64-67, asst scientist, 67-69; assoc prof nutrit & animal path, Mass Inst Technol, 69-73; prof vet path, Univ Mo, 73-76; PROF & CHMN DEPT MICROBIOL & PATH, WASH STATE UNIV, 76- Concurrent Pos: Lectr physiol, Univ Toronto, 67-69; Mary Mitchell res award, 63. Mem: Am Vet Med Asn; Am Diabetes Asn; Can Vet Med Asn; Am Inst Nutrit; Coun On Arteriosclerosis, Am Heart Asn. Res: Endocrinology; nutrition; diabetes; carcinogenesis; cardiovascular diseases; experimental pathology of nutrition and cardiovascular diseases. Mailing Add: Col of Vet Med Wash State Univ Pullman WA 99163

WILSON, ROBERT CURTIS, b Canton, Tex, Aug 31, 12; m 38; c 1. CHEMISTRY. Educ: NTex State Teachers Col, BS, 35, MS, 36; Univ Tex, PhD(org chem), 41. Prof Exp: Jr chemist, US Dept Agr, Tex, 38-40; chemist, 41-49, sect mgr, 49-57, asst mgr res & develop, 57-60, mgr, 60-64, SR TECH ASSOC, GAF CORP, 64- Mem: Am Chem Soc. Res: Photographic chemicals; photographic color formers and sensitizing dyes; organic chemical intermediates; pharmaceuticals; dyestuffs; acetylene derivatives and vinyl polymers. Mailing Add: 20 Colby Lane Cranford NJ 07016

WILSON, ROBERT E, b Norristown, Pa, Jan 16, 37; m 63; c 2. ASTROPHYSICS, X-RAY ASTRONOMY. Educ: Univ Pa, AB, 58, MS, 60, PhD(astron), 63. Prof Exp: Asst prof astron, Georgetown Univ, 63-66; Nat Res Coun sr res assoc, Inst Space Studies, 72-74; assoc prof, 66-70, PROF ASTRON, UNIV S FLA, 70-; PROF PHYSICS & ASTRON, UNIV FLA, 75- Concurrent Pos: Consult, Goddard Space Flight Ctr, NASA, 65-70; res grants, NASA, 66-69 & NSF, 70-73. Mem: Am Astron Soc; Int Astron Union; Royal Astron Soc. Res: Theory and observation of binary stars; x-ray astronomy. Mailing Add: Dept of Astron Univ of S Fla Tampa FL 33620

WILSON, ROBERT FRANCIS, b Scranton, Pa, Aug 9, 34; m 72; c 7. THORACIC SURGERY, CARDIOVASCULAR SURGERY. Educ: Lehigh Univ, BA, 57; Temple Univ, MD, 58. Prof Exp: From instr to assoc prof, 63-71, PROF SURG, SCH MED, WAYNE STATE UNIV, 71-, DIR AFFIL PROG THORACIC SURG, 71-, ASST DEAN DETROIT GEN HOSP AFFAIRS, 72-; CHIEF SECT THORACIC & CARDIOVASC SURG, HARPER HOSP, 72- Concurrent Pos: Markle scholar acad med; pres med staff, Detroit Gen Receiving Hosp, 72-74. Mem: Soc Univ Surg; Am Asn Thoracic Surg; Am Asn Surg Trauma; Am Col Surg; Am Col Chest Physicians. Res: Shock; respiratory failure; fluid and electrolytes. Mailing Add: Sch of Med Dept of Surg Wayne State Univ Detroit MI 48201

WILSON, ROBERT G, b Galesburg, Ill, Aug 30, 30; m 53; c 3. CHEMISTRY, BIOCHEMISTRY. Educ: Knox Col, AB, 52; Purdue Univ, MS, 56; Okla State Univ, PhD(chem), 61. Prof Exp: Res chemist, Nat Cancer Inst, 63-67; asst prof biochem, 67-71, ASSOC PROF BIOCHEM, COL MED NJ, 71- Concurrent Pos: NIH fel biochem, Brandeis Univ, 61-63. Mem: Am Chem Soc. Res: Initiation of RNA synthesis; effect of mutagens and carcinogens upon transcription. Mailing Add: Dept of Biochem Col of Med of NJ Newark NJ 07103

WILSON, ROBERT GRAY, b Wooster, Ohio, Apr 7, 34; m 57; c 2. ELECTRONICS, NUCLEAR PHYSICS. Educ: Ohio State Univ, BSc, 56, PhD(physics), 61. Prof Exp: Prin physicist, Battelle Mem Inst, 54-58; res asst, Res Found, Ohio State Univ, 58-60; sr physicist, NAm Aviation/Rocketdyne, 61-63; SR MEM TECH STAFF, RES LABS, HUGHES AIRCRAFT CO, 63- Mem: Am Phys Soc; Inst Elec & Electronics Eng. Res: Ion implantation; semiconductor devices and integrated circuits; electron and ion emission from surfaces; scanning acoustic microscopy; experimental low energy nuclear physics. Mailing Add: Dept Electron Device Physics Hughes Res Labs 3011 Malibu Canyon Rd Malibu CA 90265

WILSON, ROBERT HALLOWELL, b Baltimore, Md, July 30, 24; m 48; c 2. ENVIRONMENTAL HEALTH. Educ: Univ Rochester, BS, 45. Prof Exp: Jr scientist, Atomic Energy Proj, 46-51, instr indust hyg & toxicol & asst scientist, Atomic Energy Proj, 51-56, scientist, 56-62, CHIEF ENGR ATOMIC ENERGY PROJ, SCH MED & DENT, UNIV ROCHESTER, 62-, ASST PROF RADIATION BIOL & BIOPHYS, 56-, CHIEF ENVIRON HEALTH & SAFETY, CHIEF SAFETY OFFICER, UNIV, 75- Mem: AAAS; Am Nuclear Soc; Health Physics Soc; Soc Study Amphibians & Reptiles; Am Indust Hyg Asn. Res: Generation, sampling and behavior of aerosols; control of radiation hazards; air safety considerations of nuclear weapons transport and storage; environmental impact of mercury. Mailing Add: Sch of Med & Dent Univ of Rochester Rochester NY 14642

WILSON, ROBERT JAMES, b Edmonton, Alta, Feb 5, 15; m 41; c 2. BACTERIOLOGY, IMMUNOLOGY. Educ: Univ BC, BA, 35, MA, 37; Univ Toronto, MD, 42, DPH, 46. Prof Exp: Assoc hyg & prev med, 47-48, from asst prof to assoc prof, 48-71, PROF HEALTH ADMIN & MICROBIOL, UNIV TORONTO, 71- Concurrent Pos: Res assoc, Connaught Labs Ltd, 48-56, res mem, 56-57, from asst dir to assoc dir, 57-72, chmn & sci dir, 72- Res: Staphylococcal food poisoning; staphylococcus toxoid; pertussis vaccine and multiple antigens; multiple antigens contained in poliomyelitis vaccine; clinical trials of live oral poliomyelitis virus vaccine; clinical studies of penicillin. Mailing Add: Connaught Labs Ltd 1755 Steeles Ave W Willowdale ON Can

WILSON, ROBERT JOHN, b St Louis, Mo, Apr 23, 35; m 57; c 5. RADIOLOGICAL PHYSICS, NUCLEAR MEDICINE. Educ: St Mary's Univ, Tex, BS & BA, 56; Wash Univ, PhD(physics), 63. Prof Exp: Res assoc physics, Wash Univ, 63-66; res physicist, US Naval Radiol Defense Lab, San Francisco, 66-69; asst prof radiol, 69-72, ASSOC PROF NUCLEAR MED, UNIV TENN, MEMPHIS, 72- Mem: Am Asn Physicists in Med; Am Phys Soc; Soc Nuclear Med; Am Col Radiol; Radiol Soc NAm. Mailing Add: Dept of Nuclear Med Univ of Tenn Memphis TN 38163

WILSON, ROBERT LAKE, b Gallipolis, Ohio, July 2, 24; m 50; c 2. GEOLOGY. Educ: Wheaton Col, AB, 48; Univ Iowa, MS, 50; Univ Tenn, PhD, 67. Prof Exp: Asst geol, Univ Iowa, 49-50 & Univ Tenn, 50-52; area geologist, Tenn Div Geol, 52-55; from instr to assoc prof geol, 55-66, PROF GEOL & GEOG, UNIV TENN, CHATTANOOGA, 66- Concurrent Pos: NSF fac fel, 60-61. Mem: Soc Econ Paleontologists & Mineralogists; Nat Speleol Soc; Geol Soc Am; Nat Asn Geol Teachers; Am Asn Petrol Geol. Res: Paleozoic stratigraphy and sedimentation of Southern Appalachians; economic geology of the eastern United States. Mailing Add: Dept of Geol Univ of Tenn Chattanooga TN 37401

WILSON, ROBERT LEE, b Champaign, Ill, Mar 7, 17; m 40; c 4. MATHEMATICS. Educ: Univ Fla, AB, 38; Univ Wis, MA, 40, PhD(math), 47. Prof Exp: Asst math, Univ Wis, 39-41, 46-47; from instr to asst prof, Univ Tenn, 47-56; sr aerophys engr, Gen Dynamics/Convair, Tex, 56-58; PROF MATH, OHIO WESLEYAN UNIV, 58- Concurrent Pos: Adj prof, Tex Christian Univ, 56-58; vis prof & dir comput ctr, Univ Ibadan, 66-68. Mem: AAAS; Am Math Soc; Math Asn Am; Asn Comput Mach; Australian Math Soc. Res: Galois theory; numerical analysis; computing; matrix theory. Mailing Add: Dept of Math Ohio Wesleyan Univ Delaware OH 43015

WILSON, ROBERT LEE, b Washington, DC, Jan 16, 46; m 67. ALGEBRA. Educ: Am Univ, BA, 65; Yale Univ, PhD(math), 69. Prof Exp: Instr math, Courant Inst Math Sci, NY Univ, 69-71; asst prof, 71-75, ASSOC PROF MATH, RUTGERS UNIV, NEW BRUNSWICK, 75- Mem: Am Math Soc; Math Asn Am. Res: Lie algebras over fields of prime characteristic. Mailing Add: Dept of Math Rutgers Univ New Brunswick NJ 08903

WILSON, ROBERT LEE, b Peoria, Ill, Dec 16, 18; m 41; c 3. ECONOMIC GEOLOGY, MINING GEOLOGY. Educ: Colo Sch Mines, GeolE, 41; Univ Ariz, PhD(geol), 56. Prof Exp: Asst regional geologist, Bur Reclamation, US Dept Interior, 46-51; dist geologist, US Army Corps Engrs, US Dept Defense, 51-53; consult geologist, 55-56; geologist, 56-72, EXPLOR MGR, KAISER STEEL CORP, OAKLAND, 56- Mem: Geol Soc Am; Soc Econ Geol; Am Inst Mining, Metall & Petrol Eng; Asn Prof Geol Scientists. Res: Exploration methods applied to discovery of mineral deposits; genesis of mineral deposits. Mailing Add: 3153 Windsor Ct Lafayette CA 94549

WILSON, ROBERT LEE, JR, b Auburn, Ala, Jan 3, 42; m 62; c 2. MATHEMATICS. Educ: Ohio Wesleyan Univ, BA, 62; Univ Wis-Madison, MA, 63, PhD(math), 69. Prof Exp: Asst prof math, Univ Wis-Madison, 69-74; ASSOC PROF MATH, WASHINGTON & LEE UNIV, 74- Mem: Am Math Soc; Math Asn Am. Res: Combinatorics; graph theory; universal algebra; generalizations of group theory. Mailing Add: Dept of Math Washington & Lee Univ Lexington VA 24450

WILSON, ROBERT NORTON, b Walla Walla, Wash, Oct 7, 27; m 56; c 1. MATHEMATICAL PHYSICS, ATMOSPHERIC PHYSICS. Educ: Whitman Col, AB, 48; Stanford Univ, MS, 50, PhD(physics, math), 60. Prof Exp: Res scientist, Microwaves, Kane Eng Labs, Calif, 60-65 & Atmospheric Physics, Lockheed Res Lab, 65-71; RES SCIENTIST, RADIATIVE TRANSFER & HYDRODYN, MISSION RES CORP, 71- Mem: Am Phys Soc; Am Inst Physics. Res: Electromagnetic theory and experimentation; multiple quantum effect physics; non-equilibrium statistical mechanics; hydrodynamics; magnetohydrodynamics; radiation physics. Mailing Add: 16 W Mountain Dr Santa Barbara CA 93103

WILSON, ROBERT PAUL, b Revere, Mo, Dec 28, 41; m 62; c 3. BIOCHEMISTRY. Educ: Univ Mo, Columbia, BSEd, 63, MS, 65, PhD(biochem), 68. Prof Exp: Instr agr chem, Univ Mo, Columbia, 68-69; asst prof, 69-72, ASSOC PROF BIOCHEM, MISS STATE UNIV, 72- Mem: AAAS; Am Chem Soc. Res: Comparative toxicity and metabolism of ammonia; comparative purine and pyrimidine metabolism; fish biochemistry and nutrition. Mailing Add: Dept of Biochem Miss State Univ Drawer BB Mississippi State MS 39762

WILSON, ROBERT RATHBUN, b Frontier, Wyo, Mar 4, 14; m 40; c 3. PHYSICS. Educ: Univ Calif, AB, 36, PhD(physics), 40. Hon Degrees: MA, Harvard Univ, 46. Prof Exp: From instr to asst prof physics, Princeton Univ, 40-46, in tech chg isotron develop proj, 42-43; physicist, Los Alamos Sci Lab, 43-46, leader cyclotron group, 43-44, head exp res div, 44-46; assoc prof physics, Harvard Univ, 46-47; prof physics & dir lab nuclear studies, Cornell Univ, 47-67; PROF DEPT PHYSICS & ENRICO FERMI INST NUCLEAR STUDIES, UNIV CHICAGO, 67-; DIR, NAT ACCELERATOR LAB, 67- Concurrent Pos: Mem, Comt Atomic Casualties, Nat Res Coun, 48-51; exchange prof, Univ Paris, 54-55; mem, Steering Comt, Proj Sherwood, US AEC, 58- Mem: Nat Acad Sci; Am Phys Soc; Am Acad Arts & Sci; Am Philos Soc. Res: Nuclear and particle physics. Mailing Add: Nat Accelerator Lab PO Box 500 Batavia IL 60510

WILSON, ROBERT STEVEN, b Hartford, Conn, Dec 26, 42; m 66; c 3. PHYSICAL CHEMISTRY. Educ: Brown Univ, BS, 62, PhD(phys chem), 68. Prof Exp: Fel phys chem, Yale Univ, 68-69; asst prof, 69-73, ASSOC PROF PHYS CHEM, NORTHERN ILL UNIV, 73- Concurrent Pos: Res prof, Solid State Sci Div, Argonne Nat Lab, 71- Mem: Am Phys Soc; Am Chem Soc. Res: Statistical mechanics of irreversible processes; optical properties of impurity systems; critical transport properties of fluids. Mailing Add: Dept of Chem Northern Ill Univ De Kalb IL 60115

WILSON, ROBERT WARREN, b Oakland, Calif, July 26, 09; m 41; c 2. VERTEBRATE PALEONTOLOGY. Educ: Calif Inst Technol, BS, 30, MS, 32, PhD(vert paleontol), 36. Prof Exp: Asst geol, Calif Inst Technol, 30-34, Sterling res fel, 36-37, fel, 37-39; from instr to asst prof geol, Univ Colo, 39-46, Nat Res Coun fel, 46-47; assoc prof zool & assoc cur vert paleont, Univ Kans, 47-61; prof paleont & dir mus geol, SDak Sch Mines & Technol, 61-75; VIS PROF, TEX TECH UNIV, 75- Concurrent Pos: Guggenheim fel, Univ London, 56-57; Fulbright sr res scholar, Univ Vienna, 67-68. Honors & Awards: Arnold Guyot Mem Award, Nat Geog Soc, 74. Mem: Fel Geol Soc Am; Paleont Soc; Soc Vert Paleont (secy-treas, 54, pres, 55); Am Soc Mammal. Res: Tertiary mammalian faunas. Mailing Add: The Museum Tex Tech Univ Lubbock TX 79409

WILSON, ROBERT WOODROW, b Houston, Tex, Jan 10, 36; m 58; c 3. RADIO ASTRONOMY. Educ: Rice Univ, BA, 57; Calif Inst Technol, PhD(physics), 62. Prof Exp: Res fel radio astron, Calif Inst Technol, 62-63; MEM TECH STAFF, BELL LABS, 63- Mem: Am Astron Soc. Res: Problems related to the galaxy; absolute flux and background temperature measurements; millimeter-wave measurements of interstellar molecules. Mailing Add: Bell Labs PO Box 400 Holmdel NJ 07733

WILSON, ROGER EDWIN, plant ecology, deceased

WILSON, RONALD F, ecology, see 12th edition

WILSON, RONALD HARVEY, b Belle Fourche, SDak, Oct 2, 32; m 55; c 3. EXPERIMENTAL SOLID STATE PHYSICS, ENERGY CONVERSION. Educ: SDak State Univ, BS, 56, MS, 58; Rensselaer Polytech Inst, PhD(physics), 64. Prof Exp: Physicist, Flight Propulsion Lab, Gen Elec Co, 58-59, res trainee, Physics, Res Lab, 59-61, physicist, Gen Eng Lab, 61-62; teaching asst physics, Rensselaer Polytech Inst, 62-63; PHYSICIST, GEN ELEC RES & DEVELOP CTR, 63- Mem: AAAS; Am Phys Soc; Inst Elec & Electronics Eng. Res: Physics and properties of thin films; physics of semiconductor devices; semiconductor processing; imaging and storage devices. Mailing Add: Gen Elec Res & Develop Ctr PO Box 8 Schenectady NY 12301

WILSON, RONALD WAYNE, b Iowa Falls, Iowa, Aug 4, 39. BOTANY, MYCOLOGY. Educ: Iowa State Univ, BS, 61; Mich State Univ, PhD(bot), 65. Prof Exp: USPHSfel, Med Ctr, Ind Univ, 65-67; asst prof, 67-70, ASSOC PROF NATURAL SCI, MICH STATE UNIV, 70- Mem: AAAS; Am Inst Biol Sci. Res: Fungal physiology and biochemical aspects of fungal morphogenesis; general-liberal studies in science; lysine metabolism. Mailing Add: Dept of Natural Sci Mich State Univ East Lansing MI 48823

WILSON, ROY D, b Shawnee, Okla, June 6, 31; m 57; c 3. ANESTHESIOLOGY, PHYSIOLOGY. Educ: Stephen F Austin State Col, BA, 51; Baylor Univ, MD, 55; Am Bd Anesthesiol, dipl, 62. Prof Exp: Intern, Hermann Hosp, Houston, 55-56; pvt pract, 56-57; from asst prof to assoc prof anesthesiol, 63-70, PROF ANESTHESIOL, UNIV TEX MED BR GALVESTON, 70-, DIR RES & RESIDENCY TRAINING, 72- Concurrent Pos: Chief anesthesiol div, Shriners Burn Inst, Galveston Univ, 66-72; mem bd med adv & past chmn, Am Asn Inhalation Ther, 68-; mem bd trustees, Am Registry Inhalation Ther, 72-76. Mem: Fel Am Col Anesthesiol; AMA; Am Soc Pharmacol & Exp Therapeut; Am Soc Clin Pharmacol & Therapeut; Pan-Am Med Asn (sect vpres, 68-70). Res: Cardio-pulmonary physiology and pharmacology; physiopathology of burned children; pediatric anesthesiology. Mailing Add: Dept of Anesthesiol Univ of Tex Med Br Galveston TX 77550

WILSON, SHIRLEY LANE, b Barryton, Mich, Sept 7, 18. PLANT PHYSIOLOGY, PLANT PATHOLOGY. Educ: Mich State Col, BS, 50; Univ Ill, MS, 51, PhD(bot), 54. Prof Exp: Asst, Univ Ill, 51-53; asst plant physiologist, Conn Agr Exp Sta, 54-55; asst prof bot, Southern Ill Univ, 55-63; ASST PROF BOT, DRAKE UNIV, 63- Mem: Am Soc Plant Physiol; Bot Soc Am; Sigma Xi. Res: Photoperiodism; control of algal blooms; plant pigments; bioelectric currents in mimosa; flower opening rhythms. Mailing Add: Dept of Biol Drake Univ Des Moines IA 50311

WILSON, SLOAN JACOB, b Dallas, Tex, Jan 22, 10; m 48; c 2. INTERNAL MEDICINE. Educ: Wichita State Univ, AB, 31, MS, 32; Univ Kans, BS, 34, MD, 36; Am Bd Internal Med, dipl, 48. Prof Exp: From intern to asst resident, Ohio State Univ Hosp, 36-38, resident res med, 38-39, instr path, 39-40; from asst prof to assoc prof med, 46-59, PROF INTERNAL MED, UNIV KANS MED CTR, KANSAS CITY, 59- Mem: AAAS; Soc Exp Biol & Med; fel AMA; fel Am Col Physicians; Am Soc Hemat. Res: Blood hematology. Mailing Add: Univ of Kans Med Ctr 39th St & Rainbow Blvd Kansas City KS 66103

WILSON, STANLEY PORTER, b Andalusia, Ala, Sept 4, 31; m 53; c 1. POPULATION GENETICS. Educ: Auburn Univ, BS, 54, MS, 58; Okla State Univ, PhD(genetics), 61. Prof Exp: Nat Acad Sci-Nat Res Coun fel, Purdue Univ, 61-63; coordr regional poultry breeding proj, US Dept Agr, 63-65, leader, Genetics Invest, Poultry Res Br, Animal Husb Res Div, Agr Res Ctr, Md, 65-67; dir, Pioneering Res Lab, Purdue Univ, 67-75; ASSOC DIR & ASST DEAN, AGR EXP STA, AUBURN UNIV, 75- Mem: AAAS; Genetics Soc Am; Poultry Sci Asn; Am Soc Animal Sci; Am Genetic Asn. Res: Selection studies with tribolium, mice, poultry and swine; effects of mating systems on selection and genetic parameters. Mailing Add: Sch of Agr Auburn Univ Auburn AL 36830

WILSON, STEPHEN JAMES, b Philadelphia, Pa, Feb 10, 44. NUCLEAR PHYSICS. Educ: Duke Univ, BS, 66; Univ Wis-Madison, MS, 68, PhD(physics), 73. Prof Exp: FEL NUCLEAR PHYSICS, UNIV GUELPH, 74- Res: High-spin states in intermediate weight nuclei via fusion-evaporation reactions. Mailing Add: Dept of Physics Univ of Guelph Guelph ON Can

WILSON, STEPHEN ROSS, b Oklahoma City, Okla, Mar 13, 46; m 67. SYNTHETIC ORGANIC CHEMISTRY. Educ: Rice Univ, BA, 69, MA, 72, PhD(org chem), 72. Prof Exp: NIH fel org chem, Calif Inst Technol, 72-74; ASST PROF ORG CHEM, IND UNIV, BLOOMINGTON, 74- Concurrent Pos: Sigma Xi res award, Rice Univ, 72. Mem: AAAS; The Chem Soc; Am Chem Soc. Res: Development of new approaches to the synthesis of naturally occurring compounds of biological significance, and the structure elucidation and total synthesis of such substances. Mailing Add: Dept of Chem Ind Univ Bloomington IN 47401

WILSON, THOMAS CHARLES, polymer chemistry, see 12th edition

WILSON, THOMAS HASTINGS, b Philadelphia, Pa, Jan 31, 25; m 52; c 2. PHYSIOLOGY. Educ: Univ Pa, MD, 48; Sheffield Univ, 51-53, PhD(biochem), 53. Prof Exp: Instr physiol, Univ Pa, 49-50; instr biochem, Wash Univ, 56-57; assoc physiol, 57-59, from asst prof to assoc prof, 59-68, PROF PHYSIOL, HARVARD MED SCH, 68- Mem: Am Soc Biol Chem; Am Physiol Soc; Brit Biochem Soc. Res: Active transport of materials across cell membranes. Mailing Add: Dept of Physiol Harvard Med Sch Boston MA 02115

WILSON, THOMAS KENDRICK, b Highland Park, Mich, June 2, 31; m 52; c 5. PLANT MORPHOLOGY. Educ: Ohio Univ, BS, 53, MS, 55; Ind Univ, PhD(bot), 58. Prof Exp: Asst bot, Ohio Univ, 53-55 & Ind Univ, 55-58; from asst prof to assoc prof, Univ Cincinnati, 58-68; assoc prof, 68-74, PROF BOT, MIAMI UNIV, 74- Mem: AAAS; Bot Soc Am; Int Soc Plant Morphol; Int Asn Plant Taxon. Res: Comparative morphology of angiosperms; origin and phylogeny of vascular plants; evolution. Mailing Add: Dept of Bot Miami Univ Oxford OH 45056

WILSON, THOMAS LEE, b Wyoming, Ohio, Nov 4, 09; m 37; c 2. PHYSICAL CHEMISTRY. Educ: Col Wooster, BS, 30; Univ Wash, MS, 34; Univ Chicago, PhD(chem), 35. Prof Exp: Res chemist, Gen Labs, US Rubber Co, 35-42, dept head, 42-54, admin asst, Res & Develop Dept, 54-58, mgr res ctr, 58-65; asst prof chem, 66-69, chmn dept, 71-73, ASSOC PROF CHEM, MONTCLAIR STATE COL, 69-, DEAN SCH MATH & SCI, 73- Mem: AAAS; Am Chem Soc; fel Am Inst Chem. Res: Oceanography; reaction rates of gaseous decomposition; rubber and inorganic chemistry. Mailing Add: Sch of Math & Sci Montclair State Col Montclair NJ 07043

WILSON, THOMAS OLIVER, physiology, see 12th edition

WILSON, THOMAS PUTNAM, b New York, NY, Sept 4, 18; m 44; c 1. CHEMISTRY. Educ: Amherst Col, BA, 39; Harvard Univ, PhD(chem physics), 43. Prof Exp: Res chemist, Manhattan Proj, M W Kellogg Co, NJ, 43-44; group leader, Kellex Corp, 44-45; group leader, Manhattan Proj & S A M Labs, Chems & Plastics Div, 45-46, res chemist, 46-62, asst dir res, 62-72, res assoc, 72-73, CORP RES FEL, RES & DEVELOP DEPT, CHEMS & PLASTICS DIV, UNION CARBIDE CORP, 73- Mem: Am Chem Soc; Catalysis Soc. Res: Adsorption on solids; kinetics and catalysis; ethylene polymerization; catalyst development and characterization. Mailing Add: Res & Develop Dept Chem & Plastics Union Carbide Corp PO Box 8361 South Charleston WV 25303

WILSON, TIMOTHY M, b Columbus, Ohio, Aug 3, 38; m; c 2. SOLID STATE PHYSICS. Educ: Univ Fla, BS, 61, PhD(chem), 66. Prof Exp: Fel solid state physics, Univ Fla, 66-68, asst prof chem, 68-69; asst prof, 69-72, ASSOC PROF PHYSICS, OKLA STATE UNIV, 72- Concurrent Pos: Res staff mem, Solid State Div, Oak Ridge Nat Lab, 74-75. Mem: Am Phys Soc; Am Asn Physics Teachers. Res: Theoretical studies of the optical and magnetic properties of impurities and defects in crystalline solids. Mailing Add: Dept of Physics Okla State Univ Stillwater OK 74074

WILSON, VANNIE WILLIAM, JR, biochemistry, virology, see 12th edition

WILSON, VERNON EARL, b Plymouth Co, Iowa, Feb 16, 15; m 47; c 2. MEDICAL ADMINISTRATION, FAMILY MEDICINE. Educ: Univ Ill, BS, 50, MS & MD, 52. Prof Exp: Asst pharmacol, Univ Ill, 50-52; intern, Univ Hosp, Chicago, 52-53; asst prof, Sch Med, Univ Kans, 53-59, asst dean sch med, 57-59, actg dean sch med & actg dir med ctr, 59; prof pharmacol, Univ Mo-Columbia, 59-70, dean sch med & dir med ctr, 59-67, exec dir health affairs, 67-68, vpres acad affairs, 68-70; adminr, Health Serv & Ment Health Admin, Dept Health Educ & Welfare, Md, 70-73; prof community health, Univ Mo-Columbia, 73-74; VCHANCELLOR MED AFFAIRS, VANDERBILT UNIV, 74- Concurrent Pos: Exec officer, Mo State Crippled Children's Serv, 60-68; mem exec coun, Am Asn Med Cols, 61-67; mem, Am Bd Family Pract, 62-74; consult, USPHS, 65-69; coun mem med educ, AMA, 67-75; coordr, Mo Regional Med Prog, 66-68; mem coun fed relationship, Am Asn Univ, 68-70; ed, Continuing Educ for Family Physician, 73-75; mem liaison comt, Grad Med Educ, 73-75; mem bd dirs, AAAS, 75- Mem: AMA; hon mem Acad Anesthesiol. Res: Renal pharmacology; medical education; public health administration. Mailing Add: Med Ctr D-3300 Vanderbilt Univ Nashville TN 37232

WILSON, VERNON ELDRIDGE, b Roanoke, Va, July 3, 18; m 53; c 2. GENETICS, PHYTOPATHOLOGY. Educ: Colo State Univ, BS, 47, MS, 49; Iowa State Univ, PhD(agron) & PhD(phytopath), 55. Prof Exp: Proj leader & phytopathologist, USDA, Univ Idaho, 52-62, PROJ LEADER & PHYTOPATHOLOGIST, USDA, WASH STATE UNIV, 62- Concurrent Pos: Consult, WPakistan Agr Pulse Res, 71. Mem: Am Genetic Asn; Am Soc Agron; Am Phytopath Soc; Am Soc Hort Sci; Crop Sci Soc Am. Res: Developing high protein pulse crops through genetic approaches; program development for improved agricultural research. Mailing Add: Dept of Agron Wash State Univ Pullman WA 99163

WILSON, VICTOR JOSEPH, b Berlin, Ger, Dec 24, 28; m 53; c 2. NEUROPHYSIOLOGY. Educ: Tufts Col, BS, 48; Univ Ill, PhD(physiol), 53. Prof Exp: Res assoc, 56-58, from asst prof to assoc prof, 58-69, PROF PHYSIOL, ROCKEFELLER UNIV, 69- Mem: Am Physiol Soc; Soc Neurosci. Res: Organization and synaptic transmission in the central nervous system, particularly the spinal cord and brain stem. Mailing Add: Rockefeller Univ York Ave & 66th St New York NY 10021

WILSON, WALTER DAVIS, b Merced, Calif, Oct 20, 35; m 59; c 3. ATOMIC PHYSICS. Educ: Univ Calif, Berkeley, BS, 57, PhD(nuclear eng), 66. Prof Exp: Chem engr, Aerojet Gen Nucleonics, Calif, 58-59; mem tech staff high-altitude nuclear effects, Aerospace Corp, 65-69; PROF PHYSICS, CALIF POLYTECH STATE UNIV, SAN LUIS OBISPO, 69- Concurrent Pos: Tech consult, Sci Applns, Inc, Calif, 71-75. Mem: Am Inst Physics. Res: High altitude physics, particularly electromagnetic field propagation, chemistry and trapped radiation; plasma physics; nuclear reactor theory. Mailing Add: Dept of Physics Calif Polytech Univ San Luis Obispo CA 93401

WILSON, WALTER ERVIN, b Salem, Ore, Apr 1, 34. RADIOLOGICAL PHYSICS. Educ: Willamette Univ, BA, 56; Univ Wis, MS, 58, PhD(physics), 61. Prof Exp: RADIATION PHYSICIST, PAC NORTHWEST LABS, BATTELLE MEM INST, 64- Concurrent Pos: Fel, Basel Univ, 61-62 & Univ Wis, 62-64. Mem: AAAS; Radiation Res Soc; Am Asn Physicists in Med; Am Phys Soc. Res: Radiation effects and radiological sciences. Mailing Add: Pac NW Labs Battelle Mem Inst PO Box 999 Richland WA 99352

WILSON, WALTER LEROY, b Phoenixville, Pa, Sept 1, 18; m 44; c 2. PHYSIOLOGY. Educ: Pa State Teachers Col, West Chester, BS, 41; Univ Pa, PhD(zool), 49. Prof Exp: Biologist, Off Sci Res & Develop, Univ Pa, 43-44, asst instr zool, 46-47; biologist, Manhattan Proj, Columbia Univ, 44-46; instr physiol & biophys, Col Med, Univ Vt, 49-52, from asst prof to assoc prof, 52-65; PROF BIOL SCI, OAKLAND UNIV, 65- Concurrent Pos: Lectr, Middlebury Col, 52-53; mem corp, Marine Biol Lab, Woods Hole. Mem: Am Physiol Soc; Soc Gen Physiol; Am Soc Zoologists. Res: Effects of high temperature on living systems; protoplasmic viscosity changes during cell division; the release of anticoagulant substances from living cells and the inhibitory action of these anticoagulants on cell division; role of cellular cortex in stimulation and cell division. Mailing Add: Dept of Biol Oakland Univ Rochester MI 48063

WILSON, WALTER LUCIEN, JR, b Montgomery, Ala, Apr 26, 27; m 47; c 2. MATHEMATICS. Educ: Univ Ala, AB, 50, MA, 51; Univ Calif, Los Angeles, PhD(math), 59. Prof Exp: Res engr, NAm Aviation, Inc, Calif, 59-60; ASSOC PROF MATH, UNIV ALA, 60- Mem: Am Math Soc; Math Asn Am. Res: Calculus of variations; numerical analysis; linear programming. Mailing Add: Dept of Math Univ of Ala University AL 35486

WILSON, WILBOR OWENS, b Ft Towson, Okla, Dec 1, 10; m 38; c 4. AVIAN PHYSIOLOGY. Educ: Okla Agr & Mech Col, BS, 32; Kans State Col, MS, 33; Iowa State Col, PhD(animal breeding), 47. Prof Exp: Poultry caretaker, Okla Agr & Mech Col, 34-35; poultry proj supvr, Gen Mills, Inc, Mich, 35-37; asst prof poultry husb, SDak State Col, 37-41, actg head poultry dept, 41-44; res assoc, Iowa State Col, 44-46; from asst prof & asst poultry physiologist to assoc prof & assoc poultry physiologist, 46-72, chmn dept poultry husb, Univ, 65-69, PROF POULTRY HUSB, UNIV CALIF, DAVIS & POULTRY PHYSIOLOGIST, EXP STA, 72- Concurrent Pos: Collabr, US Dept Agr, 54-62; trainee, Nat Inst Neurol Dis & Blindness, Wash State Univ, 60-61; mem, Comn Physiol Effects of Environ Factors on Animals, Agr Bd, Nat Res Coun-Nat Acad Sci, 63-71. Mem: Fel AAAS; Am Physiol Soc; fel Poultry Sci Asn (pres, 69-70); Int Soc Biometeorol; Int Soc Chronobiol. Mailing Add: Dept of Avian Sci Univ of Calif Davis CA 95616

WILSON, WILBUR WILLIAM, b Ferriday, La, Jan 10, 48; m 69; c 1. PHYSICAL CHEMISTRY. Educ: Northeast La State Col, BS, 69; Univ NC, PhD(phys chem), 73. Prof Exp: ASST PROF CHEM, MISS STATE UNIV, 74- Mem: Am Chem Soc; Sigma Xi. Res: Laser light scattering by macromolecules. Mailing Add: Box 3348 Mississippi State MS 39762

WILSON, WILFRED J, b Ferndale, Calif, Mar 4, 30; m 68. EMBRYOLOGY. Educ: Sacramento State Col, AB, 52; Univ Calif, Davis, MA, 58, PhD(zool), 64. Prof Exp: Teaching asst zool, Univ Calif, Davis, 58-61, assoc, 61-63; from asst prof to assoc prof, 63-70, PROF ZOOL, SAN DIEGO STATE UNIV, 70- Concurrent Pos: Shell merit fel, Stanford Univ, 69; vis lectr, Burma, 71; Fulbright lectr, US Dept of State, Nat Taiwan Univ, 70-71. Mem: Am Soc Zool. Res: General and invertebrate biology; crustacean water balance; teaching methods in college introductory biology; early development of marine invertebrates. Mailing Add: Dept of Zool San Diego State Univ San Diego CA 92182

WILSON, WILLIAM A, physical chemistry, see 12th edition

WILSON, WILLIAM AUGUST, JR, b St Louis, Mo, July 3, 24; m 54; c 4. NEUROPSYCHOLOGY. Educ: Univ Calif, AB, 43, PhD(psychol), 56; Yale Univ, MD, 53. Prof Exp: Vis instr, psychol, Wesleyan Univ, 50-51; res assoc neurophysiol, Inst of Living, Univ Conn, 53-56, dir dept exp psychol, 56-58; vis asst prof psychol, Univ Calif, 58; asst prof, Univ Colo, 59-60; assoc prof, Bryn Mawr Col, 60-64; assoc dean grad sch, 71-72, PROF PSYCHOL, UNIV CONN, 64-, BIOBEHAV SCI, 69-

Concurrent Pos: Mem exp psychol res review comt, NIMH, 69-73, chmn, 71-73. Mem: AAAS; Am Psychol Asn; Inst Math Statist; Animal Behav Soc; Am Asn Lab Animal Sci. Res: Neurological determinants of behavior, especially visual perception and learning; function of regions of association cortex and the mechanisms of learning. Mailing Add: Dept of Psychol Univ of Conn Storrs CT 06268

WILSON, WILLIAM BUFORD, b Sulphur Springs, Tex, Mar 11, 23; m 47; c 6. MARINE BIOLOGY. Educ: Tex A&M Univ, BS, 48, MS, 50, PhD(biol oceanog), 66. Prof Exp: Marine biologist, Tex Game & Fish Comn, 48-49; oceanogr, Oceanog Off, US Dept Navy, 50-53; fishery res biologist, US Fish & Wildlife Serv, 53-62; res scientist, Tex A&M Univ, 62-63; res assoc, Fla State Univ, 63-66; sr res scientist, 66-67, asst prof marine biol, 67-69, ASSOC PROF MARINE BIOL & OCEANOG, MOODY COL MARINE SCI & MARINE RESOURCES, TEX A&M UNIV, 69- Concurrent Pos: Consult, Fla Bd Conserv, 63-66, Tex A&M Marine Lab, 64-66, USPHS, 64-65, Univ SFla, 66, Du Pont Chem Co, 67, Oceanonics, Inc, 68-72 & Dow Chem Co, 69-70; grants, USPHS, 62-65 & 66-69, Environ Protection Agency, 69-72 & NSF, 70-72. Mem: Am Soc Limnol & Oceanog; Soc Protozool; Ecol Soc Am. Res: Marine phytoplankton, ecology and fouling. Mailing Add: Moody Col Marine Sci & Resources Tex A&M Univ Bldg 311 Galveston TX 77550

WILSON, WILLIAM CHOURY, physical chemistry, see 12th edition

WILSON, WILLIAM CURTIS, b Orlando, Fla, Dec 29, 27; m 52; c 1. PLANT PHYSIOLOGY. Educ: Cornell Univ, BS, 49; Univ Fla, MAgr, 58, PhD(fruit crops), 66. Prof Exp: Asst mgr agr res, Fla Agr Res Inst, 57-61; ASSOC PLANT PHYSIOLOGIST, CITRUS EXP STA, AGR RES & EDUC CTR, UNIV FLA, 66-, ASSOC PROF, 74- Concurrent Pos: Merck & Co grant, 69; Julian C Miller award, Asn Southern Agr Workers, Inc, 66; Ciba-Geigy grants, 70-72. Mem: Am Soc Hort Sci; Int Soc Citricult. Res: Abscission chemicals to facilitate easier removal of citrus fruit to aid mechanical or hand harvesting. Mailing Add: Res & Educ Ctr Univ Fla Box 1088 Lake Alfred FL 33850

WILSON, WILLIAM D, b Pittsburgh, Pa, Nov 8, 25; m 49; c 5. PARASITOLOGY. Educ: Dickinson Col, BS, 50; Univ Kans, MA, 53; Mich State Univ, PhD(microbiol, pub health), 57. Prof Exp: Instr, Mich State Univ, 56-57; PROF BIOL, STATE UNIV NY COL ONEONTA, 57- Mailing Add: 94 Main St Unadilla NY 13849

WILSON, WILLIAM DENNIS, b New York, NY, July 20, 40; m 60; c 3. SOLID STATE PHYSICS. Educ: Queens Col, NY, BS, 63, MA, 65; City Univ New York, PhD(physics), 67. Prof Exp: Res physicist, Queens Col, NY, 67 & 68-69; fel physics, City Univ New York, 67-68; RES PHYSICIST, LIVERMORE LABS, SANDIA CORP, 69- Concurrent Pos: Consult, Lawrence Radiation Lab, Calif, 67-69. Mem: Am Phys Soc. Res: Interatomic potentials; defects in solids; hydrogen and helium in metals; diffusion. Mailing Add: Sandia Labs Livermore CA 94550

WILSON, WILLIAM EDWARD, III, b Providence, RI, June 4, 34; m 58; c 2. HYDROGEOLOGY. Educ: Harvard Univ, AB, 56; Univ Ill, PhD(geol), 63. Prof Exp: HYDROLOGIST, WATER RESOURCES DIV, US GEOL SURV, 62- Mem: Geol Soc Am; Am Water Resources Asn; Am Geol Inst; Nat Water Well Asn; Am Inst Prof Geol. Res: Areal investigations on the occurrence, movement and availability of ground water. Mailing Add: Water Resources Div US Geol Surv 500 Zack St Tampa FL 33602

WILSON, WILLIAM ENOCH, JR, b El Dorado, Ark, Jan 15, 33; m 63; c 2. ATMOSPHERIC CHEMISTRY. Educ: Hendrix Col, BA, 53; Purdue Univ, PhD(phys chem), 57. Prof Exp: Instr chem, Wis State, La Crosse, 55-56; Fulbright res fel, Inst Technol, Munich, Ger, 57-58; sr chemist, Appl Physics Lab, Johns Hopkins Univ, 58-67; assoc fel, Battelle Mem Inst, 67-71; chief atmospheric aerosol res sect, 71-75, CHIEF AEROSOL RES BR, ENVIRON PROTECTION AGENCY, 75- Concurrent Pos: Adj assoc prof, Environ Sci & Technol Dept, Sch Pub Health, Univ NC, 73- Mem: AAAS; Am Chem Soc; Am Phys Soc; Combustion Inst; Am Inst Aeronaut & Astronaut. Res: Sources, formation, dynamics, transport, removal, and effects of atmospheric aerosols; atmospheric chemistry and physics; molecular spectroscopy; chemical kinetics and thermodynamics pertinent to combustion, propulsion, and air pollution. Mailing Add: Environ Protection Agency Aerosol Res Br MD-59 Research Triangle Park NC 27711

WILSON, WILLIAM ERNEST, b Sparta, Ill, June 7, 07; m 40; c 3. BOTANY, PLANT PATHOLOGY. Educ: Wheaton Col, BS, 36; Univ Ill, MS, 39, PhD(bot), 41. Prof Exp: Asst biol, Wheaton Col, 36-37; asst bot, Univ Ill, 37-41; from instr to asst prof biol, Muskingum Col, 41-47; from asst prof to assoc prof, 47-68, PROF BOT, MIAMI UNIV, 68- Mem: Am Phytopath Soc; Am Forestry Asn. Res: Physiology of fungi, especially two species of Diplodia parasitic on corn; nutrition of Diplodia in artificial media; myxomycetes; teaching methods. Mailing Add: Dept of Bot Miami Univ Oxford OH 45056

WILSON, WILLIAM EWING, b New Orleans, La, Nov 8, 32; m 61; c 3. MOLECULAR PHARMACOLOGY. Educ: King Col, AB, 54; Univ Tenn, MS, 56, PhD(biochem), 59. Prof Exp: Res assoc biochem, Okla State Univ, 59-60, instr, 61-63; res biochemist, Radio Isotope Serv, Vet Admin Hosp, Little Rock, 63-64, chief, Biochem Sect, Southern Res Support Ctr, 64-68; asst prof biochem, Sch Med, Univ Ark, Little Rock, 63-68; res chemist, Anal & Synthetic Chem Br, 68-73, RES CHEMIST, ENVIRON TOXICOL BR, NAT INST ENVIRON HEALTH SCI, 73- Mem: AAAS; Am Chem Soc; NY Acad Sci. Res: Interaction of chemicals with membranal enzymes; molecular pharmacology of drugs and environmental agents; analytical biochemistry; solution interactions. Mailing Add: Environ Toxicol Br Nat Inst of Environ Health Sci Research Triangle Park NC 27709

WILSON, WILLIAM JEWELL, b St Joseph, Mo, Dec 14, 32; m 57; c 5. RADIOLOGY. Educ: William Jewell Col, AB, 54; Univ Mo-Columbia, MD, 58. Prof Exp: Intern, St Albans Hosp, Long Island & US Naval Hosp, 58-59; resident, Univ Mo-Columbia Hosp, 62-65, instr radiol, Univ, 64-65; asst prof cardiovasc radiol & dir dept, Univ Va, 66-67; asst prof diag radiol, 67-68; found prof radiol & chmn dept, Med Ctr, Univ Nebr, Omaha, 68-73; DIR RADIOL, LONG BEACH MEM HOSP, 73- Concurrent Pos: USPHS res fel, Univ Mo, 64-65; USPHS advan res fel cardiovasc radiol, Univ Minn, Minneapolis, 65-66; Markle scholar acad med, 67; chmn subcomt comput in diag radiol, Nat Ctr Radiol Health, 67-68; mem subcomt comput based diag radiol info, Am Bd Radiol, 71-73, chmn cardiovasc comt, Nat Bd Examr, 71-74. Mem: Radiol Soc NAm; Asn Univ Radiol; Am Col Radiol; AMA; fel Am Col Cardiol. Res: Cardiovascular radiology including physical properties of generating and analyzing radiant images; pharmacologic effects of contrast materials utilized in clinical angiography; aberrations in subsegmental arterial perfusion of organs; computer techniques in radiology. Mailing Add: Dept of Radiol Long Beach Mem Hosp Long Beach CA 90801

WILSON, WILLIAM JOHN, b Spokane, Wash, Dec 16, 39; m 63; c 2. RADIO ASTRONOMY. Educ: Univ Wash, BSEE, 61; Mass Inst Technol, MSEE, 63, PhD(elec eng), 70. Prof Exp: MEM TECH STAFF MILLIMETER-WAVE RADIO ASTRON & ELEC ENG, AEROSPACE CORP, 70- Concurrent Pos: User's Comt, Nat Radio Astron Observ, 71- Mem: Am Astron Soc; Int Union Radio Sci; Int Astron Union. Mailing Add: Aerospace Corp PO Box 92957 Los Angeles CA 90009

WILSON, WILLIAM MARK DUNLOP, b Glasgow, Scotland, Jan 23, 49; Brit citizen; m 73. ANIMAL SCIENCE, RUMINANT NUTRITION. Educ: Univ Glasgow, BSc, 71; Univ Ill, Urbana, MS, 73, PhD(animal sci), 75. Prof Exp: Asst animal sci & ruminant nutrit, Univ Ill, Urbana, 71-75; YOUNG PROF AGR, WORLD BANK, 75- Mem: Am Soc Animal Sci; Brit Soc Animal Prod. Res: Interested in all facets of animal husbandry, nutrition and science and agronomy as they relate to agriculture in developing countries. Mailing Add: World Bank 1818 H St NW Washington DC 20433

WILSON, WILLIAM PRESTON, b Fayetteville, NC, Nov 6, 22; m 51; c 5. PSYCHIATRY. Educ: Duke Univ, BS, 43, MD, 47. Prof Exp: Intern, Gorgas Hosp, CZ, 47-48; staff psychiatrist, State Hosp, Raleigh, NC, 48-49; instr psychiat & asst resident, Sch Med, Duke Univ, 49-52, resident neurol, 52, assoc psychiat & chief resident, 52-54, asst prof psychiat, 55-58; assoc prof, dir psychiat res labs & consult, Hogg Found, Med Br, Univ Tex, 58-60; assoc prof psychiat, 58-61, PROF PSYCHIAT, MED CTR, DUKE UNIV, 61-, DIR NEUROPHYSIOL LABS, 58-; DIR PSYCHIAT RES LABS & STAFF PSYCHIATRIST, VET ADMIN HOSP, 58- Concurrent Pos: Fel med, Duke Univ, 52-54; NIH fel, Montreal Neurol Inst, McGill Univ, 54-55; mem, Am Bd Qual EEG, 69-, secy-treas, 71-74. Mem: AAAS; AMA; Am Psychiat Asn; Am Psychopath Asn; Asn Res Nerv & Ment Dis. Res: Clinical psychiatry and neurochemistry; clinical and experimental neurophysiology; electroencephalography. Mailing Add: Dept of Psychiat Box 3838 Duke Univ Med Ctr Durham NC 27710

WILSON, WILLIAM RALPH, internal medicine, deceased

WILSON, WILLIAM SOLOMON, b Reynoldsburg, Ohio, Aug 9, 08. PHYSICAL CHEMISTRY. Educ: Brown Univ, ScB, 31, ScM, 34; Yale Univ, PhD(phys chem), 36. Prof Exp: Res chemist, Shell Develop Co, Calif, 37; instr chem, Blue Ridge Col, 38-40; from asst prof to assoc prof physics, Georgetown Col, 40-45, phys sci, 45-47; prof chem & chem eng, 47-72, head dept chem eng, 57-64, head dept gen sci, 64-72, EMER PROF CHEM & GEN SCI, 72- Concurrent Pos: Instr, Garth Sch, Ky, 44-47; asst prof, Univ Wyo, 48-49; acad fel, Geophys Inst, Univ Alaska, 49-50, actg dir, 51-52, assoc dir, 52-54; regional counsr, Am Asn Physics Teachers, Alaska, 64-69. Mem: Fel AAAS; Am Chem Soc; Am Phys Soc; Am Meteorol Soc; Am Soc Eng Educ. Res: Raman spectroscopy; theory of electrolytes; atomic structure; pressure effects on spectral lines; wave mechanics; geophysics; chemistry and physics of the atmosphere; science education. Mailing Add: Box 1010 Soldotna AK 99669

WILSON, WILLIAM THOMAS, b Midvale, Utah, July 22, 32; m 58; c 5. INVERTEBRATE PATHOLOGY, ENTOMOLOGY. Educ: Colo Agr & Mech Col, BS, 55; Colo State Univ, MS, 57; Ohio State Univ, PhD(entom), 67. Prof Exp: Asst entom, Colo State Univ, 55-57, from instr to asst prof, 60-65; asst, Ohio State Univ, 65-67; asst res pathobiologist, Univ Calif, Irvine, 67-68; RES ENTOMOLOGIST, AGR RES SERV, US DEPT AGR & MEM GRAD FAC, UNIV WYO, 68- Mem: Bee Res Asn; Soc Invert Path; Entom Soc Am. Res: Pathology of bacteria in the gut and hemolymph of insects; chemotherapeutic treatment of insect diseases; invertebrate microbiology; diseases and toxicology of the honey bee. Mailing Add: Bee Dis Invests Agr Res Serv US Dept of Agr Univ Sta Box 3168 Laramie WY 82071

WILSON, WOODROW, JR, b Pittsfield, Mass, May 1, 44; m 68. THEORETICAL CHEMISTRY. Educ: Temple Univ, BA, 66; Calif Inst Technol, PhD(chem), 70. Prof Exp: Fel chem, Univ Toronto, 70-72 & Argonne Nat Lab, 72-74; MEM STAFF, SCI APPLICATIONS, INC, 74- Mem: Am Phys Soc. Res: Theoretical study of reaction surfaces; gas phase kinetics. Mailing Add: Sci Applns Inc 1651 Old Meadow Rd McLean VA 22101

WILSON BELL, MARGARET DYKES, physics, see 12th edition

WILT, FRED H, b South Bend, Ind, Dec 12, 34; m 57; c 1. ZOOLOGY. Educ: Ind Univ, AB, 56; Johns Hopkins Univ, PhD(biol), 59. Prof Exp: Fel, Carnegie Inst Technol, 59-60; assoc prof biol, Purdue Univ, 60-64; assoc prof, 64-71, PROF ZOOL, UNIV CALIF, BERKELEY, 71- Concurrent Pos: NIH spec fel, 63-64; fel, Guggenheim Found, 75. Mem: Soc Develop Biol; Am Soc Cell Biol; Am Soc Zool. Mailing Add: Dept of Zool Univ of Calif Berkeley CA 94720

WILT, JAMES WILLIAM, b Chicago, Ill, Aug 28, 30; m 53; c 5. ORGANIC CHEMISTRY. Educ: Univ Chicago, AB, 49, MSc, 53, PhD(chem), 54. Prof Exp: Instr chem, Univ Conn, 55; from instr to assoc prof, 55-66, PROF CHEM, LOYOLA UNIV CHICAGO, 66-, CHMN DEPT, 70- Mem: Am Chem Soc. Res: Rearrangements in organic chemistry; reaction mechanisms; diazoalkane chemistry; free radical chemistry; chemistry of bicyclic compounds. Mailing Add: Dept of Chem Loyola Univ 6525 N Sheridan Rd Chicago IL 60626

WILT, JOHN CHARLES, b Moose Jaw, Sask, Feb 20, 20; m; c 4. BACTERIOLOGY. Educ: Univ Man, MD, 45, MSc, 50; Am Bd Path, dipl, 50; FRCP(C). Prof Exp: Demonstr, 46-53, assoc prof bact, 54-56, prof & head dept, 57-69, PROF MED MICROBIOL & HEAD DEPT, FAC MED, UNIV MAN, 69-, ASSOC DEAN FAC MED & DIR DEPT CLIN MICROBIOL, HEALTH SCI CTR, 67- Concurrent Pos: Asst pathologist, Winnipeg Gen Hosp, 46-53, dir dept bact, 67-; consult bacteriologist, Man Dept Health, Children's Hosp Winnipeg, Grace Hosp & Winnipeg Gen Hosp. Mem: Fel Am Col Physicians; Can Soc Microbiol; Can Pub Health Asn; Can Asn Med Microbiol (vpres, 57-58); Can Asn Path. Res: General microbiology, especially virology. Mailing Add: Dept of Med Microbiol Univ of Manitoba Med Col Winnipeg MB Can

WILT, JOHN ROBERT, b Cleveland, Ohio, Mar 4, 39; m 59; c 5. PHYSICAL CHEMISTRY. Educ: Mass Inst Technol, BS, 60; Univ Calif, Los Angeles, PhD(chem), 65. Prof Exp: Sr res chemist, Eastman Kodak Res Labs, 67-71; mgr photochem mat, Addressograph-Multigraph Corp, 71-74; MGR MAT SCI, ROYAL TYPEWRITER CO, 74- Mem: Am Phys Soc. Res: Spectroscopy; photochemistry; photoconductivity; triboelectric phenomena. Mailing Add: Royal Typewriter Co 150 New Park Ave Hartford CT 06106

WILT, MYRON HARDING, organic chemistry, see 12th edition

WILT, PAXTON MARSHALL, b Louisville, Ky, July 8, 42; m 67. MOLECULAR SPECTROSCOPY. Educ: Centre Col Ky, BA, 64; Vanderbilt Univ, PhD(physics), 67. Prof Exp: Asst prof, 67-72, ASSOC PROF CHEM PHYSICS, CENTRE COL KY, 72- Concurrent Pos: Consult, Res Corp Am, 67. Res: Molecular vibration-rotation spectres of small polyatomic molecules. Mailing Add: Dept of Chem Physics Centre Col of Ky Danville KY 40422

WILTBANK, JAMES N, b Eagar, Ariz, Oct 9, 24; m 51; c 9. ANIMAL

PHYSIOLOGY. Educ: Brigham Young Univ, BS, 51; Univ Wis, MS, 52, PhD, 55. Prof Exp: Coop agent, US Dept Agr-Univ Wis, 55; geneticist, Agr Res Ctr, US Dept Agr, Md, 55-57, animal physiologist, 57-58, Ft Robinson Beef Cattle Res Sta, 58-65; animal physiologist, Squibb Inst Med Res, E R Squibb & Sons, Inc, 65-67; assoc prof physiol, Colo State Univ, 67-72, prof, 72-74; PROF PHYSIOL, AGR RES STA, TEX A&M UNIV, 74- Honors & Awards: Physiol & Endocrinol Award, Am Soc Animal Sci, 75. Mem: Am Soc Animal Sci. Res: Physiology of reproduction in beef cattle. Mailing Add: Tex A&M Univ Agr Res Sta St Rt 2 Box 43C Beeville TX 78102

WILTBANK, WILLIAM JOSEPH, b Clifton, Ariz, Jan 1, 27; m 49; c 5. HORTICULTURE, PLANT PHYSIOLOGY. Educ: NMex State Univ, BSAgr, 50; Univ Fla, PhD(fruit crops), 67. Prof Exp: Instr voc agr, NMex State Dept Voc Educ, 50-53; instr hort, NMex State Univ, 53-54, ext horticulturist, 54-59; hort adv, US Agency Int Develop, Costa Rica, 59-64; res asst, 64-68, asst prof, 68-73, ASSOC PROF FRUIT CROPS, UNIV FLA, 73- Honors & Awards: Gourley Award, Am Soc Hort Sci, 71. Mem: Am Soc Hort Sci; Am Inst Biol Sci; Int Soc Hort Sci; Int Soc Citricult. Res: Physiology of plant reproduction; plant tolerance to temperature and water stress; mineral nutrition of plants. Mailing Add: Dept of Fruit Crops Univ of Fla Gainesville FL 32611

WILTON, ARTHUR CHARLES, b Carragana, Sask, Jan 18, 24; US citizen; m 57; c 3. CYTOGENETICS, PLANT BREEDING. Educ: Univ BC, BSA, 49; Univ Sask, MSc, 54; Univ Man, PhD(plant breeding), 63. Prof Exp: Asst res officer agron, Can Dept Agr, BC, 49-55; asst exp officer clover breeding, Welsh Plant Breeding Sta, 55-56; agronomist, Univ Alaska, Palmer Ctr, 57-66; asst res officer cotton breeding, Univ Calif, Davis, 66-67; overseas agr adv, Md, 67-70, turf breeder & res agronomist, 70-72, RES AGRONOMIST FORAGE BREEDING, PASTURE RES STA, UNIVERSITY PARK, AGR RES SERV, USDA, 72-; ADJ PROF SOIL PHYSICS, DEPT AGRON, COL AGR, PA STATE UNIV, 74- Mem: Genetics Soc Can; Am Soc Agron; Crop Sci Soc Am. Res: Cytogenetics and breeding of forage grasses. Mailing Add: Dept Agron Col of Agr Pa State Univ University Park PA 16802

WILTS, JAMES REED, b Marshalltown, Iowa, June 7, 23; m 50. ELECTRONIC PHYSICS. Educ: Iowa State Col, BS, 44; Calif Inst Technol, MS, 49, PhD(physics), 52. Prof Exp: Fel, Calif Inst Technol, 52-53; physicist, Gen Elec Co, NY, 53-54; sr electronics engr, Gen Dynamics/Convair, Calif, 54-55; asst prof elec eng, Mich State Univ, 55-56; assoc prof, Colo State Univ, 56-57; staff physicist, Int Bus Mach Corp, NY, 57-62; PHYSICIST, NAVAL ELECTRONICS LAB CTR, 63- Mem: Am Phys Soc; Inst Elec & Electronics Engrs. Res: Ionospheric and plasma physics; physics of fluids; statistical communication theory. Mailing Add: Naval Electronics Lab Ctr Code 4300 San Diego CA 92152

WILTSE, JEANETTE A, molecular genetics, see 12th edition

WILTSE, MILTON ADAIR, JR, b Watertown, NY, July 11, 42; m 65. GEOLOGY, GEOCHEMISTRY. Educ: Univ Pa, AB, 64; Ind Univ, MA, 66, PhD(geol), 68. Prof Exp: ASST PROF ECON GEOL, COLO SCH MINES, 68- Mem: Geol Soc Am; Mineral Soc Am; Goechem Soc; Am Inst Mining, Metall & Petrol Eng. Res: Analysis of mineral deposit environments; correlating analytical data with field observations to deduce paragenesis and environment and mode of formation; analysis of naturally occurring mineral systems; quantitative analyses correlated with crystal structure, mode of occurrence and paragenesis. Mailing Add: Dept of Geol Colo Sch of Mines Golden CO 80401

WILTSHIRE, CHARLES THOMAS, b Kansas City, Mo, Apr 5, 41; m 62; c 4. AQUATIC ECOLOGY. Educ: Culver-Stockton Col, BA, 63; Drake Univ, MA, 65; Univ Mo-Columbia, PhD(zool), 73. Prof Exp: Asst prof, 66-73, ASSOC PROF BIOL, CULVER-STOCKTON COL, 73-, CHMN DIV NATURAL SCI, 76- Mem: AAAS; Am Soc Zoologists; Am Inst Biol Sci. Res: Taxonomy and natural history of conchostracans, such as Cyzius; ecology of the middle Mississippi River. Mailing Add: Dept of Biol Culver-Stockton Col Canton MO 63435

WILZBACH, KENNETH EUGENE, organic chemistry, photochemistry, see 12th edition

WIMAN, ROBERT EDGAR, b Bozeman, Mont, Nov 24, 33; m 54; c 4. ORGANIC CHEMISTRY. Educ: Mont State Univ, BS, 57; Ore State Univ, MS, 61, PhD(org chem), 62. Prof Exp: From chemist to sr chemist, Texaco, Inc, NY, 61-65; chemist, Fiber Industs, Inc, NC, 65-66; group leader textile chem, Dan River Mills, Inc, 66-67; mgr textile res, Emery Industs, Inc, Ohio, 67-70; mgr fiber lubricants, Tanatex Chem Co, 70-72; EXEC VPRES, REG BURNETT, INC, 72- Concurrent Pos: Vis lectr, State Univ NY New Paltz, 64-65; pres, Dispersions, Inc, 72-; pres, Dyatherm, Inc, 75- Mem: Am Chem Soc; The Chem Soc; Am Asn Textile Chem & Colorists; Am Inst Chem. Mailing Add: PO Box 2052 Dalton GA 30720

WIMBER, DONALD EDWARD, b Greeley, Colo, Jan 2, 30; m 57; c 2. BOTANY, CYTOLOGY. Educ: San Diego State Col, BA, 52; Claremont Col, MA, 54, PhD(bot), 56. Prof Exp: Res assoc, Dos Pueblos Orchid Co, 54-57; res collabr, Brookhaven Nat Lab, 57-60, asst biologist, 61-63; res collabr, Royal Cancer Hosp, 60-61; assoc prof, 63-68, PROF BIOL, UNIV ORE, 68- Concurrent Pos: NIH fel, 58-61, career develop award, 66-71. Mem: Bot Soc Am; Radiation Res Soc; Am Soc Cell Biol. Res: Chromosome structure; radiation cytology; genetics and cytology of orchids; localization of gene function. Mailing Add: Dept of Biol Univ of Ore Eugene OR 97403

WIMBERLEY, STANLEY, b Detroit, Mich, Dec 22, 27. MARINE GEOLOGY, OCEANOGRAPHY. Educ: Johns Hopkins Univ, BA, 52; Univ Tex, MA, 54; Univ Southern Calif, PhD(geol), 64. Prof Exp: Res technician, Chesapeake Bay Inst, Johns Hopkins Univ, 51-52; sedimentologist, Gulf Res & Develop Co, 54-58; res assoc, Allan Hancock Found, Univ Southern Calif, 59-61; asst prof geol, Univ PR, 62-65 & Univ South Fla, 65-67; ASSOC PROF GEOL, CHAPMAN COL, 67- Mem: Geol Soc Am; Soc Econ Paleont & Mineral; Am Asn Petrol Geol; Marine Technol Soc. Res: Detrital sediments of barrier islands, beaches and continental shelf; sea floor topography. Mailing Add: Dept of Geol Chapman Col Orange CA 92666

WIMER, BRUCE MEADE, b Tuckerton, NJ, Aug 31, 22; m 50; c 3. INTERNAL MEDICINE, HEMATOLOGY. Educ: Franklin & Marshall Col, BS, 43; Jefferson Med Col, MD, 46; Am Bd Internal Med, dipl; Am Bd Hemat, dipl, 72. Prof Exp: Intern, Hosp, Jefferson Med Col, 46-47, resident internal med & hemat, 48-51; asst internal med & hemat, Guthrie Clin, Sayre, Pa, 53-59; pvt pract, Summit, NJ, 59-61; assoc med dir, Squibb Inst Med Res, 61-62; CHIEF HEMAT & ONCOL, LOVELACE BATAAN MED CTR, 62- Mem: AMA; fel Am Col Physicians; Am Soc Hemat; Int Soc Hemat. Res: Cancer immunotherapy, especially adoptive leukocyte therapy; testosterone in treatment of hematologic disorders; hematologic therapeutics; hemolytic characteristics of the McLeod phenotype. Mailing Add: Lovelace Bataan Med Ctr 5200 Gibson Blvd SE Albuquerque NM 87108

WIMER, CYNTHIA CROSBY, b Boston, Mass, Oct 23, 33; m 57; c 3. PHYSIOLOGICAL PSYCHOLOGY. Educ: Wellesley Col, BA, 55; McGill Univ, MA, 58; Rutgers Univ, PhD(psychol), 61. Prof Exp: Res asst psychoacoustics, Mass Inst Technol, 55-56; res assoc psychol, Inst Develop Studies, NY Med Col, 61 & Jackson Lab, 63-69; ASSOC RES SCIENTIST, DIV NEUROSCI, CITY OF HOPE MED CTR, 69- Mem: Am Psychol Asn; Behav Genetics Asn. Res: Psychometrics; behavior genetics; neuroanatomical correlates of behavior; structure of intellect. Mailing Add: Div of Neurosci City of Hope Med Ctr 1500 E Duarte Rd Duarte CA 91010

WIMER, DAVID CARLISLE, b Champaign, Ill, July 20, 26; m 55; c 1. ANALYTICAL CHEMISTRY, ORGANIC CHEMISTRY. Educ: Univ Ill, BS, 51. Prof Exp: From chemist to sr chemist, 51-65, group leader invest drugs res, 65-70, SR ANAL RES CHEMIST, ANAL RES DEPT, ABBOTT LABS, N CHICAGO, 70- Mem: AAAS; Am Chem Soc. Res: Acid-base interactions; non-aqueous solvent chemistry; functional group analysis, particularly organic nitrogen functions; thin layer chromatography; organic and inorganic qualitative analysis; ultraviolet absorption spectra of inorganic and organic compounds. Mailing Add: 2312 11th St Winthrop Harbor IL 60096

WIMER, LARRY THOMAS, b Stuttgart, Ark, Dec 20, 36; m 59; c 3. INSECT PHYSIOLOGY. Educ: Phillips Univ, BA, 57; Rice Inst, MA, 59; Univ Va, PhD(physiol), 63. Prof Exp: Instr biol, Northwestern Univ, 62-64; asst prof, 64-70, ASSOC PROF BIOL, UNIV SC, 70- Mem: AAAS; Am Soc Zoologists. Res: Insect developmental physiology; hormonal regulation of developmental metabolic systems. Mailing Add: Dept of Biol Univ of SC Col Arts & Sci Columbia SC 29208

WIMER, RICHARD E, b Tulare, Calif, Apr 8, 32; m 57; c 3. BEHAVIORAL GENETICS. Educ: San Jose State Col, AB, 52; Ohio Univ, MSc, 53; McGill Univ, PhD(psychol), 59. Prof Exp: Instr psychol, Douglass Col, Rutgers Univ, 58-60; sr res assoc psychiat, NY Med Col, 60-61; assoc staff scientist, Jackson Lab, 61-65, staff scientist, 65-69; SR RES SCIENTIST & HEAD BEHAV GENETICS SECT, DIV NEUROSCI, CITY OF HOPE MED CTR, 69- Mem: Am Psychol Asn; Genetics Soc Am; Am Genetic Asn; Soc Neurosci; Behav Genetics Asn. Res: Genetic variations in brain structure and correlated behavioral function. Mailing Add: City of Hope Med Ctr Div of Neurosci 1500 E Duarte Rd Duarte CA 91010

WIMETT, THOMAS FREDERICK, b Stevensville, Mont, Jan 1, 21; m 54; c 5. REACTOR PHYSICS. Educ: Mont State Col, BS, 45; Mass Inst Technol, SB, 46, PhD(physics), 53. Prof Exp: Mem staff, Radiation Lab, Mass Inst Technol, 43-45, asst, Electronics Servo Lab, 46-47, res assoc physics, Synchrotron & Magnet Labs, 47-53; MEM STAFF, LOS ALAMOS SCI LAB, 53- Mem: Am Nuclear Soc; Am Phys Soc. Res: Electronics; nuclear magnetic resonance experiments; neutron physics; prompt burst reactors. Mailing Add: Nuclear Div Los Alamos Sci Lab Los Alamos NM 87544

WIMMER, CHARLES ROBERT, chemistry, see 12th edition

WIMMER, ECKARD, b Berlin, Ger, May 22, 36; m 65; c 2. MOLECULAR BIOLOGY, VIROLOGY. Educ: Univ Göttingen, diplom chemist, 59, PhD(org chem), 62. Prof Exp: Asst org chem, Univ Göttingen, 62-64; res fel biochem, Univ BC, 64-66; res assoc molecular biol, Univ Ill, 66-68; from asst prof to assoc prof microbiol, Sch Med, St Louis Univ, 68-74; ASSOC PROF MICROBIOL, SCH BASIC HEALTH SCI, STATE UNIV NY STONY BROOK, 74- Concurrent Pos: Vis prof, Mass Inst Technol, 69. Mem: Am Soc Microbiol. Res: Molecular biology of animal and bacterial viruses; biochemistry of nucleic acids; cell biology. Mailing Add: Sch of Basic Health Sci State Univ of NY Stony Brook NY 11790

WIMS, ANDREW MONTGOMERY, b Phila, Pa, Apr 29, 35; m 59; c 4. PHYSICAL CHEMISTRY, POLYMER CHEMISTRY. Educ: Howard Univ, BS, 57, MS, 59, PhD, 67. Prof Exp: Res chemist, Nat Bur Standards, 60-68; SR RES CHEMIST, GEN MOTORS RES LAB, 69- Mem: Am Chem Soc; Am Inst Chem; Am Soc Test & Mat; Sigma Xi. Res: Characterization of polymeric materials; light scattering; phase equilibriums; thermodynamics of polymers in solution; organic analysis. Mailing Add: Anal Chem Dept Gen Motors Res Labs 12 Mile & Mound Rds Warren MI 48090

WIMSATT, WILLIAM ABELL, b Washington, DC, July 28, 17; m 40; c 6. HISTOPHYSIOLOGY, REPRODUCTIVE BIOLOGY. Educ: Cornell Univ, AB, 39, PhD(hist, embryol), 43. Prof Exp: Asst histol & embryol, Cornell Univ, 40-43; asst anat, Med Sch, Harvard Univ, 43-44, instr, 44-45; from asst prof to assoc prof, 45-50, mem bd trustees, 60-65, PROF ZOOL, CORNELL UNIV, 51- Concurrent Pos: Res collabr, Brookhaven Nat Lab, 54-59; Guggenheim fel, 62-63. Mem: Fel AAAS; Am Soc Zool; Am Asn Anat; Soc Study Reprod; Am Soc Mammal. Mailing Add: Div of Biol Sci Genetics Sect Cornell Univ Ithaca NY 14853

WINANS, EDGAR VINCENT, b Salt Lake City, Utah, Apr 23, 30; m 52; c 2. ANTHROPOLOGY. Educ: Univ Calif, Los Angeles, BA, 52, MA, 54, PhD(anthrop & sociol), 59. Prof Exp: Actg instr anthrop, Univ Calif, Los Angeles, 55-56; instr, Univ Wash, 57-59; from asst prof to assoc prof, Univ Calif, Riverside, 60-65, asst dean lett & sci, 62-65; assoc prof, 65-66, chmn dept, 70-73, PROF ANTHROP, UNIV WASH, 66- Concurrent Pos: Grant, NSF, EAfrica, 61-65; grant, NIH, 61-65; Haines Found grant, Univ Calif, Riverside, 63-64; consult, Pac State Hosp, Pomona, Calif, 63-65; proj specialist, Ford Found, 68-70; econ adv rural develop, Govt Kenya, 68-70; consult, Int Labor Orgn/UN Develop Prog, 72; consult ed, J B Lippincott Co, 72- Mem: Fel Am Anthrop Asn; fel African Studies Asn; fel Royal Anthrop Inst. Res: African social and economic structure; history of political and economic change in East Africa; economics of development in Africa, Asia and Latin America. Mailing Add: Dept of Anthrop Univ of Wash Seattle WA 98105

WINANS, SARAH SCHILLING, b Hannibal, Mo, Apr 23, 41; m 65. NEUROANATOMY, NEUROPSYCHOLOGY. Educ: Cornell Univ, BA, 63, PhD(anat), 69. Prof Exp: Instr anat, State Univ NY Downstate Med Ctr, 68-70; asst prof, 70-75, ASSOC PROF ANAT, UNIV MICH, ANN ARBOR, 75- Mem: AAAS; Am Asn Anatomists; Soc Neurosci. Res: Structure and function of the olfactory and limbic systems in the central nervous system; structure and function of the visual system. Mailing Add: Dept of Anat Univ of Mich Med Sci II Ann Arbor MI 48104

WINAWER, SIDNEY J, b New York, NY, July 9, 31. INTERNAL MEDICINE, GASTROENTEROLOGY. Educ: NY Univ, BA, 52; State Univ NY, MD, 56. Prof Exp: Asst med, Harvard Med Sch, 62-64, instr, 65-66; asst prof, 66-72, CLIN ASSOC PROF MED, MED COL, CORNELL UNIV, 72- Concurrent Pos: Fel med, Boston City Hosp, 62-64, assist physician, 65-66; NIH spec fel, 65-67; asst physician, New York Hosp, 66-; dir gastrointestinal lab, 72-; asst clinician, Sloan-Kettering Inst. Mem: Fel Am Col Physicians; Am Gastroenterol Asn; Am Soc Gastrointestinal Endoscopy; Am Fedn Clin Res; Am Col Gastroenterol. Res: Clinical investigation in gastrointestinal diseases, particularly morphology, physiology, cell proliferation and other aspects of gastritis; malabsorptive studies such as massive bowel resection; clinical and investigative aspects of gastrointestinal and liver cancer. Mailing Add: Mem Sloan-Kettering Cancer Ctr 1275 York Ave New York NY 10021

WINBORN, WILLIAM BURT, b Victoria, Tex, Oct 6, 31; m 53; c 1. ANATOMY. Educ: Univ Tex, BS, 56; La State Univ, PhD(anat), 63. Prof Exp: From instr to asst prof anat, Med Units, Univ Tenn, Memphis, 63-68; asst prof, 68-69, ASSOC PROF ANAT, UNIV TEX HEALTH SCI CTR SAN ANTONIO, 69- Mem: AAAS; Electron Micros Soc Am; Am Soc Cell Biol; Am Asn Anatomists. Res: Electron microscopic studies of Islets of Langerhans and of the gastrointestinal tract; cytochemistry of the gastrointestinal tract. Mailing Add: Dept of Anat Univ of Tex Health Sci Ctr San Antonio TX 78284

WINBURY, MARTIN M, b New York, NY, Aug 4, 18; m 42; c 2. PHARMACOLOGY, PHYSIOLOGY. Educ: Long Island Univ, BS, 40; Univ Md, MS, 42; NY Univ, PhD(physiol), 51. Prof Exp: Economist, US Bur Mines, 42-44; mem staff biochem & pharmacol, Merck Inst Therapeut Res, 44-47; pharmacologist, Div Biol Res, G D Searle & Co, 47-55; sr pharmacologist, Schering Corp, 55-58, dir dept pharmacol, 58-61, assoc dir biol res, 61; DIR, DEPT PHARMACOL, WARNER LAMBERT RES INST, 61- Concurrent Pos: Mem vis fac, Col Physicians & Surgeons, Columbia Univ. Mem: Soc Exp Biol & Med; Am Soc Pharmacol & Exp Therapeut; Am Chem Soc; Am Heart Asn; Am Col Cardiol. Res: Pharmacology and physiology of cardiovascular, coronary and autonomic agents; cardiac arrhythmias; microcirculation. Mailing Add: Dept of Pharmacol Warner Lambert Res Inst Morris Plains NJ 07950

WINCH, BRADLEY LOUIS, b Elkader, Iowa, June 17, 33; m 56; c 3. ORGANIC CHEMISTRY. Educ: Loras Col, BS, 55; Wayne State Univ, PhD(org chem), 58; William Mitchell Col Law, JD, 67. Prof Exp: Res assoc, Wayne State Univ, 60; patent asst, Parke, Davis & Co, Mich, 60-61; res assoc res admin, Wayne State Univ, 61-64; proj mgr res prod develop, Cent Res Labs, Gen Mills, Inc, Minn, 64-67, dept head polymer res, 67-68, mgr specialty polymers, 68-69, tech dir, Craft, Game & Toy Div, 69-70; mgr new ventures, Mattel, Inc, Calif, 70-72; PRES B L WINCH ASSOCS, 72- Concurrent Pos: NSF res fel, Inst Org Chem, Univ Karlsruhe, Ger, 58-59, Ger Chem Soc res fel, 59-60; off sci visitor Soviet Union, Nat Acad Sci & Soviet Acad Sci, Inst Metallo-Org Compounds, Moscow, lectr, Russia, Romania, Hungary, Poland and Czech, 63-64; educ systs consult, Behav Res Labs, Inc, Calif, 71-; consult, Nat Inst Educ; prin, Foreign Study League, 73-75; pres, Instructional Technol Enterprises, 73-; mem educ adv bd, Educ Develop Opers, 73- Mem: Am Chem Soc. Res: Licensing of technology; use of technology in new business ventures; educational product development, introduction and marketing; human interaction and self-awareness programs; diagnostic and prescriptive educational systems; career education; consulting. Mailing Add: B L Winch Assocs PO Box 1185 Torrance CA 90505

WINCH, FRED EVERETT, JR, b Mass, June 16, 14; m 39; c 3. FORESTRY, SILVICULTURE. Educ: Univ Maine, BS, 36; Cornell Univ, MS, 37. Prof Exp: Forestry specialist, Soil Conserv Serv, USDA, 38-40, farm planning technician, 40-43; from asst prof to assoc prof, 43-56, actg head dept natural resources, 72-73, PROF FORESTRY, CORNELL UNIV, 56-, ACTG ASSOC DIR FORESTRY EXTEN, 73- Mem: Soc Am Foresters. Res: Plantation establishment and early growth; maple syrup and Christmas tree production; forest recreation, resource development and conservation; land use inventory analysis and planning; forest tax impacts. Mailing Add: Dept of Natural Resources Cornell Univ Ithaca NY 14850

WINCH, JOHN E, agronomy, see 12th edition

WINCHELL, C PAUL, b Ionia, Mich, Oct 14, 21; m 46; c 3. MEDICINE. Educ: Univ Mich, MD, 45. Prof Exp: From instr to assoc prof, 51-71, PROF MED, SCH MED, UNIV MINN, MINNEAPOLIS, 71- Concurrent Pos: Consult cardiol, Vet Hosp, Minneapolis & Off Hearings & Appeals, Social Security Admin. Mem: Am Heart Asn; Am Fedn Clin Res; Am Col Physicians. Res: Cardiovascular diseases. Mailing Add: Univ Hosps Univ of Minn Minneapolis MN 55455

WINCHELL, HARRY SAUL, b Coaldale, Pa, Mar 1, 35; m 64; c 4. MEDICAL PHYSICS, NUCLEAR MEDICINE. Educ: Bucknell Univ, BA, 54; Hahnemann Med Col, MD, 58; Univ Calif, Berkeley, PhD(biophys), 61. Hon Degrees: DSc, Bucknell Univ, 72. Prof Exp: Intern, San Francisco Hosp, Univ Calif, 58-59; univ fel, Univ Calif, Berkeley, 59-61; resident, Mt Sinai Hosp, NY, 61-62; ASSOC RES PHYSICIAN, DONNER LAB, UNIV CALIF, BERKELEY, 62-, LECTR MED PHYSICS, UNIV, 66-; EXEC VPRES & DIR RES & DEVELOP, MEDI-PHYSICS, INC, 72- Concurrent Pos: NSF fel, Univ Calif, Berkeley, 59-60, NIH fel, 60-61; spec consult, Sealab II Exp, La Jolla, Calif, 65. Honors & Awards: George Von Hevesy Award, Europ Orgn Nuclear Med, 69. Mem: Fel Am Col Physicians; Am Fedn Clin Res; Soc Nuclear Med; Soc Exp Biol & Med. Mailing Add: Medi-Physics Inc 5855 Christie Ave Emeryville CA 94608

WINCHELL, HORACE, b Madison, Wis, Jan 1, 15; m 37. MINERALOGY, GEOLOGY. Educ: Univ Wis, BA & MA, 36; Harvard Univ, MA, 37, PhD(mineral, crystallog), 41. Prof Exp: Asst geologist, City Bd Water Supply, Honolulu, 38-40; res crystallographer, Hamilton Watch Co, 41-45; from instr to asst prof, 45-51, ASSOC PROF MINERAL & CUR MINERAL, PEABODY MUS, YALE UNIV, 51- Concurrent Pos: Asst, Conn Geol Natural Hist Surv, 46-52. Mem: AAAS; fel Geol Soc Am; fel Mineral Soc Am; Geochem Soc; Soc Econ Geol. Res: Physical properties of sapphire; sapphire jewel bearing; diamond dies, design and crystallography; grading of diamond powder; mineralogy of Connecticut; petrology of Oahu, Hawaii; systematic mineralogy; optical mineralogy and crystallography. Mailing Add: 2161 Yale Sta New Haven CT 06520

WINCHELL, ROBERT E, b Wichita, Kans, Sept 21, 31; m 58; c 3. MINERALOGY, CRYSTALLOGRAPHY. Educ: Stanford Univ, BS, 56; Mich Col Mining & Technol, MS, 59; Ohio State Univ, PhD(mineral), 63. Prof Exp: Jr eng geologist, Bridge Dept, Calif Div Hwys, 57-58; res scientist, AC Spark Plug Div, Gen Motors Corp, 63-66; assoc prof, 66-72, PROF GEOL & MINERAL, CALIF STATE UNIV, LONG BEACH, 72- Mem: Mineral Soc Am; Brit Mineral Soc. Res: Mineral synthesis and characterization; optical, x-ray and morphological crystallography; x-ray diffraction; electron microscopy; phase equilibrium studies; crystal chemistry, growth and structure analysis; solid state and materials science. Mailing Add: Dept of Geol Sci Calif State Univ Long Beach CA 90804

WINCHESTER, ALBERT MCCOMBS, b Waco, Tex, Apr 20, 08; m 34; c 1. BIOLOGY. Educ: Baylor Univ, AB, 29; Univ Tex, MA, 31, PhD(zool), 34. Prof Exp: Instr zool, Univ Tex, 32-34; prof, Lamar Col, 34-35; head dept biol, Ouachita Baptist Col, 35-36; head dept, Tenn State Teachers Col, 36-37; head dept biol & chmn div sci, Okla Baptist Univ, 37-43; prof zool, Baylor Univ, 43-46; head biol dept, Stetson Univ, 46-61; consult, Biol Sci Curric Study, Am Inst Biol Sci, Univ Colo, 61-62; PROF BIOL, UNIV NORTHERN COLO, 62- Concurrent Pos: Rockefeller fel; Carnegie grant, 48; AAAS vis lectr, 64-; mem, Colo Bd Exam Basic Sci, 66- Mem: AAAS; Genetics Soc Am; Am Soc Human Genetics; Am Eugenics Soc. Res: Genetics; effects of x-ray on Drosophila; induced sex ratio variation in Drosophila; radiation; human genetics. Mailing Add: Dept of Biol Univ of Northern Colo Greeley CO 80631

WINCHESTER, CLARENCE FLOYD, b Chicago, Ill, Oct 14, 01; m 43; c 1. NUTRITION. Educ: Univ Calif, BS, 24, MS, 35; Univ Mo, PhD(agr chem), 39. Prof Exp: Educr, Calif, 24-31; res scientist, Univ NH, 31-32, Univ Calif, 32-37 & Univ Mo, 37-47; res scientist, Univ Fla, 47, assoc prof nutrit & assoc biochemist, 48-49; animal physiologist, Animal & Poultry Husb Res Br, Agr Res Serv, USDA, 49-58; livestock adv, US Opers Mission, Ceylon, 58-59; nutritionist, USDA, 60-61 & US Fish & Wildlife Serv, 61-65; lectr nutrit, Howard Univ, 66-75; RETIRED. Mem: AAAS; Am Chem Soc; Am Inst Chemists; Am Inst Nutrit; Am Soc Animal Sci. Res: Supplementation of cereal diets; nutritive requirements of farm animals; continuous growth of cattle; endocrine relationships in meat and egg production; water requirements of cattle. Mailing Add: 2124 Sudbury Pl NW Washington DC 20012

WINCHESTER, JAMES ALWYN, nematology, see 12th edition

WINCHESTER, JOHN WIDMER, b Chicago, Ill, Oct 8, 29; m 58; c 1. PHYSICAL CHEMISTRY. Educ: Univ Chicago, AB, 50, SM, 52; Mass Inst Technol, PhD(chem), 55. Prof Exp: Fulbright grant, Neth, 55-56; from asst prof to assoc prof geochem, Mass Inst Technol, 56-66; assoc prof meteorol & oceanog, Univ Mich, Ann Arbor, 67-69, prof oceanog & asst dir Great Lakes Res Div, 69-70; PROF OCEANOG & CHMN DEPT, FLA STATE UNIV, 70- Concurrent Pos: Vis lectr, Lamont Geol Observ, Columbia Univ, 59; res partic, Oak Ridge Inst Nuclear Studies & Oak Ridge Nat Lab, 58, 59 & 61; Fulbright vis prof, Taipei, 62-63; vis scientist, Inst Marine Sci, Univ Alaska, 64; vis prof, La Plata, 66. Mem: Am Chem Soc; Geochem Soc; Geol Soc Am; Am Phys Soc; Am Geophys Union. Res: Activation analysis; atmospheric and marine geochemistry. Mailing Add: Dept of Oceanog Fla State Univ Tallahassee FL 32306

WINCHESTER, RICHARD ALBERT, b Denver, Colo, Nov 20, 23; m 52; c 3. AUDIOLOGY, SPEECH PATHOLOGY. Educ: Univ Denver, BA, 47, MA, 48; Univ Southern Calif, PhD(audiol, speech path), 57. Prof Exp: Resident audiol & speech path, Orthop Hosp, Los Angeles, 48-50; asst prof, Univ Denver, 50-53; res audiologist, Walter Reed Army Med Ctr, 53-54; dir hearing & speech clin, Vet Admin Hosp, San Francisco, 54-55, res audiologist, Vet Admin Regional Off, Los Angeles, 55-58; asst prof audiol, Sch Med, Temple Univ, 59-63; DIR DIV COMMUN DISORDERS & RES AUDIOLOGIST, CHILDREN'S HOSP PHILADELPHIA, 63-; ASST PROF AUDIOL, SCH MED, UNIV PA, 64- Concurrent Pos: Dir audiol & speech path, Otologic Group Philadelphia, 59-66; consult, Pa Acad Ophthal & Otolaryngol, 60-; spec lectr, Univ Md, 63-64. Mem: AAAS; Am Speech & Hearing Asn; assoc fel Am Acad Ophthal & Otolaryngol; Am Cleft Palate Asn; Am Audiol Soc. Res: Speech in deafness; auditory perception in brain injury; nonorganic deafness; deafness in otosclerosis; congenital mixed deafness; hearing patterns in vestibular disorders; central auditory functions; sound spectrography and cleft palate speech; auditory behavior in infancy and early childhood. Mailing Add: Children's Hosp 34th & Civic Ctr Blvd Philadelphia PA 19104

WINCHESTER, SAMUEL CLYDE, chemical engineering, see 12th edition

WINCKLER, JOHN RANDOLPH, b North Plainfield, NJ, Oct 27, 16; m 43; c 5. PHYSICS. Educ: Rutgers Univ, BS, 42; Princeton Univ, PhD(physics), 46. Prof Exp: Res physicist, Johns-Manville Corp, NJ, 37-42; instr physics, Palmer Lab, Princeton Univ, 46-49; from asst prof to assoc prof, 49-58, PROF PHYSICS, UNIV MINN, MINNEAPOLIS, 58- Concurrent Pos: Guggenheim fel, Meudon Observ, 65-66; mem math & phys sci div comt, NSF. Honors & Awards: Agr Res Serv Space Sci Award, 62. Mem: Fel AAAS; fel Am Phys Soc; fel Am Geophys Union; Int Acad Astronaut. Res: High speed flow of gases and shock waves; geomagnetic effects and energy spectrum of primary cosmic rays; atmospheric total radiation; solar produced cosmic rays and energetic processes in solar flares; geomagnetic storm influences on energetic particles in the magnetosphere. Mailing Add: Sch of Physics & Astron Univ of Minn Minneapolis MN 55455

WINCKLHOFER, ROBERT CHARLES, b Newark, NJ, Dec 14, 26; m 49; c 2. POLYMER PHYSICS. Educ: Columbia Univ, BS, 53. Prof Exp: Engr, Cent Res Lab, 53-61, group leader, 61-64, res supvr, 64, Fibers Div, Tech Dept, 64-68, mgr res, 68-70, tech dir heavy denier nylon, 70-72, TECH DIR ADVAN TECHNOL, FIBERS DIV, TECH DEPT, ALLIED CHEM CORP, 72- Mem: AAAS. Res: Fiber polymer physics; crystallization kinetics; polymer characterization; differential thermal analysis of polymers. Mailing Add: Fibers Div Tech Dept Allied Chem Corp PO Box 31 Petersburg VA 23803

WINDECKER, RICHARD CHASE, b Newark, NJ, Oct 20, 41; m 69. BIOPHYSICS, EXPERIMENTAL SOLID STATE PHYSICS. Educ: Ohio Wesleyan Univ, BA, 63; Univ Ill, Urbana, MS, 65, PhD(physics), 70. Prof Exp: Vis lectr physics, Univ Chiangmai, Thailand, 70-73; RES ASST BIOPHYS, UNIV GUELPH, ONT, 74- Mem: Am Phys Soc; Sigma Xi; Inst Elec & Electronics Engr. Res: Probabilistic adaptive and learning models of neural interactions. Mailing Add: Dept of Physics Univ of Guelph Guelph ON Can

WINDELL, JOHN THOMAS, b Hammond, Ind, Apr 4, 30; m 59; c 3. AQUATIC BIOLOGY. Educ: Ind Cent Col, BS, 53; Ind Univ, MA, 58, PhD(limnol), 65. Prof Exp: Teacher, Griffith High Sch, 55-58; asst prof biol, Ind Cent Col, 58-62; assoc zool, Ind Univ, Bloomington, 62-65; asst prof, Ind Univ Northwest, 65-66; assoc prof, 66-70, PROF BIOL, UNIV COLO, BOULDER, 70- Concurrent Pos: Ind Univ fac fel, 66; partic, Int Symp Biol Basis Freshwater Fish Prod, Reading, Eng, 66. Honors & Awards: Lieber Mem Teaching Award, Ind Univ, 65. Mem: AAAS; Am Soc Zoologists; Am Inst Biol Sci; Am Fisheries Soc. Res: Ecological physiology; biological basis of fish production; food consumption in fishes; conversion coefficients and the ecology of fishes. Mailing Add: Dept of Biol Univ of Colo Boulder CO 80302

WINDER, CHARLES GORDON, b Ottawa, Ont, June 13, 22; m 48; c 2. GEOLOGY. Educ: Univ Western Ont, BSc, 49; Cornell Univ, MS, 51, PhD(geol), 53. Prof Exp: Lectr, 53-56, from asst prof to assoc prof, 56-64, head dept, 65-71, PROF GEOL, UNIV WESTERN ONT, 64- Mem: Fel Geol Soc Am; fel Paleont Soc; Soc Econ Mineral & Paleont; Am Asn Petrol Geologists; fel Geol Asn Can. Res: Stratigraphy of southern Ontario; conodonts; carbonate petrology; history of geology and Canadian geologists. Mailing Add: Dept of Geol Univ of Western Ont London ON Can

WINDER, DALE RICHARD, b Marion, Ind, Aug 27, 29; m 53. SOLID STATE PHYSICS. Educ: DePauw Univ, AB, 51; Univ Nebr, MA, 54; Case Inst Technol, PhD(physics), 57. Prof Exp: Lab & teaching asst physics, DePauw Univ, 50-51, Univ Nebr, 51-54 & Case Inst Technol, 54-57; physicist, Nat Carbon Res Labs, Union Carbide Corp, 57-60; asst prof, 60-64, ASSOC PROF PHYSICS, COLO STATE UNIV, 64- Concurrent Pos: Physicist, Boulder Lab, Nat Bur Standards, 62-; Idaho Nuclear Corp-Asn Western Univs fac appointee, 67. Mem: Am Phys Soc; Am Asn Physics Teachers; Am Crystallog Asn. Res: Photoelectric effect; lattic dynamics; crystal growth; radiation damage; nuclear fuels and moderators; transport property measurements. Mailing Add: Dept of Physics Colo State Univ Ft Collins CO 80521

WINDER, WILLIAM CHARLES, b Salt Lake City, Utah, Nov 16, 14; m 39. FOOD

SCIENCE, DAIRY INDUSTRY. Educ: Utah State Univ, BS, 46, MS, 48; Univ Wis, PhD(dairy indust), 49. Prof Exp: Plant supt, Winder Dairy, Utah, 34-45; from instr to assoc prof, 49-60, PROF FOOD SCI, UNIV WIS-MADISON, 60- Mem: Am Dairy Sci Asn. Res: Effects of ultrasound on food products; physical and chemical effects of freezing and drying food products; analysis of foods. Mailing Add: Dept of Food Sci Univ of Wis Madison WI 53706

WINDHAGER, ERICH E, b Vienna, Austria, Nov 4, 28; US citizen; m 56; c 2. PHYSIOLOGY, BIOPHYSICS. Educ: Univ Vienna, MD, 54. Prof Exp: Fel biophys, Harvard Med Sch, 56-58; instr physiol, Med Col, Cornell Univ, 58-61; vis scientist, Biochem Inst, Univ Copenhagen, 61-63; from asst prof to assoc prof physiol, 63-69, PROF PHYSIOL, MED COL, CORNELL UNIV, 69-, CHMN DEPT, 73- Concurrent Pos: Career scientist, Res Coun New York, 63-71 & Irma Hirschl Found, 73-; sect ed, Am J Physiol, 69-74. Mem: Am Physiol Soc; Biophys Soc; Int Soc Nephrology; Harvey Soc; Am Soc Nephrology. Res: Renal tubular transfer of sodium chloride; electrophysiology of the nephron; micropuncture techniques and nephron function; kidney, water and electrolytes. Mailing Add: Dept of Physiol Cornell Univ Med Col New York NY 10021

WINDHAM, MICHAEL PARKS, b Houston, Tex, Sept 23, 44; m 70; c 1. MATHEMATICS. Educ: Rice Univ, BA, 66, MA & PhD(math), 70. Prof Exp: Instr math, Univ Miami, 70-71; ASST PROF MATH, UTAH STATE UNIV, 71- Mem: Am Math Soc; Math Asn Am. Res: Several complex variables; differential geometry; mathematics history. Mailing Add: Dept of Math Utah State Univ Logan UT 84321

WINDHAM, PAT MORRIS, b Roscoe, Tex, Sept 3, 20. PHYSICS. Educ: NTex State Col, BS, 47, MS, 51; Rice Inst, PhD(physics), 55. Prof Exp: Asst prof physics, Tex Tech Col, 55-56; asst prof, 56-59, ASSOC PROF PHYSICS, N TEX STATE UNIV, 59- Mem: Am Phys Soc; Am Asn Physics Teachers. Res: Low energy nuclear physics; charge exchange studies; negative ions. Mailing Add: Dept of Physics NTex State Univ Denton TX 76203

WINDHAM, RONNIE LYNN, b Jasper, Tex, Mar 2, 43; m 72. ANALYTICAL CHEMISTRY. Educ: Pan Am Univ, BA, 65; Eastern NMex Univ, MS, 69; Tex A&M Univ, PhD(anal chem), 71. Prof Exp: RES CHEMIST, JEFFERSON CHEM CO, 72- Mem: Am Chem Soc. Res: Analytical chemistry of trace metals; solvent extraction; atomic absorption spectrophotometry. Mailing Add: 11608 January Dr Austin TX 78753

WINDHAM, STEVE LEE, b Miss, Sept 19, 22; m 48; c 3. HORTICULTURE. Educ: Miss State Col, BS, 43, MS, 48; Mich State Col, PhD(hort), 53. Prof Exp: Asst, Miss State Col, 47-48; asst horticulturist 48-51, assoc horticulturist, 53-61, HORTICULTURIST, TRUCK CORPS BR, MISS AGR EXP STA, MISS STATE UNIV, 61-, ADJ ASSOC PROF HORT, 74- Concurrent Pos: Asst, Mich State Col, 51-53. Res: Vegetable crop response and utilization of nutrient elements. Mailing Add: Truck Crops Exp Sta Miss State Univ Crystal Springs MS 39059

WINDHEUSER, JOHN JOSEPH, b Bonn, Ger, Dec 2, 26; US citizen; m 51; c 4. PHYSICAL PHARMACY. Educ: Rutgers Univ, Newark, BSc, 51; Univ Wis-Madison, MS, 59, PhD(pharm), 61. Prof Exp: Res pharmacist, Ciba Pharmaceut Co, NJ, 51-57; asst prof pharm, Univ Wis-Madison, 61-63; dir pharm res & develop, Sandoz Pharmaceut, NJ, 63-66; assoc prof pharm, Univ Wis-Madison, 66-72; dir pharm res & develop, Alza Corp, 72-74; SR VPRES, INTERx RES CORP, 74- Mem: AAAS; Am Pharmaceut Asn; fel Acad Pharmaceut Sci. Res: Physical chemistry of solutions; drug interaction in solution; biopharmaceutics of cancer-chemotherapeutic agents; drug analysis. Mailing Add: INTERx Res Corp 2201 W 21st St Lawrence KS 66044

WINDHOLZ, THOMAS BELA, b Arad, Romania, Jan 10, 23; US citizen; m 48; c 2. ORGANIC CHEMISTRY. Educ: Univ Cluj, MS, 47. Prof Exp: Res chemist, Chinoin Pharmaceut Co, Budapest, 48-50; sr res chemist, Res Inst Pharmaceut Indust, 51-54, sect head, 55-56; res chemist, Res Labs, Celanese Corp Am, NJ, 57-59; sr res chemist, 60-63, sect head, 64-69, assoc dir, 70-72, DIR INT REGULATORY AFFAIRS, RES LABS, MERCK & CO, INC, 72- Mem: Am Chem Soc. Res: Synthetic organic chemistry; steroids and other natural products; medicinal chemistry, metabolism; international scientific relations. Mailing Add: Merck & Co Res Labs Rahway NJ 07065

WINDHOLZ, WALTER M, b Gorham, Kans, Apr 25, 33; m 61; c 3. MATHEMATICS. Educ: Ft Hays Kans State Col, AB, 53; Kans State Univ, MS, 58. Prof Exp: Mathematician, Thiokol Chem Corp, Utah, 61-65; MATHEMATICIAN, KAMAN SCI CORP, 65- Mem: Math Asn Am. Res: Applied mathematics; structural dynamics. Mailing Add: 1338 Whitehouse Dr Colorado Springs CO 80904

WINDHORST, DOROTHY BAKER, b Pawhuska, Okla, Mar 25, 28; c 2. DERMATOLOGY, IMMUNOLOGY. Educ: Univ Chicago, MD, 54. Prof Exp: Asst prof dermat, Univ Minn, 62-68; from asst prof to assoc prof med, Univ Chicago, 68-73; MED OFFICER, NAT CANCER INST, 72- Concurrent Pos: Mem training grants comt, Nat Inst Arthritis, Metab & Digestive Dis, 72-73; mem subcomt arsenic, Comt Med & Biol Effects of Environ Pollutants, Nat Acad Sci, 74-76. Mem: Am Asn Immunologists; Am Fedn Clin Res; Soc Invest Dermat. Res: Cancer immunotherapy; human phagocyte function; genetic defects in defense mechanisms. Mailing Add: Off of Assoc Dir for Immunol Nat Cancer Inst Bldg 10 Rm 4B17 Bethesda MD 20014

WINDISCH, RITA M, b Pittsburgh, Pa. CLINICAL CHEMISTRY. Educ: Duquesne Univ, BS, 60, PhD(chem), 64. Prof Exp: CHIEF CLIN CHEMIST PATH, MERCY HOSP, 65- Concurrent Pos: Clin prof, Sch Med Technol, Carlow Col, 65-; clin prof, Sch Med Technol, Duquesne Univ, 69-; med staff affil, Div Clin Chem, Dept Path, Mercy Hosp, 69- & Div Nuclear Path & Oncol, Dept Radiol, 74- Mem: Am Asn Clin Chemists; Am Chem Soc. Res: Diabetes, clinical chemistry and toxicology. Mailing Add: Dept of Path Mercy Hosp Locust St Pittsburgh PA 15219

WINDLAN, HAROLD MILTON, b Tonica, Ill, Aug 11, 20; m 59; c 4. DAIRY CHEMISTRY, BIOCHEMISTRY. Educ: Univ Ga, BS, 51, MS, 52; Cornell Univ, PhD, 55. Prof Exp: Food technologist, Am Cyanamid Co, 55-56, mgr food technol labs, 56-58; res chemist, Continental Baking Co, NY, 58; dir lab, Carr Baking Co, 59; vpres, 59-66, PRES, C W ENGLAND LABS, INC, 66-, DIR, 74- Mem: Am Chem Soc. Res: Food chemicals, preservation and analysis; flavor analysis. Mailing Add: C W England Labs Inc 11213 Somerset Ave Beltsville MD 20705

WINDLE, WILLIAM FREDERICK, b Huntington, Ind, Oct 10, 98; m 23; c 2. ANATOMY. Educ: Denison Univ, BS, 21; Northwestern Univ, MS, 23, PhD(anat), 26. Hon Degrees: ScD, Denison Univ, 47. Prof Exp: From asst to assoc prof anat, Med Sch, Northwestern Univ, 22-35, prof micros anat, 35-42, prof neurol & dir neurol inst, 42-46; prof anat & chmn dept, Sch Med, Univ Wash, Seattle, 46-47; prof anat & chmn dept, Sch Med, Univ Pa, 47-51, vis res prof, 51-52; sci dir, Baxter Lab, Inc, 51-53; chief, Lab Neuroanat Scis, NIH, 54-60, asst dir, Nat Inst Neurol Dis & Blindness, 60-61, chief lab perinatal physiol, NIH, San Juan, PR, 61-63; res prof rehab med & dir res, NY Univ Med Ctr, 63-71; RES PROF, DENISON UNIV, 71- Concurrent Pos: Guest investr, Cambridge Univ, 35-36; vis prof, Univ Tenn, 41; Commonwealth vis prof, Univ Louisville, 44; Harvey lectr, 45; lectr, Postgrad Assembly Gynec, Univ Southern Calif, 47; mem res adv bd, United Cerebral Palsy Asns, 55-64, human embryol & develop study sect, NIH, 55-58, sci rev comn, Div Health Res Facils & Resources, 64-67; ed, Exp Neurol, 58-75; mem hon fac med, Univ Chile, 61; prof ad honoren, Univ PR, 62-65; vis prof, Univ Calif, Los Angeles, 70-; chmn sci adv comt, Nat Paraplegia Found, 70-73. Honors & Awards: Weinstein Award, United Cerebral Palsy Asns, 57; Albert Lasker Award Basic Med Res, 68; Award, Asn Res Nerv & Ment Dis, 71; William T Wakeman Basic Res Award, Nat Paraplegia Found, 72; Paralyzed Vet Asn Am Res Award, 72. Mem: AAAS; Am Physiol Asn; Soc Exp Biol & Med; Am Asn Anat; Am Neurol Asn. Res: Embryology and histology of the nervous system; development of behavior; fetal physiology; cerebral anoxia; regeneration in the central nervous system. Mailing Add: 229 S Cherry St Granville OH 43023

WINDLER, DONALD RICHARD, b Centralia, Ill, Feb 4, 40; m 63; c 1. PLANT TAXONOMY. Educ: Southern Ill Univ, Carbondale, BS, 63, MA, 65; Univ NC, Chapel Hill, PhD(bot), 70. Prof Exp: Asst prof, 69-73, ASSOC PROF BIOL, TOWSON STATE COL, 73-, CUR HERBARIUM, 69- Mem: Int Asn Plant Taxonomists; Am Soc Plant Taxon; Torrey Bot Club; Sigma Xi. Res: Systematics of Leguminosae; flora of Maryland and Delaware; plant distribution in the Eastern United States. Mailing Add: Herbarium Towson State Col Baltimore MD 21204

WINDMUELLER, HERBERT GEORGE, b Westphalia, Ger, July 5, 31; nat US; m 58; c 2. BIOCHEMISTRY. Educ: Va Polytech Inst, BS, 52, MS, 56, PhD(biochem), 58. Prof Exp: Asst, Va Polytech Inst, 54-57; NIH fel biochem, Brandeis Univ, 58-61; BIOCHEMIST, NAT INST ARTHRITIS, METAB & DIGESTIVE DIS, 61- Mem: Am Inst Nutrit; Am Soc Biol Chemists; Am Chem Soc; Am Heart Asn. Res: Lipoproteins; lipid metabolism; nucleotides; nutrition; intestinal metabolism; isolated organ perfusion. Mailing Add: Nat Inst of Arth Met & Dig Dis Bethesda MD 20014

WINDOLPH, JOSEPH R, b Humphrey, Nebr, Nov 30, 20. MATHEMATICS. Educ: Univ Notre Dame, MS, 53. Prof Exp: From instr to assoc prof, 53-75, PROF MATH, QUINCY COL, 75-, CHMN DEPT, 60- Mem: Math Asn Am; Nat Coun Teachers Math. Res: Abstract algebra; foundations; electronics. Mailing Add: Dept of Math Quincy College Quincy IL 62301

WINDOM, HERBERT LYNN, b Macon, Ga, Apr 23, 41; m 63; c 2. OCEANOGRAPHY. Educ: Fla State Univ, BS, 63; Univ Calif, San Diego, MS, 65, PhD(earth sci), 68. Prof Exp: ASSOC PROF OCEANOG, SKIDAWAY INST OCEANOG & GA INST TECHNOL, 68- Res: Marine environmental quality; chemical oceanography; marine biochemistry of trace elements; marine sediments; environmental effects of dredging. Mailing Add: Skidaway Inst of Oceanog Box 13687 Savannah GA 31406

WINDSOR, DONALD ARTHUR, b Chicago, Ill, Mar 22, 34; m 63, 69; c 4. INFORMATION SCIENCE, INVERTEBRATE ZOOLOGY. Educ: Univ Ill, Urbana, BS, 59, MS, 60. Prof Exp: Res asst parasitol, Dept Zool, Univ Ill, Urbana, 62-64; unit leader, 66-67, doc sect chief, 67-74, INFO SCIENTIST III, RES & DEVELOP DEPT, NORWICH PHARMACAL CO, 74-; RES DIR, SciAESTHETICS INST, 69- Mem: AAAS; NY Acad Sci; Am Soc Info Sci; Med Libr Asn; Inst Info Sci. Res: Bibliometric investigations of information transfer systems in clinical development of chemotherapeutic pharmaceuticals, in biomedical specialty fields, and in the species of Hirudinea. Mailing Add: SciAesthetics Inst PO Box 604 Norwich NY 13815

WINDSOR, DONALD MONTGOMERY, b Chicago, Ill, Aug 4, 44; m 71. ANIMAL BEHAVIOR. Educ: Purdue Univ, BS, 66; Cornell Univ, PhD(animal behav), 72. Prof Exp: Fel, Orgn Trop Studies, 72-73; fel, 73-75, ZOOLOGIST, SMITHSONIAN TROP RES INST, 75- Res: Ecological and genetical factors influencing reproductive success of the wasp, Polistes versicolor, in Panama. Mailing Add: Smithsonian Trop Res Inst PO Box 2072 Balboa CZ

WINDSOR, EMANUEL, b Gloucester, Mass, Sept 2, 13; m 47; c 3. BIOCHEMISTRY. Educ: Calif Inst Technol, BS, 38, MS, 48, PhD(biochem), 51. Prof Exp: Asst biochem, Sansum Clin, 38-41; sr sci aide, Phys & Chem Mat Test, US Eng Labs, 41-43; anal chemist, Gooch Labs, 43-44; res biochemist, Huntington Mem Hosp, 47-48; asst microbiol, Calif Inst Technol, 48-49; chief clin lab, Delos Comstock X-ray & Clin Labs, 50-51; res fel biol, Calif Inst Technol, 51-53; sr biochemist, Riker Labs, 53-70; CLIN BIOCHEMIST, LOS ANGELES COUNTY-OLIVE VIEW MED CTR, 71- Mem: AAAS; Am Heart Asn; Am Asn Clin Chemists. Res: Lipid metabolism; cardiovascular disease; enzyme induction; clinical laboratory methodology; liver microsomal enzymes. Mailing Add: 6421 Nagle Ave Van Nuys CA 91401

WINDSOR, ROBERT BEACH, b Freeport, NY, July 20, 20; m 52. APPLIED PHYSICS. Educ: Hobart Col, BA, 41; Mass Inst Technol, PhD(physics), 44. Prof Exp: Mem staff, Radiation Lab, Mass Inst Technol, 43-45; physicist, Allan B Du Mont Labs, Inc, NJ, 45-46; asst prof physics, Johns Hopkins Univ, 46-50; res engr, Westinghouse Elec Corp, 50-52; chief engr, Waters Mfg Co, 52-54; chief engr, Vectron, Inc, 54-56; eng scientist, Aerospace Systs Div, Radio Corp Am, 56-58, sr eng scientist, 58-60, mgr advan instrumentation, 60-62; consult, Inst Defense Anal, 62-63; sr mem tech staff, Mitre Corp, Mass, 63-67; CONSULT ENG PHYSICIST, 68- Mem: Am Phys Soc; Am Inst Aeronaut & Astronaut; Inst Elec & Electronics Engrs. Res: Physical electronics of semiconductors; precision electronic instruments; optical and electronic systems. Mailing Add: 81 Plain Rd Wayland MA 01778

WINE, JEFFREY JUSTUS, b Pittsburgh, Pa, Feb 10, 40; m 66. NEUROSCIENCES. Educ: Univ Pittsburgh, BS, 66; Univ Calif, Los Angeles, PhD(psychol), 71. Prof Exp: NIH fel, Dept Biol Sci, 71-72, ASST PROF PSYCHOL, STANFORD UNIV, 72- Mem: Am Psychol Asn; Soc Neurosci; Animal Behav Soc; Soc Exp Biol; AAAS. Res: Neurophysiological and neuroanatomical analysis of invertebrate behavior. Mailing Add: Dept of Psychol Stanford Univ Stanford CA 94305

WINE, PAUL HARRIS, b Detroit, Mich, Mar 18, 46; m 74. CHEMICAL PHYSICS. Educ: Univ Mich, BS, 68; Fla State Univ, PhD(phys chem), 74. Prof Exp: ROBERT A WELCH FEL CHEM, UNIV TEX, DALLAS, 74- Mem: Am Chem Soc; Am Phys Soc. Res: Energy transfer; laser excited fluorescence spectroscopy; photochemistry of small molecules. Mailing Add: Inst for Chem Sci Univ of Tex Dallas Box 688 Richardson TX 75080

WINE, RUSSELL LOWELL, b Indian Springs, Tenn, Aug 17, 18; m 42; c 3. STATISTICS. Educ: Bridgewater Col, BA, 41; Univ Va, MA, 45; Va Polytech Inst, PhD(statist), 55. Prof Exp: Instr math, Univ Va, 43-45, Amherst Col, 45-46 & Univ Okla, 46-47; asst prof, Washington & Lee Univ, 47-52; assoc prof statist, Va Polytech Inst, 55-57; assoc prof, 57-60, PROF STATIST & HEAD DEPT, HOLLINS COL, 60- Mem: Fel AAAS; Inst Math Statist; Am Statist Asn; Biomet Soc; Math Asn Am.

Res: Least squares; design of experiments; multiple tests; sample surveys. Mailing Add: Dept of Statist Hollins College Hollins College VA 24020

WINEFORDNER, JAMES D, b Geneseo, Ill, Dec 31, 31; m 57; c 3. ANALYTICAL CHEMISTRY. Educ: Univ Ill, BS, 54, MS, 55, PhD(anal chem), 58. Prof Exp: Fel, Univ Ill, 58-59; from asst prof to assoc prof, 59-67, PROF CHEM, UNIV FLA, 67- Honors & Awards: Meggers Award in Spectros, 69; Am Chem Soc Award Anal Chem, 73. Mem: Soc Appl Spectros; Am Chem Soc. Res: Atomic, ionic and molecular emission; absorption; fluorescence spectroscopy; gas chromatographic detectors; trace analysis. Mailing Add: Dept of Chem Univ of Fla Gainesville FL 32611

WINEGRAD, ALBERT IRVIN, b Philadelphia, Pa, Sept 26, 26; m 58. MEDICINE. Educ: Univ Pa, MD, 52. Prof Exp: From intern to resident med, Hosp Univ Pa, 52-55; from asst instr to assoc prof, 57-71, PROF MED, SCH MED, UNIV PA, 71-, DIR, GEORGE S COX MED RES INST, 66- Concurrent Pos: Am Col Physicians res fel med, Peter Bent Brigham Hosp & USPHS fel, Harvard Med Sch, 55-57; Am Diabetes Asn res fel, 57-59; Markle scholar, 59-64. Res: Endocrinology and metabolic diseases. Mailing Add: Cox Med Res Inst Hosp of the Univ of Pa Philadelphia PA 19104

WINEGRAD, SAUL, b Philadelphia, Pa, Mar 15, 31; m 63; c 2. PHYSIOLOGY. Educ: Univ Pa, BA, 52, MD, 56. Prof Exp: Intern, Peter Bent Brigham Hosp, 56-57; sr asst surgeon, NIH, 57-59, fel, Nat Heart Inst, 59-60, surgeon, NIH, 60-61; hon res assoc, Univ Col, Univ London, 61-62; from asst prof to assoc prof physiol, 62-69, PROF PHYSIOL, SCH MED, UNIV PA, 69- Concurrent Pos: Assoc, Sch Med, George Washington Univ, 58-61; NSF sr fel, Univ Col, Univ London, 71-72. Mem: Am Physiol Soc; Soc Gen Physiol; Biophys Soc; Cardiac Muscle Soc. Res: Cardiovascular and muscle physiology. Mailing Add: Dept of Physiol Univ of Pa Sch of Med Philadelphia PA 19104

WINEK, CHARLES L, b Erie, Pa, Jan 13, 36; m 60; c 3. TOXICOLOGY, PHARMACOLOGY. Educ: Duquesne Univ, BS, 57, MS, 59; Ohio State Univ, PhD(pharmacol), 62. Prof Exp: Res assoc phytochem, Ohio State Univ, 59-62; res toxicologist, Procter & Gamble Col, 62-63; from asst prof pharmacol & toxicol to assoc prof toxicol, 63-69, PROF TOXICOL, DUQUESNE UNIV, 69-; CHIEF TOXICOLOGIST, ALLEGHENY COUNTY CORONER'S OFF, 66- Concurrent Pos: Consult, Dept Anesthesiol, St Francis Hosp, 67-; mem panel ther, Poison Control Ctrs, Dept Health; mem adv comt lab act, Pa Dept Health; mem adv bd, Drug Res Proj, Franklin Inst, Philadelphia; fac mem, Bur Narcotics & Dangerous Drugs, Police Educ Prog; adj prof, Sch Med, Univ Pittsburgh; ed at large toxicol, Marcel Dekker, Inc, New York; ed, Toxicol Newslett, Sch Pharm, Duquesne Univ; ed, Toxicol Ann, 74. Mem: Soc Toxicol; Am Acad Forensic Sci; Acad Pharmaceut Sci; Am Asn Poison Control Ctrs; Drug Info Asn. Res: Toxicity of antifungal agents; safety evaluations; rapid methods of toxicological analyses. Mailing Add: Dept of Pharmacol Duquesne Univ Pittsburgh PA 15219

WINELAND, WILLIAM CLEMARD, b Monroe, La, Dec 5, 12; m 50; c 2. PHYSICS, RESEARCH ADMINISTRATION. Educ: Univ Ky, BS, 33, MS, 35, PhD(physics), 40. Prof Exp: Asst physics, Univ Ky, 34-36; instr math & physics, Morehead State Col, 36-40, assoc prof, 40-42, head dept, 42-46; physicist, US Naval Ord Lab, 46-54, chief weapons anal div, 54-55, dep chief physics res dept, 55-57, chief, 57-58, physics prog chief, 58-64; dept dir, NATO Supreme Allied Commander, Atlantic, Antisubmarine Warfare Ctr, Italy, 64-66; assoc head physics res, US Naval Ord Lab, 66-73, assoc tech dir res, 73-75, HEAD RES & TECHNOL, NAVY SURFACE WEAPONS CTR, 75- Concurrent Pos: Lectr, Univ Md, 62-64. Honors & Awards: US Navy Meritorious Civilian Serv Award, 53, Superior Civilian Serv Award, 66. Mem: Am Phys Soc; Acoust Soc Am. Res: Magnetic measurements; underwater acoustics; antisubmarine warfare; systems analysis. Mailing Add: 10304 Eastwood Ave Silver Spring MD 20901

WINEMAN, ROBERT JUDSON, b Chicago, Ill, 1919; m 44; c 5. BIO-ORGANIC CHEMISTRY, BIOMATERIALS. Educ: Williams Col, AB, 41; Univ Mich, MS, 42; Harvard Univ, PhD(chem), 49. Prof Exp: Chemist, E I du Pont de Nemours & Co, 42-43; res chemist, Monsanto Chem Co, 49-53, res group leader, 54-60, Monsanto Res Corp, Mass, 60-61, dir, Boston Lab, 61-69; dir biomed res labs, Am Hosp Supply Corp, 69-70; ASSOC CHIEF, ARTIFICIAL KIDNEY-CHRONIC UREMIA PROG, NAT INST ARTHRITIS, METAB & DIGESTIVE DIS, 70- Concurrent Pos: Instr, Northeastern Univ, 52-53. Mem: AAAS; Am Chem Soc; Am Soc Artificial Internal Organs. Res: Organic synthesis; steroids; amino acids; sulfur compounds; medical devices. Mailing Add: Nat Inst of Hlth Bldg 31 Rm 9A05 9000 Rockville Pike Bethesda MD 20014

WINER, ALFRED D, b Lynn, Mass, Dec 24, 26; m 55; c 2. BIOCHEMISTRY. Educ: Northeastern Univ, BS, 48; Purdue Univ, MS, 50; Duke Univ, PhD(biochem), 57. Prof Exp: Instr org chem, Univ Mass, 50-51; USPHS fel, Med Nobel Inst, Sweden, 58-60; from instr to asst prof biochem, 60-65, ASSOC PROF BIOCHEM, MED CTR, UNIV KY, 65- Concurrent Pos: USPHS career develop award, 60-70; consult, Pabst Labs, Wis, 64- Mem: Am Chem Soc; Brit Biochem Soc; Am Soc Biol Chemists. Res: Mechanism of action of dehydrogenase-coenzyme complexes; hormonal effects on enzymes in spermatogenesis. Mailing Add: Dept of Biochem Univ of Ky Med Ctr Lexington KY 40506

WINER, HERBERT ISAAC, b New York, NY, Sept 19, 21; m 43; c 4. FORESTRY. Educ: Yale Univ, BA, MF, 49, PhD, 56. Prof Exp: Instr forestry, Sch Forestry, Yale Univ, 52-56, asst prof lumbering, 56-64; sci consult, Pulp & Paper Inst Can, 63-64, forester, 64-65, sr forester, 65-71, dir, Logging Res Div, 71-75, RES DIR, EASTERN DIV, FOREST ENG RES INST CAN, 75- Concurrent Pos: Mem, Int Union Forest Res Orgns. Mem: Soc Am Foresters; Can Inst Forestry; Forest Hist Soc. Res: Research management in all aspects of forest engineering; forest history. Mailing Add: Forest Eng Res Inst 245 Hymus Blvd Pointe Claire PQ Can

WINESTOCK, CLAIRE HUMMEL, b US, July 7, 32; m 56; c 1. ORGANIC CHEMISTRY. Educ: Univ Utah, BS, 52; Univ Wis, PhD(org chem), 56. Prof Exp: Res assoc biochem, Columbia Univ, 56-59; res fel chem, Univ Utah, 59, res assoc biochem, Col Med, 60-61, res instr, 61-65; grants assoc, NIH, 65-66, health scientist adminr, Nat Inst Arthritis & Metab Dis, 66-69, EXEC SECY, VIROL STUDY SECT, DIV RES GRANTS, NIH, 69- Mem: AAAS; Am Chem Soc. Res: Organic synthesis; chemistry of natural products; science administration. Mailing Add: Div of Res Grants Nat Insts of Health Bethesda MD 20014

WINET, HOWARD, b Chicago, Ill, Sept 13, 37; m 68. BIOPHYSICS. Educ: Univ Ill, BS, 59; Univ Calif, Los Angeles, MA, 62, PhD(zool), 69. Prof Exp: Res fel eng sci, 69-73, RES BIOPHYSICIST ENG SCI, CALIF INST TECHNOL, 73- Concurrent Pos: Adv, Nat Sci Comt, Calif State Comn Teacher Preparation & Licensing, 72- Mem: Am Physiol Soc; Am Soc Zoologists; Biophys Soc; Soc Exp Biol & Med; Soc Gen Physiologists. Res: Biophysical fluid mechanics of muco-ciliary systems, cell suspensions and bacterial and sperm propulsion. Mailing Add: Div of Eng Calif Inst of Technol Pasadena CA 91125

WINFIELD, ARNOLD FRANCIS, b Chicago, Ill, Sept 29, 26; m 51; c 2. BIOCHEMISTRY. Educ: Howard Univ, BS, 49. Prof Exp: Chemist, Ord Corps, US Dept Army, 52-53; chemist, 53-69, biochemist, 69-71, REGULATORY AFFAIRS ADMINR, CONSUMER & HOSP PRODS DEPT, ABBOTT LABS, 71- Mem: Am Chem Soc; Soc Indust Microbiol; Am Soc Qual Control. Mailing Add: Dept 491 Abbott Labs PO Box 68 Abbott Park North Chicago IL 60064

WINFIELD, JOHN BUCKNER, b Kentfield, Calif, Mar 19, 42; m 69; c 3. IMMUNOLOGY, RHEUMATOLOGY. Educ: Williams Col, BA, 64; Cornell Univ, MD, 68. Prof Exp: Intern internal med, New York Hosp, 68-69; staff assoc immunol, NIH, 69-71; resident, 71-73, instr, 74-75, ASSOC PROF INTERNAL MED, UNIV VA, 76- Concurrent Pos: Fel immunol, Rockefeller Univ, 73-75; fel, Arthritis Found, 73, sr investr, 76. Mem: Am Fedn Clin Res; Am Col Physicians; Am Rheumatology Asn; Am Asn Immunol. Res: Clinical immunology; auto immune diseases. Mailing Add: Rheumatol Div Dept Internal Med Univ of Va Sch of Med Charlottesville VA 22901

WINFREE, ARTHUR T, b St Petersburg, Fla, May 5, 42; m 65; c 2. BIOLOGY, BIOPHYSICS. Educ: Cornell Univ, BS, 65; Princeton Univ, PhD(biol), 70. Prof Exp: Asst prof math biol, Univ Chicago, 69-72; ASSOC PROF BIOL, PURDUE UNIV, 72- Concurrent Pos: Vis prof, Univ Sussex & Med Res Coun Lab Molecular Biol, Cambridge, Eng, 71; res career develop award, NIH, 73- Res: Chemical oscillations and waves; circadian clocks and photo-periodism; daily and seasonal temporal organization in ecosystems. Mailing Add: Dept of Biol Sci Purdue Univ West Lafayette IN 47907

WINFREY, J C, b Post, Tex, Feb 10, 27; m 47; c 1. ORGANIC CHEMISTRY, ANALYTICAL CHEMISTRY. Educ: ETex State Teachers Col, BS & MS, 49. Prof Exp: Teacher high sch, Tex, 49-51; chemist, Eagle-Picher Lead Co, 51 & Lone Star Gas Co, 51-56; res chemist, Dow Chem Co, 56-62; anal chemist, Res Dept, Signal Oil & Gas Co, 62-68; anal sect supvr res mgt, Signal Chem Co, 69-71; chief chemist, Geneva Industs, Inc, 71-73; gen mgr & corp secy, Anal Serv, Inc, 73-75; CONSULT, 75- Concurrent Pos: Mem, Adv Comt Proj, 44, Am Petrol Inst, 68-71; anal consult, Haines & Assocs, 71. Mem: Am Chem Soc; Am Inst Chemists. Res: Instrumental analytical chemistry; organic synthesis of amines and epoxides; gas chromatography; thin layer chromatography; liquid chromatography; mass spectrometry; infrared spectrometry; computer applications in analytical chemistry; spectrochemical analysis of used lube oils. Mailing Add: 5215 Georgi Lane Houston TX 77018

WING, CHARLES GODDARD, oceanography, geophysics, see 12th edition

WING, ELIZABETH S, b Cambridge, Mass, Mar 5, 32; m 57; c 2. ZOOLOGY. Educ: Mt Holyoke Col, BA, 55; Univ Fla, MS, 57, PhD(zool), 62. Prof Exp: Asst cur zoo-archaeol, 61-74, ASST PROF ANTHROP, UNIV FLA, 70-, ASSOC CUR, FLA STATE MUS, 74- Concurrent Pos: NSF grants, 61-64, 66-68, 69-73 & 75, co-investr, 61-64; Caribbean Res Prog grant, 64 & 65; Ctr Latin Am Studies res grant, 66. Mem: Am Soc Mammal; Soc Am Archaeol; AAAS. Res: Identification and analysis of faunal remains excavated from Indian sites in Southeastern United States and Latin America; prehistoric subsistence and animal domestication in the Andes. Mailing Add: Fla State Mus Univ of Fla Gainesville FL 32601

WING, GEORGE MILTON, b Rochester, NY, Jan 21, 23; m 72. APPLIED MATHEMATICS. Educ: Univ Rochester, BA, 44, MS, 47; Cornell Univ, PhD(math), 49. Prof Exp: Scientist, Los Alamos Sci Lab, Univ Calif, 45-46, mem staff, 51-58; instr math, Univ Rochester, 46-47; instr, Univ Calif, Los Angeles, 49-51, asst prof, 51-52; assoc prof, Univ NMex, 58-59; mem staff, Sandia Corp, 59-64; prof math, Univ Colo, 64-66 & Univ NMex, 66-73; vis prof, Tex Tech Univ, 75-76. Concurrent Pos: Consult, Los Alamos Sci Lab, 58-59 & 64-, Sandia Corp, 58-59, E H Plesset Assocs, 58-59 & 65-69 & Rand Corp, 58-65; mem, Panel Phys Sci & Eng, Comt Undergrad Prog Math, 63-67. Mem: AAAS; Am Math Soc; Math Asn Am; Soc Indust & Appl Math. Res: Transport theory. Mailing Add: 107 Tunyo Los Alamos NM 87544

WING, JAMES, b Highland Park, Mich, July 8, 29; m 57; c 2. NUCLEAR CHEMISTRY. Educ: Univ Tenn, BS, 51; Purdue Univ, MS, 53, PhD(chem), 56. Prof Exp: Asst chemist, Argonne Nat Lab, 55-65, assoc chemist, 65-69; chemist, Anal Div, Nat Bur Standards, 69-75; NUCLEAR CHEMIST, US NUCLEAR REGULATORY COMN, 75- Concurrent Pos: Fulbright lectr, Chinese Univ Hong Kong, 64-65. Mem: Am Phys Soc; Am Chem Soc. Res: Nuclear mass systematics; nuclear activation analysis; cross sections and mechanisms of nuclear reactions; radioactivities of new isotopes; carrier-free radiochemical separation techniques; computer automation of laboratory experiments; computer-aided information storage and retrieval; chemical safety for nuclear reactor operation. Mailing Add: US Nuclear Regul Comn NRR-TR-AAB Washington DC 20555

WING, JAMES MARVIN, b Anniston, Ala, Mar 17, 20; m 46; c 2. ANIMAL NUTRITION. Educ: Berea Col, BS, 46; Colo State Univ, MS, 48; Iowa State Univ, PhD(dairy husb), 52. Prof Exp: From asst prof to assoc prof, 51-66, PROF DAIRY SCI, UNIV FLA, 66- Mem: Am Soc Animal Sci; Am Dairy Sci Asn. Res: Nutrition; digestibility of carotenoids; nucleic acids; medicated feeds; ensilability; digestibility and consumption of herbage; climatic adaptation of cattle; optimum levels of carbohydrates for cattle; environmental quality and the animal industries; internal development in Colombia, El Salvador and Viet Nam. Mailing Add: Dairy Res Unit Univ of Fla Rte 3 Box 73 Gainesville FL 32601

WING, LARRY DEAN, b Webster City, Iowa, Aug 13, 35; m 58; c 1. WILDLIFE ECOLOGY. Educ: Univ Idaho, BS, 58, MD, 62, PhD(forestry), 70. Prof Exp: Res biologist, Wash State Univ, 62-65; asst prof wildlife ecol, 69-72, ASSOC PROF ZOOL & ENTOM, IOWA STATE UNIV, 72- Mem: AAAS; Wildlife Soc; EAfrican Wildlife Soc; Am Inst Biol Sci. Res: Effects of environmental pesticides on wildlife; mechanisms controlling the selection of plants by grazing herbivores; conservation of African wildlife. Mailing Add: Dept of Fisheries & Wildlife Biol Rm 105 Sci Bldg II Iowa State Univ Ames IA 50010

WING, LEONARD WILLIAM, zoology, deceased

WING, MERLE WESLEY, b Ft Fairfield, Maine, Aug 14, 16; m 70. ENTOMOLOGY. Educ: Univ Maine, BS, 39; Univ Minn, PhD(entom), 67. Prof Exp: Asst prof zool, NC State Univ, 42-45 & 46-51; actg asst prof, Tulane Univ, 45-46; asst prof biol, Middle Tenn State Col, 58-60; assoc prof zool, State Univ NY Col Cortland, 60-63; lectr entom, Cornell Univ, 63-67; assoc prof biol, Slippery Rock State Col, 67-69; assoc prof, 69-72, PROF BIOL, GENEVA COL, 72- Mem: AAAS; Am Soc Zoologists; Am Inst Biol Sci; Biomet Soc. Res: Systematics and evolution of social insects, especially Hymenoptera, Formicidae. Mailing Add: Dept of Biol Geneva College Beaver Falls PA 15010

WING, ROBERT EDWARD, b Bridgeport, Conn, Nov 29, 41; m 64; c 2. CHEMISTRY. Educ: Millikin Univ, BA, 63; Southern Ill Univ, Carbondale, PhD(biochem), 67. Prof Exp: Northern Regional Lab grant, Southern Ill Univ, Carbondale, 67-68; RES CHEMIST, CEREAL PROD LAB, NORTHERN REGIONAL LAB, USDA, 68- Concurrent Pos: Res assoc, Peoria Sch Med, Univ Ill Col Med, 71-; instr, Bradley Univ, 73- Mem: Am Chem Soc; Sigma Xi. Res: Water pollution of heavy metal ions; reactive carbohydrate polymers; carbohydrate thin-layer chromatography and enzyme interactions; carbohydrate slow release pesticide formulations. Mailing Add: Cereal Prod Lab USDA Northern Regional Lab Peoria IL 61604

WING, ROBERT FARQUHAR, b New Haven, Conn, Oct 31, 39; m 63; c 3. ASTRONOMY. Educ: Yale Univ, BS, 61; Univ Calif, Berkeley, PhD(astron), 67. Prof Exp: Asst prof, 67-71, ASSOC PROF ASTRON, OHIO STATE UNIV, 71- Mem: Am Astron Soc; fel Royal Astron Soc. Res: Spectroscopy and photometry of cool stars, especially Mira variables; infrared spectra; determination of chemical composition. Mailing Add: Dept of Astron Ohio State Univ Columbus OH 43210

WING, WILLIAM HINSHAW, b Ann Arbor, Mich, Jan 11, 39; m 67; c 2. ATOMIC PHYSICS, MOLECULAR PHYSICS. Educ: Yale Univ, BA, 60; Rutgers Univ, New Brunswick, MS, 62; Univ Mich, Ann Arbor, PhD(physics), 68. Prof Exp: Res staff physicist, Yale Univ, 68-70, res assoc physics, 70-72, asst prof, 72-74; ASSOC PROF PHYSICS, UNIV ARIZ, 74- Concurrent Pos: Res Corp Cottrell grant, 71; Nat Bur Stand, US Dept Com Precision Measurement grant, 74. Mem: Am Phys Soc; Inst Elec & Electronics Eng; Optical Soc Am. Res: Fundamental physical constants; simple atomic and molecular physics; lasers; physics of infinitesima. Mailing Add: Dept of Physics & Optical Sci Ctr Univ of Ariz Tucson AZ 85721

WINGARD, LEMUEL BELL, JR, b Pittsburgh, Pa, July 10, 30; m 65. PHARMACOLOGY. Educ: Cornell Univ, BChE, 53, PhD(biochem eng), 65. Prof Exp: Res engr, Jackson Lab, E I du Pont de Nemours & Co, 56-58 & Seaford Nylon Plant, 58-61; asst prof chem eng, Cornell Univ, 65-66; assoc prof, Univ Denver, 66-67; from asst prof to assoc prof chem eng, Univ, 67-75, ASSOC PROF PHARMACOL, SCH MED, UNIV PITTSBURGH, 72-, ADJ PROF CHEM ENG, UNIV, 76- Concurrent Pos: NIH spec fel, Sch Pharm, State Univ NY Buffalo, 70-72; chmn confs enzyme eng, Eng Found, 71 & 73. Mem: Fel AAAS; Am Inst Chem Engr; NY Acad Sci; Am Chem Soc; Biomed Eng Soc. Res: Kinetics and mass transfer in biological systems; pharmacokinetics; immobilized enzymes. Mailing Add: Dept of Pharmacol Univ of Pittsburgh Sch Med Pittsburgh PA 15261

WINGARD, NORMAN EDWARD, b Montpelier, Ohio, Aug 23, 36; m 65; c 2. GEOLOGY. Educ: Bowling Green State Univ, BS, 60; Mich State Univ, MS, 62, PhD(geol), 69. Prof Exp: Consult geologist, Anaconda Am Brass Co, 62-65; res geologist, Mich State Hwy Dept, 64-70; asst prof, 70-73, ASSOC PROF GEOL, WESTERN ILL UNIV, 73- Mem: AAAS; Geol Soc Am. Res: Environmental and and engineering geology; quaternary and glacial geology; sedimentology. Mailing Add: Dept of Geol Western Ill Univ Macomb IL 61455

WINGARD, PAUL SIDNEY, b Akron, Ohio, Jan 10, 30; m 53; c 5. PHYSICAL GEOLOGY. Educ: Miami Univ, AB, 52, MS, 55; Univ Ill, PhD(geol), 61. Prof Exp: Instr geol, Kans State Univ, 57-61, asst prof, 61-66; assoc prof geol & asst dean lib arts, 66-67, PROF GEOL & ASSOC DEAN COL ARTS & SCI, UNIV AKRON, 67- Mem: AAAS; Geol Soc Am. Res: Geochronology and geology of south central Maine; geology of central and southern Colorado; post-Pleistocene geology and life of northern Ohio. Mailing Add: Col of Arts & Sci Univ of Akron Akron OH 44325

WINGATE, CATHARINE L, b Boston, Mass. RADIOLOGICAL PHYSICS, BIOPHYSICS. Educ: Simmons Col, BS, 43; Harvard Univ, MA, 48; Columbia Univ, PhD(biophys), 61. Prof Exp: Res asst radiation physics, Mass Inst Technol, 43-45; sr technician, Woods Hole Oceanog Inst, 45-46; instr physics, Univ Conn, New London, 48-49; res asst med biophys, Sloan-Kettering Inst Cancer Res, 49-51; instr physics, Adelphi Col, 51-54; res scientist radiol physics, Col Physicians & Surgeons, Columbia Univ, 54-63; radiol physicist, Naval Radiol Defense Lab, 63-66; sr res scientist biophys, NY Univ, 66-67; assoc radiol physicist, Brookhaven Nat Lab, 67-70; asst prof radiol physics, Sch Med & asst dean, Sch Basic Health Sci, 70-74, res asst prof radiol physics, Sch Med, 74-75, RES ASSOC PROF RADIOL, SCH MED, STATE UNIV NY STONY BROOK, 75- Concurrent Pos: Consult, Vet Admin Hosp, Northport, NY, 71-74; physicist, Nassau County Med Ctr, East Meadow, NY, 74-; collabr, Brookhaven Nat Lab, 74- Mem: AAAS; Radiation Res Soc; Biophys Soc; Health Physics Soc; Am Asn Physicists in Med. Res: Absorbed dose at bone-soft tissue interface; ionization chamber techniques; luminescence and medical dosimetry; calcium uptake in stressed bone; delta ray dose around a charged particle beam; microdosimetry; proton Bragg peak localization for therapy. Mailing Add: Health Sci Ctr State Univ NY Stony Brook NY 11790

WINGATE, MARTIN BERNARD, b London, Eng, m; c 2. OBSTETRICS & GYNECOLOGY. Educ: Univ London, MB, BS, 48, MD, 64; FRCS, 53: FRCS(E), 55; FRCS(C), 66. Prof Exp: House surgeon orthop, St Mary's Hosp, London, 48-50 & Obstet Unit, Whittington Hosp, 50-51; demonstr anat, St Mary's Hosp Med Sch, 51-52, prosecutor, 52, sr house officer surg, Hosp, 52-53; house surgeon, Cent Middlesex Hosp, 53-54; locum surg registr, Royal Northern Hosp, 54-55; locum surg registr obstet & gynec, Middlesex Hosp & Hosp for Women, 56-58; sr registr, Southampton Gen Hosp, 58-60; first asst, St George's Hosp, 58-62; sr registr obstet & gynec, Queen Charlotte & Chelsea Hosps, 62-63; asst examr obstet & gynec, Univ Bristol, sr lectr & consult, United Bristol Hosps & Univ Bristol & examr, Cent Midwives Bd, 63-66; assoc prof obstet & gynec, Univ Man, 67-71; prof, Temple Univ, 69-71; prof obstet & gynec & pediat, Thomas Jefferson Univ, 71-75; PROF OBSTET & GYNEC & PEDIAT, ALBANY MED COL, 75- Concurrent Pos: Res fel obstet & pediat, Guy's Hosp Med Sch, 62-63. Mem: Can Med Asn; Sob Obstet & Gynaec Can; fel Am Col Obstet & Gynec. Res: Genetics of human aberrations and malignant lesions of the cervix; surgery of infertility and the transplantation of reproductive organs together with means of modification of the rejection phenomena. Mailing Add: Dept of Obstet & Gynec Albany Med Col Albany NY 12208

WINGELETH, DALE CLIFFORD, b Cleveland, Ohio, June 8, 43; m 64; c 2. INORGANIC CHEMISTRY, CLINICAL CHEMISTRY. Educ: Cleveland State Univ, BES, 66; Univ Colo, Boulder, PhD(inorg chem), 70. Prof Exp: Clin biochemist, St Joseph Hosp, Denver, 70-72; clin chemist, St Lawrence Hosp, Mich, 72; forensic chemist, Poisonlab Inc, 72, VPRES & TECH DIR, POISONLAB DIV, CHEMED CORP, 72- Mem: Am Chem Soc; The Chem Soc; Am Asn Clin Chemists; Am Acad Forensic Sci; Forensic Sci Soc. Res: Inorganic hydrides; clinical toxicology. Mailing Add: 201 Cedarbrook Rd Boulder CO 80302

WINGENDER, RONALD JOHN, b Menominee, Mich, Sept 30, 36; m 63; c 1. ANALYTICAL CHEMISTRY. Educ: Univ Wis, BS, 59, PhD(anal chem), 69; Univ Iowa, MS, 61. Prof Exp: Chemist, Forest Prod Lab, 61-64; chemist, Ansul Co, Wis, 69, mgr anal res, 69-72; SECT HEAD CHEM, INDUST BIO-TEST LABS, 72- Mem: Fel Am Inst Chemists; Am Chem Soc. Res: Proton nuclear magnetic resonance of cobalt II and nickel II aminopolycarboxylic acid and polyamine complexes; development of pesticide residue analytical procedures; pesticide analysis and metabolite determination for Food & Drug Administrations petitions for tolerance; spectroscopic identification of trace contaminants. Mailing Add: Indust Bio-Test Labs Inc Northbrook IL 60062

WINGER, MILTON EUGENE, b Mayville, NDak, Aug 28, 31; m 54; c 2. MATHEMATICS, STATISTICS. Educ: Mayville State Col, BS, 53; Univ NDak, MS, 56; Iowa State Univ, PhD(statist), 72. Prof Exp: Inspector eng, US Army Corps Engrs, 57; asst prof math, Univ NDak, 60-68; instr statist, Iowa State Univ, 68-70; ASSOC PROF MATH & STATISTICS, UNIV NDAK, 71- Concurrent Pos: Statist consult, Inst Appl Math & Statist, 74-; vis lectr, Inst Math Statist, 76-77. Mem: Math Asn Am; Am Statist Asn. Res: Stochastic modelling study for coal revenues to state. Mailing Add: 623 24th Ave S Grand Forks ND 58201

WINGERD, WINSTON HAROLD, chemistry, food technology, see 12th edition

WINGET, CARL HENRY, b Noranda, Que, Sept 28, 38; m 64; c 2. FOREST ECOLOGY, TREE PHYSIOLOGY. Educ: Univ NB, BScF, 60; Univ Wis, MSc, 62, PhD(forestry, bot), 64. Prof Exp: Res scientist, Forestry Res Lab, Dept Forestry & Rural Develop Can, 64-67; prof forestry & geod, Laval Univ, 67-73; res scientist, Can Forestry Serv, 73-75, PROG MGR, LAURENTIAN FORESTRY RES CTR, CAN FORESTRY SERV, 75- Mem: Can Inst Forestry. Res: Forest resource management; silviculture of tolerant northern hardwoods. Mailing Add: Laurentian Forest Res Ctr Box 3800 Ste Foy PQ Can

WINGET, CHARLES M, b Garden City, Kans, Dec 26, 25; c 2. AEROSPACE SCIENCES. Educ: San Francisco State Col, BA, 51; Univ Calif, PhD, 57. Prof Exp: Chemist poultry husb, Univ Calif, 51-53, res asst, 53-56, jr res poultry physiologist, 56-57, res fel, 57-59; assoc prof avian physiol, Ont Agr Col, Univ Guelph, 59-63; res scientist, 63-67, PROJ SCIENTIST, BIOSATELLITE PROJ, NASA-AMES RES CTR, MOFFETT FIELD, 67- Concurrent Pos: Nat Inst Neurol Dis & Blindness fel, 57-59; lectr, Univ Calif, Davis, 64-; adj prof physiol, Sch Med, Wright State Univ, 75- Mem: Poultry Sci Asn; Biophys Soc; Aerospace Med Asn; Int Soc Chronobiol; Am Physiol Soc. Res: Rhythms and social schedule; hypokinesis and drugs in man; bird and monkey response to change in photoperiod; physiological changes associated with aeronautical environment; biotelemetry; biorhythm data acquisition and reduction. Mailing Add: Human Studies Branch NASA Ames Res Ctr Moffett Field CA 94035

WINGET, GARY DOUGLAS, b Dayton, Ohio, Mar 27, 39; m 60; c 3. PLANT BIOCHEMISTRY, PLANT PHYSIOLOGY. Educ: Miami Univ, AB, 61, MA, 63; Mich State Univ, PhD(bot), 68. Prof Exp: Res chemist, Mound Lab, Monsanto Res Corp, 63-64; asst prof, 66-71, ASSOC PROF BIOL SCI, UNIV CINCINNATI, 71- Concurrent Pos: Vis assoc prof biochem, molecular & cell biol, Cornell Univ, 75. Mem: Am Soc Photobiol; Am Chem Soc; Am Soc Plant Physiologists. Res: Inhibitors of photosynthesis; mechanism of photophosphorylation; physiological action of phlorizin. Mailing Add: Dept of Biol Sci Univ of Cincinnati Cincinnati OH 45221

WINGFIELD, EDWARD CHRISTIAN, b Charlottesville, Va, Nov 17, 23; m 47; c 3. PHYSICS, SCIENCE ADMINISTRATION. Educ: Univ Va, BA, 46, MA, 49; Univ NC, PhD(physics), 54. Prof Exp: Asst prof physics, Univ Richmond, 49-51; physicist, Savannah River Lab, E I du Pont de Nemours & Co, 54-62; chief, Instrumentation Develop & Serv, 62-65, ASST MGR ENG OPERS, UNITED TECHNOLOGIES RES CTR, 65- Res: Instrumentation; reactor physics; simulation by computers; aerodynamic testing; management of research and development activities. Mailing Add: United Technologies Res Ctr Silver Lane East Hartford CT 06108

WINGO, CURTIS W, b Fair Grove, Mo, Aug 30, 15; m 52; c 2. ENTOMOLOGY. Educ: Southwestern Mo State Teachers Col, AB, 36; Univ Mo, MA, 39; Iowa State Col, PhD(entom), 51. Prof Exp: Assoc prof, 51-59, PROF ENTOM, UNIV MO-COLUMBIA, 59- Res: Biology and control of insect parasites of man and domestic animals. Mailing Add: 1-69 Agr Bldg Dept of Entom Univ of Mo Columbia MO 65201

WINGO, WILLIAM JACOB, b Ladonia, Tex, July 17, 18; m 40; c 2. BIOCHEMISTRY. Educ: Univ Tex, BA, 38, MA, 40; Univ Mich, PhD(biochem), 46. Prof Exp: Tutor biochem, Med Br, Univ Tex, 38-39, instr, 39-41, 45-48, lectr biochem & nutrit, 48-54, asst prof biochem, Postgrad Sch Med, 50-54, res assoc, M D Anderson Hosp Cancer Res, 48-50, assoc biochemist, 50-54; ASSOC PROF BIOCHEM, MED COL & SCH DENT, UNIV ALA, BIRMINGHAM, 54- Mem: AAAS; Am Soc Biol Chem; Am Chem Soc; Soc Exp Biol & Med; Soc Protozool. Res: Chemistry and metabolism of amino acids; growth and metabolism of ciliate Protozoa; histochemistry; apparatus development. Mailing Add: Dept of Biochem Univ of Ala Birmingham AL 35233

WINGROVE, ALAN SMITH, b Hanford, Calif, Mar 4, 39. ORGANIC CHEMISTRY. Educ: Univ Calif, Berkeley, BS, 60; Univ Calif, Los Angeles, PhD(chem), 64. Prof Exp: NSF fel, 64-65; asst prof chem, Univ Tex, Austin, 65-71; lectr & sci researcher & writer, 71-73; ASSOC PROF CHEM & CHMN DEPT, TOWSON STATE COL, 73- Mem: Am Chem Soc; The Chem Soc. Res: Chemistry of second row elements and participation in solvolysis and base-catalyzed cleavages; destabilized carbonium ions; carbenophiles; stereochemistry; synthetic methods. Mailing Add: Dept of Chem Towson State Col Baltimore MD 21204

WINHOLD, EDWARD JOHN, b Brantford, Ont, Jan 3, 28; nat US; m 51; c 3. PHYSICS. Educ: Univ Toronto, BA, 49; Mass Inst Technol, PhD(physics), 53. Prof Exp: Asst physics, Mass Inst Technol, 51-53, res staff mem, Lab Nuclear Sci, 53-54; from instr to asst prof physics, Univ Pa, 54-57; from asst prof to assoc prof, 57-69, PROF PHYSICS, RENSSELAER POLYTECH INST, 69- Concurrent Pos: Vis staff mem, Atomic Energy Res Estab, Harwell, Eng, 68-69. Mem: Am Phys Soc. Res: Experimental nuclear physics. Mailing Add: Dept of Physics Rensselaer Polytech Inst Troy NY 12181

WINICK, HERMAN, b New York, NY, June 27, 32; m 53; c 3. HIGH ENERGY PHYSICS. Educ: Columbia Univ, AB, 53, PhD(physics), 57. Prof Exp: Asst physics, Columbia Univ, 53-54, asst, Nevis Cyclotron Lab, 54-57; res assoc & instr physics, Univ Rochester, 57-59; res fel, Cambridge Electron Accelerator, Harvard Univ, 59-65, sr res assoc & lectr, 65-74; DEP DIR, STANFORD SYNCHROTRON RADIATION PROJ, STANFORD UNIV, 74- Mem: Am Phys Soc. Res: Meson scattering; bremstrahlung research; accelerator design and development; colliding beams; synchrotron radiation production and experimentation. Mailing Add: Synchrotron Radiation Proj Stanford Linear Accelerator Ctr Bldg 69 PO Box 4349 Stanford CA 94305

WINICK, MYRON, b New York, NY, May 4, 29; m 64; c 2. PEDIATRICS, NUTRITION. Educ: Columbia Univ, AB, 51; Univ Ill, Urbana, MS, 52; State Univ NY Downstate Med Ctr, MD, 56. Prof Exp: From asst resident pediat to chief resident, Med Col, Cornell Univ, 57-60; Bank Am-Giannini Found fel, Stanford Univ,

62-63, attend pediatrician & instr pediat, Med Col, 63-64; asst prof, Med Col, Cornell Univ, 64-67, dir birth defects treatment ctr, 64-71, from assoc prof to prof pediat & nutrit, 68-71; ROBERT R WILLIAMS PROF NUTRIT, PROF PEDIAT & DIR INST HUMAN NUTRIT, COL PHYSICIANS & SURGEONS, COLUMBIA UNIV, 72-, DIR, CTR NUTRIT, GENETICS & HUMAN DEVELOP, 75- Concurrent Pos: NIH spec fel, 63-64; vis prof, Univ Chile, 67; USPHS career develop award, 69-74; mem comt nutrit, brain develop & behav, Nat Acad Sci; consult, Pan-Am Health Orgn. Honors & Awards: E Mead Johnson Award Pediat Res, 70. Mem: Soc Pediat Res; Am Pediat Soc; Am Inst Nutrit; Am Soc Clin Nutrit; Am Acad Pediat. Res: Growth retardation; biochemistry of growth; malnutrition and connected cellular and biochemical effects. Mailing Add: Inst of Human Nutrit 511 W 166th St New York NY 10032

WINICOUR, JEFFREY, b Providence, RI, Apr 12, 38; m 64; c 1. THEORETICAL PHYSICS. Educ: Mass Inst Technol, BS, 59; Syracuse Univ, PhD(physics), 64. Prof Exp: Res asst, Syracuse Univ, 59-64; res physicist, Aerospace Res Labs, 64-72; ASSOC PROF PHYSICS, UNIV PITTSBURGH, 72- Mem: Am Phys Soc. Res: General relativity; equations of motion; gravitational radiation. Mailing Add: Dept of Physics Univ of Pittsburgh Pittsburgh PA 15213

WINICOV, EDITH, b Philadelphia, Pa, Aug 7, 28; m 49; c 3. ORGANIC CHEMISTRY. Educ: Univ Pa, BA, 49, MS, 51; Bryn Mawr Col, PhD(phys & org chem), 58. Prof Exp: Asst chem, Rohm & Haas Co, 49-50; demonstr, Bryn Mawr Col, 51-52; asst prof, Long Island Univ, 58-67; lectr, Hunter Col, 67-70; lit chemist, Texaco Develop Corp, 70-73; ANALYST, NEW YORK DEPT HEALTH ENVIRON HEALTH SERV, 73- Mem: AAAS; Am Chem Soc. Res: Aromatic substitution; physical properties of aromatic compounds; toxic materials; air contaminants. Mailing Add: 63-58 78th St Middle Village NY 11379

WINICOV, HERBERT, b Brooklyn, NY, Mar 14, 35; m 58; c 2. ORGANIC CHEMISTRY. Educ: Univ Pa, BA, 56; Univ Wis, PhD(chem), 61. Prof Exp: Sr chemist, 60-68, SR INVESTR, SMITH, KLINE & FRENCH LABS, 68- Mem: AAAS; Am Chem Soc. Res: Organic synthesis and process development of pharmaceuticals; antibiotics, steroids and natural products. Mailing Add: Smith Kline & French Labs 1500 Spring Garden St Philadelphia PA 19101

WINICOV, ILGA BUTELIS, b Riga, Latvia, May 16, 35; US citizen; m 58; c 2. MOLECULAR BIOLOGY. Educ: Univ Pa, AB, 56, PhD(microbiol), 71; Univ Wis, MS, 58. Prof Exp: Res asst bact genetics, Univ Wis, 58-60; res asst biochem, Fels Res Inst, 64-67; assoc, 72-74, RES ASSOC MOLECULAR BIOL, INST CANCER RES, 74- Concurrent Pos: Fel, Damon Runyon Mem Fund, 72-73. Mem: Am Soc Microbiol. Res: The post transcriptional mechanism and regulation of RNA processing in eukaryotic cells, as achieved by interaction of RNA with both structural proteins and processing enzymes. Mailing Add: Inst for Cancer Res 7701 Burholme Ave Fox Chase Philadelphia PA 19111

WINICUR, DANIEL HENRY, b New York, NY, May 6, 39; m 60; c 1. CHEMICAL PHYSICS. Educ: City Col New York, BME, 61; Univ Conn, MSME, 63; Univ Calif, Los Angeles, PhD(chem dynamics), 68. Prof Exp: Res engr, Space Systs Div, Hughes Aircraft Co, 63-64; Shell Oil fel chem dynamics, A A Noyes Lab Chem Physics, Calif Inst Technol, 68-70; ASST PROF PHYS CHEM, UNIV NOTRE DAME, 70- Mem: AAAS; Am Chem Soc; Am Phys Soc. Res: Molecular beam studies of chemical reactions; excited atomic and molecular states; energy transfer processes. Mailing Add: Dept of Chem Univ of Notre Dame Notre Dame IN 46556

WINICUR, SANDRA, b New York, NY, Oct 4, 39; m 60; c 2. CELL PHYSIOLOGY. Educ: Hunter Col, BA, 60; Univ Conn, MS, 63; Calif Inst Technol, PhD(biochem), 71. Prof Exp: ASST PROF BIOL, IND UNIV, SOUTH BEND, 70- Mem: Am Soc Cell Biol; Am Inst Biol Sci. Res: Enzymes in insect molting fluid; motility in Protozoa. Mailing Add: Dept of Biol Indiana Univ at South Bend South Bend IN 46615

WINIKOFF, BEVERLY, b New York, NY, Aug 26, 45; m 73. PUBLIC HEALTH, NUTRITION. Educ: NY Univ, MD, 71; Harvard Univ, AB, 66, MPH, 73. Prof Exp: Intern, Gen Rose Mem Hosp, Denver, 71-72; res fel, Dept Nutrit, Sch Pub Health, Harvard Univ, 73-74; prog assoc & nutrit specialist, 74-75, ASST DIR HEALTH SCI, ROCKEFELLER FOUND, 75- Mem: Am Pub Health Asn. Res: Development and implementation of nutrition policies and programs. Mailing Add: Rockefeller Found 1133 Ave of the Americas New York NY 10036

WINJUM, JACK KEITH, b Platte, SDak, Feb 5, 33; m 54; c 3. FOREST ECOLOGY. Educ: Ore State Univ, BS, 55; Univ Wash, MS, 61; Univ Mich, PhD(forest ecol), 65. Prof Exp: Forester, US Forest Serv, 55; forest technologist, Forestry Res Ctr, 58-63, regeneration ecologist, 63-73, MGR FOREST REGENERATOR RES, FORESTRY RES CTR, WEYERHAEUSER CO, 73- Mem: Soc Am Foresters; Ecol Soc Am. Res: Cone and seed yield of Douglas fir; ecology of forest nurseries; stock handling and field out planting of seedlings in the regeneration period of Douglas fir management; forest regeneration ecology. Mailing Add: Forestry Res Ctr Weyerhaeuser Co Centralia WA 98531

WINKEL, CLEVE R, b Logan, Utah, Mar 20, 32; m 55; c 6. BIOCHEMISTRY, ORGANIC CHEMISTRY. Educ: Utah State Univ, BS, 54, MS, 55; Brigham Young Univ, PhD, 70. Prof Exp: PROF CHEM, RICKS COL, 59-, CHMN DIV NATURAL SCI, 72- Res: Enzymology; enzyme mechanism; medical biochemistry. Mailing Add: Dept of Chem Ricks College Rexburg ID 83440

WINKEL, DAVID EDWARD, b Sibley, Iowa, Mar 10, 31; m 53; c 3. PHYSICAL CHEMISTRY, COMPUTER SCIENCE. Educ: Iowa State Univ, PhD(chem), 57. Prof Exp: From asst prof to assoc prof chem, 57-71, PROF CHEM & COMPUT SCI, UNIV WYO, 71-, DIR DIV COMPUT SERV, 72-, HEAD DEPT COMPUT SCI, 74- Concurrent Pos: Vis prof, Ind Univ, 62-63 & 71. Res: Nature of adsorption of gases on metals; computer architecture. Mailing Add: Div Comput Serv Univ Wyo Box 3945 Univ Sta Laramie WY 82070

WINKELMANN, FREDERICK CHARLES, b Brooklyn, NY, Apr 11, 41; m 68; c 2. HIGH ENERGY PHYSICS. Educ: Mass Inst Technol, BS, 62, PhD(physics), 68. Prof Exp: Res assoc physics, Lab Nuclear Sci, Mass Inst Technol, 68-69; res assoc, Stanford Linear Accelerator Ctr, 69-72; RES ASSOC PHYSICS, LAWRENCE BERKELEY LAB, UNIV CALIF, 72- Mem: Am Phys Soc. Res: Elementary particle research using bubble chambers and spectrometers; phenomenology of strong interactions. Mailing Add: Lawrence Berkeley Lab Univ of Calif Berkeley CA 94720

WINKELMANN, JOHN ROLAND, b Champaign, Ill; m 62; c 2. VERTEBRATE ZOOLOGY, MAMMALOGY. Educ: Univ Ill, Urbana, BS, 54; Univ Mich, Ann Arbor, MA, 60, PhD(zool), 71. Prof Exp: ASST PROF BIOL, GETTYSBURG COL, 63- Concurrent Pos: Fac fel grant for res in Mex, Gettysburg Col, 72-73. Mem: AAAS; Soc Study Evolution; Am Soc Mammal. Res: Biology of nectar-feeding bats. Mailing Add: Dept of Biol Gettysburg Col Gettysburg PA 17325

WINKELMANN, RICHARD KNISELY, b Akron, Ohio, July 12, 24; m; c 4. DERMATOLOGY. Educ: Univ Akron, BS, 47; Marquette Univ, MD, 48; Univ Minn, PhD(dermat), 56. Prof Exp: Res assoc chem, Wash Univ, 49, res assoc anat, 50; asst pub health officer, USPHS, Ala, 52-54; from instr to asst prof dermat, 56-65, PROF DERMAT, MAYO GRAD SCH MED, UNIV MINN, 65-, ASSOC PROF ANAT, 64-, CONSULT SECT DERMAT, MAYO FOUND, 56- Concurrent Pos: Instr & res assoc, Med Col, Univ Ala, 53-54; chmn dept dermat, Mayo Clin. Res: Anatomy. Mailing Add: Mayo Clin 200 SW First St Rochester MN 55901

WINKELSTEIN, WARREN, JR, b Syracuse, NY, July 1, 22; m 47; c 3. MEDICINE, EPIDEMIOLOGY. Educ: Univ NC, BA, 43; Syracuse Univ, MD, 47; Columbia Univ, MPH, 50. Prof Exp: Dist health officer, Erie County Health Dept, NY, 50-51; regional rep pub health div, Tech & Econ Mission, Mutual Security Agency, Cambodia, Laos & Viet Nam, 51-53; dir div commun dis control, Erie County Health Dept, NY, 53-56; from asst prof to prof prev med, Sch Med, State Univ NY Buffalo, 56-69, chief dept epidemiol, Chronic Dis Res Inst, 57-64; PROF EPIDEMIOL & DEAN SCH PUB HEALTH, UNIV CALIF, BERKELEY, 68- Concurrent Pos: Spec res fel, Nat Heart Inst, 56-67, career develop award, 62-68; dep health comnr, Erie County Health Dept, 59-62; mem heart dis control prog adv comt & air pollution training comt, USPHS, 62-65; mem subcomt, Nat Comt Health Statist, 65-; mem res comt, Am Heart Asn, 66-71. Mem: AAAS; fel Am Pub Health Asn; Asn Teachers Prev Med; Am Col Prev Med; Am Col Prev Med. Res: Epidemiology of cardiovascular diseases; air pollution and viral exanthemata. Mailing Add: Dept of Epidemiol Univ of Calif Sch Pub Health Berkeley CA 94720

WINKERT, JOHN WYNIA, b Brooklyn, NY, Dec 27, 29; m 51; c 4. PHYSIOLOGY, CHEMISTRY. Educ: Polytech Inst Brooklyn, BS, 51; NY Univ, MS, 57, PhD(physiol), 60. Prof Exp: Instr physiol, NY Med Col, 60-64; asst prof pharmacol, State Univ NY Buffalo, 64-67; assoc prof physiol, Meharry Med Col, 67-72; ASST PRIN SUPVR MORRIS HEPATOMA PROJ, COL MED, HOWARD UNIV, 72- Concurrent Pos: Helen Hay Whitney Found fel, NY Med Col, 61-64; NIH trainee nucleic acid & chromatin chem, Dept Obstet & Gynec, Vanderbilt Univ, 71-72. Mem: AAAS; Am Physiol Soc; NY Acad Sci; Am Inst Biol Sci; Am Soc Zool. Res: Steroid chemistry and metabolism; regulation of fibrinogen synthesis; regulation of erythropoiesis isolation and structure of erythropoietin; effect of erthropoietin in chromatin priming of RNA synthesis; toxicity of histones; role of cyclic adenosine monophosphate in hematopoiesis and lymphomas; cryopreservation of hepatomas. Mailing Add: Cancer Res Unit Howard Univ Col of Med Washington DC 20001

WINKLER, BARRY STEVEN, b New York, NY, Apr 17, 45; m 66; c 1. PHYSIOLOGY. Educ: Harpur Col, BA, 65; State Univ NY Buffalo, MA, 68, PhD(physiol), 71. Prof Exp: Instr physiol, Sch Med, State Univ NY Buffalo, 70-71; ASST PROF BIOL SCI, OAKLAND UNIV, 71- Res: Physiology of the retina; analysis of ionic and metabolite contributions to electroretinogram potentials; relationship of electroretinogram to ganglion cell activity. Mailing Add: Dept of Biol Sci Oakland Univ Rochester MI 48063

WINKLER, BERTRAM STANLEY, b New York, NY, Aug 20, 31; m 63; c 2. PHARMACOLOGY. Educ: Univ Ill, BS, 51; NY Univ, MS, 58, PhD(pharmacol), 65. Prof Exp: Res assoc pharmacol, Postgrad Med Sch, NY Univ, 59-65, instr, 65-66, asst prof, 66-70; ASST PROF BIOL, QUEENSBOROUGH COMMUNITY COL, 70- Concurrent Pos: City Col New York Res Found res fel, 71; NSF res fel, 74. Mem: AAAS. Res: Hormonal regulation of fat and carbohydrate metabolism. Mailing Add: 624 Duke St Westbury NY 11590

WINKLER, BRUCE CONRAD, b Milwaukee, Wis, Sept 25, 37; m 59; c 2. BIOCHEMISTRY. Educ: Valparaiso Univ, BA, 59; Iowa State Univ, MS, 62; Univ Okla, PhD(biochem), 67. Prof Exp: Instr chem, Cent State Univ, 62-64; fel biocm, Univ Alta, 67-69; ASST PROF BIOCHEM, KANSAS CITY COL OSTEOP MED, 69-, ACTG CHMN DEPT, 73- Res: Muscle phosphorylase; clinical chemistry, especially proteins and enzymes. Mailing Add: Dept of Biochem Kansas City Col of Osteop Med 2105 Independence Ave Kansas City MO 64124

WINKLER, CARL ARTHUR, b Virden, Man, Oct 24, 09; m 37; c 4. CHEMISTRY. Educ: Univ Man, BSc, 30, MSc, 31; McGill Univ, PhD(phys chem), 33; Oxford Univ, DPhil(chem), 39. Hon Degrees: DSc, Univ Man, 66 & Sir George Williams Univ, 72; LLD, Queen's Univ, 68. Prof Exp: Tutor chem, Univ Man, 30-31; Rhodes scholar, Oxford Univ, 33-36; asst res biologist, Nat Res Coun, 36-39; from asst prof to assoc prof, 39-55, chmn dept, 55-61, vprin, 66-69, PROF CHEM, McGILL UNIV, 55- Honors & Awards: Gov Gen's Silver Medal, 36; Chem Inst Can Medal, 58, Chem Educ Award, 70; Officer, Order of Brit Empire. Mem: Fel Royal Soc Can; fel Chem Inst Can. Res: Kinetics of chemical reactions; electrochemistry, active nitrogen. Mailing Add: Dept of Chem McGill Univ Montreal PQ Can

WINKLER, CHARLES HERMAN, JR, b Austin, Tex, May 5, 15; m 37; c 2. MICROBIOLOGY, HEALTH SCIENCES. Educ: Tex A&M Univ, BS, 34, MS, 38; Univ Wis, PhD(zool), 42. Prof Exp: Prin pub sch, Tex, 36-38; asst zool, Univ Wis, 38-42; instr bact, Sch Med, Univ Ark, 42-44, asst prof, 44-46; from asst prof to assoc prof, 46-61, asst to vpres allied health sci, 66-69, chmn div allied health sci, 67-71, PROF MICROBIOL, MED CTR, UNIV ALA, BIRMINGHAM, 61-, ASSOC DEAN SCH COMMUNITY & ALLIED HEALTH RESOURCES, 71- Mem: Fel AAAS; Am Soc Microbiol; Asn Schs Allied Health Professions. Res: Parasitology; clinical microbiology; allied health sciences. Mailing Add: Sch of Commun & Allied Hlth Res Univ of Ala Birmingham AL 35294

WINKLER, DELOSS EMMET, b Atchison, Kans, Feb 4, 14; m 41; c 2. POLYMER CHEMISTRY. Educ: Univ Kans, AB, 36, MA, 39, PhD(chem), 41. Prof Exp: Teacher high sch, Kans, 36-37; asst instr chem, Univ Kans, 37-39; chemist, Shell Develop Co, 41-70; real estate salesman, 71-72; chemist, Spinco Div, 72, SR RES CHEMIST, BECKMAN INSTRUMENTS, INC, 72- Mem: Am Chem Soc. Res: Vapor phase catalysis; plastics; rubber; oxidation of hydrocarbons; chromatographic polymers for separation of amino acids and polymers for solid state synthesis of amino acids. Mailing Add: 133 Lombardy Lane Orinda CA 94563

WINKLER, ERHARD MARIO, b Vienna, Austria, Jan 8, 21; nat US; m 53; c 2. ENVIRONMENTAL GEOLOGY. Educ: Univ Vienna, ScD, 45. Prof Exp: Asst eng geol, Univ Vienna, 40-45; sci asst geol, Vienna Tech Univ, 46-48; from instr to assoc prof, 48-73, PROF GEOL, UNIV NOTRE DAME, 73- Honors & Awards: E B Burwell Jr Award, Geol Soc Am, 75. Mem: Fel AAAS; fel Geol Soc Am. Res: Engineering geology. Mailing Add: Dept of Geol Univ of Notre Dame Notre Dame IN 46556

WINKLER, ERNST HANS, b Halle, Ger, Jan 20, 11; nat US; m 40. PHYSICS. Educ: Univ Halle, Dr Rer Nat(physics), 37. Prof Exp: Asst prof physics, Univ Halle, 37-45; aeronaut res engr, Naval Ord Lab, 46-74, AERONAUT RES ENGR, US NAVAL SURFACE WEAPONS CTR, WHITE OAK, 74- Concurrent Pos: Supvry physicist, Ger Army Res Ctr, 39-46. Res: Theoretical and experimental aerodynamics as applied to ballistic missiles. Mailing Add: 12490 Lime Kiln Rd Fulton MD 20759

WINKLER, EVA MARIA, b Ger, June 15, 15; m 40. PHYSICS. Educ: Univ Halle, Dr rer nat(physics), 41. Prof Exp: Res assoc atomic & molecular physics, Ger Army Res Ctr, Peenemünde, 41-43; res assoc atomic & molecular physics, Univ Halle, 43-44 & Ger Army Res Ctr, Kochel, 44-45; aeronaut res engr, US Naval Ord Lab, 47-56, chief hypersonics group, 56-74, high-temperature aerodyn group, 60-74, CHIEF HYPERSONICS GROUP & HIGH-TEMPERATURE AERODYN GROUP, US NAVAL SURFACE WEAPONS CTR, WHITE OAK, 74- Res: Experimental and theoretical hypersonic research; physics of high temperature gases. Mailing Add: 12490 Lime Kiln Rd Fulton MD 20759

WINKLER, HANS J S, organic chemistry, physical chemistry, see 12th edition

WINKLER, HERBERT H, b Highland Park, Mich, June 18, 39; m 61; c 1. MICROBIOLOGY, BIOCHEMISTRY. Educ: Kenyon Col, BA, 61; Harvard Univ, PhD(physiol), 66. Prof Exp: NSF fel physiol chem, Sch Med, Johns Hopkins Univ, 66-68; asst prof, 68-71, ASSOC PROF MICROBIOL, SCH MED, UNIV VA, 71- Concurrent Pos: NIH res career develop award. Mem: Am Soc Microbiol; Am Soc Biol Chemists. Res: Transport of molecules across biological membranes; biology of rickettsiae. Mailing Add: Dept of Microbiol Univ of Va Sch of Med Charlottesville VA 22901

WINKLER, LINDSAY ROBERT, zoology, see 12th edition

WINKLER, LOUIS, b Elizabeth, NJ, Sept 7, 33; m 57. ASTRONOMY. Educ: Rutgers Univ, BS, 55; Adelphi Univ, MS, 59; Univ Pa, PhD(astron), 64. Prof Exp: Engr, Am Bosch Arma Corp, 56-59; proj engr, Philco Corp, 59-64; ASST PROF, PA STATE UNIV, 64- Mem: Am Astron Soc. Res: Double stars; astronomy and astrology of early America.

WINKLER, MARVIN HOWARD, b Brooklyn, NY, Dec 17, 26; m 53; c 2. BIOPHYSICAL CHEMISTRY. Educ: NY Univ, BA, 49, MS, 51, PhD, 54. Prof Exp: Res assoc immunochem, Roswell Park Mem Inst, 54-58; asst & assoc investr, Protein Found, Inc, 58-66; asst prof biochem, Albert Einstein Col Med, 66-67; ASSOC PROF IMMUNOCHEM, DEPT MED, MT SINAI SCH MED, 67- Concurrent Pos: Res assoc, Dept Biochem, Harvard Med Sch, 62-66; asst head dept microbiol, Montefiore Hosp, Bronx, NY, 66-67; consult, Brockton Vet Admin Hosp, Mass, 63-; Protein Found, Inc, 66- Mem: AAAS; Am Chem Soc; Biophys Soc; Soc Exp Biol & Med; Am Soc Biol Chem; Am Asn Immunologists. Res: Protein small molecular interactions; immunochemistry; fluorescence. Mailing Add: Mt Sinai Sch of Med 100th St & Fifth Ave New York NY 10029

WINKLER, NORMAN WALTER, b Englewood, NJ, May 28, 35; m 59; c 2. BIOCHEMISTRY, DERMATOLOGY. Educ: Univ Rochester, AB, 57; Univ Chicago, MD, 65, PhD(biochem), 70. Prof Exp: Res asst fibrinolysis, Sloan-Kettering Inst Cancer Res, 58-59; from intern to resident internal med, Univ Chicago Hosps & Clins, 68-70; USPHS fel dermat, Med Sch, Univ Ore, 70-72; ASST PROF DERMAT, SCH MED, STATE UNIV NY BUFFALO, 72-; CHIEF DERMAT SERV, BUFFALO VET ADMIN HOSP, 73- Mem: AAAS. Res: Enzymology; membrane receptors in cutaneous disease; keratinocyte differentiation. Mailing Add: Vet Admin Hosp Rm 1001-B 3495 Bailey Ave Buffalo NY 14215

WINKLER, PAUL FRANK, b Nashville, Tenn, Nov 10, 42; m 66; c 2. ASTROPHYSICS. Educ: Calif Inst Technol, BS, 64; Harvard Univ, AM, 65, PhD(physics), 70. Prof Exp: ASST PROF PHYSICS, MIDDLEBURY COL, 69- Concurrent Pos: Environ Protection Agency res grant, 71-73; vis scientist, Mass Inst Technol, 73-74; res affil, 74-; NSF res grant, 74-75. Mem: Am Phys Soc; Am Asn Physics Teachers; Am Astron Soc. Res: Supernova remnants; galactic and extragalactic x-ray sources; atomic and molecular physics; solid waste separation technology. Mailing Add: Dept of Physics Middlebury Col Middlebury VT 05753

WINKLER, ROBERT RANDOLPH, b Washington, DC, June 16, 33; m 55; c 3. ORGANIC CHEMISTRY. Educ: Univ Md, BS, 55; Univ Mich, MS, 60, PhD(org chem), 62. Prof Exp: Phys sci aide plant indust sta, Agr Res Serv, USDA, Md, 55, chemist, 57-58; teaching asst & res fel, Univ Mich, 58-61; asst prof, 61-66, ASSOC PROF ORG CHEM, OHIO UNIV, 66- Mem: AAAS; Am Chem Soc. Res: Chemical education; mechanism and stereochemistry of carbonyl condensation reactions. Mailing Add: Dept of Chem Ohio State Univ Athens OH 45701

WINKLER, SHELDON, b New York, NY, Jan 25, 32; m 61; c 2. DENTISTRY. Educ: NY Univ, BA, 53, DDS, 56. Prof Exp: From instr to asst prof denture prosthesis, Col Dent, NY Univ, 58-68; asst prof removable prosthodont, 68-70, ASSOC PROF REMOVABLE PROSTHODONT, STATE UNIV NY BUFFALO, 70- Concurrent Pos: Dir mat res, CMP Industs, Inc, 63-65, consult, 65-66; lectr, New York Community Col, 67-68; consult, Coe Labs, Inc, Ill, 67-; consult dent auxiliary training progs, Bd Coop Educ Serv, Cheektowaga, NY, 70- Mem: Fel Am Col Dent; Am Prosthodont Soc; Am Dent Asn; Sigma Xi; Am Acad Plastics Res Dent. Res: Dental resins and alloys; preservation and embedment of specimens in methyl methacrylate; demineralization of bone; geriatric dentistry; laser radiation applications in dentistry. Mailing Add: 52 Woodbury Dr Buffalo NY 14226

WINKLER, VIRGIL DEAN, b Danvers, Ill, Feb 9, 17; m 43; c 2. GEOLOGY. Educ: Univ Ill, AB & BS, 38, MS, 39, PhD(geol), 41. Prof Exp: Instr geol, Univ Ill, 38-39; paleontologist, 41-45, chief paleontologist, 45-55, paleont coordr, 55-56, eval geologist, 56-61, eval & opers geologist, 61-63, SPEC STUDIES & EVAL GEOLOGIST, CREOLE PETROL CORP, 63- Concurrent Pos: Prof, Cent Univ Venezuela, 58-59 & 66- Mem: AAAS; Paleont Soc; Soc Econ Paleont & Mineral; Geol Soc Am; Asn Geol, Mineral & Petrol, Venezuela (vpres, 54-55, secy-treas, 59-60). Res: Paleontology of Paleozoic rocks; world-wide occurrence of oil; Mesozoic and Cenozoic stratigraphy of Venezuela. Mailing Add: Geol Lab Creole Petrol Corp Apt 889 Caracas Venezuela

WINKLEY, DONALD CHARLES, b Geneva, Ill, Apr 9, 38; m 62; c 1. INORGANIC CHEMISTRY. Educ: Northwestern Univ, BS, 61; Univ Tenn, PhD(chem), 65. Prof Exp: Res chemist inorg res & develop, 65-69, supvr metals applns, 69-71, MGR METALS APPLNS, INORG RES & DEVELOP, FMC CORP, 71- Res: Chemistry of peroxygen chemicals and applications of peroxygen chemicals in metal treating and ore processing. Mailing Add: FMC Corp PO Box 8 Princeton NJ 08540

WINN, A VERNON, b Galt, Calif, Mar 15, 15; m 40; c 2. ORGANIC CHEMISTRY. Educ: Pac Union Col, BA, 38; Univ Wash, MS, 50; Stanford Univ, PhD(chem), 59. Prof Exp: Instr sci, Auburn Acad, 41-49; instr chem, Can Union Col, 50-54; assoc prof, 54-60, chmn dept, 60-74, PROF CHEM, PAC UNION COL, 60- Concurrent Pos: Lectr, Stanford Univ, 57-58. Mem: Am Chem Soc. Res: Mechanisms of aromatic desulfonation reactions and thermal and acid catalyzed isomerization of thujone. Mailing Add: Dept of Chem Pac Union Col Angwin CA 94508

WINN, EDWARD BARRIERE, b Baltimore, Md, Dec 27, 22; m 49; c 4. SCIENCE ADMINISTRATION. Educ: Univ SC, BSEE, 46; Univ Va, MS, 47; Univ Minn, PhD(physics), 50. Prof Exp: Elec engr, Westinghouse Elec Corp, 46; asst, Univ Minn, 48-50; res physicist, Textile Fibers Dept, E I du Pont de Nemours & Co, 50-58, res supvr, 58-62, tech mgr, du Pont de Nemours Int, SA, 62-70; independent consult S P A, 74- Mem: Am Phys Soc; Am Chem Soc. Res: Physics of high polymers; textile fibers; processing and applications technology of synthetic fibers; physics of electrical insulating materials; electrical insulation technology; industrial and technical marketing; new business ventures in textile, polymers and chemicals. Mailing Add: 37d chemin des Coudriers Geneva Switzerland

WINN, GRANT SAUNDERS, b Salt Lake City, Utah, Feb 28, 11; m 35; c 4. AIR POLLUTION, BIOCHEMISTRY. Educ: Univ Utah, BA, 33, MA, 35; Purdue Univ, PhD(biochem), 40. Prof Exp: Asst chem, Univ Utah, 33-35; asst gen chem, Purdue Univ, 35-37, biochem, 38-40, Nat Res Coun fel, 40-41; indust hygienist, Delco Battery Opers Div, Gen Motors Corp, 41-46; instr chem, Univ Utah, 46-47; hyg chemist, Am Smelting & Refining Co, 47-58; dir div occup health, 58-67, chief air qual sect, 67-74, DIR BUR AIR QUAL, UTAH STATE DIV HEALTH, 74- Mem: Am Indust Hyg Asn; Am Conf Govt Indust Hygienists; Air Pollution Control Asn; State & Territorial Air Pollution Prog Adminrs Asn (pres, 70). Res: Lead metabolism; analysis of Great Salt Lake water; control of lead hazard and exposure in storage battery manufacturing; methods of evaluating extent of exposure to lead; urinary thallium and lead determination; urinary coproporphryn versus lead intoxication symptoms. Mailing Add: 3439 E Magic View Dr Salt Lake City UT 84121

WINN, HENRY JOSEPH, b Lowell, Mass, Mar 2, 27; m 53; c 6. IMMUNOLOGY. Educ: Ohio State Univ, BA, 48, MS, 50, PhD(bact), 52. Prof Exp: Fel med & bact, Ohio State Univ, 52-54; fel chem, Calif Inst Technol, 54-55; res assoc, Jackson Mem Lab, 55-57, staff scientist, 57-65; assoc immunologist, Mass Gen Hosp, 65-73; asst prof bact, 65-70, ASSOC PROF MICROBIOL & MOLECULAR GENETICS, HARVARD MED SCH, 69-; IMMUNOLOGIST, MASS GEN HOSP, 73- Mem: Am Asn Immunol. Res: Immunology of homotransplantation; immunogenetics. Mailing Add: Dept of Surg Mass Gen Hosp Boston MA 02114

WINN, HOWARD ELLIOTT, b Winthrop, Mass, May 1, 26; m 51; c 3. BIOLOGICAL OCEANOGRAPHY, BIOACOUSTICS. Educ: Bowdoin Col, AB, 48; Univ Mich, MS, 50, PhD(zool), 55. Prof Exp: Specialist, Am Mus Natural Hist, 54-55; from asst to prof zool, Univ Md, 55-65; PROF OCEANOG & ZOOL, UNIV RI, 65- Concurrent Pos: Guggenheim fel, 62-63. Mem: AAAS; Am Inst Biol Sci; Am Soc Ichthyol & Herpet; Animal Behav Soc; Am Soc Mammal. Res: Comparative animal behavior; biology of fishes; sounds in animals; behavior and sounds of whales. Mailing Add: Grad Sch of Oceanog Univ of RI Kingston RI 02881

WINN, HUDSON SUMNER, b Tokyo, Japan, Jan 8, 19; US citizen; m 42; c 3. ZOOLOGY. Educ: Ill Col, AB, 40; Northwestern Univ, PhD(zool), 50. Prof Exp: Asst prof zool, Kans State Univ, 50-52; from asst prof to assoc prof, 52-59, PROF BIOL SCI, STATE UNIV NY ALBANY, 59- Mem: AAAS; Am Soc Zoologists; Am Ornith Union. Res: Avian physiology, especially photoperiodic effects. Mailing Add: Dept of Biol Sci State Univ of NY Albany NY 12222

WINN, MARTIN, b Brooklyn, NY, Jan 25, 40; m 66; c 2. MEDICINAL CHEMISTRY. Educ: Cooper Union Univ, BChE, 61; Northwestern Univ, PhD(org chem), 65. Prof Exp: SR CHEMIST, ABBOTT LABS, 65- Mem: Am Chem Soc. Res: Pharmaceuticals; nonclassical aromatic systems; heterocycles; psychotropic drugs; antihypertensive drugs, diuretics. Mailing Add: Dept 464 Abbott Labs Research Div North Chicago IL 60064

WINNAIL, DOUGLAS SAMUEL, b Akron, Ohio, Apr 16, 42; m 69; c 2. BIOLOGY. Educ: Washington & Jefferson Col, BA, 64; Univ Miss, MS, 67, PhD(anat), 68; Ambassador Col, BA, 70. Prof Exp: Asst prof, 68-75, V CHMN DEPT JOINT SCI, AMBASSADOR COL, 74-, ASSOC PROF BIOL SCI, 75- Mem: AAAS; Am Inst Biol Sci. Res: Human ecology, stress physiology; environmental biology. Mailing Add: Ambassador College 300 W Green St Pasadena CA 91123

WINNER, BERNARD MARK, b Chicago, Ill, Sept 3, 18; m 46; c 2. ORGANIC CHEMISTRY, BIOCHEMISTRY. Educ: Ind Univ, AB, 42. Prof Exp: Chemist, Allison Div, Gen Motors Corp, 42-43; res chemist, Univ Chicago, 44-45; supvr, Monsanto Chem Co, 45; res chemist, 47-49, foreman, 49-51, foreman packing & loading, 51-53, foreman dry starch, 53-54, div mgr packaging, 54-56, shift supt, 56-58, assoc dir res, 59-69, DIR QUAL CONTROL, AM MAIZE-PROD CO, 69- Mem: Am Chem Soc; Instrument Soc Am; Am Asn Cereal Chemists; Am Soc Qual Control. Mailing Add: 100 Park Ave Calumet City IL 60409

WINNER, IRENE PORTIS, b Chicago, Ill, Apr 7, 23; m 42; c 2. CULTURAL ANTHROPOLOGY. Educ: Radcliffe Col, BA, 43; Columbia Univ, MA, 53; Univ NC, PhD(anthrop), 67. Prof Exp: Res analyst, US Off War Info, 43-46; vis instr anthrop, Wayne State Univ, 62-63; res assoc anthrop & interdiscipline, Off Provost, Brown Univ, 68-71; Ossabaw Island Proj fel, Ga, 71; vis lectr anthrop, Tufts Univ, 72; RES FEL ANTHROP, BROWN UNIV, 72- Concurrent Pos: Am Coun Learned Socs-Soc Sci Res Coun fel, Poland & USSR, 72-73; lectr anthrop, Tufts Univ, 73 & 75, Emmanuel Col, 75. Mem: Fel Am Anthrop Asn; fel Soc Appl Anthrop; Am Ethnol Soc; Am Asn Advan Slavic Studies; Int Asn Semiotic Studies. Res: Comparative peasant studies with focus on Eastern Europe; structural and theoretical anthropology; Slovene ethnicity in greater Cleveland; semiotics; aesthetic anthropology.

WINNER, ROBERT WILLIAM, b Columbus, Ohio, Apr 5, 27; m 51; c 2. ECOLOGY. Educ: Ohio State Univ, PhD(wildlife mgt), 57. Prof Exp: Instr zool, 57-59, asst prof biol, 59-65, assoc prof zool, 65-69, PROF ZOOL, MIAMI UNIV, 69- Mem: Am Soc Limnol & Oceanog; Ecol Soc Am; Int Soc Limnol. Res: Limnology; ecology of planktonic communities; effect of heavy metals on aquatic organisms and communities. Mailing Add: Dept of Zool Miami Univ Oxford OH 45056

WINNETT, GEORGE, b New York, NY, Jan 4, 23; m 48; c 2. ENVIRONMENTAL SCIENCES, ANALYTICAL CHEMISTRY. Educ: Pa State Col, BS, 47; NY Univ, MA, 51. Prof Exp: Chemist, Queens Gen Hosp, NY, 48-49; Lutheran Hosp, New York, 49-50; lab asst & teacher high schs, 50-52; control chemist, Valspar Corp, Valentine Paint Co, 52-53, res chemist, 53-55; serv chemist, Reichold Chem, Inc, 55-57; asst prof, 57-65, ASSOC PROF AGR CHEM, RUTGERS UNIV, 65- Concurrent Pos: Smithsonian Inst consult, Microanal Lab & Pesticide Monitoring Proj, Univ Tehran, Iran. Mem: AAAS; Am Chem Soc; Sigma Xi (secy, 66-69, vpres, 70, pres, 71); Asn Offs Anal Chem. Res: Methodology for microanalysis of pesticide residues in soil and foodstuffs, using gas chromatograph, spectrometers, atomic absorption and miscellaneous allied equipment. Mailing Add: Dept of Entom & Econ Zool Rutgers Univ PO Box 231 New Brunswick NJ 08903

WINNICK, CHARLES NATHAN, chemistry, see 12th edition

WINNIE, WILLIAM W, JR, b Pontiac, Mich, Jan 27, 28; m 63. GEOGRAPHY, SOCIOLOGY. Educ: Univ Fla, BS, 53, MA, 55, PhD(inter-Am studies), 56. Prof Exp: Statistician, Int Statist Prog Off, US Bur Census, 56-58; asst prof geog, La State

Univ, 58-59; prof econ geog, Nuevo Leon Univ, 59-61; Fulbright lectr, Univ Chile, 61-62; specialist social planning & programming, Dept Social Affairs, Pan Am Union, 62-63; sociologist, Eval & Coord Group, Social Develop Div, Inter-Am Develop Bank, 63-64; consult, Latin Am Ctr, Univ Calif, Los Angeles, 64-66; assoc res sociologist, 66-67; prof sociol, Colo State Univ, 67-70; Fulbright res prof econ, 70-73, RES GEOGR, FAC ECON, UNIV GUADALAJARA, 73-; GEN EXEC COORDR, W MEX SOCIAL ADVAN STUDY, 72- Concurrent Pos: Res scholar, Univ Calif, Los Angeles, assoc, Calif State Col Long Beach & lectr, Calif State Col Fullerton, 64-65; actg dir ctr Latin Am studies, Colo State Univ, 68-69; Am Philos Soc field work grant, Mex, 69-70. Mem: AAAS; Latin Am Studies Asn. Res: Modernization; population studies; relationships between local community and broader levels of society. Mailing Add: Colon 36 Ajijic Jaliso Mexico

WINNIFORD, ROBERT STANLEY, b Portland, Ore, Oct 10, 21; m 44; c 4. PHYSICAL CHEMISTRY. Educ: Ore State Col, BS, 43; Calif Inst Technol, MS, 48; Univ Tenn, PhD, 51. Prof Exp: Instr chem, Univ Tenn, 47-49; res chemist, Calif Res Corp, Standard Oil Co Calif, 51-63; asst prof, 63-67, ASSOC PROF CHEM, WHITWORTH COL, WASH, 67-, CHMN DEPT, 71- Mem: Am Chem Soc; Sigma Xi. Res: Colloid and surface chemistry; nonaqueous solutions; asphalt chemistry and rheology. Mailing Add: Dept of Chem Whitworth College Spokane WA 99218

WINNIK, MITCHELL ALAN, b Milwaukee, Wis, July 17, 43. ORGANIC CHEMISTRY, PHOTOCHEMISTRY. Educ: Yale Univ, BA, 65; Columbia Univ, PhD(org chem), 69. Prof Exp: USPHS fel, Calif Inst Technol, 69-70; asst prof, 70-75, ASSOC PROF ORG CHEM, UNIV TORONTO, 75- Mem: AAAS; Am Chem Soc; Chem Inst Can. Res: Conformational analysis of flexible chain molecules. Mailing Add: Dept of Chem Univ of Toronto Toronto ON Can

WINNINGHAM, JOHN DAVID, b Mexia, Tex, Dec 28, 40; m 63; c 1. MAGNETOSPHERIC PHYSICS. Educ: Tex A&M Univ, BS, 63, MS, 65, PhD(physics), 70. Prof Exp: From res asst to res sci asst physics, 66-71, res assoc, 71-73, RES SCIENTIST PHYSICS, UNIV TEX, DALLAS, 73- Concurrent Pos: Consult, Los Alamos Sci Lab, Univ Calif, 74- Mem: Am Geophys Union. Res: Investigation of the source and acceleration mechanisms of corpuscular fluxes that produce the aurora and concomitant physical processes by means of rocket and satellite instruments. Mailing Add: Univ of Tex at Dallas PO Box 688 Richardson TX 75080

WINNINGHOFF, FRANCIS JOSEPH, b Los Angeles, Calif, Oct 2, 36. METEOROLOGY. Educ: Univ Calif, BA, 60, MA, 61, PhD(meteorol), 68. Prof Exp: Meteorologist, Extended Forecast Div, Nat Meteorol Ctr, Nat Oceanic & Atmospheric Admin, 61-63, Anal & Forecast Div, 63-64, Meteorol Satellite Lab, 64; meteorologist, Fleet Numerical Weather Ctr, US Navy, Dept Defense, 69-70, asst prof, Naval Postgrad Sch, 70-71; meteorologist, Univ Calif, Los Angeles, 71-73; res scientist, Atmospheric Environ Serv, Can, 73-75. Mem: Am Meteorol Soc. Res: Development of dynamic initialization procedures for use in modern global numerical weather prediction models. Mailing Add: 2004 Apt C Oak St Santa Monica CA 90405

WINOCUR, JOSEPH, physics, see 12th edition

WINOGRAD, NICHOLAS, b New London, Conn, Dec 27, 45; m 68. ANALYTICAL CHEMISTRY. Educ: Rensselaer Polytech Inst, BS, 67; Case Western Reserve Univ, PhD(chem), 70. Prof Exp: Asst prof, 70-75, ASSOC PROF CHEM, PURDUE UNIV, WEST LAFAYETTE, 75- Concurrent Pos: Res grants, Res Corp, 70-72, Am Chem Soc, 70-73, NSF, 71- & Air Force Off Sci Res, 72-; Sloan fel, 74- Mem: Am Chem Soc; Electrochem Soc. Res: Studies of electrode surfaces using electron spectroscopy and internal reflection spectroscopy; measurement of fast chemical reaction rates. Mailing Add: Dept of Chem Purdue Univ West Lafayette IN 47907

WINOGRAD, SHMUEL, b Israel, Jan 4, 36; m 58; c 2. MATHEMATICS. Prof Exp: MEM STAFF, DEPT MATH, T J WATSON RES CTR, IBM CORP, 61- Concurrent Pos: McKay lectr, Univ Calif, Berkeley, 67-68; IBM fel, 72; vis prof, Israel Inst Technol, 72- Res: Computer mathematics; reliable computations; complexity of computattions. Mailing Add: Math Sci Dept T J Watson Res Ctr PO Box 218 Yorktown Heights NY 10598

WINOGRAD, TERRY ALLEN, b Takoma Park, Md, Feb 24, 46; m 68. COMPUTER SCIENCE. Educ: Colo Col, BA, 66; Mass Inst Technol, PhD(appl math), 70. Prof Exp: Instr math, Mass Inst Technol, 70-71, asst prof elec eng, 71-73; ASST PROF COMPUT SCI & LING, STANFORD UNIV, 73- Concurrent Pos: Consult, Palo Alto Res Ctr, Xerox Corp, 73-; mem, COSERS Panel Artificial Intel, NSF, 75. Mem: Asn Comput Ling. Res: Artificial intelligence; computational linguistics; cognitive modelling. Mailing Add: Dept of Comput Sci Stanford Univ Stanford CA 94305

WINOKUR, GEORGE, b Philadelphia, Pa, Feb 10, 25; m 51; c 3. PSYCHIATRY. Educ: Johns Hopkins Univ, BA, 44; Univ Md, MD, 47; Am Bd Psychiat & Neurol, dipl, 53. Prof Exp: From instr to prof psychiat, Sch Med, Wash Univ, 51-71; PROF PSYCHIAT & HEAD DEPT, COL MED, UNIV IOWA & DIR, IOWA PSYCHOPATHIC HOSP, 71- Concurrent Pos: From asst psychiatrist to assoc psychiatrist, Barnes Hosp, 55-71; attend, Malcolm Bliss Psychiat Hosp. Honors & Awards: Anna Monika Prize Res Affective Disorder; Hofheimer Prize Psychiat Res, Am Psychiat Asn. Mem: Soc Biol Psychiat; fel Am Psychiat Asn; Psychiat Res Soc; Asn Res Nerv & Ment Dis; Am Psychopath Asn. Res: Conditioning and habituation; sexual variables in psychiatric patients and controls; genetics and epidemiological studies of psychiatric diseases. Mailing Add: Iowa Psychopath Hosp 500 Newton Rd Iowa City IA 52240

WINOKUR, MORRIS, b Philadelphia, Pa, Nov 8, 10; m 36; c 3. BIOLOGY. Educ: City Col New York, BS, 31; NY Univ, MSc, 33, EdD, 38; Columbia Univ, PhD(bot), 46. Prof Exp: From tutor to instr biol, City Col New York, 33-47; from asst prof to assoc prof, 47-67, supvr div biol, 57-68, PROF BIOL & CHMN DEPT, BARUCH COL, 68-, DEAN SUMMER SESSION, 69-, ASSOC DEAN LIB ARTS, 70- Concurrent Pos: Fel, Cold Spring Harbor, 31 & Marine Biol Lab, Woods Hole, 32; vis scholar, Columbia Univ, 47-48. Mem: AAAS; Bot Soc Am; Am Soc Plant Physiol; Nat Sci Teachers Asn; Nat Asn Biol Teachers. Res: Science orientation; cytology of protozoa; comparative physiology of green alga Chlorella. Mailing Add: Dept of Biol Baruch Col New York NY 10010

WINRICH, LONNY B, b Eau Claire, Wis, July 10, 37; m 61; c 5. COMPUTER SCIENCE, APPLIED MATHEMATICS. Educ: Wis State Univ, Eau Claire, BS, 60; Univ Wyo, MS, 62; Iowa State Univ, PhD(appl math), 68. Prof Exp: Physicist, Boulder Labs, Nat Bur Standards, 60-62; mathematician, Aerospace Div, Honeywell, Inc, 62-64; instr math & comput sci, Iowa State Univ, 64-68; asst prof comput sci, Univ Mo-Rolla, 68-71; ASSOC PROF COMPUT SCI & CHMN DEPT, UNIV WIS-LA CROSSE, 71- Concurrent Pos: Lectr, Viterbo Col; consult, Dairyland Power Co. Mem: Asn Comput Mach; Math Asn Am; Soc Indust & Appl Math; Asn Educ Data Systs. Res: Numerical solution of ordinary and partial differential equations; matrix computations; programming languages; medical computing. Mailing Add: Dept of Comput Sci Univ of Wis-La Crosse La Crosse WI 54601

WINSBERG, GWYNNE ROESELER, b Chicago, Ill, Nov 28, 30; m 50; c 2. EPIDEMIOLOGY. Educ: Univ Chicago, MS, 62, PhD(biopsychol), 67. Prof Exp: Instr biol, Univ Chicago, 65-67; asst prof anat, 67-71, ASST PROF COMMUNITY HEALTH & PREV MED, MED SCH, NORTHWESTERN UNIV, CHICAGO, 71- Concurrent Pos: Lectr, Ill Col Optom, 62-67; consult, Ill Dept Ment Health, 69-72; USPHS grant, Fac Inst Med Care Orgn, Univ Mich, 73 & 74; Nat Endowment Humanities grant, Univ Pa, 74; spec asst to regional health adminr, Region V, USPHS, 74-76. Mem: Am Pub Health Asn; Am Soc Trop Med & Hyg; Asn Teachers Prev Med; Int Soc Psychosom Obstet & Gynec. Res: Social and medical epidemiology; medical care organization. Mailing Add: Dept Commun Health & Prev Med Northwestern Univ Med Sch Chicago IL 60611

WINSBERG, LESTER, b Montreal, Que, Jan 31, 21; nat US; m 49; c 4. NUCLEAR SCIENCE. Educ: Univ Chicago, BS, 42, PhD(chem), 47. Prof Exp: Jr scientist, Metall Lab, 43-45 & Los Alamos Sci Lab, 45-46; assoc physicist, Argonne Nat Lab, 47-50; sr scientist, Weizmann Inst, 50-52, 54-55 & Inst Nuclear Studies, Univ Chicago, 52-54; res scientist, Lawrence Radiation Lab, Univ Calif, 55-60; PROF PHYSICS, UNIV ILL, CHICAGO CIRCLE, 61- Concurrent Pos: Assoc chemist, Argonne Nat Lab, 60-64, consult, 64- Mem: Am Phys Soc. Res: Nuclear fission; neutron diffraction; natural radioactivity; meson interactions; high energy nuclear and heavy ion induced reactions; stopping of heavy ions; nuclear physics-high energy spallation reactions. Mailing Add: Univ of Ill Box 4348 Chicago IL 60680

WINSBERG, MORTON DANIEL, b Chicago, Ill, Aug 1, 30. GEOGRAPHY. Educ: Univ Ill, BS, 51, MS, 54; Univ Fla, PhD(geog), 58. Prof Exp: Asst prof geog, ECarolina Col, 58-62; asst prof geog & geol, Ohio Univ, 62-65; ASSOC PROF GEOG, FLA STATE UNIV, 65- Concurrent Pos: Am Coun Learned Socs grant, Arg, 60-61; Agr Develop Coun grant, 66; Fulbright lectr, Helsinki & Swedish Schs Econ, Finland, 67-68 & Univ Col, Cork, 73. Mem: Am Geog Soc; Asn Am Geog. Res: Population and settlement in Latin America and the agricultural geography of Argentina. Mailing Add: Dept of Geog Fla State Univ Tallahassee FL 32306

WINSLOW, ALFRED EDWARDS, b Clinton, Mass, Oct 8, 19; m 44; c 2. ORGANIC CHEMISTRY. Educ: Worcester Polytech Inst, BS, 41; Mass Inst Technol, PhD(org chem), 47. Prof Exp: Jr chemist, Tenn Eastman Corp, 44-45; asst, Sugar Res Found, Mass Inst Technol, 45-47; res chemist, Union Carbide Chem Co, 47-64; RES & DEVELOP CHEMIST, BORDEN CHEM CO, 64- Mem: Am Chem Soc; Am Soc Qual Control. Res: Water soluble polymers; condensation polymerizations; reactions in aqueous media; paper resins; binders; applications orientated experimental designs; resin analyses; manufacturing procedures and quality control. Mailing Add: Borden Chem Co Adhes & Chem Div 14 Johnson St Bainbridge NY 13733

WINSLOW, DAVID CLINTON, b Creston, Iowa, July 15, 14; m 36; c 2. GEOGRAPHY. Educ: Univ Okla, BA, 36; Univ Nebr, MA, 39, PhD(geog), 48. Prof Exp: Naturalist-ranger, US Nat Park Serv, 41-43; dust control adv, US Air Force, 43; soil conservationist, US Soil Conserv Serv, 43-45; vis assoc prof geog, Univ Okla, 46; asst prof, Okla State Univ, 47-50, 53-55, assoc prof, 55-57; cartogr, Res Div, Aeronaut Chart & Info Ctr, US Air Force, 57-58; PROF GEOG & REGIONAL PLANNING, INDIANA UNIV PA, 58- Concurrent Pos: Soc Sci Comt grant in aid, 48-50; ed, Geog, Geol & Meteorol Col Workbk Ser, 50-, textbks & atlas, Hq, Air Force Reserve Officers Training Corps, Air Univ, 51-53 & Okla Geogr, 53-56; Aerospace Sem scholar, 61; ed, Pa Geogr, GTU Handbk & Pa Geogr, 63-; researcher, Latin Am, 68; Europ trip sponsor, Six Eastern Univs Consortium, 70; NSF scholar, Systs Oriented Air Transp Workshop, Space Inst, Univ Tenn, 71; Sigma Xi nat lectr, 71; nat rev bd mem, US Nat Air Transp Syst, Fed Aviation Agency, 71-72; pop consult, US Dept State, 72; Phi Delta Kappa partial scholar, USSR Educ Sem, 73. Mem: Asn Am Geogrs; Am Geog Soc; Nat Coun Geog Educ; Am Antiq Soc; Am Meteorol Soc. Mailing Add: Dept of Geog Indiana Univ of Pa Indiana PA 15701

WINSLOW, DONALD J, b Lisbon Falls, Maine, Aug 31, 15; m 41; c 3. PATHOLOGY. Educ: Bates Col, BS, 37; Harvard Med Sch, MD, 41. Prof Exp: Intern surg, Boston City Hosp, 41-42; lab officer & pathologist, US Army, 42-46; pvt pract, Maine & Calif, 46-50; pathologist & chief lab, US Army, 50-52; resident pathologist, Letterman Army Hosp, 52-56; asst chief geog & infectious dis, Armed Forces Inst Path, 56-58, chief soft tissue br & head WHO Int Reference Ctr Soft Tissue Tumors, 58-60; dir dept path, Sisters' Hosp, Waterville, Maine, 60-62; chief infectious dis br, Armed Forces Inst Path, 62-67; CHIEF LAB SERVS, VET ADMIN HOSP, 67- Concurrent Pos: Res fel internal med, Cornell Serv, Bellevue Hosp, New York, 47-48. Mem: Fel Col Am Path; fel Am Soc Clin Path; Am Soc Trop Med & Hyg; Int Acad Path. Res: Infectious diseases and cancer. Mailing Add: Vet Admin Ctr Bay Pines FL 33504

WINSLOW, FIELD HOWARD, b Proctor, Vt, June 10, 16; m 45; c 3. ORGANIC CHEMISTRY. Educ: Middlebury Col, BS, 38; RI State Col, MS, 40; Cornell Univ, PhD(org chem), 43. Prof Exp: Res chemist, Manhattan Proj, Columbia Univ, 43-45; MEM TECH STAFF, BELL LABS, INC, 45- Concurrent Pos: Prof, Stevens Inst Technol, 64-67; ed, Macromolecules, 67- Mem: AAAS; Am Chem Soc. Res: Photochemistry; organic semiconductors; polymer morphology, chemical reactivity and photochemistry; deterioration and stabilization of rubbers and plastics; fluorocarbons. Mailing Add: Bell Labs Inc Murray Hill NJ 07974

WINSLOW, GEORGE HARVEY, b Washington, DC, June 21, 16; m 44; c 2. PHYSICS. Educ: Carnegie Inst Technol, BS, 38, MS, 39, DSc, 46. Prof Exp: Instr physics, Carnegie Inst Technol, 38, res physicist, 43-46; ASSOC PHYSICIST, ARGONNE NAT LAB, 46- Mem: Am Phys Soc. Res: Magnetic moments by molecular beams; high speed deformation of metals; shaped charges; solid state; attempts to find requantization of space quantized atoms at collision; alpha decay theory high temperature physical chemistry. Mailing Add: Argonne Nat Lab 9700 S Cass Ave Argonne IL 60439

WINSLOW, JOHN DURFEE, b Ft Monroe, Va, June 21, 23; m 52. GEOLOGY. Educ: Brown Univ, AB, 49; Univ Ill, PhD, 57. Prof Exp: Supvry hydrologist, Water Resources Div, US Geol Surv, 50-74; HYDROLOGIST, US DEPT INTERIOR, 74- Concurrent Pos: Eng geologist, State Geol Surv, Ind, 58-59; assoc prof, Univ Kans, 63-71; co-dir course on ground water, UNESCO, Buenos Aires, Arg, 65, course on appl geol, Medellin, Colombia, 66; int prof ground water, Pan-Am Health Orgn, Santiago, Chile, 66; mem adv panel hydrol, Arg Pampa, Nat Acad Sci, 70-71. Honors & Awards: Antarctic Serv Medal, Us Antarctic Res Prog, US Dept Interior, 69. Mem: AAAS; Geol Soc Am; Soc Econ Paleont & Mineral; Am Geophys Union; Int Asn Hydrogeol. Res: Hydrology; ground water and engineering geology. Mailing Add: US Dept Interior 230 S Dearborn St 32nd floor Chicago IL 60604

WINSLOW, JOHN HATHAWAY, b Trieste, Italy, Apr 16, 32; US citizen; m 54; c 3. CULTURAL GEOGRAPHY, HISTORICAL GEOGRAPHY. Educ: Univ Mich, AB, 54; Cambridge Univ, PhD(geog), 73. Prof Exp: Asst prof geog & anthrop, Calif State

Univ, Hayward, 60-64, assoc prof & lectr geog, 68-71; assoc prof geog, Univ Papua New Guinea, 71-75; VIS PROF, TRINITY COL, UNIV DUBLIN, IRELAND, 75- Res: Charles Darwin's geographical-geological contributions; development of Darwin's thinking; plant and animal migrations; historical geography of the Pacific, especially Melanesia; history of science; environmental problems. Mailing Add: Dept of Geog Trinity Col Univ of Dublin Dublin Ireland

WINSLOW, LEON E, b Centralia, Ill, Nov 17, 34; m 59; c 6. MATHEMATICAL ANALYSIS. Educ: Marquette Univ, BS, 56, MS, 60; Duke Univ, PhD(math), 65. Prof Exp: Asst physics, Marquette Univ, 56-57, Comput Ctr, 58-59; prin physicist, Battelle Mem Inst, 57-58; instr math, Rockhurst Col, 59-60; asst, Duke Univ, 60-64, res assoc spec projs numerical anal, 64-65; asst prof math, Rockhurst Col, 65-66; asst prof comput sci, Univ Notre Dame, 66-72; ASSOC PROF COMPUT SCI, WRIGHT STATE UNIV, 72- Mem: Math Asn Am; Am Math Soc; Asn Comput Mach. Res: Numerical, functional and systems analysis. Mailing Add: Dept of Comput Sci Wright State Univ Dayton OH 45431

WINSLOW, RICHARD EDWARD, b Troy, NY, Feb 8, 41; m 70. MATHEMATICS. Educ: Columbia Univ, BA, 63; Brandeis Univ, MA, 65, PhD(math), 70. Prof Exp: Instr math, Boston Col, 67-70; ASST PROF MATH, LOWELL STATE COL, 70- Mem: Am Math Soc. Res: Algebra; relative homological algebra. Mailing Add: Dept of Math Lowell State Col Lowell MA 01854

WINSMANN, FRED RUDOLPH, b Jersey City, NJ, Sept 24, 25; m 53; c 5. PHYSIOLOGY, BIOLOGY. Educ: Springfield Col, BS, 51; Boston State Col, MEd, 61. Prof Exp: Physiologist, Environ Protection Res Div, Dept of Army, 51-53; scientist rehab, NY Univ, 53-54; high sch teacher, NJ, 54-55 & NY, 55-56; RES BIOLOGIST, MIL ERGONOMICS DIV, RES INST ENVIRON MED, NATICK DEVELOP CTR, OFF SURGEON GEN, DEPT OF ARMY, 56- Concurrent Pos: Supvr activities, Warren Ctr for Phys Educ & Recreation, Boston Bouve Col, Northeastern Univ, 68-, lectr, Col, 72- Mem: Sigma Xi. Res: Effects of heat, cold, work and altitude on man's ability to perform; extend his tolerance and determine his susceptibility to injury in climatic and work extremes. Mailing Add: US Army Res Inst Environ Med Natick Develop Ctr Natick MA 01760

WINSOR, TRAVIS WALTER, b San Francisco, Calif, Dec 1, 14; m 39; c 2. MEDICINE. Educ: Stanford Univ, BA, 37, MD, 41. Prof Exp: Asst med, Med Sch, Tulane Univ, 41-42, instr, 42-45; clin instr, 45-47, asst clin prof, 47-63, CLIN PROF MED, MED SCH, UNIV SOUTHERN CALIF, 63-; DIR, MEM HEART RES FOUND, INC, 53- Concurrent Pos: Mem coun on circulation, Am Heart Asn & Intersoc Comn Heart Dis Resources. Mem: Soc Exp Biol & Med; fel Am Heart Asn; fel AMA; fel Am Col Physicians; Am Col Cardiol. Res: Biophysics; arteriosclerosis; hypertension. Mailing Add: 4041 Wilshire Blvd Los Angeles CA 90010

WINSTEAD, JACK ALAN, b Dixon, Ky, June 13, 32; m 56; c 3. BIOCHEMISTRY. Educ: Univ Ky, BS, 54; Okla State Univ, MS, 59; Univ Ill, PhD(chem), 64. Educ: Res officer chem, Mat Lab, US Air Force, Wright-Patterson AFB, Ohio, 54-57, asst prof, Air Force Acad, 59-62, res biochemist, Sch Aerospace Med, 64-68, res chemist, Frank J Seiler Res Lab, US Air Force Acad, 68-70, dir, Directorate Chem Sci, 70-72, dep dir, Toxic Hazards Div, Aerospace Med Res Lab, Wright-Patterson AFB, 72-75; PROF ASSOC, NAT ACAD SCI, 75- Mem: Am Soc Biol Chemists; AAAS; Radiation Res Soc; Am Chem Soc. Res: Structure and function of proteins; radiation biochemistry; organic synthesis and toxicology. Mailing Add: Nat Acad Sci 2101 Constitution Ave Washington DC 20418

WINSTEAD, JANET, b Wichita Falls, Tex, Mar 13, 32. MYCOLOGY. Educ: Midwestern Univ, BS, 53; Ohio Univ, MS, 55; Univ Tex, Austin, PhD(bot), 70. Prof Exp: Instr biol, Ky Wesleyan Col, 56-57; asst prof, Atlantic Christian Col, 57-65; ASSOC PROF BIOL, MADISON COL, VA, 69- Mem: Am Inst Biol Sci; Mycol Soc Am. Res: Monospore culture of myxomycetes. Mailing Add: Dept of Biol Madison College Harrisonburg VA 22801

WINSTEAD, JOE EVERETT, b Wichita Falls, Tex, Mar 17, 38; m 60; c 1. BOTANY, ECOLOGY. Educ: Midwestern Univ, BS, 60; Ohio Univ, MS, 62; Univ Tex, Austin, PhD(bot), 68. Prof Exp: Instr biol, Delta Col, 62; asst prof, 68-72, ASSOC PROF BIOL, WESTERN KY UNIV, 72- Mem: Ecol Soc Am; Bot Soc Am. Res: Ecotype differentiation of plant species; natural revegetation of stripmines; differentiation of wood cell types and wood anatomy; environmental physiology. Mailing Add: Dept of Biol Western Ky Univ Bowling Green KY 42101

WINSTEAD, MELDRUM BARNETT, b Lincolnton, NC, Oct 19, 26; m 59; c 3. ORGANIC CHEMISTRY. Educ: Davidson Col, BS, 46; Univ NC, MA, 49, PhD(chem), 52. Prof Exp: Instr chem, Davidson Col, 46-47; asst, Univ NC, 47-50; from asst prof to assoc prof, 52-69, PROF CHEM, BUCKNELL UNIV, 69- Concurrent Pos: Consult, Glyco Chem, Inc, 58 & Sadtler Res Labs, 61-70; vis chem assoc, Calif Inst Technol, 67-68; USPHS spec res fel, 67-68; res assoc, Lawrence Radiation Lab, Univ Calif, Berkeley, 68-69; vis scientist, Medi-Physics, Inc, Calif, 72- Honors & Awards: USPHS Award, Nat Inst Gen Med Sci, 74. Mem: Soc Nuclear Med; Sigma Xi; Am Chem Soc; Coblentz Soc. Res: Organic medicinals; preparation and scintigraphic study of pharmaceuticals containing short-lived radiocarbon-11. Mailing Add: Dept of Chem Bucknell Univ Lewisburg PA 17837

WINSTEAD, NASH NICKS, b Durham, Co, NC, June 12, 25; m 49; c 1. PLANT PATHOLOGY. Educ: NC State Col, BS, 48, MS, 51; Univ Wis, PhD(plant path), 53. Prof Exp: From asst prof to assoc prof, 53-60, PROF PLANT PATH, NC STATE UNIV, 60-, PROVOST & V CHANCELLOR, 74- Concurrent Pos: Dir, Inst Biol Sci & asst dir res, NC Agr Exp Sta, 65-67, asst provost, NC State Univ, 67-73, assoc provost, 73-74; Phillips Found internship acad admin, Ind Univ, 65-66; mem, Comt Planned Res Basic Bio-sci during manned earth-orbiting missions, Am Inst Biol Sci-NASA, 65-67; mem, Bd Dirs, Consortium Cooperating Raleigh Cols, 69-, pres, 71-73. Honors & Awards: Res Award, Sigma Xi, 61. Mem: AAAS; Am Phytopath Soc; Am Inst Biol Sci; Sigma Xi. Res: Vegetable diseases; breeding for resistance; physiology of parasitism. Mailing Add: Rm A Holliday Hall NC State Univ Raleigh NC 27607

WINSTEN, SEYMOUR, b Jersey City, NJ, June 14, 26; m 49; c 3. BIOCHEMISTRY. Educ: Rutgers Univ, AB, 48, PhD(microbiol, physiol), 56; NY Univ, MSc, 50; Am Bd Clin Chem, dipl. Prof Exp: Asst, Merck Inst Therapeut Res, 50-56; assoc microbiol, Univ Pa, 56-57; HEAD DEPT CHEM, ALBERT EINSTEIN MED CTR, 57-; ASSOC PROF BIOCHEM, SCH MED, TEMPLE UNIV, 70- Concurrent Pos: Consult, Atlantic City Hosp, Mass Rehab Hosp, Surgeon Gen US & Walson Gen Hosp; dir lab, Mass Rehab Hosp, 75- Honors & Awards: John Gunther Reinhold Award, 68. Mem: Fel Am Asn Clin Chem. Res: Clinical chemistry; immunochemistry; chemical diagnosis of disease; mycology; endocrine chemistry and its relationship to various disease processes. Mailing Add: Div of Lab Albert Einstein Med Ctr Philadelphia PA 19141

WINSTEN, WALTER ABBOTT, b New York, NY, June 3, 15; m 41; c 2. BIOCHEMISTRY. Educ: City Col New York, BS, 35; Columbia Univ, AM, 36, PhD(biochem), 39. Prof Exp: Res chemist, Int Vitamin Div, Am Home Prod Corp, 40-41, chief res chemist, 41-44; sect mgr, SAM Labs, Columbia Univ, 44-46; group leader, Schenley Distillers Corp, 46-47; chief chemist, Food Res Labs, 47-50; DIR, WINSTEN LABS, 50- Concurrent Pos: Spec assoc prof, Hofstra Univ, 60-67, prof, 67- Mem: Am Chem Soc. Res: Antibiotics, vitamins; amino acids; microbiological assay methods; antimetabolites; synthetic organic chemistry; partition chromatography; pharmaceutical product and process development; transfer RNA. Mailing Add: 671 Fairview Ave Westbury NY 11590

WINSTON, ANTHONY, b Washington, DC, Dec 5, 25; m 52; c 4. POLYMER CHEMISTRY. Educ: George Washington Univ, BS, 50; Duke Univ, MA, 52, PhD, 55. Prof Exp: Res chemist, Armstrong Cork Co, 54-59; from asst prof to assoc prof, 59-75, PROF CHEM, WVA UNIV, 75- Concurrent Pos: Res assoc, Water Res Inst, WVa Univ, 75- Mem: AAAS; Am Chem Soc. Res: Polymer synthesis and reactions; metal complexing polymers; selective chelating ion exchange resins; stereochemistry; hemiacetal equilibria; wood plastic combinations. Mailing Add: Dept of Chem WVa Univ Morgantown WV 26506

WINSTON, ARTHUR WILLIAM, b Toronto, Ont, Feb 11, 30; US citizen; m 49; c 4. PHYSICS, MATHEMATICS. Educ: Univ Toronto, BASc, 51; Mass Inst Technol, PhD(physics), 54. Prof Exp: Eng physicist, Nat Res Coun Can, 49-51; res asst, Mass Inst Technol, 51-54; sr engr, Schlumberger Well Surv Corp, 54-57; sr engr, Nat Res Corp, 57-59; chief scientist, Allied Res Assocs, Inc, 59-61; pres, Space Sci, Inc, 61-65; pres, 65-75, CHMN, IKOR, INC, 75- Concurrent Pos: Lectr, Northeastern Univ, 57-65, adj prof, 65-; dir, New Boston Bank & Trust Co & Metric Inc, 69- Mem: Am Inst Aeronaut & Astronaut; Inst Elec & Electronics Engrs; Am Geophys Union; Am Phys Soc; Am Inst Mining, Metall & Petrol Engrs. Res: Electromagnetic propagation and measurements; nuclear physics applied to geophysics; thin film technology. Mailing Add: Ikor Inc Northwest Indust Park Burlington MA 01803

WINSTON, DONALD, b Washington, DC, Apr 4, 31. GEOLOGY. Educ: Williams Col, BA, 53; Univ Tex, MA, 57, PhD(geol), 63. Prof Exp: From instr to asst prof, 61-70, ASSOC PROF GEOL, UNIV MONT, 70- Mem: AAAS; Soc Econ Paleont Mineral. Res: Sedimentary petrology, particularly carbonate petrology of Pennsylvanian rocks and modern carbonate areas; stratigraphy and sedimentation, particularly Precambrian rocks; Cambrian paleontology. Mailing Add: Dept of Geol Univ of Mont Missoula MT 59801

WINSTON, HARVEY, b Newark, NJ, Aug 11, 26; m 49; c 2. PHYSICAL CHEMISTRY. Educ: Columbia Univ, AB, 45, MA, 46, PhD(chem), 49. Prof Exp: Asst chem, Columbia Univ, 45-49; Jewett fel, Univ Calif, 49-50, instr chem, 50-51; asst prof, Univ Calif, Los Angeles, 51-52; mem tech staff, Hughes Aircraft Co, 52-58, mgr mat res lab, Semiconductor Div, 58-60; assoc dir quantum electronics lab, Quantatron, Inc, 61, vpres, Quantum Tech Labs, Inc, 61-63; mgr, 63-69, SR SCIENTIST, QUANTUM PHYSICS DEPT, HUGHES RES LABS, 69- Mem: Fel Am Phys Soc; sr mem Inst Elec & Electronics Engrs. Res: Solid state spectroscopy; lasers and laser systems; semiconductor physics and devices. Mailing Add: 1450 San Remo Dr Pacific Palisades CA 90272

WINSTON, JAMES J, b New York, NY, Mar 17, 15; m 40; c 2. FOOD CHEMISTRY. Educ: City Col New York, BS, 36. Prof Exp: Chemist, Jacobs Cereal Labs, Inc, 36-38, chief chemist, 39-49; DIR FOOD LABS, JACOBS-WINSTON LABS, INC, 50- Concurrent Pos: Dir res, Nat Macaroni Mfrs Asn, 50; lectr, City Col New York, 60. Honors & Awards: Pub Serv Award, Gov A Harriman, NY, 58. Mem: AAAS; Am Asn Cereal Chemists; fel Am Inst Chemists; Inst Food Technologists; NY Acad Sci. Res: Food chemistry, enrichment, sanitation, macaroni, egg noodles, lecithin and cereals; new product development; meat and fish technology. Mailing Add: 156 Chambers St New York NY 10007

WINSTON, JAY STEVEN, meteorology, see 12th edition

WINSTON, PAUL WOLF, b Chicago, Ill, Aug 9, 20; m 48; c 1. BIOLOGY. Educ: Univ Mass, BS, 48; Northwestern Univ, MS, 50, PhD, 52. Prof Exp: Instr, Brown Univ, 51-52; from instr to assoc prof, 52-68, PROF BIOL, UNIV COLO, BOULDER, 68- Mem: AAAS; Ecol Soc Am; Am Soc Zoologists; Soc Exp Biol; Entom Soc Am. Res: Humidity relations and water balance of terrestrial arthropods, especially cuticular control of water exchange with air; physiology of chronic exposure to heavy metals in mammals. Mailing Add: Dept of EPO Biol Univ of Colo Boulder CO 80302

WINSTON, ROLAND, b Moscow, USSR, Feb 12, 36; US citizen; m 57; c 3. EXPERIMENTAL PHYSICS, PARTICLE PHYSICS. Educ: Shimer Col, BA, 53; Univ Chicago, BS, 56, MS, 57, PhD(physics), 63. Prof Exp: Asst prof physics, Univ Pa, 63-64; asst prof, 64-71, ASSOC PROF PHYSICS, UNIV CHICAGO, 71- Concurrent Pos: Sloan Found fel, 67-69. Mem: Fel Am Phys Soc. Res: Elementary particle physics; leptonic decays of hyperons; muon physics, especially hyperfine effects in muon capture by complex nuclei; solar energy concentrators; infra-red detectors; optics of visual receptors. Mailing Add: Dept of Physics Univ of Chicago Chicago IL 60637

WINSTROM, LEON OSCAR, b Holland, Mich, Apr 8, 12; m 38; c 3. PHYSICAL CHEMISTRY, TEXTILE CHEMISTRY. Educ: Hope Col, AB, 34; Carnegie Inst Technol, MS, 37, DSc(phys chem), 38. Prof Exp: Instr, Carnegie Inst Technol, 37-38; res chemist, Nat Aniline Chem Div, Allied Chem Corp, 38-53, asst supvr, 53-57, sr scientist, 57-58, group leader, 58-64, res supvr, 64-68, sr res assoc, Spec Chem Div, 68-71; MGR RES & DEVELOP, FLOCK DIV, MALDEN MILLS, LAWRENCE, 71- Honors & Awards: Schoellkopf Medal, 66. Mem: AAAS; Am Chem Soc; Am Asn Textile Chemists & Colorists. Res/ Vapor and liquid phase hydrogenation and oxidation; ammination by reduction; recovery of organic oxidation products. Mailing Add: 9 Prospect Rd Andover MA 01810

WINTER, ALDEN RAYMOND, b Bridgeport, WVa, Sept 23, 97; m 26. POULTRY SCIENCE. Educ: Univ WVa, BA, 20; Ohio State Univ, MS, 21, PhD(chem), 28. Prof Exp: Asst nutrit invests, Ohio Exp Sta, 20-25; asst prof poultry sci, 26-33, part-time instr bact, 32-35 & 43-44, from assoc prof to prof poultry sci, 34-68, chmn dept, 55-64, EMER PROF POULTRY SCI, OHIO STATE UNIV, 68- Concurrent Pos: Res assoc prof, Exp Sta, Iowa State Col, 44-45; USAID poultry adv, India, 64-67. Mem: AAAS; Am Soc Microbiol; Poultry Sci Asn; World Poultry Sci Asn; Inst Food Technologists. Res: Microbiology, preservation and measurements of quality in poultry products. Mailing Add: 1909 Harwich Rd Columbus OH 43221

WINTER, ALEXANDER J, b Vienna, Austria, June 21, 31; nat US; m 59; c 3. IMMUNOBIOLOGY. Educ: Univ Ill, DVM, 55; Univ Wis, PhD(med & vet path), 59. Prof Exp: From asst to assoc prof vet sci, Pa State Univ, 59-63; assoc prof, 63-68, PROF VET MICROBIOL, NY STATE VET COL, CORNELL UNIV, 68- Concurrent Pos: Mem, Bacteriol & Mycol Sect, NIH, 71-75. Mem: Conf Res Workers Animal Dis; Am Soc Microbiol; AAAS. Res: Microbial and immunologic factors affecting animal reproduction; bacterial virulence mechanisms. Mailing Add: NY State Col of Vet Med Cornell Univ Ithaca NY 14853

WINTER, CHARLES ERNEST, b Colorado Springs, Colo, Sept 9, 14; m 37; c 2. MICROBIOLOGY. Educ: Colo Col, BA, 36, MA, 38; Univ Md, MS, 45, PhD(bact), 47. Prof Exp: Instr biol, China Training Inst, Kiangsu, China, 39-42, prof biol sci, 47-50; prof, Southern Jr Col, 42-43; asst prof, Washington Missionary Col, 43-45; bacteriologist, US Fish & Wildlife Serv, 45-47; from asst prof to assoc prof, 50-61, PROF MICROBIOL, LOMA LINDA UNIV, 61- Concurrent Pos: Mem, Am Asn Dent Schs. Mem: AAAS; Am Soc Microbiol; Am Pub Health Asn; Asn Am Med Cols. Res: Medical microbiology; immunology. Mailing Add: Dept of Microbiol Loma Linda Univ Loma Linda CA 92354

WINTER, CHARLES GORDON, b Hanover, Pa, Dec 28, 36; m 58; c 3. BIOCHEMISTRY. Educ: Juniata Col, BS, 58; Univ Mich, MS, 63, PhD(biochem), 64. Prof Exp: Technician, Metab Res Unit, Univ Mich, Ann Arbor, 58-60; Childs Mem Fund Med Res fel phys chem, Sch Med, Johns Hopkins Univ, 64-66; asst prof, 66-73, ASSOC PROF BIOCHEM, SCH MED, UNIV ARK, LITTLE ROCK, 73- Mem: AAAS; Am Chem Soc; Biophys Soc. Res: Mechanism of action of alkali-cation-dependent adenosinetriphosphatase; role of autolysins in chain formation by bacteria. Mailing Add: Dept of Biochem Univ of Ark for Med Sci Little Rock AR 72201

WINTER, CHESTER CALDWELL, b Cazenovia, NY, June 2, 22; m 45; c 3. MEDICINE. Educ: Univ Iowa, BA, 43, MD, 46; Am Bd Urol, dipl. Prof Exp: Asst prof surg, Sch Med, Univ Calif, Los Angeles, 58-61; PROF UROL & DIR DIV, COL MED, OHIO STATE UNIV, 61- Concurrent Pos: Mem staff, Univ Hosp, 61- & Children's Hosp, 61- Mem: Am Urol Asn; Am Col Surg; Soc Univ Urol; Soc Univ Surg; Am Asn Genito-Urinary Surg. Res: Urological surgery; renal hypertension; diagnostic isotopes in urology. Mailing Add: Div of Urol Ohio State Univ Med Ctr Columbus OH 43210

WINTER, DAVID ARTHUR, b Windsor, Ont, June 16, 30; m 58; c 3. BIOMEDICAL ENGINEERING, ELECTRICAL ENGINEERING. Educ: Queen's Univ, Ont, BSc, 53, MSc, 61; Dalhousie Univ, PhD(physiol), 67. Prof Exp: From lectr to asst prof elec eng, Royal Mil Col, Ont, 58-63; from asst prof to assoc prof, NS Tech Col, 63-69; assoc prof surg, Univ Man, 69-74, adj prof elec eng, 70-74; ASSOC PROF KINESIOLOGY, UNIV WATERLOO, 74- Concurrent Pos: Can Coun fel med, eng & sci, Dalhousie Univ, 66-68. Mem: Inst Elec & Electronics Eng; Can Med & Biol Eng Soc (pres, 70-74); Soc Photo-Optical Instrument Eng. Res: Signal processing of biological signals; medical image processing; biomechanics; locomotion studies; powered prosthetic devices. Mailing Add: Dept of Kinesiology Univ of Waterloo Waterloo ON Can

WINTER, DAVID JOHN, b Painesville, Ohio, May 2, 39; m 65; c 1. MATHEMATICS. Educ: Antioch Col, BA, 61; Yale Univ, MS, 63, PhD(math), 65. Prof Exp: Instr math, Yale Univ, 65-67; NSF fel, Univ Bonn, 67-68; from asst prof to assoc prof, 68-74, PROF MATH, UNIV MICH, ANN ARBOR, 74- Concurrent Pos: Vis assoc prof, Calif Inst Technol, 72-73. Mem: Am Math Soc. Res: Algebra. Mailing Add: Dept of Math 4200 Angell Hall Univ of Mich Ann Arbor MI 48104

WINTER, DAVID LEON, b New York, NY, Nov 10, 33; m 73; c 5. RESEARCH ADMINISTRATION. Educ: Columbia Col, AB, 55; Wash Univ, MD, 59. Prof Exp: Surg intern, Sch Med, Wash Univ, 59-60; Nat Inst Neurol Dis & Blindness fel, Baylor Univ, 60-62; med res officer, Nat Inst Neurol Dis & Blindness, 62-64; neurophysiologist, Walter Reed Army Inst Res, 64-66, chief dept neurophysiol, 66-71; dep dir life sci, Ames Res Ctr, Moffett Field, Calif, 71-74, DIR LIFE SCI, NASA HQ, DC, 74- Honors & Awards: Hans Berger Prize, Am Electroencephalog Soc, 64. Mem: AAAS; Am Physiol Soc; Soc Neurosci; Aerospace Med Asn. Res: Somatosensory systems; visceral reflexes; autonomic nervous system; psychophysiology; aerospace physiology. Mailing Add: Life Sci NASA Hq Washington DC 20546

WINTER, DONALD F, b Buffalo, NY, Oct 6, 31; m 57; c 3. APPLIED MATHEMATICS. Educ: Amherst Col, BA, 54; Harvard Univ, MA, 59, PhD(appl physics), 62. Prof Exp: Mathematician, Air Force Cambridge Res Labs, 54-56; engr, Missile Systs Lab, Sylvania Elec Prod, Inc, 56-58, eng specialist, Appl Res Lab, 58-62, sr eng specialist, 62-63; mem staff, Geo-astrophys Lab, Boeing Sci Res Labs, 63-70; assoc prof, Ctr Quantitative Sci & Dept Oceanog, 70-74, PROF OCEANOG, UNIV WASH, 74- Concurrent Pos: Vis lectr, Univ Manchester, 66-67. Res: Applied analysis; methods of mathematical physics with applications to solar system astronomy; hydrodynamical and biological processes in oceanography; growth and transport processes in biological systems. Mailing Add: Dept of Oceanog Univ of Wash Seattle WA 98195

WINTER, EDWARD H, b Poughkeepsie, NY, Aug 21, 23; c 4. ANTHROPOLOGY, ETHNOLOGY. Educ: Harvard Univ, BA, 44; Harvard Univ, MA, 49, PhD(social anthrop), 53. Prof Exp: Asst prof anthrop, Univ Ill, 55-57, assoc prof, 57-59; chmn dept sociol & anthrop, 59-65, PROF ANTHROP, UNIV VA, 59- Concurrent Pos: Sr fel anthrop, EAfrica Inst Social Res, 53-55; NSF fel field res in EAfrica, 65-66. Mem: Am Anthrop Asn; Brit Asn Social Anthrop. Res: Social structure; African societies. Mailing Add: Dept Sociol & Anthrop Univ of Va Charlottesville VA 22903

WINTER, FRANK COUNSEL, b Charlotte, NC, June 16, 22; m 43; c 3. MEDICINE. Educ: Stanford Univ, BA, 43, MD, 46; Am Bd Ophthal, dipl, 52. Prof Exp: Asst surg, Stanford Univ, 51; assoc ophthal, Johns Hopkins Univ, 52; assoc prof ophthal & chief dept, Sch Med, Univ NC, 53-55; asst clin prof surg, 55-61, from actg chief to chief div ophthal, 59-68, ASSOC CLIN PROF SURG, SCH MED, STANFORD UNIV, 61-, DIR EYE PATH LAB, 55- Concurrent Pos: Am Acad Ophthal fel, Armed Forces Inst Path, 49-50; NIH fel, Wilmer Inst, 52-53. Mem: AAAS; AMA; Asn Res Vision & Ophthal; Am Acad Ophthal & Otolaryngol. Res: Ophthalmic pathology; diabetes. Mailing Add: Dept of Surg Stanford Univ Palo Alto CA 94304

WINTER, HARRY CLARK, b New Britain, Conn, Feb 26, 41. BIOCHEMISTRY. Educ: Pa State Univ, BS, 62; Univ Wis, MS, 64, PhD(biochem), 67. Prof Exp: NSF fel cell physiol, Univ Calif, Berkeley, 67-68; asst prof biochem, Pa State Univ, 68-75; LECTR BIOL CHEM, UNIV MICH, ANN ARBOR, 75- Mem: Am Chem Soc. Res: Biological nitrogen fixation; photosynthesis; biosynthetic pathways of plants and bacteria; enzyme mechanisms. Mailing Add: Dept of Biol Chem Univ of Mich Ann Arbor MI 48104

WINTER, HENRY FRANK, JR, b Wooster, Ohio, Dec 25, 36. PHYSIOLOGY. Educ: Case Inst Technol, BSc, 58; Baylor Univ, MSc, 62, PhD(physiol, biochem, anat), 65. Prof Exp: Asst prof, 65-72, ASSOC PROF PHYSIOL, SCH DENT MED, WASH UNIV, 73- Mem: AAAS; Am Physiol Soc; Int Asn Dent Res. Res: Oral biology, neurophysiology; instrumentation for medical research; growth and development. Mailing Add: Dept of Physiol Wash Univ Sch Dent Med St Louis MO 63110

WINTER, IRWIN CLINTON, b Clinton, Okla, July 17, 10; m 38; c 4. PHARMACOLOGY. Educ: Allegheny Col, BS, 31; Northwestern Univ, MS, 33, PhD, 34; Univ Tenn, MD, 41. Prof Exp: Asst physiol chem, Med Sch, Northwestern Univ, 31-34, fel, 34-35; physiologist, Res Dept, Parke, Davis & Co, Mich, 35-36; instr physiol & pharmacol, Col Med, Baylor Univ, 36-39; assoc prof pharmacol, Sch Med, Univ Okla, 39-42; dir clin res, 46-75, med dir, 59-75, vpres med affairs, 62-75, CONSULT, G D SEARLE & CO, 75- Concurrent Pos: Mem coun arteriosclerosis, Am Heart Asn. Mem: Soc Exp Biol & Med; Am Soc Pharmacol & Exp Therapeut; AMA; Am Rheumatism Asn; Am Fedn Clin Res. Res: Liver damage and fat metabolism; physiology and pharmacology of micturition; autonomic pharmacology. Mailing Add: G D Searle & Co PO Box 5110 Chicago IL 60680

WINTER, JEANETTE, b New York, NY, Dec 19, 17. MICROBIOLOGY. Educ: Brooklyn Col, BA, 37; NY Univ, PhD(microbiol), 60. Prof Exp: Instr, 64-68, asst prof, 68-71, ASSOC PROF MICROBIOL, MED SCH, NY UNIV, 71- Mem: AAAS; Am Soc Microbiol. Res: Mechanism of competence for DNA uptake in bacterial transformation; role of nucleases in DNA integration during bacterial transformation of streptococci. Mailing Add: Dept of Microbiol NYU Med Sch 550 First Ave New York NY 10016

WINTER, JEREMY STEPHEN DRUMMOND, b Duncan, BC, Dec 11, 37; m 61; c 2. PEDIATRICS, ENDOCRINE PHYSIOLOGY. Educ: Univ BC, MD, 61; Am Bd Pediat, dipl, 67; FRCP(C), 68. Prof Exp: Intern & resident, Montreal Gen, Montreal Children's & Royal Victoria Hosp, 61-64; instr pediat, Univ Pa, 64-67; asst prof, 67-71, ASSOC PROF PEDIAT, UNIV MAN, 71-; ENDOCRINOLOGIST, HEALTH SCI CTR, WINNIPEG, 71- Concurrent Pos: NIH fel endocrinol, Children's Hosp Philadelphia, 64-67; Med Res Coun grant, Univ Man, 67-; consult, St Boniface Hosp, 67-; scientist, Queen Elizabeth II Res Found, 72. Mem: Endocrine Soc; Soc Pediat Res; Can Soc Clin Invest; Am Fedn Clin Res; Can Pediat Soc. Res: Physiology of the pituitary-gonadal axis during fetal life, childhood and puberty. Mailing Add: Health Sci Ctr 685 Bannatyne Winnipeg MB Can

WINTER, JOSEPH WOLFGANG, b Vienna, Austria, Oct 27, 15; nat US; m 41; c 2. MICROBIOLOGY. Educ: Univ Calif, BA, 47; Stanford Univ, MA, 49, PhD(bact), 50. Prof Exp: Asst, Stanford Univ, 48-50; bacteriologist, Virus Lab, State Dept Pub Health, Calif, 50-51 & Vet Admin Hosp, Oakland, Calif, 51-53; fel, New York Pub Health Res Inst, 54-56; MICROBIOLOGIST, BETH ISRAEL MED CTR, 56- Concurrent Pos: Lectr, Hunter Col, 57-65; asst prof, Mt Sinai Sch Med, 68- Mem: Am Soc Microbiol; Am Pub Health Asn; NY Acad Sci. Res: Clinical microbiology; virology; enteric bacteriology. Mailing Add: Beth Israel Med Ctr 10 Nathan Dr Perlman Pl New York NY 10003

WINTER, PETER MICHAEL, b Sverdlovsk, Russia, Aug 5, 34; US citizen; m 64; c 2. ANESTHESIOLOGY. Educ: Cornell Univ, AB, 58; Univ Rochester, MD, 62; Am Bd Anesthesiol, dipl, 72. Prof Exp: USPHS res fel, Harvard Univ, 65-66; res assoc physiol, State Univ NY Buffalo, 66-67, asst res prof anesthesiol, 67-69; assoc prof, 69-74, PROF ANESTHESIOL, SCH MED, UNIV WASH, 74- Concurrent Pos: Consult, Virginia Mason Res Ctr, Seattle, 69-; Nat Heart & Lung Inst grant, Sch Med, Univ Wash, 71-74, res career develop award, 72-77. Mem: Am Col Chest Physicians; AMA; Am Soc Anesthesiol; NY Acad Sci. Res: Respiration therapy; critical care medicine; hyperbaric physiology; oxygen toxicity. Mailing Add: Dept of Anesthesiol Univ of Wash Sch of Med Seattle WA 98195

WINTER, PHILLIP E, b Milwaukee, Wis, May 28, 35; m 56; c 3. PREVENTIVE MEDICINE, EPIDEMIOLOGY. Educ: Carroll Col, BA, 56; Wash Univ, MD, 60; Univ Calif, MPH, 64. Prof Exp: Rotating intern, Presby-St Lukes Hosp, Chicago, 60-61; fel infectious dis, Barnes Hosp Med Ctr, Univ Wash, 61-62; Med Corps, US Army, 62-, chief prev med, Army Hosp, Ft Leonard Wood, Mo, 62-63, resident, Walter Reed Army Inst Res, 64-65, chief dept epidemiol, SEATO Med Res Lab, 65-69, chief prev med div, Off Surgeon Gen, 69-72, dir US Army med component, SEATO Med Res Lab, 72-75, CHIEF RES PLANNING OFF, US ARMY MED RES & DEVELOP COMMAND, MED CORPS, US ARMY, 75- Res: Epidemiology of dengue; dengue hemorrhagic fever; malaria; melioidoses; plague; military preventive medicine; public health. Mailing Add: US Army Med Res Develop Command Washington DC 20314

WINTER, ROBERT JOHN, b Toledo, Ohio, Oct 13, 45; m 72; c 1. PEDIATRIC ENDOCRINOLOGY. Educ: Amherst Col, BA, 67; Northwestern Univ, MD, 71. Prof Exp: Intern pediat, Hartford Hosp, Conn, 71-72; resident pediat, Boston City Hosp, 72-73; fel pediat endocrinol, Johns Hopkins Univ, 73-75; ASST PROF PEDIAT ENDOCRINOL, CHILDREN'S MEM HOSP & NORTHWESTERN UNIV, CHICAGO, 75- Mem: Am Diabetes Asn. Res: Disorders of growth and of glucose homeostasis; primarily clinical research. Mailing Add: Children's Mem Hosp 2300 Children's Plaza Chicago IL 60614

WINTER, ROLAND ARTHUR EDWIN, b Reval, Estonia, Aug 29, 35; US citizen; m 59; c 3. ORGANIC CHEMISTRY, POLYMER CHEMISTRY. Educ: Stuttgart Tech Univ, Cand chem, 57; Harvard Univ, AM, 61, PhD(org chem), 65. Prof Exp: Res chemist, J R Geigy AG, Basel, Switz, 65-66, res assoc, Geigy Chem Corp, NY, 66-69, group leader, 69-70, Ciba-Geigy Corp, 70-72, SR STAFF SCIENTIST, CIBA-GEIGY CORP, 72- Mem: Am Chem Soc; Ger Chem Soc. Res: Synthetic organic chemistry; heterocyclic chemistry; high temperature polymers and plastic additives. Mailing Add: 23 Banksville Rd Armonk NY 10504

WINTER, ROLF GERHARD, b Düsseldorf, Ger, June 30, 28; nat US; m 51; c 3. NUCLEAR PHYSICS. Educ: Carnegie Inst Technol, BS, 48, MS, 51, DSc, 52. Prof Exp: Asst physics, Carnegie Inst Technol, 46-51; instr, Western Reserve Univ, 51-52, asst prof, 52-54; from asst prof to assoc prof, Pa State Univ, 54-64; chmn dept, 66-72, PROF PHYSICS, COL WILLIAM & MARY, 64- Concurrent Pos: Vis physicist & lectr, Carnegie Inst Technol, 55-56 & Oxford Univ, 61-62; indust & acad consult. Mem: Fel Am Phys Soc. Res: Beta decay; quantum theory; nuclear reactions; history of physics. Mailing Add: Dept of Physics Col of William & Mary Williamsburg VA 23185

WINTER, RUDOLPH ERNST KARL, b Vienna, Austria, Nov 27, 35; US citizen; m 64; c 3. ORGANIC CHEMISTRY. Educ: Columbia Univ, AB, 57; Johns Hopkins Univ, MA, 59, PhD(org chem), 64. Prof Exp: NIH fel chem, Karlsruhe Tech Univ, 62-63 & Harvard Univ, 63-64; asst prof org chem, Polytech Inst Brooklyn, 64-69; ASSOC PROF ORG CHEM, UNIV MO-ST LOUIS, 69- Concurrent Pos: Vis res prof, Swiss Fed Univ, Zurich, 75-76. Mem: Am Chem Soc; The Chem Soc. Res: Chemistry of naturally occurring substances, especially terpenes and sesquiterpenes; photochemical and thermal reactions; small ring compounds; four membered rings. Mailing Add: Dept of Chem Univ of Mo St Louis MO 63121

WINTER, STEPHEN SAMUEL, b Vienna, Austria, Feb 27, 26; US citizen; m 51; c 3. SCIENCE EDUCATION. Educ: Albright Col, BS, 48; Columbia Univ, PhD(phys chem), 53. Prof Exp: Res chemist, Atlas Powder Co, 52-53; asst prof chem, Northeastern Univ, 53-58; asst prof chem & educ, Univ Minn, 58-61; assoc prof educ, State Univ NY Buffalo, 61-66, prof, 66-71, dir teacher educ, 68-71; PROF EDUC & CHMN DEPT, TUFTS UNIV, 71- Concurrent Pos: NSF fac fel, Harvard Univ, 57-58, consult, Proj Physics, 64-70; consult & hon assoc prof, Nat Univ Paraguay, 65; consult, Div Sci Teaching, UNESCO, 69-71. Mem: AAAS; Am Chem Soc; Nat Sci

Teachers Asn; Nat Asn Res Sci Teaching. Res: Measurements of outcomes of science instruction; effectiveness of multi-modal teaching. Mailing Add: Lincoln Filene Ctr Tufts Univ Medford MA 02155

WINTER, STEVEN RAY, b Belvidere, Ill, Jan 16, 44; m 70. AGRONOMY, PLANT PHYSIOLOGY. Educ: Univ Ill, BS, 66, MS, 68; Purdue Univ, PhD(agron), 71. Prof Exp: ASST PROF CROP PROD, TEX A&M UNIV, 71- Mem: Am Soc Agron; Am Soc Sugar Beet Technol; Weed Sci Soc Am. Res: Production and physiology of sugar beets on the Texas high plains. Mailing Add: Tex Agr Exp Sta Tex A&M Univ Bushland TX 79012

WINTER, THOMAS C, JR, b East Grand Rapids, Mich, June 19, 34; m 56; c 2. SPACE PHYSICS, OPTICS. Educ: US Mil Acad, BS, 56; Univ Calif, Los Angeles, MA, 61, PhD(planetary & space physics), 66. Prof Exp: Officer, US Army Corps Engrs, 56-70; mem staff, Coun Environ Qual, Exec Off President, 70-72, MIL STAFF ASST TO DIR, DEFENSE RES & ENG, OFF SECY DEFENSE, 73- Concurrent Pos: Exec secy, High Energy Laser Review Group, Dept Defense Adv Comt, 73- Honors & Awards: Skylab Achievement Award, NASA, 74, Solar Sci Exp Team Group Achievement Award, 74. Mem: Sigma Xi. Res: Hydrogen geocorona; extreme ultraviolet solar spectrum. Mailing Add: 5941 Thomas Dr Springfield VA 22150

WINTER, THOMAS GREELEY, b Los Angeles, Calif, Apr 21, 27; m 61; c 4. PHYSICS. Educ: Stanford Univ, BS, 49; Cath Univ, MS, 61, PhD(physics), 63. Prof Exp: Appl engr, Westinghouse Elec Corp, 49-52; self employed, 52-56; proj engr, US Dept Defense, 56-59; from asst prof to assoc prof physics, Okla State Univ, 63-72; KISTLER PROF, UNIV TULSA, 72- Mem: AAAS; Acoust Soc Am; Am Asn Physics Teachers; Am Phys Soc. Res: Acoustic propagation in gases and soils. Mailing Add: Dept of Physics Univ of Tulsa Tulsa OK 74104

WINTER, WILLIAM KENNETH, b Manitowoc, Wis, Apr 26, 26; m 63; c 2. PHYSICS. Educ: Univ Wis, BA, 50; Kans State Col, MS, 52, PhD(physics), 56. Prof Exp: RES PHYSICIST, PHILLIPS PETROL CO, 56- Mem: AAAS; Soc Petrol Eng. Res: Molecular spectroscopy; fluid flow through porous media; reservoir simulation. Mailing Add: Res Ctr Phillips Petrol Co Bartlesville OK 74003

WINTER, WILLIAM PHILLIPS, b Uniontown, Pa, Aug 17, 38; m 60; c 2. BIOCHEMISTRY. Educ: Pa State Univ, University Park, BS, 60, MS, 62, PhD(biochem), 65. Prof Exp: Instr biochem, Pa State Univ, University Park, 63-65; res assoc, Univ Wash, 65-67, actg asst prof, 67-69; res assoc, 69-73, assr res scientist, 73-75, ASSOC RES SCIENTIST HUMAN GENETICS, MED SCH, UNIV MICH, ANN ARBOR, 75- Concurrent Pos: NIH fel, Univ Wash, 65-66, Am Cancer Soc fel, 66-67; investr, Howard Hughes Med Res Inst, 67-69. Mem: AAAS; Am Chem Soc; Am Soc Hemat. Res: Structure and function of human blood proteins; structural abnormalities in proteins in inherited and congenital disease. Mailing Add: Dept of Human Genetics Univ of Mich Med Sch Ann Arbor MI 48104

WINTERBERG, FRIEDWARDT, b Berlin, Ger, June 12, 29. THEORETICAL PHYSICS. Educ: Univ Frankfurt, MS, 53; Univ Göttingen, PhD(nuclear physics), 55. Prof Exp: Group leader theoret physics, Res Reactor, Hamburg, Ger, 55-59; asst prof plasma physics & relativity, Case Univ, 59-63; assoc prof physics, 63-68, PROF PHYSICS, UNIV NEV, RENO, 68- Mem: Am Phys Soc; hon mem Hermann Oberth Soc. Res: Neutron physics; plasma physics; magnetohydrodynamics; controlled fusion; intense relativistic electron beams; thermonuclear microexplosions; nuclear rocket propulsion; general relativity; atmospheric physics; energy research. Mailing Add: Desert Res Inst Univ of Nev Syst Reno NV 89507

WINTERBOTTOM, ROBERT, organic chemistry, see 12th edition

WINTERCORN, ELEANOR STIEGLER, b Morristown, NJ, Jan 15, 35; m 58. AUDIOLOGY, SPEECH PATHOLOGY. Educ: Rockford Col, BA, 56; Univ Wis, MS, 58; Univ Md, PhD, 69. Prof Exp: Clin instr speech path & phonetics, Rockford Col, 56-57; speech & hearing therapist, El Paso Cerebral Palsy Treatment Ctr, 58-59; audiologist, 60-66, supvr clin audiol, 66-70, ASST CHIEF AUDIOL & SPEECH PATH SERV, VET ADMIN HOSP, DC, 71- Concurrent Pos: Mem res comt hearing aid eval processes, Am Speech & Hearing Asn, 66-67; Vet Admin rep comt hearing, bioacoust & biomech, Nat Res Coun-Nat Acad Sci, 68-71; res assoc & Vet Admin rep comt hearing, Bioacoust Lab, Univ Md, 68-72; res asst prof, Univ Md, 73-; dir, Vet Admin Nat Hearing Aid Testing Prog, 75- Res: Hearing aids; speech intelligibility. Mailing Add: Vet Admin Hosp 50 Irving St NW Washington DC 20422

WINTERMOYER, JOHN PAUL, b Hedgesville, WVa, Nov 16, 13; m 45; c 4. APPLIED CHEMISTRY. Educ: Univ Md, BS, 38, MS, 40, PhD(soil chem), 42. Prof Exp: Res asst soil chem, Univ Md, 38-42; assoc anal chemist, Eastern Exp Sta, US Bur Mines, 42-46; res chemist, Naval Ord Lab, 46-52, ord engr, Res & Develop Div, Ammunition & Explosive Br, 52-60, supvry tech engr, Missile Ord Div, 60-66, SUPVY RES ENGR, ARMAMENT DIV, NAVAL AIR SYSTS COMMAND, US DEPT NAVY, WASHINGTON, DC, 66- Mem: Am Chem Soc; Am Ord Asn. Res: Pyrotechnic and chemiluminescent chemistry; chemical-biological warfare; cartridge actuated devices. Mailing Add: 5616 Ruatan St Berwyn Heights College Park MD 20740

WINTERMYER, ROBERT LEE, physical organic chemistry, see 12th edition

WINTERNHEIMER, P LOUIS, b Evansville, Ind, Feb 9, 31; m 51; c 2. BOTANY. Educ: Purdue Univ, West Lafayette, BS, 53; Univ Iowa, MS, 55; Ind Univ, Bloomington, PhD(bot), 71. Prof Exp: ASSOC PROF BIOL, UNIV EVANSVILLE, 57- Mem: Am Bot Soc. Res: Biosystematic studies of Oenothera biennis and other species. Mailing Add: Univ of Evansville PO Box 329 Evansville IN 47702

WINTERNITZ, THOMAS W, b Baltimore, Md, Nov 14, 16; m 41; c 3. ELECTROMAGNETICS, COMMUNICATIONS ENGINEERING. Educ: Univ Chicago, BS, 38; Harvard Univ, MS, 40, PhD(electromagnetics theory), 48. Prof Exp: Test engr, Western Elec Co, Ill, 40-42; engr, Bell Tel Labs, NJ, 42-45; fel electronics, Harvard Univ, 45-48; mem staff, supvr & dept head, Bell Tel Labs, 48-67, dir, Kwajalein Radar Field Sta, 67-69, DIR, SAFEGUARD RADAR LAB, BELL LABS, 69- Mem: Inst Elec & Electronics Eng. Res: Military systems and electronics; radar and digital computers. Mailing Add: Safeguard Radar Lab Bell Labs Whippany Rd Whippany NJ 07981

WINTERNITZ, WILLIAM WELCH, b New Haven, Conn, June 21, 20; m 49; c 3. MEDICINE. Educ: Dartmouth Col, AB, 42; Johns Hopkins Univ, MD, 45. Prof Exp: From instr to asst prof med & physiol, Yale Univ, 52-59; assoc prof med, 59-66, PROF MED, COL MED, UNIV KY, 66- Mem: AAAS; Endocrine Soc; AMA; Am Fedn Clin Res; Am Diabetes Asn. Res: Endocrine regulation of metabolism. Mailing Add: Rte 4 Nicholasville KY 40356

WINTERRINGER, GLEN SPELMAN, plant taxonomy, deceased

WINTERS, EARL D, b Rio Grande, Ohio, Aug 28, 37; m 60; c 1. PHYSICAL CHEMISTRY. Educ: Ohio Wesleyan Univ, BA, 59; Mass Inst Technol, PhD(phys chem), 65. Prof Exp: MEM TECH STAFF, BELL LABS, INC, 65- Mem: Am Chem Soc; Electrochem Soc; Electroplaters Soc. Res: Electrodeposition, etching and corrosion of metals. Mailing Add: R D 4 Box 312 Quakertown PA 18951

WINTERS, EDWARD PHILLIP, b Chicago, Ill, Sept 17, 26; m 56; c 2. PHARMACY. Educ: Univ Ill, BS, 51, MS, 55; Univ Fla, PhD(pharm), 59. Prof Exp: Asst pharm, Univ Ill, 51-55, instr, 55-56; instr, Univ Fla, 56-59; asst prof, Ferris Inst, 59-60; mem res staff, Pharmaceut Develop, 60-63, mgr int new prod develop, 63-67, dir int prod planning, 67-73, AREA DIR PAC & FAR EAST, ABBOTT LABS, 73- Mem: Am Pharmaceut Asn. Res: Pharmaceutical chemistry; international pharmaceutical research and development. Mailing Add: Abbott Labs 14th & Sheridan Rd North Chicago IL 60064

WINTERS, HAROLD ABRAHAM, b Chicago, Ill, Aug 22, 30; m 55. GEOGRAPHY, GEOLOGY. Educ: Northern Ill Univ, BA, 55; Northwestern Univ, MS, 57, PhD(geog), 60. Prof Exp: Asst prof earth sci, Northern Ill Univ, 59-64; assoc prof geog, Portland State Col, 64-65; assoc prof, 65-71, PROF GEOG, MICH STATE UNIV, 71-, CHMN, COMN COL GEOG, 72- Mem: Asn Am Geog; Am Geog Soc; Geol Soc Am. Res: Geomorphology; glaciation. Mailing Add: Dept of Geog Mich State Univ East Lansing MI 48823

WINTERS, HAROLD F, physics, see 12th edition

WINTERS, HARVEY, b Paterson, NJ, Aug 23, 42; m 65; c 1. MICROBIOLOGY, BIOCHEMISTRY. Educ: Fairleigh Dickinson Univ, BS, 64, MS, 66; Columbia Univ, PhD(chem biol), 71. Prof Exp: From instr to asst prof, 69-75, ASSOC PROF BIOL, FAIRLEIGH DICKINSON UNIV, 75- Honors & Awards: Roon Award, Soc Paint Technol, 73. Mem: Am Soc Microbiol; Soc Indust Microbiol; Sigma Xi. Res: Microbiology of aqueous coatings. Mailing Add: Dept of Biol Fairleigh Dickinson Univ Teaneck NJ 07666

WINTERS, JAMES CLINTON, chemistry, see 12th edition

WINTERS, JOHN CALVIN, physical chemistry, see 12th edition

WINTERS, LAWRENCE JOSEPH, b Chicago, Ill, June 11, 30; m 61; c 3. ORGANIC CHEMISTRY. Educ: Wash Univ, AB, 53; Univ Kans, PhD(chem), 59. Prof Exp: Asst chem, Univ Kans, 56-58; fel, Fla State Univ, 59-61; from asst prof to prof, Drexel Univ, 61-71, actg chmn dept, 68-69, asst dean grad sch, 69-71; PROF CHEM & CHMN DEPT, VA COMMONWEALTH UNIV, 72- Mem: Am Chem Soc. Res: Bipyridine chemistry; organic reaction mechanisms; structure-activity relationships; aliphatic nitro-compounds. Mailing Add: Dept of Chem Va Commonwealth Univ Richmond VA 23220

WINTERS, LOREN MEL, b Wichita, Kans, Aug 20, 48. ATOMIC PHYSICS. Educ: Kans State Univ, MSc, 72, PhD(physics), 74. Prof Exp: Lectr physics & math, US AID, 74-75; VIS ASST PROF PHYSICS, ECAROLINA UNIV, 75- Mem: Am Phys Soc; Am Asn Physics Teachers. Res: Investigations of atomic innershell ionization phenomena in ion-atom collisions; in particular, measurement of x-ray production cross sections and x-ray spectroscopy. Mailing Add: Dept of Physics ECarolina Univ Greenville NC 27834

WINTERS, MARY ANN, b Paterson, NJ, Nov 14, 37. BIOCHEMISTRY. Educ: Seton Hill Col, BA, 67; Univ Pittsburgh, PhD(biochem), 72. Prof Exp: Teacher elem & high schs, Pa & Ariz, 56-66; from instr to asst prof, 67-76, ASSOC PROF CHEM, SETON HILL COL, 76- Mem: Am Chem Soc. Res: Purification of nucleic acid synthesizing enzymes and the isolation and identification of necleic acids. Mailing Add: Seton Hill Col Greensburg PA 15601

WINTERS, RAY WYATT, b Takoma Park, Md, Feb 17, 42; m 67. PHYSIOLOGICAL PSYCHOLOGY. Educ: Mich State Univ, BS, 64, MA, 66, PhD(psychol), 69. Prof Exp: Asst prof, 69-74, ASSOC PROF PSYCHOL, UNIV MIAMI, 74- Concurrent Pos: NIH & NSF instnl grants, 69-, NIH grant, 72- Mem: Optical Soc Am. Res: Human psychophysical research in conjunction with animal neurophysiology, especially sensory systems and vision. Mailing Add: Dept of Psychol Univ of Miami Coral Gables FL 33124

WINTERS, ROBERT WAYNE, b Evansville, Ind, May 23, 26; m 48; c 2. PEDIATRICS, PHYSIOLOGY. Educ: Ind Univ, AB, 48; Yale Univ, MD, 52. Prof Exp: Intern pediat, Univ Calif, 52-53; from asst to chief resident, Univ NC, 54-56; res fel med, Univ NC, 56-58; asst prof physiol, Univ Pa, 58-61; assoc prof pediat, 61-62, PROF PEDIAT, COL PHYSICIANS & SURGEONS, COLUMBIA UNIV, 62- Concurrent Pos: Res fel, Univ Calif, 52-53; res fel biochem, Univ Pa, 58. Honors & Awards: E Mead Johnson Prize, 66; Borden Award, 74. Mem: Soc Pediat Res; Am Soc Clin Invest; Am Pediat Soc; Am Acad Pediat; Am Physiol Soc. Res: Renal and acid base physiology; metabolism of water and electrolytes; intravenous nutrition. Mailing Add: Babies Hosp Pediat Dept 3975 Broadway New York NY 10032

WINTERS, ROGER, b Stilson, Tex, Dec 16, 06; m 30. PHYSICS. Educ: Univ Kans, AB, 28; Univ Mo, AM, 40. Prof Exp: Instr chem & physics, Hannibal-LaGrange Col, 35-37; instr math & physics, William Jewell Col, 40-42; prof physics, 42-74, EMER PROF PHYSICS, WESTMINSTER COL, MO, 74- Res: Electronics; electricity and magnetism; chemistry; algebra. Mailing Add: Dept of Physics Westminster Col Fulton MO 65251

WINTERS, RONALD HOWARD, b Los Angeles, Calif, Apr 13, 42; m 69. PHARMACOLOGY. Educ: San Fernando Valley State Col, BA, 63; Ore State Univ, PhD(pharmacol), 69. Prof Exp: Biochemist, Riker Labs, Inc, 64-65; asst to dean undergrad studies, 72-74, from instr to asst prof, 68-73, ASSOC PROF PHARMACOL, SCH PHARM, ORE STATE UNIV, 73-, ASST DEAN, 74- Concurrent Pos: Res grants, Ore Heart Asn, 70-72 & Ore Educ Coord Coun, 70-72. Res: Cardiovascular pharmacology; anesthesia. Mailing Add: Dept of Pharmacol Ore State Univ Sch of Pharm Corvallis OR 97331

WINTERS, RONALD ROSS, b Marion, Va, June 4, 41; m 60; c 2. NUCLEAR PHYSICS, ASTROPHYSICS. Educ: King Col, AB, 63; Va Polytech Inst & State Univ, PhD(physics), 67. Prof Exp: ASSOC PROF PHYSICS, DENISON UNIV, 66- Concurrent Pos: Consult, Oak Ridge Nat Lab, 72-74; dir, Sci Semester Prog, Great Lakes Col Asn, 75- Mem: Sigma Xi; Asn Advan Physics Teaching. Res: Measurement of neutron capture cross sections; s-process nucleosynthesis; origin of the earth-moon system. Mailing Add: Dept of Physics Denison Univ Granville OH 43023

WINTERS, STEPHEN SAMUEL, b New York, NY, June 29, 20; m 43; c 2. GEOLOGY. Educ: Rutgers Univ, BA, 42; Columbia Univ, MA, 48, PhD(geol), 55. Prof Exp: Instr geol, Rutgers Univ, 48-49; from asst prof to assoc prof, 49-66, PROF GEOL, FLA STATE UNIV, 66-, DIR DIV BASIC STUDIES, 64-, DIR HONORS

PROF, 67- Mem: Geol Soc Am; Paleont Soc; Soc Econ Paleont & Mineral; Am Asn Petrol Geol. Res: Stratigraphy and invertebrate paleontology of late Paleozoic; sedimentation of Tertiary and terrace materials. Mailing Add: Dept of Geol 105 Dodd Hall Fla State Univ Tallahassee FL 32306

WINTERS, WALLACE DUDLEY, b New York, NY, June 20, 29; m 53; c 4. NEUROPHARMACOLOGY, CLINICAL PHARMACOLOGY. Educ: George Washington Univ, AB, 50; Univ Mich, Ann Arbor, MA, 52; Univ Wis-Madison, PhD(pharmacol), 54; Med Col Wis, MD, 58. Prof Exp: Asst pharmacol, Univ Mich, Ann Arbor, 51-52 & Univ Wis-Madison, 52-54; instr, Med Col Wis, 54-58; intern, Milwaukee Hosp, Wis, 58-59; Ment Health trainee neuropharmacol, Univ Calif, Los Angeles, 59-61, res pharmacologist, 61-63, assoc prof pharmacol, Sch Med, 63-68, prof pharmacol & psychiat, 68-71; PROF PHARMACOL, PSYCHIAT, FAMILY PRACT & EMERGENCY MED SERV, SCH MED, UNIV CALIF, DAVIS, 71- Concurrent Pos: Ment Health Prog rep pharmacol, Univ Calif, Los Angeles, 61-71, mem, Brain Res Inst & chmn ment health training prog, Educ Comt, 65-71, mem brain res adv comt, 70-71; mem, Preclin Psychopharmacol Res Rev Comt, 65-69. Honors & Awards: A E Bennet Award, Soc Biol Psychiat, 66. Mem: AAAS; Am Soc Pharmacol & Exp Therapeut; AMA. Res: Neuropharmacological action of central nervous system acting drugs; models of psychosis; scheme of anesthetic, excitant, hallucinogen and convulsant drug action. Mailing Add: Dept of Pharmacol Univ of Calif Sch of Med Davis CA 95616

WINTERS, WENDEL DELOS, b Herrin, Ill. VIROLOGY, IMMUNOLOGY. Educ: Univ Ill, Urbana, BS, 62, MS, 66, PhD(med microbiol), 68. Prof Exp: Res asst med microbiol, Univ Ill Col Med, 63-65, teaching asst, 65-68; vis scientist, Nat Inst Med Res, London, 68-71; asst prof surg & med microbiol, Sch Med, Univ Calif, Los Angeles, 71-75; ASSOC PROF MICROBIOL, SCH MED, UNIV TEX HEALTH SCI CTR SAN ANTONIO, 76- Concurrent Pos: Biochemist, Chicago Bd Health, 63-68; microbiologist, Presby St Lukes Hosp, Chicago, 65-68; Med Res Coun Eng grant, Nat Inst Med Res, London, 68-69; Damon Runyon Mem Fund Cancer Res fel, 70-71; consult, Vet Admin Hosp, Sepulveda, Calif, 71. Mem: Am Asn Immunologists; Am Asn Cancer Res; Am Soc Microbiol; Tissue Cult Asn; Brit Soc Gen Microbiol. Res: Mechanisms of virus assembly; immunochemistry of human cancers. Mailing Add: Dept of Microbiol Sch Med Univ Tex Health Sci Ctr San Antonio TX 78284

WINTERSCHEID, LOREN COVART, b Manhattan, Kans, Oct 5, 25; m 48; c 6. THORACIC SURGERY, CARDIOVASCULAR SURGERY. Educ: Willamette Univ, BA, 48; Univ Pa, PhD(microbiol), 62. Prof Exp: Asst surg, 57-58, from instr to assoc prof, 58-72, PROF SURG, SCH MED, DIR UNIV HOSP & ASST DEAN CLIN AFFAIRS, UNIV WASH, 72- Concurrent Pos: Resident surgeon, Affil Hosps, Univ Wash, 55-62; NIH fel, 57-60; mem bd trustees, Willamette Univ, 60-; attend surgeon, Univ & King County Hosps, Seattle, 63. Mem: AMA. Res: General surgery. Mailing Add: Dept of Surg RF-25 Univ Hosp Seattle WA 98195

WINTHROP, JOEL ALBERT, b Elizabeth, NJ, Oct 30, 42. APPLIED MATHEMATICS. Educ: Univ Calif, BA, 64, MA, 70, PhD(math), 71. Prof Exp: ASST PROF MATH, UNIV MO, 71- Mem: Am Math Soc; Math Asn Am; Soc Indust & Appl Math; Sigma Xi. Res: Digital signal processing. Mailing Add: Dept of Math Univ of Mo Columbia MO 65201

WINTHROP, STANLEY OSCAR, b Cowansville, Que, June 22, 27; m 56; c 3. ORGANIC CHEMISTRY. Educ: McGill Univ, BEng, 48; Ga Inst Technol, MS, 49; Univ Tex, PhD(org chem), 51. Prof Exp: Res chemist, Sterling-Winthrop Res Inst, 52-54; head med chem, Ayerst Res Labs, 54-64; dir res & develop, Lever Bros, Can, 64-69; sci adv, Off Sci & Technol, Can Dept Indust, Trade & Com, 69-71; DIR GEN, AIR POLLUTION CONTROL DIRECTORATE, CAN DEPT ENVIRON, 71- Mem: Fel Can Inst Chem; Can Res Mgt Asn. Res: Pharmaceuticals; nitrogen heterocycles; fats and oils; detergents; research administration. Mailing Add: Air Pollution Control Directorate Can Dept of the Environ Ottawa ON Can

WINTNER, CLAUDE EDWARD, b Princeton, NJ, Apr 8, 38; m 67; c 2. ORGANIC CHEMISTRY. Educ: Princeton Univ, AB, 59; Harvard Univ, MA, 60, PhD(chem), 63. Prof Exp: From instr to asst prof, Yale Univ, 63-68; asst prof, Swarthmore Col, 68-69; ASSOC PROF CHEM, HAVERFORD COL, 69- Mem: AAAS; Am Chem Soc; The Chem Soc. Res: Chemistry of the azo and azoxy groups; free radicals. Mailing Add: Dept of Chem Haverford Col Haverford PA 19041

WINTON, CHARLES NEWTON, b Raleigh, NC, Sept 22, 43; m 66; c 1. MATHEMATICS. Educ: NC State Univ, BS, 65; Univ NC, Chapel Hill, MA & PhD(math), 69. Prof Exp: Asst prof math, Univ SC, 69-74; ASSOC PROF MATH SCI, UNIV N FLA, 74- Mem: Am Math Soc; Math Asn Am. Res: Ring structure theory, including quotient objects and various torsion theories. Mailing Add: Dept of Math Sci Univ of NFla Jacksonville FL 32216

WINTON, LAWSON LOWELL, b Ontario, Calif, July 29, 29; m 54; c 2. FOREST PHYSIOLOGY, TISSUE CULTURE. Educ: Univ Calif, Berkeley, BS, 57; Univ Minn, MS, 63, PhD(forestry, bot), 64. Prof Exp: Asst forestry & plant physiol, Sch Forestry, Univ Calif, Berkeley, 54-59; asst forestry & bot, Univ Minn, 59-64; res fel & asst prof tree cytol & tissue culture, 64-69, RES ASSOC & ASSOC PROF FOREST BIOL, INST PAPER CHEM, 69- Mem: Int Soc Plant Morphol; Int Asn Plant Tissue Cult. Res: Cytology of polyploid seedlings of white and black spruce; cytology and tissue culture of polyploid aspen trees; clonal propagation of conifer tree species from cell and callus cultures. Mailing Add: Div of Natural Mat & Systs Inst of Paper Chem Appleton WI 54911

WINTON, LOWELL SHERIDAN, b Townville, Pa, Oct 23, 09; m 35; c 4. MATHEMATICS. Educ: Grove City Col, BS, 31; Oberlin Col, AM, 32; Duke Univ, PhD(math), 37. Prof Exp: Asst math, Duke Univ, 32-35; instr, Wofford Col, 35; PROF MATH, NC STATE UNIV, 35- Mem: Am Math Soc; Am Soc Eng Educ; Math Asn Am. Res: Linear operators; analysis; electrical apparatus; power generation; magnetism. Mailing Add: Dept of Math NC State Univ Raleigh NC 27607

WINTON, MARIANNE YVONNE, anthropology, see 12th edition

WINTON, RAYMOND SHERIDAN, b Raleigh, NC, Jan 4, 40; m 73. MOLECULAR SPECTROSCOPY. Educ: NC State Univ, BS, 62; Duke Univ, PhD(physics), 72. Prof Exp: Physicist electro-optics, US Army Electronics Command, 65-67; ASST PROF PHYSICS, COL, 72- Mem: Sigma Xi; Am Phys Soc; Math Asn Am. Res: High precision measurements in microwave molecular spectroscopy and saturation effects in molecular absorption spectra. Mailing Add: Dept of Physics Miss Col Clinton MS 39058

WINTROBE, MAXWELL MYER, b Halifax, NS, Oct 27, 01; nat US; m 27; c 2. INTERNAL MEDICINE. Educ: Univ Man, BA, 21, MD, 26, BSc, 27; Tulane Univ, PhD(internal med), 29; Am Bd Internal Med, dipl, 37. Hon Degrees: DSc, Univ Man, 58, Univ Utah, 67 & Med Col Wis, 74. Prof Exp: Intern, King George Hosp, Winnipeg, 25 & Winnipeg Gen Hosp, 25-26; Bell fel, Univ Man, 26-27; instr med, Sch Med, Tulane Univ, 27-30; instr, Sch Med, Johns Hopkins Univ, 30-35, assoc, 35-43; prof internal med, 43-70, head dept, 43-67, dir lab study hereditary & metab disorders, 45-72, dir, Cardiovasc Res & Training Inst, 69-74, DISTINGUISHED PROF INTERNAL MED, SCH MED, UNIV UTAH, 70- Concurrent Pos: Asst vis physician, Charity Hosp, New Orleans, La, 27-30; asst physician, Johns Hopkins Hosp, 30-39, assoc physician, 39-43, physician in charge clin nutrit, gastrointestinal & hemopoietic disorders, 41-43; physician in chief, Salt Lake Gen Hosp, Utah, 43-65 & Med Ctr, Univ Utah, 65-67; chief consult, Vet Admin Hosp, 46-; mem consult, AEC, 48-; spec consult to Surgeon Gen, US Army, 49, mem adv comt metab, 60, nutrit anemias adv drug reactions, WHO, spec consult nutritional anemias, 59-; mem comt revision, US Pharmacopoeia, 50-60, mem panel hemat, 61-65; mem comt res life sci, Nat Acad Sci, 66-; mem coun arthritis & metab dis & coun allergy & infectious dis, NIH, 67-70. Vis prof, Univ Colo, 55, Tufts Univ & New Eng Med Ctr, 56, Univ NC, 61, Vanderbilt Univ & Ohio State Univ, 64, Univ Okla, 65, Emory Univ & Univ Ala, 69, Univ Rochester, 70, Univ Wash, 71, Southwestern Med Col & George Washington Univ, 72, Univ Fla, 73, Wake Forest Univ, Univ Ottawa & Univ PR, 74; Univ Toronto, McGill Univ, Univ Calif, Los Angeles & Harvard Univ, 75. Fulbright lectr, India, 56; Lambie Dew oration, Univ Sydney, 58; Pfizer lectr, Australia & NZ, 58; York lectr, Univ BC, 60; Thayer lectr & vis prof, Johns Hopkins Univ & Hosp, 66; Rosenthal lectr, Mt Sinai Hosp, New York, 67; Lilly lectr, Royal Col Physicians, London, 68; Falk lectr & vis prof, St John's Mercy Hosp, St Louis, Mo, 70; lectr & vis prof, Queen's Med Ctr, Hawaii & lectr, Palm Springs Acad Med, Calif, 71; Gifford-Hill lectr & vis prof, Univ Tex Southwest Med Sch Dallas; Canfield lectr, Univ Mich & vis distinguished prof, Med Ctr, George Washington Univ, 72. Honors & Awards: Gold-Headed Cane, Univ Calif, 58; Modern Med Award Distinguished Achievement, 58; Francis G Blake Mem Award, Asn Am Physicians, 65; Phillips Mem Award, Am Col Physicians, 67; Mayo Soley Award, Western Soc Clin Res, 70; Robert H Williams Award, Asn Prof Med, 73; Kober Medal, Asn Am Physicians, 74. Mem: Nat Acad Sci; Asn Am Physicians (vpres, 63-64, pres, 64-65); master Am Col Physicians; Asn Prof Med (pres, 65-66); Am Soc Hemat (vpres, 69, pres, 71-72). Res: Hematology; clinical and experimental nutrition, especially nutritional deficiencies in swine; leukemia and related neoplastic diseases. Mailing Add: Univ of Utah Med Ctr 50 N Medical Dr Salt Lake City UT 84132

WINTTER, JOHN ERNEST, b Birmingham, Ala, Oct 31, 24; m 44; c 2. MEDICINAL CHEMISTRY. Educ: Howard Col, BS, 49; Univ Fla, MS, 50, PhD(pharmaceut chem), 52. Prof Exp: Asst pharm, Univ Fla, 50-52; assoc prof, 52-58, PROF PHARM, SCH PHARM, SAMFORD UNIV, 58-, DEAN, 72- Mem: Acad Pharmaceut Sci; Am Pharmaceut Asn. Res: Amino acids in sponge; alginic acid derivatives; sapote gum suspending properties; esters of amino acids; phthalimidoacetic acid esters; antifungal esters of thiomalic acid. Mailing Add: Sch of Pharm Samford Univ 800 Lakeshore Dr Birmingham AL 35209

WINZELER, ROBERT LEE, b Canton, Ohio, Feb 25, 40; m 61; c 2. CULTURAL ANTHROPOLOGY, ETHNOGRAPHY. Educ: Kent State Univ, BA, 63; Univ Chicago, AM, 66, PhD(anthrop), 70. Prof Exp: Instr anthrop, Northern Ill Univ, 68-69; lectr, 69-70, chmn dept, 75 & 77, ASST PROF ANTHROP, UNIV NEV, RENO, 70- Concurrent Pos: NIH res grant, 75. Mem: Am Anthrop Asn; Asn Asian Studies. Res: Religion, politics and change in Southeast Asia; ethnic relations in Southeast Asia; ethnography and ethnology of Southeast Asia. Mailing Add: Dept of Anthrop Univ of Nev Reno NV 89507

WINZENREAD, MARVIN RUSSELL, b Indianapolis, Ind, Nov 22, 37; m 60; c 2. MATHEMATICS, EDUCATION. Educ: Purdue Univ, BS, 60; Univ Notre Dame, MS, 64; Ind Univ, Bloomington, EdD(math educ), 69. Prof Exp: Teacher high sch, Ind, 60-63; from instr to asst prof math, Northwest Mo State Col, 64-67; lectr, Ind Univ Indianapolis, 69; asst prof, 69-73, ASSOC PROF MATH, CALIF STATE UNIV, HAYWARD, 73- Res: Mathematics in the inner city school. Mailing Add: Dept of Math Calif State Univ 25800 Hillary St Hayward CA 94542

WIORKOWSKI, JOHN JAMES, b Chicago, Ill, Sept 30, 43; m 66; c 1. STATISTICS. Educ: Univ Chicago, BS, 65, MS, 66, PhD(statist), 72. Prof Exp: Asst prof statist, Grad Prog Health Care Admin, US Army-Baylor Univ, 68-71; res assoc, Univ Chicago, 71-73; asst prof, Pa State Univ, 73-75; ASSOC PROF STATIST, UNIV TEX, DALLAS, 75- Concurrent Pos: Assoc dir, Statist Consult & Coop Res Ctr, Pa State Univ, 73-74, dir, 74-75; consult, Fed Energy Admin, 75. Mem: Sigma Xi; Am Statist Asn; Biomet Soc; Inst Math Statist. Res: Interest in applied statistics, specifically biostatistics, linear models, time series analysis, genetic statistics. Mailing Add: Grad Prog in Math Sci Univ of Tex Dallas PO Box 688 Richardson TX 75080

WIPF, FRANCIS LOUISE, b Iola, Wis, Aug 18, 06. CYTOLOGY. Educ: Wis State Teachers Col, Oshkosh, BE, 30; Univ Wis, PhM, 33, PhD(bot, agr bact), 39. Prof Exp: Asst biol, Wis State Teachers Col, Oshkosh, 30-32; asst vet sci, 39-43, asst genetics, 43-47, instr, 47-49, instr vet sci, 49-55, asst prof, 55-72, EMER ASST PROF VET SCI, UNIV WIS-MADISON, 72- Mem: AAAS; Genetics Soc Am; Conf Res Workers Animal Dis. Res: Root nodules; fur farm animals; cytogenetics; histopathology; microtechnique. Mailing Add: 1214 Vilas Ave Madison WI 53715

WIPKE, WILL TODD, b St Charles, Mo, Dec 16, 40; c 2. CHEMISTRY. Educ: Univ Mo-Columbia, BS, 62; Univ Calif, Berkeley, PhD(chem), 65. Prof Exp: Res fel chem, Harvard Univ, 67-69; asst prof, Princeton Univ, 69-75; ASSOC PROF CHEM, UNIV CALIF, SANTA CRUZ, 75- Concurrent Pos: NIH res fel, Harvard Univ, 68-69; NIH spec res resource grant, Princeton Univ, 70-75 & Merck prof develop grant, 70-75; consult, Merck, Sharp & Dohme, 70-; mem bd adv, Chem Abstr Serv, 70-73; dir, NATO Advan Study Inst Comput Rep & Manipulation Chem Info, 73; mem, Nat Res Coun Comt Nat Res Comput Chem, 74- Mem: Am Chem Soc; The Chem Soc; Asn Comput Mach. Res: Organic synthesis; computer assisted design of organic syntheses. Mailing Add: Dept of Chem Univ of Calif Santa Cruz CA 95064

WIRKA, HERMAN W, b Madison, Wis, July 24, 03; m 29; c 2. ORTHOPEDIC SURGERY. Educ: Univ Wis, BA, 29, MD, 30. Prof Exp: Assoc prof, 50-54, PROF ORTHOP SURG, UNIV HOSPS, SCH MED, UNIV WIS-MADISON, 54-, CHMN DEPT, 62- Concurrent Pos: Consult, St Mary's & Madison Gen Hosps, 40- & Vet Admin Hosp. Mem: Am Acad Orthop Surg; Clin Orthop Soc; Am Col Surg; Int Soc Orthop Surg & Traumatol. Mailing Add: Dept of Surg Univ of Wis Sch of Med Madison WI 53706

WIRKKALA, ROBERT, organic chemistry, see 12th edition

WIRSZUP, IZAAK, b Wilno, Poland, Jan 5, 15; US citizen; m 49; c 1. MATHEMATICS. Educ: Univ Wilno, Mag Philos, 39; Univ Chicago, PhD(math), 55. Prof Exp: Lectr math, Tech Inst, Wilno, Poland, 39-41; dir bur studies & spec statist, Cent Soc Purchase; dir Soc Anonyme des Monoprix, France, 46-49; from instr to assoc prof math, 49-65, PROF MATH, UNIV CHICAGO, 65- Concurrent Pos: Dir Surv East Europ Math Lit, proj Univ Chicago, under NSF grant, 56-; consult sch math study group, Yale Univ & Stanford Univ, 60, 61 & 66; Ford Found consult, Univ Math Progs, Colombia, SAm, 65 & 66; mem, US Comn Math Instr, 69-73; adv math, Encyclopaedia Britannica, 71- Honors & Awards: Quantrell Award, Univ

Chicago, 58. Mem: Am Math Soc; Math Asn Am. Res: Mathematical analysis; international mathematics education. Mailing Add: Dept of Math Univ of Chicago Chicago IL 60637

WIRTH, HENRY EDGAR, b Bellingham, Wash, June 4, 08; m 31; c 3. CHEMISTRY. Educ: Univ Wash, BS, 29, MS, 30, PhD(chem), 34. Prof Exp: From actg asst prof to assoc prof quant anal, NDak Col, 34-39; from asst prof to assoc prof chem, Ohio State Univ, 39-50, vchmn dept, 46-49; chmn dept, 50-65, prof, 50-72, EMER PROF CHEM, SYRACUSE UNIV, 72- Mem: Am Chem Soc. Res: Chemical oceanography; partial molal volumes of electrolytes in aqueous solution; colorimetry; physical properties of soap; low temperature research. Mailing Add: Dept of Chem Syracuse Univ Syracuse NY 13210

WIRTH, JOHN CHRISTIAN, b New York, NY; m 45; c 2. BIOCHEMISTRY, ORGANIC CHEMISTRY. Educ: Fordham Univ, PhD(org chem), 43. Prof Exp: Instr chem, Fordham Univ, 37-43, res chemist, Off Sci Res & Develop Contract, 43-44; chemist, S B Penick & Co, NY, 44-47; from asst prof to assoc prof, 47-57, PROF CHEM, ST JOHN'S UNIV, NY, 57- Mem: AAAS; Am Chem Soc; Brit Biochem Soc. Res: Chemistry of metabolic products of microorganisms. Mailing Add: Dept of Chem St John's Univ Jamaica NY 11439

WIRTH, JOSEPH GLENN, b Onawa, Iowa, Nov 19, 34; m 53; c 1. ORGANIC CHEMISTRY. Educ: Univ Wash, BS, 59; Univ Mich, Ann Arbor, MS & PhD(chem), 65. Prof Exp: Chemist, Boeing Co, 58-62; res chemist, 65-71, MGR RES & DEVELOP PROD DEVELOP, SILICONE PROD DEPT, RES & DEVELOP CTR, GEN ELEC CO, 71- Res: Organic synthesis, nitrogen heterocycles; polymer synthesis; fluorescence, organosilicon chemistry. Mailing Add: Silicone Prods Dept Gen Elec Co Waterford NY 12188

WIRTH, THOMAS HENRY, inorganic chemistry, see 12th edition

WIRTH, WILLIS WAGNER, b Dunbar, Nebr, Oct 17, 16; m 42; c 2. ENTOMOLOGY. Educ: Iowa State Col, BS, 40; La State Univ, MS, 47; Univ Calif, PhD(entom), 50. Prof Exp: Instr biol, La Polytech Inst, 40-41; ENTOMOLOGIST, SYST ENTOM LAB, USDA, WASHINGTON, DC, 49- Concurrent Pos: Sr asst sanitarian, USPHS, 42-47; Fulbright res scholar, Australia, 56-57; courtesy prof entom, Univ Fla, 65-; vis prof entom, Univ Md, College Park, 74- Mem: Fel Entom Soc Am; Soc Syst Zool. Res: Systematic entomology; taxonomy of Diptera. Mailing Add: 806 Copley Lane Silver Spring MD 20904

WIRTS, CHARLES WILMER, b Pittsburgh, Pa, June 12, 07; m 38; c 2. PHYSIOLOGY. Educ: Lafayette Col, BS, 30; Jefferson Med Col, MD, 34; Am Bd Internal Med, dipl, 46; Am Bd Gastroenterol, dipl, 48. Prof Exp: From asst prof to assoc prof med, 51-59, prof clin med, 59-72, HON PROF MED, JEFFERSON MED COL, 72- Concurrent Pos: Assoc physician, Pa Hosp, 39-66; consult, Walson Gen Hosp, US Dept Army, Ft Dix, 47-; dir postgrad inst, Philadelphia County Med Soc, 57-59; mem exam bd, Am Bd Gastroenterol, 57-60. Mem: Am Soc Gastrointestinal Endoscopy (secy-treas, Gastroscopic Soc, 55, vpres, 58-59, pres, 59-60); Am Gastroenterol Asn; AMA; fel Am Col Physicians; Am Col Gastroenterol (vpres, 54, pres, 57). Res: Gastrointestinal physiology; nutrition; malabsorption; gastrointestinal bleeding. Mailing Add: 2017 Delancey Philadelphia PA 19104

WIRTSCHAFTER, JONATHAN DINE, b Cleveland, Ohio, Apr 9, 35; m 59; c 5. OPHTHALMOLOGY, NEUROLOGY. Educ: Reed Col, BA, 56; Harvard Med Sch, MD, 60; Linfield Col, MS, 63. Prof Exp: Intern, Philadelphia Gen Hosp, 60-61; resident neurol, Good Samaritan Hosp, Portland, Ore, 61-63; resident ophthal, Johns Hopkins Hosp, 63-66; fel neurol, New York Neurol Inst, Columbia-Presby Med Ctr, New York, 66-67; from asst prof to assoc prof ophthal & neurol, 67-72, dir div ophthal, 67-74, PROF OPHTHAL & NEUROL, COL MED, UNIV KY, 72-, CHMN DEPT OPHTHAL, 74- Concurrent Pos: Attend surgeon, Vet Admin Hosp, Lexington, Ky, 67-69, consult surgeon, 69-; vis prof, Hadassah Hosp & Hebrew Univ Jerusalem, 73-74. Mem: Am Acad Neurol; fel Am Acad Ophthal & Otolaryngol; Am Asn Ophthal; fel Am Col Surg; Asn Res Vision & Ophthal. Res: Clinical neuro-ophthalmology; interactive teaching methods in ophthalmology; retinal venous pathophysiology; strabismus diagnosis and management; biology of the pterygium. Mailing Add: Dept of Ophthal Univ of Ky Med Ctr Lexington KY 40506

WIRTZ, GEORGE H, b Kohler, Wis, Apr 29, 31; m 52; c 2. IMMUNOCHEMISTRY, BIOCHEMISTRY. Educ: Univ Wis, BS, 53, MS, 56; George Washington Univ, PhD(biochem), 62. Prof Exp: Biochemist, Walter Reed Army Inst Res, 58-62; fel biol, Johns Hopkins Univ, 62-63; from asst prof to assoc prof, 63-72, PROF BIOL, MED CTR, W VA UNIV, 72- Concurrent Pos: Vis prof, Mainz Univ, 68. Mem: AAAS; Am Asn Immunol. Res: Tumor Immunochemistry. Mailing Add: Dept of Biochem WVa Univ Med Ctr Morgantown WV 26506

WIRTZ, JOHN HAROLD, b Sheboygan, Wis, Nov 13, 23; m 50. NATURAL HISTORY, VERTEBRATE ZOOLOGY. Educ: Loyola Univ, Ill, BS, 52; Univ Wyo, MS, 54; Ore State Univ, PhD, 61. Prof Exp: Asst prof, 57-70, ASSOC PROF NATURAL HIST & GEN BIOL, PORTLAND STATE UNIV, 70- Mem: AAAS; Nat Audubon Soc. Res: Visual behavior, mobility and orientation in sciurid rodents. Mailing Add: Dept of Biol Portland State Univ Portland OR 97207

WIRTZ, WILLIAM OTIS II, b Montclair, NJ, Aug 16, 37; m 72. MAMMALOGY. Educ: Rutgers Univ, BA, 59; Cornell Univ, PhD(ecol, evolutionary biol), 68. Prof Exp: Res cur, Smithsonian Inst, 62-66; ASST PROF ZOOL, POMONA COL, 68- Mem: Am Soc Mammalogists; Ecol Soc Am; Wildlife Soc; Am Soc Zoologists; Am Inst Biol Sci. Res: Mammalian population ecology and behavior, especially limiting factors in population dynamics; avian population ecology; evolution and systematics of mammals. Mailing Add: Dept of Zool Pomona Col Claremont CA 91711

WISBY, WARREN JENSEN, b Denmark, Nov 14, 22; US citizen; m 62; c 2. FISHERY BIOLOGY. Educ: Univ Wis, BA, 48, MA, 50, PhD(zool), 52. Prof Exp: Res assoc zool, Univ Wis, 52-59; assoc prof marine biol, Inst Marine Sci, Univ Miami, 59-65; dir, Nat Fisheries Ctr & Aquarium Dept of Interior, 65-72; adj prof, Inst Marine Sci, 65-72, ASSOC DEAN, SCH MARINE & ATMOSPHERIC SCI, UNIV MIAMI, 72- Mem: Am Fisheries Soc; Am Soc Zool; Animal Behav Soc. Res: Behavior and sensory physiology of marine organisms. Mailing Add: Sch of Marine & Atmospheric Sci Univ of Miami 10 Rickenbacker Causeway Miami FL 33149

WISCHMEIER, WALTER HENRY, b Lincoln, Mo, Jan 18, 11; m 47; c 2. SOIL CONSERVATION. Educ: Univ Mo, BS, 53; Purdue Univ, MS, 57. Prof Exp: Researcher, Soil & Water Conserv Res Div, Agr Res Serv, USDA, 40-61, res invests leader soil erosion, Corn Belt Br, 61-72; assoc prof agr eng, 65-75, PROF AGR ENG, PURDUE UNIV, WEST LAFAYETTE, 75-; TECH ADV WATER EROSION, NORTH CENT REGION, AGR RES SERV, USDA, 72- Honors & Awards: Superior Serv Awards, USDA, 59 & 73. Mem: Am Soc Agron; fel Soil Conserv Soc Am; Am Soc Agr Eng. Res: Soil and water conservation; quantitative relationship of soil erosion to rainfall characteristics, topographic features, management, productivity level and factor interactions; conservation farm planning; runoff and soil-loss prediction equations. Mailing Add: 2009 Indian Trail Dr West Lafayette IN 47906

WISCHNITZER, SAUL, b New York, NY, Apr 10, 30. BIOLOGY, ANATOMY. Educ: Yeshiva Univ, BA, 51; Univ Notre Dame, MS, 54, PhD(biol), 56. Prof Exp: Resident res assoc biol, Argonne Nat Lab, 56-57; from instr to asst prof anat, NY Med Col, 57-64; assoc prof, 64-68, asst to dean, 64-66, PROF BIOL, YESHIVA UNIV, 68-, ASST DEAN, 66- Concurrent Pos: USPHS spec res fel, 60-61, career res develop award, 64-69; adj prof, Hunter Col, 71- Mem: AAAS; Electron Micros Soc Am; Am Asn Anat; Am Soc Cell Biol; Royal Micros Soc. Res: Electron microscopy; fine structure of cells; oocytes; nuclear physiology. Mailing Add: Dept of Biol Yeshiva Univ New York NY 10033

WISCHOW, RUSSELL P, b Sentinel Butte, NDak, Dec 22, 29; c 2. NUCLEAR CHEMISTRY. Educ: NDak State Univ, BS, 51, MS, 52; Vanderbilt Univ, PhD(phys chem), 58. Prof Exp: Group leader, Oak Ridge Nat Lab, 52-61; sr res chemist, Callery Chem Co, 61-63; supvr chem unit, Martin Co, 63-65; asst to vpres, Nuclear Chem Div, Nuclear Fuel Serv, Inc, W R Grace & Co, 65-66; dir res & develop & asst to pres, Nuclear Fuel Serv, Inc, 66-67, asst gen mgr, 67; dir div nuclear mat safeguards, AEC, 67-70; VPRES & TREAS, E R JOHNSON ASSOCS INC, 70- Concurrent Pos: Pres, Nuclear Audit & Test Co; vpres & secy, Fuel Mgt Corp. Mem: Fel Am Inst Chem; Am Chem Soc; Am Nuclear Soc. Res: Nuclear fuel reprocessing; fission product separation; kinetics of solvent extraction; heavy element separations; high energy rocket fuel and oxidizers; radioisotope heat sources and auxiliary power generators. Mailing Add: E R Johnson & Assocs Inc 910 17th St NW Washington DC 20006

WISCOMBE, WARREN JACKMAN, applied mathematics, see 12th edition

WISDOM NORVELL EDWIN, JR, b Oklahoma City, Okla, Apr 20, 37. PHYSICAL CHEMISTRY. Educ: Univ Tex, BS, 58, PhD(phys chem), 62. Prof Exp: Staff mem, Corp Res Labs, Esso Res & hng Co, 62-69; sr res chemist, Proj Norton Electroplating Tech, 69-72, group leader, 72-75, MGR PROD DEVELOP, COATED ABRASIVE DIV, NORTON CO, 75- Honors & Awards: Silver Medal, Am Electroplaters Soc, 72. Mem: Am Chem Soc. Res: Electroplating; electrochemical kinetics; coated abrasive product design. Mailing Add: Coated Abrasive Res & Develop Dept Norton Co Troy NY 12181

WISE, BURTON LOUIS, b New York, Nov 24, 24; m 59. NEUROSURGERY. Educ: Columbia Univ, AB, 44; New York Med Col, MD, 47. Prof Exp: Clin instr neurol surg, 54, from instr to assoc prof, 55-68, vchmn dept, 65-68, ASSOC CLIN PROF NEUROSURG, SCH MED, UNIV CALIF, SAN FRANCISCO, 68-; CHIEF DEPT NEUROSCI, MT ZION HOSP & MT ZION NEUROL INST, 75- Concurrent Pos: Attend neurol surgeon, Ft Miley Vet Admin Hosp, San Francisco, 54-68 & San Francisco Gen Hosp, 58-69; consult neurosurgeon, Laguna Honda Home & Langley Porter Neuropsychiat Inst, 57-68 & Letterman Army Hosp, 58-; chief neurosurg, Mt Zion Hosp & Mt Zion Neurol Inst, 70-74. Mem: Am Asn Neurol Surg; AMA; Am Col Surg; Am Fedn Clin Res. Res: Metabolic responses to central nervous system lesions; brain stem mechanisms in salt and water homeostasis; effects of hypertonic solutions on cerebrospinal fluid pressure; neuroendocrinology; pediatric neurosurgery and hydrocephalus. Mailing Add: Mt Zion Hosp & Neurol Inst 1600 Divisadero St San Francisco CA 94115

WISE, CHARLES DAVIDSON, b Huntington, WVa, June 13, 26; m 47; c 1. INVERTEBRATE ZOOLOGY, LIMNOLOGY. Educ: Univ WVa, AB & MS, 50; Univ NMex, PhD(invert zool), 62. Prof Exp: Asst zool, Marshall Univ, 50-51; teacher high sch, WVa, 51-53; instr biol, Amarillo Col, 55-57; res scientist, Inst Marine Sci, Univ Tex, 58-60; asst biol, Univ NMex, 60-61; from asst prof to assoc prof, 61-72, PROF BIOL, BALL STATE UNIV, 72- Concurrent Pos: Ind State rep, 66-68; Ind State senator, 68-72; mem, Int Comt Recent Ostracoda, 63-73. Mem: Ecol Soc Am; Am Soc Limnol & Oceanog; Am Micros Soc; Nat Audubon Soc; Sigma Xi. Res: Ecology; biological oceanography; marine and freshwater ostracods, especially taxonomy and ecology. Mailing Add: Dept of Biol Ball State Univ Muncie IN 47306

WISE, DAVID HAYNES, b Mineral Wells, Tex, Apr 28, 45; m 67; c 1. POPULATION ECOLOGY. Educ: Swarthmore Col, BA, 67; Univ Mich, MS, 69, PhD(zool), 74. Prof Exp: Instr biol, Albion Col, 69-70; lectr zool, Univ Mich, 70-71; ASST PROF BIOL, UNIV NMEX, 74- Mem: Ecol Soc Am; Am Arachnol Soc; Sigma Xi. Res: Population dynamics and regulation of population density; community interactions, especially competition and predation; evolutionary ecology; interspecific competition and the structure of tenebrionid beetle communities. Mailing Add: Dept of Biol Univ of NMex Albuquerque NM 87131

WISE, DONALD L, b Indianapolis, Ind, May 27, 29; m 52; c 4. CELL PHYSIOLOGY. Educ: Wabash Col, AB, 51; NY Univ, MS, 54, PhD, 58. Prof Exp: Instr natural sci, Univ Chicago, 57-58; from instr to assoc prof biol, 58-66, PROF BIOL, COL WOOSTER, 66-, CHMN DEPT, 72- Concurrent Pos: Staff biologist, Comn Undergrad Educ Biol, 67-68; vis prof, George Washington Univ, 67-68. Mem: AAAS; Soc Protozool; Am Soc Zool; Am Inst Biol Sci. Res: Protozoan and cellular metabolism and physiology. Mailing Add: Dept of Biol Col of Wooster Wooster OH 44691

WISE, DONALD U, b Reading, Pa, Apr 21, 31; m 65; c 2. GEOLOGY. Educ: Franklin & Marshall Col, BS, 53; Calif Inst Technol, MS, 55; Princeton Univ, PhD(geol), 57. Prof Exp: From asst prof to prof geol, Franklin & Marshall Col, 57-68; chief scientist & dep dir, NASA Apollo Lunar Explor Off, 68-69; PROF GEOL, UNIV MASS, AMHERST, 69- Concurrent Pos: Mem planetology subcomt, Space Sci Bd, Nat Acad Sci, 64-68; consult, NASA, 64- & Pa Geol Surv, 65-66. Mem: AAAS; Geol Soc Am; Am Geophys Union. Res: Structural geology; structure and basement features of the middle Rocky Mountains; flow mechanics of rocks; structures of the Appalachian Piedmont; regional fracture analysis; lunar and planetary geology. Mailing Add: Dept of Geol Univ of Mass Amherst MA 01002

WISE, DWAYNE ALLISON, b Lewisburg, Tenn, Feb 5, 45; m 66; c 1. CYTOGENETICS. Educ: David Lipscomb Col, BA, 67; Fla State Univ, MS, 70, PhD(genetics), 72. Prof Exp: Res fel cytogenetics, Health Sci Ctr, Univ Tex, 73-74; instr zool, 74-75, RES ASSOC CELL BIOL, DUKE UNIV, 75- Concurrent Pos: NIH fel, 75-77. Mem: Am Genetics Asn; Am Soc Plant Taxonomists; Sigma Xi. Res: Investigation of the control of chromosome structure during the cell cycle and of chromosome distribution at meiosis and mitosis. Mailing Add: Dept of Zool Duke Univ Durham NC 27706

WISE, EDMUND MERRIMAN, JR, b Jersey City, NJ, Aug 10, 30; m 52; c 2. MICROBIAL BIOCHEMISTRY. Educ: Oberlin Col, BA, 52; Harvard Univ, PhD(biochem), 63. Prof Exp: Jr biologist, Parke, Davis & Co, 54-55; NIH fel, Med Sch, Tufts Univ, 64-66, from instr to asst prof molecular biol & microbiol, 65-73; SR RES MICROBIOLOGIST, BURROUGHS-WELLCOME CO, 73- Concurrent Pos: NIH fel, Med Sch, Harvard Univ, 63-64. Mem: AAAS; Am Chem Soc; Am Soc

Microbiol. Res: Control of enzyme activity; bacterial cell wall synthesis and degradation; design of enzyme inhibitors; microbial cofactor biosynthesis. Mailing Add: Microbiol Dept Wellcome Res Labs Research Triangle Park NC 27709

WISE, EDWARD NELSON, b Athens, Ohio, May 30, 15; m 36; c 2. CHEMISTRY, RESEARCH ADMINISTRATION. Educ: Ohio Univ, BS, 37, MS, 38; Univ Kans, PhD(chem), 53. Prof Exp: Teacher chem & physics, Gallia Acad High Sch, 38-42; qual control chemist, Baker & Adamson Div, Gen Chem Co, 42; supvr, Standards Lab, WVa Ord Works, 42-45; res engr graphic arts, Battelle Mem Inst, 45-47; mem staff anal instrumentation, Los Alamos Sci Lab, 47-50; tech asst, Hercules Powder Co, 51; from asst prof to assoc prof, 52-61, assocxcoordrfrom assoc coordr to coordr res, 64-72, PROF CHEM, UNIV ARIZ, 61- Mem: Am Chem Soc. Res: Electrophoretic deposition of natural and synthetic rubbers; halftone and color separation techniques in the graphic arts; electrostatic image formation and development; xerography; analytical instrumentation; automatic titrimetry and coulometric analysis. Mailing Add: Dept of Chem Univ of Ariz Tucson AZ 85721

WISE, ERNEST GEORGE, b Dunkirk, NY, Sept 6, 20; m 50; c 2. RADIATION BIOLOGY, MICROBIOLOGY. Educ: State Univ NY Col Fredonia, BEd, 42; Columbia Univ, MA, 47; Syracuse Univ, PhD(sci educ), 60. Prof Exp: Instr sci & math, State Univ NY Col New Paltz, 47-48, PROF BIOL, STATE UNIV NY COL OSWEGO, 48- Concurrent Pos: AEC-NSF acad year fel radiation biol, Cornell Univ, 64-65; AEC equip grant, 68. Mem: AAAS; Health Phys Soc; NY Acad Sci; Int Radiation Protection Asn. Res: Science education from elementary grades through the college level. Mailing Add: 90 Sixth Ave Oswego NY 13126

WISE, GARY E, b Yuma, Colo, July 30, 42; m 62; c 1. CELL BIOLOGY. Educ: Univ Denver, BA, 64; Univ Calif, Berkeley, PhD(zool), 68. Prof Exp: NIH fel cell biol, Univ Colo, Boulder, 69-71; asst prof biol struct, 72-75, ASSOC PROF BIOL STRUCT, SCH MED, UNIV MIAMI, 75- Mem: Am Soc Cell Biol; Am Asn Anatomists. Res: Origin and function of cytoplasmic membranes, especially the Golgi apparatus; localization and function of macromolecules involved in nucleocytoplasmic interactions; ultrastructure of amphibian chromatophores. Mailing Add: Dept of Biol Struct Univ of Miami Miami FL 33152

WISE, GENE, b Willard, Ohio, Apr 13, 22; m 64; c 1. ORGANIC CHEMISTRY, PHYSICAL CHEMISTRY. Educ: Capital Univ, BS, 47; Western Reserve Univ, PhD(chem), 50. Prof Exp: From asst prof to assoc prof, 50-60, PROF CHEM, VA MIL INST, 60- Concurrent Pos: Consult, Gen Tire & Rubber Co, 57-63 & Environ Sci Serv Admin, 67-68. Mem: Am Chem Soc. Res: Computer applications in chemistry; chemical kinetics; properties of polymers. Mailing Add: Dept of Chem Va Mil Inst Lexington VA 24450

WISE, GEORGE HERMAN, b Saluda, SC, July 7, 08; m 37; c 4. ANIMAL NUTRITION, ANIMAL PHYSIOLOGY. Educ: Clemson Col, BS, 30; Univ Minn, MS, 32, PhD(dairy husb), 37. Prof Exp: Asst dairy husb, Univ Minn, 33-36; assoc, Clemson Col, 37-44; from assoc prof to prof, Kans State Col, 44-47; assoc prof, Iowa State Col, 47-49; prof animal indust, 49-51, head nutrit sect, 49-66, WILLIAM NEAL REYNOLDS PROF ANIMAL SCI, NC STATE UNIV, 51- Concurrent Pos: Mem comt animal nutrit, Nat Res Coun, 51-53; consult, State Exp Sta Div, Agr Res Serv, USDA, 55-62; study leave, Univ Calif, Davis, 66-67. Honors & Awards: Award, Am Feed Mfrs Asn, 48; Borden Award, 49; Award of Honor, Am Dairy Sci Asn, 66. Mem: Am Soc Animal Sci; Am Dairy Sci Asn (vpres, 63-64, pres, 64-65); Am Inst Nutrit; Soc Nutrit Educ; Nutrit Today Soc. Res: Nutrition and physiology of animals. Mailing Add: 229 Woodburn Rd Raleigh NC 27605

WISE, GEORGE NELMS, medicine, ophthalmology, deceased

WISE, HAROLD B, b Hamilton, Ont, Feb 14, 37. SOCIAL MEDICINE, INTERNAL MEDICINE. Educ: Univ Toronto, MD, 61. Prof Exp: Physician, Prince Albert Clin, Sask, Can, 62-63; resident, Kaiser Found Hosp, San Francisco, Calif, 63-64, Montefiore Hosp & Med Ctr, Bronx, NY, 64-65; actg dir ambulatory serv & home care, Morrisania City Hosp, 65-66; dir health ctr, Dr Martin Luther King, Jr Health Ctr, 66-71; DIR ANAL & DEVELOP HEALTH TEAMS, MONTEFIORE HOSP & MED CTR, 71-; ASSOC PROF COMMUNITY HEALTH, ALBERT EINSTEIN COL MED, 70- Concurrent Pos: Milbank Mem Fund fel; Albert Einstein Col Med, 67-72; dir internship & residency prog social med, Montefiore Hosp & Med Ctr, 69-, dir inst health team develop, 72-; dir, Family Ctr Health. Mem: Nat Inst Med. Res: Research into the family and the healing processes. Mailing Add: Inst for Health Team Develop Montefiore Hosp Bronx NY 10467

WISE, HENRY, b Ciechanow, Poland, Jan 14, 19; nat US; m 43, 60; c 6. PHYSICAL CHEMISTRY. Educ: Univ Chicago, SB, 41, SM, 44, PhD(phys chem), 47. Prof Exp: Res assoc, Univ Chicago, 41-46; dir field lab, NY Univ, 46-47; scientist, Nat Adv Comt Aeronaut, Ohio, 47-49; phys chemist, Calif Inst Technol, 49-55; chmn chem dynamics dept, 55-71, SCI FEL, STANFORD RES INST, 71- Concurrent Pos: Lectr sch eng, Stanford Univ, 60-; vis prof, Israel Inst Technol, 65; mem comt motor vehicle emission, Nat Acad Sci. Mem: AAAS; Am Chem Soc; Am Phys Soc; Catalysis Soc; The Chem Soc. Res: Heterogeneous catalysis; chemical kinetics. Mailing Add: Mat Sci Stanford Res Inst Menlo Park CA 94025

WISE, HUGH EDWARD, JR, b Lafayette, Ind, Oct 12, 30. ORGANIC CHEMISTRY. Educ: Vanderbilt Univ, BA, 52; Univ Fla, PhD(chem), 61. Prof Exp: Proj leader, Tech Serv Lab, Union Carbide Corp, NY, 61-66, proj specialist, 66-68; res chemist, Nalco Chem Co, Ill, 68-70; res chemist, 70-71, FIELD SERV SUPVR, RES & DEVELOP, WASTE TREATMENT DIV, CLOW CORP, 71- Mem: Am chem Soc. Res: Environmental science and engineering; waste treatment technology. Mailing Add: Waste Treatment Div Clow Corp Box 324 Florence KY 41042

WISE, JOHN HICE, b Marysville, Pa, Nov 6, 20; m 43; c 3. PHYSICAL CHEMISTRY. Educ: Haverford Col, BS, 42; Brown Univ, PhD(chem), 47. Prof Exp: Asst chem, Brown Univ, 42-43, res chemist, Off Sci Res & Develop & Manhattan Proj, 43-46; from instr to asst prof chem, Stanford Univ, 47-53; assoc prof, 53-61, dept head, 70, PROF CHEM, WASHINGTON & LEE UNIV, 61- Concurrent Pos: Vis assoc prof, Brown Univ, 59-60; vis prof, Hollins Col, 63-64; teacher, Univ Va, 63 & 65 & Stanford Univ, 66; fac res partic, Argonne Nat Lab, 72-73. Mem: Am Chem Soc; Am Phys Soc. Res: Absorption and emission spectroscopy; infrared spectrometry; atomic and molecular structure. Mailing Add: Dept of Chem Washington & Lee Univ Lexington VA 24450

WISE, JOHN P, b Boston, Mass, Feb 9, 24; m 60; c 3. FISHERIES BIOLOGY, MARINE BIOLOGY. Educ: Suffolk Univ, AB, 50; Univ NH, MS, 53. Prof Exp: Biologist, Woods Hole Lab, Bur Com Fisheries, 53-60; Food & Agr Orgn biologist, Govt Arg, Brazil, Tunisia & Uruguay, 60-64; biologist, Southwest Fisheries Ctr, 65-73, SR RES ANALYST, NAT MARINE FISHERIES SERV, 73- Concurrent Pos: Consult, Food & Agr Orgn, 59 & 75. Mem: AAAS; fel Am Inst Fishery Res Biol. Res: Stocks of marine animals, involving studies of ecology, growth rates, mortality rates, predation and parasitology, directed at eventual exploitation by man for maximum sustained yield. Mailing Add: 2265 Wheystone St Vienna VA 22180

WISE, JOHN THOMAS, b Orangeburg, SC, Nov 29, 26; m 55; c 3. CHEMISTRY, PAPER CHEMISTRY. Educ: The Citadel, BS, 46; Purdue Univ, MS, 49. Prof Exp: Chemist, Thiokol Chem Co, 49-50; chemist, 52-57, sr chemist, 57-65, group leader res & develop, 65-69, mgr tech serv sect, 69-70, mgr lab serv, 70-73, RES COORDR, SONOCO PROD CO, 73- Concurrent Pos: Instr, Univ SC, 60. Mem: Am Chem Soc; Tech Asn Pulp & Paper Indust. Res: Physical and chemical testing; pulp and paper; paper products; utilization of waste products; organic synthesis; instrumentation. Mailing Add: Res Lab Sonoco Prod Co Hartsville SC 29950

WISE, LAWRENCE DAVID, b Canton, Ohio, Oct 13, 40; m 67. ORGANIC CHEMISTRY. Educ: Manchester Col, BA, 62; Ohio State Univ, MS, 64, PhD(org chem), 67. Prof Exp: Res scientist org chem, Goodyear Tire & Rubber Co, 67-69; RES SCIENTIST ORG CHEM, WARNER-LAMBERT RES INST, 69- Mem: Am Chem Soc. Res: Synthetic organic chemistry, particularly heterocycles directed toward drug design. Mailing Add: Warner-Lambert Res Inst 170 Tabor Rd Morris Plains NJ 07950

WISE, LOUIS M, physical chemistry, colloid chemistry, see 12th edition

WISE, LOUIS NEAL, b Slagle, La, Jan 27, 21; m 44; c 2. AGRONOMY. Educ: Northwestern State Col, BS, 42; La State Univ, BS, 46, MS, 47; Purdue Univ, PhD(agron), 50. Prof Exp: Asst, Purdue Univ, 47-50; asst prof, 50-53, from assoc agronomist to agronomist, Exp Sta, 50-66, dir regional res lab, 52-66, dean sch agr, 61-66, PROF AGRON, MISS STATE UNIV, 53-, VPRES AGR & FORESTRY, 66- Mem: Am Soc Agron. Res: Pasture production and management; seed research. Mailing Add: Miss State Univ Col Agr Dept of Agron Mississippi State MS 39762

WISE, MATTHEW NORTON, b Tacoma, Wash, Apr 2, 40; m 65. HISTORY OF PHYSICS, HISTORY OF SCIENCE. Educ: Pac Lutheran Univ, BS, 62; Wash State Univ, PhD(physics), 68. Prof Exp: Asst prof physics, Auburn Univ, 67-69 & Ore State Univ, 69-71; NSF sci fac fel hist of sci, Princeton Univ, 71-72; LECTR HIST, UNIV CALIF, LOS ANGELES, 75- Mem: Am Phys Soc; Am Asn Physics Teachers; Hist Sci Soc. Res: History of nineteenth and twentieth century physical sciences. Mailing Add: Dept of Hist 405 Hilgard Ave Univ of Calif Los Angeles CA 90024

WISE, MILTON BEE, b Newland, NC, July 17, 29; m 51; c 3. ANIMAL NUTRITION. Educ: Berea Col, BS, 51; NC State Col, MS, 53; Cornell Univ, PhD(animal nutrit), 57. Prof Exp: Lab supvr, Berea Col, 47-51; asst, NC State Col, 52-53, res assoc, 53-54; asst, Cornell Univ, 54-55, instr animal husb, 55-57; from asst prof to prof animal sci, NC State Univ, 57-70; PROF ANIMAL SCI & HEAD DEPT, VA POLYTECH INST & STATE UNIV, 70- Mem: Am Soc Animal Sci. Res: Mineral and nutrient metabolism; forage utilization; physiology of digestion. Mailing Add: Dept of Animal Sci VA Polytech Inst & State Univ Blacksburg VA 24061

WISE, PAUL HENRY, b Streetsboro, Ohio, Aug 7, 14; m 59; c 2. ORGANIC CHEMISTRY. Educ: Kent State Univ, BS, 36; Ohio State Univ, PhD(chem), 40. Prof Exp: Fel chem, Ohio State Univ, 40-41; res chemist, Goodyear Tire & Rubber Co, 41-46; res scientist, Nat Adv Comt Aeronaut, 46-57; sr res chemist, 57-58, sect head basic process res, 58-65, MGR CHEM, PROCESS RES, GOODYEAR TIRE & RUBBER CO, 65- Mem: Am Chem Soc. Res: Monomers for rubbers and plastics; aircraft fuels and lubricants; organic chemical process development. Mailing Add: Dept 455B Goodyear Tire & Rubber Co 1444 E Market St Akron OH 44316

WISE, RALEIGH WARREN, b Plainfield, NJ, Sept 30, 28; m 57. RESEARCH ADMINISTRATION, RUBBER CHEMISTRY. Educ: Univ Va, BS, 51. Prof Exp: Anal chemist, 51-53, anal res chemist, 54-56, res group leader, 56-65, res sect mgr, 65-71, group mgr, Instrument & Equip Div, 71-74, dir, 74-75, DIR TECHNOL, RUBBER CHEM DIV, MONSANTO INDUST CHEM CO, 75- Mem: Am Chem Soc; Instrument Soc Am; fel Am Inst Chemists. Res: Instrumentation; chemical and elastomer research. Mailing Add: Monsanto Indust Chem Co 260 Springside Dr Akron OH 44313

WISE, RICHARD MELVIN, b Greentown, Ohio, Sept 27, 24; m 68; c 3. ORGANIC CHEMISTRY. Educ: Mt Union Col, BS, 49; Ohio State Univ, PhD(org chem), 55. Prof Exp: Res chemist, 55-58, sr res chemist, 58-72, RES SCIENTIST, GEN TIRE & RUBBER CO, 72- Mem: Am Chem Soc. Res: Organic research; synthesis of monomers; preparation of polymerization catalysts and polymers; tire cord adhesives. Mailing Add: 2780 Wright Rd Uniontown OH 44685

WISE, ROBERT IRBY, b Barstow, Tex, May 19, 15; m 40; c 3. BACTERIOLOGY, INTERNAL MEDICINE. Educ: Univ Tex, BA, 37, MD, 50; Univ Ill, MS, 38, PhD(bact), 42; Am Bd Internal Med, dipl, 57. Prof Exp: Asst, Div Animal Genetics, Exp Sta, Univ Ill, 38-39, asst instr bact, 39-42; dir, Wichita City-County Pub Health Lab, Tex, 42-43; dir, Houston Pub Health Lab, 43; asst prof bact, Sch Med, Univ Tex, 43-46, dir bact & serol labs, Univ Hosp, 46-50; asst surgeon, USPHS Hosp, New Orleans, La, 50-51; fel med, Univ Minn, 51-53, asst prof, 53-54, asst prof med & bact, 54-55; from asst prof to assoc prof med, 55-59, Magee prof med & head dept, 59-75, EMER MAGEE PROF MED, JEFFERSON MED COL, 75-; ASST CHIEF MED STAFF, VET ADMIN HOSP, 75- Concurrent Pos: Bacteriologist, Univ Hosp, Univ Minn, 57-59; assoc mem comn streptococcal dis, Armed Forces Epidemiol Bd, 58-66; physician-in-chief, Thomas Jefferson Univ Hosp, 59-75; mem bd trustees, Magee Mem Hosp, Philadelphia, 59-75 & Drexel Univ, 66-75; mem, Greater Philadelphia Comt Med-Pharmaceut Sci, 63-75, chmn, 70-74; mem adv comt, Inter-Soc Comt Heart Dis Resources, 67-70; mem bd-adv comt registry of tissue reaction, Univs Assoc for Res & Educ on Path, Inc, 70-; mem exec comt, Int Cong Internal Med, 71. Mem: Am Fedn Clin Res; fel Am Col Physicians; Asn Am Physicians; Am Infectious Dis Soc; Am Pharmaceut Asn. Res: Infectious diseases; chemotherapy; antibiotics. Mailing Add: Vet Admin Hosp Togus ME 04330

WISE, SHERWOOD WILLING, JR, b Jackson, Miss, May 31, 41; m 65. GEOLOGY, PALEONTOLOGY. Educ: Washington & Lee Univ, BS, 63; Univ Ill, MS, 65, PhD(geol), 70. Prof Exp: NSF fel, Swiss Fed Inst Technol, 70-71; asst prof geol, 71-75, ASSOC PROF GEOL, FLA STATE UNIV, 75- Concurrent Pos: NSF res grant, 72-; res grant, Petrol Res Fund, 73-75. Honors & Awards: Outstanding Paper Award, Soc Econ Paleont & Mineral, 71. Mem Am Asn Petrol Geol; Soc Econ Paleont & Mineral; Am Micros Soc; Geol Soc Am; Swiss Geol Soc. Res: Skeletal ultrastructure; taxonomy and biostratigraphy of calcareous nannoplankton; early diagenesis of carbonate and siliceous sediment; circum-Antarctic marine geology. Mailing Add: Dept of Geol Fla State Univ Tallahassee FL 32306

WISE, WILLIAM BERNARD, b Baltimore, Md, May 21, 37; m 65; c 1. PHYSICAL CHEMISTRY. Educ: Loyola Col, Md, BS, 58; Carnegie-Mellon Univ, MS, 61, PhD(chem), 64. Prof Exp: Systs & Commun Sci fel, Carnegie-Mellon Univ, 63-64; dir, Nuclear Magnetic Resonance Specialties, 64-66; sr scientist, Arco Chem Co, 66-72; MEM FAC, DEPT CHEM, UNIV MASS, 72- Mem: Am Chem Soc; Am Crystallog

Asn; Am Inst Physics. Res: Studies in molecular structure using nuclear magnetic resonance and x-ray diffraction. Mailing Add: Dept of Chem Univ of Mass Amherst MA 01002

WISE, WILLIAM CURTIS, b Louisville, Ky, Nov 24, 40; m 63; c 2. PHYSIOLOGY, BIOPHYSICS. Educ: Transylvania Univ, AB, 63; Univ Ky, PhD(physiol & biophys), 67. Prof Exp: Physiologist, McDonnell-Douglas Corp, 67-68; asst prof, 68-73, ASSOC PROF PHYSIOL, MED UNIV SC, 73- Concurrent Pos: Koebig Trust grant physiol, Med Univ SC, 72-73; Nat Cancer Inst res career develop award, 74- Mem: AAAS; Am Physiol Soc; Aerospace Med Asn; Biophys Soc. Res: Membrane transport; neoplastic disease; renal and acid-base physiology. Mailing Add: Dept of Physiol Med Univ of SC Charleston SC 29401

WISE, WILLIAM STEWART, b Carson City, Nev, Aug 18, 33; m 55; c 3. MINERALOGY, PETROLOGY. Educ: Stanford Univ, BS, 55, MS, 58; Johns Hopkins Univ, PhD(geol), 61. Prof Exp: Instr geol, Stanford Univ, 58 & Johns Hopkins Univ, 60-61; from asst prof to assoc prof, 61-73, PROF GEOL, UNIV CALIF, SANTA BARBARA, 73- Concurrent Pos: Consult, US Geol Surv, 65-67. Mem: AAAS; Geol Soc Am; Mineral Soc Am; Mineral Asn Can; Mineral Soc Gt Brit. Res: Paragenesis of minerals, principally zeolites and associated minerals, barium silicates and phosphates. Mailing Add: Dept of Geol Sci Univ of Calif Santa Barbara CA 93106

WISEBLATT, LAZARE, b Montreal, Que, Can, June 13, 20; nat US; m 64; c 1. CHEMISTRY. Educ: McGill Univ, BEng, 48, MSc, 50; Univ Sask, PhD(chem), 52. Prof Exp: Res chemist, Int Milling Co, 52-54; cereal chemist, Am Inst Baking, 54-65; SECT MGR, RES & DEVELOP, QUAKER OATS CO, 66-, INSTRUMENT DEVELOP SPECIALIST, 73- Mem: AAAS; Am Asn Cereal Chem; Inst Food Technol. Res: Cereal chemistry; bakery foods technology; instrumentation in process monitoring quality assurance. Mailing Add: Quaker Oats Co 617 W Main St Barrington IL 60010

WISEMAN, BILLY RAY, b Sudan, Tex, Mar 28, 37; m 63; c 2. ENTOMOLOGY, HORTICULTURE. Educ: Tex Tech Col, BS, 59; Kans State Univ, MS, 61, PhD(entom), 67. Prof Exp: Res asst host plant resistance, Kans State Univ, 59-61 & 64-66; res entomologist, Southern Grains Invests, Okla, 66-67; RES ENTOMOLOGIST, SOUTHERN GRAIN INSECTS LAB, AGR RES SERV, USDA, 67- Concurrent Pos: Mem grad fac, Univ Ga & Univ Fla. Mem: Entom Soc Am; NY Acad Sci. Res: Entomological research in host plant resistance of small grains, corn, sorghum and vegetable crops and the insects attacking these crops, including feeding stimulants, deterrents, food utilization, behavior and biology. Mailing Add: Southern Grain Insects Res Lab Ga Coastal Plain Exp Sta Tifton GA 31794

WISEMAN, EDWARD H, b Portsmouth, Eng, Nov 14, 34; m 57; c 4. BIOCHEMICAL PHARMACOLOGY. Educ: Univ Birmingham, BSc, 56, PhD(org chem), 59. Prof Exp: Fel, Ohio State Univ, 60; res chemist, 61-64, supvr biochem pharmacol, 64-67, asst to res vpres, 67, mgr biochem pharmacol, 67-71, DIR PHARMACOL, PFIZER INC, 71- Mem: Am Rheumatism Asn; Am Soc Pharmacol & Exp Therapeut. Res: Biochemistry of metabolic diseases. Mailing Add: Pfizer Inc Groton CT 06341

WISEMAN, GEORGE EDWARD, b Brooklyn, NY, May 28, 18; m 45; c 3. INORGANIC CHEMISTRY, ORGANIC CHEMISTRY. Educ: St Peter's Col, BS, 40; Polytech Inst Brooklyn, PhD(chem), 56. Prof Exp: Assoc prof chem, St John's Univ, NY, 46-59; chmn dept chem, 59-66, assoc grad dean, Conolly, 66-71, PROF CHEM, LONG ISLAND UNIV, 59- Mem: Am Chem Soc; fel Am Inst Chemists. Res: Preparation, properties and structures of organoselenium compounds; heterocyclic compounds; metallic derivatives of aromatic hydrocarbons. Mailing Add: Dept of Chem Long Island Univ Brooklyn NY 11201

WISEMAN, GORDON G, b Livingston, Wis, Feb 24, 17; m 42; c 2. PHYSICS. Educ: SDak State Univ, BS, 38; Univ Kans, MS, 41, AM, 47, PhD(physics), 50. Prof Exp: Instr physics, Culver-Stockton Col, 41-43; from instr to assoc prof, 43-64, PROF PHYSICS, UNIV KANS, 64- Mem: Am Phys Soc; Am Asn Physics Teachers. Res: Dielectrics; absorption microspectrophotometry; ferroelectricity. Mailing Add: Dept of Physics & Astron Univ of Kans Lawrence KS 66044

WISEMAN, GORDON MARCY, b Winnipeg, Man, Feb 24, 34; m 56. BACTERIOLOGY. Educ: Univ Man, BSc, 56, MSc, 61; Univ Edinburgh, PhD(bact), 63, DSc, 74. Prof Exp: Demonstr, 57-59, asst prof, 65-72, ASSOC PROF BACT, FAC MED, UNIV MAN, 72- Concurrent Pos: Med Res Coun fel bact, Fac Med, Univ Man, 64 & scholar, 65- Mem: Can Soc Microbiol. Res: Neisseria gonorrhoeae. Mailing Add: Dept Med Microbiol Univ of Manitoba Winnipeg MB Can

WISEMAN, JOHN R, b Patriot, Ohio, May 4, 36; m 56; c 3. ORGANIC CHEMISTRY. Educ: Univ Colo, BS, 57; Stanford Univ, PhD(chem), 65. Prof Exp: NSF fel chem, Univ Calif, Berkeley, 64-65, lectr, 65-66, fel, 66; asst prof, 66-70, ASSOC PROF CHEM, UNIV MICH, 70- Concurrent Pos: Grants, Petrol Res Fund, 68-73; Res Corp grant, 68-69; Am Cancer Soc grant, 72-73. Mem: AAAS; Am Chem Soc; The Chem Soc. Res: strain of bicyclic bridgehead alkenes; reaction of carbonium ions; synthesis of natural products. Mailing Add: Dept of Chem Univ Mich Ann Arbor MI 48104

WISEMAN, LAWRENCE LINDEN, b Galion, Ohio, Apr 27, 44; m 64; c 2. DEVELOPMENTAL BIOLOGY. Educ: Hiram Col, AB, 66; Princeton Univ, MA, 69, PhD(biol), 70. Prof Exp: Nat Cancer Inst fel, Princeton Univ, 70-71; ASST PROF BIOL, COL WILLIAM & MARY, 71- Concurrent Pos: Vis sceintist, Human Leukemia Prog, Ont Cancer Inst, Toronto, 74-75. Mem: AAAS; Soc Develop Biol; Am Soc Zool; Am Soc Cell Biol. Res: Cell adhesion, cell movement; vertebrate embryology. Mailing Add: Dept of Biol Col of William & Mary Williamsburg VA 23185

WISEMAN, PARK ALLEN, b Amsden, Ohio, Dec 29, 18; m 42; c 1. ORGANIC CHEMISTRY. Educ: DePauw Univ, AB, 40; Purdue Univ, MA, 42, PhD(org chem), 44. Prof Exp: Asst org & phys chem, Purdue Univ, 40-42, Monsanto Chem Co res fel org chem, 44-46; res chemist, Firestone Tire & Rubber Co, Ohio, 46-47; from asst prof to assoc prof, 47-56, head dept, 65-69, PROF CHEM, BALL STATE UNIV, 56- Concurrent Pos: NSF vis prof, Tech Inst Northwestern Univ, 69-70. Mem: AAAS; Am Chem Soc. Res: High pressure oxidation of hydrocarbons; catalytic vapor-phase oxidation of hydrocarbons; fluorine chemistry; organic synthesis and natural products. Mailing Add: Dept of Chem Ball State Univ Muncie IN 47306

WISEMAN, RALPH FRANKLIN, b Washington, DC, Sept 1, 21; m 51; c 2. MICROBIOLOGY. Educ: Univ Md, BS, 49; Univ Hawaii, MS, 53; Univ Wis, PhD(bact), 56. Prof Exp: Lab asst, Nat Cancer Inst, 39-42, med bacteriologist, Nat Inst Dent Res, 49-51; asst bact, Univ Hawaii, 51-53; asst, Univ Wis, 53-55, instr, 55-56; from instr to assoc prof, 56-66, PROF MICROBIOL, UNIV KY, 66- Concurrent Pos: Vis prof, Hacettepe Univ, Turkey, 68-69 & Rega Inst Med Res, Cath Univ, Louvain, 69. Mem: Am Soc Microbiol; Asn Gnotobiotics; fel Am Acad Microbiol. Res: Intestinal microbiology; germ free-like cheracteristics in antibiotic-treated animals; animal-microbial ecosystems. Mailing Add: Morgan Sch of Biol Sci Univ of Ky Lexington KY 40506

WISEMAN, WILLIAM JOSEPH, JR, b Summit, NJ, June 16, 43; m 65; c 2. OCEANOGRAPHY. Educ: Johns Hopkins Univ, BES, 64, MS, 66, MA, 68, PhD(oceanog), 69. Prof Exp: Instr geol, Univ NH, 69-70, asst prof earth sci, 70-71; ASST PROF MARINE SCI, COASTAL STUDIES INST, LA STATE UNIV, BATON ROUGE, 71- Mem: Am Geophys Union; Inst Elec & Electronics Engrs. Res: Estuarine and nearshore circulation. Mailing Add: Coastal Studies Inst La State Univ Baton Rouge LA 70803

WISER, CYRUS WYMER, b Wartrace, Tenn, Jan 14, 23; m 45; c 3. AQUATIC ECOLOGY, PHYSIOLOGY. Educ: Harding Col, BS, 45; George Peabody Col, MA, 46; Vanderbilt Univ, PhD(biol), 56. Prof Exp: Instr biol, David Lipscomb Col, 46-49; assoc prof, Jacksonville State Teachers Col, 49-51, 53-54; vis asst prof, Vanderbilt Univ, 54-55; assoc prof, 56-61, PROF BIOL, MID TENN STATE UNIV, 61- Mem: Sigma Xi. Res: Physiology, aquatic ecology and limnology; population studies of ponds and lakes; accumulation of radioactive isotopes by aquatic organisms. Mailing Add: Dept of Biol Mid Tenn State Univ Murfreesboro TN 37130

WISER, HORACE CLARE, b Lewiston, Utah, Jan 26, 33; m 53; c 4. MATHEMATICS. Educ: Univ Utah, BA, 53, PhD(math), 61; Univ Wash, BS, 54. Prof Exp: From asst prof to assoc prof, 61-74, PROF MATH, WASH STATE UNIV, 74- Mem: Am Math Soc; Math Asn Am. Res: Undergraduate mathematics curriculum; point set topology. Mailing Add: SE 775 Derby St Pullman WA 99163

WISER, JAMES ELDRED, b Wartrace, Tenn, Dec 31, 15; m 41; c 1. ANALYTICAL CHEMISTRY. Educ: Mid Tenn State Col, BS, 38; Peabody Col, MA, 40, PhD(sci educ), 47. Prof Exp: Teacher high sch, Fla, 38-39 & Ala State Teachers Col, 40-41; teacher physics, Vanderbilt Univ, 42; teacher chem & physics, David Lipscomb Col, 42-46; PROF CHEM & HEAD DEPT CHEM & PHYSICS, MID TENN STATE UNIV, 46- Concurrent Pos: Instr, Peabody Col, 44; NSF panelist, 64, 68 & 71. Mem: AAAS; Am Chem Soc; Am Inst Chem. Res: Food chemistry; educational psychology. Mailing Add: Dept of Chem & Physics Mid Tenn State Univ Murfreesboro TN 37130

WISHART, DAVID JOHN, b Corbridge, Eng, Aug 15, 46; m 71. CULTURAL GEOGRAPHY. Educ: Univ Sheffield, Eng, BA, 67; Univ Nebr-Lincoln, MA, 68, PhD(geog), 71. Prof Exp: Asst prof geog, Univ Ariz, 71-72; asst prof, Beloit Col, 72-74; ASST PROF GEOG, UNIV NEBR-LINCOLN, 74- Mem: Asn Am Geogrs; Survival Int. Res: The frontier process on the Great Plains; the impact of the Euro-American settlement on the Native American; Indian removal and images of the Great Plains. Mailing Add: Dept of Geog Univ of Nebr Lincoln NE 68508

WISHART, FRANKLYN OGILVIE, b Can, Oct 23, 05; m 35; c 1. IMMUNOLOGY. Educ: Univ Toronto, BA, 28, MD, 31, MA, 33, DPH, 42. Prof Exp: From assoc prof to prof hyg & prev med, 44-56, prof microbiol, 56-71, EMER PROF MICROBIOL, UNIV TORONTO, 71- Mem: Hon mem Can Pub Health Asn. Res: Serology of vaccina-variola viruses and typhus fever; immunity in diptheria and tetanus. Mailing Add: 56 Servington Crescent Toronto ON Can

WISHINSKY, HENRY, b New York, NY, Feb 2, 19; m 47; c 4. BIO-ORGANIC CHEMISTRY. Educ: NY Univ, BA, 41, MS, 44; Georgetown Univ, PhD(org biochem), 51. Prof Exp: Res chemist, Wallace & Tiernan, Inc, NJ, 41-46; dir res develop, Universal Synthetics, Inc, NY, 46-49; chem dir endocrinol lab, Med Sch & Hosp, Georgetown Univ, 49-51; dir div biochem, Sinai Hosp Baltimore, 52-64; dir res lab, 64-68, VPRES RES & DEVELOP, AMES CO, MILES LABS, INC, 68- Concurrent Pos: Lectr, Sinai Sch Nursing, Baltimore, 52-64; consult biochem, Lutheran Hosp, Baltimore, 54-64, Franklin Square Hosp, Baltimore, 56-64, James L Kernan Hosp, Baltimore, 62-64 & Guerin & Kime Clin Lab, Md, 61-64. Mem: AAAS; Am Asn Clin Chem; Am Asn Poison Control Ctrs; Asn Clin Sci; Am Chem Soc. Res: Proteins, steroids, instrumentation and laboratory design. Mailing Add: Ames Co Div Miles Labs Inc Elkhart IN 46514

WISHMAN, MARVIN, organic chemistry, see 12th edition

WISHNER, LAWRENCE ARNDT, b New York, NY, Sept 7, 32; m 55; c 2. BIOCHEMISTRY. Educ: Univ Md, BS, 54, MS, 61, PhD(food chem), 64. Prof Exp: Asst dairy dept, Univ Md, 57-61, from asst prof to assoc prof chem, 61-68, chmn dept, 67-71, PROF CHEM, MARY WASHINGTON COL, 68-, ASST DEAN, 71- Concurrent Pos: Mem, Am Conf Acad Deans. Mem: Am Chem Soc; Am Oil Chem Soc; Am Inst Chem; NY Acad Sci; Sigma Xi. Res: Light-induced oxidation of milk; thermal oxidation of fats; autoxidation of tissue lipids in vivo; biological antioxidants. Mailing Add: Off of the Dean Mary Washington Col Fredericksburg VA 22401

WISHNETSKY, THEODORE, b New York, NY, July 5, 25; m 48; c 2. FOOD TECHNOLOGY. Educ: Cornell Univ, BS, 49, MS, 50; Univ Mass, PhD(food technol), 58. Prof Exp: Res assoc, NY Agr Exp Sta, Geneva, 50-54; chemist, Eastman Chem Prod, Inc, 58-62; sr scientist, Air Prod & Chem, Inc, Pa, 62-68; ASSOC PROF FOOD SCI, MICH STATE UNIV, 68- Mem: Inst Food Technol; Soc Cryobiol. Res: Cryogenics in food processing; low-temperature preservation of foods; phase change as related to functional properties of foods and biologicals; controlled atmospheres; packaging; antioxidants; food colorimetry. Mailing Add: Dept Food Sci & Human Nutrit Mich State Univ East Lansing MI 48823

WISHNICK, MARCIA M, b New York, NY, Oct 10, 38; m 60; c 1. BIOCHEMISTRY, PEDIATRICS. Educ: Barnard Col, BA, 60; NY Univ, PhD(biochem), 70, MD, 74. Prof Exp: Chemist, Lederle Labs, Am Cyanamid Co, 60-66; assoc biochem, Pub Health Res Labs, City of New York, 70-71; RES ASSOC PHARMACOL, SCH MED, NY UNIV, 71- Concurrent Pos: Resident pediat, NY Univ-Bellevue Med Ctr, 74-77. Mem: AAAS; Am Chem Soc. Res: Mechanisms of action of carbon dioxide-fixing enzymes; fatty acid biosynthetic and elongation enzymes; drug-acetylation activity in erythrocytes; inborn errors in metabolism. Mailing Add: 151 W 74th St New York NY 10023

WISMAN, EVERETT LEE, b Woodstock, Va, Oct 1, 22; m 48; c 3. POULTRY SCIENCE. Educ: Va Polytech Inst, BS, 46; Cornell Univ, MS, 49; Pa State Col, PhD(biochem, poultry husb), 52. Prof Exp: County agr agent, 47-48, PROF POULTRY SCI, VA POLYTECH INST & STATE UNIV, 52- Honors & Awards: Distinguished Serv Award, Am Sci Affiliation, AAAS, 71. Mem: Am Poultry Sci Asn; Am Inst Nutrit; AAAS. Res: Role of antibiotics, arsenicals and other feed derivatives in chick growth stimulation; evaluation of animal by-products in poultry rations. Mailing Add: Dept of Poultry Sci Va Polytech Inst & State Univ, Blacksburg VA 24061

WISMAR, BETH LOUISE, b Cleveland, Ohio, Feb 18, 29. ANATOMY, MEDICAL EDUCATION. Educ: Western Reserve Univ, BSc, 51, MSc, 57; Ohio State Univ,

PhD(anat), 61. Prof Exp: Instr embryol & histol, Col Med, 61-63, asst prof anat, Col Med & Col Arts & Sci, 63-69, ASSOC PROF ANAT, COL MED & COL ARTS & SCI, OHIO STATE UNIV, 69- Res: Comparative and human histology, especially cardiovascular and urogenital systems; audio-visual methods and programming in science education. Mailing Add: Dept of Anat Ohio State Univ Columbus OH 43210

WISMER, CHESTER AARON, phytopathology, see 12th edition

WISMER, MARCO, b Switz, Dec 26, 21; nat US; m 52; c 2. POLYMER CHEMISTRY. Educ: Swiss Fed Inst Technol, PhD, 48. Prof Exp: Res chemist, Amercoat Corp, 49-51; tech mgr plastics div, Ciba Prod Corp, 51-56; res assoc, Springdale Res Ctr, 56-62, scientist, 62-64, dir advan res dept, 64-74, VPRES RES & DEVELOP, RES CTR, PPG INDUST, INC, 74- Honors & Awards: IR-100, Indust Res, 66 & 74. Mem: Am Chem Soc; Am Soc Test & Mat; Fedn Soc Paint Technol. Res: Synthesis of epoxy resins; polyester resins; polyether polyols; chlorinated compounds; urethane technology; epoxidation thecnology; synthetic organic chemistry; polyolefin chemistry; unsaturated polyesters; radiation technology and electrodeposition. Mailing Add: 1215 Applewood Dr Gilsonia PA 15044

WISMER, ROBERT KINGSLEY, b Atlantic City, NJ, June 18, 45; m 70. X-RAY CRYSTALLOGRAPHY. Educ: Haverford Col, BS, 67; Iowa State Univ, PhD(phys chem), 72. Prof Exp: Instr chem, Iowa State Univ, 72; systs analyst comput sci, Ames Lab, Energy Res & Develop Admin, 73; asst prof chem, Luther Col, Iowa, 73-74; ASST PROF CHEM, DENISON UNIV, 74- Mem: Am Chem Soc; AAAS; Am Crystallog Asn. Res: Solution of the phase problem through deconvolution of the Patterson function, especially development of techniques which can be automated and those adaptable to small computers. Mailing Add: Dept of Chem Denison Univ Granville OH 43023

WISNER, JACKSON WARD, JR, b Baltimore, Md, Aug 7, 25; m 63; c 3. ORGANIC CHEMISTRY. Educ: Univ Vt, BS, 50, MS, 52; Western Reserve Univ, PhD(org chem), 57. Prof Exp: Chemist, 57-59, sr chemist, 59-64, res chemist, 64-72, SR RES CHEMIST, TEXACO RES CTR, 72- Mem: Am Chem Soc; Sigma Xi. Res: Evaluation of additives for lubricants; formulation of lubricants useful in automotive vehicles. Mailing Add: Texaco Res Ctr PO Box 509 Beacon NY 12508

WISNER, ROBERT JOEL, b Hannibal, Mo, Jan 18, 25; m 47; c 4. ALGEBRA. Educ: Univ Ill, BS, 48, MS, 49; Univ Wash, PhD(math), 53. Prof Exp: Assoc math, Univ Wash, 51-53, res mathematician, Pub Opinion Lab, 52-53; instr math, Univ BC, 53-54; from asst prof to assoc prof, Haverford Col, 54-60; assoc prof, Mich State Univ, Oakland, 60-63; assoc prof, 63-70, PROF MATH SCI & HEAD DEPT MATH, NMEX STATE UNIV, 70- Concurrent Pos: Consult, Burroughs Corp, 57-58; NSF fel & mem, Inst Advan Study, 59-60; ed, Rev, Soc Indust & Appl Math, 59- Mem: Am Math Soc; Soc Indust & Appl Math; Math Asn Am; Am Statist Asn; Can Math Cong. Res: Rings; Abelian groups; number theory. Mailing Add: Dept of Math NMex State Univ Las Cruces NM 88001

WISNICKI, BOLESLAW PAUL, structural engineering, see 12th edition

WISNIESKI, BERNADINE JOANN, b Baltimore, Md, Feb 26, 45. PHYSICAL BIOCHEMISTRY. Educ: Univ Md, College Park, BS, 67; Univ Calif, Berkeley, PhD(genetics), 71. Prof Exp: Damon Runyon Mem Fund fel, 71-73, Celeste Durand Rogers Mem Found fel, 73-74, actg asst prof, 74-75, ASST PROF BACT, UNIV CALIF, LOS ANGELES, 75- Concurrent Pos: Mem, Cancer Res Ctr, Univ Calif, Los Angeles, 75-, assoc mem, Molecular Biol Inst, 75- Mem: AAAS. Res: Function and physical structure of animal cell membranes; membrane alterations. Mailing Add: Dept of Bact Univ of Calif Los Angeles CA 90024

WISNIEWSKI, HENRYK MIROSLAW, b Luszkowko, Poland, Feb 27, 31; US citizen; m 54; c 2. NEUROPATHOLOGY, PATHOLOGY. Educ: Med Acad, Danzig, physician dipl, 55; Med Acad, Warsaw, Dr Med, 60, docent, 65. Prof Exp: Resident res fel, Med Acad, Gdansk, Poland, 55-58; from asst to assoc prof neuropath, head of lab & assoc dir, Inst Neuropath, Polish Acad Sci, Warsaw, 58-66; from res assoc to asst prof path, 66-69, assoc prof neuropath, 69-74, PROF NEUROPATH, ALBERT EINSTEIN COL MED, 74- Concurrent Pos: Health Res Coun New York career scientist award, 70-72; Nat Multiple Sclerosis Soc fel, 71-74; NIH fel, 72-77; vis neuropathologist, Univ Toronto, 61-62; vis scientist, lab of Neuropath, Nat Inst Neurol Dis & Blindness, 62-63; consult, Merck Labs, Rahway, NJ, 72-74; dirxdir, Demyelinating Dis Unit, Med Res Coun, Newcastle-upon-Tyne, Eng. Honors & Awards: Weil Award, Am Asn Neuropathologists, 69, Moore Award, 72. Mem: Am Asn Neuropath; Polish Asn Neuropath; Polish Asn Path; Am Asn Geront. Res: Light and ultrastructural studies of the pathological brain; experimental neuropathology; synaptic and axonal pathology; pre and senile dementia; multiple sclerosis and other human and experimental demyelinating diseases. Mailing Add: Dept of Path Albert Einstein Col of Med Bronx NY 10461

WISOTSKY, MAX J, organic chemistry, see 12th edition

WISOTZKY, JOEL, b Chicago, Ill, Feb 17, 23; m 49; c 3. DENTISTRY, DENTAL RESEARCH. Educ: Cent YMCA Col, BS, 45; Loyola Univ, DDS, 47; Univ Rochester, PhD(exp path), 56. Prof Exp: Pvt pract, 48-49; sr asst dent surgeon, Fed Correctional Inst, Tex, 49-51; fel, Univ Rochester, 51-56; res assoc exp path, Dent Med Div, Colgate-Palmolive Co, 56-59; hon assoc res specialist, Bur Biol Res, Rutgers Univ, 57-59; assoc prof res dent med, 59-63, prof med & dir dent res, 63-72, PROF ORAL BIOL, SCH DENT, CASE WESTERN RESERVE UNIV, 63-, DIR GRAD TRAINING & RES, 72- Concurrent Pos: USPHS res fel, 52-56, career develop award, 59-65. Mem: AAAS; Am Soc Exp Path; Int Asn Dent Res. Res: Cariology; aging changes in oral tissues; phosphorescence of oral structures; electro-physiology; theoretical oral biology. Mailing Add: 3407 Blanche Ave Cleveland Heights OH 44118

WISSBRUN, KURT FALKE, b Brackwede, Ger, Mar 19, 30; nat US. POLYMER SCIENCE, RHEOLOGY. Educ: Univ Pa, BS, 52; Yale Univ, MS, 53, PhD(phys chem), 56. Prof Exp: Dreyfus fel, Univ Rochester, 55-57; res chemist, 57-60, group leader, 60-62, res assoc, 62-72, SR RES ASSOC, CELANESE RES CO, SUMMIT, 72- Concurrent Pos: Adj prof chem eng, Univ Del, 74- Mem: Am Chem Soc; Soc Rheol. Mailing Add: 806 Morris Turnpike Apt 3P7 Short Hills NJ 07078

WISSEMAN, CHARLES LOUIS, JR, b Seguin, Tex, Oct 2, 20; m 41; c 4. MEDICAL MICROBIOLOGY. Educ: Southern Methodist Univ, BA, 41; Kans State Col, MS, 43; Southwestern Univ, MD, 46; Am Bd Path, dipl; Am Bd Microbiol, dipl. Prof Exp: Chief chemotherapeut res sect, Dept Virus & Rickettsial Dis, Army Med Serv Grad Sch, Walter Reed Army Med Ctr, DC, 48-54, asst chief dept, 52-54; PROF MICROBIOL & HEAD DEPT, SCH MED, UNIV MD, BALTIMORE CITY, 54-, ASST PROF MED, 57- Concurrent Pos: Instr med, Sch Med, Georgetown Univ & actg dir bact & serol labs, Univ Hosp, 50-54; dep dir comn rickettsial dis, Armed Forces Epidemiol Bd, 57-59, dir, 59-72; consult, Surgeon Gen, US Army, NIH, WHO & Pan-Am Health Orgn. Mem: Am Soc Microbiol; Infectious Dis Soc Am; Am Soc Trop Med & Hyg; Am Soc Clin Invest; Am Asn Immunol. Res: Infectious diseases; viral and rickettsial diseases; pathogenesis and immunity. Mailing Add: Dept of Microbiol Univ of Md Sch of Med Baltimore MD 21201

WISSEMAN, WILLIAM ROWLAND, b Halletsville, Tex, Nov 2, 32; m 59; c 3. PHYSICS. Educ: NC State Univ, BNuclearEng, 54; Duke Univ, PhD(physics), 59. Prof Exp: Res assoc & instr physics, Duke Univ, 59-60; mem tech staff, 60-75, BR MGR, TEX INSTRUMENTS, INC, 75- Mem: Am Phys Soc; Inst Elec & Electronics Engrs. Res: Electromagnetic wave propagation in solids; properties of semiconductors; solid state microwave sources. Mailing Add: 5747 Melshire Dr Dallas TX 75230

WISSING, THOMAS EDWARD, b Milwaukee, Wis, Aug 15, 40; m 69. FRESH WATER ECOLOGY. Educ: Marquette Univ, BS, 62, MS, 64; Univ Wis-Madison, PhD(zool), 69. Prof Exp: Asst aquatic ecol, Marquette Univ, 62-63, asst gen biol, 63-64; asst gen zool, Univ Wis, 64-65, Fed Water Pollution Control Admin trainee aquatic ecol, 65-69; asst prof zool, 69-73, ASSOC PROF ZOOL, MIAMI UNIV, 73- Mem: Am Fisheries Soc; Am Soc Limnol & Oceanog; Int Asn Ecol; Ecol Soc Am; Int Asn Theoret & Appl Limnol. Res: Fisheries biology; bioenergetics. Mailing Add: Dept of Zool Miami Univ Oxford OH 45056

WISSLER, ROBERT WILLIAM, b Richmond, Ind, Mar 1, 17; m 40; c 4. PATHOLOGY. Educ: Earlham Col, AB, 39; Univ Chicago, MS, 43, PhD(path), 46, MD, 48; Am Bd Path, dipl, 51. Hon Degrees: DSc, Earlham Col, 59, Univ Heidelberg, 73. Prof Exp: Asst chem, Earlham Col, 38-39; asst path, 41-43, from instr to prof, 43-72, chmn dept, 57-72, DONALD N PRITZKER PROF PATH, SCH MED, UNIV CHICAGO, 72- Concurrent Pos: Intern, Chicago Marine Hosp, 49-50; mem path study sect, USPHS, 57-61, consult, Surgeon Gen Path Training Comt, 63-68; mem comt path, Nat Res Coun, 58-64; consult, Armed Forces Inst Path, 61-72, chmn sci adv comt, 66-67; chmn comt path, Nat Acad Sci-Nat Res Coun, 63-; secy-treas, Am Asn Chmn Med Sch Dept Path, 63-64, pres, 67-68; chmn coun arteriosclerosis, Am Heart Asn, 65-66; vpres-dir, Univs Asn Res & Educ Path, Inc, 65, pres, 69-71; chmn ad hoc comt animal models, Artificial Heart-Myocardial Infarction Prog, Nat Heart Inst & mem Vet Admin Eval & Rev Comt, Res in Path & Lab Med, 66; vchmn bd trustees, Am Asn Accreditation on Lab Animal Care, 67, chmn, 72-74; trustee, Am Bd Path, 68-, secy, 74; mem path adv coun, Vet Admin, 70-74; mem adv comt, Life Sci Res Off, 71-; mem nat adv food comt, Food & Drug Admin, 72-74. Honors & Awards: Award of Merit, Am Heart Asn, 71. Mem: Soc Exp Biol & Med; Am Soc Exp Path (vpres, 60-61, pres, 61-62); AMA; Am Asn Path & Bact (vpres, 67, pres, 68-69); Am Asn Cancer Res. Res: Protein, lipid nutrition and metabolism; cardiovascular disease; experimental induction and regression of atherosclerosis; cellular immunological reactions including tumor immunity; immunohistochemistry of atherosclerosis; lipoprotein arterial wall cell interaction. Mailing Add: Dept of Path Univ of Chicago Chicago IL 60637

WISSNER, ALLAN, b New York, NY, Nov 14, 45. ORGANIC CHEMISTRY. Educ: Long Island Univ, BS, 67; Univ Pa, PhD(org chem), 71. Prof Exp: NIH fel chem, Cornell Univ, 72-74; RES CHEMIST, LEDERLE LABS, AM CYANAMID CO, 74- Mem: Am Chem Soc. Res: Medicinal chemistry. Mailing Add: Metab Dis Ther Res Sect Lederle Labs Pearl River NY 10965

WISSOW, LENNARD JAY, b Philadelphia, Pa, May 23, 21; m 46; c 1. CHEMISTRY. Educ: Pa State Col, BS, 42; Duke Univ, AM, 43, PhD(org chem), 45. Prof Exp: Asst instr chem & asst org chem res, Duke Univ, 43-45; res org chemist, Publicker, Inc, 45; res & develop chemist, Nat Foam Syst, Inc, Pa, 46-47; sr res & develop chemist, Merck & Co, Inc, 47-51; head develop res, Otto B May, Inc, 51-58; treas, 59-60, CHIEF CHEMIST, J & H BERGE, INC, 58-, PRES, 60- Mem: Am Chem Soc; Am Inst Chem; AAAS; Sigma Xi; NY Acad Sci. Res: Synthetic organic chemistry; fine organic chemicals and processes; pharmaceuticals; vat dyestuffs and intermediates; research and sales administration. Mailing Add: 207 Chestnut Ave N North Plainfield NJ 07063

WIST, ABUND OTTOKAR, b Vienna, Austria, May 23, 26; US citizen; m 63; c 2. THEORETICAL PHYSICS, THERMODYNAMICS. Educ: Graz Tech Univ, BS, 48; Univ Vienna, PhD(thermodyn), 51. Prof Exp: Technician physics, Vienna Tech Univ, 51-52; res & develop engr, Radiowerke Wien, Austria, 52-54 & Siemens & Halske AG, WGer, 54-58; dir res & develop precision lab instruments, Fisher Sci, Inc, 64-69; res assoc, Grad Sch Pub Health, Univ Pittsburgh, 70-72 & Dept Radiol, Sch Med, 72-74; ASST PROF, VA COMMONWEALTH UNIV, 74- Concurrent Pos: Air pollution & med consult. Mem: Sr mem Inst Elec & Electronics Engrs; Am Chem Soc; Calorimetry Conf. Res: Physical and analytical chemistry; electrical engineering; solid state devices; circuitry and reactions; reaction kinetics; catalysis; precision instrumentation in air pollution and medicine. Mailing Add: Va Commonwealth Univ 910 W Franklin St Richmond VA 23220

WISTAR, RICHARD, b NJ, Nov 2, 05; m 33; c 3. CHEMISTRY. Educ: Haverford Col, BS, 28; Harvard Univ, AM, 31, PhD, 39. Prof Exp: Instr chem, Haverford Col, 32-35 & Bennington Col, 35-38; from assoc prof to prof, 39-71, EMER PROF CHEM, MILLS COL, 71- Mem: AAAS; Am Chem Soc. Res: Physical organic chemistry; reaction kinetics and mechanisms in organic chemistry. Mailing Add: Duck Cove Inverness CA 94937

WISTENDAHL, WARREN ARTHUR, b Jersey City, NJ, Mar 12, 20; m 45; c 3. PLANT ECOLOGY. Educ: Rutgers Univ, BSc, 52, PhD(bot), 55. Prof Exp: Asst bot, Rutgers Univ, 52-55; asst prof biol, Carthage Col, 55-56, assoc prof & head dept, 56-57; from asst prof to assoc prof bot, 57-67, chmn dept, 64-69, PROF BOT, OHIO UNIV, 67-; DIR, DYSART WOODS LAB, 67- Mem: AAAS; Ecol Soc Am; Torrey Bot Club; Bot Soc Am; Am Bryol & Lichenological Soc. Res: Analyses of upland and flood plain plant communities; plant succession; ecology of bryophytes. Mailing Add: Dept of Bot Porter Hall Ohio Univ Athens OH 45701

WISTREICH, GEORGE A, b New York, NY, Aug 12, 32; m 57; c 2. MICROBIOLOGY, ELECTRON MICROSCOPY. Educ: Univ Calif, Los Angeles, AB, 57, MS, 61; Univ Southern Calif, PhD(bact), 68. Prof Exp: Res asst zool, Univ Calif, Los Angeles, 58-60, res virologist, 60-61; from instr to asst prof biol, 61-71, ASSOC PROF LIFE SCI, EAST LOS ANGELES COL, 71-, CHMN DEPT, 72-, DIR ALLIED HEALTH SCI PROGS, 68- Concurrent Pos: Aerospace consult, Garrett Corp, Calif, 66-67; lectr, Upward Bound Prog, East Los Angeles Col, 68- Mem: Am Soc Microbiol; Am Inst Biol Sci; NY Acad Sci; fel Am Inst Chem; fel Royal Soc Health. Res: Insect pathology; virology and tissue culture; cytology and cytochemistry; undergraduate education in biological sciences; electron microscopy. Mailing Add: Dept of Life Sci East Los Angeles Col Los Angeles CA 90022

WISTREICH, HUGO ERYK, b Jasto, Poland, Aug 8, 30; nat US; m 58; c 3. FOOD SCIENCE. Educ: Inst Agr Tech, France, Ingenieur, 53; Rutgers Univ, MS, 57, PhD(food tech), 59. Prof Exp: Dir res, Reliable Packing Co, 58-60, Preservaline Mfg Co, 60-63 & Dubuque Packing Co, 63-64; vpres technol, 64-75, PRES, B HELLER & CO, 75- Mem: AAAS; Am Chem Soc; Inst Food Technol; Am Meat Sci Asn; Am Soc Test & Mat. Res: Meat curing; electrical anesthesia in animals; food analysis; nutrition; food-meat biochemistry. Mailing Add: 10127 S Seeley Ave Chicago IL 60643

WISWALL, RICHARD H, JR, b Peabody, Mass, Mar 7, 16; m 46; c 5. PHYSICAL CHEMISTRY. Educ: Harvard Univ, AB, 37; Princeton Univ, PhD(chem), 41. Prof Exp: Chemist, Am Cyanamid Co, NJ, 40-43 & Union Carbide & Carbon Chem Corp, 46-49; CHEMIST, BROOKHAVEN NAT LAB, 49- Mem: Am Chem Soc; fel Am Nuclear Soc. Res: Chemistry of nuclear energy production; fluorine chemistry; fused salts; metal hydrides; energy storage. Mailing Add: 331 Beaver Dam Rd Brookhaven NY 11719

WISWELL, JOHN GORDON, b Halifax, NS, Apr 9, 19; nat US; m 49; c 2. ENDOCRINOLOGY. Educ: Dalhousie Univ, BA, 38, BSc, 40, MD, CM, 43. Prof Exp: Intern, Victoria Gen Hosp, Halifax, 42-43; asst resident med, Royal Victoria Hosp, Montreal, Que, 43-46; fel, Mass Gen Hosp, Boston, 46-48; Am Cancer Soc fel biochem, Univ Utah, 49-51; Damon Runyon Mem Fund fel med, Johns Hopkins Hosp, 51-53, asst, Sch Med, Johns Hopkins Univ, 53-54; from asst prof to assoc prof, 55-67, PROF MED, SCH MED, UNIV MD, BALTIMORE CITY, 67- Concurrent Pos: Asst chief med, Baltimore City Hosps, 55-58; asst prof med, Sch Med, Johns Hopkins Univ, 56-74; consult physician, Baltimore Vet Admin Hosp & USPHS Hosp. Mem: AAAS; Am Thyroid Asn; Endocrine Soc; Am Clin & Climat Asn; Am Fedn Clin Res. Res: Relationship of the thyroid and catecholamines; physiologic and clinical aspects of the thyroid gland. Mailing Add: Dept of Med Univ of Md Hosp Baltimore MD 21201

WIT, ANDREW LEWIS, b Oceanside, NY, Jan 18, 42; m 65; c 1. CARDIOVASCULAR PHYSIOLOGY, PHARMACOLOGY. Educ: Bates Col, BS, 63; Columbia Univ, PhD(pharmacol), 68. Prof Exp: Res physiologist, USPHS Hosp, Staten Island, NY, 68-70; assoc, 70-71, asst prof, 71-74, ASSOC PROF PHARMACOL, COL PHYSICIANS & SURGEONS, COLUMBIA UNIV, 74- Concurrent Pos: Res assoc, Rockefeller Univ, 70-71, vis asst prof, 71-74, adj assoc prof, 74-; NY Heart Asn sr investr, Columbia Univ, 71-75, Am Heart Asn grant in aid, 72-76. Mem: Am Heart Asn; Am Fedn Clin Res; Soc Gen Physiol; Am Physiol Soc. Res: Cardiac electrophysiology, pharmacology and arrhythmias. Mailing Add: Col of Physicians & Surgeons Columbia Univ Dept Pharmacol New York NY 10034

WITCHER, WESLEY, b Chatham, Va, July 9, 23; m 55; c 2. PLANT PATHOLOGY. Educ: Va Polytech Inst, BS, 49, MS, 58; NC State Col, PhD(plant path), 60. Prof Exp: Instr voc agr, Pittsylvania County Sch Bd, 49-54; asst county agent, Exten Serv, Va Polytech Inst, 54-56; asst, NC State Col, 57-60; PROF FOREST PATH, CLEMSON UNIV, 60- Mem: Am Phytopath Soc; Soc Nematol. Res: Fungus-nematode complex of tobacco; diseases of highbush blueberries; forest disease. Mailing Add: Dept of Plant Path & Physiol Clemson Univ Clemson SC 29631

WITCOFSKI, RICHARD LOU, b Peiping, China, Mar 29, 35; US citizen; m 56; c 2. MEDICAL BIOPHYSICS, NUCLEAR MEDICINE. Educ: Lynchburg Col, BS, 56; Vanderbilt Univ, MS, 60; Wake Forest Univ, PhD(anat), 67. Prof Exp: Res asst, 57-61, from instr to prof, 61-73, PROF RADIOL, BOWMAN GRAY SCH MED, 73- Mem: Soc Nuclear Med; Health Physics Soc; Radiation Res Soc; Am Asn Physicists in Med; Am Inst Ultrasonics in Med. Res: Radiation biology. Mailing Add: Dept of Radiol Bowman Gray Sch of Med Winston-Salem NC 27103

WITELSON, SANDRA FREEDMAN, b Montreal, Que, Feb 24, 40. NEUROPSYCHOLOGY. Educ: McGill Univ, BSc, 60, MSc, 62, PhD(psychol), 66. Prof Exp: Lectr psychol, Yeshiva Univ, 66; NIMH res fel, Sch Med, NY Univ, 66-68; instr, NY Med Col, 68-69; asst prof, 69-74, ASSOC PROF PSYCHOL, SCH MED, McMASTER UNIV, 74- Concurrent Pos: Ont Ment Health Found res grant, McMaster Univ, 70-77. Mem: AAAS; Can Psychol Asn; Am Psychol Asn; Acad Aphasia; Int Neuropsychol Soc. Res: Perception; cognition; language; brain function; developmental psychology. Mailing Add: Dept of Psychiat McMaster Univ Hamilton ON Can

WITHAM, ABNER CALHOUN, b Atlanta, Ga, Apr 28, 21; m 48. CARDIOLOGY. Educ: Emory Univ, AB, 42; Johns Hopkins Univ, MD, 45. Prof Exp: Intern internal med, Emory Univ, 45-46, asst resident, 48-51; res asst, Cardiovasc Lab, Brit Post-Grad Med Sch, London, 51-52; asst prof physiol, 52-55, resident med, 54-58, assoc prof, 58-64, PROF MED, MED COL GA, 64- Mem: AMA; Am Fedn Clin Res; Am Clin & Climat Asn. Res: Physiology of venoms; pulmonary circulation; congenital heart disease; vector cardiography. Mailing Add: Div of Cardiol Med Col of Ga Augusta GA 30902

WITHAM, FRANCIS H, b Waltham, Mass, Apr 26, 36; m 61; c 3. PLANT PHYSIOLOGY. Educ: Univ Mass, BS, 58, MA, 60; Ind Univ, PhD(plant physiol), 64. Prof Exp: Lectr plant physiol, Ind Univ, 63-64; asst prof, 66-70, ASSOC PROF BIOL, PA STATE UNIV, 70- Mem: Am Soc Plant Physiol. Res: Biosynthesis, chemistry and mechanism of action of cytokinins and their interaction with other plant growth substances. Mailing Add: Dept of Biol Sci Pa State Univ University Park PA 16802

WITHBROE, GEORGE LUND, b Green Bay, Wis, Dec 14, 38; m 64; c 2. ASTROPHYSICS, SOLAR PHYSICS. Educ: Mass Inst Technol, BS, 61; Univ Mich, MS, 63, PhD(astron), 65. Prof Exp: Res fel astron, 65-69, RES ASSOC ASTRON, HARVARD UNIV, 69-, LECTR, 70-; ASTROPHYSICIST, SMITHSONIAN ASTROPHYS OBSERV, 73- Concurrent Pos: Mem, Solar Physics Spacelab Working Group, NASA & team leader, Solar EUV-XUV Soft X-ray Telescope Definition Team, 75- Mem: Int Astron Union; Am Astron Soc; Am Geophys Union. Res: Interpretation of solar and stellar visible, radio and EUV radiation; determination of solar chemical abundances; temperature density structure of solar atmosphere and terrestrial atmosphere. Mailing Add: Ctr for Astrophys 60 Garden St Cambridge MA 02138

WITHER, ROSS PLUMMER, b Portland, Ore, Dec 29, 22; m 44; c 3. ORGANIC CHEMISTRY, PULP & PAPER TECHNOLOGY. Educ: Univ Ore, BS, 47, MA, 49; Stanford Univ, PhD(chem), 56. Prof Exp: SR RES CHEMIST, CENT RES DIV, CROWN ZELLERBACH CORP, 55- Mem: Am Chem Soc; Tech Asn Pulp & Paper Indust. Res: Cellulose chemistry; pulp and paper research; paper coatings research; specialty papers development. Mailing Add: Cent Res Div Crown Zellerbach Corp Camas WA 98607

WITHERELL, DONALD RAY, b Iowa City, Iowa, Nov 29, 31; m 58; c 2. ORGANIC CHEMISTRY. Educ: Coe Col, BA, 55; BS, 59, PhD(chem), 64. Prof Exp: Chemist, W S Merrell Pharmaceut Co, 59-61; SR ENGR, COLLINS RADIO CO, 63- Mem: Am Chem Soc. Res: Development of chemical processes for the fabrication of electron components. Mailing Add: 2130 Birchwood Dr NE Cedar Rapids IA 52402

WITHERELL, EGILDA DEAMICIS, b Fall River, Mass, Nov 1, 22; m 56. RADIOLOGICAL PHYSICS, NUCLEAR MEDICINE. Educ: Mass Inst Technol, SB, 44; Am Bd Radiol, dip, 53; Am Bd Health Physics, dipl, 60. Prof Exp: Mem staff physics, Radiation Lab, Mass Inst Technol, 44-45, mem staff math, Dynamic Anal & Control Lab, 46-47; asst instr chem, Northeastern Univ, 45-56; radiol physicist, Cancer Res Inst, New Eng Deaconess Hosp, Boston, Mass, 47-66 & Peter Bent Brigham Hosp, 66-67; RADIOL PHYSICIST, NEWTON WELLESLEY HOSP, NEWTON LOWER FALLS, 67- Mem: Am Col Radiol; Soc Nuclear Med; Am Asn Physicists Med; Health Physics Soc. Res: Radiological physics. Mailing Add: Newton Wellesley Hosp Newton Lower Falls MA 02162

WITHERELL, MICHAEL STEWART, b Toledo, Ohio, Sept 22, 49. ELEMENTARY PARTICLE PHYSICS. Educ: Univ Mich, BS, 68; Univ Wis, MA, 70, PhD(physics), 73. Prof Exp: Instr, 73-75, ASST PROF PHYSICS, PRINCETON UNIV, 75- Mem: Sigma Xi; Am Phys Soc. Res: Counter and spark chamber experiments in elementary particle physics. Mailing Add: Dept of Physics Princeton Univ PO Box 708 Princeton NJ 08540

WITHERS, ARNOLD MOORE, b Pueblo, Colo, May 28, 16; m 41; c 2. ANTHROPOLOGY. Educ: Univ Ariz, BA, 38, MA, 41. Prof Exp: Field archaeologist, Ariz State Mus, 39-40 & Amerind Found, Ariz, 41; ranger naturalist, Nat Park Serv, Grand Canyon, 41; asst prof, 47-57, ASSOC PROF ANTHROP, UNIV DENVER, 57- Concurrent Pos: With Amerind Found, Casas Grandes, Mex, 59-61. Mem: Fel Am Anthrop Asn; Am Soc Archaeol. Res: Archaeology of the southwest and the western plains; Mexico. Mailing Add: Dept of Anthrop Univ of Denver Denver CO 80210

WITHERS, EDWARD DONALD, b Felton, Pa, June 8, 23; m 47; c 2. ANALYTICAL CHEMISTRY. Educ: Lebanon Valley Col, BS, 44; Univ Cincinnati, MS, 48. Prof Exp: Instr chem, Lafayette Col, 48-51; res chemist, Plastics Div, Allied Chem Corp, 51-57, group leader, 57-60, tech supvr, 60-65; group leader, Celanese Res Co, NJ, 65-67, sect head, 67-72; MGR, GAF CORP, 72- Mem: Am Chem Soc. Res: Analysis and characterization of polymeric materials; instrumental methods of analysis. Mailing Add: Olde Forge East 37-5A Morristown NJ 07960

WITHERS, HUBERT RODNEY, b Stanthorpe, Australia, Sept 21, 32; m 59; c 1. RADIOBIOLOGY, RADIOTHERAPY. Educ: Univ Queensland, MB, BS, 56; Univ London, PhD(path), 65; Col Radiol Australasia, dipl, 61. Prof Exp: Intern med, Royal Brisbane Hosp, Australia, 57; registr radiother, Queensland Radium Inst, 58 & 60-61; registr path, Royal Brisbane Hosp, 59 & 62-63; vis scientist, Gray Lab, Northwood, Eng, 63-65; vis res fel radiobiol, Nat Cancer Inst, Md, 66-68; assoc prof exp radiother, 68-71, PROF EXP RADIOTHER, UNIV TEX M D ANDERSON HOSP & TUMOR INST HOUSTON, 71- Concurrent Pos: Nat Cancer Inst res grant, Univ Tex M D Anderson Hosp & Tumor Inst Houston, 68-; mem comt cellular radiation biol, Los Alamos Meson Physics Facil, 70-; mem comn hazards californium, AEC, 70-; mem radiation study sect, Nat Cancer Inst, 71-; assoc ed, Radiation Res, 73-76; assoc ed, Int J Radiation Oncol, Biol & Physics, 75- Mem: Am Soc Therapeut Radiol; Radiol Soc NAm; Brit Inst Radiol; Radiation Res Soc; Brit Asn Radiation Res. Res: Response of normal tissues to radiation. Mailing Add: Sect of Exp Radiother M D Anderson Hosp & Tumor Inst Houston TX 77030

WITHERSPOON, DON MEADE, b Lamar, SC, Mar 6, 30; m 53; c 2. MEDICINE, SURGERY. Educ: Univ Ga, DVM, 59, PhD(physiol), 70. Prof Exp: From asst prof to assoc prof med & surg, large animal surg & med, Auburn Univ, 64-67; res assoc physiol, 67-69, PROF MED & SURG, COL VET MED, UNIV GA, 69- Mem: Am Asn Equine Practitioners; Am Asn Bovine Practitioners; Am Asn Vet Study Breeding Soundness; Am Vet Med Asn. Res: Equine, bovine and swine reproduction. Mailing Add: Dept of Med & Surg Univ of Ga Col of Vet Med Athens GA 30601

WITHERSPOON, JAMES DONALD, b Springfield, Mo, Dec 19, 33; m 58; c 2. ANIMAL PHYSIOLOGY. Educ: Purdue Univ, BS, 55, MS, 60, PhD(physiol), 63. Prof Exp: From instr to asst prof biol, Western Md Col, 60-68; ASSOC PROF BIOL, SOUTHWESTERN AT MEMPHIS, 68- Concurrent Pos: Consult, Doubleday & Co, 60-62 & Narco Bio-Systs, 70- Mem: AAAS; Am Inst Biol Sci; Underwater Med Soc. Res: Aerospace biology; decompression; student laboratory experiments; audiovisual tapes; autonomic nervous system; biofeedback. Mailing Add: Dept of Biol Southwestern at Memphis Memphis TN 38112

WITHERSPOON, JOHN PINKNEY, JR, b Hamlet, NC, Feb 28, 31; m 52; c 5. RADIATION ECOLOGY, PLANT ECOLOGY. Educ: Emory Univ, BS, 52, MS, 53; Univ Tenn, PhD(bot), 62. Prof Exp: Res asst biol, Emory Univ, 55-57; health physicist, 62, ECOLOGIST, OAK RIDGE NAT LAB, 62- Mem: Ecol Soc Am; Health Physics Soc. Res: Radiological impact assessments of nuclear fuel cycle facilities. Mailing Add: 100 Wade Lane Oak Ridge TN 37830

WITHERSPOON, SAMUEL MCBRIDGE, b Marion, SC, Dec 26, 26; m 52; c 2. ANESTHESIOLOGY. Educ: Clemson Col, BS, 54; Med Col SC, MD, 52. Prof Exp: Intern, Jefferson Davis Hosp, Houston, Tex, 52-53; pvt pract, SC, 53-57; resident, Med Ctr Hosps, 57-59, instr, 59-60, assoc, 60-63, asst prof, 63-66, ASSOC PROF ANESTHESIOL, MED UNIV SC, 66- Mem: Am Soc Anesthesiol; AMA; Int Anesthesia Res Soc. Res: Effective compounds applicable in the patient with intractable pain. Mailing Add: 6 Pierates Cruze Mt Pleasant SC 29464

WITHINGTON, WILLIAM ADRIANCE, b Honolulu, Hawaii, Feb 17, 24; m 55; c 2. GEOGRAPHY OF SOUTHEAST ASIA. Educ: Harvard Col, AB, 46; Northwestern Univ, MA, 48, PhD(geog), 55. Prof Exp: From instr to asst prof geog, George Washington Univ, 46-53; sci writer, US Corps Engrs-New Eng-NY Inter-Agency Comn, 54-55; from instr to asst prof, 55-63, ASSOC PROF GEOG, UNIV KY, 63- Concurrent Pos: Vis prof geog & sociol, Fac Econ, Nommensen Univ, Sumatra, 57-59; sem mem, Urban Develop Comt, Southeast Asia Develop Adv Group, 68-71; consult, Int Develop Res Ctr, Ont, 71-72. Mem: Fel Am Geog Soc; Asn Asian Studies; Asn Am Geog; Sigma Xi; Nat Coun Geog Educ. Res: Urban development; regional, cultural and economic processes and patterns, with particular reference to Southeast Asia, especially Indonesia; urban transportational linkages in Kentucky. Mailing Add: Dept of Geog Univ of Ky Lexington KY 40506

WITHNER, CARL LESLIE, JR, b Indianapolis, Ind, Mar 3, 18; m 41; c 3. PLANT MORPHOGENETICS. Educ: Univ Ill, BA, 41; Yale Univ, MS, 43, PhD(bot), 48. Prof Exp: Asst instr bot, Yale Univ, 41-43, asst, 46-47; from instr to assoc prof biol, 48-63, dep chmn dept, 60-64, actg chmn dept, 64-65, PROF BIOL, BROOKLYN COL, 64- Concurrent Pos: Resident investr orchids, Brooklyn Bot Garden, 49-; Guggenheim fel, 61-62. Mem: Bot Soc Am; Am Soc Plant Physiol. Res: Orchids;

physiology of higher plants in relation to growth and development. Mailing Add: Dept of Biol Brooklyn Col Brooklyn NY 11210

WITHROW, ALICE PHILLIPS, b Louisville, Ky, Sept 5, 07; m 31; c 2. BIOLOGICAL CHEMISTRY, PHOTOBIOLOGY. Educ: Butler Univ, AB, 29; Univ Cincinnati, MA, 30; Purdue Univ, PhD(agr bot), 42. Prof Exp: Asst plant physiol, Purdue Univ, 43-49; plant physiologist biochem, Smithsonian Inst, 49-58; assoc prog dir sci personnel & educ, 58-66, assoc prog dir advan sci educ prog, Div Grad Educ in Sci, 66-73, PROG MGR, ALTERNATIVES IN HIGHER EDUC, DIV HIGHER EDUC SCI, NSF, 73- Mailing Add: Nat Sci Found Washington DC 20550

WITHROW, CLARENCE DEAN, b Hutchinson, WVa, Mar 6, 27; m 53; c 3. PHARMACOLOGY. Educ: Davis & Elkins Col, BS, 48; Univ Utah, MS, 55, PhD(pharmacol), 59. Prof Exp: Res instr, 59-63, from instr to asst prof, 63-70, ASSOC PROF PHARMACOL, COL MED, UNIV UTAH, 70- Mem: AAAS; Am Soc Pharmacol & Exp Therapeut. Res: Acid-base metabolism, particularly intracellular pH regulation; renal pharmacology; mineralocorticoids; polarography. Mailing Add: Dept of Pharmacol Univ of Utah Col of Med Salt Lake City UT 84132

WITHSTANDLEY, VICTOR DEWYCKOFF, III, b New York, NY, Sept 1, 21; m 58; c 3. MOLECULAR SPECTROSCOPY. Educ: Cornell Univ, BA, 50; Univ Calif, Berkeley, MA, 52; Pa State Univ, DEd(physics), 66, PhD(physics), 72. Prof Exp: Asst seismologist, Geotech Corp, Tex, 52-56; res asst underwater acoustics, Ord Res Lab, Pa State Univ, 59-62; instr math, Juniata Col, 66-67; res asst, Ctr Air Environ Studies, Pa State Univ, 69-73; PRIN SCIENTIST, SCITEK, INC, 74- Mem: Am Phys Soc; Am Geophys Union; Air Pollution Control Asn. Res: Seismic wave and underwater sound studies; magnetic anisotropies of single crystals; computer analysis of time series; optical engineering; infrared spectroscopy of molecules; remote sensing for geophysical and environmental studies. Mailing Add: 127 W Whitehall Rd State College PA 16801

WITIAK, DONALD T, b Milwaukee, Wis, Nov 16, 35; m 55; c 2. ORGANIC CHEMISTRY, MEDICINAL CHEMISTRY. Educ: Univ Wis, BS, 58, PhD(med chem), 61. Prof Exp: From asst prof to assoc prof med chem, Univ Iowa, 61-67; assoc prof, 67-71, PROF MED CHEM, COL PHARM, OHIO STATE UNIV, 71-, CHMN DEPT, 73- Mem: Am Chem Soc; Am Pharmaceut Asn; fel Acad Pharmaceut Sci. Res: Synthesis of biologically active compounds; stereo-structure activity relationships; hypolipidemic drugs; CNS drugs; carcinogenesis and anticancer agents. Mailing Add: Dept of Med Chem Ohio State Univ Col of Pharm Columbus OH 43210

WITKAMP, MARTINUS, microbial ecology, see 12th edition

WITKIN, EVELYN MAISEL, b New York, NY, Mar 9, 21; m 43; c 2. MICROBIAL GENETICS. Educ: NY Univ, AB, 41; Columbia Univ, MA, 43, PhD(zool), 47. Prof Exp: Res assoc bact genetics, Carnegie Inst, 46-49, mem staff genetics, 49-55; assoc prof med, Col Med, State Univ NY Downstate Med Ctr, 55-69, prof, 69-71; PROF BIOL SCI, DOUGLASS COL, RUTGERS UNIV, 71- Concurrent Pos: Am Cancer Soc fel, 47-49; res assoc, Carnegie Inst, 55-71, fel, 56; Waksman lectr, 59. Mem: Am Soc Microbiol; Genetics Soc Am; Am Soc Nat; Radiation Res Soc. Res: Mechanism of spontaneous and induced mutation in bacteria; genetic effects of radiation; enzymatic repair of DNA damage. Mailing Add: Dept of Biol Sci Douglass Col Rutgers Univ New Brunswick NJ 08903

WITKIN, GEORGE JOSEPH, b New York, NY, Dec 22, 15; m 40; c 2. DENTISTRY. Educ: NY Univ, AB, 38, DDS, 42; Am Bd Periodont, dipl; Am Bd Oral Path, dipl. Prof Exp: Asst, 46-47, from instr to assoc prof, 47-66, PROF PERIODONTIA, NY UNIV, 66-, DIR TEACHER TRAINING PROG COL DENT, 61-, CHMN DEPT, PERIODONTIA & ORAL MED, 66-, ASST DEAN, 67- Concurrent Pos: Res fel periodontia, NY Univ, 42; mem attend staff, Univ Hosp; spec consult, Munic Civil Serv Comn, New York & USPHS; consult, US Vet Admin. Mem: Sci Res Soc Am; Am Dent Asn; fel Am Col Dent; Am Acad Periodont; fel Am Acad Oral Med. Res: Periodontia; dental education. Mailing Add: 1 Rockefeller Plaza New York NY 10020

WITKIN, STEVEN S, b Brooklyn, NY, Oct 19, 43; m 66. MOLECULAR BIOLOGY, VIROLOGY. Educ: Hunter Col, BA, 65; Univ Conn, MS, 67; Univ Calif, Los Angeles, PhD(microbiol), 70. Prof Exp: Staff assoc, Inst Cancer Res, Columbia Univ, 72-74; ASSOC CELL BIOL, SLOAN-KETTERING INST CANCER RES, 74- Concurrent Pos: Fel, Roche Inst Molecular Biol, Nutley, NJ, 70-72. Mem: AAAS; Am Soc Microbiol. Res: Transfer and expression of genetic information; RNA tumor viruses; spermatozoa. Mailing Add: Div of Cell Biochem Sloan-Kettering Inst Cancer Res New York NY 10021

WITKIND, IRVING JEROME, b New York, NY, Mar 28, 17; m 42. ECONOMIC GEOLOGY. Educ: Brooklyn Col, BA, 39; Columbia Univ, MA, 41; Univ Colo, PhD(geol), 56. Prof Exp: GEOLOGIST, US GEOL SURV, 46- Mem: Geol Soc Am; Am Asn Petrol Geol. Res: Pleistocene geology; localization of sodium sulfate; localization of uranium minerals; laccolithic mountains of southeastern Utah and central Montana; stratigraphy and structural geology of southwestern Montana and southeastern Idaho. Mailing Add: US Geol Surv Fed Ctr Denver CO 80225

WITKOP, BERNHARD, b Freiburg, Ger, May 9, 17; nat US; m 45; c 3. ORGANIC CHEMISTRY. Educ: Univ Munich, PhD(org chem), 40, ScD, 46. Prof Exp: Privat docent, Univ Munich, 46-47; Mellon Found fel, Harvard Univ, 47-48, instr, 48-50, USPHS spec fel, 50-51; vis scientist, NIH, 51-53, chemist, 53-55, CHIEF SECT METABOLITES, NAT INST ARTHRITIS, METAB & DIGESTIVE DIS, 55-, CHIEF LAB CHEM, 57- Concurrent Pos: Mem, Nat Acad Sci-Nat Res Coun, 59-62; vis prof, Kyoto Univ, 61 & Univ Freiburg, 62; mem, Bd Int Sci Exchange, Nat Acad Sci, 75-77. Honors & Awards: Superior Serv Award, US Dept Health, Educ & Welfare, 67; Paul Karrer Gold Medal, Univ Zurich, 71. Mem: Nat Acad Sci; Am Chem Soc; Leopoldine Ger Acad Res Natural Sci. Res: Alkaloids; arrow and mushroom poisons; oxidation mechanisms; peroxides; ozonides; intermediary and labile metabolites; nonenzymatic selective cleavage of proteins and enzymes; photochemistry of amino acids and nucleotides; venoms of amphibians; biochemical mechanisms; dynamics of modified homopolynucleotides; stimulation of interferon. Mailing Add: Nat Inst of Arthritis Metab & Dig Dis Bethesda MD 20014

WITKOP, CARL JACOB, JR, b East Grand Rapids, Mich, Dec 27, 20; m 66; c 7. HUMAN GENETICS, ORAL PATHOLOGY. Educ: Mich State Col, BS, 44; Univ Mich, DDS, 49, MS, 54; Am Bd Oral Path, dipl, 57. Prof Exp: Asst dent surgeon & intern, US Marine Hosp, USPHS, Seattle, Wash, 49-50, sr asst dent surgeon, US Coast Guard Yard, Baltimore, 50; oral pathologist, Nat Inst Dent Res, 50-57, chief human genetics sect, 57-63, chief human genetics br, 63-66; PROF HUMAN & ORAL GENETICS & CHMN DIV, SCH DENT & PROF DERMAT, MED SCH, UNIV MINN, MINNEAPOLIS, 66- Concurrent Pos: Fel, Univ Mich, 52-54; consult, Children's Hosp, Washington, DC, 56-, Nat Found Congenital Malformation, 63- & Easter Seal Soc, 63-66; lectr, Schs Med & Dent, Howard Univ, 56-, Georgetown Univ & Johns Hopkins Univ; dent dir, Int Comt Nutrit Nat Develop, Chile, 60 & Paraguay, 65; chief dent sect, Inst Nutrit Cent Am, Panama & Guatemala, 64. Mem: Fel AAAS; Am Soc Human Genetics (secy, 68-70); Am Dent Asn; fel Am Acad Oral Path (vpres, 60, 72, pres elect, 73, pres, 74). Res: Albinism and pigment defects; exfoliative cytology; population isolates; congenital malformations; oral epidemiology and nutrition. Mailing Add: 9 Manitoba Rd Hopkins MN 55343

WITKOSKI, FRANCIS CLEMENT, b Scranton, Pa, Dec 19, 22; m 46; c 5. ORGANIC CHEMISTRY. Educ: Univ Scranton, BSc, 43; Bucknell Univ, MSc, 50. Prof Exp: From instr to asst prof org chem, Univ Scranton, 47-55; dir res, State Dept Hwy, Pa, 55-58, assoc dir res & testing, 58-60; secy & tech dir, Am Testing Labs, Inc, 60-63; PRES, MAT ENG & SERV CO, 63- Concurrent Pos: Mem, Comt Rigid Pavement Design & Comt Prestressed Concrete, Hwy Res Bd, Nat Acad Sci-Nat Res Coun. Mem: AAAS; Am Chem Soc; Am Soc Testing & Mat. Res: Chemistry of lignin; incinerated anthracite mine waste in asphalt pavements; engineering properties of construction materials. Mailing Add: Mat Eng & Serv Co 21st & Chestnut Sts PO Box 621 Camp Hill PA 17011

WITKOVSKY, PAUL, b Chicago, Ill, May 24, 37; m 64. SENSORY PHYSIOLOGY. Educ: Univ Calif, Los Angeles, BA, 58, MA, 60, PhD(physiol), 62. Prof Exp: NIH fel neurophysiol, Sci Res Inst, Caracas, Venezuela, 62-63; instr ophthal, Columbia Univ, 64-65, from asst prof to assoc prof physiol, 65-73; PROF ANAT SCI, STATE UNIV NY STONEY STONY BROOK, 75- Concurrent Pos: Res grants, Nat Inst Neurol Dis & Blindness, 64- & Nat Coun Combat Blindness, 66-67. Mem: AAAS; Asn Res Vision & Ophthal; Biophys Soc; Soc Neurosci. Res: Central nervous system organization of tactile sensation; neurophysiological organization of the retina. Mailing Add: Dept of Anat Sci State Univ NY Stony Brook NY 11794

WITKOWSKI, JOHN FREDERICK, b Beatrice, Nebr, Apr 30, 42; m 74; c 1. ENTOMOLOGY. Educ: Univ Nebr, BSc, 65, MSc, 70; Iowa State Univ, PhD(entom), 75. Prof Exp: Res rep agr chem, Chemagro Corp, 70-72; res assoc entom, Iowa State Univ, 72-75; ENTOMOLOGIST, UNIV NEBR, 75- Mem: Entom Soc Am. Res: The biology and chemical control of insects damaging corn and soybeans. Mailing Add: Northeast Sta Concord NE 68728

WITKOWSKI, JOSEPH THEODORE, b Ft Worth, Tex, Oct 29, 42; m 65; c 1. ORGANIC CHEMISTRY, MEDICINAL CHEMISTRY. Educ: NTex State Univ, BS, 65, MS, 66; Univ Utah, PhD(chem), 70. Prof Exp: RES CHEMIST, NUCLEIC ACID RES INST, ICN PHARMACEUT, INC, 69- Mem: Am Chem Soc; Int Soc Heterocyclic Chem. Res: Synthesis and properties of nucleosides and heterocycles; synthesis of antiviral agents. Mailing Add: ICN Pharmaceut Inc Nucleic Acid Res Inst 2727 Campus Dr Irvine CA 92664

WITKUS, ELEANOR RUTH, b New York, NY, July 11, 18. BIOLOGY. Educ: Hunter Col, BA, 40; Boston Univ, MA, 41; Fordham Univ, PhD(cytol), 44. Prof Exp: Instr zool, Marymount Col, NY, 43-44; from instr to assoc prof bot & bact, 44-74, PROF BIOL SCI & CHMN DEPT, FORDHAM UNIV, 74- Mem: Bot Soc Am; Torrey Bot Club (corresp secy, 53-56). Res: Botanical cytology. Mailing Add: Dept of Biol Sci Fordham Univ New York NY 10458

WITLIN, BERNARD, b Philadelphia, Pa, July 18, 14; m 44; c 1. BACTERIOLOGY. Educ: Univ Calif, Los Angeles, AB, 36; Philadelphia Col Pharm, MSc, 38, DSc(bact, pub health), 40; Am Bd Clin Chem, dipl. Prof Exp: Res bacteriologist, Sharp & Dohme, Inc, Pa, 38-40; dir, Barlin Labs, 40-41; bacteriologist, USPHS, US Dept Army & US Dept Navy, 41-46 & Mellon Inst, 46-50; assoc prof bact, 50-68, prof, 50-71, EMER PROF BACT, PA COL OPTOM, 71-; PROF BACT & DIR DIV BACT & PUB HEALTH, PHILA COL PHARM & SCI, 68- Concurrent Pos: Instr, Philadelphia Col Pharm & Sci, 40-41; assoc prof microbiol & pub health, Philadelphia Col Osteopath Med, 50-68; clin pathologist, Metrop Hosp, Philadelphia, 52- Mem: AAAS; Am Soc Microbiol; Am Pub Health Asn; Am Pharmaceut Asn; Sigma Xi (pres, 71-72). Res: Antiseptics and disinfectants; water purification; blood banks; serology. Mailing Add: Dept of Bact Philadelphia Col of Pharm & Sci Philadelphia PA 19104

WITMAN, EUGENE DEWALD, b New Albany, Ind, Sept 21, 10; m 38; c 1. AGRICULTURAL CHEMISTRY. Educ: Ohio State Univ, AB, 35, PhD(chem), 39. Prof Exp: Sherwin-Williams res fel, Ohio State Univ, 39-47, lectr insecticide chem, 40-47; asst dir, Agr Chem Div, Sherwin-Williams Co, 47-50; mgr agr chem develop, 50-55, asst dir mkt develop, 55-64, asst to dir mkt serv, 64-68, mgr com develop, 68-72, TECH ASST TO VPRES MKT, CHEM DIV, PPG INDUSTS, INC, 72- Res: Insecticide chemistry and development; chemical market development. Mailing Add: 744 Country Club Dr Pittsburgh PA 15228

WITMAN, GEORGE BODO, III, b Upland, Calif, July 19, 45; m 69. CELL BIOLOGY. Educ: Univ Calif, Riverside, BA, 67; Yale Univ, PhD(cellular & develop biol), 72. Prof Exp: NIH fel cell biol, Whitman Lab, Univ Chicago, 72-73; fel molecular biol, Lab Molecular Biol & Biophys, Univ Wis-Madison, 73-74; ASST PROF BIOL, PRINCETON UNIV, 74- Mem: AAAS; Am Soc Cell Biol; Int Fedn Cell Biol; Soc Protozoologists. Res: Structure, composition, function and development of cell organelles. Mailing Add: Dept of Biol Princeton Univ Princeton NJ 08540

WITMAN, ROBERT CHARLES, b Pottsville, Pa, Feb 15, 27; m 52; c 3. ORGANIC CHEMISTRY. Educ: Pa State Col, BS, 49, MS, 50; Univ Del, PhD, 53. Prof Exp: Res chemist phys org chem, Shell Chem Corp, 53-61, sr technologist prod develop, 61-63; dir com develop, 63-65, dir mkt, 65-69, mgr corp develop, 69-70, vpres & tech mgr, 70-74, RES DIR, CINCINNATI MILACRON CHEM, INC, 74- Mem: Am Chem Soc. Res: Organometallics; organic phosphines and derivatives; organic sulfur compounds. Mailing Add: 501 Reily Rd Cincinnati OH 45215

WITMER, HEMAN JOHN, b Bayonne, NJ, Apr 5, 44; m 70; c 2. VIROLOGY, BIOCHEMISTRY. Educ: Delaware Valley Col, BS, 65; Ind Univ, Bloomington, PhD(microbiol), 69. Prof Exp: NIH fel, McArdle Lab, Univ Wis-Madison, 69 & Ind Univ, Bloomington, 69-71, vis asst prof microbiol, 71, asst prof, Med Ctr, 71-72; ASST PROF BIOL SCI, UNIV ILL, CHICAGO CIRCLE, 72- Res: Controls of gene expression of coliphage T4; photodynamic action. Mailing Add: Dept of Biol Sci Univ of Ill at Chicago Circle PO Box 4348 Chicago IL 60680

WITMER, RICHARD EVERETT, b Swarthmore, Pa, July 3, 42; m 66; c 3. GEOGRAPHY. Educ: Univ Fla, BS, 62, MS, 64, PhD(geog), 67. Prof Exp: Instr geog, Fla Atlantic Univ, 65-67, res asst prof, 67-68; assoc prof, ETenn State Univ, 68-74; COORDR & PHYS SCIENTIST, GEOG PROG, US GEOL SURV, 74- Concurrent Pos: Prin investr, NASA Test Site 177, Tenn Valley, 68-; consult, Asn Am Geog, 69- & Tenn Valley Authority, 71-72. Mem: Asn Am Geog; Am Soc Photogram. Res: Applications of aircraft and spacecraft remote sensing data to earth science problems. Mailing Add: Mail Stop 710 US Geol Surv Reston VA 22092

WITMER, WILLIAM BYRON, b Clarksville, Tex, June 29, 31; m 55; c 3. INORGANIC CHEMISTRY, PHYSICAL CHEMISTRY. Educ: Tex A&M Univ, BS, 52, MS, 58, PhD(chem), 60. Prof Exp: Res chemist, Chemstrand Res Ctr, Monsanto

Co, NC, 59-64, sr res chemist, 64, supvr spec anal lab, Textiles Div, Ala, 64-65, supt tech lab, 65-68, supt qual control, Decatur Plant, 68-70, qual control mgr, Lingen Plant, Monsanto (Deutschland), GMBH, WGer, 70-73, SUPT, TECH DEPT, SAND MOUNTAIN PLANT, MONSANTO TEXTILES CO, 73- Mem: Am Chem Soc; Am Soc Qual Control. Res: Amine-halogen complexes; polymer characterizition techniques; polymer properties related to synthetic fiber production. Mailing Add: Monsanto Textiles Co Sand Mountain Plant Star Route Guntersville AL 35976

WITNAUER, LEE P, b East Amherst, NY, Jan 7, 22; m 43; c 2. PHYSICAL CHEMISTRY. Educ: Canisius Col, BS, 42; Temple Univ, AM, 48, PhD(phys chem), 53. Prof Exp: Asst chief chemist, Lake Ontario Ord Works, 42-43; res chemist, Eastern Regional Res Lab, 43-53, head sect phys testing, 53-57, prin res phys chemist, 57-60, supvry phys chemist, Prod Properties Eval Invests, 60, CHIEF LAB & RES ADMINR, PHYS CHEM PROBS, AGR COMMODITIES, PHYS CHEM LAB, EASTERN UTILIZATION RES & DEVELOP DIV, AGR RES SERV, USDA, 60- Concurrent Pos: Adj prof, Eve Col, Drexel Univ, 53- Honors & Awards: Superior Serv Award, USDA, 58. Mem: AAAS; Am Chem Soc; NY Acad Sci. Mailing Add: Eastern Regional Res Lab 600 E Mermaid Lane Philadelphia PA 19118

WITONSKY, PHIL, biochemistry, see 12th edition

WITORSCH, RAPHAEL JAY, b New York, NY, Dec 12, 41; m 64; c 2. PHYSIOLOGY. Educ: NY Univ, AB, 63; Yale Univ, MS, 65, PhD(physiol), 68. Prof Exp: USPHS trainee, Sch Med, Univ Va, 68-69, NIH fel, 69-70; ASST PROF PHYSIOL, MED COL VA, VA COMMONWEALTH UNIV, 70- Concurrent Pos: Dir first yr med curriculum, Med Col Va, 75-; co-investr, NIH Grant Breast Cancer, 75-78. Mem: Endocrine Soc; Sigma Xi. Res: Endocrinology; hormone receptors in normal and neoplastic tissues. Mailing Add: Dept of Physiol Med Col of Va Richmond VA 23298

WITRIOL, NORMAN MARTIN, b Brooklyn, NY, June 9, 40; m 66. Educ: Polytech Inst Brooklyn, BS, 61; Brandeis Univ, MA, 64, PhD(physics), 68. Prof Exp: RES PHYSICIST, PHYS SCI LAB, US ARMY MISSILE COMMAND, 68- Concurrent Pos: Asst prof, Univ Ala, Huntsville, 70- Mem: Am Phys Soc; Am Asn Physics Teachers. Res: Molecular, chemical, quantum crystal, quantum, mathematical, many-body and theoretical physics. Mailing Add: 2501 Lancelot Dr Huntsville AL 35803

WITSCHARD, GILBERT, b Morehead City, NC, Mar 13, 33; m 60; c 3. ORGANIC CHEMISTRY. Educ: Queens Col, BS, 57; Univ Pittsburgh, PhD(org chem), 63. Prof Exp: Res chemist, 63-70, SR RES CHEMIST, HOOKER CHEM CORP, NIAGARA FALLS, 70- Mem: Am Chem Soc. Res: Organo-phosphorus chemistry; fire retardance; polymer synthesis; polymers stabilization; powder coatings; polyvinyl chloride. Mailing Add: 2078 Long Rd Grand Island NY 14072

WITSCHI, HANSPETER RUDOLF, b Berne, Switz, Mar 17, 33. TOXICOLOGY. Educ: Univ Berne, MD, 60. Prof Exp: Asst path, Inst Forensic Med, Univ Berne, 61-64; res fel, Toxicol Res Unit, Med Res Coun, Eng, 65-66; res fel exp path, Univ Pittsburgh, 67-69; asst prof, 69-74, ASSOC PROF TOXICOL & PHARMACOL, FAC MED, UNIV MONTREAL, 74- Concurrent Pos: Mem drug toxicol group, Med Res Coun, Can, 71. Mem: Am Soc Pharmacol & Exp Therapeut; Soc Exp Biol & Med; Brit Biochem Soc; Soc Toxicol; Pharmacol Soc Can. Res: Experimental toxicology; biochemical pathology; interaction of drugs and toxic agents with organ function at the cellular level. Mailing Add: Dept Pharmacol Fac Med Univ Montreal PO Box 6128 Montreal PQ Can

WITSCHONKE, CHARLES RICHARD, b Milwaukee, Wis, Mar 26, 16; m 42; c 5. ENVIRONMENTAL CHEMISTRY. Educ: Kalamazoo Col, AB, 37; Brown Univ, PhD(phys chem), 41. Prof Exp: Res chemist, Brown Univ, 40-42; res chemist, US Naval Res Lab, Washington, DC, 42-43, head res & develop group, 43-45; res chemist, Calco Chem Div, 45-54, mgr phys anal sect, Res Div, NJ, 54, tech asst to dir, Res Serv Dept, Conn, 54-55, mgr phys anal res sect, 55-57, mgr refining & mining chem res, 57-61, mgr paper chem res, 61-65, mgr process chem res, 65-72, water treating & mining chem res, 72-74, SR RES CHEMIST, WATER TREATING, AM CYANAMID CO, 74- Mem: Am Chem Soc; fel Am Inst Chemists. Res: Synthesis and application of polyelectrolytes for cleaning up water and waste streams; water and waste treatment chemicals; flocculants; industrial biocides; surfactants; paper chemicals. Mailing Add: 33 Beach Dr Noroton CT 06820

WITSENHAUSEN, HANS S, b Frankfurt, Ger, May 6, 30; US citizen; m 61; c 3. APPLIED MATHEMATICS. Educ: Free Univ Brussels, ICME, 53, lic sc phys, 56; Mass Inst Technol, SM, 64, PhD(elec eng), 66. Prof Exp: Asst elec eng, Free Univ Brussels, 53-57; appln eng, European Ctr, Electronic Assoc, Inc, 57-60, sr engr, Res Div, 60-63; res asst control theory & Lincoln Lab assoc, Electronic Systs Lab, Mass Inst Technol, 63-65; MEM TECH STAFF, MATH RES CTR, BELL LABS, 66- Concurrent Pos: Vinton Hayes sr fel, Harvard Univ, 75-76. Mem: Inst Elec & Electronics Eng; Am Math Soc. Res: System theory; optimization; geometry; inequalities; information. Mailing Add: Bell Labs Murray Hill NJ 07974

WITT, ADOLF NICOLAUS, b Bad Oldesloe, Ger, Oct 17, 40; m 67; c 2. ASTROPHYSICS. Educ: Univ Hamburg, Vordiplom, 63; Univ Chicago, PhD(astrophys), 67. Prof Exp: From asst prof to assoc prof, 67-74, PROF ASTRON, UNIV TOLEDO, 74-, ASSOC DIR, RITTER OBSERV, 72- Concurrent Pos: Vis fel, Lab Atmospheric & Space Physics, Univ Colo, Boulder, 75-76. Mem: AAAS; Am Astron Soc; Int Astron Union. Res: Interstellar matter; spectral classification; photometry; astronomical instrumentation. Mailing Add: Dept of Physics & Astron Univ of Toledo Toledo OH 43606

WITT, ARTHUR, JR, fishery biology, see 12th edition

WITT, DONALD REINHOLD, b LeMars, Iowa, Apr 15, 23; m 50; c 4. CHEMISTRY. Educ: Westmar Col, BS, 48; Univ SDak, MA, 50. Prof Exp: Res chemist, 50-62, GROUP LEADER CHEM, PHILLIPS PETROL CO, 62- Mem: Am Chem Soc. Res: Polymerization of olefins and diolefins to solid polymers; catalyst development related to such reactions. Mailing Add: Res Ctr Phillips Petrol Co Bartlesville OK 74003

WITT, ENRIQUE ROBERTO, b Buenos Aires, Arg, May 10, 26; nat US; m 55; c 3. ORGANIC CHEMISTRY. Educ: Univ Buenos Aires, Lic, 51, DrChem, 53. Prof Exp: Technician, E R Squibb & Sons, Arg, 51-52; res chemist, Arg AEC, 53-55; RES ASSOC, CELANESE CHEM CO, 56- Mem: AAAS; Am Chem Soc; Water Pollution Control Fedn. Res: Synthetic lubricants; phosphorus compounds; polyester technology; environmental maintenance; anaerobic biological treatment of industrial wastes. Mailing Add: 1037 Brock Corpus Christi TX 78412

WITT, JOHN, JR, b Muskegon, Mich, Oct 5, 35; m 64. ORGANIC CHEMISTRY. Educ: Mich State Univ, BS, 57; Univ Ill, PhD(org chem), 61. Prof Exp: Res chemist, Ethyl Corp, 60-62; asst head chem process develop, 62-69, mgr spec synthesis & process develop, 69-72, MGR, SYNTHESIS DEVELOP, G D SEARLE & CO, 72- Mem: Am Chem Soc. Res: Organic synthesis; steroids; heterocyclics; amino acids. Mailing Add: G D Searle & Co PO Box 5110 Chicago IL 60680

WITT, PAUL CHANDLER, organic chemistry, deceased

WITT, PETER NIKOLAUS, b Berlin, Ger, Oct 20, 18; m 49; c 2. PHARMACOLOGY. Educ: Univ Tübingen, MD, 46. Prof Exp: Asst, Univ Tübingen, 45-49; sr asst, Univ Berne, 49-56, privat-docent, 56; from asst prof to assoc prof pharmacol, Col Med, State Univ NY Upstate Med Ctr, 56-66; DIR DIV RES, NC DEPT MENT HEALTH, 66- Concurrent Pos: Rockefeller fel, Harvard Med Sch, 52-53; Lederle med fac award, 57-59; adj prof, NC State Univ & Univ NC, Chapel Hill, 66- Honors & Awards: Buergi Award, 56. Mem: Am Soc Pharmacol & Exp Therapeut; Ger Pharmacol Soc; Swiss Pharmacol Soc. Res: Effect of drugs on web building behavior of spiders; invertebrate behavior; effect of cardioactive drugs on ion movements in heart muscle; objective testing of fine motor behavior in healthy and diseased human subjects under the influence of drugs. Mailing Add: Div of Res NC Dept Ment Health Box 7532 Raleigh NC 27611

WITT, SHIRLEY HILL, b Whittier, Calif, Apr 17, 34; c 2. BIOLOGICAL ANTHROPOLOGY. Educ: Univ Mich, BA, 65, MA, 66; Univ NMex, PhD(phys anthrop), 69. Prof Exp: Vis asst prof anthrop, Univ NC, Chapel Hill, 70-72; assoc prof anthrop, Colo Col, 72-75; REGIONAL DIR, US COMN CIVIL RIGHTS, 75- Concurrent Pos: Vis lectr, Am Anthrop Asn, 70-; consult, Off Civil Rights, US Dept Health, Educ & Welfare, 71-72; comnr, Human Rels Comn, Colorado Springs, Colo, 71-; mem planning comt, Nat Conf Civil Rights, US Comn Civil Rights, 73. Mem: AAAS; fel Am Anthrop Asn; fel Am Asn Phys Anthrop. Res: Genetic demography; biocultural interface; nutrition; native Americans; Latin Americans; Appalachia; studies of broad patterns of civil rights violations in seven mountain states region. Mailing Add: 1726 Champa Denver CO 80202

WITTBECKER, EMERSON LAVERNE, b Freeport, Ill, Feb 25, 17; m 40; c 2. POLYMER CHMISTRY, TEXTILE CHEMISTRY. Educ: Univ Ill, AB, 39; Pa State Univ, MS, 41, PhD(org chem), 42. Prof Exp: Jr res assoc, 46-52, res assoc, 52-55, res mgr pioneering res, 55-60, Orlon-Lycra res, 60-64, DIR, CAROTHERS LAB, NYLON RES DIV, TEXTILE FIBERS DEPT, E I DU PONT DE NEMOURS & CO, 64- Mem: Am Chem Soc. Res: Condensation polymers; fibers. Mailing Add: E I du Pont de Nemours & Co Exp Sta Wilmington DE 19898

WITTCOFF, HAROLD, b Marion, Ind, July 3, 18; m 46; c 2. ORGANIC CHEMISTRY. Educ: DePauw Univ, AB, 40; Northwestern Univ, PhD(org chem), 43. Prof Exp: Head chem res dept, 43-56, dir chem res, 56-68, vpres chem res & develop, 68-69, vpres & dir corp res, 69-74, SPEC ADV TO THE PRES, GEN MILLS CHEM, INC, 73-; ADJ PROF CHEM, UNIV MINN, 73- Mem: Am Chem Soc; Am Oil Chem Soc; Fedn Socs Paint Technol; Com Develop Asn; Inst Food Technol. Res: Phosphatides; polymers; protective coatings; resins and plastics; research adminstration. Mailing Add: Gen Mills Chem Inc 4620 W 77th St Minneapolis MN 55435

WITTE, DAVID L, b San Diego, Calif, 43; m 66; c 2. BIOCHEMISTRY, PATHOLOGY. Educ: St Olaf Col, BA, 65; Iowa State Univ, PhD(biochem), 71. Prof Exp: ASST PROF PATH & CLIN CHEM, UNIV IOWA, 73- Concurrent Pos: Fel clin chem, Univ Iowa, 71-73. Mem: Am Asn Clin Chem; Am Soc Clin Path. Res: Methods in clinical chemical analysis, specifically cholesterol. Mailing Add: Dept of Path Univ of Iowa Iowa City IA 52242

WITTE, JOHN JACOB, b Passaic, NJ, Mar 10, 32; m 68; c 3. PREVENTIVE MEDICINE, PEDIATRICS. Educ: Hope Col, AB, 54; Johns Hopkins Univ, MD, 59; Harvard Univ, MPH(microbiol), 66. Prof Exp: Intern & resident pediat, Johns Hopkins Univ, 59-62; med epidemiologist infectious dis, 62-65, asst chief immunization br, 66-70, chief, 70-74, DIR IMMUNIZATION DIV, CTR FOR DIS CONTROL, 74- Concurrent Pos: Consult, Adv Comt Immunizing Agents, Can, 69- & Adv Comt Epidemiol, Can, 69- Honors & Awards: Commendation Medal, USPHS, HEW, 72. Mem: Fel Am Acad Pediat; Infectious Dis Soc; Am Col Preventive Med; Am Pub Health Asn. Res: Epidemiology of communicable diseases; development and field testing of vaccines; complications of vaccine administration. Mailing Add: Ctr for Dis Control Atlanta GA 30333

WITTE, MICHAEL, b Poland, Mar 15, 11; nat US; m 40; c 4. ORGANIC CHEMISTRY, POLLUTION CHEMISTRY. Educ: Loyola Univ, Ill, BS, 37; Univ Ill, MS, 38, PhD(org chem), 41. Prof Exp: Asst chem, Univ Ill, 38-41; res chemist, Nat Aniline Div, Allied Chem Corp, 41-47; prod supvr, Gen Aniline & Film Corp, 46-54, prod mgr, NJ, 54-56; pres, Simpson Labs, Inc, 57-59; pres, Carnegies Fine Chem Div, Rexall Drug & Chem Corp, 59-60; PRES, M WITTE ASSOCS, 60- Concurrent Pos: Chem consult. Mem: AAAS; Am Chem Soc. Res: Pharmaceutical intermediates; dyestuffs manufacture; biochemistry; chemical management. Mailing Add: 420 River Rd Apt C-11 Chatham NJ 07928

WITTE, ROBERT SARNOW, solid state physics, see 12th edition

WITTEBORN, FRED CARL, b St Louis, Mo, Dec 27, 34; m 57; c 1. PHYSICS, ASTROPHYSICS. Educ: Calif Inst Technol, BS, 56; Stanford Univ, MS, 58, PhD(physics), 65. Prof Exp: Res assoc physics, Stanford Univ, 65-68; res scientist, 69-70, chief astrophysics br, 70-74, RES SCIENTIST, NASA AMES RES CTR, 75- Concurrent Pos: Vis scholar, Stanford Univ, 70- Mem: Am Phys Soc; Am Astron Soc. Res: Infrared astronomy; gravity; atomic physics; low temperature physics. Mailing Add: Astrophys Br NASA Ames Res Ctr N-245-6 Moffett Field CA 94035

WITTEKIND, RAYMOND RICHARD, b Jamaica, NY, May 9, 29; m 60. ORGANIC CHEMISTRY. Educ: Polytech Inst Brooklyn, BS, 51; Columbia Univ, AM, 55, PhD, 59. Prof Exp: Res chemist, McNeil Labs, Inc, 58-61; scientist, 61-73, SR SCIENTIST, WARNER-LAMBERT RES INST, 73- Mem: Am Chem Soc; The Chem Soc; Swiss Chem Soc. Res: Synthesis of fused ring heterocyclic compounds; stereochemistry and mechanism of organic reactions. Mailing Add: 30 Valley View Dr Morristown NJ 07960

WITTELS, BENJAMIN, b Minneapolis, Minn, Jan 22, 26; m 55; c 2. PATHOLOGY, BIOCHEMISTRY. Educ: Univ Minn, BA, 48, MD, 52; Am Bd Path, dipl, 57. Prof Exp: Assoc prof, 60-70, PROF PATH, MED CTR, DUKE UNIV, 70- Mem: Am Asn Path & Bact. Res: Cardiac metabolism; hematology. Mailing Add: Dept of Path Duke Univ Med Ctr Durham NC 27707

WITTELS, MARK C, b Minneapolis, Minn, July 14, 21; m 51; c 3. MINERALOGY, SOLID STATE PHYSICS. Educ: Univ Minn, BS, 47; Mass Inst Technol, PhD(geol, ceramic eng), 51. Prof Exp: Staff scientist, Oak Ridge Nat Lab, 51-63; sr solid state physicist, US AEC, 63-74; CHIEF, SOLID STATE PHYSICS & MAT CHEM BR, DIV PHYS RES, MAT SCI PROG, ENERGY RES & DEVELOP ADMIN, 74- Concurrent Pos: Vis prof, Wash Univ, 68-69. Mem: Mineral Soc Am; Am Crystallog Asn; Am Phys Soc. Res: Radiation effects in crystalline solids and in lunar materials;

x-ray diffraction instrumentation; crystal growth techniques; defects in crystals; stored energy in reactor-irradiated graphite. Mailing Add: Mat Sci Prog Div of Phys Res Energy Res & Develop Admin Washington DC 20545

WITTEMANN, JOSEPH KLAUS, b Heilbronn, Ger, Nov 13, 41; US citizen; m 65; c 3. DENTISTRY, PSYCHOLOGY. Educ: State Univ NY, BS, 64; Ohio State Univ, MA, 67, PhD(coun), 72. Prof Exp: Psychologist, Ohio Penitentiary, 67-68; asst prof psychol, Ohio Dominican Col, 69-72, chmn dept, 70-71; ASST PROF GEN & PREV DENT, MED COL VA, 72-, DIR OFF EDUC EVAL, PLANNING & RES, 74- Concurrent Pos: Consult alcoholism prog, Ohio Dept Health, 70-; mem ad hoc comt learning environ, Asn Am Med Schs, 73- Mem: Int Asn Dent Res; Am Educ Res Asn. Res: Career development psychology; evaluation research program; pain control and analgesia; institutional research and management. Mailing Add: Off Educ Eval & Planning & Res Med Col of Va Sch of Dent Richmond VA 23298

WITTEN, BENJAMIN, b Baltimore, Md, Oct 4, 16; m 45; c 3. ORGANIC CHEMISTRY. Educ: Johns Hopkins Univ, PhD(org chem), 40. Prof Exp: RES CHEMIST, DEPT OF DEFENSE, 40- Concurrent Pos: Asst surg, Sch Med, Johns Hopkins Univ, 55-; res assoc oncol & cell biol, Sinai Hosp, Baltimore, 59-; Secy Army res & study fel, 66. Honors & Awards: Meritorious Civilian Serv Awards, Dept of Defense, 46 & 72; Award of Merit, Am Chem Soc, 75. Mem: Am Chem Soc; Am Asn Cancer Res; NY Acad Sci. Res: Physiologically active compounds and the determination of their chemical, physical and toxicological properties. Mailing Add: 7501 Prince George Rd Baltimore Md 21208

WITTEN, GERALD LEE, b Daviess County, Mo, May 12, 29; m 51; c 3. SCIENCE EDUCATION. Educ: Kans State Teachers Col, 56, MS, 58, EdS(phys sci), 62. Prof Exp: Teacher high sch, Kans, 56-62; ASSOC PROF PHYS SCI, EMPORIA KANS STATE COL, 62- Mem: Nat Sci Teachers Asn; Am Asn Physics Teachers. Res: Development of take home laboratory exercises for high school physics and general education physical science classes. Mailing Add: Dept of Phys Sci Emporia Kans State Col Emporia KS 66801

WITTEN, LOUIS, b Baltimore, Md, Apr 13, 21; m 48; c 4. PHYSICS. Educ: John Hopkins Univ, BE, 41, PhD(physics), 51; NY Univ, BS, 44. Prof Exp: Res assoc, Proj Matterhorn, Princeton Univ, 51-53; instr fluid mech, Univ Md, 52-53; staff scientist, Lincoln Lab, Mass Inst Technol, 53-54; prin scientist, Res Inst Advan Study, Martin Marietta Corp, Md, 54-65, assoc dir, 65-68; head dept physics, 68-74, PROF PHYSICS, UNIV CINCINNATI, 68- Concurrent Pos: Adj prof, Drexel Univ, 56-68; Fulbright lectr, Weizmann Inst Sci, Israel, 64-65; trustee, Gravity Res Found, 66-, vpres, 72- Mem: AAAS; Am Phys Soc; Am Math Soc; Am Asn Physics Teachers; Am Astron Soc. Res: General theory of relativity; statistical mechanics; theory of particles and fields. Mailing Add: Dept of Physics Univ of Cincinnati Cincinnati OH 45221

WITTEN, MAURICE HADEN, b Jamesport, Mo, Dec 5, 31; m 69; c 3. PHYSICS. Educ: Kans State Teachers Col, BA, 56; Univ Nebr, Lincoln, MA, 60; Univ Iowa, PhD(sci educ, physics), 67. Prof Exp: Engr, Int Business Mach Corp, 56-57; from instr to assoc prof physics, 60-69, PROF PHYSICS, 69-, CHMN DEPT, 70- Mem: Am Asn Physics Teachers; Nat Sci Teachers Asn. Res: Nuclear emulsion techniques; physics education. Mailing Add: Dept of Physics Ft Hays Kans State Col Hays KS 67601

WITTEN, THOMAS A, b Tulsa, Okla, July 19, 16; m 43; c 9. INTERNAL MEDICINE, GASTROENTEROLOGY. Educ: Univ Tex, AB, 38, MD, 42. Prof Exp: Intern, Med Col Va, 42-43; house officer, Johns Hopkins Hosp, Baltimore, 46-47; resident, Univ Utah, 47-49; clin instr, 51-53, asst clin prof, 53-57, from asst prof to assoc prof, 57-75, PROF MED, SCH MED, UNIV COLO MED CTR, DENVER, 75- Concurrent Pos: Instr, Vet Admin Hosp, Grand Junction, Colo, 51-53, asst dir res lab, Vet Admin Hosp, Ft Logan, 51-53, asst chief med & chief gastroenterol, Vet Admin Hosp, Denver, 52-, assoc chief staff res & educ, 61-68. Mem: Fel Am Col Physicians; Am Fedn Clin Res; Am Gastroenterol Asn; AMA. Res: Studies in the effect of alcohol on metabolic processes. Mailing Add: Vet Admin Hosp Rm A501a 1055 Clermont Denver CO 80220

WITTEN, THOMAS ADAMS, JR, b Raleigh, NC, Aug 24, 44. THEORETICAL PHYSICS. Educ: Reed Col, AB, 66; Univ Calif, San Diego, PhD(physics), 71. Prof Exp: Instr physics, Princeton Univ, 71-74; foreign collabr, Comn Atomic Energy, Saclay, France, 74-75; ASST PROF PHYSICS, UNIV MICH, ANN ARBOR, 75- Mem: Am Phys Soc. Res: Renormalization scaling symmetry in extended matter; excitations of systems with long-range order. Mailing Add: Randall Lab Univ of Mich Ann Arbor MI 48109

WITTEN, THOMAS RINER, b Tazewell, Va, Apr 5, 42; m 69; c 2. EXPERIMENTAL NUCLEAR PHYSICS. Educ: Va Polytech Inst & State Univ, BS, 63, PhD(physics), 70. Prof Exp: Fel physics, Univ Victoria, 69-71; res assoc, Rice Univ, 71-75; ASST PROF PHYSICS, KENT STATE UNIV, 75- Res: Interaction of pions and nucleons with nuclei at intermediate energy from standpoint of basic physics and radiotherapy. Mailing Add: Dept of Physics Kent State Univ Kent OH 44242

WITTENBACH, VERNON ARIE, b Belding, Mich, Dec 13, 45; m 68; c 1. PLANT PHYSIOLOGY. Educ: Mich State Univ, BS, 68, MS, 70, PhD(hort), 74. Prof Exp: RES BIOLOGIST, CENT RES & DEVELOP DEPT, E I DU PONT DE NEMOURS & CO, 74- Mem: Am Soc Plant Physiologists; Am Soc Hort Sci. Res: Physiology of plant senescence, especially the mode of action of cytokinins in delarying senescence. Mailing Add: E I du Pont de Nemours & Co Cent Res & Develop Dept Exp Sta Wilmington DE 19898

WITTENBERG, ALBERT M, b Newark, NJ, Jan 3, 25; m 61; c 2. PHYSICAL ELECTRONICS. Educ: Union Col, NY, BS, 49; Johns Hopkins Univ, PhD(physics), 55. Prof Exp: MEM TECH STAFF PHYSICS, BELL LABS, 55- Mem: Am Phys Soc; Inst Elec & Electronics Eng. Res: Atomic structure of diatomic molecules; gaseous electronics; radiative heat transfer; optical spectroscopy; photoconductivity; interconnection technology. Mailing Add: 19 Exeter Rd Short Hills NJ 07078

WITTENBERG, BEATRICE A, b Berlin, Ger, Nov 6, 28; US citizen; m 54; c 3. BIOCHEMISTRY, PHYSIOLOGY. Educ: Univ Toronto, BA, 49, MA, 50; Western Reserve Univ, PhD(pharmacol), 54. Prof Exp: Res assoc physiol, Western Reserve Univ, 54-55; cancer res, Delafield Hosp, NY Univ, 55-56; ASST PROF PHYSIOL, ALBERT EINSTEIN COL MED, 64- Mem: Am Soc Biol Chem. Res: Heme proteins. Mailing Add: Dept of Physiol Albert Einstein Col of Med Bronx NY 10461

WITTENBERG, JONATHAN B, b New York, NY, Sept 19, 23; m 54; c 3. BIOCHEMISTRY, PHYSIOLOGY. Educ: Harvard Univ, BS, 45; Columbia Univ, MA, 46, PhD, 50. Prof Exp: From instr to asst prof biochem, Western Reserve Univ, 52-55; asst prof, 55, from asst prof to assoc prof physiol, 56-65, PROF PHYSIOL, ALBERT EINSTEIN COL MED, 65- Concurrent Pos: NIH res fel, 52. Res: Porphyrins; swimbladders; retia mirabilia; oxygen transport; myoglobin; hemeproteins; hemoglobin. Mailing Add: Dept of Physiol Albert Einstein Col of Med Bronx NY 10461

WITTENBERG, LAYTON JUNIOR, b Toledo, Ohio, May 6, 27; m 53; c 3. PHYSICAL INORGANIC CHEMISTRY. Educ: Ohio State Univ, BS, 49; Univ Wis, PhD(chem), 53. Prof Exp: SR RES SPECIALIST, MOUND LAB, US ENERGY RES & DEVELOP ADMIN, MONSANTO RES CORP, 53- Mem: Am Chem Soc; Am Nuclear Soc. Res: Plutonium and rare earth metallurgy and ceramics; nuclear fuels; liquid metals; fusion reactor materials. Mailing Add: Monsanto Res Corp Mound Lab Miamisburg OH 45342

WITTENBERGER, CHARLES LOUIS, b Ogallala, Nebr, June 22, 30; m 56; c 3. BACTERIOLOGY. Educ: Univ Creighton, BS, 52, MS, 54; Ind Univ, PHD(bact), 59. Prof Exp: Vis scientist, NIH, 59-61, res mcirobiologist, 61-68, CHIEF MICROBIAL PHYSIOL SECT, NAT INST DENT RES, 68- Mem: Am Acad Microbiol; Am Soc Microbiol; Am Soc Biol Chemists. Res: Microbial metabolism; electron transport in Hydrogenomonas species; intermediary metabolism of lactic acid; biochemical regulation of microbial metabolism. Mailing Add: Lab of Microbiol Nat Inst of Dent Res Bethesda MD 20014

WITTENBORN, AUGUST FERDINAND, b Guadalupe Co, Tex, Aug 11, 23; m 46; c 2. PHYSICS. Educ: Univ Tex, PhD(physics), 51. Prof Exp: Res physicist, Defense Res Lab, Univ Tex, 49-51 & 53-58; mem staff, Los Alamos Sci Lab, 51-52; sr vpres, Tracor, Inc, 58-69, pres, Tracor Comput Corp, 69-72, PRES, NAT CON-SERV, INC, 72- Mem: Acoust Soc Am. Res: Hydrodynamics and electromagnetic wave propagation. Mailing Add: Nat Con-Serv Inc 6500 Tracor Lane Box 6-13 Austin TX 78721

WITTENBROOK, LAWRENCE SETH, organic chemistry, see 12th edition

WITTER, JOHN ALLEN, b Jamestown, NY, Sept 2, 43; m 71. FOREST ENTOMOLOGY, INSECT ECOLOGY. Educ: Va Polytech Inst, BS, 65, MS, 67; Univ Minn, St Paul, PhD(entom), 71. Prof Exp: Res technician entom, Southeastern Forest Exp Sta, US Forest Serv, 66-67; res asst, Univ Minn, St Paul, 67-71, res fel, 71-72, instr forest entom, 72; ASST PROF FOREST ENTOM, SCH NATURAL RESOURCES, UNIV MICH, ANN ARBOR, 72- Mem: Entom Soc Am; Entom Soc Can; Soc Am Foresters. Res: Population dynamics of forest insects, parasites and predators of forest insects; defoliators; tent caterpillars; large aspen tortrix; sugar maple insects; life tables; insect behavior. Mailing Add: Samuel T Dana Bldg Sch of Natural Resources Univ of Mich Ann Arbor MI 48104

WITTER, LLOYD DAVID, b Chicago, Ill, May 15, 23; m 50; c 3. FOOD MICROBIOLOGY, FOOD SCIENCE. Educ: Univ Wash, Seattle, BS, 45, MS, 50, PhD(microbiol), 53. Prof Exp: Asst microbiol, Univ Wash, Seattle, 52-53; res chemist, Continental Can Co, Ill, 53-56; from asst prof to assoc prof, 56-67, PROF FOOD MICROBIOL, UNIV ILL, URBANA, 67- Concurrent Pos: Vis prof appl biochem, Univ Nottingham, Eng, 72. Mem: Fel Am Acad Microbiol; Am Soc Microbiol; Am Dairy Sci Asn; Inst Food Technol; NY Acad Sci. Res: Microbiology of food and dairy products; psychrophilic bacteria; bacterial growth on solid surfaces; heat resistance and injury of bacteria; institutional foods. Mailing Add: Dept of Food Sci Univ of Ill Urbana IL 61801

WITTER, RICHARD L, b Bangor, Maine, Sept 10, 36; m 62; c 1. POULTRY PATHOLOGY. Educ: Mich State Univ, BS, 58, DVM, 60; Cornell Univ, MS, 62, PhD, 64. Prof Exp: Res vet, Regional Poultry Res Lab, 64-75, DIR, REGIONAL POULTRY RES LAB, AGR RES SERV, USDA, 75- Concurrent Pos: Asst prof, Mich State Univ, 64-71, assoc prof, 71- Honors & Awards: Am Asn Avian Path Award, 66; Res Award, Poultry Sci Asn, 71; Res Award, Sigma Xi, 75. Mem: Am Vet Med Asn; Am Asn Avian Path; Conf Res Workers Animal Dis; Poultry Sci Asn; Sigma Xi. Res: Epizootiology and control of poultry diseases, especially virology and pathology; viral-induced neoplasia, especially Marek's disease and lymphoid leukosis of chickens. Mailing Add: Regional Poultry Res Lab 3606 E Mt Hope Rd East Lansing MI 48823

WITTERHOLT, EDWARD JOHN, b Osceola Mills, Pa, Nov 12, 35; m 57; c 5. APPLIED MATHEMATICS. Educ: Manhattan Col, BS, 57; Brown Univ, ScM, 59, PhD(appl math), 64. Prof Exp: RES PROJ MATHEMATICIAN, SCHLUMBERGER TECHNOL CORP, 63- Mem: Am Phys Soc; Acoust Soc Am. Res: Analysis of pressure and temperature phenomena in petroleum reservoirs; multi-phase fluid flow; acoustic wave propagation. Mailing Add: Schlumberger Technol Corp Ridgefield CT 06877

WITTERHOLT, VINCENT GERARD, b New York, NY, Sept 24, 32; m 54; c 6. ORGANIC CHEMISTRY. Educ: Queens Col, NY, BS, 53; Purdue Univ, PhD(org chem), 58. Prof Exp: Res chemist, Org Chem Dept, 58-59, res supvr, 69-73, sr supvr sulfur colors area, Chambers Works, 73-74, chief supvr, Azo Lab, 74-75, DIV HEAD, ORG CHEM DEPT, JACKSON LAB, CHAMBERS WORKS, E I DU PONT DE NEMOURS & CO, INC, 75- Mem: Am Chem Soc. Res: Organofluorine chemistry; heterocyclic and dye chemistry; photochemistry; fluorescence; photo stabilization; colloid chemistry; new dyes product and process development. Mailing Add: Org Chem Dept Jackson Lab E I du Pont de Nemours & Co Inc Deepwater NJ 08023

WITTERS, ROBERT DALE, b Cheyenne, Wyo, May 2, 29. PHYSICAL CHEMISTRY. Educ: Univ Colo, BA, 51; Mont State Univ, PhD(phys chem), 64. Prof Exp: Chemist, E I du Pont de Nemours & Co, 51-53; asst prof chem, State Univ NY Col Plattsburgh, 59-62 & Mont State Univ, 62-63; res fel, Harvey Mudd Col, 64-65; asst prof, 65-69, ASSOC PROF CHEM, COLO SCH MINES, 69- Mem: Am Chem Soc; Am Crystallog Asn. Res: X-ray crystallography. Mailing Add: Dept of Chem Colo Sch of Mines Golden CO 80401

WITTERS, WELDON L, b Dayton, Ohio, Dec 13, 29; m 59; c 2. REPRODUCTIVE PHYSIOLOGY, BIOLOGY. Educ: Ball State Univ, BS, 52, MS, 55; Purdue Univ, MS, 64, PhD(animal physiol), 67. Prof Exp: Teacher, High Schs, Ind, 55-67; asst prof, 67-70, ASSOC PROF ZOOL, OHIO UNIV, 70- Concurrent Pos: NSF travel grant, France, 68; res grant, Ohio Univ, 68-69; NSF grant biol educ, 70-72; consult drug educ, Ohio. Mem: Nat Asn Biol Teachers; Nat Sci Teachers Asn; Soc Study Reproduction; Am Soc Animal Sci. Res: Metabolic pathways functional in sperm; effect of metabolic inhibitors on sperm; hallucinogenic alkaloids metabolic effect on tissue; laboratory concepts taught with visuals and video tapes. Mailing Add: 47 Avon Pl Athens OH 45701

WITTES, JANET TURK, b Pittsburgh, Pa, May 12, 43; m 64; c 2. STATISTICS, BIOSTATISTICS. Educ: Radcliffe Col, BA, 64; Harvard Univ, MA, 65, PhD(statist), 70. Prof Exp: Res assoc, Sch Pub Health, Univ Pittsburgh, 70-72 & Dept Statist, George Washington Univ, 72-73; adj asst prof pub health, Dept Epidemiol, Sch Pub Health, Columbia Univ, 73-74; ASST PROF MATH, HUNTER COL, 74- Concurrent Pos: Consult, Dept Epidemiol, Sch Pub Health, Columbia Univ, 74- Mem: Biomet

Soc; Am Statist Asn. Res: Applied statistical methodology, especially biostatistical techniques. Mailing Add: Dept of Math Hunter Col 695 Park Ave New York NY 10021

WITTHOFT, JOHN, b Oneonta, NY, Oct 4, 21; m; c 3. ANTHROPOLOGY. Educ: NY State Teachers Col, BA, 44; Univ Pa, MA, 45. Prof Exp: Instr, Army Spec Training Prog, Univ Pa, 45-46; asst state anthropologist, State Mus, Pa, 48-49, state anthropologist, 48-50, dir, 55-58, state anthropologist, 59-67; lectr anthrop, 67-69, ASSOC PROF ANTHROP & RES ASSOC AM SECT, UNIV MUS, UNIV PA, 69- Mem: Am Anthrop Asn; Am Soc Archaeol. Res: Archaeology of eastern woodlands, especially New York and Pennsylvania; ethnology of eastern woodlands, especially Cherokee and Iroquois. Mailing Add: Dept of Anthrop Univ of Pa Philadelphia PA 19174

WITTHUHN, BURTON ORRIN, b Allentown, Pa, Aug 22, 34; m 61; c 1. GEOGRAPHY. Educ: Kutztown State Col, BS, 56; Pa State Univ, MEd, 62, PhD(geog), 68. Prof Exp: Teacher math, Allentown Sch Dist, 56-63; instr geog, Pa State Univ, 65, res asst, 66; asst prof, Ohio State Univ, 67-70; PROF GEOG & CHMN DEPT, EDINBORO STATE COL, 70- Concurrent Pos: Consult, Proj Africa, 69-71 & Inst Community Serv, 73- Mem: Asn Am Geogr; Nat Coun Geog Educ; African Studies Asn; Regional Sci Asn; Am Geog Asn. Res: Patterns of social change; regional resource planning and development; organization of territory; enviromental impact and decision-making. Mailing Add: Dept of Geog Edinboro State Col Edinboro PA 16444

WITTICK, JAMES JOHN, b New York, NY, Aug 17, 30; m 56; c 4. ANALYTICAL CHEMISTRY. Educ: Col Holy Cross, BS, 52; Tufts Univ, MS, 55; Univ Pa, PhD(chem), 66. Prof Exp: Chemist, Merck Sharp & Dohme Res Labs, NJ, 55-57, group leader phys & anal res, 57-60, unit head pharmaceut anal, Pa, 60-62, sr res chemist, 65-68, res fel, 68-70, ASSOC DIR QUAL CONTROL, MERCK CHEM DIV, MERCK & CO, INC, 70- Mem: Am Chem Soc. Res: Purity and structure determination of organic compounds; pharmaceutical analysis; electro-analytical chemistry; x-ray diffraction; chromatography. Mailing Add: Qual Control Dept Merck & Co Inc Merck Chem Div Rahway NJ 07065

WITTIE, LARRY DAWSON, b Bay City, Tex, Mar 9, 43; m 72; c 1. COMPUTER SCIENCE. Educ: Calif Inst Technol, BS, 66; Univ Wis-Madison, MS, 67, PhD(comput sci), 73. Prof Exp: Systs programmer, Calif Inst Technol, 63-66 & IBM Corp, Sunnyvale, 66; NASA trainee comput sci, Univ Wis-Madison, 66-69, res asst, 69-72; asst prof comput sci, Purdue Univ, 72-73; ASST PROF COMPUT SCI, STATE UNIV NY BUFFALO, 73- Mem: Asn Comput Mach; Inst Elec & Electronics Engrs; Soc Neurosci. Res: Parallel information processing in networks, with emphasis on computer architecture interconnection techniques for efficient communications among millions of microcomputers; simulation of large brain models; neural distributed memory mechanisms. Mailing Add: Dept of Comput Sci State Univ NY at Buffalo 4226 Ridge Lea Rd Amherst NY 14226

WITTIG, GERTRAUDE CHRISTA, b Glauchau, Ger, Oct 4, 28; US citizen. ZOOLOGY, INSECT PATHOLOGY. Educ: Univ Tübingen, Dr rer nat(zool, bot, biochem), 55. Prof Exp: Teacher, Musterschule Glauchau, Ger, 46-47 & Preuniv Sch Neckarsulm, 55; Ger Res Asn res assoc, Zool Inst, Univ Tübingen, 56-58; res assoc entom, Univ Calif, Berkeley, 58-59; microbiologist, Insect Path Pioneering Lab, Entom Res Div, Agr Res Serv, USDA, Md, 59-62; res microbiologist, Forestry Sci Lab, Pac Northwest Forest & Range Exp Sta, US Forest Serv, Ore, 62-68; assoc prof biol sci, 68-75, PROF BIOL SCI, SOUTHERN ILL UNIV, EDWARDSVILLE, 75- Concurrent Pos: Lalor Found fel, 58; Ger Acad Exchange Serv & Ministry of Educ Baden-Württemburg res grants, 58; Fulbright travel grant, 58-59; consult, Univ Ariz, 61; assoc prof, Ore State Univ, 62-68. Mem: AAAS; Am Soc Cell Biologists; Entom Soc Am; Electron Micros Soc Am; Soc Invert Path. Res: Biological ultrastructure; morphology and histology of insects; insect pathology, particularly histopathological and ultrastructural problems. Mailing Add: Biol Sci Southern Ill Univ Edwardsville IL 62026

WITTIG, HEINZ JOSEPH, b Munich, Ger, May 19, 21; US citizen; m 60; c 4. ALLERGY, IMMUNOLOGY. Educ: Univ Munich, MD, 51; Am Bd Pediat, dipl, 56, cert allergy, 61. Prof Exp: Instr & fel pediat allergy, Univ Rochester, 57-58; fel, NY Univ, 58-60; clin asst prof, Sch Med, Seton Hall Univ, 60-64; asst prof pediat, Sch Med, WVa Univ, 64-68; chief div pediat allergy, 68, PROF PEDIAT, COL MED, UNIV FLA, 68- Mem: Fel Am Acad Allergy; fel Am Acad Pediat; Am Thoracic Soc. Res: Clinical immunology; bronchial asthma, etiology of intrinsic form; lymphocyte stimulation with various antigens, including cancer; macrophage inhibition; bronchial challenge procedures. Mailing Add: Dept Pediat Div Pediat Allergy Univ Fla Col Med Box J-296 Gainesville FL 32610

WITTIG, KENNETH PAUL, b Pittsburgh, Pa, Aug 19, 46; m 69. INVERTEBRATE PHYSIOLOGY. Educ: St Vincent Col, BS, 68; Kent State Univ, PhD(animal physiol), 74. Prof Exp: Teacher biol & chem, St Thomas Aquinas High Sch, 68-69; ASST PROF BIOL & PHYSIOL, SIENA COL, 74- Mem: AAAS; Sigma Xi; Am Inst Biol Sci; Am Soc Zoologists. Res: Investigation of the neuroendocrine control of calcium metabolism in the crayfish. Mailing Add: Dept of Biol Siena Col Loundonville NY 12211

WITTING, HARALD LUDWIG, b Duisburg, Ger, Sept 23, 36; US citizen; m 60; c 3. PLASMA PHYSICS. Educ: Mass Inst Technol, BS & MS, 59, ScD(plasma), 64. Prof Exp: PHYSICIST, GEN ELEC RES & DEVELOP CTR, 64- Mem: AAAS; Am Phys Soc. Res: Gas discharges; electrodes; photochemical reactions. Mailing Add: Gen Elec Res & Develop Ctr PO Box 8 Schenectady NY 12305

WITTING, JAMES M, b Chicago, Ill, Jan 14, 38; m 62; c 3. PHYSICAL OCEANOGRAPHY, HYDRODYNAMICS. Educ: John Carroll Univ, BS, 59, MS, 60; Mass Inst Technol, PhD(physics), 64. Prof Exp: Assoc physicist, IIT Res Inst, 64-66, res physicist, 66; asst prof hydrodyn, Dept Geophys Sci, Univ Chicago, 66-70; phys sci adminr, Off Naval Res, 70-73, oceanogr, Phys Oceanog Prog, 72-73, HEAD, PHYS OCEANOG BR, US NAVAL RES LAB, 73- Mem: Am Phys Soc; Am Geophys Union. Res: Waves in dispersive media; water waves; undular bores; numerical modeling of oceanic circulation. Mailing Add: Phys Oceanog Br US Naval Res Lab Washington DC 20375

WITTING, LLOYD ALLEN, b Chicago, Ill, May 18, 30; m 56; c 3. NUTRITION, LIPID CHEMISTRY. Educ: Univ Ill, BS, 52, MS, 53, PhD, 56. Prof Exp: Asst food technol, Univ Ill, 52-55; assoc biochemist, Am Meat Inst Found, Chicago, 55-57; proj assoc physiol chem, Univ Wis, 57-59; med res assoc, Mendel Res Lab, Elgin State Hosp, Ill, 59-68, actg dir, 68-70, res scientist, 69-72; assoc prof food sci & technol, Tex Woman's Univ, 72-74; CONSULT, CITY OF DENTON, TEX, 75- Concurrent Pos: Asst prof, Col Med, Univ Ill, 62-72. Mem: Am Inst Nutrit; Am Oil Chem Soc; Geront Soc; Am Soc Biol Chemists; Am Soc Neurochem. Res: Polyunsaturated fatty acids; glycolipids; clinical and animal nutrition; chemistry and biochemistry of lipids; tocopherol; gas chromatography. Mailing Add: 3409 Dunes St Denton TX 76201

WITTKE, JAMES PLEISTER, b Westfield, NJ, Apr 2, 28; m 52; c 2. OPTICAL PHYSICS. Educ: Stevens Inst Technol, ME, 49; Princeton Univ, MA, 52, PhD(physics), 55. Prof Exp: Instr physics, Princeton Univ, 54-55; MEM TECH STAFF, RCA LABS, 55- Mem: Fel Am Phys Soc; Inst Elec & Electronics Eng; Optical Soc Am. Res: Microwave spectroscopy; masers; lasers; fiber optics; optical instrumentation. Mailing Add: RCA Labs Princeton NJ 08540

WITTKOWER, ANDREW BENEDICT, b London, Eng, Nov 7, 34; m 57; c 2. ATOMIC PHYSICS. Educ: McGill Univ, BSc, 55; Cambridge Univ, MSc, 59; Univ London, PhD(atomic physics), 67. Prof Exp: Proj physicist, High Voltage Eng Corp, 59-64, res physicist, 64-67, assoc dir res, 67-71; sr vpres, Extrion Corp, 71-75, MKT & ASST GEN MGR, EXTRION DIV, VARIAN ASSOC, 75- Mem: Fel Am Phys Soc; Brit Inst Physics. Res: Atomic physics applied to the development of ion accelerators. Mailing Add: Varian/Extrion Box 1226 Gloucester MA 01930 22

WITTKOWER, ERIC DAVID, b Berlin, Ger, Apr 4, 99; nat Can; m 31; c 2. PSYCHIATRY. Educ: Univ Berlin, MD, 24; FRCPS, cert psychiat, 55. Prof Exp: Lectr psychosom med, Univ Berlin, 31; mem res staff, Cent Path Lab, Maudsley Hosp, London, Eng, 33-35; Sir Halley-Stewart res fel & physician, Tavistock Clin, 35; res fel psychiat, Nat Asn Prev Tuberc, Gt Brit, 45; psychiatrist, Dermat Dept, St Bartholomew's Hosp, London, 48; from asst prof to prof psychiat, 51-72, dir transcult 56, EMER PROF PSYCHIAT, McGILL UNIV, 72- Concurrent Pos: Consult, Royal Victoria & Montreal Gen Hosps, 51-; mem expert adv panel ment health, WHO, 64-69; consult, Queen Elizabeth Hosp, 65; mem bd, Fr-Am Inst Ment Health, Paris, 73. Mem: Fel Am Asn Social Psychiat; fel Royal Col Psychiatrists; Am Psychiat Asn; Can Psychoanal Soc (pres, 66-); Int Col Psychosom Med (pres, 72). Res: Psychosomatic medicine; transcultural psychiatry. Mailing Add: Sect Transcult Psychiat Studies McGill Univ Beatty Hall Montreal PQ Can

WITTLAKE, EUGENE BISHOP, b Sheridan, Wyo, Aug 14, 12; m 47; c 1. BRYOLOGY. Educ: Augustana Col, AB, 36; Univ Iowa, MS, 38; Univ Kans, PhD(bot), 54 Prof Exp: Instr bot, Univ Ark, 46-51 & Univ Kans, 51-54; asst prof biol, Iowa State Teachers Col, 54-56; from asst prof to assoc prof, 56-59, cur mus, 59-68, PROF BOT, ARK STATE UNIV, 59-, DIR MUS, 68- Concurrent Pos: Partic, Southeast Mus Conf. Mem: Am Asn Mus. Res: Bryology and paleobotany of Arkansas. Mailing Add: Ark State Univ Mus Drawer HH State University AR 72467

WITTLE, JOHN KENNETH, b Lancaster, Pa, July 20, 39; c 1. INORGANIC CHEMISTRY, ANALYTICAL CHEMISTRY. Educ: Franklin & Marshall Col, AB, 62; Purdue Univ, Lafayette, PhD(inorg chem), 68. Prof Exp: Inorg chemist, 67-68, proj engr, 68-72, mgr polymer technol, 72-75, MGR DIELECTRIC MAT LAB, GEN ELEC CO, 75- Mem: AAAS; Am Chem Soc; The Chem Soc; Am Inst Mining, Metall & Petrol Eng. Res: Insulation systems for electrical systems; decomposition of electrical insulation under electrical stress; materials application. Mailing Add: Gen Elec Co 6901 Elmwood Ave Philadelphia PA 19142

WITTLE, LAWRENCE WAYNE, b Mt Joy, Pa, Nov 20, 41; m 64; c 2. PHYSIOLOGY. Educ: Lebanon Valley Col, BS, 63; Univ Va, PhD(biol), 68. Prof Exp: NIH fel, Inst Marine Sci, Univ Miami, 68-70; ASST PROF BIOL, ALMA COL, 70- Mem: AAAS; Am Soc Zool; Int Soc Toxinology. Res: Physiological and pharmacological properties of marine toxins. Mailing Add: Dept of Biol Alma Col Alma MI 48801

WITTLER, RUTH GRAESER, b Baltimore, Md, July 19, 18. MICROBIOLOGY. Educ: Oberlin Col, AB, 40; Johns Hopkins Univ, MS, 44; Western Reserve Univ, PhD(bact), 47; Am Bd Microbiol, dipl, 64. Prof Exp: Lab technician, St Joseph's Hosp, Baltimore, 41-42; technician, Harriet Lane Home, Johns Hopkins Univ, 42; instr immunochem, Med Sch, Western Reserve Univ, 47-51, sr instr biochem, 51-52, asst prof, 52-53; Vet Admin bacteriologist, Walter Reed Army Inst Res, 53-64, microbiologist, 64-73; RETIRED. Concurrent Pos: Jenner Mem fel, Lister Inst Prev Med, London, 49-51; adj instr, Sch Med, Univ Md, 58-73; mem tech adv comt, Am Type Cult Collection, 63-66, extramural curator Mollicutes & L-phase variants, 72-73; civilian career prog coordr, Walter Reed Army Med Ctr, 68-71, career adv, 71-73; mem subcomt nomenclature mycoplasmas, Int Mycoplasma Characterization Prog, WHO-Food & Agr Orgn, 69-73. Mem: AAAS; Am Soc Microbiol; fel Am Acad Microbiol; Soc Gen Microbiol. Res: Biology, taxonomy and type culture collections of mycoplasmatales and L-phase variants; role of L-phase of bacteria invivo. Mailing Add: 83 Bay Dr Bay Ridge Annapolis MD 21403

WITTLIFF, JAMES LAMAR, b Taft, Tex, June 15, 38; m 62; c 2. BIOCHEMISTRY. Educ: Univ Tex, Austin, BA, 61, PhD(molecular biol), 67; La State Univ, MS, 63. Prof Exp: USPHS fel biochem regulation, Oak Ridge Nat Lab, Tenn, 67-69; asst prof biochem, Sch Med & Dent, 69-74, ASSOC PROF BIOCHEM & HEAD SECT ENDOCRINE BIOCHEM, CANCER CTR, UNIV ROCHESTER, 75- Mem: AAAS; Am Chem Soc; Am Asn Cancer Res; Am Soc Biol Chemists; Endocrine Soc. Res: Hormonal control of protein and nucleic acid synthesis; role of specific hormone receptors in target cell response. Mailing Add: Dept of Biochem Sch Med & Dent Univ of Rochester Rochester NY 14642

WITTMAN, JAMES SMYTHE, III, b Ft Bragg, NC, Mar 1, 43; m 65; c 3. BIOCHEMISTRY, NUTRITION. Educ: La Col, BA, 64; Tulane Univ, PhD(biochem), 70. Prof Exp: Chemist, Southern Regional Res Lab, USDA, 64-65; org chemist, US Customs Lab, La, 65-66; asst biochemist, 68-70, SR BIOCHEMIST, HOFFMANN-LA ROCHE INC, 70- Mem: AAAS. Res: Biochemical nutrition; regulation of intermediary metabolism and protein synthesis. Mailing Add: Biochem Nutrit Dept Hoffmann-La Roche Inc Nutley NJ 07110

WITTMAN, WILLIAM F, b Pittsburgh, Pa, Oct 10, 37; m 64; c 4. ORGANIC CHEMISTRY. Educ: Carnegie Inst Technol, BS, 59; Univ Nebr, PhD(org chem), 65. Prof Exp: Sr res chemist, 64-72, PATENT LIAISON SPECIALIST & PATENT AGENT, 3M CO, 72- Mem: Am Chem Soc. Res: Pharmaceutical, agrichemical and biopolymer patents. Mailing Add: 218-1 Riker Labs 3M Co 3M Ctr St Paul MN 55101

WITTMANN, WALTER I, oceanography, meteorology, see 12th edition

WITTNER, MURRAY, b New York, NY, Apr 23, 27; m 55; c 2. PHYSIOLOGY, PARASITOLOGY. Educ: Univ Ill, ScB, 48, ScM, 49; Harvard Univ, PhD, 55; Yale Univ, MD, 61. Prof Exp: Instr path & parasitol & consult path, 56-57, PROF PATH & PARASITOL & DIR PARASITOL LABS, ALBERT EINSTEIN COL MED, 67- Concurrent Pos: Career scientist, Health Res Coun, New York, 67-; attend physician, Bronx Munic Hosp Ctr, Lincoln Hosp & Albert Einstein Col Med Hosp; dir, Trop Dis Clin, Lincoln Hosp & Bronx Munic Hosp Ctr. Mem: Soc Protozool; Am Soc Cell Biol; Am Asn Path & Bact; Am Soc Zool; Am Soc Parasitol. Res: Physiology and biochemistry of oxygen poisoning; physiology of parasites; experimental pathology. Mailing Add: Dept of Path Albert Einstein Col of Med Bronx NY 10461

WITTRY, ESPERANCE, b Marshall, Minn, Jan 13, 20. BIOLOGY. Educ: Col St

Catherine, BA, 46; Univ Notre Dame, MS, 54, PhD(biol), 60. Prof Exp: Assoc prof, 60-71, PROF BIOL, COL ST CATHERINE, 71- Res: Physiology; radiation biology; neurophysiology. Mailing Add: Dept of Biol Col of St Catherine St Paul MN 55105

WITTSELL, LAWRENCE EUGENE, b Neosho County, Kans, Mar 2, 29; m 51; c 6. SOIL SCIENCE. Educ: Brigham Young Univ, BS, 57; Kans State Univ, MS, 59, PhD(agron), 64. Prof Exp: Res asst soil mgt, Kans State Univ, 56-62; asst agronomist, Trop Res Div, United Fruit Co, 62-65; plant nutritionist, Agr Res Lab, 65-73, AGRONOMIST, PESTICIDE DEVELOP DEPT, BIOL SCI RES CTR, SHELL DEVELOP CO, 73- Mem: Am Soc Agron; Soil Sci Soc Am; Int Soc Soil Sci; Coun Agr Sci & Technol. Res: Effects of artificial compaction of the soil on growth of plants in field and greenhouse; weed control and other cultural practices in bananas; general fertilizer and plant nutrition investigations; research and development of herbicides. Mailing Add: Shell Develop Co PO Box 4248 Modesto CA 95352

WITTSON, CECIL L, b Camden, SC, Jan 14, 07; m 34; c 1. PSYCHIATRY. Educ: Univ SC, BS, 28; Med Col of SC, MD, 31; Am Bd Neurol & Psychiat, dipl, 48. Prof Exp: Staff psychiatrist, Cent Islip State Hosp, NY, 32-50; prof neurol & psychiat, 50-71, chmn dept, 50-64, dean col, 64-71, pres, Med Ctr, 68-71, EMER PROF NEUROL & PSYCHIAT, COL MED & EMER CHANCELLOR, MED CTR, UNIV NEBR, OMAHA, 72-; MED PROG DIR, HENNINGSON, DURHAM & RICHARDSON, INC, 72- Concurrent Pos: Attend psychiatrist, Out-Patient Dept, NY Psychiat Inst, 35-36; res consult, 46-; dir, Nebr Psychiat Inst, 50-64; chief clin servs, Nebr Bd Control, 51-64, dir ment health, 54-67; mem, Surgeon Gen Adv Comt Indian Health, 65-; nat adv comt alcoholism, US Dept Health, Educ & Welfare, 66- Mem: AAAS; fel Am Psychiat Asn; Asn Mil Surg US; Am Psychopath Asn; Am Asn Ment Deficiency. Res: Psychiatric selection; medical education. Mailing Add: 9651 N 29th St Omaha NE 68112

WITTSTRUCK, THOMAS ANDREW, physical chemistry, see 12th edition

WITTWER, JOHN WILLIAM, b Columbus, Ohio, Apr 35; m 59; c 3. PERIODONTOLOGY. Educ: Ohio State Univ, DDS, 59, MSc, 65. Prof Exp: Instr periodont, Ohio State Univ, 61-64; asst prof, Sch Dent, Loyola Univ Chicago, 66-69; assoc prof, 69-75, PROF PERIODONT, SCH DENT, UNIV LOUISVILLE, 75- Mem: Int Asn Dent Res; Am Acad Periodont. Res: Periodontal anatomy; periodontal immunology. Mailing Add: Health Sci Ctr Univ of Louisville Sch Dent Louisville KY 40201

WITTWER, LELAND S, b Belleville, Wis, Apr 26, 19. ANIMAL NUTRITION. Educ: Mich State Univ, BS, 52; Cornell Univ, MS, 54, PhD(dairy prod), 56. Prof Exp: Asst prof animal sci, Univ Mass, 56-58; PROF ANIMAL SCI, UNIV WIS-RIVER FALLS, 58- Mem: AAAS; Am Dairy Sci Asn; Am Soc Animal Sci. Mailing Add: Dept of Animal Sci Univ of Wis Col Agr River Falls WI 54022

WITTWER, SYLVAN HAROLD, b Hurricane, Utah, Jan 17, 17; m 38; c 4. HORTICULTURE. Educ: Utah State Agr Col, BS, 39; Univ Mo, PhD(hort), 43. Prof Exp: Asst, Univ Mo, 40-43, instr hort, 43-46; from asst prof to assoc prof, 46-51, PROF HORT, MICH STATE UNIV, 51-, DIR AGR EXP STA, 65- Concurrent Pos: Consult, Rockefeller Found, Mex, 68-69, Ford Found, Ceylon, 69- & UN Develop Prog, 71-; mem agr bd, Nat Acad Sci, 71-; chmn bd agr & renewable resources, Nat Acad Sci-Nat Res Coun, 73- Honors & Awards: Campbell Award, AAAS, 57. Mem: AAAS; Soc Develop Biol; Am Soc Hort Sci; Am Soc Plant Physiol; Bot Soc Am. Res: Physiology of reproduction in horticulture; plant growth regulators for improving fruit set and control flowering; nutrition of horticultural crops; radioisotopes in mineral nutrition of plants. Mailing Add: Agr Exp Sta Mich State Univ East Lansing MI 48824

WITTY, RALPH, b Windsor Mills, Que, Nov 19, 20; m 52; c 2. BIOCHEMISTRY, NUTRITION. Educ: McGill Univ, BS, 48, PhD(biochem), 51. Prof Exp: Res chemist, 51-56, group leader biochem, 56-68, ASST DIR RES, RES & DEVELOP LABS, CAN PACKERS LTD, 68- Mem: AAAS; Can Inst Food Sci; Nutrit Soc Can; Can Biochem Soc. Res: Animal and human fats; oils, soaps and detergents; digestion and absorption of proteins and availability of amino acids in proteins. Mailing Add: Can Packers Ltd Res Ctr 2211 St Clair Ave W Toronto ON Can

WITUCKI, JEANNETTE RENNER, b Omaha, Nebr, Feb 28, 30; m 49; c 2. ANTHROPOLOGY, ANTHROPOLOGICAL LINGUISTICS. Educ: Univ Calif, Los Angeles, BA, 60, PhD(anthrop), 66. Prof Exp: Asst prof, 67-74, ASSOC PROF ANTHROP, CALIF STATE UNIV, LOS ANGELES, 74- Concurrent Pos: Res grants, Calif State Univ Found, 68 & Sigma Xi, 71. Mem: Am Anthrop Asn; Am Ethnol Soc; Ling Soc Am; Int Ling Asn. Res: Cross-cultural and ethno-linguistic research; correlations between language and culture; ethno-semantics; cognitive anthropology; basic linguistic analysis. Mailing Add: Dept of Anthrop Calif State Univ 5151 State University Dr Los Angeles CA 90032

WITWER, BRUCE DONALD, b Toledo, Ohio, July 8, 17; m 54; c 2. RESOURCE MANAGEMENT. Educ: Purdue Univ, BSChE, 39; Columbia Univ, AM, 47; Univ Chicago, MBA, 55. Prof Exp: Physicist, Sylvania Elec Prod, Inc, 48-51; chief spec weapons, Res & Develop Command, US Air Force, 51-54, chief nuclear powered aircraft proj, Hq US AEC-Air Force Joint Off, 56-61, chief nuclear power div, Dep Chief of Staff, Res & Develop Hq, DC, 61-63, chief mgt procedures div, 63-65; sr assoc, Control Systs Res, Inc, Arlington, 66-74; SR ASSOC, KAPPA SYST, INC, 74- Concurrent Pos: Consult to hq, US Air Force, 66. Mem: AAAS; Am Inst Biol Sci. Res: Military and social science research development and demonstration; management of military health care and anti-poverty systems. Mailing Add: 5601 Chesterfield Dr Camp Springs MD 20031

WITZ, DENNIS FREDRICK, b Milwaukee, Wis, Dec 10, 38; m 71; c 1. MICROBIOLOGY. Educ: Carroll Col, Wis, BS, 61; Univ Wis-Madison, MS, 64, PhD(bact), 67. Prof Exp: RES ASSOC MICROBIOL, UPJOHN CO, 67- Mem: AAAS; Am Soc Microbiol; Am Inst Biol Sci. Res: Process of biological nitrogen fixation by microorganisms; biosynthesis of antibiotics and secondary metabolites; production of antibiotics by fermentation. Mailing Add: 1400-89-1 Upjohn Co Kalamazoo MI 49081

WITZ, GISELA, b Breslau, Ger, Mar 16, 39; US citizen. CANCER. Educ: NY Univ, BA, 62, MS, 65, PhD(phys org chem), 69. Prof Exp: Fel biochem, Sloan-Kettering Inst Cancer Res, 69-70; assoc res scientist cancer res, 70-73, RES SCIENTIST ENVIRON MED, MED CTR, NY UNIV, 73- Mem: Am Chem Soc. Res: Chemical carcinogenesis; fluorescence spectrophotometric determination of conformational states of mammalian plasma membranes; effect of tumor promoters on cell membranes. Mailing Add: Dept Environ Med NY Univ Med Ctr New York NY 10016

WITZ, KLAUS GERHARD, b Breslau, Ger, July 1, 37; US citizen; m 60; c 2. MATHEMATICS. Educ: Univ NC, BS, 58, MA, 60, PhD(math), 63. Prof Exp: Instr math, Univ NC, 62-63; from instr to asst prof, 63-69, ASSOC PROF MATH, UNIV ILL, URBANA, 69- Concurrent Pos: NSF res contracts, 64-; Dept Health, Educ & Welfare contracts, 70- Mem: AAAS; Am Math Soc. Res: Functional analysis; mathematical foundations of quantum mechanics; mathematical models in social sciences; Piagetian psychology; cognitive processes in children; cognition; structural analysis in anthropology. Mailing Add: Dept Math 273 Attgeld Hall Univ of Ill Urbana IL 61801

WITZ, SAMUEL, inorganic chemistry, see 12th edition

WITZEL, DONALD ANDREW, b Artesian, SDak, Sept 9, 26. VETERINARY PHYSIOLOGY. Educ: Univ Minn, BS, 53, DVM, 57; Iowa State Univ, MS, 65, PhD(vet physiol), 70. Prof Exp: Res vet physiol, Nat Animal Dis Lab, 61-72, VET MED OFFICER PHYSIOL & TOXICOL, VET TOXICOL & ENTOM RES LAB, AGR RES SERV, USDA, 72- Concurrent Pos: Consult, Baylor Col Med, 72-74. Mem: AAAS; Am Vet Med Asn; NY Acad Sci; Am Soc Vet Physiologists & Pharmacologists. Res: Electrophysiological studies of the visual system of domestic animals as related to toxicological problems. Mailing Add: Vet Toxicol & Entom Res Lab PO Drawer GE College Station TX 77840

WITZEL, EVERET WAYNE, b La Valle, Wis, Jan 26, 34; m 57; c 4. ANATOMY. Educ: Emmanuel Missionary Col, BA, 57; Loma Linda Univ, MS, 60, MD, 62, PhD, 72. Prof Exp: Trainee anat, Univ Chicago, 63-64; from instr to asst prof, Loma Linda Univ, 64-72; pvt pract, 72-74; DIR MED EDUC, HINSDALE HOSP, ILL, 74-, DIR FAMILY PRACT RESIDENCY, 75- Concurrent Pos: Lectr, Christian Med Col, Vellore, India, 65-67; med dir, San Bernardino Col Med & Dent Asst, 72-73. Mem: Am Acad Family Physicians; Am Acad Med Dir; Am Pub Health Asn; Soc Teachers Family Med; Asn Hosp Med Educ. Res: Physiological and ultrastructural studies in neurological sciences; physiological parameters during perfusion fixation of the rat central nervous system for electron microscopy. Mailing Add: Hinsdale Sanitarium & Hosp 120 N Oak St Hinsdale IL 60521

WITZEL, FRANK, b Amsterdam, NY, July 9, 16; m 46; c 2. ORGANIC CHEMISTRY. Educ: Union Univ, NY, BS, 46. Prof Exp: Res chemist, Beech-Nut Packing Co, 46-52, head gum develop lab, 52-53, asst mgr, Gum Base Div, 53-58, chief chemist, Beech Nut-Life Savers, Inc, 48-59, mgr prototype develop, 59-71, SR RES ASSOC, BEECH NUT-LIFE SAVERS, INC, 71- Mem: Am Chem Soc; Inst Food Technol; Am Asn Candy Technol. Res: Natural gums and related synthetic polymers; essential oils. Mailing Add: 610 S Main St Spring Valley NY 10977

WITZGALL, CHRISTOPH JOHANN, b Hindelang, Ger, Feb 25, 29; US citizen; m 64; c 3. OPERATIONS RESEARCH, NUMERICAL ANALYSIS. Educ: Univ Munich, PhD(math), 58. Prof Exp: Res assoc math, Princeton Univ, 59-60, Univ Mainz, 60-62 & Argonne Nat Lab, 62; mathematician, Nat Bur Standards, 62-66 & Boeing Co, 66-73; MATHEMATICIAN, NAT BUR STANDARDS, 73- Concurrent Pos: Vis prof, Univ Tex, Austin, 71 & Univ Würzburg, 72; assoc ed, Math Prog, 73- Mem: Asn Comput Mach; Soc Indust & Appl Math; AAAS. Res: Further development of operations research, numerical analysis and programming languages as needed for planning and systems applications. Mailing Add: Appl Math Div Nat Bur of Standards Washington DC 20234

WITZIG, FREDERICK THEODORE, b Ill, Feb 26, 25; m 51; c 3. GEOGRAPHY. Educ: Bradley Univ, BS, 49; Univ Ill, MS, 51, PhD, 57. Prof Exp: From instr to assoc prof, 53-67, chmn div soc sci, 67-74, PROF GEOG, UNIV MINN, DULUTH, 67- Concurrent Pos: Consult, State Legis Interim Study Comn Water Pollution, Minn, 59. Mem: Asn Am Geog. Res: Environmental conservation; mineral geography of Europe. Mailing Add: Dept of Geog Univ of Minn Duluth MN 55812

WITZLEBEN, CAMILLUS LEO, b Dickinson, NDak, Apr 20, 32; m 56; c 6. PATHOLOGY. Educ: Univ Notre Dame, BS, 53; St Louis Univ, MD, 57. Prof Exp: Resident path, St Louis Univ, 57-60; NSF fel, Hosp Sick Children, London, Eng, 60-61; fel, Harvard Univ, 61-62; dir labs, Children's Hosp Med Ctr Northern Calif, 64-66; dir labs, Cardinal Glennon Mem Hosp Children, St Louis, 66-73; PROF PATH & PEDIAT, UNIV PA, 73-; DIR PATH, CHILDREN'S HOSP PHILADELPHIA, 73- Concurrent Pos: Consult, San Francisco Gen Hosp, 64-66; asst prof, Univ Calif, 64-66; from asst prof to prof, St Louis Univ, 66-73; NIH res grant, 68- Mem: Am Asn Path & Bact. Res: Hepatobiliary system; pediatric disease; heavy metals. Mailing Add: 229 Cornell Ave Swarthmore PA 19081

WIXOM, ROBERT LLEWELLYN, b Philadelphia, Pa, July 6, 24; m 49; c 2. BIOCHEMISTRY. Educ: Earlham Col, AB, 47; Univ Ill, PhD(biochem), 52. Prof Exp: Asst biochem, Univ Ill, 48-52; from instr to assoc prof, Sch Med, Univ Ark, 52-64; assoc prof, 64-72, PROF BIOCHEM, SCH MED, UNIV MO-COLUMBIA, 72- Concurrent Pos: Fel, Univ Ill, 55; Lalor Found res fel, 58; NIH spec res fel, 70-71. Mem: AAAS; Am Soc Biol Chemists; Am Chem Soc; Soc Exp Biol & Chem; Am Inst Nutrit. Res: Requirements and metabolism of essential amino acids in man; nutritional aspects of glycine in the chick; biosynthesis of amino acids in microorganisms and plants; inborn errors of amino acid metabolism; intravenous alimentation of amino acid solutions. Mailing Add: Dept of Biochem Univ of Mo Columbia MO 65201

WIXSON, ELDWIN A, JR, b Winslow, Maine, Nov 30, 31; m 54; c 4. MATHEMATICS. Educ: Univ Maine, BS, 53; Colby Col, MST, 62; Temple Univ, MSEd, 62; Univ Mich, PhD, 69. Prof Exp: Teacher, High Sch, Maine, 53-54 & 57-60; TV teacher, Maine Dept Educ, 60-61; teacher, High Sch, Maine, 62-63; assoc prof math, Keene State Col, 63-65; assoc prof, 66-70, PROF MATH, PLYMOUTH STATE COL, 70-, CHMN DEPT, 66- Mem: Math Asn Am; Nat Coun Teachers Math. Res: Mathematics education, especially at the undergraduate college level. Mailing Add: Dept of Math Plymouth State Col Plymouth NH 03264

WIZENBERG, MORRIS JOSEPH, b Toronto, Ont, Apr 9, 29; US citizen; m 55; c 3. RADIOTHERAPY. Educ: Univ Toronto, MD, 53. Prof Exp: Resident obstet & gynec, Sinai Hosp, Baltimore, Md, 54-58; resident therapeut radiol, Univ Hosp, 59-61, from instr to assoc prof, Sch Med, 61-68, PROF RADIOL & HEAD DIV THERAPEUT RADIOL, UNIV MD, BALTIMORE CITY, 68- Concurrent Pos: Nat Cancer Inst spec fel, Sch Med, Univ Md, Baltimore City, 61-63; consult numerous hosps. Mem: Am Cancer Soc; fel Am Col Radiol; Am Radium Soc; Am Soc Therapeut Radiol; Radiol Soc NAm. Res: Clinical radiation therapy, radiation biology. Mailing Add: Div of Radiation Ther Univ of Md Hosp Baltimore MD 21201

WLECH, RAYMOND LEE, b Emporia, Kans, Nov 2, 43; m 64; c 1. ORGANIC CHEMISTRY, CHEMICAL KINETICS. Educ: Kans State Teachers Col, BA, 65; Iowa State Univ, PhD(org chem), 69. Prof Exp: RES CHEMIST, HERCULES RES CTR, WILMINGTON, 69- Mem: Am Chem Soc. Res: Product development; polyesters; kinetics & mechanism of hydrocarbon autoxidation. Mailing Add: 306 Ware Rd Newark DE 19711 Wilmington DE

WOBESER, GARY ARTHUR, b Regina, Sask, Feb 12, 42; m 65; c 2. FISH PATHOLOGY, WILDLIFE PATHOLOGY. Educ: Univ Toronto, BSA, 63; Univ Guelph, MSc, 66, DVM, 69; Univ Sask, PhD(vet path), 73. Prof Exp: ASSOC PROF VET PATH, WESTERN COL VET MED, UNIV SASK, SASKATOON, 73-

Concurrent Pos: Coun mem, Wildlife Dis Asn, 75-, chmn student awards comt, 76. Mem: Wildlife Dis Asn; Can Asn Vet Pathologists; Am Fisheries Soc; Can Vet Med Asn. Res: Diseases of free-living fish and wild life, and of cultured salmonid fishes, with particular emphasis on infectious, degenerative and toxic problems. Mailing Add: Dept of Vet Path Western Col of Vet Med Univ Sask Saskatoon SK Can

WOBSCHALL, DAROLD C, b Wells, Minn, Feb 24, 32; m 57; c 3. BIOPHYSICS, ELECTRICAL ENGINEERING. Educ: St Olaf Col, BA, 53; State Univ NY Buffalo, MA, 60, PhD(biophys), 66. Prof Exp: Res assoc physics, Univ Buffalo, 58-60; assoc physicist, Cornell Aeronaut Lab, 60-62; cancer res scientist, Roswell Park Mem Inst, 66-67; asst prof, 67-71, ASSOC PROF ENG & BIOPHYS, STATE UNIV NY BUFFALO, 71- Concurrent Pos: NIH spec fel, 66-67. Mem: AAAS; Am Phys Soc; Biophys Soc; Inst Elec & Electronic Eng. Res: Electrical and mechanical properties of membranes; bioengineering; electronics instrumentation. Mailing Add: 4234 Ridge Lea Buffalo NY 14226

WOBUS, REINHARD ARTHUR, b Norfolk, Va, Jan 11, 41; m 67; c 2. GEOLOGY. Educ: Wash Univ, AB, 62; Harvard Univ, MA, 63; Stanford Univ, PhD(geol), 66. Prof Exp: Asst prof, 66-72, ASSOC PROF GEOL, WILLIAMS COL, 72-; GEOLOGIST CENT ENVIRON GEOL BR, US GEOL SURV, 67- Mem: Fel Geol Soc Am. Res: Igneous and metamorphic petrology; Precambrian geology of central Colorado. Mailing Add: Dept of Geol Williams Col Williamstown MA 01267

WOCHOK, ZACHARY STEPHEN, b Philadelphia, Pa, Dec 29, 42; m 70. PLANT PHYSIOLOGY. Educ: La Salle Col, BA, 64; Villanova Univ, MS, 67; Univ Conn, PhD(plant physiol), 70. Prof Exp: Asst prof, 71-74, ASSOC PROF BIOL, UNIV ALA, 74- Concurrent Pos: NIH fel, Yale Univ, 70-71; consult, Gulf States Paper Corp, 75-76. Mem: Am Soc Plant Physiol; Bot Soc Am; Int Asn Plant Tissue Cult; Scand Soc Plant Physiol. Res: Amelioration of strip mined spoils; study of plant growth and development through in vitro methodology; cell determination in meristematic growth zones and differentiation of callus tissue grown in culture. Mailing Add: Dept of Biol Univ of Ala PO Box 1927 University AL 35486

WODARCZYK, FRANCIS JOHN, physical chemistry, see 12th edition

WODICKA, VIRGIL ORVILLE, b St Louis, Mo, Mar 5, 15; m 41; c 2. FOOD TECHNOLOGY. Educ: Wash Univ, BS, 34, MS, 35; Rutgers Univ, PhD, 56. Prof Exp: Asst, Procter & Gamble Co, 35-36; res chemist, Ralston Purina Co, 36-41, mgr cereal labs, 46-48; head nutrit res dept, Libby, McNeil & Libby, Ill, 48-51; asst to sci dir, Qm Food & Container Inst, US Dept Army, 51-52, chief animal prod div, 54-55, assoc dir, Food Div, 55-58, dir, 58-61; tech dir food elements, Hunt Food & Industs Inc, 61-65; tech dir, Hunt-Wesson Foods, Inc, Calif, 65-69, vpres, 65-69; dir, Bur Foods, US Food & Drug Admin, 70-74; CONSULT, 75- Mem: AAAS; Am Chem Soc; Am Oil Chem Soc; Am Soc Qual Control; assoc Tech Asn Pulp & Paper Indust. Res: Methods for vitamin analysis of food and feeds; vitamin stability in processing and storage of foods and feeds; cured meat pigments; dehydration; new food products, processes and packages; food quality control; food safety. Mailing Add: 1307 Norman Pl Fullerton CA 92631

WODINSKY, ISIDORE, b New York, NY, Mar 6, 19; m 42; c 2. ONCOLOGY. Educ: Brooklyn Col, BA, 39; George Washington Univ, MS, 51. Prof Exp: Biologist, Nat Cancer Inst, 46-55, sci adminr, 55-57; HEAD CANCER CHEMOTHER RES, LIFE SCI SECT, ARTHUR D LITTLE, INC, 59- Mem: Int Union Against Cancer; Am Asn Cancer Res; Int Soc Chemother; NY Acad Sci. Res: Chemotherapy of cancer carcinogenesis; biology and kinetics of experimental neoplasms; cryobiology. Mailing Add: Arthur D Little Inc Life Sci Sect 30 Memorial Dr Cambridge MA 02142

WODZINSKI, RUDY JOSEPH, b Chicago, Ill, June 12, 33; m 56; c 3. MICROBIAL BIOCHEMISTRY. Educ: Loyola Univ, Ill, BS, 55; Univ Wis, MS, 57, PhD(bact), 60. Prof Exp: Asst bact, Univ Wis, 55-60; sr res scientist, Squibb Inst Med Res, New Brunswick, NJ, 60-62; res microbiologist, Int Minerals & Chem Corp, Ill, 62-64, supvr microbial biochem, 64-68, mgr animal sci, 68-70; PROF BIOL SCI, FLA TECHNOL UNIV, 70- Mem: Am Soc Microbiol. Res: Enzymology; molecular biology; virology; microbial genetics and physiology; fermentation; available moisture requirements of microorganisms; water and waste microbiology. Mailing Add: Dept of Biol Sci Fla Technol Univ Orlando FL 32816

WOEHLER, KARLHEINZ EDGAR, b Berlin, Ger, June 5, 30; m 56; c 1. PHYSICS. Educ: Univ Bonn, BS, 53; Aachen Tech Univ, Dipl, 55; Univ Munich, PhD(physics), 62. Prof Exp: Physicist commun technol, Siemens & Halske, Ger, 55-59; res assoc plasma physics, Max-Planck Inst Physics, 59-62; asst prof physics, US Naval Postgrad Sch, 62-64; sr res assoc, Inst Plasma Physics, 64-65; asst prof physics, 65-72, PROF PHYSICS, NAVAL POSTGRAD SCH, 72-, CHMN DEPT PHYSICS & CHEM, 74- Concurrent Pos: Consult physicist, Atomics Int Div, NAm Aviation, Inc, 63-64; Nat Acad Sci res grant, Res Labs, NASA, 66; consult, Naval Electronics Lab Ctr, Calif, 68, 69 & 72. Mem: Am Phys Soc. Res: Plasma physics; general relativity and cosmology. Mailing Add: Dept of Physics & Chem Naval Postgrad Sch Monterey CA 93940

WOEHLER, MICHAEL EDWARD, b Appleton, Wis, Feb 16, 45. IMMUNOLOGY, MEDICAL MICROBIOLOGY. Educ: Northwestern Univ, BA, 67; Marquette Univ, PhD(microbiol), 71. Prof Exp: Fel biochem, Univ Ga, 71-74; MEM TECH STAFF IMMUNOL, GTE LABS, INC, 74- Mem: Am Soc Microbiol; AAAS; Sigma Xi; NY Acad Sci. Res: Immunology, specifically structure function relationships of immunoglobulins G and E antibodies. Mailing Add: GTE Labs Inc 40 Sylvan Rd Waltham MA 02154

WOELFEL, JULIAN BRADFORD, b Baltimore, Md, Dec 17, 25; m 48; c 3. DENTISTRY. Educ: Ohio State Univ, DDS, 48. Prof Exp: PROF DENT, COL DENT, OHIO STATE UNIV, 48- Concurrent Pos: Consult & Am Dent Asn res assoc, Nat Bur Stands, 57-63; consult, Vet Admin Hosp, Dayton, Ohio, 66-69 & Fed Penitentiary, Chillicothe, 65-66; pres, Carl O Boucher Prosthodontic Conf, 67- Honors & Awards: Int Asn Dent Res Award, 67. Mem: Am Prosthodont Soc; Am Dent Asn; Int Asn Dent Res; fel Am Col Dent; Acad Denture Prosthetics. Res: Prosthodontic dentistry; denture base resins; clinical evaluation of complete dentures; electromyography; jaw and denture movement; mandibular motion in three dimensions; accuracy of impression materials; soft and hard tissue and facial dimension changes beneath complete dentures during six years; computer analysis of mandibular resorption. Mailing Add: Col of Dent Ohio State Univ Columbus OH 43210

WOELKE, CHARLES EDWARD, b Seattle, Wash, Jan 8, 26; m 47; c 2. FISHERIES, MARINE ECOLOGY. Educ: Univ Wash, BS, 50, PhD(fisheries), 68. Prof Exp: Aquatic biologist, Ore Fisheries Comn, 50-51; fisheries biologist, 51-64, RES SCIENTIST FISHERIES, WASH DEPT FISHERIES, 68- Concurrent Pos: Affil prof, Univ Wash, 69- Mem: Fel Am Inst Fishery Res Biol; Nat Shellfisheries Asn. Res: Molluscan commercial shellfish; bioassays with bivalve embryos; development of in situ bioassays with bivalve larvae; development of water quality standards and criteria; biometrics and ecological systems analysis. Mailing Add: 2378 Crestline Blvd Olympia WA 98502

WOERNER, DALE EARL, b Oak Hill, Kans, Jan 15, 26; m 50; c 6. ANALYTICAL CHEMISTRY. Educ: Kans State Univ, BS, 49; Univ Ill, MS, 51, PhD(anal chem), 53. Prof Exp: Asst chem, Univ Ill, 49-53; assoc prof, Hanover Col, 53-55; from asst instr to asst prof, Kans State Univ, 55-58; from asst prof to assoc prof, 58-66, PROF CHEM, UNIV NORTHERN COLO, 66- Mem: Am Chem Soc. Res: Amperometric titrations; spectroscopy. Mailing Add: Dept of Chem Univ of Northern Colo Greeley CO 80631

WOERNER, LLOYD MARTIN, physical chemistry, see 12th edition

WOERNER, ROBERT LEO, b Evanston, Ill, Apr 21, 48. LOW TEMPERATURE PHYSICS. Educ: Mass Inst Technol, SB & SM, 71, PhD(physics), 74. Prof Exp: Fel liquid helium, Dept Physics, Mass Inst Technol, 74-75; MEM TECH STAFF PHYSICS, BELL TEL LABS, HOLMDEL, NJ, 75- Mem: Am Phys Soc. Res: Light scattering from the elementary excitations in superfluid helium. Mailing Add: Bell Tel Labs Holmdel NJ 07733

WOESSNER, DONALD EDWARD, b Milledgeville, Ill, Oct 6, 30; m 58. PHYSICAL CHEMISTRY, NUCLEAR MAGNETIC RESONANCE. Educ: Carthage Col, AB, 52; Univ Ill, PhD(chem), 57. Prof Exp: Asst phys chem, Univ Ill, 55-57, fel chem, 57-58; sr res technologist, 58-62, RES ASSOC, FIELD RES LAB, MOBIL RES & DEVELOP CORP, 62- Honors & Awards: W T Doherty Award, Am Chem Soc, 75. Mem: AAAS; Am Phys Soc; Am Chem Soc. Res: Relationships of nuclear spin relaxation times in nuclear magnetic resonance and use in determining structure and motion in physical systems. Mailing Add: Field Res Lab Mobil Res & Develop Corp Dallas TX 75221

WOESSNER, JACOB FREDERICK, JR, b Pittsburgh, Pa, May 8, 28; m 53; c 2. BIOCHEMISTRY. Educ: Valparaiso Univ, BA, 50; Mass Inst Technol, PhD(biochem), 55. Prof Exp: Asst, Mass Inst Technol, 53-55; Lilly fel natural sci, Univ Mich, 55-56; res asst prof, 56-64, assoc prof, 64-72, PROF BIOCHEM & ASSOC PROF MED, SCH MED, UNIV MIAMI, 72- Concurrent Pos: Investr, Labs Cardiovasc Res, Howard Hughes Med Inst, Fla, 56-71; vis scientist, Max Planck Inst Protein & Leather Res, 61-62; mem gen med B study sect, NIH, 71- Mem: Am Chem Soc; Geront Soc; Am Rheumatism Asn; Am Soc Biol Chemists; Biochem Soc. Res: Formation, metabolism and aging of connective tissues; proteolytic eyzymes; lysosomes and tissue resorption; arthritis. Mailing Add: Univ of Miami Sch of Med Biscayne Annex PO Box 875 Miami FL 33152

WOESSNER, RONALD ARTHUR, b Pittsburgh, Pa, Apr 27, 37; m 57; c 3. FORESTRY, GENETICS. Educ: WVa Univ, BS, 63; NC State Univ, MS, 66, PhD(forest genetics), 68. Prof Exp: Asst prof plant & forest sci, Tex A&M Univ, 68-74, assoc prof forestry, 74, assoc geneticist, Tex Forest Serv, 68-74; SUPVR RES & DEVELOP, NAT BULK CARRIERS INC, 74- Mem: Soc Am Foresters; AAAS. Res: Genetic improvement of pine and hardwood species, temperate and tropical; quantitative genetics of forest trees; selection indices; prediction of response to selection, genotype-environment interactions; inter-population crossing. Mailing Add: Nat Bulk Carriers Inc 1345 Ave of Americas New York NY 10019

WOFFORD, IRVIN MIRLE, b White Co, Ga, Dec 11, 16; m 38. AGRONOMY. Educ: Univ Ga, BSA, 48; Univ Fla, MSA, 49; Mich State Col, PhD(farm crops), 53. Prof Exp: Instr agron, Univ Fla, 49-51, asst agronomist, Exp Sta, 53-56; asst, Mich State Col, 51-53; dir agron, Southern Nitrogen Co, Inc, 56-64; MGR AGR PUB RELS, KAISER AGR CHEM, 64- Mem: Am Soc Agron. Res: Crop management and production; fertilizer studies; variety testing; date of planting; rotations; plant population studies. Mailing Add: PO Box 246 Savannah GA 31402

WOFSY, LEON, b Stamford, Conn, Nov 23, 21; m 42; c 2. CHEMISTRY, IMMUNOLOGY. Educ: City Col New York, BS, 42; Yale Univ, MS, 60, PhD(chem), 61. Prof Exp: PROF IMMUNOL, UNIV CALIF, BERKELEY, 64- Mem: Am Chem Soc. Res: Study of antibody specificity; mechanisms of cellular differentiation. Mailing Add: Dept of Bacteriol & Immunol Univ of Calif Berkeley CA 94720

WOGAN, GERALD NORMAN, b Altoona, Pa, Jan 11, 30; m 57; c 2. TOXICOLOGY. Educ: Juniata Col, BS, 51; Univ Ill, MS, 53, PhD(physiol), 57. Prof Exp: Instr physiol, Univ Ill, 56-57; asst prof, Rutgers Univ, 57-61; res assoc food toxicol, 61-62, from asst prof to assoc prof, 62-68, PROF FOOD TOXICOL, MASS INST TECHNOL, 68- Mem: AAAS; Am Inst Nutrit; Soc Toxicol; Int Soc Toxinology; Am Asn Cancer Res. Res: Chemical carcinogenesis; physiological and biochemical responses to toxic substances; isolation and characterization of toxic microbial metabolites; natural carcinogens. Mailing Add: Dept Nutrit & Food Sci Mass Inst Technol Cambridge MA 02139

WOGEN, WARREN RONALD, b Forest City, Iowa, Feb 19, 43; m 69; c 2. MATHEMATICS. Educ: Luther Col, Iowa, BA, 65; Ind Univ, Bloomington, MS, 67, PhD(math), 69. Prof Exp: ASSOC PROF MATH, UNIV NC, CHAPEL HILL, 69- Mem: Am Math Soc; Math Asn Am. Res: Operator theory and operator algebras. Mailing Add: Dept of Math Univ of NC Chapel Hill NC 27514

WOGMAN, NED ALLEN, b Spokane, Wash, Oct 25, 39; m 59; c 3. NUCLEAR CHEMISTRY, PHYSICAL CHEMISTRY. Educ: Wash State Univ, BS, 61; Purdue Univ, PhD(phys chem), 66. Prof Exp: Sr res scientist, 65-68, mgr radiol chem, 68-72, RES ASSOC, PAC NORTHWEST LAB, BATTELLE MEM INST, 72- Concurrent Pos: Lectr, Joint Ctr Grad Study, Wash State Univ/Univ of Wash/Ore State Univ, 72-; mem sci comt, Nat Coun Radiation Protection, 73- Mem: Am Chem Soc. Res: Rates and mechanisms of biological, meteorological, oceanographic and ecological processes by natural and artificial radionuclides, including development of sensitive multidimensional gamma-ray spectrometer systems for trace radionuclide measurements. Mailing Add: Battelle Northwest Labs PO Box 999 Richland WA 99352

WOHL, ARNOLD J, b Brooklyn, NY, May 5, 37; m 60; c 1. PHARMACOLOGY. Educ: Queens Col, NY, BS, 58; Albany Med Col, PhD(pharmacol), 64. Prof Exp: Supvr cardiovasc pharmacol, Schering Corp, 64-70, cardiovasc sect leader, 70-72; ASSOC DIR RES, FOSTER D SNELL INC, 72- Mem: Am Soc Pharmacol & Exp Therapeut. Res: Cardiovascular pharmacology; etiology of hypertension; interaction of antihypertensive agents with altered calcium metabolism in hypertensive vasculature; molecular orbital theory as a predictor in pharmacology. Mailing Add: Foster D Snell Inc Elizabeth NJ 07201

WOHL, BERNARD G, b Philadelphia, Pa, July 20, 16; m 41; c 2. ORGANIC CHEMISTRY, MEDICINAL CHEMISTRY. Educ: Pa State Col, BS, 40; Temple Univ, MA, 46, PhD, 49. Prof Exp: Chemist, Distillers-Seagram Corp, 40-42; med chemist, Wyeth Inc, 42-46; res chemist, Fels & Co, 46-48 & Pyrene Corp, 48-52; dir res, Lemmon Pharmacal Co, 52-59; tech dir, Philadelphia Ampoule Labs, Inc, 59-61; TECH DIR, WOHL LAB, 61- Concurrent Pos: Lectr, Pa State Univ, 46-48; secy-treas, Cunningham Distributors & Murd Co, 66- Mem: AAAS; Am Chem Soc. Res: Organic medicinal application of Mennich reaction to medicinal chemistry. Mailing Add: 1222 Vilsmeier Rd Lansdale PA 19446

WOHL, GEORGE T, b Omaha, Nebr, Sept 1, 17; m 45; c 3. RADIOLOGY. Educ: Univ Pa, AB, 38; Temple Univ, MD, 42. Prof Exp: Lectr, Div Grad Med, Univ Pa, 54-67, prof clin radiol, Sch Med, 67-72, dir div diag radiol, 69-72; DIR DIV DIAG RADIOL, LANKENAU HOSP, PHILADELPHIA, 69-; PROF RADIOL, MED COL, THOMAS JEFFERSON UNIV, 74- Concurrent Pos: Prof radiol, Med Sch, Temple Univ, 60-67; chmn dept radiol, Philadelphia Gen Hosp, 60-72. Mem: Roentgen Ray Soc; Radiol Soc NAm; fel Am Col Radiol. Res: Medical and diagnostic radiology; applications of medical uses of radioisotopes. Mailing Add: Div of Diag Radiol Hosp Philadelphia PA 19151

WOHL, PHILIP R, b 1944; US citizen; m; c 1. MATHEMATICS. Educ: Queens Col, NY, BA, 66; Cornell Univ, PhD(appl math), 71. Prof Exp: Teaching asst math, Cornell Univ, 70-71; asst prof, NY Univ, 71-72; RES ASSOC MATH, CARLETON UNIV, 72- Mem: Soc Indust & Appl Math; Math Asn Am. Res: Classical analysis and methods of ordinary and partial differential equations; perturbation methods in fluid mechanics; low Reynolds number hydrodynamics; blood flow and other flows of suspensions in tubes; biofluid mechanics; mathematical modelling in biology. Mailing Add: Dept of Math Carleton Univ Ottawa ON Can

WOHLEBER, DAVID ALAN, b Pittsburgh, Pa, Oct 1, 40; m 61; c 4. CHEMISTRY. Educ: Univ Pittsburgh, BS, 62; John Carroll Univ, MS, 67; Kent State Univ, PhD(chem), 70. Prof Exp: Chemist anal chem, Develop Lab, Standard Oil Co, Ohio, 61-63, chemist polymer chem, Res Lab, 63-66; sr res scientist phys chem, 70-74, SECT HEAD, EXTRACTIVE METALL DIV, ALCOA TECH CTR, ALUMINUM CO AM, 70- Mem: Am Chem Soc; Catalysis Soc; Sigma Xi. Res: Chlorination technolgoy; inorganic and physical chemistry of alumina refining and aluminum smelting; physical adsorption; polymer synthesis and characterization. Mailing Add: Alcoa Center PA 15069

WOHLENBERG, ERNEST HAROLD, b Liberal, Kans, Mar 12, 36; m 63; c 1. ECONOMIC GEOGRAPHY. Educ: Univ Kans, BA, 59; Univ Wash, MA, 66, PhD(geog), 70. Prof Exp: Programmer, IBM Corp, 59-60 & Boeing Co, 62-63; part-time instr, Seattle Pac Col, 66-69; ASST PROF GEOG, IND UNIV, BLOOMINGTON, 70- Mem: Asn Am Geogrs; Regional Sci Asn. Res: Regional aspects of social and economic inequalities. Mailing Add: Dept of Geog Ind Univ Bloomington IN 47401

WOHLER, JAMES RICHARD, II, b Pittsburgh, Pa, Sept 9, 40; m 75; c 1. ECOLOGY, LIMNOLOGY. Educ: Univ Pittsburgh, BS, 63, MS, 66; PhD(biol), 69. Prof Exp: Lab asst biochem pharmacol, Mellon Inst, 58-63; from asst prof to assoc prof biol, Allegheny Col, 68-75, chmn dept, 72-75; ASST VPRES, FREEPORT BRICK CO, 75- Mem: Ecol Soc Am; Int Asn Gt Lakes Res; Int Soc Theoret & Appl Limnology; Am Soc Limnol & Oceanog. Res: Aquatic plant ecology; plant physiology. Mailing Add: RD 1 Saegertown PA 16433

WOHLERS, HERBERT C, b Brooklyn, NY, Sept 27, 18; m 45; c 3. ORGANIC CHEMISTRY, RESEARCH ADMINISTRATION. Educ: Mass Inst Technol, SB, 40, PhD(org chem), 44. Prof Exp: In charge res, WIndia Chem, Ltd, Bahamas, 40-41; res dir, Mich Chem Corp, 44-48, tech dir, 48-55; chemist, Solvay Process Div, 55-56, assoc res supvr, 56-58, asst dir res, 58-59, dir, 59-61, dir, Nat Aniline Div, 61-63, dir res, Plastics Div, 63-69, DIR CHEM RES CTR, ALLIED CHEM CORP, 71- Mem: Am Chem Soc. Res: Research administration in organic and polymer chemistry. Mailing Add: Corp Res & Develop Allied Chem Corp Morristown NJ 07960

WOHLETZ, ERNEST W, b Nekoma, NDak, July 10, 07; m 35; c 2. FORESTRY. Educ: Univ Calif, BS, 30, MS, 47. Prof Exp: Jr forester, US Forest Serv, Calif, 33-34; assoc forestry, Univ Calif, 36-37; from asst prof to assoc prof forestry, 37-48, assoc dir forest wildlife & range exp sta, 48-53, dean col forestry & dir forest wildlife & range exp sta, 53-71, PROF FORESTRY, UNIV IDAHO, 48-, EMER DEAN COL FORESTRY, 72- Concurrent Pos: Dir range resources study, Pub Land Law Rev Comn, 70-72. Mem: Soc Am Foresters. Res: Forest economics. Mailing Add: Col Forest Wildlife & Range Sci Univ of Idaho Moscow ID 83843

WOHLFORD, DUANE DENNIS, b Newcastle, Ind, May 20, 37; m 66; c 3. GEOLOGY, PETROLOGY. Educ: Univ Wis, BS, 59; Univ Colo, PhD(geol), 65. Prof Exp: Asst prof geol, 64-74, ASSOC PROF EARTH SCI, STATE UNIV NY COL ONEONTA, 74- Mem: AAAS; Geol Soc Am. Res: Petrology of the Precambrian rocks of the southern Adirondack Mountains of New York. Mailing Add: Dept of Earth Sci State Univ of NY Col Oneonta NY 13820

WOHLFORT, SAM WILLIS, b Toledo, Ohio, June 8, 26; m 49; c 3. ANALYTICAL CHEMISTRY, MATERIALS SCIENCE. Educ: Univ Toledo, BS, 48; Ohio State Univ, MS, 50. Prof Exp: Asst instrumental & spectrog anal, Ohio State Univ, 50-53; chemist, Sharp-Schurtz Co, 53-56; chemist, Libbey-Owens-Ford Glass Co, 56-58, group leader anal res, 58-65; supvr anal methods & mfg processes, Milchem Inc, Tex, 65-69; supvr phys measurements dept, 69-73, SUPVR MAT TECHNOL DEPT, GOODYEAR ATOMIC CORP, 73- Mem: Am Chem Soc; Electron Micros Soc Am; Soc Appl Spectros; Am Inst Chem. Res: Electron microscopy; x-ray diffraction and fluorescence; ultraviolet, infrared and visible spectrophotometry; differential thermal analysis; thin films; corrosion protection; vacuum deposition; x-ray stress analysis; incipient failure detection; non-destructive testing; metallography. Mailing Add: Mat Technol Dept Goodyear Atomic Corp PO Box 628 Piketon OH 45661

WOHLGELERNTER, DEVORA KASACHKOFF, b Washington, DC, Apr 1, 41; m 67; c 3. MATHEMATICS. Educ: Yeshiva Univ, BA, 61, MA, 63, PhD(math), 70. Prof Exp: ASST PROF MATH, BARUCH COL, 70- Mem: Am Math Soc. Res: Polynomial approximation of functions of a complex variable. Mailing Add: Dept of Math Baruch Col New York NY 10010

WOHLGEMUTH, JOHN HAROLD, b New York, NY, Dec 20, 46; m 73. EXPERIMENTAL SOLID STATE PHYSICS. Educ: Rensselaer Polytech Inst, BS, 68, MS, 72, PhD(physics), 73. Prof Exp: Fel physics, Univ Waterloo, 73-75; FEL ELEC ENG, MOORE SCH ELEC ENG & SCI, UNIV PA, 75- Mem: Am Phys Soc; Sigma Xi. Res: Using electron beam guns to float zone crystallize large sheets of silicon for making solar cells. Mailing Add: Moore Sch of Elec Eng & Sci Univ of Pa Philadelphia PA 19174

WOHLHIETER, JOHN ANDREW, b Pittsburgh, Pa, Mar 18, 32; m 61; c 2. BIOPHYSICS. Educ: Univ Pittsburgh, BS, 54, MS, 57, PhD(biophys), 60. Prof Exp: Res assoc biophys, Univ Pittsburgh, 54-59; res biochemist, 62-69, ASST CHIEF, DEPT BACT IMMUNOL, WALTER REED ARMY INST RES, 69- Concurrent Pos: Mem genetics fac grad prog, NIH, 68-70, mem microbial chem study sect, 69-, chmn comt eval extramural prog mechanisms resistance antimicrobial agents, Nat Inst Allergy & Infectious Dis, 75. Mem: AAAS; Biophys Soc; Am Chem Soc; Am Soc Microbiol. Res: Microbial genetics; formation of intergeneric hybrids; isolation and characterization of extrachromosomal DNA; biophysical characterization of nucleic acids, enzymes and proteins; resistance to antimicrobial agents. Mailing Add: Dept of Bact Immunol Walter Reed Army Inst Res Washington DC 20012

WOHLPART, ALFRED, b New York, NY, July 30, 37; m 60; c 5. PLANT PHYSIOLOGY, MOLECULAR BIOLOGY. Educ: Colo State Univ, BS, 63; Univ Tex, Austin, PhD(molecular biol), 67. Prof Exp: Damon Runyon Mem Fund for Cancer res fel molecular biol, Univ Zurich, 67-69; ASSOC PROF BIOL, KENYON COL, 69- Mem: Am Inst Biol Sci; Bot Soc Am; Am Soc Plant Physiol. Res: Biogenesis and physiology of plant pigments such as betacyanins and betaxanthins; uptake of nucleic acids by plant cells. Mailing Add: Dept of Biol Kenyon Col Gambier OH 43022

WOHLRAB, HARTMUT, b Berlin, WGer, July 2, 41; m 67; c 1. BIOCHEMISTRY. Educ: Rensselaer Polytech Inst, BS, 62; Stanford Univ, PhD(biophys), 68. Prof Exp: NIH trainee, Johnson Res Found, Univ Pa, 68-70; asst biochem, Univ Munich, 70-72; STAFF SCIENTIST, BOSTON BIOMED RES INST, 72- Concurrent Pos: Estab investr, Am Heart Asn, 73- Mem: Geront Soc; Biophys Soc; Fedn Europ Biochem Socs; Am Aging Asn; AAAS. Res: Mitochondrial biochemistry; mitochondria in developing and aging, or senescent, tissues. Mailing Add: Div III Boston Biomed Res Inst 20 Staniford St Boston MA 02114

WOHLSCHLAG, DONALD EUGENE, b Bucyrus, Ohio, Nov 6, 18; m 43; c 3. MARINE ECOLOGY. Educ: Ind Univ, PhD, 49. Prof Exp: Asst chem & zool, Heidelberg Col, 40-41; res assoc, Univ Wis, 49; from asst prof to prof biol, Stanford Univ, 49-65; dir, Inst, 65-70, PROF ZOOL, MARINE SCI INST, UNIV TEX, 65- Concurrent Pos: Arctic fisheries researcher, Off Naval Res & Arctic Inst NAm, 52-53 & 54-55, Antarctic fishery biol researcher, 58-65, Gulf coastal fishery biol research, 65-; consult, Inst Ecol, 74-; ed, Contrib Marine Sci, Univ Tex Marine Sci Inst, 75- Mem: AAAS; Am Fisheries Soc; Am Soc Limnol & Oceanog; Am Soc Zoologists; Ecol Soc Am. Res: Ecology of fishes; metabolism and growth; population dynamics. Mailing Add: Marine Sci Inst Univ of Tex Port Aransas TX 78373

WOHLTMANN, HULDA JUSTINE, b Charleston, SC, Apr 10, 23. PEDIATRICS. Educ: Col Charleston, BS, 44; Med Col SC, MD, 49; Am Bd Pediat, dipl, 55. Prof Exp: From instr to asst prof pediat, Sch Med, Wash Univ, 53-61, USPHS spec res fel biochem, Univ, 61-63; asst prof pediat, 65-70, PROF PEDIAT, MED UNIV SC, 70- Mem: AAAS; fel Am Acad Pediat; Am Diabetes Asn; NY Acad Sci; Endocrine Soc. Res: Pediatric metabolism. Mailing Add: Dept of Pediat Med Univ of SC Charleston SC 29401

WOHNSIEDLER, HENRY PETER, b US, Jan 9, 04; m 34. PLASTICS CHEMISTRY. Educ: City Col New York, BS, 24. Prof Exp: Chemist, Res Lab, Am Cyanamid Co, 24-34, dir, Anal Lab, Plastics Div, 34-41, group leader, Cent Res Lab, 41-46, sr chemist, Basic Res Dept, 46-56, sr res chemist, Plastics & Resins Div, 56-67; sr res scientist & consult, Am Standard Corp, NJ, 67-70; INDUST CONSULT, 70- Mem: Am Chem Soc. Res: Thermosetting polymer systems and plastics; paper resins; aminotriazine resins and polyelectrolytes; cross-linking and cure reactions; morphology; materials science. Mailing Add: PO Box 3082 Noroton CT 06820

WOHNUS, JOHN FREDERICK, b Chester, Pa, July 15, 13; m 39; c 1. ZOOLOGY. Educ: Williams Col, BA, 35, MA, 37; Univ Calif, Los Angeles, PhD(zool), 41. Prof Exp: Asst biol, Williams Col, 35-37; asst zool, Univ Calif, Los Angeles, 37-40; Kelco Indust fel, Scripps Inst Oceanog, 41-43; PROF ZOOL, BENNINGTON COL, 46- Concurrent Pos: Res assoc, Columbia Univ, 57-59, Med Ctr, Dartmouth Univ, 62-63 & City of Hope Med Ctr, 68-69 & 73-74; vis prof biol, Williams Col, 73. Mem: AAAS; Am Soc Zool. Res: Embryology and microanatomy; avian blood parasites; mammalian and amphibian chromosome morphology; electrophoretic studies of the genetically inherited water soluble muscle proteins. Mailing Add: Dept of Zool Bennington Col Bennington VT 05201

WOIDICH, FRANCIS DE SALES, b Battle Creek, Mich, Mar 6, 25; m 71. MEDICINE, PHYSIOLOGY. Educ: Harvard Univ, BS, 45; George Washington Univ, MD, 52. Prof Exp: Res fel pediat, George Washington Univ, 53-54, res assoc physiol, 54-59; DIR, WASHINGTON INST BIOPHYS RES, 59- Mem: Biophys Soc; NY Acad Sci; Am Soc Photobiol; Int Acad Prev Med. Res: Biophysics; systems analysis; responses of living systems to specific force fields, including changes in consciousness and physiology in humans; principles of organization of living systems; preventive medicine and health renewal strategies. Mailing Add: 1830 Baldwin Dr McLean Va 22101

WOISARD, EDWIN LEWIS, b Newark, NJ, Jan 21, 26; m 53, 69; c 6. PHYSICS, OPERATIONS RESEARCH. Educ: Drew Univ, BA, 50; Lehigh Univ, MS, 52, PhD(physics), 59. Prof Exp: Instr physics, Moravian Col, 56-59; assoc res physicist, Res Lab, Whirlpool Corp, 59-61; proj leader, Weapons Systs Eval Div, Inst Defense Anal, DC, 61-67; exec vpres, John D Kettelle Corp, 67-70; pvt consult, 70-71; asst for mid-range objectives to dir, Systs Anal Div, 71-74, SPEC ASST NAVY NET ASSESSMENT, OFF CHIEF NAVAL OPERS, US NAVY, 74- Mem: Sigma Xi; Am Phys Soc; Opers Res Soc Am. Res: Thermoelectricity; solid state physics; microwave absorption; systems and defense analysis. Mailing Add: 1587B N Van Dorn Alexandria VA 22304

WOJCICKI, ANDREW, b Warsaw, Poland, May 5, 35; nat US; m 68. INORGANIC CHEMISTRY, ORGANOMETALLIC CHEMISTRY. Educ: Brown Univ, BS, 56; Northwestern Univ, PhD(chem), 60. Prof Exp: Asst chem, Northwestern Univ, 56-58, assoc, 59; NSF fel inorg chem, Univ Nottingham, 60-61; from asst prof to assoc prof, 61-69, PROF INORG CHEM, OHIO STATE UNIV, 69- Concurrent Pos: Vis assoc prof, Case Western Reserve Univ, 67; US sr scientist award, Humboldt Found, Ger, 75-76; Guggenheim Found fel, 76. Mem: AAAS; Am Chem Soc; The Chem Soc. Res: Synthesis and mechanism of reactions of inorganic and organometallic compounds. Mailing Add: Dept of Chem Ohio State Univ Columbus OH 43210

WOJCICKI, STANLEY G, b Warsaw, Poland, Mar 30, 37; US citizen; m 61. HIGH ENERGY PHYSICS. Educ: Harvard Univ, AB, 57; Univ Calif, Berkeley, PhD(physics), 62. Prof Exp: Physicist, Lawrence Radiation Lab, Univ Calif, 61-66; from asst prof to assoc prof, 66-74, PROF PHYSICS, STANFORD UNIV, 74- Concurrent Pos: NSF fel, 64-65; Alfred P Sloan Found fel, 68- Mem: Am Phys Soc. Res: Resonances in high energy physics. Mailing Add: Dept of Physics Stanford Univ Stanford CA 94305

WOJCIK, ANTHONY STEPHEN, b Chicago, Ill, Sept 18, 45; m 69; c 2. COMPUTER SCIENCE, ALGEBRA. Educ: Univ Ill, Urbana-Champaign, BS, 67, MS, 68, PhD(comput sci), 71. Prof Exp: ASST PROF COMPUT SCI, ILL INST TECHNOL, 71- Mem: Sigma Xi; Asn Comput Mach; Inst Elec & Electronics Engrs. Res: Switching and automata theory, especially the theory and application of multi-valued switching theory; computer architecture; reliable design of digital systems; design automation and digital systems. Mailing Add: Dept Comput Sci Ill Inst Technol 10 W 31st St Chicago IL 60616

WOJCIK, BRUNO HENRY, organic chemistry, see 12th edition

WOJCIK, JOHN F, b Ashley, Pa, Nov 12, 38; m 60; c 4. PHYSICAL CHEMISTRY. Educ: King's Col, Pa, BS, 60; Cornell Univ, PhD(phys chem), 65. Prof Exp: Asst prof chem, St Francis Col, Pa, 65-66; ASST PROF CHEM, VILLANOVA UNIV, 66- Concurrent Pos: Petrol Res Fund res grant, 66-68. Mem: Am Chem Soc. Res: Kinetics of chelation reactions; stability constants of ethylenediaminetetra acetate type complexes. Mailing Add: Dept of Chem Villanova Univ Villanova PA 19085

WOJCIK, RONALD JOSEPH, applied mathematics, computer science, see 12th edition

WOJNAR, ROBERT JOHN, b Thompsonville, Conn, Jan 29, 35; m 58; c 3. IMMUNOBIOLOGY, BIOCHEMISTRY. Educ: Univ Conn, BA, 56, MS, 60, PhD(biochem), 64. Prof Exp: NIH fel, Yale Univ, 63-65; staff scientist, Worcester Found Exp Biol, 65-68; res investr, 68-69, SR INVESTR BIOCHEM IMMUNOL, SQUIBB INST MED RES, 69- Mem: AAAS; NY Acad Sci. Res: Cellular immunology; inflammation; nucleic acids; immunopharmacology. Mailing Add: Squibb Inst for Med Res Princeton NJ 08540

WOJTASZEK, JOSEPH HENRY, b Adams, Mass, Sept 2, 37; m 58. THEORETICAL PHYSICS. Educ: Mass Inst Technol, BS, 59; Univ Rochester, PhD(physics), 65. Prof Exp: ASST PROF PHYSICS, UNIV MIAMI, 65- Mem: Am Phys Soc. Res: Particle scattering theory; Regge pole theory; electromagnetic mass differences of elementary particles. Mailing Add: Dept of Physics Univ of Miami Coral Gables FL 33124

WOJTOWICZ, JOHN ALFRED, b Niagara Falls, NY, Oct 12, 26; m 57; c 3. INDUSTRIAL CHEMISTRY. Educ: Univ Buffalo, BA, 54; Niagara Univ, MS, 66. Prof Exp: Analyst chem anal, E I du Pont de Nemours & Co, Inc, 45-50, develop analyst chem anal & process develop, 50-54, develop chemist, 54-56; res chemist inorg synthesis & chem kinetics, 56-61, sr res chemist org synthesis & process develop, 61-71, RES ASSOC INORG & ORG SYNTHESIS & PROCESS DEVELOP, OLIN CORP, 71- Mem: Am Chem Soc; Sigma Xi. Res: Organic and inorganic synthesis; process development. Mailing Add: Olin Corp 275 Winchester Ave New Haven CT 06504

WOJTOWICZ, MARIA BARBARA, biochemistry, see 12th edition

WOJTOWICZ, PETER JOSEPH, b Elizabeth, NJ, Sept 22, 31; m 53; c 3. THEORETICAL PHYSICS. Educ: Rutgers Univ, BSc, 53; Yale Univ, MS, 54, PhD(phys chem), 56. Prof Exp: MEM TECH STAFF & THEORET CHEM PHYSICIST, RCA LABS, 56- Mem: Fel Am Phys Soc. Res: Statistical mechanics; theory of liquid and solid states; theory of magnetism and properties of magnetic substances; liquid crystals. Mailing Add: RCA Labs Princeton NJ 08540

WOJTOWICZ, WESLEY JOSEPH, b Detroit, Mich, July 1, 15; m 38, 51; c 4. PHYSICAL CHEMISTRY. Educ: Lawrence Inst Technol, BChE, 37; Wayne Univ, PhD(chem), 53. Prof Exp: Prod control chemist, 36-48, vpres, 48-75, DIR RES, H A MONTGOMERY CO, 48- Mem: AAAS; Am Chem Soc; Am Soc Lubrication Eng. Res: Colloidal aggregates; tactoids; emulsions; polymers; lubricants. Mailing Add: H A Montgomery Co Inc 17191 Walter P Chrysler Freeway Detroit MI 48203

WOLBACH, DEAN, organic chemistry, air pollution, see 12th edition

WOLBACH, ROBERT ALBERT, b New York, NY, Sept 3, 30; m 74; c 6. PHYSIOLOGY. Educ: Cornell Univ, BA, 51, PhD(physiol), 54; NY Univ, MD, 61. Prof Exp: Instr physiol, Sch Med, NY Univ, 56-60; USPHS spec fel, Copenhagen Univ, 61; asst prof, 62-69, asst to dean, 72-74, ASSOC PROF PHYSIOL, COL MED, UNIV UTAH, 69- Concurrent Pos: Mem, Mt Desert Island Biol Lab; mem, Utah Task Force Health & Ment Health for Span Speaking, 70-73; mem, Utah Migrant Health Policy Bd, 73-, actg chmn, 75. Mem: AAAS; Int Soc Nephrology; Am Soc Nephrology; Am Physiol Soc. Res: Comparative physiology of excretion of vertebrates; metabolism of phosphorus, calcium and hydrogen ions; pathophysiology of cold injury. Mailing Add: Dept of Physiol Univ of Utah Col of Med Salt Lake City UT 84112

WOLBARSHT, MYRON LEE, b Baltimore, Md, Sept 18, 24; m 63; c 3. BIOPHYSICS, BIOMEDICAL ENGINEERING. Educ: St Johns Col, AB, 50; Johns Hopkins Univ, PhD(biol, biophys), 58. Prof Exp: Head physicist, Naval Med Res Inst, 58-68; PROF OPHTHALMOL & BIOMED ENG, EYE CTR, DUKE UNIV, 68- Concurrent Pos: Guest scientist, Naval Med Res Inst, 54-58; res assoc, Psychiat Inst, Med Sch, Univ Md, 54-60; res fel biol, Johns Hopkins Univ, 58-63; consult, York Hosp, Pa, 63-68; mem exec panel, Nat Res Coun Armed Forces Comt Vision, 63-; chmn eye hazards subcomt, Laser Safety Comt, Am Nat Standards Inst, 68- Mem: Am Physiol Soc; Inst Elec & Electronics Engr; Soc Gen Physiol; Optical Soc Am; Royal Soc Med. Res: Laser safety; biomedical engineering applications to ophthalmology; structure and function of sense organs, especially vision, chemoreception, mechanoreception; electrophysiology of central nervous system. Mailing Add: Dept of Ophthal Duke Univ Eye Ctr Durham NC 27710

WOLBARST, ANTHONY BRINTON, b Philadelphia, Pa, Jan 29, 44. HIGH PRESSURE PHYSICS, MAGNETIC RESONANCE. Educ: Trinity Col, Conn, BA, 66; Dartmouth Col, PhD(physics), 71. Prof Exp: Vis asst prof physics, State Univ NY Buffalo, 70-71; sr technician med physics, Intensive Care Unit, Mass Gen Hosp, 72; res fel physics, Univ Witwatersrand, 72-76; SR RES SCIENTIST PHYSICS, NAT PHYS RES LAB, COUN SCI & INDUST RES, S AFRICA, 76- Mem: Am Phys Soc. Res: High-pressure low-temperature electron nuclear double resonance spectroscopy; high-pressure high-temperature polymorphic phase transitions; color centers; group representation theory; many-body theory. Mailing Add: Div of High Pressure Physics NPRL Coun Sci & Indust Res Pretoria South Africa

WOLBERG, GERALD, b New York, NY, Aug 18, 37; m 67; c 2. IMMUNOLOGY, MICROBIOLOGY. Educ: NY Univ, BA, 58; Univ Ky, MS, 63; Tulane Univ, PhD(immunol), 67. Prof Exp: Fel immunol, Pub Health Res Inst of City of New York, 67-68, NIH fel, 68-70; MEM STAFF, WELLCOME RES LABS, BURROUGHS WELLCOME CO, 70- Mem: Am Asn Immunol; Am Soc Microbiol. Res: Pathogenic Escherichia coli; antigenicity of bovine serum albumin; passive hemagglutination techniques; antibody formation in vitro; antibody producing cells; immunosuppression; immunoactivation. Mailing Add: Burroughs Wellcome Co 3030 Cornwallis Rd Research Triangle Park NC 27709

WOLBERG, WILLIAM HARVEY, b July 10, 31; US citizen; m 55; c 6. SURGERY, ONCOLOGY. Educ: Univ Wis-Madison, BS, 53, MD, 56. Prof Exp: Intern, Ohio State Univ, 56-57; resident surg, 57-61, chmn gen surg, 72-75, from instr to assoc prof, 61-71, PROF SURG, UNIV WIS-MADISON, 71- Concurrent Pos: Consult merit rev bd oncol, Vet Admin, 72. Mem: Am Asn Cancer Res; Am Col Surg; Asn Acad Surg. Res: Nucleic acid synthesis in human tumors; immune response to human tumors. Mailing Add: Dept of Surg Univ of Wis Madison WI 53706

WOLCOTT, ARTHUR RIPATTE, b Lake City, Mich, Jan 10, 18; m 40; c 3. SOIL MICROBIOLOGY. Educ: Mich State Univ, BS, 40; Va Polytech Inst, MS, 51; Purdue Univ, PhD(plant physiol), 56. Prof Exp: Instr farm crops & soil sci, Upper Peninsula Substa, Agr Exp Sta, Univ Mich, 47-52; assoc prof soil sci, 55-65, PROF SOIL SCI, MICH STATE UNIV, 65- Concurrent Pos: Consult, Dept Army, Corps Engr, 72-73. Mem: AAAS; Soil Sci Soc Am; Am Soc Agron. Res: Pesticide-soil interactions; impact parameters of wastes in land disposal systems. Mailing Add: Dept Crop & Soil Sci Mich State Univ East Lansing MI 48824

WOLCOTT, DAMIEN, inorganic chemistry, see 12th edition

WOLCOTT, JOHN H, b Columbus, Ohio, Sept 23, 41; m 64; c 1. AEROSPACE MEDICINE. Educ: Colo State Univ, BS, 64, MS, 66, PhD(plant physiol), 68. Prof Exp: Physiol training officer, Air Force Hosp, Moody AFB, US Air Force, 69-70, physiol training officer, Hq Aerospace Med Div, Brooks AFB, 70-73; AEROSPACE PHYSIOLOGIST, ARMED FORCE INST PATH, 73- Mem: AAAS; Am Soc Plant Physiol; Aerospace Med Asn. Res: Pathway of synthesis of pyrimidine nucleotides in plants and effects of light on synthesis of these nucleotides; aerospace physiology, including hypobaric and hyperbaric medical research. Mailing Add: Aerospace Path Div Armed Forces Inst Path Washington DC 20306

WOLCOTT, MARK WALTON, b Mansfield, Ohio, Apr 16, 15; m 41; c 2. SURGERY, THORACIC SURGERY. Educ: Lehigh Univ, BA, 37; Univ Pa, MD, 41. Prof Exp: Instr surg, Grad Sch Med, Univ Pa, 50-52; assoc, Med Col Pa, 52-53; asst clin prof, Med Col Ga, 54-57; chief of surg, Vet Admin Hosp, Coral Gables, Fla, 57-64; chief of res in surg, Vet Admin Cent Off, DC, 64-70; PROF SURG, SCH MED, UNIV UTAH, 70-; CHIEF OF STAFF, VET ADMIN HOSP, SALT LAKE CITY, 70- Concurrent Pos: Assoc prof, Sch Med, Univ Miami, 57-64; asst clin prof, Med Sch, George Washington Univ, 65-70. Mem: Am Thoracic Soc; AMA; Soc Thoracic Surg; Am Col Surg; Am Col Chest Physicians. Res: Coagulation defects associated with thoracic surgery; deep hypothermia and extra corporeal circulation; tissue transplantation; clostridial infection and hyperbaric medicine. Mailing Add: 500 Foothill Blvd Salt Lake City UT 84113

WOLCOTT, ROBERT B, b Janesville, Wis, July 27, 14; m 46; c 2. DENTISTRY. Educ: Beloit Col, BS, 36; Marquette Univ, DDS, 41; Georgetown Univ, MS, 51. Prof Exp: Head oper dent, US Naval Dent Sch, 44-48 & 52-54; guestworker, Dent Mat Lab, Nat Bur Stand, 49-51; dent asst, Naval Training Ctr, Calif, 56-58; dir training div, Dent Res Labs, Md, 58 & Ill, 59-62; PROF RESTORATIVE DENT, CHMN & DIR CLIN, SCH DENT, CTR HEALTH SCI, UNIV CALIF, LOS ANGELES, 62- Concurrent Pos: Consult, Vet Admin, 66- Mem: Fel Am Col Dent; Am Dent Asn; Am Acad Restorative Dent; Int Asn Dent Res. Res: Clinical and laboratory investigations in development and utilization of dental materials. Mailing Add: Sch of Dent Univ of Calif Ctr for Health Sci Los Angeles CA 90024

WOLCOTT, ROBERT MICHAEL, b Greenville, Ky, June 24, 40; m 67; c 1. IMMUNOLOGY, BIOCHEMISTRY. Educ: Ill Wesleyan Univ, BS, 65; Vanderbilt Univ, PhD(biochem), 70. Prof Exp: ASST PROF MICROBIOL, SCH MED, UNIV ALA, BIRMINGHAM, 71- Concurrent Pos: NIH fel med, Sch Med, Univ Ala, Birmingham, 69-71. Res: Deleterious effects of environmental toxicants on the immune system. Mailing Add: Dept of Microbiol Univ of Ala Sch of Med Birmingham AL 35294

WOLD, AARON, b NY, May 8, 27; m 57; c 3. INORGANIC CHEMISTRY. Educ: Polytech Inst Brooklyn, BS, 46, MS, 48, PhD(chem), 52. Prof Exp: Res assoc chem, Univ Conn, 51-52; from instr to asst prof, Hofstra Col, 52-56; mem staff, Lincoln Lab, Mass Inst Technol, 56-63; assoc prof eng & chem, 63-67, PROF ENG & CHEM, BROWN UNIV, 67- Concurrent Pos: Ed, J Solid State Chem, 68-75; assoc ed, J Inorg Chem, 74-; consult, E I du Pont de Nemours & Co, Inc. Mem: Am Chem Soc. Res: Solid state chemistry of rare earths and transition elements. Mailing Add: Dept Chem & Div Eng Brown Univ Providence RI 02912

WOLD, DONALD C, b Fargo, NDak, Sept 24, 33; m 56; c 3. ACOUSTICS. Educ: Univ Wis-Madison, BA, 55, MA, 57; Ind Univ, Bloomington, PhD(physics), 68. Prof Exp: Lectr physics, Forman Christian Col, WPakistan, 58-63, head dept, 61-63; asst res physicist, Univ Calif, Los Angeles, 68-69; assoc prof physics, 69-74, head dept, 70-74, PROF PHYSICS & CHMN DEPT PHYSICS & ASTRON, UNIV ARK, LITTLE ROCK, 74- Mem: Am Phys Soc; Am Asn Physics Teachers; Asn Comput Mach. Res: Acoustics of speech and music. Mailing Add: Dept of Physics & Astron Univ of Ark Little Rock AR 72204

WOLD, FINN, b Stavanger, Norway, Feb 3, 28; nat US; m 53; c 2. BIOCHEMISTRY. Educ: Okla State Univ, MS, 53; Univ Calif, PhD(biochem), 56. Prof Exp: Res assoc biochem, Univ Calif, 56-57; from asst prof to assoc prof, Univ Ill, 57-66; prof biochem, Med Sch, Univ Minn, Minneapolis, 66-74; PROF BIOCHEM & HEAD DEPT, UNIV MINN, ST PAUL, 74- Concurrent Pos: Lalor res award, 58; Guggenheim fel immunochem, London, Eng, 60-61; USPHS res career develop award, 61-66; vis prof chem, Nat Taiwan Univ, 71; consult biochem & molecular biol fel rev comt, Nat Inst Gen Med Sci, 66-70, consult biochem training comt, 71-74; consult biochem & biophys res eval comt, Vet Admin, 69-71, consult res serv merit rev basic sci, 72-75; consult res personnel comt, Am Cancer Soc, 74-77; vis prof biochem, Rice Univ, 74. Mem: AAAS; Am Soc Biol Chemists; Am Chem Soc; Brit Biochem Soc. Res: Protein chemistry; physical, chemical and biological properties of proteins; relation of protein structure and function; mechanism of enzyme action. Mailing Add: Dept of Biochem Univ of Minn St Paul MN 55108

WOLD, RICHARD JOHN, b Oshkosh, Wis, Oct 23, 37; m 60; c 2. MARINE GEOPHYSICS. Educ: Univ Wis, BS, 60, PhD(geophys), 66. Prof Exp: Lectr geophys, Univ Wis, 66-67; asst prof geol & geophys, 67-70, chmn dept geol sci, 70-72, assoc dean res, Grad Sch, 72-73, dir, Gt Lakes Res Facility, 73-75; ASSOC PROF GEOL SCI, UNIV WIS-MILWAUKEE, 70- Concurrent Pos: Mem, Nat Adv Coun, Univ-Nat Oceanog Lab Syst, 74-75; assoc br chief, Off Marine Geol, Atlantic-Gulf of Mex Br, US Geol Surv, 75-76. Mem: Am Geophys Union; Soc Explor Geophys. Res: Geophysical studies in Great Lakes; geophysical and geological studies of inland lakes; marine geophysical instrumentation. Mailing Add: Dept Geol Sci Univ of Wis Milwaukee WI 53201

WOLDSETH, ROLF, b Trondheim, Norway, Jan 18, 30; US citizen; m 55; c 4. ATOMIC PHYSICS, NUCLEAR PHYSICS. Educ: Tech Univ Norway, BSc, 55; Wash Univ, St Louis, PhD(physics), 65. Prof Exp: Res physicist, Joint Estab for Nuclear Energy Res, Norway, 55-56; asst prof physics, Rensselaer Polytech Inst, 63-67 & Wake Forest Univ, 67-70; DIR APPLN LAB, KEVEX CORP, FOSTER CITY, CALIF, 70- Mem: Am Phys Soc. Res: Positron annihilation; x-ray spectra; photonuclear reactions; fast neutron induced reactions. Mailing Add: 220 Stilt Ct Foster City CA 94404

WOLEN, ROBERT LAWRENCE, b New York, NY, May 20, 28. PHARMACOLOGY. Educ: West Chester State Col, BS, 50; Univ Del, MS, 51, PhD(biochem), 60. Prof Exp: Lectr atomic energy, Oak Ridge Inst Nuclear Studies, 60-61; lab dir biochem, Res Inst, St Joseph Hosp, 61-62; res scientist, Lilly Res Labs, 62-65, RES ASSOC PHARMACOL, LILLY LAB CLIN RES, 65-; ASST PROF PHARMACOL, SCH MED, IND UNIV, INDIANAPOLIS, 73- Mem: Am Chem Soc; Sigma Xi; Am Soc Pharmacol & Exp Therapeut; Am Asn Clin Chemists. Res: Phase I clinical pharmacology including drug metabolism, pharmacokinetics, drug interactions and therapeutic drug analysis. Mailing Add: Lilly Lab for Clin Res Wishard Mem Hosp Indianapolis IN 46202

WOLF, ABNER, b New York, NY, May 14, 02; m 33, 49; c 1. NEUROPATHOLOGY. Educ: Columbia Univ, AB, 23, MD, 26. Prof Exp: From instr to asst prof path, 30-39, instr neurol, 31-35, assoc instr, 33-35, pathologist, 35-46, from asst prof to assoc prof neuropath, 39-51, PROF NEUROPATH, COL PHYSICIANS & SURGEONS, COLUMBIA UNIV, 51- Concurrent Pos: Asst attend pathologist, Presby Hosp, 31-33, attend neuropathologist, 44-; attend neuropathologist, Babies' Hosp, 32-, Vanderbilt Clin, 44- & Columbia-Presby Med Ctr, 46-; consult, Vet Admin Hosp, Kingsbridge, 46- & USPHS; asst ed, J Neuropath & Exp Neurol, 57-; mem neurol grad training grant comt, Nat Inst Neurol Dis & Blindness, 57-60, bd sci counr, 59-63. Mem: Am Soc Exp Path; Harvey Soc; Am Asn Neuropath; Int Soc Neuropath (pres, 70-74); Am Neurol Asn. Res: Infections and regeneration of the nervous system; pathogenesis of schizophrenia; sensitization of the central nervous system to neural tissue. Mailing Add: Col of Physicians & Surgeons Columbia Univ Dept of Path New York NY 10032

WOLF, ALBERT ALLEN, b Nashville, Tenn, Sept 2, 35; m 56; c 4. ELEMENTARY PARTICLE PHYSICS. Educ: Vanderbilt Univ, BA, 58, MA, 60; Ga Inst Technol, PhD(physics), 66. Prof Exp: Physicist, Aladdin Electronics Div, Aladdin Industs, Inc, 58; engr, Sperry Rand, Inc, 59-61; instr physics, Ga Inst Technol, 61-64; asst prof, 65-69, ASSOC PROF PHYSICS, DAVIDSON COL, 69- Concurrent Pos: Guest prof, Univ Ulm, WGer, 71-72. Mem: Am Phys Soc; Am Asn Physics Teachers. Res: Quantization of non-linear fields. Mailing Add: Dept of Physics Davidson Col Box 285 Davidson NC 28036

WOLF, ALFRED PETER, b New York, NY, Feb 13, 23; m 46; c 1. ORGANIC CHEMISTRY, PHYSICAL CHEMISTRY. Educ: Columbia Univ, BA, 44, MA, 48, PhD(chem), 52. Prof Exp: MEM STAFF, CHEM DEPT, BROOKHAVEN NAT LAB, 51- Concurrent Pos: Adj prof, Columbia Univ, 53-; ed, J Labelled Compounds, 65-; consult, Univ Mass, 65-66, Philip Morris, 66-, NIH, 66 & Int Atomic Energy Agency, 69-; adv, Ital Nat Res Coun, 69-; mem eval panel, Nat Bur Standards, 72- Honors & Awards: Award, Am Chem Soc, 71. Mem: Am Chem Soc; The Chem Soc; Soc Ger Chem; Soc Nuclear Med. Res: Radiopharmaceutical research and nuclear medicine; organic reaction mechanisms; chemical effects of nuclear transformations; chemistry of carbon-11, nitrogen-13, oxygen-15 and iodine-123; accelerators for nuclide production and radiopharmaceutical production. Mailing Add: PO Box 1043 Setauket NY 11733

WOLF, ARNOLD VERYL, physiology, deceased

WOLF, BENJAMIN, b Deerfield, NJ, Dec 2, 13; m 40; c 2. SOIL CHEMISTRY. Educ: Rutgers Univ, BS, 35, MS, 38, PhD(soil chem), 40. Prof Exp: Asst instr soil chem, Rutgers Univ, New Brunswick, 40-41; soil chemist, Seabrook Farms Co, 41-49; CONSULT, DR WOLF'S AGR LABS, 49- Concurrent Pos: Consult veg & floricult crops, Cent Am, SAm & Caribbean; assoc ed, Commun Soil Sci & Plant Anal, 70- Mem: AAAS; fel Soil Sci Soc Am; Am Soc Hort Sci. Res: Soil and plant analysis; herbicides. Mailing Add: Dr Wolf's Agr Labs 6861 SW 45th St Ft Lauderdale FL 33314

WOLF, BENJAMIN, b Detroit, Mich, June 27, 26; m 52; c 2. MICROBIOLOGY. Educ: Wayne State Univ, BS, 49; Univ Mich, MS, 52; Univ Pa, PhD(microbiol), 59. Prof Exp: From asst instr to asst prof, 57-69, ASSOC PROF MICROBIOL, SCH VET MED, UNIV PA, 69- Mem: Am Soc Microbiol. Res: Medical microbiology; immunology. Mailing Add: Dept of Microbiol Univ Pa Sch Vet Med Philadelphia PA 19104

WOLF, BERNARD SAUL, b New York, NY, Sept 8, 12; m 39; c 2. RADIOLOGY. Educ: NY Univ, BS, 32, MD, 36. Prof Exp: Resident path, 36-37, intern surg, 37-40, asst path, 40, resident radiol, 40-42, asst radiotherapist, 42-44, assoc radiotherapist, 47-49, CHIEF DEPT RADIOL, MT SINAI HOSP, 49-; PROF RADIOL & CHMN DEPT, MT SINAI SCH MED, 65- Concurrent Pos: Med dir, NY Opers Off, AEC, 47-49; mem, First Radiation Surv Party, Hiroshima. Mem: Roentgen Ray Soc; Am Gastroenterol Asn; fel Am Col Radiol; Am Soc Neuroradiol; Radiol Soc NAm. Res: Clinical radiology; radium and radioactive materials. Mailing Add: Mt Sinai Hosp Dept Radiol Fifth Ave & E 100th New York NY 10029

WOLF, BEVERLY, b Chicago, Ill, Jan 14, 35. MICROBIOLOGY. Educ: Univ Colo, BA, 55; Univ Calif, Los Angeles, PhD(biochem), 59. Prof Exp: Guest investr, Rockefeller Inst, 59-61; res assoc, Harvard Univ, 61-63; asst res biologist, Univ Calif, Berkeley, 65-72; vis asst prof, Mills Col, 72-73; SR SCIENTIST, CETUS CORP, 73- Res: Biochemical genetics of neurospora; genetical transformation of pneumococcus; origin and direction of desoxyribonucleic acid synthesis in Endamoeba coli; temperature sensitive DNA synthesis mutants of Escherichia coli; microbial antibiotic production. Mailing Add: Cetus Corp 600 Bancroft Way Berkeley CA 94710

WOLF, CALVIN NEWTON, organic chemistry, see 12th edition

WOLF, CAROL EUWEMA, b New Castle, Pa, June 11, 36; m 58; c 2. MATHEMATICAL LOGIC. Educ: Swarthmore Col, BA, 58; Cornell Univ, MA, 62, PhD(math), 64. Prof Exp: Asst prof math, State Univ NY Brockport, 68-75; ASST PROF MATH, IOWA STATE UNIV, 75- Mem: Am Math Soc; Math Asn Am; Asn Symbolic Logic. Res: Investigation of application of recursive function theory to understanding of brain function. Mailing Add: 3611 Ross Rd Ames IA 50010

WOLF, CHARLES TROSTLE, b West Reading, Pa, Mar 20, 30; m 53; c 3. MATHEMATICS. Educ: Millersville State Col, BS, 53; Univ Del, MS, 59. Prof Exp: Teacher, Pequea Valley High Sch, 53-56; asst math, Univ Del, 56-58; asst prof, Shippensburg State Col, 58-61; ASSOC PROF MATH, MILLERSVILLE STATE COL, 61- Mem: Math Asn Am. Res: Modern mathematics for elementary teachers. Mailing Add: Dept of Math Millersville State Col Millersville PA 17551

WOLF, CLARENCE J, physical chemistry, see 12th edition

WOLF, DALE DUANE, b Alma, Nebr, June 16, 32; m 52; c 4. AGRONOMY. Educ: Univ Nebr, BSc, 54, MSc, 57; Univ Wis, PhD(agron), 62. Prof Exp: Asst prof agron, Univ Conn, 62-67; asst prof, 67-71, ASSOC PROF AGRON, VA POLYTECH INST & STATE UNIV, 71- Mem: Am Soc Agron; Crop Sci Soc Am. Res: Forage crops; plant physiology. Mailing Add: Dept of Agron Va Polytech Inst & Sta Univ Blacksburg VA 24060

WOLF, DALE E, b Kearney, Nebr, Sept 6, 24; m 45; c 4. AGRONOMY. Educ: Univ Nebr, BSc, 43; Rutgers Univ, PhD(farm crops, weed control), 49. Prof Exp: Asst farm crops, Rutgers Univ, 46-49, assoc prof & assoc res specialist, 49-50; agronomist, USDA, 47-50; asst mgr agr chem res, 50-54, mgr, 54-56, asst dist sales mgr, 56-59, dist sales mgr, 60-64, sales mgr biochem, Del, 64-67, mgr, Planning Div, 67-68, asst dir, Agr Div, 68-70, dir, Indust Specialities Div, 71-72, dir mkt agrichem, 72-75, ASST GEN MGR, BIOCHEM DEPT, E I DU PONT DE NEMOURS & CO, INC, 75- Mem: Am Soc Agron. Res: Agricultural chemicals; weed control. Mailing Add: B-14348 E I du Pont de Nemours & Co Inc Wilmington DE 19898

WOLF, DIETER, b Sindelfingen, WGer, Nov 3, 46. THEORETICAL SOLID STATE PHYSICS, MAGNETIC RESONANCE. Educ: Univ Stuttgart, Diplom, 70, Dr rer nat(physics), 73. Prof Exp: Res assoc mat sci, Max Planck Inst Metal Res, Stuttgart, 72-74; RES ASST PROF PHYSICS, UNIV UTAH, 74- Mem: Am Phys Soc. Res: Diffusion in crystals; theory of motion-induced magnetic resonance and relaxation; atomic, molecular and defect motions in metals and ionic, molecular and liquid crystals studied by nuclear magnetic resonance. Mailing Add: Dept of Physics Univ of Utah Salt Lake City UT 84112

WOLF, DON PAUL, b Lansing, Mich, Aug 8, 39; m 67; c 2. BIOCHEMISTRY, REPRODUCTIVE BIOLOGY. Educ: Mich State Univ, BS, 61, MS, 62; Univ Wash, PhD(biochem), 67. Prof Exp: Fel biochem, Univ Geneva, 67-68 & Univ Calif, Davis, 68-71; ASST PROF OBSTET & GYNEC & BIOPHYS, UNIV PA, 71- Mem: AAAS; Am Fertil Soc; Soc Study Reproduction; Am Soc Cell Biol; Am Soc Biol Chemists. Res: Characterization of reproductive processes; the role of cortical granules in fertilization and early development; sperm excluding mechanism operative in animal ova; isolation and characterization of cervical mucin. Mailing Add: 307 Med Labs Univ Pa Div Reprod Biol Philadelphia PA 19174

WOLF, DONALD EDWIN, b Oskaloosa, Iowa, Nov 11, 12; m 45; c 5. ORGANIC CHEMISTRY. Educ: Wheaton Col, BS, 35; Univ Ill, PhD(org chem), 39. Prof Exp: Fel, Univ Ill, 39-40; res chemist, Org & Biochem Res Dept, Merck & Co, Inc, 40-66, RES CHEMIST, ANIMAL DRUG METAB DEPT, MERCK SHARP & DOHME RES LABS, 66- Mem: AAAS; Am Chem Soc. Res: Isolation of natural products; drug metabolism. Mailing Add: Animal Drug Metab Dept Merck Sharp & Dohme Res Labs Rahway NJ 07065

WOLF, DUANE CARL, b Springfield, Mo, Apr 7, 46; m 68. SOIL MICROBIOLOGY. Educ: Univ Mo-Columbia, BS, 68; Univ Calif, Riverside, PhD(soils), 73. Prof Exp: ASST PROF SOILS, UNIV MD, COLLEGE PARK, 73- Mem: Am Soc Agron; Soil Sci Soc Am; Am Soc Microbiol; AAAS. Res: Microbiology of sludge amended soils; evaluation of forest and grass buffer strips in improving the water quality of manure-polluted runoff; effect of salt fallout from cooling towers on soil. Mailing Add: Dept of Agron Univ of Md College Park MD 20742

WOLF, EDWARD D, b Quinter, Kans, May 30, 35; m 55; c 3. SURFACE PHYSICS. Educ: McPherson Col, BS, 57; Iowa State Univ, PhD(phys chem), 61. Prof Exp: Res assoc, Princeton Univ, 61-62; sr chemist, Atomics Int Div, NAm Aviation, Inc, 63-64, res specialist, 65, mem tech staff, Sci Ctr Div, 64-65; mem tech staff, 65-67, sr staff chemist, 67-72, sect head electron beam surface physics, 72-74, SR SCIENTIST, HUGHES RES LABS, MALIBU, 74- Concurrent Pos: Res assoc, Univ Calif, Berkeley, 68. Mem: Am Phys Soc; Electron Micros Soc Am; sr mem Inst Elec & Electronics Eng. Res: Ionic mobilities of high temperature inorganic liquids; magnetohydrodynamic energy conversion; field emission and scanning electron microscopy; electron beam microfabrication.

WOLF, EMIL, b Prague, Czech, July 30, 22; m 51; c 2. MATHEMATICAL PHYSICS. Educ: Bristol Univ, BSc, 45, PhD(physics), 48; Univ Edinburgh, DSc(physics), 55. Prof Exp: Res asst optics, Cambridge Univ, 48-51; res asst & lectr math physics, Univ Edinburgh, 51-54; res fel theoret physics, Univ Manchester, 54-59; assoc prof optics, 59-61, PROF PHYSICS, UNIV ROCHESTER, 61- Concurrent Pos: Guggenheim fel & vis prof, Univ Calif, Berkeley, 66-67. Mem: Fel Optical Soc Am; Fel Am Phys Soc; fel Brit Inst Physics. Res: Theoretical optics; electromagnetic theory. Mailing Add: Dept of Physics & Astron Univ of Rochester Rochester NY 14627

WOLF, ERIC W, b Frankfurt am Main, Ger, Feb 20, 22; US citizen; m 49; c 2. COMPUTER SCIENCE, SYSTEMS SCIENCE. Educ: City Col New York, BEE, 49; Ohio State Univ, MS, 51. Prof Exp: Electronic scientist, Wright Air Develop Ctr, 49-53; res engr, Lincoln Lab, Mass Inst Technol, 53-59; tech adv tech ctr, Supreme Hq Allied Powers Europe, The Hague, 59-60; dir comput prog, Data & Info Systs Div, Int Tel & Tel Corp, 61-62, tech dir, 62-64; tech dir, Naval Command Systs Support Activity, 64-70; SR SCIENTIST, BOLT BERANEK & NEWMAN INC, 70- Honors & Awards: Prof Commendation, US Air Force Res & Develop Command, 56. Mem: AAAS; Am Soc Info Sci; Asn Comput Mach; Inst Elec & Electronics Engrs; Am Inst Aeronaut & Astronaut. Res: Design and development of computer-based information systems; computer communications; computer networks; information management; man-machine interaction in natural language. Mailing Add: Bolt Beranek & Newman Inc 1701 N Ft Myer Dr Arlington VA 22209

WOLF, FRANK E, b New York, NY, Nov 15, 20; m 54. BIOLOGY. Educ: NY Univ, BS, 50, MA, 51, EdD, 56. Prof Exp: Teacher, New York Bd Educ, 50-52; teacher, Pub Sch, Del, 52-53; Fulbright teacher & supvr schs, Burma, 53-54; teacher, Pub Sch, NY, 54-55; assoc prof biol sci, Plymouth Teachers Col, 56; prof spec educ, NY State Col Teachers, Buffalo, 56-57; PROF BIOL SCI, FITCHBURG STATE COL, 57- Mem: Nat Asn Biol Teachers; Nat Sci Teachers Asn; Am Micros Soc; NY Acad Sci. Res: Development of specialized attachments and accessories to the microscope; development of polychrome stains and stainmountants. Mailing Add: Dept of Biol Sci Fitchburgh State Col Fitchburg MA 01420

WOLF, FRANK JAMES, b Xenia, Ohio, Nov 7, 16; m 42; c 3. ORGANIC CHEMISTRY. Educ: Miami Univ, AB, 38; Univ Ill, PhD(org chem), 42. Prof Exp: Sr chemist, Merck & Co, Inc, 42-59, asst dir microbiol, Merck Sharp & Dohme Res Labs, 59-70, DIR ANIMAL DRUG METAB & RADIOCHEM, MERCK SHARP & DOHME RES LABS, 70- Concurrent Pos: Asst, Nat Defense Res Comt, 41-42. Mem: Am Soc Pharmacol & Exp Therapeut; Am Chem Soc. Res: Phthalides; substituted sulfaquinoxalines; benzotriazines; antibiotics; vitamin B-12; catalysis; biochemical separations; drug metabolism; animal tissue residue; mass spectroscopy. Mailing Add: 126 E Lincoln Ave Rahway NJ 07065

WOLF, FRANK LOUIS, b St Louis, Mo, Apr 18, 24; m 47; c 4. MATHEMATICS. Educ: Wash Univ, BS, 44, MA, 48; Univ Minn, PhD(math), 55. Prof Exp: Tech supvr chem eng, Carbon & Carbide Chem Corp, Tenn, 44-46; instr math, St Cloud State Col, 49-51; from instr to assoc prof, 52-67, PROF MATH, CARLETON COL, 67-

Concurrent Pos: NSF fel, 60-61. Mem: AAAS; Math Asn Am. Res: Statistics; foundations of mathematics. Mailing Add: Dept of Math Carleton Col Northfield MN 55057

WOLF, FRANTISEK, b Prostejov, Czech, Nov 30, 04; nat US; m 45; c 1. MATHEMATICS. Educ: Masaryk Univ, PhD(anal), 28. Prof Exp: Instr, High Schs, Prague, 30-31; privat docent, Charles Univ, Prague, 38-41; instr math, Macalester Col, 41-42; from instr to prof, 42-72, EMER PROF MATH, UNIV CALIF, BERKELEY, 72- Concurrent Pos: Denis Fund fel, Cambridge Univ, 37; Govt Sweden fel, Univ Stockholn, 39-41; Fulbright fel, Free Univ Brussels, 59-60; Miller res fel, 63-64; vis prof, Aachen Tech Univ, 63; dir progs for gifted high sch students, NSF, 60-67; Tufts Univ Sch Law & Diplomacy Latin Am teaching fel math, San Carlos Univ Guatemala, 75. Mem: Am Math Soc; Math Asn Am. Res: Trigonometric expansions; summability and uniqueness theory; theory of operators, especially differential operators. Mailing Add: 181 Stonewall Rd Berkeley CA 94705

WOLF, FREDERICK TAYLOR, b Auburn, Ala, July 11, 15; m 45; c 2. BOTANY. Educ: Harvard Univ, AB, 35; Univ Wis, MA, 36, PhD(bot), 38. Prof Exp: Nat Res Coun fel, Harvard Univ, 38-39; from instr to assoc prof bot, 39-56, PROF BOT, VANDERBILT UNIV, 56- Mem: Mycol Soc Am; Bot Soc Am; Am Soc Plant Physiol; Brit Mycol Soc; Int Soc Human & Animal Mycol. Res: Mycology; physiology of fungi; plant physiology. Mailing Add: Dept of Gen Biol Vanderbilt Univ Nashville TN 37235

WOLF, GEORGE, b Vienna, Austria, June 16, 22; nat US; m 48; c 3. NUTRITION, BIOCHEMISTRY. Educ: Univ London, BSc, 44; Oxford Univ, DPhil, 47. Prof Exp: Res fel, Chester Beatty Res Inst, Royal Cancer Hosp, Eng, 47-48, Harvard Univ, 48-50 & Univ Wis, 50-51; from asst prof to assoc prof animal nutrit, Univ Ill, Urbana, 51-62; assoc prof, 62-74, PROF NUTRIT & FOOD SCI, MASS INST TECHNOL, 74- Concurrent Pos: Guggenheim fel, 58-59. Mem: Am Soc Biol Chemists; Am Inst Nutrit. Res: Intermediary metabolism of amino acids and vitamins; biosynthesis and metabolism of carnitine; metabolic function of vitamin A. Mailing Add: Dept of Nutrit & Food Sci Mass Inst of Technol Cambridge MA 02139

WOLF, GEORGE ANTHONY, JR, b East Orange, NJ, Apr 20, 14; m 39; c 2. MEDICINE. Educ: NY Univ, BS, 36; Cornell Univ, MD, 41; Am Bd Internal Med, dipl, 48. Hon Degrees: DSc, Union Univ, 65. Prof Exp: Intern med, New York Hosp, 41-42; asst, Med Col, Cornell Univ, 42-43, instr, 43, fel med res, 44-46, asst prof clin med, 49-52; prof clin med & dean, Col Med, Univ Vt, 52-61; prof med, Sch Med & vpres med & dent affairs, Tufts Univ, 61-66; dean & provost, Univ Kansas Med Ctr, Kansas City, 66-70; PROF MED, COL MED, UNIV VT, 70- Concurrent Pos: Asst resident med, New York Hosp, 42-43, resident physician, 43-44, physician to out-patients, 44, asst attend physician & dir out-patient dept, 49-52; dir, Tufts New Eng Med Ctr, 61-66; attend physician, Med Ctr Hosp, Vt, 70- Mem: Harvey Soc; fel Am Col Physicians; fel NY Acad Med; Asn Am Med Cols (past pres). Res: Internal medicine. Mailing Add: Dept of Med Univ of Vt Col Med Burlington VT 05401

WOLF, GEORGE L, b Cadjavica, Yugoslavia, July 17, 27; US citizen; m 62; c 1. VETERINARY PATHOLOGY. Educ: Univ Munich, DVM, 59; Ohio State Univ, MSc, 61, PhD(path), 65. Prof Exp: From instr to asst prof path, Ohio State Univ, 59-65; pathologist, Eastman Kodak Co, NY, 65-68; asst prof lab animal med, Col Med, Univ Cincinnati, 68-72; pathologist, Hoechst Pharmaceut, Inc, 72-73; PATHOLOGIST, MERCK SHARP & DOHME, 73- Concurrent Pos: Consult, Vet Admin, 68- Mem: Electron Micros Soc Am; Am Vet Med Asn. Res: Spontaneous and experimental canine cryptococcosis; Rocky Mountain spotted fever in monkeys. Mailing Add: Merck Sharp & Dohme West Point PA 19450

WOLF, GEORGE WILLIAM, b Newark, NJ, Jan 15, 43. ASTRONOMY. Educ: Univ Pa, BA, 65, MS, 67, PhD(astron), 70. Prof Exp: Res assoc astron, Mt John Observ, Univ Canterbury, NZ, 68-69; ASST PROF ASTRON, SOUTHWEST MO STATE UNIV, 71- Mem: Am Astron Soc. Res: Observational astronomy in the areas of linear and circular polarimetry, spectroscopy and photoelectric photometry of stars. Mailing Add: Dept of Physics & Astron Southwest Mo State Univ Springfield MO 65802

WOLF, GERALD LEE, b Sidney, Nebr, Apr 2, 38; m 64; c 2. RADIOLOGY. Educ: Univ Nebr, BS, 62, MS, 64, PhD(physiol & pharmacol), 65; Harvard Univ, MD, 68. Prof Exp: Asst prof physiol, 68-69, asst prof physiol & med, 69-71, ASST PROF PHARMACOL & RADIOL & DIR RADIOL RES, COL MED, UNIV NEBR, OMAHA, 71- Concurrent Pos: Life Ins Med res fel, 64-69; intern, Col Med, Univ Nebr, 68-69; res consult, Omaha Vet Admin Hosp, 69-, clin investr, 70-73. Mem: Am Col Radiol; fel Am Soc Clin Pharmacol & Therapeut; Asn Am Med Cols; AMA; Am Soc Pharmacol & Exp Therapeut. Res: Renal and endocrine participation in electrolyte homeostasis; physiology of the pump-perfused dog kidney. Mailing Add: Dept of Radiol Univ of Nebr Col of Med Omaha NE 68105

WOLF, HAROLD HERBERT, b Quincy, Mass, Dec 19, 34; m 57; c 2. PHARMACOLOGY. Educ: Mass Col Pharm, BS, 56; Univ Utah, PhD(pharmacol, exp psychiat), 61. Prof Exp: From asst prof to assoc prof, 61-69, KIMBERLY PROF PHARMACOL & CHMN DEPT, COL PHARM, OHIO STATE UNIV, 69- Concurrent Pos: NIMH & Nat Inst Drug Abuse res grants, 63-; Am Asn Cols Pharm vis lectr, 64-; mem pharm rev comt, NIH, 69-71; Fulbright-Hays sr scholar, Univ Sains Malaysia, 74. Mem: AAAS; Am Asn Col Pharm (pres-elect, 76); Am Pharmaceut Asn; Am Soc Pharmacol & Exp Therapeut; fel Acad Pharmaceut Sci. Res: Investigation of effects of drugs on central nervous system; psychotropics, central nervous system stimulants, anticonvulsants, narcotic analgetics, thermoregulation and animal behavior. Mailing Add: Dept of Pharmacol Ohio State Univ Col of Pharm Columbus OH 43210

WOLF, HAROLD WILLIAM, b Chicago, Ill, July 13, 21; m 44; c 4. ENVIRONMENTAL HEALTH. Educ: Univ Iowa, BS, 49, MS, 50, DrPH, 65. Prof Exp: From sanit engr to sr sanit engr, Fed Water Pollution Control Admin, USPHS, Calif, 50-68, asst chief res & develop, Water Supply & Sea Resources, Ohio, 68, dir div criteria & standards, Bur Water Hygiene, Md, 68-70; PROF CIVIL ENG, TEX A&M UNIV, 70-, DIR ENVIRON SCI ENG, 75- Concurrent Pos: Res fel, Calif Inst Technol, 60-62; lectr, Univ Calif, Los Angeles, 65-66; dir, Dallas Water Reclamation Res Ctr, 70-75; mem, Nat Drinking Water Adv Coun, 75. Mem: AAAS; Sigma Xi; Am Soc Civil Eng; Am Pub Health Asn; Am Soc Microbiol. Res: Water quality criteria; air-borne pathogens. Mailing Add: 202 Civil Eng Bldg Tex A&M Univ College Station TX 77843

WOLF, HENRY, b Munich, Ger, Aug 30, 19; mat US; m 44; c 4. APPLIED MATHEMATICS, CELESTIAL MECHANICS. Educ: Univ Toronto, BA, 46, MA, 47, PhD, Brown Univ, 50. Prof Exp: Instr math, Univ Toronto, 46-48; res assoc appl math, Brown Univ, 48-50; from asst prof to assoc prof math, Hofstra Col, 50-56; prin engr, Repub Aviation Corp, 56-57, group engr, 57-58, sr group engr, 58-59, develop engr, Appl Math Sect, 59-60, chief numerical methods, 60-63; sr scientist & secy treas, 63-74, VPRES, ANAL MECH ASSOCS, JERICHO, 74- Mem: Am Math Soc. Res: Plastic wave propagation; interior ballistics and aeroelasticity; numerical analysis. Mailing Add: Anal Mech Assocs 50 Jericho Turnpike Jericho NY 11753

WOLF, IRA KENNETH, b New York, NY, Nov 14, 42; m 64; c 1. MATHEMATICS. Educ: Tufts Univ, BA, 64; Yale Univ, MA, 66; Rutgers Univ, PhD(math), 71. Prof Exp: ASST PROF MATH, BROOKLYN COL, 71- Mem: Math Asn Am. Res: Category theory. Mailing Add: Dept of Math Brooklyn Col Brooklyn NY 11210

WOLF, JAMES STUART, b Chicago, Ill, Mar 1, 35; m 58; c 2. MEDICINE, SURGERY. Educ: Grinnell Col, AB, 57; Univ Ill, BS, 59, MD, 61. Prof Exp: USPHS res fel transplantation, 64-66, from instr to assoc prof, 67-74, PROF SURG, MED COL VA, 74-; CHIEF SURG, McGUIRE VET ADMIN HOSP, 68- Concurrent Pos: Attend surgeon, McGuire Vet Admin Hosp, 67-68. Mem: Transplantation Soc; Am Soc Nephrology; Asn Acad Surg; fel Am Col Surg; Soc Univ Surg. Res: Transplantation immunology; clinical and experimental organ transplantation. Mailing Add: Med Col of Va Dept Surg PO Box 167 MDV Sta Richmond VA 23298

WOLF, JOHN PETER, III, organic chemistry, see 12th edition

WOLF, JOSEPH ALBERT, b Chicago, Ill, Oct 18, 36. MATHEMATICS. Educ: Univ Chicago, BS, 56, MS, 57, PhD(geom), 59. Prof Exp: NSF fel math & mem, Inst Advan Study, 60-62; from asst prof to assoc prof, 62-66, PROF MATH, UNIV CALIF, BERKELEY, 66- Concurrent Pos: Alfred P Sloan res fels, 66-68, Miller Res Prof, 72. Mem: AAAS; Am Math Soc; Math Soc France; Swiss Math Soc. Res: Riemannian geometry, Lie groups and harmonic analysis. Mailing Add: Dept of Math Univ of Calif Berkeley CA 94720

WOLF, JULIUS, b Boston, Mass, Aug 15, 18; m 45; c 3. MEDICINE. Educ: Boston Univ, SB, 40, MD, 43. Prof Exp: Chief med serv, 54-72, CHIEF OF STAFF, VET ADMIN HOSP, BRONX, 72-; PROF CLIN MED, MT SINAI SCH MED, 68-, ASSOC DEAN VET ADMIN PROGS, 72- Concurrent Pos: Assoc clin prof med, Col Physicians & Surgeons, Columbia Univ, 62-68; chmn, Vet Admin Lung Cancer Study Group, 62- Res: Lung cancer chemotherapy. Mailing Add: Vet Admin Hosp 130 W Kingsbridge Rd Bronx NY 10468

WOLF, KARL HEINZ, b Ruesselsheim, Ger, Dec 9, 30; Can citizen; c 2. GEOLOGY. Educ: McMaster Univ, BSc, 57; Univ Sydney, PhD(geol), 63. Prof Exp: Geologist, Aluminum Co Can, 57-58 & Imp Oil Co, 58-59; demonstr geol, Univ Sydney, 59-62; sr demonstr, Australian Nat Univ, 63-64; Nat Res Coun Can res fel, Univ Toronto, 64-65; geologist, Del Geol Surv, 65-66; asst prof geol, Ore State Univ, 66-69; consult, Lacanex Mining Co, Mex, 69-70; ASSOC PROF GEOL, LAURENTIAN UNIV, 70-; CONSULT GEOLOGIST & CHIEF GEOL ED, WATTS, GRIFFIS & McOUAT LTD FOR SAUDI ARABIAN DIRECTORATE GEN MINERAL RESOURCES, JIDDAH, 75- Concurrent Pos: Consult, Timor Oil Pty Ltd & Amalgamated Petrol Pty Ltd, Australia, 62-63 & Cent Coast Oil Co, 63-64; co-ed, Sedimentary Geol, 67- & Earth-Sci Revs, 70- Mem: Int Asn Sedimentol. Res: Carbonate petrology; diagenesis; environments; clastic and pyroclastic rocks; sedimentary ore genesis. Mailing Add: Watts Griffis & McOuat Ltd Box 5218 Jiddah Saudi Arabia

WOLF, KENNETH EDWARD, b Chicago, Ill, Oct 22, 21; m 48; c 3. MICROBIOLOGY. Educ: Utah State Univ, BS, 51, MS, 52, PhD(fish path), 56. Prof Exp: DIR, EASTERN FISH DIS LAB, US FISH & WILDLIFE SERV, 54- Mem: AAAS; Am Soc Microbiol; Sigma Xi; Wildlife Dis Asn; NY Acad Sci. Res: Diseases of fishes, especially of viral etiology; methods of cultivation of fish cells and tissues. Mailing Add: US Fish & Wildlife Serv RD 1 Box 17A Kearneysville WV 25430

WOLF, LARRY LOUIS, b Madison, Wis, Oct 21, 38; m 65. ZOOLOGY. Educ: Univ Mich, BS, 61; Univ Calif, Berkeley, PhD(zool), 66. Prof Exp: Assoc zool, Univ Calif, Berkeley, 64-66; Elsie Binger Naumberg res fel ornith, Am Mus Natural Hist, 66-67; asst prof zool, 67-70, ASSOC PROF BIOL, SYRACUSE UNIV, 70- Mem: Am Ornith Union; Cooper Ornith Soc; Brit Ornith Union; Soc Study Evolution; Ecol Soc Am. Res: Ecological determinants of social systems, principally in birds. Mailing Add: Dept of Zool Syracuse Univ Syracuse NY 13210

WOLF, LAURENCE GRAMBOW, b New York, NY, Sept 4, 21; div; c 3. CULTURAL GEOGRAPHY, URBAN GEOGRAPHY. Educ: City Col New York, BS, 43; Columbia Univ, MA, 47; Syracuse Univ, PhD(geog), 66. Prof Exp: From instr to asst prof, 52-69, ASSOC PROF GEOG, UNIV CINCINNATI, 69- Concurrent Pos: Coordr res & cartog, Housing Anal Greater Cincinnati, Action, Inc, NY, 60; consult geog & cartog, Adv Comt Educ, Human Rels Comn, Cincinnati, 67-68; dir, Miami Valley Conurbation Base-Map Proj, Eco-Tec Found, 74-75. Mem: Asn Am Geogr; Am Geog Soc. Res: Cultural societal ecology and urban ecology, especially population redistribution; geography of energy; refinement of concepts and conceptual frameworks in these areas and their application to interdisciplinary environmental education. Mailing Add: Dept of Geog Univ of Cincinnati Cincinnati OH 45221

WOLF, LEONARD NICHOLAS, b McKeesport, Pa, Jan 3, 11; m 43; c 7. BIOLOGY, ZOOLOGY. Educ: St Vincent Col, BS, 33; Univ Pittsburgh, MS, 34, PhD(biol), 39. Hon Degrees: ScD, St Vincent Col, DSc, 73. Prof Exp: From assoc prof to prof biol, Univ Scranton, 37-69, head dept, 40-69; vpres planning & govt relations, 69-74, EXEC DIR, GREATER DEL VALLEY REGIONAL MED PROG, 74- Concurrent Pos: Chmn, State Comn Acad Facilities; mem, Coun Higher Educ & State Bd Educ, Commonwealth of Pa; chmn regional adv group, Greater Del Valley Regional Med Prog; mem, Gov Adv Health Coun. Mem: Am Soc Zool; Nat Asn Biol Teachers; Asn Am Med Cols; NY Acad Sci. Res: Animal pigmentation; sex hormones. Mailing Add: Rm 344 3401 Market St Philadelphia PA 19104

WOLF, LESLIE RAYMOND, b East Chicago, Ind, Feb 18, 49; m 72. ORGANIC CHEMISTRY, POLYMER CHEMISTRY. Educ: Lewis Univ, BA, 71; Pa State Univ, PhD(chem), 74. Prof Exp: CHEMIST, ROHM AND HAAS CO, 73- Concurrent Pos: Instr chem, Eve Div, LaSalle Col, 74- Mem: Am Chem Soc. Res: Polymer coatings and monomer research. Mailing Add: Res Labs Rohm and Haas Co Norristown & McKean Rd Spring House PA 19477

WOLF, MARK ADAM, b Phoneton, Ohio, June 25, 15; m 37; c 3. BACTERIOLOGY. Educ: Purdue Univ, BS, 37; Univ Wis, MS, 39; Am Bd Indust Hyg, dipl. Prof Exp: Asst, Univ Wis, 37-39; asst bacteriologist, Kans State Bd Health, 39-40, bacteriologist in charge br lab, 40-42; chemist, 42-58, toxicologist, 58-68, res toxicologist, 68-73, INFO SERV SUPVR & GROUP LEADER, DOW CHEM CO, 73- Mem: Am Indust Hyg Asn; Soc Toxicol. Res: Toxicology of chemicals; storage and retrieval information. Mailing Add: Dow Chem Co 1803 Bldg Midland MI 48640

WOLF, MARVIN ABRAHAM, b Syracuse, NY, Dec 26, 25; m 54; c 2. MICROMETEOROLOGY. Educ: NMex Inst Mining & Technol, BS, 51; Univ Wash, MS, 62. Prof Exp: Assoc res engr, Boeing Airplane Co, Wash, 58-60; res meteorologist, Meteorol Res Inc, Calif, 60-63; physicist, Pac Missile Range, Calif, 63-66; RES ASSOC, PAC NORTHWEST LABS, BATTELLE-MEM INST, 66- Mem: Am Meteorol Soc; Am Geophys Union; Sigma Xi. Res: Micrometeorological and mesometeorological processes which control the suspension, transport, diffusion and deposition of airborne materials from agricultural and industrial sources. Mailing Add: 1736 Horn Ave Richland WA 99352

WOLF, MICHAEL JOSEPH, b Grand Rapids, Mich, Feb 18, 09. CHEMISTRY. Educ: Univ Wis, PhD(bot), 42. Prof Exp: Dir, Post Eng Lab, US Army Eng Corps, Ft Benning, Ga, 41-44; assoc chemist, Argonne Nat Lab, 44-45; chemist, 45-61, head cereal micros & qual invests, 61-74, RES LEADER CEREAL MICROS & QUAL INVESTS, NORTHERN REGIONAL RES LAB, AGR RES SERV, USDA, 74- Mem: Am Asn Cereal Chem; Am Soc Plant Physiol; Soc Econ Bot; Am Chem Soc. Res: Plant physiology and chemistry of potato alkaloids; chemistry of plutonium and other radioactive elements; microscopy of starches; chemical composition and structure of cereal grains. Mailing Add: Cereal Micros Invest Northern Regional Res Lab Agr Res Serv USDA Peoria IL 61604

WOLF, MICHAEL WALTER, geology, see 12th edition

WOLF, NEIL STEVEN, b Brooklyn, NY, Nov 15, 37; m 59; c 2. PLASMA PHYSICS. Educ: Queens Col, NY, BS, 58; Stevens Inst Technol, MS, 60, PhD(physics), 66. Prof Exp: Res assoc physics, Space Physics Labs, G C Dewey Corp, 65-67; asst prof, 67-72, ASSOC PROF PHYSICS & COORDR DEPT SCI, DICKINSON COL, 72-, CHMN DEPT PHYSICS & ASTRON, 74- Concurrent Pos: Res Corp Frederick Cottrell res grant, 67- Mem: Am Inst Physics. Res: Plasma physics related to control of instabilities in linear discharges immersed in strong magnetic fields; study of low frequency ionization waves which affect instabilities and plasma confinement. Mailing Add: Dept of Physics Dickinson Col Carlisle PA 17013

WOLF, NORMAN SANFORD, b Kansas City, Mo, July 22, 27; m 67; c 4. EXPERIMENTAL PATHOLOGY, RADIOBIOLOGY. Educ: Kans State Univ, BS & DVM, 53; Northwestern Univ, PhD(exp path), 60; Am Col Lab Animal Med, dipl, 55. Prof Exp: Dir dept animal care, Med Sch, Northwestern Univ, 53-58; vis scientist, Pasteur Lab, Inst Radium, Paris, 60-61; consult radiation biol, Path & Physiol Sect, Biol Div, Oak Ridge Nat Lab, 61-62; res asst prof exp biol, Baylor Col Med, 62-68; ASSOC PROF PATH & MEM RADIOL SCI GROUP, SCH MED, UNIV WASH, 68- Concurrent Pos: NSF fel, Pasteur Lab, Inst Radium, Paris, 60-61; consult, Animal Quarters, Vet Admin Res Hosp, Chicago, 54-60 & Vet Admin Hosp, Seattle, 73- Mem: Radiation Res Soc; Transplantation Soc; Am Soc Exp Path; Int Soc Exp Hemat. Res: Hematopoietic regeneration and transplantation following ionizing radiation; immune competence; hemopoietic stem cell identification; control of hematopoiesis by hematopoietic organ stroma; recovery after irradiation by cells and tissues. Mailing Add: Dept of Path Univ of Wash Sch of Med Seattle WA 98195

WOLF, PAUL A, b Terre Haute, Ind, Feb 18, 12; m 35; c 7. BIOCHEMISTRY. Educ: Purdue Univ, PhD(biochem), 38. Prof Exp: Res chemist, Mich State Dept Health, 38-42; res chemist, Biochem Res Dept, Dow Chem Co, 42-70, tech specialist, Designed Prods Dept, 70-75; RETIRED. Concurrent Pos: Vis prof chem, Paine Col, 76- Mem: Fel AAAS; Soc Indust Microbiol; Am Soc Microbiol; Sigma Xi; Am Chem Soc. Res: Antimicrobial mechanisms; microbiological deterioration and its prevention. Mailing Add: Dept of Chem Paine Col Augusta GA 30901

WOLF, PAUL LEON, b Detroit, Mich, Oct 4, 28; m 52; c 3. PATHOLOGY. Educ: Wayne State Univ, BA, 48; Univ Mich, MD, 52; Am Bd Path, cert path anat & clin path, 60. Prof Exp: Intern, Detroit Receiving Hosp, 52-53; resident path, Wayne State Univ Hosps, 56-60, from asst prof to prof, 60-68; dir clin lab, Med Ctr, Stanford Univ, 68-74; PROF PATH, UNIV CALIF, SAN DIEGO, 74- Concurrent Pos: Assoc, Detroit Receiving Hosp; mem staff, Dearborn Vet Admin Hosp. Mem: AMA; Am Soc Clin Path; Am Asn Path & Bact; Col Am Path. Res: Anatomic pathology; histochemistry; immunopathology; cancer immunology. Mailing Add: Univ Hosp PO Box 3548 225 W Dickensen San Diego CA 92103

WOLF, PHILIP FRANK, b New York, NY, Apr 12, 38; m 60; c 3. PHYSICAL ORGANIC CHEMISTRY. Educ: NY Univ, BS, 60; Columbia Univ, MA, 61, PhD(org chem), 64. Prof Exp: Fel org chem, Yale Univ, 64-65; res chemist, 65-70, from proj scientist to res scientist, 70-74, GROUP LEADER ORG CHEM, UNION CARBIDE CORP, 74- Mem: Am Chem Soc. Res: Studies on oxidation-oxygen transfer mechanisms, substitution reactions of ethylene oxide, free radical telomerization, heterogeneous gas phase kinetics and Diels-Alder reactions. Mailing Add: Union Carbide Corp Tarrytown NY 10591

WOLF, PIERRE L, b Strasbourg, France, July 18, 21; MICROBIOLOGY. Educ: Univ Geneva, Lac, 46, DSc, 50. Prof Exp: Res asst, Yale Univ, 50-51; bibliogr anat, McGill Univ, 52; instr biol, Xavier Univ La, 52-57; asst prof, Canisius Col, 57-59 & Fairleigh Dickinson Univ, 59-61; asst prof biol, 61-72, ASSOC PROF BIOL SCI, STATEN ISLAND COMMUNITY COL, 72- Res: Respiratory systems and cyanhydric inhibition of respiratory systems of Pseudomonas fluorescens. Mailing Add: Dept of Biol Staten Island Community Col 715 Ocean Staten Island NY 10301

WOLF, RICHARD ALAN, b Pittsburgh, Pa, Nov 10, 39; m 71. SPACE PHYSICS, ASTROPHYSICS. Educ: Cornell Univ, BEngPhys, 62; Calif Inst Technol, PhD(nuclear astrophys), 66. Prof Exp: Res fel physics, Calif Inst Technol, 66; mem tech staff, Bel Tel Labs, 66-67; from asst prof to assoc prof space sci, 67-74, PROF SPACE PHYSICS & ASTRON, RICE UNIV, 74- Concurrent Pos: Mem, Inst Advan Study, 69 & 74-75. Mem: Am Phys Soc; Am Geophys Union. Res: Physics of the solar wind; magnetosphere and ionosphere; theoretical astrophysics. Mailing Add: Dept Space Physics & Astron Rice Univ Houston TX 77001

WOLF, RICHARD CLARENCE, b Lancaster, Pa, Nov 28, 26; m 52; c 2. PHYSIOLOGY, ENDOCRINOLOGY. Educ: Franklin & Marshall Col, BS, 50; Rutgers Univ, PhD(zool), 54. Prof Exp: Waksman-Merck fel, Rutgers Univ, New Brunswick, 54-55; Milton fel, Sch Dent Med, Harvard Univ, 55-56, USPHS fel, 56-57; asst prof physiol, Primate Lab, 57-61, assoc prof, 61-66, co-dir progs, 68-70, MEM ENDOCRINOL-REPROD PHYSIOL PROG, UNIV WIS-MADISON, 63-, DIR, 70-, PROF PHYSIOL, 66- CHMN DEPT, 71- Concurrent Pos: Mem res career award comt, NIH, 70-72, mem contract res & adv comt, 73-; mem sci adv bd, Yerkes Regional Primate Res Ctr, Emory Univ, 72-; consult, Ford Found, 72- Mem: Soc Study Reprod; Endocrine Soc; Am Physiol Soc; Brit Soc Endocrinol; Brit Soc Study Fertil. Res: Endocrinology of pregnancy. Mailing Add: Dept of Physiol Univ of Wis Serv Mem Insts 470 N Charter St Madison WI 53706

WOLF, RICHARD EDWARD, JR, microbiology, molecular genetics, see 12th edition

WOLF, RICHARD EUGENE, b Dixon, Ill, June 25, 36; m 60; c 3. ORGANIC CHEMISTRY, POLYMER CHEMISTRY. Educ: Northern Ill Univ, BSE, 57; Univ San Francisco, MS, 65; Univ Calif, Berkeley, PhD(chem), 68. Prof Exp: Sect leader org synthesis, 68-70, mgr, Long Range Res Dept, 70-74, RES SCIENTIST, DeSOTO, INC, 74- Concurrent Pos: Chmn patent comt, DeSoto, Inc, 71-, chmn basic res comt, 74- Mem: Am Chem Soc; Fedn Socs Paint Technol; Nat Micrographics Asn. Res: Photochemistry; photoconduction in organic molecules; photopolymerization; emulsion polymerization; coatings technology; water treatment. Mailing Add: DeSoto Inc 1700 S Mt Prospect Rd Des Plaines IL 60018

WOLF, ROBERT LAWRENCE, b New York, NY, Aug 7, 28; m 57; c 1. PHYSIOLOGY, BIOCHEMISTRY. Educ: Duke Univ, BS, 50, MD, 52; Am Bd Internal Med, dipl, 66. Prof Exp: Intern med & surg serv, Mt Sinai Hosp, 52-53, chief resident path, 53-54, asst resident med, 56-57, chief resident, 57-58, clin asst, Mt Sinai Hosp & Mt Sinai Sch Med, 58-65, res asst, 60-66, ASST CLIN PROF MED, MT SINAI SCH MED, 66- Concurrent Pos: Arthritis & Rheumatism Found res fel, 58-59; res fel med, Mt Sinai Hosp, 58-61; AEC byprod mat license for res, 59-; var pharmaceut corps res grants, 59-; USPHS res grants, 59-; Lupus Erhthematosus Found, Inc res grants, 61-; mem coun high blood pressure res & coun circulation, Am Heart Asn; asst attend physician, Mt Sinai Hosp, 65-; Syntex Res Found res grants, 65-; Ciba Co res grant, 66; guest lectr, Nat Univ Colombia, 66, Cath Univ Chile, 66 & Inst Med Res, Univ SAm, Buenos Aires, 66; G D Searle & Co res grant, 66-68; Health Res Coun of City New York res grant, 66-69; US partic, Int Atomic Energy Agency Symp Radioimmunoassay & Related Procedures in Clin Med & Res, Istanbul, Turkey, 73; consult ed, AMA Drug Eval, 73-74. Mem: Am Nuclear Soc; Am Physiol Soc; fel Am Col Physicians; Am Soc Internal Med; Am Col Chest Physicians. Res: Hypertension; circulatory physiology and cardiology. Mailing Add: Dept of Med Mt Sinai Sch of Med New York NY 10029

WOLF, ROBERT OLIVER, b Mansfield, Ohio, Mar 14, 25; m 70; c 1. ORAL BIOLOGY. Educ: NCent Col, BA, 50; Ohio State Univ, MA, 52, DDS, 58. Prof Exp: Res asst physiol genetics, Ohio State Univ, 52-54; res asst oral biol, Col Dent, 57-58; intern clin dent, USPHS Hosp, New Orleans, La, 58-59; investr salivary physiol & biochem, Human Genetics Br, 59-70 & Oral Med & Surg Br, 70-73, INVESTR SALIVARY PHYSIOL & BIOCHEM, LAB ORAL MED, NAT INST DENT RES, 73- Mem: Am Dent Asn; Am Soc Human Genetics; Int Asn Dent Res; Am Inst Biol Sci; Tissue Cult Asn. Res: Human and animal salivary physiology, biochemistry, enzymology and genetics, especially isoamylases; human salivary gland disease, diagnosis and treatment, especially Sjögren's syndrome. Mailing Add: Rm 2B-13 Bldg 10 Nat Inst of Dent Res Bethesda MD 20014

WOLF, ROBERT PETER, b Long Branch, NJ, Oct 27, 39; m 60; c 2. LOW TEMPERATURE PHYSICS. Educ: Mass Inst Technol, BS, 60, PhD(physics), 63. Prof Exp: From asst prof to assoc prof, 63-74, PROF PHYSICS, HARVEY MUDD COL, 74- Concurrent Pos: NSF sci faculty fel, Oxford Univ, 69-70. Res: Liquid helium; phase transitions; phonon interactions; philosophy of science. Mailing Add: Dept of Physics Harvey Mudd Col Claremont CA 91711

WOLF, ROBERT STANLEY, b New York, NY, May 14, 46; m 75. MATHEMATICAL LOGIC. Educ: Mass Inst Technol, BS, 66; Stanford Univ, MS, 67, PhD(math), 74. Prof Exp: Vis asst prof math, Univ Ore, 73-74; vis scholar, Stanford Univ, 74-75; LECTR MATH, CALIF POLYTECH STATE UNIV, SAN LUIS OBISPO, 75- Mem: Am Math Soc; Asn Symbolic Logic. Res: Continuing study of set theories with intuitionistic logic; theory of infinite games, specifically Almost-Borel games; point-set topology, mathematical biology, artificial intelligence and mathematics education. Mailing Add: Dept of Math Calif Polytech State Univ San Luis Obispo CA 93407

WOLF, STEPHEN NOLL, b Biloxi, Miss, Dec 4, 44; m 73. ACOUSTICS. Educ: Lebanon Valley Col, BS, 66; Univ Md, PhD(molecular spectros), 72. Prof Exp: RES PHYSICIST UNDERWATER ACOUSTICS, US NAVAL RES LAB, 71- Mem: Sigma Xi. Res: At-sea experimental studies of acoustic propagation of sound in shallow water. Mailing Add: Code 8120 US Naval Res Lab Overlook Ave Washington DC 20375

WOLF, STEWART GEORGE, JR, b Baltimore, Md, Jan 12, 14; m 42; c 3. CLINICAL MEDICINE, INTERNAL MEDICINE. Educ: Johns Hopkins Univ, AB, 34, MD, 38. Hon Degrees: MD, Gothenburg Univ, 68. Prof Exp: Intern med, NY Hosp, 38-39, from asst resident to resident, 39-42; from asst prof to assoc prof med, Med Col, Cornell Univ, 46-52; prof med & consult prof psychiat, Sch Med, Univ Okla, 52-67, regents prof med, psychiat, neurol & behav sci, 67-70, head dept med, 52-69, prof physiol, Sch Med, 68-70; PROF MED, UNIV TEX SYST, 70-; DIR, MARINE BIOMED INST, 70-; PROF INTERNAL MED & PHYSIOL, UNIV TEX MED BR GALVESTON, 70- Concurrent Pos: Res fel, Bellevue Hosp, 39-42; Nat Res Coun fel, Cornell Univ, 41-42; assoc vis neuropsychiatrist, Bellevue Hosp, 48-52; asst attend physician in charge psychosom clin, Cornell Univ, 46-52; mem comt psychiat, Nat Res Coun, 48-52, mem comt vet med probs, 51-52; head psychosom sect, Okla Med Res Found, 52-, head neurosci sect, 67-70; mem spec study group, Off Res & Develop, Dept Defense, 52-55; mem pharmacol & exp therapeut study sect, 56-57; mem comt prof educ, 56-63, chmn, 57-63; mem gen med study sect, NIH, 57-61, chmn gastroenterol training grant comt, 58-61; mem adv comt, Space Med & Behav Sci, NASA, 60-61 & Nat Adv Heart Coun, 61-65; mem coun ment health, AMA, 60-64; consult, Europ Off, Off Int Res, NIH, 63-64; mem adv comt admis, Nat Formulary, 65-69; mem bd regents, Nat Libr Med, 65-69, chmn, 68-69; mem comt int progs, Am Heart Asn, 65-, chmn, 65-70; mem educ & supply panel, Nat Adv Comn Health Manpower, 66-67. Honors & Awards: Award, Am Gastroenterol Asn, 42; Hofheimer Prize, Am Psychiat Asn, 53; Award of Merit, Am Heart Asn, 64; Distinguished Serv Citation, Univ Okla, 68. Mem: Am Psychosom Soc (past pres); Am Gastroenterol Asn (pres, 69-70); fel Am Col Physicians; Pavlovian Soc (pres, 66-67); Am Col Clin Pharmacol & Chemother (pres, 66-67). Res: Gastrointestinal, cardiovascular, sensory and neural physiology. Mailing Add: Marine Biomed Inst 200 University Blvd Galveston TX 77550

WOLF, THOMAS, b Sept 10, 32; US citizen; m 59; c 2. ANALYTICAL CHEMISTRY. Educ: Cambridge Univ, BA, 55, MA, 61; Univ RI, PhD(anal chem), 66. Prof Exp: Chemist, Coates Bros, Eng, 55-57; lab supvr, Wymat Corp, NJ, 57-58; chemist, Enthone Inc, Conn, 58-59 & Eltex Res Corp, RI, 59-62; sr res chemist, 66-75, RES ASSOC, COLGATE-PALMOLIVE CO, 75- Mem: Am Chem Soc; Am Soc Testing & Mat. Res: Analytical research, with particular reference to liquid chromatography, to thermal analysis and to laboratory automation. Mailing Add: Colgate-Palmolive Co 909 River Rd Piscataway NJ 08854

WOLF, THOMAS MICHAEL, b Highland Park, Mich, Dec 28, 42; c 2. GENETICS, ENVIRONMENTAL BIOLOGY. Educ: Western Mich Univ, BS, 65; Wayne State Univ, MS, 67, PhD(genetics), 72. Prof Exp: Instr, 71-72, ASST PROF BIOL, WASHBURN UNIV TOPEKA, 72- Mem: Genetics Soc Am. Res: Human karyotyping and human histocompatibility genetics. Mailing Add: Dept of Biol Washburn Univ Topeka Topeka KS 66621

WOLF, WALTER, b Frankfort, Ger, May 25, 31; US citizen; m 55; c 2. RADIOCHEMISTRY, RADIOPHARMACY. Educ: Univ of the Repub, Uruguay, BSc, 49, MS, 52; Univ Paris, PhD, 56. Prof Exp: Asst chem, Univ of the Repub, Uruguay, 51-52; asst chem, Nat Cent Sci Res, France, 55-56, attache, 56; assoc prof org chem & biochem, Concepcion Univ, 56-58; traveling fel, McGill Univ, 58; res assoc, Amherst Col, 58-59; res assoc org chem, 59-62, vis asst prof pharmaecut chem,

61-62, from asst prof to assoc prof, 62-70, chmn dept biomed chem, 70-74, PROF BIOMED CHEM, UNIV SOUTHERN CALIF, 70-, DIR, RADIOPHARM PROG, 68-; DIR RADIOPHARM SERV, LOS ANGELES COUNTY-UNIV SOUTHERN CALIF MED CTR, 70- Concurrent Pos: Consult radiopharm, Int Atomic Energy Agency & US Vet Admin. Mem: Am Chem Soc; Soc Nuclear Med; Fr Soc Biol Chem; Radiation Res Soc; Acad Pharmaceut Sci. Res: New radiopharmaceuticals, especially in clinical pharmacology; chemistry and biochemistry of organic iodo compounds; radioiodination; electron spin resonance of free radicals. Mailing Add: Radiopharm Prog Sch Pharm Univ of Southern Calif Los Angeles CA 90033

WOLF, WALTER ALAN, b New York, NY, Mar 9, 42; m 62; c 1. BIOCHEMISTRY, ORGANIC CHEMISTRY. Educ: Wesleyan Univ, BA, 62; Brandeis Univ, MA, 64, PhD(org chem), 67. Prof Exp: Fel, Mass Inst Technol, 67-70; ASST PROF BIOCHEM, COLGATE UNIV, 70- Concurrent Pos: Vis prof, State Univ NY Agr & Tech Col Morrisville, 71-72; res corp grant, Colgate Univ, 71-72. Mem: AAAS; Am Chem Soc; The Chem Soc. Res: Role of acetate in primary and secondary metabolism; vitamin B12 models; enzymology. Mailing Add: Dept of Chem Colgate Univ Hamilton NY 13346

WOLF, WALTER J, b Hague, NDak, May 2, 27; m 49; c 5. BIOCHEMISTRY. Educ: Col St Thomas, BS, 50; Univ Minn, PhD(biochem), 56. Prof Exp: Assoc chemist, 56-58, chemist, 58-61, prin chemist, 61-68, LEADER MEAL PROD RES, NORTHERN REGIONAL LAB, USDA, 68- Concurrent Pos: Assoc ed, Cereal Chem, 70-73 & Cereal Sci Today, 70-73. Mem: AAAS; Am Scn Cereal Chem; Am Chem Soc; NY Acad Sci. Res: Isolation and characterization of soybean proteins; protein interactions, denaturation and structure; food uses of soybean proteins. Mailing Add: Northern Regional Lab 1815 N University St Peoria IL 61604

WOLF, WAYNE ROBERT, b Kenton, Ohio, May 18, 43; m 71. ANALTYICAL CHEMISTRY, NUTRITION. Educ: Kent State Univ, BS, 65, PhD(chem), 69. Prof Exp: Res chemist, Aerospace Res Lab, US Air Force, Wright-Patterson AFB, Ohio, 67-71; Nat Res Coun assoc, 71-72, res chemist, Human Nutrit Res Div, Agr Res Serv, 71-75, RES CHEMIST, NUTRIENT COMPOSITION LAB, NUTRIT INST, USDA, 75- Mem: AAAS; Am Chem Soc; Am Inst Chem. Res: Instrumental analytical methodology for nutrient content of foods, trace element nutrition, biological availability of trace elements; atomic absorption spectrometry; optical emission spectrometry; gas-liquid chromatography. Mailing Add: Nutrient Compos Lab Nutrit Inst USDA Beltsville MD 20705

WOLF, WERNER PAUL, b Vienna, Austria, Apr 22, 30; m 54; c 2. PHYSICS. Educ: Oxford Univ, BA, 51, MA & DPhil(physics), 54. Hon Degrees: MA, Yale Univ, 65. Prof Exp: Res assoc physics, Clarendon Lab, Oxford Univ, 54-56, Imp Chem Industs res fel, 57-59; res fel appl physics, Harvard Univ, 56-57; univ lectr physics, Oxford Univ, 59-63; assoc prof physics & appl sci, 63-65, PROF PHYSICS & APPL SCI, YALE UNIV, 65- Concurrent Pos: Consult, Hughes Aircraft Co & E I du Pont de Nemours & Co, Inc, 57, Gen Elec Co, 60 & 66-, Mullard Res Labs, 61 & Int Bus Mach Corp, 62-66; prog comt chmn, Conf Magnetism & Magnetic Mat, 68, gen conf chmn, 71, adv comt chmn, 72; vis prof, Munich Tech Univ, 69. Mem: Am Asn Crystal Growth; fel Am Phys Soc; sr mem Inst Elec & Electronics Eng. Res: Magnetism; experimental and theoretical study of magnetic materials, especially at low temperatures; magnetic cooling, relaxation, microwave resonance, optical properties, crystal fields and anisotropy; magnetic thermal properties, critical points and magnetic phase transitions; synthesis of new materials and growth of single crystals. Mailing Add: Becton Ctr Yale Univ New Haven CT 06520

WOLFARTH, EUGENE F, b Washington, DC, June 24, 32; m 57; c 3. PHYSICAL ORGANIC CHEMISTRY. Educ: Univ Md, BS, 54; Ohio State Univ, PhD(chem), 61. Prof Exp: Res chemist res & develop command, Wright Patterson Air Force Base, Ohio, 57; asst instr chem, Ohio State Univ, 58-60; RES ASSOC, RES LABS, EASTMAN KODAK CO, 61- Mem: Am Chem Soc. Res: Kinetics and reaction mechanisms; organic synthesis; color photography, the mechanisms of development and designing new film systems. Mailing Add: Eastman Kodak Co Res Labs 1669 Lake Ave Rochester NY 14650

WOLFBAUER, CLAUDIA ANN, b St Louis, Mo, Oct 15, 46. SEDIMENTOLOGY. Educ: St Louis Univ, BS, 68; Univ Wyo, PhD(geol), 72. Prof Exp: Res assoc geol, Univ Wyo, 72-73; res geologist, Esso Prod Res Co, 73; GEOLOGIST, US GEOL SURV, 73- Concurrent Pos: Ed adv, Geol, 75- Mem: Soc Econ Paleontologists & Mineralogists; AAAS; Geol Soc Am. Res: Bentonite; oilshale; cretaceous stratigraphy of the Rocky Mountain area; lacustrine rocks; tertiary oil and gas reservoirs. Mailing Add: Box 25046 MS 912 Fed Ctr US Geol Surv Lakewood CO 80225

WOLFE, ALLAN FREDERICK, b Olyphant, Pa, Oct 22, 38; m 63; c 3. INVERTEBRATE PHYSIOLOGY, HISTOLOGY. Educ: Gettysburg Col, BA, 63; Drake Univ, MA, 65; Univ Vt, PhD(zool), 68. Prof Exp: ASST PROF BIOL, LEBANON VALLEY COL, 68- Mem: AAAS; Am Soc Zool; Am Inst Biol Sci. Res: Histology and physiology of Artemia salina; invertebrate reproductive systems. Mailing Add: Dept of Biol Lebanon Valley Col Anneville PA 17003

WOLFE, ALVIN WILLIAM, b Schuyler, Nebr, Mar 1, 28; m 52; c 3. ANTHROPOLOGY, SOCIAL ANTHROPOLOGY. Educ: Univ Nebr, AB, 50; Northwestern Univ, PhD(anthrop), 57. Prof Exp: Instr anthrop & sociol, Middlebury Col, 55-57; asst prof, Lafayette Col, 57-61; assoc prof, Wash Univ, 61-68; prof anthrop, Univ Wis-Milwaukee, 68-74; PROF ANTHROP, UNIV S FLA, 74- Mem: fel AAAS; fel Am Anthrop Asn; fel Soc Appl Anthrop (secy, 74-); fel Am Ethnol Soc; fel Royal Anthrop Inst Gt Brit & Ireland. Res: Comparative studies in social structure; urban anthropology; social networks. Mailing Add: Dept of Anthrop Univ of SFla Tampa FL 33620

WOLFE, BERNARD MARTIN, b Killdeer, Sask, Dec 31, 34; m 70. MEDICINE, BIOCHEMISTRY. Educ: Univ Sask, BA, 56; Oxford Univ, BM, BCh, 63, MA, 67; McGill Univ, MSc, 67; FRCPS(C), 68. Prof Exp: House physician & surgeon med, Guy's Hosp, London, Eng, 63-64; from jr asst resident to sr asst resident, Royal Victoria Hosp, Montreal, Que, 64-68; Med Res Coun Can centennial fel, Cardiovasc Res Inst, Med Ctr, Univ Calif, San Francisco, 68-70; asst prof, 70-73, ASSOC PROF MED, UNIV WESTERN ONT, 73-, HON LECTR BIOCHEM, 72- Concurrent Pos: Consult, Westminster Hosp, London, Ont, 70-; consult & chief endocrinol & metab, Univ Hosp, London, Ont, 72- Mem: Can Soc Clin Invest; Can Soc Endocrinol & Metab; Am Fedn Clin Res. Res: Clinical investigation of the effects of diets and drugs on lipid, carbohydrate and amino acid metabolism in man and experimental animals in the fed state. Mailing Add: 17 Metamora Crescent London ON Can

WOLFE, BERTRAM, b US, June 26, 27; m 50; c 3. NUCLEAR PHYSICS, NUCLEAR ENGINEERING. Educ: Princeton Univ, BA, 50; Cornell Univ, PhD(nuclear physics), 54. Prof Exp: Physicist, Eastman Kodak Co, 54-55; physicist, Nuclear Energy Div, Gen Elec Co, 55-56, mgr develop reactor physics, 57-59, mgr conceptual design & anal, 59-64, mgr plant eng & develop, Advan Prod Oper, 64-69; assoc dir, Pac Northwest Labs, Battelle Mem Inst, 69-70; vpres & tech dir, Wadco Corp, 70; gen mgr, Breeder Reactor Dept, 70-73, GEN MGR, FUEL RECOVERY & IRRADIATION PROD DEPT, GEN ELEC CO, 74- Concurrent Pos: Mem, Atomic Indust Forum. Mem: AAAS; Am Phys Soc; fel Am Nuclear Soc. Res: Nuclear power technology; advanced energy technology. Mailing Add: Gen Elec Co 175 Curtner Ave San Jose CA 95125

WOLFE, CALEB WROE, b Washington, DC, Oct 22, 08; m 57; c 6. GEOLOGY. Educ: River Falls State Teachers Col, BE, 30; Harvard Univ, MA, 35, PhD(mineral), 40. Prof Exp: Asst, Harvard Univ, 37-41; from asst to prof, 41-74, EMER PROF GEOL, BOSTON UNIV, 74-; PROF GEOL, SALEM STATE COL, 74- Concurrent Pos: Prod engr, Raytheon Co, 44; crystallog consult, Lincoln Labs, 61-64. Honors & Awards: Neil Miner Award, Nat Asn Geol Teachers, 74. Mem: Fel Geol Soc Am; fel Am Mineral Soc. Res: Geometrical crystallography; genetic mineralogy; mountain building geophysics. Mailing Add: Dept of Geol Salem State Col Salem MA 01970

WOLFE, CARVEL STEWART, b Minneapolis, Minn, June 11, 27; m 54; c 3. MATHEMATICS. Educ: Univ Ariz, BS, 50, MS, 51. Prof Exp: Asst math, Univ Wash, 51-53; asst prof, Shepherd State Col, 53-54; asst, Univ Md, 54-56; asst prof, 56-63, ASSOC PROF MATH, US NAVAL ACAD, 63- Mem: AAAS; Am Math Soc; Math Asn Am. Res: Numerical analysis; linear programming; tensor analysis and geometry. Mailing Add: Dept of Math US Naval Acad Annapolis MD 21402

WOLFE, CLINTON RAY, b Huntington, WVa, Oct 7, 40; m 58; c 3. ANALYTICAL CHEMISTRY. Educ: Marshall Univ, BS, 62; Univ NMex, PhD(chem), 66. Prof Exp: Asst chem, Univ NMex, 62-66; SR CHEMIST, WESTINGHOUSE RES LABS, WESTINGHOUSE ELEC CORP, 66- Res: Electrochemical methods for trace elements in environmental samples; rapid, accurate electrochemical methods for uranium and transition metals in mining solutions and thermoanalytical techniques. Mailing Add: Westinghouse Res Labs Churchill Boro Pittsburgh PA 15235

WOLFE, DAVID FRANCIS ZEKE, b Lost Creek, WVa, Dec 10, 37; m 60; c 3. PHYSICAL CHEMISTRY. Educ: WVa Wesleyan Col, BS, 60; Univ WVa, MS, 62, PhD(phys chem), 67. Prof Exp: Asst chem, Univ WVa, 60-62, part-time instr, 62-65; from instr to asst prof, 65-69, ASSOC PROF CHEM, WVA WESLEYAN COL, 69- Concurrent Pos: Vis asst prof, Univ WVa, 67 & 68. Mem: Am Chem Soc. Res: Grignard reactions of 9-phenanthrenemethyl magnesium chloride; transfer of force constants from pure to mixed halides of boron. Mailing Add: Dept of Chem WVa Wesleyan Col Box 69 Buckhannon WV 26201

WOLFE, DAVID M, b Philadelphia, Pa, Oct 27, 38; m 65; c 2. HIGH ENERGY PHYSICS. Educ: Univ Pa, BA, 59, MS, 61, PhD(physics), 66. Prof Exp: Res assoc physics, Enrico Fermi Inst, Univ Chicago, 66-69; vis asst prof physics, Univ Wash, 69-71; asst prof, 71-75, ASSOC PROF PHYSICS, UNIV NMEX, 75- Mem: Am Phys Soc; Fedn Am Scientists; Sigma Xi. Res: Experimental research in the weak interactions of elementary particles and nucleon-nucleon interaction. Mailing Add: Dept Physics & Astron Univ of NMex Albuquerque NM 87131

WOLFE, DOROTHY WEXLER, b Springfield, Ill, Aug 20, 20; m 42; c 1. MATHEMATICS. Educ: Univ Ill, BS, 41; Wayne State Univ, MA, 53; Univ Pa, PhD(math), 66. Prof Exp: Asst math, Univ Pa, 54-55; instr, Swarthmore Col, 62-64; asst prof, 65-69, ASSOC PROF MATH, WIDENER COL, 69- Mem: Am Math Soc; Math Asn Am. Res: Mathematical analysis; metric spaces. Mailing Add: Dept of Math Widener Col Chester PA 19013

WOLFE, DOUGLAS ARTHUR, b Dayton, Ohio, July 6, 39; m 59; c 3. MARINE CHEMISTRY, POLLUTION BIOLOGY. Educ: Ohio State Univ, BSc, 59, MSc, 61, PhD(physiol chem), 64. Prof Exp: Chief biogeochem prog, Radiobiol Lab, US Bur Com Fisheries, 64-70; dir estuarine res, Atlantic Estuarine Fisheries Ctr, Nat Marine Fisheries Serv, 70-75; STAFF DIR ECOL, OUTER CONTINENTAL SHELF ENVIRON ASSESSMENT PROG, NAT OCEANIC & ATMOSPHERIC ADMIN, 75- Concurrent Pos: Adj asst prof, NC State Univ, 66-70, adj assoc prof, 70-75; chief scientist I marine biol, Nuclear Ctr, PR, 69-70; consult, Panel on Zinc, Nat Acad Sci, 73-75. Mem: Estuarine Res Fedn; Am Chem Soc; Ecol Soc Am; Am Malacol Union; Am Soc Limnol & Oceanog. Res: Comparative biochemistry of carotenoids and lipids; ecology and biology of molluscs; petroleum, radioisotopes and trace metals in marine environment. Mailing Add: Environ Res Labs Nat Oceanic & Atmos Admin Boulder CO 80302

WOLFE, EDWARD W, b Brooklyn, NY, Jan 21, 36; m 56; c 5. GEOLOGY. Educ: Col Wooster, BA, 57; Ohio State Univ, PhD(geol), 61. Prof Exp: Instr geol, Col Wooster, 59-61; geologist, Pac Coast Br, Calif, 61-68, GEOLOGIST, CTR ASTROGEOL, US GEOL SURV, 68- Mem: AAAS; Geol Soc Am; Am Asn Petrol Geol; Soc Econ Paleont & Mineral; Am Geophys Union. Res: Areal and lunar geology. Mailing Add: Ctr of Astrogeol US Geol Surv 601 E Cedar Ave Flagstaff AZ 86001

WOLFE, GLEN ALAN, organic chemistry, x-ray crystallography, see 12th edition

WOLFE, GORDON A, b Chicago, Ill, Sept 8, 31; m 56; c 2. SOLID STATE PHYSICS. Educ: Ill Inst Technol, BS, 60; Univ Mo, MS, 63, PhD(physics), 67. Prof Exp: Instr physics, Univ Mo, 64-67; asst prof, 67-74, ASSOC PROF PHYSICS, SOUTHERN ORE STATE COL, 74- Mem: AAAS; Am Asn Physics Teachers. Res: Anharmonic effects in crystals. Mailing Add: Dept of Physics Southern Ore State Col Ashland OR 97520

WOLFE, HARRY BERNARD, b Vancouver, BC, Dec 29, 27; m 52; c 4. OPERATIONS RESEARCH. Educ: Univ BC, BA, 49, MA, 51; Columbia Univ, PhD(physics), 56. Prof Exp: Sr staff mem oper res, Mass, 56-63, mgr, San Francisco Opers Res Group, Calif, 63-69, SR STAFF MEM, HEALTH CARE SECT, ARTHUR D LITTLE, INC, MASS, 69- Mem: Inst Mgt Sci; Opers Res Soc Am; Can Opers Res Soc. Res: Health care; hospitals; management science, including inventory and scheduling theory and marketing analysis; nuclear physics. Mailing Add: Arthur D Little Inc 35 Acorn Park Cambridge MA 02140

WOLFE, HARRY WALTER, JR, organic chemistry, see 12th edition

WOLFE, HERBERT GLENN, b Uniontown, Kans, Mar 14, 28; m 50; c 3. DEVELOPMENTAL GENETICS. Educ: Kans State Univ, BS, 49; Univ Kans, PhD(zool), 60. Prof Exp: Assoc staff scientist physiol genetics, Jackson Lab, Maine, 60-63; from asst prof to assoc prof, 63-70, PROF PHYSIOL GENETICS, UNIV KANS, 70- Concurrent Pos: Mem genetics standards subcomt, Inst Lab Animal Resources, 65-68; NIH spec res fell, Harwell, Eng, 69-70. Mem: Soc Develop Biol; Genetics Soc Am; Sigma Xi. Res: Physiological genetics, specifically genetic control of physiological and developmental processes related to reproduction; blood proteins and hematopoiesis; pigmentation in mice. Mailing Add: Dept Physiol & Cell Biol Univ of Kans Lawrence KS 66045

WOLFE, HUGH CAMPBELL, b Parkville, Mo, Dec 18, 05; m 29; c 4. PHYSICS. Educ: Park Col, AB, 26; Univ Mich, MS, 27, PhD(physics), 29. Hon Degrees: ScD,

Park Col, 62. Prof Exp: Asst physics, Univ Mich, 26-27, instr, 27-29; Nat Res Coun fel, Calif Inst Technol, 29-31; Lorentz Found fel, State Univ Utrecht, 31-32; Heckscher res asst, Cornell Univ, 32; instr, Ohio State Univ, 32-33; from instr to assoc prof physics, City Col New York, 34-49; prof & head dept, Cooper Union, 49-60; dir publ, Am Inst Physics, 60-70, tech asst publ & info, 71-75; CONSULT, 75- Concurrent Pos: Mem comn on symbols, units & nomenclature, Int Union Pure & Appl Physics, 61-, chmn comn on publ, 66-72; mem adv panel, Int Standards Orgn, 63-; chmn comt symbols, units & terminology, Nat Acad Sci, 64-; chmn, Gordon Conf Sci Info Probs in Res, 65; mem metric adv comt, Am Nat Standards Inst, 70-; mem metric practice comt, Am Nat Metric Coun, 74- Mem: Fel Am Phys Soc; Am Fedn Am Scientists; Am Asn Physics Teachers. Res: Applications of quantum theory; development of information systems. Mailing Add: 30 Lawrence Pkwy Tenafly NJ 07670

WOLFE, JACK, b New York, NY, Sept 24, 09; m 33; c 1. MATHEMATICS, INFORMATION SCIENCE. Educ: City Col New York, BS, 30, MS, 32; NY Univ, PhD, 40. Prof Exp: From tutor to assoc prof math & infor sci, 32-57, PROF COMPUT & INFO SCI, BROOKLYN COL, 57- Concurrent Pos: Lectr, NSF Insts. Mem: Am Math Soc; Math Asn Am; Asn Comput Mach. Res: Programming for electronic computers; cryptoanalysis. Mailing Add: Dept of Info Sci Brooklyn Col Brooklyn NY 11210

WOLFE, JACK ALBERT, paleontology, see 12th edition

WOLFE, JACK MORRIS, b Damascus, Va, Dec 23, 06; wid; c 1. ANATOMY. Educ: Carson-Newman Col, BS, 28; Vanderbilt Univ, PhD(anat), 31. Prof Exp: Asst anat, Vanderbilt Univ, 29-30, from instr to asst prof, 30-36; from asst prof to prof, 36-74, chmn dept, 46-71, EMER PROF ANAT, ALBANY MED COL, 74- Mem: Am Asn Anatomists. Res: Cytology and cytochemistry of anterior pituitary gland. Mailing Add: Dept of Anat Albany Med Col Albany NY 12208

WOLFE, JAMES ALVIS, b Rogersville, Tenn, Aug 16, 31; m 68. BOTANY, ECOLOGY. Educ: Carson-Newman Col, BS, 53; Univ Tenn, Knoxville, MS, 56, PhD(bot), 67. Prof Exp: Res technician agron, Univ Fla, 56; res technician ecol, Oak Ridge Nat Lab-Union Carbide Corp, 59-60; res technician, Univ Tenn, Knoxville, 60-64; asst prof biol, Campbellsville Col, 68-72; SOIL SCIENTIST, SOIL CONSERV SERV, USDA, 73- Mem: Am Inst Biol Sci; Ecol Soc Am; Am Soc Agron; Soil Sci Soc Am. Res: Biogeochemical cycling in terrestrial ecosystems; factors influencing soil formation. Mailing Add: Volusia County Agr Ctr 3100 E New York Ave De Land FL 32720

WOLFE, JAMES F, b York, Pa, Oct 5, 36; m 59; c 2. ORGANIC CHEMISTRY. Educ: Lebanon Valley Col, BS, 58; Ind Univ, PhD(org chem), 63. Prof Exp: Teaching asst org chem, Ind Univ, 58-60; res assoc, Duke Univ, 63-64; from asst prof to assoc prof, 64-74, PROF CHEM, VA POLYTECH INST & STATE UNIV, 74- Mem: Am Chem Soc. Res: Use of multiple anions in organic synthesis; synthesis of new medicinal agents. Mailing Add: Dept of Chem Va Polytech Inst & Sta Univ Blacksburg VA 24061

WOLFE, JAMES H, b Salt Lake City, Utah, Jan 7, 22; m 56. MATHEMATICS. Educ: Univ Utah, BA, 42; Harvard Univ, MA, 43, PhD(math), 48. Prof Exp: PROF MATH, UNIV UTAH, 48- Mem: Am Math Soc. Res: Topology and integration theory; matrices. Mailing Add: Dept of Math Univ of Utah Salt Lake City UT 84112

WOLFE, JAMES LEONARD, b Milton, Fla, May 5, 40; m 63. VERTEBRATE ZOOLOGY, ANIMAL BEHAVIOR. Educ: Univ Fla, BS, 62; Cornell Univ, PhD(vert zool), 66. Prof Exp: Asst prof biol, Univ Ala, University, 66-68; asst prof zool, 68-70, ASSOC PROF ZOOL, MISS STATE UNIV, 70- Mem: Animal Behav Soc; Am Soc Mammal; Ecol Soc Am. Res: Ecology and behavior of mammals. Mailing Add: Dept of Zool Drawer Z Miss State Univ Mississippi State MS 39762

WOLFE, JAMES RICHARD, JR, b Elizabeth, NJ, Aug 6, 32; m 57; c 3. ORGANIC CHEMISTRY. Educ: Mass Inst Technol, BS, 54; Univ Calif, PhD(chem), 58. Prof Exp: RES CHEMIST, E I DU PONT DE NEMOURS & CO, INC, 57- Mem: Fel Am Inst Chem. Res: Polymers; physical organic chemistry. Mailing Add: Exp Sta Bldg 353 E I du Pont de Nemours & Co Wilmington DE 19898

WOLFE, JAMES WALLACE, b Ludlowville, NY, Apr 11, 32; m 54; c 2. NEUROPHYSIOLOGY, PSYCHOLOGY. Educ: Univ Calif, Riverside, BA, 63; Univ Rochester, PhD(psychol), 66. Prof Exp: Res psychologist, Army Med Res Lab, Ft Knox, Ky, 66-68; RES NEUROPHYSIOLOGIST, US AIR FORCE SCH AEROSPACE MED, 68- Concurrent Pos: Lectr, Univ Louisville, 67-68 & St Mary's Univ, Tex, 69- Mem: Int Brain Res Orgn; Soc Res Otolaryngol; Soc Neurosci; Aerospace Med Asn; Barany Soc. Res: Cerebellar integration of sensory information; effects of drugs on electrophysiological responses; neurophysiological control of oculomotor function. Mailing Add: c/o Chief Vestibular Function US Air Force Sch Aerospace Med Brooks AFB TX 78235

WOLFE, JAMES WILEY, veterinary medicine, see 12th edition

WOLFE, JOHN KAVANAUGH, b Washington, DC, Aug 5, 14; m 39; c 3. ORGANIC CHEMISTRY. Educ: Univ Md, BS, 36, PhD(org chem), 39. Prof Exp: Milton fel med, Harvard Med Sch & fel chem, Harvard Univ, 39-41; assoc chemist, NIH, 41-42; chemist, US Naval Res Lab, 42-46; res assoc, 46-55, mgr advan degree personnel, 55-65, consult educ rels, 65-68, MGR UNIV RELS, RES & DEVELOP CTR, GEN ELEC CO, SCHENECTADY, 68- Concurrent Pos: Asst, Peter Bent Brigham Hosp, 39-41; lectr, Univ Md, 41-42; with Nat Bur Standards, 44; consult, Off Econ Coop & Develop, Paris, 58-; chmn bd, Int Asn Exchange of Students for Technol Experience, 57- Mem: Fel AAAS; Am Chem Soc; NY Acad Sci. Res: Isolation and identification of steroids; carbohydrate and steroid synthesis; synthetic organic chemistry; fluorocarbons; electrochemical syntheses; thermal reactions; interaction between industry and universities. Mailing Add: 6 Yorkshire Terr Sherwood Forest Clifton Park NY 12065

WOLFE, JOHN NICHOLAS, plant ecology, see 12th edition

WOLFE, JON WILLARD, organic chemistry, see 12th edition

WOLFE, LAUREN GENE, b Kenton, Ohio, Nov 7, 39; m 66. PATHOLOGY. Educ: Ohio State Univ, DVM, 63, MS, 65, PhD(vet path), 68; AM Col Vet Path, dipl, 68. Prof Exp: Asst prof path, Univ Ill Med Ctr, 68-71; assoc prof, 71-74, PROF MICROBIOL, RUSH-PRESBY-ST LUKE'S MED CTR, 74- Concurrent Pos: Asst microbiologist, Presby-St Luke's Hosp, 68-71. Mem: AAAS; Am Soc Exp Path; Am Asn Immunol; Am Asn Cancer Res; Int Acad Path. Res: Experimental and comparative pathology; oncology. Mailing Add: Rush-Presby-St Luke's Med Ctr 1753 W Congress Pkwy Chicago IL 60612

WOLFE, LEONHARD SCOTT, b Auckland, NZ, Mar 3, 25; Can citizen; m 60; c 2. BIOCHEMISTRY, NEUROCHEMISTRY. Educ: Univ NZ, BSc, 47; Cambridge Univ, PhD(insect physiol, biochem), 52, ScD, 76; Univ Western Ont, MD, 58; FRCP(C), 72. Prof Exp: Jr lectr zool, Univ Canterbury, 49-50; assoc entomologist, Agr Res Inst, Can Dept Agr, 52-54; from asst prof neurochem to assoc prof neurol & neurosurg, 60-70, PROF NEUROL & NEUROSURG, MONTREAL NEUROL INST, McGILL UNIV, 70-, DIR, DONNER LAB EXP NEUROCHEM, 65- Concurrent Pos: Nat Res Coun Can med res fel, 59-60; Sister Elizabeth Kenny Found scholar, 60-61; med res assoc, Med Res Coun Can, 63-, mem grants comt neurol sci, 70-74, mem priorities selection & rev comt, 72-; hon lectr biochem, McGill Univ, 60-70, prof biochem, 71-; consult dermat res unit, Royal Victoria Hosp, Montreal, 65-67. Mem: Int Brain Res Orgn; Am Soc Biol Chemists; Can Biochem Soc; Can Physiol Soc; Int Soc Neurochem. Res: Entomology; biology and control of biting flies; insect cholinesterases; metabolism of insecticides; biochemistry and function of complex glycolipids in neurones; membranes; role of lipid anions in excitable tissues; convulsive states; biosynthesis, release and action of prostaglandins; biochemistry of lipid storage diseases and degenerative neurological diseases. Mailing Add: Montreal Neurol Inst 3801 University St Montreal PQ Can

WOLFE, MICHAEL LOWNDES, wildlife ecology, wildlife management, see 12th edition

WOLFE, PAUL JAY, b Mansfield, Ohio, Oct 2, 38; m 60; c 2. NUCLEAR PHYSICS, EXPLORATION GEOPHYSICS. Educ: Case Inst Technol, BS, 60, MS, 63, PhD(nuclear physics), 66. Prof Exp: Design engr, Lamp Div, Gen Elec Co, 60-61; asst prof physics, 66-71, chmn dept, 72-75, ASSOC PROF PHYSICS, WRIGHT STATE UNIV, 71- Mem: Soc Explor Geophysicists; Am Phys Soc; Am Asn Physics Teachers. Res: Nuclear structure physics. Mailing Add: Dept of Physics Wright State Univ Dayton OH 45431

WOLFE, PETER EDWARD, b Hammonton, NJ, Apr 27, 11. GEOLOGY. Educ: Rutgers Univ, BS, 33; Princeton Univ, MA, 40, PhD(geol), 41. Prof Exp: Asst conservationist, USDA, 35-37; asst instr geol, Princeton Univ, 37-41, res assoc, 44-45; geologist, Nfld Dept Nat Resources, 41-44; from asst prof to assoc prof geol, 45-60, PROF GEOL, RUTGERS UNIV, NEW BRUNSWICK, 60- Concurrent Pos: Vis prof, Osmania Univ, India, 58-59; Fulbright fel, 58. Mem: Fel Geol Soc Am; Sigma Xi. Res: Geomorphology; geohydrology; environmental geology; pleistocene and periglacial research, Atlantic Coastal Plain. Mailing Add: 6 Mercer St Princeton NJ 08540

WOLFE, PETER NORD, b Lakewood, Ohio, July 24, 29; m 51; c 3. PHYSICS. Educ: Ohio Wesleyan Univ, BA, 51; Ohio State Univ, MS, 52, PhD(physics), 55. Prof Exp: NSF fel, 54-55; from res physicist to mgr systs physics Dept, Res Labs, 55-72, MGR TRANSFORMER TECHNOL, WESTINGHOUSE ELEC CORP, 72- Mem: Inst Elec & Electronics Engrs. Res: Power technology; applied physics; microwave spectra. Mailing Add: Westinghouse Elec Corp Westinghouse Bldg Gateway Ctr Pittsburgh PA 15222

WOLFE, PHILIP, b San Francisco, Calif, Aug 11, 27. APPLIED MATHEMATICS. Educ: Univ Calif, AB, 48, PhD, 54. Prof Exp: Instr math, Princeton Univ, 54-57; mathematician, Rand Corp, 57-66; MEM RES STAFF, IBM RES, 66-; PROF ENG MATH, COLUMBIA UNIV, 68- Mem: Fel AAAS; Am Math Soc; Soc Indust & Appl Math; Math Prog Soc. Res: Mathematics of optimization; linear and nonlinear programming; computational complexity. Mailing Add: IBM Res PO Box 218 Yorktown Heights NY 10598

WOLFE, RALPH STONER, b New Windsor, Md, July 18, 21; m 50; c 3. MICROBIOLOGY. Educ: Bridgewater Col, BS, 42; Univ Pa, MS, 49, PhD, 53. Prof Exp: Asst instr microbiol, Univ Pa, 47-49, instr, 51-52; asst limnol, Acad Natural Sci, Pa, 49-50; from instr to assoc prof, 53-60, PROF MICROBIOL, UNIV ILL, URBANA, 61- Concurrent Pos: NSF fel, 58; Guggenheim fel, 60. Mem: Am Soc Microbiol. Res: Metabolism and physiology of bacteria. Mailing Add: Dept of Microbiol Univ Ill 127 Burrill Hall Urbana IL 61803

WOLFE, RAYMOND, b Hamilton, Ont, Apr 8, 27; m 54; c 3. SOLID STATE PHYSICS. Educ: Univ Toronto, BA, 49, MA, 50; Bristol Univ, PhD(physics), 55. Prof Exp: Physicist, Eastman Kodak Co, NY, 50-52; physicist, Gen Elec Co, Eng, 54-57; MEM TECH STAFF, BELL TEL LABS, 57- Concurrent Pos: Ed, Appl Solid State Sci. Mem: Fel Am Phys Soc. Res: Theoretical and experimental solid state physics; transport and optical properties of semiconductors and metals; thermoelectric materials and devices; magnetic materials; magnetic bubble devices. Mailing Add: Electronic Mat Lab Bell Tel Labs Murray Hill NJ 07974

WOLFE, RAYMOND GROVER, JR, b Oakland, Calif, June 1, 20; m 46; c 3. BIOCHEMISTRY. Educ: Univ Calif, AB, 42, MA, 48, PhD(biochem), 55. Prof Exp: Biochemist, Donner Lab, Univ Calif, 48-55; Nat Found Infantile Paralysis fel chem, Univ Wis, 55-56; from asst prof to assoc prof, 56-67, Lalor res fel, summer 56, PROF CHEM, UNIV ORE, 67- Concurrent Pos: Guggenheim fel, Inst Biochem, Univ Vienna, 63-64; vis prof, Bristol Univ, 70-71. Mem: AAAS; Am Chem Soc; Am Soc Biol Chem. Res: Enzyme catalytic mechanism; structure-function relationship of polymeric enzymes; enzyme kinetics and inhibition; protein structure studies, particularly in dehydrogenases. Mailing Add: Dept of Chem Univ of Ore Eugene OR 97403

WOLFE, ROBERT NORTON, b Falls City, Nebr, Feb 22, 08; m 33; c 2. PHYSICS. Educ: Parsons Col, BS, 30; Univ Rochester, MS, 32, PhD(physics), 35. Prof Exp: Physicist, Eastman Kodak Co, 35-73; RETIRED. Concurrent Pos: Pres, Rochester Coun Sci Socs, 65-67. Mem: Fel Optical Soc Am; Soc Photog Sci & Eng; Am Phys Soc. Res: Radioactivity; photographic sensitometry; microsensitometry; optical and photographic image structure. Mailing Add: 340 Cobbs Hill Dr Rochester NY 14610

WOLFE, ROGER THOMAS, b Mt Vernon, Ill, July 31, 32; m 56; c 2. ORGANIC CHEMISTRY, RESEARCH ADMINISTRATION. Educ: Bradley Univ, BS, 54; Rensselaer Polytech Inst, PhD(chem), 59; Salmon P Chase Col, JD, 69. Prof Exp: Lab asst anal chem, Bradley Univ, 52-54; asst gen & org chem, Rensselaer Polytech Inst, 54-56; res chemist, Sterling-Winthrop Res Inst, NY, 56-60, from assoc patent agent to patent agent, 60-65, patent agent, Hilton-Davis Chem Co Div, Sterling Drug Co, 65-66, asst dir res & develop, 66-69, vpres res & develop, 70-75, PATENT ATTORNEY, HILTON-DAVIS CHEM CO DIV, STERLING DRUG CO, 69-, VPRES RES ADMIN & LEGAL AFFAIRS, 75- Mem: Am Chem Soc. Res: Patents; pigments; dyes; intermediates; patent law. Mailing Add: Hilton-Davis Chem Co 2235 Langdon Farm Rd Cincinnati OH 45237

WOLFE, ROY ISRAEL, b Staszow, Poland, Nov 19, 17; Can citizen; m 49; c 4. GEOGRAPHY. Educ: McMaster Univ, BA, 40; Univ Toronto, MA, 47, PhD(geog), 56. Prof Exp: Dir visual educ, Vet Rehab Inst Toronto, 45-47; res geogr, Ont Dept Hwys, 52-67; Resources of the Future grant, 67, assoc prof geog, 67-68, PROF GEOG, YORK UNIV, 68- Concurrent Pos: Co-founder, York Univ Transport Centre, 67-68; partner, R I Wolfe Assocs, 67-; assoc, Kates, Peat, Marwick & Co, 69-70; spec lectr, Univ Toronto, 69-70; mem comn transp, Int Geog Union, 69-72, mem working group on geog of tourism & recreation, 72-; assoc ed, J Leisure Res, 70-73; dir, Ecol Res Ltd, Toronto, 71-72. Mem: Can Asn Geogr; Int Peace Res Soc; World Soc Ekistics; Regional Sci Asn. Res: Human biology; modeling recreational travel; search for a theory of surrogates; structure of the human brain. Mailing Add: Dept of Geog York Univ 4700 Keele St Downsview ON Can

WOLFE, STEPHEN JAMES, b San Francisco, Calif, Jan 30, 43. MATHEMATICS. Educ: Williams Col, BA, 65; Univ Calif, Riverside, MA, 67, PhD(math), 70. Prof Exp: Asst prof math, 70-75, ASSOC PROF MATH, UNIV DEL, 75- Mem: Am Math Soc; Math Asn Am; Inst Math Statist. Res: Probability limit theorems; infinitely divisible distribution functions; characteristic functions. Mailing Add: Dept of Math Univ of Del Newark DE 19711

WOLFE, STEPHEN LANDIS, b Sept 23, 32; m 53; c 2. CELL BIOLOGY, ELECTRON MICROSCOPY. Educ: Bloomsburg State Col, BS, 54; Ohio State Univ, MS, 59; Johns Hopkins Univ, PhD(biol), 62. Prof Exp: NIH fel zool, Univ Minn, 62-63; asst prof, 63-68, ASSOC PROF ZOOL, UNIV CALIF, DAVIS, 68- Concurrent Pos: Vis prof, Yale Univ, 73. Mem: AAAS; Am Soc Cell Biol; Electron Micros Soc Am. Res: Fine structure of chromosomes. Mailing Add: Dept of Zool Univ of Calif Davis CA 95616

WOLFE, WALTER MCILHANEY, b Baltimore, Md, Aug 15, 21; m 45; c 3. OBSTETRICS & GYNECOLOGY. Educ: Univ Md, MD, 46; Am Bd Obstet & Gynec, dipl, 57. Prof Exp: Asst prof, 65-68, actg chmn dept, 69-72, ASSOC PROF OBSTET & GYNEC, SCH MED, UNIV LOUISVILLE, 68- Concurrent Pos: Proj dir family planning, Dept Health Educ & Welfare Grant, Louisville Gen Hosp, 71- Mem: Am Col Obstet & Gynec. Res: Applications of current technological and educational techniques to community reproductive health. Mailing Add: Dept of Obstet & Gynec Univ of Louisville Sch of Med Louisville KY 40202

WOLFE, WILLIAM LOUIS, JR, b Yonkers, NY, Apr 5, 31; m 55; c 3. OPTICS, ELECTRICAL ENGINEERING. Educ: Bucknell Univ, BS, 53; Univ Mich, MS, 56, MSE, 66. Prof Exp: Asst proj engr, Sperry Gyroscope Co, 52-53; from res asst to res assoc infrared & optics, Univ Mich, 53-57, from assoc res engr to res engr, 57-66, lectr elec eng, 62-66; chief engr, Honeywell Radiation Ctr, 66-68, mgr electro-optics Dept, Honeywell, Inc, 68-69; PROF OPTICAL SCI, UNIV ARIZ, 69- Concurrent Pos: Lectr, Northeastern Univ, 68-69; mem panel of comt undersea warfare-assessment electro optics, Nat Acad Sci; mem adv comt, Army Res Off, study panel on Army Countermine Adv Comt & adv comt, Nat Bur Stand & Air Force Systs Command. Mem: Fel Optical Soc Am; sr mem Inst Elec & Electronics Engrs. Res: Optical materials for infrared use; radiometry; space navigation using star trackers; electro-optical system design. Mailing Add: Optical Sci Ctr Univ of Ariz Tucson AZ 85721

WOLFE, WILLIAM RAY, JR, b Grafton, WVa, Nov 16, 24; m 52; c 2. PHYSICAL CHEMISTRY. Educ: WVa Wesleyan Col, BS, 49; Western Reserve Univ, MS, 50, PhD(phys chem), 53. Prof Exp: Asst phys chem, Western Reserve Univ, 50-52; res chemist cent res dept, 52-60, chemist explosives dept, 60-64, chemist develop dept, 65-68, RES ASSOC PLASTICS DEPT EXP STA, E I DU PONT DE NEMOURS & CO, INC, 68- Mem: Am Chem Soc; Electrochem Soc. Res: Fused salt and aqueous electrochemistry; energy conversion; high temperature chemistry; catalysis; plastics processing. Mailing Add: Plastics Dept Exp Sta E I du Pont de Nemours & Co Inc Wilmington DE 19898

WOLFENBARGER, DAN A, b White Plains, NY, Sept 23, 34; m 59; c 3. ENTOMOLOGY. Educ: Univ Fla, BSA, 56; Iowa State Univ, MS, 57; Ohio State Univ, PhD(entom), 61. Prof Exp: Entomologist agr exp sta, Tex A&M Univ, 61-65; ENTOMOLOGIST COTTON INSECT RES, AGR RES SERV, USDA, 65- Mem: Entom Soc Am. Res: Activity of and resistance and mode of inheritance to insecticides against tobacco budworm and boll weevil; biochemical analysis of tobacco budworm and insecticide residues on cotton plant. Mailing Add: Agr Res Serv USDA PO Box 1033 Brownsville TX 78520

WOLFENBARGER, DANIEL OTIS, b Ottawa Co, Okla, June 22, 04; m 29; c 3. ENTOMOLOGY. Educ: Colo Agr Col, BS, 28; Cornell Univ, PhD(econ entom), 38. Prof Exp: Teacher high sch, Wyo, 28-29; asst dept entom & zool, SDak State Col, 29-30; asst, NY State Col Agr, Cornell, 30-33; agent & asst entomologist, USDA, 34-43; asst entomologist, Del Agr Exp Sta, 43-45; entomologist, Exp Sta, 45-75; EMER PROF ENTOM, UNIV FLA, 75- Honors & Awards: Spokesman of Year Award, Farm Chem Mag, 74. Mem: Entom Soc Am. Res: Economic entomology; dispersion of small organisms; factors affecting dispersal distance of small organisms. Mailing Add: 29220 SW 187th Ave Homestead FL 33030

WOLFENDEN, RICHARD VANCE, b Oxford, Eng, May 17, 35; US citizen; m 65; c 2. BIOCHEMISTRY. Educ: Princeton Univ, AB, 56; Oxford Univ, BA & MA, 60; Rockefeller Inst, PhD(biochem), 64. Prof Exp: Asst prof biochem, Princeton Univ, 64-70; assoc prof, 70-73, PROF BIOCHEM, SCH MED, UNIV NC, CHAPEL HILL, 73- Concurrent Pos: Mem, NSF Adv Panel Molecular Biol, 74- Mem: AAAS; Am Chem Soc; Am Soc Biol Chemists. Res: Physical organic chemistry in relation to enzyme-catalyzed reactions; analogs of intermediates in substrate transformation. Mailing Add: Dept of Biochem Univ of NC Sch of Med Chapel Hill NC 27514

WOLFENSTEIN, LINCOLN, b Cleveland, Ohio, Feb 10, 23; m 43, 57; c 3. THEORETICAL HIGH ENERGY PHYSICS. Educ: Univ Chicago, BS, 43, MS, 44, PhD(physics), 49. Prof Exp: Physicist, Nat Adv Comt Aeronaut, 44-46; instr physics, 48-49, from asst prof to assoc prof, 49-60, PROF PHYSICS, CARNEGIE-MELLON UNIV, 60- Concurrent Pos: NSF sr fel, Europ Orgn Nuclear Res, Geneva, Switz, 64-65; vis prof, Univ Mich, 70-71; Guggenheim fel, 74-75; mem physics adv comt, NSF, 74- Mem: Am Phys Soc. Res: Nuclear collisions; weak interactions. Mailing Add: Dept of Physics Carnegie-Mellon Univ Pittsburgh PA 15213

WOLFERSBERGER, MICHAEL GREGG, b Northampton, Pa, June 14, 44; m 65. PHYSIOLOGICAL CHEMISTRY. Educ: Lebanon Valley Col, BS, 66; Temple Univ, PhD(biochem), 71. Prof Exp: Res assoc biochem, Lab Exp Dermat, Albert Einstein Med Ctr, Philadelphia, 71-73, assoc mem div res, 73-75; ASST PROF, DIV NATURAL SCI & MATH, ROSEMONT COL, 74- Mem: AAAS; Am Chem Soc. Res: Metabolism of low molecular weight compounds. Mailing Add: Div of Natural Sci & Math Rosemont Col Rosemont PA 19010

WOLFF, ALBERT ELI, b New York, NY, Jan 4, 12; div; c 2. BIOSTATISTICS, EPIDEMIOLOGY. Educ: Syracuse Univ, BS, 35, MS, 36; NY Univ, MA, 53; Univ Tex, PhD(econ, bus statist), 62. Prof Exp: Prof statist, Loyola Univ, La, 47-50; asst prof, Univ NMex, 57-59; assoc prof, Tulsa Univ, 60-61 & St Michael's Col, NMex, 62-63; assoc prof, 63-70, PROF EPIDEMIOL, COMMUNITY MED, BIOSTATIST & PUB HEALTH, MED CTR, UNIV ALA, BIRMINGHAM, 70- Mem: Am Statist Asn. Res: Economic, administrative and social aspects of medicine, dentistry and related life sciences. Mailing Add: Dept of Biostatist Univ of Ala Med Ctr Birmingham AL 35294

WOLFF, ARTHUR HAROLD, b Trenton, NJ, Dec 23, 19; m 46; c 3. ENVIRONMENTAL HEALTH. Educ: Mich State Univ, DVM, 42. Prof Exp: Res investr, Commun Dis Ctr, USPHS, 46-50, res biologist, Sanit Eng Ctr, 50-58, chief training br, Div Radiol Health, 58-60; consult, UN Food & Agr Orgn, Italy, 60-61; chief radiation bio-effects prog, Nat Ctr Radiol Health, USPHS, 61-68; asst dir res, Consumer Protection & Environ Health, Dept Health, Educ & Welfare, 68-69, asst dir radiation prog, US Environ Protection Agency, 69-71; PROF ENVIRON HEALTH & CHMN DEPT, SCH PUB HEALTH, UNIV ILL MED CTR, 71- Concurrent Pos: Fel, Duke Univ & Oak Ridge Inst Nuclear Studies, 50-51; exec secy, Nat Adv Comt Radiation, 59-61; consult & mem expert panels, UN Food & Agr Orgn, 60-63 & WHO, 61-; USPHS rep, Fed Radiation Coun Expert Panels, 61-; vis lectr, Univ Mo & Univ Pa, 62-66. Mem: AAAS; Am Vet Med Asn; Am Pub Health Asn; Soc Occup & Environ Health; NY Acad Sci. Res: Radiation biology; comparative oncology; environmental toxicology. Mailing Add: Sch of Pub Health Univ of Ill Med Ctr PO Box 6998 Chicago IL 60680

WOLFF, CHARLES LAMBERT, astronomy, physics, see 12th edition

WOLFF, DAVID A, b Cleveland, Ohio, Nov 2, 34; m 58; c 2. CELL BIOLOGY, VIROLOGY. Educ: Col Wooster, AB, 56; Univ Cincinnati, MS, 60, PhD(microbiol, virol), 65. Prof Exp: Asst prof virol & microbiol, 64-67, ASSOC PROF VIROL & MICROBIOL, OHIO STATE UNIV, 67- Concurrent Pos: NIH res grant, 66-69; Am-Swiss Found Sci Exchange lectr, Switz, 70; on res leave, Univ Uppsala, Sweden, 71. Mem: AAAS; Am Soc Microbiol; Am Soc Cell Biol; Tissue Cult Asn. Res: Viral-induced cytopathic effects and relation of lysosomal enzymes; purification of virus; electron microscopy of virus infected cells; purification of lysosomes; virus interactions with synchronized cells. Mailing Add: Dept Microbiol Ohio State Univ 484 W 12th Ave Columbus OH 43210

WOLFF, DONALD JOHN, b New York, NY, Feb 23, 42. BIOCHEMISTRY, PHARMACOLOGY. Educ: Fordham Univ, BS, 63; Univ Wis, PhD(biochem), 69. Prof Exp: ASST PROF PHARMACOL, RUTGERS MED SCH, COL MED & DENT NJ, 72- Concurrent Pos: NIH trainee pediat, J P Kennedy, Jr Labs, Med Sch, Univ Wis, 68-72. Mem: AAAS. Res: Calcium-binding proteins; regulation of cyclic nucleotide metabolism by clacium ion. Mailing Add: Dept of Pharmacol Rutgers Med Sch PO Box 101 Piscataway NJ 08854

WOLFF, EMILY TOWER, b Dover, NH, Aug 23, 13. BOTANY. Educ: Temple Univ, AB, 37; Pa State Univ, MS, 38, PhD(bot), 47. Prof Exp: Res botanist, Am Cyanamid Co, 43-36; asst prof bot, Univ Ga, 47-49; instr biol, Temple Univ, 49-50; res botanist, Mistaire Labs, 51; asst prof bot, Wellesley Col, 52-56; asst prof, 56-66, ASSOC PROF BIOL, HOBART & WILLIAM SMITH COLS, 66- Mem: Fel AAAS; Bot Soc Am; Am Soc Plant Physiol; Ecol Soc Am; Am Inst Biol Sci. Mailing Add: Dept of Biol Hobart & William Smith Cols Geneva NY 14456

WOLFF, FREDERICK WILLIAM, b Berlin, Ger, Aug 21, 20. PHARMACOLOGY, MEDICINE. Educ: Univ Durham, MB, BS, 46, MD, 57. Prof Exp: House physician, Royal Victoria Infirmary, Univ Durham, 46-47; house physician, med registr & resident med officer, Southend-on-Sea Gen Hosp, Eng, 47-50; med registr, Whittington Hosp, 53-54; clin pharmacologist, Wellcome Res Inst, 55-59; asst prof med, Sch Med & Endocrine Clin, Johns Hopkins Univ, 59-63; PROF MED, SCH MED, GEORGE WASHINGTON UNIV, 65- Concurrent Pos: Sr res asst, Post-Grad Med Sch, Univ London & Whittington Hosp, 55-59; consult, Food & Drug Admin, DC & Children's Med Ctr, DC, 71- Mem: Am Diabetes Asn; Am Soc Pharmacol & Exp Therapeut; Am Fedn Clin Res; Am Heart Asn; Royal Soc Med. Res: Therapeutics; clinical pharmacology; endocrinology; diabetes; hypertension. Mailing Add: 800 Notley Rd Silver Spring MD 20904

WOLFF, GEORGE LOUIS, b Hamburg, Ger, Aug 24, 28; US citizen; m 53; c 2. GENETICS. Educ: Ohio State Univ, BS, 50; Univ Chicago, PhD(zool), 54. Prof Exp: Biologist, Nat Cancer Inst, 56-58; res assoc, Inst Cancer Res, 58-63, supvr animal colony, 58-68, asst mem, 63-72, geneticist, 68-72; chief mammalian genetics br, 72-74, CHIEF DIV MUTAGENIC RES, NAT CTR TOXICOL RES, US FOOD & DRUG ADMIN, 74- Concurrent Pos: USPHS fel, Nat Cancer Inst, 54-56; prof assoc, Nat Acad Sci, 56-57; consult, Am Asn Accreditation Lab Animal Care, 70-75; mem, Interagency Panel Environ Mutagenesis, 73-; asst prof biochem, Med Sch, Univ Ark, 73- Mem: AAAS; Am Asn Cancer Res; Am Genetic Asn; Environ Mutagen Soc; Genetics Soc. Res: Genetic and hormonal regulation of metabolism in neoplastic and normal mammalian tissues; genetic cocarcinogenesis; mouse genetics; chemical mutagenesis in the mammal. Mailing Add: Div of Mutagenic Res Nat Ctr for Toxicol Res Jefferson AR 72079

WOLFF, GUNTHER ARTHUR, b Essen, Ger, Mar 31, 18; nat US; m 45; c 2. PHYSICAL CHEMISTRY. Educ: Univ Berlin, BS, 44, MS, 45; Tech Univ, Berlin, ScD(theoret inorg chem), 48. Prof Exp: Res assoc, Fritz-Haber Inst, Ger, 44-50, sci asst head & dep chief, 50-53; consult & sr res scientist, Signal Corps Res & Develop Labs, US Dept Army, NJ, 53-60; sr group leader mat res solid state res dept, Harshaw Chem Co, Ohio, 60-63; dir mat res, Erie Tech Prod, Inc, 63-64; prin scientist, Tyco Labs, Inc, 64-70; CONSULT CHEMIST, LIGHTING RES & TECH SERV OPER, GEN ELEC CO, EAST CLEVELAND, 70- Concurrent Pos: Mem comn crystal growth, Int Union Crystallog, 66-75, mem, Am Comt Crystal Growth, 67-72. Mem: Am Asn Crystal Growth; Am Chem Soc; Electrochem Soc; fel Mineral Soc Am; fel Am Inst Chem. Res: Crystal growth and dissolution, including evaporation and etching; crystal imperfections; electroluminescence and luminescence; lamp envelopes; semiconductors and ceramics; chemical bonding; solid state chemistry and physics. Mailing Add: 3776 Northampton Rd Cleveland Heights OH 44121

WOLFF, HANS, chemistry, see 12th edition

WOLFF, IVAN A, b Louisville, Ky, Feb 10, 17; m 41; c 4. ORGANIC CHEMISTRY, RESEARCH ADMINISTRATION. Educ: Univ Louisville, BA, 37; Univ Wis, MA, 38, PhD(org chem), 40. Prof Exp: Fel biochem, Univ Wis, 40-41; asst chemist northern regional res lab bur agr chem & eng, 41-42, from assoc chemist to chemist, 43-48, unit leader, 48-54, asst head cereal crops sect, 54-58, chief indust crops lab northern utilization res & develop div, Agr Res Serv, 58-69, DIR EASTERN REGIONAL RES CTR, AGR RES SERV, USDA, 69- Concurrent Pos: Mem subcomt natural toxicants food protection comt, Nat Acad Sci-Nat Res Coun, 70-74. Honors & Awards: Super Serv Award, USDA, 57. Mem: Am Chem Soc; Am Oil Chem Soc; Soc Econ Bot (pres, 64-65); Inst Food Technol. Res: Biochemistry, nutrition, processing utilization of farm commodities, and the constituents, components and derivatives from them. Mailing Add: Eastern Regional Res Ctr Agr Res Serv USDA 600 E Mermaid Philadelphia PA 19118

WOLFF, JACOB, chemistry, see 12th edition

WOLFF, JAMES A, b New York, NY, June 19, 14; m 46; c 4. PEDIATRICS, HEMATOLOGY. Educ: Harvard Univ, AB, 35; NY Univ, MD, 40; Am Bd Pediat, dipl, 48, cert pediat hemat-oncol, 74. Prof Exp: Intern, Lenox Hill Hosp, New York, 40-42; asst resident, Boston Children's Hosp, 45-47, chief resident outpatient dept, 47; asst pediatrician, Babies Hosp, New York, 48-51, from asst attend pediatrician to assoc attend pediatrician, 51-68; from assoc prof to prof clin pediat, 61-72, PROF PEDIAT, COL PHYSICIANS & SURGEONS, COLUMBIA UNIV, 72-; ATTEND PEDIATRICIAN, BABIES HOSP, 68- Concurrent Pos: Fel pediat hemat, Harvard Med Sch, 48; prin investr children's cancer study group A, NIH, 58-; consult, Englewood Hosp, NJ, 62-, Valley Hosp, Ridgewood, 66-, St Luke's Hosp, New York & Community Hosp, Sullivan County, 69-; mem, Sub-board Pediat Hemat-Ontology, Am Bd Pediat, 72-; mem, Nat Wilms' Tumor Study Comt. Mem: Harvey Soc; Am Pediat Soc; Soc Pediat Res; Am Soc Hemat; Int Soc Hemat. Res: Pediatric hematology and oncology. Mailing Add: Babies Hosp Dept of Pediat 3975 Broadway New York NY 10032

WOLFF, JAN, b Dusseldorf, Ger, Apr 25, 25; US citizen; m 55; c 2. BIOCHEMISTRY, ENDOCRINOLOGY. Educ: Univ Calif, BA, 45, PhD(physiol, biochem), 49; Harvard Univ, MD, 53. Prof Exp: Res asst, Harvard Univ, 54-55; MED DIR RES, NIH, 55- Concurrent Pos: NSF sr fel, 58-59. Mem: Am Thyroid Asn; Am Soc Biol Chem; Endocrine Soc; Am Soc Clin Invest. Res: Biochemistry and chemistry of thyroid hormone synthesis; mechanism of hormone action, particularly thyroid hormone; properties and function of biological membranes; microtubules and secretory mechanisms. Mailing Add: Rm 8N319 NIH Clin Ctr Bethesda MD 20014

WOLFF, JOHN BRUNO, b Ger, May 5, 25; nat US; m 50; c 1. BIOPHYSICS. Educ: Hunter Col, AB, 50; Johns Hopkins Univ, MA, 51, PhD(biol), 55. Prof Exp: NIH fel, 52-54; biochemist, Smithsonian Inst, 54-58; vis scientist & res assoc, Nat Inst Arthritis & Metab Dis, 58-60, chemist, Nat Inst Neurol Dis & Blindness, 60-62, health scientist adminstr, 62-65, HEALTH SCIENTIST ADMINSTR, DIV RES GRANTS, NIH, 65- Mem: AAAS; Am Soc Biol Chem; Biophys Soc (treas, 71-79); Am Chem Soc. Res: Microbial and plant biochemistry; enzymology. Mailing Add: Westwood Bldg Room 4A07 5333 Westbard Ave Bethesda MD 20016

WOLFF, JOHN SHEARER, III, b Rochester, NY, Feb 9, 41; m 63; c 2. BIOCHEMISTRY, VIROLOGY. Educ: Wittenberg Univ, AB, 62; Univ Cincinnati, PhD(biochem), 66. Prof Exp: Asst prof biol, Fed City Col, 69-70; sr investr, 70-71, LAB HEAD MOLECULAR BIOL, JOHN L SMITH MEM FOR CANCER RES, PFIZER, INC, 71- Concurrent Pos: Nat Inst Allergy & Infectious Dis fel immunochem, Col Med, Univ Ill, 68-69. Mem: AAAS; Am Chem Soc; Am Soc Microbiol; Reticuloendothelial Soc. Res: Relationships between biochemical and biological activities of tumor viruses, especially RNA tumor viruses. Mailing Add: Pfizer Inc Mem for Cancer Res 199 Maywood Ave Maywood NJ 07607

WOLFF, MANFRED ERNST, b Berlin, Ger, Feb 14, 30; nat US; div; c 3. MEDICINAL CHEMISTRY. Educ: Univ Calif, BS, 51, MS, 53, PhD(pharmaceut chem), 55. Prof Exp: Asst, Univ Calif, 52-55; res fel, Univ Va, 55-57; sr med chemist, Smith Kline & French Labs, 57-60; from asst prof to assoc prof pharmaceut chem, 60-65, PROF PHARMACEUT CHEM, UNIV CALIF, SAN FRANCISCO, 65-, CHMN DEPT, 70- Concurrent Pos: Vis prof, Imperial Col Sci, London, 67-68. Mem: Am Chem Soc; Am Pharmaceut Asn; fel Am Acad Pharmaceut Sci. Res: Synthesis of potential anabolic, anti-inflammatory or anti-aldosterone hormone analogs; synthesis of aldosterone, of cardiac glycosides and aglycones; steroid chemistry. Mailing Add: Dept of Pharmaceut Chem Univ of Calif San Francisco CA 94143

WOLFF, MANFRED PAUL, b New York, NY, Apr 26, 38; m 62; c 2. GEOLOGY, SEDIMENTOLOGY. Educ: Hofstra Univ, BS, 61; Univ Rochester, MS, 63; Cornell Univ, PhD(geol), 67. Prof Exp: Wilson P Foss fel, Cornell Univ, 63-65, Penrose Bequest grant, 66; asst prof geol, 67-75, NSF instnl grant, 69, actg chmn dept, 71-75, ASSOC PROF GEOL, HOFSTRA UNIV, 75- Mem: Geol Soc Am; Soc Econ Paleont & Mineral; Nat Asn Geol Teachers; Int Asn Sedimentol. Res: Ancient and recent depositional environments; physical stratigraphy; coastal geomorphology; sedimentation; sedimentology; clastic and carbonate petrology; marine geology and geochemistry; clay mineralogy; sedimentology of barrier islands and beaches; geomorphology of coastal regions. Mailing Add: Dept of Geol Hofstra Univ Hempstead NY 11550

WOLFF, MARIANNE, b Berlin, Ger; US citizen; m 52; c 2. PATHOLOGY, SURGICAL PATHOLOGY. Educ: Hunter Col, BA, 48; Columbia Univ, MD, 52. Prof Exp: Intern med, Presby Hosp, New York, 52-53; asst resident lab, Mt Sinai Hosp, New York, 53-54, asst resident path, St Luke's Hosp, 54-56; from instr to asst prof surg path, Col Physicians & Surgeons, Columbia Univ, 56-70; asst surg pathologist, 68-71, ASSOC SURG PATHOLOGIST, PRESBY HOSP, NEW YORK, 71-; ASSOC PROF SURG PATH, COL PHYSICIANS & SURGEONS, COLUMBIA UNIV, 70-, ASSOC PROF CLIN SURG PATH, 75- Concurrent Pos: Resident, Presby Hosp, 56-57; from asst pathologist to assoc pathologist, Roosevelt Hosp, New York, 57-68. Mem: Fel Am Soc Clin Path. Res: Various problems in surgical pathology studies with the aid of the electron microscope. Mailing Add: Col of Physicians & Surgeons Columbia Univ Lab of Surg Path New York NY 10032

WOLFF, MILO MITCHELL, b Glen Ridge, NJ, Aug 9, 23; m 54; c 5. PHYSICS, ELECTRICAL ENGINEERING. Educ: Upsala Col, BS, 48; Univ Pa, MS, 53, PhD(physics), 58. Prof Exp: Electronic engr, Philco Corp, Pa, 49-51; lectr electronics, Community Col, Temple Univ, 52-53; instr physics, Univ Pa, 58; Univ Ky-Agency Int Develop asst prof, Bandung Tech Inst, 58-61, assoc prof, 62; res physicist, Mass Inst Technol, 63-69; prof physics, Nanyang Univ, Singapore, 70-72; mem tech staff, Aerospace Corp, Los Angeles, 72-75; CHIEF, SCI & TECHNOL SECT, ECON COMN AFRICA-UN, ADDIS ABABA, 75- Concurrent Pos: Asia Found vis prof physics, Vidyalankara Univ, Ceylon, 66-68; mem, US-Pakistan Sci Surv Team, US NSF, 74; mem methane gas panel, Nat Acad Sci, 74. Mem: Am Phys Soc. Res: Electronics; computer science; planetary optics; methods of cultural adaptation to technology in traditional societies; nuclear physics; space physics; upper atmosphere; technical development in Southeast Asia; polarized light; atmospheric and biological energy; economic development. Mailing Add: Econ Comn for Africa PO Box 3001 Addis Ababa Ethiopia

WOLFF, NIKOLAUS EMANUEL, b Munich, Ger, July 7, 21; nat US; m 54; c 3. CHEMISTRY, ELECTRONICS. Educ: Munich Tech Univ, Cand, 48; Princeton Univ, MA, 51, PhD(chem), 52. Prof Exp: Asst instr, Princeton Univ, 50-52, instr chem, 52-53; res chemist, Jackson Lab & Exp Sta, E I du Pont de Nemours & Co, 53-58; mem tech staff, Labs, David Sarnoff Res Ctr, RCA Corp, 59-63, head mat processing res, 63-66; assoc lab dir, Process Res & Develop Lab, 67-68; mgr mat progs, 68-69, mgr photoreceptor technol, 69-71, MGR MAT INFO TECHNOL GROUP, XEROX CORP, 71- Mem: Fel AAAS; Am Chem Soc; Am Phys Soc; fel Am Inst Chem; Soc Photog Sci & Eng. Res: Steroid, fluorine and polymer chemistry; organometallics; electronic properties of organic materials; chemistry of recording media; electrophotography; solid state technology and integrated circuits. Mailing Add: 63 Southern Pkwy Rochester NY 14618

WOLFF, PAUL M, meteorology, oceanography, see 12th edition

WOLFF, PETER ADALBERT, b Oakland, Calif, Nov 15, 23; m 48; c 2. PHYSICS. Educ: Univ Calif, AB, 45, PhD(physics), 51. Prof Exp: Physicist, Bell Tel Labs, Inc, NJ, 52-68, dir, Electronics Res Lab, 68-70; PROF PHYSICS, MASS INST TECHNOL, 70-, ASSOC DIR, CTR MAT SCI & ENG, 74- Concurrent Pos: Prof, Univ Calif, San Diego, 63-64. Mem: Fel Am Phys Soc. Res: Spin susceptibility of electron gas; theory of local moments in metals; plasma effects in solids; interaction of light with electrons in solids; tunable sources of infrared radiation. Mailing Add: Dept of Physics Mass Inst of Technol Cambridge MA 02139

WOLFF, PETER HARTWIG, b Krefeld, Ger, July 8, 26; US citizen; m 62; c 4. MEDICINE, PSYCHOBIOLOGY. Educ: Univ Chicago, BS, 47, MD, 50. Prof Exp: Asst psychiat, Harvard Med Sch, 56-59, instr, 58-61, assoc, 61-64, asst prof, 64-71; res assoc, 56-61, ASSOC, CHILDREN'S HOSP MED CTR, 61-, DIR RES, 64-; PROF PSYCHIAT, HARVARD MED SCH, 71- Concurrent Pos: Fel neurophysiol, Univ Chicago, 51-52; Kirby Collier Mem lectr, Rochester, NY, 63; instr, Boston Psychoanal Inst, 67-; Sandor Rado lectr, Columbia Univ, 69. Honors & Awards: Helen Sargent Prize, Menninger Found, 66; Felix & Helene Deutsch Prize, Boston Psychoanal Inst, 66. Mem: Fel Am Psychiat Asn. Res: Developmental psychobiology; biological basis of behavior. Mailing Add: Children's Hosp Med Ctr 300 Longwood Ave Boston MA 02115

WOLFF, RICHARD JAMES, b St Paul, Minn, Oct 17, 40; m 62. ASTROPHYSICS. Educ: Carleton Col, BA, 62; Univ Calif, Berkeley, PhD(physics), 67. Prof Exp: Asst prof, 67-70, asst astronr, 70-72, ASSOC ASTRONR, UNIV HAWAII, 72- Mem: Optical Soc Am. Res: Stellar spectroscopy; studies of early type stars; instrumentation for telescopes and for reduction of astronomical data. Mailing Add: Inst for Astron Univ of Hawaii 2680 Woodlawn Dr Honolulu HI 96822

WOLFF, ROBERT L, b Marion, Tex, Dec 12, 39; m 60; c 1. AGRICULTURE. Educ: Tex A&I Univ, BS, 66; Tex A&M Univ, MS, 68; La State Univ, Baton Rouge, PhD(agr), 71. Prof Exp: Asst prof agr, Tex A&I Univ, 66-72; ASST PROF AGR, SOUTHERN ILL UNIV, CARBONDALE, 72- Mem: Am Soc Agr Eng. Res: Agricultural mechanization. Mailing Add: Dept of Agr Industs Southern Ill Univ Carbondale IL 62901

WOLFF, ROGER GLEN, b Eureka, SDak, Sept 7, 32; m 59; c 2. GEOLOGY, HYDROLOGY. Educ: SDak Sch Mines & Technol, BS, 58; Univ Ill, MS, 60, PhD(geol), 61. Prof Exp: RES GEOLOGIST, WATER RESOURCES DIV, US GEOL SURV, 61- Mem: Am Geophys Union; Clay Minerals Soc; Geol Soc Am. Res: Role of confining layers in fluid and solute movement; regional tectonic stress determinations. Mailing Add: US Geol Surv Water Resources Div Reston VA 22092

WOLFF, SHELDON, b Peabody, Mass, Sept 22, 28; m 54; c 3. CYTOGENETICS, RADIOBIOLOGY. Educ: Tufts Col, BS, 50; Harvard Univ, MA, 51, PhD(biol), 53. Prof Exp: Biologist, Oak Ridge Nat Lab, 53-66, mem sr res staff, 65-66; PROF CYTOGENETICS, UNIV CALIF, SAN FRANCISCO, 66- Concurrent Pos: Mem subcomt radiobiol, Nat Acad Sci-Nat Res Coun, 61-, mem space sci bd & mem comt nuclear sci, 74-; vis prof, Univ Tenn, 62; mem comt 15 environ biol & chmn panel radiation biol of comt 15, Space Sci Bd, Nat Acad Sci, 62, mem comt postdoctoral fels div biol & agr, 62-65; consult spec facil prog, NSF, 62-64; mem exec comt, Nat Acad Sci-Nat Res Coun Space Biol Summer Study, 68, mem exec comt priorities study for NASA, 70, mem subcomt genetics effects adv comt to Environ Protection Agency, Div Med Sci, 70-; prog chmn, XIII Int Cong Genetics. Honors & Awards: E O Lawrence Award, US AEC, 73. Mem: Environ Mutagen Soc; Genetics Soc Am; Radiation Res Soc; Bot Soc Am; Am Soc Cell Biol. Res: Chromosome structure; radiation genetics and cytology; chromosome structure; genetics and cytology. Mailing Add: Lab of Radiobiol Univ of Calif San Francisco CA 94143

WOLFF, SHELDON MALCOLM, b Newark, NJ, Aug 19, 30; m 56; c 3. INFECTIOUS DISEASES, IMMUNOLOGY. Educ: Univ Ga, BS, 52; Vanderbilt Univ, MD, 57; Am Bd Internal Med, dipl. Prof Exp: Intern med, Sch Med, Vanderbilt Univ, 57-58, asst resident, 58-59; sr resident, Bronx Munic Hosp Ctr & Albert Einstein Col Med, 59-60; clin assoc, 60-62, clin investr, 62-63, sr investr, 63-65, head physiol sect, Lab Clin Invest, 64-74, CLIN DIR, NAT INST ALLERGY & INFECTIOUS DIS & CHIEF LAB CLIN INVEST, 68- Concurrent Pos: Res asst, Sch Med, Vanderbilt Univ, 56-59; lectr, Sch Med, Georgetown Univ, 62-69, clin prof, 69-; sr consult infectious dis, Nat Naval Med Ctr, Md, 70- Honors & Awards: Super Serv Award, NIH, 71. Mem: AAAS; Am Asn Immunol; Am Clin & Climat Asn; Am Fedn Clin Res; Am Soc Clin Invest. Res: Mechanisms of host responses. Mailing Add: Lab of Clin Invest Bldg 10 Nat Inst Allergy & Infect Dis Bethesda MD 20014

WOLFF, SIDNEY CARNE, b Sioux City, Iowa, June 6, 41; m 62. ASTROPHYSICS. Educ: Carleton Col, BA, 62; Univ Calif, Berkeley, PhD(astron), 66. Prof Exp: Res astronr, Lick Observ, Univ Calif, Santa Cruz, 67; asst astronr, 67-71, ASSOC ASTRONR, INST ASTRON, UNIV HAWAII, 71- Mem: Am Astron Soc; Int Astron Union. Res: Stellar spectroscopy; photoelectric photometry; magnetic stars. Mailing Add: Inst for Astron Univ of Hawaii 2680 Woodlawn Dr Honolulu HI 96822

WOLFF, STEVEN, b New York, NY, Apr 15, 43. ORGANIC CHEMISTRY. Educ: Williams Col, BA, 65; Yale Univ, PhD(org chem), 70. Prof Exp: Fel org chem, Squibb Inst Med Res, 70-71; res assoc, 71-73, ASST PROF ORG CHEM, ROCKEFELLER UNIV, 73- Mem: Am Chem Soc; The Chem Soc. Res: Mechanistic organic photochemistry. Mailing Add: Rockefeller Univ 1230 York Ave New York NY 10021

WOLFF, THEODORE ALBERT, b Philadelphia, Pa, Feb 24, 43; m 62. ENTOMOLOGY. Educ: NMex Highlands Univ, BS, 65; Univ NC, Chapel Hill, MSPH, 69; Univ Utah, PhD(biol), 76. Prof Exp: Vol biol, Peace Corps, Malaysia, 65-68; environ scientist entom, NMex Environ Improv Agency, 69-72; teaching fel biol, Univ Utah, 72-74; environ scientist, 74-76, PROG MGR RADIATION PROTECTION, N MEX ENVIRON IMPROV AGENCY, 76- Mem: Assoc Sigma Xi; Am Mosquito Control Asn. Res: Systematic studies of mountain Aedes mosquitoes; subgenus Ochlerotatus of Arizona and New Mexico. Mailing Add: Rte 1 Box 12-C Santa Cruz NM 87567

WOLFF, VERNON CLARE, JR, organic chemistry, see 12th edition

WOLFF, WILLIAM FRANCIS, b Newark, NJ, June 17, 21; m 48; c 4. ORGANIC CHEMISTRY. Educ: Yale Univ, BS, 47. Prof Exp: Res chemist, Standard Oil Co, Ind, 47-53; res chemist, Pa Salt Mfg Co, 53-54; res chemist, 54-58, sr proj chemist, 58-61, SR RES SCIENTIST, STANDARD OIL CO, IND, 61- Mem: AAAS; Am Chem Soc. Res: Hydrocarbon polymers; organic sulfur and chlorine compounds; carbons and aromatic complexes; heterogeneous catalysis. Mailing Add: Amoco Res Ctr Naperville IL 60540

WOLFF, WILLIAM I, b New York, NY. SURGERY. Educ: NY Univ, BS, 36; Univ Md, Baltimore City, MD, 40. Prof Exp: Intern med, Second Div, Bellevue Hosp, 40-

41, resident surg, 41-42, resident thoracic surg, Chest Surg Serv, 42-43; resident surg, Bronx Vet Hosp, 46-48; instr, Med Sch, Cornell Univ, 49-56; assoc prof clin surg, Sch Med, NY Univ, 56-69; PROF SURG, MT SINAI SCH MED, 69-, DIR DEPT SURG, BETH ISRAEL MED CTR, 69- Concurrent Pos: Vis surgeon, St Vincent's Hosp, 49-, consult, 58-; consult, F D Roosevelt Vet Admin Hosp, Montrose, 54- & NY Vet Admin Hosp, Manhattan, 65- Mem: AMA; Am Col Surg; Am Asn Thoracic Surg; Soc Thoracic Surg; Int Soc Surg. Res: Fiberoptic endoscopy, particularly colonoscopy, new method of removing colonic polyps and diagnosing cancer of colon; mediastinal tumors, esophageal diseases and lung cancer. Mailing Add: Beth Israel Med Ctr Dept Surg 10 Nathan D Perlman Pl New York NY 10003

WOLFGANG, ROBERT W, biology, parasitology, see 12th edition

WOLFGRAM, FREDERICK JOHN, b Los Angeles, Calif, Mar 18, 25; m 52; c 3. NEUROCHEMISTRY. Educ: Univ Calif, AB, 49; Calif Inst Technol, PhD(physiol), 52. Prof Exp: Jr res neurologist, 54-55, from asst res neurologist to assoc res neurologist, 55-64, assoc prof neurol, 64-70, PROF NEUROL, UNIV CALIF, LOS ANGELES, 70- Concurrent Pos: Nat Found Infantile Paralysis res fel, Johns Hopkins Univ, 52-54. Mem: Am Acad Neurol; Am Neurol Asn; Am Soc Neurochem; Int Soc Neurochem. Res: Myelin chemistry; demyelinating diseases; amyotrophic lateral sclerosis. Mailing Add: Reed Neurol Res Ctr Univ of Calif Sch of Med Los Angeles CA 90024

WOLFHAGEN, JAMES LANGDON, b Portland, Ore, Dec 9, 20; m 48; c 3. ORGANIC CHEMISTRY. Educ: Linfield Col, AB, 46; Univ Calif, Berkeley, PhD(chem), 51. Prof Exp: Chemist, Northwest Testing Labs, Ore, 41-42; asst prof chem, Whitworth Col, Wash, 49-52; from asst prof to assoc prof, 52-64, PROF CHEM, UNIV MAINE, ORONO, 64-, HEAD DEPT, 67- Mem: AAAS; Am Chem Soc. Res: Reaction of alkali metals with dimethylformamide; synthesis and stereochemistry of glycidic esters; epoxidation reactions; carbonium ions. Mailing Add: Dept of Chem Univ of Maine Orono ME 04473

WOLFHARD, HANS G, b Basel, Switz, Apr 2, 12; nat US; m 40; c 4. PHYSICS, PHYSICAL CHEMISTRY. Educ: Univ Göttingen, Dr rer nat, 38. Prof Exp: Res scientist, Aeronaut Res Sta, Ger, 39-46; prin sci officer, Royal Aircraft Estab, Eng, 46-53; sr prin sci officer, Rocket Propulsion Dept, 53-56; res scientist, US Bur Mines, 56-58; mgr physics dept reaction motors div, Thiokol Chem Corp, 59-63; MEM SR RES STAFF, INST DEFENSE ANAL, 63- Mem: Assoc fel Am Inst Aeronaut & Astronaut; Combustion Inst. Res: Physics of combustion; spectroscopy; applied combustion; jet propulsion; reentry physics. Mailing Add: Inst for Defense Anal 400 Army-Navy Dr Arlington VA 22202

WOLFLE, DAEL (LEE), b Puyallup, Wash, Mar 5, 06; m 29; c 3. SCIENCE POLICY, ADMINISTRATIVE SCIENCES. Educ: Univ Wash, BS, 27, MS, 28; Ohio State Univ, PhD(psychol), 31. Hon Degrees: DSc, Drexel Inst Technol, 56, Ohio State Univ, 57 & Western Mich Univ, 60. Prof Exp: Instr psychol, Ohio State Univ, 29-32; prof, Univ Miss, 32-36; examr biol sci, Univ Chicago, 36-39, from asst prof to assoc prof psychol, 38-45; exec secy, Am Psychol Asn, 46-50; dir, Comn Human Resources & Advan Training, Assoc Res Couns, 50-54; exec officer, AAAS, 54-70; actg dean archit & urban planning, 72-73, PROF PUB AFFAIRS, GRAD SCH PUB AFFAIRS, UNIV WASH, 70- Concurrent Pos: Civilian training adminr electronics, US Army Sig Corps, 41-43; tech aide, Off Sci Res & Develop, 44-46; mem or vchmn, Bd Trustees, Russell Sage Found, 61-; mem or chmn, Bd Trustees, James McKeen Cattell Fund, 62-; trustee, Pac Sci Ctr Found, 62-; mem ed adv bd, Sci Yearbk, Encycl Brittanica, 67-; mem res adv comt, Am Coun Educ, 68-73; trustee, Biosci Info Servs, 68-74; mem or chmn, Geophys Inst Adv Comt, Univ Alaska, 69-; mem manpower inst, Nat Indust Conf Bd, 70; chmn rev comt sci resources studies, NSF, 72-73; mem comt grad med educ, Asn Am Med Cols, 72-75; mem, US-USSR Joint Group Experts Sci Policy, 73-; mem comn human resources, Nat Acad Sci-Nat Res Coun, 74- Honors & Awards: Presidential Cert of Merit, President US, 48; Except Serv Medal, US Air Force, 57; Montgomery lectr, Univ Nebr, 59; Walter Van Dyke Bingham lectr, Columbia Univ, 60; Herbert S Langfeld lectr, Princeton Univ, 69. Mem: AAAS (exec officer, 54-70); Am Psychol Asn (exec secy 46-50); Am Coun Educ (secy, 66-67). Res: Education, utilization, mobility, supply and demand trends of scientific and specialized personnel. Mailing Add: Grad Sch of Pub Affairs Univ of Wash Seattle WA 98195

WOLFLE, THOMAS LEE, b Eugene, Ore, Apr 24, 36; m 62; c 2. VETERINARY MEDICINE, ANIMAL BEHAVIOR. Educ: Tex A&M Univ, BS, 59, DVM, 61; Univ Calif, Los Angeles, MA, 67, PhD(physiol psychol), 70; Am Col Lab Animal Med, dipl, 66. Prof Exp: Vet primate colony mgt, Sch Aerospace Med, Brooks AFB, Tex, 61-65, chief comp toxicol lab, Aeromed Res Lab, Holloman AFB, 65-66; chief flight environ br, Aerospace Med Res Labs, Wright-Patterson AFB, Ohio, 70-73, asst chief weapons effects br, Sch Aerospace Med, Brooks AFB, Tex, 73-75; SR VET OFFICER, NIH, 75- Concurrent Pos: Adj prof psychol, Wright State Univ, 71-73, consult, Lab Animal Med, 72-73. Mem: Am Vet Med Asn; Am Asn Lab Animal Practrs; AAAS; Am Asn Lab Animal Sci; assoc fel Col Vet Toxicol. Res: Behavioral effects of toxic-radiation environments on animals; impact of early non-specific environmental influences upon research on selected animal-disease models; identification of animal models of human disease. Mailing Add: Bldg 102 Rm 102 NIH Animal Ctr Bethesda MD 20014

WOLFMAN, EARL FRANK, JR, b Buffalo, NY, Sept 14, 26; m 46; c 3. SURGERY. Educ: Harvard Univ, BS, 46; Univ Mich, MD, 50; Am Bd Surg, dipl, 58. Prof Exp: From instr to asst prof surg, Univ Mich, 57-63, assoc prof, Med Sch, 63-66, asst to dean, 60-61, asst dean, 61-64; PROF SURG, CHMN DIV & DEPT & ASSOC DEAN, SCH MED, UNIV CALIF, DAVIS, 66- Concurrent Pos: Consult, Vet Admin Hosps, Travis AFB & Martinez, Calif, 66-; chief div surg serv, Sacramento Med Ctr, 68- Mem: Fel Am Col Surgeons; AMA; Soc Surg Alimentary Tract. Mailing Add: Dept of Surg Univ of Calif Sch of Med Davis CA 95616

WOLFORD, JACK ARLINGTON, b Brookville, Pa, Dec 5, 17; m 44; c 2. PSYCHIATRY. Educ: Allegheny Col, AB, 40; Univ Pa, MD, 43; Am Bd Psychiat, dipl, 53. Prof Exp: Intern med, Allegheny Gen Hosp, Pittsburgh, Pa, 43; resident psychiat, Warren State Hosp, 44-46, sr psychiatrist, 48-51, clin dir, 51-56; dir, Hastings State Hosp, Nebr, 56-58; asst prof psychiat, Sch Med, 58-69, chief social psychiat, Western Psychiat Inst & Clin, 58-72, dir community ment health, Retardation Ctr, 67-74, psychiatrist-in-chief & actg chmn dept psychiat, Sch Med, 72-73, PROF PSYCHIAT, SCH MED, UNIV PITTSBURGH, 69-, DIR ADULT SERV, WESTERN PSYCHIAT INST & CLIN, 72- Concurrent Pos: Resident psychiat, Psychiat Inst & Clin, Pittsburgh, 53; asst prof psychiat, Univ Nebr, 56-58; vis fac sem, Lab Community Psychiat, Harvard Univ; vis fel indust med, Mellon Inst; chmn, Governor's Task Force Commitment Procedures; mem, Governor's Adv Comt to Dept Welfare & Comn Ment Health; mem commun adv comt, NIMH; mem med adv comt, Highland Drive Vet Admin Hosp, Pittsburgh; chmn, Gov Adv Comt Ment Health & Ment Retardation, Dept Pub Welfare & Comn Ment Health, 74 & 75, Comprehensive Comt Ment Health Planning, 75 & Pa Asn Community Ment Health & Ment Retardation Ctrs, 75; vpres & pres elect, Group for Advan Psychiat, 75-77. Mem: Fel Am Psychiat Asn (vpres, 75); fel Am Col Psychiat; AMA. Res: Social and community psychiatry; urban mental health and illness. Mailing Add: Western Psychiat Inst & Clin 3811 O'Hara St Pittsburgh PA 15261

WOLFORD, JOHN HENRY, b Osgood, Ind, June 11, 36. AVIAN PHYSIOLOGY, POULTRY SCIENCE. Educ: Purdue Univ, BS, 58; Mich State Univ, MS, 60, PhD(avian physiol), 63. Prof Exp: From asst prof to assoc prof poultry sci, Mich State Univ, 63-74; PROF ANIMAL & VET SCI & CHMN DEPT, UNIV MAINE, ORONO, 74- Mem: Poultry Sci Asn; Sigma Xi; Am Soc Animal Sci. Res: Reproductive physiology of the turkey breeder hen; physiological alterations of fatty liver syndrome in laying chickens. Mailing Add: Dept of Animal & Vet Sci Univ of Maine Orono ME 04473

WOLFORD, LIONEL THOMAS, chemistry, see 12th edition

WOLFORD, RICHARD KENNETH, b Pa, Jan 3, 32. CHEMISTRY. Educ: WVa Wesleyan Col, BS, 53; Univ Ky, MS, 55, PhD(chem), 59. Prof Exp: Asst chem, Univ Ky, 53-58; chemist, Nat Bur Standards, 58-66; PHYS SCI ADMINSTR, NAT OCEAN SURV, NAT OCEANIC & ATMOSPHERIC ADMIN, 66- Mem: AAAS; Am Chem Soc. Res: Oceanography. Mailing Add: Nat Ocean Surv Nat Ocean & Atmospheric Admin Rockville MD 20852

WOLFORD, RICHARD WILSON, chemistry, deceased

WOLFOWITZ, JACOB, b Poland, Mar 19, 10; US citizen; m 34; c 2. MATHEMATICS. Educ: City Col New York, BS, 31; NY Univ, PhD(math), 42. Prof Exp: Assoc prof math statist, Univ NC, 45-46; from assoc prof to prof, Columbia Univ, 46-51; prof math, Cornell Univ, 51-70; PROF MATH, UNIV ILL, URBANA, 70- Concurrent Pos: Vis prof, Univ Calif, Los Angeles, 52-53, Univ Ill, 53, Israel Inst Technol, 57, Univ Paris, 67 & Univ Heidelberg, 69. Mem: Nat Acad Sci; fel Am Acad Arts & Sci; Am Math Soc; fel Economet Soc; fel Inst Math Statist. Res: Mathematical statistics; probability; information theory. Mailing Add: Dept of Math Univ of Ill Urbana IL 61801

WOLFRAM, THOMAS, b St Louis, Mo, July 27, 36. SOLID STATE PHYSICS. Educ: Univ Calif, Riverside, AB, 59, PhD(physics), 63; Univ Calif, Los Angeles, MA, 60. Prof Exp: Mem tech staff, Atomics Int, 60-63; mem tech staff, NAm Rockwell Sci Ctr, 63-68, group leader solid state physics, 68-72, dir physics & chem, Rockwell Int Sci Ctr, 72-74; PROF PHYSICS & CHMN DEPT, UNIV MO, COLUMBIA, 74- Concurrent Pos: Adj prof physics, Univ Calif, Riverside, 68-69. Mem: Fel Am Phys Soc. Res: Lattice dynamics; spin waves; superconductivity; electronic and optical properties; physics and chemistry of surfaces; catalysis. Mailing Add: Dept of Physics Univ of Mo Columbia MO 65201

WOLFSBERG, KURT, b Hamburg, Ger, Nov 1, 31; nat US; m 55; c 3. RADIOCHEMISTRY, NUCLEAR CHEMISTRY. Educ: St Louis Univ, BS, 53; Wash Univ, St Louis, MA, 55, PhD(chem), 59. Prof Exp: STAFF MEM RADIOCHEM GROUP, LOS ALAMOS SCI LAB, 59- Concurrent Pos: Consult aircraft nuclear propulsion comt nuclear measurements & standards, US Air Force, 56-57; mem subcomt radiochem, Nat Acad Sci, 72-; guest & Fulbright grantee, Univ Mainz, WGer, 74-75; Fulbright award, 74. Mem: AAAS; Am chem Soc; fel Am Inst Chem. Res: Emanation techniques; mass and charge distribution in fission; high temperature diffusion of fission products; lanthanide and actinide chemistry; properties of very heavy nuclides. Mailing Add: 303 Venado Los Alamos NM 87544

WOLFSBERG, MAX, b Hamburg, Ger, May 28, 28; nat US; m 57; c 1. PHYSICAL CHEMISTRY. Educ: Wash Univ, St Louis, AB, 48, PhD(chem), 51. Prof Exp: Asst chem, Wash Univ, St Louis, 48-50; assoc chemist, Brookhaven Nat Lab, 51-54, from chemist to sr chemist, 54-69; Regents' lectr, 68, PROF CHEM, UNIV CALIF, IRVINE, 69- Concurrent Pos: NSF sr fel, 58-59; vis prof chem, Cornell Univ, 63 & Ind Univ, 65; prof, State Univ NY Stony Brook, 66-69. Mem: Am Chem Soc. Res: Theoretical chemistry; isotope effects, chemical reactions, electronic structure of molecules. Mailing Add: Dept of Chem Univ of Calif Irvine CA 92664

WOLFSON, ALBERT, b New York, NY, Feb 3, 17; m 37, 71; c 2. ZOOLOGY. Educ: Cornell Univ, BS, 37; Univ Calif, PhD(zool), 42. Prof Exp: Assoc zool, Univ Calif, 42-43, instr, 43-44; from instr to assoc prof, 44-57, PROF ZOOL, NORTHWESTERN UNIV, EVANSTON, 57- Concurrent Pos: Vis scientist, Am Inst Biol Sci, 59-; NSF sr fel, Univ Tokyo, 61-62; mem steering comt, Biol Sci Curric Study Comt, 61-65; Alumni Fund lectr, 64; consult environ biol panel, NSF, 65-67; res assoc, Field Mus Natural Hist, 66-; mem bd governors, Chicago Acad Sci, 66. Honors & Awards: Brewster Mem Medal & Award, Am Ornith Union, 62. Mem: Fel AAAS; Am Soc Zoologists; Am Ornith Union (secy, 51-52); Endocrine Soc; Wilson Ornith Soc. Res: Migration, reproductive physiology, endocrines, physiology and annual cycles of birds; zoogeography; Circadian rhythms; neuroendocrinology. Mailing Add: Dept of Biol Sci Northwestern Univ Evanston IL 60201

WOLFSON, ALFRED MORTIMER, b New York, NY, May 23, 99; m 26; c 2. PLANT MORPHOLOGY. Educ: Cornell Univ, BS, 21; Univ Wis, AM, 22, PhD(plant cytol), 24. Prof Exp: Comn Relief Belg Educ Found fel, Univ Louvain & Brussels Univ, 24-26; Nat Res Coun fel, Cornell Univ, 26-27; res botanist, United Fruit Co, 27-29; prof biol & head dept, Murray State Univ, 29-69, EMER PROF BIOL, MURRAY STATE UNIV, 69- Mem: Fel AAAS; Am Inst Biol Sci. Res: Cytology and genetics of Hapaticae. Mailing Add: 310 N 14th St Murray KY 42071

WOLFSON, C JACOB, b Kingsley, Iowa, Apr 12, 38; m 64; c 1. PHYSICS. Educ: Grinnell Col, BA, 60; Univ Utah, PhD(physics), 66. Prof Exp: RES SCIENTIST, LOCKHEED PALO ALTO RES LAB, 66- Mem: Am Geophys Union. Res: Cosmic ray muon and neutrino studies; cosmic ray neutron monitor space physics. Mailing Add: Lockheed Palo Alto Res Lab 3251 Hanover St Palo Alto CA 94304

WOLFSON, EDWARD A, b New York, NY, Feb 4, 26; m 55; c 3. PREVENTIVE MEDICINE, MEDICINE. Educ: Cornell Univ, AB, 48, MNS, 49, MD, 53; Columbia Univ, MPH, 71. Prof Exp: Pvt pract internal med, 57-69; PROF PREV MED & COMMUNITY HEALTH & VCHMN DEPT, COL MED NJ, 69-, ASSOC DEAN HEALTH CARE, 72-, DIR, OFF PRIMARY HEALTH CARE EDUC, 75- Concurrent Pos: Consult, Spec Action for Drug Abuse Prev, White House, 71, Valley Hosp, Martland Hosp, East Orange Vet Admin & St Joseph's Hosps. Mem: Fel Am Col Physicians; fel Am Col Prev Med. Res: Drug abuse; delivery of health care. Mailing Add: Dept of Prev Med & Com Health Col of Med of NJ Newark NJ 07103

WOLFSON, JAMES, b Chicago, Ill, Mar 16, 43; m 71. PHYSICS. Educ: Grinnell Col, BA, 64; Mass Inst Technol, PhD(physics), 68. Prof Exp: Mem staff physics lab nuclear sci, 68-70, ASST PROF PHYSICS, MASS INST TECHNOL, 70- Mem: AAAS; Am Phys Soc. Res: Experimental high energy physics. Mailing Add: Dept of Physics Mass Inst of Technol Cambridge MA 02139

WOLFSON, JOSEPH LAURENCE, b Winnipeg, Man, July 22, 17; m 44; c 2. NUCLEAR PHYSICS. Educ: Univ Man, BSc, 42, MSc, 53; McGill Univ,

PhD(physics), 48. Prof Exp: Asst res officer physics, Atomic Energy Can, Ltd, Chalk River, 48-55; physicist, Jewish Gen Hosp, Montreal, Que, 55-58; assoc res officer, Nat Res Coun Can, 58-64; prof physics, Univ Sask, 64-74; DEAN SCI, CARLETON UNIV, OTTAWA, 74- Mem: Am Phys Soc; Can Asn Physicists. Res: Beta and gamma ray spectroscopy; experimental gravitation. Mailing Add: Carleton Univ Ottawa ON Can

WOLFSON, KENNETH GRAHAM, b New York, NY, Nov 21, 24; m 47; c 2. MATHEMATICS. Educ: Brooklyn Col, BA, 47; Johns Hopkins Univ, MA, 48; Univ Ill, PhD(math), 52. Prof Exp: From instr to assoc prof, 52-60, PROF MATH, RUTGERS UNIV, 60-, CHMN DEPT, 61- Mem: Am Math Soc; Math Asn Am. Res: Spectral theory of differential equations; rings of linear transformations; structure of rings. Mailing Add: Dept of Math Rutgers Univ New Brunswick NJ 08903

WOLFSON, LEONARD LOUIS, b Wilkes-Barre, Pa, Dec 13, 19; m 47; c 1. MICROBIOLOGY. Educ: Univ Chicago, MS, 51. Prof Exp: Res bacteriologist, Ill State Dept Pub Welfare, 52-53; head bact lab, Wilson & Co, 54-56; group leader, Nalco Chem Co, 57-66; sr res assoc, Indust Bio-Test Labs, 66-69; CORP DIR QUAL ASSURANCE, WILL ROSS, INC & PRES, WILRO SCI LABS, 69- Mem: Am Soc Microbiol; Soc Indust Microbiol; NY Acad Sci. Res: Microbiology of industrial waters; use of microbiocides; air and water pollution control; microbiology of medical devices, cosmetics and drugs. Mailing Add: 7734 N Berwyn Glendale WI 53209

WOLFSON, NANCY DOLLY, b Boston, Mass, Aug 7, 28. CELL PHYSIOLOGY, DEVELOPMENTAL PHYSIOLOGY. Educ: Univ Wis, BA, 49; Univ Tenn, MS, 51; Duke Univ, PhD(cell physiol), 60. Prof Exp: J C Childs Mem Fund fel embryol, Brussels, 60-61; res assoc bot, Univ Ill, 61-62; instr physiol, Mt Holyoke Col, 62-63; asst prof zool, 64-70, ASSOC PROF BIOL, McGILL UNIV, 70- Concurrent Pos: Am Cancer Soc grant, 63-64; NSF fel, Marine Lab, Duke Univ, 64; Nat Res Coun Can grant, 65- Res: Sulfhydryl related functions in cell division and development. Mailing Add: Dept of Biol McGill Univ Montreal PQ Can

WOLFSON, ROBERT JOSEPH, b Philadelphia, Pa, Sept 11, 29; m 53; c 2. OTOLARYNGOLOGY. Educ: Temple Univ, BA, 52, MS, 61; Hahnemann Med Col, MD, 57. Prof Exp: PROF OTOLARYNGOL & HEAD DEPT, MED COL PA, 69- Mem: Am Acad Ophthal & Otolaryngol; Am Otol Soc; Am Laryngol, Rhinol & Otol Soc; Royal Soc Med; AMA. Mailing Add: Dept of Surg Med Col of Pa Philadelphia PA 19129

WOLFSON, SEYMOUR J, b Detroit, Mich, Feb 13, 37; m 58; c 4. COMPUTER SCIENCE. Educ: Wayne State Univ, BS, 59, PhD(physics), 65; Univ Chicago, MS, 60. Prof Exp: Res assoc physics, Wayne State Univ, 63-65; sr scientist, Comput Sci Corp, 65-68; asst prof comput sci, 68-73, ASSOC PROF COMPUT SCI, WAYNE STATE UNIV, 73- Concurrent Pos: Consult, Comput Sci Corp, 68-71, Lincorp Corp, 72- & Dept Housing & Urban Develop, 75-; secy, Comput Sci Bd, 74-76. Mem: Am Phys Soc; Asn Comput Mach; Am Arbit Asn; Inst Elec & Electronics Engrs. Res: Computer networks and applications; numerical methods. Mailing Add: Dept of Comput Sci Wayne State Univ Detroit MI 48202

WOLFSON, SIDNEY KENNETH, JR, b Philadelphia, Pa, June 14, 31; m 58; c 3. NEUROSURGERY, BIOMEDICAL ENGINEERING. Educ: Univ Pa, AB, 51; Univ Chicago, MD, 58. Prof Exp: Resident surg, Univ Pa Hosp, 59-63, asst prof surg res, Sch Med, Univ Pa, 63-68; dir surg res, Michael Reese Hosp & Med Ctr, 68-71; ASSOC PROF NEUROSURG & DIR SURG RES, UNIV PITTSBURGH, 71- Concurrent Pos: Career Develop Award, Nat Heart & Lung Inst, 63; assoc prof surg, Univ Chicago, 69-71; chmn spec study sect, Nat Inst Arthritis & Metab Dis, 74-75; consult, Artificial Heart Assessment Panel, NIH, 74, Med Devices Prog, Nat Heart & Lung Inst, 73-74. Mem: Am Soc Artificial Internal Organs; Am Asn Neurol Surgeons; Soc Acad Surgeons; Soc Neurosci; Asn Advan Med Instrumentation. Res: Hypothermia and circulatory arrest; implantable biofuel cell for powering implanted devices, devices, indwelling arterial oxygen electrode. Mailing Add: Dept of Surg Montefiore Hosp 3459 5th Ave Pittsburgh PA 15213

WOLGA, GEORGE JACOB, b New York, NY, Apr 2, 31; m 59; c 3. PHYSICS. Educ: Cornell Univ, BEngPhys, 53; Mass Inst Technol, PhD(physics), 57. Prof Exp: Asst physics, Mass Inst Technol, 53-56, instr, 57-60, asst prof, 60-61; from asst prof to assoc prof, 61-68, PROF PHYSICS, CORNELL UNIV, 68- Concurrent Pos: Consult, Gen Elec, Sylvania & US Naval Res Lab, vpres-dir res, Lansing Res Corp, 64- Mem: Am Phys Soc; Inst Elec & Electronics Engrs. Res: Excited state spectroscopy; molecular physics; quantum electronics; molecular energy transfer and relaxation; laser induced chemistry; infrared tunable lasers and spectrscopy. Mailing Add: Dept of Elec Eng & Appl Physics Cornell Univ Ithaca NY 14850

WOLGAMOTT, GARY, b Alva, Okla, July 23, 40; m 62; c 2. MICROBIOLOGY, MOLECULAR BIOLOGY. Educ: Northwestern State Col, Okla, BS, 63; Okla State Univ, PhD(microbiol), 68. Prof Exp: From asst prof to assoc prof microbiol, 68-75, PROF MICROBIOL, SOUTHWESTERN OKLA STATE UNIV, 75- Concurrent Pos: NSF fel col med, Univ Iowa, 71; chmn int sci fair, Micro Judges, Am Soc Microbiol, 75. Mem: AAAS; Am Soc Microbiol. Res: Study of the mode of action of specific chemotheopeutic agents on the activities of microorganisms, including their physiology and microstructure; action of microbial hemolysins on tissue culture cells ultrastructure; virulence analysis of space flown microautoflora from astronauts. Mailing Add: Dept of Microbiol Southwestern Okla State Univ Weatherford OK 73096

WOLGEMUTH, KENNETH MARK, b Mechanicsburg, Pa, Dec 23, 43; m 70. MARINE GEOCHEMISTRY. Educ: Wheaton Col, BS, 65; Columbia Univ, MS, 69, PhD(geochem), 72. Prof Exp: ASST PROF GEOL, DICKINSON COL, 71- Concurrent Pos: Vis asst prof geol, World Campus Afloat, Chapman Col, 75. Mem: Geol Soc Am; Am Geophys Union; AAAS. Res: Trace element geochemistry of surface waters. Mailing Add: Dept of Geol Dickinson Col Carlisle PA 17013

WOLICKI, ELIGIUS ANTHONY, b Buffalo, NY, May 10, 27; m 54; c 4. NUCLEAR PHYSICS. Educ: Canisius Col, BS, 46; Univ Notre Dame, PhD(physics), 50. Prof Exp: Asst physics, Univ Notre Dame, 46-47, asst, 47-48; res assoc nuclear physics, Univ Iowa, 50-52; nuclear physicist nuclear sci div, 52-66, CONSULT & ACTG ASSOC SUPT RADIATION TECHNOL DIV, US NAVAL RES LAB, 66- Honors & Awards: Centennial of Sci Award, Univ Notre Dame, 65; Meritorious Civilian Serv Award, US Naval Res Lab, 71. Mem: Am Phys Soc; Am Nuclear Soc; Sigma Xi; Fed Prof Asn. Res: Nuclear reactions, structure and applications; electrostatic accelerators; radiation detectors; applications of nuclear radiation, nuclear techniques and ion beam accelerators. Mailing Add: 1310 Gatewood Dr Alexandria VA 22307

WOLIN, ALAN GEORGE, b New York, NY, Apr 2, 33; m 54; c 4. FOOD SCIENCE. Educ: Cornell Univ, BS, 54, MS, 56, PhD, 58. Prof Exp: Sr proj leader yeast tech div, Fleischmann Labs, Standard Brands, Inc, 58-60; head dairy tech lab, Vitex Labs Div, Nopco Chem Co, 60-62; prod improv mgr, 62-67, QUAL COORDR, M&M CANDIES DIV, MARS, INC, 67- Concurrent Pos: Ed, J Appl Microbiol, 72-; mem res comt, Nat Confectioners Asn, 75-, tech comt, Grocery Mfrs Asn, 73- & NJ Pub Health Adv Comn, 73-75. Mem: Am Dairy Sci Asn; Am Soc Microbiol; Inst Food Technol; Soc Consumer Affairs Prof; Am Chem Soc. Res: Use of radioisotopes in study of food flavors; dairy started cultures; enzymes; fermentation; antimicrobial agents in milk; food fortification; phosphatase activity of chocolate milk; natural bacterial inhibitors in raw milk; chocolate and confection research; quality assurance improvement programs and handling of regulatory aspects of food administration. Mailing Add: 44 Stonehenge Rd Morristown NJ 07960

WOLIN, HAROLD LEONARD, b Brooklyn, NY, June 22, 27; m 56; c 3. MICROBIOLOGY. Educ: Univ Calif, AB, 50, MA, 52; Cornell Univ, PhD(bact), 56; Nat Registry Clin Chem, cert. Prof Exp: Res scientist, Pac Yeast Prod Co, 56-57; instr, Hahnemann Med Col, 57-59; asst prof microbiol col med & dent, Seton Hall Univ, 59-63; CLIN ASSOC PROF MICROBIOL, COL MED & DENT, NJ, 63-; ASSOC DIR CLIN LAB, BROOKDALE HOSP CTR, BROOKLYN, NY, 63- Mem: AAAS; Am Soc Microbiol; Brit Soc Gen Microbiol; Am Asn Clin Chem. Res: Clinical microbiology and chemistry; microbial physiology. Mailing Add: Dept of Clin Labs Brookdale Hosp Ctr Brookdale Plaza Brooklyn NY 11212

WOLIN, LEE ROY, b Cleveland, Ohio, Dec 8, 27; m 50; c 3. PSYCHOLOGY, NEUROPHYSIOLOGY. Educ: Los Angeles State Col, BS, 50; Cornell Univ, PhD(psychol), 55. Prof Exp: Mem fac psychol, Sarah Lawrence Col, 55-59; res assoc develop, Child Study Ctr, Clark Univ, 59-60; res assoc, Cleveland Psychiat Inst, 60-69, dir lab neuropsychol, 69-73; ASST PROF NEUROSURG, CASE WESTERN RESERVE UNIV, 70-; DIR LAB NEUROPSYCHOL & EEG, OHIO MENT HEALTH & MENT RETARDATION RES CTR, 73-; ADMIN DIR, DRUG ABUSE TREATMENT UNIT, CLEVELAND PSYCHIAT INST, 75- Concurrent Pos: Lectr, Lakewood High Exten, Ohio State Univ, 63-66 & Bedford Exten, Cleveland State Univ, 66; lectr, Kent State Univ, 65. Mem: AAAS; Am Psychol Asn; NY Acad Sci. Res: Neurophysiology of vision; behavioral and neurophysiological manifestations of brain disfunction; perceptual, cognitive and emotional aspects of neuropsychiatric disorders perception. Mailing Add: Ohio Ment Health & Ret Res Ctr 1708 Aiken Ave Cleveland OH 44109

WOLIN, MEYER JEROME, b Bronx, NY, Nov 10, 30; m 55. MICROBIAL ECOLOGY, MICROBIAL PHYSIOLOGY. Educ: Cornell Univ, BS, 51; Univ Chicago, PhD(microbiol), 54. Prof Exp: NIH fel microbiol, Univ Minn, 54-55; fel, Univ Ill, Urbana, 55-56, from asst prof to assoc prof dairy sci, 56-67, assoc prof microbiol, 65-67, prof dairy sci & microbiol, 67-74; CHIEF RES SCIENTIST DIV LAB & RES, NY STATE HEALTH DEPT, 74- Concurrent Pos: NSF sr fel, Univ Newcastle, 64-65; mem microbial chem study sect, NIH, 71-75. Mem: AAAS; Am Chem Soc; Am Soc Microbiol; Am Dairy Sci Asn; Am Soc Biol Chem. Res: Microbial biochemistry and ecology; fermentations in anaerobic ecosystems; interspecies interactions; hydrogen metabolism; methane production. Mailing Add: 120 New Scotland Ave Albany NY 12201

WOLINSKI, LEON EDWARD, b Buffalo, NY, Apr 3, 26; m 54; c 3. ORGANIC POLYMER CHEMISTRY. Educ: Univ Buffalo, BA, 49, PhD(chem), 51. Prof Exp: Res chemist film dept, E I du Pont de Nemours & Co, 51-57, staff scientist org polymer chem, 57-59, res assoc, 59-70; CORP DIR RES, PRATT & LAMBERT CO, 70- Concurrent Pos: Lectr, Canisius Col, 55-59. Mem: Am Chem Soc; Am Inst Chem. Res: Grignard reagents; organo silicon chemistry; surface chemistry; adhesives; condensation and addition polymers; coatings; films. Mailing Add: 35 Parkview Terr Cheektowaga NY 14225

WOLINSKY, ALBERT, b Vienna, Austria, Apr 12, 13; US citizen; m 46; c 3. PHYSICS, MATHEMATICS. Educ: Univ Vienna, PhD(physics, math), 37. Prof Exp: Ins mathematician, Assicurazioni Generali, Vienna, Austria, 36-38; instr math, RI State Col, 46; instr, NY Univ, 46-51; assoc physicist, Farrand Optical Co, Inc, NY, 51-53; sr staff mem, Gen Precision Lab, Inc, NY, 53-60; staff engr, TRW Systs Group, Calif, 60-70; LECTR MATH, STATIST & COMPUT SCI, CALIF STATE UNIV, LOS ANGELES, 70-; OWNER, COMPUT INFO SYSTS, 70- Concurrent Pos: Lectr, Hunter Col, 46-49, Columbia Univ, 54-55, Univ Southern Calif, 70-71 & Calif State Col, Long Beach, 70-71; consult, Fed Aviation Agency, DC, 61. Mem: Am Math Soc; Math Asn Am; Asn Comput Mach. Res: Latent photographic image; electrostatic electron lenses; microwave antenna systems; digital computer design; air traffic control automation; mathematical theory of shift registers; associative memory-processor algorithms and design; number theory; algebra. Mailing Add: 13674 Bayliss Rd Los Angeles CA 90049

WOLINSKY, EMANUEL, b New York, NY, Sept 23, 17; m 47; c 2. INFECTIOUS DISEASES. Educ: Cornell Univ, BA, 38, MD, 41. Prof Exp: Intern med, New York Hosp, 43-44, resident, 44-45; asst dir tuberc res, Trudeau Lab, Trudeau Found, 46-56; asst prof med, 56-62, asst prof microbiol, 59-62, assoc prof, 62-68, PROF MED, SCH MED, CASE WESTERN RESERVE UNIV, 68- Concurrent Pos: Bacteriologist chg, Cleveland Metrop Hosp, 59-; former mem tuberc panel, US-Japan Coop Med Sci Prog; former mem strep & staph comn, Armed Forces Epidemiol Bd; assoc ed, Am Rev Respiratory Dis. Mem: Am Soc Microbiol; Am Thoracic Soc; Infectious Dis Soc Am. Res: Medical microbiology and pulmonary diseases; tuberculosis bacteriology and experimental chemotherapy; infectious diseases. Mailing Add: Metrop Gen Hosp Cleveland OH 44109

WOLINSKY, HARVEY, b Cleveland, Ohio, June 3, 39. MEDICINE, PATHOLOGY. Educ: Western Reserve Univ, AB, 60; Univ Chicago, MD & MS, 63, PhD(path), 67. Prof Exp: Intern, Univ Chicago Hosps, 63-64; asst resident internal med, Mt Sinai Hosp, Cleveland, Ohio, 64-65; resident chest serv, Bronx Municipal Hosp Ctr, NY, 67-68; assoc, 68-70, assoc path, 69-70, asst prof med & path, 70-73, ASSOC PROF MED, ALBERT EINSTEIN COL MED, 73-, MEM CARDIOVASC-PULMONARY-RENAL RES UNIT, 68- Concurrent Pos: USPHS res fel, Univ Chicago Hosps, 65-67; USPHS res career develop award, Nat Heart & Lung Inst, 72-77; assoc attend physician, Bronx Municipal Hosp Ctr, NY, 68-71, attend physician, 72-; mem coun arteriosclerosis, Am Heart Asn, 70- Mem: Am Thoracic Soc; Am Soc Exp Path; Fedn Am Socs Exp Biol; Am Soc Clin Invest; Harvey Soc. Res: Comparative pathology; structure, function and biochemistry of blood vessels; effects of hormonal and mechanical factors on blood vessel structure. Mailing Add: Dept of Med Albert Einstein Col of Med Bronx NY 10461

WOLINSKY, JOSEPH, b Chicago, Ill, Dec 3, 30; m 51; c 5. ORGANIC CHEMISTRY. Educ: Univ Ill, BS, 52; Cornell Univ, PhD, 56. Prof Exp: Proj assoc, Univ Wis, 56-58; from asst prof to assoc prof chem, 58-67, PROF CHEM, PURDUE UNIV, 67- Mem: Am Chem Soc. Res: Chemistry of terpenes, alkaloids and related natural products. Mailing Add: Dept of Chem Purdue Univ West Lafayette IN 47906

WOLK, COLEMAN PETER, b New York, NY, Sept 28, 36; m 65; c 1. DEVELOPMENTAL MICROBIOLOGY. Educ: Mass Inst Technol, SB & SM, 58; Rockefeller Inst, PhD(biol), 64. Prof Exp: Nat Acad Sci-Nat Res Coun res fel biol, Calif Inst Technol, 64-65; from asst prof to assoc prof, 65-74, PROF BOT, MICH STATE UNIV, 74- Mem: Am Soc Microbiol; Soc Develop Biol. Res: Physiological and biochemical basis of differentiation and morphogenesis of blue-green algae. Mailing Add: Energy Res & Develop Admin Plant Res Lab Mich State Univ East Lansing MI 48824

WOLK, ELLIOTT SAMUEL, b Springfield, Mass, Aug 5, 19; m 50; c 3. MATHEMATICS. Educ: Clark Univ, AB, 40; Brown Univ, ScM, 47, PhD(math), 54. Prof Exp: Instr math, 50-56, from asst prof to assoc prof, 56-63, chmn dept, 67-73, PROF MATH, UNIV CONN, 63- Concurrent Pos: Consult elec boat div, Gen Dynamics Corp, 55-58. Mem: Am Math Soc; Math Asn Am. Res: Partially ordered sets; general topology. Mailing Add: Dept of Math Univ of Conn Storrs CT 06268

WOLK, ROBERT GEORGE, b New York, NY, Mar 10, 31; m 56; c 5. ORNITHOLOGY. Educ: City Col New York, BS, 52; Cornell Univ, MS, 54, PhD(vert zool), 59. Prof Exp: Asst prof biol, St Lawrence Univ, 57-63; assoc prof vert morphol & behav, Adelphi Univ, 63-67; CUR LIFE SCI, NASSAU COUNTY MUS, 67- Mem: AAAS; Am Asn Mus; Am Inst Biol Sci; Am Ornith Union; Am Soc Zool. Res: Behavioral adaptations and functional morphology of birds; evolution, systematics and distribution of gulls, terns and skimmers; avian crepuscular vision. Mailing Add: Tackapausha Mus of Natural Hist Seaford NY 11783

WOLKE, RICHARD ELWOOD, b East Orange, NJ, June 2, 33; m 64; c 2. VETERINARY PATHOLOGY. Educ: Cornell Univ, BS, 55, DVM, 62; Univ Conn, MS, 66, PhD(vet path), 68. Prof Exp: Vet, Am Soc Prev Cruelty Animals, 62-63; pvt practice, 63-64; res asst path, Univ Conn, 64-68, res assoc, 68-69, res assoc icthyopath, 69-70; asst prof, 70-75, ASSOC PROF ICTHYOPATH, UNIV RI, 75- Concurrent Pos: Conn Res Comn grant, Univ Conn, 69-70; Nat Oceanic & Atmospheric Admin sea grant, Univ RI, 70- Mem: AAAS; Int Asn Aquatic Animal Med; World Maricult Soc; Am Vet Med Asn; NY Acad Sci. Res: Ichthyopathology; comparative pathology; oncology. Mailing Add: Dept of Animal Path Peckham Res Lab Univ of RI Kingston RI 02881

WOLKE, ROBERT LESLIE, b Brooklyn, NY, Apr 2, 28; m 64; c 1. NUCLEAR CHEMISTRY, RADIOCHEMISTRY. Educ: Polytech Inst Brooklyn, BS, 49; Cornell Univ, PhD, 53. Prof Exp: Res assoc nuclear chem, Enrico Fermi Inst, Univ Chicago, 53-56; nuclear chemist, Gen Atomic Div, Gen Dynamics Corp, 56-57; from asst prof to assoc prof chem, Univ Fla, 57-60; assoc prof, 60-67, PROF CHEM, UNIV PITTSBURGH, 67- Concurrent Pos: Res partic, Oak Ridge Inst Nuclear Studies, 58 & 59; vis prof, Univ PR, 70; vis prof, US AID, Univ Oriente, Venezuela, 73. Mem: AAAS; Am Chem Soc; Am Phys Soc. Res: Nuclear reactions; recoil studies, interaction of energetic ions with matter; natural radioactivity; marine radioactivity. Mailing Add: Wherrett Lab of Nuclear Chem Dept of Chem Univ of Pittsburgh Pittsburgh PA 15260

WOLKE, SARA RICHARDSON, b York, Ala, Oct 29, 51; m 73. ANALYTICAL CHEMISTRY. Educ: Miss Univ Women, BS, 73. Prof Exp: Chief chemist, 73-75, VPRES ANAL CHEM, MACMILLAN RES, LTD, 75- Res: Studies of aluminum recovery from wastes of bauxite processing plants; future plans include the detection of ethylene oxide by gas chromatography. Mailing Add: MacMillan Res Ltd 1221 Barclay Circle Box 1305 Marietta GA 30061

WOLKEN, GEORGE, JR, b Jersey City, NJ, Nov 11, 44; m 67; c 1. CHEMICAL PHYSICS, THEORETICAL CHEMISTRY. Educ: Tufts Univ, BS, 66; Harvard Univ, PhD(chem physics), 71. Prof Exp: Fel, Max Planck Inst Aerodyn, Univ Göttingen, 71-72; asst prof chem, Ill Inst Technol, 72-74; MEM STAFF, BATTELLE MEM INST, 74- Mem: Am Phys Soc. Res: Theoretical chemical kinetics, both of gas phase reactions and heterogeneous reactions. Mailing Add: Battelle Mem Inst 505 King Ave Columbus OH 43201

WOLKEN, JEROME JAY, b Pittsburgh, Pa, Mar 28, 17; m 45, 56; c 4. BIOPHYSICS. Educ: Univ Pittsburgh, BS, 46, MS, 48, PhD(biophys), 49. Prof Exp: Res fel, Mellon Inst, 43-47; res fel, Rockefeller Inst, 51-52; DIR BIOPHYS RES LAB, MELLON INST SCH SCI, CARNEGIE-MELLON UNIV, 53-, PROF BIOPHYS, 64- Concurrent Pos: AEC fel, 49-51; Am Cancer Soc fel, 51-53; asst prof sch med, Univ Pittsburgh, 53-57, from assoc prof to prof, 57-66; Nat Coun Combat Blindness fel, 57; career prof, USPHS, 62-64; guest prof, Pa State Univ, 63; vis prof, Univ Paris, 67-68, Univ Col, Univ London, 71 & Pasteur Inst, Paris, 72. Mem: Fel AAAS; fel Optical Soc Am; Am Chem Soc; Am Soc Cell Biol; Soc Gen Physiol. Res: Photosynthesis; vision and nerve excitation. Mailing Add: 5817 Elmer St Pittsburgh PA 15232

WOLKOFF, A STARK, b Uniontown, Pa, Sept 2, 21; m 56. OBSTETRICS & GYNECOLOGY, PHYSIOLOGY. Educ: Univ Scranton, BS, 43; Hahnemann Med Col, MD, 50. Prof Exp: Instr obstet & gynec, Univ Louisville, 55-56; asst prof, State Univ NY Downstate Med Ctr, 56-57; asst prof gynec & obstet, Sch Med, Univ NC, 59-64; PROF GYNEC & OBSTET & ASST PROF PHYSIOL, UNIV KANS MED CTR, KANSAS CITY, 64- Concurrent Pos: Spec res fel physiol, Sch Med, Yale Univ, 57-58; consult, Kansas City Vet Hosp & Munsen Hosp, Ft Leavenworth. Mem: Fel Am Col Surgeons; fel Am Col Obstet & Gynec; AMA; Soc Gynec Invest. Res: Fetal and placental physiology using pregnant ewe and intrauterine catheterization techniques. Mailing Add: Dept of Obstet & Gynec Univ of Kans Sch of Med Kansas City KS 66103

WOLKOFF, AARON WILFRED, b Toronto, Ont, Feb 12, 44; m 66; c 2. ANALYTICAL CHEMISTRY, ENVIRONMENTAL CHEMISTRY. Educ: Univ Toronto, BSc, 65, MSc, 67, PhD(org chem), 71. Prof Exp: Fel, Inst Environ Sci & Eng, Univ Toronto, 71-72; teaching master math & chem, Seneca Col Appl Arts & Technol, 72-73; RES SCIENTIST ENVIRON ANAL, CAN CENTRE FOR INLAND WATERS, DEPT ENVIRON, 73- Mem: Am Chem Soc; Chem Inst Can. Res: Development of new analytical methods for environmental analysis with emphasis on the applications of high pressure liquid chromatography. Mailing Add: Can Centre for Inland Waters Box 5050 Burlington ON Can

WOLKOFF, HAL NORMAN, b New York, NY, June 6, 35; m 58; c 3. PHARMACEUTICS. Educ: Columbia Univ, BS, 56; Univ Wis, MS, 60, PhD(pharm), 61. Prof Exp: Res chemist, Schering Corp, 60-62, sect head pharm, 62-65, dept mgr, 65-68, assoc dir pharmaceut res & develop, 68-70, dir, 70-73, dir pharmaceut res & develop, Schering-Plough Corp, 73-74, DIR MFG SERV, SCHERING-PLOUGH CORP, 74- Concurrent Pos: Guest lectr col pharm, Columbia Univ, 61-63. Mem: Am Pharmaceut Asn; Am Chem Soc; Soc Cosmetic Chem; Int Pharmaceut Fedn; fel Am Acad Pharmaceut Sci. Res: Factors influencing drug stability and biological availability; pharmaceutical dosage forms. Mailing Add: Schering-Plough Corp 86 Orange St Bloomfield NJ 07003

WOLL, JOHN WILLIAM, JR, b Philadelphia, Pa, May 19, 31; m 52; c 2. MATHEMATICS. Educ: Haverford Col, BS, 52; Princeton Univ, PhD, 56. Prof Exp: Instr math, Princeton Univ, 56-57; asst prof, Lehigh Univ, 57-58, Univ Calif, 58-61 & Univ Wash, 61-68; PROF MATH, WESTERN WASH STATE COL, 68- Mem: Am Math Soc. Res: Functional analysis and stochastic processes. Mailing Add: Dept of Math Western Wash State Col Bellingham WA 98225

WOLLAEGER, ERIC EDWIN, b Milwaukee, Wis, Mar 6, 10; m 38; c 4. CLINICAL MEDICINE. Educ: Dartmouth Col, AB, 31; Harvard Univ, MD, 34; Univ Minn, MS, 41; Am Bd Internal Med, dipl, 55. Prof Exp: From instr to assoc prof, 41-56, PROF MED, MAYO GRAD SCH MED, UNIV MINN, 56-; CONSULT, MAYO CLIN, 41- Concurrent Pos: Mem subspec bd gastroenterol, Am Bd Internal Med, 55-61; dir training prog gastroenterol, Mayo Grad Sch Med, Univ Minn, 60-65; pres of staff, Mayo Clin, 62. Mem: Am Soc Clin Invest; Am Gastroenterol Asn. Res: Gastroenterology, particularly the digestive and absorptive functions of the human gastrointestinal tract in health and disease. Mailing Add: 1026 Plummer Circle Rochester MN 55901

WOLLAN, DAVID STRAND, b Boston, Mass, Mar 25, 37; m 63; c 1. SOLID STATE PHYSICS. Educ: Amherst Col, AB, 59; Univ Ill, MS, 61, PhD(physics), 66. Prof Exp: Asst prof physics, Va Polytech Inst & State Univ, 66-74; PHYSICIST, US ARMS CONTROL & DISARMAMENT AGENCY, 74- Mem: AAAS; Am Phys Soc; Inst Elec & Electronics Engrs; Fedn Am Scientists. Res: Electron and nuclear magnetic resonance in solids. Mailing Add: US Arms Control & Disarmament Agency Washington DC 20451

WOLLAN, ERNEST OMAR, b Glenwood, Minn, Nov 6, 02; m 30; c 3. PHYSICS. Educ: Concordia Col, BA, 23; Univ Chicago, MS, 27, PhD(physics), 29. Hon Degrees: DSc, Concordia Col, 65. Prof Exp: Instr physics, NDak State Col, 26-27; instr, Univ Chicago, 30-32; Nat Res Coun fel, Univ Zurich, 32-33; res assoc, Univ Chicago, 33-34; asst prof, Wash Univ, St Louis, 34-38; res assoc, Univ Chicago, 39-41, asst prof, 41-42, sect chief metall lab, 42-44; physicist, Chicago Tumor Inst, 38-49; sect chief & assoc dir physics div, 44-64, dir div, 64-67, CONSULT SOLID STATE DIV, OAK RIDGE NAT LAB, 67- Honors & Awards: Wetherill Medal, Franklin Inst, 67. Mem: Fel Am Phys Soc. Res: X-ray scattering; cosmic rays; radiation physics; nuclear physics and neutron diffraction, especially magnetism. Mailing Add: 107 Oneida Lane Oak Ridge TN 37830

WOLLAN, GERHARD NORVAL, b Minn, June 27, 10; m; c 3. MATHEMATICS. Educ: Luther Col, AB, 31; Univ Iowa, MS, 36; Univ Ga, PhD(math), 52. Prof Exp: Teacher high sch, Minn, 31-35; actuarial clerk, Home Life Ins Co, 36-41; physicist, US Dept Navy, 41-43; engr, A Brothman & Assocs, 43-46; assoc prof math, Simpson Col, 46-48 & NGa Col, 48-50; asst prof, Memphis State Col, 51-53; asst prof, Ft Wayne Ctr, 53-58, ASSOC PROF MATH, PURDUE UNIV, LAFAYETTE, 58- Mem: AAAS; Am Math Soc; Math Asn Am. Res: Infinite series. Mailing Add: Dept of Math Purdue Univ Lafayette IN 47907

WOLLAN, JOHN JEROME, b Chicago, Ill, July 7, 42; m 67; c 2. LOW TEMPERATURE PHYSICS. Educ: St Olaf Col, BA, 64; Iowa State Univ, PhD(physics), 70. Prof Exp: Vis asst prof physics, Univ Ky, 70-73; Nat Res Coun assoc physics, Air Force Mat Lab, 73-74; STAFF MEM PHYSICS, LOS ALAMOS SCI LAB, 74- Mem: Am Phys Soc; Sigma Xi. Res: Development and evaluation of superconducting wire for magnetic energy transfer and storage applications for fusion power systems, in particular, static and fast pulse energy losses in superconducting wire. Mailing Add: CTR-9 MS 464 Los Alamos Sci Lab Los Alamos NM 87545

WOLLEBEN, JAMES A, b Brooklyn, NY, May 6, 35; m 59; c 6. PALEONTOLOGY, STRATIGRAPHY. Educ: Univ Southern Miss, BS, 56; La State Univ, MS, 59; Univ Tex, PhD(geol), 66. Prof Exp: Asst prof geol, Univ Mo-Columbia, 65-67; asst prof, 67-69, ASSOC PROF EARTH SCI, LA STATE UNIV, NEW ORLEANS, 69- Res: Cretaceous paleontology and stratigraphy; paleobiology of Tertiary echinoderms; biometrics. Mailing Add: Col of Sci La State Univ New Orleans LA 70122

WOLLENSAK, JOHN CHARLES, b Rochester, NY, Dec 16, 32; m 57; c 5. ORGANIC CHEMISTRY. Educ: Col Holy Cross, BS, 54; Mass Inst Technol, PhD(org chem), 58. Prof Exp: Res chemist, 58-66, SUPVR CHEM RES, ETHYL CORP, 66- Mem: Am Chem Soc. Res: Organic synthesis; organometallics. Mailing Add: Ethyl Corp 1600 W Eight Mile Rd Detroit MI 48220

WOLLER, WILLIAM HENRY, b San Antonio, Tex, Feb 28, 33; m 58; c 3. PHYSICAL PHARMACY, COSMETIC CHEMISTRY. Educ: Univ Tex, Austin, BS, 55. Prof Exp: Mfg pharmacist drug prod, 56-64, res & develop scientist drugs & cosmetics, 65-73, RES & DEVELOP MGR DRUGS & COSMETICS, TEX PHARMACAL CO, DIV WARNER LAMBERT, MORRIS PLAINS, NJ, 74- Mem: Am Pharm Asn. Res: Development of vehicles for organic peroxides for use in the treatment of dermatological disorders; screening alpha hydroxy acids for use in treatment of icthyosis. Mailing Add: 3314 Yorktown San Antonio TX 78230

WOLLERMANN, LOUIS ALBERT, food technology, see 12th edition

WOLLIN, GOESTA, b Ystad, Sweden, Oct 4, 22; nat US; m 50; c 1. OCEANOGRAPHY. Educ: Hermods Col, Sweden, Phil; Columbia Univ, MS, 53. Prof Exp: Newspaper reporter, Ystads Allehanda, 39-41; free lance writer, Swedish Newspapers, 45-50; asst, 50-56, RES CONSULT, LAMONT GEOL OBSERV, COLUMBIA UNIV, 56- Mem: AAAS; NY Acad Sci; Glaciol Soc. Res: Recent and pleistocene climates and marine sedimentation; micropaleontological research of marine sediments; the relationship between climatic changes and variations in the earth's magnetic field. Mailing Add: Snedens Landing Palisades NY 10964

WOLLMAN, HARRY, b Brooklyn, NY, Sept 26, 32; m 57; c 3. ANESTHESIOLOGY. Educ: Harvard Univ, AB, 54, MD, 58; Am Bd Anesthesiol, dipl, 64. Prof Exp: Intern med & surg, Univ Chicago Clins, 58-59; resident, Hosp Univ Pa, 59-63, assoc, Univ, 63-65, from asst prof to prof, 65-70, ROBERT DUNNING DRIPPS PROF ANESTHESIA & CHMN DEPT, SCH MED, UNIV PA, 72-, PROF PHARMACOL, 71- Concurrent Pos: NIH res trainee, 59-63; Pharmaceut Mfrs Asn fel, 60-61; consult, Vet Admin Hosp, Philadelphia, 63-64 & Valley Forge Army Hosp, 65-66; mem pharm & toxicol training grants comt, NIH, 66-68, mem anesthesia training grants comt, 71-73, mem surg A study sect, 74-; mem anesthesia drug panel drug efficacy study, Comt Anesthesia, Nat Acad Sci-Nat Res Coun, 70-71, mem comt adverse reactions to anesthesia drugs, 71-72; assoc ed, Anesthesiol, 70-75; prin investr, Anesthesia Res Ctr, Univ Pa, 72-, prog dir anesthesia res training grant, 72-, chmn comt studies involving human beings, 72-; John Harvard scholar. Honors & Awards: Detur Award. Mem: Am Physiol Soc; Sigma Xi; Asn Univ Anesthetists. Res: Circulatory physiology; cerebral blood flow and metabolism; regional blood flow during anesthesia. Mailing Add: Dept of Anesthesiol Hosp Univ of Pa Philadelphia PA 19104

WOLLMAN, SEYMOUR HORACE, b New York, NY, May 17, 15; m 44; c 3. PHYSIOLOGY. Educ: NY Univ, BS, 35, MS, 36; Duke Univ, PhD(physics), 41. Prof Exp: Physicist, Bur Ord, US Dept Navy, 42-46; BIOPHYSICIST, NAT CANCER INST, 48- Concurrent Pos: Nat Cancer Inst fel, Rockefeller Inst, 48. Mem: Am Physiol Soc; Am Thyroid Asn; Am Soc Cell Biol; Europ Thyroid Asn. Res: Thyroid gland and thyroid tumors. Mailing Add: Nat Cancer Inst Bethesda MD 20014

WOLLNER, THOMAS EDWARD, b Rochester, Minn, Dec 30, 36; m 58; c 2.

ORGANIC CHEMISTRY, POLYMER CHEMISTRY. Educ: St John's Univ, Minn, BA, 58; Wash State Univ, PhD(chem), 64. Prof Exp: Res mgr polymer res, 64-74, LAB MGR POLYMER RES, CENT RES LAB, 3M CO, 74- Mem: Am Chem Soc; Sigma Xi. Res: Pressure sensitive tapes; biosciences; heterocyclic chemistry involving sulfur & nitrogen. Mailing Add: 3M Co Cent Res Lab PO Box 33221 St Paul MN 55133

WOLLRAB, JAMES EDWARD, physical chemistry, see 12th edition

WOLLSCHLAEGER, GERTRAUD, b Muenchen, Ger, Feb 28, 24; US citizen; m 48; c 3. MEDICINE, RADIOLOGY. Educ: Univ Munich, physicum, 54, MD, 57; State Univ NY, MD, 65. Prof Exp: Instr radiol, Albert Einstein Col Med, 61-64; from asst prof to assoc prof, Sch Med, Univ Mo-Columbia, 64-71; asst chief sect neuroradiol, William Beaumont Hosp, Royal Oak, Mich, 71-72; PROF RADIOL & NEURORADIOL, SCH MED, WAYNE STATE UNIV, 73- Concurrent Pos: Res fel radiol, Albert Einstein Col Med, Yeshiva Univ, 61-62, NIH spec fel neuroradiol, 62-64; res grant, Univ Mo-Columbia, 64-71; co-dir, NIH spec fel training prog, 67-71, consult, Crippled Children's Serv, 69-71. Mem: AAAS; Asn Univ Neuroradiol; Radiol Soc NAm; Inst Elec & Electronics Eng; AMA. Res: Postmortem cerebral angiography; cerebral microangiography; microtumor-circulation. Mailing Add: 5885 Wing Lake Rd Birmingham MI 48010

WOLLSCHLAEGER, PAUL BERNHARD, b Berlin, Ger, Feb 3, 20; US citizen; m 48; c 3. MEDICINE, RADIOLOGY. Educ: Free Univ Berlin, physicum; Univ Munich, MD, 55; State Univ NY, MD, 65. Prof Exp: Fel med, Univ Munich Clin, 55-57; instr radiol, Albert Einstein Col Med, 61-64; from asst prof to prof, Sch Med, Univ Mo-Columbia, 64-71; chief sect neuroradiol, William Beaumont Hosp, Royal Oak, Mich, 71-72; PROF RADIOL & NEURORADIOL, SCH MED, WAYNE STATE UNIV, 73- Concurrent Pos: NIH spec fel, Albert Einstein Col Med, Yeshiva Univ, 61-64; res grants, Univ Mo-Columbia, 64-71; dir NIH fel training prog, 67-71; consult, Crippled Children's Serv, Univ Mo, 69-71, consult adv comt for PhD biomed eng, 70-71. Mem: AAAS; AMA; Asn Univ Radiol; Radiol Soc NAm; Inst Elec & Electronics Eng (vpres, 70-71). Res: Correlation pre- and postmortem cerebral angiography; cerebral microangiography; sonography; thermography. Mailing Add: 5885 Wing Lake Rd Birmingham MI 48010

WOLLUM, ARTHUR GEORGE, II, b Chicago, Ill, July 26, 37; m 60; c 2. SOILS, MICROBIOLOGY. Educ: Univ Minn, BS, 59; Ore State Univ, MS, 62, PhD(soils), 65. Prof Exp: Forester, Gifford Pinchot Nat Forest, USDA, 59-60, res forester, Pac Northwest Forest & Range Exp Sta, 60-61; asst soils, Ore State Univ, 64-65, asst prof, 65-67; asst prof, NMex State Univ, 67-71; ASSOC PROF SOILS, NC STATE UNIV, 71- Mem: Soc Am Foresters; Soil Sci Soc Am; Am Soc Microbiol; Am Soc Agron. Res: Microbiology of nodule-formation of nonleguminous plants; ecology of nitrogen fixing plants in forested ecosystems; nitrogen cycle in forested ecosystems; microbiology of environmental pollution. Mailing Add: Dept of Soil Sci NC State Univ Raleigh NC 27607

WOLLWAGE, JOHN CARL, b Chicago, Ill, Oct 11, 14; m 39; c 3. PAPER CHEMISTRY. Educ: Northwestern Univ, BS, 34; Lawrence Col, MS, 36, PhD(paper chem), 38. Prof Exp: Mem staff, Res Dept, Hammermill Paper Co, Pa, 36; mem staff chem res, Beveridge-Marvellum Co, Mass, 37; res chemist, Kimberly-Clark Corp, 39-40, 42, tech supt, Mill, 41-42, war prod develop, 43-45, mill mgr, 45-49, asst tech dir, 49-52, dir res, 52-55, mgr forest opers, 55-59, gen mgr creped wadding mfg processes, Consumer Prod Div, 59-62, vpres mfg, 62-68, vpres, C W Mfg, Res & Eng, 68-71, vpres corp res & eng, 71-73; V PRES RES, INST PAPER CHEM, 73- Mem: Tech Asn Pulp & Paper Indust (pres, 63-64); Am Chem Soc; Can Pulp & Paper Asn; AAAS. Res: Alum and its effect on hydrogen ion concentration of paper; flocculation of paper making fibers. Mailing Add: Inst for Paper Chem PO Box 1039 Appleton WI 54911

WOLLWAGE, PAUL CARL, b Appleton, Wis, Mar 15, 41; m 65; c 2. CARBOHYDRATE CHEMISTRY, PULP CHEMISTRY. Educ: St Olaf Col, BA, 63; Inst Paper Chem, MS, 66, PhD(chem), 69. Prof Exp: SR RES CHEMIST, ST REGIS TECH CTR, 69- Mem: Am Chem Soc; Tech Asn Pulp & Paper Indust. Res: Pulping and bleaching processes; pulp characterization; wood chemicals; papermaking additives. Mailing Add: St Regis Tech Ctr West Nyack Rd West Nyack NY 10994

WOLMA, FRED J, b Albuquerque, NMex, Dec 10, 16; m 43; c 2. MEDICINE, SURGERY. Educ: Univ Tex, BA, 40, MD, 43; Am Bd Surg, dipl, 53. Prof Exp: Intern, Med Col Va, 43-44, resident surg, 47-48; res physician, St Mary's Infirmary, Galveston, Tex, 46-47; resident surg, 48-51, from instr to assoc prof, 51-69, chief div gen surg, 67-70, PROF SURG, UNIV TEX MED BR GALVESTON, 69- Concurrent Pos: Consult, USPHS Hosp, St Mary's Hosp, Galveston, Tex, Galveston County Hosp, La Marque & Clear Creek Hosp, Webster. Mem: AMA; Am Col Surg; Am Asn Surg of Trauma. Res: Clinical medicine; peripheral vascular surgery. Mailing Add: Dept of Surg Univ of Tex Med Br Galveston TX 77550

WOLMAN, ERIC, b New York, NY, Sept 25, 31; m 63; c 2. OPERATIONS RESEARCH, SYSTEMS ENGINEERING. Educ: Harvard Univ, AB, 53, MA, 54, PhD(appl math), 57. Prof Exp: Mem tech staff, 57-66, head traffic systs anal dept, 66-68, head traffic res dept, 68-72, HEAD NETWORK ENG DEPT, BELL TEL LABS, 72- Concurrent Pos: Vis lectr, Harvard Univ, 64; mem comt fire res, Nat Acad Sci-Nat Res Coun, 66-70; ad hoc eval panel for fire prog, Nat Bur Standards, 71-74, eval panel, Inst Appl Technol, Nat Bur Standards, 74- Mem: Fel AAAS; Am Math Soc; Inst Elec & Electronics Eng; Soc Indust & Appl Math; Opers Res Soc Am. Res: Applied probability; queuing theory; teletraffic theory. Mailing Add: Bell Tel Labs Everett Rd Holmdel NJ 07733

WOLMAN, MARKLEY GORDON, b Baltimore, Md, Aug 16, 24; m 51; c 4. GEOLOGY. Educ: Johns Hopkins Univ, BA, 49; Harvard Univ, MA, 51, PhD(geol), 53. Prof Exp: GEOLOGIST, US GEOL SURV, 51-; PROF GEOG & CHMN DEPT GEOG & ENVIRON ENG, JOHNS HOPKINS UNIV, 58- Honors & Awards: Asn Am Geog Award, 72. Mem: Geol Soc Am; Am Geophys Union; Asn Am Geog. Res: River morphology; water resources. Mailing Add: Dept of Geog & Environ Eng Johns Hopkins Univ Baltimore MD 21218

WOLMAN, SANDRA R, b New York, NY, Nov 23, 33; m 63; c 2. PATHOLOGY, CYTOGENETICS. Educ: Radcliffe Col, AB, 55; NY Univ, MD, 59. Prof Exp: Intern path, Bellevue Hosp, New York, 59-60, resident, 60-63; asst assoc pathologist, Morristown Mem Hosp, 64-66; asst pathologist, Monmouth Med Ctr, 66-67; ASST PROF CLIN PATH, SCH MED, NY UNIV, 67-, ASST PROF PATH, 72- Concurrent Pos: Teaching fel path, Sch Med, NY Univ, 62-64, Nat Cancer Inst res fel oncol, 63, Children's Cancer Res Found fel, 64; asst pathologist, Bellevue Hosp, 63-64, asst vis pathologist, 67-; consult pathologist, Morristown Mem Hosp, 66-; assoc pathologist, French & Polyclin Hosps, 70-71. Mem: AAAS; Am Soc Clin Path; Am Soc Human Genetics; Am Asn Cancer Res; Tissue Cult Asn. Res: Clinical cytogenetics; tumor cytogenetics; chemical and viral carcinogenesis in vivo and in vitro. Mailing Add: Dept of Path NY Univ Sch of Med New York NY 10016

WOLMAN, WALTER, chemistry, see 12th edition

WOLMAN, WILLIAM WOLFGANG, b Ger, Aug 5, 22; nat US; m 67; c 3. MATHEMATICAL STATISTICS. Educ: City Col New York, BBA, 46; Columbia Univ, AM, 49; Univ Rochester, PhD(math), 60. Prof Exp: Statistician, State Div Housing, NY, 47-50; head statist div, Off Naval Inspector Ord, Eastman Kodak Co, 50-55; head statist methodology & reliability, Qual Control Div, Bur Ord, US Dept Navy, 55-56, statistician consult, Bur Yards & Docks, 56-60; chief statistician, Off Reliability & Systs Anal, NASA, 60-66; CHIEF TRAFFIC SYSTS DIV, OFF RES & DEVELOP, FED HWY ADMIN, US DEPT TRANSP, DC, 66- Concurrent Pos: Prof lectr, George Washington Univ, 56- Mem: Am Statist Asn; Inst Math Statist; Int Asn Statist in Phys Sci. Res: Probability theory; modern statistical methodology for problems in physical sciences; reliability theory in space technology; transportation research. Mailing Add: Traffic Systs Div Off Res & Develop Fed Hwy Admin Washington DC 20591

WOLNY, FRIEDRICH FRANZ, b Troppau, Czechoslovakia, Aug 24, 31; US citizen; m 56; c 1. ORGANIC POLYMER CHEMISTRY. Educ: Munich Tech Univ, BS, 54, MS, 55. Prof Exp: Group leader resin res, Sued-West-Chemie, WGer, 56-58, asst res & develop dir, 59-64; chemist, 65-68, group leader, 69-72, MGR RESIN RES, SCHENECTADY CHEM, INC, NY, 73- Mem: Electrochem Soc WGer; Am Chem Soc; Soc Automotive Eng. Res: Organic polymer chemistry, poly condensation products, phenol formaldehyde resins, applied in friction materials. Mailing Add: 823 Dean St Schenectady NY 12309

WOLOCHOW, HYMAN, b Richdale, Can, Feb 10, 18; nat US; m 43; c 2. MICROBIOLOGY. Educ: Univ Alta, BSc, 38, MSc, 40; Univ Calif, PhD(microbiol), 50. Prof Exp: Asst dairy bact, Univ Alta, 38-41; asst dairy bact, Univ Calif, 47-48 & plant nutrit, 48-50, asst res bacteriologist, 50-59; assoc bacteriologist, 59-73, ASSOC RES BACTERIOLOGIST, NAVAL BIOL LAB, NAVAL SUPPLY CTR, UNIV CALIF, 73- Concurrent Pos: Consult, US Dept Defense. Res: Dairy bacteriology; enzyme chemistry; medical microbiology. Mailing Add: Naval Biol Lab Naval Supply Ctr Oakland CA 94625

WOLOCK, FRED WALTER, b Whitinsville, Mass, Mar 8, 26. MATHEMATICAL STATISTICS. Educ: Holy Cross Col, BS, 47; Cath Univ, MS, 49; Va Polytech Inst, PhD(math statist), 64. Prof Exp: Instr math, Lewis Col, 53-54; instr, Iona Col, 54-56; instr, St John's Univ, NY, 56-57; instr, Worcester Polytech Inst, 57-60; asst prof math & statist, Boston Col, 64-65; ASSOC PROF MATH & STATIST, SOUTHEASTERN MASS UNIV, 65- Concurrent Pos: Consult, Nat Ctr Air Pollution Control, USPHS, 65- & Berkshire Hathaway Co, 66- Mem: Inst Math Statist; Am Statist Asn; Math Asn Am. Res: Experimental design and applications of statistics to biological and medical sciences; industrial applications of statistics. Mailing Add: Dept of Math Southeastern Mass Univ North Dartmouth MA 02747

WOLONTIS, VIDAR MICHAEL, b Helsinki, Finland, May 28, 26; nat US; m 48; c 2. MATHEMATICS. Educ: Univ Helsinki, MA, 47; Harvard Univ, PhD(math), 49. Prof Exp: Asst prof math, Univ Kans, 49-53; mem staff, Data Systs Eng Div, 65-70, MEM STAFF, BELL LABS, INC, 53-, EXEC DIR OPERS RES, 70- Mem: Am Math Soc; Asn Comput Mach; sr mem Inst Elec & Electronics Engrs. Res: Functions of a complex variable; digital computers; data processing and communications. Mailing Add: Rm 2D-406 Bell Labs Inc Holmdale NJ 07733

WOLOSHIN, HENRY JACOB, b Philadelphia, Pa, Aug 23, 13; m 48; c 3. RADIOLOGY. Educ: Temple Univ, BS, 34, MD, 38; Am Bd Radiol, dipl, 47. Prof Exp: Instr, 47-48, assoc, 48-55, from asst prof to assoc prof, 55-65, PROF RADIOL, SCH MED, TEMPLE UNIV, 65-, ASSOC, UNIV HOSP, 47- Mem: AMA; Am Col Radiol. Mailing Add: Dept of Radiol Temple Univ Hosp Philadelphia PA 19140

WOLOWYK, MICHAEL WALTER, b Rain-Amlech, Ger, Oct 26, 42; Can citizen; m 66; c 2. PHARMACOLOGY, CELL BIOLOGY. Educ: Univ Alta, BSc, 65, PhD(pharmacol), 69. Prof Exp: Asst prof clin pharm, 71-75, PROF PHARM, UNIV ALTA, 75-, HON ASST PROF PHARMACOL, 71-, SCI RES ASSOC, HOSP, 73- Concurrent Pos: Med Res Coun fel, Wellcome Res Labs, Beckenham, Eng, 69-71. Mem: Pharmacol Soc Can. Res: Physiology, pharmacology and morphology of smooth muscle and red blood cells, with particular reference to active transport and exchange of cations. Mailing Add: Dept of Clin Pharm Univ of Alta Fac of Pharm Edmonton AB Can

WOLPE, JOSEPH, b Johannesburg, SAfrica, Apr 20, 15; US citizen; m 48; c 2. PSYCHIATRY, PSYCHOLOGY. Educ: Univ Witwatersrand, MB & BCh, 39, MD, 48. Prof Exp: Lectr psychiat, Univ Witwatersrand, 50-59; prof sch med, Univ Va, 60-65; PROF PSYCHIAT SCH MED, TEMPLE UNIV & EASTERN PA PSYCHIAT INST, 65- Concurrent Pos: Fel, Ctr Advan Study Behav Sci, Stanford Univ, 56-57; ed, J Behav Ther & Exp Psychiat, 70- Mem: Am Psychiat Asn; Am Psychol Asn; Am Psychopath Asn; Asn Advan Behav Ther. Mailing Add: Temple Univ Med Sch & Eastern Pa Psychiat Inst 3300 Henry Ave Philadelphia PA 19129

WOLPERT, ARTHUR, b New York, NY, Sept 18, 32; m 54; c 5. PSYCHIATRY, CHILD PSYCHIATRY. Educ: Columbia Univ, BS, 53; Univ Md, MS, 55, PhD(pharmacol), 59, MD, 61. Prof Exp: Asst dir, Sagamore Children's Ctr, Melville, NY, 69-72; RES PSYCHIATRIST, STATE HOSP, CENT ISLIP, 72- Concurrent Pos: Consult, Islip, WIslip, Brentwood, Patchogue & Port Jefferson Pub Schs, 70-, Suffolk Ctr Emotionally Disturbed Children, 71- & Brookhaven Nat Labs, 71-; asst prof dept psychiat, Med Sch, State Univ NY Stony Brook, 71- Mem: Fel Am Psychiat Asn; fel Am Acad Child Psychiat. Res: Child psychopharmacology and electroencephalography. Mailing Add: 20 Fourth Ave Bay Shore NY 11706

WOLPERT, STEPHEN MICHAEL, polymer physics, see 12th edition

WOLPERT-DEFILIPPES, MARY KATHERINE, b Sioux City, Iowa, Dec 13, 39; m 73. PHARMACOLOGY. Educ: Creighton Univ, BS, 63; Univ Mich, Ann Arbor, MS, 66, PhD(pharmacol), 69. Prof Exp: Fel pharmacol, Sch Med, Yale Univ, 69-71; staff fel, 71-75, SR STAFF FEL PHARMACOL, NAT CANCER INST, NIH, 75- Res: Biochemical pharmacology, particularly in cancer chemotherapy; mechanisms of drug resistance, metabolism of anticancer agents and mitotic inhibitors as anticancer agents. Mailing Add: Drug Eval Br Nat Cancer Inst NIH Blair Bldg Rm 532 8300 Colesville Rd Silver Spring MD 20910

WOLPOFF, MILFORD HOWELL, b Chicago, Ill, Oct 28, 42; m 66. PHYSICAL ANTHROPOLOGY. Educ: Univ Ill, Urbana, AB, 64, PhD(anthrop), 69. Prof Exp: Asst prof anthrop, Case Western Reserve Univ, 68-70; ASSOC PROF ANTHROP, UNIV MICH, ANN ARBOR, 70- Concurrent Pos: NSF res grant, Transvaal Mus, Nat Mus Kenya, 72; Am Philos Soc equip grant, 72-73. Mem: Fel Am Asn Phys Anthrop; fel Am Anthrop Asn. Res: Paleoanthropology; human origins and evolution; evolution theory; biomechanics; computer analysis; dental variation. Mailing Add: Dept of Anthrop Univ of Mich Ann Arbor MI 48104

WOLSEY, WAYNE C, b Battle Creek, Mich, Nov 12, 36; m 65; c 2. INORGANIC CHEMISTRY. Educ: Mich State Univ, BS, 58; Univ Kans, PhD(chem), 62. Prof Exp: Sr res chemist chem div, Pittsburgh Plate Glass Co, Ohio, 62-65; asst prof chem, 65-72, ASSOC PROF CHEM, MACALESTER COL, 72- Concurrent Pos: Vis asst prof, Ariz State Univ, 71-72. Mem: AAAS; Am Chem Soc. Res: Coordination compounds; chemistry of chlorine and nitrogen compounds. Mailing Add: Dept of Chem Macalester Col St Paul MN 55105

WOLSKY, ALAN MARTIN, b Brooklyn, NY, May 17, 43; m 69. PHYSICS. Educ: Columbia Col, AB, 64; Univ Pa, MS, 65, PhD(physics), 69. Prof Exp: Nat Res Coun fel math physics, Courant Inst Math Sci, NY Univ, 70, vis mem, 71; ASST PROF PHYSICS, TEMPLE UNIV, 71- Mem: Am Phys Soc. Res: Elementary particle theory with emphasis on weak interactions and field theory. Mailing Add: Dept of Physics Temple Univ Philadelphia PA 19122

WOLSKY, ALEXANDER ALBERT, b Budapest, Hungary, Aug 12, 02; nat US; m 40; c 2. DEVELOPMENTAL BIOLOGY. Educ: Eötvös Lorand Univ, Budapest, PhD(zool), 28. Prof Exp: Asst lectr, Eötvös Lorand Univ, Budapest, 25-29; res asst, Hungarian Biol Res Inst, 29-34, res assoc, 34-39, actg dir, 39-45; prof zool & chmn dept, Eötvös Lorand Univ, Budapest, 45-49; prin sci officer, UNESCO, 48-54; prof exp embryol, Fordham Univ, 54-66; prof biol & chmn dept sci, 66-72, EMER PROF BIOL, MARYMOUNT COL, NY, 72-; ADJ PROF RADIATION BIOL, NY UNIV MED SCH, 73- Concurrent Pos: Rockefeller fel, 35-36; ed in chief, Monographs Develop Biol, 68 & Exp Biol & Med, 75. Mem: Am Soc Zool; Int Soc Develop Biol; fel Zool Soc India; Am Teilhard de Chardin Asn. Res: Physiology and genetics of development, especially the eye; physiology of regeneration; effects of morphostatic substances, radiations and ultrasound. Mailing Add: Dept of Radiol NY Univ Med Ctr 550 First Ave New York NY 10016

WOLSKY, MARIA DE ISSEKUTZ, b Kolozsvar, Romania, June 17, 16; nat US; m 40; c 2. CELL BIOLOGY. Educ: Med Univ, Budapest, MD, 43. Prof Exp: Asst pharmacol, Univ Budapest, 38-39; res assoc, Hungarian Biol Res Inst, 40-45; from instr to assoc prof, 56-67, PROF BIOL, MANHATTANVILLE COL, 67- Concurrent Pos: Res assoc, Fordham Univ, 58-62. Mem: AAAS. Res: Cell physiology; factors controlling cell division in developing systems; tissue homotransplantation; cell differentiation. Mailing Add: Dept of Biol Manhattanville Col Purchase NY 10577

WOLSKY, SUMNER PAUL, b Boston, Mass, Aug 21, 26; m 50; c 2. PHYSICAL CHEMISTRY. Educ: Northeastern Univ, BS, 47; Boston Univ, MA, 49, PhD(chem), 52. Prof Exp: Mem res staff, Raytheon Co, 52-61; dir lab phys sci, 61-74, DIR RES & DEVELOP LAB PHYS SCI, P R MALLORY & CO, INC, 74- Mem: AAAS; Am Chem Soc; Am Phys Soc; Int Elec & Electronics Eng; Electrochem Soc. Res: Batteries; semiconductors; surface physics; physical electronics; sputtering; thin films; vacuum microbalances. Mailing Add: P R Mallory & Co Inc Northwest Indust Park Third Ave Burlington MA 01803

WOLSSON, KENNETH, b Paterson, NJ, Oct 12, 33. MATHEMATICS. Educ: Brooklyn Col, BS, 54; Columbia Univ, AM, 55; NY Univ, PhD(math), 62. Prof Exp: Prin res mathematician, Repub Aviation Corp, 62-63, sr res mathematician, 63-64; ASST PROF MATH, FAIRLEIGH DICKINSON UNIV, TEANECK, 64- Mem: Am Math Soc; Math Asn Am. Res: Partial and ordinary differential equations. Mailing Add: Dept of Math 1000 River Rd Fairleigh Dickinson Univ Teaneck NJ 07666

WOLSTENCROFT, RAMON DAVID, b Chelmsford, Eng, July 25, 36; m 63; c 1. ASTROPHYSICS. Educ: Univ Col, London, BSc, 59; Cambridge Univ, PhD(astrophys), 62. Prof Exp: Jr astronr, Kitt Peak Nat Observ, 62-65; sr sci officer astron & astrophys, Royal Observ, Edinburgh, Scotland, 65-67; lectr, Univ Edinburgh, 67-68; assoc prof physics & astrophys, 68-73, PROF PHYSICS & ASTRON, UNIV HAWAII, 73- Concurrent Pos: Fulbright scholar, 62-65; mem, UK Sci Res Coun Working Group on Moon, Planets & Interplanetary Matter, 66-68. Mem: AAAS; Am Astron Soc; Royal Astron Soc; Int Astron Union. Res: Linear and circular polarimetry of zodiacal light, planets, stars, galaxies; properties of interplanetary, interstellar dust; stellar magnetic fields; continuum radiation from galactic nuclei and quasars; peculiar galaxies. Mailing Add: Inst for Astron Univ of Hawaii 2680 Woodlawn Dr Honolulu HI 96822

WOLSTENHOLME, DAVID ROBERT, b Bury, Eng, Nov 5, 37; m 63. MOLECULAR BIOLOGY, CELL BIOLOGY. Educ: Univ Sheffield, BSc, 58, PhD(genetics), 61. Prof Exp: Fel zool, Univ Wis, 61-62, res assoc, 62-63; vis lectr genetics, Univ Groningen, 63-64; res fel biol, Beermann Div, Max Planck Inst Biol, 64-67; res assoc, Whitman Lab, Univ Chicago, 67-68; assoc prof, Kans State Univ, 68-70; assoc prof, 70-72, PROF BIOL, UNIV UTAH, 72- Concurrent Pos: Res career develop award, Nat Inst Gen Med Sci, Univ Utah, 72-76. Mem: Brit Genetical Soc; Am Soc Cell Biol; Genetics Soc Am. Res: Structure of chromosomes; form and structure of extranuclear DNA. Mailing Add: Dept of Biol Univ of Utah Salt Lake City UT 84112

WOLSTENHOLME, WILLIAM ERNEST, b Paterson, NJ, Nov 5, 11; m 45. MATHEMATICS. Educ: NY Univ, BEE, 40, MS, 50. Prof Exp: Res physicist res ctr, Uniroyal Inc, NJ, 46-59, sr res physicist, 60-68; consult, Wayne Tech Serv Corp, 68-70; PVT CONSULT & TEACHER, 70- Concurrent Pos: Prin investr, US Army Contract, 64-67 & US Air Force, 66-67. Mem: Am Phys Soc; Soc Rheol; Am Chem Soc; Am Soc Test & Mat; Inst Elec & Electronics Eng. Res: High polymer physics; rheological behavior of high polymers; thermodynamical properties; correlation of mechanical with molecular properties; high speed testing of thermoplastics; stress failure of metals and plastics. Mailing Add: 953 Berdan Ave Wayne NJ 07470

WOLSZON, JOHN DONALD, b Chicago, Ill, Jan 27, 29; m 53; c 6. ANALYTICAL CHEMISTRY. Educ: Univ Ill, BS, 51; Pa State Univ, PhD(anal chem), 55. Prof Exp: Instr chem, Marshall Univ, 55-58; asst prof, Univ Mo, 58-63; ASSOC PROF CHEM, PURDUE UNIV, 63- Mem: Am Chem Soc; Water Pollution Control Fedn; Am Water Works Asn. Res: Methods of chemical analysis; water and waste water chemistry. Mailing Add: Sch of Civil Eng Purdue Univ Lafayette IN 47907

WOLTEN, GERARD MARTIN, b Berlin, Ger, Nov 7, 20; nat US; m 42; c 1. PHYSICAL CHEMISTRY, FORENSIC SCIENCE. Educ: NY Univ, BA, 42, MS, 44, PhD(phys chem), 48. Prof Exp: Dir res develop, Tempil Corp, 44-56; sr tech specialist, Atomics Int Div NAm Aviation, Inc, 56-61; staff scientist, 61-74, MGR FORENSIC SCI SECT, AEROSPACE CORP, 74- Mem: Am Chem Soc; Am Crystallog Asn; Am Phys Soc; Am Acad Forensic Sci. Res: Phase equilibria and transformations; crystallography; high temperature x-ray diffraction; machine computations; inorganic structures. Mailing Add: 22926 Cass Ave Woodland Hills CA 91364

WOLTER, FREDERICK JOHN, b Ohiowa, Nebr, Apr 5, 16; m 43; c 3. CHEMISTRY. Educ: Nebr State Teachers Col, AB, 38; Iowa State Col, PhD(phys chem), 46. Prof Exp: Res assoc Manhattan proj, Iowa State Col, 42-46; res chemist indust chem dept, E I du Pont de Nemours & Co, Inc, 46-53, res chemist sales develop, 53-62, res chemist mkt res, 62-75; RETIRED. Mem: Am Chem Soc; Chem Mkt Res Asn. Res: General industrial chemical markets and technology. Mailing Add: Indust Chem Dept E I du Pont de Nemours & Co Inc Wilmington DE 19898

WOLTER, GERHARD HERMAN, b Ger, July 26, 08; nat US; m 59; c 2. CHEMICAL PHYSICS. Educ: Friedrichs Wilhelm Univ, Berlin, BS, 31, MS, 32. Prof Exp: Jr res physicist, Osram-Studien-Gesellschaft, Ger, 36-38; res physicist, Tel & Tel, Osram, 38-43; sr res physicist, Bernhard Berghaus Lab, 43-45; instr, Teacher Training Prog Sch Admin, Berlin, 46-50; dir res, Kahl Sci Instrument Corp, Calif. 50-57; from asst prof to assoc prof physics, 57-67, prof, 67-75, EMER PROF PHYSICS, SAN DIEGO STATE UNIV, 75- Concurrent Pos: Consult, Narmco Industs, Inc Div, Telecomput Corp, 59-62; res fel, San Diego Biomed Res Inst, 61-, pres, 70-71; consult, Physics, Appl Physics & Biophys, 62- Mem: Am Asn Physics Teachers; NY Acad Sci. Res: Spectroscopy; high temperature and plasma physics; biochemistry; biophysics. Mailing Add: Dept of Physics San Diego State Univ San Diego CA 92182

WOLTER, J REIMER, b Halstenbek, Ger, May 9, 24; US citizen; m 52; c 4. OPHTHALMOLOGY, PATHOLOGY. Educ: Univ Hamburg, MD, 49. Prof Exp: Intern med, Med Sch, Univ Hamburg, 49-50, resident instr ophthal, 50-53; res assoc path, 53-56, from asst prof to assoc prof ophthal, 56-64, PROF OPHTHAL, MED SCH, UNIV MICH, ANN ARBOR, 64- Concurrent Pos: Chief ophthal serv, Vet Admin Hosp, Ann Arbor, 62-; ed, J Pediat Ophthal, 67- Mem: AMA; Am Ophthal Soc; Am Acad Ophthal & Otolaryngol; Asn Res Ophthal; Ger Ophthal Soc. Res: Clinical ophthalmology; opthalmic pathology. Mailing Add: Dept of Ophthal Univ of Mich Med Ctr Ann Arbor MI 48104

WOLTER, KARL ERICH, b New York, NY, Nov 8, 30. PLANT PHYSIOLOGY. Educ: State Univ NY Col Forestry, Syracuse, BS, 58; Univ Wis, PhD(plant physiol), 64. Prof Exp: RES PLANT PHYSIOLOGIST, FOREST PROD LAB, FOREST SERV, USDA, 63- Concurrent Pos: Vis prof, Iowa State Univ, 71-72. Mem: Am Soc Plant Physiol; Scand Soc Plant Physiol; Japanese Soc Plant Physiol. Res: Growth, differentiation and nutrition of plants, specifically tree species; host-pathogen interactions. Mailing Add: Forest Prod Lab USDA Madison WI 53705

WOLTERINK, LESTER FLOYD, b Marion, NY, July 28, 15; m 38; c 2. BIOPHYSICS. Educ: Hope Col, AB, 36; Univ Minn, MA, 40, PhD(zool), 43. Prof Exp: Lab asst, Hope Col, 34-36; asst, Univ Minn, 36-41; instr physiol, 41-45, from asst prof to assoc prof, 45-52, PROF PHYSIOL, MICH STATE UNIV, 52-, ASST EXP STA, 45- Concurrent Pos: Assoc physiologist, Argonne Nat Lab, Ill, 48; proj scientist biosatellite proj, Ames Res Ctr, NASA, 65-66; mem subcomt nitrogen oxides, Nat Acad Sci-Nat Res Coun, 73-75. Mem: Am Soc Zool; Biophys Soc; Am Physiol Soc; Radiation Res Soc; Brit Biol Eng Soc. Res: Biological rhythms and oscillatory time series. Mailing Add: Dept of Physiol Mich State Univ East Lansing MI 48824

WOLTERMANN, GERALD M, b Bellevue, Ky, Nov 19, 47. INORGANIC CHEMISTRY. Educ: Thomas More Col, AB, 69; Univ Ky, PhD(inorg chem), 73. Prof Exp: Res fel bioinorg chem, Univ Ill, 73-75; RES CHEMIST, ENGELHARD MINERALS & CHEM CORP, 75- Concurrent Pos: Res fel, Univ Ky, 73. Mem: Am Chem Soc. Res: Catalysis; especially metal and zeolite catalysis; enzymatic systems especially model systems involving coordination complexes; electron spin resonance as applied to inorganic coordination crystals. Mailing Add: Engelhard Minerals & Chem Corp Menlo Park Edison NJ 07068

WOLTERS, ROBERT JOHN, b St Louis, Mo, Nov 7, 40; m 70; c 1. PHARMACEUTICAL CHEMISTRY, PHARMACOLOGY. Educ: St Louis Col Pharm, BS, 65; NDak State Univ, MS, 68, PhD(pharmaceut chem), 71. Prof Exp: REV CHEMIST, US FOOD & DRUG ADMIN, ROCKVILLE, 71- Mem: Am Chem Soc; Am Pharmaceut Asn; Sigma Xi. Res: Synthesis of potential pharmaceutically active compounds; mescaline analogs. Mailing Add: 14427 Gunstock Ct Silver Spring MD 20906

WOLTERSDORF, OTTO WILLIAM, JR, b Philadelphia, Pa, June 19, 35; m 57; c 2. MEDICINAL CHEMISTRY. Educ: Gettysburg Col, AB, 56; Pa State Univ, MS, 59. Prof Exp: Res assoc org synthesis, 59-65, from res chemist to sr res chemist, 65-73, RES FEL, MERCK SHARP & DOHME RES LABS, 73- Mem: Am Chem Soc. Res: Organic synthesis; radioisotope synthesis. Mailing Add: 200 Dorset Way Chalfont PA 18914

WOLTHUIS, ENNO, chemistry, see 12th edition

WOLTHUIS, ROGER A, b Champaign, Ill, Mar 3, 37; m 65; c 2. CARDIOVASCULAR PHYSIOLOGY. Educ: Univ Mich, BA, 63; Mich State Univ, MS, 65, PhD(physiol), 68. Prof Exp: Res scientist, Technol, Inc, NASA, 68-71, proj leader, Cardiovasc Res Lab, Johnson Space Ctr, Technol, Inc, NASA, 71-74; RES SCIENTIST INTERNAL MED, SCH AEROSPACE MED, BROOKS AFB, 74- Concurrent Pos: Partic, Apollo & Skylab Med Experiments. Mem: AAAS; Aerospace Med Asn; Am Physiol Soc; Inst Elec & Electronics Eng. Res: Studies on man's physiological adaptation to normal and zero gravity environment as measured by orthostatic stress techniques; studies on stress testing and hypertension. Mailing Add: Dept of Internal Med/NGI Sch of Aerospace Med Brooks AFB TX 78235

WOLTJER, LODEWYK, b Holland, Apr 26, 30; m; c 1. ASTROPHYSICS. Educ: Univ Leiden, PhD(astron), 57. Prof Exp: Res assoc, Yerkes Observ, Univ Chicago, 57-58 & Fermi Inst Nuclear Studies, 60; mem, Inst Advan Study, 59; lectr, Univ Leiden, 59-60; mem, Inst Advan Study, 61; prof theoret astrophys & plasma physics, Univ Leiden, 61-64; RUTHERFURD PROF ASTRON & CHMN DEPT, COLUMBIA UNIV, 63- Concurrent Pos: Vis prof, Mass Inst Technol, 62-63 & Univ Md, 63. Mem: Int Astron Union; Am Astron Soc. Res: Galactic dynamics; radio astronomy. Mailing Add: Dept of Astron 109 Mem Libr Columbia Univ New York NY 10027

WOLTZ, FRANK EARL, b Bethlehem, Pa, Nov 29, 16; m 47; c 2. PHYSICAL CHEMISTRY. Educ: Bethany Col, WVa, BS, 38; Univ WVa, MS, 40, PhD(chem), 43; Ohio Univ, MS, 70. Prof Exp: Mat engr, Westinghouse Elec Corp, Pa, 42-44; lab mgr, Goodyear Synthetic Rubber Corp, Ohio, 44-47; res chemist, Goodyear Tire & Rubber Co, 47-50, rubber compounder, 50-53, supvr opers anal, Goodyear Atomic Corp, Piketon, 53-67, SUPT ENG DEVELOP, GOODYEAR ATOMIC CORP, PIKETON, 67- Concurrent Pos: Consult radiation adv bd, Ohio Dept Health, 74- Mem: Am Nuclear Soc. Res: Electrical insulating varnishes; analytical test methods; rubber manufacture; inert electrode systems; rubber compounding for use in tire manufacturing; fluid flow; material and energy optimization of gaseous diffusion processes; computer control of chemical processes; nuclear criticality safety. Mailing Add: 400 E Third St Waverly OH 45690

WOLTZ, SHREVE SIMPSON, b Clifton, Va, Apr 9, 24; m 47; c 2. HORTICULTURE. Educ: Va Polytech Inst, BS, 43; Rutgers Univ, PhD(soils, plant physiol), 51. Prof Exp: Dir fertilizer res, Baugh & Sons Co, 51-53; asst horticulturist, 53-62, assoc plant physiologist, 62-68, PLANT PHYSIOLOGIST, GULF COAST EXP STA, UNIV FLA, 68- Mem: Am Soc Hort Sci; Am Soc Plant Physiol; Am

Phytopath Soc; Scand Soc Plant Physiol. Res: Plant nutrition; gladiolus and chrysanthemum culture; soil fertility; physiology of disease. Mailing Add: Agr Res & Educ Ctr Bradenton FL 33505

WOLTZ, WILLIE GARLAND, b Macclesfield, NC, Oct 12, 11; m 40; c 1. AGRONOMY, SOIL FERTILITY. Educ: NC State Col, BS, 39; Cornell Univ, PhD(soil chem), 48. Prof Exp: Assoc soil technologist, Va Truck Exp Sta, 43-46; assoc prof agron, 46-51, PROF AGRON, NC STATE UNIV, 51- Concurrent Pos: Consult, R J Reynolds Tobacco Co. Honors & Awards: Dedicatee, Tobacco Sci, 72; Tobacco Man of Year Award, Tobacco Int, 71. Mem: Am Soc Agron; Soil Sci Soc Am; Tobacco Res Chemists Conf. Res: Soil fertility and mineral nutrition investigations with flue-cured tobacco. Mailing Add: 105 Country Club Dr Oxford NC 27565

WOLVEN, ANNE M, b New York, NY, Feb 2, 25. TOXICOLOGY. Educ: Hunter Col, BA, 45. Prof Exp: Group leader pharmacol, Schering Corp, NJ, 45-47; group leader, Leberco Labs, NJ, 47-52, from asst dir to assoc dir labs, 52-72; mgr toxicol, Alza Res Corp, 72-75; SR TOXICOLOGIST, SHELL CHEM CO, 75- Concurrent Pos: Round Table discussant, Gordon Conf Toxicol & Safety Eval, 70, chairperson, 76; lectr, Ctr Prof Advan, 72; consult toxicol, Nat Inst Drug Abuse, 74-75 & Ministry Health, Mex, 75; mem hazardous mat adv comt, Environ Protection Agency, 75-77. Mem: AAAS; Soc Toxicol; Soc Cosmetic Chem; Am Soc Microbiol. Res: Eye and skin irritation and absorption phenomena; pesticide toxicology. Mailing Add: Shell Chem Co 2401 Crow Canyon Rd San Ramon CA 94583

WOLVERTON, BILLY CHARLES, b Scott Co, Miss, Oct 13, 32; m 55; c 1. CHEMISTRY. Educ: Miss Col, BS, 60. Prof Exp: Res asst, Med Ctr, Univ Miss, 60-63; res chemist, US Naval Weapons Lab, 63-65; br chief res chemist, Air Force Armament Lab, 65-71; ENVIRON SCIENTIST, NAT SPACE TECHNOL LAB, BAY ST LOUIS, MISS, 71- Concurrent Pos: Mem Panel Unconventional Approaches to Aquatic Weed Control & Utilization, Nat Acad Sci, 75- Honors & Awards: Super Sci Achievement Award, Dept Navy, 65; Sci Achievement Award, Dept Air Force, 69, Spec Sci Achievement Award, 71; Sci Technol Utilization Award, Am Inst Aeronaut & Astronaut, 70; Except Sci Serv Medal, NASA, 75. Res: Vascular, aquatic plants plants as biological filtration systems for removing domestic and industrial pollutants from wastewater and the utilization of harvested plant material as renewable sources of feed, fertilizer and methane. Mailing Add: 1129 Parkwood Circle Picayune MS 39466

WOMACK, FRANCES C, b Owensboro, Ky, Mar 23, 31; m 53; c 2. GENETICS, ENZYMOLOGY. Educ: Vanderbilt Univ, BA, 52, MA, 55, PhD(biol), 62. Prof Exp: Asst prof genetics, 62-63 & 64-65, res assoc, 63-64, res assoc enzymol, 65-72, ASST PROF ENZYMOL, SCH MED, VANDERBILT UNIV, 72- Mem: Genetics Soc Am. Res: Protein structure, function and their relation to genetic information; bacteriophage genetics. Mailing Add: Dept of Microbiol Vanderbilt Univ Sch of Med Nashville TN 37203

WOMACK, HERBERT, b Mantee, Miss, May 25, 29; m 57; c 3. ENTOMOLOGY. Educ: Miss State Col, BS, 55, MS, 57. Prof Exp: Res entomologist stored prod insects br, Mkt Res & Develop, Agr Res Serv, USDA, 57-66; EXTEN ENTOMOLOGIST, UNIV GA, 66- Mem: Entom Soc Am. Res: Insects attacking stored products; cotton insects. Mailing Add: Ga Coop Exten Serv PO Box 1209 Tifton GA 31794

WOMACK, JAMES E, b Anson, Tex, Mar 30, 41; m 63; c 2. GENETICS. Educ: Abilene Christian Col, BS, 64; Ore State Univ, PhD(genetics), 68. Prof Exp: From asst prof to assoc prof biol, Abilene Christian Col, 68-73; vis scientist, 73-75, STAFF SCIENTIST, JACKSON LAB, 75- Concurrent Pos: Geneticist, Maine Genetics Counseling Ctr, 73- Mem: AAAS; Genetics Soc Am; Am Genetics Asn; Am Soc Human Genetics. Res: Radiation genetics of mammals; radiation induced isozyme mutants; mammalian developmental genetics; comparative mammalian genetics; genetics of lysosomal enzymes. Mailing Add: Jackson Lab Bar Harbor ME 04609

WOMACK, MADELYN, b Waco, Tex, Mar 13, 10. NUTRITION. Educ: Tex State Col Women, BS, 31; Univ Ill, MS, 33, PhD(biochem), 35. Prof Exp: From asst to res assoc biochem, Univ Ill, 35-47; RES CHEMIST NUTRIT INST, AGR RES SERV, USDA, 47- Mem: AAAS; Am Chem Soc; Soc Exp Biol & Med; Am Inst Nutrit. Res: Effects of interrelationships among nutrients on the nutritive value of amino acids and proteins. Mailing Add: Nutrit Inst Agr Res Serv USDA Beltsville MD 20705

WOMACK, NATHAN ANTHONY, surgery, deceased

WOMBLE, EUGENE WILSON, b High Point, NC, June 27, 31; m 59; c 4. MATHEMATICS. Educ: Wofford Col, BS, 52; Univ NC, Chapel Hill, MA, 59; Univ Okla, PhD(math), 70. Prof Exp: Teacher, Kernersville High Sch, 56-58; instr math, Wake Forest Col, 59-61; asst prof, Pfeiffer Col, 61-66; spec instr, Univ Okla, 69-70; Charles A Dana prof, 72, PROF MATH, PRESBY COL, SC, 70- Mem: Math Asn Am; Nat Coun Teachers Math. Res: Foundations of convexity; convexity structures. Mailing Add: Dept of Math Presby Col Clinton SC 29325

WOMER, WALTER DALE, b Salt Lake City, Utah, Dec 11, 31; m 55; c 2. ORGANIC CHEMISTRY. Educ: Univ Omaha, BA, 53; Univ Iowa, MS, 55, PhD(org chem), 57. Prof Exp: Off Army Ord res grant, 54-57; res chemist, E I du Pont de Nemours & Co, Inc, 57-64; sr chemist, Cent Res Labs, Minn Mining & Mfg Co, 64-67; SR ENGR, PHILLIPS FIBERS CORP, 67- Mem: Am Chem Soc. Res: Textile fibers; polymer chemistry; synthetic resins and coatings; textile engineering. Mailing Add: Phillips Fiber Corp Tech Ctr PO Box 66 Greenville SC 29605

WOMMACK, JOEL BENJAMIN, JR, b Benton, Ky, Dec 5, 42; m 67; c 2. AGRICULTURAL CHEMISTRY. Educ: David Lipscomb Col, BS, 64; Vanderbilt Univ, PhD(org chem), 68. Prof Exp: Res chemist, 68-74, RES SUPVR, BIOCHEM DEPT, EXP STA, E I DU PONT DE NEMOURS & CO, 74- Mem: Am Chem Soc. Mailing Add: 1007 Jeffrey Rd Wilmington DE 19810

WONDERLING, THOMAS FRANKLIN, b Utica, Ohio, Feb 4, 15; m 41; c 2. AGRICULTURE. Educ: Ohio State Univ, BS, 39. Prof Exp: Teacher bd educ, Ohio, 39-45; farm mgr, Tiffin State Hosp, State Dept Pub Welfare, Ohio, 45-46, farm mgr, Lima State Hosp, 46-48; supt outlying farms, Ohio Agr Exp Sta, 48-63; coordr res opers, Ohio Agr Res & Develop Ctr, 63-69, coordr res opers & phys plant, 69-74; COORDR RES OPERS & SUPVR N APPALACHIAN WATERSHED EXP STA, 74- Mailing Add: NAppalachian Watershed Exp Sta Coshocton OH 43812

WONDERS, WILLIAM CLARE, b Toronto, Ont, Apr 22, 24; m 51; c 3. GEOGRAPHY, GEOGRAPHY OF NORTHERN LANDS. Educ: Univ Toronto, BA, 46, PhD(geog), 51; Syracuse Univ, MA, 48. Prof Exp: Teaching asst geog, Syracuse Univ, 46-48; lectr, Univ Toronto, 48-53; from asst prof to assoc prof, 53-57, head dept, 57-67, PROF GEOG, UNIV ALTA, 57- Concurrent Pos: Mem Can nat comt, Int Geog Union, 53-61 & 64-72; guest prof, Uppsala Univ, 62-63; Can Coun sr res fel, Uppsala Univ, 62-63; NSF sr foreign scientist fel & vis prof, Univ Okla, 65-66; mem, Nat Adv Comt Geog Res, 65-69; res fel & Can Coun res grant, Univ Aberdeen, 70-71. Mem: Can Asn Geog (pres, 61-62); Asn Am Geog; fel Arctic Inst NAm; Sigma Xi. Res: Settlement geography, including historical aspects; Canada and western United States; northern lands, especially North America and Scandinavia; northern Scotland; historical geography of settlement by Scandinavians in central Alberta; significance of forestry in settlement in highland Scotland. Mailing Add: Dept of Geog Univ of Alta Edmonton AB Can

WONENBURGER, MARIA JOSEFA, b La Coruna, Spain, July 19, 27. MATHEMATICS. Educ: Univ Madrid, Lic math, 50, Dr(math), 60; Yale Univ, PhD(math), 57. Prof Exp: Nat Res Coun Can fel, Queen's Univ, Ont, 60-62; from asst prof to assoc prof math, Toronto Univ, 62-66; prof, State Univ NY, Buffalo, 66-67; PROF MATH, IND UNIV, BLOOMINGTON, 67- Res: Algebra. Mailing Add: Dept of Math Ind Univ Bloomington IN 47401

WONES, DAVID R, b San Francisco, Calif, July 13, 32; m 58; c 4. PETROLOGY, GEOCHEMISTRY. Educ: Mass Inst Technol, SB, 54, PhD(geol), 60. Prof Exp: Bush fel, Carnegie Inst, 57-59; geologist, US Geol Surv, 59-67; assoc prof geol, Mass Inst Technol, 66-71; chief br exp geochem & mineral, 72-76, GEOLOGIST, US GEOL SURV, 71- Mem: Mineral Soc Am; Geochem Soc; Am Geophys Union; Geol Soc Am; Brit Mineral Soc. Res: Mineral synthesis; phase equilibria of systems applicable to rocks and minerals; studies of topical field problems. Mailing Add: Stop 959 US Geol Surv Reston VA 22092

WONG, ALAN YAU KUEN, b Hong Kong, Feb 6, 37; Can citizen; m 67; c 2. PHYSICS, BIOPHYSICS. Educ: Dalhousie Univ, BSc, 62, MSc, 63, PhD(biophys), 67. Prof Exp: Res assoc comput sci, 66-68, lectr biophys, 68-71, ASST PROF BIOPHYS, DALHOUSIE UNIV, 71- Concurrent Pos: Can Heart Found fel biophys & bioeng res lab, Dalhousie Univ, 68-71, Med Res Coun Can res scholar, 71- Mem: Med & Biol Eng Soc Can; Int Orgn Med Physics; Soc Math Biol. Res: Application of sliding filament theory to the mechanics of cardiac muscle and left ventricle; excitation-contraction coupling of cardiac muscle. Mailing Add: Biophysics & Bioeng Res Lab Dalhousie Univ Halifax NS Can

WONG, ALFRED, b Macao, Portugal, Feb 4, 37; m 65. PLASMA PHYSICS. Educ: Univ Toronto, BASc, 58, MA, 59; Univ Ill, MSc, 61; Princeton Univ, PhD(plasma physics), 63. Prof Exp: Res assoc plasma physics lab, Princeton Univ, 62-64; from asst prof to assoc prof physics, 64-72, PROF PHYSICS, UNIV CALIF, LOS ANGELES, 72- Concurrent Pos: Sloan res fel, 66-68. Mem: Am Phys Soc. Res: Waves and radiation from plasmas; nonlinear phenomena. Mailing Add: Dept of Physics Univ of Calif Los Angeles CA 90024

WONG, BING KUEN, b Shanghai, China, Oct 4, 38; m 66; c 2. MATHEMATICAL ANALYSIS. Educ: Kans State Col, Pittsburg, AB, 61; Univ Ill, MA, 63, PhD(math), 66. Prof Exp: Asst prof math, Univ Western Ill, 65-66; asst prof, Rochester Inst Technol, 66-68; PROF MATH & CHMN DEPT, WILKES COL, 68- Mem: Am Math Soc; Math Asn Am. Res: Analysis. Mailing Add: Dept of Math Wilkes Col Wilkes-Barre PA 18703

WONG, CHAK-KUEN, b Macao, China; m 70. COMPUTER SCIENCE, APPLIED MATHEMATICS. Educ: Univ Hong Kong, BA, 65; Columbia Univ, MA, 66, PhD(math), 70. Prof Exp: RES STAFF MEM COMPUT SCI, T J WATSON RES CTR, IBM CORP, 69- Concurrent Pos: Vis assoc prof, Univ Ill, Urbana, 72-73. Honors & Awards: Outstanding Invention Award, IBM Corp, 71. Mem: Am Math Soc; Inst Elec & Electronics Eng; Asn Comput Mach. Res: Mathematical analysis; discrete and combinatorial mathematics; application of mathematics to computers and computing; analysis of optimum and near-optimum algorithms in computing; system impact on future computer memory technology. Mailing Add: T J Watson Res Ctr IBM Corp Yorktown Heights NY 10598

WONG, CHEUK-YIN, b Kwangtung, China, Apr 28, 41; m 66; c 2. PHYSICS. Educ: Princeton Univ, AB, 61, MA, 63, PhD(physics), 66. Prof Exp: Physicist, Oak Ridge Nat Lab, 66-68; res fel physics, Niels Bohr Inst, Copenhagen, Denmark, 68-69; PHYSICIST, OAK RIDGE NAT LAB, 69- Mem: Am Phys Soc. Res: Theoretical studies of nuclear stability, nuclear fission and nuclear reactions; dynamics of nuclear fluid. Mailing Add: Oak Ridge Nat Lab Oak Ridge TN 37830

WONG, CHI SONG, b Cheng Tak, Hunan, China, May 26, 38; m 66; c 3. MATHEMATICAL STATISTICS, OPERATOR THEORY. Educ: Nat Taiwan Univ, BS, 62; Univ Ore, MS, 66; Univ Ill, Urbana, MS, 67, PhD(functional anal), 69. Prof Exp: Tutor, Chinese Univ, Hong Kong, 62-65; asst prof math, Southern Ill Univ, Carbondale, 69-71; asst prof, 71-73, ASSOC PROF MATH, UNIV WINDSOR, 73- Concurrent Pos: Can Nat Res Coun fels, 72-76. Mem: Can Math Cong. Res: Using algebra, functional analysis, geometry and topology to characterize certain classes of self maps which have fixed points; mathematical analysis. Mailing Add: Dept of Math Univ of Windsor Windsor ON Can

WONG, CHIU MING, b Canton, China, July 8, 35; m 60; c 2. ORGANIC CHEMISTRY. Educ: Nat Taiwan Univ, BSc, 59; Univ NB, PhD(chem), 64. Prof Exp: Asst, Univ NB, 64, res fels, 64-65; res fels, Harvard Univ, 65-66; asst prof, 66-68, ASSOC PROF ORG CHEM, UNIV MAN, 68- Mem: Am Chem Soc. Res: Synthesis of hydronaphthacenic, antibiotic and antineoplastic compounds. Mailing Add: Dept of Chem Univ of Man Winnipeg MB Can

WONG, CHUEN, US citizen. PHYSICS. Educ: Chung Chi Col, Hong Kong, dipl sci, 60; Case Western Reserve Univ, PhD(physics), 67. Prof Exp: Demonstr physics, Chung Chi Col, Hong Kong, 60-63; instr, 67-70, ASST PROF PHYSICS, LOWELL TECHNOL INST, 70- Mem: Am Phys Soc. Res: Experimental solid state physics; elastic constants; semiconductors. Mailing Add: Dept of Physics Lowell Technol Inst Lowell MA 01854

WONG, CHUN-MING, b Hong Kong, Brit Crown Colony, Nov 12, 40; m 71. ORGANIC POLYMER CHEMISTRY. Educ: Univ Calif, Berkeley, BS, 65; Wayne State Univ, MS, 66; NDak State Univ, PhD(chem), 73. Prof Exp: Chemist, Inmont Corp, 66-70; RES CHEMIST, E I DU PONT DE NEMOURS & CO, INC, 74- Mem: Am Chem Soc. Res: Water dispersible organic coatings aimed at reducing pollution and saving energy. Mailing Add: E I du Pont de Nemours & Co 945 Stephenson Hwy Troy MI 48084

WONG, DAVID C, b Hong Kong, Brit Crown Colony, Sept 26, 41; m 71. SEMICONDUCTORS. Educ: Boston Univ, BA, 66; Univ Mass, Amherst, MS, 68, PhD(physics), 75. Prof Exp: Res & teaching asst, 67-75, assoc physics, Univ Mass, Amherst, 75-76; ENGR, MACHLETT LABS, 76- Mem: Am Phys Soc. Res: To study experimentally the energy levels of electrons bound to sulfur impurities deep in the forbidden gap of silicon. Mailing Add: Machlett Labs Stanford CT 06907

WONG, DAVID TAIWAI, b Hong Kong, Brit Crown Colony, Nov 6, 35; US citizen; m 63; c 3. BIOCHEMISTRY. Educ: Seattle Pac Col, BS, 60; Ore State Univ, MS, 64;

Univ Ore, PhD(biochem), 66. Prof Exp: Fel biophys chem, Univ Pa, 66-68; sr biochemist, 68-72, RES BIOCHEMIST, LILLY RES LABS, ELI LILLY & CO, 73- Mem: Biophys Soc; Am Soc Pharmacol & Exp Therapeut. Res: Synthetic chemicals which block the uptake of specific neurotransmitters investigated as potentially useful therapeutic agents for mental disorders; chemicals of microbial origin which carry metal cations across biomembranes. Mailing Add: 1640 Ridge Hill Lane Indianapolis IN 46217

WONG, DAVID YUE, b Swatow, China, Apr 16, 34; US citizen; m 60; c 2. PHYSICS. Educ: Hardin-Simmons Univ, BA, 54; Univ Md, College Park, PhD(physics), 58. Prof Exp: Theoretical physicist, Univ Calif, Berkeley, 58-60, from asst prof to assoc prof, 60-67, PROF PHYSICS, UNIV CALIF, SAN DIEGO, 67- Concurrent Pos: Alfred P Sloan fel, Univ Calif, San Diego, 63-66. Mem: Am Inst Physics; Am Phys Soc. Res: Theoretical high energy physics. Mailing Add: Dept of Physics Univ of Calif San Diego La Jolla CA 92038

WONG, DONALD TAI ON, b Honolulu, Hawaii, Nov 1, 26; m 54; c 2. IMMUNOLOGY. Educ: St Louis Univ, BS, 49; Wash Univ, PhD(microbiol), 53. Prof Exp: Res chemist, Dept Bact Microbial Chem Sect, Walter Reed Army Inst Res, Army Med Ctr, Washington, DC, 52-61; res chemist, Blood Antigen Lab, Div Animal Husb, Agr Res Ctr, Md, 61-65; RES CHEMIST, DEPT IMMUNOCHEM, DIV COMMUN DIS & IMMUNOL, WALTER REED ARMY INST RES, ARMY MED CTR, 65- Mem: AAAS; Am Chem Soc. Res: Oxidative metabolism in microorganisms; alternate pathways and carbon-2-carbon-2 condensation mechanisms; immunoglobulin specifity and structure; mechanisms involved with immediate type hypersensitivity reactions. Mailing Add: Dept Immunochem Div Commun Dis & Immun Walter Reed Army Inst Res Army Med Ctr Washington DC 20012

WONG, DOROTHY PAN, b Nanking, China, July 8, 37; US citizen; m 68. PHYSICAL CHEMISTRY. Educ: Univ Okla, BS, 57; Univ Minn, MS, 59; Case Inst Technol, PhD(phys chem), 64. Prof Exp: Res chemist, Continental Oil Co, 57; assoc chemist, Airforce Midway Lab, Univ Chicago, 59-60; asst prof phys chem, Calif State Col Fullerton, 64-65; res assoc quantum chem, Princeton Univ, 65-66; asst prof, 66-70, PROF PHYS CHEM, CALIF STATE UNIV, FULLERTON, 70- Mem: Am Chem Soc; Am Phys Soc. Res: Non-empirical quantum mechanical calculations for geometry of molecules; molecular properties and rotation barriers of nitrogen compounds and for other molecules of current chemical interest. Mailing Add: Dept of Chem Calif State Univ 800 N State College Blvd Fullerton CA 92634

WONG, EDWARD WAI CHE, physical organic chemistry, polymer chemistry, see 12th edition

WONG, HARRY YUEN CHEE, b Kapaa, Hawaii, Oct 23, 17; m 43; c 3. PHYSIOLOGY, ENDOCRINOLOGY. Educ: Okla State Univ, BS, 42; Univ Southern Calif, MS, 47, PhD(endocrinol, physiol), 50. Prof Exp: Asst physiol, Univ Southern Calif, 46-48, lab assoc anat, 48-49; assoc prof biol, Andrews Univ, 49-51; instr zool, 51-52, from asst prof to assoc prof physiol, Sch Med, 52-66, PROF PHYSIOL, SCH MED, HOWARD UNIV, 66-, DIR ENDOCRINOL & METAB, 53-; DIR BASIC ENDOCRINE RES, FREEDMEN'S HOSP, 52- Concurrent Pos: Consult, Off Surgeon Gen, US Air Force, 63-; vis prof hormone lab, II Med Clin, Univ Hamburg, 69; vis scientist, Armed Forces Inst Path, DC, 70; fel, coun arteriosclerosis, Am Heart Asn; mem, Int Cong Physiol Sci, Int Cong Pharmacol, Int Cong Hormonal Steroids & Int Cong Endocrinol; consult to chief dept med & clin sci, Brooks AFB. Honors & Awards: Citation, Food & Drug Admin, 62; Cert Outstanding Achievement, US Air Force, 69. Mem: Am Physiol Soc; Endocrine Soc; NY Acad Sci. Res: Comparative actions of male hormones and their steroids in atherosclerosis; endocrine changes in birds and man; lipid mobilizing factor; factor of stress and exercise and sex hormones in atherosclerosis; enzymes, catecholamines and hormones of heart and adrenal glands; lipid changes in chickens, man and gerbils; chemical determination of androgens. Mailing Add: Dept of Physiol & Biophys Howard Univ Washington DC 20059

WONG, HORNE RICHARD, b Hong Kong, China, Jan 9, 23; Can citizen; m 58; c 3. ENTOMOLOGY. Educ: Univ Man, BSA, 47; Mich State Univ, MS, 50; Univ Ill, PhD(entom), 60. Prof Exp: Sr agr asst, Can Dept Agr, 47-48, officer-in-charge forest insect surv, 49-52, res officer, 53-60; res officer, 61-65, RES SCIENTIST, NORTHERN FOREST RES CTR, CAN DEPT ENVIRON, 66- Concurrent Pos: Mem, Can Comt Common Names Insects, 58-70. Mem: AAAS; Entom Soc Am; Soc Syst Zool; Soc Study Evolution; Entom Soc Can. Res: Systematics, biology and phylogeny of Symphta; life history and habits of forest insects. Mailing Add: Northern Forest Res Ctr Can Dept of Environ 5320 122 St Edmonton AB Can

WONG, HOW-KIN, b Hong Kong, Oct 13, 41. GEOPHYSICS, PHYSICS. Educ: Univ Hong Kong, BSc, 63, PhD(marine physics), 68. Prof Exp: Demonstr physics, Univ Hong Kong, 64-67; asst scientist, Woods Hole Oceanog Inst, 67-71; ASSOC PROF GEOL, NORTHERN ILL UNIV, 71- Mem: Am Geophys Union; Am Geol Inst; Brit Inst Physics; AAAS. Res: Geophysical studies of the Azores-Gibraltar lineation, the African rift valley lakes and the central Mediterranean and reconstructions of their evolutionary history. Mailing Add: Dept of Geol Northern Ill Univ DeKalb IL 60115

WONG, JACOB YAU-MAN, b Hong Kong, Sept 23, 39; US citizen; m 64. SOLID STATE PHYSICS, BIONICS. Educ: Princeton Univ, AB, 62; Stanford Univ, MA, 63, PhD(physics), 68. Prof Exp: Physicist, Stanford Res Inst, 67-69; physicist, Hewlett-Packard Co, Calif, 69-72, PROJ MGR, MED ELECTRONICS DIV, HEWLETT-PACKARD CO, 72- Res: Indium arsenide antimonide electroluminescent diodes and photovoltaic detectors; medical gas analysis; respiratory monitoring and pulmonary functional testing instrumentation; blood gas analysis. Mailing Add: Hewlett-Packard Co 175 Wyman St Waltham MA 02154

WONG, JAMES CHIN-SZE, b Hong Kong, Dec 5, 40; Can citizen. MATHEMATICS. Educ: Univ Hong Kong, BA, 63; Univ BC, PhD(math), 69. Prof Exp: Nat Res Coun Can fel, McMaster Univ, 69-71; ASST PROF MATH, UNIV CALGARY, 71- Mem: Can Math Cong; Am Math Soc. Res: Functional analysis. Mailing Add: Dept of Math Univ of Calgary Calgary AB Can

WONG, JAMES SAI-WING, b Shanghai, China, May 2, 40; m 63; c 2. MATHEMATICAL ANALYSIS. Educ: Baylor Univ, BSc, 60; Calif Inst Technol, PhD, 65. Prof Exp: From asst prof to assoc prof math, Univ Alta, 64-68; assoc prof, Carnegie-Mellon Univ, 68-70; PROF MATH, UNIV IOWA, 70- Concurrent Pos: Nat Res Coun Can grant, 65-68; Defence Res Bd Can grant, 66-68; vis assoc prof math res ctr, Univ Wis, 67-68; Scaife Found fel, 68-70; US Army grant, Univ Durham, 72-75. Mem: Am Math Soc; Math Asn Am; Can Math Cong. Res: Fixed point theorems; differential equations, ordinary, partial and functional; integral equations; asymptotic theory. Mailing Add: Dept of Math Univ of Iowa Iowa City IA 52240

WONG, JAMES TENG, applied mathematics, see 12th edition

WONG, JEFFREY TZE-FEI, b Hong Kong, Aug 5, 37; Can citizen; m 61; c 3. BIOCHEMISTRY. Educ: Univ Toronto, BA, 59, PhD(biochem), 62. Prof Exp: Asst prof, 65-70, ASSOC PROF BIOCHEM, FAC MED, UNIV TORONTO, 70- Concurrent Pos: Med Res Coun Can grant, Univ Toronto, 65- Mem: Can Soc Biochem; Am Soc Biochem. Res: Enzyme kinetics and mechanism; bacterial regulations of RNA synthesis. Mailing Add: Dept of Biochem Univ of Toronto Fac of Med Toronto ON Can

WONG, JOE, b Hong Kong, Aug 8, 42; m 69; c 2. PHYSICAL CHEMISTRY, SOLID STATE SCIENCE. Educ: Univ Tasmania, BSc, 65, Hons, 66; Purdue Univ, Lafayette, PhD(phys chem), 70. Prof Exp: Anal chemist, Australian Titan Prod, Tasmania, 62-63; res asst, Electrolytic Zinc Co Australasia, 63-64 & Dept Chem, Univ Tasmania, 64-65; res chemist, Electrolytic Zinc Co Australasia, Tasmania, 66; res asst chem, Walker Lab, Rensselaer Polytech Inst, 66-67 & Purdue Univ, Lafayette, 67-70; PHYS CHEMIST, CORP RES & DEVELOP, GEN ELEC CO, 70- Concurrent Pos: Adj lectr chem, Royal Hobart Col, 66. Mem: Am Chem Soc; Royal Australian Chem Inst; Am Ceramic Soc; Electrochem Soc. Res: Molten salt chemistry; thermodynamic and spectroscopic studies; spectroscopy of simple inorganic glasses; thin films; deposition and structural characterization; impurity diffusion in semiconductors; microstructure of electronic polycrystalline ceramics. Mailing Add: Gen Elec Co Res & Develop Ctr Bldg K1 Rm 5C11 PO Box 8 Schenectady NY 12301

WONG, KAI-WAI, b Aug 7, 38; Brit citizen. PHYSICS. Educ: Duke Univ, BS, 59; Northwestern Univ, MS, 60, PhD(physics), 63. Prof Exp: Res assoc physics, Northwestern Univ, 63; res assoc, Univ Iowa, 63-64; from asst prof to assoc prof, 64-72, PROF PHYSICS, UNIV KANS, 72- Concurrent Pos: Vis assoc prof, Univ Southern Calif, 69-71; vis prof, Univ Calif, Los Angeles, 72-73. Mem: Am Phys Soc. Res: Theoretical physics; many-body problems; statistical mechanics; high-energy theory. Mailing Add: Dept of Physics Univ of Kans Lawrence KS 66045

WONG, KEITH KAM-KIN, b Hong Kong, Feb 11, 29; US citizen; m 61; c 2. BIOCHEMISTRY, PHYSIOLOGY. Educ: Southwestern at Memphis, BA, 55; Univ Tenn, MSc, 57; NY Univ, PhD(biol), 69. Prof Exp: Res asst chemother, Sloan-Kettering Inst Cancer Res, 57-58; mem staff biochem & drug metab, Worcester Found Exp Biol, 60-63; res assoc biochem pharmacol, Schering Corp, 63-66; res assoc exp hemat, NY Univ, 66-69; SR RES INVESTR DRUG METAB, SQUIBB INST MED RES, 69- Mem: AAAS; Am Chem Soc; NY Acad Sci. Res: Drug metabolism; biochemical pharmacology; biogenesis of erythropoietin; metabolism of biogenic amines; amino acid activation and transfer; protein synthesis; transformation of nucleic acid. Mailing Add: Dept of Drug Metab Squibb Inst New Brunswick NJ 08903

WONG, KIN-PING, b China, Aug 18, 41; m 68; c 2. BIOCHEMISTRY, BIOPHYSICAL CHEMISTRY. Educ: Univ Calif, Berkeley, BS, 64; Purdue Univ, PhD(biochem), 68. Prof Exp: Res fel phys biochem, Med Ctr, Duke Univ, 68-70; from asst prof to assoc prof chem, Univ SFla, Tampa, 70-75; ASSOC PROF BIOCHEM, UNIV KANS MED CTR, 75- Concurrent Pos: Am Cancer Soc res grant, Univ SFla, Tampa, 70-71, NIH biomed res grant, 71-72, Cottrell res grant, 71-, Damon Runyon cancer res grant, 72-74; USPHS res career develop award, Nat Inst Gen Med Sci, 73. Mem: AAAS; Am Chem Soc. Res: Physical biochemistry of protein biosynthesis; mechanism of protein folding; ribosome structure; physicochemical studies of ribosomal proteins; isolation and purification of proteins from endoplasmic reticulum. Mailing Add: Dept of Biochem Univ of Kans Med Ctr Kansas City KS 66103

WONG, MAURICE KING FAN, b Shanghai, China, Apr 9, 32. MATHEMATICAL PHYSICS. Educ: Univ Hong Kong, BSc, 54; Berhmans Col, AB, 58, MA, 61; Univ Birmingham, PhD(math physics), 64. Prof Exp: Res assoc, Inst Advan Studies, Dublin, 64-65; res fel res inst nat sci, Woodstock Col, Md, 65-68; asst prof, 69-72, ASSOC PROF MATH, FAIRFIELD UNIV, 72- Concurrent Pos: Fel, St Louis Univ, 69. Mem: AAAS; Am Phys Soc. Res: Lie groups; superconductivity; Mössbauer effect; elementary particles; nuclear physics; quantum theory. Mailing Add: Dept of Math Fairfield Univ Fairfield CT 06430

WONG, MICHAEL Y M, physical chemistry, see 12th edition

WONG, MING DAK, b Can citizen. BIOCHEMISTRY. Educ: Simon Fraser Univ, BSc, 69; Univ BC, PhD(biochem), 73. Prof Exp: Res asst photobiol, Biol Dept, Simon Fraser Univ, 69; teaching asst biochem, Univ BC, 70-73; FEL BIOCHEM PHARMACOL, STANFORD UNIV, 73- Res: Mechanism of steroid hormone action at the molecular level; role of glucocorticoids in cell and tissue growth and differentiation; transcriptional modifications of embryonic tissue by hormones during development. Mailing Add: Dept of Pharmacol Sch of Med Stanford Univ Stanford CA 94305

WONG, MING MING, b Singapore, Jan 3, 28; US citizen. PARASITOLOGY. Educ: Wilmington Col, Ohio, BS, 52; Ohio State Univ, MS, 53; Tulane Univ La, PhD(med parasitol), 63. Prof Exp: Med technologist, Good Samaritan Hosp, Zanesville, Ohio, 54-55; teacher, Diocesan Girls' Sch, Hong Kong, 55-56; demonstr parasitol & bact fac med, Univ Hong Kong, 56-59; teaching asst med parasitol med sch, Tulane Univ La, 59-63, NIH res fel trop med, 63-64, res assoc, 64-65; res assoc fac med, Univ Malaya, 64-65, lectr parasitol, 65-67; asst res parasitologist, 67-73, ASSOC RES PARASITOLOGIST, PRIMATE RES CTR, UNIV CALIF, DAVIS, 73- Concurrent Pos: WHO res grant, Univ Malaya, 66-67; NIH grants primate res ctr, Univ Calif, Davis, 70-75. Mem: Am Soc Parasitol; Am Soc Trop Med & Hyg; Royal Soc Trop Med & Hyg; Am Heartworm Soc; Am Soc Clin Pathologists. Res: Filariasis; primate parasitology; immunology of parasitic diseases. Mailing Add: Primate Res Ctr Univ of Calif Davis CA 95616

WONG, MORTON MIN, b Canton, China, Oct 2, 24; US citizen; m 56; c 4. ELECTROCHEMISTRY, CHEMICAL ENGINEERING. Educ: Univ Calif, BS, 51. Prof Exp: Trainee, Am Potash & Chem Corp, 51-53; researcher, 53-54, asst proj leader, 54-56, res proj leader, 56-60, res group leader, 60-62, res proj coordr, 62-71, RES SUPVR, US BUR MINES, 71- Mem: Am Inst Mining, Metall & Petrol Eng; Am Inst Chem. Res: Chemical reduction of rare earths; hydrometallurgical and electrolytical processing of lead-zinc ores; metallurgical treatment of titanium ore. Mailing Add: 2281 Riviera St Reno NV 89509

WONG, PATRICK TIN-CHOI, b Hong Kong, Mar 9, 42; m 67; c 2. GENETICS. Educ: Univ Calif, Berkeley, BA, 64; Ore State Univ, MS, 66, PhD(genetics), 69. Prof Exp: Asst genetics, Ore State Univ, 64-69; RES SCIENTIST, CITY OF HOPE MED CTR, 70- Concurrent Pos: Fel biol, Biomed Inst, City of Hope Med Ctr, 69-70. Res: Behavior genetics; insect neurophysiology. Mailing Add: Dept of Biol City of Hope Med Ctr Duarte CA 91010

WONG, PATRICK T T, b Hong Kong, June 4, 33; Can citizen; m 67; c 2. PHYSICAL CHEMISTRY. Educ: Univ NB, PhD, 67. Prof Exp: ASSOC RES OFFICER, NAT RES COUN CAN, 67- Mem: Chem Inst Can. Res: Raman spectroscopy under high pressure. Mailing Add: Div of Chem Nat Res Coun of Can Ottawa ON Can

WONG, PATRICK YUI-KWONG, b Kiangsi, China, Nov 25, 44. BIOCHEMISTRY, BIOCHEMICAL PHARMACOLOGY. Educ: Nat Taiwan Norm Univ, BSc, 67; Univ Vt, PhD(biochem), 75. Prof Exp: Fel, Med Col Wis, 74-75; INSTR PHARMACOL, COL BASIC MED SCI, UNIV TENN, MEMPHIS, 75- Mem: Am Chem Soc. Res: Control and regulation of prostaglandin synthesis and metabolism in cardiovascular disorders and inflammation process in arthritis. Mailing Add: Dept Pharmacol Col Basic Med Sci Univ of Tenn 800 Madison Ave Memphis TN 38163

WONG, PATRICK YU-PEI, b Amoy, China, Oct 27, 39; US citizen; m 67; c 3. GASTROENTEROLOGY. Educ: Univ Sydney, MBBs, 65. Prof Exp: From intern to resident, Loma Linda Univ Hosp, 67-70; NIH res fel, Harvard Med Sch, 70-72, instr med, 73-74; ASST PROF MED, ALBANY MED COL, 74- Mem: Am Fedn Clin Res; Am Asn Study Liver Dis; Am Gastroenterol Asn. Res: Study of interrelationship between the Kallibrein-Kinin and the renin angiotensin systems in normals and hypertensives. Mailing Add: Albany Med Col Albany NY 12208

WONG, PAUL WING-KON, US citizen. PEDIATRICS, BIOCHEMICAL GENETICS. Educ: Univ Hong Kong, MD, 58; Univ Man, MSc, 67; Am Bd Pediat, dipl, 64. Prof Exp: Instr pediat, Children's Mem Hosp, Northwestern Univ, Chicago, 63-64; from asst prof to prof pediat, Chicago Med Sch, 67-73; prof pediat & dir metab unit, Abraham Lincoln Sch Med, Univ Ill Med Ctr, 73-76; PROF PEDIAT & DIR GENETIC SECT, RUSH MED SCH & PRESBY-ST LUKE MED CTR, CHICAGO, 76- Concurrent Pos: USPHS res fel biochem & med genetics, Northwestern Univ, Chicago, 62-64; Children's Res Fund fel, Ment Retardation Res Unit, Royal Manchester Children's Hosp, Eng, 65-67; attend physician, Cook County Hosp, Chicago, Ill, 65-72, consult, 72-; dir infant's aid perinatal res labs & premature & newborn nurseries, Mt Sinai Hosp, 67-73; attend physician, Univ Ill Hosp & Presby-St Luke Med Ctr, 73- Mem: Am Acad Pediat; Am Fedn Clin Res; Soc Pediat Res; Am Soc Human Genetics. Res: Metabolic diseases; biochemical genetics. Mailing Add: Dept of Pediat Presby-St Luke Med Ctr Chicago IL 60612

WONG, PETER ALEXANDER, b Honan, China, Apr 9, 41; US citizen; m 66; c 1. CHEMICAL INSTRUMENTATION. Educ: Pac Union Col, BS, 62; Rensselaer Polytech Inst, PhD(chem), 69. Prof Exp: US AEC grant, Purdue Univ, 67-69; ASST PROF CHEM, ANDREWS UNIV, 69- Mem: Am Chem Soc; Fedn Am Scientists. Res: Nuclear and chemical instrumental analysis for biomedical applications. Mailing Add: Dept of Chem Andrews Univ Berrien Springs MI 49104

WONG, PUI KEI, b Canton, China, Nov 7, 35; m 67. MATHEMATICS. Educ: Pac Union Col, BS, 56; Carnegie Inst Technol, MS, 58, PhD(math), 62. Prof Exp: Instr math, Carnegie Inst Technol, 60-62; asst prof, Lehigh Univ, 62-64; from asst prof to assoc prof, 64-72, PROF MATH, MICH STATE UNIV, 72- Mem: Am Math Soc. Res: Stability and oscillation theory of differential equation; function-theoretic differential equations; non-linear boundary value problems. Mailing Add: Dept of Math Wells Hall Mich State Univ East Lansing MI 48823

WONG, ROBERT, inorganic chemistry, physical chemistry, see 12th edition

WONG, RODERICK SUE-CHEUNG, b Shanghai, China, Oct 2, 44. MATHEMATICAL ANALYSIS. Educ: San Diego State Col, AB, 65; Univ Alta, PhD(math), 69. Prof Exp: Asst prof math, 69-73, ASSOC PROF MATH, UNIV MAN, 73-, NAT RES COUN CAN GRANT, 69- Mem: Soc Indust & Appl Math. Res: Asymptotic expansions; special functions. Mailing Add: Dept of Math Univ of Man Winnipeg MB Can

WONG, RUTH (LAU), b Hong Kong, Nov 25, 25; m 52; c 4. PATHOLOGY. Educ: Lingnan Univ, MB, BS, 48. Prof Exp: Resident path, Children's Hosp, Washington, DC, 51-52; resident, Duke Univ Hosp, 52-53; resident, Michael Reese Hosp, 53-54; res asst, La Rabida Sanitarium, 55-56; res asst, Univ Chicago, 56-57; asst prof, 57-66, ASSOC PROF PATH, UNIV ILL COL MED, 66- Concurrent Pos: Fel, Michael Reese Hosp, Chicago, Ill, 54-55. Res: Surgical pathology; serotonin content of mast and enterochromaffin cells; relationship of mast cells to tissue response to inflammation, and of enterochromaffin cells to gastric physiology and pathology; medical information science; data retrieval of pathology records. Mailing Add: Dept of Path Univ of Ill Col of Med Chicago IL 60612

WONG, SAMUEL SHAW MING, b Peking, China, May 10, 37; m 67; c 2. PHYSICS. Educ: Int Christian Univ, Tokyo, BA, 59; Purdue Univ, MS, 61; Univ Rochester, PhD(theoret physics), 68. Prof Exp: Asst prof, 69-71, ASSOC PROF PHYSICS, UNIV TORONTO, 71- Mem: Am Phys Soc; Can Asn Physics. Res: Nuclear structure theory. Mailing Add: Dept of Physics Univ of Toronto Toronto ON Can

WONG, SHEK-FU, b Canton, China, Dec 5, 43; m 73. ATOMIC PHYSICS, MOLECULAR PHYSICS. Educ: Chung Chi Col, Chinese Univ Hong Kong, BSc, 65; Univ Del, PhD(physics), 72. Prof Exp: Instr physics, Chung Chi Col, Chinese Univ Hong Kong, 65-67; fel, 72-74, res staff, 74-75, ASST PROF ENG & APPL SCI, YALE UNIV, 75- Mem: Am Phys Soc. Res: Crossed-beam electron impact experiments, with particular interest in the resonant vibrational and rotational excitation in small molecules. Mailing Add: Mason Lab Dept Eng & Appl Sci Yale Univ New Haven CT 06520

WONG, SHI-YIN, b Hoiping, Canton, China, Apr 27, 41; US citizen; m 67; c 1. ORGANIC CHEMISTRY. Educ: Univ Calif, Los Angeles, BS, 64; Univ Southern Calif, PhD(chem), 68. Prof Exp: MEM TECH STAFF, HUGHES RES LAB, HUGHES AIRCRAFT CO, MALIBU, CALIF, 69- Concurrent Pos: Air Force Off Sci Res fel, Hughes Res Lab, 72-73. Res: Electrohydrodynamics of liquid crystal; molecular correlation of liquid crystal. Mailing Add: 547 Tenth St Santa Monica CA 90402

WONG, SHUE TUCK, b Seremban, Negri Sembilan, Malaysia, May 3, 34; m 66; c 2. GEOGRAPHY, WATER RESOURCES. Educ: Augustana Col, AB, 59; Yale Univ, MA, 60; Univ Chicago, PhD(geog), 68. Prof Exp: Res asst geog, Univ Chicago, 60-62; res assoc water resources, Northeastern Ill Planning Comn, 62-65; urban transportation planner, Chicago Area Transportation Study, 65-66; asst prof, 67-69, ASSOC PROF GEOG, SIMON FRASER UNIV, 70- Concurrent Pos: Pres res grant, 68-70; res award, Can Coun, 73-74; vis assoc prof, Asian Inst Technol, 76-78. Mem: Am Geophys Union; AAAS. Res: Water resources management; hydrology; applied multivariate analysis; quantitative methodology. Mailing Add: Dept of Geography Simon Fraser Univ Burnaby BC Can

WONG, SIU GUM, b San Francisco, Calif, Feb 21, 47. OPTOMETRY, PUBLIC HEALTH. Educ: Univ Calif, Berkeley, BS, 68, OD, 70, MPH, 72. Prof Exp: Res assoc community med, Sch Med, St Louis Univ, 72-73; ASST PROF OPTOM & PUB HEALTH, COL OPTOM, UNIV HOUSTON, 73- Concurrent Pos: Consult, Health Power Assoc, Inc, New Orleans, 74. Mem: Fel Am Acad Optom; Am Pub Health Asn. Res: Community optometry. Mailing Add: Dept of Optom Univ of Houston Col of Optom Houston TX 77004

WONG, STEWART, b Toronto, Ont, Jan 2, 30; m 59; c 2. PHARMACOLOGY. Educ: Univ Toronto, BA, 58, MA, 60; Purdue Univ, PhD(pharmacog), 63. Prof Exp: Assoc prof pharmacol, Col Pharm, NDak State Univ, 63-65; instr, Col Med, Univ Iowa, 65-67; sect head, Appl Sci Div, Litton Indust Inc, 68; sect head, Union Carbide Corp, 68-69; GROUP LEADER PHARMACOL, McNEIL LABS, INC, JOHNSON & JOHNSON, 69- Mem: AAAS; Am Soc Pharmacol & Exp Therapeut; NY Acad Sci. Res: Experimental diseases; arthritis; inflammation and healing of soft and skeletal tissue; diabetes mellitus; cardiovascular and metabolic diseases; autonomic nervous system; natural product biosynthesis and biological activity. Mailing Add: McNeil Labs Inc Ft Washington PA 19034

WONG, TANG-FONG FRANK, b Canton, China, Jan 21, 44; m 69; c 1. THEORETICAL HIGH ENERGY PHYSICS. Educ: Chinese Univ Hong Kong, BSc, 65; Brown Univ, PhD(physics), 70. Prof Exp: Res assoc physics, Brookhaven Nat Lab, 69-71; res fel, 71-73, vis asst prof, 73-74, ASST PROF PHYSICS, RUTGERS UNIV, NEW BRUNSWICK, 74- Mem: Am Phys Soc. Res: High energy behavior of renormalizable field theories; symmetry and symmetry breaking in field theories; strong interaction phenomenology. Mailing Add: Dept of Physics Rutgers Univ New Brunswick NJ 08903

WONG, TIN KIN, b China, Spet 18, 37; c 2. MATHEMATICS. Educ: Chu Hai Col, Hong Kong, dipl, 62; Northcote Col, Hong Kong, cert, 63; Ind Univ, Bloomington, MA, 67, PhD(math), 69. Prof Exp: Teacher math, Tun Yu Sch, Hong Kong, 63-65; asst prof, Wayne State Univ, 69-75; SR LECTR, HONG KONG POLYTECH, 75- Mem: Am Math Soc. Res: Functional analysis; operator theory. Mailing Add: Dept of Math & Sci Hong Kong Polytech Kowloon Hong Kong

WONG, VICTOR KENNETH, b San Francisco, Calif, Nov 1, 38; m 64; c 1. PHYSICS. Educ: Univ Calif, Berkeley, BS, 60, PhD(physics), 66. Prof Exp: Fel physics, Ohio State Univ, 66-67; res assoc, 67-68; lectr, 68-69, ASST PROF PHYSICS, UNIV MICH, ANN ARBOR, 69- Mem: Am Phys Soc. Res: Many-body theory; equilibrium properties of interacting bosons at low temperatures; two-band superconductors; liquid helium; superfluidity; quantum fluids; critical phenomena; surface phenomena; and dielectric formulation of Bose liquids. Mailing Add: Dept of Physics Univ of Mich Ann Arbor MI 48104

WONG, WALTER MUN-FAY, b Canton, China, June 22, 12; US citizen; m 41; c 1. DENTISTRY. Educ: St Mary's Col, Calif, AB, 34; Col Physicians & Surgeons, San Francisco, DDS, 40. Prof Exp: Clin instr prosthetic dent, 49-54, asst clin prof, 54-57, from asst clin prof to asst prof oper dent, 57-65, ASSOC PROF OPER DENT, UNIV OF THE PAC, 65- Concurrent Pos: Consult, Vet Admin Hosp, Martinez, Calif, 62- Mem: AAAS; Am Dent Asn; Asn Mil Surg US; Royal Soc Health. Res: Operative and proethetic dentistry. Mailing Add: Univ of the Pac Sch of Dent 2155 Webster St San Francisco CA 94115

WONG, WARREN JAMES, b Masterton, NZ, Oct 16, 34; m 62; c 3. ALGEBRA. Educ: Univ Otago, NZ, BSc, 55, MSc, 56; Harvard Univ, PhD(math), 59. Prof Exp: Lectr math, Univ Otago, NZ, 60-63, sr lectr, 64; assoc prof, 64-68, PROF MATH, UNIV NOTRE DAME, 68- Concurrent Pos: Vis fel, Univ Auckland, 69. Mem: Am Math Soc; Math Asn Am; Australian Math Soc. Res: Finite group theory. Mailing Add: Dept of Math Univ of Notre Dame Notre Dame IN 46556

WONG, YUEN-FAT, b Kwangtung, China, Sept 22, 35; US citizen; m 62; c 3. MATHEMATICS. Educ: Cornell Univ, PhD(math), 64. Prof Exp: ASSOC PROF MATH, DEPAUL UNIV, 64- Concurrent Pos: NSF fel, DePaul Univ, 65-67. Mem: Am Math Soc. Res: Algebraic topology. Mailing Add: Dept of Math DePaul Univ 2323 N Seminary Ave Chicago IL 60614

WONG-RILEY, MARGARET TZE TUNG, b Shanghai, China, Oct 20, 41; US citizen; m 70; c 1. NEUROANATOMY, ANATOMY. Educ: Columbia Univ, BS, 65, MA, 66; Stanford Univ, PhD(anat), 70. Prof Exp: ASST PROF ANAT & NEUROANAT, UNIV CALIF, SAN FRANCISCO, 73- Concurrent Pos: Fight for Sight fel, Univ Wis, 70-71; NIH fel, Lab Neurophysiol, Nat Inst Neurol Dis & Stroke, 72-73; Alexander Ryan Endowment Fund fel, Univ Calif, San Francisco, 74-75. Mem: AAAS; Soc Neurosci; Asn Am Anat. Res: Study of the vertebrate visual system through various anatomical as well as physiological, histochemical and radioautographic means. Mailing Add: Dept of Anat Univ of Calif San Francisco CA 94143

WOO, CHIA-WEI, b Shanghai, China, Nov 13, 37; m 60; c 3. THEORETICAL PHYSICS. Educ: Georgetown Col, BS, 56; Washington Univ, MA, 61, PhD(physics), 66. Prof Exp: Appl mathematician, Monsanto Co, Mo, 59-63; res assoc physics, Washington Univ, 66; asst res physicist, Univ Calif, San Diego, 66-68; from asst prof to assoc prof, 68-73, PROF PHYSICS, NORTHWESTERN UNIV, EVANSTON, 73-, CHMN DEPT, 74- Concurrent Pos: Lectr, Washington Univ, 59-61; consult, Argonne Nat Lab, 68-; vis assoc prof, Univ Ill, Urbana-Champaign, 70-71; Alfred P Sloan res fel, 71-73. Mem: Am Phys Soc. Res: Quantum many body theory; low temperature physics; surface physics; liquid crystals; neutron stars. Mailing Add: Dept of Physics Northwestern Univ Evanston IL 60201

WOO, CHING CHANG, b Canton, China, Dec 12, 15; US citizen; m 49; c 4. PETROLOGY, MINERALOGY. Educ: Nat Univ Peking, BS, 37; Univ Chicago, MS, 48, PhD(petrol, mineral), 52. Prof Exp: Asst stratig, paleont & struct geol, Nat Univ Chungking, 38-40; asst geologist, Geol Surv Szechwan, China, 40-42, chief paleontologist, 42-44; asst geol, Univ Chicago, 49-51; geologist & field engr, Crane Co, Ill, 52-55; chief geologist, Heavy Minerals Co, Tenn, 55-58; div geologist, SE US, Vitro Minerals Corp, 58-59, res geologist, 59-63; geologist, Washington, DC, 63-71, GEOLOGIST, US GEOL SURV, OFF MARINE GEOL, WOODS HOLE OCEANOG INST, 71- Mem: Geol Soc Am; Mineral Soc Am; Mineral Asn Can. Res: Geotectonics; sedimentary petrology of dried lake sediments; petrogenesis of metamorphic and sedimentary rocks; petrographic analysis; mineral economics and mineral benefication of ores and ore deposits, especially placer type; lunar regoliths of Apollo programs; continental shelf sediments. Mailing Add: Off of Marine Geol US Geol Surv Woods Hole Oceanog Inst Woods Hole MA 02543

WOO, GAR LOK, b Canton, China, Jan 14, 35; US citizen; m 64; c 2. ORGANIC CHEMISTRY. Educ: Univ Calif, Berkeley, BS, 59; Mass Inst Technol, PhD(org chem), 62. Prof Exp: Res chemist, 62-69, sr res chemist, 69-75, SR RES ASSOC, CHEVRON RES CO, RICHMOND, 75- Mem: Am Chem Soc. Res: Physical organic chemistry, mechanism, stereochemistry and synthesis; exploratory petrochemicals and surfactants; sulfur and organo-sulfur chemistry. Mailing Add: 200 Blackfield Dr Tiburon CA 94920

WOO, GEORGE CHI SHING, b Shanghai, China, Feb 15, 41; Can citizen; m 67; c 3. OPTOMETRY. Educ: Col Optom Ont, OD, 64; Ind Univ, Bloomington, MS, 68, PhD(physiol optics), 70. Prof Exp: Optometrist, Can Red Cross, 64-66; clin instr optom, Ind Univ, 66-67; teaching assoc physiol optics, 67-68, res asst, 68-69, res assoc & assoc instr, 69-70; asst prof, 70-74, ASSOC PROF OPTOM, UNIV WATERLOO,

74- Mem: Am Acad Optom; Am Optom Asn; Can Asn Optom. Res: Care of the partially sighted; vision and lighting. Mailing Add: Sch of Optom Univ of Waterloo Waterloo ON Can

WOO, HENRY KYI-OEN, physical chemistry, see 12th edition

WOO, JAMES T K, b Shanghai, China, June 7, 38. ORGANIC CHEMISTRY. Educ: Wabash Col, BA, 61; Univ Md, PhD(chem), 67. Prof Exp: Asst, Univ Md, 61-63, res asst, 63-66, fel, 66-67; res chemist, Dow Chem Co, Mich, 67-71; mem staff, Horizon Res Corp, 71-72; SCIENTIST, GLIDDEN DURKEE CO, 72- Mem: Am Chem Soc. Res: Polymer chemistry. Mailing Add: Dwight P Joyce Res Ctr Glidden Durkee Co Strongsville OH 44136

WOO, KWANG BANG, b Kyoto, Japan, Jan 25, 34; m 63; c 2. BIOMEDICAL ENGINEERING, ELECTRICAL ENGINEERING. Educ: Yonsei Univ, Korea, BE, 57, ME, 59; Ore State Univ, MS, 62, PhD(elec eng), 65. Prof Exp: Instr elec eng, Yonsei Univ, Korea, 59-60; res mem, Sci Res Inst, Ministry Defense, Korea, 57-60; res assoc, Ore State Univ, 65; asst prof elec eng & biomed eng, Wash Univ & res assoc, Ctr Biol Natural Systs, 66-71; SR STAFF FEL, OFF OF DEP DIR, DIV CANCER TREATMENT, NAT CANCER INST, 71- Concurrent Pos: Fel biophys, Inst Sci Technol, Univ Mich, 65-66. Mem: AAAS; Inst Elec & Electronics Eng; NY Acad Sci; Biophys Soc. Res: Control mechanisms in metabolic processes; cell cycle kinetics and cancer therapy; tumor-marker interactions; nonlinear systems analysis and computer techniques. Mailing Add: Nat Cancer Inst 9000 Rockville Pike Bethesda MD 20014

WOO, NORMAN TZU TEH, b Shanghai, China, Sept 28, 39. MATHEMATICS. Educ: Wabash Col, BA, 62; Southern Methodist Univ, MS, 64; Wash State Univ, PhD(math), 68. Prof Exp: ASSOC PROF MATH, CALIF STATE UNIV, FRESNO, 68- Mem: Am Math Soc. Res: Number theory of mathematics. Mailing Add: Dept of Math Calif State Univ Shaw & Cedar Ave Fresno CA 93710

WOO, SHIEN-BIAU, b Shanghai, China, Aug 13, 37; m 63; c 2. ATOMIC PHYSICS, MOLECULAR PHYSICS. Educ: Georgetown Col, BS, 57; Washington Univ, MA, 61, PhD(physics), 64. Prof Exp: Instr math, Univ Mo, St Louis, 61-62; res assoc physics, Joint Inst Lab Astrophys, Univ Colo, 64-66; asst prof, 66-70, ASSOC PROF PHYSICS, UNIV DEL, 70- Mem: Am Phys Soc; Am Asn Physics Teachers. Res: Inference of ion-molecule reaction cross sections from rate constants; ion velocity distributions; photodetachment of molecular negative ions. Mailing Add: Dept of Physics Univ of Del Newark DE 19711

WOOD, ALBERT ELMER, b Cape May Court House, NJ, Sept 22, 10; m 37; c 3. VERTEBRATE PALEONTOLOGY. Educ: Princeton Univ, BS, 30; Columbia Univ, MA, 32, PhD(geol), 35. Hon Degrees: MA, Amherst Col, 54. Prof Exp: Asst biol, Long Island Univ, 30-33, tutor, 33-34; from asst geologist to geologist, US Army Engrs, 36-41 & geologist, 46; from asst prof to prof, 46-70, chmn dept, 62-66, EMER PROF BIOL, AMHERST COL, 70- Concurrent Pos: Partic, Paleont Expeds, Western US, 28, 31-32 & 35; dir, Amherst Col Exped, 48, 57, 60, 63, 65 & 68; assoc cur vert paleont, Pratt Mus, Amherst Col, 48-70; NSF sr fel, Naturhistorisches Mus, Basel, Switz, 66-67. Mem: AAAS; Am Soc Mammalogists; Paleont Soc; fel Geol Soc Am; Am Soc Zoologists. Res: Rodent and lagomorph classification, paleontology and evolution. Mailing Add: 20 Hereford Ave Cape May Court House NJ 08210

WOOD, ALEX JAMES, b Vancouver, BC, Sept 3, 14; m 41; c 2. BIOCHEMISTRY, BACTERIOLOGY. Educ: Univ BC, BSA, 35, MSA, 38; Cornell Univ, PhD(bact, biochem), 40. Prof Exp: Bacteriologist, Fisheries Res Bd Can, 40-43, Defence Res Bd, Can, 43-46 & Ministry Supply, UK, 47-48; prof animal sci, Univ BC, 46-65; PROF BIOCHEM, UNIV VICTORIA, BC, 65-, HEAD DEPT, 74- Mem: AAAS; Am Chem Soc; Am Nutrit Soc; Can Biochem Soc; Can Soc Microbiol. Res: Animal growth and energetics. Mailing Add: Dept of Biochem Univ of Victoria Victoria BC Can

WOOD, BENJAMIN W, b Cardston, Alta, Nov 13, 38; US citizen; m 59; c 6. BOTANY, RANGE SCIENCE. Educ: Brigham Young Univ, BS, 63, MS, 67; Ore State Univ, PhD(range sci), 71. Prof Exp: Asst prof biol, Boise Col, 67-68; instr range mgt, Ore State Univ, 69-70, asst prof, 70-71; ASST PROF BOT & RANGE SCI, BRIGHAM YOUNG UNIV, 71- Concurrent Pos: Navajo-Kaiparowits Power Generating Sta fel, Ctr Health & Environ Studies, Brigham Young Univ, 71- Mem: Soc Range Mgt. Res: Desert ecology; wildlife habitat improvement. Mailing Add: Dept of Bot & Range Sci Brigham Young Univ Provo UT 84602

WOOD, BRUCE, b Kintnersville, Pa, Apr 9, 38; m 63; c 2. MATHEMATICS, MECHANICAL ENGINEERING. Educ: Pa State Univ, BS, 60; Univ Wyo, MS, 64; Lehigh Univ, PhD(math), 67. Prof Exp: Asst air pollution control engr, Bethlehem Steel Corp, 60-62; asst math, Univ Wyo, 63-64 & Lehigh Univ, 64-67; asst prof, 67-71, ASSOC PROF MATH, UNIV ARIZ, 71- Mem: Am Math Soc. Res: Linear approximation theory; summability theory; theory of complex variables; linear positive operators. Mailing Add: Dept of Math Univ of Ariz Tucson AZ 85721

WOOD, BURRELL LUSHA, JR, b Cosby, Tenn, Mar 30, 20; m 65. ORGANIC CHEMISTRY. Educ: Presby Col, SC, BS, 40, BA, 47; Univ Ga, MS, 42; Univ NC, PhD(chem), 52. Prof Exp: Asst chem, Univ Ga, 40-42; res chemist, Taylor-Colquitt Co, SC, 42-43; assoc prof chem, Presby Col, SC, 43-49; from assoc prof to prof, Furman Univ, 49-57, actg head dept, 53-57; prof & admis counr, NMex Inst Mining & Technol, 57-60; ed, Things of Sci & asst ed, Chemistry, Sci Serv, 60-61; sci ed specialist, Div Tech Info, AEC, 61-62, chief res info anal sect, Div Isotopes Develop, 62-64, exhibit coord, Div Spec Projs, 64-65, overseas exhibit mgr, Div Tech Info, 65-68, chief exhibits opers br, 68, asst dir exhibits, 68-72, environ nuclear educ & AV specialist, Div Nuclear Educ & Training, 72-75; SPEC ASST, OFF PUB AFFAIRS, US ENERGY RES & DEVELOP ADMIN, 75- Concurrent Pos: NSF fac fel, Univ Zurich, 57-58. Mem: Fel AAAS; Am Nuclear Soc. Res: Lossen rearrangement; chlorination; alkene addition of nitrosyl chloride; organic synthesis; thiazoles; tritium analysis; halide-contaminated ice. Mailing Add: Off of Pub Affairs US Energy Res & Develop Admin Washington DC 20545

WOOD, CALVIN DALE, b Salt Lake City, Utah, July 13, 33; m 55; c 5. NUCLEAR PHYSICS, ATOMIC PHYSICS. Educ: Univ Calif, Berkeley, AB, 57, PhD(high energy physics), 62. Prof Exp: Res asst high energy physics, Lawrence Radiation Lab, Univ Calif, Berkeley, 58-62, physicist, 61-62; asst prof physics, Univ Utah, 62-64; PHYSICIST, LAWRENCE LIVERMORE LAB, 64- Concurrent Pos: Consult, Utah Hwy Patrol Accident Invest Staff, 62-64 & Hill AFB Accident Invest Staff, 64. Mem: Am Phys Soc; Ital Phys Soc. Res: Shock hydrodynamics; neutron cross sections; high speed digital computers; nuclear processes. Mailing Add: Lawrence Livermore Lab PO Box 808 L-24 Livermore CA 94550

WOOD, CARL EUGENE, b Alice, Tex, Aug 28, 40; m 73; c 3. MARINE ECOLOGY, INVERTEBRATE ZOOLOGY. Educ: Tex A&M Univ, BS, 62, MS, 65; PhD(fisheries), 69. Prof Exp: Fishery technician, Nat Marine Fisheries Serv, 63-64; res asst limnol, Tex A&M Univ, 65-67; limnologist, Tenn Valley Authority, 67-69; ASSOC PROF INVERT & MARINE BIOL, TEX A&I UNIV, 69- Concurrent Pos: Maricult consult, Flato Corp, 73-74. Mem: Am Soc Limnol & Oceanog; Fedn Estuarine Res. Res: Marine invertebrate ecology; shrimp of the suborder Natantia systematics; primary productivity of estuaries. Mailing Add: Box 158 Dept of Biol Tex A&I Univ Kingsville TX 78363

WOOD, CAROL SAUNDERS, b Pennington Gap, Va, Feb 9, 45. MATHEMATICAL LOGIC. Educ: Randolph-Macon Womans Col, AB, 66; Yale Univ, PhD(math), 71. Prof Exp: Gast dozent math, Univ Erlangen-Nürnberg, WGer, 71-72; lectr, Yale Univ, 72-73; vis instr, 70-71, ASST PROF MATH, WESLEYAN UNIV, 73- Mem: Am Math Soc; Math Asn Am; Asn Symbolic Logic. Res: Application of model theory to algebra, in particular to fields and to differential fields. Mailing Add: Dept of Math Wesleyan Univ Middletown CT 06457

WOOD, CARROLL E, JR, b Roanoke, Va, Jan 13, 21. BOTANY. Educ: Roanoke Col, BS, 41; Univ Pa, MS, 43; Harvard Univ, PhD(biol), 49. Prof Exp: Instr biol, Harvard Univ, 49-51; from asst prof to assoc prof bot, Univ NC, 51-54; assoc cur, 54-69, CUR, ARNOLD ARBORETUM, HARVARD UNIV, 69-, PROF BIOL, 74-, MEM FAC ARTS & SCI, 71- Concurrent Pos: Lectr biol, Harvard Univ, 70-74. Mem: Am Soc Plant Taxon; Bot Soc Am; Int Asn Plant Taxon. Res: Flora of southeastern United States; biosystematics and taxonomy of flowering plants. Mailing Add: 22 Divinity Ave Cambridge MA 02138

WOOD, CHARLES, b London, Eng, Nov 6, 24; US citizen; m 50; c 3. EXPERIMENTAL SOLID STATE PHYSICS. Educ: Univ London, BSc, 51, MSc, 55, PhD(physics), 62. Prof Exp: Physicist, Gen Elec Res Labs, Eng, 51-53 & Electronic Tubes Ltd, 53-54; dep group leader, Caswell Res Lab, Plessey Co, 54-56; group supvr, Res Div, Philco Corp, Pa, 56-60; sect head solid state prod group, Kearfott Div, Gen Precision Instruments, NJ, 60-61; dir thermoelec, Intermetallic Prod, Inc, 61-63; mgr, Mat Res Dept, Xerox Corp Res Labs, NY, 63-67; head dept, 67-70, PROF PHYSICS, NORTHERN ILL UNIV, 67- Mem: Am Phys Soc; Sigma Xi. Res: Semiconductors; photoconductors; thin films; crystal growth; Hall-effect devices; thermoelectricity; electrophotography. Mailing Add: Dept of Physics Northern Ill Univ De Kalb IL 60115

WOOD, CHARLES DONALD, b Ravena, Ky, Feb 4, 25; m; c 4. PHARMACOLOGY. Educ: Univ Ky, BS, 49, MS, 50; Univ NC, PhD(physiol), 57. Prof Exp: Neurophysiologist, Nat Inst Neurol Dis & Blindness, 53-55; instr physiol & pharmacol, Med Sch, Univ Ark, 57-59, from asst prof to assoc prof pharmacol, Med Ctr, 59-67; PROF PHYSIOL & PHARMACOL, MED SCH, LA STATE UNIV, SHREVEPORT, 67-, HEAD MED COMMUN, 71- Concurrent Pos: Consult, US Naval Aerospace Med Inst, 62- Mem: Aerospace Med Asn. Res: Temporal lobe epilepsy; influence of temporal lobe structures on behavior; neuropharmacology; pulmonary edema; oxygen toxicity; neurogenic hypertension; antimotion sickness drugs; aerospace pharmacology; drugs and athletic performance. Mailing Add: Dept of Physiol & Pharmacol La State Univ Med Sch Shreveport LA 71101

WOOD, COLIN, b Farnborough, Eng, Dec 20, 23; nat US; m 56; c 2. PATHOLOGY. Educ: Univ Birmingham, MB, ChB, 46, MD, 57. Prof Exp: Sr resident, Case Western Reserve Univ, 53-54; sr registr, Bernhard Baron Inst Path, London Hosp, Eng, 54-58; from asst prof to assoc prof, 58-70, PROF PATH, SCH MED, UNIV MD, BALTIMORE CITY, 70- Mem: Am Col Path; Am Soc Clin Path; Int Acad Path. Res: Dermatopathology; surgical pathology. Mailing Add: Dept of Path Univ of Md Sch of Med Baltimore MD 21201

WOOD, CORINNE SHEAR, b Baltimore, Md, Apr 14, 25; m 46; c 4. PHYSICAL ANTHROPOLOGY, MEDICAL ANTHROPOLOGY. Educ: Univ Calif, Riverside, BA, 68, PhD(anthrop), 73. Prof Exp: Res asst med, Johns Hopkins Univ, 50-55, Sinai Hosp, Baltimore, 55-58 & Johns Hopkins Univ, 59-61; med technologist, Riverside Community Hosp, Calif, 62-71; teaching asst anthrop, Univ Calif, Riverside, 68-70; ASST PROF ANTHROP, CALIF STATE UNIV, FULLERTON, 73- Concurrent Pos: Res mem, Univ Calif Med Ctr, San Francisco, 73; lectr, Univ Calif, Riverside, 74-75, lab consult, 75. Mem: Fel Am Anthrop Asn; Am Asn Phys Anthropologists; Med Anthrop Soc. Res: Human physiological variables related to disease and disease vectors; health conditions of American Indian populations; women, nutrition and health; interrelationship of human culture and disease. Mailing Add: Dept of Anthrop Calif State Univ Fullerton CA 92631

WOOD, CRAIG ADAMS, b Rochester, NY, Jan 31, 41; m 64; c 2. MATHEMATICS. Educ: Col Wooster, BA, 62; Fla State Univ, MS, 63, PhD(math), 67. Prof Exp: Instr math, Fla State Univ, 67-68; from asst prof to assoc prof, Okla State Univ, 72-73, NASA res grant, 69-70; ASSOC PROF MATH SCI & HEAD DIV, UNIV HOUSTON, VICTORIA, 73- Mem: Am Math Soc; Math Asn Am; Asn Comput Mach. Res: Commutative ring theory in algebra; algebraic equations in numerical analysis. Mailing Add: Div of Math Sci Univ of Houston Victoria TX 77901

WOOD, DARRELL FENWICK, b Kentville, NS, May 13, 42; m 70; c 1. FOOD SCIENCE. Educ: McGill Univ, BSc, 63, MSc, 65; Univ BC, PhD(food sci), 73. Prof Exp: Res officer food sci, Can Dept Agr, 65-67; res officer, Atomic Energy Can Ltd, 67-70; RES SCIENTIST FOOD SCI, AGR CAN, 73- Mem: Can Inst Food Sci Technol; Inst Food Technologists. Res: Development of technology for reducing acidity in wine and testing new grape varieties for wine quality. Mailing Add: Agr Can Res Sta Summerland BC Can

WOOD, DARWIN LEWIS, b East Orange, NJ, July 21, 21; m 45; c 6. PHYSICS, CHEMISTRY. Educ: Princeton Univ, AB, 42; Ohio State Univ, PhD(physics, physiol), 70. Prof Exp: Physicist, Rohm and Haas Co, 42-46; fel, Univ Mich, 50-52, asst prof physics, 53-56; MEM TECH STAFF, CHEM DEPT, BELL LABS, INC, 56- Mem: Optical Soc Am. Res: Polymers; proteins; optics; spectroscopy; crystal spectra; ions in crystals. Mailing Add: Chem Dept Bell Labs Inc PO Box 261 Murray Hill NJ 07974

WOOD, DAVID, b Woodlawn, Ill, Oct 10, 28; m 58; c 3. ANALYTICAL CHEMISTRY. Educ: Univ Ill, BS, 50; Univ Wis, PhD(chem), 56. Prof Exp: Chemist, Velsicol Chem Corp, 50-52; asst chem, Univ Wis, 52-56; chemist, Spencer Kellogg & Sons, Inc, 56-59; assoc res chemist, 59-75, RES CHEMIST & GROUP LEADER, STERLING-WINTHROP RES INST, 75- Mem: Am Chem Soc. Res: Synthetic organic chemistry; synthesis of pharmaceuticals; gas chromatography. Mailing Add: Sterling-Winthrop Res Inst Rensselaer NY 12144

WOOD, DAVID ALVRA, b Flora Vista, NMex, Dec 21, 04; m 37; c 5. PATHOLOGY. Educ: Stanford Univ, AB, 26, MD, 30. Prof Exp: Asst pharmacol, Stanford Univ, 28-29, from instr to assoc prof path, 30-51; from assoc prof to prof, Sch Med, 51-72, dir, Cancer Res Inst, 51-72, EMER PROF PATH, SCH MED, RES ONCOLOGIST & EMER DIR, CANCER RES INST, UNIV CALIF, SAN FRANCISCO, 72- Concurrent Pos: Spec consult, Nat Cancer Inst, 56-63; mem sci adv bd, Armed Forces Inst Path, 57-62; co-chmn, mem adv comt & spec consult, Cancer Control Prog, Bur State Serv, USPHS, 59-63; mem clin studies panel cancer chemother, Nat Cancer

Inst, NIH, 59-64; mem US nat comt, Int Union Against Cancer, 61-64; mem histopath nomenclature & classification tumors comt, WHO, Tokyo, 69-74; mem US nat comt X, Int Cancer Cong, 70; fel selection comt, Int Agency Res Cancer, Lyon, 70-73. Honors & Awards: Am Cancer Soc Award, 50 & Distinguished Serv Award, 72; Award, Col Am Pathologists, 58; Lucy Wortham James Award, James Ewing Soc, 70. Mem: AAAS; Am Cancer Soc (pres, 56-57); Am Asn Cancer Res; Col Am Pathologists (pres, 52-55); Am Asn Cancer Insts (pres, 70-72). Res: Neoplastic diseases; dual pulmonary circulation; exfoliative cytology; evaluation of cancer education in medical and dental schools; oral contraceptives and tumors of the breast, epidemiological and morphological correlations. Mailing Add: 54 Commonwealth Ave San Francisco CA 94118

WOOD, DAVID BELDEN, b Glendale, Calif, Nov 15, 35; m 56; c 3. ASTRONOMY. Educ: Univ Calif, Berkeley, AB, 57, PhD(astron), 63. Prof Exp: Mem tech staff astron, Bellcomm, Inc, 67-69, supvr astrophys, 69-71; ADV PLANS STAFF, GODDARD SPACE FLIGHT CTR, NASA, 71- Mem: AAAS; Am Astron Soc; Royal Astron Soc. Res: Eclipsing binary stars; extragalactic research; photoelectric photometry; space astronomy; operations research; computer modeling. Mailing Add: 5108 Viking Rd Bethesda MD 20014

WOOD, DAVID ELDON, b Ada, Okla, Dec 1, 39; m 60; c 3. PHYSICAL CHEMISTRY. Educ: Okla State Univ, BS, 60; Calif Inst Technol, PhD(phys chem), 64. Prof Exp: Res chemist, Orlando Div, Martin-Marietta Corp, 63-65; res fel chem, Calif Inst Technol, 65-66; asst prof, Carnegie-Mellon Univ, 66-71; asst prof, 71-74, ASSOC PROF CHEM, UNIV CONN, STORRS, 74- Concurrent Pos: Am Chem Soc Petrol Res Fund grant, 66-68. Mem: Am Chem Soc; Am Phys Soc. Res: Electron paramagnetic resonance; studies of structure and reactions of free radicals. Mailing Add: Dept of Chem Univ of Conn Storrs CT 06268

WOOD, DAVID LEE, b St Louis, Mo, Jan 8, 31; m 60; c 2. FOREST ENTOMOLOGY, INSECT ECOLOGY. Educ: State Univ NY Col Forestry, Syracuse Univ, BS, 52; Univ Calif, Berkeley, PhD(entom), 60. Prof Exp: Asst entomologist, Boyce Thompson Inst Plant Res, 59-60; from lectr entom & asst entomologist to assoc prof entom & assoc entomologist, 60-70, PROF ENTOM & ENTOMOLOGIST, UNIV CALIF, BERKELEY, 70- Mem: AAAS; Sigma Xi; Entom Soc Am; Soc Am Foresters; Entom Soc Can. Res: Forest insect ecology and pest management; insect-host relationships, especially host selection behavior, insect pheromones and host resistance with special emphasis on bark beetles. Mailing Add: Dept of Entom Sci Univ of Calif Berkeley CA 94720

WOOD, DAVID ROY, b Mar 3, 35; US citizen; m 67. SPECTROSCOPY. Educ: Friends Univ, AB, 56; Univ Mich, MS, 58; Purdue Univ, PhD(physics), 67. Prof Exp: Instr physics, Friends Univ, 58-59; instr amth & physics, Scattergood Sch, 59-61; res assoc physics, Purdue Univ, 67; asst prof, 67-74, ASSOC PROF PHYSICS, WRIGHT STATE UNIV, 74- Mem: Optical Soc Am. Res: Experimental atomic spectroscopy; analysis of the energy level structure of the lead atom; Fabry-Perut interferometry and Zeeman effect analysis. Mailing Add: Dept of Physics Wright State Univ Dayton OH 45431

WOOD, DERICK, b Bolton, Eng, July 19, 40. COMPUTER SCIENCE. Educ: Univ Leeds, BS, 63, dipl electronic comput, 64, PhD(math), 68. Prof Exp: Comput asst, Univ Leeds, 64-68; asst res scientist, Courant Inst Math Sci, NY Univ, 68-70; asst prof, 70-74, ASSOC PROF COMPUT SCI, McMASTER UNIV, 74- Mem: Asn Comput Mach; Can Math Cong; Can Info Processing Soc. Res: Formal language theory with particular emphasis on syntax analysis and compiling. Mailing Add: Dept of Appl Math McMaster Univ Hamilton ON Can

WOOD, DON CLIFTON, b Farmington, Utah, Mar 25, 23; m 46; c 5. BIOCHEMISTRY. Educ: Brigham Young Univ, BS, 47, MS, 48; Cornell Univ, PhD(biochem), 52. Prof Exp: Biochemist, Ft Detrick, Md, 47-49 & Vet Admin Hosp, Salt Lake City, 52-60; dir biochem, St Vincent & Providence Hosp, Portland, 60-64; head res dept, Providence Hosp, 64-73; VPRES, MEDLAB COMPUT SERV, INC, 73- Concurrent Pos: Asst res prof, Col Med, Univ Utah, 52-60, assoc clin prof path, 75; res assoc surg, Med Sch, Univ Ore, 67, assoc prof, 70-73; clin chemist-consult, Cottonwood Hosp, 75. Mem: Asn Res Vision & Ophthal; Am Asn Clin Chemists; NY Acad Sci. Res: Chemistry lens protein; clinical chemistry; mechanism of action of dimethyl sulfoxide; cancer. Mailing Add: 2038 Royal Circle Salt Lake City UT 84108

WOOD, DONALD EUGENE, b St Paul, Minn, Dec 26, 30; m 54; c 2. NUCLEAR PHYSICS. Educ: Univ Nev, BS, 51; Northwestern Univ, MS, 53, PhD(physics), 56. Prof Exp: Asst physics, Northwestern Univ, 51-55; physicist, Hanford Labs, Gen Elec Co, 55-58, sr physicist, 58-63; res scientist, 63-68, mgr nuclear prod, 68-74, SR SCIENTIST, NUCLEAR SERV PROG, KAMAN SCI CORP, 74- Concurrent Pos: Lectr, Hanford Grad Ctr, Univ Wash, 58-63. Mem: Am Nuclear Soc; Am Phys Soc. Res: Reactor safety; probabilistic analysis; reliability prediction; activation analysis; reactor physics; shielding. Mailing Add: Kaman Sci Corp PO Box 7463 Colorado Springs CO 80933

WOOD, DONALD L, organic chemistry, see 12th edition

WOOD, DONALD ROY, b Keats, Kans, Apr 17, 21; m 43; c 2. AGRONOMY. Educ: Kans State Col, BS, 43; Colo State Univ, MS, 49; Univ Wis, PhD, 56. Prof Exp: From asst prof agron & asst agronomist to assoc prof agron & assoc agronomist, 47-63, PROF AGRON & AGRONOMIST, COLO STATE UNIV, 63- Concurrent Pos: Asst, Univ Wis, 50-51; res assoc, Univ Calif, 60-61; res assoc, Inst Nutrit Cent Am, Panama, 74. Mem: Fel AAAS; Genetics Soc Am; Genetics Soc Can; Am Soc Agron; Am Phytopath Soc. Res: Dry field bean breeding; dry bean disease resistance; breeding for improved nutritional value. Mailing Add: Dept of Agron Colo State Univ Ft Collins CO 80523

WOOD, EARL HOWARD, b Mankato, Minn, Jan 1, 12; m 36; c 4. PHYSIOLOGY. Educ: Macalester Col, BA, 34; Univ Minn, BS, 36, MS, 39, PhD(physiol), 40; MD, 41. Hon Degrees: DSc, Macalester Col, 50. Prof Exp: Instr physiol, Univ Minn, 39-40; instr pharmacol, Harvard Med Sch, 42; assoc prof physiol, 42-60, PROF PHYSIOL & MED, MAYO GRAD SCH MED, UNIV MINN, 42-; CONSULT, MAYO CLIN, 42- Concurrent Pos: Am Physiol Soc travel award, Int Physiol Cong, Oxford Univ; career investr, Am Heart Asn, 61-; vis scientist, Univ Bern, 65-66 & Univ Col, Univ London, 72-73. Honors & Awards: President's Cert of Merit. Mem: AAAS; Am Physiol Soc; Soc Exp Biol & Med; Am Soc Clin Invest; Am Soc Pharmacol. Res: Electrolyte metabolism of cardiac and voluntary muscle; glucose reabsorption in amphibian kidney; effect of cardiac glycoside on electrolyte metabolism; cardiopulmonary effects of gravitational and inertial forces, aerospace medicine; cardiovascular and respiratory physiology of man. Mailing Add: Mayo Grad Sch of Med Univ of Minn Rochester MN 55901

WOOD, ELWYN DEVERE, b Everett, Wash, Sept 15, 34; m 66; c 2. MARINE GEOCHEMISTRY. Educ: Western Wash State Col, BA(chem) & BA(educ), 64; Univ Wash, MS, 66; Univ Alaska, PhD(chem oceanog), 71. Prof Exp: Oceanogr, Univ Wash, 66-67 & PR Nuclear Ctr, 70-75; OCEANOGR, OUTER CONTINENTAL SHELF OFF, BUR LAND MGT, DEPT INTERIOR, 75- Mem: Am Chem Soc; Am Geophys Union; Am Soc Limnol & Oceanog; Geochem Soc. Res: Trace element chemistry in the marine environment; determination and evaluation of natural and induced radioactivity in the environment; evaluation of near-shore currents for the disposal of wastes. Mailing Add: Bur of Land Mgt Fed Bldg New Orleans LA 70334

WOOD, ERNEST HARVEY, radiology, deceased

WOOD, EUNICE MARJORIE, b Venice, Calif, Sept 5, 27. CELL BIOLOGY, ELECTRON MICROSCOPY. Educ: Rutgers Univ, New Brunswick, BS, 48; Mt Holyoke Col, MA, 50; Harvard Med Sch, PhD(anat), 68. Prof Exp: Instr zool, Wellesley Col, 50-52; lectr, Barnard Col, Columbia Univ, 53-55; res technician hemat, City of Hope Med Ctr, Calif, 57-59; res assoc, Inst Cancer & Blood Res, 59-62; instr biol, Mt Holyoke Col, 62-63; USPHS fel, Med Sch, Univ Southern Calif, 67-68; asst prof, 68-72, ASSOC PROF BIOL, CALIF STATE UNIV, LONG BEACH, 72- Concurrent Pos: Electron micros consult oncol unit, Med Sch, Univ Southern Calif, 68- Mem: Am Soc Cell Biol; Am Asn Anat. Res: Electron microscopy of hemopoietic organs; ultrastructural effects of drugs on leukemic cells; electronmicroscopic cytochemistry of human myeloma cells. Mailing Add: Dept of Biol Calif State Univ 6101 E Seventh St Long Beach CA 90840

WOOD, FERGUS JAMES, b London, Ont, May 13, 17; nat US; m 46; c 2. GEOPHYSICS. Educ: Univ Calif, AB, 38. Prof Exp: Asst astron, Univ Mich, 40-42; instr physics & astron, Pasadena City Col, 46-48 & John Muir Col, 48-49; asst prof physics, Univ Md, 49-50; assoc physicist, Appl Physics Lab, Johns Hopkins Univ, 50-55; sci ed, Encycl Americana, 55-60; aeronaut & space res scientist & sci asst to dir off space flight progs, NASA, 60-61; prog dir foreign sci info, NSF, 61-62; phys scientist, Off Dir, US Coast & Geod Surv, 62-70; phys scientist, 70-73, RES ASSOC, OFF DIR, NAT OCEAN SURV, NAT OCEANIC & ATMOSPHERIC ADMIN, ROCKVILLE, 73- Honors & Awards: Spec Achievement Award for Tidal Res, Nat Oceanic & Atmospheric Admin, 74. Res: Environmental geoscience; wind-profile studies over navy ships at sea; perigean and proxigean spring tide analysis and potential for coastal flooding; gravitational-geophysical correlations; science education, history and film documentation. Mailing Add: 10408 Sweetbriar Pkwy Silver Spring MD 20903

WOOD, FLOYD WILLIAM, b Eugene, Ore, May 31, 26; m 64; c 4. PHYSICAL METALLURGY. Educ: Univ Ore, BS, 48; Ore State Univ, MetE, 67, MMatS, 69, PhD(phys metall), 74. Prof Exp: Plant anal clerk, Pac Tel & Tel Co, 48-49; eng aid, City of Portland, Ore, 50-51; cement inspector & tester, Seattle Lab, US Bur Standards, Wash, 51-52; from physicist to res physicist, 52-75, RES SUPVR, ALBANY METALL RES CTR, US BUR MINES, 75- Mem: Am Soc Metals; Am Vacuum Soc; Am Soc Qual Control; Int Metallog Soc; Int Solar Energy Soc. Res: Development and testing of wear-resistant materials and solar-thermal selective absorbers; casting; powder metallurgy; coating; hardening; thin-film technology; metallography; metals; ceramics; intermediate phases; structural control and property modification. Mailing Add: Albany Metall Res Ctr US Bur of Mines PO Box 70 Albany OR 97321

WOOD, FORREST GLENN, b South Bend, Ind, Sept 13, 18; m 46. MARINE ZOOLOGY. Educ: Earlham Col, AB, 40; Yale Univ, MS, 50. Prof Exp: Resident biologist, Lerner Marine Lab, BWI, 50-51; cur, Marine Studios & Res Lab, Marineland, Fla, 51-63; head marine biosci facil, Naval Missile Ctr, Calif, 63-70; sr scientist & consult, Ocean Sci Dept, 70, HEAD MARINE BIOSCI PROG OFF, UNDERSEA SCI DEPT, NAVAL UNDERSEA CTR, 72- Mem: Am Soc Ichthyologists & Herpetologists; Am Soc Mammalogy; Sigma Xi; Animal Behav Soc. Res: Behavior of marine mammals, sharks. Mailing Add: 750 Albion St San Diego CA 92106

WOOD, FRANCIS A, b Perryville, Mo, Nov 17, 32; m 54; c 7. AIR POLLUTION, PHYTOPATHOLOGY. Educ: Univ Mo, BS, 55, MA, 56; Univ Minn, PhD(plant path), 61. Prof Exp: Asst prof forest path, Pa State Univ, 61-66, from assoc prof to prof plant path, 66-72, asst dir ctr air environ studies, 65-72; PROF PLANT PATH & HEAD DEPT, UNIV MINN, ST PAUL, 72- Concurrent Pos: Consult, Pa Elec Co, 62-72 & State of Pa, 66-72; USPHS grant, 66- Mem: Am Phytopath Soc; Mycol Soc Am; Air Pollution Control Asn. Res: Forest pathology; epidemiology of forest tree diseases; effects of photochemical air pollutants on trees; vascular wilt and canker diseases. Mailing Add: Dept of Plant Path Univ of Minn St Paul MN 55108

WOOD, FRANCIS C, JR, b Philadelphia, Pa, Oct 20, 28; m 58; c 2. INTERNAL MEDICINE, ENDOCRINOLOGY. Educ: Princeton Univ, AB, 50; Harvard Med Sch, MD, 54; Am Bd Internal Med, dipl, 63. Prof Exp: Intern, King County Hosp, Seattle, Wash, 54-55; resident, Vet Admin Hosp, 55-56; resident, Univ Wash Hosp, 60-61; from instr to asst prof, Sch Med, 61-68, asst prog dir, Clin Res Ctr, 62-64, prog dir, 64-70, ASSOC PROF MED, SCH MED, UNIV WASH, 68-, ASSOC DEAN, 70- Concurrent Pos: Res fel, Harvard Med Sch & Peter Bent Brigham Hosp, 58-60; chief of staff, Seattle Vet Admin Hosp, 70- Mem: Endocrine Soc; Am Diabetes Asn; fel Am Col Physicians. Mailing Add: Vet Admin Hosp 4435 Beacon Ave S Seattle WA 98108

WOOD, FRANCIS CLARK, b Wellington, SAfrica, Oct 1, 01; US citizen; m 26; c 3. MEDICINE. Educ: Princeton Univ, AB, 22; Univ Pa, MD, 26. Hon Degrees: DSc, Trinity Col, Dublin, 62, Princeton Univ, 64 & Univ Pa, 71. Prof Exp: Intern, Hosp, 26-28, from asst instr to prof med, Sch Med, 28-70, chmn dept, Hosp, 47-65, Frank Wistar Thomas prof med, Sch Med, 55, pres med bd, 61, EMER PROD MED, SCH MED, UNIV PA, 70- Concurrent Pos: Mem bd trustees, Assoc Univs, Inc, Brookhaven, 53-61; vis prof, Peter Bent Brigham Hosp, Boston, 55, Univ Ore, 57 & 63, Duke Univ, 58, Univ Ark, 64 & Northwestern Univ, Ill, 65; Hugh J Morgan vis prof, Vanderbilt Univ, 62. Mem: Am Clin & Climat Asn (pres, 56); master Am Col Physicians; Am Heart Asn; Am Soc Clin Invest; Asn Am Physicians (pres, 66). Res: Internal medicine; cardiovascular disease. Mailing Add: 216 Maloney Clin Bldg Hosp of Univ of Pa Philadelphia PA 19104

WOOD, FRANCIS EUGENE, b Kirksville, Mo, Sept 19, 32; m 55; c 4. ENTOMOLOGY. Educ: Univ Mo-Columbia, BS, 58, MS, 62; Univ Md, PhD(entom), 70. Prof Exp: Exten entomologist, Univ Mo-Columbia, 60-64, EXTEN ENTOMOLOGIST, UNIV MD, 64-, ASST PROF ENTOM, 71- Mem: Entom Soc Am. Res: Taxonomy of Coleoptera; household insects; youth entomology. Mailing Add: Dept of Entom Univ of Md College Park MD 20742

WOOD, FRANK BRADSHAW, b Jackson, Tenn, Dec 21, 15; m 45; c 4. ASTRONOMY. Educ: Univ Fla, BS, 36; Princeton Univ, MA, 40, PhD(astron), 41. Prof Exp: Res assoc, Princeton Univ, 46; Nat Res Coun fel, Steward Observ, Univ Ariz & Lick Observ, Univ Calif, 46-47; asst prof astron & asst astronr, Univ Ariz, 47-50; from assoc prof to prof astron, Univ Pa, 50-58, Flower prof & chmn dept, 58-68; PROF ASTRON & DIR OPTICAL ASTRON OBSERV, UNIV FLA, 68-, ASSOC CHMN ASTRON, 71- Concurrent Pos: Mem comts, Int Astron Union, 38 & 42, mem orgn comt & chmn comt int progs, 42, pres comt, 42, 68-71; Fulbright fel, Mt Stromlo Observ, Australian Nat Univ, 57-58; exec dir, Flower & Cook Observ, Univ Pa, 50-54, dir, 54-68; NATO sr fel sci, 73; Am Astron Soc vis lectr astron, 73- Mem: AAAS (secy, Sect Astron, 58-70); Am Astron Soc; Royal Astron Soc; hon mem Royal Astron Soc NZ. Res: Photoelectric photometry eclipsing binary stars; analysis of spectrophotometric data in the far ultraviolet as taken from the Copernicus satellite; emphasis on close double stars. Mailing Add: Dept of Physics & Astron Univ of Fla Gainesville FL 32611

WOOD, GALEN THEODORE, b Philadelphia, Pa, Feb 7, 29; m 55; c 3. NUCLEAR PHYSICS. Educ: Washington Univ, BS, 51, PhD(physics), 56. Prof Exp: Physicist, Argonne Cancer Res Hosp, Univ Chicago, 55-57; NSF res fel nuclear spectros, Inst Theoret Physics, Univ Copenhagen, 57-59; asst prof physics & res nuclear spectros, Univ Pa, 59-65, NSF res grant radioactive nuclei, 63-65; assoc physicist, Argonne Nat Lab, 65-69; ASSOC PROF PHYSICS, CLEVELAND STATE UNIV, 69- Mem: Am Phys Soc. Res: Nuclear spectroscopy of radiations from radioactive decay and nuclear reactions, decay scheme, gamma-gamma directional and polarization correlations, magnetic moments, lifetimes and nuclear magnetic hyperfine fields; electron accelerator developments; nuclear spectroscopy. Mailing Add: Dept of Physics Cleveland State Univ Cleveland OH 44115

WOOD, GARNETT ELMER, b Gloucester, Va, Feb 14, 29; m 53; c 2. BIOCHEMISTRY. Educ: Va State Col, BS, 51, MS, 56; Georgetown Univ, PhD(chem), 66. Prof Exp: Res microbiologist, Div Vet Med, Walter Reed Army Med Ctr, 56-64; RES CHEMIST, DIV CHEM & PHYSICS, FOOD & DRUG ADMIN, 65- Concurrent Pos: Lectr chem, Fed City Col, Washington, DC, 73- Mem: AAAS; Am Chem Soc; Am Oil Chemists Soc. Res: Chemistry of toxic and deleterious compounds that may arise in certain foods as a result of handling, storage and/or processing. Mailing Add: 4020 20th St NE Washington DC 20018

WOOD, GENE WAYNE, b Bedford, Va, Oct 23, 40; m 65; c 2. WILDLIFE ECOLOGY. Educ: Va Polytech Inst & State Univ, BS, 63; Pa State Univ, MS, 66, PhD(agron), 71. Prof Exp: Instr wildlife mgt, Pa State Univ, University Park, 67-71, asst prof wildlife ecol, 71-74; ASST PROF FORESTRY, BELLE W BARUCH RES INST, CLEMSON UNIV, 74- Mem: Ecol Soc Am; Wildlife Soc; Soc Am Foresters. Res: Effects of silvicultural practices on animal population and habitat; nutrient distribution in forest ecosystems. Mailing Add: Belle W Baruch Res Inst Clemson Univ Box 596 Georgetown SC 29440

WOOD, GEORGE WILLIAM, b Warrensburg, Mo, June 16, 19. ACOUSTICS, RESEARCH ADMINISTRATION. Educ: Cent Mo State Col, BS, 41; La State Univ, MS, 46. Prof Exp: Instr physics, La State Univ, 43-48; res scientist, Defense Res Lab, Univ Tex, 49-53; actg exec secy, Comt Undersea Warfare, 53-54, exec secy, 55-59, exec dir, 66-69, asst exec secy, Div Phys Sci, 68-73, exec secy, Physics Surv Comt, 69-73, spec progs officer, 73-75, EXEC SECY, OFF PHYS SCI, NAT ACAD SCI-NAT RES COUN, 75- Concurrent Pos: Vis assoc prof physics, Tulane Univ, 53-54; res assoc, Antisubmarine Warfare Res Ctr, Supreme Allied Command, Atlantic, 59-61, asst sci dir, 61-62; head int progs, Off Foreign Secy, Nat Acad Sci, 62-63; staff mem, Inst Defense Anal, 64-65. Mem: AAAS; fel Acoust Soc Am; fel Royal Soc Arts; Sigma Xi. Mailing Add: Nat Acad Sci 2101 Constitution Ave NW Washington DC 20418

WOOD, GERRY ODELL, b Oklahoma City, Okla, Nov 19, 43; m 65. INDUSTRIAL HYGIENE. Educ: Univ Okla, BSCh, 65; Univ Tex, Austin, PhD(phys chem), 69. Prof Exp: Res asst phys chem, Kerr-McGee Res Ctr, Okla, 65; fel, 69-71, STAFF MEM CHEM, LOS ALAMOS SCI LAB, UNIV CALIF, 72- Mem: Am Chem Soc. Res: Air sampling techniques; analytical methods development; chemical kinetics; photochemistry; dynamics of gas phase reactions. Mailing Add: Indust Hyg Group Los Alamos Sci Lab Univ of Calif Los Alamos NM 87544

WOOD, GILBERT CONGDON, biology, entomology, see 12th edition

WOOD, GLEN MEREDITH, b Dallas, Tex, Apr 17, 20; m 50; c 7. AGRONOMY. Educ: RI State Col, BS, 47; Rutgers Univ, MS, 48, PhD(agr), 50. Prof Exp: Asst, Rutgers Univ, 47-50; actg chmn dept, 53-55, ASSOC PROF AGRON & ASSOC AGRONOMIST, UNIV VT, 50- Concurrent Pos: Golf Course Supts Asn Am res grants, 67, 68 & 71-74; assoc prof & assoc agronomist, Wash State Univ, 69-70. Mem: Am Soc Agron; Crop Sci Soc Am; Int Turfgrass Soc; Am Forage & Grassland Coun. Res: Cold hardiness in ladino clover; physiological and environmental studies with birdsfoot trefoil and other forage crops; forage utilization by poultry; turfgrass management; shade and drouth studies with turfgrasses; application of infrared photography to turfgrass research; cold hardiness studies with forage and turfgrasses. Mailing Add: Dept of Plant & Soil Sci Univ of Vt Burlington VT 05401

WOOD, GORDON WALTER, b NS, Can, Apr 6, 33; m 56; c 2. ORGANIC CHEMISTRY, MASS SPECTROMETRY. Educ: Mt Allison Univ, BSc, 55, MSc, 56; Syracuse Univ, PhD(org chem), 62. Prof Exp: Elem sch teacher, NS, 51-52; chemist, Paints Div, Can Industs Ltd, 56-58; fel with A C Cope, Mass Inst Technol, 62-63; from asst prof to assoc prof, 63-75, PROF CHEM, UNIV WINDSOR, 75- Concurrent Pos: Vis assoc res scientist, Space Sci Lab, Univ Calif, Berkeley, 69-70. Mem: AAAS; Am Chem Soc; The Chem Soc; Am Soc Mass Spectrometry; Chem Inst Can. Res: Applications of field ionization and field desorption mass spectrometry to problems in organic and biochemistry. Mailing Add: Dept of Chem Univ of Windsor Windsor ON Can

WOOD, GWENDOLYN BILLINGS, physical chemistry, see 12th edition

WOOD, HARLAND G, b Delavan, Minn, Sept 2, 07; m 29; c 3. BIOCHEMISTRY, MICROBIOLOGY. Educ: Macalester Col, BA, 31; Iowa State Col, PhD, 35. Hon Degrees: ScD, Macalester Col, 46 & Northwestern Univ, 72. Prof Exp: From instr to asst prof bact, Iowa State Col, 36-43; assoc prof physiol chem, Univ Minn, 43-46; prof biochem & dir, 46-65, dean sci, 67-69, PROF BIOCHEM, CASE WESTERN RESERVE UNIV, 65-, UNIV PROF, 70- Concurrent Pos: Nat Res Coun fel, Univ Wis-Madison, 35-36; Fulbright fel, Univ Dunedin, 55; Commonwealth fel, Max Planck Inst, Munich, Ger, 62; mem adv coun, Life Ins Med Res Fund, 57-62; mem, NIH Training Grant Comt, 65-69; mem adv bd, Am Cancer Soc, 65-69; mem, President's Adv Comt, 67-71; mem coun, Int Union of Biochem, 67-, secy gen, 70-73; mem phys study sect, NIH, 73- Honors & Awards: Eli Lilly Award in Bact, 42; Carl Neuberg Medal, 52; Bayerischen Akademie, Ger, 63; Modern Med Award for Distinguished Achievement, 68; Lynen Lectr & Medal, 72. Mem: Nat Acad Sci; Am Acad Arts & Sci; Am Soc Biol Chemists; Am Chem Soc; Soc Am Microbiol. Res: Tracer studies with labeled compounds; structure of enzymes; mechanism of enzyme action; role of biotin B-12 and metals. Mailing Add: Dept of Biochem Case Western Reserve Univ Cleveland OH 44106

WOOD, HAROLD ARTHUR, b Jamaica, BWI, June 1, 21; nat US; m 45; c 2. GEOGRAPHY. Educ: McMaster Univ, MA, 51; Univ Toronto, PhD(geog), 58. Prof Exp: Headmaster, Col Int, Cap Haitien, Haiti, 43-mod langs, Stoney Brook Sch, NY, 47-49; from lectr to assoc prof, 50-66, PROF GEOG, McMASTER UNIV, 66- Concurrent Pos: Leader geog surv, Can Govt, Manitoba, 51, Newfoundland, 52 & St Lawrence Seaway area, 53; consult land use classification study, Orgn Am Studies, 65-66; consult, Geog & Hist, 65-, pres geog comn, 73-, pres orgn, 1st Interam sem Definition of Regions Develop Planning, 67. Mem: Asn Am Geog; Am Geog Soc; Am Soc Photogram; Can Asn Geog; Can Inst Surv. Res: Latin America; cartography; urban geography; air photo interpretation. Mailing Add: Dept of Geog McMaster Univ Hamilton ON Can

WOOD, HARRY ALAN, b Albany, NY, Apr 24, 41; m 63; c 1. PLANT VIROLOGY. Educ: Middlebury Col, AB, 63; Purdue Univ, MS, 65, PhD(plant virol), 68. Prof Exp: RES SCIENTIST, BOYCE THOMPSON INST PLANT RES, 68- Mem: Am Phytopath Soc; Am Soc Microbiol. Res: Physical and biological properties of viruses of plants and fungi with special interest in double stranded RNA viruses. Mailing Add: Boyce Thompson Inst for Plant Res 1086 N Broadway Yonkers NY 10701

WOOD, HARRY BURGESS, JR, b Washington, DC, July 30, 19; m 46; c 1. ORGANIC CHEMISTRY. Educ: Washington & Lee Univ, AB, 42; Ohio State Univ, MS, 47, PhD(chem), 50. Prof Exp: Supvr prod control, Res & Develop, Merck & Co, Inc, 42-45; res assoc & spec asst carbohydrate chem & micro org anal, Ohio State Univ, 47-48, asst gen chem, 48-49 & org chem, 49-50; res chemist chem natural prod, Nat Heart Inst, 50-53 & carbohydrate chem, Nat Inst Arthritis & Metab Dis, 53-61, CHIEF DRUG DEVELOP BR, NAT CANCER INST, NIH, 61- Mem: AAAS; Am Chem Soc; Am Soc Hort Sci; The Chem Soc; Fr Soc Therapeut Chem. Res: Carbohydrate and medicinal chemistry; synthetic and . structural chemistry of natural products; degradation and structure determinations; cancer chemotherapy. Mailing Add: Drug Develop Br Nat Cancer Inst NIH Bethesda MD 20014

WOOD, HENDERSON KINGSBERRY, b Huntington, WVa, Feb 24, 13; m 38; c 1. GENETICS, PHYSIOLOGY. Educ: Ohio Wesleyan Univ, BA, 37; Fisk Univ, MA, 40; Ind Univ, PhD(zool), 53. Prof Exp: Instr biol, Ala State Teachers Col, 40-44; from instr to asst prof, Fisk Univ, 44-48; PROF BIOL, TENN STATE UNIV, 48-, HEAD DEPT BIOL SCI, 56- Concurrent Pos: Grad consult, Tenn State Univ, 52-62. Mem: Am Genetic Asn; Nat Inst Sci. Res: Protozoan genetics; physiology of Protozoa. Mailing Add: Dept of Biol Sci Tenn State Univ Nashville TN 37203

WOOD, HENRY NELSON, b Passaic, NJ, June 24, 25; m 51; c 3. BIOCHEMISTRY. Educ: Wagner Col, BS, 49; NC State Col, MS, 51; Purdue Univ, PhD(biochem), 55. Prof Exp: Res assoc, 55-62, asst prof, 62-67, ASSOC PROF PLANT BIOL, ROCKEFELLER UNIV, 67- Mem: Harvey Soc; Am Soc Biol Chemists. Res: Biochemistry and physiology of growth processes. Mailing Add: Rockefeller Univ 66th St & York Ave New York NY 10021

WOOD, HOWARD JOHN, b Baltimore, Md, July 19, 38; m 61; c 2. ASTRONOMY. Educ: Swarthmore Col, BA, 60; Ind Univ, MA, 62, PhD(astron), 65. Prof Exp: Res asst, Sproul Observ, 57-59, Goethe Link Observ, 58-62 & Lowell Observ, 62-63; from instr to assoc prof astron, Univ Va, 64-70; staff astronr, Europ Southern Observ, Santiago, Chile, 70-75; FULBRIGHT VIS PROF, UNIV OBSERV, VIENNA, 75- Concurrent Pos: Guest investr, McDonald Observ, 59-60, Lowell Observ, 62-65 & Kitt Peak Nat Observ, 63-69; NSF grants, 66-70. Mem: Fel AAAS; Am Astron Soc; fel Royal Astron Soc. Res: Photoelectric and spectrophotometric studies of the Balmer lines in the spectra of the magnetic and related stars; Zeeman spectroscopy of magnetic stars; photometric studies of asteroids; photography of Mars. Mailing Add: Univ Observ Tuerkenschanzstr 17 A-1180 Vienna Austria

WOOD, IRWIN BOYDEN, b Concord, NH, Apr 27, 26; m; c 3. PARASITOLOGY. Educ: Univ NH, BS, 49, MS, 51; Kans State Univ, PhD(parasitol), 58. Prof Exp: Asst zoologist, Univ NH, 50; chemist, 52-54, parasitologist, 54-56, res parasitologist & group leader, Agr Div, 58-64, mgr animal res & develop, Cyanamid Int, 64-74, DIR ANIMAL PROD RES & DEVELOP, CYANAMID INT, AM CYANAMID CO, WAYNE, 74- Mem: Am Soc Parasitologists; World Asn Advan Vet Parasitologists. Res: Chemotherapy and physiology of helminths; host-parasite relations; bacterial chemotherapy; acaricides; animal health and feed product development. Mailing Add: RR 1 Box 182A Pennington NJ 08534

WOOD, JACK SHEEHAN, b St Albans, Vt, Oct 31, 31; m 58; c 2. ENVIRONMENTAL PHYSIOLOGY. Educ: Univ Maine, Orono, BS, 54; Mich State Univ, MA, 60, PhD(ecol, animal physiol), 63. Prof Exp: From asst prof to assoc prof, 63-75, PROF BIOL, WESTERN MICH UNIV, 75-, ASSOC DIR ENVIRON STUDIES, INST PUB AFFAIRS, 72- Concurrent Pos: Water qual dir, Mich SCent Planning & Develop Region, 75- Mem: AAAS; Wildlife Soc; Am Inst Biol Sci; Soc Exp Biol & Med. Res: Physiological response to adverse environmental conditions, including general systematic stress responses, reproductive inhibition and related phenomena in vertebrates. Mailing Add: Dept of Biol Western Mich Univ Kalamazoo MI 49001

WOOD, JACKIE DALE, b Picher, Okla, Feb 16, 37; m 56; c 2. PHYSIOLOGY, ZOOLOGY. Educ: Kans State Col Pittsburg, BS, 64, MS, 66; Univ Ill, PhD(physiol), 69. Prof Exp: Asst prof biol, Williams Col, 69-71; asst prof, 71-74, ASSOC PROF PHYSIOL, UNIV KANS MED CTR, KANSAS CITY, 74- Mem: AAAS; Am Inst Biol Sci; Am Soc Zool; Soc Neurosci; Am Physiol Soc. Res: Nerve and muscle physiology; comparative animal physiology; mechanisms of nervous control of intestinal muscle. Mailing Add: Dept of Physiol Univ of Kans Med Ctr Kansas City KS 66103

WOOD, JAMES, physiology, see 12th edition

WOOD, JAMES ALAN, b Richmond, Va, Sept 16, 39; m 65; c 1. MATHEMATICAL ANALYSIS. Educ: Georgetown Univ, BS, 61; Univ Va, MA, 63, PhD(math), 66. Prof Exp: From instr to asst prof math, Georgetown Univ, 65-69; asst prof, 69-72, ASSOC PROF MATH, VA COMMONWEALTH UNIV, 72- Concurrent Pos: NSF res grant, 68-70. Mem: Am Math Soc. Res: Operational calculus and dynamical systems; multiplier theory. Mailing Add: 10250 Gwynnbrook Rd Richmond VA 23235

WOOD, JAMES ALEXANDER, b Melfort, Sask, Sept 20, 16; m; c 2. PHARMACY. Educ: Univ Sask, BSP, 46; Univ Wash, Seattle, MS, 52, PhD, 58. Prof Exp: From instr to assoc prof, 46-63, PROF PHARM, UNIV SASK, 63- Mem: Can Pharmaceut Asn; Asn Fac Pharm Can. Res: Surface free energy values as parameters in the prediction of adhesive bond strengths developed by tablet film coatings; diffusion of medicaments in and from dermatologic preparations. Mailing Add: Col of Pharm Univ of Sask Saskatoon SK Can

WOOD, JAMES C, JR, b Spartanburg, SC, Aug 21, 39; m 64; c 2. SOLID STATE PHYSICS. Educ: Clemson Univ, BS, 61, MS, 63; Univ Va, PhD(physics), 66. Prof Exp: Res physicist, Cent Res Div, Am Cyanamid Co, Conn, 66-71; sr res scientist, TRW Eastern Res Lab, 71-73; HEAD DEPT SCI TEACHING PHYSICS & PHYS SCI, TRI-COUNTY TECH COL, 73- Mem: Am Phys Soc; Am Asn Physics Teachers. Mailing Add: Tri-County Tech Col Pendleton SC 29670

WOOD, JAMES DOUGLAS, b Aberdeen, Scotland, Jan 25, 30; m 56; c 3. NEUROCHEMISTRY. Educ: Aberdeen Univ, BSc, 51, PhD(biochem), 54. Prof Exp: Res officer, Can Dept Agr, 54-57; assoc scientist, Fisheries Res Bd, Can, 57-61; head biochem group, Defence Res Med Labs, Can, 61-63, head physiol chem sect, 63-68; PROF BIOCHEM & HEAD DEPT, UNIV SASK, 68- Mem: Can Physiol Soc; Am Soc Neurochem; Int Soc Neurochem; Can Biochem Soc. Res: Gamma-aminobutyric acid metabolism; oxygen toxicity. Mailing Add: Dept of Biochem Univ of Sask Saskatoon SK Can

WOOD, JAMES EDWIN, III, b Charlottesville, Va, Feb 5, 25; m 48; c 4. MEDICINE. Educ: Harvard Med Sch, MD, 49. Prof Exp: Intern med, Mass Mem Hosps, 49-50; asst, Sch Med, Boston Univ, 50-51, instr, 53-58; assoc prof, Med Col Ga, 58-64; prof, Sch Med, Univ Va, 64-69, Va Heart Asn res prof cardiol, 64-68, prof physiol & chmn dept, 65-66, assoc dean sch med, 68-69; PROF MED, SCH MED, UNIV PA, 69-, DIR MED, PA HOSP, 69- Concurrent Pos: USPHS spec res fel, 57-58; res fel, Evans Mem & Mass Mem Hosps, 50-51 & 54-56; resident, Mass Mem Hosp, 53-54, asst vis physician, 56-58; asst mem, Evans Mem & Mass Mem Hosps, 56-58; dir, Ga Heart Asn Lab Cardiovasc Res, 58-64. Mem: Am Physiol Soc; Am Soc Clin Invest; Am Heart Asn; Am Fedn Clin Res; Am Clin & Climat Asn. Res: Venous distensibility; physiological and pathological responses of the veins; peripheral vascular diseases; hypertensive diseases. Mailing Add: Pa Hosp Eighth & Spruce St Philadelphia PA 19107

WOOD, JAMES KENNETH, b Boulder, Colo, Jan 29, 42; m 66; c 2. SYNTHETIC ORGANIC CHEMISTRY. Educ: Colo State Univ, BS, 64; Kans State Col, MS, 65; Ohio State Univ, PhD(chem), 69. Prof Exp: Asst prof, 69-74, ASSOC PROF CHEM, UNIV NEBR AT OMAHA, 74- Mem: Am Chem Soc. Res: Synthesis of novel and biologically active compounds; development of new synthetic techniques and methods; development of new methods for the resolution of racemates. Mailing Add: Dept of Chem Univ of Nebr at Omaha Omaha NE 68101

WOOD, JAMES LEE, b Cordele, Ga, Sept 5, 40; m 60; c 4. INORGANIC CHEMISTRY, THERMOCHEMISTRY. Educ: Vanderbilt Univ, BA, 62, PhD(inorg chem), 66. Prof Exp: Res fel chem, Rice Univ, 65-66; asst prof, 66-69, ASSOC PROF CHEM, DAVID LIPSCOMB COL, 69- Concurrent Pos: Sr fel, Rice Univ, 71-73; Indust Res 100 Award, Indust Res Mag. Mem: Am Chem Soc. Res: Thermodynamics and reaction calorimetry; fluorine chemistry; coordination compounds. Mailing Add: Dept of Chem David Lipscomb Col Nashville TN 37203

WOOD, JAMES MANLEY, JR, b Birmingham, Ala, July 5, 27; m 53; c 3. PHYSICAL CHEMISTRY. Educ: Howard Col, BA, 47; Univ Wis, PhD, 52. Prof Exp: Res chemist, La, 52-69, RES ASSOC, ETHYL CORP, 69- Mem: Am Chem Soc. Res: Molecular spectra; electrochemistry of fused salts; high energy batteries; decomposition of organometallic compounds, vapor plating; high temperature chemistry. Mailing Add: Ethyl Corp PO Box 341 Baton Rouge LA 70821

WOOD, JAMES THORNTON, b Montclair, NJ, Apr 1, 40; m 66; c 2. MATHEMATICS. Educ: Amherst Col, BA, 61; Univ Pa, MA, 63, PhD(math), 67. Prof Exp: Instr math, Haverford Col, 65-66; from instr to asst prof, Swarthmore Col, 66-70; ASST PROF MATH, COLO COL, 70- Mem: Math Asn Am; Hist Sci Soc; Nat Coun Teachers Math. Res: Operator algebras; history of mathematics; mathematical history of quantum mechanics. Mailing Add: Dept of Math Colo Col Colorado Springs CO 80903

WOOD, JAMES W, b Seattle, Wash, Jan 22, 25; m 53; c 2. BIOLOGY. Educ: Univ Wash, BS, 50, MS, 58. Prof Exp: Aquatic biologist, Fish Comn Ore, 50-55, fish pathologist, 55-60; fish pathologist, 60-70, SUPVR FISH CULT RES, WASH STATE DEPT FISHERIES, 70- Concurrent Pos: Consult, Int Pac Salmon Fisheries Comn, 64-65, Can Dept Fisheries, 66 & Repub of Chile Dept Fisheries, 70-71. Mem: Am Fisheries Soc; Am Inst Fishery Res Biol; Wildlife Dis Asn. Res: Infectious and nutritional diseases of salmonid fishes. Mailing Add: Wash State Dept of Fisheries Univ of Wash Seattle WA 98105

WOOD, JEANIE MCMILLIN, b Spartanburg, SC, Sept 26, 39; m 68; c 2. BIOCHEMISTRY, CHEMISTRY. Educ: Converse Col, BA, 61; Univ NC, Chapel Hill, PhD(biochem), 67. Prof Exp: Instr biochem, Univ NC, Chapel Hill, 67-68; res assoc, Dept Med, Cornell Univ & Inst Muscle Dis, Inc, 68-69; res assoc, 69-70, res instr, 70-71, instr, 71-72, res asst prof myocardial biol, 72-73, ASST PROF CELL BIOPHYS, BAYLOR COL MED, 73- Concurrent Pos: USPHS grant, Univ NC, Chapel Hill, 67-68; Muscular Dystrophy Asn Am grant, Cornell Univ & Inst Muscle Dis, Inc, 68-69; Tex Med Ctr grant myocardial biol, Baylor Col Med, 69-70; Tex Heart Asn grant, 70-72; NIH grant, 75-79; mem, Int Study Group Res Cardiac Metab. Mem: Sigma Xi; Biophys Soc. Res: Effect of myocardial ischemia on the mitochondrial functions of energy production and fatty acid transport and oxidation; carnitine palmityltransferase system. Mailing Add: Dept of Cell Biophys Baylor Col of Med Houston TX 77025

WOOD, JESSE HERMON, organic chemistry, see 12th edition

WOOD, JOE GEORGE, b Victoria, Tex, Dec 8, 28; m 55; c 1. ANATOMY. Educ: Univ Houston, BS, 53, MS, 58, PhD(anat), 62. Prof Exp: Asst biol, Univ Houston, 56-58; instr anat, Dent Br, Univ Tex, 61 & Sch Med, Yale Univ, 62-63; asst prof, Sch Med, Univ Ark, 63-66; assoc prof, Univ Tex Med Sch, San Antonio, 66-70, asst dean acad develop, 67-69; dir prog neurostruct & function, 70-75, PROF NEUROBIOL & CHMN DEPT NEUROBIOL & ANAT, UNIV TEX MED SCH, HOUSTON, 70- Concurrent Pos: USPHS trainee, Sch Med, Yale Univ, 62-63; mem neuroanat vis scientist prog, USPHS, 65-66. Mem: Am Asn Anat; Soc Exp Biol & Med; Am Soc Cell Biol; Asn Am Med Cols; Histochem Soc. Res: Histochemistry and cytochemistry of neurons; histochemical and electron microscopic localization of biogenic amines and their relation to nerve function in animals under stress and drug administration. Mailing Add: Dept of Neurobiol & Anat Univ of Tex Med Sch Houston TX 77025

WOOD, JOHN ARMSTEAD, JR, b Roanoke, Va, July 28, 32; m 58; c 2. GEOCHEMISTRY. Educ: Va Polytech Inst, BS, 54; Mass Inst Technol, PhD(geol), 58. Prof Exp: Geologist, Smithsonian Astrophys Observ, 59; Am Chem Soc-Petrol Res Fund fel, Cambridge Univ, 59-60; geologist, Smithsonian Astrophys Observ, 60-62; res assoc, Enrico Fermi Inst Nuclear Studies, Univ Chicago, 62-65; GEOLOGIST, SMITHSONIAN ASTROPHYS OBSERV, 65- Concurrent Pos: Res assoc, Harvard Col Observ, 60-; mem comt surfaces & interiors of moon & planets, Space Sci Bd, Nat Acad Sci, 63; vchmn, Lunar Sample Anal Planning Team, 71-; lectr geol, Harvard Univ, 73- Honors & Awards: NASA Medal for Exceptional Sci Achievement, 73. Mem: Fel AAAS; fel Am Geophys Union; Geochem Soc; Meteoritical Soc (pres, 70-72). Res: Study of meteorites as samples of primordial planetary material; lunar petrology and geophysics; origin of the planets. Mailing Add: Smithsonian Astrophys Observ 60 Garden St Cambridge MA 02138

WOOD, JOHN CHARLES, organic chemistry, see 12th edition

WOOD, JOHN DAVID, b Galt, Ont, Feb 23, 34; m 64; c 2. GEOGRAPHY. Educ: Univ Toronto, BA, 55, MA, 58; Univ Edinburgh, PhD(geog), 62. Prof Exp: Asst lectr geog, Univ Edinburgh, 57-62; asst prof, Univ Alta, 62-65; chmn dept, Atkinson Col, York Univ, 65-72, ASSOC PROF GEOG, ATKINSON COL, YORK UNIV, 65- Concurrent Pos: Nat Res Coun Can grant, 63; Can Coun res grants, 64, 66 & 67-68. Mem: Can Asn Geog; Royal Scottish Geog Soc. Res: Migration, especially British to North America; pioneer settlement; theory of geography. Mailing Add: Dept of Geog Atkinson Col York Univ 4700 Keele St Downsview ON Can

WOOD, JOHN EDWARD, III, b Lynchburg, Va, May 20, 16; m 46. ORGANIC CHEMISTRY. Educ: Lynchburg Col, AB, 36; Mass Inst Technol, PhD(org chem), 39. Prof Exp: Res chemist, Standard Oil Develop Co, 39-41; plant technol supvr, Standard Oil Co La, 41-42, asst supvr synthetic alcohol plants, 42-45; gen foreman synthetic alcohol & olefin extraction plants, Esso Standard Oil Co, 45-49, head process eng, 50-52, asst head tech div, 52-53, head petrol prod div, 54, head chem prod div, 54-55, asst gen mgr, Chem Prod Dept, 55-56, gen mgr, 56-58; pres, Enjay Co, Inc, 58-60 & Enjay Chem Co Div, Humble Oil & Refining Co, 60-65; pres, Chem Div, 65-69, exec vpres chem & metallics, 65-72, VPRES, VULCAN MAT CO, 72- Mem: Am Chem Soc; Am Inst Chem Engrs; Soc Chem Indust; Am Inst Chemists. Res: Friedel-Crafts reaction. Mailing Add: Vulcan Mat Co PO Box 7497 Birmingham AL 35223

WOOD, JOHN GRADY, b Atlanta, Ga, Aug 1, 42. NEUROBIOLOGY. Educ: Ga State Univ, BS, 67; Emory Univ, PhD(anat), 71. Prof Exp: Fel neurobiol, Inst Animal Physiol, Cambridge, Eng, 71-73 & City of Hope Med Ctr, Duarte, Calif, 73-74; ASST PROF ANAT, CTR HEALTH SCI, UNIV TENN, 74- Concurrent Pos: Fel, Nat Multiple Sclerosis Soc, 71-72 & Multiple Sclerosis Soc Gt Brit & Northern Ireland, 72-73; independent res fel neurobiol, Friday Harbor Marine Labs, Friday Harbor, Wash, 74; Alfred P Sloan Found res fel, 76. Mem: Am Asn Anat; Soc Neurosci; Am Soc Cell Biol; Am Soc Neurochem; Am Soc Zoologists. Res: The fine structural immunocytochemical localization of carbohydrates and membrane antigens in the nervous system; the role of glycoproteins in myelin structure and function; properties of reaggregating sponge cells. Mailing Add: Anat Dept Ctr Health Sci Univ of Tenn Memphis TN 38163

WOOD, JOHN HENRY, b Calgary, Alta, Nov 18, 24; nat US; m 50; c 2. PHARMACEUTICS. Educ: Univ Man, BSc, 46, MSc, 47; Ohio State Univ, PhD(phys chem), 50. Prof Exp: Proj chemist, Colgate-Palmolive Co, 50-53; sr res assoc chem, Rensselaer Polytech Inst, 53-54; proj chemist, Colgate-Palmolive Co, 54-56, group leader, 56-57; head phys chem sect, Prod Div, Bristol-Myers Co, NJ, 57-61, head phys chem dept, 61-65, asst dir res & develop labs, 65-67, dir chem res, 67-69; PROF PHARM, MED COL VA, VA COMMONWEALTH UNIV, 69- Concurrent Pos: Mem comt rev, US Pharmacopoeia, 70-75. Mem: Am Chem Soc; Soc Rheol; Soc Cosmetic Chemists; Am Pharmaceut Asn; fel Acad Pharmaceut Sci. Res: Phase rule studies; surfactants and micellar phenomena; rheology; physical pharmacy; biopharmaceutics and pharmacokinetics; physics of tabletting. Mailing Add: 1504 Cedarbluff Dr Richmond VA 23233

WOOD, JOHN HERBERT, b Michigan City, Ind, Oct 12, 29; m 59; c 3. SOLID STATE PHYSICS. Educ: Purdue Univ, BS, 51; Mass Inst Technol, PhD(solid state physics), 58. Prof Exp: Res assoc solid state physics, Mass Inst Technol, 58-62, asst prof, 62-66, dir coop comput lab, 64-65; consult, 65-66, STAFF MEM, LOS ALAMOS SCI LAB, 66- Concurrent Pos: Res assoc, Atomic Energy Res Estab, Harwell, Eng, 74-75. Mem: Am Phys Soc. Res: Calculation of atomic wave functions and energy levels; calculation of energy band structures; calculation of molecular structure via complete neglect of differential overlap and scattered wave methods. Mailing Add: Group CMB-5 MS 730 Los Alamos Sci Lab Los Alamos NM 87544

WOOD, JOHN JACKSON, b Decatur, Tex, Mar 21, 40; m 61; c 4. CULTURAL ANTHROPOLOGY, ARCHAEOLOGY. Educ: Univ Colo, BA, 62, MA, 64, PhD(anthrop), 67. Prof Exp: ASSOC PROF ANTHROP, NORTHERN ARIZ UNIV, 66- Mem: Am Anthrop Asn; Am Asn Phys Anthrop; Soc Am Archaeol. Res: Statistical methods and computer applications; North America, Africa. Mailing Add: Dept of Anthrop Northern Ariz Univ Flagstaff AZ 86001

WOOD, JOHN KARL, b Logan, Utah, July 8, 19; m 47; c 4. PHYSICS. Educ: Utah State Agr Col, BS, 41; Pa State Col, MS, 42, PhD(physics), 46. Prof Exp: Asst petrol refining, Pa State Univ, 44-46; optical engr, Bausch & Lomb Optical Co, NY, 46-48; from asst prof to assoc prof physics, Univ Wyo, 48-56; PROF PHYSICS, UTAH STATE UNIV, 56- Concurrent Pos: NSF sci fac fel, Sweden, 66. Mem: Am Phys Soc; Am Soc Metals; Optical Soc Am. Res: Crystal orientation in metals studies by means of x-rays; Raman spectroscopy; pole figures of the effect of some cold rolling mill variables on low carbon steel; light; molecular and atomic physics; general mathematics; sound. Mailing Add: Dept of Physics Utah State Univ Logan UT 84322

WOOD, JOHN LANGILLE, b Boston, Mass, Nov 15, 13; m 38; c 2. MYCOLOGY. Educ: Univ Mass, BS, 36, MS, 47; Columbia Univ, PhD(bot, mycol), 51. Prof Exp: Instr bot, Univ Mass, 46 & Pa State Univ, 46-49; asst, Columbia Univ, 49-51; instr dermat, Sch Med, Johns Hopkins Univ & dir fungus lab, Johns Hopkins Hosp, 51-56, lectr microbiol, Sch Hyg & Pub Health, 52-54, res assoc, 54-56; asst prof bot & bact, Univ Cincinnati, 56-59; assoc prof, 59-61, PROF MARINE SCI, COL WILLIAM & MARY, 61-; ASSOC PROF MARINE SCI, UNIV VA, 63-; ASSOC DIR, VA INST MARINE SCI, 67- Concurrent Pos: NIH grant, 57-58; assoc marine biologist, Va Fisheries Lab, 59-60; asst dir & sr marine scientist, Va Inst Marine Sci, 61-67; assoc dir sch marine sci, Col William & Mary, 74- Mem: Nat Shellfisheries Asn. Res: Marine microbiology; infectious diseases of shellfish; physiology of pathogenic microorganisms; trace element nutrition; medical mycology; life histories and cytology of ascomycetes; research administration. Mailing Add: Va Inst of Marine Sci Gloucester Point VA 23062

WOOD, JOHN LEWIS, b Homer, Ill, Aug 7, 12; m 41; c 2. BIOCHEMISTRY. Educ: Univ Ill, BS, 34; Univ Va, PhD(org chem), 37. Hon Degrees: DSc, Blackburn Univ, 55. Prof Exp: Asst biochem, Med Sch, George Washington Univ, 37-38; asst, Med Col, Cornell Univ, 38-39; assoc chemist, Eastern Regional Res Lab, Bur Agr Chem & Eng, USDA, 41-42; asst biochem, Med Col, Cornell Univ, 42-44, asst prof, 44-46; assoc prof, Col Med, 46-50, prof biochem, Med Units, 50-71, head dept biochem, 50-55, chmn dept, 55-67, ALUMNI DISTINGUISHED SERV PROF BIOCHEM, UNIV TENN CTR HEALTH SCI, MEMPHIS, 71- Concurrent Pos: Finney-Howell fel, Harvard Univ, 39-41; Guggenheim fel, 54; USPHS spec res fel, 65; Nat Acad Sci-Polish Acad Sci exchange visitor, 70. Mem: AAAS; Am Chem Soc; Am Soc Biol Chem; Soc Exp Biol & Med; Am Asn Cancer Res. Res: Biochemistry of amino acids; proteins; carcinogenesis; thiocyano derivatives; sulfur compounds. Mailing Add: Dept of Biochem Univ of Tenn Ctr for Health Sci Memphis TN 38163

WOOD, JOHN MARTIN, b Huddersfield, Eng, Mar 22, 38; m 62; c 1. BIOCHEMISTRY, ORGANIC CHEMISTRY. Educ: Univ Leeds, BSc, 61, PhD(biochem), 64. Prof Exp: Lectr org chem, Leeds Col Technol, Eng, 61-63; res assoc microbiol, Univ Ill, Urbana, 64-66, from asst prof to prof biochem, 66-72;

Guggenheim fel, Oxford, Eng, 72-73; PROF BIOCHEM & ECOL & DIR, FRESHWATER BIOL INST, UNIV MINN, 74- Honors & Awards: Medal for Environ Chem, Synthetic Org Chem Mfrs Asn US, 72. Mem: Am Chem Soc; Am Soc Microbiol. Res: Mechanisms of enzymes containing transition metals as prosthetic groups; chemistry of free and bound vitamin B12; effect of side-chain substituents on cleavage of aromatic compounds by oxygenases. Mailing Add: Freshwater Biol Inst PO Box 100 Navarre MN 55391

WOOD, JOHN STANLEY, b Stoke-on-Trent, Eng, Oct 9, 36; m 62; c 2. CHEMISTRY. Educ: Univ Keele, BA, 58; Univ Manchester, PhD(chem), 62. Prof Exp: Res assoc chem, Mass Inst Technol, 62-64; lectr, Univ Southampton, 64-70; PROF CHEM, UNIV MASS, AMHERST, 70- Mem: The Chem Soc; Am Chem Soc. Res: Inorganic chemistry; x-ray crystallography; studies of stereochemistries and electron structures of inorganic compounds. Mailing Add: Dept of Chem Univ of Mass Amherst MA 01002

WOOD, JOSEPH M, b Richmond, Ind, May 2, 21; m 57; c 2. BOTANY, PALEOBOTANY. Educ: Ind Univ, BA, 53, PhD(plant morphol, paleobot), 60; Univ Mich, MSc, 56. Prof Exp: From instr to assoc prof bot & paleobot, 57-73, asst dir div biol sci, 71-75, PROF BIOL SCI, UNIV MO-COLUMBIA, 73- Mem: AAAS; Bot Soc Am; Paleont Soc; Am Asn Stratig Palynologists; Brit Palaeont Asn. Res: Paleozoic and Mesozoic plant macro/micro fossils and stratigraphic significances. Mailing Add: Div of Biol Sci Tucker Hall Univ of Mo Columbia MO 65201

WOOD, KENNETH GEORGE, b Niagara Falls, Ont, Jan 11, 24; nat US; m 48; c 3. LIMNOLOGY. Educ: Univ Toronto, BA, 47, MA, 49; Ohio State Univ, PhD(hydrobiol), 53. Prof Exp: Asst prof biol, Buena Vista Col, 53-55 & RI Col, 55-56; prof, Thiel Col, 56-65; assoc prof, 65-71, PROF BIOL, STATE UNIV NY COL FREDONIA, 71- Concurrent Pos: Sabbatical leave, Calspan Corp, NY, 72-73. Mem: Ecol Soc Am; Sigma Xi; Am Soc Limnol & Oceanog; Int Asn Theoret & Appl Limnol. Res: Ecology of aquatic animals; primary productivity; heavy metal cycling in ecosystem. Mailing Add: Dept of Biol State Univ of NY Fredonia NY 14063

WOOD, LANDLEY HARRISS, b Lynchburg, Va, Aug 22, 24; m 51; c 5. ENVIRONMENTAL PHYSIOLOGY. Educ: Col William & Mary, BS, 49; Columbia Univ, AM, 50; Cornell Univ, PhD(biol), 65. Prof Exp: Instr sociol, Winthrop Col, 50-51; pvt bus, 52-56; researcher, Bur Commercial Fisheries, US Fish & Wildlife Serv, 56-57; res asst biol, Woods Hole Oceanog Inst, 57-58, Inst Fish Res, Univ NC, 59 & Lerner Lab, Am Mus Natural Hist, 59-60; assoc marine scientist & head dept, Va Inst Marine Sci, 61-67, sr marine scientist & head dept environ physiol, 67-69; prof zool & chmn dept, Univ NH, 69-72; COORDR ENVIRON STUDIES PROG, SWEET BRIAR COL, 72- Concurrent Pos: From asst prof to assoc prof, Col William & Mary, 61-69; asst prof, Univ Va, 63-69. Mem: AAAS; Am Soc Limnol & Oceanog; Am Soc Zoologists; Animal Behav Soc; Estuarine Res Soc. Res: Physiological and behavioral effects upon marine organisms of changes in sensory and biochemical characteristics of environment. Mailing Add: Box Z Sweet Briar Col Sweet Briar VA 24595

WOOD, LAWRENCE ARNELL, b Peekskill, NY, Jan 15, 04; m 51; c 2. PHYSICS. Educ: Hamilton Col, AB, 25; Cornell Univ, PhD(physics), 32. Prof Exp: From asst to instr physics, Cornell Univ, 27-35; res physicist, 35-43, chief rubber sect, 43-62, CONSULT RUBBER, NAT BUR STANDARDS, 62- Concurrent Pos: Deleg, Int Rubber Technol Conf, London, 38, 48 & 62, Kuala Lumpur, Malaysia, 68 & Rio de Janeiro, Brazil, 74. Honors & Awards: Meritorious Serv Award, US Dept Com, 58. Mem: Fel Am Phys Soc; Am Chem Soc. Res: Semiconductors; Hall effect; blocking layer photocells; physics and technology of polymers, especially synthetic rubbers and natural rubber. Mailing Add: Polymers Div Nat Bur of Standards Washington DC 20234

WOOD, LEON S, plant pathology, see 12th edition

WOOD, LEONARD ALTON, b Gratiot Co, Mich, Aug 22, 22; m 42; c 3. GEOLOGY. Educ: Mich State Univ, BS, 46. Prof Exp: Geologist, Water Resources Div, Mich, 46-51, Tex, 52-63 & Colo, 63-67, coordr subsurface waste disposal studies, 71-75, STAFF HYDROLOGIST, WATER RESOURCES DIV, US GEOL SURV, 67- Mem: Geol Soc Am; Am Asn Petrol Geologists; Am Geophys Union; Asn Eng Geologists. Res: Occurrence of ground water; relation of ground water to surface water in Colorado. Mailing Add: 431 Blair Rd NW Vienna VA 22180

WOOD, LINDSAY WALLACE, microbial ecology, see 12th edition

WOOD, LOUIS L, b Washington, DC, July 26, 31; m 58; c 1. ORGANIC CHEMISTRY. Educ: Univ Del, BS, 53; Ohio State Univ, PhD(org chem), 59. Prof Exp: RES CHEMIST, W R GRACE & CO, WASHINGTON RES CTR, CLARKSVILLE, 58- Mem: Am Chem Soc. Res: Polymers; organic chemical synthesis; textile applications; foam technology. Mailing Add: 11760 Gainsborough Rd Rockville MD 20854

WOOD, LOWELL THOMAS, b Ada, Okla, Sept 8, 42; m 66; c 2. PHYSICS, SCIENCE EDUCATION. Educ: Univ Kans, BS, 64; Univ Tex, Austin, PhD(physics), 68. Prof Exp: Asst prof physics, Univ Tex, Austin, 68-69; ASST PROF PHYSICS, UNIV HOUSTON, 69- Concurrent Pos: NASA grant, Univ Houston, 70-73. Mem: Am Phys Soc. Res: Thermomagnetic torques in gases; solid state physics; science teacher education. Mailing Add: Dept of Physics Univ of Houston Houston TX 77004

WOOD, MARGARET GRAY, b Jamaica, NY, May 23, 18; m 50; c 3. MEDICINE, DERMATOLOGY. Educ: Univ Ala, BA, 41; Woman's Med Col Pa, MD, 48. Prof Exp: Assoc, 53-68, asst prof, 68-71, ASSOC PROF DERMAT, SCH MED, UNIV PA, 71- Concurrent Pos: Assoc, Grad Div Med Dermat, Univ Pa, 53-66, assoc prof, 66-71; asst prof, Woman's Col Pa, 58-66, vis asst prof, 66-; asst vis physician, Philadelphia Gen Hosp, 54-70, consult, 70-; consult, Philadelphia Vet Hosp, 66-; consult, Dent Sch, Univ Pa, 66-, consult, Vet Sch Med, 70- Mem: Histochem Soc; Soc Invest Dermat; AMA; Am Med Women's Asn; Am Acad Dermat. Res: Histochemistry; dermatopathology. Mailing Add: Dept of Dermat Hosp of Univ of Pa Philadelphia PA 19104

WOOD, NANCY ELIZABETH, b Martins Ferry, Ohio. SPEECH PATHOLOGY. Educ: Ohio Univ, BS, 43, MS, 47; Northwestern Univ, PhD(speech path), 52. Prof Exp: Assoc prof lang path, Case Western Reserve Univ, 52-60; consult specialist, Off Educ, Dept Health, Educ & Welfare, 60-62, chief neurol & sensory dis res, USPHS, 62-64; prof commun dis, 65-74, PROF SURG, SCH MED, UNIV SOUTHERN CALIF, 65-, DIR COMMUN DIS, 71-, PROF & RES DIR, SCH JOUR, 75- Concurrent Pos: Asst dir, Cleveland Hearing & Speech Ctr, 52-56, coordr clin serv, 56-59, dir lang dis, 59-60. Mem: AAAS; Soc Res Child Develop; fel Am Speech & Hearing Asn; fel Am Psychol Asn. Res: Language development, disorders and pathology; differential diagnosis of young children; aphasia; mental retardation; hearing loss; test design; communication science research; memory, perception and auditory processing. Mailing Add: Univ of Southern Calif 734 W Adams Blvd Los Angeles CA 90007

WOOD, NORMAN KENYON, b Perth, Ont, Dec 1, 35; m 69; c 1. ORAL PATHOLOGY. Educ: Univ Toronto, DDS, 58; Cook County Hosp, Chicago, dipl oral surg, 65; Northwestern Univ, MS, 66, PhD(oral path), 68; Am Bd Oral Surg, dipl, 70; Am Bd Oral Path, dipl, 71. Prof Exp: Pvt pract, 58-62; res assoc biol mat, Dent Sch, Northwestern Univ, 67-68; asst prof oral path, 68-70, ASSOC PROF ORAL DIAG & CHMN DEPT, DENT SCH, LOYOLA UNIV CHICAGO, 70- Concurrent Pos: Consult, Hines Vet Admin Hosp, Ill, 71- Mem: Am Acad Oral Path. Res: Oral embryology, induction of cleft palates in fetal mice; implantology, tissue compatibility studies on implant materials. Mailing Add: Loyola Univ of Chicago Dent Sch 2160 First Ave Maywood IL 60153

WOOD, NORRIS PHILIP, b Binghamton, NY, July 8, 24; m 55; c 2. BACTERIOLOGY. Educ: Hartwick Col, BS, 49; Cornell Univ, MNS, 51; Univ Pa, PhD(microbiol), 55. Prof Exp: From asst prof to assoc prof microbiol, Agr & Mech Col Tex, 55-63; from asst prof to assoc prof bact, 63-72, PROF MICROBIOL, UNIV RI, 72-, CHMN DEPT MICROBIOL & BIOPHYS, 70- Concurrent Pos: Res partic, Oak Ridge Nat Lab, 58 & 62; mem, State Adv Comt Regional Med Prog, 66- Mem: AAAS; Am Chem Soc; Am Soc Microbiol. Res: Bacterial physiology; intermediary metabolism; chemistry of microorganisms; microbial ecology. Mailing Add: Dept of Microbiol & Biophys Univ of RI Kingston RI 02881

WOOD, PAUL ALAN, geology, see 12th edition

WOOD, PAULINE J, b Springdale, Pa, Nov 7, 22. DEVELOPMENTAL BIOLOGY, HISTOLOGY. Educ: Adrian Col, BS, 51; Univ Mich, MS, 54, PhD(zool), 60. Prof Exp: Instr zool, Univ Mich, 58-59; res instr embryol, Dent Sch, Univ Wash, 59-61; asst prof zool, Knox Col, 61-62; from asst prof to assoc prof, 62-73, PROF ZOOL, UNIV DETROIT, 73- Mem: AAAS; Am Soc Zool; Am Inst Biol Sci; Reticuloendothelial Soc. Res: Phylogeny of mesenchymal and hemopoietic cells. Mailing Add: Dept of Biol Univ of Detroit 4001 W McNicholas Rd Detroit MI 48221

WOOD, PETER DOUGLAS, b London, Eng, Aug 25, 29; m 53; c 1. BIOCHEMISTRY, CHEMISTRY. Educ: Univ London, BSc, 52, MSc, 56, PhD(chem), 62, DSc, 73. Prof Exp: Chemist, Weston Res Labs, Eng, 52-55; res chemist, Imp Chem Industs Australia & NZ, 56-59; res asst chem, Univ Sask, 59; res assoc, Inst Metab Res, Oakland, Calif, 62-68; ADJ PROF MED, MED CTR, STANFORD UNIV, 69- Concurrent Pos: Fel coun arteriosclerosis, Am Heart Asn, 68; dep dir, Stanford Lipid Res Clin & Stanford Specialized Ctr Res, 71- Mem: Fel Royal Inst Chem; Am Inst Nutrit; Am Soc Clin Nutrit; Am Heart Asn; Am Oil Chem Soc. Res: Lipid chemistry, metabolism and methodology; exercise. Mailing Add: Rm S005 Stanford Univ Med Ctr Stanford CA 94305

WOOD, RANDALL DUDLEY, b Palmer, Ky, Aug 3, 36; m 59; c 1. BIOCHEMISTRY, ORGANIC CHEMISTRY. Educ: Univ Ky, BS, 59, MS, 61; Tex A&M Univ, PhD(biochem), 65. Prof Exp: Scientist, Oak Ridge Assoc Univs, 66-70; assoc prof, Stritch Sch Med, Loyola Univ Chicago & Hines Vet Admin Hosp, 70-71; assoc prof med & biochem, Sch Med, Univ Mo-Columbia, 71-76; PROF BIOCHEM, TEX A&M UNIV, COLLEGE STATION, 76- Concurrent Pos: AEC fel, 65-66. Mem: AAAS; Am Asn Cancer Res; Am Chem Soc; Am Oil Chem Soc; Am Soc Biol Chemists. Res: Lipid biochemistry and metabolism of normal, tumor and embryonic tissues; biosynthesis, metabolism and occurrence of alkyl glyceryl ethers and plasmalogens; structural and metabolic relationships between molecular species of various classes. Mailing Add: Dept of Med & Biochem Univ of Mo Med Sch Columbia MO 65201

WOOD, RAYMOND ARTHUR, b Middletown, NY, Nov 28, 24. ZOOLOGY, PARASITOLOGY. Educ: Mt St Mary's Col, Md, BS, 50; Univ Notre Dame, MS, 53, PhD, 55. Prof Exp: Instr & spec lectr anat & physiol, Ind Univ, 54-55; asst prof, Pan Am Col, 55-56; PROF ZOOL & CHMN DIV, ORANGE COUNTY COMMUNITY COL, 56- Concurrent Pos: NSF grants, Bermuda Biol Sta, 56, Comp Anat Inst, Harvard Univ, 63, Col Biol Inst, Williams Col, 66, Marine Lab, Duke Univ, 65, 67 & Marine Labs, Naples, Italy, 70; Sigma Xi grant, 57. Mem: Fel AAAS; Am Soc Parasitologists; Am Soc Zoologists; Soc Syst Zool; fel Royal Soc Trop Med & Hyg. Res: Systematics of monogenea. Mailing Add: Div of Biol & Health Sci Orange County Community Col Middletown NY 10940

WOOD, REED RALPH, b Hill City, Kans, Feb 17, 09; m 33; c 1. AGRONOMY. Educ: Colo State Univ, BS, 41. Prof Exp: Agt, USDA, 36-42; agronomist, Great Western Sugar Co, 42-60, mgr, Agr Res Sta, 60-68, dir agr serv, 68-72, mgr seed mkt develop, 72-74; CONSULT, BEET SUGAR DEVELOP FOUND, 75- Mem: AAAS; Am Soc Sugar Beet Technologists. Res: Plant breeding; chemistry. Mailing Add: 2100 Longs Peak Ave Longmont CO 80501

WOOD, REUBEN ESSELSTYN, b Lansing, Mich, Apr 1, 15. PHYSICAL CHEMISTRY. Educ: Calif Inst Technol, BS, 36, PhD(phys chem), 39; Univ Chicago, MS, 37. Prof Exp: Fel, Calif Inst Technol, 39-43; fel, Md Res Labs, 43-45; from asst prof to assoc prof, 45-58, PROF CHEM, GEORGE WASHINGTON UNIV, 58- Concurrent Pos: Pres, Sigma Press, 60-; chemist, Nat Bur Standards; consult, Off Saline Water, US Dept Interior. Mem: AAAS; Am Chem Soc; The Chem Soc. Res: Electrochemistry; chemical thermodynamics; fused salts. Mailing Add: Dept of Chem George Washington Univ Washington DC 20006

WOOD, RICHARD DAWSON, b Toledo, Ohio, Apr 28, 18; m 46; c 2. BOTANY. Educ: Ohio State Univ, BA & BSc, 40; Northwestern Univ, MS, 42, PhD(bot), 47. Prof Exp: Asst, Ohio State Univ, 38-40 & Northwestern Univ, 40-42 & 46-47; from instr to assoc prof, 47-59, PROF BOT, UNIV RI, 59- Concurrent Pos: Lectr, Marine Biol Lab, Woods Hole, 48-49, instr, 50-51; prof, Univ NH, 52; Fulbright res scholar, Australia, 61; res assoc, Narragansett Marine Lab. Honors & Awards: Darbaker Award, 66. Mem: AAAS; Bot Soc Am; Brit Phycol Soc; Phycol Soc Am (treas, 52-54); Am Soc Plant Taxonomists. Res: Aquatic plant ecology; taxonomy of marine and freshwater algae; monograph of Characeae. Mailing Add: Dept of Bot Univ of RI Kinston RI 02881

WOOD, RICHARD ELLET, b Farmington, Utah, Mar 3, 28; m 48; c 5. NUCLEAR PHYSICS. Educ: Univ Utah, BS, 52, PhD(physics), 55. Prof Exp: Res assoc neutron cross sects, Brookhaven Nat Lab, 53-54; nuclear engr, Gen Elec Co, 55-56, supvr low power test opers, 56-59, supvr initial eng test opers, 59-60, supvr anal, Idaho Test Sta, Air Craft Nuclear Propulsion Dept, 60-61, physicist, Atomic Power Equip Dept, 61-62, mgr, Idaho Eng, Nuclear Mat & Propulsion Oper, Idaho Test Sta, 62-68; chief nuclear eng br, Idaho Opers Off, AEC, 68-74; DIR, REACTOR SUPPORT DIV, US ENERGY RES & DEVELOP ADMIN, 74- Mem: Am Nuclear Soc. Res: Nuclear engineering and neutron physics. Mailing Add: US Energy Res & Develop Admin 550 Second St Idaho Falls ID 83401

WOOD, RICHARD FROST, b Lebanon, Tenn, June 6, 31; m; c 3. PHYSICS. Educ: Fordham Univ, BS, 53; Ohio State Univ, MS, 56, PhD(physics), 59. Prof Exp: Res specialist, Opers Res, NAm Aviation, Inc, 58-60; asst prof physics, Univ Fla, 60-62; RES SCIENTIST & HEAD THEORY SECT, SOLID STATE DIV, OAK RIDGE

NAT LAB, 62- Concurrent Pos: Vis prof, Univ Uppsala, 60-61, NSF fel, 61-62. Mem: AAAS; fel Am Phys Soc. Res: Theoretical solid state physics; lattice defects; lattice dynamics; optical properties and electronic structure of solids. Mailing Add: Solid State Div Oak Ridge Nat Lab Oak Ridge TN 37830

WOOD, RICHARD LEE, b Ft Dodge, Iowa, Sept 8, 30; m 55; c 3. VETERINARY MICROBIOLOGY. Educ: Univ Mo-Columbia, DVM, 61; Iowa State Univ, MS, 66, PhD(vet microbiol), 70. Prof Exp: RES VET, NAT ANIMAL DIS CTR, AGR RES SERV, USDA, 61- Mem: Am Vet Med Asn; Am Soc Microbiol; Conf Res Workers Animal Dis. Res: Epizootiology and pathogenesis of swine erysipelas and streptococcal lymphadenitis of swine. Mailing Add: Nat Animal Dis Ctr PO Box 70 Ames IA 50010

WOOD, RICHARD LYMAN, b Allamore, Tex, Jan 2, 29; m 51; c 2. CYTOLOGY. Educ: Linfield Col, BA, 50; Univ Wash, PhD(zool), 57. Prof Exp: From instr to asst prof anat, Univ Wash, 59-64; assoc prof, Univ Minn, Minneapolis, 64-70; prof, Sch Med, Univ Miami, 70-74; PROF ANAT, SCH MED, UNIV SOUTHERN CALIF, 74- Concurrent Pos: NIH fel, Univ Wash, 57-59; NIH res grant, Dept Biol Struct, Univ Wash, 60-64 & Dept Anat, Univ Minn, 64-70. Mem: AAAS; Am Soc Cell Biol; Am Asn Anat; Electron Micros Soc Am; Soc Develop Biol. Res: Fine structure; cellular anatomy by electron microscopy; animal cytology and histology. Mailing Add: Dept of Anat Univ Southern Calif Sch of Med Los Angeles CA 90033

WOOD, ROBERT CHARLES, b Lakewood, Ohio, May 7, 29. BACTERIOLOGY. Educ: Lehigh Univ, BA, 51, MS, 52; Univ Md, PhD, 55. Prof Exp: Sr res microbiologist, Wellcome Res Labs, Burroughs Wellcome & Co, 56-60; asst prof microbiol, Sch Med, George Washington Univ, 60-64; asst prof, 64-65, ASSOC PROF MICROBIOL, UNIV TEX MED BR GALVESTON, 65- Concurrent Pos: Res fel bact physiol, Univ Pa, 55-56. Mem: AAAS; Am Soc Microbiol; Brit Soc Gen Microbiol. Res: Chemotherapy; mechanisms of action of drugs; folic acid metabolism; comparative aspects of bacterial physiology; membrane permeability; bacterial genetics. Mailing Add: Dept of Microbiol Univ of Tex Med Br Galveston TX 77550

WOOD, ROBERT E, b Philadelphia, Pa, May 16, 38; m 62; c 2. OPERATIONS RESEARCH. Educ: Ga Inst Technol, BS, 60, MS, 62, PhD(physics), 65. Prof Exp: From asst prof to assoc prof physics, Emory Univ, 64-75; SR OPERS RES ANALYST, SOUTHERN RWY SYST, 75- Concurrent Pos: Consult, Ga Inst Technol, 65, fel chem, 66; consult, Allied Gen Nuclear Serv, 73-75 & Aston Co, 75- Mem: AAAS; Am Phys Soc; Am Asn Physics Teachers. Res: Res: Application of models to transportation industry. Mailing Add: Southern Rwy Syst 125 Spring St Rm 815 Atlanta GA 30303

WOOD, ROBERT HEMSLEY, b Brooklyn, NY, May 8, 32; m 56; c 2. PHYSICAL CHEMISTRY. Educ: Calif Inst Technol, BS, 53; Univ Calif, PhD(chem), 57. Prof Exp: From instr to assoc prof, 57-70, chmn dept, 69-71, PROF CHEM, UNIV DEL, 70- Mem: Am Chem Soc. Res: Solution thermodynamics; electrolytes, non-electrolytes, non-aqueons, and high temperature. Mailing Add: Dept of Chem Univ of Del Newark DE 19711

WOOD, ROBERT MANNING, b Bronxville, NY, May 13, 38; m 64; c 1. NUCLEAR PHYSICS. Educ: Princeton Univ, AB, 60; Univ Wis, PhD(physics), 64. Prof Exp: Res assoc physics, Univ Wis, 64-66; ASST PROF PHYSICS, UNIV GA, 66- Mem: Am Phys Soc; Am Asn Physics Teachers. Res: Low energy experimental nuclear physics; fast neutron spectroscopy; neutron polarization phenomena. Mailing Add: Dept of Physics Univ of Ga Athens GA 30601

WOOD, ROBERT WINFIELD, b Detroit, Mich, Dec 29, 31; m 59; c 3. RADIATION BIOPHYSICS. Educ: Univ Detroit, BS, 53; Vanderbilt Univ, MA, 55; Cornell Univ, PhD(biophys), 61. Prof Exp: Radiol physicist, AEC, 62-73; MGR PHYS & TECHNOL PROGS, DIV BIOMED & ENVIRON RES, US ENERGY RES & DEVELOP ADMIN, 73- Concurrent Pos: Nat Inst Gen Med Sci fel, 61-62. Mem: Am Soc Nuclear Med; Health Physics Soc; Radiation Res Soc; Am Asn Physicists in Med. Res: Electron spin resonance, aromatic hydrocarbon negative ions and irradiated biological compounds; radiological physics; dosimetry; biomedical instrumentation. Mailing Add: Div of Biomed & Environ Res US Energy Res & Develop Admin Washington DC 20545

WOOD, RONALD MCFARLANE, b New York, NY, Oct 11, 15; m 43; c 2. MEDICAL MICROBIOLOGY. Educ: McGill Univ, BSc, 39; Johns Hopkins Univ, PhD(biol), 49; Am Bd Med Microbiol, dipl, 62. Prof Exp: Microbiologist, Royal Victoria Hosp, 39-40; res microbiologist, Ayerst McKinna & Harrison, Ltd, 40-41; from instr to assoc prof ophthal, Sch Med, Johns Hopkins Univ, 49-63; CHIEF MICROBIAL DIS LAB, STATE DEPT PUB HEALTH, CALIF, 64- Concurrent Pos: Asst prof, Sch Hyg & Pub Health, Johns Hopkins Univ, 51-56; assoc prof, Sch Med, Univ Fla, 63-64. Honors & Awards: Barnett Cohen Award, 62. Mem: AAAS; fel Royal Soc Health; fel Am Pub Health Asn; Am Soc Microbiol; Nat Tuberc & Respiratory Dis Asn. Res: Immunology; public health; infectious diseases. Mailing Add: Microbial Dis Lab State of Calif Dept of Pub Health Berkeley CA 94704

WOOD, ROY KELLUM, b Augusta, Ark; m 41; c 3. WILDLIFE MANAGEMENT. Educ: Univ Ark, BS, 39; Va Polytech Inst, MS, 43. Prof Exp: Coordr fed aid projs, State Game & Fish Comn, Ark, 41-45; biologist, US Fish & Wildlife Serv, 45-52, regional supvr river basin studies, 52-59; mem staff, US Study Comn Southeast River Basins, 59-66; chief div water resources studies, Bur Outdoor Recreation, US Dept Interior, Washington, DC, 66-67, regional dir, Atlanta, 67-73; ACQUISITION OFFICER, GA STATE DEPT NATURAL RESOURCES, 73- Res: Fish and wildlife management of impounded waters; game management; ecology. Mailing Add: 875 Woodstock Rd Roswell GA 30075

WOOD, SARALUE, b Wehadkee, Ala, June 26, 34. NUCLEAR PHYSICS. Educ: Mercer Univ, AB, 55; Univ Fla, MS, 59. Prof Exp: Instr math, Mercer Univ, 56-57; teaching asst physics, Univ Fla, 57-59; asst prof, 59-66, actg chmn dept, 70-72, ASSOC PROF PHYSICS, AUSTIN PEAY STATE UNIV, 66-, CHMN DEPT, 72- Concurrent Pos: NSF acad year exten grants, 64-68; dir, NSF Physics Prog for Teachers, 69-75. Mem: Fel AAAS; Am Asn Physics Teachers; Am Phys Soc. Res: Low energy nuclear reactions using low atomic mass number targets; preparation and examination of targets for use in nuclear reactions; health physics technology; teaching of physics in high schools and colleges. Mailing Add: Dept of Physics Austin Peay State Univ Clarksville TN 37040

WOOD, SCOTT EMERSON, b Ft Collins, Colo, Apr 9, 10; m 36; c 1. PHYSICAL CHEMISTRY. Educ: Univ Denver, BS, 30, MS, 31; Univ Calif, PhD(chem), 35. Prof Exp: Res assoc chem, Mass Inst Technol, 35-40; from instr to asst prof, Yale Univ, 40-48; from assoc prof to prof, 48-75, admin officer & vchmn dept, 60-62, actg assoc dean for res, 62-64, EMER PROF CHEM, ILL INST TECHNOL, 75- Concurrent Pos: Consult, Argonne Nat Lab, 60-; vis prof, Univ Col, Dublin, 66-67. Mem: AAAS; Am Chem Soc; Am Phys Soc. Res: Vapor pressures of nonaqueous solutions; density and coefficient of expansion of solutions; index of refraction; theory and thermodynamics of nonaqueous solutions; chromatography; paper electrochromatography; fused salts. Mailing Add: 1124 Community Dr La Grange Park IL 60525

WOOD, SHERWIN FRANCIS, b Lake Placid, NY, June 23, 08; m 33; c 2. PROTOZOOLOGY, MEDICAL ENTOMOLOGY. Educ: Univ Calif, Los Angeles, AB, 31; Univ Calif, MA, 33, PhD(zool), 35. Prof Exp: Reader biol, Univ Calif, Los Angeles, 30-31, tech asst zool, Univ Calif, Berkeley, 31-34, asst, 33-35; from assoc to instr biol, 35-38, instr life sci, 38-43 & 46-65, PROF BIOL, LOS ANGELES CITY COL, 65- Concurrent Pos: AAAS grant, 39; USPHS grant, 50; Southern Calif Acad Sci grant, 57; mem, Int Union Conserv Nature & Natural Resources. Mem: AAAS; Soc Protozool; Entom Soc Am; Am Soc Trop Med & Hyg; Cooper Ornith Soc. Res: Chagas' disease and insect vectors; vertebrate blood parasites. Mailing Add: 1015 N Alexandria Ave Los Angeles CA 90029

WOOD, SPENCER HOFFMAN, b Portland, Ore, Nov 18, 38; m 61; c 3. GEOLOGY. Educ: Colo Sch Mines, GE, 64; Calif Inst Technol, MS, 70, PhD(geol), 75. Prof Exp: Seismol engr, Geophys Serv, Mobil Oil Corp, 64-65, geophysicist, Mobil Oil Libya, Ltd, 65-68; INSTR GEOL, OCCIDENTAL COL, 74- Mem: Geol Soc Am; Am Geophys Union; Sigma Xi; Am Quaternary Asn. Res: Geomorphology; late quaternary stratigraphy and geochronology of the Sierra Nevada Region; tephrachronology, obsidian hydration-rind dating. Mailing Add: Dept of Geol Occidental Col Los Angeles CA 90041

WOOD, STEPHEN CRAIG, b Cleveland, Ohio, Sept 28, 42; m 67; c 3. PHYSIOLOGY. Educ: Kent State Univ, BS, 64, MA, 66; Univ Ore, PhD(physiol), 70. Prof Exp: Nat Res Coun res assoc physiol, Submarine Med Res Lab, Groton, Conn, 70-71; asst prof zoophysiol, Aarhus Univ, 71-72; asst prof physiol, Southern Ill Univ, Edwardsville, 72-74; ASST PROF PHYSIOL, SCH MED, UNIV NMEX, 74- Concurrent Pos: Sr vis res fel, Marine Biol Lab, Plymouth, Eng, 72; Danish Natural Sci Res Coun grant, Comoro Islands Coelacanth Exped, 72. Mem: Am Physiol Soc; Underseas Med Soc; Am Soc Zoologists. Res: Diving physiology; comparative physiology of respiration and blood gas transport; metabolism and function of red blood cells; environmental physiology. Mailing Add: Dept of Physiol Univ of NMex Sch of Med Albuquerque NM 87131

WOOD, STEPHEN LANE, b Logan, Utah, July 2, 24; m 47; c 3. ENTOMOLOGY. Educ: Utah State Univ, BS, 46, MS, 48; Univ Kans, PhD(entom), 53. Prof Exp: High sch instr, Utah, 48-50; asst instr biol, Univ Kans, 50-53; syst entomologist, Can Dept Agr, 53-56; from asst prof to assoc prof, 56-68, PROF ZOOL & ENTOM, BRIGHAM YOUNG UNIV, 68- Concurrent Pos: Ed, Great Basin Naturalist, 72- Mem: AAAS; Entom Soc Am; Coleopterists' Soc. Res: Systematics of Scolytidae. Mailing Add: Dept of Zool Brigham Young Univ Provo UT 84602

WOOD, SUMNER, JR, medicine, pathology, deceased

WOOD, THOMAS DONALD, physical chemistry, see 12th edition

WOOD, THOMAS HAMIL, b Atlanta, Ga, June 22, 23; m 51; c 3. BIOPHYSICS. Educ: Univ Fla, BS, 46; Univ Chicago, PhD(biophys), 53. Prof Exp: Res assoc, Univ Chicago, 53; from asst prof to assoc prof, 53-63, PROF PHYSICS, UNIV PA, 63- Concurrent Pos: NSF sr fel, Inst Radium, Paris, 61-62: vis prof, Univ Leicester, 67-68. Mem: Biophys Soc; Radiation Res Soc; Genetics Soc Am. Res: Effects of radiations on microorganisms; influence of temperature and protective agents; cellular freezing; bacterial conjugation; genetic recombination. Mailing Add: Dept of Physics Univ of Pa Philadelphia PA 19104

WOOD, THOMAS ROSS, b Portland, Colo, May 7, 16; m 39; c 3. BIOCHEMISTRY, HEALTH SCIENCES. Educ: Univ Denver, BS, 37; Univ Ill, PhD(biochem), 40. Prof Exp: Fel, Harvard Univ, 40-42, fel biochem, Med Sch, 42; Nutrit Found sr fel, Univ Pittsburgh, 42-43; res chemist, Gen Foods Corp, 43-46 & Merck & Co, Inc, 46-50; res chemist, Pharmaceut Div, 50-52, dir, 52-74, SITE MGR, STINE LAB, E I DU PONT DE NEMOURS & CO, INC, 74- Mem: Am Chem Soc; Am Soc Biol Chemists; NY Acad Sci. Res: Amino acid metabolism; parathyroid hormone chemistry; fat metabolism; phytochemistry; vitamin B12; new and unidentified vitamins; pharmaceuticals. Mailing Add: Stine Lab E I du Pont de Nemours & Co Inc Newark DE 19711

WOOD, TIMOTHY SMEDLEY, b Port Washington, NY, Dec 4, 42; m 64; c 2. ECOLOGY. Educ: Earlham Col, AB, 64; Univ Colo, PhD(zool), 71. Prof Exp: ASST PROF AQUATIC ECOL, WRIGHT STATE UNIV, 71- Mem: AAAS; Am Soc Zoologists; Am Micros Soc; Int Bryozool Asn; Am Soc Limnol & Oceanog. Res: Aquatic community ecology; structure and function of animal colonies, with emphasis on the Ectoprocta. Mailing Add: Dept of Biol Sci Wright State Univ Dayton OH 45431

WOOD, VAN EARL, b New York, NY, May 25, 33; m 58; c 2. PHYSICS. Educ: Union Col, BS, 55; Case Inst Technol, MS, 59, PhD(physics), 61. Prof Exp: Fel, 60-73, SR PHYSICIST, COLUMBUS LABS, BATTELLE MEM INST, 73- Mem: Am Phys Soc. Res: Theoretical solid-state physics and materials science; related applied mathematics. Mailing Add: Columbus Labs Battelle Mem Inst Columbus OH 43201

WOOD, WALTER ABBOTT, b Hoosick Falls, NY, Mar 23, 07; m 30; c 2. GEOGRAPHY. Educ: Am Geog Soc, cert, 32. Hon Degrees: DSc, Univ Alaska, 55. Prof Exp: Head dept explor & field res, Am Geog Soc, NY, 34-42; dir NY off, Arctic Inst NAm, 47-58; pres, Am Geog Soc, 57-67; chmn, Am Geog Soc, 67-70, CHMN, ARCTIC INST N AM-AM GEOG SOC ICEFIELD RANGES RES PROJ, CAN, 60-, MEM BD GOV, ARCTIC INST N AM & AM GEOG SOC, 72- Concurrent Pos: Mem expeds, Kashmir Himalaya, 29, Panama & Guatemala, 31 & Colombia, SAm, 36; mem, Louise A Boyd Exped, EGreenland, 33; leader, Wood Yukon Exped, Am Geog Soc, 35, 36, 39 & 41; mem, Am Mus Natural Hist Exped, Grand Canyon, 37, Cabot Colombian Exped, 39 & US Army Alaskan Test Exped, 42; US deleg, Int Photogram Cong, Italy, 38; proj leader, Snow Cornice, Arctic Inst NAm, Alaska & Can, 48-51; proj dir, McCall Glacier Sta, Int Geophys Year, Alaska, 56-57; pres, Explorers Club, 67-71. Honors & Awards: Explorers Medal, 72; Charles P Daly Medal, Am Geog Soc, 74. Mem: Fel AAAS; fel Am Geog Soc; Asn Am Geogr; fel Arctic Inst NAm; Am Geophys Union. Res: Geographical exploration; mapping; photogrammetry. Mailing Add: PO Box EEE Southampton NY 11968

WOOD, WALTER RAYMOND, anthropology, see 12th edition

WOOD, WARREN WILBUR, b Pontiac, Mich, Apr 9, 37; m 61; c 1. GEOCHEMISTRY, HYDROLOGY. Educ: Mich State Univ, BS, 59, MS, 62, PhD(geol), 69. Prof Exp: Geologist II, Mich Hwy Dept, 62-63; RES HYDROLOGIST, US GEOL SURV, 64- Concurrent Pos: Adj prof geosci, Tex Tech Univ, 73- Mem: Am Geophys Union; Geochem Soc; Int Asn Geochem & Cosmochem; Int Asn Sci Hydrol. Res: Geochemistry as applied to the hydrologic cycle; research on geochemistry of artificial recharge including solute transport and permeability changes associated with chemical and biological activity. Mailing Add: US Geol Surv PO Box 3355 Lubbock TX 79410

WOOD, WILLIAM BAINSTER, b Surry Co, NC, Feb 7, 31; m 52; c 4. MEDICINE. Educ: Univ NC, BS, 53, MD, 56. Prof Exp: Intern, NC Mem Hosp, Chapel Hill, 56-57, resident internal med, 57-59; assoc physiol, George Washington Univ, 61-63; from instr to asst prof, 63-68, ASSOC PROF MED, SCH MED, UNIV NC, CHAPEL HILL, 68- Concurrent Pos: Fel chest dis, Sch Med, Univ NC, 59-60; attend physician, consult & dir pulmonary lab, NC Mem Hosp, Chapel Hill, 63-; attend physician, Gravely Sanatorium, 63-; sr scientist, Wrightsville Maine-Biomed Lab, 66-; hon sr lectr, Cardio Thoracic Inst, London, Eng, 73-74; dir med educ & res, Eastern NC Hosp, 74-, med dir, Wilson, NC, 75- Mem: Am Thoracic Soc; AMA; Am Fedn Clin Res; Am Soc Internal Med. Res: Pulmonary diseases; morphology of lung in disease; mechanics and control of respiration; effects of hyperbaric atmospheres on respiration; respiratory physiology in deep sea diving; immunological and hypersensitivity lung diseases. Mailing Add: Sch of Med Univ of NC Chapel Hill NC 27514

WOOD, WILLIAM BARRY, III, b Baltimore, Md, Feb 19, 38; m 61; c 2. BIOCHEMISTRY, GENETICS. Educ: Harvard Univ, AB, 59; Stanford Univ, PhD(biochem), 64. Prof Exp: Air Force Off Sci Res, Nat Acad Sci-Nat Res Coun fel molecular biol, Univ Geneva, 63-64; from asst prof to assoc prof, 65-70, PROF BIOL, CALIF INST TECHNOL, 70- Concurrent Pos: Nat Acad Sci-US Steel Award Molecular Chem, 69; Guggenheim fel, Dept Molecular, Cellular & Develop Biol, Univ Colo, Boulder, 75-76. Mem: Nat Acad Sci; AAAS; Am Soc Biol Chemists; Soc Develop Biol. Res: Genetic control of macromolecular assembly; morphogenesis, structure and function of bacterial viruses. Mailing Add: Biol Div Calif Inst of Technol Pasadena CA 91109

WOOD, WILLIAM BOOTH, b New York, NY, Sept 9, 22. PHYSIOLOGY. Educ: Ft Hays Kans State Col, MS, 51; Univ Kans, PhD(physiol, anat), 59. Prof Exp: Lab asst physiol, Univ Kans, 55-56, asst, 56-58, res assoc, 58, asst instr, 58-59; from instr to assoc prof, 59-73, PROF PHARMACOL, COL MED, UNIV TENN, MEMPHIS, 73- Mem: AAAS; assoc Am Physiol Soc; Am Soc Pharmacol & Exp Therapeut. Res: Respiratory physiology and pharmacology of cardiovascular system. Mailing Add: Dept of Pharmacol Univ of Tenn Ctr for Health Sci Memphis TN 38163

WOOD, WILLIAM GARDNER, biochemistry, see 12th edition

WOOD, WILLIAM HULBERT, b Red Creek, NY, Sept 26, 11; m 55; c 1. GEOLOGY, MATHEMATICS. Educ: Mass Inst Technol, SB, 34; Univ Ariz, PhD, 56. Prof Exp: Asst prof, 55-70, ASSOC PROF GEOL, UNIV SOUTHWEST LA, 70- Mem: Geol Soc Am. Res: Statistical geology; lower Paleozoic stratigraphy; environmental geology; sedimentology. Mailing Add: Dept of Geol Univ of Southwestern La Lafayette LA 70501

WOOD, WILLIAM OTTO, b Oklahoma City, Okla, Apr 7, 25; m 50; c 1. CELL BIOLOGY, BIOPHYSICS. Educ: Southern Methodist Univ, BS, 50. Prof Exp: Indust bacteriologist, Joseph E Seagrams & Sons, Inc, Ky, 50-53; phys chem scientist, Bur Mines, US Dept Interior, Ore, 53-54; serologist, Animal Dis Eradication Div, USDA, Ore, 54-58 & Md, 58-59, tissue cult res biologist, Animal Dis & Parasite Res Div, Md, 59-61, tissue cult res cytologist, Nat Animal Dis Lab, Iowa, 61-64; res biologist tissue cult, Res Labs, Biophys Lab, 64-69, sci adminr, Res/Chem Labs, Edgewood Arsenal, Aberdeen Proving Ground, US Dept Army, 69-74; RETIRED. Honors & Awards: Serv Award, USDA, 61; Serv Award, US Dept Army, 68, Sustained Super Performance Award, 73, Cert Achievement, 74, Cert Appreciation, 74, Lett Appreciation, 74. Mem: Tissue Cult Asn; Am Soc Cell Biol. Res: Comparative studies in tissue culture preservation; in vitro studies of regenerating liver cells wounded by missiles; wounding and hyperbaric oxygenation; culture of cells from cloisonne goat kidneys. Mailing Add: 1300 Locust Ave Bel Air MD 21014

WOOD, WILLIAM WAYNE, b Terry, Mont, Nov 1, 24; m 46; c 5. STATISTICAL MECHANICS. Educ: Mont State Col, BS, 47; Calif Inst Technol, PhD(chem), 51. Prof Exp: Staff mem, 50-58, group leader, 58-71, STAFF MEM, LOS ALAMOS SCI LAB, 71- Concurrent Pos: Vis prof, Univ Colo, 69-70. Mem: AAAS; fel Am Phys Soc. Res: Theory of optical activity; detonation. Mailing Add: Los Alamos Sci Lab Los Alamos NM 87544

WOOD, WILLIS A, b Johnson City, NY, Aug 6, 21; m 47; c 3. MICROBIOLOGY, BIOCHEMISTRY. Educ: Cornell Univ, BS, 47; Ind Univ, PhD, 50. Prof Exp: From asst prof to assoc prof dairy bact, Univ Ill, 50-58; prof agr chem, 58-61, chmn dept biochem, 68-74, PROF BIOCHEM, MICH STATE UNIV, 61- Concurrent Pos: Dir, Gilford Instrument Labs; chmn res inst & spec profs adv comt, Nat Inst Dent Res. Honors & Awards: Lilly Award, Am Chem Soc, 55. Mem: Am Chem Soc; Am Soc Microbiol; Am Soc Biol Chemists. Res: Chemical activities of microorganisms; amino acid metabolism; enzymology; protein structure; instrumentation. Mailing Add: Rm 410 Dept of Biochem Mich State Univ East Lansing MI 48824

WOODALL, DAVID MONROE, b Perryville, Ark, Aug 2, 45; m 66. PLASMA PHYSICS. Educ: Hendrix Col, BA, 67; Columbia Univ, MS, 68; Cornell Univ, PhD(appl physics), 75. Prof Exp: Nuclear engr, Westinghouse Nuclear Energy Systs, 68-70; ASST PROF MECH ENG, UNIV ROCHESTER, 74- Mem: Am Phys Soc; Am Nuclear Soc; Am Soc Mech Engrs. Res: Plasma physics, laser-plasma interaction experiments; x-ray imaging. Mailing Add: Dept of Mech & Aero Sci Univ of Rochester Rochester NY 14627

WOODALL, WILLIAM ROBERT, JR, b Augusta, Ga, May 29, 45; m 66; c 2. AQUATIC ECOLOGY, ENVIRONMENTAL BIOLOGY. Educ: Univ Ga, BS, 67, MS, 69, PhD(entom), 72. Prof Exp: Ecol consult, 72, ENVIRON SPECIALIST, GA POWER CO, 72- Mem: Ecol Soc Am; NAm Benthological Soc; Soc Power Indust Biologists. Res: Thermal effects; ecology of large rivers, their floodplains, oxbow lakes and swamps; invasion of exotic bivalves; entrainment and impingement of aquatic organisms; larval fish and invertebrate drift; nutrient cycling. Mailing Add: Ga Power Co Environ Ctr 791 Dekalb Industrial Way Decatur GA 30030

WOODARD, GEOFFREY, b Big Rapids, Mich, Nov 16, 15; m 48; c 5. PHARMACOLOGY. Educ: George Washington Univ, BS, 39, PhD, 51. Prof Exp: From jr pharmacologist to pharmacologist, US Food & Drug Admin, 39-52, chief pharmacodyn br, 52-57; PRES, WOODARD RES CORP, 57-, CHMN, 74- Mem: AAAS; Am Soc Pharmacol & Exp Therapeut; Soc Toxicol; NY Acad Sci. Res: Toxicity and pharmacology of drugs; metabolism of drugs and chemicals; bioassay. Mailing Add: Woodard Res Corp PO Box 405 Herndon VA 22070

WOODARD, HELEN QUINCY, b Detroit, Mich, Aug 8, 00. BIOCHEMISTRY, RADIOBIOLOGY. Educ: Stetson Univ, BS, 20; Columbia Univ, AM, 21, PhD(chem), 25. Prof Exp: Res chemist, 25-56, assoc biochemist, 56-66, EMER ASSOC BIOCHEMIST, MEM CTR CANCER & ALLIED DIS, 66-; CONSULT, SLOAN-KETTERING INST, 68- Concurrent Pos: Asst mem, Sloan-Kettering Inst, 48-60, assoc mem, 60-68; asst prof, Med Col, Cornell Univ, 52-64, assoc prof, 64-68. Mem: AAAS; Am Chem Soc; Radiation Res Soc; Am Asn Cancer Res; Health Physics Soc. Res: Chemical effects of gamma and x-rays; serum and tissue phosphatases; clinical biochemistry in bone disease; effects of radiation on bone; metabolism of bone-seeking isotopes; biological effects of radiation. Mailing Add: Sloan-Kettering Inst 1275 York Ave New York NY 10021

WOODARD, HENRY HERMAN, JR, b Salisbury, Mass, Dec 18, 25; m 49; c 2. GEOLOGY. Educ: Dartmouth Col, AB, 47, MA, 49; Univ Chicago, PhD, 55. Prof Exp: Geologist, US Geol Surv, 47-49 & State Develop Comn, Maine, 50-51; assoc prof geol, 53-66, chmn div natural sci & math, 62-66, PROF GEOL & CHMN DEPT, BELOIT COL, 66- Concurrent Pos: Geologist, Geol Surv Nfld, 55 & US Geol Surv, 57. Mem: Geol Soc Am; Geochem Soc; Nat Asn Geol Teachers; Am Geophys Union. Res: Petrology and structural geology of igneous rocks; diffusion in naturally occurring silicates; geology of Newfoundland and Boulder batholith; structure and petrology of southwestern Maine; contact alteration associated with tertiary stocks in central Colorado; marine sanidines from Ordovician of Wisconsin. Mailing Add: Dept of Geol Beloit Col Beloit WI 53511

WOODARD, JAMES CARROLL, b Birmingham, Ala, Nov 19, 33. NUTRITION, COMPARATIVE PATHOLOGY. Educ: Auburn Univ, DVM, 58; Mass Inst Technol, PhD(nutrit, path), 65. Prof Exp: Parasitologist, Fla Vet Diag Lab, Fla Livestock Bd, 58-59; instr vet path, Auburn Univ, 61-62; res assoc, Mass Inst Technol, 63-65; NIH fel, 65-66, asst prof, 66-70, ASSOC PROF PATH, COL MED, UNIV FLA, 70- Mem: Am Vet Med Asn; Int Acad Path; Am Inst Nutrit; Am Soc Exp Path; NY Acad Sci. Res: Biochemical pathology, biochemical relationships of disease to microscopic pathology. Mailing Add: Dept of Path Univ of Fla Col of Med Gainesville FL 32611

WOODARD, JOHN, b New Bedford, Mass, Jan 14, 17; m 49; c 4. BIOLOGY. Educ: Univ Va, BS, 42, MS, 49; Univ Wis, PhD(bot), 54. Prof Exp: Res assoc zool, 54-55, Nat Cancer Inst fel, 56-57, res assoc zool, 57-70, ASSOC PROF BIOL & RES ASSOC, UNIV CHICAGO, 70- Res: Cytochemistry and electron microscopy of nucleoproteins. Mailing Add: Dept of Biol Whitman Lab Univ of Chicago Chicago IL 60637

WOODARD, MARIE W, b Jersey City, NJ, Nov 15, 21; m 48; c 5. BIOCHEMISTRY. Educ: Dunbarton Col, BA, 43; Georgetown Univ, MS, 46. Prof Exp: Statist clerk, Chem Warfare Serv, US Dept Army, 38; from asst to assoc chemist, Nat Bur Stand, 43-44; tech aide, Nat Res Coun, 44-45; pharmacologist, US Food & Drug Admin, 45-48; TOXICOLOGIST & VPRES, WOODARD RES CORP, 57- Honors & Awards: Cert Meritorious Serv, Nat Res Coun, 45. Mem: Soc Toxicol; Am Asn Lab Animal Sci (secy-treas, 62-67); NY Acad Sci; Int Primatol Soc. Res: Safety evaluation of drugs, pesticides and cosmetics following administration to animals; laboratory animal care and management. Mailing Add: 12310 Pinecrest Rd Herndon VA 22070

WOODARD, RALPH EMERSON, b Nelsonville, Ohio, May 27, 21; m 43; c 3. RESEARCH ADMINISTRATION, REACTOR PHYSICS. Educ: Wittenberg Col, BA, 53; Oak Ridge Sch Reactor Technol, dipl, 57; Indust Col Armed Forces, dipl, 70; George Washington Univ, MSBA, 71. Prof Exp: Res physicist, Wright Air Develop Ctr, US Dept Air Force, 53-56, gen physicist, Oak Ridge Nat Lab, 56-57, nuclear physicist, Wright Air Develop Ctr, 57-61, gen engr res & tech plans, Aeronaut Systs Div, 61-63, supvry physicist & dep dir gen physics res lab, Aerospace Res Labs, 63-71, phys sci adminr, 71-75, PHYS SCI ADMINR, AIR FORCE WRIGHT AERONAUT LABS, 75- Concurrent Pos: Instr, Wittenberg Col, 57-58. Mem: AAAS; Am Phys Soc; Am Nuclear Soc. Res: Laboratory management; plasma physics; solid state physics; mathematics; chemistry; fluid mechanics; metallurgy and ceramics; energy conversion; research management. Mailing Add: 305 Gordon Rd Springfield OH 45504

WOODARD, ROBERT LOUIS, b South Bristol, NY, June 16, 17; m 44; c 2. SCIENCE EDUCATION, ASTRONOMY. Educ: Syracuse Univ, BS, 39; State Univ NY Col Geneseo, MS, 56; Cornell Univ, PhD(sci educ), 63. Prof Exp: Boy's worker, Woods Run Settlement House, Pa, 39-41; supt parks, Greensboro, NC, 46-47; self-employed, 47-49; munic engr, Naples, NY, 49-51; elem & jr high sch teacher, NY, 53-57; assoc prof, 60-64, PROF PHYS SCI & ASTRON, INDIANA UNIV PA, 64-, DIR INSTNL RES, 67- Mem: Nat Asn Res Sci Teaching; Asn Instnl Res; Am Asn Higher Educ. Mailing Add: Off of Instnl Res Indiana Univ Indiana PA 15701

WOODBREY, JAMES CALVIN, b Standish, Maine, Oct 16, 34; m 56; c 3. POLYMER CHEMISTRY, POLYMER SCIENCE. Educ: Univ Maine, BS, 56; Mich State Univ, PhD(phys chem, physics, org chem), 60. Prof Exp: Sr res chemist, Res Div, W R Grace & Co, Md, 60-61; res chemist, Plastics Div, Fund & Explor Res, Mass, 61-65, res specialist & proj leader, Cent Res Dept, Mo, 65-67, sr res specialist & proj leader, New Enterprise Div, 67-69, SCI FEL & RES PROJ MGR, EXPLOR RES, NEW ENTERPRISE DIV, MONSANTO CO, 69- Mem: AAAS; Am Chem Soc; Am Phys Soc; Am Inst Chemists. Res: Syntheses, molecular and solid-state structures, and chemical, physical and mechanical properties of macromolecular systems and composites; magnetic resonance and vibrational spectros; thermodynamics. Mailing Add: Monsanto Co Res Ctr 800 N Lindbergh Blvd St Louis MO 63166

WOODBRIDGE, CASPAR LIGON, JR, physics, see 12th edition

WOODBRIDGE, CYRIL GORDON, b Vancouver, BC, Oct 23, 15; nat US; m 48; c 2. HORTICULTURE. Educ: Univ BC, BA, 35; Univ Wash, MSc, 38; State Col Wash, PhD(chem), 48. Prof Exp: Chemist, Can Dept Agr, 35-40 & 45-54; assoc prof, 54-61, PROF HORT & HORTICULTURIST, WASH STATE UNIV, 61- Concurrent Pos: Fulbright award, 62-63. Mem: AAAS; Am Chem Soc. Res: Mineral nutrition and physiological disorders of fruit trees, small fruits and vegetables; effect of pesticides on chemical composition of plants. Mailing Add: Dept of Hort Wash State Univ Pullman WA 99163

WOODBRIDGE, DAVID DAVIS, b Seattle, Wash, Jan 29, 22; m 50; c 3. PHYSICS, METEOROLOGY. Educ: Univ Wash, BS, 49; Ore State Univ, MS, 51, PhD(physics), 56. Prof Exp: Asst physics, Ore State Univ, 51-54; asst prof, Colo Sch Mines, 54-57; chief space environ br, Res Projs Lab, Ballistic Missile Agency, US Dept Army, 57-60, dir exp progs, Requirement & Plans Div, Rocket & Guided Missile Agency, 60-61; staff scientist, Chance Vought, Tex, 61-62; dir res & head physics dept, 62-69, PROF SCI EDUC & PHYSICS, DIR UNIV CTR POLLUTION RES & HEAD DEPT SCI EDUC, FLA INST TECHNOL, 69- Concurrent Pos: Consult, High Altitude Observ, Univ Colo, 56-57; prof, Exten Div, Univ Ala, 57-61; vis lectr, Washington Univ, 60; pollution consult, Envirolab, Inc, Fla, 71- Mem: Am Meteorol Soc; Am Inst Aeronaut & Astronaut; Am Asn Physics Teachers; Am Geophys Union; Solar Energy Soc. Res: Pollution abatement; irradiation of waste water; physics of atmosphere and space; effects of solar storms on high-level circulation patterns; extra low frequency radiation; energy conversion. Mailing Add: 1856 Washington Ave Melbourne FL 32935

WOODBRIDGE, JOSEPH ELIOT, b Philadelphia, Pa, July 15, 21; m 49; c 6. CLINICAL CHEMISTRY, PHYSICAL CHEMISTRY. Educ: Princeton Univ, PhD(chem), 48. Prof Exp: Chemist, Manhattan Proj, 44; from res chemist to group leader, Atlantic Refining Co, 46-60; dir res, Hartman-Leddon Co, 60-66; res dir, Sadtler Res Labs, Inc, Pa, 66-68; dir clin res, Worthington Biochem Corp, NJ, 68-71; VPRES DIAG PROD, PRINCETON BIOMEDIX INC, 71- Mem: Am Chem Soc; Am Asn Clin Chemists; Sigma Xi. Res: Synthetic detergents; petrochemicals; clinical reagents; mass spectrometry. Mailing Add: 84 Bayard Lane Princeton NJ 08540

WOODBRIDGE, MARGARET YOUNG, mathematics, see 12th edition

WOODBURN, HENRY MILTON, b Lockport, NY, May 30, 02; m 33; c 3. ORGANIC CHEMISTRY. Educ: Univ Buffalo, BS, 23; Northwestern Univ, MS, 25; Pa State Univ, PhD(org chem), 31. Prof Exp: From instr to prof chem, 23-72, actg chmn dept, 43-45, head dept, 45-56, dean grad sch arts & sci, 53-66, distinguished serv prof, 72, EMER PROF CHEM, STATE UNIV NY BUFFALO, 72- Mem: Am Chem Soc. Res: Tertiary alcohols; reaction of cyanogen with organic compounds; reaction of trifluoroacetonitrile with hydrogen-containing functional groups; chemistry of oxamidines, cyanoformadinines, oxaldiimidates and cyanoforminidates; chemical literature. Mailing Add: 93 Lehn Springs Dr Williamsville NY 14221

WOODBURN, MARGY JEANETTE, b Pontiac, Ill, Sept 5, 28. FOOD SCIENCE, MICROBIOLOGY. Educ: Univ Ill, Urbana, BS, 50; Univ Wis, Madison, MS, 56, PhD(exp foods), 59. Prof Exp: Instr foods & nutrit, Univ Wis, Madison, 56-57; from assoc prof to prof, Purdue Univ, 59-69; PROF & HEAD DEPT FOODS & NUTRIT, SCH HOME ECON, ORE STATE UNIV, 69- Concurrent Pos: Nat Res Coun res assoc, Ft Detrick, Md, 68. Mem: Am Dietetic Asn; Am Home Econ Asn; Am Pub Health Asn; Am Soc Microbiol; Inst Food Technol. Res: Food microbiology; staphylococcal enterotoxins; foodborne pathogenic bacteria. Mailing Add: Dept of Foods & Nutrit Sch of Home Econ Ore State Univ Corvallis OR 97331

WOODBURNE, MICHAEL O, b Ann Arbor, Mich, Mar 8, 37; m 60; c 2. VERTEBRATE PALEONTOLOGY, STRATIGRAPHY. Educ: Univ Mich, BS, 58, MS, 60; Univ Calif, Berkeley, PhD(paleont), 66. Prof Exp: Mus technician vert paleont, Univ Calif, Berkeley, 62-65, mus scientist, 65-66; res assoc, Princeton Univ, 66; from lectr to asst prof, 66-71, ASSOC PROF GEOL, UNIV CALIF, RIVERSIDE, 71- Mem: Soc Vert Paleont; Am Soc Mammal; Paleont Soc; Soc Study Evolution; Geol Soc Am. Res: Mammalian paleontology, including Australian marsupials; biostratigraphy and paleontology of the Mojave Desert. Mailing Add: Dept of Geol Sci Univ of Calif Riverside CA 92502

WOODBURNE, RUSSELL THOMAS, b London, Ont, Nov 2, 04; US citizen; m 34; c 3. ANATOMY. Educ: Univ Mich, AB, 32, MA, 33, PhD(anat), 35. Prof Exp: From instr to prof, 36-74, chmn dept, 58-73, EMER PROF ANAT, MED SCH, UNIV MICH, ANN ARBOR, 75- Mem: Am Asn Anat (secy-treas, 64-72, pres, 73-74); Can Asn Anat. Res: Structure of mammalian midbrain; pleura; blood vessels of pancreas, liver, urinary bladder, ureter and urethra. Mailing Add: Dept Anat 5810 Med Sci II Bldg Univ of Mich Med Sch Ann Arbor MI 48104

WOODBURY, DIXON MILES, b St George, Utah, Aug 6, 21; m 45; c 3. PHARMACOLOGY. Educ: Univ Utah, BS, 42, MS, 45; Univ Calif, PhD(zool), 48. Prof Exp: Asst zool, Univ Calif, 44-47; res instr physiol & pharmacol, 47-50, asst res prof, 50-53, assoc res prof pharmacol, 53-59, assoc prof, 59-61, PROF PHARMACOL, COL MED, UNIV UTAH, 61-, CHMN DEPT, 72- Concurrent Pos: USPHS grant, Col Med, Univ Utah, 47-50; NIH res career award, 62-; Nat Neurol Res Found scientist, Univ Utah, 58; mem pharmacol training comt, NIH, 61-65; mem adv panel on epilepsies, Surgeon Gen, USPHS, 66-69; mem prog proj comt A, Nat Inst Neurol Dis & Stroke, 67-71 & ad hoc comt anticonvulsants, 68-; Boerhaave prof, State Univ Leiden, 68; mem neurobiol merit rev bd, Vet Admin, 72-74; mem prof adv bd & nat bd dirs, Epilepsy Found Am, 73-; distinguished res prof, Univ Utah, 73-74. Mem: Am Soc Pharmacol & Exp Therapeut; Endocrine Soc; Soc Neurosci; Am Soc Neurochem; Int Soc Neurochem. Res: Brain electrolyte metabolism; convulsive disorders; endocrinology; effects of hormones and therapeutic gases on the nervous system; neurochemistry; development of the nervous system; blood-brain barrier; metabolism of drugs. Mailing Add: Dept of Pharmacol Univ of Utah Col of Med Salt Lake City UT 84132

WOODBURY, ELTON NORRIS, b Beverly, Mass, Dec 7, 13; m 39; c 3. ENTOMOLOGY. Educ: Ohio Wesleyan Univ, AB, 36; Ohio State Univ, PhD(entom), 39. Prof Exp: Crop Protection Inst investr, Purdue Univ, 39-42; with dept zool & entom, Pa State Col, 42-43; entomologist, 43-53, supvr, Agr Chem Lab, 53-68, MGR PESTICIDE CONTROL SERV, SYNTHETICS DEPT, HERCULES INC, 69- Mem: Entom Soc Am. Res: Insect toxicology. Mailing Add: Hercules Inc Synthetics Dept 900 Market St Wilmington DE 19899

WOODBURY, ERIC JOHN, b Washington, DC, Feb 9, 25; m 46; c 3. PHYSICS. Educ: Calif Inst Technol, BS, 47, PhD(physics), 51. Prof Exp: Asst, Calif Inst Technol, 47-51; mem tech staff, 51-53, group head & sr staff engr, Electronics Dept, Guided Missile Lab, 53-60, sr staff physicist, Radar & Missile Electronics Lab, 61-62, sr scientist, 62-63, asst dept mgr, Laser Develop Dept, 63-66, mgr laser dept, 66-69, mgr laser develop dept, 69-72, MGR TACTICAL LASER SYSTS LAB, HUGHES AIRCRAFT CO, 72- Concurrent Pos: Mem indust adv comt, Tech Educ Res Ctr, 72-75. Mem: Am Phys Soc; fel Inst Elec & Electronics Engrs; Sigma Xi. Res: Noise in electronic systems; application of solid state to electronic devices; electronic systems for use in missiles and satellites; experimental nuclear physics; optical masers; laser systems and components. Mailing Add: 18621 Tarzana Dr Tarzana CA 91356

WOODBURY, GEORGE WALLIS, JR, b Oct 13, 37; US citizen; m 60; c 2. PHYSICAL CHEMISTRY. Educ: Univ Idaho, BS, 59; Univ Minn, PhD(phys chem), 64. Prof Exp: Res assoc chem, Univ Minn, 64-65 & Cornell Univ, 65-66; asst prof, 66-70, ASSOC PROF CHEM, UNIV MONT, 70- Mem: Am Phys Soc. Res: Statistical mechanics of cooperative phenomena. Mailing Add: Dept of Chem Univ of Mont Missoula MT 59801

WOODBURY, HENRY HUGH, b Paterson, NJ, Sept 24, 28; m 55; c 7. PHYSICS. Educ: Calif Inst Technol, BS, 49, PhD(physics), 53. Prof Exp: PHYSICIST, RES & DEVELOP CTR, GEN ELEC CO, 53- Mem: Fel Am Phys Soc. Res: Solid state physics, particularly research and development on semiconducting materials. Mailing Add: Res & Develop Ctr Gen Elec Co Box 8 Schenectady NY 12301

WOODBURY, JOHN F L, b London, Eng, Dec 22, 18; Can citizen; m 43; c 3. MEDICINE. Educ: Dalhousie Univ, BSc, 39, MD, CM, 43; Royal Col Physicians Can, cert, 52; FRCP(C), 72. Prof Exp: Assoc prof, 46-49, PROF MED & DIR RHEUMATIC DIS UNIT, DALHOUSIE UNIV, 69- Mem: Fel Am Col Physicians; Am Rheumatism Asn; Can Rheumatism Asn; Can Med Asn. Res: Etiology and immunology of arthritis. Mailing Add: Suite 8-016 Victoria Gen Hosp Halifax NS Can

WOODBURY, JOHN WALTER, b St George, Utah, Aug 7, 23; m 49; c 4. PHYSIOLOGY, BIOPHYSICS. Educ: Univ Utah, BS, 43, MS, 47, PhD(physiol), 50. Prof Exp: Lab asst physics, Univ Utah, 42-43; staff mem, Radiation Lab, Mass Inst Technol, 43-45; res asst physiol, Univ Utah, 45-47; from instr to asst prof physiol, Sch Med, Univ Wash, 50-57, from assoc prof to prof physiol & biophys, 57-73; PROF PHYSIOL, COL MED, UNIV UTAH, 73- Mem: AAAS; Am Physiol Soc; Biophys Soc; Inst Elec & Electronics Eng. Res: Electrophysiology of excitable tissues; ion transport through membranes; characteristics of anion channels; anion permeability. Mailing Add: Dept of Physiol 4C208 Univ Med Ctr Salt Lake City UT 84132

WOODBURY, LOWELL ANGUS, b St George, Utah, Oct 11, 10; m 36; c 3. BIOSTATISTICS. Educ: Univ Utah, BS, 33, MS, 34; Univ Mich, PhD(zool), 40. Prof Exp: Res assoc anat, Sch Med, Univ Utah, 45-46, asst prof physiol, 46-52; chief biostatistician, Atomic Bomb Casualty Comn, 52-58; statistician, WHO, 59-74; ASSOC RES PROF, DIV RADIOBIOL, COL MED, UNIV UTAH, 74- Mem: Am Statist Asn; Biomet Soc; Am Physiol Soc. Res: Bone micromorphommetry; stereology; epidemiology of convulsive disorders; clinical drug trials; health information systems. Mailing Add: Div of Radiobiol Univ of Utah Med Ctr Salt Lake City UT 84132

WOODBURY, MAX ATKIN, b St George, Utah, Apr 30, 17; m 47; c 4. BIOMATHEMATICS, COMPUTER SCIENCE. Educ: Univ Utah, BS, 39; Univ Mich, MS, 40, PhD(math), 48. Prof Exp: Instr math, Univ Mich, 47-49; Off Naval Res grant & mem, Inst Advan Study, 49-50; res assoc math & econ, Princeton Univ, 50-52; assoc prof statist, Univ Pa, 52-54; prin investr, Logistics Res Proj, George Washington Univ, 54-56; res prof math, Col Eng, NY Univ, 56-62, prof exp neurol, Med Ctr, 63-65; PROF BIOMATH, MED CTR, DUKE UNIV, 66-, PROF COMPUT SCI, 71- Concurrent Pos: Gov & indust consult, 51-; mem opers res adv coun, New York, 64-68; mem diag radiol adv group, Nat Cancer Inst, 74-77; sr fel, Ctr Study Aging & Human Develop, 75-; mem several nat comts weather modification & NIH & Food & Drug Admin study sects. Mem: Fel AAAS; fel Inst Math Statist; fel Am Statist Asn; Classification Soc; Biomet Soc. Res: Computing; statistics; models in biology and medicine; quantitative models of information about biomedical systems require mathematics, probability and basic sciences for formulation, statistics of estimation of parametrics and testing, computing and numerical analysis for calculation. Mailing Add: Biomath Duke Univ Med Ctr PO Box 3200 Durham NC 27710

WOODBURY, NATHALIE FERRIS SAMPSON, b Humboldt, Ariz, Jan 25, 18; m 48. CULTURAL ANTHROPOLOGY, ETHNOGRAPHY. Educ: Barnard Col, AB, 39. Prof Exp: Instr anthrop, Brooklyn Col, 44-45; asst prof anthrop & geog, Eastern NMex Col, 45-46; instr anthrop, Univ Ariz, 46-47; asst archaeologist, Zaculeu Proj, United Fruit Co, Guatemala, 48-50; lectr anthrop, Barnard Col, 52-58, actg exec officer, Dept Anthrop, 54-56, asst dean studies, 56-58; res assoc anthrop, Ariz State Mus, 58-64; res assoc, US Nat Mus, Smithsonian Inst, 64-70; secy, 70-75, MEM EXEC BD, AM ANTHROP ASN, 75- Concurrent Pos: Assoc ed, Am Antiquity, 58-62; res assoc, Mus Northern Ariz, 58-71; Wenner-Gren Found grant, Am Anthrop Asn. 69-70; ed, Guide to Depts of Anthrop, Am Anthrop Asn, 69-; assoc ed, Am Anthropologist, 73-75. Mem: Fel Am Anthrop Asn; fel Soc Appl Anthrop; Soc Med Anthrop; Am Ethnol Soc (secy, 60-66); Soc Am Archaeol (treas, 65-69). Res: American Indian ethnology; American archaeology; history of anthropology and folklore. Mailing Add: Am Anthrop Asn 1703 New Hampshire Ave NW Washington DC 20009

WOODBURY, RICHARD BENJAMIN, b West Lafayette, Ind, May 16, 17; m 48. ANTHROPOLOGY. Educ: Harvard Univ, BS, 39, MA, 42, PhD(anthrop), 49. Prof Exp: Archeologist, Zaculeu Proj, United Fruit Co, Guatemala, 47-50; assoc prof anthrop, Univ Ky, 50-52 & Columbia Univ, 52-58; Social Sci Res Coun res grant, NMex, 58-59; res assoc prof anthrop, Univ Ariz, 59-63; cur, US Nat Mus, Smithsonian Inst, 63-65, chmn, Off Anthrop, 65-67, cur, US Nat Mus, 67-69; chmn dept anthrop, 69-73, actg assoc provost & dean grad sch, 73-74, PROF ANTHROP, UNIV MASS, AMHERST, 69- Concurrent Pos: Staff mem, Awatovi Exped, Peabody Mus, Harvard Univ, 38-; Wenner-Gren Found res grant anthrop, NMex, 54-55; mem div anthrop & psychol, Nat Res Coun, 54-57; ed, J Soc Am Archaeol, 54-58; NSF res grant, 59-63; mem bd dirs, Human Rels Area Files, Inc, Conn, 64-69; ed-in-chief, Am Anthropologist, 75-78. m: Fel AAAS; fel Am Anthrop Asn; Soc Am Archaeol (treas, 53-54, pres, 58-59. Res: Archeology of the Southwestern United States and Mexico and Central America; preindustrial systems of water control and utilization in arid regions. Mailing Add: Dept of Anthrop Machmer Hall Univ of Mass Amherst MA 01002

WOODBURY, ROBERT ARTHUR, b Pittsburg, Kans, Sept 1, 04; m 31; c 3. PHARMACOLOGY, PHYSIOLOGY. Educ: Univ Kans, BS, 24, MS, 28, PhD(physiol, pharmacol), 31; Univ Chicago, MD, 34. Prof Exp: Asst instr physiol & pharmacol, Univ Kans, 27-29, instr, 29-31; instr physiol, YMCA Col, 31-32; instr biol, Cent YMCA Col, 32-33; intern, Gen Hosp, Kansas City, 33-34; from asst prof physiol & pharmacol to prof pharmacol & head dept, Univ Ga, 34-47; PROF PHARMACOL & CHMN DEPT, UNIV TENN, MEMPHIS, 47- Mem: Am Physiol Soc; Am Soc Pharmacol & Exp Therapeut; Swiss Asn Physiol & Pharmacol. Res: Drugs influencing autonomic nervous system; posterior pituitary hormones; dysmenorrhea; cardiovascular and uterine physiology and pharmacology. Mailing Add: Dept of Pharmacol Univ of Tenn Med Units Memphis TN 38163

WOODBY, LAUREN G, b Beaverton, Mich, Sept 26, 14; m 39; c 5. MATHEMATICS. Educ: Cent Mich Univ, BS, 34; Univ Mich, MA, 40, PhD, 52. Prof Exp: High sch teacher, Mich, 34-41 & CZ, 41-42; instr math, CZ Jr Col, 46-51; assoc prof, Mankato State Col, 52-54; from assoc prof to prof, Cent Mich Univ, 54-61; specialist, US Off Educ, 61-66; PROF MATH, MICH STATE UNIV, 66- Concurrent Pos: NSF fac fel, 58-59; fel teacher training, NY Univ, 69-70; coordr, US presentation, 2nd Int Cong Math Educ, 72; prog mgr, NSF, 73-74, consult math educ, 74-75. Mem: Nat Coun Teachers Math; Math Asn Am. Res: School mathematics curriculum; teacher education. Mailing Add: Dept of Math Mich State Univ East Lansing MI 48823

WOODCOCK, ALFRED HERBERT, b Atlanta, Ga, Sept 7, 05; m 41; c 3. OCEANOGRAPHY. Hon Degrees: DSc, Long Island Univ, 61. Prof Exp: Technician, Woods Hole Oceanog Inst, 31-42, res assoc, 42-46, oceanogr, 46-63; RES ASSOC GEOPHYS, HAWAII INST GEOPHYS, 63- Mem: AAAS; fel Am Meteorol Soc; assoc Am Geophys Union. Res: Marine meteorology; air-sea interaction; sea-salt nuclei in marine atmospheres; cloud physics. Mailing Add: Hawaii Inst of Geophys Univ of Hawaii Honolulu HI 96822

WOODCOCK, CHARLES MARTIN, b Newark, NJ, Aug 9, 20. FOOD SCIENCE. Educ: Univ Mass, BSc, 42; Ohio State Univ, MSc, 47. Prof Exp: Instr food freezing & preserv, Ohio State Univ, 46-47; proj leader, Packaging Res Sect, Cent Labs, 47-53, head, 53-58, lab mgr, Post Div, 58-72, SR RES SPECIALIST, TECH CTR, GEN FOODS CORP, 72- Mem: Sigma Xi. Res: Packaging of foodstuffs including papers, films, foils, glass and shipping containers; product and process improvement of cereals and beverage powders. Mailing Add: Tech Ctr Gen Foods Corp Tarrytown NY 10591

WOODCOCK, CHRISTOPHER LEONARD FRANK, b Essex, Eng, July 9, 42; m 64;

c 3. CELL BIOLOGY, BOTANY. Educ: Univ Col, Univ London, BSc, 63, PhD(bot), 66. Prof Exp: Res fel biophys, Univ Chicago, 66-67; res fel bot, Harvard Univ, 67-69, lectr biol, 69-72; asst prof, 72-75, ASSOC PROF ZOOL, UNIV MASS, AMHERST, 75- Mem: Am Soc Cell Biol. Res: Cell ultrastructure and function; information processing in cells; chromatin structure and function. Mailing Add: Dept of Zool Univ of Mass Amherst MA 01002

WOODFIN, BEULAH MARIE, b Chicago, Ill, June 22, 36. BIOCHEMISTRY. Educ: Vanderbilt Univ, BA, 58; Univ Ill, Urbana, MS, 60, PhD(biochem), 63. Prof Exp: Res assoc biochem, Univ Mich, 63-66, instr, 66-67; ASST PROF BIOCHEM, SCH MED, UNIV N MEX, 67- Concurrent Pos: USPHS fel, Univ Mich, 63-65. Mem: AAAS; Am Chem Soc; NY Acad Sci. Res: Correlation of activity and structure of enzymes. Mailing Add: Dept of Biochem Univ of NMex Sch of Med Albuquerque NM 87131

WOODFORD, VERNON RICH, JR, b Beaufort, SC, Nov 22, 20; Can citizen; m 43; c 6. BIOCHEMISTRY. Educ: Col Charleston, BS, 42; McGill Univ, MSc, 48, PhD(biochem), 58. Prof Exp: Lab asst biol, Col Charleston, 40-42; demonstr biochem, McGill Univ, 48-50; asst prof, 50-62, ASSOC PROF BIOCHEM, UNIV SASK, 62- Mem: Fel Chem Inst Can; Can Biochem Soc. Res: Metabolism of brain; amine metabolism. Mailing Add: Dept of Biochem Univ of Sask Saskatoon SK Can

WOODGATE, BRUCE EDWARD, b Eastbourne, Sussex, Eng, Feb 19, 39; m 65; c 2. ASTROPHYSICS. Educ: Univ London, BSc, 61, PhD(astron), 65. Prof Exp: From res asst to res assoc physics, Univ Col, Univ London, 65-71; sr res assoc, Columbia Univ, 71-74, assoc dir Astrophys Lab, 72-74; sci systs analyst, 74-75, ASTROPHYSICIST, GOODARD INST SPACE STUDIES, NASA, 75- Concurrent Pos: Mem working group, NASA Outlook for Space Study, 74. Mem: Am Astron Soc; Fel Royal Astron Soc. Res: X-ray astronomy; astronomy of galaxies and their interstellar medium; solar physics and solar-terrestrial relations; remote sensing of earth resources. Mailing Add: Code 683 Goddard Space Flight Ctr NASA Greenbelt MD 20771

WOODHALL, BARNES, b Rockport, Maine, Jan 22, 05; m 28; c 2. NEUROSURGERY. Educ: Williams Col, AB, 26; Johns Hopkins Univ, MD, 30. Hon Degrees: DSc, Williams Col, 70. Prof Exp: Intern surg, Johns Hopkins Hosp, 30-31, asst resident surgeon & neurosurgeon, 31-36, resident surg, 36-37, instr, 31-37; asst prof surg chg neurosurg, Sch Med, 37-47, prof neurosurg, 47-70, dean sch med, 60-63, vprovost univ, 60-67, assoc provost, 67-68, asst pres, 68, chancellor, 69-71, JAMES B DUKE PROF NEUROSURG, SCH MED, DUKE UNIV, 70- Honors & Awards: Distinguished physician, Vet Admin, 71. Mem: Am Surg Asn; Am Acad Neurol Surg (pres, 44); Am Asn Neurol Surg (pres, Harvey Cushing Soc, 63); Soc Neurol Surg (pres, 64). Res: Experimental studies in peripheral nerve pathology; experimental brain tumors; chemotherapy of brain tumors. Mailing Add: Duke Univ Med Ctr Baker House 142 Durham NC 27710

WOODHAM, DONALD W, b Jacksonville, Ala, July 9, 29; m 54; c 1. ANALYTICAL CHEMISTRY. Educ: Jacksonville State Univ, BS, 54; Am Inst Chemists, cert, 73. Prof Exp: Chem aide, Phosphate Develop Works, Tenn Valley Authority, 54-55, control chemist, 55-57; res chemist, Pesticide Chem Br, Entom Res Div, 57-65, res chemist, Plant Pest Control Div, Miss, 65-67, supvry chemist, 67-70, chemist in-chg plant pest control prog, Animal & Plant Health Inspection Serv, 70-75, LAB SUPVR, PINK BOLLWORM REARING FACIL, ANIMAL & PLANT HEALTH INSEPCTION SERV, AGR RES SERV, USDA, 75- Mem: AAAS; Am Chem Soc; Entom Soc Am; Am Inst Chemists; Sigma Xi. Res: Analysis of phosphorous intermediates used in the synthesis of nerve gas; pesticide residues in crops; animal products and miscellaneous materials; analysis of environmental samples for pesticide and herbicide residues; development of new methods for the analysis of pesticides and pesticide residues; confirmation methods for identification purposes; methods utilized in the rearing, irradiation and shipment of pink bollworm moths. Mailing Add: Pink Bollworm Rearing Facil 4125 E Broadway Phoenix AZ 85040

WOODHOUR, ALLEN F, b Newark, NJ, Feb 21, 30; m 55; c 1. VIROLOGY. Educ: St Vincent Col, AB, 52; Cath Univ Am, MS, 54, PhD(bact), 56. Prof Exp: Bacteriologist, Walter Reed Army Inst Res, 56-57; res assoc virol, Charles Pfizer & Co, Inc, 57-60; res assoc, 60-64, asst dir, Dept Virus Dis, 64-66, dir viral vaccine res, 66-73, asst dir virus & cell biol res, 68-73, SR DIR & ASST AREA HEAD VIRUS & CELL BIOL RES, MERCK INST THERAPEUT RES, 74- Mem: AAAS; Am Asn Immunol; Sigma Xi; Int Asn Biol Stand. Res: Use of adjuvants in immunology; metabolizable vegetable oil water-in-oil adjuvant; respiratory viruses for vaccine development; development of bacterial vaccines. Mailing Add: Merck Inst of Therapeut Res West Point PA 19486

WOODHOUSE, BERNARD LAWRENCE, b Norfolk, Va, Aug 14, 36; m 64; c 2. PHARMACOLOGY. Educ: Howard Univ, BS, 58, MS, 63, PhD(pharmacol), 73. Prof Exp: Instr zool, A&T Col NC, 63-64; from instr to asst prof, 64-73, ASSOC PROF BIOL, SAVANNAH STATE COL, 73- Concurrent Pos: Hoffmann-La Roche res grant & NIH res grant, 75- Res: Study of the mechanism of the antihypertensive effects of beta adrenergic blocking drugs on various species of animals. Mailing Add: Dept of Biol Savannah State Col Savannah GA 31404

WOODHOUSE, EDWARD JOHN, b Norwich, Eng, Apr 22, 39. CHEMISTRY, TOXICOLOGY. Educ: Univ Nottingham, BSc, 61, PhD(chem), 64. Prof Exp: Res assoc inorg chem, Ore State Univ, 64-67; sr chemist, 67-74, PRIN CHEMIST, MIDWEST RES INST, 74- Mem: AAAS; Am Chem Soc; NY Acad Sci; Int Asn Forensic Toxicol. Res: Radiochemistry; solvent extration; analysis of drugs in body fluids and dosage form; marijuana chemistry and metabolism. Mailing Add: Midwest Res Inst 425 Volker Blvd Kansas City MO 64110

WOODHOUSE, JOHN CRAWFORD, b New Bedford, Mass, Apr 29, 98; m 27; c 2. CHEMISTRY. Educ: Dartmouth Col, BA, 21, MA, 23; Harvard Univ, AM, 24, PhD(chem), 27. Hon Degrees: MA, Dartmouth Col, 61. Prof Exp: Instr chem, Dartmouth Col, 21-23; from asst to instr, Harvard Univ, 23-25; consult, 26-28; chemist, Ammonia Dept, E I du Pont de Nemours & Co, 28-42, asst dir chem div, Grasselli Chem Dept, 42, dir, 43-44, dir tech div, 44-50, dir atomic energy div, 50-62; CONSULT ATOMIC ENERGY & CHEM, 62-; COUNR & SECY BD, COL MARINE STUDIES, UNIV DEL, 72- Concurrent Pos: Overseer, Thayer Sch Eng, Dartmouth Col, 56, trustee, Col, 60; consult mem steering comt, Panel Tech Adv, US Dept Defense, 58. Mem: Am Chem Soc; Am Nuclear Soc; Am Inst Chem Engrs; Brit Soc Chem Indust. Res: Corrosion; gaseous adsorption; high pressure; organic and inorganic polymers; biological chemotherapy; nuclear metallurgy and nuclear power; use of nuclear isotopes; marine studies. Mailing Add: 21 Woodbrook Circle Wilmington DE 19810

WOODHOUSE, WILLIAM WALTON, JR, b White Oak, NC, May 24, 10; m 34; c 2. SOILS. Educ: NC State Col, BS, 32, MS, 41; Cornell Univ, PhD(soils), 48. Prof Exp: From asst agronomist to assoc agronomist, 36-45, res assoc prof, 41-50, PROF AGRON, NC STATE UNIV, 51- Mem: Am Soc Agron; Soil Sci Soc Am. Res: Fertilization and management of pastures and forage plants; use of cover in the control of water on and in the soils; dune stabilization; land reclamation; marsh building. Mailing Add: Dept of Soil Sci NC State Univ Raleigh NC 27607

WOODHULL, ANN MCNEAL, b Orange, NJ, Oct 20, 42. BIOPHYSICS, NEUROBIOLOGY. Educ: Swarthmore Col, BA, 64; Univ Wash, PhD(physiol, biophys), 72. Prof Exp: NIH fel physiol & biophys, Univ Wash, 72; ASST PROF BIOL, DIV NATURAL SCI, HAMPSHIRE COL, 72- Concurrent Pos: Lectr neurobiol, Harvard Univ, 75. Mem: Biophys Soc. Res: Chemical identity and molecular pharmacology of the neurotoxin of the salamander Notophthalmus viridescens. Mailing Add: Div of Natural Sci Hampshire Col Amherst MA 01002

WOODIN, HOWARD EUGENE, plant ecology, see 12th edition

WOODIN, TERRY STERN, b New York, NY, Dec 25, 33; m 54; c 5. BIOCHEMISTRY. Educ: Alfred Univ, BA, 54; Univ Calif, Davis, MA, 65, PhD(biochem), 67. Prof Exp: Res assoc biochem, Univ Calif, Davis, 67-68; adj asst prof, Univ Nev, Reno, 68-69, asst prof, 69-72; ASST PROF BIOCHEM, HUMBOLDT STATE UNIV, 72- Mem: AAAS; Am Chem Soc; Sigma Xi; Am Soc Plant Physiologists. Res: Sulfate metabolism in fungi; glutathione reductase; chorismate metabolism in plants; thermophilic fungi; microbial degradation of pesticides. Mailing Add: Dept of Chem Humboldt State Univ Arcata CA 95521

WOODIN, WILLIAM GRAVES, b Dunkirk, NY, July 22, 14; m 40; c 2. ALLERGY. Educ: Cornell Univ, AB, 36, MD, 39. Prof Exp: Intern & asst resident med, Univ Hosps Cleveland, 39-41; asst, Med Col, Cornell Univ, 46, instr, 47-48; from instr to asst prof, 48-54, clin assoc prof, 54-63, CLIN PROF MED, STATE UNIV NY UPSTATE MED CTR, 63-, DIR ALLERGY CLIN, 48- Concurrent Pos: Fel, Roosevelt Hosp, New York, 47-48; consult hosps, 50- Mem: AMA; fel Am Acad Allergy; NY Acad Sci. Mailing Add: 109 S Warren St Syracuse NY 13202

WOODIN, WILLIAM HARTMAN, III, b New York, NY, Dec 16, 25; m 48; c 4. ZOOLOGY. Educ: Univ Ariz, BA, 50; Univ Calif, MA, 56. Prof Exp: From assoc dir to dir, 53-71, EMER DIR, ARIZ-SONORA DESERT MUS, 72- Mem: Fel AAAS. Res: Herpetology; taxonomy; ecology; desert ecology. Mailing Add: 3600 N Larrea Lane Tucson AZ 85715

WOODING, FRANK JAMES, b Pontiac, Ill, Feb 1, 41; m 64; c 3. AGRONOMY. Educ: Univ Ill, Urbana, BS, 63; Kans State Univ, MS, 66, PhD(agron), 70. Prof Exp: Res assoc plant physiol, Pa State Univ, 69-70; asst prof agron, 70-75, ASSOC PROF AGRON, INST AGR SCI, UNIV ALASKA, 75- Mem: Am Soc Agron; Crop Sci Soc Am; Soil Sci Soc Am. Res: Plant nutrition; plant growth under arctic and subarctic conditions; crop physiology; crop production. Mailing Add: Inst of Agr Sci Univ of Alaska Fairbanks AK 99701

WOODING, WILLIAM MINOR, b Waterbury, Conn, Aug 24, 17; m 40; c 2. EXPERIMENTAL STATISTICS, CHEMISTRY. Educ: Polytech Inst Brooklyn, BChemE, 53. Prof Exp: Analyst inorg chem, Scovill Mfg Co, Conn, 36-40; technician, Am Cyanamid Co, 41-46, chemist, 46-51, res chemist, 51-56, coordr personnel admin serv, 56-57; asst chief chemist, Revlon, Inc, NY, 57-61, assoc res dir, 61-65; assoc res dir, 65-67, DIR TECH SERV, CARTER PROD DIV, CARTER-WALLACE, INC, 67-, DIR STATIST SERV, 75- Mem: AAAS; Am Chem Soc; fel Am Inst Chemists; fel Am Soc Qual Control; fel Soc Cosmetic Chem. Res: Experimental design and applied statistics, principally in medical and biological fields. Mailing Add: Cranbury NJ

WOODINGS, ERIC T, microbiology, see 12th edition

WOODLAND, BERTRAM GEORGE, b Mountain Ash, Wales, Apr 4, 22; m 52; c 2. STRUCTURAL GEOLOGY. Educ: Univ Wales, BSc, 42; Univ Chicago, PhD(geol), 62. Prof Exp: Exp officer, Ministry Home Security, Gt Brit Air Ministry, 43-46; res asst mineral surv, Ministry Town & Country Planning, 46-49; asst res officer, Ministry Housing & Local Govt, 49-54; from instr to asst prof geol, Univ Mass, 54-56; asst prof, Mt Holyoke Col, 56-58; assoc cur, Chicago Natural Hist Mus, 58-62; CUR, FIELD MUS NATURAL HIST, 63- Concurrent Pos: Consult, Petrol Brasileiro Depex, Rio de Janeiro, Brazil, 55-56. Mem: Fel Geol Soc London; Brit Geol Asn. Res: Metamorphism; igneous rocks; cone-in-cone structure; tectonics and microstructures. Mailing Add: Dept of Geol Field Mus of Natural Hist Chicago IL 60605

WOODLAND, DOROTHY JANE, b Warren, Ohio, Sept 20, 08. PHYSICAL CHEMISTRY. Educ: Col Wooster, BS, 29; Ohio State Univ, MSc, 30, PhD(chem), 32. Prof Exp: Asst, Col Wooster, 28-29 & Ohio State Univ, 29-32; from instr to asst prof chem, Wellesley Col, 32-38; assoc prof & head dept, Western Col, 38-42, prof, 42-44; prof & head dept, 44-74, EMER PROF CHEM, JOHN BROWN UNIV, 74- Mem: Am Chem Soc. Res: Surface energy; relation between radius of curvature of droplets and surface energy. Mailing Add: Dept of Chem John Brown Univ Siloam Springs AR 72761

WOODLAND, JOHN TURNER, b Melrose, Mass, June 26, 23. ZOOLOGY. Educ: Boston Univ, AB, 45, AM, 46; Harvard Univ, MA, 49, PhD(zool), 50. Prof Exp: From asst to instr biol, Boston Univ, 45-47; asst prof, Lebanon Valley Col, 50-52; assoc prof & chmn dept, Richmond Prof Inst, Col William & Mary, 52-55; asst prof, Northeastern Univ, 55-58; asst prof biol sci, Mass State Col, Salem, 58-63, assoc prof, 63-67, PROF BIOL, BOSTON STATE COL, 67- Concurrent Pos: NSF fac fel, 59-60. Mem: Am Soc Microbiol. Res: Insect embryology and histology; physiology of bacterial spores. Mailing Add: Dept of Biol Boston State Col Boston MA 02115

WOODLAND, WILLIAM CHARLES, b Highland Park, Mich, Nov 22, 19; m 44; c 3. PHYSICAL CHEMISTRY. Educ: Col Wooster, BA, 41; Carnegie Inst Technol, MS, 49, DSc(phys chem), 50. Prof Exp: Instr org microanal, NY Univ, 49-51; chemist, Jackson Lab, 51, Chambers Works, NJ, 51-64, color specialist, Washington Works, Parkersburg, WVa, 64-72, SR CHEMIST, WASHINGTON WORKS, E I DU PONT DE NEMOURS & CO, INC, PARKERSBURG, 72- Res: Thermoplastic resins; color technology; specialized analytical chemistry. Mailing Add: 9 Ashwood Dr Vienna WV 26101

WOODLEY, CHARLES LEON, b Montgomery, Ala, Jan 22, 44; m 66; c 1. BIOCHEMISTRY. Educ: Univ Ala, BS, 66, MS, 68; Univ Nebr, PhD(chem), 72. Prof Exp: Res assoc biochem, Univ Nebr, Lincoln, 71-74; ASST PROF BIOCHEM, UNIV MISS MED CTR, 74- Mem: Sigma Xi. Res: Control of protein synthesis initiation and translation in eukaryotic and viral systems. Mailing Add: Dept of Biochem Univ of Miss Med Ctr Jackson MS 39216

WOODLEY, ROBERT EARL, b Portland, Ore, Nov 22, 25; m 52; c 2. PHYSICAL CHEMISTRY. Educ: Ore State Univ, BS, 49; Univ Wash, MS, 51; Ore State Univ, PhD(chem), 65. Prof Exp: Engr, Gen Elec Co, 51-57; res fel chem, Ore State Univ, 57-60; sr engr, Gen Elec Co, 60-65; sr scientist, Battelle Mem Inst, 65-70; SR SCIENTIST, WESTINGHOUSE-HANFORD CO, 70- Res: Investigation of oxygen potential, composition relationships in mixed uranium-plutonium oxide nuclear fuels; other aspects of ceramic nuclear fuel chemistry, including the kinetics of oxidation and reduction. Mailing Add: Westinghouse-Hanford Co PO Box 1970 Richland WA 99352

WOODLEY, WILLIAM LEE, meteorology, see 12th edition

WOODMAN, DANIEL RALPH, b Portland, Maine, Apr 20, 42; m 67. VIROLOGY. Educ: Univ Maine, Orono, BS, 64; Univ Md, College Park, MS, 66, PhD(microbiol), 72. Prof Exp: Asst microbiol, Univ Md, College Park, 64-67; naval officer virol, Deseret Test Ctr, Salt Lake City, 67-69; instr, Univ Col, Univ Md, 70; exec officer, US Naval Unit, Ft Detrick, Md, 71-74, HEAD VIROL DIV, NAVAL MED RES INST, 74- Mem: Am Soc Microbiol; Asn Mil Surgeons US; Am Soc Trop Med & Hyg; AAAS. Res: Animal virology with a special interest in viral immunology and chemotherapeutics; arbovirus replication and pathogenicity. Mailing Add: Dept of Microbiol Naval Med Res Inst NNMC Bethesda MD 20014

WOODMAN, JAMES NELSON, b Ft Worth, Tex, May 26, 36; m 69. SILVICULTURE, FOREST PHYSIOLOGY. Educ: Univ Wash, BS, 59, PhD(silvicult), 68; Univ Calif, Berkeley, MF, 63. Prof Exp: Forester, US Forest Serv, 63-64; res assoc, Col Forest Resources, Univ Wash, 64-68; SILVICULTURIST, FORESTRY RES CTR, WEYERHAEUSER CO, 68-, PROJ LEADER INTENSIVE STAND CULT RES, 73- Concurrent Pos: Exchange scientist, NZ Forest Serv-Forest Res Inst, 75-76. Mem: Soc Am Foresters; Int Union Forestry Res Orgn; Int Asn Plant Physiologists. Res: Intensive measurement of photosynthesis rates and principal environmental factors within large forest grown trees; correlation and prediction of various stand growth parameters from representative sample measurements of forest environment; development of thinning, fertilization and irrigation prescriptions for major western tree species. Mailing Add: Weyerhaeuser Co PO Box 420 Centralia WA 98531

WOODMAN, RICHARD J, b Bromley, Eng, May 7, 33; m 61; c 3. BIOCHEMISTRY. Educ: Univ London, BSc, 54, MSc, 55, PhD(neurochem), 61. Prof Exp: Res assoc biochem, Sch Path, Oxford Univ, 61-62, Imp Cancer Res Fund, London, Eng, 62-64, Roswell Park Mem Inst, 64-65, Sch Med, Stanford Univ, 65-67 & Baltimore City Hosps, 67-68; SR INVESTR, CANCER CHEMOTHER-DRUG EVAL DIV, MICROBIOL ASSOC, INC, 68- Res: Anionic macromolecules of tumor and normal cell surfaces and their detection in ambient fluids. Mailing Add: Microbiological Associates Inc 5221 River Rd Bethesda MD 20016

WOODMANSEE, ROBERT ASBURY, b Cincinnati, Ohio, Mar 6, 26; m 49; c 3. MARINE ECOLOGY. Educ: Univ Miami, BS, 48, MS, 49; Western Reserve Univ, PhD(zool), 52. Prof Exp: Asst zool, Western Reserve Univ, 49-52, technician biochem, 52; oceanogr, US Naval Hydrog Off, 52-54; from asst prof to prof biol, Miss Southern Col, 54-64; assoc marine scientist, Va Inst Marine Sci, 64-65; from assoc prof to prof biol, Univ SAla, 65-72; HEAD ECOL SECT, GULF COAST RES LAB, 72- Mem: Am Soc Limnol & Oceanog; Ecol Soc Am. Res: Ecology of plankton; seasonal distribution and daily vertical migrations of zooplankton; distribution of planktonic diatoms in estuary. Mailing Add: Ecol Sect Gulf Coast Res Lab PO Box AG Ocean Springs MS 39564

WOODMANSEE, ROBERT GEORGE, b Albuquerque, NMex, Sept 11, 41; m 63; c 1. RANGE ECOLOGY, FOREST ECOLOGY. Educ: Univ NMex, BS, 67, MS, 69; Colo State Univ, PhD(forest ecol & soils), 72. Prof Exp: Fel grassland ecol, 72-74, SR RES ECOLOGIST, NATURAL RESOURCE ECOL LAB, COLO STATE UNIV, 74-, ASST DIR GRASSLAND BIOME, 75- Mem: Ecol Soc Am; AAAS; Am Inst Biol Sci; Am Soc Agron; Soc Range Mgt. Res: Field experimentation and simulation of modeling of nutrient cycling in grassland and forest ecosystems. Mailing Add: Natural Resources Ecol Lab Colo State Univ Ft Collins CO 80523

WOODRIFF, RAY ALAN, b Pueblo, Colo, Jan 22, 09; m 38; c 3. CHEMISTRY, SPECTROSCOPY. Educ: Univ Ore, BS, 33; Ore State Col, MS, 35; Univ Colo, PhD(chem), 42. Prof Exp: Asst chem, Ore State Col, 33-34; Univ Colo, 35-39; from instr to assoc prof, 39-53, PROF CHEM, MONT STATE UNIV, 53- Concurrent Pos: NSF grants, 68-73 & 75. Mem: Am Chem Soc; Soc Appl Spectros. Res: Emission spectroscopy; analytical chemistry; inorganic chemistry; magnetohydrodynamics; theory of mountain building; challenge of ultimate survival; nature of light; valence theory; development of nonflame atomic absorption using a high temperature furnace. Mailing Add: Dept of Chem Mont State Univ Bozeman MT 59715

WOODRIFF, ROGER L, b Bozeman, Mont. MATHEMATICS. Educ: Mont State Univ, BS, 64; Univ Wis-Milwaukee, MS, 65. Prof Exp: Asst math, Mont State Univ, 63-64; instructing asst, Univ Wis-Milwaukee, 64-65; vis instr, Mid East Tech Univ, Ankara, 65-67; asst prof, Humboldt State Col, 67-70; PROF MATH, MENLO COL, 70- Mem: Am Math Soc; Math Asn Am. Res: Algebraic topology. Mailing Add: Dept of Math Menlo Col Menlo Park CA 94025

WOODRING, JAY PORTER, b Philipsburg, Pa, Sept 29, 32; m 55; c 2. ENVIRONMENTAL PHYSIOLOGY. Educ: Pa State Univ, BS, 54; Univ Minn, MS, 58, PhD(entom & zool), 60. Prof Exp: From instr to assoc prof, 60-70, PROF ZOOL, LA STATE UNIV, BATON ROUGE, 71- Concurrent Pos: Humboldt scholar, 67-68. Res: Functional morphology, physiology and ecology of mites, millipedes, insects and other invertebrates that occur in the litter. Mailing Add: Dept of Zool La State Univ Baton Rouge LA 70803

WOODRING, WENDELL PHILLIPS, b Reading, Pa, June 13, 91; m 18. GEOLOGY. Educ: Albright Col, AB, 10; Johns Hopkins Univ, PhD(geol), 16. Hon Degrees: DSc, Albright Col, 52. Prof Exp: Field asst, US Geol Surv, 13-16, geologist, 19-27; prof invert paleont, Calif Inst Technol, 27-30; geologist, US Geol Surv, 30-61; RES ASSOC, SMITHSONIAN INST, 61- Concurrent Pos: Geologist, Sinclair Cent Am Oil Corp, 17-18; geologist in chg, Haitian Geol Surv, 20-24; geol explor, Haiti, 20-21, 41-42 & Cuba, 42-43; paleontologist, Trop Oil Co, Colombia, 22. Honors & Awards: Penrose Medal, Geol Soc Am, 49; Thompson Medal, Nat Acad Sci, 67. Mem: Nat Acad Sci; AAAS; fel Geol Soc Am (pres, 53); Paleont Soc (pres, 48); Am Philos Soc. Res: Invertebrate paleontology; tertiary mollusks; stratigraphy; areal geology. Mailing Add: Nat Mus of Natural Hist E-507 Washington DC 20560

WOODROOF, JASPER GUY, b Mountville, Ga, May 23, 00; m 25; c 3. HORTICULTURE. Educ: Univ Ga, BSA, 22, MSA, 26; Mich State Univ, PhD(hort), 32. Prof Exp: Asst horticulturist, Exp Sta, Univ Ga, 22-26, assoc horticulturist, 26-32, pres, Abraham Baldwin Agr Col, 32; conservationist, Resettlement Admin, Wash, 33-36; adminr, Farm Security Admin, 36-38; horticulturist, Exp Sta, 38-40, food technologist & head dept food technol, 40-67, chmn div food sci, 50-67, Alumni Found distinguished prof, 58-67, ALUMNI FOUND DISTINGUISHED EMER PROF FOOD SCI, EXP STA, UNIV GA, 67- Concurrent Pos: Consult, lectr & author; mem staff, Study Foods for Shelter Storage, Off Civil & Defense Mobilization, US Dept Defense; sci adv coun, Refrig Res Found, 50; consult to Foreign Agr Serv, London, Eng, 65. Mem: Fel AAAS; fel Am Soc Heat, Refrig & Air Conditioning Engrs; Inst Food Technologists. Res: Anatomy and morphology of pecans; preservation of fruits, vegetables and meats by freezing; cold storage of peanuts and pecans and their products; stability of military rations at widely different environments; foods for shelter storage. Mailing Add: Dept of Food Sci Univ of Ga Exp Sta Experiment GA 30212

WOODROW, DONALD L, b Washington, Pa, Nov 25, 35; m 60; c 2. STRATIGRAPHY, SEDIMENTOLOGY. Educ: Pa State Univ, BS, 57; Univ Rochester, MS, 50, PhD(geol), 65. Prof Exp: From asst prof to assoc prof, 65-75, PROF GEOL, HOBART & WILLIAM SMITH COLS, 75- Concurrent Pos: NSF res grant, 68-75; Res Corp res grant, 69-75; vis res geologist, Univ Reading, 71-72; consult, var indust orgns, 74- Mem: Geol Soc Am; Am Asn Petrol Geologists; Soc Econ Paleontologists & Mineralogists; Int Asn Sedimentol. Res: Sedimentology of Paralic and lake sediments; Upper Devonian stratigraphy of the Appalachian Plateau. Mailing Add: Dept of Geosci Hobart & William Smith Cols Geneva NY 14456

WOODRUFF, CALVIN WATTS, b Conn, July 16, 20; m 50; c 3. MEDICINE. Educ: Yale Univ, BA, 41, MD, 44. Prof Exp: From instr to assoc prof pediat, Sch Med, Vanderbilt Univ, 50-60; PROF PEDIAT, SCH MED, UNIV MO-COLUMBIA, 65- Concurrent Pos: Markle scholar med sci, Vanderbilt Univ, 52-58. Mem: Soc Pediat Res; Am Inst Nutrit; Am Pediat Soc; Am Soc Clin Nutrit. Mem: Nutritional anemias; metabolic aspects of growth. Mailing Add: Dept of Child Health Univ of Mo Sch of Med Columbia MO 65201

WOODRUFF, CHARLES MARSH, JR, b Columbia, Tenn, Aug 26, 44; m 73. PHYSICAL GEOLOGY. Educ: Vanderbilt Univ, BA, 66, MS, 68; Univ Tex, Austin, PhD(geol), 73. Prof Exp: Geologist, Tenn Div Geol, 69-70; RES SCIENTIST GEOL, BUR ECON GEOL, UNIV TEX, AUSTIN, 72- Concurrent Pos: Geol consult, Coastal Mgt Prog, Tex Gen Land Off, 74-76. Mem: Geol Soc Am; Soc Mining Engrs; AAAS; Sigma Xi. Res: Quaternary research in carbonate terranes and subhumid areas; environmental geology, especially the assessment of impacts between man and natural systems. Mailing Add: Bur of Econ Geol Univ of Tex Univ Sta Box X Austin TX 78712

WOODRUFF, CLARENCE MERRILL, b Kansas City, Mo, Apr 8, 10; m 37; c 2. AGRONOMY. Educ: Univ Mo, BS, 32, MA, 39, PhD, 53. Prof Exp: Lab technician, Exp Sta, Univ Mo, 32-34; supt in chg soil conserv exp sta, USDA, 34-38; from instr to assoc prof soils, 38-57, chmn dept, 67-70, PROF SOILS, EXP STA, COL AGR, UNIV MO-COLUMBIA, 57- Mem: Am Soc Agron; Soil Sci Soc Am; assoc Am Geophys Union. Mailing Add: Dept of Agron Univ of Mo Columbia MO 65201

WOODRUFF, DAVID SCOTT, b Penrith, Eng, June 12, 43; Australian citizen; m 72. ECOLOGY, MEDICAL PARASITOLOGY. Educ: Univ Melbourne, BSc, 65, PhD(zool), 73. Prof Exp: Tutor biol, Trinity Col, Univ Melbourne, 66-69; Frank Knox fel biol, Harvard Univ, 69-71, Alexander Agassiz lectr biogeog, 72, res fel biol, Mus Comp Zool, 73-74; ASST PROF BIOL, PURDUE UNIV, 74- Concurrent Pos: Lectr ecol, Comn Exten Courses, Harvard Univ, 72-74. Mem: Soc Study Evolution; Am Soc Ichthyol & Herpetol; Soc Syst Zool; Ecol Soc Am; Soc Exp & Descriptive Malacol. Res: Genetic variation, population ecology and evolutionary biogeography of land snails; hybridization and evolutionary ecology of anuran amphibians; biological control of schistosomiasis. Mailing Add: Dept of Biol Sci Purdue Univ West Lafayette IN 47907

WOODRUFF, EDYTHE PARKER, b Bellwood, Ill, Jan 15, 28; m 50; c 2. TOPOLOGY. Educ: Univ Rochester, BA, 48, MS, 52; Rutgers Univ, New Brunswick, MS, 67; State Univ NY Binghamton, PhD(math), 71. Prof Exp: ASST PROF MATH, TRENTON STATE COL, 71- Mem: Am Math Soc; Math Asn Am. Res: Topology of Euclidean 3-space; monotone decompositions, P-lifting; crumpled cubes. Mailing Add: 11 Fairview Ave East Brunswick NJ 08816

WOODRUFF, HAROLD BOYD, b Bridgeton, NJ, July 22, 17; m 42; c 2. MICROBIOLOGY. Educ: Rutgers Univ, BS, 39, PhD(microbiol), 42. Prof Exp: Asst soil microbiol, Rutgers Univ, 38-42; res microbiologist, Merck Sharp & Dohme Res Labs, 42-46, head res sect, Microbiol Dept, 47-49, from asst dir to dir, 49-57, dir microbiol & natural prod res dept, 57-69, exec dir biol sci, Merck Inst Therapeut Res, 69-73, EXEC ADMINR, MERCK SHARP & DOHME RES LABS, MSD (JAPAN) CO, LTD, 73- Concurrent Pos: Lectr, US Off Educ; ed, Appl Microbiol, 53-62; pres bd dirs, Am Soc Microbiol Found, 72; mem bd trustees, Biol Abstracts, 72-, treas, 74-; mem sci adv comt, Charles F Kettering Res Lab, 72- Honors & Awards: Charles Thom Award, Soc Indust Microbiol, 73. Mem: Am Soc Microbiol (treas, 64-70); Soc Indust Microbiol (pres, 54-56); Am Chem Soc; Am Acad Microbiol; Brit Soc Gen Microbiol. Res: Antibiotics; physiology of microorganisms; production of chemicals by microorganisms; analytical procedures using microorganisms; isolation of natural products. Mailing Add: Merck Sharp & Dohme Res Labs Rahway NJ 07065

WOODRUFF, JAMES DONALD, b Baltimore, Md, June 20, 12; m 39; c 3. OBSTETRICS & GYNECOLOGY. Educ: Dickinson Col, BS, 33; Johns Hopkins Univ, MD, 37. Prof Exp: From instr to assoc prof gynec, 42-60, from assoc prof to prof gynec & obstet, 60-75, assoc prof path, 63-75, RICHARD W TeLINDE PROF GYNEC & PATH, SCH MED, JOHNS HOPKINS UNIV, 75-, HEAD GYNEC PATH LAB, JOHNS HOPKINS HOSP, 51- Concurrent Pos: Chief gynecologist, Md Gen Hosp, 51-58 & Hosp for Women of Md, 58-62. Mem: Int Soc Study Vulvar Dis (pres, 73-75); Am Asn Obstet & Gynec (secy); Am Gynec Soc; fel Am Col Obstet & Gynec. Res: Gynecologic pathology; study of functional activity of ovarian neoplasms and vulvar disease. Mailing Add: Johns Hopkins Hosp 601 N Broadway Baltimore MD 21205

WOODRUFF, JAMES F, b Detroit, Mich, Feb 11, 20; m 49; c 4. GEOGRAPHY. Educ: Univ Mich, PhD(geog), 52. Prof Exp: Asst prof geog, Univ Ga, 52-56 & San Diego State Col, 56-60; assoc prof geog & geol, Grad Sch, 61-64, PROF GEOG, UNIV GA, 64- Mem: Asn Am Geog; Am Geol Soc. Res: Geomorphology and historical geography. Mailing Add: Dept of Geog Univ of Ga Athens GA 30601

WOODRUFF, JOHN H, JR, b Barre, Vt, Dec 14, 11; m 50; c 1. RADIOLOGY. Educ: Univ Vt, BS, 35, MD, 38. Prof Exp: Intern, US Marine Hosp, 38-39; resident radiol, Mary Fletcher Hosp, 39-40 & Royal Victoria Hosp, 40-41; asst prof, Univ Vt, 42-44; pvt pract, Calif, 46-50; clin instr, 50-52, asst prof, 52-54, ASSOC CLIN PROF RADIOL, UNIV CALIF, LOS ANGELES, 54-; CHIEF RADIOL, US VET ADMIN HOSP, SEPULVEDA, 71- Concurrent Pos: Chief radiologist, Los Angeles, County Harbor Gen Hosp, 52-65; consult, Terminal Island Fed Prison, 57-65 & Long Beach Vet Admin Hosp, 59-68; radiologist, Univ Calif Med Ctr, 65-67; clin prof radiologic serv, Univ Calif, Irvine, 67-68, clin prof radiologic sci, 68-70; chief radiol, San Fernando Vet Admin Hosp, 68-71. Mem: Am Roentgen Ray Soc; Radiol Soc NAm; AMA; Am Col Radiol. Res: Radiologic aspects of diseases of kidneys, lungs, gastrointestinal tract and of trauma to the abdomen and its contents. Mailing Add: 2618 Palos Verdes Dr W Palos Verdes Estates CA 90274

WOODRUFF, JOSEPH FRANKLIN, b Columbus, Ohio, Aug 8, 13; m 38; c 2.

INSTRUMENTATION, MATERIALS SCIENCE. Educ: Capital Univ, BS, 35. Prof Exp: High sch instr, Ohio, 36-41; analyst, 41-43, spectrochem analyst, 43-45, supv spectrochemist, 45-68, MGR X-RAY & SPECTROCHEM LAB, RES & TECHNOL, ARMCO STEEL CORP, 68- Honors & Awards: Award of Merit, Am Soc Testing & Mat, 68, H V Churchill Award, 74. Mem: Fel Am Soc Testing & Mat. Res: Analytical chemistry; x-ray diffraction; electron microprobe auger spectroscopy for quality control of steel plastics, graphite fibers. Mailing Add: Res & Technol Armco Steel Corp Middletown OH 45043

WOODRUFF, LAURENCE CLARK, b Kingman, Kans, Aug 5, 02; m 25; c 1. INSECT PHYSIOLOGY. Educ: Univ Kans, AB, 24, AM, 30; Cornell Univ, PhD(entom), 34. Prof Exp: Jr entomologist, Bur Entom, USDA, 24-28; instr entom, Univ Kans, 28-30; instr biol, Cornell Univ, 30-34; from asst prof to assoc prof entom, 34-42, resistrar, 42-46, dean men, 47-53, prof biol, 50-72, dean students, 53-67, EMER PROF BIOL, UNIV KANS, 72- Mem: Fel Entom Soc Am; Nat Asn Biol Teachers. Res: Growth in insects; insect nutrition. Mailing Add: 2 Westwood Rd Lawrence KS 66044

WOODRUFF, MARVIN WAYNE, b New York, NY, Mar 9, 28; m 54; c 3. UROLOGY. Educ: Columbia Univ, BS, 51; NY Univ, MD, 55; Am Bd Urol, dipl, 64. Prof Exp: Intern surg, Bellevue Hosp, New York, 55-56, asst resident, 56-57; resident urol, Univ Mich Hosp, 57-58; clin instr, Univ Mich, 58-60; chief, Dept Urol, Roswell Park Mem Inst, 60-68; PROF SURG, ALBANY MED COL, 68-, DIR DIV UROL, ALBANY MED CTR HOSP, 68- Concurrent Pos: Dir clin studies sect, Kidney Dis Inst, NY State Dept Health, 68- Honors & Awards: Award, Am Urol Asn, 60. Mem: Am Urol Asn; fel Am Col Surg. Res: Estrogen-androgen antagonism of prostatic growth; renal handling of contrast media as determined by stop flow analysis; dialysis in infancy; renal excretion mechanism of nitrofurantoin. Mailing Add: 9 Loudon Heights North Loudonville NY 12211

WOODRUFF, RICHARD EARL, b Elk River, Idaho, July 3, 29; m 54; c 3. HORTICULTURE, PLANT PHYSIOLOGY. Educ: Wash State Univ, BS, 52, MS, 57; Mich State Univ, PhD, 59. Prof Exp: Asst hort, Tree Fruit Exp Sta, USDA, 54-55, Wash State Univ, 55-57 & Mich State Univ, 57-59; horticulturist, USDA, Tex, 59-60 & United Fruit Co, NJ, 60-63; head dept plant physiol, Tela RR Co, LaLima, Honduras, 63-70; plant physiologist, Inter-Harvest Inc, 70-72; DIR RES, TRANSFRESH CORP, 72- Mem: Am Soc Plant Physiol; Am Soc Hort Sci. Res: Post harvest physiology and handling of fruit. Mailing Add: TransFresh Corp PO Box 1788 Salinas CA 93901

WOODRUFF, RICHARD IRA, b Glen Ridge, NJ, Aug 19, 40; m 62; c 3. DEVELOPMENTAL BIOLOGY, REPRODUCTIVE BIOLOGY. Educ: Ursinus Col, BS, 62; West Chester State Col, MEd, 65; Univ Pa, PhD(biol), 72. Prof Exp: Teacher high sch, 62-66; from instr to assoc prof, 66-72, PROF BIOL, WEST CHESTER STATE COL, 72- Concurrent Pos: Res fel, Univ Pa, 72- Mem: AAAS; Am Soc Zool. Res: Developmental biology; electrophysiological events during egg formation. Mailing Add: Dept of Biol West Chester State Col West Chester PA 19380

WOODRUFF, RICHARD L, b Bakersfield, Calif, Oct 12, 25; m 50; c 3. PHYSICAL CHEMISTRY. Educ: Univ Calif, BS, 47. Prof Exp: Asst res chemist, Calif Res Corp, Stand Oil Co, Calif, 47-51; res technologist, Shell Oil Co, 51-55, res group leader, 55-64, chief res engr, 64-65, spec engr, 65-66, res dir, Shell Chem Co, NJ, 66-71, mgr prod develop-plastics, 71-72, res & develop coordr automotive lubricants, Shell Oil Co, 72-75, RES & DEVELOP COORDR INDUST LUBRICANTS, SHELL OIL CO, 75- Mem: Soc Automotive Engrs. Res: Petroleum and petrochemical products and processes. Mailing Add: Shell Oil Co One Shell Plaza Box 2463 Houston TX 77002

WOODRUFF, ROBERT ARNOLD, JR, b Rochester, Mich, May 19, 34; m 56; c 2. PSYCHIATRY. Educ: Harvard Univ, AB, 56, MD, 60. Prof Exp: Intern med, Mary Imogene Bassett Hosp, Cooperstown, NY, 60-61; asst resident psychiat, 61-63, chief resident, 65-66, from instr to assoc prof, 66-74, PROF PSYCHIAT, DIR PSYCHIAT RESIDENCY TRAINING & DIR PSYCHIAT IN-PATIENT SERV, SCH MED, WASH UNIV, 74- Concurrent Pos: USPHS spec res fel, 67-69; USPHS grant, 71- Mem: AAAS; AMA; Royal Col Psychiat; Am Psychiat Asn; Psychiat Res Soc. Res: Clinical psychiatric research; epidemiology and the somatic therapies. Mailing Add: Renard Hosp 4940 Audubon Ave St Louis MO 63110

WOODRUFF, ROBERT EUGENE, b Kennard, Ohio, July 20, 33; m 54; c 2. ENTOMOLOGY. Educ: Ohio State Univ, BSc, 56; Univ Fla, PhD(entom), 67. Prof Exp: Entomologist, Ky State Health Dept, 56-58; ENTOMOLOGIST, PLANT INDUST DIV, FLA DEPT AGR, 58- Concurrent Pos: Mem, Orgn Trop Studies, NSF, Costa Rica, 64; ed, Coleopterists Bull, 71- Mem: Entom Soc Am; Soc Syst Zool; Asn Trop Biol. Res: Systematic entomology; taxonomy, ethology, ecology of beetles of the family Scarabaeidae, especially myrmecophilous and termitophilous species. Mailing Add: Div of Plant Indust State Dept of Agr Box 1269 Gainesville FL 32601

WOODRUFF, ROBERT WILSON, b Lagrange, Ohio, May 5, 25; m 50; c 2. POLYMER PHYSICS. Educ: Oberlin Col, AB, 47; Univ Rochester, PhD(optics), 54. Prof Exp: Res physicist, 54-66, SR RES PHYSICIST, E I DU PONT DE NEMOURS & CO, INC, 66- Mem: Optical Soc Am; Soc Photog Sci & Eng. Res: Soft x-ray spectroscopy; vacuum evaporation techniques; optical-mechanical relationships in polymers; graphic arts reproduction systems; evaluation of photosensitive systems. Mailing Add: 11 Fairview Ave East Brunswick NJ 08816

WOODRUFF, RODGER KING, b Melrose Park, Ill, Nov 9, 40; m 63; c 3. METEOROLOGY, PHYSICAL OCEANOGRAPHY. Educ: Ore State Univ, BS, 63, MS, 67. Prof Exp: Commissioned officer, US Coast & Geod Surv, Environ Sci Serv Admin, 63-69; res scientist meteorol, 69-72, mgr atmospheric physics sect, 72-74, MGR APPL METEOROL SECT, PAC NORTHWEST LABS, BATTELLE MEM INST, 74- Mem: Am Meteorol Soc; Air Pollution Control Asn. Res: Atmospheric diffusion, transport, deposition, resuspension and turbulence; industrial siting and aircraft operations. Mailing Add: Appl Meteorol Sect Pac Northwest Labs Richland WA 99352

WOODRUFF, RONNY CLIFFORD, b Greenville, Tex, Mar 12, 43; m 65. GENETICS. Educ: ETex State Univ, BS, 66, MS, 67; Utah State Univ, PhD(zool), 71. Prof Exp: NIH reproduction & develop training grant, Univ Tex, Austin, 71-73, asst prof zool, 73-74; SR RES ASST GENETICS, UNIV CAMBRIDGE, 74- Mem: Genetics Soc Am; Genetical Soc. Res: Structure and function of the genetic material of higher organisms, particularly Drosophila melanogaster. Mailing Add: Dept of Genetics Univ of Cambridge Cambridge England

WOODRUFF, SAMUEL AMOS, physical chemistry, see 12th edition

WOODRUFF, TRUMAN OWEN, b Salt Lake City, Utah, May 26, 25; m 48. THEORETICAL SOLID STATE PHYSICS. Educ: Harvard Univ, AB, 47, BA, 50; Calif Inst Technol, PhD(physics), 55. Prof Exp: Res assoc physics, Univ Ill, 54-55; physicist, Res Lab, Gen Elec Co, 55-62; PROF PHYSICS, MICH STATE UNIV, 62- Concurrent Pos: Vis prof, Univ Ariz, 67; Fulbright fel, Univ Pisa, 68-69. Mem: Fel Am Phys Soc. Res: Quantum mechanics; solid state physics, especially the quantum theory of solids; electronic structure of defects in solids; quantum statistical mechanics and kinetics of the excitations of complex systems. Mailing Add: Dept of Physics Mich State Univ East Lansing MI 48824

WOODRUFF, WILLIAM LEE, b Seward, Nebr, Oct 21, 38; m 63; c 3. REACTOR PHYSICS. Educ: Nebr Wesleyan Univ, BA, 60; Univ Nebr, MS, 64; Tex A&M Univ, PhD(nuclear eng), 70. Prof Exp: Lab asst & technician physics, Nebr Wesleyan Univ, 60-63, vis lectr, 63-64; instr, Univ Omaha, 64-66; NUCLEAR ENGR, ARGONNE NAT LAB, 68- Concurrent Pos: Inst Atomic Energy, Brazil, 75- Mem: Am Nuclear Soc; Sigma Xi. Res: Methods and computer code development and physics analysis for design and safety of liquid metal and gas-cooled fast breeder reactors. Mailing Add: Argonne Nat Lab 9700 S Cass Ave Argonne IL 60439

WOODS, ALAN CHURCHILL, JR, b Baltimore, Md, July 1, 18; m 44; c 4. SURGERY. Educ: Princeton Univ, AB, 40; Johns Hopkins Univ, MD, 43; Am Bd Surg, dipl, 51. Prof Exp: Intern & asst resident surgeon, Johns Hopkins Hosp, 44-45; resident surg, Henry Ford Hosp, 45-56; from asst resident surgeon to resident surgeon, Johns Hopkins Hosp, 48-49, surgeon & surgeon chg outpatient dept, 50; asst, 49, from instr to asst prof, 49-67, ASSOC PROF SURG, JOHNS HOPKINS UNIV, 67- Concurrent Pos: William Stewart Halsted fel surg, Johns Hopkins Univ, 49-50. Mem: AMA; fel Am Col Surgeons; Soc Head & Neck Surg. Res: Abdominal, head and neck surgery. Mailing Add: Dept of Surg Johns Hopkins Univ Baltimore MD 21205

WOODS, ALEXANDER HAMILTON, b Tuxedo, NY, July 26, 22; m 56; c 2. IMMUNOLOGY. Educ: Harvard Univ, BS, 44; Johns Hopkins Univ, MD, 52; Am Bd Internal Med, dipl, 60. Prof Exp: Instr med, Sch Med, Duke Univ, 55-56; asst prof med & microbiol, Med Ctr, Univ Okla, 58-64; PROF IMMUNOL & ASSOC PROF INTERNAL MED, COL MED, UNIV ARIZ, 64-; ASSOC CHIEF STAFF, VET ADMIN HOSP, TUCSON, 70- Concurrent Pos: Res fel biochem, Duke Univ, 56-58; clin investr, Vet Admin Hosp, Oklahoma City, 59-61; dir res, Vet Admin Hosp, Tucson, 64-70. Mem: AAAS; Am Chem Soc; NY Acad Sci; Brit Biochem Soc. Res: Immunochemistry; hematology; cancer chemotherapy. Mailing Add: Vet Admin Hosp Tucson AZ 85713

WOODS, ALFRED DAVID BRAINE, b St John's, Nfld, July 16, 32; m 54; c 3. SOLID STATE PHYSICS, LOW TEMPERATURE PHYSICS. Educ: Dalhousie Univ, BSc, 53, MSc, 55; Univ Toronto, PhD(low temperature physics), 57. Prof Exp: Res fel low temperature physics, Univ Toronto, 57-58; RES OFFICER SOLID STATE PHYSICS, ATOMIC ENERGY CAN LTD, 58-, HEAD NEUTRON & SOLID STATE PHYSICS BR, 71- Mem: Am Phys Soc; Can Asn Physicists. Res: Dynamics of condensed matter using inelastic neutron scattering. Mailing Add: Atomic Energy of Can Ltd Chalk River ON Can

WOODS, ALVIN EDWIN, b Murfreesboro, Tenn, Mar 17, 34; m 59; c 2. FOOD CHEMISTRY, BIOCHEMISTRY. Educ: Mid Tenn State Univ, BS, 56; NC State Univ, MS, 58, PhD(food flavor), 62. Prof Exp: From asst prof to assoc prof, 61-69, PROF CHEM, MID TENN STATE UNIV, 69- Concurrent Pos: Consult, Off Int Res, NIH, 65-67. Mem: Am Chem Soc; fel Am Inst Chemists. Res: Enzyme stereochemistry and kinetics; food flavors, metal ions in biological systems; organo phosphorous compounds. Mailing Add: Dept of Chem Mid Tenn State Univ Murfreesboro TN 37130

WOODS, CECIL LAMBORN, b Covington, Ky, July 5, 03; m 25; c 3. MATHEMATICS. Educ: Emmanuel Missionary Col, Andrews, AB, 25; Univ Cincinnati, MS, 38; Ohio State Univ, PhD(math), 46. Prof Exp: Instr, Hinsdale Acad, 25-28; instr math, Washington Missionary Col, 28-31, head premed dept, Training Inst, China, 31-37; instr math, Emmanuel Missionary Col, Andrews, 38-47; prof, 47-71, EMER PROF MATH, PAC UNION COL, 71- Concurrent Pos: Asst, Army Specialized Training Prog, Ohio State Univ, 42-44. Mem: Am Math Soc; Math Asn Am. Res: Physics; convex functions. Mailing Add: PO Box 747 Angwin CA 94508

WOODS, CHARLES WILLIAM, b Akron, Ohio, June 1, 28; m 61; c 2. ORGANIC CHEMISTRY. Educ: Ohio State Univ, BS, 51; Univ Md, PhD(org chem), 58. Prof Exp: Chemist, E I du Pont de Nemours & Co, Ky, 57-58; CHEMIST, ENTOM RES DIV, USDA, 59- Mem: AAAS; Am Chem Soc. Res: Synthesis of chemosterilants for insects; synthesis of radioactive organic compounds. Mailing Add: ARC Rm 113 Bldg 306 Beltsville MD 20705

WOODS, CLIFTON, III, b Mecklenburg Co, NC, Aug 28, 44. PHYSICAL INORGANIC CHEMISTRY. Educ: NC Cent Univ, BS, 66; NC State Univ, MS, 69, PhD(chem), 71. Prof Exp: Asst prof chem, Univ Fla, 71-73, Bowling Green State Univ, 73-74; ASST PROF CHEM, UNIV TENN, 74- Mem: Am Chem Soc; Sigma Xi. Res: Use of nuclear magnetic resonance, infrared and Raman spectroscopies in the determination of structures and conformational changes of some organosilicon and germanium compounds. Mailing Add: Dept of Chem Univ of Tenn Knoxville TN 37916

WOODS, CLYDE M, b Seattle, Wash, Nov 12, 32; m 60; c 4. ANTHROPOLOGY. Educ: San Jose State Col, BA, 60, MS, 61; Stanford Univ, MA, 63, PhD(anthrop), 68. Prof Exp: ASST PROF ANTHROP, UNIV CALIF, LOS ANGELES, 68- Mem: Fel Am Anthrop Asn; assoc Soc Appl Anthrop. Res: Modernization; cultural dynamics; economic anthropology; peasant societies; Middle and Latin America. Mailing Add: Dept of Anthrop Univ of Calif 405 Hilgard Ave Los Angeles CA 90024

WOODS, DALE, b Stone Co, Mo, Nov 1, 22. MATHEMATICS. Educ: Southwest Mo State Col, BS, 44; Okla State Univ, MS, 50, EdD, 61. Prof Exp: High sch teacher, Mo, 41-44 & Ill, 44-45; high sch prin, Nebr, 46-47; instr math, NDak State Univ, 47-51, Tex Western Col, 51-52 & Southern Miss Univ, 52-53; res mathematician, Halliburton Res Lab, 53-54; asst prof math, Memphis State Univ, 54-57 & Idaho State Univ, 57-59; assoc prof, 59-61, PROF MATH, NORTHEAST MO STATE UNIV, 61-, HEAD DIV MATH, 65- Concurrent Pos: Asst, Okla State Univ, 49-50, 57-58 & 60-61. Mem: Math Asn Am; Am Math Soc; Nat Coun Teachers Math. Res: Differential gemoetry and equations; theory of numbers. Mailing Add: Div of Math Northeast Mo State Univ Kirksville MO 63501

WOODS, EUGENE FRANCIS, pharmacology, see 12th edition

WOODS, FRANK ROBERT, b Mt Vernon, NY, June 20, 16; m 42; c 3. HYDRODYNAMICS, GAS DYNAMICS. Educ: NY Univ, BA, 41, MS, 47, PhD(physics), 55. Prof Exp: Instr physics, Univ NH, 48-53, from asst prof to assoc prof, 53-57; physicist, Boeing Airplane Co, 57-58; assoc prof physics, Mont State Col, 58-63; lectr aerospace eng & eng physics, Univ Va, 63-73, sr scientist, 63-69, prin scientist, Dept Aerospace Eng & Eng Physics, 69-73; MASTER, HILL SCH, POTTSTOWN, PA, 73- Mem: Am Phys Soc. Res: Hydrodynamics; scattering; theoretical hydrodynamics. Mailing Add: The Hill Sch Pottstown PA 19464

WOODS, FRANK WILSON, b Covington, Va, Apr 1, 24; m 48; c 5. FOREST ECOLOGY, ENVIRONMENTAL SCIENCES. Educ: NC State Col, BS, 49; Univ Tenn, MS, 51, PhD, 57. Prof Exp: Instr, Univ Tenn, 52; res forester, Southern Forest Exp Sta, US Forest Serv, 53-58; asst prof silvicult, Duke Univ, 58-64, assoc prof forest ecol, 64-69; PROF FORESTRY, UNIV TENN, KNOXVILLE, 69- Concurrent Pos: Vis res partic, Oak Ridge Inst Nuclear Studies, 64; vis res assoc, Univ Calif, Los Angeles, 65; NSF-Soc Am Foresters vis lectr, 66; consult, Orgn Econ Develop, Portugal, 69- & Oak Ridge Nat Lab, 71- Mem: Soc Am Foresters; Ecol Soc Am; Int Asn Trop Biologists; fel AAAS; Int Asn Ecologists. Res: Ecology and physiology of root systems; effects of environment on tree growth. Mailing Add: Univ of Tenn Dept of Forestry PO Box 1071 Knoxville TN 37901

WOODS, GEORGE THEODORE, b Tyro, Kans, Aug 21, 24; m 48; c 3. VETERINARY MEDICINE, PUBLIC HEALTH. Educ: Kans State Univ, DVM, 46; Univ Calif, MPH, 59; Purdue Univ, MS, 60. Prof Exp: Inspector animal dis eradication, Ill State Dept Agr, 46-47; supvr lab animal med, Med Sch, Northwestern Univ, 47-48; vet, 48-49; asst prof vet path & hyg, 49-59, assoc prof vet microbiol & pub health, 59-66, PROF VET MICROBIOL, PUB HEALTH & RES, COL VET MED, UNIV ILL, URBANA, 66- Concurrent Pos: Trainee, USPHS, 58-59; mem adv comt, Nat Specific Pathogen Free Swine Cert Agency; collabr, USDA. Mem: Am Vet Med Asn; Am Pub Health Asn; Conf Pub Health Vets; Asn Teachers Vet Pub Health & Prev Med (pres). Res: Preventive veterinary medicine; epidemiology; viral respiratory diseases of cattle and swine; bovine myxovirus para-influenza 3. Mailing Add: Dept of Vet Path & Hyg Univ of Ill Col of Vet Med Urbana IL 61801

WOODS, GERALDINE PITTMAN, b West Palm Beach, Fla; m 45; c 3. NEUROEMBRYOLOGY. Educ: Howard Univ, BS, 42; Radcliffe Col, MA, 43, PhD(neuroembryol), 45. Prof Exp: Instr biol, Howard Univ, 45-46; SPEC CONSULT, NAT INST GEN MED SCI, NIH, 69- Concurrent Pos: Mem, Nat Adv Coun, Gen Med Sci Inst, NIH, 64-68; chmn, Defense Adv Comt Women in Serv, 68; mem bd trustees, Howard Univ, 68-, chmn, 75-; mem, Gen Res Support Prog Adv Comt, Div Res Resources, NIH, 70-73; mem bd trustees, Calif Mus Found of Calif Mus Sci & Indust, 71-; mem air pollution manpower develop adv comt, Environ Protection Agency, 73-75; mem bd dirs, Robert Wood Johnson Health Policy Fels, Inst of Med Nat Acad Sci, 73-, Nat Comn Cert Physicians Assts, 74-; mem bd trustees, Atlanta Univ, 74- & mem Calif Post Sec Educ Comn, 74- Honors & Awards: Spec Award, President's Coun on Youth Opportunity. Mem: Inst of Med of Nat Acad Sci; AAAS; Fedn Am Scientists; Nat Inst Sci; Asn Advan Colored People. Res: Encouraging the participation of minorities in the regular programs at the National Institutes of Health and developing two programs that would further assist colleges with minorities move into research and research training to project more participation in health and scientific careers. Mailing Add: 12065 Rose Marie Lane Los Angeles CA 90049

WOODS, JAMES E, b Chicago, Ill, May 7, 26. ENDOCRINOLOGY, HISTOCHEMISTRY. Educ: Univ Ill, BS, 51; DePaul Univ, MS, 57; Loyola Univ, Ill, PhD(anat), 64. Prof Exp: Res asst endocrinol, VioBin Corp, 51-52; res asst, Michael Reese Med Ctr, Chicago, 53-54; res assoc histochem & endocrinol, Stritch Sch Med, Loyola Univ Chicago, 64-66; asst prof, 66-73, ASSOC PROF HISTOCHEM & ENDOCRINOL, DePAUL UNIV, 73- Mem: AAAS; Am Soc Zool; Soc Study Reproduction. Res: Reproductive endocrinology of adult and embryonic domestic fowl. Mailing Add: Dept of Biol Sci DePaul Univ Chicago IL 60614

WOODS, JAMES W, b Graham, Tex, July 27, 23; m 67; c 1. MEDICAL EDUCATION. Educ: Johns Hopkins Univ, BA, 49, PhD(physiol), 54. Prof Exp: Instr sci, Sch Nursing, Johns Hopkins Hosp, 51-54; res assoc, Postgrad Med Sch, Univ London, 56-57; from instr to asst prof physiol, Sch Med, Johns Hopkins Univ, 57-68; assoc prof physiol, 68-71, PROF PHYSIOL & BIOPHYS, HEALTH SCI CTR, UNIV OKLA, 71-, ASSOC PROF PHARMACOL & DIR MULTIDISCIPLINARY LABS, 68- Concurrent Pos: Am Cancer Soc fel med res, Johns Hopkins Univ, 55 & Univ London, 55-56; USPHS sr fel, 57-62 & career develop award, 62-67; vis prof, Sch Basic Med Sci, Univ Ill, Urbana-Champaign, 74-75. Mem: AAAS; Asn Am Med Cols; Asn Multidiscipline Educ Health Sci; Nat Soc Med Res; Endocrine Soc. Res: Central nervous system mechanisms in homeostasis and emotional behavior; neuroendocrinology; instructional design and methodology; curriculum evaluation; temperature regulation and mechanisms involved in fever; physiology of the thyroid gland. Mailing Add: Multidisciplinary Labs Univ of Okla Health Sci Ctr Oklahoma City OK 73190

WOODS, JAMES WATSON, b Lewisburg, Tenn, Feb 20, 18; m 44; c 3. MEDICINE. Educ: Univ Tenn, BA, 39; Vanderbilt Univ, MD, 43; Am Bd Internal Med, dipl. Prof Exp: Instr, Sch Med, Univ Pa, 47-48; from asst prof to assoc prof, 53-64, PROF MED, SCH MED, UNIV NC, CHAPEL HILL, 64- Concurrent Pos: NIH spec fel, Dept Med, Vanderbilt Univ, 67-68; fel, Coun Clin Cardiol & mem med adv bd, Coun High Blood Pressure, Am Heart Asn; consult, Social Security Admin Bur Hearings & Appeals, Dept Health, Educ & Welfare. Mem: AAAS; Am Fedn Clin Res; Am Clin & Climat Asn; fel Am Col Physicians. Res: Cardiology; pathogenesis of hypertension. Mailing Add: Dept of Med Univ of NC Sch of Med Chapel Hill NC 27514

WOODS, JIMMIE DALE, b Albuquerque, NMex, Oct 8, 33; m 56; c 3. MATHEMATICS, STATISTICS. Educ: US Coast Guard Acad, BS, 55; Trinity Col, Conn, MS, 63; Univ Conn, PhD(math statist), 68. Prof Exp: US Coast Guard, 51-, from instr to assoc prof math, US Coast Guard Acad, 60-68, asst head dept, 64-68, PROF MATH & HEAD DEPT, US COAST GUARD ACAD, 68- Mem: Am Statist Asn; Am Math Soc; Math Asn Am. Res: Application of matrix techniques to statistical distribution theory; mathematical modeling in management science. Mailing Add: RD 2 Box 136 Wheeler Rd Stonington CT 06378

WOODS, JOE DARST, b Knoxville, Iowa, Jan 2, 23; m 45; c 2. INORGANIC CHEMISTRY. Educ: Cent Col, Iowa, BS, 44; Iowa State Univ, MS, 50, PhD(chem), 54. Prof Exp: From instr to assoc prof, 52-64, PROF CHEM, DRAKE UNIV, 64-, CHMN DEPT, 69- Concurrent Pos: Fulbright-Hays lectr, St Louis Univ, Philippines, 67-68. Mem: AAAS; Am Chem Soc. Res: Mechanism of decomposition of chlorates; radioactive tracers; oxidation of metals; dissolution of metals in acids. Mailing Add: 4107 Ardmore Rd Des Moines IA 50310

WOODS, JOHN PRICE, b San Antonio, Tex, Feb 25, 03; m 25; c 5. GEOPHYSICS. Educ: Univ Tex, BS, 25, MS, 30, PhD(physics), 32. Prof Exp: Asst, Univ Tex, 29-33; asst dir geophys lab, Shell Oil Co, 33-42; group leader radio res lab, Harvard Univ, 42-44; dir geophys lab, Atlantic Ref Co, 44-66; PROF PHYSICS & MATH, ALASKA METHODIST UNIV, 66- Mem: Soc Explor Geophys; Seismol Soc Am; Am Geophys Union. Res: Exploration geophysics; electronic circuits. Mailing Add: Dept of Physics Alaska Methodist Univ Anchorage AK 99504

WOODS, JOSEPH FRANCIS, solid state physics, see 12th edition

WOODS, JOSEPH JAMES, b Camden, NJ, June 21, 43; m 68; c 2. PHYSIOLOGY. Educ: St Jospeh's Col, Pa, BS, 65; Rutgers Univ, PhD(physiol), 71. Prof Exp: ASST PROF BIOL, QUINNIPIAC COL, 70- Concurrent Pos: NIH co-investr, 71-74. Res: Application of the principles of muscle biophysics to muscular exercise; muscle activity and oxygen consumption during varying rates of positive and negative work. Mailing Add: Dept of Biol Quinnipiac Col PO Box 304 Hamden CT 06518

WOODS, LAUREN ALBERT, b Aurora Co, SDak, Spet 10, 19; m 44; c 3. PHARMACOLOGY. Educ: Dakota Wesleyan Univ, BA, 39; Iowa State Col, PhD(org chem), 43; Univ Mich, MD, 49. Prof Exp: Asst, Nat Defense Res Comt, Iowa State Col, 43-44; from instr to prof pharmacol, Univ Mich, 46-60, actg chmn dept, 56; prof & head dept, Col Med, Univ Iowa, 60-70; PROF PHARMACOL & VPRES HEALTH SCI, MED COL VA, VA COMMONWEALTH UNIV, 70- Mem: AAAS; Am Chem Soc; Am Soc Pharmacol & Exp Therapeut; NY Acad Sci. Res: Metabolism of drugs; chemical structure; biological activity relationships; compounds affecting the central nervous system; mechanisms of development of tolerance and physical dependence to narcotics; radioactive tracer studies; histochemical distribution of drugs. Mailing Add: Box 606 Med Col of Va Va Commonwealth Univ Richmond VA 23219

WOODS, LLOYD LANDER, chemistry, see 12th edition

WOODS, LOREN PAUL, b Poseyville, Ind, Aug 4, 13; m 37, 67; c 4. ICHTHYOLOGY. Educ: Earlham Col, AB, 36. Prof Exp: Asst field naturalist, Roosevelt Wildlife Mem, Syracuse, 36; guide-lectr, Field Mus Natural Hist, 38-41, asst cur fishes, 41-46; assoc cur, US Nat Mus, DC, 46-48; CUR FISHES, FIELD MUS NATURAL HIST, 48- Mem: Am Soc Ichthyologists & Herpetologists (asst secy, 47, vpres, 57); Soc Vert Paleont; Soc Syst Zool. Res: Tropical marine fishes; morphology, taxonomy and ecology of fishes. Mailing Add: Div of Fishes Field Mus of Natural Hist Chicago IL 60605

WOODS, MARIBELLE, b Albany, Ga, Aug 16, 19. PHARMACOLOGY. Educ: Univ Chattanooga, BS, 42; Yale Univ, MS, 48. Prof Exp: Bioassayer, Chattanooga Med Co, 42-48, pharmacologist, 48-66; PHARMACOLOGIST, CHATTEM DRUG & CHEM CO, 66- Concurrent Pos: Pharmacologist, Brayten Pharmaceut Co, 48-73. Mem: Am Asn Lab Animal Sci; NY Acad Sci. Res: Laxative action of senna; premenstrual tension; uterine physiology; antacids. Mailing Add: 311 Guild Dr Chattanooga TN 37421

WOODS, MARK WINTON, b Takoma Park, Md, Oct 15, 08; m 33; c 2. CYTOLOGY, PHYSIOLOGY. Educ: Univ Md, BS, 31, MS, 33, PhD(cytol), 36. Prof Exp: Asst bot & plant path, Univ Md, 34-36, instr, 36-37, from asst prof to assoc prof plant path, 34-47; biologist, Lab Biochem, Nat Cancer Inst, 47-73; RETIRED. Mem: AAAS; Am Phytopath Soc; Am Soc Biol Chemists; Am Asn Cancer Res; NY Acad Sci. Res: Virus diseases; cytology, metabolism and cytochemistry of cancer. Mailing Add: 10718 Brookside Dr Sun City AZ 85351

WOODS, MARY, b Webster Groves, Mo, Dec 22, 23. INORGANIC CHEMISTRY. Educ: Rosary Col, AB, 45; Univ Ill, MA, 47; Univ Wis, PhD(chem), 61. Prof Exp: Teacher, Trinity High Sch, Ill, 49-53; from instr to assoc prof chem, 53-73, PROF CHEM, ROSARY COL, 73- Concurrent Pos: Consult, Argonne Nat Lab, 75- Mem: Am Chem Soc. Res: Complex ion equilibria and kinetics; kinetics of redox reactions of actinide ions in solution. Mailing Add: Dept of Chem Rosary Col River Forest IL 60305

WOODS, PHILIP SARGENT, b Concord, NH, Nov 25, 21; m 46; c 2. CELL BIOLOGY, MOLECULAR BIOLOGY. Educ: Mich State Univ, BS, 47; Univ Wis, PhD(cytol), 52. Prof Exp: Head cellular res sect, Dept Radiobiol, US Army Med Res Lab, Ky, 52-53; USPHS fel, Columbia Univ, 53-55; mem staff biol dept, Brookhaven Nat Lab, 55-61; assoc prof biol sci, Univ Del, 61-67; PROF BIOL, QUEEN'S COL, NY, 67- Mem: Am Soc Cell Biol; Sigma Xi; Electron Micros Soc Am. Res: Biosynthesis of macromolecules; nucleic acid and protein metabolism within cells; autoradiography with tritium-labeled precursors; fine structure of cells. Mailing Add: Dept of Biol Queen's Col Flushing NY 11367

WOODS, RAYMOND DOUGLAS, b Evangeline, La, Sept 14, 10. GEOLOGY. Educ: Univ Tex, BA, 31, MA, 34. Prof Exp: Tutor geol, Univ Tex, 31-34; paleontologist, Humble Oil & Refining Co, 34-42, sr paleontologist, 42-49, sr geologist, 49-56, res coordr, 56-61; mgr geol div, Jersey Prod Res Co, 61-62; asst chief geol res, Humble Oil & Refining Co, 63-64; asst mgr geol div, Esso Prod Res Co, 65; asst chief geologist, Humble Oil & Refining Co, 66-72, CONSULT, EXXON CO USA, 73- Concurrent Pos: Instr, Univ Houston, 36-42; lab instr, La State Univ, 46. Mem: Geol Soc Am; Soc Econ Paleontologists & Mineralogists; Am Asn Petrol Geologists. Res: Stratigraphy; micropaleontology. Mailing Add: 2701 Westheimer Houston TX 77006

WOODS, ROBERT CLAUDE, b Atlanta, Ga, Mar 24, 40; m 63; c 2. PHYSICAL CHEMISTRY. Educ: Ga Inst Technol, BS, 61; Harvard Univ, AM, 62, PhD(phys chem), 65. Prof Exp: Instr chem, US Naval Acad, 65-67; asst prof, 67-73, ASSOC PROF CHEM, UNIV WIS-MADISON, 73- Mem: Am Chem Soc; Am Phys Soc. Res: Microwave spectroscopy of molecules with internal rotors and of transient species present in electrical discharges and related theoretical problems. Mailing Add: Dept of Chem Univ of Wis Madison WI 53706

WOODS, ROBERT JAMES, b London, Eng, Feb 8, 28; m 58; c 4. RADIATION CHEMISTRY. Educ: Univ London, BSc, 49, PhD(org chem), 51; Imp Col London, dipl, 51; FRIC. Prof Exp: Nat Res Coun Can fel, Prairie Regional Lab, Sask, 51-53; Univ NZ res fel, Victoria, NA, 53-54; res assoc, Univ Sask, 55-62; sr res fel, Royal Mil Col Sci, Eng, 62-63; asst prof chem, 63-65, ASSOC PROF CHEM, UNIV SASK, 65- Concurrent Pos: Vis prof, Saclay Nuclear Res Ctr, France, 72-73. Mem: Chem Inst Can; fel The Chem Soc. Res: Radiation chemistry of organic compounds, both pure and in aqueous solution. Mailing Add: Dept of Chem & Chem Eng Univ of Sask Saskatoon SK Can

WOODS, ROBERT MCDILL, JR, physics, see 12th edition

WOODS, ROGER DAVID, b Los Angeles, Calif, Mar 28, 24; m 52; c 3. THEORETICAL PHYSICS, COMPUTER SCIENCE. Educ: Univ Redlands, AB, 45; Univ Calif, Los Angeles, MA, 49, PhD, 54. Prof Exp: Res physicist, Univ Calif, Los Angeles, 54; asst prof physics, Univ Miami, 54-61; assoc prof, Univ Redlands, 61-65; assoc prof, 65-72, PROF PHYSICS, SAN BERNARDINO VALLEY COL, 72- Mem: AAAS; Am Phys Soc. Res: Nucleon-nuclei scattering; molecular structure; electron structure of atoms; energy bands in solids; generalized theory of gravitation; visible, ultraviolet and infrared spectroscopy; scientific computer application. Mailing Add: 16 S Ash St Redlands CA 92373

WOODS, ROY ALEXANDER, b Columbia, Mo, Oct 31, 13; m 34; c 1. PHYSICS, ELECTRONICS. Educ: Lincoln Univ, Mo, AB, 34; Boston Univ, AM, 46 & 48, EdD, 60. Prof Exp: Teacher high schs, Mo, 34-41; instr electronics & radar technician, US Navy, 42-45; prof physics & chmn div natural sci, Va State Col, 48-68; PROF PHYSICS, NORFOLK STATE COL, 68- Concurrent Pos: Electronic engr, Lab Electronics Res & Develop, Mass, 52-53, 59-60. Mem: Am Asn Physics Teachers.

Res: Instrumentation for naval radar. Mailing Add: Dept of Physics Norfolk State Col Norfolk VA 23504

WOODS, SHERWYN MARTIN, b Des Moines, Iowa, June 25, 32; m 71. PSYCHIATRY, PSYCHOANALYSIS. Educ: Univ Wis, BS, 54, MD, 57; Am Bd Psychiat & Neurol, dipl, 65. Prof Exp: Intern, Philadelphia Gen Hosp, 57-58; resident, Univ Hosps, Univ Wis Med Sch, 58-61, instr psychiat, Med Sch, 61; from asst prof to assoc prof, 63-74, PROF PSYCHIAT, SCH MED, UNIV SOUTHERN CALIF, 74-, DIR GRAD EDUC, 63-, DIR STUDENT PSYCHIAT SERV, 66- Concurrent Pos: NIMH career teacher award, 64-66; clin assoc, Southern Calif Psychoanal Inst, 64-68, mem, 68-, instr, 69-; examr, Am Bd Psychiat & Neurol, 67-; consult, Calif Dept Ment Hyg, 70-; pres, Am Asn Dir Psychiat Residency Training, 74- Mem: Fel Am Psychiat Asn; Asn Advan Psychother; fel Am Col Psychiat. Res: Medical education; human sexuality; psychotherapy. Mailing Add: Dept of Psychiat Div Grad Educ Univ Southern Calif Sch of Med Los Angeles CA 90033

WOODS, STUART B, b Pathlow, Sask, Apr 26, 24; m 49; c 2. SOLID STATE PHYSICS. Educ: Univ Sask, BA, 44, MA, 48; Univ BC, PhD, 52. Prof Exp: Res officer physics, Nat Res Coun Can, 52-59; assoc prof, 59-65, PROF PHYSICS, UNIV ALTA, 65-, ASSOC DEAN FAC GRAD STUDIES & RES, 74- Concurrent Pos: Vis prof, Univ Bristol, 66-67 & Univ St Andrews, Scotland, 73-74. Mem: Am Phys Soc; Can Asn Physicists. Res: Low temperature and solid state physics, chiefly experimental transport properties in solids. Mailing Add: Dept of Physics Univ of Alta Edmonton AB Can

WOODS, THOMAS STEPHEN, b Florence, Ala, Dec 13, 44; m 66; c 1. ORGANIC CHEMISTRY. Educ: Auburn Univ, BS, 67; Univ Ill, PhD(org chem), 71. Prof Exp: Res chemist, Div Med Chem, Walter Reed Army Inst Res, 72-74; RESCHEMIST, E I DU PONT DE NEMOURS & CO, BIOCHEM DEPT, 74- Mem: Am Chem Soc. Res: Synthesis of novel organic compounds of biological utility; heterocyclic chemistry; organosulfur and selenium chemistry; theories of tautomerism and resonance; organic photochemistry. Mailing Add: 1919 Floral Dr N Graylyn Crest Wilmington DE 19810

WOODS, WALTER RALPH, b Grant, Va, Dec 2, 31; m 53; c 2. ANIMAL NUTRITION. Educ: Murray State Univ, BS, 54; Univ Ky, MS, 55; Okla State Univ, PhD, 57. Prof Exp: Instr, Okla State Univ, 56-57; from asst prof to assoc prof animal husb, Iowa State Univ, 57-62; from assoc prof to prof, Univ Nebr-Lincoln, 62-71; HEAD DEPT ANIMAL SCI, PURDUE UNIV, WEST LAFAYETTE, 71- Mem: Am Soc Animal Sci; Am Inst Nutrit; Am Dairy Sci Asn. Res: Protein and nonprotein nitrogen utilization in beef cattle and sheep; energy utilization as influenced by diet composition and processing. Mailing Add: Dept of Animal Sci Purdue Univ West Lafayette IN 47907

WOODS, WARREN WHITNEY, b Los Angeles, Calif, Nov 3, 18; m 43; c 3. PETROLEUM CHEMISTRY. Educ: Stanford Univ, AB, 43. Prof Exp: Asst, Stanford Univ, 42-45; res chemist, Basic & Theoret Lab, Socony-Vacuum Oil Co, 45-51; res group leader colloid chem, 51-58, mgr petrol prod div, 58-75, SR RES FEL, RES & DEVELOP DEPT, CONTINENTAL OIL CO, 75- Res: Greases; mineral oil additives; surfactants; petroleum products. Mailing Add: Res & Develop Dept Continental Oil Co Ponca City OK 74601

WOODS, WENDELL DAVID, b Liberal, Kans, Dec 5, 32; m 64; c 2. BIOCHEMISTRY, PHYSIOLOGY. Educ: Univ Mo, BS, 58, MS, 59, PhD(biochem), 65. Prof Exp: Res asst, 63-66, from instr to asst prof, 66-73, ASSOC PROF OPHTHAL, SCH MED, EMORY UNIV, 73- Concurrent Pos: NIH grant vision, Emory Univ, 72- Mem: AAAS; Asn Res Ophthal; Am Chem Soc; Am Sci Affil. Res: Biochemical mechanisms relating cyclic-adenosine monophosphate and prostaglandins to fluid dynamics in the eye. Mailing Add: Lab for Ophthal Res Emory Univ Atlanta GA 30322

WOODS, WILLIAM GEORGE, b Superior, Wis, Dec 21, 31; m 52; c 3. PHYSICAL ORGANIC CHEMISTRY. Educ: Univ Calif, Los Angeles, BS, 53; Calif Inst Technol, PhD, 57. Prof Exp: Asst, Calif Inst Technol, 53-55; res chemist, Res Lab, Gen Elec Co, 56-58; sr scientist, 58-74, MGR AGR RES & DEVELOP, US BORAX RES CORP, 74- Mem: Am Chem Soc. Res: Carbonium ion and pyrolysis mechanism; semi-inorganic polymer systems; semi-empirical molecular orbital calculations; nuclear magnetic resonance spectroscopy; organometallic and organic synthesis; herbicide synthesis and metabolism; direct agricultural research programs. Mailing Add: US Borax Res Corp 412 Crescent Way Anaheim CA 92801

WOODSIDE, DONALD G, b Pittsburgh, Pa, Apr 28, 27; Can citizen; m 53; c 3. ORTHODONTICS. Educ: Dalhousie Univ, BSc, 48, DDS, 52; Univ Toronto, MSc, 56; FRCD(C), 67. Prof Exp: Assoc, 55-58, asst prof & actg head dept, 58-59, assoc prof, 59-61, actg head dept, 61-62, assoc dean, 70-73, PROF ORTHOD & HEAD DEPT, FAC DENT, UNIV TORONTO, 62- Concurrent Pos: Consult, Ont Hosp Crippled Children & Burlington Orthod Res Ctr, 62. Mem: Am Asn Orthod; Europ Orthod Soc; Edward H Angle Soc; Charles Tweed Found. Res: Human growth and development; study of human mandibular growth distance, velocity curves and growth direction to establish predicability factors. Mailing Add: Dept of Orthod Univ of Toronto Fac of Dent Toronto ON Can

WOODSIDE, EUGENE EMERSON, microbiology, biochemistry, deceased

WOODSIDE, GILBERT LLEWELLYN, b Curwensville, Pa, Feb 9, 09; m 34; c 2. DEVELOPMENTAL BIOLOGY. Educ: DePauw Univ, AB, 32; Harvard Univ, MA, 33, PhD(biol), 36. Hon Degrees: DSc, Univ Mass, 65. Prof Exp: Asst zool, Harvard Univ, 33-36; from asst prof to prof biol, Univ Mass, 36-64, head dept zool, 48-61, dean grad sch, 50-62, provost, 61-64; assoc dir extramural progs, 64-67, assoc dir Inst, 67-75, DEP DIR, NAT INST CHILD HEALTH & HUMAN DEVELOP, 75- Concurrent Pos: Spec res fel, Nat Cancer Inst, 57. Mem: Fel AAAS; Am Soc Zoologists; Electron Micros Soc Am; Soc Develop Biol; Am Soc Cell Biol. Res: Embryonic induction; auxin in animal embryos; effect of hormones on development; growth of chick embryos in tissue culture; embryonic mortality as influenced by nutrition; chemotherapy of mouse tumors; electron microscopy of mouse lung tissue. Mailing Add: Nat Inst of Child Health & Human Develop Bethesda MD 20014

WOODSIDE, KENNETH HALL, b Northampton, Mass, June 18, 38; m 60; c 2. PHYSIOLOGY, BIOCHEMISTRY. Educ: Oberlin Col, AB, 59; Univ Rochester, PhD(biochem), 69. Prof Exp: Res assoc physiol, 68-70, ASST PROF PHYSIOL & HEAD MULTIDISCIPLINE LABS, COL MED, PA STATE UNIV, 70- Mem: Am Physiol Soc; AAAS; Am Chem Soc. Res: Chemistry and biology of glucagon; hormonal and non-hormonal regulation of liver protein biosynthesis and degradation. Mailing Add: Dept of Physiol Hershey Med Ctr Pa State Univ Hershey PA 17033

WOODSIDE, WILLIAM, b Ft William, Ont, July 5, 31; m 53; c 3. PHYSICS, MATHEMATICS. Educ: Queen's Univ, Belfast, BSc, 51, MSc, 59, DSc, 62. Prof Exp: Asst Royal Naval Sci Serv, Baldock, Eng, 51 & Toronto, 53; master physics, Ridley Col, Can, 52-55; res officer, Nat Res Coun, Can, 54-58; res physicist, Gulf Res & Develop Co, 58-60; master math, Ridley Col, 60-66; asst prof, 66-70, ASSOC PROF MATH, QUEEN'S UNIV, ONT, 70- Mem: Can Asn Physicists; Brit Inst Physics. Res: Heat transfer and fluid flow in porous media. Mailing Add: Dept of Math Queen's Univ Kingston ON Can

WOODSON, BERNARD ROBERT, b Richmond, Va, Oct 13, 23; m 52; c 4. ALGOLOGY. Educ: Va State Col, BS, 45; Howard Univ, MS, 48; Mich State Univ, PhD(bot), 58. Prof Exp: From instr to assoc prof, 52-60, chmn dept, 66-71, PROF BIOL, VA STATE COL, 60-, DEAN SCI & TECHNOL, 73- Concurrent Pos: Grants in aid, Va Acad Sci, 55-56, Richmond Area Univ Ctr & Am Acad Sci, 58; asst dir, NSF Inst, 58-, dir undergrad res partic prog, 62-67 & 71-72, dir acad yr inst, 68-71; consult, NIH, Moton Col Serv Bur & Environ Protection Agency, 74- Honors & Awards: NIH Citation of Serv, 72. Mem: AAAS; Phycol Soc Am; Ecol Soc Am; Nat Asn Biol Teachers; Am Soc Pub Adminrs. Res: Population dynamic of algae; physiology, taxonomy and ecology of algae; pollution ecology. Mailing Add: 20007 Roosevelt Ave Colonial Heights VA 23834

WOODSON, BRUCE ALAN, biochemistry, virology, see 12th edition

WOODSON, JOHN HODGES, b Hartford, Conn, May 25, 33; m 60; c 3. PHYSICAL CHEMISTRY, THEORETICAL CHEMISTRY. Educ: Wesleyan Univ, BA, 55; Northwestern Univ, PhD(phys chem), 59. Prof Exp: Asst prof chem, Wesleyan Univ, 59-61; from asst prof to assoc prof, 61-70, PROF CHEM, SAN DIEGO STATE UNIV, 70- Mem: Am Chem Soc; Sigma Xi. Res: Valence theory; kinetics of oscillating reactions; ion-selective electrodes. Mailing Add: Dept of Chem San Diego State Univ San Diego CA 92182

WOODSON, PAUL BERNARD, b St Louis, Mo, Jan 20, 49; div. NEUROBIOLOGY. Educ: Univ Calif, San Diego, BA, 70, PhD(neurosci), 75. Prof Exp: SCHOLAR NEUROBIOL, DEPT PSYCHIAT, UNIV CALIF, SAN DIEGO, 75- Mem: Soc Neurosci. Res: Mechanisms and modulations of synaptic transmission; correlations of synaptic and behavioral plasticities; pharmacology of synaptic transmission; synaptic tolerance phenomena; membrane fluidity as a physiological regulation point in synaptic transmission. Mailing Add: Dept of Psychiat Sch of Med Univ of Calif BSB 2056 LaJolla CA 92093

WOODSTOCK, LOWELL WILLARD, b Harvey, Ill, July 6, 31; m 61; c 2. PLANT PHYSIOLOGY, BIOCHEMISTRY. Educ: Univ Ill, BS, 54; Univ Wis, PhD(bot), 59. Prof Exp: Nat Cancer Inst fel, Royal Bot Garden, Scotland, 59-61 & Univ Wis, 61-62; res assoc, Argonne Nat Lab, 62-63; res plant physiologist, 63-67, leader seed qual invests, 67-73, RES PLANT PHYSIOLOGIST, AGR RES SERV, USDA, 73- Concurrent Pos: Am Seed Res Found grant, 69-72; intergovt employee exchange act trainee, Ore State Univ, 73-74. Mem: AAAS; Bot Soc Am; Am Soc Plant Physiol; Asn Off Seed Anal; Sigma Xi. Res: Photosynthesis in unicellular algae; growth and differentiation of cells in vascular plants; metabolism and physiology of seeds relative to germination, dormancy, storage and vigor. Mailing Add: Agr Res Ctr-West Bldg 049 USDA Agr Res Serv Beltsville MD 20705

WOODWARD, ARTHUR AMOS, JR, physiology, see 12th edition

WOODWARD, ARTHUR EUGENE, b Los Angeles, Calif, Oct 16, 25; m 52; c 1. PHYSICAL CHEMISTRY. Educ: Occidental Col, BA, 49, MA, 50; Polytech Inst Brooklyn, PhD(phys chem), 53. Prof Exp: Asst Occidental Col, 50; US Govt grantee chem, Cath Univ Louvain, 53-54; res fel, Harvard Univ, 54-55; asst prof, Pa State Univ, 55-59, from asst prof to assoc prof physics, 59-64; assoc prof chem, 64-66, PROF CHEM, CITY COL NEW YORK, 67- Concurrent Pos: Guggenheim fel, Queen Mary Col, London, 62-63. Mem: Am Chem Soc; Am Phys Soc; NY Acad Sci. Res: Low temperature dynamic mechanical properties; nuclear magnetic resonance of high polymers; polymer crystals. Mailing Add: Dept of Chem City Col of New York New York NY 10031

WOODWARD, CLARE K, b Houston, Tex, Dec 10, 41; m 67. BIOCHEMISTRY. Educ: Smith Col, BA, 63; Rice Univ, PhD(biol), 67. Prof Exp: Fel genetics, Univ Minn, St Paul, 67-68; fel phys chem, Univ Minn, Minneapolis, 68-70, asst prof lab med, Med Sch, 70-72; ASST PROF BIOCHEM, UNIV MINN, ST PAUL, 72- Mem: Biophys Soc; Am Soc Biol Chemists. Res: Protein chemistry. Mailing Add: Dept of Biochem Univ of Minn St Paul MN 55101

WOODWARD, DAVID ALFRED, b Leamington Spa, Eng, Aug 29, 42; m 66; c 2. CARTOGRAPHY. Educ: Univ Wales, BA, 64; Univ Wis-Madison, MA, 67, PhD(geog), 70. Prof Exp: DIR, HERMON DUNLAP SMITH CTR HIST CARTOG, NEWBERRY LIBR, 72- Concurrent Pos: Travel grants, Am Coun Learned Soc, 71, 73 & 75; gen ed, Studies Hist Discoveries, 69-76; fel, William Andrews Clark Mem Libr & exchange fel, Brit Acad, 73; assoc prof geog, Northwestern Univ, 73-; lectr, Hist Cartog, Univ Chicago, 74- Mem: Asn Am Geogrs; Am Cong Surv & Mapping; Int Soc Hist Cartog; Printing Hist Soc London; Royal Geog Soc. Res: History of map printing; development of cartography of the Great Lakes region; cartographic methodology. Mailing Add: Newberry Libr 60 W Walton St Chicago IL 60610

WOODWARD, DAVID HARVEY, physics, mathematics, see 12th edition

WOODWARD, DAVID LEE, b Sioux City, Iowa, Feb 14, 40; m 59; c 2. PHARMACOLOGY. Educ: Utah State Univ, BS, 64; Univ Utah, PhD(pharmacol), 68. Prof Exp: Asst prof, Fac Dent, Univ Man, 69-70; res assoc, 70, ASSOC DIR, HOECHST-ROUSSEL PHARMACEUT INC, 70- Concurrent Pos: NIH trainee pharmacol, Univ Man, 68-69. Mem: AAAS; NY Acad Sci. Res: Cardiovascular-renal pharmacology. Mailing Add: Dept of Pharmacol Bldg L Hoechst-Roussel Pharmaceut Inc Somerville NJ 08876

WOODWARD, DAVID WILLCOX, b Oxford, NY, July 24, 13; m 38; c 1. ORGANIC CHEMISTRY. Educ: Amherst Col, AB, 34; Harvard Univ, PhD(org chem), 37. Prof Exp: Res chemist, Chem Dept, 37-51, res supvr, Photo Prod Dept, 51-54, res mgr, 45-68, LAB DIR, PHOTO PROD DEPT, E I DU PONT DE NEMOURS & CO, INC, 68- Mem: Am Chem Soc; Soc Photog Sci & Eng. Res: Photography; furane and polymer chemistry; cyanides; nitriles; photochemistry; photopolymerization; dyes. Mailing Add: 103 Taylor Lane Kennett Square PA 19348

WOODWARD, DONALD JAY, b Detroit, Mich, Aug 1, 40; m 63; c 3. PHYSIOLOGY. Educ: Univ Mich, BS, 62, PhD(physiol), 66. Prof Exp: From instr to asst prof, 66-74, ASSOC PROF PHYSIOL, DEPT PHYSIOL & CTR BRAIN RES, SCH MED & DENT, UNIV ROCHESTER, 74- Concurrent Pos: Nat Inst Neurol Dis & Stroke teacher-investr trainee, Univ Rochester, 71- Mem: Am Physiol Soc; Neurosci Soc. Res: Developmental neurobiology; cerebellar neurophysiology and neuropharmacology. Mailing Add: Dept of Physiol Univ of Rochester Sch of Med & Dent Rochester NY 14642

WOODWARD, DOW OWEN, b Logan, Utah, Dec 1, 31; m 56; c 4. MOLECULAR

BIOLOGY. Educ: Utah State Univ, BS, 56; Yale Univ, MS, 57, PhD(bot), 59. Prof Exp: Mem res staff radiobiol, Aerospace Med Ctr, US Air Force, 59-62; assoc prof, 62-74, PROF BIOL, STANFORD UNIV, 74- Mem: Genetics Soc Am; Am Soc Biol Chem. Res: Biochemical genetics in microorganisms; membrane structure and function; cytoplasmic inheritance; biological rhythms; enzymology. Mailing Add: Dept of Biol Stanford Univ Stanford CA 94305

WOODWARD, EDWARD ROY, b Chicago, Ill, Sept 6, 16; m 39; c 2. SURGERY. Educ: Grinnell Col, BA, 38; Univ Chicago, MD, 42; Am Bd Surg, dipl, 51. Prof Exp: Asst resident, Univ Clins, Univ Chicago, 47-49, instr & resident, Grad Sch Med, 49-50 & 51-52, instr, 52-53; from asst prof to assoc prof, Med Ctr, Univ Calif, Los Angeles, 53-57; PROF SURG & HEAD DEPT, COL MED, UNIV FLA, 57- Concurrent Pos: Douglas Smith fel surg, Univ Chicago, 46-47; Markle scholar, 52-57. Mem: Am Physiol Soc; Soc Univ Surg; Am Surg Asn; Am Gastroenterol Asn; fel Am Col Surgeons. Res: Physiology of gastrointestinal tract. Mailing Add: Dept of Surg Univ of Fla Gainesville FL 32610

WOODWARD, ERVIN CHAPMAN, JR, b Long Beach, Calif, Apr 8, 23; m 49; c 2. RADIATION PHYSICS. Educ: Univ Calif, PhD(physics), 52. Prof Exp: PHYSICIST, LAWRENCE LIVERMORE LAB, UNIV CALIF, 52- Mem: Am Phys Soc. Res: Spectroscopy; hyperfine structure; nuclear spin and isotope shift; high speed optics. Mailing Add: Lawrence Livermore Lab Univ of Calif PO Box 808 Livermore CA 94550

WOODWARD, FRED ERSKINE, b Boston, Mass, Aug 20, 21; m 44; c 4. SURFACE CHEMISTRY, ORGANIC POLYMER CHEMISTRY. Educ: Dartmouth Col, AB, 43; Univ Ill, MA, 44, PhD(org chem), 46. Prof Exp: Chemist, Gen Aniline & Film Corp, Pa, 46-48, sr chemist, Dyestuff & Chem Div, 48-54, mgr indust sect, Antara Tech Dept, 54-58, asst prog mgr appln res, 58-62; dir res & develop, Nopco Chem Co, NJ, 62-69; PRES, SURFACE CHEMISTS FLA, INC, 69- Mem: Am Chem Soc; Am Oil Chem Soc; Soc Lubrication Eng; Am Inst Mining & Metall Engrs; Am Asn Textile Chemists & Colorists. Res: Surface active agents; lubrication; defoaming; detergency; synthesis and applications research; surface chemistry of antimicrobial agents; mechanisms of clay dewatering; iodophors. Mailing Add: Surface Chemists of Fla Inc Palm Beach Int Airport Bldg 541 West Palm Beach FL 33406

WOODWARD, GLENN JONES, b Milton, Ore, Nov 8, 08; m 36; c 1. INORGANIC CHEMISTRY, BIOCHEMISTRY. Educ: Whitman Col, AB, 30; Univ Ore, MA, 32; State Col Wash, PhD(chem), 51. Prof Exp: Instr chem, George Fox Col, 34-35; teacher high sch, Ore, 35-37 & Wash, 37-40; instr chem, Multnomah Col, 40-42; control chemist, McKesson & Robbins, Inc, 42-43; from instr to prof, 43-74, EMER PROF CHEM, WHITMAN COL, 74- Mem: Am Chem Soc; NY Acad Sci. Res: Fungicidal properties of certain phenol derivatives; sulfur metabolism of penicillium molds; colorimetric methods of analysis. Mailing Add: Dept of Chem Whitman Col Walla Walla WA 99362

WOODWARD, HARRY W, b Norwich, Norfolk, Eng, Feb 2, 23; Can citizen; m 51; c 3. PETROLEUM GEOLOGY. Educ: Queen's Univ, Ont, BSc, 49; Univ Wis, PhD(geol), 53. Prof Exp: Sr staff geologist, Gulf Oil Can Ltd, 53-66; adminr oil & gas sect, 66-68, DIR OIL & MINERAL DIV, DEPT INDIAN & NORTHERN AFFAIRS, GOVT CAN, 68- Res: Resource management; surface structure and subsurface geology of Western Canada; resource development of Northern Canada. Mailing Add: 2220 Louisiana Ave Ottawa ON Can

WOODWARD, HUBERT EDMUND, b Penarth, SWales, June 20, 98; m 41; c 2. CHEMISTRY. Educ: Univ Sask, BSc & MSc, 25; Univ Pittsburgh, PhD(chem), 27. Prof Exp: Instr chem, Univ Pittsburgh, 27-28, asst prof, 28-33; asst prof med res, Univ Toronto, 33-41, dominion analyst, 41-47; supt food & drugs lab, Can Dept Nat Health & Welfare, 47-50, regional dir, 50-64, consult food & drugs, 64-75; RETIRED. Mem: AAAS; fel Chem Inst Can. Res: Biochemistry; microchemistry. Mailing Add: 14 Sandalwood Pl Don Mills ON Can

WOODWARD, J GUY, b Carleton, Mich, Nov 19, 14; m 45; c 3. PHYSICS. Educ: NCent Col, BA, 36; Mich State Col, MS, 38; Ohio State Univ, PhD(physics), 42. Prof Exp: Asst physics, Mich State Col, 36-39; asst physics, Ohio State Univ, 39-42; res physicist, Mfg Co, RCA Corp, 42, res engr, RCA Labs, 42-72, RES FEL, RCA LABS, 72- Honors & Awards: Emile Berliner Award, Audio Eng Soc, 68. Mem: Fel AAAS; Acoust Soc Am; Audio Eng Soc (pres, 71-72); fel Inst Elec & Electronics Engrs. Res: Physical optics; room and music acoustics; radio interference from motor vehicles; underwater sound; electroacoustic transducers; high fidelity phonograph systems; viscometry; video magnetic tape recording. Mailing Add: RCA Labs Princeton NJ 08540

WOODWARD, JAMES CRAWFORD, b Lennoxville, Que, Apr 7, 10; m 35; c 1. ANIMAL NUTRITION. Educ: McGill Univ, BSA, 30; Cornell Univ, MS, 32, PhD(animal nutrit), 34. Prof Exp: Jr chemist, Can Dept Agr, 34-47, asst chemist, 38-40, chemist, 46-47, prin chemist, 48-49, chief chem div, 49-55, assoc dir, Exp Farms Serv, 55-59, asst dir-gen, Res Br, 59-62, assoc dir-gen, 62-68, asst dep minister res, 68-74; PVT CONSULT, 74- Honors & Awards: Silver Acorn, Boy Scouts Can, 73. Mem: Nutrit Soc Can; fel Chem Inst Can; fel Agr Inst Can (pres, 57-58). Res: Animal and plant nutrition; soils. Mailing Add: 21 Findlay Ave Ottawa ON Can

WOODWARD, JAMES D, b Scottsville, Ky, Nov 25, 36; m 56; c 3. DENTISTRY, PROSTHODONTICS. Educ: Univ Louisville, DMD, 65. Prof Exp: Pub health dentist, Ky State Dept Health, 65-67; pvt dent pract, Ky, 67-69; asst prof prosthodontics, Col Dent, Univ Ky, 69-75; ASSOC PROF PROSTHODONTICS, COL DENT, UNIV OKLA, 75- Concurrent Pos: Consult, Okla State Med Examr, Okla State Hwy Patrol & Fed Aviation Admin. Mem: Am Dent Asn; Am Asn Dent Schs; Am Soc Forensic Odontol; Am Acad Forensic Sci. Mailing Add: Col of Dent Univ of Okla PO Box 26901 Oklahoma City OK 73190

WOODWARD, JAMES KENNETH, b Anderson, Mo, Feb 5, 38; m 60; c 2. PHARMACOLOGY. Educ: Southwest Mo State Col, BS & AB, 60; Univ Pa, PhD(pharmacol), 67. Prof Exp: Pharmacologist, Stine Lab, E I du Pont de Nemours & Co, Inc, Del, 63-67, res pharmacologist, 67-71; head cardiovasc-autonomic dis res, 71-72, head cardiovasc-respiratory dis res, 72-74, HEAD PHARMACOL DEPT, MERRELL-NAT LABS, 74- Concurrent Pos: Lectr, Sch Med, Univ Pa, 67-71. Mem: NY Acad Sci. Res: Cardiovascular-autonomic pharmacology, especially etiology, pathology and treatment of hypertension; role of central nervous system and autonomic nervous system in control of cardiovascular function. Mailing Add: Pharmacol Dept Merrell-Nat Labs 110 Amity Rd Cincinnati OH 45215

WOODWARD, JOHN MORRILL, b Concord, NH, Oct 16, 18; m 42; c 3. BACTERIOLOGY. Educ: Univ NH, BS, 41; Mass State Col, MS, 43; Univ Kans, PhD(bact, biochem), 49. Prof Exp: Asst bact, Univ Kans, 46-47; asst prof, Univ Maine, 49-51; assoc prof, 51-57, PROF BACT, UNIV TENN, KNOXVILLE, 57- Concurrent Pos: Consult, St Mary's Hosp, 51-54, Baptist Hosp, 55-59, Children's Hosp, 57- & Biol Div, Oak Ridge Nat Lab, 59-63. Mem: AAAS; Am Soc Microbiol; Am Acad Microbiol; Can Soc Microbiol. Res: Staphylococcal toxins; immunity to tularemia in laboratory animals; host-parasite relationships in tularemia; radiation injury and bacterial invasion of animal tissues; serological grouping of the serratia; new techniques in antibiotic sensitivity testing; pseudomonas infection in laboratory animals; bacterial endotoxins and host response to infection. Mailing Add: Dept of Microbiol Univ of Tenn Knoxville TN 37916

WOODWARD, KENT THOMAS, b Cleveland, Ohio, Dec 11, 23; m 49; c 3. RADIOTHERAPY. Educ: Clemson Univ, BS, 47; Univ SC, MD, 47; Univ Rochester, PhD(radiation biol), 66. Prof Exp: Intern, Boston City Hosp, Boston, Mass, 47-48; intern res med, Georgetown Univ Hosp, DC, 48-49; intern chest disease, Fitzsimons Army Hosp, Denver, Colo, 49, 50-51; intern med, Walter Reed Army Hosp, DC, 51-52; staff mem biomed res, Los Alamos Sci Lab, 52-56; chief biophys dept, Walter Reed Army Inst Res, 56-60, dir div nuclear med, 62-68; prog dir radiation, Nat Cancer Inst, 68-69; fel radiation ther, Univ Tex, M D Anderson Hosp & Tumor Inst Houston, 69-72; assoc prof, Med Univ SC, 72-73; ASSOC PROF RADIOL, UNIV PA, 73- Mem: AAAS; Am Med Asn; Radiation Res Soc; Health Physics Soc; fel Am Col Physicians. Res: Biological effects ionizing radiation on normal tissues and tumors. Mailing Add: Dept of Radiol Univ of Pa Philadelphia PA 19104

WOODWARD, LEE ALBERT, b Omaha, Nebr, Apr 22, 31; m 52; c 4. STRUCTURAL GEOLOGY. Educ: Univ Mont, BA, 58, MS, 59; Univ Wash, PhD(geol), 62. Prof Exp: Geologist, US Bur Reclamation, 58 & Pan Am Petrol Corp, 62-63; instr geol, Olympic Col, 63-65; from asst prof to assoc prof, 65-73, PROF GEOL, UNIV NMEX, 73-, CHMN DEPT, 70- Concurrent Pos: NSF fel, 58-61; Monsanto fel, 61-62; Univ NMex res allocations comt grant, 65-67. Mem: Geol Soc Am; Am Asn Petrol Geologists. Res: Regional tectonics of western United States. Mailing Add: Dept of Geol Univ of NMex Albuquerque NM 87106

WOODWARD, LEROY ALBERT, b Hartford, Conn, Nov 22, 16; m 42; c 2. PHYSICS. Educ: Ga Inst Technol, BS, 43; Univ Mich, MS, 47. Prof Exp: Mem sci staff, US Navy Underwater Sound Lab, Columbia Univ, 44; contract physicist, David Taylor Model Basin, US Dept Navy, DC, 45; from instr to asst prof physics, Ga Inst Technol, 47-51, assoc prof physics & res physicist, Ext S Exp Sta, 51-55; res physicist, Scripto, Inc, 55-58, dir res, 58-60; res asst prof physics, Eng Exp Sta, 60-63, asst prof, 63-65, ASSOC PROF PHYSICS, GA INST TECHNOL, 65- Mem: Royal Astron Soc Can; Am Asn Physics Teachers. Res: Optics and optical microscopy. Mailing Add: 834 Oakdale Rd NE Atlanta GA 30307

WOODWARD, RALPH STANLEY, b Williston, SC, Oct 1, 14; m 38; c 2. HORTICULTURE. Educ: Clemson Col, BS, 36; La State Univ, MS, 47. Prof Exp: Inspector, USDA, 36; tech asst, Univ Ark, 36-39; asst prof hort, La Polytech Inst, 39-42; prod supt, Fed Rubber Reserve Comn, 43-44; asst exten horticulturist, Univ, 45-47, asst hort & supt, N La Exp Sta, 47-55, assoc, 55-58, PROF HORT & SUPT, N LA EXP STA, LA STATE UNIV, 58- Mem: Am Soc Hort Sci. Res: General agriculture; tree ripened peaches. Mailing Add: North La Exp Sta Calhoun LA 71225

WOODWARD, RAY R, b Willow Creek, Mont. GENETICS. Educ: Mont State Col, BS, 39, MS, 47; Univ Minn, PhD(genetics), 53. Prof Exp: Asst animal husbandman, NMont Br Sta, USDA, 39-42, animal husbandman, Range Livestock Exp Sta, Animal Husb Res Div, Agr Res Serv, 46-60; dir beef cattle breeding, 60-74, BEEF CONSULT, AM BREEDERS SERV, 74- Honors & Awards: Outstanding Contrib to Am Agr Award, Fed Land Bank, 67; Beef Res Pioneer Award, Beef Improv Fedn, 74. Mem: Am Soc Animal Sci; Soc Range Mgt. Res: Animal breeding, especially in field record of performance in beef cattle. Mailing Add: Box 1195 Bozeman MT 59715

WOODWARD, ROBERT BURNS, b Boston, Mass, Apr 10, 17; m 38, 46; c 4. ORGANIC CHEMISTRY. Educ: Mass Inst Technol, BS, 36, PhD(chem), 37. Hon Degrees: Many degrees for US & foreign univs. Prof Exp: Fel, 37-38, mem, Soc Fels, 38-40, from instr to prof chem, 41-53, Morris Loeb prof, 53-60, DONNER PROF SCI, HARVARD UNIV, 60- Concurrent Pos: Consult, Polaroid Corp, 42, Comt Med Res, 44-45, War Prod Bd, 44-45 & Chas Pfizer & Co, Inc, 51-; dir, Woodward Res Inst, Univ Basel, 63-; mem bd gov, Weizmann Inst Sci, 68-; mem bd dirs, Ciba-Geigy Ltd, Basel, 70- Numerous hon lect, US & abroad, 48- Honors & Awards: Nobel Prize in Chem, 65; Scott Medal, Franklin Inst & City of Philadelphia, 45; Ledlie Prize, Harvard Univ, 55; Res Corp Award, 55; Baekeland Medal, Am Chem Soc, 55, Nichols Medal, 56, Synthetic Org Chem Award, 57, Richards Medal, 58, Adams Medal, 61, Kirkwood Medal, 65, Willard Gibbs Medal, 67 & Arthur C Cope Award, 73; Davy Medal, The Chem Soc, 59; Pius XI Gold Medal, Pontifical Acad Sci, 61; Priestley Medallion, Dickinson Col, 62; Gold Medal, Synthetic Org Chem Mfrs Asn, 62; Stas Medal, Belg Chem Soc, 62; Nat Medal of Sci, 64; Lavoisier Medal, Chem Soc France, 68; Order of the Rising Sun Japan, 70; Hanbury Mem Medal, Pharmaceut Soc Gt Brit, 70; Pierre Bruylants Medal, Cath Univ Louvain, 70; Sci Achievement Award, AMA, 71. Mem: Nat Acad Sci; fel Am Acad Arts & Sci; Am Philos Soc; hon mem Ger Chem Soc; hon mem The Chem Soc. Res: Chemistry of natural products. Mailing Add: Dept of Chem Harvard Univ 12 Oxford St Cambridge MA 02138

WOODWARD, STEPHEN COTTER, b Atlanta, Ga, July 19, 35; m 57; c 2. PATHOLOGY. Educ: Emory Univ, MD, 59. Prof Exp: Pathologist, Georgetown Univ Hosp, 64-68; asst prof, 64-68, ASSOC PROF PATH, SCH MED, GEORGETOWN UNIV, 68-; PATHOLOGIST, HUNTER LAB, SIBLEY HOSP, 68- Concurrent Pos: Consult, Children's Hosp, 64-, Vet Admin Hosp, 65- & comt skeletal syst, Div Med Sci, Nat Res Coun, 68-; attend pathologist, DC Gen Hosp, 67- Mem: Col Am Path; Am Soc Clin Path; Am Soc Exp Path; AMA. Res: Fibroplasia; collagen elaboration; effects of endocrine and vulnerary agents upon wound repair; quality control methods for clinical laboratories. Mailing Add: Sch of Med & Dent Georgetown Univ Washington DC 20016

WOODWARD, THEODORE ENGLAR, b Westminster, Md, Mar 22, 14; m 38; c 4. MICROBIOLOGY, MEDICINE. Educ: Franklin & Marshall Col, BS, 34; Univ Md, MD, 38; Am Bd Internal Med, dipl. Hon Degrees: DSc, Western Med Col, 50 & Franklin & Marshall Col, 54. Prof Exp: Asst prof med, Sch Med & Univ Hosp, 46-48, assoc prof med & dir sect infectious dis, Sch Med, 48-54, PROF MED & HEAD DEPT, SCH MED, UNIV MD, BALTIMORE CITY, 54- Concurrent Pos: Instr, Sch Med, Johns Hopkins Univ, 46-48, lectr, 48-54; attend physician, Vet Admin Hosp, 46-48, consult, 48-; consult, State Health Dept, Md, 50; mem comt int ctrs med res & training, USPHS, 61-62; mem US adv comt, US-Japan Coop Med Sci Prog, 65- Mem: Am Soc Clin Invest; AMA; Am Clin & Climat Asn; Am Asn Physicians; fel Am Col Physicians. Res: Infectious and rickettsial diseases; enteric diseases including typhoid fever; internal medicine. Mailing Add: Dept of Med Univ of Md Sch of Med Baltimore MD 21201

WOODWARD, VAL WADDOUPS, b Preston, Idaho, July 26, 27; m 47, 67; c 3. GENETICS. Educ: Utah State Univ, BS, 50; Kans State Univ, MS, 50; Cornell Univ, PhD(genetics), 53. Prof Exp: NIH fel & guest assoc biologist, Brookhaven Nat Lab, 53-55; assoc prof genetics, Kans State Univ, 55-58; prof biol & chmn dept, Univ Wichita, 58-61; assoc prof, Rice Univ, 61-67; PROF GENETICS, UNIV MINN, ST

PAUL, 67- Concurrent Pos: Fel, Birmingham Univ, 62-63. Mem: AAAS; Am Soc Cell Biol; Genetics Soc Am. Res: Gene-enzyme transport; cell wall; neurospora; self-assembly of membrane and other organelle proteins. Mailing Add: 442 Biosci Univ of Minn St Paul MN 55101

WOODWARD, WILLIAM MOONEY, b Hartford, Conn, Sept 19, 16; m 42; c 2. NUCLEAR PHYSICS. Educ: Columbia Univ, AB, 38; Princeton Univ, PhD(physics), 42. Prof Exp: Physicist, Manhattan Proj, Los Alamos Sci Lab, 43-46; asst prof physics, Mass Inst Technol, 46-48; assoc prof, 48-58, PROF PHYSICS, CORNELL UNIV, 58- Mem: Am Phys Soc. Res: Electron scattering; high energy physics. Mailing Add: Dept of Physics Lab of Nuclear Studies Cornell Univ Ithaca NY 14850

WOODWELL, GEORGE MASTERS, b Cambridge, Mass, Oct 23, 28; m 55; c 4. ECOLOGY, BOTANY. Educ: Dartmouth Col, AB, 50; Duke Univ, AM, 56, PhD(bot), 58. Prof Exp: From asst bot to assoc prof, Univ Maine, 57-61; sr ecologist, Brookhaven Nat Lab, 61-75; DIR ECOSYSTS CTR, MARINE BIOL LAB, 75- Concurrent Pos: Assoc, Conserv Found, 58-61; lectr, Sch Forestry, Yale Univ, 67-; adj assoc prof, NY Univ, 67-71; founding mem bd trustees, Environ Defense Fund, 67-68 & 73-; mem, Inst Ecol, 70-71, Natural Resources Defense Coun, 70-, World Wildlife Fund, 70- & Live Sci Adv Comt, NASA, 71-; chmn, Suffolk County Coun Environ Qual, 72; mem environ adv panel, US Senate Comt Pub Works. Honors & Awards: NY Bot Garden Sci Award, 75. Mem: Fel AAAS; Ecol Soc Am (vpres, 67); Radiation Res Soc; Am Inst Biol Sci; Brit Ecol Soc. Res: Structure, function and development of terrestrial and marine ecosystems; environmental cycling of nutrients, radioactive isotopes and organic compounds, especially pesticides; ecological effects of ionizing radiation. Mailing Add: Ecosysts Ctr Marine Biol Lab Woods Hole MA 02543

WOODWICK, KEITH HARRIS, b Tappen, NDak, Jan 4, 27; m 49; c 3. INVERTEBRATE ZOOLOGY. Educ: Jamestown Col, BS, 49; Univ Wash, MS, 51; Univ Southern Calif, PhD(zool), 55. Prof Exp: Asst zool, Univ Wash, 49-51; instr, Univ Southern Calif, 54, asst, 54-55; from instr to assoc prof zool, 55-66, chmn dept, 65-69, PROF ZOOL, CALIF STATE UNIV, FRESNO, 66-, COORDR MARINE SCI, 68- Mem: AAAS; Soc Syst Zool. Res: Systematics and larval development of polychaetes; Enteropneusta. Mailing Add: Dept of Biol Calif State Univ Fresno CA 93740

WOODWORTH, CURTIS WILMER, b Reading, Pa, Aug 30, 42; m 64; c 2. ORGANIC CHEMISTRY. Educ: Albright Col, BS, 64; Princeton Univ, PhD(chem), 69. Prof Exp: Res chemist, Am Cyanamid Co, Bound Brook, 68-73, group leader, 73-75, DEPT HEAD, LEDERLE LABS, AM CYANAMID CO, PEARL RIVER, 75- Mem: Am Chem Soc. Res: Structure-reactivity relationships; process research and development on pharmaceuticals and fine chemicals. Mailing Add: Lederle Labs Pearl River NY 10965

WOODWORTH, HAROLD CYRIL, microbiology, immunology, see 12th edition

WOODWORTH, ROBERT CUMMINGS, b Cambridge, Mass, Nov 11, 30; m 52; c 3. BIOCHEMISTRY. Educ: Univ Vt, BS, 53; Pa State Univ, PhD(chem), 57. Prof Exp: Res chemist, Nat Inst Allergy & Infectious Dis, 56-60; from instr to assoc prof, 61-75, PROF BIOCHEM, COL MED, UNIV VT, 75- Concurrent Pos: USPHS fel, Clin Chem Lab, Malmo Gen Hosp, Sweden, 60-61; USPHS spec fel & vis prof, Inorg Chem Lab, Oxford Univ, 68-69; Fogarty Int sr fel, Oxford Univ, 76-77. Mem: Am Chem Soc; Am Soc Biol Chemists; Biochem Soc; AAAS. Res: Protein structure-function; nature of iron-binding proteins; mechanisms of metal binding and release; role of bound anions, physiological function and structure. Mailing Add: Dept of Biochem Univ of Vt Given Med Bldg Burlington VT 05401

WOODWORTH, ROBERT HUGO, b Boston, Mass, Feb 24, 02; m 27; c 3. BIOLOGY. Educ: Mass Col, BS, 24; Harvard Univ, AM, 27, PhD(biol), 28. Prof Exp: Instr biol, Williams Col, 24 & 26; tutor biol, Harvard Univ, 27-35, from instr to asst prof bot, 27-35, cur bot garden, 29-35; chmn div sci, 36-46, mgr col farm, 42-45, PROF SCI, BENNINGTON COL, 35- Concurrent Pos: Vis prof, Hiram Col, 55-56, Univ Fla, 62-67, Williams Col, 63-64 & Southwestern at Memphis, 71-72. Mem: AAAS. Res: Chromosome number in birches and related species; biomicrocinematography. Mailing Add: Div of Sci Bennington Col Bennington VT 05201

WOODWORTH, WAYNE LEON, b Liberal, Kans, Jan 31, 40; m 62; c 3. MATHEMATICS. Educ: Kans State Univ, BS, 62, MS, 63; Iowa State Univ, PhD(math), 68. Prof Exp: From instr to asst prof, 64-72, ASSOC PROF MATH & CHMN DEPT, DRAKE UNIV, 72- Mem: Math Asn Am; Am Math Soc. Res: Measure and integration theory; functional analysis. Mailing Add: Dept of Math Drake Univ Des Moines IA 50311

WOODY, A-YOUNG MOON, b Pyungyang, Korea, Mar 7, 34; US citizen; m 65; c 2. CHEMISTRY, BIOCHEMISTRY. Educ: Univ Calif, Berkeley, BS, 59; Cornell Univ, PhD(biochem), 64. Prof Exp: Res assoc chem, Cornell Univ, 64-65; res assoc microbiol, Univ Ill, Urbana, 65-66; res assoc biochem, 67-69; res assoc develop biol, Dept Zool, Ariz State Univ, 72-74. Mem: Am Chem Soc. Res: Structure and function of proteins and enzymes. Mailing Add: 2607 Shadow Ct Ft Collins CO 80521

WOODY, CHARLES DILLON, b Brooklyn, NY, Feb 6, 37; m 59; c 2. NEUROPHYSIOLOGY. Educ: Princeton Univ, AB, 57; Harvard Med Sch, MD, 62. Prof Exp: Intern med, Strong Mem Hosp, Univ Rochester, 62-63; resident, Boston City Hosp, Mass, 63-64; res assoc, Lab Neurophysiol, NIH, 64-67, res officer, Lab Neural Control, 68-71; assoc prof anat, physiol & psychiat, 71-75, ASSOC PROF ANAT & PSYCHIAT, MENT RETARDATION CTR, NEUROPSYCHIAT INST, UNIV CALIF, LOS ANGELES, 76- Concurrent Pos: Harvard Moseley fel & Nat Acad Sci exchange fel neurophysiol, Inst Physiol, Czech Acad Sci, 67-68; res fel neurol, Harvard Med Sch, 63-64. Honors & Awards: Nightingale Prize, Brit Biol Eng Soc & Int Fedn Med Electronics & Biol Eng, 69. Mem: AAAS; Am Physiol Soc; Soc Neurosci; Biomed Eng Soc. Res: Neurophysiology of learning and memory; neurophysiology of learned motor performance; electrophysiologic data analysis by linear filter techniques employing digital computers. Mailing Add: Ment Retardation Ctr Univ of Calif Neuropsychiat Inst Los Angeles CA 90024

WOODY, CHARLES OWEN, JR, b Somerville, Tenn, Oct 28, 30; m 60; c 4. REPRODUCTIVE PHYSIOLOGY. Educ: Miss State Univ, BS, 57, MS, 59; NC State Univ, PhD(animal sci), 63. Prof Exp: Trainee endocrinol, Univ Wis-Madison, 63, proj assoc reproductive physiol, 64-68; ASSOC PROF ANIMAL INDUSTS, UNIV CONN, 68- Mem: Am Soc Animal Sci; Brit Soc Study Fertility; Soc Study Reproduction. Res: Corpus luteum physiology; testis growth and function. Mailing Add: Dept of Animal Industries Univ of Conn Storrs CT 06268

WOODY, NORMAN COOPER, b Winnfield, La, June 15, 14; m 40. PEDIATRICS. Educ: Centenary Col, BS, 36; La State Univ, MD, 41; Am Bd Pediat, dipl, 48. Prof Exp: Intern, Charity Hosp, New Orleans, La, 41-42; resident pediat, Tulane Serv, 45-47; instr, Sch Med, Tulane Univ, 47-48; pvt pract, Tex, 48-59; from asst prof to assoc prof, 59-64, PROF PEDIAT, SCH MED, TULANE UNIV, 64- Concurrent Pos: Lederle med fac award, 62-65; asst vis physician, Charity Hosp, New Orleans, 59-67. Mem: Am Pediat Soc. Res: Human parasitology; neonatology; inborn metabolic errors. Mailing Add: Tulane Univ Sch of Med New Orleans LA 70112

WOODY, ROBERT WAYNE, b Newton, Iowa, Dec 5, 35; m 65; c 2. PHYSICAL CHEMISTRY, BIOCHEMISTRY. Educ: Iowa State Col, BS, 58; Univ Calif, Berkeley, PhD(chem), 62. Prof Exp: Res assoc phys chem, Cornell Univ, 62-64, Nat Inst Gen Med Sci fel, 63-64; asst prof, Univ Ill, Urbana, 64-70; from assoc prof to prof, Ariz State Univ, 70-75; PROF PHYS CHEM, COLO STATE UNIV, 75- Mem: AAAS; Am Chem Soc; Am Soc Biol Chemists; Biophys Soc. Res: Optical properties of molecules; structure of proteins; interaction of small molecules with proteins. Mailing Add: Dept of Biochem Colo State Univ Ft Collins CO 80523

WOODYARD, JACK RAMON, b Los Angeles, Calif, Oct 18, 44; m 65; c 3. ATOMIC PHYSICS. Educ: Colo State Univ, BS, 66; Univ Nev, Reno, MS, 68, PhD(physics), 74. Prof Exp: PHYSICIST, RENO METALL RES CTR, US BUR MINES, 71- Concurrent Pos: Adj asst prof, Univ Nev, Reno, 75- Mem: Am Phys Soc; Am Asn Physics Teachers; AAAS. Res: Ab initio atomic transition probabilities for rare gas configurations are being calculated by a method which takes into account: L-S term dependence, spin-orbit interaction and configuration interaction. Mailing Add: Dept of Physics Univ of Nev Reno NV 89507

WOODYARD, JAMES DOUGLAS, b San Antonio, Tex, Oct 8, 38. ORGANIC CHEMISTRY. Educ: Tex Christian Univ, BA, 61, MA, 63, PhD(chem), 67. Prof Exp: NASA fel Univ Ill, Chicago, 66-67; asst prof, 67-74, ASSOC PROF CHEM, W TEX STATE UNIV, 74- Concurrent Pos: Robert A Welch grant, 68. Mem: Am Chem Soc; The Chem Soc. Res: Reaction of carbenes and stereochemistry of carbene reactions; triplet state of organic molecules. Mailing Add: Dept of Chem WTex State Univ Canyon TX 79015

WOODYARD, JAMES ROBERT, b Pittsburgh, Pa, July 18, 36; m 60; c 4. PHYSICS. Educ: Duquesne Univ, BEd, 60; Univ Del, MS, 62, PhD(physics), 66. Prof Exp: AEC fel, Oak Ridge Nat Lab, 65-67; asst prof physics, Univ Ky, 67-69; mem tech staff, Gen Tel & Electronics Labs, 69-71; asst prof physics, Univ Hartford, 71-72; asst prof, Trenton State Col, 72-75; ASSOC PROF, DIV SCI & TECHNOL, COL LIFELONG LEARNING, WAYNE STATE UNIV, 75- Mem: AAAS; Am Phys Soc; Am Vacuum Soc; Sigma Xi; Am Asn Physics Teachers. Res: Particle surface interactions; ultra high vacuum technique; instructional methods. Mailing Add: Col of Lifelong Learning Wayne State Univ Detroit MI 48202

WOODYARD, WILLIAM T, b St Joseph, Mo, Apr 12, 19; m 42; c 6. ORGANIC CHEMISTRY. Educ: Univ Mo, BS, 50, MA, 51; Univ Denver, PhD(higher educ), 65. Prof Exp: US Air Force, 41-, from instr to assoc prof chem, US Mil Acad, 51-54, prof & head dept, US Air Force Acad, 54-65, chief scientist, Europ Off Aerospace Res, 65-67, DEAN FAC, US AIR FORCE ACAD, 68- Mem: Am Chem Soc. Res: Chemical education. Mailing Add: Off of Dean of Fac US Air Force Academy CO 80840

WOOFTER, HARVEY DARRELL, b Glenville, WVa, Jan 31, 23; m 44. WEED SCIENCE. Educ: WVa Univ, BS, 43; Ohio State Univ, MS, 49, PhD(agron), 53. Prof Exp: County agr agt, WVa, 43-44 & 46-48; asst agron, Ohio State Univ, 48-49; res agronomist, Chem Corps Biol Warfare Labs, US Dept Army, Md, 49-54; field res rep, Pittsburgh Coke & Chem Co, 54-56; field res rep, Chemagro Corp, 56-62, asst supvr field res, 62-64; mgr field res, Diamond Alkali Co, 64-66; mgr prod develop, Ciba Agrochem Co, Fla, 66-68; mgr, Fla Res Sta, Velsicol Chem Corp, 68-72, MGR AGROCHEM STA, HOFFMAN-LA ROCHE, INC, 72- Mem: Am Soc Agron; Entom Soc Am; Weed Sci Soc Am; Aquatic Plant Mgt Soc. Res: Weed control; development of new pesticides, fungicides, nematocides, bactericides, defoliants, desiccants and plant growth regulators. Mailing Add: Hoffman-La Roche Sci Inc PO Box X Vero Beach FL 32960

WOOL, IRA GOODWIN, b Newark, NJ, Aug 22, 25; m 50; c 2. BIOCHEMISTRY. Educ: Syracuse Univ, AB, 49; Univ Chicago, MD, 53, PhD(physiol), 54. Prof Exp: Intern med, Beth Israel Hosp, Boston, Mass, 54-55, asst resident, 55-56; from asst prof to assoc prof physiol, 57-65, assoc prof biochem, 64-65, PROF BIOCHEM, UNIV CHICAGO, 65-, A J CARLSON PROF BIOL SCI, 73- Concurrent Pos: Fel physiol, Harvard Univ, 56; Commonwealth Fund fel, Univ Chicago, 56-57; vis scientist, Dept Biochem, Cambridge Univ, 60-61; vis prof, Wayne State Univ, 64, 66 & Fla State Univ, 66; Bernstein lectr, Beth Israel Hosp, Harvard Med Sch, 64; Lederle sci lectr, Lederle Labs, 66; vis fac mem, Mayo Grad Sch Med, 66; vis prof, Rutgers Univ, 67; ed, Vitamins & Hormones. Honors & Awards: Ginsburg Award, Univ Chicago, 52; mem overseas panel, J Biochem, 73-; Alexander von Humboldt spec.fel, Fed Repub Ger, 73-74; vis res scientist, Max-Planck Inst Molecular Genetics, Berlin, Ger, 73-74. Mem: AAAS; Am Physiol Soc; Brit Biochem Soc; Am Soc Biol Chemists; Am Soc Cell Biol. Res: Regulation of protein synthesis; structure and function of animal ribosomes; mechanism of hormone action. Mailing Add: Dept of Biochem Univ of Chicago Chicago IL 60637

WOOLAVER, LAWRENCE BRENTON, inorganic chemistry, analytical chemistry, see 12th edition

WOOLCOTT, WILLIAM STARNOLD, b Coffeyville, Kans, Apr 14, 22; m 46; c 2. VERTEBRATE ZOOLOGY. Educ: Austin Peay State Col, BS, 47; Peabody Col, MA, 48; Cornell Univ, PhD(vert zool), 55. Prof Exp: Asst prof biol, Carson-Newman Col, 49-53; assoc prof, 55-67, PROF BIOL, UNIV RICHMOND, 67- Mem: Am Soc Zool; Am Soc Ichthyol & Herpet; Soc Syst Zool. Res: Morphological and ecological aspects of fishes. Mailing Add: Dept of Biol Univ of Richmond Richmond VA 23173

WOOLDRIDGE, DAVID DILLEY, b Seattle, Wash, Mar 12, 27; m 48, 70; c 5. FOREST SOILS, FOREST HYDROLOGY. Educ: Univ Wash, BS, 50, PhD(forestry), 61. Prof Exp: Forester, Rayonier Inc, 50-52, res forester, 53-56; res forester, Forest Hydrol Lab, US Forest Serv, Wash, 56-68; ASSOC PROF FOREST HYDROL, COL FOREST RESOURCES, UNIV WASH, 68- Concurrent Pos: Asst prof, Univ Wash, 65-68; consult, King County Flood Control Div, 66-; consult local eng firms & US Corps Eng; res grants, Off Water Resources & Res & US Forest Serv. Mem: Soc Am Foresters; Soil Sci Soc Am; Am Geophys Union. Res: Soil chemistry and physics in relation to soil erosion; hydrology of forest watersheds and influences on management on water yield and quality; urban hydrology; environmental quality. Mailing Add: Col of Forest Resources Univ of Wash Seattle WA 98195

WOOLDRIDGE, DAVID PAUL, b Terre Haute, Ind, Dec 25, 31; m 56. ENTOMOLOGY. Educ: Ind Univ, BS, 56, PhD(zool), 62. Prof Exp: Asst prof biol, Wilkes Col, 62-63; asst prof zool, Southern Ill Univ, 63-67; assoc ed, Biol Abstr, 67-68; asst prof biol, 68-72, ASSOC PROF BIOL, PA STATE UNIV, 72- Mem: Entom Soc Am; Coleopterist's Soc. Res: Taxonomy and biochemistry of aquatic Coleoptera;

effects of water pollution on aquatic Coleoptera. Mailing Add: 810 E Prospect Ave North Wales PA 19454

WOOLDRIDGE, ELIZABETH TAYLOR, b Salem, Ky, Sept 6, 08; m 37; c 1. MATHEMATICS. Educ: Murray State Univ, AB, 31; Peabody Col, MA, 41; Univ Nebr, PhD(ed psychol), 64. Prof Exp: Teacher, High Sch, Ky, 28-29 & 36-42, Elem Schs, 33-35; High Sch, La, 42-45 & Nebr, 45-46; teacher English, Wayne State Col, 46-47, teacher math, 47-57; asst statist, Univ Nebr, 58-59; assoc prof math, Wayne State Col, 59-62; PROF MATH, UNIV N ALA, 62- Concurrent Pos: NSF vis scientist lectr, 66- Mem: Math Asn Am. Res: Statistics; factorial study of changes in ability patterns of students in college algebra. Mailing Add: Dept of Math Univ N Ala Florence AL 35630

WOOLDRIDGE, GENE LYSLE, b Randalia, Iowa, Apr 16, 24; m 45; c 5. ATMOSPHERIC SCIENCES, PHYSICS. Educ: Upper Iowa Col, BS, 44; Mankato State Col, MS, 61; Colo State Univ, PhD(atmospheric sci), 70. Prof Exp: Instr physics, Rochester State Jr Col, Minn, 61-62; instr, Mankato State Col, 62-64, asst prof atmospheric sci, 65-67; ASSOC PROF ATMOSPHERIC SCI, UTAH STATE UNIV, 70- Mem: Am Meteorol Soc. Res: Mesoscale circulations and transport processes; mesoscale-macroscale and mesoscale-microscale energy interactions and mechanisms. Mailing Add: Dept of Soil Sci & Biometeorol Utah State Univ Logan UT 84322

WOOLES, WALLACE RALPH, b Lawrence, Mass, Mar 8, 31; m 51; c 5. PHYSIOLOGY, PHARMACOLOGY. Educ: Boston Col, BS, 58, MS, 61; Univ Tenn, PhD(physiol), 63. Prof Exp: From instr to assoc prof pharmacol, Med Col Va, 63-70; PROF PHARMACOL & ASSOC VCHANCELLOR HEALTH AFFAIRS, E CAROLINA UNIV, 70- Mem: AAAS; Soc Toxicol; Int Soc Res Reticuloendothelial Systs; Am Soc Pharmacol & Exp Therapeut. Res: Radiation injury and lipid metabolism; alcohol and lipid metabolism; reticuloendothelial system and drug metabolism. Mailing Add: Sch of Med ECarolina Univ Greenville NC 27834

WOOLEVER, PATRICIA S, b Dana Point, Calif, 38. INSECT MORPHOLOGY. Educ: Cent State Col, Okla, BS, 58; Univ Calif, Berkeley, PhD(entom), 65. Prof Exp: Asst specialist insect morphol, Univ Calif, Berkeley, 63-68; asst prof biol, Cent State Univ, Okla, 69-70; NATURE PHOTOGRAPHER, 71-; RES ENTOMOLOGIST, UNIV CALIF, BERKELEY, 75- Concurrent Pos: Asst specialist insect morphol, Univ Calif, Berkeley, 73-74. Mem: AAAS. Res: Electron microscopy and histology of insect tissues; sporozoans in cockroach Malpighian tubules; neurometamorphosis and compound eye development in Galleria; neuromorphology in cricket circadian rhythms. Mailing Add: 201 Wellman Univ of Calif Div of Entom Berkeley CA 94720

WOOLEY, BENNIE CECIL, microbiology, biochemistry, see 12th edition

WOOLF, C R, b Cape Town, SAfrica, Jan 26, 25; Can citizen; m 52; c 3. PULMONARY DISEASES, INTERNAL MEDICINE. Educ: Univ Cape Town, BSc, 44, MB, ChB, 47, MD, 51; FRCP(C), 57. Prof Exp: Jr asst path, Univ Cape Town, 49, jr asst med, 50-51; house physician, Brompton Hosp, London, Eng & London Chest Hosp, 52-53; resident chest serv, Bellevue Hosp, New York, 53-54; res assoc, Ont Heart Found, 55-63; from asst prof to assoc prof, 63-73, PROF MED, UNIV TORONTO, 73-, DIR TRI-HOSP RESPIRATORY SERV, 63- Concurrent Pos: Francis Esther res fel cardio-respiratory, Univ Toronto, 54-55; AMA, Med Res Coun Can, Ont Heart Found, Ont Tuberc Asn & Can Tuberc Asn grants; sr physician, Toronto Gen Hosp, 58- Mem: Am Thoracic Soc; Am Col Chest Physicians; Can Thoracic Soc. Res: Respiratory physiology; cardiac dyspnea; role of oxygen in the regulation of respiration; effects of smoking; surgical treatment of emphysema. Mailing Add: Respiratory Div Off Toronto Gen Hosp Toronto ON Can

WOOLF, CHARLES MARTIN, b Salt Lake City, Utah, Aug 23, 25; m 50; c 4. GENETICS. Educ: Univ Utah, BS, 48, MS, 49; Univ Calif, PhD(genetics), 54. Prof Exp: Asst geneticist, Lab Human Genetics, Univ Utah, 50-51, dir, 57-61, res instr genetics, Univ, 53-55, asst prof, 55-59, assoc prof, 59-61; dean col lib arts, 73-75, assoc prof genetics, 61-64, PROF ZOOL, ARIZ STATE UNIV, 64-, VPRES GRAD STUDIES & RES & DEAN GRAD COL, 76- Mem: Am Soc Human Genetics (treas, 61-63); Genetics Soc Am. Res: Genetics of congenital malformation; consanguinity and genetic effects; drosophila behavior and developmental genetics. Mailing Add: Dept of Zool Ariz State Univ Tempe AZ 85281

WOOLF, CYRIL, organic chemistry, see 12th edition

WOOLF, MICHAEL A, b Detroit, Mich, Feb 3, 38; m 60; c 2. LOW TEMPERATURE PHYSICS. Educ: Harvard Col, AB, 59; Univ Calif, Berkeley, PhD(physics), 64. Prof Exp: Mem tech staff physics, Bell Tel Labs, 64-66; asst prof, Univ Calif, Los Angeles, 66-71; CONSULT, 71-; ASSOC PROF PHYSICS, HAMPSHIRE COL, 75- Concurrent Pos: Alfred P Sloan fel, 67-69. Res: Tunneling in superconductors; light scattering in liquid helium. Mailing Add: Watson Rd Ashfield MA 01330

WOOLF, WILLIAM BLAUVELT, b New Rochelle, NY, Sept 18, 32. MATHEMATICS. Educ: Pomona Col, BA, 53; Claremont Col, MA, 55; Univ Mich, PhD(math), 60. Prof Exp: Instr math, Mt San Antonio Jr Col, 54 & Univ Mich, 54-59; from instr to assoc prof, Univ Wash, 59-68; staff assoc, 68-69, ASSOC SEC & DIR ADMIN, AM ASN UNIV PROFS, 69- Concurrent Pos: Fulbright res scholar, Univ Helsinki, 63-64. Mem: Am Math Soc; AAAS; Math Asn Am. Res: Functions of a complex variable; mathematics education. Mailing Add: Am Asn of Univ Profs Suite 500 One Dupont Circle Washington DC 20036

WOOLFENDEN, GLEN EVERETT, b Elizabeth, NJ, Jan 23, 30; m 54; c 3. ORNITHOLOGY. Educ: Cornell Univ, BS, 53; Univ Kans, MA, 56; Univ Fla, PhD(zool), 60. Prof Exp: Instr biol, Univ Fla, 59-60; from instr to assoc prof zool, 60-70, PROF BIOL, UNIV SOUTH FLA, TAMPA, 70- Concurrent Pos: Res Soc Am res grant, 61; consult, Encephalitis Res Ctr, Tampa, 61-64; res assoc, Am Mus Natural Hist, 70- Mem: Am Ornith Union; Wilson Ornith Soc; Cooper Ornith Soc; Brit Ornith Union. Res: Behavior; systematics; paleontology; ecology. Mailing Add: Dept of Biol Univ of South Fla Tampa FL 33620

WOOLFOLK, CLIFFORD ALLEN, b Riverside, Calif, June 5, 35; m 57; c 2. MICROBIOLOGY. Educ: Univ Calif, Riverside, BA, 57; Univ Wash, Seattle, MSc, 59, PhD(microbiol), 63. Prof Exp: Res asst microbiol, Univ Wash, 57-59 & 62-63; USPHS fel enzymol lab biochem, Nat Heart Inst, 63-65; asst prof, 65-68, ASSOC PROF MICROBIOL, UNIV CALIF, IRVINE, 68- Mem: AAAS; Am Soc Microbiol; Brit Soc Gen Microbiol; Am Chem Soc; Am Soc Biol Chem. Res: Microbial physiology; hydrogenase and hydrogenase mediated reduction of inorganic compounds; cumulative feedback inhibition of glutamine synthetase from Escherichia coli; bacterial purine oxidizing enzymes. Mailing Add: Dept of Molecular Biol & Biochem Univ of Calif Irvine CA 92664

WOOLFOLK, ROBERT WILLIAM, b Riverside, Calif, Feb 9, 37; m 68; c 2. PHYSICAL CHEMISTRY. Educ: Univ Calif, Riverside, BS, 58; Univ Calif, Berkeley, PhD(phys chem), 64. Prof Exp: Staff scientist, United Tech Ctr Div, United Aircraft Corp, 63-65; PHYS CHEMIST & ASST DIR PROG DEVELOP, STANFORD RES INST, 65- Mem: AAAS; Am Chem Soc; Am Phys Soc; Am Inst Aeronaut & Astronaut. Res: Shock wave phenomenon; gas phase reaction kinetics; photochemistry; fluorine chemistry; explosive sensitivity; nonideal explosions; combustion and energy programs. Mailing Add: Stanford Res Inst-Washington 1611 N Kent St Arlington VA 22209

WOOLFORD, ROBERT GRAHAM, b London, Ont, Apr 14, 33; m 55; c 2. ORGANIC CHEMISTRY. Educ: Univ Western Ont, BSc, 55, MSc, 56; Univ Ill, PhD(org chem), 59. Prof Exp: From asst prof to assoc prof chem, 59-68, assoc dean sci, 67-75, PROF CHEM, UNIV WATERLOO, 68- Concurrent Pos: Res fel chem, Univ Ill, 56-58. Mem: Fel Chem Inst Can. Res: Electroorganic chemistry of halogenated carboxylic acids; synthesis of polymers. Mailing Add: Dept of Chem Univ of Waterloo Waterloo ON Can

WOOLFSON, ARNOLD PETER, b Toronto, Ont, Mar 14, 36; m 63; c 1. CULTURAL ANTHROPOLOGY. Educ: Univ Toronto, BA, 58, MA, 61; State Univ NY Buffalo, PhD(anthrop), 67. Prof Exp: Instr Eng, Univ Buffalo, 63-64; instr Eng & anthrop, State Univ NY Buffalo, 64-65; asst prof anthrop, Wayne State Univ, 66-70; asst prof, 70-72, ASSOC PROF ANTHROP, UNIV VT, 72- Concurrent Pos: Co-chmn standing comt ling & cognition, Coun Anthrop & Educ, 72, mem, Coun. Mem: Fel AAAS; fel Am Anthrop Asn; Ling Soc Am. Res: Value orientations of bicultural and bilingual children in Northeastern Vermont. Mailing Add: Dept of Anthrop Univ of Vt Burlington VT 05401

WOOLLARD, GEORGE PRIOR, b Savannah, Ga, Dec 20, 08; m 37; c 9. GEOPHYSICS. Educ: Ga Inst Technol, BS, 32, MS, 34; Princeton Univ, AM, 35, PhD(struct geol), 37. Hon Degrees: PhD, Univ Uppsala, 73; DSc, Univ Wis, 75. Prof Exp: Asst, Ga Inst Technol, 32-34 & Princeton Univ, 35-36; lectr geophys, Rutgers Univ, 39; Nat Res Coun fel, Lehigh Univ, 39-40; Guggenheim fel, Princeton Univ, 40-41, 47-48, lectr geophys, 41 & 48; res group leader, Oceanog Inst, Woods Hole, 42-47, assoc geophys & oceanog, 47-57; from assoc prof to prof geophys & eng geol, Univ Wis, 48-63, dir geophys & polar res ctr, 59-63; PROF GEOPHYS, UNIV HAWAII & DIR INST GEOPHYS, 63- Concurrent Pos: Trustee, Bermuda Biol Sta, 39-47; mem study group V, Int Union Geod & Geophys, 56-, mem study group VI, 70-, mem upper mantle comt; chmn gravity & crustal struct, US Comt, Int Geophys Year, 57-60; mem adv comt sci & technol, Gov Hawaii, 64-; mem oceanog adv comt, 66-; mem, President's Adv Comt Oceanog, 70-71; mem upper mangle comt, Nat Acad Sci, chmn geod comt, Space Sci Bd; fel, World Acad Arts & Sci; mem, Pan Am Inst Geog & Hist Geophys Comn, 71. Honors & Awards: Bowie Medal, Am Geophys Union, 73. Mem: Fel Geol Soc Am; Soc Explor Geophys; Am Asn Petrol Geol; fel Am Geophys Union (vpres, 60-63, pres, 64-66); Europ Asn Explor Geophys. Res: Relation of gravity and magnetic anomalies and seismic refraction measurements to geology and the earth's crustal structure; application of geology and geophysics to engineering problems; world's gravity field. Mailing Add: Hawaii Inst of Geophys Univ of Hawaii Honolulu HI 96822

WOOLLETT, ALBERT HAINES, b Oxford, Miss, Jan 23, 30; m 64; c 3. PHYSICS. Educ: Univ Miss, BA, 49, MS, 50; Univ Okla, PhD(physics), 56. Prof Exp: Asst prof physics, Fisk Univ, 56-57; instr, Reed Col, 57-59; assoc prof, High Point Col, 59-61; asst prof, Va Polytech Inst & State Univ, 61-63; asst prof, 63-72, ASSOC PROF PHYSICS, MEMPHIS STATE UNIV, 72- Mem: Am Asn Physics Teachers; Optical Soc Am; assoc mem Am Astron Soc. Res: Infrared and Raman spectroscopy; physical optics. Mailing Add: Dept of Physics Memphis State Univ Memphis TN 38111

WOOLLETT, RALPH STORER, b Newport, RI, Dec 30, 17. ACOUSTICS. Educ: Mass Inst Technol, SB, 39; Univ Conn, MS, 60, PhD, 67. Prof Exp: PHYSICIST, NAVAL UNDERWATER SYSTS CTR, 47- Mem: Acoust Soc Am; Inst Elec & Electronics Eng. Res: Magnetostriction; piezoelectricity; electromechanical transducers. Mailing Add: Naval Underwater Systs Ctr New London CT 06320

WOOLLEY, DONALD GRANT, b Magrath, Alta, Dec 12, 25; m 47; c 5. AGRONOMY, PHYSIOLOGY. Educ: Utah State Univ, BSc, 51, MSc, 56; Iowa State Univ, PhD(crop physiol), 59. Prof Exp: Asst prof agron, Iowa State Univ, 59-60; res officer, Can Dept Agr, 60-63; assoc prof, 63-67, PROF AGRON, IOWA STATE UNIV, 67- Mem: Am Soc Agron; Am Soc Crop Sci. Res: Crop physiology and climatology. Mailing Add: 1816 Bel Air Dr Ames IA 50010

WOOLLEY, DOROTHY ELIZABETH SCHUMANN, b Wapakoneta, Ohio, Feb 2, 29; m 50; c 3. PHYSIOLOGY, PHARMACOLOGY. Educ: Bowling Green State Univ, BS, 50; Ohio State Univ, MS, 56; Univ Calif, Berkeley, PhD(physiol), 61. Prof Exp: NSF fel, Univ Calif, Berkeley, 61-62, asst res physiologist, 62-65, lectr physiol, 63-64; from asst prof to assoc prof physiol & environ toxicol, 65-74, PROF ANIMAL PHYSIOL, UNIV CALIF, DAVIS, 74- Concurrent Pos: Lectr, Sch Med, Univ Calif, San Francisco, 60-65. Mem: AAAS; Am Physiol Soc; Endocrine Soc; Am Soc Pharmacol & Exp Therapeut. Res: Effects of hormones, drugs and neurotoxins on brain electrical activity, neurochemistry and behavior in rats and monkeys. Mailing Add: Dept of Animal Physiol Univ of Calif Davis CA 95616

WOOLLEY, EARL MADSEN, b Richfield, Utah, Apr 10, 42; m 66; c 5. PHYSICAL CHEMISTRY, ANALYTICAL CHEMISTRY. Educ: Brigham Young Univ, BS, 66, PhD(phys chem), 69. Prof Exp: Nat Res Coun Can fel, Univ Lethbridge, 69-70; ASSOC PROF ANAL & PHYS CHEM, BRIGHAM YOUNG UNIV, 70- Mem: Am Chem Soc. Res: Acid-base equilibria in aqueous-organic mixtures; hydrogen bonding in solutions; calorimetric and thermal methods of analysis. Mailing Add: Dept of Chem Brigham Young Univ Provo UT 84602

WOOLLEY, GEORGE WALTER, b Osborne, Kans, Nov 9, 04; m 36; c 3. GENETICS. Educ: Iowa State Col, BS, 30; Univ Wis, MS, 31, PhD(genet), 35. Prof Exp: Asst genet, Univ Wis, 31-35; res assoc, Jackson Mem Lab, 36-49, mem bd dirs, 37-49, vpres, 43-47, asst dir & sci adminr, 47-49; chief div steroid biol, Sloan-Kettering Inst, 49-58, chief div human tumor exp chemother, 58-61, chief div & head sect tumor biol, 61-66; head biol sci sect, 66-67, HEALTH SCIENTIST ADMINR & GENET PROG ADMINR, NAT INST GEN MED SCI, 66- Concurrent Pos: Fel, Univ Wis, 35-36; mem, Int Genet Cong, NY, 36 & Edinburgh, 39, Int Cancer Cong, Atlantic City, 39 & St Louis, 47, sect biol, Nat Res Coun, 54-57, bd trustees, Dalton Schs, Inc, NY, 56- & Mt Desert Island Biol Labs; vis assoc, Sloan-Kettering Inst Cancer Res, 48-49, assoc scientist, 66-; vis res assoc, Jackson Mem Lab, 49-53; prof, Sloan-Kettering Div, Med Col, Cornell Univ, 51-66; consult, Nat Cancer Inst, 56-60. Mem: Fel AAAS; Genet Soc Am; Am Asn Cancer Res; Am Soc Human Genet; fel NY Acad Sci. Res: Basic and human genetics; cancer; endocrinology. Mailing Add: Apt 336 Kenwood Pl 5301 Westbard Circle Washington DC 20016

WOOLLEY, JOSEPH TARBET, b Denver, Colo, Jan 25, 23; m 52; c 3. PLANT PHYSIOLOGY. Educ: Utah State Agr Col, BS, 48, MS, 52; Univ Calif, PhD(plant physiol), 56. Prof Exp: Asst prof, 56-65, ASSOC PROF PLANT PHYSIOL, UNIV

ILL, URBANA, 65-; PLANT PHYSIOLOGIST, AGR RES SERV, USDA, 56- Mem: Am Soc Plant Physiol; Am Soc Agron; Soc Exp Biol. Res: Physics of water in plants and soils; reflection of light by leaves; soil-plant-atmosphere interactions. Mailing Add: Dept of Agron Univ of Ill Urbana IL 61801

WOOLLEY, LEGRAND H, b Salt Lake City, Utah, Apr 22, 31; m 53; c 4. ORAL PATHOLOGY, DENTISTRY. Educ: Univ Mo-Kansas City, DDS, 58; Univ Ore, MS, 66; Am Bd Oral Path, dipl, 73. Prof Exp: Asst prof path, Dent Sch, 66-70, sr clin instr dent & oral med, 67-71, ASSOC PROF PATH, DENT SCH, UNIV ORE, 70-, ASSOC PROF DENT, MED SCH, 71- Concurrent Pos: Am Cancer Soc fel, 66-68; consult, Fairview Hosp, Salem, Ore, 66-68. Mem: Fel Am Acad Oral Path; Am Dent Asn. Res: Effects of use of electrical dental equipment on dogs with artificial cardiac pacemakers. Mailing Add: Univ of Ore 611 SW Campus Dr Portland OR 97201

WOOLLEY, PAUL VINCENT, b Kansas City, Mo, Mar 9, 09; m 38; c 2. PEDIATRICS. Educ: Univ Kans, AB, 30; Univ Idaho, MS, 31; Harvard Univ, MD, 35. Prof Exp: Asst dir labs, Commonwealth of Mass, 38-39; from fel to assoc prof pediat, Univ Ore, 39-46; PROF PEDIAT & CHMN DEPT, SCH MED, WAYNE STATE UNIV, 46- Concurrent Pos: Dir, Variety Club Growth & Develop Clin; pediatrician-in-chief, Children's Hosp Mich, 46-; emer consult, Surgeon Gen, US Dept Air Force & Detroit Receiving, Herman Kieffer & Sinai Hosps; trustee, Child Res Ctr Mich. Mem: Soc Pediat Res; Am Pediat Soc; Am Acad Pediat. Res: Bacteriology; infectious diseases; abnormal growth and development. Mailing Add: Dept of Pediat Wayne State Univ Sch of Med Detroit MI 48201

WOOLLEY, TYLER ANDERSON, b Los Angeles, Calif, Apr 3, 18; m 43; c 3. ZOOLOGY. Educ: Univ Utah, BS, 39, MS, 41; Ohio State Univ, PhD(entom), 48. Prof Exp: Sr asst comp anat, Univ Utah, 38-39; asst zool, Ohio State Univ, 46-47, asst instr, 47-48; from asst prof to assoc prof, 48-58, PROF ZOOL, COLO STATE UNIV, 58- Res: Acarology; Oribatei; taxonomy; biology; invertebrate zoology. Mailing Add: Dept of Zool & Entom Colo State Univ Ft Collins CO 80521

WOOLRIDGE, DEAN EVERETT, b Chickasha, Okla, May 30, 13; m 36; c 3. PHYSICS. Educ: Univ Okla, BA, 32, MS, 33; Calif Inst Technol, PhD(physics), 36. Prof Exp: Mem tech staff, Bell Tel Labs, Inc, 36-46; co-dir res & develop lab, Hughes Aircraft Corp, 46-51, dir, 51-52, vpres, 52-53; pres, Ramo-Woolridge Corp, 53-58, pres & dir, Thompson Ramo Woolridge, Inc, 58-62, dir, TRW, Inc, 62-70; INDEPENDENT RESEARCHER, 70- Concurrent Pos: Res assoc, Calif Inst Technol, 62- Honors & Awards: Hackett Award, 55; Distinguished Serv Citation. Univ Okla, 60; Westinghouse Award, AAAS, 63. Mem: Nat Acad Sci; AAAS; fel Am Phys Soc; fel Am Acad Arts & Sci; fel Am Inst Aeronaut & Astronaut. Res: Separation of isotopes; systems analysis; analog computers; magnetic recording; physical electronics; neurobiology; science writing. Mailing Add: 4545 Via Esperanza Santa Barbara CA 93110

WOOLRIDGE, ROBERT LEONARD, b Garretson, SDak, Oct 13, 19; m 46; c 1. MEDICAL BACTERIOLOGY, IMMUNOLOGY. Educ: Univ SDak, BA, 41; Univ Chicago, MSc, 43; Keio Univ, Japan, DSc(microbiol), 60. Prof Exp: Res asst path, Sch Med, Univ Chicago, 43-48; microbiologist, US Naval Med Res Unit 4, 48-59 & Unit 2, 59-66; health scientist adminr, US-Japan Coop Med Sci Prog, NIH, 66-68, chief, Pac Off, Japan, 68-71, Dept Health, Educ & Welfare health scientist liaison officer to Dept State, 71-72, CHIEF PREV BR, CANCER CONTROL PROG, NAT CANCER INST, 72- Concurrent Pos: Res assoc, Univ Chicago, 49-52; asst clin prof, Univ Wash, 61-; tech consult, WHO, Geneva, 62-66. Mem: Fel AAAS; Am Soc Microbiol; Asia-Pac Acad Ophthal. Res: Vaccine development; immunological prophylaxis; laboratory diagnosis of respiratory viruses, arboviruses and the trachomainclusion conjunctivitis agents; epidemiology and chemoprophylaxis studies in trachomatous children in the Far East. Mailing Add: Blair Bldg 614 Nat Cancer Inst Bethesda MD 20014

WOOLSEY, CLINTON NATHAN, b Brooklyn, NY, Nov 30, 04; m 42; c 3. NEUROPHYSIOLOGY. Educ: Union Univ, NY, AB, 28; Johns Hopkins Univ, MD, 33. Prof Exp: Asst physiol, Sch Med, Johns Hopkins Univ, 33-34, instr, 34-39, from assoc to assoc prof physiol, 41-48; Charles Sumner Slichter res prof neurophysiol, 48-75, dir lab neurophysiol, 60-73, EMER PROF NEUROPHYSIOL, UNIV WIS-MADISON, 75-, BIOMED UNIT COORDR, WAISMAN CTR, 73- Concurrent Pos: Rockefeller Found fel, Johnson Found, Univ Pa, 38-39; fel neuromuscular physiol, Johns Hopkins Univ, 39-41; James Arthur lectr, Am Mus Nat Hist, New York, 52; Hines lectr, Emory Univ, 61; Bishop lectr, Wash Univ, 61; hon lectr, Albany Med Col, 66; von Monakow lectr, Univ Zurich, 68; J Hughlings Jackson Mem lectr, Montreal Neurol Inst, 71; Donald D Matson lectr, Harvard Med Sch, 75; hon mem fac med, Univ Chile. Spec consult, Ment Health Study Sect, NIH, 52-53 & neurol study sect, 53-56 & 57-58; mem nat adv coun, Nat Inst Neurol Dis & Blindness, 58-62, spec consult neurol prog proj comt, 63-67, mem bd sci counr, 65-69; mem div med sci, Nat Res Coun, 52-58 & Inter-Coun Adv Comt Career Res Professorships, 60-61; mem med exchange mission pharmacol & physiol of nerv syst, USSR, 58; US organizer, Int Brain Res Orgn-UNESCO vis sem, Chile, 66; mem, Nat Acad Sci-Colciencias Panel Grad Educ & Res Biol, Colombian Univs, 72. Honors & Awards: Medalist, Univ Brussels, 68. Mem: Nat Acad Sci; Am Physiol Soc; hon mem Am Neurol Asn; assoc Am Asn Neurol Surg; Soc Neurosci. Res: Cerebral localization; sensory and motor. Mailing Add: Dept of Neurophysiol Univ of Wis Madison WI 53706

WOOLSEY, GERALD BRUCE, b Brooks, Ga, Aug 16, 37; m 60; c 3. PHYSICAL CHEMISTRY. Educ: Univ SC, BS, 60, PhD(phys chem), 67. Prof Exp: Chemist, Tenn Corp, Cities Serv Co, 60-63; res chemist, 67-69, develop supvr, 69-71, res supvr, 71-74, MFG SUPVR, E I DU PONT DE NEMOURS & CO, INC, 74- Mem: Am Chem Soc. Res: Thermodynamics of solutions; hydration in solutions of concentrated electrolytes; amide solutions; polyester films. Mailing Add: 500 Poe Ave Worthington OH 43085

WOOLSEY, MARION ELMER, b Croft, Kans, July 27, 19; m 40; c 3. MICROBIOLOGY. Educ: Univ Tex, Austin, BA, 64, MA, 66, PhD(microbiol), 68. Prof Exp: Res assoc immunol, Univ Tex, Austin, 68, asst prof microbiol, 68-70; ASST PROF MICROBIOL, UNIV TULSA, 70- Concurrent Pos: NIH fel, Univ Tex, Austin, 68. Mem: Am Soc Microbiol. Res: Basic and applied research in medical microbiology, immunology and immunochemistry. Mailing Add: Dept of Life Sci Univ of Tulsa Tulsa OK 74104

WOOLSEY, NEIL FRANKLIN, b Tieton, Wash, Apr 30, 35; m 56; c 5. ORGANIC CHEMISTRY. Educ: Univ Portland, BS, 57; Univ Wis, PhD(org chem), 62. Prof Exp: NSF fel, Imp Col, London, 61-63; res assoc org chem, Iowa State Univ, 63-65; asst prof chem, 65-70, ASSOC PROF CHEM, UNIV NDAK, 70- Mem: Am Chem Soc; The Chem Soc. Res: Diazoketone reactions; organic photochemistry; synthesis of polyhedranes; coal chemistry. Mailing Add: Dept of Chem Univ of NDak Grand Forks ND 58201

WOOLSEY, ROBERT S, b Chicago, Ill, May 30, 31; m 67; c 1. MEDICINE, NEUROLOGY. Educ: St Louis Univ, BS, 53, MD, 57. Prof Exp: Intern med, St Louis Univ Hosp, 57-58; resident neurol, Univ Mich Hosp, 58-61; from instr to assoc prof, 62-75, PROF NEUROL, ST LOUIS UNIV, 75- Concurrent Pos: Fel neuropath, Col Physicians & Surgeons, Columbia Univ, 61-62. Mem: AMA; Asn Res Nerv & Ment Dis; Am Acad Neurol; Am Electroencephalog Soc. Mailing Add: Wohl Mem Ment Health Inst 1221 S Grand Blvd St Louis MO 63104

WOOLSEY, THOMAS ALLEN, b Baltimore, Md, Apr 17, 43; m 69; c 2. NEUROANATOMY. Educ: Univ Wis-Madison, BS, 65; Johns Hopkins Univ, MD, 69. Prof Exp: Intern surg, Barnes Hosp St Louis, Mo, 69-70; ASST PROF ANAT & NEUROBIOL, MED SCH, WASH UNIV, 71- Concurrent Pos: NIH fel anat, Med Sch, Wash Univ, 70-71, Nat Inst Neurol Dis & Stroke grant, 72- Mem: Am Asn Anat; Soc Neurosci; AAAS. Res: Structure of somatosensory areas of the cerebral cortex. Mailing Add: Dept of Anat & Neurobiol Wash Univ Med Sch St Louis MO 63110

WOOLSON, EDWIN ALBERT, b Takoma Park, Md, Oct 2, 41; m 61; c 3. PESTICIDE CHEMISTRY. Educ: Univ Md, BS, 63, MS, 66, PhD(soil chem), 69. Prof Exp: Analyst soil testing, Univ Md, 62; phys sci aide, USDA, 62-63, chemist, 63-65, anal chemist, 65-67, RES CHEMIST, USDA, 67- Honors & Awards: USDA Superior Serv Award, 74. Mem: Am Chem Soc; Am Soc Agron; Soil Sci Soc Am; Weed Sci Soc Am. Res: Behavior and fate of arsenic, herbicides and insecticides in soil and water; method development for pesticides in soils, plants and water; toxicity of pesticides to plants; impurities in pesticides; bioaccumulation. Mailing Add: Bldg 050 Agr Res Ctr-West USDA Beltsville MD 20705

WOONTON, GARNET ALEXANDER, b London, Ont, July 9, 06; m 34; c 1. PHYSICS, MAGNETIC RESONANCE. Educ: Univ Western Ont, BA, 25, MA, 31. Hon Degrees: DSc, Univ Western Ont, 55. Prof Exp: Demonstr physics, Univ Western Ont, 31-34; asst & fel, 34-39, from assoc prof to prof, 39-48, MacDonald prof & chmn dept, 55-68, dir, Eaton Electronics Res Lab, 50-69, HON PROF PHYSICS, McGILL UNIV, 69-; HON PROF FAC SCI, LAVAL UNIV, 73-; HON PROF PHYSICS, UNIV WESTERN ONT, 74- Concurrent Pos: Titular prof, Fac Sci & dir, Ctr Res Atoms & Molecules, Laval Univ, 69-73; res fel, Ctr Nuclear Studies, Grenoble, France, 68-69. Honors & Awards: Medal Achievement Physics, Can Asn Physicists. Mem: Can Asn Physicists (pres, 48-49); fel Inst Elec & Electronics Engrs; fel Royal Soc Can. Res: Magnetic resonance of gases; electron paramagnetic resonance; microwave techniques applied to physical measurements; spectroscopy. Mailing Add: Apt 1105 1231 Richmond St London ON Can

WOOSLEY, RAYMOND LEON, b Roundhill, Ky, Oct 2, 42; m 63; c 2. CLINICAL PHARMACOLOGY. Educ: Western Ky State Univ, BS, 64; Univ Louisville, PhD(pharmacol), 67; Univ Miami, MD, 73. Prof Exp: Sr pharmacologist, Meyer Labs, Inc, 68-69, dir res pharmaceut, 69-71; from intern to resident med, Vanderbilt Univ Hosp, 73-74, FEL CLIN PHARMACOL, VANDERBILT UNIV, 74- Concurrent Pos: NIH fel pharmacol, Med Sch, Univ Louisville, 67-68, lectr, 69-71; instr med & clin pharmacol, Vanderbilt Univ, 75. Mem: Soc Exp Biol & Med; Am Soc Pharmacol & Exp Therapeut; fel Am Col Clin Pharmacol. Res: Lipid metabolism and its relation to atherosclerotic vascular disease; zinc metabolism and its importance in protein synthesis and tissue repair; clinical pharmacology of new antiarrhythmic drugs. Mailing Add: Dept of Clin Pharmacol Vanderbilt Univ Nashville TN 37232

WOOSLEY, ROYCE STANLEY, b Caneyville, Ky, June 17, 34; m 59; c 2. ORGANIC CHEMISTRY. Educ: Western Ky Univ, BS, 56; Univ Conn, MS, 59; Ohio Univ, PhD(org chem), 67. Prof Exp: Sect leader tech serv, Olin Mathieson Chem Corp, 58-62; from asst prof to assoc prof, 66-74, PROF CHEM, WESTERN CAROLINA UNIV, 74-, CHMN DEPT, 69- Mem: Am Chem Soc. Res: Thermal rearrangement reactions of unsaturated hydrocarbons. Mailing Add: Dept of Chem Western Carolina Univ Cullowhee NC 28723

WOOSTER, HAROLD ABBOTT, b Hartford, Conn, Jan 3, 19; m m 41, 68; c 4. INFORMATION SCIENCE. Educ: Syracuse Univ, AB, 39; Univ Wis, MS, 41, PhD(physiol chem), 43. Prof Exp: Asst, Toxicity Lab, Univ Chicago, 43-46; res assoc, Pepper Lab, Univ Pa, 46-47; sr fel, Mellon Inst, 47-56; dir res commun, Air Force Off Sci Res, 56-59, chief info sci div, 59-62, dir info sci, 62-70; chief res & develop br, 70-74, SPEC ASST PROG DEVELOP, LISTER HILL NAT CTR BIOMED COMMUN, NAT LIBR MED, 74- Concurrent Pos: Exec secy panel info sci & tech, Comt Sci & Tech Info, Off Dir Defense Res & Eng, 65-66; adj instr, Grad Sch Libr Sci, Drexel Inst Technol, 67. Mem: AAAS; Am Soc Info Sci; Am Chem Soc. Res: Biomedical communications; computer-aided instruction; medical television; information storage and retrieval; library automation. Mailing Add: Lister Hill Nat Ctr Biomed Commun Nat Libr of Med 8600 Rockville Pike Bethesda MD 20014

WOOSTER, WARREN SCRIVER, b Westfield, Mass, Feb 20, 21; m 48; c 3. OCEANOGRAPHY. Educ: Brown Univ, BSc, 43; Calif Inst Technol, MS, 47; Univ Calif, PhD, 53. Prof Exp: Asst res oceanogr, Scripps Inst, Univ Calif, 51-58, assoc res oceanogr, 58-61; dir off oceanog, UNESCO, 61-63; prof oceanog, Scripps Inst Oceanog, Univ Calif, 63-73, chmn grad dept, 67-69; PROF OCEANOG & DEAN, ROSENSTIEL SCH MARINE & ATMOSPHERIC SCI, UNIV MIAMI, 73- Sea, Concurrent Pos: Dir invests, Coun Hydrobiol Invests, Peru, 57-58; vpres, Int Coun Explor Sea, 74- Mem: Am Soc Limnol & Oceanog; fel Am Geophys Union; Arctic Inst NAm; fel Am Meteorol Soc. Res: Descriptive oceanography of the Pacific Ocean; physical, chemical and fishery oceanography; ocean affairs. Mailing Add: Univ of Miami Rosenstiel Sch 4600 Rickenbacker Causeway Miami FL 33149

WOOTEN, ARTHUR LEE, organic chemistry, see 12th edition

WOOTEN, BENJAMIN ALLEN, b Opelike, Ala, Apr 11, 17; m 40. PHYSICS. Educ: Univ Ala, BA, 37; Columbia Univ, MA, 42, PhD, 58. Prof Exp: Instr physics, Columbia Univ, 44-46; assoc prof, Southwestern at Memphis, 47-57; from asst prof to assoc prof, 57-61, PROF PHYSICS, WORCESTER POLYTECH INST, 61- Concurrent Pos: Asst, Nevis Cyclotron Labs, Columbia Univ, 52-58. Mem: AAAS; Am Phys Soc; Am Asn Physics Teachers. Res: Nuclear physics. Mailing Add: Dept of Physics Worcester Polytech Inst Worcester MA 01609

WOOTEN, FRANK THOMAS, b Fayetteville, NC, Sept 24, 35; m 62; c 3. BIOMEDICAL ENGINEERING. Educ: Duke Univ, BS, 57, PhD(elec eng), 64. Prof Exp: Sr engr, Electronic Res Lab, Corning Glass Works, 64-67; res engr, 67-71, mgr biomed eng, 71-75, EXEC ASST TO PRES, RES TRIANGLE INST, 75- Mem: Asn Advan Med Instrumentation; Inst Elec & Electronics Eng. Res: Medical instrumentation; analysis of pulmonary sound; technology transfer. Mailing Add: Res Triangle Inst PO Box 12194 Research Triangle Park NC 27709

WOOTEN, FREDERICK (OLIVER), b Linwood, Pa, May 16, 28; m 52; c 2. SOLID STATE PHYSICS. Educ: Mass Inst Technol, BS, 50; Univ Del, PhD(chem), 55. Prof Exp: Staff physicist, All Am Eng Co, 54-57; res chemist, Lawrence Livermore Lab, 57-72; vchmn dept appl sci, 72-73, PROF APPL SCI, UNIV CALIF, DAVIS, 72-, CHMN DEPT, 73- Concurrent Pos: Vis prof, Drexel Univ, 64; lectr, Univ Calif,

Davis, 65-72; vis prof, Chalmers Univ Technol, Sweden, 67-68; consult, Lawrence Livermore Lab, 72- Mem: AAAS; Am Phys Soc. Mailing Add: Dept of Appl Sci Univ of Calif Davis CA 95616

WOOTEN, JEAN W, b Douglasville, Ga, Jan 11, 29; m 52; c 1. BOTANY. Educ: NGa Col, BS, 46; Fla State Univ, MS, 64, PhD(bot), 68. Prof Exp: Asst prof bot, Iowa State Univ, 68-74; ASST PROF BIOL, UNIV SOUTHERN MISS, 74- Mem: Am Soc Plant Taxonomists; Soc Study Evolution; Int Asn Plant Taxon; Bot Soc Am. Res: Evolution and systematics in aquatic vascular plants. Mailing Add: Dept of Biol Univ of Southern Miss Hattiesburg MS 39401

WOOTEN, WILLIS CARL, JR, b Homerville, Ga, Mar 9, 22; m 49; c 3. POLYMER CHEMISTRY. Educ: Univ NC, PhD(chem), 50. Prof Exp: Sr res chemist, 51-60, admin asst, 60-65, DIV HEAD, RES LAB, TENN EASTMAN CO, 65- Mem: Am Chem Soc. Res: Synthetic fibers and plastics; organic chemistry. Mailing Add: Res Lab Tenn Eastman Co Kingsport TN 37662

WOOTTEN, MICHAEL JOHN, b Somersham, Eng, Mar 2, 44; m 66; c 2. PHYSICAL CHEMISTRY, WATER CHEMISTRY. Educ: Univ Leicester, BS, 66, PhD(phys chem), 69. Prof Exp: Res fel chem, Carnegie-Mellon Univ, 69-71; sr res fel phys chem, Univ Southampton, Eng, 71-73; sr engr, Belg, 73-75, SR ENGR, WESTINGHOUSE RES & DEVELOP CTR, PITTSBURGH, 75- Mem: Nat Asn Corrosion Engrs; Royal Inst Chem, Eng; Electrochem Soc Eng. Res: The physical chemistry of aqueous solutions at high temperatures and pressures with particular reference to power generation. Mailing Add: Westinghouse Res & Develop Ctr Beulah Rd Pittsburgh PA 15235

WOOTTON, DONALD MERCHANT, b Paonia, Colo, Apr 13, 16; m 41; c 1. PARASITOLOGY, ZOOLOGY. Educ: Santa Barbara State Col, BA, 41; Univ Wash, MS, 43; Stanford Univ, PhD(biol), 49. Prof Exp: Actg instr zool & bot, Santa Barbara State Col, 41-42, instr zool, 49-52; asst prof, Univ Calif, Santa Barbara, 52-56; PROF BIOL, CALIF STATE UNIV, CHICO, 57- Mem: Am Soc Parasitol. Res: Helminths; branchipods; limnology. Mailing Add: Dept of Biol Sci Calif State Univ Chico CA 95926

WOOTTON, JAMES CHARLES, organic chemistry, see 12th edition

WOOTTON, JOHN FRANCIS, b Penn Yan, NY, May 31, 29; m 59; c 4. BIOCHEMISTRY. Educ: Cornell Univ, BS, 51, MS, 53, PhD(biochem), 60. Prof Exp: Clin chemist, Clifton Springs Sanitarium & Clin, NY, 56; Nat Found fel chem, Univ Col London, 60-62; from asst prof to assoc prof physiol chem, 62-70, PROF PHYSIOL CHEM, NY STATE COL VET MED, CORNELL UNIV, 70- Concurrent Pos: Hon res asst, Univ London, 62; vis scientist, Lab Molecular Biol, Med Res Coun, Cambridge, Eng, 69-70. Mem: Am Chem Soc. Res: Enzymology; proteolytic enzymes; relationship of enzyme structure to function. Mailing Add: NY State Col of Vet Med C126C Cornell Univ Ithaca NY 14853

WOOTTON, PETER, b Peterborough, Eng, Apr 30, 24; m 47; c 3. MEDICAL PHYSICS, RADIOLOGY. Educ: Univ Birmingham, BSc, 44. Prof Exp: Physicist, Res & Develop Labs, Farrow's Br, Reckitt & Coleman Ltd, 44-48; radiation physicist, Royal Infirmary, Glasgow, Scotland, 48-51; instr radiol physics, Univ Tex M D Anderson Hosp, 51-53; radiation physicist, Tumor Inst, Swedish Hosp, Seattle, Wash, 53-64; from asst prof to assoc prof radiol, 64-72, PROF RADIOL, UNIV WASH, 72- Concurrent Pos: Instr, Med Sch, Univ Ore, 54-60, Univ Seattle, 56 & Penrose Cancer Hosp, Colorado Springs, Colo, 56-57; clin asst prof & radiation physicist, Univ Wash, 59-64; mem tech adv bd radiation control, Dept Health, Wash, 62-, US Nat Comt Med Physics, 64-, comt radiation ther studies, Hyperbaric Oxygen Steering Comt, Nat Cancer Inst, 65- & sci comt, 25, Nat Coun Radiation Protection & Measurements, 67- Mem: Am Asn Physicists in Med; Soc Nuclear Med; Am Col Radiol; Am Soc Therapeut Radiol; fel Brit Inst Physics. Res: Applications of radiation physics in medicine, especially dosimetry of all types of ionizing radiations and the effects of physical parameters such as high pressure oxygen and pulsed radiation in radiobiology; fast neutron therapy. Mailing Add: NN158 Univ Hosp RC-08 Univ of Wash Seattle WA 98195

WOPSCHALL, ROBERT HAROLD, b Glendale, Calif, May 1, 40; m 62; c 2. PHYSICAL CHEMISTRY. Educ: Harvey Mudd Col, BS, 62; Univ Wis, PhD(phys chem), 67. Prof Exp: RES CHEMIST, E I DU PONT DE NEMOURS & CO, INC, 66- Mem: Am Chem Soc. Res: Stationary electrode polarography; adsorption of electroactive species on electrodes and coupled chemical reactions; photopolymerization; electroless deposition. Mailing Add: Photo Prod Dept E I du Pont de Nemours & Co Inc Wilmington DE 19898

WORCESTER, JOHN LANG, nuclear physics, atomic physics, see 12th edition

WORDEN, DAVID GILBERT, b Minneapolis, Minn, Mar 9, 24; m 47; c 2. SOLID STATE PHYSICS, PLASMA PHYSICS. Educ: Earlham Col, AB, 50; Iowa State Univ, PhD(physics), 56. Prof Exp: Physicist, Gen Elec Res Lab, 56-61; mgr surface physics sect, Electro-Optical Systs, Inc, 61-65, electron device res sect, 65-66, electron & image device dept, 66-67; chmn dept physics, 67-68, PROF PHYSICS & ACAD VPRES, NDAK STATE UNIV, 68- Concurrent Pos: Assoc part-time prof, Calif State Col, Los Angeles, 65-67. Mem: AAAS; Am Phys Soc; Am Asn Physics Teachers. Res: Conduction in thin films; electron emission; ion production on surfaces; physical and chemical adsorption; thermionic energy conversion; thermionic electron microscopy; gas lasers; low-energy gaseous electrical discharges in magnetic fields. Mailing Add: Admin Bldg NDak State Univ Fargo ND 58102

WORDEN, EARL FREEMONT, JR, b Portsmouth, NH, Nov 30, 31; m 60; c 1. PHYSICAL CHEMISTRY. Educ: Univ NH, BS, 53, MS, 55; Univ Calif, PhD(chem), 59. Prof Exp: CHEMIST, LAWRENCE LIVERMORE LAB, UNIV CALIF, 58- Mem: Am Chem Soc; Sigma Xi; Optical Soc Am; Soc Appl Spectros. Res: Atomic and molecular optical spectroscopy; laser isotope separation; molecular lasers. Mailing Add: Lawrence Livermore Lab Univ of Calif Livermore CA 94550

WORDEN FREDERIC GARFIELD, b Syracuse, NY, Mar 22, 18; m 44; c 5. PSYCHIATRY. Educ: Dartmouth Col, AB, 39; Univ Chicago, MD, 42. Prof Exp: Intern med, Johns Hopkins Hosp, 42-43, from house officer to resident psychiat, 46-50; instr, Med Sch, Johns Hopkins Univ, 50-52; clin dir, Sheppard & Enoch Hosp, Towson, Md, 52-53; from asst prof to prof psychiat, Med Sch, Univ Calif, Los Angeles, 53-71, from assoc res psychiatrist to res psychiatrist, 56-61; prog dir, Neurosci Res Prog, 69-71, exec dir, 71-74, PROF PSYCHIAT, MASS INST TECHNOL, 69-, DIR NEUROSCI RES PROG, 74- Concurrent Pos: Career investr, NIMH, 56-61, mem behav sci & exp psychol study sects, 57-61, ment health career award comt, 61-66 & res scientist & mem develop rev comt, 71-; mem comt brain sci, Nat Res Coun, 71-74; mem bd dirs fund for res in psychiat, Neurosci Res Prog, Mass Inst Technol, 73- Mem: Am Psychiat Asn; Am Psychoanal Asn; AMA; NY Acad Sci; fel Am Acad Arts & Sci. Res: Relationships between brain function and behavior, especially auditory electrophysiology of attention, habituation and learning. Mailing Add: 45 Hilltop Rd Weston MA 02193

WORDEN, JOHN LORIMER, JR, b South Bend, Ind, Dec 11, 07; m 33; c 4. PUBLIC HEALTH ADMINISTRATION. Educ: Univ Notre Dame, BS, 28; St Bonaventure Univ, MS, 37, PhD(physiol), 57. Prof Exp: Asst prof biol & premed adv, Col St Thomas, 30-33; prof histol & embryol, 33-73, head dept biol, 36-47, registr, Grad Sch, 40-47, dir physiol, 50-73, EMER PROF BIOL, ST BONAVENTURE UNIV, 73-; COORDR EMERGENCY MED SERV DIV, CATTARAUGUS COUNTY HEALTH DEPT, 73- Mem: Sigma Xi; AAAS; Am Chem Soc; Am Soc Cell Biol; Tissue Cult Asn. Res: Histology; embryology; cell and tissue culture; emergency medical services. Mailing Add: 141 S Barry St Olean NY 14760

WORDEN, LEONARD RUSSELL, b Summit, NJ, June 6, 37; m 58; c 3. FAMILY MEDICINE. Educ: Kalamazoo Col, BA, 59; Univ Kans, PhD(org chem), 63; Univ Mich, Ann Arbor, med, 73- Prof Exp: From instr to asst prof chem, Kalamazoo Col, 63-73, dir health sci prog, 72-73. Concurrent Pos: Res consult, Kalamazoo Spice Extraction Co, 63-73. Mem: AMA; Am Acad Family Physicians; Nat Asn Residents & Interns. Res: Chemistry of naturally occurring and synthetic oxygen heterocyclic compounds; structure and synthesis of hop bittering principles. Mailing Add: Univ of Mich Med Sch Ann Arbor MI 48104

WORDEN, RALPH EDWIN, b Berwyn, Nebr, Dec 23, 17; m 43; c 9. PHYSICAL MEDICINE. Educ: Univ Nebr, BS, 41, MD, 44; Univ Minn, MS, 49. Prof Exp: Instr phys med & rehab, Col Med, Univ Minn, 49-51; from asst prof to prof, Col Med, Ohio State Univ, 51-58; prof phys med, 58-67, PROF REHAB, DEPT MED, SCH MED, UNIV CALIF, LOS ANGELES, 67-, PROF, DEPT PSYCHIAT, 74- Concurrent Pos: Co-consult, University Heights Hosp, Bel Air, Calif. Mem: AMA. Res: Physiological effects of physical agents; functional anatomy; alcoholism; rehabilitation. Mailing Add: Rehab Ctr Univ of Calif Sch of Med Los Angeles CA 90024

WORDEN, SIMON PETER, b Mt Clemens, Mich, Oct 21, 49. SOLAR PHYSICS. Educ: Univ Mich, BS, 71; Univ Ariz, PhD(astron), 75. Prof Exp: Res asst astron, Kitt Peak Nat Observ, 71-75; ASTROPHYSICIST, SACRAMENTO PEAK OBSERV, AIR FORCE GEOPHYS LAB, 75- Mem: Am Astron Soc; Royal Astron Soc. Res: The study of large-scale convective motions on the sun and observation and interpretation of stellar phenomenon related to solar surface activity. Mailing Add: Sacramento Peak Observ Sunspot NM 88349

WORF, DOUGLAS LOWELL, b Bowbells, NDak, Oct 20, 16; m 48; c 3. ENVIRONMENTAL SCIENCES. Educ: Univ Toledo, BS, 38; Georgetown Univ, PhD(biochem), 53. Prof Exp: Fel org mat, Mellon Inst, 41-44; res analyst toxicol, Off Naval Res, 46-52; biochemist, AEC, 52-58; dir, Life Support Systs, NASA, 58-62; res dir environ pollution, Alaska Water Lab, USPHS, 65-67; SCI ADMINR, ENVIRON PROTECTION AGENCY, 71- Concurrent Pos: Mem, Interagency Comt US Partic Int Biol Yr, 67-68, Off Sci & Technol, President's Sci & Adv Comt, Environ Qual Comt Res Panel. Mem: Am Chem Soc; Soc Environ Geochem & Health. Res: Administration of environmental pollution research including indoor air pollution, toxicology of chemicals and metals, health aspects of pesticides, auto emissions, administration of research grants and contracts. Mailing Add: 109 Perth Ct Cary NC 27511

WORF, GAYLE L, b Garden City, Kans, Nov 17, 29; m 52; c 2. PLANT PATHOLOGY. Educ: Kans State Univ, BS, 51, MS, 53; Univ Wis, PhD(plant path), 61. Prof Exp: County agt, Kans State Univ, 55-58; plant pathologist, Iowa State Univ, 61-63; PLANT PATHOLOGIST, UNIV WIS-MADISON, 63- Mem: Am Phytopath Soc. Res: Interpreting current research in plant pathology and analyzing its application to field situations; diagnostic procedures; economical appraisal of disease outbreaks; effective control programs. Mailing Add: 285 Russell Labs Univ of Wis Madison WI 53706

WORGUL, BASIL VLADIMIR, b New York, NY, June 30, 47; m 69; c 2. CELL BIOLOGY. Educ: Univ Miami, BS, 69; Univ Vt, PhD(zool), 74. Prof Exp: Staff assoc, 74-75, NIH FEL & RES ASSOC CELL BIOL, DEPT OPHTHAL, COL PHYSICIANS & SURGEONS, COLUMBIA UNIV, 75- Mem: Sigma Xi; AAAS; Asn Res Vision & Ophthal; Am Soc Cell Biol. Res: The cytopathomechanism of radiation cataractogenesis; control of growth and differentiation in ocular epithelia. Mailing Add: Col of Physicians & Surgeons Columbia Univ 630 W 168th St New York NY 10032

WORK, HENRY HARCUS, b Buffalo, NY, Nov 11, 11; m 45; c 4. PEDIATRICS, PSYCHIATRY. Educ: Hamilton Col, AB, 33; Harvard Univ, MD, 37; Am Bd Pediat, dipl, 47; Am Bd Psychiat & Neurol, dipl, 50, cert child psychiat, 60. Prof Exp: Psychiat serv adv, US Children's Bur, 47-49; assoc prof pediat & psychiat, Univ Louisville, 49-55; from assoc prof to prof psychiat, Univ Calif, Los Angeles, 55-72; DEP MED DIR, AM PSYCHIAT ASN, 72- Mem: Am Psychiat Asn; fel Am Orthopsychiat Asn; Am Pub Health Asn; Am Acad Pediat. Res: Identification of child with mother and other relatives. Mailing Add: Am Psychiat Asn 1700 18th St NW Washington DC 20009

WORK, JAMES LEROY, b Lancaster, Pa, Feb 6, 35; m 55; c 2. POLYMER CHEMISTRY, PHYSICAL CHEMISTRY. Educ: Franklin & Marshall Col, BA, 62; Univ Del, PhD(phys chem), 70. Prof Exp: RES SCIENTIST, ARMSTRONG CORK CO, LANCASTER, 52- Mem: Am Chem Soc. Res: Structure-property relationships in polymers and polymeric composites. Mailing Add: 845 Paramount Ave Lampeter PA 17537

WORK, JOE BOEHM, chemistry, see 12th edition

WORK, PHIL STRATFORD, biochemistry, see 12th edition

WORK, RAY ALEXANDER, III, physical inorganic chemistry, see 12th edition

WORK, RICHARD NICHOLAS, b Ithaca, NY, Aug 7, 21; m 48; c 3. POLYMER PHYSICS. Educ: Cornell Univ, BA, 42, MS, 44, PhD(appl physics), 49. Prof Exp: Asst physics, Cornell Univ, 44-46, 48-49; physicist, Nat Bur Standards, 49-51; res assoc, Plastics Lab, Princeton Univ, 51-56; from asst prof to assoc prof physics, Pa State Univ, 56-65; from asst dean to assoc dean, Col Lib Arts, 65-73, PROF PHYSICS, ARIZ STATE UNIV, 65- Concurrent Pos: Consult, Thiokol Chem Corp, 54-55, White Sands Missile Range, 55-60, Sandia Corp, 65-69 & Acushnet Co, 74- Mem: Fel AAAS; fel Am Phys Soc; Am Asn Physics Teachers. Res: Relaxation phenomena and transitions in polymers; theory of dielectrics; electrical breakdown; instrumentation. Mailing Add: 413 E Geneva Dr Tempe AZ 85282

WORK, ROBERT WYLLIE, b Chicago, Ill, July 10, 07; m 34. POLYMER CHEMISTRY. Educ: Univ Ill, BS, 29; Cornell Univ, PhD, 32. Prof Exp: Chemist, Swift & Co, Ill, 29; asst chem, Cornell Univ, 29-32; res engr, Gen Elec Co, 33-41;

chief chemist, Celanese Corp Am, 41-43, dir phys res, Res Lab, 43-52, asst mgr res, 52-54, asst mgr opers, Textile Div, 54-55, asst tech dir, 55-56; mgr tech res, Chemstrand Res Ctr, Inc, 57-64; dir textile res & prof textiles, 64-73, EMER PROF TEXTILES, NC STATE UNIV, 73- Concurrent Pos: Mem, Textile Res Inst. Mem: Fel AAAS; Am Chem Soc; Fiber Soc; fel Brit Textile Inst. Res: Man-made and spider fibers and textiles. Mailing Add: 109 David Clark Labs NC State Univ Raleigh NC 27607

WORK, STEWART D, b Chicago, Ill, Oct 17, 37; m 59; c 4. ORGANIC CHEMISTRY. Educ: Oberlin Col, AB, 59; Duke Univ, PhD(org chem), 63. Prof Exp: Fel, Duke Univ, 63 & Purdue Univ, 63-64; from asst prof to assoc prof, 64-73, PROF CHEM, EASTERN MICH UNIV, 73- Mem: Am Chem Soc. Res: Base-catalyzed condensation reactions; organo-silicon chemistry. Mailing Add: Dept of Chem Eastern Mich Univ Ypsilanti MI 48197

WORK, TELFORD HINDLEY, b Selma, Calif, July 11, 21; m; c 3. BIOLOGY, EPIDEMIOLOGY. Educ: Stanford Univ, AB, 42, MD, 46; Univ London, dipl, 49; Johns Hopkins Univ, MPH, 52. Prof Exp: Res assoc filariasis, Fiji Islands Colonial Med Serv, 49-51; staff mem virus labs, Rockefeller Found, NY, 52; mem, US Naval Med Res Unit 3, Egypt, 53-54; staff mem virus res ctr, Rockefeller Found, Poona, India, 54-55, dir, 55-58; staff mem virus labs, Rockefeller Found, NY, 58-60; chief virus & rickettsia sect, Commun Dis Ctr, USPHS, 61-67; prof infectious & trop dis, 66-69, prof immunol & microbiol, 69-71, PROF INFECTIOUS & TROP DIS, MED MICROBIOL & IMMUNOL & PREV & SOCIAL MED, MED CTR, UNIV CALIF, LOS ANGELES, 71-, VCHMN SCH PUB HEALTH, 66-, HEAD DIV INFECTIOUS & TROP DIS, 69- Concurrent Pos: Mem expert comt virus dis, WHO, 60-; dir, WHO Regional Arbovirus Lab, 61-66; assoc mem comn viral infections, Armed Forces Epidemiol Bd, 61-; Pan-Am Health Orgn consult, Arg, 62, Venezuela, 64, Jamaica, 66; regents lectr, Univ Calif, Los Angeles, 63; consult, US Army Engrs Medico-Ecol Study, Atlantic-Pac Interoceanic Canal Studies, 67- Mem: AAAS; Cooper Ornith Soc; Am Ornith Union; Am Soc Trop Med & Hyg; Am Pub Health Asn. Res: Medical ecology, epidemiology and medicine of virus infections, especially arthropod-borne viruses; tropical medicine. Mailing Add: Ctr for the Health Sci Univ of Calif Med Ctr Los Angeles CA 90024

WORKER, GEORGE F, JR, b Ordway, Colo, June 1, 23; m 49; c 4. AGRONOMY, BOTANY. Educ: Colo State Univ, BS, 49; Univ Nebr, MS, 53. Prof Exp: Asst county agent, Nebr, 49-51, asst in agron, 53; SPECIALIST IN AGRON & SUPT, IMP VALLEY FIELD STA, UNIV CALIF, DAVIS, 53- Mem: Am Soc Agron. Res: Grain sorghum production, plant function, breeding and adaption to desert climate; adaption of other field crops such as barley, sugar beets and flax to southwestern desert areas. Mailing Add: Imp Valley Field Sta 1004 E Holton Rd El Centro CA 92243

WORKMAN, ERWIN FRANKLIN, JR, b Greensboro, NC, Aug 24, 46; m 71. BIOCHEMISTRY. Educ: Duke Univ, BA, 68; Tex A&M Univ, PhD(biochem), 75. Prof Exp: Res asst biochem, Tex A&M Univ, 72-75; RES ASSOC BIOCHEM, DENT RES CTR, UNIV NC, CHAPEL HILL, 75- Mem: Sigma Xi. Res: Study of normal hemostatic mechanisms with emphasis on platelet function. Mailing Add: 204 Barington Hills Rd Chapel Hill NC 27514

WORKMAN, GARY LEE, b Birmingham, Ala, Apr 21, 40; m 67; c 1. PHYSICAL CHEMISTRY, MOLECULAR SPECTROSCOPY. Educ: Col William & Mary, BS, 64; Univ Rochester, PhD(phys chem), 69. Prof Exp: Res fel chem, Ohio State Univ, 69-70; Nat Acad Sci res assoc, Marshall Space Flight Ctr, NASA, 70-72; dir, PBR Electronics, Inc, 72-76; PROF PHYS, ATHENS COL, 76- Mem: Am Phys Soc; Am Chem Soc; Instrument Soc Am; Am Soc Nondestructive Testing. Res: Systems interfacing; electrooptics applications; solar photochemistry; reaction kinetics; spectroscopy. Mailing Add: Dept of Phys Athens Col PO Box 158 Athens AL 35611

WORKMAN, JOHN PAUL, b Salem, Ore, Feb 18, 43; m 64; c 2. RANGE MANAGEMENT, AGRICULTURAL ECONOMICS. Educ: Univ Wyo, BS, 65; Utah State Univ, MS, 67, PhD(range econ), 70. Prof Exp: Asst prof range resource econ, 70-75, ASSOC PROF RANGE ECON, UTAH STATE UNIV, 75- Mem: AAAS; Soc Range Mgt; Am Agr Econ Asn. Res: Economics of range utilization, range improvement and range livestock production. Mailing Add: Dept of Range Sci Utah State Univ Logan UT 84322

WORKMAN, MARCUS ORRIN, b Canton, Ohio, Sept 20, 40. INORGANIC CHEMISTRY. Educ: Manchester Col, BA, 62; Ohio State Univ, PhD(inorg chem), 66. Prof Exp: Teaching assoc chem, Ohio State Univ, 65-66; res assoc, Northwestern Univ, 66-67; ASST PROF CHEM, UNIV VA, 67- Concurrent Pos: Adv Res Projs Agency fel, 66-67. Mem: Am Chem Soc; The Chem Soc. Res: Coordination complexes of transition metals with polydentate ligands; complexes with oxide and sulfoxide donor atoms; complexes of lanthanides and actinides. Mailing Add: Dept of Chem Univ of Va Charlottesville VA 22901

WORKMAN, MILTON, b Chicago Heights, Ill, Oct 1, 20; m 49; c 6. PLANT PHYSIOLOGY. Educ: Colo Agr & Mech Col, BS, 50; Univ Calif, PhD(plant physiol), 54. Prof Exp: Asst veg crops, Univ Calif, 50-54; from instr to assoc prof hort, Purdue Univ, 54-66; PROF HORT, COLO STATE UNIV, 66- Res: Pre and post harvest physiology. Mailing Add: Dept of Hort Colo State Univ Ft Collins CO 80521

WORKMAN, PETER L, b Providence, RI, May 19, 37; m 67. GENETICS, ANTHROPOLOGY. Educ: Univ Calif, Davis, BS, 57, PhD(genetics), 62. Prof Exp: Staff fel, Nat Inst Arthritis & Metab Dis, 62-63; NIH fel, Dept Genetics, Cambridge Univ, 63-65; asst res biologist, Univ Calif, Davis, 65-67; res biologist, Human Genetics Br, Nat Inst Dent Res, 67-70; assoc prof pediat, Mt Sinai Sch Med, 70-71; assoc prof anthrop, Univ Mass, Amherst, 71-75; PROF ANTHROP & CHMN DEPT, UNIV N MEX, 75- Concurrent Pos: Lectr, Mt Sinai Sch Med, 71-75; Nat Inst Child Health & Human Develop career develop award, 72-75. Mem: Soc Study Human Biol; Brit Eugenics Soc; Am Soc Human Genetics; Am Soc Phys Anthrop. Res: Population and human genetics; child growth, development and behavior. Mailing Add: Dept of Anthrop Univ of NMex Albuquerque NM 87131

WORKMAN, RALPH BURNS, b Omaha, Nebr, June 25, 24; m 51; c 4. ECONOMIC ENTOMOLOGY, VEGETABLE CROPS. Educ: Colo State Univ, BS, 51, MS, 52; Ore State Univ, PhD(entom), 58. Prof Exp: Res asst entom, Colo State Univ, 52-55; asst entomologist, 58-67, ASSOC ENTOMOLOGIST, AGR RES CTR, UNIV FLA, 67- Mem: Entom Soc Am. Res: Economic entomology; biology and control of cruciferous and potato insects. Mailing Add: Agr Res Ctr Univ of Fla Hastings FL 32045

WORKMAN, WESLEY RAY, b Mich, Feb 1, 26; m 48; c 4. ORGANIC CHEMISTRY. Educ: Mich State Univ, BS, 49, MS, 50; Univ Minn, PhD(org chem), 54. Prof Exp: Sr chemist, 54-68, mgr photog sci & photo chem, 68-72, dir imaging, Res Div, 72-74, DIR SYSTS RES LAB, CENT RES LABS, 3 M CO, 74- Mem: Am Chem Soc; Soc Photog Scientists & Engrs. Res: Photochemistry. Mailing Add: 1533 Conway St Paul MN 55106

WORKMAN, WILLIAM EDWARD, b Richmond, Va, May 13, 41; m 59; c 2. ENVIRONMENTAL GEOLOGY. Educ: Univ Va, BS, 62, MS, 64; Univ Tex, Austin, PhD(geol), 68. Prof Exp: Asst prof geol, Albion Col, 68-73; geoscientist, Palmer & Baker Engrs, Inc, 73-75; SUPVRY ENVIRON GEOLOGIST, US CORPS ENGRS, MOBILE, 75- Concurrent Pos: Sigma Xi-Sci Res Soc Am grants in aid of res, 63 & 66; geologist C, Va Div Mineral Resources, Charlottesville, 63-; consult, Palmer & Baker Engrs, Inc, 75-; environ consult to var pvt firms, 75- Mem: Sigma Xi; Geol Soc Am. Res: Regional metamorphism in Llano Uplift, Texas; coastal erosion; engineering geology relative to coastal processes.

WORKMAN, WILLIAM GLENN, b Sheridan, Wyo, Mar 19, 47; m 72; c 1. AGRICULTURAL ECONOMICS. Educ: Univ Wyo, BS, 69; Utah State Univ, MA, 72, PhD(resource econ), 76. Prof Exp: Res asst econ, Utah State Univ, 69-72; ASST PROF ECON, UNIV ALASKA, 73- Mem: Am Econ Asn. Res: Economic analysis of allocation of public funds in the provision of outdoor recreation opportunities; economic analysis of producing vegetables under controlled environment conditions in northern latitudes. Mailing Add: Dept of Agr Econ Univ of Alaska Fairbanks AK 99701

WORL, RONALD GRANT, b Dunlap, Iowa, Jan 16, 38; m 60; c 3. ECONOMIC GEOLOGY. Educ: Utah State Univ, BS, 60; Univ Wyo, MA, 63, PhD(geol), 68. Prof Exp: Teaching asst, Univ Wyo, 62-64; asst prof petrol & mineral deposits, Colo State Univ, 66-67; geologist, Cent Mineral Resources, 67-73, GEOLOGIST, SAUDI ARABIAN PROJ, US GEOL SURV, 73- Mem: Soc Econ Geologists. Res: Distribution of minerals and elements in mineral deposits that formed at or near surface; structural petrology of migmatites; fluorine deposits, geologic and geochemical perspectives and commodity aspects; geology and economic aspects of massive sulfide, sedimentary copper and hydrothermal gold deposits of the Arabian Shield. Mailing Add: US Geol Surv c/o Am Embassy APO New York NY 09697

WORLEY, DAVID EUGENE, b Cadiz, Ohio, Aug 6, 29; m 68; c 2. PARASITOLOGY. Educ: Col Wooster, AB, 51; Kans State Univ, MS, 55, PhD(parasitol), 58. Prof Exp: Assoc res parasitologist, Parke, Davis & Co, 58-62; from asst prof to assoc prof, 62-72, PROF VET RES LAB, MONT STATE UNIV, 72- Mem: Am Soc Parasitol; Am Micros Soc; Wildlife Dis Asn. Res: Zoology; helminthology, including chemotherapy of parasitic infections and helminth life cycles. Mailing Add: Vet Res Lab Mont State Univ Bozeman MT 59715

WORLEY, JOHN DAVID, b Texarkana, Tex, Dec 10, 38; m 63; c 4. BIOPHYSICAL CHEMISTRY. Educ: Hendrix Col, BA, 60; Univ Okla, PhD(phys chem), 64. Prof Exp: NIH fel biophys chem, Northwestern Univ, 64-66; asst prof chem, Univ Cincinnati, 66-70; ASST PROF CHEM, ST NORBERT COL, 70- Mem: Am Chem Soc. Res: Protein structure and denaturation in solution; solutions of nonelectrolytes; hydrogen bonding. Mailing Add: Dept of Chem St Norbert Col West De Pere WI 54178

WORLEY, JOSEPH FRANCIS, b Washington, DC, Oct 10, 28; m 58. PLANT PATHOLOGY. Educ: George Washington Univ, BS, 55, MS, 58, PhD(bot), 65. Prof Exp: Bact technician, Am Type Cult Collection, 54-56; PLANT PATHOLOGIST, PLANT HORMONE LAB, AGR RES CTR, USDA, 56- Mem: AAAS; Am Phytopath Soc; Bot Soc Am; Am Soc Plant Physiologists. Res: Morphology and ultrastructure of diseased plants. Mailing Add: Plant Hormone Lab Agr Res Ctr USDA Beltsville MD 20705

WORLEY, RAY EDWARD, b Robbinsville, NC, May 4, 32; m 55; c 3. HORTICULTURE, AGRONOMY. Educ: NC State Col, BS, 54, MS, 58; Va Polytech Inst, PhD(agron), 61. Prof Exp: Asst field crops, NC State Col, 56-58; instr agron, Va Polytech Inst, 58-61; asst horticulturist, 61-72, ASSOC HORTICULTURIST, GA COASTAL PLAIN EXP STA, 72- Mem: Am Soc Hort Sci. Res: Pecan tree nutrition, management and physiology; vegetable and forage nutrition and physiology. Mailing Add: Hort Dept Ga Coastal Plain Exp Sta Tifton GA 31794

WORLEY, RICHARD DIXON, b Little Rock, Ark, Dec 24, 26; m 51; c 2. PHYSICS. Educ: Hendrix Col, BS, 49; Univ Ark, MA, 51; Univ Chicago, MS, 60; Univ Calif, Berkeley, PhD(physics), 63. Prof Exp: Nuclear physicist, Wright-Patterson AFB, Ohio, 51-54; nuclear engr, Douglas Air Craft Co, Calif, 57-59; sr physicist, Lawrence Radiation Lab, Univ Calif, 63-70; PHYSICIST, MASON & HANGER, PANTEX PLANT, SILAS MASON CO, 70- Mem: Am Phys Soc. Res: Characteristic x-ray production by ion bombardment of both polycrystalline and single-crystal targets; atomic beams and the hyperfine interaction; high explosive research and development. Mailing Add: 3710 Huntington Dr Amarillo TX 79109

WORLEY, ROBERT DUNKLE, b Trenton, NJ, Jan 24, 25; m 50; c 3. PHYSICS. Educ: Williams Col, AB, 49; Columbia Univ, AM, 51, PhD, 55. Prof Exp: From asst to lectr, Columbia Univ, 50-54; MEM STAFF, BELL LABS, INC, 54-, SUPVR, 60- Mem: Am Phys Soc; Acoust Soc Am; NY Acad Sci. Res: Heat capacities of superconductors; sound transmission in the ocean; operations research; sonar systems. Mailing Add: 17 Knollwood Trail Brookside NJ 07926

WORLEY, SMITH, JR, b Columbia, SC, June 14, 24; m 47; c 3. AGRONOMY. Educ: Clemson Col, BS, 49; Univ Ariz, MS, 53; La State Univ, PhD, 58. Prof Exp: Asst plant breeder, Univ Ariz, 49-55; res assoc agron, NC State Col, 55-56 & La State Univ, 56-58; RES AGRONOMIST, SOUTHERN REGION, AGR RES SERV, USDA, 58- Honors & Awards: Service Award, Am Soc Testing & Mat, 74. Mem: Am Soc Qual Control; Am Soc Testing & Mat. Res: Genetics of cotton quality; fiber technology; statistics; instrumentation for evaluation of components of fiber quality; textile quality control. Mailing Add: US Cotton Fiber Res Lab Univ Tenn Agr Campus Knoxville TN 37916

WORLEY, WILL J, b Gibson City, Ill, Aug 2, 19; m 54; c 3. THEORETICAL MECHANICS, APPLIED MECHANICS. Educ: Univ Ill, BS, 43, MS, 45, PhD, 52. Prof Exp: From instr to assoc prof, 43-60, PROF THEORET & APPL MECH, UNIV ILL, URBANA, 60-, PROJ DIR, NASA PROJ, 63- Concurrent Pos: Consult, Magnavox Co, Ind, 56-58, Ill, 58-65, Chris-Kaye Mfg Co, Ill, 57, A O Smith Corp, Wis, 59 & attorneys, Ill, 59-; proj dir, Wright Air Develop Ctr Nonlinear Mech Proj, 59-61; consult, IBM Corp, NY, 60-61 & 63-64, Calif, 61. Mem: Acoust Soc Am; Am Soc Mech Eng. Res: Acoustical noise reduction; mechanical vibrations and nonlinear mechanics; static and dynamic behavior of plates and shells; optimum structural design; mechanical properties of materials; synthesis in system design; failure investigation of components and systems; systems approach to prevention and analysis of system failure. Mailing Add: 306e Talbot Lab Univ of Ill Urbana IL 61801

WORLOCK, JOHN M, b Kearney, Nebr, Feb 15, 31; div; c 2. SOLID STATE PHYSICS. Educ: Swarthmore Col, BA, 53; Cornell Univ, PhD(physics), 62. Prof Exp: NSF fel, 62-63; actg asst prof physics, Univ Calif, Berkeley, 63-64; MEM TECH STAFF, BELL LABS, INC, 64- Mem: Am Phys Soc. Res: Lattice dynamics; transport and optical properties of non-metallic crystals; light scattering; phase transitions. Mailing Add: Bell Labs Inc Holmdel NJ 07733

WORMAN, JAMES JOHN, b Allentown, Pa, Feb 17, 40; m 61; c 4. ORGANIC CHEMISTRY. Educ: Moravian Col, BS, 61; NMex Highlands Univ, MS, 64; Univ Wyo, PhD, 68. Prof Exp: Instr chem, Moravian Col, 62-63; res asst, Univ Wyo, 65-67; asst prof, 67-72, ASSOC PROF CHEM, S DAK STATE UNIV, 72- Mem: Am Chem Soc; The Chem Soc. Res: Theoretical and experimental organic photochemistry, including syntheses and reaction of large ring nitrogen heterocycles. Mailing Add: Dept of Chem SDak State Univ Brookings SD 57006

WORMSER, HENRY C, b Strasbourg, France, Sept 10, 36; US citizen; m 63; c 2. PHARMACEUTICAL CHEMISTRY. Educ: Temple Univ, BSc, 59, MSc, 61; Univ Wis, PhD(pharmaceut chem), 65. Prof Exp: Asst prof, 65-71, ASSOC PROF PHARMACEUT CHEM, WAYNE STATE UNIV, 71- Mem: Am Chem Soc; Am Pharmaceut Asn. Res: Synthesis of model compounds to be used in study of drug-enzyme or drug-receptor site interactions in an effort to determine specific mechanisms of drug activity. Mailing Add: Wayne State Univ Col Pharm & Allied Hlth Profns Detroit MI 48202

WORNARDT, WALTER WILLIAM, JR, b Milwaukee, Wis, Mar 21, 34; m 63; c 4. PALEONTOLOGY. Educ: Univ Wis-Madison, BS, 56, MS, 58; Univ Calif, Berkeley, PhD(paleont), 63. Prof Exp: Consult paleont, Calif, 64-65; res geologist, Esso Prod Res Co, Tex, 65-67; SR RES GEOLOGIST, UNION OIL CO CALIF, 67- Mem: Geol Soc Am; Am Asn Petrol Geol; Soc Econ Paleont & Mineral; Am Asn Stratig Palynologists. Res: Biostratigraphy, biogeography and paleoecology of diatoms and silicoflagellates, especially fossil marine species from the world's Cenozoic rocks; zonation of Neogene strata. Mailing Add: Union Oil Co of Calif PO Box 76 Brea CA 92621

WOROBEY, WALTER, solid state physics, see 12th edition

WOROCH, EUGENE LEO, b Kenosha, Wis, Mar 18, 22; m 49; c 4. MEDICINAL CHEMISTRY. Educ: Univ Wis, BS, 44, MS, 45, PhD(org chem), 48; Univ Chicago, MBA, 71. Prof Exp: Proj dir natural prod, Wis Alumni Res Found, 48-49; group leader & consult, Bjorksten Labs, 49; res assoc natural prod, Mayo Clin, 49-51; group leader, Glidden Co, 51-58; head dept org chem res, 58-75, DIR DIV ANTIBIOTIC RES, ABBOTT LABS, 75- Mem: Am Chem Soc; AAAS; Am Soc Microbiol. Res: Natural products; steroids; antibiotics; peptides and structural chemistry. Mailing Add: Abbott Labs 1400 Sheridan Rd North Chicago IL 60064

WORONICK, CHARLES LOUIS, b Meriden, Conn, Dec 4, 30. BIOCHEMISTRY, CLINICAL CHEMISTRY. Educ: Univ Conn, BS, 53; Univ Calif, Berkeley, MS, 55; Univ Wis, PhD(biochem), 59. Prof Exp: Asst prof chem, Brown Univ, 62-66; assoc non-clin investr path, Pa Hosp, 66-68; biochemist, Med Res Labs, Dept Med, Hartford Hosp, 68-73; ASST PROF LAB MED, SCH MED, UNIV CONN, FARMINGTON, 73- Concurrent Pos: NSF fel, 59-61; fel enzyme chem, Nobel Med Inst, Stockholm, 59-62; Nat Cancer Inst fel, 61-62; assoc path, Med Sch, Univ Pa, 66-68. Mem: Am Asn Clin Chem; Am Chem Soc; NY Acad Sci. Res: Enzyme kinetics; equilibria and mechanisms; fluorometry of enzyme complexes; immunochemistry. Mailing Add: Dept of Lab Med Sch of Med Univ of Conn Health Ctr Farmington CT 06032

WORRALL, JOHN GATLAND, b Cleethorpes, Eng, May 22, 38; Can citizen. DENDROLOGY. Educ: Univ Durham, BSc, 59; Yale Univ, MF, 64, PhD(forestry), 69. Prof Exp: ASSOC PROF FORESTRY, UNIV BC, 68- Res: Environmental control of cambial activity; breakage of dormancy seeds and plants. Mailing Add: Fac of Forestry Univ of BC Vancouver BC Can

WORRALL, PAUL MICHAEL, b Dursley, Eng, Jan 17, 35; m 65; c 2. MEDICAL RESEARCH. Educ: Univ Birmingham, Eng, BS, 56, MB ChB, 59; Royal Col Surgeons & Physicians, MRCS LRCP, 60. Prof Exp: Dep med dir, Lederle Labs, Eng, 64-65; var, Bristol Labs, NY, 65-69; proj officer med, Army Investigational Drug Rev Bd, US Army, 69-71; dir med res, Bristol-Myers Int Div, 71-73; dir med res, 73-75, MED DIR, MEAD JOHNSON CO, 75- Mem: Am Rheumatism Asn; AAAS; Brit Med Asn. Res: Management of drug development area (human use) in a major pharmaceutical company. Mailing Add: Mead Johnson Co Evansville IN 47721

WORRALL, WINFIELD SCOTT, b Cheltenham, Pa, Jan 12, 21; m 49; c 2. ORGANIC CHEMISTRY. Educ: Haverford Col, BS, 42; Harvard Univ, MA, 47, PhD(chem), 49. Prof Exp: Res chemist, Monsanto Chem Co, 50-54; from instr to assoc prof chem, Trinity Col, Conn, 54-64; assoc prof chem, 64-68, ASSOC PROF GEN SCI STUDIES, STATE UNIV NY COL PLATTSBURGH, 68- Res: Steroids; heterocyclics. Mailing Add: Dept of Gen Studies State Univ of NY Col Plattsburgh NY 12901

WORRELL, ALBERT CADWALLADER, b Philadelphia, Pa, May 14, 13; m 37; c 3. FOREST ECONOMICS, RESOURCE ECONOMICS. Educ: Univ Mich, BSF, 35, MF, 35, PhD(forest econ), 53. Prof Exp: Asst forester, Soil Conserv Serv, USDA, 35-40; dist forester, Va Forest Serv, 44-47; from asst prof to assoc prof forest econ, Univ Ga, 47-55; from assoc prof to prof, 55-67, EDWIN W DAVIS PROF FOREST POLICY, YALE UNIV, 67- Concurrent Pos: Forest economist, UN Econ Comn Latin Am, 60-61; guest prof forest econ, Univ Freiburg, 70. Mem: Soc Am Foresters. Res: Renewable natural resource economics and policy. Mailing Add: Sch of Forestry & Environ Studies Yale Univ 120 High St New Haven CT 06511

WORRELL, FRANCIS TOUSSAINT, b Hartford, Conn, Apr 19, 15; m 48; c 2. PHYSICS. Educ: Univ Mich, BSE, 36; Univ Pittsburgh, MS, 40, PhD(physics), 41. Prof Exp: Mem staff, Physicists Res Co, Mich, 36-37; asst, Univ Pittsburgh, 37-41; instr, Univ Tenn, 41-42; staff mem, Radiation Lab, Mass Inst Technol, 42-46; res assoc, Inst Metals, Univ Chicago, 46-47; asst prof physics, Rensselaer Polytech Inst, 47-55; assoc prof, DePauw Univ, 55-58; Fulbright lectr, Al-Hikma Univ Baghdad, 58-59; assoc prof, Beloit Col, 59-60; staff mem, Lincoln Lab, Mass Inst Technol, 60-63; PROF PHYSICS, UNIV LOWELL, 63- Concurrent Pos: Vis lectr, Univ Bristol, 72-73. Mem: AAAS; Am Phys Soc; Am Asn Physics Teachers. Res: Design of experiments; structure of materials; thermionic emission; atmospheric optics. Mailing Add: Dept of Physics & Appl Physics Univ Lowell Lowell MA 01854

WORRELL, JAY H, b Manchester, NH, July 14, 38; m 59; c 3. PHYSICAL INORGANIC CHEMISTRY. Educ: Univ NH, BS, 61, MS, 63; Ohio State Univ, PhD(inorg chem), 66. Prof Exp: Res assoc chem, State Univ NY Stony Brook, 66-67; asst prof, 67-72, ASSOC PROF CHEM, UNIV S FLA, TAMPA, 72- Concurrent Pos: Res Corp NY res grant, 68- Mem: Am Chem Soc. Res: Preparation, properties and theory of coordination compounds; stereochemistry and inorganic reaction kinetics for ligand substitution and oxidation reduction processes. Mailing Add: Dept of Chem Univ of SFla Tampa FL 33620

WORRELL, JOHN MAYS, JR, b El Paso, Tex, Oct 3, 33; m 66; c 1. MATHEMATICS, MEDICINE. Educ: Univ Tex, BA, 54, MD, 57, PhD(math), 61. Prof Exp: Intern med, Denver Gen Hosp, Colo, 57-58; instr math, Univ Tex, 58-59; NSF fel, 61-62; mem tech staff math res, Sandia Corp, NMex, 62-72; PROF MATH, OHIO UNIV, 72- Concurrent Pos: Consult clin med & biomed sci. Mem: Am Math Soc; Am Med Asn. Res: Problems having topological character; clinical medicine; biological processes. Mailing Add: Dept of Math Ohio Univ Athens OH 45701

WORRELL, LEE FRANK, b Orleans, Ind, Feb 27, 13; m 34; c 1. PHARMACEUTICAL CHEMISTRY. Educ: Purdue Univ, BS, 35, PhD(pharmaceut chem), 40. Prof Exp: Assoc prof pharm, Drake Univ, 38-42; from instr to assoc prof, Univ Mich, 42-60; dean col pharm, 62-66, PROF PHARMACEUT CHEM, UNIV TEX, AUSTIN, 60- Mem: AAAS; Am Chem Soc; Am Pharmaceut Asn. Res: Analytical pharmaceutical chemistry. Mailing Add: Univ of Tex Col of Pharm Austin TX 78712

WORRELL, WAYNE L, b Rock Island, Ill, Oct 25, 37; m 68; c 2. HIGH TEMPERATURE CHEMISTRY. Educ: Mass Inst Technol, BS, 59, PhD(metall), 63. Hon Degrees: MA, Univ Pa, 71. Prof Exp: Fel metall, Univ Calif, Berkeley, 63-64, lectr, 64-65; from asst prof to assoc prof, 65-74, PROF METALL, UNIV PA, 74- Concurrent Pos: Chmn comt high temp sci & technol, Nat Acad Sci-Nat Res Coun, 74-77; consult to indust & govt orgns; vis prof, Dept Chem, Univ Calif, Berkeley, 75-76. Mem: AAAS; Am Inst Mining, Metall & Petrol Eng; Electrochem Soc; Am Soc Metals; Am Ceramic Soc. Res: Thermodynamics and kinetics of high-temperature reactions; thermodynamics of metallic alloys; corrosion at elevated temperatures; high-temperature materials chemistry; solid state electrochemistry. Mailing Add: Dept Metall & Mat Sci Univ of Pa Philadelphia PA 19174

WORREST, ROBERT CHARLES, b Hartford, Conn, July 6, 35; m 57; c 1. PHOTOBIOLOGY, MARINE ECOLOGY. Educ: Williams Col, BA, 57; Wesleyan Univ, MA, 64; Ore State Univ, PhD(radiation biol), 75. Prof Exp: Teacher, Canterbury Sch, Conn, 57-59, Belmont Hill Sch, Mass, 59-71; instr biol, 71-72, trainee environ health, 72-75, RES ASSOC MARINE ECOL, ORE STATE UNIV, 75- Mem: Sigma Xi; AAAS; Am Soc Photobiol; Radiation Res Soc; Am Soc Limnol & Oceanog. Res: Assessment of the impact of increased solar mid-ultraviolet radiation upon marine ecosystems. Mailing Add: Dept of Gen Sci Ore State Univ Corvallis OR 97331

WORSFOLD, RICHARD JOHN, b Torquay, Eng, Apr 22, 40; m 65; c 3. GEOLOGY. Educ: Univ Exeter, BSc, 62; Univ Birmingham, PhD(geol), 67. Prof Exp: Geologist, Brit Antarctic Surv, 62-68; STAFF GEOLOGIST, INT NICKEL CO CAN, LTD, 68- Honors & Awards: Polar Medal, Gt Brit, 72. Mem: Can Inst Mining & Metall; Geol Asn Can. Res: Metamorphic petrology; geological setting of nickeliferous deposits. Mailing Add: Can Nickel Co Can Ltd Hwy 17 W Coppercliff ON Can

WORSHAM, ARCH DOUGLAS, b Culloden, Ga, Feb 22, 33; m 56; c 2. WEED SCIENCE, PLANT PHYSIOLOGY. Educ: Univ Ga, BSA, 55, MS, 57; NC State Univ, PhD(crop sci), 61. Prof Exp: From exten asst prof to exten assoc prof, 60-67, assoc prof, 67-69, PROF CROP SCI, NC STATE UNIV, 69- Concurrent Pos: Res grants, 66-74. Honors & Awards: Outstanding Publ Award, Weed Sci Soc Am, 75. Mem: Weed Sci Soc Am; Am Inst Biol Sci; Am Soc Plant Physiologists. Res: Pesticides; crop science; basic and applied research in weeds in agronomic crops and non-tillage crop production. Mailing Add: Dept of Crop Sci NC State Univ Raleigh NC 27607

WORSHAM, HERBERT J, JR, b Lynchburg, Va, Jan 22, 35; m 57; c 3. NUCLEAR PHYSICS. Educ: Va Polytech Inst, BS, 56. Prof Exp: From physicist to sr physicist, 59-69, mgr reactor opers, Nuclear Serv Dept, 69-74, sect mgr, 74, RESIDENT PROJ MGR, NUCLEAR SERV DEPT, BABCOCK & WILCOX CO, 74- Concurrent Pos: Supvr, Lynchburg Pool Reactor, 65-69. Mem: Am Nuclear Soc. Res: Instrumentation, particularly, in-core and operation of reactors using advanced control techniques; operator training, start-up and testing of power reactors. Mailing Add: Babcock & Wilcox Co PO Box 1260 Lynchburg VA 24505

WORSHAM, JAMES ESSEX, JR, b Newport News, Va, Apr 29, 25. PHYSICAL CHEMISTRY, BIOMEDICAL ENGINEERING. Educ: Univ Richmond, BS, 47; Vanderbilt Univ, MS, 49; Duke Univ, PhD(chem), 53. Prof Exp: Instr high sch, Va, 48-50; assoc prof chem, Hampden-Sydney Col, 53-54; from asst prof to assoc prof, 54-67, PROF CHEM, UNIV RICHMOND, 67-, DIR ACAD COMPUT, 74- Concurrent Pos: Res assoc, Mass Inst Technol, 59-60; consult, Dept Med, Med Col, Va, 63-, adj res prof med, 63- Mem: Am Chem Soc; Sigma Xi. Res: Molecular structure; neutron diffraction; solutions; biomedical instrumentation; control of circulation. Mailing Add: Dept of Chem Univ of Richmond Richmond VA 23173

WORSHAM, WALTER CASTINE, b Turbeville, SC, Aug 17, 38; m 59; c 2. TEXTILE CHEMISTRY. Educ: Col Charleston, BS, 61; Univ NC, PhD(phys chem), 66. Prof Exp: Chemist, Fiber Industs, Inc, 66-67; chemist, 67-75, TECH MGR, EMERY INDUSTS, INC, MAULDIN, 75- Mem: Am Chem Soc; Am Asn Textile Chemists & Colorists. Res: Kinetics of photochemical reactions; chemistry of textile processing. Mailing Add: 101 Silver Pine Ct Greer SC 29651

WORSING, ROBERT A, applied mathematics, see 12th edition

WORSLEY, THOMAS RAYMOND, biostratigraphy, oceanography, see 12th edition

WORT, DENNIS JAMES, b Eng, July 19, 06; m 36; c 2. PLANT PHYSIOLOGY. Educ: Univ Sask, BSc, 32, MSc, 34; Univ Chicago, PhD, 40. Prof Exp: Instr biol & bot, Nutana Collegiate Inst, 28-45; assoc prof, 45-47, PROF PLANT PHYSIOL, UNIV BC, 47- Concurrent Pos: Lectr, Univ Sask, 43-45; Nuffield Res fel, Oxford Univ, 59. Mem: Fel AAAS; Am Soc Plant Physiol; Am Soc Agron; Bot Soc Am; Can Soc Plant Physiol. Res: Chemical control of plant growth; plant biochemistry. Mailing Add: Dept of Bot Univ of BC Vancouver BC Can

WORTH, CARLETON RUSSELL, mathematics, see 12th edition

WORTH, DONALD CALHOUN, b Brooklyn, NY, Oct 20, 23; m 46; c 4. NUCLEAR PHYSICS. Educ: Carnegie Inst Technol, BS, 44; Yale Univ, MS, 48, PhD(physics), 49. Prof Exp: Instr physics, Berea Col, 50-52, assoc prof, 52-53; from asst prof to assoc prof, 54-60, PROF PHYSICS, INT CHRISTIAN UNIV, TOKYO, 60-, DEAN COL LIBERAL ARTS, 70- Concurrent Pos: Vis asst prof, Ala Polytech Inst, 51; vis assoc prof, Univ Chicago, 53-54; NSF fel, Univ Va, 58-59; vis prof, Univ Wis, 63-64 & State Univ NY Stony Brook, 68-69. Mem: AAAS; Am Phys Soc; Am Asn Physics Teachers. Res: Low energy nuclear polarization studies, especially in nucleon-nucleon scattering; physics education. Mailing Add: Dept Physics Int Christian Univ 10-2 Osawa 3-Chome Mitaka-shi Tokyo Japan

WORTH, JAMES JUDSON BLACKLEY, b Charleston, WVa, Dec 11, 27; m 52; c 3. GEOPHYSICS, METEOROLOGY. Educ: Pa State Univ, BS, 52; Univ Mich, MS, 57. Prof Exp: Liaison off pres off, Pa State Univ, 51-52; indust engr, Armstrong Cork

Co, 52-56; res assoc meteorol group, Eng Res Inst, Univ Mich, 56-57; engr, Commun Dept, Bendix Systs Div, Bendix Corp, 57-58, mgr eng group, 58-59, head meteorol & geophys dept, 59-62, asst gen mgr & dir environ sci, 69-70, dir indust develop, 70-71, V PRES, RESEARCH TRIANGLE INST, 71- Concurrent Pos: Am del, Int Geophys Conf, Helsinki, Finland, 60; mem atmospheric sci new tech comt, Nat Acad Sci, 61-62; adj asst prof, Dept Geol, Univ NC, 64-65, mem vis coun, 64-69, mem water resources res inst, 65-70; ad hoc sci adv comt geophys, US Army, 67-68. Mem: Am Meteorol Soc; Am Geophys Union; Royal Meteorol Soc. Res: Engineering meteorology; micrometeorology; physical oceanography; atmospheric chemistry; environmental sciences. Mailing Add: Res Triangle Inst PO Box 12194 Research Triangle Park NC 27709

WORTH, ROBERT MCALPINE, b Kiangsi Prov, China, Aug 27, 24; US citizen; m 51; c 2. EPIDEMIOLOGY. Educ: Univ Calif, Berkeley, BA, 50, PhD(epidemiol), 62; Univ Calif, San Francisco, MD, 54; Harvard Univ, MPH, 58. Prof Exp: Intern med, Southern Pac Hosp, San Francisco, 54-55; resident family pract, San Mateo Community Hosp, 55-56; physician leprosy, Hawaii Dept Health, 56-57, asst chief chronic dis, 57, health officer, 58-60; PROF EPIDEMIOL, SCH PUB HEALTH, UNIV HAWAII, MANOA, 63- Concurrent Pos: Hooper Found res fel, Univ Hong Kong, 61-63; vis prof, Sch Med, Univ Calif, 69-70. Mem: Am Pub Health Asn; Am Epidemiol Soc. Res: Leprosy epidemiology and control; disease survey methodology; automation medical record systems for epidemiologic and quality control purposes; public health in modern China. Mailing Add: Univ of Hawaii Sch of Pub Health 1960 East-West Rd Honolulu HI 96822

WORTH, ROY EUGENE, b Broxton, Ga, Mar 24, 38; m 65; c 1. FUNCTIONAL ANALYSIS. Educ: Univ Ga, BS, 60, MA, 62, PhD(math), 68. Prof Exp: Asst prof math, WGa Col, 63-68; asst prof, 68-71, ASSOC PROF MATH, GA STATE UNIV, 71- Mem: Am Math Soc; Math Asn Am. Res: Banach semilinear spaces and semialgebras; applications of optimal control theory in economics, health care, etc. Mailing Add: Dept of Math Ga State Univ Atlanta GA 30303

WORTH, WILLIAM SUTHERLAND, b Cleveland, Ohio, Apr 7, 26; m 54; c 3. NEUROSCIENCES, NUTRITION. Educ: Univ Ohio, BS, 51; Univ Wyo, MS, 52. Prof Exp: Res physiologist, Dept Chem, Univ Colo, 54-57 & Med Res & Nutrit Lab, US Dept Army, Colo, 57-63; dir, Lipid Res Lab, Vet Admin Hosp, Denver, 63-67; res physiologist, Dept Pharmacol, 67-75, MEM FAC NEUROL CHEM, DEPT NEUROL, UNIV COLO MED CTR, DENVER, 75- Concurrent Pos: Fulbright-Hays scholar, Nat Res Ctr, Univ Alexandria, 65-66; mem staff, Apollo Biomed Exp Anal & Integration, Martin Marietta Corp, Colo, 67-70. Mem: Soc Nuclear Med; Soc Exp Biol & Med; Am Physiol Soc; Soc Neurosci; Int Soc Neurochem. Res: Neurochemistry, biogenic amines; metabolism and nutrition; international health. Mailing Add:

WORTHEN, HOWARD GEORGE, b Provo, Utah, Dec 21, 25; m 50; c 5. PEDIATRICS, BIOCHEMISTRY. Educ: Brigham Young Univ, AB, 47; Northwestern Univ, MD, 51; Univ Minn, PhD, 61. Prof Exp: From instr to asst prof pediat, Univ Minn, 56-62; assoc prof, Med Col, Cornell Univ, 62-65; PROF PEDIAT, UNIV TEX HEALTH SCI CTR DALLAS, 65- Mem: Am Soc Exp Path; Electron Micros Soc Am; Harvey Soc; Soc Pediat Res; Am Soc Cell Biol. Res: Renal disease; histochemistry; electron microscopy. Mailing Add: Dept of Pediat Univ of Tex Health Sci Ctr Dallas TX 75235

WORTHEN, LEONARD ROBERT, b Woburn, Mass, Dec 28, 25; m 55; c 3. MICROBIOLOGY. Educ: Mass Col Pharm, BS, 50; Temple Univ, MS, 52; Univ Mass, PhD(bact), 57. Prof Exp: Instr pharmacol, Sch Nursing, Holyoke Hosp, 55-57; asst prof pharm, 57-63, assoc prof pharmacog, 63-70, PROF PHARMACOG, UNIV RI, 70-, DIR ENVIRON HEALTH SCI PROG, 72- Mem: Am Soc Microbiol; Am Pharmaceut Asn; Am Soc Pharmacog; AAAS; Am Pub Health Asn. Res: Fungal metabolites, particularly antibiotics and other metabolites of medicinal importance; natural products from marine sources. Mailing Add: Col of Pharm Univ of RI Kingston RI 02881

WORTHINGTON, CHARLES ROY, b Penola, Australia, May 17, 25; m 59; c 3. BIOPHYSICS. Educ: Univ Adelaide, PhD(physics), 55. Prof Exp: Res assoc crystallog, Polytech Inst Brooklyn, 55-57; staff mem biophys, Biophys Res Unit, Med Res Coun, King's Col, Univ London, 58-61; from asst prof to assoc prof physics, Univ Mich, Ann Arbor, 61-69; prof chem & physics, 69-72, PROF PHYSICS & BIOL, MELLON INST SCI, CARNEGIE-MELLON UNIV, 72- Concurrent Pos: Consult, Nat Bur Stand, DC, 57-58. Mem: Biophys Soc; Am Crystallog Asn. Res: Membrane structure; molecular organization of biological systems and theories of biological mechanisms; x-ray biophysics and microscopy. Mailing Add: Mellon Inst of Sci Carnegie-Mellon Univ Pittsburgh PA 15213

WORTHINGTON, EDWARD ARTHUR, b New Orleans, La, Nov 28, 20; m 46; c 3. PHYSICAL CHEMISTRY. Educ: Loyola Univ, BS, 41; St Louis Univ, MS, 43; Univ Pittsburgh, PhD(chem), 48. Prof Exp: Lectr inorg chem, Univ Pittsburgh, 46-47, instr, 47-48; sr res chemist, Allied Chem & Dye Corp, 48-51; res supvr, Kaiser Aluminum & Chem Corp, 51-58, mgr res, 58-69, dir res serv, 69-75; DIR RES & DEVELOP, GRAIN PRODS INC, KRAUSE MILLING CO, 75- Mem: Am Chem Soc. Res: Electrolysis of aqueous solutions and fused salts; extractive metallurgy; thermodynamics of high temperature reactions; process development. Mailing Add: Res & Develop Grain Prods Inc Krause Milling Co PO Box 1156 Milwaukee WI 53201

WORTHINGTON, HARVEY ROBERT, JR, physics, see 12th edition

WORTHINGTON, JAMES BRIAN, b Sandwich, Ill, Nov 29, 43; m 65; c 1. ANALYTICAL CHEMISTRY. Educ: Augustana Col, BA, 65; Purdue Univ, Lafayette, PhD(anal chem), 70. Prof Exp: Res chemist, 70-72, GROUP LEADER ENVIRON ANAL, DIAMOND SHAMROCK CORP, 72- Mem: Am Chem Soc. Res: Kinetic methods of analysis; design of chemical instrumentation; real time computer automation; environmental related analyses. Mailing Add: T R Evans Res Ctr Diamond Shamrock Corp Painesville OH 44077

WORTHINGTON, JOHN THOMAS, III, b Catonsville, Md, Feb 1, 24; m 49; c 2. HORTICULTURE. Educ: Univ Md, BS, 49, MS, 53. Prof Exp: Biol aide, Bur Plant Indust, 49-51, from jr horticulturist to horticulturist, 51-67, sr res horticulturist, 67-74, PRIN HORTICULTURIST, MKT RES DIV, AGR MKT SERV, USDA, 74- Mem: Am Soc Hort Sci. Res: Fruits and vegetables, especially storage, market diseases and field studies on nursery stock, particularly strawberry plants; quality evaluation of fresh fruits and vegetables by light transmission methods. Mailing Add: 4600 Drexel Rd College Park MD 20740

WORTHINGTON, LAWRENCE VALENTINE, b London, Eng, Mar 6, 20; US citizen; m 52; c 2. OCEANOGRAPHY. Prof Exp: Hydrographic technician, 46-50, res assoc phys oceanog, 50-58, phys oceanogr, 58-63, sr scientist, 63-74, CHMN, DEPT PHYS OCEANOG, WOODS HOLE OCEANOG INST, 74- Mem: Am Geophys Union; AAAS. Res: General circulation of ocean and deep water problems. Mailing Add: 540 Woods Hole Rd Woods Hole MA 02543

WORTHINGTON, RICHARD DANE, b Houston, Tex, Sept 20, 41; m 64. MORPHOLOGY, HERPETOLOGY. Educ: Univ Tex, Austin, BA, 63; Univ Md, MS, 66, PhD(zool), 68. Prof Exp: USPHS trainee, Univ Chicago, 68-69; asst prof biol, 69-74, ASSOC PROF BIOL SCI, UNIV TEX, EL PASO, 74- Mem: Soc Syst Zool; Soc Study Evolution; Ecol Soc Am; Am Soc Ichthyologists & Herpetologists; Soc Study Amphibians & Reptiles. Res: Evolutionary biology of caudate amphibians; functional morphology of salamander skeletal system; evolutionary morphology; lizard ecology. Mailing Add: Dept of Biol Sci Univ of Tex El Paso TX 79968

WORTHINGTON, ROBERT EARL, b Kingston, Ga, Jan 2, 29; m 60; c 3. LIPID CHEMISTRY, FOOD SCIENCE. Educ: Berry Col, BSA, 52; NC State Col, MS, 55; Iowa State Univ, PhD(biochem), 62. Prof Exp: Res instr agr & biol chem, NC State Col, 55-56; assoc biochem & biophys, Iowa State Univ, 56-61, asst prof animal sci, 61-64; ASST PROF FOOD SCI, GA STA, UNIV GA, 64- Mem: AAAS; Am Chem Soc; Am Oil Chem Soc. Res: Lipid chemistry of foods. Mailing Add: Dept Food Sci Ga Sta Univ of Ga Experiment GA 30212

WORTHINGTON, THOMAS KIMBER, b New York, NY, Aug 4, 47. LOW TEMPERATURE PHYSICS. Educ: Franklin & Marshall Col, BA, 69; Wesleyan Univ, PhD(physics), 75. Prof Exp: FEL PHYSICS, RUTGERS UNIV, 75- Mem: Am Inst Physics. Res: Specific heat and thermal conductivity of granular aluminum films. Mailing Add: Dept of Physics Rutgers Univ New Brunswick NJ 08903

WORTHINGTON, WARD CURTIS, JR, b Savannah, Ga, Aug 8, 25; m 47; c 2. ANATOMY. Educ: The Citadel, BS, 52; Med Col SC, MD, 52. Prof Exp: Intern surg, Boston City Hosp, Mass, 52-53; instr anat, Sch Med, Johns Hopkins Univ, 53-56; asst prof, Col Med, Univ Ill, 56-57; from asst prof to assoc prof, 57-66, asst dean, 66-69, PROF ANAT, MED UNIV SC, 66-, CHMN DEPT, 69-, ASSOC DEAN, 70-, ACTG VPRES ACAD AFFAIRS, 75- Concurrent Pos: NIH spec fel, Dept Human Anat, Oxford Univ, 64-65. Mem: AAAS; Am Asn Anat; Endocrine Soc; Am Physiol Soc; Microcirc Soc. Res: Anatomy and physiology of pituitary circulation; histology; neuroendocrinology. Mailing Add: Med Univ of SC 80 Barre St Charleston SC 29401

WORTHMAN, ROBERT PAUL, b Chicago, Ill, Aug 3, 19; m 46; c 2. VETERINARY ANATOMY. Educ: Kans State Univ, DVM, 43; Iowa State Univ, MS, 53. Prof Exp: Asst prof vet anat, Wash State Univ, 46-49 & Iowa State Univ, 49-53; asst prof, 53-72, PROF VET ANAT, WASH STATE UNIV, 72- Mem: Am Vet Med Asn; Am Asn Anat; Am Asn Vet Anatomists (pres, 75); World Asn Vet Anatomists (vpres, 75). Res: Veterinary surgical and functional anatomy; techniques of teaching museum specimen preparation. Mailing Add: Dept of Anat Wash State Univ Pullman WA 99163

WORTIS, MICHAEL, b New York, NY, Sept 28, 36; m 64; c 2. THEORETICAL PHYSICS, SOLID STATE PHYSICS. Educ: Harvard Univ, BA, 58, MA, 59, PhD(physics), 63. Prof Exp: Miller fel physics, Univ Calif, Berkeley, 62-64; NSF fel, Fac Sci, Univ Paris, 64-65; vis prof, Pakistan Atomic Energy Comn, WPakistan, 65; res asst prof, 66, from asst prof to assoc prof, 66-73, PROF PHYSICS, UNIV ILL, URBANA, 73- Concurrent Pos: A P Sloan Found fel, Univ Ill, Urbana, 67-69; Ford Found consult, Univ Islamabad, WPakistan, 71. Mem: Am Phys Soc. Res: Statistical physics; magnetic phenomena; phase transitions; phase transitions and related phenomena in strongly coupled and complex systems. Mailing Add: Dept of Physics Univ of Ill Urbana IL 61801

WORTMAN, BERNARD, b Brooklyn, NY, Apr 23, 24; m 52; c 3. BIOCHEMISTRY. Educ: Syracuse Univ, AB, 48; Univ Tex, MA, 51; Ohio State Univ, PhD(physiol), 55. Prof Exp: From res asst to res asst prof ophthal, Sch Med, Wash Univ, 55-65; res assoc prof, Albany Med Col, 65-66; res chemist, Food & Drug Admin, Washington, DC, 66-69; SCIENTIST ADMINR, NAT EYE INST, 69- Concurrent Pos: NIH spec fel, 60-61; estab investr, Am Heart Asn, 61-65; vis scientist, Am Physiol Soc, 63; consult to lab serv, Vet Admin Hosp, Albany, 65-66; res assoc prof, Med Sch, George Washington Univ, 67-70. Mem: Am Physiol Soc; Soc Gen Physiol; Am Soc Cell Biol; Am Chem Soc. Res: Cell physiology; general metabolism of cornea; biosynthesis of sulfated mucopolysaccharides in cornea; biochemistry of connective tissue; macromolecular biochemistry. Mailing Add: Nat Eye Inst Bethesda MD 20014

WORTMAN, LEO STERLING, JR, b Quinlan, Okla, Apr 3, 23; m 49; c 4. PLANT BREEDING. Educ: Okla State Univ, BS, 43; Univ Minn, MS, 48, PhD(plant genetics), 50. Prof Exp: Geneticist in-chg corn breeding, Rockefeller Found, 50-54; head dept plant breeding, Pineapple Res Inst, Hawaii, 55-60; asst dir, Int Rice Res Inst, Rockefeller Found, 60-64; dir, Pineapple Res Inst, Hawaii, 64-65; dir agr sci, 66-70, VPRES, ROCKEFELLER FOUND, 70- Concurrent Pos: Mem agr bd, Nat Acad Sci, 67-71, adv comt biol & med sci, 71-72, bd sci & technol int develop, 73-; vis comt, Dept Nutrit & Food Sci, Mass Inst Technol, 71-74; adv comt res, NSF, 73-74; adv panel agr & rural develop, World Bank, 73-75; chmn, Nat Acad Sci Plant Studies Deleg to People's Repub of China, 74; pres, Int Agr Develop Serv, 75- Honors & Awards: Int Serv Agron Award, Am Soc Agron, 75; Joseph C Wilson Award for Significant Achievement & Contrib in Int Affairs, Rochester Comt for UN, Xerox Corp & Univ Rochester, 75. Mem: Fel AAAS; Am Soc Agron; fel Philippine Asn Advan Sci. Res: Agricultural research and education. Mailing Add: Rockefeller Found 1133 Ave of the Americas New York NY 10036

WORTMAN, ROGER MATTHEW, b Chicago, Ill, Dec 11, 32; m. PHYSICS. Educ: Univ Chicago, PhD(phys chem), 57. Prof Exp: Res physicist, Carnegie Inst Technol, 57-61; res scientist, Douglas Aircraft Co, Inc, 61-63; criminalist, Oakland Police Dept, Calif, 63-67; CHEMIST, SCI LIAISON & ADV GROUP, US ARMY, 67- Res: Field emission microscopy and radiation damage of surfaces; aerospace devices; scientific criminal investigation; military technology; research administration. Mailing Add: 3100 Twin Ct Bowie MD 20715

WORTS, GEORGE FRANK, JR, b Toledo, Ohio, Apr 24, 16; m 50; c 1. GEOLOGY. Educ: Stanford Univ, BS, 39. Prof Exp: Geologist, Ground Water Br, 40-51, geologist in charge, Long Beach, 52-56, dist geologist, Sacramento, 56-58, br area chief, Pac Coast Area, 58-62, DIST CHIEF, NEV, US GEOL SURV, 62- Mem: Fel Geol Soc Am; Am Geophys Union; Asn Eng Geol. Res: National and international hydrology, especially of arid regions; ground-water resources, particularly quantitative analysis, water quality, coastal hydrology and water management; direction of complex hydrologic studies and applied research. Mailing Add: PO Box 2222 Carson City NV 98701

WORZEL, JOHN LAMAR, b West Brighton, NY, Feb 21, 19; m 41; c 4. GEOPHYSICS. Educ: Lehigh Univ, BS, 40; Columbia Univ, MA, 48, PhD, 49. Prof Exp: Res assoc, Woods Hole Oceanog Inst, 40-46; geodesist, Columbia Univ, 47-49, res assoc geol, 48-49, instr, 49-51, asst prof, 51-52, assoc prof geophys, 52-57, prof, 57-72, asst dir, Lamont Geol Observ, 51-62, actg dir, 63, assoc dir, 64-72; PROF GEOPHYS & DEP DIR EARTH & PLANETARY SCI DIV, MARINE BIOMED

INST, UNIV TEX MED BR GALVESTON, 72-, DIR GEOPHYS LAB, MARINE SCI INST, 74- Concurrent Pos: Geophys consult, Off Naval Res, 50; Guggenheim Found fel, 63; chmn spec study group, Int Union Geod & Geophys. Mem: AAAS; Seismol Soc Am; Soc Explor Geophys; Am Geophys Union; Am Asn Petrol Geologists. Res: Gravity and seismic observations at sea; underwater photography; oceanography. Mailing Add: Geophys Lab Marine Sci Inst Univ of Tex 700 The Strand Galveston TX 77550

WOS, JOHN DAVID, bio-organic chemistry, see 12th edition

WOS, LAWRENCE THOMAS, mathematics, see 12th edition

WOSILAIT, WALTER DANIEL, b Racine, Wis, Feb 4, 24; m 48; c 1. PHARMACOLOGY, BIOCHEMISTRY. Educ: Wabash Col, BA, 49; Johns Hopkins Univ, PhD(biol), 53. Prof Exp: Jr instr pharmacol, Western Reserve Univ, 53-56; asst prof, State Univ NY Downstate Med Ctr, 56-63, assoc prof, 63-65; assoc prof, 65-66, PROF PHARMACOL, SCH MED, UNIV MO-COLUMBIA, 66- Concurrent Pos: USPHS res grant, State Univ NY Downstate Med Ctr, 56-65; USPHS res grant, Sch Med, Univ Mo-Columbia, 65- Mem: Am Soc Biol Chemists; Am Soc Pharmacol & Exp Therapeut; Am Soc Clin Pharmacol & Therapeut; Soc Exp Biol & Med; Harvey Soc. Res: Anticoagulant interactions. Mailing Add: Dept of Pharmacol Univ Mo Sch of Med Columbia MO 65201

WOSKE, HARRY MAX, b Reading, Pa, Feb 26, 24; m 72; c 7. INTERNAL MEDICINE, CARDIOLOGY. Educ: Columbia Univ, AB, 45; Long Island Col Med, MD, 48. Prof Exp: Instr med, Sch Med, Univ Pa, 55-57, assoc, 57-64, from adj asst prof to adj assoc prof, 64-69, assoc prof med & assoc dean, 69-73; assoc dean, 74, PROF MED & CHIEF MED SERV, COL MED & DENT NJ-RUTGERS MED SCH, 73-, ACTG DEAN, 75- Concurrent Pos: NIH fel cardiol, Sch Med, Univ Pa, 55-57. Mem: Fel Am Col Physicians; fel Am Col Cardiol; fel NY Acad Sci. Res: Exercise projects. Mailing Add: Dept of Med Raritan Valley Hosp Green Brook NJ 08812

WOSTMANN, BERNARD STEPHAN, b Amsterdam, Neth, Nov 6, 18; US citizen; m 46; c 5. BIOCHEMISTRY, NUTRITION. Educ: Univ Amsterdam, BS, 40, MS, 45, DSc, 48. Prof Exp: Instr org chem, Univ Amsterdam, 43, instr biochem, 45-58, lectr, 48-50, sci off, 48-55; from asst prof to assoc prof, 55-65, PROF BIOCHEM, UNIV NOTRE DAME, 65- Concurrent Pos: Asst dir, Neth Inst Nutrit, 48-55; Rockefeller res fel, 50-51. Mem: AAAS; Am Inst Nutrit; Soc Exp Biol & Med; Asn Gnotobiotics (pres, 66-68); NY Acad Sci. Res: Biochemical background of host-contaminant relationship; role of intestinal flora in nutrition. Mailing Add: Dept of Microbiol Lobund Lab Univ of Notre Dame Notre Dame IN 46556

WOTHERSPOON, NEIL, b New York, NY, Oct 24, 30; m 54; c 1. PHYSICAL CHEMISTRY, INSTRUMENTATION. Educ: Polytech Inst Brooklyn, BS, 52, PhD, 57. Prof Exp: Res scientist, Radiation & Solid State Lab, NY Univ, 57-68; ASST PROF BIOPHYS & BIOENG, MT SINAI SCH MED, 68- Mem: Am Chem Soc. Res: Instrumentation for physics, chemistry and biomedical sciences, including optical electronic and computer techniques for automated analysis; analog and digital data acquisition, interfacing and processing. Mailing Add: Mt Sinai Sch of Med New York NY 10029

WOTIZ, HERBERT HENRY, b Vienna, Austria, Oct 8, 22; nat US; m 47; c 3. ORGANIC CHEMISTRY, BIOCHEMISTRY. Educ: Providence Col, ScB, 44; Yale Univ, PhD(org chem), 51. Prof Exp: Instr, 51-53, asst prof, 53-55, assoc res prof, 55-61, assoc prof, 61-63, PROF BIOCHEM, SCH MED, BOSTON UNIV, 63-; RES ASSOC, R D EVANS MEM HOSP, 55- Concurrent Pos: Res fel, R D Evans Mem Hosp, 50-55; USPHS sr res fel, 60-64 & res career develop fel, 65-69. Mem: Am Chem Soc; Am Soc Biol Chem; Endocrine Soc; Am Asn Cancer Res. Res: Steroid metabolism and analysis; mechanism of hormone action, action of impeding estrogens. Mailing Add: Dept of Biochem Boston Univ Sch of Med Boston MA 02118

WOTIZ, JOHN HENRY, b Moravska Ostrava, Czech, Apr 12, 19; nat US; m 45; c 3. ORGANIC CHEMISTRY. Educ: Furman Univ, BS, 41; Univ Richmond, MS, 43; Ohio State Univ, PhD(org chem), 48. Prof Exp: Asst chem, Univ Richmond, 41-43 & Ohio State Univ, 43-44 & 46-47; from instr to assoc prof, Univ Pittsburgh, 48-57; prof chem & chmn dept, Marshall Univ, 62-67; PROF CHEM & CHMN DEPT, SOUTHERN ILL UNIV, 67- Concurrent Pos: With Fed Security Agency, 44; Nat Acad Sci exchange prof, USSR, 69, Bulgaria, Hungary, Poland, Rumania & Yugoslavia, 72; mem int activ comt, Am Chem Soc, 75- Mem: Am Chem Soc. Res: Propargylic rearrangement; radical ions from vicinal diamines; institutional research in eastern socialist southeast Asian and Pacific Ocean countries. Mailing Add: Dept of Chem Southern Ill Univ Carbondale IL 62901

WOTMAN, STEPHEN, b Pelham, NY, Aug 5, 31; m 60; c 2. DENTISTRY. Educ: Univ Pa, DDS, 56. Prof Exp: Res asst stomatol, 62-66, asst clin prof dent & commun health, 68-70, ASSOC PROF PREV DENT & ASST DEAN ADMIN AFFAIRS, SCH DENT & ORAL SURG, COLUMBIA UNIV, 70- Concurrent Pos: Nat Inst Dent Res grant, 68-73; co-investr, Hypertension Ctr, 70-75. Mem: AAAS; Int Asn Dent Res. Res: Salivary composition, control and relationships to systemaic disease; cystic fibrosis; hypertension; digitalis effects; hormonal influence; Parkinson's disease as well as the dental phenomena; plaque, caries and calculus as related to saliva. Mailing Add: Sch of Dent & Oral Surg Columbia Univ New York NY 10034

WOTSCHKE, DETLEF, b Treuenbrietzen, Ger, Apr 14, 44; m 69. COMPUTER SCIENCE, SYSTEMS SCIENCE. Educ: Univ Braunschweig, Vordiplom, 67, Diplom, 69; Univ Calif, Los Angeles, PhD(syst sci), 74. Prof Exp: Mgr safety eng, Tebera Tech Res & Consult, Braunschweig, 65-70; teaching & res assoc syst sci, Univ Calif, Los Angeles, 71-74; ASST PROF COMPUT SCI, PA STATE UNIV, 74- Concurrent Pos: Ger Acad Exchange Serv fel, 70; chmn arrangements, Eighth Ann Symp Theory Comput, Asn Comput Mach, 75-76. Mem: Asn Comput Mach; Ger Soc Comput Sci. Res: Formal languages; automata theory; theory of grammars and parsing; stochastic systems. Mailing Add: Dept of Comput Sci Pa State Univ University Park PA 16802

WOTT, JOHN ARTHUR, b Fremont, Ohio, Apr 10, 39; m 59; c 3. HORTICULTURE. Educ: Ohio State Univ, BS, 61; Cornell Univ, MS, 66, PhD(hort), 68. Prof Exp: Instr, Coop Exten Serv, Ohio State Univ, 61-64; res asst hort, Cornell Univ, 64-68; asst prof, 68-73, ASSOC PROF HORT, PURDUE UNIV, WEST LAFAYETTE, 73- Concurrent Pos: County exten agent 4-H, Wood County, Bowling Green, Ohio, 61-64. Mem: Am Soc Hort Sci; Am Hort Soc; Int Plant Propagators Soc. Res: Nutrition of cuttings during propagation; application of nutrient mist; horticultural problems of homeowners, such as foliage plants, annuals and perennials; herbicides in annual flowers; morphological and physiological response of woody plants to flooding. Mailing Add: Dept of Hort Purdue Univ West Lafayette IN 47906

WOTTON, ROBERT MOORE, b New York, NY, Nov 13, 09; m 40; c 2. ANATOMY. Educ: Columbia Univ, AB, 38, AM, 39; Univ Pittsburgh, PhD, 51. Prof Exp: Lab instr parasitol, NY Med Col, 39; instr anat, Essex Col Med & Surg, 45-46; instr biol, Long Island Univ, 46-48; asst prof zool, Duquesne Univ, 48-51 & Univ Pittsburgh, 51-56; assoc prof anat & zool, Univ Nebr, 56-60; assoc prof anat, Sch Med, La State Univ, 60-63; ASSOC PROF ZOOL & PHYSIOL, UNIV NEBR-LINCOLN, 63- Mem: Am Asn Anatomists; Am Soc Cell Biol. Res: Cytophysiology of fats. Mailing Add: 2641 N 65th St Lincoln NE 68507

WOUK, ARTHUR, b New York, NY, Mar 25, 24; m 44; c 2. MATHEMATICS. Educ: City Col New York, BS, 43; Johns Hopkins Univ, MA, 47, PhD(math), 51. Prof Exp: Instr math, Johns Hopkins Univ, 47-50 & Queens Col, NY, 50-52; mathematician, Proj Cyclone, Reeves Instrument Corp, 52-54; supvr math anal sect, Missile Systs Lab, Sylvania Elec Prod, Inc, 54-56, sr eng specialist, Appl Res Lab, Sylvania Electronic Systs Div, Gen Tel & Electronics Corp, 56-62; vis prof, Math Res Ctr, Univ Wis, 62-63; from assoc prof to prof appl math, Northwestern Univ, Evanston, 63-72; PROF COMPUT SCI & CHMN DEPT, UNIV ALTA, 72- Concurrent Pos: Ed, Commun, Asn Comput Mach, 58-64; consult, Appl Res Lab, Sylvania Electronic Systs Div, Gen Tel & Electronics Corp & Argonne Nat Lab, 63-69; ed, SIAM Rev, Soc Indust & Appl Math, 63- Mem: Am Math Soc; Soc Indust & Appl Math; Opers Res Soc Am; Asn Comput Mach. Res: Optimal control theory; numerical and functional analysis. Mailing Add: Dept of Comput Sci Univ of Alta Edmonton AB Can

WOURMS, JOHN BARTON, b New York, NY, Apr 30, 37; m 72. CELL BIOLOGY, DEVELOPMENTAL BIOLOGY. Educ: Fordham Univ, BS, 58, MS, 60; Stanford Univ, PhD(biol), 66. Prof Exp: Am Cancer Soc fel, Harvard Univ, 66-68; Nat Res Coun Can grant & asst prof cell develop biol, Dept Biol, McGill Univ, 68-71, res assoc, Dept Path, 71-72; ASSOC RES SCIENTIST, NY OCEAN SCI LAB, 72- Concurrent Pos: Vis lectr, Biol Labs, Harvard Univ, 71-72. Mem: Am Soc Cell Biol; Am Soc Ichthyologists & Herpetologists; Soc Develop Biol; Int Soc Develop Biologists; Marine Biol Asn UK. Res: Cell differentiation; cell ultrastructure; reproduction and development of fishes and marine invertebrates; oogenesis; ultrastructure and chemistry of extra-cellular matrices; biology of annual fishes; evolutionary biology; elasmobranch biology; marine biomedical resources. Mailing Add: PO Box 467 Abraham's Path Amagansett NY 11930

WOUTERS, LOUIS FRANCIS, nuclear physics, see 12th edition

WOYCHIK, JOHN HENRY, b Scranton, Pa, Mar 30, 30; m 53; c 2. BIOCHEMISTRY, PHYSIOLOGY. Educ: Univ Scranton, BS, 53; Univ Tenn, MS, 55, PhD(biochem), 57. Prof Exp: Chemist, Northern Regional Lab, 57-63, prin chemist, Eastern Mkt & Res Div, 63-74, CHIEF, DAIRY LAB, EASTERN REGIONAL RES CTR, USDA, 74- Mem: Am Chem Soc; Am Dairy Sci Asn; Am Soc Biol Chemists; Inst Food Technologists. Res: Isolation and characterization of milk and cereal proteins; glycoproteins; basic research on milk components and dairy product development. Mailing Add: Eastern Regional Res Ctr USDA 600 E Mermaid Lane Philadelphia PA 19118

WOYSKI, MARGARET SKILLMAN, b West Chester, Pa, July 26, 21; m 48; c 4. GEOLOGY. Educ: Wellesley Col, BA, 43; Univ Minn, MS, 45, PhD(geol), 46. Prof Exp: Instr geol, Univ Minn, 46; geologist, Mo Geol Surv, 46-48; instr geol, Univ Wis, 48-52; lectr, Calif State Col, Long Beach, 63-67; from asst prof to assoc prof, 67-74, PROF GEOL & CHMN DEPT EARTH SCI, CALIF STATE UNIV, FULLERTON, 74- Concurrent Pos: Lectr geol, Calif State Univ, Fullerton, 66-67. Mem: Mineral Soc Am; Nat Asn Geol Teachers; Geol Soc Am. Res: Intrusive rocks of central Minnesota; Precambrian sediments of Missouri; laboratory manuals for historical and physical geology. Mailing Add: Dept of Earth Sci Calif State Univ 800 N State College Blvd Fullerton CA 92634

WOZENCRAFT, PAUL, b Columbus, Ohio, May 23, 09. PATHOLOGY. Educ: Univ Cincinnati, AB, 34, MD, 39. Prof Exp: Intern, Cincinnati Gen Hosp, 38-39; assoc prof, 49-61, PROF PATH, COL MED, UNIV CINCINNATI, 61- Concurrent Pos: Fel, Mayo Found, 39-41 & 46-47; fel, Mem Hosp, New York, 47-48; attend pathologist, Cincinnati Gen Hosp, 49-; consult pathologist, Vet Admin, 50- Mem: Am Asn Clin Path; Int Acad Path. Res: General and surgical pathology. Mailing Add: Dept of Path Cincinnati Gen Hosp Cincinnati OH 45229

WOZNIAK, WAYNE THEODORE, b Chicago, Ill, Oct 13, 45; m 71. CHEMISTRY, SPECTROSCOPY. Educ: Ill Benedictine Col, BS, 67; Fla State Univ, PhD(phys inorg chem), 71. Prof Exp: Res assoc phys chem, Princeton Univ, 71-73; res assoc chem physics, Univ Ill, 73-75; RES ASSOC BIOPHYS, AM DENT RES INST, 75- Concurrent Pos: Instr, Univ Ill, 74. Mem: Am Chem Soc; AAAS; Soc Appl Spectros. Res: Flourescence spectroscopy of dental materials and calcified tissue; optical changes in tissue due to aging; resonance Raman spectroscopy; applications of vibrational spectroscopy to biological systems and organometallic chemistry. Mailing Add: Am Dent Asn Res Inst 211 E Chicago Ave Chicago IL 60611

WOZNICK, BENJAMIN JOSEPH, b Jersey City, NJ, Aug 8, 36; m 59; c 1. THEORETICAL PHYSICS, COMPUTER SCIENCE. Educ: Mass Inst Technol, BS, 57, PhD(physics), 63. Prof Exp: Sr scientist, Avco Everett Res Lab, 62-65; sr scientist, Geosci Inc, Ampex Corp, 65-70, vpres, 68-70; proj dir, Trans Technol Inc, 70-72; dir opers, 72-73, vpres, 73-75, PRIN COMPUT SCIENTIST ENVIRON RES & TECHNOL, COMPUT LIBR SERV INC, 75- Mem: Am Phys Soc; Inst Elec & Electronics Engrs; Asn Comput Mach. Res: Applications of computers to scientific calculations, data communication and processing. Mailing Add: 259 Marlborough St Boston MA 02116

WRAGG, LAURENCE EDWARD, b Bowmanville, Ont, Aug 12, 21; m 56 & 63; c 6. ANATOMY, FAMILY MEDICINE. Educ: McMaster Univ, BA, 46; Univ Toronto, MA, 51; Univ Wis, PhD(anat), 55; Northwestern Univ, MD, 62. Prof Exp: Asst wildlife conserv, Ont Res Comn, 46-47; asst prof biol, Mt Allison Univ, 48-50; asst endocrinol, Roscoe B Jackson Lab, 50-51; asst anat, Univ Wis, 54-55 & dent br, Univ Tex, 55-58; instr, Northwestern Univ, 58-61; intern, Spokane, Wash, 62-63; from asst prof to assoc prof, Med Sch, Northwestern Univ, 63-71; gen pract, Calif, 71- Concurrent Pos: Gen pract residency, Contra Costa County Hosp, 71-72; chief of staff, Eskaton Colusa Healthcare Ctr, 75. Mem: AAAS; Am Asn Anat. Res: Pineal physiology; development of the face. Mailing Add: 124 E Webster St Colusa CA 95932

WRAIGHT, AARON JOSEPH, geography, physical geography, see 12th edition

WRAIGHT, COLIN ALLEN, b London, Eng, Nov 27, 45. BIOPHYSICS. Educ: Univ Bristol, BSc, 67, PhD(biochem), 71. Prof Exp: Fel biophys, State Univ Leiden, 71-72; assoc, Cornell Univ, 72-74; asst prof biol, Univ Calif, Santa Barbara, 74-75; ASST PROF BOT & BIOPHYS, UNIV ILL, URBANA-CHAMPAIGN, 75- Mem: Am Soc Photobiol. Res: Membrane functions and mechanisms of electron and ion transport in biological energy conservation. Mailing Add: Depts of Bot & Biophys 255 Morrill Hall Univ of Ill Urbana IL 61801

WRANGELL, LEWIS J, b Milwaukee, Wis, July 30, 14; m 40; c 1. ANALYTICAL

CHEMISTRY. Educ: Marquette Univ, BS, 36, MS, 39. Prof Exp: Chemist, Res Labs, Allis-Chalmers Mfg Co, Wis, 42-45, chemist-in-chg anal lab, 45-59, res chemist, 59-69; chief chem eng & sci, Geo J Meyer Mfg, 69-70; indust consult, 70-72; ANAL CHEMIST, REDUCTION SYSTS DIV, ALLIS-CHALMERS CORP, 72- Mem: Fel Am Inst Chemists; Am Chem Soc. Res: Analytical chemistry of metallic materials with solids and gases at elevated temperatures; analysis of unusual and new compounds; analysis of reduced iron ore. Mailing Add: 2325 N 86th St Wauwatosa WI 53226

WRASIDLO, WOLFGANG JOHANN, b Beuthen, Ger, Oct 31, 38; US citizen; m; c 2. POLYMER CHEMISTRY. Educ: San Diego State Univ, BA, 63; Univ Nürnberg, MS, 67. Prof Exp: Sr res chemist, Whittaker Corp, 63-65; res chemist, US Naval Ord Lab, 67-69; staff mem polymers, Sci Res Labs, Boeing Co, 69-71; STAFF CHEMIST, UOP, 71- Honors & Awards: Civilian Serv Award, US Naval Ord Lab, 70. Mem: Am Chem Soc. Res: Solid state polymers; synthesis and characterization of high temperature polymers; mechanical, thermal and morphological behavior of polymers; reverse osmosis membranes. Mailing Add: UOP 4901 Morena Blvd San Diego CA 92117

WRATHALL, CARL RICHARD, biochemical genetics, see 12th edition

WRATHALL, DONALD PRIOR, b Pittsburgh, Pa, Mar 9, 36; m 62; c 4. PHYSICAL CHEMISTRY. Educ: Brigham Young Univ, BA, 64, PhD(phys chem), 68. Prof Exp: NIH fel, Yale Univ, 67-68; SR RES CHEMIST, EASTMAN KODAK CO, 68- Res: Thermodynamics of reactions at solid-liquid interfaces and with macromolecules in solution. Mailing Add: Res Labs Eastman Kodak Co 1669 Lake Ave Rochester NY 14650

WRATHALL, JAY W, b Salt Lake City, Utah, May 12, 33; m 69; c 2. INORGANIC CHEMISTRY. Educ: Brigham Young Univ, BS, 57, MS, 59; Ohio State Univ, PhD(inorg chem), 62. Prof Exp: Asst prof chem, Univ Calif, Berkeley, 62-64 & Univ Hawaii, 64-69; assoc prof, Church Col Hawaii, 69-73, prof & div chmn, 73-74, PROF CHEM & DIV CHMN, BRIGHAM YOUNG UNIV, HAWAII CAMPUS, 74- Mem: Am Chem Soc. Res: Coordination chemistry; reactions of coordinated ligands; biological activity of transition metal complexes. Mailing Add: Brigham Young Univ Hawaii Campus 55-220 Kulanui St Laie HI 96762

WRATHALL, JEAN REW, b Brooklyn, NY, Dec 3, 42; m 60; c 2. GENETICS, CELL BIOLOGY. Educ: Univ Utah, BS, 64, PhD(genetics, molecular biol), 69. Prof Exp: Asst prof biol, State Univ NY Col Geneseo, 69-70; from instr to asst prof genetics, Med Col, Cornell Univ, 70-74; res assoc, 74-75, ASST PROF ANAT, MED COL, GEORGETOWN UNIV, 75- Concurrent Pos: Damon Runyon Mem Fund fel, Cornell Univ, 71- Mem: AAAS; Soc Develop Biol; Soc Cell Biol; Tissue Cult Asn. Res: Control of differentiated function in eukaryotic cells in culture; abnormal functions in malignant cells in culture; 5-bromodeoxyuridine suppression of melanin synthesis and tumorigenicity of melanoma cells. Mailing Add: Dept of Anat Georgetown Univ Med Col Washington DC 20007

WRATTEN, CRAIG CHARLES, biochemistry, see 12th edition

WRAY, GRANVILLE WAYNE, b Elk City, Okla, Dec 16, 41; m 65. CELL BIOLOGY, BIOCHEMISTRY. Educ: Phillips Univ, BS, 63; Okla State Univ, MS, 65; Univ Tex, PhD(cell biol), 70. Prof Exp: Res asst biochem, Okla State Univ, 63-65; res asst, Univ Tex M D Anderson Hosp & Tumor Inst, 65-66; ASST PROF CELL BIOL, BAYLOR COL MED, 72- Concurrent Pos: Damon Runyon Cancer res fel, McArdle Lab Cancer Res, Univ Wis-Madison, 71-72. Honors & Awards: Mike Hogg Award, 68. Mem: AAAS; Am Chem Soc; Am Soc Cell Biol. Res: Isolation, morphology and biochemistry of the mammalian metaphase chromosome. Mailing Add: Dept of Cell Biol Baylor Col of Med Houston TX 77025

WRAY, JAMES DAVID, b Norton, Kans, Oct 3, 36; c 1. ASTRONOMY. Educ: Univ MNex, BS, 59; Univ Cincinnati, MS, 62; Northwestern Univ, PhD(astron), 66. Prof Exp: Res assoc meteoritics, Univ NMex, 62-64, dir, Inst Meteoritics, 66-67; asst prof astron, Northwestern Univ, 67-72; RES SCIENTIST ASTRON, UNIV TEX, AUSTIN, 72- Concurrent Pos: Consult, Boller & Chivens Div, Perkin-Elmer Corp, 73- Mem: Int Astron Union; Am Astron Soc. Res: Space astronomy, ultraviolet stellar spectroscopy; extra-galactic research, surface distribution of color in galaxies; digital image processing; numerical data base management and applications. Mailing Add: Dept of Astron Univ of Tex Austin TX 78712

WRAY, JOE D, b Conway, Ark, Sept 30, 26; m 51; c 5. PEDIATRICS, PUBLIC HEALTH. Educ: Stanford Univ, BA, 47, MD, 52; Univ NC, MPH, 67; Am Bd Pediat, dipl. Prof Exp: Intern, Charity Hosp La, New Orleans, 51-52; intern & resident pediat, Grace-New Haven Community Hosp, Conn, 54-56; chief resident, Hacettepe Children's Hosp, Ankara, Turkey, 56-58, assoc pediatrician, 58-61; vie prof pediat, Fac Med, Univ Valle, Colombia, 61-66; MEM ROCKEFELLER FOUND FIELD STAFF BIOMED SCI, RAMATHIBODI HOSP MED SCH, MIHIDOL UNIV, THAILAND, 67- Concurrent Pos: Mem field staff biomed sci, Rockefeller Found, NY, 60- Mem: Fel Am Acad Pediat; Am Pub Health Asn; Int Epidemiol Asn; Int Union Nutrit Sci. Res: Preschool child nutrition; growth and development; delivery of health services to children in developing countries; family planning. Mailing Add: Rockefeller Found GPO Box 2453 Bangkok Thailand

WRAY, JOE WILLIE, b Cedartown, Ga, Oct 27, 12. MATHEMATICS. Educ: Univ Ga, BS, 39; Univ NC, MS, 44; Univ Ill, MS, 48, PhD(math), 52. Prof Exp: Instr math, NC State Col, 42-44; prof, Flora Macdonald Col, 44-46, instr, NC State Col, 46-47; asst, Univ Ill, 47-50; asst prof, Univ Idaho, 50-56; asst prof, 56-59, ASSOC PROF MATH, GA INST TECHNOL, 59- Mem: Am Math Soc. Res: Analysis; nonanalytic functions. Mailing Add: Dept of Math Ga Inst of Technol Atlanta GA 30332

WRAY, JOHN LEE, b Charleston, WVa, July 10, 25; m 52; c 1. GEOLOGY. Educ: Univ WVa, BS, 50, MS, 51; Univ Wis, PhD(geol), 56. Prof Exp: Teaching asst geol, Univ WVa, 48-51; geologist, WVa State Hwy Dept, 51-53; res asst chem, Univ Wis, 53-56; RES ASSOC, RES CTR, MARATHON OIL CO, 56- Concurrent Pos: Adj prof geol, Colo Sch Mines, 70- Mem: Geol Soc Am; Paleont Soc; Am Asn Petrol Geol; Soc Econ Paleont & Mineral. Res: Paleontology; fossil algae; carbonate sedimentation, petrology and geochemistry. Mailing Add: Marathon Oil Co Res Ctr PO Box 269 Littleton CO 80120

WRAY, KURT LEO, physical chemistry, see 12th edition

WRAY, VIRGINIA LEE POLLAN, b Grove, Okla, Mar 20, 40; m 65. BIOCHEMISTRY. Educ: Okla State Univ, BS, 62, MS, 66; Univ Tex Grad Sch Biomed Sci Houston, PhD(biochem), 70. Prof Exp: Res asst biochem virol, Col Med, Baylor Univ, 66; ASST PROF CELL BIOL, BAYLOR COL MED, 73- Concurrent Pos: Fel, McArdle Lab Cancer Res, Univ Wis-Madison, 70-72. Mem: Am Soc Cell Biol; Am Chem Soc. Res: Membrane biochemistry; composition and function of nuclear and plasma membranes. Mailing Add: Dept of Cell Biol Baylor Col of Med Houston TX 77030

WREDE, DON EDWARD, b Cincinnati, Ohio, Feb 13, 30; m 50; c 5. RADIOLOGICAL PHYSICS, BIOPHYSICS. Educ: Univ Cincinnati, BS, 57, MS, 60, PhD(biophys), 63. Prof Exp: Tech engr, Gen Elec Co, Ohio, 48-54; lectr physics, Univ Cincinnati, 58-62; assoc prof & chmn dept, Heidelberg Col, 62-66; Fulbright lectr, Cheng Kung Univ, Taiwan, 66-68; asst prof radiation med, 69-72, ASSOC PROF RADIATION MED & CHIEF CLIN PHYSICIST, A B CHANDLER MED CTR, UNIV KY, LEXINGTON, 72-, RADIOL PHYSICIST, 69- Concurrent Pos: NIH res fel, Radioisotope Lab, Gen Hosp, Univ Cincinnati, 68-69; med physics liaison, Australia & Taiwan, 74-; vis lectr, Bowling Green State Univ, 63-64; vis scientist, NSF, 69- Mem: Radiation Res Soc; Am Asn Physicists in Med (secy-treas, 71-); Radiol Soc NAm; Am Col Radiol. Res: Medical radiation dosimetry, particularly the study of irregular shaped fields from teletherapy devices; chemical protection from radiation; mathematical models of biological systems, particularly the endocrine. Mailing Add: Dept of Radiation Med Univ of Ky A B Chandler Med Ctr Lexington KY 40506

WREDE, JAMES ALLEN, organic chemistry, see 12th edition

WREDE, ROBERT C, JR, b Cincinnati, Ohio, Oct 19, 26; m 48; c 3. MATHEMATICS. Educ: Miami Univ, Ohio, BS, 49, MA, 50; Ind Univ, PhD(math), 56. Prof Exp: Instr math, Miami Univ, Ohio, 50-51; from instr to assoc prof, 55-63, PROF MATH, CALIF STATE UNIV, SAN JOSE, 63- Concurrent Pos: Consult phys & res lab, Int Bus Mach Corp, Calif, 56-58; Hunter's Point Radiation Lab, 60. Mem: Am Math Soc; Math Asn Am; Tensor Soc. Res: Relativity theory; differential geometry; vector and tensor analysis. Mailing Add: Dept of Math McQuarry Hall Calif State Univ San Jose CA 95114

WREN, FRANK LYNWOOD, b Yorkville, Tenn, Sept 28, 94; m 21; c 2. MATHEMATICS. Educ: Univ of the South, AB, 15; George Peabody Col, MA, 25; Univ Chicago, PhD(math), 30. Prof Exp: Teacher high sch, Tenn, 15-17; asst math, Johns Hopkins Univ, 17; teacher high sch, Tenn, 19-24; asst statist, George Peabody Col, 25, from assoc prof to prof math, 27-59, head dept, 27-59, JULIA A SEARS EMER PROF MATH, GEORGE PEABODY COL, 59-; EMER PROF MATH, CALIF STATE UNIV, NORTHRIDGE, 75- Concurrent Pos: Vis prof, Univ Calif, Los Angeles, 58; lectr NSF Insts; consult, Indian Springs Sch; prof math, Calif State Univ, Northridge, 59-75; Southwestern Regional Lab Educ Res & Develop Rev panelist, NSF, 60-61. Mem: Fel AAAS; Math Asn Am. Res: Calculus of variations; trigonometry; arithmetic; teaching of arithmetic and secondary mathematics; basic algebraic concepts. Mailing Add: 8956 Etiwanda Ave Northridge CA 91324

WREN, HENRY K, b Loganton, Pa, July 26, 26; m 61. ORGANIC CHEMISTRY. Educ: Fremont Col, BS, 56; Milan Polytech Inst, PhD(chem), 60; Am Inst Chemists, PC-A. Prof Exp: Head, Resin Dept, Perfection Paint & Chem Co, 60-61; chemist, Am Art Clay Co, 61-62; head sect, Sherwin-Williams Res Ctr, 62-64; dir res, Baltimore Paint & Chem Corp, 64-66; mat engr, RCA Magnetic Prod, 67; chemist, Naval Avionics Facil, Ind, 67-69 & Wren Res & Develop Corp, Mich, 69-70; mem staff, Sargent Paint, 70-75; CHIEF CHEMIST TRADE SALES, PERFECTION PAINT & COLOR CO, 75- Concurrent Pos: Teacher, Purdue Univ, Indianapolis, 68-69 & Ind Cent Col, 68-69. Mem: AAAS; fel Am Inst Chemists; Am Chem Soc; Info Coun Fabric Flammability. Res: Polymers; organic synthesis. Mailing Add: 8521 Lamira Lane Indianapolis IN 46234

WRENCH, JOHN WILLIAM, JR, mathematics, see 12th edition

WRENN, MCDONALD EDWARD, b New York, NY, Apr 16, 36; m 67; c 2. ENVIRONMENTAL HEALTH, RADIOLOGICAL HEALTH. Educ: Princeton Univ, AB, 58; NY Univ, MS, 62, PhD(nuclear eng, environ health), 67. Prof Exp: Res scientist, 62-67, from instr to asst prof, 67-72, ASSOC PROF ENVIRON MED, MED CTR, NY UNIV, 72-; BIOMED SCIENTIST RADIOBIOL, DIV BIOMED & ENVIRON RES, US ENVIRON RES & DEVELOP ADMIN, 73- Concurrent Pos: Mem sci comt environ radiation measurements & sci comt maximum permissible concentrations for occupation exposures, Nat Coun Radiation Protection & Measurements, 71-; chmn N-13 comt, Am Nat Standards Inst, 72- Mem: Radiation Res Soc; Health Physics Soc; fel Am Pub Health Asn; Am Inst Biol Sci; NY Acad Sci. Res: Biological effects of environmental agents on man and animals, particularly radiations and radioactive materials; environmental cycling and transport of trace and radioactive elements; mammalian metabolism of actinides and development of environmental radiation detection instruments. Mailing Add: Dept of Environ Med NY Univ Med Ctr New York NY 10016

WRENN, SAMUEL NATHANIEL, b Vance Co, NC, Dec 15, 04; wid. ORGANIC CHEMISTRY. Educ: Duke Univ, AB, 27, AM, 29; Pa State Col, PhD(org chem), 35. Prof Exp: Asst chem, Duke Univ, 27-29; asst, Pa State Col, 29-34, instr exten div, 34-36; asst prof, The Citadel, 30-31; asst prof, 36-42, actg dean, Univ Jr Col, 38-39, from assoc prof to prof, 42-70, EMER PROF CHEM, GEORGE WASHINGTON UNIV, 70- Mem: AAAS; Am Chem Soc; fel Am Inst Chemists. Res: Aliphatic and synthetic organic chemistry; organic nitrogen compounds; tertiary compounds; polymerized olefins. Mailing Add: 3130 Fifth St N Arlington VA 22201

WRENSHALL, GERALD ALFRED, b North Battleford, Sask, Apr 3, 12; m 42. MEDICINE. Educ: Univ Sask, BSc, 33, MSc, 35; Yale Univ, PhD(nuclear physics), 39; Univ Toronto, MA, 47, PhD(carbohydrate metab), 51. Prof Exp: Radon pumpman, Sask Cancer Comn, 35-36; lectr physics, McMaster Univ, 39-43; asst, 43-46, res assoc, 46-47, from asst prof to assoc prof, 47-60, PROF MED RES, UNIV TORONTO, 60- Mem: Am Phys Soc; Am Diabetes Asn; NY Acad Sci; Can Physiol Soc; Can Asn Physicists. Res: Scattering of alphaparticles and protons by light nuclei; x-ray filter design; applications of radioactive isotopic tracers to rate measurements in physiological systems; extractable insulin of pancreas; insulin, glucagon and glucose turnover during running. Mailing Add: Charles H Best Inst Univ of Toronto Toronto ON Can

WRIEDE, PETER ARTUR, b Grossenhain, Ger, Jan 27, 44; US citizen; m 67; c 2. ORGANIC CHEMISTRY. Educ: Columbia Univ, AB, 65, MA, 66, PhD(chem), 69. Prof Exp: CHEMIST, E I DU PONT DE NEMOURS & CO, INC, 69- Mem: AAAS; Am Chem Soc. Res: Photochemistry; photochemical pathways for reaction in solution and in the solid state. Mailing Add: E I du Pont de Nemours & Co Exp Sta Wilmington DE 19898

WRIEDT, HENRY ANDERSON, b Melbourne, Australia, Feb 6, 28; nat US; m 53 & 60; c 3. PHYSICAL CHEMISTRY. Educ: Univ Melbourne, BMetE, 49; Mass Inst Technol, ScD(metall), 54. Prof Exp: Asst metall, Mass Inst Technol, 49-53; technologist, Appl Res Lab, 53-55, scientist, Edgar C Bain Lab Fund Res, 55-66, sr scientist, 66-72, SR SCIENTIST, RES LAB, US STEEL CORP, 72- Mem: Am Inst Mining, Metall & Petrol Eng. Res: Physical chemistry of metals; chemical effects of crystal imperfections; thermodynamics of metallic systems. Mailing Add: US Steel Corp Res Lab Monroeville PA 15146

WRIGHT, ALAN CARL, b Bangor, Maine, Aug 16, 39; m 64; c 2. CHEMISTRY. Educ: Univ Maine, Orono, BS, 61; Univ Fla, PhD(chem), 66. Prof Exp: Res chemist, Cent Res Div, Am Cyanamid Co, 66-69; ASST PROF CHEM, EASTERN CONN STATE COL, 70- Mem: Am Chem Soc; AAAS. Res: Development of new laboratory experiments for organic chemistry involving compounds of biological interest; synthesis of organic fluorine compounds. Mailing Add: Dept of Phys Sci Eastern Conn State Col Willimantic CT 06226

WRIGHT, ALDEN HALBERT, b Missoula, Mont, Apr 23, 42; m 67; c 2. TOPOLOGY. Educ: Dartmouth Col, BA, 64; Univ Wis-Madison, PhD(math), 69. Prof Exp: Vis asst prof math, Univ Utah, 69-70; asst prof, 70-74, ASSOC PROF MATH, WESTERN MICH UNIV, 74- Mem: Am Math Soc. Res: Geometric topology of manifolds; monotone mappings; topology of 3-manifolds. Mailing Add: Dept of Math Western Mich Univ Kalamazoo MI 49008

WRIGHT, ALLEN KENT, b Richmond, Va, Feb 9, 39; m 59; c 2. BIOPHYSICS, BIOMETRY. Educ: Randolph-Macon Col, BA, 61; Med Col Va, PhD(biophys), 69. Prof Exp: ASST PROF BIOMET & BIOCHEM, MED UNIV SC, 69- Res: Molecular biophysics and development of models to account for experimental observations, specifically in area to rotational diffusion models. Mailing Add: Dept of Biomet Med Univ of SC 80 Barre St Charleston SC 29401

WRIGHT, ANDREW, b Edinburgh, Scotland, Jan 28, 35; m 57; c 2. MOLECULAR BIOLOGY. Educ: Univ Edinburgh, BSc, 57, PhD(biochem), 60. Prof Exp: Fel biochem, Univ Minn, 60-62; fel biol, Mass Inst Technol, 63-67; ASSOC PROF MOLECULAR BIOL, MED SCH, TUFTS UNIV, 67- Mem: Fedn Am Socs Exp Biol. Res: Structure and function of bacterial cell membranes and mechanisms of bacteriophage infection. Mailing Add: Dept of Molecular Biol Tufts Univ Med Sch Boston MA 02111

WRIGHT, ANN ELIZABETH, b Mooringsport, La, Feb 20, 22; m 40; c 1. RADIOLOGICAL PHYSICS, BIOPHYSICS. Educ: Univ Houston, BA, 65; Univ Tex M D Anderson Hosp & Tumor Inst Houston, MS, 67, PhD(radiol physics), 70. Prof Exp: Instr radiol, Baylor Col Med, 68-70, asst prof radiol physics, 70-75; ASSOC PROF RADIOL, UNIV TEX MED BR GALVESTON, 75- Concurrent Pos: Consult physicist radiother, Vet Admin Hosp, Houston. Mem: Am Inst Physics; Am Asn Physicist in Med (treas, 73-75). Res: Radiological physics measurements using solid state devices; interaction of radiation with molecules and penetration in inhomogeneous mediums. Mailing Add: Dept of Radiol Univ of Tex Med Br Galveston TX 77550

WRIGHT, ANTHONY AUNE, b Los Angeles, Calif, Jan 4, 43; m 65. ANIMAL BEHAVIOR, PSYCHOPHYSICS. Educ: Stanford Univ, BA, 65; Columbia Univ, MA, 70, PhD(psychol), 71. Prof Exp: Instr psychol, Columbia Univ, 69-71; asst prof, Univ Tex, Austin, 71-72, ASST PROF NEURAL SCI, SENSORY SCI CTR, UNIV TEX HEALTH SCI CTR, HOUSTON, 72- Mem: Psychonomic Soc; Asn Res Vision & Ophthal; Sigma Xi. Res: Animal sensory processes; discrimination learning; theoretical psychophysics; color vision. Mailing Add: Sensory Sci Ctr UT Health Sci Ctr Tex Med Ctr 6420 Lamar Fleming Ave Houston TX 77030

WRIGHT, ARCHIBALD NELSON, b Toronto, Ont, May 22, 32; m 55; c 3. PHYSICAL CHEMISTRY. Educ: McGill Univ, BSc, 53, PhD, 57. Prof Exp: Grace Chem fel with Prof F S Dainton, Univ Leeds, 57-59; res fel with Prof C A Winkler, McGill Univ, 59-63; phys chemist, 63-68, mgr photochem br, Chem Lab, 68-72, mgr reactions & processes br, 72-73, MGR PLANNING & RESOURCES, MAT SCI & ENG, RES & DEVELOP CTR, GEN ELEC CO, 73- Mem: Am Chem Soc; Am Phys Soc; The Chem Soc; NY Acad Sci. Res: Gas phase kinetics, especially reactions to hydrogen and nitrogen atoms and excited nitrogen molecules; anionic polymerization; reactions of clean metal surfaces; photolysis; photopolymerization. Mailing Add: Gen Elec Co Res & Develop Ctr Schenectady NY 12301

WRIGHT, ARTHUR GILBERT, b Carthage, Ill, Jan 22, 09; m. ZOOLOGY. Educ: Carthage Col, AB, 32; Univ Ill, MS, 47. Prof Exp: Zoologist, Ill State Mus, 33-37; Rockefeller fel & mus asst, Biol Sect, Buffalo Mus Sci, 37-38; cur, Ill State Mus, 39-53 & exhibits, State Mus Fla, 53-60; supv exhibits specialist, Nat Park Serv, 61-63; asst chief, Off Exhibits, 63-69, sr museologist, 69-72, spec asst to dir, 72-73, writer ed, Off Exhibits Cent, 74-75, CONSULT, NAT MUS NATURAL HIST, SMITHSONIAN INST, 75-; ASSOC PROF LECTR, DEPT ANTHROP, GEORGE WASHINGTON UNIV, 72- Mem: AAAS; Am Asn Mus; Philos Sci Asn. Res: Museum exhibit planning. Mailing Add: 12201 Galway Dr Silver Spring Md 20904

WRIGHT, ARTHUR WILLIAM, b Windsor Locks, Conn, July 20, 94; m 29 & 63. PATHOLOGY. Educ: Harvard Univ, AB, 17, MD, 23; Am Bd Path, dipl, 39. Prof Exp: Intern path & bact, Boston City Hosp, 23-24, asst, 24-25; asst prof path, Sch Med, Vanderbilt Univ, 25-29; dir, Bender Hyg Lab, NY, 29-34; prof path & bact & head dept, 34-56, prof path & head dept, 56-59, prof path, 59-68, EMER PROF PATH, ALBANY MED COL, 68- Concurrent Pos: Lectr, Albany Med Col, 29-34; pathologist & bacteriologist-in-chief, Albany Med Ctr Hosp, 34-56, pathologist-in-chief, 56-59, sr pathologist, 59-; secy, NY State Bd Med Exam, 61-65; mem, Nat Bd Med Exam, 66-74. Mem: Fel Am Soc Clin Path; fel AMA; fel Am Pub Health Asn; Am Asn Cancer Res; fel NY Acad Sci. Res: Cancer; spontaneous tumors; aging changes in tissue. Mailing Add: 642 Western Ave Albany NY 12203

WRIGHT, BARBARA EVELYN, b Pasadena, Calif, Apr 6, 26; m 51; c 3. BIOCHEMISTRY. Educ: Stanford Univ, PhD(microbiol), 51. Prof Exp: Biologist, Nat Heart Inst, 53-61; res assoc, Huntington Labs, Mass Gen Hosp, 61-67; RES DIR, BOSTON BIOMED RES INST, 67- Concurrent Pos: Nat Res Coun fel, Carlsberg Lab, Copenhagen, Denmark, 50-51, Childs Mem Fund fel, 51-52; tutor, Harvard Univ & assoc prof, Harvard Med Sch; ed ann rev microbiol, Exp Mycol & J Bact; Found for Microbiol lectr, 70-71. Mem: Am Soc Biol Chem; Soc Gen Physiol; NY Acad Sci. Res: Biochemical basis of differentiation in the slime mold; kinetic modelling of metabolic networks. Mailing Add: Boston Biomed Res Inst 20 Staniford St Boston MA 02114

WRIGHT, BARTON ALLEN, b Bisbee, Ariz, Dec 21, 20; m 49; c 2. ANTHROPOLOGY. Educ: Univ Ariz, BA, 52, MA, 54. Prof Exp: State archaeol asst, Town Creek Indian Mound, NC, 49-51; archaeologist, Am-Ind Found, Dragoon, Ariz, 52-55; CUR, MUS NORTHERN ARIZ, 55- Concurrent Pos: Nat Endowment for Arts fel, Mus Northern Ariz, 72-73. Mem: Soc Am Archaeol; Am Asn Mus. Res: Kachinas; general Hopi ethnology; arts and crafts. Mailing Add: Mus of Northern Ariz PO Box 1389 Flagstaff AZ 86001

WRIGHT, BILL C, b Waterford, Miss, June 15, 30; m 59; c 3. SOIL SCIENCE. Educ: Miss State Univ, BS, 52, MS, 56; Cornell Univ, PhD(soil sci), 59. Prof Exp: From asst prof to assoc prof soil sci, Miss State Univ, 59-64; assoc soil scientist, 64-73, AGR PROJ LEADER, ROCKEFELLER FOUND, ANKARA, TURKEY, 73- Mem: Int Soc Soil Sci; Am Soc Agron; Soil Sci Soc Am; Indian Soc Agron; Indian Soil Sci Soc. Res: Soil-phosphorus reactions products; phosphate components of fertilizers; evaluation techniques for fertilizers in field and laboratory; fertilizer use and cultural management of cereal crops in India; wheat production in areas of low rainfall. Mailing Add: Rockefeller Found 111 W 50th St New York NY 10020

WRIGHT, BRADFORD LAWRENCE, b Colfax, Wash, May 10, 40; m 61; c 2. PHYSICS. Educ: Reed Col, BA, 61; Mass Inst Technol, PhD(physics), 67. Prof Exp: Asst prof physics, Mass Inst Technol, 67-70 & Middlebury Col, 70-74; MEM STAFF, LOS ALAMOS SCI LAB, 74- Mem: Am Phys Soc; Am Asn Physics Teachers. Res: Plasma physics; gaseous electronic. Mailing Add: MS 648 Los Alamos Sci Lab Los Alamos NM 87545

WRIGHT, BRUCE STANLEY, b Quebec, Que, Sept 17, 12; m 37; c 2. WILDLIFE ECOLOGY. Educ: Univ New Brunswick, BSc, 36; Univ Wis, MSc, 47. Prof Exp: Res asst forestry, Dom Forest Serv, 37-39; in chg waterfowl surv, Ducks Unlimited, Can, 45-47; DIR WILDLIFE RES, WILDLIFE MGT INST, 47-; ADJ PROF WILDLIFE BIOL, UNIV NB, FREDERICTON, 71- Concurrent Pos: Res fel biol, Univ NB, Fredericton, 47-71; Can Nat Res Coun grants, 49- Honors & Awards: John Pearce Mem Award, Wildlife Soc, 71. Mem: Am Ornith Union; Wildlife Soc; Can Soc Wildlife & Fishery Biol. Res: Survival of the cougar, Felis concolor, in the northeast; breeding ecology of the black duck, Anas rubripes; effects of pesticides on woodcocks, Philohela minor Gmelin. Mailing Add: Northeastern Wildlife Sta Univ of NB Fredericton NB Can

WRIGHT, BYRON TERRY, b Waco Tex, Oct 19, 17; wid; c 3. NUCLEAR PHYSICS. Educ: Rice Univ, BA, 38; Univ Calif, Berkeley, PhD(physics), 41. Prof Exp: Physicist, Navy Radio & Sound Lab, Calif, 41-42 & Manhattan Dist, Calif, Tenn & NMex, 42-46; from asst prof to assoc prof, 46-56, PROF PHYSICS, UNIV CALIF, LOS ANGELES, 56- Concurrent Pos: Fulbright res scholar, 56-57; Guggenheim fel, 63-64; Ford Found fel, Europ Orgn Nuclear Res, 63-64. Mem: Am Phys Soc. Res: Accelerators; nuclear structure. Mailing Add: Dept of Physics Univ of Calif Los Angeles CA 90024

WRIGHT, CHARLES CATHBERT, b Hanford, Calif, Dec 18, 19; m 41; c 3. CHEMISTRY. Educ: Univ Calif, Los Angeles, BA, 41. Prof Exp: Chemist, A R Maas Chem Co, 46-47, res chemist, 47-48; independent consult, 48-49; gen mgr, Oilwell Res, Inc, 49-50, pres, 50-72, chmn bd, 72-74; TECH ADVISOR, ENG DEPT, ARABIAN AM OIL CO, 74- Concurrent Pos: Lectr, Univ Southern Calif, 57-; pres, Am Coun Independent Labs, 64-66; dir water technol, Rhodes Corp, 66-68. Mem: Am Chem Soc; Am Water Works Asn; Am Inst Chemists; Am Soc Testing & Mat; Soc Petrol Eng. Res: Water treatment of subsurface injection; oilwell drilling fluids and the chemistry of oilwell production; oilfield corrosion control. Mailing Add: Arabian Am Oil Co Dhahran Saudi Arabia

WRIGHT, CHARLES DEAN, b Yankton, SDak, June 25, 30; m 52; c 4. POLYMER CHEMISTRY. Educ: Augustana Col, SDak, BA, 52; Univ Minn, PhD(org chem), 56. Prof Exp: Instr org chem, Univ Minn, 55-56; res chemist, 56-64, MGR RES & NEW PROD GROUPS, ADHESIVES, COATINGS & SEALERS DIV, MINN MINING & MFG CO, 64- Mem: Am Chem Soc. Res: Addition of carbenes to indenes; stereospecific polymers; oxidizers for rocket fuels; general polymer chemistry; adhesive compounding and testing; adhesion; new business development; relationships of science and Christianity. Mailing Add: 2579 Elm Dr White Bear Lake MN 55110

WRIGHT, CHARLES GARY, b Okmulgee, Okla, Apr 25, 43. NEUROANATOMY, OTORHINOLARYNGOLOGY. Educ: Univ Okla, BS, 66; Ind Univ, PhD(neurosci), 70. Prof Exp: Fel neuroanat, Dept Anat & Kresge Hearing Res Inst, Univ Mich, 71-73, lectr anat, Depts Anat & Otorhinolaryngol, 73-76; ASST PROF ANAT, UNIV TEX, DALLAS, 76- Mem: Sigma Xi. Res: Anatomy and pathology of the inner ear and central auditory and vestibular pathways. Mailing Add: Callier Ctr for Commun Disorders Univ of Tex Dallas TX 75235

WRIGHT, CHARLES GERALD, b Boynton, Pa, June 12, 30; m 53; c 1. ENTOMOLOGY. Educ: Univ Md, BS, 51, MS, 53; NC State Univ, PhD(entom), 58. Prof Exp: Entomologist, Wilson Pest Control, 58-63; from asst prof to assoc prof, 63-75, PROF ENTOM, NC STATE UNIV, 75- Concurrent Pos: Vchmn NC struct pest control comt. Mem: Entom Soc Am. Res: Urban and industrial entomology; cockroaches; rodent control. Mailing Add: Box 5215 NC State Univ Raleigh NC 27607

WRIGHT, CHARLES HUBERT, b Appleton City, Mo, Oct 30, 22; m 52; c 5. ANALYTICAL CHEMISTRY. Educ: Univ Mo, PhD(chem), 52. Prof Exp: Instr anal chem, Univ Mo, 52-54; chemist, US Radium Corp, 54-58; anal group leader, Spencer Chem Co, 58-66; supvr anal sect, Gulf Res & Develop Co, 66-71; SR RES CHEMIST, PITTSBURG & MIDWAY COAL MINING CO, 71- Mem: AAAS; Am Chem Soc. Res: Analytical chemistry of fertilizers, herbicides, polymers and fuels. Mailing Add: Pittsburg & Midway Coal Mining Co 9009 W 67th St Merriam KS 66202

WRIGHT, CHARLES JOSEPH, b Montour Falls, NY, May 27, 38. ORGANIC CHEMISTRY. Educ: Univ Rochester, BS, 60; Mass Inst Technol, MS, 62. Prof Exp: Res chemist, 64-70, SR RES CHEMIST, RES LABS, EASTMAN KODAK CO, 70- Mem: Am Chem Soc. Res: Synthetic organic chemistry; organic sulfur chemistry; chemical modification of gelatin. Mailing Add: Res Labs Eastman Kodak Co Kodak Park Rochester NY 14650

WRIGHT, CHARLES MILTON, plant pathology, see 12th edition

WRIGHT, CHARLES R B, b Lincoln, Nebr, Jan 11, 37; m 55; c 2. ALGEBRA. Educ: Univ Nebr, BA, 56, MA, 57; Univ Wis, PhD(math), 59. Prof Exp: Res fel math, Calif Inst Technol, 59-60, instr, 60-61; from asst prof to assoc prof, 61-72, PROF MATH, UNIV ORE, 72-, ASSOC DEAN, COL LIBERAL ARTS, 73- Mem: Am Math Soc; Math Asn Am. Res: Finite groups. Mailing Add: Dept of Math Univ of Ore Eugene OR 97403

WRIGHT, CHARLES V, JR, b Webb City, Mo, Sept 14, 23; m 44; c 2. SANITARY ENGINEERING, PUBLIC HEALTH. Educ: Univ Tex A&M Col, BSCE, 48, MEng, 50; Univ Calif, MPH, 55. Prof Exp: Engr II, Mo Hwy Dept, 48-49; from jr staff engr to sanit eng dir, USPHS, 50-66; assoc prof civil eng & pub health, Univ Mo-Columbia, 67-72; DEP ADMINR REGION VII, ENVIRON PROTECTION AGENCY, 72- Concurrent Pos: Lectr, Sch Pub Health & Prev Med, Univ Wash, 55-57. Mem: Am Pub Health Asn; Am Water Works Asn; Water Pollution Control Fedn; Nat Soc Prof Eng. Res: Solid waste disposal; water treatment.

WRIGHT, CHRISTINE GERDA, b Wurzen, EGer, Mar 3, 40; US citizen; m 65; c 2. MOLECULAR BIOLOGY, STRUCTURAL CHEMISTRY. Educ: Ind Univ, BS, 65; Univ Calif, San Diego, PhD(chem), 69. Prof Exp: From lab asst to technician biochem, Lederle Labs, 59-63; NIH fel molecular biol, MRC Lab Molecular Biol, Cambridge, Eng, 69-71; RES ASSOC BIOCHEM, PRINCETON UNIV, 71- Concurrent Pos: Fel, Nat Cancer Inst, 74-75. Res: Investigation into the structure and function of lectins, in particular wheat germ agglutinin, using x-ray diffraction methods. Mailing Add: Dept of Biochem Sci Frick Lab Princeton Univ Princeton NJ 08540

WRIGHT, CLARENCE PAUL, b Cliffside, NC, Apr 15, 39. GENETICS. Educ: Lenoir-Rhyne Col, BS, 62; Univ Utah, MS, 65, PhD(genetics), 68. Prof Exp: ASST PROF BIOL, WESTERN CAROLINA UNIV, 68- Mem: Genetics Soc Am. Res: Developmental genetics. Mailing Add: Dept of Biol Western Carolina Univ Cullowhee NC 28723

WRIGHT, CLAUDE-STARR, b Laurens, SC, Mar 7, 17; m 58. HEMATOLOGY. Educ: Univ SC, BS, 39; Med Col SC, MD, 42; Am Bd Internal Med, dipl, 50. Prof Exp: Intern, Hosp of Richland County, SC, 42-43; asst resident internal med, Barnes Hosp, St Louis, Mo, 43-44; chief resident hematol, Ohio State Univ, 46-47, from instr to assoc prof, 47-55; assoc prof, 55-58, dir dept continuing educ, 58-64, PROF MED & HEAD DIV HEMAT, MED COL GA, 58- Concurrent Pos: Res Found fel med, Ohio State Univ, 44-45, Baruch fel med res, 45-46, Stone fel, 46-47. Mem: AAAS; Am Soc Hemat; Am Soc Clin Invest; fel Am Col Physicians; fel Int Soc Hemat. Res: Erythropoietin; cancer chemotherapy; immunohematology; splenic physiology and pathology. Mailing Add: Div of Hemat Med Col of Ga Augusta GA 30902

WRIGHT, CYNTHIA ROSEMAN, geology, see 12th edition

WRIGHT, DAVID ANTHONY, b Baltimore, Md, Aug 19, 41; m 62; c 3. DEVELOPMENTAL BIOLOGY, GENETICS. Educ: Univ Md, College Park, BS, 63; Univ Ill, Urbana, MS, 65; Wash Univ, PhD(biol), 68. Prof Exp: NIH fel med genetics, Univ Tex M D Anderson Hosp & Tumor Inst Houston, 68-70; asst prof biol, Univ Tex Grad Sch Biomed Sci Houston, 70-75; ASSOC BIOLOGIST & ASST PROF BIOL, UNIV TEX CANCER CTR, M D ANDERSON HOSP & TUMOR INST, 75- Mem: AAAS; Am Soc Zoologists; Soc Develop Biol. Res: Patterns and control of gene expression during embryogenesis, especially of enzyme phenotypes in nuclear-cytoplasmic hybrids in amphibians. Mailing Add: Dept of Biol M D Anderson Hosp Houston TX 77025

WRIGHT, DAVID ARTHUR, physical chemistry, see 12th edition

WRIGHT, DAVID FRANKLIN, b Quincy, Mass, Feb 19, 29; m 52; c 1. PHYSICAL CHEMISTRY. Educ: Tufts Univ, BS, 51; Ohio State Univ, PhD(chem), 57. Prof Exp: PROF BASIC SCI & CHMN DEPT, MASS MARITIME ACAD, 69- Mem: Am Chem Soc. Res: Low temperature thermodynamics; clathrates of hydroquinone. Mailing Add: Dept of Basic Sci Mass Maritime Acad Buzzards Bay MA 02532

WRIGHT, DAVID GRANT, b Am Fork, Utah, Aug 21, 46; m 70; c 2. TOPOLOGY. Educ: Brigham Young Univ, BA, 70; Univ Wis-Madison, MA, 72, PhD(math), 73. Prof Exp: Lectr math, Univ Wis-Madison, 74; FEL MATH, MICH STATE UNIV, 74- Mem: Am Math Soc. Res: Geometric topology including piecewise linear topology and topological embeddings in manifolds. Mailing Add: Dept of Math Mich State Univ East Lansing MI 48824

WRIGHT, DAVID LAVERNE, b Ostrander, Minn, Sept 11, 36. PEDIATRICS. Educ: Univ Minn, BS, 59, MD, 66, PhD, 68; Am Bd Pediat, dipl. Prof Exp: Staff pediatrician, 70-71, chief dept pediat, 71-72, dir med educ & hosp dep dir, 72-75, CHIEF DEPT PEDIAT, USPHS HOSP, NEW ORLEANS, 75- Concurrent Pos: USPHS fel pediat, Med Sch, Univ Minn, 68-70; vis scientist, Charity Hosp New Orleans, 70-; clin asst prof pediat, Med Sch, Tulane Univ, 72-; vis staff, Crippled Childrens Hosp, New Orleans, 73- Mem: AMA; Am Acad Pediat. Res: Experimental diabetes. Mailing Add: USPHS Hosp 210 State St New Orleans LA 70118

WRIGHT, DAVID LEE, b Mattoon, Ill, Dec 1, 49; m 69; c 1. MATHEMATICS. Educ: David Lipscomb Col, BA, 71; Columbia Univ, MS, 73, PhD(math), 75. Prof Exp: ASST PROF MATH, WASH UNIV, 75- Mem: Sigma Xi; Am Math Soc. Res: Behavior of polynomial algebras, their automorphisms, their stable structure. Mailing Add: Dept of Math Wash Univ St Louis MO 63130

WRIGHT, DENNIS CHARLES, b Flint, Mich, Oct 14, 39. NEUROSCIENCES. Educ: Univ Mich, Ann Arbor, BA, 60; Univ Calif, Berkeley, PhD(psychol), 69. Prof Exp: Asst prof, 68-73, ASSOC PROF PSYCHOL, UNIV MO-COLUMBIA, 73- Concurrent Pos: Assoc investr, Dalton Res Ctr, 68- Mem: Asn Psychophysiol Study Sleep. Res: Neurophysiology of sleep; biochemistry and electrophysiology of learning and memory; state-dependent learning. Mailing Add: Dept of Psychol Univ of Mo Columbia MO 65201

WRIGHT, DONALD DECKER, chemistry, deceased

WRIGHT, DONALD LEE, analytical chemistry, see 12th edition

WRIGHT, DONALD N, b Provo, Utah, Dec 6, 35; m 57; c 4. BACTERIOLOGY, BIOCHEMISTRY. Educ: Univ Utah, BS, 58; Iowa State Univ, PhD(bact), 64. Prof Exp: Chief bacteriologist & head serol div, Philadelphia Naval Hosp, Pa, 60-62; res bacteriologist, Naval Biol Lab, Calif, 64-69; assoc prof, 69-74, PROF MICROBIOL, BRIGHAM YOUNG UNIV, 74- Mem: AAAS; Am Soc Microbiol; Brit Soc Gen Microbiol; Am Oil Chem Soc. Res: Physiology of microorganisms, particularly growth, inhibition, taxonomy and aerosol behavior of the Mycoplasma; biochemical responses and growth rate control of microorganisms as a function of their environment. Mailing Add: Dept of Microbiol Brigham Young Univ Provo UT 84602

WRIGHT, EDWARD RAY, analytical chemistry, see 12th edition

WRIGHT, EDWIN T, b Ione, Wash, Oct 25, 22; m 46; c 2. DERMATOLOGY. Educ: La Sierra Col, BS, 43; Loma Linda Univ, MD, 47. Prof Exp: Intern, Los Angeles County Gen Hosp, 46-47; resident dermat & syphil, Vet Admin Ctr Hosp, Los Angeles, 50-53; res microbiologist, 54, assoc clin prof, 60-67, ASSOC PROF MED & DERMAT, SCH MED, UNIV CALIF, LOS ANGELES, 67-; DIR TRAINING & CHIEF DERMAT SECT, VET ADMIN CTR, 55- Mem: Fel Am Col Physicians; fel Am Acad Dermat; Soc Invest Dermat; fel NY Acad Sci; fel Royal Soc Med. Res: Dermatopathology, mycology and biochemistry of the skin. Mailing Add: Dept of Dermat Univ of Calif Sch of Med Los Angeles CA 90024

WRIGHT, ELISABETH MURIEL JANE, b Ottawa, Ont, July 28, 26; m 58; c 2. MATHEMATICS. Educ: Univ Toronto, BA, 49, MA, 50, cert, 51; Wash Univ, PhD(educ), 57. Prof Exp: Specialist schs, Ont, 51-55; asst prof educ, Wash Univ, 57-61, asst prof math, 59-61; from asst prof to assoc prof, 63-72, PROF MATH, CALIF STATE UNIV, NORTHRIDGE, 72- Concurrent Pos: Consult, Santa Barbara Schs, 61-62 & Minn Nat Lab, State Dept Educ, 62-67. Mem: AAAS; Math Asn Am; Am Educ Res Asn. Res: Psychological problems in mathematics education. Mailing Add: Dept of Math Calif State Univ Northridge CA 91324

WRIGHT, ELLEN, textiles, see 12th edition

WRIGHT, EVERETT JAMES, b Meriden, Conn, Sept 20, 29; m 51; c 1. ORGANIC CHEMISTRY. Educ: Hobart Col, BS, 51; Univ Del, MS, 54, PhD(chem), 57. Prof Exp: Chemist, Olin Industs, 51-52; RES CHEMIST, E I DU PONT DE NEMOURS & CO, 57- Mem: Am Chem Soc. Res: Organic nitrogen heterocycles; aliphatic nitrogen compounds; polymerization; textile chemicals; application techniques; fibers and fabrics; personnel and industrial relations. Mailing Add: 2615 Bardell Dr Wilmington DE 19808

WRIGHT, FARROLL TIM, b Hume, Mo, June 24, 41; m 65; c 2. STATISTICS, MATHEMATICS. Educ: Univ Mo, AB, 63, AM, 64, PhD(statist), 68. Prof Exp: Asst prof math, Univ Mo-Rolla, 68-69; asst prof, 69-71, ASSOC PROF STATIST, UNIV IOWA, 71- Mem: Inst Math Statist. Res: Rates of convergence in law of large numbers; equivalent invariant measures; isotonic regresson. Mailing Add: Dept of Statist Univ of Iowa Iowa City IA 52240

WRIGHT, FRANCES WOODWORTH, b Providence, RI, Apr 30, 97. ASTRONOMY. Educ: Brown Univ, AB, 19, AM, 20; Radcliffe Col, PhD(astron), 58. Prof Exp: From instr to prof math & astron, Elmira Col, 20-27; asst, Observ, Harvard Univ, 31-32 & eclipsing binaries on Harvard photog, Princeton Univ, 33-37; asst, Harvard Univ, 37-42, teaching fel navig, 42-46, asst meteor proj & exec secy dept astron, 46-63, lectr astron, 58-67, HON ASSOC, HARVARD COL OBSERV, 68-; ASTRONR, SMITHSONIAN ASTROPHYS OBSERV, 61- Concurrent Pos: Consult, 59-61. Mem: AAAS; Am Astron Soc; Int Astron Union; Am Inst Navig. Res: Variable stars; eclipsing binaries; stellar photometry; radiant paths of photographic meteors; extraterrestrial dust; navigation. Mailing Add: Smithsonian Astrophys Observ Cambridge MA 02138

WRIGHT, FRANCIS HOWELL, b New York, NY, Jan 30, 08; m 69; c 2. PEDIATRICS. Educ: Haverford Col, BS, 29; Johns Hopkins Univ, MD, 33; Am Bd Pediat, dipl, 41. Prof Exp: Asst pediat, Johns Hopkins Univ, 34-35; asst, Columbia Univ, 35-36, instr, 36-38; from asst prof to prof, Sch Med, Univ Chicago, 40-73, chmn dept, 46-62; EXEC SECY, AM BD PEDIAT, 69- Concurrent Pos: Fel bact & path, Rockefeller Inst, 38-40; mem, Am Bd Pediat, 62-67, from vpres to pres, 64-67. Mem: AAAS; Am Pediat Soc; Soc Pediat Res (vpres, 52); Am Acad Pediat. Res: Virology; infant care; psychologic adjustment of infants and children. Mailing Add: Am Bd of Pediat Mus 57th St & Lake Shore Dr Chicago IL 60637

WRIGHT, FRANCIS STUART, b Pittsfield, Mass, Feb 24, 29; m 58; c 4. PEDIATRIC NEUROLOGY. Educ: Univ Mass, BS, 51; Univ Rochester, MD, 55; Am Bd Pediat, dipl, 64; Am Bd Psychiat & Neurol, dipl & cert neurol, 66, cert child neurol, 72. Prof Exp: Asst prof, 63-68, ASSOC PROF PEDIAT & NEUROL, MED CTR, UNIV MINN, MINNEAPOLIS, 68- Concurrent Pos: Fel pediat, Med Ctr, Univ Minn, Minneapolis, 56-58, Nat Inst Neurol Dis & Blindness spec fel pediat neurol, 60-63; consult, Minneapolis Pub Sch Syst, 65-; assoc, Grad Fac, Univ Minn, 67- Res: Developmental electrophysiology. Mailing Add: Dept of Neurol Univ of Minn Med Ctr Minneapolis MN 55455

WRIGHT, FRED BOYER, b Roanoke, Va, Dec 14, 25; m 48; c 2. MATHEMATICS. Educ: Univ NC, BA, 47, MA, 48; Univ Chicago, PhD(math), 53. Prof Exp: Instr math, Univ NC, 48-49; sr mathematician adv bd simulation, Univ Chicago, 53-54, consult systs res; from instr to prof math, Tulane Univ, 54-68; PROF MATH & CHMN DEPT, UNIV NC, CHAPEL HILL, 68- Concurrent Pos: Vis prof, Cambridge Univ, 58-59 & Northwestern Univ, 63; Sloan Found fel, 58-62; chmn comt regional develop, Nat Acad Sci-Nat Res Coun, 69-71, pregrad fel panel, 71-72; mem math adv panel, NSF, 71-; actg chmn opers res curric, Univ NC, Chapel Hill, 72. Mem: Am Math Soc; Math Asn Am; London Math Soc. Res: Algebra; functional analysis; topological algebra. Mailing Add: Buttons Lane Chapel Hill NC 27514

WRIGHT, FRED MARION, b Aurora, Ill, Sept 29, 23; m 47. MATHEMATICS. Educ: Denison Univ, BA, 44; Northwestern Univ, MS, 49, PhD(math), 53. Prof Exp: Instr math, Denison Univ, 47; asst, Northwestern Univ, 47-51; from instr to assoc prof, 53-64, PROF MATH, IOWA STATE UNIV, 64- Concurrent Pos: Vis asst prof, Univ Mich, 57-58. Mem: Am Math Soc. Res: Continued fractions and function theory. Mailing Add: Dept of Math Iowa State Univ Ames IA 50010

WRIGHT, FREDERICK FENNING, b Princeton, NJ, Mar 16, 34. MARINE GEOLOGY, OCEANOGRAPHY. Educ: Columbia Univ, BS, 59, AM, 61; Univ Southern Calif, PhD(geol), 67. Prof Exp: Teaching asst geol, Columbia Univ, 59-61; res asst, Univ Southern Calif, 61-65; eng geologist, Div Water Resources, State of Calif, 65-66; asst prof marine sci, Inst Marine Sci, Univ Alaska, Fairbanks, 66-72, asst prof oceanog & exten oceanogr, Marine Adv Prog, Anchorage, 72-74; oceanog consult, 74-75; dir, Alaska Coastal Mgt Prog, Off Gov, State of Alaska, 75; ALASKA OCS RES MGT OFFICER, OUTER CONTINENTAL SHELF ENVIRON ASSESSMENT PROG, OFF GOV, STATE OF ALASKA & NAT OCEANIC & ATMOSPHERIC AGENCY, 75- Concurrent Pos: Asst dir, NSF Inst Oceanog, Marine Lab, Tex A&M Univ, 63; Geol Soc Am Penrose res grant, 64-66. Mem: Geol Soc Am; Am Geophys Union; Am Asn Petrol Geol; Am Soc Limnol & Oceanog; Arctic Inst NAm. Res: Inshore oceanography and sedimentation in subarctic; fisheries oceanography; coastal resource management and planning. Mailing Add: Box 537 Douglas AK 99824

WRIGHT, FREDERICK HAMILTON, b Washington, DC, Dec 2, 12; m 47; div. PHYSICS. Educ: Haverford Col, BA, 34; Calif Inst Technol, PhD(physics), 48. Prof Exp: Aerodynamicist, Douglas Aircraft Co, Inc, 40-46; res engr, Jet Propulsion Lab, Calif Inst Technol, 46-59; div mgr & prog mgr, Space Gen Corp, 59-69; MEM STAFF, AEROJET ELECTROSYSTS CO, AZUSA, 69- Mem: Am Inst Aeronaut & Astronaut; Am Phys Soc; Inst Elec & Electronic Engrs. Res: Sensor systems; space science; stars in the infrared. Mailing Add: 515 Palmetto Dr Pasadena CA 91105

WRIGHT, FULTON WATKINS, JR, physics, see 12th edition

WRIGHT, GEORGE BUFORD, b Dallas, Tex, Apr 4, 26; m 53; c 4. SOLID STATE PHYSICS. Educ: Univ Tex, BS, 49; Mass Inst Technol, PhD(physics), 60. Prof Exp: Engr, Int Bus Mach Corp, 49-51 & Radio Corp Am, 52-55; staff assoc, Lincoln Lab, Mass Inst Technol, 58-60, staff physicist, 60-62, asst leader solid state physics group, 62-71; mem staff phys prog, Off Naval Res, 71-72; BATCHELOR PROF ELEC ENG, STEVENS INST TECHNOL, 72- Concurrent Pos: Exchange prof, Solid Physics Lab, Univ Paris VI, 71. Mem: AAAS; fel Am Phys Soc; Optical Soc Am; Inst Elec & Electronics Engrs. Res: Optical properties of solids; Raman scattering; luminescence; magnetoplasmas; ultraviolet reflectivity; band structure; lattice dynamics; excitons; quantum electronics. Mailing Add: Dept of Elec Eng Stevens Inst of Technol Hoboken NJ 07030

WRIGHT, GEORGE CARLIN, b Shawsville, Md, Mar 3, 26; m 55; c 3. SYNTHETIC ORGANIC CHEMISTRY. Educ: Johns Hopkins Univ, BE, 47; Univ Del, MS, 55, PhD(chem), 58. Prof Exp: Chemist, Glenn L Martin Co, Md, 47-50; res chemist, Gen Aniline & Film Corp, Pa, 50; chemist, Army Chem Ctr, 51-52; res chemist, Gen Aniline & Film Corp, Pa, 52-53; SR RES CHEMIST, NORWICH PHARMACAL CO, 58- Mem: Am Chem Soc. Res: Polymerization and development chemistry; degradation of carbamates; organic synthesis; hydantoins; nitrofurans; hydrazidines; quinolines. Mailing Add: Norwich Pharmacal Co Norwich NY 13815

WRIGHT, GEORGE EDWARD, b Milwaukee, Wis, Oct 21, 41; m 65; c 3. MEDICINAL CHEMISTRY. Educ: Univ Ill, Chicago, BS, 63, PhD(chem), 67. Prof Exp: Sr res asst chem, Univ Durham, 66-68; asst prof med chem, Sch Pharm, Univ Md, Baltimore, 68-74; ASSOC PROF PHARMACOL, UNIV MASS MED SCH, 74- Mem: Am Chem Soc; The Chem Soc; Int Soc Heterocyclic Chem. Res: Synthesis and structure of heterocyclic compounds; nuclear magnetic resonance spectroscopy; DNA polymerase inhibitors. Mailing Add: 298 Highland St Worcester MA 01602

WRIGHT, GEORGE F, b Council Bluffs, Iowa, Feb 23, 04; m 37; c 3. CHEMISTRY. Educ: Iowa State Col, BS, 29, PhD(chem), 32. Prof Exp: Nat Res Coun fel, Harvard Univ & Univ Vienna, 33-35; lectr chem, McGill Univ, 35-37; from asst to assoc, 38-40, PROF CHEM, UNIV TORONTO, 41- Concurrent Pos: Mem comt explosives, Nat Res Coun Can, 40-45. Honors & Awards: Medal of Freedom; Order Brit Empire. Mem: Am Chem Soc; Chem Inst Can (pres, 64-65). Res: Chemistry of furans; organometallic compounds; lignins; explosives; stereochemistry; x-ray diffraction; electrical polarization; tobacco smoke carcinogens; far-infrared spectroscopy; pharmaceuticals. Mailing Add: Dept of Chem Univ of Toronto Toronto ON Can

WRIGHT, GEORGE GREEN, b Ann Arbor, Mich, Aug 17, 16; m 57; c 3. IMMUNOLOGY, MICROBIOLOGY. Educ: Olivet Col, BA, 36; Univ Chicago, PhD(bact), 41. Prof Exp: Instr immunol, Univ Chicago, 41-42; vis investr, NIH, 46-48; med microbiologist, Ft Detrick, Md, 48-71; ASST DIR BIOL LABS, MASS STATE DEPT PUB HEALTH, 71- Concurrent Pos: Logan fel Univ Chicago, 41-42; Nat Res Coun fel, Calif Inst Technol, 42-43, fel immunol, 43-46; US Secy Army res fel, Oxford Univ, 57-58. Mem: Soc Exp Biol & Med; Am Soc Microbiol; Am Asn Immunol. Res: Elaboration, isolation and characterization of microbial antigens and toxins; production of biologics; general immunology and serology; immunity in anthrax; serum proteins. Mailing Add: Biologic Labs 375 South St Boston MA 02130

WRIGHT, GEORGE JOSEPH, b Allendale, Ill, June 4, 31; m 53; c 2. DRUG METABOLISM, BIOCHEMISTRY. Educ: Univ Ill, Urbana, BS, 52, PhD(animal nutrit & biochem), 60. Prof Exp: Med researcher human biochem, L B Mendel Res Lab, Elgin State Hosp, Ill, 60-62; biochemist toxicol & indust hyg, Biochem Res Lab, Dow Chem Co, Midland, Mich, 52-67; DEPT HEAD DRUG METAB, MERRELL-NAT LABS, DIV RICHARDSON-MERRELL, INC, 67- Concurrent Pos: Lectr chem, Eve Col, Univ Cincinnati, 70-; consult clin and preclin pharmacol, Walter Reed Army Inst Res. Mem: AAAS; Soc Toxicol; Am Soc Pharmacol & Exp Therapeut. Res: Drug disposition; bioavailability; biopharmaceutics; pharmacokinetics; enzymology; toxicology. Mailing Add: Dept of Drug Metab Merrell-Nat Labs Cincinnati OH 45215

WRIGHT, GEORGE LEONARD, JR, b Ludington, Mich, Feb 8, 37; m 64; c 2. IMMUNOCHEMISTRY, MICROBIOLOGY. Educ: Albion Col, BA, 59; Mich State Univ, MS, 62, PhD(microbiol, path, biochem), 66. Prof Exp: Fel immunol & immunochem, Sch Med, George Washington Univ, 66-67, asst prof microbiol & immunochem, 67-74; MEM FAC, DEPT MICROBIOL & CELL BIOL, EASTERN VA MED SCH, 74- Concurrent Pos: Sigma Xi award sci res, 66; spec consult, Vet Admin Hosp, Wilmington, Del, 66-; deleg & rapporteur, Tuberc Panel, US-Japan Coop Med Sci Prog, Tokyo, 68 & 70; Hartford Found grant, 68-; NIH grant, 72-74; spec consult, Beckman Instruments; consult, Washington, DC Vet Admin & Children's Hosps; mem subcomt, US-Japan Med Sci Prog Stand Mycobact Antigens. Mem: AAAS; Nat Tuberc & Respiratory Dis Asn; Am Soc Microbiol; NY Acad Sci; Reticuloendothelial Soc. Res: Separation, isolation and immunobiological characterization of specific mycobacterial antigens and human tumor associated antigens; development of gradient polyacrylamide electrophoretic techniques for the separation and isolation of serum and tissue proteins. Mailing Add: Dept of Microbiol & Cell Biol Eastern Va Med Sch 358 Mowbray Arch Norfolk VA 23507

WRIGHT, GORDON PRIBYL, b Crosby, Minn, May 18, 38; m 61; c 1. OPERATIONS RESEARCH, STATISTICS. Educ: Macalester Col, BA, 60; Univ Mass, MA, 63; Case Western Reserve Univ, PhD(opers res), 67. Prof Exp: Computer analyst, Conn Gen Life Ins Co, 60-61; statistician, Goodyear Tire & Rubber Co, 63-64; asst opers res, Case Inst Technol, 64-67; asst prof, Northwestern Univ, 67-70; ASSOC PROF STATIST, PURDUE UNIV, WEST LAFAYETTE, 70- Concurrent Pos: Prin investr, NSF Res Grant, 68- Mem: Opers Res Soc Am; Inst Mgt Sci; Soc Indust & Appl Math. Res: Mathematical optimization theory, especially nonlinear and dynamic programming and inventory theory. Mailing Add: Dept of Indust Admin Purdue Univ West Lafayette IN 47907

WRIGHT, GRANT MACLACHLAN, geology, deceased

WRIGHT, HARRY TUCKER, JR, b Louisville, Ky, July 21, 29. PEDIATRICS, EPIDEMIOLOGY. Educ: Wake Forest Col, BS, 51, MD, 55; Univ Calif, Berkeley, MPH, 75. Prof Exp: From instr to assoc prof, 63-73, PROF PEDIAT, UNIV SOUTHERN CALIF, 73- Concurrent Pos: Teaching fel pediat, Case Western Reserve Univ, 58-59; NIH res training grant, 61-64. Mem: Am Acad Pediat; Am Fedn Clin Res; Soc Pediat Res; Am Pediat Soc; Am Soc Microbiol. Res: Clinical and laboratory studies of cytomegalovirus and other herpes viruses; central nervous system syndromes of viral etiology and newborn infections; virology. Mailing Add: Childrens Hosp of Los Angeles PO Box 54700 Terminal Annex Los Angeles CA 90054

WRIGHT, HARVEL AMOS, b Mayflower, Ark, July 6, 33; m 54; c 2. MATHEMATICS. Educ: Ark State Teachers Col, BS, 54; Univ Ark, MA, 56; Univ Tenn, PhD, 67. Prof Exp: Teacher high sch, Ark, 55-56; instr math & physics, Ark State Teachers Col, 56-58; instr math, Univ Tenn, 58-62; physicist, 62-70, health physicist, 70-74, CHIEF, BIOL & RADIATION PHYSICS SECT, OAK RIDGE NAT LAB, 74- Concurrent Pos: Consult, Oak Ridge Nat Lab, 60-62. Mem: Math Asn Am; Am Phys Soc; Health Physics Soc; Radiation Res Soc; AAAS. Res: Interaction of radiation with matter; dosimetry of ionizing radiation; plasma physics; theory of real variable. Mailing Add: Health Phys Div Bldg 4500S Oak Ridge Nat Lab Oak Ridge TN 37830

WRIGHT, HASTINGS KEMPER, b Boston, Mass, Aug 22, 28; m 54; c 4. MEDICINE, PHYSIOLOGY. Educ: Harvard Univ, AB, 50, MD, 54. Prof Exp: Asst prof surg, Western Reserve Univ, 62-66; assoc prof, 66-72, PROF SURG, SCH MED, YALE UNIV, 72- Concurrent Pos: Crile fel surg, Western Reserve Univ, 61-62. Mem: AAAS; Soc Univ Surg; Am Fedn Clin Res; Am Gastroenterol Asn; Am Surg Asn. Res: Gastrointestinal fluid and electrolyte absorption. Mailing Add: Dept of Surg Yale-New Haven Hosp 333 Cedar St New Haven CT 06510

WRIGHT, HENRY ALBERT, b Modesto, Calif, June 1, 35; m 61; c 3. RANGE MANAGEMENT, APPLIED STATISTICS. Educ: Univ Calif, Davis, BS, 57; Utah State Univ, MS, 62, PhD(range mgt), 64. Prof Exp: Exten specialist range mgt, Univ Calif, Davis, 57-58; range aid, Intermountain Forest & Range Exp Sta, Boise, Idaho, 60-64, res assoc, 64-67; from asst prof to assoc prof, 67-74, PROF RANGE MGT, TEX TECH UNIV, 74- Concurrent Pos: Mem grad coun, Tex Tech Univ, 71-74; educ comt, Southwest Interagency Fire Coun, 71- Mem: Soc Range Mgt; Ecol Soc Am; Weed Sci Soc Am. Res: Fire ecology particularly developing prescription techniques for burning rangeland communities and studying the effect of fire on their total ecosystem. Mailing Add: Dept of Range & Wildlife Mgt Tex Tech Univ Lubbock TX 79409

WRIGHT, HERBERT EDGAR, JR, b Malden, Mass, Sept 13, 17; m 43; c 5. PALEOECOLOGY. Educ: Harvard Univ, AB, 39, AM, 41, PhD(geol), 43. Hon Degrees: DSc, Trinity Col, Dublin, 66. Prof Exp: Instr geol, Brown Univ, 46-47; from asst prof to prof, 47-74, REGENTS' PROF GEOL, ECOL & BOT, UNIV MINN, MINNEAPOLIS, 74-, DIR LIMNOL RES CTR, 63- Concurrent Pos: Geologist, US Geol Surv, DC, 46-47; mem Boston Col-Fordham archaeol exped, 47; geologist, Minn Geol Surv, 48-63; Wenner-Gren fel, 51; geologist, US Geol Surv, DC, 52-53; mem archaeol exped, Oriental Inst, 51, 54-55, 60, 63, 64 & 70. Mem: Geol Soc Am; Ecol Soc Am; Am Soc Limnol & Oceanog; Am Quaternary Asn (pres, 71-72); Glaciol Soc. Res: Geomorphology; Pleistocene geology and paleoecology; vegetation history; paleolimnology. Mailing Add: Dept of Geol Univ of Minn Minneapolis MN 55455

WRIGHT, HERBERT FESSENDEN, b Worcester, Mass, July 19, 17; m 41; c 1. MEDICINAL CHEMISTRY. Educ: Oberlin Col, AB, 40; Cornell Univ, MS, 42, PhD(org chem), 44; Am Inst Chem, cert. Prof Exp: Asst chem, Cornell Univ, 42-43, Off Sci Res & Develop antimalarial proj, 43-44; res chemist, Lever Bros Co, Mass, 44-45; res assoc, Mass Inst Technol, 45-46; instr chem, Tufts Col, 46-48; res chemist, Arthur D Little, Inc, 48-49; instr chem, Yale Univ, 49-52; sr res chemist, Olin Industs, 52-54; CONSULT CHEM, 54-; PROF BIOL & PHYS SCI, UNIV NEW HAVEN, 66-, CHMN DEPT, 67-, DIR ENVIRON STUDIES, 70- Concurrent Pos: Pres & res dir, W Elsworth Co, Inc, 59-64; asst prof biol & phys sci, Univ New Haven, 64-65, assoc prof, 6S-66; dir, Univ Res Insts Conn, 67- Mem: AAAS; Am Chem Soc; NY Acad Sci; fel Am Inst Chem; Am Inst Biol Sci. Res: Synthetic drugs and vitamin A, organic ortho silicate esters; polymers; fungicides; antimetabolites; organic electrode reactions; organic synthetic methods; molecular biology; forensic science. Mailing Add: Dept of Sci & Biol Univ New Haven 300 Orange Ave West Haven CT 06516

WRIGHT, HERBERT N, b Berwyn, Ill, May 23, 28; m 52; c 3. PSYCHOACOUSTICS, AUDIOLOGY. Educ: Grinnell Col, BA, 50; Ind Univ, MA, 53; Northwestern Univ, PhD(audiol), 58. Prof Exp: Res assoc otolaryngol, Northwestern, 56-58; res assoc blind mobility, Auditory Res Ctr, 60-61; from res instr to res assoc prof otolaryngol, 61-63, ASSOC PROF OTOLARYNGOL, STATE UNIV NY UPSTATE MED CTR, 63- Concurrent Pos: Res fel psychoacoustics, Cent Inst Deaf, St Louis, Mo, 58-60 & Syracuse Univ, 61-63. Mem: AAAS; Am Speech & Hearing Asn; Acoust Soc Am; Psychonomic Soc; Int Soc Audiol. Res: Psychoacoustics of individual differences, especially with reference to disorders of the auditory system; temporal factors of audition. Mailing Add: Dept of Otolaryngol & Commun Sci State Univ of NY Upstate Med Ctr Syracuse NY 13210

WRIGHT, HOWARD EDWARDS, JR, b Petersburg, Va, June 29, 13; m 41; c 3. ORGANIC CHEMISTRY. Educ: Hampden-Sydney Col, BS, 35; Univ Richmond, MS, 37. Prof Exp: Res chemist, Am Tobacco Co, 39-43; res & develop chemist, US Naval Ord Lab, Md, 43-46; RES CHEMIST, AM TOBACCO CO, 46-, ASST PROCESS DEVELOP MGR, 70- Concurrent Pos: Supvr tobacco basic res, Am Tobacco Co, 65-68, asst mgr basic mat res, 69-70. Mem: Am Chem Soc; fel Am Inst Chemists; Phytochem Soc NAm (treas, 66-70). Res: Organic and biochemical investigations of tobacco leaf and smoke composition and processes. Mailing Add: Dept of Res & Develop Am Tobacco Co Box 799 Hopewell VA 23860

WRIGHT, HOWARD OLIVER, marine biology, animal behavior, see 12th edition

WRIGHT, IAN GLAISBY, b Fredericton, NB, Sept 15, 35; m 58; c 3. SYNTHETIC ORGANIC CHEMISTRY. Educ: Univ BC, BA, 57; Univ NB, Fredericton, MSc, 59; Univ Wis-Madison, PhD(org chem), 65. Prof Exp: Fel with A I Scott, Univ BC, 63-65; sr org chemist, 65-73, SR ORG CHEMIST PROCESS RES & DEVELOP DIV, LILLY RES LABS, ELI LILLY & CO, 73- Mem: AAAS; Am Chem Soc; The Chem Soc. Res: Chemistry and biological activity of cephalosporin antibiotics; chemistry of natural products from plant and microbiological sources. Mailing Add: Process Res & Develop Div Eli Lilly & Co Indianapolis IN 46206

WRIGHT, JAMES A, plant breeding, see 12th edition

WRIGHT, JAMES ARTHUR, b Toronto, Ont, Dec 29, 41; m 70. GEOPHYSICS. Educ: Univ Toronto, BASc, 64, MA, 65, PhD(physics), 68. Prof Exp: Nat Res Coun Can fel, Brunswick Tech Univ, 68-69; asst prof, 69-74, ASSOC PROF PHYSICS, MEM UNIV NFLD, 74- Concurrent Pos: Mem geomagnetism subcont, Assoc Comt Geod & Geophys, Nat Res Coun Can, 71-73. Mem: Can Asn Physicists; Geol Soc Can; Am Geophys Union; Soc Explor Geophysicists. Res: Geomagnetism and the earth's interior; exploration geophysics. Mailing Add: Dept of Physics Mem Univ of Nfld St Johns NF Can

WRIGHT, JAMES EDWARD, b Little Rock, Ark, Sept 6, 46; m 66; c 2. EXERCISE PHYSIOLOGY. Educ: Fairleigh Dickinson Univ, BS, 69; Miss State Univ, PhD(zool), 73. Prof Exp: Asst prof biol sci, Simon's Rock Early Col, 73-75; NIH FEL & ASST RES PHYSIOLOGIST EXERCISE & ENVIRON PHYSIOL, INST ENVIRON STRESS, UNIV CALIF, SANTA BARBARA, 75- Mem: AAAS; Am Col Sports Med. Res: Longitudinal and cross sectional investigation of functional changes produced by weight training, concentrating on fluid shifts and cardiovascular parameters; exercise- and heat-induced body fluid shifts in normal adult males; peripheral blood flow under negative pressure stress; effects of 96 hour fasting on normal adults. Mailing Add: Inst of Environ Stress Univ of Calif Santa Barbara CA 93106

WRIGHT, JAMES ELBERT, b Kerrville, Tex, Oct 7, 40; m 62; c 2. MEDICAL ENTOMOLOGY, BACTERIOLOGY. Educ: Tex A&M Univ, BS, 63; Ohio State Univ, PhD(med entom, bact), 66; Am Registry Cert Entom, cert. Prof Exp: Res asst mosquitoes entom, Ohio State Univ, 63-66, res assoc, 66; RES ENTOMOLOGIST, VET TOXICOL & ENTOM RES LAB, AGR RES SERV, USDA, 66- Mem: Fel AAAS; Entom Soc Am; Am Chem Soc; Soc Invert Path; Am Mosquito Control Asn. Res: Insect juvenile and molting hormones interrelationships; insect biochemistry; mosquito biology; diapause mechanisms; livestock arthropods; biting flies; stable fly and horn fly biology; studies on temperature and photoperiod; biology of livestock ticks; insect growth regulators. Mailing Add: Vet Toxicol & Entom Res Lab Agr Res Serv USDA PO GE College Station TX 77840

WRIGHT, JAMES EVERETT, JR, b Deepstep, Ga, Apr 28, 23; m 48; c 3. GENETICS. Educ: Univ Ga, BS, 46; Cornell Univ, PhD(genetics), 50. Prof Exp: From instr to assoc prof, 49-60, PROF GENETICS, PA STATE UNIV, 60- Mem: AAAS; Genetics Soc Am; Am Fisheries Soc; Am Genetic Asn. Res: Genetics and breeding of fishes. Mailing Add: 202 Buckhout Lab Pa State Univ University Park PA 16802

WRIGHT, JAMES FRANCIS, b Philadelphia, Pa, Jan 18, 24; m 55; c 3. COMPARATIVE PATHOLOGY, TOXICOLOGY. Educ: Univ Pa, VMD, 51; Univ Calif, Davis, PhD(comp path), 69. Prof Exp: Vet in-chg animal quarantine, Plum Island Animal Dis Lab, USDA, NY, 54-57; vet, Nat Zool Park, Smithsonian Inst, 57-62; res scientist, USPHS, US Air Force Radiobiol Lab, Univ Tex, Austin, 62-64, Yerkes Regional Primate Res Ctr, Emory Univ, 64-65 & Radiobiol Lab, Univ Calif, Davis, 65-69; chief toxicol studies sect, Twinbrook Res Lab, 69-73, CHIEF PATH STUDIES SECT, EXP BIOL LAB, NAT ENVIRON RES CTR, ENVIRON PROTECTION AGENCY, 73- Concurrent Pos: Panel mem interagency bd, US Civil Serv Exam. Mem: Am Vet Med Asn; Am Asn Zoo Vets; Am Asn Lab Animal Sci; Wildlife Dis Asn; Radiation Res Soc. Res: Laboratory animal medicine; pathology and toxicology; diseases of wild animals; comparative radiation pathology. Mailing Add: Exp Biol Lab Nat Envir Res Ctr MO-72 Environ Protect Agency Research Triangle Park NC 27711

WRIGHT, JAMES LOUIS, b Pleasant Grove, Utah, Aug 28, 34; m 58; c 6. AGRONOMY, AGRICULTURAL METEOROLOGY. Educ: Utah State Univ, BS, 59, MS, 61; Cornell Univ, PhD(soil physics), 65. Prof Exp: SOIL SCIENTIST, AGR RES SERV, USDA, 65-; ASSOC PROF RES CLIMATOL & AFFIL PROF SOILS, UNIV IDAHO, 70- Mem: Am Soc Agron; Soil Sci Soc Am; Sigma Xi. Res: Microclimate investigations to determine rate of evaporation of water from agricultural crops using energy balance and micrometeorological approaches. Mailing Add: Snake River Conserv Res Ctr Rte 1 Box 186 Kimberly ID 83341

WRIGHT, JAMES MALCOLM, physical chemistry, see 12th edition

WRIGHT, JAMES P, astrophysics, see 12th edition

WRIGHT, JAMES R, b Riversdale, NS, July 19, 16; m 42; c 2. SOIL CHEMISTRY. Educ: McGill Univ, BSc, 40; Mich State Univ, MS, 48, PhD(soil chem), 53. Prof Exp: Asst chemist, NS Dept Agr, 40-41, asst prov chemist, 45-50; head soil genesis sect, Soil Res Inst, 51-61, DIR, KENTVILLE RES STA, CAN DEPT AGR, 61- Concurrent Pos: Lectr, NS Agr Col, 45-50; prof, Acadia Univ, 65- Mem: Am Soc Agron; Soil Sci Soc Am; fel Chem Inst Can; Can Soc Soil Sci (pres, 61-62); Agr Inst Can. Res: Chemical nature of soil organic matter; soil genesis, fertility and plant nutrition. Mailing Add: Kentville Res Sta Kentville NS Can

WRIGHT, JAMES REUBEN, biological chemistry, see 12th edition

WRIGHT, JAMES ROSCOE, b White Hall, Md, July 7, 22; m 50; c 2. ORGANIC CHEMISTRY. Educ: Md State Col Salisbury, BSEd, 46; Wash Col, BS, 48; Univ Del, MS, 49, PhD(chem), 51; Harvard Univ, prog mgt develop, 67. Prof Exp: Res chemist, Southwest Res Inst, 51-52 & Chevron Res Corp, Stand Oil Co Calif, 52-60; from chemist to sr chemist, 60-66, phys sci adminstr, 66-72, dir, Ctr for Bldg Technol, 72-74, DEP DIR, INST APPL TECHNOL, NAT BUR STANDARDS, 74- Concurrent Pos: Asst prof, Trinity Univ, Tex, 51-52; Com Sci & Tech fel, US Patent Off, 64-65; pres, Int Union Testing & Res Labs Mat & Struct, 71-72, mem bur, 74-78 & del, 75-79; mem bldg res adv bd, 75-76. Honors & Awards: Gold Medal Award, Dept Com, 75. Mem: Am Chem Soc; Am Soc Testing & Mat. Res: Organosilicon compounds; radiation effects on organic lubricants; emulsification of petroleum products; photochemical stability of organic building materials; application of performance concept in building research, codes and standards. Mailing Add: Nat Bur of Standards Bldg 225 Rm B-115 Washington DC 20234

WRIGHT, JAMES SHERMAN, b Seattle, Wash, Aug 27, 40; m 63; c 1. THEORETICAL CHEMISTRY. Educ: Stanford Univ, BS, 62; Univ Calif, Berkeley, PhD(chem), 68. Prof Exp: Nat Ctr Sci Res, France grant, Fac Sci, Orsay, France, 69-70; ASST PROF CHEM, CARLETON UNIV, 70- Res: Chemistry. Mailing Add: Dept of Chem Carleton Univ Ottawa ON Can

WRIGHT, JANE COOKE, b New York, NY, Nov 30, 19; m 47; c 2. MEDICINE. Educ: Smith Col, AB, 42; NY Med Col, MD, 45. Hon Degrees: DrMedSci, Women's Med Col Pa, 65; DSc, Denison Univ, 71. Prof Exp: Sch physician, New York City Dept Health, 49; clinician, Cancer Res Found, Harlem Hosp, 49-52, dir, 52-55; from instr to asst prof res surg, Postgrad Med Sch, NY Univ, 55-61, adj assoc prof, 61-67, dir cancer chemother serv res, 55-67; PROF SURG & ASSOC DEAN, NY MED COL, 67- Concurrent Pos: Clin asst vis physician, Harlem Hosp, 49, asst vis physician, 49-55; asst vis physician, 4th Surg Div, Bellevue Hosp, 55-59, assoc vis physician, 60-; asst attend physician, Univ Hosp, 56- & Manhattan Vet Admin Hosp, 63-; mem courtesy staff, Midtown Hosp, 62-; consult, Health Ins Plan, 62-, Boulevard Hosp, 63-, St Luke's Hosp, Newburgh, NY, 64- & St Vincent's Hosp, New York, 66-; mem, President's Comn Heart Dis, Cancer & Stroke, 64, Nat Adv Cancer Coun, 66-70 & Nat Coun Negro Women, 56 & 63; attend surg, Flower-Fifth Ave Hosps, Metrop Hosp & Bird S Coler Mem Hosp, 67-; consult oncol, Wyckoff Heights Hosp, 69-; attend physician, Gen Surg Sect, Chemother, Grassland Hosp, 71-74. Honors & Awards: Mademoiselle Award, 52; Spence Chapin Award, 58; Spirit Achievement Award, 65. Mem: AMA; Am Soc Clin Oncol (secy-treas, 64-); Am Asn Cancer Res; Nat Med Asn; NY Acad Sci. Res: Cancer chemotherapy. Mailing Add: NY Med Col 106th St at Fifth Ave New York NY 10029

WRIGHT, JEFFREY LAWSON CAMERON, b Clarkston, Scotland, Nov 11, 42; m 64; c 1. BIOLOGICAL CHEMISTRY. Educ: Glasgow Univ, BSc, 64, PhD(chem), 67. Prof Exp: Fel biol chem, Nat Res Coun Can, 67-68; res asst bio-org chem, Sci Res Coun, 68-69; dept asst enzymol, Oxford Univ, 69-71; res assoc biol chem, Dalhousie Univ, 72-74; RES OFFICER BIOL CHEM, NAT RES COUN CAN, 74- Mem: Chem Inst Can. Res: Application of nuclear magnetic resonance spectroscopy to the study of biosynthetic and metabolic processes and problems of structure and stereochemistry; analysis of macromolecular structures via selective multinuclear biosynthetic enrichment. Mailing Add: Atlantic Regional Lab 1411 Oxford St Halifax NS Can

WRIGHT, JEROME J, b Aurora, Nebr, Apr 2, 14; m 41; c 1. GEOLOGY, HYDROGEOLOGY. Educ: Univ Nebr, BSc, 41, MSc, 47; Univ Ariz, PhD(geol, hydrol), 64. Prof Exp: Geologist, Nebr Surv, 41-42 & 46-47 & Calif Co, 47-56; chief geologist, Coronet Oil Co, 56-60; res assoc geol & hydrol, Univ Ariz, 60-64, from asst prof to assoc prof, 64-74; CHIEF HYDROLOGIST, METROP UTILITIES MGT AGENCY, 74- Concurrent Pos: Geologist, US Geol Surv, 46-47; mem, Int Geother Coun; asst dean, Col Earth Sci, Univ Ariz, 72-73. Mem: Fel AAAS; Am Asn Petrol Geol; Am Water Resources Asn; Int Asn Hydrogeol. Res: Hydrogeology; borehole geophysics; geothermal occurrences; groundwater resources; water quality; wastewater reclamation and recharge. Mailing Add: 7711 E Lee Tucson AZ 85715

WRIGHT, JOE CARROL, b Benton, Ark, Feb 6, 33; m 53; c 4. ORGANIC CHEMISTRY. Educ: Ouachita Baptist Univ, BS, 54; Univ Ark, MS, 62, PhD(org mech), 66. Prof Exp: Asst prof chem, Mobile Col, 64-66; assoc prof, 66-72, PROF CHEM, HENDERSON STATE UNIV, 72-, DEAN SCH SCI, 69- Mem: Am Chem Soc; The Chem Soc. Res: Isotope effect studies in organic reaction mechanism determinations. Mailing Add: Off of the Dean Sch Sci Henderson State Univ Arkadelphia AR 71923

WRIGHT, JOHN BRENTON, b Providence, RI, Nov 7, 16; m 42; c 2. MEDICINAL CHEMISTRY. Educ: Providence Col, BSc, 40; Columbia Univ, MA, 43, PhD(org chem), 45. Prof Exp: Res chemist, Columbia Univ, 44-45; res chemist, 46-54, RES ASSOC, UPJOHN CO, 54- Concurrent Pos: Vis res assoc, Harvard Univ, 64-65. Mem: Am Chem Soc; Soc Heterocyclic Chem. Res: Antiviral, antidiabetic and anti-inflammatory compounds; reactions of acetals; Mannich reactions; synthetic medicinal and heterocyclic chemistry; anti-asthma. Mailing Add: Dept 7244-25-6 Upjohn Co Henrietta St Kalamazoo MI 49006

WRIGHT, JOHN CLIFFORD, b Livingston, Mont, Jan 29, 19; m 44; c 2. LIMNOLOGY. Educ: Mont State Univ, BS, 41; Ohio State Univ, PhD, 50. Prof Exp: From instr to assoc prof, 49-61, PROF BOT, MONT STATE UNIV, 61-, ASSOC COORDR, CTR ENVIRON STUDIES, 71- Concurrent Pos: NSF sr fel, 59-60; exec secy, XV Int Cong Limnol, 61-62; dir, Ctr Environ Studies, Mont State Univ, 66-71. Mem: AAAS; Ecol Soc Am; Am Soc Limnol & Oceanog; Int Asn Theoret & Appl Limnol. Res: Oceanography; ecology. Mailing Add: Dept of Bot & Microbiol Mont State Univ Bozeman MT 59715

WRIGHT, JOHN COLLINS, b WVa, Aug 5, 27; m; c 2. ORGANIC CHEMISTRY. Educ: WVa Wesleyan Col, BS, 48; Univ Ill, PhD(org chem), 51. Prof Exp: Res chemist, Hercules Powder Co, 51-57; prof chem, WVa Wesleyan Col, 57-64; asst prog dir, NSF, 64-65; fel, Univ Mich, 65-66; dean col arts & sci, Northern Ariz Univ, 66-70, PROF CHEM & DEAN COL ARTS & SCI, WVA UNIV, 70- Concurrent Pos: Hon res assoc, Univ Col, Univ London, 62-63. Mem: Am Chem Soc. Mailing Add: Dept of Chem WVa Univ Morgantown WV 26506

WRIGHT, JOHN CUSHING, b Lakewood, Ohio, May 25, 47. ANIMAL BEHAVIOR. Educ: Wittenberg Univ, BA, 70; Miami Univ, MA, 72, PhD(exp psychol), 76. Prof Exp: Instr psychol, Miami Univ, 74-75; ASST PROF PSYCHOL, BEREA COL, 75- Concurrent Pos: Dir, Animal Lab, Behav Genetics Lab, Berea Col, 75- Mem: Animal Behav Soc; Behav Genetics Asn; Sigma Xi. Res: The early development of communication and intraspecific social relationships in Canis familiaris; agonistic behavior and the development of social structure in several genetically inbred strains of mice in a semi-natural setting. Mailing Add: Berea Col CPO 2056 Berea KY 40403

WRIGHT, JOHN FOWLER, b London, Ont, July 28, 21; US citizen; m 49; c 2. PHYSICAL CHEMISTRY. Educ: Univ Western Ont, BSc, 45, MSc, 46. Prof Exp: SR RES ASSOC, RES LABS, EASTMAN KODAK CO, 46- Mem: Am Chem Soc. Res: Polymer synthesis; surface treatments of polymers; colloid chemistry; surface active agents. Mailing Add: 94 Chestnut Hill Dr Rochester NY 14617

WRIGHT, JOHN JAY, b Torrington, Conn, July 10, 43; m 65; c 2. ATOMIC PHYSICS. Educ: Worcester Polytech Inst, BS, 65; Univ NH, PhD(physics), 70. Prof Exp: Fel lasers, Joint Inst Lab Astrophys, Colo, 69-70; ASST PROF PHYSICS, UNIV NH, 70- Mem: Am Inst Physics; Am Phys Soc; Am Asn Physics Teachers. Res: Optical pumping; liquid crystals. Mailing Add: Dept of Physics Univ of NH Durham NH 03824

WRIGHT, JOHN MARLIN, b Long Beach, Calif, Feb 2, 37; m 58; c 2. CHEMICAL INSTRUMENTATION, MAGNETIC RESONANCE. Educ: Calif Inst Technol, BS, 60; Harvard Univ, PhD(chem), 67. Prof Exp: Res fel chem, Harvard Univ, 67-70; assoc specialist, 70-73, SPECIALIST & LECTR CHEM, UNIV CALIF, SAN DIEGO, 73- Mem: Am Chem Soc; AAAS. Res: Application of nuclear magnetic resonance and mass spectrometry to chemical problems; development of new instrumental techniques. Mailing Add: Dept of Chem B-017 Univ Calif San Diego La Jolla CA 92037

WRIGHT, JOHN RICKEN, b Batesville, Ark, Jan 3, 39; m 64; c 1. BIOINORGANIC CHEMISTRY. Educ: Ark State Univ, BS, 60; Univ Miss, MS, 67, PhD(chem), 71. Prof Exp: Res assoc, Dept Bot, Wash Univ, 67-68; NIH fel, Fla State Univ, 72-73; ASST PROF CHEM, DEPT PHYS SCI, SOUTHEASTERN OKLA STATE UNIV, 73- Mem: Sigma Xi. Res: The metabolism and transport of biological forms of copper; the interaction of copper with membranes in normal and pathological states. Mailing Add: Dept of Phys Sci Southeastern Okla State Univ Durant OK 74701

WRIGHT, JOHNIE ALGIE, b Atlas, Ala, Feb 5, 24; m 49; c 2. HORTICULTURE. Educ: Tenn Polytech Inst, BS, 47; Iowa State Col, MS, 49; La State Univ, PhD, 60. Prof Exp: Asst prof hort, Tenn Polytech Inst, 49-53; assoc prof, 53-60, PROF HORT, LA TECH UNIV, 60-, ASSOC DEAN, COL LIFE SCI, 72- Concurrent Pos: Asst, La State Univ, 60. Mem: Am Soc Hort Sci. Res: Nutrient culture of roses; methods of watering greenhouse roses; container nursery stock studies. Mailing Add: Dept of Hort La Tech Univ PO Box 5136 Ruston LA 71270

WRIGHT, JON ALAN, b Tacoma, Wash, Jan 18, 38; m 62; c 2. THEORETICAL PHYSICS, PARTICLE PHYSICS. Educ: Calif Inst Technol, BS, 59; Univ Calif, Berkeley, PhD(physics), 65. Prof Exp: Physicist, Aerospace Corp, 65; res assoc physics, Univ Calif, San Diego, 65-67; ASSOC PROF PHYSICS, UNIV ILL, URBANA, 67- Mem: Am Inst Physics. Res: Theoretical elementary particle physics. Mailing Add: Dept of Physics Univ of Ill Urbana IL 61801

WRIGHT, JONATHAN WILLIAM, b Spokane, Wash, June 29, 16; m 40; c 1. FOREST GENETICS. Educ: Univ Idaho, BS, 38; Harvard Univ, MF, 39, AM, 41, PhD(genetics), 42. Prof Exp: Instr forestry, Purdue Univ, 42-45; geneticist, Northeastern Forest Exp Sta, US Forest Serv, Morris Arboretum, 46-57; PROF FORESTRY, MICH STATE UNIV, 57- Concurrent Pos: Assoc ed, Silvae Genetica, 56- Mem: Soc Am Foresters. Res: Interspecific hybridization and geographic variation in Pinus and Picea; genetic statistics. Mailing Add: 2034 Yuma Trail Okemos MI 48864

WRIGHT, JOSEPH WILLIAM, JR, b Indianapolis, Ind, Oct 30, 16; m 42, 65; c 2. MEDICINE. Educ: Univ Mich, AB, 38, MD, 42. Prof Exp: Intern, Univ Mich Hosp, 42-43; resident, Ind Univ Hosp, 46-48; ASSOC PROF OTOLARYNGOL, SCH MED, IND UNIV, INDIANAPOLIS, 58- Concurrent Pos: Fel surg, Northwestern Univ, 55-56; consult, Off Surgeon Gen, US Army, 49-; chmn otolaryngol serv, Community Hosp, 57-, mem bd trustees; mem, Ind State Hearing Comn; pres, Wright Inst Otol; consult, Crossroad Rehab Ctr, Indianapolis, Ind, & Indianapolis Speech & Hearing Ctr; consult mem, St Vincent's Hosp; chmn, legis commun comt, Am Coun Otolaryngol, 75-; mem, Otosclerosis Study Group, 75- Mem: AMA; Am Acad Ophthal & Otolaryngol; Am Laryngol, Rhinol & Otol Soc; Pan-Am Med Asn; Int Col

Surg. Res: Bioceramics as applied to otology. Mailing Add: 5506 E 16th St Indianapolis IN 46218

WRIGHT, KENNETH A, b Timmins, Ont, Apr 7, 36; m 59; c 3. ZOOLOGY, PARASITOLOGY. Educ: Univ Toronto, BA, 58, MA, 60; Rice Univ, PhD(parasitol, cytol), 62. Prof Exp: Res zoologist, Univ Calif, Riverside, 62-63, asst res nematologist, Univ Calif, Davis, 63-64; res scientist parasitol & cytol, Ont Res Found, 64-66; from asst prof to assoc prof, Sch Hyg, 66-75, ASSOC PROF MICROBIOL & PARASITOL, FAC MED, UNIV TORONTO, 75- Concurrent Pos: Assoc ed, Can Soc Zoologists, 74- Mem: Am Soc Parasitol; Soc Nematol; Can Soc Zoologists; Can Soc Microscopists. Res: Cytology of parasites, especially of nematodes. Mailing Add: Dept of Microbiol & Parasitol Fac Med Univ Toronto Toronto ON Can

WRIGHT, KENNETH ARTHUR, b Elsie, Mich, June 16, 05; m 39; c 2. NUCLEAR PHYSICS, ELECTRONICS. Educ: Cent Mich Univ, BS, 34; Univ Mich, MS, 39. Prof Exp: Prin & teacher pub schs, Mich, 28-42; from asst prof to prof, 42-74, EMER PROF PHYSICS, CENT MICH UNIV, 74- Mem: Am Asn Physics Teachers. Res: Decay schemes of elements. Mailing Add: Dept of Physics Cent Mich Univ Mt Pleasant MI 48858

WRIGHT, KENNETH HAROLD, b NDak, Apr 21, 21; m 48; c 2. ENTOMOLOGY. Educ: Univ Wash, BS, 48; Duke Univ, MS, 50. Prof Exp: Res forest entomologist, Exp Sta, 48-67, ASST DIR DIV FOREST PROTECTION RES, PAC NORTHWEST FOREST & RANGE EXP STA, US FOREST SERV, 67- Mem: Soc Am Foresters; Entom Soc Am. Res: Forest insects; tree killing bark beetles. Mailing Add: Pac NW Forest & Range Exp Sta US Forest Serv PO Box 3141 Portland OR 97208

WRIGHT, KENNETH JAMES, b Pittsburgh, Pa, Aug 26, 39; m 66; c 1. PHYSICAL INORGANIC CHEMISTRY, ENVIRONMENTAL SCIENCES. Educ: Portland State Univ, BS, 62; Univ Idaho, PhD(inorg chem), 72. Prof Exp: Anal & res chemist, Harvey Aluminum Corp, Ore, 63-66; INSTR CHEM, N IDAHO COL, 71-, CHMN DIV PHYS SCI, 72- Concurrent Pos: NSF trainee chem, Univ Idaho, 67, Nat Defense Educ Act fel, 68-70. Mem: Am Chem Soc; Sigma Xi. Res: Environmental effects of air and water pollutants; energy resources and conservation; science education methodologies; analytical chemistry of environmental pollutants. Mailing Add: Div of Phys Sci N Idaho Col Coeur d'Alene ID 83814

WRIGHT, KENNETH OSBORNE, b Ft George, BC, Nov 1, 11; m 37 & 70; c 1. ASTROPHYSICS. Educ: Univ Toronto, BA, 33, MA, 34; Univ Mich, PhD(astron), 40. Prof Exp: Asst astron, Univ Toronto, 33-34; asst, 36-40, astronomer, 40-60, asst dir, 60-66, DIR, DOM ASTROPHYS OBSERV, 66- Concurrent Pos: Lectr, Univ BC, 43-44; spec lectr, Univ Toronto, 60-61; res asst, Mt Wilson & Palomar Observ, 62; chmn assoc comt astron, Nat Res Coun Can, 71-74. Honors & Awards: Gold Medal, Royal Astron Soc Can, 33. Mem: Int Astron Union; Am Astron Soc; Royal Astron Soc Can (pres, 64-66); fel Royal Soc Can; Can Astron Soc. Res: Stellar radial velocities; observations of stellar line intensities; curves of growth; stellar atmospheres, peculiar A-type stars; observation and analysis of atmospheres of giant eclipsing systems. Mailing Add: Dom Astrophys Observ 5071 W Saanich Rd Victoria BC Can

WRIGHT, LAUREN ALBERT, b New York, NY, July 9, 18. GEOLOGY. Educ: Univ Southern Calif, AB, 40, MS, 43; Calif Inst Technol, PhD(geol), 51. Prof Exp: From jr geologist to asst geologist, US Geol Surv, 42-46; assoc geologist, State Div Mines, Calif, 47-51, sr mining geologist, 51-54, supv mining geologist, 54-61; PROF GEOL, PA STATE UNIV, 61- Mem: Fel Geol Soc Am; Am Soc Econ Geologists; Am Inst Mining, Metall & Petrol Eng. Res: Geologic occurence, origin and economics of industrial minerals; geology of Death Valley region in California; structural geology. Mailing Add: Dept of Geosci Pa State Univ University Park PA 16802

WRIGHT, LEMUEL DARY, b Nashua, NH, Mar 1, 13; m 41; c 5. BIOCHEMISTRY. Educ: Univ NH, BS, 35, MS, 36; Ore State Col, PhD(biochem), 40. Prof Exp: Asst biochem, Pa State Univ, 36-37; asst nutrit chemist, Ore State Col, 37-40; fel, Univ Tex, 40-41; instr biochem, Sch Med, Univ WVa, 41-42; res biochemist, Med Res Div, Merck Sharp & Dohme Div, 42-47, dir nutrit res, 47-50, dir microbiol chem, 50-56; PROF NUTRIT, CORNELL UNIV, 56- Mem: Am Chem Soc; Am Soc Biol Chemists; Soc Exp Biol & Med; Am Inst Nutrit. Res: Vitamin B complex; microbiological methods of assay; bacterial growth factors; biogenesis and metabolism of pyrimidines. Mailing Add: Div Nutrit Sci Cornell Univ Ithaca NY 14850

WRIGHT, LEO MILFRED, geology, see 12th edition

WRIGHT, LEON WENDELL, b Los Angeles, Calif, July 16, 23; m 50; c 4. PHYSICAL CHEMISTRY. Educ: Univ Calif, Los Angeles, BS, 46, MS, 47; Univ Del, PhD(chem), 51. Prof Exp: Instr chem, Mont State Col, 47-49 & Univ Del, 49-51; res chemist, Houdry Process Corp, 51-56; from res chemist to sr res chemist, Atlas Chem Indust, Inc, 56-67, prin chemist, Chem Res Dept, 67-70, SUPVR, ORG & PROCESS RES GROUP, ICI US, INC, 70- Mem: Am Chem Soc; Sigma Xi. Res: Catalytic hydrogenation and hydrogenolysis; hetero and homogeneous activation of hydrogen; isomerization of polyhydric alcohols. Mailing Add: Chem Res Dept ICI US Inc Concord Pike & New Murphy Rd Wilmington DE 19897

WRIGHT, LOUIS EDGAR, b Buras, La, Oct 18, 40; m 69; c 1. THEORETICAL PHYSICS, NUCLEAR PHYSICS. Educ: La State Univ, BS, 61; Duke Univ, PhD(physics), 66. Prof Exp: Res assoc physics, Duke Univ, 66; prog mgr theoret physics, US Army Res Off-Durham, 66-69; asst prof physics, Duke Univ, 69-70; asst prof, 70-73, ASSOC PROF PHYSICS, OHIO UNIV, 73- Concurrent Pos: Vis physicist, Inst Theoret Physics, Univ Frankfurt, 68-69. Mem: Am Phys Soc. Res: Theoretical investigations of the electromagnetic structure of the atomic nucleus. Mailing Add: Dept of Physics Ohio Univ Athens OH 45701

WRIGHT, MADISON JOHNSTON, b Washington, DC, Apr 9, 24; m 54; c 4. AGRONOMY. Educ: Univ NC, BA, 47; Univ Wis, MS, 50, PhD(agron, bot), 52. Prof Exp: Asst prof agron, Univ Wis, 52-59; assoc prof, 59-68, chmn dept, 70-75, PROF AGRON, CORNELL UNIV, 68- Mem: Am Soc Agron; Crop Sci Soc Am; Am Soc Animal Sci; Am Inst Biol Sci. Res: Crop management. Mailing Add: Dept of Agron Cornell Univ Ithaca NY 14853

WRIGHT, MARGARET RUTH, b Rochester, NY, Mar 24, 13. ZOOLOGY. Educ: Univ Rochester, AB, 34, MS, 38; Yale Univ, PhD(zool), 46. Prof Exp: Asst, Univ Rochester, 36-38; histol technician, Med Sch, Yale Univ, 38-41, asst, Osborn Zool Lab, 42-43; instr biol, Middlebury Col, 43-46; from instr to assoc prof, 46-59, PROF ZOOL, VASSAR COL, 59- Concurrent Pos: Mem exped, Alaska, 36; grant, Nat Inst Neurol Dis & Blindness, 54-64; Vassar Col fels, 54-55, 63-64 & 68-69; cur, Natural Hist Mus, Vassar Col. Mem: AAAS; Am Soc Zool; NY Acad Sci; Ecol Soc Am. Res: Limnology and biogeography; experimental morphology; trophic action in sensory systems. Mailing Add: Dept of Biol Vassar Col Poughkeepsie NY 12601

WRIGHT, MARION IRENE, b Cranston, RI, June 17, 22. GEOGRAPHY. Educ: RI Col, EdB, 44; Clark Univ, MA, 46. Prof Exp: Chmn dept soc sci, RI Col, 53-70, PROF GEOG, RI COL, 53- Mem: Asn Am Geog. Res: Areas of Europe and Africa; metropolitan growth. Mailing Add: Dept of Geog RI Col 600 Mount Pleasant Providence RI 02908

WRIGHT, MARTIN, b Tex, Feb 12, 12; m 39; c 3. MATHEMATICS. Educ: Univ Tex, BA, 35, MA, 37; Rice Inst, PhD, 56. Prof Exp: Teacher, Pub Schs, Tex, 36-42; head dept, 56-68, PROF MATH, UNIV HOUSTON, 38- Concurrent Pos: Dir, NSF Inst; consult, Gen Dynamics/Convair, 55. Mem: Math Asn Am. Res: Various mean in the summation of divergent series; asymptotic Dirichlet series in a strip. Mailing Add: Dept of Math Univ of Houston 3801 Cullen Blvd Houston TX 77004

WRIGHT, MARY LOU, b Milford, Mass, Dec 4, 34. DEVELOPMENTAL BIOLOGY. Educ: Col Our Lady Elms, BS, 57; Univ Detroit, MS, 66; Univ Mass, PhD(develop biol), 72. Prof Exp: TEACHING & RES BIOL, COL OUR LADY ELMS, 58- Concurrent Pos: Sigma Xi Res Grant, 73. Mem: Soc Develop Biol; Am Soc Zoologists; Tissue Cult Asn. Res: Investigation of the influence of thyroxine and prolactin on cell population kinetics in Amphibian tadpole limb epidermis. Mailing Add: Col of Our Lady of Elms Chicopee MA 01013

WRIGHT, MAURICE MORGAN, b Assiniboia, Sask, July 29, 16; m 45; c 3. ELECTROCHEMISTRY. Educ: Univ BC, BA & BASc, 38; Princeton Univ, MA, 48, PhD(chem), 52. Prof Exp: Res chemist, 38-45, RES CHEMIST, COMINCO, LTD, 49- Concurrent Pos: Indust fel, Nat Res Coun Can, 49-52. Mem: Electrochem Soc; Am Chem Soc; Chem Inst Can; Nat Asn Corrosion Engrs. Res: Electrowinning and refining of metals; electrolytic hydrogen; ammonia synthesis; physical methods of analysis; deuterium separation, analysis and exchange reactions; reactive metals; anodic films; corrosion; protective coatings; lead-acid battery chemistry. Mailing Add: 1454 Willowdown Rd Oakville ON Can

WRIGHT, MICHAEL JAMES, mathematics, see 12th edition

WRIGHT, NORMAN SAMUEL, b BC, Dec 8, 20; m 49. PLANT PATHOLOGY. Educ: Univ BC, BSA, 44, MSA, 46; Univ Calif, PhD(plant path), 51. Prof Exp: Plant pathologist, Sci Serv, Plant Path Lab, 46-60, HEAD PLANT PATH SECT, RES STA, CAN DEPT AGR, 60- Mem: Potato Asn Am; Can Phytopath Soc; Agr Inst Can. Res: Potato viruses. Mailing Add: Agr Can Res Sta 6660 NW Marine Dr Vancouver BC Can

WRIGHT, OSCAR LEWIS, b Murphysboro, Ill, Apr 21, 17; m 39; c 3. ORGANIC CHEMISTRY, PHYSICAL CHEMISTRY. Educ: Southern Ill Univ, BEd, 38; Univ Mo, BS, 47, PhD(org & phys chem), 49. Prof Exp: Teacher pub sch, Ill, 39-40; anal chemist, Wis Steel Co, 40-41 & E I du Pont de Nemours & Co, 41-43; assoc prof, Southwest Mo State Col, 49-50; head dept chem, Col of Emporia, 50-52; org chemist, Continental Oil Co, 52-55, Pittsburgh Coke & Chem Co, 55-58 & MSA Res Corp Div, Mine Safety Appliance Co, 58-61; prof chem & chmn div sci & math, Rockhurst Col, 61-67; PROF CHEM & CHMN DEPT, NORTHEAST LA UNIV, 67- Res: Organic reactions of ion-exchange agents; steric factors in aromatic electrophilic substitution; organometallics; gas-solid reaction systems; effect of solvents in aromatic electrophilic substitution reactions. Mailing Add: Dept of Chem Northeast La Univ Monroe LA 71201

WRIGHT, PAUL ALBERT, b Nashua, NH, June 15, 20; m 43; c 3. REPRODUCTIVE ENDOCRINOLOGY. Educ: Bates Col, SB, 41; Harvard Univ, AM, 42, PhD(endocrinol), 44. Prof Exp: Asst biol, Harvard Univ, 42-43, instr, 44-45; instr zool & physiol, Univ Wash, 45-46; instr biol sci, Boston Univ, 46-47; from instr to assoc prof zool, Univ Mich, 47-58; assoc prof, 58-62, endocrinologist, Agr Exp Sta, 58-68, chmn dept, Univ, 63-69, PROF ZOOL, UNIV NH, 62- Mem: AAAS; Am Soc Zoologists (treas, 65-68); Am Physiol Soc; Soc Exp Biol & Med; Soc Study Reproduction. Res: Ovulation in the frog; physiology of melanophores of Amphibia; blood sugar studies in lower vertebrates; control of corpus luteum life by the uterus. Mailing Add: Dept of Zool Spaulding Bldg Univ of NH Durham NH 03824

WRIGHT, PAUL ERNEST, organic chemistry, pharmaceutical chemistry, see 12th edition

WRIGHT, PAUL MCCOY, b Alfalfa Co, Okla, Sept 11, 04; m 30; c 3. PHYSICAL CHEMISTRY. Educ: Wheaton Col, Ill, BS, 26; Ohio State Univ, MS, 28, PhD(phys chem), 30. Prof Exp: Asst chem, Wheaton Col, Ill, 25-26; asst gen chem, Ohio State Univ, 26-28, asst phys chem, 28-29; from asst prof to prof chem, 29-70, actg chmn dept chem & geol, 39-40, chmn dept geol, 40-59, chmn dept chem, 40-69, dir field camp, SDak, 46-52 & 60, EMER PROF CHEM, WHEATON COL, ILL, 70- Mem: AAAS; Am Chem Soc; fel Am Inst Chemists. Res: Dimensions of vapor particles; eutectics of explosive mixtures; equilibria of glycerol esters; radiochemistry; color centers in rose quartz by electron paramagnetic resonance. Mailing Add: 717 N Washington St Wheaton IL 60187

WRIGHT, PETER HEDLEY, b Edinburgh, Scotland, Sept 6, 21; c 2. PHARMACOLOGY, ENDOCRINOLOGY. Educ: Univ Liverpool, BSc, 42, MSc, 48, MB, ChB, 52, MD, 60. Prof Exp: Intern, Alder Hey Children's Hosp, Liverpool, Eng, 52-53; registr path, Liverpool Stanley Hosp, 53-54; registr clin path, Postgrad Sch Med London, 54-55; lectr & sr lectr chem path, Guy's Hosp Med Sch, London, 55-63; assoc prof, 63-69, PROF PHARMACOL, SCH MED, IND UNIV, INDIANAPOLIS, 69- Concurrent Pos: Rockefeller Found traveling fel, Sch Med, Northwestern Univ, 60-61. Mem: Am Diabetes Asn; Endocrine Soc; Am Physiol Soc; Brit Med Asn; Brit Biochem Soc. Res: Assay of insulin in blood; production, assay, and characterization of antibodies to insulin; induction of diabetes and estimation of insulin secretion with anti-insulin serum; other protein hormones and their antibodies; metabolic disorders of hormonal origin. Mailing Add: Dept of Pharmacol Ind Univ Med Ctr Indianapolis IN 46202

WRIGHT, PHILIP LINCOLN, b Nashua, NH, July 9, 14; m 39; c 3. ZOOLOGY. Educ: Univ NH, BS, 35, MS, 37; Univ Wis, PhD(zool), 40. Prof Exp: From instr to assoc prof, 39-51, chmn dept, 56-69 & 70-71, PROF ZOOL, UNIV MONT, 51- Concurrent Pos: Ed, Gen Notes & Rev, J Mammal, 66-67; sabbatical leave, Africa, 70. Mem: Am Soc Mammalogists; Am Soc Zoologists; Wildlife Soc; Soc Study Reproduction; Am Ornith Union. Res: Reproductive cycles of birds and mammals; especially Mustelidae. Mailing Add: Dept of Zool Univ of Mont Missoula MT 59801

WRIGHT, RICHARD DONALD, b Jersey City, NJ, July 2, 38; m 62; c 1. CELL BIOLOGY. Educ: Univ NH, BA, 64; Tulane Univ, MS, 69, PhD(biol), 70. Prof Exp: Researcher biochem, Dept Exp Med & Biochem, Sch Med, Univ Ore, 65-66; fel & lectr biol, Univ Mass, 70-71; ASST PROF BIOL, SANGAMON STATE UNIV, 71- Concurrent Pos: Adj asst prof, Sch Med, Southern Ill Univ, 72-; tech E M consult, Affil Labs Inc, 75- Mem: Sigma Xi; Am Soc Parasitologists. Res: Cell surface specializations; morphological and chemical aspects of human infective yeast organisms during development; acanthocephalan cellular biology. Mailing Add: Sangamon State Univ Springfield IL 62703

WRIGHT, RICHARD T, b Haddonfield, NJ, June 28, 33; m 61; c 3. AQUATIC ECOLOGY. Educ: Rutgers Univ, AB, 59; Harvard Univ, PhD(biol), 63. Prof Exp: NSF fel, Inst Limnol, Univ Uppsala, 63-65; PROF BIOL, GORDON COL, 65- Concurrent Pos: NSF res grants, 66-71 & 73-77; NSF sci fac fel, Ore State Univ, 69-70. Mem: Ecol Soc Am; Am Soc Limnol & Oceanog; Am Sci Affil; Am Soc Microbiologists. Res: Dissolved organic matter of natural waters and the microorganisms using, transforming and producing it. Mailing Add: Dept of Biol Gordon Col Wenham MA 01984

WRIGHT, ROBERT ANDERSON, b El Paso, Tex, July 22, 33; m 55; c 4. BOTANY. Educ: NMex State Univ, BS, 55, MS, 60; Univ Ariz, PhD(bot), 65. Prof Exp: PROF BIOL, W TEX STATE UNIV, 64- Mem: AAAS; Biomet Soc; Ecol Soc Am; Soc Study Evolution; Brit Ecol Soc. Res: Vegetation changes; statistical ecology; net primary productivity. Mailing Add: Dept of Biol WTex State Univ Canyon TX 79015

WRIGHT, ROBERT CRUNN, b Detroit, Mich, Nov 21, 37. URBAN GEOGRAPHY, EDUCATION. Educ: Eastern Mich Univ, BA, 60, MA, 65; Clark Univ, PhD(geog), 72. Prof Exp: Teacher geog, sci & math, Romulus Pub Schs, Mich, 60-68; instr geog, Mich Lutheran Col, 68-69; ASST PROF GEOG, WAYNE STATE UNIV, 72- Mem: Asn Am Geogr; Nat Coun Geog Educ. Res: Electoral patterns in urban areas; political geography of school districts. Mailing Add: Dept of Geog Wayne State Univ Detroit MI 48202

WRIGHT, ROBERT DEAN, b Menasha, Wis, Mar 30, 09; m 34; c 3. PUBLIC HEALTH. Educ: Univ Wis, BA & MA, 33; Wash Univ, MD, 35; Johns Hopkins Univ, MPH, 40; cert pub health & prev med, 50. Prof Exp: Dist health officer, Mo State Bd Health, 37-38; mem off med educ, Div Sanit Reports & Statist, USPHS, 38-39 & Sch Hyg & Pub Health, Johns Hopkins Univ, 39-40; dir, Mobile Clin Demonstration, Ga, 40-41; venereal dis consult, NC State Bd Health, 41-44; med officer-in-chg venereal dis, Rapid Treat Hosp, WVa, 44-46; res officer, Venereal Dis Res Lab, US Marine Hosp, 46-50, dir res & prof educ, Div Venereal Dis Control, 50-51; prof social & environ med & chmn dept, Univ Va, 51-56, clin prof prev med, 56-63; prof pub health admin, 63-69, PROF INT HEALTH, SCH HYG & PUB HEALTH, JOHNS HOPKINS UNIV, 69- Concurrent Pos: World Health Study fel, Eng & Europe, 51; dep med dir region III, USPHS, DC, 56-58, asst dir, Off Voc Rehab, 59-62, med consult, Div Int Health, 62-; sr res officer, AID, 62-63; prof community health & chmn dept, Col Med, Univ Lagos, 63-69. Mem: Fel Am Pub Health Asn; Am Col Prev Med. Res: Anatomy; penicillin therapy of syphilis; public health methods. Mailing Add: Dept of Int Health Johns Hopkins Univ Sch of Hyg Baltimore MD 21205

WRIGHT, ROBERT HAMILTON, b Vancouver, BC, Dec 26, 06; m 31; c 3. PHYSICAL CHEMISTRY. Educ: Univ BC, BA, 28; McGill Univ, MSc, 30, PhD(chem), 31. Hon Degrees: DSc, Univ NB, 73. Prof Exp: From asst prof to assoc prof chem, Univ NB, 31-41, prof phys chem, 41-46; head div chem, BC Res Coun, 46-62, head olfactory response invest, 62-71; CONSULT, FOOD & AGR ORGN, 72- Mem: Entom Soc Am; fel Chem Inst Can; Can Entom Soc. Res: Fundamental and applied odor studies; physical basis of odors; insect attractants and repellents. Mailing Add: 6822 Blenheim St Vancouver BC Can

WRIGHT, ROBERT JAMES, b Bridgewater, Va, Dec 16, 18; m 46; c 4. GEOLOGY. Educ: Denison Univ, BA, 40; Columbia Univ, MA, 42, PhD(geol), 47. Prof Exp: Field geologist, US Geol Surv, Washington, DC, 42-44; lectr geol, Columbia Univ, 46-47; asst prof, St Lawrence Univ, 47-49; staff geologist, US AEC, NY, 49-54, chief geol br, Colo, 51-54; supvr explor, Climax Uranium Co, 54-58; mgr western explor, Am Metal Climax, Inc, 59-65, vpres, AMAX Explor, Inc, 65-68, mgr overseas mining activities group, AMAX, Inc, 68-73; consult geologist, Roan Consolidated Mines Ltd, Zambia, 73-75; CHIEF GEOLOGIST, NUCLEAR FUEL CYCLE & PROD DIV, ENERGY RES & DEVELOP ADMIN, 75- Mem: Fel Geol Soc Am; Soc Econ Geol; Am Inst Mining, Metall & Petrol Eng. Res: Economic geology. Mailing Add: Nuclear Fuel Cycle & Prod Div Energy Res & Develop Admin Washington DC 20545

WRIGHT, ROBERT L, b Buckhannon, WVa, Sept 7, 30; m 52; c 1. ORGANIC CHEMISTRY, RUBBER CHEMISTRY. Educ: WVa Wesleyan Col, BS, 52. Prof Exp: Anal trainee, 52, res chemist, 55-64, sr res chemist, 64-75, RES SPECIALIST, MONSANTO CO, 75- Mem: Am Chem Soc. Res: Process development; exploratory synthesis in the field of rubber chemicals; product development of adhesive systems. Mailing Add: 3288 Dowling Dr Akron OH 44313

WRIGHT, ROBERT PAUL, b Hartford, Conn, Jan 31, 43; m 65; c 2. PALEOECOLOGY, INVERTEBRATE PALEONTOLOGY. Educ: Univ Conn, BA, 65; Univ Mich, MS, 67, PhD(geol), 71. Prof Exp: Res asst paleont, Univ Mich, 65-70, field asst stratig, 65, teaching fel geol & paleont, 67-69; instr, 70-71, ASST PROF GEOL, OHIO STATE UNIV, 71- Mem: Paleont Soc; Soc Econ Paleontologists & Mineralogists; Int Comn Palezoic Microflora. Res: Paleontology and stratigraphy of the Jurassic system in the United States western interior; Chitinozoa zoology and biostratigraphy of Devonian strata in the United States midwest. Mailing Add: Ohio State Univ 1465 Mt Vernon Ave Marion OH 43302

WRIGHT, ROBERT RAYMOND, b Buffalo, NY, May 20, 31; m 60; c 3. SCIENCE POLICY. Educ: Syracuse Univ, BS, 57; Wayne State Univ, MA, 61, EdD, 64. Prof Exp: Asst dir instnl res, Wayne State Univ, 60-66; dir instnl res, State Univ NY Cent Staff, 66-72; consult univ sci & data eval, NSF, 72-73; dir instnl res, State Univ NY Cent Staff, 73-74; STAFF ASSOC & HEAD SCI INDICATORS UNIT, NSF, 74- Concurrent Pos: Chmn, Data Standards Comt, Asn Instnl Res, 67-69, mem, Comt Access to Fed Data, 72-73; mem, Comt Fed Reporting, Nat Asn Col & Univ Bus Officers, 67-69. Mem: AAAS; Asn Instnl Res; Soc Social Studies Sci. Res: Economics of research and development; sociology of science. Mailing Add: 1098 Larkspur Terr Rockville MD 20850

WRIGHT, ROBERT W, b Auburn, NY, Aug 2, 32; m 55. PHYSICAL ORGANIC CHEMISTRY, POLYMER CHEMISTRY. Educ: NY State Col Forestry, BS, 59, PhD(chem), 64. Prof Exp: Chemist, Owens-Ill, Inc, Ohio, 64-65, res chemist, 65-66, sect leader polymer chem, 66-69; GROUP DIR SYNTHETIC MAT, CORP RES CTR, INT PAPER CO, 69- Mem: Am Chem Soc; Tech Asn Pulp & Paper Indust; Soc Plastics Engrs; Coun Agr & Chemurgic Res. Res: Plastics processing; polymer coatings and adhesives; photo-curable coatings and inks; health and medical devices. Mailing Add: Corp Res & Develop Div Int Paper Co Tuxedo NY 10987

WRIGHT, RUFUS WILLIAM, b Adair, Okla, Sept 3, 14; m 46; c 4. PHYSICS. Educ: Southwestern Col, Kans, BA, 36; Univ Wis, PhD(physics), 40. Prof Exp: Res physicist, Corning Glass Works, 40-41; res assoc, Radiation Lab, Mass Inst Technol, 41-45; physicist, 45-71, ASSOC SUPT ELECTRONICS TECHNOL DIV, US NAVAL RES LAB, WASHINGTON, DC, 71- Mem: Am Phys Soc; Sigma Xi. Res: Thermionic emission; dielectrics; ferromagnetism. Mailing Add: 1144 Westmoreland Rd Alexandria VA 22308

WRIGHT, RUSSELL EMERY, b Muscatine, Iowa, June 19, 39; m 63; c 3. MEDICAL ENTOMOLOGY. Educ: Iowa State Univ, BSc, 63, MS, 66; Univ Wis-Ann Arbor, PhD(entom), 69. Prof Exp: Asst prof, 69-75, ASSOC PROF ENTOM, UNIV GUELPH, 75- Concurrent Pos: Mem, Can Comt Biting Flies, 74- Mem: Entom Soc Am; Am Mosquito Control Asn. Res: Behavior, biology and control of insect pests of livestock; mosquitoes and blackflies; arthropod borne viruses. Mailing Add: Dept of Environ Biol Univ of Guelph Guelph ON Can

WRIGHT, SEWALL, b Melrose, Mass, Dec 21, 89; m 21; c 3. GENETICS, EVOLUTION. Educ: Lombard Col, BS, 11; Univ Ill, MS, 12; Harvard Univ, ScD(zool), 15. Hon Degrees: ScD, Univ Rochester, 42, Yale Univ, 49, Harvard Univ, 51, Knox Col, 57, Western Reserve Univ, 58, Univ Chicago, 59, Univ Ill, 61, Univ Wis, 65; LLD, Mich State Col, 55. Prof Exp: Sr animal husbandman, Animal Husb Div, Bur Animal Indust, USDA, 15-25; from assoc prof to prof, 26-37, Ernest D Burton distinguished serv prof, 38-54, EMER PROF ZOOL, UNIV CHICAGO, 55- Concurrent Pos: Mem exec comt, Div Biol & Agr, Nat Res Coun, 31-33; Hitchcock prof, Univ Calif, 43; Fulbright prof, Univ Edinburgh, 49-50; mem comt biol effects of radiation, Nat Acad Sci, 55-; Leon J Cole prof genetics, Univ Wis-Madison, 55-60, emer prof, 60-; pres, Int Cong Genetics, Montreal, 58. Honors & Awards: Weldon Mem Medal, Oxford Univ, 47; Elliot Medal, Nat Acad Sci, 47, Kimber Award, 56; Lewis Prize, Am Philos Soc, 50; Nat Medal Sci, 66. Mem: Nat Acad Sci; AAAS; Am Acad Arts & Sci; Am Soc Naturalists (treas, 29-32, pres, 52); Am Philos Soc. Res: Genetics of guinea pig and populations; theory of evolution; mathematical theory of population genetics. Mailing Add: 3905 Council Crest Madison WI 53711

WRIGHT, SYDNEY COURTENAY, b Vancouver, BC, Oct 16, 23; m 48; c 3. PHYSICS. Educ: Univ BC, BA, 43; Univ Calif, PhD(physics), 49. Prof Exp: NSF fel, 49-50, res assoc, 50-55, asst prof, 55, assoc prof, 57-68, PROF PHYSICS, ENRICO FERMI INST, UNIV CHICAGO, 69- Concurrent Pos: Consult, Brookhaven Nat Lab, 53 & Argonne Nat Lab, 57-60. Mem: Am Phys Soc. Res: Experimental particle physics; particle accelerator design. Mailing Add: Dept of Physics Univ Chicago 5801 S Ellis Ave Chicago IL 60637

WRIGHT, THEODORE RICHARD, b Washington, DC, Mar 5, 15; m 41; c 1. PLANT PATHOLOGY. Educ: Pa State Col, BS, 37. Prof Exp: Pathologist & asst county agent, Ala, 40-43 & 46; from asst plant pathologist to assoc plant pathologist, Bur Plant Indust, USDA, 46-52, assoc plant pathologist, Qual Maintenance & Improv, Agr Mkt Serv, 52-55; pathologist, Standard Fruits, Inc, 55-65; AGR RES TECHNICIAN, QUAL MAINTENANCE RES, AGR RES SERV, USDA, 65- Mem: Am Phytopath Soc. Res: Post harvest diseases; fruits and vegetables. Mailing Add: PO Box 1844 Wenatchee WA 98801

WRIGHT, THEODORE ROBERT FAIRBANK, b Kodaikanal, India, Apr 10, 28; US citizen; m 51. DEVELOPMENTAL GENETICS. Educ: Princeton Univ, AB, 49; Wesleyan Univ, MA, 54, PhD(zool), 59. Prof Exp: Asst prof biol, Johns Hopkins Univ, 59-65; assoc prof, 65-75, PROF BIOL, UNIV VA, 75- Concurrent Pos: Mem genetics study sect, NIH, 72-74; fel, Max Planck Inst Biol, 75-76. Mem: Fel AAAS; Genetics Soc Am; Soc Develop Biol; Am Soc Zoologists. Res: Developmental genetics of embryonic mutants in Drosophila; biochemical genetics of Drosophila. Mailing Add: Dept of Biol Univ of Va Charlottesville VA 22901

WRIGHT, THOMAS DODSON, b Temple, Tex, Mar 10, 36; m 58; c 2. ZOOLOGY, LIMNOLOGY. Educ: Univ Tex, BA, 61; Univ Wis, MS, 64, PhD(zool), 68. Prof Exp: Spec investr pesticides, Wis Dept Natural Resources, 67-68, sr biologist, 68; res assoc thermal pollution, Dept Water Resources, Cornell Univ, 68-69; asst prof wetlands & assoc marine scientist, Va Inst Marine Sci, 69-70; asst prof, 70-73, ASSOC PROF BIOL SCI, MICH TECHNOL UNIV, 73- Mem: Am Fisheries Soc; Am Soc Ichthyologists & Herpetologists. Res: Water pollution; ichthyology; bioengineering. Mailing Add: Dept of Biol Sci Mich Technol Univ Houghton MI 49931

WRIGHT, THOMAS L, b Chicago, Ill, July 26, 35; m 58; c 2. GEOLOGY, PETROLOGY. Educ: Pomona Col, AB, 57; Johns Hopkins Univ, PhD(geol), 61. Prof Exp: Geologist, Washington, DC, 61-64, staff geologist, Hawaiian Volcano Observ, Hawaii Nat Park, 64-69, geologist, Geol Div, Washington, DC, 69-74, WITH US GEOL SURV, RESTON, VA, 74- Mem: Mineral Soc Am; Geochem Soc; Am Geophys Union. Res: Igneous petrology; petrology and mineralogy of Hawaiian basalt; study of crystallization of basalt in the lava lakes of Kilauea volcano; chemical and stratigraphic study of the basalts of the Columbia River plateau. Mailing Add: US Geol Surv 959 Nat Ctr Reston VA 22092

WRIGHT, THOMAS PAYNE, b Ft Worth, Tex, Dec 23, 43; m 66; c 2. PHYSICS, PLASMA PHYSICS. Educ: St Bonaventure Univ, BS, 66; NMex State Univ, MS, 68, PhD(physics), 69. Prof Exp: STAFF MEM PHYSICS, SANDIA LABS, 69- Mem: Am Phys Soc. Res: Plasma waves and instabilities; electromagnetic theory; theory of kinetic equations; laser-plasma interaction; statistical mechanics; relativistic electron beams. Mailing Add: Sandia Labs PO Box 5800 Albuquerque NM 87115

WRIGHT, THOMAS PERRIN, JR, b Great Falls, SC, June 23, 39; m 61; c 2. TOPOLOGY. Educ: Davidson Col, AB, 60; Univ Wis, MA, 63, PhD(math), 67. Prof Exp: ASSOC PROF MATH, FLA STATE UNIV, 67- Concurrent Pos: NSF grant, Fla State Univ, '67-72. Mem: Am Math Soc. Res: Topology of manifolds. Mailing Add: 1808 Skyland Dr Tallahassee FL 32303

WRIGHT, WALTER EUGENE, b Terre Haute, Ind, July 16, 24; m 51; c 5. BIOPHARMACEUTICS. Educ: Purdue Univ, BS, 48, MS, 50, PhD(pharmaceut chem), 53. Prof Exp: Sr biochemist, 53-65, res scientist, 65-69, RES ASSOC, ELI LILLY & CO, 69- Mem: Am Chem Soc; NY Acad Sci; Am Soc Microbiol. Res: Intestinal and drug absorption; active transport; study of the absorption, metabolism and excretion of new medicinal agents in experimental animals. Mailing Add: 7553 N Audubon Rd Indianapolis IN 46250

WRIGHT, WAYNE GORDON, b Yankton, SDak, June 13, 35; m 57; c 3. WEED SCIENCE, ENTOMOLOGY. Educ: SDak State Univ, BS, 57, MS, 61. Prof Exp: Asst mgr state seed lab, SDak State Univ, 57-62, mgr found seed stock div, 63-64, in charge of weed res, 64-66; regional tech specialist agr herbicides, Dow Chem Co, 67-71, field group leader, Dow Chem USA, 71-74, PROD TECH SPECIALIST, DOW CHEM USA, 74- Mem: Weed Sci Soc Am. Res: Seed technology; crop production; weed control in crops and non-crop areas; brush control; harvest aid defoliation; insect and disease control in crops and ornamentals. Mailing Add: Dow Chem USA PO Box 1706 Midland MI 48640

WRIGHT, WAYNE MITCHELL, b Sanford, Maine, July 12, 34; m 59; c 4. ACOUSTICS. Educ: Bowdoin Col, AB, 56; Harvard Univ, MS, 57, PhD(appl physics), 61. Prof Exp: Res fel, Harvard Univ, 61-62; from asst prof to assoc prof, 62-75, PROF PHYSICS & CHMN DEPT, KALAMAZOO COL, 75- Mem: Acoust Soc Am; Am Asn Physics Teachers. Res: Physical acoustics; ultrasonics; experimental studies of finite-amplitude sound phenomena in air. Mailing Add: Dept of Physics Kalamazoo Col Kalamazoo MI 49007

WRIGHT, WELLESLEY HORTON, b Los Angeles, Calif, Apr 26, 32; m 55; c 2. DENTISTRY, PERIODONTOLOGY. Educ: Univ Wash, DDS, 59, BA, 61; Baylor Univ, MS, 63. Prof Exp: Dent intern, USPHS, Tex, 59-60, staff dent off, 60-61; clin instr & res assoc dent, Baylor Univ, 61-63; staff periodontologist, USPHS, NY, 63-64; vis lectr, Royal Dent Col, Denmark, 64-66; hon staff mem, 66; asst prof, Sch Dent Med, Univ Pa, 66-68; assoc prof periodont, Sch Dent, 68-75, assoc clin prof surg, Sch Med, 69-75, PROF PERIODONT & DIR GRAD PERIODONT, SCH DENT, UNIV ORE, 75-, CLIN PROF SURG, SCH MED, 75- Concurrent Pos: Fulbright scholar, Royal Dent Col, Denmark, 64-66; master clinician, Marion County Res Soc, 69-70 & Medford Periodont Res Soc, 69-71. Mem: AAAS; Sigma Xi; fel Am Col Dent; fel Int Col Dent; Am Dent Asn. Res: Epidemiology of periodontal disease; healing of periodontal surgical procedures; quantitation of blood loss during periodontal surgical procedures; immunology and periodontal disease. Mailing Add: Univ of Ore Dent Sch 611 S W Campus Dr Portland OR 97201

WRIGHT, WILBUR HERBERT, b Kansas City, Mo, Feb 28, 20; m 43; c 2. PHYSICS. Educ: Oberlin Col, AB, 42; Rutgers Univ, PhD(physics), 52. Prof Exp: From asst to instr physics, Rutgers Univ, 47-52; asst prof, Univ NH, 52-56; assoc prof, 56-60, PROF PHYSICS, COLO COL, 60- Concurrent Pos: NSF fac fel, Stanford Univ, 59-60. Mem: Am Phys Soc; Am Asn Physics Teachers. Res: Superconductivity; applied mathematics; measurement of small magnetic susceptibilities; non-equilibrium thermodynamics. Mailing Add: Dept of Physics Colo Col Colorado Springs CO 80903

WRIGHT, WILLIAM BLYTHE, JR, b Washington, DC, Sept 29, 18; m 48; c 2. PHARMACEUTICAL CHEMISTRY. Educ: Univ Va, BS, 39; Univ Mich, PhD(chem), 42. Prof Exp: Resin res chemist, Rohm and Haas Co, Pa, 42-47; pharmaceut res chemist, Bound Brook Labs, 47-55, PHARMACEUT RES CHEMIST, LEDERLE LABS, AM CYANAMID CO, 55- Mem: Am Chem Soc. Res: Coatings; plywood adhesives; ion exchange resins; pharmaceuticals. Mailing Add: 18 Clinton Pl Woodcliff Lake NJ 07675

WRIGHT, WILLIAM ERSKINE, nuclear physics, see 12th edition

WRIGHT, WILLIAM HERBERT, III, b Newton, Mass, Feb 13, 43. STRUCTURAL GEOLOGY. Educ: Middlebury Col, BA, 65; Ind Univ, Bloomington, MA, 67; Univ Ill, Urbana, PhD(geol), 70. Prof Exp: Explor geologist, Chevron Oil Co, Colo, 66; asst prof, 69-72, ASSOC PROF GEOL, CALIF STATE COL, SONOMA, 72- Mem: AAAS; Geol Soc Am; Am Geophys Union. Res: Folding, metamorphic structures; structural evolution of mountain belts; structure and geologic history of Calaveras Formation, Sierra Nevada, California. Mailing Add: Dept of Geol Calif State Col Sonoma Rohnert Park CA 94928

WRIGHT, WILLIAM LELAND, b Darbyville, Ohio, Aug 14, 30; m 52; c 3. WEED SCIENCE. Educ: Ohio Univ, BS, 53, MS, 57; Purdue Univ, PhD(plant physiol), 64. Prof Exp: Plant physiologist, Dow Chem Co, Tex, 57-58; from plant physiologist to sr plant physiologist, 58-65, head plant sci res, 65-72, prod plans adv, Elanco Prod Co, 72-73, REGULATORY SERV ADV, ELI LILLY & CO, 73- Mem: AAAS; Am Soc Plant Physiologists; Weed Sci Soc Am. Res: Chemical weed control; plant growth regulation, insecticides and aquatic weed control; fate of herbicides in environment; product planning; regulatory affairs. Mailing Add: 1410 Bowman Dr Greenfield IN 46140

WRIGHT, WILLIAM RAY, b Iola, Wis, Agu 16, 41; m 63. SOIL MORPHOLOGY. Educ: Wis State Univ-River Falls, BS, 66; Univ Md, College Park, MS, 69, PhD(soils), 72. Prof Exp: ASST PROF SOIL SCI, UNIV RI, 72- Mem: Am Soc Agron; Soil Sci Soc Am; Soil Conserv Soc Am. Res: Soil genesis, classification and land use. Mailing Add: Dept of Plant & Soil Sci Univ of RI Kingston RI 02881

WRIGHT, WILLIAM REDWOOD, b Philadelphia, Pa, Sept 17, 27; m 56; c 3. PHYSICAL OCEANOGRAPHY. Educ: Princeton Univ, BA, 50; Univ RI, MS, 65, PhD(oceanog), 70. Prof Exp: Teacher, RI Pvt Sch, 50-52; pub info officer, 60-62, res asst phys oceanog, 62-70, ASST SCIENTIST, WOODS HOLE OCEANOG INST, 70- Concurrent Pos: Reporter, Auburn Citizen-Advertiser, NY, 52-54 & Providence J, RI, 54-60. Mem: AAAS; Am Geophys Union; Am Soc Limnol & Oceanog. Res: Deep circulation of the world oceans; coastal circulation. Mailing Add: Box 54 Woods Hole MA 02543

WRIGHT, WILLIAM ROBERT, b Cleveland, Ohio, Nov 24, 28. THEORETICAL PHYSICS. Educ: Harvard Univ, AB, 51, MA, 52, PhD(physics), 57. Prof Exp: Asst prof physics, Univ Kans, 57-61; assoc prof, 61-62, head dept, 62-66, PROF PHYSICS, UNIV CINCINNATI, 62- Concurrent Pos: Vis mem dept physics, Oxford Univ, 67-68. Mem: AAAS; Am Phys Soc; Am Asn Physics Teachers. Res: Solid state effects in nuclear orientation; Green's function techniques in magnetism; variational methods in statistical mechanics. Mailing Add: Dept of Physics Univ of Cincinnati Cincinnati OH 45221

WRIGHT, WILLIAM WYNN, b Baltimore, Md, Aug 13, 23; m 45; c 5. CHEMISTRY. Educ: Loyola Col, Md, BS, 44; Georgetown Univ, MS, 46, PhD(biochem), 48. Prof Exp: Chemist, Nat Bur Stand, Washington, DC, 45; chief antibiotic chem br, 45-55, dir antibiotic control labs, 55-57, dir antibiotic res, 57-64, dep dir div antibiotics & insulin cert, 64-69, dep dir pharmaceut res & testing, 69-71, dir div drug biol, 71-75, DEP ASSOC DIR PHARMACEUT RES & TESTING, BUR DRUGS, US FOOD & DRUG ADMIN, 75- Concurrent Pos: Mem, WHO Expert Panels on Antibiotics, 60-75 & Biol Stand, 75- Honors & Awards: Superior Serv Award, US Dept Health, Educ & Welfare, 64; Award of Merit, US Food & Drug Admin, 71. Mem: Fel AAAS; Am Chem Soc; fel Asn Off Anal Chemists; NY Acad Sci; fel Acad Pharmaceut Sci. Res: Antibiotic testing by chemical, physical, microbial and biological methods; absorption, excretion, distribution and tissue residues of antibiotics; bacterial susceptibility; pharmaceutical analysis. Mailing Add: 1301 Dilston Pl Silver Spring MD 20903

WRIGHTON, MARK STEPHEN, b Jacksonville, Fla, June 11, 49; m 68. PHOTOCHEMISTRY, INORGANIC CHEMISTRY. Educ: Fla State Univ, BS, 69; Calif Inst Technol, PhD(chem), 72. Prof Exp: ASST PROF CHEM, MASS INST TECHNOL, 72- Concurrent Pos: Petrol Res Fund starter grant, Mass Inst Technol, 72-75, Res Corp grant, 72- Honors & Awards: Herbert Newby McCoy Award, Calif Inst Technol, 72. Mem: AAAS; Am Chem Soc. Res: Excited state processes in transition metal containing molecules. Mailing Add: Dept of Chem 6-331 Mass Inst of Technol Cambridge MA 02139

WRIGHTSON, JOHN MOWLL, organic chemistry, see 12th edition

WRIGLEY, ARTHUR NELSON, b Philadelphia, Pa, Apr 24, 14; m 44; c 2. ORGANIC CHEMISTRY. Educ: Haverford Col, AB, 37, MS, 43; Temple Univ, PhD(chem), 58. Prof Exp: Instr, Fla Prep Sch, 39-41 & Country Day, Pa, 41-42; asst, Princeton Univ, 43; chemist, Eastern Regional Res Lab, 43-61, HEAD PLASTICS INVESTS, ANIMAL FAT PROD LAB, USDA, 61- Mem: AAAS; Am Chem Soc; Am Oil Chemists' Soc. Res: Allyl ethers; oxidation; copolymers; coatings; detergents; periodic table; urethane foams; internal plasticization; polyblends. Mailing Add: Eastern Regional Res Ctr US Dept of Agr Philadelphia PA 19118

WRIGLEY, CHARLES YONGE, physics, see 12th edition

WRIGLEY, WALTER, b Brockton, Mass, Mar 26, 13; m 41; c 2. PHYSCIS. Educ: Mass Inst Technol, SB, 34, ScD(appl physics), 41. Prof Exp: Physicist, Interchem Corp, 34-37 & United Color & Pigment Co, 37; asst aircraft instruments, Mass Inst Technol, 38-40; proj engr, Sperry Gyroscope Co, 40-46; dep dir instrumentation lab, 46-56, assoc prof aeronaut eng, 51-56, prof instrumentation & astronaut & ed dir, C S Draper Lab, 56-75, EMER PROF INSTRUMENTATION & ASTRONAUT, MASS INST TECHNOL, 75- Concurrent Pos: Mem adv coun, US Air Force Acad, 66-, chmn, 70-, mem adv group, Aeronaut Res & Develop, 65-69; mem bd acad visitors, New Eng Aeronaut Inst, 70. Honors & Awards: Superior Achievement Award, Inst Navig, 73. Mem: AAAS; fel Am Inst Aeronaut & Astronaut; Am Phys Soc; Optical Soc Am; Int Acad Astronaut. Res: Aircraft and marine instruments for navigation and control; vibration measuring equipment; design of optical instruments; weapon systems; inertial guidance. Mailing Add: 93 Grand View Ave Wollaston MA 02170

WRIST, PETER ELLIS, b Mirfield, Eng, Oct 9, 27; m 55; c 3. PHYSICS, MATHEMATICS. Educ: Univ Cambridge, BA, 48, MA, 52; Univ London, MSc, 52. Prof Exp: Res physicist, Brit Paper & Board Indust Res Asn, 49-52 & Que NShore Paper Co, Can, 52-56; res physicist, 56-60, assoc dir res, Cent Res Labs, 60-62, dir res, 62-65, mgr res & eng, 65-68, VPRES, MEAD CORP, 68- Concurrent Pos: Chmn bd gov, Nat Coun Paper Indust Air & Stream Improv, Inc. Honors & Awards: Smith Medal, 54; Weldon Medal, 56. Mem: Tech Asn Pulp & Paper Indust; NY Acad Sci; Brit Inst Physics. Res: Fluid mechanical behavior of fibre suspensions; high speed paper manufacture; filtration and associated problems. Mailing Add: 3390 Sunny Crest Lane Dayton OH 45419

WRISTERS, HARRY (JAN), b Helmond, Neth, July 18, 39; US citizen; m 60; c 2. ORGANIC CHEMISTRY. Educ: Hope Col, AB, 60; Ohio Univ, PhD(org chem), 64. Prof Exp: Res chemist, Esso Res & Eng Co, 64-72, SR STAFF CHEMIST, EXXON CHEM CO, 72- Mem: Am Chem Soc; The Chem Soc. Res: Pyrolysis of non-conjugated alkadiynes; Ziegler-Natta catalysis; polymer morphology. Mailing Add: Exxon Chem Co PO Box 4255 Baytown TX 77520

WRISTON, JOHN CLARENCE, JR, b Boston, Mass, Aug 12, 25; m 45; c 4. BIOCHEMISTRY. Educ: Univ Vt, BS, 48; Columbia Univ, PhD(biochem), 53. Prof Exp: Nat Found Infantile Paralysis fel & instr biochem, Sch Med, Univ Colo, 53-55; from asst prof to assoc prof, 55-69, PROF CHEM, UNIV DEL, 69- Mem: Am Soc Biol Chemists. Res: Protein chemistry; structure and mechanism of action of L-asparaginase. Mailing Add: Dept of Chem Univ of Del Newark DE 19711

WRISTON, ROBERT S, physics, see 12th edition

WROBEL, JOSEPH JUDE, b Chicago, Ill, Mar 18, 47; m 70; c 1. CHEMICAL PHYSICS. Educ: Loyola Univ, Chicago, BS, 68; Univ Fla, PhD(chem physics), 76. Prof Exp: RES PHYSICIST, EASTMAN KODAK CO RES LABS, 75- Mem: Am Chem Soc. Res: Image storage and display systems. Mailing Add: Eastman Kodak Co Bldg 81 Kodak Park Rochester NY 14650

WROBEL, JOSEPH STEPHEN, b Syracuse, NY, Aug 15, 39. SOLID STATE PHYSICS. Educ: Syracuse Univ, BS, 61, MS, 64, PhD(physics), 67. Prof Exp: Res asst physics, Syracuse Univ, 61-66; MEM TECH STAFF, TEX INSTRUMENTS INC, 66- Mem: AAAS; Am Phys Soc. Res: Semiconductor materials; photoconductivity; infrared physics. Mailing Add: Tex Instruments Inc PO Box 5936 MS 118 Dallas TX 75222

WROBEL, STANLEY JOHN, JR, biochemistry, organic chemistry, see 12th edition

WROBEL, THEODORE FRANK, b New York, NY, Jan 15, 42; m 68; c 1. SOLID STATE PHYSICS. Educ: Calif State Univ, BS, 68, MS, 71. Prof Exp: Res asst solid state physics, Gen Atomic Co, 68-70, sr physicist, 70-73; STAFF PHYSICIST, IRT CORP, 73- Res: Radiation calorimetry, including both high and low energy photons and electrons; radiation-stimulated failure mechanisms in semiconductor devices and development of medical applications of fiber optic systems. Mailing Add: IRT Corp 7650 Convoy Ct San Diego CA 92138

WROGEMANN, KLAUS, b Berlin, Ger, Dec 8, 40; m 67; c 2. BIOCHEMISTRY, MEDICINE. Educ: Univ Marburg, MD, 66; Univ Man, PhD(biochem), 69. Prof Exp: Intern surg & med, Med Hosp, Hanover, Ger, 70; asst prof, 70-74, ASSOC PROF BIOCHEM, FAC MED, UNIV MAN, 74- Concurrent Pos: Muscular Dystrophy Asn Can res grant, Fac Med, Univ Man, 71-74 & 74-77; Can Heart Found res grant, 74-75 & 75-76. Mem: Can Biochem Soc; Am Chem Soc; Int Study Group Res Cardiac Metab. Res: Metabolism of normal and dystrophic heart and skeletal muscle; protein synthesis in genetic diseases. Mailing Add: Dept of Biochem Fac of Med Univ of Man Winnipeg MB Can

WROLSTAD, RONALD EARL, b Oregon City, Ore, Feb 5, 39. FOOD SCIENCE, AGRICULTURAL CHEMISTRY. Educ: Ore State Univ, BS, 60; Univ Calif, Davis, PhD(agr chem), 64. Prof Exp: Grad scientist protein chem sect, Unilever Res Lab, Eng, 64-65; res assoc, 65-66, asst prof, 66-71, ASSOC PROF FOOD SCI, ORE STATE UNIV, 71- Concurrent Pos: Sabbatical leave, Plant Dis Div, Dept Sci & Indust Res, Auckland, NZ, 72-73. Mem: Inst Food Technol; Am Chem Soc. Res: Flavor chemistry; chemistry of natural products; anthocyanin pigments; flavonoids; food chemistry. Mailing Add: Dept of Food Sci Ore State Univ Corvallis OR 97331

WRONA, WLODZIMIERZ STEFAN, b Deszno, Poland, Oct 16, 12; m 44; c 3. GEOMETRY. Educ: Jagiellonian Univ, MPh, 34, PhD(math), 45; Acad Mining & Metall, Cracow, docent, 49. Prof Exp: Asst lectr math, Acad Mining & Metall, Cracow, 32-45, from asst prof to assoc prof, 45-60; assoc prof, Warsaw Tech Univ, 60-65; prof, Univ Ghana, Legon, 65-66, Warsaw Tech Univ, 66-69 & Ahmadu Bello Univ, Nigeria, 69-70; PROF MATH, CALIF STATE UNIV, HAYWARD, 70- Concurrent Pos: Brit Coun scholar, Univ Leeds, 47-48. Honors & Awards: S Zaremba Award, Polish Math Soc, 50. Mem: Am Math Soc; Polish Math Soc; London Math Soc. Res: Metric differentiable manifolds; properties of curvature; Lie groups. Mailing Add: Dept of Math Calif State Univ 25800 Hillary St Hayward CA 94542

WRONSKI, CHRISTOPHER ROMAN, b Warsaw, Poland, Mar 2, 39; US citizen; m 63; c 4. SOLID STATE PHYSICS. Educ: Univ London, BS, 60, PhD(physics) & DIC, 63. Prof Exp: Mem staff physics, Cent Res Labs, Minn Mining & Mfg Co, 63-66; MEM TECH STAFF PHYSICS, RCA LABS, 66- Mem: Am Phys Soc. Res: Semiconductor and insulator junctions with particular emphasis on photovoltaic devices in both photoelectric and solar cell applications. Mailing Add: RCA Labs Princeton NJ 08540

WRONSKI, JOSEPH PETER, b Minneapolis, Minn, July 15, 14; m 38; c 1. PHYSICS.

Educ: Univ Minn, BCh, 40, MS, 47. Prof Exp: Chemist, Minneapolis Munic Waterworks, 40-41; res engr, Twin Cities Ord Plant, 41-44; physicist, Minn Mining & Mfg Co, 44-47; sr physicist, Shell Oil Co, Tex, 47-48; res engr, Battelle Mem Inst, 48-56; contract adminstr, Bettis Atomic Power Div, Westinghouse Elec Corp, 56-59; MGR ANAL SERV, R/M FRICTION MAT CO, 59- Mem: AAAS; Assoc Am Phys Soc; Sigma Xi. Res: Infrared and emission spectroscopy; x-ray diffraction; operations analysis; materials properties. Mailing Add: R/M Friction Mat Co 123 E Stiegel St Manheim PA 17545

WROTENBERY, PAUL TAYLOR, b Pollok, Tex, Apr 24, 34; m 54; c 2. DATA PROCESSING. Educ: Univ Tex, BS, 58, MA, 62, PhD(physics, chem), 64. Prof Exp: Res scientist, Defense Res Labs, Tex, 58; sr scientist & proj dir, Tracor Inc, 58-64; sci consult, Int Bus Mach Corp, Washington, DC, 64-68; dir comput serv dept, Tracor Inc, 68, sr vpres, Tracor Comput Corp, 68-70; pres & chmn bd, United Systs Int, 70-74; PRES, EQUIMATIC CO & DIR, INFORMATICS INC, 74- Res: Surfaces and solid-liquid interfaces; semiconductor electrolyte interface properties; signal processing; digital filtering; underwater sound; acoustics; digital control theory analysis; management information systems; systems analysis. Mailing Add: 3411 Monte Vista Austin TX 78731

WRUCKE, CHESTER THEODORE, JR, b Portland, Ore, Oct 24, 27; m 54; c 3. GEOLOGY. Educ: Stanford Univ, BS, 51, MS, 52, PhD, 66. Prof Exp: GEOLOGIST, US GEOL SURV, 52- Mem: Geol Soc Am. Res: Petrology of igneous and metamorphic rocks; structural geology. Mailing Add: US Geol Surv 345 Middlefield Rd Menlo Park CA 94025

WU, ALFRED CHI-TAI, b Chekiang, China, Jan 24, 33; m 67; c 2. THEORETICAL PHYSICS. Educ: Wheaton Col, BS, 55; Univ Md, PhD(physics), 60. Prof Exp: Mem, Inst Advan Study, 60-62; asst prof, 62-66, ASSOC PROF PHYSICS, UNIV MICH, ANN ARBOR, 66- Concurrent Pos: John Simon Guggenheim Mem Found fel, 68-69. Mem: Am Phys Soc. Res: Quantum field theory; particle physics. Mailing Add: Dept of Physics Univ of Mich Ann Arbor MI 48104

WU, CHANGSHENG, b Liaoning, Manchuria, China, Oct 3, 23; nat US; m 55; c 2. EXPLORATION GEOPHYSICS. Educ: Nat Southwestern Assoc Univs, China, 44; Univ Tex, BS, 47; Rice Univ, MA, 63, PhD(geophys), 66. Prof Exp: Seismologist, United Geophys Co, 47-51; party chief, Tex Seismog Co, 52-53; rev geophysicist, Precision Explor Co, 53-55 & Ralph E Fair, Inc, 55-56; seismic prospecting agent, Tech Assistance Admin, UN, 57-60; asst geophys, Rice Univ, 62-64; res geophysicist, Western Geophys Co, 66-69, sr res geophysicist, 69-74, SR STAFF SCIENTIST, WESTERN GEOPHYS CO, LITTON INDUSTS, INC, 74- Mem: Soc Explor Geophysicists. Res: Elastic wave propagation. Mailing Add: 6522 Redding Rd Houston TX 77036

WU, CHENG-TSU, b China, Oct 9, 22; US citizen; m 51; c 1. GEOGRAPHY OF CHINA, GEOGRAPHY. Educ: Nat Cent Univ, Taiwan, BSc, 45; Ohio State Univ, MA, 51; Clark Univ, PhD(geog), 58. Prof Exp: ASSOC PROF GEOG, HUNTER COL, 57- Mem: Asn Am Geog; Asn Asian Studies; Am Geog Soc; NY Acad Sci. Res: Chinese and Chinatowns in America as a social-geographical study. Mailing Add: Dept of Geol & Geog Hunter Col 695 Park Ave New York NY 10021

WU, CHENG-WEN, b Taipei, Taiwan, June 19, 38; m 63; c 3. BIOPHYSICS, BIOCHEMISTRY. Educ: Nat Taiwan Univ, MD, 64; Case Western Reserve Univ, PhD(biochem), 69. Prof Exp: Assoc phys biochem, Cornell Univ, 69-71; asst prof, 72-74, ASSOC PROF BIOPHYS, ALBERT EINSTEIN COL MED, 75- Concurrent Pos: NIH spec fel biophys, Yale Univ, 71-72; Am Cancer Soc res grant, Albert Einstein Col Med, 72-75, NIH res grant, 72-79 & res career develop award, 72-77. Mem: AAAS; Am Chem Soc; Am Soc Biol Chem; Biophys Soc; NY Acad Sci. Res: Regulation and mechanism of gene expression; carcinogenesis; optical studies of nucleic acid and protein interaction; fast reactions in biological systems; absorption and emission spectroscopy. Mailing Add: Dept of Biophys Albert Einstein Col of Med Bronx NY 10461

WU, CHIEN-SHIUNG, b Shanghai, China, May 29, 12; nat US; m 42; c 1. PHYSICS. Educ: Nat Cent Univ, China, BS, 34; Univ Calif, PhD, 40. Hon Degrees: DSc, Princeton Univ, 58, Smith Col, 59, Goucher Col, 60, Rutgers Univ, 61, Yale Univ, 67. Prof Exp: PROF PHYSICS, COLUMBIA UNIV, 57- Concurrent Pos: Res Award, Res Corp, 59; Award, Am Asn Univ Women, 60; Comstock Award, Nat Acad Sci, 64. Mem: Nat Acad Sci; hon fel Royal Soc Edinburgh; Am Phys Soc; Chinese Acad Sci. Res: Nuclear physics; non-conservation of parity in beta decay. Mailing Add: Dept of Physics Columbia Univ New York NY 10027

WU, CHII-HUEI, b Taichung, Taiwan, May 14, 41; US citizen; m 65; c 1. MEDICAL MICROBIOLOGY. Educ: Chung-Hsing Univ, BS, 63; Okla State Univ, MS, 68, PhD(microbiol), 70. Prof Exp: NIH & USPHS fel, Dept Biol, Yale Univ, 70-72; SR MED MICROBIOLOGIST, LAB, FAXTON HOSP, 72- Mem: Sigma Xi; Am Soc Microbiol. Res: Regulatory control mechanisms governing the synthesis of bacterial enzymes and medical microbiology and immunology. Mailing Add: Clin Path Lab Faxton Hosp 1676 Sunset Ave Utica NY 13502

WU, CHING KUEI, b Hopei, China, Feb 26, 19; m 38; c 4. BIOLOGY, GENETICS. Educ: Cath Univ, Peiping, BS, 41, MS, 43; Northern Ill Univ, cert advan study, 63; Brown Univ, PhD(biol), 65. Prof Exp: Teacher high sch, China, 43-45; asst prof biol, Prov Med Col Honan, China, 4S-47; dean study high sch, Normal Univ China, 47-55; instr biol, Univ Taiwan, 55-61; res asst, Wash Univ, St Louis, 61-63; res assoc, Brown Univ, 63-65; asst prof, 65-69, ASSOC PROF BIOL, ADRIAN COL, 69- Mem: AAAS; Genetics Soc Am; Am Genetic Asn. Res: Radiation genetics and chemical mutagens, especially behavior of chromosomes in Drosophila melanogaster. Mailing Add: Dept of Biol Adrian Col Adrian MI 49221

WU, CHING-HSONG, b Taipei, Taiwan, Apr 22, 39; m 67; c 2. CHEMICAL KINETICS. Educ: Nat Taiwan Univ, BS, 62; NMex Highland Univ, MS, 65; Univ Calif, Berkeley, PhD(chem), 69. Prof Exp: SR RES CHEMIST, SCI RES LAB, FORD MOTOR CO, DEARBORN, 69- Mem: Am Chem Soc. Res: Gas phase kinetics; free radical reactions; mechanism of smog formation; energy transfer; carbon dioxide laser induced chemical reactions. Mailing Add: 38945 Worchestor Dr Westland MI 48185

WU, CHING-YONG, US citizen. INDUSTRIAL ORGANIC CHEMISTRY. Educ: Nat Taiwan Univ, BS, 55; Univ Pittsburgh, PhD(chem), 61. Prof Exp: Nat Res Coun Can fel chem, 63-65; fel, Mellon Inst Sci, 65-67; res chemist, 67-71, SR RES CHEMIST, GULF RES & DEVELOP CO, 71- Mem: Am Chem Soc; Catalysis Soc. Res: Physical organic chemistry and petrochemical research; homogeneous and heterogeneous catalysis, polymer synthesis and new polymerization. Mailing Add: Gulf Res & Develop Co PO Drawer 2038 Pittsburgh PA 15230

WU, CHIN-HUA SHIH, chemistry, see 12th edition

WU, CHISUNG, b Chishan, Taiwan, Jan 1, 32; m 57, 72; c 2. ORGANIC CHEMISTRY. Educ: Nat Taiwan Univ, BSc, 55; Case Western Reserve Univ, PhD(org chem), 60. Prof Exp: Res chemist, Harris-Intertype Corp, 59-61; res chemist, Chem & Plastics Opers Div, Union Carbide Corp, 61-65, proj scientist, 65-75, PROJ LEADER, UNION CARBIDE CAN LTD, 76- Mem: Am Chem Soc. Res: Organic synthesis; organophosphorus chemistry; polymer synthesis; organometallic chemistry; catalysis. Mailing Add: Union Carbide Can Ltd 10555 Metropolitan Blvd Montreal East PQ Can

WU, CHUNG, b Foochow, China, Dec 13, 19; nat US; m 50; c 4. BIOCHEMISTRY. Educ: Fukien Christian Univ, China, BS, 41; Univ Mich, MS, 48, PhD(biol chem), 52. Prof Exp: Res asst, Mayo Clin, 52-56; from instr to asst prof, 56-65, ASSOC PROF BIOL CHEM, MED SCH, UNIV MICH, ANN ARBOR, 65- Mem: Am Soc Biol Chem; Am Chem Soc; Am Asn Cancer Res. Res: Mechanisms of enzyme action; enzymology of cancer; metabolic controls. Mailing Add: 5685 Kresge Bldg Univ of Mich Med Sch Ann Arbor MI 48104

WU, CHUNG PAO, b Kwantung, China, May 15, 42; Hong Kong citizen; m 71; c 2. SOLID STATE ELECTRONICS. Educ: Yale Univ, BS, 65, MS, 66, PhD(physics), 68. Prof Exp: Res physicist, Electron Accelerator Lab, Yale Univ, 68-70; asst prof physics, Nanyang Univ, 70-72; MEM TECH STAFF SOLID STATE ELECTRONICS, SARNOFF RES CTR, RCA CORP, 73- Mem: Am Phys Soc; Inst Elec & Electronics Engrs. Res: Development of methods which accurately control the generation and implantation of ions in solids; the study and characterization of ion implantation techniques for semiconductor device fabrication. Mailing Add: David Sarnoff Res Ctr RCA Corp Princeton NJ 08540

WU, DAISY YEN, b Shanghai, China, June 12, 02; US citizen; m 24; c 5. NUTRITION. Educ: Ginling Col, China, BA, 21; Columbia Univ, MA, 23; Chinese-French Acad, China, dipl, 44; UN Lang Training Course, dipl, 63. Prof Exp: Asst biochem, Peking Union Med Col, 23-24; res assoc, Med Col Ala, 49-53; tech assoc nutrit, Food Conserv Div, UNICEF, 60-64; assoc pub health nutrit, Inst Human Nutrit, Columbia Univ, 64-71; RES ASSOC NUTRIT, ST LUKE'S HOSP CTR, 71- Mem: Am Inst Nutrit; fel Am Pub Health Asn; fel Royal Soc Health; Sigma Xi. Res: Proteins and amino acids; metabolic studies in man; vegetarian diets and dietaries; development, design and administration of an information retrieval system. Mailing Add: 449 E 14th St New York NY 10009

WU, ELLEN LEM, b Shanghai, China, Dec 6, 30; m 54; c 2. PHYSICAL CHEMISTRY. Educ: Carleton Col, BA, 54; Univ Minn, PhD(phys chem), 62. Prof Exp: Sr res chemist, Appl Res Div, 62-72, RES ASSOC, PROCESS RES & TECHNOL SERV DIV, MOBIL RES & DEVELOP CORP, 72- Mem: Am Chem Soc. Res: Catalysis research; physicochemical methods employed to elucidate nature of catalysts used in hydrocarbon conversion processes. Mailing Add: Res Dept Mobil Res & Develop Corp Paulsboro NJ 08066

WU, EN SHINN, b Kwangtung, China, Apr 20, 43. CHEMICAL PHYSICS. Educ: Nat Taiwan Univ, BS, 65; Cornell Univ, PhD(appl physics), 72. Prof Exp: Res assoc chem, Syracuse Univ, 72-74; ASST PROF PHYSICS, UNIV MD BALTIMORE COUNTY, 74- Mem: Am Phys Soc; Biophys Soc; Sigma Xi. Res: General properties of simple fluids and fluid mixtures, in particular, their thermodynamic behaviors near the critical points; the experimental techniques employed are primarily light-scattering and small angle X-ray scattering. Mailing Add: Dept of Physics Univ Md Baltimore County Baltimore MD 21228

WU, FA YUEH, b China, Jan 5, 32; m 63; c 2. THEORETICAL PHYSICS. Educ: Chinese Naval Col, BS, 54; Nat Tsing Hua Univ, MS, 59; Wash Univ, PhD(physics), 63. Prof Exp: Res assoc physics, Wash Univ, 63; asst prof, Va Polytech Inst, 63-67; from asst prof to assoc prof, 67-75, PROF PHYSICS, NORTHEASTERN UNIV, 75- Concurrent Pos: Sr Fulbright res fel, Australian Nat Univ, 73. Mem: Am Phys Soc. Res: Many body theory; theory of quantum liquids; statistical mechanics; solid state theory. Mailing Add: Dept of Physics Northeastern Univ Boston MA 02115

WU, FELICIA YING-HSIUEH, b Taipei, Taiwan, Feb 27, 39; US citizen; m 63; c 3. BIOPHYSICS, BIOCHEMISTRY. Educ: Nat Taiwan Univ, BS, 61; Univ Minn, MS, 63; Case Western Reserve Univ, PhD(org chem), 69. Prof Exp: Med technician biochem, US Naval Med Res Unit 2, Taipei, Taiwan, 63-65; res assoc biochem, Sect Biochem & Molecular Biol, Cornell Univ, 69-71; res assoc pharmacol, Yale Univ, 71; res fel, 72, assoc, 72-73, INSTR BIOPHYS, ALBERT EINSTEIN COL MED, YESHIVA UNIV, 73- Mem: Biophys Soc; Am Chem Soc; The Chem Soc. Res: Regulation and mechanism of gene transcription; carcinogenesis; optical studies of nucleic acid and protein interaction; fast reactions in biological systems; absorption and emission spectroscopy. Mailing Add: Dept of Biophys Albert Einstein Col of Med Bronx NY 10461

WU, FRANCIS TAMING, b Shanghai, China, May 27, 36; m 66. GEOPHYSICS. Educ: Nat Taiwan Univ, BS, 59; Calif Inst Technol, PhD(geophys), 66. Prof Exp: Asst prof geophys, Boston Col, 68-69; asst prof, 70-72, ASSOC PROF GEOPHYS, STATE UNIV NY, BINGHAMTON, 72- Concurrent Pos: Adv, Chinese Earthquake Res Ctr, 72. Mem: AAAS; Am Geophys Union; Seismol Soc Am. Res: Faulting as a dynamic phenomenon, its seismic radiation, rate of growth and driving mechanism. Mailing Add: Dept of Geol State Univ of NY Binghamton NY 13901

WU, HENRY CHI-PING, b Fenghua, Chekiang, China, May 21, 35; m 65; c 3. BIOCHEMISTRY. Educ: Nat Taiwan Univ, MD, 60; Harvard Univ, PhD(biochem), 66. Prof Exp: Jane Coffin Childs Fund fel, Mass Inst Technol, 66-67, Med Found fel, 67-69; asst prof, 69-74, ASSOC PROF MICROBIOL, HEALTH CTR, UNIV CONN, 74- Mem: AAAS; Am Soc Microbiol; Am Soc Biol Chemists. Res: Biochemistry of cell division in bacteria; biochemical and genetic studies of cell surface. Mailing Add: Dept of Microbiol Univ of Conn Health Ctr Farmington CT 06032

WU, HSIN-I, b Tokyo, Japan, May 25, 37; m 64; c 2. MATHEMATICAL PHYSICS. Educ: Tunghai Univ, Taiwan, BS, 60; Univ Mo, MS, 64, PhD(physics), 67. Prof Exp: Asst prof, 67-72, ASSOC PROF PHYSICS, SOUTHEAST MO STATE UNIV, 72- Mem: Am Asn Physics Teachers; Am Crystallog Asn. Res: Small angle x-ray scattering theory; calculating the chord distribution function; theory of x-ray study of membranes. Mailing Add: Dept of Physics Southeast Mo State Univ Cape Girardeau MO 63701

WU, HUNG-HSI, b Hong Kong, May 25, 40; US citizen; m 67. MATHEMATICS. Educ: Columbia Col, AB, 61; Mass Inst Technol, PhD(math), 63. Prof Exp: Res assoc math, Mass Inst Technol, 63-64; mem, Inst Advan Study, 64-65; from asst prof to assoc prof, 65-73, PROF MATH, UNIV CALIF, BERKELEY, 73- Concurrent Pos: Alfred P Sloan fel, 71-73. Mem: Am Math Soc. Res: Differential geometry; complex manifolds. Mailing Add: Dept of Math Univ of Calif Berkeley CA 94720

WU, JANG-MEI GLORIA, b Hangchow, Repub China, 48. ANALYTICAL MATHEMATICS. Educ: Nat Taiwan Univ, BS, 70; Univ Ill, Urbana, MS, 71,

PhD(math), 74. Prof Exp: Vis lectr math, Univ Ill, Urbana, 75-76, VIS ASST PROF MATH, IND UNIV, BLOOMINGTON, 76- Mem: Am Math Soc. Res: Complex function theory. Mailing Add: Dept of Math Ind Univ Bloomington IN 47401

WU, JIA-HSI, b Formosa, July 6, 26; m 56; c 1. PLANT PHYSIOLOGY, VIROLOGY. Educ: Univ Taiwan, BA, 50; Cornell Univ, MS, 52; Wash Univ, PhD(bot), 58. Prof Exp: Instr plant physiol, Univ Taiwan, 52-55; fel, Univ Wis, 58-59; asst botanist, Univ Calif, Los Angeles, 59-63; asst prof plant physiol, Tex Tech Col, 63-65; asst biologist, Univ Calif, San Diego, 65-66; asst prof cell physiol, 66-72, ASSOC PROF CELL , PHYSIOL, CALIF STATE POLYTECH UNIV, POMONA, 72- Concurrent Pos: NSF grants, 64-68 & 70-72. Res: Cell physiology. Mailing Add: Dept of Biol Sci Calif State Polytech Univ 3801 W Temple Ave Pomona CA 91768

WU, LILIAN SHIAO-YEN, b Peiking, China, July 6, 47. APPLIED MATHEMATICS. Educ: Univ Md, BS, 68; Cornell Univ, MS, 72, PhD(appl math), 74. Prof Exp: RES STAFF APPL MATH, THOMAS J WATSON RES CTR, IBM CORP, 73- Concurrent Pos: Vis scientist, Marine Biol Lab, 75- Mem: Soc Indust & Appl Math. Res: Population dynamics; game theory. Mailing Add: IBM Thomas J Watson Res Ctr Box 218 Yorktown Heights NY 10598

WU, MING TSUNG, b Taiwan, Rep China, Oct 9, 41; m 68; c 1. CHEMISTRY, FOOD SCIENCE. Educ: Nat Taiwan Univ, BS, 64, MS, 67; Utah State Univ, PhD(plant nutrit & biochem), 71. Prof Exp: Res assoc food sci, Utah State Univ, 71-72; res assoc food sci, Univ Ga, 72-73; RES ASST PROF FOOD SCI, UTAH STATE UNIV, 73- Mem: Am Soc Plant Physiol; Am Soc Microbiol; Inst Food Technol; Sigma Xi; Soc Econ Bot. Res: Food storage; postharvest physiology; mycotoxins; plant growth regulators; biochemistry and toxicology of potato toxicants. Mailing Add: Dept of Nutrit & Food Sci Utah State Univ Logan UT 84322

WU, MING-CHI, b Nantou, Taiwan, Nov 13, 40; US citizen; m 68; c 2. BIOCHEMISTRY. Educ: Nat Taiwan Univ, BS, 63; Univ Wis-Madison, MS, 68, PhD(biochem), 70. Prof Exp: Res fel physiol chem, Sch Med, Johns Hopkins Univ, 69-71; res assoc biochem, Sch Med, Univ Pittsburgh, 71-73; RES ASSOC HEMAT, HOWARD HUGHES MED INST, 74-, ASST PROF MED, SCH MED, UNIV MIAMI, 75- Mem: Sigma Xi. Res: Biochemistry of cultured cancer cells and hemapoiesis in the bone marrow. Mailing Add: 1207 Placetas Ave Coral Gables FL 33146

WU, MU TSU, b Changhwa, Taiwan, Oct 25, 29; US citizen; m 57; c 4. ORGANIC CHEMISTRY, MEDICINAL CHEMISTRY. Educ: Nat Taiwan Univ, BS, 51; Univ Md, PhD(pharmaceut chem), 61; Univ Tohoku, Japan, DSc(chem), 61. Prof Exp: Res chemist, Ord Res Inst, 51-58; res assoc pharmaceut chem, Univ Md, 58-62, assoc res prof, 64-65; res assoc chem, Univ NH, 62-64; sr res chemist, 65-72, RES FEL, MERCK & CO, INC, 72- Mem: AAAS; Am Chem Soc; Am Inst Chemists. Res: Synthetic organic and medicinal chemistry. Mailing Add: 35 Lance Dr Clark NJ 07066

WU, NING GAU, applied mechanics, heat transfer, see 12th edition

WU, PEI-HSING LIN, biochemistry, nutrition, see 12th edition

WU, RAY J, b Peking, China, Aug 14, 28; nat US; m 56; c 2. BIOCHEMISTRY. Educ: Univ Ala, BS, 50; Univ Pa, PhD, 55. Prof Exp: Asst instr biochem, Univ Pa, 51-55, Damon Runyon fel cancer res, 55-57; from asst to assoc, Pub Health Res Inst New York, 57-61, assoc mem, 61-66; assoc prof, 66-72, PROF BIOCHEM, CORNELL UNIV, 72- Mem: Am Soc Biol Chem. Res: DNA sequence analysis; cancer research; enzymology. Mailing Add: Dept of Biochem Wing Hall Cornell Univ Ithaca NY 14850

WU, RICHARD LI-CHUAN, b Tainan, Taiwan, Aug 21, 40; m 68; c 1. PHYSICAL CHEMISTRY. Educ: Nat Cheng Kung Univ, Taiwan, BS, 63; Univ Kans, PhD(chem), 71. Prof Exp: Res chemist, Aerospace Res Labs, Wright-Patterson Air Forc Base, 71-75; RES ASST PROF CHEM, WRIGHT STATE UNIV, 75- Mem: Am Chem Soc; Am Soc Mass Spectrometry; Sigma Xi. Res: Mass Spectrometry; high temperature chemistry; chemical kinetics; vaporization processes; ion-molecule reactions; thermodynamics. Mailing Add: 1571 Sycamore Ave Dayton OH 45432

WU, RODNEY TA-CHUAN, physical chemistry, see 12th edition

WU, ROY SHIH-SHYONG, b Shanghai, China, Nov 15, 44; US citizen. CELL BIOLOGY, BIOCHEMISTRY. Educ: Univ Calif, Berkeley, AB, 67; Albert Einstein Col Med, Bronx, NY, PhD(biochem), 72. Prof Exp: NIH fel develop biol, Dept Zool, Univ Calif, Berkeley, 72-74; fel cell biol, Children's Hosp, Oakland, Calif, 74-75; SCIENTIST CELL BIOL, BIOTECH RES LAB, INC, ROCKVILLE, MD, 75- Mem: AAAS; Am Chem Soc; Am Soc Biol Chemists; Soc Develop Biol. Res: Turnover rates of specific ribonucleic acids in mammary gland explants and epithelial cells; screening of new materials for anti-oncogenic virus related properties. Mailing Add: Biotech Res Lab 12601 Twinbrook Pkwy Rockville MD 20852

WU, SHIRLEY SHAO-NING, b China, May 9, 47; Repub China citizen. BIOCHEMISTRY. Educ: Nat Taiwan Univ, BS, 69; Univ Notre Dame, PhD(microbiol), 75. Prof Exp: RES ASSOC BIOCHEM, NC STATE UNIV, 74- Concurrent Pos: Fel, NC State Univ, 74-75. Res: Isolation and identification of a stimulating factor in mouse adipose glyceride biosynthesis system. Mailing Add: Dept of Biochem NC State Univ Raleigh NC 27607

WU, SHY-HSIEN, polymer chemistry, chemical engineering, see 12th edition

WU, SING-CHOU, b China, June 2, 36; m 64; c 2. STATISTICS, ECONOMETRICS. Educ: Nat Taiwan Univ, BA, 59; Utah State Univ, MS, 66; Colo State Univ, PhD(statist), 70. Prof Exp: Economist, Bank of China, 61-63; programmer, Comput Ctr, Utah State Univ, 65-66; asst statist, Colo State Univ, 66-69; asst prof, 69-72, ASSOC PROF STATIST, CALIF STATE POLYTECH UNIV, SAN LUIS OBISPO, 72- Mem: AAAS; Am Statist Asn; Inst Math Statist. Res: Design of experiment; statistical computation. Mailing Add: Dept of Comput Sci & Statist Calif Polytech State Univ San Luis Obispo CA 93401

WU, SOUHENG, b Tainan, China, Jan 16, 36; m 65; c 1. PHYSICAL CHEMISTRY. Educ: Nat Taiwan Univ, BS, 58; Univ Kans, PhD(chem), 65. Prof Exp: Asst researcher, Taiwan Sugar Corp, 58-61; res chemist, 65-69, staff chemist, 69-71, RES ASSOC, E I DU PONT DE NEMOURS & CO, INC, 71- Mem: Am Chem Soc. Res: Polymer physics; interfacial sciences; adhesion; coatings. Mailing Add: Exp Sta E I du Pont de Nemours & Co Inc Wilmington DE 19898

WU, SZU HSIAO ARTHUR, b Ho-Fei, China, Sept 15, 19; nat US; m 56; c 1. PHYSIOLOGY. Educ: Nat Cent Univ, China, BSc, 41; Ore State Col, MSc, 49, PhD(animal breeding), 52. Prof Exp: Instr, Oberlin Mem Sch, 44-45; asst, 47-52, instr physiol of reproduction, 52-57, asst prof animal sci, 57-66, ASSOC PROF ANIMAL SCI, ORE STATE UNIV, 67- Mem: Am Dairy Sci Asn; Am Soc Animal Sci; Soc Study Reproduction; Electron Micros Soc Am. Res: Physiology of reproduction; physiology and ultrastructure of germ cells. Mailing Add: Dept of Animal Sci Ore State Univ Corvallis OR 97331

WU, TAI TE, b Shanghai, China, Aug 2, 35; m 66. BIOPHYSICS, APPLIED MATHEMATICS. Educ: Univ Hong Kong, MB & BS, 56; Univ Ill, Urbana, BS, 58; Harvard Univ, SM, 59, PhD(eng), 61. Prof Exp: Res fel struct mech, Harvard Univ, 61-63; asst prof eng, Brown Univ, 63-65; res assoc biol chem, Harvard Med Sch, 65-66; from asst prof to assoc prof biomath, Med Col, Cornell Univ, 67-70; from assoc prof to prof physics & eng sci, 70-74, PROF BIOCHEM & MOLECULAR BIOL & ENG SCI, NORTHWESTERN UNIV, 74- Concurrent Pos: Res scientist, Hydronaut, Md, 62; res fel biol chem, Harvard Med Sch, 64; chmn comt biophys & mem comt biomed eng, Northwestern Univ, 72- Honors & Awards: Res Career Develop Award, NIH, 74. Mem: AAAS; Am Phys Soc; Am Soc Microbiol; Biophys Soc; NY Acad Sci. Res: Three-dimensional structures of macromolecules, especially those of antibodies; bacterial evolution; biochemistry; molecular biology. Mailing Add: Dept of Biochem & Eng Sci Northwestern Univ Evanston IL 60201

WU, TAI TSUN, b Shanghai, China, Dec 1, 33; m 67. PHYSICS. Educ: Univ Minn, BS, 53; Harvard Univ, SM, 54, PhD(appl physics), 56. Prof Exp: Jr fel, Soc Fels, 56-59, from asst prof to assoc prof, 59-66, GORDON McKAY PROF APPL PHYSICS, HARVARD UNIV, 66- Concurrent Pos: Mem, Inst Advan Study, 58-59, 60-61 & 62-63; vis prof & NSF sr fel, Rockefeller Univ, 66-67; Guggenheim Mem Found fel, Deutsches Elektronen-Synchotron, Hamburg, Ger, 70-71. Mem: Am Phys Soc; Inst Elec & Electronics Eng. Res: Electromagnetic theory; statistical mechanics; elementary particles. Mailing Add: Rm 308 Gordon McKay Lab of Appl Sci Harvard Univ Cambridge MA 02138

WU, TAI WING, b Hong Kong; Can citizen. BIOCHEMISTRY, BIOCHEMICAL ENGINEERING. Educ: Chinese Univ Hong Kong, BSc, 66; Univ Toronto, MSc, 68, PhD(biochem), 71. Prof Exp: Med Res Coun Can fel develop biol, 71-73; SR RES CHEMIST CLIN BIOCHEM & BIOCHEM ENG, EASTMAN KODAK CO RES LABS, 73- Concurrent Pos: Referee publ, Can J Biochem, Nat Res Coun Can, 73-; referee papers, Biochemistry, J Biol Chem, 73- Mem: Can Biol Socs; AAAS. Res: Mechanistic investigation of diverse biochemical phenomena of fundamental and practical interest; design and testing of totally novel techniques for clinical analyses. Mailing Add: Eastman Kodak Co 1669 Lake Ave Rochester NY 14650

WU, TING KAI, b Nanking, China, Aug 5, 37; m 66; c 2. CHEMISTRY. Educ: Mt St Mary's Col, Md, BS, 60; Columbia Univ, MA, 61, PhD(phys chem), 65. Prof Exp: Res assoc, Columbia Univ, 65; res chemist, 65-72, SR RES CHEMIST, E I DU PONT DE NEMOURS & CO, INC, 72- Concurrent Pos: Vis assoc prof, Nat Sci Coun China, Nat Taiwan Univ, 70-71. Mem: AAAS; Am Chem Soc; Am Inst Physics. Res: Structure and properties of molecules and macromolecules; spectroscopic analyses of molecular structure; polymer characterization. Mailing Add: Plastics Dept Exp Sta E I du Pont de Nemours & Co Inc Wilmington DE 19898

WU, TSE CHENG, b Hong Kong, Aug 21, 23; nat US; m 63; c 3. ORGANIC CHEMISTRY, POLYMER CHEMISTRY. Educ: Yenching Univ, China, BS, 46; Univ Ill, MS, 48; Iowa State Univ, PhD(org chem), 52. Prof Exp: From res asst to res assoc, Iowa State Univ, 48-53; res chemist, Textile Fibers Dept, E I du Pont de Nemours & Co, Inc, 53-60; res chemist, Silicone Prod Dept, Gen Elec Co, 60-71; SR RES CHEMIST, ABCOR, INC, 71- Mem: Am Chem Soc. Res: Monomer and polymer syntheses; acrylic, vinyl, polyester, polyamide and silicone polymers; rubber, fiber, plastic, emulsion and microencapsulation technology; biomaterials and biomedical implantation devices; organometallic and organosilicon compounds. Mailing Add: 21 Berkeley St Melrose MA 02176

WU, TSU MING, b Taipei, Taiwan, Dec 18, 36; m; c 2. PHYSICS. Educ: Univ Taiwan, BS, 59; Univ Pa, PhD(physics), 66. Prof Exp: Fel physics, Case Western Reserve Univ, 66-68; asst prof, 68-71, ASSOC PROF PHYSICS, STATE UNIV NY, BINGHAMTON, 71- Mem: Am Phys Soc. Res: Many-body problems in solid state physics, especially superconductivity and magnetism. Mailing Add: Dept of Physics State Univ of NY Binghamton NY 13901

WU, WEN-LI, b Szuchuan, China, Nov 13, 45; m 72; c 1. POLYMER PHYSICS. Educ: Nat Taiwan Univ, BS, 67; Mass Inst Technol, MS, 69, PhD(polymer physics), 72. Prof Exp: Res assoc polymer physics, Mass Inst Technol, 72-73; SR RES ENGR, MONSANTO TEXTILES CO, 73- Mem: Am Chem Soc. Res: Relations among the polymer morphology, its processing condition and its physical properties. Mailing Add: Tech Ctr Monsanto Textiles Co Pensacola FL 32575

WU, WILLIAM GAY, b Portland, Ore, Feb 5, 31; m 57; c 3. MEDICAL MICROBIOLOGY, IMMUNOLOGY. Educ: Ore State Univ, BS, 49, MS, 61; Univ Utah, PhD(immunol, cell biol), 62. Prof Exp: Lab asst soil microbiol, Ore State Univ, 57-58, res asst vet microbiol, 58-59; res asst immunol, Univ Utah, 59-62; from asst prof to assoc prof, 62-70, PROF MICROBIOL, CALIF STATE UNIV, SAN FRANCISCO, 70-, CHMN DEPT, 67- Concurrent Pos: Res Corp grants, 64-65; NSF grants, 65-67. Mem: AAAS; Am Soc Microbiol; Am Asn Immunol. Res: Natural resistance mechanisms; cellular immunity; acute disease mechanisms. Mailing Add: Dept of Microbiol San Francisco State Univ 1600 Holloway Ave San Francisco CA 94132

WU, YAO HUA, b Soochow, China, July 16, 20; nat US; m 50; c 2. ORGANIC CHEMISTRY. Educ: Chiao Tung Univ, BS, 43; Univ Nebr, MS, 48, PhD(org chem), 51. Prof Exp: Pharmaceut chemist, Int Chem Works, Shanghai, 43-47; res chemist, Smith-Dorsey Co, 51-53; res chemist, 53-60, sr res fel, 60-70, DIR CHEM RES, MEAD JOHNSON & CO, 70- Mem: Am Chem Soc. Res: Synthetic pharmaceuticals. Mailing Add: Mead Johnson Res Ctr Evansville IN 47721

WU, YING VICTOR, b Peiping, China, Nov 1, 31; nat US; m 60; c 1. PHYSICAL CHEMISTRY. Educ: Univ Ala, BS, 53; Mass Inst Technol, PhD(phys chem), 58. Prof Exp: Asst phys chem, Mass Inst Technol, 53-57; res assoc chem, Cornell Univ, 58-61; res chemist, 61-64, PRIN CHEMIST, NORTHERN REGIONAL LAB, USDA, 64- Mem: AAAS; Am Chem Soc; Am Asn Cereal Chemists; Am Soc Biol Chemists; Inst Food Technologists. Res: Physical chemistry of protein protein structure, cereal protein concentrates and isolates; optical rotatory disperson; hydrogen ion equilibria. Mailing Add: Northern Regional Lab US Dept of Agr 1815 N University St Peoria IL 61604

WU, YUNG-CHI, b Canton, China, Oct 3, 23; US citizen; m 45; c 2. THERMODYNAMICS. Educ: Sun Yat-Sen Univ, BS, 47; Univ Houston, MS, 52; Univ Chicago, PhD(chem), 57. Prof Exp: Chemist, Res & Develop Lab, Portland Cement Asn, 57-62, Watson Res Ctr, Int Bus Mach Corp, 63-66 & Oak Ridge Nat Lab, 66-67; CHEMIST, NAT BUR STANDARDS, 67- Mem: AAAS; Am Chem Soc; Am Inst Aeronaut & Astronaut. Res: Electrolyte solutions; thermodynamics; biotechnology of heat transfer. Mailing Add: Nat Bur of Standards Washington DC 20234

WUBBELS, GENE GERALD, b Lanesboro, Minn, Sept 21, 42; m 67; c 2. CHEMISTRY. Educ: Hamline Univ, BS, 64; Northwestern Univ, PhD(chem), 68. Prof Exp: ASSOC PROF ORG & BIOL CHEM & CHMN DEPT CHEM, GRINNELL COL, 68- Concurrent Pos: Petrol Res Fund grant, 71-73; res assoc, State Univ NY Buffalo, 74-75; res grant, Am Chem Soc-Petrol Res Fund, 74-76. Mem: Am Chem Soc. Res: Catalytic mechanisms for photosubstitution, photoreduction and photoaddition reactions of aromatic compounds. Mailing Add: Dept of Chem Grinnell Col Grinnell IA 50112

WUCHTER, RICHARD B, b Wadsworth, Ohio, July 21, 37; m 72. ORGANIC CHEMISTRY. Educ: Western Reserve Univ, AB, 59; Cornell Univ, PhD(org chem), 63. Prof Exp: Asst org chem, Cornell Univ, 60-62; group leader process res, 63-70, GROUP LEADER POLLUTION CONTROL CHEM SYNTHESIS, FLUID PROCESS LAB, ROHM AND HAAS CO, 70- Mem: Am Chem Soc. Res: Monomer synthesis and process development; plastics; modifiers for plastics; fibers; pollution control and ion exchange syntheses. Mailing Add: Fluid Process Lab Rohm and Haas Co Morristown & McKean Rd Spring House PA 19477

WUDL, FRED, organic chemistry, see 12th edition

WUEHRMANN, ARTHUR H, b Bayonne, NJ, Jan 20, 14; m 39; c 3. DENTISTRY. Educ: Tufts Univ, DMD, 37, AB, 60. Prof Exp: Lab instr path, Sch Dent Med, Tufts Univ, 36-37, from instr to assoc prof clin dent, 38-51, from actg dir to dir div grad & postgrad studies, 48-51; from assoc dean to actg dean sch dent, 56-61, PROF DENT, SCH DENT, UNIV ALA, BIRMINGHAM, 51-; DIR DENT RADIOL RES TRAINING PROG, NAT INST DENT RES, 62- Concurrent Pos: Consult, New Eng Area Vet Admin, 45-51, Pan-Am Sanit Bur, Pan-Am Health Orgn, 62-63, Ala Vet Admin Hosps, 52-, dent dept, Ft Benning, Ga, 55-, bur radiol health, USPHS, 62- & div dent health, 69-; mem, Nat Adv Comt on Radiation, USPHS, 58-61, nat training grants comt, Nat Inst Dent Res, 61-65 & Nat Ctr for Radiol Health training grant comt, USPHS, 64-68, chmn, 66-68. Mem: Am Dent Asn; Int Asn Dent Res; fel Am Col Dent; Am Pub Health Asn; Am Acad Dent Radiol. Res: Dental radiology interpretation; radiological health and radiation biology, specifically related to effects of x-radiation on bone. Mailing Add: Sch of Dent Univ of Ala Birmingham AL 35233

WUELLNER, JAMES ALBERT, organic chemistry, see 12th edition

WUENSCHEL, PAUL CLARENCE, b Erie, Pa, May 13, 21; m 50; c 6. GEOPHYSICS. Educ: Colo Sch Mines, GeolEngr, 44; Columbia Univ, PhD(geol), 55. Prof Exp: Res assoc, Columbia Univ, 46-52; dir geol & geophys res, Res, Inc, 52-55; res assoc, 55-74, GEOPHYSICIST, GULF RES & DEVELOP CO, 74- Mem: Soc Explor Geophys; Seismol Soc Am; Acoust Soc Am; Am Geophys Union; Europ Asn Explor Geophys. Res: Seismology; potential, electrical and seismic methods of geophysical exploration. Mailing Add: 128 Marian Ave Glenshaw PA 15116

WUEPPER, KIRK DEAN, b Bay City, Mich, Mar 18, 38; m 67; c 3. DERMATOLOGY. Educ: Univ Mich, MD, 63; Am Bd Dermat, dipl, 70. Prof Exp: USPHS fel, Scripps Clin & Res Found, 68-70, assoc dermat & path, 70-72; ASSOC PROF DERMAT, MED SCH, UNIV ORE, 72- Concurrent Pos: Res career develop award, 71- Honors & Awards: Charles W Burr Award for Res, 64. Mem: Am Asn Immunol; Am Soc Exp Path; Soc Invest Dermat; Am Fedn Clin Res; Am Acad Dermat. Mailing Add: Dept of Dermat Univ of Ore Med Sch Portland OR 97201

WUERKER, RALPH FREDERICK, b Los Angeles, Calif, Jan 18, 29; m 53; c 3. PHYSICS. Educ: Occidental Col, BA, 51; Stanford Univ, PhD(physics), 60. Prof Exp: Engr, AiRes, Inc, 56-58; mem tech staff, Res Lab, Ramo-Wooldridge, Inc, 58-59; mem sr staff, Res Lab, Space Technol Labs, Inc, 60-61; mem sr staff, Quantatron Corp, 61-62; MEM PROF STAFF, TRW SYSTS GROUP, TRW INC, REDONDO BEACH, 63- Concurrent Pos: Consult, Lawrence Livermore Labs, 73- Honors & Awards: Res Soc Award, TRW Systs, 66. Mem: Am Asn Physics Teachers; Am Phys Soc; Optical Soc Am. Res: Holography and coherent optics; plasma particle resonances; electrooptics; lasers; physical optics; plasma physics; electron and general experimental physics. Mailing Add: 4036 Via Pima Palos Verdes Estates CA 90274

WUEST, PAUL JOSEPH, b Philadelphia, Pa, Feb 26, 37; m 61; c 4. PLANT PATHOLOGY. Educ: Pa State Univ, BS, 58, PhD(plant path), 63. Prof Exp: Asst, 58-63, from asst prof to assoc prof exten div, 64-74, PROF PLANT PATH EXTEN DIV, PA STATE UNIV, 74- Concurrent Pos: Fel, Univ Guelph, 70-71. Mem: Am Phytopath Soc; Can Phytopath Soc; Soc Nematol. Res: Diseases of the commercial mushroom; soil treatment and disease occurrence; fungicide tolerance; epidemiology. Mailing Add: Dept of Plant Path Pa State Univ University Park PA 16802

WUESTHOFF, MICHAEL TORREY, organic chemistry, see 12th edition

WUJEK, DANIEL EVERETT, b Bay City, Mich, Oct 26, 39; m 66. BOTANY. Educ: Cent Mich Univ, BS, 61, MA, 62; Univ Kans, PhD(bot), 66. Prof Exp: From asst prof to assoc prof bot, Wis State Univ, La Crosse, 66-68; assoc prof, 68-73, PROF BOT, CENT MICH UNIV, 73- Concurrent Pos: Wis State Univ fac grant, 66-68; NSF grant, 71-72; Cent Mich Univ fac res grant, 71-75. Honors & Awards: Dimond Award, Bot Soc Am, 75. Mem: AAAS; Bot Soc Am; Am Micros Soc; Phycol Soc Am; Int Phycol Soc. Res: Algal life history studies and electron microscopy, including ecology and relations to water quality. Mailing Add: Dept of Biol Cent Mich Univ Mt Pleasant MI 48858

WUKELIC, GEORGE EDWARD, b Steubenville, Ohio, Sept 17, 29; m 55; c 3. ATMOSPHERIC PHYSICS, SPACE PHYSICS. Educ: WVa Univ, AB, 52. Prof Exp: Prin physicist, Battelle Mem Inst, 52-60, sr physicist, 60-72, ASSOC CHIEF FLIGHT SYSTS TECHNOL & APPLNS RES DIV, PHYSICS & METALL DEPT, BATTELLE-COLUMBUS LABS, 72- Concurrent Pos: US Air Force rep, Comt on Exten to the Standard Atmosphere, Int Civil Aviation Orgn, 59-61. Mem: Am Geophys Union. Res: Space geophysics; remote sensing applications; environmental earth resource surveys. Mailing Add: Physics & Metall Dept Battelle-Columbus Labs Columbus OH 43201

WULBERT, DANIEL ELIOT, b Chicago, Ill, Dec 17, 41; m 63. MATHEMATICS. Educ: Knox Col, Ill, BA, 63; Univ Tex, MA, 64, PhD(math), 66. Prof Exp: Vis asst prof math, Univ Lund, 66-67; asst prof, Univ Wash, 67-74; ASSOC PROF MATH, UNIV CALIF, SAN DIEGO, 74- Concurrent Pos: Fel, Univ Lund, 66-67; NSF res grant, 68- Mem: Am Math Soc. Res: Approximation theory; functional analysis. Mailing Add: Dept of Math Univ of Calif at San Diego La Jolla CA 92037

WULF, OLIVER REYNOLDS, b Norwich, Conn, Apr 22, 97; m 22. METEOROLOGY, PHYSICS. Educ: Worcester Polytech Inst, BS, 20; Am Univ, MS, 22; Calif Inst Technol, PhD, 26. Prof Exp: Jr chemist, USDA, 20-22; asst, Calif Inst Technol, 23-26, Nat Res Coun fel chem, 26-27; Nat Res Coun fel, Univ Calif, 27-28; from assoc chemist to sr physicist, Bur Chem & Soils, USDA, 28-39; meteorologist, US Weather Bur, 39-63, res meteorologist, 63-65; res meteorologist, Inst Atmospheric Sci, Environ Sci Serv Admin, US Dept Com, 65-67; res assoc, 45-67, emer res assoc, 67-74, EMER SR RES ASSOC PHYS CHEM, CALIF INST TECHNOL, 74- Concurrent Pos: Guggenheim fel, Univs Berlin & Göttingen, 32-33; fel & res assoc, Univ Chicago, 41-45. Mem: Nat Acad Sci; AAAS; fel Am Phys Soc; Am Meteorol Soc; Am Chem Soc. Res: Chemical kinetics; photochemistry; molecular structure; band spectra; atmospheric and solar physics; geomagnetism; meteorology of the upper atmosphere; general circulation. Mailing Add: Noyes Lab of Chem Physics Div of Chem & Chem Eng Calif Inst of Technol Pasadena CA 91125

WULF, RONALD JAMES, b Davenport, Iowa, July 24, 28; m 59; c 3. PHARMACOLOGY. Educ: Univ Iowa, BS, 50, MS, 57; Purdue Univ, PhD(biochem), 64. Prof Exp: Res chemist, John Deere & Co, 50-52; asst pharmacol, Univ Iowa, 54-57; res pharmacologist, Lederle Labs, Am Cyanamid Co, NY, 57-61; teaching asst biochem, Purdue Univ, 61-64; assoc prof pharmacol, Univ Conn, 64-70; ASST RES DIR PHARMACOL & TOXICOL, CARTER WALLACE INC, 70- Mem: AAAS; Am Chem Soc. Res: Biochemical pharmacology; mechanisms of drug action; chemically induced fibrinolysis; inhalation toxicology. Mailing Add: Carter Wallace Inc Cranbury NJ 08512

WULFERS, THOMAS FREDERICK, b Cape Girardeau, Mo, Oct 4, 39; m 63. ORGANIC CHEMISTRY. Educ: Univ St Louis, BS, 61; Washington Univ, MA, 63; Univ Chicago, PhD(chem), 65. Prof Exp: Res chemist, Shell Oil Co, Mo, 65-72; RES CHEMIST, ATLANTIC RICHFIELD CO, ILL, 72- Mem: Am Chem Soc. Res: Petroleum products research. Mailing Add: 2921 Greenwood Rd Hazel Crest IL 60429

WULFF, BARRY LEE, b Mt Kisco, NY, Feb 17, 40; m 66; c 1. ECOLOGY, MARINE BIOLOGY. Educ: State Univ NY Cortland, BS, 65; Col William & Mary, MA, 68; Ore State Univ, PhD(bot), 70. Prof Exp: ASST PROF BIOL, EASTERN CONN STATE COL, 70- Mem: AAAS; Am Soc Limnol & Oceanog; Phycol Soc Am; Am Inst Biol Sci. Res: Structure and function of benthic aquatic plant communities; marine algae. Mailing Add: Dept of Biol Eastern Conn State Col Willimantic CT 06226

WULFF, CLAUS ADOLF, b Hamburg, Ger, July 20, 38; US citizen; m 60; c 2. PHYSICAL CHEMISTRY. Educ: Cornell Univ, AB, 59; Mass Inst Technol, PhD(phys chem), 62. Prof Exp: Fel, Inst Sci & Technol, Univ Mich, 62-63; asst prof chem, Carnegie Inst Technol, 63-65; from asst prof to assoc prof, 65-73, PROF CHEM, UNIV VT, 73- Concurrent Pos: Am-Scand Found fel, Univ Lund, 71-72. Mem: Am Chem Soc; The Chem Soc; Calorimetry Conf. Res: Low-temperature and solution calorimetry. Mailing Add: Dept of Chem Univ of Vt Burlington VT 05401

WULFF, JOHN LELAND, b Oakland, Calif, Mar 19, 32; m 68; c 3. MATHEMATICS. Educ: Sacramento State Col, AB, 54; Univ Calif, Davis, MA, 57, PhD(math), 66. Prof Exp: From instr to assoc prof, 55-68, chmn dept, 68-71, PROF MATH, CALIF STATE UNIV, SACRAMENTO, 69- Res: Measure theory and integration. Mailing Add: Dept of Math Calif State Univ Sacramento CA 95819

WULFF, VERNER JOHN, b Essen, Ger, Aug 16, 16; m 42; c 3. NEUROPHYSIOLOGY. Educ: Wayne Univ, AB, 38; Northwestern Univ, MA, 40; Univ Iowa, PhD(zool), 42. Prof Exp: Asst zool & comp vert anat, Wayne Univ, 36-38; asst zool, embryol & endocrinol, Northwestern Univ, 38-40; asst, Univ Iowa, 40-42, instr & assoc mammalian physiol, 46-47; asst prof mammalian neurophysiol, Univ Ill, 47-51; prof zool & chmn dept, Syracuse Univ, 51-60; ASSOC DIR RES, MASONIC MED RES LAB, 61- Mem: AAAS; Am Soc Zool; Soc Exp Biol & Med. Res: Electrophysiology of retinae; relation between photochemical events and retinal action potential. Mailing Add: Masonic Med Res Lab Bleeker St Utica NY 13501

WULFMAN, CARL E, b Detroit, Mich, Nov 29, 30; m 52; c 4. THEORETICAL PHYSICS, THEORETICAL CHEMISTRY. Educ: Univ Mich, BS, 53; Univ London, PhD(org chem), 57. Prof Exp: Instr chem, Univ Tex, 56-57; assoc prof, Defiance Col, 57-61; chmn dept, 61-74, PROF PHYSICS, UNIV OF THE PAC, 61- Concurrent Pos: Vis mem, Ctr Theoret Studies, Coral Gables, Fla, 67; NSF sci fac fel, Oxford Univ, 67-68; vis prof, Japan Soc Promotion Sci, 74-75. Mem: AAAS; Am Chem Soc; Am Phys Soc; Am Asn Physics Teachers. Res: Transformation properties of dynamical equations; continuous groups; applications to chemical kinetics, atomic and molecular quantum mechanics. Mailing Add: Dept of Physics Univ of the Pac Stockton CA 95204

WULFMAN, DAVID SWINTON, b Detroit, Mich, Sept 1, 34; m 61; c 3. SYNTHETIC ORGANIC CHEMISTRY, PHYSICAL ORGANIC CHEMISTRY. Educ: Univ Mich, Ann Arbor, BS, 56; Dartmouth Col, AM, 58; Stanford Univ, PhD(chem), 62; Alliance Francaise, Paris, France, IVe, French, 74. Prof Exp: Res asst chem, Univ Mich, 54-56; sr develop engr, Hercules Inc, Utah, 61-63; asst prof, 63-66, ASSOC PROF CHEM, UNIV MO-ROLLA, 66- Concurrent Pos: Consult, Dept Chem, Stanford Univ, 69; lectr, Wash Univ, 70 & Chem Soc France, 75; res assoc, Nat Ctr Sci Res, Teachers' Training Col, Paris, 74-75. Mem: Am Chem Soc; The Chem Soc; Chem Inst Can. Res: Synthesis and study of theoretically important molecules; homogeneous catalysis; kinetics of photochemical processes. Mailing Add: Dept of Chem Univ of Mo Rolla MO 65401

WULLSTEIN, LEROY HUGH, b Nampa, Idaho, Nov 23, 31; m 56; c 1. BOTANY, MICROBIOLOGY. Educ: Univ Utah, BS, 57; Ore State Univ, MS, 61, PhD(microbiol), 64. Prof Exp: Asst prof soil sci, Univ BC, 64-66; from asst prof to assoc prof, 66-75, PROF BOT, UNIV UTAH, 75- Concurrent Pos: Sr Fulbright res fel nitrogen fixation, Ireland, 72-73; consult, Brookhaven Labs, 74- Mem: Am Soc Microbiol; Am Chem Soc; Soil Sci Soc Am. Res: Inorganic nitrogen transformations; denitrification; nitrogen fixation in the rhizosphere and root nodule induction. Mailing Add: Dept of Bot Univ of Utah Salt Lake City UT 84112

WUN, CHUN KWUN, b Canton, China, Feb 15, 40; m 64; c 2. BIOLOGY. Educ: Chung Chi Col, Chinese Univ Hong Kong, BS, 64; Springfield Col, MS, 69; Univ Mass, Amherst, MS, 71, PhD(lipid chem), 74. Prof Exp: Asst educ officer sci, Educ Dept Hong Kong, 65-66; from asst instr to instr biol, Springfield Col, 68-70; res asst lipid chem, Univ Mass, Amherst, 70-73; asst prof biol, Springfield Col, 73-74; FEL, DEPT ENVIRON SCI, UNIV MASS, AMHERST, 75- Concurrent Pos: Adj asst prof biol, Springfield Col, 75- Res: Lipid metabolism, particularly triglyceride synthesis and its control in mycobacterium smegmatis. Mailing Add: Dept of Environ Sci Univ of Mass Amherst MA 01002

WUNDER, BRUCE ARNOLD, b Monterey Park, Calif, Feb 10, 42; m 63; c 2. PHYSIOLOGICAL ECOLOGY, VERTEBRATE ZOOLOGY. Educ: Whittier Col, BA, 63; Univ Calif, Los Angeles, PhD(vert zool), 68. Prof Exp: NIH fel, Inst Arctic Biol, Univ Alaska, 68-69; ASST PROF ZOOL, COLO STATE UNIV, 69- Concurrent Pos: Small mammal ecologist, Biol Res Assocs, Inc & consult, Thorne Ecol Inst, 72- Mem: AAAS; Am Soc Zoologists; Am Soc Mammalogists; Am Inst Biol Sci; Ecol Soc

Am. Res: Temperature regulation and energetics; water balance and mechanisms of evaporative water loss, particularly in vertebrates; feeding strategies and distribution patterns in vertebrates. Mailing Add: Dept of Zool & Entom Colo State Univ Ft Collins CO 80523

WUNDER, CHARLES COOPER, b Pittsburgh, Pa, Oct 2, 28; m 62; c 3. PHYSIOLOGY, BIOPHYSICS. Educ: Washington & Jefferson Col, AB, 49; Univ Pittsburgh, MS, 52, PhD(biophys), 54. Prof Exp: Asst biophys, Univ Pittsburgh, 49-51; assoc physiol, 54-56, asst prof, 56-63, assoc prof physiol & biophys, 63-71, PROF PHYSIOL & BIOPHYS, UNIV IOWA, 71- Concurrent Pos: NIH res career develop award, 61-66; vis scientist & NIH spec fel, Mayo Clin, 66-67. Mem: Soc Exp Biol & Med; Am Soc Zoologists; Soc Develop Biol; Biophys Soc; Am Physiol Soc. Res: Environmental biophysics of growth and function; gravitational biology. Mailing Add: Dept of Physiol & Biophys Univ of Iowa Iowa City IA 52240

WUNDER, WILLIAM W, b Lake Park, Iowa, June 4, 30; m 60; c 3. POPULATION GENETICS, BIOMETRICS. Educ: Iowa State Univ, BS, 58; Mich State Univ, MS, 64, PhD(dairy cattle breeding), 67. Prof Exp: Asst prof exten dairy sci, Univ Ky, 67-68; asst prof, 68-74, ASSOC PROF ANIMAL SCI, IOWA STATE UNIV, 74- Mem: Am Dairy Sci Asn. Res: Effects of environment on milk production and statistical methods of adjusting for or removing these effects in selecting dairy cattle for higher milk production. Mailing Add: Dept of Animal Sci 123 Kildee Hall Iowa State Univ Ames IA 50010

WUNDERLICH, BERNHARD, b Brandenburg, Ger, May 28, 31; nat US; m 53; c 2. POLYMER CHEMISTRY. Educ: Univ Frankfurt, BSc, 54; Northwestern Univ, PhD, 57. Prof Exp: Instr chem, Northwestern Univ, 57-58; from instr to asst prof, Cornell Univ, 58-63; assoc prof, 63-65, PROF CHEM, RENSSELAER POLYTECH INST, 65- Concurrent Pos: Consult, E I du Pont de Nemours & Co, Inc. Honors & Awards: Mettler Award for Thermal Anal, 71. Mem: Am Chem Soc; NAm Thermal Anal Soc; Int Confedn Thermal Anal; fel Am Phys Soc; NY Acad Sci. Res: Physical chemistry of the solid state of high polymers; transitions of high polymers at elevated temperatures and high pressures. Mailing Add: 211 Winter St Exten Troy NY 12180

WUNDERLICH, FRANCIS J, b Philadelphia, Pa, Mar 9, 38; m 62; c 5. PHYSICAL CHEMISTRY. Educ: Villanova Univ, BS, 59; Georgetown Univ, PhD(chem), 64. Prof Exp: Res asst chem, Villanova Univ, 57-58, instr, 59; instr, Georgetown Univ, 59-61, res assoc, 61-63; fel molecular physics, 63-65; from asst prof to assoc prof physics & chem, Col Virgin Islands, 65-69; ASST PROF PHYSICS, VILLANOVA UNIV, 69- Concurrent Pos: Dir, NSF Grant, Undergrad Sci Equip Prog, Col Virgin Islands, 65-67, dir, Etelman Astron Observ, 66-67. Honors & Awards: Award, Am Inst Chemists, 59. Mem: Am Chem Soc; The Chem Soc. Res: Theoretical molecular physics; gas phase free radicals; laser-induced gas phase reactions. Mailing Add: Dept of Physics Villanova Univ Villanova PA 19085

WUNDERLICH, MARVIN C, b Decatur, Ill, May 8, 37; m 60; c 2. MATHEMATICS. Educ: Concordia Teachers Col, Ill, BS, 59; Univ Colo, PhD(math), 64. Prof Exp: Asst prof math, State Univ NY, Buffalo, 64-67; assoc prof, 67-72, PROF MATH, NORTHERN ILL UNIV, 72- Concurrent Pos: NSF res grant, 66-; vis, Univ Nottingham, 72-73. Mem: Am Math Soc; Asn Comput Mach. Res: Number theory; computing mathematics. Mailing Add: Dept of Math Northern Ill Univ De Kalb IL 60115

WUNDERMAN, IRWIN, b New York, NY, Apr 24, 31; m 51; c 3. ELECTROOPTICS, ELECTRONIC INSTRUMENTATION. Educ: City Col New York, BSEE, 52; Univ Southern Calif, MSEE, 56; Stanford Univ, EEE, 61, PhD(elec eng), 64. Prof Exp: Jr engr draftsman, Lockheed-Calif Co, 52, jr engr, 52-53, res engr, 53-56; lab sect leader, Hewlett Packard Co, 56-61, co-founder, Hewlett Packard Assocs, 61-65, lab mgr, Hewlett Packard Corp Labs, 65-67; pres & gen mgr, Cintra Inc, Cintra Physics Int, 67-71; PVT RESEARCHER, 71- Honors & Awards: Outstanding Invited Paper, Solid State Circuits Conf, Inst Elec & Electronics Engrs, 65; Nat Commendation Letter, 68. Mem: AAAS; sr mem Inst Elec & Electronics Engrs; Optical Soc Am; Am Inst Physics; Sigma Xi. Res: Electrooptics instrumentation; radiometry; photometry; optoelectronic solid state devices and circuits; computer architecture and systems; modeling of nonlinear physical systems; physics of optics and quanta; wave/particle dilemma; photons, wavavs and the classical origin of quanta theory in mathematical physics. Mailing Add: 655 Eunice Ave Mountain View CA 94040

WUNSCH, CARL ISAAC, b New York, NY, May 5, 41; m 70; c 1. OCEANOGRAPHY. Educ: Mass Inst Technol, SB, 62, PhD(geophys), 66. Prof Exp: Lectr oceanog, 66-67, from asst prof to assoc prof, 67-75, PROF PHYS OCEANOG, MASS INST TECHNOL, 75- Concurrent Pos: Vis sr investr, Cambridge Univ, 69 & 74-75. Honors & Awards: James B Macelwane Award, Am Geophys Union, 71. Mem: Am Geophys Union; Royal Astron Soc. Res: Time series analysis; internal waves; sea level, mixing processes, circulation dynamics; ocean and solid earth tides. Mailing Add: Dept of Earth & Planetary Sci Mass Inst of Technol Cambridge MA 02139

WUNZ, PAUL RICHARD, JR, b Erie, Pa, Oct 18, 23; m 48; c 3. ORGANIC CHEMISTRY. Educ: Pa State Col, BS, 44, MS, 47; Univ Del, PhD(chem), 50. Prof Exp: Instr chem, Univ Del, 50; asst prof chem & head dept, Augsburg Col, 50-51; res chemist, Nopco Chem Co, 51-53; res chemist & group leader, Callery Chem Co, 53-57; from asst prof to assoc prof chem, Geneva Col, 57-65; chmn dept, 65-73, PROF CHEM, INDIANA UNIV PA, 65- Mem: AAAS; Am Chem Soc; Sigma Xi. Res: Synthetic organic chemistry; organometallic compounds; pharmaceuticals, steroids; heterocyclic compounds. Mailing Add: Dept of Chem Indiana Univ of Pa Indiana PA 15701

WURDACK, JOHN J, b Pittsburgh, Pa, Apr 28, 21; m 59; c 2. BOTANY. Educ: Univ Pittsburgh, BS, 42; Univ Ill, BS, 49; Columbia Univ, PhD, 52. Prof Exp: Asst bot, Univ Pittsburgh, 42; tech asst bot, NY Bot Garden, 49-52; from asst cur to assoc cur, 52-60, assoc cur, Div Phanerogams, 60-63, CUR BOT, NAT MUS NATURAL HIST, SMITHSONIAN INST, 63- Concurrent Pos: Mem exped, Venezuela, 50-59 & Peru, 62. Mem: Am Soc Plant Taxonomists; Torrey Bot Club; Int Asn Plant Taxon. Res: Taxonomy of Melastomataceae and flowering plants of northern South America. Mailing Add: Dept Bot Nat Mus of Natural Hist Smithsonian Inst Washington DC 20560

WURST, GLEN GILBERT, b Mt Holly, NJ, Apr 17, 45. GENETICS. Educ: Juniata Col, BS, 67; Univ Pittsburgh, PhD(biol), 75. Prof Exp: Teaching asst biol, Univ Pittsburgh, 67-71, teaching fel, 71-75; ASST PROF BIOL, ALLEGHENY COL, 75- Res: Structural and functional aspects of pleiotropic genes in Drosophila Melanogaster. Mailing Add: Dept of Biol Allegheny Col Meadville PA 16335

WURSTER, DALE E, b Sparta, Wis, Apr 10, 18; m 44; c 2. PHARMACY. Educ: Univ Wis, BS, 42, PhD, 47. Prof Exp: From instr to prof pharm, Univ Wis, 47-71; prof pharm & pharmaceut chem & dean col pharm, NDak State Univ, 71-72; PROF PHARM & DEAN COL PHARM, UNIV IOWA, 72- Concurrent Pos: Am Asn Cols Pharm-NSF vis scientist, 63-66; consult, USPHS, 66-72; mem revision comt, US Pharmacopoeia, 61-72. Honors & Awards: Res Achievement Award, Am Pharmaceut Asn, 65. Mem: Acad Pharmaceut Sci; Am Pharmaceut Asn; hon mem Rumanian Soc Med Sci. Res: Physical factors influencing dissolution kinetics; diffusion kinetics in biological membranes, drug release mechanisms from pharmaceutical systems, percutaneous absorption, air-suspension coating and granulating technique. Mailing Add: Col of Pharm Univ of Iowa Iowa City IA 52240

WURSTER, RICHARD T, b Riverside, NJ, Mar 5, 36; m 57; c 3. HORTICULTURE. Educ: Rutgers Univ, BS, 59; Univ Calif, Davis, MS, 60; Cornell Univ, PhD(veg crops), 64. Prof Exp: Res asst dept veg crops, Univ Calif, Davis, 59-60 & Cornell Univ, 60-64; agr adv hort, Ahwaz Agr Col, Molla-Sani, Iran, 64-66; horticulturist, Plant Indust Sta, USDA, 66-67; from asst prof to assoc prof hort, Int Progs, WVa Univ, 67-73; DIR OUTREACH & TRAINING PROG, INT POTATO CTR, PERU, 73- Concurrent Pos: US Agency Int Develop-WVa Univ contract & lectr, Dept Crop Sci & Prod, Makerere Univ, Uganda, 67-73. Mem: Am Soc Hort Sci; Int Soc Hort Sci. Res: Tropical and subtropical horticultural crops, particularly those crops which can provide the protein requirements of populations in developing countries; potato productivity increase in tropical and subtropical regions. Mailing Add: Int Potato Ctr Apartado 5969 Lima Peru

WURSTER, WALTER HERMAN, b Ger, Aug 17, 25; nat US. PHYSICS. Educ: Univ Buffalo, BA, 50, PhD, 57. Prof Exp: Res physicist, Cornell Aeronaut Lab, Inc, 55-61, prin physicist, 61-72, staff scientist, 72-74; MEM STAFF, CALSPAN CORP, 74- Mem: Am Phys Soc. Res: Beta ray spectroscopy; absorption and emission spectroscopy of high temperature gases in infrared and ultraviolet; shock tube techniques and instrumentation. Mailing Add: Calspan Corp PO Box 235 Buffalo NY 14221

WURSTER-HILL, DORIS HADLEY, b Washington, DC, Sept 9, 32. CYTOGENETICS. Educ: George Washington Univ, BS, 54; Stanford Univ, MA, 56, PhD(biol), 58. Prof Exp: Res assoc endocrinol, 62-65, from res assoc & instr to res assoc & asst prof cytogenetics, 67-72, ASST PROF CYTOGENETICS, DARTMOUTH MED SCH, 72- Concurrent Pos: USPHS trainee cytogenetics, Dartmouth Med Sch, 65-67; consult, Vet Admin Hosp, White River Jct, Vt, 69- Mem: AAAS; Am Soc Human Genetics; Am Soc Mammalogists; Ecol Soc Am. Res: Clinical and comparative mammalian cytogenetics. Mailing Add: Dept of Path Dartmouth Med Sch Hanover NH 03755

WURTELE, MORTON GAITHER, b Harrodsburg, Ky, July 25, 19; m 42; c 2. DYNAMIC METEOROLOGY. Educ: Harvard Univ, BS, 40; Univ Calif, Los Angeles, MA, 44, PhD, 53. Prof Exp: Res asst prof meteorol, Mass Inst Technol, 53-58; assoc prof, 58-64, vchmn dept, 69-72, PROF METEOROL, UNIV CALIF, LOS ANGELES, 64-, CHMN DEPT, 72- Concurrent Pos: Fulbright grants, Univ Sorbonne, 49 & Hebrew Univ, Israel, 65; NATO sr fel, 62; consult, Atmospheric Sci Lab, White Sands Missile Range, 65- Mem: Am Meteorol Soc. Res: Small- and medium-scale atmospheric motions; sound propagation; atmospheric-ocean interaction. Mailing Add: Dept of Meteorol Univ of Calif Los Angeles CA 90024

WURTELE, ZIVIA SYRKIN, b New York, NY, Apr 1, 21; m 42; c 2. STATISTICS, ECONOMETRICS. Educ: Hunter Col, BA, 40; Univ Calif, Los Angeles, MA, 44; Columbia Univ, PhD(math statist), 54. Prof Exp: From asst res statistician to assoc res statistician, Univ Calif, Los Angeles, 59-65; consult econ & math models, Syst Develop Corp, 65-74; CONSULT STATIST, SCI APPLN, INC, 74- Concurrent Pos: Res fel econ, Harvard Univ, 56-58. Mem: Inst Math Statist; NY Acad Sci. Res: Mathematical models of social science processes. Mailing Add: 432 E Rustic Rd Santa Monica CA 90403

WURTH, MICHAEL JOHN, b Highland Park, Ill, Mar 31, 37; m 65; c 1. ORGANIC CHEMISTRY, PHOTOGRAPHY. Educ: Lake Forest Col, BA, 64; Northwestern Univ, Evanston, PhD(org chem), 69. Prof Exp: ASSOC PROF CHEM, OKLAHOMA CITY UNIV, 68- Mem: Am Chem Soc. Res: Synthesis of new bicyclic heterocyclic compounds; photographic chemistry, especially emulsions and developing agents. Mailing Add: Dept of Chem Oklahoma City Univ Oklahoma City OK 73106

WURTMAN, JUDITH JOY, b Brooklyn, NY, Aug 4, 37; m 59; c 2. NUTRITION. Educ: Wellesley Col, BA, 59; Harvard Univ, MAT, 60; George Washington Univ, PhD(cell biol), 71. Prof Exp: Asst prof biol & nutrit, Newton Col, 72-74; ASST PROF NUTRIT, SIMMON GRAD SCH EDUC, MASS INST TECHNOL, 74-, FEL, DEPT NUTRIT, 74- Concurrent Pos: Consult nutrit educ, Newton Pub Schs, 72-74; NIH fel, 74-76. Mem: Soc Nutrit Educ; Am Dietetic Asn; Nutrit Today. Res: Effect of nutritional factors on growth and development of the neonate, especially in the development of behavior regulating food intake, and the development of obesity. Mailing Add: Dept of Nutrit Mass Inst of Technol Cambridge MA 02139

WURTMAN, RICHARD JAY, b Philadelphia, Pa, Mar 9, 36; m 59; c 2. ENDOCRINOLOGY, NEUROBIOLOGY. Educ: Univ Pa, AB, 56; Harvard Med Sch, MD, 60. Prof Exp: Intern & asst resident med, Mass Gen Hosp, 60-62; res assoc, Lab Clin Sci, NIMH, 62-64, med res officer, 65-67; assoc prof, 67-70, PROF ENDOCRINOL & METAB, MASS INST TECHNOL, 70- Concurrent Pos: Josiah Macy, Jr Found fel, Mass Inst Technol, 60-62; res fel endocrinol, Mass Gen Hosp, 64-65; clin asst med, Mass Gen Hosp, 65-; vis lectr, Am Chem Soc, 66; lectr, Harvard Med Sch, 69-; mem, Preclin Psychopharmacol Study Sect, NIMH, 71-75; Am Inst Biol Sci Adv panel, Biosci Prog, NASA; res adv bd, Parkinson's Dis Foster Elting Bennett lectr, Am Neurol Asn, 74; assoc, Neurosci Res Prog. Honors & Awards: Am Therapeut Soc Prize, 66; Soc Biol Psychiat Prize, 66; John Jacob Abel Award, Am Soc Pharmacol Exp Therapeut, 68; Alvarenga Prize & Lect, Col Physicians Philadelphia, 70; Ernst Oppenheimer Prize, Endocrine Soc, 73; Louis B Flexner lectr, 75. Mem: Am Physiol Soc; Am Soc Pharmacol & Exp Therapeut; Am Soc Neurochem; Am Soc Biol Chem; Am Soc Clin Invest. Res: Neuroendocrinology; neuropharmacology; biological rhythms; pineal gland; catecholamines; amino acid metabolism; effects of nutrition on brain; biological effects of light; L-Dopa. Mailing Add: Dept of Nutrit & Food Sci Lab of Neuroendocrine Regulation Mass Inst of Technol 56-245 Cambridge MA 02139

WURTZ, ROBERT HENRY, b St Louis, Mo, Mar 28, 36; m 58; c 2. NEUROPHYSIOLOGY, NEUROPSYCHOLOGY. Educ: Oberlin Col, AB, 58; Univ Mich, Ann Arbor, PhD(physiol psychol), 62. Prof Exp: Res assoc neurophysiol, Wash Univ, 62-65; res fel, Lab Neurophysiol, NIH, 65-66, RES SCIENTIST, LAB NEUROBIOL, NIMH, 66- Mem: Am Physiol Soc; Am Psychol Asn; Soc Neurosci. Res: Neurophysiological basis of behavior, specifically the physiology of vision and learning. Mailing Add: Lab of Neurobiol Nat Inst of Ment Health Bethesda MD 20014

WURZBURG, OTTO BERNARD, b Grand Rapids, Mich, Aug 1, 15; m 40; c 6. CARBOHYDRATE CHEMISTRY. Educ: Univ Mich, BS, 38, MS, 39. Prof Exp: Chemist & supvr cent control, Nat Starch Prod, 39-44, res chemist & supvr starch res, 45-55, assoc res dir, 56-68, V PRES RES, STARCH DIV, NAT STARCH & CHEM

CORP, 68- Concurrent Pos: Mem bd dirs, Customaize Inc, 74- Mem: Am Chem Soc; Inst Food Technologists; Am Asn Cereal Chemists. Res: Starch; carbohydrates; industrial applications. Mailing Add: RR 1 Box 45 Felmly Rd Whitehouse Station NJ 08889

WUSKELL, JOSEPH P, b New York, NY, Nov 14, 38. CHEMISTRY. Educ: Univ Conn, BA, 60; Univ Minn, PhD(chem), 67. Prof Exp: Chemist, Merck Sharp & Dohme Res Labs, 60-62; sr chemist, Ott Chem Co, Corn Prod Co, 67-68; GROUP LEADER, QUAKER OATS CO, 68- Mem: Am Chem Soc; The Chem Soc. Res: Organic synthesis and reaction mechanisms. Mailing Add: Quaker Oats Co 617 W Main St Barrington IL 60010

WUSSOW, GEORGE C, b Milwaukee, Wis, Mar 10, 23; m 49; c 3. ORAL SURGERY, ORAL PATHOLOGY. Educ: Marquette Univ, DDS, 49. Prof Exp: From instr to assoc prof, 53-67, PROF ORAL SURG, MARQUETTE UNIV, 67-, CHMN DEPT, 61-, LECTR, SCH DENT HYG, 70- Concurrent Pos: Attend oral surg, Vet Admin Ctr, Wood, Wis & mem consult staff, Milwaukee County Gen Hosp, 57-; consult, Great Lakes Naval Hosp, 70-; mem adv bd, Milwaukee Area Tech Col, 70- Mem: Am Soc Oral Surg; Am Acad Oral Path; fel Am Col Dent; Int Asn Oral Surg; fel Royal Soc Health. Res: Clinical evaluation of proteolytic enzymes in the management of impacted mandibular third molars. Mailing Add: Dept of Oral Surg Marquette Univ Sch of Dent Milwaukee WI 53233

WUST, CARL JOHN, b Providence, RI, July 2, 28; m 51; c 5. IMMUNOLOGY. Educ: Providence Col, BS, 50; Brown Univ, MSc, 53; Ind Univ, PhD(microbiol), 57. Prof Exp: Electron microscopist, Ind Univ, 53-55; NIH fel, Yale Univ, 57-59; biochemist, Biol Div, Oak Ridge Nat Lab, 59-70; assoc prof, 70-74, PROF MICROBIOL, UNIV TENN, KNOXVILLE, 74- Mem: AAAS; Am Soc Microbiol; Am Asn Immunologists; Sigma Xi; Soc Exp Biol & Med. Res: Antibody biosynthesis; protein synthesis; RNA in the immune response; leukemia antigens; cell-mediated immunity. Mailing Add: Dept of Microbiol Univ of Tenn Knoxville TN 37916

WUTHIER, ROY EDWARD, b Rushville, Nebr, Nov 11, 32; m 56; c 2. BIOCHEMISTRY. Educ: Univ Wyo, BS, 54; Univ Wis, MS, 58, PhD, 60. Prof Exp: Asst biochem, Univ Wis, 55-60; res fel, Forsyth Dent Ctr, Harvard Med Sch, 60-63, asst mem staff & assoc biol chem, 63-69; assoc prof biochem, Depts Orthop Surg & Biochem, Col Med, Univ Vt, 69-75; PROF CHEM, COL ARTS & SCI & COORDR BIOCHEM, COL MED, UNIV SC, 75- Mem: AAAS; Am Soc Biol Chem; Am Chem Soc; Orthop Res Soc; Int Asn Dent Res. Res: Mechanism of calcification; lipid and membrane involvement in calcification; metabolism of calcifying tissue. Mailing Add: Dept of Chem Phys Sci Ctr Univ of SC Columbia SC 29208

WUU, TING-CHI, b Salt County, China, Sept 27, 34; US citizen; m 62; c 3. BIOCHEMISTRY. Educ: Nat Taiwan Univ, BSc, 58, MSc, 60; McGill Univ, PhD(biochem), 67. Prof Exp: Prof asst biochem res, McGill Univ, 67-69; asst prof, 69-74, ASSOC PROF BIOCHEM, MED COL OHIO, 75- Mem: AAAS. Res: Structure and function of peptides and proteins; brain, peptides and proteins; isolation and characterization of hormone-binding proteins of neurohypophysis; biological, chemical and physical properties of neurophysins and neurosecretory granules. Mailing Add: Dept of Biochem Med Col of Ohio Toledo OH 43614

WYANDT, HERMAN EDWIN, JR, b Port Angeles, Wash, Feb 28, 39. CYTOGENETICS, MEDICAL GENETICS. Educ: Western Wash State Col, BA, 62; Ore State Univ, MS, 65, PhD(cell biol), 72. Prof Exp: From res asst to res assoc med & cytogenetics, Med Sch, 68-74, INSTR MED GENETICS, HEALTH SCI CTR, UNIV ORE, 74- Concurrent Pos: Basil O'Connor starter res grant, Nat Found March Dimes, 73; Med Res Found Ore res grant, 74. Mem: AAAS. Res: Basic and clinical research in human cytogenetics; chromosome structure and function. Mailing Add: Univ of Ore Health Sci Ctr PO Box 574 Portland OR 97207

WYANT, GORDON MICHAEL, b Frankfurt, Ger, Mar 28, 14; m; c 5. ANESTHESIOLOGY. Educ: Univ Bologna, MD, 38; Royal Col Physicians & Surgeons Eng, dipl, 45; Am Bd Anesthesiol, dipl, 53; Royal Col Physicians & Surgeons Can, dipl anesthesiol, 52, FRCP(C), 63. Prof Exp: Asst prof anesthesia, Col Med, Univ Ill, 50-53; asst prof surg & head div anesthesia, Stritch Sch Med, Loyola Univ, Ill, 53-54; PROF ANESTHESIA, UNIV SASK, 54- Mem: Am Soc Anesthesiol; fel Am Geriat Soc; Am Geog Soc; fel Am Col Anesthesiol; fel Int Col Anesthetists. Res: Related clinical and basic sciences of anesthesia. Mailing Add: Dept of Anesthesia Univ of Sask Col of Med Saskatoon SK Can

WYANT, JAMES CLAIR, b Morenci, Mich, July 31, 43; m 71. OPTICS. Educ: Case Inst Technol, BS, 65; Univ Rochester, MS, 67, PhD(optics), 68. Prof Exp: Optical engr & head, Optical Eng Sect, Itek Corp, 68-74; ASSOC PROF OPTICS, OPTICAL SCI CTR, UNIV ARIZ, 74- Concurrent Pos: Guest ed, J Optical Eng, Soc Photo-Optical Instrumentation Engrs, 74. Mem: Am Inst Physics; Optical Soc Am; Soc Photo-Optical Instrumentation Engrs. Res: Interferometry; holography; optical testing; optical processing; optical properties of the atmosphere; active optics. Mailing Add: Optical Sci Ctr Univ of Ariz Tucson AZ 85721

WYATT, ANDY JACK, b Rogers, Ark, Dec 8, 26; m 57; c 4. ANIMAL BREEDING. Educ: Univ Ark, BSA, 49; Iowa State Univ, MS, 51, PhD(poultry breeding genetics), 53. Prof Exp: Geneticist, Ghostley's Poultry Farms, Inc, 53-64, vpres breeding & res, 64-68; geneticist, Babcock Poultry Farm, Inc, 68-72, V PRES RES & DEVELOP, BABCOCK SWINE, INC, 72- Concurrent Pos: Chmn, Nat Comt Random Sample Poultry Testing, 64-65. Mem: Poultry Sci Asn; Biomet Soc. Res: Poultry breeding and genetics of poultry populations with special emphasis on characteristics of economic merit. Mailing Add: Babcock Swine Inc Box 388 Rochester MN 55901

WYATT, BENJAMIN WOODROW, b Farrar, Ga, Dec 24, 16; m 48; c 4. CHEMISTRY. Educ: Southwestern Univ, Tex, BS, 37; Univ Tex, MA, 40, PhD(org chem), 43. Prof Exp: Tutor chem, Univ Tex, 38-43; assoc mem & asst to patent agent, 43-50, patent agent, 50-61, from asst dir to assoc dir, 61-74, DIR, PATENT DIV, STERLING-WINTHROP RES INST, 74- Mem: Am Chem Soc. Res: Organic chemistry. Mailing Add: Sterling-Winthrop Res Inst Rensselaer NY 12144

WYATT, COLEN CHARLES, b Geneva, NY, Dec 10, 27; m 48; c 5. HORTICULTURE. Educ: Cornell Univ, BS, 53. Prof Exp: Sr res horticulturist, H J Heinz Co, 54-64; univ horticulturist & asst dir maintenance, Bowling Green State Univ, 65-66; plant breeder, Libby McNeill & Libby, Ohio, 67-72; PLANT BREEDER, PETO SEED CO, 72- Mem: Am Hort Soc; Am Forestry Asn. Res: Breeding tomatoes for disease resistance and mechanical harvest; breeding vegetable crops. Mailing Add: PO Box 1255 Rte 4 Woodland CA 95695

WYATT, ELLIS JUNIOR, b Norton, Kans, Oct 30, 30; m 53; c 3. PARASITOLOGY, INVERTEBRATE ZOOLOGY. Educ: Lewis & Clark Col, BS, 57; Ore State Univ, MS, 61, PhD(zool), 71. Prof Exp: Aquatic biologist, Ore State Fish Comn, 57, 58-60 & 61; asst prof biol, Cent Ore Col, 61-65 & 67-68; aquatic biologist, Ore State Fish Comn, 68-71; asst prof, 71-72, ASSOC PROF BIOL & CHMN DEPT, HAMLINE UNIV, 72- Concurrent Pos: Lab teaching asst, Ore State Univ, 57. Mem: AAAS; Am Soc Parasitologists; Am Soc Zoologists; Am Fisheries Soc. Res: Parasitic protozoa of fresh water fishes; bacteriology, helminthology, mycology, therapeutics and toxicology of fresh water fishes; ecology of fish parasitism. Mailing Add: Dept of Biol Hamline Univ St Paul MN 55104

WYATT, GERARD ROBERT, b Palo Alto, Calif, Sept 3, 25; m 51; c 3. INSECT PHYSIOLOGY, BIOCHEMISTRY. Educ: Univ BC, BA, 45; Cambridge Univ, PhD(natural sci), 50. Prof Exp: Sci officer, Insect Path Res Inst, Can Dept Agr, Ont, 50-54; asst prof biochem, Yale Univ, 54-60, from assoc prof to prof biol, 60-73; PROF BIOL & HEAD DEPT, QUEEN'S UNIV, ONT, 73- Concurrent Pos: Guggenheim Mem fel, 54. Mem: Am Soc Biol Chemists. Res: Composition of nucleic acids; biochemistry and physiology of insects; composition of insect hemolymph; carbohydrate metabolism and its regulation; physiology of development; actions of insect hormones, especially ecdysone and juvenile hormone. Mailing Add: Dept of Biol Queen's Univ Kingston ON Can

WYATT, JEFFREY RENNER, b Hampton, Va, Jan 1, 46; m 69. CHEMISTRY. Educ: Univ Calif, Riverside, AB, 63; Northwestern Univ, Evanston, PhD(chem), 71. Prof Exp: Teaching assoc chem, Univ Kans, 71-72; Nat Res Coun res fel, 72-73, RES CHEMIST, NAVAL RES LAB, 73- Mem: Am Chem Soc; Am Soc Mass Spectrometry. Res: Mass spectrometry; chemical dynamics; flame chemistry; ion-molecule reactions. Mailing Add: Naval Res Lab Code 6110 Washington DC 20390

WYATT, JOHN POYNER, b Winnipeg, Man, Feb 25, 16; m 45; c 4. PATHOLOGY. Educ: Univ Man, MD, 38. Prof Exp: Coroner's pathologist, City of Toronto, Can, 46-49; prof path, Sch Med, St Louis Univ, 49-67; prof path & head dept, Med Col, Univ Man, 67-73; PROF PATH, COL MED, UNIV KY, 73- Concurrent Pos: Fel path, Banting Inst, Univ Toronto, 39-41; Littauer fel, Harvard Med Sch, 45-46; lectr, Univ Toronto, 46-49; consult, Silicosis Bd, Ont Prov Govt, 46-49; chief pathologist, St Louis County, Mo, 49-67. Mem: Am Asn Path & Bact; Col Am Path; Path Soc Great Brit & Ireland; Int Acad Path. Res: Iron metabolism, particularly ferrodynamics of iron storage; chronic lung disease, particularly pneumoconioses and emphysema; virology, especially salivary gland virus. Mailing Add: Tobacco & Health Res Inst 109 Kincaid Hall Univ of Ky Lexington KY 40506

WYATT, PHILIP JOSEPH, b Los Angeles, Calif, Apr 16, 32; m 57; c 3. BIOPHYSICS, MICROBIOLOGY. Educ: Univ Chicago, BA, 52, BS, 54; Univ Ill, MS, 56; Fla State Univ, PhD, 59. Prof Exp: Staff mem, Los Alamos Sci Lab, 59; prin scientist, Aeronutronic Div, Ford Motor Co, 59-62; dir adv planning, Plasmadyne Corp, 62-63; mem tech staff, DRC Inc, 63-67; sr sci specialist, EG&G, Inc, 67-68; PRES & CHMN, SCI SPECTRUM, INC, 68- Mem: Fel Am Phys Soc; Optical Soc Am; Am Soc Microbiol; Sigma Xi. Res: Light scattering studies of microorganisms; development of new assay and identification techniques using resonance scattering; antibiotic susceptibility testing; development of light scattering instrumentation; bioassays for antineoplastic drugs. Mailing Add: PO Box 3003 1216 State Santa Barbara CA 93105

WYATT, RAYMOND L, b Salisbury, NC, Nov 23, 26; m 52; c 1. PLANT MORPHOLOGY, PLANT TAXONOMY. Educ: Wake Forest Col, BS, 48; Univ NC, MA, 54, PhD, 56. Prof Exp: Instr biol, Mars Hill Col, 48-52 & Univ NC, 55-56; from asst prof to assoc prof, 56-75, lectr, NSF Insts Sci Teachers, 59-60, PROF BIOL, WAKE FOREST UNIV, 75- Mem: AAAS. Res: Embryology of Asarum; floral morphology and phylogeny of Aristolochiaceae and Annonaceae; survival of American chestnut in North Carolina. Mailing Add: Dept of Biol Wake Forest Univ Reynolds Sta Winston-Salem NC 27109

WYATT, RICHARD J, b Los Angeles, Calif, June 5, 39; m; c 2. PSYCHOPHARMACOLOGY. Educ: Johns Hopkins Univ, BA, 61, MD, 64; Am Bd Psychiat & Neurol, dipl. Prof Exp: Intern pediat, Western Reserve Univ Hosp, 64-65; resident psychiat, Mass Ment Health Ctr, Boston, 65-67; clin assoc, Sect Psychophysiol Sleep, Adult Psychiat Br, 67-68, clin assoc, Lab Clin Psychobiol, 68-69, res psychiatrist, Lab Clin Psychopharmacol, 69-71, ACTG CHIEF, LAB CLIN PSYCHOPHARMACOL, NAT INST MENT HEALTH, 72- Concurrent Pos: Teaching asst, Harvard Univ, 65-67; assoc prof psychiat, Med Ctr, Stanford Univ, 73-74; clin prof, Med Ctr, Duke Univ, 75. Honors & Awards: Harry Solomon Res Award, Mass Ment Health Ctr, Boston, 68; A E Bennett Award Clin Psychiat Res, Soc Biol Psychiat, 71; Psychopharmacol Award, Am Psychol Asn, 72. Mem: Am Psychiat Asn; Soc Psychophysiol Study Sleep; Psychiat Res Soc; Soc Biol Psychiat; Am Col Neuropsychopharmacol. Mailing Add: St Elizabeths Hosp WAW Bldg Rm 536 Washington DC 20032

WYATT, ROBERT EUGENE, b Chicago, Ill, Nov 11, 38; m 64; c 1. THEORETICAL CHEMISTRY. Educ: Ill Inst Technol, 61; Johns Hopkins Univ, MA, 63, PhD(chem), 65. Prof Exp: NSF fel chem, Keele Univ, 65-66 & Harvard Univ, 66-67; asst prof, 67-72, ASSOC PROF CHEM, UNIV TEX, AUSTIN, 72- Mem: AAAS; Am Chem Soc; Am Phys Soc. Res: Internal rotation barriers; scattering theory; theoretical chemical kinetics. Mailing Add: Dept of Chem Univ of Tex Austin TX 78712

WYATT, ROGER DALE, b Albemarle, NC, Apr 16, 48; m 68; c 2. MICROBIOLOGY, POULTRY SCIENCE. Educ: NC State Univ, BS, 70, MS, 72, PhD(micorbiol), 74. Prof Exp: ASST PROF, DEPT POULTRY SCI, UNIV GA, 74- Mem: Poultry Sci Asn; Am Soc Microbiol. Res: Biological effects of dietary mycotoxins on poultry and evaluation of antifungal compounds for use in grain and poultry feeds. Mailing Add: Dept of Poultry Sci Univ of Ga Athens GA 30602

WYATT, STANLEY PORTER, JR, b Medford, Mass, Apr 20, 21; m 48; c 3. ASTRONOMY. Educ: Dartmouth Col, AB, 42; Harvard Univ, AM, 48, PhD(astron), 50. Prof Exp: Instr astron, Univ Mich, 50-53; from asst prof to assoc prof, 53-61, PROF ASTRON, UNIV ILL, URBANA, 61- Mem: Int Astron Union; Am Astron Soc. Res: Galactic and extragalactic astronomy; interplanetary physics. Mailing Add: Dept of Astron Univ of Ill Urbana IL 61801

WYBLE, D O, b Jackson, Mich, Nov 26, 16; m 48; c 6. GEOPHYSICS. Educ: Eastern Mich Univ, AB, 38; Mich Technol Univ, BS, 42, MS, 50; Pa State Univ, PhD(geophys), 58. Prof Exp: Teacher, High Sch, Mich, 38-40; instr physics, Mich Technol Univ, 47-50; asst, Pa State Univ, 55-57; assoc prof, 57-62, PROF PHYSICS, MICH TECHNOL UNIV, 62- Concurrent Pos: Consult, Columbia Explor, Inc, 59. Mem: Soc Explor Geophys; Am Geophys Union. Res: Physical properties of rock; analog simulation. Mailing Add: Dept of Physics Mich Technol Univ Houghton MI 49931

WYCKOFF, DELAPHINE GRACE ROSA, b Beloit, Wis, Sept 11, 06; m 42. MICROBIOLOGY, BACTERIOLOGY. Educ: Univ Wis, PhB, 27, PhM, 28, PhD(bact), 38. Prof Exp: From instr to asst prof bact, NDak Agr Col, 28-37; from instr to assoc prof bact & bot, 38-57, prof bact, 57-72, EMER PROF BACT, WELLESLEY COL, 72- Concurrent Pos: Consult, Traveling Sci Teachers Prog, Oak Ridge Inst Nuclear Studies & Biol Sci Curric Study, Am Inst Biol Sci. Mem: AAAS;

Am Soc Microbiol; Am Acad Microbiol; Brit Soc Gen Microbiol. Res: Physiological variation; induced mutations in actinomycetes; bactericidal agents; antibiotics from actinomycetes; biochemical activities of yeasts; marine halophilic bacteria; soil microbiology. Mailing Add: 78 Cedar St Newington CT 06111

WYCKOFF, HAROLD ORVILLE, b Traverse City, Mich, Apr 26, 10; m 40; c 2. PHYSICS. Educ: Univ Wash, BS, 34, PhD(physics), 40. Prof Exp: Jr physicist, Nat Bur Stand, 41-42, from asst physicist to assoc physicist, 42-43; expert consult, Ninth US Army Air Force, Europe, 43-45; physicist & asst chief, X-ray Sect, Nat Bur Stand, 45-49, chief, 49-53, chief radiation physics lab, 53-66; dep sci dir, Armed Forces Radiobiol Res Inst, 66-71; consult, Bur Radiol Health, 71-74; CONSULT RADIATION PHYSICS, 74- Concurrent Pos: Secy & mem, Int Comn Radiation Units & Measurements, 53-69, chmn, 69-; mem comt, Int Comn Radiol Protection, 53-69; mem adv comt, Health Physics Div, Oak Ridge Nat Lab, 62-69; mem bd, Nat Coun Radiation Protection & Measurements, 64-70; consult, Fed Aviation Agency, 71-74. Honors & Awards: Bronze Star, 46; Silver Medal, Dept Com, 52, Gold Medal, 60; Gold Medal, Radiol Soc NAm, 63; Exceptional Civilian Serv Medal, Defense Atomic Support Agency, 71; Gold Medal, XIII Int Cong Radiol, 73. Mem: Health Physics Soc; fel Am Phys Soc; Radiation Res Soc; assoc fel Am Col Radiol; Radiol Soc NAm (treas, 70-). Res: Radiation physics; radiation protection and measurement. Mailing Add: 4108 Montpelier Rd Rockville MD 20853

WYCKOFF, HAROLD WINFIELD, b Niagara Falls, NY, Dec 3, 26; m 55; c 3. MOLECULAR BIOPHYSICS. Educ: Antioch Col, BS, 49; Mass Inst Technol, PhD(biophys), 55. Prof Exp: Res assoc biol, Mass Inst Technol, 55; NIH fel, Cambridge Univ, 56; res physicist, Am Viscose Corp, Pa, 57-63; ASSOC PROF MOLECULAR BIOPHYS, YALE UNIV, 63- Mem: Am Crystallog Asn; Biophys Soc. Res: Structure and function of biological macromolecules, especially enzymes, as determined by x-ray diffraction analysis. Mailing Add: Dept of Molecular Biophys & Biochem Yale Univ Box 2166 Yale Station New Haven CT 06520

WYCKOFF, JAMES M, b Niagara Falls, NY, July 3, 24; m 47, 68; c 5. NUCLEAR PHYSICS. Educ: Antioch Col, BS, 48; Univ Rochester, MS, 52. Prof Exp: Electronics technician, Airborne Instruments Lab, 46-47; asst physics, Antioch Col, 47-48; res asst, Univ Rochester, 48-51; physicist, Nat Bur Stand, 51-67; health physicist, Stanford Linear Accelerator Ctr, 67-68; physicist, 68-71, COORDR, RADIATION SAFETY PROG, NAT BUR STAND, 71- Concurrent Pos: Mem comt 22, Nat Coun Radiation Protection & Measurement, 66-72, mem off telecommun policy side effects subcomt, 72-; mem, Interagency Comt Fed Guid Occup Exposures to Ionizing Radiation, 74- Mem: Am Phys Soc. Res: Detection of high energy x-rays; measurement of attenuation coefficients; induced radioactivity and development of on-line computer system; application of measurements to safe and effective use of x-rays; radioactivity, ultraviolet light, lasers, electromagnetic and ultrasonic radiation sources; development of measurement systems adequate to the protection of those near such radiation sources. Mailing Add: Nat Bur of Stand Metrol Bldg Rm B368 Washington DC 20234

WYCKOFF, KENNETH KEITH, organic chemistry, see 12th edition

WYCKOFF, PETER HINES, b Brooklyn, NY, June 1, 13; m 42; c 1. GEOPHYSICS. Educ: Carnegie Inst Technol, BS, 36; Calif Inst Technol, MS, 37. Prof Exp: Res engr, Res Lab, Westinghouse Elec Corp, 37-41, 46-48; chief atmospheric physics lab, Air Force Cambridge Res Labs, 48-56, aerosol physics lab, 56-58, aerophys lab, 58-60; asst dir physics res, Armour Res Found, 61-63; asst dir staff opers, IIT Res Inst, 63-64; prog dir weather modification, NSF, 64-72; RETIRED. Concurrent Pos: Dir prog 4, Oper Greenhouse, Eniwetok, 51; mem, Tech Panel Rocketry & Exec Comt Rocket & Satellite Res, US Nat Comt Int Geophys Yr, Nat Acad Sci; mem panel, Atmosphere Res & Develop Bd, 52-54; mem, Bd Dirs, US Civil Serv Exam, 50-61; chief, Exp Physics Lab, Geophys Res Directorate, Watson Labs, US Air Force, 41-46. Mem: Fel AAAS; Optical Soc Am; Am Phys Soc; Am Meteorol Soc; assoc fel Am Inst Aeronaut & Astronaut. Res: Upper air and cloud physics. Mailing Add: 3066 Valley Lane Falls Church VA 22044

WYCKOFF, RALPH D, geophysics, deceased

WYCKOFF, RALPH WALTER GRAYSTONE, b Geneva, NY, Aug 9, 97; m; c 3. PHYSICAL CHEMISTRY, BIOPHYSICS. Educ: Hobart Col, BS, 16; Cornell Univ PhD(chem), 19. Hon Degrees: MD, Masaryk Univ, Brno, 47; ScD, Univ Strasbourg, 52, Hobart Col, 75. Prof Exp: Instr anal chem, Cornell Univ, 17-19; phys chemist, Carnegie Inst Geophys Lab, 19-27; assoc mem subdiv biophys, Rockefeller Inst, 27-38; with Lederle Labs, 38-40, assoc res dir, 40-42; tech dir, Reichel Labs, 42-43; lectr, Univ Mich, 43-45; sr scientist, NIH, 45, scientist dir, 46-52, sci attache, Am Embassy, Eng, 52-54, biophysicist, 54-59; PROF PHYSICS & BACT, UNIV ARIZ, 59- Concurrent Pos: Res assoc, Calif Inst Technol, 21-22; dir res, Nat Ctr Sci Res, France, 58; ed, Exp Cell Res. Honors & Awards: Medal, Pasteur Inst. Mem: Nat Acad Sci; AAAS; Am Chem Soc; fel Am Phys Soc; Electron Micros Soc Am (past pres). Res: Structure of crystals; effect of radiation on cells; development of air-driven ultracentrifugation of proteins and viruses; electron microscopy; purification of viruses and macromolecules; ultra-soft x-rays; ultrastructure of fossils. Mailing Add: Bldg 81 Dept of Physics Univ of Ariz Tucson AZ 85721

WYCKOFF, ROBERT CUSHMAN, b Peoria, Ill, Feb 7, 13; m 42. PHYSICS, BIOPHYSICS. Educ: Univ Ill, BS, 35, MS, 36. Prof Exp: Prof physics & head dept, Buena Vista Col, 38-47; instr, Univ Okla, 47-51; head sci & technol unit, Off Naval Intel, Europe, 51-57, sr ord analyst, 57-58; chief sci & technol br, Army Correlation Ctr, US Signal Intel Agency, Va, 58-60; staff engr, Librascope Div, Gen Precision, Inc, 60-61; head oper anal group, Res Anal Sect, Space Sci Div, Jet Propulsion Lab, Calif Inst Technol, 61-72; SR SCIENTIST, ELECTRO-OPTICAL SYSTS, VISTA LAB, 72- Concurrent Pos: Ballistician, Des Moines Ord Plant, 42-43. Mem: AAAS; Am Phys Soc. Res: Spectroscopy; interferometry; high vacuum techniques; trace elements in biological tissue; military and naval scientific and technical intelligence; digital computer systems analysis. Mailing Add: Vista Lab PO Box 5608 Pasadena CA 91107

WYCKOFF, SUSAN, b Santa Cruz, Calif, Mar 18, 41; m 67. ASTRONOMY. Educ: Mt Holyoke Col, AB, 62; Case Inst Technol, PhD(astron), 67. Prof Exp: Fel, Inst Sci & Technol, Univ Mich, 67-68; asst prof physics, Albion Col, 68-70; res assoc astron, Univ Kans, 70-72 & Wise Observ, Tel Aviv Univ, 72-75; PRIN RES FEL, ROYAL GREENWICH OBSERV, 75- Mem: Am Astron Soc; Int Astron Union. Res: Spectroscopy of long-period variable stars, cool stars and comets; observational astrophysics. Mailing Add: Royal Greenwich Observ Herstmonceux Castle Hailsham England

WYCOFF, HARLAND DEWITT, b St Paul, Minn, Dec 28, 17; m 42; c 4. BIOCHEMISTRY. Educ: Univ Wis, BS, 42, MS, 48, PhD(biochem), 52. Prof Exp: Asst prof agr chem, Univ Idaho, 50-53; from asst prof to assoc prof, 53-62, PROF BIOCHEM, MED COL GA, 62- Mem: AAAS; Am Chem Soc. Res: Physiological chemistry; protein and lipid metabolism; blood coagulation. Mailing Add: Dept of Cell & Molecular Biol Med Col of Ga Augusta GA 30902

WYCOFF, SAMUEL JOHN, b Berry, Ala, Feb 25, 29. PUBLIC HEALTH, DENTISTRY. Educ: Univ Ala, Tuscaloosa, BS, 50; Univ Ala, Birmingham, DMD, 54; Univ Mich, Ann Arbor, MPH, 59; Am Bd Dent Pub Health, dipl, 65. Prof Exp: Staff dent officer, Div Dent Health, USPHS, Washington, DC, 59-63, regional prog dir, Boston, Mass, 63-65; assoc prof prev dent & community health & chmn dept, Sch Dent Loyola Univ Chicago, 67-69; PROF PREV DENT & COMMUNITY HEALTH & CHMN DEPT, SCH DENT, UNIV SAN FRANCISCO, 69- Concurrent Pos: Fels, Univ Calif, San Francisco, 69-; consult, Coun Dent Educ, Am Dent Asn, 68- & Calif State Health Manpower Coun, 70-; mem comt acad affairs, Am Asn Dent Schs, 71-; mem tech adv comt, Calif State Dept Pub Health, 72-; mem, Pierre Fouchard Acad. Honors & Awards: H Trendley Dean Citation, 65. Mem: Am Dent Asn; fel Am Pub Health Asn; Am Asn Hosp Dent; fel Am Col Dent; Int Dent Fedn. Res: Epidemiology of oral disease; studies on health care delivery system and health manpower needs. Mailing Add: Univ of Calif Sch of Dent 532 Parnassus Ave San Francisco CA 94122

WYDER, JOHN ERNEST, b Grand Forks, BC, Jan 3, 38; m 60; c 4. GEOPHYSICS, GEOLOGY. Educ: Univ BC, BASc, 61; Univ Sask, MSc, 64, PhD(geophys geol), 68. Prof Exp: Res scientist, Can Dept Energy, Mines & Resources, 61-71; chief geophysicist, 71-75, GEN MGR, KENTING EXPLOR SERV LTD, 75- Mem: Soc Explor Geophys; Europ Asn Explor Geophys; Geol Asn Can; Can Inst Mining & Metall. Res: Mining, groundwater, permafrost and engineering geophysics with emphasis on electrical and borehold geophysical techniques. Mailing Add: Kenting Explor Serv Ltd 524 11th Ave SW Calgary AB Can

WYDEVEN, THEODORE, b Wausau, Wis, Jan 18, 36; m 63; c 2. PHYSICAL CHEMISTRY. Educ: Marquette Univ, BS, 58; Univ Wash, PhD(phys chem), 64. Prof Exp: RES SCIENTIST ENVIRON CONTROL SYSTS, NASA AMES RES CTR, 64- Mem: AAAS; Am Chem Soc; The Chem Soc. Res: Research and development of advanced environmental control systems for purifying water, recycling oxygen and controlling atmospheric trace contaminants. Mailing Add: NASA Ames Res Ctr Moffett Field CA 94035

WYDOSKI, RICHARD STANLEY, b Nanticoke, Pa, Feb 3, 36; m 59; c 4. FISHERIES, ZOOLOGY. Educ: Bloomsburg State Col, BS, 60; Pa State Univ, MS, 62, PhD(zool), 65. Prof Exp: Teaching asst zool, Pa State Univ, 61-65; fisheries biologist, Bur Com Fisheries, US Fish & Wildlife Serv, 65-66; asst prof fisheries & asst leader coop fishery unit, Ore State Univ, 66-69; assoc prof fisheries & asst leader coop fishery unit, Univ Wash, 69-73; ASSOC PROF WILDLIFE SCI & LEADER COOP FISHERY UNIT, UTAH STATE UNIV, 73- Mem: Am Fisheries Soc; Am Inst Fishery Res Biol. Res: Aquatic ecology; reponses of fish populations to alterations of the aquatic environment; fish behavior and habit requirements; socio-economic studies as related to fish management. Mailing Add: 279 Natural Resources-Zool UMC 52 Utah State Univ Logan UT 84322

WYE, EDWIN JAMES, b Toronto, Ont, Oct 8, 18; m 43. BIOCHEMISTRY. Educ: Univ Toronto, BA, 49. Prof Exp: Chemist, W R Drynen Nutrit Labs, Can Canners Ltd, 49-50; chemist, Res Labs, Can Breweries Ltd, 50-58; tech assoc, J E Siebel Sons Co, Ill, 58-60; sr res asst, 60-62, res assoc, 62-72, ASST DIR, CONNAUGHT LABS LTD, 72- Res: Protein and flavonoid compounds of brewing materials; beer stability; pancreatic hormones; starch chemistry; blood serum proteins; blood coagulation. Mailing Add: Connaught Labs Ltd 1755 Steeles Ave Willowdale ON Can

WYGANT, JAMES CALVIN, b Guys Mills, Pa, Oct 3, 23; m 48; c 3. ORGANIC CHEMISTRY. Educ: Allegheny Col, BS, 48; Univ Mich, MS, 49, PhD(chem), 52. Prof Exp: RES CHEMIST, MONSANTO CO, 52- Mem: Am Chem Soc. Res: Product and process research and development; organic synthesis. Mailing Add: Monsanto Co 800 N Lindbergh Blvd St Louis MO 63166

WYGANT, NOEL DARWIN, b Roanoke, Ind, Apr 18, 08; m 33; c 3. FORESTRY. Educ: Purdue Univ, BSF, 32; Syracuse Univ, PhD(forest entom), 40. Prof Exp: Jr entomologist, Forest Insect Lab, Bur Entom & Plant Quarantine, USDA, 35-37, from asst entomologist to sr entomologist, 37-53, PRIN ENTOMOLOGIST, ROCKY MT FOREST & RANGE EXP STA, US FOREST SERV, 53- Concurrent Pos: Mem fac forest & wood sci, Colo State Univ, 69-74. Mem: AAAS; Soc Am Foresters. Res: Forest entomology; shelter-belt insects in Great Plains; bark beetles and methods of control; forest insect problems in central and southern Rocky Mountain region. Mailing Add: 5435 Mohawk Rd Littleton CO 80120

WYKES, ARTHUR ALBERT, b Boston, Mass, May 21, 23; m 56; c 2. BIOCHEMICAL PHARMACOLOGY, TOXICOLOGY. Educ: Univ Ill, BS, 45; Univ Wis, MS, 49; Purdue Univ, PhD(pharmaceut chem, pharmacol, biochem), 57. Prof Exp: Res asst biochem, Res Div, Armour & Co, 45-46; res asst biochem, Univ Wis, 47-49; teaching asst pharmaceut chem, Purdue Univ, 49-51; res biochemist, Armour & Co, 51-52; sr res biochemist, Res Div, Int Minerals & Chem Corp, 52-53; sr res biochemist, Res Dept, Baxter Labs, Inc, 53-55; res asst pharmaceut chem, Purdue Univ, 55-57; res biochemist, Chem Pharmacol Sect, Abbott Labs, 57-61; chief & supvry res pharmacologist, biochem pharmacol sect, US Air Force Sch Aerospace Med, 61-67; sr pharmacologist, Chem & Life Sci Labs, Res Triangle Inst, 67-68; PHARMACOLOGIST & SR DRUG & TOXICOL LIT SPECIALIST, NAT LIBR MED, 68- Concurrent Pos: Assoc clin prof, Div Pharmacol & Physiol, Med Ctr, Duke Univ, 67-69; vis assoc prof pharmacol, Milton S Hershey Med Ctr, 71-76. Mem: Fel AAAS; Am Chem Soc; Int Soc Biochem Pharmacol; Soc Toxicol; Drug Info Asn. Res: Drug effects on biological, enzyme and metabolic systems at cellular and subcellular levels, especially as influenced by environmental agents, husbandry factors, other drugs and toxic chemicals; drug interactions; neurobiochemistry; pharmacology; enzymology; central nervous system drugs; drugs of abuse; biogenic amines; psychopharmacologic drugs and agents; drug, toxic agent and animal models; biomedical literature and computerized data. Mailing Add: Spec Info Serv Toxicol Info Prog Nat Libr of Med NIH Bethesda MD 20014

WYKHUIS, WALTER ARNOLD, b Oostburg, Wis, Apr 6, 11; m 50. DENTISTRY. Educ: Calvin Col, BA, 32; Chicago Col Dent Surg, DDS, 36. Prof Exp: From instr to asst prof prosthodont, Chicago Col Dent Surg, 38-43; assoc prof, Sch Dent, Univ Tenn, 43-45 & Univ Ore, 45-46; mem staff prof & sales dept, Dentists Supply Co NY, 46-55; assoc prof prosthodont, Sch Dent, Univ Minn, 55-56; ASSOC PROF PROSTHODONT, SCH DENT, UNIV WASH, 56- Concurrent Pos: Consult, USPHS Hosp, 58-67, Vet Admin Hosps, Seattle, 58-67, Am Lake, 59-67 & Walla Walla, 62-67. Mem: Am Dent Asn. Res: Prosthodontics; epoxy resin denture base material. Mailing Add: Dept of Prosthodont Univ of Wash Sch of Dent Seattle WA 98105

WYKLE, ROBERT LEE, b Belmont, NC, Mar 17, 40; m 71. BIOCHEMISTRY. Educ: Teacher physics, biol & gen sci, Waynesville High Sch, 63-65; teaching asst biochem, Med Sch, Univ Tenn, 65-67; fel, 68-70, assoc scientist, 70-71, SCIENTIST, OAK RIDGE ASSOC UNIVS, 76- Res: Biochemistry and function of lipids in normal and neoplastic cells with special interest in ether-linked lipids. Mailing Add: Med & Health Sci Div Oak Ridge Assoc Univs PO Box 117 Oak Ridge TN 37830

WYKOFF, DALE EMERSON, b Lakewood, Ohio, Jan 12, 27; m 50; c 2. PARASITOLOGY. Educ: Va Mil Inst, BS, 49; Tulane Univ, MS, 51, PhD(parasitol), 58; Bernard Nocht Tropeninstitut, Hamburg, dipl, 60; Am Bd Microbiol, dipl. Prof Exp: US Army, 49-, parasitologist, 406 Med Lab, Tokyo, Japan, 51-54, Walter Reed Army Inst Res, 54-57, US Army in Europe Med Lab, Landstuhl, Ger, 58-61, SEATO Med Res Lab, Bangkok, Thailand, 62-65, Makerere Univ Col, Uganda, 66-69 & Walter Reed Army Med Ctr, 69-71, chief res planning off, US Army Med Res & Develop Command, Washington, DC, 71-73, CHIEF US ARMY MED RES UNIT, NAIROBI, KENYA, 73- Res: Resource management for broad international military medical research. Mailing Add: US Army Med Res Unit Nairobi Dept of State Washington DC 20520

WYKOFF, MATTHEW HENRY, b Kewanee, Ill, Apr 11, 23; m 52; c 5. EXPERIMENTAL SURGERY, PHYSIOLOGY. Educ: Iowa State Univ, DVM, 46; Univ Mo, MS, 60. Prof Exp: Pvt pract, 46-50 & 52-57; asst prof vet anat, Univ Mo, 57-61; assoc prof physiol, Comp Animal Res Lab, Univ Tenn-AEC, 61-72; PRIN SCIENTIST VET SURG, ETHICON RES FOUND, 72- Res: Effects of ionizing irradiation on central nervous and cardiovascular systems; placental transfer; prenatal effects of ionizing irradiation; healing. Mailing Add: Ethicon Res Found Bridgewater NJ 08807

WYLD, GARRARD ERNEST ALFRED, analytical chemistry, see 12th edition

WYLD, HENRY WILLIAM, JR, b Portland, Ore, Oct 16, 28; m 55; c 3. PHYSICS. Educ: Reed Col, BA, 49; Univ Chicago, MS, 52, PhD(physics), 54. Prof Exp: Instr physics, Princeton Univ, 54-57; from asst prof to assoc prof, 57-63, PROF PHYSICS, UNIV ILL, URBANA, 63- Concurrent Pos: Consult, Space Tech Labs, Inc, Calif, 57-63; NSF sr fel, Oxford Univ, 63-64; Guggenheim fel, Europ Coun Nuclear Res, 71. Mem: Am Phys Soc. Res: Theoretical high energy and plasma physics. Mailing Add: Dept of Physics Univ of Ill Urbana IL 61801

WYLEN, HERBERT E, b Philadelphia, Pa, Mar 16, 33; m 57; c 3. THEORETICAL PHYSICS. Educ: Univ Del, BS, 61; Bryn Mawr Col, MA, 65, PhD(physics), 69. Prof Exp: Asst prof physics, Union Col, NY, 66-75; DEAN, WASHINGTON & JEFFERSON COL, 75- Mem: Am Phys Soc; Am Asn Physics Teachers. Res: Nuclear astrophysics; cosmology. Mailing Add: Off of Dean Washington & Jefferson Col Washington PA 15301

WYLER, OSWALD, b Scuol, Grisons, Switz, Apr 2, 22; m 60; c 3. MATHEMATICS. Educ: Swiss Fed Inst Technol, dipl, 47, Dr sc math, 50. Prof Exp: Asst inst geophys, Swiss Fed Inst Technol, 46-50; lectr math, Northwestern Univ, 51-53; from asst prof to assoc prof, Univ NMex, 53-65; PROF MATH, CARNEGIE-MELLON UNIV, 65- Mem: Am Math Soc; Math Asn Am; Swiss Math Soc. Res: Categorical algebra; categorical topology; theory of convergence spaces. Mailing Add: Dept of Math Carnegie-Mellon Univ Pittsburgh PA 15213

WYLIE, AUBREY EVANS, b Carthage, Ark, Nov 21, 17; m 49; c 1. FORESTRY. Educ: Colo State Univ, BS, 46, MF, 47; State Univ NY Col Environ Sci & Forestry, PhD(wood sci), 50. Prof Exp: Instr assoc prof wood prod, Col Environ Sci & Forestry, State Univ NY, 50-56; prof, Mich State Univ, 56-68; vpres, Freeman Corp, 68-70; FOREST SCIENTIST FORESTRY, COOP STATE RES SERV, USDA, 70- Mem: Soc Am Foresters; Forest Prod Res Soc. Res: Institutional and national planning of forestry research; review of institutional research programs; research budgeting, coordination of federal and state research. Mailing Add: USDA Rm 427W Admin Bldg Coop State Res Serv Washington DC 20250

WYLIE, CLARENCE RAYMOND, JR, b Cincinnati, Ohio, Sept 9, 11; m 35, 58; c 2. GEOMETRY. Educ: Wayne State Univ, BA & BS, 31; Cornell Univ, MS, 32, PhD(math), 34. Prof Exp: Instr & asst prof math, Ohio State Univ, 34-46; prof & head dept & actg dean col eng, US Air Force Inst Technol, 46-48; prof, Univ Utah, 48-69, head dept, 48-67; prof, 69-71, chmn dept, 70-76, KENAN PROF MATH, FURMAN UNIV, 71- Concurrent Pos: Consult, Gen Elec Co, NY, 37 & Briggs Mfg Co, Mich, 41; mech engr, Wright Field Propellor Lab, Ohio, 43-46 & Aero Prod Div, Gen Motors Corp, 45-47; lectr, Educ Prog, Union Carbide Corp, 65- Mem: Fel AAAS; Am Math Soc; Math Asn Am; Am Soc Eng Educ. Res: Projective geometry, especially line geometry; applied mathematics, especially mechanical vibrations. Mailing Add: Dept of Math Furman Univ Greenville SC 20613

WYLIE, DOUGLAS WILSON, b Saskatoon, Sask, Nov 12, 26; nat US; m 51; c 4. PHYSICS. Educ: Univ NB, BSc, 47; Dalhousie Univ, MSc, 49; Univ Conn, PhD(solid state physics), 62. Prof Exp: Asst, Brown Univ, 49-50; instr, Univ NB, 50-51; from instr to prof, Univ Maine, 51-68; chmn dept, 68-73, PROF PHYSICS, WESTERN ILL UNIV, 68- Mem: AAAS; Am Phys Soc; Am Asn Physics Teachers; Can Asn Physicists. Res: Solid state physics; radiation damage; electron spin resonance. Mailing Add: Dept of Physics Western Ill Univ Macomb IL 61455

WYLIE, EDWIN J, b Cincinnati, Ohio, Oct 13, 18; m 45; c 3. SURGERY. Educ: Pomona Col, AB, 39; Harvard Med Sch, MD, 43. Prof Exp: From asst prof to assoc prof, 58-67, PROF SURG, SCH MED, UNIV CALIF, SAN FRANCISCO, 67-, VCHMN DEPT, 59-, CHIEF VASCULAR SURG, 55- Concurrent Pos: Consult, Vet Admin Hosp, Ft Miley, Calif, 51- Mem: Soc Vascular Surg; Soc Univ Surg; Am Surg Asn; Am Col Surg; Int Cardiovasc Soc. Res: Vascular surgery. Mailing Add: Dept of Surg San Francisco Med Ctr Univ of Calif Sch of Med San Francisco CA 94122

WYLIE, RICHARD MICHAEL, b Louisville, Ky, June 17, 34; m 69. BIOLOGY, NEUROPHYSIOLOGY. Educ: Harvard Univ, BA, 56, MA, 58, PhD(biol), 62. Prof Exp: Fel neurophysiol, Univ Utah, 62-65, res assoc, 65-66; res assoc, Rockefeller Univ, 66-69; RES PHYSIOLOGIST, DEPT NEUROPHYSIOL, WALTER REED ARMY INST RES, 69- Mem: Soc Neurosci; assoc mem Am Physiol Soc. Res: Biophysics of sensory mechanisms; mechanisms of sensory discrimination in central nervous systems; integration in sensory and motor systems. Mailing Add: Dept of Neurophysiol Walter Reed Army Inst of Res Walter Reed Army Med Ctr Washington DC 20012

WYLIE, WILLIAM DICKEY, b Carthage, Ark, Nov 18, 14; m 40; c 4. ENTOMOLOGY. Educ: Univ Ark, BSA, 37; Cornell Univ, PhD(econ entom), 41. Prof Exp: Asst entom, Cornell Univ, 37-42; entomologist, US Sugar Corp, 42-45 & Everglades Exp Sta, Univ Fla, 46-47; asst prof, Univ & asst entomologist, Exp Sta, 47-49, assoc prof & assoc entomologist, 49-57, PROF ENTOM, UNIV & ENTOMOLOGIST, EXP STA, UNIV ARK, FAYETTEVILLE, 57- Mem: Entom Soc Am. Res: Biology and control of insects of fruit and vegetable crops. Mailing Add: Dept of Entom Univ of Ark Fayetteville AR 72701

WYLLER, ARNE AUGUST, astronomy, see 12th edition

WYLLIE, GILBERT ALEXANDER, b Saltcoats, Scotland, Jan 11, 28; US citizen; m 57; c 3. BIOLOGY, ECOLOGY. Educ: Col Idaho, BS, 58; Sacramento State Col, MA, 60; Purdue Univ, PhD(ecol), 63. Prof Exp: Assoc prof biol, WTex State Univ, 63-65; asst prof, 65-66, ASSOC PROF BIOL, BOISE STATE UNIV, 66- Mem: Ecol Soc Am. Res: Effects of environment on morphology, distribution, behavior of invertebrates and lower vertebrates. Mailing Add: Dept of Biol Boise State Univ Boise ID 83701

WYLLIE, MALCOLM ROBERT JESSE, b Cape Town, SAfrica, July 31, 19; nat US; m 47; c 2. GEOPHYSICS, ENGINEERING. Educ: Univ Cape Town, BSc, 39; Oxford Univ, DPhil(electrochem), 43, DSc, 58. Prof Exp: Metallurgist, Union Steel Corp, SAfrica, 40; fel, Johns Hopkins Univ, 46-47; electrochemist, Gulf Res & Develop Co, 47-49, head petrophys sect, 49-55, asst div dir geol & geochem, 55, asst div dir reservoir mech div, 55-64, dir explor & prod dept, 64-67, from vpres to pres admin, 67-70, from exec vpres to pres, Gulf Oil Co-Eastern Hemisphere, 70-75, CHMN BD, GULF OIL CO-EASTERN HEMISPHERE, 75- Concurrent Pos: Mem staff, Admiralty Res & Develop, India. Honors & Awards: Gold Medal Achievement Award, Soc Prof Well Log Analysts. Mem: Am Chem Soc; Am Inst Mining, Metall & Petrol Engrs; Soc Prof Well Log Analysts; Brit Inst Petrol; The Chem Soc. Res: Well logging, electrochemistry of ion-exchange resins; fluid flow in porous media; acoustical velocity in rocks; petroleum reservoir engineering. Mailing Add: Gulf Oil Corp Gulf Bldg Pittsburgh PA 15230

WYLLIE, PETER JOHN, b London, Eng, Feb 8, 30; m 56; c 3. GEOLOGY, GEOCHEMISTRY. Educ: Univ St Andrews, BSc, 52 & 55, PhD, 58. Hon Degrees: DSc, Univ St Andrews, 74. Prof Exp: Geologist, Brit NGreenland Exped, 52-54; asst lectr geol, Univ St Andrews, 55-56; asst geochem, Pa State Univ, 56-58, asst prof, 58-59; res fel, Leeds Univ, 59-60, lectr exp petrol, 60-61; assoc prof petrol, Pa State Univ, 61-65; master of col & assoc dean phys sci div, 72-73, PROF PETROL & GEOCHEM, UNIV CHICAGO, 65- Concurrent Pos: Mem, Int Comn Exp Petrol at High Pressures & Temperatures, 71-; ed, J Geol. Honors & Awards: Polar Medal for Geol Surv & Explor in Greenland; Award, Mineral Soc Am, 65. Mem: Geol Soc Am; Geochem Soc; Am Geophys Union; Mineral Soc Am; Brit Mineral Soc. Res: Igneous and metamorphic petrology; experimental petrology; high pressure studies on hydrothermal systems; application of phase equilibrium studies to igneous and metamorphic petrology. Mailing Add: Hinds Geophys Lab 5734 S Ellis Ave Chicago IL 60637

WYLLIE, THOMAS DEAN, b Hinsdale, Ill, Dec 4, 28; m 50; c 3. PLANT PATHOLOGY. Educ: San Diego State Col, AB, 52; Univ Minn, MS, 57, PhD(plant path), 60. PLANT PATH, UNIV MO-COLUMBIA, 60- Mem: Am Phytopath Soc. Res: Physiology of host-parasite interactions; mycotoxin and mycotoxioses research; ecological relationships of non-specific soil borne pathogenic fungi on the soybean. Mailing Add: Dept of Plant Path 108 Waters Hall Univ of Mo Columbia MO 65201

WYLY, LEMUEL DAVID, JR, b Seneca, SC, Aug 9, 16; m 38; c 2. NUCLEAR PHYSICS. Educ: The Citadel, BS, 38; Univ NC, MA, 39; Yale Univ, PhD(physics), 49. Prof Exp: Instr physics, Ga Sch Technol, 39-41, asst prof, 46; asst instr, Yale Univ, 46-48; from assoc prof to prof, 49-58, REGENTS' PROF PHYSICS, GA INST TECHNOL, 58- Concurrent Pos: Consult, Oak Ridge Nat Lab, 52- Mem: Fel Am Phys Soc. Res: Nuclear energy levels; proportional and scintillation and solid state detectors; decay schemes of radioactive isotopes and from neutron capture. Mailing Add: Sch of Physics Ga Inst of Technol Atlanta GA 30332

WYMA, RICHARD J, b Grand Rapids, Mich, June 25, 36; m 64; c 2. INORGANIC CHEMISTRY. Educ: Hope Col, AB, 58; Univ Mich, MS, 60, PhD(phys chem), 64. Prof Exp: Asst prof chem, Geneva Col, 64-69; ASSOC PROF CHEM, IND UNIV-PURDUE UNIV, INDIANAPOLIS, 70- Mem: Am Chem Soc; Soc Appl Spectros. Res: Application of molecular spectroscopy to structure determination and to bounding theories of inorganic systems. Mailing Add: Dept of Chem 1201 E 38th St Ind Univ-Purdue Univ Indianapolis IN 46205

WYMAN, BOSTWICK FRAMPTON, b Aiken, SC, Aug 22, 41. MATHEMATICS. Educ: Mass Inst Technol, SB, 62; Univ Calif, Berkeley, MA, 64, PhD(math), 66. Prof Exp: Instr math, Princeton Univ, 66-68; asst prof, Stanford Univ, 68-72; ASSOC PROF MATH, OHIO STATE UNIV, 72- Concurrent Pos: Vis asst prof, Univ Oslo, 70-71. Mem: Am Math Soc; Math Asn Am. Res: Algebraic number theory; class field theory and ramification; algebraic system theory. Mailing Add: Dept of Math Ohio State Univ Columbus OH 43210

WYMAN, DONALD, b Templeton, Calif, Sept 18, 03; m 27; c 4. HORTICULTURE. Educ: Pa State Col, BSA, 26; Cornell Univ, MSA, 31, PhD(hort), 35. Prof Exp: Instr, Pa State Univ, 27-29; investr, Cornell Univ, 29-31, instr, 31-35; horticulturist, 36-70, EMER HORTICULTURIST, ARNOLD ARBORETUM, HARVARD UNIV, 70- Honors & Awards: Coleman Award, Am Asn Nurserymen, 49 & 51; NY Hort Soc Distinguished Serv Award, 60; Garden Club Fedn Am Medal of Honor, 65; Veitch Medal, Royal Hort Soc, 69; George Robert White Medal of Honor, Mass Hort Soc, 70; Arthur Hoyt Scott Gold Medal, Swarthmore Col, 71; L H Bailey Medal, Am Hort Soc, 71. Mem: Am Soc Hort Sci (vpres, 52-53); Am Hort Soc (pres, 61-62); Am Asn Bot Gardens & Arboretums; fel Nat Recreation & Park Asn. Res: Ornamental horticulture; plant propagation; winter hardiness; selection of best varieties of woody plants for landscape use. Mailing Add: 59 Jericho Rd Weston MA 02193

WYMAN, DONALD PAUL, organic chemistry, polymer chemistry, see 12th edition

WYMAN, GEORGE MARTIN, b Budapest, Hungary, Oct 13, 21; nat US; m 51; c 1. ORGANIC CHEMISTRY. Educ: Cornell Univ, AB, 41, MS, 43, PhD(org chem), 44. Prof Exp: Res chemist, Gen Chem Co, 44-45, Gen Aniline & Film Corp, 45-49 & Nat Bur Standards, 49-54; chief spectros sect, Qm Res & Develop Ctr, US Dept Army, 54-57, sci adv, European Res Off, 57-60, DIR CHEM DIV, US ARMY RES OFF, 60- Concurrent Pos: Adj prof chem, Univ NC, Chapel Hill, 73- Mem: Am Chem Soc. Res: Spectrophotometry; cis-trans isomerization of conjugated compounds; organic photochemistry; excited state chemistry of dyes. Mailing Add: US Army Res Off Box 12211 Research Triangle Park NC 27709

WYMAN, HAROLD ROBERTSON, b Yarmouth, NS, Nov 12, 05; wid; c 3. CHEMISTRY. Educ: Dalhousie Univ, BSc, 27; McGill Univ, MSc, 30. Prof Exp: Demonstr chem, McGill Univ, 27-28; asst chemist, Halifax Refinery, Imp Oil, Ltd, 30-38; consult chemist, 38-64; pres, Wyman & West Ltd, 64-73; LAB MGR, CALEB BRETT MARITIMES LTD, 73- Mem: AAAS; Am Chem Soc; fel Chem Inst Can; Marine Chem Asn. Res: Surface energy relationships; chemical analytical methods, especially identification of oil sources, detection and identification of drugs. Mailing Add: 3232 Barrington St Halifax NS Can

WYMAN, JEFFRIES, b West Newton, Mass, June 21, 01; m 54; c 2. MOLECULAR BIOLOGY. Educ: Harvard Univ, AB, 23; Univ London, PhD, 27. Prof Exp: From instr to assoc prof zool, Harvard Univ, 28-51; sci adv, US Embassy, Paris, 51-54; dir, UNESCO Sci Off MidE, 55-58; GUEST PROF, BIOCHEM INST, UNIV ROME &

ISTITUTO REGINA ELENA, 60- Concurrent Pos: Past secy gen, European Molecular Biol Orgn. Mem: Nat Acad Sci; Am Acad Arts & Sci. Mailing Add: Ist Fisioterapici Ospitalieri Ist Regina Elena Viale Viale Regina Elena 291 Rome 00161 Italy

WYMAN, JOHN E, b Amsterdam, NY, Feb 20, 31; m 52; c 4. CHEMISTRY. Educ: Univ Mich, BS, 52, Purdue Univ, MS, 55, PhD, 56. Prof Exp: Res chemist, Linde Co, Union Carbide Corp, 56-58, res chemist, Union Carbide Chem Co, 58-59; res chemist, Spec Proj Dept, Monsanto Chem Co, 59-60, res group leader, Monsanto Res Corp, 60-65; MEM SCI STAFF, ITEK CORP, 65- Mem: Am Chem Soc. Res: Photochemistry; complex transition element organometallic chemistry; metal carbonyls; propellant, explosive and inorganic chemistry; graphic arts, film and paper coatings. Mailing Add: 191 Grove St Lexington MA 02173

WYMAN, MAX, b Can, Apr 14, 16; m 40; c 1. MATHEMATICS. Educ: Univ Alta, BSc, 37; Calif Inst Technol, PhD(math), 40. Prof Exp: Munitions gauge inspector, Nat Res Coun Can, 40-41; lectr math, Univ Sask, 41-42; munitions gauge inspector, Nat Res Coun Can, 42-43; lectr math, 43-45, from asst prof to prof, 45-74, head dept math, 61-64, dean sci, 63-64, acad vpres, 64-69, pres, 69-74, UNIV PROF MATH, UNIV ALTA, 74- Concurrent Pos: Chmn, Alta Human Rights Comn, 73-76. Mem: Math Asn Am; fel Royal Soc Can; NY Acad Sci. Res: Relativity theory and asymptotics. Mailing Add: 836 Educ Bldg Univ of Alta Edmonton AB Can

WYMAN, MILTON, b Cleveland, Ohio, Oct 11, 30; m 56; c 2. VETERINARY MEDICINE, OPHTHALMOLOGY. Educ: Ohio State Univ, DVM, 63, MS, 64; Am Col Vet Ophthal, dipl. Prof Exp: Res assoc ophthal, Cols Med & Vet Med, 62-64, instr vet ophthal, Col Vet Med, 64-66, from asst prof to assoc prof vet ophthal & med, 66-73, PROF VET OPHTHAL & MED, COL VET MED, OHIO STATE UNIV, 73-, CHIEF SMALL ANIMAL SERV, 72-, ASSOC PROF OPHTHAL, COL MED, 72- Mem: Am Soc Vet Ophthal; Am Vet Med Asn; Am Asn Vet Clin. Res: Congenital ocular defects in dogs and their relationship to man; glaucoma in the basset hound; ocular fundus anomaly in collies; medical application of soft contact lenses in animals and man. Mailing Add: 2615 Carriage Lane Powell OH 43065

WYMAN, ROBERT J, b Syracuse, NY, June 8, 40. NEUROSCIENCES, COMPARATIVE PHYSIOLOGY. Educ: Harvard Univ, AB, 60; Univ Calif, Berkeley, MA, 63, PhD(biophys), 65. Prof Exp: Math analyst, Tech Res Group, Inc, 59; NSF res fel appl sci, Calif Inst Technol, 66; asst prof biol, 66-70, ASSOC PROF BIOL, YALE UNIV, 71- Mem: Soc Neurosci. Res: Nervous control of motor activity in insects and vertebrates; application of computers to mathematical analysis of neural data; central nervous generation of motor output patterns; genetics of Drosophila nervous system. Mailing Add: Dept of Biol Yale Univ New Haven CT 06520

WYMAN, STANLEY M, b Cambridge, Mass, Aug 3, 13; m; c 4. MEDICINE, RADIOLOGY. Educ: Harvard Univ, AB, 35, MD, 39. Prof Exp: Radiologist, Mass Gen Hosp, 47-68; asst clin prof radiol, Harvard Med Sch, 54-75; VIS RADIOLOGIST, MASS GEN HOSP, 68-; ASSOC CLIN PROF RADIOL, HARVARD MED SCH, 75- Concurrent Pos: Consult, US Navy, 57-73. Honors & Awards: Silver Medal, Roentgen Ray Soc, 52; Gold Medal, Am Col Radiol, 72; Gold Medal, Radiol Soc NAm, 74. Mem: Radiol Soc NAm (pres, 68); Roentgen Ray Soc; AMA; Am Col Radiol (pres, 71). Res: Cardiovascular radiology. Mailing Add: 575 Mt Auburn St Cambridge MA 02139

WYMER, RAYMOND GEORGE, b Colton, Ohio, Oct 1, 27; m 48; c 4. RESEARCH ADMINISTRATION. Educ: Memphis State Col, BS, 50; Vanderbilt Univ, MA & PhD, 53. Prof Exp: Mem staff, Oak Ridge Nat Lab, 53-56; assoc prof, Ga Inst Technol, 56-58; chief nuclear chem, Indust Reactor Labs, 58-59; res chemist, 59-62, sect chief, 62-73, ASSOC DIR CHEM TECHNOL DIV, OAK RIDGE NAT LAB, 73- Mem: Fel Am Inst Chemists; Sigma Xi; Am Nuclear Soc. Res: Colloid, radiation, transuranium element and complex ion chemistry; kinetics; nuclear fuel cycle. Mailing Add: Chem Technol Div Oak Ridge Nat Lab Oak Ridge TN 37830

WYMORE, CHARLES ELMER, b Iowa, Jan 2, 32; m 52; c 3. INORGANIC CHEMISTRY. Educ: Cent Col, Iowa, BA, 53; Univ Ill, PhD(chem), 57. Prof Exp: Res chemist, 56-63, prof leader, 63-68, group leader, 68-71, res mgr, 71-74, RES DIR, DOW CHEM CO, 74- Mem: Am Chem Soc. Res: Coordination and organometallic chemistry; homogeneous catalysis; organic chemistry; heterogeneous catalysis. Mailing Add: 2601 Lambros Dr Midland MI 48640

WYNDER, ERNST LUDWIG, b Ger, Apr 30, 22; nat US. PREVENTIVE MEDICINE, EPIDEMIOLOGY. Educ: NY Univ, BA, 43; Wash Univ, BS & MD, 50. Prof Exp: Intern, Georgetown Univ Hosp, 50; asst prof prev med, Grad Sch Med Sci, Med Col, Cornell Univ, 54-56, assoc prof, 56-69; PRES & MED DIR, AM HEALTH FOUND, 69- Concurrent Pos: Asst, Sloan-Kettering Inst Cancer Res, 52-54, assoc, 54-60, assoc mem, 60-69, assoc scientist, 69-71; jr asst resident, Mem Hosp for Cancer & Allied Dis, 51-52, sr asst resident, 52-54, clin asst physician, 54-64, asst attend physician, 64-69, consult epidemiologist, 69-; clin vis asst, James Ewing Hosp, 54-64, asst vis physician, 64-68; mem, Task Force Lung Cancer, Tobacco Working Group, 67-; mem, Nat Cancer Plan, 71; ed, Prev Med J, 72. Mem: AMA; Am Asn Cancer Res; Am Pub Health Asn; NY Acad Sci. Res: Environmental factors affecting major chronic disease development, preventive medical aspects. Mailing Add: Am Health Found 1370 Ave of the Americas New York NY 10019

WYNGAARD, JOHN C, b Madison, Wis, Dec 4, 38; m 65; c 2. FLUID DYNAMICS. Educ: Univ Wis-Madison, BSc, 61, MSc, 62; Pa State Univ, PhD(mech eng), 67. Prof Exp: Res physicist, Air Force Cambridge Res Labs, 67-75; PHYSICIST, WAVE PROPAGATION LAB, NAT OCEANIC & ATMOSPHERIC ADMIN, 75- Concurrent Pos: Vis assoc prof atmospheric sci, Univ Wash, 73. Mem: Am Meteorol Soc; Am Phys Soc; Sigma Xi. Res: The structure and dynamics of the mean and turbulent components of the lower atmosphere and their parameterization for use in mesoscale meteorological models. Mailing Add: 3835 Lakebriar Dr Boulder CO 80302

WYNGAARDEN, JAMES BARNES, b East Grand Rapids, Mich, Oct 19, 24; m 46; c 5. BIOCHEMISTRY, METABOLISM. Educ: Univ Mich, MD, 48. Prof Exp: Asst pharmacol, Med Sch, Univ Mich, 46-48; mem med house staff, Mass Gen Hosp, 48-52; vis investr, Pub Health Res Inst New York, 53; investr, Nat Heart Inst, 53-54 & Nat Inst Arthritis & Metab Dis, 54-56; assoc prof med & biochem, Sch Med, Duke Univ, 56-61, prof med & assoc prof biochem, 61-65; prof med & chmn dept, Univ Pa, 65-67; FREDERIC M HANES PROF MED & CHMN DEPT, DUKE UNIV, 67- Concurrent Pos: Dalton scholar med res, Mass Gen Hosp, 51; consult, Vet Admin Hosp, Durham, NC; consult, Off Sci & Technol, Exec Off of the President, 66-72; mem adv comt biol & med, AEC, 67-69; mem adv bd, Howard Hughes Med Inst, 69-; mem bd sci counrs, Nat Inst Arthritis, Metab & Digestive Dis, 71-74; mem, Nat Res Coun, 71- & President's Sci Adv Comt, 72-73; mem exec comn, Assembly of Life Sci, 72- Mem: Inst of Med of Nat Acad Sci; Am Soc Clin Invest; Am Soc Biol Chem; Am Rheumatism Asn; Am Fedn Clin Res. Res: Control of purine synthesis; purine metabolism in normal and gouty man; metabolism of iodine and steroids; oxalate synthesis; inborn errors of metabolism. Mailing Add: Dept of Med Duke Univ Med Ctr Durham NC 27706

WYNKOOP, RAYMOND, b Bethayres, Pa, Aug 3, 16; m 42; c 4. CHEMICAL ENGINEERING, CHEMISTRY. Educ: Worcester Polytech Inst, BS, 42; Princeton Univ, PhD(chem eng), 48. Prof Exp: Proj engr, Publicker Commercial Alcohol Co, Pa, 42-45; assoc prof chem eng, Tufts Univ, 48; sr design engr, Standard Oil Co, Ind, 48-52; sr chem engr, Nat Distillers Prod Corp, 52-56; mgr, Catalytic Construct Corp, 57-58; asst to pres, Houdry Process Corp, 56-57; asst dir res petrochem, 58-59, mgr basic res, 59-61, dir com develop, res & eng, 61-64, dir patent dept, 64-67, admin dir, 67-69, sci adv res & eng, 69-70, DIR CORP RES & DEVELOP, SUN OIL CO, 70- Mem: AAAS; Soc Automotive Eng; Am Chem Soc. Res: Heat transfer; thermodynamics; heat transfer and kinetics in tubular reactors during hydrogenation of ethylene over copper magnesia catalyst; heterogeneous reaction kinetics; solvent extraction. Mailing Add: Sun Res & Develop Co Box 1135 Marcus Hook PA 19061

WYNN, CHARLES MARTIN, SR, b New York, NY, May 8, 39; m 66; c 3. ORGANIC CHEMISTRY, ACADEMIC ADMINISTRATION. Educ: City Col New York, BChE, 60; Univ Mich, MS, 63, PhD(chem), 65. Prof Exp: Instr gen chem, Univ Mich, 65-67; US Peace Corps lectr chem, Malayan Teachers' Col, 67-69; from asst prof to assoc prof phys sci, 69-74, asst to provost, 74-75, PROF PHYS SCI, OAKLAND COMMUNITY COL, 74-, CHMN DEPT, 69- Mem: Am Chem Soc; Am Asn Higher Educ; Am Educ Sci Asn. Res: Structural directivity in diene synthesis. Mailing Add: Dept of Phys Sci Oakland Community Col Farmington MI 48024

WYNN, JACK THOMAS, b Hawkinsville, Ga, Aug 16, 40; m 66; c 3. ANTHROPOLOGY. Educ: Ga State Col, BA, 68; Univ Mo-Columbia, MA, 71, PhD(anthrop), 75. Prof Exp: Teaching asst anthrop, Univ Mo-Columbia, 71-74, teaching asst archaeol, 74-75; ASST PROF ANTHROP, MISS STATE UNIV, 75- Mem: Am Anthrop Asn; Soc Am Archaeol; Latin Am Anthrop Group. Res: Archaeology and ethnology of South America and Mesoamerica; peasant and prehistoric trade systems and settlement patterns; cultural evolution and ceramic chronology in Latin America; prehistory of the southeastern United States. Mailing Add: Dept of Anthrop PO Drawer GN Miss State Univ Mississippi State MS 39762

WYNN, JAMES ELKANAH, b Pennington Gap, Va, Feb 7, 42; m 64; c 1. MEDICINAL CHEMISTRY, ANALYTICAL CHEMISTRY. Educ: Va Commonwealth Univ, BS, 64, PhD(med chem), 69. Prof Exp: Res fel med & anal chem, Med Col Va, 69; asst prof, 69-73, ASSOC PROF MED CHEM, COL PHARM, UNIV SC, 73- Concurrent Pos: Comn prod scholar grant, Col Pharm, Univ SC, 70-71, lectr, Proj Upward Bound, 70- Mem: Am Chem Soc; Am Pharmaceut Asn; Am Asn Cols Pharm; Sigma Xi. Res: Organic chemistry; cancer chemotherapeutic agents of the alkylating type; synthesis, testing and correlation of activity with physical parameters; mechanism of dimenthyl sulfoxide interaction with isolated enzyme systems; synthesis of agents for urolithiasis treatment. Mailing Add: Col of Pharm Univ of SC Columbia SC 29208

WYNN, JOHN HEYWARD, JR, bacterial physiology, see 12th edition

WYNN, RALPH MATTHEW, b New York, NY, Nov 1, 30. OBSTETRICS & GYNECOLOGY. Educ: Harvard Univ, AB, 51; NY Univ, MD, 54; Am Bd Obstet & Gynec, dipl. Prof Exp: From instr to assoc prof obstet & gynec, State Univ NY Downstate Med Ctr, 61-68; PROF OBSTET & GYNEC & HEAD DEPT, UNIV ILL MED CTR, 68- Concurrent Pos: USPHS res fel, 62-63; mem study sect, NIH; mem, Nat Med Comn Planned Parenthood-World Pop. Honors & Awards: Purdue Frederick Res Award, Am Col Obstet & Gynec, 67. Mem: Fel Am Asn Obstet & Gynec; fel Am Gynec Soc; Am Asn Anat; Asn Profs Gynec & Obstet; Soc Gynec Invest. Res: Electron microscopy; comparative anatomy of placenta; ultrastructure of placenta; fetal membranes; endometrium. Mailing Add: Dept of Obstet & Gynec Univ of Ill at the Med Ctr Chicago IL 60680

WYNN, ROBERT WALTER, chemistry, see 12th edition

WYNN, WILLARD KENDALL, JR, b Raleigh, NC, Mar 28, 32; m 63. PLANT PATHOLOGY. Educ: NC State Univ, BS, 55; Univ Fla, PhD(plant path), 63. Prof Exp: Assoc plant pathologist, Boyce Thompson Inst Plant Res, 63-68; ASSOC PROF PLANT PATH, UNIV GA, 68- Mem: Am Phytopath Soc; Am Plant Physiol. Res: Physiology of uredospore germination and rust infection. Mailing Add: Dept of Plant Path Univ of Ga Athens GA 30601

WYNNE, ELMER STATEN, b El Paso, Tex, Oct 23, 17; m 38; c 2. MEDICAL MICROBIOLOGY. Educ: Univ Tex, BA, 38, MA, 44, PhD(bact), 48; Am Bd Microbiol, cert microbiol & bact. Prof Exp: Asst bact, Univ Tex, 38-39, tutor, 39-42, instr, 46, res assoc, 46-48, res bacteriologist, M D Anderson Hosp & Tumor Inst, 50-58, assoc prof microbiol, Dent Br, Univ, 58-59; asst prof, Univ Okla, 48-50; bacteriologist, US Air Force Sch Aerospace Med, 59-60, res prof bact & chief microbiol, 60-67, sr microbiologist, 68-69; ASSOC PROF MED LAB TECHNOL, ST PHILLIPS'S COL, 70- Mem: AAAS; Am Soc Microbiol; fel Am Acad Microbiol. Res: Enteric bacteriology; bacterial antagonism; physiology of Clostridium spore germination; microbiological aspects of cancer research; aerospace microbiology; hand disinfection. Mailing Add: 4826 Hershey Dr San Antonio TX 78220

WYNNE, JOHNNY CALVIN, b Williamston, NC, May 17, 43; m 61; c 2. PLANT BREEDING, PLANT GENETICS. Educ: NC State Univ, BS, 65, MS, 68, PhD(crop sci), 74. Prof Exp: Instr, 68-74, ASST PROF CROP SCI, NC STATE UNIV, 74- Mem: Am Soc Agron; Am Peanut Res & Educ Asn. Res: Improvement of cultivated peanuts through breeding for higher productivity, disease resistance, insect resistance and better quality; evolution and genetics of subspecific groups in peanuts. Mailing Add: Dept of Crop Sci NC State Univ Raleigh NC 27607

WYNNE, KENNETH JOSEPH, b Rumford, RI, Jan 17, 40; m 67; c 2. INORGANIC CHEMISTRY, POLYMER CHEMISTRY. Educ: Providence Col, BS, 61; Univ Mass, Amherst, MS & PhD(chem), 65. Prof Exp: Fel inorg chem, Univ Calif, Berkeley, 65-67; asst prof, Univ Ga, 67-73, SCI OFFICER, OFF NAVAL RES, 73- Mem: AAAS; Am Chem Soc. Res: Synthetic inorganic and organometallic chemistry; coordination chemistry; inorganic ring systems; fluorine chemistry. Mailing Add: Off of Naval Res Chem Prog 800 N Quincy St Arlington VA 22217

WYNNE, LYMAN CARROLL, b Lake Benton, Minn, Sept 17, 23; m 47; c 5. PSYCHIATRY, PSYCHOLOGY. Educ: Harvard Med Sch, MD, 47; Harvard Univ, PhD(soc psychol), 58. Prof Exp: Intern med, Peter Bent Brigham Hosp, Boston, 47-48; resident, Mass Gen Hosp, 51; psychiatrist, Lab Socio-Environ Studies, NIMH, Md, 52-54, Adult Psychiat, 54-71, chief family studies sect, 57-67, chief adult psychiat br, 61-71; PROF PSYCHIAT & CHMN DEPT, SCH MED & DENT, UNIV ROCHESTER, 71-; PSYCHIATRIST-IN-CHIEF, STRONG MEM HOSP, 71- Concurrent Pos: Mem fac, Wash Sch Psychiat, 56-71; mem fac, Wash Psychoanal Inst, 60-71, teaching analyst, 66-71; consult & collab investr, WHO, 65-; mem-at-large, Div Behav Sci, Nat Res Coun, 69-72; mem rev comt career develop awards, NIMH, 72-76; vis lectr, Am Univ Beirut. Honors & Awards: Commendation medal, USPHS, 65, Meritorious Serv Medal, 66; Fromm-Reichmann Award, Am Acad Psychoanal, 66; Hofheimer Prize, Am Psychiat Asn, 66; Salmon lectr, 73; Stanley R Dean Res Award, Am Col Psychiatrists, 76. Mem: Am Psychiat Asn; Am Psychosom Soc; Am Orthopsychiat Asn; Am Psychoanal Asn; Am Psychopath Asn. Res: Family research and therapy; schizophrenia; cross-cultural studies; child development. Mailing Add: Dept of Psychiat Univ Rochester Sch Med & Dent Rochester NY 14642

WYNNE, MICHAEL, b St Louis, Mo, Feb 4, 40. MARINE PHYCOLOGY. Educ: Washington Univ, AB, 62; Univ Calif, Berkeley, PhD(bot), 67. Prof Exp: NSF fel, Univ Wash, 67-69; asst prof bot, 69-72, ASSOC PROF BOT, UNIV TEX, AUSTIN, 72- Concurrent Pos: Instr, Hopkins Marine Sta, 68, Friday Harbor Labs, 70 & Marine Biol Lab, 71 & 72. Mem: Brit Phycol Soc; Japan Phycol Soc; Phycol Soc Am (secy, 73-75, vpres, 76); Bot Soc Am; Int Phycol Soc. Res: Culturing and experimental studies of brown and red algae; systematics of marine algae. Mailing Add: Dept of Bot Univ of Tex Austin TX 78712

WYNNE-EDWARDS, HUGH ROBERT, b Montreal, Que, Jan 19, 34; m 56, 72; c 2. GEOLOGY. Educ: Aberdeen Univ, BSc, 55; Queen's Univ, Ont, MA, 57, PhD(geol), 59. Prof Exp: Tech officer geol, Geol Surv Can, 58-59; lectr, Queen's Univ, Ont, 59-61, from asst prof to assoc prof, 61-68, prof geol & head dept geol sci, 68-72; PROF GEOL & HEAD DEPT GEOL SCI, UNIV BC, 72- Concurrent Pos: Vis fel, Aberdeen Univ, 65-66 & Univ Witwatersrand, 72; pres, Can Geosci Coun, 74. Mem: Fel Geol Soc Am; fel Geol Asn Can; Mineral Asn Can; Can Inst Mining & Metall; Royal Soc Can. Res: Structure and petrology of plutonic and metamorphic rocks; deformation of rocks at low rates of strain; tectonics of the Grenville Province. Mailing Add: Dept of Geol Sci Univ of BC Vancouver BC Can

WYNNE-ROBERTS, CAROLINE ROSALES, b London, Eng, US citizen. RHEUMATOLOGY, ELECTRON MICROSCOPY. Educ: Soc Apothecaries, London, LMSSA, 60; London Univ, Eng, MB, BS, 61. Prof Exp: Intern & resident med, Med Ctr, Univ Mich, Ann Arbor, 63-67; fel rheumatology, Rackham Arthritis Res Unit, Univ Mich, 67-69; res assoc path-electron micros, Univ Colo & Vet Admin Hosp, 69-71; from instr to asst prof med & rheumatology, Univ Pittsburgh, 71-75; ASSOC PROF RHEUMATOLOGY & MED, SCH MED, SOUTHERN ILL UNIV, SPRINGFIELD, 75- Concurrent Pos: Instr path-electron micros, Dept Path, Univ Colo Med Ctr, 69-71; chief, Electron-Micros Unit, Dept of Path, Vet Admin Hosp, Pittsburgh, 71-75. Mem: Am Rheumatism Asn; Am Soc Cell Biol; Electron Micros Soc Am; AAAS; Am Fedn Clin Res. Res: Ultrastructure rheumatic disease tissues, especially synovium, muscle and skin; mucopolysaccharide of cartilage; steroid uptake/metabolism tissue cultured fibroblasts; rheumatic and normal human. Mailing Add: Sch of Med Southern Ill Univ Dept Med/Rheumatology Box 3926 Springfield IL 62708

WYNSTON, LESLIE K, b San Diego, Calif, Jan 5, 34; m 63. BIOCHEMISTRY. Educ: San Diego State Col, BS, 55; Univ Calif, Los Angeles, MS, 58, PhD(physiol chem), 60. Prof Exp: Instr biochem, Med Sch, Northwestern Univ, 60-61; lectr, Med Sch, Univ Calif, San Francisco, 61-63; USPHS fel, Max Planck Inst Protein & Leather Res, 63-65; asst prof chem, 65-69, ASSOC PROF CHEM, CALIF STATE UNIV, LONG BEACH, 69- Concurrent Pos: Co-prin investr, USPHS-NIH Res Grant, 60-61; supvr, Metab Res Lab, Chicago Wesley Mem Hosp, 60-61; consult, NAm Aviation, Inc, 65-67; vis prof, Univ Zurich, 71-72. Mem: AAAS; Am Chem Soc; NY Acad Sci. Res: Protein purification and characterization; chemical isolation procedures; chromatographic and electrophoretic methods; endocrinology. Mailing Add: Dept of Chem Calif State Univ Long Beach CA 90840

WYNSTRA, JOHN, b Grand Rapids, Mich, June 9, 17; m 46; c 3. ORGANIC POLYMER CHEMISTRY. Educ: Univ Mich, BS, 39, MS, 40, PhD(org chem), 43. Prof Exp: Res chemist, Bakelite Corp, 43-51, group leader, 51-56, sect head, 56-65, res assoc, Res & Develop Dept, Chem & Plastics Opers Div, 65-70, RES FEL, RES & DEVELOP DEPT, UNION CARBIDE CORP, 70- Mem: Am Chem Soc; Am Inst Chemists. Res: Condensation polymers; thermoset polymer systems; polyesters; epoxy resins; phenolic resins. Mailing Add: Union Carbide Corp Bldg 200 Bound Brook NJ 08805

WYNVEEN, ROBERT ALLEN, b Baldwin, Wis, July 24, 39; m 64; c 4. HEALTH PHYSICS, MEDICAL PHYSICS. Educ: Univ Wis-River Falls, BS, 61; Rutgers Univ, MS, 63, PhD(radiation biophysics), 72. Prof Exp: Health physicist, Argonne Nat Lab, 63-65; RADIOL HEALTH PHYSICIST, RUTGERS MED SCH, RUTGERS UNIV, 65- Concurrent Pos: Radiol health physics consult, Colgate-Palmolive Res Ctr, 68-, Warner-Lambert Res Ctr, 72-, Ortho Diag & Pharmaceut, Inc, 74- & Fusion Energy Corp, 75- Mem: Am Asn Physicists Med; Nat Health Physics Soc. Res: Immediate and transient effects of radiation, especially ionizing, microwave and laser, on biological systems' functions with emphasis on cellular energy production and active transport across membranes. Mailing Add: Dept Rad Safety 3572 Doolittle Bldg Rutgers Univ Busch Campus New Brunswick NJ 08903

WYON, JOHN BENJAMIN, b London, Eng, May 3, 18; m 46; c 2. EPIDEMIOLOGY, DEMOGRAPHY. Educ: Cambridge Univ, BA, 40, MB, BCh, 42; Harvard Univ, MPH, 53. Prof Exp: Med officer, Friends Ambulance Unit, Ethiopia, 43-45; med missionary to India from Church Missionary Soc, London, 47-52; res assoc epidemiol, 53-58, instr, 58-60, res fel, 60-61, res assoc pop studies, 61-62, asst prof, 62-66, lectr pop studies & sr res assoc, Ctr Pop Studies, 66-71, SR LECTR POP STUDIES, SCH PUB HEALTH, HARVARD UNIV, 71- Concurrent Pos: Field dir, India-Harvard-Ludhiana Pop Study & asst prof, Christian Med Col, Ludhiana, India, 53-60. Mem: Am Pub Health Asn; fel Royal Soc Med; Int Union Sci Study Pop. Res: Population control; internal medicine; field research on births, deaths and migrations in urban United States and in rural developing countries; development of local education units to demonstrate implications of population changes. Mailing Add: Dept of Pop Sci Harvard Univ Sch of Pub Health 665 Huntington Ave Boston MA 02115

WYRICK, PRISCILLA BLAKENEY, b Greensboro, NC, Apr 28, 40. BACTERIOLOGY. Educ: Univ NC, Chapel Hill, BS, 62, MS, 67, PhD(bact), 71. Prof Exp: Technician clin microbiol, NC Mem Hosp, 62-64, asst supvr, 64-65, supvr in chg mycol & mycobact, 65-66; ASST PROF BACT, SCH MED, UNIV NC, CHAPEL HILL, 73- Concurrent Pos: Med Res Coun fel, Nat Inst Med Res, London, Eng, 71-73; consult, Dept Hosp Labs, NC Mem Hosp, 73- Mem: Am Soc Microbiol; Brit Soc Gen Microbiol. Res: Bacterial L-forms; bacterial ultrastructure; pathogenesis of Chlamydia; medical microbiology. Mailing Add: Dept of Bact Univ of NC Sch of Med Chapel Hill NC 27514

WYRICK, RONALD EARL, b Kansas City, Mo, Nov 4, 44; m 66; c 1. BIOCHEMISTRY, ALLERGY. Educ: Calif State Col, Stanislaus, BA, 68; Univ Calif, Davis, PhD(biochem), 74. Prof Exp: INDUST RES ALLERGY, HOLLISTER-STIER LABS, SUBSID CUTTER LABS, WASH, 74- Res: Elucidation of allergy mechanisms to provide research directions for potential new treatments for the allergic condition. Mailing Add: Hollister-Stier Labs Box 3145 Terminal Annex Spokane WA 99220

WYROBEK, ANDREW JULIUS, US citizen. MEDICAL BIOPHYSICS, ANIMAL GENETICS. Educ: Univ Notre Dame, BS, 70; Univ Toronto, PhD(med biophys), 75. Prof Exp: SR STAFF BIOPHYSICIST, DIV BIOMED & ENVIRON RES, LAWRENCE LIVERMORE LAB, UNIV CALIF, 75- Mem: Environ Mutagen Soc. Res: Effects of mutagens, carcinogens and teratogens on the genetics of sperm production in mammals with emphasis on the development of adequate testing systems for environmental pollutants. Mailing Add: Biomed Div L-523 Lawrence Livermore Lab PO Box 808 Livermore CA 94550

WYRTKI, KLAUS, b Tarnowitz, Ger, Feb 7, 25; m 53; c 2. PHYSICAL OCEANOGRAPHY. Educ: Kiel Univ, PhD(phys oceanog), 50. Prof Exp: Scientist, Ger Hydrographic Inst, Hamburg, 50-51; res fel oceanog, Kiel Univ, 51-54; scientist, Inst Marine Res, Djakarta, 54-57; scientist, Int Hydrographic Bur, Monaco, 58; res officer, Commonwealth Sci & Indust Res Orgn, Australia, 58-61; res oceanogr, Scripps Inst, Univ Calif, 61-64; PROF OCEANOG, UNIV HAWAII, 64- Concurrent Pos: Ed, Atlas Phys Oceanog for Int Indian Ocean Exped, 65-; chmn, NPac Exp, 74- Mem: AAAS; Am Geophys Union; Am Meteorol Soc. Res: General circulation of the oceans; water masses; equatorial circulation; air-sea energy exchange; ocean-atmosphere interaction. Mailing Add: Dept of Oceanog Univ of Hawaii Honolulu HI 96822

WYSE, BENJAMIN DELANEY, b Columbia, SC, July 20, 27; m 52; c 2. ORGANIC CHEMISTRY. Educ: Erskine Col, AB, 48; Vanderbilt Univ, MA, 51; Univ SC, PhD(chem), 57. Prof Exp: Instr math, Erskine Col, 49-51; chemist, Celanese Corp, SC, 51-53; chemist, Tech Sect, 56-61, SR RES CHEMIST, TECH SECT, E I DU PONT DE NEMOURS & CO, INC, 61- Mem: Am Chem Soc. Res: Acrylic polymerization processes and reaction mechanisms; isocyanate chemistry; solution and melt spinning processes of elastomers and polyamides. Mailing Add: Tech Sect E I du Pont de Nemours & Co Inc Chattanooga TN 37401

WYSE, FRANK OLIVER, b Milwaukee, Wis, Apr 22, 30; m 67; c 3. MATHEMATICS. Educ: Harvard Univ, AB, 52; Princeton Univ, AM, 55; Ore State Univ, PhD(math), 64. Prof Exp: Instr math, Lehigh Univ, 55-58; from instr to asst prof, Ore State Univ, 58-70; prof & chmn dept, Talladega Col, 70-73; ASSOC PROF MATH, CLARK COL, 73- Concurrent Pos: Asst prof, Cleveland State Univ, 66-70. Mem: Am Math Soc. Res: Algebra; topology. Mailing Add: 2307 Greenglade Rd Atlanta GA 30345

WYSE, GORDON ARTHUR, b San Jose, Calif, July 12, 40; m 63; c 3. ZOOLOGY, NEUROPHYSIOLOGY. Educ: Swarthmore Col, BA, 61; Univ Mich, MA, 63, PhD(zool), 67. Prof Exp: From instr to asst prof, 66-72, ASSOC PROF ZOOL, UNIV MASS, AMHERST, 72- Concurrent Pos: Nat Inst Neurol Dis & Stroke res grant, 69-; vis scholar, Stanford Univ, 72-73. Mem: AAAS; Brit Soc Exp Biol; Soc Gen Physiol; Am Soc Zoologists; Am Soc Mammal. Res: Comparative neurophysiology; neural integration of central and sensory information to control rhythmic and other behavior patterns; mechanoreception, propioception and chemoreception. Mailing Add: Dept of Zool Univ of Mass Amherst MA 01002

WYSE, ROGER EARL, b Wauseon, Ohio, Apr 22, 43. PLANT PHYSIOLOGY, BIOCHEMISTRY. Educ: Ohio State Univ, BSAgr, 65; Mich State Univ, MS, 67, PhD(crop sci), 69. Prof Exp: Fel, Mich State Univ, 69-70; PLANT PHYSIOLOGIST, AGR RES SERV, USDA, 70- Mem: AAAS; Am Soc Plant Physiol; Am Soc Agron; Am Soc Crop Sci. Res: Post harvest physiology of sugarbeets; oligosaccharide metabolism and mechanism of sucrose storage in beet roots; biochemical methods of testing for superior breeding lines. Mailing Add: USDA Agr Res Serv Crops Res Lab Utah State Univ Logan UT 84321

WYSHAK, GRACE, b Boston, Mass. BIOSTATISTICS. Educ: Smith Col, BA, 49; Harvard Univ, MSHyg, 56; Yale Univ, PhD(biomet), 64. Prof Exp: Res assoc epidemiol, Harvard Univ, 56-60; instr math, Albertus Magnus Col, 64-65; ASSOC PROF BIOMET, YALE UNIV, 65- Concurrent Pos: NIH res career develop award, 68-72; consult, NIH, 70-, Vet Admin Coop Studies Ctr, 72- & Radcliffe Inst Prog Health Care, 75. Mem: Sigma Xi; Am Statist Asn; Biomet Soc; Am Epidemiol Asn; Int Epidemiol Asn. Res: Inheritance of twinning; biometric and epidemiologic methods; statistical methods in virology; statistical applications in psychiatry. Mailing Add: 32 Commonwealth Ave Chestnut Hill MA 02167

WYSOCKI, ALLEN JOHN, b Chicago, Ill, Dec 22, 36; m 63; c 2. INDUSTRIAL ORGANIC CHEMISTRY. Educ: Loyola Univ Chicago, BS, 58; Northwestern Univ, Evanston, PhD(org chem), 63. Prof Exp: Res chemist, IIT-Res Inst, 62-64; DIV RES MGR, SOAP & HOUSEHOLD PROD DIV, ARMOUR-DIAL, INC, PHOENIX, 64- Mem: Am Chem Soc. Res: Development of soaps, detergents and other household products. Mailing Add: 4135 N 57th Way Phoenix AZ 85018

WYSOLMERSKI, THERESA, b West Rutland, Vt, Oct 25, 32. ZOOLOGY, ECOLOGY. Educ: Col St Rose, BS, 59; Univ Notre Dame, MS, 61; Rutgers Univ, New Brunswick, PhD(zool, ecol), 73. Prof Exp: Teacher, St John's Cath Acad, 56-59; instr chem, 59-60, asst prof biol, 61-66, ASSOC PROF BIOL, COL ST ROSE, 66- Concurrent Pos: Col rep, Hudson Valley Mohawk League Consortium, 69-70. Mem: Ecol Soc Am; Am Inst Biol Sci; Sigma Xi. Res: Distribution of animal populations and their energy impact on forest floors. Mailing Add: Col of St Rose Div Natural Sci 432 Western Ave Albany NY 12203

WYSONG, DAVID SERGE, b Glasgow, Ky, Apr 20, 34; m 58; c 5. PLANT PATHOLOGY. Educ: Colo State Univ, BS, 58, MS, 61; Univ Ill, PhD(plant path), 64. Prof Exp: ASSOC PROF PLANT PATH & EXTEN PLANT PATHOLOGIST, UNIV NEBR-LINCOLN, 64- Mem: Am Phytopath Soc. Res: Practical application of plant pathology. Mailing Add: 305 Plant Industry Bldg Univ of Nebr Lincoln NE 68503

WYSS, MAX, b Zurich, Switz, Sept 10, 39; m 70; c 2. SEISMOLOGY. Educ: Swiss Fed Inst Technol, Dipl, 64; Calif Inst Technol, MS, 68, PhD(geophysics), 70. Prof Exp: Res scientist geophysics, Univ Calif, San Diego, 70; res scientist seismology, Lamont-Doherty Geol Observ, Columbia Univ, 70-72, res assoc, 72-73; asst prof, 73-75, ASSOC PROF GEOL, UNIV COLO BOULDER, 75- Concurrent Pos: Ed, Pure & Appl Geophysics, 74; vis prof geophysics, Univ Karlsruhe, Ger, 75. Mem: Am Geophys Union; Seismol Soc Am; Geol Soc Am. Res: Earthquake source mechanism; earthquake predictions; magnetism of rocks. Mailing Add: CIRES Univ of Colo Boulder CO 80302

WYSS, ORVILLE, b Medford, Wis, Sept 10, 12; m 41; c 3. MICROBIOLOGY. Educ: Univ Wis, BS, 37, MS, 38, PhD(bact), 41. Prof Exp: Asst bact, Univ Wis, 37-41; res bacteriologist, Wallace & Teirnan Prod, 41-45; assoc prof bact, 45-48, chmn dept microbiol, 59-69, PROF BACT, UNIV TEX, AUSTIN, 48-, CHMN DEPT MICROBIOL, 75- Concurrent Pos: Fulbright grant, Univ Sydney, 71. Mem: AAAS; Am Chem Soc; Am Soc Microbiol (pres, 65); Am Soc Biol Chemists; Am Acad Microbiol. Res: Bacterial physiology and genetics; microbial survival. Mailing Add: Dept of Microbiol Univ of Tex Austin TX 78712

WYSS, WALTER, b Matzendorf, Switz, Mar 26, 38; m 61; c 3. MATHEMATICS, PHYSICS. Educ: Swiss Fed Inst Technol, dipl phys, 61, Dr Sc Nat(math physics), 65. Prof Exp: Instr physics, Swiss Fed Inst Technol, 61-66; instr, Princeton Univ, 66-68; asst prof math & physics, 68-71, ASSOC PROF MATH & PHYSICS, UNIV COLO, BOULDER, 71- Concurrent Pos: Swiss Nat Found stipend, Princeton Univ, 66-68; res fel, Univ Colo, Boulder, 69-70, NSF res grant, 70-72. Mem: Am Math Soc. Res: Axiomatic theory of quantized fields; general relativity; functional analysis; infinite parameter lie groups. Mailing Add: 2810 Iliff Boulder CO 80303

WYSSBROD, HERMAN ROBERT, b Louisville, Ky, Oct 17, 41; m 63. PHYSIOLOGY, BIOPHYSICS. Educ: Univ Louisville, BEE, 63, PhD(physiol), 68. Prof Exp: Asst prof physiol, 72-73, ASSOC PROF PHYSIOL & BIOPHYS, MT SINAI SCH MED, 74- Concurrent Pos: NIH res career develop award, Mt Sinai Sch Med, 72-77; asst prof biophys chem, Mt Sinai Grad Sch Biol Sci, City Univ New York, 68-73; vis asst prof, Rockefeller Univ, 71-73, vis assoc prof, 74- Mem: AAAS; Am Physiol Soc; Am Chem Soc; Soc Exp Biol & Med; NY Acad Sci. Res: Conformation-function relationships of biologically active peptides; transmembrane transport. Mailing Add: Dept of Physiol & Biophys Mt Sinai Sch of Med New York NY 10029

WYSTRACH, VERNON PAUL, b St Paul, Minn, May 8, 19; m 49; c 4. ORGANIC CHEMISTRY. Educ: Univ Minn, BCh, 41; Univ Rochester, PhD(org chem), 44. Prof Exp: Res chemist, 44-52, group leader, Res Div, 52-54, group leader, Basic Res Dept, 54-59, mgr chem synthesis sect, Contract Res Dept, 59-66, mgr appl res sect, 66-67, mgr prod res sect, Chem Dept, Cent Res Div, 67-72, prof mgr prod develop dept, Chem Res Div, 72-74, EMPLOYMENT SUPVR, AM CYANAMID CO, 74- Concurrent Pos: Am Cyanamid Co grant, Univ Cambridge, 61-62. Mem: Am Chem Soc; The Chem Soc. Res: Cyanamide derivatives and nitrogen heterocycles; organic phosphorus compounds; rocket propellants and explosives; chemistry of adhesion; fire retardants. Mailing Add: Am Cyanamid Co 1937 W Main St Stamford CT 06904

WYSZECKI, GUNTER, b Tilsit, Ger, Nov 8, 25; Can citizen; m 54; c 2. PHYSICS, MATHEMATICS. Educ: Tech Univ Berlin, Dipl Ing, 51, Dr Ing(math), 53. Prof Exp: Fulbright scholar, Nat Bur Standards, DC, 53-54; physicist, Fed Inst Mat Test, Ger, 54-55; PRIN RES OFFICER & HEAD OPTICS SECT, DIV PHYSICS, NAT RES COUN CAN, 55- Concurrent Pos: Chmn colorimetry comt, Int Comn Illum, 63-75; adj prof, Sch Optom, Fac Sci, Univ Waterloo, 69- Mem: Fel Optical Soc Am; fel Illum Eng Soc. Res: Colorimetry; photometry; color vision. Mailing Add: Div of Physics Nat Res Coun Ottawa ON Can

WYTTENBACH, CHARLES RICHARD, b South Bend, Ind, Jan 28, 33; m 59; c 3. DEVELOPMENTAL BIOLOGY. Educ: Ind Univ, AB, 54, MA, 56; Johns Hopkins Univ, PhD(biol), 59. Prof Exp: From instr to asst prof anat, Univ Chicago, 59-66; asst prof zool, 66-70, assoc prof physiol & cell biol, 70-75, PROF PHYSIOL & CELL BIOL, UNIV KANS, 75- Concurrent Pos: Managing ed, Univ Kans Sci Bull, 68-74; mem corp, Marine Biol Lab, Woods Hole, Mass. Mem: AAAS; Soc Develop Biol; Am Soc Zoologists; Soc Neurosci. Res: Developmental biology, particularly growth and morphogenesis in colonial hydroids; vertebrate neurogenesis. Mailing Add: Dept of Physiol & Cell Biol Univ of Kans Lawrence KS 66045

X

XAVIER, K S, b Kerala, India, June 1, 30; m 58; c 3. BOTANY, MICROBIOLOGY. Educ: Univ Madras, BA, 56; Wayne State Univ, MS, 61, PhD(biol), 67. Prof Exp: Demonstr bot, S H Col, Univ Madras, 56-59; asst biol, Wayne State Univ, 60-65; instr, Detroit Inst Technol, 65-66; asst prof, 66-69, ASSOC PROF BIOL, ADRIAN COL, 70- Concurrent Pos: NSF partic grant, Univ Tex, Austin, 69; NSF exten grant, Adrian Col, 70-71, col res grant, Alton Jones Cell Sci Ctr, 71. Mem: AAAS; Am Inst Biol Sci; Tissue Cult Asn; Electron Micros Soc Am. Res: Pollen morphology and development; pollen physiology ultrastructure and plant tissue culture. Mailing Add: Dept of Biol Adrian Col Adrian MI 49221

XINTARAS, CHARLES, b New Bedford, Mass, Sept 5, 28; m 57; c 2. INDUSTRIAL HEALTH. Educ: Harvard Univ, AB, 52; Univ Cincinnati, ScD(indust health), 64. Prof Exp: Res chemist, Filtrol Corp, 52-55; eng inspector, Los Angeles County, 55-59; pub health adv, 59-62, pharmacologist brain res, 62-74, ASST CHIEF BEHAV MOTIVATION FACTORS BR, INST OCCUP SAFETY & HEALTH, USPHS, 71- Concurrent Pos: Consult behav toxicol, WHO; mem, Permanent Comn & Int Asn Occup Health, Soc Neurosci. Res: Behavioral toxicology; behavioral and neurophysiological indicators for the monitoring and early detection of potential industrial health and safety problems. Mailing Add: Nat Inst Occup Safety & Health 1014 Broadway Cincinnati OH 45202

Y

YABLON, ISADORE GERALD, b Montreal, Que, May 30, 33; m 62; c 2. ORTHOPEDIC SURGERY. Educ: McGill Univ, BSc, 54; Univ Toronto, MD, 58. Prof Exp: Instr orthop, McGill Univ, 67-71; asst prof, 71-73, ASSOC PROF ORTHOP, BOSTON UNIV, 73- Concurrent Pos: Vis surgeon orthop, Univ Hosp, 71- & Boston City Hosp, 71- Mem: Am Acad Orthop Surg; Orthop Res Soc; Can Orthop Soc; Can Orthop Res Soc; fel Royal Col Physicians & Surgeons Can. Res: Developing technique of joint homografting for clinical application; methods to prevent graft rejection. Mailing Add: Univ Hosp 75 E Newton St Boston MA 02118

YABLONSKI, MICHAEL EUGENE, b Minneapolis, Minn, July 11, 40; m 62; c 4. MEDICINE, PHYSIOLOGY. Educ: Univ Minn, BS, 65, MD, 67, PhD(physiol), 73. Prof Exp: Intern surg, Albert Einstein Col Med, 68; resident physician, Naval Air Develop Ctr, 71-73; resident ophthal, Univ Minn Hosps, 73- Res: Transport physiology; transcapillary exchange; physiology of the eye and vision. Mailing Add: 7315 Ridgeway Rd Minneapolis MN 55427

YABLONSKY, HARVEY ALLEN, b New York, NY, Nov 24, 33; m 64; c 2. PHYSICAL CHEMISTRY. Educ: Brooklyn Col, BS, 54, MA, 58; Stevens Inst Technol, MS, 57, PhD(phys chem), 64; Am Inst Chem, cert. Prof Exp: Res chemist, NY State Dept Health, 55; lectr chem, Brooklyn Col, 55-56; teaching asst, Stevens Inst Technol, 56-59; lectr, Hunter Col, 60-63; asst prof, US Merchant Marine Acad, 63-64; head dept phys chem, Bristol Myers Prod Div, 64-69; PROF CHEM, DEPT PHYS SCI, KINGSBOROUGH COL, 69- Concurrent Pos: Res biochemist, Messinger Res Found, 56-58; lectr, Hunter Col, 63-64 & Rutgers Univ, 67-69; independent consult. Mem: Am Chem Soc; fel Am Inst Chem. Res: Kinetics of redox systems; structure of solutions; sorption at surfaces; complex ion chemistry; fiber and powder rheology; piezoelectricity of biological materials; nonuniform surface photometry; pharmacokinetics. Mailing Add: Dept of Phys Sci Kingsborough Col Brooklyn NY 11235

YACHNIN, STANLEY, b New York, NY, June 28, 30; m 60; c 2. INTERNAL MEDICINE, HEMATOLOGY. Educ: NY Univ, MD, 54; Am Bd Internal Med, dipl, 62. Prof Exp: House officer, Peter Bent Brigham Hosp, Boston, Mass, 54-55, jr asst resident med, 55-56, sr resident, 60-61; from asst prof to assoc prof, 61-69, PROF MED, SCH MED, UNIV CHICAGO, 69-, HEAD SECT HEMAT-ONCOL, 66- Concurrent Pos: USPHS res fel, Peter Bent Brigham Hosp & Harvard Med Sch, 58-60; Markle scholar acad med, 63-68. Mem: AAAS; Am Soc Clin Invest; Am Fedn Clin Res; Am Soc Hemat; Asn Am Physicians. Res: Hemolytic anemias; complement; lymphocyte transformation; mitogenic proteins; paroxysmal nocturnal hemoglobinuria. Mailing Add: Dept of Med Univ of Chicago Sch of Med Chicago IL 60637

YACKEL, JAMES W, b Sanborn, Minn, Mar 6, 36; m 60; c 2. MATHEMATICS. Educ: Univ Minn, BA, 58, MA, 60, PhD(math), 64. Prof Exp: John Wesley Young res instr math, Dartmouth Col, 64-66; asst prof, 66-70, ASSOC PROF MATH, PURDUE UNIV, 70- Mem: Am Math Soc; Inst Math Statist; Math Asn Am. Res: Stochastics processes and graph theory; probability theory; combinatorial theory. Mailing Add: Dept of Math Purdue Univ West Lafayette IN 47906

YACOWITZ, HAROLD, b New York, NY, Feb 17, 22; m 41; c 3. NUTRITION. Educ: Cornell Univ, BS, 47, MNS, 48, PhD(animal nutrit), 50. Prof Exp: Assoc res biochemist, Parke, Davis & Co, 50-51; from asst prof to assoc prof poultry nutrit, Ohio State Univ, 51-55, assoc prof, Agr Exp Sta, 51-55; head nutrit res dept, Squibb Inst Med Res, 55-59; dir appl res, Nopco Chem Co, 59-61; RES ASSOC, FAIRLEIGH DICKINSON UNIV, 61- Concurrent Pos: Nutrit consult, 61- Mem: Am Chem Soc; NY Acad Sci; Poultry Sci Asn; Am Soc Animal Sci; Am Inst Nutrit. Res: Vitamin B-12 microbiological assays; vitamin interrelationships; antibiotic absorption and effects of dietary antibiotics on chicks and hens; vitamin requirements; antifungal agents; calcium and fat metabolism in man and animals; atherosclerosis. Mailing Add: 221 Second Ave Piscataway NJ 08854

YADAV, KAMALESHWARI PRASAD, b Burhiatikar, India, Jan 5, 37; m 57; c 2. ANALYTICAL BIOCHEMISTRY. Educ: Univ Bihar, BSc, 59; Univ Mo, MS, 61, PhD(biochem), 66. Prof Exp: Instr animal husb, Ranchi Agr Col, Bihar, 59-61; res asst agr chem, Univ Mo, 62-66; res biochemist, 66-70, SR BIOCHEMIST, FALSTAFF BREWING CORP, 70- Mem: Fel Am Inst Chemists; Am Soc Brewing Chem; Am Chem Soc. Res: Brewing and fermentation. Mailing Add: Falstaff Brewing Corp 1620 Shenandoah St Louis MO 63104

YADAV, RAGHUNATH P, b Kanpur, India, Jan 2, 35; m 64; c 2. ENTOMOLOGY. Educ: Agra Univ, BS, 56, MS, 58; La State Univ, PhD(entom), 64. Prof Exp: From asst prof to assoc prof entom, 64-74, ASSOC PROF BIOL, SOUTHERN UNIV, BATON ROUGE, 74- Mem: Entom Soc Am. Res: Improvement of artificial diet media for rearing of sugarcane borer; development of laboratory techniques for the detection of sugarcane borer resistance to insecticides. Mailing Add: Dept of Biol Sci Southern Univ Baton Rouge LA 70813

YADAVALLI, SRIRAMAMURTI VENKATA, b Secunderabad, India, May 12, 24; US citizen; m 52. PHYSICS, ELECTRICAL ENGINEERING. Educ: Andhra Univ, India, BS, 42, MS, 45; Univ Calif, MS, 49, PhD(elec eng), 53. Prof Exp: Officer in charge, Physics & Chem Labs, Eng Res Dept, Hyderabad, India, 46-48; asst elec eng, Univ Calif, 49-52, lectr, 52, res engr, Inst Eng Res, 52-53; mem tech staff, Gen Elec Co, 53-58, consult engr, 59; sr math physicist, 59-67, staff scientist, 67-68, STAFF SCIENTIST ENG SCI, STANFORD RES INST, 68- Concurrent Pos: Lectr, Exten, Univ Calif, 57-58; consult, Raytheon Co, 59-62, Litton Indust, 62-64, McGraw Hill Book Co, 62-70 & Rand Corp, 71- Mem: AAAS; Inst Elec & Electronic Eng; Am Inst Aeronaut & Astronaut; Am Phys Soc; Soc Eng Sci; NY Acad Sci. Res: Stochastic processes; electron and plasma physics; statistical and mathematical physics; electrohydrodynamics; optics; biomathematics. Mailing Add: Stanford Res Inst Menlo Park CA 94025

YADEN, SENKALONG, b Nagaland, India, Apr 21, 35. ZOOLOGY. Educ: Wilson Col, Bombay, India, BSc, 56; Univ Bombay, MSc, 58; Univ Minn, PhD(zool), 65. Prof Exp: ASSOC PROF BIOL, JARVIS CHRISTIAN COL, 67- Mem: AAAS; NY Acad Sci; Am Inst Biol Sci. Res: Ecology of freshwater organisms. Mailing Add: Dept of Biol Jarvis Christian Col US Hwy 80 Hawkins TX 75765

YAEGER, JAMES AMOS, b Chicago, Ill, Aug 10, 28; m 52; c 4. HISTOLOGY. Educ: Ind Univ, DDS, 52, MS, 55; Univ Ill, PhD(anat), 59. Prof Exp: Instr anat & clin dent, 55-57; from asst prof to prof histol & head dept, Col Dent, Univ Ill, Chicago, 59-68; PROF ORAL BIOL & HEAD DEPT, SCH DENT MED, UNIV CONN, 68- Concurrent Pos: Resident assoc, Argonne Nat Lab, 63-64. Mem: Am Soc Cell Biol; Am Asn Anat; Int Asn Dent Res. Res: Mechanisms of hard tissue formation and destruction. Mailing Add: Dept of Oral Biol Univ of Conn Health Ctr Farmington CT 06032

YAEGER, ROBERT GEORGE, b Rochester, NY, Oct 25, 17; m 53; c 3. PARASITOLOGY. Educ: Univ Rochester, AB, 50; Univ Tex, MA, 52; Tulane Univ, PhD(parasitol, biochem), 55. Prof Exp: Instr bact & parasitol, Med Sch, Univ Tex, 52-54; from instr to assoc prof, 55-70, PROF PARASITOL & MED, SCH MED, TULANE UNIV, 70- Mem: Am Soc Parasitol; Am Soc Trop Med & Hyg; Soc Exp Biol & Med; Am Inst Nutrit; Soc Protozool. Res: Nutritional and immunological relationships of host and parasite; culture methods; Chagas' disease; other parasitic protozoa which infect man. Mailing Add: Dept of Parasitol Tulane Med Ctr New Orleans LA 70112

YAFET, YAKO, b Istanbul, Turkey, Jan 2, 23; nat US; m 49; c 3. SOLID STATE PHYSICS. Educ: Tech Univ Istanbul, ME, 45; Univ Calif, PhD(physics), 52. Prof Exp: Res assoc physics, Univ Ill, 52-54; physicist, Westinghouse Elec Corp Res Labs, 54-60; MEM TECH STAFF, BELL TEL LABS, INC, 60- Mem: Fel Am Phys Soc. Res: Electronic properties of semiconductors and metals; theoretical solid state physics. Mailing Add: Bell Tel Labs Inc PO Box 261 Rm 1d-334 Murray Hill NJ 07974

YAFFE, LEO, b US, July 6, 16; Can citizen; m 45; c 2. RADIOCHEMISTRY. Educ: Univ Man, BSc, 40, MS, 41; McGill Univ, PhD, 43. Prof Exp: Proj leader nuclear chem & tracer res, Atomic Energy Can, Ltd, 43-52; spec lectr radiochem, 52-54, assoc prof, 54-58, chmn dept chem, 65-72, MACDONALD PROF CHEM, McGILL UNIV, 58-, VPRIN ADMIN & PROF FAC, 74- Concurrent Pos: Dir res, Int Atomic Energy Agency, Vienna, Austria, 63-65; res collabr, Brookhaven Nat Lab. Res: Nuclear chemistry; fission studies. Mailing Add: Dept of Chem McGill Univ Montreal PQ Can

YAFFE, ROBERTA, b Chelsea, Mass, Jan 2, 44. ORGANIC CHEMISTRY. Educ: Bryn Mawr Col, AB, 65; Mass Inst Technol, PhD(chem), 70. Prof Exp: Teaching asst org chem, Mass Inst Technol, 65-66; SR CHEMIST, BEACON RES LABS, TEXACO, INC, 69- Mem: Am Chem Soc; Sigma Xi. Res: Lubricant additives and technology. Mailing Add: Texaco Beacon Res Labs PO Box 509 Beacon NY 12508

YAFFE, RUTH POWERS, b Duluth, Minn, June 4, 27; wid; c 2. RADIOCHEMISTRY. Educ: Macalester Col, BA, 48, PhD(phys chem), 51. Prof Exp: AEC fel radiochem, Ames Lab, 51-52; chemist, Oak Ridge Nat Lab, 52-53; instr chem, Univ Tenn, 55-56; from asst prof to assoc prof, 57-66, PROF CHEM, SAN JOSE STATE UNIV, 66- Mem: Am Nuclear Soc; Am Chem Soc; Health Physics Soc. Res: Chemistry of ruthenium; environmental soil and water analysis for radionuclides. Mailing Add: Dept of Chem San Jose State Univ San Jose CA 95192

YAFFE, SUMNER J, b Boston, Mass, May 9, 23; m; c 4. PEDIATRICS, PHARMACOLOGY. Educ: Harvard Univ, AB, 45, MA, 50; Univ Vt, MD, 54; Am Bd Pediat, dipl, 60. Prof Exp: From intern to sr asst resident, Children's Hosp, Boston, 54-56; exchange resident, St Mary's Hosp, London, Eng, 56-57; from instr to asst prof pediat, Stanford Univ, 59-63; assoc prof, State Univ NY Buffalo, 63-66, prof pediat, 66-75, assoc chmn dept, 69-75; PROF PEDIAT & PHARMACOL, UNIV PA, 75-; HEAD DIV CLIN PHARMACOL, CHILDREN'S HOSP PHILADELPHIA, 75- Concurrent Pos: Res fel metab, Children's Hosp, Boston, 57-59; teaching fel pediat, Harvard Med Sch, 56; Fulbright scholar, St Mary's Hosp, London, Eng, 56-57; Am Heart Asn adv res fel, 60; Lederle med fac award, 62; attend pediatrician, Palo Alto-Stanford Hosp, 59, dir newborn nursery serv, 60, prof dir clin res ctr premature infants, 62; dir pediat renal clin, Stanford Med Ctr, 60; attend pediatrician, Children's Hosp, Buffalo, 63-75; prog dir, Clin Res Ctr for Children, 63; prog consult, Nat Inst Child Health & Human Develop, 63, mem training grant comt, 63, mem reprod biol comt, 65; on leave, Dept Pharmacol, Karolinska Inst, Stockholm, Sweden, 69-70. Mem: Soc Pediat Res; Am Acad Pediat; Am Soc Clin Pharmacol & Therapeut; Am Pediat Soc; Am Soc Pharmacol & Exp Therapeut. Res: Pediatric clinical pharmacology; neonatal, perinatal, fetal and pediatric pharmacology; developmental pharmacology; drug metabolism; drug disposition in sick infants and children; bilirubin metabolism and binding albumin. Mailing Add: Div of Clin Pharmacol Children's Hosp 34th & Civic Ctr Blvd Philadelphia PA 19104

YAGER, BILLY JOE, b Cameron, Tex, Dec 16, 32; m 53; c 4. PHYSICAL ORGANIC CHEMISTRY. Educ: Southwest Tex State Col, BS, 53; Tex A&M Univ, MS, 60, PhD(chem), 62. Prof Exp: Instr chem, Tex A&M Univ, 61-62; from asst prof to assoc prof, 62-71, PROF CHEM, SOUTHWEST TEX STATE UNIV, 71- Mem: Am Chem Soc. Res: Solvent effects upon saponification rate constants; effect of solvent composition upon activity of reactants. Mailing Add: Dept of Chem Southwest Tex State Univ San Marcos TX 78666

YAGER, JAMES DONALD, JR, b Milwaukee, Wis, Dec 29, 43; m 68; c 1. CELL BIOLOGY, CHEMICAL CARCINOGENESIS. Educ: Marquette Univ, BS, 65; Univ Conn, PhD(develop biol), 71. Prof Exp: ASST PROF BIOL SCI, DARTMOUTH COL, 74- Concurrent Pos: Nat Cancer Inst fel oncol, McArdle Lab Cancer Res, Med Ctr, Univ Wis-Madison, 71-74; Nat Cancer Inst grant, 75-; Milheim Found res grant, 75. Mem: AAAS; Am Asn Cancer Res; Am Soc Cell Biologists; Sigma Xi. Res: Oncology. Mailing Add: Dept of Biol Sci Dartmouth Col Hanover NH 03755

YAGER, PHILIP MARVIN, b Los Angeles, Calif, Aug 5, 38; m 73; c 2. EXPERIMENTAL HIGH ENERGY PHYSICS. Educ: Univ Calif, Berkeley, BA, 61; Univ Calif, San Diego, MS, 64, PhD(physics), 71. Prof Exp: From lectr to asst prof physics, 68-72, ASSOC PROF PHYSICS, UNIV CALIF, DAVIS, 75- Mem: AAAS; Am Phys Soc. Res: Hadronic interactions at high energy. Mailing Add: Dept of Physics Univ of Calif Davis CA 95616

YAGER, ROBERT EUGENE, b Carroll, Iowa, Apr 13, 30; m 55; c 2. SCIENCE EDUCATION, PLANT PHYSIOLOGY. Educ: State Col Iowa, BS, 50; Univ Iowa, MS, 53, PhD, 57. Prof Exp: Res asst plant physiol, 55-56, from instr to assoc prof sci educ, 56-67, PROF SCI EDUC, UNIV IOWA, 67- Concurrent Pos: Dir sec sci training prog, NSF, 59-, dir, In-Serv Inst, 61-, dir, Summer Inst, 63-, dir acad yr prog for sci supv, summer inst in-serv teachers, coop col sch sci prog & undergrad pre-serv teacher educ prog; ed, Jour Nat Asn Res Sci Teaching, 64- Mem: Nat Asn Res Sci Teaching; Nat Sci Teachers Asn; Nat Asn Biol Teachers. Res: Chemical control of abscission processes; effect of student interaction upon learning process; teacher affects upon learning outcomes; development of science curricula. Mailing Add: Dept of Sci Educ Univ of Iowa Iowa City IA 52240

YAGER, ROBERT H, b Somerset, Va, Oct 4, 13; m 39; c 1. VETERINARY MEDICINE. Educ: Univ Pa, VMD, 39; Am Bd Vet Pub Health, dipl; Am Col Lab Animal Med, dipl; Am Bd Med Microbiol, dipl. Prof Exp: Vet Corps, US Army, 39-65, chief vet sect, Second Serv Command Lab, 41-44, chief vet br, 18th Med Gen Lab, 44-46, commanding officer, 46-47, asst dir vet div, Army Med Dept Res & Grad Sch, 47-48, vis investr, Rockefeller Inst, 48-49, dir vet div, Army Med Serv Grad Sch, 49-54, staff vet, US Forces, Austria, 54-55 & Europe Commun Zone, 55-57, asst to commanding officer, Army Med Unit, Ft Detrick, 57-60, dir div vet med, Walter Reed Army Inst Res, 60-65; exec secy, Inst Lab Animal Resources, Nat Acad Sci, 65-76; RETIRED. Mem: Soc Exp Biol & Med; Am Vet Med Asn; Am Asn Immunol. Res: Etiology, pathogenesis, ecology, pathology, immunologic and public health aspects of diseases of animals transmissible to man; tularemia and anthrax; laboratory animal medicine. Mailing Add: Nac Res Coun Nat Acad of Sci Washington DC 20418

YAGI, FUMIO, b Seattle, Wash, July 14, 17; m 54. MATHEMATICS. Educ: Univ Wash, BS, 38, MS, 41; Mass Inst Technol, PhD(math), 43. Prof Exp: Fel, Inst Advan Study, 43; instr, Univ Wash, 46-49, asst prof, 49-53; mathematician, Ballistic Res Labs, Aberdeen Proving Ground, 53-56; sr res engr, Jet Propulsion Lab, Calif Inst Technol, 56-58, res specialist, 58-63; appl mathematician, 63-66, mem systs anal staff, 66-67, GROUP SUPVR SYSTS ANAL, GRUMMAN AIRCRAFT ENG CORP, BETHPAGE, 67- Concurrent Pos: Lectr, Univ Md, 56, Univ Calif, Los Angeles, 57-61 & Adelphi Univ, 63-64 & 66; adj prof, C W Post Col, Long Island Univ, 66- Mem: Am Math Soc; Am Phys Soc. Res: Analysis; space trajectory and guidance studies; systems performance and error analysis. Mailing Add: 74 Oakdale Dr Centerport NY 11721

YAGI, HARUHIKO, b Sendai, Japan, June 27, 39; m 69; c 2. ORGANIC CHEMISTRY, ELECTROCHEMISTRY. Educ: Tohoku Univ, Japan, MS, 65, PhD(systhesis of isoquinoline alkaloid), 68. Prof Exp: Asst org synthesis, Pharmaceut Inst, Tohoku Univ, Japan, 68-69; res asst electrochem, Univ Conn, 69-70; res assoc chem kinetics, Johns Hopkins Univ, 70-71; VIS SCIENTIST DRUG METABOLISM, LAB CHEM, NAT INST ARTHRITIS & METAB DIS, NIH, 71- Concurrent Pos: Fel, Univ Conn, 69-70 & Johns Hopkins Univ, 70-71. Mem: Pharmaceut Soc Japan; Am Chem Soc. Mailing Add: Lab Chem Rm 230 Bldg 4 NIAMD NIH Bethesda MD 20014

YAGUCHI, MAKOTO, b Yokohama, Japan, Oct 19, 30; m 60; c 1. PROTEIN CHEMISTRY, AGRICULTURAL CHEMISTRY. Educ: Tokyo Univ Agr, BAgr, 53; Univ Calif, Davis, MS, 57, PhD(agr chem), 63. Prof Exp: Res assoc protein chem, Purdue Univ, 63-64; res food technologist, Univ Calif, Davis, 64-65; asst res officer, 65-69, ASSOC RES OFFICER, DIV BIOL SCI, NAT RES COUN CAN, 70- Mem: AAAS; Am Dairy Sci Asn; Inst Food Technol; Agr Chem Soc Japan; Can Biochem Soc. Res: Vitamin B-12 in milk; column chromatography of milk proteins; milk lipase; bacterial flagellin; kappa-casein; glycoproteins and lipopolysaccharides; plant cell culture; amino acid sequence of ribosomal proteins. Mailing Add: Div of Biol Sci Nat Res Coun Ottawa ON Can

YAHIA, JACK, solid state physics, see 12th edition

YAHIKU, PAUL Y, b Honolulu, Hawaii, July 25, 38; m 65; c 1. STATISTICS. Educ: Pac Union Col, BA, 61; Univ Calif, Los Angeles, PhD(biostatist), 67. Prof Exp: Asst prof, 66-74, ASSOC PROF BIOSTATIST, LOMA LINDA UNIV, 74- Mem: Am Statist Asn; Biomet Soc. Res: Statistical methodology. Mailing Add: Dept of Biostatist Loma Linda Univ Loma Linda CA 92354

YAHIRO, ARTHUR T, microbiology, biochemistry, see 12th edition

YAHNER, JOSEPH EDWARD, b Chicago, Ill, June 16, 31; m 58; c 3. SOIL CLASSIFICATION. Educ: Purdue Univ, BS, 54; Ore State Univ, MS, 61, PhD(soils), 63. Prof Exp: Agronomist, Purdue Univ-Brazil Proj, US AID Contract, Vicosa, Brazil, 63-67, EXTEN AGRONOMIST, PURDUE UNIV, WEST LAFAYETTE, 67-, ASSOC PROF AGRON, 70- Mem: Am Soc Agron; Soil Conserv Soc Am. Res: Use of soil maps and soil information in land use planning; urban or nonagricultural land use; waste disposal on land. Mailing Add: Dept of Agron Purdue Univ West Lafayette IN 47907

YAHR, CHARLES CORBIN, b Carlinville, Ill, June 23, 25; m 49; c 4. CULTURAL GEOGRAPHY, GEOGRAPHY OF SOUTH ASIA. Educ: Ill State Univ, BS, 49, MS, 50; Univ Ill, PhD(geog), 56. Prof Exp: Ed asst geog, Silver Burdett Co, 52-54; PROF GEOG, CALIF STATE UNIV, SAN DIEGO, 55- Concurrent Pos: Consult, Teledyne-Ryan Aeronautical Corp, 67-68; vis prof geog, Ill State Univ, 68-69 & Univ Wis-Oshkosh, 75-76. Mem: Asn Am Geog; Asn Asian Studies. Res: Geography of southern and southeastern Asia. Mailing Add: Dept of Geog San Diego State Univ San Diego CA 92182

YAHR, MELVIN DAVID, b New York, NY, Nov 18, 17; m 48; c 4. NEUROLOGY. Educ: NY Univ, AB, 39, MD, 43; Am Bd Psychiat & Neurol, dipl, 49. Prof Exp: Res asst, Col Physicians & Surgeons, Columbia Univ, 48-50, instr, 50-51, assoc, 51-53, from asst prof to assoc prof clin neurol, 53-70, from asst dean to assoc dean, 59-73, Merritt prof clin neurol, 70-73; HENRY P & GEORGETTE GOLDSCHMIDT PROF NEUROL & CHMN DEPT, MT SINAI SCH MED, 73- Concurrent Pos: Nat Res Coun res assoc & asst neurologist, Neurol Inst, Presby Hosp, 48-50, asst attend neurologist, 50-53, assoc attend, 53-61, attend, 61-; asst adj neurol serv, Lenox Hill Hosp, 48-49, adj, 49-54, assoc, 54-60; asst neurologist, Montefiore Hosp, 49-51; mem neurol study sect, NIH, 50-; consult, USPHS, 52-55 & Neuro-Psychiat Inst, NJ, 52-59; med dir, Parkinson's Dis Found, 58-73; mem comt drug ther in neurol, Nat Inst Neurol Dis & Blindness, 59- Mem: Am Epilepsy Soc; Am Neurol Asn (secy-treas); Asn Am Med Cols; Asn Res Nerv & Ment Dis; fel Am Acad Neurol. Res: Cause and treatment of epilepsy; cerebro-vascular diseases; Parkinsonism and multiple sclerosis. Mailing Add: Dept of Neurol Mt Sinai Sch of Med New York NY 10029

YAKAITIS, RONALD WILLIAM, b Baltimore, Md, Oct 13, 41; m 68; c 2. ANESTHESIOLOGY. Educ: Loyola Col, BS, 63; Univ Md, MD, 67. Prof Exp: Staff instr anesthesia, US Naval Hosp, Oakland, Calif, 71-73; asst prof, Med Univ SC, 73-75; ASST PROF ANESTHESIA, UNIV ARIZ MED CTR, 75- Concurrent Pos: Consult anesthesia, Univ Calif, San Francisco, 72-73, Vet Admin Hosp, Charleston, SC, 73-75 & Vet Admin Hosp, Tucson, Ariz, 75- Mem: Am Soc Anesthesiologists; Int Anesthesia Res Soc; Am Col Physicians; Am Med Asn. Res: Pulmonary ultramicroscopic and biochemical changes due to oxygen toxicity; cardiovascular drug pharmacokinetics during acid-base imbalance; new techniques for intraoperative anesthetic management. Mailing Add: Dept of Anesthesiol Univ of Ariz Med Ctr Tucson AZ 85724

YAKAITIS-SURBIS, ALBINA ANN, b Harvey, Ill, Feb 16, 23; m 67. ANATOMY. Educ: Univ Chicago, BS, 45, MS, 49; Univ Minn, PhD(anat), 55. Prof Exp: Instr biol, Univ Akron, 46-47; instr, Univ Minn, 47-49, asst anat, 49-54; res assoc path, Med Sch, Univ Mich, 59-61; ASST PROF ANAT, SCH MED, UNIV MIAMI, 61- Concurrent Pos: George fel, Detroit Inst Cancer Res, 56-58; res assoc, Vet Admin Hosp, Miami, 64- Mem: Am Soc Hemat; Am Asn Anat. Res: Oncology; electron microscopy; pathology. Mailing Add: Dept of Biol Structure Univ of Miami Sch of Med Miami FL 33152

YAKATAN, GERALD JOSEPH, b Philadelphia, Pa, May 20, 42; m 64; c 2. PHARMACY. Educ: Temple Univ, BS, 63, MS, 65; Univ Fla, PhD(pharmaceut sci), 71. Prof Exp: ASST PROF PHARM, UNIV TEX, AUSTIN, 72-, ASST DIR DRUG DYNAMICS INST, 75- Mem: Am Pharmaceut Asn; Acad Pharmaceut Sci. Res: Pharmacokinetics; biopharmaceutics. Mailing Add: Col of Pharm Univ of Tex Austin TX 78712

YAKEL, HARRY L, b Brooklyn, NY, July 24, 29. X-RAY CRYSTALLOGRAPHY. Educ: Polytech Inst Brooklyn, BS, 49; Calif Inst Technol, PhD(chem), 52. Prof Exp: Fel chem, Calif Inst Technol, 52-53; GROUP LEADER METALS & CERAMICS, OAK RIDGE NAT LAB, 53- Mem: AAAS; Am Chem Soc; Am Crystallog Asn; Mineral Soc Am; Sigma Xi. Res: Structural studies of solids using x-ray diffraction methods. Mailing Add: Oak Ridge Nat Lab PO Box X Oak Ridge TN 37830

YAKUBIK, JOHN, b Fords, NJ, Sept 23, 28; m 52; c 2. PHARMACEUTICAL CHEMISTRY. Educ: Rutgers Univ, BS, 49; Purdue Univ, MS, 50, PhD(pharmaceut chem), 52. Prof Exp: Res assoc, Squibb Inst Med Res, 52-55; from scientist to sr scientist, Schering Corp, 55-61, mgr pharmaceut develop dept, 61-65, dir sci liaison, Schering Labs, 65-71, dir new prod planning, 71-74, SR DIR CORP PROD DEVELOP, SCHERING CORP, BLOOMFIELD, 74- Mem: AAAS; Soc Cosmetic Chem; Am Pharmaceut Asn; NY Acad Sci. Res: Pharmaceutical product development; pharmaceutical and medicinal chemistry; pharmacology. Mailing Add: Bloomfield NJ

YALE, CHARLES E, b Aurora, Ill, Mar 21, 25; m 48; c 4. MEDICINE, SURGERY. Educ: Univ Ill, Urbana, BS, 49; Case Western Reserve Univ, MD, 55; Univ Cincinnati, DSc(surg), 61. Prof Exp: Resident surg, Univ Cincinnati, 56-62, instr, 62-64; from asst prof to assoc prof, 64-72, PROF SURG, MED SCH, UNIV WIS-MADISON, 72-, VCHMN DEPT, 73- Concurrent Pos: NIH grant, Univ Wis-Madison, 66-71; clin investr surg, Cincinnati Vet Admin Hosp, 62-64; attend surgeon, Univ Wis Hosps, 64-, dir gnotobiotic lab, Univ Wis-Madison, 65- Mem: AAAS; Asn Gnotobiotics (pres, 70-71); Asn Acad Surg; Am Col Surg; AMA; Am Soc Microbiol. Res: Gastrointestinal surgery; intestinal obstruction and strangulation; wound healing; surgical infections and septic shock; gnotobiotics. Mailing Add: Dept of Surg Univ of Wis Med Sch Madison WI 53706

YALE, FRANCIS GAYMON, b Yale, Iowa, Aug 24, 09; m 45; c 5. SCIENCE EDUCATION. Educ: Colo State Col Educ, AB, 32, MA, 41; Columbia Univ,

EdD(sci educ), 52. Prof Exp: Prin & teacher, High Schs, Colo, 32-41; instr, Kans Wesleyan Univ, 41-43; asst prof natural sci, Univ Denver, 46-51; from asst prof to assoc prof phys sci, 52-70, PROF SCI EDUC, ARIZ STATE UNIV, 70- Concurrent Pos: Consult sci prog, Jr Cols & Pub Schs, Ariz, 58- Mem: AAAS. Res: Natural science, especially general teacher education. Mailing Add: Dept of Physics Ariz State Univ Tempe AZ 85281

YALE, HARRY LOUIS, b Chicago, Ill, Dec 18, 13; m 43; c 3. ORGANIC CHEMISTRY. Educ: Univ Ill, BSc, 37; Iowa State Col, PhD(org chem), 40. Prof Exp: Res chemist, Nat Defense Res Comt, 40-41 & Shell Develop Co, 41-45; res chemist, 46-67, SR RES FEL, SQUIBB INST MED RES, 67- Honors & Awards: Lasker Award. Mem: Am Chem Soc; NY Acad Sci; Swiss Chem Soc. Res: Organometallic compounds; quinoline derivatives; furan; high temperature oxidation and chlorination of olefins; chelate compounds; explosives; antituberculous drugs; diuretics; ataractic agents. Mailing Add: Dept of Org Chem Squibb Inst for Med Res Princeton NJ 08540

YALE, IRL KEITH, b Billings, Mont, Mar 13, 39. MATHEMATICS. Educ: Univ Mont, BA, 60; Univ Calif, Berkeley, PhD(math), 66. Prof Exp: Instr math, Univ Mont, 64-65; asst prof, Morehouse Col, 66-67; asst prof, 67-73, ASSOC PROF MATH, UNIV MONT, 73- Mem: Am Math Soc; Math Asn Am. Res: Functional and harmonic analysis. Mailing Add: Dept of Math Univ of Mont Missoula MT 59801

YALE, PAUL B, b Geneva, NY, Apr 28, 32; m 53; c 5. MATHEMATICS. Educ: Univ Calif, Berkeley, BA, 53; Harvard Univ, MA & PhD(math), 59. Prof Exp: Asst prof math, Oberlin Col, 59-61; from asst prof to assoc prof, 61-74, PROF MATH, POMONA COL, 74- Concurrent Pos: NSF sci fac fel, 67-68. Mem: Am Math Soc; Math Asn Am; Asn Comput Mach. Res: Geometry; symmetry; group theory; computer graphics; numerical analysis. Mailing Add: Dept of Math Pomona Col Claremont CA 91711

YALE, SEYMOUR HERSHEL, b Chicago, Ill, Nov 27, 20; m 43; c 2. RADIOLOGY. Educ: Univ Ill, BS, 44, DDS, 45. Prof Exp: Asst clin dent, Col Dent, 48-49, from instr to asst prof, 49-54, assoc prof radiol & head dept, 56-57, prof, 57, mem fac grad col, 60, admin asst to dean col dent, 61-63, from asst dean to actg dean, 63-65, DEAN COL DENT, UNIV ILL MED CTR, 65- Concurrent Pos: Res consult, Hines Vet Admin Hosp, Ill, 59; consult, West Side Vet Admin Hosp, Chicago, 61-, dent proj sect, Nat Inst Radiol Health, 61-, div radiol health, Bur State Serv, 62- & Vet Admin Res Hosp, Chicago, 63-; mem sect comt dent film specifications, US Am Stand Inst, 59-, subcomt 16, Nat Comt Radiation Protection, 63-, gen res support adv comt, Dept Health, Educ & Welfare, 65-, Mayor's Comt Heart, Cancer & Stroke, Chicago & comt dent care ment ill, Ill State Dept Ment Health. Mem: AAAS; Am Dent Asn; Am Acad Dent Radiol; fel Am Col Dent; Int Asn Dent Res. Res: Morphology; radiographic anatomy; radiation biology and control. Mailing Add: PO Box 6998 Chicago IL 60680

YALKOVSKY, RALPH, b Chicago, Ill, Oct 11, 17. OCEANOGRAPHY. Educ: Univ Chicago, BS, 46, MS, 55, PhD(geol), 56. Prof Exp: Geologist, US Corps Engrs, Wash, 49-50; jr engr, State Div Hwys, Calif, 50-52; jr engr geophys comput, Western Geophys Co, 52-54; assoc geol engr, Crane Co, Ill, 55-56; asst prof geol, Mont State Univ, 56-61 & State Univ NY Col New Paltz, 61-62; from asst prof to assoc prof, 62-65, PROF GEOL, BUFFALO STATE UNIV COL, 65- Concurrent Pos: Res grant, Mont State Univ, 59-60 & 60-61; vis investr, Archives of Indies, Naval Mus Madrid, 59-60 & 68, Spanish Inst Oceanog, Spain, 59-60 & Royal Spanish Acad Hist, 68; vis scientist, US Coast Geodetic Surv ship Discoverer, 69. Mem: Fel AAAS; Nat Asn Geol Teachers; Am Geophys Union; NY Acad Sci; Brit Inst Int & Comp Law. Res: Marine geology and geochemistry of marine sediments; water resources; history of science; international law of the sea; public policy. Mailing Add: Dept of Gen Sci Buffalo State Univ Col Buffalo NY 14222

YALKOWSKY, SAMUEL HYMAN, b New York, NY, Dec 5, 42; m 62; c 2. PHARMACEUTICAL CHEMISTRY. Educ: Columbia Univ, BS, 65; Univ Mich, Ann Arbor, MS, 68, PhD(pharm chem), 69. Prof Exp: STAFF SCIENTIST PHARMACY RES, UPJOHN CO, 69- Res: Physical chemistry of surfaces and micelles; solubility and related phenomena. Mailing Add: 1327 Banbury Rd Kalamazoo MI 49001

YALL, IRVING, b Chicago, Ill, Jan 31, 23. MICROBIAL PHYSIOLOGY. Educ: Brooklyn Col, BA, 48; Univ Mo, MA, 51; Purdue Univ, PhD(bact), 55. Prof Exp: Asst bact, Univ Mo, 49-51; res fel, Purdue Univ, 54-56, resident res assoc biochem, Argonne Nat Lab, 56-57; from asst prof microbiol to assoc prof, 57-68, PROF MICROBIOL & MED TECHNOL, UNIV ARIZ, 68- Mem: Fel AAAS; Am Soc Microbiol; Am Chem Soc. Res: Phosphorus metabolism in wastewaters; intermediary metabolism of microorganisms; control of nucleic acids. Mailing Add: Dept of Microbiol Univ of Ariz Tucson AZ 85721

YALMAN, RICHARD GEORGE, b Indianapolis, Ind, Apr 16, 23; m 44; c 2. INORGANIC CHEMISTRY. Educ: Harvard Univ, BS, 43, MA, 47, PhD(chem), 49. Prof Exp: Jr chemist, Monsanto Chem Co, 44; res chemist & sr group leader, Mound Lab, 49-50; from asst prof to assoc prof chem, 50-61, chmn dept, 58 & 61-66, PROF CHEM, ANTIOCH COL, 61-, CHMN DEPT, 72- Concurrent Pos: Consult, Signal Corps, Air Force Off Sci Res Projs, Antioch, 50-58, Monsanto Co, 55 & Kettering Res Lab, 67-71; pres, Mad River Chem Co, 67-; consult, Yellow Springs Instrument Co, 71-72. Mem: AAAS; Am Chem Soc; The Chem Soc; Indian Chem Soc; Soc Chem Indust. Res: Kinetics; metal complexes; high temperature synthesis; metalloporphines; hydrothermal properties of oil shale. Mailing Add: Dept of Chem Antioch Col Yellow Springs OH 45387

YALOW, ABRAHAM AARON, b Syracuse, NY, Sept 18, 19; m 43; c 2. PHYSICS, MEDICAL BIOPHYSICS. Educ: Syracuse Univ, AB, 39; Univ Ill, MS, 42, PhD(physics), 45. Prof Exp: Asst physics, Syracuse Univ, 39-41 & Univ Ill, 41-42; asst physicist, Nat Defense Res Comt, 43; asst physics, Univ Ill, 44-45; asst engr, Fed Telecommun Labs, NY, 45-46; asst prof physics, NY State Maritime Col, 47-48; from instr to assoc prof, 48-66, PROF PHYSICS, COOPER UNION, 66- Concurrent Pos: Consult physicist, Montefiore Hosp, 46- Mem: AAAS; Am Phys Soc; Am Asn Physics Teachers; assoc fel Am Col Radiol; Fedn Am Scientists. Res: Neutron resonance absorption and scattering; microwave transmission; medical applications of radioactive isotopes; Mössbauer effect. Mailing Add: Dept of Physics Cooper Union Sch of Eng & Sci New York NY 10003

YALOW, ROSALYN SUSSMAN, b New York, NY, July 19, 21; m 43; c 2. MEDICAL PHYSICS. Educ: Hunter Col, AB, 41; Univ Ill, MS, 42, PhD(physics), 45; Am Bd Radiol, dipl. Prof Exp: Asst physics, Univ Ill, 41-43; instr, 44; asst engr, Fed Telecommun Lab, NY, 45-46; lectr & asst prof physics, Hunter Col, 46-50; physicist & asst chief radioisotope serv, 50-70, actg chief radioisotope serv, 68-70, CHIEF RADIOIMMUNOASSAY REFERENCE LAB, VET ADMIN HOSP, 69-, CHIEF NUCLEAR MED SERV, 70-, SR MED INVESTR, 72- Concurrent Pos: Consult, Vet Admin Hosp, Bronx, 47-50; mem, President's Study Group Careers for Women, 66-; res prof, Dept Med, Mt Sinai Sch Med, City Univ New York, 68-74, Distinguished Serv Prof, 74-; mem endocrinol study sect, NIH, 69-72 & bd sci coun, Nat Inst Arthritis & Metab Dis, 72-75; consult, subcomt human applns radioactive mat, New York Dept Health, 72- Honors & Awards: Middleton Award, 60; Lilly Award, Am Diabetes Asn, 61; Fed Woman's Award, 61; Van Slyke Award, Am Asn Clin Chem, 68; Gairdner Found Int Award, 71; Am Col Physicians Award, 71; Koch Award, Endocrine Soc, 72; sustaining mem lect award, Asn Mil Surgeons, 75; A Cressy Morrison Award, Natural Sci, NY Acad Sci, 75; Vet Admin Except Serv Award, 75; AMA Sci Achievement Award, 75; Am Asn Clin Chem Award, 75; Albion O Bernstein, MD Award, Med Soc, State NY, 74. Mem: Nat Acad Sci; Am Physiol Soc; Radiation Res Soc; Am Asn Physicists in Med; fel NY Acad Sci. Res: Medical use of radioisotopes, radioimmunoassay and radiation chemistry. Mailing Add: Vet Admin Hosp 130 W Kingsbridge Rd Bronx NY 10468

YAM, LUNG TSIONG, b Canton, China, Apr 10, 36; US citizen; m 64; c 2. INTERNAL MEDICINE. Educ: Nat Taiwan Univ, MD, 60 Prof Exp: From instr to asst prof med, Sch Med, Tufts Univ, 67-72; assoc hemat, Scripps Clin & Res Found, 72-74; ASSOC PROF MED, UNIV LOUISVILLE, 74-; CHIEF HEMAT-ONCOL, VET ADMIN HOSP, 74- Concurrent Pos: Res assoc, New Eng Med Ctr Hosp, 68-72, head cytol & histochem, 70-72. Mem: Am Soc Hemat; Am Soc Histochem & Cytochem. Res: Use of morphologic approach to study problems related to hematology-oncology; use of cytochemistry, immunochemistry and electrophoresis for isoenzymes to identify the origin of normal and neoplastic cells. Mailing Add: Vet Admin Hosp 800 Zorn Ave Louisville KY 40206

YAMADA, ESTHER V, b London, Ont, July 19, 23; m 53; c 1. CHEMISTRY. Educ: Univ Western Ont, BSc, 45, PhD(biochem), 51; McGill Univ, MSc, 47. Prof Exp: Nat Cancer Inst fel, Univ Western Ont, 52-55; res assoc, Karolinska Inst, Sweden, 55-57; res fel, NIH, 57-59; lectr, 59-60, asst prof, 60-66, ASSOC PROF BIOCHEM, FAC MED, UNIV MAN, 66- Mem: AAAS; Can Biochem Soc; Chem Inst Can. Res: Nucleic acid metabolism; enzymes of nucleotide and nucleic acid metabolism in differentiating and developing tissues. Mailing Add: Dept of Biochem Univ of Man Fac of Med Winnipeg MB Can

YAMADA, MASAAKI, b Japan, Aug 9, 42; m 71; c 2. PLASMA PHYSICS. Educ: Univ Tokyo, Japan, BS, 66, MS, 68; Univ Ill, PhD(physics), 73. Prof Exp: Res asst physics, Univ Ill, 69-73; res assoc, 73-75, RES STAFF PLASMA PHYSICS, PLASMA PHYSICS LAB, PRINCETON UNIV, 75- Mem: Am Phys Soc; Phys Soc Japan. Res: Experimental studies of basic plasma physics, including microinstabilities and transport properties of plasmas; statistical physics view of plasmas. Mailing Add: Plasma Physics Lab PO Box 451 Princeton Univ Forrestal Campus Princeton NJ 08540

YAMADA, RYUJI, b Hiroshima City, Japan, Jan 3, 32; m m 60; c 2. EXPERIMENTAL HIGH ENERGY PHYSICS. Educ: Hiroshima Univ, Japan, BS, 54; Univ Tokyo, Japan, MS, 56, PhD(physics), 62. Prof Exp: Res assoc high energy physics, Inst Nuclear Study, Univ Tokyo, Japan, 56-68; res assoc, Brookhaven Nat Lab, 63-65; res assoc accelerator physics, Cornell Univ, 65-66; accelerator consult, 67, PHYSICIST HIGH ENERGY PHYSICS, FERMI NAT ACCELERATOR LAB, 68- Mem: Am Phys Soc; Phys Soc Japan. Res: Development of super conducting magnets for energy doubler project; high energy experiments using existing 400 GeV Proton Synchrotron. Mailing Add: 1228 Modaff St Naperville IL 60540

YAMADA, SYLVIA BEHRENS, b Hamburg, Ger, May 7, 46; m 75. POPULATION ECOLOGY. Educ: Univ BC, BSc, 68, MSc, 71; Univ Ore, PhD(marine ecol), 74. Prof Exp: BIOLOGIST FISHERIES, PAC BIOL STA FISHERIES & MARINE SERV, DEPT OF ENVIRON CAN, 74- Mem: Ecol Soc Am. Res: Incorporation and retension of trace elements in salmon tissue; ecology of intertidal prosobranchs, littorina sitkana, littorina scutulata and littorina planaxis. Mailing Add: Pac Biol Sta Fisheries & Marine Serv Dept of Environ Nanaimo BC Can

YAMADA, YOSHIKAZU, b Honokaa, Hawaii, May 20, 15; m 50; c 4. ORGANIC CHEMISTRY. Educ: Univ Hawaii, BS, 37; Univ Mich, MS, 38; Purdue Univ, PhD(chem), 50. Prof Exp: Res chemist, Davidson Corp, Ill, 50-53; from sr proj engr to res engr, Mergenthaler Linotype Co, NY, 53-59; prin res chemist, Bell & Howell Co, Ill, 59-60, prin res chemist, Res Labs, Calif, 60-72; PRES, YAMADA-GRAPHICS CORP, COSTA MESA, 72- Mem: AAAS; Am Chem Soc. Res: Photosensitive systems; graphic media. Mailing Add: 6151 Sierra Bravo Rd Irvine CA 92715

YAMAGISHI, FREDERICK GEORGE, b Reno, Nev, Sept 14, 43; m 68; c 2. ORGANIC CHEMISTRY. Educ: Univ Calif, Los Angeles, BS, 65, PhD(org chem), 72; Calif State Col, MS, 67. Prof Exp: Fel, Dept Physics, Univ Pa, 72-73, res assoc org chem, 73-74; MEM TECH STAFF ORG CHEM, HUGHES RES LAB, HUGHES AIRCRAFT CO, 74- Mem: Am Chem Soc. Res: Organic conductors; one-dimensional materials; conducting polymers. Mailing Add: 3011 Malibu Canyon Rd Malibu CA 90265

YAMAGUCHI, MASATOSHI, b San Leandro, Calif, Mar 12, 18; m 42; c 3. PLANT PHYSIOLOGY, BIOCHEMISTRY. Educ: Univ Calif, BS, 40, PhD(agr chem), 50. Prof Exp: Prin lab technician, 41 & 46-50, instr, 50-52, asst olericulturist, 50-58, assoc olericulturist, 58-64, lectr, 64-73, PROF VEG CROPS, UNIV CALIF, DAVIS, 73-, OLERICULTURIST, 64- Concurrent Pos: Fulbright res scholar, 59-60; vis prof, Univ Man, 67-68. Mem: Am Chem Soc; Am Soc Hort; Int Soc Hort Sci; Am Soc Plant Physiol; Can Soc Plant Physiol. Res: Chemical constituents and quality of vegetables; physiological disorders of vegetable crops; biochemistry and physiology of vegetable fruit development and ripening. Mailing Add: Dept of Veg Crops Univ of Calif Davis CA 95616

YAMAGUCHI, SHOGO, b Gresham, Ore, Mar 29, 16; m 53; c 3. PLANT PHYSIOLOGY. Educ: Univ Calif, Los Angeles, BA, 40, MA, 48, PhD(bot sci), 54. Prof Exp: Sr lab technician citrus path, Univ Calif, Riverside, 53-54, asst res botanist, Univ Calif, Davis, 54-65; ASSOC PROF BIOL, TUSKEGEE INST, 66- Mem: AAAS; Bot Soc Am; Weed Sci Soc Am. Res: Translocation and distribution pattern of herbicides in plants by the use of autoradiography. Mailing Add: Dept of Biol Tuskegee Institute AL 36088

YAMAKAWA, KAZUO ALAN, physics, see 12th edition

YAMAMOTO, HARRY Y, b Honolulu, Hawaii, Nov 26, 33; m 57; c 2. BIOCHEMISTRY, FOOD TECHNOLOGY. Educ: Univ Hawaii, BS, 55; Univ Ill, MS, 58; Univ Calif, Davis, PhD(biochem), 62. Prof Exp: Asst prof, 61-64, assoc prof, 65-70, PROF FOOD SCI, UNIV HAWAII, MANOA, 70- Concurrent Pos: USPHS spec fel, Charles F Kettering Res Lab, 68-69. Honors & Awards: Samuel Cate Prescott Award, Inst Food Technologists, 69. Mem: Am Chem Soc; Am Soc Plant Physiol; Inst Food Technologists. Res: Carotenoid function; photosynthesis; food biotechnology. Mailing Add: Dept of Food Sci Univ of Hawaii Honolulu HI 96822

YAMAMOTO, JOE, b Los Angeles, Calif, Apr 18, 24; m 47; c 2. PSYCHIATRY.

Educ: Univ Minn, BS, 46, MB, 48, MD, 49. Prof Exp: Asst prof psychiat, Sch Med, Univ Okla, 55-58, asst prof, 58-61; from asst prof to assoc prof psychiat, Sch Med, 58-66, mem fac, Psychoanal Inst, 66-69, PROF PSYCHIAT, SCH MED, UNIV SOUTHERN CALIF, 69- Concurrent Pos: Clin dir, Adult Outpatient Psychiat Clin, Los Angeles County-Univ Southern Calif Med Ctr. Mem: AAAS; fel Am Acad Psychoanal; fel Am Psychiat Asn; fel Am Col Psychiatrists. Res: Clinical and preventive psychiatry. Mailing Add: Dept of Psychiat Univ of Southern Calif Los Angeles CA 90033

YAMAMOTO, KEITH ROBERT, b Des Moines, Iowa, Feb 4, 46. MOLECULAR BIOLOGY. Educ: Iowa State Univ, BSc, 68; Princeton Univ, PhD(biochem sci), 73. Prof Exp: Res asst biochem, Dept of Biochem & Biophys, Iowa State Univ, 67-68; NIH trainee biochem sci, Princeton Univ, 68-73; fel biochem, Lab Gordon M Tomkins, 73-75, ASST PROF BIOCHEM, UNIV CALIF, SAN FRANCISCO, 76- Concurrent Pos: Fel, Helen Hay Whitney Found, 73-76; fel consult, Found Res Hereditary Dis, 74. Mem: Am Soc Microbiol. Res: Mechanisms of gene regulation in eukaryotic cells. Mailing Add: Dept of Biochem & Biophys Univ of Calif San Francisco CA 94143

YAMAMOTO, MASANOBU, biochemistry, see 12th edition

YAMAMOTO, NOBUTO, b Tagawa City, Japan, Apr 25, 25; m 54; c 3. MICROBIOLOGY, BIOPHYSICS. Educ: Kurume Inst Technol, BS, 47; Kyushi Univ, MS, 53; Nagoya Univ, PhD(bact), 58. Prof Exp: Asst prof, Sch Med, Gifu Univ, 58-62; vis scientist molecular biol, NIH, 62-63; from asst prof to assoc prof microbiol, 63-71, PROF MICROBIOL, FELS RES INST & SCH MED, TEMPLE UNIV, 71- Concurrent Pos: Vis researcher virol, Inst Cancer Res, Philadelphia, 59-61; vis scientist, Dept Bact, Ind Univ, 61-62. Mem: Am Soc Microbiol; Am Asn Cancer Res. Res: Virology; genetics; cancer research; molecular biology. Mailing Add: Fels Res Inst Temple Univ Sch of Med Philadelphia PA 19140

YAMAMOTO, RICHARD, b Wapato, Wash, May 27, 27; m 50; c 7. VETERINARY MICROBIOLOGY. Educ: Univ Wash, BS, 52; Univ Calif, MA, 55, PhD(microbiol), 57. Prof Exp: Asst specialist vet pub health, Univ Calif, 57-58, asst res microbiologist, 59; asst prof vet serol & asst vet serologist, Ore State Univ, 59-61; asst microbiologist, 62-64, lectr, 64-67, assoc prof, 67-70, assoc microbiologist, 64-70, PROF & MICROBIOLOGIST, UNIV CALIF, DAVIS, 70- Honors & Awards: Tom Newman Int Award, 67; Nat Turkey Fedn Res Award, 70. Mem: Am Soc Microbiol; Poultry Sci Asn; Am Asn Avian Path; Animal Health Asn; World Poultry Sci Asn. Res: Microbiology, immunology and epidemiology of avian diseases. Mailing Add: Dept of Epidemiol & Prev Med Univ of Calif Davis CA 95616

YAMAMOTO, RICHARD KUMEO, b Honolulu, Hawaii, June 29, 35; m 61; c 3. PHYSICS. Educ: Mass Inst Technol, SB, 57, PhD(physics), 63. Prof Exp: Res staff physics, 63-64, from instr to assoc prof, 64-75, PROF PHYSICS, MASS INST TECHNOL, 75- Mem: Am Phys Soc. Res: High energy nuclear physics; bubble chamber techniques and automated scanning and measuring of bubble chamber film. Mailing Add: Dept of Physics Mass Inst of Technol Cambridge MA 02139

YAMAMOTO, RICHARD SUSUMU, b Honolulu, Hawaii, May 15, 20; m 46; c 1. BIOCHEMISTRY. Educ: Univ Hawaii, AB, 46; George Washington Univ, AM, 49; Johns Hopkins Univ, ScD, 54. Prof Exp: Res assoc biochem, Johns Hopkins Univ, 54-55; biochemist, Lab Nutrit & Endocrinol, Nat Inst Arthritis & Metab Dis, 55-62; biochemist, Biol Br, 62-70, BIOCHEMIST, EXP PATH BR, ETIOLOGY, 70- Mem: AAAS; Am Inst Nutrit; Soc Exp Biol & Med; Am Asn Cancer Res; Am Asn Clin Chemists. Res: Vitamin B12; nutritional obesity; lipid metabolism; chemical carcinogenesis; nutrition and endocrines in carcinogenesis. Mailing Add: Rm 3-C-22 Bldg 37 Nat Cancer Inst Bethesda MD 20014

YAMAMOTO, ROBERT TAKAICHI, b Hawaii, May 26, 27. ENTOMOLOGY. Educ: Univ Hawaii, BA, 53; Univ Ill, MS, 55, PhD, 57. Prof Exp: Res assoc entom, Univ Ill, 57-60; mem staff, Entom Res Div, USDA, 60-65; ASSOC PROF ENTOM, NC STATE UNIV, 65- Mem: Entom Soc Am. Res: Insect physiology and behavior. Mailing Add: Dept of Entom NC State Univ PO Box 5126 Raleigh NC 27607

YAMAMOTO, SACHIO, b Petaluma, Calif, Dec 12, 32; m 58; c 3. MARINE CHEMISTRY, PHYSICAL CHEMISTRY. Educ: Univ Calif, Berkeley, BS, 55; Iowa State Univ, PhD(phys chem), 59. Prof Exp: Res chemist, Calif Res Corp, 59-63; res chemist, US Naval Radiol Defense Lab, 63-69, res chemist, Naval Undersea Ctr, 69-71, SUPVRY RES CHEMIST, NAVAL UNDERSEA CTR, 71- Mem: Am Chem Soc; AAAS. Res: Environmental sciences; trace metal analysis; gas solubility; x-ray fluorescence analysis. Mailing Add: 3725 Notre Dame Ave San Diego CA 92122

YAMAMOTO, TATSUZO, b Hardieville, Alta, Feb 8, 28. MICROBIOLOGY. Educ: Univ Alta, BSc, 52, MSc, 55; Yale Univ, PhD(virol), 61. Prof Exp: Fel microbiol, Univ Toronto, 61-62; from asst prof to assoc prof virol, 62-74, PROF MICROBIOL, UNIV ALTA, 74- Concurrent Pos: Consult, Govt Can, 72- Mem: AAAS; Am Soc Microbiol; Can Soc Microbiol; Electron Micros Soc Am. Res: Replication and structure of viruses; sporulation in Bacillus subtilis; microbial diseases of fish. Mailing Add: Dept of Microbiol Univ of Alta Edmonton AB Can

YAMAMOTO, TOMOKO, b Tokyo, Japan, Nov 7, 43. BIOPHYSICS. Educ: Tokyo Metrop Univ, Japan, BS, 66; Bradley Univ, BS, 68; Univ Mich, PhD(biophysics), 74. Prof Exp: Biophysics trainee, Univ Mich, 68-72; assoc biophysics, Div Biol Sci, 73-75, ASSOC CHEM, DEPT CHEM, CORNELL UNIV, 75- Mem: Biophys Soc Japan; Am Soc Photobiol. Res: Resonance Raman spectroscopy of heme and iron-sulfur proteins; photochemical and photoelectrical phenomena in photosynthesis. Mailing Add: Dept of Chem Cornell Univ Ithaca NY 14853

YAMAMOTO, WILLIAM SHIGERU, b Cleveland, Ohio, Sept 22, 24; m 65; c 3. PHYSIOLOGY, COMPUTER SCIENCE. Educ: Park Col, AB, 45; Univ Pa, MD, 49. Hon Degrees: MS, Univ Pa, 71. Prof Exp: Instr physiol, Sch Med, Univ Pa, 52-53, assoc, 55-57, instr biostatist & asst prof physiol, 57-66, prof physiol, 66-70; prof physiol & biomath, Sch Med, Univ Calif, 70-71; PROF CLIN ENG & CHMN DEPT, SCH MED, GEORGE WASHINGTON UNIV, 71- Concurrent Pos: Mem study sect, NIH, 63-65, mem Nat Adv Coun Res Resources, 71-75; consult, Health Care Tech Div, Nat Ctr Health Serv Res & Develop, '68- Mem: AAAS; Am Physiol Soc; Asn Comput Mach; Biomed Eng Soc. Res: Computer applications in health services; physiology of respiratory regulation by carbon dioxide homeostasis. Mailing Add: Dept of Clin Eng George Washington Univ Med Ctr Washington DC 20037

YAMAMOTO, Y LUCAS, b Hokkaido, Japan, Jan 19, 28; Can citizen; m 58; c 3. NEUROSURGERY, NUCLEAR MEDICINE. Educ: Hokkaido Univ, BSc, 48, MD, 52; Yokohama Nat Univ, PhD(radiobiol), 61. Prof Exp: Intern med, Int Cath Hosp, Tokyo, Japan, 53; resident neurosurg, Med Ctr, Georgetown Univ, 54-58; res assoc nuclear med & radiobiol, Med Dept, Brookhaven Nat Lab, 58-61; res assoc neurosurg res, 61-68, ASST PROF NEUROL & NEUROSURG, MONTREAL NEUROL INST, MCGILL UNIV, 68- Concurrent Pos: Mem, Am Bd Nuclear Med, 73- Mem: Soc Nuclear Med; Radiation Res Soc; Am Col Nuclear Physicians; Can Neurosurg Soc. Res: Neurological science; cerebral circulation. Mailing Add: Montreal Neurol Inst 3801 University St Montreal PQ Can

YAMAMOTO, YASUSHI STEPHEN, b Topaz, Utah, Aug 6, 43. ORGANIC CHEMISTRY, PHOTOGRAPHIC CHEMISTRY. Educ: Univ Wis, BS, 65; Pa State Univ, PhD(org chem), 71. Prof Exp: SR RES CHEMIST, EASTMAN KODAK CO RES LABS, 71- Mem: Am Chem Soc. Res: Investigating synthesis of color couplers for photographic applications; mechanisms and rates of coupling. Mailing Add: Eastman Kodak Co Res Labs 1669 Lake Ave Rochester NY 14650

YAMAMURA, HENRY ICHIRO, b Seattle, Wash, June 25, 40; m 64; c 1. NEUROCHEMISTRY, NEUROPHARMACOLOGY. Educ: Univ Wash, BS, 64, MS, 68, PhD(pharmacol), 69. Prof Exp: Pharm intern, Seattle, Wash, 60-64, staff pharmacist, 64-66; spec lectr pharmacol, Seattle Pac Col, 66-67; ASST PROF, DEPT PHARMACOL, COL MED, UNIV ARIZ, 75- Concurrent Pos: NIMH spec fel pharmacol, Sch Med, Johns Hopkins Univ, 72-75; NIMH Res Scientist Develop Awardee, 75- Mem: AAAS; Soc Neurosci; Am Soc Neurochem; Am Soc Pharmacol & Exp Therapeut. Res: Release and uptake of brain neurohumoral transmitters; demonstration of brain neurotransmitter receptors. Mailing Add: Dept Pharmacol Col of Med Univ of Ariz Tucson AZ 85724

YAMAMURA, STANLEY SATOSHI, analytical chemistry, inorganic chemistry, see 12th edition

YAMANAKA, WILLIAM KIYOSHI, b Kauai, Hawaii, Mar 19, 31; m 58; c 3. NUTRITION, COMMUNITY HEALTH. Educ: Univ Hawaii, BS, 55; Univ Calif, PhD(nutrit), 69. Prof Exp: Res assoc nutrit, Children's Hosp of East Bay, Oakland, Calif, 59-64; res asst, Univ Calif, Berkeley, 64-69; assoc prof nutrit, Univ Mo-Columbia, 69-75; MEM FAC, UNIV WASH, 75- Concurrent Pos: Nutrit adv, Delta Area Head Start, Mo, 69-74; adv, Sickle Cell Anemia Adv Bd, Mo, 74. Mem: AAAS; Am Dietetic Asn; Soc Nutrit Educ; Am Soc Clin Path. Res: Role of lipids in cardiovascular disease; effect of protein malnutrition of growing mammalian organisms; applied nutrition programs in developing countries. Mailing Add: DL-10 Univ of Wash Seattle WA 98195

YAMANE, GEORGE M, b Honolulu, Hawaii, Aug 9, 24; m 51; c 3. ORAL MEDICINE, ORAL PATHOLOGY. Educ: Haverford Col, AB, 46; Univ Minn, Minneapolis, DDS, 50, PhD(oral path), 63. Prof Exp: Asst chem & zool, Univ Hawaii, 43-44; asst oral path & diag, Univ Minn, Minneapolis, 51-53; asst prof oral path, Univ Ill, Chicago, 57-59; asst prof oral path, Univ Wash, 59-63, dir tissue lab, 60-63; prof oral diag, med & roentgenol & chmn div, Univ Minn, Minneapolis, 63-70; PROF ORAL DIAG & RADIOL & CHMN DEPT, COL DENT NJ, COL MED & DENT NJ, 70- Concurrent Pos: Consult, Children Orthop Hosp & Med Ctr, Seattle, 60-63, Vet Admin Hosps, American Lake, Wash, 62-63 & Minneapolis, 64-70 & div dent health, Wyo State Bd Health, 66-70. Mem: Fel AAAS; fel Am Acad Oral Path; fel Int Col Dent; Int Asn Dent Res; fel Am Col Dent. Res: Bone tissue formation; radiobiology; magnesium metabolism; psychosomatic etiology of oral lesions. Mailing Add: Dept of Oral Diag & Radiol Col of Med & Dent of NJ Newark NJ 07103

YAMANOUCHI, TAIJI, b Tokyo, Japan, Aug 16, 31; m 61; c 2. PHYSICS. Educ: Tokyo Univ Ed, BS, 53, MS, 55; Univ Rochester, PhD(physics), 60. Prof Exp: Res assoc physics, Univ Rochester, 60-65, sr res assoc, 65-69; PHYSICIST, NAT ACCELERATOR LAB, 69- Mem: Am Phys Soc. Res: Experimental particle physics. Mailing Add: Nat Accelerator Lab PO Box 500 Batavia IL 60510

YAMASHIRO, STANLEY MOTOHIRO, b Honolulu, Hawaii, Nov 26, 41; m 64; c 2. BIOMEDICAL ENGINEERING. Educ: Univ Southern Calif, BS, 64, MS, 66, PhD(elec eng), 70. Prof Exp: Mem tech staff elec eng, Hughes Aircraft Co, 64-70; res assoc, 70-71, asst prof, 71-74, ASSOC PROF BIOMED ENG, UNIV SOUTHERN CALIF, 74- Res: Cardiopulmonary physiology; application of control theory and computer technology to biological systems. Mailing Add: Dept of Biomed Eng Univ of Southern Calif Los Angeles CA 90007

YAMASHIROYA, HERBERT MITSUGI, b Honolulu, Hawaii, Sept 14, 30; m 47; c 4. MICROBIOLOGY, VIROLOGY. Educ: Univ Hawaii, BA, 53; Univ Ill, Chicago, MS, 62, PhD(microbiol), 65; Registry of Medical Technologists, cert. Prof Exp: Med technician supvr clin lab, Atomic Bomb Casualty Comn, Nat Acad Sci-Nat Res Coun, Hiroshima, Japan, 56-58; assoc scientist, IIT Res Inst, 64-65, res scientist, 65-68, sr scientist, 68-71; ASST PROF PATH, UNIV ILL COL MED, 71-, ASST DIR HOSP LAB, 71- Mem: AAAS; Am Soc Microbiol; Am Soc Clin Path. Res: Tissue culture and its application to viruses and rickettsiae; tissue culture nutritional studies; immunologic techniques and the immune mechanism; diagnostic procedures in microbiology. Mailing Add: Dept of Path Univ of Ill at the Med Ctr Chicago IL 60680

YAMAUCHI, HIROSHI, b Honolulu, Hawaii, Mar 26, 23; m 60; c 2. THEORETICAL PHYSICS. Educ: Univ Hawaii, BS, 47; Harvard Univ, MA, 48, PhD(physics), 50. Prof Exp: From instr to asst prof physics, Colby Col, 50-54; theoret physicist, Nuclear Develop Assoc, Inc, NY, 54-55; from asst prof to assoc prof math, Univ Hawaii, 55-65; assoc prof math & physics, 66-69, PROF MATH & PHYSICS, CHAMINADE COL HONOLULU, 69- Mem: Am Phys Soc. Res: Foundations of quantum mechanics; quantum field theory; mathematical physics; theoretical nuclear physics. Mailing Add: Dept Math Chaminade Col 3140 Waialae Ave Honolulu HI 96816

YAMAUCHI, MASANOBU, b Maui, Hawaii, Mar 3, 31; m 58; c 2. INORGANIC CHEMISTRY. Educ: Univ Hawaii, BA, 53; Univ Mich, MS, 58, PhD(chem), 61. Prof Exp: Asst prof chem, Univ NMex, 60-65; from asst prof to assoc prof, 65-74, PROF CHEM, EASTERN MICH UNIV, 74- Mem: Am Chem Soc. Res: Chemistry of boron hydrides and related compounds. Mailing Add: Dept of Chem Eastern Mich Univ Ypsilanti MI 48197

YAMAUCHI, TOSHIO, b Newell, Calif, Feb 13, 45. MEDICAL GENETICS. Educ: Northwestern Univ, BA, 66, PhD(biol sci), 72. Prof Exp: Fel med genetics, 72-74, RES ASSOC MED GENETICS, UNIV TEX SYST CANCER CTR, M D ANDERSON HOSP & TUMOR INST, 74- Mem: AAAS. Res: Study of enzymes which are involved in the metabolism and activation of chemical carcinogens and the development of systems to study these enzymes in the human population. Mailing Add: Dept of Biol Univ Tex Syst Cancer Ctr M D Anderson Hosp 6723 Bertner Ave Houston TX 77025

YAMAZAKI, HIROSHI, b Hokkaido, Japan, Sept 5, 31; m 61; c 3. BIOCHEMISTRY. Educ: Hokkaido Univ, BS, 54, MS, 56; Univ Wis, PhD(biochem), 60. Prof Exp: Proj assoc biochem, Univ Wis, 61-63, res assoc, 63-65; res officer biol, Atomic Energy Can Ltd, 65-67; asst prof, 67-70, ASSOC PROF BIOL, CARLETON UNIV, 70- Res: Control mechanisms of RNA and protein biosynthesis in bacteria. Mailing Add: Dept of Biol Carleton Univ Ottawa ON Can

YAMAZAKI, WILLIAM TOSHI, b San Francisco, Calif, May 10, 17; m 42; c 3. CEREAL CHEMISTRY. Educ: Univ Calif, BS, 39, MS, 41; Ohio State Univ, PhD(agr chem), 50. Prof Exp: CHEMIST IN CHG SOFT WHEAT QUAL LAB, AGR RES SERV, USDA, OHIO AGR RES & DEVELOP CTR, 63- Concurrent Pos: Adj prof agron, Ohio State Univ & Ohio Agr Res & Develop Ctr, 57- Mem: AAAS; Am Asn Cereal Chem; Am Chem Soc. Res: Chemical and physical basis for processing quality in soft wheat and soft wheat flour. Mailing Add: USDA Agr Res Serv Ohio Agr Res & Develop Ctr Wooster OH 44691

YAMBERT, PAUL ABT, b Toledo, Ohio, May 15, 28; m 50; c 5. CONSERVATION. Educ: Univ Mich, BS, 50, MS, 51, MA, 55, PhD(conserv), 60. Prof Exp: Field scout exec, Boys Scouts of Am, 51-52; teacher, Ann Arbor High Sch, Mich, 53-57; prof natural resources, Wis State Univ, Stevens Point, 57-69; dean outdoor labs, 69-74, PROF, LITTLE GRASSY OUTDOOR LAB, PINE HILLS FIELD RES STA, SOUTHERN ILL UNIV, CARBONDALE, 74- Concurrent Pos: Vis prof, Univ Mich, 63. Res: International resource development; demography; environmental quality; outdoor recreation. Mailing Add: Little Grassy Outdoor Lab Southern Ill Univ Carbondale IL 62901

YAMDAGNI, RAGHAVENDRA, b Aligarh, India, June 30, 41. PHYSICAL CHEMISTRY, MASS SPECTROMETRY. Educ: Allahabad Univ, BSc, 57, MSc, 59, PhD, 65. Prof Exp: Res assoc, Cornell Univ, 65-68; res assoc mass spectrom, Univ Alta, 68-75; PROF ASSOC, DEPT CHEM, UNIV CALGARY, 75- Mem: Am Soc Mass Spectrom. Res: Ion molecule reactions; thermodynamic and reaction kinetics studies. Mailing Add: Dept of Chem Univ of Calgary Calgary AB Can

YAMIN, MICHAEL, b New York, NY, Aug 24, 28. PHYSICAL CHEMISTRY, COMPUTER SCIENCES. Educ: Polytech Inst Brooklyn, BS, 49; Yale Univ, PhD(chem), 52. Prof Exp: Fel chem, Yale Univ, 52-53; fel glass sci, Mellon Inst, 33-56; MEM STAFF, BELL LABS, INC, 57- Res: Semiconductors and semiconductor devices; insulating thin films; computer aids to design. Mailing Add: Bell Labs Inc Murray Hill NJ 07974

YAMIN, SAMUEL PETER, b New York, NY, July 26, 38; m 65; c 1. PARTICLE PHYSICS. Educ: Mass Inst Technol, SB, 60; Univ Pa, MS, 61, PhD(physics), 66. Prof Exp: Res assoc physics, Brookhaven Nat Lab, 66-69; ASST PROF PHYSICS, RUTGERS UNIV, NEW BRUNSWICK, 69- Mem: Am Phys Soc. Res: Experimental elementary particle physics; musical acoustics. Mailing Add: Dept of Physics Rutgers Univ New Brunswick NJ 08903

YAMINS, JACOB LOUIS, b Fall River, Mass, Jan 8, 14; m 48; c 1. FOOD SCIENCE. Prof Exp: Res assoc, Res Lab Org Chem, Mass Inst Technol, 39-42; sr chemist, Nat Fireworks, Inc, Mass, 43-46; sr chemist, Biochem Div, Interchem Corp, NJ, 47-48; res chemist & proj leader, Res Labs Div, Nat Dairy Prod Corp, 48-49, asst to vpres & dir res, 50-54, asst to pres, 54-58; dept head fundamental studies, Res & Develop Div, Am Sugar Co, NY, 58-65, dir sci develop, 65-67; VIS PROF FOOD SCI, RUTGERS UNIV, NEW BRUNSWICK, 67-; CONSULT, 69- Concurrent Pos: Adv chem dept, Adelphi Col, 50-; abstractor, Chem Abstrs. Mem: Fel AAAS; Am Chem Soc; Sigma Xi; Soc Chem Indust; Solar Energy Soc. Res: Protein hydrolysis and isolation of amino acids; vitamin syntheses; preparation of primary explosives; tall oil; sterol isolation and syntheses; tocopherols; flavors; syntheses of long chain surface active agents and bactericides; antioxidants; carbohydrates; food product development; nutrition; single cell proteins; packaging. Mailing Add: PO Box 150 Freeport NY 11520

YAN, JOHNSON FAA, b Amoy, China, May 21, 34; m 70. PHYSICAL CHEMISTRY. Educ: Nat Taiwan Univ, BS, 59; Kent State Univ, MS, 65, PhD(chem), 67. Prof Exp: Chief lab, Hwa Ming Pulp & Paper Manufactory, Chu-nan, Taiwan, 59-62; res assoc & fel, Cornell Univ, 67-69; DEVELOP ASSOC PULP & PAPER, BOWATERS CAROLINA CORP, 69- Mem: AAAS; Am Chem Soc. Res: Physical chemistry, surface and colloid; polymers; biopolymers; pulp and paper. Mailing Add: Bowaters Carolina Corp PO Box 7 Catawba SC 29704

YAN, MAXWELL MENUHIN, organic chemistry, see 12th edition

YAN, TUNG-MOW, b Keelung, Taiwan, Nov 27, 36; m 64; c 2. HIGH ENERGY PHYSICS. Educ: Nat Taiwan Univ, BS, 60; Nat Tsinghua Univ, Taiwan, MS, 62; Harvard Univ, PhD(physics), 68. Prof Exp: Res assoc physics, Stanford Linear Accelerator Ctr, 68-70, vis scientist, 73-74; asst prof, 70-75, ASSOC PROF PHYSICS, CORNELL UNIV, 76- Concurrent Pos: Alfred P Sloan Found fel, 74-76. Mem: Am Phys Soc. Res: Structure of elementary particles and properties of quantum field theories. Mailing Add: Lab of Nuclear Studies Cornell Univ Ithaca NY 14853

YANAGIMACHI, RYUZO, b Sapporo, Japan, Aug 27, 28; m 60. REPRODUCTIVE BIOLOGY. Educ: Hokkaido Univ, BSc, 53, DSc(biol), 60. Prof Exp: Res scientist, Worcester Found Exp Biol, 60-64; lectr biol, Hokkaido Univ, 64-66; from asst prof to assoc prof, 66-73, PROF ANAT, SCH MED, UNIV HAWAII, 73- Mem: Soc Study Reproduction; Am Asn Anat; Am Soc Cell Biol; Brit Soc Study Fertil. Res: Biology of reproduction, particularly biology and physiology of gametes and early development of mammals. Mailing Add: Dept of Anat & Reproductive Biol Univ of Hawaii Sch of Med Honolulu HI 96822

YANAI, HIDEYASU STEVE, b Tokyo, Japan, Feb 26, 28; nat US; m 56; c 2. ANALYTICAL CHEMISTRY. Educ: Tokyo Agr Col, BS, 47; Calif State Polytech Col, BS, 53; Univ Minn, PhD(phys chem), 58. Prof Exp: Sr scientist, 58-69, anal lab head, Res Div, 69-73, ANAL RES PROJ LEADER, RES DIV, ROHM AND HAAS CO, 73- Mem: Am Crystallog Asn; Am Chem Soc. Res: X-ray crystallography; small angle x-ray scattering; polymer physics. Mailing Add: Rohm and Haas Co Res Div PO Box 219 Bristol PA 19007

YANARI, SAM SATOMI, b Gilcrest, Colo, May 27, 23; m 51; c 3. IMMUNOLOGY, BIOCHEMISTRY. Educ: Univ Chicago, BS, 48, PhD, 52. Prof Exp: Res assoc, Univ Chicago, 52-53; res assoc, Armour & Co, 53-56; res assoc, Minn Mining & Mfg Co, 56-64; head, Biochem Dept, Armour Pharmaceut Co, Ill, 65-69; res dir & vpres, Wilson Labs, 69-72; HEAD, RES LAB, DIV ALLERGY & CLIN IMMUNOL, HENRY FORD HOSP, DETROIT, 72- Mem: Am Chem Soc; Am Fedn Clin Res; Am Asn Clin Chem. Res: Mechanism of enzyme action; protein chemistry and structure; proteolytic enzymes; absorption and fluorescence spectroscopy; hormones; immunology. Mailing Add: 13328 Wales St Huntington Woods MI 48070

YANCEY, JOEL A, organic chemistry, see 12th edition

YANCHICK, VICTOR A, b Joliet, Ill, Dec 3, 40; m 63; c 2. PHARMACY. Educ: Univ Iowa, BS, 62, MS, 66; Purdue Univ, PhD(pharm), 68. Prof Exp: Instr pharm, Purdue Univ, 66-68; asst prof, 68-72, actg asst dean, 72-74, ASSOC PROF PHARM, UNIV TEX, AUSTIN, 72-, ASST DEAN ACAD AFFAIRS, COL PHARM, 74- Honors & Awards: Parenteral Drug Asn Res Awards, 66 & 68. Mem: Am Pharmaceut Asn; Am Soc Hosp Pharmacists; Acad Pharmaceut Sci. Res: Parenteral drugs, primarily inactivation by other agents; drug interactions and incompatibilities. Mailing Add: Col of Pharm Univ of Tex Austin TX 78712

YANDERS, ARMON FREDERICK, b Lincoln, Nebr, Apr 12, 28; m 48; c 2. GENETICS. Educ: Nebr State Col, AB, 48; Univ Nebr, PhD(zool), 53. Prof Exp: Res assoc genetics, Oak Ridge Nat Lab & Northwestern Univ, 53-54; biophysicist, US Naval Radiol Defense Lab, 55-58; assoc geneticist, Argonne Nat Lab, 58-59; assoc prof zool, Mich State Univ, 59-65, prof & asst dean col natural sci, 65-69; DEAN COL ARTS & SCI, UNIV MO-COLUMBIA, 69- Concurrent Pos: Vis scientist, Commonwealth Sci & Indust Res Orgn, Canberra, Australia, 66; mem bd dir & consult, Assoc Midwestern Univs, 66-68; mem bd trustees, Argonne Univs Asn, 68-, vpres, 71-72, interim pres, 72-73, chmn bd trustees, 73- Mem: AAAS; Am Soc Zoologists; Genetics Soc Am; Soc Study Evolution; Am Soc Naturalists. Res: Drosophila cytogenetics. Mailing Add: Col of Arts & Sci Univ of Mo Columbia MO 65201

YANEY, PERRY PAPPAS, b Columbus, Ohio, July 28, 31; m 61; c 3. PHYSICS. Educ: Univ Cincinnati, EE, 54, MS, 57, PhD(physics), 63. Prof Exp: Design engr, Baldwin Piano Co, Ohio, 54-55; res physicist, St Eloi Corp, 55-59; physicist, Electronics & Ord Div, Avco Corp, 59-62; Univ Cincinnati res fel, Wright-Patterson AFB, 62-63; assoc res physicist, Res Inst, 63-65, asst prof physics, Univ, 65-69, ASSOC PROF PHYSICS, UNIV DAYTON, 69- Concurrent Pos: Consult, Optical Spectros, Owens-Corning Fiberglas Corp & Reheis Chem Co, 74; consult & vis scientist, Univ Southern Calif, 75. Mem: AAAS; Am Phys Soc; Optical Soc Am; Inst Elec & Electronics Engrs; Am Asn Physics Teachers. Res: Optical and electrical properties of ions in crystals; Raman spectroscopy; lasers and their applications; electroluminescence in solids; electrooptical instrumentation techniques. Mailing Add: Dept of Physics Univ of Dayton Dayton OH 45469

YANG, CHAO-CHIH, b Changsha, China, Dec 17, 28; m 57; c 1. COMPUTER SCIENCE. Educ: Chinese Naval Col Technol, BS, 53; Nat Chiao Tung Univ, MS, 62; Northwestern Univ, MS, 64, PhD(elec eng), 66. Prof Exp: Asst prof elec eng & info sci, Wash State Univ, 66-67; prof comput sci, Nat Chiao Tung Univ, 67-71; scientist, Int Bus Mach, San Jose Res Lab, 71-72; ASSOC PROF INFO SCI, UNIV ALA, BIRMINGHAM, 72- Mem: Inst Elec & Electronics Engrs; Asn Comput Mach; Sigma Xi. Res: Design, implementation and analysis of computer algorithms related to theory of computation, operating systems principles. Mailing Add: Dept of Info Sci Univ of Ala Univ Sta Birmingham AL 35294

YANG, CHAO-HUI, b Taichung, Taiwan, Aug 20, 28; m 63; c 2. MATHEMATICS. Educ: Nat Taiwan Univ, BS, 51; Univ Mich, MA, 55; Univ Cincinnati, PhD(math), 58. Prof Exp: Res fel, Inst Math Sci, NY Univ, 58-59, assoc res scientist, 59-61; lectr math, Rutgers Univ, 61-64; PROF MATH, STATE UNIV NY COL ONEONTA. Mem: Am Math Soc; Math Asn Am. Res: Maximal binary matrices; integral equations; integrability of trigonometric series; combinatorial and functional analyses. Mailing Add: Dept of Math State Univ of NY Col Oneonta NY 13820

YANG, CHARLES (YU-DI), b Peiking, China, Apr 13, 31; US citizen; m 61; c 2. PLANT PATHOLOGY, PHYSIOLOGY. Educ: Taiwan Prov Col Agr, BS, 55; Univ Wis-Madison, MS, 61, PhD(plant path, bot), 64. Prof Exp: Proj assoc fungal physiol, Univ Wis-Madison, 64-65; asst prof plant path, Univ Ky, 65-72; PLANT PATHOLOGIST & HEAD DEPT PLANT PATH, ASIAN VEG RES & DEVELOP CTR, TAIWAN, 72- Concurrent Pos: NSF fel, Dept Plant Path, Univ Wis-Madison, 64-65. Mem: AAAS; Am Phytopath Soc; Mycol Soc Am; Am Soc Microbiol; Brit Soc Gen Microbiol. Res: Fungal physiology; rust and other soybean diseases; pathogen studies of plant root and vegetable diseases; soil and aquatic microflora; mycotoxins; growth and differentiation of Phycomycetes; electron microscopy. Mailing Add: Asian Veg Res & Develop Ctr PO Box 42 Tainan Taiwan

YANG, CHEN NING, b Hofei, Anhwei, China, Sept 22, 22; m 50; c 3. PHYSICS. Educ: Southwest Assoc Univ, China, BSc, 42; Univ Chicago, PhD(physics), 48. Hon Degrees: DSc, Princeton Univ, 58, Polytechnic Inst Brooklyn, 65 & Univ Wroclaw, 75. Prof Exp: Instr physics, Univ Chicago, 48-49; mem, Inst Advan Study, 49-66; EINSTEIN PROF PHYSICS & DIR INST THEORET PHYSICS, STATE UNIV NY STONY BROOK, 66- Honors & Awards: Nobel Prize in Physics, 57; Einstein Award, 57. Mem: Nat Acad Sci; Am Philos Soc; Am Phys Soc. Res: Theoretical physics. Mailing Add: Inst for Theoret Physics State Univ of NY Stony Brook NY 11790

YANG, CHIA HSIUNG, b Peikang, Taiwan, Sept 24, 40; US citizen; m 69; c 2. NUCLEAR PHYSICS. Educ: Tunghai Univ, Taiwan, BSc, 62; Tsing Hua Univ, Taiwan, MSc, 65; Washington Univ, MA, 67, PhD(physics), 71. Prof Exp: Asst prof, 71-75, ASSOC PROF PHYSICS, SOUTHERN UNIV, 75- Concurrent Pos: NASA res grant, 72- Mem: Am Phys Soc; Sigma Xi. Res: Microscopic study of isotropic superfluidity of neutron star matter by Yang and Clark method which combines Bardeen-Cooper-Schrieffer and correlated basis function theories. Mailing Add: Southern Br PO Box 9393 Southern Univ Baton Rouge LA 70813

YANG, CHUI-HSU (TRACY), b Hunan, China, Nov 19, 38; US citizen; m 64; c 2. RADIATION BIOPHYSICS. Educ: Tunghai Univ, BS, 59; North Tex State Univ, MS, 64; Univ Ill, PhD(biophys), 67. Prof Exp: Appointee biol, Argonne Nat Lab, 67-69; BIOPHYSICIST, LAWRENCE BERKELEY LAB, UNIV CALIF, 69- Mem: Am Inst Biol Sci; Biophys Soc; Radiation Res Soc; Sigma Xi; AAAS. Res: Effects of radiation on membrane, development and longevity; mechanisms and kinetics of recovery; space biology; mechanism of aging; responses of cultured mammalian cells to heavy ions and other environmental factors; tumor induction by radiation and virus. Mailing Add: Bldg 74B Rm 127 Univ of Calif Lawrence Berkeley Lab Berkeley CA 94720

YANG, CHUNG SHU, b Peking, China, Aug 8, 41; m 66; c 2. BIOCHEMISTRY. Educ: Nat Taiwan Univ, BS, 62; Cornell Univ, PhD(biochem), 67. Prof Exp: Fel biochem, Scripps Clin & Res Found, 67-69; res assoc, Yale Univ, 69-71; asst prof, 71-75, ASSOC PROF BIOCHEM, COL MED NJ, 75- Honors & Awards: Fac Res Award, Am Cancer Soc; Future Leaders Award, Nutrit Found. Mem: Teratology Soc; Am Chem Soc. Res: Mechanisms of biological oxygenation and carcinogen activation, structure and function of phosphoprotein phosphatase. Mailing Add: Dept of Biochem Col of Med & Dent of NJ Newark NJ 07103

YANG, CHUNG-CHUN, b Kiang-su, China, Nov 21, 42; m 67; c 3. PURE MATHEMATICS, APPLIED MATHEMATICS. Educ: Nat Taiwan Univ, BS, 64; Univ Wis-Madison, MS, 66, PhD(math), 69. Prof Exp: Res assoc math, Mich State Univ, 69-70; RES MATHEMATICIAN, NAVAL RES LAB, 70- Honors & Awards: Res Publ Award, Naval Res Lab, 73. Mem: Am Math Soc; Japanese Math Soc. Res: Factorization theory in the function theory of one complex variable; applications of theory of meromorphic functions to some physical and engineering problems. Mailing Add: Naval Res Lab Washington DC 20375

YANG, CHUNG-TAO, b Pingyang, China, May 4, 23; m 57; c 3. MATHEMATICS. Educ: Chekiang Univ, BS, 46; Tulane Univ, PhD, 52. Prof Exp: Asst math, Chekiang

Univ, 46-48; asst, Nat Acad Sci, China, 48-49; instr, Nat Taiwan Univ, 49-50; res assoc, Univ Ill, 52-54; mem staff, Inst Advan Study, 54-56; from asst prof to assoc prof, 56-61, PROF MATH, UNIV PA, 61- Mem: Am Math Soc; Math Asn Am. Res: General and algebraic topology; topological and differential transformation groups. Mailing Add: Dept of Math Univ of Pa Philadelphia PA 19174

YANG, DA-PING, genetics, cell biology, see 12th edition

YANG, DAVID CHIH-HSIN, b Hsinchiang, China, Jan 8, 47; m 71. BIOCHEMISTRY. Educ: Nat Taiwan Univ, BSc, 68; Yale Univ, PhD(biochem), 73. Prof Exp: Res assoc, The Rockefeller Univ, 73-75; ASST PROF BIOCHEM, GEORGETOWN UNIV, 75- Mem: Am Chem Soc; AAAS. Res: Conformational analysis of protein and nucleic acid; fluorescence spectroscopy; chemical modification of nucleic acid. Mailing Add: Dept of Chem Georgetown Univ Washington DC 20057

YANG, DOMINIC TSUNG-CHE, b Tainan, Taiwan, Oct 9, 33; US citizen; m 62; c 2. ORGANIC CHEMISTRY. Educ: St Benedict's Col, Kans, BS, 59; Univ Ga, PhD(org chem), 69. Prof Exp: Chemist, Nalco Chem Co, Ill, 59-64; instr toxicol & NIH res grant, VanderVanderbilt Univ, 68-70; asst prof org chem, 70-74, ASSOC PROF ORG CHEM, UNIV ARK, LITTLE ROCK, 74- Mem: Am Chem Soc; The Chem Soc. Res: Isolation and structure elucidation of fungal metabolites; organic synthesis. Mailing Add: Dept of Chem Univ of Ark Little Rock AR 72204

YANG, DOROTHY CHUAN-YING, b Shanghai, China, May 27, 18; US citizen; div; c 2. PEDIATRICS, NEUROLOGY. Educ: St Johns Univ, China, MD, 45; New York Med Col, MMSc, 50; Am Bd Pediat, dipl, 52. Prof Exp: Intern, St Luke's Hosp, Shanghai, China, 44-45; med resident, Govt Hosp, Free China, 45-46; resident pediat, Children's Ctr, New York, 47-48; asst resident, New Eng Hosp Women & Children, Boston, Mass, 48-49; asst resident, Syracuse Med Ctr, 49-50; instr pediat, New York Med Col, 52-54, assoc, 54-55, asst clin prof, 55-56, asst prof, 56-60; res neurol, Children's Hosp Philadelphia, 61-62; assoc prof pediat, New York Med Col, 64-69, asst prof neurol, 66-69; CLIN ASSOC PROF PEDIAT, STATE UNIV NY DOWNSTATE MED CTR, 69-; ASSOC DIR STANLEY S LAMM INST DEVELOP DIS, LONG ISLAND COL HOSP, 69- Concurrent Pos: Teaching & res fel, New York Med Col, 50-52; NIH spec fel pediat neurol, 60-63; asst pediatrician, Flower & Fifth Ave Hosp, New York, 54-60; pediatrician, Collab Study Neurol Dis & Blindness, NIH, 57-60, neurologist, 64-69; asst vis pediatrician, Metrop Hosp, New York, 58-60; spec training electroencephalog, Grad Hosp, Univ Pa, 62-63; assoc attend pediatrician, Flower & Fifth Ave & Metrop Hosps, 64-69, asst attend neurologist, 66-69; assoc vis pediatrician, Kings County Hosp Ctr, 71-; consult pediat neurol, St John's Episcopal Hosp, Brooklyn, NY, 72- Mem: Fel Am Acad Pediat; Child Neurol Soc; Am Acad Neurol; Am Med Electroencephalog Asn. Mailing Add: 110 Amity St Brooklyn NY 11201

YANG, GRACE L, b Queichow, China; m 64; c 2. STATISTICS. Educ: Univ Calif, Berkeley, MA, 63, PhD(statist), 66. Prof Exp: Asst prof, 69-74, ASSOC PROF STATIST, UNIV MD, COLLEGE PARK, 74- Mem: Inst Math Statist. Res: Mathematical statistics; biostatistics. Mailing Add: Dept of Math Univ of Md College Park MD 20742

YANG, HOYA Y, b Amoy, China, June 3, 12; nat US; m 46; c 2. FOOD TECHNOLOGY. Educ: Nanking Univ, BS, 36; Ore State Col, MS, 40, PhD(food technol), 44. Prof Exp: Asst prof, 43-47, ASSOC PROF FOOD TECHNOL, ORE STATE UNIV, 48- Mem: Am Chem Soc; Am Soc Enol; Inst Food Technologists; Am Soc Microbiol. Res: Food fermentation; food additives; food enzymes; food and food product analysis. Mailing Add: Dept of Food Sci & Technol Ore State Univ Corvallis OR 97331

YANG, JEN TSI, b Shanghai, China, Mar 18, 22; m 49; c 2. BIOPHYSICAL CHEMISTRY. Educ: Nat Cent Univ, China, BS, 44; Iowa State Univ, PhD(biophys chem), 52. Prof Exp: Asst anal chem, Nat Cent Univ, China, 46-47; res assoc protein chem, Iowa State Univ, 52-54; res fel polypeptide & protein chem, Harvard Univ, 54-56; res chemist, Am Viscose Corp, 56-59; assoc prof biochem, Dartmouth Med Sch, 59-60; assoc prof, 60-64, PROF BIOCHEM, UNIV CALIF, SAN FRANCISCO, 64- Concurrent Pos: Guggenheim fel, 59-60. Mem: Am Chem Soc; Am Soc Biol Chemists; Biophys Soc. Res: Physical chemistry of biopolymers. Mailing Add: Cardiovasc Res Inst Rm 831 HSW Univ of Calif San Francisco CA 94143

YANG, JEONG SHENG, b Taiwan, July 11, 34; m 60; c 3. TOPOLOGY. Educ: Taiwan Normal Univ, BS, 58; Univ Ala, Tuscaloosa, MA, 63; Univ Miami, PhD(math), 67. Prof Exp: Asst math, Taiwan Normal Univ, 57-62; asst prof, La State Univ, New Orleans, 66-68; vis asst prof, Univ Miami, 68-69; asst prof, 69-72, ASSOC PROF MATH, UNIV SC, 72- Mem: Am Math Soc; Math Asn Am. Res: Function spaces; topological groups; transformation groups. Mailing Add: Dept of Math & Comput Sci Univ of SC Columbia SC 29208

YANG, JOHN YUN-WEN, b Changsha, China, May 19, 30; US citizen; m 58; c 3. ENVIRONMENTAL CHEMISTRY. Educ: St Benedict's Col, Kans, BS, 52; Univ Kans, PhD(chem), 57. Prof Exp: Res assoc recoil carbon-14, Brookhaven Nat Lab, 57-59; sr chemist, US Naval Radiol Defense Lab, Calif, 59-60; res specialist, Atomics Int Div, NAm Aviation, Inc, 60-62, mem tech staff, Sci Ctr, 62-67; sr res scientist, Western NY Nuclear Res Ctr, 67-72; PRIN CHEMIST, ENVIRON SYSTS DEPT, CALSPAN CORP, 72- Concurrent Pos: Adj assoc prof, State Univ NY Buffalo, 70- Mem: Am Chem Soc; Am Nuclear Soc; NY Acad Sci. Res: Environmental air and water chemistry; physiocochemical methods of wastewater treatment; air pollution technology; radiation and isotope applications; radiation photochemistry; electrochemical processes. Mailing Add: Calspan Corp Environ Systs Dept PO Box 235 Buffalo NY 14221

YANG, JULIE CHI-SUN, b Peking, China, June 10, 28. INORGANIC CHEMISTRY. Educ: Tsing Hua Univ, China, BS, 49; Ind Univ, MA, 52; Univ Ill, PhD, 55. Prof Exp: Asst chem, Ind Univ, 50-52; res chemist, Res Ctr, Johns-Manville Corp, 55-59, sr res chemist, 59-67, res assoc, 67-72; sr group leader, 72-74, RES MGR, CONSTRUCT PROD DIV, W R GRACE & CO, 75- Concurrent Pos: Mem, Hwy Res Bd, Nat Acad Sci-Nat Res Coun. Mem: Am Chem Soc; Am Ceramic Soc; Am Concrete Inst. Res: Inorganic silicate and cement chemistry; synthesis; properties; structures; material research. Mailing Add: W R Grace & Co 62 Whittemore Ave Cambridge MA 02140

YANG, KEI-HSIUNG, b Taiwan, Dec 10, 40; US citizen; m 71; c 2. QUANTUM OPTICS, MEDICAL PHYSICS. Educ: Nat Taiwan Univ, BS, 64; Univ Notre Dame, MS, 67; Univ Calif, Berkeley, PhD(physics), 74. Prof Exp: Teaching asst, Univ Notre Dame, 65-67; teaching asst, Univ Calif, Berkeley, 67-69; mem tech staff res, Bell Tel Labs, Murray Hill, NJ, 69; res asst, Lawrence Berkeley Lab, 69-73; MEM STAFF, RES & DEVELOP CTR, GEN ELEC CORP, 73- Mem: Am Phys Soc. Res: Continuous tunable coherent vacuum ultraviolet source; medical x-ray devices. Mailing Add: Res & Devel Ctr Gen Elec Corp K-1 5C31 PO Box 8 Schenectady NY 12301

YANG, MAN-CHIU, b Hankow, China, Aug 16, 46; m 71; c 1. BIOCHEMISTRY. Educ: Chinese Univ Hong Kong, BSc, 70; Univ Nebr, PhD(chem), 74. Prof Exp: RES ASSOC, STATE UNIV NY BUFFALO, 74- Mem: Biophys Soc; Am Chem Soc. Res: Study of the intermediates and enzymic systems in oxidative phosphorylation and photophosphorylation. Mailing Add: 168 Acheson Hall State Univ NY Buffalo NY 14214

YANG, MARK CHAO-KUEN, b Tsuchuan, China, Dec 14, 42; m 68; c 1. STATISTICS. Educ: Nat Taiwan Univ, BS, 64; Univ Wis, MS, 67, PhD(statist), 70. Prof Exp: Asst prof, 70-75, ASSOC PROF STATIST, UNIV FLA, 75- Concurrent Pos: Consult, Redstone Arsenal, US Army Command, Ala, 72-73 & Offshore Power Co, Fla, 73-74. Mem: Am Statist Asn; Inst Math Statist; AAAS. Res: Applied probability; stochastic processes; time series analysis. Mailing Add: Dept of Statist Nuclear Sci Bldg Univ of Fla Gainesville FL 32601

YANG, MEILING-TSAI, physical chemistry, analytical chemistry, see 12th edition

YANG, NIEN-CHU, b Shanghai, China, May 1, 28; nat US; m 54; c 3. CHEMISTRY. Educ: St John's Univ, China, BS, 48; Univ Chicago, PhD, 52. Prof Exp: Res assoc, Mass Inst Technol, 52-55; res fel, Harvard Univ, 55-56; from asst prof to assoc prof, 56-63, PROF CHEM, UNIV CHICAGO, 63- Concurrent Pos: Alfred P Sloan fel, 60-64. Mem: AAAS; Am Chem Soc; The Chem Soc. Res: Photochemistry; free radicals; synthesis of natural products and carcinogenesis. Mailing Add: Dept of Chem Univ of Chicago Chicago IL 60637

YANG, OVID Y H, b Korea; US citizen. PATHOLOGY, IMMUNOLOGY. Educ: Yonsei Univ, Korea, MD, 50; Univ Ottawa, PhD, 62. Prof Exp: DIR CLIN LAB, PARK PLACE HOSP, PORT ARTHUR, TEX, 69- Mem: Am Soc Exp Path; Int Acad Path; Am Soc Clin Path; Col Am Path; AMA. Res: Chemical carcinogenesis; cancer immunology. Mailing Add: 3807 Platt Port Arthur TX 77640

YANG, PAULINE YUN-WO CHEN, biochemistry, physical chemistry, see 12th edition

YANG, SEN-LIAN, b Taipei, Taiwan, Jan 10, 38; Chinese citizen; m 66; c 2. OBSTETRICS & GYNECOLOGY. Educ: Nat Taiwan Univ, MD, 63. Prof Exp: Resident obstet & gynec, Nat Taiwan Univ Hosp, 64-68; res fel immunol, US Naval Res Unit 2, 68-70; Ford Found fel reproductive biol & immunobiol, Dept Obstet & Gynec, Univ Pa, 70-71; resident obstet & gynec, Chicago Lying-in Hosp, Univ Chicago, 72-73; instr, 73-74, ASST PROF OBSTET & GYNEC, UNIV CHICAGO, 74- Concurrent Pos: Vis staff obstet & gynec, Nat Taiwan Univ Hosp, 68-70; fel reproductive biol & immunobiol, Chicago Lying-in Hosp, Univ Chicago, 73-74. Mem: Int Fedn Gynec & Obstet; Am Fertil Soc; Soc Study Reproduction; Am Col Obstetricians & Gynecologists; Asn Obstet & Gynec Repub China. Res: Immunobiology of reproductive medicine. Mailing Add: Dept of Obstet & Gynec Chicago Lying-in Hosp Univ Chicago 5841 S Maryland Ave Chicago IL 60637

YANG, SHANG-FA, b Tainan, Formosa, Nov 10, 32; m 65; c 2. PLANT PHYSIOLOGY. Educ: Nat Taiwan Univ, BS, 56, MS, 58; Utah State Univ, PhD(plant biochem), 62. Prof Exp: Fel, Univ Calif, Davis, 62-63; res assoc biochem, NY Univ Med Ctr, 63-64 & Univ Calif, San Diego, 64-66; from asst biochemist to assoc biochemist, 66-74, BIOCHEMIST, UNIV CALIF, DAVIS, 74-, LECTR BIOCHEM, 66- Concurrent Pos: NSF res grant, 67-; Environ Protection Agency res grant, 71-; vis prof, Univ Konstanz, 74; NIH res grant, 75- Honors & Awards: Campbell Award, Am Inst Biol Sci, 69. Mem: AAAS; Am Soc Biol Chemists; Am Soc Plant Physiol; Am Soc Hort Sci. Res: Biosynthesis and hormonal action of ethylene; postharvest biochemistry of fruits and vegetables; biochemical effects of sulfur dioxide on vegetation. Mailing Add: Dept of Veg Crops Univ of Calif Davis CA 95616

YANG, SHAW-MING, b Tainan, Taiwan, May 10, 33; m 66; c 1. MICROBIOLOGY. Educ: Nat Taiwan Univ, BSA, 57, MSA, 59; Univ Wis-Madison, PhD(plant path), 65. Prof Exp: Res asst plant path, Nat Taiwan Univ, 55-57 & Univ Wis-Madison, 62-65; fel food microbiol, Nat Res Coun Can, 65-67; res scientist plant path, Can Dept Agr, 67-68; RES PLANT PATHOLOGIST, USDA, 68- Mem: Am Phytopath Soc; Mycol Soc Am. Res: Bacterial and fungal diseases of sugarcane; disease control; nitrogen fixation. Mailing Add: USDA Sugarcane Field Sta PO Box 470 Houma LA 70360

YANG, SHIANG-PING, b Hankow, China, Mar 5, 19; nat US; m 60. NUTRITION. Educ: Nat Cent Univ, China, BS, 42; Iowa State Univ, MS, 49, PhD(nutrit), 56. Prof Exp: Animal husbandman, Nan-An Dairy Farms, China, 42-45; tech trainee, USDA, 45-46; sr animal husbandman, Chinese Nat Relief & Rehab Admin, 46-47; res assoc nutrit, Iowa State Univ, 49-56; chemist, Mead Johnson Res Ctr, 56-57; asst prof food & nutrit, Purdue Univ, 57-62; assoc prof, Va Polytech Inst, 62; prof, La State Univ, Baton Rouge, 63-69; PROF FOOD & NUTRIT & CHMN DEPT, TEX TECH UNIV, 69- Concurrent Pos: Fulbright lectr & vis prof, Nat Taiwan Univ, 59-60. Mem: Am Soc Animal Sci; Am Chem Soc; Am Inst Nutrit; Inst Food Technologists; Am Dietetic Asn. Res: Protein metabolism; nutritional improvement of dietary proteins; factors influencing the qualities of food. Mailing Add: Dept of Food & Nutrit Tex Tech Univ Lubbock TX 79409

YANG, SHUNG-JUN, b Tintsin, China, Jan 13, 34; US citizen; m 64; c 3. CELL BIOLOGY. Educ: Taiwan Univ, BS, 55; Univ Toronto, MS, 58; NC State Col, PhD(genetics), 62. Prof Exp: Fel cytogenetics, Baylor Col Med, 62-63, instr, 63-64; res assoc radiation biol, Sch Med, Stanford Univ, 64-68; res assoc cell biol, Albert Einstein Col Med, 73-74; RADIATION BIOLOGIST, METHODIST HOSP, BROOKLYN, NY, 74- Concurrent Pos: Clin asst prof radiation biol, State Univ NY Downstate Med Ctr, 75- Res: Proliferation kinetics of mammalian cells in culture and tumors in vivo; immunology studies of malignant diseases. Mailing Add: Dept of Radiother Methodist Hosp 506 Sixth St Brooklyn NY 11215

YANG, TIEN WEI, ecology, see 12th edition

YANG, TSU-JU (THOMAS), b Fengshang, Taiwan, Repub China, Aug 14, 32; m 61; c 3. PATHOBIOLOGY, IMMUNOBIOLOGY. Educ: Nat Taiwan Univ, BVM, 55; Ministry Exam, Taipei, Taiwan, DVM, 59; McGill Univ, PhD(immunol), 71. Prof Exp: Assoc mem immunol, Academia Sinica Inst Zool, 61-64; res assoc cytogenetics, Dept Animal Biol, Univ Pa, 64-66; res fel immunol, Dept Microbiol, Univ Minn, 66-67; demonstr immunol, McGill Univ, 68-71; from asst prof to assoc prof immunol, Univ Tenn Mem Res Ctr, 71-75; ASSOC PROF PATHOBIOL, UNIV CONN, STORRS, 75- Mem: Am Asn Cancer Res; Am Soc Cell Biol; Am Soc Microbiol; AAAS. Res: Tumor-host interaction in canine transmissible sarcoma and Marek's lymphoma; granulopoiesis control mechanism in canine cyclic neutropenia; mode of action of membrane reactive agents: antibodies and lectins. Mailing Add: Dept of Pathobiol Univ of Conn Storrs CT 06268

YANG, WEN-KUANG, b Taiwan, China, Oct 19, 36; m 63; c 2. BIOCHEMISTRY, MEDICINE. Educ: Nat Taiwan Univ, BM, 62; Tulane Univ, PhD(biochem), 66; Educ Coun Foreign Med Grad, cert, 62. Prof Exp: Vis investr enzym, 66-68, staff biochemist, 68-73, GROUP LEADER, BIOL DIV, OAK RIDGE NAT LAB, 73- Concurrent Pos: Res fel nutrit & metab, Sch Med, Tulane Univ, 63-66; Damon Runyon Mem res fel cancer, 66-68; lectr biomed sci, Oak Ridge Biomed Grad Sch, Univ Tenn, 69- Mem: Am Asn Biol Chem; Am Asn Cancer Res; Geront Soc; Formosan Med Asn. Res: Protein biosynthesis in cancer tissues; host cell-RNA oncogenic virus interaction; enzymology of DNA synthesis in mammalian normal and tumor cells; isoaccepting transfer RNA of mammalian tissues. Mailing Add: Carcinogenesis Prog Biol Div Oak Ridge Nat Lab PO Box Y Oak Ridge TN 37830

YANG, WILLIAM CHI TSU, b Peiping, China, Nov 30, 22; nat US. PHYSIOLOGY. Educ: Soochow Univ, BS, 42; Univ Southern Calif, MS, 50, PhD, 56. Prof Exp: Instr zool, Soochow Univ, 47-48; lectr biol, 54-55, from instr to asst prof pharmacol, 56-64, ASSOC PROF PHARMACOL, UNIV SOUTHERN CALIF, 64- Res: Cell physiology; metabolism and culture of protozoa; metabolic activities of mitochondria. Mailing Add: Dept of Pharmacol Univ of Southern Calif Los Angeles CA 90033

YANG, WON-TACK, b Pusan, Korea, May 4, 24; m 45; c 3. MARINE BIOLOGY. Educ: Univ Miami, BS, 55, MS, 57, PhD(marine biol), 67. Prof Exp: Res instr water pollution, Marine Lab, Univ Miami, 57-58; chief aquacult, Cent Fisheries Exp Sta, Korea, 58-61; res scientist, Inst Marine Sci, Univ Miami, 64-68, assoc prof marine sci, Sch Marine & Atmospheric Sci, 68-75; FISHERY BIOLOGIST, FOOD & AGR ORGN, UN, 75- Concurrent Pos: Inst Sea Grant Prog fel brachyuran crab cult, Univ Miami, 70-; consult, Oak Ridge Nat Lab, 70-71, Ralston Purina Co, 71-72 & Los Roques Sci Found, Venezuela, 72- Mem: AAAS; Marine Biol Asn UK; World Maricult Soc; Am Fisheries Soc. Res: General marine aquaculture, with specialization in shrimp and crab larval mass-culture and farming process. Mailing Add: 2821 SW 124th Ct Miami FL 33175

YANICK, NICHOLAS SAMUEL, b Oakburn, Man, Dec 4, 07; nat US; m 53. PHYSICAL CHEMISTRY. Educ: Univ Man, BSc, 30, MSc, 32; NY Univ, PhD(chem), 35. Prof Exp: Chemist, Mathieson Alkali Works, Va, 35-36; res chemist, US Gypsum Co, Ill, 36-38; chief chemist, Wahl-Henius Inst, Ill, 38-45; scientist, Metall Labs, Univ Chicago, 45; chief chemist, Chapman & Smith Co, 45-48; res chemist, John F Jelke Co, 49-51; sr scientist, Res & Develop Div, Kraftco Corp, 51-70; TECH CONSULT, 70- Mem: Am Chem Soc; Inst Food Technologists. Res: Colloid-chemical aspects of food products; solubilities; physico-chemical properties of organic compounds; winterization; separation of fatty acids and fats; evaluation of edible proteins; vanilla; evaluation and formulation of retail and institutional, nutritive and dietetic food products; food ingredients application; food technology; development of industrial food products. Mailing Add: 1643 David Dr Escondido CA 92026

YANIV, SHLOMO STEFAN, b Poznan, Poland, Sept 11, 31; US citizen; m 59; c 2. HEALTH PHYSICS, RADIOLOGICAL PHYSICS. Educ: Israel Inst Technol, BS, 54, Ingenieur, 55; Univ Pittsburgh, MS, 65, DSc(radiation health), 69. Prof Exp: Radiation protection engr, Israel AEC, 58-62; univ health physicist, Grad Sch Pub Health, Univ Pittsburgh, 63-67, asst prof health physics & asst prof radiol, Sch Med, 69-72; sr health physicist, Prod Stand Br, Directorate Regulatory Stand, US AEC, DC, 72-75; TECH ASST TO DIR, DIV SAFEGUARDS FUEL CYCLE & ENVIRON RES, US NUCLEAR REGULATORY COMN, WASHINGTON, DC, 75- Concurrent Pos: Am Cancer Soc grant, Univ Pittsburgh, 71-72; radiol physicist, Montefiore Hosp, 67-72. Mem: Health Physics Soc; Am Asn Physicists in Med; Soc Nuclear Med. Res: Dose reduction in diagnostic radiology; environmental aspects of radionuclides use; application of short life radionuclides in nuclear medicine; radiation dosimetry; effects of population exposure to ionizing radiation. Mailing Add: 11216 Broad Green Dr Potomac MD 20854

YANIV, SIMONE LILIANE, b France, May 17, 38; US citizen; c 2. PSYCHOACOUSTICS, BIOACOUSTICS. Educ: Univ Pittsburgh, BS, 66, MS, 68, PhD(noise control psychoacoust), 72. Prof Exp: Noise pollution consult, Allegheny County Health Dept, 72-73; bioacoust scientist effects noise on people, US Environ Protection Agency, Noise Abatement & Control Off, 73-75; RES PSYCHOACOUSTICIAN, NAT BUR STAND, 75- Honors & Awards: Spec Achievement Award, US Dept Com, Nat Bur Stand, 76. Mem: Acoust Soc Am; Nat Acad Sci; Nat Res Coun. Res: Identification of factors affecting human response to noise encountered in and around buildings for the purpose of developing measurement techniques and criteria for rating acoustical spaces in doors and new noise control techniques. Mailing Add: Nat Bur of Stand Bldg 226 Rm A-313 Washington DC 20234

YANIV, ZOHARA, b Tel-Aviv, Israel, Apr 6, 37; m 64; c 2. PLANT PHYSIOLOGY, PHYTOPATHOLOGY. Educ: Hebrew Univ, Israel, MSc, 62; Columbia Univ, PhD(bot), 67. Prof Exp: Asst fungus physiol, Hebrew Univ, Israel, 61-62; asst biochem, Albert Einstein Col Med, 62-63; asst bot, Columbia Univ, 64-65; asst plant physiol physiol, 65-67; Herman Frasch Found fel plant biochem, 67-69, ASST PLANT BIOCHEMIST, BOYCE THOMPSON INST PLANT RES, 69- Mem: AAAS; Am Soc Plant Physiol; Am Inst Biol Sci; Am Soc Photobiol; Am Phytopath Soc. Res: Protein synthesis in plants; phytochromes; DNA synthesis and the control of differentiation; physiology of parasitism in plants. Mailing Add: Boyce Thompson Inst for Plant Res 1086 N Broadway Yonkers NY 10701

YANKAUER, ALFRED, b New York, NY, Oct 12, 13; m 48; c 3. PUBLIC HEALTH. Educ: Dartmouth Col, BA, 34; Harvard Univ, MD, 38; Columbia Univ, MPH, 47. Prof Exp: Dist health officer, New York City Dept Health, 48-50; asst prof prev med & pub health, Med Col, Cornell Univ, 48-50 & Sch Med, Univ Rochester, 50-52; dir bur maternal & child health, NY State Dept Health, 52-61; regional adv maternal & child health, Pan-Am Health Orgn, WHO, 61-66; sr lectr maternal & child health, 66-70, SR LECTR HEALTH SERV ADMIN, SCH PUB HEALTH, HARVARD UNIV, 70-; PROF COMMUNITY MED, MED SCH, UNIV MASS, WORCESTER, 73- Concurrent Pos: Dir maternal & child health serv, Health Bur, Rochester, NY, 50-52; lectr, Albany Med Col, 52-61; WHO vis prof, Madras Med Col, India, 57-59; ed, Am J Pub Health, 75- Mem: Fel Am Pub Health Asn; fel Am Acad Pediat. Res: Maternal and child health; school health; social medicine; health care. Mailing Add: Dept of Community & Family Med Univ of Mass Med Sch Worcester MA 01604

YANKEE, ERNEST WARREN, b Hayward, Calif, Nov 18, 43; m 65; c 2. ORGANIC CHEMISTRY, MEDICINAL CHEMISTRY. Educ: La Sierra Col, BA, 65; Univ Calif, Los Angeles, PhD(org chem), 70. Prof Exp: RES ASSOC ORG CHEM, UPJOHN CO, 70- Mem: Am Chem Soc. Res: Synthesis and structure-activity relationships of prostaglandins. Mailing Add: Upjohn Co Kalamazoo MI 49001

YANKEE, RONALD AUGUST, b Franklin, Mass, May 24, 34. MEDICINE. Educ: Tufts Univ, BS, 56; Yale Univ, MD, 60. Prof Exp: Intern med, Univ Va, 60-61; resident, Univ Mich, 62-63; sr investr, Nat Cancer Inst, 63-75; MEM STAFF, SIDNEY FARBER CANCER CTR, 75- Mem: Transplantation Soc; Am Soc Hemat. Res: Bone marrow transplantation; histocompatibility; platelet transfusion therapy; cancer chemotherapy. Mailing Add: Sidney Farber Canter Ctr 35 Binney St Boston MA 02115

YANKEELOV, JOHN ALLEN, JR, b Elizabeth, NJ, Dec 9, 32; m 61; c 5. BIOCHEMISTRY, ORGANIC CHEMISTRY. Educ: Marietta Col, BS, 54; Cornell Univ, PhD(org chem), 59. Prof Exp: Res assoc enzym, Brookhaven Nat Lab, 59-61; asst prof, Rockefeller Inst, 61-63; asst ,prof, 63-68, ASSOC PROF BIOCHEM, SCH MED, UNIV LOUISVILLE, 68- Mem: AAAS; Am Chem Soc; Am Soc Biol Chem. Res: Protein structure and function; biochemical methodology. Mailing Add: Dept of Biochem Sch of Med Univ Louisville Health Sci Ctr Louisville KY 40201

YANKELL, SAMUEL L, b Bridgeton, NJ, July 4, 35; m 58; c 3. BIOCHEMISTRY, PHYSIOLOGY. Educ: Ursinus Col, BS, 56; Rutgers Univ, MS, 57, PhD, 60. Prof Exp: Instr, Georgian Court Col, 60; sr res biochemist, Colgate-Palmolive Co, 60-63; head dept biochem & pharmacol, Smith, Miller & Patch, 63-65; sr pharmacologist, Menley & James Res Labs, 66-67; sect head biol labs, Smith Kline & French Inter-Am Corp, 67-69; head dept biol sci, Menley & James Labs, 69-74; RES ASSOC, SCH DENT MED, UNIV PA, 74- Mem: Am Chem Soc; NY Acad Sci; Int Asn Dent Res; Am Soc Pharmacol & Exp Therapeut. Res: Digestion and adsorption; dermatology; dental research. Mailing Add: Univ of Pa Sch of Dent Med 4001 Spruce St Philadelphia PA 19174

YANKO, JOHN ALEXIS, b Monessen, Pa, May 24, 15; m 42; c 2. POLYMER SCIENCE, RUBBER CHEMISTRY. Educ: Geneva Col, BS, 37; Western Reserve Univ, MS, 41, PhD(phys chem), 42. Prof Exp: Instr phys chem, Fenn Col, 41-42; res chemist, Res Ctr, 42-56, proj leader, 56-60, sect leader, 60-66, SR RES ASSOC, B F GOODRICH CO, 66- Mem: Am Chem Soc, Res: Dielectric constants; binary molten tin alloys; high polymers; polymerization; fractionation of polymers; effect of structure and molecular weight on physical properties of elastomers, fibers and plastics; colloidal chemistry; detergents; latex and sponge rubber. Mailing Add: 8440 Wiese Rd Brecksville OH 41141

YANKO, WILLIAM HARRY, b Monessen, Pa, Jan 6, 19; m 42; c 2. RADIOCHEMISTRY, ORGANIC CHEMISTRY. Educ: Geneva Col, BS, 40; Pa State Univ, PhD(org chem), 44. Prof Exp: Org res chemist, Cent Res Lab, Monsanto Chem Co, 43-46; org res chemist, Clinton Labs, Tenn, 46-47; radiochem res group leader, Cent Res Lab, Monsanto Chem Co, 47-60, GROUP LEADER, MONSANTO RES CORP, DAYTON, 61- Mem: Am Chem Soc; NY Acad Sci. Res: Synthetic antimalarials; anticancer drugs; radioisotopic synthesis. Mailing Add: 5612 Royalwood Dr Centerville OH 45429

YANKWICH, PETER EWALD, b Los Angeles, Calif, Oct 20, 23; m 45; c 3. PHYSICAL CHEMISTRY. Educ: Univ Calif, BS, 43, PhD(chem), 45. Prof Exp: Res chemist, Radiation Lab, Univ Calif, 44-46, instr, chem, 47-48; from asst prof to assoc prof, 48-57, PROF CHEM, UNIV ILL, URBANA, 57- Concurrent Pos: NSF fel, Calif Inst Technol & Brookhaven Nat Lab, 60-61. Mem: Fel AAAS; Sigma Xi; Am Chem Soc; fel Am Phys Soc. Res: Chemical kinetics; isotope effects. Mailing Add: Dept of Chem Univ of Ill Urbana IL 61801

YANNONI, COSTANTINO SHELDON, b Boston, Mass, May 20, 35; m 66; c 3. MAGNETIC RESONANCE. Educ: Harvard Col, AB, 57; Columbia Univ, MA, 60, PhD(chem), 67. Prof Exp: Res chemist, Union Carbide Res Inst, 66-67; res chemist, Watson Res Ctr, 67-71, RES CHEMIST, RES LAB, IBM CORP, 71- Mem: Am Chem Soc. Mailing Add: Res Lab Dept K34 IBM Corp Monterey & Cottle Rds San Jose CA 95193

YANNONI, NICHOLAS, b Boston, Mass, Aug 3, 27; m 55; c 4. PHYSICAL CHEMISTRY. Educ: Boston Univ, BA, 54, PhD(chem), 61. Prof Exp: Res fel chem, Mellon Inst, 54-55; staff scientist, Device Develop Corp, 60-61; physicist, 61-64, chief energetics br, 64-73, CHIEF OPTO-ELECTRONIC PHYSICS BR, AIR FORCE CAMBRIDGE RES LABS, 74- Honors & Awards: Am Inst Chemists Medal, 54. Mem: Sigma Xi; Am Crystallog Asn; Am Chem Soc; Am Phys Soc. Res: Crystal structure analysis; optics. Mailing Add: Air Force Cambridge Res Labs LQS L G Hanscom AFB Bedford MA 01731

YANO, FLEUR BELLE, US citizen. THEORETICAL PHYSICS. Educ: Columbia Univ, BS, 54; Univ Southern Calif, MA, 58; Univ Rochester, PhD(physics), 66. Prof Exp: From asst prof to assoc prof, 64-73, PROF PHYSICS, CALIF STATE UNIV, LOS ANGELES, 73- Concurrent Pos: Res Corp grant, Calif State Univ, Los Angeles & State Univ Groningen, 72-73. Mem: Am Phys Soc; Fedn Am Sci. Res: Theoretical nuclear physics; radiative muon capture by complex nuclei; exchange currents in nuclear physics. Mailing Add: Dept of Physics Calif State Univ Los Angeles CA 90032

YANOF, HOWARD MERAR, b Chicago, Ill, June 11, 33; m 55; c 3. BIOPHYSICS, PHYSIOLOGY. Educ: Ohio State Univ, BSc, 53, MSc, 56; Univ Calif, Berkeley, PhD(biophys), 61. Prof Exp: Res assoc biophys, Dept Surg Res, Michael Reese Hosp, Chicago, 60-63; asst prof physiol, Sch Med, St Louis Univ, 63-67, proj dir, Med Ctr Comput Lab, 65-67; sr scientist & asst physiol studies, NASA Electronics Res Ctr, 67; assoc prof physiol, Med Col Va, 67-68; ASSOC PROF PHYSIOL, MED COL OHIO, 68- Concurrent Pos: Schweppe Found career fel, 61-62; NIH res grant, 61-63 & 64-77, career develop award, 65-67; Am Heart Asn grants, 68-71; adj assoc prof, Univ Toledo, 68- Mem: Biophys Soc; Am Physiol Soc; Soc Neurosci; Asn Schs Allied Health Professions. Res: Biophysics of circulation; biomedical instrumentation, especially development of non-invasive biomedical instruments; computer science; cerebral vascular blood flow. Mailing Add: 5224 Kearsdale Rd Toledo OH 43623

YANOFSKY, CHARLES, b New York, NY, Apr 17, 25; m 49; c 3. MOLECULAR BIOLOGY. Educ: City Col New York, BS, 48; Yale Univ, MS, 50, PhD(microbiol), 51. Prof Exp: Res asst microbiol, Yale Univ, 51-54; asst prof, Sch Med, Western Reserve Univ, 54-58; assoc prof biol, 58-61, PROF BIOL, STANFORD UNIV, 61- Concurrent Pos: Lederle med fac award, 55-57; Am Heart Asn career investr, 69- Honors & Awards: Eli Lilly Award, 59; US Steel Found Award, 64; Howard Taylor Ricketts Award, 66; Lasker Med Res Award, 71; Nat Acad Sci Award Microbiol, 72. Mem: Nat Acad Sci; AAAS; Am Acad Arts & Sci; Am Soc Microbiol; Genetics Soc Am. Mailing Add: Dept of Biol Sci Stanford Univ Stanford CA 94305

YANOWITCH, MICHAEL, b Minsk, Russia, Feb 1, 23; nat US; m 52; c 2. APPLIED MATHEMATICS. Educ: Cooper Union, BSE, 43; NY Univ, MS, 50, PhD(math), 53. Prof Exp: Elec engr, Philco Corp, 43-46; instr elec eng, Polytech Inst Brooklyn, 48-49; asst, Inst Math Sci, NY Univ, 50-52, assoc res scientist, 57-58; sr mathematician, Reeves Instrument Corp, 52-57; assoc prof math, 58-62, PROF MATH, ADELPHI UNIV, 62- Concurrent Pos: Consult, Surv Bur Corp, 59-60 & Grumman Aircraft Eng Corp, 61-63; vis scientist, Nat Ctr Atmospheric Res, 65-66. Mem: AAAS; Am Math Soc; Soc Indust & Appl Math; Math Asn Am. Res: Asymptotics; wave motion; atmospheric waves. Mailing Add: Dept of Math Adelphi Univ Garden City NY 11530

YANTIS, RICHARD P, b Westerville, Ohio, July 1, 32; m 59; c 2. MATHEMATICS, OPERATIONS RESEARCH. Educ: US Naval Acad, BS, 54; Univ NC, MA, 62; Ohio State Univ, PhD(indust eng), 66. Prof Exp: US Air Force, 54-74, intel officer, 54-56, instr navig, 56-60, from instr to assoc prof math, US Air Force Acad, 62-70, assoc prof opers res, US Air Force Inst Technol, 71-74; TEACHER, COLUMBUS ACAD, 74- Concurrent Pos: Proj leader underground coal mining proj, Battelle Mem Inst, 74-75, consult, 75- Mem: Inst Mgt Sci; Sigma Xi; Am Inst Indust Eng; Inst Mgt Sci. Res: Linear programming; integer linear programming. Mailing Add: 265 Storington Rd Westerville OH 43081

YAO, ALDEN, food technology, chemistry, see 12th edition

YAO, JERRY SHI KUANG, b Peiping, China, Oct 12, 25; m 46; c 2. PHOTOGRAPHIC CHEMISTRY. Educ: Peking Univ, BS, 46; Mont State Col, PhD, 60. Prof Exp: Fel biochem, Mont State Col, 60-61; asst prof org & gen chem, Wis State Col, Stevens Point, 61-62; assoc prof, Dubuque, 62-63; res chemist, 63-66, TECH SPECIALIST, PHOTO & REPRODUCTIVE DIV, GAF CORP, 66- Mem: Am Chem Soc. Res: Heterocyclic chemistry in relation to photography. Mailing Add: GAF Corp Res & Develop Ctr Charles St Binghamton NY 13902

YAO, JOE, b Antung, China, Feb 11, 30; m 68. WOOD SCIENCE, WOOD CHEMISTRY. Educ: Chung Hsing Univ, Taiwan, BS, 54; Mont State Univ, MS, 58; NC State Univ, PhD(wood sci, wood technol), 65. Prof Exp: Asst prof, 63-70, ASSOC PROF WOOD SCI & TECHNOL, MISS STATE UNIV, 70-; ASSOC WOOD TECHNOLOGIST, MISS FOREST PROD UTILIZATION LAB, 70- Concurrent Pos: Asst forester, Miss Agr Exp Sta, 63-67; asst wood technologist, Miss Forest Prod Utilization Lab, 67-70; vis prof, Nat Chung Hsing Univ, Taiwan, 73-74. Mem: AAAS; Soc Wood Sci & Technol; Forest Prod Res Soc; Am Inst Chemists; NY Acad Sci. Res: Wood particleboard properties; wood capillary structure; water diffusion in wood; shrinkage and related properties; low grade hardwood utilization; utilization of recycled wood fiber material. Mailing Add: Dept of Wood Sci & Technol Miss State Univ Mississippi State MS 39762

YAO, KENNETH TSOONG-SIEU, b Shanghai, China, Dec 30, 14; nat US; m 54; c 8. GENETICS, CYTOLOGY. Educ: Nat Cent Univ, China, BS, 36, MS, 41; Univ Edinburgh, PhD, 47. Prof Exp: Coop agt dairy breeding, Ohio State Univ, 48-51; res collabr animal genetics, Agr Res Ctr, USDA, 51-54; asst prof poultry genetics & breeding, Univ Nebr, 54-61, res fel, Dept Zool & Physiol, 61-63; GENETICIST, BUR RADIOL HEALTH, US DEPT HEALTH, EDUC & WELFARE, 63- Mem: AAAS; Genetics Soc Am; Tissue Cult Asn; Am Genetic Asn; Radiation Res Soc. Res: Radiation genetics. Mailing Add: Bur of Radiol Health 12709 Twinbrook Pkwy Rockville MD 20852

YAO, SHANG JEONG, b Canton, China, June 6, 34; US citizen. CHEMICAL PHYSICS, BIOMEDICAL SCIENCES. Educ: Taipei Inst Technol, Taiwan, Dipl chem eng, 55; Univ Ore, MA, 61; Univ Minn, Minneapolis, PhD, 66. Prof Exp: Asst prof phys sci, Wilbur Wright Campus, Chicago City Col, 68-69; assoc surg res, Michael Reese Hosp & Med Ctr, Chicago, 69-71; ASST PROF NEUROL SURG, SCH MED, UNIV PITTSBURGH, 71- Concurrent Pos: Robert A Welch Found fel theoret chem, Tex A&M Univ, 66-67; fel, Northwestern Univ, Evanston, 67-68; instr, Univ Chicago, 70-71; sr res chemist, Montefiore Hosp, Pittsburgh, Pa, 71- Mem: Am Phys Soc; Am Soc Artificial Internal Organs; Soc Neurosci. Res: Irreversible thermodynamics; bioenergetics; quantum theory of enzyme specificity; implantable energy sources; neuroscience. Mailing Add: Dept of Neurol Surg Univ of Pittsburgh Pittsburgh PA 15261

YAO, YORK-PENG EDWARD, b Canton, China, Sept 11, 37; m 65; c 4. THEORETICAL HIGH ENERGY PHYSICS. Educ: Univ Calif, Berkeley, BS, 60; Harvard Univ, MA, 63, PhD(physics), 64. Prof Exp: Assoc mem natural sci, Inst Advan Study, 64-66; asst prof physics, 66-72, ASSOC PROF PHYSICS, UNIV MICH, ANN ARBOR, 72- Mem: Am Phys Soc. Res: Quantum field theory; elementary particle physics. Mailing Add: Dept of Physics Univ of Mich Ann Arbor MI 48104

YAO, YUNG FANG YU, b Kweiyang, China, Apr 1, 27; m 55. PHYSICAL CHEMISTRY. Educ: Chekiang Univ, China, BS, 48; Lehigh Univ, MS, 54, PhD(chem), 56. Prof Exp: Asst, Res Inst Appl Chem, China, 48-49; res assoc, Ord Res Inst of China, Formosa, 49-52; asst, Lehigh Univ, 52-56, res assoc, 56-59; res assoc, Polytech Inst Brooklyn, 59-61; RES SCIENTIST, SCI LAB, FORD MOTOR CO, 61- Res: Polymer structures; chemisorption of polar molecules on metals and metal oxides; ion exchange; diffusion in ionic solids; catalysis and its application to automotive emission control. Mailing Add: Sci Lab Ford Motor Co Dearborn MI 48121

YAP, FUNG YEN, b Jamaica, WI, Oct 12, 33. COMPUTER SCIENCES, SCIENCE EDUCATION. Educ: Brandeis Univ, BA, 58; Johns Hopkins Univ, PhD(physics), 67. Prof Exp: Asst prof physics, Wilson Col, 66-74, chmn dept, 72-74; SR DIGITAL PROGRAMMER, COMPUT SCI TECHNICOLOR ASSOCS, 74- Concurrent Pos: Proj dir, NSF Award for Purchase of Instnl Sci Equip for Physics Dept, Wilson Col, 68-70. Mem: Am Phys Soc; Am Asn Physics Teachers. Res: Analysis of errors in x-ray spectroscopy; precision measurement of x-ray wavelengths; determination of nuclear decay schemes; environmental pollution. Mailing Add: Goddard Space Flight Ctr Bldg 23 Rm E-309 Greenbelt MD 20770

YAP, WILLIAM TAN, b Amoy, China, Aug 10, 34; US citizen; m 69; c 1. PHYSICAL CHEMISTRY. Educ: Mass Inst Technol, BS, 56, MS, 58, PhD(phys chem), 64. Prof Exp: Res chemist, Res Ctr, Hercules, Inc, 64-69; vis assoc biophys chem, NIH, 69-71; RES CHEMIST, NAT BUR STAND, 72- Res: Biophysical chemistry; solution properties of proteins and other macromolecules. Mailing Add: 6204 Mori St McLean VA 22101

YAPEL, ANTHONY FRANCIS, JR, b Soudan, Minn, Aug 14, 37; m 60; c 3. PHYSICAL CHEMISTRY, BIOPHYSICAL CHEMISTRY. Educ: St John's Univ, Minn, BA, 59; Univ Minn, Minneapolis, PhD(phys chem), 67. Prof Exp: RES SPECIALIST, MINN MINING & MFG CO, 66- Mem: Am Chem Soc; AAAS. Res: Fast reaction, temperature-jump relaxation and enzyme kinetics; physical chemistry of membranes; reverse osmosis phenomena; structure-activity correlations on biological systems; column chromatography; drug delivery systems. Mailing Add: Cent Res Lab Minn Mining & Mfg Co St Paul MN 55101

YAPHE, WILFRED, b Lachine, Que, July 9, 21; m 46; c 3. BACTERIOLOGY. Educ: McGill Univ, BSc, 49, PhD(agr bact), 52. Prof Exp: Assoc res officer, Atlantic Regional Lab, Nat Res Coun Can, 52-66; assoc prof microbiol & immunol, 66-72, PROF MICROBIOL & IMMUNOL, McGILL UNIV, 72- Mem: Am Soc Microbiol; Can Soc Microbiol. Res: Marine microbiology; marine algae; bacterial decomposition of agar, carrageenin and other algal polysaccharides. Mailing Add: Dept of Microbiol & Immunol McGill Univ Montreal PQ Can

YAQUB, ADIL MOHAMED, b Jordan, Jan 19, 28; nat US; m 51; c 2. ALGEBRA. Educ: Univ Calif, AB, 50, MA, 51, PhD(math), 55. Prof Exp: Asst, Univ Calif, 50-55; from instr to asst prof math, Purdue Univ, 55-60; assoc prof, 60-67, PROF MATH, UNIV CALIF, SANTA BARBARA, 67- Mem: Am Math Soc; Math Asn Am. Res: Algebraic structures; ring theory; number theory. Mailing Add: Dept of Math Univ of Calif Santa Barbara CA 93106

YAQUB, JILL COURTANEY DONALDSON SPENCER, b Almondsbury, Eng, Dec 17, 31; m 59. GEOMETRY. Educ: Oxford Univ, BA, 53, MA, 57, PhD(math), 60. Prof Exp: Asst lectr math, Royal Holloway Col, Univ London, 56-59; instr, Wash Univ, 60-61; asst prof, Tufts Univ, 61-63; from asst to assoc prof, 63-75, PROF MATH, OHIO STATE UNIV, 75- Concurrent Pos: Fel, Alexander von Humboldt Found, 70-71. Mem: Am Math Soc. Res: Non-Desarguesian planes, inversive planes and their automorphism groups. Mailing Add: Dept of Math Ohio State Univ Columbus OH 43210

YARBOROUGH, VICTOR ANTHONY, chemistry, see 12th edition

YARBOROUGH, WILLIAM WALTER, JR, b Tylertown, Miss, Jan 6, 45; m 68. PLASMA PHYSICS. Educ: Univ Chattanooga, AB, 67; Vanderbilt Univ, PhD(physics), 74. Prof Exp: ASST PROF PHYSICS, PRESBY COL, 74- Mem: Am Asn Physics Teachers; Am Inst Physics. Res: Low energy theta pinch devices, particularly losses from such devices. Mailing Add: Dept of Math-Physics Presby Col Clinton SC 29325

YARBRO, CLAUDE LEE, JR, b Jackson, Tenn, Sept 26, 22; m 51; c 3. ECOLOGY. Educ: Lambuth Col, BA, 43; Univ NC, PhD(biochem), 54. Prof Exp: Actg prof math & physics, Lambuth Col, 46-47; instr physics, Union Col, Tenn, 48; instr biochem, Vanderbilt Univ, 49-51; asst, Univ NC, 51-54, res assoc, 54-57, instr, 54-60; biologist, Biol Br, Res & Develop Div, US AEC, 60-67, biol scientist, Res Contracts Br, Lab & Univ Div, 67-72, BIOL SCIENTIST, RES & DEVELOP BR, RES & TECHNOL SUPPORT DIV, US ENERGY RES & DEVELOP ADMIN, OAK RIDGE OPERS, 72- Mem: Ecol Soc Am; AAAS; fel Am Inst Chemists; Int Oceanog Found. Res: Phospholipid chemistry and metabolism; mechanism of renal calculus; formation and physical chemistry of calcium phosphate; ecological succession on sandstone bluffs. Mailing Add: 147 Alger Rd Oak Ridge TN 37830

YARBROUGH, ARTHUR C, JR, b Mitchell, Ga, June 12, 28. ORGANIC CHEMISTRY, SCIENCE EDUCATION. Educ: Ga Southern Col, BS, 49; George Peabody Col, MA, 52. Prof Exp: Pub sch instr, Ga, 49-51 & 52-53; asst prof chem, Emory at Osford, 53-55; instr, George Peabody Col, 56-57; ASSOC PROF CHEM, TOWSON STATE COL, 57- Concurrent Pos: Consult var high & jr high schs, Md, 57-, Univ Md, 63 & Baltimore County Fire Dept, 67- Mem: Am Chem Soc. Res: Reactions of transition elements. Mailing Add: Dept of Chem Towson State Col Baltimore MD 21204

YARBROUGH, CHARLES GERALD, b Lumberton, NC, Oct 13, 39; m 60; c 2. ZOOLOGY. Educ: Wake Forest Univ, BS, 61, MA, 63; Univ Fla, PhD(zool), 70. Prof Exp: From instr to asst prof biol, 64-75, ASSOC PROF BIOL, CAMPBELL COL, 75- Concurrent Pos: Chapman res grant, Am Mus Natural Hist, 64; AEC proj ecol researcher, Battelle Mem Inst, 68; vis grad prof, Univ Va, 75. Mem: Am Ornith Union; Cooper Ornith Soc. Res: Influence of physical factors, nutrients and heavy metals on biotic communities; metabolism and temperature regulation in vertebrates; ecological implications of energetics in animals. Mailing Add: Dept of Biol Campbell Col Buies Creek NC 27506

YARBROUGH, GEORGE GIBBS, b Houston, Tex, Jan 20, 43; m 64; c 2. NEUROPHARMACOLOGY. Educ: Univ Houston, BS, 68; Vanderbilt Univ, PhD(pharmacol), 72. Prof Exp: Fel physiol, Univ Man, 72-73; from lectr to asst prof, Univ Sask, 73-75; RES FEL NEUROPSYCHOPHARMACOL, MERCK INST THERAPEUT RES, 75- Concurrent Pos: Med Res Coun Can scholar, Univ Sask, 74-75. Mem: Soc Neurosci. Res: Physiology and pharmacology of synaptic transmission in the mammalian central nervous system. Mailing Add: Merck Sharp & Dohme Res Labs West Point PA 19486

YARBROUGH, HENRY FLOYD, JR, b Richmond, Va, Jan 21, 24; m 46; c 2. BACTERIOLOGY. Educ: Univ Ill, BS, 50, MS, 51, PhD(bact), 53. Prof Exp: RES ASSOC, MOBIL RES & DEVELOP CORP, 53- Mem: Am Soc Microbiol; Soc Indust Microbiol; Sigma Xi. Res: Petroleum microbiology. Mailing Add: Mobil Res & Develop Corp Res Lab PO Box 900 Dallas TX 75221

YARBROUGH, JOHN ALONZO, b Texarkana, Ark, Oct 17, 05; m 33; c 2. BOTANY, BIOLOGY. Educ: Okla Baptist Univ, AB, 25; Univ Okla, MS, 31; Univ Iowa, PhD(bot), 34. Prof Exp: Teacher, High Sch, Okla, 26-30; prof biol, Southwest Baptist Col, 34-35; from asst prof to assoc prof bot, Baylor Univ, 35-43; prof biol, 43-74, EMER PROF BIOL, MEREDITH COL, 74-; RES ASSOC, NC STATE UNIV, 45- Mem: AAAS; Bot Soc Am. Res: Morphology of foliage leaves; anatomy of plant organs; seedling morphology; foliage leaf development. Mailing Add: Dept of Biol Meredith Col Raleigh NC 27611

YARBROUGH, KAREN MARGUERITE, b Memphis, Tenn, Mar 4, 38. GENETICS, MICROBIOLOGY. Educ: Miss State Univ, BS, 61, MS, 63; NC State Univ, PhD(genetics), 67. Prof Exp: Asst prof biol, 67-70, assoc prof microbiol & biol, 70-71, ASSOC PROF MICROBIOL & DIR INST GENETICS, UNIV SOUTHERN MISS, 72- Mem: Genetics Soc Am; Am Genetic Asn; Sigma Xi; NY Acad Sci. Res: Population genetics; human genetics; dermatoglyphics. Mailing Add: PO Box 421 Univ of Southern Miss Hattiesburg MS 39401

YARBROUGH, LYNN DOUGLAS, b Ft Worth, Tex, July 17, 30; m 61; c 3. COMPUTER SCIENCE. Educ: Rice Inst, BA, 53; Univ Ill, MS, 55. Prof Exp: Comput prog analyst, MacDonnell Aircraft Corp, 55; from sr engr to prin scientist, Space & Info Systs Div, NAm Aviation Inc, 55-65; asst dir, Comput Ctr, Harvard Univ, 65-67; SR SCIENTIST, ARCON CORP, 67- Mem: Asn Comput Mach. Res: Design and development of computer operating systems, computer languages and computer graphic systems. Mailing Add: 128 Simonds Rd Lexington MA 02173

YARBROUGH, MARY ELIZABETH, b Raleigh, NC, May 23, 04. NUTRITION. Educ: Meredith Col, AB, 26; NC State Col, MS, 27; Duke Univ, PhD(nutrit), 41. Prof Exp: From instr to assoc prof chem, 29-38, actg prof, 38-39, prof, 40-74, EMER PROF CHEM & PHYSICS, MEREDITH COL, 74- Mem: AAAS; Am Chem Soc; Am Inst Chem. Res: Vitamins C and A. Mailing Add: Coop Educ Meredith Col Raleigh NC 27611

YARD, ALLAN STANLEY, b Rocktown, NJ, Nov 18, 27; m 57; c 2. PHARMACOLOGY. Educ: Rutgers Univ, BS, 52; Med Col Va, PhD(pharmacol, biochem), 56. Prof Exp: Asst prof pharmacol, Rutgers Univ, 55-56; asst prof, Med Col Va, 56-60; sr scientist, Ortho Res Found, 60-63, Ortho res fel, 63-67; group chief, 67-73, dir acad & govt liaison, 73-75, ASST DIR, DRUG REGULATORY AFFAIRS, HOFFMANN-LA ROCHE INC, 75- Concurrent Pos: Mem coadj fac, Rutgers Univ, 63- Mem: AAAS; Am Soc Pharmacol & Exp Therapeut; Soc Study Reproduction; Am Chem Soc; NY Acad Sci. Res: Drug metabolism; effect of drugs on liver metabolism; synthesis of hydrazino compounds; biochemistry and pharmacology of the oviduct, uterus and fertility control agents. Mailing Add: Res Labs Hoffmann-La Roche Inc Nutley NJ 07110

YARDLEY, DARRELL GENE, b Gorman, Tex, Apr 15, 48; m 72; c 1. POPULATION GENETICS. Educ: Univ Tex, Austin, BA, 71, MA, 72; Univ Ga, PhD(zool), 75. Prof Exp: Instr human anat & physiol, Univ Ga, 75; ASST PROF POP GENETICS, CLEMSON UNIV, 75- Mem: Soc Study Evolution. Res: Laboratory, field and theoretical approaches to genetic mechanisms of evolution. Mailing Add: Dept of Zool Clemson Univ Clemson SC 29631

YARDLEY, JAMES THOMAS, III, b Taft, Calif, May 15, 42; m 66. PHYSICAL CHEMISTRY. Educ: Rice Univ, BA, 64; Univ Calif, Berkeley, PhD(chem), 67. Prof Exp: Asst prof, 67-70, ASSOC PROF CHEM, UNIV ILL, URBANA, 74- Concurrent Pos: Dreyfus Found teacher-scholar award, 70-75; Alfred P Sloan fel, 72-73. Mem: Am Phys Soc. Res: Molecular spectroscopy; vibrational energy transfer; molecular lasers; molecular dynamics. Mailing Add: Dept of Chem Univ of Ill Urbana IL 61801

YARDLEY, JOHN HOWARD, b Columbia, SC, June 7, 26; m 52; c 3. PATHOLOGY. Educ: Birmingham-Southern Col, AB, 49; Johns Hopkins Univ, MD, 53; Am Bd Path, dipl, 59. Prof Exp: Intern internal med, Vanderbilt Univ Hosp, 53-54; from instr to assoc prof, 54-72, PROF PATH, SCH MED, JOHNS HOPKINS UNIV, 72- Concurrent Pos: From asst resident to resident, Johns Hopkins Hosp, 54-58. Mem: Am Asn Path & Bact; Int Acad Path; Am Soc Exp Path; Am Gastroenterol Asn. Res: Gastrointestinal diseases; electron microscopy. Mailing Add: Dept of Path Johns Hopkins Hosp Baltimore MD 21205

YARE, ROBERT SAYLES, biochemistry, food science, see 12th edition

YARGER, DOUGLAS NEAL, b Omaha, Nebr, July 13, 37; m 60; c 4. ATMOSPHERIC PHYSICS, METEOROLOGY. Educ: Iowa State Univ, BS, 59; Univ Ariz, MS, 62, PhD(meteorol), 67. Prof Exp: Asst prof, 67-71, ASSOC PROF METEOROL & CLIMATOL, IOWA STATE UNIV, 71- Mem: Am Meteorol Soc; Optical Soc Am. Res: Radiative transfer in the atmosphere; aerosol physics. Mailing Add: 313 Curtiss Hall Iowa State Univ Ames IA 50010

YARGER, FREDERICK LYNN, b Lindsey, Ohio, Mar 8, 25; m 48; c 2. PHYSICS. Educ: Capital Univ, BSc, 50; Ohio State Univ, MSc, 53, PhD(physics), 60. Prof Exp: Res asst physics, Los Alamos Sci Lab, 52, mem staff, 53-55 & 56; sr engr, Columbus Div, NAm Aviation, Inc, Ohio, 56-58; res asst physics, Ohio State Univ, 58-60; supvry physicist, Nat Bur Standards, Colo, 60-64; sci specialist, Edgerton, Germeshausen & Grier, Inc, Nev, 64-65; sr res physicist, Falcon Res & Develop Co, Colo, 65-66; ASSOC PROF PHYSICS, NMEX HIGHLANDS UNIV, 66- Concurrent Pos: Vis staff mem, Los Alamos Sci Lab, 74-76. Mem: Am Phys Soc; Optical Soc Am; Sigma Xi; Mex Phys Soc. Res: High pressure equations of state; molecular spectroscopy. Mailing Add: Dept of Physics & Math NMex Highlands Univ Las Vegas NM 87701

YARGER, HAROLD LEE, b Ypsilanti, Mich, Mar 15, 40; m 65; c 3. EXPLORATION GEOPHYSICS. Educ: Antioch Col, BS, 62; State Univ NY Stony Brook, MA, 65, PhD(physics), 68. Prof Exp: Instr physics, State Univ NY Stony Brook, 67-68; res assoc, Northwestern Univ, 68-69; res assoc, Univ, 69-70, RES ASSOC GEOPHYS, KANS GEOL SURV, UNIV KANS, 70- Concurrent Pos: NASA res contracts, 72-74 & 73-75; US Geol Surv res contract, 74- Mem: AAAS; Am Geophys Union; Am Phys Soc; Soc Explor Geophys. Res: Gravity and magnetics; remote sensing; airplane and satellite imagery for exploration and management of earth resources; high energy physics; bubble chamber work. Mailing Add: Kans Geol Surv Univ of Kans Lawrence KS 66044

YARIAN, DEAN ROBERT, b Warsaw, Ind, Oct 12, 33; m 54; c 7. ORGANIC CHEMISTRY. Educ: DePauw Univ, BA, 55; Univ Wash, PhD(chem), 60. Prof Exp: Asst lectr, Univ Wash, 58-59; sr chemist, Cent Res Lab, 60-65, sr chemist, Paper Prod Div, 65-70, CHEMIST SPECIALIST, PAPER PROD LAB, MINN MINING & MFG CO, ST PAUL, 70- Mem: Am Chem Soc. Res: Organic synthesis; imaging chemistry; colloids; paper chemistry; statistics. Mailing Add: PO Box 57 Afton MN 55001

YARINGTON, CHARLES THOMAS, JR, b Sayre, Pa, Apr 26, 34; m 63; c 3. OTORHINOLARYNGOLOGY. Educ: Princeton Univ, AB, 56; Hahnemann Med Col, MD, 60; Am Bd Otolaryngol, dipl, 65. Prof Exp: Rotating intern, Rochester Gen Hosp, 60-61; res gen surg, Dartmouth Med Sch Affil Hosps, 61-62; asst otolaryngol, Sch Med, Univ Rochester, 62-63, instr, 63-65; chief, Eye, Ear, Nose & Throat Serv, US Army Hosp, Ft Carson, Colo, 65-67; asst prof, Sch Med WVa Univ, 67-68; from assoc prof to prof otorhinolaryngol & chmn dept, Col Med, Univ Nebr, Omaha, 68-74; CLIN PROF OTOLARYNGOL, UNIV WASH, 74-; HEAD & NECK SURGEON, DEPT OTOLARYNGOL, MASON CLIN, 74- Concurrent Pos: Res otolaryngol, Sch Med, Univ Rochester, 62-65; consult, Colo State Hosp, 65-67 & Vet Admin Hosp, WVa, 67-68; chief, Vet Admin Hosp, Omaha & Univ Nebr Hosp, Omaha, 68-74; staff physician, Mason Clin & Virginia Mason Hosp, 74-; mem, Deafness Res Found & Am Coun Otolaryngol. Honors & Awards: Prof Doctor Ignacio Barraquer Mem Award, 68. Mem: Fel Am Acad Facial Plastic & Reconstruct Surg; fel Am Acad Ophthal & Otolaryngol; fel Am Col Chest Physicians; fel Am Rhinol Soc; fel Int Broncho-Esophagol Soc. Res: Clinical studies in the pathology and therapy of congenital and neoplastic defects in the head and neck; histopathology of salivary gland disease. Mailing Add: Mason Clin 1118 Ninth Ave Seattle WA 98101

YARINSKY, ALLEN, b Brooklyn, NY, May 6, 29; m 52; c 3. PARASITOLOGY, CHEMOTHERAPY. Educ: City Col New York, BS, 51; Columbia Univ, MS, 53; Univ NC, MS, 57, PhD, 61. Prof Exp: Med bacteriologist, Ft Detrick, Md, 54-56; res scientist, New York Dept Health, 61-65; res biologist, 66-69, HEAD PARASITOL DEPT, STERLING-WINTHROP RES INST, 69- Mem: Am Soc Parasitol; Am Soc Trop Med & Hyg; Royal Soc Trop Med & Hyg. Res: Research on Clostridium botulinum type E toxin; immunological relationships of experimental Trichinella spirals infections; laboratory diagnosis of Protozoan and helminth infections; chemotherpy of parasitic infections. Mailing Add: Sterling-Winthrop Res Inst Rensselaer NY 12144

YARIS, ROBERT, b New York, NY, Oct 16, 35; m 64. PHYSICAL CHEMISTRY. Educ: Univ Calif, Los Angeles, BS, 58; Univ Wash, PhD(phys chem), 62. Prof Exp: Res assoc phys chem, Univ Minn, 62-64; from asst prof to assoc prof chem, 64-70, PROF CHEM, WASH UNIV, 70- Concurrent Pos: Alfred P Sloan fel, 66- Mem: Am Phys Soc. Res: Theoretical and quantum chemistry; time-dependent perturbation theory; many body theory. Mailing Add: Dept of Chem Wash Univ St Louis MO 63130

YARMOLINSKY, MICHAEL BEZALEL, b New York, NY, Jan 18, 29; m 62; c 1. MOLECULAR BIOLOGY. Educ: Harvard Univ, AB, 50; Johns Hopkins Univ, PhD(biol), 54. Prof Exp: Instr pharmacol, Col Med, NY Univ, 54-55; res assoc, McCollum-Pratt Inst, Johns Hopkins Univ, 58-61, asst prof, 61-63; res chemist, NIH, 64-74; DIR RES, INST MOLECULAR BIOL, NAT CTR SCI RES, PARIS, 74- Concurrent Pos: NSF fel, Pasteur Inst, Paris, 63-64. Mem: Am Soc Biol Chemists. Res: Protein biosynthesis and its regulation; interactions between temperate bacteriophage and its host. Mailing Add: Inst of Molecular Biol Nat Ctr of Sci Res Paris France

YARMUSH, DAVID LEON, b New York, NY, June 10, 28. APPLIED MATHEMATICS. Educ: Harvard Univ, BA, 49; Princeton Univ, PhD(math), 59. Prof Exp: Mathematician chem & radiation labs, Army Chem Ctr, Md, 52-54; asst math, Princeton Univ, 54-56; mathematician, Tech Res Group, Inc, 56-67; RES SCIENTIST, COURANT INST, NY UNIV, 65- Mem: Am Math Soc. Res: Dynamic theory of games and mathematical economics; radiation transport theory; structural vibrations and sound radiation; theorem-proving by computer; computer programming of deduction procedures. Mailing Add: Courant Inst 251 Mercer St New York NY 10012

YARNALL, JOHN LEE, b Elizabeth, NJ, Jan 27, 32; m 53; c 3. INVERTEBRATE ZOOLOGY. Educ: Univ Mont, BS, 53, MA, 62; Stanford Univ, PhD(biol), 72. Prof Exp: Asst prof, 69-72, ASSOC PROF BIOL, HUMBOLDT STATE UNIV, 72- Mem: AAAS; Am Soc Zoologists. Res: Invertebrate functional morphology and behavior, especially locomotion and feeding; cephalopod molluscs. Mailing Add: Dept of Biol Humboldt State Univ Arcata CA 95521

YARNELL, CHARLES FREDERICK, b Napoleon, Ohio, Sept 22, 39; m 68; c 3. PHYSICAL CHEMISTRY. Educ: Defiance Col, BS, 61; Mass Inst Technol, PhD(phys chem), 65. Prof Exp: STAFF MEM LEAD-ACID BATTERIES, BELL LABS, 65- Mem: Electrochem Soc. Res: Materials research on the positive and negative plate of the lead-acid battery. Mailing Add: Bell Labs 600 Mountain Ave Murray Hill NJ 07974

YARNELL, JOHN LEONARD, b Topeka, Kans, Mar 1, 22; m; c 4. PHYSICS. Educ: Univ Kans, AB, 47, AM, 49; Univ Minn, PhD(physics), 52. Prof Exp: Asst instr physics & math, Univ Kans, 47-49; asst, Univ Minn, 49-51; staff mem, Physics Div, 52-65, GROUP LEADER, PHYSICS DIV, LOS ALAMOS SCI LAB, 65- Mem: Fel Am Phys Soc; Am Nuclear Soc. Res: Lattice dynamics; neutron diffraction; reactors; cryogenics; solid state physics. Mailing Add: Los Alamos Sci Lab Box 1663 Los Alamos NM 87544

YARNELL, RICHARD ASA, b Boston, Mass, May 11, 29; m; c 4. ANTHROPOLOGY, ETHNOBOTANY. Educ: Duke Univ, BS, 50; Univ NMex, MA, 58; Univ Mich, PhD(anthrop), 63. Prof Exp: From instr to assoc prof anthrop, Emory Univ, 62-71, Am Acad Arts & Sci fel, 63; assoc prof anthrop, 71-75, assoc chmn dept, 73-75, PROF ANTHROP, UNIV NC, CHAPEL HILL, 75- Mem: AAAS; Am Anthrop Asn; Soc Am Archaeol; Soc Econ Bot; Am Soc Ethnohist. Res: Analysis of archaeological plant remains; evolution of plant domestication; aboriginal plant utilization; cultural ecology; economic botany. Mailing Add: Dept of Anthrop Univ of NC Chapel Hill NC 27514

YARNELLE, JOHN E, b Ft Wayne, Ind, Apr 23, 10; m 33. MATHEMATICS. Educ: Williams Col, Mass, AB, 32; Univ Chicago, SM, 47; Univ Pittsburgh, PhD(math), 55. Prof Exp: Instr, Forman Sch, Conn, 32-34, asst head master, 34-44; dir residence systs, Univ Chicago, 44-45, dir student activities, 46; PROF MATH, HANOVER COL, 47- Concurrent Pos: Lectr, NSF, 61; mem panel on suppl pub sch math study group, 64-67. Mem: Am Math Soc; Math Asn Am. Res: Measure theory. Mailing Add: Dept of Math Hanover Col Hanover IN 47243

YARNS, DALE A, b Jackson, Minn, July 9, 30; m 54; c 2. ANIMAL PHYSIOLOGY. Educ: Univ Minn, BS, 56; SDak State Col, MS, 58; Univ Md, PhD(animal sci), 64. Prof Exp: Res asst dairy husb, SDak State Univ, 56-58; lab technician animal sci, Univ Calif, Davis, 58-61; animal husbandryman, Beef Cattle Br, USDA, 61-64; assoc res physiol, Animal Med Ctr, 64-66; asst prof, 66-72, ASSOC PROF PHYSIOL & CHMN DEPT BIOL, WAGNER COL, 72- Res: Comparative cardiac and ruminant physiology. Mailing Add: Dept of Biol Wagner Col Staten Island NY 10301

YAROSEWICK, STANLEY J, b Epping, NH, Sept 10, 39; m 64; c 2. ATOMIC PHYSICS, SPECTROSCOPY. Educ: Univ NH, BS, 61; Clarkson Col Technol, MS, 63, PhD(physics), 66. Prof Exp: Asst prof physics, Clarkson Col Technol, 66-69; assoc prof, 69-74, PROF PHYSICS, WEST CHESTER STATE COL, 74- Mem: Am Asn Physics Teachers. Res: Atomic emission spectra. Mailing Add: Dept of Physics West Chester State Col West Chester PA 19380

YARRANTON, GEORGE ANTHONY, b London, Eng, July 3, 40; m 62; c 1. PLANT ECOLOGY, PLANT MORPHOLOGY. Educ: Univ London, BSc, 61; Univ Exeter, PhD(bot), 65. Prof Exp: Sci officer, Scottish Marine Biol Asn, 63-65; asst prof, Scarborough Col, 65-72, ASSOC PROF BOT, UNIV TORONTO, 72- Concurrent Pos: Grants, Nat Res Coun, 65- Mem: Am Bryol Soc; Can Bot Soc; Brit Ecol Soc. Res: Mathematical models of vegetation; plant succession; ecological strategy and population dynamics of plants; quantitative analysis of morphological variation in plants, community dynamics; forest regeneration. Mailing Add: Dept of Bot Univ of Toronto Toronto ON Can

YARUS, MICHAEL J, b Pikeville, Ky, Mar 2, 40; m 62; c 2. MOLECULAR BIOLOGY, BIOCHEMISTRY. Educ: Johns Hopkins Univ, BA, 60; Calif Inst Technol, PhD(biophys), 66. Prof Exp: USPHS & NIH fels biochem, Stanford Univ, 65-67; asst prof, 67-74, ASSOC PROF MOLECULAR, CELLULAR & DEVELOP BIOL, UNIV COLO, BOULDER, 74- Concurrent Pos: USPHS & NIH grant, 68- Mem: AAAS. Res: Minute viruses; transfer DNA; mammalian embryogeny. Mailing Add: Dept of Chem Univ of Colo Boulder CO 80302

YARWOOD, CECIL EDMUND, b Sumas, Wash, Sept 16, 07; m 36; c 5. PLANT PATHOLOGY. Educ: Univ BC, BSA, 29; Purdue Univ, MS, 31; Univ Wis, PhD(plant path), 34. Prof Exp: Plant dis investr, Div Bot, Can Dept Agr, 28-29; asst bot, Purdue Univ, 29-32; agent div forage crops & dis, Bur Plant Indust, USDA, 29-34; assoc plant path, Exp Sta, 34-35, instr plant path univ & jr plant pathologist, Exp Sta, 35-40, asst prof & asst plant pathologist, 40-46, assoc prof & assoc plant pathologist, 46-50, prof plant path & plant pathologist, 50-75, EMER PROF PLANT PATH, UNIV CALIF, BERKELEY, 75- Concurrent Pos: Guggenheim Mem fel, Univ Cambridge, 57-58 & Miller Inst Basic Res, 63-64; exchange scientist, Acad-Sci USA-USSR, 75. Mem: Fel AAAS; fel Am Phytopath Soc; Bot Soc Am; Am Soc Plant Physiol; Mycol Soc Am. Res: Obligate parasitism; diurnal periodicity; acquired immunity; predisposition; transmission of viruses; latency of viruses; selective accumulation of chemicals; heat therapy; translocated heat injury; physiological adaptation to heat; viruses in fungi. Mailing Add: Dept of Plant Path Univ of Calif Berkeley CA 94720

YARWOOD, EVANGELINE ALDERMAN, zoology, see 12th edition

YARZABAL, LUIS ALBERTO, b Melo, Uruguay, Jan 30, 38; m 62; c 3. MEDICAL MICROBIOLOGY Educ: Univ Montevideo, Med Dr, 64. Prof Exp: Assoc prof internal med, Fac Med Montevideo, 69-72; dir mycol div, Int Bact Santiago, Chile, 72-73; assoc prof microbiol, Fac Med Panama, 73-74; SR RESEARCHER IMMUNOL, FAC MED LILLE, FRANCE, 74- Concurrent Pos: Consult, Pan Am Health Orgn, 71-73. Mem: Uruguyan Soc Microbiol; Int Soc Human & Animal Mycol. Res: Immunology of mycosis-characterization of fungal antigens, studies on mechanisms of host immune reactions, serology. Mailing Add: Ctr Immunol & Parasitic Biol Pasteur Inst 20 Blvd Louis XIV 59012 Lille France

YASHON, DAVID, b Chicago, Ill, May 13, 35; c 3. NEUROSURGERY. Educ: Univ Ill, BSM, 58, MD, 60; FRCS(C), 69. Prof Exp: Instr neurosurg, Univ Chicago, 65-66; asst prof, Case Western Reserve Univ, 66-69; ASSOC PROF NEUROSURG, OHIO STATE UNIV, 69- Mem: Cong Neurol Surg; Am Asn Neurol Surg; Soc Univ Surg; Am Acad Neurol; Asn Acad Surg. Res: Cerebral physiology and metabolism during circulatory deficiency; spinal cord injury and metabolic effects. Mailing Add: 410 W Tenth Columbus OH 43212

YASKO, RICHARD N, b Conemaugh, Pa, Aug 29, 35; m 64. NUCLEAR PHYSICS. Educ: Pa State Univ, BS, 57, MS, 61, PhD(physics), 63. Prof Exp: Fel nuclear physics, Argonne Nat Lab, 63-64; asst prof physics, Villanova Univ, 64-67; ADV PHYSICIST, IBM CORP, ENDICOTT, 68- Mem: Am Phys Soc; Inst Elec & Electronics Eng; Am Vacuum Soc. Res: Surface physics; semiconductor device physics; ion implantation, spreading resistance diffusion profiling and capacitance; voltage testing of MO5 devices. Mailing Add: 1010 Knoll Dr Endwell NY 13760

YASMINEH, WALID GABRIEL, b Amman, Jordan, Jan 21, 31; US citizen; m 60; c 3. BIOCHEMISTRY. Educ: Am Univ Cairo, BSc, 53; Univ Minn, Minneapolis, MSc, 63, PhD(biochem), 66. Prof Exp: From jr scientist to assoc scientist pediat, 59-65, asst prof lab med, 67-72, ASSOC PROF LAB MED, SCH MED, UNIV MINN, MINNEAPOLIS, 72- Concurrent Pos: Grad Sch grant, Sch Med, Univ Minn, Minneapolis, 68- Mem: Am Chem Soc. Res: Mammalian constitutive heterochromatin and repetitive DNA, nature, origin, function and relation to disease. Mailing Add: 2057 Woodbridge St Paul MN 55113

YASPAN, ARTHUR, b Youngstown, Ohio, Feb 2, 18; m 44; c 1. MATHEMATICS, OPERATIONS RESEARCH. Educ: Univ Chicago, MS, 38; Case Inst Technol, PhD(opers res), 61. Prof Exp: Res asst physics, Sonar Anal Sect, Div War Res, Columbia Univ, 44-47; instr math, Western Reserve Univ, 52-54; res asst opers res, Case Inst Technol, 56-58; asst prof math, Polytech Inst Brooklyn, 63-67, assoc prof, 67-72; ASSOC PROF MATH, YORK COL, CITY UNIV NEW YORK, 74- Res: Applied probability, with special reference to gambling theory. Mailing Add: 83-19 141st St Jamaica NY 11435

YASSO, WARREN E, b New York, NY, Oct 19, 30; m 57; c 3. GEOLOGY, GEOMORPHOLOGY. Educ: Brooklyn Col, BS, 57; Columbia Univ, MA, 61, PhD(geomorphol), 64. Prof Exp: Instr earth sci, Adelphi Univ, 61-64; asst prof geol, Va Polytech Inst, 64-66; ASSOC PROF SCI EDUC, TEACHERS COL, COLUMBIA UNIV, 66- Mem: AAAS; Geol Soc Am; Int Asn Sedimentol; Nat Asn Geol Teachers; Nat Sci Teachers Asn. Res: Coastal and continental shelf geological processes; curriculum research in earth sciences. Mailing Add: 528 Franklin Turnpike Ridgewood NJ 07450

YASUDA, HIROTSUGU, b Kyoto, Japan, Mar 24, 30; m 68; c 2. POLYMER CHEMISTRY, PHYSICAL CHEMISTRY. Educ: Kyoto Univ, BS, 53; State Univ NY Col Environ Sci & Forestry, MS, 59, PhD(polymer & phys chem), 61. Prof Exp: Fel, State Univ NY Col Environ Sci & Forestry, Syracuse, 61; chemist, Camille Dreyfus Lab, Res Triangle Inst, 61-63; res assoc, Ophthalmic Plastic Lab, Mass Eye & Ear Infirmary, 63-64 & Cedars-Sinai Med Ctr, 64-65; guest scientist, Royal Inst Technol, Sweden, 65-66; head membrane & med polymer sect, Camille Dreyfus Lab, 66-75, MGR POLYMER DEPT, RES TRIANGLE INST, 75- Mem: Am Chem Soc; AAAS. Res: Preparation and characterization of polymers; transport phenomena through polymer membrane; biomedical application of polymers, membrane technology, plasma polymerization and surface modifications. Mailing Add: Res Triangle Inst PO Box 12194 Research Triangle Park NC 27709

YASUDA, STANLEY K, b Pahoa, Hawaii, Jan 7, 31; m 55; c 3. ANALYTICAL CHEMISTRY. Educ: Park Col, BA, 53; Kans State Univ, MS, 55, PhD(anal chem), 57. Prof Exp: STAFF MEM, LOS ALAMOS SCI LAB, 57- Concurrent Pos: Sr analyst, Chemagro Corp, 64. Mem: Fel Am Inst Chemists; Am Chem Soc. Res: Microanalytical methods for analysis of explosive and non-explosive materials, utilizing wet and instrumental techniques. Mailing Add: Los Alamos Sci Lab PO Box 1663 Los Alamos NM 87545

YASUMURA, SEIICHI, b New York, NY, Sept 28, 32; m 63; c 1. ENDOCRINOLOGY. Educ: Occidental Col, AB, 58; Univ Cincinnati, PhD(anat), 62. Prof Exp: Instr, 64-66, ASST PROF PHYSIOL, STATE UNIV NY DOWNSTATE MED CTR, 66- Concurrent Pos: Fel, State Univ Groningen, 62-63; NSF fel, State Univ NY Downstate Med Ctr, 63-64. Mailing Add: Dept of Physiol State Univ NY Downstate Med Ctr New York NY 11203

YASUNOBU, KERRY T, b Seattle, Wash, Nov 21, 25; m 52; c 1. BIOCHEMISTRY. Educ: Univ Wash, PhD(biochem), 54. Prof Exp: Res scientist, Univ Tex, 54-55; res assoc, Med Sch, Univ Ore, 55-58; asst prof chem, 58-62, assoc prof biochem, 62-64, PROF BIOCHEM, UNIV HAWAII, MANOA, 64- Concurrent Pos: NSF sr fel, 63-64; NIH sr fel, 71-72. Mem: Am Chem Soc; Am Soc Biol Chemists. Res: Enzymology, especially oxidative, heme-enzymes and proteolytic enzymes. Mailing Add: Dept Biochem-Biophys Univ of Hawaii at Manoa Honolulu HI 96822

YATES, ALBERT CARL, b Memphis, Tenn, Sept 29, 41; m 62; c 2. THEORETICAL PHYSICAL CHEMISTRY. Educ: Memphis State Univ, BS, 65; Ind Univ, Bloomington, PhD(chem physics), 68. Prof Exp: Res assoc chem, Univ Southern Calif, 68-69; from asst prof to assoc prof, Ind Univ, Bloomington, 69-74; ASSOC PROF CHEM & ASSOC UNIV DEAN FOR GRAD EDUC & RES, UNIV CINCINNATI, 74- Mem: Am Phys Soc; Am Chem Soc. Res: Collisions of charged particles with atomic and molecular systems; heavy-particle collisions; photo-absorption processes. Mailing Add: Dept of Chem Univ of Cincinnati Cincinnati OH 45221

YATES, ALFRED RANDOLPH, b Guelph, Ont, July 3, 28; m 52; c 4. FOOD MICROBIOLOGY. Educ: Univ Toronto, BSA, 50, MSA, 54; Univ Nottingham, PhD(microbiol), 65. Prof Exp: Control chemist, Cow & Gate, Ltd, 55-57; res officer food microbiol, Microbiol Div, Can Dept Agr, 57-59; teacher high schs, Ont, 59-61; res officer food microbiol, Plant Res Inst, 61-62, RES SCIENTIST, FOOD RES INST, CAN DEPT AGR, 62- Concurrent Pos: Food microbiol expert, Food & Agr Orgn UN, Food Technol Ctr, Malaysia, 72-73. Mem: Am Soc Microbiol; Can Inst Food Sci & Technol. Res: Dairy microbiology; physiology and ecology of bacteria and moulds. Mailing Add: Food Res Inst Res Br Can Dept Agr Cent Exp Farm Ottawa ON Can

YATES, ANN MARIE, b Ogdensburg, NY, Sept 29, 40. ANALYTICAL

CHEMISTRY. Educ: St Lawrence Univ, BS, 62; Ariz State Univ, PhD(chem), 66. Prof Exp: Asst prof anal chem, Ariz State Univ, 66-67; NIH res assoc inorg chem, Univ Pittsburgh, 67-68; dir labs, Chemalytics Inc, 68-74; MEM CHEM FAC, MARICOPA COUNTY COMMUNITY COL DIST, 76- Mem: AAAS; Am Chem Soc. Res: Microanalytical techniques.

YATES, CLAIRE HILLIARD, b Cornwall, Ont, Mar 27, 20; m 42; c 3. ANALYTICAL CHEMISTRY, PHARMACEUTICAL CHEMISTRY. Educ: Sir George Williams Col, BSc, 46; McGill Univ, PhD(biochem), 51. Prof Exp: Prod chemist, Charles E Frosst & Co, Montreal, 41-47; res chemist, 48-69, ANAL UNIT HEAD, PHARM RES DEPT, MERCK FROSST LABS, 69- Concurrent Pos: Lectr, Sir George Williams Col, 48-61. Mem: Chem Inst Can. Res: Steroids; synthesis of radioactive organic substances; analyses of drug formulations; stability of drugs in dosage forms. Mailing Add: 994 Second Ave Verdun PQ Can

YATES, FRANCIS EUGENE, b Pasadena, Calif, Feb 26, 27; m 49; c 5. PHYSIOLOGY. Educ: Stanford Univ, BA, 47, MD, 51. Prof Exp: Intern, Philadelphia Gen Hosp, 50-51; instr physiol, Harvard Med Sch, 55-57, assoc, 57-59, asst prof, 59-60; from assoc prof to prof, Stanford Univ, 60-69, actg exec head dept, 64-69; PROF BIOMED ENG, GRAD CTR ENG SCI, UNIV SOUTHERN CALIF, 69- Concurrent Pos: Res fel physiol, Harvard Univ, 53-55; Markle scholar med sci, Harvard Med Sch, 59; mem, physiol training comt, Nat Inst Gen Med Sci, 64-70, mem med scientist training prog comt, 71-73; vis prof, Stanford Univ, 69-; sect ed endocrinol & metab, Am J Physiol, 69-74; managing ed, Annals Biomed Eng, 71-74; consult prin scientist, Alza Corp; mem sci info prog adv comt, Nat Inst Neurol & Communicative Dis & Stroke, 76-; managing ed, Am J Physiol, Regulation, Integration on Adaptation, 76- Honors & Awards: Upjohn Award, Endocrine Soc, 62. Mem: AAAS; Biomed Eng Soc (pres, 74-75); Am Physiol Soc; Endocrine Soc; NY Acad Sci. Res: Metabolism and inactivation of adrenal cortical hormones; analysis of endocrine feedback systems. Mailing Add: Dept of Biomed Eng Sci Hall 375 Univ of Southern Calif Los Angeles CA 90007

YATES, GEORGE KENNETH, b Chicago, Ill, Sept 24, 25; m 52; c 4. SPACE PHYSICS. Educ: Harvard Univ, AB, 50; Univ Chicago, MS, 55, PhD(physics), 64. Prof Exp: RES PHYSICIST, AIR FORCE CAMBRIDGE RES LABS, 64- Mem: Am Phys Soc; Am Geophys Union. Res: Physics of the near space environment, especially particle radiation. Mailing Add: AFCRL-PHE Air Force Cambridge Res Labs L G Hanscom AFB Bedford MA 01731

YATES, HARRIS OLIVER, b Paducah, Ky, Apr 14, 34; m 54; c 3. BIOLOGY. Educ: David Lipscomb Col, BA, 56; George Peabody Col, MA, 57; Vanderbilt Univ, PhD(biol), 65. Prof Exp: Instr, 57-59, from asst prof to assoc. prof, 63-68, PROF BIOL & CHMN DEPT, DAVID LIPSCOMB COL, 68- Res: Experimental plant taxonomy. Mailing Add: Dept of Biol David Lipscomb Col Nashville TN 37203

YATES, HARRY ORBELL, III, b Camden, NJ, May 15, 31; m 58; c 3. FOREST ENTOMOLOGY. Educ: Univ Maine, BS, 54; Duke Univ, MF, 58; Ohio State Univ, PhD(entom), 64. Prof Exp: Entomologist, US Forest Serv, Ga, 58-60, Ohio, 60-63, ENTOMOLOGIST, US FOREST SERV, GA, 63- Mem: Entom Soc Am. Res: Terminal feeding moths of the genus Rhyacionia; influence of pine oleoresin on insect attack; seed and cone insects. Mailing Add: Forestry Sci Lab Carlton St Athens GA 30602

YATES, JEROME DOUGLAS, b Center Point, Ark, Jan 5, 35; m 69; c 3. POULTRY NUTRITION. Educ: Univ Ark, BSA, 58, MS, 59; Mich State Univ, PhD(poultry nutrit), 64. Prof Exp: Nutrit technician, Mich State Univ, 59-63; res assoc nutrit & food sci, 64-68, RES SCIENTIST, CAMPBELL SOUP CO, 69- Mem: Poultry Sci Asn; World Poultry Sci Asn. Res: Poultry nutrition emphasizing the influence of nutrients and other dietary components on quality of poultry meat; mineral and amino acid nutrition. Mailing Add: Campbell Inst for Agr Res Fayetteville AR 72701

YATES, JOHN THOMAS, JR, b Winchester, Va, Aug 3, 35; m 58; c 2. PHYSICAL CHEMISTRY, SURFACE CHEMISTRY. Educ: Juniata Col, BS, 56; Mass Inst Technol, PhD, 60. Prof Exp: Res assoc chem, Mass Inst Technol, 60; instr & asst prof, Antioch Col, 60-63; Nat Res Coun-Nat Bur Standards res assoc, 63-65; staff mem, Phys Chem Div, 65-74, ACTG CHIEF, SURFACE PROCESSES & CATALYSIS SECT, NAT BUR STANDARDS, 74- Concurrent Pos: Consult, Westgate Labs, Ohio, 63; vis examr, Swarthmore Col, 66; sr vis fel, Univ EAnglia, 70-71 & 72; trustee, Am Vacuum Soc, 74; Sherman Fairchild scholar, Calif Inst Technol, 75. Honors & Awards: Silver Medal, Dept of Com, 73. Mem: Am Chem Soc; Am Phys Soc; Am Vacuum Soc. Res: Spectra of adsorbed molecules; metal carbonyls; heterogeneous catalysis; kinetics of adsorption and desorption; isotopic mixing in the chemisorbed layer; electron impact studies of adsorbed species; electronic properties of the chemisorbed layer. Mailing Add: Sur Processes & Catalysis Sect Nat Bur of Standards Washington DC 20234

YATES, KEITH, b Preston, Eng, Oct, 22, 28; Can citizen; m 53; c 3. PHYSICAL ORGANIC CHEMISTRY. Educ: Univ BC, BA, 56, MSc, 57, PhD(org chem), 59; Oxford Univ, DPhil(phys chem), 61. Prof Exp: From asst prof to assoc prof chem, 61-67, asst dean, Sch Grad Studies, 67-70, PROF CHEM, UNIV TORONTO, 67-, CHMN DEPT, 74- Concurrent Pos: Ed, Can J Chem, 74- Res: Physical and theoretical organic chemistry; acidity functions and reaction mechanisms. Mailing Add: Dept of Chem Univ of Toronto Toronto ON Can

YATES, KENNETH PIDCOCK, b Bucks Co, Pa, Jan 5, 20; m 46; c 2. CHEMICAL PHYSICS. Educ: Col Wooster, BA, 41; Ohio State Univ, MA, 43, PhD(physics), 45. Prof Exp: Asst math, Col Wooster, 39-41; asst physics, Ohio State Univ, 41-42, instr, 42-44, res assoc antenna lab, Res Found, 44-46; res physicist, Pure Oil Co, 46-47, group leader physics res, 47-49, supvr & sr res physicist, Physics Sect, 49-60, res assoc, 61-65; res assoc, Res Ctr, Union Oil Co Calif, 66-75; RETIRED. Mem: Am Phys Soc; Am Chem Soc. Res: Molecular chemical physics; analytical spectroscopies; x-ray and radiotracers applications and safety; industrial electrostatics; electrical physics; infrared spectroscopy; magnetic susceptibility; instrumentation; catalyst research. Mailing Add: 3651 Coronado Dr Fullerton CA 92635

YATES, LELAND MARSHALL, b Stevensville, Mont, Feb 11, 15; m; c 4. PHYSICAL CHEMISTRY. Educ: Mont State Univ, BA, 38, MA, 40; Wash State Univ, PhD(chem), 55. Prof Exp: Instr chem & physics, Custer County Jr Col, 40-42 & 45-47; instr chem, Univ Mont, 47-49; asst, Wash State Univ, 49-51; from instr to assoc prof, 51-71, PROF CHEM, UNIV MONT, 71- Mem: Am Chem Soc. Res: Equilibrium constants and thermodynamics of complex ions; analysis for small concentration of ions. Mailing Add: Dept of Chem Univ of Mont Missoula MT 59801

YATES, PAUL C, physical chemistry, inorganic chemistry, see 12th edition

YATES, PETER, b Wanstead, Eng, Aug 26, 24; m 50; c 3. ORGANIC CHEMISTRY. Educ: Univ London, BSc, 46; Dalhousie Univ, MSc, 48; Yale Univ, PhD(chem), 51. Prof Exp: Instr chem, Yale Univ, 51-52; from instr to asst prof, Harvard Univ, 52-60;

PROF CHEM, UNIV TORONTO, 60- Concurrent Pos: Sloan Found fel, 57-60; Merck Sharp & Dohme lectr, Chem Inst Can, 63; vis prof, Yale Univ, 66; con- sult, Ciba-Geigy Pharmaceut Div. Honors & Awards: Centennial Medal, Govt Can, 67. Mem: Am Chem Soc; The Chem Soc; Chem Inst Can; Royal Soc Can. Res: Structural, synthetic and mechanistic organic chemistry, including natural and photochemical products, aliphatic diazo and heterocyclic compounds. Mailing Add: Dept of Chem Univ of Toronto Toronto ON Can

YATES, RICHARD ALAN, biochemistry, see 12th edition

YATES, RICHARD LEE, b Red Oak, Iowa, June 15, 31; m 61; c 2. MATHEMATICS. Educ: Fla Southern Col, BS, 52; Univ Fla, MS, 54, PhD(math), 57. Prof Exp: Asst math, Univ Fla, 52-56; asst prof, Univ Houston, 57-60; from asst prof to assoc prof, Kans State Univ, 60-67; assoc prof & chmn math sect, 67-70, prof & acad dean, 70-75, EXEC ASST TO CHANCELLOR, PURDUE UNIV, 75- Mem: Math Asn Am. Res: Classical number theory; modern algebra; lattice theory. Mailing Add: Purdue Univ Calumet Campus Hammond IN 46323

YATES, ROBERT DOYLE, b Birmingham, Ala, Feb 28, 31; m 57; c 2. CYTOLOGY. Educ: Univ Ala, BS, 54, MS, 56, PhD(anat), 60. Prof Exp: Instr gross anat & neuroanat, Univ Tex Med Br, Galveston, 61-64, from asst prof to prof microanat, 64-70; PROF ANAT & CHMN DEPT, SCH MED, TULANE UNIV, LA, 72- Concurrent Pos: Fel, Med Ctr, Univ Ala, 61-62; NIH career res develop award, 64. Mem: AAAS; Am Soc Cell Biol; Am Asn Anat; Am Soc Neuropath. Res: Electron microscopic studies of reversible alterations in the organelles and inclusions of cells subjected to experimentally induced stresses. Mailing Add: Dept of Anat Sch of Med Tulane Univ New Orleans LA 70112

YATES, ROBERT EDMUNDS, b Bisbee, Ariz, Aug 15, 26; m 47; c 4. PHYSICAL CHEMISTRY. Educ: Univ Ariz, BS, 48, MS 49; Mich State Univ, PhD(phys chem), 52. Prof Exp: Asst chem, Mich State Univ, 49-51; res engr, Dow Chem Co, 52-58; res chemist, Aerojet-Gen Corp, 58-61 & Rocket Power, Inc, 61-65; res chemist, Aerojet-Gen Corp, Calif, 66-67, chem specialist, 67-71; PHYSICIST, McCLELLAN AFB, SACRAMENTO, 71- Mem: Am Chem Soc. Res: Boron, fluorine and high temperature chemistry; thermodynamics and spectroscopy. Mailing Add: 7313 Pine Grove Way Folsom CA 95630

YATES, SHELLY GENE, b Altus, Okla, Feb 29, 32; m 54; c 4. CHEMISTRY. Educ: Southwestern State Col, Okla, BS, 56; Okla State Univ, MS, 58. Prof Exp: RES CHEMIST, NORTHERN REGIONAL RES LAB, AGR RES SERV, USDA, 58- Mem: Am Chem Soc. Res: Natural products, mycotoxins; isolation and characterization. Mailing Add: 5619 N Plaza Dr Peoria IL 61614

YATES, STEVEN WINFIELD, b Memphis, Mo, Apr 19, 46; m 75. NUCLEAR CHEMISTRY, RADIOCHEMISTRY. Educ: Univ Mo-Columbia, BS, 68; Purdue Univ, Lafayette, PhD(chem), 73. Prof Exp: Res asst chem, Purdue Univ, Lafayette, 71-73; fel, Argonne Nat Lab, 73-75; ASST PROF CHEM, UNIV KY, 75- Mem: Am Chem Soc; Am Phys Soc. Res: Lavel structures of transitional and deformed nuclei; heavy-ion reactions; inelastic scattering and transfer reactions; nuclear isomerism and high-spin phenomena; neutron induced reactions; x-ray fluorescence. Mailing Add: Dept of Chem Univ of Ky Lexington KY 40506

YATES, VANCE JOSEPH, b Smithville, Ohio, Oct 25, 17; m 42; c 4. VETERINARY VIROLOGY. Educ: Ohio State Univ, BSc, 40, DVM, 49; Univ Wis, PhD, 60. Prof Exp: Instr high sch, Ohio, 40-41; asst prof animal path, 49-50, assoc prof & assoc res prof, 51-55, PROF ANIMAL PATH & RES PROF, UNIV RI, 55-, HEAD DEPT, 51- Concurrent Pos: Mem temp staff, Rockefeller Found, 63-64; vis prof dept exp biol, Baylor Col Med, 71-72. Mem: Am Vet Med Asn; Am Asn Avian Path. Res: Avian virology and pathology; oncogenicity of avian adenoviruses. Mailing Add: Dept of Animal Path Univ of RI Kingston RI 02881

YATES, WESLEY ROSS, b Omaha, Nebr, July 19, 30; m 56; c 3. PHYSICAL CHEMISTRY. Educ: Univ Omaha, BA, 56; Univ Hawaii, PhD(chem), 64. Prof Exp: SR RES SCIENTIST, TECH CTR, AVERY PRODS CORP, 68- Mem: Am Chem Soc. Res: Kinetics of free radical reactions; surface and colloid chemistry; mechanical properties of polymers; polymer analytical chemistry. Mailing Add: 8537 Keokuk Ave Canoga Park CA 91306

YATES, WILLARD F, JR, b Findlay, Ohio, June 20, 30; m 65; c 1. PLANT TAXONOMY, CYTOGENETICS. Educ: Eastern Ill Univ, BS, 58; Ind Univ, MA, 60, PhD(bot), 67. Prof Exp: Instr biol, Cumberland Col, 60-62; asst prof, Ball State Univ, 65-67; ASSOC PROF BOT, BUTLER UNIV, 67- Mem: AAAS; Bot Soc Am; Am Soc Plant Taxon; Torrey Bot Club; Am Inst Biol Sci. Res: Plant cytotaxonomy; phytochemistry. Mailing Add: Dept of Bot Butler Univ Indianapolis IN 46208

YATSU, EIJU, b Uchihara, Japan, July 10, 20; m 46; c 3. MINERALOGY, GEOLOGY. Educ: Tokyo Bunrika Univ, BSc, 45; Univ Tokyo, DSc, 57. Prof Exp: Asst prof earth sci, Chuo Univ, Tokyo, 54-57, prof eng geol, 58-66; vis prof geomorphology, Univ Ottawa, 66-67; assoc prof, 67-69, PROF GEOMORPHOLOGY, UNIV GUELPH, 69- Concurrent Pos: Vis assoc prof, La State Univ, 65-66. Mem: Am Geophys Union; Clay Minerals Soc; Geochem Soc. Res: Weathering of rocks and minerals; clay minerals; mechanical behaviors of unconsolidated rocks. Mailing Add: Dept of Geog Univ of Guelph Guelph ON Can

YATSU, FRANK MICHIO, b Los Angeles, Calif, Nov 28, 32; m 55; c 1. NEUROLOGY. Educ: Brown Univ, AB, 55; Case Western Reserve Univ, MD, 59. Prof Exp: From asst prof to assoc prof neurol, Univ Calif Med Ctr, San Francisco, 67-75, vchmn dept, 73-75; PROF NEUROL & CHMN DEPT, UNIV ORE HEALTH SCI CTR, 75- Concurrent Pos: Chief neurol serv, San Francisco Gen Hosp, 69-75; mem cardiovasc A res study comt, Am Heart Asn, 74-77; mem neurol disorders prog, Proj A Rev Comt, Nat Inst Neurol & Commun Disorders & Stroke, NIH, 75-79; mem adv coun, Epilepsy Ctr of Ore, 75- Mem: Am Acad Neurol; Am Neurol Asn; Am Soc Neurochem; Int Soc Neurochem. Res: Brain ischemia and atherosclerosis. Mailing Add: Dept of Neurol Univ of Ore Health Sci Ctr Portland OR 97201

YATSU, LAWRENCE Y, b Pasadena, Calif, Aug 2, 25; m 54; c 2. PLANT PHYSIOLOGY. Educ: Mich State Univ, BS, 49; Univ Calif, MS, 50; Cornell Univ, PhD, 60. Prof Exp: Chemist, Strong, Cobb & Co, 54-55; res assoc, Cornell Univ, 55-60; plant physiologist, Field Lab Tung Invest, USDA, 60-61, RES CHEMIST, SOUTHERN REGIONAL RES CTR, USDA, 61- Concurrent Pos: Adj assoc prof biol, Tulane Univ, 73- Mem: AAAS; Am Chem Soc; Bot Soc Am; Am Soc Plant Physiol; Am Inst Biol Scientists. Res: Cell biology; biochemistry. Mailing Add: 7611 Dalewood Rd New Orleans LA 70126

YATVIN, MILTON B, b New Brunswick, NJ, Nov 12, 30; m 52; c 3. PHYSIOLOGY, RADIOBIOLOGY. Educ: Rutgers Univ, BS, 52, MS, 54, PhD(endocrinol, reproductive physiol), 62. Prof Exp: Instr dairy sci, Rutgers Univ, 55-56; lectr reproductive physiol, Univ PR, 57-59; from instr to assoc prof, 63-71, PROF RADIOBIOL, MED SCH, UNIV WIS-MADISON, 71- Concurrent Pos: NIH fel endocrinol, Rutgers Univ, 62-63. Mem: Biophys Soc; Radiation Res Soc; Soc Exp Biol & Med; Am Physiol Soc. Res: Cell damage and repair after exposure to ionizing radiation; nucleic acid-membrane relationships in cells. Mailing Add: Depts of Radiol & Path Med Sch Univ of Wis Madison WI 53706

YAU, SHING-TUNG, b Kwuntung, China, Apr 4, 49. MATHEMATICS. Educ: Univ Calif, Berkeley, PhD(math), 71. Prof Exp: Fel, Inst Advan Study, Princeton Univ, 71-72; asst prof math, State Univ NY, Stony Brook, 72-73; vis asst prof, 73-74, ASSOC PROF MATH, STANFORD UNIV, 74- Mem: Am Math Soc. Res: Differential geometry. Mailing Add: Dept of Math Stanford Univ Stanford CA 94305

YAU, WALLACE WEN-CHUAN, b Shianghai, China, Feb 20, 37; US citizen; m 61; c 2. ANALYTICAL CHEMISTRY, POLYMER PHYSICS. Educ: Nat Taiwan Univ, BS, 59; Univ Mass, PhD(phys chem), 66. Prof Exp: RES CHEMIST ANAL CHEM, CENT RES & DEVELOP DEPT, E I DU PONT DE NEMOURS & CO, INC, 65- Mem: Am Chem Soc. Res: Studies of polymer structures and properties using chromatographic, optical and mechanical characterization techniques. Mailing Add: Cent Res & Develop Dept E I du Pont de Nemours & Co Wilmington DE 19898

YAVERBAUM, SIDNEY, b New York, NY, Jan 28, 23; m 67; c 2. MEDICAL MICROBIOLOGY, IMMUNOCHEMISTRY. Educ: Univ Pa, PhD(med microbiol), 52. Prof Exp: Asst, Univ Pa, 51-52, res assoc, 52-53; fel microbiol, Boyce Thompson Inst Plant Res, 54-55; res med bacteriologist, Bio-Detection Br, Phys Defense Div, US Dept Army, Ft Detrick, 55-70; SR RES BIOLOGIST, CORNING GLASS WORKS, 70- Mem: AAAS; Am Soc Microbiol; Am Chem Soc; Sigma Xi; NY Acad Sci. Res: Cytology of yeasts, fungi and bacteria; genetics and nutrition of bacteria; assay of fungicides; physiology of aerobic sporeforming bacteria; biochemical composition of microorganisms; radioactive antibodies; solid-phase radioimmunoassay; immobilized enzyme research; automated instrumentation for microbiology. Mailing Add: Bio-Org Technol Corning Glass Works Corning NY 14830

YAVIN, AVIVI I, b Haifa, Israel, June 20, 28; m 51; c 2. PHYSICS. Educ: Hebrew Univ, Jerusalem, MSc, 54; Univ Wash, PhD(physics), 58. Prof Exp: Teacher physics & math, Hebrew Gym, Jerusalem, 51-53; asst physics grad sch, Univ Wash, 54-58; res assoc, Univ Ill, Urbana, 58-59, res asst prof, 59-60; physicist, Israeli AEC, 60-62; from asst prof to prof physics, Univ Ill, Urbana, 62-70; PROF PHYSICS, TEL-AVIV UNIV, 69-, DEAN FAC EXACT SCI, 71- Concurrent Pos: Consult cyclotron, Hebrew Univ, Jerusalem & Univ Grenoble, 60-62; nucleus res, Saclay, France & Los Alamos, NMex. Mem: Fel Am Phys Soc. Res: Nuclear research with cyclotrons, betatrons, linacs and reactors; accelerator design. Mailing Add: Dept of Physics & Astron Tel-Aviv Univ Tel-Aviv Israel

YAVORSKY, JOHN MICHAEL, b Renovo, Pa, June 11, 19. WOOD TECHNOLOGY. Educ: State Univ NY, BS, 42, MS, 47, PhD(wood eng), 55. Prof Exp: Res assoc & assoc prof wood utilization, State Univ NY Col Forestry, Syracuse, 48-56; chief wood utilization sect, Forestry Div, Food & Agr Orgn, UN, Rome, Italy, 57-63, proj mgr forestry proj, Lima, Peru, 63-67, sr proj officer, UN Spec Fund, NY, 67; PROF FORESTRY, STATE UNIV NY COL ENVIRON SCI & FORESTRY, SYRACUSE, 67-, DEAN SCH CONTINUING EDUC, 73- Mem: Forest Prod Res Soc; Soc Am Foresters; Soc Wood Sci & Technol. Res: Planning and supervision of continuing education activities in forestry and forest products technology; world forestry aspects of forest industries development. Mailing Add: State Univ of NY Col of Environ Sci & Forestry Syracuse NY 13210

YAW, KATHERINE EMILY, b Ft Smith, Ark, Oct 16, 14. MICROBIOLOGY. Educ: Univ Mich, AB, 36, MS, 37; Yale Univ, PhD(microbiol), 48. Prof Exp: Sr res scientist, Parke-Davis & Co, 38-44 & Brookhaven Nat Lab, 48-50; asst prof biol, Univ Del, 50-59; prof, Clarion State Col, 59-60; assoc prof, 60-65, PROF BIOL, WASHINGTON COL, 65- Mem: AAAS; Am Soc Microbiol. Res: Infectious diseases; immunology. Mailing Add: Dept of Biol Washington Col Chestertown MD 21620

YAWGER, ERNEST STANLEY, JR, food bacteriology, see 12th edition

YCAS, MARTYNAS, b Voronezh, Russia, Dec 10, 17; nat US; m 45; c 3. BIOLOGY. Educ: Univ Wis, BA, 47; Calif Inst Technol, PhD(embryol), 50. Prof Exp: Instr, Univ Wash, 50-51; biologist pioneering res labs, Qm Corps, US Dept Army, 51-56; asst prof microbiol, Sch Med, 56-65, PROF MICROBIOL, SCH MED, STATE UNIV NY UPSTATE MED CTR, 65- Res: Biochemical evolution; theoretical biology. Mailing Add: 109 Croyden Rd Syracuse NY 13224

YEADON, DAVID ALLOU, b New Orleans, La, Nov 10, 20; m 49; c 2. CHEMISTRY. Educ: Loyola Univ, La, BS, 40; Univ Detroit, MS, 42. Prof Exp: Asst & lab instr, Univ Detroit, 40-42; chemist, Gelatin Prod Corp, Mich, 42-43; chemist, Esso Standard Oil Co, La, res chemist, Esso Labs, 43-50; res chemist, Southern Regional Res Lab, Naval Store Div, USDA, 50; res chemist, Alpine Corp, Miss, 50-53; res chemist, Oilseed Crops Lab, 53-65, RES CHEMIST, COTTON FINISHES LAB, SOUTHERN REGIONAL RES LAB, USDA, 65- Honors & Awards: Superior Serv Awards, USDA, 67 & 73. Mem: Am Chem Soc; Sigma Xi; Am Asn Textile Chemists & Colorists. Res: Synthetic rubber; hydrocarbons; polymers, resins and coatings; chemistry, synthesis and applications of fats and oils; modifications to improve utilization of cotton; fire retardant cotton textiles. Mailing Add: 1460 Pressburg St New Orleans LA 70122

YEAGER, CHARLES LEVANT, b Rose City, Mich, Sept 12, 07; m 34. NEUROLOGY, PSYCHIATRY. Educ: Emanuel Missionary Col, Andrew, BS, 29; Col Med Evangelists, MD, 34; Univ Minn, MS, 40, PhD(neurol, psychiat), 44. Prof Exp: Dir clin psychiat & electroencephalog lab, Vet Admin Hosp, Waco, Tex, 46-47; lectr, 47-48, asst prof, 48-53, from asst clin prof to clin prof, 53-66, PROF PSYCHIAT IN RESIDENCE, SCH MED, UNIV CALIF, SAN FRANCISCO, 66-, PSYCHIATRIST & DIR ELECTROENCEPHALOG LAB, 47- Concurrent Pos: Consult, Vet Admin Hosp, 47-, Surgeon Gen, US Air Force, 49-57, Mt Zion Hosp, 51- & Marin Gen Hosp, 56-; dir electroencephalog labs, Calif State Hosps, 53- Mem: Am Electroencephalog Lab; Soc Exp Biol & Med; Sci Res Soc Am; Am Psychiat Asn; AMA. Res: Electroencephalography; neurophysiology; behavior. Mailing Add: Dept of Psychiat Univ of Calif Sch of Med San Francisco CA 94122

YEAGER, ERNEST BILL, b Orange, NJ, Sept 26, 24. ELECTROCHEMISTRY. Educ: Montclair State Col, BA, 45; Western Reserve Univ, MS, 46, PhD(phys chem), 48. Prof Exp: Asst physics & phys chem, 45-47, from instr to assoc prof chem, 48-58, actg chmn dept, 64-65, chmn dept, 69-72, chmn fac senate, 72-73, PROF CHEM, CASE WESTERN RESERVE UNIV, 58- Concurrent Pos: Consult, Union Carbide Corp, 55- & Gen Motors Corp, 70; mem comt undersea warfare, Nat Acad Sci-Nat Res Coun, 63-73; rep mem phys sci div, Nat Res Coun, 69-73; mem comn electrochem, Int Union Pure & Appl Chem; mem underwater sound adv group, Off Naval Res, 72-74. Honors & Awards: Acoust Soc Am Award, 56; Cert of Commendation, US Navy, 73. Mem: Fel AAAS; fel Acoust Soc Am (vpres, 67-68); Am Chem Soc; Int Soc Electrochem (vpres, 67-68, pres, 70-71); Electrochem Soc (vpres, 62-64, pres, 64-65). Res: Ultrasonics; electrode kinetics; electrolytes; relaxation spectroscopy. Mailing Add: Dept of Chem Case Western Reserve Univ Cleveland OH 44106

YEAGER, HOWARD LANE, b Pittsburgh, Pa, Dec 24, 43. ANALYTICAL CHEMISTRY. Educ: Univ Pittsburgh, BS, 65; Univ Wis-Madison, MS, 67; Univ Alta, PhD(chem), 69. Prof Exp: Lectr anal chem, Univ Wis, 69-70; ASST PROF CHEM, UNIV CALGARY, 70- Mem: AAAS; Chem Inst Can. Res: Ionic solvation in nonaqueous solvents; development of analytical methods utilizing nonaqueous solvents. Mailing Add: Dept of Chem Univ of Calgary Calgary AB Can

YEAGER, JOHN FREDERICK, b Orange, NJ, Jan 3, 27; m 57; c 4. CHEMISTRY. Educ: NJ State Teachers Col, Montclair, BA, 49; Western Reserve Univ, MS, 51, PhD(phys chem), 53. Prof Exp: Res chemist, Nat Carbon Co, 53-60, res group leader, Consumer Prod Div, 60-65, res tech mgr primary batteries, 65-67, tech mgr Leclanche cells, 67-72, dir battery develop lab, Battery Prod Div, 72-75, DIR TECHNOL, BATTERY PROD DIV, UNION CARBIDE CORP, 76- Mem: Inst Electrochem Soc; Electrochem Soc. Res: Electrochemistry; plating; batteries; fuel cells. Mailing Add: Union Carbide Corp Btry Prod Div PO Box 6056 Cleveland OH 44101

YEAGER, SANDRA ANN, b Philadelphia, Pa, Jan 4, 39. ORGANIC CHEMISTRY. Educ: Thiel Col, AB, 60; Univ NH, MS, 63, PhD(org chem), 68. Prof Exp: Asst prof chem, Hudson Valley Community Col, NY, 62-64; res asst, Children's Cancer Res Found, Boston, Mass, 68-69; asst prof chem, Pa State Univ, Mont Alto, 69-73; asst prof org & biochem, Wilson Col, 73-74; ASSOC PROF ORG & BIOCHEM, MILLERSVILLE STATE COL, 74- Mem: Am Chem Soc; World Future Soc; Nutrit Today Soc; Sigma Xi. Res: Analysis of biochemically important substances using varied chromatographic techniques. Mailing Add: Millersville State Col Millersville PA 17551

YEAGER, VERNON LEROY, b Williston, NDak, Nov 20, 26; m 47; c 4. ANATOMY. Educ: Minot State Col, BS, 49; Univ NDak, PhD(anat), 55. Prof Exp: Teacher sci, Garrison High Sch, 49-51; prof anat, Univ NDak, 55-67; assoc prof, St Louis Univ, 67-68; Rockefeller Found vis prof, Mahidol Univ, Thailand, 68-71; PROF ANAT, ST LOUIS UNIV, 71- Concurrent Pos: NSF fel, Northwestern Univ, Chicago, 60; NIH grant, Univ NDak, 60-67. Mem: Am Asn Anat; Sigma Xi; Am Soc Cell Biol. Res: Pathology of connective tissues. Mailing Add: Dept of Anat St Louis Univ Sch of Med St Louis MO 63104

YEAKEL, ALLEN EGGER, b Fair Oaks, Pa, Mar 21, 23; c 4. ANESTHESIOLOGY. Educ: Carnegie Inst Technol, BS, 44; Univ Pa, MD, 51. Prof Exp: Pvt pract, Pa, 53-59; from instr to prof anesthesia, Med Ctr, WVa Univ, 61-70; PROF ANESTHESIA & CHMN DEPT, HERSHEY MED CTR, PA STATE UNIV, 70- Mem: Am Soc Anesthesiol; Int Anesthesia Res Soc. Mailing Add: Dept of Anesthesia Pa State Univ Hershey Med Ctr Hershey PA 17033

YEAKEY, ERNEST LEON, b Sikeston, Mo, Aug 5, 34; m 61; c 2. ORGANIC CHEMISTRY. Educ: Southeast Mo State Col, BS, 56; State Univ, Iowa, PhD(org chem), 60. Prof Exp: Res chemist, 60-67, SUPVR EXPLOR RES, JEFFERSON CHEM CO, 67- Mem: Am Chem Soc. Res: Hydrogenation of nitriles to amines; reductive amination of alcohols to amines; synthetic routes to alpha olefins; catalytic synthesis of ethyleneamines. Mailing Add: Jefferson Chem Co PO Box 4128 N Austin Sta Austin TX 78751

YEANDLE, STEPHEN SAFFORD, b Bayonne, NJ, Sept 28, 29; m 66. BIOPHYSICS. Educ: Cornell Univ, BA, 51; Johns Hopkins Univ, PhD, 57. Prof Exp: NIH fel, Yale Univ, 58-59; asst prof physics, George Washington Univ, 59-65; BIOPHYSICIST, NAVAL MED RES INST, 65- Mem: Biophys Soc. Res: Biophysics of visual excitation; neurophysiology; biological applications of thermodynamics. Mailing Add: 4800 Oxford St Garrett Park MD 20766

YEARDLEY, NELSON PAUL, b Parkersburg, WVa, Sept 18, 11; m 40; c 2. MATHEMATICS. Educ: La State Univ, BS, 36, MS, 38; Lehigh Univ, MA, 40; Univ Cincinnati, PhD(math), 49. Prof Exp: Asst prof math, The Citadel, 42-44; asst prof, Univ of the South, 44-46; instr, Univ Ky, 46-47; asst prof, Purdue Univ, 49-51; asst prof, Iowa State Univ, 51-52; PROF MATH, THIEL COL, 52- Res: Laguerre series in the complex plane. Mailing Add: Dept of Math Thiel Col Greenville PA 16125

YEARGERS, EDWARD KLINGENSMITH, b Houma, La, Apr 27, 38; m 75; c 2. BIOPHYSICS. Educ: Ga Inst Technol, BS, 60; Emory Univ, MS, 62; Mich State Univ, PhD(biophys), 66. Prof Exp: US AEC res fel radiation physics, Oak Ridge Nat Lab, 66-67; NIH res fel theoret chem, Czech Acad Sci, 67; asst prof biol, 68-72, ASSOC PROF BIOL, GA INST TECHNOL, 72- Mem: Biophys Soc. Res: Molecular biophysics; protein structure; microwave irradiation of proteins. Mailing Add: Sch of Biol Ga Inst Technol Atlanta GA 30332

YEARIAN, HUBERT JOSE, physics, see 12th edition

YEARIAN, MASON RUSSELL, b Lafayette, Ind, July 5, 32; m 56, 65; c 3. NUCLEAR PHYSICS, HIGH ENERGY PHYSICS. Educ: Purdue Univ, BS, 54; Stanford Univ, MS, 56, PhD(physics), 61. Prof Exp: Res assoc physics, Univ Pa, 59-61; from asst prof to assoc prof, 61-71, PROF PHYSICS, STANFORD UNIV, 71-, DIR HIGH ENERGY PHYSICS LAB, 73- Mem: Am Phys Soc. Res: Electron scattering from nuclei and nucleons; nucleon form factors; charge distribution in nuclei; high energy particle physics. Mailing Add: Dept of Physics Stanford Univ Stanford CA 94305

YEARIAN, WILLIAM C, b Lake Village, Ark, May 20, 37; m 60; c 1. ENTOMOLOGY. Educ: Univ Ark, Fayetteville, BS, 60, MS, 61; Univ Fla, PhD(entom), 66. Prof Exp: From asst prof to assoc prof, 65-74, 74- PROF ENTOM, UNIV ARK, FAYETTEVILLE, Mem: Soc Invert Path; Entom Soc Am. Res: Forest entomology; applied insect pathology. Mailing Add: Dept of Entom Univ of Ark Fayetteville AR 72701

YEARICK, ELISABETH STELLE, b Spokane, Wash, July 1, 13. NUTRITION, BIOCHEMISTRY. Educ: Univ Wis, BS, 34, MS, 35; Univ Iowa, PhD(nutrit), 60; Am Bd Nutrit, cert human nutrit. Prof Exp: Asst dir dietetics, Duke Hosp, 48-53, asst prof, Duke Univ, 53; assoc prof nutrit, Univ Iowa, 53-57; assoc prof, WVa Univ, 60-66; PROF NUTRIT & FOODS, ORE STATE UNIV, 66- Mem: AAAS; Am Chem Soc; Am Dietetic Asn; Am Home Econ Asn. Res: Relationship of diet to serum glucose, lipids and amino acids. Mailing Add: Sch of Home Econ Ore State Univ Corvallis OR 97331

YEAROUT, PAUL HARMON, JR, b Coeur d'Alene, Idaho, Sept 15, 24; m 49. MATHEMATICS. Educ: Reed Col, BA, 49; Univ Wash, MS, 58, PhD, 61. Prof Exp: Instr math, Reed Col, 51-52; instr exten ctr, Portland State Col, 52-55; asst prof, Knox Col, 59-62; assoc prof 62-67, PROF MATH, BRIGHAM YOUNG UNIV, 67- Concurrent Pos: Consult sch med, Univ Wash, 59-64. Mem: Math Asn Am. Res: Abstract algebra; theory and structure of groups and rings. Mailing Add: 328 TMCB Dept of Math Brigham Young Univ Provo UT 84602

YEARY, ROGER A, b Cleveland, Ohio, Apr 26, 32; m 53; c 4. VETERINARY PHARMACOLOGY. Educ: Ohio State Univ, DVM, 56; Am Col Lab Animal Med, dipl; Am Bd Vet Toxicol, dipl. Prof Exp: Staff sr toxicologist, Charles Pfizer & Co, Inc, Conn, 60-61; chief toxicol sect, Lakeside Labs Div, Colgate-Palmolive Co, Wis, 61-65; exten vet toxicologist, Coop Exten Serv, 65-67, assoc prof vet physiol & pharmacol, 67-72, PROF VET PHYSIOL & PHARMACOL, COL VET MED, OHIO STATE UNIV, 72- Concurrent Pos: Toxicol consult, Lakeside Labs Div, Colgate-Palmolive Co, 65-67 & Minn Mining & Mfg Co, 67; NIH res grants, toxicol study sect, 73-77. Mem: AAAS; Am Asn Lab Animal Sci; Am Soc Pharmacol & Exp Therapeut; Am Vet Med Asn; Soc Toxicol. Res: Drug toxicology; perinatal pharmacology. Mailing Add: Dept of Vet Physiol & Pharmacol Ohio State Univ 1900 Coffey Rd Columbus OH 43210

YEATES, MAURICE, b UK, May 24, 38; m 62; c 2. URBAN GEOGRAPHY, ECONOMIC GEOGRAPHY. Educ: Univ Reading, BA, 60; Northwestern Univ, MA, 62, PhD(geog), 63. Prof Exp: Asst prof geog, Univ Fla, 63-65; from asst prof to assoc prof, 65-70, PROF GEOG, QUEEN'S UNIV, ONT, 70-, HEAD DEPT, 73- Concurrent Pos: Partic NSF Conf, Northwestern Univ, 64 & Univ Cincinnati, 68; consult, Area Develop Agency, Dept Indust, Ottawa, 66-68; Area Develop Agency fel, 67-68; Can Coun grant, 69-71, res & travel grant, 71-72; Queen's res award, 69 & 69-71; Ministry State Urban Affairs fel, 72-74; consult, Llewellyn-Davies, Weeks, Forester-Walker & Bor. Honors & Awards: Partic Award, Asn Am Geogr, 64. Res: Macro-urban problems; regional development; internal structure of cities. Mailing Add: Dept of Geog Queen's Univ Kingston ON Can

YEATMAN, CHRISTOPHER WILLIAM, b Port Pirie, Australia, Aug 6, 27; m 54; c 4. FOREST GENETICS. Educ: Univ Adelaide, BSc, 51; Australian Sch Forestry, Canberra, dipl, 51; Yale Univ, MF, 57, PhD(forest genetics), 66. Prof Exp: Forester, Dept Woods & Forests, SAustralia, 51; forest officer, Forestry Comn Gt Brit, 51-53; asst, Petawawa Forest Exp Sta, 53-54, res forest officer, 54-66, RES SCIENTIST, PETAWAWA FOREST EXP STA, 66- Concurrent Pos: Mem & secy, Comt Forest Tree Breeding Can, 55-66, exec secy, 66-68. Mem: Biomet Soc; Can Inst Forestry; Genetics Soc Can. Res: Silviculture; plantation establishment; tree breeding and forest genetics; genecology. Mailing Add: Petawawa Forest Exp Sta Can Forestry Serv Chalk River ON Can

YEATMAN, HARRY CLAY, b Ashwood, Tenn, June 22, 16; m 49; c 2. ZOOLOGY. Educ: Univ NC, AB, 39, MA, 42, PhD(zool), 53. Prof Exp: Asst zool, Univ NC, 39-42, instr, 47-50; from asst prof to assoc prof biol, 50-59, PROF BIOL, UNIV OF THE SOUTH, 59-, CHMN DEPT, 71- Concurrent Pos: Consult, US Nat Mus, 48-, Woods Hole Oceanog Inst, 60- & SEATO, US Army, Thailand, 66-; mem adv group, Tech Aqua Biol Sta. Mem: AAAS; Soc Syst Zool; Am Soc Limnol & Oceanog; Am Soc Ichthyologists & Herpetologists; Am Ornith Union. Res: Limnology; taxonomy and ecology of freshwater and marine copepods. Mailing Add: Dept of Biol Univ of the South Sewanee TN 37375

YEATMAN, JOHN NEWTON, b Washington, DC, Apr 30, 20; m 54; c 2. FOOD SCIENCE. Educ: Univ Md, BS, 44; Univ Calif, Los Angeles, MS, 48. Prof Exp: Plant physiologist, USDA, 44-47; plant physiologist, Chem Corps, US Dept Army, Ft Detrick, 48-53; res food technol & leader, Qual Eval Invest, Agr Res Serv, USDA, 54-68, dir color res lab, Mkt Qual Res Div, 68-71; res food technol, Bur Foods, Div Food Technol, Food & Drug Admin, 71-75; CONSULT FOOD STANDS-QUAL EVAL, 75- Mem: Inst Food Technol; Inter-Soc Color Coun; Am Soc Test & Mat. Res: Research and development of methods for standards improvement by objectively measuring by physical and chemical means identity and quality factors in processed fruits and vegetables and their products; food standards and quality evaluation; instruments and inspection lighting. Mailing Add: 11106 Cherry Hill Rd Adelphi MD 20783

YEATS, FREDERICK TINSLEY, b Gadsden, Ala, Apr 4, 42; m 69; c 2. BOTANY. Educ: Miss Col, BS, 64; Univ Miss, MS, 67; Univ SC, PhD(biol), 71. Prof Exp: ASSOC PROF BIOL, HIGH POINT COL, 69- Mem: Sigma Xi; Am Inst Biol Sci. Res: Developmental morphology in fern gametophytes; embryology in the genus Smilax. Mailing Add: Dept of Biol High Point Col High Point NC 27262

YEATS, ROBERT SHEPPARD, b Miami, Fla, Mar 30, 31; m 52; c 5. GEOLOGY. Educ: Univ Fla, AB, 52; Univ Wash, MS & PhD(geol), 58. Prof Exp: Exploitation engr, Shell Oil Co, 58-62, sr prod geologist, 62-64, sr geologist, 64-67; assoc prof geol, 67-71, PROF GEOL, OHIO UNIV, 72- Concurrent Pos: Consult, F Beach Leighton & Assocs, Calif & Energy Resources Br, US Geol Surv, 75; co-chief scientist, Deep Sea Drilling Proj, 73-75. Mem: AAAS; fel Geol Soc Am; Am Geophys Union; Am Asn Petrol Geol. Res: Structural evolution of Pacific continental margin of Americas; application of plate tectonics to petroleum accumulation; petrology of deep ocean basalts. Mailing Add: Dept of Geol Ohio Univ Athens OH 45701

YEATS, RONALD BRADSHAW, b Newcastle-upon-Tyne, Eng, Mar 17, 41; m 62; c 2. ORGANIC CHEMISTRY. Educ: Univ Durham, BSc, 62, PhD(org chem), 65. Prof Exp: Mayo fel, Univ Western Ont, 65-67, lectr org chem, 67-68; asst prof, 68-74, ASSOC PROF ORG CHEM & CHMN DEPT, BISHOP'S UNIV, 74-; ASST RESEARCHER, UNIV HOSP CTR, UNIV SHERBROOKE, 70- Mem: The Chem Soc; Chem Inst Can. Res: Terpene synthesis; synthetic aspects of organic photochemistry; mechanism of solvolysis reactions; microbiological oxidation reactions. Mailing Add: Dept of Chem Bishop's Univ Lennoxville PQ Can

YEATTS, FRANK RICHARD, b Altoona, Pa, Mar 5, 36; m 60; c 2. THEORETICAL PHYSICS. Educ: Pa State Univ, BS, 58; Univ Ariz, MS, 63, PhD(physics), 64. Prof Exp: Asst prof physics, 64-67, ASSOC PROF PHYSICS, COLO SCH MINES, 67- Mem: Am Geophys Union; Seismol Soc Am; Am Asn Physics Teachers. Res: Theoretical geophysics. Mailing Add: Dept of Physics Colo Sch of Mines Golden CO 80401

YEATTS, LEROY BROUGH, JR, b Newport, Pa, Mar 28, 22; m 49; c 3. PHYSICAL INORGANIC CHEMISTRY. Educ: Lebanon Valley Col, BS, 43; Cornell Univ, MS, 45, PhD, 48. Prof Exp: Asst prof chem, Lafayette Col, 48-52; CHEMIST, OAK RIDGE NAT LAB, 52- Mem: Am Chem Soc. Res: High temperature aqueous electrolyte solutions; tobacco smoke chemistry. Mailing Add: 633 Lake Shore Dr Kingston TN 37763

YEDINAK, PETER DEMERTON, b Bath, NY, Aug 20, 39; m 63; c 6. THEORETICAL PHYSICS, SOLID STATE PHYSICS. Educ: Union Col, BS, 62; Clark Univ, MA, 67, PhD(chem physics), 68. Prof Exp: Asst prof, 67-72, ASSOC PROF PHYSICS, WESTERN MD COL, 72- Concurrent Pos: Consult, Aberdeen Res & Develop Ctr, Ballistic Res Labs, 69- Mem: AAAS; Am Inst Physics; Am Asn Physics Teachers; Am Phys Soc; NY Acad Sci. Res: Theoretical lattice dynamics; determination of transition rates for acoustical energy absorption by molecules in a solid containing isotopic defects. Mailing Add: Dept of Physics Western Md Col Westminster MD 21157

YEDLOUTSCHNIG, RONALD JOHN, b Centralia, Wash, Jan 14, 30; m 56; c 2. VETERINARY MEDICINE. Educ: Wash State Univ, DVM, 54. Prof Exp: Pvt pract, 57-60; field vet, Animal Health Div, USDA, 60-65, foreign animal dis epidemiologist, 65-66, trainee exotic dis, 66-67, PRIN VET, PLUM ISLAND ANIMAL DIS LAB, USDA, 67- Res: Exotic animal diseases. Mailing Add: Plum Island Animal Dis Lab PO Box 848 Greenport NY 11944

YEE, KANE SHEE-GONG, b Kwangtung, China, Mar 26, 34; US citizen; m 62; c 2. APPLIED MATHEMATICS. Educ: Univ Calif, Berkeley, BS, 57, MS, 58, PhD(appl math), 63. Prof Exp: Asst prof math, Univ Fla, 66-68; assoc prof, 68-73, PROF MATH, KANS STATE UNIV, 73- Concurrent Pos: Consult, Lawrence Livermore Lab, Univ Calif, 66-, grant, 70- Mem: Soc Indust & Appl Math. Res: Mathematical physics; numerical solution to partial differential equations. Mailing Add: Dept of Math Kans State Univ Manhattan KS 66502

YEE, SINCLAIR SHEE-SING, b China, Jan 20, 37; US citizen; m 61; c 2. BIOENGINEERING, MICROELECTRONICS. Educ: Univ Calif, Berkeley, BS, 59, MS, 61, PhD(elec eng), 65. Prof Exp: Res engr, Lawrence Livermore Lab, 64-66; from asst prof to assoc prof elec eng, 66-74, PROF ELEC ENG & DIR MICROTECHNOL LAB, UNIV WASH, 74- Concurrent Pos: Consult, Lawrence Livermore Lab, 66-; NIH spec res fels, Case Western Reserve Univ, 72-73 & Bioeng Ctr, Univ Wash, 73-74. Mem: Am Phys Soc; Inst Elec & Electronics Eng. Res: Semiconductor physics and devices; bioinstrumentation; microelectronic devices. Mailing Add: Dept of Elec Eng Univ of Wash Seattle WA 98105

YEE, TIN BOO, b Canton, China, Feb 25, 15; US citizen. CHEMISTRY, CERAMICS. Educ: Ark State Col, BS, 38; Univ Ark, MS, 40; Univ Ill, AM, 50, PhD(chem), 54. Prof Exp: Chemist, Chem Warfare Serv, Huntsville Arsenal, Ala, 42-45; asst chemist, State Geol Surv, Ill, 45-55; RES CHEMIST, REDSTONE ARSENAL, 55- Concurrent Pos: Instr, Exten, Univ Ala, 60; vis res scientist, Union Indust Res Inst, Taiwan, 70-71. Mem: Am Chem Soc; Am Ceramic Soc. Res: Crystal growth studies by the thin film method in ceramic materials; solid rocket propellants; mutations in flowers and plants by radiations; material research in microelectronics. Mailing Add: Syst Design Prototype Develop Lab Missile Res & Develop Command Redstone Arsenal AL 35809

YEE, TUCKER TEW, b Toyshun, Canton, China, Mar 9, 36; US citizen; m 68; c 1. ORGANIC CHEMISTRY, PHYSICAL ORGANIC CHEMISTRY. Educ: Knox Col, Ill, BA, 60; Univ Mass, PhD(org chem), 64. Prof Exp: Res assoc, Princeton Univ, 64-65; res chemist, Eastern Lab, E I du Pont de Nemours & Co, Inc, 65-67; sr res chemist, Arco Chem Co Div, Atlantic Richfield Co, Pa, 67-72; RES CHEMIST, NAVAL WEAPONS CTR, 72- Mem: Sigma Xi; Am Chem Soc. Res: Chemistry of nitrogen containing heterocycles; radiochemical tracer technique; general organic syntheses; organic polymer syntheses and polymer applications. Mailing Add: 908 Sylvia Ave Ridgecrest CA 93555

YEEND, WARREN ERNEST, b Colfax, Wash, May 14, 36; m 64; c 1. GEOLOGY. Educ: Wash State Univ, BS, 58; Univ Colo, MS, 61; Univ Wis, PhD(geol), 65. Prof Exp: GEOLOGIST, US GEOL SURV, 65- Honors & Awards: Spec Achievement Award, US Geol Surv, 75. Mem: Geol Soc Am. Res: Surficial geology in an area of oil shade interest in western Colorado; geomorphology; gold bearing gravels of the Sierra Nevada; economic geology; engineering geology along the proposed Trans-Alaska pipeline; mapping and copper resource evaluation in southern Arizona. Mailing Add: US Geol Surv 345 Middlefield Rd Menlo Park CA 94025

YEH, BILLY KUO-JIUN, b Foochow, China, Aug 28, 37; m 65; c 3. CARDIOVASCULAR DISEASES, CLINICAL PHARMACOLOGY. Educ: Nat Taiwan Univ, MD, 61; Univ Okla, MS, 63; Columbia Univ, PhD(pharmacol), 67. Prof Exp: Intern med, Nat Taiwan Univ Hosp, 60-61; resident path, Med Ctr, Univ Okla, 63; teaching asst, Col Physicians & Surgeons, Columbia Univ, 64-67; asst resident, Emory Univ Affil Hosps, 68-69; staff cardiologist & chief sect clin pharmacol, Div Cardiol, Mt Sinai Med Ctr, Miami Beach, 69-71; asst prof med, 70-73, asst prof pharmacol, 72-73, CLIN ASST PROF MED, SCH MED, UNIV MIAMI, 73-; ASSOC DIR, DIV CLIN INVEST, MIAMI HEART INST, MIAMI BEACH, 73- Concurrent Pos: Fel, Univ Okla, 62-63; fel pharmacol, Col Physicians & Surgeons, Columbia Univ, 63-64; fel med, Sch Med, Emory Univ & Grady Mem Hosp, 67-68. Mem: Fel Am Col Physicians; fel Am Col Cardiol; Am Soc Pharmacol & Exp Therapeut; Am Physiol Soc; Am Fedn Clin Res. Mailing Add: 7110 SW 148th Terr Miami FL 33158

YEH, GEORGE CHIAYOU, b Kagi, Taiwan, Oct 3, 26; US citizen; m 57; c 4. PHYSICAL CHEMISTRY, CHEMICAL PHYSICS. Educ: Taiwan Univ, BSc, 50; Univ Tokyo, DEng, 53; Univ Toronto, MSc, 55, PhD(phys chem), 57. Prof Exp: Lectr, Japanese Engrs Union, Tokyo, 51-53; assoc prof chem eng, Auburn Univ, 57-61; assoc prof, 61-63, vpres acad affairs, 73, PROF CHEM ENG, VILLANOVA UNIV, 63-, DIR RES & PATENT AFFAIRS, 74- Concurrent Pos: Prin investr res grants, Petrol Res Fund, 63-65 & NASA, 65- Honors & Awards: Achievement Award, United Inventors & Scientists Am, 74. Mem: Fel Am Inst Chemists; Am Inst Chem Engrs; Am Chem Soc; Japanese Soc Chem Eng. Res: Reactions at interfaces; interfacial phenomena; catalysis; quantum mechanics; solid state chemistry; electrochemical analysis; liquids separation; gas separation; energy conversion; vapor engine. Mailing Add: Dept of Chem Eng Villanova Univ Villanova PA 19085

YEH, HERMAN JIA-CHAIN, b Taipei, Taiwan, Nov 15, 39. PHYSICAL CHEMISTRY. Educ: Cheng-Kung Univ, Taiwan, BS, 63; Univ Mass, Amherst, PhD(chem), 68. Prof Exp: Fel chem, Univ Mass, 68-69; vis fel, Lab Chem, 70-71, staff fel, 72-74, SR STAFF FEL, NAT INST ARTHRITIS, METABOLISM & DIGESTIVE DIS, 74- Mem: Am Chem Soc. Res: Nuclear magnetic resonance spectroscopy. Mailing Add: Lab of Chem Bldg 4 Rm B2-33 Nat Inst Arthritis Metabolism & Digestive Dis Bethesda MD 20014

YEH, HSIN-YANG, b Hsin-Chu, Formosa, Jan 1, 30; US citizen; m 67; c 2. MATHEMATICAL PHYSICS, PARTICLE PHYSICS. Educ: Nat Taiwan Univ, BS, 52; Kyushu Univ, MS, 57; Univ NC, Chapel Hill, PhD(theoret physics), 60. Prof Exp: Res physicist nuclear data group, Nat Acad Sci-Nat Res Coun, Washington, DC, 60-61; sr scientist, Edgerton, Germeshausen & Grier, Inc, 61-63; PROF PHYSICS, MOORHEAD STATE UNIV, 66- Mem: Am Phys Soc; NY Acad Sci; Int Soc Gen Relativity & Gravitation. Res: Quantum field theory; nuclear physics; group theory. Mailing Add: Dept of Physics & Astron Moorhead State Univ Moorhead MN 56560

YEH, HSU-CHI, b Taipei, Taiwan, Sept 30, 40; US citizen; m 66; c 2. ENVIRONMENTAL SCIENCES, MECHANICAL ENGINEERING. Educ: Nat Taiwan Univ, BS, 63; Univ Minn, MS, 67, PhD(mech eng), 72. Prof Exp: Teaching asst mech eng, Univ Minn, 64-65, res asst mech eng & aerosol physics, 65-72; res assoc, 72-73, RES SCIENTIST AEROSOL PHYSICS, INHALATION TOXICOL RES INST, LOVELACE FOUND, 73- Mem: Health Physics Soc; AAAS; Sigma Xi. Res: Aerosol science and technology; inhalation toxicity associated with inhaled aerosols and the particle deposition in mammalian lungs including mammalian airway morphometry. Mailing Add: Inhalation Toxicol Res Inst Lovelace Found PO Box 5890 Albuquerque NM 87115

YEH, JAMES JUI-TIN, b Tainan, Formosa, Apr 26, 27; US citizen. MATHEMATICS. Educ: Taiwan Univ, BS, 50; Univ Minn, MA, 54, PhD(math), 57. Prof Exp: Instr math inst technol, Univ Minn, 57-58 & Mass Inst Technol, 58-60; asst prof, Univ Rochester, 60-64; vis mem, Courant Inst Math Sci, NY Univ, 64-65; assoc prof, 65-68, PROF MATH, UNIV CALIF, IRVINE, 68- Mem: Am Math Soc. Res: Integration in function spaces; functional analysis; stochastic processes. Mailing Add: Dept of Math Univ of Calif Irvine CA 92664

YEH, KUO-CHEN, b Taiwan, Oct 1, 35; m 67; c 2. ORGANIC CHEMISTRY. Educ: Taiwan Norm Univ, BS, 58; Univ Ala, Tuscaloosa, MS, 64, PhD(chem), 67. Prof Exp: Fel, State Univ NY Albany, 67-68; SR GROUP LEADER, BLOCK DRUG CO, 68- Mem: AAAS; Am Chem Soc. Res: Organic synthesis, reaction mechanism. Mailing Add: 11 Seneca Rd Cranford NJ 07016

YEH, KWAN-NAN, b Taichung, Taiwan, Feb 27, 38; m 65; c 2. TEXTILE CHEMISTRY. Educ: Nat Taiwan Univ, BS, 61; Tulane Univ, La, MS, 65; Univ Ga, PhD(chem), 70. Prof Exp: Cotton Found res assoc fel, Nat Bur Standards, 70-72, res chemist, 72-73; ASST PROF TEXTILES & CONSUMER ECON, UNIV MD, COLLEGE PARK, 73- Mem: AAAS; Am Chem Soc; Am Asn Textile Chemists & Colorists. Res: Thermodynamics; thermochemistry; textile and polymer flammability; mechanism of flame retardant action; combustion of polymers. Mailing Add: Dept of Textiles & Consumer Econ Univ of Md College Park MD 20742

YEH, NOEL KUEI-ENG, b Malacca, Malaysia, Dec 15, 37; m 65; c 2. PARTICLE PHYSICS. Educ: Williams Col, BA, 61; Yale Univ, MS, 62, PhD(physics), 66. Prof Exp: Res assoc physics, Nevis Labs, Columbia Univ, 66-68, instr univ, 68-69; asst prof, 69-73, ASSOC PROF PHYSICS, STATE UNIV NY BINGHAMTON, 73- Mem: Am Phys Soc. Res: Properties of mesons and hyperons; resonances; weak and strong interactions of elementary particles. Mailing Add: Dept of Physics State Univ of NY Binghamton NY 13901

YEH, RAYMOND T, b Hunan, China, Nov 5, 37; m 66; c 1. COMPUTER SCIENCE. Educ: Univ Ill, BS, 61, MA, 63, PhD(math), 66. Prof Exp: Asst prof comput sci, Pa State Univ, 66-69; assoc prof, 69-74, PROF COMPUT SCI & ELEC ENG, UNIV TEX, AUSTIN, 74-, CHMN DEPT COMPUT SCI, 75- Concurrent Pos: Ed transactions software eng, Inst Elec & Electronics Eng. Mem: Am Math Soc; Asn Comput Mach; Inst Elec & Electronics Eng. Res: Software engineering; computation theory; programing methodology. Mailing Add: Dept of Comput Sci Univ of Tex Austin TX 78712

YEH, SAMUEL D J, b Kunming, China, Apr 23, 26; US citizen; m 59; c 2. BIOCHEMISTRY. Educ: Nat Defense Med Ctr, Shanghai, MD, 48; Johns Hopkins Univ, ScD(biochem), 60. Prof Exp: Instr med, Nat Defense Med Ctr, 48-53; asst resident, Lutheran Hosp, Md, 53-54; from instr to asst prof biochem, Sch Hyg & Pub Health, Johns Hopkins Univ, 60-63; instr, Med Col, Cornell Univ, 63-69; asst attend physician, Mem Hosp, New York City, 69-70 & 72-75; ASST PROF MED, MED COL, CORNELL UNIV, 69-; ASSOC ATTEND PHYSICIAN, MEM HOSP, NEW YORK CITY, 75- Concurrent Pos: Assoc, Sloan-Kettering Inst, 63-71 & 72- Mem: AAAS; Am Chem Soc; Am Inst Nutrit; Soc Nuclear Med; Am Col Nuclear Physicians. Res: Interrelationship between nutrients; intestinal absorption and marginal deficiencies; role of protein synthesis on metabolic functions; tumor localizing radionuclides. Mailing Add: 303 E 71st St Apt 4A New York NY 10021

YEH, SHU-YUAN, b Kwangtung, China, June 26, 26; US citizen; m 57; c 2. PHARMACEUTICAL CHEMISTRY, PHARMACOLOGY. Educ: Nat Defense Med Ctr, Taiwan, BS, 51; Univ Iowa, MS, 57, PhD(pharmaceut chem), 59. Prof Exp: Pharmacist, Army, Navy & Air Force Hosp, Taiwan, 51-52; clin, Off of President, Taiwan, 53-55; res asst pharm, Univ Iowa, 55-59, res assoc med, 59-63, from res instr to res asst prof drug metab, Col Med, 63-70; asst prof, Univ Ky, 71; PHARMACOLOGIST, ADDICTION RES CTR, NAT INST DRUG ABUSE, 72- Mem: AAAS; Am Pharmaceut Asn; Acad Pharmaceut Sci; Am Soc Pharmacol & Exp Med; NY Acad Sci. Res: Development of methods for detection, isolation and identification of drug and its metabolites in the biological fluids. Mailing Add: Addiction Res Ctr Nat Inst Drug Abuse PO Box 12390 Lexington KY 40511

YEH, YIN, b Chungking, China, Nov 1, 38; US citizen; m 61; c 2. QUANTUM ELECTRONICS, CHEMICAL PHYSICS. Educ: Mass Inst Technol, BS, 60; Columbia Univ, PhD(physics), 65. Prof Exp: Res assoc physics, Columbia Radiation Lab, Columbia Univ, 65-66; Lawrence Radiation Lab fel, Lawrence Livermore Lab, 66-67, sr physicist, 67-72; ASSOC PROF APPL SCI, UNIV CALIF, DAVIS, 72- Concurrent Pos: Lectr, St Mary's Col, Calif, 67-68; lectr, Univ Calif, Davis, 71-72; consult, Lawrence Livermore Lab, 72- Mem: AAAS; Am Phys Soc. Res: Application of laser spectroscopy to study dynamic phenomena in chemical physics, biophysics and solid state physics. Mailing Add: Dept of Appl Sci Univ of Calif Davis CA 95616

YEHLE, CLIFFORD OMER, b Syracuse, NY, Apr 9, 41; m 62; c 3. GENETICS, BIOCHEMISTRY. Educ: Syracuse Univ, BS, 62, MS, 65, PhD(microbiol), 67. Prof Exp: ASST PROF BIOCHEM & GENETICS, ROCKEFELLER UNIV, 74- Concurrent Pos: Am Cancer Soc fel genetics, Sch Med, Stanford Univ, 69-74. Mem: Am Soc Microbiol. Res: Differential expression of bacteriophage genomes in vegetative and sporulating Bacillus subtilis; nucleotide sequence relationships between phage and host genomes; DNA synthesis and genetic recombination in phage-infected cells. Mailing Add: Rockefeller Univ New York NY 10021

YEISER, ANDREW STURM, b Omaha, Nebr, Aug 5, 25; m 54; c 2. COMPUTER SCIENCE. Educ: Univ Calif, Berkeley, AB, 47. Prof Exp: Physicist, Univ Calif, Berkeley, 47-49; partner & gen mgr, Bay Instruments, 49; physicist, Western Regional Res Lab, USDA, 50-54; mem tech staff, Hughes Aircraft Co, 54-57; pres & gen mgr, Bytronics Corp, 57-58; proj mgr digital comput design, Ramo-Wooldridge Div, TRW, Inc, 58-63; mgr resource mgt systs, Autonetics Div, NAm Aviation, Inc, Calif, 63-68; pres, Gen Systs Industs, Inc, 68-72; PRES, ANDREW S YEISER & ASSOCS, 72- Concurrent Pos: Lectr, Univ Calif, Los Angeles, 50-55 & Univ Calif, Irvine, 55-; mgt info systs consult, UN Indust Develop Orgn, Turkey, 74- Mem: Am Phys Soc; Inst Elec & Electronics Eng; Soc Rheology. Res: Digital computer design and applications; management information systems; resources management; mathematical modeling; operations research; systems engineering; instrumentation and control systems. Mailing Add: Andrew S Yeiser & Assocs 302 Cleveland Dr Huntington Beach CA 92648

YELENOSKY, GEORGE, b Vintondale, Pa, July 20, 29; m 63; c 3. PLANT PHYSIOLOGY. Educ: Pa State Univ, BS, 55, MS, 58; Duke Univ, DF, 63. Prof Exp: Res forester, Northeastern Forest Exp Sta, US Forest Serv, 55-56 & 58-61; Int Shade

Tree Conf res fel, 63-64; PLANT PHYSIOLOGIST, CITRUS INVESTS, USDA, 64- Mem: Am Soc Plant Physiol; Am Soc Hort Sci. Res: Forestry; soil aeration and tree growth; cold hardiness of citrus trees; cryobiology membership. Mailing Add: USDA Agr Res Serv 2120 Camden Rd Orlando FL 32803

YELIN, ROBERT EMIL, inorganic chemistry, textile chemistry, see 12th edition

YELLEN, JAY, b New York, NY, Dec 17, 48; m 71; c 1. ALGEBRA. Educ: Polytech Inst Brooklyn, BS, 69, MS, 71; Colo State Univ, PhD(math), 75. Prof Exp: Teaching asst math, Polytech Inst Brooklyn, 69-71; teaching asst, Colo State Univ, 71-75; ASST PROF MATH, ALLEGHENY COL, 75- Mem: Am Math Soc. Res: Group representation theory, particularly groups of central type; algebraic coding theory. Mailing Add: Dept of Math Allegheny Col Meadville PA 16335

YELLIN, ABSALOM MOSES, b Tel-Aviv, Israel, July 25, 36; US citizen; m 66; c 2. PSYCHOPHYSIOLOGY, CHILD PSYCHIATRY. Educ: Univ Del, BA, 65, MA, 68, PhD(psychol), 70. Prof Exp: Fel psychol, Univ Calif, Los Angeles Sch Med, 69-71; asst res psychologist, Neuropsychiat Inst, Univ Calif, Los Angeles, 71-72; asst prof, Dept Psychiat, Univ Calif, Davis, 72-74; ASST PROF PSYCHOL, UNIV MINN, 74- Mem: Am Psychol Asn; AAAS; NY Acad Sci. Res: Attention and information processing; hyperkinesis and other disorders involving attentional deficits biorhythms. Mailing Add: Div Child & Adolescent Psychiat Univ Minn Box 95 Mayo Bldg Minneapolis MN 55455

YELLIN, EDWARD L, b Brooklyn, NY, July 2, 27; m 48; c 3. BIOENGINEERING, CARDIOVASCULAR PHYSIOLOGY. Educ: Colo State Univ, BS, 59; Univ Ill, MS, 61, PhD(mech eng), 64. Prof Exp: Res assoc surg, 65-66, assoc surg & physiol, 66-68, asst prof, 68-72, ASSOC PROF SURG & PHYSIOL, ALBERT EINSTEIN COL MED, 72- Concurrent Pos: NIH sr fel cardiovasc tech, Univ Wash, 64-65; mem coun basic sci, Am Heart Asn. Mem: Am Asn Univ Prof; Am Soc Mech Eng; Asn Advan Med Instrumentation; Am Soc Artificial Internal Organs; NY Acad Med. Res: Application of fluid mechanics, particularly turbulence to cardiovascular physiology; natural and artificial heart valves; cardiac dynamics. Mailing Add: 38 Lakeside Dr New Rochelle NY 10801

YELLIN, HERBERT, b New York, NY, May 27, 35; m 63; c 5. HEALTH SCIENCES, EXPERIMENTAL NEUROLOGY. Educ: City Col New York, BA, 56; Univ Calif, Los Angeles, PhD(anat & neurophysiol), 66. Prof Exp: Cytologist, Div Labs, Cedars of Lebanon Hosp, Los Angeles, 57-58; cytologist radiation path, Armed Forces Inst Path, US Army, 58-60; staff fel exp neurol, Nat Inst Neurol Dis & Blindness, NIH, 66-68; res physiologist trophic nerve functions, Nat Inst Neurol Dis & Stroke, NIH, 68-73; res physiologist neuronal develop & regeneration, Nat Inst Neurol & Communicative Dis & Stroke, NIH, 73-76; GRANTS ASSOC SCI ADMIN, DIV RES GRANTS, NIH, 76- Concurrent Pos: Lectr anat, Sch Med & Dent, Georgetown Univ, 67-75; councilman assembly scientists, Nat Inst Neurolog & Communicative Dis & Stroke, NIH, 73-75. Mem: Am Asn Anatomists; Am Physiol Soc; Soc Neurosci. Res: Sensorimotor characteristics of posture and locomotion; interrelationships of neurons and skeletal muscle. Mailing Add: Grants Assocs Prog Div of Res Grants NIH Bethesda MD 20014

YELLIN, STEVEN JOSEPH, b San Francisco, Calif, Dec 27, 41. EXPERIMENTAL HIGH ENERGY PHYSICS. Educ: Calif Inst Technol, BS, 63, PhD(physics), 71. Prof Exp: Physicist, Deutsches Elektronen-Synchrotron, 71-73; ASST PROF & ASST RES PHYSICIST, UNIV CALIF, SANTA BARBARA, 73- Res: Electromagnetic interactions in elementary particle physics. Mailing Add: Dept of Physics Univ of Calif Santa Barbara CA 93106

YELLIN, TOBIAS O, b Tel Aviv, Israel, Aug 22, 34; US citizen; m 66; c 2. BIOCHEMISTRY, PHARMACOLOGY. Educ: Philadelphia Col Pharm, BS, 59, MS, 62; Univ Del, PhD(biochem), 66. Prof Exp: Sr scientist, Abbott Labs, Ill, 66-71; sr pharmacologist, Smith Kline & French Labs, Pa, 71-73; head gastroenterol sect, Rorer Res Labs, Pa, 73-75; HEAD GASTROENTEROL UNIT, DEPT PHARMACOL, ICI UNITED STATES INC, WILMINGTON, DEL, 75- Concurrent Pos: Vis instr pharmacol, Med Col Pa, 73- Mem: AAAS; Am Chem Soc; Am Soc Pharmacol & Exp Therapeut. Res: Gastrointestinal pharmacology and biochemistry; histamine and histamine antagonists. Mailing Add: 4 Waterford Way Wallingford PA 19086

YELLIN, WILBUR, b Passaic, NJ, Nov 1, 32; m 70; c 2. SPECTROSCOPY, PHYSICAL INORGANIC CHEMISTRY. Educ: Rutgers Univ, BSc, 54; Cornell Univ, PhD, 62. Prof Exp: Chemist, Nat Starch Corp, 54; asst chem, Cornell Univ, 54-58, res asst, 58-59; sr res chemist, Ozalid Div, Gen Aniline & Film Corp, 59-62; res chemist, 62-70, SECT HEAD, MIAMI VALLEY LABS, PROCTER & GAMBLE CO, 70- Mem: AAAS; Soc Appl Spectros; Am Chem Soc. Res: Application of spectroscopic and physical measurement techniques to chemical problems and structure determinations; hydration of biological materials; aqueous complex ions; Raman, infrared, nuclear magnetic resonance and mass spectrometry. Mailing Add: Miami Valley Labs Procter & Gamble Co Cincinnati OH 45247

YELON, ARTHUR MICHAEL, b New York, NY, Apr 15, 34; m 58; c 2. SOLID STATE PHYSICS. Educ: Cornell Univ, BA, 55; Case Inst Technol, MS, 59, PhD(physics), 61. Prof Exp: Asst physics, Case Inst Technol, 55-57, instr, 57-61; assoc mem res staff, Res Ctr, Int Bus Mach Corp, 61-62, mem res staff, 62-63; mem res staff, Lab Electrostatics & Metal Phys, Grenoble, France, 63-66; assoc prof appl sci, Yale Univ, 66-73; assoc prof eng physics, 73-74, PROF ENG PHYSICS, POLYTECH SCH MONTREAL, 74- Mem: Am Phys Soc; Am Vacuum Soc; Am Asn Physics Teachers; Inst Elec & Electronics Eng; Can Asn Physicists. Res: Structure and properties of thin metallic films; ferromagnetism; superconductivity; surface physics; electron tunneling; electronic properties of polymers. Mailing Add: Dept Eng Physics Polytech Sch Montreal Post Box 6079 Sta A Montreal PQ Can

YELON, WILLIAM B, b Brooklyn, NY, Aug 23, 44; m 66; c 2. EXPERIMENTAL SOLID STATE PHYSICS. Educ: Haverford Col, BA, 65; Carnegie Mellon Univ, MS, 67, PhD(physics), 70. Prof Exp: Fel physics, Brookhaven Nat Lab, 70-72; res physicist, Inst Laue-Langevin, 72-75; SR RES SCIENTISLEADER NEUTRON SCATTERING, UNIV RES REACTOR, 75- & ASSOC PROF PHYSICS, UNIV MO-COLUMBIA, 75- Mem: Am Phys Soc. Res: Neutron scattering; studies of phase transitions; dynamics and statics of nearly one and two dimensional systems; gamma ray diffraction. Mailing Add: Univ Mo Res Reactor Res Park Columbia MO 65201

YELTON, CHESTLEY LEE, b Butler, Ky, Nov 29, 09; m 39; c 2. ORTHOPEDIC SURGERY. Educ: Transylvania Col, AB, 32; Univ Louisville, MD, 37; Am Bd Orthop Surg, dipl, 52. Prof Exp: Chief dept orthop surg, Lloyd Noland Hosp, Fairfield, Ala, 52-63; PROF ORTHOP SURG & CHMN DEPT, MED COL ALA, 63- Mem: Fel Am Col Surg; fel Am Acad Orthop Surg; fel Am Orthop Foot Soc; Am Asn Surg of Trauma; Int Soc Orthop Surg & Traumatol. Mailing Add: 619 S 19th St Birmingham AL 35233

YELTON, DAVID BAETZ, b Cincinnati, Ohio, Jan 16, 45. MICROBIOLOGY. Educ: Mass Inst Technol, BS, 66; Univ Mass, Amherst, MS, 69, PhD(microbiol), 71. Prof Exp: Instr cell biol, Med Sch, Univ Md, 72-73; ASST PROF MICROBIOL, MED CTR, W VA UNIV, 73- Mem: Am Soc Microbiol; Tissue Cult Asn. Res: Tumor virology; cell biology. Mailing Add: Dept of Microbiol Med Ctr W Va Univ Morgantown WV 26506

YEMMA, JOHN JOSEPH, b Youngstown, Ohio, July 14, 34; m 56; c 6. CYTOCHEMISTRY. Educ: Youngstown State Univ, BS, 61; George Peabody Col, MA, 64; Pa State Univ, PhD(biol chem), 71. Prof Exp: Instr biol, Pa State Univ, 65-70; asst prof, 71-74, ASSOC PROF BIOL, YOUNGSTOWN STATE UNIV, 74- Mem: AAAS; Am Bot Soc. Res: Genetic regulation and differentiation in the fungi. Mailing Add: Dept of Biol Youngstown State Univ Youngstown OH 44503

YEN, CHEN-WAN LIU, b Tainan, Taiwan, Jan 26, 32; US citizen; m 58; c 2. AEROSPACE SCIENCE. Educ: Nat Taiwan Univ, BS, 54; Mass Inst Technol, PhD(physics), 64. Prof Exp: Tech staff space sci, KMS Technol, 68-72; MEM TECH STAFF SPACE SCI, JET PROPULSION LAB, CALIF INST TECHNOL, 72- Mailing Add: 867 Marymount Lane Claremont CA 91711

YEN, DAVID HSEIN-YAO, b Tsingtao, Shantung, Apr 18, 34; m 64; c 3. APPLIED MATHEMATICS, ENGINEERING MECHANICS. Educ: Nat Taiwan Univ, BS, 56; Mich State Univ, MS, 61; NY Univ, PhD(math), 66. Prof Exp: Asst prof civil eng & mech, Bradley Univ, 61-62; from asst prof to assoc prof, 65-72, PROF MECH & MATH, MICH STATE UNIV, 72- Concurrent Pos: Vis scientist, Inst Comp Appln Sci & Eng, NASA Langley Res Ctr, 74. Mem: Am Math Soc; Am Soc Mech Engrs; Soc Indust Appl Math; Am Phys Soc. Res: Analytical and numerical studies of linear and nonlinear wave phenomena in various continuous media; methods of applied mathematics. Mailing Add: Dept of Math Mich State Univ East Lansing MI 48824

YEN, ELIZABETH HSI, b Shanghai, China, Oct 4, 27; m 57; c 2. MATHEMATICAL STATISTICS. Educ: Univ Taiwan, BA, 52; Univ Minn, MA, 56, PhD(math statist), 63. Prof Exp: Asst statistician, NC State Univ, 62-63, assoc statistician, 63-64; res asst prof math, Statist Lab, Cath Univ, 64-66; asst prof math statist, Columbia Univ, 66-69; STAFF STATIST CONSULT, GRUMMAN DATA SYSTS CORP, 69- Mem: AAAS; Inst Math Statist. Res: Theoretical and applied statistics; non-parametric statistical inference; computer performance evaluation. Mailing Add: Grumman Data Systs Corp 1111 Stewart Ave Bethpage NY 11714

YEN, LEWIS C, b Luho, Kiangsu, China, June 21, 30; m 61; c 2. PHYSICAL CHEMISTRY, CHEMICAL ENGINEERING. Educ: Taiwan Col Eng, BS, 53; Univ Fla, MS, 57; Univ Tex, PhD(chem eng, phys chem), 62. Prof Exp: Jr chem engr, Chinese Petrol Corp, Taiwan, 54-55; sr eng analyst, Phillips Petrol Co, 61-65, eng consult, 65-66; mgr chem eng data, 66-72, MGR CHEM ENG, LUMMUS CO, 72- Mem: Am Chem Soc; Am Inst Chem Eng. Res: Thermodynamics; phase equilibria; separations; kinetics. Mailing Add: 1 Tuxedo Dr Livingston NJ 07039

YEN, PETER KAI JEN, b China, Feb 10, 22. DENTISTRY, BIOPHYSICS. Educ: West China Union Univ, DDS, 47; Harvard Univ, DMD, 54. Prof Exp: Intern children's dent, Forsyth Dent Infirmary, 49-50; asst oral path, 54-57, instr dent med, 57-61, clin assoc, 61-67, asst prof orthod, 67-70, ASSOC PROF ORTHOD, SCH DENT MED, HARVARD UNIV, 70- Concurrent Pos: Fel orthod, Forsyth Dent Infirmary, 50-52 & 55-57. Mem: Fel AAAS; Am Dent Asn; Am Asn Orthod; NY Acad Sci; Int Asn Dent Res. Res: Cranio-facial growth and development. Mailing Add: Harvard Sch of Dent Med 188 Longwood Ave Boston MA 02115

YEN, SAMUEL SHOW-CHIH, b Peking, China, Feb 22, 27; m 58; c 3. REPRODUCTIVE ENDOCRINOLOGY. Educ: Chee-Loo Univ, China, BS, 49; Univ Hong Kong, MD, 54; Am Bd Obstet & Gynec, dipl, 66. Prof Exp: Intern med, Queen Mary Hosp, Hong Kong, 54-55; resident obstet & gynec, Johns Hopkins Hosp, 56-60; instr, Johns Hopkins Univ, 58-60; chief dept, Guam Mem Hosp, 60-62; asst prof, Sch Med, Case Western Reserve Univ, 62-67, assoc prof reprod biol, 67-70; assoc dir obstet, Univ Hosps of Cleveland, 68-70; PROF OBSTET & GYNEC, UNIV CALIF, SAN DIEGO, 70-, CHMN DEPT, 72- Concurrent Pos: Teaching & res fels, Harvard Med Sch, 62. Mem: Endocrine Soc; Am Diabetes Asn; Am Soc Gynec Invest; fel Am Col Obstet & Gynec. Res: Obstetrics and gynecology. Mailing Add: Dept of Obstet & Gynec Univ of Calif Sch of Med La Jolla CA 92037

YEN, TEH FU, b Kunming, China, Jan 9, 27; US citizen; m 59. ENVIRONMENTAL SCIENCE, BIOCHEMICAL ENGINEERING. Educ: Cent China Univ, BS, 47; WVa Univ, MS, 53; Va Polytech, PhD(org chem), 56. Prof Exp: Asst, Cent China Univ, 47-48, Yunnan Univ, 48-49 & WVa Univ, 50-53; sr res chemist, Res Div, Goodyear Tire & Rubber Co, 55-59; fel petrol chem, Mellon Inst, 59-65, sr fel, Carnegie-Mellon Univ, 65-68; assoc prof, Calif State Univ, Los Angeles, 68-69; ASSOC PROF BIOCHEM, DEPT MED, UNIV SOUTHERN CALIF, 69-, DEPT CHEM ENG, 70-, ENVIRON ENG PROG, 71- Mem: Sr mem Am Chem Soc; Am Inst Physics; Am Inst Chem Eng; Am Soc Artificial Internal Organs; fel The Chem Soc; fel Am Petrol Inst; fel Am Inst Chem. Res: Structure of large molecules by physical methods; biogeoorganic chemistry; geomicrobiology, fossil fuels science and technology, asphaltenes; biochemical energy conversion, energy and environment; biomaterials; heavy metals in environments; new energy sources. Mailing Add: Dept of Chem Eng Univ of Southern Calif University Park Los Angeles CA 90007

YEN, TERENCE TSIN TSU, b Shanghai, China, May 2, 37; m 64; c 3. BIOCHEMISTRY, GENETICS. Educ: Nat Taiwan Univ, BS, 58; Univ NC, PhD(biochem, genetics), 66. Prof Exp: RES SCIENTIST, RES LABS, ELI LILLY & CO, 65- Mem: AAAS; Genetics Soc Am; Sigma Xi; Am Soc Biol Chem; Am Chem Soc. Res: Biochemical defects of metabolic diseases in higher organisms, obesity, diabetes and hypertension. Mailing Add: Lilly Res Labs Indianapolis IN 46206

YEN, WILLIAM MAO-SHUNG, b Nanking, China, Apr 5, 35; m 68. SOLID STATE PHYSICS. Educ: Univ Redlands, BS, 56; Washington Univ, PhD(physics), 62. Prof Exp: Res assoc physics, Washington Univ, 62 & Stanford Univ, 62-65; from asst prof to assoc prof physics, 65-72, PROF PHYSICS, UNIV WIS-MADISON, 72- Concurrent Pos: Consult cent res dept, Varian Assocs, Calif, 63-65; vis prof inst solid state physics, Univ Tokyo, 71-72; sr vis fel, Stanford Univ, 74-75; physicist, Lawrence Livermore Lab, 74-75, consult, 75- Mem: AAAS; Am Phys Soc; Optical Soc Am. Res: Solid state spectroscopy; spectroscopy in magnetic materials; quantum electronics; laser spectroscopy; ultraviolet spectra of solids. Mailing Add: Dept of Physics Sterling Hall Univ of Wis Madison WI 53706

YENCHA, ANDREW JOSEPH, b Pa, July 3, 38; m 67. PHYSICAL CHEMISTRY, ATMOSPHERIC CHEMISTRY. Educ: Univ Calif, Berkeley, BS, 63, Univ Calif, Los Angeles, PhD(chem), 68. Prof Exp: Res asst chem, Dow Chem Co, Calif, 60-62; res asst, Univ Calif, Berkeley, 62-63, teaching asst, Los Angeles, 64-65, res asst, 65-68; fel, Yale Univ, 68-70; ASST PROF, STATE UNIV NY ALBANY, 70- Mem: AAAS; Am Inst Physics; Am Chem Soc; NY Acad Sci. Res: Molecular beam and molecular spectroscopy. Mailing Add: Dept of Chem State Univ of NY Albany NY 12222

YENDOL, WILLIAM G, b Pomona, Calif, Feb 22, 31; m 59; c 2. ENTOMOLOGY. Educ: Calif State Polytech Univ, BS, 53; Purdue Univ, MS, 57, PhD(entom), 64. Prof Exp: Tech rep agr chem, L H Butcher Chem Co, 58-59; entomologist, Lake States Forest Exp Sta, 63-65; ASSOC PROF ENTOM RES, PESTICIDE LAB, PA STATE UNIV, UNIVERSITY PARK, 65- Mem: Entom Soc Am; Soc Invert Path; Int Orgn Biol Control; Am Inst Biol Sci. Res: Insect pathology; biological control; biochemistry. Mailing Add: Pesticide Lab Dept of Entom Pa State Univ University Park PA 16802

YENGOYAN, ARAM A, b Fresno, Calif, Sept 14, 35; m 61; c 2. ANTHROPOLOGY. Educ: Fresno State Col, AB, 56; Univ Calif, Los Angeles, MA, 58; Univ Chicago, PhD(anthrop), 63. Prof Exp: Instr anthrop, Univ Okla, 62-63; from asst prof to assoc prof, 63-73, PROF ANTHROP, UNIV MICH, ANN ARBOR, 73- Concurrent Pos: Res fels, Rackham Found & Australian Inst Aboriginal Studies, 66-67; Fulbright-Hays fel, Capiz, Philippines, 69-70; sr fel, Pop Inst, East-West Ctr, Univ Hawaii, 72; Rockefeller-Ford Found Pop Policy grant, Philippines & Turkey, 74-75. Mem: Fel Am Anthrop Asn; Am Ethnol Asn. Res: Social anthropology; human and cultural ecology, especially demographic aspects; South East Asia and Australia. Mailing Add: Dept of Anthrop Univ of Mich Ann Arbor MI 48104

YENI-KOMSHIAN, GRACE HELEN, b Beirut, Lebanon, Jan 29, 36. PSYCHOPHYSIOLOGY. Educ: Am Univ Beirut, BA, 57; Cornell Univ, MS, 62; McGill Univ, PhD(psychol), 65. Prof Exp: Res assoc psychol, Ctr Appl Ling, Washington, DC, 64-65, consult, 65-66; instr med psychol, Sch Med, 66-68, Nat Inst Neurol Dis & Stroke spec fel, Med Insts, 69-71, instr otolaryngol, Sch Med, 72-73, ASST PROF OTOLARYNGOL, SCH MED, JOHNS HOPKINS UNIV, 73- Concurrent Pos: Consult, Ford Found, Manila, Philippines, 68. Mem: AAAS; Am Psychol Asn; Acoust Soc Am; Soc Res Child Develop. Res: Speech perception and production; neurophysiology of hearing; speech and brain functions; language acquisition; cognitive development. Mailing Add: Dept of Otolaryngol Traylor 420 Johns Hopkins Univ Sch of Med Baltimore MD 21205

YENNIE, DONALD ROBERT, b Paterson, NJ, Mar 4, 24; m 50; c 2. THEORETICAL HIGH ENERGY PHYSICS. Educ: Stevens Inst Technol, ME, 45; Columbia Univ, PhD(physics), 51. Prof Exp: Instr physics, Stevens Inst Technol, 46-47; mem, Inst Advan Study, 51-52; from instr to asst prof, Stanford, 52-57; from assoc prof to prof, Univ Minn, Minneapolis, 57-64; PROF PHYSICS, CORNELL UNIV, 64- Concurrent Pos: NSF sr fel, 60-61; mem prog adv comt, Stanford Linear Accelerator Ctr, 65-68, vis scientist, 70-71. Mem: Am Phys Soc; Am Asn Physics Teachers. Res: Quantum field theory; theory of high energy electromagnetic interactions; theory of Lamb shift and hyperfine splitting; renormalization theory; infrared divergence in quantum electrodynamics. Mailing Add: Newman Lab Cornell Univ Ithaca NY 14853

YENSEN, ARTHUR ERIC, b Nampa, Idaho, Oct 13, 44; m 66. ECOLOGY. Educ: Col Idaho, BS, 66; Ore State Univ, MA, 71; Univ Ariz, PhD(zool), 73. Prof Exp: ASST PROF BIOL, MILLSAPS COL, 73- Mem: Ecol Soc Am; Soc Study Evolution; Soc Syst Zool; Cooper Ornith Soc; Am Entom Soc. Res: Community structure including species diversity, predator-prey relationships, competition, niche utilization and the invasibility of communities; biosystematics and evolution of Throscidae. Mailing Add: Dept of Biol Millsaps Col Jackson MS 39210

YENTSCH, CHARLES SAMUEL, b Louisville, Ky, Sept 13, 27. MARINE BIOLOGY. Educ: Univ Louisville, BS, 50; Fla State Univ, MS, 53. Prof Exp: Asst marine biol, Fla State Univ, 52-53; asst biol oceanog, Univ Wash, 53-55; res assoc marine ecol, Woods Hole Oceanog Inst, 55-67; assoc prof oceanog, Nova Univ, 67-69, assoc prof marine biol, 69-71; PROF MARINE SCI & DIR MARINE STA, UNIV MASS, AMHERST, 71- Mem: Am Soc Limnol & Oceanog; Phycol Soc Am. Res: Marine phytoplankton ecology. Mailing Add: Marine Sta Univ of Mass at Amherst Gloucester MA 01930

YEN-WATSON, BELINDA R S, b Szechuen, China, Sept 2, 41; US citizen; m 67. IMMUNOLOGY. Educ: Southern Ill Univ, Carbondale, BS, 62; Univ Ark, Fayetteville, MS, 66, PhD(immunol), 71. Prof Exp: Am Heart Asn fel, 71-72, Arthritis Found res fel, 72-73; res assoc clin immunol, Sch Med, Case Western Reserve Univ, 73-74; RES FEL, CLEVELAND CLIN FOUND, 74- Mem: AAAS; Am Soc Microbiol. Res: Clinical immunology, currently inflammatory responses in humans and animals; regulation of immune response. Mailing Add: Cleveland Clin 9500 Euclid Ave Cleveland OH 44106

YEO, RICHARD RUSSELL, botany, agronomy, see 12th edition

YEOMAN, FREDERICK ALBERT, organic chemistry, see 12th edition

YEOMAN, LYNN CHALMERS, b Evanston, Ill, May 17, 43; m 66; c 2. BIOCHEMISTRY. Educ: DePauw Univ, BA, 65; Univ Ill, Champaign, PhD(biochem), 70. Prof Exp: Instr, 72-73, ASST PROF PHARMACOL, BAYLOR COL MED, 73- Concurrent Pos: NIH training grant & univ fel, Baylor Col Med, 70-72. Mem: Am Asn Cancer Res; Am Chem Soc. Res: Structural and physical chemistry of the proteins of the nucleus and the nucleolus; non-histone chromosomal proteins and gene regulation. Mailing Add: Dept of Pharmacol Baylor Col of Med Houston TX 77025

YEOMANS, DONALD KEITH, b Rochester, NY, May 3, 42; m 70; c 1. ASTRONOMY. Educ: Middlebury Col, BS, 64; Univ Md, MS, 67, PhD(astron), 70. Prof Exp: Sr math analyst astron, Bendix Field Eng Corp, 70-72; tech supvr, Comput Sci Corp, 72-75; SR ENGR ASTRON, JET PROPULSION LAB, CALIF INST TECHNOL, 76- Concurrent Pos: Fel, Univ Md, 70; consult, NASA's Comets & Asteroids Sci Working Group, 73-75. Mem: Int Astron Union; Am Astron Soc; Astron Soc Pac. Res: Comet and asteroid orbit determination; interplanetary mission analysis; celestial mechanics; history of science. Mailing Add: Jet Propulsion Lab 4800 Oak Grove Dr Pasadena CA 91103

YEOWELL, DAVID ARTHUR, b London, Eng, Jan 3, 37; m 64. PHARMACEUTICAL CHEMISTRY. Educ: Bristol Univ, BSc, 58, PhD, 61. Prof Exp: Swiss Nat Fund fel org chem, Univ Zurich, 61-62; Imp Chem Indust res fel biogenetics, Univ Liverpool, 62-64; sr org chemist, Wellcome Res Labs, 64-68, sr develop chemist, Chem Develop Labs, 68-71, head develop res, 71-73, MGR CHEM DEVELOP LABS, BURROUGHS WELLCOME CO, 73- Mem: Am Chem Soc; The Chem Soc. Res: Fingerprinting routes to drugs; cost analysis; synthetic organic chemistry; structure determinations; pharmaceutical research; heterocyclic compounds, especially pyrimidine derivatives; continuous processes. Mailing Add: Chem Develop Labs Burroughs Wellcome Co 3030 Cornwallis Rd Research Triangle Park NC 27709

YERANSIAN, JAMES A, b Chicago, Ill, Apr 8, 28; m 50; c 5. ANALYTICAL CHEMISTRY. Educ: Cornell Univ, BA, 48; Adelphi Univ, MS, 55. Prof Exp: Chemist, Nat Dairy Res Labs, Inc, 48-55; assoc chem, 55-58, proj leader food res, 58-60, sr chemist, 60-64, group leader coffee res, 64-65, sr group leader, 65-66, sr group leader, Flavor & Prod Develop, 66-68, SUPVR CORP ANAL LAB, FLAVOR & PROD DEVELOP, GEN FOODS CORP RES, 68- Mem: AAAS; Am Chem Soc. Res: Food analysis; coffee technology; analytical instrumentation and automation. Mailing Add: Anal Lab Gen Foods Tech Ctr White Plains NY 10625

YERBY, ALONZO SMYTHE, b Augusta, Ga, Oct 14, 21; m 43; c 3. PUBLIC HEALTH, HEALTH ADMINISTRATION. Educ: Univ Chicago, BS, 41; Meharry Med Col, MD, 46; Harvard Univ, MPH, 48; Am Bd Prev Med, dipl, 53. Prof Exp: Asst to dir food res inst, Univ Chicago, 42-43; intern, Coney Island Hosp, Brooklyn, NY, 46-47; health officer-in-training, New York City Dept Health, 47-48; field med officer, UN Int Refugee Orgn, US Zone, Ger, 48-49, dep chief med affairs, Off US High Comnr Ger, 49-50; assoc med dir, Health Ins Plan Greater New York, 50-54, consult, 54; regional med consult, Off Voc Rehab, 54-57; dep comnr med affairs, NY State Dept Soc Welfare, 57-60; exec dir med servs, New York City Dept Health & med welfare adminr, Dept Welfare, 60-65 & coordr welfare servs, Dept Hosps, 64-65, comnr, Dept Hosps, 65-66; head dept, 66-75, PROF HEALTH SERV ADMIN, SCH PUB HEALTH, HARVARD UNIV, 66- Concurrent Pos: Staff physician, Sidney Hillman Health Ctr, New York, 51-53; adj asst prof admin med, Sch Pub Health, Columbia Univ, 60-66; lectr, Yale Univ, 65-66; mem, Surg Gen Adv Comt Urban Health Affairs, USPHS & task force on orgn community health servs, Nat Comn Health Servs, 63-66; mem summer study sci & urban develop, Dept Housing & Urban Develop & President's Off Sci & Tech, 66; adv comt on rels with state health agencies, Dept Health, Educ & Welfare, 66; mem, President's Nat Adv Comn Health Manpower, 66-67; mem nat pub health training coun, NIH, 69-71, consult, 71-, mem bd adv, John E Fogarty Int Ctr, 72-76, vis prof, 75; WHO consult health manpower training, Govt Sierra Leone, WAfrica, 70; mem coun, Nat Inst Med, 70-73. Mem: Fel Am Pub Health Asn; fel Am Col Prev Med; Am Pub Welfare Asn; NY Acad Med. Res: Health service administration; public health practice. Mailing Add: Dept of Health Serv Admin Harvard Univ Sch of Pub Health Boston MA 02115

YERG, DONALD G, b Lewistown, Pa, Mar 4, 25; m 48; c 3. METEOROLOGY. Educ: Pa State Univ, BS, 46, MS, 47, PhD(meteorol), 53. Prof Exp: Asst agr eng, Univ Calif, 47-48; instr math, Univ Alaska, 48-50; consult, Tech Info Div, Library of Cong, 51-52; asst ionosphere res, Pa State Univ, 52-53; lectr physics, Univ PR, 53-55; assoc prof, 55-60, PROF PHYSICS & DEAN GRAD SCH, MICH TECHNOL UNIV, 60- Mem: AAAS; Am Meteorol Soc; Am Geophys Union. Res: Meteorology of ionospheric regions; micrometeorology; biometeorology. Mailing Add: Grad Sch Mich Technol Univ Houghton MI 49931

YERG, RAYMOND A, b Jersey City, NJ, Apr 4, 17; m 46; c 3. AEROSPACE MEDICINE, OCCUPATIONAL MEDICINE. Educ: Seton Hall Col, BS, 38; Georgetown Univ, MD, 42; Harvard Univ, MPH, 55. Prof Exp: Comdr, 1st Missile Div, Vandenberg AFB, US Air Force, Calif, 59-61, dep bioastronaut Air Force Eastern Test Range, 61-65, comdr, Aerospace Med Res Lab, Wright-Patterson AFB, Ohio, 65-68, chief sci & tech div, Hq US Air Force, The Pentagon, 68-72; CORP MED DIR, AM CAN CO, 72- Honors & Awards: AMA Spec Aerospace Med Citation, 62; Air Force Asn Meritorious Award Support Mgt, 64. Mem: Fel Aerospace Med Asn; fel Am Col Prev Med; NY Acad Sci; Am Acad Occup Med; Am Occup Med Asn. Mailing Add: 160 Eden Rd Stamford CT 06907

YERGANIAN, GEORGE, b New York, NY, June 14, 23; m 50; c 2. BIOLOGY. Educ: Mich State Univ, BS, 47; Harvard Univ, PhD(biol), 50. Prof Exp: Instr bot, Univ Minn, 50-51; AEC fel, Brookhaven Nat Lab, 51-52 & Boston Univ, 52-53; USPHS res fel, Boston Univ & Children's Cancer Res Found, 53-54; RES ASSOC, CHILDREN'S CANCER RES FOUND, CHILDREN'S HOSP MED CTR, 54-; RES ASSOC PATH, HARVARD UNIV, 54- Mem: Genetics Soc Am; Radiation Res Soc; Am Soc Human Genetics; Am Asn Cancer Res; Am Soc Cell Biologists. Res: Mammalian cytogenetics with emphasis on dwarf species of hamsters. Mailing Add: Children's Cancer Res Found 35 Binney St Boston MA 02115

YERGER, RALPH WILLIAM, b Reading, Pa, July 31, 22; m 54; c 4. SYSTEMATIC ICHTHYOLOGY. Educ: Pa State Univ, BS, 43, MS, 47; Cornell Univ, PhD(zool), 50. Prof Exp: Teacher high sch, Pa, 43; instr biol, Pa State Univ, 47-48; instr nature study, Reading Mus, 48; from asst prof to assoc prof zool, 50-61, actg head dept, 53-55, PROF ZOOL, FLA STATE UNIV, 61-, ASSOC CHMN UNDERGRAD STUDIES, 75- Mem: Am Soc Ichthyologists & Herpetologists; Am Inst Biol Scientist; Soc Syst Zool; Asn Trop Biol; Am Fisheries Soc. Res: Taxonomy, ecology and distribution of fresh and salt water fishes of the southeastern United States, Central America and the Caribbean. Mailing Add: Dept of Biol Sci Fla State Univ Tallahassee FL 32306

YERGEY, ALFRED L, III, b Philadelphia, Pa, Sept 17, 41; m 63; c 3. ION OPTICS, CHEMICAL KINETICS. Educ: Muhlenberg Col, BS, 63; Pa State Univ, PhD(chem), 67. Prof Exp: Res fel chem, Rice Univ, 67-69; chemist, Esso Res & Eng Co, NJ, 69-71; SR SCIENTIST, SCI RES INSTRUMENTS CORP, 71- Mem: AAAS; Am Chem Soc; Air Pollution Control Asn. Res: Ion/electron optics; chemical ionization kinetics; applications of mass spectrometry; heterogeneous reaction kinetics; non-isothermal method; stable isotope applications to clinical situations; quadrupole mass spectrometry. Mailing Add: Sci Res Instruments Corp 6707 Whitestone Rd Baltimore MD 21207

YERGIN, PAUL FLOHR, b New York, NY, Apr 21, 23; m 47; c 2. NUCLEAR PHYSICS. Educ: Union Univ, NY, BS, 44; Columbia Univ, MA, 49, PhD, 53. Prof Exp: Asst electronics res, Gen Elec Co, NY, 42; asst gen physics lab, Union Univ, NY, 42-43; physicist radiation lab, Columbia Univ, 44-45, mem sci staff, 45-52; instr physics, Univ Pa, 52-55, asst prof, 55-56; from asst prof to assoc prof, 56-74, PROF PHYSICS, RENSSELAER POLYTECH INST, 74- Mem: Am Phys Soc; Am Asn Physics Teachers. Res: Photonuclear reactions; neutron cross sections. Mailing Add: Dept of Physics Rensselaer Polytech Inst Troy NY 12181

YERICK, ROGER EUGENE, b Kingsville, Tex, July 6, 32. ANALYTICAL CHEMISTRY. Educ: Tex Col Arts & Indust, BS, 53. Prof Exp: Res asst anal chem, Iowa State Univ, 53-57; asst prof chem, Tex Col Arts & Indust, 57-58; from asst prof to assoc prof, 58-65, PROF CHEM, LAMAR UNIV, 65-, DEAN COL SCI, 74- Concurrent Pos: Educ consult, Spec Training Div, Oak Ridge Assoc Univs, 62- Mem: AAAS; Am Chem Soc. Res: Analytical chemistry of chelates; analytical radiochemistry; analytical applications of liquid scintillation counting techniques. Mailing Add: Dept of Chem Lamar Univ Beaumont TX 77705

YERMANOS, DEMETRIOS M, b Thessaloniki, Greece, June 29, 21; US citizen; m 55; c 3. GENETICS, PLANT BREEDING. Educ: Univ Thessaloniki, MS, 47; Iowa State Univ, MS, 52; Univ Calif, Davis, PhD(genetics), 60. Prof Exp: Agronomist, UN Relief & Rehab Admin, Greece, 45-47 & Econ Coop Admin, 53-55; asst specialist, Univ Calif, Davis, 56-60; from asst prof to assoc prof agron, 61-72, vchmn dept plant sci, 72-74, PROF AGRON, UNIV CALIF, RIVERSIDE, 72- Concurrent Pos: Res grants, NSF & USPHS, 63-65, Dept Health, Educ & Welfare, 71-74. Mem: Am Soc Agron. Res: Genetics and plant breeding of oil crops. Mailing Add: Dept of Plant Sci Univ of Calif Riverside CA 92502

YERUSHALMY, JACOB, biometry, deceased

YESAIR, DAVID WAYNE, b Newbury, Mass, Sept 9, 32; m 54; c 3. BIOCHEMISTRY. Educ: Univ Mass, BS, 54; Cornell Univ, PhD(biochem), 58. Prof Exp: Asst biochem, Cornell Univ, 55-57, res assoc, 58; res biochemist, Lederle Labs Div, Am Cyanamid Co, 59-61; NSF fel, Reading, Eng, 61-62; consult biochemist, 62-74, DIR BIOCHEM & BIOCHEM PHARMACOL, ARTHUR D LITTLE, INC, 74- Concurrent Pos: Nat Cancer Inst spec res award, Inst Org Chem, Gif-sur-Yvette, France, 71-72; lectr, Mass Inst Technol, 72- Mem: Am Chem Soc; NY Acad Sci; Am Asn Cancer Res; Am Soc Pharmacol & Exp Therapeut; fel Am Inst Chemists. Res: Lipid biochemistry; cancer, obesity and diabetes; metabolism and mode of action of drugs affecting lipid metabolism; isolation, characterization and metabolic action of biologically active agents. Mailing Add: Dept of Biochem Arthur D Little Inc Acorn Park Cambridge MA 02140

YESNER, RAYMOND, b Columbus, Ga, Apr 18, 14; m 47; c 3. PATHOLOGY. Educ: Harvard Univ, AB, 35; Tufts Col, MD, 41; Yale Univ, MA, 72. Prof Exp: Intern & res, Beth Israel Hosp, Boston, Mass, 41-44; pathologist & chief lab serv, Vet Admin Hosp, Newington, 47-53; from asst clin prof to assoc clin prof, 49-64, assoc prof, 64-72, assoc dean, 69-74, PROF PATH, MED SCH, YALE UNIV, 72-; CHIEF PATHOLOGIST, VET ADMIN HOSP, WEST HAVEN, 74- Concurrent Pos: Consult pathologist, Coop Study of Prostate, Vet Admin, chmn path panel, Lung Cancer Chemother Study Group, 58-, mem path res eval comt, sr physician, 71-74; pathologist & chief lab serv, Vet Admin Hosp, West Haven, 53-74, chief staff, 69-74. Mem: AMA; Am Asn Path & Bact; Col Am Path; Int Acad Path. Res: Changes in blood viscosity; liver, lung and gastrointestinal disease; carcinoma of prostate and bladder. Mailing Add: Vet Admin Hosp W Spring St West Haven CT 06516

YESSIK, MICHAEL JOHN, b Webster, Mass, Nov 22, 41; m 70. SOLID STATE PHYSICS. Educ: Williams Col, BA, 62; Syracuse Univ, PhD(solid state sci), 66; Univ Cambridge, MA, 67. Prof Exp: NATO fel, Cavendish Lab, Univ Cambridge, 66-67, NSF fel, 67-68; sr res scientist physics, Sci Res Staff, Ford Motor Co, 68-75; MGR PROCESS DEVELOP, PHOTON SOURCES, INC, 75- Mem: Am Asn Physics Teachers; Am Phys Soc; Inst Elec & Electronics Engrs. Res: Electronic and magnetic properties of metals and alloys; physical properties of high temperature ceramics; materials processing using high-power lasers. Mailing Add: Photon Sources Inc 37100 Plymouth Rd Livonia MI 48150

YETHON, ANDREW EDWARD, organic chemistry, see 12th edition

YETT, FOWLER REDFORD, b Johnson City, Tex, Oct 18, 19; m 45; c 3. APPLIED MATHEMATICS. Educ: Univ Tex, BS, 43, MA, 52; Iowa State Univ, PhD(appl math), 55. Prof Exp: Chemist & chem engr, Manhattan Proj, 43-45; owner camera supply co, 46-49; instr math, Iowa State Univ, 52-55; asst prof, Long Beach State Col, 55-56 & Univ Tex, 56-65; chmn dept math, 65-68, PROF MATH, UNIV S ALA, 65- Mem: Am Math Soc; Math Asn Am. Res: Nonlinear differential equations. Mailing Add: Dept of Math Univ of SAla Mobile AL 36608

YEUNG, SHIU FONG, b Nanking, China, Feb 12, 36. APPLIED MATHEMATICS. Educ: Samford Univ, BA, 57; Univ Fla, MS, 60, PhD(math), 62; Va Polytech Inst, MS, 66. Prof Exp: Asst prof math, Va Polytech Inst, 61-67; res assoc, Res Div, 67-73, SR SYST ANALYST, FOSTER WHEELER CORP, 73- Concurrent Pos: Res assoc, Nat Inst Dent Health, 66-67. Mem: Am Math Soc; Am Soc Mech Engrs. Res: Mathematical elasticity; data reduction and data base analysis, system analysis for forecasting and cost optimization, numerical structural analysis. Mailing Add: 29 Radtke Rd RD 3 Dover NJ 07801

YEVICH, JOSEPH PAUL, b McKees Rocks, Pa, Sept 20, 40; m 64; c 2. ORGANIC CHEMISTRY, MEDICINAL CHEMISTRY. Educ: Carnegie-Mellon Univ, BS, 62, MS, 67, PhD(org chem), 69. Prof Exp: Chemist, Gulf Res & Develop Corp, 62-65; sr scientist, 69-74, SR INVESTR, MEAD JOHNSON & CO, 74- Mem: AAAS; Am Chem Soc; Sigma Xi. Res: Design and synthesis of potential medicinal agents. Mailing Add: RR 5 Box 586 Newburgh IN 47630

YEVICH, PAUL PETER, b Berwick, Pa, June 16, 24; m 53. HISTOPATHOLOGY. Educ: Pa State Univ, BA, 49. Prof Exp: Histologist chem warfare lab med directorate, Army Chem Ctr, US Dept Defense, 50-54, histopathologist, 54-60; res histopathologist physiol sect, Div Occup Health, US Dept Health, Educ & Welfare, 61-66; res biologist, Invert Sect, 67-71, RES TEAM LEADER, INVERT SECT, NAT MARINE WATER QUAL LAB, ENVIRON PROTECTION AGENCY, 71- Mem: Soc Invert Path. Res: Comparative histology and pathology; toxic effects of various compounds on cells and tissues of various species of animals; invertebrate histology and effects of toxic compounds on invertebrates; histopathologic effects of oil pollutants on marine life. Mailing Add: Environ Res Labs West Kingston RI 02892

YEVICK, GEORGE JOHANNUS, b Berwick, Pa, Apr 24, 22; m 45. PHYSICS. Educ: Mass Inst Technol, BSc, 42, DSc(physics), 47. Hon Degrees: MEng, Stevens Inst Technol, 58. Prof Exp: Staff mem, Radiation Lab, Mass Inst Technol, 44-46; from asst prof to assoc prof, 48-57, PROF PHYSICS, STEVENS INST TECHNOL, 57- Mem: Am Phys Soc; Soc Photo-Optical Instrument Engr; NY Acad Sci. Res: Theory of elementary particles; dynamical theory of many interacting particles; causal theory of quantum mechanics; control of thermonuclear fusion. Mailing Add: Dept of Physics Stevens Inst of Technol Hoboken NJ 07030

YFF, PETER, b Chicago, Ill, Mar 8, 24; m 51; c 4. MATHEMATICS. Educ: Roosevelt Univ, BS, 47; Univ Chicago, MS, 48; Univ Ill, PhD(math), 57. Prof Exp: Lectr math, Roosevelt Univ, 48-50; asst prof, Am Univ Beirut, 51-55, chmn dept, 52-55; asst, Univ Ill, 55-57; asst prof, Fresno State Col, 57-58; assoc prof, 58-64, PROF MATH, AM UNIV BEIRUT, 64- Mem: Am Math Soc; Math Asn Am; Edinburgh Math Soc. Res: Theory of groups; projective and Euclidean geometries. Mailing Add: Dept of Math Am Univ of Beirut Beirut Lebanon

YGUERABIDE, JUAN, b Laredo, Tex, Oct 9, 35; m 56; c 4. BIOCHEMISTRY, BIOPHYSICS. Educ: St Mary's Univ, Tex, BS, 57; Univ Notre Dame, PhD(phys chem), 62. Prof Exp: Res assoc, Radiation Lab, Univ Notre Dame, 61-63; mem res staff, Sandia Corp, NMex, 63-68; res assoc biochem, Stanford Univ, 68-69; lectr & res assoc biochem & biophys, Yale Univ, 69-72; ASSOC PROF BIOL, UNIV CALIF, SAN DIEGO, 72- Concurrent Pos: Fel, Radiation Lab, Univ Notre Dame, 62-63; consult prof chem, Univ NMex, 66-68; sci consult, Sandia Corp, NMex, 68-69. Mem: AAAS; Biophys Soc. Res: Structure, conformation and function of proteins and biological membranes; nanosecond fluorescence spectroscopy; mathematical physics. Mailing Add: Dept of Biol Univ of Calif at San Diego La Jolla CA 92037

YIANNIOS, CHRIST NICHOLAS, b Molai, Greece, Oct 10, 24; m 62; c 2. SYNTHETIC ORGANIC CHEMISTRY. Educ: Nat Univ Athens, BS, 51; Univ Ga, MS, 55. Prof Exp: Res chemist, Allied Chem Corp, 56-57, Cedars Lebanon Hosp, Los Angeles, Calif, 58-59 & Western Lacquer Corp, 59-60; res chemist, Olin Mathieson Chem Corp, 61-65, sr res chemist, 65-69; consult self-employed, 69-72, SR RES CHEMIST, CARBOLABS, INC, 73- Mem: Am Chem Soc; Sigma Xi. Res: Organic chemistry of sulfur; synthesis of biologically active compounds; synthesis of new chemicals for agricultural and pharmaceutical purposes. Mailing Add: Carbolabs Inc Fairwood Rd Bethany CT 06525

YIANNOS, PETER N, b Olympia, Greece, Nov 27, 32; US citizen; m 62; c 2. PHYSICAL CHEMISTRY, ENGINEERING. Educ: Univ Mo-Rolla, BS, 56; Lawrence Univ, MS, 58, PhD(phys chem), 60. Prof Exp: From res group leader to sr res group leader, Scott Paper Co, 60-65, sect head, 65-66, mgr pioneering res, 66-67; assoc prof eng, PMC Cols, 67-69; mgr paper res, 69-73, DIR PAPER RES, SCOTT PAPER CO, 73- Concurrent Pos: Instr, Tech Develop Prog, Scott Paper Training Course, 64-69; lectr eng, Eve Div, PMC Cols, 66-67; tech consult, Scott Paper Co, 67-69; adj prof, Widener Col, 72- Honors & Awards: Albert Award, Tech Asn Pulp & Paper Indust. Mem: Am Chem Soc; Am Inst Chem Eng; Tech Asn Pulp & Paper Indust. Res: Molecular forces and surface phenomena; fibers and fiber bonding; wood technology and pulping; mechanical properties of fibers and sheet assemblies; materials science. Mailing Add: Scott Paper Co Res Div Philadelphia PA 19113

YIELDING, K LEMONE, b Auburn, Ala, Mar 25, 31; m 73; c 4. MOLECULAR BIOLOGY, MEDICINE. Educ: Ala Polytech Inst, BS, 49; Univ Ala, MS, 52, MD, 54. Prof Exp: Intern, Med Ctr, Univ Ala, 54-55; clin assoc, NIH, 55-57; resident, USPHS Hosp, 57-58; asst prof med, Georgetown Univ, 58-64; PROF BIOCHEM, ASSOC PROF MED & CHIEF LAB MOLECULAR BIOL, MED CTR, UNIV ALA, BIRMINGHAM, 64- Concurrent Pos: Sr investr, Nat Inst Arthritis & Metab Dis, 58-64; consult, USPHS, 64- Mem: AAAS; Soc Exp Biol & Med; Am Soc Biol Chem. Res: Molecular basis for biological regulation, including both genetic mechanisms and control of enzyme activity; elucidation of disease mechanisms and drug action in molecular terms. Mailing Add: Dept of Molecular Biol Med Ctr Univ of Ala Birmingham AL 35294

YIH, ROY YANGMING, b Changsha, China, Oct 5, 31; nat US; m 60; c 3. AGRICULTURAL CHEMISTRY. Educ: Nat Taiwan Univ, BS, 56; Univ SC, MS, 59; Rutgers Univ, PhD(plant physiol, biochem), 63. Prof Exp: Sr scientist, 62-71, lab head, 72-73, PROJ LEADER, ROHM AND HAAS CO, SPRINGHOUSE, 73- Mem: Am Chem Soc; Weed Sci Soc Am. Res: Synthesis and evaluation of compounds as herbicides and plant growth regulators in order to discover new marketable products. Mailing Add: 94 Windover Lane Doylestown PA 18901

YIM, GEORGE KWOCK, WAH, b Honolulu, Hawaii, Jan 7, 30; m 52; c 5. PHARMACOLOGY. Educ: Univ Iowa, BS, 52, MS, 54, PhD(pharmacol), 56. Prof Exp: Instr pharmacol, Univ Iowa, 55-56; from asst prof to assoc prof, 56-70, PROF PHARMACOL, PURDUE UNIV, WEST LAFAYETTE, 70- Concurrent Pos: USPHS career develop award, 61-66; NIH spec fel, 66-67. Mem: AAAS; Am Pharmaceut Asn; Am Soc Pharmacol & Exp Therapeut; Soc Exp Biol & Med. Res: Action of cholinergic and tremorgenic substances on the central nervous system; pharmacology of central transmitter substances. Mailing Add: Dept of Pharmacol Purdue Univ Sch of Pharm Lafayette IN 47907

YIN, BARBARA HSIN-HSIN, b Hangchow, China, Nov 11, 43; US citizen; m 67; c 2. MATHEMATICAL STATISTICS. Educ: Nat Taiwan Univ, BS, 64; Wayne State Univ, MA, 67, PhD(math statist), 71. Prof Exp: Teaching asst math, Wayne State Univ, 65-67; mgt scientist, Burroughs Corp, 67-69; teaching asst math, Wayne State Univ, 69-71; ASST PROF MATH, OCCIDENTAL COL, 71- Mem: Opers Res Soc Am; Inst Math Statist. Res: Queueing theory; optimal stopping rule. Mailing Add: Dept of Math Occidental Col Los Angeles CA 90041

YIN, FAY HOH, b Peking, China, Mar 10, 32; US citizen; m 59; c 2. VIROLOGY. Educ: Univ Wis-Madison, BA, 54, MS, 55, PhD(biochem), 60. Prof Exp: Res asst biochem, Univ Wis-Madison, 60; res assoc virol, Dept Path, Univ Pa, 63-65; RES CHEMIST, CENT RES DEPT, E I DU PONT DE NEMOURS & CO, INC, 66- Mem: Am Soc Microbiol. Res: Biochemical studies of arbovirus and picornavirus replication. Mailing Add: Cent Res Dept E I du Pont de Nemours & Co Inc Wilmington DE 19898

YING, KUANG LIN, b Kiangsu, China, June 12, 27; Can citizen; m 55; c 3. GENETICS, CYTOGENETICS. Educ: Nat Taiwan Univ, BSc, 52; Univ Sask, PhD(cytol, genetics), 61. Prof Exp: Sr specialist plant breeding & genetics, Sino-Am Joint Comn Rural Reconstruct, 61-64; Med Res Coun Can res assoc, 64-67, asst prof, 67-73, ASSOC PROF HUMAN CYTOGENETICS, DEPT PEDIAT, UNIV SASK, 73- Concurrent Pos: Assoc prof, Grad Sch, Nat Taiwan Univ, 62-63. Mem: Genetics Soc Can; Am Soc Human Genetics; Tissue Cult Asn. Res: Human cytogenetics; mammalian cytogenetics; prenatal detection of genetic disorders. Mailing Add: Dept of Pediat Univ of Sask Saskatoon SK Can

YING, SEE CHEN, b Shanghai, China, Apr 4, 41; m 68. SOLID STATE PHYSICS. Educ: Univ Hong Kong, BSc, 63 & 64; Brown Univ, PhD(physics), 68. Prof Exp: Res assoc physics, Brown Univ, 68-69; asst res scientist, Univ Calif, San Diego, 69-71; asst prof, 71-75, ASSOC PROF PHYSICS, BROWN UNIV, 75- Concurrent Pos: Res fel, A P Sloan Found, 72. Mem: Am Phys Soc. Res: Theoretical solid state physics; electronic processes at metallic surfaces; quantum fluids. Mailing Add: Dept of Physics Brown Univ Providence RI 02912

YINGER, JOHN MILTON, b Quincy, Mich, July 25, 16; m 41; c 3. ANTHROPOLOGY. Educ: DePauw Univ, AB, 37; La State Univ, AM, 39; Univ Wis, PhD(sociol, anthrop), 42. Prof Exp: From instr to assoc prof sociol & anthrop, Ohio Wesleyan Univ, 41-47; assoc prof, 47-52, PROF SOCIOL & ANTHROP, OBERLIN COL, 52- Concurrent Pos: Vis prof, Univ Mich, 58, Univ Wash, 59, Univ Hawaii, 61 & Wayne State Univ, 71; res grant, Oberlin Col, 63-64; Guggenheim fel & sr specialist, East-West Ctr, Univ Hawaii, 68-69. Honors & Awards: Anisfeld-Wolf Award, Sat Rev-Anisfeld-Wolf Comt, 59. Mem: Am Sociol Asn (secy, 71-74, pres-elect, 75-76, pres, 76-77); Am Anthrop Asn; Soc Sci Study Relig; Sigma Xi. Res: Racial and ethnic relations; sociology and anthropology of religion; field theory. Mailing Add: 272 Oak Oberlin OH 44074

YINGST, HARVEY AUSTIN, b Somerset Co, Pa, July 29, 38; div. PHYSICAL INORGANIC CHEMISTRY. Educ: Ohio State Univ, BS, 61; Univ Cincinnati, PhD(inorg chem), 65. Prof Exp: Asst prof chem, Hollins Col, 65-68; res assoc, Brigham Young Univ, 68-70; ASST PROF CHEM, WILSON COL, 70- Concurrent Pos: Res with A E Martell, Tex A&M Univ, 67-68. Mem: AAAS; Am Chem Soc; The Chem Soc. Res: Bonding in complex ions; electrolyte solution theory; metal chelate equilibria, especially thermochemistry in aqueous solutions. Mailing Add: Dept of Chem Wilson Col Chambersburg PA 17201

YINGST, RALPH EARL, b Lebanon, Pa, Aug 5, 29; m 64. INORGANIC CHEMISTRY. Educ: Univ Chicago, AB, 50; Lebanon Valley Col, BS, 55; Univ Pittsburgh, PhD(chem), 64. Prof Exp: Instr chem, Johnstown Col, 61-63; asst prof, 64-70, ASSOC PROF CHEM, YOUNGSTOWN STATE UNIV, 70- Mem: AAAS; Am Chem Soc. Res: Coordination compounds of metals with pyridine and substituted pyridines, especially those containing olefinic linkages, such as 2-vinylpyridine and 2-

allylpyridine; optically active metal complexes, especially cobalt. Mailing Add: Dept of Chem Youngstown State Univ Youngstown OH 44555

YIP, CECIL CHEUNG-CHING, b Hong Kong, June 11, 37; m 60; c 2. BIOCHEMISTRY, ENDOCRINOLOGY. Educ: McMaster Univ, BSc, 59; Rockefeller Univ, PhD(biochem, endocrinol), 63. Prof Exp: Res assoc endocrinol, Rockefeller Univ, 63-64; from asst prof to assoc prof, 64-74, PROF ENDOCRINOL, BANTING & BEST DEPT MED RES, C H BEST INST, UNIV TORONTO, 74- Concurrent Pos: Med Res Coun Can med res scholar, 67-71. Mem: AAAS; Am Soc Biol Chem; Can Biochem Soc. Res: Biosynthesis of proinsulin and insulin; hormone-receptor interaction. Mailing Add: C H Best Inst Univ of Toronto Toronto ON Can

YIP, GEORGE, b Oakland, Calif, Nov 14, 26; m 53; c 1. BIOCHEMISTRY, FOOD TECHNOLOGY. Educ: Univ Calif, Berkeley, BS, 51; Georgetown Univ, MS, 59. Prof Exp: Chemist, Nat Canners Asn, Calif, 51-52; chemist, Div Food Chem, 55-56, res chemist pesticides, 56-63, sect chief herbicides & plant growth regulators, 63-71, chief, Biochem Technol Br, 71-72, CHIEF, INDUST CHEM CONTAMINANT BR, DIV CHEM TECHNOL, FOOD & DRUG ADMIN, 72- Mem: Am Chem Soc; Asn Off Anal Chem. Res: Methods of analysis for industrial chemical contaminants in foods; identification of unknown contaminants including degradation products. Mailing Add: Div of Chem Technol Food & Drug Admin 200 C St SW Washington DC 20204

YIP, KWOK LEUNG, b Canton, People's Repub China, Sept 23, 44; m 72; c 1. SOLID STATE PHYSICS. Educ: Chung Chi Col, Chinese Univ Hong Kong, BSc, 65; Providence Col, MS, 70; Lehigh Univ, PhD(physics), 73. Prof Exp: Teacher physics, King's Col, Hong Kong, 66-68; res assoc physics, Univ Ill, Urbana, 73-75; ASSOC SCIENTIST PHYSICS, WEBSTER RES CTR, XEROX CORP, 75- Mem: Am Phys Soc. Res: Electronic structure of semiconductors, semimetals and insulators; electronic properties of impurities and defects in these materials; electronic structure, lattice vibrations and electron-vibration interactions in molecular and organic solids; physics of ligand binding to heme proteins. Mailing Add: Webster Res Ctr Xerox Corp Webster NY 14580

YIP, LILY CHUNG, b Canton, China, Aug 10, 37; US citizen; m 62; c 4. BIOCHEMISTRY, MICROBIOLOGY. Educ: Taiwan Prov Normal Univ, BSc, 58; Univ Cincinnati, PhD, 65. Prof Exp: Instr biochem, Sch Med, Ind Univ, 67-69; res assoc, 70-73, ASSOC, SLOAN-KETTERING INST, 73- Concurrent Pos: Fel, Ind Univ, 65-67. Mem: Am Chem Soc. Res: Enzymes involved in purine metabolic pathway; control mechanism of the metabolic process. Mailing Add: Sloan-Kettering Inst 410 E 68th St New York NY 10021

YIP, MORRIS CHUK MING, biochemistry, neurobiology, see 12th edition

YIP, PATRICK CHEUNG-YUM, b Hong Kong, Apr 21, 43; Can citizen; m 70. APPLIED MATHEMATICS, NUCLEAR PHYSICS. Educ: Mem Univ Nfld, BSc, 65; McMaster Univ, PhD(theoret physics), 70. Prof Exp: Fel physics, 70, ASST PROF APPL MATH, McMASTER UNIV, 70- Mem: Can Asn Physicists. Res: Nuclear structure theory; mathematical theory of communication. Mailing Add: Dept of Appl Math McMaster Univ Hamilton ON Can

YIP, RODERICK WING, b Vancouver, BC, Oct 12, 37; m 63; c 3. PHYSICAL CHEMISTRY, ORGANIC CHEMISTRY. Educ: Univ BC, BSc, 60; Univ Western Ont, BSc, 61, PhD(chem), 65. Prof Exp: Jr res fel chem, Univ Sheffield, 65-66; asst res officer, Div Pure Chem, 66-71, ASSOC RES OFFICER, DIV CHEM, NAT RES COUN CAN, 71- Res: Photophysics; photochemistry and photobiology. Mailing Add: Div of Chem Nat Res Coun of Can Ottawa ON Can

YNTEMA, CHESTER LOOMIS, b Holland, Mich, Sept 1, 04; m 34. ANATOMY. Educ: Hope Col, AB, 26; Yale Univ, PhD(zool), 30. Prof Exp: Instr anat, Sch Med, Univ Pa, 30-34; from instr to asst prof, Med Col, Cornell Univ, 34-46; from assoc prof to prof, Col Med, Syracuse Univ, 46-50; prof, 50-74, actg dean, Sch Grad Studies, 69-73, EMER PROF ANAT, COL MED, STATE UNIV NY UPSTATE MED CTR, 74- Concurrent Pos: Mem Marine Biol Lab, Woods Hole, 40-; mem consult comt anat sci training, Nat Inst Gen Med Sci, NIH, 62-66. Mem: Am Soc Nat; Harvey Soc; Am Asn Anat; Int Inst Embryol. Res: Embryology of nervous system in amphibian and chick; development of the ear in the amphibia; embryology of the turtle. Mailing Add: Dept of Anat State Univ of NY Upstate Med Ctr Syracuse NY 13210

YNTEMA, GEORGE BUSEY, b Champaign, Ill, Oct 23, 26. PHYSICS. Educ: Swarthmore Col, AB, 48; Yale Univ, MS, 49, PhD(physics), 52. Prof Exp: Res assoc physics, Univ Ill, 52-54; instr, Cornell Univ, 54-57; res physicist, Opers Res, Inc, Md, 57-61; prin scientist, 61-67, SR PRIN SCIENTIST, UNITED AIRCRAFT RES LABS, EAST HARTFORD, 67- Res: Cryogenics, including Type II superconductivity; superconducting magnets; cryotrons, magneto-resistance, superfluid helium, laser theory, irreversible statistical mechanics; signal detection theory; devices which learn; operations research. Mailing Add: United Aircraft Res Lab East Hartford CT 06108

YNTEMA, JAN LAMBERTUS, b Neth, Oct 5, 20; m 48; c 4. PHYSICS. Educ: Free Univ, Amsterdam, NatPhilDrs, 48, DrPhysics, 52. Prof Exp: Res assoc physics, Princeton Univ, 49-52; asst prof, Univ Pittsburgh, 52-55; assoc physicist, 55-68, SR PHYSICIST, ARGONNE NAT LAB, 68- Mem: Am Phys Soc. Res: Radioactivity; gases at high temperatures; nuclear physics. Mailing Add: Argonne Nat Lab Argonne IL 60439

YNTEMA, MARY KATHERINE, b Urbana, Ill, Jan 20, 28. MATHEMATICS. Educ: Swarthmore Col, BA, 50; Univ Ill, AM, 61, PhD(math), 65. Prof Exp: Teacher, Am Col Girls, Istanbul, 50-54 & Columbus Sch Girls, Ohio, 54-57; programmer, Lincoln Lab, Mass Inst Technol, 57-58; teacher high sch, Mont, 59-60; asst prof math, Univ Ill, Chicago, 65-67; asst prof comput sci, Pa State Univ, 67-71; asst prof, 71-72, ASSOC PROF MATH, SANGAMON STATE UNIV, 72-, COORDR MATH SYSTS PROG, 75- Mem: Am Math Soc; Asn Comput Mach; Math Asn Am. Res: Automata; context-free languages. Mailing Add: Dept of Math Systs Sangamon State Univ Springfield IL 62708

YOAKUM, ANNA MARGARET, b Loudon, Tenn, Jan 13, 33. ANALYTICAL CHEMISTRY, PHYSICAL METALLURGY. Educ: Maryville Col, AB, 54; Univ Fla, MS, 56, PhD(anal chem), 60. Prof Exp: Supvr control lab, Greenback Indust, Inc, 56-59; sr res chemist, Chemstrand Res Ctr, Inc, 60-64; mem res staff, Oak Ridge Nat Lab, 64-69; EXEC V PRES & LAB DIR, STEWART LABS, INC, 67- Mem: Am Chem Soc; Soc Appl Spectros; NY Acad Sci; Am Soc Test & Mat; fel Am Inst Chem. Res: Analytical chemistry and trace analysis; research and method development in emission, flame, atomic absorption, x-ray fluorescence and infrared spectroscopy. Mailing Add: Stewart Labs Inc 5815 Middlebrook Pike Knoxville TN 37921

YOCH, DUANE CHARLES, b Parkston, SDak, Nov 4, 40; m 64; c 2. MICROBIOLOGY, BIOCHEMISTRY. Educ: SDak State Univ, BS, 63, MS, 65; Pa State Univ, PhD(microbiol), 68. Prof Exp: Asst res microbiologist, 68-69, ASSOC

SPECIALIST, AGR EXP STA, DEPT CELL PHYSIOL, UNIV CALIF, BERKELEY, 70- Mem: Am Soc Microbiol. Res: Bioenergetics; investigations of electron transport coupled to nitrogenase in bacteria; nigrogen fixation; iron-sulfur proteins. Mailing Add: Dept of Cell Physiol Univ of Calif 251 Hilgard Hall Berkeley CA 94720

YOCHELSON, ELLIS LEON, b Washington, DC, Nov 14, 28; m 50; c 3. INVERTEBRATE PALEONTOLOGY. Educ: Univ Kans, BS, 49, MS, 50; Columbia Univ, PhD, 55. Prof Exp: Asst, Univ Kans & Columbia Univ, 50-52; PALEONTOLOGIST, US GEOL SURV, 52- Concurrent Pos: Mem, Nat Res Coun, 57-71; treas, Int Paleont Asn, 71- Mem: AAAS; Paleont Soc (pres, 76); Soc Syst Zool (secy, 62-65). Res: Systematics and evolution of Paleozoic gastropods; phylogeny of Mollusca, especially early Paleozoic major taxa. Mailing Add: Rm E-317 US Nat Mus Washington DC 20560

YOCHELSON, LEON, b Buffalo, NY, July 23, 17; m 42; c 3. PSYCHIATRY. Educ: Univ Buffalo, AB, 38, MD, 42. Prof Exp: Clin prof psychiat, 59-69, PROF PSYCHIAT & BEHAV SCI, SCH MED, GEORGE WASHINGTON UNIV, 59- Concurrent Pos: Guest lectr, Catholic Univ, 48-58; mem fac, Wash Sch Psychiat, 50-58; consult, Vet Admin, 52- & NIMH, 54-; supv & training analyst, Wash Psychoanal Inst, 59-; chmn prof assocs, Psychiat Inst Washington, DC, chmn inst, chmn found, 66-; chmn comt psychiat hosps, Am Fedn Hosps, 74-; consult, Off Voc Rehab, Dept Health, Educ & Welfare. Mem: Fel Am Psychiat Asn; Am Psychoanal Asn. Res: Psychoanalysis. Mailing Add: Psychiat Inst of Wash 2141 K St NW Washington DC 20037

YOCHIM, JEROME M, b Chicago, Ill, Feb 23, 33; m 57; c 2. ENDOCRINOLOGY, PHYSIOLOGY. Educ: Univ Ill, BS, 55, MS, 57; Purdue Univ, PhD(biol sci), 60. Prof Exp: NIH fel anat, Col Med, Univ Ill, 60-62; from asst prof to assoc prof physiol, 62-70, PROF PHYSIOL, UNIV KANS, 71- Concurrent Pos: NIH career develop award, 71-76. Mem: AAAS; Am Soc Zoologists; Am Physiol Soc; Soc Study Reprod; Endocrine Soc. Res: Physiology of reproduction. Mailing Add: Dept of Physiol & Cell Biol Univ of Kans Lawrence KS 66045

YOCKEY, HUBERT PALMER, b Alexandria, Minn, Apr 15, 16; m 46; c 3. THEORETICAL BIOLOGY. Educ: Univ Calif, Berkeley, AB, 38, PhD(physics), 42. Prof Exp: Jr physicist, Nat Defense Res Comt, Calif, 41-42; physicist radiation lab, Univ Calif, 42-44; sr physicist, Tenn Eastman Corp, 44-46; group leader irradiation physics, NAm Aviation, Inc, 46-52; chief nuclear physics, Convair Div, Gen Dynamics Corp, Tex, 52-53; asst dir health & physics div, Oak Ridge Nat Lab, Tenn, 53-59; mgr res & develop div, Aerojet-Gen Nucleonics Corp, 59-62 & Hughes Res Labs, 63-64; CHIEF REACTOR BR, ABERDEEN PROVING GROUND, 64- Concurrent Pos: Consult, Oak Ridge Nat Lab, 60. Mem: Am Phys Soc; Am Nuclear Soc; Radiation Res Soc; Health Phys Soc. Res: Information theory in biology; solid state physics; pulsed reactors. Mailing Add: Reactor Br Ballistics Res Labs Aberdeen Proving Ground MD 21005

YOCOM, CHARLES FREDERICK, b Logan, Iowa, Oct 21, 14; m 39; c 3. WILDLIFE MANAGEMENT. Educ: Iowa State Univ, BS, 39; Wash State Univ, MS, 42, PhD, 49. Prof Exp: Lab technician, Iowa Coop Wildlife Res Unit, 39-40; game biologist, State Dept Game, Wash, 42-47; from instr to asst prof game mgt, Wash State Univ, 47-53; assoc prof wildlife mgt & head game mgt, 53-56, chmn div natural resources, 56-60, coordr, 60-61, PROF WILDLIFE MGT, HUMBOLDT STATE UNIV, 58- Concurrent Pos: Sabbatical leave, NZ, Australia & other foreign countries, 68. Mem: Wildlife Soc; Cooper Ornith Soc; assoc Am Ornith Union. Res: Wildlife conservation; ornithology; mammalogy; aquatic biology; ecology; scientific illustrations. Mailing Add: Sch of Natural Resources Humboldt State Univ Arcata CA 95521

YOCOM, PERRY NIEL, b Auburn, Maine, Sept 27, 30; m 62; c 3. INORGANIC CHEMISTRY. Educ: Pa State Univ, BS, 54; Univ Ill, PhD, 58. Prof Exp: Mem tech staff, 57-70, RES GROUP HEAD, DAVID SARNOFF RES CTR, RCA CORP, 70- Mem: AAAS; Am Chem Soc; Electrochem Soc; Am Inst Chemists; Mineral Soc Am. Res: Chemistry of fused salts; crystal growth; defects in solids; rare earth phase chemistry; luminescence. Mailing Add: David Sarnoff Res Ctr RCA Corp Princeton NJ 08540

YOCUM, CHARLES FREDRICK, b Storm Lake, Iowa, Oct 31, 41; m 63; c 1. BIOCHEMISTRY. Educ: Iowa State Univ, BS, 63; Ind Univ, PhD(biochem), 71. Prof Exp: Biochemist protein chem, ITT Res Inst, 63-68; NIH Fel biochem, Cornell Univ, 71-73; ASST PROF BIOL, UNIV MICH, 73- Mem: Am Chem Soc; Am Soc Plant Physiologists; AAAS; Biophys Soc. Res: Mechanisms of photosynthetic electron transport and energy transduction in chloroplasts and blue-green algae. Mailing Add: Dept Cellular & Molecular Biol Div of Biol Sci Univ of Mich Ann Arbor MI 48104

YOCUM, CONRAD SCHATTE, b Swarthmore, Pa, Mar 29, 19; m 46; c 3. PLANT PHYSIOLOGY. Educ: Col William & Mary, BS, 40; Univ Md, MS, 47; Stanford Univ, PhD, 52. Prof Exp: Asst marine biol, Va Fisheries Lab, 46-47; asst plant physiol, Hopkins Marine Sta, 47-48; instr plant physiol, Harvard Univ, 52-55; asst prof, Cornell Univ, 55-61; assoc prof, 61-64, PROF PLANT PHYSIOL, UNIV MICH, ANN ARBOR, 64- Mem: Am Soc Plant Physiol; Bot Soc Am. Res: Photosynthesis; respiration; tropisms; nitrogen fixation. Mailing Add: Dept of Bot Univ of Mich Ann Arbor MI 48104

YOCUM, RONALD H, organic chemistry, polymer chemistry, see 12th edition

YODAIKEN, RALPH EMILE, b Johannesburg, SAfrica, Aug 25, 25; US citizen. PATHOLOGY, ENDOCRINOLOGY. Educ: Univ Witwatersrand, MB & BCh, 56. Prof Exp: Lectr path, Univ Witwatersrand, 58-63; assoc pathologist, Buffalo Gen Hosp, NY, 63-67; assoc prof path, Sch Med, Univ Cincinnati, 68-71; PROF PATH & ASSOC PROF MED, SCH MED, EMORY UNIV, 71- Concurrent Pos: Chief electron micros, Vet Admin Hosp, 71- Mem: Electron Micros Soc Am; Am Diabetes Asn; fel Am Col Path; fel Royal Micros Soc. Res: Vascular pathology with special reference to diabetes. Mailing Add: Atlanta Vet Admin Hosp 1670 Clairmont Rd Decatur GA 30033

YODER, CLAUDE H, b West Reading, Pa, Mar 16, 40; m 66; c 2. INORGANIC CHEMISTRY. Educ: Franklin & Marshall Col, BA, 62; Cornell Univ, PhD(chem), 66. Prof Exp: Asst prof chem, 66-74, ASSOC PROF & CHMN DEPT, FRANKLIN & MARSHALL COL, 74- Concurrent Pos: Dreyfus Found teacher scholar, 71. Mem: AAAS; Am Chem Soc. Res: Bonding in organometallic compounds. Mailing Add: Dept of Chem Franklin & Marshall Col Lancaster PA 17604

YODER, DAVID LEE, b Bellefontaine, Ohio, June 23, 36; m 61; c 2. PLANT PATHOLOGY, SOIL MICROBIOLOGY. Educ: Goshen Col, BA, 60; Mich State Univ, MS, 68, PhD(plant path), 71. Prof Exp: PLANT PATHOLOGIST, HUNT-WESSON FOODS, INC, 71- Res: Soil-borne plant diseases; disease of tomatoes; soil fungistasis. Mailing Add: Hunt-Wesson Foods Inc Davis CA 95616

YODER, DAVID SMILEY, pharmaceutical chemistry, see 12th edition

YODER, DONALD MAURICE, b Elkhart Co, Ind, Jan 3, 20; m 45; c 3. PLANT PATHOLOGY. Educ: Wabash Col, BA, 42; Cornell Univ, PhD(plant path), 50. Prof Exp: Sr fel biol res div, Union Carbide Chem Co, 50-54, head div, 54-61, mem staff tech develop, Agr Chem Div, 61-67, sr analyst mkt res & technol deleg UAR, Union Carbide Tech Serv Co, 67-68; mgr prod develop agr chem, 68-74, MGR REGIST & TOXICOL, BASF-WYANDOTTE CORP, 74- Concurrent Pos: Fel, Boyce Thompson Inst Plant Res, 50. Mem: Am Chem Soc; Am Phytopath Soc; Am Inst Biol Sci. Res: Evaluation of organic chemicals for agricultural uses; agricultural chemicals and food; pesticide regulations. Mailing Add: Agr Chem Dept BASF-Wyandotte Corp PO Box 181 Parsippany NJ 07054

YODER, HARRY WHITAKER, JR, b Chicago, Ill, Aug 2, 28; m 50; c 3. VETERINARY MEDICINE. Educ: Univ Ill, BS, 50; Iowa State Col, DVM, 54. Prof Exp: Asst prof vet res, Vet Res Inst, Iowa State Univ, 57-65; RES VET, SOUTHEAST POULTRY RES LAB, USDA, 65- Mem: Am Vet Med Asn; Am Asn Avian Path. Res: Poultry respiratory diseases, especially pleuropneumonia-like organism infections. Mailing Add: USDA SE Poultry Res Lab 934 College Station Rd Athens GA 30601

YODER, HATTEN SCHUYLER, b Cleveland, Ohio, Mar 20, 21; m 59; c 2. PETROLOGY. Educ: Univ Chicago, SB, 41, cert, 42 & 46; Mass Inst Technol, PhD(petrol), 48. Prof Exp: DIR, CARNEGIE INST WASHINGTON GEOPHYS LAB, 48- Concurrent Pos: Vis prof, Calif Inst Technol, 58, Univ Tex, 64, Univ Colo, 66 & Univ Cape Town, 67. Honors & Awards: Columbia Univ Bicentennial Medal, 54; Mineral Soc Am Award, 54; Day Medal, Geol Soc Am, 62; Arthur L Day Prize, Nat Acad Sci, 72; A G Werner Medal, Ger Mineral Soc, 72. Mem: Nat Acad Sci; Mineral Soc Am (pres, 72); Geol Soc Am; Am Geophys Union (pres, Volcanology, Geochem & Petrol Sect, 61-64); Geochem Soc. Res: Experimental petrology; phase equilibria in mineral systems; piezochemistry; properties of minerals at high pressure and high temperature; hydrothermal mineral synthesis. Mailing Add: Carnegie Inst Wash Geophys Lab 2801 Upton St NW Washington DC 20008

YODER, JOHN L, b Holden, Mo, Nov 10, 25; m 51; c 4. DENTISTRY. Educ: Cent Mo State Col, BS, 51, Univ Mo-Kansas City, DDS, 55; Univ Iowa, MS, 61. Prof Exp: From instr to asst prof, 59-64, ASSOC PROF CROWN & BRIDGE, COL DENT, UNIV IOWA, 64-, ACTG HEAD DEPT, 69- Concurrent Pos: USPHS teacher trainee, 59-61, instnl grant, 66. Mem: Am Dent Asn; Am Inst Oral Biol; Am Acad Hist Dent. Res: Fixed partial prosthesis, especially occlusion; preventive dentistry. Mailing Add: Col of Dent Univ of Iowa Iowa City IA 52240

YODER, JOHN MENLY, b Ft Wayne, Ind, Oct 4, 31; m 60; c 2. ENDOCRINOLOGY, IMMUNOCHEMISTRY. Educ: Purdue Univ, BS, 53, PhD(animal physiol), 61. Prof Exp: Res biochemist, Ames Co Div, 61-72, SR RES SCIENTIST, AMES CO DIV, MILES LABS, INC, 72- Mem: Am Chem Soc; NY Acad Sci. Res: Plant and animal physiology; silage fermentation; protein purification; characterization of proteins and polysaccharides by immunochemistry; gonadotropins; antibody production. Mailing Add: Ames Res Lab 819 McNaughton St Elkhart IN 46514

YODER, JULIAN CLIFTON, b Danville, Va, May 29, 10; m 37; c 2. GEOGRAPHY. Educ: Appalachian State Teachers Col, BS, 33; Peabody Col, MA, 38; Univ NC, PhD(geog), 49. Prof Exp: Instr geog, 33-38, assoc prof, 38-49, prof & chmn dept social studies, 59-65, prof geog & geol & chmn dept, 65-72, prof geog, 72-74, EMER PROF GEOG, APPALACHIAN STATE UNIV, 74- Mem: Asn Am Geogr; Nat Coun Geog Educ; Am Geog Soc. Res: Mountain, political and economic geography. Mailing Add: 118 Highland Ave Boone NC 28607

YODER, LEVON LEE, b Middlebury, Ind, June 22, 36; m 60; c 1. ELEMENTARY PARTICLE PHYSICS. Educ: Goshen Col, BA, 58; Univ Mich, MA, 61, PhD, 63. Prof Exp: Asst prof physics & chmn dept, Millikin Univ, 63-65; assoc prof, 65-71, PROF PHYSICS, ADRIAN COL, 71-, CHMN DEPT, 65- Mem: AAAS; Am Asn Physics Teachers. Res: High energy and cosmic ray physics. Mailing Add: 2499 Sword Hwy Adrian MI 49221

YODER, NEIL RICHARD, b Wichita, Kans, Mar 27, 37; m 68. PARTICLE PHYSICS. Educ: Kans State Teachers Col, BA, 59; Pa State Univ, PhD(physics), 69. Prof Exp: Instr physics, Mich State Univ, 65-67; SR RES ASSOC, UNIV MD, COLLEGE PARK, 67- Mem: Am Phys Soc. Res: Phenomenological analysis of moderate energy nucleon-nucleon data; application of computers to on-line analysis of nuclear physics experimental data. Mailing Add: Dept of Physics & Astron Univ of Md College Park MD 20742

YODER, OLEN CURTIS, b Fairview, Mich, Jan 26, 42; m 67. PLANT PATHOLOGY. Educ: Goshen Col, BA, 64; Mich State Univ, MS, 68, PhD(plant path), 71. Prof Exp: ASST PROF PLANT PATH, CORNELL UNIV, 71- Concurrent Pos: USDA res grant, 72-75; Rockefeller Found res grant, 74. Mem: AAAS; Am Phytopath Soc; Am Soc Plant Physiol; Int Soc Plant Path; Sigma Xi. Res: Physiology of plant disease; host-specific fungal toxins; control of post-harvest diseases. Mailing Add: Dept of Plant Path Cornell Univ Ithaca NY 14850

YODER, PAUL RUFUS, JR, b Huntingdon, Pa, Feb 6, 27; m 48; c 4. OPTICS. Educ: Juniata Col, BS, 47; Pa State Univ, MS, 50. Prof Exp: Assoc prof physics & math, Bridgewater Col, 50-51; physicist, US Army Frankford Arsenal, 51-61; proj engr, 61-67, MGR OPTICAL SYSTS DEPT, PERKIN-ELMER CORP, 67- Mem: Fel Optical Soc Am; Soc Photo-Optical Instrumentation Engrs. Res: Design and development of specialized optical instrumentation. Mailing Add: 9 Bhasking Ridge Rd Wilton CT 06897

YODER, ROBERT E, b Richmond, Va, Jan 1, 30; m 51; c 3. HEALTH PHYSICS, INDUSTRIAL HYGIENE. Educ: Appalachian State Teachers Col, BS, 51; Harvard Univ, ScD(radiol health), 63. Prof Exp: Jr health physicist, Oak Ridge Nat Lab, 54-57; from instr to asst prof health physics, Sch Pub Health, Harvard Univ, 57-66; health physicist, Lawrence Livermore Lab, Univ Calif, 65-72; ASST DIR NUCLEAR FACIL, DIV OPER SAFETY, US AEC, 72- Concurrent Pos: Consult radiation safety, Mass, 58-64; dir health safety & environ, Rockwell Int, Rocky Flats Plant, Colo. Mem: AAAS; Health Physics Soc. Res: Aerosol technology; basic properties of aerosols; physics and chemistry of small particles and their influence on health; radiation protection program administration in nuclear research institutions. Mailing Add: US AEC Germantown MD 20767

YODER, WAYNE ALVA, b Grantsville, Md, July 6, 43; m 67; c 1. INVERTEBRATE ZOOLOGY, ENTOMOLOGY. Educ: Goshen Col, BA, 65; Mich State Univ, MS, 71, PhD(zool), 72. Prof Exp: Asst prof, 72-75, ASSOC PROF BIOL, FROSTBURG STATE COL, 75- Mem: Am Inst Biol Sci; Am Soc Zoologists; Entom Soc Am; Sigma Xi. Res: Systematic and ecological studies of the mites associated with silphid beetles. Mailing Add: Dept of Biol Frostburg State Col Frostburg MD 21532

YODH, GAURANG BHASKAR, b Ahmedabad, India, Nov 24, 28; nat US; m 54; c 3. PHYSICS. Educ: Univ Bombay, BSc, 48; Univ Chicago, MS, 51, PhD(physics), 55. Prof Exp: Instr physics, Stanford Univ, 54-56; res fel, Tata Inst Fundamental Res, India, 57-58; res physicist, Carnegie Inst Technol, 58-59, asst prof physics, 59-61; assoc prof, 61-64, PROF PHYSICS, UNIV MD, COLLEGE PARK, 65-, PROF ASTRON, 74- Concurrent Pos: Consult, US Naval Res Lab, DC, 65- & Argonne Nat Lab, 66-; vis prof, Univ Ariz, 66-67. Mem: Am Phys Soc. Res: Experimental and phenomenological study of high energy interactions of elementary particles. Mailing Add: Dept of Physics Univ of Md College Park MD 20742

YOELI, MEIR, b Kaunas, Lithuania, Aug 20, 12; m 46; c 3. TROPICAL MEDICINE, PARASITOLOGY. Educ: Univ Kaunas, Lithuania, MSc, 32; Univ Basel, MD, 39. Prof Exp: Sci asst, Malaria Res Sta, Hebrew Univ, Israel, 34-38, lectr parasitol & trop med, 49-56; assoc prof prev med, 56-68, PROF PREV MED, SCH MED, NY UNIV, 68- Concurrent Pos: Vis prof, Univ Pa; consult infectious & parasitic dis, Willowbrook State Sch; mem int expert panel malaria, WHO. Honors & Awards: Wallach Prize, Ministry Health, Israel, 55; Laveran Medal, 65; Maimonides Award, 66. Mem: Am Soc Parasitol; Am Soc Trop Med & Hyg. Res: Tropical diseases; host-parasite relationship; immunological aspects of insect born diseases; geographical-pathology of infectious zoonosis. Mailing Add: Dept of Prev Med NY Univ Med Ctr New York NY 10016

YOESTING, CLARENCE C, b Apr 5, 12; US citizen; m 40; c 2. PHYSICS, SCIENCE EDUCATION. Educ: Cent State Col, Okla, BS, 36; Univ Okla, MEd, 47, EdD(sci educ), 65. Prof Exp: Teacher sci, Lacy Schs, Okla, 36-38, Loyal Schs, 38-40, Newkirk, Okla, 41-42 & Ponca Mil Acad, 45-47; prin & teacher, Tonkawa, Okla, 47-51; counr & teacher, Northeast High Sch, Oklahoma City, 51-61; PROF PHYSICS, CENT STATE UNIV OKLA, 61- Mem: Nat Sci Teachers Asn; Am Inst Physics; Am Asn Physics Teachers. Mailing Add: Dept of Phys Sci Cent State Univ Edmond OK 73034

YOGORE, MARIANO G, JR, b Iloilo City, Philippines, Dec 29, 21; m 45; c 7. PARASITOLOGY, PUBLIC HEALTH. Educ: Univ Philippines, MD, 45; Johns Hopkins Univ, MPH, 48, DrPH, 57; Philippine Bd Prev Med & Pub Health, dipl, 56. Prof Exp: From instr to prof parasitol, Univ Philippines, 45-67; res assoc & assoc prof, 67-69, RES ASSOC & PROF PARASITOL, UNIV CHICAGO, 69- Concurrent Pos: USPHS res fel, Dept Microbiol, Univ Chicago, 59-61; mem, Nat Res Coun Philippines, 57- Mem: Am Soc Trop Med & Hyg. Res: Immunity to parasitic diseases with special interest in schistosomiasis. Mailing Add: Dept of Microbiol Univ of Chicago Chicago IL 60637

YOH, PHILIP, astronomy, space science, see 12th edition

YOHE, CLEON RUSSELL, b New York, NY, July 8, 41; m 66; c 1. ALGEBRA. Educ: Univ Pa, AB, 62; Univ Chicago, MS, 63, PhD(math), 66. Prof Exp: Asst prof, 66-71, ASSOC PROF MATH, WASH UNIV, 71- Mem: AAAS; Am Math Soc. Res: Structure theory of rings, specifically structure of rings of endomorphisms of modules over commutative noetherian rings. Mailing Add: Dept of Math Wash Univ St Louis MO 63130

YOHE, DANIEL CHARLES, b Urbana, Ill, Nov 19, 40; m 62; c 3. PHYSICAL CHEMISTRY, SCIENCE EDUCATION. Educ: DePauw Univ, BA, 61; Case Western Reserve Univ, MS, 63, PhD(phys chem), 67. Prof Exp: Asst prof chem, Cornell Col, Iowa, 67-71; asst prog dir, Instructional Sci Equip Prog, Undergrad Educ in Sci, 72-73, PROG MGR, PRE-COL EDUC IN SCI DIV, NSF, 73- Mem: Am Chem Soc. Mailing Add: Precol Educ in Sci NSF Washington DC 20550

YOHE, GAIL ROBERT, b Burlington, Iowa, Feb 8, 04; m 31; c 2. ORGANIC CHEMISTRY. Educ: Cornell Col, BA, 25; Univ Ill, MS, 27, PhD(org chem), 29. Prof Exp: Res chemist, Goodyear Tire & Rubber Co, Ohio, 29-30; from asst prof to assoc prof chem, Ohio Wesleyan Univ, 30-37; from assoc chemist to sr chemist, 37-72, head div coal chem, 45-72, EMER SR CHEMIST, ILL STATE GEOL SURV, 72- Mem: Am Chem Soc. Res: Chemical nature of coal. Mailing Add: 313 E Washington St Urbana IL 61801

YOHE, JAMES MICHAEL, b Delaware, Ohio, June 8, 36; m 61; c 3. MATHEMATICS, COMPUTER SCIENCE. Educ: DePauw Univ, BA, 57; Univ Wis-Madison, MS, 62, PhD(math), 67. Prof Exp: Asst prof math, Math Res Ctr, Univ Wis-Madison, 67-68; asst prof, Pa State Univ, 68-69; proj assoc, 69-71, ASST DIR, MATH RES CTR, UNIV WIS-MADISON, 71-, ASST PROF COMPUT SCI, 73- Concurrent Pos: Lectr, Univ Wis, 71-72. Mem: Am Math Soc; Math Asn Am; Asn Comput Mach. Res: Computer systems programming; computer arithmetic; topology of 3-dimensional manifolds with emphasis on link theory. Mailing Add: Math Res Ctr Univ of Wis 610 Walnut St Madison WI 58706

YOHN, DAVID STEWART, b Shelby, Ohio, June 7, 29; m 50; c 5. MICROBIOLOGY. Educ: Otterbein Col, BS, 51; Ohio State Univ, MS, 53, PhD, 57; Univ Pittsburgh, MPH, 60. Prof Exp: Res assoc, Univ Pittsburgh, 56-60, asst res prof microbiol, Grad Sch Pub Health, 60-62; asst res prof, State Univ NY Buffalo, 62-71; assoc cancer res scientist, Roswell Park Mem Inst, 62-69; PROF VIROL, OHIO STATE UNIV, 69-, DIR OHIO STATE UNIV CANCER RES CTR, 73- Concurrent Pos: Consult, Nat Cancer Inst, 70-; mem med & sci adv bd & bd trustees, Leukemia Soc Am, 71-; secy gen, Int Asn Comp Res Leukemia & Related Dis, 74- Mem: AAAS; Am Soc Microbiol; Am Asn Cancer Res. Res: Mammalian and oncogenic viruses; virus host-cell relationships; tumor immunology. Mailing Add: Ohio State Univ Cancer Res Ctr Suite 357 1580 Cannon Dr Columbus OH 43210

YOHO, CLAYTON W, b Glen Dale, WVa, Dec 4, 24; m 49; c 3. ORGANIC CHEMISTRY. Educ: W Liberty State Col, BSc, 49; Univ Pittsburgh, MSc, 51, PhD(org chem), 57. Prof Exp: Jr fel org res, Mellon Inst Indust Res, 51-54; process develop chemist, Merck & Co, Inc, 57-60; res supvr org res, 65-71, STAFF RES SUPVR, JOHNSON WAX, 71-, SR CHEMIST, 60- Mem: AAAS; Am Chem Soc. Res: Process development work involving vitamins B-1, B-12, and gibrel; organic synthesis work in the areas of adhesives, insect repellents, insect attractants and insecticides; product development of oral hygiene products. Mailing Add: 419 William St Racine WI 53402

YOHO, JAMES GIBSON, b Brownsville, Pa, Sept 3, 21; m 43; c 1. FOREST ECONOMICS. Educ: Univ Ga, BS, 47; NY Univ, MF, 48; Mich State Univ, PhD, 56. Prof Exp: Asst forest mgt, Col Forestry, NY Univ, 47-48; asst prof forest econ & mgt, Stephen F Austin State Col, 48-51; asst forestry & forest econ, Mich State Univ, 51-53; from asst prof to assoc prof, Iowa State Univ, 53-57; from assoc prof to prof, Duke Univ, 57-68, dir grad studies, Sch Forestry, 62-68; ASST DIR WOODLANDS, INT PAPER CO, 69- Concurrent Pos: Consult, Southeastern Forest Exp Sta, US Forest Serv, 57-59; ed, J Forestry, 61-68; res assoc, Resources for the Future, 64-65, 65; partic, Guest Prog Culture & Sci, Fed Repub Ger, 65; Fulbright lectr, Lincoln Col, Canterbury, 68; partic, Commonwealth Land Use & Develop Symp, Oxford Univ; consult, Govt Southern Resource Anal Comt, NC & Legis of NY. Mem: Am Econ

Asn; Am Agr Econ Asn; Soc Am Foresters. Res: Forest land, policy and production economics. Mailing Add: Woodlands Int Paper Co 220 E 42nd St New York NY 10017

YOHO, ROBERT OSCAR, b Solsberry, Ind, Sept 29, 13; m 34; c 3. PUBLIC HEALTH EDUCATION. Educ: Ind Univ, AB, 34, MA, 38, HSD, 57. Prof Exp: High sch teacher, Ind, 35-41; health educ consult, 41-45, dir div pub health educ, 45-68, dir bur pub health educ, rec & statist, 46-68, ASST STATE HEALTH COMNR, IND STATE BD HEALTH, 68- Concurrent Pos: Instr, Med Sch, Ind Univ, Indianapolis & Butler Univ. Mem: Am Pub Health Asn; Am Asn Health, Phys Educ & Recreation. Res: Interrelation of services provided by rehabilitation agencies of Indiana. Mailing Add: 2318 N Fisher Ave Indianapolis IN 46224

YOHO, TIMOTHY PRICE, b Nov 8, 41. DEVELOPMENTAL BIOLOGY, ENTOMOLOGY. Educ: West Liberty State Col, BS, 67; WVa Univ, PhD(develop biol & entomol), 72. Prof Exp: Fel res, WVa Univ, 72-74; ASST PROF BIOL, LOCK HAVEN COL, 74- Mem: Sigma Xi; Entomol Soc Am; Am Inst Biol Sci. Res: Electron microscopy; biochemistry; electrophysiology to study the photodynamic effect of light on dye-fed insects; death in visible light-exposed insects caused by food, drug and cosmetic dyes; danger to consumer and new insecticide development. Mailing Add: Dept of Biol Lock Haven Col Lock Haven PA 17745

YOKE, JOHN THOMAS, b New York, NY, Feb 27, 28; m 56; c 3. INORGANIC CHEMISTRY. Educ: Yale Univ, BS, 48; Univ Minn, MS, 50, PhD(chem), 54. Prof Exp: Res chemist, Procter & Gamble Co, 56-58; instr chem, Univ NC, 58-59; asst prof, Univ Ariz, 59-64; assoc prof, 64-70, PROF CHEM, ORE STATE UNIV, 70- Mem: Am Chem Soc. Res: Inorganic synthesis; coordination chemistry; group V compounds; chemical binding; oxidation of ligands; catalysis. Mailing Add: Dept of Chem Ore State Univ Corvallis OR 97331

YOKLEY, ORANGE EDWARD, organic chemistry, analytical chemistry, see 12th edition

YOKLEY, PAUL, JR, b Mitchellville, Tenn, Aug 3, 23; m 52; c 2. ZOOLOGY. Educ: George Peabody Col, BS, 49, MA, 50; Ohio State Univ, PhD(zool), 68. Prof Exp: Instr biol, 50-53, asst prof zool, 53-68, PROF ZOOL, UNIV NORTH ALA, 68- Concurrent Pos: Sci consult, Colbert Co Schs, 67-68; Tenn Game & Fish Comn res grant, 69-72; fisheries scientist, Am Fisheries Soc; consult, Tenn Valley Authority, 71- Honors & Awards: Res Award, Asn Southeastern Biologists, 70; State Conserv Educr Year, Ala Wildlife Fedn, 72. Mem: Soc Syst Zool; Am Malacol Union. Res: Life history and ecology of freshwater mussels; ecology of the freshwater mussels in the Tennessee River. Mailing Add: Dept of Biol Univ of NAla Box 5153 Florence AL 35630

YOKOSAWA, AKIHIKO, b Kofu, Japan, Nov 19, 27; US citizen; m 57; c 3. HIGH ENERGY PHYSICS. Educ: Tohoku Univ, Japan, BS, 51; Univ Cincinnati, MS, 53; Ohio State Univ, PhD(nuclear physics), 57. Prof Exp: Assoc prof physics, Ill State Univ, 57-59; physicist, 59-70, SR PHYSICIST, ARGONNE NAT LAB, 70- Mem: Fel Am Phys Soc. Res: Elementary particle physics. Mailing Add: Argonne Nat Lab 9700 S Cass Ave Argonne IL 60439

YOKOYAMA, HENRY, agricultural chemistry, plant biochemistry, see 12th edition

YOKOYAMA, HISAKO OGAWA, histochemistry, see 12th edition

YOKOYAMA, KATSUYUKI, plant physiology, biochemistry, see 12th edition

YOKOYAMA, MITSUO, b Fukuoka, Japan, Mar 6, 27; m 55. IMMUNOLOGY, IMMUNOHEMATOLOGY. Educ: Juntendo Med Sch, Tokyo, MD, 50; Tokyo Med & Dent Univ, DMSc, 58. Prof Exp: Intern, Japanese Red Cross Cent Hosp, 51; staff, United Nat Blood Bank, 406 Med Gen Lab, Tokyo, 51-53; res asst, Tokyo Med & Dent Univ, 54-56, chief consult, 56-58, lectr, 59; res assoc, NIH, Japan, 58-59; vis scientist, NIH, US, 59-62; assoc res physician & vis res asst prof, Sch Med, Univ Calif, 62-63; assoc prof genetics, Univ Hawaii, 64-65, clin assoc prof med & genetics, 66-72; vpres sci affairs & dir labs, Kallestad Labs, 73-74; ASSOC PROF PATH & MICROBIOL, COL OF MED & COL BASIC MED SCI, UNIV ILL &HEAD CLIN IMMUNOL, UNIV HOSP LAB, 74- Concurrent Pos: Dir res, Kuakini Med Res Inst, 65-72; consult hemat, Tripler Army Med Ctr, Honolulu, Hawaii, 72. Res: Blood group; human genetics; clinical immunology. Mailing Add: Clin Immunol Univ Hosp Lab Univ of Ill Med Ctr Chicago IL 60680

YOLDAS, BULENT ERTURK, b Turkey, Feb 19, 38; US citizen. CERAMICS, GLASS TECHNOLOGY. Educ: Ohio State Univ, BCerE, 63, MS, 64, PhD(glass & refractory), 66. Prof Exp: Sr engr mat sci, Owens-Ill Tech Ctr, 66-74; SR SCIENTIST MAT SCI, WESTINGHOUSE RES LABS, 74- Mem: Am Ceramic Soc; Sigma Xi. Res: Coating technology for electronic and consumer products; formation of glass and ceramic materials by chemical polymerization; high surface area; catalytic materials; porous ceramic, metal-organic compounds. Mailing Add: Res & Develop Ctr Westinghouse Elec Corp Pittsburgh PA 15235

YOLE, RAYMOND WILLIAM, b Middlesbrough, Eng, Feb 21, 27; Can citizen; m 57; c 4. STRATIGRAPHY, SEDIMENTOLOGY. Educ: Univ New Brunswick, BSc, 47; Johns Hopkins Univ, MA, 58; Univ BC, PhD(geol), 65. Prof Exp: Geologist, Calif Standard Co, Alta, 47-51, asst to vpres explor, 51-53, dist stratigr, 53-56; asst prof, 63-67, chmn dept, 67-70, ASSOC PROF GEOL, CARLETON UNIV, 67- Concurrent Pos: Vis res geologist, Univ Reading, 70-71; assoc ed, Can Soc Petrol Geol, 74- Mem: Can Soc Petrol Geol; Petrol Soc; Can Inst Mining & Metall; fel Geol Asn Can; Am Asn Petrol Geol. Res: Petroleum geology; Permian stratigraphy, Canadian cordillera; Paleozoic stratigraphy and sedimentology. Mailing Add: Dept of Geol Carleton Univ Ottawa ON Can

YOLLES, SEYMOUR, b Brooklyn, NY, Oct 12, 14; m 40; c 2. CHEMISTRY. Educ: Brooklyn Col, BS, 35; Univ Miami, MS, 48; Univ NC, PhD(inorg chem), 51. Prof Exp: Anal chemist, US Testing Co, NJ, 37-38; tutor chem, Brooklyn Col, 38; res chemist, Colgate-Palmolive-Peet Co, NJ, 38-42, Ridbo Labs, 42-43 & Alrose Chem Co, RI, 43-45; consult chemist, 45-47; from instr to asst prof chem, Univ Miami, 47-52; staff chemist, Fabric & Finishes Dept, E I du Pont de Nemours & Co, Inc, 52-67; PROF CHEM, UNIV DEL, 67- Concurrent Pos: AEC asst, Univ NC, 50-51; adj prof, Univ Del, 56-67; consult, E I du Pont de Nemours & Co, Inc, 68- Mem: Am Chem Soc. Res: Organic and inorganic chemistry; reactions in liquid sulphur dioxide; quinoxalines; zirconium and hafnium; boron-phosphorus and silicon-phosphorus polymers; inorganic polymers; long time sustained release compositions for morphine antagonists. Mailing Add: Dept of Chem Univ of Del Newark DE 19711

YOLLES, STANLEY FAUSST, b New York, NY, Apr 19, 19; m 42; c 2. MEDICINE, PSYCHIATRY. Educ: Brooklyn Col, AB, 39; Harvard Univ, AM, 40; NY Univ, MD, 50; Johns Hopkins Univ, MPH, 57. Prof Exp: Parasitologist, Sector Malaria Lab, US Dept Army, 41-42, assoc dir, 42-44; intern, USPHS Hosp, Staten Island, NY, 50-51, resident psychiat, Lexington, Ky, 51-54; staff psychiatrist, Ment Health Study Ctr, NIMH, 54-55, from assoc dir to dir, 55-60, from assoc dir to dep dir extramural progs, 60-63, from dep dir to dir, NIMH, 63-70, assoc adminr for ment health, US Dept Health, Educ & Welfare, 68-70; PROF PSYCHIAT & BEHAV SCI & CHMN DEPT, STATE UNIV NY STONY BROOK, 71- Concurrent Pos: Clin prof psychiat, George Washington Univ, 67-; spec consult, NY City Bd Educ; mem, Prof Adv Bd, Int Comt Against Ment Illness, 68-, Nat Adv Panel, Am Jewish Comt, 69-, Expert Adv Panel Ment Health, WHO, 69-, Med Adv Comt, Am Joint Distribution Comt, 70-; assoc ed bd, J Dis Dis Nerv Syst, 68-; trustee, NY Sch Psychiat; sr consult, Southside Hosp, Bay Shore, South Oaks Hosp, Amityville & Nassau County Med Ctr, NY. Mem: AAAS; fel Am Psychiat Asn; fel Am Pub Health Asn; fel Am Col Psychiat; fel NY Acad Sci. Res: Community mental health; mental health asministration; epidemiology of mental health. Mailing Add: 2 Soundview Ct Stony Brook NY 11790

YOLLICK, BERNARD LAWRENCE, b Toronto, Ont, Mar 24, 22; nat US; m 47; c 2. ANATOMY, SURGERY. Educ: Univ Toronto, MD, 45; Am Bd Surg, dipl, 57; Am Bd Otolaryngol, dipl, 67. Prof Exp: Instr anat, Col Med, Univ Sask, 47-49; clin asst prof, Col Med, Baylor Univ, 54-67; asst prof, Univ Tex Dent Sch, 54-61; lectr surg, Univ Tex Postgrad Sch Med, 57-67; ASST PROF OTOLARYNGOL, UNIV TEX HEALTH SCI CTR DALLAS, 67- Concurrent Pos: Fel surg, Am Cancer Soc, 53-54; consult, Houston Pulmonary Cytol Proj, 59 & Vet Admin Hosp, Dallas. Mem: Soc Human Genetics; Am Asn Anat; AMA; fel Am Col Surg; Am Soc Head & Neck Surg. Res: Induction of bone tumors in animals using heavy metals; experimental surgery in animals. Mailing Add: 4229 Bobbitt Dr Dallas TX 75229

YONCE, LLOYD ROBERT, b Roscoe, Mont, Sept 27, 24; m 48; c 2. PHYSIOLOGY. Educ: Mont State Col, BS, 49; Oregon State Univ, MS, 52; Univ Mich, PhD(physiol), 55. Prof Exp: Instr physiol, Ore State Univ, 51-52; instr, Med Sch, Univ Mich, 55-56; asst prof, 57-61, ASSOC PROF PHYSIOL, SCH MED, UNIV NC, CHAPEL HILL, 61- Concurrent Pos: Fel, Med Col Ga, 56-57; USPHS spec fel physiol, Univ Gothenburg. Mem: Am Phys Soc; Am Microcirculation Soc. Res: Cardiovascular physiology; neurophysiology; physiology of diving animals. Mailing Add: Dept of Physiol Univ of NC Sch of Med Chapel Hill NC 27514

YONDA, ALFRED WILLIAM, b Cambridge, Mass, Aug 10, 19; m 49; c 4. MATHEMATICS. Educ: Univ Ala, BS, 52, MA, 54. Prof Exp: Mathematician, Rocket Res, Redstone Arsenal, Ala, 53 & US Army Ballistic Res Labs, Aberdeen Proving Ground, Md, 54-56; instr math, Temple Univ, 56-57; assoc scientist, Res & Adv Develop Div, Avco Corp, 57-59; sr proj mem tech staff, Radio Corp Am, 59-66; mgr comput anal & prog dept, Raytheon Corp, Bedford, 66-70, prin engr, Missile Systs Div, 70-73; MGR SYSTS ANAL & PROG DEPT, EASTERN DIV, GTE SYLVANIA, NEEDHAM, MASS, 73- Mem: Nat Soc Prof Engrs; Am Math Soc; Math Asn Am; Asn Comput Mach; NY Acad Sci. Res: Computer systems analysis; simulation; communications systems analysis; numerical analysis. Mailing Add: 12 Sunset Dr Medway MA 02053

YONENAKA, HIDEO H, b San Jose, Calif, July 7, 27; m 65; c 2. MARINE MICROBIOLOGY. Educ: Univ Calif, Berkeley, BA, 51; Univ Southern Calif, MA, 55, PhD(bact, biochem), 61. Prof Exp: Fel, Scripps Clin & Res Found, 61-62; vis lectr embryol & parasitol, Univ Calif, Santa Barbara, 62-63; assoc prof microbiol, 63-72, PROF MICROBIOL, SAN FRANCISCO STATE UNIV, 72- Concurrent Pos: Vis scientist, Woods Hole Oceanog Inst, 71. Mem: Am Soc Microbiol. Res: Marine microbial ecology and physiology; bacteriology; biochemistry. Mailing Add: Dept of Biol San Francisco State Univ San Francisco CA 94132

YONETANI, TAKASHI, b Kagawa-ken, Japan, Aug 6, 30; US citizen; m 58; c 1. BIOCHEMISTRY, BIOPHYSICS. Educ: Osaka Univ, BA, 53, PhD(biochem), 61. Prof Exp: Res fel biochem, Johnson Found, Univ Pa, 58-61; Swedish Med Res Coun res fel, Nobel Med Inst, Stockholm, 62-64; from asst prof to assoc prof phys biochem, 64-68, PROF PHYS BIOCHEM, JOHNSON RES FOUND, UNIV PA, 68- Concurrent Pos: USPHS career develop award, 67-72. Mem: AAAS; Am Soc Biol Chemists; Am Chem Soc; Biophys Soc. Res: Purification, crystallization and characterization of cytochrome oxidase, alcohol dehydrogenase and cytochrome c peroxidase; determination of structure and function of these enzymes by spectrophotometry, electron spin resonance and x-ray diffraction techniques; heart disease; artificial hemoglobin; metalloporphyrin synthesis. Mailing Add: Johnson Res Found Univ Pa 37th & Hamilton Walk Philadelphia PA 19174

YONG, FOOK CHOY, b Malaysia. BIOLOGICAL CHEMISTRY. Educ: Nanyang Univ, BS, 61; Univ Man, MS, 64; Ore State Univ, PhD(chem), 68. Prof Exp: Res assoc, 68-70, vis asst prof, 70-75, ASST PROF BIOCHEM, DIV RES, STATE UNIV NY ALBANY, 75- Concurrent Pos: Lectr, Univ Malaya, Malaysia, 73-75. Mem: Am Chem Soc. Res: Electron transport and oxidative phosphorylation; porphyrins, hemes and cytochrome chemistry; oxygen metabolism; protein chemistry. Mailing Add: Dept of Chem State Univ NY Albany NY 12222

YONG, MAN SEN, b Sumatra, Indonesia, Mar 29, 41; Can citizen; m 65; c 2. PHARMACOLOGY. Educ: Nat Taiwan Univ, BSc, 62; Univ Alta, MSc, 64, PhD(pharmacol), 68. Prof Exp: ASST PROF PHARMACOL, McGILL UNIV, 71- Concurrent Pos: Med Res Coun Can fel, McGill Univ, 68-71, Can Heart Found res scholar, 71- Mem: Pharmacol Soc Can. Res: Adrenergic receptor mechanism; disposition and metabolism of sympathomimetic amines in vascular tissue; mechanism of cardiovascular hypersensitivity to noradrenaline. Mailing Add: Dept of Pharmacol & Therapeut McGill Univ Montreal PQ Can

YONGE, KEITH A, b London, Eng, June 22, 10; m 47; c 4. PSYCHIATRY. Educ: McGill Univ, MD, CM, 48; Univ London, dipl psychol med, 52. Prof Exp: Mem staff psychiat, Med Res Coun, Eng, 51-52; dir, Ment Health Clin, Can, 52-54; from asst prof to assoc prof psychiat, Univ Sask, 55-57, clin dir, Univ Hosp, 55-57; prof & head dept, 55-75, EMER PROF PSYCHIAT, UNIV ALTA, 75- Concurrent Pos: Can Coun leave fel, 71-72. Mem: Psychiat Asn; Can Med Asn; Can Psychiat Asn; Can Ment Health Asn. Res: Basic and clinical psychiatry; phenomenology of depression; cognitive effects of cannabis; nature of human aggression. Mailing Add: Kingscote Rd RR 2 Cobble Hill BC Can

YONGUE, WILLIAM HENRY, b Charlotte, NC, Aug 21, 26; m 48; c 1. PROTOZOOLOGY, AQUATIC ECOLOGY. Educ: Johnson C Smith Univ, BS, 49; Univ Mich, MS, 62; Va Polytech Inst & State Univ, PhD(zool), 72. Prof Exp: Head dept sci, West Charlotte High Sch, 59-70; from instr to asst prof, 70-73, ASSOC PROF ZOOL, VA POLYTECH INST & STATE UNIV, 73- Concurrent Pos: Res assoc & proj scientist, Univ Mich Biol Sta, 69-75. Mem: Soc Protozoologists; Am Inst Biol Sci; AAAS; Nat Asn Biol Teachers. Res: The ecology of freshwater protozoans using polyurethane foam substrates for sampling and as microhabitats. Mailing Add: Dept of Biol Va Polytech Inst & State Univ Blacksburg VA 24061

YONKE, THOMAS RICHARD, b Kankakee, Ill, Nov 30, 39; m 63; c 3. ENTOMOLOGY, SYSTEMATICS. Educ: Loras Col, BS, 62; Univ Wis, Madison,

MS, 64, PhD(entom), 67. Prof Exp: Instr entom, Univ Wis, Madison, 66-67; asst prof, 67-71, ASSOC PROF ENTOM, UNIV MO, COLUMBIA, 71- Mem: Entom Soc Am; Soc Syst Zool; Entom Soc Can. Res: Biology and taxonomy of Hemiptera; taxonomy of immature Heteroptera. Mailing Add: Dept of Entom Univ of Mo Columbia MO 65201

YONKMAN, FREDRICK FRANCIS, pharmacology, deceased

YONUSCHOT, GENE R, b Brooklyn, NY, Oct 29, 36; m 57; c 2. BIOCHEMISTRY. Educ: Univ Mo-Columbia, PhD(biochem), 69. Prof Exp: Instr biochem, Univ NC, 69-71; ASST PROF BIOCHEM, GEORGE MASON UNIV, 71- Res: Acidic chromosomal proteins and t-RNA in relation to the control of cell division and differentiation. Mailing Add: Dept of Biochem George Mason Univ Fairfax VA 22030

YOO, BONG YUL, b Pusan, Korea, June 30, 35; Can citizen. PLANT PHYSIOLOGY, CELL BIOLOGY. Educ: Seoul Nat Univ, BSc, 58; Okla State Univ, MSc, 61; Univ Calif, Berkeley, PhD(bot), 65. Prof Exp: Nat Res Coun Can fel, 65-66; asst prof, 66-71, ASSOC PROF BIOL, UNIV NB, FREDERICTON, 71- Mem: Am Soc Cell Biol; Am Soc Plant Physiol; Bot Soc Am; Can Soc Cell Biol. Res: Biogenesis of plant cell organelles. Mailing Add: Dept of Biol Univ of NB Fredericton NB Can

YOO, TAI-JUNE, b Seoul, Korea, Mar 7, 35; US citizen; m 63; c 3. IMMUNOLOGY, INTERNAL MEDICINE. Educ: Seoul Nat Univ, MD, 59; Univ Calif, Berkeley, PhD(med physics), 63. Prof Exp: Teaching asst biophys, Div Med Physics, Univ Calif, Berkeley, 60-61; res asst, Lawrence Radiation Lab, 61-63; asst in med, Sch Med, Wash Univ, 63-66; sr cancer res scientist, Roswell Park Mem Inst, 66-68; res prof biol, Niagara Univ, 68-69; asst prof, 72-75, ASSOC PROF MED, COL MED, UNIV IOWA, 75- Concurrent Pos: NIH fel immunol, Wash Univ, 65-66; from intern to asst resident, Barnes Hosp, St Louis, Mo, 63-65; assoc resident, Sch Med, NY Univ, 68-69; clin investr, Vet Admin Hosp, Iowa City, Iowa, 72- Mem: AAAS; Biophys Soc; Am Asn Immunol; NY Acad Sci; Am Fedn Clin Res. Res: Molecular and cellular biology of immune phenomena; structure and function of antibody active site; interaction of ligand and protein; regulation in immune response; mechanism of action of immunopotentiator; immunologic and allergic disorders. Mailing Add: Div of Allergy & Immunol Univ of Iowa Col of Med Iowa City IA 52240

YOOD, BERTRAM, b Bayonne, NJ, Jan 6, 17; m 44; c 3. MATHEMATICAL ANALYSIS. Educ: Yale Univ, BS, 38, PhD(math), 47; Calif Inst Technol, MS, 39. Prof Exp: From instr to asst prof math, Cornell Univ, 47-53; from asst prof to prof, Univ Ore, 53-72; PROF MATH, PA STATE UNIV, 72- Concurrent Pos: Vis assoc prof, Univ Calif, 56-57; vis res assoc, Yale Univ, 58-59; mem, Inst Adv Study, 61-62. Mem: Am Math Soc. Res: Banach algebra; Banach spaces; analysis. Mailing Add: Dept of Math Pa State Univ University Park PA 16802

YOON, CHAI HYUN, b Korea, Aug 14, 20; nat US; m 53; c 2. GENETICS. Educ: Alma Col, AB, 50; Ohio State Univ, PhD(genetics), 53. Prof Exp: Res assoc genetics, Ohio State Univ, 53-55, instr, 55-56; asst prof biol, Lycoming Col, 56-58; from asst prof to assoc prof genetics, 59-66, PROF GENETICS, BOSTON COL, 66- Mem: AAAS; Genetics Soc Am; Am Genetic Asn. Res: Neuromuscular mutations in mice. Mailing Add: Dept of Biol Boston Col Chestnut Hill MA 02167

YOON, DO YEUNG, b Inchon, Korea, Jan 22, 47; m 71; c 3. POLYMER CHEMISTRY. Educ: Seoul Nat Univ, BS, 69; Univ Mass, MS, 71, PhD(polymer sci), 73. Prof Exp: Res assoc chem, Stanford Univ, 73-75; RES SCIENTIST POLYMER MAT, RES LAB, IBM CORP, 75- Mem: Am Chem Soc; Am Phys Soc. Res: Conformational statistics and conformation-dependent properties of polymers; morphology and properties of polymers. Mailing Add: Res Lab K42/282 IBM Corp 5600 Cottle Rd San Jose CA 95193

YOON, JONG SIK, b Suwon, Korea, Jan 25, 37; US citizen; m 62; c 3. GENETICS, EVOLUTION. Educ: Yonsei Univ, Korea, BS, 61; Univ Tex, Austin, MA, 64, PhD(genetics), 65. Prof Exp: Res scientist assoc IV, Univ Tex, Austin, 62-65; res assoc oncol, M D Anderson Hosp, Univ Tex, 66-68; asst prof genetics & cytol, Yonsei Univ, Korea, 68-71; res scientist IV & V genetics, 71-74, instr cell biol, 74-75, RES SCIENTIST GENETICS, UNIV TEX, AUSTIN, 75- Honors & Awards: Young Scientist Award, Int Union Against Cancer, 70. Mem: Genetics Soc Am; Soc Study Evolution; AAAS; Sigma Xi. Res: Cytogenetics; mutation; oncogenetics; radiation genetics; genome organization, speciation and evolution of Drosophila and other species; genic balance between euchromatin and heterochromatin of chromosomes. Mailing Add: Dept of Zool Univ of Tex Austin TX 78712

YORAN, CALVIN S, organic chemistry, see 12th edition

YORK, CARL MONROE, JR, b Macon, Ga, July 2, 25; m 52; c 3. PHYSICS. Educ: Univ Calif, Berkeley, AB, 46, MA, 50, PhD(physics), 51. Prof Exp: Fulbright fel physics, Univ Manchester, 51-52; res fel, Calif Inst Technol, 52-54; asst prof, Univ Chicago, 54-59; Ford Found & Guggenheim Found fel, Europ Orgn Nuclear Res, Geneva, 59-60; from assoc prof to prof, Univ Calif, Los Angeles, 60-69, asst chancellor res, 65-69, assoc dean grad div, 63-65; tech asst basic sci, Off Sci & Technol, Exec Off of the President, 69-72; vchancellor acad affairs, Univ Denver, 72-74; CONSULT, 74- Concurrent Pos: Consult, Argonne Nat Lab, 57-61, TRW Systs, Calif, 61-, Film Assocs, Calif, 62-, Lawrence Radiation Lab, 66, NSF, 72-, Fedn Rocky Mountain States, 74-75, Colo Energy Res Inst, 75- & Calif Energy Resource, Conserv & Develop Comn, 75- Mem: AAAS; Am Phys Soc. Res: Cosmic rays; elementary particles; positive sigma hyperon; high energy accelerator design; pion-nucleon scattering and production; muon decay and interactions; photo-production of pions; regional and state energy policies and plans; federal budgets for basic science and national science policy. Mailing Add: 679 Carlston Ave Oakland CA 94610

YORK, CHARLES JAMES, b Calif, Sept 28, 19; m 44; c 4. VIROLOGY, BACTERIOLOGY. Educ: Univ Calif, AB, 43; Ohio State Univ, DVM, 48; Cornell Univ, PhD(virol, bact), 50. Prof Exp: Asst bact, Univ Calif, 41-43, bacteriologist, Vet Sci Dept, 43-44; sr bacteriologist, Med Res Dept, Ohio State Univ, 44-46, bacteriologist, Vet Col, 46; res assoc, Vet Virus Inst, Cornell Univ, 48-52; dir virus res lab, Pitman-Moore Co, Ind, 52-63; prof vet sci & head dept vet res lab, Mont State Univ, 63-65; dir inst comp biol, Zool Soc San Diego, 65-70; ASSOC PROF COMP PATH, MED SCH, UNIV CALIF, SAN DIEGO, 67- Concurrent Pos: Mem WHO. Mem: Soc Exp Biol & Med; Am Vet Med Asn; Tissue Cult Asn; US Animal Health Asn; Am Pub Health Asn. Res: Virus diseases of man and animals; comparative medicine for animal research models. Mailing Add: Med Sch Dept Path Univ of Calif at San Diego La Jolla CA 92037

YORK, CHRISTOPHER LAFAYETTE, ecology, see 12th edition

YORK, DEREK H, b Yorkshire, Eng, Aug 12, 36; m 61; c 1. GEOPHYSICS. Educ: Oxford Univ, BA, 57, DPhil(physics), 60. Prof Exp: Lectr, 60-62, from asst prof to assoc prof, 62-74, PROF PHYSICS, UNIV TORONTO, 74- Concurrent Pos: Chmn subcomt isotope geophys & mem comt geod & geophys, Nat Res Coun Can, 67.

Mem: Am Geophys Union; Can Asn Physicists. Res: Isotopic geophysics; temporal evolution of continents; reversals of earth's magnetic field. Mailing Add: Dept of Physics Univ of Toronto Toronto ON Can

YORK, DONALD GILBERT, b Shelbyville, Ill, Oct 28, 44; m 66; c 1. ASTROPHYSICS. Educ: Mass Inst Technol, BA, 66; Univ Chicago, PhD(astrophysics), 70. Prof Exp: From res asst to res assoc, 70-72, RES STAFF ASTROPHYSICS, PRINCETON UNIV, 72- Mem: Am Astron Soc. Res: Determination physical properties of interstellar gas and dust, using ultraviolet and visual spectroscopic techniques in our own galaxy as well as distant galactic systems. Mailing Add: Princeton Univ Observ Princeton NJ 08540

YORK, DONALD HAROLD, b Moose Jaw, Sask, Jan 30, 44; m 65; c 2. NEUROPHYSIOLOGY, NEUROSCIENCE. Educ: Univ BC, BSc, 65, MSc, 66; Monash Univ, Australia, PhD(neurophysiol), 69. Prof Exp: Asst prof physiol, Queen's Univ, Ont, 68-75; ASSOC PROF PHYSIOL, SCH MED, UNIV MO-COLUMBIA, 75- Concurrent Pos: Med Res Coun Can grant, Queen's Univ, Ont, 69-72; scholar, Med Res Coun Can, 70. Mem: AAAS; Can Physiol Soc; Am Physiol Soc; Soc Neurosci; Pharmacol Soc Can. Res: Motor control; basal ganglia and movement; synaptic transmission in central nervous system. Mailing Add: Dept of Physiol Univ of Mo Sch of Med Columbia MO 65201

YORK, E TRAVIS, JR, soil science, agronomy, see 12th edition

YORK, GEORGE KENNETH, II, b Tucson, Ariz, July 1, 25; m 47; c 5. MICROBIOLOGY Educ: Stanford Univ, AB, 50; Univ Calif, PhD(microbiol), 60. Prof Exp: Asst bacteriologist, Nat Canners Asn, 51-53; actg asst prof, 58-60, asst prof food sci & technol, 60-66, EXTEN MICROBIOLOGIST, UNIV CALIF, DAVIS, 66- Mem: Am Soc Microbiol; Inst Food Technol. Res: Food microbiology; intermediary metabolism of microbes; thermomicrobiology; modes of inhibition of microbes by chemicals; treatment and disposal of waste. Mailing Add: Dept of Food Sci & Technol Univ of Calif Davis CA 95616

YORK, GEORGE WILLIAM, b St Louis, Mo, Sept 26, 45; m 68; c 2. LASERS. Educ: St Louis Univ, BS, 67; Univ Mo, Rolla, MS, 69, PhD(physics), 71. Prof Exp: Res assoc physics, Joint Inst Lab Astrophys, 72-74; MEM STAFF PHYSICS, LOS ALAMOS SCI LAB, 74- Mem: Am Phys Soc; Laser Inst Am. Res: Development of high power gas lasers utilizing preionized electrical discharges in metal vapor system. Mailing Add: Los Alamos Sci Lab Lab L-2 MS-552 Los Alamos NM 87545

YORK, HERBERT FRANK, b Rochester, NY, Nov 24, 21; m 47; c 3. PHYSICS, SCIENCE POLICY. Educ: Univ Rochester, AB, 42, MS, 43; Univ Calif, PhD, 49. Hon Degrees: DSc, Case Western Reserve Univ, 60; LLD, Univ San Diego, 64; DrHumL, Claremont Grad Sch, 74. Prof Exp: Asst physics, Univ Rochester, 42-43; physicist, Radiation Lab, Univ Calif, 43-54, assoc dir, 54-58, dir, Livermore Lab, 54-58, asst prof physics, Univ, 51-54; dir advan res projs div, Inst Defense Anal & chief scientist, Advan Res Projs Agency, US Dept Defense, 58, dir defense res & eng, Off Secy Defense, 58-61; chancellor, 61-64 & 70-72, grad dean, 69-70 & 72-73, PROF PHYSICS, UNIV CALIF, SAN DIEGO, 65- Concurrent Pos: Mem sci adv bd, US Air Force, 53-57, ballistic missile adv comt, Secy Defense, 55-58 & sci adv panel, US Army, 56-58; mem, President's Sci Adv Comt, 57-58 & 64-67, vchmn, 65-67; mem gen adv comt, US Arms Control & Disarmament Agency, 61-69; mem bd trustees, Aerospace Corp, 62- & Inst Defense Anal, 65 & 67-; Guggenheim fel, 72-73; mem continuing comt, Conf Sci & World Affairs, 73- Mem: Am Phys Soc; Am Astronaut Soc; Am Inst Aeronaut & Astronaut; Inst Elec & Electronics Eng; Int Acad Astronaut. Res: Science and public affairs; disarmament problems. Mailing Add: 6110 Camino de la Costa La Jolla CA 92037

YORK, JAMES WESLEY, JR, b Raleigh, NC, July 3, 39; m 61; c 2. THEORETICAL PHYSICS. Educ: NC State Univ, BS, 62, PhD(physics), 66. Prof Exp: Asst prof physics, NC State Univ, Raleigh, 65-68; res assoc physics, Princeton Univ, 68-69, lectr, 69-70, asst prof, 70-73; ASSOC PROF PHYSICS, UNIV NC, CHAPEL HILL, 73- Honors & Awards: Third Prize, Gravity Res Found, Mass, 75. Mem: Am Phys Soc; Am Asn Physics Teachers; AAAS. Res: Gravitation and relativity; mathematical, astrophysical and quantum theoretic aspects. Mailing Add: Dept of Physics & Astron Univ of NC Chapel Hill NC 27514

YORK, JOHN LYNDAL, b Morton, Tex, Aug 14, 36; m 58; c 2. BIOCHEMISTRY, PHYSICAL ORGANIC CHEMISTRY. Educ: Harding Col, BS, 58; Johns Hopkins Univ, PhD(biochem), 62. Prof Exp: NIH trainee, 62-64; biochemist, Stanford Res Inst, 64-65; asst prof biochem, Med Units, Univ Tenn, Memphis, 65-68; ASSOC PROF BIOCHEM, SCH MED, UNIV ARK, LITTLE ROCK, 68- Concurrent Pos: Vis prof, Karolinska Inst, Stockholm, 74-75; fel, Swedish Med Res Coun, 75. Mem: Am Soc Biol Chem; Am Chem Soc. Res: Nature of the active site and mechanisms of action of non-heme iron proteins; Mössbauer effect; mechanisms of oxidation and oxygenation. Mailing Add: Dept of Biochem Univ of Ark Med Col Little Rock AR 72201

YORK, JOHN OWEN, b Parkin, Ark, July 11, 23; m 49; c 2. PLANT BREEDING, PLANT GENETICS. Educ: Miss State Univ, BS, 48, MS, 50; Tex A&M Univ, PhD(plant breeding), 62. Prof Exp: From instr to assoc prof, 52-68, PROF AGRON, UNIV ARK, FAYETTEVILLE, 68- Mem: Am Soc Agron; Am Genetic Asn. Res: Hybrid corn and grain sorghum breeding and production. Mailing Add: Dept of Agron Univ of Ark Fayetteville AR 72701

YORK, OWEN, JR, b Evansville, Ind, Oct 18, 27; m 48; c 2. ORGANIC CHEMISTRY. Educ: Evansville Col, BA, 48; Univ Ill, MA, 50, PhD(org chem), 52. Prof Exp: Asst, Univ Ill, 48-52; res chemist, Hercules Powder Co, 52-56; head dept chem, Ill Wesleyan Univ, 56-60; res supvr, W R Grace & Co, 60-61; assoc prof, 61-64, chmn dept, 64-73; reader & table leader, Advan Placement Chem, 64-72, PROF CHEM, KENYON COL, 64-, CHIEF READER, ADVAN PLACEMENT CHEM, 72- Concurrent Pos: NSF fel, Stanford Univ, 68-69. Mem: AAAS; Am Chem Soc; The Chem Soc. Res: Grignard reaction; oxidation; electrophilic substitution; isomerization of aromatic acids; synthetic photochemistry; catalysis. Mailing Add: Dept of Chem Kenyon Col Gambier OH 43022

YORK, ROBERT JOSEPH, inorganic chemistry, see 12th edition

YORK, SHELDON STAFFORD, b New Haven, Conn, Oct 29, 43; m 68; c 2. BIOPHYSICAL CHEMISTRY. Educ: Bates Col, BS, 65; Stanford Univ, PhD(biochem), 71. Prof Exp: Am Cancer Soc res fel chem, Calif Inst Technol, 70-72; ASST PROF CHEM, UNIV DENVER, 72- Mem: Am Chem Soc. Res: Protein - DNA recognition processes, specifically the conformational changes within the lac repressor protein which affect its ability to bind to the lac operator. Mailing Add: Dept of Chem Univ of Denver Denver CO 80210

YORKE, JAMES ALAN, b Peking, China, Aug 3, 41; US citizen; m 63; c 2. APPLIED MATHEMATICS, BIOMATHEMATICS. Educ: Columbia Univ, AB, 63; Univ Md,

College Park, PhD(math), 66. Prof Exp: From res assoc to res assoc prof, 66-72, RES PROF MATH, INST FLUID DYNAMICS & APPL MATH, UNIV MD, COLLEGE PARK, 72- Concurrent Pos: NSF grant, 70-75. Mem: AAAS; Am Math Soc; Soc Indust & Appl Math. Res: Qualitative ordinary differential equations; applications in epidemiology. Mailing Add: Inst Fluid Dynamics & Appl Math Univ of Md College Park MD 20740

YORTON, JOAN BANNISTER, b Chicago, Ill, Aug 6, 33; m 66; c 2. ORGANIC CHEMISTRY. Educ: Univ Fla, BS, 54, MS, 57, PhD(chem), 59. Prof Exp: Asst scientist & chem teacher, Oak Ridge Inst Nuclear Studies, 59-63, scientist, 63-65; fel chem, 65-66, INSTR PHYS SCI, UNIV FLA, 67- Mem: Am Chem Soc; Am Asn Univ Profs; Nat Sci Teachers Asn. Res: Synthesis of derivatives of piperazine; configurational relationships of 1,2,2-triphenylethanol and 1,2,2-triphenyethylamine; radio chemical criterion for configurational relationship. Mailing Add: 12Rte 1 Box 306B Alachua FL

YOS, DAVID ALBERT, b Trenton, NJ, Apr 24, 23; m 46; c 2. BOTANY, HISTORY OF BIOLOGY. Educ: NY Univ, AB, 48; Univ Mo, MA, 52; Univ Iowa, PhD(bot, plant anat), 60. Prof Exp: Instr biol, Burlington Col, 52-59; asst prof bot, Univ Wis-Green Bay, 60-62; teacher, High Sch, NJ, 63; from asst prof to assoc prof biol, 63-73, PROF BIOL, EASTERN NMEX UNIV, 73- Mem: AAAS. Res: Fluorescence microscopy of plant tissues; development of periderm; microtechnique; history of microscopy and photomicrography. Mailing Add: Dept of Biol Eastern NMex Univ Portales NM 88130

YOS, JERROLD MOORE, b Clinton, Iowa, Jan 1, 30; m 60; c 3. THEORETICAL PHYSICS, Educ: Univ Nebr, AB, 52, MS, 54, PhD(physics), 56. Prof Exp: SR SCIENTIST, SYSTS DIV, AVCO CORP, 57- Mem: Am Phys Soc. Res: Electromagnetics; atomic physics; kinetic theory of gases; electrical discharges. Mailing Add: Avco Systs Div 201 Lowell St Wilmington MA 01887

YOSHIDA, AKIRA, b Okayama, Japan, May 10, 24; m 54. BIOCHEMISTRY, GENETICS. Educ: Univ Tokyo, MS, 47, DSc, 54. Prof Exp: From instr to asst prof chem, Univ Tokyo, 51-54, assoc prof chem & biochem, 54-60; res assoc biochem, Univ Pa, 61-63; res chemist, NIH, 63-65; res prof med genetics, Univ Wash, 64-72; DIR DEPT BIOCHEM GENETICS, CITY OF HOPE MED CTR, 72- Concurrent Pos: Rockefeller Found Int scholar, 55-56. Mem: AAAS; Am Soc Biol Chem; Int Soc Hemat; Soc Human Genetics. Res: Study on the changes of protein structure and properties of the enzymes due to mutation; regulatory mechanism of gene action. Mailing Add: Dept of Biochem Genetics City of Hope Med Ctr Duarte CA 91010

YOSHIDA, SHIRO, b Tokyo, Japan, Dec 24, 23; m; c 2. NUCLEAR PHYSICS. Educ: Univ Tokyo, BSc, 49, PhD(physics), 55. Prof Exp: Res assoc physics, Inst Fundamental Physics, Kyoto Univ, 52-56; assoc prof, Inst Nuclear Study, Univ Tokyo, 56-63; prof, Osaka Univ, 63-68; PROF PHYSICS, RUTGERS UNIV, NEW BRUNSWICK, 68- Concurrent Pos: Res fel math physics, Univ Birmingham, 54-56; vis prof physics, Univ Pittsburgh, 59-61; vis prof, Rutgers Univ, New Brunswick, 67-68. Mem: Am Phys Soc. Res: Theoretical nuclear structure physics, especially direct reactions, collective models and isobaric analog states. Mailing Add: Dept of Physics Rutgers Univ New Brunswick NJ 08903

YOSHIDA, TAKESHI, b Fukuoka, Japan, July 24, 38; m 64; c 2. IMMUNOPATHOLOGY. Educ: Univ Tokyo, MD, 63, Dr Med Sci, 70. Prof Exp: Res mem immunol, Dept Tuberc, NIH, Tokyo, 64-71; asst prof, Dept Path, State Univ NY Buffalo, 71-74; asst prof, 74-75, ASSOC PROF PATH, DEPT PATH, UNIV CONN HEALTH CTR, 75- Concurrent Pos: Res assoc, Dept Path, NY Univ Med Ctr, 67-68; vis assoc, Lab Immunol, Nat Inst Allergy & Infectious Dis, Bethesda, 68-69, vis scientist, 71; Buswell fel, State Univ NY Buffalo , Buffalo, 71-74, asst dir, Ctr Immunol, 73-74; Nat Inst Allergy & Infectious Dis res career develop award, 75. Mem: Am Asn Immunologists; Am Soc Exp Path; Am Asn Univ Pathologists. Res: Mechanisms of cell-mediated immunity, biological and physicochemical characterizations of effector molecules produced in vitro by stimulated lymphocytes, and in vivo activities of lymphokines. Mailing Add: Dept of Path Univ of Conn Health Ctr Farmington CT 06032

YOSHIHARA, HIDEO, physics, see 12th edition

YOSHIKAWA, HERBERT HIROSHI, b South Dos Palos, Calif, May 13, 29; m 60. SOLID STATE PHYSICS. Educ: Univ Chicago, PhB, 48, MS, 51; Univ Pa, PhD(physics), 58. Prof Exp: Asst, Proj Big Ben, Univ Pa, 53-56 & dept physics, 56-58; from physicist to sr physicist, Hanford Labs, Gen Elec Co, 58-64, actg mgr, 64; res assoc, Pac Northwest Labs, Battelle Mem Inst, 65, from actg mgr to mgr, 64-70; mgr irradiation anal sect, 70-74, MGR MAT ANAL & APPLN SECT, HANFORD ENG DEVELOP LAB, WESTINGHOUSE-HANFORD CO, 74- Concurrent Pos: Chmn panel in-pile dosimetry, Int Atomic Energy Agency, 64; chmn, Radiation Damage Subgroup, Working Group on Reactive Radiation Measurements, Int Atomic Energy Agency, 70- Mem: AAAS; Am Phys Soc; Am Nuclear Soc. Res: Radiation effects on solids; dosimetry for radiation damage studies; radiation effects to materials for fission and fusion power reactor applications. Mailing Add: Hanford Eng Develop Lab PO Box 1970 Richland WA 99352

YOSHIMINE, MASAO, organic chemistry, see 12th edition

YOSHIMOTO, CARL MASARU, b Honolulu, Hawaii, Apr 27, 22; m 57; c 2. ENTOMOLOGY. Educ: Iowa Wesleyan Col, BA, 50; Kans State Univ, MS, 52; Cornell Univ, PhD(entom), 55. Prof Exp: Entomologist, Entom Res Div, USDA, 55-57 & BP Bishop Mus, 58-69; SR RES SCIENTIST, DEPT ENVIRON, CAN FORESTRY SERV, 69- Concurrent Pos: Affil fac, Grad Sch, Univ Hawaii, 64-69; NSF grant, Brit Mus, London, Eng, 67-68. Mem: Entom Soc Am; Entom Soc Can. Res: Taxonomy of Hymenoptera; insect behavior and dispersal; zoogeography. Mailing Add: Biosysts Res Inst Cent Exp Farm Ottawa ON Can

YOSHIMURA, SEI, b Hilo, Hawaii, May 24, 22. CHEMOTHERAPY, VIROLOGY. Educ: Univ Hawaii, BS, 46. Prof Exp: Bacteriologist, Honolulu Health Dept, 47-49; lab technician, Hektoen Inst, Ill, 50-52; bacteriologist, Vet Admin Hosp, Minneapolis, Minn, 52-54 & Mt Sinai Hosp, Chicago, Ill, 54-55; virologist, Nepera Chem Co, Warner-Chillcott Co, 55-56; sect head microbiol, Grove Labs Div, Bristol-Myers Co, 56-65; SR RES ASST MICROBIOL, MERRELL NAT-LABS, RICHARDSON-MERRELL INC, 65- Res: Viral chemotherapy; medical microbiology; general toxicology; analgesics. Mailing Add: Infectious Dis Res Dept Merrell-Nat Labs 110 E Amity Rd Cincinnati OH 45215

YOSHINAGA, KOJI, b Yokohama, Japan, Mar 20, 32; m 61. ENDOCRINOLOGY. Educ: Univ Tokyo, BSc, 55, MSc, 57, PhD(agr), 60. Prof Exp: Trainee physiol reproduction, Worcester Found Exp Biol, 61-64; vis scientist, Agr Res Coun Unit Reproduction Physiol & Biochem, Cambridge, 64-66; staff scientist, Worcester Found Exp Biol, 66-69; res assoc, 69, asst prof, 69-72, ASSOC PROF ANAT, HARVARD MED SCH, 72- Concurrent Pos: Pop Coun fel, Worcester Found Exp Biol, 62-63; Pop Coun fel, Agr Res Coun Unit Reproduction Physiol & Biochem, Cambridge, 64-65, Lalor Found fel, 65-66. Mem: Am Asn Anat; Endocrine Soc; Soc Study Reproduction; Am Physiol Soc; Soc Study Fertil. Res: Endocrinology and physiology of reproduction in female animals, especially the mechanisms involved in ovo-implantation, ovarian function and relationship between the egg development and hormone action. Mailing Add: Harvard Med Sch 45 Shattuck St Boston MA 02115

YOSHINO, TIMOTHY PHILLIP, b Turlock, Calif, Apr 5, 48. PARASITOLOGY, IMMUNOBIOLOGY. Educ: Univ Calif, Santa Barbara, BA, 70, MA, 71, PhD(biol), 75. Prof Exp: Res assoc parasitol, Univ Calif, Santa Barbara, 71-73; RES ASSOC INVERT IMMUNOL, LEHIGH UNIV, 75- Concurrent Pos: USPHS res fel, Nat Inst Allergy & Infectious Dis, 75. Mem: Am Soc Parasitologists; Soc Invert Path; Am Soc Zoologists; AAAS; Sigma Xi. Res: Seralogical and cellular mechanisms of internal defense in the pelecypod and gastropod molluscs. Mailing Add: Inst Pathobiol Ctr Health Sci Bldg 17 Lehigh Univ Bethlehem PA 18015

YOSIM, SAMUEL JACK, b St Petersburg, Fla, Apr 21, 20; m 49; c 2. PHYSICAL CHEMISTRY. Educ: Univ Fla, BS, 48; Univ Chicago, MS, 49, PhD(phys chem), 52. Prof Exp: Assoc chemist, Argonne Nat Lab, 51-52; res chemist, 52-56, mgr pollution technol unit, 56-75, MGR PHYS CHEM UNIT, CHEM GROUP, ATOMICS INT DIV, ROCKWELL INT, 75- Res: Molten salts, high temperature chemistry; nuclear chemistry; fuel chemistry. Mailing Add: 23812 Killion Woodland Hills CA 91364

YOSS, KENNETH M, b Hudson, Iowa, Jan 13, 26; m 55; c 3. ASTRONOMY. Educ: Univ Mich, BS, 48, MS, 50, PhD(astron), 53. Prof Exp: Asst prof astron & physics, Wilson Col, 52-53; from asst prof to assoc prof, La State Univ, 53-59; assoc prof astron, Mt Holyoke Col, 59-64; PROF ASTRON, UNIV ILL, URBANA, 64- Mem: Am Astron Soc; Int Astron Union. Res: Spectrophotometry of objective prism and slit spectra; spectral and luminosity classification. Mailing Add: Univ of Ill Observ Urbana IL 61801

YOSS, ROBERT EUGENE, b Spooner, Wis, Nov 28, 24; m 47; c 3. NEUROANATOMY, NEUROLOGY. Educ: Univ Tenn, MD, 48; Univ Mich, MS & PhD(neuroanat), 52. Prof Exp: From instr to asst prof anat, Med Sch, Univ Mich, 49-54; from asst prof to assoc prof, 57-70, PROF NEUROL, MAYO GRAD SCH MED, UNIV MINN, 70, CONSULT, MAYO CLIN, 57- Concurrent Pos: Fel neruol, Mayo Grad Sch Med, Univ Minn, 55-57. Mem: Am Asn Anat. Res: Anatomy of spinal cord; narcolepsy. Mailing Add: Mayo Grad Sch of Med Univ of Minn Rochester MN 55901

YOST, DON M, b Tedrow, Ohio, Oct 30, 93; m 17; c 2. INORGANIC CHEMISTRY. Educ: Univ Calif, BS, 23; Calif Inst Technol, PhD, 26. Prof Exp: Fel, 26-28, asst prof chem 29-35, assoc prof inorg chem, 35-42, prof, 42-64, EMER PROF INORG CHEM, CALIF INST TECHNOL, 64- Concurrent Pos: Int Ed Bd fel, Sweden & Ger, 28-29; div mem & sect chmn, Off Sci Res & Develop, 40-43; mem adv comt to dir res, AEC, 50-; Little vis prof, Mass Inst Techno.l, 53. Honors & Awards: Presidental Cert of Merit, 48. Mem: Nat Acad Sci; AAAS; fel Am Phys Soc; Am Acad Arts & Sci. Res: Physical chemistry; rates of chemical reactions; Raman spectra; gas equilibria; chemical effects of x-rays; spectroscopy; chemistry of platinum metals; electrode potentials of rare elements; artificial radioactivity; low temperature thermodynamics; fluorine; microwaves; rare earths; mathematics. Mailing Add: 1270 Cardova St Apt 9 Pasadena CA 91106

YOST, EARL KNISELY, JR, analytical statistics, mathematical statistics, see 12th edition

YOST, FRANCIS LORRAINE, b Punxsutawney, Pa, Dec 18, 08; m 54. THEORETICAL PHYSICS. Educ: Univ Ky, BS, 29, MS, 31; Univ Wis, PhD(physics), 36. Prof Exp: Asst physics, Univ Wis, 31-36 & Purdue Univ, 36-37; physicist, US Rubber Co, Mich, 37-42; sr tech aide, Div 17, Nat Defense Res Comt, 43-44; Off Sci Res & Develop, Eng & France, 44-45, Washington, 45-46; chief math anal subdiv, US Naval Ord Lab, 46-47 & weapons anal div, 50-54; assoc prof physics, Ill Inst Technol, 47-51; prof physics, Univ Ky, 54-74, head dept, 54-65; RETIRED. Concurrent Pos: Vis prof, Univ Indonesia, 56-58. Mem: Am Phys Soc; Am Asn Physics Teachers. Res: Nuclear physics; physical properties of rubber; naval ordnance; guided missile lethalities. Mailing Add: 1320 E Cooper Dr Lexington KY 40502

YOST, GEORGE PALMER, high energy physics, see 12th edition

YOST, HENRY THOMAS, JR, b Baltimore, Md, Jan 22, 25; m 48; c 1. RADIATION GENETICS. Educ: Johns Hopkins Univ, AB, 47, PhD, 51. Hon Degrees: MA, Amherst Col, 65. Prof Exp: From instr to assoc prof, 51-65, PROF BIOL, AMHERST COL, 65- Concurrent Pos: NIH fel, Chester Beatty Res Inst, London, Eng, 62-63 & 69-70. Mem: Fel AAAS; Soc Develop Biol; NY Acad Sci; Bot Soc Am; Genetics Soc Am. Res: Investigation of the structure of the chromosome and other particulates; radiation biology. Mailing Add: 75 N East St Amherst MA 01002

YOST, JOHN FRANKLIN, b Brodbecks, Pa, Mar 21, 19; m 43; c 2. ORGANIC CHEMISTRY. Educ: Western Md Col, BS, 43; Johns Hopkins Univ, AM, 48, PhD(org chem), 50. Hon Degrees: DS, Western Md Col, 64. Prof Exp: Jr instr, Johns Hopkins Univ, 46-50; res chemist, Synthetic Rubber & Plastics Lab, US Rubber Co, 43-44 & 46; agr chemist & group leader, Agr Chem Labs, 50-57, mgr tech dept, Phosphate & Nitrate Div, 57, dir plant indust develop, Agr Div, 58-62, dir prod develop & govt registrn, 62-66, DIR AGR RES & DEVELOP, INT DEPT, AM CYANAMID CO, 66- Mem: Am Chem Soc. Res: Synthetic organic chemistry; agricultural product analysis, formulation and registration; pesticides, fertilizers, feed additives and veterinary products; research administration. Mailing Add: Int Dept Am Cyanamid Co PO Box 400 Princteon NJ 08540

YOST, ROBERT STANLEY, b Pottsville, Pa, Jan 24, 21; m 43; c 2. ORGANIC CHEMISTRY. Educ: Pa State Univ, BS, 42; Duke Univ, PhD(org chem), 48. Prof Exp: Chemist, Hercules Powder Co, 42-43; lab instr chem, Duke Univ, 43-44, chemist, 44-46; chemist, 47-49, head process res group, Redstone Res Labs, 59-68, CHEMIST, RES LABS, ROHM AND HAAS CO, 68- Mem: Am Chem Soc. Res: Organic synthesis including synthetic resins and explosives; process research. Mailing Add: Res Labs Rohm & Haas Co Spring House PA 19477

YOST, WILLIAM A, b Dallas, Tex, Sept 21, 44; m 69; c 1. PSYCHOACOUSTICS, PSYCHOPHYSICS. Educ: Colo Col, BA, 66; Ind Univ, Bloomington, PhD(psychol), 70. Prof Exp: NSF fel, Univ Calif, San Diego, 70-71; asst prof speech psychol, 71-74, ASSOC PROF PSYCHOL, UNIV FLA, 74- Concurrent Pos: Mem, Comt Acoust Stand, 72-75. Mem: Acoust Soc Am; Am Psychol Asn; Sigma Xi; AAAS. Res: Binaural hearing; pitch perception; speech perception; auditory sensitivity and discrimination in young children; noise pollution. Mailing Add: Rm 36 Arts & Sci Bldg IASCP Univ of Fla Gainesville FL 32601

YOST, WILLIAM JACQUE, b Fairview, WVa, Dec 23, 13; m 38, 66; c 2. PHYSICS. Educ: Furman Univ, AB, 35, BS, 36; Syracuse Univ, MA, 38; Brown Univ,

PhD(physics), 40. Prof Exp: Instr math, Furman Univ, 35-36 & physics, Syracuse Univ, 40-41; res physicist, Field Res Dept, Magnolia Petrol Co, 41-44, group leader, 44-46, res assoc, 46-52, tech adv to res dir, 52-54, dir res, Marathon Oil Co, 54-73; MGT CONSULT, 73- Mem: AAAS; Am Phys Soc; Soc Explor Geophys. Res: Theoretical quantum mechanics; infrared and ultraviolet spectroscopy; electromagnetic phenomena in conducting media; x-ray and electron diffraction; electromagnetic and elastic wave propagation theory. Mailing Add: PO Box 746 Estes Park CO 80517

YOST, WILLIAM LASSITER, b Washington, DC, Mar 14, 23; m 51; c 4. ORGANIC CHEMISTRY. Educ: Univ Va, BS, 44, MS, 47, PhD(chem), 49. Prof Exp: AEC fel biol Sci, Calif Inst Technol, 49-50, NIH fel, 50-51; res chemist, US Naval Ord Test Sta, 51-52; sr chemist, Ciba Pharmaceut Co, 52-65, head appl math, 65-67, dir chem develop & appl math, 67-69; PRES, UNION DATA CORP, 69- Mem: AAAS; Am Chem Soc; Asn Comput Mach; NY Acad Sci. Res: Application of mathematics and computer techniques to the solution of problems arising in medical and pharmaceutical research. Mailing Add: 208 Schooley's Mountain Rd Long Valley NJ 07853

YOTIS, WILLIAM WILLIAM, b Almyros, Greece, Jan 17, 30; US citizen; m 57; c 3. MICROBIOLOGY. Educ: Wayne State Univ, BS, 54, MS, 56; Northwestern Univ, PhD(microbiol), 60. Prof Exp: Asst microbiol, Wayne State Univ, 54-56; clin bacteriologist, Univ Hosp, Univ Mich, 56-57; from instr to assoc prof, 60-72, interim chmn, 64-65, PROF MICROBIOL, MED SCH, LOYOLA UNIV CHICAGO, 73- Concurrent Pos: Res grants, NIH, 60-, Eli Lilly Res Labs, 65-66, Syntex Res Labs, 66- & Upjohn Co, 66-; consult, Am Type Cult Collection, 66- Mem: AAAS; Am Med Asn; Am Soc Microbiol; Brit Soc Gen Microbiol; NY Acad Sci. Res: Investigations on mechanisms of microbial pathogenicity; nonspecific host defense mechanisms; bactericidal activity of body fluids; staphylococcal host-parasite relationship; hormonal influence on infection; bacterial physiology; isotochophoresis and scanning isoelectric focusing of medically important proteins. Mailing Add: Loyola Univ Med Sch 2160 S First Ave Maywood IL 60153

YOUDIN, MYRON, b Stamford, Conn, Oct 14, 18; m 43; c 2. BIOMEDICAL ENGINEERING. Educ: NY Univ, BEE, 40, MEE, 48. Prof Exp: Chief engr, Amperex Electronic Prod Inc, 39-49; vpres eng, Anton Electronic Labs, 49-61; vpres & chief engr, EON Corp, 61-66; biomed engr, Mt Sinai Sch Med, 66-67; SR RES SCIENTIST, INST REHAB MED, NY UNIV MED CTR, 67- Concurrent Pos: Adj prof med technol, C W Post Col, 67- Mem: NY Acad Sci; NY Acad Med; Inst Elec & Electronics Engrs; Biomed Eng Soc; Soc Nuclear Med. Res: Development of instrumentation for the measurement of distributed cerebral blood flow in humans; development of locomotion and environmental equipment for quadriplegic and other severely handicapped patients. Mailing Add: 54-46 186th St Flushing NY 11365

YOUKER, JAMES EDWARD, b Cooperstown, NY, Nov 13, 28; m 63; c 2. RADIOLOGY. Educ: Colgate Univ, AB, 50; Univ Buffalo, MD, 54; Am Bd Radiol, dipl, 60. Prof Exp: Asst prof radiol, Med Col Va, 61-63; from asst prof to assoc prof, Univ Calif, San Francisco, 64-68; PROF RADIOL & CHMN DEPT, MILEAUKEE COUNTY GEN HOSP, WIS, 68- Concurrent Pos: NIH res fel, Allmanna Sjukhuset, Malmo, Sweden, 62 & 63; attend radiologist, Proj HOPE, Indonesia, 58; USPHS grant dir, Training Radiologist & Technician Teams in Mammography, & co-dir, Training Prog Cardiovasc Radiol, 65-68. Mem: Radiol Soc NAm; Am Col Radiol; Asn Univ Radiol; Int Soc Lymphology. Res: Pulmonary function changes with lymphographic contrast media; pathology of congenital heart disease; chylous ascites and the spectrum of the disease. Mailing Add: Dept of Radiol Milwaukee County Gen Hosp Milwaukee WI 53202

YOUKER, JOHN, b Auburn, NY, Sept 7, 43; m 67; c 2. PHYSICAL CHEMISTRY. Educ: Rensselaer Polytech Inst, BS, 65, PhD(phys chem), 69. Prof Exp: Chemist, Coated Abrasive & Tape Div, Norton Co, 68-69; ASST PROF CHEM, HUDSON VALLEY COMMUNITY COL, 69- Mem: Am Chem Soc. Res: Spectroscopy; analytical chemistry; environmental science. Mailing Add: Dept of Chem Hudson Valley Community Col Troy NY 12180

YOUMANS, ANNE STEWART, b Springfield, Mo, Sept 26, 16; m 45. MEDICAL MICROBIOLOGY, TUBERCULOSIS. Educ: Stanford Univ, AB, 38; Northwestern Univ, MS, 43, PhD(bact, immunol), 46; Am Bd Med Microbiol, cert pub health & lab microbiol. Prof Exp: Asst, 46-49, from instr to assoc prof, 49-71, PROF MICROBIOL, MED SCH, NORTHWESTERN UNIV, 71- Honors & Awards: Pasteur Award, Ill Soc Microbiol, 70. Mem: Fel AAAS; fel Am Acad Microbiol; Am Asn Immunol; Am Soc Microbiol; Reticuloendothelial Soc. Res: Medical microbiology; tuberculosis both chemotherapy and immunity; a specific RNA vaccine for tuberculosis; host-parasite relationships; cellular immunity and its relationship to delayed hypersensitivity; metabolism of the tuberclebacillus. Mailing Add: Dept of Microbiol Northwestern Univ Med Sch Chicago IL 60611

YOUMANS, GUY PARRY, b St Anthony, Idaho, Feb 21, 08; m 45. BACTERIOLOGY. Educ: Univ Wash, BS, 31, MS, 32, PhD(bact, immunol), 35; Northwestern Univ, MD, 43; Am Bd Path, dipl, 55. Prof Exp: From instr to assoc prof microbiol, 35-49, chmn dept, 49-75, PROF MICROBIOL, MED SCH, NORTHWESTERN UNIV, 49-, PROF IMMUNOL, 75- Mem: AAAS; Am Soc Microbiol; Soc Exp Biol & Med; Am Thoracic Soc; fel Am Acad Microbiol. Res: Tuberculosis; medical bacteriology; immunology. Mailing Add: Dept of Microbiol Med Sch Northwestern Univ Chicago IL 60611

YOUMANS, HUBERT LAFAY, b Lexsy, Ga, Aug 2, 25; m 51; c 1. ANALYTICAL CHEMISTRY. Educ: Emory Univ, AB, 49, MS, 50; La State Univ, PhD(anal chem), 61. Prof Exp: Chemist, Savannah River Plant, E I du Pont de Nemours & Co, SC, 52-57; develop chemist, Sucrochem Div, Colonial Sugars Co, Lab, 57-58; asst prof chem, Ft Hays Kans State Col, 61-64; res chemist, Atlas Chem Indust, Inc, Del, 64-67; ASSOC PROF CHEM, WESTERN CAROLINA UNIV, 67- Mem: Am Chem Soc. Res: Absorptiometry; analytical separations; communications for chemistry students. Mailing Add: Box 375 A Cullowhee NC 28723

YOUMANS, JULIAN RAY, b Baxley, Ga, Jan 2, 28; m 54; c 3. NEUROSURGERY. Educ: Emory Univ, BS, 49, MD, 52; Univ Mich, MS, 55, PhD(neuroanat), 57; Am Bd Neurol Surg, dipl, 60. Prof Exp: From asst prof to assoc prof neurosurg, Sch Med, Univ Miss, 59-63; from assoc prof to prof, Med Col SC, 63-67, chief div, 63-67; PROF NEUROSURG & CHIEF DEPT, SCH MED, UNIV CALIF, DAVIS, 67- Mem: AMA; Am Acad Neurol; Am Asn Surg of Trauma; Am Col Surg; Am Asn Automotive Med. Res: Physiology of cerebral blood flow. Mailing Add: Dept of Neurol Surg Univ of Calif Sch of Med Davis CA 95616

YOUMANS, WILLIAM BARTON, b Cincinnati, Ohio, Feb 3, 10; m 32; c 3. PHYSIOLOGY. Educ: Western Ky State Col, BS, 32, MA, 33; Univ Wis, PhD(animal med), 38; Univ Ore, MD, 44. Prof Exp: Instr biol, Western Ky State Col, 32-35; from asst to instr physiol, Univ Wis, 35-38; from instr to prof, Med Sch, Univ Ore, 38-52, head dept, 45-52; chmn dept, 52-71, PROF PHYSIOL, SCH MED, UNIV WIS-MADISON, 52- Concurrent Pos: USPHS spec fel, 61-62; intern, Henry Ford Hosp, 44-45; mem physiol study sect, USPHS, 52-56, mem physiol training comt, 58-62. Mem: Am Physiol Soc; Soc Exp Biol & Med; Am Soc Pharmacol & Exp Therapeut. Res: Innervation of intestine; gastrointestinal motility; visceral reflexes; cardiac innervation, neurohormones; angiotensin. Mailing Add: Dept of Surg Sch of Med Univ of Wis Madison WI 53706

YOUNATHAN, EZZAT SAAD, b Deirut, Egypt, Aug 25, 22; nat US; m 58; c 2. BIOCHEMISTRY. Educ: Univ Cairo, BSc, 44; Fla State Univ, MA, 53, PhD, 55. Prof Exp: Chemist, Govt Labs, Egypt, 44-50; Seagrams' Int Training Prog fel, 50-51; res asst, Fla State Univ, 51-55, asst prof, 58-59; res assoc, Col Med, Univ Ill, 55-57; asst prof biochem, Sch Med, Univ Ark, Little Rock, 59-63, assoc prof & actg head dept, 63-66; NIH spec fel & vis prof, Inst Enzyme Res, Univ Wis, 66-67; PROF BIOCHEM, LA STATE UNIV, BATON ROUGE, 68- Mem: Am Soc Biol Chem; Am Chem Soc; Asn Am Med Cols; Soc Exp Biol & Med; NY Acad Sci. Res: Mechanism of enzyme action; protein chemistry; control of intermediary metabolism; diabetogenic substances. Mailing Add: Dept of Biochem La State Univ Baton Rouge LA 70803

YOUNATHAN, MARGARET TIMS, b Clinton, Mass, Apr 25, 26; m 58; c 2. FOOD SCIENCE, NUTRITION. Educ: Southern Miss Univ, BA, 46, BS, 50; Univ Tenn, MS, 51; Fla State Univ, PhD(food sci, nutrit), 58. Prof Exp: Instr food & nutrit, Ore State Univ, 51-55; Qm Food & Container Inst res fel, Fla State Univ, 58-59; instr pediat, Sch Med, Univ Ark, Little Rock, 62-65, asst prof, 65-68; ASSOC PROF FOOD & NUTRIT, LA STATE UNIV, BATON ROUGE, 71- Concurrent Pos: Consult, Ark State Health Dept, 62-68. Mem: Inst Food Technol; Am Home Econ Asn; Am Dietetic Asn. Res: Heme pigments; antioxidants; lipid oxidation; infant and child nutrition. Mailing Add: Dept of Food & Nutrit La State Univ Baton Rouge LA 70803

YOUNCE, GORDON BALDWIN, b San Francisco, Calif, Feb 6, 33; m 57; c 3. STRUCTURAL GEOLOGY, PHYSICAL OCEANOGRAPHY. Educ: Yale Univ, BS, 55; Cornell Univ, PhD(geol), 70. Prof Exp: Explor geologist, Mobil Oil Co, 62-65 & Humble Oil Co, 65-67; ASST PROF GEOL, RUTGERS UNIV, CAMDEN, 70-, ACTG CHMN DEPT, 73- Concurrent Pos: Rutgers Univ grant, Bonavista Bay, Nfld, 71-72. Mem: Am Asn Petrol Geologists; Geol Soc Am. Res: Structural geology of Bonavista Bay, Newfoundland; structure of Delaware Basin of West Texas; salt water intrusion and mixing in Damlico Sound, Delaware Bay. Mailing Add: Dept of Geol Rutgers Univ Camden NJ 08102

YOUNG, AINSLIE THOMAS, JR, b Norman, Okla, June 3, 43; m 68. POLYMER CHEMISTRY, PHOTOCHEMISTRY. Educ: Memphis State Univ, BS, 66, MS, 68; Univ Ky, PhD(phys chem), 71. Prof Exp: Fel, Univ Calif, Berkeley, 71-73; fel, La State Univ, Baton Rouge, 73-74; RES CHEMIST POLYMER CHEM, CTR RES LAB, MEAD CORP, 74- Mem: Am Chem Soc; Sigma Xi; AAAS. Res: Physical chemical and chemical kinetic studies of photo-polymerization reactions. Mailing Add: Ctr Res Lab Mead Corp Eighth & Hickory Sts Chillicothe OH 45601

YOUNG, ALLAN CHARLES, b Can, May 23, 11; nat US; m 46; c 1. PHYSICS, PHYSIOLOGY. Educ: Univ BC, BA, 30, MA, 32; Univ Toronto, PhD(physics), 34. Prof Exp: Tech develop engr, Northern Elec Co, 34-36; fel, Univ Rochester, 36-37, instr physiol, 38, Rockefeller fel, 38-40; physicist, Nat Res Coun, 40-45; head div physics, BC Res Coun, 45-49; res assoc physiol, 49-51, from asst prof to assoc prof physiol & biophys, 51-60, PROF PHYSIOL & BIOPHYS, UNIV WASH, 60- Mem: Am Phys Soc; Am Physiol Soc. Res: Electricity; magnetism; industrial physics; nerve and muscle physiology; respiration; control systems. Mailing Add: Dept of Physiol & Biophys Univ of Wash Seattle WA 98195

YOUNG, ALLEN MARCUS, b Ossining, NY, Feb 23, 42. ECOLOGY. Educ: State Univ NY New Platz, BA, 64; Univ Chicago, PhD(zool), 68. Prof Exp: Orgn Trop Studies, Inc fel for study in Costa Rica, Univ Chicago, 68-70; asst prof biol, Lawrence Univ, Appleton, Wis, 70-75; CUR & HEAD DEPT INVERTEBRATE ZOOL, MILWAUKEE PUB MUS, 75- Concurrent Pos: NSF fel for study in Costa Rica, Lawrence Univ, 71-74; NSF res grant, 75-; res assoc, Nat Mus Costa Rica, 75- Mem: AAAS; Ecol Soc Am; Lepidop Soc; Asn Trop Biol. Res: Population biology and behavior of neotropical Lepidoptera and Cicadidae; ecology of laboratory populations of Tribolium. Mailing Add: Dept of Invertebrate Zool Milwaukee Pub Mus 800 W Wells St Milwaukee WI 53233

YOUNG, ALVIN, organic chemistry, textile chemistry, see 12th edition

YOUNG, ANDREW TIPTON, b Canton, Ohio, Apr 4, 35; m 54, 63, 68; c 2. ASTRONOMY. Educ: Oberlin Col, BA, 55; Harvard Univ, MA, 57, PhD(astron), 62. Prof Exp: Res fel, Observ, Harvard Univ, 60-65, lectr astron, Univ, 62-65; asst prof, Univ Tex, 65-67; mem tech staff, Aerospace Corp, 67-68 & Jet Propulsion Lab, 68-73; res scientist, 73-75, VIS ASST PROF ASTRON, TEX A&M UNIV, 75- Concurrent Pos: Co-investr TV experiments, Mariner Mars, 69 & 71. Mem: Fel AAAS; Am Astron Soc; fel Royal Astron Soc; Optical Soc Am. Res: Observational astronomy; astronomical photometry and instrumentation; photomultipliers; scintillation; planetary physics. Mailing Add: Dept of Physics Tex A&M Univ College Station TX 77843

YOUNG, ARCHIE RICHARD, II, b Camden, NJ, June 8, 28; m 51; c 7. INORGANIC CHEMISTRY. Educ: Lincoln Univ, Pa, AB, 49; Univ Pa, MS, 50, PhD(phys chem), 55. Prof Exp: Instr chem, Ft Valley State Col, 50-51 & Va Union Univ, 51-52; assoc prof, Tenn State Univ, 54-56; from res chemist to sr res chemist, Reaction Motors Div, Thiokol Chem Corp, NJ, 56-63, supvr inorg synthesis, 63-67; sr chemist, 67-68, SR RES CHEMIST, EXXON RES & ENG CO, LINDEN, 68- Mem: AAAS; Am Chem Soc. Res: Aluminum hydride and inorganic fluorine chemistry; fluorocarbons; fuel cells; solid state synthesis; ferrites. Mailing Add: 18 Franklin Pl Montclair NJ 07042

YOUNG, ARNOLD E, organic chemistry, see 12th edition

YOUNG, ARTHUR, b New York, NY, Jan 4, 40; m 60; c 1. ASTRONOMY. Educ: Allegheny Col, BS, 60; Ind Univ, MA, 65, PhD(astron), 67. Prof Exp: From asst prof to assoc prof, 67-74, PROF ASTRON, SAN DIEGO STATE UNIV, 74- Concurrent Pos: Acad consult, Spitz Labs, 63- Mem: Am Astron Soc. Res: Spectroscopy; galactic kinematics. Mailing Add: Dept of Astron San Diego State Univ San Diego CA 92115

YOUNG, ARTHUR WESLEY, b Shenandoah, Iowa, May 14, 04; m 29; c 3. AGRONOMY, SOILS. Educ: Iowa State Col, BS, 29, MS, 30, PhD(soil bact), 32. Prof Exp: Instr agr bact, Univ Tenn, 32-34; prof agron, Panhandle Agr & Mech Col, 34-35; from assoc prof to emer prof agron, Tex Tech Univ, 35-69, from actg head dept to head dept, 37-69; CONSULT, 69- Concurrent Pos: Consult, Bunge y Born, Argentina, 68-71; independent agr consult, 69-; chmn, State Seed & Plant Bd, Tex; res consult, Dept of Agr, Lubbock Christian Col, 75- Honors & Awards: Agr Chem Award, W Tex Agr Chem Inst, 69; Gerold W Thomas Outstanding Agriculturalist Award, Tex Tech Univ, 70. Mem: AAAS; Am Soc Agron; Soil Sci Soc Am; hon mem Int Crop Improve Asn (vpres, 58-59, pres, 60-61). Res: Soil bacteriology, chemistry and fertility; direction of microbial studies on influence of organic matter and fertilizer to selected soils. Mailing Add: 3305 45th St Lubbock TX 79413

YOUNG, AUSTIN HARRY, b Brighton, Mass, Oct 25, 28; m 58; c 2. PHYSICAL CHEMISTRY. Educ: Tufts Univ, BS, 50; Univ Wis, PhD(chem), 59. Prof Exp: Develop chemist, Fabrics & Finishes Dept, EI du Pont de Nemours & Co, 50-51, supvr, Explosives Dept, 51-52; res asst, Army Chem Ctr, Edgewood, Md, 53-54; from res chemist to sr res chemist, 58-69, RES ASSOC, A E STALEY MFG CO, 69- Mem: AAAS; Am Chem Soc; Am Asn Cereal Chem. Res: Starch, colloid, polymer and radiation chemistry; kinetics; polymer synthesis and characterization; chromatography; thermal analysis; rheology; radiotracer techniques; graphic arts; health physics. Mailing Add: Res & Develop Div AE Staley Mfg Co Decatur IL 62525

YOUNG, BERNARD THEODORE, b Tarentum, Pa, Apr 13, 30; m 55; c 5. PHYSICS. Educ: Slippery Rock State Col, BS, 52; Tex A&M Univ, MS, 61, PhD(physics), 64. Prof Exp: Res & develop engr, Tex Div, Dow Chem Co, 56-59; res asst physics, Tex A&M Univ, 60-63; from assoc prof to prof, Sam Houston State Col, 63-68, dir, 65-68; assoc dean, 68-70, GRAD DEAN, ANGELO STATE UNIV, 70- Concurrent Pos: Mem elem transparency proj, Tex Educ Agency, 66-67. Mem: Am Phys Soc. Res: Molecular spectroscopy; atomic structure. Mailing Add: Off of Grad Dean Angelo State Univ San Angelo TX 76901

YOUNG, BEVERLEY GEORGE, b Windsor, Ont, Jan 30, 28; m 57; c 2. PHYSICS. Educ: Western Ont Univ, BSc, 53, MSc, 54; McMaster Univ, PhD(physics), 58. Prof Exp: RES SCIENTIST, DEFENCE RES ESTAB OTTAWA, DEPT NAT DEFENCE CAN, 58- Mem: Can Asn Physicists. Res: Infrared spectrometry; remote sensing; image interpretation. Mailing Add: 13 Riverbrook Rd Ottawa ON Can

YOUNG, BING-LIN, b Honan, China, Feb 3, 37; m 64; c 1. HIGH ENERGY PHYSICS. Educ: Nat Taiwan Univ, BS, 59; Univ Minn, PhD(physics), 66. Prof Exp: Res assoc physics, Ind Univ, 66-68 & Brookhaven Nat Lab, 68-70; asst prof, 70-74, ASSOC PROF PHYSICS, IOWA STATE UNIV, 74- Concurrent Pos: Assoc physicist, Ames Lab, US AEC, 70-74, physicist, 74- Mem: Am Phys Soc. Res: Theoretical physics of the elementary particles. Mailing Add: Dept of Physics Iowa State Univ Ames IA 50010

YOUNG, BOBBY GENE, b Cape Girardeau, Mo, Aug 3, 29; m 52; c 2. MICROBIOLOGY, GENETICS. Educ: Southeast Mo State Col, BS, 51; Johns Hopkins Univ, PhD(biol), 65. Prof Exp: Biologist, Nat Cancer Inst, 54-56, chemist, 56-62; teaching asst, Johns Hopkins Univ, 62-64; staff fel, Nat Cancer Inst, 65-66, res microbiologist, Div Biologics Standards, NIH, 66-69; assoc prof, 69-74, actg chmn dept, 70-71, PROF MICROBIOL, UNIV MD, COLLEGE PARK, 74-, CHMN DEPT, 71- Mem: AAAS; Am Soc Microbiol. Res: Microbial genetics; cancer research, virology and interferon aspects of malignancy. Mailing Add: Dept of Microbiol Univ of Md College Park MD 20742

YOUNG, BRUCE ARTHUR, b Sydney, Australia, Jan 16, 39; m 65; c 2. AGRICULTURE, ANIMAL PHYSIOLOGY. Educ: Univ New Eng, Australia, BRurSc, 62, MRurSc, 65, PhD(physiol), 69. Prof Exp: Asst prof, 68-72, ASSOC PROF ANIMAL PHYSIOL, UNIV ALTA, 72- Mem: Am Soc Animal Sci; Can Soc Animal Sci; Agr Inst Can. Res: Environmental physiology of animals; adaptation to physical environment; livestock production and energy metabolism in cold climates. Mailing Add: Dept of Animal Sci Univ of Alta Edmonton AB Can

YOUNG, BRUCE C, b Brawley, Calif, Aug 12, 31; m 58; c 1. GEOGRAPHY. Educ: Univ Calif, Los Angeles, BA, 53, MA, 57, PhD(geog), 68. Prof Exp: PROF GEOG, SANTA MONICA COL, 57-, CHMN DEPT EARTH SCI, 71- Mem: Asn Am Geogr; Am Geog Soc. Mailing Add: Dept of Earth Sci Santa Monica Col Santa Monica CA 90405

YOUNG, BYRON ARLEN, physics, see 12th edition

YOUNG, CHARITY LOUISE, b Chicago, Ill, Oct 4, 46. CELL BIOLOGY. Educ: Conn Col Women, BA, 68; Georgetown Univ, PhD(biol), 76. Prof Exp: Instr, 75-76, ASST PROF BIOL, ST JOSEPH'S COL, PHILADELPHIA, 76- Res: Ultrastructural and physiological aspects of fish pathology; biochemical aspects of fish microbiol pathogens. Mailing Add: Dept of Biol St Joseph's Col Philadelphia PA 19131

YOUNG, CHARLES ALBERT, b Dodge Center, Minn, Oct 11, 11; m 50; c 2. INDUSTRIAL ORGANIC CHEMISTRY. Educ: Purdue Univ, BS, 33; Univ Notre Dame, MS, 34, PhD(org chem), 36. Prof Exp: Asst chem, Univ Notre Dame, 33-36; staff chemist, Jackson Lab, 36-60, STAFF CHEMIST, EXP STA, E I DU PONT DE NEMOURS & CO, INC, 60- Mem: Am Chem Soc. Res: Synthetic rubber; organic water repellants; adhesives; dye synthesis and application; industrial and automotive finishes. Mailing Add: 111 Beech Lane Forest Brook Glen Wilmington DE 19804

YOUNG, CHARLES EDWARD, b Petrolia, Ont, June 1, 41. CHEMICAL PHYSICS. Educ: Univ Toronto, BSc, 63; Univ Calif, Berkeley, PhD(chem), 66. Prof Exp: Res assoc chem, Mass Inst Technol, 66-68; mem tech staff, Bell Labs, 68-70; CHEMIST, ARGONNE NAT LAB, 70- Mem: Am Chem Soc; Am Phys Soc. Res: Reaction and excitation cross sections for supra thermal collisions of atoms and simple molecules studied by accelerated beam techniques, chemiluminescence and laser induced fluorescence. Mailing Add: Chem Div Argonne Nat Lab Argonne IL 60439

YOUNG, CHARLES GILBERT, b Fawn Grove, Pa, Feb 25, 30; m 59; c 3. OPTICS. Educ: Elizabethtown Col, BS, 52; Univ Conn, MA, 56, PhD(physics), 61. Prof Exp: Physicist, Brookhaven Nat Lab, 56 & Navy Electronics Lab, Calif, 57; instr physics, Conn Col, 57-59; from res asst to instr, Univ Conn, 59-62; consult, 61-62, sr physicist, 62-64, mgr systs res dept, 65-70, gen mgr laser prod dept, 70-73, CHIEF PHYSICIST, AM OPTICAL CO, 64-, DIR PROD DEVELOP, 73- Mem: Sr mem Inst Elec & Electronics Eng; Optical Soc Am; Am Phys Soc. Res: General research and development management of new products and new technology in lenses frames and materials, optical vacuum coatings, plastic laminates, electro-optical devices, polarizers and pilot line and production work. Mailing Add: 17 Southwood Rd Storrs CT 06268

YOUNG, CHARLES WESLEY, b Enid, Okla, Dec 2, 29; m 52; c 3. DAIRY HUSBANDRY. Educ: Okla State Univ, BS, 56; NC State Univ, MS, 58, PhD(animal indust), 61. Prof Exp: From asst prof to assoc prof, 60-69, PROF DAIRY HUSB, UNIV MINN, ST PAUL, 69- Mem: Am Dairy Sci Asn. Res: Systems of breeding in dairy cattle; relationship between size and production in dairy cattle; economics of selection for milk yield in dairy cattle. Mailing Add: Dept of Animal Sci Univ of Minn St Paul MN 55108

YOUNG, CHARLES WILLIAM, b Denver, Colo, Nov 19, 30; m 63; c 3. INTERNAL MEDICINE, CANCER. Educ: Columbia Univ, AB, 52; Harvard Univ, MD, 56. Prof Exp: Intern med, Second Med Div, Bellevue Hosp, New York, 56-57, resident, Second Med Div, Bellevue Hosp & Mem Hosp, 57-59; from res assoc to assoc, 62-71, ASSOC MEM, SLOAN-KETTERING INST CANCER RES, 71-; ASST PROF MED, MED COL CORNELL UNIV, 66- Concurrent Pos: Fel, Mem Hosp & Sloan-Kettering Inst Cancer Res, 59-60. Mem: AAAS; Am Asn Cancer Res; Am Soc Clin Oncol; Am Fedn Clin Res; NY Acad Sci. Res: Oncology; biochemical pharmacology; embryology. Mailing Add: Sloan-Kettering Inst Cancer Res 410 E 68th St New York NY 10021

YOUNG, CHARLOTTE MARIE, b Minneapolis, Minn, Aug 19, 10. NUTRITION. Educ: Univ Minn, BS, 35; Iowa State Univ, MS, 37, PhD(human nutrit), 40. Hon Degrees: DSc, Syracuse Univ, 73. Prof Exp: Asst nutrit, Univ Minn, 34-35 & Iowa State Univ, 36-40; instr, Mich State Univ, 40-42; from instr to prof, 42-74, secy grad sch nutrit, 52-74, secy grad fac, 71-74, EMER PROF MED NUTRIT, CORNELL UNIV, 74- Concurrent Pos: WHO consult, Inst Nutrit, Cent Am & Panama; US Opers Mission-Int Coop Admin consult, Peru, 61 & 63; Cooper mem lectr; Lydia J Roberts Mem lectr, 72; consult, Nat Heart & Lung Inst, 73-; consult, USDA, 74- Honors & Awards: Dorr Medal; Borden Award; Copher Award, Am Dietetic Asn, 72. Mem: Fel Am Pub Health Asn; Am Home Econ Asn; Am Dietetic Asn; Am Inst Nutrit; Am Soc Clin Nutrit. Res: Dietary study techniques; nutritional status; basal metabolism; weight control; body composition; food habits; frequency of feeding studies. Mailing Add: 110 Warren Rd Ithaca NY 14850

YOUNG, CLYDE THOMAS, b Durham, NC, Aug 22, 30; m 55; c 5. FOOD SCIENCE. Educ: NC State Univ, BS, 52, MS, 55; Okla State Univ, PhD(food sci), 70. Prof Exp: Instr chem & math, NC State Univ, 54-56; asst head claims dept, Anderson-Clayton & Co, Ga, 58-59; res chemist, Ga Inst Technol, 59-60; ASST RES CHEMIST, GA STA, UNIV GA, 60- Mem: Am Peanut Res & Educ Asn; Am Chem Soc; Am Oil Chemists Soc; Inst Food Technologists. Res: Major investigations in changes and variations in biochemical constituents of peanuts—factors affecting aroma, flavor, color, maturation and protein during development, harvesting, curing, storage, and roasting. Mailing Add: Dept of Food Sci Ga Sta Experiment GA 30212

YOUNG, DALE W, b Perry, Utah, Apr 23, 18; m 42. PLANT PHYSIOLOGY. Educ: Utah State Univ, BS, 42, MS, 50; Iowa State Univ, PhD(plant physiol), 53. Prof Exp: Agronomist, USDA, 48-51; plant physiologist, Rohm and Haas Co, 53-60; supvr pesticide eval, Hooker Chem Corp, 60-66; mem field res staff, Chemagro, 66-67, mgr biol res, Kansas City, Mo, 68-72; MGR COM DEVELOP, GULF OIL CHEM CO, MERRIAM, 73- Mem: Weed Sci Soc Am; Entom Soc Am. Res: Chemicals to increase food production. Mailing Add: 9639 Delmar Overland Park KS 66207

YOUNG, DANIEL TEST, b Kansas City, Mo, Aug 21, 23; m 49; c 3. INTERNAL MEDICINE. Educ: Guilford Col, BS, 46; Harvard Med Sch, MD, 50. Prof Exp: From intern to asst resident med, Jefferson Hosp, Philadelphia, Pa, 50-52; from asst resident to chief resident, NC Mem Hosp, 52-54; from instr to assoc prof, 54-71, PROF MED, SCH MED, UNIV NC, CHAPEL HILL, 72- Mem: AAAS; AMA; Am Heart Asn. Res: Cardiology; muscle physiology. Mailing Add: Dept of Med Univ of NC Chapel Hill NC 27514

YOUNG, DAVID A, b Carmel, Calif, Sept 26, 42. PHYSICAL CHEMISTRY. Educ: Pomona Col, BA, 64; Univ Chicago, PhD(chem), 67. Prof Exp: PHYSICIST, LAWRENCE LIVERMORE LAB, 67- Mem: Am Chem Soc; Am Phys Soc. Res: Statistical mechanics of gases, liquids and solids. Mailing Add: Lawrence Livermore Lab PO Box 808 Livermore CA 94550

YOUNG, DAVID ALLAN, b Wilkinsburg, Pa, May 26, 15; m 34; c 1. INSECT TAXONOMY. Educ: Louisville Univ, AB, 39; Cornell Univ, MS, 42; Univ Kans, PhD(entom), 51. Prof Exp: Instr, Louisville Univ, 46-48; asst, Univ Kans, 48-49; entomologist, Insect Identification & Parasite Introd Sect, Entom Res Br, Agr Res Serv, USDA, 50-57; assoc prof, 57-61, PROF ENTOM, NC STATE UNIV, 61- Mem: Entom Soc Am. Res: Insect taxonomy; taxonomy of the Auchenorrhynchous Homoptera; reclassification of Cicadellinae (Homoptera: Auchenorrhyncha). Mailing Add: Dept of Entom NC State Univ Raleigh NC 27607

YOUNG, DAVID BRUCE, b Pittsburgh, Pa, Mar 13, 45; m 65; c 3. PHYSIOLOGY. Educ: Univ Colo, BA, 67; Ind Univ, PhD(physiol), 72. Prof Exp: Instr, 72-74, ASST PROF PHYSIOL, SCH MED, UNIV MISS, 74- Concurrent Pos: NIH trainee, Sch Med, Univ Miss, 72-74. Res: Fluid and electrolyte balance control mechanisms; hypertension. Mailing Add: Dept of Physiol Univ of Miss Med Ctr Jackson MS 39216

YOUNG, DAVID CALDWELL, b Memphis, Tenn, June 18, 24; m 55; c 3. ORGANIC CHEMISTRY. Educ: Davidson Col, BS, 46; Univ Fla, MS, 48, PhD(chem), 50. Prof Exp: Chemist, Edgar C Britton Res Lab, 50-53, group leader, 53-62, patents coord, 62-65, asst to dir, 65-70, asst to dir chem biol res, 70-74, ASST TO DIR PHARMACEUT RES & DEVELOP, DOW CHEM CO, 74- Mem: AAAS; Am Chem Soc. Res: Leuckart reaction; phthalaldehydic acid; organic research and process development; chemical patents. Mailing Add: 1223 Holyrood St Midland MI 48640

YOUNG, DAVID EDWARD, inorganic chemistry, see 12th edition

YOUNG, DAVID K, marine biology, ecology, see 12th edition

YOUNG, DAVID MARSHALL, b Minot, ND, Aug 26, 42; m 63; c 2. PATHOLOGY. Educ: Colo State Univ, DVM, 66; Ohio State Univ, MS, 67, PhD(comp path), 70. Prof Exp: COMP PATHOLOGIST, LAB TOXICOL, NAT CANCER INST, 73- Concurrent Pos: Sr staff fel, NIH, 70-73; consult pathologist, Statutory Adv Comt, Food & Drug Admin, Dept Health, Educ & Welfare, 71; mem fac, Found Advan Educ in the Sci, NIH, 72- Honors & Awards: C L Davis Jour Award, C L Davis Found Advan Vet Path, 73. Mem: AAAS; Int Acad Path; Am Soc Exp Path; Soc Pharmacol & Environ Path; Am Asn Cancer Res. Res: Comparative pathology; endocrine pathology; cancer; mineral metabolism; orthopedic pathology; animal models of human disease; toxicology. Mailing Add: Lab of Toxicol Nat Cancer Inst Bethesda MD 20014

YOUNG, DAVID MATHESON, b London, Eng, Apr 19, 28; m 53; c 6. ENVIRONMENTAL CHEMISTRY. Educ: Univ London, BSc, 48, PhD(chem), 49. Prof Exp: Asst lectr phys chem, St Andrews Univ, 49-51; res assoc, Amherst Col, 51-52; res fel, Nat Res Coun Can, 52-53; Royal Mil Col, Can, 53-54 & 55-56; Humboldt Scholar, Phys Chem Inst, Munich, 54-55; res chemist, 56-58, supvr res & develop lab, 58-65, asst res mgr, 65-72, MGR CHEM RES & DEVELOP, DOW CHEM CAN, LTD, 72- Mem: Chem Inst Can. Res: Physical adsorption of gases; boron chemistry; gas chromatography; heterogeneous catalysis; photochemistry. Mailing Add: 542 Highbury Park Sarnia ON Can

YOUNG, DAVID MONAGHAN, JR, b Boston, Mass, Oct 20, 23; m 49; c 3. MATHEMATICS. Educ: Webb Inst Naval Archit, BS, 44; Harvard Univ, MA, 47, PhD(math), 50. Prof Exp: Instr & res assoc math, Harvard Univ, 50-51; mathematician, Aberdeen Proving Ground, 51-52; assoc prof math, Univ Md, 52-55; mgr math anal dept, Ramo-Wooldridge Corp, 55-58; prof math & dir comput ctr, 58-

70, PROF MATH & COMPUT SCI & DIR CTR NUMERICAL ANAL, UNIV TEX, AUSTIN, 70- Mem: Am Math Soc; Soc Indust & Appl Math; Math Asn Am; Asn Comput Mach. Res: Numerical analysis, especially the numerical solution of partial differential equations by finite difference methods; high-speed computing. Mailing Add: Ctr Numerical Anal Univ of Tex Austin TX 78712

YOUNG, DAVID PARIS, b St Louis, Mo, Sept 28, 37; m 60; c 5. SCIENCE EDUCATION. Educ: Park Col, AB, 59; Univ Kans, PhD(chem), 63. Prof Exp: Asst prof, 63-66, chmn dept, 68-74, ASSOC PROF CHEM, MARYVILLE COL, 66-, DIR PROJ ON FUTURISTICS, 74- Concurrent Pos: NSF sci fac fel, Cornell Univ, 70-71. Mem: AAAS; Am Chem Soc. Res: Impact of scientific knowledge on society; futuristics; biomedical ethics. Mailing Add: Box 2894 Maryville Col Maryville TN 37801

YOUNG, DAVID ROSS, b San Diego, Calif, Nov 19, 37; m 68; c 1. OCEANOGRAPHY. Educ: Pomona Col, BA, 60; Scripps Inst Oceanog, PhD(oceanog), 70. Prof Exp: SR ENVIRON SPECIALIST OCEANOG, SOUTHERN CALIF COASTAL WATER RES PROJ, 70- Mem: AAAS. Res: Introduction, transport, fate and ecological effects of trace elements and synthetic organic compounds in the marine environment. Mailing Add: Southern Calif Coastal Water Res 1500 E Imperial Hwy El Segundo CA 90245

YOUNG, DAVIS ALAN, b Abington, Pa, Mar 5, 41; m 65; c 2. PETROLOGY, MINERALOGY. Educ: Princeton Univ, BSE, 62; Pa State Univ, MS, 65; Brown Univ, PhD(geol), 69. Prof Exp: Asst prof geol, NY Univ, 68-74; ASSOC PROF GEOL, UNIV NC, WILMINGTON, 74- Concurrent Pos: NSF instnl grant, NY Univ, 69-70. Mem: AAAS; Mineral Soc Am. Res: Precambrian igneous and metamorphic geology of New Jersey and southeastern Pennsylvania; petrology of syenites. Mailing Add: Dept of Geol Univ of NC Wilmington NC 28403

YOUNG, DELANO VICTOR, b Honolulu, Hawaii, Nov 17, 45; m 70; c 1. BIOCHEMISTRY. Educ: Stanford Univ, BS, 67; Columbia Univ, PhD(biochem), 73. Prof Exp: Fel cell biol, Salk Inst Biol Studies, 73-75; ASST PROF CHEM, BOSTON UNIV, 75- Mem: AAAS; Sigma Xi. Res: Peptide growth factors from serum and other biological sources required by transformed eucaryotic cells and the mechanisms by which these factors regulate growth in such cells. Mailing Add: Dept of Chem Boston Univ 675 Commonwealth Ave Boston MA 02215

YOUNG, DENNIS LEE, b St Louis, Mo, Jan 22, 44; div. MATHEMATICAL STATISTICS. Educ: St Louis Univ, BS, 65; Purdue Univ, MS, 67, PhD(statist), 70. Prof Exp: Asst prof statist, NMex State Univ, 70-75; ASSOC PROF STATIST, ARIZ STATE UNIV, 75- Mem: AAAS; Inst Math Statist; Am Statist Asn. Res: Multivariate statistical analysis. Mailing Add: Dept of Math Ariz State Univ Tempe AZ 85281

YOUNG, DEWALT SECRIST, b Medford, Ore, Aug 17, 15; m 41; c 1. ORGANIC CHEMISTRY. Educ: Cornell Col, AB, 36; Duke Univ, MS, 38, PhD(chem), 40. Prof Exp: Res chemist, Gen Chem Co, 39-44; res chemist, 44-72, COORDR APPL RES, TENN EASTMAN CORP, 72- Mem: Am Chem Soc. Res: Fluorine chemistry; hydroquinone derivatives; photographic chemicals. Mailing Add: Tenn Eastman Co PO Box 511 Kingsport TN 37662

YOUNG, DONALD ALCOE, b Fredericton, NB, Oct 21, 29; m 55; c 3. GENETICS, PLANT BREEDING. Educ: McGill Univ, BSc, 52; Univ Wis, MS, 54, PhD(genetics), 57. Prof Exp: Res officer, 57-66, sect head potato breeding, 66-73, PROG MGR, CAN DEPT AGR, 73- Concurrent Pos: Mem, Work Planning Comt Potato Breeding, 59-; chmn, Work Planning Comt Potato Texture, 64- Mem: Potato Asn Am; Can Soc Hort Sci; Genetics Soc Can. Res: Potato breeding; sample selection as related to potato quality; disease resistance. Mailing Add: Can Dept of Agr Res Sta PO Box 280 Fredericton NB Can

YOUNG, DONALD C, b Paducah, Ky, Feb 25, 33; m 51; c 4. INORGANIC CHEMISTRY, AGRICULTURAL CHEMISTRY. Educ: Univ Calif, Riverside, BA, 61, PhD(inorg chem), 66. Prof Exp: From res asst petrochem to sr res chemist, 53-69, res assoc, 69-74, SR RES ASSOC FERTILIZER CHEM, UNION OIL CO, 74- Concurrent Pos: Res scholar, Univ Calif, Riverside, 66-68. Mem: Am Chem Soc. Res: Chemistry and technology of polyphosphoric acid; transition-metal complexes in homogeneous catalysis; carborane chemistry; transition metal complexes of dicarbollide ion; fertilizer and soil chemistry. Mailing Add: 309 W Jacaranda Pl Fullerton CA 92632

YOUNG, DONALD CHARLES, b Fremont, Ohio, June 29, 44; m 68; c 2. ANALYTICAL CHEMISTRY. Educ: Harvard Univ, AB, 66; Univ NC, Chapel Hill, PhD(anal chem), 71. Prof Exp: Res assoc chem, Purdue Univ, 71-72; ASST PROF CHEM, OAKLAND UNIV, 72- Mem: Am Chem Soc; Sigma Xi. Res: Coordination chemistry; molecular structure and conformation in solution; catalysis by coordination; ligand electrosynthesis. Mailing Add: Dept of Chem Oakland Univ Rochester MI 48063

YOUNG, DONALD EDWARD, b Lake Zurich, Ill, June 13, 22; m 47; c 3. HIGH ENERGY PHYSICS. Educ: Ripon Col, BA, 46; Univ Minn, MS, 51, PhD(nucleon scattering), 59. Prof Exp: Asst physics, Univ Minn, 49-53; physicist, Labs, Gen Mills Co, 53-60; head, Physics Div, Midwestern Univs Res Asn, 60-67; prof physics, Univ Wis, 67; PHYSICIST, NAT ACCELERATOR LAB, 67- Mem: Am Phys Soc. Res: Proton linear accelerator design; high energy particle accelerators design and operation; magnetic field measurements; nuclear physics and radioactivity; proton-proton scattering; dosimetry; radiation damage. Mailing Add: Accelerator Div Fermi Nat Accelerator Lab PO Box 500 Batavia IL 60501

YOUNG, DONALD MACKEY, chemistry, see 12th edition

YOUNG, DONALD REEDER, b Logan, Utah, July 21, 21; m 46; c 4. PHYSICS. Educ: Utah State Agr Col, BA, 42; Mass Inst Technol, PhD(physics), 49. Prof Exp: Mem staff, Radiation Lab, Mass Inst Technol, 42-45, asst, Insulation Lab, 45-49; tech engr, 49-52, proj engr, 52-61, sr engr & mgr device & mat characterization, 61-71, RES STAFF MEM, IBM CORP, 71- Mem: Fel Am Phys Soc. Res: Electrical breakdown; ferroelectric materials; superconductors; semiconductors. Mailing Add: IBM Res Ctr Yorktown Heights NY 10598

YOUNG, DONALD RUDOLPH, b Chicago, Ill, Oct 29, 23; m 55; c 2. PHYSIOLOGY, METABOLISM. Educ: Loyola Univ, Ill, BS, 49; Univ Calif, Berkeley, MA, 51, PhD(physiol), 54. Prof Exp: Res physiologist poultry nitrit, Univ Calif, Berkeley, 54-55, asst prof physiol, 55; nutritionist, Qm Food & Container Inst, 55-57, head performance lab, 58-61, actg chief nutrit br, 61-62; res scientist, Biotechnol Div, 62-72; CHIEF ENVIRON PHYSIOL, BIOMED RES DIV, AMES RES CTR, NASA, 72- Concurrent Pos: Mem panel environ physiol, Dept Army, 57-62 & biomed exp working group, NASA, 63-65; lectr, Stanford Univ, 63- Mem: AAAS; Am Physiol Soc; Soc Exp Biol & Med; Aerospace Med Asn; NY Acad Sci. Res: Inert gases; nutritional requirements during stress conditions; methematical models of fat metabolism; carbohydrate turnover rate. Mailing Add: Biomed Res Div Ames Res Ctr NASA Moffett Field CA 94035

YOUNG, DONALD STIRLING, b Belfast, Northern Ireland, Dec 17, 33. CLINICAL PATHOLOGY. Educ: Aberdeen Univ, MB & ChB, 57; Univ London, PhD(chem path), 62. Prof Exp: Lectr mat med, Aberdeen Univ, 58-59; resident chem path, Royal Postgrad Med Sch London, 62-64; vis scientist, 65-67, CHIEF CLIN CHEM, NIH, 67- Concurrent Pos: Chmn bd ed, Clin Chem, Am Asn Clin Chemists, 73- Honors & Awards: Gerard B Lambert Award, Gerard B Lambert Awards Orgn, 75. Mem: AAAS; Brit Asn Clin Biochem; Am Asn Clin Chem; Acad Clin Lab Physicians & Scientists. Res: Clinical chemistry, development of high resolution analytical techniques and applications of computers to improve quality and usefulness of laboratory data. Mailing Add: NIH Bldg 10 Rm 4N-309 Bethesda MD 20014

YOUNG, EDMOND GROVE, b Govans, Md, Oct 29, 17; m 46; c 3. FLUORINE CHEMISTRY. Educ: Univ Md, BS, 38, PhD(org chem), 43. Prof Exp: Asst chem, Univ Md, 38-43; chemist, Sharples Chem, Inc, Mich, 43-44; res chemist, E I du Pont de Nemours & Co, Inc, 44-48; tech sales, Kinetics Chem, Inc, 48-49, sales mgr aerosol propellants, 49-50; sales develop, Kinetic Chem Div, 50-52, mgr kinetic chem div, 52-57, mgr develop conf, 57-68, mgr develop conf & govt liaison, Develop Dept, 68-73, MGR BUS DEVELOP, CENT RES & DEVELOP DEPT, E I DU PONT DE NEMOURS & CO, INC, 73- Concurrent Pos: Mem, Franklin Inst. Mem: AAAS; Am Chem Soc; Com Develop Asn; Soc Chem Indust. Res: Reaction of metallo-organics; chemistry of fluorinated compounds; commercial chemical development and marketing. Mailing Add: Cent Res & Develop Dept E I du Pont de Nemours & Co Inc Tenth & Market Sts Wilmington DE 19898

YOUNG, EDWARD JOSEPH, b Roselle, NJ, Feb 18, 23; m 55; c 2. GEOCHEMISTRY, MINERALOGY. Educ: Rutgers Univ, BS, 48; Mass Inst Technol, MS, 50, PhD(geol), 54. Prof Exp: GEOLOGIST, US GEOL SURV, 52- Mem: Am Geol Soc; Mineral Soc Am; Geochem Soc; Mineral Asn Can. Res: Petrology; geochemistry of apatite. Mailing Add: US Geol Surv Fed Ctr Denver CO 80225

YOUNG, ELDRED EMSLY, organic chemistry, see 12th edition

YOUNG, ELEANOR ANNE, b Houston, Tex, Oct 8, 25. NUTRITION. Educ: Incarnate World Col, BA, 47; St Louis Univ, MEd, 55; Univ Wis, PhD(nutrit), 68. Prof Exp: Asst prof foods & Nutrit, Incarnate World Col, 53-63 & 68-72; sr res assoc, Univ Tex Health Sci Ctr, Dept Med, 68-72; ASSOC PROF FOODS & NUTRIT, INCARNATE WORD COL, 72-; ASST PROF MED, UNIV TEX HEALTH SCI CTR, DEPT MED, SAN ANTONIO, 72- Concurrent Pos: Consult, Audie Murphy Vet Admin Hosp, San Antonio, 73- Mem: Am Inst Nutrit; Am Bd Human Nutrit; Am Soc Clin Nutrit; Am Dietetic Asn; Am Pub Health Asn. Res: Metabolic response and feeding-fasting intervals in man; metabolism of intravenously infused maltose; lactose intolerance; nutritional adaptations after partial small bowel resections. Mailing Add: Univ Tex Health Sci Ctr 7703 Floyd Curl San Antonio TX 78284

YOUNG, ELIZABETH BELL, b Franklinton, NC, July 2, 29; m. SPEECH PATHOLOGY, AUDIOLOGY. Educ: NC Col Durham, AB, 48, MA, 50; Ohio State Univ, PhD(speech sci, speech & hearing ther), 59. Prof Exp: Chmn dept English, Barber-Scotia Col, 48-53; chmn dept speech, Talladega Col, 53-55; asst prof, Va State Col, 56-57; prof speech correction & dir speech clin, Fla A&M Univ, 59; chmn dept English & speech, Fayetteville State Col, 59-63; asst prof speech path, Col Dent, Howard Univ, 63-64; chmn dept English & lang, Md State Col, 65-66; asst prof speech path, undergrad & grad prog & supvr speech & audiol training clin prog, 66-69, PROF SPEECH PATH, GRAD SCH & SUPVR SPEECH CLIN, CATH UNIV AM, 69- Mem: Am Speech & Hearing Asn. Res: Pathology of speech and hearing mechanism and speech science; observations in the field of speech and hearing pathology and science. Mailing Add: 8104 W Beach Dr NW Washington DC 20012

YOUNG, ELTON THEODORE, b Brush, Colo, May 1, 40; m 62; c 2. MOLECULAR BIOLOGY. Educ: Univ Colo, BA, 62; Calif Inst Technol, PhD(biophys), 67. Prof Exp: Fel molecular biol, Univ Geneva, 67-69; asst prof, 69-75, ASSOC PROF BIOCHEM & GENETICS, UNIV WASH, 75- Res: Regulation of transcription and translation in bacteriophage. Mailing Add: Dept of Biochem Univ of Wash SJ-70 Seattle WA 98195

YOUNG, EUGENE FREDERICK, spectroscopy, instrumentation, see 12th edition

YOUNG, EUTIQUIO CHUA, b Del Gallego, Philippines, July 17, 32; m 61; c 3. MATHEMATICAL ANALYSIS. Educ: Far Eastern Univ, Manila, BS, 54; Univ Md, MA, 60, PhD(math), 62. Prof Exp: Asst prof math, Univ Conn, 61-62; head dept math, Far Eastern Univ, Manila, 62-64; assoc prof, De la Salle Col, Manila, 64-65; from asst prof math to assoc prof, 65-74, PROF MATH, FLA STATE UNIV, 74- Mem: Math Asn Am; Am Math Soc. Res: Cauchy problems, uniqueness of solutions of boundary value problems and comparison theorems for partial differential equations. Mailing Add: Dept of Math Fla State Univ Tallahassee FL 32306

YOUNG, EVAN JOHNSON, b Akron, Ohio, Oct 7, 18; m 44; c 5. RUBBER CHEMISTRY. Educ: Manchester Col, AB, 41; Ohio State Univ, MSc, 44, PhD(org chem), 47. Prof Exp: Asst, Ohio State Univ, 41-47; asst, Off Sci Res & Develop & Manhattan Proj, 42-44; assoc chemist, Clinton Labs, Tenn, 44-45; res chemist, WVa, 47-52, from group leader to sr group leader, 52-68, ADMIN MGR, MONSANTO CO, OHIO, 68- Mem: AAAS; Am Inst Chem; Am Chem Soc. Res: Chemistry of organic phosphates and organic fluorides; radiochemistry; isolation of radioactive isotopes; petroleum chemicals and lubricants; rubber chemicals. Mailing Add: Monsanto Co 260 Springside Dr Akron OH 44313

YOUNG, EVIE FOUNTAIN, JR, b Baton Rouge, La, Sept 9, 28; m 57; c 3. PLANT BREEDING, AGRONOMY. Educ: La State Univ, BS, 51, MS, 53; Okla State Univ, PhD(plant breeding, genetics), 64. Prof Exp: Agronomist, Ibec Res Inst, 53-55 & NAR Farm, 55-56; asst agronomist, La State Univ, 56-59; agronomist, 59-63, RES AGRONOMIST, AGR RES SERV, USDA, 63- Concurrent Pos: Instr, Okla State Univ, 59-63. Res: Genetics. Mailing Add: 10601 N Loop Rd El Paso TX 79927

YOUNG, FRANCIS ALLAN, b Utica, NY, Dec 29, 18; m 45; c 2. PSYCHOPHYSIOLOGY. Educ: Tampa Univ, BS, 41; Case Western Reserve Univ, MA, 45; Ohio State Univ, PhD(psychol), 49. Prof Exp: From instr to assoc prof, 48-61, PROF PSYCHOL, WASH STATE UNIV, 61-, DIR PRIMATE RES CTR, 57- Concurrent Pos: Nat Acad Sci-Nat Res Coun sr fel physiol psychol, 56-57; vis res prof, Med Sch, Univ Ore, 63-64; actg asst dir, Regional Primate Res Ctr, Univ Wash, 66-68; vis res prof, Med Sch, Univ Uppsala, 71; mem, Conf Prof & Soc Issues Psychol. Mem: AAAS; Am Psychol Asn; Am Acad Optom; Asn Res Vision & Ophthal; Animal Behav Soc. Res: Vision and audition including comparative studies within primates; classical and instrumental conditioning; sexual behavior and genetics of behavior in primates. Mailing Add: Primate Res Ctr Wash State Univ Pullman WA 99163

YOUNG, FRANK COLEMAN, b Roanoke, Va, June 10, 35; m 59; c 3. NUCLEAR PHYSICS. Educ: Johns Hopkins Univ, BA, 57; Univ Md, PhD(nuclear physics), 62. Prof Exp: NSF res fel, US Naval Res Lab, 62-63; from asst prof to assoc prof physics, Univ Md, 63-72; RES PHYSICIST, US NAVAL RES LAB, 72- Honors & Awards: Res Publ Award, US Naval Res Lab, 72. Mem: Am Phys Soc; Inst Elec & Electronics Engr; Nuclear & Plasma Sci Soc. Res: Experimental nuclear physics; application of nuclear techniques to studies of hot dense plasmas. Mailing Add: 100 Mel Mara Dr Oxon Hill MD 20021

YOUNG, FRANK E, b Mineola, NY, Sept 1, 31; m 56; c 5. PATHOLOGY, MICROBIOLOGY. Educ: State Univ NY, MD, 56; Western Reserve Univ, PhD(microbiol), 62. Prof Exp: From intern to resident path, Univ Hosps, Cleveland, Ohio, 56-60; from instr to assist prof, Western Reserve Univ, 62-65; from assoc mem to mem, Depts Microbiol & Exp Path, Scripps Clin & Res Found, 65-70; PROF MICROBIOL, PATH, RADIATION BIOL & BIOPHYS & CHMN DEPT MICROBIOL, SCH MED & DENT, UNIV ROCHESTER, 70- Concurrent Pos: Am Cancer Soc res grant, 62-; NIH res grants, 65-, training grant, 70-; NSF res grant, 70-72; fac res assoc, Am Cancer Soc, 62-70; assoc prof, Univ Calif, San Diego, 67-70; dir clin microbiol labs, Strong Mem Hosp, 70-; dir, Health Dept Labs, Monroe County, 70- Mem: AAAS; Am Soc Microbiol; Am Soc Biol Chem; Am Soc Exp Path; Am Asn Path & Bact. Res: Mechanism of deoxyribonucleic and mediated transformation of bacterial and animal cells; regulation of bacterial cell surface; pathobiology of Neisseria gonorrhoeae. Mailing Add: Dept Microbiol Sch Med & Dent Univ of Rochester Rochester NY 14642

YOUNG, FRANK EVANS, b Colorado Springs, Colo, June 17, 14; m 40; c 4. PHYSICAL CHEMISTRY. Educ: Colo Col, AB, 36, MA, 38; Univ Calif, PhD(chem), 41. Prof Exp: Asst chem, Wash Univ, St Louis, 36-37 & Univ Calif, 38-41; res chemist, Standard Oil Co, Calif, 41-42; instr chem & actg head dept, Col Puget Sound, 42-43; res chemist, US Bur Mines, 43-45; res chemist, Western Regional Res Lab, Bur Agr & Indust Chem, USDA, 45-53 & Western Utilization Res Br, Agr Res Serv, 53-56; from assoc prof to prof phys chem, Calif State Polytech Col, 56-65; Fulbright lectureship, Khonkaen Univ, Thailand, 65-66; UNESCO tech adv chem, Ceylon Col Technol, 66-69, UN Develop Prog adv mat sci, 69-71; PROF CHEM, UNIV KEBANGSAAN, MALAYSIA, 73- Concurrent Pos: Prof from Univ Ky Contract Team, Inst Technol Bandung, Indonesia, 60-62. Honors & Awards: Superior Serv Award, USDA, 55. Mem: Am Chem Soc; fel Am Inst Chem. Res: Calorimetry; heat capacity; heat of solution and formation; phase equilibria; oxidation of hydrocarbons; science teaching materials and apparatus; electrochemistry; electrolytic solutions. Mailing Add: Dept of Chem Univ of Kebangsaan Kuala Lumpur Malaysia

YOUNG, FRANK GLYNN, b New York, NY, Dec 29, 16; m 41. CHEMISTRY. Educ: Dartmouth Col, AB, 37; Columbia Univ, PhD(org chem), 41. Prof Exp: Asst chem, Columbia Univ, 37-41; res chemist, 41-55, group leader radiation & isotope chem, 55-60, chem physics, Parma Res Labs, 60-63, SR RES SCIENTIST, UNION CARBIDE CORP, 63- Mem: Am Chem Soc; Am Soc Testing & Mat. Res: Mechanisms of catalytic processes; isotopic tracer studies; heterogeneous catalysis and surfaces. Mailing Add: Tech Ctr Union Carbide Corp PO Box 8361 South Charleston WV 25303

YOUNG, FRANK HOOD, b Baltimore, Md, Dec 31, 39; m 61; c 4. MATHEMATICS. Educ: Haverford Col, BA, 61; Univ Pa, MA, 63, PhD(math), 68. Prof Exp: Instr math, Temple Univ, 65-68; asst prof, 68-74, ASSOC PROF MATH, KNOX COL, ILL, 74- Mem: Am Math Soc; Math Asn Am. Res: Algebra. Mailing Add: Dept of Math Knox Col Galesburg IL 61401

YOUNG, FRANK NELSON, JR, b Oneonta, Ala, Nov 2, 15; m 43; c 2. BIOLOGY. Educ: Univ Fla, BS, 38, MS, 40, PhD(biol), 42. Prof Exp: Asst prof biol, Univ Fla, 46-49; from asst prof to assoc prof, 49-62, PROF ZOOL, IND UNIV, BLOOMINGTON, 62- Concurrent Pos: Guggenheim fel, 60-61; fel, La State Univ, 63. Mem: Am Soc Zool; Soc Study Evolution. Res: Taxonomy and ecology of aquatic Coleoptera; medical entomology; speciation and extinction of animals; land snails of genus Liguus. Mailing Add: Dept of Zool Ind Univ Bloomington IN 47401

YOUNG, FRANKLIN, b Beijing, China, Feb 1, 28; nat US. NUTRITION, BIOCHEMISTRY. Educ: Mercer Univ, AB, 51; Univ Fla, BSA, 52, MAgr, 54, PhD(nutrit), 60. Prof Exp: Asst vet sci, Univ Fla, 54-60, fel biochem, 60-61; res assoc, Bowman Gray Sch Med, 61-65, res instr prev med & assoc biochem, 65-66; ASSOC PROF FOOD & NUTRIT SCI, UNIV HAWAII, 66- Mem: Am Inst Nutrit. Res: Atherosclerosis, lipid metabolism and human nutrition. Mailing Add: Dept of Food & Nutrit Sci Univ of Hawaii 2500 Dole St Honolulu HI 96822

YOUNG, FRANKLIN ALDEN, JR, b Harrisburg, Pa, Mar 14, 38; m 59; c 3. MATERIALS SCIENCE. Educ: Univ Fla, BIE, 60, MSE, 63; Univ Va, DSc(mat sci), 68. Prof Exp: Instr metall, Clemson Univ, 63-65, asst prof mat eng, 68-70; assoc prof, 70-75, PROF DENT MAT & CHMN DEPT, COL DENT MED, MED UNIV S C, 75- Mem: Am Soc Metals; Int Asn Dent Res. Res: Mechanical properties of alloys; materials of medicine and dentistry. Mailing Add: Col of Dent Med Med Univ of S C 80 Barre St Charleston SC 29401

YOUNG, FREDERICK, b Towaco, Colo, July 2, 32; m 52; c 7. PHYSICS. Educ: Univ NMex, BS, 61, MS, 63, PhD(physics), 72. Prof Exp: Res physicist, Air Force Weapons Lab, Kirtland AFB, NMex, 63-65; res assoc high energy neutron physics, Univ NMex, 65-71; staff mem x-ray spectros, 71-72, STAFF MEM LASER FUSION, LOS ALAMOS SCI LAB, UNIV CALIF, 72- Concurrent Pos: Mem, Nat Res Coun, 72- Mem: Am Phys Soc. Res: High energy gamma ray physics; solar wind physics; high energy neutron physics; plasma physics in exploding wire systems and other high energy pulsed power systems; satellite instrumentation; design and physics of high energy vacuum systems. Mailing Add: Los Alamos Sci Lab Group L-4 PO Box 1663 Los Alamos NM 87545

YOUNG, FREDERICK GRIFFIN, b Niagara Falls, Ont, Nov 7, 40; m 63; c 2. STRATIGRAPHY. Educ: Queen's Univ, Ont, BSc, 63; McGill Univ, MSc, 64, PhD(geol), 70. Prof Exp: Geologist, Hudson's Bay Oil & Gas Co, 64-66; RES SCIENTIST STRATIG, INST SEDIMENTARY & PETROL GEOL, GEOL SURV CAN, 69- Mem: Soc Econ Paleont & Mineral; Can Soc Petrol Geologists. Res: Physical stratigraphy; lithofacies analyses; geology of Upper Precambrian and Cambrian; clastic sedimentation; trace fossils; Mesozoic and Cenozoic geology of Mackenzie Delta area. Mailing Add: Inst Sedimentary & Petrol Geol Geol Surv Can 3303 33rd St Calgary AB Can

YOUNG, FREDERICK HARRIS, mathematics, see 12th edition

YOUNG, FREDERICK WALTER, JR, b Hebron, Va, Sept 13, 24; m 50; c 2. PHYSICAL CHEMISTRY. Educ: Hampden-Sydney Col, BS, 44; Univ Va, PhD(chem), 50. Prof Exp: Instr, Hampden-Sydney Col, 44-46; res assoc, 50-51, chemist, Solid State Div, 56-69, ASSOC DIR SOLID STATE DIV, OAK RIDGE NAT LAB, 69- Concurrent Pos: Res assoc, Univ Va, 51-56. Mem: Fel AAAS; Am Crystallog Asn; fel Am Phys Soc; Am Asn Crystal Growth. Res: Chemical properties of metal surfaces; observations of dislocations in metals; radiation damage in metals. Mailing Add: Gallaher Ferry Rd Rte 5 Lenoir City TN 37771

YOUNG, GAIL SELLERS, JR, b Chicago, Ill, Oct 3, 15; m 34, 68; c 1. MATHEMATICS. Educ: Univ Tex, BA, 39, PhD(math), 42. Prof Exp: Tutor pure math, Univ Tex, 38-39, instr, 39-42; from instr to asst prof math, Purdue Univ, 42-47; from asst prof to prof, Univ Mich, 47-59; prof, Tulane Univ, La, 59-70, head dept, 61-70; chmn dept, 70-76, PROF MATH, UNIV ROCHESTER, 70- Concurrent Pos: Mem, Nat Res Coun, 65-68 & 70-73, Conf Bd Math Sci, 66-70; chmn, US Comn Math Educ & Steering Coun, The Rochester Plan, 75- Mem: Fel AAAS; Am Math Soc; Math Asn Am (1st vpres, 66-68, pres, 70-71). Res: Topology; analysis; differential equations. Mailing Add: 276 Bonnie Brae Ave Rochester NY 14618

YOUNG, GALE, b Baroda, Mich, Mar 5, 12; m 49; c 2. NUCLEAR PHYSICS. Educ: Milwaukee Sch Eng, BS, 33; Univ Chicago, BS & MS, 36. Prof Exp: Asst math biophys res, Univ Chicago, 36-40; head dept math & physics, Olivet Col, 40-42; physicist, Manhattan Dist Proj, Univ Chicago, 42-46; physicist, Clinton Labs, Tenn, 46-48; tech dir, Nuclear Develop Assocs, 48-55, vpres, Nuclear Develop Corp Am, 55-61; div vpres, United Nuclear Corp, 61-62; asst dir, 62-71, CONSULT, OAK RIDGE NAT LAB, 71- Concurrent Pos: Mem sci adv bd, US Dept Air Force, 54-58. Mem: Soc Indust & Appl Math; Math Asn Am. Res: Nuclear reactors; applied mathematics. Mailing Add: Oak Ridge Nat Lab Oak Ridge TN 37831

YOUNG, GEORGE ANTHONY, b New York, NY, Nov 8, 19; m 49; c 4. METEOROLOGY. Educ: NY Univ, BS, 48, MS, 49, PhD(meteorol), 65. Prof Exp: Res asst meteorol, NY Univ, 49-50; RES ASSOC, NAVAL SURFACE WEAPONS CTR, 50- Mem: Am Meteorol Soc. Res: Micrometeorology; hydrodynamics; turbulence; underwater explosions; environmental effects. Mailing Add: 3611 Janet Rd Silver Spring MD 20906

YOUNG, GEORGE JAMISON, b Hornell, NY, Aug 31, 25; m 46; c 3. PHYSICAL CHEMISTRY. Educ: Rensselaer Polytech Inst, BS, 50; Lehigh Univ, MS, 52, PhD(phys chem), 54. Prof Exp: Instr phys chem, Lehigh Univ, 54-55; fel, Mellon Inst, 55-56; from asst prof to assoc prof, Pa State Univ, 56-58; prof chem, Alfred Univ, 58-61; pres, Surface Processes Corp, 61-69; DIR CORP ENG, PITNEY-BOWES, INC, 69- Mem: The Chem Soc. Res: Engineering science; operations analysis; research administration. Mailing Add: Pitney-Bowes Inc 69 Walnut St Stamford CT 06904

YOUNG, GEORGE ROBERT, b Monmouth, Ill, Mar 9, 25; m 46; c 4. BIOLOGICAL CHEMISTRY. Educ: Univ Ind, BS, 49, PhD(biol chem), 56; Northwestern Univ, MS, 52. Prof Exp: Fel biol chem, Univ Ind, 55-56; instr, Dent Br, Univ Tex, 56-57, asst prof, 57-62; assoc prof biochem & nutrit, Sch Dent, 62-68, coord grad studies, 67-72, PROF BIOCHEM & NUTRIT, SCH DENT, UNIV MO-KANSAS CITY, 68-, CHMN DEPT, 62-, PROF MED, SCH MED, 73- Mem: Am Chem Soc; Int Asn Dent Res. Res: Collagen and collagenase; invasiveness of cells; solubility of tooth enamel; nutrition and periodontal metabolism. Mailing Add: Dept Biochem & Nutrit Sch Dent Univ of Mo 650 E 25th St Kansas City MO 64108

YOUNG, GILBERT FLOWERS, JR, b Mayesville, SC, Sept 23, 22. NEUROLOGY, PEDIATRICS. Educ: Col Charleston, BA, 42; Univ NC, MA, 46, Med Col SC, MD, 47; Am Bd Pediat, dipl, 58; Am Bd Psychiat & Neurol, dipl, 66, cert neurol with spec competence child neurol, 69. Prof Exp: Intern, Med Col Va Hosp, 47-48; instr pharmacol, Univ NC, 48-49; resident pediat, Roper Hosp, Charleston, SC, 51-53; pvt pract, 53-56; resident child develop, Children's Hosp, Columbus, Ohio, 56-57; from asst prof to assoc prof pediat, 57-71, from asst prof to assoc prof neurol, 60-71, PROF NEUROL & PEDIAT, MED UNIV SC, 71- Concurrent Pos: Resident, Mass Gen Hosp, 60-63; consult, Med Univ SC Hosp & Roper Hosp, Charleston. Mem: AMA; Am Acad Pediat; Am Acad Neurol. Res: Child neurology and development; congenital encephalopathies. Mailing Add: Med Univ Hosp 80 Barre St Charleston SC 29401

YOUNG, GRANT MCADAM, b Glasgow, Scotland, Aug 23, 37; m 60; c 3. STRATIGRAPHY, SEDIMENTOLOGY. Educ: Glasgow Univ, BSc, 60, PhD(geol), 67. Prof Exp: Res demonstr geol, Univ Wales, 62-63; from lectr to asst prof, 63-70, ASSOC PROF GEOL, UNIV WESTERN ONT, 70- Concurrent Pos: Nat Res Coun Can & Geol Surv Can grants, Univ Western Ont, 65- Mem: Soc Econ Paleont & Mineral; Geol Asn Can. Res: Stratigraphy and sedimentation of Precambrian supracrustal rocks; glaciogenic rocks in global correlation; Huronian rocks of Ontario and Upper Precambrian rocks of Arctic Canada. Mailing Add: Dept of Geol Univ of Western Ont London ON Can

YOUNG, HARLAND HARRY, b Portland, Ore, July 29, 08; m 32; c 2. ORGANIC CHEMISTRY. Educ: Reed Col, BA, 29; Mass Inst Technol, PhD(org chem), 32. Prof Exp: From anal chemist to res chemist, Swift & Co, 24-36, in charge new prod develop div, 36-39, asst to chief chemist, 39-41, asst chief chemist, 41-46, from asst dir res to dir res, 46-70; PRES, RES ADV SERV, INC, 71- Mem: Am Chem Soc; Am Leather Chem Asn. Res: Adhesives; food fats; industrial oils; soap; glycerin; colloids; emulsions; detergents; proteins; fibers; nonwoven fabrics; resins; polymers; byproduct utilization. Mailing Add: 4724 Wolf Rd Western Springs IL 60588

YOUNG, HAROLD, b New York, NY, Feb 2, 23; m 45; c 2. BIOCHEMISTRY. Educ: Cornell Univ, BS, 47, MS, 48. Prof Exp: Plant mgr, Cooperdale Dairy Co, 47; microbiologist & biochemist, Res Labs, 48-55, chief chemist, Consol Prod Div, 55-57, gen prod mgr & chief chemist, 57-58, chemist & whey technologist, Kraft Foods Div, 58-62, prod mgr indust prod, 62-64, prod mgr indust plants, 64-68, nat production prod mgr, 68-70, prod mgr indust plants, 70-75, STAFF ADMIN PRODUCTION MGR, KRAFT FOODS DIV OF KRAFTCO CORP, NAT DAIRY PROD CORP, CHICAGO, 75- Mem: Am Chem Soc; Soc Indust Microbiol; Am Dairy Sci Asn. Res: Fermentation products from dairy by-products; utilization of processing by-products for foods and animal feeds; production technology on fermentation products and processing by-products. Mailing Add: 1011 Pershing Ave Wheaton IL 60187

YOUNG, HAROLD EDLE, b Arlington, Mass, Sept 4, 17; m 43; c 4. FORESTRY. Educ: Univ Maine, BS, 37; Duke Univ, MF, 46, PhD(tree physiol), 48. Prof Exp: Field asst, US Forest Serv, 37-40; from instr to assoc prof, 48-61, PROF FORESTRY, UNIV MAINE, 61- Concurrent Pos: Fulbright res scholar, Norway, 63-64; vis appointment, Dept Forestry, Australian Nat Univ, 68-69; consult, Off Opers Anal, US Air Force, 51-59. Honors & Awards: Hitchcock Award, Forest Prod Res Soc Am, 74. Mem: Fel AAAS; Soc Am Foresters; Ecol Soc Am; Am Soc Plant Physiol; corresp mem Soc Forestry Finland. Res: Tree physiology and growth phenomena; soils. Mailing Add: Complete Tree Inst Univ of Maine Orono ME 04473

YOUNG, HAROLD HENRY, b Malone, NY, Sept 11, 27; m 51; c 5. ANALYTICAL CHEMISTRY. Educ: St Michael's Col, Vt, BS, 51. Prof Exp: Chem tech, Works Lab, Gen Elec Co, 51-52; shift supvr, Ind Ord Works, E I du Pont de Nemours & Co, Inc, Ind, 52-54, lab supvr, Savannah River Plant, SC, 54-57; tech reviewer, Div Civilian Appln, Oak Ridge, Tenn, 57-58, non-destructive testing specialist, Div Isotopes

Develop, Washington, DC, 58-59, isotopes training specialist, 59-62, nuclear educ & training specialist, 62-73, educ & training specialist, Div Biomed & Environ Res, 73-75, SR TRAINING COORDR, DIV UNIV & MANPOWER DEVELOP PROGS, US AEC, US ENERGY RES & DEVELOP ADMIN, WASHINGTON, DC, 75- Mem: Am Nuclear Soc. Res: Quality control of nitrocellulose, plutonium and special nuclear materials; radioisotope and radiation applications. Mailing Add: Div Univ & Manpower Develop Progs US Energy Res & Develop Admin Washington DC 20545

YOUNG, HAROLD WILLIAM, horticulture, plant breeding, see 12th edition

YOUNG, HARRISON HURST, JR, b Drumright, Okla, Sept 24, 19; m 42; c 2. PHYSICAL CHEMISTRY. Educ: Princeton Univ, AB, 40; Columbia Univ, PhD(chem), 50. Hon Degrees: Juris Dr, Fordham Univ, 74. Prof Exp: Asst, Nat Defense Res Comt, 41-45; instr chem, Williams Col, Mass, 47-50; res group leader, Westvaco Div, Food Mach & Chem Corp, 50-52, res sect mgr, 52-56, asst to dir res, Westvaco Mineral Prod Div, 56-58, mgr detergent applns res, Inorg Res & Develop Dept, 59-62, tech recruitment mgr, Chem Div, 62-70, MEM STAFF CHEM GROUP, PATENT & LICENSING DEPT, FMC CORP, 70-, PATENT ATTORNEY, 75- Mem: Am Chem Soc. Res: Reaction kinetics; industrial inorganic chemicals; detergent applications; agricultrual pesticides. Mailing Add: FMC Corp 2000 Market St Philadelphia PA 19103

YOUNG, HARRY CURTIS, JR, b East Lansing, Mich, Sept 6, 18; m 43; c 4. PLANT PATHOLOGY. Educ: Ohio State Univ, BSc, 40; Univ Minn, MSc, 43, PhD, 49. Prof Exp: Asst, Div Plant Path, Exp Sta, Univ Minn, 40-43 & 46-47; asst prof plant path, NY State Agr Exp Sta, Geneva, 47-50; assoc prof, 50-56, PROF PLANT PATH, OKLA STATE UNIV, 56- Concurrent Pos: Fel, Guggenheim Mem Found, 61; sr postdoctoral fel, Fulbright-Hays Prog, 69. Mem: Am Phytopath Soc; Am Soc Agron. Res: Pathology of elm, tomato, cucumber, fruit nursery stock, cereal and forage crops; breeding for disease resistance in cereal crops. Mailing Add: Dept of Bot & Plant Path Okla State Univ Stillwater OK 74074

YOUNG, HENRY Y, organic chemistry, see 12th edition

YOUNG, HERBERT LEWIS, b Detroit, Mich, July 7, 25; m 49; c 4. INDUSTRIAL ORGANIC CHEMISTRY. Educ: DePaul Univ, BS, 49, MS, 53; Purdue Univ, PhD(chem), 56. Prof Exp: Res Chemist, Del, 56-71, STAFF SCIENTIST, BACCHUS WORKS, HERCULES INC, UTAH, 71- Mem: Am Chem Soc; Am Inst Chem; Soc Advan Mat & Process Eng. Res: Resins for adhesive bonding; solid state reactions; nitrogen chemistry; explosives and propellants. Mailing Add: Bacchus Works Hercules Inc Magna UT 84044

YOUNG, HO LEE, b Canton, China, July 15, 20; nat US; m 49. PHYSIOLOGY. Educ: Lingnan Univ, BS, 43; Univ Calif, PhD, 54. Prof Exp: Asst gen biol & comp anat, Nat Med Col Shanghai, China, 44-49; student res physiologist, Univ Calif, 52-54, jr res physiologist, 54-56; res assoc radiol, Stanford Univ, 56-57; asst res physiologist, Med Ctr, Univ Calif, San Francisco, 57-63; res scientist, NASA Ames Res Ctr, 63-72; chief chem sect, 72-74, chief air sect, 74-76, chief lab br, 74, CHIEF WATER SECT, US ENVIRON PROTECTION AGENCY, 76- Concurrent Pos: San Francisco Heart Asn sr res fels, 60-62. Mem: Am Physiol Soc; Am Soc Cell Biol; Am Soc Microbiol; Aerospace Med Asn. Res: Cell physiology; cell particulates; cellular and fat metabolism; enzyme systems; protein synthesis; radiation effect on cells; electrolyte transport in skeletal muscle and isolated cells; effect of toxic materials and stresses on biochemical and physiological process of living systems. Mailing Add: 5978 Greenridge Rd Castro Valley CA 94546

YOUNG, HONG YIP, b Wailuku, Hawaii, Nov 27, 10; m 37; c 3. AGRICULTURAL CHEMISTRY. Educ: Univ Hawaii, BS, 32, MS, 33. Prof Exp: Sci aide, Pineapple Res Inst, 33-40, from jr chemist to chemist, 40-67; from assoc agronomist to agronomist, 67-75, EMER AGRONOMIST, AGR EXP STA, COL TROP AGR, UNIV HAWAII, 75- Concurrent Pos: Vis scientist, Int Rice Res Inst, 65-66; consult, Indian Agr Res Inst, 68. Mem: AAAS; Am Chem Soc. Res: Plant, soil, hormone and pesticide residue analysis; mineral nutrition of plants. Mailing Add: 676 Hakaka St Honolulu HI 96816

YOUNG, HOWARD FREDERICK, b Fond du Lac, Wis, Nov 13, 18; m 42; c 3. ZOOLOGY. Educ: Univ Wis, BA, 46, MA, 47, PhD, 50. Prof Exp: Instr, Univ Ark, 50, asst prof, 50-53; assoc prof biol, Western Ill State Col, 53-55; from asst prof to assoc prof, 55-63, PROF BIOL, UNIV WIS, LA CROSSE, 63- Concurrent Pos: NSF grant, 59-60. Mem: Wilson Ornith Soc; Am Ornith Union. Res: Ornithology; ecology; behavior; population. Mailing Add: Dept of Biol Univ of Wis La Crosse WI 54601

YOUNG, HOWARD SETH, b Birmingham, Ala, July 7, 24; m 45; c 7. PHYSICAL CHEMISTRY. Educ: Birmingham Southern Col, BS, 42; Brown Univ, PhD(chem), 48. Prof Exp: Chemist, 44-46 & 48-51, sr res chemist, 51-62, from res assoc to sr res assoc, 63-70, HEAD ENG RES DIV, TENN EASTMAN CORP, 70- Mem: Am Chem Soc. Res: Inorganic chemistry; catalysis. Mailing Add: 1909 E Sevier Ave Kingsport TN 37664

YOUNG, HUGH DAVID, b Ames, Iowa, Nov 3, 30; m 60; c 2. THEORETICAL PHYSICS. Educ: Carnegie Inst Technol, BS, 52, MS, 53, PhD(physics), 59, Carnegie-Mellon Univ, BFA, 72. Prof Exp: From instr to asst prof physics, 56-65, head dept natural sci, 62-74, ASSOC PROF PHYSICS, CARNEGIE-MELLON UNIV, 65- Mem: Am Phys Soc; Am Asn Physics Teachers. Res: Meson theory; new teaching materials for introductory college physics courses; administrative work in science education. Mailing Add: Dept of Physics Carnegie-Mellon Univ Pittsburgh PA 15213

YOUNG, IN MIN, b Seoul, Korea, July 13, 26; US citizen; m 53; c 3. AUDIOLOGY, PSYCHOACOUSTICS. Educ: Yonsei Univ, MD, 48; Jefferson Med Col, MSc, 66. Prof Exp: Instr otolaryngol, Sch Med, Yonsei Univ, 54-59; resident, Newark Eye & Ear Infirmary, NJ, 59-60; res audiologist, 60-65, from asst prof to assoc prof, 65-72, PROF AUDIOL, JEFFERSON MED COL, 72- Concurrent Pos: Dir, Hearing & Speech Ctr, Thomas Jefferson Univ Hosp, Philadelphia, 72-, audiologist, Affil Staff, 73-; consult otolaryngol, US Naval Hosp, Philadelphia; United Fund grant, 73-76. Mem: Acoust Soc Am; Am Speech & Hearing Asn; Am Neurotol Soc; Am Audiol Soc. Res: Auditory threshold and suprathreshold adaptation; Bekesy audiometry and marking. Mailing Add: Jefferson Med Col Thomas Jefferson Univ Philadelphia PA 19107

YOUNG, IRVING, b New York, NY, Aug 15, 22; m 48; c 3. MEDICINE, PATHOLOGY. Educ: Johns Hopkins Univ, AB, 43, MD, 46. Prof Exp: Asst pathologist, Kings County Hosp, Brooklyn, NY, 49-51; assoc dir labs, Div Path, 52-71, CHMN DIV LABS, ALBERT EINSTEIN MED CTR, 71-; CLIN ASSOC PROF PATH, SCH MED, TEMPLE UNIV, 75- Mem: AMA. Res: Immunomorphologic correlation; histochemistry of chromosomes; surgical pathology. Mailing Add: Div of Labs Albert Einstein Med Ctr Philadelphia PA 19141

YOUNG, IRVING GUSTAV, b Brooklyn, NY, Dec 10, 19; m 41; c 2. ANALYTICAL CHEMISTRY, PHYSICAL CHEMISTRY. Educ: City Col New York, BS, 39; Polytech Inst Brooklyn, MS, 50; Temple Univ, PhD, 67. Prof Exp: Res asst, Bellevue Hosp, Columbia Univ, 39-42; asst chemist, Picatinny Arsenal, 42-44; battery technologist, US Elec Mfg Corp, 44-51; sr res chemist, Int Resistance Co, 51-56, sr res scientist, 59-64; chief chemist, Transition Metals & Chem Co, 56-57; asst res scientist, Leeds & Northrup Co, 57-59; res fel, Temple Univ, 64-65; chemist, Advan Technol Staff, Indust Div, Honeywell, Inc, 65-74, DEVELOP SUPVR, HONEYWELL POWER SOURCES CTR, 74- Mem: Am Chem Soc; Instrument Soc Am; Air Pollution Control Asn; Water Pollution Control Fedn; Am Soc Test & Mat. Res: Electrochemistry; process analyzers; air and water pollution. Mailing Add: 22 Four Leaf Rd Levittown PA 19056

YOUNG, J LOWELL, b Perry, Utah, Dec 13, 25; m 50; c 4. SOIL BIOCHEMISTRY. Educ: Brigham Young Univ, BS, 53; Ohio State Univ, PhD(soils), 56. Prof Exp: Asst agron, Agr Exp Sta, Ohio State Univ, 53-56, fel agr biochem, Univ, 56-57; chemist, Agr Res Serv, USDA, 57-60; asst prof, 57-63, ASSOC PROF SOILS, ORE STATE UNIV, 63-; RES CHEMIST, AGR RES SERV, USDA, 60- Concurrent Pos: Assoc ed, Soil Sci Soc Am publ, 75-78. Mem: AAAS; Am Soc Agron; Soil Sci Soc Am; Clay Minerals Soc; Int Asn Study Clays. Res: Chemistry of soil nitrogen and organic matter; amino acids of soils, organic matter, root exudates, endomycorrhiza and stream waters; amorphous soil colloids, mineral-organic complexes and suspended sediments; D-amino acids and higher plants. Mailing Add: Dept of Soil Sci Ore State Univ Corvallis OR 97331

YOUNG, JACK ELWOOD, comparative endocrinology, comparative physiology, see 12th edition

YOUNG, JACK PHILLIP, b Huntington, Ind, Oct 28, 29; m 55; c 5. ANALYTICAL CHEMISTRY. Educ: Ball State Teachers Col, BS, 50; Univ Ind, PhD(chem), 55. Prof Exp: CHEMIST, UNION CARBIDE NUCLEAR CO, 55- Mem: AAAS; Am Chem Soc. Res: Actinide chemistry; spectroscopy of solutions, molten salts, solid state compounds; laser spectroscopy; photoionization studies. Mailing Add: 100 Westlook Circle Oak Ridge TN 37830

YOUNG, JAMES ALBERT, range ecology, genetics, see 12th edition

YOUNG, JAMES ARTHUR, JR, b Tacoma, Wash, Feb 12, 21; m 43; c 2. PHYSICS, ELECTRICAL ENGINEERING. Educ: Calif Inst Technol, BS, 43; Univ Wash, PhD(physics), 53. Prof Exp: Res engr rocket instrumentation, Jet Propulsion Lab, Calif Inst Technol, 46-47; teaching fel physics, Univ Wash, 47-53; CO-DIR, RADIO RES LAB, BELL TEL LABS, HOLMDEL, 53- Mem: Inst Elec & Electronics Engrs; Am Phys Soc. Res: Communications research; encoding, modulation transmission and switching of information signals, particularly for high radio frequency and optical media. Mailing Add:

YOUNG, JAMES CHRISTOPHER F, b Charlottetown, PEI, Apr 1, 40; m 66; c 2. ANALYTICAL CHEMISTRY, ENTOMOLOGY. Educ: Mt Allison Univ, BSc, 60; McMaster Univ, MSc, 62; Mass Inst Technol, PhD(org chem), 71. Prof Exp: Teacher sci, Kitchener-Waterloo Col & Voc Sch, 62-64; lectr chem, Waterloo Lutheran Univ, 64-66; RES SCIENTIST, CHEM & BIOL RES INST, AGR CAN, 72- Concurrent Pos: Rockefeller Found fel, NY Col Forestry, Syracuse Univ, 71-72. Mem: AAAS; Chem Inst Can; Am Chem Soc. Res: Natural product chemistry; chemicals used in animal communication; insect pheromones and hormones and feeding stimulants; organic chemistry analytical methodology. Mailing Add: Agr Can Chem & Biol Res Inst Ottawa ON Can

YOUNG, JAMES EDWARD, b Wheeling, WVa, Jan 18, 26; m 48; c 1. PARTICLE PHYSICS, NUCLEAR PHYSICS. Educ: Howard Univ, BS, 46, MS, 49; Mass Inst Technol, MS, 51, PhD(physics), 53. Prof Exp: Instr physics, Hampton Inst, 46-49; fel acoustics, Mass Inst Technol, 53-55; staff mem physics, Los Alamos Sci Lab, 57-70; PROF PHYSICS, INST THEORET PHYSICS, MASS INST TECHNOL, 70- Concurrent Pos: Shell B P fel, Univ Southampton, 56; consult, Gen Atomics, Calif, 57-58; Nat Acad Sci-Nat Res Coun & Ford fel, Bohr Inst, Copenhagen, 61-62; vis assoc prof, Univ Minn, 64; res asst, Oxford Univ, 65-66; visitor, Inst Theoret Physics, Mass Inst Technol, 68-69. Mem: Am Phys Soc. Res: Pion physics at intermediate energies; three-body relativistic models; multipheripheral equations for hadrons at high energies. Mailing Add: Ctr for Theoret Physics Mass Inst of Technol Cambridge MA 02139

YOUNG, JAMES GEORGE, b Milwaukee, Wis, July 18, 26; m 54; c 3. INDUSTRIAL PHARMACY. Educ: Univ Wis, BS, 48, MS, 49; Univ NC, PhD(pharm), 52. Prof Exp: Asst prof pharmaceut chem, Med Col Va, 51-54 & Univ Tenn, 56-58; sr chemist, Riker Labs, Inc, 58-60, dir prod develop, 60-72; DIR DEVELOP, G D SEARLE & CO, 72- Mem: AAAS; Am Chem Soc; Am Pharmaceut Asn; NY Acad Sci. Res: Pharmaceutical aerosol formulation; drug stabilization; general pharmaceutical development. Mailing Add: GD Searle & Co PO Box 5110 Chicago IL 60680

YOUNG, JAMES HOWARD, b Norfolk, Va, May 9, 24; m 50; c 1. MATHEMATICS. Educ: Univ Va, BA, 48; Duke Univ, MA, 51. Prof Exp: Instr physics, Norfolk Div, Col William & Mary, 47-51; instr math, Johns Hopkins Univ, 55-69, res scientist, Inst Coop Res, 51-69; SR RES ANALYST, FALCON RES & DEVELOP, 69- Mem: Am Math Soc. Res: Weapons systems analysis; military operations research. Mailing Add: 106 Regester Ave Baltimore MD 21212

YOUNG, JAMES ORVILLE, dairy technology, chemistry, see 12th edition

YOUNG, JAMES ROBERT, biological chemistry, deceased

YOUNG, JAMES ROGER, b Fordland, Mo, June 14, 23; m 45; c 3. PHYSICS. Educ: Park Col, AB, 46; Univ Mo, BA, 49, PhD(physics), 52. Prof Exp: Res physicist, Res Lab, 51-63, mgr advan develop vacuum prod, 63-64, mgr eng, 64-66, MGR PLASMA LIGHT SOURCES, RES & DEVELOP CTR, GEN ELEC CO, SCHENECTADY, 74- Mem: Am Vacuum Soc (treas, 73-); Am Phys Soc; Am Inst Physics. Res: Vacuum physics; physical electronics. Mailing Add: Schenectady NY

YOUNG, JANICE EDITH, b Princeton, NJ, Jan 14, 46; m 72. ZOOLOGY, COMPARATIVE PHYSIOLOGY. Educ: Groucher Col, AB, 67; Northwestern Univ, Evanston, PhD(biol), 70. Prof Exp: Asst prof biol, Keuka Col, NY, 70-73 & Springfield Col, Mass, 73-74. Mem: AAAS; Am Soc Zool; Am Inst Biol Sci. Res: Crustacean endocrinology, especially the control of the reproductive phenomena in male decapods and photoperiodism. Mailing Add: 1433 S Watson Rd Mt Pleasant MI 48858

YOUNG, JANIS DILLAHA, b Little Rock, Ark, July 12, 27; m 56; c 2. BIOCHEMISTRY, IMMUNOCHEMISTRY. Educ: Hendrix Col, BA, 49; Univ Okla, MS, 51; Univ Calif, Berkeley, PhD(biochem), 60. Prof Exp: Biochemist, Armour Labs, Chicago, Ill, 51-54; instr chem, Colby Col, 54-55; NIH trainee virol, Univ Calif, Berkeley, 59-61; assoc res scientist, Lab Med Entom, Kaiser Found Res Inst, 61-70;

ASSOC RES BIOCHEMIST, SPACE SCI LAB & ADJ ASSOC PROF IMMUNOL, DEPT BACT & IMMUNOL, UNIV CALIF, BERKELEY, 71- Mem: Am Soc Biol Chem; Am Asn Immunol; Am Chem Soc. Res: Chemistry and enzymatic study of peptides and proteins; relating structure of proteins to their immunological activity using peptide synthesis. Mailing Add: Univ of Calif Space Sci Lab 1414 S Tenth St Richmond CA 94804

YOUNG, JAY ALFRED, b Huntington, Ind, Sept 8, 20; m 42, 62; c 18. PHYSICAL CHEMISTRY. Educ: Univ Ind, BS, 39; Oberlin Col, AM, 40; Univ Notre Dame, PhD, 50. Prof Exp: Chief chemist, Asbestos Mfg Co, Ind, 40-42; ord engr, US War Dept, DC, 42-44; from instr to prof chem, King's Col, Pa, 49-69; vis prof, Carleton Univ, 69-70; Hudson Prof Chem, Auburn Univ, 70-75; PROF CHEM, FLA STATE UNIV, 75- Concurrent Pos: Mem, Adv Coun Col Chem. Honors & Awards: Centennial of Sci Award, Univ Notre Dame, 65; Excellence in Teaching Award, Mfg Chem Asn, 71. Mem: AAAS; Am Chem Soc. Res: The relationship between the art and the science of teaching, especially the proper function of laboratory work. Mailing Add: Dept of Chem Fla State Univ 415 N Monroe Tallahassee FL 32301

YOUNG, JAY MAITLAND, b Louisville, Ky, Nov 26, 44. PHYSICAL BIOCHEMISTRY, ENZYMOLOGY. Educ: Vanderbilt Univ, BA, 66; Yale Univ, MS, 67, MPh, 68, PhD(chem), 71. Prof Exp: ASST PROF CHEM, BRYN MAWR COL, 70- Concurrent Pos: NIH fel, Oxford Univ, 71-72; vis scientist, Inst Cancer Res, Philadelphia, Pa, 75-76. Mem: AAAS; Am Chem Soc. Res: Mechanisms of enzymatic reactions; nuclear magnetic resonance; chemical modifications of proteins. Mailing Add: Dept of Chem Bryn Mawr Col Bryn Mawr PA 19010

YOUNG, JERRY H, b Fitzhugh, Okla, Aug 4, 31; m 52; c 1. ENTOMOLOGY. Educ: Okla State Univ, BS, 55, MS, 56; Univ Calif, Berkeley, PhD(parasitol), 59. Prof Exp: PROF ENTOM, OKLA STATE UNIV, 59- Mem: Entom Soc Am; Entom Soc Can. Res: Cotton insect control; mite morphology; Hymenoptera taxonomy; integrated and biological control of cotton insects. Mailing Add: Dept of Entom Okla State Univ Stillwater OK 74074

YOUNG, JERRY WESLEY, b Mulberry, Tenn, Aug 19, 34; m 59; c 2. ANIMAL NUTRITION. Educ: Berry Col, BS, 57; NC State Univ, MS, 59, PhD(animal nutrit), 63. Prof Exp: Res asst animal nutrit, NC State Univ, 57-63; USPHS fel biochem, Inst Enzyme Res, Univ Wis, 63-65; asst prof animal nutrit, 65-70, assoc prof animal sci & biochem, 70-74, PROF ANIMAL SCI & BIOCHEM, IOWA STATE UNIV, 74- Mem: Am Chem Soc; Am Dairy Sci Asn; Am Inst Nutrit; Am Soc Animal Sci. Res: Volatile fatty acid metabolism in ruminants; mechanism and control of gluconeogenesis and interrelationships with lipid metabolism; lipid absorption in the bovine; control of milk fat synthesis. Mailing Add: Dept of Animal Sci Iowa State Univ Ames IA 50010

YOUNG, JOEL EDWARD, biochemistry, see 12th edition

YOUNG, JOHN A, b Washington, DC, July 4, 39; m 62; c 2. METEOROLOGY. Educ: Miami Univ, BA, 61; Mass Inst Technol, PhD(meteorol), 66. Prof Exp: NSF fel, Univ Oslo, 66; from asst prof to assoc prof, 66-75, NSF res grant, 71-75, PROF METEOROL, UNIV WIS-MADISON, 75- Concurrent Pos: Vis assoc prof, Mass Inst Technol, 73-74; mem Global Atmospheric Res Prog, Nat Acad Sci, 73-75. Mem: AAAS; Am Meteorol Soc. Res: Dynamic meteorology; numerical modeling; planetary boundary layer; geophysical fluid dynamics. Mailing Add: Dept of Meteorol Univ of Wis Madison WI 53706

YOUNG, JOHN ADAMS, chemistry, see 12th edition

YOUNG, JOHN ALBION, JR, b Newport, RI, Aug 29, 09; m 37. PETROLEUM GEOLOGY. Educ: Brown Univ, PhB, 32, ScM, 34; Harvard Univ, PhD(geol), 46. Prof Exp: Asst geol, Brown Univ, 32-37; asst paleont, Harvard Univ, 37-39; instr geol, Mich State Col, 39-44; geologist, Sun Oil Co, 44-46; asst prof geol, Syracuse Univ, 46-47; geologist, Sun Oil Co, 47-50, staff geologist, 50-70; vpres, secy, dir & dir oil & gas div, Tax Shelter Adv Serv, Inc, Pa, 70-73; VPRES, WORLD RESOURCES, 73-; SR VPRES, OMNI-EXPLOR, INC, NARBERTH, 74- Mem: Geol Soc Am; Am Soc Petrol Geologists. Res: Petroleum geology; stratigraphy; subsurface geology. Mailing Add: PO Box 362 Hunters Lane Devon PA 19333

YOUNG, JOHN CANNON, b Salt Lake City, Utah, June 27, 28; m 56; c 4. GEOLOGY. Educ: Univ Utah, BS, 50, MS, 53; Princeton Univ, PhD(geol), 60. Prof Exp: Explor geologist, Standard Oil Co Calif, 50-54 & 59-61; PROF GEOL & CHMN DEPT, HUMBOLDT STATE UNIV, 61- Mem: AAAS; Am Asn Petrol Geologists; Nat Asn Geol Teachers; Geol Soc Am. Res: Regional stratigraphy and structural geology in the western United States; geomorphology; tectonics in Klamath Mountains. Mailing Add: Dept of Geol Humboldt State Univ Arcata CA 95521

YOUNG, JOHN COLEMAN, b Leesville, La, July 13, 42; m 64; c 1. STATISTICS. Educ: Northwestern State Univ, BA, 64, MS, 65; Southern Methodist Univ, PhD(statist), 71. Prof Exp: Asst instr math, Northwestern State Univ, 64-65; asst prof, McNeese State Univ, 67-69; instr statist, Southern Methodist Univ, 69-71; asst prof, 71-75, ASSOC PROF MATH, McNEESE STATE UNIV, 75- Mem: Am Statist Asn. Res: Multivariate analysis; discrimination, analysis of variance and goodness of Fit test for multivariate populations. Mailing Add: Dept of Math Sci McNeese State Univ Lake Charles LA 70601

YOUNG, JOHN DAVIS, b Harrisonburg, Va, July 9, 21; m 46; c 3. ORGANIC CHEMISTRY. Educ: Univ NC, BS, 43; Univ Ill, PhD(org chem), 47. Prof Exp: Asst org chem, Univ Ill, 43-46, chemist, Off Rubber Reserve, 46-47; CHEMIST, E I DU PONT DE NEMOURS & CO, INC, 47- Mem: Am Chem Soc; Soc Plastics Engrs; Soc Automotive Engrs. Res: Synthesis of substituted styrene monomers for polymerization studies; design development with plastic materials in automotive industry. Mailing Add: E I du Pont de Nemours & Co Inc 26300 Northwestern Hwy Southfield MI 48076

YOUNG, JOHN FALKNER, b Tyler, Tex, Apr 3, 40. PHARMACODYNAMICS, TOXICOLOGY. Educ: NTex State Univ, BA, 63; Univ Houston, BS, 66; Univ Fla, MS, 69, PhD(pharmaceut res), 73. Prof Exp: RES BIOLOGIST TERATOLOGY, NAT CTR TOXICOL RES, 73- Honors & Awards: Commendable Serv Award, Nat Ctr Toxicol Res, Food & Drug Admin, 74. Mem: Am Pharmaceut Asn; Soc Appl Spectros. Res: Application of the principles of pharmacokinetics to teratogenic research; development of the analytical procedures used to quantitate the chemicals from biological fluids and tissues; simulation of data on hybrid computer. Mailing Add: Teratology Div Nat Ctr for Toxicol Res Jefferson AR 72079

YOUNG, JOHN H, b Shamokin, Pa, Aug 16, 40. PHYSICS. Educ: Gettysburg Col, BA, 62; Univ NH, MS, 64; Clark Univ, PhD, 69. Prof Exp: Asst physics, Clark Univ, 64-66; ASST PROF PHYSICS, UNIV ALA, BIRMINGHAM, 70- Mem: Am Phys Soc. Res: Description of the gravitational field of a rotating mass in the general theory by exact means; two body problem in general relativity; three body nucleon problem. Mailing Add: Dept of Physics Univ of Ala Univ Sta Birmingham AL 35294

YOUNG, JOHN KIGER, b Rock Springs, Wyo, May 3, 08; m 33; c 5. DENTISTRY. Educ: Stanford Univ, AB, 31, AM, 34; Col Physicians & Surgeons San Francisco, DDS, 40; Am Bd Oral Path, dipl, 52. Prof Exp: Asst prof bact, path & histol, 31-42, from assoc prof to prof bact & path, 42-70, prof path & histol & chmn dept, 70-74, EMER PROF PATH, UNIV OF THE PAC, 74- Concurrent Pos: Asst, Stanford Univ, 33-36, asst, Sch Med, 43-. Mem: Assoc Am Dent Asn; fel Am Col Dent; fel Am Acad Oral Path. Res: Cytopathology of poliomyelitis; bacteriophage behavior; sedimentation rates in peridontal disease; histopathology of treated teeth; oral tumors and cysts. Mailing Add: Dept of Path & Histol Univ of the Pac Sch of Dent San Francisco CA 94115

YOUNG, JOHN PAUL, b Baltimore, Md, Dec 20, 23; m 53; c 2. OPERATIONS RESEARCH. Educ: Univ Md, BS, 50; Johns Hopkins Univ, DEng(opers res), 62. Prof Exp: Indust engr, Westinghouse Elec Corp, 50-54; sr analyst opers res, 54-59, instr, 59-62, lectr, 62-64, asst dir opers res div, Hosp, 62-68, from asst prof to assoc prof pub health admin & opers res, 64-69, assoc provost, 68-72, actg vpres & provost, 70-71, PROF PUB HEALTH ADMIN & OPERS RES, SCH HYG & PUB HEALTH, JOHNS HOPKINS UNIV, 69- Concurrent Pos: USPHS grant; consult, USPHS, 62-, mem health serv res study sect, 66-69, mem health care systs study sect, 69-72; consult, Vet Admin, 62-, Hosp Coun of Md, 65-, Am Hosp Asn, 66- & CSF, Ltd, 75-; bd trustees, Chesapeake Res Consortium, 72- Mem: Fel AAAS; Opers Res Soc Am; Am Pub Health Asn; Inst Mgt Sci; fel Royal Soc Health. Res: Operations research in health services. Mailing Add: Stebbins Bldg Johns Hopkins Univ Baltimore MD 21205

YOUNG, JOHN WILLIAM, b Toronto, Ont, Nov 16, 12; nat US; m 39; c 1. MATHEMATICS, COMPUTER SCIENCE. Educ: Univ Fla, AB, 34, BS, 37, MA, 40, PhD(math), 52. Prof Exp: Prin & sch teacher, Fla, 34-42; instr math, Univ Fla, 46-52; appl sci rep, Int Bus Mach Corp, Ga, 54-55; mathematician & comput prog consult, Res Comput Ctr, NY, 55-57 & Missile Test Ctr, Radio Corp Am, 57-58; head anal & info processing, Res Div, Radiation, Inc, 59-60, head eng comput serv, 61-68, mem sr staff, 68-71; mem sr specialist staff, Data Systs Div, Martin Marietta Corp, Fla, 72-75; VIS ASSOC PROF COMPUT SCI, FLA TECHNOL UNIV, 75- Mem: Nat Coun Teachers Math; Math Asn Am; Asn Comput Mach. Res: Applications of electronic digital computers; computer performance measurement and evaluation; simulation of computer systems. Mailing Add: PO Box 1661 Melbourne FL 32901

YOUNG, JON NATHAN, b Hibbing, Minn, May 30, 38; m 61; c 2. ANTHROPOLOGY, RESOURCE MANAGEMENT. Educ: Univ Ariz, BA, 60, PhD(anthropol), 67; Univ Ky, MA, 62. Prof Exp: Teaching asst anthropol, Univ Ariz, 66-67; archeologist, 67-72, EXEC ORD CONSULT CULT RES MGT, NAT PARK SERV, US DEPT OF INTERIOR, 73- Concurrent Pos: Bd dir, Mus San Carlos Apache, 69- & Anthropol Related Sci & Art, Kokopelli Sessions Inc, 71- Mem: Fel Am Anthropol Asn; fel AAAS; fel Royal Anthropol Inst; Soc Am Archaeol. Res: Archeology and ethnology of the southwestern United States; federal historic preservation laws, policies and procedures. Mailing Add: 6970 Camino Namara Tucson AZ 85715

YOUNG, JOSEPH HARDIE, b Salt Lake City, Utah, Aug 11, 18; m 66; c 1. BIOLOGY. Educ: Stanford Univ, PhD(biol), 54. Prof Exp: Asst biol, Stanford Univ, 49-52, instr, 52-54; from instr to asst prof zool, Tulane Univ, 54-59; from asst prof to assoc prof biol, 59-65, PROF BIOL, SAN JOSE STATE UNIV, 65-, CHMN DEPT, 66- Mem: AAAS; Entom Soc Am; Am Soc Zool. Res: Insect embryology; arthropod morphology; marine biology. Mailing Add: Dept of Biol Sci San Jose State Univ San Jose CA 95112

YOUNG, JOSEPH MARVIN, b Marshall, Tex, Oct 16, 19; m 42; c 4. PATHOLOGY, ANATOMY. Educ: Harvard Univ, BS, 43; Johns Hopkins Univ, MD, 45. Prof Exp: Intern surg, Johns Hopkins Hosp, 45-46; resident surg, 46-51, resident path, 51-54, CHIEF LAB SERV, VET ADMIN HOSP, 54-; PROF PATH, UNIV TENN MED UNITS, MEMPHIS, 62- Honors & Awards: Spec Commendation Award, Chief Med Dir, Vet Admin Cent Off, 67. Mem: AAAS; Col Am Path; Am Asn Path & Bact; Am Soc Exp Path. Res: Joint reactions and spread of cancer. Mailing Add: Vet Admin Hosp Lab Serv 1030 Jefferson Ave Memphis TN 38104

YOUNG, JOSEPH ORAN, b Wilton, Conn, July 22, 14; m 39; c 5. BOTANY. Educ: Hobart Col, BS, 27; Brown Univ, MSc, 39; Univ Chicago, PhD(bot), 41. Prof Exp: Head agr res dept, Libby, McNeill & Libby, 45-58; PROF HORT, UNIV NEBR-LINCOLN, 58- Concurrent Pos: Consult prof, Inst Colombiano Agropecorio, Bogota, Colombia, 70-72 & Kabul Univ, 75- Mem: AAAS; fel Am Soc Hort Sci. Res: Breeding nutrition, development, pathology and storage of canning crops; vegetable crops. Mailing Add: Dept of Hort Univ of Nebr Lincoln NE 68583

YOUNG, KEITH PRESTON, b Buffalo, Wyo, Aug 18, 18; m 49; c 3. PALEONTOLOGY, ENVIRONMENTAL GEOLOGY. Educ: Univ Wyo, BA, 40, MA, 42; Univ Wis, PhD(stratig), 48. Prof Exp: From asst prof to assoc prof, 48-58, PROF GEOL, UNIV TEX, AUSTIN, 58- Mem: Paleont Soc; Geol Soc Am; Soc Econ Paleont & Mineral; Am Asn Petrol Geol; Am Inst Prof Geologists. Res: Cretaceous stratigraphy of Texas; cephalopods and paleontology. Mailing Add: Dept of Geol Sci Univ of Tex Austin TX 78712

YOUNG, KENNETH CHRISTIE, b Rochester, NY, Nov 9, 41. CLOUD PHYSICS. Educ: Ariz State Univ, BS, 65; Univ Chicago, MS, 67, PhD(geophys), 73. Prof Exp: Fel, Nat Ctr Atmospheric Res, 73-74; ASST PROF ATMOSPHERIC SCI, UNIV ARIZ, 74- Mem: Am Meteorol Soc. Res: Precipitation processes in clouds; ice phase nucleation; hail suppression; orographic precipitation and its enhancement; numerical simulations of microphysical processes in clouds. Mailing Add: Inst of Atmospheric Physics Univ of Ariz Tucson AZ 85721

YOUNG, KENNETH KONG, b Vancouver, BC, Mar 19, 37; m 67. HIGH ENERGY PHYSICS. Educ: Univ Wash, BSc, 59; Univ Pa, PhD(physics), 65. Prof Exp: Res assoc physics, Univ Mich, 65-67; asst prof, 67-70, ASSOC PROF PHYSICS, UNIV WASH, 70- Mem: Am Phys Soc. Res: Experimental weak interactions; time reverse invariance; particle physics. Mailing Add: Dept of Physics Univ of Wash Seattle WA 98195

YOUNG, LAURENCE CHISHOLM, b Göttingen, Ger, July 14, 05; m 34; c 6. PURE MATHEMATICS. Educ: Cambridge Univ, ScD, 39. Prof Exp: Lectr, Cambridge Univ & Univ London, 36-38; prof pure math, Univ Cape Town, 38-49; prof math, Univ, 49-68, distinguished res prof, 68-73, PROF MATH, MATH RES CTR, UNIV WIS-MADISON, 73- Concurrent Pos: Vis prof, Ohio State Univ, 47-48 & Ind Univ, 60-61; mem, Inst Advan Study, 47-48; lectr, Europe, 58-59. Mem: Am Math Soc; Royal Astron Soc; London Math Soc. Res: Calculus of variations; prime ends; theory of the integral. Mailing Add: Dept of Math Univ of Wis Madison WI 53706

YOUNG, LAWRENCE EUGENE, b Waterville, Ohio, Mar 18, 13; m 40; c 4. MEDICINE. Educ: Ohio Wesleyan Univ, BA, 35; Univ Rochester, MD, 39; Am Bd Internal Med, dipl. Hon Degrees: DSc, Ohio Wesleyan Univ, 67. Prof Exp: From intern to asst resident med, Strong Mem Hosp, 39-41, asst bact, Sch Hyg & Pub Health, Johns Hopkins Univ & Hosp, 41-42; chief resident, Strong Mem Hosp, 42-43; instr med, 43-44, 46-47, from asst prof to assoc prof, 48-57, Charles A Dewey prof & chmn dept, 57-74, ALUMNI DISTINGUISHED SERV PROF MED, SCH MED & DENT, UNIV ROCHESTER, 74-, DIR PROG INTERNAL MED, UNIV ROCHESTER ASSOC HOSPS, 74- Concurrent Pos: Buswell fel, 46-47; physician-in-chief, Strong Mem Hosp, 57-74; mem comt blood, Nat Res Coun, 51-53; hemat study sect mem, USPHS, 53-57. Mem: Am Soc Clin Invest; Asn Am Physicians (vpres, 72-73, pres, 73-74); regent Am Fedn Clin Res; Asn Profs Med (pres, 66-67); fel Int Soc Hemat. Res: Hematology; patient care. Mailing Add: Box 702 Univ of Rochester Rochester NY 14642

YOUNG, LEE ADAMS, physics, chemical physics, see 12th edition

YOUNG, LEONA GRAFF, b New York, NY, Dec 22, 36; m 58; c 3. PHYSIOLOGY. Educ: Bryn Mawr Col, BA, 58; Univ SC, MS, 60; Emory Univ, PhD(physiol), 67. Prof Exp: Instr physiol, 67-68, USPHS fel biochem, 69-71, instr physiol, 71-72, ASST PROF PHYSIOL, SCH MED, EMORY UNIV, 72- Concurrent Pos: USPHS res grant, 72- Mem: AAAS; Am Soc Cell Biol. Res: Mammalian spermatozoan motility; spermatogenesis; fertility; infertility; contraception; contractile proteins. Mailing Add: Dept of Physiol Emory Univ Sch of Med Atlanta GA 30322

YOUNG, LEONARD M, b Dallas, Tex, Oct 20, 35. GEOLOGY. Educ: Rice Univ, BA, 57; Univ Okla, MS, 60; Univ Tex, Austin, PhD(geol), 68. Prof Exp: Asst prof, 67-74, ASSOC PROF GEOL, NORTHEAST LA UNIV, 74- Mem: Soc Econ Paleontologists & Mineralogists; Nat Asn Geol Teachers. Res: Carbonate and terrigenous sedimentary petrology; sedimentary processes; textural parameters; sedimentary structures; fluid inclusion paleotemperatures. Mailing Add: Dept of Geol Northeast La Univ Monroe LA 71201

YOUNG, LEWIS BREWSTER, b Los Angeles, Calif, Feb 25, 43; m 66. ORGANIC CHEMISTRY. Educ: Univ Calif, Riverside, BA, 64; Iowa State Univ, PhD(org chem), 68. Prof Exp: NIH fel, Univ Colo, Boulder, 68-70; SR RES CHEMIST, MOBIL CHEM CO, 70- Mem: Am Chem Soc; The Chem Soc; Sigma Xi. Res: Oxidation mechanisms; homogeneous catalysis; heterogeneous catalysis; ion-molecule reactions; catalysis by zeolites. Mailing Add: Mobil Chem Co PO Box 240 Edison NJ 08817

YOUNG, LIONEL WESLEY, b New Orleans, La, Mar 14, 32; m 57; c 3. PEDIATRICS, RADIOLOGY. Educ: St Benedict's Col, Kans, BS, 53; Howard Univ, MD, 57. Prof Exp: From sr instr to asst prof radiol, 65-69, asst prof pediat, 66-69, assoc prof radiol & pediat, Med Ctr, Univ Rochester, 69-75; PROF RADIOL & PEDIAT, HEALTH CTR, UNIV PITTSBURGH, 75- Concurrent Pos: Nat Cancer Inst traineeship grant radiation ther, Med Sch, Univ Rochester, 59-60; Children's Bur, Dept Health, Educ & Welfare fel pediat radiol, Sch Med, Univ Cincinnati, 63-65; abstractor radiol, Am J Roentgenol & Radium Ther & Nuclear Med, 65-; clin consult, NY State Dept Health, 67-75; mem radiol training comt, Nat Inst Gen Med Sci, NIH, 71-73; pediat radiol consult, Comt Prof Self-Eval & Continuing Educ, Am Col Radiol, 72- Honors & Awards: Spec Paper Award, Soc Pediat Radiol, 69; Gold Cert, Nat Med Asn, 70. Mem: Radiol Soc NAm; Am Roentgen Ray Soc; Soc Pediat Radiol; Am Col Radiol; Asn Univ Radiol. Res: Magnification radiography and tomagraphy in pediatric radiology; radiology of renal hypoplasias and dysplasias; duodenal, pancreatic and renal injury from blunt trauma; skeletal dysplasias and metabolic bone disease in childhood. Mailing Add: Dept of Radiol Children's Hosp of Pittsburgh Pittsburgh PA 15213

YOUNG, LLOYD MARTIN, b Merricourt, NDak, Nov 9, 42; m 66; c 2. EXPERIMENTAL NUCLEAR PHYSICS. Educ: Univ NDak, BS, 65, MS, 66; Univ Ill, PhD(physics), 72. Prof Exp: Res assoc, 72-74, ASST PROF PHYSICS, UNIV ILL, 74- Mem: Am Phys Soc. Res: Development of an electron accelerator having a 100% duty factor using a superconducting linac through which the beam is recirculated several times. Mailing Add: Physics Res Lab Univ Ill Champaign IL 61820

YOUNG, LOUIS LEE, b El Paso, Tex, Nov 22, 41. FOOD SCIENCE. Educ: Tex A&M Univ, BS, 64, MS, 67, PhD(poultry prod technol), 70. Prof Exp: Res asst poultry nutrit, Tex Agr Exp Sta, 64-67; res fel poultry prod technol, Tex A&M Univ, 67-68; res assoc, Tex Agr Exp Sta, 68-71; RES FOOD TECHNOLOGIST, POULTRY PROD TECHNOL, RUSSELL RES CTR, AGR RES SERV, US DEPT AGR, 71- Mem: AAAS; Poultry Sci Asn; Inst Food Technol; Sigma Xi. Res: Food chemistry and microbiology; poultry processing; recovery and utilization of protein from food processing waste. Mailing Add: Russell Res Ctr Agr Res Serv US Dept of Agr PO Box 5677 Athens GA 30604

YOUNG, LOUISE GRAY, applied physics, see 12th edition

YOUNG, M WHARTON, b Spartanburg, SC, Oct 24, 04. NEUROANATOMY. Educ: Howard Univ, BS, 26, MD, 30; Univ Mich, PhD, 34. Prof Exp: From asst prof to prof, 36-73, EMER PROF NEUROANAT, COL MED, HOWARD UNIV, 73-; PROF ANAT, COL MED, UNIV MD, BALTIMORE CITY, 73- Mem: AAAS. Res: Hypertensive deafness; phonoreception and photoreception; needle puncture treatment of deafness; radio theory of the ear, an electronic concept of hearing; anatomical similarities in the visual and the auditory systems. Mailing Add: Dept of Anat Univ of Md Col of Med Baltimore MD 21201

YOUNG, MAHLON GILBERT, b Texarkana, Tex, Nov 25, 19; m 40; c 4. CHEMISTRY. Prof Exp: Prod chemist, Fansteel Metall Corp, 40-42; control chemist pharmaceut, Abbott Labs, 45-52, opers supvr, Radio Pharmaceut Div, 52-59; group leader low-level radio anal, Nuclear Sci & Eng Corp, 59-60; appln engr, 60-66, sr appln chemist, Pa, 66-68, INSTRUMENT SPECIALIST, FISHER SCI CO, 68- Mem: Am Chem Soc; fel Am Inst Chemists. Res: Synthesis of tagged organics; szilard-chalmers separations; instrumental analysis; nucleonics. Mailing Add: 3704 Southampton Ct Raleigh NC 27604

YOUNG, MARGARET CLAIRE, b Austin, Tex, Sept 23, 43; m 64; c 2. ANATOMY, PHYSIOLOGY. Educ: Univ Tex, BA, 64; Univ Tex Med Br Galveston, PhD(physiol), 69. Prof Exp: Res technician II physiol, 64-65, res assoc physiol, 69-70, instr anat, 70-71, ASST PROF ANAT, UNIV TEX MED BR, GALVESTON, 71- Concurrent Pos: Jeanne B Kempner fel, Univ Tex Med Sch, San Antonio, 69-70. Mem: AAAS; Am Asn Anat. Res: Autoradiographic evidence of leucocytic participation in nervous system injury; autoradiographic study of pathways of nerve fibers in and out of spinal cord. Mailing Add: Dept of Anat Univ of Tex Med Br Galveston TX 77550

YOUNG, MARTIN DUNAWAY, b Moreland, Ga, July 4, 09; m 38; c 2. PARASITOLOGY, MALARIOLOGY. Educ: Emory Univ, BS, 31, MS, 32; Johns Hopkins Univ, ScD(parasitol), 37; Am Bd Med Microbiol, dipl. Hon Degrees: DSc, Emory Univ, 63 & Mich State Univ, 75. Prof Exp: Jr zoologist, NIH, 37-40, dir malaria res lab, 41-50, in charge imported malaria studies, 43-46, dir, Malaria Sur Liberia, 48, sanitarian, 44-48, sr scientist, 48-50, scientist dir, 50-64, head sect epidemiol, Lab Trop Dis, Nat Inst Allergy & Infectious Dis, 50-61, asst chief lab parasite chemother, 61-62, assoc dir extramural res, 62-64; dir, Gorgas Mem Lab, 64-74, dir res, Gorgas Mem Inst, 72-74; RES PROF PARASITOL, COL VET MED & DEPT IMMUNOL & PROF MED MICROBIOL, COL MED, UNIV FLA, 74- Concurrent Pos: Vis prof, ETenn State Teachers Col, 39; lectr, Meharry Med Sch SC, 60 & Med Sch, Univ Panama, 64-69; vis prof, Ala Med Ctr, 65-; clin prof, Sch Med, La State Univ, 67-; consult, Int Coop Admin, India, 57, WHO, Rumania, 61 & CZ Dept, 64-; mem expert adv panel malaria, WHO, 50-; mem malaria adv panel, Pan Am Health Orgn, 57-68; mem Columbia Bd Health, SC, 60-61; mem malaria & parasitic dis comns, Armed Forces Epidemiol Bd, Dept Defense, 65-73; hon res assoc, Smithsonian Inst, 66-; res assoc, Gorgas Mem Lab, 74- Honors & Awards: Rockefeller Pub Serv Award, 53; Darling Medal & Prize, WHO, 63; Order of Manuel Amador Guerrero, Govt Panama, 74; Cert Merit, Gorgas Mem Inst Trop & Prev Med, Inc, 74; Gorgas Medal, Asn Mil Surgeons of US, 74. Mem: AAAS; Am Soc Parasitol (pres, 65); Am Soc Trop Med & Hyg (pres, 52); Royal Soc Trop Med & Hyg; Australian Soc Parasitol. Res: Malaria parasitology, epidemiology and treatment; parasitic protozoa, helminths; parasitic diseases, especially biology, epidemiology and treatment. Mailing Add: 8421 NW Fourth Pl Gainesville FL 23601

YOUNG, MARVIN KENDALL, JR, b Tulia, Tex, Mar 28, 24; m 48; c 3. BIOCHEMISTRY. Educ: McMurry Col, BS, 46; Univ Tex, MA, 50, PhD(chem), 62; cert, Nat Registry Clin Chem & Am Bd Clin Chem. Prof Exp: Teacher high sch, Tex, 47-48; res scientist, Clayton Found Biochem Inst, Tex, 50-51; biochemist, Surg Res Unit, Brooke Army Med Ctr, Ft Sam Houston, Tex, 51-56, chief physiol sect, 56-57; clin chemist, Woman's Hosp, Detroit, Mich, 62-65, chmn dept biochem, Hutzel Hosp, 65-66; chmn biochem, Clin Path Labs, 66-73; dir, Health & Environ Technol, Inc, 73-74; DIR, BIO-SCI LABS, CHICAGO BR, 74- Concurrent Pos: Affil instr, Sch Med, Wayne State Univ, 63-66; consult, Holy Cross, Seton & St David Hosps, Austin, 66-, spec lectr, Col Pharm, Univ Tex, Austin. Mem: AAAS; Am Chem Soc; fel Am Asn Clin Chemists; fel Am Inst Chemists; NY Acad Sci. Res: Body fluid distribution; intermediary metabolism of carbohydrates; enzymology; methodology in clinical chemistry, including spectrophotometric, radioisotopic, gas chromatographic and atomic absorption spectrometric methods. Mailing Add: 679 Glen Haven Glen Ellyn IL 60137

YOUNG, MAURICE DURWARD, b North Vancouver, BC, Sept 20, 12; m 54; c 2. PEDIATRICS, CARDIOLOGY. Educ: Cambridge Univ, BA, 33, MA & MB, BCh, 38; FRCP(C), 48. Prof Exp: House physician & officer, London Hosp, 37-40; resident med officer, Warleywoods Hosp, 40-44; sr intern, Children's Mem Hosp, Montreal, Que, 46-47; from asst prof to assoc prof, 53-69, PROF PEDIAT, UNIV BC, 69-, ASSOC DEAN FAC MED, 74- Concurrent Pos: Fel pediat, Johns Hopkins Univ, 48-49; asst physician, Cardiac Clin, Johns Hopkins Hosp, 48-49; mem sr staff, Vancouver Gen Hosp; Queen's hon surgeon, 58-60. Mem: Am Heart Asn; Am Acad Pediat; Can Pediat Soc; Can Med Asn; Can Cardiovasc Soc. Res: Pediatric cardiology. Mailing Add: Dept of Pediat Univ of BC Vancouver BC Can

YOUNG, MICHAEL WARREN, b Miami, Fla, Mar 28, 49. GENETICS. Educ: Univ Tex, Austin, BA, 71, PhD(genetics), 75. Prof Exp: NIH RES FEL BIOCHEM, STANFORD MED SCH, 75- Mem: Genetics Soc Am. Res: Organization and function of DNA sequences in eukaryotes. Mailing Add: Dept of Biochem Stanford Med Sch Stanford CA 94305

YOUNG, MORRIS NATHAN, b Lawrence, Mass, July 20, 09; m 48; c 2. OPHTHALMOLOGY, SURGERY. Educ: Mass Inst Technol, BS, 30; Harvard Univ, MA, 31; Columbia Univ, MD, 35; Am Bd Ophthal, dipl. Prof Exp: Intern, Queen's Gen Hosp, NY, 35-37; resident ophthal, Harlem Eye & Ear Hosp, 38-40; asst flight surgeon, Maxwell Field, Ala, 41-42, sr eye, ear, nose & throat officer, Walter Reed Gen Hosp, Washington, DC, 42, sr eye, ear, nose & throat serv, 69th Sta Hosp, 42-44, chief eye, ear, nose & throat sect, 235th Gen Hosp, 44-45, med officer & mem staff, 301st Logistical Support Brigade, 66, dep comdr, 343rd Gen Hosp, 66-67, dep comdr & chief prof serv, 307th Gen Hosp, 67-69, staff med officer, 818th Hosp Ctr, 69; DIR OPHTHAL & ATTEND, BEEKMAN DOWNTOWN HOSP, NEW YORK CITY, 69- Concurrent Pos: Ophthalmologist & auth, 45-; attend & prof, Fr & Polyclin Med Sch & Health Ctr, 63-; med adv, Dir Selective Serv, NY Dist, 65-; consult ophthalmologist, Beth Israel Med Ctr, New York City, 72- Honors & Awards: Grand Hospitaler, Knights of Malta, Order of St John of Jerusalem, 72; Order of Lafayette. Mem: AMA; Pan-Am Med Asn; Contact Lens Asn Ophthal; Am Acad Ophthal & Otolaryngol; Acad Comp Med. Res: Medicine; mnemonics and art of memory; science exhibits; illusion practices. Mailing Add: 170 Broadway New York NY 10038

YOUNG, NANCY LIZOTTE, b Rumford, Maine. BIOCHEMISTRY. Educ: Antioch Col, BS, 59; Purdue Univ, PhD(develop biol), 74. Prof Exp: FEL BIOCHEM, PURDUE UNIV, 74- Res: Regulation of 3-hydroxy-3-methylglutaryl coenzyme A reductase in leukocytes. Mailing Add: Dept of Biochem Purdue Univ West Lafayette IN 47907

YOUNG, NELSON FORSAITH, b Everett, Wash, Oct 17, 14; m 42; c 4. BIOCHEMISTRY. Educ: Univ Wash, BS, 36; NY Univ, PhD(biochem), 45. Prof Exp: Res chemist, Mem Hosp, 40-43; asst, Sloan-Kettering Inst, 45-48; from asst prof to assoc prof clin biochem, 48-56, PROF CLIN PATH, MED COL VA, 56-, LECTR BIOCHEM, 70- Mem: Am Soc Clin Path; Am Chem Soc. Res: Renal function; protein metabolism; radiation effects. Mailing Add: Dept of Clin Path Med Col of Va Box 696 Richmond VA 23298

YOUNG, NORTON BRUCE, b Renton, Wash, Aug 11, 26; m 50; c 3. SPEECH PATHOLOGY, AUDIOLOGY. Educ: Univ Wash, BS, 50, MA, 53; Purdue Univ, PhD(speech path, audiol), 57. Prof Exp: Sr clinician speech path & audiol, Seattle Speech & Hearing Ctr, 54-55; asst prof, Univ, 55-60, instr audiol, Med Sch, 60-64, from instr to assoc prof audiol & pediat, 64-73, assoc prof speech path & audiol, Crippled Children's Div, 70-73, PROF PEDIAT, MED SCH & PROF SPEECH PATH & AUDIOL, CRIPPLED CHILDREN'S DIV, UNIV ORE, 73- Mem: Am Speech & Hearing Asn. Res: Pedo-audiology; psychoacoustics; organic disorders of speech; clinical audiology. Mailing Add: Dept of Speech Univ of Ore Med Sch Portland OR 97201

YOUNG, ORSON WHITNEY, b Raymond, Alta, Oct 9, 04; US citizen; m 32; c 4. ZOOLOGY. Educ: Univ Utah, BA, 32, MA, 33; Univ Mich, PhD(limnol), 42. Prof Exp: Prof zool, 36-74, head dept, 36-67, chmn life sci div, 67-68, EMER PROF ZOOL, WEBER STATE COL, 74- Mem: AAAS; Am Nature Study Soc. Res: Limnology of Periphyton in streams and in Douglas Lake, Michigan; invertebrate zoology; mosquito control. Mailing Add: Dept of Zool Weber State Col Ogden UT 84403

YOUNG, OTIS BIGELOW, b Dodge Center, Minn, Nov 17, 99; m 28; c 2. PHYSICS. Educ: Wabash Col, AB, 21; Univ Ill, AM, 23, PhD(physics), 28. Prof Exp: Instr, High Schs, Ind & Ill, 21-25; asst physics, Univ Ill, 25-28; from assoc prof to prof physics &

astron, 28-75, chmn dept, 38-53, dir atomic & capacitor res, 53-75, EMER PROF PHYSICS & ASTRON, SOUTHERN ILL UNIV, CARBONDALE, 75- Concurrent Pos: Head dept physics, McKendree Col, 28-29; coordr civilian pilot training & war training serv, Civil Aeronaut Admin, 39-43; dir, Vet Info Bur, 44-45; physicist, Sangamo Elec Co, 49-; dir res, Univ Chicago, 53-; comnr & secy, Southern Ill Airport Auth. Mem: Am Phys Soc; Am Asn Physics Teachers. Res: Dielectric constants of gases and electrolytic solutions; electrical discharges in gases; cooperative dielectric, capacitor and atomic research; cosmic rays; low energy elementary and fundamental particles; magnetic monopole. Mailing Add: Heritage Hills RD 1 Carbondale IL 62901

YOUNG, PAUL ANDREW, b St Louis, Mo, Oct 3, 26; m 49; c 10. ANATOMY. Educ: St Louis Univ, BS, 47, MS, 53; Univ Buffalo, PhD(anat), 57. Prof Exp: From asst to instr anat, Univ Buffalo, 53-57; from asst prof to assoc prof, 57-72, actg chmn dept, 69-73, PROF ANAT, SCH MED, ST LOUIS UNIV, 72-, CHMN DEPT, 73-, ASSOC PROF NEUROANAT IN NEUROL & PSYCHIAT, 67- Mem: AAAS; Am Asn Anat; Soc Neurosci; Electron Micros Soc Am. Res: Neuroanatomy, especially mammalian forebrain centers, basal ganglia, thalamus, cerebral cortex; degenerative and axoplasmic flow techniques; electron microscopy of nervous tissue; pathology of nutritional encephalopathy. Mailing Add: Dept of Anat St Louis Univ St Louis MO 63104

YOUNG, PAUL MCCLURE, b Seaman, Ohio, Feb 13, 16; m 42; c 2. MATHEMATICS. Educ: Miami Univ, AB, 37; Ohio State Univ, MA, 39, PhD(math), 41. Prof Exp: From instr to asst prof math, Miami Univ, 41-47; from assoc prof to prof, Kans State Univ, 47-62, assoc dean sch arts & sci, 56-62; vpres acad affairs, Univ Ark, 62-66; EXEC DIR, MID-AM STATE UNIVS ASN, 66-; PROF MATH & VPRES UNIV DEVELOP, KANS STATE UNIV, 70- Mem: AAAS; Am Math Soc; Math Asn Am. Res: Analysis; approximation of functions by integral means; characterization of integral means. Mailing Add: Kans State Univ Manhattan KS 66506

YOUNG, PAUL RUEL, b St Marys, Ohio, Mar 16, 36; c 2. COMPUTER SCIENCES, MATHEMATICAL LOGIC. Educ: Antioch Col, BS, 59; Mass Inst Technol, PhD(math), 63. Prof Exp: Asst prof math, Reed Col, 63-66; from asst prof to assoc prof, 66-72, PROF MATH & COMPUT SCI, PURDUE UNIV, LAFAYETTE, 72- Concurrent Pos: NSF fel, Stanford Univ, 65-66; vis prof, Univ Calif, Berkeley, 72-73. Mem: Asn Comput Mach; Am Math Soc; Asn Symbolic Logic. Res: Theory of computational complexity and theory of algorithms. Mailing Add: Div of Math Sci Purdue Univ Lafayette IN 47907

YOUNG, PETER CHUN MAN, b Hong Kong, Dec 19, 36; Can citizen; m 67; c 2. BIOCHEMISTRY. Educ: McGill Univ, BS, 61, MS, 63, PhD(biochem), 67. Prof Exp: Res assoc endocrinol, St Michael's Hosp, 67-71; ASST PROF ENDOCRINOL, MED CTR, IND UNIV, INDIANAPOLIS, 71- Mem: AAAS; Can Biochem Soc; NY Acad Sci. Res: Biochemistry of steroid hormones; reproductive endocrinology. Mailing Add: Dept of Obstet & Gynec Ind Univ Med Ctr Indianapolis IN 46202

YOUNG, PHILIP ROSS, b Emory, Va, Oct 24, 40. ANALYTICAL CHEMISTRY. Educ: Emory & Henry Col, BS, 62; Va Polytech Inst & State Univ, MS, 71, PhD(chem), 76. Prof Exp: CHEMIST POLYMER RES, NASA LANGLEY RES CTR, 62- Mem: Am Chem Soc. Res: Using liquid and gel permeation chromatography techniques to characterize and evaluate new monomers and high performance polymers for aerospace applications. Mailing Add: NASA Langley Res Ctr Mail Stop 226 Hampton VA 23665

YOUNG, PHILLIP D, b Ottawa, Ill, Oct 18, 36; m; c 5. ETHNOLOGY. Educ: Univ Ill, Urbana, BA, 61, PhD(anthrop), 68. Prof Exp: Asst prof, 66-70, ASSOC PROF ANTHROP, UNIV ORE, 70- Mem: Am Anthrop Asn; Am Soc Ethnohist. Res: Social anthropology; linguistics; Latin America; ethnohistory of Central America; revitalization movements; colonization; ritual and symbolism among native peoples of Central and South America. Mailing Add: Dept of Anthrop Univ of Ore Eugene OR 97403

YOUNG, PHILLIP GAFFNEY, b Beeville, Tex, July 21, 37; m 60; c 3. NUCLEAR PHYSICS. Educ: Univ Tex, Austin, BS, 61, MA, 62; Australian Nat Univ, PhD(nuclear physics), 65. Prof Exp: Res fel nuclear physics, Australian Nat Univ, 65-66; res fel, 66-68, MEM STAFF NUCLEAR CROSS SECT EVAL, LOS ALAMOS SCI LAB, 68- Mem: Am Phys Soc; Am Nuclear Soc. Res: Low energy nuclear physics; neutron-particle and charged-particle cross sections and polarization. Mailing Add: Group T-2 Los Alamos Sci Lab Los Alamos NM 87544

YOUNG, POH-SHIEN, b Chekiang, China, Jan 18, 26; m 59; c 3. PARTICLE PHYSICS. Educ: Nat Chi-Nan Univ, China, BS, 49; Okla State Univ, MS, 57; Univ Calif, Berkeley, PhD(physics), 65. Prof Exp: Asst physics, Okla State Univ, 55-57; nuclear physicist, Admiral Corp, Ill, 57 & Nuclear Chicago Corp, 57-59; instr physics, Univ Ill, Chicago, 59-60 & Oakland City Col, Calif, 60-62; asst, Univ Calif, Berkeley, 62-65, physicist, Lawrence Radiation Lab, 65-66; asst prof, 66-70, ASSOC PROF PHYSICS, MISS STATE UNIV, 70- Mem: Am Phys Soc. Res: Cosmic ray and reactor physics. Mailing Add: Dept of Physics Box 5167 Miss State Univ Mississippi State MS 39762

YOUNG, RALPH ALDEN, b Arickaree, Colo, July 14, 20; m 42; c 2. SOIL FERTILITY. Educ: Colo State Univ, BS, 42; Kans State Univ, MS, 47; Cornell Univ, PhD(agron), 53. Prof Exp: Instr soils, Kans State Univ, 47-48; asst prof soils, Univ & asst soil scientist, Exp Sta, 48-50 & 53-55, from assoc prof & assoc soil scientist to prof soils & soil scientist, 55-62; vis prof, Univ Calif, Davis, 62-63; chmn dept agr biochem & soil sci, 63-65, chmn div plant, soil & water sci, 65-75, PROF SOIL SCI, UNIV & SOIL SCIENTIST, NEV AGR EXP STA, UNIV NEV, RENO, 63-, ASSOC DIR, NEV AGR EXP STA, 75- Mem: Fel AAAS; Soil Sci Soc Am; Am Soc Agron; Int Soc Soil Sci. Res: Fertilizer-water interactions; soil as a waste treatment system. Mailing Add: Nev Agr Exp Sta Univ of Nev Reno NV 89507

YOUNG, RALPH HOWARD, b Berkeley, Calif, Mar 22, 42; m 68; c 1. CHEMICAL PHYSICS, SOLID STATE PHYSICS. Educ: Calif Inst Technol, BS, 64; Stanford Univ, PhD(chem physics), 68. Prof Exp: Lectr chem, Stanford Univ, 68; asst prof, Jackson State Col, 68-70; lectr, Univ Calif, Riverside, 70; SR RES CHEMIST, CHEM PHYSICS LAB, EASTMAN KODAK CO, 71- Res: Photoconduction in organic solids; quantum chemistry; quantum axiomatics. Mailing Add: Chem Physics Lab Eastman Kodak Co Res Labs Rochester NY 14650

YOUNG, RALPH WALDO, b Ft Wayne, Ind, Nov 5, 08; m 40. MATHEMATICS. Educ: Ball State Teachers Col, AB, 31; Ind Univ, MS, 33; Lehigh Univ, AM, 48; Univ Fla, DEd, 51. Prof Exp: Teacher high sch, Ind, 31-49; instr math, Univ Fla, 49-51; prof & head dept, Henderson State Teachers Col, 51-52; from chmn prog to prof math, chmn depts math & physics & gen subjects & dean arts & sci, Ind Inst Technol, 52-60, dean fac, 60-67; chmn dept math, 67-72, dean faculties, 72-74, CHMN DEPT SCI, TRI STATE COL, 72-, EMER DEAN FAC, 74- Concurrent Pos: Instr, Lehigh Univ, 48-49. Mem: Math Asn Am. Res: Teacher education. Mailing Add: 525 E South St Angola IN 46703

YOUNG, RAYMOND A, b Buffalo, NY, Mar 14, 45; m 62; c 2. WOOD CHEMISTRY. Educ: State Univ NY, BS, 66, MS, 68; Univ Wash, PhD(wood chem), 73. Prof Exp: Process supvr paper prod, Kimberly-Clark Corp, 68-69; Textile Res Inst fel, Princeton Univ, 73-74, staff scientist textiles, 74-75; ASST PROF FORESTRY, UNIV WIS-MADISON, 75- Mem: Tech Asn Pulp & Paper Indust; Fiber Soc. Res: Chemical modification of cellulose and high yield pulp fibers; bonding of cellulose fibers in blended flexible fiber composites such as nonwovens and synthetic papers. Mailing Add: Dept of Forestry Univ of Wis Madison WI 53706

YOUNG, RAYMOND HINCHCLIFFE, JR, b Pennsauken, NJ, Nov 22, 28; m 53; c 4. CLAY MINERALOGY. Educ: Pa Mil Col, BS, 53; Univ Maine, MS, 55, PhD(org chem), 61. Prof Exp: Asst chemG U chem, Univ Maine, 53-54, res asst org chem, 54-55, instr chem, 55-60; res chemist, Bircham Bend Plant, Monsanto Co, 60-70; RES CHEMIST, FREEPORT KAOLIN CO, 70- Honors & Awards: A K Doolittle Award, Am Chem Soc, 72. Mem: AAAS; Am Chem Soc. Res: Surface modified kaolins; flame retardants; light scattering of pigments; computer science. Mailing Add: Freeport Kaolin Co PO Box 337 Gordon GA 31031

YOUNG, REUBEN B, b Wilmington, NC, Apr 2, 30; m 53; c 3. MEDICINE, PEDIATRICS. Educ: Med Col Va, BS, 53, MD, 57. Prof Exp: Instr pediat, Univ Pa, 61-63; from asst prof to assoc prof, 63-71, PROF PEDIAT, MED COL VA, 71-, VCHMN DEPT & PROF GENETICS, 75- Concurrent Pos: NIH grant, 65-71. Mem: Am Acad Pediat; Endocrine Soc; Am Diabetes Asn. Res: Pediatric endocrinology; catecholamine metabolism in children; hypoglycemia in children. Mailing Add: Dept of Pediat Med Col of Va Box 65 Richmond VA 23219

YOUNG, RICHARD A, b Providence, RI, Aug 5, 40; m 64. GEOLOGY. Educ: Cornell Univ, BA, 62; Wash Univ, PhD(geol), 66. Prof Exp: Asst prof earth sci, 66-72, ASSOC PROF GEOL SCI, STATE UNIV NY COL GENESEO, 72- Concurrent Pos: Prin investr geol mapping proj using Apollo Mission photog, NASA Contract, 72-75; vis faculty mem, Univ Canterbury, 72. Mem: AAAS; Geol Soc Am. Res: Cenozoic geology, including sedimentation, geomorphology, glacial geology and vulcanism; lunar geology. Mailing Add: Dept of Geol Sci State Univ NY Col Geneseo NY 14454

YOUNG, RICHARD ACCIPITER, b Pittsburgh, Pa, Mar 24, 42; m 68; c 2. SOLID STATE PHYSICS. Educ: Lehigh Univ, BS, 64; Univ Chicago, PhD(physics), 68. Prof Exp: Res assoc, 68-69, asst prof, 69-74, Alfred P Sloan fel, 71-73, ASSOC PROF PHYSICS, UNIV ARIZ, 74- Concurrent Pos: Vis asst prof, Univ Calif, Berkeley, 72- Mem: Am Phys Soc. Res: Electronic transport properties of metals. Mailing Add: Dept of Physics Univ of Ariz Tucson AZ 85721

YOUNG, RICHARD EDWARD, b Los Angeles, Calif, Aug 20, 38; m 63; c 2. BIOLOGICAL OCEANOGRAPHY. Educ: Pomona Col, BA, 60; Univ Southern Calif, MS, 64; Univ Miami, PhD(oceanog), 68. Prof Exp: Res asst oceanog, Inst Marine Sci, Univ Miami, 65-68, res scientist, 68; asst prof zool, Ohio Wesleyan Univ, 68-69; asst prof oceanog, 69-74, ASSOC PROF OCEANOG, UNIV HAWAII, MANOA, 74- Mem: AAAS; Marine Biol Asn UK. Res: Cephalopod, deep-sea and invertebrate biology. Mailing Add: Dept of Oceanog Univ of Hawaii at Manoa Honolulu HI 98622

YOUNG, RICHARD EVANS, b Trinidad, Colo, Aug 21, 43; m 75; c 1. DYNAMIC METEOROLOGY, GEOPHYSICS. Educ: Univ Calif, Berkeley, BS, 66; Univ Calif, Los Angeles, MS, 69, PhD(geophys & space physics), 72. Prof Exp: Aerospace engr, NASA, Ames Res Ctr, Moffett Field, Calif, 66-67; fel, Nat Ctr Atmospheric Res, 72-73; RES GEOPHYSICIST, UNIV CALIF, LOS ANGELES, 73- Concurrent Pos: Consult, Terestrial Planets Sci Adv Group, Jet Propulsion Lab, 76-77. Mem: Am Astron Soc; AAAS. Res: Dynamics of planetary atmospheres, especially at the moment that of Venus; generation of magnetic fields by flow motions in stars and planetary bodies; thermal convection in planetary interiors. Mailing Add: NASA Ames Res Ctr Space Sci Div Moffett Field CA 94035

YOUNG, RICHARD L, b Rushville, Ill, Nov 22, 32; m 53; c 3. ORGANIC CHEMISTRY. Educ: Univ Ill, BSc, 54; Brown Univ, PhD(org chem), 59. Prof Exp: Res chemist, Res Inst Med & Chem, 58-61; res assoc biochem, Col Agr, Univ Wis, 61-63; asst prof agr biochem, Univ Hawaii, 64-66; GROUP LEADER ORG CHEM, NEW ENG NUCLEAR CORP, 66- Mem: Am Chem Soc. Res: Peptide synthesis; tritium and carbon-14 radiochemicals. Mailing Add: New Eng Nuclear Corp 549 Albany St Boston MA 02118

YOUNG, RICHARD LAWRENCE, b Proctor, Vt, Mar 12, 25; m 57; c 3. ANALYTICAL BIOCHEMISTRY, NEUROCHEMISTRY. Educ: Cath Univ Am, AB, 49; St Louis Univ, PhD(biochem), 56. Prof Exp: Trainee res & develop, SK&F Lab, 49-51, sr res scientist biochem, 56-59; Nat Inst Neurol Dis & Blindness spec trainee neurochem, Inst Living, 59-60, spec fel, Sch Med, Washington Univ, 60-62; res assoc, Inst Living, 62-66; sr biochemist pharmacol & biochem, 66-73, asst chief clin biochem, 73-74, RES FEL BIOCHEM, HOFFMANN-LA ROCHE INC, 74- Mem: AAAS; Am Chem Soc; Am Soc Neurochem. Res: Development and use of micromethods for investigating the disposition, metabolism and target effects of therapeutic agents in vivo; quantitative histochemistry, radioisotopes, radioimmunoassay; bioavailability and pharmacokinetics. Mailing Add: Dept of Biochem & Drug Metab Hoffmann-La Roche Inc Nutley NJ 07110

YOUNG, RICHARD S, b Southampton, NY, Mar 6, 27; m 55; c 3. EXOBIOLOGY, DEVELOPMENTAL BIOLOGY. Educ: Gettysburg Col, AB, 48; Fla State Univ, PhD(zool), 55. Hon Degrees: ScD, Gettysburg Col, 66. Prof Exp: Res scientist biochem, Food & Drug Admin, 56-58; chief special fields br, Army Ballistic Missile Agency, Ala, 58-60; chief space biol, 60-61, chief exobiol div, Ames Res Ctr, 61-67, CHIEF BIOSCI DIV, NASA, WASHINGTON, DC, 67- Mem: Int Soc Study Origin of Life (secy gen, 72); AAAS. Res: Science administration of national program in the areas of the origin and evolution of life; chemical evolution; organic geochemistry; search for extraterrestrial life; planetary exploration; comparative planetology; space and gravitational biology. Mailing Add: 7927 Falstaff Rd McLean VA 22101

YOUNG, RICHARD WAIN, b Albany, NY, Dec 15, 29; m 55; c 4. ANATOMY. Educ: Antioch Col, BA, 56; Columbia Univ, PhD(anat), 59. Prof Exp: From asst prof to assoc prof, 60-68, PROF ANAT, SCH MED, UNIV CALIF, LOS ANGELES, 68- Concurrent Pos: NSF fel anat, Univ Bari & Caroline Inst, Sweden, 59-60; Markle scholar med sci, 62-67; guest investr anat, Dept Biol, Saclay Nuclear Res Ctr, France, 66-67; mem, Jules Stein Eye Inst, Univ Calif, Los Angeles. Honors & Awards: Fight for Sight Res Citation, 69; Friedenwald Award, Asn Res in Vision & Opthalmol, 76. Mem: Am Asn Anat; Am Soc Cell Biol; Asn Res Vision & Opthal. Res: Cell biology; radioisotope studies of ocular tissues. Mailing Add: Dept of Anat Univ of Calif Sch of Med Los Angeles CA 90024

YOUNG, ROBERT, JR, b Sept 14, 30; US citizen; m 55; c 3. VETERINARY MEDICINE. Educ: Univ Calif, BS, 53, MS, 56, DVM, 61. Prof Exp: Vet, 61-67, dept head vet med, 67-69, DEPT HEAD DRUG DEVELOP, SHELL DEVELOP CO, 69- Mem: Am Vet Med Asn; Am Soc Animal Sci; Indust Vet Med Asn. Res: Drug development and secondary screening; growth regulators; anthelmintics and ectoparasiticides. Mailing Add: Shell Develop Co PO Box 4248 Modesto CA 95352

YOUNG, ROBERT A, b New York, NY, June 8, 29; m 57, 69; c 1. PHYSICS, PHYSICAL CHEMISTRY. Educ: Univ Wash, BS, 51, PhD(physics), 59. Prof Exp: Res asst, Univ Wash, 53-59; engr, Boeing Airplane Co, 59-60; physicist, Stanford Res Inst, 60-62, sr physicist, 62-67, chmn atmospheric chem physics dept, 67-68; PROF PHYSICS, YORK UNIV, 68- Concurrent Pos: Consult, Dept Physics, Univ Wash, 59-60; vis fel atmospheric physics, Joint Inst Lab Astrophys, Univ Colo, 66-67; chmn & pres, Intra-Space Int, 72-75; vis fel, Lab Space Physics, Univ Colo, 74-75; dir res, Xonics, Inc, 75- Mem: AAAS; Am Phys Soc; Am Geophys Union. Res: Energy transfer; atomic and molecular processes and spectra; rocket experimentation; baloon, aircraft stratospheric measurements. Mailing Add: Dept of Physics York Univ Downsview ON Can

YOUNG, ROBERT ALAN, b St Cloud, Minn, Jan 24, 21; m 48; c 3. SOLID STATE PHYSICS, CRYSTALLOGRAPHY. Educ: Polytech Inst Brooklyn, PhD(physics), 59. Prof Exp: Res asst prof & res physicist, Ga Inst Technol, 51-53; instr physics, Polytech Inst Brooklyn, 53-57; res assoc prof physics & head diffraction lab, 57-63, PROF PHYSICS & HEAD CRYSTAL PHYSICS BR, GA INST TECHNOL, 64- Concurrent Pos: Co-ed, J Appl Crystallog, 67-69, ed, 70-; mem, US Nat Comt Crystallog, 69-74. Mem: AAAS; fel Brit Inst Physics; French Soc Mineral & Crystallog; fel Am Phys Soc; Am Crystallog Asn (treas, 68-71, vpres, 72, pres, 73). Res: Crystal physics; x-ray, neutron and electron diffraction theory and applications. Mailing Add: Sch of Physics & Eng Exp Sta Ga Inst of Technol Atlanta GA 30332

YOUNG, ROBERT BRUCE, b Mt Sterling, Ky, Jan 28, 16; m 46; c 3. CHEMISTRY. Educ: Univ Ky, AB, 38, MS, 40; Mich State Col, PhD(phys chem), 44. Prof Exp: Asst, Univ Ky, 38-40; asst, Mich State Col, 40-42; res chemist, Plastics Div, Gen Elec Co, 42-51, appln engr, Chem Mat Dept, 51-53, supvr methylon develop group, 53-54, supvr adv & develop group, 54-57, wire enamels & varnishes, 57-62; tech dir magnet wire div, 62-71, DIR POLYMER TECHNOL, CORP RES, ESSEX INT, INC, 72- Mem: Fel AAAS; Am Chem Soc. Res: Chromatography; electrode potential measurements; solution phenomena; reaction rate studies; synthetic coating resins; wire enamels and varnishes. Mailing Add: Essex Int/Mellon Inst 4400 Fifth Ave Pittsburgh PA 15213

YOUNG, ROBERT CARL, b San Antonio, Tex, Oct 19, 31; c 4. THEORETICAL PHYSICS, COMPUTER SCIENCES. Educ: Baylor Univ, BS & MS, 53; Rice Univ, MA, 57, PhD(physics), 60. Prof Exp: Sr theoret physicist, Nat Reactor Testing Sta, Atomic Energy Div, Phillips Petrol Co, 60-64; prof physics, Houston Baptist Col, 64-69; theoret physicist, Columbia Sci Res Inst, Houston, 69-71; SCIENTIST, AEROJET NUCLEAR CO, 71- Mem: Am Phys Soc. Res: Theoretical problems in low energy nuclear physics and applications of mathematical physics. Mailing Add: Aerojet Nuclear Co 550 Second St Idaho Falls ID 83401

YOUNG, ROBERT ELLSWORTH, b Stroud, Okla, Jan 11, 08; m 30; c 3. HORTICULTURE. Educ: Okla Agr & Mech Col, BS, 30; Ohio State Univ, MS, 31. Prof Exp: Asst res prof, 31-37 & 47-49, assoc res prof, 49-54, RES PROF VEG CROPS, EXP STA, UNIV MASS, 54- Res: Breeding vegetable crops; fertilizer research with vegetable crops. Mailing Add: Suburban Exp Sta Univ of Mass Waltham MA 02154

YOUNG, ROBERT GLEN, economic geology, see 12th edition

YOUNG, ROBERT HAYWARD, b Andover, NB, Mar 18, 40; m 65; c 2. PHOTOCHEMISTRY, ORGANIC POLYMER CHEMISTRY. Educ: Mt Allison Univ, BSc, 62, MSc, 63; Mich State Univ, PhD(org chem), 67. Prof Exp: Asst prof org chem, Georgetown Univ, 67-73; PROJ SCIENTIST, UNION CARBIDE CORP, 73- Mem: Am Chem Soc; Chem Inst Can. Res: Organic chemistry and photochemistry of singlet oxygen; polymer chemistry, adhesive bonding fundamentals. Mailing Add: Union Carbide Corp One River Rd Bound Brook NJ 08805

YOUNG, ROBERT JOHN, b Calgary, Alta, Feb 10, 23; m 50; c 2. ANIMAL NUTRITION. Educ: Univ BC, BSA, 50; Cornell Univ, PhD(animal nutrit), 53. Prof Exp: Asst poultry nutrit, Cornell Univ, 50-53; res assoc, Banting & Best Dept Med Res, Univ Toronto, 53-56; res biochemist, Int Minerals & Chem Corp, Ill, 56-58 & Procter & Gamble Co, Ohio, 58-60; assoc prof animal nutrit & poultry husb, 60-65, PROF ANIMAL NUTRIT & HEAD DEPT POULTRY SCI, NY STATE COL AGR & LIFE SCI, CORNELL UNIV, 65- Mem: PoulPoultry Sci Asn; Am Inst Nutrit. Res: Mineral metabolism; energy value of fats and fatty acids; nutrition and metabolism of protein and amino acids. Mailing Add: Dept of Poultry Sci NY State Col Agr & Life Sci Cornell Univ Ithaca NY 14850

YOUNG, ROBERT M, b Brooklyn, NY, Sept 10, 39. BIOCHEMISTRY, BIOLOGY. Educ: Brooklyn Col, BS, 60, MA, 65; Univ Pittsburgh, PhD(biochem), 71. Prof Exp: Sr res asst clin enzymol, Beth Israel Med Ctr, 60-65; asst prof, 70-75, ASSOC PROF BIOL & CHMN DEPT, RI COL, 75- Concurrent Pos: Lectr, Brooklyn Col, 62-65; res asst, Creedmoor State Hosp, 63-64. Mem: AAAS; Am Inst Biol Sci; Am Soc Microbiol; Am Soc Zoologists. Res: Ribosomal structure and function; protein and RNA synthesis; biosynthesis of folates and pteridines; use of enzymes for the diagnosis of cancer. Mailing Add: Dept of Biol RI Col Providence RI 02908

YOUNG, ROBERT SPENCER, b Charlottesville, Va, Apr 14, 22; m 45; c 2. GEOLOGY. Educ: Univ Va, BA, 50, MA, 51; Cornell Univ, PhD(geol), 54. Prof Exp: Field geologist, Va State Geol Surv, 52-53, stratigrapher-struct geologist, 53-56; regional geologist & br mgr, Roland F Beers, Inc, 56-59; from asst prof to assoc prof geol, Univ Va, 59-68, asst dean col, 60-64; PRES, N AM EXPLOR INC, 64- Concurrent Pos: Lectr, Harpur Col, State Univ NY, 54; partner, Beers & Young, 60-63; pres, Mem: Geol Soc Am; Am Asn Petrol Geologists; Soc Explor Geophysicists; Am Inst Mining, Metall & Petrol Engrs; Soc Econ Geologists. Res: Evaluation of energy resources of Gulf Coastal states; economic and structural studies of the Paleozoics in the Appalachians of the eastern United States; geological, geophysical and geochemical explorations for sulfide deposits in the United States. Mailing Add: NAm Explor Inc PO Box 7584 Charlottesville VA 22906

YOUNG, ROBERT WILLIAM, b Mansfield, Ohio, May 11, 08; m 47; c 3. PHYSICS, ACOUSTICS. Educ: Ohio Univ, BS, 30; Univ Wash, PhD(physics), 34. Prof Exp: Physicist, C G Conn, Ltd, 34-42; physicist, Div War Res, Univ Calif, 42-46, res assoc, Marine Phys Lab, 46; physicist, US Navy Electronics Lab, 46-67, tech consult, Naval Undersea Ctr, 67-70, SR SCIENTIST & CONSULT ACOUST, NAVAL UNDERSEA CTR, 70- Mem: AAAS; fel Acoust Soc Am (vpres, 53-54, pres, 60-61). Res: Acoustics of wind musical instruments; piano strings and tuning; underwater ambient and ship noise and sound propagation; acoustical standards; techniques for analyzing sound; architectural acoustics; community noise measurement and description. Mailing Add: 1696 Los Altos Rd San Diego CA 92109

YOUNG, ROGER GRIERSON, b Moose Jaw, Sask, Dec 18, 20; nat US; m 54; c 3. COMPARATIVE BIOCHEMISTRY. Educ: Univ Alta, BSc, 43, MSc, 48; Univ Ore, PhD(chem), 52. Prof Exp: Rockefeller asst, Cornell Univ, 51-52, Geer res fel, 52-53; assoc biochemist, Ethicon, Inc, 53-55; asst prof, 55-60, ASSOC PROF ENTOM, CORNELL UNIV, 60- Concurrent Pos: Vis prof, Col Agr, Univ Philippines, 68-69. Mem: AAAS; Entom Soc Am; Am Chem Soc. Mailing Add: Dept of Entom Cornell Univ Ithaca NY 14850

YOUNG, ROGER HOLLEY, plant physiology, plant biochemistry, see 12th edition

YOUNG, RONALD JEROME, b Hong Kong, Aug 10, 32; m 62; c 2. BIOCHEMISTRY, MOLECULAR BIOLOGY. Educ: Univ Sydney, BSc, 54; Univ NSW, PhD, 58. Prof Exp: Fel, Univ Wis, 58-62; lectr biochem, Monash Univ, 63-67; fel, Univ Calif, 67-70; assoc prof biol sci, Drexel Univ, 70-73; ASSOC PROF REPRODUCTIVE BIOL, MED CTR, CORNELL UNIV, 73- Mem: The Chem Soc. Res: Biochemistry of fertilization; reproductive biology. Mailing Add: Cornell Univ Med Ctr 1300 York Ave New York NY 10021

YOUNG, ROY ALTON, b McAlister, NMex, Mar 1, 21; m 50; c 2. PLANT PATHOLOGY. Educ: NMex State Univ, BS, 41; Iowa State Univ, MS, 42, PhD(plant path), 48. Prof Exp: From asst prof to assoc prof bot & plant path, 48-53, head dept, 58-66, dean res, 66-70, actg pres, Univ, 69-70, PROF BOT & PLANT PATH, ORE STATE UNIV, 53-, VPRES RES & GRAD STUDIES, 70- Concurrent Pos: Mem, Comn Undergrad Educ in Biol Sci, 63-67; mem exec coun, Nat Gov Coun Sci & Technol, 70-; consult, State Exp Sta, USDA, mem exec comt, Study Probs Pest Control: A Technol Assessment, Nat Acad Sci-Nat Acad Eng, 72-; mem US nat comt man & biosphere, UNESCO, 73-; mem comt to rev int biol prog, Nat Acad Sci-Nat Res Coun, 74-; mem bd dirs, Pac Power & Light Co, 74- & Boyce Thompson Inst Plant Res, 75-; chmn, Comt Environ & Energy, 75. Mem: Fel AAAS; fel Am Phytopath Soc. Res: Diseases of ornamentals and potatoes; soil-borne diseases; chemical control of plant diseases; administration. Mailing Add: Res & Grad Studies Ore State Univ Corvallis OR 97331

YOUNG, ROY E, b Lincoln, Nebr, Apr 9, 18; m 50; c 2. PLANT PHYSIOLOGY. Educ: Univ Calif, Los Angeles, PhD(biochem), 60. Prof Exp: Assoc biochemist, Univ Calif, Los Angeles, 60-66; asst prof biochem, 66-71, ASSOC PROF PLANT PHYSIOL, UNIV CALIF, RIVERSIDE, 71- Concurrent Pos: Consult, Scripps Inst Oceanog, Univ Calif, San Diego, 65-66; sr fel, Dept Sci & Indust Res, Auckland, NZ, 66-67. Mem: AAAS; Am Chem Soc; Am Soc Plant Physiologists; Am Inst Biol Sci; Am Soc Hort Sci. Res: Biochemistry of fruit ripening; dark carbon dioxide fixation; phosphate metabolism in ripening fruit; changes in mitochondria associated with ripening; extraction and characterization of isoenzymes of ripening fruit; ethylene production. Mailing Add: Dept of Plant Sci Univ of Calif Riverside CA 92502

YOUNG, RUSSELL DAWSON, b Huntington, NY, Aug 17, 23; m 54; c 4. OPTICS, SURFACE SCIENCE. Educ: Rensselaer Polytech Inst, BS, 53; Pa State Univ, MS, 56, PhD(physics), 59. Prof Exp: Res assoc, Pa State Univ, 59-61; mem staff, 61-73, CHIEF OPTICS & MICROMETROLOGY SECT, NAT BUR STANDARDS, 73- Honors & Awards: Edward Uhler Condon Award, Nat Bur Standards, 74. Mem: Am Phys Soc; Optical Soc Am. Res: Physical characterization of surfaces; micrometrology. Mailing Add: Optics & Micrometrology Sect Nat Bur of Standards Washington DC 20234

YOUNG, RUTH STEUART, b Hereford, Tex, Jan 5, 26; m 56; c 3. BIOLOGY, BIOCHEMISTRY. Educ: Col Ozarks, BS, 48; Univ Ark, MA, 52. Prof Exp: Lab asst biol, Col Ozarks, 46-48; high sch teacher, Ark, 48-51; asst zool, Univ Ark, 51-52; demonstr biol, Bryn Mawr Col, 52-55; asst biochem, Med Ctr, Univ Ark, 56-62; RESEARCHER, 62- Mem: Soc Exp Biol & Med. Res: Metabolic effects of vitamin deficiencies; taxonomy and life history studies of Myriapoda. Mailing Add: 23 Nob Hill Cove Little Rock AR 72205

YOUNG, SANFORD TYLER, b Chicago, Ill, Apr 14, 36; m 61; c 2. ORGANIC CHEMISTRY, ANALYTICAL CHEMISTRY. Educ: Univ Ill, BS, 58; Univ Rochester, PhD(org chem), 63. Prof Exp: Res chemist, 62-68, sr res chemist, 68-74, RES ASSOC, AGR CHEM DIV, FMC CORP, 74- Mem: AAAS; Am Chem Soc. Res: Synthesis of biologically active materials and organic analytical chemistry. Mailing Add: Agr Chem Div FMC Corp Middleport NY 14105

YOUNG, SETH YARBROUGH, III, b Victoria, Miss, June 8, 41; m 61; c 3. ENTOMOLOGY, VIROLOGY. Educ: Miss State Univ, BS, 63; Auburn Univ, PhD(entom), 67. Prof Exp: Res entomologist, Stored Prod Res Lab, Mkt Qual Res Div, USDA, Ga, 66-67; asst prof, 67-71, ASSOC PROF ENTOM, UNIV ARK, FAYETTEVILLE, 71- Mem: Entom Soc Am; Soc Invert Path. Res: Insect virology. Mailing Add: 2740 Loxley Fayetteville AR 72701

YOUNG, SHARON CLAIRENE, b Elk City, Okla, Aug 3, 42. ZOOLOGY. Educ: Bethany Nazarene Col, BS, 64; Okla State Univ, MS, 65, PhD(entom), 69. Prof Exp: From asst prof to assoc prof, 68-73, PROF BIOL, BETHANY NAZARENE COL, 73- Mem: AAAS; Sigma Xi. Res: Plant resistance to insects; thrips resistance in peanuts. Mailing Add: Dept of Biol Bethany Nazarene Col Bethany OK 73008

YOUNG, SIMON NESBITT, b Godalming, Eng, Feb 2, 45; m 69; c 1. NEUROSCIENCES. Educ: Oxford Univ, BA, 67; London Univ, MS, 68, PhD(biochem), 71. Prof Exp: Res asst biochem, Inst Neurol, London Univ, 68-71; fel, 71-75, ASST PROF NEUROCHEM, DEPT PSYCHIAT, McGILL UNIV, 75- Concurrent Pos: Fel, Que Med Res Coun, 72-73; J B Collip fel, McGill Univ, 73-75. Mem: Can Biochem Soc. Res: Investigation of tryptophan metabolism and brain 5 hydroxytryptamine synthesis in experimental animals and in man. Mailing Add: Dept of Psychiat McGill Univ 1033 Pine Ave W Montreal PQ Can

YOUNG, STEPHEN DEAN, b Pasadena, Calif, Apr 11, 42; m 63. INVERTEBRATE PHYSIOLOGY, MARINE BIOLOGY. Educ: Reed Col, BA, 64; Univ Calif, Los Angeles, MA, 66, PhD(zool), 69. Prof Exp: Res assoc, Dent Res Ctr, Univ NC, 69-71; asst prof zool, Ind Univ Northwest, 71-73; asst res biologist, Univ Calif, San Diego, 73-75; ASST PROF BIOL SCI, ST JOHN'S UNIV, NY, 75- Concurrent Pos: Adj asst prof, Northwest Ctr Med Educ, Ind Univ, 72- Mem: AAAS; Am Soc Zoologists. Res: Structural macromolecule and skeleton biochemistry in invertebrates; coelenterate biology including reef ecology and symbiosis; comparative physiology of invertebrates; cnidarian collagen biochemistry. Mailing Add: Dept of Biol Sci St John's Univ Jamaica NY 11439

YOUNG, STEPHEN JAMES, b Long Beach, Calif, Feb 28, 42; m 65. ATMOSPHERIC PHYSICS. Educ: Univ Alaska, College, BS, 64, PhD(physics), 68. Prof Exp: Sr res asst prof, Geophys Inst, Univ Alaska, 67-68; res scientist, US Army Cold Regions Res & Eng Lab, 68-70; MEM TECH STAFF, AEROSPACE

CORP, 71- Mem: Optical Soc Am. Res: Molecular physics and spectroscopy; radiative transfer. Mailing Add: Chem & Physics Lab Aerospace Corp Box 95085 Los Angeles CA 90045

YOUNG, STEVEN BURR, botany, evolutionary biology, see 12th edition

YOUNG, STEVEN WILFORD, b Chicago, Ill, Apr 13, 48; m 72. SEDIMENTARY PETROLOGY. Educ: Albion Col, AB, 70; Ind Univ, AM, 74, PhD(sedimentary petrol), 75. Prof Exp: ASST PROF GEOL, UNIV MINN, DULUTH, 75- Mem: Geol Soc Am; Sigma Xi; Soc Econ Paleontologists & Mineralogists. Res: Provenance and petrology of sands and sandstones. Mailing Add: Dept of Geol Univ of Minn Duluth MN 55812

YOUNG, STUART, b Haslington, Eng, Dec 30, 25; nat US; m 53. VETERINARY PATHOLOGY. Educ: Royal Vet Col, Univ London, MRCVS, 48; Royal Sch Vet Studies, Univ Edinburgh, DVSM, 51; Mich State Univ, MS, 54; Univ Calif, Davis, PhD(path), 63. Prof Exp: Asst vet invest officer, Vet Invest Lab, North of Scotland Col Agr, 49-55; from asst pathologist to assoc pathologist, Vet Res Lab, Mont State Col, 55-61; assoc prof path, Univ Minn, 63-64; assoc prof path, 64-72, dir NIH grad training prog, 65-72, PROF PATH, COL VET MED & BIOMED SCI, COLO STATE UNIV, 72- Concurrent Pos: NIH spec res fel, Cambridge Univ, 59 & Univ Bern, Switz, 70-71. Mem: AAAS; Am Asn Neuropath; Am Vet Med Asn; Conf Res Workers Animal Dis. Res: Metabolic and developmental disorders of central nervous system; pathology of chronic, progressive pneumonopathies; pathogenesis of nutritional myopathies. Mailing Add: Dept of Path Colo State Univ Col Vet Med & Biomed Sci Ft Collins CO 80521

YOUNG, SUE ELLEN, b Port Arthur, Tex, Nov 28, 39; m 72. OPHTHALMOLOGY. Educ: Univ Tex, BA, 61; Univ Tex, Galveston, MD, 69. Prof Exp: From intern to resident ophthal, Univ Tex, Houston, 69-73; fel neuro ophthal, Johns Hopkins Hosp, 73-74; ASST PROF OPHTHAL, UNIV TEX CANCER SYST, 74- Concurrent Pos: Fel, Columbia Presby Hosp, 74; clin instr consult, Univ Tex Med Br, Houston, 74- Mem: AMA. Res: Effect of immunotherapy on prognosis of ocular melanoma; effect of chemotherapeutic agents on retinal function and the prevention of irradiation keratitis by topical L-cysteine. Mailing Add: 405 Piney Pt Houston TX 77024

YOUNG, SYDNEY SZE YIH, b Fukien, China, Nov 8, 24; m 54; c 1. QUANTITATIVE GENETICS, POPULATION GENETICS. Educ: Nantung Univ, BAgrSc, 47; Sydney Tech Col, FSTC, 51; Univ NSW, MSc, 56, DSc(genetics), 66; Univ Sydney, PhD(genetics), 59. Prof Exp: From res scientist to prin res scientist, Div Animal Genetics, Commonwealth Sci & Indust Res Orgn, 59-67; PROF GENETICS, OHIO STATE UNIV, 67- Concurrent Pos: Commonwealth Sci & Indust Res Orgn overseas fel, 63-64. Mem: Biomet Soc; Genetics Soc Am. Res: Theoretical and experimental quantitative and population genetics; animal breeding; biostatistics. Mailing Add: Acad Fac of Genetics Ohio State Univ Columbus OH 43210

YOUNG, THEODORE R, physics, see 12th edition

YOUNG, THOMAS EDWARD, b Chaves Co, NMex, June 15, 28; m 60; c 2. NUCLEAR PHYSICS. Educ: Rice Univ, PhD(physics), 58. Prof Exp: Asst prof physics, Pac Col, Calif, 58-60 & Trinity Univ, Tex, 60-61; PHYSICIST, AEROJET NUCLEAR CO, 61- Mem: Am Asn Physics Teachers; Am Phys Soc. Res: Neutron reactions and low energy charged particle reactions; radiation shielding. Mailing Add: Aerojet Nuclear Co PO Box 1845 Idaho Falls ID 83401

YOUNG, THOMAS EDWIN, b Manheim, Pa, Sept 7, 24; m 45; c 1. ORGANIC CHEMISTRY. Educ: Lehigh Univ, BS, 49, MS, 50; Univ Ill, PhD(chem), 52. Prof Exp: Asst chem, Lehigh Univ, 49-50; res chemist, E I du Pont de Nemours & Co, 52-55; asst prof chem, Antioch Col, 55-58; from asst prof to assoc prof, 58-66, PROF CHEM, LEHIGH UNIV, 66- Mem: Am Chem Soc. Res: Heterocyclic chemistry; indoles; structure and reactivity of heteroaromatic compounds; organosulfur chemistry. Mailing Add: 1952 Pinehurst Rd Bethlehem PA 18015

YOUNG, THOMAS WILBUR, b Washington, Ind, Nov 26, 05; m 36; c 1. HORTICULTURE, SOILS. Educ: Purdue Univ, BS, 30; Univ Fla, MS, 34; Cornell Univ, PhD, 42. Prof Exp: Assoc horticulturist, Citrus Exp Sta, Univ Fla, 42-49, PROF HORT, UNIV FLA, 42-, HORTICULTURIST, AGR RES & EDUC CTR, HOMESTEAD & INST FOOD & AGR SCI, UNIV, 54- Concurrent Pos: Prod mgr, Am Fruit Growers, Fla, 49-54. Mem: Am Soc Hort Sci. Res: Citrus, mango, lychee nutrition; soil moisture relations; variations in mineral concentrations in avocado and mango leaves as an aid to more efficient fertilizer practices. Mailing Add: Agr Res & Educ Ctr Inst Food & Agr Sci Univ of Fla Homestead FL 33030

YOUNG, ULYSSES SIMPSON, b Philadelphia, Pa, June 10, 13. ANTHROPOLOGY. Educ: WVa State Univ, AB, 36; Univ Pa, MA, 39; Univ Md, PhD(anthrop, sociol), 65. Prof Exp: Res asst anthrop, Fisk Univ, 40-41; PROF ANTHROP, BOWIE STATE COL, 45-, DIR EVE SCH, 65-, HEAD DEPT ANTHROP & SOCIOL, 67- Concurrent Pos: Consult, Arundel City Housing Surv Comn, 65; rep, Am Asn Cols Teachers Educ; Ford Found travel grant, New Eng Anthrop Soc. Mem: Fel Am Anthrop Asn. Res: Cultural tradition of midwifery in the Mississippi Delta. Mailing Add: Dept of Anthrop & Sociol Bowie State Col Bowie MD 20715

YOUNG, VERNON ROBERT, b Rhyl, Wales, Nov 15, 37; m 66; c 1. NUTRITION, BIOCHEMISTRY. Educ: Univ Reading, BSc, 59; Cambridge Univ, dipl agr, 60; Univ Calif, Davis, PhD(nutrit), 65. Prof Exp: Lectr nutrit biochem, 65-66, asst prof physiol chem, 66-72, ASSOC PROF NUTRIT BIOCHEM, MASS INST TECHNOL, 72- Res: Protein and clinical nutrition; muscle protein metabolism. Mailing Add: Dept of Nutrit & Food Sci Mass Inst of Technol Cambridge MA 02139

YOUNG, VICTOR JAY, physics, see 12th edition

YOUNG, VIOLA MAE, b Allegan, Mich, Oct 9, 15; m; c 2. MICROBIOLOGY. Educ: Mich State Col, BS, 36; Univ Ill, MS, 43; Loyola Univ, Ill, PhD, 53. Prof Exp: Technician, Ill Res & Educ Hosp, Chicago, 37-43; instr bact, Univ Chicago Med Sch, 43-45; bacteriologist & parasitologist, Mt Sinai Hosp & Res Found, Chicago, 45-47; res assoc, Sch Trop Med, PR, 47-48; parasitologist, Hektoen Inst, Cook County Hosp, Ill, 48-54; supvry bacteriologist, Dept Bact, Walter Reed Army Inst Res, 54-61; chief microbiol serv, Clin Path Dept, Clin Ctr, NIH, 61-68; CHIEF RES MICROBIOL, BALTIMORE CANCER RES CTR, CLIN BR, NAT CANCER INST, 68- Concurrent Pos: Asst supv bacteriologist, State Hosp Serv, Ill, 43-44; dir lab, Presby Hosp, San Juan, PR, 47-48; lectr, Loyola Univ, Ill, 49-54; chmn, Subcomt Enterobacteriaceae, 64-67; mem fac, Rackham Grad Sch, Univ Mich, 69-; mem, Am Bd Microbiol. Mem: Am Soc Microbiol; Am Soc Trop Med & Hyg; Soc Exp Biol & Med; NY Acad Sci; Sigma Xi. Res: All infectious agents causing diarrhea; ecology of intestinal tract; Pseudomonas aeruginosa; host-parasite relationships; normal antibodies and immunological response; infection prevention in cancer patients; interrelationships among microorganisms; opportunistic infections. Mailing Add: Baltimore Cancer Res Ctr 22 S Greene St Baltimore MD 21201

YOUNG, WARREN MELVIN, b Massillon, Ohio, Dec 30, 37; m 60; c 2. ASTRONOMY. Educ: Case Inst Technol, BS, 60; Ohio State Univ, MS, 61, PhD(astron), 71. Prof Exp: ASSOC PROF ASTRON & PLANETARIUM DIR, YOUNGSTOWN STATE UNIV, 62- Mem: Am Astron Soc; Int Soc Planetarium Educ. Res: Spectrum binary stars; photometry. Mailing Add: Dept of Physics & Astron Youngstown State Univ Youngstown OH 44503

YOUNG, WEI, b China, Feb 10, 19; nat US; m 49; c 1. BIOPHYSICS. Educ: Cath Univ, China, BS, 43; Univ Calif, PhD(biophys), 57. Prof Exp: Asst physiol, Inst Physiol, China, 43-45; instr, Nat Med Col Shanghai, 45-49; asst med physics, Univ Calif, 52-54, res biophysicist, Donner Lab, 57-63, biophysicist, Biomed Div, Lawrence Radiation Lab, 63-70; BIOPHYSICIST, NASA AMES RES CTR, 71- Mem: AAAS; Am Physiol Soc; Soc Cryobiol; NY Acad Sci. Res: Effects of chemical and physical factors on in situ enzyme kinetics; regression and quantitation of atherosclerosis. Mailing Add: NASA Ames Res Ctr Moffett Field CA 94035

YOUNG, WESLEY O, b Nampa, Idaho, Oct 7, 25; m 50; c 2. DENTISTRY. Educ: Northwest Nazarene Col, AB, 46; Univ Ore, DMD, 47; Univ Mich, Ann Arbor, MPH, 51; Am Bd Dent Pub Health, dipl, 57. Prof Exp: Pvt pract, Idaho, 47-50; dir dent health sect, Idaho Dept Health, 51-53; sr dent surgeon, Div Dent Resources, USPHS, 53-55; chief dent health sect, Idaho Dept Health, 55-63, head cleft palate rehab prog, 56-63, dir child health div, 60-63; prof community dent & chmn dept, Col Dent, Univ Ky, 63-71; PROF COMMUNITY DENT, SCH DENT, UNIV ALA, BIRMINGHAM, 72- Concurrent Pos: Consult, West Interstate Comn for Higher Educ, 56-58 & Idaho State Univ, 57-59; staff mem, Comn Surv Dent in US, Am Coun Educ, Ill, 60; lectr, dept dent hyg, Idaho State Univ, 62-63; exam mem, Am Bd Dent Pub Health, 66-70, pres, 70-71; mem, Nat Traineeship Adv Comt, USPHS, DC, 62, Nat Adv Coun Health, Bur Health Manpower, 67- & health serv res study sect, NIH, 64. Honors & Awards: Sippy Mem Award, Am Pub Health Asn, 60. Mem: Am Dent Asn; Am Soc Dent for Children (secy-treas, 59-60, vpres, 61, pres, 63-64); fel Am Col Dent; fel Am Pub Health Asn; Int Asn Dent Res. Res: Preventive dentistry; health service research; dental public health. Mailing Add: Dept of Community Dent Univ of Ala Sch of Dent Birmingham AL 35233

YOUNG, WILLIAM ALLEN, b St Marys, Ohio, May 21, 30. PETROLEUM GEOCHEMISTRY. Educ: Miami Univ, Ohio, BA, 52; Ohio State Univ, MSc, 54, PhD(chem), 57. Prof Exp: Res chemist, Carter Oil Co, 57-58, Jersey Prod Res Co, 58-64, Esso Prod Res Co, 64-66, sr res specialist, 66-72, res assoc, 72-75, SR RES ASSOC, EXXON PROD RES CO, 75- Mem: AAAS; Am Asn Petrol Geol; Am Chem Soc. Res: Petroleum and organic geochemistry; clay-organic-water interactions; hydrogeology. Mailing Add: Exxon Prod Res Co PO Box 2189 Houston TX 77001

YOUNG, WILLIAM ANTHONY, b Cleveland, Ohio, Feb 10, 23; m 54; c 2. PHYSICAL CHEMISTRY, ENGINEERING MANAGEMENT. Educ: Univ Wash, Seattle, BS, 49, MS, 53. Prof Exp: Res chemist, Am Marietta Co, 53-54; chemist, Thermodyn Sect, Nat Bur Standards, 54-55; test engr, Douglas Aircraft Co, 55-56; res engr, Atomics Int Div, NAm Rockwell Corp, 56-57, sr res chemist, 57-68, mem tech staff, 68-71; mgr anal develop & testing lab, Nuclear Energy Div, 72-74, MGR ANAL TECHNOL DEVELOP, NUCLEAR FUEL DEPT, GEN ELEC CO, 74- Mem: Am Chem Soc; Am Phys Soc; Am Nuclear Soc; Sigma Xi; fel Am Inst Chem. Res: Metal hydrides; solid state chemistry; high temperature heterogeneous reactions; diffusion in solids; reaction kinetics; thermophysical properties; molecular structure; analytical chemistry; computer applications. Mailing Add: M/C K51 Gen Elec Co PO Box 780 Wilmington DE 28401

YOUNG, WILLIAM DONALD, JR, b Glen Ridge, NJ, Nov 2, 38. BACTERIOLOGY, SYSTEMATICS. Educ: Fairleigh Dickinson Univ, BS, 60, MS, 74. Prof Exp: Res asst parasitol, NY Med Col, 61-62; sci asst clin chem, Walter Reed Army Inst Res, 62-64; SCIENTIST BACT, WARNER LAMBERT RES INST, MORRIS PLAINS, 64- Mem: Am Soc Microbiol. Res: Development of rapid biochemical tests for use in diagnostic bacteriology; use of computer technology in the identification of bacterial cultures. Mailing Add: 68 Gates Ave Montclair NJ 07042

YOUNG, WILLIAM GLENN, JR, b Washington, DC, Feb 26, 25; m 52; c 4. THORACIC SURGERY. Educ: Duke Univ, MD, 48; Am Bd Surg, dipl, 58; Am Bd Thoracic Surg, dipl, 58. Prof Exp: Resident surg, Dake Hosp, 56-57; from asst prof to assoc prof, 57-63, PROF SURG, MED CTR, DUKE UNIV, 63- Concurrent Pos: Mem sr surg staff, Duke Hosp, 57-; attend surgeon, Vet Admin Hosp, Durham, NC, 57-; consult, Watts Hosp, Durham, 58- & Womack Army Hosp, Ft Bragg, 59- Mem: Soc Vascular Surg; Soc Univ Surg; AMA; Am Asn Thoracic Surg; fel Am Col Surg. Res: Cardiovascular surgery, application of moderate and profound hypothermia in cardiovascular surgery. Mailing Add: Dept of Surg Duke Univ Med Ctr Durham NC 27706

YOUNG, WILLIAM GOULD, b Colorado Springs, Colo, July 30, 02; m 26. CHEMISTRY. Educ: Colo Col, BA, 24, MA, 25; Calif Inst Technol, PhD, 29. Hon Degrees: DSc, Colo Col, 62; LLD, Univ of the Pac, 66 & Univ Calif, 72. Prof Exp: Asst, Coastal Lab, Div Plant Biol, Carnegie Inst, 25-27; asst, Am Petrol Inst, 27-28; Nat Res Coun fel, Stanford Univ, 29-30; instr, 30-31, from asst prof to prof chem, 31-70, head dept chem, 40-48, dean div phys sci, 46-57, fac res lectr, 57, vchancellor, 57-70, EMER PROF CHEM & EMER VCHANCELLOR, UNIV CALIF, LOS ANGELES, 70- Honors & Awards: Tolman Medal, Am Chem Soc, 61, Award Chem Educ, 63, Priestley Medal, 67. Mem: Nat Acad Sci; Am Chem Soc. Res: Atmospheric oxidation; chemistry of plant pigments; carbohydrates; preparation and configuration of stereoisomeric compounds; thermal decomposition of alcohols; molecular rearrangements; mechanism of substitution reactions; allylic organo-metallic compounds. Mailing Add: 5036 Avenida Del Sol Laguna Hills CA 92653

YOUNG, WILLIAM IRVING, b Vineland, NJ, Dec 15, 39; m 67; c 2. MEDICAL GENETICS. Educ: Pa State Univ, BS, 61; Univ Minn, MS, 66, PhD(human genetics), 74. Prof Exp: Geneticist, Minn Dept Health, 65-68; NIMH trainee behav genetics, Univ Minn, 68-73; ASST PROF CLIN GENETICS & BEHAV SCI, DEPT PSYCHIAT & BEHAV SCI, EASTERN VA MED SCH, 74- Mem: Am Soc Human Genetics; Behav Genetics Asn. Res: Behavioral genetics; genetic factors in Parkinson's disease; hyperkinetic syndrome; alcoholism. Mailing Add: Dept of Psychiat & Behav Sci Eastern Va Med Sch Norfolk VA 23501

YOUNG, WILLIAM JOHNSON, II, b Lynn, Mass, Dec 10, 25; m 50; c 2. GENETICS, ANATOMY. Educ: Amherst Col, BA, 50, MA, 52; Johns Hopkins Univ, PhD(biol), 56. Prof Exp: Res assoc biol, Johns Hopkins Univ, 56-57, from asst prof to assoc prof anat, Sch Med, 57-66, assoc prof biophys, 66-68; PROF ANAT & CHMN DEPT, COL MED, UNIV VT, 68- Mem: Genetics Soc Am; Am Soc Human Genetics; Am Genetic Asn; Am Asn Anat. Res: Drosophila biochemical genetics; cytogenetics. Mailing Add: Dept of Anat Univ of Vt Col of Med Burlington VT 05401

YOUNG, WILLIAM LEWIS, III, organic chemistry, inorganic chemistry, see 12th edition

YOUNG, WILLIAM PAUL, b Spokane, Wash, Oct 11, 13; m 42; c 3. SURGERY. Educ: Univ Wis, BS, 37, MS, 39, MD, 41. Prof Exp: Intern, Res Hosp, Kansas City, Mo, 41-42; res surg, Univ Wis Hosps, 46-49, instr surg, Univ, 50; chief surg serv, Southeast Fla State Tuberc Hosp, Lantana, 51; chief surg serv, Vet Admin Hosp, Madison, Wis, 52; from asst prof to assoc prof, 53-56, PROF SURG, CARDIOVASC SURG SECT, MED SCH, UNIV WIS-MADISON, 56- Concurrent Pos: USPHS trainee, Univ Wis Hosps, 49-50; consult, Vet Admin. Mem: AMA; Am Heart Asn; Am Col Surg; Am Asn Thoracic Surg. Res: Cardiovascular surgery; pulmonary hypertension associated with congenital heart disease; homografts; myocardial revascularization and studies of anticoagulation techniques. Mailing Add: Univ of Wis Madison WI 53706

YOUNG, WILLIAM ROBERT, organic chemistry, see 12th edition

YOUNG, WILLIAM ROBERT, b East Rochester, NY, Oct 20, 26; m 49; c 3. ECONOMIC ENTOMOLOGY. Educ: Univ Rochester, BA, 51; Cornell Univ, PhD(entom), 55. Prof Exp: From asst entomologist to entomologist, 55-75, AGR PROJ LEADER & FOUND REP TO THAILAND, ROCKEFELLER FOUND, 75- Mem: Entom Soc Am. Res: Economic entomology; pests of cereal crops; biology; ecology; population studies; host plant resistance to insects. Mailing Add: Rockefeller Found GPO Box 2453 Bangkok Thailand

YOUNG, WILLIAM STANLEY, agronomy, see 12th edition

YOUNGBERG, CHESTER THEODORE, b Seattle, Wash, Mar 26, 17; m 41; c 4. FOREST SOILS. Educ: Wheaton Col, BS, 41; Univ Mich, MF, 47; Univ Wis, PhD(soils), 51. Prof Exp: Asst soils, Univ Wis, 47-51, forest soils specialist, Weyerhaeuser Timber Co, 51-52; assoc prof soils, Ore State Univ, 52-57; forestry specialist, Monsanto Chem Co, 57-58; PROF FOREST SOILS, ORE STATE UNIV, 58- Concurrent Pos: Exchange prof, NC State Univ, 69-70. Mem: Soc Am Foresters; fel Am Soc Agron; Soil Sci Soc Am; Sigma Xi. Res: Soil-vegetation relationships; forest humus; symbiotic nitrogen fixation in non-leguminous plants; tree nutrition; slope-stability. Mailing Add: Dept of Soil Sci Ore State Univ Corvallis OR 97331

YOUNGBLOOD, BETTYE SUE, b Powhatan, Ala, Dec 6, 26. ORGANIC CHEMISTRY. Educ: Auburn Univ, BS, 46; Univ Ala, MS, 49, PhD(chem), 57. Prof Exp: High sch teacher, Ala, 46-50; instr chem, Univ Miss, 50-52; high sch teacher, Ala, 56-57; asst ed chem nomenclature, Chem Abstr Serv, 57-59, assoc ed, 59-62; asst prof, 62-65, PROF CHEM, JACKSONVILLE STATE UNIV, 65- Mem: AAAS; Am Chem Soc; fel Am Inst Chemists. Res: Reaction mechanisms of aliphatic sulfonyl chlorides and derivatives; organic nomenclature—steroids and alkaloids. Mailing Add: Dept of Chem Jacksonville State Univ Jacksonville AL 36265

YOUNGBLOOD, DAVE HARPER, b Waco, Tex, Oct 30, 39. PHYSICS. Educ: Baylor Univ, BS, 61; Rice Univ, MA, 63, PhD(physics), 65. Prof Exp: Fel, Argonne Nat Lab, 65-67; ASSOC PROF PHYSICS, CYCLOTRON INST, TEX A&M UNIV, 67- Mem: Am Phys Soc. Res: Nuclear spectroscopy and reaction theory; nuclear physics. Mailing Add: Cyclotron Inst Tex A&M Univ College Station TX 77843

YOUNGDALE, GILBERT ARTHUR, b Detroit, Mich, Jan 15, 29; m 56; c 5. ORGANIC CHEMISTRY. Educ: Univ Detroit, BS, 54, MS, 56; Wayne State Univ, PhD(org chem), 59. Prof Exp: RES ASSOC MED CHEM, UPJOHN CO, 59- Mem: Am Chem Soc. Res: Prostaglandins for fertility control; anti-ulcer, blood pressure regulation. Mailing Add: Fertil Res Upjohn Co Kalamazoo MI 49001

YOUNGER, DANIEL H, b Flushing, NY, Sept 30, 36; m 65; c 3. MATHEMATICS. Educ: Columbia Univ, AB, 57, BS, 58, MS, 59, PhD(elec eng), 63. Prof Exp: Sloan fel, Princeton Univ, 63-64; res engr, Res & Develop Ctr, Gen Elec Co, NY, 64-67; assoc prof, 68-75, PROF MATH, UNIV WATERLOO, 75- Concurrent Pos: Managing ed, J Combinatorial Theory, 68-75. Res: Graph theory, especially minimax theory of directed graphs; algorithms; use of computer in combinatorial mathematics. Mailing Add: Dept of Combinatorics Univ of Waterloo Waterloo ON Can

YOUNGGREN, NEWELL A, b River Falls, Wis, Mar 15, 15; m 41; c 2. BIOLOGY. Educ: River Falls State Col, BE, 37; Univ Wis, MPh, 40; Univ Colo, PhD, 56. Prof Exp: Asst prof biol, Northland Col, 46-48; asst prof, Bradley Univ, 48-55; asst prof, Univ Colo, 55-60, chmn dept, 58-60; head dept biol sci, 68-74, PROF BIOL SCI, UNIV ARIZ, 61- Concurrent Pos: Inst dir, NSF, 58-60. Mem: Fel AAAS; Am Inst Biol Sci; Nat Asn Biol Teachers. Res: Cellular biology; biology education; slime mold physiology. Mailing Add: Dept of Biol Sci Univ of Ariz Tucson AZ 85721

YOUNGKEN, HEBER WILKINSON, JR, b Philadelphia, Pa, Aug 13, 13; m 42; c 2. PHARMACOGNOSY. Educ: Bucknell Univ, AB, 35; Mass Col Pharm, BS, 38; Univ Minn, MS, 40, PhD, 42. Prof Exp: Asst biol & pharmacog, Mass Col Pharm, 35-39; asst pharmacog, Col Pharm, Univ Minn, 39-42; from instr to prof, Univ Wash, 45-57; PROF PHARMACOG & DEAN COL PHARM, UNIV R I, 57-, PROVOST, HEALTH SCI AFFAIRS, 69- Concurrent Pos: Chmn, Plant Sci Seminar, 51; Nat Adv Coun Ed Health prof, 64- Honors & Awards: E L Newcomb Res Award, 53. Mem: Fel AAAS; Am Pharmaceut Asn. Res: Plant chemistry and pharmacology of plant constituents; effects of growth hormones and radiation on medicinal plant growth and constituents; biosynthesis of drug plant glycosides and alkaloids. Mailing Add: Dept of Pharmacog Univ of RI Kingston RI 02881

YOUNGLAI, EDWARD VICTOR, b Trinidad, WI, July 15, 40; Can citizen; m 70; c 2. BIOCHEMISTRY, REPRODUCTIVE PHYSIOLOGY. Educ: McGill Univ, BSc, 64, PhD(biochem), 67. Prof Exp: Res asst biochem, McGill Univ, 64-67; asst prof, 70-75, ASSOC PROF OBSTET & GYNEC, McMASTER UNIV, 75- Concurrent Pos: Fel exp med, McGill Univ, 67-68; Med Res Coun Can fel vet clin studies, Cambridge Univ, 68-70; Med Res Coun Can scholar, 70-75 & res grants, McMaster Univ, 70- Mem: Endocrine Soc; Brit Soc Study Fertil; Brit Soc Endocrinol; Soc Study Reproduction; Can Soc Clin Chem. Res: Control of gonadal function; gonadal steroid biosynthesis and metabolism; secretion of hormones. Mailing Add: Dept of Obstet & Gynec McMaster Univ Health Sci Ctr Hamilton ON Can

YOUNGLOVE, JAMES NEWTON, b Coleman, Tex, Dec 16, 27; m 49; c 3. MATHEMATICS. Educ: Univ Tex, BA, 51, PhD(math), 58. Prof Exp: Instr math, Univ Tex, 52-58; asst prof, Univ Mo, 58-65; assoc prof, 65-71, PROF MATH, UNIV HOUSTON, 71- Mem: Am Math Soc; Math Asn Am. Res: Point set topology. Mailing Add: 5431 Willowbend Houston TX 77004

YOUNGMAN, ARTHUR L, b Chicago, Ill, Oct 24, 37; m 63; c 1. BOTANY. Educ: Mont State Univ, BA, 59; Western Reserve Univ, MS, 61; Univ Tex, PhD(bot), 65. Prof Exp: ASST PROF BIOL, WICHITA STATE UNIV, 65- Mem: Ecol Soc Am; Am Inst Biol Sci. Res: Physiological ecology of vascular plants; environmental impact of industrial activity on terrestrial ecosystems. Mailing Add: Dept of Biol Wichita State Univ Wichita KS 67208

YOUNGMAN, EDWARD AUGUST, b Fresno, Calif, May 17, 25; m 47; c 3. ORGANIC CHEMISTRY, POLYMER CHEMISTRY. Educ: Univ Wash, BS, 48, PhD(chem), 52. Prof Exp: Chemist, Shell Develop Co, Calif, 52-58, res supvr, 58-66, head plastics & resins res dept, 66-67, dir plastics technol Ctr, Shell Chem Co, NJ, 67-69, DIR PHYS SCI, BIOL SCI RES CTR, SHELL DEVELOP CO, CALIF, 69- Mem: Am Chem Soc; Soc Chem Indust. Res: Physical science of biologically active molecules, especially their synthesis and application. Mailing Add: 300 Durham Lane Modesto CA 95350

YOUNGMAN, VERN E, b Valley, Nebr, Sept 11, 28; m 54. AGRONOMY. Educ: Univ Nebr, BS, 55, MS, 57; Washington State Univ, PhD(agron), 62. Prof Exp: Instr agron, Univ Nebr, 56-58; from instr to asst prof, Washington State Univ, 58-67; ASSOC PROF AGRON, COLO STATE UNIV, 67- Mem: Am Soc Agron; Soc Econ Bot. Res: Physiology, ecology and management of crop plants. Mailing Add: Dept of Agron Colo State Univ Ft Collins CO 80521

YOUNGMANN, CARL ERNST, b Denver, Colo, Nov 9, 43. CARTOGRAPHY, GEOGRAPHY. Educ: Midland Col, AB, 65; Univ Kans, MA, 68, PhD(geog), 72. Prof Exp: Asst prof cartog, Ohio State Univ, 70-73; ASST PROF CARTOG, UNIV WASH, 73- Concurrent Pos: Geogr, Comput Graphics, Geog Info Syst, Battelle Mem Inst, 71-74; consult, Regional Plan Asn, NY, 72-75. Mem: Am Cong Surg & Mapping; Asn Am Geogr; Asn Comput Mach. Res: Applications of automation to cartography, including data analysis and organization; interactive computer graphics and cartographic animation. Mailing Add: Dept of Geog DP-10 Univ of Wash Seattle WA 98195

YOUNGNER, JULIUS STUART, b New York, NY, Oct 24, 20; m 43; c 2. MEDICAL MICROBIOLOGY. Educ: NY Univ, AB, 39; Univ Mich, MS, 41, ScD(bact), 44; Am Bd Med Microbiol, dipl. Prof Exp: From asst to instr bact, Univ Mich, 41-44; asst path, Manhattan Proj, Univ Rochester, 45-46; instr, Univ Mich, 46-47; sr asst scientist, Nat Cancer Inst, 47-49; asst res prof virol & bact, Sch Med, 49-56, assoc prof microbiol, 56-60, PROF MICROBIOL, SCH MED, UNIV PITTSBURGH, 60-, CHMN DEPT, 66- Concurrent Pos: Vis prof, Nat Univ Athens, 63; mem virol & rickettsial study sect, NIH, 66-70; mem comn influenza, Armed Forces Epidemiol Bd, 70-73; mem bd sci counr, Nat Inst Allergy & Infectious Dis, 70-74; nat lectr, Found Microbiol, 72-73. Honors & Awards: Educ Award, E I du Pont de Nemours & Co, 74. Mem: AAAS; Am Acad Microbiol; Brit Soc Gen Microbiol; Am Soc Microbiol; Infectious Dis Soc Am. Res: Replication and properties of animal viruses; cellular and host resistance to virus infection; persistent viral infections. Mailing Add: Dept of Microbiol Univ of Pittsburgh Sch of Med Pittsburgh PA 15261

YOUNGNER, PHILIP GENEVUS, b Nelson, Minn, July 13, 20; m 47; c 4. PHYSICS. Educ: St Cloud State Col, BS, 44; Univ Wis, MS, 47, PhD(physics), 58. Prof Exp: Pub sch teacher, Minn, 39-41; radium technician, Wis Gen Hosp, 46-47; instr physics, Exten, Univ Wis, 47-49; from asst prof to assoc prof, 49-59, PROF PHYSICS, ST CLOUD STATE UNIV, 59-, CHMN DEPT, 60- Mem: Am Phys Soc; Am Asn Physics Teachers. Res: Molecular spectra; solar energy; salt water conversion; atmospheric electricity. Mailing Add: Dept of Physics St Cloud State Univ St Cloud MN 56301

YOUNGNER, VICTOR BERNARR, b Nelson, Minn, Apr 19, 22; m 44; c 2. AGRONOMY, PLANT ECOLOGY. Educ: Univ Minn, BS, 48, PhD(hort), 52. Prof Exp: Res asst hort, Univ Minn, 48-52; plant geneticist, Ferry Morse Seed Co, 52-55; from asst prof to assoc prof ornamental hort, Univ Calif, Los Angeles, 55-65; assoc prof agron, 65-68, chmn dept, 68-70, PROF AGRON, UNIV CALIF, RIVERSIDE, 68- Mem: AAAS; Bot Soc Am; fel Am Soc Agron; Crop Sci Soc Am; Sigma Xi. Res: Genetics, breeding and ecology of grasses; ecology and physiology of chaparral plants; renovation and recycling of waste water. Mailing Add: Dept of Plant Sci Univ of Calif Riverside CA 92502

YOUNGQUIST, MARY JOSEPHINE, b Fullerton, NDak, Oct 12, 30. Educ: Univ Minn, BA, 57; Mass Inst Technol, PhD(org chem), 61. Prof Exp: Res fel org chem, Univ Minn, 61-63; sr chemist, 63-74, TECH STAFF ASST, EASTMAN KODAK CO, 75- Mem: Am Chem Soc. Res: Technical editing; training coordinating; photographic chemistry. Mailing Add: Bldg 59 Eastman Kodak Co Res Labs Rochester NY 14650

YOUNGQUIST, RUDOLPH WILLIAM, b Minneapolis, Minn, Aug 10, 35; m 59; c 3. FOOD BIOCHEMISTRY. Educ: Univ Minn, BChem, 57; Iowa State Univ, MS, 60, PhD(biochem), 62. Prof Exp: RES BIOCHEMIST, PROCTER & GAMBLE CO, 62- Mem: AAAS; Am Chem Soc; Am Asn Cereal Chemists. Res: Starch and protein biochemistry. Mailing Add: Procter & Gamble Co Miami Valley Labs PO Box 39175 Cincinnati OH 45239

YOUNGQUIST, WALTER, b Minneapolis, Minn, May 5, 21; m 43; c 4. GEOLOGY. Educ: Gustavus Adolphus Col, BA, 42; Univ Iowa, MS, 43, PhD(geol), 48. Prof Exp: Asst geol, Univ Iowa, 42-43; jr geologist, Groundwater Div, US Geol Surv, Iowa, Va & La, 43-44; asst paleont, Univ Iowa, 45-47; asst prof geol, Univ Idaho, 48-51; geologist, Int Petrol Co, Peru, 51-52, sr geologist, 52-53, chief spec studies sect, 53-54; prof geol, Univ Kans, 54-57; from assoc prof to prof, Univ Ore, 57-66; consult, Minerals Dept, Humble Oil & Refining Co, 66-73; GEOTHERMAL RESOURCES CONSULT, EUGENE WATER & ELEC BD, 73- Concurrent Pos: Mem, Geothermal Resources Coun. Mem: Fel AAAS; Geol Soc Am; Am Asn Petrol Geologists; Am Inst Prof Geologists. Res: Geology and economics of mineral resources; petroleum geology; geothermal resources. Mailing Add: PO Box 5501 Eugene OR 97405

YOUNGS, ROBERT LELAND, b Pittsfield, Mass, Feb 10, 24; m 49; c 5. FORESTRY. Educ: State Univ NY, BS, 48; Univ Mich, MWT, 50; Yale Univ, PhD(forestry), 57. Prof Exp: Forest prod technologist, Forest Prod Lab, 51-66, proj leader fundamental properties, 58-64, chief div solid wood prod res, 64-66, dir div forest prod & eng res, 67-70, dir, Southern Forest Exp Sta, 70-72, assoc dep chief res, 72-75, DIR, FOREST PROD LAB, US FOREST SERV, 75- Honors & Awards: Wood Award, Forest Prod Res Soc, 57. Mem: AAAS; Soc Am Foresters; Soc Wood Sci & Technol (secy-treas, 58-59, vpres, 60-61, pres, 62-63); Forest Prod Res Soc; Am Forestry Asn. Res: Basic physical and mechanical properties of wood and related factors. Mailing Add: Forest Prod Lab PO Box 5130 Madison WI 53705

YOUNGS, VERNON LEROY, b Roseglen, NDak, Oct 1, 24; m 48; c 2. ORGANIC CHEMISTRY. Educ: Jamestown Col, BS, 48; NDak State Univ, MS, 62, PhD(chem), 65. Prof Exp: Instr, Columbus High Sch, NDak, 48-55 & NDak Sch Forestry, 55-60; res asst, NDak State Univ, 61-65; res cereal technologist, Agr Res Serv, 65-70, CHEMIST IN-CHG NAT OAT QUAL LAB, USDA, 70-; PROF AGRON, UNIV WIS-MADISON, 75- Concurrent Pos: Assoc prof agron, Univ Wis-Madison, 70-75. Mem: AAAS; Am Inst Chemists; Am Chem Soc; Am Asn Cereal Chemists; Inst Food Technologists. Res: Chemistry of cereal crops, particularly oats and wheat. Mailing Add: 6018 Piping Rock Rd Madison WI 53711

YOUNGSTOM, KARL ARDEN, b Akron, Iowa, Sept 15, 08; m 33; c 2. RADIOLOGY, ANATOMY. Educ: Univ Kans, AB, 30, MA, 32, PhD(anat), 37; Duke Univ, MD, 44. Prof Exp: Asst instr bact, Univ Kans, 30-31, from asst instr to instr anat, 31-37; from instr to assoc, Sch Med, Duke Univ, 37-43; from asst prof to prof radiol, 53-71, PROF ANAT & RADIOL, MED CTR, UNIV KANS, 71- Concurrent Pos: Consult, Vet Admin Hosp, Kansas City, Mo, 58-, Wadsworth, Kans, 59- Mem: AAAS; Radiol Soc NAm; Am Asn Anat; AMA; fel Am Col Radiol. Res: Developmental neuroanatomy and physiology; neuroradiology; angiography; methods for presenting radiological anatomy; effects of ultrasound on embryonic tissue; essential hypertension. Mailing Add: Dept of Anat Univ of Kans Med Ctr Kansas City KS 66103

YOUNGSTROM, RICHARD EARL, b Durham, NC, Sept 11, 43; m 66; c 2. ORGANIC CHEMISTRY, RADIOCHEMISTRY. Educ: Duke Univ, BS, 65; Wash Univ, MA, 69. Prof Exp: From assoc scientist steroid chem to scientist radiochem, 69-70, SR SCIENTIST RADIOCHEM, SCHERING CORP, 71- Mem: Am Chem Soc; AAAS. Mailing Add: Natural Prod Res Dept Schering Corp 60 Orange St Bloomfield NJ 07003

YOUNKIN, STUART G, b US, Jan 16, 12; m 43; c 5. PLANT PATHOLOGY. Educ: Iowa State Univ, BS, 36, MS, 39; Cornell Univ, PhD, 43. Prof Exp: From asst plant pathologist & geneticist to plant pathologist & geneticist, 43-52, asst to dir res, 52-53, dir agr res, 53-62, VPRES AGR RES, CAMPBELL SOUP CO, 62-, PRES, CAMPBELL INST AGR RES, 66- Concurrent Pos: Mem agr bd, Nat Acad Sci-Nat Res Coun, 62-68, pres, Agr Res Inst, 64-65; mem panel world food supply, President's Sci Adv Comt, 66-67. Mem: Soc Econ Botanists; Am Soc Hort Sci; Am Phytopath Soc. Res: Virus diseases of potatoes; vegetable disease control; breeding of tomatoes and peppers. Mailing Add: Agr Res Campbell Soup Co Campbell Pl Camden NJ 08101

YOUNT, DAVID EUGENE, b Prescott, Ariz, June 5, 35; m 62, 74; c 2. ELEMENTARY PARTICLE PHYSICS. Educ: Calif Inst Technol, BS, 57; Stanford Univ, MS, 59, PhD(physics), 63. Prof Exp: From instr to asst prof physics, Princeton Univ, 62-64; NSF fel, Linear Accelerator Lab, Orsay, France, 64-65; res assoc, Stanford Linear Accelerator Ctr, 65-69; assoc prof, 69-72, PROF PHYSICS, UNIV HAWAII, HONOLULU, 72- Concurrent Pos: 3M Co fel, Princeton Univ, 63; dir, Hawaii Topical Conf Particle Physics, 71. Mem: Am Phys Soc; Undersea Med Soc. Res: Experimentation in positron scattering, leptonic K-meson decay, hadronic photon absorption and photoproduction of mesons; instrumentation for particle beam monitors, spark chambers, streamer chambers and multiwire proportional chambers. Mailing Add: Dept of Physics & Astron Univ of Hawaii Honolulu HI 96822

YOUNT, ERNEST H, b Lincolnton, NC, Feb 23, 19; m 42; c 3. MEDICINE. Educ: Univ NC, BA, 40; Vanderbilt Univ, MD, 43. Prof Exp: Asst med, Univ Chicago, 45-48; from instr to assoc prof, 48-54, chmn dept, 52-72, PROF MED, BOWMAN GRAY SCH MED, 54- Concurrent Pos: Consult, Oak Ridge Inst Nuclear Studies, 50-58; mem dean's comt, Vet Admin Hosp, Salisbury, 54-63; mem, Nat Bd Med Exam, 58-61, chmn, 61. Mem: Am Fedn Clin Res; Am Soc Internal Med; Am Col Physicians; Am Diabetes Asn; Asn Prof Med. Res: Malaria; adrenal and thyroid function; diabetes. Mailing Add: Dept of Med Bowman Gray Sch of Med Winston-Salem NC 27103

YOUNT, RALPH GRANVILLE, b Indianapolis, Ind, Mar 25, 32; m 57; c 3. BIOCHEMISTRY. Educ: Wabash Col, AB, 54; Iowa State Univ, PhD, 58. Prof Exp: Res assoc enzymol, Brookhaven Nat Lab, 58-60; from asst prof chem & asst chemist to assoc prof & assoc chemist, 60-72, PROF BIOCHEM, WASH STATE UNIV, 72-, CHMN BIOCHEM/BIOPHYS PROG, 73- Concurrent Pos: NIH spec fel, Sch Med, Univ Pa & vis prof, Johnson Found, 69-70. Mem: AAAS; Am Soc Biol Chemists; Am Chem Soc; NY Acad Sci; Biophys Soc. Res: Mechanism of enzyme action as it applies to contractile proteins. Mailing Add: Dept of Chem Wash State Univ Pullman WA 99163

YOUNTS, SANFORD EUGENE, b Lexington, NC, Aug 29, 30; m 54; c 1. SOIL SCIENCE, AGRONOMY. Educ: NC State Univ, BS, 52, MS, 55; Cornell Univ, PbD(agron), 57. Prof Exp: Asst prof soils, Univ Md, 57-58; agronomist, Am Potash Inst, 58-60; assoc prof soils, NC State Col, 60-64; regional dir, Am Potash Inst, 64-67, vpres, 67-69; assoc dean col agr & dir rural develop ctr, 69-72, VPRES SERV, UNIV GA, 72- Mem: AAAS; Am Soc Agron; Int Soc Soil Sci. Res: Soil fertility and crop physiology; root growth of field crops as influenced by fertilizer and lime placement; chloride nutrition of corn; potash requirements of forage crops; nitrogen sources for turf; copper nutrition of wheat, corn and soybeans. Mailing Add: Univ of Ga 300 Old College Athens GA 30601

YOURNO, JOSEPH DOMINIC, b Utica, NY, July 14, 36; m 60; c 2. MOLECULAR BIOLOGY, MICROBIAL GENETICS. Educ: Kenyon Col, AB, 58; Johns Hopkins Univ, PhD(biochem), 64; Univ Miami, MD, 74. Prof Exp: Geneticist, Brookhaven Nat Lab, NY, 65-72; INSTR & RESIDENT CLIN PATH, STATE UNIV NY UPSTATE MED CTR, 74- Res: Gene-protein studies; effect of mutation on protein primary structure in bacteria; gene fusion; carcinogens as mutagens; diagnostic enzymology. Mailing Add: State Univ NY Upstate Med Ctr 750 E Adams St Syracuse NY 13210

YOURTEE, JOHN ASHBY, b Brownsville, Md, Aug 27, 13; m 40; c 2. CHEMISTRY. Educ: Univ Md, BS, 33, PhD(org chem), 43. Prof Exp: Res chemist, Sylvania Indust Corp, Va, 34-40, patent searcher, 40-43, res chemist, 43-46; chief chemist, Plastics Dept, Sylvania Div, Am Viscose Corp, 46-53, dir qual control, 53-55, res coordr, Film Div, 55-56, tech supt, Marcus Hook Plant, 56-65, sr proj coordr, Am Viscose Div, 65-69, FOOD & DRUG ADMIN COORDR, AM VISCOSE DIV, FMC CORP, MARCUS HOOK, PA, 69- Concurrent Pos: Asst, Univ Md, 40-43. Mem: AAAS; Am Soc Qual Control; Am Chem Soc; fel Am Inst Chemists; Am Soc Testing & Mat. Res: Cellulose and derivatives; lacquers and coating; synthetic resins and plastics; special printing inks for regenerated cellulose; method for preparation of succindialdehyde. Mailing Add: 108 Hilltop Rd Hilltop Manor Wilmington DE 19809 Marcus Hook PA

YOURTEE, LAWRENCE KARN, b Brunswick, Md, Mar 6, 17; m 41; c 1. ORGANIC CHEMISTRY. Educ: Washington Col, Md, BS, 37; Ga Inst Technol, MS, 39; Univ Tex, PhD(chem), 48. Prof Exp: Instr chem, Ga Inst Technol, 40-42 & Univ Tex, 46-47; asst prof, Univ Tenn, 47-48; assoc prof, 48-57, chmn dept, 57-71, CHILDS PROF CHEM, HAMILTON COL, 58- Mem: Am Chem Soc. Res: Synthesis and properties of heterocyclic nitrogen compounds; Pfitzinger reaction; use of ion-exchange resins in organic synthesis and separations. Mailing Add: Dept of Chem Hamilton Col Clinton NY 13323

YOUSE, BEVAN K, b Markle, Ind, Apr 5, 27; m 58; c 1. MATHEMATICAL ANALYSIS. Educ: Auburn Univ, BS, 49; Univ Ga, MS, 52. Prof Exp: Instr math, Memphis State Univ, 52-53 & Univ Ga, 53-54; asst prof, 54-67, ASSOC PROF MATH, EMORY UNIV, 67- Concurrent Pos: NSF fac fel, 60-61. Mem: AAAS; Am Math Soc; Math Asn Am. Res: Mathematical analysis. Mailing Add: Dept of Math Emory Univ Atlanta GA 30322

YOUSE, HOWARD RAY, b Bryant, Ind, May 22, 15; m 42. BOTANY. Educ: DePauw Univ, BA, 37; Ore State Col, MS, 42; Purdue Univ, PhD(bot), 51. Prof Exp: From instr to assoc prof, 40-55, PROF BOT, DePAUW UNIV, 55-, HEAD DEPT BOT & BACT, 73- Mem: AAAS; Bot Soc Am. Res: Pollen grains; seed germination. Mailing Add: Dept of Bot DePauw Univ Greencastle IN 46135

YOUSEF, MOHAMED KHALIL, b Cairo, Egypt, Aug 19, 35; US citizen; m 63; c 2. ENVIRONMENTAL PHYSIOLOGY. Educ: Ain Shams Univ, Cairo, BSc, 59; Univ Mo-Columbia, MS, 63, PhD(environ physiol), 66. Prof Exp: Res assoc environ physiol, Univ Mo-Columbia, 66-67; vis asst prof, Inst Arctic Biol, Univ Alaska, 67-68; asst prof, Lab Environ Patho-physiol, Desert Res Inst, Univ Nev, 68-70; from asst prof to assoc prof, 70-74, PROF BIOL & PHYSIOL, UNIV NEV, LAS VEGAS, 74-, COORDR HEALTH PREPROF PROG, 73- Mem: Int Soc Biometeorol; Am Physiol Soc; Soc Exp Biol & Med; Endocrine Soc; Am Col Sports Med. Res: Physiological adaptations to desert, mountain and arctic environments; role of the respiratory, cardiovascular and endocrine systems in adaptation; comparative thermoregulation during rest and exercise under different environments; comparative adaptations of organisms to various stressful environments; emphasis is on the role of cardiovascular, respiratory and endocrine systems. Mailing Add: Dept of Biol Sci Univ of Nev Las Vegas NV 89154

YOUSSEF, MARY NAGUIB, US citizen. STATISTICS. Educ: Univ Cairo, BS, 58; Columbia Univ, MA, 64; Stanford Univ, MS, 67; Ore State Univ, PhD(statist), 70. Prof Exp: Asst instr celestial mech, Cairo Univ, 58-67; mem tech staff & researcher, Oper Res Ctr, Inst Nat Planning, Cairo, 63-65; mem tech staff syst anal, Bell Tel Labs, 70-76, SR ENGR STATIST ANAL, AM BELL INT INC, AM TEL & TEL CO, 76- Res: Modeling and analyzing queuing systems; development of approximate analytic solutions for unsolved queuing problems; methods for projecting telecommunication traffic in a special environment. Mailing Add: Am Bell Int Inc PO Box 5000 South Plainfield NJ 07080

YOUSSEF, NABIL NAGUIB, b Cairo, Egypt, Oct 19, 37; US citizen; m 63; c 3. MORPHOLOGY, CELL BIOLOGY. Educ: Ain Shams Univ, Cairo, BSc, 58; Utah State Univ, MS, 64, PhD(zool), 66. Prof Exp: Asst instr entom & zool, Ain Shams Univ, Cairo, 58-60; from res asst to res assoc, 64-68, asst prof, 68-75, ASSOC PROF ZOOL, UTAH STATE UNIV, 75- Concurrent Pos: USDA grant, 66-68. Mem: AAAS; fel Royal Entom Soc London; Entom Soc Am; Soc Protozoologists. Res: Fine structure of Protozoa and Insecta with special emphasis on morphogeneses of normal and abnormal tissues induced by drugs or pathogens. Mailing Add: Dept of Biol Utah State Univ Logan UT 84322

YOUSTEN, ALLAN A, b Racine, Wis, Nov 9, 36; m 62; c 2. MICROBIAL PHYSIOLOGY. Educ: Univ Wis, BS, 58; Cornell Univ, MS, 60, PhD, 63. Prof Exp: Microbial biochemist, Int Minerals & Chem Corp, 65-69; NIH spec fel, Univ Wis, 69-71; ASST PROF MICROBIOL, VA POLYTECH INST & STATE UNIV, 71- Mem: AAAS; Am Soc Microbiol; Soc Indust Microbiol. Res: Physiology and structure of microorganisms; bacterial spore formation and germination; bacterial insect pathogens. Mailing Add: Dept of Biol Va Polytech Inst & State Univ Blacksburg VA 24061

YOUTCHEFF, JOHN SHELDON, b Newark, NJ, Apr 16, 25; m 50; c 5. ASTROPHYSICS. Educ: Columbia Univ, AB & BS, 50; Univ Calif, Los Angeles, PhD, 54. Prof Exp: Dir test staff, US Naval Air Missile Test Ctr, 50-53; opers analyst, Advan Electronics Ctr, Gen Elec Co, 53-56, functional engr, Missile & Space Div, 56-60, consult engr, 60-63, mgr advan reliability concepts oper, 63-72; mgr reliability & maintainability, Litton Industs, 72-73; PROG DIR, US POSTAL SERV HQ, WASHINGTON, DC, 73- Mem: Fel AAAS; sr mem Inst Elec & Electronics Engrs; assoc fel Am Inst Aeronaut & Astronaut; Am Soc Mech Engrs; sr mem Am Astron Soc. Res: Operations analysis; advanced systems planning; aerospace and environmental systems. Mailing Add: 543 Midland Ave Berwyn PA 19312

YOUTSEY, KARL JOHN, b Chicago, Ill, May 6, 39; m 69; c 2. PHYSICS. Educ: Loyola Univ, Chicago, BS, 61; Ill Inst Technol, MS, 65, PhD(physics), 68. Prof Exp: Physicist, Physics Dept, 61-64, physicist, Mat Sci Lab, 68-73, dir mat sci, Corp Res, 73-75, DIR PROD & PROCESS DEVELOP, WOLVERINE DIV, UOP, INC, 75- Mem: Am Phys Soc; Am Soc Metals. Res: Electronic and physical properties of ceramics; fuel cell technology; solar thermal energy systems; thin film technology; laboratory and industrial automation systems and design. Mailing Add: Wolverine Div UOP Inc PO Box 2202 Decatur AL 35601

YOUTZ, BYRON LEROY, b Burbank, Calif, Nov 10, 25; m 51; c 3. NUCLEAR PHYSICS. Educ: Calif Inst Technol, BS, 48; Univ Calif, Berkeley, PhD(physics), 53. Prof Exp: Res physicist, Lawrence Radiation Lab, Univ Calif, 50-53; asst prof physics, Am Univ Beirut, 53-56, actg chmn dept, 55-56; from asst prof to prof, Reed Col, 56-68, actg pres, Col, 67-68; prof physics & acad vpres, State Univ NY Col Old Westbury, 68-70; PROF PHYSICS, DIV SCI, EVERGREEN STATE COL, 70- Concurrent Pos: Guest, Japanese Phys Educ Soc & Asia Found Physics Curricula, 61 & 66; mem steering comt, Phys Sci Study Comn, 61-; consult, Educ Serv Inc, Mass, 61-; prin lectr, Sem Sec Sch Physics Curricula, Salisbury, Fedn Rhodesia & Nyasaland, 63 & Sem Advan Topics in Phys Sci Study Comt Physics, Santiago, Chile, 64; mem adv comt, Boston Univ Math Proj, 73- Mem: AAAS; Am Phys Soc; Am Asn Physics Teachers. Res: Astrophysics; nuclear structures; energy sources. Mailing Add: 6113 Buckhorn NW Olympia WA 98502

YOVANOVITCH, DRASKO D, b Belgrade, Yugoslavia, May 24, 30; US citizen; m 54; c 2. HIGH ENERGY PHYSICS. Educ: Belgrade Univ, BSc, 52; Univ Chicago, MSc, 56, PhD(physics), 59. Prof Exp: Res assoc physics, Enrico Fermi Inst, Univ Chicago, 59-60; asst physicist, Univ Calif, San Diego, 60-62; assoc physicist, High Energy Physics Div, Argonne Nat Lab, 62-72; ASSOC PHYSICIST, NAT ACCELERATOR LAB, 72- Mem: Fel Am Phys Soc. Mailing Add: Nat Accelerator Lab PO Box 500 Batavia IL 60510

YOVITS, MARSHALL CLINTON, b Brooklyn, NY, May 16, 23; m 52; c 3. COMPUTER SCIENCES. Educ: Union Col, BS, 44, MS, 48; Yale Univ, MS, 50, PhD(physics), 51. Prof Exp: Physicist, Nat Adv Comt for Aeronaut, Langley Field, Va, 44-46; instr physics, Union Col, 46-48; instr, Yale Univ, 48-50; sr physicist, Appl Physics Lab, Johns Hopkins Univ, 51-56; physicist, Off Naval Res, 56, head info systs br, 56-62, dir naval anal group, 62-66; PROF COMPUT & INFO SCI & CHMN DEPT, OHIO STATE UNIV, 66- Honors & Awards: Outstanding Performance Award, US Navy, 61, Superior Civilian Serv Award, 64. Mem: Inst Elec & Electronics Eng; Asn Comput Mach; Am Soc Eng Educ. Res: Information systems; management information; self-organizing systems; information science. Mailing Add: Dept of Comput & Info Sci Ohio State Univ 2036 Neil Ave Mall Columbus OH 43210

YOW, FRANCIS WAGONER, b Asheville, NC, May 1, 31; m 49; c 2.

EMBRYOLOGY. Educ: Western Carolina Univ, BS, 55; Emory Univ, MS, 56, PhD(protozool), 58. Prof Exp: Asst prof biol, Western Carolina Col, 58-60; from asst prof to assoc prof, 60-65, chmn dept, 63-74, PROF BIOL, KENYON COL, 65- Concurrent Pos: Consult-examr, NCent Asn Cols & Univs, 71- Mem: AAAS; Soc Protozoologists; Am Soc Zoologists; Soc Develop Biol. Res: Morphogenesis in ciliate protozoa; radiation biology and nutrition of invertebrates; nucleic acid synthesis; inhibition of cellular activities by radiation and chemical means. Mailing Add: Dept of Biol Kenyon Col Gambier OH 43022

YOW, MARTHA DUKES, b Talbotton, Ga, Jan 15, 22; m 44; c 3. PEDIATRICS. Educ: Univ SC, BS, 40, MD, 43; Am Bd Pediat, dipl. Prof Exp: Instr bact, 49-50, instr pediat, 50-52, from instr to asst assoc prof, 55-69, PROF PEDIAT, BAYLOR COL MED, 69-, DIR PEDIAT INFECTIOUS DIS SECT, 64- Concurrent Pos: Res fel pediat, Baylor Col Med, 50-52, Jones fel, 55-; NIH grant; mem bd sci coun, Nat Inst Allergy & Infectious Dis. Mem: Am Fedn Clin Res; Am Soc Microbiol; Soc Pediat Res; Infectious Dis Soc Am; Am Acad Pediat. Res: Infectious diseases; applied virology. Mailing Add: Dept of Pediat Baylor Col of Med Houston TX 77025

YOWELL, HOWARD LOGAN, organic chemistry, see 12th edition

YOZWIAK, BERNARD JAMES, b Youngstown, Ohio, July 5, 19; m 43; c 4. MATHEMATICS. Educ: Marietta Col, AB, 40; Univ Pittsburgh, MS, 51, PhD(math), 61. Prof Exp: Clerk, Youngstown Sheet & Tube Co, 40-41; high sch prin & teacher, Ohio, 41-42; civilian instr, US Army Air Forces Tech Training Command, Ill & Wis, 42-44; clerk, Youngstown Sheet & Tube Co, 44-45; high sch prin & teacher, Ohio, 45-47; from asst prof to assoc prof, 47-63, chmn dept, 66-71, PROF MATH, YOUNGSTOWN STATE UNIV, 63-, DEAN COL ARTS & SCI, 71- Mem: AAAS; Math Asn Am; Sigma Xi. Res: Summability methods. Mailing Add: 2080 S Schenley Ave Youngstown OH 44511

YPHANTIS, DAVID ANDREW, b Boston, Mass, July 14, 30; m 53; c 5. BIOPHYSICS, BIOCHEMISTRY. Educ: Harvard Univ, AB, 52; Mass Inst Technol, PhD(biophys), 55. Prof Exp: Am Cancer Soc fel, Mass Inst Technol, 55-56; from asst biophysicist to assoc biophysicist, Argonne Nat Lab, 56-58; from asst prof to assoc prof biochem, Rockefeller Univ, 58-65; prof biol, State Univ NY Buffalo, 65-68, prof biophys & chmn dept biol, 67-68; PROF BIOL, UNIV CONN, 68- Concurrent Pos: Consult, Argonne Nat Lab, 58-62 & NIH, 67- Mem: AAAS; Am Chem Soc; Biophys Soc; Am Soc Biol Chem. Res: Physical biochemistry; protein physical chemistry; ultracentrifugation. Mailing Add: Biol Sci Group Univ of Conn Storrs CT 06268

YU, ALBERT TZENG-TYNG, b Taiwan, June 22, 40; m 64; c 2. PLANT BREEDING. Educ: Nat Chung-Hsing Univ, Taiwan, BS, 63; Univ Calif, Davis, MS, 70, PhD(genetics), 72. Prof Exp: Res assoc plant breeding, Chung-Hsing Univ, 65-66; res geneticist, Univ Calif, Davis, 72; plant breeder, Niagara Seeds, FMC Corp, 72-74; PLANT BREEDER, PETOSEED CO, INC, GEORGE BALL CORP, 74- Mem: Sigma Xi. Res: Plant breeding programs for summer squash and peppers, both hot and sweet, including disease resistance. Mailing Add: Petoseed Co Rte 4 PO Box 1255 Woodland CA 95695

YU, ARTHUR J, b Shanghai, China, July 31, 30; US citizen; m 58; c 3. POLYMER CHEMISTRY. Educ: Univ Pa, AB, 52, MS, 54; PhD(org chem), 57; Univ Conn, MBA, 75. Prof Exp: Res chemist polymer, Am Viscose Corp, 56-59; sr res chemist, Thiokol Chem Corp, 59-61, supvr, 61-67, supvr, 67-69, MGR POLYMERS, STAUFFER CHEM CO, 69- Concurrent Pos: Consult, Foreign Info Serv, NSF, 59-60. Mem: Am Chem Soc. Res: Structure, morphology and performance relationships of polymeric materials; cost-benefit study of polymer as a material; input-output analysis of plastic industry. Mailing Add: Eastern Res Ctr Stauffer Chem Co Dobbs Ferry NY 10522

YU, BYUNG PAL, b Ham Hung, Korea, June 27, 31; US citizen; m 59; c 1. BIOCHEMISTRY, CELL PHYSIOLOGY. Educ: Mo State Univ, BS, 60; Univ Ill, PhD(lipid chem), 65. Prof Exp: From res instr to res asst prof, Med Col Pa, 65-68, from asst prof to assoc prof, 68-73; PROF PHYSIOL, UNIV TEX MED SCH SAN ANTONIO, 73- Concurrent Pos: Am Diabetes Asn res & career develop award. Mem: Am Oil Chem Soc; Am Physiol Soc. Res: Biochemical study of leucocyte lysosomes; platelet membrane; calcium transport in skeletal microsomes. Mailing Add: Dept of Physiol Univ of Tex Health Sci Ctr San Antonio TX 78284

YU, CHIA-NIEN, b Shanghai, China, Aug 5, 31; US citizen; m 66; c 3. ORGANIC CHEMISTRY. Educ: Univ Ill, Urbana, BS, 58; Univ Mich, MS, 59. Prof Exp: Res chemist, 59-67, RES SCIENTIST, NORWICH PHARMACAL CO, 67- Res: Synthesis of organic compounds for biological screenings. Mailing Add: Norwich Pharmacal Co Norwich NY 13815

YU, DAVID U L, b Hong Kong, Aug 27, 40; US citizen; m 65; c 2. NUCLEAR PHYSICS, PARTICLE PHYSICS. Educ: Seattle Pac Col, BSc, 61; Univ Wash, PhD(theoret physics), 64. Prof Exp: Res assoc theoret physics, Stanford Univ, 64-66; Brit Sci Res Coun fel physics, Univ Surrey, 66-67; from asst prof to assoc prof, Seattle Pac Col, 67-74; MEM STAFF, INFONET, COMPUT SCI CORP, 74- Mem: Am Phys Soc; Am Asn Physics Teachers. Res: Nuclear structure and reactions; elementary particle physics. Mailing Add: Infonet Comput Sci Corp 9841 Airport Blvd Los Angeles CA 90045

YU, GEORGE CHINSHIH, b Hupei, China, Oct 30, 37; US citizen; m 62; c 2. STATISTICS. Educ: Univ Nebr-Lincoln, BS, 60; Univ Md, College Park, PhD(statist), 69. Prof Exp: Actuarial analyst, Metrop Life Ins Co, 62-65; asst prof math, State Univ NY Albany, 69-73; SR STATISTICIAN, HOFFMANN-LA ROCHE, INC, 73- Concurrent Pos: Fac fel, State Univ NY Albany, 70, NSF res grant, 71-72; res assoc, Univ Calif, Berkeley, 72-73. Mem: Inst Math Statist; Am Statist Asn; Biomet Soc. Res: Non-parametric methods; time series analysis; statistical quality control; biostatistics; biometrics. Mailing Add: Res Statist Hoffmann-La Roche Inc Nutley NJ 07110

YU, GRACE WEI-CHI HU, b Feb 10, 37; US citizen; m 62; c 1. PLANT PHYSIOLOGY, CELL BIOLOGY. Educ: Nat Taiwan Univ, BS, 59; Wash State Univ, MS, 63; Duke Univ, PhD(plant physiol), 67. Prof Exp: Res assoc plant physiol, Duke Univ, 66-68; lectr bot, 68, res assoc plant physiol, 68-71, asst res biologist, Neuropsychiat Inst, 71-72, MENT HEALTH TRAINING PROG FEL, BRAIN RES INST, SCH MED, UNIV CALIF, LOS ANGELES, 72-, RES ASSOC HEMAT & ONCOL, DEPT PEDIAT, 75- Mem: AAAS; Am Soc Plant Physiologists. Res: Plant physiology, especially ion transport; developmental biology. Mailing Add: 30303 Via Borica Palos Verdes Peninsula CA 90274

YU, GRETA Y, b Canton, China, Jan 12, 17; nat US. PHYSICS, OPERATIONS RESEARCH. Educ: Sun Yat-Sen Univ, BS, 38; Univ Ill, Univ MS, 40; Univ Cincinnati, PhD(physics), 43. Prof Exp: Spectroscopist, Wright Aeronaut Corp, 43-45; physicist, US Naval Ord Plant, 45-49; res physicist, Cornell Aeronaut Lab, Inc, 50-56; sr tech specialist & tech staff, NAm Aviation, Inc, 56-57, sr mem tech staff, NAm Rockwell Corp, 67-71; TECH CONSULT ECON RES, STATE OF OHIO, 71-, STAFF SCIENTIST, OHIO POWER SITING COMN, 74- Mem: Am Phys Soc; Opers Res Soc Am; Sigma Xi. Res: Energy and environment research; nuclear and fossil power siting and evaluation; applied physics. Mailing Add: 601 Fairway Blvd Columbus OH 43213

YU, HYUK, b Kapsan, Korea, Jan 20, 33; m 63; c 1. PHYSICAL CHEMISTRY, POLYMER CHEMISTRY. Educ: Seoul Nat Univ, BS, 55; Univ Southern Calif, MS, 58; Princeton Univ, PhD(phys chem), 62. Prof Exp: Res chemist, Nat Bur Stand, 63-67; asst prof, 67-69, ASSOC PROF CHEM, UNIV WIS-MADISON, 69- Concurrent Pos: Res assoc, Dartmouth Col, 62-63; Fulbright lectr, 72; consult, Nat Bur Stand & Eastman Kodak Co, NY. Mem: AAAS; Am Chem Soc; Am Phys Soc. Res: Phase equilibria of biological macromolecules and membranes; polymer solution characterizations; syntheses of macromolecules. Mailing Add: Dept of Chem Univ of Wis Madison WI 53706

YU, JAMES CHUN-YING, b Hunan, China, Oct 14, 40; US citizen; m 65; c 2. ACOUSTICS, FLUID DYNAMICS. Educ: Nat Taiwan Univ, BSc, 62; Syracuse Univ, MSc, 68, PhD(mech eng), 71. Prof Exp: Instr mech eng, Syracuse Univ, 70-71; asst res prof, 71-75, ASSOC RES PROF ACOUST, GEORGE WASHINGTON UNIV, 75- Concurrent Pos: NASA res grant, Langley Res Ctr, 71- Mem: Am Inst Aeronaut & Astronaut; Acoust Soc Am. Res: Sound generation from fluid flows; acoustic measurements and instrumentation; turbulent flows. Mailing Add: Joint Inst Acoust & Flight Sci MS-169 NASA Langley Ctr Hampton VA 23365

YU, LEEPO CHENG, b Shanghai, China, June 25, 39; m 65; c 1. BIOPHYSICS. Educ: Brown Univ, BS, 63; Univ Md, PhD(physics), 69. Prof Exp: Res assoc muscle physiol, Brown Univ, 69-72; staff fel, 73-75, SR STAFF FEL MUSCLE X-RAY DIFFRACTION, NAT INST ARTHRITIS, METAB & DIGESTIVE DIS, 75- Mem: Am Phys Soc. Res: X-ray diffraction of striated vertebrate muscle; theoretical modelling of force generation. Mailing Add: Lab of Phys Biol Nat Inst of Arthritis Metab & Digestive Dis Bethesda MD 20014

YU, MANG CHUNG, b Hong Kong, Mar 4, 39; US citizen; m 66. ANATOMY, NEUROANATOMY. Educ: St Edward's Univ, BS, 63, MS, 66, PhD(anat), 70. Prof Exp: Fels, State Univ NY Buffalo, 70-72; ASST PROF ANAT, COL MED & DENT NJ, 72- Concurrent Pos: Consult neuroanat, Vet Admin Hosp, East Orange, NJ, 72- Mem: AAAS; Am Asn Anat. Res: Neurobiology; neuropathology. Mailing Add: Dept of Anat Col of Med & Dent of NJ Newark NJ 07103

YU, MING-HO, b Kaohsiung, Taiwan, May 22, 28; m 56; c 3. ENVIRONMENTAL HEALTH, NUTRITION. Educ: Nat Taiwan Univ, BS, 53; Utah State Univ, MS, 64, PhD(nutrit & biochem), 67. Prof Exp: Res asst agr chem, Taiwan Agr Res Inst, 54-55; asst res fel chem, Inst Chem Acad Sinica, 59-62; fel, Utah State Univ, 67 & Univ Alta, 67-68; vis asst prof plant biochem, 69-70, lectr environ biol, 70-71, asst prof, 71-74, ASSOC PROF ENVIRON BIOL, HUXLEY COL, WESTERN WASH STATE COL, 74- Mem: AAAS; Am Pub Health Asn; Int Soc Fluoride Res. Res: Fluoride effects on the physiology and biochemistry of animals and plants; effects of pollutants on health; vitamin C metabolism. Mailing Add: Huxley Col Western Wash State Col Bellingham WA 98225

YU, MING-HUNG, b Hsinwu, Taiwan, Feb 5, 36; m 68; c 2. PLANT CYTOGENETICS. Educ: Nat Chung-Hsing Univ, Taiwan, BS, 58; Tex A&M Univ, MS, 67; Iowa State Univ, PhD(plant breeding, cytogenetics), 72. Prof Exp: Agronomist, Taiwan Prov Dept Agr & Forestry, 60-65; geneticist, plant breeding, O M Scott & Sons Co, ITT Res Div, 72-75; RES AGRONOMIST, CYTOGENETICS, AGR RES SERV, USDA, 75- Concurrent Pos: Consult, Beet Sugar Develop Found, 75. Honors & Awards: R D & E Award, O M Scott & Sons Co, 72. Mem: Am Soc Agron; Crop Sci Soc Am; Am Soc Sugar Beet Technol. Res: Transferral of nematode and disease resistance from wild Beta species to the cultivated sugar beets. Mailing Add: USDA Agr Res Serv PO Box 5098 Salinas CA 93901

YU, NAI-TENG, b Pingtung, Formosa, Aug 19, 39; m 66. BIOPHYSICAL CHEMISTRY. Educ: Nat Taiwan Univ, BS, 63; NMex Highlands Univ, MS, 66; Mass Inst Technol, PhD(phys chem), 69. Prof Exp: Res chemist, Arthur D Little, Mass, 66; res assoc chem, Mass Inst Technol, 69-70; asst prof, 70-75, ASSOC PROF CHEM, GA INST TECHNOL, 75- Concurrent Pos: Res Corp res grant, Ga Inst Technol, 71-72, USPHS res grant, 71-78. Mem: AAAS; Am Chem Soc; Biophys Soc; Asn Res Vision & Ophthal. Res: Laser Raman spectroscopy of biopolymers; temperature-jump relaxation kinetics; mechanisms of cataract lens formation. Mailing Add: Sch of Chem Ga Inst of Technol Atlanta GA 30332

YU, PAO-LO, b Shanghai, China, Feb 1, 24; US citizen; m 59; c 3. GENETICS, BIOSTATISTICS. Educ: Nanking Univ, BS, 47; Univ Minn, Minneapolis, MS, 54; NC State Univ, PhD(statist), 61. Prof Exp: Res asst exp statist, NC State Univ, 56-58; statistician, John L Smith Mem Cancer Res, Chas Pfizer & Co, 58-64; asst prof, 64-68, ASSOC PROF MED GENETICS, SCH MED, IND UNIV, INDIANAPOLIS, 68- Mem: Am Statist Asn; Biomet Soc; Am Soc Human Genetics. Res: Determination of the genetics of quantitative traits; development of statistical methods for genetic study. Mailing Add: Dept of Med Genetics Ind Univ Med Ctr Indianapolis IN 46202

YU, PAUL N, b Kiangsi, China, Nov 17, 15; nat US; m 44; c 4. INTERNAL MEDICINE, CARDIOLOGY. Educ: Nat Med Col Shanghai, China, MD, 39; London Sch Trop Med & Hyg, 46, dipl, 47; Am Bd Internal Med, dipl, 56, cert cardiovasc dis, 57. Prof Exp: Instr med, asst resident & chief resident physician, Cen Hosp, Chunking, China, 40; asst resident physician, Hosp, 47, from instr to prof med, 48-69, SARA McCORT WARD PROF MED, SCH MED, UNIV ROCHESTER, 69-, HEAD CARDIOL UNIT & PHYSICIAN, HOSP, 63- Concurrent Pos: Hochstetter fel, Sch Med, Univ Rochester, 48-54; consult, State Depts Health & Social Welfare, NY, 55-, Genesee Hosp, Rochester, 57-, Vet Admin Hosp, Bath, 59-, Highland Hosp, Rochester, Frederick Thompson Mem Hosp, Canandaigua & Newark Community Hosps & St Mary's Hosp; from asst to sr assoc physician, Univ Rochester Hosp, 52-63, founding fel coun clin cardiol, Am Heart Asn. Mem: Fel Am Col Physicians; sr mem Am Fedn Clin Res; Am Physiol Soc; Asn Univ Cardiol; Am Heart Asn (pres, 72-73). Res: Pulmonary circulation hemodynamics; electrocardiography. Mailing Add: Dept of Med Univ of Rochester Sch of Med Rochester NY 14642

YU, PETER YOUND, b Shanghai, China, Sept 8, 44; m 71; c 1. EXPERIMENTAL SOLID STATE PHYSICS, SEMICONDUCTORS. Educ: Univ Hong Kong, BSc, 66 & 67; Brown Univ, PhD(physics), 72. Prof Exp: Fel, Univ Calif, Berkeley, 71-72, lectr physics, 72-73; RES STAFF MEM, THOMAS J WATSON RES CTR, IBM CORP, 73- Mem: Am Phys Soc; Sigma Xi. Res: Optical properties and light scattering of semiconductors. Mailing Add: Thomas J Watson Res Ctr IBM Corp Yorktown Heights NY 10598

YU, RILEY CHAOPING, b Chekaing, China, Apr 4, 36; m 65; c 1. NEUROBIOLOGY, GENETICS. Educ: Nat Taiwan Univ, BSc, 62, MSc, 64; Univ Man, PhD(genet), 69. Prof Exp: Res instr neurobiol, Sch Med, Wash Univ, 70-72;

ASST PROF ANAT, UNIV TEX MED BR, GALVESTON, 72- Concurrent Pos: Res fel human cytogenet, Med Col, Cornell Univ & NY Blood Ctr, 69-70; res fel neurobiol, Col Physicians & Surgeons, Columbia Univ, 70; Nat Multiple Sclerosis Soc fel, Sch Med, Wash Univ, 70-72. Mem: AAAS; Genetics Soc Am; Am Soc Cell Biol; Genetics Soc Can; Biol Coun Can. Res: Genetic studies of nervous tissue in culture system and experimental pathology of myelin sheath in nervous system. Mailing Add: Tissue Cult Lab Dept of Anat Univ of Tex Med Br Galveston TX 77550

YU, ROBERT KUAN-JEN, b China, Jan 27, 38; m 72. BIOCHEMISTRY, NEUROCHEMISTRY. Educ: Tunghai Univ, Taiwan, BS, 60; Univ Ill, Urbana, PhD(chem), 67. Prof Exp: Res assoc neurol biochem, Albert Einstein Col Med, 68-69; from instr to asst prof, 69-75, ASSOC PROF NEUROL BIOCHEM, MED SCH, YALE UNIV, 75- Concurrent Pos: NIH fel, Albert Einstein Col Med, 67-68. Mem: AAAS; Am Chem Soc; Am Soc Neurochem; Int Soc Neurochem. Res: Chemistry and metabolism of sphingolipids in the central nervous system, parasynpathetic nervous system and body fluids; sphingolipidosis; ionic properties of lipids in solution and membrane. Mailing Add: Dept of Neurol Yale Sch of Med New Haven CT 06510

YU, RUEY JIIN, b Hsin-chu, Taiwan, Mar 23, 32; m 59; c 3. CLINICAL PHARMACOLOGY. Educ: Nat Taiwan Univ, BSc, 56, MSc, 60; Univ Ottawa, PhD(org chem), 65. Prof Exp: Lectr chem, Nat Univ Taiwan, 61-62; Nat Res Coun Can fel, 65-67; asst prof, 67-73, ASSOC PROF DERMAT, SKIN & CANCER HOSP, TEMPLE UNIV, 73- Mem: Soc Invest Dermat. Res: Dermatopharmacology for skin disorders, such as psoriasis, acne and ichthyosis. Mailing Add: Skin & Cancer Hosp 3322 N Broad St Philadelphia PA 19140

YU, SHIH-AN, b Hupei, China, May 10, 27; m 55. BOTANY. Educ: Nat Taiwan Univ, BS, 50; Univ NH, MS, 56, PhD(hort), 59. Prof Exp: Res assoc forage breeding, Mich State Univ, 59-64; lectr bot, Univ Mich, Ann Arbor, 65-66, res assoc, 66-67; asst prof, 67-70, ASSOC PROF BIOL, EASTERN MICH UNIV, 70- Res: Plant breeding and genetics. Mailing Add: Dept of Biol Eastern Mich Univ Ypsilanti MI 48197

YU, SHIU YEH, b Formosa, China, June 1, 26; nat US; m 60. BIOCHEMISTRY, ORGANIC CHEMISTRY. Educ: Provincial Col Agr, China, BS, 51; Okla State Univ, MS, 56; St Louis Univ, PhD, 63. Prof Exp: Chemist & res assoc, Indust Res Inst, Formosa, 51-52; res asst, Okla State Univ, 52-56; res assoc, Inst Exp Path, Jewish Hosp, St Louis, 56-60; fel, Med Sch, St Louis Univ, 61-62; BIOCHEMIST, VET ADMIN HOSP, JEFFERSON BARRACKS, 63- Concurrent Pos: Clin biochem consult, St Louis State Hosp & Sch; res asst prof, Sch Med, Washington Univ, 72-; instr, Forest Park Community Col, 73- Mem: Geront Soc; Electron Micros Soc Am; Am Soc Exp Path; Am Heart Asn; Brit Biochem Soc. Res: Immunology; biochemistry of arteriosclerosis; structure of elastin and chemistry of elastase; mechanism of delayed hypersensitivity; mechanism of antibody formation. Mailing Add: Vet Admin Hosp Jefferson Barracks St Louis MO 63125

YU, TERRY TA-JEN, b Yen-Jer, Jin-Ching, Nov 23, 40; Chinese citizen. ORGANIC CHEMISTRY, MEDICINAL CHEMISTRY. Educ: Cheng Kung Univ, Taiwan, BSc, 65; Univ NB, PhD(org chem), 70. Prof Exp: Asst chem, Univ NB, 66-70; res assoc, Col Pharm, 70-75, RES SCHOLAR NUCLEAR MED, UNIV MICH, ANN ARBOR, 75- Mem: Am Chem Soc; Chem Inst Can. Res: Total synthesis of alkaloid; synthesis compound for anticancer; radiopharmacy; organic synthesis; radiopharmaceutic chemistry; nuclear medicine. Mailing Add: Nuclear Med Univ of Mich Ann Arbor MI 48104

YU, TS'AI-FAN, b Shanghai, China, Oct 24, 11; nat US. MEDICINE, METABOLISM. Educ: Ginling Col, China, BA, 32; Peiping Union Med Col, China, MD, 36. Prof Exp: From intern to chief resident med, Peiping Union Med Col, 35-40; instr med, Col Physicians & Surgeons, Columbia Univ, 50-56, assoc, 56-59, asst prof, 60-66; assoc prof, 66-73, RES PROF, MT SINAI SCH OF MED, 73- Concurrent Pos: Res fel med, Col Physcians & Surgeons, Columbia Univ, 47-50. Mem: Am Physiol Soc; Am Soc Pharmacol & Exp Therapeut; Harvey Soc; AMA; Am Rheumatism Asn. Res: Calcium and phosphorous metabolism in osteomalacia; purine metabolism and gout. Mailing Add: Mt Sinai Hosp 11 E 100th St New York NY 10029

YU, WAI-MAO, applied physics, electrical engineering, see 12th edition

YUAN, EDWARD LUNG, b China, July 2, 29; US citizen; m 54; c 2. PHYSICAL CHEMISTRY. Educ: Univ Calif, Berkeley, BS, 51; Univ Wis-Madison, PhD(phys chem), 54. Prof Exp: Res assoc high temperature kinetics, Univ Wis, 54-55; res chemist, 55-62, staff chemist, 63-68, res assoc, 68-72, RES FEL HIGH TEMPERATURE KINETICS, E I DU PONT DE NEMOURS & CO, INC, 72- Mem: Am Chem Soc. Res: Electronics and high temperature materials; polymer chemistry; high temperature polymers; film formation mechanisms; electrical and electronics materials technology. Mailing Add: E I du Pont de Nemours & Co 3500 Grays Ferry Ave Philadelphia PA 19146

YUAN, LUKE CHIA LIU, b Changtehfu, China, Apr 5, 12; US citizen; m 42; c 1. PHYSICS. Educ: Yenching Univ, BS, 32, MS, 34; Calif Inst Technol, PhD(physics), 40. Prof Exp: Asst physics, Yenching Univ, 32-34; asst, Calif Inst Technol, 37-40, fel, 40-42; res physicist, RCA Labs, 42-46; res assoc, Princeton Univ, 46-49; SR PHYSICIST, BROOKHAVEN NAT LAB, UPTON, 49- Concurrent Pos: Guggenheim fel, 58; vis physicist, Europ Orgn Nuclear Res, 72-76. Honors & Awards: Achievement Award, Chinese Inst Elec Eng, 62. Mem: Fel Am Phys Soc; Acad Sinica; NY Acad Sci. Res: High energy physics; super energy accelerator and particle detection systems; cosmic rays; radio direction finding; frequency modulation radar systems. Mailing Add: 15 Claremont Ave New York NY 10027

YUAN, TZU-LIANG, b Ningpo, China, May 27, 22; nat US; m 60; c 2. SOILS. Educ: Nat Univ Chekiang, China, BSc, 45; Ohio State Univ, MSc, 52, PhD(agron), 55. Prof Exp: Asst instr soils, Tat Nat Univ Chekiang, China, 45; instr, Tax Univ, Nat Univ, Kweichow, 46-48; asst soils chemist, Taiwan Sugar Exp Sta, 48-51; res asst, Ohio Agr Exp Sta, 52-55; from asst prof to assoc prof, 55-70, PROF SOIL CHEM, AGR EXP STA, UNIV FLA, 70- Mem: Fel AAAS; Soil Sci Soc Am; Clay Minerals Soc; NY Acad Sci; Int Soil Sci Soc. Res: Soil-forming processes; chemical nature of soils; soil properties in relation to plant nutrition. Mailing Add: Dept of Soil Sci Univ of Fla Gainesville FL 32611

YUAN, WILLIAM JEN CHUN, b Kiangsu Prov, Repub of China; US citizen. MATHEMATICAL STATISTICS, MEDICAL STATISTICS. Educ: Cheng Kung Univ, Taiwan, BA, 62; Wayne State Univ, MA, 66; Univ Calif, Berkeley, PhD(statist), 74. Prof Exp: Statistician & programmer, Texaco, Inc, 66-67; opers res analyst, Teknekron, Inc, 68-70; from teaching asst to teaching assoc statist, Univ Calif, Berkeley, 67-73; lectr statist, Univ Calif, Davis, 73-74; ASST PROF STATIST, STATE UNIV NY STONY BROOK, 74- Mem: Sigma Xi; Inst Math Statist; Am Statist Asn. Res: Comparisons of statistical methods, nonparametric statistics, data analysis, biostatistics, analysis of categorized data, design of experiment, and sampling. Mailing Add: Dept of Appl Math & Statist State Univ of NY Stony Brook NY 11794

YUCEOGLU, YUSUF ZIYA, b Cesme, Turkey, Mar 28, 19; m 47. INTERNAL MEDICINE, CARDIOLOGY. Educ: Istanbul Univ, MD, 44. Prof Exp: Intern, Istanbul Univ Hosp, 43-44; resident internal med, Ankara Univ Hosp, 47-50, spec asst, 50-52; fel cardiol, Mt Sinai Hosp, New York, 53-55; res fel, Cardiopulmonary Lab, Maimonides Hosp, 55-57; NY Heart Asn fel, 55-57; from res assoc to assoc, Maimonides Hosp, 57-60, from asst attend physician to assoc attend physician, 61-67, dir cardiopulmonary lab, 66-67; ASSOC PROF MED, NEW YORK MED COL, 67- Concurrent Pos: From instr to asst prof med, State Univ NY Downstate Med Ctr, 55-67; asst attend physician, Kings County Hosp, 66-67; assoc attend physician, Flower & Fifth Ave Hosps, 67- & Metrop Hosps, 67- Honors & Awards: Cert Honor, Am Col Angiol & Int Col Angiol, 65. Mem: AMA; fel Am Col Cardiol; fel Am Col Angiol. Res: Cardiac physiology; vectorcardiography. Mailing Add: Div of Cardiol Metrop Hosp 1901 First Ave New York NY 10029

YUDELSON, JOSEPH SAMUEL, b Philadelphia, Pa, July 20, 25; m 52; c 4. POLYMER CHEMISTRY, PHOTOCHEMISTRY. Educ: Univ Pittsburgh, BS, 50; Ill Inst Technol, PhD(chem), 55. Prof Exp: Res chemist, 54-57, sr res chemist, 57-60, res assoc phys chem, 60-70, SR LAB HEAD SOLID STATE PHOTOSCI LAB, EASTMAN KODAK RES LAB, 70- Mem: Assoc Am Chem Soc; Soc Photog Scientists & Engrs; Electrochem Soc. Res: Physical chemistry of hydrophilic polymers; photographic chemistry; nonaqueous solvents; inorganic photochemistry. Mailing Add: Eastman Kodak Co 1669 Lake Ave Rochester NY 14650

YUDOWITCH, KENNETH LOUIS, b Hartford, Conn, Mar 21, 20; m 41; c 5. PHYSICS, OPERATIONS RESEARCH. Educ: Trinity Col, Conn, BS, 43; Univ Mo, PhD(physics), 48. Prof Exp: Instr physics, Trinity Col, 43-44; res asst, Manhattan Proj, Columbia Univ, 44-45; res asst, Univ Mo, 45-47, O M Stewart fel, 47-48; from asst prof to assoc prof physics, Fla State Univ, 48-51; supvr physics of solids, Ill Inst Technol, 51-53; chmn infantry proj, Opers Res Off, Johns Hopkins Univ, 53-59, chief tactics div, 59-61; Europ liaison officer, Res Anal Corp, 61-63, chief field exp div, 63-64; analyst, Weapons Systs Eval Group, Inst Defnese Anal, 64-65; tech dir field exp, Data Dynamics Corp, 65-66; mgr opers anal dept, Stanford Res Inst, 66-69; PRES, OPERS RES ASSOCS, 69- Concurrent Pos: Lectr natural sci, San Jose State Univ. Mem: Mil Opers Res Soc (pres). Mailing Add: Opers Res Assocs 20800 Valley Green Dr Cupertino CA 95014

YUEN, PO SANG, b Tsing Tao, China, Apr 5, 41; m 68; c 2. POLYMER CHEMISTRY, ORGANIC CHEMISTRY. Educ: Baylor Univ, BS, 64; Univ Mich, PhD(org chem), 68. Prof Exp: Res chemist, Stauffer Chem Co, 68-72; pres, Tecin Mat Appln Inc, 72-74; EXEC DIR, CHINATOWN MANPOWER PROJ INC, 74- Res: Kinetic studies of solvolysis using polymers as catalysts; polymer development. Mailing Add: 2028 Haviland Ave Bronx NY 10472

YUEN, TED GIM HING, b Canora, Sask, Dec 21, 33; m 58; c 4. EXPERIMENTAL PATHOLOGY. Educ: Andrews Univ, BA, 56; Univ Southern Calif, PhD(exp path), 69. Prof Exp: Fel exp path, Univ Southern Calif, 69-71; RES ASSOC EXP PATH, HUNTINGTON INST APPL MED RES, 72- Concurrent Pos: Res assoc, Dept Path, Univ Southern Calif, 75-76. Res: Ultrastructural study of the effects of electrical stimulation of the brain. Mailing Add: Huntington Inst of Appl Med Res 734 Fairmount Ave Pasadena CA 91105

YUHAS, JOHN M, b Passaic, NJ, Aug 10, 40; m 64; c 2. RADIOBIOLOGY, CANCER. Educ: Univ Scranton, BS, 62; Univ Md, MS, 64, PhD(radiation biol), 66. Prof Exp: Assoc staff scientist, Jackson Lab, 66-69; biologist, Oak Ridge Nat Lab, 69-76; ASSOC DIR, CANCER RES & TREATMENT CTR, UNIV NMEX, 76- Mem: AAAS; Am Asn Cancer Res; Radiation Res Soc; Genetics Soc Am; NY Acad Sci. Res: Mammalian radiobiology; radioprotective drugs; experimental radiotherapy; toxicology; tumor immunology; carcinogenesis; oncogenic viruses. Mailing Add: Cancer Res & Treatment Ctr Univ of NMex Albuquerque NM 87131

YUHAS, JOSEPH GEORGE, b Cleveland, Ohio, Aug 26, 38; m 60; c 4. ANIMAL BEHAVIOR, ZOOLOGY. Educ: Ohio State Univ, BSc Agr, 60, BScEd & MSc, 62, PhD(zool), 70. Prof Exp: Res assoc wildlife, Ohio State Univ, 68-69; asst prof biol, Defiance Col, 69-75, chmn dept natural systs, 72-75; ASSOC PROF LIFE SCI & DIR, CTR FOR LIFE SCI, ST FRANCIS COL, 75- Mem: Animal Behav Soc; Am Inst Biol Sci. Res: Effects of environmental contamination on behavior; innate behavior and evolution; endocrine control of behavior; induced ovulation and delayed implantation. Mailing Add: Dept of Life Sci St Francis Col Biddeford ME 04005

YUILL, ROBERT STANLEY, b Grayling, Mich, Sept 1, 37; m 63; c 2. URBAN GEOGRAPHY, ECONOMIC GEOGRAPHY. Educ: Antioch Col, BS, 60; Univ Mich, MA, 64, PhD(geog), 69. Prof Exp: Instr geog, Univ Mich, 66-67; asst prof, State Univ NY Buffalo, 67-71; chmn dept, 71-75, ASSOC PROF GEOG, CENT MICH UNIV, 71- Mem: Asn Am Geog; Regional Sci Asn. Res: Urban spatial growth; remote sensing of urban areas; urban systems. Mailing Add: Dept of Geog Cent Mich Univ Mt Pleasant MI 48858

YUILL, THOMAS MACKAY, b Berkeley, Calif, June 14, 37; m 60; c 1. VIROLOGY, ECOLOGY. Educ: Utah State Univ, BS, 59; Univ Wis, MS, 62, PhD(vet sci), 64. Prof Exp: Lab officer virol, Walter Reed Army Inst Res, 64-66, med biologist, SEATO Med Res Lab, 66-68; asst prof, 68-72, ASSOC PROF VET SCI, UNIV WIS-MADISON, 72- Mem: Am Soc Microbiol; Wildlife Soc; Wildlife Dis Asn; Am Soc Trop Med & Hyg; Royal Soc Trop Med & Hyg. Res: Arbovirus epizootiology; wildlife diseases, especially duck plague in waterfowl; virus ecology. Mailing Add: Dept of Vet Sci Univ of Wis Madison WI 53706

YUK, JAMES PETER, organic chemistry, analytical chemistry, see 12th edition

YULE, HERBERT PHILLIP, b Chicago, Ill, Apr 17, 31; m 61; c 2. NUCLEAR CHEMISTRY, COMPUTER SCIENCE. Educ: Univ Chicago, PhD(nuclear chem), 60. Prof Exp: Res chemist, Calif Res Corp, 57-61; staff mem, Gen Atomic Div, Gen Dynamics Corp, Calif, 62-66; assoc prof activation anal, Tex A&M Univ, 66-69, assoc prof comput sci, 67-69; res chemist, Nat Bur Standards, 69-72; staff consult, 72-74, MGR COMPUT SERV, NUS CORP, 74- Concurrent Pos: Adj assoc prof biochem, Baylor Col Med, 67-69; lectr, Nat Bur Standards, 68 & NATO, 70; consult, Nat Bur Standards, 74-75. Mem: Am Nuclear Soc; Am Chem Soc. Res: Activation analysis, especially computer techniques; numerical analysis of data; computer and system programming. Mailing Add: NUS Corp 4 Research Pl Rockville MD 20850

YULE, THOMAS J, b Chicago, Ill, Nov 21, 40; m 66. NUCLEAR PHYSICS, REACTOR PHYSICS. Educ: John Carroll Univ, BS, 62; Univ Wis-Madison, MS, 64, PhD(physics), 68. Prof Exp: ASST PHYSICIST, ARGONNE NAT LAB, 68- Mem: Am Phys Soc. Res: Nuclear research instrumentation. Mailing Add: Div of Reactor Physics Argonne Nat Lab Bldg 316 Argonne IL 60439

YUN, KWANG-SIK, b Seoul, Korea, July 27, 29; m 60; c 1. PHYSICAL CHEMISTRY. Educ: Seoul Nat Univ, BS, 52; Univ Cincinnati, PhD(chem), 61. Prof Exp: Res assoc, Inst Molecular Physics, Univ Md, 60-63; Nat Res Coun Can fel, 63-65; scientist, Nat

Ctr Atmospheric Res, Colo, 65-67; ASSOC PROF CHEM, UNIV MISS, 67- Mem: Am Chem Soc. Res: Kinetic theory of gases and liquids; gas phase chemical kinetics and chemical reactions of atmospheric gases. Mailing Add: Dept of Chem Univ of Miss University MS 38677

YUN, SEUNG SOO, b Korea, Mar 1, 31; US citizen; m 57; c 3. ACOUSTICS. Educ: Clark Univ, AB, 57; Brown Univ, MSc, 61, PhD(physics), 64. Prof Exp: Asst physicist, Ore Regional Primate Res Ctr, 63-65; res physicist, IIT Res Inst, 65-67; asst prof, 67-70, ASSOC PROF PHYSICS, OHIO UNIV, 70- Mem: Acoust Soc Am. Res: Physical acoustics and ultrasonics; absorption and dispersion of ultrasound in liquid and solid; critical phenomena. Mailing Add: Dept of Physics Ohio Univ Athens OH 45701

YUN, YOUNG MOK, b Chung Song Co, Korea, Sept 22, 31; m 66; c 2. ENTOMOLOGY. Educ: Wash State Univ, BS, 61; Ore State Univ, BS, 62; Mich State Univ, MS, 64, PhD(entom), 67. Prof Exp: ENTOMOLOGIST, AGR RES CTR, GREAT WESTERN SUGAR CO, 67- Mem: Entom Soc Am; Am Soc Sugar Beet Technologists. Res: Biology, ecology, and control of insects, nematodes, and diseases affecting sugar beets. Mailing Add: Agr Res Ctr Great Western Sugar Co Longmont CO 80501

YUND, MARY ALICE, b Xenia, Ohio, Feb 12, 43; m 66. DEVELOPMENTAL BIOLOGY. Educ: Knox Col, BA, 65; Harvard Univ, MA, 67, PhD(biol), 70. Prof Exp: NIH fel, Univ Calif, Berkeley, 70-72, NIH trainee, 72-73, res geneticist, 73; asst prof biol, Wayne State Univ, 74-75; ASST RES GENETICIST, UNIV CALIF, BERKELEY, 75- Concurrent Pos: NSF res grant, 75. Mem: Sigma Xi; Soc Develop Biol; Genetics Soc Am; Am Soc Zoologists; AAAS. Res: Hormonal control of gene activity in differentiation of imaginal discs of Drosophila melanogaster; mechanism of steroid hormone action. Mailing Add: Dept of Genetics Univ of Calif Berkeley CA 94720

YUND, RICHARD ALLEN, b Ill, Dec 14, 33; m 57; c 2. MINERALOGY. Educ: Univ Ill, PhD(geol), 60. Prof Exp: Fel, Geophys Lab, Carnegie Inst, 60-61; from asst prof to assoc prof, 61-68, PROF GEOL, BROWN UNIV, 68- Mem: Mineral Soc Am; Geochem Soc; Am Geophys Union. Res: Kinetics and mechanisms of mineral reactions. Mailing Add: Dept of Geol Sci Brown Univ Providence RI 02912

YUNE, HEUN YUNG, b Seoul, Korea, Feb 1, 29; US citizen; m 56; c 3. RADIOLOGY, SURGERY. Educ: Severence Union Med Col, MD, 56; Am Bd Radiol, dipl, 64; Korean Bd Radiol, dipl, 65. Prof Exp: Resident gen surg, Presby Med Ctr, Korea, 56-60; resident radiol, Vanderbilt Univ Hosp, 60-63, instr, Univ, 62-64; chief radiologist, Presby Med Ctr, Korea, 64-66; from asst prof to assoc prof radiol, Vanderbilt Univ, 66-71; PROF RADIOL, IND UNIV, INDIANAPOLIS, 71- Concurrent Pos: Consult radiologist, Indianapolis Vet Admin Hosp, 71 & Wishard Mem Hosp, Indianapolis, 75- Honors & Awards: Magna cum Laude Award, Radiol Soc NAm, 69; Cert of Merit, AMA & Am Roentgen Ray Soc, 70, 72, 73 & 74; Silver Medal, Am Roentgen Ray Soc, 71, Bronze Medal & Cert Merit, 75; Cert of Merit, Radiol Soc NAm, 75. Mem: Radiol Soc NAm; Am Roentgen Ray Soc; Am Col Radiol; Asn Univ Radiol; Int Soc Lymphology. Res: Vascular radiology; tumor angiography, angiography in trauma, angiography in endocrine disorder and lymphangiography; eye, ear, nose and throat radiology; head and neck tomography and contrast radiography in head and neck. Mailing Add: Dept of Radiol Ind Univ Med Ctr Indianapolis IN 46202

YUNGBLUTH, THOMAS ALAN, b Warren, Ill, Dec 12, 34. GENETICS, PLANT BREEDING. Educ: Univ Ill, BS, 56; Univ Minn, PhD(genetics), 66. Prof Exp: Asst prof, 66-72, ASSOC PROF BIOL, WESTERN KY UNIV, 72- Mem: Am Soc Agron; Crop Sci Soc Am. Mailing Add: Dept of Biol Western Ky Univ Bowling Green KY 42101

YUNGHANS, WAYNE N, b Lakewood, Ohio, Dec 10, 45; m 69; c 2. CYTOLOGY. Educ: Heidelberg Col, BS, 67; Purdue Univ, MS, 69, PhD(cytol), 74. Prof Exp: Res asst microbiol, US Army, 69-71; ASST PROF CYTOL, STATE UNIV NY COL FREDONIA, 74- Honors & Awards: State Univ NY Res Found Award, 75. Mem: AAAS; Am Soc Plant Physiologists. Res: Isolation and purification of cellular membranes including plasma membranes, Golgi apparatus and endoplasmic reticulum; membranes characterized for enzyme activities and protein kinases and phosphoproteins. Mailing Add: Dept of Biol State Univ Ny Col Fredonia NY 14063

YUNGUL, SULHI HASAN, b Istanbul, Turkey, Oct 20, 19; nat US; m 47; c 3. EXPLORATION GEOPHYSICS. Educ: Mont Sch Mines, BS, 43; Calif Inst Technol, MS, 44, GeophEngr, 45; Tex A&M Univ, PhD, 62. Hon Degrees: Geophys Engr, Mont Col Mineral Sci & Technol, Engr, 71. Prof Exp: From geophysicist to chief geophysicist, Mining Res & Explor Inst, Govt Turkey, 45-53, chief geophysicist, Etibank, 53-55; SR RES ASSOC, CHEVRON OIL FIELD RES CO, STANDARD OIL CO CALIF, 55- Concurrent Pos: Vis lectr, Univ Calif, Riverside, 72 & Calif State Univ, Fullerton, 75. Mem: Soc Explor Geophys; Am Geophys Union; Europ Asn Explor Geophys. Res: Solid earth exploration geophysics, especially gravity, magnetic and electrical; tectonophysics. Mailing Add: Chevron Oil Field Res Co PO Box 446 La Habra CA 90631

YUNICE, ANDY ANIECE, b Rahbeh-Akkar, Lebanon, Jan 2, 25; US citizen; m 59; c 3. BIOCHEMISTRY, ENVIRONMENTAL SCIENCE. Educ: Am Univ Beirut, BS, 49; Wayne State Univ, MS, 58; Univ Okla, PhD(environ health), 70. Prof Exp: Instr, Am Boys Sch Tripoli, Lebanon, 49-53; prin high sch, 53-54; res asst med, Wash Univ, 58-67, res instr, 67-70; RES BIOCHEMIST & DIR TRACE METAL RES LABS, VET ADMIN HOSP, 62- Concurrent Pos: Asst prof & mem grad fac, Sch Med, Univ Okla, 70-; hosp grant & co-prin investr, NIH grant, Vet Admin Hosp, 70- Mem: AAAS; Am Fedn Clin Res; Geront Soc; Am Chem Soc; Sigma Xi. Res: Trace metal metabolism in post-alcoholic cirrhosis and cardiovascular diseases, particularly hypertension in human beings. Mailing Add: Renal Div Dept of Med Vet Admin Hosp 921 NE 13th St Oklahoma City OK 73104

YUNICK, ROBERT P, b Schenectady, NY, Oct 27, 35; m 59; c 3. ORGANIC POLYMER CHEMISTRY. Educ: Union Col, NY, BS, 47; Rensselaer Polytech Inst, PhD(org chem), 61. Prof Exp: Res chemist, Olefins Div, Union Carbide Chem Corp, 61-63; mgr, 63-72, DIR RES, W HOWARD WRIGHT RES CTR, SCHENECTADY CHEM, INC, 72- Mem: Am Chem Soc; Am Ornithologists Union. Res: Organic synthesis of intermediates for resin synthesis; synthesis of phenolic, hydrocarbon, resorcinolic, polyester and polyesterimide resins; synthesis of high-temperature polymers. Mailing Add: Schenectady Chem Inc 2750 Balltown Rd Schenectady NY 12309

YUNIS, ADEL A, b Rahbeh, Lebanon, Mar 17, 30; m 59; c 3. INTERNAL MEDICINE. Educ: Am Univ Beirut, BA, 50, MD, 54. Prof Exp: Clin fel hemat, Washington Univ, 57-58, res fel, 58-59, res assoc biochem, 59-61, from instr to asst prof med, Med Sch, 61-64; from asst prof to assoc prof & dir hemat, 64-68, PROF MED & BIOCHEM, SCH MED, UNIV MIAMI, 68-, DIR DIV HEMAT, HOWARD HUGHES LABS HEMAT RES, 68- Concurrent Pos: Am Leukemia Soc scholar, 61-66; USPHS res career develop award, 66-71. Mem: Am Fedn Clin Res; Am Soc Hemat; Am Soc Exp Path; Am Soc Clin Invest; Int Soc Hemat. Res: Comparative studies on glycogen phosphorylase; enzymes of glycogen metabolism in white blood cells; mechanism of action of bone marrow toxins and the pathogenesis of chloramphenicol-induced blood dyscrasias. Mailing Add: Univ of Miami Sch of Med PO Box 875 Biscayne Annex Miami FL 33152

YUNIS, EDMOND, b Sincelejo, Colombia, Aug 8, 29; US citizen; m 65; c 1. MEDICINE, PATHOLOGY. Educ: Nat Univ Colombia, MD, 54. Prof Exp: Resident anat path, Univ Kans, 55-57; resident clin path, 57-59, from instr to prof lab med, 60-71, dir blood bank, 61-68, PROF LAB MED & PATH, UNIV HOSPS, UNIV MINN, MINNEAPOLIS, 71-, DIR DIV IMMUNOL, 66- Mem: Am Soc Exp Path; Am Asn Immunol. Res: Antigenicity in cells and animals; immunological capacity in animals related to thymus; transplantation immunology. Mailing Add: Univ of Minn Hosp Box 198 Mayo Minneapolis MN 55455

YUNIS, JORGE J, b Sincelejo, Colombia, Oct 5, 33; US citizen. GENETICS, PATHOLOGY. Educ: Inst Simon Araujo, Columbia, BS, 51; Cent Univ Madrid, MD, 56, Dr MD, 57. Prof Exp: Intern, Prov Hosp, Barranquilla, Colombia, 57-58, resident internal med, 58-59; resident clin & anat path, 59-62, from instr to assoc prof lab med, 62-69, dir grad studies lab med & path, 69-74, PROF LAB MED & PATH, HEAD DIV MED GENETICS, UNIV MINN, MINNEAPOLIS, 69- Concurrent Pos: Fel lab med, Univ Minn, Minneapolis, 62-63; chmn human genetics comt health sci, Univ Minn, Minneapolis, 72- Mem: Am Soc Human Genetics; Am Soc Cell Biol; Am Asn Path & Bact; Acad Clin Lab Physicians & Sci; Am Soc Hemat. Res: Fine structure and molecular organization of chromosomes; chromosome defects in man. Mailing Add: Div of Med Genetics Box 198 Mayo Univ of Minn Hosps Minneapolis MN 55455

YUNKER, CONRAD ERHARDT, b Matawan, NJ, Dec 22, 27; m 58; c 4. ZOOLOGY, PARASITOLOGY. Educ: Univ Md, BS, 52, MS, 54, PhD, 58. Prof Exp: Staff mem med zool, US Naval Med Res Unit, Egypt, 55-57; res assoc zool, Univ Md, 58, asst prof, 58-59; entomologist, Entom Res Inst, Can Dept Agr, 59-60; entomologist, Mid Am Res Unit, NIH, 60-61, SCIENTIST DIR, ROCKY MOUNTAIN LAB, NAT INST ALLERGY & INFECTIOUS DIS, 61- Concurrent Pos: Consult, Nat Inst Allergy & Infectious Dis, 59; mem, Bolivian Hemorrhagic Fever Comn, 63. Honors & Awards: Award, Inst Acarology, Ohio State Univ, 72; Sigrid Juselius Found, Helsinki, Finland, 75. Mem: Fel AAAS; Soc Syst Zool; Am Soc Zool; Am Soc Parasitol; Am Soc Trop Med & Hyg. Res: Systematic acarology; medical entomology; arthropod tissue culture; arthropod-borne viruses. Mailing Add: Rocky Mountain Lab Hamilton MT 59840

YUNKER, MARTIN HENRY, b Milton Junction, Wis, Dec 28, 28; m 53; c 2. PHARMACY. Educ: Univ Wis, BS, 51, MS, 53, PhD(phys pharm), 57. Prof Exp: Tech asst to mgr pharmaceut prod, Merck, Sharp & Dohme Div, Merck, Inc, 57-58, supvr granulation dept, 58-59, supvr qual control, 59-61, res assoc pharmaceut res, 61-63; SR RES PHARMACIST, ABBOTT LABS, 63- Mem: Am Pharmaceut Asn. Res: Pharmaceutical research, including tablet formulations; biopharmaceutics; in vitro drug dissolution; preformulation characterization of drugs. Mailing Add: Dept 493 Abbott Labs North Chicago IL 60064

YUNKER, WAYNE HARRY, b Corvallis, Ore, Jan 8, 36; m; c. PHYSICAL INORGANIC CHEMISTRY. Educ: Ore State Col, BS, 57; Univ Wash, PhD(chem), 61. Prof Exp: Res scientist chem, Gen Elec Co, 63-65; sr res scientist, Battelle Mem Inst, 65-70; SR RES SCIENTIST CHEM, WESTINGHOUSE HANFORD CO, 70- Mem: Am Chem Soc; Sigma Xi. Mailing Add: Westinghouse Hanford Co PO Box 1970 Richland WA 99352

YURA, HAROLD T, theoretical physics, see 12th edition

YURCHENCO, JOHN ALFONSO, b San Juan, Arg, Feb 22, 15; nat US; m 44; c 4. MEDICAL MICROBIOLOGY. Educ: Albion Col, BA, 41; Johns Hopkins Univ, ScD(bact), 49. Prof Exp: Chief div microbiol, Eaton Labs, Inc, 48-55; head dept chemother, Squibb Inst Med Res, 55-56; SR RES SCIENTIST, WYETH LABS, INC, 56- Res: Chemotherapy bacterial infections; bacterial pathogenicity and virulence; low temperature stability of infectious bacterial pools; immunology. Mailing Add: Wyeth Labs Inc Radnor PA 19088

YUREK, GERALD G, b San Francisco, Calif, May 7, 35; m 59; c 1. BIOMEDICAL ENGINEERING, ELECTRICAL ENGINEERING. Educ: Calif Inst Technol, BS, 56; Stanford Univ, MS, 57, Engr & MS, 61, PhD(physiol), 64. Prof Exp: Elec engr, Instrument Sect, NIH, 57-59, develop engr, Instrument Eng & Develop Br, 61-63, SR INVESTR, LAB TECH DEVELOP, NAT HEART INST, 63- Concurrent Pos: Vpres, Joint Comt Eng in Med & Biol, 66, treas, 67-68, gen chmn ann conf, 70. Mem: AAAS; Asn Advan Med Instrumentation; Inst Elec & Electronics Engrs; Biomed Eng Soc. Res: Instrument and methods development for biochemical analysis; microchemical analysis instrumentation; cardiac anaphylaxis; microimmunochemistry; artificial circulatory support devices. Mailing Add: Lab of Tech Develop Nat Heart Inst Bethesda MD 20014

YURKIEWICZ, WILLIAM J, b Bloomsburg, Pa, Sept 21, 39; m 65; c 2. INSECT PHYSIOLOGY, BIOCHEMISTRY. Educ: Bloomsburg State Col, BS, 60; Bucknell Univ, MS, 62; Pa State Univ, PhD(entom), 65. Prof Exp: Asst entom, Pa State Univ, 63-65; res entomologist, USDA, Ga, 65-66; PROF BIOL, MILLERSVILLE STATE COL, 66- Mem: Entom Soc Am. Res: Insect flight physiology; neutral lipid and phospholipid composition and metabolism in insects; neural control of insect flight. Mailing Add: Dept of Biol Millersville State Col Millersville PA 17551

YURKOWSKI, MICHAEL, b Sask, Can, Sept 1, 28; m 57; c 3. BIOCHEMISTRY, NUTRITION. Educ: Univ Sask, BSA, 51, MSc, 59; Univ Guelph, PhD(nutrit), 68. Prof Exp: Anal chemist, Western Potash Corp Ltd, Sask, 51-52; indust chemist, Cereal & Oilseed Processing, Sask Wheat Pool, Can, 52-56; anal chemist, Plant Prod Div, Can Dept Agr, Ont, 59-60; food & drug directorate, Can Dept Nat Health & Welfare, Man, 60-62; res scientist marine lipid biochem, Halifax Lab, 62-65, RES SCIENTIST FISHERIES RES BD, CAN DEPT ENVIRON, NUTRIT BIOCHEM, FRESHWATER INST, FISHERIES SERV, 68- Mem: Chem Inst Can; Can Inst Food Sci & Technol. Res: Applied and basic nutrition of fish and freshwater organisms; lipid biochemistry of freshwater organisms; odours in freshwater and freshwater fish. Mailing Add: Freshwater Inst Fisheries Serv Can Dept Environ 501 Univ Crescent Winnipeg MB Can

YURKSTAS, A ALBERT, b Lawrence, Mass, Nov 25, 21; m 49; c 3. DENTISTRY. Educ: Univ Mass, BS, 47; Tufts Univ, DMD, 49, MS, 56. Prof Exp: Res assoc oral physiol, 49-54, from asst prof to assoc prof prosthetics, 54-60, PROF PROSTHETICS & CHMN DEPT COMPLETE DENTURE PROSTHESIS, SCH DENT MED, TUFTS UNIV, 60-, DIR CLINS, 65- Concurrent Pos: Consult, USPHS Hosp, Brighton, Mass & Vet Admin Hosps, Brockton & West Roxbury, Mass & Manchester,

NH, 60- Mem: Am Col Dent; Am Acad Oral Med; Int Asn Dent Res. Res: Denture prosthetics; oral physiology. Mailing Add: Tufts Univ Sch of Dent Med 136 Harrison Ave Boston MA 02111

YUROW, HARVEY WARREN, b New York, NY, Feb 14, 32; m 56; c 3. ANALYTICAL CHEMISTRY. Educ: Queens Col (NY), BS, 54; Pa State Univ, PhD(anal chem), 60. Prof Exp: Dept Defense fel, Rutgers Univ, 59-60; RES CHEMIST, EDGEWOOD ARSENAL, 60- Mem: Am Chem Soc; Sigma Xi. Res: Trace analysis of organic compounds via chromogen formation; structure-activity relationships for physiologically active compounds; organic analysis via chemiluminescence. Mailing Add: 3801 Maryland Ave Abingdon MD 21009

YUSHOK, WASLEY DONALD, b Woodbine, NJ, Mar 11, 20; m 50. BIOCHEMISTRY, CANCER. Educ: Rutgers Univ, BS, 41, MS, 43; Cornell Univ, PhD(chem embryol), 50. Prof Exp: Asst, Rutgers Univ, 41-43 & Cornell Univ, 46-49; asst to ed handbk biol data, Nat Res Coun, 50; res assoc, Univ Tex Med Br, 50-52; res biochemist, Biochem Res Found, 52-59, head div cancer biochem, 59-66; ASSOC MEM DIV BIOCHEM, INST FOR CANCER RES, 66- Mem: AAAS; Am Chem Soc; Am Asn Cancer Res; NY Acad Sci. Res: Regulation of metabolism and enzymes in cancer cells; nucleotide, protein and carbohydrate metabolism. Mailing Add: Inst for Cancer Res 7701 Burholme Ave Philadelphia PA 19111

YUSKA, HENRY, b Brooklyn, NY, Nov 7, 14; m 44; c 2. ORGANIC CHEMISTRY. Educ: City Col New York, BS, 35; Polytech Inst Brooklyn, MS, 39; Univ Ill, PhD(org chem), 42. Prof Exp: Res chemist, Jewish Hosp, Brooklyn, NY, 35-39; asst chem, Univ Ill, 41; res org chemist, Barrett Div, Allied Chem & Dye Corp, 42-43; resin group leader, Interchem Corp, 43-60, dir dept org chem, Cent Res Labs, 60-63; tech dir, Sun Chem Corp, 63-66; PROF CHEM, BROOKLYN COL, 66- Concurrent Pos: Instr, Eve Div, Hunter Col, 43-46. Mem: Am Chem Soc. Res: Synthetic resins; organic synthesis of monomers; biochemistry of lead poisoning; analytical methods for blood chemistry. Mailing Add: Dept of Chem Brooklyn Col Bedford Ave & Ave H Brooklyn NY 11210

YUSPA, STUART HOWARD, b Baltimore, Md, July 19, 41; m 65; c 2. CANCER. Educ: Johns Hopkins Univ, BS, 62; Univ Md, MD, 66; Am Bd Internal Med, dipl, 72. Prof Exp: Intern internal med, Hosp Univ Pa, 66-67; res assoc cancer, Nat Cancer Inst, 67-70; resident internal med, Hosp Univ Pa, 70-72; SR INVESTR CANCER, NAT CANCER INST, 72- Concurrent Pos: Mem biol models segment, Carcinogenesis Prog, Nat Cancer Inst, 72- Mem: Am Asn Cancer Res; AAAS. Res: Determine mechanisms whereby chemicals initiate or promote malignant transformation of epithelial cells by using cell culture model systems. Mailing Add: Nat Cancer Inst In Vitro Pathogenesis Sect Exp Pathol Br Bethesda MD 20014

YUSTER, PHILIP HAROLD, b Fargo, NDak, Nov 7, 17; m 47; c 2. RADIATION PHYSICS. Educ: NDak Col, BS, 39; Wash Univ, PhD(phys chem), 49. Prof Exp: Jr chemist, Panama Canal, 42-43; asst, Manhattan Dist, Univ Chicago, 43-44; asst, Los Alamos Sci Lab, 44-45; asst, Wash Univ, 45-49; assoc chemist, 49-58, SR CHEMIST, ARGONNE NAT LAB, 58- Mem: Am Phys Soc. Res: Mass spectroscopy; photochemistry; luminescence; color centers. Mailing Add: S831 Washington St Downers Grove IL 60515

YU-SUN, CLARE CHUAN CHANG, b Nanking, China, Aug 16, 26; nat US; m 53; c 2. MYCOLOGY, GENETICS. Educ: Seton Hill Col, BA, 50; Columbia Univ, MA, 51, PhD(bot), 53. Prof Exp: From asst prof to assoc prof, 57-64, PROF BIOL, UNIV ALBUQUERQUE, 64- Concurrent Pos: NSF fac fel & res fel, Calif Inst Technol, 63-64; vis scientist, Cambridge Univ, Eng, 73-74. Mem: AAAS; Mycol Soc Am; Bot Soc Am; Genetics Soc Am. Res: Genetic studies of Ascobolus immersus. Mailing Add: Dept of Biol St Joseph Place NW Univ of Albuquerque Albuquerque NM 87120

YUWILER, ARTHUR, b Mansfield, Ohio, Apr 4, 27; m 50; c 3. BIOCHEMISTRY. Educ: Univ Calif, Los Angeles, BS, 50, PhD(biochem), 56. Prof Exp: Asst chem, Univ Calif, Los Angeles, 50-51, asst physiol chem, 52-54, asst chem, 54-56, res biochemist, 57-59; res neurobiochemist, Vet Admin, Calif, 56-57; res assoc & dir labs & biochem sect, Schizophrenia & Psychopharmacol Res Proj, Ypsilanti State Hosp & Univ Mich, 59-62; asst prof biochem, 64-70, ASSOC PROF PSYCHIAT, UNIV CALIF, LOS ANGELES, 70-; CHIEF NEUROBIOCHEM RES, VET ADMIN BRENTWOOD HOSP, 62- Concurrent Pos: Res biochemist, Ment Health Res Inst, Univ Mich, 59-62; mem, Brain Res Inst, Univ Calif, Los Angeles, 65-; mem career develop award comt, MINH, 70-76 & Calif State Ment Health Adv Comt, 70- Mem: Am Soc Neurochem; Int Soc Neurochem; Am Col Neuropsychopharmacol; Soc Biol Psychiat; Am Soc Biol Chemists. Res: Amino acids and proteins; enzymes; biochemistry of schizophrenia; intermediary metabolism of neurohumors; stress. Mailing Add: 20620 Clarendon Woodland Hills CA 91364

Z

ZABARA, JACOB, b Philadelphia, Pa, May 8, 32; m 70; c 1. PHYSIOLOGY, NEUROPHYSIOLOGY. Educ: Johns Hopkins Univ, BS, 53; Univ Pa, MS, 58, PhD(physiol), 59. Prof Exp: USPHS fel, Univ Pa, 59-60; instr pharmacol, Dartmouth Col, 60-61; instr pharmacol, Univ Pa, 61-63; USPHS spec fel biomath, 63-64; instr physiol, Univ Pa, 64-65, assoc pharmacol, 65-67, assoc physiol, 65-67; asst prof, 67-72, ASSOC PROF PHYSIOL, GRAD SCH, TEMPLE UNIV, 72- Mem: Soc Neurosci; Aerospace Med Asn; Undersea Med Soc; Am Asn Anat; Int Asn Cybernet. Res: Neurophysiology and cybernetics. Mailing Add: Dept of Physiol & Biophys Temple Univ Grad Sch Philadelphia PA 19140

ZABEL, CARROLL WAYNE, b Deer Creek, Minn, Oct 5, 20; m 43; c 3. PHYSICS. Educ: Lawrence Col, BA, 46; Mass Inst Technol, PhD(physics), 49. Prof Exp: Mem staff, Microwave Radar, Mass Inst Technol, 42-46; mem staff nuclear physics, Los Alamos Sci Lab, 49-65; assoc prof physics & assoc dean col arts & sci, 65-67, dir res & assoc dean grad sch, 67-71, PROF PHYSICS, UNIV HOUSTON, 70- Concurrent Pos: Secy nuclear cross sect adv group, AEC, 52-57, adv comt reactor safeguards, 65-69; mem, Atoms for Peace Mission, NZ, 58; corp vis comt, Dept Nuclear Eng, Mass Inst Technol, 63-66. Mem: AAAS: Am Phys Soc; Am Asn Physics Teachers; NY Acad Sci. Res: Nuclear engineering and physics. Mailing Add: Dept of Physics Univ of Houston 3801 Cullen Blvd Houston TX 77004

ZABEL, LOWELL WALLACE, b Valley City, NDak, Apr 11, 13; m 36; c 3. PHYSICS. Educ: Lawrence Col, AB, 35. Prof Exp: Physicist, Kimberly Clark Corp, 36-42, Wilcox Elec Co, 42-45 & Eastern Air Lines & TACA Airways, 45-46; physicist, Kimberly Clark Corp, 46-58, supt instrumentation lab, 58-62, tech supt, Coosa River Newsprint Div, 62-67; assoc prof, 67-68, L C CALDER PROF PULP & PAPER TECHNOL, UNIV MAINE, ORONO, 68- Mem: Tech Asn Pulp & Paper Indust. Res: Instrumentation; laboratory and process control. Mailing Add: Dept of Chem Eng Univ of Maine Orono ME 04473

ZABEL, NORMAN RALPH, physics, see 12th edition

ZABEL, ROBERT ALGER, b Boyceville, Wis, Mar 11, 17; m 44; c 5. FOREST PATHOLOGY. Educ: Univ Minn, BS, 38; State Univ NY, MS, 41, PhD(bot), 48. Prof Exp: Lab aide, Lake States Forest Exp Sta, US Forest Serv, 40; from asst prof to assoc prof forest path, 47-55, head dept bot & forest path, 55-65, assoc dean biol sci & undergrad instr, 65-66, PROF FOREST PATH, STATE UNIV NY COL ENVIRON SCI & FORESTRY, 55-, ASSOC DEAN BIOL SCI & INSTR, 66-, VPRES ACAD AFFAIRS, 70- Mem: AAAS; Soc Am Foresters; Forest Prod Res Soc; Am Phytopath Soc; Am Inst Biol Sci. Res: Forest products deterioration; wood decays; lumber stains; evaluation of preservatives; wood durability evaluations; root and heart rots; toxicants and slimicides. Mailing Add: State Univ NY Col Environ Sci Forestry Syracuse NY 13210

ZABETAKIS, MICHAEL GEORGE, b Francis, Pa, Mar 3, 24; m 46; c 2. PHYSICAL CHEMISTRY, SAFETY ENGINEERING. Educ: Washington & Jefferson Col, BS, 43, MA, 45; Univ Pittsburgh, PhD(chem), 56. Prof Exp: Asst chem, Washington & Jefferson Col, 42-43, instr physics, 44-45 & 46-48; instr physics, Univ Pittsburgh, 48-49; physicist, US Bur Mines, 49-55, chief br gas explosions, 55-65, res dir, Health & Safety Res & Testing Ctr, Pa, 65-67; prof math & chmn dept, Washington & Jefferson Col, 67-71; res supvr, US Bur Mines, 71-73; dep asst adminr, Educ & Training, 73-75, SUPT, NAT MINE HEALTH & SAFETY ACAD, MINING ENFORCEMENT & SAFETY ADMIN, 75- Honors & Awards: Distinguished Serv Award, US Dept Interior, 74. Mem: Am Soc Safety Engrs; Sigma Xi; Combustion Inst; Am Chem Soc; Am Phys Soc. Res: Academic administration; flammability characteristics of gases and vapors; combustion; ignition; flame propagation; cryogenics; instrumentation. Mailing Add: 207 Queen St Beckley WV 25801

ZABIK, MATTHEW JOHN, b South Bend, Ind, Aug 22, 37; m 58; c 1. CHEMISTRY, TOXICOLOGY. Educ: Purdue Univ, Lafayette, BS, 59; Mich State Univ, MS, 62, PhD(org chem), 65. Prof Exp: Asst prof, 65-69, ASSOC PROF PESTICIDE RES & ENTOM, MICH STATE UNIV, 69- Mem: Am Chem Soc. Res: Photochemistry of pesticides and toxicology of pesticides. Mailing Add: Dept of Entom Mich State Univ East Lansing MI 48893

ZABIN, BURTON ALLEN, b Chicago, Ill, Mar 18, 36; m 72. INORGANIC CHEMISTRY, ANALYTICAL CHEMISTRY. Educ: Univ Ill, BS, 57; Purdue Univ, PhD(inorg chem), 62. Prof Exp: Res assoc chem, Stanford Univ, 62-63; dir res, 63-72, DIV MGR CHEM, BIO-RAD LABS, 72- Mem: Am Chem Soc. Res: Separations chemistry, including ion exchange resins, gel filtration materials and other column chromatographic materials. Mailing Add: Bio-Rad Labs 32nd & Griffin Richmond CA 94804

ZABIN, IRVING, b Chicago, Ill, Nov 13, 19; m 42; c 3. BIOCHEMISTRY, MOLECULAR BIOLOGY. Educ: Univ Ill, BS, 40; Univ Chicago, PhD(biochem), 49. Prof Exp: Res assoc biochem, Univ Chicago, 49-50; res assoc biol chem, 50-51, from lectr to assoc prof, 51-64, PROF BIOL CHEM, SCH MED, UNIV CALIF, LOS ANGELES, 64- Concurrent Pos: Nat Multiple Sclerosis Soc scholar, 59-60; Guggenheim fel, 67-68; NATO sr fel sci, 75; vis prof, Pasteur Inst, Paris, 59-60 & 67-68 & Imp Col London, 75. Mem: AAAS; Am Soc Biol Chem; Am Soc Microbiol. Res: Protein structure, synthesis and control. Mailing Add: Dept of Biol Chem Univ of Calif Sch of Med Los Angeles CA 90024

ZABINSKI, ROSE MARIE C, b Chicago, Ill, Mar 2, 45. CLINICAL CHEMISTRY, TOXICOLOGY. Educ: Loyola Univ Chicago, BS, 68; Univ Ill, Urbana, MS, 70, PhD(biochem), 72. Prof Exp: Fel biochem, Med Sch, Northwestern Univ, 72-74; CLIN CHEMIST, MASON-BARRON LAB, INC, DAMON CORP, 74- Mem: Am Chem Soc; AAAS; Sigma Xi. Res: Development of new test procedures in clinical chemistry and toxicology. Mailing Add: Mason-Barron Lab Inc 4720 W Montrose Ave Chicago IL 60641

ZABKA, GEORGE GENE, botany, see 12th edition

ZABLOCKA-ESPLIN, BARBARA, b Warsaw, Poland, Jan 5, 25; m 64; c 4. NEUROPHARMACOLOGY, NEUROPHYSIOLOGY. Educ: Med Acad, Warsaw, dipl, 52, MD, 61. Prof Exp: Asst prof pharmacol, Med Acad, Warsaw, 52-55, sr asst prof, 55-61; Riker Int fel, Col Med, Univ Utah, 61-62, res assoc, 62-65, asst res prof, 65-68; ASST PROF PHARMACOL & THERAPEUT, McGILL UNIV, 68- Mem: Am Soc Pharmacol & Exp Therapeut; Can Pharmaceut Asn. Res: Central excitatory and depressant drugs; central transmitter substances. Mailing Add: 3 Parkside Pl Montreal PQ Can

ZABLOW, LEONARD, b New York, NY, Sept 3, 27; m 50; c 2. BIOPHYSICS, NEUROPHYSIOLOGY. Educ: Calif Inst Technol, BSc, 48; Columbia Univ, MA, 50. Prof Exp: Res worker biochem, Worcester Found Exp Biol, 51-52; res asst neurol, 52-60, res assoc neurol, 60-73, SR STAFF ASSOC NEUROL, COL PHYSICIANS & SURGEONS, COLUMBIA UNIV, 73- Concurrent Pos: Lectr, Polytech Inst Brooklyn, 59-66. Mem: Fel AAAS; Biophys Soc. Res: Implications of scalp electroencephalograms for cortical source localization; spatial electroencephalogram analysis; focal generator size in clinical and experimental epilepsy; electromyography. Mailing Add: Dept of Neurol Columbia Univ Col of Physicians & Surgeons New York NY 10032

ZABOLOTNY, ERNEST, physical chemistry, see 12th edition

ZABOR, JOHN WILLIAM, b Cleveland, Ohio, Jan 27, 15; m 40; c 3. PHYSICAL CHEMISTRY. Educ: Hiram Col, AB, 36; Brown Univ, MS, 38; Univ Rochester, PhD(phys chem), 40. Prof Exp: Instr chem, Williams Col, 40-42; chem res assoc, Cent Lab, Nat Defense Res Comt, Northwestern Univ, 42-45; dir res activated carbon div, Pittsburgh Coke & Chem Co, 45-48, dir res & develop, 48-54, asst to vpres chem, 54-56; dir res, Res & Eng Div, Wyandotte Chem Corp, 56-58, dir res div, 58-63, vpres res & develop, 63-65; vpres res & develop, US Gypsum Co, 65-74; EXEC VPRES, APOLLO CHEM CORP, 74- Concurrent Pos: Trustee, Hiram Col; past mem bd dirs, Indust Res Inst; mem subcomt adv coun & consult, Chem Corps, US Army, 51-52. Honors & Awards: Presidential Cert Merit. Mem: AAAS; Am Chem Soc; Am Inst Chem; Asn Res Dirs; Newcomen Soc. Res: Activated carbon; industrial chemicals and polymers; building materials. Mailing Add: Apollo Chem Corp 35 South Jefferson Rd Whippany NJ 07981

ZABOROWSKI, LEON MICHAEL, b Chicago, Ill, Nov 1, 42; m 64; c 2. ENVIRONMENTAL CHEMISTRY. Educ: Univ Wis-River Falls, BS, 66; Univ Idaho, PhD(chem), 70. Prof Exp: ASST PROF CHEM, UNIV WIS-RIVER FALLS, 70-, DIR CONTINUING EDUC & EXTEN, 74- Mem: Am Chem Soc. Res: Improvement in chemical education; chemical problems and the average citizen; fluorine chemistry. Mailing Add: Dept of Chem Univ of Wis River Falls WI 54022

ZABORSKI, BOGDAN, b Warsaw, Poland, Apr 5, 01; nat Can; m 28; c 2. GEOGRAPHY. Educ: Warsaw Tech Univ, PhD, 25; Cracow Tech Univ, Dozent, 30.

Prof Exp: Mem staff, Dept Geog, Warsaw Tech Univ, 24-27, prof geog, 38-39, actg prof, Cracow Tech Univ, 30-33, Polish Free Univ, 34-37; head geog sect & cartog sect & cartog printing off, Exiled Polish Govt, Eng, 42-45, head sect Polish univ students, Comt Educ Poles in Great Brit & prof geog, Polish Univ Col, Eng, 46-48; assoc prof, McGill Univ, 48-57; prof geog & head inst geog, Univ Ottawa, Ont, 57-66; PROF GEOG, SIR GEORGE WILLIAMS UNIV, 66- Concurrent Pos: Geogr, Planning Off, Union Polish Mt Lands, 34-39; cartographer & head cartog off, Polish Inst Study Nat Probs, 34-39. Res. Geography of the Soviet Union; urban and rural settlements in Canada; cultural pattern of population of Eurasia. Mailing Add: Dept of Geog George Williams Campus of Concordia Univ Montreal PQ Can

ZABORSKY, OSKAR RUDOLF, b Neuwalddorf, Czech, Oct 6, 41; US citizen; m 68; c 1. BIOLOGICAL CHEMISTRY. Educ: Philadelphia Col Pharm & Sci, BSc, 64; Univ Chicago, PhD(chem), 68. Prof Exp: NIH fel, Harvard Univ, 68-69; sr res chemist, Corp Res Labs, Esso Res & Eng Co, Linden, NJ, 69-74; PROG MGR ENZYME TECHNOL, RENEWABLE RESOURCES PROG, NSF, 74- Mem: AAAS; Am Chem Soc. Res: Chemical modification of enzymes; immobilized enzymes; enzyme technology. Mailing Add: Renewable Resources Prog Nat Sci Found Washington DC 20550

ZABRANSKY, RONALD JOSEPH, b Little Ferry, NJ, Mar 18, 35; m 58; c 3. CLINICAL MICROBIOLOGY, CLINICAL BACTERIOLOGY. Educ: Rutgers Univ, BS, 56; Ohio State Univ, MS, 61, PhD(microbiol), 63; Am Bd Med Microbiol, dipl, 69. Prof Exp: Microbiologist, Battelle Mem Inst, 59-60; teaching asst med microbiol bact, Ohio State Univ, 60-61, res asst, 61-63; assoc consult microbiol, Mayo Clin, 63-64, consult, 64-69; DIR DIV MICROBIOL, MT SINAI MED CTR, MILWAUKEE, 69- Concurrent Pos: Asst clin prof, Dept Microbiol, Med Col Wis, 70-74; assoc adj prof microbiol, Med Col Wis, 75- Mem: AAAS; fel Am Acad Microbiol; Am Pub Health Asn; Am Soc Microbiol; Am Soc Clin Path. Res: Pyocine and serologic typing of Pseudomonas aeruglinosa; isolation and antibiotic testing of anaerobic bacteria. Mailing Add: Div Microbiol Mt Sinai Med Ctr 950 N 12th St Milwaukee WI 53201

ZABRISKIE, FRANKLIN ROBERT, b New York, NY, Dec 21, 33. ASTRONOMY. Educ: Princeton Univ, BSE, 55, MSE, 57, PhD(astron), 60. Prof Exp: Asst prof astron, Wesleyan Univ, 60-66; ASSOC PROF ASTRON, PA STATE UNIV, UNIVERSITY PARK, 66- Mem: Am Astron Soc; Int Astron Union; Am Geophys Union. Res: Design of instruments for optical telescopes; photoelectric stellar classification; studies of long period variable stars. Mailing Add: Davey Lab Dept of Astron Pa State Univ University Park PA 16802

ZABRISKIE, JOHN LANSING, JR, b Auburn, NY, June 8, 39; m 63; c 2. ORGANIC CHEMISTRY. Educ: Dartmouth Col, BA, 61; Univ Rochester, PhD(chem), 66. Prof Exp: Sr chemist process res, Merck & Co, Inc, NJ, 65-68, tech asst to exec dir, animal sci res, 69-70, sr mgr pharmacol qual control, Merck Sharp & Dohme, 70-72, mgr pharmaceut mfg, 72-74, DIR PHARMACEUT MFG, MERCK SHARP & DOHME, 74- Mem: Pharmaceut Mfrs Asn. Res: Organic synthesis and reaction mechanisms. Mailing Add: Merck Sharp & Dohme West Point PA 19486

ZABUSKY, NORMAN J, b New York, NY, Jan 4, 29; m 54; c 3. APPLIED MATHEMATICS, COMPUTER SCIENCE. Educ: City Col New York, BEE, 51; Mass Inst Technol, MS, 53; Calif Inst Technol, PhD(physics), 59. Prof Exp: Res asst, Servo-Mech Lab, Mass Inst Technol, 51-53; mem tech staff guided missile simulation & control, Raytheon Mfg Co, 53-59; NSF fel, Plasma Physics, Max Planck Inst Physics & Astrophys, 59-60; vis res assoc, Plasma Physics Lab, Princeton Univ, 60-61; mem tech staff, Bell Tel Labs, Inc, 61-63, supvr plasma & computational physics, 63-68, head computational physics res dept, 68-76; PROF MATH & ENG, UNIV PITTSBURGH, 76- Concurrent Pos: Dir, Int Sch Nonlinear Math & Physics, Ger, NATO, 66; mem comt support res in math sci, Nat Acad Sci, 66-; J S Guggenheim fel, Oxford Univ, Math Inst & Weizmann Inst Sci, 71-72. Mem: AAAS; fel Am Phys Soc; fel NY Acad Sci. Res: Computational mathematics approach for nonlinear physical and chemical systems. Mailing Add: Dept of Math Univ of Pittsburgh Pittsburgh PA 15260

ZACCHEI, ANTHONY GABRIEL, b Philadelphia, Pa, Mar 31, 40; m 63; c 3. DRUG METABOLISM, MASS SPECTROMETRY. Educ: Villanova Univ, BS, 62, MS, 65; Univ Minn, Minneapolis, PhD(biochem), 68. Prof Exp: Res assoc pharmacol, 61-64, sr res pharmacologist, 68-70, res fellow, 70-73, SR RES FELLOW DRUG METABOLISM, MERCK INST THERAPEUT RES, 73- Concurrent Pos: Vis asst prof, Inst Lipid Res, Baylor Col Med, 72. Mem: AAAS; Am Chem Soc; Am Soc Mass Spectrometry. Res: Investigations into the physiological disposition of new drug products including absorption, excretion and metabolic studies. Mailing Add: Drug Metab Dept 805 Merck Inst for Therapeut Res West Point PA 19486

ZACHARIAS, DAVID EDWARD, b Philadelphia, Pa, May 16, 26; m 68; c 3. X-RAY CRYSTALLOGRAPHY, ORGANIC CHEMISTRY. Educ: Temple Univ, AB, 53, AM, 54; Univ Pittsburgh, PhD(x-ray crystallog), 69. Prof Exp: From jr chemist to sr chemist, Smith Kline & French Labs, Pa, 54-71; RES ASSOC, MOLECULAR STRUCT LAB, INST FOR CANCER RES, 71- Mem: AAAS; Am Crystallog Asn; Am Chem Soc; The Chem Soc; Am Inst Chem. Res: Synthesis of heterocyclic compounds; single crystal x-ray structure determination of organic and biologically important compounds; x-ray powder diffraction. Mailing Add: Molecular Struct Lab Inst for Cancer Res Philadelphia PA 19111

ZACHARIAS, JERROLD REINACH, b Jacksonville, Fla, Jan 23 05; m 27; c 2. PHYSICS, SCIENCE EDUCATION. Educ: Columbia Univ, AB, 26, AM, 27, PhD(physics), 32. Prof Exp: Tutor physics, City Col New York, 29-30; instr, Hunter Col, 31-36, asst prof, 36-40; staff mem, Radiation Labs, Mass Inst Technol, 41-45; head eng div, Los Alamos Sci Lab, 45; prof physics, 46-66, dir lab nuclear sci, 46-56, inst prof, 66-70, dir educ res ctr, 68-72, EMER INST PROF & EMER PROF PHYSICS, MASS INST TECHNOL, 70- Concurrent Pos: Dir, Sprague Elec Co; trustee, Educ Develop Ctr. Mem: Nat Acad Sci; fel Am Phys Soc; fel Am Acad Arts & Sci. Res: Electric and magnetic shapes of atomic nuclei; atomic clocks; military technology; education reform. Mailing Add: 32 Clifton St Belmont MA 02178

ZACHARIAS, LEONA RUTH, b New York, NY, Feb 6, 07; m 27; c 2. HUMAN DEVELOPMENT. Educ: Columbia Univ, AB, 27, MA, 28, PhD(anat, embryol), 37. Prof Exp: Asst, Am Mus Natural Hist, 28-30; asst neurol, Col Physicians & Surgeons, Columbia Univ, 29-32, instr anat, physiol & embryol, 36-45, instr, dept optom, Univ, 44-46; instr biol & vert embryol, Hunter Col, 32-33; lectr opthal res, 46-65, res assoc obstet & gynec, 65-69, PRIN ASSOC OBSTET & GYNEC, HARVARD MED SCH, 69- Concurrent Pos: Biologist, Mass Eye & Ear Infirmary, 46-65; assoc biologist, Vincent Mem Hosp, Mass Gen Hosp, 67-; res assoc, Mass Inst Technol. Mem: Am Asn Anat; Endocrine Soc. Res: Experimental embryology; pituitary innervation; blindness in premature infants; growth and sexual development of adolescent girls. Mailing Add: Dept of Gynec Mass Gen Hosp Fruit St Boston MA 02114

ZACHARIASEN, FREDRIK, b Chicago, Ill, June 14, 31; m 57. THEORETICAL PHYSICS. Educ: Univ Chicago, PhB, 50, BS, 51; Calif Inst Technol, PhD(physics), 56. Prof Exp: Instr physics, Mass Inst Technol, 55-56, jr res physicist, Univ Calif, 56-57; res assoc physics, Stanford Univ, 57-58, asst prof, 58-60; from asst prof to assoc prof, 60-65, PROF PHYSICS, CALIF INST TECHNOL, 65- Concurrent Pos: Consult, Rand Corp, 56-; Sloan Found fel, 60-64; consult, Los Alamos Sci Lab, 61-; Inst Defense Anal, 61-; Guggenheim Found fel, 70-71. Res: High energy particle physics. Mailing Add: Dept of Physics Calif Inst of Technol Pasadena CA 91109

ZACHARIASEN, FREDRIK WILLIAM HOULDER, b Langesund, Norway, Feb 5, 06; nat US; m 30; c 2. PHYSICS. Educ: Univ Oslo, MS, 26, PhD(physics), 28. Prof Exp: Asst prof physics, Univ Oslo, 28-30; from asst prof to prof, 30-62, Ernest DeWitt Burton Distinguished Serv prof, 62-74, chmn dept, 45-49, 56-59, dean div phys sci, 59-62, ERNEST DeWITT BURTON DISTINGUISHED SERV PROF, UNIV CHICAGO, 74- Concurrent Pos: Rockefeller Found fel, Univ Manchester, 28-29; Guggenheim fel, 35; sr physicist, Manhattan Proj, 43-45. Mem: Nat Acad Sci; fel Am Phys Soc; Mineral Soc Am; Am Crystallog Asn; for mem Norweg Acad Sci & Letters. Res: X-rays and crystal structure. Mailing Add: Dept of Physics Univ of Chicago Chicago IL 60637

ZACHARIUS, ROBERT MARVIN, b New York, NY, Mar 21, 20; m 44; c 4. PLANT BIOCHEMISTRY. Educ: NY Univ, BA, 43; Univ Colo, MA, 48; Univ Rochester, PhD(plant physiol), 53. Prof Exp: Asst chem, Univ Colo, 47-48; asst bot, Cornell Univ, 51-52; res chemist, Gen Cigar Co, Inc, Pa, 52-54; biochemist, Eastern Utilization Res Br, 54-57, biochemist, Eastern Utilization Res & Develop Div Pa, 57-71, biochemist, Plant Prod Lab, Eastern Regional Res Ctr, 71-74, RES CHEMIST, EASTERN REGIONAL RES CTR, AGR RES SERV, USDA, 74- Concurrent Pos: Lectr, St Joseph's Col, Pa. Mem: AAAS; Biochem Soc; Am Chem Soc; Am Soc Plant Physiol. Res: Non-protein nitrogen compounds of plants; plant proteins; nicotiana alkaloids; ion-exchange and chromatographic methods; electrophoresis; antimetabolites; plant metabolism; proteolytic inhibitors; plant-parasite interactions, stress physiology. Mailing Add: USDA Agr Res Serv Eastern Regional Res Ctr Philadelphia PA 19118

ZACHARUK, R Y, b Yorkton, Sask, May 1, 28; m 52; c 2. INSECT MORPHOLOGY, INSECT PATHOLOGY. Educ: Univ Sask, BSA, 50, MSc, 55; Univ Glasgow, PhD(histochem, physiol), 61. Prof Exp: Res officer entom, Res Sta, Can Dept Agr, 50-63; assoc prof biol, 63-65, chmn dept, 65-67, PROF BIOL, UNIV REGINA, 67- Concurrent Pos: Agr Inst Can fel, 59-61. Mem: AAAS; Soc Invert Path; Electron Micros Soc Am; Can Soc Entom; Can Soc Zool. Res: Sense organ ultrastructure; neurophysiology; entomophagous fungi; histochemistry. Mailing Add: Dept of Biol Univ of Regina Regina SK Can

ZACHER, ALBERT RICHARD, b Woodland, Calif, Apr 21,. 40; m 68. HIGH ENERGY PHYSICS. Educ: Calif Inst Technol, BS, 62; Princeton Univ, PhD(physics), 68. Prof Exp: Res assoc comput sci, Washington Univ, 68-69, physics, 69, asst prof, 69-75; MEM STAFF, ASTRONIX, INC, 75- Mem: AAAS; Am Phys Soc. Res: Experimental physics of elementary particles; computer image processing for physics and biomedical applications. Mailing Add: Astronix Inc 1314 Hanley Industrial Ct St Louis MO 63144

ZACHMANOGLOU, ELEFTHERIOS CHARALAMBOS, b Thessaloniki, Greece, Mar 19, 34; US citizen. APPLIED MATHEMATICS. Educ: Rensselaer Polytech Inst, BAeroE, 56, MS, 57; Univ Calif, Berkeley, PhD(appl math), 62. Prof Exp: From asst prof to assoc prof, 62-70, PROF MATH, PURDUE UNIV, LAFAYETTE, 70- Concurrent Pos: Fulbright res grant, Univ Rome, 65-66. Mem: Am Math Soc. Res: Partial differential equations; wave propagation. Mailing Add: Dept of Math Purdue Univ Lafayette IN 47907

ZACK, NEIL RICHARD, b Canton, Ohio, Apr 26, 47. INORGANIC CHEMISTRY, FLUORINE CHEMISTRY. Educ: Rensselaer Polytech Inst, BS, 69; Marshall Univ, MS, 70; Univ Idaho, PhD(chem), 74. Prof Exp: Fel chem, Utah State Univ, 74-75; fel, Univ Idaho, 75-76; SR CHEMIST, ALLIED CHEM CORP, 76- Mem: Am Chem Soc; Sigma Xi. Res: Inorganic heterocyclics; fluorine containing derivatives of catenated sulfur compounds. Mailing Add: 828 Crestmont Idaho Falls ID 83401

ZACKS, SHELEMYAHU, b Tel Aviv, Israel, Oct 15, 32; m 55; c 2. MATHEMATICAL STATISTICS, STATISTICAL ANALYSIS. Educ: Hebrew Univ, Israel, BA, 55; Israel Inst Technol, MSc, 60; Columbia Univ, PhD(indust eng), 62. Prof Exp: Sr lectr statist, Israel Inst Technol, 63-65; prof, Kans State Univ, 65-68; prof, Univ NMex, 68-70; PROF MATH & STATIST, CASE WESTERN RESERVE UNIV, 70-, CHMN DEPT, 74- Concurrent Pos: Fel statist, Stanford Univ, 62-63; consult, Inst Mgt Sci & Eng, George Washington Univ, 67- Mem: Fel Am Statist Asn; fel Inst Math Statist. Res: Optimal design of sequential experiments; statistical adaptive processes; analysis of contingency tables; manpower forecasting for large military organizations; stochastic control. Mailing Add: Dept of Math & Statist Case Western Reserve Univ Cleveland OH 44106

ZACKS, SUMNER IRWIN, b Boston, Mass, June 29, 29; m 53; c 3. PATHOLOGY. Educ: Harvard Univ, BA, 51, MD, 55; Am Bd Path, dipl, 61. Hon Degrees: MA, Univ Pa. Prof Exp: From intern to asst resident path, Mass Gen Hosp, Boston, 55-58; from asst prof to assoc prof, 62-71, PROF PATH, SCH MED, UNIV PA, 71-; NEUROPATHOLOGist, PA HOSP, 61-, ASSOC DIR, AYER LAB, 64- Concurrent Pos: Asst path, Pa Hosp, 61-64. Honors & Awards: Hektoen Bronze Medal, AMA, 61. Mem: Histochem Soc (secy, 65-69); Am Soc Exp Path; Am Soc Cell Biol; Am Asn Neuropath; fel Col Am Path. Res: Fine structure of neuromuscular junctions, normal and in disease; fine structure pathology of muscle; molecular pathology of endotoxins. Mailing Add: Ayer Lab Pa Hosp Eighth & Spruce Sts Philadelphia PA 19107

ZACZEK, NORBERT MARION, b Baltimore, Md, Aug 15, 36. ORGANIC CHEMISTRY. Educ: Loyola Col, Md, BS, 58; Carnegie-Mellon Univ, PhD(org chem), 62. Prof Exp: From instr to assoc prof, 62-71, PROF CHEM, LOYOLA COL (MD), 71- Mem: AAAS; Am Chem Soc; The Chem Soc. Res: Metal hydride reduction of unsaturated ketoamides. Mailing Add: Dept of Chem Loyola Col Baltimore MD 21210

ZADEH, NORMAN, b New York, NY, Apr 17, 50. OPERATIONS RESEARCH. Educ: Univ Calif, Berkeley, BA, 70, PhD(oper res), 72. Prof Exp: Res staff mem comput sci, Thomas J Watson Res Ctr, IBM Corp, 72-73; ASST PROF OPERS RES, COLUMBIA UNIV, 75- Res: Combinatorial optimization; game theory; international financing. Mailing Add: 310 Seeley W Mudd Bldg Columbia Univ New York NY 10027

ZADNIK, VALENTINE EDWARD, b Cleveland, Ohio, Feb 13, 34; m 58; c 4. GEOLOGY, ENGINEERING GEOLOGY. Educ: Western Reserve Univ, BA, 57; Univ Ill, MS, 58, PhD(geol, civil eng), 60. Prof Exp: Res geologist, Jersey Prod Res Co, 64-65; sr res geologist, Esso Prod Res Co, 65-66; res geologist, US Army Res Off, 66-74; STAFF GEOLOGIST, OFF ENERGY RESOURCES, US GEOL SURV, 74- Concurrent Pos: Fel int affairs, Princeton Univ, 70-71; mem comt rock mech & comt seismol, Nat Acad Sci. Mem: Asn Eng Geol. Res: Petroleum exploration; carbonate rock petrography; solid-earth geophysics, particularly seismology, gravity, geomagnetism and geodesy. Mailing Add: 105 S Park Dr Arlington VA 22204

ZADO, FRANJO M, b Velika, Yugoslavia, Mar 28, 34; m 57; c 1. ANALYTICAL CHEMISTRY. Educ: Univ Zagreb, BS, 57, PhD(inorg anal), 59. Prof Exp: Res chemist inorg chem, Rudjer Boskovic Inst, Zagreb, Yugoslavia, 59-64; Int Atomic Energy Agency fel, Univ Ill, 64-65, NSF res assoc, 65-66; res chemist anal chem, Rudjer Boskovic Inst, Zagreb, 67-69; MEM RES STAFF, WESTERN ELEC RES CTR, 70- Mem: Am Chem Soc. Res: Trace gas analysis; monitoring systems development; chemical characterization of soldering fluxes. Mailing Add: Western Elec Res Ctr PO Box 900 Princeton NJ 08540

ZADOFF, LEON NATHAN, b Passaic, NJ, Aug 6, 23; m 44; c 2. PHYSICS. Educ: Cooper Union, BChE, 48; NY Univ, PhD(physics), 58. Prof Exp: Chem engr, Flintkote Co, 48; petrol chemist, Paragon Oil Co, 48-49; protein chemist, Botany Mills, Inc, 49; jr chem engr, City Fire Dept, New York, 49-52; physicist, NY Naval Shipyard, 52-54; proj supvr reactor physics, Ford Instrument Co, 54-58; assoc scientist, Repub Aviation Corp, 58-64, specialist physicist, Repub Aviation Div, Fairchild-Hiller Corp, NY, 64-72; CHIEF SCIENTIST, EMS DEVELOP CORP, FARMINGDALE, 72- Concurrent Pos: Lectr, Adelphi Col, 58-; adj assoc prof, NY Inst Technol, 72- Mem: AAAS; Am Phys Soc; Am Inst Aeronaut & Astronaut; NY Acad Sci. Res: Theoretical, plasma and reactor physics; electromagnetic wave propagation; quantum mechanics. Mailing Add: 17 Spruce St Merrick NY 11566

ZADROZNY, MITCHELL G, b Chicago, Ill, Dec 23, 23. GEOGRAPHY. Educ: Ill State Univ, BS, 47; Univ Chicago, SM, 49, PhD(geog), 56. Prof Exp: Geog analyst, Geog Br, Gen Hq, US Army, Tokyo, Japan, 50-52; lectr geog, Univ Chicago, 52-55; PROF GEOG, WRIGHT CITY COL, 55- Concurrent Pos: Dir res, Cambodia-Laos Proj, Univ Chicago Subcontract, 54-55; bd mem, President's Water Pollution Control Adv Bd, 72-75. Mem: Asn Am Geog. Res: Water pollution; ocean dumping. Mailing Add: 4158 N McVicker Ave Chicago IL 60634

ZADUNAISKY, JOSE ATILIO, b Rosario, Arg, July 15, 32; m 54; c 2. PHYSIOLOGY, BIOPHYSICS. Educ: Univ Buenos Aires, MD, 56. Prof Exp: Instr, Inst Physiol, Med Sch, Univ Buenos Aires, 52-56; Arg Nat Res Coun estab investr, Dept Biophys, Med Sch, Univ Buenos Aires, 60-63; assoc prof physiol & dir res, Dept Opthal, Sch Med, Univ Louisville, 64-67; assoc prof ophthal & physiol, Sch Med, Yale Univ, 67-73; PROF PHYSIOL & EXP OPHTHAL, MED SCH, NY UNIV, 73- Concurrent Pos: Arg Res Coun fel biochem, Univ Col, Dublin, 58-59 & Inst Biol Chem, Copenhagen, 59-60; USPHS grants, 62- Mem: Am Physiol Soc; Biophys Soc; Asn Res Vision & Ophthal; Soc Gen Physiol; Arg Soc Biophys. Res: Transport and permeability of biological membranes, especially epithelial tissues. Mailing Add: Dept of Physiol NY Univ Med Sch New York, NY 10016

ZAEHRINGER, MARY VERONICA, b Philadelphia, Pa, May 27, 11. FOODS. Educ: Temple Univ, BS, 46; Cornell Univ, MS, 48, PhD(foods), 53. Prof Exp: Asst food res, Cornell Univ, 46-48, 51-53; from instr to asst prof home econ, Mont State Col, 48-50; res prof home econ, 53-72, res prof food sci, 72-73, RES PROF BACT & BIOCHEM, AGR EXP STA, UNIV IDAHO, 73- Concurrent Pos: Res fel, Inst Storage & Processing Agr Produce, State Agr Univ, Wageningen, 67-68. Mem: Am Asn Cereal Chem; Am Chem Soc; Inst Food Technol; Potato Asn Am; Am Soc Hort Sci. Res: Potato texture; quality of food products. Mailing Add: 614 Ash St Moscow ID 83843

ZAERR, JOE BENJAMIN, b Los Angeles, Calif, Sept 9, 32; m 54; c 3. FOREST PHYSIOLOGY. Educ: Univ Calif, Berkeley, BS, 54, PhD(plant physiol), 64. Prof Exp: Res assoc, Crops Res Div, Agr Res Serv, USDA, Md, 64-65; asst prof, 65-71, ASSOC PROF FORESTRY, ORE STATE UNIV, 71-; VPRES, PMS INSTRUMENT CO, CORVALLIS, 67- Mem: AAAS; Am Soc Plant Physiol; Soc Am Foresters; Scand Soc Plant Physiol. Res: Plant growth regulators; forest regeneration; bioelectrical potential in plants; root growth. Mailing Add: Forestry Sch Ore State Univ Corvallis OR 97331

ZAFFARANO, DANIEL JOSEPH, b Cleveland, Ohio, Dec 16, 17; m 46; c 6. NUCLEAR PHYSICS. Educ: Case Inst Technol, BS, 39; Ind Univ, MS, 48, PhD(physics), 49. Prof Exp: Tech liaison, Nat Carbon Co, 40-45; contract liaison, Appl Physics Lab, Johns Hopkins Univ, 45-46; from assoc prof to prof, 49-67, chmn dept physics, 61-71, DISTINGUISHED PROF PHYSICS, IOWA STATE UNIV, 67-, VPRES FOR RES & GRAD DEAN, 71- Concurrent Pos: Sci liaison officer, US Off Naval Res, London, 57-58. Mem: Fel Am Phys Soc. Res: Experimental nuclear physics. Mailing Add: 201 Beardshear Iowa State Univ Ames IA 50010

ZAFFARONI, ALEJANDRO, b Montevideo, Uruguay, Feb 27, 23; m 46; c 2. BIOCHEMISTRY. Educ: Univ Montevideo, BS, 41; Univ Rochester, PhD(biochem), 49. Hon Degrees: DSc, Univ Rochester, 72. Prof Exp: Dir biol res, Syntex SA, Mex, 51-54, dir res & develop, 54-56, vpres, 56-61, pres, Syntex Labs, Inc, exec vpres, Syntex Corp & pres, Syntex Res Ctr, Calif, 61-68; PRES & DIR RES, ALZA CORP, 68-; PRES, DYNAPOL, 72- Concurrent Pos: Hon prof, Univ Montevideo, 59. Mem: Am Chem Soc; Soc Exp Biol & Med; Endocrine Soc; Am Soc Biol Chemists; Am Soc Microbiol. Res: Biochemistry and pharmacology of steroid hormones. Mailing Add: Alza Corp 950 Page Mill Rd Palo Alto CA 94304

ZAFIRATOS, CHRIS DAN, b Portland, Ore, Nov 18, 31; m 57; c 4. NUCLEAR PHYSICS. Educ: Lewis & Clark Col, BS, 57; Univ Wash, PhD(physics), 62. Prof Exp: Res instr nuclear physics, Univ Wash, 62; staff mem, Los Alamos Sci Lab, 62-64; asst prof physics, Ore State Univ, 64-68; from asst prof to assoc prof, 68-72, PROF PHYSICS, UNIV COLO, 72- Concurrent Pos: Chmn nuclear physics lab, Univ Colo, 74-76. Mem: Am Phys Soc. Res: Nuclear reactions; neutron and nuclear structure physics. Mailing Add: Dept of Physics Univ of Colo Boulder CO 80302

ZAFIRIOU, OLIVER C, organic chemistry, see 12th edition

ZAFRAN, MISHA, b Berlin, Ger, Aug 10, 49; US citizen. MATHEMATICAL ANALYSIS. Educ: Univ Calif, Riverside, BS, 68, PhD(math), 72. Prof Exp: Mem math res, Inst Advan Study, 72-73; ASST PROF MATH, STANFORD UNIV, 73- Mem: Am Math Soc. Res: Interrelationships between harmonic analysis, spectral theory and interpolation of operators. Mailing Add: Dept of Math Stanford Univ Stanford CA 94305

ZAGAR, WALTER T, b Brooklyn, NY, Oct 15, 28; m 51; c 4. PHYSICAL CHEMISTRY, POLYMER CHEMISTRY. Educ: Manhattan Col, BS, 50; Fordham Univ, MS, 55, PhD(phys chem), 58. Prof Exp: Chemist, Dextran Corp, 52-53; instr chem, Notre Dame Col, NY, 55-56; instr, Manhattan Col, 56-57; sr scientist, Polymer Chem Div, W R Grace & Co, 57-66; mgr plastics div, Allied Chem Corp, 66; mgr polymer develop, Chemplex Corp, 66-67; sr res chemist, Plastics Div, NJ, 67-72, La, 72-73, TECH SUPVR, ALLIED CHEM CORP, LA, 73- Mem: Soc Plastics Engrs; Am Chem Soc; NY Acad Sci. Res: Polymer development in area of polyethylene blends and additive systems for polymers, especially antioxidants, antistats and ultraviolet absorbers. Mailing Add: 5013 Parkhurst Dr Baton Rouge LA 70816

ZAGON, IAN STUART, b New York, NY, Mar 28, 43; m 64. NEUROBIOLOGY, PROTOZOOLOGY. Educ: Univ Wis-Madison, BS, 65; Univ Ill, Urbana, MS, 69; Univ Colo, Denver, PhD(anat), 72. Prof Exp: Asst prof biol struct, Med Sch, Univ Miami, 72-74; ASST PROF ANAT, MILTON S HERSHEY MED SCH, PA STATE UNIV, 74- Mem: Soc Neurosci; Am Asn Anat; Soc Protozool; Tissue Culture Asn; Am Soc Cell Biol. Res: Developmental neurobiology, focusing on normal and abnormal cerebellar development; effects of nitroso compounds and narcotic analgesics on cerebellar morphogenesis. Mailing Add: Dept of Anat Hershey Med Ctr Pa State Univ Hershey PA 17033

ZAGOR, HERBERT IVAN, physics, see 12th edition

ZAHALSKY, ARTHUR C, b New York, NY, Oct 31, 30; m 67; c 3. BIOCHEMISTRY, PARASITOLOGY. Educ: McGill Univ, BSc, 52; NY Univ, PhD(biol), 63. Prof Exp: Res assoc & staff mem, Haskins Labs, 58-66; res collabr, Brookhaven Nat Labs, 68-74; asst prof microbiol, Queens Col (NY), 66-69; assoc prof biochem, doctoral prog, City Univ New York, 69-71; PROF LAB BIOCHEM, PARASITOL, SOUTHERN ILL UNIV, EDWARDSVILLE, 71- Mem: AAAS; Am Soc Parasitol; Am Zool Soc; Biochem Soc; Royal Soc Trop Med Hyg. Res: Chemotherapy and immunology of trypanosomiasis; mechanisms of action of trypanocides; host immune response to parasites. Mailing Add: Lab Biochem Parasitol Biol Sci Southern Ill Univ Edwardsville IL 62026

ZAHARIS, JOHN LOUIS, b Westbrook, Maine, Nov 24, 28; m 50; c 9. MEDICAL ENTOMOLOGY. Educ: Univ Okla, BS, 55, MS, 58; Kans State Univ, PhD(entom), 60. Prof Exp: Asst zool, Univ Okla, 55-57; asst entom, Kans State Univ, 57-60, asst prof, 60; asst prof biol, head dept & chmn sci div, Benedictine Heights Col, 60-61; assoc prof biol, Upper Iowa Univ, 61-62; asst prof, Loyola Col, Md, 62-64, assoc prof & chmn dept, 64-68; CHMN DEPT BIOL, MIAMI-DADE COMMUNITY COL, SOUTH CAMPUS, 68- Concurrent Pos: Am Coun Educ acad admin fel, 72-73; Southeastern regional dir, Nat Task Force Two-Year Col Biologists, Am Inst Biol Sci. Mem: AAAS; Entom Soc Am; Am Inst Biol Sci. Res: Microbiology of insects. Mailing Add: Dept of Biol 11011 S W 104th St Miami-Dade Community Col S Campus Miami FL 33156

ZAHARKO, DANIEL SAMUEL, b New Westminster, BC, Nov 3, 30; US citizen; m 59; c 3. PHARMACOLOGY, PHYSIOLOGY. Educ: Univ BC, BPE, 53; Univ Ill, MS, 54, PhD(physiol), 63. Prof Exp: From instr to asst prof pharmacol, Ind Univ, Bloomington, 63-68; USPHS res fel, 68-70, PHARMACOLOGIST, LAB CHEM PHARMACOL, NAT CANCER INST, 70- Mem: AAAS; Am Soc Pharmacol & Exp Therapeut. Res: Environmental physiology; insulin metabolism, secretion and effects; drug effects on cell membranes; cancer chemotherapy; pharmacokinetics. Mailing Add: Lab of Chem Pharmacol Nat Cancer Inst Bethesda MD 20014

ZAHL, PAUL ARTHUR, b Bensenville, Ill, Mar 20, 10; m 46; c 2. NATURAL SCIENCE, EXPERIMENTAL BIOLOGY. Educ: NCent Col, AB, 32; Harvard Univ, AM, 34, PhD(biol), 36. Hon Degrees: DSc, NCent Col, 72. Prof Exp: Asst comp anat & histol, Harvard Univ, 34-36, Parke Davis & Co res fel endocrinol, 36-37; staff physiologist, 37-46, assoc dir & secy corp, 46-58, RES ASSOC, HASKINS LABS, INC, NEW YORK, 58-; SR SCIENTIST, NAT GEOG SOC, 58- Concurrent Pos: Mem var expeds, 37-67; res assoc, Union Col, NY, 37-39; guest investr, Mem Hosp, New York, 39-41; res assoc, Am Mus Natural Hist, 51-64; adj prof biol, Fordham Univ, 64-; USPHS & Dept Health, Educ & Welfare grants. Mem: Am Soc Zoologists; Soc Exp Biol & Med; Ecol Soc Am; Radiol Soc NAm; Am Asn Cancer Res. Res: Experimental biology; natural history; exploration. Mailing Add: Nat Geog Soc 17th & M Sts NW Washington DC 20036

ZAHLER, RAPHAEL, b New York, NY, June 14, 45; m 71; c 1. MATHEMATICS. Educ: Harvard Univ, AB, 66; Univ Chicago, SM, 67, PhD(math), 70. Prof Exp: Lectr math, Univ Calif, Berkeley, 70-72; asst prof, 72-75, ASSOC PROF MATH, DOUGLASS COL, RUTGERS UNIV, 75- Concurrent Pos: Prin investr, NSF res grant, 72- Mem: Am Math Soc. Res: Algebraic and differential topology; mathematical biology; applied mathematics. Mailing Add: Dept of Math Douglass Col Rutgers Univ New Brunswick NJ 08903

ZAHLER, STANLEY ARNOLD, b New York, NY, May 28, 26; m 52; c 3. MICROBIAL GENETICS. Educ: NY Univ, AB, 48; Univ Chicago, MS, 49, PhD, 52. Prof Exp: Instr gen bact, Northern Ill Col Optom, 51; USPHS fel bact, Univ Ill, 52-54; instr microbial genetics, Univ Wash, 54-57, asst prof, 57-58; instr, Med Ctr, WVa Univ, 59; Instr, 59-64, ASSOC PROF MICROBIAL GENETICS, CORNELL UNIV, 64- Concurrent Pos: Consult, Gen Elec Corp, 63-66; USPHS spec fel, Scripps Clin & Res Found, 66-67. Mem: AAAS; Am Soc Microbiol; Genetics Soc Am; Brit Soc Gen Microbiol. Res: Bacteriophages; microbial genetics; developmental biology; metabolic controls. Mailing Add: Sect of Genetics Cornell Univ Ithaca NY 14853

ZAHLER, WARREN LEIGH, b Springville, NY, June 28, 41; m 68; c 2. BIOLOGICAL CHEMISTRY. Educ: Alfred Univ, BA, 63; Univ Wis-Madison, MS, 66, PhD(biochem), 68. Prof Exp: NIH fel molecular biol, Vanderbilt Univ, 67-71, res assoc, 71-72; ASST PROF BIOCHEM, UNIV MO-COLUMBIA, 72- Mem: Am Oil Chem Soc; Am Chem Soc; Soc Study Reproduction. Res: Isolation and characterization of sperm membranes. Study of the function of sperm membranes and membrane bound enzymes in reproduction. Mailing Add: Dept of Biochem Univ of Mo 105 Schweitzer Hall Columbia MO 65201

ZAHM, HARRY VICTOR, chemistry, see 12th edition

ZAHN, KENNETH CHARLES, organic chemistry, inorganic chemistry, see 12th edition

ZAHND, HUGO, b Berne, Switz, May 16, 02; nat US; m 26; c 2. BIOCHEMISTRY. Educ: NY Univ, BS, 26; Columbia Univ, AM, 29, PhD(biochem), 33. Prof Exp: Tutor chem, 28-34, from instr to prof, 34-72, EMER PROF CHEM, BROOKLYN COL, 72- Mem: Fel AAAS; Am Chem Soc; Hist Sci Soc; fel Am Inst Chem; NY Acad Sci. Res: Labile sulfur in proteins; quantitative inorganic and organic analysis; history of chemistry; chromatography as applied to the fields of alkaloids, amino acids and proteins. Mailing Add: 84A Enfield Court Ridge NY 11961

ZAHNLEY, JAMES CURRY, b Manhattan, Kans, Apr 16, 38; m 65; c 2. BIOCHEMISTRY. Educ: Kans State Univ, BS, 58; Purdue Univ, MS, 62, PhD(biochem), 63. Prof Exp: Res assoc enzym, Syntex Inst Molecular Biol, Calif, 63-64, assoc biochem, 64-65; asst res biochemist, NIH grant, 65-66; RES CHEMIST, WESTERN REGIONAL RES CTR, AGR RES SERV, USDA, 66- Mem: AAAS; Am Chem Soc. Res: Enzymology; protein biochemistry; food biochemistry. Mailing Add: Western Regional Res Ctr US Dept of Agr 800 Buchanan St Berkeley CA 94710

ZAIDEL, ERAN, b Kibbutz Yagur, Israel, Jan 23, 44; US citizen; m 65; c 2. NEUROPSYCHOLOGY. Educ: Columbia Univ, AB, 67; Calif Inst Technol, MSc, 68, PhD(psychobiol), 73. Prof Exp: RES FEL PSYCHOBIOL, CALIF INST TECHNOL, 73- Mem: Acad Aphasia. Res: Neurolinguistics and psycholinguistics; cognitive and developmental psychology; epistemology and the philosophy of science, of mind and of language. Mailing Add: 156-29 Div of Biol Calif Inst Technol Pasadena CA 91125

ZAIDI, SYED AMIR ALI, b Lahore, Pakistan, Apr 15, 35; m 62. NUCLEAR PHYSICS. Educ: Punjab Univ, BSc, 56; Univ Göttingen, 57-58; Univ Heidelberg, dipl physics, 60, PhD(physics), 64. Prof Exp: Vis res scientist, Max Planck Inst Nuclear Physics, 64-66; asst prof, 66-68, ASSOC PROF PHYSICS, UNIV TEX, AUSTIN, 68-, ASSOC DIR CTR NUCLEAR STUDIES, 67- Mem: Fel Am Phys Soc. Res: Isobaric analogue resonances; nuclear structure studies using shell model description of reaction theory; heavy ion induced reactions; elementary particle physics. Mailing Add: Dept of Physics Univ of Tex Austin TX 78712

ZAIDINS, CLYDE STEWART, physics, see 12th edition

ZAIDMAN, SAMUEL, b Bucharest, Romania, Sept 4, 33; m 62; c 4. MATHEMATICS. Educ: Univ Bucharest, Lic, 55; Univ Paris, Dr d'Etat, 70. Prof Exp: Asst math, Univ Bucharest, 55-59; vis prof, Univ Milan, 61-64; PROF MATH, UNIV MONTREAL, 64- Concurrent Pos: Vis prof, Univ Geneva, 66-68. Mem: Am Math Soc. Res: Partial differential equations; operator theory; almost-periodic functions and equations. Mailing Add: Dept of Math Univ of Montreal Box 6128 Montreal PQ Can

ZAIKA, LAURA LARYSA, b Kharkow, Ukraine, b June 23, 38; US citizen. FOOD CHEMISTRY. Educ: Drexel Inst Tech, BS, 60; Univ Pa, PhD(org chem), 64. Prof Exp: RES CHEMIST MICROBIOL INVESTS, AGR RES SERV, USDA, 64- Mem: Am Chem Soc; Inst Food Technologists. Res: 2-aryl benzimidazoles; 1, 2, 3-benzotriazines; meat flavor investigations; chromatography; microbial metabolites; fermented meat products. Mailing Add: 5023 N Rosehill St Philadelphia PA 19120

ZAITLIN, MILTON, b Mt Vernon, NY, Apr 2, 27; m 51; c 4. PLANT VIROLOGY. Educ: Univ Calif, BS, 49; Calif, Los Angeles, PhD(plant physiol), 54. Prof Exp: Res officer, Commonwealth & Indust Res Orgn, Canberra, Australia, 54-58; asst prof hort, Univ Mo, 58-60; asst agr biochem, Univ Ariz, 60-62, from assoc prof to prof agr biochem & plant path, 66-73; PROF PLANT PATH, CORNELL UNIV, 73- Concurrent Pos: Guggenheim & Fulbright fels, 66-67; assoc ed, Virol, 66-71, ed, 72- Mem: Soc Gen Microbiol; Am Asn Univ Profs; Am Soc Plant Physiologists; Am Phytopath Soc. Res: Plant viruses; physiology of plant virus disease; molecular basis of plant-virus infections. Mailing Add: Dept of Plant Path Cornell Univ Ithaca NY 14853

ZAJAC, ALFRED, b Vienna, Austria, Feb 18, 17; US citizen; m 50; c 2. PHYSICS. Educ: Univ St Andrews, BSc, 46, Hons, 48; NY Univ, MS, 52; Polytech Inst Brooklyn, PhD(physics), 57. Prof Exp: From instr to assoc prof, 55-74, PROF PHYSICS & CHMN DEPT, ADELPHI UNIV, 74- Concurrent Pos: NSF res grant, 62-64. Mem: Am Phys Soc; Am Asn Physics Teachers; Am Crystallog Asn. Res: Crystal perfection, thermal motion and anomalous transmission of x-rays. Mailing Add: Dept of Physics Adelphi Univ Garden City NY 11530

ZAJAC, BARBARA ANN, b Fountain Springs, Pa, Mar 15, 37; m 57; c 1. VIROLOGY, ELECTRON MICROSCOPY. Educ: Univ Pa, BA, 58, PhD(microbiol), 67. Prof Exp: Assoc pediat, virol, Univ Pa, 69-70; asst prof, 70-75, ASSOC PROF MICROBIOL, MED COL PA, 75- Concurrent Pos: NIH fel, Div Virus Res, Children's Hosp Philadelphia, Pa, 67-69; grants, Res Corp & Anna Fuller Fund, 72-73, Damon Runyon Mem Fund, 73-75 & Nat Cancer Inst, 73-76. Mem: AAAS; Am Soc Microbiol; Electron Micros Soc Am; NY Acad Sci. Res: Human tumor viruses, their host-parasite relationships and electron microscopic analysis. Mailing Add: Dept of Microbiol Med Col of Pa Philadelphia PA 19129

ZAJAC, FELIX EDWARD, III, b Baltimore, Md, Dec 4, 41; m 62; c 2. NEUROPHYSIOLOGY, BIOMEDICAL ENGINEERING. Educ: Rensselaer Polytech Inst, BEE, 62; Stanford Univ, MS, 65, PhD(neurosci), 68. Prof Exp: Staff assoc, Lab Neural Control, Nat Inst Neurol Dis & Stroke, 68-70; asst prof, 70-73, ASSOC PROF ELEC ENG, UNIV MD, COLLEGE PARK, 73-, DIR BIOMED RES LAB, 71- Mem: AAAS; Am Physiol Soc; Soc Neurosci; Inst Elec & Electronics Eng. Res: Neural control and biomechanics of animal movement with emphasis on cat locomotion and jumping. Mailing Add: Dept of Elec Eng Univ of Md College Park MD 20742

ZAJAC, IHOR, b Lwiw, Ukraine, May 26, 31; US citizen; m 57; c 1. MEDICAL MICROBIOLOGY, VIROLOGY. Educ: Univ Pa, BA, 58; Hahnemann Med Col, MS, 60, PhD(microbiol), 64. Prof Exp: Asst microbiol, Hahnemann Med Col, 64-65, instr, 65; asst prof, Jefferson Med Col, 65-71; assoc sr investr, 71-75, SR INVESTR, SMITH KLINE & FRENCH LABS, 75- Concurrent Pos: Vis lectr, Med Col Pa, 72- Mem: Am Soc Microbiol; Tissue Cult Asn; Brit Soc Gen Microbiol; Soc Exp Biol & Med. Res: Enteroviruses; cell-virus interactions; cell membrane; interferon; bacterial and viral chemotherapy; anaerobic bacteria. Mailing Add: L-37 Res & Develop Smith Kline & French Lab Philadelphia PA 19101

ZAJAC, WALTER WILLIAM, JR, b Central Falls, RI, July 19, 34; m 59; c 6. ORGANIC CHEMISTRY. Educ: Providence Col, BS, 55; Va Polytech Inst, MS, 57, PhD(chem), 60. Prof Exp: From asst prof to assoc prof, 59-72, PROF CHEM, VILLANOVA UNIV, 72- Concurrent Pos: Fel, Univ Alta, 65-66. Mem: Am Chem Soc; The Chem Soc. Res: Reduction of organic compounds; reactions of nitriles; synthesis and mechanism of ring closure reactions in homocyclic and heterocyclic systems; conformation of heterocyclic compounds. Mailing Add: Dept of Chem Villanova Univ Villanova PA 19085

ZAJACEK, JOHN GEORGE, b Allentown, Pa, May 8, 36; m 64; c 3. ORGANIC CHEMISTRY. Educ: Lehigh Univ, BA, 58; Cornell Univ, PhD(org chem), 62. Prof Exp: MGR PETROCHEM RES, ARCO CHEM CO DIV, ATLANTIC-RICHFIELD CO, 62- Concurrent Pos: Instr, Drexel Inst Tech, 66- Mem: Am Chem Soc. Res: Oxidation of hydrocarbons; epoxidation of olefins; reactions of carbon monoxide; reduction of aromatic nitro compounds; metal catalyzed reactions. Mailing Add: Res Dept Arco Chem Co 500 S Ridgeway Ave Glenolden PA 19087

ZAJCEW, MYKOLA, b Pisky, Ukraine, Nov 5, 94; nat US; m 18. CHEMISTRY. Educ: Polytech Univ Prague, Czech, ChE, 24. Hon Degrees: DSc, Ukrainian Tech Univ, Ger, 57. Prof Exp: From asst prof to assoc prof chem, Ukrainian Tech Univ, Czech, 24-35; prof chem technol, head dept & rector, Regensburg & Munich, Ger, 45-51; mgr, Extraction Merz Plant, Czech, 35-36; chief chemist, Kosmos Factories, 37-45; res chemist, 51-65, VIS RESEARCHER, ENGELHARD INDUSTS, INC, 66- Honors & Awards: Cert of Recognition, Chem Abstr Serv, Am Chem Soc, 71. Mem: Am Oil Chemists Soc; NY Acad Sci. Res: Catalytic reactions of fats, especially hydrogenation of fatty oils with precious metal catalysts. Mailing Add: 147 Norwood St Newark NJ 07106

ZAJIC, JAMES EDWARD, b Wichita, Kans, Mar 8, 28; m 52; c 3. MICROBIOLOGY, LAW. Educ: Univ Kans, BA, 51; Univ Wis, MS, 53; Univ Calif, PhD(microbiol), 56; Okla City Univ, JD, 67. Prof Exp: Fel biochem, Argonne Nat Lab, 56-57; res microbiologist, Grain Processing Corp, Iowa, 57-61; group leader biochem, Kerr McGee Industs, 61-66; assoc prof biochem, 67-70, PROF BIOCHEM ENG, UNIV WESTERN ONT, 70-, ASST DEAN FAC ENG SCI, 72- Mem: Am Chem Soc; Soc Indust Microbiol; Am Soc Microbiol; Am Inst Chem Eng; Biomed Eng Soc. Res: Industrial fermentations; continuous fermentations and related engineering topics; environmental engineering; biochemistry of hydrocarbons; biogeochemistry. Mailing Add: Fac of Eng Sci Univ of Western Ont London ON Can

ZAK, BENNIE, b Detroit, Mich, Sept 29, 19; m 46; c 3. CLINICAL BIOCHEMISTRY. Educ: Wayne State Univ, BS, 48, PhD(chem), 52. Prof Exp: Asst prof clin chem, 57-60, assoc prof path, 60-65, PROF PATH, SCH MED, WAYNE STATE UNIV, 65-; MED LAB ANALYST & JR ASSOC PATH, DETROIT RECEIVING HOSP, 51- Concurrent Pos: Res technician, Detroit Receiving Hosp, 50-51. Mem: Am Chem Soc; Am Asn Clin Chem. Res: Spectrophotometric methods; rapid agar gel electrophoretic separations instrumentation in clinical chemistry; phosphate and phosphomonoesterase analysis and automation. Mailing Add: Dept of Path Wayne State Univ Col of Med Detroit MI 48021

ZAK, BRATISLAV, b Washington, DC, Feb 3, 19. FOREST PATHOLOGY. Educ: Pa State Univ, BS, 41; Duke Univ, MF, 49, DF, 54. Prof Exp: With southeastern forest exp sta, US Forest Serv, 46-62, PRIN PLANT PATHOLOGIST, FORESTRY SCI LAB, PAC NORTHWEST FOREST & RANGE EXP STA, US FOREST SERV, 62- Mem: Am Phytopath Soc; Mycol Soc Am; Brit Mycol Soc. Res: Characterization and classification of mycorrhizae of Pacific Northwest forest trees; role of mycorrhizae in root disease. Mailing Add: Forestry Sci Lab 3200 Jefferson Way Corvallis OR 97331

ZAK, FREDERICK GERARD, b Vienna, Austria, Feb 13, 15; nat US; m 58; c 2. PATHOLOGY. Educ: Univ Basel, MD, 39; Am Bd Path, dipl, 48, cert clin path, 52. Prof Exp: Resident path, Mt Sinai Hosp, 43-44, asst pathologist, 45-46; res asst dermatopath, NY Skin & Cancer Unit, NY Univ, 47-52; lectr path, Col Med, State Univ NY Downstate Med Ctr, 52-55, assoc clin prof, 55-64; prof, NY Med Col, 64-68; CLIN PROF STATE UNIV NY DOWNSTATE MED CTR, 68-; CHIEF ANAT PATH, METHODIST HOSP, 68- Concurrent Pos: Fel exp surg, Med Sch, Johns Hopkins Univ, 40-41; Dazian Found fel, Mt Sinai Hosp, New York, 44-45; fel dermatopath, NY Skin & Cancer Unit, NY Univ, 47-52; res asst, Mt Sinai Hosp, 46-56, assoc attend pathologist, 57-64; pathologist & dir labs, Knickerbocker Hosp, 49-51; chief clin lab, Regional Off, Vet Admin, NY, 50-53; assoc pathologist, Wyckoff Heights Hosp, 52-53; neuropathologist, Kings County Hosp, NY, 52-53; pathologist & dir labs, North Shore Hosp, Manhasset, 53-56; sr pathologist, Vet Admin Hosp, Ft Hamilton, NY, 55-56; consult pathologist, Hosp for Joint Dis, Polyclin Hosp & Wyckoff Heights Hosp; res assoc otolaryngic path, Mt Sinai Hosp, 63-71; lectr otolaryngic path, Mt Sinai Med Sch, 71-75; consult pathologist, Health Ins Plan of Greater NY, State Dept Health, Albany, NY, St Joseph's Hosp, Yonkers, NY & Northern Westchester Hosp, Mt Kisco, NY. Mem: Fel AMA; Am Asn Path & Bact; Am Asn Cancer Res; fel Col Am Path; Int Acad Path. Res: Pathology of tumors, vascular diseases and skin; experimental pathology, especially cancer research; otolaryngeal pathology. Mailing Add: Methodist Hosp 506 Sixth St Brooklyn NY 11215

ZAKAIB, DANIEL D, b Montreal, Que, Apr 1, 25; m 48; c 2. PETROLEUM CHEMISTRY, ANALYTICAL CHEMISTRY. Educ: Montreal Tech Inst, dipl chem, 46; Sir George Williams Univ, BSc, 53. Prof Exp: Supvr, Montreal Refinery Lab, Brit Am Oil Co, 48-55, asst refinery chemist, 55-58, anal technologist, Head Off Toronto, 58-63, coordr, Anal Res Labs, Brit Am Res & Develop Co, 63-66; mgr anal & chem res, Res & Develop Dept, Gulf Oil Can, Ltd, 66-69, supvr petrol chem sect, Phys Sci Div, Gulf Res & Develop Co, 69-71, DIR TECH OPERS, RES & DEVELOP DEPT, GULF OIL CAN, LTD, 71- Mem: Am Chem Soc; Am Soc Testing & Mat; Can Asn Appl Spectros; fel Chem Inst Can. Res: Technical administration of industrial research facility. Mailing Add: Gulf Oil Can Ltd Res & Develop Dept 2489 N Sheridan Way Sheridan Park ON Can

ZAKHARY, RIZKALLA, b Assiut, Egypt, Sept 5, 24; m 66; c 2. HUMAN ANATOMY. Educ: Cairo Univ, BS, 49, MS, 54; Tulane Univ, PhD(anat), 64. Prof Exp: Technician & res asst biochem, US Naval Med Res Unit 3, Cairo, Egypt, 50-56; instr biol chem & gen sci, Am Univ Cairo, 57-60; asst prof anat & physiol, Sch Dent, Loyola Univ, La, 64-67; asst prof anat, 67-70, ASSOC PROF ANAT, SCH DENT, UNIV SOUTHERN CALIF, 70- Concurrent Pos: NIH grant, Tulane Univ, 65-67. Mem: AAAS; Am Asn Anatomists. Res: Hypothermia, academic and applied aspects; cryobiology; stress and hypothermia. Mailing Add: Dept of Anat Univ of Southern Calif Sch Dent Los Angeles CA 90007

ZAKI, ABD EL-MONEIM EMAM, b Cairo, UAR, Dec 18, 33; US citizen. HISTOLOGY, ORAL BIOLOGY. Educ: Cairo Univ, BChD, 55, DDR, 58; Ind Univ, MSD, 62; Univ Ill, PhD(anat), 69. Prof Exp: Dent surgeon, Demonstration & Training Ctr, Qualyub, UAR, 55-59; from instr to asst prof, Fac Dent, Cairo Univ, 62-67; res assoc, Col Dent, 67-70, asst prof histol, Col Dent & lectr, Col Med, 70-72, assoc prof histol, Col Dent & Sch Basic Med Sci, Med Ctr, 72-75, PROF HISTOL, COL DENT, UNIV ILL, 75- Concurrent Pos: UAR govt spec mission mem grad study, US, 59-62. Mem: AAAS; Am Asn Anat; Electron Micros Soc Am; Int Asn Dent Res. Res: Cellular control of mineralization using an amphibian odontogenic model and cytological aspects of odontogenesis. Mailing Add: Dept of Histol Col of Dent Univ of Ill at the Med Ctr Chicago IL 60612

ZAKI, MAHFOUZ H, b Cairo, Egypt, Apr 14, 24; US citizen; m 70. ENVIRONMENTAL MEDICINE, PUBLIC HEALTH. Educ: Cairo Univ, MB, ChB, 49; Univ Alexandria, MPH, 58; Columbia Univ, DrPH, 62. Prof Exp: Intern med & surg, Kasr-El-Eini Univ Hosp, Cairo, 50-51; med officer, Abu-Sidhom Health Ctr, Minia, 51-56; instr prev med, High Ints Pub Health, 57-59; staff mem edpidemiol, Sch Pub Health, Columbia Univ, 59-62; asst prof prev med, 62-68, ASSOC PROF ENVIRON MED & COMMUNITY HEALTH, STATE UNIV NY DOWNSTATE MED CTR, 68- Concurrent Pos: Grants, Health Res Coun City New York, 64-68 & tuberc br, USPHS & Peace Corps; Peace Corps physician & prog tech adv, US State Dept, Afghanistan, 70-71; dir pub health, Suffolk County, New York; adj prof pub health, City Univ New York, 73-; lectr, Sch Pub Health & Admin Med, Columbia Univ, 73- Mem: Fel Am Pub Health Asn; fel Royal Soc Trop Med & Hyg; Am Soc Trop Med & Hyg; fel Am Col Chest Physicians; Am Statist Asn. Res: Etiology of sarcoidosis; prevalence of infection with typical and atypical strains of Mycobacterium tuberculosis and of infection with Histoplasma capsulatum in Afghanistan. Mailing Add: Dept of Health Servs H Lee Dennison Bldg Hauppauge NY 11787

ZAKRISKI, PAUL MICHAEL, b Amsterdam, NY, July 12, 40; m 64; c 2. ANALYTICAL CHEMISTRY. Educ: Univ Rochester, AB, 62; Univ Cincinnati,

PhD(org chem), 67. Prof Exp: Sr res chemist, 66-74, SECT LEADER, RES DIV, B F GOODRICH CO, BRECKSVILLE, 74- Mem: Am Chem Soc; Am Soc Mass Spectrometry. Res: Mass spectrometry for structure elucidation; high speed chromatography. Mailing Add: 8329 Wyatt Rd Broadview Heights OH 44147

ZAKRZEWSKA, BARBARA (MRS M BOROWIECKI), b Warsaw, Poland, Nov 20, 24; nat US; m 71; c 1. GEOGRAPHY, GEOLOGY. Educ: Ind Univ, BA, 56, MA, 57; Univ Wis, PhD(geog), 62. Prof Exp: Instr Polish, US Air Force Lang Sch, Indiana, 55-56; from asst prof to assoc prof, 60-69, PROF GEOG, UNIV WIS-MILWAUKEE, 69- Concurrent Pos: Univ res grants, Univ Wis-Madison, 60, 63 & 66, Int Res & Exchange Bd res fel, 70; vis prof, Ind Univ, 63-64; Nat Defense Educ Act grant, 67. Mem: AAAS; Int Geog Union; Am Asn Advan Slavic Studies; Asn Am Geog; Am Geog Soc. Res: Physical geography; geomorphology; glacial geology; terrain form on loess surfaces; geomorphology of stream valleys; quantitative land form analysis; terrain types of the Great Plains; geography of East Central Europe. Mailing Add: Dept of Geog Univ of Wis Milwaukee WI 53201

ZAKRZEWSKI, GEORGE ANTHONY, physical inorganic chemistry, see 12th edition

ZAKRZEWSKI, RICHARD JEROME, b Hamtramck, Mich, Nov 5, 40; m 66; c 2. VERTEBRATE PALEONTOLOGY. Educ: Wayne State Univ, BS, 63; Univ Mich, MS, 65, PhD(vert paleont), 68. Prof Exp: NSF fel geol, Idaho State Univ-Los Angeles County Mus, 68-69; asst prof, 69-74, ASSOC PROF EARTH SCI, FT HAYS KANS STATE COL, 74-, DIR STERNBERG MEM MUS, 73- Mem: Soc Vert Paleont; Paleont Soc; Am Soc Mammal; Soc Syst Zool; Am Quaternary Asn. Res: Fossil mammals, particularly rodents; late Cenozoic stratigraphy. Mailing Add: Sternberg Mem Mus Ft Hays Kans State Col Hays KS 67601

ZAKRZEWSKI, SIGMUND FELIX, b Buenos Aires, Arg, Sept 15, 19; m 56; c 1. BIOCHEMISTRY. Educ: Univ Hamburg, MS, 52, PhD(biochem), 54. Prof Exp: Res asst, Sch Med, Western Reserve Univ, 52-53; res asst, Sch Med, Yale Univ, 53-56; sr cancer res scientist, 56-61, assoc cancer res scientist, 61-71, PRIN CANCER RES SCIENTIST, DEPT EXP THERAPEUT, ROSWELL PARK MEM INST, 71- Concurrent Pos: Res prof, Dept Chem, Niagara Univ; assoc prof, Dept Pharmacol, State Univ NY Buffalo, Roswell Park Div. Mem: AAAS; Am Chem Soc; Am Asn Cancer Res; Am Soc Biol Chemists; NY Acad Sci. Res: Cancer chemotherapy; metabolism of folic acid and folic acid antagonist. Mailing Add: 260 Lakewood Pkwy Buffalo NY 14226

ZALAR, FRANK VICTOR, b Cleveland, Ohio, Dec 23, 40; m 70. ORGANIC CHEMISTRY, POLYMER CHEMISTRY. Educ: John Carroll Univ, BS, 62; Ohio State Univ, PhD(chem), 66. Prof Exp: RES CHEMIST, LUBRIZOL CORP, 66- Mem: Am Chem Soc; Am Soc Lubrication Engrs. Res: Lubricant additive chemistry. Mailing Add: Lubrizol Corp 29400 Lakeland Blvd Wickliffe OH 44092

ZALAY, ETHEL SUZANNE, b Budapest, Hungary, Sept 1, 19; m 46; c 3. ORGANIC CHEMISTRY. Educ: Univ Sci Budapest, Hungary, PhD, 44. Prof Exp: Owner, Dr Somody Lab, 45-51; org chemist, Fine Chem Producing Union, 51-54; owner, Dr Somody Lab, 54-56; org chemist, Textile Res Inst, 57; res chemist, Biol Res Lab, Philadelphia Gen Hosp, 57-58; res chemist, 58-69, ASSOC RES CHEMIST, STERLING-WINTHROP RES INST, STERLING DRUG, INC, 69- Mem: Am Chem Soc. Res: Fine organic chemicals; phospholipids; pharmaceuticals; heterocyclic chemistry. Mailing Add: Sterling-Winthrop Res Inst Columbia Turnpike Rensselaer NY 12144

ZALESKI, WITOLD ANDREW, b Pyzdry, Poland, Apr 4, 20; Can citizen; m 48; c 4. MEDICINE. Educ: Univ Edinburgh, MB, ChB, 46; Royal Col Physciians & Surgeons, Ireland, dipl psychol med, 52; Royal Col Physicians & Surgeons Can, cert psychiat, 62; Univ Sask, MD, adeundem, 64; FRCP(C), 72; Royal Col Psychiat, cert, 73. Prof Exp: Dep supt & consult psychiatrist, Ment Retardation Insts, Regional Hosp Bd Eng, Birmingham, 54-58; clin dir, Sask Training Sch, Mosse Jaw, Can, 58-67; assoc prof, 67-73, PROF PEDIAT, UNIV SASK, 73-, DIR ALVIN BUCKWOLD MENT RETARDATION UNIT, UNIV HOSP, 67- Concurrent Pos: Vis consult psychiat, Univ Sask Hosp, 62-67. Mem: Fel Am Asn Ment Deficiency; Am Acad Ment Retardation; Can Med Asn; Can Psychiat Asn. Res: Etiology and prevention of mental retardation; inborn errors of metabolism; chromosomal anomalies; behavioral programs for the retarded; delivery of services in mental retardation. Mailing Add: Alvin Buckwold Ctr Univ of Sask Hosp Saskatoon SK Can

ZALEWSKI, EDMUND JOSEPH, b Schenectady, NY, July 23, 31; m 58; c 5. ORGANIC POLYMER CHEMISTRY. Educ: Union Col, BS, 64. Prof Exp: From technician polyester to group leader polymer, 50-67, MGR POLYMER, SCHENECTADY CHEM INC, 67- Mem: Am Chem Soc; Soc Plastics Engrs. Res: Development of organic and heterocyclic polymers exhibiting excellent mechanical properties coupled with chemical and thermal resistance for use as electrical insulation. Mailing Add: 2761 Maida Lane Schenectady NY 12306

ZALEWSKI, EDWARD FRANCIS, chemical physics, see 12th edition

ZALIK, SARA E, b Mex, May 23, 39; m 66; c 2. DEVELOPMENTAL BIOLOGY, CELL BIOLOGY. Educ: Nat Univ Mex, BS, 59; Univ Ill, PhD(anat), 63. Prof Exp: NIH int fel, Biol Div, Oak Ridge Nat Lab, 63-64; asst prof cell biol, Ctr Res & Advan Studies, Nat Polytech Inst, Mex, 64-66; asst prof, 66-72, ASSOC PROF ZOOL, UNIV ALTA, 72- Mem: AAAS; Can Soc Cell Biol; Am Soc Zool; Soc Develop Biol; Am Soc Cell Biol. Res: Cell differentiation and metaplasia; cell surface and its role in differentiation. Mailing Add: Dept of Zool Univ of Alta Edmonton AB Can

ZALIK, SAUL, b Ratcliffe, Sask, May 11, 21; m 66; c 2. PLANT PHYSIOLOGY, BIOCHEMISTRY. Educ: Univ Man, BSA, 43, MSc, 48; Purdue Univ, PhD(plant physiol), 52. Prof Exp: Lectr plant sci, Univ Man, 48-49; from asst prof to assoc prof, 52-64, PROF PLANT PHYSIOL & BIOCHEM, UNIV ALTA, 64- Mem: AAAS; Am Soc Plant Physiol; Can Biochem Soc. Res: Metabolism of lipids; nucleic acids and proteins in relation to plant differentiation and development. Mailing Add: Dept of Plant Sci Univ of Alta Edmonton AB Can

ZALIPSKY, JEROME JAROSLAW, b Ukraine; US citizen; c 2. ANALYTICAL CHEMISTRY. Educ: St Joseph's Col, BS, 58, MS, 62; Univ Pa, PhD(anal chem), 70. Prof Exp: From chemist to group leader anal chem, Nat Drug Co, Richardson-Merrill Inc, 58-70; from sr scientist to group leader phys chem, 70-74, SECT HEAD PHYS & MICROANAL CHEM, WILLIAM H RORER INC, 74- Mem: Am Chem Soc. Res: Chemical structure elucidation of new drug substance; kinetics; characterization of hydrolysis products; analytical and physical profile of drug substance. Mailing Add: William H Rorer Inc 500 Virginia Dr Ft Washington PA 19034

ZALKIN, ALLAN, physical chemistry, crystallography, see 12th edition

ZALKIN, HOWARD, b New York, NY, Dec 31, 34; m 66; c 1. BIOCHEMISTRY. Educ: Univ Calif, Davis, BS, 56, MS, 59, PhD(biochem), 61. Prof Exp: Res assoc chem, Harvard Univ, 61-62; res assoc biochem, Pub Health Res Inst New York, 62-64; res assoc, Col Physicians & Surgeons, Columbia Univ, 64-66; from asst prof to assoc prof, 66-72, PROF BIOCHEM, PURDUE UNIV, LAFAYETTE, 72- Concurrent Pos: Fels, NSF, 61-63, USPHS, 63-64 & USPHS fel biol sci, Stanford Univ, 72-73. Mem: Am Soc Biol Chemists. Res: Regulation of enzyme activity, allosteric enzymes; structure and function of glutamine amidotransferases. Regulation of tryptophan biosynthesis. Mailing Add: Dept of Biochem Purdue Univ Lafayette IN 47907

ZALKOW, LEON HARRY, b Millen, Ga, Nov 27, 29; m 71; c 1. ORGANIC CHEMISTRY. Educ: Ga Inst Technol, BCE, 52, PhD(chem), 56. Prof Exp: Res fel, Wayne State Univ, 55-56, 57-59; res chemist, E I du Pont de Nemours & Co, 56-57; asst prof chem, Okla State Univ, 59-62, assoc prof, 62-65; assoc prof, 65-69, PROF CHEM, GA INST TECHNOL, 69- Concurrent Pos: Prof & head dept chem, Univ of the Negev, 70-72. Mem: AAAS; Am Chem Soc; The Chem Soc. Res: Natural products; conformational analysis; chemistry of bicyclic azides. Mailing Add: Dept of Chem Ga Inst of Technol Atlanta GA 30332

ZALL, ROBERT ROUBEN, b Lowell, Mass, Dec 6, 25; m 49; c 3. FOOD SCIENCE. Educ: Univ Mass, BS, 49, MS, 50; Cornell Univ, PhD(food sci), 68. Prof Exp: Lab dir dairy prod, Grandview Dairies, Inc, NY, 50-51, mgr, Butter & Cheese Div, 51-53, mgr, Condensed Milks & Powder Div, 53-57, gen mgr corp, 57-66; dir res & prod, Crowley Food Co, 68-71; ASSOC PROF FOOD SCI, CORNELL UNIV, 71- Concurrent Pos: Environ Protection Agency pollution abatement demonstration grant, Whey Fractionation Plant, Crowley Foods Co, 69-72, proj dir, 71- Honors & Awards: Cert of Appreciation, Environ Protection Agency, 75. Mem: Inst Food Technol; Am Soc Agr Eng; Int Asn Milk, Food & Environ Sanit. Res: Detergents as inhibitors in food; reusing cleaning fluids to reduce consumption of energy and chemicals; reclamation and renovation of food wastes; membrane filtration processing; utilization of whey fractions in foods. Mailing Add: Dept Food Sci 118 Stocking Hall Cornell Univ Ithaca NY 14853

ZALLEN, EUGENIA MALONE, b Camp Hill, Ala, July 18, 32; m 59. FOOD SCIENCE. Educ: Auburn Univ, BS, 53; Purdue Univ, MS, 60; Univ Tenn, PhD(food sci), 74. Prof Exp: Dietetic intern, Med Ctr, Duke Univ, 53-54, assoc dietitian, 54-57; asst chief dietary, Univ Hosp, Emory Univ, 57-58; from instr to asst prof food & nutrit, Auburn Univ, 62-66; instr food, nutrit & inst admin, Univ Md, 67-72; researcher dairy sci, Okla State Univ, 72-73; researcher, Univ Tenn, 73-74; ASSOC PROF & DIR SCH HOME ECON, UNIV OKLA, 74- Concurrent Pos: Consult, Head Start Day Care Ctrs, Ala, 65-66; field reader, Bur Res, Dept Health Educ & Welfare, DC, 66-; consult, Univ Consult, Inc, Ala, 68; consult, Optimal Systs, Inc, Ga, 69-; pres, Acad World, Inc, Consults, 75- Mem: Am Home Econ Asn; Am Dietetic Asn; Inst Food Technol; Sigma Xi; Soc Nutrit Educ. Res: Institution administration; quality factors in quantity food production; curriculum development for dietetics; systems analysis and automatic data processing for volume food service. Mailing Add: Sch of Home Econ Univ of Okla Norman OK 73069

ZALLEN, HAROLD, pharmaceutical chemistry, health physics, see 12th edition

ZALLEN, RICHARD, b New York, NY, Jan 1, 37; m 64; c 2. SOLID STATE PHYSICS. Educ: Rensselaer Polytech Inst, BS, 57; Harvard Univ, AM, 59, PhD(solid state physics), 64. Prof Exp: Res assoc, Res Div, Raytheon Co, 58-59; res asst solid state physics, Harvard Univ, 59-64, res fel, 64-65; STAFF MEM, XEROX RES LABS, 65- Concurrent Pos: Vis assoc prof, Israel Inst Technol, 71-72. Mem: Fel Am Phys Soc. Res: Optical properties of solids; lattice vibrations and electronic structure; layer crystals and amorphous solids; intermolecular interactions; Raman scattering; phase transitions; pressure effects; solid state theory. Mailing Add: Xerox Res Labs 800 Phillips Rd Bldg 114 Webster NY 14580

ZALOKAR, MARKO, b Ljubljana, Yugoslavia, July 14, 18; nat US; m 51; c 2. DEVELOPMENTAL GENETICS. Educ: Univ Ljubljana, dipl, 40; Univ Geneva, DSc(zool), 44. Prof Exp: Asst biol, Univ Geneva, 42-44, instr genetics, 46-47; fel, Calif Inst Technol, 47-49; asst prof, Univ Wash, 49-51; vis scientist, NIH, 52-54; res assoc, Wesleyan Univ, 54-55; vis scientist, Yale Univ, 55-61; assoc res biologist, Univ Calif, San Diego, 61-66, lectr biol, Univ Calif, Davis, 66-67 & Univ Calif, San Diego, 67-68; DIR RES, CTR MOLECULAR GENETICS, GIF-SUR-YVETTE, FRANCE, 68- Mem: AAAS; Genetics Soc Am; Soc Develop Biol. Res: Lens regeneration; anatomy and development of Drosophila; biochemical genetics of Neurospora; autoradiography; protein and ribonucleic acid biosynthesis; developmental genetics of Drosophila. Mailing Add: Ctr for Molecular Genetics Nat Ctr of Sci Res 91 Gif-sur-Yvette France

ZALUBAS, ROMUALD, b Pandelys, Lithuania, July 20, 11; nat US; m 39; c 1. ASTROPHYSICS. Educ: Kansas State Univ, MA, 36; Georgetown Univ, PhD(astrophys), 55. Prof Exp: Asst astron, Vilnius State Univ, 40-44; dir sec sch, Ger, 45-49; instr math, Nazareth Col (NY), 49-51; instr, Georgetown Univ, 52-57; PHYSICIST, NAT BUR STANDARDS, 55- Mem: AAAS; Am Astron Soc; Optical Soc Am. Res: Description and analysis of atomic spectra; thorium wavelength standards. Mailing Add: Nat Bur of Standards Washington DC 20234

ZALUCKY, THEODORE B, b Beleluja, West-Ukraine, Apr 11, 19; US citizen; m 46; c 2. PHARMACY, PHARMACEUTICAL CHEMISTRY. Educ: Univ Vienna, MPharm, 42, DSc Nat(pharmaceut chem), 45; Ill, Chicago, BS, 52. Prof Exp: Anal chemist, Control Lab, Chicago Pharm Co, 52-53; res chemist, Food Prod Lab, Scholl Mfg Co, Inc, 53-55; asst prof pharmaceut chem, 55-63, assoc prof pharm & pharmaceut chem, 63-72, PROF PHARM, COL PHARM, HOWARD UNIV, 72- Concurrent Pos: AEC grant, 62-63. Mem: Am Chem Soc; Am Pharmaceut Asn; Am Asn Cols Pharm; Acad Pharmaceut Sci. Res: Chemistry of morphine alkaloids, epoxy ethers and some benzolypiperidines; structure-chromogenic activity relationship of phenolic compounds with Ehrlich reagent; isolation and structure of some new anhalonium alkaloids; spiro-compounds containing geranium and organo-metallic compounds. Mailing Add: Col of Pharm Howard Univ Washington DC 20001

ZALUSKY, RALPH, b Pawtucket, RI, Oct 11, 31; m 58; c 3. INTERNAL MEDICINE, HEMATOLOGY. Educ: Brown Univ, AB, 53; Boston Univ, MD, 57; Am Bd Internal Med, dipl, 64. Prof Exp: From intern to sr resident med, Duke Univ Hosp, 57-62; asst prof, 66-69, ASSOC PROF MED, MT SINAI SCH MED, 69-, ACTG CHIEF HEMAT, MT SINAI HOSP, 72- Concurrent Pos: USPHS fel, Thorndike Mem Lab, Harvard Univ, 59-61; res assoc med, Mt Sinai Hosp, 64-65, asst attend hematologist, 65-66. Mem: Am Fedn Clin Res; Am Soc Clin Nutrit. Res: Interrelationships between vitamin B-twelve and folic acid metabolism; sodium and potassium membrane transport; abnormal hemoglobins; erythropoietin physiology. Mailing Add: Div of Hemat Mt Sinai Hosp 100th St & Fifth Ave New York NY 10029

ZAM, STEPHEN G, III, b Toledo, Ohio, Nov 3, 32. PARASITOLOGY. Educ: Georgetown Univ, BS, 54; Catholic Univ, MS, 56; Univ Southern Calif, PhD(biol sci), 66. Prof Exp: Asst prof biol, Loyola Univ (Calif), 66; asst prof parasitol, 66-71,

ASSOC PROF ZOOL, UNIV FLA, 71- Concurrent Pos: Consult, Marineland, 67- Mem: Am Soc Parasitol; Am Soc Trop Med & Hyg; Int Soc Parasitol. Res: Nematode physiology including biochemistry of nematode egg hatching and larval molting; immunology to helminth infections. Mailing Add: Div of Biol Sci Univ of Fla Rm 611 Life Sci Bldg Gainesville FL 32601

ZAMBERNARD, JOSEPH, b Sept 5, 30; US citizen. CYTOLOGY. Educ: Univ Ala, AB, 53, MS, 56; Tulane Univ, PhD(cytol), 64. Prof Exp: Fel virol & immunol, Sch Med, Univ Colo, Denver, 64-66, asst prof anat, 66-72; ASSOC PROF ANAT, ALBANY MED COL, 72- Mem: Electron Micros Soc Am; Am Soc Cell Biol; Am Asn Anatomists; Tissue Cult Asn. Res: Virology and immunology; ultrastructure. Mailing Add: Dept of Anat Albany Med Col Albany NY 12208

ZAMBITO, ARTHUR JOSEPH, b Rochester, NY, Sept 7, 14; m 42; c 7. ORGANIC CHEMISTRY. Educ: Univ Mich, BS, 40, MS, 41, PhD(org chem), 47. Prof Exp: From res chemist to sr res chemist, 41-64, sect leader, Res & Develop Lab, 64-75, SR RES FEL, MERCK & CO INC, 75- Mem: Am Chem Soc. Res: Synthesis of pharmaceuticals; synthesis and isolation of amino acids; preparation of parenteral solutions and emulsions; synthesis of anticancer agents. Mailing Add: Res & Develop Lab Merck & Co Inc Lawrence St Rahway NJ 07065

ZAMBONI, LUCIANO, b South Dona diPiave, Italy, Sept 6, 29; m 57; c 2. PATHOLOGY, ELECTRON MICROSCOPY. Educ: Univ Rome, MD, 55, dipl gastroenterol, 58. Prof Exp: Instr path, Univ Rome, 55-59; asst resident anat, Univ Calif, Los Angeles, 59-60; instr, McGill Univ, 60-61; asst resident path, Karolinska Inst, Sweden, 61-63; asst resident, 63-65, PROF PATH, UNIV CALIF, LOS ANGELES, 65-; CHIEF DEPT PATH & HEAD ELECTRON MICROS, LOS ANGELES COUNTY HARBOR GEN HOSP, 63- Mem: Electron Micros Soc Am; Am Soc Cell Biol; Am Fertil Soc; Soc Study Reproduction; Ital Med Asn. Res: Reproductive biology; ultrastructural studies on embryogenesis, early reproduction and fertilization. Mailing Add: Dept of Path Harbor Gen Hosp Torrance CA 90509

ZAMCHECK, NORMAN, b Lynn, Mass, June 14, 17; m 74; c 4. GASTROENTEROLOGY, CLINICAL PATHOLOGY. Educ: Harvard Univ, AB, 39, MD, 43. Prof Exp: Asst, Fatigue Lab, Harvard Univ, 39; med res asst, Beth Israel Hosp, Boston, 39-42; resident path, Mass Gen Hosp & resident med, Peter Bent Brigham Hosp, 43; res fel exp path, Harvard Med Sch, 46-48; res fel med, Thorndike Mem Lab, Boston City Hosp, 48-51; clin assoc, 57-67, asst clin prof, 67-70, ASSOC CLIN PROF MED, HARVARD MED SCH, 70-; CHIEF, MALLORY GASTROINTESTINAL LAB, BOSTON CITY HOSP, 51- Concurrent Pos: Res assoc, Sch Med, Boston Univ, 51-; exec officer, Mallory Found, Boston, 59-70; dir, Leary Labs, 63-70; mem, President's Nat Cancer Plant & adv comt, Colo-rectal Cancer Prog, 71-72; vis physician, Harvard Med Serv, Boston City Hosp, 72- Mem: Am Gastroenterol Asn; Am Asn Cancer Res; Am Soc Exp Path; Am Soc Clin Nutrit; Am Fedn Clin Res. Res: Pathobiology of digestive disorders; laboratory medicine. Mailing Add: Boston City Hosp 784 Massachusetts Ave Boston MA 02118

ZAME, ALAN, b New York, NY, Aug 16, 41. MATHEMATICS. Educ: Calif Inst Technol, BS, 62; Univ Calif, Berkeley, PhD(math), 65. Prof Exp: ASSOC PROF MATH, UNIV MIAMI, 65- Mem: Am Math Soc; Am Math Asn. Res: Number theory; functional and combinatorial analysis. Mailing Add: Dept of Math Univ of Miami Coral Gables FL 33124

ZAME, WILLIAM ROBIN, b Long Beach, NY, Nov 4, 45. MATHEMATICS. Educ: Calif Inst Technol, BS, 65; Tulane Univ, MS, 67, PhD(math), 70. Prof Exp: Evans instr math, Rice Univ, 70-72; asst prof, State Univ NY Buffalo, 72-75; ASSOC PROF MATH, TULANE UNIV, 75- Mem: Am Math Soc; Math Asn Am. Res: Several complex variables; Banach algebras, C - algebras. Mailing Add: Dept of Math Tulane Univ New Orleans LA 70118

ZAMECNIK, PAUL CHARLES, b Cleveland, Ohio, Nov 22, 12; m 36; c 3. MEDICINE. Educ: Dartmouth Col, AB, 33; Harvard Univ, MD, 36. Hon Degrees: Dr, Univ Utrecht, 66; DSc, Columbia Univ, 71. Prof Exp: Resident med, C P Huntington Mem Hosp, Boston, 36-37; intern, Univ Hosps, Cleveland, 38-39; Moseley traveling fel from Harvard Univ, Carlsberg Lab, Copenhagen, 39-40; Finney-Howell fel, Rockefeller Inst, 41-42; from instr to assoc prof, 42-56, COLLIS P HUNTINGTON PROF ONCOL MED & DIR J C WARREN LABS, HARVARD MED SCH, 56- Concurrent Pos: Physician, Mass Gen Hosp, 56- Honors & Awards: James Ewing Award, 63; Borden Award, 65; John Collins Warren Triennial Prize, 46 & 50; Am Cancer Soc Nat Award, 68; Passano Award, 70. Mem: Nat Acad Sci; Asn Am Physicians; Am Acad Arts & Sci; Am Soc Biol Chemists; Am Asn Cancer Res (pres, 64-65). Res: Protein metabolism of normal and malignant tissues; cancer; radioactive isotope tracers. Mailing Add: Huntington Labs Mass Gen Hosp Boston MA 02114

ZAMEL, NOE, b Rio Grande, Brazil, Apr 2, 35; m 59; c 3. RESPIRATORY PHYSIOLOGY. Educ: Col Med, Fed Univ Rio Grande do Sul, Brazil, MD, 58. Prof Exp: From instr to assoc prof med, Col Med, Fed Univ Rio Grande do Sul, Brazil, 62-70; assoc prof med & dir respiratory physiol, Col Med, Univ Nebr, Omaha, 70-72; ASSOC PROF MED, FAC MED & DIR RESPIRATORY PHYSIOL, TRIHOSP RESPIRATORY SERV, UNIV TORONTO, CAN, 72- Concurrent Pos: Res fel respiratory physiol, Cardiovasc Res Inst, Univ Calif, San Francisco, 69. Honors & Awards: Miguel Couto Award, Col Med, Fed Univ Rio Grande do Sul, 58; Cecile Lehman Mayer Award, Am Col Chest Physicians, 69. Mem: Am Col Chest Physicians; Am Thoracic Soc; Am Fedn Clin Res; Can Soc Clin Invest; Sigma Xi. Res: Development of new tests of lung mechanics for early detection of chronic obstructive lung diseases; effects of cigarette smoking in its reversibility after cessation on lung mechanics. Mailing Add: 207 Torresdale Ave Willowdale ON Can

ZAMENHOF, PATRICE JOY, b Santa Rosa, Calif, Apr 2, 34; m 61. MOLECULAR GENETICS. Educ: Univ Calif, Berkeley, AB, 56, PhD(microbiol), 62. Prof Exp: Res assoc biochem, Col Physicians & Surgeons, Columbia Univ, 62-64; asst prof, 64-71, ASSOC PROF BIOL CHEM, SCH MED, UNIV CALIF, LOS ANGELES, 71- Concurrent Pos: USPHS res grants, Univ Calif, Los Angeles, 65-68 & 70-, career develop award, Nat Inst Gen Med Sci, 67, mem cancer ctr, 75- Mem: Am Soc Microbiol; Genetics Soc Am; Am Soc Biol Chemists. Res: Mutagenic mechanisms; genetic instability in microorganisms; mutator genes; repair of genetic damage; functional interactions of inactive mutant proteins. Mailing Add: Dept of Biol Chem Univ of Calif Sch of Med Los Angeles CA 90024

ZAMENHOF, STEPHEN, b Warsaw, Poland, June 12, 11; nat US; m 61. BIOCHEMISTRY. Educ: Warsaw Polytech Sch, Dr Tech Sci, 36; Columbia Univ, PhD(biochem), 49. Prof Exp: Res assoc biochem, Columbia Univ, 50-51, from asst prof to assoc prof, 51-64; PROF MICROBIOL GENETICS & BIOL CHEM, SCH MED, UNIV CALIF, LOS ANGELES, 64- Concurrent Pos: Guggenheim fel, 58-59. Mem: Am Soc Biol Chem; Am Chem Soc; Am Soc Microbiol; Genetics Soc Am; Am Acad Microbiol. Res: Microbial genetics; nucleic acids; growth hormone; prenatal brain development. Mailing Add: Dept of Microbiol & Immunol Univ of Calif Sch of Med Los Angeles CA 90024

ZAMICK, LARRY, b Winnipeg, Man, Mar 15, 35; m 66; c 2. NUCLEAR PHYSICS. Educ: Univ Man, BSc, 57; Mass Inst Technol, PhD(physics), 61. Prof Exp: Instr physics, Princeton Univ, 62-65, res assoc, 65-66; assoc prof, 66-70, PROF PHYSICS, RUTGERS UNIV, 70- Mem: Fel Am Phys Soc. Res: High energy deuteron-nucleus scattering; nuclear structure studies with the shell model. Mailing Add: Dept of Physics Rutgers Univ New Brunswick NJ 08903

ZAMIKOFF, IRVING IRA, b Toronto, Ont, Feb 13, 43; m 67; c 2. DENTISTRY, PROSTHODONTICS. Educ: Univ Toronto, DDS, 67; Univ Mich, MS, 70; Am Bd Prosthodont, dipl, 72. Prof Exp: ASSOC PROF PROSTHODONT, SCH DENT, LA STATE UNIV, NEW ORLEANS, 70- Concurrent Pos: Vis dentist, Charity Hosp, New Orleans, 70- Mem: Am Dent Asn; Int Asn Dent Res; Can Dent Asn; Am Col Prosthodont. Mailing Add: Dept of Prosthodont La State Univ Med Ctr New Orleans LA 70119

ZANCA, PETER, b New York, NY, Oct 29, 08; m 39; c 5. RADIOLOGY. Educ: Univ Rome, MD, 35; Am Bd Radiol, cert, 51. Prof Exp: Intern, St Joseph's Hosp, Lancaster, Pa, 36-37; radiol resident, Am Oncol Hosp, Philadelphia, 37-38; chief x-ray serv, William Beaumont Gen Hosp, El Paso, Tex, 47-49; radiol resident, Oliver Gen Hosp, Augusta, Ga, 50 & Letterman Gen Hosp, San Francisco, 50-51; chief x-ray serv, Camp Cook Army Hosp, 51-52 & Osaka Army Hosp, 53-54; chief radiol serv, Tokyo Army Hosp, 54-55 & Ft Dix Army lHosp, 55-58; consult, 4th Army Surgeon & dir, X-ray Sch Technol, Ft Sam Houston, Tex, 58-63; chmn dept radiol, Bexar County Hosp, 63-67; chmn dept, 67-74, PROF RADIOL, UNIV TEX MED SCH SAN ANTONIO, 67- Concurrent Pos: Fel radiol, NY Univ, 46; clin prof, Med Sch, Baylor Univ, 60-63; consult, San Antonio State Tuberc Hosp, 63-; radiol consult, Far East Command, 65; dir x-ray sch technol, St Phillips Col, Tex, 71-; consult, Vet Admin Hosp, San Antonio, 73-; chmn radiol dept, Brooke Gen Hosp, 58-63; vchief staff, Cancer Therapy Res Ctr, San Antonio, 74-75. Mem: Am Col Radiol; Radiol Soc NAm; AMA; AAAS; Am Asn Univ Profs. Res: Pulmonary diseases and pancreas; mammography and cancer of the breast. Mailing Add: 130 Wyndale Dr San Antonio TX 78209

ZAND, ROBERT, b New York, NY, Jan 7, 30; m 52; c 3. BIOPHYSICAL CHEMISTRY, NEUROCHEMISTRY. Educ: Univ Mo, BS, 51; Polytech Inst Brooklyn, MS, 54; Brandeis Univ, PhD(chem), 61. Prof Exp: Res chemist, Irvington Varnish & Insulator, 53; assoc res biophysicist, 63-73, asst prof, 68-73, ASSOC PROF BIOCHEM & RES BIOPHYSICIST, UNIV MICH, ANN ARBOR, 73- Concurrent Pos: NIH fel, Harvard Med Sch, 61-63; fel, Brandeis Univ, 61-63; ODOL Found lectr, Univ Buenos Aires, 72. Mem: Am Chem Soc; Biophys Soc; Am Soc Neurochem; Am Soc Biol Chem. Res: Conformation of proteins and small molecules by spectroscopic methods; use of amino acid analogs to determine stereochemical requirements of amino acyl synthetase and transport proteins; mechanism of the brain toxicity of bilirubin in the neonate. Mailing Add: Biophys Res Div Univ of Mich Inst of Sci & Tech Ann Arbor MI 48109

ZAND, SIAVOSH MARC, b Tehran, Iran, Dec 9, 38. HYDROLOGY, ENVIRONMENTAL SCIENCES. Educ: Univ Tehran, BCE, 61; Univ Del, MCE, 66; Univ Calif, Berkeley, PhD(hydraul), 69. Prof Exp: Consult, Resources for Future, Inc, 67-68; systs hydrologist, Automated Environ Systs, Inc, NY, 68-69; RES HYDROLOGIST, US GEOL SURV, 69- Mem: Am Soc Civil Engrs; Am Geophys Union. Res: Hydraulics. Mailing Add: US Geol Surv 345 Middlefield Rd Menlo Park CA 94025

ZANDER, ARLEN RAY, b Shiner, Tex, Dec 12, 40; m 64; c 1. NUCLEAR PHYSICS, ATOMIC PHYSICS. Educ: Univ Tex, Austin, BS, 64; Fla State Univ, PhD(nuclear physics), 70. Prof Exp: Res physicist, Phillips Petrol Co, 64-65; asst prof, 70-74, ASSOC PROF PHYSICS & EXTERNAL GRANTS COORDR, E TEX STATE UNIV, 74- Mem: Am Phys Soc; Am Asn Physics Teachers. Res: Direct nuclear reaction mechanisms; experimental fast neutron activation studies; x-ray fluorescence studies utilizing charged particle accelerators. Mailing Add: Dept of Physics E Tex State Univ Commerce TX 75428

ZANDER, DONALD VICTOR, b Bellingham, Wash, Feb 14, 16; m 45; c 3. AVIAN PATHOLOGY. Educ: Univ Calif, BS, 41, PhD(comp path), 53; Colo State Univ, MS, 45, DVM, 50. Prof Exp: Asst poultry husb, Univ Calif, 37-41, asst specialist & lectr vet med, 50-53, asst prof, 53-55; lab instr bact, Colo State Univ, 48; dir res lab, Heisdorf & Nelson Farms, Inc, 55-71, vpres, 61-72, DIR HEALTH RES, H&N INC, 72- Mem: Poultry Sci Asn; Am Vet Med Asn; Am Asn Avian Path (pres, 65-66); World Poultry Sci Asn. Res: Avian diseases; pathology, diagnosis and epizootiology; poultry husbandry and nutrition; pathogen-free poultry. Mailing Add: 14720 Redmond-Woodinville Rd Redmond WA 98052

ZANDER, HELMUT A, b Bautzen, Ger, Oct 25, 12; nat US; m 42; c 4. DENTISTRY. Educ: Univ Würzburg, DMd, 34; Northwestern Univ, DDS, 38, MS, 40. Hon Degrees: Dr, Univ Zurich, 72. Prof Exp: Asst children's dent, Northwestern Univ, 38-40, res assoc, 40-42; asst prof clin dent, Dent Sch, Tufts Col, 42-47, from assoc prof to prof oral pediat, 47-51; prof periodont, Sch Dent, Univ Minn, 51-57; clin assoc prof dent, 66-72, PROF CLIN DENT & DENT RES, SCH MED & DENT, UNIV ROCHESTER, 72-; HEAD DEPT PERIODONT, EASTMAN DENT CTR, 57- Mem: Am Soc Dent for Children; Soc Exp Biol & Med; Am Dent Asn; Am Acad Periodont; Int Asn Dent Res. Res: Pulp pathology; caries etiology; preventive dentistry; dental histology; antibiosis; periodontology. Mailing Add: Eastman Dent Ctr 800 Main St E Rochester NY 14603

ZANDER, VERNON EMIL, b Toledo, Wash, Feb 3, 39; m 66; c 1. MATHEMATICS. Educ: Univ Wash, BS, 61; Catholic Univ, MS, 65, PhD(math), 69. Prof Exp: Mathematician, NIH, 61-66; asst prof, 68-72, ASSOC PROF MATH, W GA COL, 72- Mem: Am Math Soc; Math Asn Am. Res: Development theory for integration of vector-valued functions on product spaces and integration of infinite products of vector-valued functions from Orlicz spaces. Mailing Add: Dept of Math W Ga Col Carrollton GA 30117

ZANDLER, MELVIN E, b Wichita, Kans, Nov 28, 37; m 59; c 4. PHYSICAL CHEMISTRY. Educ: Friends Univ, BA, 60; Univ Wichita, MS, 63; Ariz State Univ, PhD(phys chem), 66. Prof Exp: Asst prof, 66-75, ASSOC PROF CHEM, WICHITA STATE UNIV, 75- Concurrent Pos: Petrol Res Fund res grant, 68-70. Mem: AAAS; Am Chem Soc; Sigma Xi. Res: Theory of liquids and liquid mixtures; statistical thermodynamics; semi-empirical molecular orbital calculations; generation and optimization of reactive potential energy surfaces. Mailing Add: Dept of Chem Wichita State Univ Wichita KS 67208

ZANDY, HASSAN F, b Tehran, Iran, Mar 11, 12; US citizen; m 43; c 3. PHYSICS, SPECTROSCOPY. Educ: Univ Birmingham, BSc, 35, MSc, 49; Univ Tehran, PhD(physics, 53. Hon Degrees: PhD, Univ Tehran, 52. Prof Exp: Instr physics, Univ

Tehran, 37-46, asst prof, 50-53; lectr, Univ Leicester, 47-50; Fulbright fel, Brooklyn Polytech Inst, 53-54; PROF PHYSICS, UNIV BRIDGEPORT, 54- Concurrent Pos: Mem vis scientist prog physics, NSF, 59-; NSF fac fel, 63-64, res grant plasma res, Univ Bridgeport, 72-73. Mem: Am Asn Physics Teachers. Res: High vacuum technique; measurement of temperature of hot plasmas by x-ray spectroscopy. Mailing Add: Dept of Physics Univ of Bridgeport Bridgeport CT 06602

ZANEVELD, JACQUES RONALD VICTOR, b Leiderdorp, Neth, July 12, 44; US citizen; m 71. Educ: Old Dom Univ, BS, 64; Mass Inst Technol, SM, 66; Ore State Univ, PhD(oceanog), 71. Prof Exp: RES ASSOC OCEANOG, ORE STATE UNIV, 71- Mem: Am Geophys Union; Optical Soc Am. Res: Theoretical and experimental relationships between light attenuation and scattering properties of the ocean and the dynamic properties of suspended and dissolved materials. Mailing Add: Sch of Oceanog Ore State Univ Corvallis OR 97331

ZANEVELD, JACQUES SIMON, b The Hague, Netherlands, Dec 9, 09; US citizen; m 39; c 2. BIOLOGICAL OCEANOGRAPHY, PHYCOLOGY. Educ: State Univ Leiden, BS, 36, MS, 38, PhD(bot, algae), 41. Prof Exp: Cur fungi, Nat Herbarium, Leiden, Netherlands, 38-42; lectr biol, Gymnasium Haganum, 42-48, 52-54; phycologist, Lab Invest Sea, Djakarta, Indonesia, 48-52; dir marine biol, Caribbean Marine Biol Inst, Curacao, Netherlands Antilles, 54-59; prof biol & chmn dept, 59-65, prof oceanog & dir dept, 65-68, SAMUEL L & FAY M SLOVER PROF OCEANOG, OLD DOM UNIV, 68- Concurrent Pos: Mem tech comt seaweeds, Indo-Pac Fisheries Coun, Food & Agr Orgn, UN, 49-52, chmn, Caribbean Fisheries Sem, 59; mem, Int Oceanog Found, 60- & Zaneveld Glacier, Antarctica, 69; vis prof, Univ PR, 57; Rockefeller Found res grant, 58-59; vis prof, Va Inst Marine Sci, 60 & Marine Lab, Duke Univ, 62; prin investr, NSF res grant, 61-64 & dir undergrad res partic prog, 61-70, prin investr, NSF Antarctic Res Prog, 63-68. Mem: AAAS; Am Inst Biol Sci; Phycol Soc Am; Int Phycol Soc; Royal Netherlands Bot Soc. Res: Economy, ecology, taxonomy and distribution of benthic marine algae, especially effects of environmental factors upon horizontal and vertical zonation of marine algae in North Sea, Indonesian Archipelago, Caribbean Sea, East Coast, US and Ross Sea, Antarctica. Mailing Add: Dept of Oceanog Old Dom Univ Box 6173 Norfolk VA 23508

ZANEVELD, LOURENS JAN DIRK, b The Hague, Netherlands, Mar 22, 42; US citizen; m 65. BIOCHEMISTRY, MEDICINE. Educ: Old Dom Col, BSc, 63; Univ Ga, DVM, 67, MS, 68, PhD(biochem), 70. Prof Exp: Res assoc biochem, Univ Ga, 69-71; asst prof obstet & gynec & res assoc, Univ Chicago, 71-74; sci adv & chief pop res ctr, IIT Res Inst, 74-75; ASSOC PROF PHYSIOL, UNIV OF ILL MED CTR, 75- Mem: Am Soc Exp Path; Soc Study Reproduction. Res: Reproduction; biochemistry of male and female genital tract secretions and spermatozoa. Mailing Add: Dept of Physiol Box 6998 Univ of Ill at the Med Ctr Chicago IL 60680

ZANGER, MURRAY, b New York, NY, May 5, 32; m 62; c 2. PHYSICAL ORGANIC CHEMISTRY. Educ: City Col New York, BS, 53; Univ Kans, PhD(org chem), 59. Prof Exp: Fel chem, Univ Wis, 59-60; res chemist, Marshall Lab, E I du Pont de Nemours & Co, 60-64; asst prof, 64-67, ASSOC PROF ORG CHEM, PHILADELPHIA COL PHARM & SCI, 67- Concurrent Pos: USPHS res grant phenothiazine chem, 66-69. Mem: AAAS; Am Chem Soc; Franklin Inst. Res: Organophosphorus compounds; cyclopropane and phenothiazine chemistry; carbenes; fluorcarbon and anionic polymers; sulfa drugs; organic mechanisms; nuclear magnetic resonance spectroscopy. Mailing Add: Dept of Chem 43rd St & Kingsessing Ave Philadelphia PA 19104

ZANGERL, RAINER, b Winterthur, Switz, Nov 19, 12; nat US; m 37; c 2. VERTEBRATE PALEONTOLOGY. Educ: Univ Zurich, PhD(paleont), 36. Prof Exp: Res partner, Prof H R Schinz, 36-37; guest researcher comp anat, Harvard Univ, 37-38; prof vet anat & head dept, Sch Vet Med, Middlesex Univ, 38-39; instr zool & comp morphol, Univ Detroit, 39-42; asst prof comp anat, Univ Notre Dame, 43-45; cur fossil reptiles, Chicago Natural Hist Mus, 45-63, CHIEF CUR DEPT GEOL, FIELD MUS NATURAL HIST, 63- Concurrent Pos: Lectr vert paleont, Univ Chicago, 47-; adj prof anat, Univ Ill Col Med; mem Smithsonian Coun, Smithsonian Inst, 70-; mem vert paleont expeds, Chicago Natural Hist Mus, Carnegie Mus & Zool Mus, Zurich. Mem: Swiss Soc Paleont. Res: Comparative anatomy; vertebrate paleontology; marine paleoecology. Mailing Add: Dept of Geol Field Mus of Natural Hist Chicago IL 60605

ZANJANI, ESMAIL DABAGHCHIAN, b Resht, Iran, Dec 23, 38; m 63; c 3. HEMATOLOGY, PHYSIOLOGY. Educ: NY Univ, BA, 64, MS, 66, PhD(hemat), 69. Prof Exp: From asst to res assoc exp hemat, NY Univ, 65-70; asst prof med & physiol, 70-74, ASSOC PROF PHYSIOL, MT SINAI SCH MED, 74- Mem: AAAS; Harvey Soc; Am Soc Hemat; Am Soc Zool; NY Acad Sci. Res: Experimental hematology; hemopoietic stimulating factor; mechanisms of blood cell production and release; renal involvement in erythropoiesis; erythropoiesis in submammalian species; fetal erythropoiesis. Mailing Add: Dept of Physiol Mt Sinai Sch of Med New York NY 10029

ZANKEL, KENNETH L, b New York, NY, Mar 29, 30; m 55; c 1. BIOPHYSICS. Educ: Rutgers Univ, BS, 51; Fla State Univ, MS, 55; Mich State Univ, PhD(physics), 58. Prof Exp: Asst res prof physics, Mich State Univ, 58-59; asst prof, Univ Ore, 59-63; Fulbright fel, Univ Heidelberg, 63-64; fel, Calif Inst Technol, 64-66; res fel, Sect Genetics Develop & Physiol, Cornell Univ, 66-69; scientist, Res Inst For Advan Studies, 69-75; MEM STAFF, ENVIRON TECHNOL CTR, MARTIN MARIETTA LABS, RES & DEVELOP CTR, 75- Mem: Fel Acoust Soc Am; Am Asn Physics Teachers. Res: Biophysics; photobiology. Mailing Add: Environ Technol Ctr Res Develop Ctr Martin Marietta Labs 1450 S Rolling Rd Baltimore MD 21227

ZANNONI, VINCENT G, b New York, NY, June 12, 28. BIOCHEMISTRY, PHARMACOLOGY. Educ: City Col New York, BS, 51; George Washington Univ, MS, 56, PhD(biochem), 59. Prof Exp: Biochemist, Goldwater Mem Hosp, New York, 51-54 & Nat Heart Inst, 54-56; res chemist, Nat Inst Arthritis & Metab Dis, 57-63; from asst prof to prof biochem pharmacol, Sch Med, NY Univ, 63-74; PROF PHARMACOL, MED SCH, UNIV MICH, ANN ARBOR, 74- Mem: AAAS; Am Soc Biol Chemists; Am Soc Pharmacol & Exp Therapeut; NY Acad Sci. Res: Inborn errors of metabolism; mechanisms of reactions; amino acid metabolism; enzymology; biochemical pharmacology. Mailing Add: Dept of Pharmacol Univ of Mich Med Sch Ann Arbor MI 48109

ZANNUCCI, JOSEPH SALVATORE, organic chemistry, analytical chemistry, see 12th edition

ZANOWIAK, PAUL, b Little Falls, NJ, July 11, 33; m 57; c 4. PHARMACEUTICS. Educ: Rutgers Univ, BS, 54, MS, 57; Univ Fla, PhD(pharm), 59. Prof Exp: Instr pharm, Col Pharm, Univ Fla, 58-59; res & develop chemist, Noxell Corp, Md, 59-64; from asst prof to assoc prof pharmaceut, Sch Pharm, WVa Univ, 64-71; PROF PHARMACEUT & CHMN DEPT, TEMPLE UNIV, 71- Concurrent Pos: Am Found Pharm Educ fel; Mead-Johnson res grant. Honors & Awards: Lunsford Richardson Pharm Award. Mem: AAAS; Soc Cosmetic Chem; Am Pharmaceut Asn; Acad Pharmaceut Sci; Am Soc Hosp Pharmacists. Res: Design and evaluation of dosage forms, especially those used in topical medication. Mailing Add: Dept of Pharm 3223 N Broad St Temple Univ Sch of Pharm Philadelphia PA 19140

ZANZUCCHI, PETER JOHN, b Syracuse, NY, Apr 21, 41; m 67. ANALYTICAL CHEMISTRY. Educ: Le Moyne Col, NY, BS, 63; Univ Ill, Urbana, MS, 65, PhD(chem), 67. Prof Exp: STAFF CHEMIST, DAVID SARNOFF RES CTR, RCA CORP, 67- Mem: Am Chem Soc; Optical Soc Am; Am Soc Testing & Mat. Res: Measurement of the optical properties of inorganic materials, particularly semiconductor materials in the wavelength range 0.2 to 200 micrometers. Mailing Add: David Sarnoff Res Ctr RCA Corp Princeton NJ 08540

ZAO, ZANG Z, b Soochow, China, Aug 24, 16; US citizen; m 55; c 2. CARDIOLOGY. Educ: Med Col Düsseldorf, MD, 41. Prof Exp: Instr cardiol, Univ Tex Med Br, 57-58; instr med, Sch Med, Tulane Univ, 59-60; asst prof, Sch Med, Univ Miami, 60-62; dir vectorcardiogram lab, Degoesbriand Hosp, Col Med, Univ Vt, 62-67; head EKG & vectocardiogram labs, Cox Heart Inst, Ohio, 67-72; MEM STAFF, J ELECTROCARDIOL, 72- Concurrent Pos: Nat Heart Inst spec res fel, 59-61, res grant, 63-66; Vt Heart Asn res grant, 66; founder & first ed, J Electrocardiol; bd chmn & co-chmn dept res, Res in Electrocardiol, Inc, 68- Honors & Awards: 5th Interam Cong Cardiol Achievement Award. Mem: Sr mem Am Fedn Clin Res; fel Am Geriat Soc; fel Am Col Cardiol; Am Col Angiol. Res: Fundamental and clinical research in electrocardiology with special reference to electrocardiography and vectorcardiography. Mailing Add: J of Electrocardiol PO Box 923-B Pac Beach Sta San Diego CA 92109

ZAPHYR, PETER ANTHONY, b Wheeling, WVa, Sept 4, 26; m 56; c 2. MATHEMATICS. Educ: Bethany Col, WVa, BS, 48; Univ WVa, MS, 49; Univ Pittsburgh, PhD(math), 57. Prof Exp: Instr math, Univ WVa, 49-50; asst, Ill Inst Technol, 50-51 & Univ Pittsburgh, 51-52; analyst, 52-61, mgr digital anal & comput sect, 61-65, asst to dir, Anal Dept, 65-69, mgr eng comput systs, Nuclear Energy Systs, 69-73, MGR, ENG COMPUT SERV, POWER SYSTS COMPUT CTR, WESTINGHOUSE ELEC CORP, PITTSBURGH, 73- Mem: Asn Comput Mach; Opers Res Soc Am. Res: Administration of industrial computing services and advanced applications of computers in engineering science, manufacturing and management. Mailing Add: Power Systs Comput Ctr Westinghouse Elec Corp 1310 Beulah Rd Pittsburgh PA 15235

ZAPISEK, WILLIAM FRANCIS, b Morris, NY, Mar 29, 35; m 59; c 3. BIOCHEMISTRY, DEVELOPMENTAL BIOLOGY. Educ: Syracuse Univ, BA, 60; Univ Conn, MS, 65, PhD(biochem), 67. Prof Exp: Fel biochem, Los Alamos Sci Lab, 67-68; asst prof, 68-72, ASSOC PROF BIOCHEM, CANISIUS COL, 72- Mem: AAAS; Am Chem Soc; Soc Develop Biol. Res: Characterization of low molecular weight methylated ribonucleic acid species; ascorbic acid sulfate metabolism. Mailing Add: Dept of Chem Canisius Col Buffalo NY 14208

ZAPLATYNSKY, ISIDOR, physical metallurgy, see 12th edition

ZAPOLSKY, HAROLD SAUL, b Chicago, Ill, Dec 24, 35; m 62; c 2. THEORETICAL PHYSICS. Educ: Shimer Col, AB, 54; Cornell Univ, PhD(physics), 62. Prof Exp: Nat Acad Sci-Nat Res Coun res assoc physics, Goddard Inst Space Studies, New York, 62-63; res assoc, Univ Md, College Park, 63-65, asst prof, 65-70; from assoc prog dir to prog dir theoret physics, NSF, 70-73; PROF PHYSICS & CHMN DEPT, RUTGERS UNIV, 73- Mem: Am Phys Soc; Am Astron Soc. Res: Quantum electrodynamics; astrophysics; general relativity. Mailing Add: Dept of Physics Rutgers Univ New Brunswick NJ 08903

ZAPP, JOHN ADAM, JR, b Pittsburgh, Pa, June 22, 11; m 34; c 2. BIOCHEMISTRY. Educ: Haverford Col, BS, 32; Univ Pa, MA, 34, PhD(physiol chem), 38. Prof Exp: Instr res med, Univ Pa, 38-43; regional gas officer, Civilian Defense Region, US Off Civilian Defense, 43-44; tech aide, Nat Defense Res Comt, 44-46; asst, 46-48, asst dir, Haskell Lab, 48-52, DIR HASKELL LAB TOXICOL & INDUST MED, E I DU PONT DE NEMOURS & CO, 52- Concurrent Pos: Porter fel, 40-43; Girvin fel, 41-42; consult res comt, Res Found, Univ Del, 49-, comt acceptable concentrations of toxic dusts & gases, Am Nat Stands Inst, 57-, adv comt short-term inhalation limits, Commonwealth of Pa, 64-67, US Army Munitions Command Adv Group, 65-72 & panel biol & indust tech, comt res life sci, Nat Acad Sci-Nat Res Coun; mem bd trustees, Indust Health Found, 62- & Permanent Comt & Int Asn Occup Health, 64- Mem: AAAS; Soc Toxicol (pres, 67-68); Indust Hyg Asn; Am Soc Pharmacol & Exp Therapeut; Am Chem Soc. Res: Nitrogen extractives in mammalian muscle; chemical action of insulin; investigations of the toxic action of chemicals and techniques of industrial preventive medicine. Mailing Add: Haskell Lab E I du Pont de Nemours & Co Wilmington DE 19898

ZAPSALIS, CHARLES, b Lowell, Mass, Sept 22, 22; m 48. FOOD SCIENCE, CHEMISTRY. Educ: Springfield Col, BS, 52; Univ Mass, PhD(food sci, chem), 63. Prof Exp: Teacher, Jr High Sch, Mass, 52-53 & high, NY, 54-55; head sci dept high sch, Mass, 56-60; instrumental chemist, Beechnut Life Savers, Inc, 63, res mgr, 63-65; from asst prof to assoc prof, 65, PROF CHEM, FRAMINGHAM STATE COL, 65-, CHMN DEPT, 66- Mem: Am Chem Soc; Inst Food Technol. Res: Anthocyanins, chemical identification; pesticide methodology and characterization of tea components by gas chromatography; characterization of amino acids and polypeptides. Mailing Add: Dept of Chem Framingham State Col Framingham MA 01701

ZAR, JACOB L, b New York, NY, Mar 16, 17; m 39; c 2. LASERS. Educ: City Col New York, BS, 38; Stevens Inst Technol, MSME, 43; NY Univ, PhD(physics), 51. Prof Exp: Proj engr, Thomas A Edison, Inc, 40-47; pres, Nat Radiac, Inc, 50-56; staff engr, Repub Aviation Corp, 56-57; dir exp develop, Airborne Accessories Corp, 57-59; sr staff scientist, Gen Precision, Inc, 59-60; res mgr, Schlumberger Well Surv Corp, 60-62; PRIN RES SCIENTIST, AVCO EVERETT RES LAB, 62- Mem: Am Phys Soc; Am Inst Aeronaut & Astronaut. Res: Superconductivity; magnetic materials; laser applications; electron physics; system design, especially electromechanical systems; laser optical systems, windows and mirrors. Mailing Add: Avco Everett Res Lab Revere Beach Pkwy Everett MA 02149

ZAR, JERROLD HOWARD, b Chicago, Ill, June 28, 41; m 67; c 2. ECOLOGY, PHYSIOLOGY. Educ: Northern Ill Univ, BS, 62; Univ Ill, MS, 64, PhD(zool), 67. Prof Exp: Res assoc physiol ecol, Univ Ill, 67-68; asst prof, 68-71, ASSOC PROF ECOL & BIOSTATIST, NORTHERN ILL UNIV, 71- Mem: AAAS; Am Inst Biol Sci; Am Ornith Union; Am Soc Zoologists; Am Statist Asn. Res: Animal ecology; physiological ecology; environmental assessment; biostatistical analysis. Mailing Add: Dept of Biol Northern Ill Univ De Kalb IL 60115

ZARAFONETIS, CHRIS JOHN DIMITER, b Hillsboro, Tex, Jan 6, 14; m 43; c 1. INTERNAL MEDICINE. Educ: Univ Mich, BA, 36, MS, 37, MD, 41; Am Bd Internal Med, dipl, 50. Prof Exp: Externe, Simpson Mem Inst, Univ Mich, 40-41; intern, Boston City Hosp, 41-42; asst prof internal med & res assoc, Univ Mich, 47-

50; assoc prof med, Sch Med, Temple Univ, 50-55, clin prof, 55-57, prof clin & res med, 57-60, chief hemat sect, Univ Hosp, 50-60; PROF INTERNAL MED & DIR, SIMPSON MEM INST, MED SCH, UNIV MICH, ANN ARBOR, 60- Concurrent Pos: Res fel internal med, Med Sch, Univ Mich, 46-47; consult, Dept Defense, Directorate Res, Develop, Testing & Eval, Dept Army, Sci Adv Panel, Vet Admin Hosp, Mich & hist unit, Dept Surgeon Gen, US Dept Army; mem bd, Med in Pub Interest, Inc; asst ed, Am J Med Sci, 51-60. Honors & Awards: Typhus Comn Medal, US; Order of Ismail, Egypt. Mem: Fel Am Col Physicians; Am Soc Clin Pharmacol & Therapeut (vpres, 65-66, pres, 68-69); fel Int Soc Hemat; hon mem Agr Med Asn; Am Therapeut Soc (pres, 68-69). Res: Histoplasmosis; infectious mononucleosis; lymphogranuloma and herpes viruses; rickettsial disease; potassium para-aminobenzoate acid in collagen and bullous disorders and conditions with excess fibrosis; blood dyscrasias; lipid mobilizer hormone and dyscrasias. Mailing Add: Dept of Internal Med Univ of Mich Med Ctr Ann Arbor MI 48104

ZARATZIAN, VIRGINIA LOUIS, b Highland Park, Mich, Nov 15, 18. PHARMACOLOGY, TOXICOLOGY. Educ: Univ Mich, BS, 42 & 46, MS, 49; Wayne State Univ, PhD(Pharmacol), 56. Prof Exp: Res assoc neuropharmacol, Col Med, Univ Ill, 56-59; pharmacologist, US Food & Drug Admin, 59-61; res pharmacologist, USPHS, 61-63; survey pharmacologist, US Army, 63-66; pharmacologist, USDA, 66-68; PHARMACOLOGIST, PSYCHOPHARMACOL RES BR, NIMH, 68- Concurrent Pos: Nat Acad Sci chem consult, Nat Adv Ctr Toxicol, 57-59; assoc prof, Univ Cincinnati, 62-63; vis prof, Sch Med, Tex Tech Univ, 72. Mem: AAAS; Am Chem Soc; Soc Toxicol. Res: Psychotropic drugs adverse reactions; environmental health; toxicology of economic poisons and air pollutants. Mailing Add: NIMH Psychopharmacol Res Br 5600 Fishers Lane Rockville MD 20852

ZARCARO, ROBERT MICHAEL, b Springfield, Mass, Mar 4, 42; m 64; c 3. GENETICS, DEVELOPMENTAL BIOLOGY. Educ: Providence Col, BA, 64, MS, 66; Brown Univ, PhD(biol), 71. Prof Exp: Asst prof, 66-75, ASSOC PROF BIOL, PROVIDENCE COL, 75- Mem: AAAS; Genetics Soc Am. Res: Role of sulfhydryl compounds in mammalian pigmentation; genetic regulation of the multiple molecular forms of tyrosinase; role of protyrosinase in regulating melanogenesis. Mailing Add: Dept of Biol Providence Col Providence RI 02918

ZARCO, ROMEO MORALES, b Caloocan, Philippines, Oct 7, 20; m 51; c 3. IMMUNOCHEMISTRY, PUBLIC HEALTH. Educ: Univ Philippines, MD, 43; Johns Hopkins Univ, MPH, 54. Prof Exp: Physician internal med, Philippine Gen Hosp, 43-44; from instr to prof microbiol, Univ Philippines, 46-67; assoc dir biochem res, Cordis Corp, 67-73, DIR, CORDIS LABS, 73- Prof Exp: USPHS fel, 60-61; vis investr, Howard Hughes Med Inst, 61-62 & 64-67; asst prof, Univ Miami. Honors & Awards: Philippine Med Asn Res Award, 58. Mem: AAAS; Am Asn Immunol; NY Acad Sci. Res: Immunology of infectious diseases, including typhoid, cholera, leprosy and influenza; complement-anti-complementary factors from snake venom and immunosuppression and the separation and purification of the nine components of complement of human and guinea pig serum. Mailing Add: Cordis Labs 2140 N Miami Ave Miami FL 33127

ZARDECKI, ANDRZEJ, b Warsaw, Poland, Aug 26, 42; m 66; c 1. QUANTUM OPTICS. Educ: Univ Warsaw, BSc, 64, MSc, 64; Polish Acad Sci, DSc, 68. Prof Exp: From asst to asst prof physics, Warsaw Tech Univ, 64-73; ASST PROF PHYSICS, LAVAL UNIV, 73- Concurrent Pos: Fel, Laval Univ, 70-72. Mem: Can Asn Physicists; Am Phys Soc. Res: Functional techniques in the optical coherence theory with applications to photocount statistics problems; neoclassical radiation theory; pulse propagation in laser media; light scattering. Mailing Add: Dept of Physics Laval Univ Quebec PQ Can

ZARE, RICHARD NEIL, b Cleveland, Ohio, Nov 19, 39; m 63; c 3. CHEMICAL PHYSICS. Educ: Harvard Univ, BA, 61, PhD(chem phys), 64. Prof Exp: Fel, Harvard, 64; res assoc & fel, Joint Inst Lab Astrophys, Univ Colo, 64-65; asst prof chem, Mass Inst Technol, 65-66; asst prof physics, Univ Colo, 66-67, from asst prof to assoc prof physics & chem, 67-69; PROF CHEM, COLUMBIA UNIV, 69-, MEM, RADIATION LAB, 72- Concurrent Pos: Mem, Joint Inst Lab Astrophys, Univ Colo, 66-67, fel, 67-69, non-resident fel, 70; Alfred P Sloan res fel, 67-69; consult, Aeronomy Lab, Nat Oceanic & Atmospheric Admin, Radio Standards Physics Div, Nat Bur Standards, 68-; Higgins prof nat sci, Columbia Univ, 75- Mem: Nat Acad Sci; AAAS; fel Am Phys Soc; The Chem Soc; Am Chem Soc. Res: Problems associated with molecular photodissociation, molecular fluorescence and molecular chemiluminescence; application of lasers to chemical problems. Mailing Add: Dept Chem Box 308 Havemeyer Hall Columbia Univ New York NY 10027

ZAREM, ABE MORDECAI, b Chicago, Ill, Mar 7, 17; m 41; c 3. ELECTROOPTICS, ENGINEERING MANAGEMENT. Educ: Ill Inst Technol, BS, 39; Calif Inst Technol, MS, 40, PhD(elec eng), 44. Hon Degrees: LLD, Univ Calif, Santa Cruz, 67 & Ill Inst Technol, 68. Prof Exp: Group mgr elec eng, US Naval Ord Test Sta, 45-48; assoc dir, Stanford Res Inst, 48-56; pres & chmn bd, Electro-Optical Systems, Inc, Div Xerox Corp, 56-67, sr vpres & dir corp develop, Xerox Corp, 67-69, mgt & eng consult, 69-75, CHMN & CHIEF EXEC OFFICER, XEROX DEVELOP CORP, XEROX CORP, 76- Concurrent Pos: Mem adv coun, Sch Eng, Stanford Univ, 66-; mem comt 5 yr eng, Harvey Mudd Col, 67, mem eng clin adv bd, 69; mem vis comt, Div Eng & Appl Sci, Calif Inst Technol, 69- Mem: Am Inst Aeronaut & Astronaut; Inst Elec & Electronics Engrs. Res: Determination of the role of socio-biological factors on the development of innovative attitudes, creative thinking and motivational behavioral patterns. Mailing Add: Xerox Develop Corp 9200 Sunset Blvd Suite 700 Los Angeles CA 90069

ZAREM, HARVEY A, b Savannah, Ga, Feb 13, 32; m 58; c 3. PLASTIC SURGERY. Educ: Yale Univ, BA, 53; Columbia Univ, MD, 57. Prof Exp: Assoc prof surg, Univ Chicago, 66-73; PROF SURG & CHIEF DIV PLASTIC SURG, MED SCH, UNIV CALIF, LOS ANGELES, 73- Concurrent Pos: Markle scholar, Markle Found, 68. Mem: Plastic Surg Res Coun; Soc Univ Surgeons; Am Cleft Palate Asn; Soc Head & Neck Surgeons; Microcirc Soc. Res: Microvasculature; microsurgery. Mailing Add: Dept of Surg Univ of Calif Med Ctr Los Angeles CA 90024

ZAREMSKY, BARUCH, b Cleveland, Ohio, Sept 21, 26; m 51; c 3. ORGANOMETALLIC CHEMISTRY, PLASTICS CHEMISTRY. Educ: Western Reserve Univ, BS, 48, MS, 50, PhD(org chem), 54. Prof Exp: RES CHEMIST, FERRO CHEM DIV, FERRO CORP, BEDFORD, 53- Mem: Am Chem Soc. Res: Additives for polyvinyl chloride polypropylene, polycarbonates and polyesters; specialist in synthesis of organoten's. Mailing Add: 1708 Beaconwood South Euclid OH 44121

ZARGER, THOMAS GORDON, b Chambersburg, Pa, Oct 18, 18; m 47; c 1. FORESTRY, HORTICULTURE. Educ: Pa State Univ, BS, 40 & 49. Prof Exp: From jr forestry aide to forestry aide, Div Forestry, Fisheries & Wildlife Develop, Tenn Valley Authority, 40-43, from jr botanist to botanist, 43-48, staff forester, 49-65, forester, 65-66, STAFF FORESTER, RECLAMATION, REVEGETATION & TREE IMPROV, DIV FORESTRY, FISHERIES & WILDLIFE DEVELOP, TENN VALLEY AUTHORITY, 66- Mem: Am Soc Hort Sci; Soc Am Foresters. Res: Selection, propagation and testing of improved nut and other tree crop species; standing hardwood timber tree grades and vigor classes; forest tree improvement; coal surface mine reclamation techniques; wild land vegetation establishment. Mailing Add: Reclam Revegetat & Tree Improv Forestry Fish & Wildlife Develop Tenn Valley Authority Norris TN 37828

ZARING, WILSON MILES, b Shelbyville, Ky, Nov 9, 26; m 50; c 2. MATHEMATICS. Educ: Ky Wesleyan Col, AB, 50; Univ Ky, MA, 52, PhD(math), 55. Prof Exp: From instr to asst prof, 55-63, ASSOC PROF MATH, UNIV ILL, URBANA, 63- Mem: Am Math Soc; Math Asn Am. Res: Analysis. Mailing Add: Dept of Math Univ of Ill Urbana IL 61801

ZARISKI, OSCAR, b Kobrin, USSR, Apr 24, 99; nat US; m 24; c 2. MATHEMATICS. Educ: Univ Rome, Dr Math, 24. Hon Degrees: MA,l Harvard Univ, 47; DSc, Col Holy Cross, 49, Brandeis Univ, 65. Prof Exp: Int Educ Bd fel, Univ Rome, 25-27; Johnston scholar, Johns Hopkins Univ, 27-29, assoc math, 29-32, from assoc prof to prof, 32-45; vis prof, Univ Sao Paulo, 45; res prof, Univ Ill, 46-47; lectr, 40-41, prof, 47-61, Dwight Parker Robinson prof, 61-69, EMER DWIGHT PARKER ROBINSON PROF MATH, HARVARD UNIV, 69- Concurrent Pos: Vis mem, Inst Advan Study, 35-36, 39 & 60-61; lectr, Moscow, 36; Guggenheim fel, 39-40, Inst Hautes Etudes, Paris, 61 & 67. Honors & Awards: Cole Prize, Am Math Soc, 44; Nat Medal of Sci, 65. Mem: Nat Acad Sci; AAAS; Am Math Soc (pres, 69-70); Am Philos Soc; Math Asn Am. Res: Algebraic geometry; modern algebra; topology. Mailing Add: Dept of Math Harvard Univ Cambridge MA 02138

ZARKOWER, ARIAN, b Tarnopol, Poland, Oct 10, 29; Can citizen; m 60; c 2. VETERINARY MEDICINE, IMMUNOLOGY. Educ: Ont Vet Col, DVM, 56; Univ Maine, MS, 60; Cornell Univ, PhD(immunochem), 65. Prof Exp: Dist vet, NB Prov Vet Serv, 56-57; self-employed vet, 57-58; res asst animal path, Univ Maine, 58-70; res officer, Animal Dis Res Inst, Can Dept Agr, 60-62; res asst immunochem, Cornell Univ, 62-65; asst prof, 65-70, ASSOC PROF VET SCI & MEM STAFF, CTR AIR ENVIRON STUDY, 70- Mem: Can Vet Med Asn; Am Soc Microbiol; Can Soc Immunol. Res: Experimental pathology; immune response and resistance in animals to infections. Mailing Add: Dept of Vet Sci Pa State Univ University Park PA 16802

ZARNEGAR, BIZHAN MOHAMMED, organic chemistry, see 12th edition

ZAROOGIAN, GERALD E, microbiology, biochemistry, see 12th edition

ZAROSLINSKI, JOHN F, b Chicago, Ill, Sept 12, 25; m 51; c 2. PHARMACOLOGY, BIOCHEMISTRY. Educ: Univ Chicago, PhB, 49; Loyola Univ, Ill, PhD(pharmacol), 65. Prof Exp: Chemist, Armour Pharmaceut Co, 51-53; from pharmacologist to sr pharmacologist, Baxter Lab Inc, 53-58; sci dir, 58-65, VPRES RES & DEVELOP, ARNAR-STONE LABS, INC, DIV AM HOSP SUPPLY CORP, 65- Concurrent Pos: Prof lectr, Stritch Sch Med, Loyola Univ Chicago, 65-; lectr, Chicago Col Osteopathic Med, 69-; res consult, US Vet Hosp, Hines, Ill. Mem: Acad Pharmaceut Sci; Am Pharmaceut Asn; Am Chem Soc; NY Acad Sci; Am Soc Pharmacol & Exp Therapeut. Res: Pharmaceutical development; biochemical pharmacology; protein binding of drugs; evaluation of hypnotic drugs; pharmaceutical development and introduction of Intropin (dopamine) into therapy for treatment of shock in humans. Mailing Add: Arnar-Stone Labs Inc 601 E Kensington Rd Mt Prospect IL 60056

ZARRELLA, WILLIAM MICHAEL, b Fitchburg, Mass, June 5, 24; m 45; c 6. PHYSICAL CHEMISTRY, ORGANIC CHEMISTRY. Educ: Col Holy Cross, BS, 45; Clark Univ, MS, 47. Prof Exp: Res assoc org geochem, Mellon Inst, 47-49, jr fel, 49-53, fel, 53-57; sect head, 57-61, sr res chemist, 61-65, res assoc, 65-66, sect supvr geochem, 66-71, POLLUTION ADV, PROCESS SCI DEPT, GULF RES & DEVELOP CO, 71- Concurrent Pos: Mem geochem adv comt, Am Petrol Inst, 66-71. Mem: Am Chem Soc; Geochem Soc; Am Asn Petrol Geol. Res: Petroleum geochemistry; chemistry of the origin of petroleum; composition of petroleum; solubility and diffusion of hydrocarbons in water; chemistry of air and water contaminants; pollution abatement methods. Mailing Add: Process Sci Dept PO Drawer 2038 Gulf Res & Develop Co Pittsburgh PA 15230

ZARRUGH, LAURA HOFFMAN, b Chicago, Ill, Mar 30, 46. MEDICAL ANTHROPOLOGY. Educ: Univ Calif, Berkeley, BA, 68, MA, 69, PhD(anthrop), 74. Prof Exp: Res assoc, 74-75, LECTR ANTHROP, DEPT PSYCHIAT, SCH MED, UNIV CALIF, SAN FRANCISCO, 75- Mem: Am Anthrop Asn; Soc Med Anthrop. Res: Mexican and Mexican-American ethnography; migration studies; middle class American family life, health and culture. Mailing Add: Community Serv & Res Prog Dept of Psychiat Univ of Calif San Francisco CA 94143

ZARTMAN, DAVID LESTER, b Albuquerque, NMex, July 6, 40; m 63; c 2. GENETICS. Educ: NMex State Univ, BS, 62; Ohio State Univ, MS, 66, PhD(cytogenetics), 68. Prof Exp: Asst prof, 68-72, ASSOC PROF DAIRY SCI, N MEX STATE UNIV, 72- Concurrent Pos: NIH fel, 73. Mem: Fel AAAS; Genetics Soc Am; Am Inst Biol Sci. Res: Radiation genetics; sex and fertility control; reproductive disorders. Mailing Add: Dairy Dept Box 3149 NMex State Univ Las Cruces NM 88003

ZARTMAN, ROBERT EUGENE, b Lancaster, Pa, May 19, 36; m 75; c 7. GEOCHRONOLOGY. Educ: Pa State Univ, BS, 57; Calif Inst Technol, MS, 59, PhD(geol), 63. Prof Exp: Fel geol, Calif Inst Technol, 62; GEOLOGIST, ISOTOPE GEOL BR, US GEOL SURV, 62- Concurrent Pos: Vis assoc, Calif Inst Technol, 71-72; chmn working group on radiogenic isotopes of the Int Asn of Volcanology & Chem of the Earth's Interior, 73- Mem: Geol Soc Am; Mineral Soc Am; Am Geophys Union. Res: Determination of geologic age by the potassium-argon, rubidium-strontium and uranium-thorium-lead radiometric methods; study of geological processes by use of natural isotopic tracer systems. Mailing Add: Isotope Geol Br US Geol Surv Denver CO 80225

ZASADA, ZIGMOND ANTHONY, b Schenectady, NY, May 1, 09; m 37; c 1. FORESTRY. Educ: State Univ NY, BS, 31. Prof Exp: Forester, Chippewa Nat Forest, US Forest Serv, 33-45, proj leader, Lake States Forest Exp Sta, 45-51, res ctr leader, 51-61, res forester, DC, 61-63, asst dir, NCent Forest Exp Sta, Minn, 63-67; RES ASSOC, CLOQUET FORESTRY CTR, COL FORESTRY, UNIV MINN, 67- Mem: Soc Am Foresters. Res: Economics of forest management and utilization; silviculture; mechanized timber harvesting. Mailing Add: 1015 Third Ave NW Grand Rapids MN 55744

ZASKE, DARWIN ERHARD, b Wadena, Minn, Mar 20, 49. CLINICAL PHARMACOLOGY. Educ: Univ Minn, BS, 72, PharmD, 73. Prof Exp: CLIN PHARMACOLOGIST, ST PAULRAMSEY HOSP & MED CTR, 73-; ASST PROF PHARMACOL, UNIV MINN, 75- Concurrent Pos: Instr pharmacol, Univ Minn, 73-74. Mem: Am Burn Asn; Am Soc Hosp Pharm. Res: Clinical application of drug-kinetic principles with the goal being patient individualization of drug therapy to provide more optimal patient therapy. Mailing Add: St Paul-Ramsey Hosp & Med Ctr St Paul MN 55101

ZASLOWSKY, JOEL ALAN, organic chemistry, see 12th edition

ZASSENHAUS, HANS J, b Coblenz, Ger, May 28, 12, nat Can; m 42; c 3. MATHEMATICS. Educ: Univ Hamburg, Dr rer nat, 34, Dr habil, 38. Hon Degrees: MA, Glasgow Univ, 49; Dr, Univ Ottawa, 66 & McGill Univ, 74. Prof Exp: Asst instr, Univ Rostock, 34-36; asst, Univ Hamburg, 36-40; diaeten-docent math, Univ Hamburg, 40-47, assoc prof, 47-49; prof, McGill Univ, 49-59; prof, Univ Notre Dame, 59-64; RES PROF MATH, OHIO STATE UNIV, 64- Concurrent Pos: Vis prof math, Univ Calif, Los Angeles, 70 & Warwick Univ, Eng, 72; Fairchild scholarship, Calif Inst Technol, 74-75. Honors & Awards: Jeffery-Williams Lect Award, Can Math Cong, 74. Mem: Math Asn Am; Am Math Soc; Can Math Cong; fel Royal Soc Can; Sigma Xi. Res: Group theory; Lie algebra; number theory; geometry of numbers; applied mathematics. Mailing Add: 942 Spring Grove Lane Worthington OH 43085

ZATKO, DAVID A, b North Tonawanda, NY, Nov 12, 40; m 66; c 1. INORGANIC CHEMISTRY. Educ: Colgate Univ, BA, 62; Univ Wis, PhD(chem), 66. Prof Exp: Asst prof, 67-74, ASSOC PROF CHEM, UNIV ALA, 74- Mem: Am Chem Soc. Res: Analytic inorganic chemistry; coordination chemistry of silver II and silver III, palladium II, substituted ferrocenes; photoelectron spectroscopy; free radical ligands; oxidation-reductions in nonaqueous solvents. Mailing Add: Dept of Chem Univ of Ala University AL 35486

ZATUCHNI, GERALD IRVING, b Philadelphia, Pa, Oct 5, 35; m 58; c 3. OBSTETRICS & GYNECOLOGY. Educ: Temple Univ, AB, 54, MD, 58, MSc, 65. Prof Exp: Instr obstet & gynec, Med Sch, Temple Univ, 65-66; dir family planning, Pop Coun, Inc, 66-69; adv, Govt of India, 69-71; Pop Coun, Inc consult family planning & obstet, WHO, 71-73; ADV, GOVT IRAN, 73- Mem: AAAS; Am Fedn Clin Res; Am Fertil Soc; Am Col Obstet & Gynec. Res: Human reproductive research; contraceptive development; family planning; population study and research. Mailing Add: Population Coun 245 Park Ave New York NY 10017

ZATUCHNI, JACOB, b Philadelphia, Pa, Oct 8, 20; m 45; c 4. INTERNAL MEDICINE, CARDIOVASCULAR DISEASE. Educ: Temple Univ, AB, 41, MD, 44, MS, 50. Prof Exp: Clin prof med, 61-65, PROF MED, SCH MED & HOSP, TEMPLE UNIV, 65-, DIR DEPT MED, 74-; CHIEF SECT CARDIOL, EPISCOPAL HOSP, PHILADELPHIA, 69- Concurrent Pos: Teaching chief med, Episcopal Hosp, 58-67; fel coun clin cardiol, Am Heart Asn, 63- Mem: AAAS; Am Thoracic Soc; AMA; fel Am Col Physicians. Res: Cardiovascular diseases. Mailing Add: 1240 Valley Rd Villanova PA 19085

ZATZ, LESLIE M, b Schenectady, NY, Nov 2, 28; m 53; c 3. RADIOLOGY. Educ: Union Col, NY, BS, 48; Albany Med Col, MD, 52; Univ Pa, MMS, 59. Prof Exp: Intern, Univ Chicago Clins, 52-53; resident radiol, Hosp Univ Pa, 55-58; from instr radiol to assoc prof, 59-72, actg dir diag radiol, 66-67, PROF RADIOL, SCH MED, STANFORD UNIV, 72- Concurrent Pos: NIH spec res fel, Postgrad Med Sch, Univ London, 65-66; consult, Vet Admin Hosp, Palo Alto, Calif, 60-72, chief radiol, 72- Mem: Am Col Radiol; Asn Univ Radiol; AMA; Am Soc Neuroradiol. Res: New diagnostic radiologic methods. Mailing Add: Dept of Radiol Vet Admin Hosp Palo Alto CA 94304

ZATZ, MARION M, b New York, NY, Feb 10, 45. IMMUNOLOGY. Educ: Barnard Col, BA, 65; Cornell Univ, PhD(immunol & microbiol), 70. Prof Exp: Fel immunol, Hosp Spec Surg, 70-71; Damon Runyan fel biochem, Albert Einstein Col Med, 71-72; instr microbiol, Sch Med, Yale Univ, 72-73, asst prof microbiol & path, 73-77, dir, Tissue Typing Lab, 72-74; GUEST WORKER, IMMUNOL BR, NAT CANCER INST, 74- Mem: Am Asn Immunologists. Res: Investigation of spontaneous leukemogenesis in AKR-J mice; investigation of genetic basis of resistance to growth of lymphoma. Mailing Add: Immunol Br Nat Cancer Inst Bethesda MD 20014

ZATZICK, MICHAEL RAYMOND, b Bridgeport, Conn, Aug 16, 28; m 64; c 2. NUCLEAR PHYSICS. Educ: Univ Conn, BA, 55, MA, 56; Brown Univ, PhD(physics), 62. Prof Exp: Asst instr astron & physics, Univ Conn, 54-56; geophysicist, Air Force Cambridge Res Ctr, 56; asst instr physics, Brown Univ, 59-62, US AEC, res assoc, 62-63; sr nuclear physicist, Solid State Radiations, Inc, 63-72, vpres res & develop, SSR Instruments Co, 70-75; V PRES & GEN MGR, NDC SYSTS, INC, 75- Concurrent Pos: Consult, Radioisotope Div, Vet Mem Hosp, Providence, RI, 60-61. Mem: AAAS; Am Phys Soc. Res: Density and microstructure studies of polymers with low energy gamma rays; density profile measurements of upper atmosphere; radiation sensors; dosimetry; radiation damage at low, intermediate and high intensity; single and multiple channel photon counting instrumentation for optical and mass spectroscopy. Mailing Add: NDC Systs Inc 1237 Shamrock Ave Monrovia CA 91016

ZATZKIS, HENRY, b Holzminden, Ger, Apr 7, 15; nat US; m 51; c 2. THEORETICAL PHYSICS, APPLIED MATHEMATICS. Educ: Ohio State Univ, BS, 42; Ind Univ, MS, 44; Syracuse Univ, PhD(physics), 50. Prof Exp: Instr physics, Ind Univ, 42-44; instr, Univ NC, 44-46; instr math, Syracuse Univ, 46-51; instr, Univ Conn, 51-53; from asst prof to prof, 53-71, chmn dept, 60-71, DIS- TINGUISHED PROF MATH, NEWARK COL ENG, 71- Res: Theory of relativity; heat conduction; acoustics. Mailing Add: 5 Elliott Pl West Orange NJ 07052

ZATZMAN, MARVIN LEON, b Philadelphia, Pa, Aug 6, 27; m 51; c 2. PHYSIOLOGY. Educ: City Col New York, BS, 50; Ohio State Univ, MS, 52, PhD(physiol), 55. Prof Exp: Asst prof physiol, Ohio State Univ, 55-56; asst prof, 56-61, assoc prof, 61-73, PROF PHYSIOL, MED CTR, UNIV MO-COLUMBIA, 73- Mem: AAAS; Biophys Soc; Int Soc Nephrology; Am Physiol Soc. Res: Renal, cardiovascular, hibernation. Mailing Add: Dept of Physiol Univ of Mo Med Ctr Columbia MO 65201

ZAUDER, HOWARD L, b New York, NY, Sept 13, 23; m 53; c 2. ANESTHESIOLOGY. Educ: Univ Vt, AB, 47, MS, 49; Duke Univ, PhD(physiol, pharmacol), 52; NY Univ, MD, 55. Prof Exp: Res assoc pharmacol, Univ Vt, 51-53; from asst prof to assoc prof anesthesiol, Albert Einstein Col Med, Yeshiva Univ, 58-67; PROF ANESTHESIOL & CHMN DEPT, UNIV TEX MED SCH SAN ANTONIO & PROF PHARMACOL, UNIV TEX HEALTH SCI CTR SAN ANTONIO, 68- Mem: AAAS; Am Soc Pharmacol & Exp Therapeut; fel Am Col Anesthesiol. Res: Pharmacology of anesthetic agents; effects of radiation on response to anesthesia; pulmonary physiology; clinical application of gas chromatography. Mailing Add: Univ of Tex Med Sch 7703 Floyd Curl Dr San Antonio TX 78229

ZAUDERER, BERT, b Vienna, Austria, Mar 8, 37; US citizen; m 61; c 3. MAGNETOHYDRODYNAMICS, ENERGY CONVERSION. Educ: City Col New York, BME, 58; Mass Inst Technol, SM, 60, ScD(plasma physics), 62. Prof Exp: Res engr, 61-67, group leader, 67-70, MGR MAGNETOHYDRODYN PROGS, SPACE SCI LAB, GEN ELEC CO, 70- Concurrent Pos: Chmn prog comt symp eng aspects of magnetohydrodyn, 75-76. Mem: Am Inst Aeronaut & Astronaut; Am Phys Soc; Am Soc Mech Engrs. Res: Magnetohydrodynamic power generation for central station power uses. Mailing Add: Valley Forge Space Ctr Gen Elec Co PO Box 8555 Philadelphia PA 19101

ZAUGG, HAROLD ELMER, b Chicago, Ill, Feb 27, 16; m 40; c 3. ORGANIC CHEMISTRY. Educ: Oberlin Col, AB, 37; Univ Minn, PhD(org chem), 41. Prof Exp: Asst org chem, Univ Minn, 38-40; res chemist, 41-56, res scientist, 56-59, res fel, 59-72, SR RES FEL, ABBOTT LABS, 72- Concurrent Pos: Fel, Purdue Univ, 58; chmn, Gordon Conf Org Reactions & Processes, 61; mem med study sect, NIH, 64-68; vis prof, Univ Southern Calif, 66. Mem: Am Chem Soc. Res: Central nervous system drugs; organic syntheses and reaction mechanisms; solvent effects; chemistry of delocalized anions; amidoalkylations; medicinal chemistry of the cannabinoids. Mailing Add: Abbott Labs 1400 Sheridan Rd North Chicago IL 60064

ZAUGG, WALDO S, b LaGrande, Ore, Dec 13, 30; m 53; c 4. BIOCHEMISTRY. Educ: Brigham Young Univ, BA, 58, PhD(biochem), 61. Prof Exp: Fel, Enzyme Inst, Univ Wis, 61-62; fel, Charles F Kettering Res Lab, 62-63, staff scientist, 63-65; BIOCHEMIST, WESTERN FISH NUTRIT LAB, US FISH & WILDLIFE SERV, 65- Concurrent Pos: Asst prof, Antioch Col, 63-65; USPHS grant, 64-65. Mem: Am Soc Biol Chemists. Res: Oxidation-reduction reactions and bioenergetics in photosynthetic bacteria, plants, mammals and poikilotherms; physiology and biochemistry of anadromous fishes. Mailing Add: Western Fish Nutrit Lab Cook WA 98605

ZAUGG, WAYNE E, b West Bend, Iowa, Feb 9, 38; m 61; c 2. PHYSICAL CHEMISTRY. Educ: Walla Walla Col, BS, 61; Univ Wash, PhD(phys chem), 65. Prof Exp: Asst prof chem, Walla Walla Col, 66-67; asst prof, 67-74, asst prof biophys, 70-74, ASSOC PROF CHEM, LOMA LINDA UNIV, LA SIERRA CAMPUS, 74- Mem: Am Chem Soc. Res: Spin-label studies of biological structures and processes; polyene antifugal agent; lipid interactions in membrane systems. Mailing Add: Dept of Chem Loma Linda Univ La Sierra Campus Riverside CA 92505

ZAUKELIES, DAVID AARON, b Detroit, Mich, May 22, 25; m 56; c 2. PHYSICAL CHEMISTRY. Educ: Mich State Univ, BS, 46; Northwestern Univ, PhD(phys chem), 50. Prof Exp: Asst, Northwestern Univ, 46-49, res assoc phys chem, 49-50; res physicist, Dow Chem Co, 51-54; prof chem, Lee Col (Tenn), 54-55; res physicist, Chemstrand Corp, 55-60, Chemstrand Res Ctr, Inc, NC, 60-61, from assoc scientist to scientist, 61-70; scientist, 70-71, SCI FEL, TECH CTR, MONSANTO TEXTILES CO, 71- Mem: Am Chem Soc; Am Phys Soc. Res: Formation of fibers; deformation of solid polymers; triboelectrification; friction; fiber tenacity and fatigue; sorbtion in polymers; chromatography; instrumentation; dislocation in crystalline polymers; polyaromatic fibers, cords and composites; carpet fibers. Mailing Add: Tech Ctr Monsanto Textiles Co PO Box 1507 Pensacola FL 32502

ZAUNER, CHRISTIAN WALTER, b July 21, 30; m 57; c 3. EXERCISE PHYSIOLOGY, PULMONARY PHYSIOLOGY. Educ: West Chester State Col, BS, 56; Syracuse Univ, MS, 57; Southern Ill Univ, PhD(phys educ), 63. Prof Exp: Asst prof exercise physiol & res, Temple Univ, 63-65; assoc prof exercise physiol & res, 65-71, assoc prof med, 70-71, PROF EXERCISE PHYSIOL, RES & MED, UNIV FLA, 71- Concurrent Pos: Univ Fla fac develop grant & Thordgray Mem Fund, Dept Clin Physiol, Malmo Gen Hosp, Sweden, 71-72. Honors & Awards: Serv Citation, Coun Nat Coop in Aquatics, 72. Mem: Am Physiol Soc; Am Asn Health, Phys Educ & Recreation. Res: Lipid metabolism; exercise and training effects on lipids, work capacity, pulmonary and respiratory function; child athletes; exercise and training effects on circulation. Mailing Add: Col of Phys Educ Univ of Fla Gainesville FL 32611

ZAUSTINSKY, EUGENE MICHAEL, b Battle Creek, Mich, Oct 19, 26. GEOMETRY. Educ: Univ Calif, Los Angeles, AB, 48; Univ Southern Calif, AM, 54, PhD(math), 57. Prof Exp: Asst, Univ Southern Calif, 52-54, lectr, 54-57; asst prof math, San Jose State Col, 57; asst prof, Univ Calif, Santa Barbara, 58-61, asst prof, Univ Calif, Berkeley, 61-63; ASSOC PROF MATH, STATE UNIV NY STONY BROOK, 63- Concurrent Pos: Vis prof, Kent State Univ, 67 & Rockefeller Univ, 73. Mem: Am Math Soc; Math Asn Am; Math Soc France; Danish Math Soc; Swiss Math Soc. Res: Differential geometry and topology; synthetic differential geometry. Mailing Add: Dept of Math State Univ of NY Stony Brook NY 11794

ZAVADA, JOHN MICHAEL, JR, b Passaic, NJ, May 9, 42. THEORETICAL PHYSICS. Educ: Catholic Univ Am, Ba, 64; NY Univ, MS, 65, PhD(physics), 71. Prof Exp: RES PHYSICIST, PITMAN-DUNN LAB, FRANKFORD ARSENAL, 72- Concurrent Pos: NSF presidential intern appointment, Frankford Arsenal, 72-73. Mem: Am Phys Soc; Sigma Xi. Res: Light scattering from slightly rough surfaces; optical surface waves; quantum electrodynamics. Mailing Add: Pitman-Dunn Lab Frankford Arsenal Philadelphia PA 19137

ZAVARIN, EUGENE, b Sombor, Yugoslavia, Feb 21, 24; nat US; m 56; c 5. ORGANIC CHEMISTRY. Educ: Univ Göttingen, dipl, 49; Univ Calif, Berkeley, PhD(org chem), 54. Prof Exp: Asst, Univ Calif, Berkeley, 52-53, sr lab technician, Forest Prod Lab, 52-54, asst specialist, 54-56, asst chemist, 56-62, assoc forest prod chemist, 62-68, FOREST PROD CHEMIST, FOREST PROD LAB, UNIV CALIF, 68- Concurrent Pos: NIH fel, Inst Org Chem, Gif-sur-Yvette, France, 63. Mem: Am Chem Soc; Forest Prod Res Soc; Phytochem Soc NAm; Int Acad Wood Sci. Res: Chemosystematics; chemistry of natural products; chemistry of nonconventional wood bonding. Mailing Add: Forest Prod Lab Univ of Calif 1301 S 46th St Richmond CA 94804

ZAVIST, ALGERD FRANK, b Chicago, Ill, July 2, 21; m 51; c 4. CHEMISTRY. Educ: Roosevelt Col, BS, 43; Univ Chicago, MS, 49, PhD(chem), 50. Prof Exp: Lab technician, Infilco, Inc, 40-41; plant chemist, W H Barber Co, 41-42; instr, Wilson Jr Col, 49; chemist, Res & Develop, Gen Elec Co, 50-54, supvr chem eng processes, 54-58, supvr chem eng & insulation develop, Mat & Processes Lab, NY, 58-62, mgr chem lab, Major Appliance Labs, 62-68, MGR LABS, MAJOR APPLIANCE LABS, GEN ELEC CO, 68- Mem: Am Chem Soc; Soc Plastics Eng; The Chem Soc. Res: Reactions of free radicals in solution; physical organic and high polymer chemistry; relationships between molecular structure and electrical and physical properties. Mailing Add: Major Appliance Labs Gen Elec Co Bldg 35-1001 Appl Pk Louisville KY 40225

ZAVISTOSKI, JAMES GREGORY, inorganic chemistry, see 12th edition

ZAVISZA, DANIEL MAXMILLIAN, b Hazardville, Conn, Nov 28, 38; m 63; c 3. PHYSICAL ORGANIC CHEMISTRY. Educ: Col of the Holy Cross, BS, 60; Clark Univ, PhD(phys org chem), 66. Prof Exp: Res chemist, 66-74, SR RES CHEMIST, AM CYANAMID CO, 74- Res: Organic polymer research and development; protein chemistry; condensation polymers. Mailing Add: Am Cyanamid Co 1937 W Main St Stamford CT 06904

ZAVITKOVSKI, JAROSLAV, forest management, see 12th edition

ZAVITSANOS, PETROS D, b Spartochori-Lefkas, Greece, July 26, 31; US citizen; m 57; c 4. PHYSICAL CHEMISTRY. Educ: Univ Calif, Berkeley, AB, 55; Univ Ill, Urbana, MS, 57, PhD(phys chem), 59. Prof Exp: Mem tech staff, Bell Tel Labs, Pa, 59-60; phys chemist, Space Sci Lab, Missile & Space Div, 61-66, group leader aerothermochem, 66-72, CONSULT-TECHNOL ENG, RE-ENTRY & ENVIRON SYSTS DIV, GEN ELEC CO, 72- Mem: Am Chem Soc. Res: Solid state chemistry and isotope effects; semiconductors; high temperature chemistry; mass spectrometry; heterogeneous reaction kinetics; chemi-ionization; laser applications. Mailing Add: Re-entry & Environ Systs Div Gen Elec Co PO Box 8555 Philadelphia PA 19101

ZAVITSAS, ANDREAS ATHANASIOS, b Athens, Greece, July 14, 37; US citizen; m 59; c 1. PHYSICAL ORGANIC CHEMISTRY. Educ: City Col New York, BS, 59; Columbia Univ, MA, 61, PhD(chem), 62. Prof Exp: Res assoc chem, Brookhaven Nat Lab, 62-64; res chemist, Monsanto Co, Mass, 64-67; from asst prof to assoc prof, 67-73, PROF CHEM, LONG ISLAND UNIV, BROOKLYN CTR, 73-, GRAD DEAN, 75- Concurrent Pos: Lectr, New Sch Soc Res, 61-64. Mem: Am Chem Soc; NY Acad Sci. Res: Organic free-radical chemistry; deuterium isotope effects; phenolic resin. Mailing Add: Dept of Chem Long Island Univ Brooklyn Ctr Brooklyn NY 11201

ZAVODNI, JOHN J, b Gallitzin, Pa, June 17, 43; m 65. PHYSIOLOGY. Educ: St Francis Col (Pa), BS, 64; Pa State Univ, PhD(physiol), 68. Prof Exp: Instr biol, St Francis Col (Pa), 64-65; asst prof, 68-75, ASSOC PROF ZOOL & CHMN DEPT SCI, PA STATE UNIV, McKEESPORT CAMPUS, 75- Mem: NY Acad Sci; Am Asn Sex Educ & Coun; Sex Educ & Info Coun US. Res: Effects of exposure to increased oxygen tensions on the endocrine system; enforcement of water pollution. Mailing Add: Dept of Zool Pa State Univ McKeesport PA 15132

ZAVON, MITCHELL RALPH, b Woodhaven, NY, May 9, 23; m 47; c 4. MEDICINE. Educ: Boston Univ, MD, 49. Prof Exp: From asst prof to clin prof indust med, Kettering Lab, Col Med, Univ Cincinnati, 55-71; assoc dir, Huntingdon Res Ctr, 71-74; ASSOC DIR, MED DIV & MED DIR, ETHYL CORP, 74- Concurrent Pos: Dir occup health serv, Cincinnati Health Dept, 55-61, asst health comnr, 61-74; consult, USPHS, 57-59, 66-69, Joint Congressional Comt Atomic Energy, 58-59, Louisville-Jefferson County Health Dept, 59-60 & USDA, 63-69; exec coordr, Miami Valley Proj, Ohio, 68-71. Res: Radiation protection; biological effects of agricultural chemicals; occupational health. Mailing Add: Med Div Ethyl Corp 451 Florida Baton Rouge LA 70801

ZAVORTINK, THOMAS JAMES, b Ravenna, Ohio, May 27, 39. ENTOMOLOGY, BIOLOGY. Educ: Kent State Univ, BS, 61; Univ Calif, Los Angeles, MA, 63, PhD(zool), 67. Prof Exp: Asst res zoologist entom, Univ Calif, Los Angeles, 68-75; MEM FAC, DEPT BIOL, UNIV SAN FRANCISCO, 75- Concurrent Pos: Consult, Southeast Asia Mosquito Proj, Smithsonian Inst, 68-71. Mem: Entom Soc Am. Res: Systematics and biology of mosquitoes and bees. Mailing Add: Dept of Biol Univ of San Francisco San Francisco CA 94117

ZAWADZKI, JOSEPH FRANCIS, b Withee, Wis, May 30, 35; m 70; c 2. ORGANIC CHEMISTRY. Educ: Northland Col, BA, 57; Loyola Univ (Ill), MS, 60, PhD(org chem), 62. Prof Exp: Res assoc org chem, Univ Chicago, 62-64; RES INVESTR CHEM PROCESS RES, SEARLE LABS, 64- Mem: Am Chem Soc. Res: Synthetic organic chemistry; reaction mechanisms; molecular rearrangements. Mailing Add: Searle Labs PO Box 5110 Chicago IL 60680

ZAWESKI, EDWARD F, b Jamesport, NY, Nov 2, 33; m 65; c 3. ORGANIC CHEMISTRY. Educ: Fordham Univ, BS, 55; Iowa State Univ, PhD(org chem), 59. Prof Exp: CHEMIST, RES LABS, ETHYL CORP, 59- Mem: Am Chem Soc. Res: Organo-metallic, olefin oxidation and antioxidant chemistry. Mailing Add: Res Labs Ethyl Corp 1600 W Eight Mile Rd Ferndale MI 48220

ZAWISTOWSKI, EDWARD ARTHUR, organic chemistry, see 12th edition

ZAWOISKI, EUGENE JOSEPH, b Plains, Pa, Aug 17, 27; m 49; c 5. PHYSIOLOGY. Educ: Temple Univ, AB, 49, MA, 51; Jefferson Med Col, PhD(physiol), 63. Prof Exp: Asst histol, Temple Univ, 49-50; res asst path, Sharp & Dohme, Inc, 51, res assoc toxicol, 51-55; res assoc gastrointestinal physiol, Merck Inst Therapeut Res, Pa, 55-62, res assoc toxicol & teratology, 62-65; from instr to asst prof physiol, 65-71, ASSOC PROF PHYSIOL, JEFFERSON MED COL, 71- Concurrent Pos: Co-adj lectr pharmacol, Cooper Hosp Sch Nursing, 67-74 & Rutgers Univ, physiol, 68, 70, 72. Mem: Am Physiol Soc. Res: Gastrointestinal physiology and pharmacology; teratology; mammalian physiology. Mailing Add: Dept of Physiol Jefferson Med Col Philadelphia PA 19107

ZAYAS, ESTHER LEIJO DE, nutrition, see 12th edition

ZAYE, DAVID FRANCIS, b Toledo, Ohio. INFORMATION SCIENCE, ANALYTICAL CHEMISTRY. Educ: Univ Toledo, BS, 62, MS, 64; Univ Hawaii, PhD(anal chem), 68. Prof Exp: From assoc ed to sr assoc ed anal chem, 68-74, SR ED, CHEM ABSTR SERV, AM CHEM SOC, 74- Mem: Am Chem Soc. Res: Scientific vocabulary management processes and information transfer techniques relating to numerical data. Mailing Add: Chem Abstr Serv Ohio State Univ Columbus OH 43210

ZBARSKY, SIDNEY HOWARD, b Vonda, Sask, Feb 19, 20; m 44; c 3. BIOCHEMISTRY. Educ: Univ Sask, BA, 40; Univ Toronto, MA, 42, PhD(biochem), 46. Prof Exp: Res officer, Biol & Med Res Br, Atomic Energy Proj, 46-48; asst prof physiol chem, Univ Minn, 48-49; assoc prof, 49-62, PROF BIOCHEM, UNIV BC, 62- Concurrent Pos: Killam sr fel, Univ BC, 72-73. Mem: AAAS; Am Chem Soc; Can Physiol Soc; Can Biochem Soc (vpres, 66-67, pres, 67-68); Am Soc Biol Chemists. Res: Detoxication mechanisms; metabolism of British anti-lewisite, purines and pyrimidines; nucleases and nucleic acid enzymes in the intestinal mucosa. Mailing Add: Dept of Biochem Univ of BC Vancouver BC Can

ZBINOVSKY, VLADIMIR, biochemistry, see 12th edition

ZBORALSKE, F FRANK, b Fall Creek, Wis, Aug 2, 32; m 58; c 4. RADIOLOGY. Educ: Marquette Univ, MD, 58. Prof Exp: Intern, St Joseph's Hosp, Milwaukee, 58-59; resident, Milwaukee County Gen Hosp, 59-62; from instr to asst prof radiol, Sch Med, Marquette Univ, 62-65; actg asst prof, Med Ctr, Univ Calif, San Francisco, 64, from asst prof to assoc prof radiol & dir exp radiol lab, 65-67; chief sect gastrointestinal radiol, 66-67; assoc prof radiol, 67-72, PROF RADIOL, SCH MED, STANFORD UNIV, 72-, DIR DIAG RADIOL, 67- Concurrent Pos: Radiologist, Milwaukee County Gen Hosp, 62-65; James Picker Found scholar radiol, 62-65; attend physician & consult, Vet Admin Hosp, Wood, Wis, 65; co-dir NIH res training grants diag radiol, Med Ctr, Univ Calif, San Francisco, 65-67 & dir res training grant, Sch Med, Stanford Univ, 67-; consult, Vet Admin Hosp, Palo Alto, Calif, 68 & Santa Clara Valley Med Ctr, San Jose, 68- Mem: Asn Univ Radiologists; assoc Am Gastroenterol Asn. Res: Esophageal motility. Mailing Add: Dept of Radiol Stanford Univ Sch of Med Stanford CA 94305

ZDANIS, RICHARD ALBERT, b Baltimore, Md, July 15, 35; m 55; c 2. PHYSICS. Educ: Johns Hopkins Univ, AB, 57, PhD(physics), 60. Prof Exp: Res assoc physics, Princeton Univ, 60-61, instr, 61-62; from asst prof to assoc prof, 62-69, PROF PHYSICS, JOHNS HOPKINS UNIV, 69-, ASSOC PROVOST, 75- Mem: Am Phys Soc; AAAS. Res: Experimental elementary particle research. Mailing Add: Dept of Physics Johns Hopkins Univ Baltimore MD 21218

ZDANUK, EDWARD JOSEPH, physical chemistry, see 12th edition

ZDERIC, JOHN ANTHONY, b San Jose, Calif, Jan 5, 24; m 49; c 1. PHARMACEUTICAL CHEMISTRY. Educ: San Jose State Col, AB, 50; Stanford Univ, MS, 52, PhD(org chem), 55. Prof Exp: Squibb fel, Wayne State Univ, 55-56; res chemist, Syntex Corp, 56-59, asst dir chem res, 59-61, dir labs, Syntex Inst Molecular Biol, 61-62, dir corp planning div, 64-66, vpres com develop, Syntex, Int, Mex, 66-70, asst corp vpres, Syntex Corp, 67-70, vpres, Syntex Labs, Inc, Calif, 70-73, VPRES ADMIN & TECH AFFAIRS, SYNTEX RES, 73- Concurrent Pos: Mem staff, Swiss Fed Inst Technol, 62-64; mem bd govs, Syva, 74- Mem: Am Chem Soc; Swiss Chem Soc. Res: Raney nickel catalyzed hydrogenolyses; macrocylic antibiotics; steroidal hormones; nucleocides and nucleotides. Mailing Add: Com Develop Syntex Labs Inc 3401 Hillview Ave Palo Alto CA 94304

ZDUNKOWSKI, WILFORD G, b Driesen, Ger, May 4, 29; US citizen; m 51; c 2. PHYSICAL METEOROLOGY. Educ: Univ Utah, BS, 58, MS, 59; Univ Munich, DSc, 62. Prof Exp: Res asst atmospheric fluoride, Univ Utah, 57-58; res meteorologist, Intermountain Weather, Inc, 58-63; from asst prof to assoc prof, 63-72, PROF METEOROL, UNIV UTAH, 72- Concurrent Pos: Res meteorologist, Univ Mainz, 59-61 & Meteorol Inst, Univ Munich, 61-62. Mem: Am Meteorol Soc; Am Geophys Union. Res: Atmospheric radiation. Mailing Add: Dept of Meteorol Univ of Utah Salt Lake City UT 84112

ZEALEY, MARION EDWARD, b Augusta, Ga, Mar 26, 13; m 48; c 4. BIOCHEMISTRY, MICROBIOLOGY. Educ: Paine Col, AB, 34; Atlanta Univ, MS, 40; Univ Minn, PhD, 60. Prof Exp: Instr chem, Miles Col, 36-43; assoc prof, Paine Col, 43-44; assoc prof biochem, 48-59, USPHS fel, 63-65, PROF MICROBIOL, MEHARRY MED COL, 65- Mem: AAAS; Am Chem Soc. Res: Protein denaturation; x-ray effects; nucleic acids; cell biology. Mailing Add: Dept of Microbiol Meharry Med Col Nashville TN 37208

ZEBOLSKY, DONALD MICHAEL, b Chicago, Ill, Aug 20, 33; m 57; c 8. PHYSICAL CHEMISTRY. Educ: Northwestern Univ, BA, 56; Kans State Univ, PhD(phys chem), 63. Prof Exp: Chemist, Baxter Labs, 56-57; asst prof phys chem, Creighton Univ, 62-63; asst prof, Northern Ill Univ, 63-64; asst prof, 64-68, ASSOC PROF PHYS CHEM, CREIGHTON UNIV, 68- Mem: Am Chem Soc; NY Acad Sci. Res: Thermodynamics, kinetics and polarography of transition metal-ion chelate formation with molecules related to imidazole; molecules containing sulfur as a ligand site. Mailing Add: Dept of Chem Creighton Univ Omaha NE 68131

ZEBOUNI, NADIM H, b Beirut, Lebanon, Apr 14, 28; m; c 2. SOLID STATE PHYSICS. Educ: Univ Paris, BS, 53; Nat Sch Advan Telecommun, France, MS, 55; La State Univ, PhD(physics), 61. Prof Exp: Eng del to Mid East, Co Gen TSF, 55-57; teaching asst, 57-58, asst prof, 60-65, ASSOC PROF PHYSICS, LA STATE UNIV, BATON ROUGE, 65-, ASTRON, 74- Mem: Am Phys Soc. Mailing Add: Dept of Physics La State Univ Baton Rouge LA 70803

ZEBOVITZ, EUGENE, b Chicago, Ill, Feb 24, 26; m 51; c 5. VIROLOGY. Educ: Roosevelt Univ, BS, 49; Univ Chicago, MS, 52, PhD(microbiol), 55. Prof Exp: Microbiologist, US Army Biol Labs, Ft Detrick, Md, 55-58; microbiologist, Universal Foods Corp, Wis, 58-62; microbiologist, US Army Biol Labs, 62-70; microbiologist, Naval Med Res Inst, Nat Naval Med Ctr, 70-74; HEALTH SCIENTIST ADMINR BIOL SCI, DIV RES GRANTS, NIH, 74- Mem: AAAS; Am Soc Microbiol; Sigma Xi; Soc Exp Biol & Med. Res: Mechanism of virus replication; molecular biology; viral genetics; microbial physiology; viral and bacterial nutrition. Mailing Add: Div of Res Grants Nat Inst of Health Bethesda MD 20014

ZECH, ARTHUR CONRAD, b Julian, Nebr, Aug 24, 27; m 59; c 1. AGRONOMY, BIOCHEMISTRY. Educ: Univ Nebr, Lincoln, BS, 58; Kans State Univ, MS, 59, PhD(agron, biochem), 62. Prof Exp: Sr agronomist, 61-65, res scientist animal nutrit, 65-69, mgr agr sci res, 69-75, MGR PLANT SCI RES, FARMLAND INDUSTS, INC, 75- Mem: Am Soc Agron; Am Soc Animal Sci; Poultry Sci Asn. Res: Animal nutrition. Mailing Add: Farmland Industs Inc 3315 N Oak Trafficway Kansas City MO 64116

ZECH, JOHN DAVID, organic chemistry, see 12th edition

ZECHIEL, LEON NORRIS, b Wilmington, Del, Sept 23, 23; m 46; c 4. ASTROPHYSICS, OPTICS. Educ: DePauw Univ, AB, 48; Ohio State Univ, MA, 51. Prof Exp: Res assoc, Res Found, Ohio State Univ, 53-58, assoc supvr & asst to dir, 58-59; sect head, GPL Div, Gen Precision, Inc, NY, 59-63, prin scientist, 63-68; mgr data mgt systs, Sanders Assocs, Inc, NH, 68-72; sr systs engr, NCR-Postal Systs Div, SC, 72-74; PROG MGR, DAYTON RES DIV, HOBART CORP, 74- Res: Reentry and space vehicle guidance and navigation; aeronautical charting; optical and infrared instrumentation; stellar photography; air navigation; reconnaissance and surveillance techniques; computerized data management systems; postal systems automation; automated weighing and package labelling systems. Mailing Add: 156 Old Salem Rd Dayton OH 45415

ZECHMAN, FREDERICK WILLIAM, JR, b Mar 16, 28; m 50; c 2. PHYSIOLOGY. Educ: Otterbein Col, BS, 49; Univ Md, MS, 51; Duke Univ, PhD, 56. Prof Exp: Asst zool, Univ Md, 49-51; biologist & asst to head biol br, US Off Naval Res, 51-53; instr physiol, Duke Univ, 53-57; from asst prof to assoc prof, Miami Univ, 57-61; from asst prof to assoc prof, 61-68, PROF PHYSIOL & CHMN DEPT PHYSIOL & BIOPHYS, MED CTR, UNIV KY, 68- Concurrent Pos: Consult biophys br, Aerospace Med Lab, Wright-Patterson AFB, 60-61. Mem: AAAS; Am Physiol Soc; Aerospace Med Asn; Soc Exp Biol & Med. Res: Respiratory regulation and mechanics; prolonged and periodic acceleration; effects of lower body negative pressure and posture change; mechanical, reflex and subjective responses to added airflow resistance; bedrest; exercise. Mailing Add: Dept of Physiol & Biophys Univ of Ky Med Ctr Lexington KY 40506

ZECHMANN, ALBERT W, b Sioux City, Iowa, Aug 21, 34; m 65; c 1. MATHEMATICS. Educ: Iowa State Univ, BS, 56, MS, 59, PhD(appl math), 61. Prof Exp: ASST PROF MATH, UNIV NEBR-LINCOLN, 61- Mem: Math Asn Am; Am Math Soc. Res: Solution of visco-elastic problems; unification of the theory of partial differential equations; study of Cauchy problem for elliptic equations. Mailing Add: Dept of Math 840 Old Hall Univ of Nebr Lincoln NE 68508

ZEDECK, MORRIS SAMUEL, b Brooklyn, NY, Jan 25, 40; m 61; c 3. PHARMACOLOGY, BIOCHEMISTRY. Educ: Long Island Univ, BS, 61; Univ Mich, PhD(pharmacol), 65. Prof Exp: Asst prof pharmacol, Sch Med, Yale Univ, 67-68; ASSOC PHARMACOL, SLOAN-KETTERING INST CANCER RES, 68-; ASST PROF PHARMACOL, GRAD SCH MED SCI, SLOAN-KETTERING DIV, CORNELL UNIV, 69- Concurrent Pos: Fel pharmacol, Sch Med, Yale Univ, 65-67. Mem: Am Asn Cancer Res; Am Soc Pharmacol & Exp Therapeut. Res: Effects of potential cancer chemotherapeutic agents upon the synthesis of nucleic acids, protein and induced enzymes; mechanism of action studies and preclinical toxicology studies of cancer chemotherapeutic agents; cellular control and regulation; chemical carcinogenesis. Mailing Add: Sloan-Kettering Inst for Cancer Res 410 E 68th St New York NY 10021

ZEDEK, MISHAEL, b Kaunas, Lithuania, July 16, 26; m 56; c 2. MATHEMATICS. Educ: Hebrew Univ, Israel, MSc, 52; Harvard Univ, PhD(math), 56. Prof Exp: Asst, Hebrew Univ, Israel, 52-53; asst, Harvard Univ, 53-55; instr math, Univ Calif, 56-58; from asst prof to assoc prof, 58-67, PROF MATH, UNIV MD, COLLEGE PARK, 68- Mem: Am Math Soc; Math Asn Am; London Math Soc. Res: Mathematical analysis; complex analysis; interpolation and approximation. Mailing Add: Dept of Math Univ of Md College Park MD 20742

ZEDLER, JOY BUSWELL, b Sioux Falls, SDak, Oct 15, 43; m 65; c 2. ECOLOGY. Educ: Augustana Col (SDak), BS, 64; Univ Wis, Madison, MS, 66, PhD(bot), 68. Prof Exp: Asst bot, Univ Wis, Madison, 64-66, fel, 66-67, res asst, 67-68; instr, Univ Mo, Columbia, 68-69; lectr, 69-72, ASST PROF BIOL, SAN DIEGO STATE UNIV, 72- Mem: Am Soc Limnol & Oceanog; Ecol Soc Am; Estuarine & Brackish-Water Sci Asn. Res: Salt marsh community structure; Algal community ecology. Mailing Add: Dept of Biol San Diego State Univ San Diego CA 92182

ZEDLER, PAUL HUGO, b Milwaukee, Wis, June 22, 41; m 65; c 2. ECOLOGY. Educ: Univ Wis, Milwaukee, BS, 63; Univ Wis, Madison, MS, 66; PhD(bot), 68. Prof Exp: Arboretum botanist, Univ Wis, 64-68; fel forestry, Univ Mo, Columbia, 68-69; ASST PROF BIOL, SAN DIEGO STATE UNIV, 69- Mem: AAAS; Brit Ecol Soc; Ecol Soc Am. Res: Structural characteristics of plant communities; nutrient relations within plant communities; plant populations. Mailing Add: Dept of Biol San Diego State Univ San Diego CA 92182

ZEE, ANTHONY, b China; m 71. THEORETICAL HIGH ENERGY PHYSICS. Educ: Princeton Univ, AB, 66; Harvard Univ, AM, 68, PhD(physics), 70. Prof Exp: Mem physics, Inst Advan Study, 70-72; asst prof, Rockefeller Univ, 72-73; ASST PROF PHYSICS, PRINCETON UNIV, 73- Concurrent Pos: A P Sloan Found fel, 73- Res: Interaction between hadronic constituents and the structure of weak interaction, with emphasis on quantum field theoretic approaches. Mailing Add: Dept of Physics Princeton Univ Princeton NJ 08540

ZEE, PAULUS, b Amsterdam, Netherlands, July 2, 28; US citizen; m 57; c 4. PEDIATRICS, BIOCHEMISTRY. Educ: Univ Amsterdam, MD, 54; Tulane Univ, PhD(biochem), 65. Prof Exp: Resident pediat, Children's Mercy Hosp, Kansas City, Mo, 56-58; asst prof pediat & physiol, Univ Tenn, 64-68, ASSOC PROF PEDIAT & PHYSIOL, UNIV TENN, 68-; MEM, ST JUDE CHILDREN's RES HOSP, 64- Mem: Am Oil Chem Soc; AMA; Am Acad Pediat; Am Inst Nutrit; Am Soc Clin Nutrit. Res: Lipid metabolism of the newborn; pediatric nutrition. Mailing Add: St Jude Children's Res Hosp PO Box 318 Memphis TN 38101

ZEE, YUAN CHUNG, b Shanghai, China, Aug 29, 35; m 66. VIROLOGY. Educ: Univ Calif, Berkeley, AB, 57, MA, 59, PhD(comp path), 66, Univ Calif, Davis, DVM, 63. Prof Exp: Res bacteriologist, Virol Div, Naval Biol Lab, Univ Calif, Berkeley, 63-66, from asst prof to assoc prof, 66-74, PROF VET MICROBIOL & CHMN DEPT, UNIV CALIF, DAVIS, 74- Mem: Am Vet Med Asn; Am Soc Microbiol. Res: Biological properties of animal viruses; mechanisms of virus replication and electron microscopy. Mailing Add: Dept of Vet Microbiol Univ of Calif Sch of Vet Med Davis CA 95616

ZEE-CHENG, KWANG YUEN, b Kashan, Chekiang, China, Sept 2, 25; m 49; c 4. ORGANIC CHEMISTRY, MEDICINAL CHEMISTRY. Educ: China Tech Inst, BS, 45; NMex Highlands Univ, MS, 57; Univ Tex, Austin, PhD(org chem), 63. Prof Exp: Res chemist & chem engr, Taiwan Pulp & Paper Corp, 46-56; asst org chem, NMex Highlands Univ, 56-57; res scientist & teaching asst, Univ Tex, Austin, 57-59; assoc chemist, Midwest Res Inst, 59-61; Welch Found fel, Univ Tex, Austin, 61-62; sr chemist, Celanese Corp Am, Tex, 62-65; sr chemist, 65-71, PRIN CHEMIST BIOL SCI DIV, MIDWEST RES INST, 71- Mem: Am Chem Soc; Sigma Xi; AAAS. Res: Physical chemistry; chemical engineering; synthesis; identification; reaction mechanism of organic compounds; heterocyclic chemistry; cancer chemotherapy. Mailing Add: Biol Sci Div Midwest Res Inst 425 Volker Blvd Kansas City MO 64110

ZEEK, WILLIAM CHARLES, b Dover, NJ, Oct 7, 19. CHEMISTRY. Educ: Syracuse Univ, BS, 43, PhD, 54. Prof Exp: Chemist, Bausch & Lomb Optical Co, NY, 43-44; phys chemist, Frankford Arsenal, US Dept Army, 51-55; chemist, Syracuse Univ, 55-56; res engr, Crosley Div, Avco Mfg Co, 56-57; chemist, Ray-O-Vac Div, Elec Storage Battery Co, Wis, 57-58, head, Solid State Physics Div, Norberg Res Ctr, 58-63; asst to tech dir, 63-68, MGR TECH SERV, C TENNANT SONS & CO, 68- Mem: Electrochem Soc; Hist Sci Soc; Am Crystallog Asn. Res: X-ray structure; nonferrous metallurgy; power source research and development. Mailing Add: C Tennant Sons & Co 100 Park Ave New York NY 10017

ZEEVAART, JAN ADRIAAN DINGENIS, b Baarland, Neth, Jan 5, 30; m 56; c 1. PLANT PHYSIOLOGY. Educ: State Agr Univ Wageningen, BSc, 53, MSc, 55, PhD(plant physiol), 58. Prof Exp: Asst plant physiol, State Agr Univ Wageningen, 55-58; res fel, Calif Inst Technol, 60-63; assoc prof, McMaster Univ, 63-65; assoc prof, 65-70, PROF PLANT PHYSIOL, MICH STATE UNIV, 70- Concurrent Pos: Guggenheim fel, Milstead Lab Chem Enzymol, Sittingbourne Res Ctr, 73-74. Mem: AAAS; Am Soc Plant Physiol; Soc Develop Biol; corresp Royal Dutch Acad Sci. Res: Physiology of flower formation; plant development as regulated by growth substances; environmental physiology. Mailing Add: Energy Res & Develop Admin Plant Res Lab Mich State Univ East Lansing MI 48824

ZEFFREN, EUGENE, b St Louis, Mo, Nov 21, 41; m 64; c 1. BIO-ORGANIC CHEMISTRY. Educ: Wash Univ, AB, 63; Univ Chicago, MS, 65, PhD(org chem), 67. Prof Exp: Res chemist enzym, 67-71, group leader, 71-74, SECT HEAD, TOILET GOODS DIV, WINTON HILL TECH CTR, PROCTER & GAMBLE CO, 74- Res: Mechanism of enzyme action; model systems for enzymic catalysis; chemistry of hair keratins; protein structure. Mailing Add: Toilet Goods Div W Hill Tech Ctr Procter & Gamble Co Cincinnati OH 45224

ZEFTEL, LEO, b Providence, RI, Aug 7, 25; m 49; c 3. ORGANIC CHEMISTRY. Educ: Brown Univ, ScB, 49; Univ Rochester, PhD(chem), 51. Prof Exp: Chemist, Jackson Lab, Org Chem Dept, 51-57 & Plants Tech Sect, 57-59, supvr, 59-66, div head, Process Dept, 66-70, chief supvr chem-tech, 70-73, chief supvr prod, 73-75,

SUPT, PROD CONTROL CHAMBERS WORKS, E I DU PONT DE NEMOURS & CO, INC, DEEPWATER, NJ, 75- Mem: Am Chem Soc. Res: Surfactants; intermediates; rubber chemicals; textile adjuvants; petroleum additives; polymers; dyes; ultraviolet absorbers; process development. Mailing Add: 4619 Sylvanus Dr Rockwood Hills Wilmington DE 19803

ZEGARELLI, EDWARD VICTOR, b Utica, NY, Sept 9, 12; m 39; c 4. DENTISTRY. Educ: Columbia Univ, AB, 34, DDS, 37; Univ Chicago, MS, 42; Am Bd Oral Med, dipl, 56. Prof Exp: Asst dent, 37-38, from instr to asst prof, 38-47, head diag & roentgenol, 47-57, prof, 57-58, actg dean, 73-74, EDWIN S ROBINSON PROF DENT, SCH DENT & ORAL SURG, COLUMBIA UNIV, 58-, DIR DIV STOMATOL, 58-, DEAN SCH DENT & ORAL SURG, 74-; DIR DENT SERV, COLUMBIA-PRESBY MED CTR, 74- Concurrent Pos: Dent alumni res award, Columbia Univ, 63; mem univ coun, Columbia Univ, 59-62, cancer coordr & chmn comt dent res, Sch Dent & Oral Surg; mem coun dent therapeut, Am Dent Asn, 63-69, vchmn, 68-69, consult, 69-, chmn coun dent mat & devices, 70-; mem, NY Bd Dent Exam, 63-, pres, 70-71; attend dent surgeon, Columbia-Presby Med Ctr & Delafield Inst Cancer Res; cent off consult & dentist in residence, Vet Admin, DC; police surgeon, New York Police Dept; chmn comt exam, NE Regional Bd Dent Examr, 69- & joint panel drugs in dent, Nat Acad Sci-Nat Res Coun-Food & Drug Admin; consult, EORange, Kingsbridge & Montrose Vet Admin Hosps, Grasslands, Phelps Mem & Vassar Bros Hosps & USPHS; consult, Nat Naval Dent Ctr, Bethesda, Md; mem, NY State Health Res Coun, 75. Honors & Awards: Austin Sniffin Medal Honor, Dent Soc, NY, 61, Jarvie-Burkhardt Medal Honor, 70; Award of Merit, Am Asn Dent Exam, 71; Samuel J Miller Medal, Am Acad Oral Med, 76. Mem: AAAS; Am Cancer Soc; fel Am Col Dent; hon mem Dent Soc Guatemala; hon fel Acad Gen Dent. Res: Diseases of the mouth and jaws, especially diagnosis; pharmacotherapeutics of oral diseases. Mailing Add: Sch of Dent & Oral Surg Columbia Univ New York NY 10032

ZEGURA, STEPHEN LUKE, b San Francisco, Calif, July 2, 43; m 65; c 1. HUMAN BIOLOGY, BIOLOGICAL ANTHROPOLOGY. Educ: Stanford Univ, BA, 65; Univ Wis-Madison, MS, 69, PhD(human biol), 71. Prof Exp: Asst prof anthrop, NY Univ, 71-72; ASST PROF ANTHROP, UNIV ARIZ, 72- Concurrent Pos: NY Univ career develop grant, Smithsonian Inst, 72. Mem: AAAS; Am Asn Phys Anthrop; Classification Soc; Am Anthrop Asn; Soc Study Human Biol. Res: Multivariate statistics; biological distance; Eskimology; population genetics. Mailing Add: Dept of Anthrop Univ of Ariz Tucson AZ 85721

ZEHNA, PETER W, mathematical statistics, see 12th edition

ZEHNER, DAVID MURRAY, b Philadelphia, Pa, Aug 7, 43; m 67. SURFACE PHYSICS. Educ: Drexel Inst Technol, BS, 66; Brown Univ, PhD(physics), 71. Prof Exp: Res asst physics, Brown Univ, 67-71; RES PHYSICIST SURFACE PHYSICS, OAK RIDGE NAT LAB, 71- Mem: Sigma Xi; Am Phys Soc; Am Vacuum Soc. Res: Investigation of surface properties of solids using surface sensitive spectroscopic techniques employing electrons, photons and ions as scattering probes; surface damage, chemisorption and catalytic phenomena. Mailing Add: Solid State Div PO Box X Oak Ridge Nat Lab Oak Ridge TN 37830

ZEHNER, LEE RANDALL, b Lansdowne, Pa, Mar 15, 47; m 73. ORGANIC CHEMISTRY. Educ: Univ Pa, BS, 68; Univ Minn, PhD(org chem), 73. Prof Exp: RES CHEMIST ORG CHEM, ARCO CHEM, ATLANTIC RICHFIELD CO, 73- Mem: Am Chem Soc. Res: Homogeneous and heterogeneous catalysis; carbonylations; petrochemicals. Mailing Add: Arco Chem Res & Develop 500 S Ridgeway Ave Glenolden PA 19036

ZEHR, ELDON IRVIN, b Manson, Iowa, June 25, 35; m 57; c 3. PLANT PATHOLOGY. Educ: Goshen Col, BA, 60; Cornell Univ, MS, 65, PhD(plant path), 69. Prof Exp: Asst prof, 69-73, ASSOC PROF PLANT PATH, CLEMSON UNIV, 73- Mem: Am Phytopath Soc; AAAS. Res: Plant diseases caused by bacteria; diseases of apples, peaches and grapes. Mailing Add: Dept of Plant Path & Physiol Clemson Univ Clemson SC 29631

ZEHR, FLOYD JOSEPH, b Lowville, NY, June 28, 29; m 57; c 4. EXPERIMENTAL PHYSICS. Educ: Eastern Mennonite Col, BS, 54; Goshen Col, BA, 57; Syracuse Univ, MS, 61 & 63, PhD(physics), 67. Prof Exp: Teacher jr high sch, PR, 54-56 & sr high sch, NY, 57-59; asst prof, 65-72, head dept, 69-71, ASSOC PROF PHYSICS, WESTMINSTER COL, PA, 72- Concurrent Pos: Researcher, Argonne Nat Lab, 71-72. Mem: Am Asn Physics Teachers. Res: Measurement of interatomic potential between lithium ions and helium atoms and lithium ions and hydrogen atoms; nuclear decay studies of LU-174 and Tm-174; neutron capture-gamma ray decay studies. Mailing Add: Dept of Physics Westminster Col New Wilmington PA 16142

ZEHRUNG, WINFIELD SCOTT, III, b Oil City, Pa, July 4, 31; m 57; c 3. ORGANIC CHEMISTRY. Educ: Allegheny Col, BS, 53, MS, 54; Univ Buffalo, PhD(chem), 57. Prof Exp: Instr chem, Allegheny Col, 54-57; res chemist, Yerkes Res Lab, E I du Pont de Nemours & Co, 57-62, lab supvr, Yerkes Film Plant, 62-65, staff scientist, Yerkes Res Lab, 65-67, tech rep venture develop sect, Film Dept, 67-71; plant mgr, Specialty Converters, Inc, 71-74, TECH DIR, SPECIALTY COMPOSITES CORP, 74- Mem: Am Chem Soc. Res: Nitrogen heterocycles; organic chemistry of cyanogen; vinyl polymers; pigmentation; color theory and applications; business analysis; urethane foam technology. Mailing Add: Specialty Composites Corp 650 Dawson Dr Newark DE 19713

ZEI, DINO, b Chicago, Ill, Aug 20, 27; m 49; c 2. EXPERIMENTAL PHYSICS, HISTORY OF SCIENCE. Educ: Beloit Col, BS, 50; Univ Wis, MS, 52, PhD, 57, MA, 72. Prof Exp: Physicist, Nat Bur Standards, 50; instr physics, Beloit Col, 52-53; asst prof, Milton Col, 53-54; assoc prof, St Cloud State Col, 55-57; PROF PHYSICS & CHMN DEPT, RIPON COL, 57- Mem: AAAS; Am Asn Physics Teachers; Am Phys Soc; Optical Soc Am; Hist Sci Soc. Mailing Add: Dept of Physics Ripon Col Ripon WI 54971

ZEIDLER, JAMES ROBERT, b Carlinville, Ill, Dec 1, 44; m 68. SOLID STATE PHYSICS. Educ: Macmurray Col, BA, 66; Mich State Univ, MS, 68; Univ Nebr, Lincoln, PhD(physics), 72. Prof Exp: Asst physics, Mich State Univ, 66-68; asst, Univ Nebr, Lincoln, 68-73, res assoc & instr optics, 73-74; PHYSICIST SIGNAL PROCESSING UNDERWATER COMMUN, NAVAL UNDERSEA CTR, 74- Mem: Am Acoust Soc; Inst Elec & Electronic Engrs. Res: Adaptive signal processing techniques; underwater acoustic communications; spectral estimation theory; digital signal processing. Mailing Add: Code 2522 Naval Undersea Ctr San Diego CA 92132

ZEIDMAN, BENJAMIN, b New York, NY, Oct 6, 31; m 56; c 1. NUCLEAR PHYSICS. Educ: City Col New York, BS, 52; Washington Univ, PhD, 57. Prof Exp: ASSOC PHYSICIST, ARGONNE NAT LAB, 57- Concurrent Pos: Ford Found fel, Niels Bohr Inst, Copenhagen, Denmark, 63-64. Mem: AAAS; Am Phys Soc. Res: Nuclear reactions, scattering, spectroscopy and structure. Mailing Add: Argonne Nat Lab Bldg 203 Argonne IL 60439

ZEIDMAN, IRVING, b Camden, NJ, Mar 17, 18; m 53; c 2. PATHOLOGY. Educ: Univ Pa, AB, 37, MD, 41. Prof Exp: From instr to assoc prof, 46-66, PROF PATH, SCH MED, UNIV PA, 66- Res: Cancer; chemical factors in cell adhesiveness; method of measuring surface area; effect of hyaluronidase on spread of tumors; transpulmonary passage of tumor cells; spread of cancer in lymphatic system. Mailing Add: Dept of Path Univ of Pa Sch of Med Philadelphia PA 19104

ZEIGEL, ROBERT FRANCIS, b Washington, DC, June 22, 31; m 57; c 2. VIROLOGY, CYTOLOGY. Educ: Eastern Ill Univ, BS, 53; Harvard Univ, AM, 55, PhD(biol), 59. Prof Exp: Res biologist, Nat Cancer Inst, 59-66; ASSOC CANCER RES SCIENTIST, ROSWELL PARK MEM INST, 66-, ASSOC PROF MICROBIOL, ROSWELL PARK MEM INST DIV, STATE UNIV NY BUFFALO, 68-, DIV REP, DEPT MICROBIOL, 70- Concurrent Pos: Consult, Nat Cancer Inst, 69-70. Mem: AAAS; Am Soc Cell Biol; Electron Micros Soc Am; Am Soc Zool. Res: Fine structural studies of mode of synthesis of oncogenic viral agents; search for viral agents and their association with human neoplasia. Mailing Add: Roswell Park Mem Inst Buffalo NY 14203

ZEIGER, HERBERT J, b Bronx, NY, Mar 16, 25; m 54; c 3. PHYSICS. Educ: City Col New York, BS, 44; Columbia Univ, MA, 48, PhD, 52. Prof Exp: Union Carbide & Carbon Corp fel, Columbia Univ, 52-53; RES PHYSICIST, LINCOLN LAB, MASS INST TECHNOL, 53- Mem: Fel Am Phys Soc. Res: Solid state and molecular physics; masers and lasers. Mailing Add: 167 Pond Brook Rd Chestnut Hill MA 02167

ZEIGER, WILLIAM NATHANIEL, b Highland Park, Mich, Sept 7, 46; m 73. NATURAL PRODUCTS CHEMISTRY. Educ: Wayne State Univ, BA, 69; Univ Pa, PhD(chem), 72. Prof Exp: NIH trainee, 72-74, ASST MEM, MONELL CHEM SENSES CTR, 72- Mem: Am Chem Soc. Res: Food and flavor chemistry. Mailing Add: McCormick & Co R&D Labs 204 Wright Ave Hunt Valley MD 21031

ZEIGLER, JAMES EDGAR, zoology, see 12th edition

ZEIGLER, JOHN MILTON, b St Augustine, Fla, May 23, 22; m 46; c 3. GEOLOGY. Educ: Univ Colo, BA, 47; Harvard Univ, PhD(geol), 54. Prof Exp: Oil geologist, Calif Co, 48-50; assoc scientist, Woods Hole Oceanog Inst, 54-67; prof marine sci, Univ PR, Mayaguez, 67-71; PROF MARINE SCI, COL WILLIAM & MARY, 71-; ASSOC PROF, UNIV VA, 71-; ASST DIR & HEAD PHYS SCI, VA INST MARINE SCI, 71- Concurrent Pos: Lectr, Univ Chicago, 64-; consult coastal processes & usage; mem bd trustees, Chesapeake Bay Consortium. Res: Beach processes; shallow water oceanography; coastal planning and management. Mailing Add: Va Inst of Marine Sci Gloucester Point VA 23062

ZEIGLER, JOSEPH ALTON, b Savannah, Ga, July 29, 41; m 68; c 2. MEDICAL MICROBIOLOGY, ELECTRON MICROSCOPY. Educ: NGa Col, BS, 64; Univ Ga, MS, 71, PhD(microbiol), 73. Prof Exp: RES ASST PROF MICROBIOL & ELECTRON MICROSCOPY, MED RES INST, FLA INST TECHNOL, 73- Mem: Sigma Xi; Am Soc Microbiol; Am Leptospirosis Res Conf. Res: Microbial pathogenesis; ultrastructure, enzymology, immunology of spirochete infections. Mailing Add: Fla Inst Technol Med Res Inst 7725 W New Haven Ave Melbourne FL 32901

ZEIGLER, ROYAL KEITH, b Kans, Dec 3, 19; m 43; c 4. MATHEMATICS, APPLIED STATISTICS. Educ: Ft Hays Kans State Col, BS, 41; Univ Nev, MS, 46; Univ Iowa, PhD(math statist), 49. Prof Exp: Assoc prof math, Bradley Univ, 49-51; statistician, AEC, 51-52; statistician, Theoret Physics Div, 52-67, group leader statist serv, 68-75, STAFF MEM, LOS ALAMOS SCI LAB, 75- Mem: Fel AAAS; fel Am Statist Asn. Res: Sampling theory. Mailing Add: 165 Chamisa St Los Alamos NM 87544

ZEIKUS, J GREGORY, b Rahway, NJ, Oct 2, 45; m 67; c 2. MICROBIOLOGY. Educ: Univ SFla, BA, 67; Ind Univ, Bloomington, MA, 68, PhD(microbiol), 70. Prof Exp: NIH fel microbiol, Univ Ill, Champaign, 70-72; ASST PROF BACT, UNIV WIS-MADISON, 72- Mem: Sigma Xi; Am Soc Microbiol; AAAS. Res: Microbial physiology and ecology; biodegradation of lignin; microbial methance formation. Mailing Add: Dept of Bact Univ of Wis Madison WI 53705

ZEILER, ANDREW GEORGE, organic chemistry, see 12th edition

ZEINER, FREDERICK NEYER, b Finley, NDak, Mar 20, 17; m 42. ZOOLOGY. Educ: Univ Denver, BS, 38; Ind Univ, PhD(zool), 42. Prof Exp: Asst zool, Ind Univ, 38-40; from asst prof to prof, 46-74, EMER PROF ZOOL, UNIV DENVER, 74- Concurrent Pos: Res consult, Martin Co, 59 & 61; mem gov bd, Am Inst Biol Sci, 67-71. Mem: AAAS; Am Soc Zool; Am Inst Biol Sci. Res: Pituitary-ovarian relationships during pregnancy; space physiology. Mailing Add: 1417 SElizabeth St Denver CO 80210

ZEINER, HELEN MARSH, b Big Timber, Mont, Oct 5, 12; m 42. BOTANY, ECOLOGY. Educ: Western Reserve Univ, AB, 33, MA, 35; Ind Univ, PhD, 44. Prof Exp: Teacher high sch, Ohio, 35-38; lectr bot, Ind Univ, 38-42; asst prof, Univ Denver, 46-49 & 58-65; CONSULT, DENVER BOT GARDENS, 65-, HON CUR HERBARIUM, 72- Res: Ecology; conservation. Mailing Add: 1417 SElizabeth St Denver CO 80210

ZEISS, HAROLD HICKS, b Evansville, Ind, Aug 16, 17; m 72; c 4. ORGANIC CHEMISTRY. Educ: Ind Univ, BS, 38; Columbia Univ, AM, 47, PhD(chem), 49. Prof Exp: Res chemist, Sunbeam Elec Mfg Co, 40-42; sr chemist, Ridbo Labs, 42-47, dir res, 47-48; from instr to asst prof chem, Yale Univ, 49-55; res assoc, Monsanto Co, 55-61, PRES & DIR, MONSANTO RES SA, 61- Concurrent Pos: Assoc prof, Univ Cincinnati, 58-61; NSF sr fel, Univ Munich & Univ Heidelberg, 60-61; adj prof, Univ Zurich, 65-68; indust prof, Ind Univ, 68- Mem: AAAS; Am Chem Soc; The Chem Soc; Ger Chem Soc; fel NY Acad Sci. Res: Stereochemical problems; mechanisms of organic oxidation reactions; structural elucidation and synthesis of naturally occurring terpenes; organometallic chemistry. Mailing Add: Monsanto Res SA Eggbühlstrasse 36 CH-8050 Zurich Switzerland

ZEIT, WALTER, b Medford, Wis, Dec 18, 98; m 20; c 2. ANATOMY, HISTOLOGY. Educ: Marquette Univ, BS, 25, MS, 27, PhD(human anat), 39. Prof Exp: Asst, 21-23, from instr to prof, 33-68, chmn dept, 47-67, from asst dean to assoc dean, 53-68, EMER PROF ANAT, MED COL WIS, 68- Mem: AAAS; Am Asn Anat. Res: Tissue responses to physical forces; structure of muscle tissue. Mailing Add: 150 SSecond St Medford WI 54451

ZEITLIN, BENJAMIN RAPHAEL, b New York, NY, Dec 6, 15; m 38; c 3. TOXICOLOGY, PHARMACOLOGY. Educ: City Col New York, BS, 37; Brooklyn Col, MS, 41. Prof Exp: Pharmacologist, McKesson & Robbins Pharmaceut Co, 47-50; res assoc pub health, Med Col, Cornell Univ, 50-52; res specialist toxicol, 52-68, SR RES SPECIALIST, CORP RES DEPT, GEN FOODS CORP, WHITE PLAINS, 68- Concurrent Pos: NIH grants chem carcinogenesis res, 64- Mem: Soc Toxicol. Res: Toxicology methodology; chemical carcinogenesis; food additive regulation and standards. Mailing Add: 8 Wood Lane Suffern NY 10901

ZEITLIN, JOEL LOEB, b Los Angeles, Calif, July 9, 42; m 72. MATHEMATICS. Educ: Univ Calif, Los Angeles, BA, 63, MA, 66, PhD(math), 69. Prof Exp: Asst prof math, Wash Univ, 69-72; prof, Cath Univ Valparaiso, 72-73; ASST PROF MATH, CALIF STATE UNIV, NORTHRIDGE, 73- Concurrent Pos: Fulbright Hays scholar, 72-73. Mem: Math Asn Am; Am Math Soc. Res: Lie groups; representation theory; special functions; geometry. Mailing Add: Dept of Math Calif State Univ Northridge CA 91324

ZEITZ, LOUIS, b Lakewood, NJ, Jan 22, 22; m 46; c 2. BIOPHYSICS, PHYSICS. Educ: Univ Calif, Berkeley, AB, 48; Stanford Univ, PhD(biophys), 62. Prof Exp: Res asst x-ray instrumentation, Appl Res Labs, Montrose, Calif, 51-52, res physicist, 52-56; res assoc physics, Univ Redlands, 56-58; res assoc, Biophys Lab, Stanford Univ, 58-59; res assoc, 62-63, ASSOC, BIOPHYS DIV, SLOAN-KETTERING INST, 63-; ASSOC PROF, SLOAN-KETTERING DIV, GRAD SCH MED, CORNELL UNIV, 69- Concurrent Pos: Nat Inst Child Health & Develop res grant, Sloan-Kettering Inst, 65-67; asst prof, Sloan-Kettering Div, Grad Sch Med, Cornell Univ, 65-69. Mem: AAAS; Biophys Soc; Am Phys Soc. Res: X-ray spectrochemical analysis; trace elements in living systems; mechanisms of radiation effects on cell development. Mailing Add: Biophys Div Sloan-Kettering Inst 410 E 68th St New York NY 10021

ZEIZEL, EUGENE PAUL, b Waterbury, Conn, Feb 26, 42. HYDROLOGY, WATER RESOURCES. Educ: Fla State Univ, BS, 64; Univ Nev, MS, 67; Va Polytech Inst & State Univ, PhD(civil eng), 74. Prof Exp: Res asst hydrol, Desert Res Inst, 67-68; scientist water resources, Hittman Assocs Inc, 73-75; PROJ DIR WATER POLLUTION, TETON COUNTY 208 PLANNING AGENCY, 75- Mem: Geol Soc Am; Am Water Res Asn. Res: Waste treatment management planning. Mailing Add: Teton County 208 Planning Agency PO Box 1727 Jackson WY 83001

ZELAC, RONALD EDWARD, b Chicago, Ill, Jan 22, 41; m 61; c 2. RADIOLOGICAL HEALTH, RADIOLOGICAL PHYSICS. Educ: Univ Ill, Urbana, BS, 62, MS, 64; Univ Mich, Ann Arbor, MS, 65; Univ Fla, PhD(environ sci), 70; Am Bd Health Physics, cert, 71. Prof Exp: Res asst solid state physics, Coord Sci Lab, Univ Ill, Urbana, 63-64; chief health physicist, IIT Res Inst, 65-68; ASST PROF RADIATION BIOL & UNIV RADIATION SAFETY OFFICER, TEMPLE UNIV, 70- Concurrent Pos: Radiation physicist, Mercy Med Ctr, Chicago, 67-68; consult, Wyeth Labs, Pa, 71- & Presby-Univ Pa Med Ctr, 74- Mem: Health Physics Soc; Am Asn Physicists in Med. Res: Radiation dosimetry and radiological safety in research and health sciences. Mailing Add: Radiation Safety Off Health Sci Ctr Temple Univ Philadelphia PA 19140

ZELANO, ANTHONY JOSEPH, physical chemistry, see 12th edition

ZELAZNY, LUCIAN WALTER, b Bristol, Conn, May 8, 42; m 62; c 4. SOIL MINERALOGY. Educ: Univ Vt, BS, 64, MS, 66; Va Polytech Inst, PhD(soil chem), 70. Prof Exp: Asst prof soils & asst soil chemist, Univ Fla, 70-75; ASSOC PROF SOIL MINERAL, VA POLYTECH INST & STATE UNIV, 75- Concurrent Pos: Nat Acad Sci-Nat Res Coun grant. Mem: Am Soc Agron; Clay Mineral Soc; AAAS. Res: The effect of deicing compounds on vegetation and water supplies; chemical, physical and mineralogical analysis of soils.Mailing Add: Dept of Agron Smyth Hall Va Polytech Inst & State Univ Blacksburg VA 24061

ZELDES, HENRY, b New Britain, Conn, June 11, 21; m 49; c 3. PHYSICAL CHEMISTRY. Educ: Yale Univ, BS, 42, MS, 44, PhD(phys chem), 47. Prof Exp: Res chemist, Sam Labs, Columbia Univ, 44-45, Carbide & Carbon Chem Corp, 45-46 & Clinton Labs, Monsanto Chem Corp, Tenn, 47; RES CHEMIST, OAK RIDGE NAT LAB, UNION CARBIDE CHEM CO, 48- Mem: Am Chem Soc; Am Phys Soc. Res: Radiochemistry; thermodynamics; electrolyte chemistry; nuclear and electron spin resonance. Mailing Add: 111 Lewis Lane Oak Ridge TN 37830

ZELDIN, MARTEL, b New York, NY, Aug 11, 37; m 58; c 4. INORGANIC CHEMISTRY. Educ: Queens Col, NY, BS, 59; Brooklyn Col, MA, 62; Pa State Univ, PhD(chem), 66. Prof Exp: Chemist, Interchem Corp, 60-62; proj scientist, Union Carbide Corp, 66-68; asst prof, 68-72, ASSOC PROF CHEM, POLYTECH INST BROOKLYN, 72- Concurrent Pos: Consult, Ramapo Sch Syst 2, 72- Mem: Am Chem Soc. Res: Chemistry of group III and group IV elements. Mailing Add: Dept of Chem Polytech Inst Brooklyn 333 Jay St Brooklyn NY 11201

ZELDIN, MICHAEL HERMEN, b Philadelphia, Pa, Mar 25, 38; m 61; c 3. CELL BIOLOGY, PLANT PHYSIOLOGY. Educ: Franklin & Marshall Col, BS, 59; Temple Univ, MA, 61, PhD(biol), 65. Prof Exp: Res assoc biol, Brandeis Univ, 65-67; asst prof, Tufts Univ, 67-74; SR RES FEL, HARVARD UNIV, 74- Concurrent Pos: NIH fel, 65-67. Mem: AAAS; Am Chem Soc; Am Soc Plant Physiol. Res: Biochemistry of development; regulation of organelle formation; interaction of cellular organelles; electrophysiology; structure and function of membranes. Mailing Add: Biol Labs Harvard Univ Cambridge MA 02138

ZELDIS, LOUIS JENRETTE, b Philadelphia, Pa, Feb 21, 12; m 41; c 4. PATHOLOGY. Educ: Syracuse Univ, BA, 37; Univ Rochester, MD, 43. Prof Exp: Intern, Strong Mem Hosp, Rochester, 43-44; instr path, Sch Med & Dent, Univ Rochester, 44-45; assoc prof, Sch Med, Emory Univ, 47-48; pathologist, Div Path, Brookhaven Nat Lab, 49-52; PROF PATH, SCH MED, UNIV CALIF, LOS ANGELES, 52-, ASST DEAN SCH MED, 72- Mem: AAAS; Am Asn Exp Path; Am Asn Path & Bact. Res: Hormonal factors in human mammary cancer; radioactive and heavy isotope tracer studies in tissue protein synthesis; obstetric-gynecologic and radiation pathology. Mailing Add: Dept of Path Univ of Calif Sch of Med Los Angeles CA 90024

ZELEN, MARVIN, b New York, NY, June 21, 27; m 50; c 2. BIOMETRY, MATHEMATICAL STATISTICS. Educ: City Col New York, BS, 49; Univ NC, MA, 51; Am Univ, PhD(statist), 57. Prof Exp: Mathematician, Nat Bur Standards, 52-61; head math statist & appl math sect, Nat Cancer Inst, 63-67; PROF STATIST, STATE UNIV NY BUFFALO, 67-, DIR STATIST LAB, 71- Concurrent Pos: Vis assoc prof, Univ Calif, Berkeley, 58; assoc prof math, Univ Md, 60-61; permanent mem math res ctr, Univ Wis, 60-63; sr Fulbright scholar, Imp Col & Sch Hyg & Trop Med, Univ London, 65-66; consult, Nat Cancer Inst. Honors & Awards: Distinguished Achievement Award, Wash Acad Sci, 67. Mem: Biomet Soc; Am Statist Asn; Inst Math Statist; Royal Statist Soc. Res: Probability and mathematical statistics; model building in biomedical sciences; statistical planning of scientific experiments; clinical trials in cancer. Mailing Add: 44 Chaumont Dr Williamsville NY 14221

ZELENIK, JOHN SLOWKO, b Ft William , Ont, Feb 21, 11; US citizen; m; c 2. OBSTETRICS & GYNECOLOGY. Educ: Lake Forest Col, BA, 32; Univ Ill, BS, 38, MD, 40; Univ Colo, MS, 52; Am Bd Obstet & Gynec, dipl, 53. Prof Exp: Chmn dept obstet & gynec, Fitzsimons Gen Hosp, US Army, Denver, Colo, 59-64 & Tripler Gen

Hosp, Honolulu, Hawaii, 64-66; prof, Col Med, Univ Ill, 66-68; PROF OBSTET & GYNEC, SCH MED, VANDERBILT UNIV, 68-, ACTG CHMN DEPT, 74- Mem: AAAS; AMA; Am Col Obstet & Gynec; Am Fertil Soc; Am Soc Cytol. Res: Hemolytic disease of the newborn; hyaline membrane disease; oral and injectable contraception. Mailing Add: Dept of Obstet & Gynec Vanderbilt Univ Sch of Med Nashville TN 37232

ZELENKA, JERRY STEPHEN, b Cleveland, Ohio, Jan 27, 36; m 58; c 3. ELECTRONIC ENGINEERING, OPTICS. Educ: Univ Mich, Ann Arbor, BS, 58, MS, 59, PhD(elec eng), 66. Prof Exp: Engr, Res Div, Bendix Corp, 59-61; res asst radar & optical processing, Univ Mich, Ann Arbor, 61-62, res assoc, 62-65, from assoc res engr to res engr, 65-72; RES ENGR, ENVIRON RES INST MICH, 72- Concurrent Pos: Lab instr, Univ Mich, Ann Arbor, 58-59, lectr, 72; consult, Westinghouse Elec Corp, 69-72, Gen Motors Corp & IBM Corp, 71- Mem: Inst Elec & Electronics Engrs; Optical Soc Am. Res: Systems analysis pertaining to coherent radars and to coherent optical processors. Mailing Add: Environ Res Inst of Mich PO Box 618 Ann Arbor MI 48107

ZELENKA, PEGGY SUE, b Joplin, Mo, Oct 4, 42; m 66. DEVELOPMENTAL BIOLOGY. Educ: Rice Univ, BA, 64; Johns Hopkins Univ, PhD(biophys), 71. Prof Exp: Fel pediat, Johns Hopkins Sch Med, 71-72; staff fel develop biol, Nat Inst Child Health & Human Develop, 72-75; SR STAFF FEL DEVELOP BIOL, NAT EYE INST, 75- Mem: Soc Develop Biol. Res: Biochemical mechanisms of cellular differentiation during embryonic development; specifically those changes occurring in developing embryonic chick lenses. Mailing Add: Nat Eye Inst Bldg 6 Rm 210 Bethesda MD 20014

ZELENY, WILLIAM BARDWELL, b Minneapolis, Minn, Mar 14, 34; m 60; c 2. THEORETICAL PHYSICS. Educ: Univ Md, BS, 56; Syracuse Univ, MS, 58, PhD(physics), 60. Prof Exp: Lectr physics, Univ Sydney, 60-62; asst prof, 62-65, ASSOC PROF PHYSICS, NAVAL POSTGRAD SCH, 65- Concurrent Pos: Consult, Data Dynamics, Inc, 65-67. Mem: Am Phys Soc. Res: Theoretical astrophysics and elementary particle physics; quantum field theory. Mailing Add: Dept of Physics Naval Postgrad Sch Monterey CA 93940

ZELEZNICK, LOWELL D, b Milwaukee, Wis, Feb 1, 35; m 61; c 1. IMMUNOLOGY, ALLERGY. Educ: Univ Ill, Chicago, BS, 56, PhD(biochem), 61. Prof Exp: Res assoc biochem, Upjohn Co, 64-68; head biochem sect, 68-72, DIR ALLERGY DEPT, ALCON LABS INC, 72- Concurrent Pos: Ciba fel microbiol, Ciba Pharmaceut Co, 60-63; fel molecular biol, Albert Einstein Col Med, 63-64; USPHS fel, 64-65; adj prof, Tex Christian Univ. Mem: AAAS; Am Chem Soc; Am Soc Biol Chem; Am Acad Allergy. Res: Drug metabolism; biosynthesis and structure of lipopolysaccharides and bacterial cell walls; immunology-allergy research. Mailing Add: Allergy Dept Alcon Labs Inc PO Box 1959 Ft Worth TX 76101

ZELEZNIK, PAULINE, b Chicago, Ill, Jan 9, 29. INORGANIC CHEMISTRY, PHYSICAL CHEMISTRY. Educ: Nazareth Col, Mich, BS, 49; Marquette Univ, MS, 55; Univ Notre Dame, PhD(chem), 64. Prof Exp: Instr, 54-60, ASSOC PROF CHEM, NAZARETH COL, 64- Mem: Am Chem Soc. Res: Radiation chemistry; bioinorganic chemistry. Mailing Add: Dept of Chem Nazareth Col Nazareth MI 49074

ZELEZNY, WILLIAM FRANCIS, b Rollins, Mont, Sept 5, 18; m 49. PHYSICAL CHEMISTRY. Educ: Mont State Col, BS, 40; Mont Sch Mines, MS, 41; Univ Iowa, PhD(phys chem), 51. Prof Exp: Chemist, Anaconda Copper Mining Co, 44-46; instr metall & phys chem, Univ Iowa, 48-49; aeronaut res scientist, Nat Adv Comt Aeronaut, 51-54; asst metallurgist, Div Indust Res, State Col Wash, 54-57; sr scientist, Atomic Energy Div, Phillips Petrol Co, Idaho, 57-66; asst metallurgist, Idaho Nuclear Corp, 66-70; STAFF MEM, LOS ALAMOS SCI LAB, 70- Mem: Am Chem Soc; Am Soc Metals; Am Inst Mining, Metall & Petrol Eng. Res: Kinetics of reaction at high temperatures; x-ray diffraction and spectroscopy; crystal structure; microprobe analysis of irradiated nuclear fuels. Mailing Add: Los Alamos Sci Lab PO Box 1663 Los Alamos NM 87544

ZELIGMAN, ISRAEL, b Baltimore, Md, July 24, 13; m 43; c 3. DERMATOLOGY. Educ: Johns Hopkins Univ, AB, 33; Columbia Univ, MedScD, 42. Prof Exp: From instr to asst prof dermat, Univ Md, 46-56; asst, 46-50, from instr to asst prof, 50-63, ASSOC PROF DERMAT, SCH MED, JOHNS HOPKINS UNIV, 63- Concurrent Pos: Pvt pract. Mem: Soc Invest Dermat; AMA; Am Acad Dermat; Am Dermat Asn. Res: Relationship of porphyrins to dermatoses; dermalogic allergy. Mailing Add: 101 W Read St Baltimore MD 21201

ZELIKOFF, STEVEN BARRY, b Brooklyn, NY, Dec 26, 37; m 59; c 3. APPLIED STATISTICS, BUSINESS ADMINISTRATION. Educ: City Col New York, BBA, 59; Univ Chicago, MBA, 60; Univ Pa, PhD(bus opers), 68. Prof Exp: Sr engr, Philco Corp, 60-64; ASSOC PROF STATIST, PHILADELPHIA COL TEXTILES & SCI, 64- Concurrent Pos: Consult, Dept of Army, 68-69. Res: Obsolescence of engineering personnel; improvement of business forecasting techniques. Mailing Add: Dept of Statist Philadelphia Col of Textiles & Sci Philadelphia PA 19144

ZELINSKI, ROBERT PAUL, b Chicago, Ill, Jan 13, 20; m 45; c 4. POLYMER CHEMISTRY, RUBBER CHEMISTRY. Educ: DePaul Univ, BS, 41; Northwestern Univ, PhD(org chem), 45. Prof Exp: Asst, Northwestern Univ, 41-42; from instr to prof chem & chmn dept, DePaul Univ, 43-55; group leader, Rubber Synthesis Br, 55-61, from sect mgr to mgr, 61-75, MGR, POLYMER RES BR, RES DIV, PHILLIPS PETROL CO, 75- Concurrent Pos: Asst, Northwestern Univ, 43-44. Mem: Am Chem Soc; Sigma Xi; Soc Plastics Engrs. Res: Synthesis of plastics and rubbers. Mailing Add: Rte 1 Box 517A Bartlesville OK 74003

ZELINSKY, DANIEL, b Chicago, Ill, Nov 22, 22; m 45; c 3. ALGEBRA. Educ: Univ Chicago, SB, 41, SM, 43, PhD(math), 46. Prof Exp: Instr math, Univ Chicago, 43-44; asst, Appl Math Group, Columbia Univ, 44-45; Nat Res Coun fel, Univ Chicago, 46, instr math, 46-47; Nat Res Coun fel, Inst Advan Study, 47-49; from asst prof to assoc prof, 49-60, PROF MATH, NORTHWESTERN UNIV, EVANSTON, 60-, CHMN DEPT, 75- Concurrent Pos: Mem exec comt, Nat Res Coun, 66-67; ed jour, Am Math Soc, 61-64. Mem: AAAS; Am Math Soc; Math Asn Am. Res: Rings; homological algebra. Mailing Add: Dept of Math Northwestern Univ Evanston IL 60201

ZELINSKY, WILBUR, b Chicago, Ill, Dec 21, 21; m 44; c 2. CULTURAL GEOGRAPHY, POPULATION GEOGRAPHY. Educ: Univ Calif, Berkeley, BA, 44; Univ Wis-Madison, MA, 46; Univ Calif, Berkeley, PhD(geog), 53. Prof Exp: Map draftsman, R M Wilmotte & Co, Washington, DC, 42-43 & J McA Smiley, NY, 44-45; terrain analyst, US Corps Engrs, Ger, 46; asst prof geog, Univ Ga, 48-52; proj assoc, Univ Wis, 52-54; indust analyst, Chesapeake & Ohio Rwy Co, Detroit, 54-59; prof geog, Southern Ill Univ, Carbondale, 59-63; PROF GEOG, PA STATE UNIV, 63- Concurrent Pos: Mem comn pop geog, Int Geog Union, 64-72; NSF grant, Pa State Univ, 65-68. Honors & Awards: Meritorious Contrib Award, Asn Am Geogr, 66. Mem: Asn Am Geogr (from vpres to pres, 71-73); Am Geog Soc; Pop Asn Am; Am Name Soc; Int Union Scientific Study Pop. Res: Population geography; cultural, social, and historical geography of Anglo-America; geography and social policy. Mailing Add: Dept of Geog 403-A Deike Bldg Pa State Univ University Park PA 16802

ZELITCH, ISRAEL, b Philadelphia, Pa, June 18, 24; m 45; c 3. BIOCHEMISTRY, PLANT PHYSIOLOGY. Educ: Pa State Univ, BS, 47, MS, 48; Univ Wis, PhD(biochem), 51. Prof Exp: Nat Res Coun fel, Col Med, NY Univ-Bellevue Med Ctr, 51-52; asst biochemist, 52-54, assoc biochemist, 54-60, biochemist, 60-74, HEAD DEPT BIOCHEM, CONN AGR EXP STA, 63-, S W JOHNSON DISTINGUISHED SCIENTIST, 74- Concurrent Pos: Lectr, Yale Univ, 58-; Guggenheim fel, Oxford Univ, 60; panel mem, NSF, 62-64 & Physiol Chem Study Sect, NIH, 66-70; Regents lectr, Univ Calif, Riverside, 71. Mem: Am Chem Soc; Am Soc Biol Chem; Am Soc Plant Physiol; Brit Biochem Soc; The Chem Soc. Res: Plant biochemistry; photosynthesis; respiration; plant productivity. Mailing Add: Dept of Biochem Conn Agr Exp Sta PO Box 1106 New Haven CT 06504

ZELL, HOWARD CHARLES, b Philadelphia, Pa, Feb 11, 22; m 52. ORGANIC CHEMISTRY. Educ: St Joseph's Col, BS, 43; Univ Del, MS, 51; Univ Pa, PhD(org chem), 64. Prof Exp: Chemist, Publicker Industs, Inc, 43-47; res assoc org chem, Merck Sharp & Dohme Res Lab, 48-65; assoc scientist, 65-67, sr scientist, 67-69, PRIN SCIENTIST, ETHICON INC, 69- Mem: AAAS; Am Chem Soc; fel Am Inst Chemists; NY Acad Sci. Res: Medicinal and polymer chemistry. Mailing Add: Ethicon Inc Somerville NJ 08876

ZELL, LAROY W, b SDak, Feb 14, 10; m 39; c 2. CYTOLOGY, PLANT TAXONOMY. Educ: Northern State Teachers Col, BS, 31; Univ Iowa, MS, 38, PhD(bot), 53. Prof Exp: Teacher high sch, SDak, 31-36, prin, 33-36, teacher & coach high sch, Nebr, 36-37, Ill, 37-41; US auditor, Govt Accounting, War Ord Air Corps, Wis, 41-45; instr biol, Mankato State Col, 45-49; instr bot, Univ Iowa, 49-51; prof cell biol & chmn dept biol sci, 51-73, EMER PROF BIOL, MANKATO STATE COL, 73- Mem: Am Inst Biol Sco; Bot Soc Am. Res: Effects of ultra centrifugal forces on yeast and radish plant cells; distribution of fern species of southern Minnesota. Mailing Add: Dept of Biol Sci Mankato State Col Mankato MN 56001

ZELLE, MAX ROMAINE, b Alleman, Iowa, Oct 11, 15; m 37; c 2. GENETICS. Educ: Iowa State col, BS, 37, PhD(genetics), 40. Prof Exp: Asst prof animal husb, Purdue Univ, 41-44; biologist, NIH, 46-48; prof bact, Cornell Univ, 48-58; chief, Biol Br, Div Biol & Med, US AEC, 58-60, asst dir biol sci, 60-61; dir ctr radiol sci & prof genetics, Univ Wash, 61-62; dir div biol & med res, Argonne Nat Lab, 62-69; PROF RADIOL & RADIATION BIOL & CHMN DEPT, COLO STATE UNIV, 69- Concurrent Pos: Consult, Oak Ridge Nat Lab, 46-57; geneticist, US AEC, 50-52 & 57-58. Mem: Fel AAAS; Genetics Soc Am; Am Soc Microbiol; Radiation Res Soc Am; Am Soc Nat. Res: Genetics of microorganisms; irradiation effects on bacteria and bacteriophages; radiation biology. Mailing Add: Dept of Radiol & Radiation Biol Colo State Univ Ft Collins CO 80521

ZELLER, EDWARD JACOB, b Peoria, Ill, Nov 6, 25. GEOCHEMISTRY. Educ: Univ Ill, AB, 46; Univ Kans, MA, 48; Univ Wis, PhD(geol), 51. Prof Exp: Asst, State Geol Surv, Ill, 45-46; asst instr gen geol, Univ Kans, 45-46; proj assoc, USAEC contract, Wis, 51-56; prof geol & prin investr, USAEC res contract, 56-69, PROF GEOL, PHYSICS & ASTRON, UNIV KANS, 69- Concurrent Pos: Mem NSF US Antarctic Res Prog, 59-63; NSF sr fel, Physics Inst, Univ Berne, 60-61; guest scientist, Brookhaven Nat Lab, 65-66; US Air Force res contract, 67-68; prin investr, NASA contract, 72 & Oak Ridge Nat Lab contraxt, 72; del, 19th & 20th Int Geol Cong. Mem: AAAS; Geol Soc Am; Geochem Soc; Am Geophys Union; Int Asn Geochem & Cosmochem. Res: Thermoluminescence and electron spin resonance in geologic materials; radiation effects from nuclear waste; chemical interactions of fast protons in solid targets; lunar and asteroidal weathering; aerosols and planetary albedo; paleoclimatology. Mailing Add: Space Technol Labs Univ of Kans 2291 Irving Hill Dr-Campus W Lawrence KS 66044

ZELLER, FRANK JACOB, b Chicago, Ill, Dec 6, 27; m 52; c 3. REPRODUCTIVE ENDOCRINOLOGY. Educ: Univ Ill, BS, 51, MS, 52; Ind Univ, PhD(zool), 57. Prof Exp: Instr zool, Bryan Col, 52-53; asst, 53-56, res assoc, 56-57, from instr to asst prof, 57-66, ASSOC PROF ZOOL, IND UNIV, BLOOMINGTON, 66- Mem: Am Soc Zool; Soc Study Reprod. Res: Reproduction in birds; effects of gonadotrophins and sex hormones on the anterior pituitary gland and gonads. Mailing Add: Dept of Zool Ind Univ Bloomington IN 47401

ZELLER, MARY CLAUDIA, b Mansfield, Ohio, Dec 1, 10. MATHEMATICS. Educ: DePaul Univ, AB, 35; Univ Mich, MA, 40, PhD(math), 44. Prof Exp: From instr to asst prof, 41-50, dean, 53-69, assoc dir three year degree prog, 72-73, PROF MATH, COL ST FRANCIS, ILL, 50-, DIR GRANTS & SPEC PROJECTS, 75- Concurrent Pos: Consult, Dept Higher Educ, Nat Cath Educ Asn, 71-72. Mailing Add: Col of St Francis 500 Wilcox St Joliet IL 60435

ZELLER, MICHAEL EDWARD, b San Francisco, Calif, Oct 8, 39; m 60; c 2. HIGH ENERGY PHYSICS. Educ: Stanford Univ, BS, 61; Univ Calif, Los Angeles, MS, 63, PhD(physics), 68. Prof Exp: Res asst physics, Univ Calif, Los Angeles, 63-68, res fel, 68-69; from instr to asst prof, 69-75, ASSOC PROF PHYSICS, YALE UNIV, 75- Mem: Am Phys Soc. Res: Polarization phenomena in the kaon-nucleon interaction; high energy, strong interaction polarization phenomena. Mailing Add: J W Gibbs Lab Yale Univ New Haven CT 06520

ZELLEY, WALTER GAUNTT, b Camden, NJ, Oct 1, 21; m 46; c 3. ELECTROCHEMISTRY. Educ: Univ Pa, BS, 42. Prof Exp: Chemist, Lake Ont Ord Works, 42-43; chemist, Burlington Reduction Works, 43-44, res engr, Res Lab, 44-65, sr scientist, 65-67, eng assoc, 67-71, SECT HEAD, ALCOA TECH CTR, ALUMINUM CO AM, 71- Mem: Am Electroplaters Soc; Sigma Xi; Tech Asn Graphic Arts. Res: Chemical and electrochemical surface finishing of aluminum; application of aluminum in the graphic arts and packaging; organic coatings for aluminum. Mailing Add: Alcoa Tech Ctr Alcoa Center PA 15069

ZELLMER, DAVID LOUIS, b Portland, Ore, June 12, 42; m 66. ANALYTICAL CHEMISTRY. Educ: Univ Mich, BSChem, 64; Univ Ill, MS, 66, PhD(anal chem), 69. Prof Exp: Asst prof, 69-74, ASSOC PROF CHEM, CALIF STATE UNIV, FRESNO, 74- Mem: Am Chem Soc; Meteoritical Soc. Res: Application of radiochemical and electrochemical techniques to the study of semiconducting electrode materials; analysis of extraterrestrial materials; instrumentation automation. Mailing Add: Dept of Chem Calif State Univ Fresno CA 93710

ZELLNER, BENJAMIN HOLMES, b Forsyth, Ga, Apr 16, 42; wid; c 2. ASTRONOMY. Educ: Ga Inst Technol, BS, 64; Univ Ariz, PhD(astron), 70. Prof Exp: Res asst atmospheric physics, Ga Inst Technol, 63-64; res asst, 67-70, RES ASSOC ASTRON, UNIV ARIZ, 70- Concurrent Pos: Res fel, Observ Paris, Meudon, France, 72-73. Mem: Am Astron Soc. Res: Astronomical polarimetry; light scattering by circumstellar and interstellar material and at the surfaces of airless solarsystem bodies. Mailing Add: Lunar & Planetary Lab Univ of Ariz Tucson AZ 85721

ZELLNER, CARL NAEHER, b Brooklyn, NY, Mar 4, 10; m 36; c 1. ORGANIC CHEMISTRY. Educ: Princeton Univ, AB, 31, MA, 32, PhD(org chem), 34. Prof Exp: Res chemist, Merck & Co, Inc, 34-37, Tidewater Assoc Oil Co, 37-55 & Celanese Corp, NJ, 55-68; CHEM CONSULT, 68- Mem: AAAS; Am Chem Soc. Res: Conversion of petroleum hydrocarbons to industrial chemicals; exploratory polymer research. Mailing Add: Box 41 RD 2 River Rd New Hope PA 18938

ZELLWEGER, HANS ULRICH, b Lugano, Switz, June 19, 09; m 40; c 2. PEDIATRICS. Educ: Univ Zurich, MD, 34. Prof Exp: Resident, Kanton Hosp, Lucerne, Switz, 34-37; mem staff, Albert Schweitzer Hosp, Lambarene, Gabun, 37-39; asst prof & resident pediat, Children's Hosp, Zurich, 40-51; prof & head dept, Am Univ Beirut, 51-57 & 58-59; res prof, 57-58, PROF PEDIAT, COL MED, UNIV IOWA, 59- Concurrent Pos: Gen secy, Int Cong Pediat, Zurich, 50. Mem: Am Acad Neurol; Am Acad Cerebral Palsey (pres, 74-75); hon mem Lebanese Pediat Soc; hon mem Austrian Pediat Soc; cor mem Swiss Pediat Soc. Res: Pediatric neurology; genetics; cytogenetics. Mailing Add: Dept of Pediat Univ Hosps Iowa City IA 52240

ZELMAN, ALLEN, b Los Angeles, Calif, Feb 12, 38; m 72; c 2. BIOPHYSICS, BIOENGINEERING. Educ: Univ Calif, Berkeley, BA, 64, PhD(biophys), 71. Prof Exp: Asst prof biophys, Meharry Med Col, 72-75; ASST PROF BIOPHYS & BIOMED ENG, RENSSELAER POLYTECH INST, 75- Mem: Biophys Soc; Electrochem Soc; Am Soc Artificial Internal Organs; AAAS. Res: Artifical kidney development; hemo-and peritoneal dialysis; membrane transport; complete characterization of coefficient matrix, alteration of charge on biological membranes and mosaic membranes; concentration feedback and automatic regulation control systems. Mailing Add: Ctr for Bioeng Rensselaer Polytech Inst Troy NY 12181

ZELMANOWITZ, JULIUS MARTIN, b New York, NY, Feb 20, 41; m 62; c 1. MATHEMATICS, ALGEBRA. Educ: Harvard Univ, AB, 62; Univ Wis-Madison, MS, 63, PhD(math), 66. Prof Exp: Instr math, Univ Wis-Madison, 66; asst prof, 66-73, ASSOC PROF MATH, UNIV CALIF, SANTA BARBARA, 73- Concurrent Pos: Vis asst prof, Univ Calif, Los Angeles, 69-70, vis assoc prof, 73-74; vis asst prof, Carnegie-Mellon Univ, 70-71, Sarah Mellon Scaife Found fel, 71; NSF res grant, 72 & 73. Mem: Math Asn Am; Am Math Soc. Res: Algebra, rings and modules. Mailing Add: Dept of Math Univ of Calif Santa Barbara CA 93106

ZELNICK, ERNEST, b Brooklyn, NY, July 10, 21; m 47; c 3. AUDIOLOGY. Educ: Brooklyn Col, BA, 42, MA, 68; St John's Univ, NY, LLB, 42; City Univ New York, PhD(audiol), 69; NY Univ, LLM, 74. Prof Exp: PRES, PROF HEARING AIDS SERV, 51-; PRES, NOISE ABATEMENT-NOISE POLLUTION CONSULT, 65- Mem: Int Soc Audiol; Acoust Soc Am; Am Pub Health Asn; Am Bar Asn; Am Audiol Soc. Res: Comparison of monotic and dichotic modes of listening as to effects on speech intelligibility for individuals suffering from sensori-neural hearing disorders which are bilaterally symmetrical. Mailing Add: 705 Flatbush Ave Brooklyn NY 11225

ZELSON, CARL, b Philadelphia, Pa, Oct 1, 05; m 34; c 2. PEDIATRICS. Educ: Univ Wis, BS, 28; Wash Univ, MD, 30. Prof Exp: From instr to prof, New York Med Col, 55-74; RETIRED. Concurrent Pos: Maternal & Child Health Serv grant, Metrop Hosp, New York, 71-74; consult newborn serv, Metrop Hosp, New York, 72-74; mem pediat staff, Danbury Hosp, Conn, 74- Mem: Fel Am Acad Pediat. Res; Effects of maternal drug addiction on the newborn infant. Mailing Add: 71 Bennetts Farm Rd Ridgefield CT 06877

ZELSON, PHILIP RICHARD, b Long Beach, Calif, Sept 3, 45; m 67; c 2. BIOCHEMISTRY. Educ: Northwestern Univ, DDS, 70; Univ Rochester, PhD(biochem), 75. Prof Exp: ASST PROF ORAL MED, SCH DENT MED, UNIV PA, 74-; RES ASSOC BIOCHEM TASTE, VET ADMIN, 75- Concurrent Pos: Res assoc biochem taste, Monell Chem Senses Ctr, 74-75. Mem: AAAS. Res: Elucidation of the molecular mechanisms involved in gustation. Mailing Add: Monell Chem Senses Ctr 3500 Market St Philadelphia PA 19104

ZELTMANN, ALFRED HOWARD, b Brooklyn, NY, Dec 25, 21; m 51; c 3. PHYSICAL CHEMISTRY. Educ: State Col Wash, BS, 48; Univ NMex, PhD(chem), 52, MS, 61. Prof Exp: STAFF MEM PHYS CHEM, LOS ALAMOS SCI LAB, UNIV CALIF, 46- Mem: Am Chem Soc. Res: Chemical kinetics; radiation chemistry; high vacuum; preparation and chemistry of gaseous hydrides; nuclear magnetic resonance; complex ions; laser spectroscopy; laser isotope separation. Mailing Add: 393 Venado St Los Alamos NM 87544

ZELTMANN, EUGENE W, b Chicago, Ill, June 26, 40. PHYSICAL CHEMISTRY. Educ: Beloit Col, BA, 62; Johns Hopkins Univ, MA, 64, PhD(chem), 67. Prof Exp: Nuclear chemist, Knolls Atomic Power Lab, Gen Elec Co, NY, 67-70; Alfred E Smith fel in NY State, 70-71; asst to dir power div, NY State Pub Serv Comn, 71-72; MGR ENVIRON PLANNING, GAS TURBINE DIV, GEN ELEC CO, 72- Mem: Am Chem Soc. Res: Chemical kinetics; fission track imaging analyses for purpose of determining presence of minute amounts of fissionable materials. Mailing Add: Gas Turbine Div Bldg 500 Rm 228 Gen Elec Co One River Rd Schenectady NY 12301

ZEMACH, CHARLES, b Los Angeles, Calif, Sept 15, 30; m 58; c 2. ELEMENTARY PARTICLE PHYSICS. Educ: Harvard Univ, BA, 51, PhD(physics), 55. Prof Exp: Nat Sci fel, 55-56; instr physics, Univ Pa, 56-57; res assoc, Univ Calif, Berkeley, 57-58, from asst prof to prof, 58-71; officer & spec asst for technol, US Arms Control & Disarmament Agency, 71-74, MEM POLICY PLANNING STAFF, STATE DEPT, 74- Concurrent Pos: Alfred P Sloan Found fel, 59-63; Guggenheim Found fel, 66-67. Mem: Am Phys Soc. Res: Thermal neutron diffraction; quantum electrodynamics; strong interactions of elementary particles. Mailing Add: Pol Plan Staff New State Dept Bldg 21st & Virginia NW Washington DC 20451

ZEMACH, RITA, b Paterson, NJ, Apr 3, 26; m 47; c 2. STATISTICS. Educ: Columbia Univ, BA, 47; Mich State Univ, MS, 61, PhD(statist, probability), 65. Prof Exp: Instr statist & probability, Mich State Univ, 65-66, asst prof systs sci, 66-72; assoc prof biostatist, Univ Mich, Ann Arbor, 72-73; CHIEF STATIST RES, MICH DEPT PUB HEALTH, 73- Concurrent Pos: Mem health care technol study sect, Dept Health, Educ & Welfare, 72-76. Mem: AAAS; Am Statist Asn; Inst Math Statist; Opers Res Soc Am; Am Pub Health Asn; Biomet Soc. Res: Health service and resource statistics; program evaluation and measurement methods. Mailing Add: Dept of Pub Health 3500 NLogan St Lansing MI 48914

ZEMAN, FRANCES JUNE, b Cleveland, Ohio, Mar 5, 25. NUTRITION. Educ: Western Reserve Univ, BS, 46, MS, 57; Ohio State Univ, PhD(nutrit), 63. Prof Exp: Dietitian, Cleveland City Hosp, Ohio, 49-51, teaching dietitian, Sch Nursing, 51-57; asst prof home econ, Kent State Univ, 57-64; from asst prof to assoc prof, 64-74, PROF NUTRIT, UNIV CALIF, DAVIS, 74- Mem: AAAS; assoc Am Physiol Soc; Am Inst Nutrit. Res: Nutrition in reproduction. Mailing Add: Dept of Nutrit Univ of Calif Davis CA 95616

ZEMAN, WOLFGANG, b Stuttgart, Ger, Apr 14, 21; nat US; m 52; c 3. NEUROPATHOLOGY. Educ: Univ Tubingen, MD, 45. Prof Exp: Resident neuropath, Ger Res Inst Psychiat, Munich, 47-49; resident neurol, Neurol Inst, Univ Hamburg, 49-51; res assoc neuropath, Neuropsychiat Inst, Univ Mich, 51-52; instr path, Univ Ark, 53-54; asst prof, Univ Pittsburgh, 55-56; asst prof psychiat & path, Ohio State Univ, 56-58, assoc prof psychiat & asst prof neurol, 58-60; assoc prof path, 60-65, PROF PATH, MED CTR, IND UNIV, INDIANAPOLIS, 65- Concurrent Pos: Consult, Vet Admin Hosps, Pittsburgh, Pa, 55-56, Chillicothe, Ohio, 57-60 & Indianapolis, Ind, 60-; guest biologist, Brookhaven Nat Lab, 58- Mem: AAAS; Am Asn Path & Bact; Am Asn Neuropath; Radiation Res Soc; NY Acad Sci. Res: Radiation injury of nervous tissues; degenerative diseases of nervous system; slow virus diseases. Mailing Add: Div of Neuropath Ind Univ Med Ctr Indianapolis IN 46202

ZEMANEK, JOSEPH, JR, b Blessing, Tex, Jan 1, 28; m 50; c 2. ACOUSTICS. Educ: Univ Tex, BS, 49; Southern Methodist Univ, MS, 57; Univ Calif, Los Angeles, PhD(physics), 62. Prof Exp: Test engr, Gen Elec Co, 49-50; jr technologist, Field Res Lab, Magnolia Petrol Co, 51-53, from res technologist to sr res technologist, 53-58, sr res technologist, Mobil Res & Develop Corp, 61-67, RES ASSOC ACOUSTIC WELL LOGGING, MOBIL RES & DEVELOP CORP, 67- Mem: Sigma Xi; Acoust Soc Am. Res: Acoustic wave propagation in isotropic media, attenuation and velocity measurements; acoustic well logging, development of instrumentation and methods; wave propagation in boreholes. Mailing Add: Field Res Lab Mobil Res & Develop Corp Dallas TX 75221

ZEMANIAN, ARMEN HUMPARTSOUM, b Bridgewater, Mass, Apr 16, 25; m 58; c 4. APPLIED MATHEMATICS. Educ: City Col New York, BEE, 47; NY Univ, MEE, 49, Eng ScD, 53. Prof Exp: Tutor elec eng, City Col New York, 47-48; engr, Maintenance Co, 48-52; from instr to assoc prof elec eng, NY Univ, 52-62; chmn dept, 67-68 & 71-74, PROF APPL MATH & STATIST, STATE UNIV NY STONY BROOK, 62- Concurrent Pos: Res fel math inst, Univ Edinburgh, 68-69; consult, All-Tronics, Inc & Anaesthesia Assocs, NY; managing ed, Siam Rev, Soc Indust & Appl Math, 69-71, ed-in-chief publs, 70-74; vis scholar, Food Res Inst, Stanford Univ, 75-76; NSF fac fel sci, 75-76. Mem: Soc Indust & Appl Math; Am Math Soc; Math Asn Am; fel Inst Elec & Electronic Engrs. Res: Mathematical systems theory; distribution theory; integral transforms; electrical network theory. Mailing Add: Dept of Appl Math & Statist State Univ of NY Stony Brook NY 11794

ZEMANSKY, MARK WALDO, b New York, NY, May 5, 00; m 32; c 2. PHYSICS. Educ: City Col New York, BS, 21; Columbia Univ, AM, 22, PhD(physics), 27. Prof Exp: Tutor, 22-25, from instr to prof, 25-66, chmn dept, 56-59, EMER PROF PHYSICS, CITY COL NEW YORK, 67- Concurrent Pos: Nat Res Coun fels, Princeton Univ, 28-30 & Kaiser Wilhelm Inst, Berlin, 30-31; mem staff, Cryogenic Lab, Columbia Univ, 46-59; lectr, New Sch Soc Res, 47-58; mem tech adv comt, Nat Bur Standards, 55-60; exec officer physics doctoral prog, City Univ New York, 63-66. Honors & Awards: Oersted Medal, Am Asn Physics Teachers, 56. Mem: Fel Am Phys Soc; Am Asn Physics Teachers (pres, 56, exec secy, 67-70). Res: Resonance radiation in medal vapors; collision broadening of spectral lines; thermodynamics; low temperature phenomena. Mailing Add: 736 Rutland Ave Teaneck NJ 07666

ZEMANY, PAUL DANIEL, physical chemistry, see 12th edition

ZEMBRODT, ANTHONY RAYMOND, b Covington, Ky, Jan 2, 43; m 66; c 2. ANALYTICAL CHEMISTRY, PHYSICAL CHEMISTRY. Educ: Thomas More Col, BA, 65; Ohio Univ, PhD(phys chem), 70. Prof Exp: SCIENTIST ANAL CHEM, DRACKETT CO, BRISTOL MYERS CO, CINCINNATI, 69- Concurrent Pos: Lectr, NKy State Col, 72- Mem: Am Chem Soc; NAm Thermal Anal Soc. Res: Thermal analysis of polymers; instrumental techniques for analyses. Mailing Add: 1004 Park Lane Covington KY 41011

ZEMJANIS, RAIMUNDS, b Petrhoff, Russia, Sept 16, 18; US citizen; m 42; c 3. REPRODUCTIVE PHYSIOLOGY. Educ: Royal Vet Col Sweden, DVM, 48; Univ Minn, PhD, 57. Prof Exp: Teaching asst anat, Univ Latvia, 38-40; chief vet, A I Ctr Ostergotland, Sweden, 48-51; instr, 52-57, assoc prof, 57-59, PROF VET OBSTET, COL VET MED, UNIV MINN, ST PAUL, 59-, HEAD DEPT VET OBSTET & GYNEC, 71-, MEM STAFF, VET HOSP, 70- Concurrent Pos: Agency Int Develop consult, Colombia, SAm, 62 & Jamaica, 65; NIH spec fel, 68-69. Mem: Am Vet Med Asn; Soc Study Reproduction; Am Asn Vet Clinicians. Res: Animal reproduction and infertility; growth actors and enzyme inhibitors and growth of Vibrio fetus; histology and electron microscopy of the bovine endometrium and normal and degenerate monkey testes. Mailing Add: Dept of Vet Obstet Univ of Minn St Paul MN 55108

ZEMKE, WARREN T, b Fairmont, Minn, Oct 9, 39; m 68; c 2. PHYSICAL CHEMISTRY, QUANTUM CHEMISTRY. Educ: St Olaf Col, BA, 61; Ill Inst Technol, PhD(chem), 69. Prof Exp: From instr to asst prof, 66-74, ASSOC PROF CHEM, WARTBURG COL, 74- Mem: AAAS; Am Chem Soc. Res: Molecular electronic wavefunctions; molecular spectroscopy and structure. Mailing Add: Dept of Chem Wartburg Col Waverly IA 50677

ZEMLICKA, JIRI, b Prague, Czech, July 31, 33; m 61; c 2. ORGANIC CHEMISTRY, BIOCHEMISTRY. Educ: Charles Univ, Prague, RNDr, 56; Czech Acad Sci, PhD(org chem), 59. Prof Exp: Res asst anal biochem, Inst Food Technol, Prague, 56; res scientist, Inst Org Chem & Biochem, Czech Acad Sci, 59-68; vis scientist, 68-69, RES SCIENTIST, MICH CANCER FOUND, 70-; ASSOC PROF BIOCHEM, SCH MED, WAYNE STATE UNIV, 71- Mem: Am Chem Soc. Res: Chemistry of nucleic acids; protein biosynthesis. Mailing Add: Mich Cancer Found 110 E Warren Ave Detroit MI 48201

ZEMLIN, WILLARD R, b Two Harbors, Minn, July 20, 20; m 54; c 2. SPEECH & HEARING SCIENCES. Educ: Univ Minn, BA, 57, MS, 60, PhD(speech), 61. Prof Exp: DIR SPEECH & HEARING RES LAB, UNIV ILL, URBANA-CHAMPAIGN, 61- Concurrent Pos: Consult, Lincoln State Sch, 65-67. Mem: Am Speech & Hearing Asn; Acoust Soc Am. Res: Anatomy and physiology of normal and pathological speech and hearing mechanisms. Mailing Add: Speech & Hearing Res Lab Univ of Ill Champaign IL 61820

ZEMMER, JOSEPH LAWRENCE, JR, b Biloxi, Miss, Feb 23, 22; m 50; c 3. ALGEBRA. Educ: Tulane Univ, BS, 43, MS, 47; Univ Wis, PhD(math), 50. Prof Exp: From asst prof to assoc prof, 50-61, PROF MATH, UNIV MO-COLUMBIA, 61-, CHMN DEPT, 67-70 & 73- Concurrent Pos: Fulbright lectr, Osmania Univ, India, 63-64. Mem: Am Math Soc; Math Asn Am; Can Math Cong. Res: Partially ordered algebraic systems; projective planes and their coordinatizing algebraic systems, especially nearfields and Veblen-Wedderburn systems. Mailing Add: Dept of Math Univ of Mo Columbia MO 65201

ZEMON, STANLEY ALAN, b Detroit, Mich, June 16, 30; m 67. PHYSICS. Educ: Harvard Univ, AB, 52; Columbia Univ, AM, 58, PhD(physics), 64. Prof Exp: Res asst physics, Columbia Univ, 58-62, res scientist, 63-64, res assoc, 64-65; MEM TECH STAFF, GTE LABS, INC, 65- Mem: Am Phys Soc; Inst Elec & Electronics Engrs. Res: Superconductivity; low temperature physics; microwaves; acoustoelectric effect; ultrasonics; semiconductors; Brillouin scattering; lasers; acoustic surface waves; nonlinear acoustics; optical guided waves; nonlinear optics. Mailing Add: Waltham Res Ctr GTE Labs Inc 40 Sylvan Rd Waltham MA 02154

ZEMP, JOHN WORKMAN, b Camden, SC, Sept 28, 31; m 58; c 4. BIOCHEMISTRY. Educ: Col Charleston, BS, 53; Med Col SC, MS, 54; Univ NC, PhD(biochem), 66. Prof Exp: Asst prof biochem, Ctr Res Pharmacol & Toxicol, Sch Med, Univ NC, Chapel Hill, 67-69; from asst prof to assoc prof biochem, 69-73, actg dean, Col Med, 74, actg vpres acad affairs, 74-75, COORDR RES, MED UNIV S C, 72-, ASSOC DEAN ACAD AFFAIRS, 73-, PROF BIOCHEM, 74- Concurrent Pos: NSF res fel, 66-67. Honors & Awards: Coker Award, Univ NC, Chapel Hill, 66. Mem: AAAS; Am Chem Soc; Soc Neurosci; Soc Exp Biol & Med; NY Acad Sci. Res: Brain function and biochemistry; developmental neurobiology; nutrition; biochemical pharmacology. Mailing Add: Dept of Biochem Med Univ of SC Charleston SC 29401

ZEN, E-AN, b Peking, China, May 31, 28; nat US. GEOLOGY, PETROLOGY. Educ: Cornell Univ, BA, 51; Harvard Univ, AM, 52, PhD(geol), 55. Prof Exp: Res fel, Oceanog Inst, Woods Hole, 55-56, res assoc, 56-58; vis asst prof, Univ NC, 58-59; RES GEOLOGIST, US GEOL SURV, 59- Mem: Nat Acad Sci; fel AAAS; fel Geol Soc Am; fel Mineral Soc Am; Mineral Asn Can. Res: Phase equilibrium of sedimentary and metamorphic rocks; stratigraphy and structure of Taconic allochthon. Mailing Add: US Geol Surv Stop 959 Nat Ctr Reston VA 22092

ZENCHELSKY, SEYMOUR THEODORE, b New York, NY, July 6, 23. ANALYTICAL CHEMISTRY. Educ: NY Univ, BA, 44, MS, 47, PhD(chem), 52. Prof Exp: From instr to assoc prof, 51-64, PROF CHEM, RUTGERS UNIV, NEW BRUNSWICK, 64- Mem: Am Chem Soc; Soc Appl Spectros; NY Acad Sci. Res: Instrumentation; nonaqueous titrations; spectrophotometry; atmospheric aerosols; organic aerosols. Mailing Add: Dept of Chem Rutgers Univ Sch of Chem New Brunswick NJ 08903

ZENDER, MICHAEL J, b Austin, Minn, Feb 27, 39. NUCLEAR PHYSICS. Educ: St John's Univ, Minn, BA, 61; Vanderbilt Univ, PhD(physics), 66. Prof Exp: From asst prof to assoc prof & chmn dept, 66-72, PROF PHYSICS, CALIF STATE UNIV, FRESNO, 73- Mem: Am Asn Physics Teachers. Res: Low energy nuclear, beta and gamma-ray spectroscopy; study of the Auger effect by using a post acceleration Geiger counter in conjunction with very thin counter windows; radiation safety. Mailing Add: Dept of Physics Calif State Univ Fresno Shaw & Cedar Ave Fresno CA 93740

ZENER, CLARENCE MELVIN, b Indianapolis, Ind, Dec 1, 05; m 31; c 4. PHYSICS. Educ: Stanford Univ, AB, 26; Harvard Univ, PhD(physics), 29. Prof Exp: Sheldon traveling fel, Ger, 29-30; Nat Res Coun fel, Princeton Univ, 30-32; fel, Bristol Univ, 32-34; instr physics, Washington Univ, 35-37 & City Col New York, 37-40; assoc prof, State Col Wash, 40-42; from physicist to prin physicist, Watertown Arsenal, 42-45; prof physics, Univ Chicago, 45-51; assoc dir res labs, Westinghouse Elec Corp, 51-56, dir res labs, 56-62, dir sci, 62-65; dean col sci, Tex A&M Univ, 66-68; UNIV PROF PHYSICS, CARNEGIE-MELLON UNIV, 68- Mem: Nat Acad Sci; fel Am Phys Soc. Res: Theoretical physics and engineering. Mailing Add: Carnegie-Mellon Univ Pittsburgh PA 15213

ZENGER, DONALD HENRY, b Little Falls, NY, July 27, 32; m 57; c 2. GEOLOGY. Educ: Union Col, NY, BS, 54; Dartmouth Col, MA, 59; Cornell Univ, PhD(geol), 62. Prof Exp: Instr, 62-64, asst prof, 64-69, ASSOC PROF GEOL, POMONA COL, 69- Concurrent Pos: Grants, Geol Soc Am, 63-65 & Am Chem Soc, 66-69. Mem: Geol Soc Am; Paleont Soc; Am Asn Geol Teachers; Soc Econ Paleont & Mineral. Res: Silurian and Devonian stratigraphy, paleontology and sedimentary petrology of New York and California. Mailing Add: Dept of Geol Pomona Col Claremont CA 91711

ZENISEK, CYRIL JAMES, b Cleveland, Ohio, Feb 6, 26; m 56; c 3. ZOOLOGY. Educ: Ohio State Univ, BSc, 49, MSc, 55, PhD(zool), 63. Prof Exp: From asst prof to assoc prof, 60-66, PROF BIOL, INDIANA UNIV PA, 66- Mem: Soc Syst Zool; Am Soc Ichthyol & Herpet; Ecol Soc Am; Soc Study Amphibians & Reptiles; Herpetologists League. Res: Taxonomy and distribution of amphibians. Mailing Add: Dept of Biol Ind Univ of Pa Indiana PA 15701

ZENITZ, BERNARD LEON, b Baltimore, Md, Mar 26, 17; m 56; c 2. MEDICINAL CHEMISTRY. Educ: Univ Md, BS, 37, PhD(pharmaceut chem), 43. Prof Exp: Sr res chemist, Frederic Stearns & Co, Mich, 44-47; SR RES CHEMIST & LAB HEAD, STERLING-WINTHROP RES INST, 47- Honors & Awards: Col Medal Gen Excellence & Pharmacog Prize, 37. Mem: Am Chem Soc; NY Acad Sci. Res: Thymus gland extracts; sympathomimetic amines; quartenary ammonium salts; synthetic detergents; sterols; preparation of B-cyclohexylalkylamines; halogen ring substituted propadrines; coronary dilators; local anesthetics; tranquilizers; antioxidants; antiinflammatory agents; antiobesity drugs; bronchodilators. Mailing Add: Sterling-Winthrop Res Inst Columbia Turnpike Rensselaer NY 12144

ZENKER, NICOLAS, b Paris, France, Dec 3, 21; nat US; m 52; c 4. PHARMACEUTICAL CHEMISTRY. Educ: Cath Univ Louvain, Cand, 48; Univ Calif, MA, 53, PhD(pharmaceut chem), 58. Prof Exp: Biochemist, Mt Zion Hosp, San Francisco, 53-60; asst prof pharmaceut chem, 60-63, assoc prof, 63-69, PROF MED CHEM & HEAD DEPT, UNIV MD, BALTIMORE CITY, 69- Mem: Am Chem Soc. Res: Synthesis, mode of action and metabolism of metabolic analogues. Mailing Add: Dept of Medicinal Chem Col of Pharm Univ of Md 636 W Lombard St Baltimore MD 21201

ZENNER, WALTER P, b Nurnberg, Ger, Oct 18, 33; US citizen; m 66. ANTHROPOLOGY. Educ: Northwestern Univ, BS, 55; Columbia Univ, MA, 58, PhD, 65. Prof Exp: Instr sociol & anthrop, Lake Forest Col, 63-65, asst prof, 65-66, ASSOC PROF ANTHROP, STATE UNIV NY ALBANY, 66- Concurrent Pos: Res assoc, Northwestern Univ proj, Israel, 67-68; consult, Am Jewish Comt, 67. Mem: Am Anthrop Asn; Am Ethnol Soc. Res: Middle Eastern ethnology; culture change, inter-group relations and comparative religion. Mailing Add: Dept of Anthrop State Univ NY Albany NY 12222

ZENTMYER, DAVID TAYLOR, b Tyrone, Pa, June 1, 15; m 42; c 2. ORGANIC CHEMISTRY. Educ: Juniata Col, BS, 37; Univ Pa, MS, 40, PhD(org chem), 48. Prof Exp: Asst instr chem, Univ Pa, 38-42; res chemist, 47-59, mgr plastics flooring res, 59-60, gen mgr floor prod res, 60-68, ASST DIR RES, FUNDAMENTAL RES & APPLN, ARMSTRONG CORK CO, 68- Mem: Am Chem Soc. Res: Synthetic plastics; plastic flooring. Mailing Add: Armstrong Cork Co Lancaster PA 17604

ZENTMYER, GEORGE AUBREY, (JR), b North Platte, Nebr, Aug 9, 13; m 41; c 3. PHYTOPATHOLOGY. Educ: Univ Calif, Los Angeles, AB, 35; Univ Calif, MS, 36, PhD(plant path), 38. Prof Exp: Asst plant path, Univ Calif, 36-37; asst pathologist, Div Forest Path, Bur Plant Indust, USDA, 37-40 & Conn Exp Sta, 40-44; asst pathologist, 44-48, assoc plant pathologist, 48-55, fac res lectr, 64, chmn dept plant path, 68-73, PLANT PATHOLOGIST, UNIV CALIF RIVERSIDE, 55-, PROF PLANT PATH, 63- Concurrent Pos: Consult, Pineapple Res Inst, 61, Trust Territory, Pac Islands, 64 & 66, Rockefeller Found, Colombia, 67, Australian Govt, 68 & US Agency Int Develop, Ghana & Nigeria, 69, NSF panel, 71-; Guggenheim fel, 64-65; mem, Nat Res Coun, 68-73; NATO sr sci fel, Eng, 71; mem var comts, Nat Acad Sci-Nat Res Coun; assoc ed, Annual Rev Phytopath, 70- Honors & Awards: Award of Merit, Am Phytopath Soc, Caribbean, 72. Mem: Fel AAAS (pres elect, Pac Div, 73-74, pres, 74-75); fel Am Phytopath Soc (secy, 59-62, pres, 65-66); Int Soc Plant Path; Mycol Soc Am; Asn Trop Biol. Res: Biology and physiology of root pathogens, especially Phytophthora; chemotaxis; chemotherapy; fungicides; diseases of avocado, cacao, other tropicals and subtropicals. Mailing Add: Dept of Plant Path Univ of Calif Riverside CA 92502

ZENTNER, THOMAS GLENN, b Rowena, Tex, Aug 6, 26; m 52; c 5. PULP CHEMISTRY, PAPER CHEMISTRY. Educ: Tex A&M Univ, BS, 48; Inst Paper Chem, MS, 50, Lawrence Col, PhD(chem), 52. Prof Exp: Mem staff, Gardner Div, Diamond Nat Corp, 52-59, dir res & develop, 59-60; dir forest prod oper, Olin Mathieson Chem Corp, 60-66; vpres res & develop, Olinkraft, Inc, La, 66-68, rep, Forest Prod Div, Olin Corp, West Monroe, 68-75, V PRES RES, OLINKRAFT, INC, 74- Mem: Am Chem Soc; corp mem Tech Asn Pulp & Paper Indust. Res: Paperboard manufacture; coating; converting for packaging; graphic arts. Mailing Add: Olinkraft Inc PO Box 488 West Monroe LA 71291

ZENZ, CARL, b Vienna, Austria, Feb 1, 23; US citizen; m 47; c 3. OCCUPATIONAL MEDICINE. Educ: Jefferson Med Col, MD, 49; Univ Cincinnati, ScD, 57. Prof Exp: Intern, Ausbury Hosp, Minneapolis, Minn, 49-50; gen pract, Minn & Wis, 50-52; resident occup med, 56-57, chief clin serv, 57-62, chief physician, 62-63, med dir, 63-65, dir med & hyg serv, 65-70, DIR MED SERV, ALLIS-CHALMERS CORP, 70- Mem: AAAS; Am Occup Med Asn; Am Indust Hyg Asn; fel Am Pub Health Asn; fel Am Acad Occup Med (past pres). Mailing Add: Allis-Chalmers Corp PO Box 512 Milwaukee WI 53201

ZEOLI, HAROLD WILSON, b Boston, Mass, Aug 24, 10; m 40; c 5. MATHEMATICS, PHYSICS. Educ: Mass State Teachers Col Bridgewater, BS, 40; Boston Univ, AM, 44. Prof Exp: Instr physics, Boston Univ, 43-44; teacher high sch, Mass, 44-48; assoc prof, 48-62, PROF MATH, CENT MICH UNIV, 62-, DIR COMPUT SCI CTR, 68- Mem: Am Phys Soc. Res: Molecular spectra; differential equations; usefulness of lighter-than-air aircraft as space vehicles; scientific computer programming. Mailing Add: Dept of Math Cent Mich Univ Mt Pleasant MI 48858

ZEPF, THOMAS HERMAN, b Cincinnati, Ohio, Feb 13, 35. PHYSICS. Educ: Xavier Univ, Ohio, BS, 57; St Louis Univ, MS, 60, PhD(physics), 63. Prof Exp: From asst prof to assoc prof, 62-75, from actg chmn dept to chmn dept, 63-73, PROF PHYSICS, CREIGHTON UNIV, 75- Concurrent Pos: Vis prof physics, St Louis Univ, 73-74. Mem: AAAS; Am Phys Soc; Am Asn Physics Teachers. Res: Surface physics; electron emission; x-ray diffraction; ultra-high vacuum techniques. Mailing Add: Dept of Physics Creighton Univ Omaha NE 68178

ZEPPA, ROBERT, b New York, NY, Sept 17, 24; m 52; c 2. THORACIC SURGERY. Educ: Columbia Univ, AB, 48; Yale Univ, MD, 52. Prof Exp: Intern, Med Ctr, Univ Pittsburgh, 52-53; asst resident surg, Sch Med, Univ NC, 53-56, thoracic resident, 56-57, instr thoracic surg, 58-60, from asst prof surg to assoc prof, 60-65; co-chmn dept, 66-71, CHMN DEPT SURG, SCH MED, UNIV MIAMI, 71-, PROG SURG & PHARMACOL, 65- Concurrent Pos: USPHS career trainee, Sch Med, Wash Univ, 56-58; Markle scholar med sci, 59-64; assoc dir clin res unit, NC Mem Hosp, 61-65; chief surg serv, Vet Admin Hosp, Miami, Fla, 71- Mem: Am Col Surg; Am Surg Asn; Soc Surg Alimentary Tract; Soc Univ Surg. Res: Biologically active amines; portal hypertension; gastrointestinal physiology. Mailing Add: Univ of Miami Sch of Med PO Box 520875 Miami FL 33152

ZERBE, JOHN IRWIN, b Hegins, Pa, June 4, 26; m 51; c 3. WOOD SCIENCE, WOOD TECHNOLOGY. Educ: Pa State Univ, BS, 51; State Univ NY, MS, 53, PhD(wood tech), 56. Prof Exp: Asst wood tech, State Univ NY, 51-56; res asst prof housing res, Univ Ill, 56-58; mgr, Govt Specifications & Standards Dept, Nat Forest Prod Asn, 58-59, asst vpres, Tech Serv, 59-70; DIR FOREST PROD & ENG RES, FOREST SERV, US DA, 70- Mem: Forest Prod Res Soc; Soc Wood Sci & Technol. Res: Mechanical properties of wood. Mailing Add: 2009 Hermitage Ave Silver Spring MD 20902

ZERBY, CLAYTON DONALD, b Cleveland, Ohio, Jan 27, 24; m 49; c 3. PHYSICS, MECHANICAL ENGINEERING. Educ: Case Western Reserve Univ, BS, 50; Univ Tenn, MS, 56, PhD(physics), 60. Prof Exp: Engr, Oak Ridge Nat Lab, 50-54, group leader physics, 54-63, mgr physics & eng, Defense & Space Systs Dept, 63-66, mgr dept, 66-67, gen mgr, Korad Laser Dept, 67-71, pres, Ocean Systs, Inc, 71-73, Domsea Farms, Inc, 72-74, tech serv mgr, 74-76, DIR OFF WASTE ISOLATION, NUCLEAR DIV, UNION CARBIDE CORP, 76- Concurrent Pos: Lectr, Univ Tenn, 61-63. Mem: Am Phys Soc; Am Nuclear Soc. Res: Theory of electromagnetic interactions; Monte Carlo methods; nuclear weapons effects; space vehicle radiation shielding; shielding against high energy particles; interaction of high energy particles with complex nuclei; nuclear waste terminal storage. Mailing Add: Nucelar Div PO Box Y Union Carbide Corp Oak Ridge TN 37830

ZERLA, FREDRIC JAMES, b Wheeling, WVa, Feb 23, 37; m 66; c 6. MATHEMATICS. Educ: Col Steubenville, BA, 58; Fla State Univ, MS, 60, PhD(math), 67. Prof Exp: Asst prof, 63-72, from asst chmn dept to actg chmn dept, 69-74, ASSOC PROF MATH, UNIV S FLA, 72-, UNDERGRAD ADV, DEPT MATH, 74- Mem: Math Asn Am. Res: Derivations in algebraic fields; field theory. Mailing Add: Dept of Math Univ of SFla Tampa FL 33620

ZERLIN, STANLEY, b New York, NY, Sept 15, 29; m 59; c 2. PSYCHOPHYSIOLOGY, ELECTROPHYSIOLOGY. Educ: City Col New York, BS, 51; Columbia Univ, MA, 53; Western Reserve Univ, PhD(audition), 59. Prof Exp: Asst proj dir, Cleveland Hearing & Speech Ctr, 57-58; proj dir, Auditory Res Lab, Vet Admin, 59-63, res scientist, 59-62, res dir, 62-63; res assoc auditory processes, Cent Inst for Deaf, 63-65; res assoc auditory evoked responses, Houston Speech & Hearing Ctr, 65-67; ASSOC PROF & RES ASSOC AUDITORY PROCESSES, OTOLARYNGOL LABS, UNIV CHICAGO, 67- Mem: AAAS; Acoust Soc Am; Am Speech & Hearing Asn; Int Soc Audiol. Res: Auditory electrophysiology, evoked response; cochlear processes; binaural interaction. Mailing Add: ENT Sect Dept of Surg Univ of Chicago 950 E 59th St Chicago IL 60637

ZERNIKE, FRITS, b Groningen, Neth, Dec 27, 30; m 62; c 3. PHYSICS. Educ: State Univ Groningen, MSc, 56. Prof Exp: Proj dir, Block Assocs, Inc, Mass, 56-61; physicist, Mithras, Inc, 61-62; sr res physicist, Perkin-Elmer Corp, Norwalk, Ct, 62-

74; CONSULT SCIENTIST, PHILIPS LABS, BRIARCLIFF MANOR, NY, 74- Mem: Am Phys Soc; Inst Elec & Electronics Eng. Res: Optics, particularly integrated optics; nonlinear optics; lasers; optical instruments. Mailing Add: Philips Labs 345 Scarborough Rd Briarcliff Manor NY 10510

ZERNOW, LOUIS, b Brooklyn, NY, Dec 27, 16; m 40; c 4. PHYSICS. Educ: Cooper Union, BChE, 38; Johns Hopkins Univ, PhD(physics), 53. Prof Exp: Mem staff, Ballistic Res Labs, Aberdeen Proving Ground, Ord Dept, US Dept Army, 40-51, chief, Rocket & Ammunition Br, 51-53 & Detonation Physics Br, 53-55; dir res & mgr ord res div, Aerojet-Gen Corp, 55-63; PRES, SHOCK HYDRODYNAMICS, DIV WHITTAKER CORP, 63- Concurrent Pos: Consult, US Dept Army. Honors & Awards: Meritorious Civilian Serv Award, US Dept Army, 44. Mem: Am Inst Aeronaut & Astronaut; Am Inst Mining, Metall & Petrol Eng; Soc Mfg Eng; Am Phys Soc; Acoust Soc Am. Res: Detonation and aerosol physics; explosives; high strain-rate behavior of materials; shock waves in solids; effects of super pressure on solids; ordnance systems and explosive metal forming. Mailing Add: 1103 E Mountain View Ave Glendora CA 91740

ZEROKA, DANIEL, b Plymouth, Pa, June 22, 41; m 67; c 2. THEORETICAL CHEMISTRY. Educ: Wilkes Col, BS, 63; Univ Pa, PhD(theoret chem), 66. Prof Exp: NSF fel statist mech, Yale Univ, 66-67; asst prof, 67-74, ASSOC PROF CHEM, LEHIGH UNIV, 74- Mem: Am Phys Soc; Am Chem Soc. Res: Quantum chemistry; statistical mechanics; magnetic resonance. Mailing Add: Dept of Chem Lehigh Univ Bethlehem PA 18015

ZERONIAN, SARKIS HAIG, b Manchester, Eng, June 30, 32; m 70. TEXTILE CHEMISTRY. Educ: Univ Manchester, BScTech, 53, MScTech, 55, PhD(cellulose chem), 62. Prof Exp: Res officer cellulose chem, Brit Cotton Indust Res Asn, Manchester, Eng, 58-60; res fel, Inst Paper Chem, 62-63; sr res fel nonwoven fabrics, Manchester Col Sci & Technol, Eng, 63-66; res assoc cellulose chem, Columbia Cellulose Co, BC, Can, 66-68; asst prof, 68-72, ASSOC PROF TEXTILE CHEM, UNIV CALIF, DAVIS, 72- Mem: Am Chem Soc; Am Asn Textile Chem & Colorists; Brit Textile Inst; Fiber Soc. Res: Chemical and physical properties of natural and man-made fibers; properties of textile finishes. Mailing Add: Div of Textiles & Clothing Univ of Calif Davis CA 95616

ZERWEKH, CHARLES EZRA, JR, b Galveston, Tex, Aug 24, 22; m 50; c 3. ORGANIC CHEMISTRY. Educ: Univ Houston, BS, 44. Prof Exp: Chemist, Humble Oil & Refining Co, 44-45, from asst res physicist to res physicist, 45-49, from patent coordr to sr patent coordr, 49-55, supvry patent coordr, 55-57, head tech info sect, Res & Develop Div, 57-63; mgr, Records Mgt Div, Stand Oil Co, NJ, 63-66; tech info specialist, Esso Res & Eng Co, NJ, 66-67; MGR TECH INFO CTR, POLAROID CORP, CAMBRIDGE, 67- Mem: AAAS; Am Chem Soc. Res: Technical information, especially patents, machine indexing, literature searching, technical files and information retrieval as applied to petrochemicals, petroleum processing and photography. Mailing Add: 14 Herrick St Winchester MA 01890

ZERZAVY, FREDERICK M, b Brno, Moravia, Dec 15, 18; US citizen; m 53; c 4. OBSTETRICS & GYNECOLOGY, PUBLIC HEALTH ADMINISTRATION. Educ: Masaryk Univ, Brno, 36-39; Univ Zagreb, MD, 42; Johns Hopkins Univ, MPH, 60, DrPH, 66; Am Bd Obstet & Gynec, dipl, 58. Prof Exp: Mem staff med surg & pediat, State & Univ Hosps, Yugoslavia, 42-43, health officer pub health & hosps, 43-44 & contract surgeon, 45-46; asst obstet & gynec, Med Sch, Univ Zagreb, 46-51; intern, Long Island Col Hosp, 51-52; resident, Hosp for Women of Md, Baltimore, 52-55; instr, Univ Md, 55-57; lectr, Johns Hopkins Univ, 63-66; consult, Fla State Bd Health, 67-68; asst prof obstet & gynec, 68-69, actg chmn dept community med, 68-70, ASSOC PROF COMMUNITY MED & SECY DEPT, CHICAGO MED SCH, 68- Concurrent Pos: Consult, Baltimore City Hosps, 56-67. Mem: Asn Teachers Prev Med; AMA; fel Am Col Obstet & Gynec; fel Am Pub Health Asn; fel Int Col Surgeons. Res: Population dynamics. Mailing Add: Dept of Community Med Chicago Med Sch Chicago IL 60612

ZESCHKE, RICHARD HERMAN, b Chicago, Ill, July 7, 41; m 64; c 2. FAMILY MEDICINE, IMMUNOLOGY. Educ: Univ Ill, Urbana, BS, 64; Univ Ill Med Ctr, PhD(anat), 70; State Univ NY Buffalo, MD, 75. Prof Exp: ASST PROF IMMUNOL, MED SCH, STATE UNIV NY BUFFALO, 70- Res: Cellular, parasitic and diagnostic immunology; immunity to infection. Mailing Add: Ctr for Immunol State Univ of NY Buffalo NY 14214

ZETLER, BERNARD DAVID, b New York, NY, Aug 27, 15; m 40; c 2. OCEANOGRAPHY. Educ: Brooklyn Col, BA, 36. Prof Exp: Jr math scientist, Hydrography Off, US Dept Navy, Washington, DC, 38; computer, US Coast & Geod Surv, 38-39, mathematician, 39-53, oceanogr, 53-56, chief, Currents & Oceanog Br, 56-60 & Oceanog Anal Br, 60-63, res group officer oceanog, 63-65, actg dir phys oceanog lab, Inst Oceanog, Environ Sci Serv Admin, Md, 65-68, dir phys oceanog lab, Atlantic Oceanog Labs, Fla, 68-72; RES OCEANOG, INST GEOPHYS & PLANETARY PHYSICS, UNIV CALIF, SAN DIEGO, 72- Concurrent Pos: Assoc prof lectr, George Washington Univ, 65-68; mem tsunami comt, Int Union Geod & Geophys. Mem: Am Geophys Union; Int Asn Geod. Res: Seismic sea waves; tides; currents; earth tides. Mailing Add: Inst of Geophys & Planetary Phys Univ of Calif San Diego A-025 La Jolla CA 92093

ZETLMEISL, MICHAEL JOSEPH, b Baltimore, Md, Feb 26, 42; m 71; c 2. ELECTROCHEMISTRY. Educ: Spring Hill Col, BS, 66; Marquette Univ, MS, 67; St Louis Univ, PhD(chem), 71. Prof Exp: RES CHEMIST & PROJ LEADER CORROSION & ELECTROCHEM, PETROLITE CORP, 71- Mem: Am Chem Soc; Sigma Xi. Res: Electrochemistry of high temperature melts as related to corrosion of metals in gas turbines and boilers. Mailing Add: Petrolite Corp 369 Marshall Ave St Louis MO 63119

ZETTEL, LARRY JOSEPH, b Detroit, Mich, Sept 12, 44. MATHEMATICS. Educ: Univ Detroit, BS, 65; Mich State Univ, MS, 66, PhD(math), 70. Prof Exp: ASST PROF MATH & CHMN DEPT, MUSKINGUM COL, 69- Mem: Am Math Soc; Math Asn Am. Res: Non-associative algebras; computer usage in undergraduate instruction. Mailing Add: Dept of Math Muskingum Col New Concord OH 43762

ZETTL, ANTON J, b Gakovo, Yugoslavia, Apr 25, 35; US citizen; m 64; c 2. MATHEMATICS. Educ: Ill Inst Technol, BS, 59; Univ Tenn, MA, 62, PhD(math), 64. Prof Exp: Assoc prof math, La State Univ, Baton Rouge, 68-69; assoc prof, 69-73, PROF MATH, NORTHERN ILL UNIV, 73- Concurrent Pos: NASA res grants, 65-67; res mem, Math Res Ctr, Univ Wis, 67-68; vis res fel, Univ Dundee, Scotland, 74-75; Brit Sci Res Coun res grant, 74-75; NSF res grant, 74-75. Mem: Am Math Soc; Math Asn Am. Res: Boundary value problems associated with ordinary differential equations; linear differential operators. Mailing Add: Dept of Math Northern Ill Univ De Kalb IL 60115

ZETTLEMOYER, ALBERT CHARLES, b Allentown, Pa, July 13, 15; m 40; c 5. COLLOID CHEMISTRY, SURFACE CHEMISTRY. Educ: Lehigh Univ, BS, 36, MS, 38; Mass Inst Technol, PhD(phys chem), 41. Hon Degrees: DSc, Clarkson Univ, 65; LLD, China Acad, Taiwan, 74. Prof Exp: Instr, Mass Inst Technol, 40-41; res chemist, Armstrong Cork Co, 41; from instr to prof, 41-60, res dir, Nat Printing Ink Res Inst, 46-68, vpres res & dir ctr surface & coatings res, 66-69, DISTINGUISHED PROF CHEM, LEHIGH UNIV, 60-, PROVOST & VPRES, 69- Concurrent Pos: Chmn, Gordon Conf Chem Interfaces, 55, mem bd trustees, Gordon Conf, 60-63; co-ed, J Colloid & Interface Sci; ed, Advances in Colloid & Interface Sci. Honors & Awards: Mattiello Award, 57; Ault Award, 60; Bond Award, 61; Hillman Award, 66; Kendall Award, 68; Welch Lectr, 72. Mem: AAAS; Am Chem Soc; Am Oil Chem Soc; fel Am Inst Chem; fel NY Acad Sci. Res: Surface chemistry and adsorption; gasses and liquids on solids; heterogeneous nucleation; flow properties of suspensions and molten resins; printability and printing quality of printing inks; heterogenous catalysis; plastic flooring. Mailing Add: Sinclair Lab Lehigh Univ Bethlehem PA 18015

ZETTLER, FRANCIS WILLIAM, b Easton, Pa, Aug 13, 38; m 61; c 2. PLANT PATHOLOGY, ENTOMOLOGY. Educ: Pa State Univ, BS, 61; Cornell Univ, MS, 64, PhD(plant path), 66. Prof Exp: Asst prof, 66-73, ASSOC PROF PLANT PATH, UNIV FLA, 73- Concurrent Pos: USDA contract, 74- Mem: Am Phytopath Soc; Entom Soc Am. Res: Transmission of plant viruses by aphids and other arthropods; general pathology and plant virology; virus diseases of ornamental plants; research involving edible root crops in the family Araceae. Mailing Add: Dept of Plant Path Univ of Fla Gainesville FL 32601

ZETTNER, ALFRED, b Laibach, Yugoslavia, Nov 21, 28; US citizen; m 59; c 2. CLINICAL PATHOLOGY. Educ: Graz Univ, MD, 54. Prof Exp: Asst prof path & clin path, Yale Univ, 63-67, assoc prof clin path, 67-68, dir dept clin micros, 63-68; PROF PATH & HEAD DIV CLIN PATH, UNIV CALIF, SAN DIEGO, 68- Concurrent Pos: NIH trainee clin path, Yale Univ, 61-63. Mem: AAAS; Am Soc Clin Path; Am Fedn Clin Res; Acad Clin Lab Physicians & Sci; Am Asn Clin Chem. Res: Competitive binding assays; folates in human serum. Mailing Add: Univ Hosp Dept of Path 225 W Dickinson St San Diego CA 92103

ZETTS, JOHN STEPHEN, b Youngstown, Ohio, June 14, 42. PHYSICS. Educ: Youngstown State Univ, BS, 64; Mich State Univ, MS, 66, PhD(physics), 71. Prof Exp: Asst prof physics, Youngstown State Univ, 70-73; ASST PROF PHYSICS & CHMN DEPT, UNIV PITTSBURGH, JOHNSTOWN, 73- Mem: Am Asn Physics Teachers. Res: Solid state physics; point defects in metals. Mailing Add: Dept of Physics Univ of Pittsburgh Johnstown PA 15904

ZEVOS, NICHOLAS, b Manchester, NH, June 24, 32; m 66; c 2. PHYSICAL CHEMISTRY. Educ: St Anselm's Col, BA, 54; Univ NH, PhD(chem), 63. Prof Exp: Instr chem, Univ NH, 63-64; res fel radiation chem, Sloan Kettering Inst, 64-66; fel, Cornell Univ, 66-68; asst prof, 68-75, ASSOC PROF CHEM, STATE UNIV NY COL POTSDAM, 75- Mem: AAAS; Am Chem Soc. Res: Gas phase kinetics; radiation and photochemistry. Mailing Add: Dept of Chem State Univ of NY Col Potsdam NY 13676

ZEY, ROBERT L, b California, Mo, Sept 19, 32; m 60; c 2. ORGANIC CHEMISTRY. Educ: Cent Methodist Col, AB, 54; Univ Nebr, MS, 59, PhD(chem), 61. Prof Exp: Res chemist, Mallinckrodt Chem Works, 61-65; from asst prof to assoc prof, 65-72, PROF CHEM, CENT MO STATE UNIV, 72- Mem: Am Chem Soc. Res: Cinnolines; aziridinones; x-ray contrast media. Mailing Add: Rte 1 Warrensburg MO 64093

ZEYA, HASAN ISMAIL, b Motihari, India, Jan 7, 31; m 57; c 5. IMMUNOLOGY, BACTERIOLOGY. Educ: Univ Bihar, ISc, 49, MD, 54; Univ NC, PhD(bact), 64. Prof Exp: Asst prof bact, Sch Med, Univ NC, Chapel Hill, 68-73; ASST PROF MED, BOWMAN GRAY SCH MED, 73- Concurrent Pos: USPHS fel, 64-66; Leukemia Soc spec grant, 66-68, scholar, 68-73. Mem: AAAS; Reticuloendoendothelial Soc; Am Soc Microbiol; NY Acad Sci; Am Fedn Clin Res. Res: Role of polymorphonuclear leukocytes in innate immunity; discovery of biologically active base proteins from granules of polymorphonuclear cells. Mailing Add: Dept of Med Bowman Gray Sch of Med Winston-Salem NC 27103

ZEYEN, RICHARD JOHN, b Mankato, Minn, Jan 17, 43. PLANT PATHOLOGY, VIROLOGY. Educ: Mankato State Col, BS, 65, MS, 67; Univ Minn, St Paul, PhD(plant path), 70. Prof Exp: Asst gen biol, Mankato State Col, 65-67; res assoc, 70-73, ASST PROF ELECTRON MICROS, MINN AGR EXP STA, UNIV MINN, ST PAUL, 73- Mem: Am Inst Biol Sci; Am Phytopath Soc; AAAS; Am Inst Biol Sci. Res: Electron microscopy; phytoarboviruses; virus transmission; histopathology of plant tissues. Mailing Add: Electron Micros Lab Dept of Plant Path Univ of Minn St Paul MN 55101

ZEZULKA, ALLISON YATES, b Inglewood, Calif, Oct 30, 47. NUTRITION. Educ: Univ Calif, Los Angeles, BS, 68, MS, 70; Univ Calif, Berkeley, PhD(nutrit), 74. Prof Exp: Intern dietetics, Vet Admin Hosp, Wadsworth, Los Angeles, 69-70; LECTR NUTRIT, UNIV CALIF, BERKELEY, 74-, PROG DIR DIETETICS, 75- Mem: Am Dietetic Asn; Am Pub Health Asn; Soc Nutrit Educ; Sigma Xi. Res: Human requirements for sulfur and sulfur containing amino acids; soy protein quality; nutritional requirements of the elderly; methods of amino acid analysis. Mailing Add: Dept of Nutrit Sci Univ of Calif 119 Morgan Hall Berkeley CA 94720

ZFASS, ALVIN MARTIN, b Norfolk, Va, Mar 30, 31; m 63; c 1. GASTROENTEROLOGY. Educ: Univ Va, BA, 53; Med Col Va, MD, 57. Prof Exp: Intern med, Bellevue-Cornell Med Ctr, 57-58; resident, Manhattan Vet Hosp, 58-60; mem, Sloan-Kettering Cancer Ctr, 60-61; from instr to asst prof med, 63-69, ASSOC PROF MED, MED COL VA, 69- Concurrent Pos: Fel gastroenterol, Manhattan Vet Hosp, 61-63; consult, McGuire Vet Hosp, 63-73, St Mary's Hosp, Richmond Mem Hosp & St Lukes Hosp. Mem: Am Gastroenterol Asn; Am Soc Gastroenterol Endoscopy. Res: Smooth muscle physiology of the esophagus. Mailing Add: Med Col of Va Richmond VA 23219

ZGANJAR, EDWARD F, b Virginia, Minn, July 31, 38; m 60; c 2. NUCLEAR PHYSICS. Educ: St John's Univ, Minn, BA, 60; Vanderbilt Univ, MA, 63, PhD(physics), 66. Prof Exp: AEC fel, Nat Reactor Testing Sta, 65-66; asst prof, 66-70, ASSOC PROF PHYSICS, LA STATE UNIV, BATON ROUGE, 70- Mem: Am Phys Soc. Res: Low energy nuclear spectroscopy; beta and gamma ray spectroscopy. Mailing Add: Dept of Physics & Astron La State Univ Baton Rouge LA 70803

ZHIVADINOVICH, MILKA RADOICICH, b San Francisco, Calif, Oct 7, 09; m 48. INORGANIC CHEMISTRY. Educ: Univ Belgrade, MChem Eng, 34; Sorbonne, DSc(chem), 39. Prof Exp: Instr inorg chem, Sch Eng, Univ Belgrade, 38-47, gen chem, 47-49, asst prof inorg chem, 49-57; asst prof, San Jose State Col, 58; asst prof, San Francisco State Col, 59-61, assoc prof phys sci, 61-64; asst prof gen & inorg chem, 64-70, PROF CHEM, CALIF STATE UNIV, HAYWARD, 70- Concurrent Pos: Vis prof, Carthage Col, 54-56 & Earlham Col, 56-57. Mem: Am Chem Soc. Mailing Add: Dept of Chem 25800 Hillary St Calif State Univ Hayward CA 94542

ZIANCE, RONALD JOSEPH, b Gallitzin, Pa, May 6, 42; m 66. PHARMACOLOGY. Educ: Univ Pittsburgh, BS, 65, PhD(pharmacol), 70. Prof Exp: Res assoc pharmacol, Sch Med, Univ Colo, 70-71; ASST PROF PHARMACOL, SCH PHARM, UNIV GA, 71- Res: Neuropharmacology; effect of centrally acting drugs on neurotransmitter mechanisms. Mailing Add: Dept of Pharmacol Univ of Ga Sch of Pharm Athens GA 30601

ZIAUDDIN, SYED, b Mysore, India, Oct 25, 19; m 46; c 2. ATMOSPHERIC PHYSICS, SPACE PHYSICS. Educ: Univ Mysore, BSc, 41; Aligarh Muslim Univ, India, MSc, 45; Ill Inst Technol, MSEE, 49; Univ Sask, PhD(space physics), 60. Prof Exp: Lectr physics, Univ Mysore, 41-50, asst prof, 51-62; prof & chmn dept, Bangalore Univ, 63-66; ASSOC PROF PHYSICS, LAURENTIAN UNIV, 66- Concurrent Pos: Nat Res Coun Can fel, 61-62. Mem: Can Asn Physicists; Am Geophys Union; Am Asn Physics Teachers; Am Geog Soc; NY Acad Sci. Res: Atmospheric physics, especially the lower ionosphere; radio techniques; undergraduate teaching of physics. Mailing Add: Dept of Physics Laurentian Univ Sudbury ON Can

ZIBOH, VINCENT AZUBIKE, b Warri, Nigeria, Apr 21, 32; m 62; c 3. BIOCHEMISTRY. Educ: Doane Col, AB, 58; St Louis Univ, PhD(biochem), 62. Prof Exp: Res fel neurochem, Ill State Psychiat Inst, Chicago, 62-64; lectr chem path, Med Sch, Univ Ibadan, 64-67; res assoc dermat, 67-69, ASST PROF DERMAT & BIOCHEM, SCH MED, UNIV MIAMI, 69- Concurrent Pos: WHO fel clin chem, Bispebjerg Hosp, Copenhagen, Denmark, 66. Mem: AAAS; Brit Biochem Soc; Am Chem Soc; Soc Invest Dermat. Res: Biochemistry of lipids and steroids; regulation of lipogenesis from glucose in skin; biosynthesis and biochemical basis of prostaglandin action in the skin. Mailing Add: Dept of Dermat Univ of Miami Sch of Med Miami FL 33136

ZICCARDI, ROBERT JOHN, b New York, NY, Dec 25, 46; m 69. BIOCHEMISTRY, IMMUNOLOGY. Educ: Univ Conn, BA, 68; Univ Calif, Los Angeles, PhD(biochem), 73. Prof Exp: RES ASSOC MOLECULAR IMMUNOL, SCRIPPS CLIN & RES FOUND, 73- Concurrent Pos: Weinberger fel, Nat Inst Allergy & Infectious Dis, 74- Res: Complement protein structure, function and interaction with biological membranes. Mailing Add: Dept of Molecular Immunol Scripps Clin & Res Found La Jolla CA 92307

ZICHIS, JOSEPH, b Lowell, Mass, Apr 13, 06; m 32. BIOCHEMISTRY, VIROLOGY. Educ: Mich State Univ, BS, 32, MS, 33, PhD(biochem), 36. Prof Exp: Res bacteriologist, Frederick Stearns & Co, 34-36; dir res biol labs, Pitman Moore Co, 36-39; res virologist, Ill State Health Dept, 39-42, asst chief bacteriologist, 42-48; res biochemist, Northwestern Univ, 41-43; owner, Markham Labs, Ill, 48-71; PRES, MARKHAM RES INST, 71- Mem: Am Pub Health Asn; Am Asn Immunol; Am Soc Microbiol. Res: Sanitation; antiseptics; acid base balance; virology and biochemistry related to prevention, epidemiology, therapy and diagnosis of viral infections; production of viral, rickettsial and bacterial diagnostic products; isolation of cellular and blood antigens. Mailing Add: PO Box 48 Rancho Santa Fe CA 92067

ZICK, WARREN HOWARD, agronomy, plant physiology, see 12th edition

ZICKER, ELDON LOUIS, b Milwaukee, Wis, Mar 12, 20; m 54; c 5. SOIL SCIENCE. Educ: Univ Wis, BS, 48, PhD(soils), 55. Prof Exp: Instr soil surv, Kans State Univ, 49; instr geol & soils, Yakima Valley Col, 59-64; INSTR SOILS, CALIF STATE UNIV, CHICO, 64-, DEAN SCH AGR & HOME ECON, 72- Mem: Am Soc Agron; Soil Sci Soc Am; Brit Soc Soil Sci. Res: Soil genesis and development; world food and environment problems. Mailing Add: Sch of Agr & Home econ Calif State Univ Chico First & Normal St Chico CA 95926

ZIDEK, JAMES VICTOR, b Acme, Alta, Sept 26, 39; m 61. MATHEMATICAL STATISTICS. Educ: Univ Alta, BSc, 61, MSc, 63; Stanford Univ, PhD(math statist), 67. Prof Exp: Lectr probability & statist, Univ Alta, 62-63; asst prof, 67-70, ASSOC PROF PROBABILITY & STATIST, UNIV BC, 70- Mem: Inst Math Statist; Statist Sci Asn Can; Can Statist Soc; Can Math Cong; Am Statist Asn. Res: Probability theory; statistical decision theory. Mailing Add: Dept of Math Univ of BC Vancouver BC Can

ZIEBARTH, TIMOTHY DEAN, b Glendive, Mont, June 10, 46; m 66; c 2. ORGANIC CHEMISTRY, PHYSICAL ORGANIC CHEMISTRY. Educ: Mont State Univ, BS, 69; Ore State Univ, PhD(org chem), 73. Prof Exp: Fel org chem, Univ Colo, 72-74; CHIEF CHEMIST, HAUSER LABS, 74- Mem: Am Chem Soc. Res: Photochemistry of allylic amines; novel methods of polymer analysis; forensic chemistry. Mailing Add: Hauser Labs 5680 Central Ave PO Box G Boulder CO 80302

ZIEBER, GEORGE HENRY, b Lang, Sask, Apr 11, 30; m 57; c 4. URBAN GEOGRAPHY. Educ: Univ Alta, BEd, 57, BA, 58; Univ Toronto, MA, 61; Univ Alta, PhD(geog), 71. Prof Exp: Teacher geog, Milton Dist High Sch, Ont, 61-62; geogr-analyst, Ont Dept Hwys, 62-64; lectr geog, Lethbridge Jr Col, 64-67; actg chmn dept, 67-68 & 72-73, asst prof, 67-74, ASSOC PROF GEOG, UNIV LETHBRIDGE, 74- Mem: Can Asn Geog; Asn Am Geog. Res: Office location on an inter and intra city basis and the decision-making process in location; neighborhood grocery and convenience chain stores. Mailing Add: Dept of Geog Univ of Lethbridge Lethbridge AB Can

ZIEBUR, ALLEN DOUGLAS, b Shawano, Wis, May 1, 23; m 49; c 2. MATHEMATICS. Educ: Univ Wis, PhD, 50. Prof Exp: From instr to assoc prof math, Ohio State Univ, 51-61; assoc prof, 61-63, PROF MATH, STATE UNIV NY BINGHAMTON, 63- Mem: Am Math Soc; Math Asn Am; Soc Indust & Appl Math. Res: Differential equations. Mailing Add: Dept of Math State Univ of NY Binghamton NY 13901

ZIEF, MORRIS, b Boston, Mass, July 8, 14; m 48; c 2. INDUSTRIAL CHEMISTRY. Educ: Harvard Univ, BA, 37; Boston Univ, MA, 38, PhD(org chem), 41. Prof Exp: Asst, Boston Univ, 37-41; asst to Prof Michael, Harvard Univ, 41-42; Hoffmann-La Roche fel, Mass Inst Technol, 42-43, res assoc, 43-46; res assoc sugar res found, Eastern Regional Res Lab, Bur Agr & Indust Chem, USDA, 46-47, proj dir, 47-50; SR CHEMIST, J T BAKER CHEM CO, 50- Mem: Am Chem Soc. Res: Preparation, containment and analysis of ultrapure materials; preparation of high-purity standards and reagents; fractional solidification; clean room technology. Mailing Add: J T Baker Chem Co Phillipsburg NJ 08865

ZIEGEL, KENNETH DAVID, b Cleveland, Ohio, Mar 15, 35; m 60; c 2. PHYSICAL CHEMISTRY. Educ: Case Inst, BS, 56; Polytech Inst Brooklyn, PhD(polymer chem), 66. Prof Exp: Res engr, Esso Res & Eng Co, 56-60; res scientist, Radiation Appln Inc, 60-62; RES SCIENTIST, E I DU PONT DE NEMOURS & CO, INC, 62- Concurrent Pos: Exchange scientist, Slovak Acad Sci, Czech, 72. Mem: AAAS; Am Chem Soc. Res: Diffusion of gases through polymeric membranes; radiation induced graft copolymerization to normally inert surfaces; adhesion; rheology of composites. Mailing Add: Elastomer Chem Dept Exp Sta E I du Pont de Nemours & Co Inc Wilmington DE 19898

ZIEGER, HERMAN ERNST, b Philadelphia, Pa, May 17, 35; m 60; c 3. ORGANIC CHEMISTRY. Educ: Muhlenberg Col, BS, 56; Pa State Univ, PhD(org chem), 61. Prof Exp: Fulbright scholar, Univ Heidelberg, 60-61; from instr to assoc prof, 61-73, PROF CHEM, BROOKLYN COL, 73- Concurrent Pos: Vis assoc, Calif Inst Technol, 67-68; Alexander von Humboldt Found fel, 74-75; mem doctoral fac, City Univ New York. Mem: AAAS; Am Chem Soc. Res: Organolithium chemistry; aryne chemistry; radical anion processes. Mailing Add: Dept of Chem Brooklyn Col Brooklyn NY 11210

ZIEGLER, ALAN CONRAD, b Galveston, Tex, Dec 10, 29; m 55; c 2. MAMMALOGY. Educ: Univ Calif, Berkeley, AB, 57, PhD(zool), 67. Prof Exp: VERT ZOOLOGIST, BISHOP MUS, 68- Concurrent Pos: Mem adv comt land vert, Hawaii Dept Agr, 68-; mammalogist mem, Hawaii Animal Species Adv Comn, 70-; affil grad fac zool, Univ Hawaii, 71-, lectr, 73. Mem: Am Soc Mammal; Australian Soc Mammal. Res: Mammalian evolution; systematics and ecology of the New Guinea vertebrates; analysis of archaeological faunal remains. Mailing Add: Bishop Mus PO Box 6037 Honolulu HI 96818

ZIEGLER, ALFRED M, b Boston, Mass, Apr 23, 38; m 64; c 2. PALEONTOLOGY, STRATIGRAPHY. Educ: Bates Col, BSc, 59; Oxford Univ, DPhil(paleont), 64. Prof Exp: Fel paleont, Calif Inst Technol, 64-66; asst prof, 66-72, ASSOC PROF PALEONT, UNIV CHICAGO, 72- Concurrent Pos: NSF grant, 72-74. Mem: Geol Soc Am; Brit Palaeontol Asn. Res: Paleontology, paleoecology and stratigraphy of the Silurian age deposits of the British Isles, Norway and eastern North America. Mailing Add: Dept of Geophys Sci Univ of Chicago Chicago IL 60637

ZIEGLER, DANIEL, b Quinter, Kans, July 6, 27; m 52; c 4. BIOCHEMISTRY. Educ: St Benedict's Col, Kans, BS, 49; Loyola Univ, Ill, PhD(biochem), 55. Prof Exp: Fel enzyme chem, Inst Enzyme Res, Univ Wis, 55-59, asst prof, 59-61; assoc prof, 61-68, PROF CHEM, UNIV TEX, AUSTIN, 68-, MEM STAFF, CLAYTON FOUND BIOCHEM INST, 61- Concurrent Pos: Estab investr, Am Heart Asn, 60-65. Mem: AAAS; Am Chem Soc; Am Soc Biol Chem. Res: Synthesis of protein hormones; mammalian mixed-function drug oxidases; flavoproteins of the electron transport system. Mailing Add: Clayton Found Biochem Inst Dept of Chem Univ of Tex Austin TX 78712

ZIEGLER, DEWEY KIPER, b Omaha, Nebr, May 31, 20; m 54; c 3. NEUROLOGY. Educ: Harvard Univ, BA, 41, MD, 45. Prof Exp: Assoc neurol, Col Physicians & Surgeons, Columbia Univ, 53-55; asst prof, Med Sch, Univ Minn, 55-56; assoc prof, 58-63, PROF NEUROL, UNIV KANS MED CTR, KANSAS CITY, 63-, CHMN DEPT, 73- Concurrent Pos: Consult, US Fed Hosp, Springfield, Mo & Vet Admin Hosp, Kansas City, Mo, 64-; clin prof neurol, Med Sch, Univ Mo-Kansas City; dir, Am Bd Psychiat & Neurol, 74- Mem: AMA; Soc Neurosci; Am Neurol Asn; fel Am Col Physicians; Am Epilepsy Soc; fel Am Acad Neurol. Res: Epidemiology and natural history of cerebrovascular disease; neurophysiological and biochemical basis of migraine. Mailing Add: Dept of Neurol Univ of Kans Med Ctr Kansas City KS 66103

ZIEGLER, FREDERICK DIXON, b Waynesboro, Pa, Oct 14, 33. BIOCHEMISTRY, PHYSIOLOGY. Educ: Shippensburg State Col, BS, 57; Pa State Univ, MS, 60, PhD(zool), 62. Prof Exp: Res trainee virol, Pa State Univ, 62-63; res assoc biochem, State Univ NY Buffalo, 63-65; asst prof, McMaster Univ, 65-69; res scientist, 69-73, SR RES SCIENTIST, DIV LABS & RES, NY STATE DEPT HEALTH, 73- Mem: Am Chem Soc; Can Biochem Soc. Res: Metabolism; coagulation. Mailing Add: NY State Dept of Health New Scotland Ave Albany NY 12201

ZIEGLER, FREDERICK EDWARD, b Teaneck, NJ, Mar 29, 38; m 62; c 2. ORGANIC CHEMISTRY. Educ: Fairleigh Dickinson Univ, BS, 60; Columbia Univ, MA, 61, PhD(chem), 64. Prof Exp: Eugene Higgins fel, Columbia Univ, 60-61, NIH fel, 61-64; NSF fel, Mass Inst Technol, 64-65; asst prof, 65-70, ASSOC PROF CHEM, YALE UNIV, 70- Mem: Am Chem Soc. Res: Organic synthetic methods and natural products synthesis. Mailing Add: Sterling Chem Lab Yale Univ New Haven CT 06520

ZIEGLER, GEORGE ELLIOTT, b Chicago, Ill, Apr 27, 07; m 35; c 1. EXPERIMENTAL PHYSICS. Educ: Univ Chicago, BS, 29, MS, 30, PhD(x-ray crystallog), 32. Prof Exp: From instr to asst prof physics, Cent YMCA Col, 32-36; asst prof, Armour Inst Technol, 36-39, x-ray physicist, Res Found, 35-39; chmn physics res, Ill Inst Technol, 40-44, sci adv, 44-45; exec scientist, Midwest Res Inst, 45-48, dir, 48-51; dir res, Zonolite Co, 51-58; CONSULT, 58-; PRES, THERMAL CONDUCTS, INC, WASH, 60- Mem: Am Phys Soc; Am Chem Soc; Optical Soc Am; Am Asn Physics Teachers; Am Ord Asn. Res: X-ray crystallography structure analysis; administration of industrial research; hydrophil balance studies on high molecular weight ketones; heat and cold underground pipe distribution systems. Mailing Add: 901 Osceola Ave Winter Park FL 32789

ZIEGLER, GEORGE WILLIAM, JR, b Cleveland, Ohio, Oct 12, 16; c 2. PHYSICAL CHEMISTRY. Educ: Monmouth Col, BS, 39; Mass Inst Technol, cert, 44; Ohio State Univ, PhD, 50. Prof Exp: Asst physics, Monmouth Col, 37-39; analyst, Grasselli Chem Div, E I du Pont de Nemours & Co, 39-40, tech supvr, 40-42, asst to chief plant technologist, 42-43; leader high temperature calorimetry group, Res Found, Ohio State Univ, 46-50; group leader process develop, Int Resistance Co, 50-51, sect head, 51-53, head dept, 53; res chemist & mem staff, Mat Control Div, Curtis Pub Co, 53-55; dir spec projs lab, 55-60, from proj engr to sr proj engr, 60-70, MGR SPEC PROD, AMP, INC, 70- Concurrent Pos: Assoc prof, Dickinson Col, 55-59. Mem: Am Chem Soc; Am Phys Soc; Am Soc Metals; Inst Elec & Electronics Eng; Inst Mgt Sci. Res: High temperature thermodynamics and kinetics; instrumentation; metallurgy; conducting films; heat transfer; color correction theory; photoengraving; radio frequency connectors; aluminum building wire connectors. Mailing Add: Circle Dr RD 8 Box 162 Carlisle PA 17013

ZIEGLER, HROLFE READ, b Can, Dec 28, 04; nat US; m 33; c 3. SURGERY. Educ: Univ Toronto, MD, 29, MS, 36; Am Bd Surg, dipl, 42; FRCS(C). Prof Exp: Asst prof anat & surg, 42-48, assoc prof surg, 48-67, clin assoc prof surg, 67-69, ASSOC PROF ANAT, SCH MED & DENT, UNIV ROCHESTER, 67- Mem: Fel Am Col Surgeons; Int Soc Surgeons. Mailing Add: Sch of Med & Dent Univ of Rochester Rochester NY 14627

ZIEGLER, JAMES FRANCIS, b Apr 20, 39; US citizen; m 69. NUCLEAR PHYSICS. Educ: Yale Univ, BS, 57, MS, 65, PhD(nuclear physics), 67. Prof Exp: Mem res staff, 67-69, DIR HIGH ENERGY ACCELERATOR LAB, RES CTR, IBM CORP, 69- Mem: Am Phys Soc. Res: Solid state analysis using nuclear physics techniques; ion-induced x-rays; nuclear backscattering; nuclear channeling; ion-implantation; optical microcircuits. Mailing Add: Res Ctr IBM Corp PO Box 218 Yorktown Heights NY 10598

ZIEGLER, JOHN BENJAMIN, b Rochester, NY, Jan 2, 17; m 46; c 3. SYNTHETIC ORGANIC CHEMISTRY. Educ: Univ Rochester, BS, 39; Univ Ill, MS, 40, PhD(org chem), 46. Prof Exp: Jr chemist, Merck & Co, Inc, NJ, 40-43; asst chem, Univ Ill, 43-46; res chemist, J T Baker Chem Co, NJ, 46-48; assoc chemist, Develop Div, Ciba Pharmaceut Co, 48-52, supvr develop res labs, 52-62, mgr process res, 62-70, dir chem develop, Pharmaceut Div, Ciba-Geigy Corp, 70-75, DIR PROCESS RES, PHARMACEUT DIV, CIBA-GEIGY CORP, 75- Mem: Am Chem Soc; Sigma Xi; Lepidop Soc (treas, 50-53); NY Acad Sci; fel Am Inst Chem. Res: Research and development of processes for synthesis of medicinal chemicals. Mailing Add: 64 Canoe Brook Pkwy Summit NJ 07901

ZIEGLER, JOHN HENRY, JR, b Altoona, Pa, Nov 10, 24; m 49; c 3. MEAT SCIENCE, ANIMAL SCIENCE. Educ: Pa State Univ, BS, 50, MS, 52, PhD(animal indust), 65. Prof Exp: Soil conservationist, USDA, 49; asst animal husb, Pa State Univ, 50-52; nutritionist, Near's Food Co, NY, 52-54; instr animal indust, 54-65, asst prof meat sci, 65-68, ASSOC PROF MEAT SCI, PA STATE UNIV, UNIVERSITY PARK, 68- Mem: Am Soc Meat Sci; Am Soc Animal Sci; Inst Food Technol. Res: Meat animal carcass evaluation and utilization; basic adipose tissue anatomy and physiology; meat product development. Mailing Add: 15 Meats Lab Pa State Univ University Park PA 16802

ZIEGLER, LOUIS WILLIAM, b Buffalo, NY, May 17, 07; m 31; c 4. HORTICULTURE. Educ: Univ Fla, BS, 30, MS, 50, PhD, 55. Prof Exp: Asst entomologist, Agr Exp Sta, Univ Fla, 30-31 & Com Sales & Res, 31-37; consult, 37-39; prod mgr, Holly Hill Fruit Prod, Inc, 39-47; from asst prof to assoc assoc prof hort, 47-58, PROF FRUIT CROPS, UNIV FLA, 58- Res: Production problems with citrus. Mailing Add: 625 NW22nd St Gainesville FL 32603

ZIEGLER, MANDELL STANLEY, organic chemistry, see 12th edition

ZIEGLER, MICHAEL ROBERT, b York, Pa, Oct 31, 42; m 64; c 2. MATHEMATICAL ANALYSIS. Educ: Shippensburg State Col, BS, 64; Univ Del, MS, 67, PhD(math), 70. Prof Exp: Instr math, Univ Del, Georgetown Exten, 69-70; fel, Univ Ky, 70-71; ASST PROF MATH, MARQUETTE UNIV, 71- Mem: Am Math Soc; Math Asn Am. Res: Complex analysis; geometric function theory; univalent functions. Mailing Add: Dept of Math Marquette Univ Milwaukee WI 53233

ZIEGLER, PAUL FOUT, b Baltimore, Md, Dec 3, 16; m 45; c 2. CHEMISTRY. Educ: Otterbein Col, BS, 39; Univ Cincinnati, MS, 47, PhD(chem), 63. Prof Exp: Analyst, Am Rolling Mills, Ohio, 39-40, chemist, 40-46; instr chem, Ala Polytech Inst, 49-52, asst prof, 52-58, ASSOC PROF CHEM, AUBURN UNIV, 58- Mem: Am Chem Soc. Res: Analytical and organic chemistry; synthesis and rearrangement. Mailing Add: Dept of Chem Auburn Univ Auburn AL 36830

ZIEGLER, PETER, b Vienna, Austria, Mar 26, 22; m 48. ORGANIC CHEMISTRY, BIOCHEMISTRY. Educ: Sir Geo Williams Col, BSc, 44; McGill Univ, PhD(biochem), 51. Prof Exp: Chemist, Frank W Horner, Ltd, 44-48; asst, McGill Univ, 48-51; group leader, 51-65, ASST DIR RES, CAN PACKERS, LTD, 66- Mem: Am Chem Soc; NY Acad Sci; fel Chem Inst Can. Res: Fine chemicals; pharmaceuticals. Mailing Add: Res & Develop Labs Can Packers Ltd Toronto ON Can

ZIEGLER, RICHARD JAMES, b Norristown, Pa, May 30, 43; m 67. VIROLOGY. Educ: Muhlenberg Col, BS, 65; Temple Univ, PhD(microbiol), 70. Prof Exp: Instr microbiol, Med Sch, Temple Univ, 69; res assoc genetics, Rockefeller Univ, 70-71; ASST PROF MICROBIOL, MED SCH, UNIV MINN, DULUTH, 71- Mem: AAAS; Tissue Cult Asn; Am Soc Microbiol; Asn Am Med Cols. Res: Effects of neurotropic viral multiplication on central nervous system function and effects of herpes simplex virus replication on primary nerve tissue culture cells. Mailing Add: Dept of Microbiol Univ of Minn Med Sch Duluth MN 55812

ZIEGLER, ROBERT G, b The Dalles, Ore, Apr 24, 24; m 53; c 3. INORGANIC CHEMISTRY. Educ: Ore State Univ, BA, 48, MS, 51; Univ Tenn, PhD, 69. Prof Exp: Control chemist, Barium Prod Ltd, 48-49; res chemist, Nitrogen Div, Allied Chem Corp, 52-57; from asst prof to assoc prof, 57-70, head dept, 69-74, PROF CHEM, LINCOLN MEM UNIV, 70- Mem: AAAS; Am Chem Soc; Am Sci Affil. Res: Analytical and coordination chemistry; fertilizer technology. Mailing Add: Dept of Chem Lincoln Mem Univ Harrogate TN 37752

ZIEGLER, THERESA FRANCES, b Budapest, Hungary; nat US; wid; c 1. CHEMISTRY. Educ: Eötvös Lorand Univ, Budapest, BS, 46, PhD(phys chem), 49. Prof Exp: Res engr, State Biochem Res Inst, Hungary, 49-50; chemist, Steel Plant Lab, 53-56; radiochemist, Nat Health & Labor Inst, 56-57; RADIOCHEMIST, STAMFORD RES LABS, AM CYANAMID CO, 57- Mem: Am Chem Soc; Am Nuclear Soc; NY Acad Sci. Res: Radiosyntheses and radio-tracerwork in the following fields—agricultural chemicals, herbicides, pesticides and insecticides; pharmaceuticals; technology of plastics, fibers and papers; surface coatings; biological membranes; industrial hygiene, toxicity studies; autoradiography, experience with sealed sources. Mailing Add: 1452 Riverbank Rd Stamford CT 06903

ZIEGLER, WILLIAM ARTHUR, b St Louis, Mo, Feb 10, 24; m 51; c 2. ANALYTICAL CHEMISTRY. Educ: Univ Ill, AB, 48, MS, 49, PhD(anal chem), 52. Prof Exp: Teaching asst, Univ Ill, 48-52; chemist, Mallinckrodt Chem Works, 52, supvr, 52-55, asst mgr anal dept, Uranium Div, 55-60, mgr anal dept, 60-66, tech asst to dir qual control, 66-67; tech specialist, Nuclear Div, Kerr-McGee Corp, 67-69, mgr qual assurance, 69-70; MGR ANAL SECT, EASTERN RES CTR, STAUFFER CHEM CO, 71- Mem: Am Chem Soc. Res: Sampling; uranium chemistry. Mailing Add: 61 Pocconock Trail New Canaan CT 06840

ZIEGRA, SUMNER ROOT, b Deep River, Conn, Feb 13, 23; m 45; c 2. PEDIATRICS. Educ: Univ Vt, BS, 45; Yale Univ, MD, 47; Am Bd Pediat, dipl, 57. Prof Exp: Fel pediat, Sch Med, NY Univ, 49-51; asst prof pediat, State Univ NY Downstate Med Ctr, 56-60; assoc prof, Jefferson Med Col, 60-63; from assoc prof to prof, Hahnemann Med Col, 63-70; prof, Col Med, Thomas Jefferson Univ, 70-72; PROF PEDIAT, MED COL PA, 73-, CHMN DEPT, 75- Concurrent Pos: Dir div B, Dept Pediat, Philadelphia Gen Hosp, 63-70, coordr, Hahnemann Div, 66-67; dir dept pediat, Lankenau Hosp, 70-72. Res: Infectious diseases. Mailing Add: 344 Hillview Rd Malvern PA 19355

ZIELEN, ALBIN JOHN, b Chicago, Ill, Dec 22, 25. PHYSICAL CHEMISTRY, INORGANIC CHEMISTRY. Educ: Miami Univ, BA, 50; Univ Calif, PhD(chem), 53. Prof Exp: Asst chem, Univ Calif, 51-52; chemist, Radiation Lab, Univ Calif, 52-53; ASSOC CHEMIST, ARGONNE NAT LAB, 53- Mem: Am Chem Soc. Res: Computer programming; neptunium solution chemistry; electrochemistry; reaction kinetics; complex ions. Mailing Add: Bldg D-200 Argonne Nat Lab 9700 SCass Ave Argonne IL 60439

ZIELEZNY, MARIA ANNA, b Kaczkowizna, Poland, Sept 20, 39; US citizen; m 69. BIOSTATISTICS. Educ: Univ Warsaw, MS, 62; Univ Calif, Los Angeles, PhD(biostatist), 71. Prof Exp: Res asst statist, Math Inst, Polish Acad Sci, Warsaw, 62-65; clin asst prof, 71-73, ASST PROF BIOSTATIST, DEPT SOCIAL & PREV MED, SCH MED, STATE UNIV NY BUFFALO, 73- Mem: Am Statist Asn; Am Pub Health Asn; Biomet Soc. Res: Statistical applications in particular to medicine, development of statistical methods in evaluation, discriminant analysis, measures of association, techniques for qualitative data. Mailing Add: Dept of Social & Prev Med State Univ of NY Sch of Med Buffalo NY 14214

ZIELEZNY, ZBIGNIEW HENRYK, b Knurow, Poland, Jan 11, 30; m 69. MATHEMATICAL ANALYSIS. Educ: Wroclaw Univ, Masters, 54, PhD(math), 59; Polish Acad Sci, docent, 64. Prof Exp: Adj math, Wroclaw Tech Univ, 58-61; adj, Polish Acad Sci, 61-64, docent, 64-69; vis prof, Univ Kiel, 69-70; assoc prof, 70-71, PROF MATH, STATE UNIV NY BUFFALO, 71- Mem: Am Math Soc. Res: Analysis, functional analysis; differential equations; existence and regularity of solutions of convolution equations in various spaces of distributions. Mailing Add: Dept of Math State Univ of NY 4246 Ridge Lea Rd Amherst NY 14226

ZIELINSKI, THERESA JULIA, b Brooklyn, NY. THEORETICAL CHEMISTRY. Educ: Fordham Univ, BS, 63, MS, 68, PhD(chem), 73. Prof Exp: ASST PROF CHEM, COL MT ST VINCENT, 72- Mem: Am Chem Soc. Res: Theoretical chemical study of drugs. Mailing Add: Col Mt St Vincent Riverdale NY 10471

ZIELINSKI, WALTER L, JR, b Staten Island, NY, May 31, 32; m 71; c 2. ANALYTICAL CHEMISTRY. Educ: Wagner Col, BS, 54; NC State Univ, MS, 59; Georgetown Univ, PhD, 72. Prof Exp: Res chemist, Div Chem & Foods, Va Dept Agr, 59-61; res chemist, Hazleton Labs, 61-62; res chemist, Div Chem & Foods, Va Dept Agr, 62-63; res chemist, Res Ctr, Union Bag-Camp Paper Corp, Ga, 63-64; staff scientist, Bionetics Res Labs, Inc, Va, 64-66, sr scientist, 67-68; sr chemist, Melpar, Inc, 68; staff res chemist, Anal Chem Div, Nat Bur Stand, 68-72; sr scientist, Litton-Bionetics Fed Contract Res Ctr, 72-76; MEM STAFF, FREDERICK CANCER RES CTR, NAT CANCER INST, 76- Mem: Am Chem Soc; Coblentz Soc. Res: Analytical methods in cancer research; method development in analytical gas chromatography; thermodynamics of solute-solvent interaction in gas chromatography; development of ion-exchange and liquid chromatographic systems; structural analysis of copolymers and ion-exchangers by infrared spectroscopy; development of particle metrology. Mailing Add: Frederick Cancer Res Ctr Nat Cancer Inst Frederick MD 21701

ZIELKE, HORST RONALD, b Pscinno, Poland, June 7, 42; US citizen; m 67; c 2. BIOCHEMISTRY, CELL BIOLOGY. Educ: Univ Ill, BS, 64; Mich State Univ, PhD(biochem), 68. Prof Exp: Res assoc, AEC Plant Res Lab, Mich State Univ, 68-71; res fel, Genetics Unit, Mass Gen Hosp, 71-73; ASST PROF PEDIAT RES, SCH MED, UNIV MD, 73- Concurrent Pos: Monsanto fel, 69. Mem: Am Soc Cell Biol. Res: Mutagenesis and enzymology of cultured mammalian cells. Mailing Add: Dept of Pediat Res Sch of Med Univ of Md Baltimore MD 21201

ZIELKE, RICHARD CARL, crop physiology, crop science, see 12th edition

ZIEMAN, CLAYTON MELVIN, physics, deceased

ZIEMAN, JOSEPH CROWE, JR, b Mobile, Ala, June 9, 43; m 67. MARINE ECOLOGY, BIOLOGICAL OCEANOGRAPHY. Educ: Tulane Univ, BS, 65; Univ Miami, MS, 68, PhD(marine sci), 70. Prof Exp: Res asst thermal pollution, Inst Marine Sci, Univ Miami, 68-70; fel syst ecol, Inst Ecol, Univ Ga, 70-71; ASST PROF ENVIRON SCI, UNIV VA, 71- Mem: Ecol Soc Am; Am Soc Limnol & Oceanog; Sigma Xi; AAAS; Int Asn Aquatic Vascular Plant Biologists. Res: Comparative studies of tropical and temperate interface zones, seagrasses, coral reefs, mangroves and salt marshes; production, colonization, succession and recovery from disturbances; simulation modeling of growth and succession. Mailing Add: Dept of Environ Sci Univ of Va Charlottesville VA 22903

ZIEMBA, FRANCIS PAUL, physics, see 12th edition

ZIEMER, PAUL L, b Toledo, Ohio, June 28, 35; m 58; c 4. HEALTH PHYSICS. Educ: Vanderbilt Univ, MS, 59; Purdue Univ, PhD(bionucleonics), 62; Am Bd Health Physics, dipl, 65. Prof Exp: Physicist, US Naval Res Lab, Wash, 57; health physicist, Oak Ridge Nat Lab, 59; from asst prof to assoc prof health physics, 62-69, PROF HEALTH PHYSICS, PURDUE UNIV, WEST LAFAYETTE, 69-, ASSOC HEAD BIONUCLEONICS, 71-, RADIOL CONTROL OFFICER, 59- Concurrent Pos: Consult, Harrison Steel Castings Co, Ind, 62-66, satellite div, Union Carbide Corp, 66-67, Calif Nuclear, Inc, 66-68, Breed Radium Inst, 68-70, Allison Div, Gen Motors Corp, 69, Mobil Field Res Labs, 70-74 & Midwest Radiation Protection, Inc, 71-; mem panel examr, Am Bd Health Physics, 69-71; mem sci comt, Nat Coun Radiation Protection, 73- Honors & Awards: Lederle Pharm Fac Award, 64; Elda E Anderson Award, Health Physics Soc, 71. Mem: Fel AAAS; Health Physics Soc (pres, 75); Am Nuclear Soc; Int Radiation Protection Asn. Res: Uptake, retention and excretion of inhaled radionuclides; radiation dosimetry. Mailing Add: Dept of Bionucleonics Purdue Univ West Lafayette IN 47907

ZIEMER, WILLIAM P, b Manitowoc, Wis, Mar 26, 34; m 57; c 3. MATHEMATICS. Educ: Univ Wis, BS, 56, MS, 57; Brown Univ, PhD(math), 61. Prof Exp: ASSOC PROF MATH, IND UNIV, BLOOMINGTON, 61- Mem: Am Math Soc; Math Asn Am. Res: Geometric analysis; area theory; surface theory; differential geometry. Mailing Add: Dept of Math Ind Univ Bloomington IN 47405

ZIENIUS, RAYMOND HENRY, b Montreal, Que, Nov 8, 34; m 70. ANALYTICAL CHEMISTRY. Educ: McGill Univ, BSc, 56, PhD, 59. Prof Exp: Asst chemist, Dom Tar & Chem Co, Ltd, 55; res chemist, Cent Res Lab, Can Industs Ltd, 59-67; asst prof, 67-72, ASSOC PROF CHEM, CONCORDIA UNIV, LOYOLA CAMPUS, 72- Mem: Fel Chem Inst Can; fel Am Chem Soc. Res: Spectroscopy; chromatography and other analytical methods. Mailing Add: Dept of Chem Concordia Univ Loyola Campus Montreal PQ Can

ZIENTY, FERDINAND B, b Chicago, Ill, Mar 21, 15; m 45; c 2. CHEMISTRY. Educ: Univ Ill, BS, 35; Univ Mich, MS, 36, PhD(pharmaceut chem), 38. Prof Exp: Res chemist, 38-40, res group leader, 40-47, from asst dir res to dir res, 47-60, adv org chem res, 60-64, MGR RES & DEVELOP, FOOD & FINE CHEM, MONSANTO CO, 64- Concurrent Pos: Vpres res, George Lueders & Co, Subsid Monsanto Co, 68-70. Mem: Fel AAAS; Am Chem Soc; Am Pharmaceut Asn; Am Inst Chem Eng; fel NY Acad Sci. Res: Antispasmodics; sulfa drugs; ethylenediamine derivatives; organic heterocycles, including imidazole and thiophene chemistry; organic acids and anhydrides; nucleophilic substitutions; nucleophilicity of thiols; catalytic oxidations; flavors. Mailing Add: 850 Rampart Dr Warson Woods MO 63122

ZIENTY, MITCHELL FRANK, b Chicago, Ill, Feb 12, 13. ORGANIC CHEMISTRY. Educ: Univ Ill, PhC, 33; Univ Mich, BS, 38, MS, 39, PhD(org chem), 41. Prof Exp: Res chemist, Upjohn Co, 41-47; res chemist, Sumner Chem Co, Inc, 47-49, mgr labs,

49-52, asst res dir, 52-70, ASSOC RES DIR, MARSCHALL DIV, MILES LABS, INC, 70- Mem: AAAS; Am Chem Soc; NY Acad Sci. Res: Organic syntheses; enzymes; polymers; catalysis. Mailing Add: Marschall Div Miles Labs Inc 820 McNaughton St Elkhart IN 46514

ZIER, ROBERT EUGENE, theoretical physics, see 12th edition

ZIERING, ALBERT, b New York, NY, Nov 21, 15; m 52; c 2. ORGANIC CHEMISTRY. Educ: NY Univ, BS, 37, MS, 39, PhD(chem), 42. Prof Exp: RES CHEMIST, HOFFMANN-LA ROCHE, INC, 42- Mem: Am Chem Soc. Res: Medicinal chemistry; analgesics; anti-inflammatory agents; hypotensives. Mailing Add: Hoffmann-La Roche Inc 21 Oakley Terr Nutley NJ 07110

ZIERING, SIGI, theoretical physics, see 12th edition

ZIERLER, KENNETH LEVIE, b Baltimore, Md, Sept 5, 17; m 41; c 5. MEDICINE, PHYSIOLOGY. Educ: Johns Hopkins Univ, AB, 36; Univ Md, MD, 41. Prof Exp: Fel med, Sch Med, Johns Hopkins Univ, 46-48, from instr to prof med, 48-72; dir, Inst Muscle Dis, Inc, 72-73; PROF PHYSIOL & MED, SCH MED, JOHNS HOPKINS UNIV, 73- Concurrent Pos: Asst physician, Outpatient Dept, Johns Hopkins Hosp, 46-53, physician, 53-55, physician-in-charge phys ther dept, 50-57, chemist, 57-58, hosp physician, 53-72 & 73-, assoc prof environ med, Sch Hyg & Pub Health, Johns Hopkins Univ, 54-59; prof physiol, Sch Med, Johns Hopkins Univ, 69-72; adj prof, Rockefeller Univ, 72-73; adj prof, Med Sch, Cornell Univ, 72-73. Mem: Am Physiol Soc; Am Soc Clin Invest; Asn Am Physicians. Res: Muscle metabolism and function; hormonal action; biomembranes; circulation; tracer kinetics; insulin, water and electrolytes. Mailing Add: Dept of Physiol Johns Hopkins Univ Sch of Med Baltimore MD 21205

ZIERLER, NEAL, b Baltimore, Md, Sept 17, 26. MATHEMATICS. Educ: Johns Hopkins Univ, AB, 45; Harvard Univ, AM, 49, PhD(math), 59. Prof Exp: Mathematician-physicist, Ballistic Res Labs, Aberdeen Proving Ground, Md, 51-52; mem staff, Instrumentation Lab, Mass Inst Technol, 52-54 & Lincoln Lab, 54-60; res group supvr, Jet Propulsion Lab, Calif Inst Technol, 60-61; sr scientist, Acron Corp, Mass, 61-62; mem staff, Lincoln Lab, Mass Inst Technol, 62; sub-dept head, Mitre Corp, Mass, 62-65; MEM STAFF COMMUN RES DIV, INST DEFENSE ANAL, 65- Mem: Am Math Soc; Math Asn Am; sr mem Inst Elec & Electronics Eng. Res: Algebra; mathematical foundations of quantum mechanics; coding and decoding of information; computer applications. Mailing Add: Inst for Defense Anal Thanet Rd Princeton NJ 08540

ZIETLOW, JAMES PHILIP, b Chicago, Ill, Dec 15, 21; m 52; c 3. PHYSICS. Educ: De Paul Univ, BS, 48; Ill Inst Technol, MS, 49, PhD(physics), 55. Prof Exp: Sr res physicist, Res & Develop Labs, Pure Oil Co, 51-56; prof physics & math, NMex Highlands Univ, 56-65, head dept physics & math, 56-63, grad dean, 63-64, head dept physics & math, 64-65; PROF PHYSICS, WESTERN MICH UNIV, 65-, ASSOC DEAN COL ARTS & SCI, 69- Mem: Am Phys Soc; Coblentz Soc. Res: Infrared, Raman, ultraviolet and mass spectroscopies. Mailing Add: Col of Arts & Sci Western Mich Univ Kalamazoo MI 49008

ZIETZ, JOSEPH RICHARD, JR, organic chemistry, see 12th edition

ZIEVE, LESLIE, b Minneapolis, Minn, Aug 6, 15; m 41; c 1. MEDICINE. Educ: Univ Minn, MA, 39, MD, 43, PhD(med), 52; Am Bd Internal Med, dipl, 51. Prof Exp: Resident med, Med Sch, Univ Minn, Minneapolis, 46-49, from instr to prof, 49-71; dir spec cancer lab, 61-72, STAFF PHYSICIAN, VET ADMIN HOSP, MINNEAPOLIS, 49-, ASSOC CHIEF STAFF FOR RES, 69-, DISTINGUISHED PHYSICIAN, 72- Concurrent Pos: Chief radioisotope unit, Vet Admin Hosp, Minneapolis, 50-72; mem exec comt, Grad Sch, Univ Minn, Minneapolis, 65-68; ed, J Lab & Clin Med, 67-70. Mem: Am Col Physicians. Res: Diseases of the liver and pancreas; lipid alterations in disease. Mailing Add: Bldg 13N Vet Admin Hosp 54th St & 48th Ave S Minneapolis MN 55417

ZIFF, MORRIS, b New York, NY, Nov 19, 13; m 40; c 2. INTERNAL MEDICINE. Educ: NY Univ, BS, 34, PhD(chem), 37, MD, 48. Prof Exp: Asst chem, NY Univ, 34-39; asst biochem, Col Physicians & Surgeons, Columbia Univ, 39-41, vis scholar, 41-44; instr & lectr, NY Univ, 44, adj asst prof, 48-50, from asst prof med to assoc prof, 54-58; PROF INTERNAL MED, SOUTHWEST MED CTR, UNIV TEX HEALTH SCI CTR DALLAS, 58- Concurrent Pos: Instr, City Col New York, 41-44; from intern to asst resident, Bellevue Hosp, 48-50, chmn clin res sect study group rheumatic dis, 52-58; consult, USPHS, 55-63. Honors & Awards: Heberden Medal, Heberden Soc London, 64. Mem: Am Chem Soc; Harvey Soc; Am Soc Clin Invest; Am Rheumatism Asn (pres, 65-66); Am Col Physicians. Res: Chemistry of connective tissue; rheumatic diseases; immunology. Mailing Add: Dept Int Med Univ Tex Southwest Med Sch 5323 Harry Hines Blvd Dallas TX 75235

ZIFFER, HERMAN, b New York, NY, Feb 22, 30; m 55; c 3. ORGANIC CHEMISTRY. Educ: City Col New York, BS, 51; Ind Univ, MA, 53; Univ Ore, PhD(chem), 55. Prof Exp: Res chemist, Nat Aniline Div, Allied Chem Corp, 55-59; RES CHEMIST, NIH, 59- Mem: Am Chem Soc; The Chem Soc. Res: Photochemistry; use of optical rotatory dispersion and other physical measurements for structure determination. Mailing Add: Bldg 2 Rm B1-06 NIH Bethesda MD 20014

ZIFFER, JACK, b New York, NY, Dec 2, 18; m 42; c 2. MICROBIOLOGY. Educ: Brooklyn Col, BS, 40; Univ London, PhD(biochem), 50. Prof Exp: Sr res microbiologist, Schenley Res Inst, 42-46; sr res fermentation chemist, A E Staley Co, 46-48; sr res microbiol chemist, Res Labs, Pabst Brewing Co, 50-55, head dept microbiol, Pabst Labs, 55-60, dir microbiol div, P-L Biochem, Inc, 60-67, vpres microbiol div, 67-76, vpres & tech dir, Premier Malt Prod, Inc, 69-76; PROF MICROBIOL, TEL-AVIV UNIV, 76- Mem: Am Chem Soc; Am Phytopath Soc; Soc Invert Path; Am Soc Testing & Mat. Res: Phytoactin; phytostreptin; penicillin; streptomycin; streptothricin; bacitracin; fungal amylase, glucoamylase, protease; bacterial amylase, gumase, protease; vitamin B12; riboflavin; gluconic acid; lactic acid; fungal anthraquinone pigments; cellulose decomposition; plant diseases; microbial insecticides; trans-N-deoxyribosylase; antimycin; septic enzymes; malt. Mailing Add: Dept of Microbiol Tel-Aviv Univ Tel Aviv Israel

ZIFFREN, SIDNEY EDWARD, b Rock Island, Ill, Aug 30, 12. SURGERY. Educ: Univ Ill, Chicago, BS, 34, MD, 36; Am Bd Surg, dipl. Prof Exp: From intern to resident, 36-42, asst, 47, from instr to assoc prof, 48-53, actg head dept, 69-71, PROF SURG, COL MED, UNIV IOWA, 53-, CHMN DIV TRAUMA & RECONSTRUCTIVE SURG, 65-, HEAD DEPT SURG, 72- Concurrent Pos: Chief surgeon, Shanghai Gen Hosp & consult surgeon, Shanghai Munic Hosps, 46-47; consult surgeon, Vet Admin Hosp, Iowa City, Iowa, 55- Mem: Fel Am Col Surg; Am Geriat Soc; Soc Exp Biol & Med; Am Asn Surg Trauma; Am Burn Asn. Res: Geriatrics; burn injuries; enzymes; pancreatitis. Mailing Add: Dept of Surg Univ Hosp Iowa City IA 52242

ZIGHELBOIM, JACOB, b Chernowitz, Rumania, Jan 2, 46; Venezuelan citizen; m 67; c 3. IMMUNOBIOLOGY. Educ: Col Moral & Luces, BS, 63; Univ Cent Venezuela, MD, 69. Prof Exp: From intern to resident internal med, Beilinson Hosp, Tel Aviv Sch Med, 70-72; fel immunobiol, Dept Microbiol & Immunol, 72-74, resident, Sch Med, 74-75, CLIN FEL HEMATOL & ONCOL, SCH MED, UNIV CALIF, LOS ANGELES, 75- Mem: Am Fedn Clin Res. Res: Identification, isolation and characterization of human leukemia antigens; develop immunoassay for measuring circulating leukemia antigenics; mechanism of action of specific and nonspecific immunotherapy in acute leukemia in man. Mailing Add: Dept of Microbiol & Immunol Univ of Calif Ctr for Health Sci Los Angeles CA 90024

ZIGMAN, SEYMOUR, b Far Rockaway, NY, Nov 21, 32; m 54; c 1. BIOCHEMISTRY. Educ: Cornell Univ, BA, 54; Rutgers Univ, MS, 56, PhD(biochem), 59. Prof Exp: Fel biochem of the eye, Mass Eye & Ear Infirmary, 59-61, res assoc, 61-62; from instr to asst prof, 62-70, ASSOC PROF BIOCHEM OF THE EYE, SCH MED & DENT, UNIV ROCHESTER, 70- Mem: Am Chem Soc; Am Soc Biol Chemists; Am Soc Photobiol; Asn Res Vision & Ophthal. Res: Chemistry and metabolism of nucleic acids and proteins of the lens and cornea of the eye as related to ocular disorders; effects of near ultraviolet light on the eye lens and retina. Mailing Add: Univ Rochester Sch Med & Dent 260 Crittenden Blvd Rochester NY 14642

ZIGMOND, MICHAEL JONATHAN, b Waterbury, Conn, Sept 1, 41. NEUROPHARMACOLOGY, PSYCHOPHARMACOLOGY. Educ: Carnegie-Mellon Univ, BS, 63; Univ Chicago, PhD(biopsychol), 68. Prof Exp: Teaching asst psychol, Carnegie-Mellon Univ, 62-63; res assoc, Mass Inst Technol, 67-69, instr, 69-70; ASST PROF BIOL & PSYCHOL, UNIV PITTSBURGH, 70- Concurrent Pos: Nat Inst Ment Health grantee, Univ Pittsburgh, 70-; mem, Neuropsychol Res Rev Comt, Nat Inst Ment Health, 74-79; Nat Inst Ment Health res career development awardee, 75-80. Mem: AAAS; Am Soc Neurochem; Soc Neurosci; Am Psychol Asn; Am Soc Pharmacol & Exp Therapeut. Res: Interactions between brain neurochemistry, behavior and environment; control of biogenic amine metabolism. Mailing Add: Dept Life Sci 571 Crawford Hall Univ of Pittsburgh Pittsburgh PA 15260

ZIGMOND, RICHARD ERIC, b Willimantic, Conn, May 9, 44. NEUROBIOLOGY, NEUROENDOCRINOLOGY. Educ: Harvard Col, BA, 66; Rockefeller Univ, PhD(neurobiol), 71. Prof Exp: Fel physiol psychol, Rockefeller Univ, 71-72; fel neurochem, Univ Cambridge, 72-75; ASST PROF PHARMACOL, HARVARD MED SCH, 75- Concurrent Pos: Tutor biochem sci, Harvard Col, 75- Mem: Soc Neurosci; Brit Pharmacol Soc; Animal Behav Soc; Am Soc Zoologists; Sigma Xi. Res: Regulation of the levels of enzymes involved in the synthesis of neurotransmitters; identification and characterization of hormone-sensitive cells in the brain. Mailing Add: Dept of Pharmacol Harvard Med Sch Boston MA 02115

ZIHLMAN, ADRIENNE LOUELLA, b Chicago, Ill, Dec 29, 40. PHYSICAL ANTHROPOLOGY. Educ: Univ Colo, Boulder, BA, 62; Univ Calif, Berkeley, PhD(anthrop), 67. Prof Exp: Asst prof anthrop, 67-73, chmn dept, 75-77, ASSOC PROF ANTHROP, OAKES COL, UNIV CALIF, SANTA CRUZ, 73- Concurrent Pos: Wenner-Gren Found Anthrop Res grants, Transvaal Mus, Pretoria, SAfrica, Univ Witwatersrand, Anthrop Inst, Zurich & Med Sch, Makerere Univ, Uganda, 69, Nat Mus Kenya, Nairobi & Transvaal Mus, 74. Mem: AAAS; Am Asn Phys Anthrop; Am Anthrop Asn; Am Soc Mammal. Res: Locomotor behavior and anatomy of primates; reconstruction of behavior and anatomy of fossil hominids; ape evolution and human origins. Mailing Add: Oakes Col Univ of Calif Santa Cruz CA 95064

ZIKAKIS, JOHN PHILIP, b Piraeus, Greece, Sept 16, 33; US citizen; m 58; c 1. BIOLOGICAL SCIENCES, BIOCHEMISTRY. Educ: Univ Del, BA, 65, MS, 67, PhD(biol, biochem), 70. Prof Exp: Res asst nutrit, Stine Lab, E I du Pont de Nemours & Co, Inc, 59-61; res assoc biochem genetics, 68-70, asst prof, 70-75, ASSOC PROF BIOCHEM GENETICS, UNIV DEL, 75- Concurrent Pos: Sci consult, Fedn Am Socs Exp Biol, 75. Mem: Am Chem Soc; Am Dairy Sci Asn; Am Inst Biol Sci; AAAS; Am Soc Animal Sci. Res: Various studies with xanthine oxidase as it may relate to atherosclerosis; immunological and nutritional studies with xanthine oxidase; biochemical genetic studies on milk and blood protein; polymorphisms. Mailing Add: Dept of Animal Sci & Agr Biochem Univ of Del Newark DE 19711

ZILBER, JOSEPH ABRAHAM, b Boston, Mass, July 27, 23; m 54; c 3. MATHEMATICS. Educ: Harvard Univ, AB, 43, MA, 46, PhD, 63. Prof Exp: Instr math, Columbia Univ, 48-50 & Johns Hopkins Univ, 50-55; asst prof, Univ Ill, 55-56; lectr, Northwestern Univ, 56-57; res assoc, Brown Univ, 57-62; res assoc, Yale Univ, 62; asst prof, 62-64, ASSOC PROF MATH, OHIO STATE UNIV, 64- Concurrent Pos: Assoc ed, Math Rev, Am Math Soc, 57-62. Mem: AAAS; Am Math Soc; Math Asn Am. Res: Algebraic topology; category theory. Mailing Add: Dept of Math Ohio State Univ Columbus OH 43210

ZILCH, KARL T, b St Louis, Mo, Nov 14, 21; m 50; c 7. ORGANIC CHEMISTRY. Educ: Univ Mo, AB, 43, MA, 47, PhD(chem), 49. Prof Exp: Asst chem, Univ Mo, 47-49; res chemist, Northern Utilization Res Br, USDA, 49-55; res chemist & group leader, 55-59, res sect head, 59-61, TECH DIR, EMERY INDUSTS, INC, 61- Concurrent Pos: Instr chem, Bradley Univ, 50-51. Mem: Am Chem Soc; Am Oil Chem Soc; Swiss Chem Soc. Res: Synthesis and processing carboxylic acids; reactions and end use application of carboxylic acids. Mailing Add: 7682 Pine Glen Dr Cincinnati OH 45224

ZILCZER, PAUL, physics, mathematics, deceased

ZILE, MAIJA HELENE, b Latvia, Aug 3, 29; nat US; m 55; c 3. BIOCHEMISTRY. Educ: Univ Md, BS, 54; Univ Wis, MS, 56, PhD(biochem), 59. Prof Exp: Res fel biochem, Univ Wis, 59 & Harvard Univ, 59-61; RES ASSOC BIOCHEM, UNIV WIS-MADISON, 61- Mem: Sigma Xi; Am Inst Nutrit. Res: Metabolism and function of vitamin A; function of vitamin A in cell proliferation and differentiation; cell cycle studies; function of vitamins and hormones at molecular level. Mailing Add: 5101 Odana Rd Madison WI 53711

ZILINSKAS, BARBARA ANN, b Waltham, Mass, Sept 21, 47. BIOCHEMISTRY. Educ: Framingham State Col, BA, 69; Univ Ill, Urbana, MS, 70, PhD(biol), 75. Prof Exp: Lab technician biol, Univ Mass Environ Exp Sta, 68-69; NASA fel, Univ Ill, Urbana, 69-72, res & teaching asst, 73-74; fel, Smithsonian Radiation Biol Lab, 75; ASST PROF BIOCHEM, COOK COL, RUTGERS UNIV, 75- Mem: Am Soc Plant Physiologists; Biophys Soc; Am Soc Photobiol; AAAS; Sigma Xi. Res: Structure-function relationships of biological membranes, especially photosynthetic membranes; biochemistry and biophysics of the photosynthetic light reactions. Mailing Add: Dept of Biochem & Microbiol Cook Col Box 231 Rutgers Univ New Brunswick NJ 08903

ZILKE, SAMUEL, b Chatfield, Man, Nov 4, 14. AGRONOMY, ECOLOGY. Educ: Univ Sask, BA, 48, BSEd, 49, BSAgr, 53, MS, 54; SDak State Univ, PhD(agron), 67. Prof Exp: Res officer & res asst plant ecol, Univ Sask, 50-55; res asst bot, SDak State Univ, 57-59 & agron & weed sci, 59-61; instr agr, Exten & Col Div, Alta Agr Col, 66-

70; agr consult, 70-71; technician, Can Wildlife Serv, 71-72; AGR CONSULT, 72- Concurrent Pos: Lectr, Col Agr, Univ Sask, 52-55, consult fertilizers & herbicides, 66-71. Mem: Agr Inst Can; Ecol Soc Am; Can Soc Soil Sci; Am Inst Biol Sci; Agron Soc Am. Res: Ecological life histories of plants and their physiology under field conditions; effect of variable soil moisture and temperature on seeds; response of field and grass crops to fertilizer and herbicides. Mailing Add: Box 147 Springside Saskatchewan SK Can

ZILKEY, BRYAN FREDERICK, b Manitou, Man, Apr 14, 41; m 64; c 3. PLANT SCIENCE, BIOCHEMISTRY. Educ: Univ Man, BSA, 62, MSc, 63; Purdue Univ, Lafayette, PhD(plant physiol), 69. Prof Exp: RES SCIENTIST TOBACCO, RES BR, CAN DEPT AGR, 69- Mem: Can Soc Plant Physiol; Am Soc Plant Physiol; Can Fedn Biol Socs. Res: Lipid and carbohydrate metabolism and biosynthesis in germinating and developing castor bean endosperm; biochemistry and physiology of tobacco growth; tobacco smoke chemistry and biological properties. Mailing Add: Res Sta Can Dept of Agr Delhi ON Can

ZILL, LEONARD PETER, b Portland, Ore, Oct 9, 20; m 52; c 2. BIOCHEMISTRY. Educ: Ore State Col, BS, 42, MS, 44; Ind Univ, PhD(biochem), 50. Prof Exp: Asst chem, Ore State Col, 42-44; asst, Forest Prod Lab, US Forest Serv, 42-44; res assoc, Ind Univ, 47-50; biochemist, Oak Ridge Nat Lab, 50-57; sr biochemist, Res Inst Advan Studies, 57-63; chief biol adaptation br, 63-67, mem lunar sample anal planning team, 68-70, CHIEF PLANETARY BIOL DIV, AMES RES CTR, NASA, 67-, SR RES SCIENTIST, 74- Mem: AAAS; Am Chem Soc; Am Soc Biol Chemists; Sigma Xi; Brit Biochem Soc. Res: Growth factors; wood and pulp chemistry; biochemistry of paramecia; chemistry and biochemistry of carbohydrates; photosynthesis; plant lipids. Mailing Add: Planetary Biol Div Ames Res Ctr NASA Moffett Field CA 94035

ZILLER, STEPHEN A, JR, b Kansas City, Mo, Nov 2, 38; m 61; c 4. NUTRITION, TOXICOLOGY. Educ: Rockhurst Col, BA, 61; St Louis Univ, PhD(biochem), 67. Prof Exp: Res biochemist, Res Div, 67-69, res nutritionist, 69-70, nutritionist, Food Prod Develop Div, 71-74, SECT HEAD FOOD SAFETY & NUTRIT, FOOD PROD DEVELOP DIV, PROCTER & GAMBLE CO, 74- Mem: AAAS; Am Chem Soc. Res: Metabolism of sterols and bile acids; drug metabolism; protein nutrition; food safety and nutrition. Mailing Add: Food Prod Develop Div Procter & Gamble Co 6071 Center Hill Rd Cincinnati OH 45224

ZILLER, WOLF GUNTHER, b Ger, Sept 2, 10; nat US; m 50. MYCOLOGY, FOREST PATHOLOGY. Educ: Univ BC, BScF, 49; Univ Toronto, MA, 52, PhD(mycol), 55. Prof Exp: Res officer forestry, Can Dept Agr, 49-65; res scientist, 66-73, EMER RES SCIENTIST, PAC FOREST RES CTR, CAN DEPT ENVIRON, 73- Res: Taxonomy and etiology of foliage diseases of conifers caused by rusts, species of Hypodermataceae and fungi associated with these pathogens. Mailing Add: 38 Kingham Pl Victoria BC Can

ZILLINSKY, FRANCIS JOHN, b Medicine Hat, Alta, July 16, 17; m 44; c 3. PLANT BREEDING. Educ: Univ Sask, BSA, 49, MSc, 50; Iowa State Univ, PhD(genetics, plant breeding), 54. Prof Exp: Cerealist, Cent Exp Farm, Genetics & Plant Breeding Inst, Res Br, Can Dept Agr, 50-65, head cereal crops sect, Res Sta, 65-67, HEAD TRITICALE IMPROV, INT MAIZE & WHEAT IMPROV CTR, MEX, 68- Mem: Am Soc Agron; Can Soc Agron; Agr Inst Can; Genetics Soc Can. Res: Interspecific hybridization among avena species; development of triticale as a food crop for the developing nations. Mailing Add: CIMMYT Londres 40 PO Box 6-641 Mexico 6 DF Mexico

ZILSEL, PAUL RUDOLPH, b Vienna, Austria, May 6, 23; nat US; div; c 2. THEORETICAL PHYSICS. Educ: Col Charleston, BS, 43; Univ Wis, MA, 45; Yale Univ, PhD(phyiscs), 48. Prof Exp: Asst theoret physics, Yale Univ, 47-48; fel, Duke Univ, 48-49; asst prof physics, Colo Agr & Mech Col, 49-50; instr, Univ Conn, 50-51, asst prof, 51-54; sr lectr, Israel Inst Technol, 54-56; asst prof, McMaster Univ, 56-58; vis lectr, 58-60, assoc prof, 60-63, PROF PHYSICS, CASE WESTERN RESERVE UNIV, 63- Concurrent Pos: NSF sci fac fel, Princeton Univ, 64-65; vis prof, Univ Wash, 71. Mem: AAAS; Am Phys Soc. Res: Theory of quantum liquids, especially superfluids; statistical mechanics, especially phase transitions. Mailing Add: Dept of Physics Case Western Reserve Univ Cleveland OH 44106

ZILVERSMIT, DONALD BERTHOLD, b Hengelo, Holland, July 11, 19; nat US; m 45; c 3. PHYSIOLOGICAL CHEMISTRY. Educ: Univ Calif, BS, 40, PhD(physiol), 48. Prof Exp: Clin demonstr, Dent Sch, Univ Calif, 46-48; from instr to prof physiol, Med Units, Univ Tenn, 48-66; PROF, DIV NUTRIT SCI & SECT BIOCHEM, MOLECULAR & CELL BIOL, DIV BIOL SCI, CORNELL UNIV, 66- Concurrent Pos: Consult, NIH, 55-; ed, J Lipid Res, 59-61; Am Heart Asn career investr, 59-; guest prof, State Univ Leiden, 61-62; vis fel exp path, Australian Nat Univ, 66; vis prof biochem, Mass Inst Technol, 72-73; mem coun arteriosclerosis, Am Heart Asn. Mem: AAAS; Am Chem Soc; Soc Exp Biol & Med; Am Inst Nutrit. Res: Lipid metabolism; lipoproteins; membrane biochemistry; phagocytosis; arteriosclerosis; use of isotopes in metabolic work. Mailing Add: Div of Nutrit Sci Cornell Univ Ithaca NY 14853

ZILZ, MELVIN LEONARD, b Detroit, Mich, Apr 15, 32; m 57; c 3. CELL BIOLOGY, BIOCHEMISTRY. Educ: Concordia Teachers Col, Ill, BS, 53; Univ Mich, Ann Arbor, MS & MA, 64; Wayne State Univ, PhD(biol), 70. Prof Exp: Teacher pvt sch, Ill, 53-57; instr high sch, Mich, 57-65, chmn dept sci, 58-65; asst prof biol, 65-72, chmn dept natural sci, 71-72, ASSOC PROF BIOL & REGISTR ADMIS, CONCORDIA SR COL, 72- Concurrent Pos: Instr, Mich Lutheran Col, 62-63. Mem: AAAS; Nat Asn Biol Teachers. Res: Cellular research, especially cell division and the anaphase movement of chromosomes. Mailing Add: Dept of Biol Concordia Sr Col Ft Wayne IN 46825

ZIMAR, FRANK, chemistry, see 12th edition

ZIMBELMAN, ROBERT GEORGE, b Keenesburg, Colo, Sept 4, 30; m 52; c 4. REPRODUCTIVE ENDOCRINOLOGY. Educ: Colo State Univ, BS, 52; Univ Wis, MS, 57, PhD(endocrinol), 60. Prof Exp: Instr genetics, Univ Wis, 57-60; res assoc, Upjohn Co, 60-64, head vet biol res, 65, head reproduction & physiol res, Agr Prod Div, 65-71, RES MGR REPRODUCTION & PHYSIOL RES, AGR PROD DIV, UPJOHN CO, 71- Mem: Am Soc Animal Sci; Soc Study Reproduction; Brit Soc Study Fertil. Res: Control of time of breeding of domestic animals; prevention of estrus in dogs; corpus luteum function in cattle and sheep; pregnancy maintenance in cattle; steroids in newborn animals; anabolic agents for domestic animals; regulatory aspects of use of hormones in food-producing animals. Mailing Add: Animal Health Res & Develop Upjohn Co Kalamazoo MI 49001

ZIMBRICK, JOHN DAVID, b Dickinson, NDak, Sept 18, 38; m 62; c 2. RADIATION BIOPHYSICS, RADIATION CHEMISTRY. Educ: Carleton Col, BA, 60; Univ Kans, MS, 62, PhD(radiation biophys), 67. Prof Exp: Asst physicist, IIT Res Inst, 62-64; chief environ studies sect, Health Serv Lab, US AEC, Idaho, 67-68; Nat Inst Gen Med Sci fel lab nuclear med & radiation biol, Univ Calif, Los Angeles, 68, US AEC fel, 68-69; asst prof radiation biophys, 69-73, ASSOC PROF RADIATION BIOPHYS, UNIV KANS, 73- Mem: Health Physics Soc; Radiation Res Soc; Biophys Soc. Res: In vivo studies on DNA base damage induced by gamma radiation; electron spin resonance spectroscopy of radicals produced in biomolecules by radiation; application of electron spin resonance to cancer detection and treatment. Mailing Add: Dept of Radiation Biophys Univ of Kans Nuclear Reactor Ctr Lawrence KS 66045

ZIMDAHL, ROBERT LAWRENCE, b Buffalo, NY, Feb 28, 35; m 56; c 3. AGRONOMY, WEED SCIENCE. Educ: Cornell Univ, BS, 56, MS, 66; Ore State Univ, PhD(agron), 68. Prof Exp: Res assoc agron, Cornell Univ, 63-64; asst prof weed sci, 68-72, ASSOC PROF WEED SCI, COLO STATE UNIV, 72- Mem: Weed Sci Soc Am; Am Chem Soc; Am Soc Agron. Res: Herbicide degradation in soil; environmental pollution by pesticides and heaheavy metals. Mailing Add: Dept of Bot & Plant Path Colo State Univ Ft Collins CO 80521

ZIMERING, SHIMSHON, b Kishinev, Romania, July 6, 33; m 66; c 2. MATHEMATICAL ANALYSIS. Educ: Univ Geneva, BSc, 56, license in math, 58; Free Univ Brussels, PhD(math, physics), 65. Prof Exp: Res asst, Weizmann Inst Sci, 58-59; res fel, Battelle Inst, Geneva, Switz, 61-66; hon res assoc math & Battelle Inst, Geneva, Switz fel, Harvard Univ, 66-67; mem advan studies ctr, Battelle Inst, Geneva, 67-68; ASSOC PROF MATH, OHIO STATE UNIV, 68- Concurrent Pos: Deleg, Int Conf Peaceful Uses Atomic Energy, Geneva, 64; NSF res grant, Ohio State Univ, 73-74. Mem: Am Math Soc. Res: Real analysis, summability and transform theory; boundary value problems; solid state physics. Mailing Add: Dept of Math Ohio State Univ Columbus OH 43210

ZIMM, BRUNO HASBROUCK, b Kingston, NY, Oct 31, 20; m 44; c 2. BIOPHYSICAL CHEMISTRY, POLYMER CHEMISTRY. Educ: Columbia Univ, AB, 41, MS, 43, PhD(chem), 44. Prof Exp: Asst chem, Columbia Univ, 41-44; res assoc & instr, Polytech Inst Brooklyn, 44-46; from instr to asst prof, Univ Calif, 46-49, assoc prof, 50-52; res assoc, Gen Elec Co, 51-60; PROF CHEM, UNIV CALIF, SAN DIEGO, 60- Concurrent Pos: Vis lectr, Harvard Univ, 50-51; vis prof, Yale Univ, 60. Honors & Awards: Leo Hendrik Baekland Award, Am Chem Soc, 57; Bingham Medal, Soc Rheol, 60; High Polymer Physics Prize, Am Phys Soc, 63. Mem: Nat Acad Sci; Am Chem Soc; Am Phys Soc; Soc Rheol; Am Soc Biol Chemists. Res: Theory of macromolecular solutions; properties and structure of high polymers and biological macromolecules. Mailing Add: Dept of Chem Univ of Calif at San Diego La Jolla CA 92093

ZIMM, GEORGIANNA GREVATT, b Jersey City, NJ, Nov 5, 17; m 44; c 2. GENETICS. Educ: Columbia Univ, BA, 40; Univ Pa, MA, 42; Univ Calif, Berkeley, PhD(zool), 50. Prof Exp: Teaching asst biol, Univ Del, 40-42; teaching asst zool, Barnard Col, Columbia Univ, 43-45, lectr, 45-46; teaching asst, Univ Calif, Berkeley, 46-50; res asst genetics & bibliogr, 68-75, RES ASSOC BIOL, UNIV CALIF, SAN DIEGO, 75- Mem: AAAS; Genetics Soc Am. Res: Mutants and cytogenetics of Drosophila. Mailing Add: Dept of Biol Univ of Calif at San Diego La Jolla CA 92093

ZIMMACK, HAROLD LINCOLN, b Chicago, Ill, Feb 12, 25; m 56; c 3. INSECT PATHOLOGY. Educ: Eastern Ill Univ, BS, 51; Iowa State Univ, MS, 53, PhD(entom), 56. Prof Exp: Asst, Iowa State Univ, 51-56; prof biol, Eastern Ky Univ, 56-63; PROF ZOOL, BALL STATE UNIV, 63- Concurrent Pos: Sigma Xi & Ind Acad Sci res grants-in-aid, 67-68. Mem: Am Soc Zoologists; Entom Soc Am; AAAS; Soc Invert Path. Res: Rapid screening of potential insect pathogen through physiological studies. Mailing Add: Dept of Biol Ball State Univ 2000 University Muncie IN 47306

ZIMMER, ALBERT MICHAEL, b Salzgitter-Lebenstedt, Ger, Sept 29, 45; US citizen; m 62. NUCLEAR MEDICINE, BIONUCLEONICS. Educ: Univ of the Pac, BS, 68; Purdue Univ, MS, 72, PhD(bionucleonics), 74. Prof Exp: Pharmacist, Thrifty Drug Stores, Inc, 68-69; asst prof radiol & nuclear pharmacist, Med Col Wis, Milwaukee County Med Complex, 73-75; ASST PROF MED RADIOL & NUCLEAR PHARMACISTS, UNIV ILL MED CTR, 75- Concurrent Pos: Consult, Union Carbide, NY, Radiopharmaceuticals, 74-75; comt mem, Nuclear Pharm Sect Am Pharmaceut Asn, 75. Mem: Soc Nuclear Med; Health Physics Soc; Am Pharmaceut Asn; Sigma Xi. Res: Development of new radiopharmaceuticals; the kinetics involved of current radiopharmaceuticals; quality control procedures of radiopharmaceuticals. Mailing Add: Univ of Ill Med Ctr Nuclear Med 840 S Wood St Chicago IL 60612

ZIMMER, ARTHUR JAMES, b St Louis, Mo, May 12, 14; m 40; c 1. PHARMACEUTICAL CHEMISTRY. Educ: St Louis Col Pharm, BS, 40; Wash Univ, MS, 43, PhD(chem), 46. Prof Exp: From instr to assoc prof, 41-55, PROF CHEM, ST LOUIS COL PHARM, 55- Concurrent Pos: Biochemist, Snodgras Lab, City Hosp, 60-; mem, US Pharmacopeial Conv, 70. Mem: Am Chem Soc; Am Pharmaceut Asn. Res: Instrumental analysis of pharmaceutical compounds. Mailing Add: 10038 Sheldon Dr St Louis MO 63137

ZIMMER, DAVID E, b Neoga, Ill, Sept 25, 35; m 56; c 2. PLANT PATHOLOGY, PLANT BREEDING. Educ: Eastern Ill Univ, BS, 57; Purdue Univ, MS, 59, PhD(plant path), 61. Prof Exp: RES PLANT PATHOLOGIST, USDA, 61- Mem: AAAS; Am Phytopath Soc; Crop Sci Soc Am. Res: Genetics of parasitism with special emphasis on obligate parasites; inheritance and nature of disease resistance in oilseed crops and the improvement of oil-seed crops through disease resistance breeding. Mailing Add: Waldron Hall NDak State Univ Fargo ND 58103

ZIMMER, GEORGE P, b Toledo, Ohio, Sept 17, 14; m 56; c 3. BIOLOGY, SCIENCE EDUCATION. Educ: Univ Toledo, BS, 37, BEd, 41; Columbia Univ, MS, 47, EdD, 50. Prof Exp: Instr natural sci, Univ Toledo, 47-49 & Columbia Univ, 49-50; assoc prof, 50-72, PROF BIOL & COORDR HEALTH SCI, STATE UNIV NY COL FREDONIA 72- Concurrent Pos: Res assoc sci ed, Harvard Univ, 61-62. Mem: Fel AAAS. Res: Microbiology. Mailing Add: Dept of Biol State Univ NY Col Fredonia NY 14063

ZIMMER, HANS, b Berlin, Ger, Feb 5, 21; m 46; c 1. ORGANIC CHEMISTRY. Educ: Tech Univ, Berlin, Cand, 47, Dipl, 49, DrIng, 50. Prof Exp: From asst prof to assoc prof, 54-61, PROF CHEM, UNIV CINCINNATI, 61- Concurrent Pos: Vis prof, Univ Mainz, 66-67, Univ Bonn, 67 & Univ Bern, 71; ed, Methodicum Chemicum; consult, Lithium Corp Am & Matheson, Coleman & Bell, 71-; ed, Ann Reports Inorg & Gen Syntheses, 72- Mem: Fel AAAS; Am Chem Soc; Ger Chem Soc. Res: Synthetic and metal organic chemistry. Mailing Add: Dept of Chem Univ of Cincinnati Cincinnati OH 45221

ZIMMER, JAMES GRIFFITH, b Lynbrook, NY, Apr 10, 32; m 71; c 3. PREVENTIVE MEDICINE, COMMUNITY HEALTH. Educ: Cornell Univ, BA, 53; Yale Univ, MD, 57; London Sch Hyg & Trop Med, dipl trop pub health, 66; Am Bd Internal Med,dipl, 65. Prof Exp: Intern internal med, Grace-New Haven Community Hosp, Conn, 57-58; resident, Strong Mem Hosp, Rochester, NY, 58-60; asst chief dermat, Walter Reed Army Inst Res, 61-63; from sr instr to asst prof prev med &

community health, 63-67, actg chmn dept prev med, 68-69, ASSOC PROF PREV MED & COMMUNITY HEALTH, SCH MED & DENT, UNIV ROCHESTER, 68- Concurrent Pos: Milbank fac fel, Univ Rochester, 64-71; pres, exec dir, Med Serv Int, Inc, 68-70; pres, Genesee Valley Med Found, 70-; med dir, Regional Utilization & Med Rev Proj, 71- Mem: Am Fedn Clin Res; Am Pub Health Asn; Int Epidemiol Asn; fel Am Col Prev Med; Royal Soc Trop Med & Hyg. Res: Community health; medical care research, especially in areas of utilization and quality of care review and care of chronically ill and aged. Mailing Add: Dept of Prev Med Univ of Rochester Med Ctr Rochester NY 14642

ZIMMER, LOUIS GEORGE, b Marseilles, Ill, Nov 30, 26; m 48; c 2. GEOLOGY. Educ: Augustana Col, BA, 50; Univ Iowa, MS, 52. Prof Exp: Subsurface geologist, Ohio Oil Co, 52-57; dist geologist, J M Huber Corp, 57-62; PARTNER, MAGAW & ZIMMER, 62- Res: Petroleum geology. Mailing Add: Magaw & Zimmer Suite 404 Park/Harvey Ctr Bldg Oklahoma City OK 73102

ZIMMER, MARTIN F, b Metz, France, Apr 25, 29; US citizen; m. THERMODYNAMICS, EXPLOSIVES. Educ: Univ Munich, BS, 55, MS, 58; Munich Tech Univ, PhD(chem technol), 61. Prof Exp: Head fuel res lab, Ger Aeronaut Res Inst, Munich, 60-62; chemist, Naval Ord Sta, 62, head thermodyn br, 62-70; DIR HIGH EXPLOSIVE RES & DEVELOP LAB, EGLIN AFB, 70- Mem: Combustion Inst. Res: Thermodynamics and combustion characteristics of fuels and oxidizers; detonation physics and explosive related phenomena; scientific and administrative management. Mailing Add: Rte 1 Box 177 Niceville FL 32578

ZIMMER, RUSSEL LEONARD, b Springfield, Ill, Nov 7, 31. INVERTEBRATE BIOLOGY. Educ: Blackburn Col, AB, 53; Univ Wash, MS, 56, PhD(zool), 64. Prof Exp: Instr, 60-63, vis asst prof, 63-64, asst prof, 64-68, ASSOC PROF ZOOL, UNIV SOUTHERN CALIF, 68-, RESIDENT DIR, SANTA CATALINA MARINE BIOL LAB, 68- Concurrent Pos: Mem, Orgn Trop Studies. Mem: AAAS; Am Soc Zool; Soc Syst Zool; Marine Biol Asn UK. Res: Reproductive biology; larval development, metamorphosis and systematics of minor invertebrate phyla, especially Phoronida and other lophophorates. Mailing Add: Dept of Biol Univ of Southern Calif Los Angeles CA 90007

ZIMMER, WILLIAM FREDERICK, JR, b Glouster, Ohio, June 4, 23; m 44; c 6. ORGANIC POLYMER CHEMISTRY. Educ: Ohio State Univ, BSc, 48, MSc, 49, PhD(chem), 52. Prof Exp: Org chemist, Res Lab, Durez Plastics & Chems, Inc, 52-55; res supvr, Hooker Chem Corp, 55-59, mgr polymer res, 59-62; mgr fiber res, Behr-Manning Div, Norton Co, 63-64, group leader chem appln res & develop, 64-68, asst dir res, Grinding Wheel Div, 68-74, RES ASSOC GRINDING WHEEL DIV, NORTON CO, 74- Mem: Am Chem Soc. Res: Organofluorine chemistry; plastics and polymer chemistry and applications; fiber technology; abrasive systems and materials research; new product research and development. Mailing Add: 28 Wallbridge Rd Paxton MA 01612

ZIMMERBERG, HYMAN JOSEPH, b New York, NY, Sept 7, 21; m 43; c 3. MATHEMATICS. Educ: Brooklyn Col, BA, 41; Univ Chicago, MS, 42, PhD(math), 45. Prof Exp: Instr math, Univ Chicago, 42-45 & NC State Col, 45-46; from instr to assoc prof math, 46-60, dir, NSF Undergrad Res Partic Prog, 62-69, 71-73, PROF MATH, RUTGERS UNIV, NEW BRUNSWICK, 60- Mem: Am Math Soc; Math Asn Am; Sigma Xi; Am Asn Univ Prof. Res: Boundary value problems; linear integro-differential-boundary-parameter problems. Mailing Add: Dept of Math Rutgers Univ New Brunswick NJ 08903

ZIMMERER, ROBERT P, b Sheboygan, Wis, Dec 7, 29; m 56; c 3. PLANT PHYSIOLOGY, MICROBIOLOGY. Educ: Univ Wis, BS, 54; Cornell Univ, MS, 61; Pa State Univ, PhD(bot), 66. Prof Exp: Asst to sales mgr, Stauffer Chem Co, 55-56; asst plant mgr, Hopkins Agr Chem Co, 56-57; chemist, Marathon Div, Am Can Co, 57-59; asst bot, Cornell Univ, 59-61; from instr to assoc prof biol, 61-74, PROF BIOL & CHMN DEPT, JUNIATA COL, 74- Concurrent Pos: Res assoc, Hershey Med Ctr, Pa State Univ, 70; vis prof, Univ Maine, 71; consult, USDA, 72-74 & J C Blair Mem Hosp, 74- Mem: Am Soc Plant Physiol; Bot Soc Am; Am Micros Soc; Am Soc Microbiol. Res: Biogenesis and function of biological membranes; biological regulatory mechanisms. Mailing Add: Dept of Biol Juniata Col Huntingdon PA 16652

ZIMMERER, ROBERT W, b Brooklyn, NY, May 21, 29; m 60; c 2. EXPERIMENTAL PHYSICS, INSTRUMENTATION. Educ: Worcester Polytech Inst, BS, 51; NY Univ, MS, 55; Univ Colo, PhD(physics), 60. Prof Exp: Design engr, Hazeltine Electronics Corp, 51-55; asst physics, Univ Colo, 55-57, res assoc, 57-60; physicist, Nat Bur Stand, 60-66; chief scientist, Wm Ainsworth & Sons, Inc, Colo, 66-69; PRES, SCIENTECH, INC, 69- Concurrent Pos: Dept Com sci fel, 65-66; phys sci consult, US Army Fitzsimons Gen Hosp, 69- Mem: Sr mem Inst Elec & Electronics Engrs; Am Phys Soc. Res: Microwave spectroscopy of gases; microwave generation and propagation at very short wavelengths; microwave power measurement; mass measurement by new methods; introduction and adaptation of scientific knowledge into industrial manufacturing process; lung physiology. Mailing Add: 2300 Kenwood Dr Boulder CO 80303

ZIMMERING, STANLEY, b New York, NY, Apr 14, 24; m 51; c 3. GENETICS. Educ: Brooklyn Col, AB, 47; Columbia Univ, AM, 49; Univ Mo, PhD(zool), 53. Prof Exp: Lectr biol & res assoc, Univ Rochester, 53-55; asst prof, Trinity Col, Conn, 55-59; res exec zool, Ind Univ, 59-62; assoc prof biol, 62-66, PROF BIOL, BROWN UNIV, 66- Res: Segregation mechanisms; radiation genetics; chemical mutagenesis. Mailing Add: Div of Biol & Med Sci Brown Univ Providence RI 02912

ZIMMERMAN, ARTHUR MAURICE, b New York, NY, May 24, 29; m 53; c 3. PHYSIOLOGY, CELL BIOLOGY. Educ: NY Univ, BA, 50, MS, 54, PhD(cell physiol), 56. Prof Exp: Technician, NY Univ, 51-52; res assoc, NY Univ & Marine Biol Lab, Woods Hole, 55-56; Lalor res fel, Marine Biol Lab, Woods Hole, 56; Nat Cancer Inst res fel, Univ Calif, 56-58; from instr to asst prof pharmacol, Col Med, State Univ NY Downstate Med Ctr, 58-64; PROF ZOOL, UNIV TORONTO, 64-, ASSOC CHMN GRAD AFFAIRS, 75- Concurrent Pos: Mem corp, Marine Biol Lab, Woods Hole. Mem: Soc Gen Physiol; Am Soc Cell Biol (treas, 74-78); Can Soc Cell Biol (pres, 76). Res: Cell division; mechanism of cytokinesis and karyokinesis; nuclear-cytoplasmic interrelations; physiological effects of temperature and pressure; amoeboid movement; physicochemical aspects of protoplasmic gels; cell cycle studies and drug action of cells. Mailing Add: Dept of Zool Univ of Toronto Toronto ON Can

ZIMMERMAN, BARRY, b New York, NY, Jan 14, 38; m 62; c 3. ORGANIC CHEMISTRY. Educ: Brooklyn Col, BSc, 59, AM, 61; Fordham Univ, PhD(org chem), 67. Prof Exp: Lectr chem, Brooklyn Col, 59-60; instr, Bronx Community Col, 62-66; proj leader chem & plastics, 66-72, MKT MGR CELLULAR & ELASTOMER MAT, UNION CARBIDE CORP, NEW YORK, 72- Mem: Am Chem Soc; The Chem Soc; NY Acad Sci. Res: Synthesis and application of new chemical species in rubber processing; cure accelerators; silanes; inorganic bonding in elastomer matricies; insecticide synthesis; microencapsulation; physiochemical properties of polyelectrolytes and other colloids; polyurethane synthesis and catalysis. Mailing Add: 5 Tara Dr Pomona NY 10970

ZIMMERMAN, BEN GEORGE, b Newark, NJ, July 1, 34; m 60; c 2. PHARMACOLOGY. Educ: Columbia Univ, BS, 56; Univ Mich, PhD(pharmacol), 60. Prof Exp: Pharmacologist, Lederle Labs, Am Cyanamid Co, 60-61; res fel, Cardiovasc Labs, Col Med, Univ Iowa, 61-63; from asst prof to assoc prof pharmacol, 63-72, PROF PHARMACOL, UNIV MINN, MINNEAPOLIS, 72- Concurrent Pos: Mem coun high blood pressure res, Coun Circulation & Coun Basic Sci, Am Heart Asn. Mem: Am Soc Pharmacol & Exp Therapeut; Soc Exp Biol & Med. Res: Vascular effects of angiotensin and other pressor agents; influence of the sympathetic nervous system on various vascular beds; vascular role of prostaglandins. Mailing Add: 105 Millard Hall Univ of Minn Minneapolis MN 55455

ZIMMERMAN, BURKE KISLING, b Grand Forks, NDak, July 4, 36. MOLECULAR BIOLOGY. Educ: Harvard Col, AB, 58; Stanford Univ, PhD(biophys), 62. Prof Exp: NIH fel, Univ Chicago, 62-64; staff biophysicist, Biol Div, Oak Ridge Nat Lab, 64-66; asst prof biochem, Mich State Univ, 66-68; assoc res biologist, Univ Calif, Santa Cruz, 68-69; ASST PROF BIOPHYS, SCH MED, JOHNS HOPKINS UNIV, 69- Mem: Biophys Soc. Res: Molecular biology of DNA; cellular differentiation in plants and animal cells. Mailing Add: Dept of Biophys Johns Hopkins Univ Sch of Med Baltimore MD 21205

ZIMMERMAN, C DUANE, b Mayville, Wis, Oct 23, 35. COMPUTER SCIENCE. Educ: Andrews Univ, BA, 57; Univ Minn, Minneapolis, MS, 60, PhD(comput sci), 69. Prof Exp: Instr math, Southern Missionary Col, 61-63; mathematician, Control Data Corp, 64-66; asst prof comput sci, Univ Minn, Minneapolis, 69-70; ASST PROF BIOMATH, LOMA LINDA UNIV, 70- Mem: Asn Comput Mach. Res: Computer-medical applications. Mailing Add: Dept of Biomath Loma Linda Univ Loma Linda CA 92354

ZIMMERMAN, CRAIG ARTHUR, b Painesville, Ohio, Mar 22, 37; m 62; c 2. BOTANY, PHYSIOLOGICAL ECOLOGY. Educ: Baldwin-Wallace Col, BS, 60; Univ Mich, MS, 62 & 64, PhD(bot), 69. Prof Exp: From instr to asst prof biol, Centre Col Ky, 67-74; environ specialist, Spindletop Res, Inc, 74-75; ASSOC PROF BIOL & CHMN DEPT, AURORA COL, 75- Mem: Am Inst Biol Sci; AAAS; Ecol Soc Am; Sigma Xi; Crop Sci Soc Am. Res: Causes, characteristics, and evolution of weed plants; comparing the biology of the weed with that of related cultivars and narrow endemics; biological indicators of stream pollution. Mailing Add: Dept of Biol Aurora Col Aurora IL 60507

ZIMMERMAN, DALE A, b Imlay City, Mich, June 7, 28; m 50; c 1. ORNITHOLOGY, ECOLOGY. Educ: Univ Mich, BS, 50, MS, 51, PhD, 56. Prof Exp: From asst prof to assoc prof, 57-72, PROF BIOL, WESTERN NMEX UNIV, 72- Concurrent Pos: Mem expeds, Africa, 61, 63, 65 & 66. Mem: Am Ornith Union; Wilson Ornith Soc; Cooper Ornith Soc; Brit Ornith Union. Res: Taxonomy and ecology of birds and plants. Mailing Add: Dept of Biol Sci Western NMex Univ Silver City NM 88061

ZIMMERMAN, DANIEL HILL, b Los Angeles, Calif, June 3, 41; m 63; c 3. BIOCHEMISTRY, IMMUNOLOGY. Educ: Emory & Henry Col, BS, 63; Univ Fla, MS, 66, PhD(biochem), 69. Prof Exp: Jr staff fel biochem, Nat Inst Arthritis, Metab & Digestive Dis, 69-71, sr staff fel, 71-73; CELLULAR IMMUNOLOGIST, RES & DEVELOP DEPT, ELECTRO NUCLEONICS LABS INC, 73- Res: Synthesis and secretion of proteins; differentiation of antibody producing cells. Mailing Add: Electro Nucleonics Labs Inc 4921 Auburn Ave Bethesda MD 20014

ZIMMERMAN, DEAN R, b Compton, Ill, July 2, 32; m 53; c 2. ANIMAL NUTRITION. Educ: Iowa State Univ, BS, 54, PhD(swine nutrit), 60. Prof Exp: Res assoc nutrit, Univ Notre Dame, 60-62; asst prof biol, Wartburg Col, 62-65; assoc prof animal sci, Purdue Univ, 65-67; ASSOC PROF ANIMAL SCI, IOWA STATE UNIV, 67- Mem: Am Soc Animal Sci; Am Inst Nutrit. Res: Nutrition research, especially compensatory growth and development; nutrition-disease interrelationships; bioavailability of amino acids. Mailing Add: Dept of Animal Sci R337 Kildee Iowa State Univ Ames IA 50011

ZIMMERMAN, DON CHARLES, b Fargo, NDak, Feb 27, 34; m 58; c 1. BIOLOGICAL CHEMISTRY, PLANT PHYSIOLOGY. Educ: NDak State Univ, BS, 55, MS, 59, PhD(biochem), 64. Prof Exp: RES CHEMIST, AGR RES SERV, USDA, 59-; ASSOC PROF BIOCHEM, NDAK STATE UNIV, 69- Concurrent Pos: Asst prof, NDak State Univ, 64-69. Mem: AAAS; Am Chem Soc; Am Soc Plant Physiol. Res: Metabolism of unsaturated fatty acids; their oxidation by lipoxidase. Mailing Add: Dept of Biochem NDak State Univ Fargo ND 58102

ZIMMERMAN, DONALD NATHAN, b Somerset, Pa, Dec 30, 32. PHYSICAL INORGANIC CHEMISTRY. Educ: Univ Md, College Park, BS, 61; Pa State Univ, MEd, 63; WVa Univ, PhD(chem), 69. Prof Exp: Teacher pub sch, Pa, 62-65; assoc prof, 69-73, PROF CHEM, IND UNIV PA, 73- Mem: AAAS; Am Chem Soc; Am Inst Physics; The Chem Soc. Res: Application of magnetochemical techniques to the determination of the structure of coordination compounds. Mailing Add: Dept of Chem Ind Univ Pa Indiana PA 15701

ZIMMERMAN, EARL GRAVES, b Detroit, Mich, Feb 15, 43; m 75; c 2. VERTEBRATE BIOLOGY, POPULATION GENETICS. Educ: Ind State Univ, Terre Haute, BS, 65; Univ Ill, Urbana, MS, 67, PhD(zool), 70. Prof Exp: Asst mammal comp anat, Univ Ill, Urbana, 65-69, asst cytogenetics, 69-70; asst prof mammal cytogenetics, 70-75, fac res grant, 70-72, ASSOC PROF MAMMAL CYTOGENETICS, N TEX STATE UNIV, 75- Honors & Awards: Jackson Award, Am Soc Mammal, 70. Mem: Am Soc Mammal; Soc Syst Zool; Soc Study Evolution. Res: Cytogenetics, population genetics and evolution of mammals. Mailing Add: Dept of Biol NTex State Univ Denton TX 76203

ZIMMERMAN, EDWARD JOHN, b Waynetown, Ind, July 12, 24; m 45; c 2. THEORETICAL PHYSICS, PHILOSOPHY OF SCIENCE. Educ: Univ Kans, AB, 45; Univ Ill, MS, 47, PhD(physics), 51. Prof Exp: Res assoc nuclear physics, Univ Ill, 50-51; from asst prof to assoc prof physics, 51-60, chmn dept, 62-66, PROF PHYSICS, UNIV NEBR-LINCOLN, 60- Concurrent Pos: Vis prof, Hamburg Univ, 57-58; NSF sci fac fel philos sci, Cambridge Univ, 66-67. Mem: Fel Am Phys Soc; Am Asn Physics Teachers; Philos Sci Asn; Brit Soc Philos Sci. Res: Foundations of physics; quantum mechanics. Mailing Add: Dept of Physics Univ of Nebr Lincoln NE 68508

ZIMMERMAN, ELMER LEROY, b Washington Co, Pa, Feb 5, 21; m 45; c 4. PHYSICS. Educ: Washington & Jefferson Col, AB, 42; Syracuse Univ, MA, 44; Ohio State Univ, PhD(physics), 50. Prof Exp: Physicist, Tenn Eastman Corp, 44-45, Oak Ridge Nat Lab, 50-55 & Nuclear Develop Corp Am, 55-58; supvr critical exp unit, Atomics Int Div, NAm Aviation, Inc, 58-60; vpres & tech dir, Solid State Radiations, Inc, 60-62; group leader radiation effects & reactor opers, Atomics Int Div, NAm

Aviation, Inc, 62-64; asst mgr weapons effects dept, Solid State Physics Lab, 64-67, mgr electronic & electro-magnetic effects dept, Vulnerability & Hardness Lab, 67-69, MEM STAFF, OPERS RES DEPT, TRW SYSTS, 69- Mem: Am Phys Soc; Am Nuclear Soc. Res: Nuclear spectroscopy and resonance; reactor physics and instrumentation; critical experiments; semiconductor radiation detectors; radiation effects and transport; ballistic missile systems engineering. Mailing Add: 22650 MacFarlane Dr Woodland Hills CA 91364

ZIMMERMAN, ELWOOD CURTIN, b Spokane, Wash, Dec 8, 12; m 41. ENTOMOLOGY. Educ: Univ Calif, BS, 36; Univ London, PhD(zool) & DIC, 56. Prof Exp: Field entomologist, Bishop Mus, Honolulu, Hawaii, 34-35, asst cur collections, 35-36, from asst entomologist to entomologist, 36-45, cur entom, 46-50, res assoc, 51-61, entomologist, 61-73; SR RES FEL, COMMONWEALTH SCI & INDUST RES ORGN, AUSTRALIA, 73- Concurrent Pos: Lectr, Univ Hawaii, 36-37, 40-41, entomologist, 58-61; assoc entomologist, Hawaiian Sugar Planters Exp Sta, 46-54; Fulbright adv researcher, Eng, 49, res & lectr, Eng, 49-51, Denmark & Sweden, 50; hon assoc, Brit Mus Natural Hist, 51- Entomologist, Mangarevan Exped Southeast Polynesia, 34, Lapham Fijian Exped, 38, Samoan Exped, 40 & NSF Eng, 54-56, 58, 66-67 & 69-73. Mem: AAAS; Australian Entom Soc; Entom Soc Am; Soc Syst Zool; Asn Trop Biol. Res: Systematic entomology; biogeography; evolution on islands; insular life; Curculionidae of Australia and Indo-Pacific; insects of Oceania and Hawaii. Mailing Add: CSIRO Box 1700 Canberra Australia

ZIMMERMAN, EMERY GILROY, b Los Angeles, Calif, June 23, 39; m 67; c 3. ANATOMY, MEDICINE. Educ: Pomona Col, BA, 61; Baylor Univ, MD, 67, PhD(anat), 71. Prof Exp: Intern med, Methodist Hosp, Houston, 68-69; instr anat, Col Med, Baylor Univ, 69-70; surgeon, Addiction Res Ctr, NIMH, Ky, 70-72; ASSOC PROF ANAT, SCH MED, UNIV CALIF, LOS ANGELES, 72- Mem: Am Physiol Soc; Endocrine Soc; Soc Exp Biol & Med; AMA; Am Soc Clin Pharmacol & Therapeut. Res: Neuroendocrinology, physiology, neuropharmacology. Mailing Add: Dept of Anat Univ of Calif Sch of Med Los Angeles CA 90024

ZIMMERMAN, ERNEST FREDERICK, b New York, NY, June 2, 33; m 57; c 2. PHARMACOLOGY. Educ: George Washington Univ, BS, 56, MS, 58, PhD(pharmacol), 61. Prof Exp: Res asst pharmacol, George Washington Univ, 56-58; fel biol, Mass Inst Technol, 60-62; from instr to asst prof pharmacol, Sch Med, Stanford Univ, 62-68; assoc prof res pediat, 68-71, PROF RES PEDIAT, COL MED, UNIV CINCINNATI, 71-, ASSOC PROF RES PHARMACOL, 68-, DIR DIV FETAL PHARMACOL, INST DEVELOP RES, CHILDREN'S HOSP RES FOUND, 68-, DIR GRAD PROG DEVELOP BIOL, 71- Mem: AAAS; Am Soc Biol Chem; Teratol Soc; Am Soc Pharmacol & Exp Therapeut. Res: Teratology; developmental biology. Mailing Add: Inst Develop Res Children's Hosp Res Found Elland & Bethesda Aves Cincinnati OH 45229

ZIMMERMAN, EUGENE MUNRO, b New Haven, Conn, June 27, 40; m 71. VIROLOGY. Educ: Yale Univ, BA, 60; Wesleyan Univ, MA, 62; Univ Md, College Park, PhD(microbiol), 68. Prof Exp: Microbiologist, Ft Detrick, Md, 69-70; asst proj dir, Microbiol Assocs Inc, Md, 70-73; SR SCIENTIST, LITTON-BIONETICS INC, 73- Concurrent Pos: Consult, Mt Sinai Sch Med, 75- Mem: AAAS; Tissue Cult Asn; Am Soc Microbiol. Res: Biology of oncogenic herpesviruses and oncarnaviruses; treatment and prophylaxis of leukemia in animal models. Mailing Add: Litton-Bionetics Inc 5516 Nicholson Lane Kensington MD 20795

ZIMMERMAN, GARY ALAN, b Seattle, Wash, Oct 19, 38; m 60; c 2. CLINICAL CHEMISTRY. Educ: Calif Inst Technol, BS, 60; Univ Wis, PhD(org chem), 65. Prof Exp: Asst prof chem, 64-69, ASSOC PROF CHEM, SEATTLE UNIV, 69-, DIR CLIN CHEM, 68-, DEAN SCH SCI & ENG, 73- Concurrent Pos: Lectr, Univ Wash, 65 & vis sci prog, Pac Sci Ctr, Seattle, 66-68; consult, Gordon Res Conf, 66-70 & Swed Hosp & Med Ctr, Seattle, 68-; vis prof chem, Univ Idaho, 73. Mem: AAAS; Am Chem Soc; Am Asn Clin Chemists; The Chem Soc; Soc Ger Chem. Res: Clinical applications of enzymatic assays; trace metal analyses; lipoproteins; isoenzymes. Mailing Add: Sch of Sci & Eng Seattle Univ Seattle WA 98122

ZIMMERMAN, GEORGE LANDIS, b Hershey, Pa, Aug 27, 20; m 53. PHYSICAL CHEMISTRY. Educ: Swarthmore Col, AB, 41; Univ Chicago, PhD, 49. Prof Exp: Res chemist sam labs, Manhattan Dist Proj, Columbia, 42-46; instr, Mass Inst Technol, 49-51; asst prof, 51-55, assoc prof, 55, PROF CHEM, BRYN MAWR COL, 55- Mem: Am Chem Soc. Res: Molecular spectroscopy. Mailing Add: Dept of Chem Bryn Mawr Col Bryn Mawr PA 19010

ZIMMERMAN, GEORGE OGUREK, b Katowice, Poland, Oct 20, 35; US citizen; m 64. LOW TEMPERATURE PHYSICS. Educ: Yale Univ, BS, 58, MS, 59, PhD(physics), 63. Prof Exp: Res asst physics, Yale Univ, 59-62, res assoc, 62-63; from asst prof to assoc prof physics, 63-73, assoc chmn dept, 72-73, PROF PHYSICS & CHMN DEPT, BOSTON UNIV, 73- Concurrent Pos: Vis scientist, Nat Magnet Lab, Mass Inst Technol, 65-; assoc physicist, Univ Calif, San Diego, 73. Mem: AAAS; NY Acad Sci; Am Phys Soc. Res: Low temperature phenomena, cryogenics, specifically pertaining to liquid and solid helium three; investigation of paramagnetic phenomena; investigation of phase transitions. Mailing Add: Dept of Physics Boston Univ 111 Cummington St Boston MA 02215

ZIMMERMAN, HARRY MARTIN, b Vilna Prov, Russia, Sept 28, 01; nat US; m 30. PATHOLOGY. Educ: Yale Univ, BS, 24, MD, 27. Hon Degrees: LHD, Yeshiva Univ, 58. Prof Exp: Ives fel, Sch Med, Yale Univ, 27-29, from asst prof path to assoc prof, 30-43; from assoc clin prof to prof, Col Physicians & Surgeons, Columbia Univ, 46-64; prof, 64-74, EMER PROF PATH, ALBERT EINSTEIN COL MED, 74- Concurrent Pos: Assoc pathologist, New Haven Hosp, Conn, 33-43; consult, Bristol Hosp, Conn, 38-43; consult, US Naval Hosp, St Albans, 46-49, Seton Hosp, 49-55, Beth Israel Hosp, 49-71, Armed Forces Inst Path, 49-71, Long Island Jewish Hosp, 54- & Vassar Bros Hosp, Poughkeepsie, NY, 61-; sr consult, Vet Admin Hosp, Bronx, 46-71; chief lab div, Montefiore Hosp, 46-73; Middleton Goldsmith lectr, NY Acad Med, 64. Honors & Awards: Golden Hope Chest Award, Nat Multiple Sclerosis Soc, 65; Max Weinstein Award, United Cerebral Palsy Found, 72. Mem: Fel Am Soc Clin Path; Am Asn Path & Bact; Am Neurol Asn; Am Asn Neuropath (pres, 44); fel Col Am Path. Res: Neuropathology; demyelinating diseases; brain tumors. Mailing Add: 200 Cabrini Blvd New York NY 10033

ZIMMERMAN, HENRY B, b Brooklyn, NY, May 6, 17; m 43; c 2. CHEMISTRY, MATHEMATICS. Educ: Brooklyn Col, BA, 43; Polytech Inst Brooklyn, MS, 52. Prof Exp: Anal chemist, Coupland Labs, 41-43; res chemist, Seydell Chem Co, 46-50; pharmaceut res chemist, Schering Corp, 50-53; sr pharmaceut chemist, Ayerst Labs Div, Am Home Prod Corp, 53-55; dir pharmaceut develop, Nepera Chem Co, Warner-Lambert Pharmaceut Co, 55-57; coordr res & develop, 57-61, dir new prod develop, 61-66, dir tech serv, 66-68, vpres tech serv, 68-73, VPRES REGULATORY AFFAIRS & NEW PROD COORD, CARTER-WALLACE, INC, 73- Mem: Am Pharmaceut Asn; Drug Info Asn; Int Acad Law & Sci. Res: New product development; research administration; regulatory affairs. Mailing Add: Carter-Wallace Inc Cranbury NJ 08512

ZIMMERMAN, HOWARD ELLIOT, b New York, NY, July 5, 26; m 50; c 2. CHEMISTRY. Educ: Yale Univ, BS, 50, PhD(chem), 53. Prof Exp: Nat Res Coun fel chem, Harvard Univ, 53-54; from instr to asst prof, Northwestern Univ, 54-60; assoc prof, 60-61, PROF CHEM, UNIV WIS-MADISON, 61- Concurrent Pos: Mem grants comt, Res Corp, 66-72. Honors & Awards: Photochem Award, Am Chem Soc, 75 & James Flack Norris Award Phys-Org Chem, 76. Mem: Am Soc Photobiol; The Chem Soc; Ger Chem Soc; Am Chem Soc. Res: Organic, physical-organic, synthetic organic chemistry; photochemistry; theoretical organic chemistry; photobiology; reaction mechanisms; stereochemistry; unusual organic phenomena and species. Mailing Add: Dept of Chem Univ of Wis Madison WI 53706

ZIMMERMAN, HOWARD KARL, JR, chemistry, see 12th edition

ZIMMERMAN, HYMAN JOSEPH, b Rochester, NY, July 14, 14; m 43; c 4. PHYSIOLOGY, METABOLISM. Educ: Univ Rochester, AB, 36; Stanford Univ, MA, 38, MD, 42. Prof Exp: Intern, Stanford Univ Hosp, 42-43; resident med, Gallinger Munic Hosp & Med Div, George Washington Univ, 46-48, clin instr, Sch Med, 48-51; asst prof, Col Med, Univ Nebr, 51; chief med serv, Vet Admin Hosp, Omaha, 51-53; clin assoc prof med, Col Med, Univ Ill, 53-57; prof & chmn dept, Chicago Med Sch, 57-65; PROF MED, GEORGE WASHINGTON UNIV, 65-; CHIEF MED SERV, VET ADMIN HOSP, DC, 68- Concurrent Pos: Asst chief med serv, Vet Admin Hosp, DC, 48-49, dir liver & metab res lab, 65-68; chief med serv, Vet Admin West Side Hosp, Chicago, 53-65; chmn dept med, Mt Sinai Hosp, 57-65; prof, Sch Med, Boston Univ, 68-69; lectr, Sch Med, Tufts Univ, 68-69; clin prof med, Sch Med, Howard Univ & Sch Med, Georgetown Univ. Mem: Am Soc Clin Invest; Endocrine Soc; Am Diabetes Asn; Am Fedn Clin Res; fel Am Col Physicians. Res: Physiology of the liver. Mailing Add: 7913 Charleston Ct Bethesda MD 20034

ZIMMERMAN, IRWIN DAVID, b Philadelphia, Pa, Oct 31, 31; m 57; c 4. NEUROPHYSIOLOGY, BIOPHYSICS. Educ: Univ Del, BA, 59; Univ Wash, PhD(physiol), 66. Prof Exp: Asst prof, 66-72, ASSOC PROF PHYSIOL & BIOPHYS, MED COL PA, 72- Concurrent Pos: NSF grant, 68-70; NIH grant, 72- Mem: Am Physiol Soc; Biophys Soc; Soc Neurosci. Res: Neural processing and coding of sensory information; membrane phenomena; development and aging of cental nervous system. Mailing Add: Dept of Physiol & Biophys Med Col of Pa Philadelphia PA 19129

ZIMMERMAN, JACK MCKAY, b New York, NY, Feb 4, 27; m 53; c 2. SURGERY. Educ: Princeton Univ, AB, 49; Johns Hopkins Univ, MD, 53; Univ Kansas City, MA, 63. Prof Exp: From intern to resident surg, Johns Hopkins Hosp, 53-59; assoc, Sch Med, Univ Kans, 59-60, asst prof, 60-65; ASSOC PROF SURG, JOHNS HOPKINS UNIV, 65- Concurrent Pos: Asst, Johns Hopkins Hosp, 54-59, surgeon, 65-, Halsted fel surg path, Univ, 55-56, instr, 58-59; staff surgeon, Vet Admin Hosp, Kansas City, 59-60, chief surg serv, 60-65; consult, Sch Dent, Univ Kans, 60; chief surg, Church Home & Hosp, 65-; consult, Vet Admin Hosp, Baltimore. Mem: AMA; Am Col Surg; Am Fedn Clin Res; Asn Am Med Cols; Soc Univ Surg. Res: Thoracic surgery; cardiovascular physiology, especially mechanisms of protection against cerebral injury following circulatory occlusion; wound healing and infections; care of advanced malignancy; medical education. Mailing Add: 100 N Broadway Baltimore MD 21231

ZIMMERMAN, JAMES JOSEPH, b Yankton, SDak, Feb 11, 33; m 61. PHARMACY, PHARMACEUTICAL CHEMISTRY. Educ: Univ of the Pac, BS, 61; Univ Calif, San Francisco, PhD(pharmaceut chem), 69. Prof Exp: Asst prof, 68-72, ASSOC PROF PHYS MED CHEM, SCH PHARM, TEMPLE UNIV, 72- Concurrent Pos: Res grant-in-aid, Temple Univ, 70-71. Honors & Awards: Lederle Award, Lederle Labs, 71. Mem: AAAS; Acad Pharmaceut Sci. Res: Substituent effects in alkaline and enzymatic hydrolysis of drug esters; linear free energy and isergonic relationships; drug stability. Mailing Add: Dept of Pharm 3223 N Broad St Temple Univ Sch of Pharm Philadelphia PA 19140

ZIMMERMAN, JAMES KENNETH, b Nelson, Nebr, Aug 23, 43. BIOCHEMISTRY. Educ: Univ Nebr, Lincoln, BS, 65; Northwestern Univ, Evanston, PhD(biochem), 69. Prof Exp: NIH fel, Univ Va, 69-71; ASST PROF BIOCHEM, CLEMSON UNIV, 71- Mem: AAAS; Am Chem Soc; Biophys Soc. Res: Associating protein systems; analytical gel chromatography by direct scanning; analytical ultracentrifugation; computer simulations. Mailing Add: Dept of Biochem Clemson Univ Clemson SC 29631

ZIMMERMAN, JAMES ROSCOE, b Norwood, Ohio, July 12, 28; m 50; c 3. ZOOLOGY, ENTOMOLOGY. Educ: Hanover Col, AB, 53; Ind Univ, MA, 55, PhD(zool), 57. Prof Exp: Asst prof zool, Univ Wichita, 57-58; assoc prof biol, Ind Cent Col, 58-61, actg chmn dept, 59-61; from asst prof to assoc prof, 61-68, PROF BIOL, N MEX STATE UNIV, 68-, HEAD DEPT, 74- Mem: AAAS; Am Soc Zool; Entom Soc Am; Soc Systs Zool; Am Entom Soc. Res: Taxonomy and ecology of aquatic Coleoptera. Mailing Add: Dept of Biol NMex State Univ Las Cruces NM 88003

ZIMMERMAN, JAY ALAN, b Philadelphia, Pa, Mar 1, 45; m 72. MAMMALIAN PHYSIOLOGY. Educ: Franklin & Marshall Col, AB, 67; Rutgers Univ, PhD(zool), 75. Prof Exp: ASST PROF BIOL, ST JOHN'S UNIV, NY, 75- Mem: Sigma Xi; Geront Soc; Soc Study Reproduction; AAAS. Res: Adaptive mechanisms of organ-system function and biochemistry during aging. Mailing Add: Dept of Biol St John's Univ Jamaica NY 11439

ZIMMERMAN, JOHN F, b Monticello, Iowa, June 22, 37; m 59; c 2. PHYSICAL CHEMISTRY, ANALYTICAL CHEMISTRY. Educ: Univ Iowa, BS, 59; Univ Kans, PhD(chem), 64. Prof Exp: Asst prof, 63-68, ASSOC PROF CHEM, WABASH COL, 68- Res: Electroanalytical measurement of diffusion coefficients; use of specific ion glass electrodes in the study of ion-pair formation; development of instrumentation for undergraduate laboratory instruction. Mailing Add: Dept of Chem Wabash Col Crawfordsville IN 47933

ZIMMERMAN, JOHN GORDON, b Brooklyn, NY, May 31, 16; m 39; c 4. PHYSICAL CHEMISTRY, INORGANIC CHEMISTRY. Educ: Univ Pa, BS, 37, MS, 39; Georgetown Univ, PhD(chem), 71. Prof Exp: Instr gen sci, Monmouth Jr Col, 38-42; shift supvr org chem prod, Ala Ord Works, 42-43; instr chem, Drexel Inst, 43-44; develop chemist org chem, Publicker Industs, Inc, 44-47; chmn sci div, St Helena exten, Col William & Mary, 47-48; asst prof, gen & phys chem, Westminster Col, 48-51; from asst prof to assoc prof, 51-72, PROF CHEM, US NAVAL ACAD, 72- Mem: Am Chem Soc. Res: Kinetics of substitution reactions of transition metal complexes. Mailing Add: Dept of Chem US Naval Acad Annapolis MD 21402

ZIMMERMAN, JOHN HARVEY, b St Paul, Minn, Feb 11, 45; m 69. MEDICAL ENTOMOLOGY. Educ: Concordia Col, Minn, BA, 67; Univ Del, Newark, MS, 69; Mich State Univ, ELansing, PhD(entom), 75. Prof Exp: ENTOMOLOGIST, NAVY ENVIRON & PREV MED UNIT, 75- Mem: Sigma Xi; Entom Soc Am; Am Mosquito Control Asn. Res: The ecology, transmission and control of arthropod-borne diseases of medical and veterinary importance. Mailing Add: Naval Med Res Unit No 5 Ethiopia APO New York NY 09319

ZIMMERMAN, JOHN LESTER, b Hamilton, Ont, Feb 17, 33; US citizen; m 55; c 3. ECOLOGY. Educ: Mich State Univ, BS, 53, MS, 59; Univ Ill, PhD(zool), 63. Prof Exp: Asst prof zool, 63-68, ASSOC PROF BIOL, KANS STATE UNIV, 68- Concurrent Pos: Mem, Audio-Tutorial Cong; sci adv environ protection, Atlantic-Richfield Co, Calif, 74-75. Mem: Ecol Soc Am; Am Ornith Union; Wilson Ornith Soc; Cooper Ornith Soc; Am Soc Zool. Res: Bioenergetics and niche requirements of birds, physiological ecology. Mailing Add: Div of Biol Kans State Univ Manhattan KS 66506

ZIMMERMAN, JOHN RICHMAN, b Eureka, Kans, Sept 11, 20; m 42; c 2. PHYSICS. Educ: Kans State Teachers Col, Emporia, AB, 42; Ohio State Univ, PhD(physics), 49. Prof Exp: Physicist, Naval Res Lab, 43-45; asst prof physics, Univ Colo, 49-53; sr res technologist, MAgnolia Petrol Co, 53-55, tech assoc, 55-56; sect supvr, Mobil Oil Corp, 56-66; prof physics & chmn dept, Southern Ill Univ, Carbondale, 66-74, chmn grad coun, 72-73, asst dean, Col Sci, 74-75; DEAN, COL SCI & TECHNOL, E TEX STATE UNIV, 74- Concurrent Pos: Consult, Magnolia Petrol Co, 52-53 & Wilmad Glass Co, 67- Mem: Am Phys Soc; Am Soc Test & Mat; Am Chem Soc; Am Inst Mining, Metall & Petrol Eng. Res: magnetic resonance; molecular adsorption; electronics; geochemistry. Mailing Add: Col of Sci & Technol ETex State Univ Commerce TX 75428

ZIMMERMAN, JOSEPH, b New York, NY, Aug 9, 21; m 48; c 3. PHYSICAL CHEMISTRY, POLYMER CHEMISTRY. Educ: City Col New York, BS, 42; Columbia Univ, AM, 47, PhD(chem), 50. Prof Exp: Res chemist, Carothers Res Lab, 50-53, res assoc, 53-62, res fel, 62-64, res mgr, Carothers Res Lab, 64-71, RES MGR, INDUST FIBERS DIV, E I DU PONT DE NEMOURS & CO, INC, 71- Mem: Am Chem Soc; Sigma Xi. Res: Polymer and fiber research, especially polyamides, polyesters and aramids. Mailing Add: 1121 Grinnell Rd Green Acres Wilmington DE 19803

ZIMMERMAN, LEO M, b Toluca, Ill, Oct 27, 98; m 28; c 2. SURGERY. Educ: Ind Univ, AB, 19; Univ Chicago, MD, 22; Am Bd Surg, dipl, 38. Prof Exp: Asst prof surg, Northwestern Univ, 24-28; chmn dept, 48-68, PROF HIST & PHILOS MED & CHMN DEPT, CHICAGO MED SCH, 68- Mem: AMA; Am Thyroid Asn; Am Asn Hist Med; fel Am Col Surgeons; fel Royal Soc Med. Res: Surgical repair of hernia; embolism; nodular goiter and carcinoma; history of surgery. Mailing Add: 5000 East End Ave Chicago IL 60615

ZIMMERMAN, LEONARD NORMAN, b Brooklyn, NY, Sept 13, 23; m 46; c 3. BACTERIOLOGY. Educ: Cornell Univ, BS, 48, MS, 49, PhD(bact), 51. Prof Exp: Asst, Cornell Univ, 48-51; from asst prof to assoc prof, 51-61, PROF BACT, PA STATE UNIV, UNIVERSITY PARK, 61- Mem: AAAS; Am Soc Microbiol. Res: Bacterial genetics and regulatory mechanisms. Mailing Add: S-233 Frear Lab Pa State Univ University Park PA 16802

ZIMMERMAN, LESTER J, b Conway, Kans, July 16, 18; m 49; c 4. AGRONOMY, MATHEMATICS. Educ: Goshen Col, BA, 47; Purdue Univ, MS, 50, PhD(soil fertil), 56; Univ Ill, MA, 61. Prof Exp: Instr chem & math, 47-49, asst prof, 50-53, PROF MATH, GOSHEN COL, 55- Concurrent Pos: Soil survr, Soil Conserv Serv, USDA, 56-58; vis prof, Univ Zambia, 68-69. Mem: AAAS; Nat Coun Teachers Math; Soil Sci Soc Am; Soil Conserv Soc Am; Am Sci Affiliation. Res: Manganese; plant nutrition. Mailing Add: Dept of Math Goshen Col Goshen IN 46526

ZIMMERMAN, LORENZ EUGENE, b Washington, DC, Nov 15, 20; m 45, 59; c 6. PATHOLOGY. Educ: George Washington Univ, AB, 43, MD, 45; Am Bd Path, dipl, 52. Prof Exp: Assoc prof, 54-63, CLIN PROF PATH & OPHTHAL, SCH MED, GEORGE WASHINGTON UNIV, 63-; CHIEF OPHTHAL PATH, ARMED FORCES INST PATH, 53- Concurrent Pos: Lectr, Johns Hopkins Univ, 59- Mem: Am Soc Clin Path; AMA; Asn Res Vision & Ophthal; Am Acad Ophthal & Otolaryngol; Int Acad Path. Res: Pathology of diseases of the eye and ocular adnexa. Mailing Add: Ophthal Path Div Armed Forces Inst of Path Washington DC 20306

ZIMMERMAN, LORRAINE MAY, b Merrill, Wis. ANTHROPOLOGY. Educ: Case Western Reserve Univ, BA, 66; Wayne State Univ, PhD(anthrop), 73. Prof Exp: Teaching fel anthrop, Wayne State Univ, 68-70, instr, 72-73; asst prof, Lawrence Univ, 73-74; VIS ASST PROF ANTHROP, UNIV WIS-PARKSIDE, 74- Mem: AAAS; fel Am Anthrop Asn; Asn Social Anthrop Oceania. Res: Land tenure among the Amish; genetic dwarfism, migration and urbanization and politics among the Buang of Papua New Guinea. Mailing Add: Div of Social Sci Univ of Wis-Parkside Kenosha WI 53140

ZIMMERMAN, MORRIS, b New York, NY, June 22, 25; m 49; c 2. BIOCHEMISTRY. Educ: City Col New York, BS, 45; NY Univ, MA, 49; Ind Univ, PhD(chem), 55. Prof Exp: Res chemist, Lederle Labs, Am Cyanamid Co, 46-47; res chemist, Wyeth, Inc, 49-51; res asst, Purdue Univ, 51-52 & Ind Univ, 52-55; sr res chemist, R J Reynolds Tobacco Co, 55-57; res fel, 57-72, SR RES FEL, MERCK SHARP & DOHME RES LABS, RAHWAY, 72- Mem: AAAS; Am Soc Biol Chem; Am Soc Cell Biol. Res: Antibiotics; protein chemistry; nicotine metabolism; cancer research; biochemical control mechanisms; tissue culture; cell biology; large scale enzyme isolation, enzyme chemistry and mechanisms; proteases, especially enzymology and role in disease. Mailing Add: 104 Johnston Dr Watchung NJ 07060

ZIMMERMAN, NORMAN, biochemistry, cancer, see 12th edition

ZIMMERMAN, PETER DAVID, b June 15, 41; US citizen; m 67; c 1. NUCLEAR PHYSICS, METEORITICS. Educ: Stanford Univ, BS, 63, PhD(physics), 69; Lund Univ, Sweden, Filosofie Licentiat, 67. Prof Exp: Res fel physics, Deutsches Electronen Synchrotron, 69-71; adj asst prof physics & planetary sci, Univ Calif, Los Angeles, 71-73; res assoc physics, Fermi Nat Accelerator Lab, 73-74; ASST PROF PHYSICS, LA STATE UNIV, 74- Concurrent Pos: Res affil, Mass Inst Technol, 75- Mem: Am Phys Soc; Meteoritical Soc. Res: Electron scattering experiments from nuclei, principally at large energy loss; origin of stone meteorites; orbital mechanics of large manned satellites. Mailing Add: Dept of Physics & Astron La State Univ Baton Rouge LA 70803

ZIMMERMAN, R ERIK, b Newark, NJ, Oct 29, 41; m 71; c 3. ASTROPHYSICS. Educ: Pomona Col, BA, 63; Univ Calif, Los Angeles, MA, 66, PhD(astron), 70. Prof Exp: Asst prof astron, Mich State Univ, 68-71; ASSOC PROF ASTRON, NEWARK STATE COL, 71- Mem: Am Astron Soc; Am Inst Physics; Royal Soc Can. Res: Stellar structure and evolution. Mailing Add: Dept of Earth & Planetary Environ Newark State Col Union NJ 07083

ZIMMERMAN, RICHARD E, b Columbus, Ohio, Jan 1, 25; m 44; c 2. PHYSICS. Educ: Univ SDak, BS, 47; Purdue Univ, MS, 50. Prof Exp: Physicist, Los Alamos Sci Lab, 50-53; analyst, Opers Res Off, Johns Hopkins Univ, 53-61; sr analyst, Res Anal Corp, 61-72; DEP DIR, GAMING & SIMULATIONS DEPT, OPERS ANAL DIV, GEN RES CORP, 72- Honors & Awards: Lanchester Prize, Opers Res Soc Am & Johns Hopkins Univ, 56. Mem: Opers Res Soc Am. Res: Computer simulation of military operations and war gaming. Mailing Add: Opers Anal Div Gen Res Corp McLean VA 22101

ZIMMERMAN, RICHARD HALE, b Bowling Green, Ohio, Apr 11, 34; m 66; c 2. PLANT PHYSIOLOGY. Educ: Mich State Univ, BS, 56; Rutgers Univ, MS, 59, PhD(hort), 62. Prof Exp: Silviculturist, Tex Forest Serv, 62-64; PLANT PHYSIOLOGIST, PLANT INDUST STA, AGR RES SERV, USDA, 65- Honors & Awards: J H Gourley Award, Am Soc Hort Sci, 72. Mem: Bot Soc Am; Int Plant Propagators Soc; AAAS; Am Soc Hort Sci. Res: Juvenility and flower initiation in fruit trees and other woody plants; effects of growth regulators on plant growth and development. Mailing Add: Fruit Lab Agr Res Ctr USDA Beltsville MD 20705

ZIMMERMAN, ROBERT A, b Lafayette, Ind, Nov 20, 26; m 47; c 2. MICROBIOLOGY, EPIDEMIOLOGY. Educ: Purdue Univ, BS, 48; Ore State Univ, MS, 57, PhD(microbiol), 60. Prof Exp: Food & drug inspector, Ind State Bd Health, 48-49; sr sanitarian, Benton County Health Dept, Ore, 49-60; heart dis control officer, USPHS, NDak, 60-62, asst chief med sect, Lab Br, 62-63, CHIEF STREPTOCOCCAL INVESTS, ECOL INVEST PROG, NAT COMMUN DIS CTR, USPHS, 63- Concurrent Pos: Mem coun rheumatic fever & congenital heart dis, Am Heart Asn; proj officer, Rheumatic Fever Proj, Ministry of Health, Cairo, United Arab Repub. Mem: Sigma Xi. Res: Investigations of the epidemiology of streptococcal diseases and their nonsuppurative sequelae; acute rheumatic fever and acute glomerulonephritis. Mailing Add: USPHS Streptococcal Dis Sect PO Box 551 Ft Collins CO 80521

ZIMMERMAN, ROBERT LYMAN, b La Grande, Ore, Dec 30, 35; m 57; c 5. PHYSICS. Educ: Univ Ore, BA, 58; Univ Wash, PhD(physics), 63. Prof Exp: Physicist, Lawrence Radiation Lab, Univ Calif, Berkeley, 64-66; asst prof, 66-74, res assoc, 70-74, ASSOC PROF PHYSICS, UNIV & ASSOC PROF, INST THEORET SCI, UNIV ORE, 74- Mem: Am Phys Soc. Res: Quantum field theory; elementary particle physics; gravitation; astrophysics. Mailing Add: Dept of Physics Univ of Ore Eugene OR 97403

ZIMMERMAN, SARAH E, b Indianapolis, Ind, Oct 29, 37. IMMUNOLOGY, BIOCHEMISTRY. Educ: Ind Univ, AB, 59, MA, 61; Wayne State Univ, PhD(biochem), 69. Prof Exp: Res asst biochem, Sch Med, Wayne State Univ, 68-69; res assoc chem, Ind Univ, Bloomington, 69-71; res assoc microbiol, 71-73, RES ASST, DEPT CLIN PATH, SCH MED, IND UNIV, INDIANAPOLIS, 73- Mem: Sigma Xi; Am Soc Microbiol; Am Chem Soc. Res: Structure and specificity of antibody; relation of antibody specificity to genetic markers of antibody molecule; genetic control of immune response; in vitro tests for cell-mediated immunity; immunological diagnosis of bacterial and fungal infections. Mailing Add: Dept of Clin Path Ind Univ Sch of Med Indianapolis IN 46202

ZIMMERMAN, SELMA BLAU, b New York, NY, Apr 1, 30; m 53; c 3. EMBRYOLOGY, PHYSIOLOGY. Educ: Hunter Col, BA, 50; NY Univ, MS, 54, PhD, 58. Prof Exp: Asst, NY Univ, 53-55; res assoc pharmacol, Col Med, State Univ NY Downstate Med Ctr, 60-61; instr biol, Hunter Col, 61-64 & York Univ, Ont, 65-66; res assoc zool, Univ Toronto, 66-69; asst prof, 69-74, ASSOC PROF NATURAL SCI, GLENDON COL, YORK UNIV, 74- Res: Pigment cell physiology; mechanisms of cell division; drug action on the peripheral circulation. Mailing Add: Natural Sci Div Glendon Col York Univ Toronto ON Can

ZIMMERMAN, SHELDON BERNARD, b New York, NY, Nov 7, 26; m 50; c 1. MICROBIOLOGY. Educ: City Col New York, BS, 48; Long Island Univ, MS, 65; NY Univ, PhD(biol), 71. Prof Exp: Chemist pharmaceut, Vitamin Corp Am, 49-51; develop microbiologist pharmaceut, Schering Corp, 51-63; SR RES MICROBIOLOGIST PHARMACEUT, MERCK INST THERAPEUT RES, 63- Mem: Am Soc Microbiol; NY Acad Sci. Res: Discovery and mode of action of antibiotics; microbial physiology; microbial ecology; structure-activity relationships of antibiotics; microbial chemotherapeutics. Mailing Add: Merck Inst Therapeut Res 50 G-102 Rahway NJ 07065

ZIMMERMAN, STANLEY DEAN, b Sao Paulo, Brazil, Aug 31, 29; m 51; c 4. RUBBER CHEMISTRY, ELECTRON MICROSCOPY. Educ: Ouachita Baptist Col, BA & BS, 49; Univ Tex, PhD, 53. Prof Exp: Chemist, Dow Chem Co, 53-58; res chemist, Petrol Chem, Inc, 58-60 & Cities Serv Res & Develop Co, 60-63; res chemist, Columbian Carbon Co, 64-70, res chemist, Cities Serv Co, 71-73, SR RES CHEMIST, PHYSICS LAB, CITIES SERV CO, CRANBURY, NJ, 73- Mem: Am Chem Soc; Electron Micros Soc Am. Res: Allylic compounds; oxidation processes; reactions of chlorinated compounds; use of gas chromatographic techniques; polymers and copolymers of butadiene; cationic polymerizations; flame retardants; electron microprobe; image analysis. Mailing Add: 554 Hammond Dr Morrisville PA 19067

ZIMMERMAN, STEVEN B, b Chicago, Ill, June 5, 34; m 56. BIOCHEMISTRY. Educ: Univ Ill, BS, 56, MS, 57; Stanford Univ, PhD(biochem), 61. Prof Exp: Nat Found res fel, 61-62; RES CHEMIST, NAT INST ARTHRITIS, METAB & DIGESTIVE DIS, 64- Mem: Am Soc Biol Chem. Res: Nucleic acid synthesis and structure; mechanism of enzyme action. Mailing Add: Nat Inst Arth Metab & Dig Dis 9000 Rockville Pike Bethesda MD 20014

ZIMMERMAN, STUART O, b Chicago, Ill, July 27, 35; m 59; c 1. MATHEMATICAL BIOLOGY. Educ: Univ Chicago, BA, 54, PhD(math biol), 64. Prof Exp: Res assoc, Univ Chicago, 63-65, from instr to asst prof math biol, 65-67; ASSOC PROF BIOMATH & ASSOC MEM, UNIV TEX DENT SCI INST HOUSTON, 67-, HEAD DEPT BIOMATH, UNIV TEX M D ANDERSON HOSP & TUMOR INST, 68-, PROF BIOMATH & BIOMATHEMATICIAN, 72- Concurrent Pos: Consult, Ill State Dent Soc, 61- & Am Dent Asn, 66-; assoc mem, Univ Tex Grad Sch Biomed Sci Houston, 68-70, mem, 70-; actg dir common res comput facil, Univ Tex M D Anderson Hosp & Tumor Inst Houston, 68-73; chmn exec bd, Univ Tex Houston Educ & Res Comput Ctr, 73- Mem: AAAS; Biophys Soc; Int Asn Dent Res; Asn Comput Mach; Soc Math Biol. Res: Biomedical computing; mathematical modeling; image processing; computer karyotyping; information management systems; design and analysis of cancer and dental clinical trials. Mailing Add: M D Anderson Hosp Dept Biomath 6723 Bertner Ave Houston TX 77025

ZIMMERMAN, THOMAS PAUL, b Plainfield, NJ, Sept 3, 42; m 66; c 2. BIOCHEMICAL PHARMACOLOGY. Educ: Providence Col, BS, 64; Brown Univ, PhD(biochem), 69. Prof Exp: Nat Inst Neurol Dis & Stroke res fel biol & med sci, Brown Univ, 69-71; RES BIOCHEMIST, WELLCOME RES LABS, BURROUGHS WELLCOME & CO, USA, INC, 71- Res: Nucleic acids-purification, chemical and physical properties; purine metabolism and the mode of action of purine antimetabolites; radioimmunoassay methodology for cyclic nucleotides. Mailing Add: Dept Exp Ther Wellcome Res Labs 3030 Cornwallis Rd Research Triangle Park NC 27709

ZIMMERMAN, TOMMY LYNN, b Lima, Ohio, July 23, 43; m 67; c 2. SOIL CONSERVATION. Educ: Ohio State Univ, BS, 66; Pa State Univ, MS, 69, PhD(agron), 73. Prof Exp: From instr to asst prof agron, Delaware Valley Col, 71-75; ASST PROF AGRON & HEAD SOIL & WATER MGT TECHNOL, AGR TECH INST, OHIO STATE UNIV, 75- Mem: Am Soc Agron; Soil Sci Soc Am; Soil Conserv Soc Am; Coun Agr Technol. Mailing Add: Agr Tech Inst Ohio State Univ Wooster OH 44691

ZIMMERMAN, WALTER BRUCE, b Evergreen Park, Ill, Nov 27, 33; m 55; c 3. SOLID STATE PHYSICS. Educ: Andrews Univ, BA, 55; Mich State Univ, MS, 57, PhD(physics), 60. Prof Exp: Asst physics, Mich State Univ, 58-60; sr physicist solar energy conversion, Gen Dynamics Astronaut, 60-62; from asst prof to assoc prof physics, Andrews Univ, 62-69; ASSOC PROF PHYSICS & DIR COMPUT SERV, IND UNIV, SOUTH BEND, 69-, CHMN DEPT PHYSICS, 74- Mem: Am Phys Soc; Am Asn Physics Teachers; Am Vacuum Soc. Res: Changes that occur in infrared absorption spectrum and lattice constant of lithium hydride as its isotopic composition is varied; magnetic effect in biological processes; agglutination of red blood cells in a magnetic field; solar energy conversion by cadmium sulfide films. Mailing Add: Dept of Physics Ind Univ 1825 Northside Blvd South Bend IN 46615

ZIMMERMAN, WILLIAM FREDERICK, b Chicago, Ill, July 7, 38; m 64; c 2. CELL BIOLOGY. Educ: Princeton Univ, BA, 60, PhD(biol), 66. Prof Exp: Instr biol, Princeton Univ, 64-66; asst prof, 66-72, ASSOC PROF BIOL, AMHERST COL, 72- Concurrent Pos: NSF res grants, 66-70; NIH spec fel, 69-70; Nat Eye Inst res grants, 70-77; vis res fel, Univ Nijmegen, 73-74. Mem: AAAS; Asn for Res Vision & Ophthalmol. Res: Mechanism of light and temperature entrainment of circadian rhythms in insects; action spectra, carotenoid metabolism; cellular biochemistry of visual cycles and photoreceptor cell renewal. Mailing Add: Dept of Biol Amherst Col Amherst MA 01002

ZIMMERMANN, BERNARD, b St Paul, Minn, June 26, 21; m 49; c 2. SURGERY. Educ: Harvard Univ, MD, 45; Univ Minn, PhD, 53; Am Bd Surg, dipl. Prof Exp: House officer, Boston City Hosp, Mass, 45-46; head surg res facil, Naval Med Res Inst, Nat Naval Med Ctr, Bethesda, Md, 46-48; fel surg, Med Sch, Univ Minn, 48-53, from instr to prof surg, 53-60; head dept, 60-73, PROF SURG, MED CTR, WVA UNIV, 60- Concurrent Pos: Am Cancer Soc scholar cancer res, 53-56; cancer coord, Sch Med, Univ Minn, 56-60; mem, USPHS Cancer Chemother Study Sect, 59-, interim comt, Orgn Cancer Coords, 57-60, bd dirs, St Paul Inst & Sci Mus, Minn, 56-60, adv comt inst res grants, Am Cancer Soc, 62-66 & prog proj rev comt, Inst Metab & Arthritis, 64-68. Mem: Endocrine Soc; Soc Exp Biol & Med; Soc Univ Surg; Am Soc Exp Path; Halsted Soc (secy-treas, 63-66, pres). Res: Surgical physiology; experimental diabetes; electrolyte balance in surgery; endocrine aspects of malignancy; adrenal function in surgery; experimental oncology. Mailing Add: Dept of Surg WVa Univ Med Ctr Morgantown WV 26506

ZIMMERMANN, CHARLES EDWARD, b Juneau, Wis, Nov 15, 30; m 56; c 4. PLANT PHYSIOLOGY. Educ: Univ Wis, BS, 53; Ore State Univ, MS, 62. Prof Exp: Technician, Wash State Univ, 56-59; res agronomist, USDA, Ore, 59-62, plant physiologist, Agr Res Serv, Ore State Univ, 62-70, RES PLANT PHYSIOLOGIST, AGR RES SERV, USDA, 70- Mem: Am Soc Plant Physiol; Am Soc Agron; Crop Sci Soc Am; Am Soc Brewing Chemists. Res: Production and quality of hops; relate gibberellins to morphogenic and physiologic changes in hops. Mailing Add: Irrig Agr Res & Exten Ctr Box 30 Prosser WA 99350

ZIMMERMANN, EUGENE ROBERT CHARLES, b New York, NY, Mar 12, 24; m 47; c 3. ORAL PATHOLOGY, MICROBIOLOGY. Educ: Univ Md, DDS, 48; Am Univ, BS, 54; Univ Tex, MA, 62. Prof Exp: Intern dent, US Marine Hosp, Seattle, Wash, 48-49, clinician, 49-50; dent resident epidemiol & biomet, NIH, 50-54; instr oral diag, oral path & pub health, 54-55, asst prof histol, path & pub health, 55-57, assoc prof path & pub health, 57-59, PROF ORAL PATH & CHMN DEPT, BAYLOR COL DENT, 59- Concurrent Pos: Consult, Vet Admin Hosps, Dallas & Temple, Tex, US Army Hosps, Ft Bliss & Ft Hood, Tex & Ft Sill, Okla & USPHS Hosp, Ft Worth, Tex. Mem: Am Dent Asn; Am Pub Health Asn; Am Inst Oral Biol; Am Acad Oral Path; Royal Soc Health. Res: Fluoridation; ultrastructure of oral lesions; effect of geography and race upon gingival disease in children; role of adenoviruses in diseases of the mouth; adaptation of gingival epithelium to tissue culture technique; effects of aging and smoking habits on oral tissues. Mailing Add: 9830 Ash Creek Dr Dallas TX 75228

ZIMMERMANN, MARTIN HULDRYCH, b Bülach, Switz, Nov 27, 26; m 53; c 2. PLANT ANATOMY, PLANT PHYSIOLOGY. Educ: Swiss Fed Inst Technol, dipl, 51, DSc(biol), 53. Hon Degrees: MA, Harvard Univ, 70. Prof Exp: Asst plant physiol, Swiss Fed Inst Technol, 51-54; lectr forest physiol & forest physiologist, Cabot Found, 54-70, actg dir, Harvard Forest, 67-68, 69-70, CHARLES BULLARD PROF FORESTRY, HARVARD UNIV & DIR HARVARD FOREST, 70- Concurrent Pos: Mem, Orgn Trop Studies; Guggenheim fel, Swiss Fed Inst Technol, 68-69. Mem: AAAS; Am Soc Plant Physiol; Am Acad Arts & Sci; Int Asn Wood Anat; Int Acad Wood Sci. Res: Structure and physiology of trees and arborescent monocotyledons; translocation of nutrients in plants. Mailing Add: Harvard Forest Petersham MA 01366

ZIMMERMANN, ROBERT ALAN, b Philadelphia, Pa, July 17, 37. MOLECULAR BIOLOGY, BIOCHEMISTRY. Educ: Amherst Col, BA, 59; Mass Inst Technol, PhD(biophysics), 64. Prof Exp: Vis scientist biochem, Acad Sci USSR, 65-66; res fel microbiol, Med Sch, Harvard Univ, 66-69; res assoc molecular biol, Univ Geneva, 70-73; ASSOC PROF MICROBIOL & BIOCHEM, UNIV MASS, AMHERST, 73- Concurrent Pos: Helen Hay Whitney Found fel, 68; sr fel, Europ Molecular Biol Orgn, 71; adv, WHO, 75-; NIH res career develop award, 75. Mem: AAAS. Res: Structure, function and biosynthesis of ribosomes in bacteria and yeast; RNA-protein interaction; primary and secondary structure of ribosomal RNA; photochemical labeling; properties of mutationally-altered ribosomes and their components. Mailing Add: Dept of Biochem Univ of Mass Amherst MA 01002

ZIMMERMANN, WILLIAM, JR, b Philadelphia, Pa, Oct 28, 30; m 62; c 3. PHYSICS. Educ: Amherst Col, AB, 52; Calif Inst Technol, PhD(physics), 58. Prof Exp: Fulbright fel, Neth, 58-59; lectr, 59-61, from asst prof to assoc prof, 61-70, PROF PHYSICS, UNIV MINN, MINNEAPOLIS, 70- Concurrent Pos: NSF sr fel, Finland, 67-68. Mem: Am Phys Soc; Am Asn Physics Teachers. Res: Low temperature physics; superfluid helium; liquid helium-3/helium-4 mixtures. Mailing Add: Sch of Physics & Astron Univ of Minn Minneapolis MN 55455

ZIMMERMANN, WILLIAM JOHN, b Mankato, Minn, May 7, 24; m 47; c 3. PARASITOLOGY. Educ: Mankato State Col, BS, 47; Iowa State Univ, MS, 52, PhD(parasitol), 55. Prof Exp: Asst zool, Iowa State Univ, 51-53, res assoc, Vet Med Res Inst, 54-55, from asst prof to assoc prof, 55-68, PROF PARASITOL PATH, VET MED RES INST, IOWA STATE UNIV, 68- Mem: Am Soc Parasitol; Conf Res Workers Animal Dis; Wildlife Dis Asn. Res: Epidemiology and epizootiology of trichiniasis and toxoplasmosis; internal parasites of domestic animals. Mailing Add: Vet Med Res Inst Iowa State Univ Ames IA 50011

ZIMMERSCHIED, WILFORD JOHN, b Pleasant Hill, Mo, Feb 22, 17; m 44; c 1. CHEMISTRY. Educ: Monmouth Col, BS, 41; Ohio State Univ, MS, 44, PhD(chem), 46. Prof Exp: Asst, Res Found, Ohio State Univ, 42-44; res chemist, Stand Oil Co, Ind, 46-60, RES CHEMIST, AMOCO CHEM CORP, 61- Concurrent Pos: Assoc fac mem, Ind Univ Northwest, 65-68. Mem: Am Chem Soc. Res: Petroleum processes; petrochemicals; aromatic acids. Mailing Add: Amoco Chem Corp Box 400 Naperville IL 60540

ZIMMT, WERNER SIEGFRIED, b Berlin, Ger, Sept 21, 21; nat US; m 47; c 3. POLYMER CHEMISTRY, POLLUTION CHEMISTRY. Educ: Univ Chicago, PhB & BS, 47, MS, 49, PhD(org chem), 51. Prof Exp: Res chemist, E I du Pont de Nemours & Co, Inc, 51-60, chem assoc, Marshall Lab, 60-62, res assoc, 62-69, RES FEL, MARSHALL LAB, E I DU PONT DE NEMOURS & CO, INC, 69- Honors & Awards: George B Heckel Award, Nat Paint & Coatings Asn, 73. Mem: Am Chem Soc. Res: Free radical chemistry; synthesis and mechanism; paint technology; role of paint solvents in air pollution. Mailing Add: E I du Pont de Nemours & Co Inc Marshall Lab Box 3886 Philadelphia PA 19146

ZIMNY, MARILYN LUCILE, b Chicago, Ill, Dec 12, 27. ANATOMY. Educ: Univ Ill, BA, 48; Loyola Univ, Ill, MS, 51, PhD(anat), 54. Prof Exp: Asst anat, Med Sch, Loyola Univ, Ill, 51-53; from asst prof to assoc prof, 54-64, PROF ANAT, LA STATE UNIV MED CTR, NEW ORLEANS, 64- Concurrent Pos: Vis prof, Sch Med, Univ Costa Rica, 61 & 62; mem, Inst Arctic Biol, Univ Alaska, 66. Mem: Am Asn Anat; Am Physiol Soc; Electron Micros Soc Am; Orthop Res Soc; Am Soc Zool. Res: Subcellular structure and cell physiology related to metabolic adjustments during hibernation and hypothermia; orthopedic research. Mailing Add: Dept of Anat La State Univ Med Ctr New Orleans LA 70112

ZIMPFER, PAUL (ELLSWORTH), b Columbus, Ohio, June 30, 10; m 37; c 2. BACTERIOLOGY. Educ: Ohio State Univ, PhD(plant physiol), 37. Prof Exp: PROF BIOL & CHMN DEPT, CAPITAL UNIV, 35- Mem: AAAS; Am Pub Health Asn. Res: Microbiology of foods and packaging materials; gnotobiotics. Mailing Add: Dept of Biol Capital Univ 2199 East Main St Columbus OH 43209

ZIMRING, LOIS JACOBS, b Chicago, Ill, Nov 19, 23; div; c 1. CHEMICAL PHYSICS. Educ: Univ Chicago, BS, 45, MS, 49, PhD(phys chem), 64. Prof Exp: Instr chem, Morgan Park Jr Col, 49-51; lectr phys sci, Univ Chicago, 59-61, instr, 61-64; asst prof chem, Univ Minn, 64-66; from asst prof to assoc prof natural sci, 66-72, PROF NATURAL SCI, MICH STATE UNIV, 72- Mem: Am Chem Soc. Res: Ultraviolet spectra of conjugated systems; crystal spectra of transition metal halides; solid state mixed alum systems. Mailing Add: Dept of Natural Sci Univ Col Mich State Univ East Lansing MI 48823

ZINDEL, HOWARD CARL, b Grand Rapids, Mich, Mar 20, 15; m 35; c 6. POULTRY HUSBANDRY. Educ: Mich State Univ, BS, 37, MS, 41, DEd(admin), 53. Prof Exp: Dist 4H Club agent, 41, exten specialist poultry sci, 46-53, PROF & CHMN DEPT POULTRY SCI, MICH STATE UNIV, 53- Mem: Poultry Sci Asn. Res: Poultry science, especially waste management, hygiene, management, housing and lighting. Mailing Add: Dept of Poultry Sci Mich State Univ East Lansing MI 48824

ZINDER, NORTON DAVID, b New York, NY, Nov 7, 28; m 49; c 2. MOLECULAR GENETICS. Educ: Columbia Univ, BA, 47; Univ Wis, MS, 49, PhD(med microbiol), 52. Prof Exp: Asst, 52-58, assoc prof, 58-64, PROF MICROBIAL GENETICS, ROCKEFELLER UNIV, 64- Honors & Awards: Eli Lilly Award Microbiol, 62; US Steel Award Molecular Biol, 66. Mem: Nat Acad Sci; AAAS; Genetics Soc Am; Am Soc Microbiol; Am Soc Biol Chem. Res: Virology; protein biosynthesis. Mailing Add: Rockefeller Univ New York NY 10021

ZINDLER, RICHARD EUGENE, b Benton Harbor, Mich, Mar 5, 27; m 58; c 2. MATHEMATICS, OPERATIONS RESEARCH. Educ: Mich State Univ, BS & MS, 49, PhD(math), 56. Prof Exp: Asst math, Mich State Univ, 49-52; asst prof, 52-57, assoc prof, 57-63, PROF ENG RES, DEPT INDUST ENG & ORD RES LAB, PA STATE UNIV, 63- Mem: AAAS; Am Math Soc; Acoust Soc Am; Opers Res Soc Am; Asn Comput Mach. Res: Weapon system analysis and synthesis; primate behavior. Mailing Add: Dept of Indust & Mgt Systs Eng Pa State Univ State Col PA 16801

ZINGARO, JOSEPH S, b Mt Morris, NY, Mar 5, 28; m 52; c 4. SCIENCE EDUCATION, CHEMISTRY. Educ: State Univ NY Col Geneseo, BS, 51; Syracuse Univ, MS, 55 & 62, PhD, 66. Prof Exp: Teacher & chmn dept sci, Vernon-Verona-Sherrill Cent Sch, 51-58; PROF CHEM, STATE UNIV NY COL BUFFALO, 58- Concurrent Pos: NSF inst grants, 66-75. Mem: AAAS; Am Chem Soc. Res: Science teaching, especially chemistry teaching; electrical conductance and thermodynamic functions as they relate to solutions. Mailing Add: Dept of Chem State Univ of NY Col Buffalo NY 14222

ZINGARO, RALPH ANTHONY, b Brooklyn, NY, Oct 27, 25; m 50; c 2. INORGANIC CHEMISTRY. Educ: City Col New York, BS, 46; Univ Kans, MS, 49, PhD, 50. Prof Exp: Sr res chemist, Eastman Kodak Co, 50-52; asst prof, Univ Ark, 52-53; res chemist, Am Cyanamid Co, 53-54; from asst prof to assoc prof, 54-64, PROF CHEM, TEX A&M UNIV, 64- Concurrent Pos: NIH spec fel, 68-69; Fulbright lectr, Univ Buenos Aires, 72. Mem: Am Chem Soc; NY Acad Sci. Res: Chemistry and biochemistry of selenium and arsenic. Mailing Add: Dept of Chem Tex A&M Univ College Station TX 77843

ZINGESER, MAURICE ROY, b Birmingham, Ala, Mar 17, 21; m 47; c 3. ANATOMY, ORTHODONTICS. Educ: New York Univ, AB, 42; Columbia Univ, DDS, 46; Tufts Univ, MS, 50; Am Bd Orthod, dipl, 63. Prof Exp: Intern surg, New York Polyclin Hosp, 46-48; clin assoc orthod, Tufts Univ, 48-50; dent surgeon, USPHS, 50-52; res assoc growth & develop, Dent Sch, Univ Ore, 54-55; VIS SCIENTIST ANTHROP, ORE REGIONAL PRIMATE RES CTR, 63- Concurrent Pos: Guest lectr, Tufts, Boston & Georgetown Univs, 60-69; contrib, Int Cong Anthrop & Ethnol Sci, 68; chmn craniofacial biol sect, Int Cong Primatol, 72; guest lectr, Hebrew Univ & Univ London, 72. Mem: Am Dent Asn; Am Asn Orthod; Int Asn Dent Res; Int Dent Fedn; Am Asn Phys Anthrop. Res: Primate odontology and craniology; especially form-functional correlations and their adaptive significance; elucidating functionally relevant aspects of craniometry for formulation of clinically useful guidelines. Mailing Add: Ore Regional Primate Res Ctr Beaverton OR 97005

ZINGESSER, LAWRENCE H, b Portchester, NY, Dec 27, 30; m; c 3. RADIOLOGY. Educ: Syracuse Univ, AB, 51; Chicago Med Sch, MD, 55. Prof Exp: From intern med to resident, Grad Hosp, Univ Pa, 55-57; resident radiol, Grace-New Haven Hosp, Yale Univ, 59-62; from asst prof to assoc prof, 65-73, PROF RADIOL, ALBERT EINSTEIN COL MED, 73-, ATTEND, 68- Concurrent Pos: NIH spec fel neuroradiol, Albert Einstein Col Med, 62-64, Nat Inst Neurol Dis & Stroke grant cerebral blood flow, 66-69; asst attend, Bronx Munic Hosp Ctr, NY, 63-65. Mem: Am Soc Neuroradiol; Asn Univ Radiol; Radiol Soc NAm; Am Col Radiol; Am Heart Asn. Res: Regional cerebral blood flow in neurologic disease states; neuroradiology. Mailing Add: Dept of Radiol Albert Einstein Col of Med Bronx NY 10461

ZINGG, WALTER, b Kloten, Switz, Mar 29, 24; Can citizen; m 49; c 4. BIOENGINEERING. Educ: Univ Zurich, MD, 50; Univ Man, MSc, 52; FRCS(C), 58. Prof Exp: Lectr physiol, Univ Man, 56-57, lectr surg, 57-61, asst prof, 61-64; from asst prof to assoc prof, 64-75, mem inst biomed eng, 72-75, PROF SURG & ASSOC DIR INST BIOMED ENG, UNIV TORONTO, 75-, SR SCIENTIST, HOSP SICK CHILDREN, 68- Concurrent Pos: Assoc scientist, Hosp Sick Children, 64-68; consult, Ont Crippled Children's Ctr, Toronto, 65- & Ont Vet Col, Univ Guelph, 70- Mem: Am Soc Artificial Internal Organs; Can Physiol Soc; Can Soc Clin Invest; fel Am Col Surg; fel Am Col Cardiol. Res: Surgical research; biomaterials; biomechanics; fetal surgery; artificial pancreas. Mailing Add: Hosp for Sick Children 555 University Ave Toronto ON Can

ZINGMARK, RICHARD G, b San Francisco, Calif, July 4, 41; m 62; c 4. ALGOLOGY. Educ: Humboldt State Col, BA, 64, MA, 65; Univ Calif, Santa Barbara, PhD(biol), 69. Prof Exp: NSF fel, Marine Lab, Duke Univ, 69-70; ASST PROF BIOL & MARINE SCI, UNIV SC, 70-, RES ASSOC MARINE SCI, BELLE W BARUCH COASTAL RES INST, 70- Concurrent Pos: Consult, Environ Res Ctr, 74- Mem: Phycol Soc Am; Int Phycol Soc; Brit Phycol Soc; Am Soc Limnol & Oceanog. Res: Sexual reproduction of dinoflagellates; effects of toxic materials on the growth and reproduction of phytoplankton; phytoplankton ultrastructure; phytoplankton life histories. Mailing Add: Dept of Biol Univ of SC Columbia SC 29208

ZINGULA, RICHARD PAUL, b Cedar Rapids, Iowa, May 31, 29; m 53; c 2. PALEONTOLOGY. Educ: Iowa State Univ, BS, 51; La State Univ, MS, 53, PhD(geol), 58. Prof Exp: Assoc geologist, Humble Oil & Refining Co, 54-60, supvry paleontologist, 60-69, sr prof geologist, 69-72; res paleontologist, Imp Oil Ltd, 72-74; SR PROF GEOLOGIST, EXXON CO USA, 74- Mem: Geol Soc Am; Paleont Soc Am; Soc Econ Paleontologists & Mineralogists; Am Asn Petrol Geologists. Res: Micropaleontology; stratigraphy. Mailing Add: Exxon Co USA PO Box 2180 Houston TX 77001

ZINK, FRANK W, b Pullman, Wash, June 17, 23; m 46; c 2. PLANT BREEDING. Educ: Univ Calif, Davis, BS, 47, MS, 48. Prof Exp: Asst specialist, 48-53, assoc specialist, 54-60, specialist, 61-69, RES SPECIALIST VEG CROPS, CALIF AGR EXP STA, UNIV CALIF, DAVIS, 69-, LECTR, COL AGR, 74- Concurrent Pos: Shell Develop grant, 60-61; Veg Growers Asn grant, 60-67; Melon Growers Asn grant, 72-75. Mem: Am Soc Hort Sci; Am Phytopath Soc; Am Soc Agron; Int Soc Hort Sci. Res: Lettuce breeding, disease resistance; melon breeding for mechanization and disease resistance. Mailing Add: Dept of Veg Crops Univ of Calif Davis CA 95616

ZINK, ROBERT EDWIN, b Minneapolis, Minn, Nov 16, 28; m 50; c 3. MATHEMATICS. Educ: Univ Minn, BA, 49, MA, 51, PhD(math), 53. Prof Exp: Asst math, Univ Minn, 49-53; instr, Purdue Univ, 53-54; lectr, George Washington Univ, 55-56; from asst prof to assoc prof math, 56-66, asst head dept, 65-69, asst dean grad sch, 69-72, PROF MATH, PURDUE UNIV, LAFAYETTE, 66- Concurrent Pos: Vis prof, Wabash Col, 61-62 & dept math, Univ Calif, Irvine, 68-69. Mem: Am Math Soc; Math Asn Am. Res: Theory of measure and integration; theory of functions of a real variable. Mailing Add: Div of Math Sci Purdue Univ Lafayette IN 47907

ZINKE, OTTO HENRY, b Webster Groves, Mo, Aug 13, 26; m 55; c 3. PHYSICS. Educ: Wash Univ, AB, 50, AM, 53, PhD, 56. Prof Exp: Salesman, Nuclear Consults Corp, 52-53; mem res staff, Linde Co Union Carbide Corp, 56-57; asst prof physics, Univ Mo, 57-59; from asst prof to assoc prof, 59-69, PROF PHYSICS, UNIV ARK, FAYETTEVILLE, 69- Res: Transient phenomena in plasmas and metals. Mailing Add: Dept of Physics Univ of Ark Fayetteville AR 72701

ZINKE, PAUL JOSEPH, b Los Angeles, Calif, Nov 10, 20; m 47; c 2. FORESTRY, SOIL SCIENCE. Educ: Univ Calif, BS, 42, MS, 52, PhD(soil sci), 56. Prof Exp: Forester, Tongass Nat Forest, US Forest Serv, 42-43, res forester, Calif Forest & Range Exp Sta, 46-56; ASSOC PROF FORESTRY & SOIL SCI, UNIV CALIF, BERKELEY, 57- Concurrent Pos: In charge, Calif Soil Veg Surv, USDA-US Forest Serv, 59-61; adv, Appl Sci Res Corp Thailand, 67-; adv, Radar Nat Resource Inventory Amazon Basin, Brazil, 71-; mem comt study defoliation effects in SEAsia, Nat Acad Sci, 71- Mem: AAAS; Soc Am Foresters; Soil Sci Soc Am; Soil Conserv Soc Am; Soc Range Mgt. Res: Forest influences and environment; forest soils; soil morphology; soil-vegetation relationships; plant ecology. Mailing Add: Sch of Forestry & Conserv 145 Mulford Hall Univ of Calif Berkeley CA 94720

ZINKEL, DUANE FORST, b Manitowoc, Wis, Aug 11, 34; m 61; c 5. ORGANIC CHEMISTRY. Educ: Univ Wis, BS, 56, PhD(biochem), 61. Prof Exp: RES CHEMIST, FOREST PROD LAB, US FOREST SERV, 61- Mem: Am Chem Soc; Sigma Xi. Res: Softwood extractives and derived products; structure determination; analytical development and analysis; biosynthesis. Mailing Add: Forest Prod Lab US Forest Serv PO Box 5130 Madison WI 53705

ZINKHAM, WILLIAM HOWARD, b Uniontown, Md, May 23, 24; m 52; c 2. PEDIATRICS. Educ: Johns Hopkins Univ, AB, 44, MD, 47. Prof Exp: From instr to asst prof, 56-61, ASSOC PROF PEDIAT, SCH MED, JOHNS HOPKINS UNIV, 61- Mem: Soc Pediat Res. Res: Hematology; metabolism of normal and abnormal erythrocytes. Mailing Add: Dept of Pediat Johns Hopkins Univ Sch of Med Baltimore MD 21218

ZINMAN, WALTER GEORGE, b New York, NY, Nov 9, 29; m 55; c 2. CHEMISTRY. Educ: Rensselaer Polytech Inst, BChE, 51; Harvard Univ, PhD(chem), 55. Prof Exp: Res chemist missile & space vehicle div, Gen Elec Co, 56-59; prin res & develop engr, Repub Aviation Corp, 59-65; res group leader, Polytech Inst Brooklyn, 65-66; propulsion engr, Grumman Aircraft Eng Corp, 66-70, consult, 70-72; asst to dir res, Surface Activation Corp, NY, 72-73; CHEM CONSULT, 73- Mem: Am Chem Soc; Am Phys Soc. Res: Reaction of dissociated gases with solids; gaseous detonations; homogeneous kinetics; fluid mechanics and reacting flows; foundations thermodynamics. Mailing Add: 8 Coventry Rd Syosset NY 11791

ZINN, DALE WENDEL, b Parkersburg, WVa, m 54; c 2. ANIMAL HUSBANDRY. Educ: WVa Univ, MS, 56. Prof Exp: Asst prof animal husb, NMex State Univ, 57-61; assoc prof, 61-71, PROF ANIMAL SCI, TEX TECH UNIV, 71-, CHMN DEPT, 69- Mem: Am Soc Animal Sci; Inst Food Technol. Res: Production and quality factors affecting quantity and quality of meat and meat products. Mailing Add: Dept of Animal Sci Tex Tech Univ Lubbock TX 79409

ZINN, DONALD JOSEPH, b New York, NY, Apr 19, 11; m 41; c 2. INVERTEBRATE ZOOLOGY. Educ: Harvard Univ, SB, 33; Univ RI, MS, 37; Yale Univ, PhD(zool), 42. Prof Exp: Dir, Bass Biol Lab, Fla, 33-35; tech asst, Ro-Lab, Conn, 38-39; asst, Osborn Zool Lab, Yale Univ, 40-41; naturalist, Marine Biol Lab, Woods Hole Oceanog Inst, 45-46; from instr to prof zool, 46-74, actg chmn dept, 60-62, chmn dept, 62-65 & 73-74, EMER PROF ZOOL, UNIV RI, 74- Concurrent Pos: Chief mosquito control proj, Pine Orchard Asn, Conn, 40; dir, Agassiz Mem Exped, Penikese Island, 47; res assoc, Narragansett Marine Lab, 55-74; deleg, Int Cong Zool, London, 58; co-ed, Psammonalia, 66-; mem, President's Adv Panel Timber & The Environ, 71-74; mem aquatic ecol sect, Int Biol Prog; ecol consult, US Plywood-Champion Papers; mem corp, Bermuda Biol Sta & Marine Biol Lab, Woods Hole Oceanog Inst; pres, Nat Wildlife Fedn, 68-71; trustee, New Eng Natural Resources Ctr, 72; mem shore erosion adv panel, US Army Corps Engrs, 74- Mem: Fel AAAS; Soc Syst Zool; Ecol Soc Am; Am Fisheries Soc; Am Soc Limnol & Oceanog. Res: Ecology and taxonomy of marine beaches intertidal interstitial fauna and flora; littoral benthos and fouling organisms; tunicate and entomostracan taxonomy; histological techniques with micrometazoa; conservation education; conservation of natural resources. Mailing Add: PO Box 589 Falmouth MA 02541

ZINN, GARY WILLIAM, b Oxford, WVa, Sept 20, 44; m 67; c 1. FOREST ECONOMICS. Educ: WVa Univ, BS, 66; State Univ NY Col Forestry, Syracuse Univ, MS, 68, PhD(forest econ), 72. Prof Exp: Instr forestry econ, State Univ NY Col Forestry, Syracuse Univ, 71; ASST PROF FOREST MGT, W VA UNIV, 72- Mem: Soc Am Foresters; Am Econ Asn. Res: Economic contributions of forest-based activity to regions; regional development; forest and natural resource policy; forest land use and management planning. Mailing Add: Div of Forestry WVa Univ Morgantown WV 26506

ZINN, JOHN, b Brooklyn, NY, Feb 28, 28; m 54; c 3. APPLIED PHYSICS. Educ: Cornell Univ, AB, 49; Univ Calif, Berkeley, PhD(phys chem), 58. Prof Exp: Chemist, Catalin Corp Am, 49-50 & M W Kellogg Co, 52-54; assoc chem, Univ Calif, 55-56, MEM STAFF, LOS ALAMOS SCI LAB, UNIV CALIF, 57- Concurrent Pos: Asst prof dept aeronaut eng, Univ Colo, 67-68. Mem: Am Phys Soc. Res: Theoretical research in atmospheric physics and chemistry; atmospheric effects of nuclear explosions. Mailing Add: Los Alamos Sci Lab PO Box 1663 Los Alamos NM 87545

ZINN, ROBERT JAMES, b Chicago, Ill, Aug 4, 46. ASTRONOMY. Educ: Case Inst Technol, BS, 68; Yale Univ, PhD(astron), 74. Prof Exp: FEL ASTRON, HALE OBSERVS, CARNEGIE INST WASHINGTON, 74- Mem: Am Astron Soc. Res: Stellar evolution; the chemical compositions of globular cluster stars, variable stars, and the stellar populations of the galaxies of the Local Group. Mailing Add: Hale Observs 813 Santa Barbara St Pasadena CA 91101

ZINN, WALTER HENRY, b Kitchener, Ont, Dec 10, 06; nat US; m 33, 66; c 2. PHYSICS. Educ: Queen's Univ, Ont, BA, 27, MA, 29; Columbia Univ, PhD(physics), 34. Hon Degrees: DSc, Queen's Univ, Ont, 57. Prof Exp: Asst physics, Queen's Univ, Ont, 27-28; asst, Columbia Univ, 31-32; from instr to asst prof, City Col New York, 32-41; physicist, Metall Lab, Manhattan Dist, Univ Chicago, 42-46; dir, Argonne Nat Lab, 45-56; vpres, Combustion Eng, Inc, 59-71; RETIRED. Concurrent Pos: Spec consult, Joint Cong Comt Atomic Energy, 56; spec mem, President's Sci Adv Comt; pres, Gen Nuclear Eng Corp, 56-64. Honors & Awards: Spec commendation, US AEC, 56, Enrico Fermi Award, 69; Cert of Merit, Am Power Conf, 57; Atoms for Peace Award, 60. Mem: Nat Acad Sci; AAAS; fel Am Phys Soc; Am Nuclear Soc (pres, 55). Res: Nuclear physics and reactor development. Mailing Add: 1155 Ford Lane Dunedin FL 33528

ZINNEMAN, HORACE HELMUT, b Frankfurt, Ger, Oct 7, 10; m 41; c 1. INTERNAL MEDICINE. Educ: Univ Vienna, MD, 37. Prof Exp: Assoc prof, 59-68, PROF INTERNAL MED, MED SCH, UNIV MINN, MINNEAPOLIS, 68- Concurrent Pos: Chief sect immunol & microbiol, Minneapolis Vet Admin Hosp; consult, Hennepin County Gen Hosp & St Paul Ramsey Gen Hosp. Mem: AMA; Am Asn Immunol; Soc Exp Biol & Med; Am Col Physicians. Res: Immunology; protein chemistry. Mailing Add: Dept of Internal Med US Vet Admin Hosp Minneapolis MN 55417

ZINNER, DORAN DAVID, b Akron, Ohio, Dec 24, 18; m 43; c 3. DENTISTRY, MICROBIOLOGY. Educ: Ohio State Univ, BA, 40, DDS, 44. Prof Exp: Asst prof microbiol, 57-70, PROF ORAL BIOL & DIR DIV, SCH MED, UNIV MIAMI, 67- Concurrent Pos: Chief dent staff, United Cerebral Palsy Clin, 48-59; dir dent res, Nat Children's Cardiac Hosp, Miami, 52-66; res assoc, Dept Med Res, Vet Admin Hosp, Coral Gables, 58-64, consult, Dept Dent, 60-; fel dent res, Univ Miami, 59-60; res assoc, Variety Children's Res Found, Miami, 61-70; res consult, Sch Dent, Univ PR, 66; consult, Sch Dent, Univ Ill, 67; dent adv & consult, Pan-Am Health Orgn, 67. Mem: AAAS; Am Soc Microbiol; Am Dent Asn; Am Fedn Clin Res; fel Am Pub Health Asn. Res: Oral microbial infections; oral pathology; peridontia; etiology and epidemiology of dental caries. Mailing Add: Div Oral Biol Univ of Miami PO Box 520875 Biscayne Annex Miami FL 33152

ZINNES, HAROLD, b New York, NY, Apr 7, 29; m 53; c 3. ORGANIC CHEMISTRY. Educ: Rutgers Univ, BS, 51, MS, 52; Univ Mich, PhD, 55. Prof Exp: Res assoc, E R Squibb & Co, 56-58; scientist, 58-63, sr scientist, 63-68, SR RES ASSOC, WARNER- LAMBERT RES INST, 68- Mem: Am Chem Soc; Am Pharmaceut Asn. Res: Medicinals; natural products; antibiotics; heterocycles; indoles; benzothiazines. Mailing Add: Warner-Lambert Res Inst 170 Tabor Rd Morris Plains NJ 07950

ZINNES, IRVING ISADORE, b Can, Mar 21, 16; nat US; m 43; c 2. THEORETICAL PHYSICS, MOLECUALR PHYSICS. Educ: City Col New York, BS, 40; NY Univ, PhD(physics), 52. Prof Exp: Physicist, Plastic Res, Grosvenor Labs, 40-41 & Degaussing Sect, US Navy, 42-43; asst magnetron, Columbia Univ, 43-45; proj engr, DeMornay-Budd, 46-47; instr, Newark Col Eng, 52-53; from asst prof to assoc prof physics, Univ Okla, 53-62; ASSOC PROF PHYSICS, FORDHAM UNIV, 62- Concurrent Pos: Res assoc, Univ Iowa, 58; Am Philos Soc fel, Princeton Univ, 60-62; vis prof, Inst Theoret Physics, Univ Geneva, 68-70. Mem: Am Phys Soc. Res: Angular correlation of gamma rays; quantum field theory; general scattering theory; molecular structure. Mailing Add: Dept of Physics Fordham Univ Bronx NY 10458

ZINS, GERALD RAYMOND, b New Ulm, Minn, Oct 23, 32; m 56; c 3. PHARMACOLOGY. Educ: SDak State Univ, BS, 54; Univ Chicago, PhD(pharmacol), 58. Prof Exp: Res assoc & instr, Univ Chicago, 58-59; res scientist, Upjohn Co, 59-66; vis assoc prof, State Univ NY Upstate Med Ctr, 66-67; sr scientist, 67-68, HEAD CARDIOVASC RES, UPJOHN CO, 68- Mem: AAAS; Am Soc Pharmacol & Exp Therapeut; Am Soc Nephrol. Res: Renal and cardiovascular pharmacology; renal prostaglandins; drug metabolism; hypertension; intermediary carbohydrate metabolism. Mailing Add: Upjohn Co 301 Henrietta St Kalamazoo MI 49001

ZINSER, EDWARD JOHN, b Toronto, Ont, Mar 13, 41; m 66; c 2. ANALYTICAL CHEMISTRY. Educ: Univ Toronto, BSc, 65, PhD(anal chem), 69. Prof Exp: Asst,

Univ Toronto, 65-68; fel, Queen's Univ, Ont, 69-70; res chemist, Marshall Lab, Pa, 70-71, res chemist, Exp Sta, Wilmington, Del, 72, sr prod specialist, Wynnewood, Pa, 72-73, tech objectives mgr, Finishes Div, Wilmington, Del, 73-74, RES SUPVR, MARSHALL RES & DEVLOP LAB, E I DU PONT DE NEMOURS & CO, PHILADELPHIA, PA, 74- Concurrent Pos: Nat Res Coun Can scholar, Univ Toronto, 68-70. Mem: Am Chem Soc; Chem Inst Can; Fedn Soc Coatings Technol. Res: Environmental analytical chemistry; electrochemistry; atomic absorption; spectroscopy; pigment dispersion; organic coatings; fluorocarbon coatings. Mailing Add: Marshall Res & Develop Lab PO Box 3886 E I Du Pont de Nemours & Co Inc Philadelphia PA 19146

ZINSSER, HANS HANDFORTH, surgery, deceased

ZINSSER, HARRY FREDERICK, b Pittsburgh, Pa, May 1, 18; m 43; c 4. CARDIOLOGY. Educ: Univ Pittsburgh, BS, 37, MD, 39. Prof Exp: Teaching fel internal med, Sch Med, Univ Pittsburgh, 40-42, asst instr, 46-47; fel, 47-48, from asst instr to assoc, 48-51, asst prof clin med, 51-53, asst prof med, 53-55, assoc prof clin med, 55-58, assoc prof med, 58-68, PROF MED, SCH MED, UNIV PA, 69-, PROF CARDIOL, DIV GRAD MED, 63-, DIR CARDIOL, GRAD HOSP, 63-, CHMN DEPT MED, 71- Mem: Am Fedn Clin Res; Am Heart Asn; Am Soc Clin Invest; Am Clin & Climat Asn; Asn Univ Cardiol. Res: Cardiovascular diseases. Mailing Add: Univ of Pa Grad Hosp 19th & Lombard Sts Philadelphia PA 19146

ZINTEL, HAROLD ALBERT, b Akron, Ohio, Dec 27, 12; m 38; c 5. SURGERY. Educ: Univ Akron, BS, 34; Univ Pa, MD, 38, DSc(med), 46. Hon Degrees: DSc, Univ Akron, 56. Prof Exp: Asst biol, Univ Akron, 33-34; asst path, Med Sch, Univ Pa, 35-36, asst chief med officer, Hosp, 39-40, from asst instr surg to instr, Med Sch, 40-47, asst prof, Med Sch & Grad Sch Med, 47, lectr, Grad Sch Nursing, 47-48, asst prof, Grad Sch Med, 50-54, assoc prof, Med Sch, 51-52, prof clin surg, 52-54; clin prof, Col Physicians & Surgeons, Columbia Univ, 54-69; ASST DIR, AM COL SURGEONS, 69-; PROF SURG, MED SCH, NORTHWESTERN UNIV, CHICAGO, 70- Concurrent Pos: Consult, Camden Munic Hosp, 48-54, Children's Hosp of Philadelphia, 51-54 & Vet Admin Hosp, Philadelphia, 53-54; attend surgeon & dir surg, St Luke's Hosp Ctr, 54-69; consult, Off Surgeon Gen, US Army, 58-67. Mem: AAAS; Soc Vascular Surg; Soc Univ Surgeons (pres, 55-56); Am Geriat Soc; Am Cancer Soc. Res: Antibiotics; wound healing; peripheral vascular hypertension; antiseptics; carcinoma of head of pancreas; nutrition; portal hypertension; trauma; computers in medicine; allied health manpower; continuing medical education. Mailing Add: Col of Surgeons 55 E Erie St Chicago IL 60611

ZIOCK, KLAUS OTTO H, b Herchen, Ger, Feb 4, 25; nat US; m 52; c 4. EXPERIMENTAL PHYSICS. Educ: Univ Bonn, Dipl, 49, Dr rer nat, 56. Prof Exp: Physicist, E Leybold's Nachfolger, Ger, 50-55; res assoc, Univ Bonn, 56-58; res assoc physics, Yale Univ, 58-60, from asst prof to assoc prof, 60-72, actg dir, Va Assoc Res Ctr, 62-64, PROF PHYSICS, UNIV VA, 72- Concurrent Pos: Vis scientist, Europ Orgn Nuclear Res, 69-70. Mem: Am Phys Soc. Res: Nuclear physics; physics of elementary particles; atomic physics. Mailing Add: Dept of Physics Univ of Va Charlottesville VA 22901

ZIOLO, RONALD F, b Philadelphia, Pa, Aug 16, 44; m 67; c 2. INORGANIC CHEMISTRY, PHYSICAL CHEMISTRY. Educ: Univ Calif, Los Angeles, BS, 66; Temple Univ, PhD(chem), 70. Prof Exp: Res fel chem, Calif Inst Technol, 71-72; SCIENTIST CHEM, XEROX CORP, 72- Mem: AAAS; Am Chem Soc; Am Crystallog Asn. Res: Transition metal chemistry; organo-tellurium and selenium compounds; x-ray crystallography; chemistry of high surface area materials; magnetic and optical properties of materials. Mailing Add: Webster Res Ctr Bldg 114 Xerox Corp Webster NY 14580

ZIONY, JOSEPH ISRAEL, b Los Angeles, Calif, Apr 6, 35; m 61; c 3. GEOLOGY. Educ: Univ Calif, Los Angeles, AB, 56, MA, 59, PhD(geol), 66. Prof Exp: Geologist, Mil Geol Br, DC, 57-59, Fuels Br, Calif, 59-60, Southwestern Br, 65-69 & Eng Geol Br, 69-73, DEP GEOL, OFF EARTHQUAKE STUDIES, US GEOL SURV, RESTON, VA, 73- Mem: Geol Soc Am; Asn Eng Geol. Res: Earthquake hazards assessment; structural and engineering geology; geology in land-use planning; regional and economic geology of the Great Basin; recency of faulting in coastal California. Mailing Add: US Geol Surv Mail Stop 905 Nat Ctr Reston VA 22092

ZIPF, THEODORE FRANCIS, physics, see 12th edition

ZIPFEL, CHRISTIE LEWIS, b Detroit, Mich, Oct 2, 41; m 64. LOW TEMPERATURE PHYSICS. Educ: Vassar Col, AB, 63; Univ Mich, MS, 65, PhD(physics), 69. Prof Exp: Instr physics, State Univ NY Stony Brook, 69-72; asst prof, Towson State Col, 72-74; RES ASSOC PHYSICS, BELL LABS, MURRAY HILL, 74- Mem: Am Phys Soc. Res: Electrons on the surface of liquid helium. Mailing Add: 31 Ridgedale Ave Madison NJ 07940

ZIPFEL, GEORGE G, JR, b Richmond, Va, Dec 23, 38; m 64. ACOUSTICS. Educ: Mass Inst Technol, BS, 60, BEE, 61; Univ Fla, MSE, 62; Univ Mich, PhD(physics), 68. Prof Exp: Res assoc, Inst Theoret Physics, State Univ NY Stony Brook, 68-71; Nat Res Coun res assoc, US Naval Res Lab, 71-73; MEM TECH STAFF, BELL LABS, 73- Mem: AAAS; Acoust Soc Am; Am Phys Soc. Res: Statistical field theory; scattering theory; particle physics; applied classical field theory; statistical field theory. Mailing Add: Bell Labs Whippany Rd Whippany NJ 07981

ZIPKIN, ISADORE, biochemistry, deceased

ZIPORIN, ZIGMUND ZANGWILL, b New York, NY, Dec 30, 15; m 38; c 2. BIOCHEMISTRY. Educ: NY Univ, BA, 35; Georgetown Univ, MS, 51, PhD(biochem), 53. Prof Exp: Biochemist, US Food & Drug Admin, 48-54; biochemist, Med Res & Nutrit Lab, Fitzsimons Gen Hosp, US Dept Army, 54-75; RETIRED. Concurrent Pos: Secy Army res fel, 59-60; fac affiliate, Colo State Univ, 66; vis lectr, Univ Colo, 67. Mem: AAAS; Am Chem Soc; Am Inst Nutrit; NY Acad Sci; Brit Biochem Soc. Res: Bone, vitamin and drug metabolism; nutrition. Mailing Add: 12735 E 13th Ave Aurora CO 80010

ZIPP, ADAM PETER, b Baltimore, Md, Nov 19, 47; m 74. BIOPHYSICAL CHEMISTRY, BIOPHYSICS. Educ: Loyola Col, BS, 69; Princeton Univ, MA, 71, PhD(chem), 73. Prof Exp: NIH FEL BIOPHYS, DEPT PHARMACEUT CHEM, UNIV CALIF MED CTR, SAN FRANCISCO, 74- Mem: Sigma Xi; AAAS; Biophys Soc. Res: Physical chemistry of proteins; application of physical chemical techniques to the study of biomedical problems; physical chemical studies of normal and diseases erythrocytes. Mailing Add: Dept of Pharmaceut Chem Univ of Calif Med Ctr San Francisco CA 94143

ZIPP, ARDEN PETER, b Dolgeville, NY, July 14, 38; m 62; c 2. PHYSICAL INORGANIC CHEMISTRY. Educ: Colgate Univ, AB, 60; Univ Pa, PhD, 64. Prof Exp: Asst prof chem, Drew Univ, 64-66; from asst prof to assoc prof, 66-73, PROF CHEM, STATE UNIV NY COL CORTLAND, 73- Mem: Am Chem Soc. Res: Non-aqueous solutions; inorganic chemistry of sulfur; transition metal complexes with sulfur containing ligands; oxidation reduction reactions of transition metal ions; bio-inorganic chemistry. Mailing Add: Dept of Chem State Univ of NY Col Cortland NY 13045

ZIPPIN, CALVIN, b Albany, NY, July 17, 26; m 64; c 2. BIOSTATISTICS, EPIDEMIOLOGY. Educ: State Univ NY, AB, 47; Johns Hopkins Univ, ScD(biostatist), 53. Prof Exp: Res asst statist, Sterling-Winthrop Res Inst, 47-50; res asst biostatist, Johns Hopkins Univ, 50-53; instr, Sch Pub Health, Univ, 53-55, from asst res biostatistician to res biostatistician, Cancer Res Inst, 55-67, asst prof prev med, 58-60, lectr, 60-67, lectr path, 61-67, PROF EPIDEMIOL, CANCER RES INST, DEPT INT HEALTH & DEPT PATH, SCH MED, UNIV CALIF, SAN FRANCISCO, 67- Concurrent Pos: Consult, US Naval Biol Lab, 55-66 & Letterman Gen Hosp, 58-; vis assoc prof, Stanford Univ, 62; NIH spec fel, London Sch Hyg & Trop Med, 64-65; temporary adv, WHO, 69, 72 & 74; Eleanor Roosevelt Int Cancer fel, Univ London, 75. Mem: Am Pub Health Asn; Am Statist Asn; Asn Teachers Prev Med; Biomet Soc; fel Royal Statist Soc. Res: Teaching of biostatistics; biometry and epidemiology in cancer research. Mailing Add: Cancer Res Inst Univ of Calif Med Ctr San Francisco CA 94143

ZIPSER, DAVID, b New York, NY, May 31, 37; m 65; c 2. MOLECULAR GENETICS. Educ: Cornell Univ, BS, 58; Harvard Univ, PhD(biochem), 63. Prof Exp: Prof biol, Columbia Univ, 65-69; INVESTR GENETICS, COLD SPRING HARBOR LAB, 70- Res: Lactose operon function in bacteria. Mailing Add: Cold Spring Harbor Lab Box 100 Cold Spring Harbor NY 11724

ZIPSER, EDWARD JERRY, meteorology, see 12th edition

ZIPURSKY, ALVIN, b Winnipeg, Man, Sept 27, 30; m 53; c 3. PEDIATRICS, PHYSIOLOGY. Educ: Univ Man, MD, 53; Royal Col Physicians Can, cert, 57, FRCP(C), 60. Prof Exp: Res fel biochem, Univ Man, 56-61; chmn dept, 67-72, PROF PEDIAT, McMASTER UNIV, 67- Concurrent Pos: Attend physician, Winnipeg Gen & Children's Hosps, 57-67; from lectr pediat to assoc prof, Univ Man, 58-67; med dir blood transfusion serv, Man Div, Can Red Cross Soc, 60-66; dir clin invest & res unit, Children's Hosp, 65-66; affil, St Joseph's Hosp & Chedoke Hosps, 67-; mem staff, Hamilton Civic Hosp & McMaster Univ Med Ctr, 69- Mem: Soc Pediat Res; Can Soc Clin Invest; Am Soc Hemat; Can Soc Hemat; Am Pediat Soc. Res: Physiology of normal and leukemic leukocytes; hematology of the newborn; pathogenesis and prevention of isoimmunization. Mailing Add: Dept of Pediat McMaster Univ Hamilton ON Can

ZIRAKZADEH, ABOULGHASSEM, b Isfahan, Iran, Feb 7, 22; m 51; c 1. MATHEMATICS. Educ: Univ Teheran, BS, 44; Univ Mich, MS, 49; Okla State Univ, PhD(math), 53. Prof Exp: Instr math, Okla State Univ, 52-53, Univ Colo, 53-54 & Wash State Univ, 54-55; asst prof, Univ Teheran, 55-56; asst prof, 57-64, ASSOC PROF MATH, UNIV COLO, BOULDER, 64- Mem: Am Math Soc; Math Asn Am. Res: Geometry; convexity. Mailing Add: Dept of Math Univ of Colo Boulder CO 80302

ZIRIN, HAROLD, b Boston, Mass, Oct 7, 29; m 57; c 2. ASTRONOMY. Educ: Harvard Univ, AB, 50, MA, 51, PhD(astrophys), 53. Prof Exp: Physicist, Rand Corp, 52-53; instr astron, Harvard Univ, 53-55; mem sr res staff, High Altitude Observ, Univ Colo, 55-64; PROF ASTROPHYS, CALIF INST TECHNOL, 64-; STAFF MEM, HALE OBSERV, 64- Concurrent Pos: Sloane fel, 58-60; Guggenheim fel, 61; dir, Big Bear Solar Observ, 69- Mem: Am Astron Soc; Am Geophys Union. Res: Solar physics; stellar spectroscopy; interstellar matter; geophysics. Mailing Add: Dept of Astrophys Calif Inst of Technol Pasadena CA 91125

ZIRINO, ALBERTO, b Berlin, Ger, Jan 7, 41; m 64; c 3. CHEMICAL OCEANOGRAPHY, POLAROGRAPHY. Educ: Univ Calif, Los Angeles, BS, 63; San Diego State Col, MS, 66; Univ Wash, PhD(oceanog), 70. Prof Exp: Res assoc oceanog, Univ Wash, 67-70; OCEANOGRAPHER, NAVAL UNDERSEA CTR, 70- Concurrent Pos: Nat Res Coun assoc, 70; instr, Grossmont Col, 70-75. Res: Voltammetric detection of trace metals in seawater; speciation of zinc, cadmium, lead, copper seawater; electrochemistry. Mailing Add: Chem & Marine Environ Group Naval Undersea Ctr San Diego CA 92132

ZIRKER, JACK BERNARD, b New York, NY, July 19, 27; m 51; c 3. SOLAR PHYSICS. Educ: City Col New York, BME, 49; Harvard Univ, PhD(astron), 56. Prof Exp: Mech eng labs, Radio Corp Am, 49-53, astrophysicist, Sacramento Peak Observ, 56-64; ASTROPHYSICIST & PROF PHYSICS, UNIV HAWAII, 64- Concurrent Pos: Consult, NASA, 68-; mem, Astron Adv Panel, NSF, 73- Mem: Am Astron Soc. Res: Physics of the outer atmosphere of the sun. Mailing Add: Dept of Physics & Astron Univ of Hawaii Honolulu HI 96822

ZIRKES, AL, b New York, NY, Dec 10, 35; m 57; c 3. RADIOLOGICAL HEALTH. Educ: City Col New York, BS, 56; Univ Calif, cert, 64. Prof Exp: Phys chemist, Res Div, Curtiss-Wright, 56-57; assoc scientist, Bettis Field, Westinghouse Elec Corp, 57-59; health physicist & indust hygienist, Marquardt Corp, 59-61; safety engr, Wyle Labs, 61-62; proj mgr nuclear & aerospace safety, Tracerlab Div, Lab Electronics, Calif, 62-71; MGR ICN FILM BADGE SERV, NUCLEAR ACCESSORIES & CHROMATOGRAPHY PROD, ICN LIFE SCI GROUP, ICN PHARMACEUT INC, 71- Mem: Am Chem Soc; Am Nuclear Soc; Health Physics Soc; Am Soc Safety Eng; Am Indust Hyg Asn. Res: Health physics; safety engineering; nuclear project management. Mailing Add: ICN Life Sci Group 26201 Miles Rd Cleveland OH 44128

ZIRKIND, RALPH, b New York, NY, Oct 20, 18; m 40; c 3. PHYSICS. Educ: City Col New York, BS, 40; Ill Inst Technol, MS, 46; Univ Md, College Park, PhD(physics), 59. Hon Degrees: DSc, Univ RI, 68. Prof Exp: Tech asst metall, Naval Inspector Ord, US Dept Navy, 41-42, physicist, 42-45, physicist, Bur Aeronaut, 45-52, chief physicist, 52-60; physicist, Advan Res Projs Agency, US Dept Defense, 60-64; prof, Aerospace Eng, Polytech Inst Brooklyn, 64-70; prof elec eng, Univ RI, 70-72; physicist, Advan Projs Res Agency, US Dept Defense, 72-74; PRIN SCIENTIST, GEN RES CORP, McLEAN, VA, 74- Concurrent Pos: Consult, Jet Propulsion Lab, 64-; prof, Univ RI, 73- Honors & Awards: Dept of Defence & Navy Meritorious Civilian Serv Awards. Mem: Am Phys Soc. Res: Optical and radiation physics; atmospheric sciences; optical physics; lasers, atmospheric physics. Mailing Add: 820 Hillsboro Dr Silver Spring MD 20902

ZIRKLE, CHARLES LEON, organic chemistry, see 12th edition

ZIRKLE, RAYMOND ELLIOTT, b Springfield, Ill, Jan 9, 02; m 24; c 2. BIOPHYSICS. Educ: Univ Mo, AB, 28, PhD(bot), 32. Prof Exp: Asst bot, Univ Mo, 28-30, instr, 30-32; Nat Res Coun fel, Johnson Found, Univ Pa, 32-34, Johnson Pound fel med physics, 34-38, lectr biophys, 36-38, instr exp radiol, 37-38, assoc, 38-40; asst prof biol, Bryn Mawr Col, 38-40; prof bot, Ind Univ, 40-44; prof, 44-48, prof radiobiol, 48-59, prof biophys, 59-67, dir inst radiobiol & biophys, 45-48, chmn comt biophys, 54-64, chmn dept biophys, 64-66, EMER PROF BIOPHYS, UNIV CHICAGO, 67- Concurrent Pos: Prin biologist, Manhattan Dist, Metall Lab, Chicago & Clinton Labs, Tenn, 42-46; Hitchcock prof, Univ Calif, 51. Mem: Nat Acad Sci; AAAS; Am Philos Soc; Biophys Soc; Radiation Res Soc (pres, 52-53). Res: Mechanisms of mitosis; irradiation of small fractions of individual cells; comparative biological effects of ultraviolet and various ionizing radiations. Mailing Add: 4675 W Red Rock Dr Perry Park SWDC Larkspur CO 80118

ZISCHKE, DELORIS PALMQUIST, parasitology, invertebrate zoology, see 12th edition

ZISCHKE, JAMES ALBERT, b Sioux Falls, SDak, Sept 18, 34; m 61; c 1. PARASITOLOGY, INVERTEBRATE ZOOLOGY. Educ: Univ Wis, BS, 57; Univ SDak, MA, 60; Tulane Univ, PhD(parasitol), 66. Prof Exp: Instr, 63-65, asst prof, 65-69, ASSOC PROF BIOL, ST OLAF COL, 69- Concurrent Pos: Partic, AEC Res Prog, PR Nuclear Ctr, 67; Duke Univ res fel, Cent Univ Venezuela, 67-68; NSF fac fel, Univ Miami, 71-72. Mem: Am Soc Zool; Am Soc Parasitol. Res: Trematode taxonomy, life histories and reproduction; schistosome dermatitis; molluscan physiology; marine littoral ecology. Mailing Add: Dept of Biol St Olaf Col Northfield MN 55057

ZISKIN, MARVIN CARL, b Philadelphia, Pa, Oct 1, 36; m 60; c 3. BIOMEDICAL ENGINEERING. Educ: Temple Univ, AB, 58, MD, 62; Drexel Inst, MSBmE, 65. Prof Exp: Intern, West Jersey Hosp, Camden, 62-63; NIH fel, Drexel Inst, 63-65; NASA fel theoret biophys, 65; instr radiol & res assoc diag ultrasonics, Hahnemann Med Col, 65-66; from asst prof to assoc prof radiol & med physics, 68-76, PROF RADIOL & MED PHYSICS, MED SCH, TEMPLE UNIV, 76-, CHMN COMT BIOPHYS & BIOENG, 74- Concurrent Pos: Lectr biomed eng, Drexel Univ, 65-71, adj assoc prof, 71-; NSF fel analog & digital electronics, 72; mem comt on sci & arts, Franklin Inst, 72-; mem bd dirs, Inst Ultrasonics in Med. Mem: Inst Ultrasonics in Med; Am Heart Asn; Inst Elec & Electronics Eng; Soc Photo-Optical Instrument Eng; NY Acad Sci. Res: Biomathematics; diagnostic ultrasonics; thermography; image processing; vision; hearing; information processing in the nervous system. Mailing Add: Dept of Radiol Temple Univ Med Sch Philadelphia PA 19140

ZISKIND, MORTON MOSES, b Syracuse, NY, Oct 15, 14; m 55. INTERNAL MEDICINE. Educ: City Col New York, BSS, 36; Tulane Univ, BS, 39, MD, 42; Am Bd Internal Med, dipl, 51; Am Bd Pulmonary Dis, dipl, 57. Prof Exp: From asst prof to assoc prof, 51-60, PROF MED, SCH MED, TULANE UNIV, 60- Concurrent Pos: Sr vis physician, Charity Hosp of La, New Orleans, consult physician, 71-; attend physician, Vet Admin Hosp, 52-71, consult physician, 72- Mem: Am Thoracic Soc; Am Col Chest Physicians; Am Col Physicians. Res: Suppurative diseases of lungs; pulmonary diseases; occupational pulmonary disease. Mailing Add: 1700 Perdido St New Orleans LA 70112

ZISMAN, WILLIAM ALBERT, b Albany, NY, Aug 21, 05; m 35; c 1. PHYSICAL CHEMISTRY. Educ: Mass Inst Technol, BS, 27, MS, 28; Harvard Univ, PhD(physics), 32. Hon Degrees: DSc, Clarkson Col Technol, 65. Prof Exp: Res assoc geophys, Harvard Univ, 31-33; statistician, Div Econ & Statist, Pub Works Admin, 33-35; mgt div, Resettlement Admin, 35-38; vis res scientist, Carnegie Inst Geophys Lab, 38-39; physicist, 39-42, head lubrication br, 47-53, surface chem br, 53-57, supt chem div, 56-68, CHAIR OF SCI IN CHEM PHYSICS, NAVAL RES LAB, WASHINGTON, DC, 69- Concurrent Pos: Lectr, Am Univ & Georgetown Univ. Honors & Awards: Hillebrand Award, Washington Chem Soc, 54; Distinguished Civilian Serv Awards, US Dept Navy, 54 & Dept Defense, 64; Carbide & Carbon Award, Am Chem Soc, 55, Kendall Award, 63; Am Soc Lubrication Eng Nat Award, 61; Capt Robert Dexter Conrad Award, Dept Navy, 68; Mayo D Hersey Award, Am Soc Mech Eng, 69; Matiello Award, Fedn Socs Paint Technol, 71. Mem: Am Chem Soc; Am Phys Soc; Am Soc Lubrication Eng; NY Acad Sci. Res: Surface and colloidal chemistry; monomolecular films; surface potentials; friction, lubrication, adhesion, corrosion inhibition, adsorption, wetting and surface properties of polymers. Mailing Add: 200 E Melbourne Ave Silver Spring MD 20901

ZISSIS, GEORGE JOHN, b Lebanon, Ind, Dec 31, 22; m 54; c 4. OPTICS. Educ: Purdue Univ, BS, 46, MS, 50, PhD(physics), 54. Prof Exp: Instr eng physics, Purdue Univ, 46-50, 52-54; assoc scientist, Atomic Power Div, Westinghouse Elec Corp, 54-55; mem spec air defense study, Off Naval Res, 57; alt head infrared lab, Inst Sci & Technol, 57-64, head lab, 64-69, chief scientist, Willow Run Labs & Tech Mgr, Infrared & Optics Div, Inst Sci & Technol, 69-73, CHIEF SCIENTIST, INFRARED & OPTICS DIV, ENVIRON RES INST MICH, UNIV MICH, 73- Concurrent Pos: Vis lectr, Univ Mich, 61-62, lectr, Dept Elec Eng, 70-72, adj prof electronics & comput eng, 73-; mem staff, Res Eng, Support Div, Inst Defense Anal, DC, 62-64, consult, 64-; consult, Army Res Off, 64-; mem comt space prog earth observ, Nat Res Coun-Nat Acad Sci, chmn, 69-72; adv, Div Earth Sci, US Geol Surv; ed-in-chief, J Remote Sensing of the Environ, 71- Mem: Fel AAAS; Optical Soc Am. Res: High resolution spectroscopy; infrared; radiometry; optical radiation physics; precision measurements. Mailing Add: 1549 Stonehaven Rd Ann Arbor MI 48104

ZISSON, JAMES, b Marlboro, Mass, June 13, 21; m 52; c 3. ORGANIC CHEMISTRY. Educ: Northeastern Univ, BS, 43; Boston Univ, MBA, 52. Prof Exp: Foundry chemist, Gen Elec Co, Mass, 41-43; chem engr, Manhattan Dist Proj, Los Alamos Sci Lab, 45-46; chemist, Res Dept, Stand Oil Co (Ind), 46-48, asst personnel supvr, 48-50, 52-54, asst to admin dir res, 54-56, admin supvr supply & transportation, 57-58; asst mgr, Amoco Chem Corp, 58-61, mgr, 61-66; purchasing agent raw mat, 66-68, mgr purchasing & traffic, 68, construct prod div, 69-71, DIR PURCHASING & TRAFFIC, DEWEY & ALMY CHEM DIV, W R GRACE & CO, 71- Mem: AAAS; Nat Asn Purchasing Mgrs; Am Chem Soc. Res: Synthesis of hydrocarbons. Mailing Add: 26 Saddle Club Rd Lexington MA 02173

ZITARELLI, DAVID EARL, b Chester, Pa, Aug 12, 41; m 66. MATHEMATICS. Educ: Temple Univ, BA, 63, MA, 65; Pa State Univ, PhD(math), 70. Prof Exp: ASST PROF MATH, TEMPLE UNIV, 70- Mem: Am Math Soc; Math Asn Am. Res: History of mathematics; algebraic theory of semigroups. Mailing Add: Dept of Math Temple Univ Philadelphia PA 19122

ZITNAK, AMBROSE, b Bratislava, Czech, Dec 30, 22; nat Can; m 50; c 3. PLANT BIOCHEMISTRY, FOOD TECHNOLOGY. Educ: Slovak Inst Tech, Czech, BSA, 46; Univ Alta, MSc, 53, PhD, 55. Prof Exp: Res officer, Agr Res Inst, Czech, 46-47; asst forage chem, Agr Exp Sta, Swiss Fed Inst Technol, 47-48; asst plant biochem, Univ Alta, 51-55, chief analyst feed & soil chem, 55-57; ASSOC PROF HORT BIOCHEM, UNIV GUELPH, 57- Mem: Can Inst Food Technol; Can Soc Plant Physiol; Can Soc Hort Sci; Can Geog Soc; Agr Inst Can. Res: Solanaceous glycoalkaloids; plant growth substances; technology of fruit and vegetable preservation; heat-browning of potato products; physiology of the potato tuber; cyanogenic glucosides. Mailing Add: Dept of Hort Sci Univ of Guelph Guelph ON Can

ZITOMER, FRED, b Chicago, Ill, Oct 19, 33; m 57; c 3. ANALYTICAL CHEMISTRY. Educ: Park Col, BA, 58; Kans State Univ, MS, 60, PhD(anal chem), 63. Prof Exp: Asst org synthesis, Park Col, 56-58; asst anal chem, Kans State Univ, 58-63; anal chemist, Res Lab, Celanese Corp, NJ, 63-64; res chemist, Agr Res Ctr, Dow Chem Co, Mich, 64-66; res chemist, Res Lab, 66-67, anal group leader, 67-71, supvr anal dept, 71-74, MGR ANAL CHEM & PHYS TESTING, CELANESE RES CO, 74- Mem: Am Chem Soc; Am Soc Mass Spectrometry. Res: Mass spectrometry; gas chromatography; trace analysis; analytical methods development; flammability, nonwovens, rubber, polymer degradation mechanisms; nuclear magnetic resonance; infrared; electron spectroscopic chemical analysis; liquid chromatography; electron spin resonance; thermal analysis. Mailing Add: Celanese Res Co PO Box 1000 Summit NJ 07901

ZITRIN, ARTHUR, b NY, Apr 10, 18; m 42; c 2. PSYCHIATRY. Educ: City Col New York, BS, 38; NY Univ, MS, 41, MD, 45. Prof Exp: Instr physiol, Hunter Col, 48-49; from clin asst psychiat to clin instr, 49-54, from asst clin prof to assoc prof, 54-67, PROF PSYCHIAT, SCH MED, NY UNIV, 67- Concurrent Pos: Pvt pract; sr psychiatrist, Bellevue Hosp, 50-, asst dir psychiat div, 54-55, dir, 55-69. Mem: Am Psychiat Asn; Am Psychoanal Asn; Asn Res Nerv & Ment Dis. Res: Physiological psychology; psychoanalysis; clinical psychiatry. Mailing Add: 56 Ruxton Rd Great Neck NY 11023

ZITRON, NORMAN RALPH, b New York, NY, May 22, 30. APPLIED MATHEMATICS. Educ: Cornell Univ, AB, 52; NY Univ, MS, 56, PhD(math), 59. Prof Exp: Lab asst physics, NY Univ, 52-53, res asst, Courant Inst Math Sci, 54-58; tech assoc appl physics, Harvard Univ, 58-59; res assoc eng, Brown Univ, 59-60; asst prof eng sci, Fla State Univ, 60-61; asst prof math, Res Ctr, Univ Wis, 61-62; ASSOC PROF MATH, PURDUE UNIV, 62- Concurrent Pos: Fulbright sr res scholar, Tech Univ Denmark, 64-65; assoc res mathematician, Radiation Lab, Univ Mich, 65-66; sr res fel, Univ Surrey, 68; res assoc, Univ Calif, Berkeley, 71; vis scholar, Stanford Univ, 72. Mem: Am Math Soc. Res: Propagation, diffraction and scattering of electromagnetic, acoustic and elastic waves; asymptotoic solutions of partial differential equations; transportation; operations research. Mailing Add: Div of Math Sci Purdue Univ Lafayette IN 47907

ZITTEL, HERMAN EDWARD, analytical chemistry, see 12th edition

ZITTER, ROBERT NATHAN, b New York, NY, Oct 3, 28; m 64. SOLID STATE PHYSICS, QUANTUM ELECTRONICS. Educ: Univ Chicago, BA, 50, MS, 52 & 60, PhD(physics), 62. Prof Exp: Mathematician, US Naval Proving Grounds, 52; res physicist, Chicago Midway Labs, 52-60; mem tech staff, Bell Tel Labs, 62-67; PROF PHYSICS, SOUTHERN ILL UNIV, CARBONDALE, 67- Mem: Am Phys Soc. Res: Semiconductors and semimetals; photoconductivity; gaseous lasers; infrared detection. Mailing Add: Dept of Physics Southern Ill Univ Carbondale IL 62901

ZITTER, THOMAS ANDREW, b Saginaw, Mich, Dec 30, 41; m 66; c 3. PLANT VIROLOGY. Educ: Mich State Univ, BS, 63, PhD(plant path), 68. Prof Exp: ASST PROF & ASST PLANT PATHOLOGIST, AGR RES & EDUC CTR, UNIV FLA, 68- Mem: Am Phytopath Soc; Entom Soc Am. Res: Isolation and identification of vegetable viruses; establishment of plant-vector-virus relationships; determination of epidemiology and control of virus diseases. Mailing Add: Agr Res & Educ Ctr Univ of Fla PO Drawer A Belle Glade FL 33430

ZITZEWITZ, PAUL WILLIAM, b Chicago, Ill, June 5, 42; m 66; c 2. EXPERIMENTAL ATOMIC PHYSICS. Educ: Carleton Col, BA, 64; Harvard Univ, AM, 65, PhD(physics), 70. Prof Exp: Scholar physics, Univ Western Ont, 70-72; res fel & sr physicist, Corning Glass Works, 72-73; ASST PROF PHYSICS, UNIV MICH, DEARBORN, 73- Mem: Am Phys Soc; Am Asn Physics Teachers; Sigma Xi. Res: Positrons; positron interactions in solids; positronium; atom-surface interactions; atomic spectroscopy; fundamental constants. Mailing Add: Dept of Natural Sci Univ of Mich Dearborn MI 48128

ZIVANOVIC, SRBISLAV V, electrical engineering, physics, see 12th edition

ZIVNUSKA, JOHN ARTHUR, b San Diego, Calif, July 10, 16. FOREST ECONOMICS, ACADEMIC ADMINISTRATION. Educ: Univ Calif, Berkeley, BS, 38, MS, 40; Univ Minn, St Paul, PhD(agr econ), 47. Prof Exp: Instr forestry, Univ Minn, 46-47; from instr to assoc prof , Sch Forestry & Conserv, Univ Calif, Berkeley, 48-59; PROF FORESTRY, SCH FORESTRY & CONSERV, UNIV CALIF, BERKELEY, 59-, DEAN, 65- Concurrent Pos: Fulbright lectr, Agr Col Norway, Vollebekk, 54-55; consult, Econ Comn Asia & Far East, 57; mem forestry res adv comt, USDA, 64-70, mem agr res policy adv comt, 71-72; pres, Asn State Cols & Univ Forestry Res Orgns, 71-72. Mem: Soc Am Foresters; Am Soc Range Mgt. Res: Demand for timber; economics of forest industries; forest taxation; international forestry. Mailing Add: Sch of Forestry & Conserv 145 Mulford Hall Univ of Calif Berkeley CA 94720

ZIZZA, FRANK, b Palermo, Italy, Dec 11, 28; nat US; m 59; c 3. PATHOLOGY. Educ: Univ Palermo, MD, 53. Prof Exp: Asst physiol, Sch Med, Univ Palermo, 53-54; Fulbright scholar, 54; intern, Mercy Hosp, Denver, Colo, 54-55; resident lab med, Univ Colo, 55-56, res fel med, 56-57, Am Heart Asn res fel, 57-58 & adv res fel, 58-59; instr physiol, Sch Med, Tulane Univ, 59-60; resident path, Mem Hosp, Danville, Va, 61-62; sr asst res, Med Ctr, Georgetown Univ, 62-63; asst chief lab serv, Vet Admin Hosps, Atlanta, Ga, 63-64 & chief lab serv, Tuskegee, Ala, 65-66; CLIN PATHOLOGIST, DOCTORS HOSP, 66-, ASSOC DIR LABS, 72- Concurrent Pos: Consult pathologist, Le Roy Hosp, New York; consult pathologist centralized lab serv, Health Ins Plan of Greater NY. Mem: AMA; fel Col Path. Res: Enzyme mechanisms; blood volume measurement; distribution and metabolism of iodine-131 labeled plasma albumins; capillary permeability; iron metabolism; nosocomial infections. Mailing Add: 170 East End Ave New York NY 10028

ZLATKIS, ALBERT, b Mar 27, 24; nat US; m 47; c 3. ANALYTICAL CHEMISTRY & ORGANIC CHEMISTRY. Educ: Univ Toronto, BASc, 47, MASc, 48; Wayne State Univ, PhD(chem), 52. Prof Exp: Demonstr chem eng, Univ Toronto, 47-48; instr, Wayne State Univ, 49-52, res assoc, 52-53; res chemist, Shell Oil Co, Tex, 53-55; from instr to assoc prof chem, 54-62, chmn dept, 58-62, PROF CHEM, UNIV HOUSTON, 62- Concurrent Pos: Adj prof, Baylor Col Med, 72- Honors & Awards: Award Chromatog, Am Chem Soc, 73. Mem: Am Chem Soc. Res: Gas chromatography and mass spectrometry of biological metabolites; clinical chemistry; environmental studies; chemistry of flavors and natural products. Mailing Add: Dept of Chem Univ of Houston Houston TX 77004

ZLETZ, ALEX, b Detroit, Mich, Mar 28, 19; m 48; c 4. PHYSICAL CHEMISTRY, ORGANIC CHEMISTRY. Educ: Wayne Univ, BS, 40, MS, 48; Purdue Univ, PhD(chem), 50. Prof Exp: Electroplating chemist, Auto City Plating Co, 41; res chemist, Stand Oil Co, 50-54, group leader, 54-61, GROUP LEADER, AM OIL CO, 61- Concurrent Pos: Guest scientist, Free Radicals Proj, Nat Bur Stand, 57. Mem: Am Chem Soc; The Chem Soc; Am Soc Lubrication Engrs. Res: High vacuum; extreme pressure; aryl borons; polyolefins; liquid rocket fuels; free radical and hydrocarbon chemistry; catalysis; fluids and lubricants; greases; railway diesel lubricating oil. Mailing Add: 1004 Mill St Apt 110 Naperville IL 60540

ZLOBEC, SANJO, b Brezicani, Yugoslavia, Nov 16, 40; m 65; c 1. APPLIED MATHEMATICS. Educ: Univ Zagreb, BSc, 63, MSc, 67; Northwestern Univ, PhD(appl math), 70. Prof Exp: Res engr, Northwestern Univ, 68-69, lectr math, 69-70; Nat Res Coun Can grant, 70-72, ASST PROF MATH, MCGILL UNIV, 70- Concurrent Pos: Vis asst prof, Univ Zagreb, 71-72. Mem: Soc Indust & Appl Math; Am Math Soc. Res: Optimization theory and applications; applied functional analysis; numerical analysis. Mailing Add: Dept of Math McGill Univ PO Box 6070 Sta A Montreal PQ Can

ZLOCHOWER, ISAAC AARON, b New York, NY, Mar 10, 38; m 65; c 5. PHYSICAL CHEMISTRY. Educ: Brooklyn Col, BS, 59; Columbia Univ, MA, 63, PhD(chem physics), 66. Prof Exp: Res asst, Sch Mines, Columbia Univ, 65-67, res assoc, 67; chemist, Esso Res & Eng Co, NJ, 67-68, res chemist, 68-70, sr res chemist, 70-74, RES ASSOC FIBER GLASS DIV, PPG INDUSTS, INC, 74- Concurrent Pos: Lectr chem eng, Univ Pittsburgh, 75. Res: Surface chemistry of glass fibers, reinforced cements, leached fiber catalysts; automobile exhaust conversion; characterization of liquid membranes. Mailing Add: 2205 Shady Ave Pittsburgh PA 15217

ZLOT, WILLIAM LEONARD, b New York, NY, June 20, 29; m 56; c 2. MATHEMATICS. Educ: City Col New York, BS, 50; Columbia Univ, MA, 52, MBA, 53, PhD(math educ), 57, MA, 59. Prof Exp: Lectr math, City Col New York, 53-57, instr, 57-59; assoc prof, Paterson State Col, 59-61; from asst prof to assoc prof math educ, Yeshiva Univ, 61-67; assoc prof math, City Col New York, 67-69; PROF MATH EDUC, DIV SCI & MATH EDUC, SCH EDUC, NY UNIV, 69- Mem: Am Math Soc; Math Asn Am. Mailing Add: Div Sci & Math Educ Sch of Educ NY Univ 32 Washington Pl New York NY 10003

ZMIJEWSKI, CHESTER MICHAEL, b Buffalo, NY, June 3, 32; m 54; c 4. IMMUNOLOGY. Educ: Univ Buffalo, BA, 55, MA, 57, PhD(immunol), 60; Millard Fillmore Hosp, cert med tech, 55. Prof Exp: Asst bact & immunol, Sch Med, Univ Buffalo, 55-58, instr & res fel, 60-61; asst prof clin path & dir blood bank, Med Col Va, 61-63; from asst prof immunol to assoc prof, Sch Med, Duke Univ, 63-70; dir transplantation immunol, Ortho Res Found, 70-73; DIR TRANSPLANTATION IMMUNOL, WILLIAM PEPPER LAB, HOSP UNIV PA, 73-, ASSOC PROF PATH, SCH MED, UNIV, 75- Concurrent Pos: Lectr, Approved Sch Med Technol, 59-61; consult blood bank & serol labs, Millard Fillmore Hosp, Buffalo, 60-61; mem ad hoc subcomt stand adv panel collab res in transplantation & immunol, NIH, 64; immunologist, Yerkes Regional Primate Res Ctr, 65; assoc res prof, Sch Med, State Univ NY Buffalo; chmn area comt immunohemat & blood banking, Nat Comt Clin Lab Stand. Mem: Affil Royal Soc Med; Am Asn Immunol; NY Acad Sci; Int Soc Blood Transfusion. Res: Immunohematology; immunogenetics; histocompatibility testing for human homo-transplantation; cancer research and tissue culture. Mailing Add: William Pepper Lab Hosp Univ of Pa Philadelphia PA 19104

ZMOLEK, WILLIAM G, b Toledo, Iowa, July 3, 21; m 45; c 5. ANIMAL SCIENCE. Educ: Iowa State Col, BS, 44, MS, 51. Prof Exp: County exten dir, 44-47, PROF ANIMAL SCI, AGR EXTEN SERV, IOWA STATE UNIV, 63-, EXTEN LIVESTOCK SPECIALIST, 48- Concurrent Pos: Agr rep, Newton Nat Bank, 48. Mem: Am Soc Animal Sci. Mailing Add: Kildee Hall 109 Iowa State Univ Ames IA 50010

ZMUDA, ALFRED JOSEPH, physics, see 12th edition

ZOBEL, BRUCE JOHN, b Calif, Feb 11, 20; m 41; c 4. FOREST GENETICS. Educ: Univ Calif, BS, 43, MF, 49, PhD(forest genetics), 51. Prof Exp: Asst to logging engr, Pac Lumber Co, 43-44; sr lab asst, Univ Calif, 46-49; silviculturist, Tex Forest Serv, 51-56; assoc prof forest genetics, 56-58, prof, 58-62, EDWIN F CONGER DISTINGUISHED PROF FORESTRY, NC STATE UNIV, 62- Honors & Awards: Biol Res Award, Am Soc Foresters, 68; Res Award, Tech Asn Pulp & Paper Asn, 73, Gold Medal, 75; Oliver Max Gardner Award, 72; Outstanding Exten Serv Award, NC State Univ, 73. Mem: Fel Am Soc Foresters; fel Tech Asn Pulp & Paper Asn; fel Int Acad Wood Sci. Res: Silviculture. Mailing Add: Sch of Forest Resources NC State Univ Raleigh NC 27607

ZOBEL, CARL RICHARD, b Pittsburgh, Pa, Aug 29, 28; m 55; c 2. MOLECULAR BIOPHYSICS, BIOPHYSICAL CHEMISTRY. Educ: Purdue Univ, BS, 51; Univ Rochester, PhD(phys chem), 54. Prof Exp: Res assoc physics, Univ Mich, 54-56; asst prof chem, Am Univ Beirut, 56-59; res assoc biophys, Johns Hopkins Univ, 59-62; asst porf, 62-68, ASSOC PROF BIOPHYS SCI, SCH MED, STATE UNIV NY BUFFALO, 68- Concurrent Pos: Du Pont fel infrared spectros, 55-56. Mem: AAAS; Electron Micros Soc Am; Biophys Soc; Am Soc Cell Biol. Res: Electron microscopy, particularly of macromolecules and ordered complexes of macromolecules; development of staining methods for electron microscopy; structure and function of motile systems, particularly striated muscle; physico-chemical studies of macromolecules and their interactions. Mailing Add: Dept of Biophys Sci Sch of Med State Univ NY 4234 Ridge Lea Rd Buffalo NY 14226

ZOBEL, DONALD BRUCE, b Salinas, Calif, July 17, 42; m 66; c 2. PLANT ECOLOGY. Educ: NC State Univ, BS, 64; Duke Univ, MA, 66, PhD(bot), 68. Prof Exp: Asst prof, 68-75, ASSOC PROF BOT, ORE STATE UNIV, 75- Mem: Ecol Soc Am. Res: Physiological plant ecology; water relations of plants. Mailing Add: Dept of Bot & Plant Path Ore State Univ Corvallis OR 97331

ZOBEL, HENRY FREEMAN, b Ft Scott, Kans, Mar 13, 22; m 44; c 3. PHYSICAL CHEMISTRY, ANALYTICAL CHEMISTRY. Educ: Univ Ill, Champaign, BS, 50, MS, 51. Prof Exp: Chemist, Northern Regional Lab, Ill, 51-67; SECT LEADER STRUCT PROPERTIES, MOFFETT TECH CTR, CPC INT, INC, 67- Concurrent Pos: Corn Industs res fel, Northern Regional Lab, Ill, 65-67; mem sci adv comt, Am Inst Baking. Mem: Am Chem Soc; Am Asn Cereal Chem. Res: Physical properties and structure of natural polymers such as granular starches, starch derived products and synthetic polymers. Mailing Add: Moffett Tech Ctr CPC Int Inc Box 345 Argo IL 60501

ZOBEL, HERBERT LAWRENCE, b Chicago, Ill, Nov 17, 24; c 2. GEOGRAPHY OF THE UNITED STATES & CANADA. Educ: Ill State Univ, BS, 48, MS, 49; Northwestern Univ, MA, 60; Univ Mich, PhD(geog & educ), 64. Prof Exp: Teacher soc sci, Chicago Pub Schs, 49-50; asst phys geog, Northwestern Univ, 51-52; asst prof & supvry teacher geog & soc sci, Western Ill Univ, 52-53; teacher unified studies, Cleveland Sch, Skokie, 54-55; instr geog & earth sci, Joliet Twp High Sch & Jr Col, 55-57; from asst prof to assoc prof geog & educ, Eastern Mich Univ, 57-64; res assoc geog, Educ Res Coun Am, 64-66; ASSOC PROF GEOG, KENT STATE UNIV, 66- Concurrent Pos: Lectr geog, Case Western Reserve Univ, 67-69. Mem: Nat Coun Geog Educ; Asn Am Geog. Res: Improvement of instructional materials and methods in geography, earth science and social sciences at all grade levels and the college and university level; improvement of teaching of these areas of specialization at all academic levels. Mailing Add: Dept of Geog Kent State Univ Kent OH 44242

ZOBELL, CLAUDE E, b Provo, Utah, Aug 22, 04; m 30, 46; c 2. MICROBIOLOGY. Educ: Utah State Univ, BS, 27, MS, 29; Univ Calif, PhD(bact), 31. Prof Exp: Prin pub sch, Idaho, 24-26; asst bact, Exp Sta, Utah State Univ, 27-28, instr col, 28-29; asst, Hooper Found, 30-31, from instr to prof microbiol, 32-72, in charge biol prog, 36-37, asst dir, 37-48, chmn div marine biol, 57-60, EMER PROF MICROBIOL, SCRIPPS INST OCEANOG, UNIV CALIF, 72- Concurrent Pos: Assoc, Univ Wis, 38-39; assoc oceanog, Inst Woods Hole, 39; spec fel, Rockefeller Found, 47-48; consult chmn, Comt Geomicrobiol, Nat Res Coun, 46-56; lectr, Princeton Univ, 48; mem polar microbiol comt, Am Soc Microbiol-Nat Res Coun, 57- Mem: AAAS; Am Soc Microbiol; Am Soc Limnol & Oceanog (pres, 49); Brit Soc Gen Microbiol; Soc Indust Microbiol. Res: Biofouling; deep-sea biology; petroleum microbiology; barobiology. Mailing Add: Scripps Inst of Oceanog A-002 Univ of Calif San Diego La Jolla CA 92093

ZOBLER, LEONARD, b New York, NY, Feb 26, 17; m 58; c 2. GEOGRAPHY, HYDROLOGY. Educ: State Col Wash, BS, 40, MS, 43; Columbia Univ, PhD(geog), 53. Prof Exp: Res assoc soils, Agr Exp Sta, Wash, 40-41; soil surveyor, Indian Serv, US Dept Interior, 41-43, 46; soil scientist, Soil Conserv Serv, USDA, 47-51; lectr geog, Columbia Univ, 51-53; instr geol & geog, Hunter Col, 53-54; instr geol, Brooklyn Col, 54-55; from asst prof to assoc prof geol & geog, 55-73, PROF GEOL & GEOG, BARNARD COL, COLUMBIA UNIV, 55- Concurrent Pos: Sci teaching grant, NSF, 75. Mem: AAAS; fel Am Geog Soc; Asn Am Geog; Soil Sci Soc Am; Soil Conserv Soc Am. Res: Land use planning; water resources; environmental impacts; raw material supplies. Mailing Add: Dept of Geog Barnard Col Columbia Univ New York NY 10027

ZOBRISKY, STEVE EDWARD, animal husbandry, see 12th edition

ZOEBISCH, OSCAR CORNELIUS, b Albany, Minn, Dec 16, 19; m 45; c 4. PLANT GENETICS, PLANT PHYSIOLOGY. Educ: Univ Minn, BS, 46, MS & PhD(plant genetics), 50. Prof Exp: Asst prof agr & asst agronomist, Univ Hawaii, 50-52; asst lab mgr pineapple div, Libby, McNeill & Libby, 52-56, asst plantation mgr, 56-57, asst dir res, Fla Citrus Div, 57-58, mgr agr res, Eastern Div Labs, 58-60, asst dir agr res, Eastern & Can Div & mgr cent agr res, 60-65; DIR DEVELOP & SERV, NEW ENG STATES, BIOL DEPT, E I DU PONT DE NEMOURS & CO, INC, 65- Mem: Am Soc Agron; Am Entom Soc; Am Phytopath Soc; AAAS. Res: Research and development of pineapple, citrus and vegetable crops. Mailing Add: E I du Pont de Nemours & Co Inc 308 E Lancaster Ave Wynnewood PA 19096

ZOGG, CARL A, b Belleville, Ill, Feb 9, 27; m 52; c 6. MEDICAL PHYSIOLOGY, NUTRITION. Educ: Univ Ill, BS, 49, MS, 60, PhD(nutrit in dairy sci), 62. Prof Exp: Milk sanitarian, Dressel Young Dairy, 49-58; asst dairy sci, Univ Ill. 58-62, res assoc physiol reprod, 62-63; asst prof physiol, 63-71, ASSOC PROF PHYSIOL & PHARMACOL, SCH MED, UNIV N DAK, 71- Res: Ruminant nutrition and physiology; male reproductive physiology; physiology of lactation of simple stomached animals; hyperbaric physiology, nutrition and microbiology; nutritional studies conducted on subjects exposed to hyperbaric helium-molecular oxygen environmental conditions. Mailing Add: Dept of Physiol Univ of NDak Sch of Med Grand Forks ND 58201

ZOGLIO, MICHAEL ANTHONY, b Providence, RI, June 27, 36; m 59; c 3. PHARMACEUTICAL CHEMISTRY. Educ: Univ RI, BSc, 58; Univ Minn, PhD(med chem), 66. Prof Exp: Mgr pharm res, Sandoz Wander Inc, 64-70; dir pharmaceut develop, Hoechst Pharmaceut Inc, 70-72; DIR PHARMACEUT RES & DEVELOP, MERRELL NAT LABS, 72- Concurrent Pos: Adj asst prof biopharmaceut, Col Pharm, Univ Cincinnati, 74- Mem: Fel Am Pharmaceut Asn; AAAS; Am Asn Col Pharm. Res: Drug stability; drug absorption; physicochemical characterization of pharmaceuticals; physical aspects of dosage form processing. Mailing Add: 6835 Stonington Rd Cincinnati OH 45230

ZOGRAFI, GEORGE, b New York, NY, Mar 13, 36; m 57; c 3. PHARMACEUTICAL CHEMISTRY, SURFACE CHEMISTRY. Educ: Columbia Univ, BS, 56; Univ Mich, MS, 58, PhD(pharmaceut chem), 61. Prof Exp: Asst prof pharm, Columbia Univ, 61-64; from asst prof to assoc prof, Univ Mich, 64-72; PROF PHARM, UNIV WIS-MADISON, 72- Concurrent Pos: Am Found Pharmaceut Educ Pfeiffer Mem res fel, Utrecht Univ, 70-71. Mem: AAAS; Am Pharmaceut Asn; fel Am Acad Pharmaceut Sci; Am Chem Soc. Res: Physical chemical basis for therapeutic activity of drugs; interfacial activity of drugs; lipids and proteins emphasizing structure and function of biological membranes. Mailing Add: Sch of Pharm Univ of Wis-Madison Madison WI 53706

ZOLBER, KATHLEEN KEEN, b Walla Walla, Wash, Dec 9, 16; m. NUTRITION. Educ: Walla Walla Col, BS, 41; Wash State Univ, MA, 61; Univ Wis, PhD(food systs admin), 68. Prof Exp: Food serv dir, Walla Walla Col, 41-50, mgr col store, 51-59, from asst prof to assoc prof foods & nutrit, 59-64; assoc prof nutrit, 64-71, PROF NUTRIT, LOMA LINDA UNIV, 72-, DIR DIETETIC INTERNSHIP, 67-, DIR DIETETIC EDUC, 71- Mem: Am Dietetic Asn; Am Mgt Asn; Am Pub Health Asn; Am Home Econ Asn; Soc Nutrit Educ. Res: Productivity in food service systems; role of technicians in heath care professions. Mailing Add: Box 981 Loma Linda CA 92354

ZOLLA-PAZNER, SUSAN BETH, b Chicago, Ill, Feb 25, 42. IMMUNOLOGY. Educ: Stanford Univ, BA, 63; Univ Calif, San Francisco, PhD(microbiol), 67. Prof Exp: NIH fel, 67-69, ASST PROF PATH, MED SCH, NY UNIV, 69-; RES MICROBIOLOGIST, MANHATTAN VET ADMIN HOSP, 69- Mem: AAAS; Am Asn Immunol. Res: Regulation of the immune response. Mailing Add: Dept of Path NY Univ Sch of Med New York NY 10016

ZOLLER, WILLIAM H, b Cedar Rapids, Iowa, Mar 3, 43; m 69; c 2. NUCLEAR CHEMISTRY. Educ: Univ Alaska, BS, 65; Mass Inst Technol, PhD(nuclear chem), 69. Prof Exp: Tech asst, Inst Geophys, Univ Hawaii, 65; res asst, Arthur A Noyes Nuclear Chem Ctr, Mass Inst Technol, 65-69, res assoc, 69; res assoc, Inst Geophys, Univ Hawaii, 69-70; asst prof nuclear & environ chem, 70-74, ASSOC PROF NUCLEAR & ENVIRON CHEM, UNIV MD, COLLEGE PARK, 74- Mem: AAAS; Am Phys Soc; Am Meteorol Soc; Am Chem Soc; Am Geophys Union. Res: Nuclear phenomenon and environmental chemical problems, especially with respect to air and water pollution; instrumental neutron and photon activation analysis of air pollutants; atmospheric chemical studies in Antarctica and the Arctic. Mailing Add: Dept of Chem Univ of Md College Park MD 20742

ZOLLINGER, JOSEPH LAMAR, b Salt Lake City, Utah, July 9, 27; m 52; c 7. ORGANIC CHEMISTRY. Educ: Univ Utah, BS, 51, PhD(chem), 54. Prof Exp: Res org chemist, M W Kellogg Co, 54-57; RES SPECIALIST, 3M CO, 57- Mem: Am Chem Soc. Res: Organic fluorine compounds; elastomers; fluoronitrogen compounds;

polymers; silicon compounds. Mailing Add: 3M Co 2301 Hudson Rd St Paul MN 55119

ZOLLINGER, ROBERT MILTON, b Millersport, Ohio, Sept 4, 03; m 29; c 2. SURGERY. Educ: Ohio State Univ, BS, 25, MD, 27; Am Bd Surg, dipl. Hon Degrees: Dr, Univ Lyons, 67. Prof Exp: Asst lab surg res, Lakeside Hosp, Cleveland, 27; intern, Peter Bent Brigham Hosp, Boston, 28-29; instr, Western Reserve Univ, 29-32, dir lab surg, 30-31, demonstr, 30-32; from instr surg to asst prof, Harvard Med Sch, 32-46; prof surg & chmn dept, Col Med & chief surg serv, Health Ctr, 46-74, EMER PROF SURG & EMER CHMN DEPT, OHIO STATE UNIV, 74- Concurrent Pos: Asst resident surgeon, Lakeside Hosp, 29-32; resident, Peter Bent Brigham Hosp, 32-34, from assoc to sr assoc, 34-40; surgeon, 40-46; ed-in-chief, Am J Surg, 58-; spec consult, Clin Res Training Comt, Nat Inst Gen Med Sci, chmn surg training comt, 67-; mem surg hist adv bd, US Vet Admin. Honors & Awards: Mod Med Award Distinguished Achievement, 67. Mem: Soc Univ Surgeons (pres, 47); Soc Surg Alimentary Tract; AMA; fel Am Surg Asn (pres, 65); Am Cancer Soc. Mailing Add: Ohio State Univ Hosp Columbus OH 43210

ZOLLWEG, JOHN ALLMAN, b Rochester, NY, July 3, 42; m 67; c 2. PHYSICAL CHEMISTRY. Educ: Oberlin Col, AB, 64; Cornell Univ, PhD(phys chem), 69. Prof Exp: NSF fel, Mass Inst Technol, 68-70; ASST PROF CHEM, UNIV MAINE, ORONO, 70- Res: Study of fluid behavior by optical mixing spectroscopy; experimental and theoretical study of phase equilibrium near critical points. Mailing Add: Dept of Chem Aubert Hall Univ of Maine Orono ME 04473

ZOLLWEG, ROBERT JOHN, b Medina, NY, Aug 1, 24; m 46; c 2. EXPERIMENTAL PHYSICS. Educ: Northwestern Univ, BS, 49, MS, 50; Cornell Univ, PhD(physics), 55. Prof Exp: Asst, Cornell Univ, 50-54; sr physicist, 54-68, ADV PHYSICIST, RES LABS, WESTINGHOUSE ELEC CORP, 68- Mem: Am Phys Soc. Res: Solid state optics; photoemission and optical absorption; electron reflection from metal surfaces; optical properties of plasmas; thermionic energy conversions; arc discharges. Mailing Add: Westinghouse Res Labs Pittsburgh PA 15235

ZOLMAN, JAMES F, b Dayton, Ohio, Aug 1, 36; m 56; c 3. NEUROPSYCHOLOGY, PSYCHOPHARMACOLOGY. Educ: Denison Univ, BS, 58; Univ Calif, Berkeley, PhD(psychol), 63. Prof Exp: NIMH fel, Dept Psychiat & Brain Res Inst, Univ Calif, Los Angeles, 62-64; asst prof, 64-69, ASSOC PROF PHYSIOL & BIOPHYS, MED CTR, UNIV KY, 69- Concurrent Pos: Found Fund fel res psychiat, Inst Psychiat, London, Eng, 71-72. Mem: AAAS; Am Psychol Asn; Psychonomic Soc; Soc Neurosci; Int Soc Develop Psychobiol. Res: Developmental psychopharmacology. Mailing Add: Dept of Physiol & Biophys Univ of Ky Med Ctr Lexington KY 40506

ZOLNOWSKI, DENNIS RONALD, b Buffalo, NY, Mar 9, 43; m 71. NUCLEAR PHYSICS. Educ: Canisius Col, NY, BS, 64; Univ Notre Dame, PhD(physics), 71. Prof Exp: Res assoc, 71-73, RES SCIENTIST PHYSICS, CYCLOTRON INST, TEX A&M UNIV, 74- Concurrent Pos: Welch Found fel, 71-72. Mem: Am Phys Soc. Res: Nuclear structure; in-beam gamma ray and conversion electron experiments; isomeric states; radioactivity. Mailing Add: Cyclotron Inst Tex A&M Univ College Station TX 77840

ZOLOTOR, LAURENCE ARTHUR, b Kansas City, Mo, Nov 30, 41; m 60; c 4. VIROLOGY. Educ: Univ Mo-Kansas City, BS, 64, MS, 66; Kans State Univ, PhD(virol), 70. Prof Exp: Fel virol, Baylor Col Med, 70-72; res assoc chem carcinogenesis; Vet Admin Hosp, Univ Minn, 72-74; MICROBIOLOGIST VIROL, NAT CTR TOXICOL RES, 74- Mem: Am Soc Microbiol. Res: Development of normal and tumor bladder cell lines to be used in toxicological and virus-chemical carcinogen interaction systems. Mailing Add: Dept of Microbiol & Immunol Nat Ctr Toxicol Res Jefferson AR 72079

ZOLTAI, TIBOR, b Gyor, Hungary, Oct 17, 25; m 50; c 3. MINERALOGY, CRYSTALLOGRAPHY. Educ: Univ Toronto, BASc, 55; Mass Inst Technol, PhD(mineral), 59. Prof Exp: From asst prof to assoc prof, 59-64, chmn dept geol & geophys, 63-71, PROF MINERAL, UNIV MINN, MINNEAPOLIS, 64- Mem: Geol Soc Am; Mineral Soc Am; Am Crystallog Asn; Mineral Asn Can. Res: Crystal structures and crystal chemistry of minerals. Mailing Add: Dept of Geol & Geophys Univ of Minn Minneapolis MN 55455

ZOLTEWICZ, JOHN A, b Nanticoke, Pa, Dec 5, 35; m 65; c 3. ORGANIC CHEMISTRY. Educ: Princeton Unib, AB, 57, PhD(org chem), 60. Prof Exp: NATO fel, Univ Munich, 60-61; Shell Corp Fund fel, Brown Univ, 61-62, NIH fel, 62-63; from asst prof to assoc prof chem, 63-73, PROF CHEM, UNIV FLA, 73- Mem: Am Chem Soc. Res: Heterocyclic chemistry; kinetics; rates and mechanism of hydrogen-deuterium exchange in heterocycles; radical-anion heteroaromatic nucleophilic substitution; thiamine model studies; covalent amination. Mailing Add: Dept of Chem Univ of Fla Gainesville FL 32611

ZOLTON, RAYMOND PETER, b Jersey City, NJ, May 21, 39; m 60; c 3. HEMATOLOGY. Educ: Newark Col Eng, BS, 63; Univ Del, MS, 67; Purdue Univ, PhD(biochem), 72. Prof Exp: Engr dyes, E I du Pont de Nemours & Co, Inc, 63-68; res assoc, Wayne State Univ, 72-74; SR SCIENTIST COAGULATION, ORTHO DIAG INC, 74- Concurrent Pos: Mem thrombosis coun, Am Heart Asn. Res: Development of diagnostic test for coagulation and fibrinolysis parameters. Mailing Add: Ortho Diag Inc Raritan NJ 08869

ZOMBECK, MARTIN VINCENT, b Peekskill, NY, Aug 14, 36; m 63; c 2. PHYSICS. Educ: Mass Inst Technol, BS, 57, PhD(physics), 69. Prof Exp: Observer, Smithsonian Astrophys Observ, 61-64; res asst physics, Mass Inst Technol, 64-69; proj scientist, Am Sci & Eng, Inc, 69-73; head physics & instrumentation, Damon Corp, 73-74; RES ASSOC, HARVARD COL OBSERV, 74- Mem: Am Phys Soc; Am Astron Soc. Res: Solar and stellar x-ray astronomy from rockets and space observatories. Mailing Add: Harvard Col Observ 60 Garden St Cambridge MA 02138

ZOMLEFER, JACK, polymer chemistry, organic chemistry, see 12th edition

ZOMPA, LEVERETT JOSEPH, b Lawrence, Mass, May 31, 38; m 66; c 6. INORGANIC CHEMISTRY. Educ: Merrimack Col, BS, 59; Col Holy Cross, MS, 60; Boston Col, PhD(inorg chem), 64. Prof Exp: Asst prof chem, Boston Col, 64-65; fel, Mich State Univ, 65-66; asst prof, 66-70, ASSOC PROF CHEM, UNIV MASS, BOSTON, 70-, CHMN DEPT, 73- Mem: Am Chem Soc; The Chem Soc. Res: Stability of transition metal complexes of stereorestrictive amines and amino acids. Mailing Add: 6 Kathleen Dr Andover MA 01810

ZOMZELY-NEURATH, CLAIRE ELEANORE, b Newark, NJ, July 4, 24; m 63. NEUROBIOLOGY. Educ: Columbia Univ, BS, 50; Harvard Univ, MS, 56, PhD(biochem, nutrit), 58. Prof Exp: Res technician, Columbia Univ, 50-53; res technician obstet & gynec, Sch Med, NY Univ, 53-54; Fulbright fel, Lab Comp Biochem, Col France, 58-59; res assoc med, Sch Med, NY Univ, 59-60; res biochemist, Monsanto Chem Co, Mo, 60-62; asst res biochemist, Sch Med, Univ Calif, Los Angeles, 62-70; ASST MEM BIOCHEM, ROCHE INST MOLECULAR BIOL, 70- Concurrent Pos: Adj prof, Col Med & Dent NJ. Mem: AAAS; Am Physiol Soc; Am Soc Biol Chem; Am Soc Neurochem; Int Soc Neurochem. Res: Regulation of cerebral protein synthesis; effects of environmental and behavioral factors on cerebral protein and nuclei acid synthesis. Mailing Add: Dept of Biochem Roche Inst of Molecular Biol Nutley NJ 07110

ZON, GERALD, b Buffalo, NY, Apr 3, 45. ORGANIC CHEMISTRY. Educ: Canisius Col, BS, 67; Princeton Univ, PhD(org chem), 71. Prof Exp: Res assoc chem, Ohio State Univ, 71-72, NIH fel, 72-73; ASST PROF CHEM, CATH UNIV AM, 73- Concurrent Pos: Res assoc, Mid-Atlantic Res Inst, 75- Mem: Am Chem Soc; Fedn Am Supporting Sci & Technol. Res: Synthesis, stereochemistry and mechanism of anticancer drugs; novel reactions of organometalloid systems. Mailing Add: Dept of Chem Cath Univ of Am Washington DC 20064

ZOOK, HARRY DAVID, b Milroy, Pa, Feb 8, 16; m 37; c 2. ORGANIC CHEMISTRY, ACADEMIC ADMINSTRATION. Educ: Pa State Col, BS, 38, PhD(org chem), 42; Northwestern Univ, MS, 39. Prof Exp: Asst org chem, Northwestern Univ, 38-39; asst chem, 39-41, from instr to assoc prof, 41-60, asst to vpres res, 65-70, PROF CHEM, PA STATE UNIV, UNIVERSITY PARK, 60-, ASST VPRES RES & ASSOC DEAN GRAD SCH, 70- Concurrent Pos: Vis lectr, Stanford Univ, 62. Mem: Am Chem Soc; The Chem Soc. Res: Kinetics and mechanism of organic reactions. Mailing Add: Pa State Univ 128 Willard University Park PA 16802

ZOOK, JAMES DAVID, solid state physics, see 12th edition

ZORACH, TIMOTHY, vertebrate ecology, see 12th edition

ZORDAN, THOMAS ANTHONY, b Rockford, Ill, Nov 21, 43. PHYSICAL CHEMISTRY. Educ: Northern Ill Univ, BS, 65; Univ Louisville, PhD(phys chem), 69. Prof Exp: Res chemist, Gulf Res & Develop Co, 69-72; sr engr, Nuclear Ctr, 72-74, MGR SAFEGUARDS EVAL, WESTINGHOUSE ELEC CORP, 74- Mem: Am Chem Soc; Calorimetry Conf; Am Soc Testing & Mat; Am Nuclear Soc. Res: Thermodynamics of solutions and pure substances; theory of the liquid state; calorimetry of chemical reactions; nuclear power reactor safety. Mailing Add: Nuclear Ctr Westinghouse Elec Corp Box 355 Pittsburgh PA 15230

ZORN, BICE SECHI, b Cagliari, Italy, May 20, 28; m 55. HIGH ENERGY PHYSICS. Educ: Univ Cagliari, Sardinia, PhD(physics), 52. Prof Exp: Asst physics, Padova, 52-56; physics assoc, Brookhaven Nat Lab, 56-62; res asst prof physics, 62-65, asst prof, 65-68, res assoc prof, 68-72, ASSOC PROF PHYSICS, UNIV MD, COLLEGE PARK, 72- Mem: Am Phys Soc; Ital Phys Soc. Res: High energy experimental particle physics using nuclear emulsions and bubble chambers as detectors of particles. Mailing Add: Dept of Physics Univ of Md College Park MD 20740

ZORN, GUS TOM, b Ada, Okla, June 18, 24; m 55. HIGH ENERGY PHYSICS. Educ: Okla State Univ, BS, 48; Univ NMex, MS, 52; Univ Padua, PhD(physics), 54. Prof Exp: Res assoc physics, Brookhaven Nat Lab, 54-56, assoc physicist, 56-62; assoc prof physics, 62-72, PROF PHYSICS, UNIV MD, COLLEGE PARK, 72- Concurrent Pos: Vis scientist, Max Planck Inst Physics, 58. Mem: Fel am Phys Soc; Ital Phys Soc. Res: Collective ion acceleration; electron ring accelerator for heavy ions; experimental high energy and elementary particle physics, using nuclear emulsions, bubble chambers, counters and spark chambers. Mailing Add: Dept of Physics & Astron Univ of Md College Park MD 20742

ZORN, JENS CHRISTIAN, b Halle, Ger, June 19, 31; US citizen; m 54; c 2. ATOMIC PHYSICS. Educ: Miami Univ, AB, 55; Yale Univ, MS, 57, PhD(physics), 61. Prof Exp: Engr, Sarkes Tarzian, Inc, Ind, 53-54; res asst physics, Univ Tubingen, 55-56; consult, Sch Med, Yale Univ, 59-61, instr, 61-62; from asst prof to assoc prof, 62-69, PROF PHYSICS, UNIV MICH, ANN ARBOR, 70- Concurrent Pos: Vis prof, Univ Puebla, 64-65; Phoenix fac fel, Univ Mich, 65-66. Mem: Fel Am Phys Soc; Am Asn Physics Teachers. Res: Atomic and molecular structure; atomic beams; microwave spectroscopy; laboratory astrophysics; space physics; scientific manpower administration. Mailing Add: Randall Lab of Physics Univ of Mich Ann Arbor MI 48104

ZORN, RALPH ALLAN, b Chicago, Ill, Aug 1, 23; m 53; c 1. INDUSTRIAL MICROBIOLOGY, ANALYTICAL CHEMISTRY. Educ: Univ Wis, BS, 49, MS, 50. Prof Exp: Fermentation supvr vitamin B12, Western Condensing Co, 50-51; chief chemist fermentation prods, Grain Processing Corp, 51-53, res microbiologist fermentation prods, 53-63, enzodex lab supvr corn syrup dextrose, 63-65, process develop fermentation scale, 65-67; RES/QUAL CONTROL LAB SUPVR INGREDIENTS & FEED SUPPL ANAL, KENT FEEDS, DIV GRAINS PROCESSING CORP, 67- Mailing Add: Kent Feeds Inc 1600 Oregon St Muscatine IA 52761

ZORNETZER, STEVEN FRANK, b New York, NY, Jan 21, 45; m 69. PSYCHOBIOLOGY, NEUROSCIENCES. Educ: State Univ NY Stony Brook, BA, 66; Univ Wis-Madison, MA, 67; Univ Calif, Irvine, PhD(biol), 71. Prof Exp: Instr biol sci, Univ Calif, Irvine, 70-71; ASST PROF NEUROSCI, COL MED, UNIV FLA, 71- Mem: AAAS; Soc Neurosci. Res: Neurobiology of memory, including the acquisition, storage and maintenance of information in the nervous system. Mailing Add: Dept of Neurosci Univ of Fla Col of Med Gainesville FL 32601

ZORY, PETER STEPHEN, JR, b Syracuse, NY, Oct 9, 36; m 61; c 2. PHYSICS. Educ: Syracuse Univ, BS, 58; Carnegie-Mellon Univ, PhD(physics), 64. Prof Exp: Physicist, Gyroscope Div, Sperry Rand Corp, 64-66, sr physicist, 66-68; MEM PROF STAFF, T J WATSON RES CTR, IBM CORP, 68- Mem: Inst Elec & Electronics Eng; Optical Soc Am. Res: Laser research. Mailing Add: T J Watson Res Ctr IBM Corp PO Box 218 Yorktown Heights NY 10598

ZORZOLI, ANITA, b New York, NY, Dec 27, 17. BIOCHEMISTRY, PHYSIOLOGY. Educ: Hunter Col, AB, 38; Columbia Univ, AM, 40; NY Univ, PhD(biol), 45. Prof Exp: Asst zool, Columbia Univ, 40-42; asst instr biol, NY Univ, 44-45; res asst path, Sch Med, Wash Univ, 45-46, instr biochem, Sch Dent, 46-48, res assoc pharmacol, Sch Med, 48-49, asst prof biochem, Sch Dent, 48-52; asst prof, Southern Ill Univ, 52-55; from assoc prof to prof biol, 55-73, PROF BIOL, JOHN GUY VASSAR CHAIR NATURAL HIST, VASSAR COL, 73- Concurrent Pos: Mem corp, Marine Biol Lab, Woods Hole. Mem: Fel AAAS; fel Geront Soc (vpres, 65-66); Am Physiol Soc; Int Asn Geront; Am Aging Asn. Res: Biochemistry of aging in mice. Mailing Add: Dept of Biol Vassar Col Poughkeepsie NY 12601

ZOSS, ABRAHAM OSCAR, b South Bend, Ind, Feb 17, 17; m 39; c 3. INDUSTRIAL CHEMISTRY. Educ: Univ Notre Dame, BSChE, 38, MS, 39, PhD(org chem), 41. Prof Exp: Asst, Univ Notre Dame, 39-41; res chemist, Gen Aniline & Film Corp, NJ,

41-43, Pa, 43-47, dept chemist, NJ, 47-49, area supt, 49-51, prod mgr, 51-54, tech mgr, 54-55, plant mgr, 55-57; mgr mfg admin, Chem Div, Minn Mining & Mfg Co, 57-58, div prod mgr, 58-60; vpres, Photek, Inc, 60-62; asst corp tech dir, Celanese Corp, 62-65, corp tech dir, 65-66, corp dir com develop, 66-69, vpres, Tenneco Chem, Inc, NY, 69-71, vpres corp dev, Universal Oil Prod Co, 71-72; GROUP VPRES, ENGELHARD MINERALS & CHEM CORP, 72- Concurrent Pos: Mem field info agency, Off Tech Serv, US Dept Com, Ger, 46; mem, Textile Res Inst Centennial of Sci Award, Univ Notre Dame, 65. Mem: AAAS; Am Chem Soc; Am Inst Chem Eng; Com Develop Asn; NY Acad Sci. Res: Acetylene and high polymer chemistry; petrochemicals; synthetic fibers; plastics; metal carbonyls and powders; coatings; forest products; catalytic chemistry. Mailing Add: 206 Emerson Lane Berkeley Heights NJ 07922

ZOTTOLA, EDMUND ANTHONY, b Gilroy, Calif, June 25, 32; m 60; c 4. FOOD SCIENCE. Educ: Ore State Univ, BS, 54, MS, 58; Univ Minn, St Paul, PhD(dairy tech), 64. Prof Exp: Res asst food sci, Ore State Univ, 56-58; res fel food sci & industs, Univ Minn, St Paul, 58-64; bacteriologist, Nat Dairy Prod Corp, Ill, 64-65; microbiologist, Nodaway Valley Foods, Iowa, 65-66; assoc prof food sci & industs, 66-72, PROF FOOD SCI & NUTRIT, UNIV MINN, ST PAUL, 72-, EXTEN FOOD MICROBIOLOGIST, 66- Mem: Inst Food Technol; Am Dairy Sci Asn; Int Asn Milk, Food & Environ Sanit; Am Soc Microbiol. Res: Spoilage and pathogenic microorganisms in food; thermal destruction of microorganisms; airborne microorganisms; food plant and equipment sanitation; detection of microorganisms in raw and processed foods. Mailing Add: Dept of Food Sci & Nutrit Univ of Minn St Paul MN 55101

ZOTTOLI, ROBERT, b Boston, Mass, Apr 17, 39; m 62; c 1. INVERTEBRATE ZOOLOGY. Educ: Bowdoin Col, BA, 60; Univ NH, MS, 63, PhD(zool), 66. Prof Exp: Asst biol, Univ NH, 61-63; from asst prof to assoc prof, 65-75, PROF BIOL, FITCHBURG STATE COL, 75- Mem: AAAS; Am Soc Zool. Res: Natural history of polychaetous annelid worms. Mailing Add: Fitchburg State Col Fitchburg MA 01420

ZOTTOLI, STEVEN JAYNES, b Boston, Mass, Aug 28, 47; c 2. NEUROBIOLOGY. Educ: Bowdoin Col, BA, 69; Univ Mass, Amherst, MS, 72, PhD(zool), 76. Prof Exp: Lectr cell physiol, Col Our Lady of the Elms, 73-74; NAT INST NEUROL & COMMUN DIS & STROKE FEL, RES INST ALCOHOLISM, NIH, 75- Mem: Am Soc Zoologists. Res: Neurophysiological, morphological and behavioral studies of Mauthner cell function in teleosts. Mailing Add: Res Inst on Alcoholism 1021 Main St Buffalo NY 14203

ZOUMAS, BARRY LEE, b Reading, Pa, July 9, 42; m 66; c 3. NUTRITION, FOOD SCIENCE. Educ: Kutztown State Col, BS, 64; Pa State Univ, MS, 66, PhD(nutrit), 69. Prof Exp: Sr scientist nutrit, Mead Johnson Res Labs, 69-71; MGR NUTRIT SCI, HERSHEY FOODS CORP, 71- Concurrent Pos: Adj asst prof behav sci, M S Hershey Med Sch, Pa State Univ, 75- Mem: AAAS; Inst Food Technologists; Am Chem Soc. Res: Nutritional composition of foods; nutritional properties of carbohydrates, the effect of nutrition on growth and development. Mailing Add: Hershey Foods Corp PO Box 54 Hershey PA 17033

ZOUROS, ELEFTHERIOS, b Lesbos, Greece, Aug 31, 39; m 68; c 1. POPULATION GENETICS. Educ: Agr Col Athens, BSc, 63, PhD(biol), 68; Univ Chicago, PhD(biol), 72. Prof Exp: Res assoc biol, Agr Col Athens, 65-68; fel pop biol, Univ Chicago, 69-73; ASST PROF BIOL, DALHOUSIE UNIV, 73- Concurrent Pos: Scholar, Greek Nat Found Scholars, 66; Ford Found fel, 69. Mem: Genetics Soc Am; Am Soc Naturalists; Genetics Soc Can; Soc Study Evolution. Res: Genetic basis of the evolutionary process. Mailing Add: Dept of Biol Dalhousie Univ Halifax NS Can

ZOVKO, CARL T, physical chemistry, see 12th edition

ZSCHEILE, FREDERICK PAUL, JR, b Burlington, Kans, May 11, 07; m 33; c 3. PLANT PHYSIOLOGY, PHYTOPATHOLOGY. Educ: Univ Calif, BS, 28, PhD(plant physiol), 31. Prof Exp: Nat Res Coun fel biol, Univ Chicago, 31-33, asst pediat, 33-34, res assoc chem, 34-37, res assoc bot, 44-46; from asst prof agron & asst chemist to assoc prof & assoc chemist, Purdue Univ, 37-44; from assoc prof to prof, 46-74, from assoc biochemist to biochemist, Exp Sta, 46-74, EMER PROF AGRON, COL AGR, UNIV CALIF, DAVIS, 74- Concurrent Pos: Guggenheim fel, 58-59. Mem: AAAS; Am Chem Soc; Am Soc Plant Physiol; fel Am Inst Chem; Am Phytopath Soc. Res: Spectrophotometry of plant pigments; preservation of carotene and vitamin A; biochemical nature of disease resistance in plants; phytotron design; gas chromatography of amino acids; lysine in food plants. Mailing Add: 236 B St Davis CA 95616

ZSCHIRNT, HANS HEINZ, physics, see 12th edition

ZSIGMOND, ELEMER K, b Budapest, Hungary, May 16, 30; US citizen. ANESTHESIOLOGY. Educ: Univ Budapest, MD, 55. Prof Exp: Intern med, Clins, Med Sch, Univ Budapest, 54-55; resident internal med, Sztalinvarosi Korhaz, Hungary, 55-56; cardiol res, Balatonfüred, Hungary, 56-57; intern med, Allegheny Gen Hosp, Pittsburgh, Pa, 60-61, resident anesthesiol, 61-63, dir anesthesiol res lab, 66-68; PROF ANESTHESIOL, MED CTR, UNIV MICH, ANN ARBOR, 68- Concurrent Pos: Res fel anesthesiol, Allegheny Gen Hosp, Pittsburgh, Pa, 63-66. Mem: Am Soc Anesthesiol; Int Anesthesia Res Soc. Res: Plasmacholinesterase studies to determine susceptibility to drugs used in anesthesia; malignant hyperpyrexia determinations. Mailing Add: Dept of Anesthesiol Univ Hosp Ann Arbor MI 48104

ZSIGRAY, ROBERT MICHAEL, b Glen Rogers, WVa, Mar 22, 39; m 70; c 4. MICROBIOLOGY. Educ: Miami Univ, AB, 61; Georgetown Univ, MS, 67, PhD(biol), 69. Prof Exp: Microbiologist, Wis State Lab Hyg, 61; res technologist genetics, US Army Biol Lab, Ft Detrick, 62-64, Nat Res Coun res assoc, 68-70; ASST PROF MICROBIOL, UNIV NH, 70- Mem: AAAS; Am Soc Microbiol. Res: Entry of exogenous DNA into cells. Mailing Add: Dept of Microbiol Univ of NH Durham NH 03824

ZSOTER, THOMAS, b Budapest, Hungary, Dec 27, 22; m 53; c 2. INTERNAL MEDICINE. Educ: Univ Budapest, MD, 47; FRCP(C). Prof Exp: Resident med, Univ Budapest, 47-49, instr internal med, 49-51; asst prof, Univ Szeged, 50-54, assoc prof med, 54-57; head circulation lab, Ayerst, McKenna & Harrison, Ltd, 57-60; ASSOC PROF PHARMACOL, UNIV TORONTO, 66-, ASSOC PROF MED, 68- Concurrent Pos: French Govt fel, Paris, 48-49; dir, Div Clin Pharmacol, Toronto Western Hosp, Ont, 66- Mem: Fel Am Col Physicians; Can Pharmacol Soc; Can Fedn Biol Soc. Res: Circulation; pathophysiology; congestive heart failure; hemodynamics in experimental valvular defects; microscopic circulation; cardiovascular pharmacology. Mailing Add: Dept of Pharmacol Univ of Toronto Toronto ON Can

ZUBAY, ELI ALAN, b Minneapolis, Minn, Apr 18, 15; m 41; c 3. STATISTICS, ACTUARIAL SCIENCE. Educ: Univ Minn, BS, 35, MA, 39; Iowa State Univ, PhD(statist), 50. Prof Exp: Prin, Dysart High Sch, 37-40; instr math, Osceola Jr Col, 40-41; prof actuarial sci, Drake Univ, 46-58; PROF ACTUARIAL SCI & MATH, GA STATE UNIV, 58-, VPRES ACAD AFFAIRS, 71- Mem: Am Acad Actuaries; Am Statist Asn; Math Asn Am; Am Risk & Ins Asn. Res: Applications of operations research to insurance. Mailing Add: Ga State Univ Univ Plaza Atlanta GA 30303

ZUBAY, GEOFFREY, b Chicago, Ill, Nov 15, 31. BIOLOGY. Educ: Univ Chicago, PhB, 49, MS, 52; Harvard Univ, PhD(phys chem), 57. Prof Exp: NSF res fel molecular biol, King's Col, Univ London, 57-59, NIH fel, 59-60; res assoc, Rockefeller Inst, 60-61; asst biochemist, Brookhaven Nat Lab, 61-63; ASSOC PROF MOLECULAR BIOL, COLUMBIA UNIV, 63- Res: Physics; molecular biology; chemistry. Mailing Add: 109 Low Mem Libr Columbia Univ New York NY 10027

ZUBECK, ROBERT BRUCE, b Minneapolis, Minn, Oct 12, 44; m 69. LOW TEMPERATURE PHYSICS, APPLIED PHYSICS. Educ: Univ Minn, BPhys, 66; Stanford Univ, MS, 68, PhD(appl physics), 73. Prof Exp: Consult, Stanford Res Inst, 69-70; RES ASSOC APPL PHYSICS, STANFORD UNIV, 73- Res: High resolution low temperature calorimetry; magnetic flux pinning in high field superconductors; electronbeam co-deposition techniques; growth morphology and crystallographic ordering of A-15 superconductors; synthesis of new superconductors. Mailing Add: Microwave Lab Stanford Univ Stanford CA 94305

ZUBECKIS, EDGAR, b Latvia, Dec 27, 02; nat Can. FOOD TECHNOLOGY. Educ: Acad Agr Jelgava, DrAgr, 44. Prof Exp: Privat docent, Fruit Tech, Univ Riga, 36-38; docent, Acad Agr Jelgava, 40-44; prof agr & fruit tech, Univ Bonn, 47-49; res scientist, Ont Res Coun, 53-55 & Hort Res Inst, Ont Dept Agr, 55-68; CONSULT, 68- Mem: NY Acad Sci; AAAS; Can Inst Food Sci & Technol. Mailing Add: Apt 406 185 Stephen Dr Toronto ON Can

ZUBER, MARCUS STANLEY, b Gettysburg, SDak, Jan 10, 12; m 41; c 1. AGRONOMY. Educ: SDak State Univ, BS, 37; Iowa State Univ, MS, 40, PhD(plant breeding), 50. Prof Exp: Agent corn invests, Div Cereal Crops & Dis, 37-42, assoc agronomist & in charge corn breeding, Div Cereal Crops, 46-50, res agronomist & in charge, 50-73, RES LEADER, CORN BREEDING, DIV CEREAL CROPS, MO AGR EXP STA, USDA, 73-; PROF AGRON, UNIV MO-COLUMBIA, 56- Mem: AAAS; Am Soc Agron. Res: Corn breeding and genetics; physiology; insect and disease resistance; cereal chemistry. Mailing Add: Dept of Agron 109 Curtis Hall Univ of Mo-Columbia Columbia MO 65201

ZUBER, WILLIAM HENRY, JR, b Memphis, Tenn, Sept 26, 37; m 59; c 3. PHYSICAL CHEMISTRY. Educ: Memphis State Univ, BS, 60; Univ Ky, PhD(phys chem), 64. Prof Exp: Asst prof chem, Murray State Univ, 64-66; asst prof, 66-70, ASSOC PROF CHEM, MEMPHIS STATE UNIV, 70- Mem: Am Chem Soc. Res: Nonaqueous solution chemistry. Mailing Add: Dept of Chem Memphis State Univ Memphis TN 38111

ZUBIN, JOSEPH, b Raseiniai, Luthuania, Oct 9, 00; nat US; m 34; c 3. BIOMETRICS, PSYCHOLOGY. Educ: Johns Hopkins Univ, AB, 21; Columbia Univ, PhD(educ psychol), 32; Am Bd Prof Psychol, dipl, 50. Hon Degrees: MD, Univ Lund, 72. Prof Exp: Asst educ psychol, Teachers Col, Columbia Univ, 30-31, instr psychomet, Col Physicians & Surgeons, 32-33; instr educ psychol, City Col New York, 34-36; asst psychologist, Ment Hosp Surv Comt, Nat Comt Ment Hyg, 36-38; assoc res psychologist, NY State Psychiat Inst & Hosp, 38-56; prin res scientist biomet, 56-60, DIR BIOMET RES UNIT, NY STATE DEPT MENT HYG, 56-, CHIEF PSYCHIAT RES, 60-; EMER PROF PSYCHOL & SPEC LECTR PSYCHOL & PSYCHIAT, COLUMBIA UNIV, 69- Concurrent Pos: From instr to asst prof psychiat, Col Physicians & Surgeons, Columbia Univ, 39-50, adj prof, 50-56, prof, Univ, 56-69; co-ed, publ, Am Psychopath Asn, 44-71; instr, Postdoctoral Inst, Am Psychol Asn, 50, 53, 55, 56, 60, 61, 64, 65 & 67; mem study sect psychopharmacol, NIMH, 62-65; mem study sect develop behav sci, NIH, 66-70; adj prof, Queens Col, 70-; prof lectr, NY Sch Psychiat, Manhattan State Hosp, 71-; consult, Vet Admin & NIMH. Honors & Awards: Paul H Hoch Lect Award, 68. Mem: Psychomet Soc; fel Am Psychol Asn; Am Asn Ment Deficiency; Am Psychopath Asn (pres, 51-52); Am Col Neuropsychopharmacol (pres, 71-72). Res: Experimental psychopathology. Mailing Add: Biomet Res Unit NY State Dept Ment Hyg 722 W 168th St New York NY 10032

ZUBKOFF, PAUL LEON, b Niagara Falls, NY, Nov 24, 34; m 60; c 2. BIOLOGICAL OCEANOGRAPHY, BIOCHEMISTRY. Educ: Univ Buffalo, BA, 56; George Washington Univ, MS, 58; Cornell Univ, PhD(biochem), 62. Prof Exp: Res asst biochem, George Washington Univ, 56-58; res asst, Cornell Univ, 58-61; res biol chemist, Univ Calif, Los Angeles, 61-63; NIH trainee & res assoc biophys, Mass Inst Technol, 63-66; asst prof biochem, Ohio State Univ, 66-70; SR MARINE SCIENTIST & HEAD ENVIRON PHYSIOL, VA INST MARINE SCI, 70-; ASSOC PROF MARINE SCI, COL WILLIAM & MARY & UNIV VA, 70- Mem: Atlantic Estuarine Res Soc; Am Chem Soc; Am Nuclear Soc; Am Soc Limnol & Oceanog; Int Asn Gt Lakes Res. Res: Dynamics of aquatic ecosystems; comparative biochemistry of mass molecular and metabolism of marine invertebrates. Mailing Add: Dept of Environ Physiol Va Inst of Marine Sci Gloucester Point VA 23062

ZUBLER, EDWARD GEORGE, b Lackawanna, NY, Mar 12, 25; m 50; c 3. PHYSICAL CHEMISTRY. Educ: Canisius Col, BS, 49; Univ Notre Dame, PhD(phys chem), 53. Prof Exp: Res phys chemist, 53-65, tech leader, 65-72, RES ADV, GEN ELEC CO, 72- Concurrent Pos: Lectr, Fenn Col, 60-65 & Cleveland State Univ, 65-66. Mem: Am Chem Soc. Res: Gas chemistry; high temperature gas-metal reactions; mass spectrometry; high and ultra high vacuum techniques. Mailing Add: Lighting Res & Tech Serv Oper Lab Gen Elec Co Number 1310 Nela Pk Cleveland OH 44112

ZUBRISKI, JOSEPH CAZIMER, b Goodman, Wis, July 7, 19; m 46; c 3. SOIL SCIENCE. Educ: Univ Wis, BS, 47, MS, 48, PhD(soils), 51. Prof Exp: From asst prof to assoc prof, 51-63, PROF SOILS, N DAK STATE UNIV, 63- Mem: Soil Sci Soc Am; Am Soc Agron. Res: soil fertility; effect of plant population and fertilizers on yield and quality of seeds; soil phosphorus. Mailing Add: Dept of Soils NDak State Univ Fargo ND 58102

ZUBROD, CHARLES GORDON, b New York, NY, Jan 22, 14; m 40; c 5. CANCER. Educ: Col of the Holy Cross, AB, 36; Columbia Univ, MD, 40. Hon Degrees: DSc, Col of the Holy Cross, 69. Prof Exp: Intern, Cent Islip State Hosp, NY, 40-41 & Jersey City Hosp, NY, 41-42; intern & asst resident med, Presby Hosp, New York, 42-43; instr med, Sch Med, Johns Hopkins Univ, 46-49, asst prof med & pharmacol, 49-53; assoc prof med & dir res, Dept Med, St Louis Univ, 53-54; chief gen med br, Nat Cancer Inst, 54-55, clin dir, 55-61, chmn med bd, 57-58, dir int res, 61-65, sci dir chemothe , 65-72, dir div cancer treatment, 72-74; PROF ONCOL & CHMN DEPT, PROF MED & DIR COMPREHENSIVE CANCER CTR, SCH MED, UNIV MIAMI, 74- Concurrent Pos: Roche res fel chemother bact dis, Johns Hopkins Hosp, 46-49; mem, Mt Desert Island Biol Lab; mem & mem exec comt, Lerner Marine Lab, Bimini. Mem: Am Soc Clin Invest; Am Soc Pharmacol & Exp Therapeut; Am Asn Cancer Res; Asn Am Physicians; Am Soc Hemat. Res: Pharmacology, especially of cancer chemotherapeutic agents; marine biology. Mailing Add: Univ of Miami Comp Cancer Ctr 1400 NW Tenth Ave PO Box 520875 Miami FL 33152

ZUBRZYCKI, LEONARD JOSEPH, b Camden, NJ, Feb 25, 32; m 54; c 1. MEDICAL MICROBIOLOGY. Educ: Temple Univ, AB, 53, PhD(med microbiol), 58. Prof Exp: Sr scientist, Wyeth, Inc, 58-61; ASSOC PROF MICROBIOL, SCH MED, TEMPLE UNIV, 61- Mem: AAAS; Am Soc Microbiol; Brit Soc Gen Microbiol. Res: Bacterial genetics. Mailing Add: Dept of Microbiol Temple Univ Philadelphia PA 19140

ZUCCARELLI, ANTHONY JOSEPH b New York, NY, Aug 11, 44; m 68; c 1. MOLECULAR BIOLOGY, MOLECULAR GENETICS. Educ: Cornell Univ, BS, 66; Loma Linda Univ, MS, 68; Calif Inst Technol, PhD(biophys), 74. Prof Exp: Fel molecular biol, Am Cancer Soc, 74-76; ASST PROF BIOL, LOMA LINDA UNIV, 76- Mem: Am Soc Microbiol; Res: Investigation of the early stages of infection with the single-stranded DNA bacteriophage phi chi 174, particularly the mechanisms of initiation and synthesis of the complementary strand in parental replicative forms. Mailing Add: Dept of Biol Loma Linda Univ Loma Linda CA 92354

ZUCCARELLO, WILLIAM A, b Trenton, NJ, July 29, 28; m 66. ENDOCRINOLOGY, BIOCHEMISTRY. Educ: Rutgers Univ, BS, 50, MS, 51; Univ Wis, PhD, 57. Prof Exp: Pharmacologist, Carroll Dunham Smith Pharmacol Co, 58-60, head pharmacol dept, NJ, 60-63; sr endocrinologist, 63-67, sr investr biochem, 67-74, CLIN DATA COORDR, SMITH KLINE & FRENCH LABS, 74- Res: Adrenal and reproductive physiology; anti-inflammatory agents; diuretic-anti hypertensive agents; neuropharmacology; medical applications of enzymes; anesthetic drugs; adrenal-thyroid-gonadal interrelationships; inhibitors of adrenal steroid biosynthesis; immunological aspects of arthritis; hypothalamic releasing factors. Mailing Add: Smith Kline & French Labs 1500 Spring Garden St Philadelphia PA 19101

ZUCK, DONALD ANTON, b Hafford, Sask, Dec 27, 18; nat US; m 44, 58; c 3. PHARMACEUTICAL CHEMISTRY. Educ: Univ Alta, BSc, 48; Univ Wis, MS, 50, PhD(pharm), 52. Prof Exp: Lectr, Sch Pharm, Univ BC, 48-50; from pharmaceut chmeist to sr pharmaceut chemist, Eli Lilly & Co, 52-65; PROF PHARM, COL PHARM, UNIV SASK, 65- Concurrent Pos: Examr, Pharm Exam Bd Can, 72-; sci ed, Can J Pharmaceut Sci, 72-; mem, Drug Qual Assessment Comt, 74- Mem: Am Pharmaceut Asn. Res: Physical chemistry as applied to pharmaceutical problems. Mailing Add: 103 Baldwin Crescent Saskatoon SK Can

ZUCK, ROBERT KARL, b Rochester, NY; m 38; c 4. BOTANY. Educ: Oberlin Col, AB, 37; Univ Tenn, MS, 39; Univ Chicago, PhD(bot), 43. Prof Exp: Asst plant pathologist, Exp Sta, Univ Tenn, 38-41; asst bot, Univ Chicago, 41-43; instr biol, Evansville Col, 43-44; asst plant pathologist, Bur Plant Indust, Soils & Agr Eng, USDA, Md, 44-46; from asst prof to assoc prof bot, 46-55, PROF BOT, DREW UNIV, 55-, CHMN DEPT, 47- Concurrent Pos: Consult ecol probs, Tex Eastern Transmission Corp, 69- Mem: Fel AAAS; Am Phytopath Soc; Bot Soc Am; Mycol Soc Am; Asn Trop Biol. Res: Mycology; biological transmutations (possible); crown gall production without bacterial contact. Mailing Add: Dept of Bot Drew Univ Madison NJ 07940

ZUCK, THOMAS FRANK, b Cleveland, Ohio, Dec 13, 33; m 61; c 2. PATHOLOGY. Educ: Carleton Col, BA, 55; Yale Univ, LLD, 58; Hahnemann Med Col, MD, 63. Prof Exp: Asst to chief, Dept Path, Fitzsimons Army Med Ctr, 68-71, asst chief, 71-74, actg chief, 72; CHIEF, DEPT SURG, LETTERMAN ARMY INST RES, 74- Concurrent Pos: Asst clin prof path, Sch Med, Univ Colo, Denver, 73-75; mem working group, Adv Comn Blood Dis & Resources, Nat Heart & Lung Inst, 75. Mem: Am Soc Hemat. Res: Investigations of preparation, storage, transfusion, and clinical effects of components of blood and blood substitutes, clinical coagulation problems with reference to modification of derangements by transfusion and/or drug therapy. Mailing Add: Dept of Surg Letterman Army Inst of Res Presidio of San Francisco CA 94129

ZUCKER, ALEXANDER, b Zagreb, Yugoslavia, Aug 1, 24; nat US; m 53; c 3. NUCLEAR PHYSICS. Educ: Univ Vt, BA, 48; Yale Univ, MS, 48, PhD(physics), 50. Prof Exp: Res asst physics, Yale Univ, 48-50; physicist, Oak Ridge Nat Lab, 50-53, sr physicist, 53-70, assoc div dir, 60-70; exec dir, Environ Studies Bd, Nat Acad Sci-Nat Acad Eng, 70-72; dir heavy ion proj, 72-74, ASSOC DIR, OAK RIDGE NAT LAB, 72-; PROF PHYSICS, UNIV TENN, 69- Concurrent Pos: Guggenheim fel & Fulbright res scholar, 66-67; del, Pugwash Conf, 71; mem comt nuclear sci & nuclear physics panel, Physics Surv, Nat Res Coun; US del peaceful uses of atomic energy, USSR. Mem: Fel AAAS; fel Am Phys Soc. Res: Nuclear reactions with heavy ions; medium energy nuclear physics, including scattering and polarization of protons; few-nucleon interactions; high-current electronuclear machines; AVF and heavy ion cyclotrons; environment and public policy. Mailing Add: Oak Ridge Nat Lab Oak Ridge TN 37830

ZUCKER, DONALD, analytical chemistry, see 12th edition

ZUCKER, IRVING, b Montreal, Que, Oct 2, 40; m 63; c 2. BIOLOGICAL RHYTHMS, NEUROENDOCRINOLOGY. Educ: McGill Univ, BSc, 61; Univ Chicago, PhD(biopsychol), 64. Prof Exp: Res assoc reprod physiol, Ore Regional Primate Res Ctr, 64-65; vis scientist, Sch Med, Univ Wis, 65; res assoc reprod physiol, Ore Regional Primate Res Ctr, 66; from asst prof to assoc prof psychol, 66-74, PROF PSYCHOL, UNIV CALIF, BERKELEY, 74- Mem: AAAS; Animal Behav Soc; Neurosci Soc. Res: Seasonal reproductive cycles; Circadian clocks; behavioral endocrinology. Mailing Add: Dept of Psychol Univ of Calif Berkeley CA 94720

ZUCKER, IRVING H, b Bronx, NY, July 13, 42; m 70; c 1. PHYSIOLOGY. Educ: City Col New York, BS, 65; Univ Mo-Kansas City, MS, 67; New York Med Col, PhD(physiol), 72. Prof Exp: USPHS fel, 72-73, ASST PROF PHYSIOL, UNIV NEBR MED CTR OMAHA, 73- Mem: Am Physiol Soc; Am Heart Asn. Res: Cardiovascular receptors and the neural control of blood volume. Mailing Add: Dept of Physiol & Biophys Univ of Nebr Med Ctr Omaha NE 68105

ZUCKER, JOSEPH, b New York, NY, Apr 11, 28; m 53; c 3. SOLID STATE PHYSICS. Educ: Univ Miami, BS, 51; NY Univ, MS, 55, PhD(physics), 61. Prof Exp: Res assoc elec eng, Res Div, Col Eng, NY Univ, 51-55; engr, Sylvania Elec Prod, Inc, 55-57, sr engr, 57-59, res engr, 59-63, adv res engr, 63-65, eng specialist. 65-66, adv eng specialist, 66-68, MEM TECH STAFF, GEN TEL & ELECTRONICS LABS, 68- Concurrent Pos: Lectr, Polytech Inst Brooklyn, 60-65, adj prof, 65-72; adj prof, Hofstra Univ, 65-68. Honors & Awards: IR 100 Award, Indust Res Mag, 65. Mem: AAAS; NY Acad Sci; Am Phys Soc; Inst Elec & Electronics Eng; Sigma Xi. Res: Transport properties of bulk semiconductors; optical probing of acoustoelectric interactions in piezoelectric semiconductors; microwave detectors, modulators and harmonic generators; photoelastic and electrooptic effect; acoustic surface wave devices; light emitting diodes; integrated optical devices; optical communication systems. Mailing Add: Gen Tel & Electronics Labs Inc 40 Sylvan Rd Waltham MA 02154

ZUCKER, LOIS MASON, biochemistry, deceased

ZUCKER, MARJORIE BASS, b New York, NY, June 10, 19; m 38; c 4. PHYSIOLOGY, HEMATOLOGY. Educ: Vassar Col, AB, 39; Columbia Univ, PhD(physiol), 44. Prof Exp: Instr physiol, Col Physicians & Surgeons, Columbia Univ, 42-44, res asst, 45-49; from asst prof to assoc prof, Col Dent, NY Univ, 49-55; assoc med, Sloan-Kettering Inst Cancer Res, 55-63; assoc prof path, 63-71, PROF PATH, SCH MED, NY UNIV, 71- Concurrent Pos: Asst res dir, Eastern Div Res Lab, Am Nat Red Cross, 63-70; mem hemat study sect, NIH, 70-74 & Int Comt Haemostasis & Thrombosis, 70- Mem: Am Physiol Soc; Soc Exp Biol & Med; Am Soc Hemat; Int Soc Hemat; Int Soc Thrombosis & Haemostasis. Res: Platelets and blood coagulation. Mailing Add: Dept of Path NY Univ Med Ctr New York NY 10016

ZUCKER, MARTIN SAMUEL, b New York, NY, Mar 15, 30; m 58; c 3. PLASMA PHYSICS, RADIATION PHYSICS. Educ: Cornell Univ, BEngPhys, 52; Univ Wis, MSc, 53, PhD(nuclear physics), 61. Prof Exp: Asst physicist, 58-62, consult, Radiation Div, Nuclear Eng Dept, 61-63, ASSOC PHYSICIST, BROOKHAVEN NAT LAB, 63- Mem: AAAS; Am Phys Soc; Am Asn Physics Teachers; Am Nuclear Soc. Res: Fast neutron polarization; electrical effects of nuclear radiations on matter; direct conversion of energy to electricity; statistical mechanics; chemical physics; accelerator development; scientific applications of computers. Mailing Add: Brookhaven Nat Lab Upton NY 11973

ZUCKER, MELVIN JOSEPH, b Charleston, SC, May 6, 29; m 58; c 2. SOLID STATE PHYSICS. Educ: Brooklyn Col, BS, 51; Rutgers Univ, PhD(physics), 57. Prof Exp: Physicist, Airborne Instruments Lab, Cutler-Hammer, Inc, 57-59; mem tech staff, Semiconductor Div, Hughes Aircraft Co, 58-62; mem tech staff, Am-Standard Corp, NJ, 62-68; CHMN DEPT MATH & PHYSICS, MERCER COUNTY COMMUNITY COL, NJ, 69- Mem: Am Phys Soc. Res: Impurities in superconductors; semiconductor devices; paramagnetic resonance; peizoresitivity studies in semiconductors. Mailing Add: Dept of Math & Physics PO Box B Mercer Community Col Trenton NJ 08690

ZUCKER, MILTON L, plant physiology, see 12th edition

ZUCKER, ROBERT MARTIN, b New York, NY, May 13, 43. BIOPHYSICS, BIOLOGY. Educ: Univ Calif, Los Angeles, BS, 65, MS, 66, PhD(biophys), 70. Prof Exp: Res asst biophys, Lab Nuclear Med & Radiation Biol, 66-70; res, Max Planck Inst Protein & Leather Res, 70-72; ASSOC SCIENTIST BIOPHYS & BIOL, PAPANICOLAOU CANCER RES INST, 72- Concurrent Pos: Mem, Breast Cancer Task Force, Nat Cancer Inst, 73. Mem: Am Chem Soc. Res: Hemoglobin development; erythroid cell development; cancer chemotherapy; tumor growth; leukemia. Mailing Add: 1155 NW 14th St PO Box 6188 Miami FL 33123

ZUCKERBERG, HYAM L, b New York, NY, Dec 5, 37; m 64; c 1. MATHEMATICS. Educ: Yeshiva Univ, BA & BHL, 59, MA, 61, PhD(math), 63. Prof Exp: Res mathematician, Davidson Lab, Stevens Inst Technol, 63- Mem: Am Math Soc. Res: Conformal mapping; potential theory; Hilbert spaces; topology; Bergman kernel function. Mailing Add: Dept of Math Long Island Univ Brooklyn Ctr Bedford Ave & Ave H Brooklyn NY 11201

ZUCKER-FRANKLIN, DOROTHEA, b Berlin, Ger, Aug 9, 29; US citizen; m 56; c 1. CELL BIOLOGY. Educ: Hunter Col, BA, 52; New York Med Col, MD, 56. Prof Exp: Intern med, Philadelphia Gen Hosp, Pa, 56-57; resident, Montefiore Hosp, New York, 57-59, USPHS res fel hemat, 59-61; USPHS res fel electron micros, 61-63, from asst prof to assoc prof, 63-74, PROF MED, SCH MED, NY UNIV, 74- Concurrent Pos: Attend physician, Bellevue Hosp, New York, 63-; attend physician, Univ Hosp, New York, 63-; spec consult path training comt, Nat Inst Gen Med Sci; USPHS res career scientist award, 66-76; Nat Inst Arthritis & Metab Dis res grant; assoc ed, Blood, 64-75 & J Reticuloendothelial Soc, 65-73. Mem: Am Fedn Clin Res; Am Soc Hemat; Am Soc Exp Path; Am Soc Clin Invest; Am Asn Physicians. Res: Hematology, including white blood cells, coagulation of blood, and platelets; immunology; electron microscopy. Mailing Add: Dept of Med NY Univ Sch of Med New York NY 10016

ZUCKERMAN, BENJAMIN MICHAEL, b New York, NY, Aug 16, 43; m 68. ASTRONOMY. Educ: Mass Inst Technol, SB & SM, 63; Harvard Univ, PhD(astron), 68. Prof Exp: Asst prof, 68-71, ASSOC PROF PHYSICS & ASTRON, UNIV MD, COLLEGE PARK, 71- Concurrent Pos: Alfred P Sloan res fel, 72-74; consult, Jet Propulsion Lab. Honors & Awards: Helen B Warner Prize, Am Astron Soc, 75. Mem: Int Astron Union; Int Union Radio Sci; Am Astron Soc. Res: Interstellar medium and radio astronomy, especially spectral line radio astronomy. Mailing Add: Dept of Physics & Astron Univ of Md College Park MD 20742

ZUCKERMAN, BERNARD, b Camden, NJ, Aug 18, 31; m 51; c 2. PHOTOGRAPHIC CHEMISTRY. Educ: Drexel Inst Technol, BS, 60; Univ Pa, MS, 63. Prof Exp: Res physicist, Atlantic Ref Co, 60-61; chemist, Missile & Space Div, Gen Elec Co, 61-63; scientist, 63-71, SR SCIENTIST, POLAROID CORP, 71- Honors & Awards: Jour Award, Soc Photog Sci & Eng, 67. Mem: Soc Photog Sci & Eng. Res: Structural analysis of organic compounds and catalysts; solid state properties of electroluminescent materials; spectral sensitization processes in silver halides; photographic materials and systems. Mailing Add: Polaroid Corp 1265 Main St Bldg 4 Waltham MA 02154

ZUCKERMAN, BERT MERTON, b New York, NY, Mar 26, 24; m 50; c 2. PLANT PATHOLOGY. Educ: NC State Col, BS, 48; State Univ NY, MS, 49; Univ Ill, PhD(plant path), 54. Prof Exp: Asst plant pathologist, Ill Natural Hist Surv, 51-54; PROF PLANT PATH, LAB EXP BIOL, UNIV MASS, 55- Mem: Am Phytopath Soc. Res: Phytonemotology; physiology of parasitism. Mailing Add: Dept of Plant Path Univ of Mass Amherst MA 01002

ZUCKERMAN, ISRAEL, b St Louis, Mo, Oct 10, 24; m 54; c 2. MATHEMATICS. Educ: City Col New York, BBA, 46; Brooklyn Col, MA, 58; Rutgers Univ, PhD(math), 63. Prof Exp: Instr math, Brooklyn Col, 56-58 & Rutgers Univ, 59-63; asst prof, Queens Col, 63-65 & Vassar Col, 65-66; ASSOC PROF MATH, LONG ISLAND UNIV, BROOKLYN CTR, 66- Mem: Math Asn Am; Am Math Soc. Res: Differential algebra; ring theory. Mailing Add: Dept of Math Long Island Univ Brooklyn Ctr Brooklyn NY 11201

ZUCKERMAN, JEROLD J, b Philadelphia, Pa, Feb 29, 36; m 59; c 5. INORGANIC CHEMISTRY, ORGANOMETALLIC CHEMISTRY. Educ: Univ Pa, BS, 57; Harvard Univ, AM, 59, PhD(chem), 60; Cambridge Univ, MA & PhD(chem), 62. Prof Exp: USPHS fel, 60-62; asst prof chem, Cornell Univ, 62-68; assoc prof, 68-71, dir res, 72-73, PROF CHEM, STATE UNIV NY ALBANY, 71- Concurrent Pos: NIH res grants, 63-68; NSF grants, 64-77; Advan Res Proj Agency res grants, 65-68; consult, Union Carbide Corp, 65-67; Res Corp res grant, 68-72; State Univ NY Res Found res grant, 69-75; Am Chem Soc Petrol Res Fund res grant, 69-72; sr fel award, Alexander von Humboldt Found, 73; vis prof, Tech Univ, Berlin, Ger, 73; mem int adv bd, Int Comn Appln Mössbauer Effect, 75; pres bd trustees, Brunswick Common Sch Dist, Rensselaer County, NY; regional ed, J Inorg & Nuclear Chem Letters;

chmn ad hoc panel, Mössbauer Spectros, Nat Acad Sci-Nat Res Coun. Mem: AAAS; Am Chem Soc; The Chem Soc; Sigma Xi; Am Asn Univ Prof. Res: Fourth group organometallic chemistry, direct synthesis from elements or oxides; structure and bonding by infrared, nuclear magnetic resonance techniques; heterocycles with unusual heteroatoms; application of tin-119 Mössbauer spectroscopy to chemical problems. Mailing Add: Dept of Chem State Univ of NY Albany NY 12222

ZUCKERMAN, LEO, b Brooklyn, NY, July 3, 17; m 42; c 3. LABORATORY MEDICINE. Educ: Brooklyn Col, BA, 42. Prof Exp: Res asst biochem, E R Squibb & Sons, 42-52; res assoc, Ortho Res Found, 52-58, supvr blood prod & biol mfg, Ortho Pharmaceut Corp, 58-62, mgr fractionation dept, Ortho Diag, 62-72, mfg dir biochem prod, 72-73, MFG DIR, ORTHO DIAG INC, 73- Mem: Am Chem Soc; Am Inst Chem; Am Asn Clin Chem; Soc Cryobiol; NY Acad Sci. Res: Coagulation; immunology; serology; chromatography; electrophoresis; lyophilization; protein isolation and characterization; clinical chemisty; cryobiology; blood fractionation. Mailing Add: Serol Prod Ortho Diag Inc Raritan NJ 08869

ZUCKERMAN, MARTIN MICHAEL, b Brooklyn, NY, June 27, 34; m 60. MATHEMATICAL LOGIC. Educ: Brandeis Univ, BA, 55; Brown Univ, MA, 60; Yeshiva Univ, PhD(math), 67. Prof Exp: Mathematician, Int Elec Co, 61-63; instr math, NY Univ, 63-66; asst prof, Hunter Col, 67-68; asst prof, 68-72, ASSOC PROF MATH, CITY COL NEW YORK, 72- Mem: Am Math Soc; Asn Symbolic Logic; Math Asn Am. Res: Set theory. Mailing Add: Dept of Math Convent Ave & 138th St City Col of New York New York NY 10031

ZUCKERMAN, PAUL ROBERT, mathematics, see 12th edition

ZUCKERMAN, SAMUEL, b New York, NY, Oct 22, 15; m 38; c 2. ORGANIC CHEMISTRY. Educ: City Col, BS, 37; Polytech Inst Brooklyn, MS, 42, PhD, 50. Prof Exp: Chemist, 36-50, tech dir & plant mgr, Brooklyn Div, 50-60, VPRES, H KOHNSTAMM & CO, INC, 59- Honors & Awards: Medal Award, Soc Cosmetic Chem, 70. Mem: Am Chem Soc; Soc Cosmetic Chem. Res: Organic synthesis of dyestuffs; chemical microscopy; cosmetic colors for camouflage; certified food, drug and cosmetic colors. Mailing Add: H Kohnstamm & Co Inc 161 Ave of the Americas New York NY 10013

ZUCKERMANN, MARTIN JULIUS, b Berlin, Ger, July 7, 36; m 60; c 4. PHYSICS. Educ: Oxford Univ, BA, 60, PhD(phyiscs), 64. Prof Exp: Fel, Univ Chicago, 64-65; asst prof physics, Univ Va, 65-67; lectr, Imp Col, Univ London, 67-69; ASSOC PROF PHYSICS, McGILL UNIV, 69- Mem: Am Inst Physics. Res: Theoretical solid state physics; investigation into superconductivity and magnetism in disordered systems; weather physics. Mailing Add: Eaton Lab McGill Univ PO Box 6070 Montreal PQ Can

ZUCLICH, JOSEPH ANTHONY, physical chemistry, biophysics, see 12th edition

ZUECH, ERNEST A, b Frontenac, Kans, Nov 17, 34; m 56; c 3. ORGANIC CHEMISTRY. Educ: Kans State Col, Pittsburg, BS, 55, MS, 56; Iowa State Univ, PhD(org chem), 60. Prof Exp: Asst org chem, Ohio State Univ, 60-61; chemist, 61-68, SECT MGR, PHILLIPS PETROL CO, 68- Mem: Am Chem Soc. Res: Organometallic chemistry. Mailing Add: 1317 Harned Dr Bartlesville OK 74003

ZUEHLKE, CARL WILLIAM, b Bonduel, Wis, Oct 28, 16; m 44; c 2. ANALYTICAL CHEMISTRY. Educ: Univ Wis, BS, 38; Univ Mich, MS, 40, PhD(anal chem), 42. Prof Exp: Chief chemist, Methods Lab, Gen Chem Div, Allied Chem & Dye Corp, 42-45, E St Louis Works, 45-46, asst mgr, Chem Control Div, 46-48; res assoc, 48-61, asst head chem div, Res Labs, 61-68, HEAD METHODS RES & TECH SERV DIV, RES LABS, EASTMAN KODAK CO, 68- Concurrent Pos: Chmn, Gordon Res Conf Anal Chem, 67. Mem: AAAS; Am Mgt Asn; Am Chem Soc. Res: Analytical chemistry of germanium; analysis for micro amounts of mercury. Mailing Add: 646 Oakridge Dr Rochester NY 14617

ZUEHLKE, RICHARD WILLIAM, b Milwaukee, Wis, June 17, 33; m 55; c 3. PHYSICAL CHEMISTRY, MARINE CHEMISTRY. Educ: Lawrence Col, BS, 55; Univ Minn, PhD(chem), 60. Prof Exp: From instr to asst prof chem, Lawrence Univ, 58-68; chmn dept chem, 68-73, acad liaison officer, 73-74, REMINGTON PROF CHEM, UNIV BRIDGEPORT, 68- Concurrent Pos: Consult, Kimberly-Clark Corp, 60-62; NSF fac fel, Univ Pittsburgh, 66-67; consult, United Illum Co, 69-71 & Wooster Davis & Cifelli Chem Specialties Corp, 70-73. Mem: AAAS; Am Chem Soc; fel Am Inst Chem. Res: Chemical education; computer simulation and modeling; marine surface chemistry. Mailing Add: Dept of Chem Univ of Bridgeport Bridgeport CT 06602

ZUELZER, WOLF W, b Berlin, Ger, May 24, 09; m 38; c 2. PATHOLOGY, PEDIATRICS. Educ: Prague Univ, MD, 35; Wayne State Univ, MD, 43. Prof Exp: Instr path, 41-44, asst prof path & pediat, 45, actg chmn dept pediat, 45-46, PROF PEDIAT RES, SCH MED, WAYNE STATE UNIV, 74-; DIR, CHILD RES CTR MICH, 55- Concurrent Pos: Pathologist, Children's Hosp Mich, 40-74, hematologist, 43-74, dir labs, 45-74, actg med dir, 45-46, assoc pediatrician-in-chief, 46, dir res, 47; ed, Blood, 51-61 & 67, Am J Dis of Children, 53-63, Transfusion, 60- & Pediatrics, 64-; pres & co-med dir, Mich Community Blood Ctr, 55-; mem study sect human embryol & develop, NIH, 59-63 & 68-72, chmn, 69-72; mem nat bd trustees med & sci adv comt, Leukemia Soc Am, 69-74. Honors & Awards: Mead Johnson Award, Am Acad Pediat, 48. Mem: AAAS; Am Soc Hemat; Soc Human Genetics; Soc Pediat Res (vpres, 55, pres, 56); Am Pediat Soc. Res: Pediatric pathology; hematology; human genetics. Mailing Add: Child Res Ctr of Mich 3901 Beaubien Blvd Detroit MI 48201

ZUFFANTI, SAVERIO, organic chemistry, see 12th edition

ZUG, GEORGE R, b Carlisle, Pa, Nov 16, 38; m 60; c 2. HERPETOLOGY, MORPHOLOGY. Educ: Albright Col, BS, 60; Univ Fla, MS, 63; Univ Mich, PhD(zool), 68. Prof Exp: Instr zool, Univ Mich, 68; asst cur herpet, 69-73, CUR HERPET, MUS NATURAL HIST, SMITHSONIAN INST, 73- Mem: Soc Study Evolution; Soc Study Amphibians & Reptiles; Soc Syst Zool; Am Soc Ichthyol & Herpet. Res: Systematics and evolution of reptiles and amphibians, particularly chelonians; morphology of reptiles and amphibians and functional relationships; locomotion of vertebrates. Mailing Add: Div of Reptiles & Amphibians US Nat Mus Smithsonian Inst Washington DC 20560

ZUGIBE, FREDERICK T, b Garnerville, NY, May 28, 28; m 51; c 7. FORENSIC MEDICINE, CARDIOVASCULAR DISEASES. Educ: St Francis Col, BS, 52; Univ Chicago, MS, 59, PhD(anat), 60; WVa Univ, MD, 68. Prof Exp: Res histologist, Lederle Labs, Am Cyanamid Co, 50-52, chemist, 53-55; asst ophthal, Col Physicians & Surgeons, Columbia Univ, 55; res histochemist in chg atherosclerosis sect, Vet Admin Hosp, Downey, Ill, 56-60; dir cardiovasc res, Vet Admin Hosp, Pittsburgh, 61-69; CHIEF MED EXAMR, ROCKLAND COUNTY, NY, 69- Concurrent Pos: Adj prof, Duquesne Univ; asst res prof, Sch Med, Univ Pittsburgh, 61-69; adj assoc prof path, Columbia Univ, 69-; consult path, ABC Labs; consult, Technicon Corp, Sirchie Fingerprint Labs, Vet Admin Comt Connective Tissue, Skeletal & Muscles, dir, Angelus Path Lab; consult path, Police Surgeon & Ambulance Corp, physician; fel coun arteriosclerosis & coun thrombosis, Am Heart Asn. Honors & Awards: Physicians Recognition Award, AMA, 71-74 & 74-77; Shields Law Enforcement Award, 73; Distingusihed Serv Award, Rockland County & Am Legion Award for Serv in Nicaragua During Earthquake, 73; Am Heart Asn Serv Recognition Award, 74. Mem: Histochem Soc; fel Am Heart Asn; NY Acad Sci; fel Am Acad Forensic Sci; fel Am Col Cardiol. Res: Atherosclerosis and aging research; carbohydrate, lipid and enzyme histochemistry; ultramicrohistochemistry; cardiovascular research; histochemistry and forensic science research. Mailing Add: One Angelus Dr Garnerville NY 10923

ZUIDEMA, GEORGE DALE, b Holland, Mich, Mar 8, 28; m 53; c 4. SURGERY. Educ: Hope Col, AB, 49; Johns Hopkins Univ, MD, 53; Am Bd Surg, dipl. Hon Degrees: DSc, Hope Col, 69. Prof Exp: Intern surg, Mass Gen Hosp, 53-54, asst resident, 54 & 57-58, chief resident, 59; from asst prof to prof, Med Sch, Univ Mich, 60-64; PROF SURG & DIR DEPT, SCH MED, JOHNS HOPKINS UNIV & SURGEON-IN-CHIEF HOSP, 64- Concurrent Pos: Fel, Harvard Med Sch, 59; attend surgeon, Ann Arbor Vet Admin Hosp, 60-64; USPHS sr res fel, 61, career develop award, 63; Markle scholar acad med, 61-66; consult, Walter Reed Army Med Ctr, Sinai & Baltimore City Hosps & clin ctr, NIH; asst ed, J Surg Res, 60, ed, 66-, mem, Inst Med, Nat Acad Sci. Honors & Awards: Russel Award, Univ Mich, 63; hon fel, Royal Col Surg, Ireland, 72. Mem: Fel Am Col Surgeons; Asn Am Med Cols; Soc Univ Surgeons; Am Surg Asn; Am Soc Clin Surg. Res: Cardiovascular and acceleration physiology; space medicine; gastrointestinal and hepatic physiology. Mailing Add: Dept of Surg Johns Hopkins Hosp Baltimore MD 21205

ZUKAS, DANUTE, b Lithuania, Mar 26, 41; US citizen; m 62; c 3. ORGANIC CHEMISTRY. Educ: NY Univ, BA, 62, MS, 65, PhD(org chem), 68. Prof Exp: Fel biochem, Dent Sch, NY Univ, 67-68; ASST PROF CHEM, SOMERSET COUNTY COL, 68- Mem: Am Chem Soc; The Chem Soc. Res: Development of audio-tutorial curricula in nursing chemistry and in non-science major chemistry courses. Mailing Add: 23 Round Top Rd Warren NJ 07060

ZUKEL, JOHN WILLIAM, b Northampton, Mass, Feb 7, 16; m 47; c 2. ENTOMOLOGY. Educ: Univ Mass, BS, 37; Univ Idaho, MS, 39; Iowa State Col, PhD(entom), 44. Prof Exp: Agent, Bur Entom & Plant Quarantine, USDA, 39-41; asst sanitarian, USPHS, DC, 43-46; entomologist, US Rubber Co, 46-64, SR RES ASSOC, UNIROYAL INC, 64- Res: Insect toxicology; anopheline mosquito biology and control; plant growth regulants; antioxidants. Mailing Add: 90 Todd St Hamden CT 06518

ZUKEL, WILLIAM JOHN, b Northampton, Mass, June 8, 22. CARDIOVASCULAR DISEASES. Educ: Univ Mass, BS, 43; Hahnemann Med Col, MD, 47; London Sch Hyg & Trop Med, dipl pub health, 61. Prof Exp: Intern & resident med, Newton-Wellesley Hosp, Newton, Mass, 47-49; from asst med officer in chg to med officer in chg, Newton Heart Prog, USPHS, Mass, 49-51, from asst chief to actg chief, Heart Dis Control Prog, DC, 51-52, asst med, Mass Gen Hosp, 52-53, asst med, Albany Med Col, 53-55, chief oper res, Heart Dis Control Prog, DC, 55-57, asst dir, Nat Heart Inst, 57-58, prog asst, Off Surgeon Gen, 58-60, assoc dir collab studies, Nat Heart Inst, 62-67, assoc dir epidemiol & biomet, 67-69, ASSOC DIR CLIN APPLNS & PREV, DIV HEART & VASCULAR DIS, NAT HEART & LUNG INST, 69- Concurrent Pos: USPHS fel cardiol, NY State Dept Health, 53-55; assoc clin prof, Sch Med, George Washington Univ, 63-72, assoc clin prof epidemiol & environ health, 72- Mem: AMA; fel Am Heart Asn; fel Am Pub Health Asn; fel Am Col Cardiol. Res: Clinical trials, prevention and epidemiological studies. Mailing Add: Rm C809 Nat Heart & Lung Inst 7910 Woodmont Ave Bethesda MD 20014

ZUKER, MICHAEL, b Montreal, Que, Apr 1, 49. BIOMATHEMATICS. Educ: McGill Univ, BSc, 70; Mass Inst Technol, PhD(probability theory), 74. Prof Exp: ASST RES OFF BIOMATH, NAT RES COUN CAN, 74- Concurrent Pos: Sessional lectr, Carleton Univ, 75-76. Mem: Am Math Soc; Can Math Cong. Res: Mathematics, usually satistical, combinatorial and numerical methods in the design and analysis of experiments in the biological sciences and for modeling biological systems. Mailing Add: Nat Res Coun 100 Sussex Dr Ottawa ON Can

ZUKOSKI, CHARLES FREDERICK, b St Louis, Mo, Jan 26, 26; m 53; c 3. SURGERY. Educ: Univ NC, AB, 47; Harvard Med Sch, MD, 51. Prof Exp: From intern surg to resident, Roosevelt Hosp, New York, 51-54; resident, Univ Ala Hosp, 55-58, instr, Univ, 58-59; res fel, Med Col Va, 59-61; from asst prof to assoc prof, Sch Med, Vanderbilt Univ, 61-68; assoc prof, Univ NC, Chapel Hill, 68-69; PROF SURG, COL MED, UNIV ARIZ, 69- Concurrent Pos: Nat Inst Neurol Dis & Blindness spec trainee, 59-61; Nat Inst Allergy & Infectious Dis spec fel, 66-67. Mem: Am Col Surg; Soc Univ Surg; Am Surg Asn; Am Soc Exp Path; Transplantation Soc. Res: Homotransplantation; renal allografts; experimental and clinical research. Mailing Add: Dept of Surg Univ of Ariz Col of Med Tucson AZ 85724

ZUKOTYNSKI, STEFAN, b Warsaw, Poland, Feb 26, 39; m 68. SOLID STATE PHYSICS, ELECTRICAL ENGINEERING. Educ: Univ Warsaw, Mag, 61, PhD(physics), 66. Prof Exp: Nat Res Coun Can fel, Univ Alta, 66-68; asst prof, 68-72, ASSOC PROF ELEC ENG, UNIV TORONTO, 72- Concurrent Pos: Consult, Elec Eng Consociates Ltd, 71- Mem: Can Asn Physicists; Am Phys Soc; Inst Elec & Electronics Eng. Res: Electrical and optical properties of semiconductors. Mailing Add: Dept of Elec Eng Univ of Toronto Toronto ON Can

ZUKOWSKI, LUCILLE PINETTE, b Millinocket, Maine, Nov 2, 16; m 55; c 1. MATHEMATICS. Educ: Colby Col, BA, 37; Syracuse Univ, MA, 43. Prof Exp: Assoc prof, 45-71, PROF MATH, COLBY COL, 71-, CHMN DEPT, 70- Concurrent Pos: Teaching fel math, Univ Mich, 54-55; vis prof, Robert Col, Istanbul, 65-66 & Iranzamin, Tehran, Iran, 72-73. Mem: Am Math Soc; Math Asn Am. Res: Algebra; analysis; geometry. Mailing Add: 16 Cherry Hill Dr Waterville ME 04901

ZULALIAN, JACK, b New York, NY, Apr 21, 36. METABOLISM. Educ: Queens Col, NY, BS, 57; Purdue Univ, Lafayette, PhD(chem), 62. Prof Exp: Res assoc natural prod biosynthesis, Korman Res Labs, Albert Einstein Med Ctr, Philadelphia, 62-66; SR RES CHEMIST, METAB, AGR DIV, AM CYANAMID CO, 66- Mem: Am Chem Soc. Res: Organic synthesis; metabolism of organic compounds designed for use in agriculture as herbicides; pesticides and animal health; radiotracter synthesis. Mailing Add: Metab Agr Div Am Cyanamid Co PO Box 400 Princeton NJ 08540

ZULESKI, FRANCIS ROBERT, biochemistry, see 12th edition

ZULL, JAMES E, b North Branch, Mich, Sept 29, 39; m 61, 68; c 3. BIOCHEMISTRY, CELLULAR BIOLOGY. Educ: Houghton Col, BA, 61; Univ Wis, MS, 63, PhD(biochem), 66. Prof Exp: Fel biochem, Univ Wis, 65-66; asst prof, 66-72, ASSOC PROF BIOL, CASE WESTERN RESERVE UNIV, 72- Concurrent Pos: NIH career develop award, 71-76. Mem: Am Chem Soc; NY Acad Sci. Res:

Membrane-hormone interactions; structure and function of biomembranes; hormone mechanisms. Mailing Add: Dept of Biol & Biochem Case Western Reserve Univ Cleveland OH 44106

ZULLO, VICTOR AUGUST, b San Francisco, Calif, July 24, 36; m 72. INVERTEBRATE PALEONTOLOGY, ZOOLOGY. Educ: Univ Calif, AB, 58, MA, 60, PhD(paleont), 63. Prof Exp: Fel systs-ecol prog, Woods Hole Marine Biol Lab, 62-63, resident systematist, 63-67, asst dir prog, 64-66; assoc cur, Dept Geol, Calif Acad Sci, 67-70, chmn, 68-70; dir prog environ sci, 71, PROF GEOL, UNIV NC, WILMINGTON, 71- Mem: AAAS; Am Soc Zool; Soc Syst Zool; Soc Vert Paleont; Paleont Soc. Res: Systematics, evolution and biogeography of Cirripedia; systematics and paleontology of Cenozoic faunas, especially mollusks and echinoids. Mailing Add: Dept of Earth Sci Univ of NC Wilmington NC 28401

ZUMAN, PETR, b Prague, Czech, Jan 13, 26; m 51; c 2. ELECTROCHEMISTRY. Educ: Charles Univ, Prague, RNDr(chem), 50; Czech Acad Sci, DrSc, 62; Univ Birmingham, DSc, 68. Prof Exp: Head org polarography div J Heyrovsky Inst Polarography, Czech Acad Sci, Prague, 50-68; PROF CHEM, CLARKSON COL TECHNOL, 70- Concurrent Pos: Sr vis fel, Univ Birmingham, 66-70; distinguished vis prof, Brooklyn Polytech Inst, 67; co-chmn org div, Int Electrochem Soc, 67; assoc mem comn electroanal chem, Int Union Pure & Appl Chem, 71; consult, Xerox Corp, 72; Theophilu Redwood lectureship, The Chem Soc, 75. Honors & Awards: Benedetti-Pichler Award, Am Microchem Soc, 75. Mem: Am Chem Soc; The Chem Soc; Electrochem Soc; Int Electrochem Soc. Res: Use of polarography and other electrochemical and optical methods for study of reactivity, equilibria, kinetics and mechanisms of reactions of organic compounds. Mailing Add: Dept of Chem Clarkson Col of Technol Potsdam NY 13676

ZUMBERGE, JAMES HERBERT, b Minneapolis, Minn, Dec 27, 23; m 47; c 4. QUATERNARY GEOLOGY. Educ: Univ Minn, BA, 46, PhD(geol), 50; Grand Valley State Col, LLD, 70; Nebr Wesleyan Univ, LHD, 72. Prof Exp: Instr geol, Duke Univ, 46-47 & Univ Minn, 48-40; from instr to prof, Univ Mich, 50-62; pres, Grand Valley State Col, 62-68; dean, Col Earth Sci, Univ Ariz, 68-72; chancellor, Univ Nebr, Lincoln, 72-75; PRES, SOUTHERN METHODIST UNIV, 75- Concurrent Pos: Chief geologist, Ross Ice Shelf Proj, Int Geophys Year, 57-58; chmn comt polar res, Nat Acad Sci, 72-; US del, Sci Comt Antarctic Res, 72-; mem, Nat Sci Bd, 74- Honors & Awards: Antarctic Serv Medal, 66; named for him, Cape Zumberge, Antarctica, NSF, 62. Mem: AAAS; fel Geol Soc Am; Am Geolphys Union; Soc Econ Geol; Int Glaciol Soc. Res: Glacial geology of Great Lakes states; ground water geology; geomorphology of glaciated regions; environmental problems of extractive industries; climatic changes; physical characteristics and recent history of the Ross Ice Shelf, Antarctica. Mailing Add: Off of the Pres Southern Methodist Univ Dallas TX 75275

ZUMBRUNN, JOHN ROBERT, b Denver, Colo, Mar 18, 42; m 68. MATHEMATICS. Educ: Princeton Univ, AB, 64; Univ Calif, Berkeley, MA, 66, PhD(math), 68. Prof Exp: Actg instr math, Univ Calif, Berkeley, 67-68; J F Ritter instr, Columbia Univ, 68-71; ASST PROF MATH, HUNTER COL, 71- Concurrent Pos: NSF fel, Columbia Univ, 68-71; City Univ New York Res Found fel, City Univ New York, 72-73. Mem: Am Math Soc. Res: Differential geometry; complex manifolds; several complex variables. Mailing Add: Dept of Math Hunter Col 69th & Lexington New York NY 10021

ZUMBRUNNEN, CHARLES EDWARD, b Grafton, WVa, Oct 29, 21; m 46; c 3. DENTISTRY. Educ: WVa Wesleyan Col, BS, 43; Northwestern Univ, Chicago, DDS, 45; Univ NC, Chapel Hill, MPH, 64. Prof Exp: Pvt dent pract, Huntington, WVa, 48-51 & 54-63; DIR, BUR DENT PUB HEALTH, NH DEPT HEALTH & WELFARE, 64- Concurrent Pos: Instr, Sch Dent Med, Tufts Univ, 69-; prof, NH Tech Inst, Concord, 70-71. Mem: Asn State & Territorial Dent Dirs; Am Col Dent; Am Dent Asn. Res: Dental health education methodology in elementary schools. Mailing Add: NH Dept of Health & Welfare 61 S Spring St Concord NH 03301

ZUMDAHL, STEVEN STANFORD, b Freeport, Ill, Nov 28, 42; m 63; c 1. INORGANIC CHEMISTRY. Educ: Wheaton Col, Ill, BS, 64; Univ Ill, Champaign, PhD, 68. Prof Exp: ASST PROF INORG CHEM, UNIV COLO, BOULDER, 68- Mem: Am Chem Soc; The Chem Soc. Res: Properties of coordination complexes with emphasis on the study of the kinetics of the substitution reactions of these complexes. Mailing Add: Dept of Chem Univ of Colo Boulder CO 80302

ZUMMO, NATALE, b Independence, La, Nov 10, 31; m 54; c 3. PLANT PATHOLOGY. Educ: Southeastern La Col, BS, 56; La State Univ, MS, 58, PhD(plant path), 60. Prof Exp: Res plant pathologist, Sugar Cane Field Sta, 60-66, RES PLANT PATHOLOGIST, US SUGAR CROPS FIELD STA, USDA, 66- Mem: Am Phytopath Soc. Res: Diseases of sugar cane and sweet sorghum; insect transmission of virus diseases. Mailing Add: USDA Sugar Crops Field Sta Rte 10 Box 151 Meridian MS 39301

ZUMOFF, BARNETT, b Brooklyn, NY, June 1, 26; m 51; c 3. MEDICINE. Educ: Columbia Univ, AB, 45; Long Island Col Med, MD, 49. Prof Exp: Res fel, Sloan-Kettering Inst, 55-57, from asst to assoc, 57-61; asst prof med, 65-71, ASSOC PROF MED, ALBERT EINSTEIN COL MED, 71-; ATTEND PHYSICIAN & ASST DIR CLIN RES CTR, MONTEFIORE HOSP, 61- Concurrent Pos: Mem coun arteriosclerosis, Am Heart Asn. Mem: Am Soc Clin Invest; Am Physiol Soc; Aerospace Med Asn; Asn Mil Surg US; Am Fedn Clin Res. Res: Human steroid metabolism; cholesterol metabolism and atherosclerosis; radioisotope tracer studies in man; human renal physiology; uric acid metabolism. Mailing Add: Dept of Oncol Montefiore Hosp & Med Ctr Bronx NY 10467

ZUMSTEG, FREDRICK C, JR, b Mansfield, Ohio, Apr 24, 43; m 68. EXPERIMENTAL SOLID STATE PHYSICS. Educ: Univ Ill, BS, 64; Univ Rochester, PhD(physics), 72. Prof Exp: Res assoc, Cornell Univ, 71-73; RES PHYSICIST, E I DU PONT DE NEMOURS & CO, INC, 73- Mem: Am Phys Soc. Res: Development and characterization of new electrooptic and nonlinear optic materials. Mailing Add: Cent Res Dept Exp Sta E I du Pont de Nemours & Co Inc Wilmington DE 19898

ZUMWALT, LLOYD ROBERT, b Richmond, Calif, Sept 4, 14; m 60; c 2. PHYSICAL CHEMISTRY. Educ: Univ Calif, BS, 36; Calif Inst Technol, PhD(phys chem), 39. Prof Exp: Asst, Calif Inst Technol, 36-39, Noyes fel, 39-41; res chemist, Shell Develop Co, 41-42; sr chemist, Oak Ridge Nat Lab, 46-48; dir, Western Div, Tracerlab, Inc, 48-56; vpres, Nuclear Sci & Eng Corp, 56-57; res staff mem, Gen Atomic Div, Gen Dynamics CorP, 57-60; sr res adv, 60-67; PROF NUCLEAR ENG, NC STATE UNIV, 67- Concurrent Pos: Consult, Gen Atomic Co, 57-, Los Alamos Sci Lab & Brookhaven Nat Lab, 73- Mem: Am Chem Soc; fel Am Nuclear Soc. Res: Fission product and tritium diffusion and sorption in materials; high temperature and nuclear reactor chemistry; effect of radiaiton on materials. Mailing Add: Dept of Nuclear Eng NC State Univ Box 5636 State Univ Sta Raleigh NC 27607

ZUND, JOSEPH DAVID, b Ft Worth, Tex, Apr 27, 39. MATHEMATICS, MATHEMATICAL PHYSICS. Educ: Agr & Mech Col Tex, BA & MS, 61; Univ Tex, PhD(math), 64. Prof Exp: Res assoc, Southwest Ctr, Advan Studies, Tex, 64 & 65; from asst prof to assoc prof math, NC State Univ, 65-69; assoc prof, Va Polytech Inst, 69-70; assoc prof math sci, 70-71, PROF MATH & MATH SCI, N MEX STATE UNIV, 72- Concurrent Pos: Res assoc, Inst Field Physics, Univ NC, Chapel Hill, 64-65, vis lectr, Dept Math, 65; vis prof, Cambridge Univ, 68-70. Mem: Am Math Soc; Am Phys Soc; Math Soc France; Ital Math Union; London Math Soc. Res: Differential and projective geometry; general relativity; electromagnetic theory. Mailing Add: Dept of Math Sci NMex State Univ Las Cruces NM 88003

ZUNG, JOSEPH T, b Vietnam, Mov 1, 30. PHYSICAL CHEMISTRY, QUANTUM CHEMISTRY. Educ: Univ Urbino, BA, 51 & 54, MA, 52 & 56; Univ Cincinnati, PhD(chem), 60. Prof Exp: NSF fel & res assoc chem, Univ Rochester, 60-61; from asst prof to assoc prof, Col William & Mary, 61-68; PROF CHEM & SR INVESTR, GRAD CTR CLOUD PHYSICS RES, UNIV MO-ROLLA, 68- Concurrent Pos: US Army Edgewood Arsenal res grant, 63-67. Mem: AAAS; Am Chem Soc. Res: Molecular structure of small molecules; Hartree-Fock wave functions for atoms and molecules; organometallic compounds; cloud dynamics and meteorological applications; evaporation and condensation of liquids and liquid particles. Mailing Add: Dept of Physics Univ of Mo Rolla MO 65401

ZUNG, WILLIAM WEN-KWAI, b Shanghai, China; US citizen. PSYCHIATRY, PSYCHOPHARMACOLOGY. Educ: Univ Wis, BS, 49; Union Theol Sem, NY, BD, 52; Trinity Univ, MS, 56; Univ Tex Med Br Galveston, MD, 61. Prof Exp: Instr, 65-66, assoc, 66-67, from asst prof to assoc prof, 67-72, PROF PSYCHIAT, MED CTR, DUKE UNIV, 73- Concurrent Pos: Clin investr, Vet Admin Hosp, Durham, NC, 65-67; NIMH res career develop award, 67-72; consult, Ctr Studies Suicide Prev, NIMH, 70-71, Career Develop Rev Div, Vet Admin Cent Off, 72- & Res Task Force, NIMH, 72-; consult pub health serv, Bur Drugs, Food & Drug Admin, 73-75 & mem geriat adv panel, 73-75. Honors & Awards: Award for Excellence, Am Acad Gen Pract, 69. Mem: Asn Psychophysiol Study Sleep; Am Asn Suicidol; Am Psychiat Asn; Acad Psychosom Med; fel Int Col Psychosom Med. Res: Affective disorders, especially depressive disorders, anxiety disorders and suicide; psychopharmacology of depression, neurophysiological aspects, including sleep disturbances in psychiatric disorders; biometric approach to psychopathology; mathematical models of psychiatric illness. Mailing Add: Vet Admin Hosp 508 Fulton St Durham NC 27705

ZUNKER, HEINZ OTTO HERMANN, b Berlin, Ger, May 27, 24; m 63; c 2. PATHOLOGY. Educ: Free Univ Berlin, MD, 54; Am Bd Path, cert, 66. Prof Exp: From intern to resident internal med, Free Univ Berlin, 54-56; head pharmacol res, Pharmaceut Co, 57-59; res assoc pharmacol, Columbia Univ, 60-61, assoc path, Col Physicians & Surgeons, 65-68, asst prof, 68; chief path & dir labs, Deaconess Hosp, Evansville, Ind, 68-73; assoc prof path & dir clin path, Col Med, Univ S Fla, 74-75; CHIEF PATHOLOGIST & DIR LABS, BEAUMONT MED SURG HOSP, BEAUMONT, 75- Concurrent Pos: Vis fel path, Columbia-Presby Med Ctr, 62-65; asst attend pathologist, Presby Hosp & consult pathologist, Harlem Hosp, New York, 66-68; assoc prof allied health sci, Univ Evansville & Ind State Univ, Evansville, 72-73. Mem: Col Am Path; Am Soc Clin Path; NY Acad Sci. Res: Pharmacology; experimental pathology; electron microscopy; clinical chemistry. Mailing Add: Beaumont Med Surg Hosp PO Box 5817 Beaumont TX 77702

ZUPKO, ARTHUR GEORGE, b Yonkers, NY, Nov 22, 16; m 47; c 1. PHARMACOLOGY. Educ: Univ Fla, BS, 42; Purdue Univ, MS, 48, PhD(pharmacol), 49. Prof Exp: Malariologist, CZ Authority, 34-36; res assoc biophys & chem, Burroughs-Wellcome & Co, 36-37; asst pharmacol, Purdue Univ, 46-48; from asst prof to prof, St Louis Col Pharm, 49-55; from assoc dean to dean, 55-60, provost, 60-73, PRES, BROOKLYN COL PHARM, LONG ISLAND UNIV, 73- Mem: AAAS; Am Pharmaceut Asn; NY Acad Sci. Res: Anticholinergics on sweating; insecticide toxicology; hypertension; alcohol metabolism; drug interactions. Mailing Add: Brooklyn Col of Pharm 600 Lafayette Ave Brooklyn NY 11216

ZURAW, EDWARD A, bacteriology, see 12th edition

ZURAWSKI, VINCENT RICHARD, JR, b Irvington, NJ, June 10, 46; m 68; c 1. BIOCHEMISTRY, IMMUNOLOGY. Educ: Montclair State Col, BA, 68; Purdue Univ, PhD(chem), 73. Prof Exp: Res assoc biochem, Purdue Univ, 74; RES FEL MED VIROL, IMMUNOCHEM & CELL BIOL, HARVARD MED SCH & CARDIAC BIOCHEM LAB, MASS GEN HOSP, BOSTON, 75- Mem: AAAS; Am Chem Soc. Res: Immunochemical and immunobiological research aimed at producing large amounts of antibody of predetermined specificity both in vivo and in vitro for structural studies and therapeutic applications. Mailing Add: Harvard Med Sch & Cardiac Biochem Lab Mass Gen Hosp Boston MA 02114

ZU RHEIN, GABRIELE MARIE, b Munich, Ger, Apr 5, 20. PATHOLOGY, NEUROPATHOLOGY. Educ: Univ Munich, MD, 53. Prof Exp: Pathologist, Munic Children's Hosp, Munich, 45-48; asst chief lab serv, 98th Gen Hosp, US Army, Munich, 48-54; from instr to assoc prof path & neuropath, 54-69, PROF PATH & NEUROPATH, MED SCH, UNIV WIS-MADISON, 69- Concurrent Pos: Pathologist, State of Wis Cent Colony & Training Sch, 59- & Henry & Lucy Moses Res Labs, Montefiore Hosp, New York, 64-65; consult, Vet Admin Hosp, Madison, 61- Mem: Am Acad Neurol; Am Asn Neuropath (vpres, 72-73); Electron Micros Soc Am. Res: Morphology of viral diseases in laboratory animals or man as studied by light or electron microscopy; degenerative diseases of the nervous system; ultrastructure of virus infected tissue cultures. Mailing Add: Dept of Path Univ of Wis Med Sch Madison WI 53706

ZURMUHLE, ROBERT W, b Lucerne, Switz, Nov 27, 33; US citizen. NUCLEAR PHYISCS. Educ: Univ Zurich, PhD(physics), 60. Prof Exp: Res assoc, 61-63, asst prof, 63-67, ASSOC PROF PHYSICS, UNIV PA, 67- Mem: Am Phys Soc. Res: Nuclear structure and nuclear reactions. Mailing Add: Dept of Physics Univ of Pa Philadelphia PA 19104

ZUSEVICS, JURIS ANDREW, soil science, chemistry, see 12th edition

ZUSI, RICHARD LAURENCE, b Winchester, Mass, Jan 27, 30; m 53; c 3. ZOOLOGY. Educ: Northwestern Univ, BA, 51; Univ Mich, MS, 53, PhD(zool), 59. Prof Exp: Lab asst dendrol, Univ Mich, 52-53; instr zool, Univ Maine, 58-61, asst prof, 61-63; RES SCIENTIST, DIV BIRDS, NAT MUS NATURAL HIST, SMITHSONIAN INST, 63- Mem: AAAS; Cooper Ornith Soc; Wilson Ornith Soc; Am Ornithologists Union; Brit Ornithologists Union. Res: Ornithology, especially functional anatomy, behavior, evolution and classification of birds. Mailing Add: Div of Birds Nat Mus of Natural Hist Washington DC 20560

ZUSMAN, FRED SELWYN, b Boston, Mass, July 24, 31; m 54; c 2. MATHEMATICS. Educ: Harvard Univ, AB, 52, MA, 54. Prof Exp: Analyst, Nat Security Agency, 52-54; sr mathematician appl physics lab, Johns Hopkins Univ, 55-61; sr scientist, Opers Res Inc, 62; dir comput lab, Nat Biomed Res Found, 63; dir comput ctr, Opers Res Inc, 63-67, sr scientist, 67-69; vpres, Sci Mgt Systs, Inc, 69-72, pres, 72; PRIN COMPUT SCIENTIST, OPERS RES INC, 72- Concurrent Pos: Lectr, Sch Hyg & Pub Health, Johns Hopkins Univ, 60-63. Mem: Asn Comput Mach; Pattern Recognition Soc; Opers Res Soc Am. Res: Operations research; computer technology; system simulation; medical applications of computers. Mailing Add: 200 E Indian Spring Dr Silver Spring MD 20901

ZUSMAN, JACK, b Brooklyn, NY, Jan 6, 34; m 55; c 4. PSYCHIATRY, PUBLIC HEALTH. Educ: Columbia Univ, AB, 55, MPH, 66; Ind Univ, MA, 56; Albert Einstein Col Med, MD, 60. Prof Exp: Intern, USPHS Hosp, New Orleans, La, 60-61, epidemic intel serv officer, Communicable Dis Ctr, 61-62, ment health career develop officer, NIMH, 62-66, staff psychiatrist, Epidemiol Studies Br, 66-67, chief, Ctr Epidemiol Studies, 67-68; assoc prof psychiat, 68-71, dir div community psychiat, 69-74, PROF PSYCHIAT, STATE UNIV NY BUFFALO, 71- Concurrent Pos: Trainee community psychiat, Columbia Univ, 64-66; adj prof law & psychiat, 74- Mem: AMA; Am Acad Psychiat & Law; Am Psychiat Asn. Res: Social factors which influence the course of mental illness and methods of their control; interaction of law and psychiatry; methods of organizing and providing medical care. Mailing Add: Dept of Psychiat State Univ of NY Buffalo NY 14215

ZUSPAN, FREDERICK PAUL, b Richwood, Ohio, Jan 20, 22; m 43; c 5. OBSTETRICS & GYNECOLOGY. Educ: Ohio State Univ, BA, 47, MD, 51. Prof Exp: Chief dept obstet & gynec, McDowell Mem Hosp, Ky, 56-58, chief clin serv, 57-58; asst prof obstet & gynec, Sch Med, Western Reserve Univ, 59-60; prof & chmn dept, Med Col Ga, 60-66; Joseph Bolivar DeLee prof & chmn dept, Univ Chicago, 66-74; CHMN DEPT OBSTET & GYNEC, OHIO STATE UNIV, 74- Concurrent Pos: Oglebay fel obstet & gynec, Sch Med, Western Reserve Univ, 58-60; consult obstetrician, Inst Res Hypnosis, 65-; assoc examr & dir, Am Bd Obstet & Gynec, 65-; gynecologist-in-chief, Chicago Lying-In Hosp, 66-74; pres, Barren Found, 74-; founding ed, J Reprod Med; ed, Am J Obstet & Gynec & Current Concepts in Obstet & Gynec; consult ed, Acta Cytologica, Exerpta Medica, Obstet & Gynec Surv, Hypertension, J Obstet & Gynec (Mex) & J Reprod Med. Res: Human reproductive physiology; epinephrine and norephinephrine in the obstetric patient. Mailing Add: Dept of Obstet & Gynec Ohio State Univ Columbus OH 43210

ZUSY, DENNIS, b Milwaukee, Wis, Dec 21, 28. ECOLOGY, PHILOSOPHY OF SCIENCE. Educ: Aquinas Inst, MA, 52 & 56; Northwestern Univ, MS, 64, PhD(biol), 67. Prof Exp: Asst prof philos, St Xavier Col, Ill, 56-62; asst prof biol, 71-75, ASSOC PROF BIOL, CLARKE COL, 75-; PROF PHILOS OF SCI, AQUINAS INST THEOL, 67- Concurrent Pos: Vis asst prof biol, Concordia Teachers Col, Ill, 67-71. Mem: AAAS; Am Inst Biol Sci; Sigma Xi. Res: Biological rhythms; freshwater ecology; philosophical implications and history of scientific concepts. Mailing Add: Aquinas Inst of Theol 2570 Asburg St Dubuque IA 52001

ZUTTY, NATHAN LEWIS, physical chemistry, organic chemistry, see 12th edition

ZUZACK, JOHN W, b St Louis, Mo, Sept 24, 38; m 62; c 2. ORGANIC CHEMISTRY. Educ: St Louis Univ, BS, 61, MS, 64, PhD(org chem), 67. Prof Exp: Lab instr freshman & org chem, St Louis Univ, 61-65, res asst med chem, 65-66; ASST PROF MED CHEM, ST LOUIS COL PHARM, 66- Mem: Am Chem Soc. Res: Leukemia chemotherapy; structure activity relationships in rickettsiostatic and analgesic agents. Mailing Add: Dept of Chem & Pharmaceut Chem St Louis Col of Pharm St Louis MO 63110

ZUZOLO, RALPH C, b Italy, Sept 5, 29; US citizen. CELL PHYSIOLOGY. Educ: NY Univ, BA, 56, MS, 60, PhD(biol), 65. Prof Exp: ASSOC PROF BIOL, CITY UNIV NEW YORK, 64-, SUPVR, DEPT BIOL, SCH GEN STUDIES, 72- Concurrent Pos: NASA grant, NY Univ, 66-, adj asst prof, NY Univ, 74- Mem: AAAS; Am Inst Physics; Soc Appl Spectros; Sigma Xi. Res: Microsurgery, the application of the laser as a microsurgical tool and the effect of lasers irradiation on cells; effect of chemical carcinogens and bisulfite on nucleic acid in living cells. Mailing Add: Dept of Biol City Univ New York Convent Ave at 138th St New York NY 10031

ZVAIFLER, NATHAN J, b Newark, NJ, Nov 26, 27; m 52; c 3. MEDICINE, IMMUNOLOGY. Educ: Haverford Col, BS, 48; Jefferson Med Col, MD, 52. Prof Exp: Resident med, Univ Mich, 55-58, instr, Med Sch, 58-59; from instr to prof, Georgetown Univ, 60-70; PROF MED, UNIV CALIF, SAN DIEGO, 70- Concurrent Pos: NIH fel arthritis, Univ Mich, 58-60. Res: Arthritis. Mailing Add: Dept of Med Univ of Calif Hosp 225 W Dickinson St San Diego CA 92103

ZVEJNIEKS, ANDREJS, b Rauna, Latvia, Jan 6, 22; m 51. ORGANIC CHEMISTRY. Educ: Latvia Univ, BS, 43, MS, 44; Royal Inst Technol, Sweden, 55. Prof Exp: Res engr, Liljeholmens Stearinfabriks, Inc, Sweden, 45-48, sec leader, 48-56; qual control supvr, Conn Adamant Plaster Co, 56; tech adv to plant mgr, Chem Div, Gen Mills, Inc, 57-58, tech dir, Petrol Chem Dept, 58-60, dir appln res, 60-62, sr scientist, 62; pres, AZ Prod, Inc, Fla, 62-72; PRES & CHIEF EXEC OFF, AZS CORP, 72- Concurrent Pos: Mem, Hwy Res Bd, Nat Acad Sci-Nat Res Coun, 59- Mem: Am Chem Soc; Am Inst Mining, Metall & Petrol Eng. Res: Ammonium compounds; ore flotation reagents; petroleum chemicals. Mailing Add: 2337 Christopher Walk Atlanta GA 30327

ZVENGROWSKI, PETER DANIEL, b New York, NY, Sept 8, 39. MATHEMATICS. Educ: Rensselaer Polytech Inst, BS, 59; Univ Chicago, MS, 60, PhD(math), 65. Prof Exp: Asst prof math, Univ Ill, Urbana, 64-70; asst prof, 70-72, chmn div pure math, 72-74, ASSOC PROF MATH, UNIV CALGARY, 72- Res: Algebraic topology and homotopy theory. Mailing Add: Dept of Math Univ of Calgary Calgary AB Can

ZWAAN, JOHAN, b Gorinchem, Neth, Sept 28, 34; m 60; c 2. DEVELOPMENTAL BIOLOGY, HUMAN ANATOMY. Educ: Univ Amsterdam, MedDrs, 60, Dr(embryol), 63. Prof Exp: From asst anat to head asst, Lab Anat & Embryol, Univ Amsterdam, 58-63; fel pediat, Sch Med, Johns Hopkins Univ, 63-64; from asst prof to assoc prof, Sch Med, Univ Va, 64-71; ASSOC PROF ANAT, HARVARD MED SCH, 71- Concurrent Pos: Lectr, Acad Phys Educ, Amsterdam, 62-63; res assoc ophthal, Children's Hosp Med Ctr, Boston, 71- Mem: AAAS; Am Asn Anatomists; Soc Develop Biol; Am Soc Cell Biol; Asn Res Vision & Ophthal. Res: Chemical and morphological changes in differentiation of vertebrate cells; developmental genetics; teratology; normal and abnormal development of the visual system. Mailing Add: Eye Res Lab Children's Hosp Med Ctr Boston MA 02115

ZWADYK, PETER, JR, b Kansas City, Kans, Apr 3, 41; m 63; c 4. MICROBIOLOGY. Educ: Univ Kans, BS, 62; Univ Iowa, MS, 66, PhD(microbiol), 71. Prof Exp: Microbiologist, Sci Assocs, 62-63; scientist microbiol, Mead Johnson & Co, 66-69; asst prof path & microbiol, 71-75, ASST PROF MICROBIOL, DUKE UNIV, 71-, ASSOC PROF PATH, 75-; CHIEF MICROBIOL, VET ADMIN HOSP, DURHAM, 71- Mem: Am Soc Microbiol. Res: Mechanism of pathogenecity; antibiotics. Mailing Add: Vet Admin Hosp Fulton St & Erwin Rd Durham NC 27705

ZWANZIG, FRANCES RYDER, b South Amboy, NJ, Oct 22, 29; m 53; c 2. CHEMISTRY. Educ: Columbia Univ, BA, 51; Yale Univ, MS, 53, PhD(chem), 56. Prof Exp: Res asst physiol chem, Sch Med, Johns Hopkins Univ, 55-58; asst ed, Rev Mod Physics, Am Phys Soc, Washington, DC, 69-73. Mem: Am Chem Soc. Res: Technical editorial writing. Mailing Add: 5314 Sangamore Rd Bethesda MD 20016

ZWANZIG, ROBERT WALTER, b Brooklyn, NY, Apr 9, 28; m 53; c 2. CHEMICAL PHYSICS. Educ: Polytech Inst Brooklyn, BS, 48; Univ Southern Calif, MS, 50; Calif Inst Technol, PhD(chem), 52. Prof Exp: Res fel theoret chem, Yale Univ, 51-54; asst prof chem, Johns Hopkins Univ, 54-58; phys chemist, Nat Bur Stand, 58-66; RES PROF, INST FLUID DYNAMICS & APPL MATH & INST MOLECUALR MOLECULAR PHYSICS, UNIV MD, COLLEGE PARK, 66- Concurrent Pos: Sherman Fairchild scholar, Calif Inst Technol, 74-75. Honors & Awards: Peter Debye Award Phys Chem, Am Chem Soc, 76. Mem: Nat Acad Sci; Am Chem Soc; Am Phys Soc; Am Acad Arts & Sci. Res: Theoretical chemical physics; statistical mechanics; theory of liquids and gases. Mailing Add: Inst for Fluid Dynamics & Appl Math Univ of Md College Park MD 20740

ZWANZIGER, DANIEL, b New York, NY, May 20, 35. THEORETICAL PHYSICS. Educ: Columbia Univ, BA, 55, PhD(physics), 60. Prof Exp: NSF fel physics, Univ Calif, Berkeley, 60-61, lectr, 61-62; scientist, Univ Rome, 62-63; vis, Saclay Ctr Nuclear Studies, France, 63-65; vis scientist, 65-67, assoc prof physics, 67-74, PROF PHYSICS, NY UNIV, 74- Mem: Am Phys Soc. Res: Quantum electrodynamics; scattering theory; elementary particle symmetries; mathematical physics. Mailing Add: Dept of Physics NY Univ 4 Washington Pl New York NY 10003

ZWARICH, RONALD JAMES, b Kamloops, BC, Apr 26, 36. PHYSICAL CHEMISTRY. Educ: Univ BC, BSc, 63, PhD(chem), 69. Prof Exp: ASST PROF CHEM, STATE UNIV NY ALBANY, 71- Mem: Am Chem Soc. Res: Molecular spectroscopy; vibrational structure and electronic properties of aromatic and heterocyclic compounds. Mailing Add: Dept of Chem State Univ of NY Albany NY 12222

ZWART, PHILIP BERNARD, applied mathematics, operations research, see 12th edition

ZWARUN, ANDREW ALEXANDER, b Pidvolochyska, Ukraine, Feb 9, 43; US citizen; m 67; c 2. MICROBIOLOGY, SOIL SCIENCE. Educ: Ohio State Univ, BSc, 65, MSc, 67; Univ Ky, PhD(soil microbiol), 70. Prof Exp: Asst prof agron, Univ Md, Eastern Shore, 70-71; chief microbiologist, Johnston Labs, Inc, 71-74; RES MICROBIOLOGIST, BETZ LABS, INC, 74- Mem: Am Soc Agron; Am Soc Microbiol. Res: Automation of microbial and biochemical procedures; trace metal toxicities; microbial activity in soils and water; industrial water treatment. Mailing Add: Betz Labs Inc Trevose PA 19047

ZWECKER, WILLIAM R, b Vienna, Austria, Apr 19, 12; US citizen; m 46; c 2. ORGANIC CHEMISTRY. Educ: Vienna Tech Univ, MChemE, 34; Dresden Tech Univ, PhD(cellulose chem), 37. Prof Exp: PRES & CONSULT ENGR, ULTRA CELLULOSE INC, 46- Concurrent Pos: Chmn war conf paper & packaging, US Army & Navy, 42. Mem: Soc Am Mil Eng. Res: Cellulose chemistry; new products; domestic and overseas market analysis; commercial cigarette papers; aircraft insulation and plastic helmet lines. Mailing Add: 101 Main St Bethany WV 26032

ZWEIFACH, BENJAMIN WILLIAM, b New York, NY, Nov 27, 10; m 37; c 3. PHYSIOLOGY, BIOENGINEERING. Educ: City Col New York, BS, 31; NY Univ, MS, 33, PhD(anat), 36. Prof Exp: Asst biol, NY Univ, 31-33, asst anat, Col Med, 35-36, res assoc, 38-45; res assoc med, Med Col, Cornell Univ, 45-47, asst prof physiol, 47-52; assoc prof biol, NY Univ, 52-55, from assoc prof path to prof, Sch Med, 55-66; PROF BIOENG, UNIV CALIF, SAN DIEGO, 66- Concurrent Pos: Charlton res fel, Med Sch, Tufts Col, 37-38; ed, Josiah Macy, Jr Found Conf Factors Regulating Blood Pressure, 48-53; mem cardiovasc study sect, USPHS & subcomt shock, Nat Res Coun; estab investr, Am Heart Asn, 55-60; career investr, Health Res Coun, New York, 60-66. Mem: Am Physiol Soc; Soc Exp Biol & Med; Histochem Soc; Microcirc Soc (secy-treas, 55-60); fel NY Acad Sci. Mailing Add: Dept of Bioeng Univ of Calif San Diego La Jolla CA 92037

ZWEIFEL, GEORGE, b Rapperswil, Switz, Oct 2, 26; m 53; c 2. ORGANIC CHEMISTRY. Educ: Swiss Fed Inst Technol, Dr sc tech, 55. Prof Exp: Res asst carbohydrate chem, Univ Edinburgh, 55-56; res fel, Univ Birmingham, 56-58; res assoc boron chem, Purdue Univ, 58-63; assoc prof chem, 63-72, PROF CHEM, UNIV CALIF, DAVIS, 72- Mem: Am Chem Soc. Res: Chemistry of natural products; utilization of organoboranes and organoalanes in organic syntheses. Mailing Add: Dept of Chem Univ of Calif Davis CA 95616

ZWEIFEL, PAUL FREDERICK, b New York, NY, June 21, 29; m 50, 60, 67; c 4. THEORETICAL PHYSICS, NUCLEAR SCIENCE. Educ: Carnegie Inst Technol, BS, 48; Duke Univ, PhD, 54. Prof Exp: Asst physicist, Chem Lab, Am Brake Shoe Co, 48; teacher, Malcom Gordon Sch, NY, 48-49; asst, Duke Univ, 50-52; res assoc, Knolls Atomic Power Lab, Gen Elec Co, 53-56, mgr theoret physics, 56-57; consult physicist, 57-58; assoc prof nuclear eng, Univ Mich, Ann Arbor, 58-60, prof, 60-68; from prof to univ prof, 68-75, DISTINGUISHED UNIV PROF PHYSICS & NUCLEAR ENG, VA POLYTECH INST & STATE UNIV, 75- Concurrent Pos: Mem adv comt reactor physics, Atomic Energy Comn, 57-64; vis prof, Middle East Tech Univ, Ankara, 64-65; consult, indust orgns & govt labs; vis prof, Rockefeller Univ, 74-75; fel, J S Guggenheim Mem Found, 74-75; consult, Los Alamos Sci Lab, 75- & Nuclear Regulatory Comn, 76. Honors & Awards: E O Lawrence Award, 72. Mem: Am Phys Soc; Am Nuclear Soc; Fedn Am Sci (secy, 57-58); Am Math Soc. Res: Mathematical physics; neutron transport theory and nuclear energy. Mailing Add: Dept of Physics Va Polytech Inst & State Univ Blacksburg VA 24061

ZWEIFEL, RICHARD GEORGE, b Los Angeles, Calif, Nov 5, 26; m 56; c 3. HERPETOLOGY. Educ: Univ Calif, Los Angeles, BA, 50, PhD(zool), 54. Prof Exp: From asst cur to assoc cur, 54-65, CUR HERPET, AM MUS NATURAL HIST, 65-, CHMN DEPT, 68- Mem: Am Soc Ichthyol & Herpet; Soc Study Evolution; Ecol Soc Am; Soc Study Amphibians & Reptiles. Res: Ecology and systematics of amphibians and reptiles. Mailing Add: Dept of Herpet Am Mus of Natural Hist Cent Park W at 79th St New York NY 10024

ZWEIFLER, ANDREW J, b Newark, NJ, Feb 2, 30; m 54; c 5. INTERNAL MEDICINE. Educ: Haverford Col, AB, 50; Jefferson Med Col, MD, 54. Prof Exp: Intern, Mt Sinai Hosp, New York, 54-55; resident internal med, 57-60, Nat Heart & Lung Inst fel, 60-63, from instr to assoc prof, 60-72, PROF INTERNAL MED, MED CTR, UNIV MICH, ANN ARBOR, 72- Concurrent Pos: Vis prof, Meharry Med Col, 67-68; fel coun arteriosclerosis & coun thrombosis, Am Heart Asn. Mem: Am Fedn Clin Res. Res: Thrombosis; vascular disease; hypertension. Mailing Add: Dept of Internal Med Univ of Mich Med Ctr Ann Arbor MI 48104

ZWEIG, ARNOLD, b New York, NY, Mar 20, 34; m 57; c 4. PHYSICAL ORGANIC

CHEMISTRY. Educ: Polytech Inst Brooklyn, BS, 55; Northwestern Univ, PhD(chem), 60. Prof Exp: Res chemist, Nalco Chem Co, 55-57; res chemist, 60-65, sr res chemist, 65-66, GROUP LEADER PHOTOCHEM, AM CYANAMID CO, 66- Concurrent Pos: Lectr, Stamford Div, Univ Conn, 65-66. Mem: Am Chem Soc; Am Soc Photobiol. Res: Electronic structure of organic molecules; organic synthesis; ultraviolet spectroscopy; structure property relationships; molecular orbital theory; photochemistry; electrochemistry; ion-radical chemistry. Mailing Add: Chem Res Div Am Cyanamid Co 1937 W Main St Stamford CT 06901

ZWEIG, GEORGE, b Moscow, USSR, May 20, 37; US citizen; m 59; c 2. HIGH ENERGY PHYSICS, NEUROSCIENCES. Educ: Univ Mich, BS, 59; Calif Inst Technol, PhD(physics), 63. Prof Exp: Nat Acad Sci-Nat Res Coun fel high energy physics, Europ Orgn Nuclear Res, 63-64; from asst prof to assoc prof physics, 64-67, PROF PHYSICS, CALIF INST TECHNOL, 67- Concurrent Pos: Sloan Found fel, 66-68. Mem: Am Phys Soc. Mailing Add: Lauritsen Lab Calif Inst of Technol Pasadena CA 91125

ZWEIG, GUNTER, b Hamburg, Ger, May 12, 23; nat US; m 49; c 2. PESTICIDE CHEMISTRY. Educ: Univ Md, BS, 44, PhD(biochem), 52. Prof Exp: Lab instr, Univ Md, 48-50; biochemist, Borden Co, 51-53 & C F Kettering Found, 53-57; from assoc chemist to supv chemist, Pesticide Residue Res Lab, Univ Calif, Davis, 57-64, lectr entom, Univ, 57-65; dir life sci div, Syracuse Univ Res Corp, 65-72, dir, Int Indust Develop Ctr, 72-73; CHIEF CHEM BR, CRITERIA & EVAL DIV, OFF PESTICIDE PROGS, ENVIRON PROTECTION AGENCY, 73- Concurrent Pos: Rothschild fel, Weizmann Inst, 63-64. Mem: AAAS; Am Chem Soc. Res: Methods of residue analysis of pesticides; metabolism and mechanism of action of pesticides. Mailing Add: 2111 Jefferson Davis Hwy Arlington VA 22202

ZWEIG, HANS JACOB, b Essen, Ger, Jan 10, 27; US citizen; m 69; c 3. MATHEMATICAL STATISTICS, OPERATIONS RESEARCH. Educ: Univ Rochester, AB, 49; Brown Univ, MA, 51; Stanford Univ, PhD(statist), 63. Prof Exp: Res phys, Eastman Kodak Co, 51-60; mgr optics technol, IBM Corp, 64-68; vpres res & develop, Credit Data, TRW Inc, 68-70; ASSOC PROF OPERS RES, NAVAL POSTGRAD SCH, 70- Concurrent Pos: Consult opers res, Systs Explor, 72-73. Mem: Fel Optical Soc Am; Inst Math Statist; NY Acad Sci. Res: Optical physics; systems analysis, especially computers and communications. Mailing Add: Dept of Opers Res & Admin Naval Postgrad Sch Monterey CA 93940

ZWEIMAN, BURTON, b New York, NY, June 7, 31; m 62; c 2. MEDICINE. Educ: Univ Pa, AB, 52, MD, 56. Prof Exp: Intern, Mt Sinai Hosp, New York, 56-57; resident med, Hosp Univ Pa, 57-58, assoc, Univ, 63-67, from asst prof to assoc prof, 68-75, PROF MED, SCH MED, UNIV PA, 75- Concurrent Pos: Allergy Found Am fels, Sch Med, Univ Pa, 59-60 & Col Med, NY Univ, 60-61. Mem: Am Asn Immunol; Am Col Physicians; Am Acad Allergy; Am Fedn Clin Res. Res: Cellular and clinical immunology; hypersensitivity inflammatory responses; collagen diseases. Mailing Add: Hosp of the Univ of Pa 3400 Spruce St Philadelphia PA 19104

ZWEMER, THOMAS J, b Mishawaka, Ind, Mar 23, 25; m 49; c 3. ORTHODONTICS. Educ: Univ Ill, DDS, 50; Northwestern Univ, MSD, 54. Prof Exp: Instr pedodont, Marquette Univ, 50-52, from asst prof to assoc prof oral rehab, 52-58; from asst prof to assoc prof orthod, Sch Dent, Loma Linda Univ, 58-66, chmn dept, 60-66; PROF DENT & ASSOC DEAN CLIN SCI, SCH DENT, MED COL GA, 66- Concurrent Pos: Mem attend staff, Wood's Vet Admin Hosp, 54-56; consult, Cerebral Palsy Clin, 54-58; chief dent serv, Milwaukee Children's Hosp, 55-58. Mem: Am Asn Orthod; Am Dent Asn; Am Col Dent; Int Asn Dent Res. Res: Health care delivery systems; physical anthropology. Mailing Add: Sch of Dent Med Col of Ga Augusta GA 30902

ZWERDLING, SOLOMON, b New York, NY, Jan 31, 22; m 44; c 3. PHYSICS, PHYSICAL CHEMISTRY. Educ: Drew Univ, BA, 43; John Hopkins Univ, MA, 44; Columbia Univ, MA, 47, PhD(chem), 52. Prof Exp: Instr chem, Johns Hopkins Univ, 43-44; asst prof naval sci & tactics, Columbia Univ, 46, asst chem, 46-49; sr res chemist, Lever Bros Co, 51-52; div staff physicist, Lincoln lab, Mass Inst Technol, 52-63, staff physicist, Div Sponsored Res, Ctr Mat Sci & Eng, 63-68; res scientist, Douglas Advan Res Labs, Calif, 68-70, mgr res, Solid State Sci Dept, McDonnell Douglas Res Labs, 70-74; DIR SOLAR ENERGY PROG, ARGONNE NAT LAB, 75- Mem: Am Phys Soc; Optical Soc Am; Soc Appl Spectros. Res: Infrared absorption of solids; infrared magneto-optical effects in semiconductors at liquid helium temperature; physics of electronic energy band structure; excitons in semiconductors; far infrared spectroscopy and detectors; metal physics, strength and corrosion; solar energy conversion; systems engineering. Mailing Add: Argonne Nat Lab Bldg 362-C181 Argonne IL 60439

ZWERLING, ISRAEL, b New York, NY, June 12, 17; m 40; c 2. PSYCHIATRY. Educ: City Col New York, BS, 37, MS, 38; Columbia Univ, PhD(psychol), 47; State Univ NY, MD, 50. Prof Exp: High sch instr, NY, 39-42; instr psychol, City Col New York, 47-49 & Hunter Col, 49-50; from asst prof to prof psychiat, Albert Einstein Col Med, 54-73, exec chmn psychiat, 68-71; dir, Bronx State Hosp, 66-73; PROF MENT HEALTH SCI & CHMN DEPT, HAHNEMANN MED COL, 73- Concurrent Pos: Lectr & psychiatrist-in-chg alcohol clin, State Univ NY, 54-55. Mem: Am Psychosom Soc; Am Psychol Asn; Am Psychoanal Asn. Res: Social and community psychiatry. Mailing Add: Dept Ment Health Sci Hahnemann Med Col Philadelphia PA 19102

ZWERMAN, PAUL JOSEPH, b Kimball, Ohio, Apr 26, 11; m 35; c 1. CONSERVATION. Educ: Ohio State Univ, BS, 31, MS, 38, PhD(soils), 49. Prof Exp: Soil scientist, Soil Conserv Serv, USDA, 34-50; assoc prof soil & water conserv, 50-68, PROF SOIL & WATER CONSERV, CORNELL UNIV, 68- Mem: Soil Sci Soc Am; Am Soc Agr Eng; Soil Conserv Soc Am; Sigma Xi; fel Am Soc Agron. Res: Environmental quality; relating soil and management to the problem of waste disposal; soil and water conservation; drainage and tillage. Mailing Add: Dept of Agron Cornell Univ Ithaca NY 14850

ZWIBEL, HARRY SAMUEL, theoretical physics, see 12th edition

ZWICK, DAAN MARSH, b New York, NY, July 28, 22; m 48; c 3. PHOTOGRAPHIC CHEMISTRY, PHYSICS. Educ: Univ Vt, BSChem, 43. Prof Exp: Instr physics, Univ Vt, 43-44; res chemist photog chem, 44-56, res assoc photog sci, 56-73, SR LAB HEAD COLOR PHYSICS, EASTMAN KODAK CO RES LABS, 73- Concurrent Pos: Soc Motion Picture & TV Engrs fel, 62. Honors & Awards: Kalmus Gold Medal, Soc Motion Picture & TV Engrs, 72 Mem: Soc Motion Picture & TV Engrs; Soc Photog Scientists & Engrs; Brit Kinematic, Sound & TV Soc. Res: Problems in the image structure of photographic materials; color reproduction with photographic materials. Mailing Add: 621 Manitou Rd Hilton NY 14468

ZWICK, EARL J, b Canton, Ohio, May 20, 31. MATHEMATICS. Educ: Kent State Univ, BS, 53, MS, 57; Ohio State Univ, PhD(math ed), 64. Prof Exp: Teacher pub schc, Ohio, 53-61; instr math, Ohio State Univ, 62-63; assoc prof, 63-72, PROF MATH, IND STATE UNIV, TERRE HAUTE, 72- Mem: Math Asn Am. Res: Teaching methods. Mailing Add: Dept of Math Ind State Univ Terre Haute IN 47809

ZWICK, ROBERT WARD, b Seattle, Wash, Sept 22, 22; m 48; c 2. ENTOMOLOGY. Educ: Univ Wash, BS, 47; Wash State Univ, MS, 51, PhD(entom), 62. Prof Exp: Res biologist, Fisheries Res Inst, 47-50; entomologist, 10th Naval Dist, San Juan, PR, 53-57 & Eng Sect, US Army Caribbean, CZ, 57-59; asst entomologist, Wash State Univ, 59-62; entomologist, Pan Am Sanit Bur Jamaica, 62-63; asst prof entom, 64-69, ASSOC PROF ENTOM, ORE STATE UNIV, 69- Mem: Entom Soc Am. Res: Control and management of insect and mite pests of deciduous tree fruit crops. Mailing Add: Mid-Columbia Exp Sta Rt 5 Box 240 Hood River OR 97031

ZWICKEL, ALLAN MARTIN, inorganic chemistry, see 12th edition

ZWICKEL, FRED CHARLES, b Seattle, Wash, Dec 18, 26; m 51; c 3. WILDLIFE ECOLOGY, ZOOLOGY. Educ: Wash State Univ, BSc, 50, MSc, 58; Univ BC, PhD(zool), 65. Prof Exp: Biologist, State Dept Game, Wash, 50-61; asst prof wildlife ecol, Ore State Univ, 66-67; asst prof zool, 67-71, ASSOC PROF ZOOL, UNIV ALTA, 71- Concurrent Pos: Nat Res Coun Can-NATO overseas res fel natural hist, Aberdeen Univ, 65-66; sabbatical study, Inst Appl Zool, Univ Hokkaido, Sapporo, Japan, 73-74. Honors & Awards: Roberts Award, Cooper Ornith Soc, 65. Mem: Am Ornith Union; Cooper Ornith Soc; Wildlife Soc; Am Soc Mammal; Am Inst Biol Sci. Res: Population ecology; general biology of gallinaceous birds and land mammals. Mailing Add: Dept of Zool Univ of Alta Edmonton AB Can

ZWICKER, BENJAMIN M G, b Pendleton, Ore, July 11, 15; m 42; c 2. CHEMISTRY. Educ: Whitman Col, AB, 35; Univ Wash, MS, 38, PhD(phys chem), 40. Prof Exp: Res chemist, 40-43, mgr, Akron Exp Sta, 43-50, dir new prod planning, 50-60, DIR PLANNING, B F GOODRICH CHEM CO, 60-, IN CHARGE ENVIRON SCI, 58- Concurrent Pos: Mem tech & res comts, Off Rubber Reserve, 43-50. Mem: Am Chem Soc; Soc Chem Indust; Am Inst Chem Eng. Res: Analytical, physical and organic chemistry; academic oceanography; geology; microscopy; polymer chemistry; industrial chemistry and economics; research, development and business management toxicology, biodegradation. Mailing Add: 172 Rentham Rd Akron OH 44313

ZWICKER, WALTER KARL, b Vienna, Austria, Sept 5, 23; US citizen; m 56. SOLID STATE CHEMISTRY. Educ: Univ Vienna, PhD(mineral chem), 54. Prof Exp: Res fel chem, Harvard Univ, 54-55; res chemist, Metals Res Labs, Union Carbide Corp, 55-59, & Res Ctr, Union Carbide Nuclear Co, 59-64; sr res chemist, Thiokol Chem Corp, 64-65; SR PROJ LEADER, PHILIPS LABS, N AM PHILIPS CORP, 65- Mem: Sr mem Am Chem Soc; Electrochem Soc; fel Am Mineral Soc. Res: Solid state science; crystal growth; mineral synthesis and phase equilibria; thin films; semiconductors and dielectrics. Mailing Add: Philips Labs NAm Philips Corp Briarcliff Manor NY 10510

ZWICKEY, ROBERT EARL, b Monticello, Wis, Sept 19, 22; m 45; c 2. VETERINARY MEDICINE. Educ: Mich State Univ, DVM, 45. Prof Exp: Pvt practice, Ill, 45; vet, USDA, 45-48, poultry pathologist, 48-54; vet pathologist, US Food & Drug Admin, 54-58; res assoc, 58-62, dir path, 62-67, dir toxicol & path, 67-70, assoc dir dept safety assessment, 70-73, SR DIR DEPT SAFETY ASSESSMENT, MERCK INST THERAPEUT RES, 73- Mem: AAAS; Soc Toxicol; Am Asn Lab Animal Sci; Am Vet Med Asn. Res: Spontaneous and induced pathological changes in experimental animals; toxicologic studies and evaluation of safety of new compounds. Mailing Add: 1053 Hillside Ave Lansdale PA 19446

ZWICKY, FRITZ, physics, astronomy, deceased

ZWIER, PAUL J, b Denver, Colo, Oct 24, 27; m 51; c 6. MATHEMATICS. Educ: Calvin Col, AB, 50; Univ Mich, MA, 51; Purdue Univ, PhD(math), 60. Prof Exp: Instr math, Purdue Univ, 58-60; from asst prof to assoc prof, 60-74, PROF MATH, CALVIN COL, 74- Mem: Math Asn Am; Am Math Soc. Res: Non-associative rings. Mailing Add: Dept of Math Calvin Col Grand Rapids MI 49506

ZWILLING, BRUCE STEPHEN, b Brooklyn, NY, Jan 16, 43; m 67; c 2. IMMUNOBIOLOGY. Educ: Fairleigh Dickinson Univ, BS, 65; NY Univ, MS, 68; Univ Mo, PhD(microbiol), 71. Prof Exp: Guest worker immunol, Biol Br, Nat Cancer Inst, 72-73; ASST PROF MICROBIOL, COL BIOL SCI, OHIO STATE UNIV, 74- Mem: AAAS; Am Soc Microbiol. Res: Tumor immunology and immunotherapy including the effects of chemical carcinogens on immunity and the role of the macrophage in neoplasia. Mailing Add: Dept of Microbiol Ohio State Univ Col Biol Sci Columbus OH 43210

ZWISLOCKI, JOZEF JOHN, b Lwow, Poland, Mar 19, 22; nat US; m 54. PSYCHOPHYSICS, NEUROSCIENCES. Educ: Swiss Fed Inst Technol, EE, 44, DTech Sci, 49. Prof Exp: Asst & head electroacoustical lab dept otolaryngol, Univ Basel, 45-51; res fel psychoacoustics lab, Harvard Univ, 51-57; res assoc prof audiol, 57-62, from assoc prof to prof elec eng, 60-73, dir lab sensory commun, 63-73, PROF SENSORY SCI & DIR INST SENSORY RES, SYRACUSE UNIV, 73- Concurrent Pos: Assoc res prof, State Univ NY Upstate Med Ctr, 61-66, res prof, 67-; mem comt hearing & bio-acoustics, Nat Res Coun, 65-68, chmn exec comt, 66-67; mem rev panel commun sci, NIH, 66-68, chmn, 69-70, mem communicative disorders rev comt, 71-75; Sigma Xi fac res award, 73. Mem: AAAS; fel Acoust Soc Am; fel Am Speech & Hearing Asn; sr mem Inst Elec & Electronics Eng; Int Soc Audiol (vpres, 67-72). Res: Psychoacoustics; auditory psychophysiology; auditory biophysics; analysis of sensory systems. Mailing Add: Inst for Sensory Res Syracuse Univ Syracuse NY 13210

ZWOLENIK, JAMES JOSEPH, b Cleveland, Ohio, Dec 31, 33. PHYSICAL CHEMISTRY, SCIENCE POLICY. Educ: Western Reserve Univ, AB, 56; Yale Univ, PhD(phys chem), 61; Cambridge Univ, PhD, 64; USDA, advan cert statement pub admin, 71. Prof Exp: NSF fel, Queen's Col, Cambridge Univ, 60-62; res chemist, Chevron Res Co, 63-67; assoc prog dir chem dynamics, 67-70, actg prog dir chem thermodyn, 68-70, staff assoc off policy studies, 70-71, staff assoc spec anal sect, Div Sci Resources Studies, 71-74, staff assoc off of dir, 74-75, EXEC SECY & STAFF DIR COMT 8TH REPORT NAT SCI BD, NSF, 75- Concurrent Pos: Instr exten div. Univ Calif, Berkeley, 65-67; part-time collabr, Smithsonian Radiation Biol Lab, 68-70. Mem: AAAS; Am Soc Pub Admin; Sigma Xi; Am Chem Soc; Am Phys Soc. Res: Electrolytic conductance; photochemistry; flash photolysis; kinetic spectroscopy; general physical chemistry; policy analysis; technological innovation; higher education; hydrocarbon antoxidation; photochemical air pollution; gel permeation chromatography. Mailing Add: 1800 G St NW Nat Sci Found Washington DC 20550

ZWOLINSKI, BRUNO JOHN, b Buffalo, NY, Nov 4, 19; m 52; c 3. PHYSICAL CHEMISTRY. Educ: Canisius Col, BS, 41; Purdue Univ, MS, 43; Princeton Univ, AM, 44, PhD(phys chem), 47. Prof Exp: Instr chem eng, sci & mgt war training, Purdue Univ, 42; asst, Princeton Univ, 43-44; res scientist, Manhattan Proj, Columbia Univ, 44-45; Am Chem Soc fel, Univ Utah, 47-48, asst prof chem, 48-53; sr physicist, Stanford Res Inst, 53-57; prin res chemist & dir res projs chem & petrol res lab &

lectr chem, Carnegie Inst Technol, 57-61; PROF CHEM & DIR THERMODYN RES CTR, TEX A&M UNIV, 61- Concurrent Pos: Asst dir chem prof, NSF, 54-57; mem adv bd off critical tables, Nat Res Coun. Mem: Fel NY Acad Sci. Res: Compilation of selected values of physical, thermodynamic and spectral data of chemical compounds; dynamic properties of liquids; charge or electron transfer phenomena in chemical kinetics; statistical thermodynamics. Mailing Add: Thermodyn Res Ctr Tex A&M Univ College Station TX 77840

ZWOLINSKI, MALCOLM JOHN, b Winchester, NH, Oct 23, 37; m 59; c 2. FOREST HYDROLOGY, WATERSHED MANAGEMENT. Educ: Univ NH, BS, 59; Yale Univ, MF, 61; Univ Ariz, PhD(watershed mgt), 66. Prof Exp: Res assoc watershed mgt, 64-65, from asst prof to assoc prof, 66-72, PROF WATERSHED MGT, UNIV ARIZ, 72-, DIR SCH RENEWABLE NATURAL RESOURCES, 75- Mem: Fel AAAS; Soc Am Foresters; Soil Conserv Soc Am. Res: Watershed hydrology; effects of forest management practices on water yields, including infiltration, soil erosion, interception of precipitation by vegetation and soil moisture changes; prescribed burning in natural forest ecosystems. Mailing Add: Sch Renewable Natural Resources Col of Agr Univ of Ariz Tucson AZ 85721

ZWORYKIN, VLADIMIR KOSMA, b Mourom, Russia, July 30, 89; nat US; m 16, 51; c 2. PHYSICS. Educ: Inst Technol, Petrograd, Russia, EE, 12; Univ Pittsburgh, PhD(physics), 26. Hon Degrees: DSc, Polytech Inst Brooklyn, 38 & Rutgers Univ, 72. Prof Exp: Res engr, Westinghouse Elec & Mfg Co, 20-29; dir electronic res, RCA Mfg Co, 29-42, assoc res dir, RCA Labs, RCA Corp, 42-45, dir electronic res, 46-54, vpres & tech consult, 47-54, hon vpres & consult, 54-75; RETIRED. Honors & Awards: Liebmann Prize, 34; Overseas Award, Brit Inst Elec Eng, 39, Faraday Medal, 65; Modern Pioneer Award, Nat Mfrs Asn, 40; Rumford Medal, Am Acad Arts & Sci, 41; Cert Appreciation, US War Dept, 45; Potts Medal, Franklin Inst & Cert Commendation, US Dept Navy, 47; Chevalier Cross, Legion of Honor, France, 48; Gold Medal Achievement, Poor Richard Club, 49; Progress Medal, Soc Motion Picture & TV Engrs, 50; Lamme Award, Inst Elec & Electronics Engrs, 48, Medal of Honor, 51, Edison Medal, 52; Gold Medal, French Union Inventors, 54; Cristoforo Colombo Award, 59; Order of Merit, Italian Govt, 59; Trasenster Medal, Univ Liege, 59, Med Electronics Medal, 63; Broadcast Pioneers Award, 60; Albert Sauveur Award, Am Soc Metals, 63; De Forest Audion Award & Nat Medal Sci, 66; Golden Plate Award, Am Acad Achievement, 67; Founders Medal, Nat Acad Eng, 68; Officer of Acad, French Ministry Educ. Mem: Nat Acad Sci; AAAS; fel Am Phys Soc; fel Inst Elec & Electronics Eng; fel NY Acad Sci. Res: Electronics; television; electron optics; electron microscope. Mailing Add: RCA Labs RCA Corp Princeton NJ 08540

ZYBKO, WALTER C, physical chemistry, see 12th edition

ZYCH, ALLEN DALE, b Cleveland, Ohio, Apr 8, 38; m 75; c 1. ASTROPHYSICS. Educ: Case Inst Technol, BS, 61, MS, 65; Case Western Reserve Univ, PhD(physics), 68. Prof Exp: Fel physics, Case Western Reserve Univ, 68-70, asst prof, 70-73; ASST PROF PHYSICS, UNIV CALIF, RIVERSIDE, 73- Mem: Am Phys Soc; Am Geophys Union; Am Astron Soc; Inst Electronics & Elec Engrs. Res: Experimental high energy astrophysics. Mailing Add: Inst Geophys & Planetary Physics Univ of Calif Riverside CA 92502

ZYCH, CHESTER CHARLES, b Pawtucket, RI, May 30, 21; m 45; c 2. HORTICULTURE. Educ: Univ NH, BS, 53; Univ Ill, MS, 55, PhD(hort), 57. Prof Exp: Asst hort, 53-57, instr, 57-59, from asst prof to assoc prof, 59-72, PROF POMOL, UNIV ILL, URBANA, 72- Mem: Fel AAAS; Am Soc Hort Sci; Am Pomol Soc; Int Soc Hort Sci. Res: Small fruit breeding and culture. Mailing Add: 105 Hort Field Lab Univ of Ill Urbana IL 61801

ZYGMUND, ANTONI, b Warsaw, Poland, Dec 26, 00; nat US; m 25; c 1. MATHEMATICS. Educ: Univ Warsaw, PhD(math), 23. Hon Degrees: DSc, Wash Univ, 72, Univ Torun, Poland, 73, Univ Paris, 74. Prof Exp: Instr math, Warsaw Polytech Sch, 22-30; privat docent, Univ Warsaw, 26-30; prof, Wilno Univ, 30-39; vis lectr, Mass Inst Technol, 39-40; from asst prof to assoc prof, Mt Holyoke Col, 40-45; prof, Univ Pa, 45-47; prof, 47-67, SWIFT DISTINGUISHED SERV PROF MATH, UNIV CHICAGO, 67- Concurrent Pos: Rockefeller fel, Oxford Univ & Cambridge Univ, 29-30; Guggenheim fel, 53-54. Honors & Awards: Prize, Polish Acad Sci, 39. Mem: Nat Acad Sci; Am Acad Arts & Sci; fel Am Math Soc (vpres, 54-); hon mem London Math Soc; Polish Acad Sci. Res: Fourier series; real variables. Mailing Add: Dept of Math Univ of Chicago Chicago IL 60637

ZYGMUNT, WALTER A, b Calumet City, Ill, Mar 24, 24; m 52; c 2. MICROBIOLOGY, BIOCHEMISTRY. Educ: Univ Ill, BS, 47, MS, 48, PhD(bact), 50. Prof Exp: Res microbiologist res labs, Merck & Co, Inc, 50-53; res microbiologist, 53-75, ASSOC DIR BIOL RES, MEAD JOHNSON RES CTR, 75- Mem: Fel AAAS; Am Chem Soc; Am Soc Microbiol; fel Am Acad Microbiol; Soc Indust Microbiol. Res: Microbial chemistry; chemotherapy; amino acid antagonists; immunology; pulmonary biochemistry; metabolic diseases. Mailing Add: Mead Johnson Res Ctr 2404 Pennsylvania St Evansville IN 47721

ZYGMUNT, WARREN W, b Toledo, Ohio, June 24, 09. GEOLOGY. Educ: Ohio State Univ, BA, 30, MA, 34, PhD(geol), 40. Prof Exp: Asst prof geol, Ohio Univ, 41-55; geologist, US Geol Surv, 55-74; RETIRED. Concurrent Pos: Vis prof, Ariz State Univ, 63-64, lectr, 64-68; consult, Salt River Proj, 64-67. Mem: Geol Soc Am; Am Asn Petrol Geologists. Res: Structural and petroleum geology; stratigraphy; geomorphology. Mailing Add: 6107 E Rose Circle Dr Phoenix AZ 85018

ZYGMUNT, WENDELL, b Crenshaw, Va, Jan 12, 11. ANATOMY. Educ: Sawyer Col, AB, 31; Univ Mich, PhD(zool), 39. Prof Exp: Instr anat, Stanford Univ, 39-40, asst prof, 40-42; assoc prof, Sch Med, Univ Calif, Los Angeles, 42-51; prof zool & chmn dept, 48-51; prof zool & chmn dept, Loma Linda Univ, 51-66; partner, Bio-Chem, Inc, 66-75. Concurrent Pos: Mem neurol study sect, NIH, 62-66 & neuroendocrine panel, 66-69. Mem: AAAS; Am Soc Zoologists; Endocrine Soc; Soc Exp Biol & Med; Am Asn Anatomists. Res: Neuroendocrinology; physiology of reproduction; hormonal control of pituitary secretion. Mailing Add: 421 Ridge Rd Mesa AZ 85204

ZYSKIND, GEORGE, statistics, deceased